LIST OF SECTIONS

METALS HANDBOOK®

Desk Edition

Edited by
Howard E. Boyer
Timothy L. Gall

AMERICAN SOCIETY FOR METALS
Metals Park, Ohio 44073

Metals Handbook is a collective effort involving thousands of technical specialists. It brings together in one book a wealth of information from world-wide sources to help scientists, engineers, and technicians solve current and long-range problems.

Great care is taken in the compilation and production of this volume, but it should be made clear that no warranties, express or implied, are given in connection with the accuracy or completeness of this publication, and no responsibility can be taken for any claims that may arise.

Nothing contained in the Metals Handbook shall be construed as a grant of any right of manufacture, sale, use, or reproduction, in connection with any method, process, apparatus, product, composition, or system, whether or not covered by letters patent, copyright, or trademark, and nothing contained in the Metals Handbook shall be construed as a defense against any alleged infringement of letters patent, copyright, or trademark, or as a defense against any liability for such infringement.

Comments, criticisms, and suggestions are invited, and should be forwarded to the American Society for Metals, Metals Park, Ohio 44073.

Library of Congress Catalog Card Number: 84-71465
ISBN: 0-87170-188-X
SAN: 204-7586

Project Director: Timothy L. Gall
ASM Staff Assistants: Joan L. Tomsic and Judith S. Gibbs
Editorial and production coordination by Carnes Publication Services, Inc.
    President: William J. Carnes
    Project Manager: Craig W. Kirkpatrick

PRINTED IN THE UNITED STATES OF AMERICA

Foreword

THE METALS HANDBOOK is not a static document. Metals technology progresses steadily, often dramatically, and with each new technological insight the need for reliable engineering data increases. For 61 years the Metals Handbook has been dynamic in its response to the continuing need for technical information. When knowledge expanded, the Handbook expanded; new formats were developed and additional subjects considered. What began as a small loose-leaf collection of data sheets developed into a multivolume series of casebound reference books — each volume a thorough, comprehensive and authoritative treatise on the subject to which it is devoted. Now in the 9th Edition, with seven volumes in print and three others in preparation, the Metals Handbook has become the standard reference for the world metallurgical community.

Although no one disputes the necessity for the multivolume set, many praise the convenience of the ever-popular 1948 edition — the last edition to be contained in one volume. The usefulness of an easily accessible, convenient and practical first reference to all of metalworking technology is evident from the hundreds of requests the Society receives yearly for a revision of the "red book." It was in response to this persistent demand that the Metals Handbook Committee, after 36 years, decided once again to pull it all together in this, the Metals Handbook Desk Edition.

The job of doing so seemed overwhelming. A total of 18 Metals Handbook volumes from both the 8th and 9th Editions stood as a reservoir of metalworking information. Instrumental in condensing this material into one volume was Howard E. Boyer. Having worked on the Handbook staff for more than 20 years, first as Managing Editor to Dr. Taylor Lyman, the distinguished editor of the 1948 edition and creator of the monumental 8th Edition, and then as editor after Dr. Lyman's death, in 1973, Howard Boyer knows the content of the Metals Handbook better than anyone else alive today. One does not exaggerate when saying that without his editorial experience, skill and commitment this book would not have been possible. Equally important to the project was William J. Carnes, who was responsible for turning the manuscript into a book. All aspects of copy editing, typesetting, proofing, printing and binding were under his expert control and direction. Aiding these gentlemen were numerous section editors, review committees and a dedicated editorial and production staff. By building upon the work of thousands of past Handbook participants, they have once again given the world metalworking community a single reference to bind us all together. On behalf of the American Society for Metals and its members, we extend our appreciation to these participants.

However, we view the Desk Edition as only a beginning. To remain viable it must remain dynamic. It is our sincere hope that within the next five years a new group of participants will step forward to improve upon this work with a revised and updated version of the Metals Handbook Desk Edition.

M. BRIAN IVES
President

EDWARD L. LANGER
Managing Director

Preface

THE METALS HANDBOOK DESK EDITION represents a fresh attack on the challenge of providing an authoritative reference volume about metals. As the technical world becomes more complex, engineers are finding their areas of expertise increasingly specialized, making it more difficult to see the whole of metals technology. When faced with questions outside their specialty, they are often at a loss as to where to begin their research. This problem provided the objective for this Desk Edition: to provide a single authoritative first reference to all of metals technology.

Under the guidance of the ASM Metals Handbook Committee, an outline was established for a single, comprehensive Desk Edition of the Metals Handbook series. Following this outline, the book is divided into four major parts: General Information; Properties and Selection; Processing; and Testing and Inspection.

General Information contains a glossary of 3000 terms and a collection of common engineering tables. It also features articles on crystal structure, phase diagrams, and metals engineering.

Properties and Selection contains articles on all of the standard industrial alloys. The compositions and mechanical properties of thousands of alloys are presented. Service and processing characteristics are covered in detail as they relate to the selection procedure.

Processing. Unlike the processing information in Part II, the information in Part III takes a how-to approach. All processes are covered, from extractive metallurgy to recycling of metals. The information is presented in enough detail to enable readers to solve many common production problems.

Testing and Inspection offers practical information on the topics of failure analysis, nondestructive testing, mechanical testing, metallography, fractography, and quality control.

Where possible, the four parts follow the structure and organization of the existing 8th and 9th Editions of Metals Handbook, thereby facilitating easy cross referencing to the larger works.

Work began on the Desk Edition with material currently in existence. The first step involved condensing 8th and 9th Edition Handbook volumes to a size appropriate for a one-volume book. This seemingly overwhelming task was undertaken by a single person, editor Howard E. Boyer, thereby ensuring an element of consistency to the compilation. The articles as condensed by the editor served two functions: to define the nature and scope of the material to be presented and to provide a framework around which to add additional material. Upon completion, these "core" articles were submitted to subject editors for further review and revision.

As one would expect, the articles underwent substantial revision in the hands of the subject editors, to reflect advances in technology. Some articles — Aluminum, Zinc, Lead, Precious Metals, Machining, and Failure Analysis — were totally rewritten. In other instances, the technical review process identified gaps in coverage of important topics, making it necessary to commission totally new articles. The articles so commissioned were Extractive Metallurgy, Iron and Steelmaking, Recycling of Metals and Alloys, and Mechanical Testing. This procedure produced a thoroughly reviewed and amended document covering all aspects of metals technology. In all, more than 40 percent of the material in these pages was specifically developed for the Desk Edition.

The challenge of converting this material into book form was accepted by Carnes Publication Services. Manuscripts flowed into their offices from the subject editors steadily for more than a year. Two individuals have gained the distinction of having read the Desk Edition from cover to cover: Craig W. Kirkpatrick, project manager on the Carnes staff, who single-handedly copy edited manuscript for the entire text; and Marjorie R. Hyslop, who compiled the more than 13,000 entries in the Desk Edition index. Now a consultant on information science and technology, Marjorie is known by many ASM members for her instrumental role in establishing Metadex, ASM's electronic database. Their efforts provided this Desk Edition with a consistency of approach and an attention to detail that will benefit readers in the use of the book.

*   *   *   *   *

Although the index usually occupies the very last pages of most books, in the Desk Edition it is followed by a short appendix designed to introduce you to a personal-computer software package called MetalSelector. Developed by ASM to supplement this book, this software package provides five functions important to metals engineers:

- It offers a sortable materials-properties database of more than 500 common industrial alloys.
- It converts common engineering units.
- It has a graphics program that allows you to plot lines and fit curves.
- It is a word processor that allows you to write reports.
- Finally, it offers a gateway into Metadex and the world of on-line information retrieval.

In the Desk Edition, the last pages are really just the beginning.

TIMOTHY L. GALL
Project Director

Contents

Part IV. Testing and Inspection

Part I
GENERAL INFORMATION

1 GLOSSARY OF METALLURGICAL TERMS AND ENGINEERING TABLES

Glossary of Terms Related to Metals and Metalworking

By the ASM Committee on Definitions of Metallurgical Terms*

An accurate understanding of technical terminology is of inestimable value to anyone who wishes to obtain information from technical literature. Not only do special words exist that have no meaning outside the technical language, but also many words with well-known literal meanings have entirely different and unrelated meanings when used as part of the technical language. This glossary clarifies more than 3000 specialized technical terms encountered in metallurgical literature. The list includes terms from: materials science; physical metallurgy; heat treating; extractive metallurgy; casting, forging, machining, forming, welding and joining; metal cleaning and finishing; electrometallurgy; powder metallurgy; mechanical testing, inspection and quality control; and metallography, fractography and failure analysis. Terms whose meanings are the same for both technical and nontechnical usage, as well as terms having universal scientific meanings, are largely excluded from this glossary.

Many cross references to preferred terms, alternative terms and closely related terms have been included; these cross references are printed in italics. Also, terms that are obsolete or otherwise inappropriate for use in current writings are so indicated. Many terms can have more than one meaning in metallurgical literature; alternative meanings are identified by parenthetical numbers preceding each alternative. Whenever possible, a general or generic meaning is given before a specific or specialized meaning, but there is no special significance to the order in which alternative meanings are given.

The definitions in this glossary have been edited with an emphasis on clarity. They are compatible with (although not necessarily identical to) definitions pub-lished by ANSI, ASQC, ASTM, AWS, SAE and other organizations. Often, when a difference in wording exists, it arises from the need for this glossary to serve readers in many branches of technology — including laymen, businessmen and students as well as scientists, engineers and technicians — whereas definitions published elsewhere frequently are intended for specialists and for use in codes and other legal documents.

A

A_{cm}, A_1, A_3, A_4. Same as *Ae_{cm}, Ae_1, Ae_3, Ae_4*.

abrasion. A roughening or scratching of a surface due to *abrasive wear*. On aluminum parts, also known as a rub mark or traffic mark.

abrasive. (1) A hard substance used for *grinding, honing, lapping, superfinishing, polishing*, pressure blasting or *barrel finishing*. It includes natural materials such as garnet, emery, corundum and diamond, and electric-furnace products like aluminum oxide, silicon carbide and boron carbide. (2) Hard particles, such as rocks, sand or fragments of certain hard metals, that wear away a surface when they move across it under pressure.

abrasive belt. A coated abrasive product, in the form of a belt, used in production grinding and polishing.

abrasive blasting. A process for cleaning or finishing by means of an abrasive directed at high velocity against the workpiece.

abrasive disk. (1) A grinding wheel that is mounted on a steel plate, with the exposed flat side being used for grinding. (2) A disk-shaped, coated abrasive product.

abrasive wear. The removal of material from a surface by hard particles sliding or rolling across the surface under pressure. The particles may be loose or may be part of another surface in contact with the surface being worn. Contrast with *adhesive wear*.

Ac_{cm}, Ac_1, Ac_3, Ac_4. Defined under *transformation temperature*.

accuracy. The closeness of approach of a measurement to the true value of the quantity measured. Since the true value cannot actually be measured, the most probable value from the available data, critically considered for sources of error, is used as "the truth." Contrast with *precision*.

acicular ferrite. A highly substructured nonequiaxed ferrite formed upon continuous cooling by a mixed diffusion and shear mode of transformation that begins at a temperature slightly higher than the transformation temperature range for upper bainite. It is distinguished from bainite in that it has a limited amount of carbon available; thus, there is only a small amount of carbide present.

acicular ferrite steels. Those steels having a microstructure consisting of either acicular ferrite or a mixture of acicular and equiaxed ferrite.

acid. A chemical substance that yields hydrogen ions (H^+) when dissolved in water. Compare with *base* (3).

acid bottom and lining. The inner bottom and lining of a melting furnace, consisting of materials like sand, siliceous rock or silica brick that give an acid reaction at the operating temperature.

acid copper. (1) Copper electrodeposited from an acid solution of a copper salt, usually copper sulfate. (2) The solution referred to in (1).

*H. E. BOYER, *Chairman*, Consultant; J. W. BARR, Aluminum Association; F. W. BOULGER, Battelle-Columbus Laboratories; P. B. BURGESS, Hayes-Albion Corp.; J. R. CUTHILL, National Bureau of Standards; C. C. DICK, Lindberg Heat Treating Co.; J. L. DOSSETT, Lindberg Heat Treating Co.; F. L. EWALD, Budd Co.; I. J. FEINBERG, National Bureau of Standards; M. FIELD, Metcut Research Associates, Inc.; R. GIBALA, Case Western Reserve University; W. C. HARMON, Republic Steel Corp.; T. J. HUGHEL, General Motors Corp.; K. J. HUMBERSTONE, American Tank & Fabricating Co.; C. G. INTERRANTE, National Bureau of Standards; F. L. JAMIESON, Steel Co. of Canada; F. LA QUE, Consultant; J. B. LONG, Tin Research Institute; P. T. LOVEJOY,

Allegheny Ludlum Steel Co.; W. S. LYMAN, Copper Development Association, Inc.; R. C. MCMASTER, Ohio State University; J. T. MICHALAK, United States Steel Corp.; T. J. MOORE, National Aeronautics and Space Administration; F. OGBURN, National Bureau of Standards; B. R. QUENEAU, retired from United States Steel Corp.; G. O. RATLIFF, Shore Metal Treating, Inc.; K. H. ROLL, Metal Powder Industries Federation; M. T. ROWLEY, American Foundrymen's Society; J. A. SIMMONS, National Bureau of Standards; F. SPEIGHT, American Welding Society; H. TURNER, McDonnell Aircraft Co.; C. F. WALTON, Iron Castings Society; W. G. WOOD, Kolene Corp.; D. BENJAMIN, *Secretary*, Senior Editor, Metals Handbook.

acid embrittlement. A form of *hydrogen embrittlement* that may be induced in some metals by acid treatment.

acid steel. Steel melted in a furnace with an *acid bottom and lining* and under a slag containing an excess of an acid substance such as silica.

activation. The changing of a passive surface of a metal to a chemically active state. Contrast with *passivation*.

activation energy. The energy required for initiating a metallurgical reaction — for example, plastic flow, diffusion, chemical reaction. The activation energy may be calculated from the slope of the line obtained by plotting the natural log of the reaction rate versus the reciprocal of the absolute temperature.

activity. A measure of the chemical potential of a substance, where chemical potential is not equal to concentration, that allows mathematical relations equivalent to those for ideal systems to be used to correlate changes in an experimentally measured quantity with changes in chemical potential.

addition agent. A substance added to a solution for the purpose of altering or controlling a process. Examples: wetting agents in acid pickles; brighteners or antipitting agents in plating solutions; inhibitors.

adhesion. Force of attraction between the molecules (or atoms) of two different phases. Contrast with *cohesion*.

adhesive bonding. A materials joining process in which an adhesive, placed between faying surfaces, solidifies to bond the surfaces together.

adhesive wear. The removal of material from a surface by the welding together and subsequent shearing of minute areas of two surfaces that slide across each other under pressure. In advanced stages, may lead to *galling* or *seizing*. Contrast with *abrasive wear*.

adjustable bed. Bed of a press designed so that the die space height can be varied conveniently.

Ae$_{cm}$, Ae$_1$, Ae$_3$, Ae$_4$. Defined under *transformation temperature*.

age hardening. Hardening by aging, usually after rapid cooling or cold working. See *aging*.

age softening. Spontaneous decrease of strength and hardness that takes place at room temperature in certain strain-hardened alloys, especially those of aluminum.

aging. A change in the properties of certain metals and alloys that occurs at ambient or moderately elevated temperatures after hot working or a heat treatment (quench aging in ferrous alloys, natural or artificial aging in ferrous and nonferrous alloys) or after a cold working operation (strain aging). The change in properties is often, but not always, due to a phase change (precipitation), but never involves a change in chemical composition of the metal or alloy. See also *age hardening, artificial aging, interrupted aging, natural aging, overaging, precipitation hardening, precipitation heat treatment, progressive aging, quench aging, step aging, strain aging*.

air bend die. Angle-forming dies in which the metal is formed without striking the bottom of the die. Metal contact is made at only three points in the cross section: the nose of the male die and the two edges of a V-shape die opening.

air bending. Bending in an *air bend die*.

air classification. The separation of metal powder into particle-size fractions by means of an air stream of controlled velocity; an application of the principle of *elutriation*.

air-hardening steel. A steel containing sufficient carbon and other alloying elements so as to harden fully during cooling in air or other gaseous media from a temperature above its transformation range. This term should be restricted to steels that are capable of being hardened by cooling in air in fairly large sections, about 2 in. or more in diameter. Same as self-hardening steel.

air-lift hammer. A type of gravity drop hammer where the ram is raised for each stroke by an air cylinder. Since length of stroke may be controlled, ram velocity and thus energy delivered to the workpiece may be varied.

alclad. Composite wrought product comprised of an aluminum alloy core having on one or both surfaces a metallurgically bonded aluminum or aluminum alloy coating that is anodic to the core and thus electrically protects the core against corrosion.

alkali metal. A metal in group IA of the periodic system — namely, lithium, sodium, potassium, rubidium, cesium and francium. They form strongly al-

kaline hydroxides; hence, the name.

alkaline cleaner. A material blended from alkali hydroxides and such alkaline salts as borates, carbonates, phosphates or silicates. The cleaning action may be enhanced by the addition of surface-active agents and special solvents.

alkaline earth metal. A metal in group IIA of the periodic system — namely, beryllium, magnesium, calcium, strontium, barium and radium — so called because the oxides or "earths" of calcium, strontium and barium were found by the early chemists to be alkaline in reaction.

alligatoring. The longitudinal splitting of flat slabs in a plane parallel to the rolled surface. Also called fishmouthing.

allotriomorphic crystal. A crystal whose lattice structure is normal but whose external surfaces are not bounded by regular crystal faces; rather, the external surfaces are impressed by contact with other crystals or another surface such as a mold wall, or are irregularly shaped because of nonuniform growth. Compare with *idiomorphic crystal*.

allotropy. A near synonym for *polymorphism*, generally restricted to description of polymorphic behavior in elements, terminal phases, and alloys whose behavior closely parallels that of the predominant constituent element.

allowance. The specified difference in limiting sizes (minimum clearance or maximum interference) between mating parts, as computed arithmetically from the specified dimensions and tolerances of each part.

alloy. A substance having metallic properties and being composed of two or more chemical elements of which at least one is a *metal*.

alloying element. An element which is added to a metal (and which remains within the metal) to effect changes in properties.

alloy plating. The codeposition of two or more metallic elements.

alloy powder. A powdered metal in which each particle is composed of the same alloy.

alloy steel. Steel containing specified quantities of alloying elements (other than carbon and the commonly accepted amounts of manganese, copper, silicon, sulfur and phosphorus) within the limits recognized for constructional alloy steels, added to effect changes in mechanical or physical properties.

all-position electrode. In arc welding, a filler-metal electrode for depositing weld metal in the flat, horizontal, overhead and vertical positions.

all-weld-metal test specimen. A test specimen wherein the portion being tested is composed wholly of weld metal.

alpha ferrite. See *ferrite*.

alpha iron. The body-centered cubic form of pure iron, stable below 910 °C (1670 °F).

alsifer. A deoxidizer (20 Al, 40 Si, 40 Fe) used for steel.

alternate-immersion test. A corrosion test in which the specimens are intermittently immersed in and removed from a liquid medium at definite time intervals.

Alumel. A nickel-base alloy containing about 2.5 Mn, 2 Al and 1 Si used chiefly as a component of pyrometric thermocouples.

aluminizing. Forming of an aluminum or aluminum alloy coating on a metal by hot dipping, hot spraying or diffusion.

aluminum bomb. A bomb-shaped container used in determining the oxygen content in liquid steel.

amalgam. An alloy of mercury with one or more other metals.

amorphous. Not having a crystal structure; noncrystalline.

amphoteric. Possessing both acidic and basic properties.

anchorite. A zinc-iron phosphate coating for iron and steel.

anelasticity. The property of solids by virtue of which strain is not a single-value function of stress in the low-stress range where no permanent set occurs.

angle of bite. In rolling of metals where all the force is transmitted through the rolls, the maximum attainable angle between the roll radius at the first contact and the line of roll centers. Operating angles less than the angle of bite are called contact angles or rolling angles.

angle of nip. In rolling, the *angle of bite*. In roll, jaw or gyratory crushing, the entrance angle formed by the tangents at the two points of contact between the

working surfaces and the (assumed) spherical particle to be crushed.

angstrom (unit). A unit of linear measurement equal to 10^{-10} m, or 0.1 nm, sometimes used to express small distances such as interatomic distances and some wavelengths.

anion. A negatively charged ion; it flows to the anode in electrolysis.

anisotropy. The characteristic of exhibiting different values of a property in different directions with respect to a fixed reference system in the material.

annealing. A generic term denoting a treatment, consisting of heating to and holding at a suitable temperature followed by cooling at a suitable rate, used primarily to soften metallic materials, but also to simultaneously produce desired changes in other properties or in microstructure. The purpose of such changes may be, but is not confined to: improvement of machinability, facilitation of cold work, improvement of mechanical or electrical properties, and/or increase in stability of dimensions. When the term is used unqualifiedly, full annealing is implied. When applied only for the relief of stress, the process is properly called *stress relieving* or stress-relief annealing.

In ferrous alloys, annealing usually is done above the upper critical temperature, but the time-temperature cycles vary widely both in maximum temperature attained and in cooling rate employed, depending on composition, material condition, and results desired. When applicable, the following commercial process names should be used: *black annealing, blue annealing, box annealing, bright annealing, cycle annealing, flame annealing, full annealing, graphitizing, in-process annealing, isothermal annealing, malleablizing, orientation annealing, process annealing, quench annealing, spheroidizing, subcritical annealing*.

In nonferrous alloys, annealing cycles are designed to: (a) remove part or all of the effects of cold working (recrystallization may or may not be involved); (b) cause substantially complete coalescence of precipitates from solid solution in relatively coarse form; or (c) both, depending on composition and material condition. Specific process names in commercial use are *final annealing, full annealing, intermediate annealing, partial annealing, recrystallization annealing*, stress-relief annealing, *anneal to temper*.

annealing carbon. Fine, apparently amorphous carbon particles formed in white cast iron and certain steels during prolonged annealing. Also called temper carbon.

annealing twin. A *twin* formed in a crystal during recrystallization.

anneal to temper. A final partial anneal that softens a cold worked nonferrous alloy to a specified level of hardness or tensile strength.

anode. The electrode where electrons leave an operating system such as a battery, an electrolytic cell, an x-ray tube or a vacuum tube. In the first of these, it is negative; in the other three, positive. In a battery or electrolytic cell, it is the electrode where oxidation occurs. Contrast with *cathode*.

anode compartment. In an electrolytic cell, the enclosure formed by a diaphragm around the anodes.

anode copper. Special-shaped copper slabs, resulting from the refinement of *blister copper* in a reverberatory furnace, used as anodes in electrolytic refinement.

anode corrosion. The dissolution of a metal acting as an anode.

anode effect. The effect produced by polarization of the anode in electrolysis. It is characterized by a sudden increase in voltage and a corresponding decrease in amperage due to the anode becoming virtually separated from the electrolyte by a gas film.

anode efficiency. *Current efficiency* at the anode.

anode film. (1) The portion of solution in immediate contact with the anode, especially if the concentration gradient is steep. (2) The outer layer of the anode itself.

anode mud. Deposit of insoluble residue formed from the dissolution of the anode in commercial electrolysis. Sometimes called anode slime.

anode polarization. See *polarization*.

anodic cleaning. *Electrolytic cleaning* in which the work is the anode. Also called reverse-current cleaning.

anodic coating. A film on a metal surface resulting from an electrolytic treatment at the anode.

anodic pickling. *Electrolytic pickling* in which the work is the anode.

anodic protection. Imposing an external electrical potential to protect a metal from corrosive attack. (Applicable only to metals that show active-passive behavior.) Contrast with *cathodic protection*.

anodizing. Forming a *conversion coating* on a metal surface by anodic oxidation; most frequently applied to aluminum.

anolyte. The electrolyte adjacent to the anode in an electrolytic cell.

antiferromagnetic material. A material wherein interatomic forces hold the elementary atomic magnets (electron spins) of a solid in alignment, a state similar to that of a *ferromagnetic material* but with the difference that equal numbers of elementary magnets (spins) face in opposite directions and are antiparallel, causing the solid to be weakly magnetic, that is, paramagnetic, instead of ferromagnetic.

antipitting agent. An *addition agent* for electroplating solutions to prevent the formation of pits or large pores in the electrodeposit.

anvil. (1) In drop forging, the base of the hammer into which the *sow block* and lower die part are set. (2) A block of steel upon which metal is forged.

anvil cap. Same as *sow block*.

apparent density. (1) The weight per unit volume of a metal powder, in contrast to the weight per unit volume of the individual particles. (2) The weight per unit volume of a porous solid, where the unit volume is determined from external dimensions of the mass. Apparent density is always less than the true density of the material itself.

approach distance. The linear distance, in the direction of feed, between the point of initial cutter contact and the point of full cutter contact.

Ar$_{cm}$, Ar$_1$, Ar$_3$, Ar$_4$, Ar', Ar''. Defined under *transformation temperature*.

arbitration bar. A test bar, cast with a heat of material, used to determine chemical composition, hardness, tensile strength, and deflection and strength under transverse loading in order to establish the state of acceptability of the casting.

arbor. (1) In machine grinding, the spindle on which the wheel is mounted. (2) In machine cutting, a shaft or bar for holding and driving the cutter. (3) In founding, a metal shape embedded in green sand or dry sand cores to support the sand or the applied load during casting.

arbor press. A machine used for forcing arbors or mandrels into drilled or bored parts preparatory to turning or grinding. Also used for forcing bushings, shafts or pins into or out of holes.

arbor-type cutter. A cutter having a hole for mounting on an arbor and usually having a keyway for a driving key.

arc blow. The swerving of an electric arc from its normal path because of magnetic forces.

arc brazing. A brazing process in which the heat required is obtained from an electric arc.

arc cutting. A group of cutting processes that melt the metals to be cut with the heat of an arc between an electrode and the base metal. See *metal-arc cutting, gas tungsten-arc cutting, plasma-arc cutting*.

arc furnace. A furnace in which material is heated either directly by an electric arc between an electrode and the work or indirectly by an arc between two electrodes adjacent to the material.

arc gouging. An arc cutting procedure used to form a bevel or groove.

arc melting. Melting metal in an electric arc furnace.

arc of contact. The portion of the circumference of a grinding wheel or cutter touching the work being processed.

arc time. The time the arc is maintained in making an arc weld. Also known as *weld time*.

arc voltage. The voltage across any electric arc — for example, across a welding arc.

arc welding. A group of welding processes that fuse metals together by heating them with an arc, with or without the application of pressure and with or without the use of filler metal.

artifact. A feature of artificial character (such as a scratch or a piece of dust on a metallographic specimen) that can be erroneously interpreted as a real feature. In inspection, an artifact often produces a *false indication*.

artificial aging. Aging above room temperature. See *aging*. Compare with *natural aging*.

athermal transformation. A reaction that proceeds without benefit of thermal fluctuations — that is, thermal activation is not required. Such reactions are

diffusionless and can take place with great speed when the driving force is sufficiently high. For example, many martensitic transformations occur athermally on cooling, even at relatively low temperatures, because of the progressively increasing driving force. In contrast, a reaction that occurs at constant temperature is an *isothermal transformation*; thermal activation is necessary in this case and the reaction proceeds as a function of time.

atmospheric riser. A riser that uses atmospheric pressure to aid feeding. Essentially a *blind riser* into which a small core or rod protrudes, the function of the core or rod being to provide an open passage so that the molten interior of the riser will not be under a partial vacuum when metal is withdrawn to feed the casting, but will always be under atmospheric pressure. Often called Williams riser.

atomic fission. The breakup of the nucleus of an atom in which the combined weight of the fragments is less than that of the original nucleus, the difference being converted to a very large energy release.

atomic hydrogen welding. An arc welding process that fuses metals together by heating them with an electric arc maintained between two metal electrodes enveloped in a stream of hydrogen. Shielding is provided by the hydrogen, which also carries heat by molecular dissociation and subsequent recombination. Pressure may or may not be used and filler metal may or may not be used. (This process is now of limited industrial significance.)

atomic number. The number of protons in an atomic nucleus, which determines the individuality of the atom as a chemical element.

atomic percent. The number of atoms of an element in a total of 100 representative atoms of a substance.

atomization. The dispersion of a molten metal into small particles by a rapidly moving stream of gas or liquid.

attenuation. The fractional decrease of the intensity of an energy flux, including the reduction of intensity resulting from geometrical spreading, absorption and scattering.

attritious wear. Wear of abrasive grains in grinding such that the sharp edges gradually become rounded. A grinding wheel that has undergone such wear usually has a glazed appearance.

ausforming. Hot deformation of metastable austenite within controlled ranges of temperature and time that avoids formation of nonmartensitic transformation products.

austempering. A heat treatment for ferrous alloys in which a part is quenched from the austenitizing temperature at a rate fast enough to avoid formation of ferrite or pearlite and then held at a temperature just above M$_s$ until transformation to bainite is complete.

austenite. A solid solution of one or more elements in face-centered cubic iron. Unless otherwise designated (such as nickel austenite), the solute is generally assumed to be carbon.

austenitic grain size. The size attained by the grains of steel when heated to the austenitic region; may be revealed by appropriate etching of cross sections after cooling to room temperature.

austenitic steel. An alloy steel whose structure is normally austenitic at room temperature.

austenitizing. Forming austenite by heating a ferrous alloy into the transformation range (partial austenitizing) or above the transformation range (complete austenitizing). When used without qualification, the term implies complete austenitizing.

autofrettage. Prestressing a hollow metal cylinder by the use of momentary internal pressure exceeding the yield strength.

autogenous weld. A fusion weld made without the addition of filler metal.

automatic brazing. Brazing with equipment that performs the brazing operation without constant observation and adjustment by a brazing operator. The equipment may or may not load and unload the work.

automatic press. A press in which the work is fed mechanically through the press in synchronism with the press action. An automation press is an automatic press that, in addition, is provided with built-in electrical and pneumatic control equipment.

automatic welding. Welding with equipment that performs the welding operation without adjustment of the controls by an operator. The equipment may or may not load and unload the work. Compare with *machine welding*.

automation press. See *automatic press*.

autoradiography. An inspection technique in which

radiation spontaneously emitted by a material is recorded photographically. The radiation is emitted by radioisotopes that are (*a*) produced in a metal by bombarding it with neutrons, (*b*) added to a metal such as by alloying, or (*c*) contained within a cavity in a metal part. The technique serves to locate the position of the radioactive element or compound.

auxiliary anode. In electroplating, a supplementary anode positioned so as to raise the current density on a certain area of the cathode and thus obtain better plate distribution.

Avogadro's number. The number of atoms (or molecules) in a mole of substance, which equals 6.02252×10^{23} per mole.

axial rake. For angular (not helical) flutes, the angle between a plane containing the tooth face and the axial plane through the tooth point. See sketch accompanying *face mill*.

axial relief. The relief or clearance behind the end cutting edge of a milling cutter.

axial runout. For any rotating element, the total variation from a true plane of rotation, taken in a direction parallel to the axis of rotation. Compare with *radial runout*.

axis of weld. A line through the length of a weld perpendicular to the cross section at its geometric center.

B

back draft. A reverse taper on a casting pattern or a forging die that prevents the pattern or forged stock from being removed from the cavity.

back extrusion. See *backward extrusion*.

backfire. The recession of a flame into the tip of a torch followed by immediate reappearance or complete extinction of the flame. See *flashback*.

backhand welding. Welding in which the back of the principal hand (torch or electrode hand) of the welder

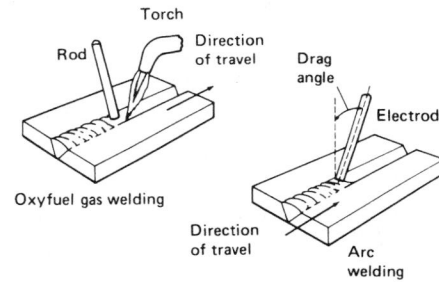

Backhand welding

faces the direction of travel. It has special significance in oxyfuel gas welding in that the flame is directed backward toward the weld bead, which provides *postheating*. Compare with *forehand welding*.

backing. (1) In grinding, the material (paper, cloth or fiber) that serves as the base for coated abrasives. (2) In welding, a material placed under or behind a joint to enhance the quality of the weld at the root. It may be a metal backing ring or strip; a pass of weld metal; or a nonmetal such as carbon, granular flux or a protective gas.

backlash. Lost motion, play or movement in moving parts such that the driving element (as a gear) can be reversed for some angle or distance before working contact is again made with a driven element.

backoff. A rapid withdrawal of a grinding wheel or cutting tool from contact with a workpiece.

back rake. The angle on a single-point turning tool corresponding to axial rake in milling. It is the angle measured between the plane of the tool face and the reference plane and lies in a plane perpendicular to the axis of the work material and the base of the tool. See sketch accompanying *single-point tool*.

backstep sequence. A longitudinal welding sequence

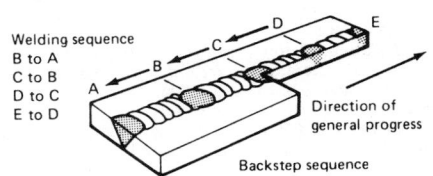

Backstep sequence

in which the direction of general progress is opposite to that in which the individual increments are welded.

backward extrusion. Same as indirect extrusion. See *extrusion*.

back weld. A weld deposited at the back of a single-groove weld.

baghouse. A chamber containing bags for filtering solids out of gases.

bail. Hoop or arched connection between the crane hook and ladle or between crane hook and mold trunnions.

bainite. A metastable aggregate of ferrite and cementite resulting from the transformation of austenite at temperatures below the pearlite range but above M_s. Bainite formed in the upper part of the bainite transformation range has a feathery appearance; bainite formed in the lower part of the range has an acicular appearance resembling that of tempered martensite.

baking. (1) Heating to a low temperature in order to remove gases. (2) Curing or hardening surface coatings such as paints by exposure to heat. (3) Heating to drive off moisture, as in baking of sand cores after molding.

balance. (1) (dynamic) Condition existing where the principal inertial axis of a body coincides with its rotational axis. (2) (static) Condition existing where the center of gravity of a body lies on its rotational axis.

ball burnishing. (1) Same as *ball sizing*. (2) Removing burrs and polishing small stampings and small machined parts by *tumbling* in the presence of metal balls.

ball mill. A machine consisting of a rotating hollow cylinder partly filled with metal balls (usually hardened steel or white cast iron) or sometimes pebbles; used to pulverize crushed ores or other substances such as pigments.

ball sizing. Sizing and finishing a hole by forcing a ball of suitable size, finish and hardness through the hole or by using a burnishing bar or broach consisting of a series of spherical lands of gradually increasing size coaxially arranged. Also called *ball burnishing*, and sometimes ball broaching.

banded structure. A segregated structure consisting of alternating nearly parallel bands of different composition, typically aligned in the direction of primary hot working.

band mark. An indentation in carbon steel sheet or strip caused by external pressure on the packaging band around cut lengths or coils; it may occur in handling, transit or storage.

bands. (1) Hot rolled steel strip, usually produced for rerolling into thinner sheet or strip. Also known as hot bands or band steel. (2) See *electron bands*.

bar. (1) An obsolete unit of pressure equal to 100 kPa. (2) An elongated rolled metal product that is relatively thick and narrow; most bars have simple, uniform cross sections such as rectangular, square, round, oval or hexagonal. Also known as barstock.

bare electrode. A filler-metal arc welding electrode in the form of a wire or rod having no coating other than that incidental to the drawing of the wire or to its preservation.

bar folder. A machine in which a folding bar or wing is used to bend a metal sheet whose edge is clamped between the upper folding leaf and the lower stationary jaw into a narrow, sharp, close and accurate fold along the edge. It is also capable of making rounded folds such as those used in wiring. A universal folder is more versatile in that it is limited to width only by the dimensions of the sheet.

bark. The decarburized layer just beneath the scale that results from heating steel in an oxidizing atmosphere.

Barkhausen effect. The sequence of abrupt changes in magnetic induction occurring when the magnetizing force acting on a ferromagnetic specimen is varied.

barrel cleaning. Mechanical or electrolytic cleaning of metal in rotating equipment.

barrel finishing. Improving the surface finish of metal objects or parts by processing them in rotating equipment along with abrasive particles that may be suspended in a liquid.

barreling. Convexity of the surfaces of cylindrical or conical bodies, often produced unintentionally during upsetting or as a natural consequence during compression testing.

barrel plating. Plating articles in a rotating container, usually a perforated cylinder that operates at least partially submerged in a solution.

barstock. Same as *bar*.

basal plane. A plane perpendicular to the principal axis (*c* axis) in a tetragonal or hexagonal structure.

base. (1) The surface on which a single-point tool rests when held in a tool post. Also known as heel. See sketch accompanying *single-point tool*. (2) In forging—see *anvil*. (3) A chemical substance that yields hydroxyl ions (OH^-) when dissolved in water.

base bullion. Crude lead containing recoverable silver, with or without gold.

base metal. (1) The metal present in the largest proportion in an alloy; brass, for example, is a copper-base alloy. (2) The metal to be brazed, cut, soldered or welded. (3) After welding, that part of the metal which was not melted. (4) A metal that readily oxidizes, or that dissolves to form ions. Contrast with *noble metal* (2).

basic bottom and lining. The inner bottom and lining of a melting furnace, consisting of materials such as crushed burned dolomite, magnesite, magnesite bricks or basic slag that give a basic reaction at the operating temperature.

basic steel. Steel melted in a furnace with a *basic bottom and lining* and under a slag containing an excess of a basic substance such as magnesia or lime.

basin. Same as *pouring basin*.

basis metal. The original metal to which one or more coatings are applied.

batch. See *lot*.

Bauschinger effect. For both single-crystal and polycrystalline metals, any change in stress-strain characteristics that can be ascribed to changes in the microscopic stress distribution within the metal, as distinguished from changes caused by strain hardening. In the narrow sense, the process whereby plastic deformation in one direction causes a reduction in yield strength when stress is applied in the opposite direction.

Bayer process. A process for extracting alumina from bauxite ore before the electrolytic reduction. The bauxite is digested in a solution of sodium hydroxide, which converts the alumina to soluble aluminate. After the "red mud" residue has been filtered out, aluminum hydroxide is precipitated, filtered out and calcined to alumina.

beach marks. Progression marks on a fatigue fracture surface that indicate successive positions of the advancing crack front. The classic appearance is of irregular elliptical or semielliptical rings, radiating outward from one or more origins. Beach marks (also known as clamshell marks or tide marks) are typically found on service fractures where the part is loaded randomly, intermittently, or with periodic variations in mean stress or alternating stress.

beaded flange. A flange reinforced by a low ridge, used mostly around a hole.

beading. Raising a ridge or projection on sheet metal.

bead weld. See preferred term *surfacing weld*.

bearing stress. The shear load on a mechanical joint (such as a pinned or riveted joint) divided by the effective bearing area. The effective bearing area of a riveted joint, for example, is the sum of the diameters of all rivets times the thickness of the loaded member.

bed. (1) The stationary portion of a press structure that usually rests on the floor or foundation, forming the support for the remaining parts of the press and the pressing load. The *bolster* and sometimes the lower die are mounted on the top surface of the bed. (2) For machine tools, the portion of the main frame that supports the tools, the work, or both.

Beilby layer. A layer of metal disturbed by mechanical working presumed to be without regular crystalline structure (amorphous); originally applied to grain boundaries.

bel. A unit denoting the ratio of power levels of signals or sound. The number of bels may be given as the common logarithm of the ratio of powers:

$$n = \log (p_1/p_2)$$

where p_1 and p_2 are the initial and final power levels.

belt grinding. Grinding with an *abrasive belt*.

bench press. Any small press that can be mounted on a bench or table.

bend allowance. The length of the arc of the neutral axis between the tangent points of a bend.

bend angle. The angle through which a bending operation is performed.

bender. Term denoting a die impression, tool or mechanical device designed to bend forging stock to conform to the general configuration of die impressions to be subsequently used.

bending brake. A *press brake* used for bending.

bending moment. The algebraic sum of the couples or the moments of the external forces, or both, to the left or right of any section on a member subjected to bending by couples or transverse forces, or both.

bending rolls. Two or three rolls with an adjustment for imparting a desired curvature in sheet or strip metal.

bend radius. (1) The inside radius of a bent section. (2) The radius of a tool around which metal is bent during fabrication.

bend tangent. A tangent point at which a bending arc ceases or changes.

bend test. A test for determining relative ductility of metal that is to be formed (usually sheet, strip, plate or wire) and for determining soundness and toughness of metal (after welding, for example). The specimen is usually bent over a specified diameter through a specified angle for a specified number of cycles.

beneficiation. Concentration or other preparation of ore for smelting.

bentonite. A colloidal claylike substance derived from the decomposition of volcanic ash composed chiefly of the minerals of the montmorillonite family. Western bentonite is slightly alkaline; southern bentonite is usually slightly acidic.

bessemer process. A process for making steel by blowing air through molten pig iron contained in a refractory lined vessel so as to remove by oxidation most of the carbon, silicon and manganese. This process is essentially obsolete in the United States.

beta ray. A ray of electrons emitted during the spontaneous disintegration of certain atomic nuclei.

beta structure. A Hume-Rothery designation for structurally analogous body-centered cubic phases (similar to beta brass) or electron compounds that have ratios of three valence electrons to two atoms. Not to be confused with a beta phase on a constitution diagram.

Betts process. A process for the electrolytic refining of lead in which the electrolyte contains lead fluosilicate and fluosilicic acid.

bevel. See preferred term, *corner angle*, and also sketch accompanying *face mill*.

bevel angle. The angle formed between the prepared edge of a member and a plane perpendicular to the surface of the member.

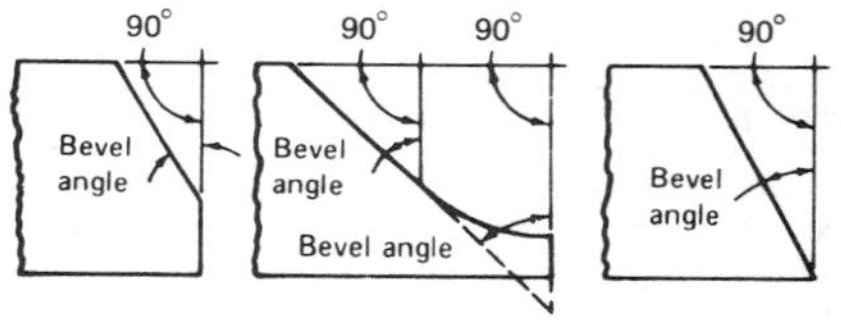

bevel flanging. Same as *flaring*.

biaxiality. In a *biaxial stress* state, the ratio of the smaller to the larger principal stress.

biaxial stress. A state of stress in which only one of the *principal stresses* is zero, the other two usually being in tension.

billet. (1) A solid semifinished round or square product that has been hot worked by forging, rolling or extrusion; usually smaller than a *bloom*. (2) A general term for wrought starting stock used in making forgings or extrusions.

billet mill. A primary rolling mill used for making steel billets.

binary alloy. An alloy containing only two component elements.

binder. (1) In founding, a material, other than water, added to foundry sand to bind the particles together, sometimes with the use of heat. (2) In powder metallurgy, a cementing medium: either a material added to the powder to increase the green strength of the compact, which is expelled during sintering; or a material (usually of relatively low melting point) added to a powder mixture for the specific purpose of cementing together powder particles that alone would not sinter into a strong body.

bipolar electrode. An *electrode* in an electrolytic cell that is not mechanically connected to the power supply, but is so placed in the electrolyte, between the anode and cathode, that the part nearer the anode becomes cathodic and the part nearer the cathode becomes anodic. Also called intermediate electrode.

bipolar field. A longitudinal magnetic field that creates two magnetic poles within a piece of material. Compare with *circular field*.

biscuit. (1) An upset blank for drop forging. (2) A small cake of primary metal (such as uranium made from uranium tetrafluoride and magnesium by bomb reduction). Compare with *derby* and *dingot*.

black annealing. Box annealing or pot annealing of ferrous alloy sheet, strip or wire. See *box annealing*.

blackheart malleable. See *malleable cast iron*.

blacking. Carbonaceous materials such as plumbago, graphite or powdered carbon used in coating pouring ladles, molds, runners, pig beds.

black light. Electromagnetic radiation not visible to the human eye. The portion of the spectrum generally used in fluorescent inspection falls in the ultraviolet region between 330 and 400 nm, with the peak at 365 nm.

black oxide. A black finish on a metal produced by immersing it in hot oxidizing salts or salt solutions.

blade-setting angle. See preferred term, *cone angle*.

blank. (1) In forming, a piece of sheet material, produced in cutting dies, that is usually subjected to further press operations. (2) A pressed, presintered or fully sintered powder metallurgy compact, usually in the unfinished condition and requiring cutting, machining or some other operation to produce the final shape. (3) A piece of stock from which a forging is made; often called a *slug* or *multiple*.

blank carburizing. Simulating the carburizing operation without introducing carbon. This is usually accomplished by using an inert material in place of the carburizing agent, or by applying a suitable protective coating to the ferrous alloy.

blankholder. The part of a drawing or forming die that holds the workpiece against the draw ring to control metal flow.

blanking. Producing desired shapes from metal to be used for forming or other operations, usually by punching.

blank nitriding. Simulating the nitriding operation without introducing nitrogen. This is usually accomplished by using an inert material in place of the nitriding agent, or by applying a suitable protective coating to the ferrous alloy.

blast furnace. A shaft furnace in which solid fuel is burned with an air blast to smelt ore in a continuous operation. Where the temperature must be high, as in the production of pig iron, the air is preheated. Where the temperature can be lower, as in smelting of copper, lead and tin ores, a smaller furnace is economical, and preheating of the blast is not required.

blasting. Cleaning or finishing metals by impingement with abrasive particles moving at high speed and usually carried by gas or liquid or thrown centrifugally from a wheel.

blemish. A nonspecific quality control term designating an imperfection that mars the appearance of a part but does not detract from its ability to perform its intended function.

blending. In powder metallurgy, the thorough intermingling of powders of the same nominal composition (not to be confused with *mixing*).

blind riser. A *riser* that does not extend through the top of the mold.

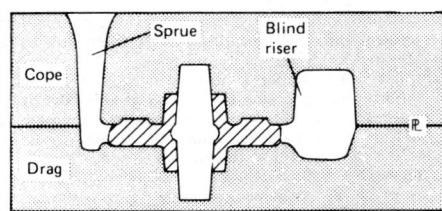

blister. A raised area, often dome shaped, resulting from (*a*) loss of adhesion between a coating or deposit and the basis metal or (*b*) delamination under the pressure of expanding gas trapped in a metal in a near subsurface zone. Very small blisters may be called pinhead blisters or pepper blisters.

blister copper. An impure intermediate product in the refining of copper, produced by blowing copper *matte* in a converter, the name being derived from the large blisters on the cast surface that result from the liberation of SO_2 and other gases.

block brazing. An obsolete brazing process in which the joint was heated using hot blocks.

blocker. The impression in the dies (often one of a series of impressions in a single die set) that imparts to the forging an intermediate shape, preparatory to forging of the final shape. Also called blocking impression.

blocker-type forging. A forging that approximates the general shape of the final part with relatively generous finish allowance and radii. Such forgings are sometimes specified to reduce die costs where only a few forgings are desired and the cost of machining each part to final shape is not excessive.

blocking. In forging, a preliminary operation performed in closed dies, usually hot, to position metal properly so that in the finish operation the dies will be filled correctly.

blocking impression. Same as *blocker*.

block sequence. A welding sequence in which separated lengths of a continuous multiple-pass weld are partly or completely built up in cross section before intervening lengths are deposited. Compare with *cascade sequence*.

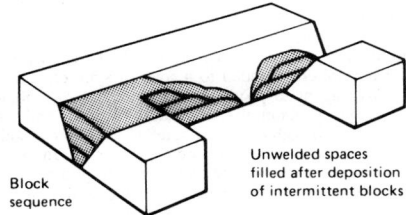

Block sequence

Unwelded spaces filled after deposition of intermittent blocks

bloom. (1) A semifinished hot rolled product, rectangular in cross section, produced on a blooming mill. See also *billet*. For steel, the width of a bloom is not more than twice the thickness, and the cross-sectional area is usually not less than about 230 cm² (36 in.²). Steel blooms are sometimes made by forging. (2) A visible exudation or efflorescence on the surface of an electroplating bath. (3) A bluish fluorescent cast to a painted surface caused by deposition of a thin film of smoke, dust or oil. (4) A loose, flowerlike corrosion product that forms when certain metals are exposed to a moist environment.

bloomer. The mill or other equipment used in reducing steel ingots to blooms.

blooming mill. A primary rolling mill used to make blooms.

blotter. In grinding, a disk of compressible material, usually blotting-paper stock, used between the grinding wheel and its flanges to avoid concentrated stresses.

blowhole. A hole in a casting or a weld caused by gas entrapped during solidification.

blowpipe. A welding or cutting torch.

blue annealing. Heating hot rolled ferrous sheet in an open furnace to a temperature within the transformation range and then cooling in air, in order to soften the metal. The formation of a bluish oxide on the surface is incidental.

blue brittleness. Brittleness exhibited by some steels after being heated to some temperature within the range of about 200 to 370 °C (400 to 700 °F), particularly if the steel is worked at the elevated temperature. Killed steels are virtually free of this kind of brittleness.

blue dip. A solution containing a mercury compound, once widely used to deposit mercury on a metal by immersion, usually prior to silver plating.

bluing. Subjecting the scale-free surface of a ferrous alloy to the action of air, steam or other agents at a suitable temperature, thus forming a thin blue film of oxide and improving the appearance and resistance to corrosion. NOTE: This term is ordinarily applied to sheet, strip or finished parts. It is used also to denote the heating of springs after fabrication in order to improve their properties.

board hammer. A type of forging hammer in which the upper die and ram are attached to "boards" that are raised to the striking position by power-driven rollers and let fall by gravity. See *drop hammer*.

bolster. A plate to which dies may be fastened, the assembly being secured to the top surface of a press bed. In mechanical forging, such a plate is also attached to the ram.

bond. (1) In grinding wheels and other relatively rigid abrasive products, the material that holds the abrasive grains together. (2) In welding, brazing or soldering, the junction of joined parts. Where filler metal is used, it is the junction of the fused metal and the heat-affected base metal. (3) In an adhesive bonded or diffusion bonded joint, the line along which the faying surfaces are joined together.

book mold. A split permanent mold hinged like a book.

bore. A hole or cylindrical cavity produced by a single-point or multipoint tool other than a drill.

boring. A machining method using single-point tools on internal surfaces of revolution.

bort. Industrial diamond.

bosh. (1) The section of a blast furnace extending upward from the tuyeres to the plane of maximum diameter. (2) A lining of quartz that builds up during the smelting of copper ores and that decreases the diameter of the furnace at the tuyeres. (3) A tank, often with sloping sides, used for washing metal parts or for holding cleaned parts.

boss. A relatively short protrusion or projection from the surface of a forging or casting, often cylindrical in shape.

bottom board. A flat base for holding the flask in making sand molds.

bottom drill. A flat-ended twist drill used to convert a cone at the bottom of a drilled hole into a cylinder.

bottoming tap. A tap with a *chamfer* of 1 to 1½ threads in length.

bottom pipe. An oxide-lined fold or cavity at the butt end of a slab, bloom or billet; formed by folding the end of an ingot over on itself during primary rolling. Bottom pipe is not *pipe*, in that it is not a shrinkage cavity, and in that sense, the term is a misnomer. Bottom pipe is similar to *extrusion pipe*. It is normally discarded when the slab, bloom or billet is cropped following primary reduction.

bowing. Deviation from flatness.

box annealing. Annealing a metal or alloy in a sealed container under conditions that minimize oxidation. In box annealing a ferrous alloy, the charge is usually heated slowly to a temperature below the transformation range, but sometimes above or within it, and is then cooled slowly; this process is also called close annealing or pot annealing. See *black annealing*.

boxing. Continuing a fillet weld around a corner as an extension of the principal weld. Also called an end return.

brake. A device for bending sheet metal to a desired angle.

brale. A diamond penetrator of specified spheroconical shape used with a Rockwell hardness tester. This penetrator is used for the A, C, D and N scales for testing hard metals.

brass. An alloy consisting mainly of copper (over 50%) and zinc, to which smaller amounts of other elements may be added.

braze welding. A method of welding by using a filler metal having a liquidus above 450 °C (840 °F) and below the solidus of the base metals. Unlike *brazing*, in braze welding, the filler metal is not distributed in the joint by capillary attraction.

brazing. A group of welding processes that join solid materials together by heating them to a suitable temperature and using a filler metal having a liquidus above 450 °C (840 °F) and below the solidus of the base materials. The filler metal is distributed between the closely fitted surfaces of the joint by capillary attraction.

brazing alloy. See preferred term *brazing filler metal*.

brazing filler metal. A nonferrous filler metal used in *brazing* and *braze welding*.

brazing sheet. Brazing filler metal in sheet form or flat-rolled metal clad with brazing filler metal on one or both sides.

breakdown. (1) An initial rolling or drawing operation, or a series of such operations, for the purpose of reducing a casting or extruded shape prior to the finish reduction to desired size. (2) A preliminary press-forging operation.

breaking stress. Same as *fracture stress* (1).

breaks. Creases or ridges usually in "untempered" or in aged material where the yield point has been exceeded. Depending on the origin of the breaks, they may be termed *cross breaks*, *coil breaks*, edge breaks or *sticker breaks*.

bridge die. A two-section extrusion die capable of producing tubing or intricate hollow shapes without the use of a separate mandrel. Metal separates into two streams as it is extruded past a bridge section, which is attached to the main die section and holds a stub mandrel in the die opening; the metal then is rewelded by extrusion pressure before it enters the die opening. Compare with *porthole die*.

bridging. (1) Premature solidification of metal across a mold section before the metal below or beyond solidifies. (2) Solidification of slag within a cupola at or just above the tuyeres. (3) Welding or mechanical locking of the charge in a downfeed melting or smelting furnace. (4) In powder metallurgy, the formation of arched cavities in a powder mass. (5) In soldering, an unintended solder connection between two or more conductors, either securely or by mere contact. Also called a crossed joint or solder short.

bright annealing. Annealing in a protective medium to prevent discoloration of the bright surface.

bright dip. A solution that produces, through chemical action, a bright surface on an immersed metal.

brightener. An agent or combination of agents added to an electroplating bath to produce a lustrous deposit.

bright finish. A high-quality finish produced on ground and polished rolls. Suitable for electroplating.

bright plate. An electrodeposit that is lustrous in the as-plated condition.

bright range. The range of current densities, other conditions being constant, within which a given electroplating bath produces a bright plate.

Brillouin zones. See *electron bands*.

Brinell hardness test. A test for determining the hardness of a material by forcing a hard steel or carbide ball of specified diameter into it under a specified load. The result is expressed as the Brinell hardness number, which is the value obtained by dividing the applied load in kilograms by the surface area of the resulting impression in square millimetres.

brinelling. Evenly spaced dents in a raceway of a rolling-element bearing that occur when the bearing assembly is subjected to a force or impact load great enough to cause the rolling elements to indent the raceway surface. Also called true brinelling. Compare with *false brinelling*.

brittle crack propagation. A very sudden propagation of a crack with the absorption of no energy except that stored elastically in the body. Microscopic examination may reveal some deformation even though it is not noticeable to the unaided eye.

brittle fracture. Separation of a solid accompanied by little or no macroscopic plastic deformation. Typically, brittle fracture occurs by rapid crack propagation with less expenditure of energy than for ductile fracture.

brittleness. The quality of a material that leads to crack propagation without appreciable plastic deformation.

broach. A bar-shaped cutting tool provided with a series of cutting edges or teeth that increase in size or change in shape from the starting to finishing end. The tool cuts in the axial direction when pushed or pulled and is used to shape either holes or outside surfaces.

bronze. A copper-rich copper-tin alloy with or without small proportions of other elements such as zinc and phosphorus. By extension, certain copper-base alloys containing considerably less tin than other alloying elements, such as manganese bronze (copper-zinc plus manganese, tin and iron) and leaded tin bronze (copper-lead plus tin and sometimes zinc). Also, certain other essentially binary copper-base alloys containing no tin, such as aluminum bronze (copper-aluminum), silicon bronze (copper-silicon) and beryllium bronze (copper-beryllium). Also, trade designations for certain specific copper-base alloys that are actually brasses, such as architectural bronze (57 Cu, 40 Zn, 3 Pb) and commercial bronze (90 Cu, 10 Zn).

bronzing. (1) Applying a chemical finish to copper or copper-alloy surfaces to alter the color. (2) Plating a copper-tin alloy on various materials.

brush anodizing. An *anodizing* process similar to *brush plating*.

brush plating. Plating with a concentrated solution or gel held in or fed to an absorbing medium, pad or brush carrying the anode (usually insoluble). The brush is moved back and forth over the area of the cathode to be plated.

brush polishing (electrolytic). A method of *electropolishing* in which the electrolyte is applied with a pad or brush in contact with the part to be polished.

buckle. (1) A local waviness in metal bar or sheet, usually transverse to the direction of rolling. (2) An indentation in a casting resulting from expansion of molding sand into the mold cavity.

buckling. Producing a bulge, bend, bow, kink or other wavy condition by compressively stressing a beam, column, plate, bar or sheet.

Bucky diaphragm. An x-ray scatter-reducing device originally intended for medical radiography but also applicable to industrial radiography in some circumstances. Thin strips of lead, with their widths held parallel to the primary radiation, are used to absorb scattered radiation preferentially; the array of strips is in motion during exposure, to prevent formation of a pattern on the film.

buffer. A substance whose purpose is to maintain a constant hydrogen-ion concentration in water solutions, even where acids or alkalis are added. Each buffer has a characteristic limited range of pH over which it is effective.

buffing. Developing a lustrous surface by contacting the work with a rotating *buffing wheel*.

buffing wheel. Buff sections assembled to the required face width for use on a rotating shaft between flanges. Sometimes called a buff.

buff section. A number of fabric, paper or leather disks with concentric center holes held together by various types of sewing to provide degrees of flexibility or hardness. These sections are assembled to make wheels for polishing.

builder. A material, such as an alkali, a buffer or a water softener, added to a soap or synthetic surface-active agent to produce a mixture having enhanced detergency. Examples: (1) alkalis—caustic soda, soda ash and trisodium phosphate; (2) *buffers*—sodium metasilicate and borax; and (3) water softeners—sodium tripolyphosphate, sodium tetraphosphate, sodium hexametaphosphate and ethylene diamine tetra-acetic acid.

buildup. Excessive electrodeposition that occurs on high-current-density areas, such as corners or edges.

buildup sequence. The order in which weld beads are deposited, generally designated in cross section as shown in the accompanying illustration.

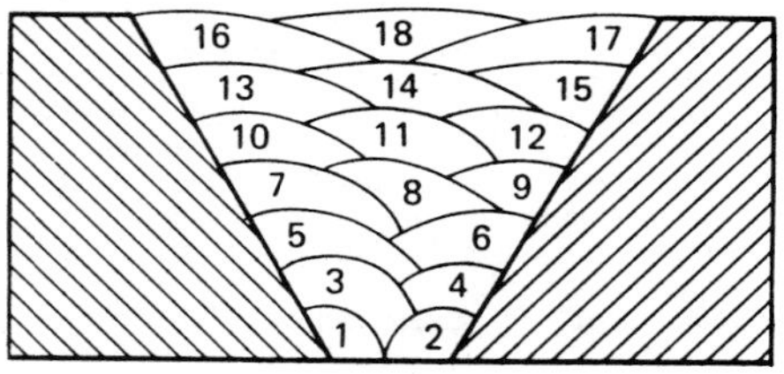

Buildup sequence

built-up edge. Chip material adhering to the tool face adjacent to the cutting edge during cutting.

bulging. Expanding the walls of a cup, shell or tube with an internally expanded segmented punch or a punch composed of air, liquids or semiliquids such as waxes, rubber and other elastomers.

bull block. A machine with a power-driven revolving drum for cold drawing wire through a drawing die as the wire winds around the drum.

bulldozer. A horizontal machine, usually mechanical, having two bull gears with eccentric pins, two connecting links to a ram, and dies to perform bending, forming and punching of narrow plate and bars. Railroad car sills are formed with a bulldozer.

bullion. (1) A semirefined alloy containing sufficient precious metal to make recovery profitable. (2) Refined gold or silver, uncoined.

bull's-eye structure. The microstructure of malleable or ductile cast iron when graphite nodules are surrounded by a ferrite layer in a pearlitic matrix.

bumper. A machine used for packing molding sand in a flask by repeated jarring or jolting.

bumping. (1) Forming a dish in metal by means of many repeated blows. (2) Forming a head. (3) Setting the seams on sheet metal parts. (4) Ramming sand in a flask by repeated jarring and jolting.

burned deposit. A dull, nodular electrodeposit resulting from excessive current density.

burned-on sand. A mixture of sand and cast metal adhering to the surface of a casting. In some instances, may resemble *metal penetration*.

burning. (1) Permanently damaging a metal or alloy by heating to cause either incipient melting or intergranular oxidation. See *overheating*. (2) In grinding, getting the work hot enough to cause discoloration or to change the microstructure by tempering or hardening.

burnishing. Smoothing surfaces through frictional contact between the work and some hard pieces of material such as hardened metal balls.

burn-off. (1) Unintentional removal of an autocatalytic deposit from a nonconducting substrate, during subsequent electroplating operations, owing to the application of excessive current or a poor contact area. (2) Removal of volatile lubricants such as metallic stearates from metal powder compacts by heating immediately prior to sintering. (3) See *melting rate*.

burr. (1) A turned-over edge on work resulting from cutting, punching or grinding. (2) A rotary tool having teeth similar to those on hand files.

burring. Same as *deburring*.

bushing. A bearing or guide.

buster. A pair of shaped dies used to combine preliminary forging operations such as edging and blocking, or to loosen the scale.

butler finish. A semilustrous metal finish composed of fine, uniformly distributed parallel lines, usually produced with a soft abrasive wheel; similar in appearance to the traditional hand-rubbed finish on silver.

buttering. A form of surfacing in which one or more layers of weld metal are deposited on the groove face of one member (for example, a high-alloy weld deposit on steel base metal that is to be welded to a dissimilar base metal). The buttering provides a suitable transition weld deposit for subsequent completion of the butt weld.

butt joint. A joint between two abutting members lying approximately in the same plane. A welded butt joint may contain a variety of grooves. See *groove weld*.

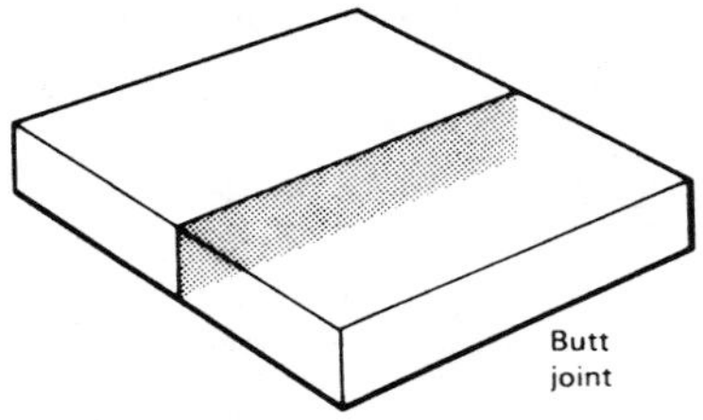

button. (1) A globule of metal remaining in an assaying crucible or cupel after fusion has been completed. (2) That part of a weld that tears out in destructive testing of a spot, seam or projection welded specimen.

butt seam welding. See *seam welding*.

butt welding. Welding a butt joint.

C

cake. (1) A copper or copper alloy casting, rectangular in cross section, used for rolling into sheet or strip. (2) A coalesced mass of unpressed metal powder.

calcination. Heating ores, concentrates, precipitates or residues to decompose carbonates, hydrates or other compounds.

calomel electrode (calomel half cell). A secondary reference electrode of the composition: $Pt/Hg\text{-}Hg_2Cl_2/KCl$ solution. For $1.0N$ KCl solution, its potential versus a hydrogen electrode at 25 °C and one atmosphere is +0.281 V.

calorizing. Imparting resistance to oxidation to an iron or steel surface by heating in aluminum powder at 800 to 1000 °C (1470 to 1830 °F).

camber. (1) Deviation from edge straightness, usually referring to the greatest deviation of side edge from a straight line. (2) Sometimes used to denote crown in rolls where the center diameter has been increased to compensate for deflection caused by the rolling pressure.

cam press. A mechanical press in which one or more of the slides are operated by cams; usually a double-action press in which the blankholder slide is operated by cams through which the dwell is obtained.

canning. (1) A dished distortion in a flat or nearly flat surface, sometimes referred to as oil canning. (2) Enclosing a highly reactive metal within a relatively inert one for the purpose of hot working without undue oxidation of the active metal.

capillary attraction. The combined force of adhesion and cohesion that causes liquids, including molten metals, to flow between very closely spaced solid surfaces, even against gravity.

capped steel. A type of steel similar to rimmed steel, usually cast in a bottle-top ingot mold, in which the application of a mechanical or a chemical cap renders the rimming action incomplete by causing the top metal to solidify. The surface condition of capped steel is much like that of rimmed steel, but certain other characteristics are intermediate between those of *rimmed steel* and those of *semikilled steel*.

capping. Partial or complete separation of a powder

metallurgy compact into two or more portions by cracks that originate near the edges of the punch faces and that proceed diagonally into the compact.

carbide. A compound of carbon with one or more metallic elements.

carbide tools. Cutting or forming tools, usually made from tungsten, titanium, tantalum, or niobium carbides, or a combination of them, in a matrix of cobalt, nickel, or other metals. Carbide tools are characterized by high hardnesses and compressive strengths and may be coated to improve wear resistance.

carbon dioxide welding. *Gas metal-arc welding* using carbon dioxide as the shielding gas.

carbon edges. Carbonaceous deposits in a wavy pattern along the edges of a sheet or strip; also known as snaky edges.

carbon electrode. A carbon or graphite rod used in carbon-arc equipment, such as in carbon-arc welding or cutting torches.

carbon equivalent. (1) For cast iron, an empirical relationship of the total carbon, silicon and phosphorus contents expressed by the formula:

$$CE = TC + \tfrac{1}{3}(Si + P)$$

(2) For rating of weldability:

$$CE = C + \frac{Mn}{6} + \frac{Cr + Mo + V}{5} + \frac{Ni + Cu}{15}$$

carbonitriding. A case hardening process in which a suitable ferrous material is heated above the lower transformation temperature in a gaseous atmosphere of such composition as to cause simultaneous absorption of carbon and nitrogen by the surface and, by diffusion, create a concentration gradient. The process is completed by cooling at a rate that produces the desired properties in the workpiece.

carbonization. Conversion of an organic substance into elemental carbon. (Should not be confused with *carburization*.)

carbon potential. A measure of the ability of an environment containing active carbon to alter or maintain, under prescribed conditions, the carbon level of the steel. NOTE: In any particular environment, the carbon level attained will depend on such factors as temperature, time and steel composition.

carbon restoration. Replacing the carbon lost in the surface layer from previous processing by carburizing this layer to substantially the original carbon level. Sometimes called recarburizing.

carbon steel. Steel having no specified minimum quantity for any alloying element (other than the commonly accepted amounts of manganese, silicon and copper) and containing only an incidental amount of any element other than carbon, silicon, manganese, copper, sulfur and phosphorus.

carbonyl powder. A metal powder prepared by the thermal decomposition of a metal carbonyl.

carburizing. Absorption and diffusion of carbon into solid ferrous alloys by heating, to a temperature usually above Ac_3, in contact with a suitable carbonaceous material. A form of *case hardening* that produces a carbon gradient extending inward from the surface, enabling the surface layer to be hardened either by quenching directly from the carburizing temperature or by cooling to room temperature, then reaustenitizing and quenching.

carburizing flame. A gas flame that will introduce carbon into some heated metals, as during a gas welding operation. A carburizing flame is a *reducing flame*, but a reducing flame is not necessarily a carburizing flame.

cascade sequence. A welding sequence in which a continuous multiple-pass weld is built up by depositing weld beads in overlapping layers, usually laid in a *backstep sequence*. Compare with *block sequence*.

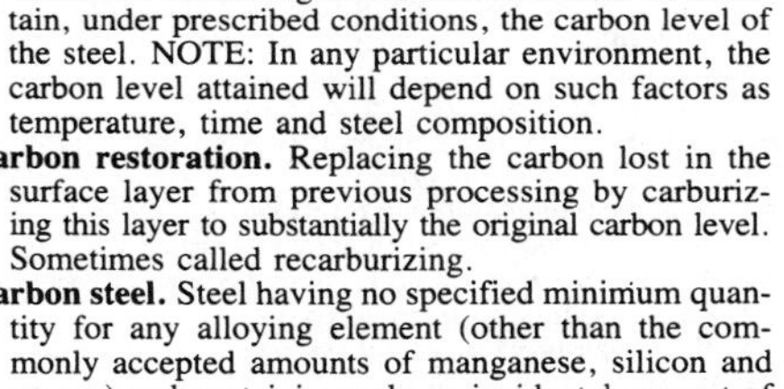

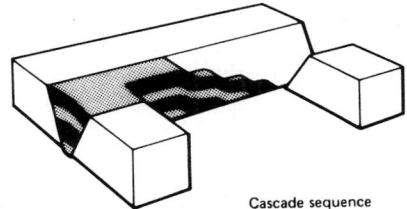

Cascade sequence

case. That portion of a ferrous alloy, extending inward from the surface, whose composition has been altered so that it can be *case hardened*. Typically considered to be the portion of the alloy (*a*) whose composition has been measurably altered from the original composition, (*b*) that appears dark on an etched cross section, or (*c*) that has a hardness, after hardening, equal to or greater than a specified value. Contrast with *core* (2).

case hardening. A generic term covering several processes applicable to steel that change the chemical composition of the surface layer by absorption of carbon, nitrogen, or a mixture of the two and, by diffusion, create a concentration gradient. The processes commonly used are *carburizing* and *quench hardening*; *cyaniding*; *nitriding*; and *carbonitriding*. The use of the applicable specific process name is preferred.

cassette. A lighttight holder, used to contain radiographic films during exposure to x-rays or gamma rays, that may or may not contain intensifying or filter screens, or both. A distinction is often made between a cassette, which has positive means for ensuring contact between screens and film and is usually rigid, and an exposure holder, which is rather flexible.

CASS test. Abbreviation for *copper-accelerated salt-spray test*.

cast. See *die proof*.

cast-alloy tool. A cutting tool made by casting a cobalt-base alloy and used at machining speeds between those for high speed steels and sintered carbides.

casting. (1) An object at or near finished shape obtained by solidification of a substance in a mold. (2) Pouring molten metal into a mold to produce an object of desired shape.

casting copper. Fire-refined tough pitch copper usually cast from melted secondary metal into ingot bars only, and used for making foundry castings but not wrought products.

casting shrinkage. (1) Liquid shrinkage—the reduction in volume of liquid metal as it cools to the liquidus. (2) Solidification shrinkage—the reduction in volume of metal from beginning to end of solidification. (3) Solid shrinkage—the reduction in volume of metal from the solidus to room temperature.

casting strains. Strains in a casting caused by *casting stresses* that develop as the casting cools.

casting stresses. Residual stresses set up when the shape of a casting impedes contraction of the solidified casting during cooling.

cast iron. A generic term for a large family of cast ferrous alloys in which the carbon content exceeds the solubility of carbon in austenite at the eutectic temperature. Most cast irons contain at least 2% carbon, plus silicon and sulfur, and may or may not contain other alloying elements. For the various forms *gray cast iron, white cast iron, malleable cast iron* and *ductile cast iron*, the word "cast" is often left out, resulting in "gray iron," "white iron," "malleable iron" and "ductile iron," respectively.

cast steel. Steel in the form of *castings*.

cast structure. The metallographic structure of a *casting* evidenced by shape and orientation of grains and by segregation of impurities.

catalyst. A substance capable of changing the rate of a reaction without itself undergoing any net change.

catastrophic failure. Sudden failure of a component or assembly that frequently results in extensive secondary damage to adjacent components or assemblies.

cathode. The electrode where electrons enter an operating system such as a battery, an electrolytic cell, an x-ray tube or a vacuum tube. In the first of these, it is positive; in the other three, negative. In a battery or electrolytic cell, it is the electrode where reduction occurs. Contrast with *anode*.

cathode compartment. In an electrolytic cell, the enclosure formed by a diaphragm around the cathode.

cathode copper. Copper deposited at the cathode in electrolytic refining.

cathode efficiency. *Current efficiency* at the cathode.

cathode film. The portion of solution in immediate contact with the cathode during electrolysis.

cathodic cleaning. *Electrolytic cleaning* in which the work is the cathode.

cathodic pickling. *Electrolytic pickling* in which the work is the cathode.

cathodic protection. Partial or complete protection of a metal from corrosion by making it a cathode, using either a galvanic or an impressed current. Contrast with *anodic protection*.

catholyte. The electrolyte adjacent to the cathode in an electrolytic cell; in a divided cell, the portion on the cathode side of the diaphragm.

cation. A positively charged ion; it flows to the cathode in electrolysis.

cationic detergent. A detergent in which the *cation* is the active part.

caustic cracking. A form of *stress-corrosion cracking* most frequently encountered in carbon steels or iron-chromium-nickel alloys that are exposed to concentrated hydroxide solutions at temperatures of 200 to 250 °C (400 to 480 °F).

caustic dip. A strongly alkaline solution into which metal is immersed for etching, for neutralizing acid or for removing organic materials such as greases or paints.

cavitation. The formation and instantaneous collapse of innumerable tiny voids or cavities within a liquid subjected to rapid and intense pressure changes. Cavitation produced by ultrasonic radiation is sometimes used to effect violent localized agitation. Cavitation caused by severe turbulent flow often leads to *cavitation damage*.

cavitation damage. Erosion of a solid surface through the formation and collapse of cavities in an adjacent liquid.

cavitation erosion. See preferred term, *cavitation damage*.

cell feed. The material supplied to the cell in the electrolytic production of metals.

cementation. Introduction of one or more elements into the outer portion of a metal object by means of diffusion at high temperature.

cement copper. Impure copper recovered by *chemical deposition* when iron (most often shredded steel scrap) is brought into prolonged contact with a dilute copper sulfate solution.

cemented carbide. A solid and coherent mass made by pressing and sintering a mixture of powders of one or more metallic carbides and a much smaller amount of a metal, such as cobalt, to serve as a binder.

cementite. A compound of iron and carbon, known chemically as iron carbide and having the approximate chemical formula Fe_3C. It is characterized by an orthorhombic crystal structure. When it occurs as a phase in steel, the chemical composition will be altered by the presence of manganese and other carbide-forming elements.

center drilling. Drilling a short, conical hole in the end of a workpiece—a hole to be used to center the workpiece for turning on a lathe.

centering plug. A plug fitting both spindle and cutter to ensure concentricity of the cutter mounting.

centerless grinding. Grinding the outside or inside of a workpiece mounted on rollers rather than on centers. The workpiece may be in the form of a cylinder or the frustum of a cone.

centrifugal casting. A casting made by pouring metal into a mold that is rotated or revolved.

ceramic tools. Cutting tools made from fused, sintered or cemented metallic oxides.

cereal. An organic *binder*, usually corn flour.

cermet. A powder metallurgy product consisting of ceramic particles bonded with a metal.

C-frame press. Same as *gap-frame press*.

CG iron. Same as *compacted graphite cast iron*.

chafing fatigue. Fatigue initiated in a surface damaged by rubbing aginst another body. See *fretting*.

chain-intermittent fillet welding. Depositing a line of intermittent fillet welds on each side of a member at a joint so that the increments on one side are essentially opposite those on the other. Contrast with *staggered-intermittent fillet welding*.

chamfer. (1) A beveled surface to eliminate an otherwise sharp corner. (2) A relieved angular cutting edge at a tooth corner.

chamfer angle. (1) The angle between a reference surface and the bevel. (2) On a milling cutter, the angle between a beveled surface and the axis of the cutter.

chamfering. Making a sloping surface on the edge of a member. Also called beveling. See *bevel angle*.

chaplet. Metal support that holds a core in place within a mold; molten metal solidifies around a chaplet and fuses it into the finished casting.

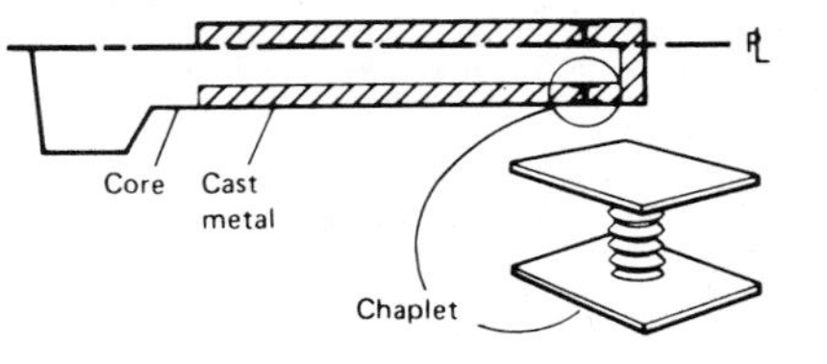

characteristic radiation. High-intensity single-wave-length x-rays, characteristic of the element emitting the rays, that appear in addition to continuous "white" radiation whenever the element is bombarded with electrons whose energy exceeds a specific critical value that is different for each element.

charge. (1) The materials fed into a furnace. (2) Weights of various liquid and solid materials put into a furnace during one feeding cycle.

charging. (1) For a lap, impregnating the surface with fine abrasive. (2) Placing materials into a furnace.

Charpy test. A pendulum-type single-blow impact test in which the specimen, usually notched, is supported at both ends as a simple beam and broken by a falling pendulum. The energy absorbed, as determined by the subsequent rise of the pendulum, is a measure of impact strength or notch toughness. Contrast with *Izod test*.

chase. To make a series of cuts each, except the first, following in the path of the cut preceding it, as in chasing a thread.

chatter. In machining or grinding, (1) a vibration of the tool, wheel or workpiece producing a wavy surface on the work and (2) the finish produced by such vibration.

checked edges. Sawtooth edges seen after hot rolling and/or cold rolling.

checkers. In a chamber associated with a metallurgical furnace, bricks stacked openly so that heat may be absorbed from the combustion products and later transferred to incoming air when the direction of flow is reversed.

checks. Numerous, very fine cracks in a coating or at the surface of a metal part. Checks may appear during processing or during service and are most often associated with thermal treatment or thermal cycling. Also called check marks, checking, *heat checks*.

cheek. The intermediate section of a flask that is used between the *cope* and the *drag* when molding a shape that requires more than one parting plane.

chelating agent. A substance used in metal finishing to control or eliminate certain metallic ions present in undesirable quantities.

chemical deposition. The precipitation or plating-out of a metal from solutions of its salts through the introduction of another metal or reagent to the solution.

chemically precipitated powder. Metal powder produced as a precipitate by chemical displacement.

chemical machining. Removing metal stock by controlled selective chemical dissolution.

chemical metallurgy. See *process metallurgy*.

chemical polishing. Improving the surface luster of a metal by chemical treatment.

chevron pattern. A fractographic pattern of radial marks (shear ledges) that look like nested letters "V"; sometimes called a herringbone pattern. Chevron patterns are typically found on brittle fracture surfaces in parts whose widths are considerably greater than their thicknesses. The points of the chevrons can be traced back to the fracture origin.

chill. (1) A metal or graphite insert embedded in the surface of a sand mold or core or placed in a mold cavity to increase the cooling rate at that point. (2) White iron occurring on a gray or ductile iron casting, such as the chill in the wedge test. Compare with *inverse chill*.

chill time. See *quench time*.

Chinese script. The angular microstructural form suggestive of Chinese writing and characteristic of the constituents α(Al-Fe-Si) and α(Al-Fe-Mn-Si) in cast aluminum alloys. A similar microstructure is found in cast magnesium alloys containing silicon as Mg_2Si.

chip breaker. (1) Notch or groove in the face of a tool parallel to the cutting edge, designed to break the continuity of the chips. (2) A step formed by an adjustable component clamped to the face of the cutting tool.

chipping. (1) Removing seams and other surface imperfections in metals manually with a chisel or gouge, or by a continuous machine, before further processing. (2) Similarly, removing excessive metal.

chips. Pieces of material removed from a workpiece by cutting tools or by an abrasive medium.

chlorination. (1) Roasting ore in contact with chlorine or a chloride salt to produce chlorides. (2) Removing dissolved gases and entrapped oxides by passing chlorine gas through molten metal such as aluminum and magnesium.

chromadizing. Improving paint adhesion on aluminum or aluminum alloys, mainly aircraft skins, by treatment with a solution of chromic acid. Also called chromodizing or chromatizing. Not to be confused with *chromating* or *chromizing*.

chromate treatment. A treatment of metal in a solution of a hexavalent chromium compound to produce a *conversion coating* consisting of trivalent and hexavalent chromium compounds.

chromating. Performing a *chromate treatment*.

Chromel. (1) A 90Ni-10Cr alloy used in thermocouples. (2) A series of nickel-chromium alloys, some with iron, used for heat-resistant applications.

chrome pickle. (1) Producing a chromate *conversion coating* on magnesium for temporary protection or for a paint base. (2) The solution that produces the conversion coating.

chromizing. A surface treatment at elevated temperature, generally carried out in pack, vapor or salt bath, in which an alloy is formed by the inward diffusion of chromium into the base metal.

chuck. A device for holding work or tools on a machine so that the part can be held or rotated during machining or grinding.

chucking lug. A projection forged or cast onto a part to act as a positive means of driving or locating the part during machining.

circle grinding. Either *cylindrical grinding* or *internal grinding*, the preferred terms.

circle shear. A shearing machine with two rotary disk cutters mounted on parallel shafts driven in unison and equipped with an attachment for cutting circles where the desired piece of material is inside the circle. It cannot be employed to cut circles where the desired material is outside the circle.

circular field. The magnetic field that (a) surrounds a nonmagnetic conductor of electricity, (b) is completely contained within a magnetic conductor of electricity or (c) both exists within and surrounds a magnetic conductor. Generally applied to the magnetic field within any magnetic conductor resulting from a current being passed through the part or through a section of the part. Compare with *bipolar field*.

clad metal. A composite metal containing two or three layers that have been bonded together. The bonding may have been accomplished by co-rolling, welding, casting, heavy chemical deposition or heavy electroplating.

clamshell marks. Same as *beach marks*.

classification. (1) The separation of ores into fractions according to size and specific gravity, generally in accordance with Stokes' law of sedimentation. (2) Separation of a metal powder into fractions according to particle size.

clay. An earthy or stony mineral aggregate consisting essentially of hydrous silicates of alumina, and which is plastic when sufficiently pulverized and wetted, rigid when dry, and vitreous when fired at a sufficiently high temperature. Clay minerals most commonly used in the foundry are montmorillonites and kaolinites.

cleanup allowance. See *finish allowance*.

clearance. (1) The gap or space between two mating parts. (2) Space provided between the relief of a cutting tool and the surface that has been cut.

clearance angle. The angle between a plane containing the flank of the tool and a plane passing through the cutting edge in the direction of relative motion between the cutting edge and the work. See sketches accompanying *face mill* and *single-point tool*.

clearance fit. Any of various classes of fit between mating parts where there is a positive allowance (gap) between the parts, even when they are made to the respective extremes of individual tolerances that ensure the tightest fit between the parts. Contrast with *interference fit*.

cleavage. Splitting (fracture) of a crystal on a crystallographic plane of low index.

cleavage fracture. A fracture, usually of a polycrystalline metal, in which most of the grains have failed by cleavage, resulting in bright reflecting facets. It is one type of *crystalline fracture* and is associated with low-energy brittle fracture. Contrast with *shear fracture*.

cleavage plane. A characteristic crystallographic plane or set of planes on which cleavage fracture easily occurs.

climb cutting. Analogous to *climb milling*.

climb milling. Milling in which the cutter moves in the direction of feed at the point of contact.

clip and shave. In forging, a dual operation in which one cutting surface in the clipping die removes the *flash* and then another shaves and sizes the piece.

close annealing. Same as *box annealing*.

closed-die forging. See *impression-die forging*.

closed dies. Forging or forming impression dies designed to restrict the flow of metal to the cavity within the die set, as opposed to open dies, in which there is little or no restriction to lateral flow.

closed pass. A pass of metal through rolls where the bottom roll has a groove deeper than the bar being rolled and the top roll has a collar fitting into the groove, thus producing the desired shape free from *flash* or fin.

close-tolerance forging. A forging held to unusually close dimensional tolerances. Often, little or no machining is required after forging.

cloudburst treatment. A form of *shot peening*.

cluster mill. A rolling mill in which each of the two working rolls of small diameter is supported by two or more backup rolls.

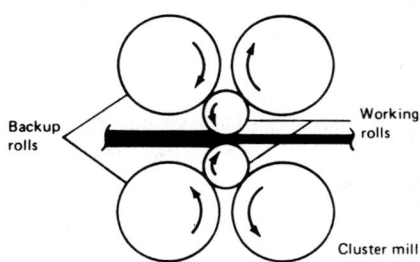

Backup rolls

Working rolls

Cluster mill

CO_2 welding. See *carbon dioxide welding*.

coalesced copper. Massive oxygen-free copper made by briquetting ground, brittle cathode copper, then sintering the briquets in a pressurized reducing atmosphere, followed by hot working.

coalescence. (1) The union of particles of a dispersed phase into larger units, usually effected at temperatures below the fusion point. (2) In welding, brazing or soldering, the union of two or more components into a single body, which usually involves melting of a filler metal or of the base metal.

coarsening. An increase in grain size, usually, but not necessarily, by *grain growth*.

coated abrasive. An abrasive product (sandpaper, for example) in which a layer of abrasive particles is firmly attached to a paper, cloth or fiber backing by means of glue or synthetic-resin adhesive.

coated electrode. See preferred term, *lightly coated electrode*.

coaxing. Improvement of the fatigue strength of a specimen by the application of a gradually increasing stress amplitude, usually starting below the fatigue limit.

coefficient of elasticity. Same as *modulus of elasticity*.

coercive force. The magnetizing force that must be applied in the direction opposite to that of the previous magnetizing force in order to reduce magnetic flux density to zero; thus, a measure of the magnetic retentivity of magnetic materials.

cogging mill. A *blooming mill*.

coherency. The continuity of lattice of precipitate and parent phase (solvent) maintained by mutual strain and not separated by a phase boundary.

coherent precipitate. A crystalline precipitate that forms from solid solution with an orientation that maintains continuity between the crystal lattice of the precipitate and the lattice of the matrix, usually accompanied by some strain in both lattices. Because the lattices fit at the interface between precipitate and matrix, there is no discernible phase boundary.

cohesion. Force of attraction between the molecules (or atoms) within a single phase. Contrast with *adhesion*.

cohesive strength. (1) The hypothetical stress causing tensile fracture without plastic deformation. (2) The stress corresponding to the forces between atoms. (3) Same as *technical cohesive strength*. (4) Same as *disruptive strength*.

coil breaks. Creases or ridges in sheet or strip that appear as parallel lines across the direction of rolling, and that generally extend the full width of the sheet or strip.

coining. (1) A closed-die squeezing operation, usually performed cold, in which all surfaces of the work are confined or restrained, resulting in a well-defined imprint of the die upon the work. (2) A *restriking* operation used to sharpen or change an existing radius or profile. (3) The final pressing of a sintered powder metallurgy compact to obtain a definite surface configuration (not to be confused with *re-pressing* or *sizing*).

coin silver. An alloy containing 90% silver, with copper being the usual alloying element.

cold-chamber machine. A *die-casting* machine in which the metal chamber and plunger are not heated.

cold extrusion. See *extrusion*.

cold heading. Working metal at room temperature in such a manner that the cross-sectional area of a portion or all of the stock is increased.

cold inspection. A visual (usually final) inspection of forgings for visible imperfections, dimensions, weight, and surface condition at room temperature. The term may also be used to describe certain nondestructive tests such as magnetic-particle, dye-penetrant and sonic inspection.

cold lap. Wrinkled markings on the surface of an ingot, caused by incipient freezing of the surface while the liquid is still in motion; results from insufficient pouring temperature. See also *cold shut* (1).

cold mill. A mill for cold rolling of sheet or strip.

cold pressing. Forming a powder metallurgy *compact* at a temperature low enough to avoid *sintering*, usually room temperature. Contrast with *hot pressing*.

cold rolled sheets. A mill product produced from a hot rolled pickled coil that has been given substantial cold reduction at room temperature. The resulting product usually requires further processing to make it suitable for most common applications. The usual end product is characterized by improved surface, greater uniformity in thickness and improved mechanical properties compared with hot rolled sheet.

cold shortness. Brittleness that exists in some metals at temperatures below the recrystallization temperature.

cold shot. A portion of the surface of an ingot or casting showing premature solidification; caused by splashing of molten metal onto a cold mold wall during pouring.

cold shut. (1) A discontinuity that appears on the surface of cast metal as a result of two streams of liquid meeting and failing to unite. (2) A lap on the surface of a forging or billet that was closed without fusion during deformation. (3) Freezing of the top surface of an ingot before the mold is full.

cold treatment. Exposing to suitable subzero temperatures for the purpose of obtaining desired conditions or properties such as dimensional or microstructural stability. When the treatment involves the transformation of retained austenite, it is usually followed by tempering.

cold trimming. Removing flash or excess metal from a forging in a trimming press when the forging is at room temperature.

cold welding. A solid-state welding process in which pressure is used at room temperature to produce coalescence of metals with substantial deformation at the weld. Compare *hot pressure welding, diffusion welding,* and *forge welding*.

cold work. Permanent strain in a metal accompanied by strain hardening.

cold working. Deforming metal plastically under conditions of temperature and strain rate that induce strain hardening. Usually, but not necessarily, conducted at room temperature. Contrast with *hot working*.

collapsibility. The requirement that a sand mold or core break down under the pressure and temperature of casting in order to avoid hot tears, or to facilitate the separation of sand and casting.

collet. A split sleeve used to hold work or tools during machining or grinding.

color buffing. Producing a final high luster by buffing. Sometimes called *coloring*.

coloring. Producing desired colors on metal by a chemical or electrochemical reaction. See also *color buffing*.

columnar structure. A coarse structure of parallel elongated grains formed by unidirectional growth, most often observed in castings, but sometimes seen in structures resulting from diffusional growth accompanied by a solid-state transformation.

combination die. (1) A die-casting die having two or more different cavities for different castings. (2) For forming, see *compound die*.

combination mill. An arrangement of a continuous mill for roughing and a *guide mill* or *looping mill* for shaping.

combined carbon. The part of the total carbon in steel

or cast iron that is present as other than *free carbon*.

combined cyanide. The cyanide of a metal-cyanide complex ion.

combined stresses. Any state of stress that cannot be represented by a single component of stress; that is, one that is more complicated than simple tension, compression or shear.

comminution. (1) Breaking up or grinding an ore into small fragments. (2) Reducing metal to powder by mechanical means.

commutator-controlled welding. Spot or projection welding in which several electrodes, in simultaneous contact with the work, function progressively under the control of an electrical commutating device.

compact. An object produced by the compression of metal powder, generally while confined in a die, with or without the inclusion of nonmetallic constituents. See also *compound compact* and *composite compact*.

compacted graphite cast iron. Cast iron having a graphite shape intermediate between the flake form typical of gray cast iron and the spherical form of fully spherulitic ductile cast iron. Also known as CG iron or vermicular iron, compacted graphite cast iron is produced in a manner similar to that for ductile cast iron, but using a technique that inhibits the formation of fully spherulitic graphite nodules.

complete fusion. Fusion that has occurred over the entire base-metal surfaces exposed for welding.

complexing agent. A substance that is an electron donor and that will combine with a metal ion to form a soluble complex ion.

complex ion. An ion that may be formed by the addition reaction of two or more other ions.

component. (1) One of the elements or compounds used to define a chemical (or alloy) system, including all phases, in terms of the fewest substances possible. (2) One of the individual parts of a vector as referred to a system of coordinates.

composite compact. A powder metallurgy *compact* consisting of two or more adhering layers of different metals or alloys with each layer retaining its original identity.

composite electrode. A welding electrode made from two or more distinct components, at least one of which is filler metal. A composite electrode may exist in any of various physical forms, such as stranded wires, filled tubes or covered wire.

composite joint. A joint in which welding is used in conjunction with mechanical joining.

composite material. A heterogeneous, solid structural material consisting of two or more distinct components that are mechanically or metallurgically bonded together (such as a *cermet*, or boron wire embedded in a matrix of epoxy resin).

composite plate. An electrodeposit consisting of layers of at least two different compositions.

composite structure. A structural member (such as a panel, plate, pipe or other shape) that is built up by bonding together two or more distinct components, each of which may be made of a metal, alloy, nonmetal or *composite material*. Examples of composite structures include: honeycomb panels, clad plate, electrical contacts, sleeve bearings, carbide-tipped drills or lathe tools, and weldments constructed of two or more different alloys.

compound compact. A powder metallurgy *compact* consisting of mixed metals, the particles of which are joined by pressing or sintering, or both, with each metal particle retaining substantially its original composition.

compound die. Any die so designed that it performs more than one operation on a part with one stroke of the press, such as blanking and piercing, where all functions are performed simultaneously within the confines of the particular blank size being worked.

compressibility. In powder metallurgy, the reciprocal of the *compression ratio* where a compact is made following a procedure in which the die, the pressure and the pressing speed are specified.

compression ratio. In powder metallurgy, the ratio of the volume of the loose powder to the volume of the compact made from it.

compressive strength. The maximum compressive stress that a material is capable of developing, based on original area of cross section. If a material fails in compression by a shattering fracture, the compressive strength has a very definite value. If a material does not fail in compression by a shattering fracture, the value obtained for compressive strength is an arbitrary value depending upon the degree of distortion

that is regarded as indicating complete failure of the material.

concave fillet weld. A fillet weld having a concave face.

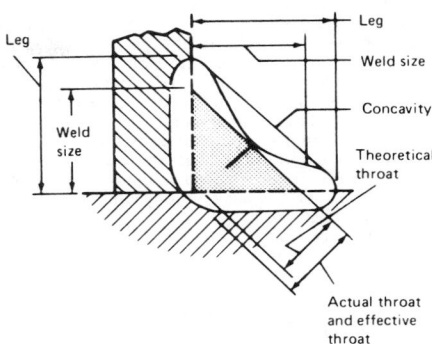

Concave fillet weld

concentration. A process for enrichment of an ore in valuable mineral content by separation and removal of waste material, or *gangue*.

concentration polarization. That part of the total polarization that is caused by changes in the activity of the potential-determining components of the electrolyte.

concurrent heating. Using a second source of heat to supplement the primary heat in cutting or welding.

conditioning heat treatment. A preliminary heat treatment used to prepare a material for a desired reaction to a subsequent heat treatment. For the term to be meaningful, the exact heat treatment must be specified.

cone angle. The angle that the cutter axis makes with the direction along which the blades are moved for adjustment, as in adjustable-blade reamers where the base of the blade slides on a conical surface.

congruent melting. An isothermal or isobaric melting in which both the solid and liquid phases have the same composition throughout the transformation.

congruent transformation. An isothermal or isobaric phase change in which both of the phases concerned have the same composition throughout the process.

constantan. A group of copper-nickel alloys containing 45 to 60% copper with minor amounts of iron and manganese and characterized by relatively constant electrical resistivity irrespective of temperature; used in resistors and thermocouples.

constituent. (1) One of the ingredients that make up a chemical system. (2) A phase or a combination of phases that occurs in a characteristic configuration in an alloy microstructure.

constitution diagram. A graphical representation of the temperature and composition limits of phase fields in an alloy system as they actually exist under the specific conditions of heating or cooling (synonymous with phase diagram). A constitution diagram may be an equilibrium diagram, an approximation to an equilibrium diagram or a representation of metastable conditions or phases. Compare with *equilibrium diagram*.

constraint. Any restriction that limits the transverse contraction normally associated with a longitudinal tension, and that hence causes a secondary tension in the transverse direction; usually used in connection with welding. Contrast with *restraint*.

consumable electrode. A general term for any arc-welding electrode made chiefly of filler metal. Use of specific names such as *covered electrode*, bare electrode, flux-cored electrode and *lightly coated electrode* is preferred.

consumable-electrode remelting. A process for refining metals in which an electric current passes between an electrode made of the metal to be refined and an ingot of the refined metal, which is contained in a water-cooled mold. As a result of the passage of electric current, droplets of molten metal form on the electrode and fall to the ingot. The refining action occurs from contact with the atmosphere, vacuum or slag through which the drop falls. See *electroslag remelting* and *vacuum arc remelting*.

contact fatigue. Cracking and subsequent pitting of a surface subjected to alternating Hertzian stresses such as those produced under rolling contact or combined rolling and sliding. The phenomenon of contact fatigue is encountered most often in rolling-element bearings or in gears, where the surface stresses are high due to the concentrated loads and are repeated many times during normal operation.

contact plating. A metal plating process wherein the plating current is provided by galvanic action between the work metal and a second metal, without the use of an external source of current.

contact potential. The potential difference at the junction of two dissimilar substances.

contact scanning. In ultrasonic inspection, a planned systematic movement of the beam relative to the object being inspected, the search unit being in contact with and coupled to this object by a thin film of coupling material.

container. The chamber into which an ingot or billet is inserted prior to extrusion. The container for backward extrusion of cups or cans is sometimes called a die.

continuous casting. A casting technique in which a cast shape is continuously withdrawn through the bottom of the mold as it solidifies, so that its length is not determined by mold dimensions. Used chiefly to produce semifinished mill products such as billets, blooms, ingots, slabs and tubes. See also *strand casting*.

continuous mill. A rolling mill consisting of a number of strands of synchronized rolls (in tandem) in which metal undergoes successive reductions as it passes through the various stands.

continuous phase. In an alloy or portion of an alloy containing more than one phase, the phase that forms the matrix in which the other phase or phases are present as isolated units.

continuous precipitation. Precipitation from a supersaturated solid solution in which the precipitate particles grow by long-range diffusion without recrystallization of the matrix. Continuous precipitates grow from nuclei distributed more or less uniformly throughout the matrix. They usually are randomly oriented, but may form a *Widmanstätten structure*. Also called general precipitation. Compare with *discontinuous precipitation, localized precipitation*.

continuous weld. A weld extending continuously from one end of a joint to the other or, where the joint is essentially circular, completely around the joint. Contrast with *intermittent weld*.

contour forming. See *stretch forming, tangent bending, wiper forming*.

contour machining. Machining of irregular surfaces, such as those generated in tracer turning, tracer boring and *tracer milling*.

contour milling. Milling of irregular surfaces. See *tracer milling*.

controlled cooling. Cooling from an elevated temperature in a predetermined manner, to avoid hardening, cracking, or internal damage, or to produce desired microstructure or mechanical properties.

controlled-pressure cycle. A forming cycle during which the hydraulic pressure in the forming cavity is controlled by an adjustable cam that is coordinated with the punch travel.

conventional forging. A forging characterized by design complexity and tolerances that fall within the broad range of general forging practice.

conventional milling. Milling in which the cutter moves in the direction opposite to the feed at the point of contact.

conventional strain. See *strain*.

conventional stress. See *stress*.

conversion coating. A coating consisting of a compound of the surface metal, produced by chemical or electrochemical treatments of the metal. (Examples are chromate coatings on zinc, cadmium, magnesium and aluminum, and oxides and phosphate coatings on steel.)

converter. A furnace in which air is blown through a bath of molten metal or matte, oxidizing the impurities and maintaining the temperature through the heat produced by the oxidation reaction.

convex fillet weld. A fillet weld having a convex face.

coolant. In metal cutting, the preferred term is *cutting fluid*.

cooling curve. A curve showing the relation between time and temperature during the cooling of a material.

cooling stresses. Residual stresses resulting from nonuniform distribution of temperature during cooling.

cooling table. Same as *hot bed*.

coordination number. (1) Number of atoms or radi-

cals coordinated with the central atom in a complex covalent compound. (2) Number of nearest neighboring atoms to a selected atom in crystal structure.

cope. The upper or topmost section of a flask, mold or pattern.

copper-accelerated salt-spray test. An accelerated corrosion test for some electrodeposits and for anodic coatings on aluminum. Often referred to as CASS test.

copper brazing. A term improperly used to denote joining with a copper-base filler metal. See preferred terms *brazing* and *braze welding*.

copperhead. A reddish spot in a porcelain enamel coating caused by iron pickup during enameling, iron oxide left on poorly cleaned basis metal, or burrs on iron or steel basis metal that protrude through the coating and are oxidized during firing.

core. (1) A specially formed material inserted in a mold to shape the interior or other part of a casting that cannot be shaped as easily by the pattern. (2) In a ferrous alloy prepared for *case hardening,* that portion of the alloy that is not part of the *case.* Typically considered to be the portion that (*a*) appears light on an etched cross section, (*b*) has an essentially unaltered chemical composition, or (*c*) has a hardness, after hardening, less than a specified value.

core blower. A machine for making foundry cores using compressed air to blow and pack the sand into the core box.

cored bar. A powder metallurgy *compact* of bar shape, the interior of which has been melted by passage of electricity.

core forging. (1) Displacing metal with a punch to fill a die cavity; (2) the product of such an operation.

core rod. The part of a die used to produce a hole in a powder metallurgy *compact*.

coring. (1) A condition of variable composition between the center and surface of a unit of microstructure (such as a dendrite, grain, carbide particle); results from nonequilibrium solidification, which occurs over a range of temperature. (2) A central cavity at the butt end of a rod extrusion, sometimes called *extrusion pipe*.

corner angle. On face milling cutters, the angle between an angular cutting edge of a cutter tooth and the axis of the cutter, measured by rotation into an axial plane. See sketch accompanying *face mill*.

corner joint. A joint between two members located approximately at right angles to each other in the form of an "L."

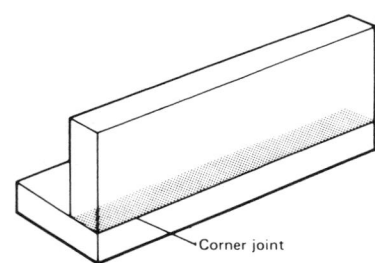

Corner joint

corona. In spot welding, an area sometimes surrounding the nugget at the faying surfaces, where solid-state welding occurs. Corona contributes variably to over-all bond strength, depending on the size of the corona and the degree of solid-state bonding achieved.

Corrodkote test. An accelerated corrosion test for electrodeposits.

corrosion. The deterioration of a metal by chemical or electrochemical reaction with its environment.

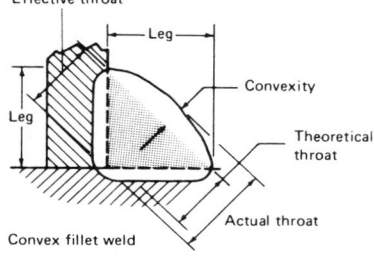

Convex fillet weld

corrosion embrittlement. The severe loss of ductility of a metal resulting from corrosive attack, usually intergranular and often not visually apparent.

corrosion fatigue. Cracking produced by the combined action of repeated or fluctuating stress and a corrosive environment.

corrugating. Forming sheet metal into a series of straight parallel alternate ridges and grooves by using a rolling mill equipped with matched roller dies or by using a press brake equipped with specially shaped punch and die.

corrugations. Transverse ripples caused by a variation in strip shape during hot or cold reduction.

corundum. Natural abrasive of the aluminum oxide type that has higher purity than emery.

Cottrell process. Removal of solid particulates from gases with electrostatic precipitation.

coulometer. An electrolytic cell arranged to measure the quantity of electricity by the chemical action produced in accordance with Faraday's law.

counterblow hammer. A forging hammer in which both the *ram* and the *anvil* are driven simultaneously toward each other by air or steam pistons.

counterboring. Drilling or boring a flat-bottomed hole, often concentric with other holes.

counterlock. A jog in the mating surfaces of dies to prevent lateral die shifting from side thrusts developed in forging irregularly shaped pieces.

countersinking. Forming a flaring depression around the top of a hole for deburring, for receiving the head of a fastener or for receiving a center.

coupling. The degree of mutual interaction between two or more elements resulting from mechanical, acoustical or electrical linkage.

coupon. A piece of metal from which a test specimen is to be prepared—often an extra piece (as on a casting or forging) or a separate piece made for test purposes (such as a test weldment).

covalent bond. A bond between two or more atoms resulting from the completion of shells by the sharing of electrons.

covered electrode. A composite filler-metal welding electrode consisting of a bare wire or a metal-cored electrode plus a covering sufficient to provide a layer of slag on deposited weld metal. The covering often contains materials that provide shielding during welding, deoxidizers for the weld metal, and arc stabilization; it may also contain alloying elements or other additives for the weld metal.

cover half. The stationary half of a die-casting die.

covering power. The ability of a solution to give satisfactory plating at very low current densities, a condition that exists in recesses and pits. This term suggests an ability to cover, but not necessarily to build up, a uniform coating, whereas *throwing power* suggests the ability to obtain a coating of uniform thickness on an irregularly shaped object.

"C" process. See *Croning process*.

crank press. A mechanical press the slides of which are actuated by a crankshaft.

crater. (1) In machining, a depression in a cutting tool face eroded by chip contact. (2) In arc welding, a depression at the termination of a bead or in the weld pool beneath the electrode.

crater crack. A crack, often star shaped, that forms in the crater of a weld bead, usually during cooling after welding.

creep. Time-dependent strain occurring under stress. The creep strain occurring at a diminishing rate is called primary creep; that occurring at a minimum and almost constant rate, secondary creep; and that occurring at an accelerating rate, tertiary creep.

creep limit. (1) The maximum stress that will cause less than a specified quantity of creep in a given time. (2) The maximum nominal stress under which the creep strain rate decreases continuously with time under constant load and at constant temperature. Sometimes used synonymously with *creep strength*.

creep recovery. Time-dependent strain after release of load in a creep test.

creep-rupture test. Same as *stress-rupture test*.

creep strength. (1) The constant nominal stress that will cause a specified quantity of creep in a given time at constant temperature. (2) The constant nominal stress that will cause a specified rate of secondary creep at constant temperature.

crevice corrosion. A type of concentration-cell corrosion; corrosion caused by the concentration or depletion of dissolved salts, metal ions, oxygen or other gases, and such, in crevices or pockets remote from the principal fluid stream, with a resultant building up of differential cells that ultimately cause deep pitting.

crimping. Forming relatively small corrugations in order to: (1) set down and lock a seam, (2) create an arc in a strip of metal, or (3) reduce an existing arc or diameter.

critical cooling rate. The rate of continuous cooling required to prevent undesirable transformation. For steel, it is the minimum rate at which austenite must be continuously cooled to suppress transformations about the M_s temperature.

critical current density. In an elecrolytic process, a current density at which an abrupt change occurs in an operating variable or in the nature of an electrodeposit or electrode film.

critical point. (1) The temperature or pressure at which a change in crystal structure, phase or physical properties occurs. Same as *transformation temperature*. (2) In an equilibrium diagram, that specific value of composition, temperature or pressure, or combinations thereof, at which the phases of a heterogeneous system are in equilibrium.

critical shear stress. The shear stress required to cause slip in a designated slip direction on a given slip plane. It is called the critical resolved shear stress if the shear stress is induced by tensile or compressive forces acting on the crystal.

critical strain. The strain just sufficient to cause *recrystallization*; because the strain is small, usually only a few percent, recrystallization takes place from only a few nuclei, which produces a recrystallized structure consisting of very large grains.

critical temperature. (1) Synonymous with *critical point* if the pressure is constant. (2) The temperature above which the vapor phase cannot be condensed to liquid by an increase in pressure.

critical temperature ranges. Synonymous with *transformation ranges*, which is the preferred term.

Croning process. A *shell molding* process utilizing a phenolic resin binder. Sometimes referred to as "C" process.

crop. (1) An end portion of an ingot that is cut off as scrap. (2) To shear a bar or billet.

cross breaks. Same as *coil breaks*.

cross-country mill. A rolling mill in which the mill stands are so arranged that their tables are parallel with a transfer (or crossover) table connecting them. Such a mill is used for rolling structural shapes, rails and any special form of bar stock not rolled in the ordinary bar mill.

crossed joint. See *bridging* (5).

cross forging. Preliminary working of forging stock in flat dies to develop mechanical properties, particularly in the center portions of heavy sections.

cross rolling. Rolling of sheet or plate so that the direction of rolling is about 90° from the direction of a previous rolling.

cross-roll straightener. A machine having paired rolls of special design for straightening round bars or tubes, the pass being made with the work parallel to the axes of the rolls.

cross-wire weld. A weld made at the junction between crossed wires or bars.

crown. (1) A contour on a sheet or roll where the thickness or diameter increases from edge to center. (2) The top section of a press structure where the cylinders and other working parts may be mounted. Also called dome, head or top platen.

crucible. A vessel or pot, made of a refractory substance or of a metal with a high melting point, used for melting metals or other substances.

crush. (1) Buckling or breaking of a section of a casting mold due to incorrect register when the mold is closed. (2) An indentation in the surface of a casting due to displacement of sand when the mold was closed.

crush forming. Shaping a grinding wheel by forcing a rotating metal roll into its face so as to reproduce the desired contour.

crushing test. (1) A radial compressive test applied to tubing, sintered-metal bearings or other similar products for determining radial crushing strength (maximum load in compression). (2) An axial compressive test for determining quality of tubing, such as weld soundness in welded tubing.

crystal. A solid composed of atoms, ions or molecules arranged in a pattern that is repetitive in three dimensions.

crystalline fracture. A pattern of brightly reflecting crystal facets on the fracture surface of a polycrys-

talline metal, resulting from cleavage fracture of many individual crystals. Contrast with *fibrous fracture, silky fracture*.

crystallization. (1) The separation, usually from a liquid phase on cooling, of a solid crystalline phase. (2) Sometimes erroneously used to explain fracturing that actually has occurred by fatigue.

crystal orientation. See *orientation*.

cubic plane. A plane perpendicular to any one of the three crystallographic axes of the cubic (isometric) system; the *Miller indices* are {100}.

cup. (1) Sheet-metal part, the product of the first step in deep drawing. (2) Any cylindrical part or shell closed at one end.

cupellation. Oxidation of molten lead containing gold and silver to produce lead oxide, thereby separating the precious metals from the base metal.

cup fracture (cup-and-cone fracture). A mixed-mode fracture, often seen in tensile-test specimens of a ductile material, where the central portion undergoes *plane-strain* fracture and the surrounding region undergoes *plane-stress* fracture. It is called a cup fracture (or cup-and-cone fracture) because one of the mating fracture surfaces looks like a miniature cup—that is, it has a central depressed flat-face region surrounded by a shear lip; the other fracture surface looks like a miniature truncated cone.

cupola. A cylindrical vertical furnace for melting metal, especially cast iron, by having the charge come in contact with the hot fuel, usually metallurgical coke.

cupping. (1) The first step in deep drawing. (2) Fracture of severely worked rods or wire where one end has the appearance of a cup and the other that of a cone.

Curie temperature. The temperature of magnetic transformation below which a metal or alloy is ferromagnetic and above which it is paramagnetic.

curling. Rounding the edge of sheet metal into a closed or partly closed loop.

current decay. In spot, seam or projection welding, the controlled reduction of the welding current from its peak amplitude to a lower value to prevent excessively rapid cooling of the weld nugget.

current efficiency. The proportion of current used in a given process to accomplish a desired result; in electroplating, the proportion used in depositing or dissolving metal.

cushion. Same as *die cushion*.

cut. (1) In castings, a rough spot or area of excess metal caused by erosion of the mold or core surface by metal flow. (2) In powder metallurgy, same as fraction.

cut-and-carry method. Stamping method wherein the part remains attached to the strip or is forced back into the strip to be fed through the succeeding stations of a progressive die.

cut edge. A mechanically sheared edge obtained by slitting, shearing or blanking.

cutoff wheel. A thin abrasive wheel for severing or slotting any material or part.

cutting down. Removing roughness or irregularities of a metal surface by abrasive action.

cutting edge. The leading edge of a cutting tool (such as a lathe tool, drill or milling cutter) where a line of contact is made with the work during machining. See sketch accompanying *single-point tool*.

cutting fluid. A fluid used in metal cutting to improve finish, tool life or dimensional accuracy. On being flowed over the tool and work, the fluid reduces friction, the heat generated and tool wear, and prevents galling. It conducts the heat away from the point of generation and also serves to wash the *chips* away.

cutting speed. The linear or peripheral speed of relative motion between the tool and workpiece in the principal direction of cutting.

cutting tip. The part of a cutting torch from which gas issues.

cyanide copper. Copper electrodeposited from an alkali-cyanide solution containing a complex ion made up of univalent copper and the cyanide radical; also, the solution itself.

cyanide slimes. Finely divided metallic precipitates that are formed when precious metals are extracted from their ores using cyanide solutions.

cyaniding. A case hardening process in which a ferrous material is heated above the lower transformation range in a molten salt containing cyanide to cause simultaneous absorption of carbon and nitrogen at the surface and, by diffusion, create a concentration gradient. *Quench hardening* completes the process.

cycle annealing. An annealing process employing a

predetermined and closely controlled time-temperature cycle to produce specific properties or microstructures.

cylindrical grinding. Grinding the outer cylindrical surface of a rotating part.

cylindrical land. *Land* having zero relief.

D

damping capacity. The ability of a material to absorb vibration (cyclical stresses) by internal friction, converting the mechanical energy into heat.

dangler. The flexible electrode used in *barrel plating* to conduct current to the work.

daylight. The maximum clear distance between the pressing surfaces of a hydraulic press when the surfaces are in their usable open position. Where a bolster is supplied, it shall be considered the pressing surface. See also *shut height*.

dc casting. Same as *direct chill casting*.

dead center. (1) A stationary center to hold rotating work. (2) Either of the two points in the path of a moving crank or connecting rod that lie at the ends of its stroke.

dead roast. A *roasting* process for complete elimination of sulfur. Also known as sweet roast.

dead soft. A *temper* of nonferrous alloys and some ferrous alloys corresponding to the condition of minimum hardness and tensile strength produced by *full annealing*.

deburring. Removing burrs, sharp edges or fins from metal parts by filing, grinding, or rolling the work in a barrel containing abrasives suspended in a suitable liquid medium. Sometimes called burring.

decalescence. A phenomenon, associated with the transformation of alpha iron to gamma iron on the heating (superheating) of iron or steel, revealed by the darkening of the metal surface owing to the sudden decrease in temperature caused by the fast absorption of the latent heat of transformation. Contrast with *recalescence*.

decarburization. Loss of carbon from the surface layer of a carbon-containing alloy due to reaction with one or more chemical substances in a medium that contacts the surface.

decomposition potential. The minimum potential difference necessary to decompose the electrolyte of a cell.

deep drawing. Forming deeply recessed parts by forcing sheet metal to undergo plastic flow between dies, usually without substantial thinning of the sheet.

deep etching. Severe *macroetching*.

defect. A departure of any *quality characteristic* from its intended (usually specified) level that is severe enough to cause the product or service not to fulfill its anticipated function. According to ANSI standards, defects are classified according to severity:

Very serious defects lead directly to severe injury or catastrophic economic loss.

Serious defects lead directly to significant injury or significant economic loss.

Major defects are related to major problems with respect to anticipated use.

Minor defects are related to minor problems with respect to anticipated use.

defective. A quality control term describing a unit of product or service containing at least one *defect*, or having several lesser imperfections that, in combination, cause the unit not to fulfill its anticipated function. NOTE: The term *defective* is not synonymous with *nonconforming* (or rejectable) and should be applied only to those units incapable of performing their anticipated functions.

deformation bands. Parts of a crystal that have rotated differently during deformation to produce bands of varied orientation within individual grains.

degasifier. A substance that can be added to molten metal to remove soluble gases that might otherwise be occluded or entrapped in the metal during solidification.

degassing. Removing gases from liquids or solids.

degreasing. Removing oil or grease from a surface. See also *vapor degreasing*.

degrees of freedom. The number of independent variables (such as temperature, pressure or concentration within the phases present) that may be altered at will without causing a phase change in an alloy system at equilibrium; or the number of such variables that must be fixed arbitrarily to define the system completely.

delayed yield. A phenomenon involving a delay in time between the application of a stress and the occurrence of the corresponding yield-point strain.

delta ferrite. See *ferrite.*

dendrite. A crystal that has a treelike branching pattern, being most evident in cast metals slowly cooled through the solidification range. See accompanying illustration.

dendritic powder. Particles of metal powder, usually of electrolytic origin, having typical pine-tree structure.

denickelification. Corrosion in which nickel is selectively leached from nickel-containing alloys. Most commonly observed in copper-nickel alloys after extended service in fresh water.

density ratio. The ratio of the determined density of a powder metallurgy compact to the absolute density of metal of the same composition, usually expressed as a percentage.

deoxidized copper. Copper from which cuprous oxide has been removed by adding a *deoxidizer,* such as phosphorus, to the molten bath.

deoxidizer. A substance that can be added to molten metal to remove either free or combined oxygen.

deoxidizing. (1) The removal of oxygen from molten metals by use of suitable *deoxidizers.* (2) Sometimes refers to the removal of undesirable elements other than oxygen by the introduction of elements or compounds that readily react with them. (3) In metal finishing, the removal of oxide films from metal surfaces by chemical or electrochemical reaction.

depolarization. A decrease in the *polarization* of an electrode.

depolarizer. A substance that produces *depolarization.*

deposition efficiency. In welding, the ratio of the weight of deposited weld metal to the net weight of electrodes consumed, exclusive of stubs.

deposition sequence. The order in which increments of weld metal are deposited.

depth of cut. The thickness of material removed from a workpiece in a single machining pass.

depth of fusion. In welding, the distance that fusion extends into the base metal or into a previous pass.

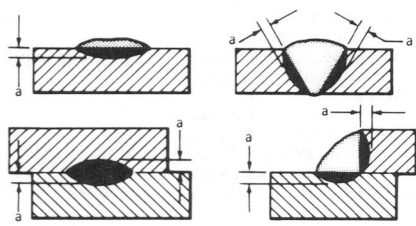

a = depth of fusion

depth of penetration. See *joint penetration* and *root penetration.*

derby. A massive piece (intermediate in size, extending to more than 100 lb, and usually cylindrical) of primary metal made by bomb reduction (such as uranium from uranium tetrafluoride reduced with magnesium). Compare with *biscuit* and *dingot.*

descaling. Removing the thick layer of oxides formed on some metals at elevated temperatures.

deseaming. Analogous to *chipping,* the surface imperfections being removed by gas cutting.

detergent. A chemical substance, generally used in aqueous solution, that removes *soil.*

detritus. Wear debris.

developed blank. A blank that requires little or no trimming when formed.

dewaxing. Removing the expendable wax pattern from an investment mold by heat or solvent.

dezincification. Corrosion in which zinc is selectively leached from zinc-containing alloys. Most commonly found in copper-zinc alloys containing less than 85% copper after extended service in water containing dissolved oxygen.

diamagnetic material. A material whose specific permeability is less than unity and is therefore repelled weakly by a magnet. Compare with *ferromagnetic material, paramagnetic material.*

diamond boring. Precision boring with a shaped diamond (but not with other tool materials).

diamond pyramid hardness test. See *Vickers hardness test.*

diamond tool. (1) A diamond, shaped or formed to the contour of a single-point cutting tool, for use in precision machining of nonferrous or nonmetallic materials. (2) Sometimes, an insert made from multicrystalline diamond compacts.

diamond wheel. A grinding wheel in which crushed and sized industrial diamonds are held in a resinoid, metal or vitrified bond.

diaphragm. (1) A porous or permeable membrane separating anode and cathode compartments of an electrolytic cell from each other or from an intermediate compartment. (2) Universal die member made of rubber or similar material used to contain hydraulic fluid within the forming cavity and to transmit pressure to the part being formed.

dichromate treatment. A chromate *conversion coating* produced on magnesium alloys in a boiling solution of sodium dichromate.

didymium. A natural mixture of the rare-earth elements praseodymium and neodymium, often given the quasichemical symbol Di.

die. A tool, usually containing a cavity, that imparts shape to solid, molten or powdered metal primarily because of the shape of the tool itself. Used in many press operations (including blanking, drawing, forging and forming), in die casting and in forming green powder metallurgy compacts. Die-casting and powder metallurgy dies are sometimes referred to as *molds.*

die block. A block, usually of tool steel, into which the desired impressions are sunk, formed or machined and from which forgings or die castings are made.

die body. The stationary or fixed part of a powder pressing die.

die casting. (1) A casting made in a die. (2) A casting process wherein molten metal is forced under high pressure into the cavity of a metal mold.

die clearance. Clearance between a mated punch and die; commonly expressed as clearance per side. Also called clearance, punch-to-die clearance.

die cushion. A press accessory located beneath or within a *bolster* or *die block* to provide an additional motion or pressure for stamping operations; actuated by air, oil, rubber or springs, or by a combination thereof.

die forging. A forging whose shape is determined by impressions in specially prepared dies.

die forming. Shaping of solid or powdered metal by forcing it into or through the cavity in a die.

die holder. A plate or block, on which the die block is mounted, having holes or slots for fastening to the bolster or the bed of the press.

die insert. A removable liner or part of a die body or punch.

die layout. The transfer of drawing or sketch dimensions to templates or die surfaces for use in sinking dies.

die life. The productive life of a die impression, usually expressed as the number of units produced before the impression has worn beyond permitted tolerances.

die lines. Lines or markings on formed, drawn or extruded metal parts caused by imperfections in the surface of the die.

die lubricant. A lubricant applied to working surfaces of dies and punches to facilitate drawing, pressing, stamping and/or ejection. In powder metallurgy, the die lubricant is sometimes mixed into the powder before pressing into a compact.

die match. The condition where dies, after having been set up in a press or other equipment, are in proper alignment relative to each other.

die opening. In flash or upset welding, the distance between the electrodes, usually measured with the parts in contact before welding has commenced or immediately upon completion of the cycle but before upsetting.

die proof. A casting of the die impression made to confirm the exactness of the impression. Also called cast.

die radius. The radius on the exposed edge of a drawing die, over which the sheet flows in forming drawn shells.

die scalping. Removing surface layers from bar, rod, wire or tube by drawing through a sharp-edged die to eliminate minor surface defects.

die set. A tool or tool holder consisting of a die base and punch plate for the attachment of a die and punch, respectively.

die shift. A condition requiring correction where, after dies have been set up in the forging equipment, displacement of a point in one die from the corresponding point in the opposite die occurs in a direction parallel to the fundamental parting line of the dies.

die sinking. Forming or machining a depressed pattern in a die.

die welding. Forge welding between dies.

differential coating. A coated product having a specified coating on one surface and a significantly lighter coating on the other surface (such as a hot dip galvanized product or electrolytic tin plate).

differential flotation. Separating a complex ore into two or more valuable minerals and *gangue* by *flotation.* Also called selective flotation.

differential heating. Heating that intentionally produces a temperature gradient within an object such that, after cooling, a desired stress distribution or variation in properties is present within the object.

diffusion. (1) Spreading of a constituent in a gas, liquid or solid, tending to make the composition of all parts uniform. (2) The spontaneous movement of atoms or molecules to new sites within a material.

diffusion aid. A solid filler metal sometimes used in *diffusion welding.*

diffusion bonding. See preferred terms *diffusion welding, diffusion brazing.*

diffusion brazing. A brazing process that joins two or more components by heating them to suitable temperatures and by using a filler metal or an *in situ* liquid phase. The filler metal may be distributed by capillary attraction or may be placed or formed at the faying surfaces. The filler metal is diffused with the base metal to the extent that the joint properties are changed to approach those of the base metal.

diffusion coating. Any process whereby a basis metal or alloy is either: (1) coated with another metal or alloy and heated to a sufficient temperature in a suitable environment or (2) exposed to a gaseous or liquid medium containing the other metal or alloy, thus causing diffusion of the coating or of the other metal or alloy into the basis metal with resultant changes in the composition and properties of its surface.

diffusion coefficient. A factor of proportionality representing the amount of substance diffusing across a unit area through a unit concentration gradient in unit time.

diffusion welding. A high-temperature solid-state welding process that permanently joins faying surfaces by the simultaneous application of pressure and heat. The process does not involve macroscopic deformation, melting, or relative motion of parts. A solid filler metal (diffusion aid) may or may not be inserted between the faying surfaces.

digging. A sudden erratic increase in cutting depth, or in the load on a cutting tool, caused by unstable conditions in the machine setup. Usually the machine is stalled, or either the tool or the workpiece is destroyed.

dilatometer. An instrument for measuring the linear expansion or contraction in a metal resulting from changes in such factors as temperature and allotropy.

dimple rupture. A fractographic term describing ductile fracture that occurs through the formation and coalescence of microvoids along the fracture path. The fracture surface of such a ductile fracture appears dimpled when observed at high magnification and usually is most clearly resolved when viewed in a scanning electron microscope.

dimpling. (1) Stretching a relatively small, shallow indentation into sheet metal. (2) In aircraft, stretching thin metal into a conical flange for use with a countersunk head rivet.

dingot. An oversized *derby* (possibly a ton or more) of a metal produced in a bomb reaction (such as uranium from uranium tetrafluoride reduced with magnesium). For these metals the term "ingot" is reserved for massive units produced in vacuum melting and casting. See *biscuit* and *derby.*

dinking. Cutting of nonmetallic materials or light-gage soft metals by using a hollow punch with a knifelike edge acting against a wooden fiber or resiliently mounted metal plate.

dip brazing. Brazing by immersing the assembly to be joined in a bath of hot molten chemicals or hot metal. A molten chemical bath may provide brazing flux; molten metal, the filler metal.

diphase cleaning. Removing *soil* by an emulsion that produces two phases in the cleaning tank: a solvent phase and an aqueous phase. Cleaning is effected by both solvent action and emulsification.

dip plating. Same as *immersion plating*.

direct-arc furnace. An electric-arc furnace in which the metallic charge is one of the poles of the arc.

direct chill casting. A continuous method of making ingots for rolling or extrusion by pouring the metal into a short mold. The base of the mold is a platform that is gradually lowered while the metal solidifies, the frozen shell of metal acting as a retainer for the liquid metal below the wall of the mold. The ingot is usually cooled by the impingement of water directly on the mold or on the walls of the solid metal as it is lowered. The length of the ingot is limited by the depth to which the platform can be lowered; therefore, it is often called semicontinuous casting.

direct-current cleaning. Same as *cathodic cleaning*.

direct extrusion. See *extrusion*.

directional property. Property whose magnitude varies depending on the relation of the test axis to a specific direction within the metal. The variation results from preferred orientation or from fibering of constituents or inclusions.

directional solidification. Solidification of molten metal in such a manner that feed metal is always available for that portion that is just solidifying.

direct quenching. (1) Quenching carburized parts directly from the carburizing operation. (2) Also used for quenching pearlitic malleable parts directly from the malleablizing operation.

discontinuity. Any interruption in the normal physical structure or configuration of a part, such as cracks, laps, seams, inclusions or porosity. A discontinuity may or may not affect the usefulness of the part.

discontinuous precipitation. Precipitation from a supersaturated solid solution in which the precipitate particles grow by short-range diffusion, accompanied by recrystallization of the matrix in the region of precipitation. Discontinuous precipitates grow into the matrix from nuclei near grain boundaries, forming cells of alternate lamellae of precipitate and depleted (and recrystallized) matrix. Often referred to as cellular or nodular precipitation. Compare with *continuous precipitation, localized precipitation*.

discontinuous yielding. Nonuniform plastic flow of a metal exhibiting a yield point in which plastic deformation is inhomogeneously distributed along the gage length. Under some circumstances, it may occur in metals not exhibiting a distinct yield point, either at the onset of or during plastic flow.

dishing. Forming a shallow concave surface, the area being large compared to the depth.

disk grinding. Grinding with the flat side of an abrasive disk or segmented wheel.

dislocation. A linear imperfection in a crystalline array of atoms. Two basic types are recognized: an edge dislocation corresponds to the row of mismatched atoms along the edge formed by an extra, partial plane of atoms within the body of a crystal; a screw dislocation corresponds to the axis of a spiral structure in a crystal, characterized by a distortion that joins normally parallel planes together to form a continuous helical ramp (with a pitch of one interplanar distance) winding about the dislocation. Most prevalent is the so-called mixed dislocation, which is the name given to any combination of an edge dislocation and a screw dislocation.

disordering. Forming a lattice arrangement in which the solute and solvent atoms of a solid solution occupy lattice sites at random. See also *ordering, superlattice*.

dispersing agent. A material that increases the stability of a suspension of particles in a liquid medium by deflocculation of the primary particles.

disruptive strength. The stress at which a metal fractures under hydrostatic tension.

distortion. Any deviation from an original size, shape or contour that occurs because of the application of stress or the release of residual stress.

disturbed metal. The cold worked metal layer formed at a polished surface during the process of mechanical grinding and polishing.

divided cell. A cell containing a diaphragm or other means for physically separating the *anolyte* from the *catholyte*.

divorced eutectic. A metallographic appearance in which the two constituents of a eutectic structure appear as massive phases rather than the finely divided mixture characteristic of normal eutectics. Often, one of the constituents of the eutectic is continuous with and indistinguishable from an accompanying proeutectic constituent.

domain. A substructure in a ferromagnetic material within which all the elementary magnets (electron spins) are held aligned in one direction by interatomic forces; if isolated, a domain would be a saturated permanent magnet.

doré silver. Crude silver containing a small amount of gold, obtained after removing lead in a cupelling furnace. Same as doré bullion and doré metal.

double-acting hammer. A forging hammer in which the ram is raised by admitting steam or air into a cylinder below the piston, and the blow intensified by admitting steam or air above the piston on the downward stroke.

double-action die. A die designed to perform more than one operation in a single stroke of the press.

double-action forming. Forming or drawing in which more than one action is achieved in a single stroke of the press.

double-action mechanical press. A press having two independent parallel movements by means of two slides, one moving within the other. The inner slide or plunger is usually operated by a crankshaft, whereas the outer or blankholder slide, which dwells during the drawing operation, is usually operated by a toggle mechanism or by cams.

double aging. Employment of two different aging treatments to control the type of precipitate formed from a supersaturated matrix in order to obtain the desired properties. The first aging treatment, sometimes referred to as intermediate or stabilizing, is usually carried out at a higher temperature than the second.

double-bevel groove weld. A groove weld in which the joint edge of one member is beveled from both sides.

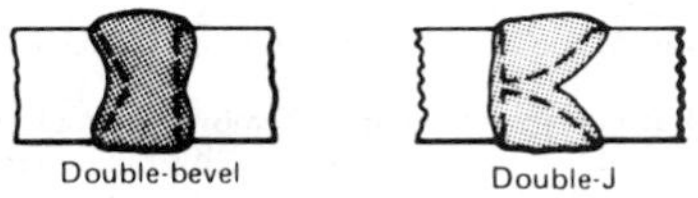

double-J groove weld. A groove weld in which the joint edge of one member is in the form of two J's, one from either side.

double salt. A compound of two salts that crystallize together in a definite proportion.

double tempering. A treatment in which a quench-hardened ferrous metal is subjected to two complete tempering cycles, usually at substantially the same temperature, for the purpose of ensuring completion of the tempering reaction and promoting stability of the resulting microstructure.

double-U groove weld. A groove weld in which each joint edge is in the form of two J's or two half-U's, one from either side of the member.

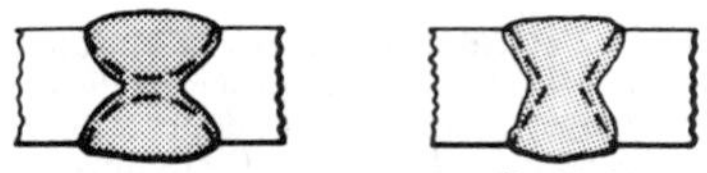

double-V groove weld. A groove weld in which each joint edge is beveled from both sides.

double-welded joint. A butt, edge, tee, corner or lap joint in which welding has been done from both sides.

down cutting. See preferred term, *climb cutting*.

downgate. Same as *sprue*.

downhand welding. See *flat-position welding*.

down milling. See preferred term, *climb milling*.

down slope time. In resistance welding, time associated with current decrease using *slope control*.

downsprue. Same as *sprue*.

Dow process. A process for the production of magnesium by electrolysis of molten magnesium chloride.

draft. (1) An angle or taper on the surface of a pattern, core box, punch or die (or of the parts made with them) that makes it easier to remove the parts from a mold or die cavity, or to remove a core from a casting. (2) The change in cross section that occurs during rolling or cold drawing.

drag. The bottom section of a flask, mold or pattern.

drag angle. In welding, the angle between the axis of the electrode or torch and a line normal to the plane of the weld when welding is being done with the torch positioned ahead of the weld puddle. See sketch accompanying *backhand welding*.

drag-in. Water or solution carried into another solution by the work and its associated handling equipment.

dragout. Solution carried out of a bath by the work and its associated handling equipment.

drag technique. A method used in manual arc welding wherein the electrode is in contact with the assembly being welded without being in short circuit. The electrode is usually used without oscillation.

drawability. A measure of the workability of a metal subject to a drawing process. A term usually expressed to indicate a metal's ability to be deep drawn.

draw bead. (1) A bead or offset used for controlling metal flow. (2) Riblike projections on draw rings or holddown surfaces for controlling metal flow.

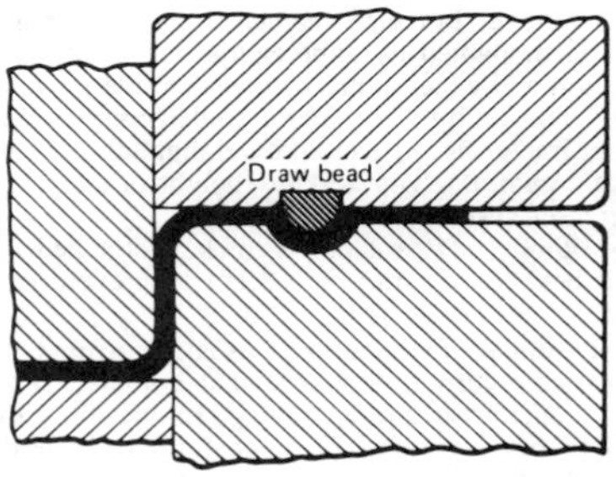

drawbench. The stand that holds the die and draw head used in drawing of wire, rod and tubing.

draw forging. See *radial forging*.

draw forming. A method of curving bars, tubes or rolled or extruded sections, in which the stock is bent around a rotating form block. Stock is bent by clamping it to the form block, then rotating the form block while the stock is pressed between the form block and a pressure die held against the periphery of the form block. Contrast with *wiper forming*.

draw head. Set of rolls or dies mounted on a drawbench for forming a section from strip, tubing or solid stock. See *Turk's-head rolls*.

drawing. (1) Forming recessed parts by forcing the plastic flow of metal in dies. (2) Reducing the cross section of bar stock, wire or tubing by pulling through a die. (3) A misnomer for tempering (see *temper*).

drawing compound. A substance applied to prevent *pickup* and *scoring* during drawing or pressing operations by preventing metal-to-metal contact of the work and die. Also known as *die lubricant*.

drawing out. A stretching operation resulting from forging a series of upsets along the length of the workpiece.

draw marks. See *scoring, galling, pickup, die lines*.

drawn shell. An article formed by drawing sheet metal into a hollow structure having a predetermined geometrical configuration.

draw plate. A circular plate with a hole in the center contoured to fit a forming punch, used to support the blank during the forming cycle.

draw radius. The radius at the edge of a die or punch over which the work is drawn.

draw ring. A ring-shaped die part over the inner edge of which the metal is drawn by the punch.

dresser. A tool used for *truing* and *dressing* a grinding wheel.

dressing. Cutting, breaking down or crushing the surface of a grinding wheel to improve its cutting ability and accuracy.

drift. (1) A flat piece of steel of tapering width used to remove taper shank drills and other tools from their holders. (2) A tapered rod used to force mismated holes into line for riveting or bolting. Sometimes called a drift pin.

drill. A rotary end-cutting tool used for making holes; it has one or more cutting lips and an equal number of helical or straight flutes for passage of chips and admission of cutting fluid.

drive fit. A type of *force fit*.

drop. A casting imperfection due to a portion of the sand dropping from the cope or other overhanging section of the mold.

drop forging. A shallow forging made in impression dies, usually with a drop hammer.

drop hammer. A forging hammer that depends on gravity for its force.

dross. The scum that forms on the surface of molten metal largely because of oxidation but sometimes because of the rising of impurities to the surface.

dry cyaniding. (obsolete) Same as *carbonitriding.*

dry sand mold. A casting mold made of sand and then dried at 100 °C (212 °F) or above before being used. Contrast with *green sand mold.*

ductile cast iron. A *cast iron* that has been treated while molten with an element such as magnesium or cerium to induce the formation of free graphite as nodules or spherulites, which imparts a measurable degree of ductility to the cast metal. Also known as nodular cast iron, spherulitic graphite cast iron and SG iron.

ductile crack propagation. Slow crack propagation that is accompanied by noticeable plastic deformation and that requires energy to be supplied from outside the body.

ductile fracture. Fracture characterized by tearing of metal accompanied by appreciable gross plastic deformation and expenditure of considerable energy.

ductility. The ability of a material to deform plastically without fracturing, measured by elongation or reduction of area in a tensile test, by height of cupping in an Erichsen test or by other means.

dummy block. In extrusion, a thick unattached disk placed between the ram and billet to prevent overheating of the ram.

dummy cathode. (1) A cathode, usually corrugated to give variable current densities, that is plated at low current densities to preferentially remove impurities from a plating solution. (2) A substitute cathode that is used during adjustment of operating conditions.

dummying. Plating with *dummy cathodes.*

duplex coating. See *composite plate.*

duplexing. Any two-furnace melting or refining process. Also called duplex melting or duplex processing.

duplicating. In machining and grinding, reproducing a form from a master with an appropriate type of machine tool, utilizing a suitable tracer or program-controlled mechanism.

duralumin. (obsolete) A term formerly applied to the class of age-hardenable aluminum-copper alloys containing manganese, magnesium or silicon.

Durville process. A casting process that involves rigid attachment of the mold in an inverted position above the crucible. The melt is poured by tilting the entire assembly, causing the metal to flow along a connecting *launder* and down the side of the mold.

dusting. Applying a powder, such as sulfur to molten magnesium, or graphite to a mold surface.

duty cycle. For electric welding equipment, the percentage of time that current flows during a specified period. In arc welding, the specified period is 10 min.

dynamic creep. Creep that occurs under conditions of fluctuating load or fluctuating temperature.

E

earing. Formation of scallops (ears) around the top edge of a drawn part, caused by directional differences in the properties of the sheet metal used.

eccentric press. A mechanical press in which the eccentric and strap are used to move the slide, rather than a crankshaft and connection.

ECM. An abbreviation for *electrochemical machining.*

eddy-current testing. An electromagnetic nondestructive testing method in which eddy-current flow is induced in the test object. Changes in flow caused by variations in the object are reflected into a nearby coil or coils where they are detected and measured by suitable instrumentation.

edge dislocation. See *dislocation.*

edge joint. A joint between the edges of two or more parallel or nearly parallel members.

edger. In forging, the portion of a die that generally distributes the metal in portions required for the shape to be forged, usually a gathering operation. A rolling edger shapes the stock into various solids of revolution; a ball edger forms a ball.

edge strain. Transverse strain lines or Lüders lines located from 25 to 300 mm (1 to 12 in.) in from the edges of cold rolled steel sheet or strip.

edging. (1) In forming, reducing the flange radius by retracting the forming punch a small amount after the stroke but prior to releasing the pressure. (2) In forging, removing flash that is directed upward between dies, usually accomplished in a lathe. (3) In rolling, working of metal where the axis of the roll is parallel to the thickness dimension. Also called edge rolling.

EDM. An abbreviation for *electrical discharge machining.*

effective rake. The angle between a plane containing a tooth face and the axial plane through the tooth point as measured in the direction of chip flow through the tooth point. Thus, it is the rake resulting from both cutter configuration and direction of chip flow.

ejector. A device mounted in such a way that it removes or assists in removing a formed part from a die.

ejector half. The movable half of a die-casting die containing the ejector pins.

ejector rod. A rod used to push out a formed piece.

elastic aftereffect. Time-dependent recovery, toward original dimensions, after the load has been reduced or removed from an elastically or plastically strained body. See *anelasticity.*

elastic constants. Factors of proportionality that describe elastic responses of a material to applied forces, including *modulus of elasticity* (either in tension, compression or shear), *Poisson's ratio, compressibility* and bulk modulus.

elastic deformation. A change in dimensions directly proportional to and in phase with an increase or decrease in applied force.

elastic hysteresis. A misnomer for an anelastic strain that lags a change in applied stress, thereby creating energy loss during cyclic loading. More properly termed *mechanical hysteresis.*

elasticity. Ability of a solid to deform in direct proportion to and in phase with increases or decreases in applied force.

elastic limit. The maximum stress to which a material may be subjected without any permanent strain remaining upon complete release of stress.

elastic modulus. Same as *modulus of elasticity.*

elastic ratio. *Yield point* divided by *tensile strength.*

elastic strain. Same as *elastic deformation.*

elastic strain energy. See *strain energy.*

elastic waves. Mechanical vibrations in an elastic medium.

electrical discharge machining. Removal of stock from an electrically conductive material by rapid, repetitive spark discharge through a dielectric fluid flowing between the workpiece and a shaped electrode. Often abbreviated EDM. Variations of the process include electrical discharge grinding and electrical discharge wire cutting.

electrical disintegration. Metal removal by an electrical spark acting in air. It is not subject to precise control, the most common application being the removal of broken tools such as taps and drills; hence the shop name "tap buster."

electrochemical corrosion. Corrosion that is accompanied by a flow of electrons between cathodic and anodic areas on metallic surfaces.

electrochemical equivalent. The weight of an element, compound, radical or ion involved in a specified electrochemical reaction during the passage of a unit quantity of electricity.

electrochemical machining. Removal of stock from an electrically conductive material by anodic dissolution in an electrolyte flowing rapidly through a gap between the workpiece and a shaped electrode. Often abbreviated ECM. Variations of the process include electrochemical deburring and electrochemical grinding.

electrochemical series. Same as *electromotive series.*

electrode. (1) In arc welding, a current-carrying rod that supports the arc between the rod and work, or

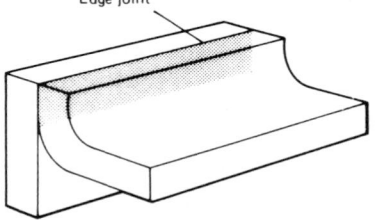

Edge joint

between two rods as in twin carbon-arc welding. It may or may not furnish filler metal. See *bare electrode, covered electrode* and *lightly coated electrode.* (2) In resistance welding, a part of a resistance welding machine through which current and, in most instances, pressure are applied directly to the work. The electrode may be in the form of a rotating wheel, rotating roll, bar, cylinder, plate, clamp, chuck or modification thereof. (3) An electrical conductor for leading current into or out of a medium.

electrode cable. Same as *electrode lead.*

electrode deposition. The weight of weld-metal deposit obtained from a unit length of electrode.

electrode force. The force between electrodes in spot, seam and projection welding.

electrode lead. The electrical conductor between the source of arc welding current and the electrode holder.

electrodeposition. The deposition of a substance on an electrode by passing electric current through an electrolyte. *Electroplating (plating), electroforming, electrorefining* and *electrowinning* result from electrodeposition.

electrode potential. The potential of a *half cell* as measured against a standard reference half cell.

electrode skid. In spot, seam or projection welding, the sliding of an electrode along the surface of the work.

electroforming. Making parts by electrodeposition on a removable form.

electrogalvanizing. The electroplating of zinc upon iron or steel.

electrogas welding. A process for *vertical-position welding* in which molding shoes confine the molten weld metal. Welding may be done by either *gas metal-arc welding* or *flux-cored arc welding.*

electroless plating. A process in which metal ions in a dilute aqueous solution are plated out on a substrate by means of autocatalytic chemical reduction.

electrolysis. Chemical change resulting from the passage of an electric current through an *electrolyte.*

electrolyte. (1) An ionic conductor. (2) A liquid, most often a solution, that will conduct an electric current.

electrolytic brightening. Same as *electropolishing.*

electrolytic cell. An assembly, consisting of a vessel, electrodes and an electrolyte, in which electrolysis can be carried out.

electrolytic cleaning. Removing soil from work by *electrolysis,* the work being one of the electrodes. The electrolyte is usually alkaline.

electrolytic copper. Copper that has been refined by electrolytic deposition, including cathodes that are the direct product of the refining operation, refinery shapes cast from melted cathodes, and, by extension, fabricators' products made therefrom. Usually when this term is used alone, it refers to electrolytic tough pitch copper without elements other than oxygen being present in significant amounts.

electrolytic deposition. Same as *electrodeposition.*

electrolytic grinding. A combination of grinding and machining wherein a metal-bonded abrasive wheel, usually diamond, is the cathode in physical contact with the anodic workpiece, the contact being made beneath the surface of a suitable electrolyte. The abrasive particles produce grinding and act as nonconducting spacers permitting simultaneous machining through electrolysis.

electrolytic machining. Controlled removal of metal by use of an applied potential and a suitable electrolyte to produce the shapes and dimensions desired.

electrolytic pickling. *Pickling* in which electric current is used, the work being one of the electrodes.

electrolytic powder. Metal powder produced by electrolytic deposition or by pulverization of an electrodeposit, or from metal made by electrodeposition.

electrolytic protection. See preferred term, *cathodic protection.*

electrometallurgy. Industrial recovery or processing of metals and alloys by electric or electrolytic methods.

electromotive force. Electrical potential; voltage.

electromotive series. A series of elements arranged according to their *standard electrode potentials.* In corrosion studies, the analogous but more practical *galvanic series* of metals is generally used. The relative positions of a given metal are not necessarily the same in the two series.

electron bands. Energy states for the free electrons in a metal, as described by use of the band theory (zone theory) of electron structure. Also called Brillouin zones.

electron-beam cutting. A cutting process that uses the

heat obtained from a concentrated beam composed primarily of high-velocity electrons, which impinge upon the workpieces to be cut; it may or may not use an externally supplied gas.

electron-beam machining. Removing material by melting and vaporizing the workpiece at the point of impingement of a focused high-velocity beam of electrons. The machining is done in high vacuum to eliminate scattering of the electrons due to interaction with gas molecules.

electron-beam microprobe analyzer. An instrument for selective analysis of a microscopic component or feature in which an electron beam bombards the point of interest in a vacuum at a given energy level. Scanning of a larger area permits determination of the distribution of selected elements. The analysis is made by measuring the wavelengths and intensities of secondary electromagnetic radiation resulting from the bombardment.

electron-beam welding. A welding process that produces coalescence of metals with the heat obtained from a concentrated beam composed primarily of high-velocity electrons impinging upon the surfaces to be joined.

electron compound. An intermediate phase on a *constitution diagram,* usually a binary phase, that has the same crystal structure and the same ratio of valence electrons to atoms as those of intermediate phases in several other systems. An electron compound is often a solid solution of variable composition and good metallic properties. Occasionally, an ordered arrangement of atoms is characteristic of the compound, in which case the range of composition is usually small. Phase stability depends essentially on electron concentration and crystal structure and has been observed at valence-electron-to-atom ratios of $^3/_2$, $^{21}/_{13}$ and $^7/_4$.

electrophoresis. Transport of charged colloidal or macromolecular materials in an electric field.

electroplating. Electrodepositing a metal or alloy in an adherent form on an object serving as a cathode.

electropolishing. (1) A technique commonly used to prepare metallographic specimens, in which a high polish is produced by making the specimen the anode in an electrolytic cell, where preferential dissolution at high points smooths the surface. (2) A variation of *chemical machining* wherein electrolytic deplating promotes chemical cutting, especially at surface irregularities.

electrorefining. Using electric or electrolytic methods to convert impure metal to purer metal, or to produce an alloy from impure or partly purified raw materials.

electroslag remelting. A *consumable-electrode remelting* process in which heat is generated by the passage of electric current through a conductive slag. The droplets of metal are refined by contact with the slag. Sometimes abbreviated ESR.

electroslag welding. A fusion welding process in which the welding heat is provided by passing an electric current through a layer of molten conductive slag contained in a pocket formed by molding shoes that bridge the gap between the members being welded. The resistance heated slag not only melts filler-metal electrodes as they are fed into the slag layer, but also provides shielding for the massive weld puddle characteristic of the process.

electrostrictive effect. The reversible interaction, exhibited by some crystalline materials, between an elastic strain and an electric field. The direction of the strain is independent of the polarity of the field. Compare with *piezoelectric effect.*

electrotinning. Electroplating tin on an object.

electrotyping. The production of printing plates by electroforming.

electrowinning. Recovery of a metal from an ore by means of electrochemical processes.

elongation. In tensile testing, the increase in the gage length, measured after fracture of the specimen within the gage length, usually expressed as a percentage of the original gage length.

elutriation. Separation of metal powder into particle-size fractions by means of a rising stream of gas or liquid.

embossing. Raising a design in relief against a surface.

embossing die. A die used for producing embossed designs.

embrittlement. Reduction in the normal ductility of a metal due to a physical or chemical change. Examples include *blue brittleness, hydrogen embrittlement* and *temper brittleness.*

emery. An impure mineral of the corundum or aluminum oxide type used extensively as an abrasive before the development of electric-furnace products.

emf. An abbreviation for *electromotive force.*

emissivity. Ratio of the amount of energy or of energetic particles radiated from a unit area of a surface to the amount radiated from a unit area of an ideal emitter under the same conditions.

emulsion. A dispersion of one liquid phase in another.

emulsion cleaner. A cleaner consisting of organic solvents dispersed in an aqueous medium with the aid of an emulsifying agent.

enantiotropy. The relation of crystal forms of the same substance in which one form is stable above a certain temperature and the other form is stable below that temperature. Ferrite and austenite are enantiotropic in ferrous alloys, for example.

end clearance angle. See *clearance angle,* and also sketches accompanying *face mill* and *single-point tool.*

end cutting-edge angle. The angle of concavity between the face cutting edge and the face plane of the cutter. It serves as relief to prevent the face cutting edges from rubbing in the cut. See sketches accompanying *face mill* and *single-point tool.*

end mark. A roll mark caused by the end of a sheet marking the roll during hot or cold rolling.

end milling. A method of machining with a rotating peripheral and end cutting tool. See also *face milling.*

end-quench hardenability test. A laboratory procedure for determining the hardenability of a steel or other ferrous alloy; widely referred to as the Jominy test. Hardenability is determined by heating a standard specimen above the upper critical temperature, placing the hot specimen in a fixture so that a stream of cold water impinges on one end, and, after cooling to room temperature is completed, measuring the hardness near the surface of the specimen at regularly spaced intervals along its length. The data are normally plotted as hardness versus distance from the quenched end.

end relief. Defined by sketch accompanying *single-point tool.*

endurance limit. The maximum stress below which a material can presumably endure an infinite number of stress cycles. If the stress is not completely reversed, the value of the mean stress, the minimum stress or the stress ratio also should be stated. Compare with *fatigue limit.*

endurance ratio. The ratio of the *endurance limit* for completely reversed flexural stress to the tensile strength of a given material.

entry mark (exit mark). A slight corrugation caused by the entry or exit rolls of a *roller leveling* unit.

epitaxy. Growth of an electrodeposit or vapor deposit in which the orientations of the crystals in the deposit are directly related to crystal orientations in the underlying crystalline substrate.

epsilon structure. A Hume-Rothery designation for structurally analogous close-packed phases or electron compounds like $CuZn_3$ that have ratios of seven valence electrons to four atoms. Not to be confused with the epsilon phase on a constitution diagram.

equiaxed grain structure. A structure in which the grains have approximately the same dimensions in all directions.

equilibrium. A dynamic condition of physical, chemical, mechanical or atomic balance which appears to be a condition of rest rather than one of change.

equilibrium diagram. A graphical representation of the temperature, pressure and composition limits of phase fields in an alloy system as they exist under conditions of complete equilibrium. In metal systems, pressure is usually considered constant.

Erichsen test. A cupping test in which a piece of sheet metal, restrained except at the center, is deformed by a cone-shaped spherical-end plunger until fracture occurs. The height of the cup in millimetres at fracture is a measure of the ductility of the metal.

erosion. Destruction of metals or other materials by the abrasive action of moving fluids, usually accelerated by the presence of solid particles or matter in suspension. When *corrosion* occurs simultaneously, the term "erosion-corrosion" is often used.

erosion-corrosion. See *erosion.*

etchant. A chemical substance or mixture used for *etching.*

etch cleaning. Removing soil by dissolving away some of the underlying metal.

etch cracks. Shallow cracks in hardened steel containing high residual surface stresses, produced on etching in an embrittling acid.

etch figures. Characteristic markings produced on crystal surfaces by chemical attack, usually having facets that are parallel to low-index crystallographic planes.

etching. (1) Subjecting the surface of a metal to preferential chemical or electrolytic attack in order to reveal structural details for metallographic examination. (2) Chemically or electrochemically removing tenacious films from a metal surface to condition the surface for a subsequent treatment, such as painting or electroplating.

eutectic. (1) An isothermal reversible reaction in which a liquid solution is converted into two or more intimately mixed solids on cooling, the number of solids formed being the same as the number of components in the system. (2) An alloy having the composition indicated by the eutectic point on an equilibrium diagram. (3) An alloy structure of intermixed solid constituents formed by a eutectic reaction.

eutectic carbide. Carbide formed during freezing as one of the mutually insoluble phases participating in the eutectic reaction of ferrous alloys.

eutectic melting. Melting of localized microscopic areas whose composition corresponds to that of the eutectic in the system.

eutectoid. (1) An isothermal reversible reaction in which a solid solution is converted into two or more intimately mixed solids on cooling, the number of solids formed being the same as the number of components in the system. (2) An alloy having the composition indicated by the eutectoid point on an equilibrium diagram. (3) An alloy structure of intermixed solid constituents formed by a eutectoid reaction.

exfoliation. A type of corrosion that progresses approximately parallel to the outer surface of the metal, causing layers of the metal to be elevated by the formation of corrosion product.

expanding. A process used to increase the diameter of a cup, shell or tube. See *bulging.*

expansion fit. An *interference* or *force fit* made by placing a cold (subzero) inside member into a warmer outside member and allowing an equalization of temperature.

explosion welding. A solid-state welding process effected by a controlled detonation, which causes the parts to move together at high velocity.

explosive forming. Shaping of metal parts wherein the forming pressure is generated by an explosive charge.

extensometer. An instrument for measuring changes in length caused by application or removal of a force. Commonly used in tension testing of metal specimens.

extractive metallurgy. The branch of *process metallurgy* dealing with the *winning* of metals from their ores. Compare with *refining.*

extra hard. A *temper* of nonferrous alloys and some ferrous alloys characterized by values of tensile strength and hardness about one-third of the way from those of *full hard* to those of *extra spring* temper.

extra spring. A *temper* of nonferrous alloys and some ferrous alloys corresponding approximately to a cold worked state above *full hard* beyond which further cold work will not measurably increase strength or hardness.

extruded hole. A hole formed by a punch that first cleanly cuts a hole and then is pushed farther through to form a flange with an enlargement of the original hole.

extrusion. Conversion of an ingot or billet into lengths of uniform cross section by forcing metal to flow plastically through a die orifice. In direct extrusion (forward extrusion), the die and ram are at opposite ends of the extrusion stock, and the product and ram travel in the same direction. Also, there is relative motion between the extrusion stock and the *container.* In indirect extrusion (backward extrusion), the die is at the ram end of the stock and the product travels in the direction opposite that of the ram, either around the ram (as in impact extrusion of cylinders such as cases for dry cell batteries) or up through the center of a hollow ram.

Impact extrusion is the process (or resultant product) in which a punch strikes a slug (usually unheated) in a confining die. The metal flow may be either between punch and die or through another opening. Impact extrusion of unheated slugs is often called cold extrusion. See also *Hooker process,* in which a pierced slug is used.

A stepped extrusion is a single product having one or more abrupt changes in cross section. It is produced by stopping extrusion to change dies. Often, such an extrusion is made in a complex die having a die section that can be freed from the main die and allowed to ride out with the product when extrusion is resumed.

extrusion billet. A metal slug used as *extrusion stock*.

extrusion defect. See preferred term, *extrusion pipe*.

extrusion ingot. A cast metal slug used as *extrusion stock*.

extrusion pipe. A central oxide-lined discontinuity that occasionally occurs in the last 10 to 20% of an extruded bar. It is caused by the oxidized outer surface of the billet flowing around the end of the billet and into the center of the bar during the final stages of extrusion. Also called *coring*.

extrusion stock. A rod, bar or other section used to make extrusions.

eyeleting. Displacing material about an opening in sheet or plate so that a lip protruding above the surface is formed.

F

face. In a lathe tool, the surface against which the chips bear as they are formed. See sketch accompanying *single-point tool*.

face mill. See definition of nomenclature in accompanying sketch.

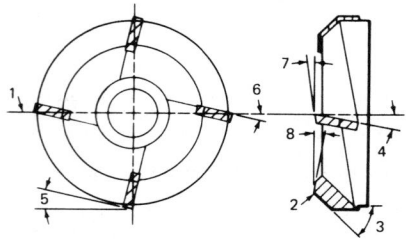

1. Reference plane
2. Tooth point
3. Corner angle (bevel)
4. Axial rake (positive)
5. Peripheral clearance angle
6. Radial rake (negative)
7. End clearance angle
8. End cutting edge angle

Face mill

face milling. Milling a surface that is perpendicular to the cutter axis.

face of weld. The exposed surface of an arc or gas weld on the side from which the welding was done. See sketch accompanying *fillet weld*.

face-type cutters. Cutters that can be mounted directly on and driven from the machine spindle nose.

facing. (1) In machining, generating a surface on a rotating workpiece by the traverse of a tool perpendicular to the axis of rotation. (2) In founding, special sand placed against a pattern to improve the surface quality of the casting. (3) For abrasion resistance, see preferred term *hard facing*.

fagot. In forging work, a bundle of iron bars that will be heated and then hammered and welded to form a single bar.

failure. A general term used to imply that a part in service (*a*) has become completely inoperable, (*b*) is still operable but is incapable of satisfactorily performing its intended function, or (*c*) has deteriorated seriously, to the point that it has become unreliable or unsafe for continued use.

false bottom. An *insert* put in either member of a die set to increase the strength and improve the life of the die.

false brinelling. Evenly spaced depressions in a raceway of a rolling-element bearing caused by fretting that occurs when the bearing is subjected to vibration while it is not rotating. Compare with *brinelling*.

false indication. In nondestructive inspection, an *indication* that may be interpreted erroneously as an *imperfection*. See also *artifact*.

false wiring. Same as *curling*.

fatigue. The phenomenon leading to fracture under repeated or fluctuating stresses having a maximum value less than the tensile strength of the material. Fatigue fractures are progressive, beginning as minute cracks that grow under the action of the fluctuating stress.

fatigue life. The number of cycles of stress that can be sustained prior to failure under a stated test condition.

fatigue limit. The maximum stress that presumably leads to fatigue fracture in a specified number of stress cycles. If the stress is not completely reversed, the value of the mean stress, the minimum stress or the stress ratio also should be stated. Compare with *endurance limit*.

fatigue notch factor (K_f). The ratio of the fatigue strength of an unnotched specimen to the fatigue strength of a notched specimen of the same material and condition; both strengths are determined at the same number of stress cycles.

fatigue notch sensitivity (q). An estimate of the effect of a notch or hole on the fatigue properties of a material; measured by $q = (K_f - 1)/(K_t - 1)$. A material is said to be fully notch sensitive if q approaches a value of 1.0; it is not notch sensitive if the ratio approaches 0. K_f is the *fatigue notch factor*, and K_t is the *stress-concentration factor*, for a specimen of the material containing a notch or hole of a given size and shape.

fatigue ratio. The *fatigue limit* under completely reversed flexural stress divided by the tensile strength for the same alloy and condition.

fatigue strength. The maximum stress that can be sustained for a specified number of cycles without failure, the stress being completely reversed within each cycle unless otherwise stated.

fatigue-strength reduction factor (K_f). The ratio of the fatigue strength of a member or specimen with no stress concentration to the fatigue strength with stress concentration. K_f has no meaning unless the stress range and the shape, size and material of the member or specimen are stated.

fatigue striations. Parallel lines frequently observed in electron microscope fractographs of fatigue fracture surfaces. The lines are transverse to the direction of local crack propagation; the distance between successive lines represents the advance of the crack front during one cycle of stress variation.

faying surface. The surface of a piece of metal (or a member) in contact with another to which it is joined or is to be joined.

feed. The rate at which a cutting tool or grinding wheel advances along or into the surface of a workpiece, the direction of advance depending on the type of operation involved.

feeder (feeder head, feedhead). A *riser*.

feeding. (1) Conveying metal stock or workpieces to a location for use or processing, such as wire to a consumable electrode, strip to a die, or workpieces to an assembler. (2) In casting, providing molten metal to a region undergoing solidification, usually at a rate sufficient to fill the mold cavity ahead of the solidification front and to make up for any shrinkage accompanying solidification.

feed lines. Linear marks on a machined or ground surface that are spaced at intervals equal to the *feed* per revolution or per stroke.

ferrimagnetic material. A material that macroscopically has properties similar to those of a *ferromagnetic material* but that microscopically also resembles an antiferromagnetic material in that some of the elementary magnetic moments are aligned antiparallel. If the moments are of different magnitudes, the material may still have a large resultant magnetization.

ferrite. (1) A solid solution of one or more elements in body-centered cubic iron. Unless otherwise designated (for instance, as chromium ferrite), the solute is generally assumed to be carbon. On some equilibrium diagrams, there are two ferrite regions separated by an austenite area. The lower area is alpha ferrite; the upper, delta ferrite. If there is no designation, alpha ferrite is assumed. (2) In the field of magnetics, substances having the general formula:

$$M^{++}O \cdot M_2^{+++}O_3$$

the trivalent metal often being iron.

ferrite banding. Parallel bands of free ferrite aligned in the direction of working. Sometimes referred to as ferrite streaks.

ferrite number. An arbitrary, standardized value designating the ferrite content of an austenitic stainless steel weld metal. This value directly replaces percent ferrite or volume percent ferrite and is determined by the magnetic test described in AWS A4.2.

ferrite streaks. Same as *ferrite banding*.

ferritic malleable. See *malleable cast iron*.

ferritizing anneal. A treatment given as-cast gray or ductile (nodular) iron to produce an essentially ferritic matrix. For the term to be meaningful, the final

microstructure desired or the time-temperature cycle used must be specified.

ferroalloy. An alloy of iron that contains a sufficient amount of one or more other chemical elements to be useful as an agent for introducing these elements into molten metal, especially into steel or cast iron.

ferrograph. An instrument used to determine the size distribution of wear particles in lubricating oils of mechanical systems.

ferromagnetic material. A material that in general exhibits the phenomena of hysteresis and saturation, and whose permeability is dependent on the magnetizing force. Microscopically, the elementary magnets are aligned parallel in volumes called *domains*. The unmagnetized condition of a ferromagnetic material results from the over-all neutralization of the magnetization of the domains to produce zero external magnetization. Compare with *paramagnetic material, diamagnetic material, ferrimagnetic material*.

fiber. (1) The characteristic of wrought metal that indicates *directional properties* and is revealed by etching of a longitudinal section or is manifested by the fibrous or woody appearance of a fracture. It is caused chiefly by extension of the constituents of the metal, both metallic and nonmetallic, in the direction of working. (2) The pattern of preferred orientation of metal crystals after a given deformation process, usually wiredrawing. See *preferred orientation*.

fiber stress. Local stress through a small area (a point or line) on a section where the stress is not uniform, as in a beam under a bending load.

fibrous fracture. A fracture whose surface is characterized by a dull gray or silky appearance. Contrast with *crystalline fracture*.

fibrous structure. (1) In forgings, a structure revealed as laminations, not necessarily detrimental, on an etched section or as a ropy appearance on a fracture. It is not to be confused with silky or ductile fracture of a clean metal. (2) In wrought iron, a structure consisting of slag fibers embedded in ferrite. (3) In rolled steel plate stock, a uniform, fine-grained structure on a fractured surface, free of laminations or shaletype discontinuities. As contrasted with (1), above, it is virtually synonymous with silky or ductile fracture.

filamentary shrinkage. A fine network of shrinkage cavities, occasionally found in steel castings, that produces a radiographic image resembling lace.

file hardness. Hardness as determined by the use of a file of standardized hardness on the assumption that a material that cannot be cut with the file is as hard as, or harder than, the file. Files covering a range of hardnesses may be employed.

filler. A material used to increase the bulk of a product without adding to its effectiveness in functional performance.

filler metal. Metal added in making a brazed, soldered or welded joint.

fillet. (1) A radius (curvature) imparted to inside meeting surfaces. (2) A concave cornerpiece used on foundry patterns.

fillet weld. A weld, approximately triangular in cross section, joining two surfaces essentially at right angles to each other in a lap, tee or corner joint.

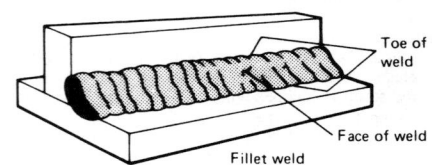

Toe of weld

Face of weld

Fillet weld

final annealing. An imprecise term used to denote the last anneal given to a nonferrous alloy prior to shipment.

fineness. A measure of the purity of gold or silver expressed in parts per thousand.

fines. (1) The product that passes through the finest screen in sorting crushed or ground material. (2) Sand grains that are substantially smaller than the predominating size in a batch or lot of foundry sand. (3) The portion of a metal powder composed of particles that are smaller than a specified size, currently less than 44 μm. See also *superfines*.

fine silver. Silver with a fineness of 999; equivalent to a minimum content of 99.9% Ag with the remaining content unrestricted.

finish. (1) Surface condition, quality or appearance of

a metal. (2) Stock on a forging or casting to be removed in finish machining.

finish allowance. The amount of excess metal surrounding the intended final configuration of a formed part; sometimes called forging envelope, machining allowance, or cleanup allowance.

finished steel. Steel that is ready for the market and has been processed beyond the stages of billets, blooms, sheet bars, slabs and wire rods.

finish grinding. The final grinding action on a workpiece, of which the objectives are surface finish and dimensional accuracy.

finishing die. The die used to make the final impression on a forging. Sometimes called finisher.

finishing temperature. The temperature at which *hot working* is completed.

finish machining. A machining process analogous to *finish grinding*.

fire-refined copper. Copper that has been refined by the use of a furnace process only, including refinery shapes and, by extension, fabricators' products made therefrom. Usually, when this term is used alone it refers to fire-refined tough pitch copper without elements other than oxygen being present in significant amounts.

fire scale. Intergranular copper oxide remaining below the surface of silver-copper alloys that have been annealed and pickled.

fir-tree crystal. A type of *dendrite*.

fish eyes. Areas on a steel fracture surface having a characteristic white crystalline appearance.

fishmouthing. Same as *alligatoring*.

fishscale. A scaly appearance in a porcelain enamel coating in which the evolution of hydrogen from the basis metal (iron or steel) causes loss of adhesion between the enamel and basis metal. Individual scales are usually small, but have been observed in sizes up to 25 mm or more in diameter. The scales are somewhat like blisters that have cracked partway around the perimeter but still remain attached to the coating around the rest of the perimeter; if detached completely, it is one form of *pop-off*.

fishtail. (1) In roll forging, the excess trailing end of a forging. It is often used, before being trimmed off, as a tong hold for a subsequent forging operation. (2) In hot rolling or extrusion, the imperfectly shaped trailing end of a bar or special section that must be cut off and discarded as mill scrap.

fit. The amount of clearance or interference between mating parts is called actual fit. Fit is the preferable term for the range of clearance or interference that may result from the specified limits on dimensions (limits of size). Refer to ANSI standards.

fixed-feed grinding. Grinding in which the wheel is fed into the work, or vice versa, by given increments or at a given rate.

fixed-position welding. Welding in which the work is held in a stationary position.

fixture. A positioning device used to hold the workpiece only.

flake powder. Metal powder in the form of flat or scalelike particles that are relatively thin.

flakes. Short, discontinuous internal fissures in ferrous metals attributed to stresses produced by localized transformation and decreased solubility of hydrogen during cooling after hot working. In a fracture surface, flakes appear as bright silvery areas; on an etched surface, they appear as short, discontinuous cracks. Also called shatter cracks or snowflakes.

flame annealing. Annealing in which the heat is applied directly by a flame.

flame cleaning. Cleaning metal surfaces of scale, rust, dirt and moisture by use of a gas flame.

flame hardening. A process for hardening the surfaces of hardenable ferrous alloys in which an intense flame is used to heat the surface layers above the upper transformation temperature, whereupon the workpiece is immediately quenched.

flame spraying. *Thermal spraying* in which a coating material is fed into an oxyfuel gas flame, where it is melted. Compressed gas may or may not be used to atomize the coating material and propel it onto the substrate.

flame straightening. Correcting distortion in metal structures by localized heating with a gas flame.

flank. The end surface of a tool that is adjacent to the cutting edge and below it when the tool is in a horizontal position, as for turning. See sketch accompanying *single-point tool*.

flank wear. The loss of relief on the flank of the tool

behind the cutting edge due to rubbing contact between the work and the tool during cutting; measured in terms of linear dimension behind the original cutting edge.

flapping. In copper refining, hastening oxidation of molten copper by striking through the slag-covered surface of the melt with a *rabble* just before the bath is poled.

flare test. A test applied to tubing, involving tapered expansion over a cone. Similar to *pin expansion test*.

flaring. (1) Forming an outward acute-angle flange on a tubular part. (2) Forming a flange by using the head of a hydraulic press.

flash. (1) In forging, excess metal forced out between the upper and lower dies. (2) In casting, a fin of metal that results from leakage between mating mold surfaces. (3) In resistance butt welding, a fin formed perpendicular to the direction of applied pressure.

flashback. The recession of a flame into or in back of the interior of a torch. See *backfire*.

flash butt welding. See *flash welding*.

flash extension. Portion of flash remaining after trimming. Flash extension is measured from the intersection of the draft and flash at the body of the forging to the trimmed edge of the stock.

flashing. In flash welding, the heating portion of the cycle, consisting of a series of rapidly recurring localized short circuits followed by molten metal expulsions, during which time the surfaces to be welded are moved one toward the other at a predetermined speed.

flash land. Relief at the parting line of a set of closed-die forging dies that is designed either to restrict or to encourage growth of flash, whichever is required to ensure complete filling of the finishing impression.

flash line. The line of location of flash formed around a forging or casting.

flash plate. A very thin final electrodeposited film of metal.

flash welding. A resistance welding process that joins metals by first heating abutting surfaces by passage of an electric current across the joint, then forcing the surfaces together by the application of pressure. Flashing and upsetting are accompanied by expulsion of metal from the joint.

flask. A metal or wood frame used for making and holding a sand mold. The upper part is called the cope; the lower, the drag.

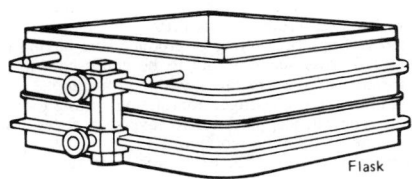

Flask

flat-die forging. Forging metal between flat or simple-contour dies by repeated strokes and manipulation of the workpiece. Also known as open-die forging, hand forging or smith forging.

flat drill. A rotary end-cutting tool constructed from a flat piece of material, provided with suitable cutting lips at the cutting end.

flat edge trimmer. A machine for trimming notched edges on shells. The slide is cam driven so as to obtain a brief dwell at the bottom of the stroke, at which time the die, sometimes called a shimmy die, oscillates to trim the part.

flat-position welding. Welding from the upper side,

the face of the weld being horizontal. Also called downhand welding.

flattening. (1) A preliminary operation performed on forging stock so as to position the metal for a subsequent forging operation. (2) Removing irregularities or distortion in sheets or plates by a method such as *roller leveling* or *stretcher leveling*.

flattening test. A quality test for tubing in which a specimen is flattened to a specified height between parallel plates.

flat wire. A roughly rectangular or square mill product, narrower than *strip*, in which all surfaces are rolled or drawn without any previous slitting, shearing or sawing.

flaw. A nonspecific term often used to imply a crack-like discontinuity. See preferred terms *discontinuity, imperfection, defect*.

flexible cam. An adjustable pressure-control cam of spring steel strips used to obtain varying pressure during a forming cycle.

flex roll. A movable jump roll designed to push up against a sheet as it passes through a roller leveler. The flex roll can be adjusted to deflect the sheet any amount up to the roll diameter.

flex rolling. Passing sheets through a flex-roll unit to minimize yield-point elongation so as to reduce the tendency for *stretcher strains* to appear during forming.

floating die. (1) A die mounted in a die holder or a punch mounted in its holder, such that a slight amount of motion compensates for tolerance in the die parts, the work or the press. (2) A die mounted on heavy springs to allow vertical motion in some trimming, shearing and forming operations.

floating plug. In tube drawing, an unsupported mandrel that locates itself at the die inside the tube, causing a reduction in wall thickness while the die is effecting a reduction in outside diameter.

floppers. On metals, lines or ridges that are transverse to the direction of rolling and generally confined to the section midway between the edges of a coil as rolled.

flospinning. Forming cylindrical, conical and curvilinear shaped parts by power spinning over a rotating mandrel.

flotation. The concentration of valuable minerals from ores by agitation of the ground material with water, oil and flotation chemicals. The valuable minerals are generally wetted by the oil, lifted to the surface by clinging air bubbles and then floated off.

flowability. A characteristic of a foundry sand mixture that enables it to move under pressure or vibration so that it makes intimate contact with all surfaces of the pattern or core box.

flow brazing. Brazing by pouring hot molten nonferrous filler metal over a joint until the brazing temperature is attained. The filler metal is distributed in the joint by capillary action.

flow brightening. Melting of an electrodeposit, followed by solidification, especially of tin plate.

flow lines. (1) Texture showing the direction of metal flow during hot or cold working. Flow lines often can be revealed by etching the surface or a section of a metal part (see macrograph at foot of page). (2) In mechanical metallurgy, paths followed by minute volumes of metal during deformation.

flow stress. The uniaxial true stress at the onset of plastic deformation in a metal.

fluidity. The ability of liquid metal to run into and fill a mold cavity.

fluorescence. The emission of characteristic electromagnetic radiation by a substance as a result of the absorption of electromagnetic or corpuscular radiation having a greater unit energy than that of the flu-

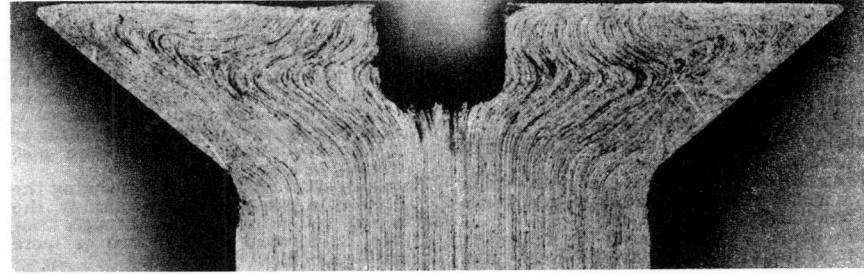

orescent radiation. It occurs only so long as the stimulus responsible for it is maintained.

fluorescent magnetic-particle inspection. Inspection with either dry magnetic particles or those in a liquid suspension, the particles being coated with a fluorescent substance to increase the visibility of the indications.

fluorescent penetrant inspection. Inspection using a fluorescent liquid that will penetrate any surface opening; after the surface has been wiped clean, the location of any surface flaws may be detected by the fluorescence, under ultraviolet light, of back-seepage of the fluid.

fluoroscopy. An inspection procedure in which the radiographic image of the subject is viewed on a fluorescent screen, normally limited to low-density materials or thin sections of metals because of the low light output of the fluorescent screen at safe levels of radiation.

flute. (1) As applied to drills, reamers and taps, the channels or grooves formed in the body of the tool to provide cutting edges and to permit passage of cutting fluid and chips. (2) As applied to milling cutters and hobs, the chip space between the back of one tooth and the face of the following tooth.

fluting. (1) Forming longitudinal recesses in a cylindrical part, or radial recesses in a conical part. (2) A series of sharp parallel kinks or creases occurring in the arc when sheet metal is roll formed into a cylindrical shape.

flux. (1) In metal refining, a material used to remove undesirable substances, like sand, ash or dirt, as a molten mixture. It is also used as a protective covering for certain molten metal baths. Lime or limestone is generally used to remove sand, as in iron smelting; sand, to remove iron oxide in copper refining. (2) In brazing, cutting, soldering or welding, material used to prevent the formation of, or to dissolve and facilitate removal of, oxides and other undesirable substances.

flux-cored arc welding. An arc welding process that joins metals by heating them with an arc between a continuous tubular filler-metal electrode and the work. Shielding is provided by a flux contained within the consumable tubular electrode. Additional shielding may or may not be obtained from an externally supplied gas or gas mixture. See also *electrogas welding.*

flux density. In magnetism, the number of *flux lines* per unit area passing through a cross section at right angles. It is given by $B = \mu H$, where μ and H are permeability and magnetic-field intensity, respectively.

flux lines. Imaginary lines used as a means of explaining the behavior of magnetic and other fields. Their concept is based on the pattern of lines produced when magnetic particles are sprinkled over a permanent magnet. Sometimes called magnetic lines of force.

flux-oxygen cutting. Oxygen cutting with the aid of a flux.

fly ash. A finely divided siliceous material formed during the combustion of coal, coke or other solid fuels.

fly cutting. Cutting with a single-tooth milling cutter.

flying shear. A machine for cutting continuous rolled products to length that does not require a halt in rolling, but rather moves along the runout table at the same speed as the product while performing the cutting, then returns to the starting point in time to cut the next piece.

fog quenching. Quenching in a fine vapor or mist.

foil. Metal in sheet form less than 0.15 mm (0.006 in.) thick.

fold. Same as *lap.*

follow board. A board contoured to a pattern to facilitate the making of a sand mold.

follow die. A *progressive die* consisting of two or more parts in a single holder, used with a separate lower die to perform more than one operation (such as piercing and blanking) on a part at two or more stations.

foot press. A small press with low capacity actuated by foot pressure on a treadle.

force fit. Any of various interference fits between parts assembled under various amounts of force.

forehand welding. Welding in which the palm of the principal hand (torch or electrode hand) of the welder faces the direction of travel. It has special significance in oxyfuel gas welding in that the flame is directed ahead of the weld bead, which provides *preheating.* Contrast with *backhand welding.*

forgeability. Term used to describe the relative ability of material to flow under a compressive load without rupture.

forge delay time. In spot, seam or projection welding, the time between the start of the welding, current or weld interval and the application of forging pressure.

forge welding. Solid-state welding in which metals are heated in a forge (in air) and then welded together by applying pressure or blows sufficient to cause permanent deformation at the interface.

forging. Plastically deforming metal, usually hot, into desired shapes with compressive force, with or without dies.

forging billet. A wrought metal slug used as *forging stock.*

forging envelope. See *finish allowance.*

forging ingot. A cast metal slug used as *forging stock.*

forging machine. A type of forging equipment, related to the mechanical press, in which the main forming energy is applied horizontally to the workpiece, which is held by dies. Commonly called upsetter or header.

forging plane. In forging, the plane that includes the principal die face and that is perpendicular to the direction of ram travel. When parting surfaces of the dies are flat, the forging plane coincides with the parting line. Contrast *parting plane.*

forging range. Temperature range in which a metal can be forged successfully.

forging rolls. A machine used in *roll forging.* Also called gap rolls.

forging stock. A rod, bar or other section used to make forgings.

formability. The relative ease with which a metal can be shaped through plastic deformation. See *drawability.*

form block. Tooling, usually the male part, used for forming sheet-metal contours, being generally employed in the rubber-pad process.

form cutter. Any cutter, profile sharpened or cam relieved, shaped to produce a specified form on the work.

form die. A die used to change the shape of a blank with minimum plastic flow.

form grinding. Grinding with a wheel having a contour on its cutting face that is a mating fit to the desired form.

forming. Making a change, with the exception of shearing or blanking, in the shape or contour of a metal part without intentionally altering its thickness.

form-relieved cutter. A cutter so relieved that by grinding only the tooth face the original form is maintained throughout its life.

form rolling. Hot rolling to produce bars having contoured cross sections; not to be confused with *roll forming* of sheet metal or with *roll forging.*

form tool. A single-edge, nonrotating cutting tool, circular or flat, that produces its inverse or reverse form counterpart upon a workpiece.

forward extrusion. Same as direct extrusion. See *extrusion.*

foundry. A commercial establishment or building where metal castings are produced.

four-high mill. A type of rolling mill, commonly used for flat-rolled mill products, in which two large-diameter backup rolls are employed to reinforce two smaller working rolls, which are in contact with the product. Either the working rolls or the backup rolls may be driven. Compare with *two-high mill, cluster mill.*

four-point press. A press whose slide is actuated by four connections and four cranks, eccentrics, or cyl-

inders, the chief merit being to equalize the pressure at the corners of the slides.

fraction. In powder metallurgy, the portion of a powder sample that lies between two stated particle sizes. Synonymous with cut.

fractography. Descriptive treatment of fracture, especially in metals, with specific reference to photographs of the fracture surface. Macrofractography involves photographs at low magnification; microfractography, photographs at high magnification.

fracture mechanics. See *linear elastic fracture mechanics.*

fracture stress. (1) The maximum principal true stress at fracture. Usually refers to unnotched tensile specimens. (2) The (hypothetical) true stress that will cause fracture without further deformation at any given strain.

fracture test. Test in which a specimen is broken and its fracture surface is examined with the unaided eye or with a low-power microscope to determine such factors as composition, grain size, case depth or soundness.

fracture toughness. See *stress-intensity factor.*

fragmentation. The subdivision of a grain into small discrete crystallites outlined by a heavily deformed network of intersecting slip as a result of cold working. These small crystals or fragments differ from one another in orientation and tend to rotate to a stable orientation determined by the slip systems.

freckling. A type of segregation revealed as dark spots on a macroetched surface of a consumable-electrode vacuum arc remelted alloy.

free carbon. The part of the total carbon in steel or cast iron that is present in elemental form as graphite or temper carbon. Contrast with *combined carbon.*

free ferrite. Ferrite that is formed directly from the decomposition of hypoeutectoid austenite during cooling, without the simultaneous formation of cementite. Also called proeutectoid ferrite.

free fit. Any of various clearance fits for assembly by hand and free rotation of parts. See *running fit.*

free machining. Pertains to the machining characteristics of an alloy to which one or more ingredients have been introduced to give small broken chips, lower power consumption, better surface finish and longer tool life; among such additions are sulfur or lead to steel, lead to brass, lead and bismuth to aluminum, and sulfur or selenium to stainless steel.

freezing range. That temperature range between *liquidus* and *solidus* temperatures in which molten and solid constituents coexist.

fretting. A type of wear that occurs between tight-fitting surfaces subjected to cyclic relative motion of extremely small amplitude. Usually, fretting is accompanied by corrosion, especially of the very fine wear debris. Also referred to as fretting corrosion, *false brinelling* (in rolling-element bearings), friction oxidation, *chafing fatigue,* molecular attrition and wear oxidation.

fretting fatigue. Fatigue fracture that initiates at a surface area where fretting has occurred.

friction welding. A solid-state process in which materials are welded by the heat obtained from rubbing together of surfaces that are held against each other under pressure.

full annealing. An imprecise term that denotes an annealing cycle designed to produce minimum strength and hardness. For the term to be meaningful, the composition and starting condition of the material and the time-temperature cycle used must be stated.

full-automatic plating. Electroplating in which the work is automatically conveyed through the complete cycle.

full center. Mild waviness down the center of a sheet or strip.

fuller. In preliminary forging, the portion of a die that reduces the cross-sectional area between the ends of the stock and permits the metal to move outward.

full hard. A *temper* of nonferrous alloys and some ferrous alloys corresponding approximately to a cold worked state beyond which the material can no longer be formed by bending. In specifications, a full hard temper is commonly defined in terms of minimum hardness or minimum tensile strength (or, alternatively, a range of hardness or strength) corresponding to a specific percentage of cold reduction following full annealing. For aluminum, a full hard temper is equivalent to a reduction of 75% from *dead soft;* for austenitic stainless steels, a reduction of about 50 to 55%.

furnace brazing. A mass-production *brazing* process

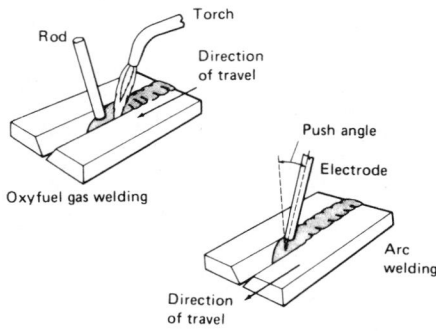

Torch

Rod

Direction of travel

Oxyfuel gas welding

Push angle

Electrode

Arc welding

Direction of travel

Forehand welding

in which the filler metal is preplaced on the joint, then the entire assembly is heated to brazing temperature in a furnace. Usually, a protective furnace atmosphere is required, and wetting of the joint surfaces is accomplished without using a brazing flux.

fusion. A change of state from solid to liquid; melting.

fusion face. A surface of the base metal that will be melted during welding.

fusion welding. Any welding process in which filler metal and base metal (substrate), or base metal only, are melted together to complete the weld.

fusion zone. In a weldment, the area of base metal melted as determined on a cross section through the weld.

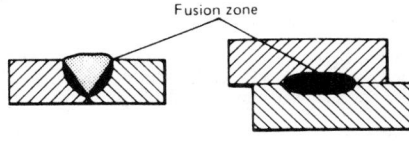
Fusion zone

G

gag. A metal spacer inserted so as to render a floating tool or punch inoperative.

gage. (1) The thickness (or diameter) of sheet or wire. The various standards are arbitrary and differ, ferrous from nonferrous products and sheet from wire. (2) An instrument used to measure thickness or length. (3) An aid for visual inspection that enables the inspector to determine more reliably whether the size or contour of a formed part meets dimensional requirements.

gage length. The original length of that portion of the specimen over which strain, change of length and other characteristics are measured.

gagger. An irregularly shaped piece of metal used for reinforcement and support in a sand mold.

galling. A condition whereby excessive friction between high spots results in localized welding with subsequent spalling and further roughening of the rubbing surface(s) of one or both of two mating parts.

galvanic cell. A cell in which chemical change is the source of electrical energy. It usually consists of two dissimilar conductors in contact with each other and with an electrolyte, or of two similar conductors in contact with each other and with dissimilar electrolytes.

galvanic corrosion. Corrosion associated with the current of a galvanic cell consisting of two dissimilar conductors in an electrolyte or two similar conductors in dissimilar electrolytes. Where the two dissimilar metals are in contact, the resulting reaction is referred to as couple action.

galvanic series. A series of metals and alloys arranged according to their relative electrode potentials in a specified environment. Compare with *electromotive series*.

galvanize. To coat a metal surface with zinc using any of various processes.

galvanneal. To produce a zinc-iron alloy coating on iron or steel by keeping the coating molten after hot-dip galvanizing until the zinc alloys completely with the basis metal.

gamma iron. The face-centered cubic form of pure iron, stable from 910 to 1400 °C (1670 to 2550 °F).

gamma ray. Short-wavelength electromagnetic radiation, similar to x-rays but of nuclear origin, with a range of wavelengths from about 10^{-14} to 10^{-10} m.

gamma structure. A Hume-Rothery designation for structurally analogous phases or electron compounds that have ratios of 21 valence electrons to 13 atoms; generally, a large complex cubic structure. Not to be confused with gamma phase on a constitution diagram.

gang milling. Milling with several cutters mounted on the same arbor or with workpieces similarly positioned for cutting either simultaneously or consecutively during a single setup.

gang slitter. A machine with a number of pairs of rotary cutters spaced on two parallel shafts, used for slitting sheet metal into strips or for trimming the edges of sheets.

gangue. The worthless portion of an ore that is separated from the desired part before smelting is commenced.

gap. The root opening in a weld joint.

gap-frame press. A general classification of presses in which the uprights or housings are made in the form of a letter "C," thereby making three sides of the die space accessible.

gas cyaniding. A misnomer for *carbonitriding*.

gas holes. Holes in castings or welds that are formed by gas escaping from molten metal as it solidifies. Gas holes may occur individually, in clusters, or distributed throughout the solidified metal.

gas metal-arc welding. A process for welding metals together by heating them with an arc between a continuous filler-metal electrode and the work. Shielding is obtained entirely from an externally supplied gas or gas mixture. Some methods of this process are called MIG or CO_2 welding. See also *electrogas welding, pulsed power welding*.

gas plating. Same as *vapor plating*.

gas pocket. A cavity caused by entrapped gas.

gas porosity. Fine holes or pores within a metal that are caused by entrapped gas or by evolution of dissolved gas during solidification.

gas-shielded arc welding. Arc welding in which the arc and molten metal are shielded from the atmosphere by a stream of gas, such as argon, helium, argon-hydrogen mixtures or carbon dioxide.

gassing. (1) Absorption of gas by a metal. (2) Evolution of gas from a metal during melting operations or on solidification. (3) Evolution of gas from an electrode during electrolysis.

gas tungsten-arc cutting. An arc-cutting process in which metals are severed by melting them with an arc between a single tungsten (nonconsumable) electrode and the work. Shielding is obtained from a gas or gas mixture.

gas tungsten-arc welding. A fusion welding process in which metals are joined by heating them with an electric arc between a nonconsumable tungsten electrode and the work. A gas or gas mixture shields the arc and the weld puddle. Pressure may or may not be applied to the joint, and filler metal may or may not be added. Sometimes referred to as TIG welding.

gas welding. See preferred term, *oxyfuel gas welding*.

gate. The portion of the runner in a mold through which molten metal enters the mold cavity. Sometimes the generic term is applied to the entire network of connecting channels that conduct metal into the mold cavity.

gated pattern. A *pattern* that includes not only the contours of the part to be cast, but also the *gates*.

gathering. A forging operation that increases the cross section of part of the stock; usually a preliminary operation.

gathering stock. Any operation whereby the cross section of a portion of the forging stock is increased beyond its original size.

geared press. A press whose main crank or eccentric shaft is connected by gears to the driving source.

ghost lines. Lines running parallel to the rolling direction that appear in a panel when it is stretched. These lines may not be evident unless the panel has been sanded or painted. (Not to be confused with *leveler lines*.)

gibs. Guides that ensure the proper restrained motion of the slide, usually being adjustable to compensate for wear.

glass electrode. A glass membrane electrode used to measure pH or hydrogen-ion activity.

glazing. Dulling the abrasive grains in the cutting face of a wheel during grinding.

glide. (1) Same as *slip*. (2) A noncrystallographic shearing movement, such as of one grain over another.

globular transfer. In consumable-electrode arc welding, a type of metal transfer in which molten filler metal passes across the arc as large droplets. Compare with *spray transfer, short-circuiting transfer*.

gold filled. Covered on one or more surfaces with a layer of gold alloy to form a clad metal. By commercial agreement, a quality mark showing the quantity and fineness of gold alloy may be affixed, indicating the actual proportional weight and karat fineness of the gold alloy cladding. For example, "$1/10$ 12K Gold Filled" means that the article consists of base metal covered on one or more surfaces with a gold alloy of 12-karat fineness comprising $1/10$th part by weight of the entire metal in the article. No article having a gold alloy coating of less than 10-karat fineness may have any quality mark affixed. No article having a gold alloy portion of less than $1/20$th by weight may be marked "Gold Filled," but may be marked

"Rolled Gold Plate" provided that the proportional fraction and fineness designation precedes. These standards do not necessarily apply to watch cases.

gooseneck. See *hot chamber machine*.

G-P zone. A *Guinier-Preston zone*.

grain. An individual crystal in a polycrystalline metal or alloy; it may or may not contain twinned regions and subgrains.

grain-boundary corrosion. Same as *intergranular corrosion*. See also *interdendritic corrosion*.

grain-fineness number. A weighted average grain size of a granular material. The AFS grain fineness number is calculated with prescribed weighting factors from the standard screen analysis.

grain flow. Fiberlike lines appearing on polished and etched sections of forgings, which are caused by orientation of the constituents of the metal in the direction of working during forging. Grain flow produced by proper die design can improve required mechanical properties of forgings.

grain growth. An increase in the average size of the grains in polycrystalline metal, usually as a result of heating at elevated temperature.

grain refiner. A material added to a molten metal to induce a finer-than-normal grain size in the final structure.

grain size. (1) For metals, a measure of the areas or volumes of grains in a polycrystalline material, usually expressed as an average when the individual sizes are fairly uniform. In metals containing two or more phases, grain size refers to that of the matrix unless otherwise specified. Grain size is reported in terms of number of grains per unit area or volume, in terms of average diameter, or as a grain-size number derived from area measurements. (2) For grinding wheels, see preferred term, *grit size*.

granular fracture. A type of irregular surface produced when metal is broken, characterized by a rough, grainlike appearance as differentiated from a smooth silky, or fibrous, type. It can be subclassified into transgranular and intergranular forms. This type of fracture is frequently called *crystalline fracture*, but the inference that the metal broke because it "crystallized" is not justified, because all metals are crystalline when in the solid state. Contrast with *fibrous fracture, silky fracture*.

granular powder. Particles of metal powder having approximately equidimensional nonspherical shapes.

granulated metal. Small pellets produced by pouring liquid metal through a screen or by dropping it onto a revolving disk, and, in both instances, chilling with water.

granulation. Production of coarse metal particles by pouring the molten metal through a screen into water or by agitating the molten metal violently during its solidification.

graphitic carbon. Free carbon in steel or cast iron.

graphitic corrosion. Corrosion of gray iron in which the iron matrix is selectively leached away, leaving a porous mass of graphite behind; it occurs in relatively mild aqueous solutions and on buried pipe and fittings.

graphitic steel. Alloy steel made so that part of the carbon is present as graphite.

graphitization. Formation of graphite in iron or steel. Where graphite is formed during solidification, the phenomenon is called primary graphitization; where formed later by heat treatment, secondary graphitization.

graphitizing. Annealing a ferrous alloy in such a way that some or all of the carbon is precipitated as graphite.

gravity hammer. A class of forging hammer wherein energy for forging is obtained by the mass and velocity of a freely falling ram and the attached upper die. Examples: *board hammers* and *air-lift hammers*.

gravity segregation. Variable composition of a casting or ingot caused by settling out of heavy constituents, or rising of light constituents, before or during solidification.

gray cast iron. A *cast iron* that gives a gray fracture due to the presence of flake graphite. Often called gray iron.

green compact. An unsintered powder metallurgy compact.

green density. Same as *pressed density*.

green rot. A form of high-temperature attack on stainless steels, nickel-chromium alloys and nickel-chromium-iron alloys subjected to simultaneous oxidation and carburization. Basically, attack occurs first by precipitation of chromium as chromium carbide, then by oxidation of the carbide particles.

green sand. A naturally bonded sand, or a compounded molding sand mixture, that has been "tempered" with water and that is used while still moist.

green sand core. (1) A *core* made of *green sand* and used as rammed. (2) A sand core that is used in the unbaked condition.

green sand mold. A casting mold composed of moist prepared molding sand. Contrast with *dry sand mold.*

grindability. Relative ease of grinding, analogous to *machinability.*

grindability index. A measure of the grindability of a material under specified grinding conditions, expressed in terms of volume of material removed per unit volume of wheel wear.

grinding. Removing material from a workpiece with a grinding wheel or abrasive belt.

grinding burn. See *burning* (2).

grinding cracks. Shallow cracks formed in the surfaces of relatively hard materials because of excessive grinding heat or the high sensitivity of the material. See *grinding sensitivity.*

grinding fluid. *Cutting fluid* used in grinding.

grinding oil. An oil-type grinding fluid; it may contain additives, but not water.

grinding relief. A groove or recess located at the boundary of a surface to permit the corner of the wheel to overhang during grinding.

grinding sensitivity. Susceptibility of a material to surface damage such as *grinding cracks;* it can be affected by such factors as hardness, microstructure, hydrogen content and residual stress.

grinding stress. *Residual stress,* generated by grinding, in the surface layer of work. It may be tensile or compressive, or both.

grinding wheel. A cutting tool of circular shape made of abrasive grains bonded together.

grit blasting. Abrasive blasting with small irregular pieces of steel, malleable cast iron or hard nonmetallic materials.

grit size. Nominal size of abrasive particles in a grinding wheel, corresponding to the number of openings per linear inch in a screen through which the particles can just pass. Sometimes, but inadvisedly, called *grain size.*

grizzly. A set of parallel bars (or grating) used for coarse separation or screening of ores, rock or other material.

groove angle. The total included angle of the groove between parts to be joined. Thus, the sum of two bevel angles, either or both of which may be zero degrees.

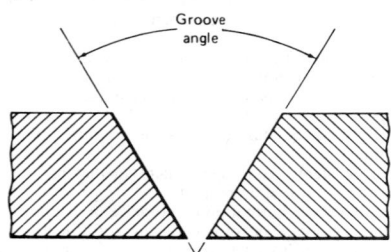

groove face. The portion of a surface or surfaces of a member included in a groove. See sketch accompanying *root of joint.*

groove weld. A weld made in the groove between two members. The standard types are square, single-bevel, single flare-bevel, single flare-V, single-J, single-U, single-V, double-bevel, double flare-bevel, double flare-V, double-J, double-U and double-V.

gross porosity. In weld metal or in a casting, pores, gas holes or globular voids that are larger and in much greater numbers than those obtained in good practice.

ground connection. In arc welding, a device used for attaching the work lead (ground cable) to the work.

growth. In cast iron, a permanent increase in dimensions resulting from repeated or prolonged heating at temperatures above 480 °C (900 °F) due either to graphitizing of carbides or to oxidation.

guard. (1) A device, often made of sheet metal or wire screening, that prevents accidental contact with moving parts of machinery. (2) In electroplating, same as *robber.*

Guerin forming. A trade-name process. See *rubber-pad forming.*

guided bend test. A test in which the specimen is bent to a definite shape by means of a jig.

guide mill. A small hand mill with several stands in a train and with guides for the work at the entrance to the rolls.

Guinier-Preston (G-P) zone. A small precipitation domain in a supersaturated metallic solid solution. A G-P zone has no well-defined crystalline structure of its own and contains an abnormally high concentration of solute atoms. The formation of G-P zones constitutes the first stage of precipitation and is usually accompanied by a change in properties of the solid solution in which they occur.

gun drill. A drill, usually with one or more flutes and with coolant passages through the drill body, used for deep hole drilling.

gutter. The clearance around the land of a forging die which provides space for the flash without trapping it in the dies.

H

habit plane. The plane or system of planes of a crystalline phase along which some phenomenon such as twinning or transformation occurs.

half cell. An electrode immersed in a suitable electrolyte, designed for measurements of electrode potential.

half hard. A *temper* of nonferrous alloys and some ferrous alloys characterized by tensile strength about midway between those of *dead soft* and *full hard* tempers.

Hall process. A commercial process for winning aluminum from alumina by electrolytic reduction of a fused bath of alumina dissolved in cryolite.

hammer forging. Forging in which the work is deformed by repeated blows. Compare with *press forging.*

hammering. Beating metal sheet into a desired shape either over a form or on a high-speed mechanical hammer and a similar anvil to produce the required dishing or thinning.

hammer welding. *Forge welding* by hammering.

hand brake. A small manual folding machine designed to bend sheet metal, similar in design and purpose to a *press brake.*

hand forging. Same as *flat-die forging.*

handling breaks. Irregular *breaks* caused by improper handling of sheets during processing. These breaks result from bending or sagging of the sheets during handling.

Hansgirg process. A process for producing magnesium by reduction of magnesium oxide with carbon.

hard chromium. Chromium electrodeposited for engineering purposes (such as to increase the wear resistance of sliding metal surfaces) rather than as a decorative coating. It is usually applied directly to basis metal and is customarily thicker than a decorative deposit, but not necessarily harder.

hard drawn. An imprecise term applied to drawn products, such as wire and tubing, that indicates substantial cold reduction without subsequent annealing. Compare with *light drawn.*

hardenability. The relative ability of a ferrous alloy to form martensite when quenched from a temperature above the upper critical temperature. Hardenability is commonly measured as the distance below a quenched surface at which the metal exhibits a specific hardness (50 HRC, for example) or a specific percentage of martensite in the microstructure.

hardener. An alloy rich in one or more alloying elements that is added to a melt to permit closer control of composition than is possible by addition of pure metals, or to introduce refractory elements not readily alloyed with the base metal. Sometimes called *master alloy* or rich alloy.

hardening. Increasing hardness by suitable treatment, usually involving heating and cooling. When applicable, the following more specific terms should be used: *age hardening, case hardening, flame hardening, induction hardening, precipitation hardening* and *quench hardening.*

hard facing. Depositing filler metal on a surface by welding, spraying or braze welding to increase resistance to abrasion, erosion, wear, galling, impact or cavitation damage.

hard head. A hard, brittle, white residue obtained in refining of tin by liquation, containing, among other things, tin, iron, arsenic and copper. Also, a refractory lump of ore only partly smelted.

hardness. Resistance of metal to plastic deformation, usually by indentation. However, the term may also refer to stiffness or temper, or to resistance to scratching, abrasion or cutting. Indentation hardness may be measured by various hardness tests, such as *Brinell, Rockwell* and *Vickers.*

hard surfacing. Same as *hard facing.*

hard temper. Same as *full hard* temper.

Haring cell. A four-electrode cell for measurement of electrolyte resistance and electrode polarization during electrolysis.

Hartmann lines. Same as *Lüders lines.*

H-band steel. Alloy steel produced to specified limits of hardenability; the chemical composition range may be slightly different from that of the corresponding grade of ordinary alloy steel.

header. See *upsetter.*

heading. Upsetting wire, rod or bar stock in dies to form parts that usually contain portions that are greater in cross-sectional area than the original wire, rod or bar.

healed-over scratch. A scratch that occurred in an earlier mill operation and was partially masked in subsequent rolling. It may open up during forming.

hearth. The bottom portions of certain furnaces, such as blast furnaces, air furnaces, and other reverberatory furnaces, that support the charge and sometimes collect and hold molten metal.

heat-affected zone. That portion of the base metal that was not melted during brazing, cutting or welding, but whose microstructure and mechanical properties were altered by the heat.

heat check. A pattern of parallel surface cracks that are formed by alternate rapid heating and cooling of the extreme surface metal, sometimes found on forging dies and piercing punches. There may be two sets of parallel cracks, one set perpendicular to the other.

heat-resisting alloy. An alloy developed for very-high-temperature service where relatively high stresses (tensile, thermal, vibratory or shock) are encountered and where oxidation resistance is frequently required.

heat time. In multiple-impulse or seam welding, the time that the current flows during any one impulse.

heat tinting. Coloration of a metal surface through oxidation by heating to reveal details of the microstructure.

heat treatable alloy. An alloy that can be hardened by heat treatment.

heat treating film. A thin coating or film, usually an oxide, formed on the surface of a metal during heat treatment.

heat treatment. Heating and cooling a solid metal or alloy in such a way as to obtain desired conditions or properties. Heating for the sole purpose of hot working is excluded from the meaning of this definition.

heel. Synonymous with *base* (1). See also sketch accompanying *single-point tool.*

hemming. Forming of an edge by bending the metal back on itself.

HERF. A common abbreviation for *high-energy-rate forging* or *high-energy-rate forming.*

herringbone pattern. Same as *chevron pattern.*

Heyn stresses. Same as *microscopic stresses.*

high-conductivity copper. Copper that, in the annealed condition, has a minimum electrical conductivity of 100% *IACS* as determined in accordance with ASTM methods of testing.

high-energy-rate forging. Producing forgings at extremely high ram velocities resulting from the sudden release of a compressed gas against a free piston. Forging is usually completed in one blow. Also known as HERF processing, high-velocity forging, high-speed forging.

high-energy-rate forming. A group of special forming processes in which metal undergoes deformation at high velocity, usually at least ten times the velocity of 0.2 to 6 m/s (0.5 to 20 ft/s) achieved in conventional forming. Commonly abbreviated HERF. Explosive forming, electrohydraulic forming and electromagnetic forming are the most common HERF processes.

high-frequency resistance welding. A resistance welding process that produces coalescence of metals with the heat generated from the resistance of the workpieces to a high-frequency alternating current in the 10 to 500 kHz range and the rapid application of an upsetting force after heating is substantially completed. The path of the current in the workpiece is controlled by use of the proximity effect (the feed current follows closely the return current conductor).

highlighting. Buffing or polishing selected areas of a complex shape to increase the luster or change the color of those areas.

high residual phosphorus copper. Deoxidized copper with residual phosphorus present in amounts (usually 0.013 to 0.04%) generally sufficient to decrease appreciably the conductivity of the copper.

hindered contraction. Contraction where the shape will not permit a casting to contract in certain regions in keeping with the coefficient of expansion.

hitch feed. Feed performed by a reciprocating head or slide carrying a gripper shoe that clamps the stock during the feeding movement and releases it on the return stroke.

hob. A rotary cutting tool with its teeth arranged along a helical thread, used for generating gear teeth or other evenly spaced forms on the periphery of a cylindrical workpiece. The hob and the workpiece are rotated in timed relationship to each other while the hob is fed axially or tangentially across or radially into the workpiece. Hobs should not be confused with multiple-thread milling cutters, rack cutters, and similar tools, where the teeth are not arranged along a helical thread.

hogging. Machining a part from bar stock, plate or a simple forging in which much of the original stock is removed.

holddown. A plate, a ring or fingers used to hold work stationary during forming, blanking, piercing or shearing.

holding furnace. A small furnace into which molten metal can be transferred to be held at the proper temperature until it can be used to make castings.

hold time. In resistance welding, the time during which pressure is applied to the work after the current ceases.

hole flanging. Forming an integral collar around the periphery of a previously formed hole. See *extruded hole*.

holidays. Discontinuities in a coating (such as porosity, cracks, gaps and similar flaws) that allow areas of basis metal to be exposed to any corrosive environment that contacts the coated surface.

homogeneous carburizing. Use of a carburizing process to convert a low-carbon ferrous alloy to one of uniform and higher carbon content throughout the section.

homogenizing. Holding at high temperature to eliminate or decrease chemical segregation by diffusion.

honing. A low-speed finishing process used chiefly to produce uniform high dimensional accuracy and fine finish, most often on inside cylindrical surfaces. In honing, very thin layers of stock are removed by simultaneously rotating and reciprocating a bonded abrasive stone or stick that is pressed against the surface being honed with lighter force than is typical of grinding.

hook. Concavity in a tooth face giving a variation in *rake* at different points along the tooth face.

Hooker process. Extrusion of a hollow billet or cup through an annulus formed by the die aperture and the mandrel or pilot to form a tube or long cup.

Hooke's law. Stress is proportional to strain. The law holds only up to the proportional limit.

Hoopes process. An electrolytic refining process for aluminum, using three liquid layers in the reduction cell.

horizontal-position welding. (1) Making a fillet weld on the upper side of the intersection of a vertical surface and a horizontal surface. (2) Making a horizontal groove weld on a vertical surface.

horizontal-rolled-position welding. Topside welding of a butt joint connecting two horizontal pieces of rotating pipe.

horn. In a resistance welding machine, a cylindrical arm or beam that transmits the electrode pressure and usually conducts the welding current.

horn press. A mechanical press equipped with or arranged for a cantilever block or horn that acts as the die or support for the die, used in forming, piercing, setting down, or riveting hollow cylinders and odd-shaped work.

horn spacing. The distance between adjacent surfaces of the horns of a resistance welding machine.

hot bed. An area adjacent to the *runout table* where hot rolled metal is placed to cool. Sometimes called the cooling table.

hot chamber machine. A *die casting* machine in which the metal chamber under pressure is immersed in the molten metal in a furnace. The chamber is sometimes called a gooseneck, and the machine is sometimes called a gooseneck machine.

hot-cold working. (1) A high-temperature thermomechanical treatment consisting of deforming a metal above its transformation temperature and cooling fast enough to preserve some or all of the deformed structure. (2) A general term synonymous with *warm working*.

hot crack. A crack formed in a cast metal because of internal stress developed on cooling following solidification. A hot crack is less open than a *hot tear* and usually exhibits less oxidation and decarburization along the fracture surface.

hot dip coating. A metallic coating obtained by dipping the basis metal into a molten metal.

hot extrusion. Extrusion at elevated temperature that does not cause strain hardening. See also *extrusion*.

hot forming. See *hot working*.

hot isostatic pressing. A process for simultaneously heating and forming a powder metallurgy compact in which metal powder, contained in a sealed flexible mold, is subjected to equal pressure from all directions at a temperature high enough for sintering to take place.

hot isostatic pressure welding. A diffusion welding method that produces coalescence of materials by heating and applying hot inert gas under pressure.

hot mill. A production line or facility for hot rolling of metals.

hot press forging. Plastically deforming metals between dies in presses at temperatures high enough to avoid strain hardening.

hot pressing. Forming a powder metallurgy compact at a temperature high enough to effect concurrent *sintering*.

hot pressure welding. A solid-state welding process that produces coalescence of materials with heat and application of pressure sufficient to produce macrodeformation of the base material. Vacuum or other shielding media may be used. See also *forge welding* and *diffusion welding*. Compare with *cold welding*.

hot quenching. An imprecise term used to cover a variety of quenching procedures in which a quenching medium is maintained at a prescribed temperature above 70 °C (160 °F).

hot rod. Same as *wire rod*.

hot shortness. A tendency for some alloys to separate along grain boundaries when stressed or deformed at temperatures near the melting point. Hot shortness is caused by a low-melting constituent, often present only in minute amounts, that is segregated at grain boundaries.

hot tear. A fracture formed in a metal during solidification because of *hindered contraction*. Compare with *hot crack*.

hot top. (1) A reservoir, thermally insulated or heated, that holds molten metal on top of a mold for feeding of the ingot or casting as it contracts on solidifying, thus preventing formation of pipe or voids. See accompanying sketch. (2) A refractory-lined steel or iron casting that is inserted into the tip of the mold and is supported at various heights to feed the ingot as it solidifies.

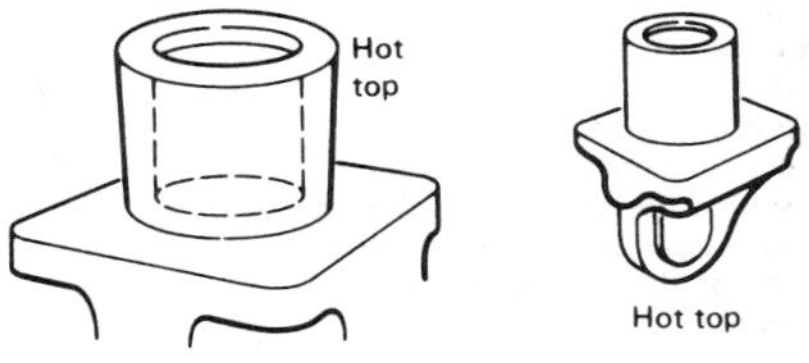

hot trimming. Removing flash or excess metal from a hot part (such as a forging) in a trimming press.

hot working. Deforming metal plastically at such a temperature and strain rate that recrystallization takes place simultaneously with the deformation, thus avoiding any strain hardening.

hubbing. Producing die cavities by pressing a male master plug, known as a hub, into a block of metal.

Hull cell. A special electrodeposition cell giving a range of known current densities for test work.

humidity test. A corrosion test involving exposure of specimens at controlled levels of humidity and temperature. Contrast with *salt-fog test*.

hydraulic press. A press in which fluid pressure is used to actuate and control the ram.

hydride descaling. *Descaling* by the action of a hydride in a fused alkali.

hydrogen brazing. A term sometimes used to denote brazing in a hydrogen-containing atmosphere, usually in a furnace; use of the appropriate process name is preferred.

hydrogen damage. A general term for the embrittlement, cracking, blistering and hydride formation that can occur when hydrogen is present in some metals.

hydrogen embrittlement. A condition of low ductility in metals resulting from the absorption of hydrogen.

hydrogen loss. The loss in weight of metal powder or of a *compact* caused by heating a representative sample for a specified time and temperature in a hydrogen atmosphere. Broadly, a measure of the oxygen content of the sample, when applied to materials containing only such oxides as are reducible with hydrogen and no hydride-forming element.

hydrogen overvoltage (in electroplating). Overvoltage associated with the liberation of hydrogen.

hydrogen-reduced powder. Metal powder produced by hydrogen reduction of a compound.

hydrometallurgy. Industrial *winning* or *refining* of metals using water or an aqueous solution.

hydrostatic tension. Three equal and mutually perpendicular tensile stresses.

hypereutectic alloy. In an alloy system exhibiting a *eutectic*, any alloy whose composition has an excess of alloying element compared with the eutectic composition, and whose equilibrium microstructure contains some eutectic structure.

hypereutectoid alloy. In an alloy system exhibiting a *eutectoid*, any alloy whose composition has an excess of alloying element compared with the eutectoid composition, and whose equilibrium microstructure contains some eutectoid structure.

hypoeutectic alloy. In an alloy system exhibiting a *eutectic*, any alloy whose composition has an excess of base metal compared with the eutectic composition, and whose equilibrium microstructure contains some eutectic structure.

hypoeutectoid alloy. In an alloy system exhibiting a *eutectoid*, any alloy whose composition has an excess of base metal compared with the eutectoid composition, and whose equilibrium microstructure contains some eutectoid structure.

hysteresis, magnetic. The lag of the magnetization of an iron or steel specimen behind any cyclic variation of the applied magnetizing field.

I

IACS. International annealed copper standard; a standard reference used in reporting electrical conductivity. The conductivity of a material, in %IACS, is equal to 1724.1 divided by the electrical resistivity of the material in nΩ·m.

idiomorphic crystal. An individual crystal that has grown without restraint so that the habit planes are clearly developed. Compare with *allotriomorphic crystal*.

immersion cleaning. Cleaning in which the work is immersed in a liquid solution.

immersion coating. A coating produced in a solution by chemical or electrochemical action without the use of external current.

immersion plating. Depositing a metallic coating on a metal immersed in a liquid solution, without the aid of an external electric current. Also called dip plating.

impact energy. The amount of energy required to fracture a material, usually measured by means of an *Izod test* or *Charpy test*. The type of specimen and test conditions affect the values and therefore should be specified.

impact extrusion. See *extrusion*.

impact line. A blemish on a drawn sheet-metal part caused by a slight change in metal thickness. The mark is called an impact line when it results from impact of the punch on the blank; it is called a recoil line when it results from transfer of the blank from the die to the punch during forming, or from a reaction to the blank being pulled sharply through the draw ring.

impact strength. Same as *impact energy*.

impact test. A test to determine the behavior of materials when subjected to high rates of loading, usually in bending, tension or torsion. The quantity measured is the energy absorbed in breaking the specimen by a single blow, as in *Charpy* and *Izod* tests.

imperfection. (1) When referring to the physical condition of a part or metal product, any departure of a quality characteristic from its intended level or state. The existence of an imperfection does not imply nonconformance (see *nonconforming*), nor does it have any implication as to the usability of a product or service. An imperfection must be rated on a scale of severity, in accordance with applicable specifications, to establish whether or not the part or metal product is of acceptable quality. (2) Generally, any departure from an ideal design, state or condition. (3) In crystallography, any deviation from an ideal space lattice.

impregnation. (1) Treatment of porous castings with a sealing medium to stop pressure leaks. (2) The process of filling the pores of a sintered compact, usually with a liquid such as a lubricant. (3) The process of mixing particles of a nonmetallic substance in a matrix of metal powder, as in diamond-impregnated tools.

impression-die forging. A forging that is formed to the required shape and size by machined impressions in specially prepared dies that exert three-dimensional control on the workpiece.

impurities. Elements or compounds whose presence in a material is undesirable.

inclinable press. A press that can be inclined to facilitate handling of the formed parts. See *open-back inclinable press*.

inclusions. Particles of foreign material in a metallic matrix. The particles are usually compounds (such as oxides, sulfides or silicates), but may be of any substance that is foreign to (and essentially insoluble in) the matrix.

indentation. In a spot, seam or projection weld, the depression on the exterior surface of the base metal.

indentation hardness. The resistance of a material to indentation. This is the usual type of hardness test, in which a pointed or rounded indenter is pressed into a surface under a substantially static load.

indication. In inspection, a response to a nondestructive stimulus that implies the presence of an *imperfection*. The indication must be interpreted to determine if (*a*) it is a true indication or a *false indication* and (*b*) whether or not a true indication represents an unacceptable deviation.

indicator. A substance that, through some visible change such as color, indicates the condition of a solution or other material as to the presence of free acid, alkali or other substance.

indirect-arc furnace. An electric-arc furnace in which the metallic charge is not one of the poles of the arc.

indirect extrusion. See *extrusion*.

induction brazing. *Brazing* in which the required heat is generated by subjecting the workpiece to electromagnetic induction.

induction furnace. An ac electric furnace in which the primary conductor is coiled and generates, by electromagnetic induction, a secondary current that develops heat within the metal charge.

induction hardening. A surface-hardening process in which only the surface layer of a suitable ferrous workpiece is heated by electromagnetic induction to a temperature above the upper critical temperature and immediately quenched.

induction heating. Heating by combined electrical resistance and hysteresis losses induced by subjecting a metal to the varying magnetic field surrounding a coil carrying alternating current.

induction melting. Melting in an *induction furnace*.

induction welding. Welding in which the required heat is generated by subjecting the workpiece to electromagnetic induction.

inert anode. An anode that is insoluble in the electrolyte under the conditions prevailing in the electrolysis.

infiltration. The process of filling the pores of a sintered or unsintered powder metallurgy compact with a metal or alloy of lower melting point.

ingate. Same as *gate*.

ingot. A casting of simple shape, suitable for hot working or remelting.

ingot iron. Commercially pure iron.

inhibitor. A substance that retards some specific chemical reaction. Picking inhibitors retard the dissolution of metal without hindering the removal of scale from steel.

inoculation. The addition of a material to molten metal to form nuclei for crystallization.

insert. (1) A part formed from a second material, usually a metal, which is placed in the molds and appears as an integral structural part of the final casting. (2) A removable portion of a die or mold.

insert die. A relatively small die which contains part or all of the impression of a forging, and which is fastened to a master die block.

inserted-blade cutters. Cutters having replaceable blades that are either solid or tipped and are usually adjustable.

intercept method. A quantitative metallographic technique in which the desired quantity (such as grain size or amount of precipitate) is expressed as the number of times per unit length a straight line on a metallographic image crosses particles of the feature being measured.

intercommunicating porosity. In a sintered powder metallurgy compact, a type of porosity in which individual pores are connected in such a way that a fluid may pass from one pore to another throughout the entire compact.

intercrystalline. Between the crystals, or grains, of a metal.

interdendritic corrosion. Corrosive attack that progresses preferentially along interdendritic paths. This type of attack results from local differences in composition, such as coring commonly encountered in alloy castings.

interface. A surface that forms the boundary between phases or systems.

interfacial tension. The contractile force of an interface between two phases.

interference. The difference in lateral dimensions at room temperature between two mating components before assembly by expansion, shrinking or press fitting. Can be expressed in absolute or in relative terms.

interference fit. Any of various classes of fit between mating parts where there is nominally a negative or zero allowance between the parts, and where there is either part interference or no gap when the mating parts are made to the respective extremes of individual tolerances that ensure the tightest fit between the parts. Contrast with *clearance fit*.

intergranular corrosion. Corrosion occurring preferentially at grain boundaries, usually with slight or negligible attack on the adjacent grains. See also *interdendritic corrosion*.

intermediate annealing. Annealing wrought metals at one or more stages during manufacture and before final treatment.

intermediate electrode. Same as *bipolar electrode*.

intermediate phase. In an alloy or a chemical system, a distinguishable homogeneous phase whose composition range does not extend to any of the pure components of the system.

intermetallic compound. An intermediate phase in an alloy system, having a narrow range of homogeneity and relatively simple stoichiometric proportions; the nature of the atomic binding can be of various types, ranging from metallic to ionic.

intermittent weld. A weld in which the continuity is broken by recurring unwelded spaces.

internal friction. The conversion of energy into heat by a material subjected to fluctuating stress. In free vibration, the internal friction is measured by the *logarithmic decrement*.

internal grinding. Grinding an internal surface such as that inside a cylinder or hole.

internal oxidation. Preferential *in situ* oxidation of certain components or phases within the bulk of a solid alloy accomplished by diffusion of oxygen into the body; a form of *subsurface corrosion*.

internal stress. See preferred term, *residual stress*.

interpass temperature. In a multipass weld, the lowest temperature of a *pass* before the succeeding one is commenced.

interrupted aging. Aging at two or more temperatures, by steps, and cooling to room temperature after each step. See *aging*, and compare with *progressive aging* and *step aging*.

interrupted-current plating. Plating in which the flow of current is discontinued for periodic short intervals to decrease anode polarization and elevate the *critical current density*. It is most commonly used in cyanide copper plating.

interrupted quenching. A quenching procedure in which the workpiece is removed from the first quench at a temperature substantially higher than that of the quenchant and is then subjected to a second quenching system having a cooling rate different from that of the first.

interstitial solid solution. A solid solution in which the solute atoms occupy positions that do not correspond to lattice points of the solvent. Contrast with *substitutional solid solution*.

intracrystalline. Within or across the crystals or grains of a metal; same as transcrystalline and transgranular.

inverse chill. A condition in an iron casting in which the interior is comprised of chilled or white iron while the surfaces are either mottled or contain free graphite.

inverse segregation. Segregation in cast metal in which an excess of lower-melting constituents occurs in the earlier freezing portions, apparently the result of liquid metal entering cavities developed in the earlier-solidified metal.

investment casting. (1) Casting metal into a mold produced by surrounding (investing) an expendable pattern with a refractory slurry that sets at room temperature, after which the wax, plastic or frozen-mercury pattern is removed through the use of heat. Also called precision casting or lost-wax process. (2) A part made by the investment casting process.

investment compound. A mixture of a graded refractory filler, a binder and a liquid vehicle, used to make molds for *investment casting*.

ion. An atom, or group of atoms, that has gained or lost one or more outer electrons and thus carries an electric charge. Positive ions, or cations, are deficient in outer electrons. Negative ions, or anions, have an excess of outer electrons.

ion exchange. The reversible interchange of ions between a liquid and solid, with no substantial structural changes in the solid.

ionic bond. A bond between two or more atoms that is the result of electrostatic attractive forces between positively and negatively charged ions.

ionic crystal. A crystal in which atomic bonds are *ionic bonds*. This type of atomic linkage, also known as (hetero) polar bonding, is characteristic of many compounds (sodium chloride, for instance).

ionization chamber. An enclosure containing two or more electrodes surrounded by a gas capable of conducting an electric current when it is ionized by x-rays or other ionizing rays. It is commonly used for measuring the intensity of such radiation.

iron casting. A part made of *cast iron*.

ironing. Thinning the walls of hollow articles by drawing them between a punch and a die.

iron-powder electrode. A welding electrode with a covering containing up to about 50% iron powder, some of which becomes part of the deposit.

irradiation. The exposure of a material in a field of radiation; the cumulative exposure.

isostatic pressing. A process for forming a powder metallurgy compact by applying pressure equally from all directions to metal powder contained in a sealed flexible mold. See also *hot isostatic pressing*.

isothermal annealing. Austenitizing a ferrous alloy and then cooling to and holding at a temperature at which austenite transforms to a relatively soft ferrite carbide aggregate.

isothermal transformation. A change in phase that takes place at a constant temperature. The time required for transformation to be completed, and in some instances the time delay before transformation begins, depends on the amount of supercooling below (or superheating above) the equilibrium temperature for the same transformation.

isotope. One of several different nuclides of an element having the same number of protons in their nuclei and therefore the same atomic number, but differing in the number of neutrons and therefore in atomic weight.

isotropy. The quality of having identical properties in all directions.

Izod test. A pendulum-type single-blow impact test in which the specimen, usually notched, is fixed at one end and broken by a falling pendulum. The energy absorbed, as measured by the subsequent rise of the pendulum, is a measure of impact strength or notch toughness. Contrast with *Charpy test*.

J

jig. A device used to hold a workpiece in place and simultaneously guide the tool in a cutting operation.

jig boring. Boring with a single-point tool where the work is positioned upon a table that can be located so as to bring any desired part of the work under the tool. Thus, holes can be accurately spaced. This type of boring can be done on milling machines or jig borers.

jig grinding. Analogous to *jig boring,* where the holes are ground rather than machined.

joggle. An offset in a flat plane consisting of two parallel bends at the same angle but in opposite directions.

joint. The location where two or more members are to be or have been fastened together mechanically or by brazing or welding.

joint efficiency. The strength of a welded joint expressed as a percentage of the strength of the unwelded base metal.

joint penetration. The minimum depth to which a groove or flange weld extends from its face into the joint, exclusive of reinforcement. Joint penetration may include *root penetration.*

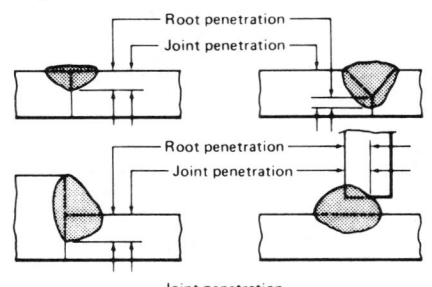

Joint penetration

Jominy test. See *end-quench hardenability test.*

K

keel block. A standard test casting, for steel and other high-shrinkage alloys, consisting of a rectangular bar that resembles the keel of a boat, attached to the bottom of a large riser, or shrinkhead. Keel blocks that have only one bar are often called Y-blocks; keel blocks having two bars, double keel blocks. Test specimens are machined from the rectangular bar, and the shrinkhead is discarded.

Kellering. A shop term. See preferred term, *tracer milling.*

kerf. The space that was occupied by material removed during cutting.

keyhole specimen. A type of specimen containing a hole-and-slot notch, shaped like a keyhole, usually used in impact bend tests. See *Charpy test* and *Izod test.*

killed steel. Steel treated with a strong deoxidizing agent such as silicon or aluminum in order to reduce the oxygen content to such a level that no reaction occurs between carbon and oxygen during solidification.

kiln. A large furnace used for baking, drying or burning firebrick or refractories, or for calcining ores or other substances.

kish. Free graphite that forms in molten hypereutectic cast iron as it cools. In castings, the kish may segregate toward the cope surface, where it lodges at or immediately beneath the casting surface.

knockout. (1) A mechanism for freeing formed parts from a die used for stamping, blanking, drawing, forging or heading operations. (2) A partly pierced hole in a sheet metal part, where the slug remains in the hole and can be forced out by hand if a hole actually is needed. (3) Removal of sand cores from a casting. (4) Jarring of an investment casting mold to remove the casting and investment from the flask.

Knoop hardness. Microhardness determined from the resistance of metal to indentation by a pyramidal diamond indenter, having edge angles of 172° 30′ and 130°, making a rhombohedral impression with one long and one short diagonal.

knuckle-joint press. A heavy short-stroke press in which the slide is directly actuated by a single toggle joint that is opened and closed by a connection and crank. It is used for embossing, coining, sizing, heading, swaging and extruding.

knurling. Impressing a design into a metallic surface, usually by means of small, hard rollers that carry the corresponding design on their surfaces.

Kroll process. A process for production of metallic titanium by reduction of titanium tetrachloride with a more active metal such as magnesium, yielding titanium in the form of granules or powder.

L

ladle. A receptacle used for transferring and pouring molten metal.

laminate. (1) A composite metal, usually in the form of sheet or bar, composed of two or more metal layers so bonded that the composite metal forms a structural member. (2) To form a metallic product of two or more bonded layers.

lamination. (1) A type of discontinuity with separation or weakness generally aligned parallel to the worked surface of a metal. May be the result of pipe, blisters, seams, inclusions or segregation elongated and made directional by working. Laminations may also occur in powder metallurgy compacts. (2) In electrical products such as motors, a blanked piece of electrical sheet that is stacked up with several other identical pieces to make a stator or rotor.

lancing. (1) A press operation in which a single-line cut is made in strip stock without producing a detached slug. Chiefly used to free metal for forming, or to cut partial contours for blanked parts, particularly in progressive dies. (2) A misnomer for *oxyfuel gas cutting.*

land. (1) For profile-sharpened milling cutters, the relieved portion immediately behind the cutting edge. (2) For reamers, drills and taps, the solid section between the flutes. (3) On punches, the portion adjacent to the nose that is parallel to the axis and of maximum diameter.

lap. A surface imperfection, with the appearance of a seam, caused by hot metal, fins or sharp corners being folded over and then being rolled or forged into the surface but without being welded.

lap joint. A joint made between two overlapping members.

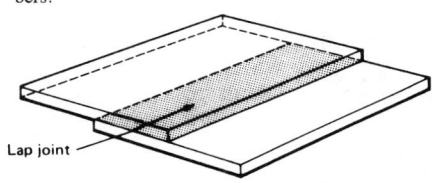

Lap joint

lapping. Finishing a surface by abrasion with an object, usually made of copper, lead, cast iron or close-grained wood, having very fine abrasive particles rolled into its surface.

laser. A device that emits a concentrated beam of electromagnetic radiation (light). Laser beams are used in metalworking to melt, cut or weld metals; in less concentrated form they are sometimes used to inspect metal parts.

laser-beam cutting. A cutting process that severs materials with the heat obtained by directing a beam from a *laser* against a metal surface. The process can be used with or without an externally supplied shielding gas.

laser-beam machining. Removing material by melting and vaporizing the workpiece at the point of impingement of a highly focused beam of coherent monochromatic light (a *laser* beam).

laser-beam welding. A welding process that joins metal parts using the heat obtained by directing a beam from a *laser* onto the weld joint.

latent heat. Thermal energy absorbed or released when a substance undergoes a phase change.

lateral extrusion. An operation in which the product is extruded sideways through an orifice in the container wall.

lateral runout. Same as *axial runout.*

lattice constant. See *lattice parameter.*

lattice parameter. The length of any side of a unit cell of a given crystal structure; if the lengths are unequal, all unequal lengths must be given.

launder. (1) A channel for conducting molten metal. (2) A box conduit conveying particles suspended in water.

lay. Direction of predominant surface pattern remaining after cutting, grinding, lapping or other processing.

leaching. Extracting an element or compound from a solid alloy or mixture by preferential dissolution in a suitable liquid.

lead. (1) The axial advance of a helix in one complete turn. (2) The slight bevel at the outer end of a face cutting edge of a face mill.

lead angle. In cutting tools, the helix angle of the flutes.

lead burning. A misnomer for welding of lead.

lead proof. See *die proof.*

leakage field. The magnetic field that leaves or enters a magnetized part at a magnetic pole.

ledeburite. The eutectic of the iron-carbon system, the constituents of which are austenite and cementite. The austenite decomposes into ferrite and cementite on cooling below Ar_1.

left-hand cutting tool. A cutter all of whose flutes twist away in a counterclockwise direction when viewed from either end.

leg of a fillet weld. (1) Actual: The distance from the root of the joint to the toe of the fillet weld. See accompanying sketch. (2) Nominal: The length of a side of the largest right triangle that can be inscribed in the cross section of the weld. See sketches accompanying *concave fillet weld* and *convex fillet weld.*

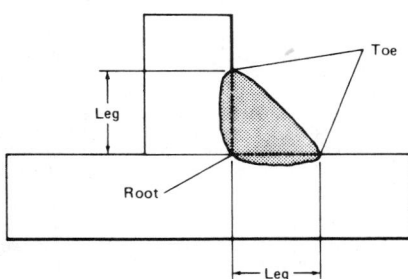

Leg of a fillet weld

leveler lines. Lines on sheet or strip running transverse to the direction of roller leveling; may be seen on stoning or light sanding after leveling (but before drawing). Usually can be removed by moderate stretching.

leveling. Flattening of rolled sheet, strip or plate by reducing or eliminating distortions. See *stretcher leveling* and *roller leveling.*

leveling action. Action exhibited by a plating solution yielding a plated surface smoother than the basis metal.

levigation. (1) Separation of fine powder from coarser material by forming a suspension of the fine material in a liquid. (2) A means of classifying a material as to particle size by the rate of settling from a suspension.

levitation melting. An *induction melting* process in which the metal being melted is suspended by the electromagnetic field and is not in contact with a container.

light drawn. An imprecise term, applied to drawn products such as wire and tubing, that indicates a lesser amount of cold reduction than for *hard drawn* products.

lightly coated electrode. A filler-metal electrode used in arc welding, consisting of a metal wire with a light coating, usually of metal oxides and silicates, applied subsequent to the drawing operation primarily for stabilizing the arc. Contrast with *covered electrode.*

light metal. One of the low-density metals, such as aluminum, magnesium, titanium, beryllium or their alloys.

limiting current density. The maximum current density that can be used to obtain a desired electrode reaction without undue interference such as from polarization.

lineage structure. (1) Deviations from perfect alignment of parallel arms of columnar dendrite as a result of interdendritic shrinkage during solidification from a liquid. This type of deviation may vary in orientation from one area to another from a few minutes to as much as two degrees of arc. (2) A type of substructure consisting of elongated subgrains.

linear elastic fracture mechanics. A method of fracture analysis that can determine the stress (or load) required to induce fracture instability in a structure containing a cracklike flaw of known size and shape. See *stress-intensity factor.*

linear strain. See *strain.*

liner. (1) The slab of coating metal that is placed on the core alloy and is subsequently rolled down to clad sheet as a composite. (2) In extrusion, a removable alloy steel cylindrical chamber, having an outside longitudinal taper firmly positioned in the container or main body of the press, into which the billet is placed for extrusion.

line reaming. Simultaneous reaming of coaxial holes in various sections of a workpiece with a reamer having cutting faces or piloted surfaces with the desired alignment.

lip. For a *milling cutter,* the material included between a relieved land and a tooth face.

lip angle. (1) For a *milling cutter,* the included angle between a tooth face and a relieved land. (2) Defined by sketch accompanying *single-point tool.*

liquation. Partial melting of an alloy, usually as a result of coring or other compositional heterogeneities.

liquation temperature. The lowest temperature at which partial melting can occur in an alloy that exhibits the greatest possible degree of segregation.

liquid honing. Producing a finely polished finish by directing an air-ejected chemical emulsion containing fine abrasives against the surface to be finished.

liquid penetrant inspection. A type of nondestructive inspection that locates discontinuities that are open to the surface of a metal by first allowing a penetrating dye or fluorescent liquid to infiltrate the discontinuity, removing the excess penetrant, and then applying a developing agent that causes the penetrant to seep back out of the discontinuity and register as an indication. Liquid penetrant inspection is suitable for both ferrous and nonferrous materials, but is limited to the detection of open surface discontinuities in nonporous solids.

liquid phase sintering. *Sintering* a powder metallurgy compact under conditions that maintain a liquid metallic phase within the compact during all or part of the sintering schedule. The liquid phase may be derived from a component of the green compact or may be infiltrated into the compact from an outside source.

liquid shrinkage. See *casting shrinkage.*

liquidus. In a constitution or equilibrium diagram, the locus of points representing the temperatures at which the various compositions in the system begin to freeze on cooling or finish melting on heating. See also *solidus.*

liquor finish. A smooth, bright finish characteristic of wet-drawn wire. Formerly produced by using liquor from fermented grain mash as a drawing lubricant.

live center. A lathe or grinder center that holds, yet rotates with, the work. It is used in either the headstock or tailstock of a machine to prevent wear and reduce the driving torque.

loading. (1) In cutting, building up of a cutting tool back of the cutting edge by undesired adherence of material removed from the work. (2) In grinding, filling the pores of a grinding wheel with material from the work, usually resulting in a decrease in production and quality of finish. (3) In powder metallurgy, filling of the die cavity with powder.

loam. A molding material consisting of sand, silt and clay, used over brickwork or other structural backup material for making massive castings, usually of iron or steel.

local action. Corrosion due to the action of "local cells"—that is, *galvanic cells* resulting from inhomogeneities between adjacent areas on a metal surface exposed to an electrolyte.

local cell. A *galvanic cell* resulting from inhomogeneities between areas on a metal surface in an electrolyte. The inhomogeneities may be of physical or chemical nature in either the metal or its environment.

local current density. Current density at a point or on a small area.

localized precipitation. Precipitation from a supersaturated solid solution similar to *continuous precipitation,* except that the precipitate particles form at preferred locations, such as along slip planes, grain boundaries or incoherent twin boundaries.

locational fit. A clearance or interference *fit* intended for locating mating parts.

lock. In forging, a condition where the flash line is not entirely in one plane. Where two or more plane changes occur, it is called a compound lock. Where a lock is placed in the die to compensate for die shift caused by a steep lock, it is called a counterlock.

logarithmic decrement (log decrement). The natural logarithm of the ratio of successive amplitudes of vibration of a member in free oscillation. It is equal to one-half the specific damping capacity.

longitudinal direction. The principal direction of flow in a worked metal.

longitudinal field. A magnetic field that extends within a magnetized part from one or more poles to one or more other poles and that is completed through a path external to the part.

long transverse. See *transverse.*

looping mill. An arrangement of hot rolling stands such that a hot bar, while being discharged from one stand, is fed into a second stand in the opposite direction.

loose metal. Refers to an area in a formed panel that is not stiff enough to hold its shape; may be confused with *oil canning.*

lost-wax process. An *investment casting* process in which a wax pattern is used.

lot. A finite quantity of a given product manufactured under production conditions that are considered uniform. Often used to describe a finite quantity of product submitted for inspection as a single group. For a bulk product (such as a chemical or powdered metal), the term "batch" is often used synonymously with lot.

lower punch. The lower part of a die, which forms the bottom of the die cavity and which may or may not move in relation to the die body; usually movable in a forging die.

low-hydrogen electrode. A covered arc welding electrode that provides an atmosphere around the arc and molten weld metal that is low in hydrogen.

low-residual-phosphorus copper. Deoxidized copper with residual phosphorus present in amounts (usually 0.004 to 0.012%) generally too small to decrease appreciably the electrical conductivity of the copper.

low shaft furnace. A short shaft-type blast furnace used to produce pig iron and ferroalloys from low-grade ores, using low-grade fuel. The air blast is often enriched with oxygen. Also used for making a variety of other products such as alumina, cementmaking slags and ammonia synthesis gas.

lubricant. Any substance used to reduce friction between two surfaces in contact.

Lüders lines. Elongated surface markings or depressions caused by localized plastic deformation that results from discontinuous (inhomogeneous) yielding. Also known as Lüders bands, Hartmann lines, Piobert lines or *stretcher strains.*

luster finish. A bright as-rolled finish, produced on ground rolls; it is suitable for decorative painting or plating, but usually must undergo additional surface preparation after forming.

lute. (1) A mixture of fireclay used to seal cracks between a crucible and its cover or between container and cover when heat is to be applied. (2) To seal with clay or other plastic material.

M

machinability. The relative ease of machining a metal.

machinability index. A relative measure of the machinability of an engineering material under specified standard conditions.

machine forging. Forging performed in upsetters or horizontal forging machines.

machine welding. Welding with equipment that performs under the continual observation and control of a welding operator. The equipment may or may not load and unload the work. Compare with *automatic welding.*

machining. Removing material from a metal part, usually using a cutting tool, and usually using a power-driven machine.

machining allowance. See *finish allowance.*

machining stress. *Residual stress* caused by machining.

macroetching. *Etching* a metal surface to accentuate gross structural details (such as grain flow, segregation, porosity or cracks) for observation by the unaided eye or at a magnification of ten diameters or less.

macrograph. A graphic reproduction of the surface of a prepared specimen at a magnification not exceeding ten diameters. When photographed, the reproduction is known as a photomacrograph.

macroscopic. Visible at magnifications up to ten diameters.

macroscopic stresses. Residual stresses that vary from tension to compression in a distance (presumably many times the grain size) that is comparable to the gage length in ordinary strain measurements, hence, detectable by x-ray or dissection methods.

macroshrinkage. Isolated, clustered or interconnected voids in a casting that are detectable macroscopically. Such voids are usually associated with abrupt changes in section size and are caused by feeding that is insufficient to compensate for solidification shrinkage.

macrostress. Same as *macroscopic stress.*

macrostructure. The structure of metals as revealed by macroscopic examination of the etched surface of a polished specimen.

magnesite wheel. A grinding wheel bonded with magnesium oxychloride.

magnetically hard alloy. A ferromagnetic alloy capable of being magnetized permanently because of its ability to retain induced magnetization and magnetic poles after removal of externally applied fields; an alloy with high coercive force. The name is based on the fact that the quality of the early permanent magnets was related to their hardness.

magnetically soft alloy. A ferromagnetic alloy that becomes magnetized readily upon application of a field and that returns to practically a nonmagnetic condition when the field is removed; an alloy with the properties of high magnetic permeability, low coercive force and low magnetic hysteresis loss.

magnetic-analysis inspection. A nondestructive method of inspection to determine the existence of variations in magnetic flux in ferromagnetic materials of constant cross section, such as might be caused by discontinuities and variations in hardness. The variations are usually indicated by a change in pattern on an oscilloscopic screen.

magnetic-particle inspection. A nondestructive method of inspection for determining the existence and extent of surface cracks and similar imperfections in ferromagnetic materials. Finely divided magnetic particles, applied to the magnetized part, are attracted to and outline the pattern of any magnetic-leakage fields created by discontinuities.

magnetic pole. The area on a magnetized part at which the magnetic field leaves or enters the part. It is a point of maximum attraction in a magnet.

magnetic separator. A device used to separate magnetic from less-magnetic or nonmagnetic materials. The crushed material is conveyed on a belt past a magnet.

magnetic writing. In magnetic-particle inspection, a *false indication* caused by contact between a magnetized part and another piece of magnetic material.

magnetizing force. A force field, resulting from the flow of electric currents or from magnetized bodies, that produces magnetic induction.

magnetostriction. The characteristic of a material that is manifest by strain when it is subjected to a magnetic field; or the inverse. Some iron-nickel alloys expand; pure nickel contracts.

malleability. The characteristic of metals that permits plastic deformation in compression without rupture.

malleable cast iron. A cast iron made by prolonged annealing of *white cast iron* in which decarburization or graphitization, or both, take place to eliminate some or all of the cementite. The graphite is in the form of temper carbon. If decarburization is the predominant reaction, the product will exhibit a light fracture surface, hence, "whiteheart malleable;" otherwise, the fracture surface will be dark, hence, "blackheart malleable." Ferritic malleable has a predominantly ferritic matrix; pearlitic malleable may contain pearlite, spheroidite or tempered martensite depending on heat treatment and desired hardness.

malleablizing. Annealing *white cast iron* in such a way that some or all of the combined carbon is transformed into graphite or, in some instances, so that part of the carbon is removed completely.

mandrel. (1) A blunt-ended tool or rod used to retain the cavity in a hollow metal product during working. (2) A metal bar around which other metal may be cast, bent, formed or shaped. (3) A shaft or bar for holding work to be machined. (4) A form, such as a mold or matrix, used as a cathode in electroforming.

Mannesmann mill. Mill used in *Mannesmann process.*

Mannesmann process. A process used for piercing tube billets in making seamless tubing. The billet is rotated between two heavy rolls mounted at an angle and is forced over a fixed mandrel.

manual welding. Welding wherein the entire welding operation is performed and controlled by hand.

maraging. A precipitation-hardening treatment applied to a special group of iron-base alloys to precipitate one or more intermetallic compounds in a matrix of essentially carbon-free martensite. NOTE: The first developed series of maraging steels contained, in addition to iron, more than 10% nickel and one or more supplemental hardening elements. In this series, aging is done at 480 °C (900 °F).

margin. The cylindrical portion of the *land* of a drill that is not cut away to provide clearance.

marquenching. See *martempering.*

martempering. (1) A hardening procedure in which an austenitized ferrous workpiece is quenched in an appropriate medium whose temperature is maintained substantially at the M_s of the workpiece, held in the medium until its temperature is uniform throughout—but not long enough to permit bainite to form—and then cooled in air. The treatment is frequently followed by tempering. (2) When the process is applied to carburized material, the controlling M_s temperature is that of the case. This variation of the process is frequently called marquenching.

martensite. A generic term for microstructures formed by diffusionless phase transformation in which the parent and product phases have a specific crystallographic relationship. Martensite is characterized by an acicular pattern in the microstructure in both ferrous and nonferrous alloys. In alloys where the solute atoms occupy interstitial positions in the martensitic lattice (such as carbon in iron), the structure is hard and highly strained; but where the solute atoms occupy substitutional positions (such as nickel in iron), the martensite is soft and ductile. The amount of high-temperature phase that transforms to martensite on cooling depends to a large extent on the lowest temperature attained, there being a rather distinct beginning temperature (M_s) and a temperature at which the transformation is essentially complete (M_f).

martensite range. The temperature interval between M_s and M_f.

martensitic transformation. A reaction that takes place in some metals on cooling, with the formation of an acicular structure called *martensite*.

mash resistance seam welding. Resistance *seam welding* in which the weld is made in a lap joint, the thickness at the lap being reduced plastically to approximately the thickness of one of the lapped parts.

masking tape. A tape used as a *resist* for stopping-off purposes.

master alloy. An alloy, rich in one or more desired addition elements, that is added to a melt to raise the percentage of a desired constituent.

match. A condition in which a point in one forging-die half is aligned properly with the corresponding point in the opposite die half within specified tolerance.

matched edges. Two edges of a forging-die face that are machined exactly at 90° to each other, and from which all dimensions are taken in laying out the die impression and aligning the dies in the forging equipment.

match lines. Same as *matched edges*.

match plate. A plate of metal or other material on which patterns for metal casting are mounted (or formed as an integral part) so as to facilitate molding. The pattern is divided along its *parting plane* by the plate.

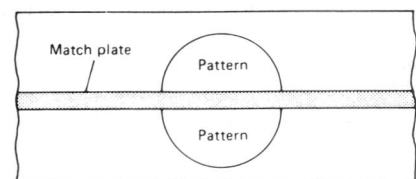

matrix. (1) The principal phase or aggregate in which another constituent is embedded. (2) In electroforming, a form used as a cathode.

matte. An intermediate product of *smelting;* an impure metallic sulfide mixture made by melting a roasted sulfide ore, such as an ore of copper, lead or nickel.

matte dip. An etching solution used to produce a dull finish on metal.

matte finish. (1) A dull texture produced by rolling sheet or strip between rolls that have been roughened by blasting. (2) A dull finish characteristic of some electrodeposits, such as cadmium or tin. Also written mat finish.

McQuaid-Ehn test. A test for revealing grain size after heating into the austenitic temperature range. Eight standard McQuaid-Ehn grain sizes are used for rating structures, No. 8 being finest, No. 1 coarsest.

mean stress. (1) In fatigue loading, the algebraic mean of the maximum and minimum stress in one cycle. Also called the steady stress component. (2) In any multiaxial stress system, the algebraic mean of three *principal stresses;* more correctly called mean normal stress.

mechanical equation of state. Any equation relating stress, strain, strain rate and temperature that is based on the concept that the instantaneous value of any one of these quantities is a single-valued function of the others, regardless of the prior history of the deformation.

mechanical hysteresis. Energy absorbed in a complete cycle of loading and unloading within the elastic limit and represented by the closed loop of the stress-strain curves for loading and unloading. Sometimes referred to as elastic, but more properly, mechanical.

mechanical metallurgy. The science and technology dealing with the behavior of metals when subjected to applied forces; often considered to be restricted to plastic working or shaping of metals.

mechanical plating. Plating wherein fine metal powders are peened onto the work by *tumbling* or other means.

mechanical press. A press whose slide is operated by a crank, eccentric, cam, toggle links or other mechanical device.

mechanical properties. The properties of a material that reveal its elastic and inelastic behavior when force is applied, thereby indicating its suitability for mechanical applications; for example, modulus of elasticity, tensile strength, elongation, hardness, and fatigue limit. Compare with *physical properties*.

mechanical testing. Determination of *mechanical properties*.

mechanical twin. A *twin* formed in a crystal by simple shear under external loading.

mechanical working. Subjecting metal to pressure, exerted by rolls, hammers or presses, in order to change the metal's shape or physical properties.

melting point. The temperature at which a pure metal, compound or eutectic changes from solid to liquid; the temperature at which the liquid and the solid are in equilibrium.

melting range. The range of temperature over which an alloy other than a compound or eutectic changes from solid to liquid; the range of temperature from *solidus* to *liquidus* at any given composition on a constitution diagram.

melting rate. In electric arc welding, the weight or length of electrode melted in a unit of time. Sometimes called melt-off rate or burn-off rate.

melt-off rate. See *melting rate*.

merchant mill. (obsolete) A mill, consisting of a group of stands of three rolls each arranged in a straight line and driven by one power unit, used to roll rounds, squares or flats of smaller dimensions than would be rolled on a bar mill.

mesh. The screen number of the finest screen of a specified standard screen scale through which almost all the particles of a powder sample will pass. Also called mesh size.

metal. (1) An opaque lustrous elemental chemical substance that is a good conductor of heat and electricity and, when polished, a good reflector of light. Most elemental metals are malleable and ductile and are, in general, denser than the other elemental substances. (2) As to structure, metals may be distinguished from nonmetals by their atomic binding and electron availability. Metallic atoms tend to lose electrons from the outer shells, the positive ions thus formed being held together by the electron gas produced by the separation. The ability of these "free electrons" to carry an electric current, and the fact that this ability decreases as temperature increases, establish the prime distinctions of a metallic solid. (3) From the chemical viewpoint, an elemental substance whose hydroxide is alkaline. (4) An *alloy*.

metal-arc cutting. Any of a group of arc cutting processes in which metals are severed by melting them with the heat of an arc between a metal electrode and the work.

metal-arc welding. Any of a group of arc welding processes in which metals are fused together using the heat of an arc between a metal electrode and the work. Use of the specific process name is preferred.

metal inert-gas welding. *Gas metal-arc welding* using an inert gas such as argon as the shielding gas.

metal leaf. Thin metal sheet, usually thinner than foil, and traditionally produced by beating rather than by rolling.

metallic bond. The principal bond between metal atoms, which arises from the increased spatial extension of valence-electron wave functions when an aggregate of metal atoms is brought close together. See *covalent bond, ionic bond*.

metallic glass. A noncrystalline metal or alloy, commonly produced by drastic supercooling of a molten alloy, by electrodeposition, or by vapor deposition. Also called amorphous alloy.

metallizing. (1) Forming a metallic coating by atomized spraying with molten metal or by *vacuum deposition*. Also called spray metallizing. (2) Applying an electrically conductive metallic layer to the surface of a nonconductor.

metallograph. An optical instrument designed for both visual observation and photomicrography of prepared surfaces of opaque materials at magnifications ranging from about 25 to about 2000 diameters. The instrument consists of a high-intensity illuminating source, a microscope and a camera bellows. On some instruments, provisions are made for examination of specimen surfaces with polarized light, phase contrast, oblique illumination, darkfield illumination and customary brightfield illumination.

metallography. The science dealing with the constitution and structure of metals and alloys as revealed by the unaided eye or by such tools as low-powered magnification, optical microscopy, electron microscopy and diffraction or x-ray techniques.

metallurgical coke. A coke, usually low in sulfur, having a very high compressive strength at elevated temperatures; used in metallurgical furnaces not only as fuel, but also to support the weight of the charge.

metallurgy. The science and technology of metals and alloys. Process metallurgy is concerned with the extraction of metals from their ores and with refining of metals; physical metallurgy, with the physical and mechanical properties of metals as affected by composition, processing and environmental conditions; and mechanical metallurgy, with the response of metals to applied forces.

metal penetration. A surface condition in castings in which metal or metal oxides have filled voids between sand grains without displacing them.

metal spraying. Coating metal objects by spraying molten metal against their surfaces. See *thermal spraying, flame spraying*.

metastable. Refers to a state of pseudoequilibrium that has a higher free energy than the true equilibrium state.

M_f temperature. For any alloy system, the temperature at which martensite formation on cooling is essentially finished. See *transformation temperature* for the definition applicable to ferrous alloys.

microfissure. A crack of microscopic proportions.

micrograph. A graphic reproduction of the surface of a prepared specimen, usually etched, at a magnification greater than ten diameters. If produced by photographic means it is called a photomicrograph (not a microphotograph).

microhardness. The hardness of a material as determined by forcing an indenter such as a Vickers or Knoop indenter into the surface of the material under very light load; usually, the indentations are so small that they must be measured with a microscope. Capable of determining hardnesses of different microconstituents within a structure, or of measuring steep hardness gradients such as those encountered in case hardening.

microprobe. See preferred term, *electron-beam microprobe analyzer*.

microradiography. The technique of passing x-rays through a thin section of an alloy in contact with a fine-grained photographic film and then viewing the radiograph at 50 to 100× to observe the distribution of alloying constituents and voids.

microscopic. Visible only at magnifications greater than ten diameters.

microscopic stresses. Residual stresses that vary from tension to compression in a distance (presumably approximating the grain size) that is small compared with the gage length in ordinary strain measurements. They are not detectable by dissection methods, but can sometimes be measured from line shift or line broadening in an x-ray diffraction pattern.

microsegregation. *Segregation* within a grain, crystal or small particle. See *coring*.

microshrinkage. A casting imperfection, not detectable microscopically, consisting of interdendritic voids. Microshrinkage results from contraction during solidification where the opporunity to supply filler material is inadequate to compensate for shrinkage. Alloys with wide ranges in solidification temperature are particularly susceptible.

microstress. Same as *microscopic stress*.

microstructure. The structure of a metal as revealed by microscopic examination of the etched surface of a polished specimen.

middling. A product intermediate between concentrate and tailing and containing enough of a valuable mineral to make retreatment profitable.

migration. Movement of entities (such as electrons, ions, atoms, molecules, vacancies and grain boundaries) from one place to another under the influence of a driving force (such as an electrical potential or a concentration gradient).

MIG welding. See *metal inert-gas welding*.

mil. One thousandth of an inch (0.001 in.).

mild steel. *Carbon steel* with a maximum of about 0.25% C.

mill. (1) A factory where metals are hot worked, cold worked, or melted and cast into standard shapes suitable for secondary fabrication into commercial products. (2) A production line, usually of four or more *stands*, for hot rolling metal into standard shapes such as bar, rod, plate, sheet or strip. (3) A single machine for hot rolling, cold rolling or extruding metal; examples include *blooming mill, cluster mill, four-high mill*, and *Sendzimer mill*. (4) A shop term for *milling cutter*. (5) A machine or group of machines for grinding or crushing ores and other minerals; see *ball mill, milling* (2).

mill edge. The normal edge produced in hot rolling. This edge is customarily removed when hot rolled sheets are further processed into cold rolled sheets.

Miller indices. A system for identifying planes and directions in any crystal system by means of sets of integers. The indices of a plane are related to the intercepts of that plane with the axes of a unit cell; the indices of a direction, to the multiples of lattice parameter that represent the coordinates of a point on a line parallel to the direction and passing through the arbitrarily chosen origin of a unit cell.

mill finish. A nonstandard (and typically nonuniform) surface finish on mill products that are delivered without being subjected to a special surface treatment (other than a corrosion-preventive treatment) after the final working or heat treating step.

milling. (1) Removing metal with a *milling cutter*. (2) The mechanical treatment of material, as in a *ball mill*, to produce particles or alter their size or shape, or to coat one component of a powder mixture with another.

milling cutter. A rotary cutting tool provided with one or more cutting elements, called teeth, which intermittently engage the workpiece and remove material by relative movement of the workpiece and cutter.

mill product. Any commercial product of a *mill*.

mill scale. The heavy oxide layer formed during hot fabrication or heat treatment of metals.

mineral dressing. Physical and chemical concentration of raw ore into a product from which a metal can be recovered at a profit.

minimized spangle. A hot dip galvanized coating of very small grain size, which makes the spangle less visible when the part is subsequently painted.

minimum bend radius. The minimum radius over which a metal product can be bent to a given angle without fracture.

minus sieve. The portion of a sample of a granular substance (such as metal powder) that passes through a standard sieve of specified number. Contrast with *plus sieve*.

mischmetal. A natural mixture of rare-earth elements (atomic numbers 57 through 71) in metallic form. It contains about 50% cerium, the remainder being principally lanthanum and neodymium.

mismatch. Error in register between forged surfaces formed by opposing dies.

misrun. A casting not fully formed, resulting from the

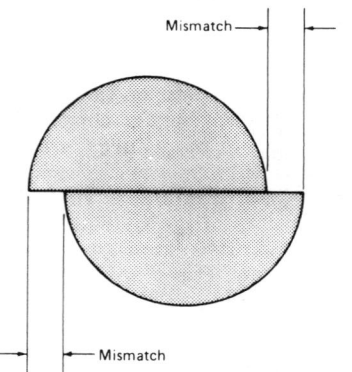

metal solidifying before the mold is filled.

mixed dislocation. See *dislocation*.

mixing. In powder metallurgy, the thorough intermingling of powders of two or more different materials (not *blending*).

mixing chamber. The part of a torch or furnace burner in which gases are mixed.

modification. Treatment of molten hypoeutectic (8 to 13% Si) or hypereutectic (13 to 19% Si) aluminum-silicon alloys to improve mechanical properties of the solid alloy by refinement of the size and distribution of the silicon phase. Involves additions of small percentages of sodium or strontium (hypoeutectic alloys) or of phosphorus (hypereutectic alloys).

modulus of elasticity. A measure of the rigidity of metal. Ratio of stress, below the proportional limit, to corresponding strain. Specifically, the modulus obtained in tension or compression is Young's modulus, stretch modulus or modulus of extensibility; the modulus obtained in torsion or shear is modulus of rigidity, shear modulus or modulus of torsion; the modulus covering the ratio of the mean normal stress to the change in volume per unit volume is the bulk modulus. The tangent modulus and secant modulus are not restricted within the proportional limit; the former is the slope of the stress-strain curve at a specified point; the latter is the slope of a line from the origin to a specified point on the stress-strain curve. Also called elastic modulus and coefficient of elasticity.

modulus of rigidity. See *modulus of elasticity*.

modulus of rupture. Nominal stress at fracture in a bend test or torsion test. In bending, modulus of rupture is the bending moment at fracture divided by the section modulus. In torsion, modulus of rupture is the torque at fracture divided by the polar section modulus.

modulus of strain hardening. See preferred term, *rate of strain hardening*.

Mohs' scale. A scratch hardness test for determining comparative hardness using ten standard minerals from talc (the softest) to diamond (the hardest).

mold. (1) A form made of sand, metal or other material that contains the cavity into which molten metal is poured to produce a casting of definite shape and outline. (2) Same as *die*.

molding machine. A machine for making sand molds by mechanically compacting sand around a pattern.

molding press. A press used to form powder metallurgy *compacts*.

mold jacket. Wood or metal form that is slipped over a sand mold for support during pouring.

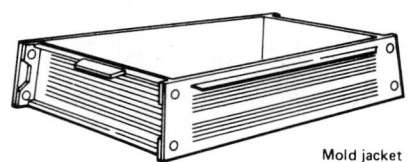

Mold jacket

mold wash. An aqueous or alcoholic emulsion or suspension of various materials used to coat the surface of a mold cavity.

Mond process. A process for extracting and purifying nickel. The main features consist of forming nickel carbonyl by reaction of finely divided reduced metal with carbon monoxide, then decomposing the nickel carbonyl, to deposit purified nickel on small nickel pellets.

monotectic. An isothermal reversible reaction in a binary system, in which a liquid on cooling decomposes into a second liquid of a different composition and a solid. It differs from a *eutectic* in that only one of the two products of the reaction is below its freezing range.

monotron hardness test. A method of determining the *indentation hardness* of a metal by measuring the load required to force a spherical penetrator into the metal to a specified depth. Now obsolete.

monotropism. The ability of a solid to exist in two or more forms (crystal structures), but in which one form is the stable modification at all temperatures and pressures. Ferrite and martensite are a monotropic pair below AC_1 in steels, for example. May also be spelled monotrophism.

mosaic structure. In crystals, a substructure in which neighboring regions have only slightly differing orientations.

M$_s$ temperature. For any alloy system, the temperature at which martensite starts to form on cooling. See *transformation temperature* for the definition applicable to ferrous alloys.

mulling. Mixing sand and clay particles with water by kneading, rolling, rubbing or stirring.

multiaxial stresses. Any stress state in which two or three principal stresses are not zero.

multiple. A piece of stock cut from a longer *mill product* to provide the exact amount of material needed for a single workpiece.

multiple-impulse welding. Spot, projection or upset welding with more than one impulse of current during a single machine cycle. Sometimes called pulsation welding.

multiple-pass weld. A weld made by depositing filler metal with two or more successive passes.

multiple-slide press. A press with individual slides, built into the main slide or connected to individual eccentrics on the main shaft, that can be adjusted so as to give variations in length of stroke and in timing.

multiple spot welding. Spot welding in which several spots are made during one complete cycle of the welding machine.

N

native metal. (1) Any deposit in the earth's crust consisting of uncombined metal. (2) The metal in such a deposit.

natural aging. Spontaneous aging of a supersaturated solid solution at room temperature. See *aging*, and compare with *artificial aging*.

natural strain. See *strain*.

necking. (1) Reducing the cross-sectional area of metal in a localized area by stretching. (2) Reducing the diameter of a portion of the length of a cylindrical shell or tube.

necking down. Localized reduction in area of a specimen during tensile deformation.

necking strain. Same as *uniform strain*.

negative rake. Describes a tooth face in rotation whose cutting edge lags the surface of the tooth face. See sketch accompanying *face mill*.

network structure. A structure in which one constituent occurs primarily at the grain boundaries, thus partially or completely enveloping the grains of the other constituents.

Neumann band. *Mechanical twin* in ferrite.

neutral flame. A gas flame in which there is no excess of either fuel or oxygen in the inner flame. Oxygen from ambient air is used to complete the combustion of CO_2 and H_2 produced in the inner flame.

neutron. Elementary nuclear particle that has a mass approximately the same as that of a hydrogen atom and that is electrically neutral; its mass is 1.008 986 mass units.

neutron embrittlement. Embrittlement resulting from bombardment with neutrons, usually encountered in metals that have been exposed to a neutron flux in the core of a reactor. In steels, neutron embrittlement is evidenced by a rise in the ductile-to-brittle transition temperature.

nibbling. Contour cutting of sheet metal by use of a rapidly reciprocating punch that makes numerous small cuts.

nitriding. Introducing nitrogen into the surface layer of a solid ferrous alloy by holding at a suitable temperature (below Ac_1 for ferritic steels) in contact with a nitrogenous material, usually ammonia or molten cyanide of appropriate composition. Quenching is not required to produce a hard case.

nitrocarburizing. Any of several processes in which both nitrogen and carbon are absorbed into the surface layers of a ferrous material at temperatures below the lower critical temperature and, by diffusion, create a concentration gradient. Nitrocarburizing is done mainly to provide an antiscuffing surface layer and to improve fatigue resistance. Compare with *carbonitriding*.

noble metal. (1) A metal whose potential is highly positive relative to the hydrogen electrode. (2) A metal with marked resistance to chemical reaction, particularly to oxidation and to solution by inorganic acids. The term as often used is synonymous with *precious metal*. Contrast with *base metal* (4).

noble potential. The potential for the passive state, if the metal can exist in both the active and passive states in a given medium.

no-draft forging. A forging with extremely close tolerances and little or no draft, requiring a minimum of machining to produce the final part. Mechanical properties can be enhanced by close control of grain flow and retention of surface material in the final component.

nodular cast iron. See preferred term, *ductile cast iron*.

nodular powder. Irregular particles of a metal powder that have knotted, rounded or other similar shapes.

nominal stress. See *stress*.

nonconforming. A quality control term describing a unit of product or service that does not meet normal acceptance criteria for the specific product or service. A nonconforming unit is not necessarily *defective*.

nondestructive inspection. Inspection by methods that do not destroy the part nor impair its serviceability.

nondestructive testing. Same as *nondestructive inspection*, but implying use of a method in which the part is stimulated and its response measured quantitatively or semiquantitatively.

nonmetallic inclusions. See *inclusions*.

normalizing. Heating a ferrous alloy to a suitable temperature above the transformation range and then cooling in air to a temperature substantially below the transformation range.

normal segregation. Concentration of alloying constituents that have low melting points in those portions of a casting that solidify last. Compare with *inverse segregation*.

normal stress. See *stress*.

nose radius. The radius of the rounded portion of the cutting edge of a tool. See sketch accompanying *single-point tool*.

nosing. Closing in the end of a tubular shape to a desired curved contour.

notch acuity. Relates to the severity of the stress concentration produced by a given notch in a particular structure. If the depth of the notch is very small compared with the width (or diameter) of the narrowest cross section, the acuity may be expressed as the ratio of the notch depth to the notch root radius. Otherwise, the acuity is defined as the ratio of one-half the width (or diameter) of the narrowest cross section to the notch root radius.

notch brittleness. Susceptibility of a material to brittle fracture at points of stress concentration. For example, in a notch tensile test, the material is said to be notch brittle if the *notch strength* is less than the tensile strength of an unnotched specimen. Otherwise, it is said to be notch ductile.

notch depth. The distance from the surface of a test specimen to the bottom of the notch. In a cylindrical test specimen, the percentage of the original cross-sectional area removed by machining an annular groove.

notch ductile. See *notch brittleness*.

notch ductility. The percentage reduction in area after complete separation of the metal in a tensile test of a notched specimen.

notching. Cutting out various shapes from the edge of a strip, blank or part.

notching press. A mechanical press used for notching internal and external circumferences and also for notching along a straight line. These presses are equipped with automatic feeds, because only one notch is made per stroke.

notch rupture strength. The ratio of applied load to original area of the minimum cross section in a stress-rupture test of a notched specimen.

notch sensitivity. A measure of the reduction in strength of a metal caused by the presence of stress concentration. Values can be obtained from static, impact or fatigue tests.

notch sharpness. See *notch acuity*.

notch strength. The maximum load on a notched tensile-test specimen divided by the minimum cross-sectional area (the area at the root of the notch). Also called notch tensile strength.

nucleation. The initiation of a phase transformation at discrete sites, the new phase growing on nuclei. See *nucleus* (1).

nucleus. (1) The first structurally stable particle capable of initiating recrystallization of a phase or the growth of a new phase, and possessing an interface with the parent matrix. The term is also applied to a foreign particle that initiates such action. (2) The heavy central core of an atom, in which most of the mass and the total positive electric charge are concentrated.

nugget. (1) A small mass of metal, such as gold or silver, found free in nature. (2) The weld metal in a spot, seam or projection weld.

O

octahedral plane. In cubic crystals, a plane with equal intercepts on all three axes.

offal. The material trimmed from blanks or formed panels.

offhand grinding. Grinding where the operator manually forces the wheel against the work, or vice versa. It often implies casual manipulation of either grinder or work to achieve the desired result. Dimensions and tolerances frequently are not specified, or are only loosely specified; the operator relies mainly on visual inspection to determine how much grinding should be done. Contrast with *precision grinding*.

offset. The distance along the strain coordinate between the initial portion of a stress-strain curve and a parallel line that intersects the stress-strain curve at a value of stress that is used as a measure of *yield strength*. It is used for materials that have no obvious *yield point*. A value of 0.2% is commonly used.

off time. In resistance welding, the time that the electrodes are off the work. This term is generally applied where the welding cycle is repetitive.

oil canning. Same as *canning*.

oilstone. A natural or manufactured abrasive stone, generally impregnated with oil, used for sharpening keen-edged tools.

Olsen ductility test. A cupping test in which a piece of sheet metal, restrained except at the center, is deformed by a standard steel ball until fracture occurs. The height of the cup (in thousandths of an inch) at time of fracture is a measure of the ductility.

open-back inclinable press. A vertical crank press that can be inclined so that the bed will have an inclination generally varying from 0° to 30°. The formed parts slide off through an opening in the back. It is often called an OBI press.

open-die forging. Same as *flat-die forging*.

open dies. See *closed dies*.

open-gap upset welding. A form of forge welding in which the weld interfaces are heated with a fuel gas flame, then forced into intimate contact by the application of force. Not to be confused with *upset welding*, which is a resistance welding process.

open-hearth furnace. A reverberatory melting furnace with a shallow hearth and a low roof. The flame passes over the charge on the hearth, causing the charge to be heated both by direct flame and by radiation from the roof and sidewalls of the furnace. In ferrous industry, the furnace is regenerative.

open rod press. A hydraulic press in which the slide is guided by vertical, cylindrical rods (usually four) that also serve to hold the crown and bed in position.

operating stress. The stress to which a structural unit is subjected in service.

optical pyrometer. An instrument for measuring the temperature of heated material by comparing the intensity of light emitted with a known intensity of an incandescent lamp filament.

orange peel. A surface roughening in the form of a pebble-grained pattern that occurs when a metal of unusually coarse grain size is stressed beyond its elastic limit. Also called pebbles and alligator skin.

ordering. Forming a *superlattice*.

ore. A natural mineral that may be mined and treated for the extraction of any of its components, metallic or otherwise, at a profit.

ore dressing. Same as *mineral dressing*.

orientation. Arrangement in space of the axes of a crystal with respect to a chosen reference or coordinate system. See also *preferred orientation*.

oscillating die press. A small high-speed press in which the die and punch move horizontally with the strip during the working stroke. Through a reciprocating motion, the die and punch return to their original positions to begin the next stroke.

overaging. Aging under conditions of time and temperature greater than those required to obtain maximum change in a certain property, so that the property is altered in the direction of the initial value. See *aging*.

overbending. Bending metal through a greater arc than that required in the finished part, to compensate for springback.

overdraft. A condition wherein a metal curves upward on leaving the rolls because of the higher speed of the lower roll.

overhauling. Cutting surface layers from castings or slabs to remove scale and surface imperfections. Sometimes called scalping or slab milling.

overhead-drive press. A mechanical press with the driving mechanism mounted in or on the crown or upper parts of the uprights.

overhead-position welding. Welding that is performed from the underside.

overheating. Heating a metal or alloy to such a high temperature that its properties are impaired. When the original properties cannot be restored by further heat treating, by mechanical working or by a combination of working and heat treating, the overheating is known as *burning*.

overlap. (1) Protrusion of weld metal beyond the toe, face or root of a weld. (2) In resistance seam welding, the area in a given weld remelted by the succeeding weld.

oversize powder. Particles of a powdered metal coarser than the maximum permitted by a given specification for particle size.

overstressing. (1) In fatigue testing, cycling at a stress level higher than that used at the end of the test.

overvoltage. The difference between the actual electrode potential when appreciable electrolysis begins and the reversible electrode potential.

oxidation. (1) A reaction in which there is an increase in valence resulting from a loss of electrons. Contrast with *reduction*. (2) A corrosion reaction in which the corroded metal forms an oxide; usually applied to reaction with a gas containing elemental oxygen, such as air.

oxidized surface (on steel). Surface having a thin, tightly adhering, oxidized skin (from straw to blue in color), extending in from the edge of a coil or sheet. Sometimes called annealing border.

oxidizing agent. A compound that causes oxidation, thereby itself becoming reduced.

oxidizing flame. A gas flame produced with excess oxygen in the inner flame.

oxyacetylene cutting. An *oxyfuel gas cutting* process in which the fuel gas is acetylene.

oxyacetylene welding. An *oxyfuel gas welding* process in which the fuel gas is acetylene.

oxyfuel gas cutting. Any of a group of processes used to sever metals by means of chemical reaction between hot base metal and a fine stream of oxygen. The necessary metal temperature is maintained by gas flames resulting from combustion of a specific fuel gas such as acetylene, hydrogen, natural gas or propane. See also *oxygen cutting*.

oxyfuel gas welding. Any of a group of processes used to fuse metals together by heating them with gas flames resulting from combustion of a specific fuel gas such as acetylene, hydrogen, natural gas or propane. The process may be used with or without the application of pressure to the joint, and with or without adding any filler metal.

oxygen cutting. Metal cutting by directing a fine stream of oxygen against a hot metal. The chemical reaction between oxygen and the base metal furnishes heat for localized melting, hence, cutting.

oxygen deficiency. A form of *crevice corrosion* in which galvanic corrosion proceeds because oxygen is prevented from diffusing into the crevice.

oxygen-free copper. Electrolytic copper free from cuprous oxide, produced without the use of residual metallic or metalloidal deoxidizers.

oxygen gouging. Oxygen cutting in which a chamfer or groove is formed.

oxygen lance. A length of pipe used to convey oxygen, either to the point of cutting in oxygen-lance cutting, or beneath the surface of the melt in a steelmaking furnace.

oxyhydrogen cutting. An *oxyfuel gas cutting* process in which the fuel gas is hydrogen.

oxyhydrogen welding. An *oxyfuel gas welding* process in which the fuel gas is hydrogen.

oxynatural gas cutting. An *oxyfuel gas cutting* process on which the fuel gas is natural gas.

oxynatural gas welding. An *oxyfuel gas welding* process in which the fuel gas is natural gas.

oxypropane cutting. An *oxyfuel gas cutting* process in which the fuel gas is propane.

oxypropane welding. An *oxyfuel gas welding* process in which the fuel gas is propane.

P

packing material. Any material in which powder metallurgy compacts are embedded during the presintering or sintering operation.

pack rolling. Hot rolling a pack of two or more sheets of metal; scale prevents their being welded together.

pancake forging. A rough forged shape, usually flat, that may be obtained quickly with a minimum of tooling. It usually requires considerable machining to attain finish size.

paramagnetic material. A material whose specific permeability is greater than unity and is practically independent of the magnetizing force. Compare with *diamagnetic material, ferromagnetic material.*

Parkes process. A process used to recover precious metals from lead and based on the principle that if 1 to 2% Zn is stirred into the molten lead, a compound of zinc with gold and silver separates out and can be skimmed off.

partial annealing. An imprecise term used to denote a treatment given cold worked material to reduce its strength to a controlled level or to effect stress relief. To be meaningful, the type of material, the degree of cold work, and the time-temperature schedule must be stated.

particle size. The controlling lineal dimension of an individual particle, such as of a powdered metal, as determined by analysis with screens or other suitable instruments.

particle-size distribution. The percentage, by weight or by number, of each fraction into which a powder sample has been classified with respect to sieve number or particle size. Preferred usage: "particle-size distribution by weight," or "particle-size distribution by frequency."

parting. (1) In the recovery of precious metals, the separation of silver from gold. (2) The zone of separation between cope and drag portions of the mold or flask in sand casting. (3) A composition sometimes used in sand molding to facilitate the removal of the pattern. (4) Cutting simultaneously along two parallel lines or along two lines that balance each other in side thrust. (5) A shearing operation used to produce two or more parts from a stamping.

parting line. (1) The intersection of the parting plane of a casting mold or the parting plane between forging dies with the mold or die cavity. (2) A raised line or projection on the surface of a casting or forging that corresponds to said intersection.

parting plane. (1) In forging, the dividing plane between dies. Contrast with *forging plane.* (2) In casting, the dividing plane between mold halves.

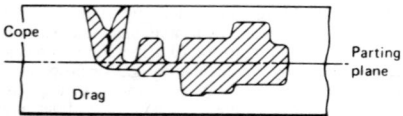

parting sand. Fine sand for dusting on sand mold surfaces that are to be separated.

parts former. A type of upsetter designed to work on short billets instead of bars and tubes, usually for cold forging.

pass. (1) A single transfer of metal through a *stand* of rolls. (2) The open space between two grooved rolls through which metal is processed. (3) The weld metal deposited in one trip along the axis of a weld.

passivation. The changing of a chemically active surface of a metal to a much less reactive state. Contrast with *activation.*

passivity. A condition in which a piece of metal, because of an impervious covering of oxide or other compound, has a potential much more positive than that of the metal in the active state.

patenting. In wiremaking, a heat treatment applied to medium-carbon or high-carbon steel before drawing of wire or between drafts. This process consists of heating to a temperature above the transformation range and then cooling to a temperature below Ae₁ in air or in a bath of molten lead or salt.

patent leveling. Same as *stretcher leveling.*

pattern. (1) A form of wood, metal or other material around which molding material is placed to make a mold for casting metals. (2) A full-scale reproduction of a part used as a guide in cutting.

Pattinson process. A process for separating silver from lead, in which the molten lead is slowly cooled so that crystals poorer in silver solidify out and are removed, leaving the melt richer in silver.

pearlite. A metastable lamellar aggregate of ferrite and cementite resulting from the transformation of aus-

tenite at temperatures above the bainite range.

pearlitic malleable. See *malleable cast iron.*

pebbles. Same as *orange peel.*

peeling. The detaching of one layer of a coating from another, or from the basis metal, because of poor adherence.

peening. Mechanical working of metal by hammer blows or shot impingement.

penetrant. A liquid with low surface tension used in *liquid penetrant inspection* to flow into surface openings of parts being inspected.

penetrant inspection. See preferred term, *liquid penetrant inspection.*

penetration. (1) In founding, an *imperfection* on a casting surface caused by metal running into voids between sand grains; usually referred to as *metal penetration.* (2) In welding, the distance from the original surface of the base metal to that point at which fusion ceased. See *joint penetration.*

penetration hardness. Same as *indentation hardness.*

percussion welding. Resistance welding in which abutting surfaces are heated by an intense spark between them, welding being consummated by applying a hammerlike blow during or immediately after the electrical discharge.

perforating. Piercing holes of desired shapes arranged in a definite pattern in sheets, blanks or formed parts.

periodic reverse. Pertains to periodic changes in direction of flow of the current in electrolysis. It applies to the process and also to the machine that controls the time for both directions.

peripheral clearance angle. See *clearance angle,* and also sketch accompanying *face mill.*

peripheral milling. Milling a surface parallel to the axis of the cutter.

peripheral speed. See preferred term, *cutting speed.*

peritectic. An isothermal reversible reaction in which a liquid phase reacts with a solid phase to produce a single (and different) solid phase on cooling.

peritectoid. An isothermal reversible reaction in which a solid phase reacts with a second solid phase to produce a single (and different) solid phase on cooling.

permanent mold. A metal, graphite or ceramic mold (other than an ingot mold) of two or more parts that is used repeatedly for the production of many *castings* of the same form. Liquid metal is poured in by gravity.

permanent set. Plastic deformation that remains on release of the stress that produces the deformation.

permeability. (1) In founding, the characteristics of molding materials that permit gases to pass through them. "Permeability number" is determined by a standard test. (2) In powder metallurgy, a property measured as the rate of passage under specified conditions of a liquid or gas through a compact. (3) A general term used to express various relationships between magnetic induction and magnetizing force. These relationships are either "absolute permeability," which is a change in magnetic induction divided by the corresponding change in magnetizing force, or "specific (relative) permeability," the ratio of the absolute permeability to the permeability of free space.

pewter. Any of various alloys in which tin is the chief constituent; especially an alloy of tin and lead formerly used for domestic utensils.

pH. The negative logarithm of the hydrogen-ion activity; it denotes the degree of acidity or basicity of a solution. At 25 °C (77 °F), 7.0 is the neutral value. Decreasing values below 7.0 indicate increasing acidity; increasing values above 7.0, increasing basicity.

phase. A physically homogeneous and distinct portion of a material system.

phase diagram. See as *constitution diagram.*

phosphating. Forming an adherent phosphate coating on a metal by immersion in a suitable aqueous phosphate solution. Also called phosphatizing.

phosphorized copper. General term applied to copper deoxidized with phosphorus. The most commonly used deoxidized copper.

photoelasticity. An optical method for evaluating the magnitude and distribution of stresses, using a transparent model of a part, or a thick film of photoelastic material bonded to a real part.

photomacrograph. See *macrograph.*

photomicrograph. See *micrograph.*

photon. The smallest possible quantity of an electromagnetic radiation that can be characterized by a definite frequency.

physical metallurgy. The science and technology deal-

ing with the properties of metals and alloys, and of the effects of composition, processing and environment on those properties.

physical properties. Properties of a metal or alloy that are relatively insensitive to structure and can be measured without the application of force; for example, density, electrical conductivity, coefficient of thermal expansion, magnetic permeability and lattice parameter. Does not include chemical reactivity. Compare with *mechanical properties.*

physical testing. Determination of *physical properties.*

pickle liquor. A spent acid-pickling bath.

pickle patch. A tightly adhering oxide or scale coating not properly removed during *pickling.*

pickle stain. Discoloration of metal due to chemical cleaning without adequate washing and drying.

pickling. Removing surface oxides from metals by chemical or electrochemical reaction.

pickoff. An automatic device for removing a finished part from the press die after it has been stripped.

pickup. Transfer of metal from tools to part or from part to tools during a forming operation. See *galling.*

Pidgeon process. A process for production of magnesium by reduction of magnesium oxide with ferrosilicon.

piezoelectric effect. The reversible interaction, exhibited by some crystalline materials, between an elastic strain and an electric field. The direction of the strain depends on the polarity of the field or vice versa. Compare with *electrostrictive effect.*

pig. A metal casting used in remelting.

pig iron. (1) High-carbon iron made by reduction of iron ore in the blast furnace. (2) Cast iron in the form of *pigs.*

Pilger tube-reducing process. See *tube reducing.*

pinchers. Surface disturbances that result from rolling processes and that ordinarily appear as fernlike ripples running diagonally to the direction of rolling.

pinch pass. A pass of sheet material through rolls to effect a very small reduction in thickness.

pinch trimming. Trimming the edge of a tubular part or shell by pushing or pinching the flange or lip over the cutting edge of a stationary punch or over the cutting edge of a draw punch.

pine-tree crystal. A type of *dendrite.*

pin expansion test. A test for determining the ability of tubes to be expanded or for revealing the presence of cracks or other longitudinal weaknesses, made by forcing a tapered pin into the open end of a tube.

pinhead blister. See *blister.*

pinhole porosity. Porosity consisting of numerous small gas holes distributed throughout the metal; found in weld metal, castings and electrodeposited metal.

pinion. The smaller of two mating gears.

Piobert lines. Same as *Lüders lines.*

pipe. (1) The central cavity formed by contraction in metal, especially ingots, during solidification. See accompanying sketch. (2) An imperfection in wrought or cast products resulting from such a cavity. (3) See

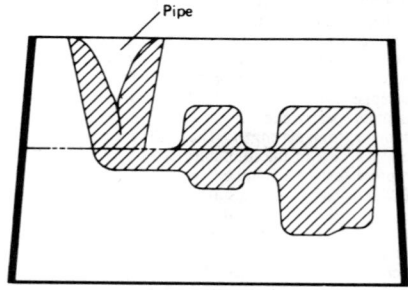

extrusion pipe. (4) A tubular metal product, cast or wrought.

pipe tap. A tap for making internal *pipe threads* within pipe fittings or holes.

pipe threads. Internal or external machine threads, usually tapered, of a design intended for making pressure-tight mechanical joints in piping systems.

pitch. See *set.*

pitting. Forming small sharp cavities in a metal surface by nonuniform electrodeposition or by corrosion.

planchet. A metal disk with milled edges, ready for coining.

plane strain. The stress condition in linear elastic fracture mechanics in which there is zero strain in a direction normal to both the axis of applied tensile stress and the direction of crack growth (i.e., parallel to the crack front); most nearly achieved in loading thick plates along a direction parallel to the plate surface. Under plane-strain conditions, the plane of fracture instability is normal to the axis of the principal tensile stress.

plane stress. The stress condition in linear elastic fracture mechanics in which the stress in the thickness direction is zero; most nearly achieved in loading very thin sheet along a direction parallel to the surface of the sheet. Under plane-stress conditions, the plane of fracture instability is inclined 45° to the axis of the principal tensile stress.

planimetric method. A method of measuring grain size in which the grains within a definite area are counted.

planing. Producing flat surfaces by linear reciprocal motion of work and the table to which it is attached, relative to a stationary single-point cutting tool.

planishing. Producing a smooth surface finish on metal by a rapid succession of blows delivered by highly polished dies or by a hammer designed for the purpose, or by rolling in a planishing mill.

plasma-arc cutting. An arc cutting process that severs metals by melting a localized area with heat from a constricted arc and removing the molten metal with a high-velocity jet of hot, ionized gas issuing from the plasma torch.

plasma-arc welding. An arc welding process that produces coalescence of metals by heating them with a constricted arc between an electrode and the workpiece (transferred arc) or the electrode and the constricting nozzle (nontransferred arc). Shielding is obtained from hot, ionized gas issuing from an orifice surrounding the electrode and may be supplemented by an auxiliary source of shielding gas, which may be an inert gas or a mixture of gases. Pressure may or may not be used, and filler metal may or may not be supplied.

plasma spraying. A *thermal spraying* process in which the coating material is melted with heat from a plasma torch that generates a nontransferred arc (defined in *plasma-arc welding*); molten coating material is propelled against the basis metal by the hot, ionized gas issuing from the torch.

plaster molding. Molding wherein a gypsum-bonded aggregate flour in the form of a water slurry is poured over a pattern, permitted to harden, and, after removal of the pattern, thoroughly dried. This technique is used to make smooth nonferrous castings of accurate size.

plastic deformation. Deformation that does or will remain permanent after removal of the load that caused it.

plastic flow. Same as *plastic deformation*.

plasticity. The ability of a metal to deform nonelastically without rupture.

plate. A flat-rolled metal product of some minimum thickness and width arbitrarily dependent on the type of metal.

platen. (1) Face of a bolster, slide or ram to which a tool assembly is attached. (2) A part of a resistance welding, mechanical testing or other machine with a flat surface to which dies, fixtures, backups or electrode holders are attached and that transmits pressure or force.

plates. Flat particles of metal powder having considerable thickness. Contrast with *flake powder*.

plating. Forming an adherent layer of metal on an object; often used as a shop term for *electroplating*.

plating rack. A fixture used to hold work and conduct current to it during electroplating.

plating range. The current-density range over which a satisfactory electroplate can be deposited.

platinum black. A finely divided form of platinum of a dull black color, usually but not necessarily produced by reduction of salts in an aqueous solution.

plug. (1) A rod or mandrel over which a pierced tube is forced. (2) A rod or mandrel that fills a tube as it is drawn through a die. (3) A punch or mandrel over which a cup is drawn. (4) A protruding portion of a die impression for forming a corresponding recess in the forging. (5) A false bottom in a die. Also called a "peg."

plug tap. A *tap* with *chamfer* extending from three to five threads.

plug weld. A circular weld made by either arc or gas welding through one member of a lap or tee joint. If a hole is used, it may be only partly filled. Neither a fillet-welded hole nor a spot weld is to be construed as a plug weld.

plumbago. A special quality of powdered graphite used to coat molds and, in a mixture with clay, to make crucibles.

plunge grinding. Grinding wherein the only relative motion of the wheel is radially toward the work.

plus sieve. The portion of a sample of a granular substance (such as metal powder) retained on a standard sieve of specified number. Contrast with *minus sieve*.

plymetal. Sheet consisting of bonded layers of dissimilar metals.

P/M. The acronym for *powder metallurgy*.

point angle. In general, the angle at the point of a cutting tool. Most commonly, the included angle at the point of a twist drill, the general-purpose angle being 118°.

pointing. (1) Reducing the diameter of wire, rod or tubing over a short length at the end by swaging or hammer forging, turning or squeezing to facilitate entry into a drawing die and gripping in the drawhead. (2) The operation in automatic machines of chamfering or rounding the threaded end or the head of a bolt.

Poisson's ratio. The absolute value of the ratio of the transverse strain to the corresponding axial strain, in a body subjected to uniaxial stress; usually applied to elastic conditions.

poke welding. Same as *push welding*.

polar bond. See *ionic bond*.

polarization. A change in the potential of an electrode during electrolysis, such that the potential of an anode becomes more noble, and that of a cathode more active, than their respective reversible potentials. Often accomplished by formation of a film on the electrode surface.

pole. (1) A means of designating the orientation of a crystal plane by stereographically plotting its normal. For example, the north pole defines the equatorial plane. (2) Either of the two regions of a permanent magnet or electromagnet where most of the lines of induction enter or leave.

pole figure. A stereographic projection representing the statistical average distribution of poles of a specific crystalline plane in a polycrystalline metal, with reference to an external system of axes. In an isotropic metal, that is, in one having a completely random distribution of orientations, the pole density is stereographically uniform; preferred orientation is shown by an increased density of poles in certain areas.

poling. A step in the fire refining of copper to reduce the oxygen content to tolerable limits by covering the bath with coal or coke and thrusting green wood poles below the surface. There is a vigorous evolution of reducing gases, which combine with the oxygen contained in the metal.

polishing. Smoothing metal surfaces, often to a high luster, by rubbing the surface with a fine abrasive, usually contained in a cloth or other soft lap. Results in microscopic flow of some surface metal together with actual removal of a small amount of surface metal. May be extended to include *electropolishing*. Contrast with *buffing*, *burnishing*.

polycrystalline. Pertaining to a solid composed of many crystals.

polymorphism. A general term for the ability of a solid to exist in more than one form. In metals, alloys and similar substances, this usually means the ability to exist in two or more crystal structures, or in an amorphous state and at least one crystal structure. See also *allotropy, enantiotropy, monotropism*.

pop-off. Loss of small portions of a porcelain enamel coating. The usual cause is outgassing of hydrogen or other gases from the basis metal during firing, but pop-off also may occur because of oxide particles or other debris on the surface of the basis metal. Usually, the pits are minute and cone shaped, but when pop-off is the result of severe *fishscale* the pits may be much larger and irregular.

pores. (1) Small voids in the body of a metal. (2) Minute cavities in a powder metallurgy compact, sometimes intentional. (3) Minute perforations in an electroplated coating.

porosity. Fine holes or pores within a metal.

porthole die. A multiple-section extrusion die capable of producing tubing or intricate hollow shapes without the use of a separate mandrel. Metal is extruded in separate streams through holes in each section and is rewelded by extrusion pressure before it leaves the die. Compare with *bridge die*.

positioned weld. A weld made in a joint that has been oriented to facilitate making the weld.

positive rake. Describes a tooth face in rotation whose cutting edge leads the surface of the tooth face. See sketch accompanying *face mill*.

postheating. Heating weldments immediately after welding, for tempering, for stress relieving, or for providing a controlled rate of cooling to prevent formation of a hard or brittle structure.

pot. (1) A vessel for holding molten metal. (2) The electrolytic reduction cell used to make such metals as aluminum from a fused electrolyte.

pot annealing. Same as *box annealing*.

pot die forming. Forming products from sheet or plate through the use of a hollow die and internal pressure which causes the preformed workpiece to assume the contour of the die.

poultice corrosion. A term used in the automotive industry to describe the corrosion of vehicle body parts due to the collection of road salts and debris on ledges and in pockets that are kept moist by weather and washing.

pouring. Transferring molten metal from a furnace or a ladle to a mold.

pouring basin. A basin on top of a mold which receives the molten metal before it enters the sprue or downgate.

powder. Particles of a solid characterized by small size, nominally within the range from 0.1 to 1000 μm.

powder lubricant. An agent mixed with or incorporated in a powder to facilitate pressing and ejection of a powder metallurgy compact.

powder metallurgy. The art of producing metal powders and of utilizing metal powders for production of massive materials and shaped objects.

powder metallurgy forging. Plastically deforming a powder metallurgy compact or preform into a fully dense finished shape using compressive force; usually done hot, and usually within closed dies.

power reel. A reel that is driven by an electric motor or some other source of power, used to wind or coil strip or wire as it is drawn through a continuous normalizing furnace, through a die, or through rolls— as in certain types of cold mills in which the work rolls are not driven.

precharge. In forming, the pressure introduced into the cavity prior to forming of the part.

precious metal. One of the relatively scarce and valuable metals: gold, silver and the platinum-group metals.

precipitation hardening. Hardening caused by precipitation of a constituent from a supersaturated solid solution. See also *age hardening* and *aging*.

precipitation heat treatment. *Artificial aging* in which a constituent precipitates from a supersaturated solid solution.

precision. The closeness of approach of each of a number of similar measurements to the arithmetic mean, the sources of error not necessarily being considered critically. *Accuracy* demands precision, but precision does not ensure accuracy.

precision casting. A metal casting of reproducible accurate dimensions, regardless of how it is made.

precision grinding. Machine grinding to specified dimensions and low *tolerances*. Contrast with *offhand grinding*.

precoat. (1) In investment casting, a special refractory slurry applied to a wax or plastic expendable pattern to form a thin coating that serves as a desirable base for application of the main slurry. (2) To make the thin coating. (3) The thin coating itself.

precoated metal products. Mill products that have a metallic, organic or conversion coating applied to their surfaces before they are fabricated into parts.

preferred orientation. A condition of a polycrystalline aggregate in which the crystal orientations are not random, but rather exhibit a tendency for alignment with a specific direction in the bulk material, commonly related to the direction of working; also called *texture*.

preforming. (1) The initial pressing of a metal powder to form a compact that is to be subjected to a subsequent pressing operation other than coining or sizing. Also preliminary shaping of a refractory metal compact after presintering and before final sintering. (2) Preliminary forming operations, especially for impression-die forging.

preheating. Heating before some further thermal or mechanical treatment. For tool steel, heating to an intermediate temperature immediately before final austenitizing. For some nonferrous alloys, heating to a high temperature for a long time, in order to homogenize the structure before working. In welding and related processes, heating to an intermediate temperature for a short time immediately before welding, brazing, soldering, cutting or thermal spraying.

presintering. Heating a powder metallurgy compact to a temperature lower than the normal temperature for final sintering, usually to increase ease of handling or forming or to remove a lubricant or binder before sintering.

press. A machine tool having a stationary bed and a slide or ram that has reciprocating motion at right angles to the bed surface, the slide being guided in the frame of the machine.

press brake. An open-frame single-action press used to bend, blank, corrugate, curl, notch, perforate, pierce or punch sheet metal or plate.

pressed density. The density of an unsintered powder metallurgy compact. Sometimes called *green density*.

press fit. An interference or *force fit* made through the use of a *press*.

press forging. *Forging* metal, usually hot, between dies in a press.

pressing. (1) In metalworking, the product or process of shallow drawing of sheet or plate. (2) Forming a powder-metal part with compressive force.

pressing area. The clear distance (left to right) between housings, stops, gibs, gibways or shoulders of strain rods, multiplied by the total distance from front to back on the bed of a *press*. Sometimes called *working area*.

pressing crack. A rupture in a green powder metallurgy compact that develops during ejection of the compact from the die; see also *capping* and *lamination*. Sometimes referred to as a *slip crack*.

pressure casting. (1) Making castings with pressure on the molten or plastic metal, as in injection molding, *die casting*, *centrifugal casting*, and cold-chamber pressure casting. (2) A casting made with pressure applied to the molten or plastic metal.

pressure gas welding. An oxyfuel gas welding process that produces coalescence simultaneously over the entire area of abutting surfaces by heating them with gas flames obtained from combustion of a fuel gas with oxygen and by application of pressure, without the use of filler metal.

primary creep. See *creep*.

primary crystal. The first type of crystal that separates from a melt on cooling.

primary current distribution. The current distribution in an electrolytic cell that is free of polarization.

primary metal. Metal extracted from minerals and free of reclaimed metal scrap. Compare with *secondary metal, native metal*.

primary mill. A mill for rolling ingots or the rolled products of ingots to blooms, billets or slabs. This type of mill is often called a *blooming mill* and sometimes a cogging mill.

primes. Metal products, principally sheet and plate, of the highest quality and free from blemishes or other visible imperfections.

principal stresses. The normal stresses on three mutually perpendicular planes on which there are no shear stresses.

prismatic plane. In noncubic crystals, any plane that is parallel to the principal axis (*c* axis).

process annealing. An imprecise term denoting various treatments used to improve workability. For the term to be meaningful, the condition of the material and the time-temperature cycle used must be stated.

process metallurgy. The science and technology of winning metals from their ores and purifying metals; sometimes referred to as chemical metallurgy. Its two chief branches are *extractive metallurgy* and *refining*.

process tolerance. The dimensional variations of a part characteristic of a specific process, once the setup is made.

profiling. Any operation that produces an irregular contour on a workpiece, for which a tracer or template-controlled duplicating equipment usually is employed.

progressive aging. Aging by increasing the temperature in steps or continuously during the aging cycle. See *aging*, and compare with *interrupted aging* and *step aging*.

progressive die. A die in which two or more sequential operations are performed at two or more positions, the work being moved from station to station.

progressive forming. Sequential forming at consecutive stations either with a single die or with separate dies.

projection welding. Resistance welding process similar to spot welding, but in which the welds are localized at projections, embossments or intersections.

proof. Any reproduction of a die impression in any material, frequently a lead or plaster cast. See *die proof*.

proof load. A predetermined load, generally some multiple of the service load, to which a specimen or structure is submitted before acceptance for use.

proof stress. (1) The stress that will cause a specified small permanent set in a material. (2) A specified stress to be applied to a member or structure to indicate its ability to withstand service loads.

proportional limit. The maximum stress at which strain remains directly proportional to stress.

pseudobinary system. (1) A three-component or ternary alloy system in which an intermediate phase acts as a component. (2) A vertical section through a ternary diagram.

pseudocarburizing. See *blank carburizing*.

pseudonitriding. See *blank nitriding*.

puckering. Wrinkling or buckling in a drawn shell in an area originally inside the draw ring.

pull cracks. In a casting, cracks that are caused by residual stresses produced during cooling, and that result from the shape of the object.

pulsation welding. Sometimes used as a synonym for *multiple-impulse welding*.

pulsed-power welding. Any arc welding process in which the power is cyclically varied to give short-duration pulses of either voltage or current that are significantly different from the average value.

pulverization. Synonymous with *comminution*.

punch. (1) The movable tool that forces material into the die in powder molding and most forming operations. (2) The movable die in a trimming press or a forging machine. (3) The tool that forces the stock through the die in rod and tube extrusion and forms the internal surface in can or cup extrusion.

punching. Producing a hole by die shearing, in which the shape of the hole is controlled by the shape of the punch and its mating die; piercing. Multiple punching of small holes is called *perforating*.

punch press. (1) In general, any mechanical press. (2) In particular, an endwheel gap-frame press with a fixed bed, used in piercing.

punch radius. The radius on the end of the punch that first contacts the work, sometimes called *nose radius*.

punch-to-die clearance. See *die clearance*.

push angle. The angle between a welding electrode and a line normal to the face of the weld when the electrode is pointing forward along the weld joint. See sketch accompanying *forehand welding*.

push bench. Equipment used for drawing moderately heavy-gage tubes by cupping sheet and forcing it through a die by pressure exerted against the inside bottom of the cup.

pusher furnace. A type of continuous furnace in which parts to be heated are periodically charged into the furnace in containers, which are pushed along the hearth against a line of previously charged containers, thus advancing the containers toward the discharge end of the furnace, where they are removed.

push fit. A loosely defined fit similar to a *snug fit*.

push welding. Spot or projection welding in which the force is applied manually to one electrode, and the work or a backing bar takes the place of the other electrode.

pyramidal plane. In noncubic crystals, any plane that intersects all three axes.

pyrometallurgy. High-temperature *winning* or *refining* of metals.

pyrometer. A device for measuring temperatures above the range of liquid thermometers.

Q

quality. (1) The totality of features and characteristics of a product or service that bear on its ability to satisfy a given need (fitness-for-use concept of quality). (2) Degree of excellence of a product or service (comparative concept). Often determined subjectively by comparison against an ideal standard or against similar products or services available from other sources. (3) A quantitative evaluation of the features and characteristics of a product or service (quantitative concept).

quality characteristic. Any dimension, mechanical property, physical property, functional characteristic or appearance characteristic that can be used as a ba-

sis for measuring the quality of a unit of product or service.

quantitative metallography. Determination of specific characteristics of a microstructure by making quantitative measurements on micrographs or metallographic images. Quantities so measured include volume concentration of phases, grain size, particle size, mean free path between like particles or secondary phases, and surface-area-to-volume ratios of microconstituents, particles or grains.

quarter hard. A *temper* of nonferrous alloys and some ferrous alloys characterized by tensile strength about midway between those of *dead soft* and *half hard* tempers.

quasibinary system. In a ternary or higher-order system, a linear composition series between two substances each of which exhibits congruent melting, wherein all equilibriums, at all temperatures or pressures, involve only phases having compositions occurring in the linear series, so that the series may be represented as binary on a phase diagram.

quench-age embrittlement. Embrittlement of low-carbon steel evidenced by a loss of ductility on aging at room temperature following rapid cooling from a temperature below the lower critical temperature.

quench aging. Aging induced by rapid cooling after *solution heat treatment*.

quench annealing. Annealing an austenitic ferrous alloy by *solution heat treatment* followed by rapid quenching.

quench cracking. Fracture of a metal during quenching from elevated temperature. Most frequently observed in hardened carbon steel, alloy steel or tool steel parts of high hardness and low toughness. Cracks often emanate from fillets, holes, corners or other stress raisers and result from high stresses due to the volume changes accompanying transformation to martensite.

quench hardening. (1) Hardening suitable alpha-beta alloys (most often certain copper or titanium alloys) by solution treating and quenching to develop a martensite-like structure. (2) In ferrous alloys, hardening by austenitizing and then cooling at a rate such that a substantial amount of austenite transforms to martensite.

quenching. Rapid cooling. When applicable, the following more specific terms should be used: *direct quenching, fog quenching, hot quenching, interrupted quenching, selective quenching, spray quenching* and *time quenching*.

quench time. In resistance welding, the time from the finish of the welding operation to the beginning of tempering. Also called chill time.

quill. (1) A hollow or tubular shaft, designed to slide or revolve, carrying a rotating member within itself. (2) Removable spindle projection for supporting a cutting tool or grinding wheel.

R

rabbit ear. Recess in the corner of a die to allow for wrinkling or folding of the blank.

rabble. A hoelike bladed tool or similar device used for stirring molten metal.

radial draw forming. Forming metals by simultaneous application of tangential stretch and radial compression forces, the operation being done gradually by tangential contact with the die member. This type of forming is characterized by very close dimensional control.

radial forging. A process utilizing two or more moving anvils or dies for producing shafts with constant or varying diameters along their length or tubes with internal or external variations in diameter; also known as draw forging or rotary swaging.

radial marks. Lines on a fracture surface that radiate from the fracture origin and are visible to the unaided eye or at low magnification. Radial lines result from the intersection and connection of brittle fractures propagating at different levels. Also called shear ledges. See also *chevron pattern*.

radial rake. The angle between the tooth face and a radial line passing through the cutting edge in a plane perpendicular to the cutter axis. See sketch accompanying *face mill*.

radial runout. For any rotating element, the total variation from true radial position, taken in a plane perpendicular to the axis of rotation. Compare with *axial runout*.

radiation damage. A general term for the alteration of

properties of a material arising from exposure to ionizing radiation (penetrating radiation) such as x-rays, gamma rays, neutrons, heavy-particle radiation, or fission fragments in nuclear fuel material.

radiation dose. Accumulated exposure to ionizing radiation during a specified period of time.

radiation energy. The energy of a given photon or particle in a beam of radiation, often expressed in electron volts.

radiation gage. An instrument for measuring the intensity and quantity of ionizing radiation.

radiation intensity. In general, the quantity of radiant energy at a specified location passing perpendicularly through unit area in unit time. It may be given as number of particles or photons per square centimetre per second, or in energy units as $J/m^2 \cdot s$ or Rhm.

radiation monitoring. The continuous or periodic measurement of the intensity of radiation received by personnel or present in any particular area.

radiation quality. A term describing roughly the spectrum of radiation produced by a radiation source, with respect to its penetrating power or its suitability for a given application.

radioactive element. An element that has at least one isotope that undergoes spontaneous nuclear disintegration to emit positive alpha particles, negative beta particles, or gamma rays.

radioactive tracer element. A radioactive isotope of an element used to study the movement and behavior of atoms by observing the distribution and intensity of radioactivity.

radioactivity. The spontaneous nuclear disintegration with emission of corpuscular or electromagnetic radiation.

radiograph. A photographic shadow image resulting from uneven absorption of penetrating radiation in a test object.

radiography. A method of nondestructive inspection in which a test object is exposed to a beam of x-rays or gamma rays and the resulting shadow image of the object is recorded on photographic film placed behind the object. Internal discontinuities are detected by observing and interpreting variations in the image caused by differences in thickness, density or absorption within the test object. Variations of radiography include electron radiography, *fluoroscopy*, neutron radiography.

radioisotope. An isotope that emits ionizing radiation during its spontaneous decay.

rake. The angular relationship between the tooth face, or a tangent to the tooth face at a given point, and a given reference plane or line. See sketches accompanying *face mill* and *single-point tool*.

ram. The moving member of a hammer, machine, or press to which a tool is fastened.

ramming. Packing sand, refractory or other material into a compact mass.

ramoff. A casting imperfection resulting from the movement of sand away from the pattern because of improper ramming.

random sequence. A longitudinal welding sequence wherein the weld-bead increments are deposited at random to minimize distortion.

range. In inspection, the difference between the highest and lowest values of a given *quality characteristic* within a single *sample*.

rare earth metal. One of the group of 15 chemically similar metals with atomic numbers 57 through 71, commonly referred to as the lanthanides.

ratcheting. Progressive cyclic inelastic deformation (growth, for example) that occurs when a component or structure is subjected to a cyclic secondary stress superimposed on a sustained primary stress. The process is called thermal ratcheting when cyclic strain is induced by cyclic changes in temperature, and isothermal ratcheting when cyclic strain is mechanical in origin (even though accompanied by cyclic changes in temperature).

ratchet marks. Lines on a fatigue fracture surface that result from the intersection and connection of fatigue fractures propagating from multiple origins. Ratchet marks are parallel to the over-all direction of crack propagation and are visible to the unaided eye or at low magnification.

rate of strain hardening. Rate of change of true *stress* with respect to true *strain* in the plastic range.

rattail. A surface imperfection on a casting, occurring as one or more irregular lines, caused by expansion of sand in the mold. Compare with *buckle* (2).

RE. Abbreviation for *rare earth* (elements).

reamed extrusion ingot. A cast hollow extrusion ingot

that has been machined to remove the original inside surface.

reamer. A rotary cutting tool with one or more cutting elements called teeth, used for enlarging a hole to desired size and contour. It is supported principally by the metal around the hole it cuts.

recalescence. A phenomenon, associated with the transformation of gamma iron to alpha iron on cooling (supercooling) of iron or steel, that is revealed by the brightening (reglowing) of the metal surface owing to the sudden increase in temperature caused by the fast liberation of the latent heat of transformation. Contrast with *decalescence*.

recarburize. (1) To increase the carbon content of molten cast iron or steel by adding carbonaceous material, high-carbon pig iron or a high-carbon alloy. (2) To carburize a metal part to return surface carbon lost in processing; also known as *carbon restoration*.

recess. A groove or depression in a surface.

reclaim rinse. A nonflowing rinse used to recover *dragout*.

recoil line. See *impact line*.

recovery. (1) Reduction or removal of work-hardening effects, without motion of large-angle grain boundaries. (2) The proportion of the desired component obtained by processing an ore, usually expressed as a percentage.

recrystallization. (1) Formation of a new, strain-free grain structure from that existing in cold worked metal, usually accomplished by heating. (2) The change from one crystal structure to another, as occurs on heating or cooling through a critical temperature.

recrystallization annealing. Annealing cold worked metal to produce a new grain structure without phase change.

recrystallization temperature. The approximate minimum temperature at which complete recrystallization of a cold worked metal occurs within a specified time.

recuperator. Equipment for transferring heat from gaseous products of combustion to incoming air or fuel. The incoming material passes through pipes surrounded by a chamber through which the outgoing gases pass.

red mud. A residue, containing a high percentage of iron oxide, obtained in purifying bauxite in the production of alumina in the *Bayer process*.

redrawing. Drawing metal after a previous cupping or drawing operation.

reducing agent. A substance that causes reduction. See *reduction* (3).

reducing flame. A gas flame produced with excess fuel in the inner flame.

reduction. (1) In cupping and deep drawing, a measure of the percentage decrease from blank diameter to cup diameter, or of diameter reduction in redrawing. (2) In forging, rolling and drawing, either the ratio of the original to final cross-sectional area or the percentage decrease in cross-sectional area. (3) A reaction in which there is a decrease in valence resulting from a gain in electrons. Contrast with *oxidation*.

reduction cell. A pot or tank in which either a water solution of a salt or a fused salt is reduced electrolytically to form free metals or other substances.

reduction of area. (1) Commonly, the difference, expressed as a percentage of original area, between the original cross-sectional area of a tensile test specimen and the minimum cross-sectional area measured after complete separation. (2) The difference, expressed as a percentage of original area, between original cross-sectional area and that after straining of the specimen.

reeding. The operation of forming serrations and corrugations by coining or embossing.

reel. (1) A spool or hub for coiling or feeding wire or strip. (2) To straighten and planish a round bar by passing it between contoured rolls.

reel breaks. Transverse breaks or ridges on successive inner laps of a coil that result from crimping of the lead end of the coil into a gripping segmented mandrel. Also called reel kinks.

reference plane. (1) The plane that contains the cutter axis and the point of the cutting edge. See sketch accompanying *face mill*. (2) A plane from which measurements are made.

refining. The branch of *process metallurgy* dealing with the purification of crude or impure metals. Compare with *extractive metallurgy*.

reflector sheet. A clad product consisting of a facing layer of high-purity aluminum capable of taking a high polish, for reflecting heat or light, and a base of com-

mercially pure aluminum or an aluminum-manganese alloy, for strength and formability.

reflowing. Melting of an electrodeposit followed by solidification. The surface has the appearance and physical characteristics of a hot dipped surface (especially tin or tin alloy plates). Also called flow brightening.

refractory. (1) A material of very high melting point with properties that make it suitable for such uses as furnace linings and kiln construction. (2) The quality of resisting heat.

refractory alloy. (1) A heat-resistant alloy. (2) An alloy having an extremely high melting point. See *refractory metal*. (3) An alloy difficult to work at elevated temperatures.

refractory metal. A metal having an extremely high melting point; for example, tungsten, molybdenum, tantalum, niobium (columbium), chromium, vanadium and rhenium. In the broad sense, this term refers to metals having melting points above the range for iron, cobalt and nickel.

regenerator. Same as *recuperator* except that the gaseous products of combustion heat brick checkerwork in a chamber connected to the exhaust side of the furnace while the incoming air and fuel are being heated by the brick checkerwork in a second chamber, connected to the entrance side. At intervals, the gas flow is reversed so that incoming air and fuel contact hot checkerwork while that in the second chamber is being reheated by exhaust gases.

regulus. The impure button, globule or mass of metal formed beneath the slag in the smelting and reduction of ores. The name was first applied by alchemists to metallic antimony because it readily alloyed with gold.

rejectable. See preferred term, *nonconforming*.

reliability. A quantitative measure of the ability of a product or service to fulfill its intended function for a specified period of time.

relief. The result of the removal of tool material behind or adjacent to the cutting edge to provide clearance and prevent rubbing (heel drag). See sketch accompanying *single-point tool*.

relief angle. The angle formed between a relieved surface and a given plane tangent to a cutting edge or to a point on a cutting edge. See sketch accompanying *single-point tool*.

relieving. Buffing or other abrasive treatment of the high points of an embossed metal surface to produce highlights that contrast with the finish in the recesses.

remanence. The magnetic induction remaining in a magnetic circuit after removal of the applied magnetizing force. Sometimes called remanent induction.

re-pressing. The application of pressure to a previously pressed and sintered powder metallurgy compact, usually for the purpose of improving some physical property.

residual elements. Elements present in an alloy in small quantities, but not added intentionally.

residual field. Same as *residual magnetic field*.

residual magnetic field. The magnetic field that remains in a part after the magnetizing force has been removed.

residual method. Method of *magnetic-particle inspection* in which the particles are applied after the magnetizing force has been removed.

residual stress. Stress present in a body that is free of external forces or thermal gradients.

resilience. (1) The amount of energy per unit volume released on unloading. (2) The capacity of a metal, by virtue of high yield strength and low elastic modulus, to exhibit considerable elastic recovery on release of load.

resinoid wheel. A grinding wheel bonded with a synthetic resin.

resist. (1) A material applied to a part of a cathode or plating rack to render the surface nonconductive. (2) A material applied to a part of the surface of an article to prevent reaction of metal from that area during chemical or electrochemical processes. (3) A material applied to prevent flow of brazing filler metal into unwanted areas.

resistance brazing. Brazing by resistance heating, the joint being part of the electrical circuit.

resistance soldering. Soldering in which the joint is heated by electrical resistance. Filler metal is either face fed into the joint or preplaced in the joint.

resistance welding. Welding with resistance heating and pressure, the work being part of the electrical circuit. Examples: resistance *spot welding*, resistance *seam welding*, *projection welding* and *flash butt welding*.

resistance welding die. The part of a resistance welding machine, usually shaped to the work contour, in which the parts being welded are held and which conducts the welding current.

resolution. The ability to separate closely related items of data or physical features using a given test method; also a quantitative measure of the degree to which they can be discriminated.

restraint. Any external mechanical force that prevents a part from moving to accommodate changes in dimensions due to thermal expansion or contraction. Often applied to weldments made while clamped in a fixture. Compare with *constraint*.

restriking. (1) Striking a trimmed but slightly misaligned or otherwise faulty forging one or more blows to improve alignment, improve surface condition, maintain close tolerances, increase hardness or effect other improvements. (2) A sizing operation in which coining or stretching is utilized to correct or alter profiles and to counteract distortion.

resultant field. The magnetic field which is the result of two or more magnetizing forces impressed on the same area of a magnetizable object. Sometimes called vector field.

resultant rake. The angle between the tooth face and an axial plane through the tooth point measured in a plane perpendicular to the cutting edge. The resultant rake of a cutter is a function of three other angles: radial rake, axial rake and corner angle. See sketch accompanying *face mill*.

retentivity. The capacity of a material to retain a portion of the magnetic field set up in it after the magnetizing force has been removed.

retort. A vessel used for distillation of volatile materials, as in separation of some metals and in destructive distillation of coal.

reverberatory furnace. A furnace with a shallow hearth, usually nonregenerative, having a roof that deflects the flame and radiates heat toward the hearth or the surface of the charge.

reverse-current cleaning. Same as *anodic cleaning*.

reverse drawing. *Redrawing* in a direction opposite to that of the original drawing.

reverse flange. A flange made by shrinking, as opposed to one formed by stretching.

reverse polarity. Direct-current arc welding circuit arrangement in which the electrode is connected to the positive terminal. Contrast with *straight polarity*.

reverse redrawing. A second drawing operation in a direction opposite to that of the original drawing.

rheology. The science of deformation and flow of matter.

rheotropic brittleness. That portion of the brittleness characteristic of non-face-centered cubic metals, when tested in the presence of a stress concentration or at low temperatures or high strain rates, that may be eliminated by prestraining under milder conditions.

riddle. A sieve used to separate foundry sand or other granular materials into various particle-size grades or to free such a material of undesirable foreign matter.

rigging. The engineering design, layout and fabrication of pattern equipment for producing castings; including a study of the casting solidification program, feeding and gating, risering, skimmers and fitting flasks.

right-hand cutting tool. A cutter all of whose flutes twist away in a clockwise direction when viewed from either end.

rimmed steel. A low-carbon steel containing sufficient iron oxide to give a continuous evolution of carbon monoxide while the ingot is solidifying, resulting in a case or rim of metal virtually free of voids. Sheet and strip products made from rimmed steel ingots have very good surface quality.

ring and circle shear. A cutting or shearing machine with two rotary-disk cutters driven in unison and equipped with a circle attachment for cutting inside circles or rings from sheet metal, where it is impossible to start the cut at the edge of the sheet. One cutter shaft is inclined to the other to provide cutting clearance so that the outside section remains flat and usable. See *circle shear*.

ringing. The audible or ultrasonic tone produced in a mechanical part by shock, and having the natural frequency or frequencies of the part. The quality, amplitude or decay rate of the tone may sometimes be used to indicate quality or soundness. See also *sonic testing, ultrasonic testing*.

ring riser. A riser block with openings matching those in the press bed.

ring rolling. The process of shaping weldless rings from pierced disks or thick-walled, ring-shaped blanks between rolls that control wall thickness, ring diameter, height and contour.

rinsability. The relative ease with which a substance can be removed from a metal surface with a liquid such as water.

riser. A reservoir of molten metal connected to a casting to provide additional metal to the casting, required as the result of shrinkage before and during solidification.

riser blocks. (1) Plates or pieces inserted between the top of a press bed or bolster and the die to decrease the height of the die space. (2) Spacers placed between bed and housings to increase *shut height* on a four-piece tie-rod straight-side press.

river pattern. A term used in fractography to describe a characteristic pattern of cleavage steps running parallel to the local direction of crack propagation on the fracture surfaces of grains that have separated by cleavage.

riveting. Joining of two or more members of a structure by means of metal rivets, the unheaded end being upset after the rivet is in place.

roasting. Heating an ore to effect some chemical change that will facilitate smelting.

robber. An extra cathode or cathode extension that reduces the current density on what would otherwise be a high-current-density area on work being electroplated.

Rochelle copper. (1) A copper electrodeposit obtained from copper cyanide plating solution to which Rochelle salt (sodium potassium tartrate) has been added for grain refinement, better anode corrosion and cathode efficiency. (2) The solution from which a Rochelle copper electrodeposit is obtained.

rock candy fracture. A fracture that exhibits separated-grain facets, most often used to describe intergranular fractures in large-grained metals.

rocking shear. A type of guillotine shear that utilizes a curved blade to shear sheet metal progressively from side to side by a rocker motion.

Rockrite tube-reducing process. See *tube reducing*.

Rockwell hardness test. An indentation hardness test based on the depth of penetration of a specified penetrator into the specimen under certain arbitrarily fixed conditions.

rod mill. (1) A *hot mill* for rolling rod. (2) A mill for fine grinding, somewhat similar to a *ball mill*, but employing long steel rods instead of balls to effect grinding.

roll bending. Curving sheets, bars and sections by means of rolls. See *bending rolls*.

roll compacting. Progressive compacting of metal powders by the use of a rolling mill.

rolled gold. Same as *gold filled* except that the proportion of gold alloy to the weight of the entire article may be less than $^1/_{20}$th. Fineness of the gold alloy may not be less than 10K.

roller leveler breaks. Obvious transverse *breaks* usually about 3 to 6 mm ($^1/_8$ to $^1/_4$ in.) apart caused by the sheet fluting during roller leveling. These will not be removed by stretching.

roller leveler lines. Same as *leveler lines*.

roller leveling. *Leveling* by passing flat stock through a machine having a series of small-diameter staggered rolls that are adjusted to produce repeated reverse bending.

roller stamping die. An engraved roller used for impressing designs and markings on sheet metal.

roll flattening. Flattening of sheets that have been rolled in packs by passing them separately through a two-high cold mill, there being virtually no deformation. Not to be confused with *roller leveling*.

roll forging. Forging with rotating dies that are not full round, the desired shape—either straight or tapered—being produced by a groove in the dies.

roll forming. Forming of flat-rolled metal by use of power-driven rolls whose contours determine the shape of the product. Roll forming is used extensively to make metal window frames, drapery rods and similar products from metal strip. The term is sometimes used to describe power spinning.

rolling. Reducing the cross-sectional area of metal stock, or otherwise shaping metal products, through the use of rotating rolls.

rolling mills. Machines used to decrease the cross-sectional area of metal stock and produce certain desired shapes as the metal passes between rotating rolls mounted in a framework comprising a basis unit called a *stand*. Cylindrical rolls produce flat shapes, grooved rolls produce rounds, squares and structural shapes. Among rolling mills may be listed the billet mill, blooming mill, breakdown mill, plate mill, sheet mill, slabbing mill, strip mill and temper mill.

roll resistance spot welding. Process for making separated resistance spot welds with one or more rotating circular electrodes. The rotation of the electrodes may or may not be stopped during the making of a weld.

roll straightening. Straightening of metal stock of various shapes by (1) passing it through a series of staggered rolls, the rolls usually being in horizontal and vertical planes; or (2) reeling in two-roll straightening machines.

roll table. A conveyor table where rolls furnish the contact surface.

roll threading. Making threads by rolling the piece between two grooved die plates, one of which is in motion, or between rotating grooved circular rolls.

roll welding. Solid-state welding in which metals are heated, then welded together by applying pressure, with rolls, sufficient to cause deformation at the faying surfaces. See also *forge welding*.

root crack. A crack in either the weld or heat-affected zone at the root of a weld.

root face. The portion of a weld groove face adjacent to the root of the joint.

root of joint. The portion of a weld joint where the members are closest to each other before welding. In cross section, this may be a point, a line or an area.

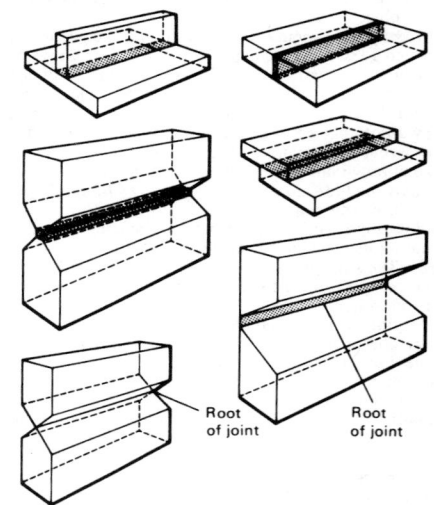

Root of joint   Root of joint

root of weld. The points, as shown in cross section, at which the weld bead intersects the base-metal surfaces either nearest to or coincident with the *root of joint*.

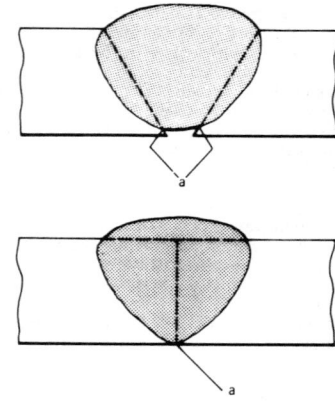

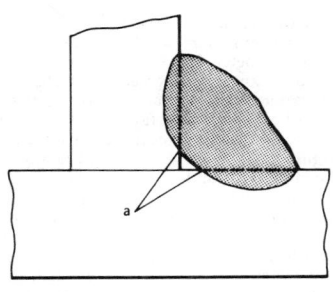

a = root of weld

root opening. In a weldment, the separation between the members at the *root of joint* prior to welding.

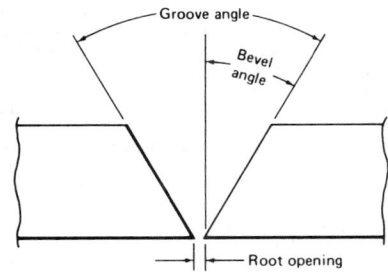

root pass. The first bead of a *multiple-pass weld,* laid in the *root of joint.*

root penetration. The depth that a weld extends into the *root of joint,* measured on the centerline of the root cross section. See sketch accompanying *joint penetration.*

rosebuds. Concentric rings of distorted coating, giving the effect of an opened rosebud. Noted only on *minimized spangle.*

rosette. (1) Rounded configuration of microconstituents arranged in whorls or radiating from a center. (2) Strain gages arranged to indicate at a single position strains in three different directions.

rotary forging. A process in which the workpiece is pressed between a flat anvil and a swiveling die with a conical working face; the platens move toward each other during forging.

rotary furnace. A circular furnace constructed so that the hearth and workpieces rotate around the axis of the furnace during heating.

rotary shear. A sheet-metal-cutting machine with two rotating-disk cutters mounted on parallel shafts driven in unison.

rotary swager. A swaging machine consisting of a power-driven ring that revolves at high speed, causing rollers to engage cam surfaces and force the dies to deliver hammerlike blows upon the work at high frequency. Both straight and tapered sections can be produced.

rouge finish. A highly reflective finish produced with rouge or other very fine abrasive; similar in appearance to the bright polish or mirror finish on sterling silver utensils.

rough grinding. Grinding without regard to finish, usually to be followed by a subsequent operation.

roughing stand. The first stand of rolls through which the reheated billet passes, or the last stand in front of the finishing rolls.

rough machining. Machining without regard to finish, usually to be followed by a subsequent operation.

roughness. Relatively finely spaced surface irregularities, the heights, widths and directions of which establish the predominant surface pattern.

roughness-width cutoff. The maximum width (in inches) of surface irregularities to be included in the measurement of roughness height.

rubber blanket. A sheet of rubber or other resilient material used as an auxiliary tool in forming.

rubber forming. Forming wherein rubber or another resilient material is used as a functional die part. Processes in which rubber is employed only to contain the hydraulic fluid are not classified as rubber forming.

rubber-pad forming. A forming operation for shallow parts wherein a pad of rubber or other resilient material is attached to the press slide and becomes the mating die for a punch, or group of punches, that has been placed on the press bed or plate. Also known as the Guerin process.

rubber wheel. A grinding wheel made with a rubber bond.

rub mark. See *abrasion.*

runner. (1) A channel through which molten metal flows from one receptacle to another. (2) The portion of the gate assembly of a casting that connects the sprue with the gate(s). (3) Parts of patterns and finished castings corresponding to the portion of the gate assembly described in (2).

runner box. A distribution box that divides molten metal into several streams before it enters the mold cavity.

running fit. Any *clearance fit* in the range used for parts that rotate relative to each other. Actual values of clearance resulting from stated shaft and hole tolerances are given in ANSI standards.

runout. (1) The unintentional escape of molten metal from a mold, crucible or furnace. (2) An imperfection in a casting caused by the escape of metal from the mold. (3) See *axial runout* and *radial runout.*

runout table. A *roll table* used to receive a rolled or extruded section.

rust. A corrosion product consisting of hydrated oxides of iron. Applied only to ferrous alloys.

S

sacrificial protection. Reduction of the extent of corrosion of a metal in an electrolyte by coupling it to another metal that is electrochemically more active in the environment.

saddling. Forming a seamless ring by forging a pierced disk over a mandrel (or saddle).

sag. An increase or decrease in the section thickness of a casting caused by insufficient strength of the mold sand of the cope or of the core.

salt fog test. An accelerated corrosion test in which specimens are exposed to a fine mist of a solution usually containing sodium chloride but sometimes modified with other chemicals. For testing details see ASTM B117.

salting out. Precipitating a substance in a solution by adding a second substance, usually a salt, without any chemical reaction such as a double decomposition taking place.

salt spray test. More properly, *salt fog test.*

sample. One or more units of a product (or a relatively small quantity of a bulk material) withdrawn from a *lot* or process stream and then tested or inspected to provide information about the properties, dimensions or other quality characteristics of the lot or process stream. Not to be confused with *specimen.*

sand. A granular material naturally or artificially produced by the disintegration or crushing of rocks or mineral deposits. In casting, the term denotes an aggregate, with an individual particle (grain) size of 0.06 to 2 mm ($^1/400$ to $^1/12$ in.) in diameter, that is largely free of finer constituents such as silt and clay, which are often present in natural sand deposits. The most commonly used foundry sand is silica; however, zircon, olivine, chromite, alumina and other crushed ceramics are used for special applications.

sand blasting. Abrasive blasting with sand. See *blasting,* and compare with *shot blasting.*

sand control. Testing and regulation of the chemical, physical and mechanical properties of foundry sand mixtures and their components.

sand hole. A pit in the surface of a sand casting resulting from a deposit of loose sand in the mold cavity.

sandwich rolling. Rolling two or more strips of metal in a pack, sometimes to form a roll-welded composite.

saponification. The alkaline hydrolysis of fats whereby a soap is formed; more generally, hydrolysis of an ester by an alkali with the formation of an alcohol and a salt of the acid portion.

satin finish. A diffusely reflecting surface finish on metals, lustrous but not mirrorlike. One type is a *butler finish.*

saw gumming. In saw manufacture, grinding away of punch marks or milling marks in the gullets (spaces between the teeth) and, in some cases, simultaneous sharpening of the teeth; in reconditioning of worn saws, restoration of the original gullet size and shape.

sawing. Cutting a workpiece with a band, blade or circular disk having teeth.

scab. An imperfection consisting of a thin, flat piece of metal attached to the surface of a sand casting. A sand scab usually is separated from the casting proper by a thin layer of sand and is joined to the casting along one edge. An erosion scab is similar in appearance to a *cut* or wash.

scale pit. (1) A surface depression formed on a forging due to scale remaining in the dies during the forging operation. (2) A pit in the ground in which scale (such as that carried off by cooling water from rolling mills) is allowed to settle out as one step in the treatment of effluent waste water.

scaling. (1) Forming a thick layer of oxidation products on metals at high temperature. (2) Depositing water-insoluble constituents on a metal surface, as in cooling tubes and water boilers.

scalped extrusion ingot. A cast, solid or hollow extrusion ingot that has been machined on the outside surface.

scalping. Removing surface layers from ingots, billets or slabs. See *die scalping.*

scarfing. Cutting surface areas of metal objects, ordinarily by using an oxyfuel gas torch. The operation permits surface imperfections to be cut from ingots, billets or the edges of plate that are to be beveled for butt welding. See *chipping.*

scarf joint. A butt joint in which the plane of the joint is inclined with respect to the main axis of the members.

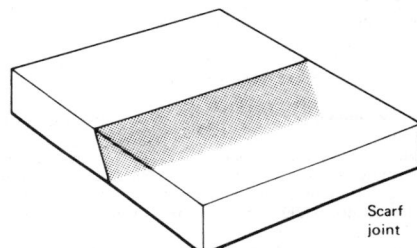

Scarf joint

Scleroscope test. A hardness test in which the loss in kinetic energy of a falling metal "tup," absorbed by indentation upon impact of the tup on the metal being tested, is indicated by the height of rebound.

scorification. Oxidation, in the presence of fluxes, of molten lead containing precious metals, to partly remove the lead in order to concentrate the precious metals.

scoring. (1) Marring or scratching of a smooth surface; most often caused by sliding contact with a mating member having a hard projection or embedded particle on its surface. (2) Reducing the thickness of a material along a line to purposely weaken it.

scouring. (1) A wet or dry cleaning process involving mechanical scrubbing. (2) A wet or dry mechanical finishing operation, using fine abrasive and low pressure, carried out by hand or with a cloth or wire wheel to produce satin or butler-type finishes.

scrap. (1) Products that are discarded because they are defective or otherwise unsuitable for sale. (2) Discarded metallic material, from whatever source, that may be reclaimed through melting and refining.

scratch hardness. The hardness of a metal determined by the width of a scratch made by drawing a cutting point across the surface under a given pressure.

screen. (1) One of a set of sieves, designated by the size of the openings, used to classify granular aggregates such as sand, ore or coke by particle size. (2) A perforated sheet placed in the gating system of a mold to separate dirt from the molten metal.

screw dislocation. See *dislocation.*

screw press. A press whose slide is operated by a screw rather than by a crank or other means.

screw stock. Free-machining bar, rod or wire.

scruff. A mixture of tin oxide and iron-tin alloy formed as dross on a tin-coating bath.

scuffing. A form of *adhesive wear* that produces superficial scratches or a high polish on the rubbing surfaces. It is observed most often on inadequately lubricated parts.

sea coal. Finely ground coal, used as an ingredient in molding sands.
sealing. (1) Closing pores in anodic coatings to render them less absorbent. (2) Plugging leaks in a casting by introducing thermosetting plastics into porous areas and subsequently setting the plastic with heat.
seal weld. Any weld used primarily to obtain tightness and prevent leakage.
seam. On a metal surface, an unwelded fool or lap that appears as a crack, usually resulting from a discontinuity.
seam welding. (1) Arc or resistance welding in which a series of overlapping spot welds is produced with rotating electrodes or rotating work, or both. (2) Making a longitudinal weld in sheet metal or tubing.
season cracking. Cracking resulting from the combined effects of corrosion and internal stress. A term usually applied to stress-corrosion cracking of brass.
secant modulus. See *modulus of elasticity.*
secondary creep. See creep.
secondary metal. Metal recovered from scrap by remelting and refining.
segment die. A die made of parts that can be separated for ready removal of the workpiece. Synonymous with split die.
segregation. Nonuniform distribution of alloying elements, impurities or microphases.
seizing. The stopping of a moving part by a mating surface as a result of excessive friction caused by *galling.*
Sejournet process. See *Ugine-Sejournet process.*
selective heating. Intentionally heating only certain portions of a workpiece.
selective leaching. Corrosion in which one element is preferentially removed from an alloy, leaving a residue (often porous) of the elements that are more resistant to the particular environment. See also *decarburization, denickelification, dezincification, graphitic corrosion.*
selective quenching. Quenching only certain portions of an object.
self-diffusion. Thermally activated movement of an atom to a new site in a crystal of its own species, as, for example, a copper atom within a crystal of copper.
self-hardening steel. See preferred term, *air-hardening steel.*
semiautomatic plating. *Plating* in which prepared cathodes are mechanically conveyed through the plating baths, with intervening manual transfers.
semiconductor. An electronic conductor whose resistivity at room temperature is in the range from 10^{-7} to 1 $\Omega \cdot m$ and in which the conductivity increases with increasing temperature over some temperature range.
semifinisher. An impression in a forging die that only approximates the finish dimensions of the forging. Semifinishers are often used to extend die life of the finishing impression, ensure proper control of grain flow during forging, and assist in obtaining desired tolerances. Also called semifinishing impression.
semifinishing. Preliminary operations performed prior to finishing.
semikilled steel. Steel that is incompletely deoxidized and contains sufficient dissolved oxygen to react with the carbon to form carbon monoxide and thus offset solidification shrinkage.
semipermanent mold. A permanent mold in which sand cores are used.
Sendzimir mill. A type of cluster mill with small-diameter working rolls and larger-diameter backup rolls, backed up by bearings on a shaft mounted eccentrically so that it can be rotated to increase the pressure between the bearings and backup rolls.
sensitivity. The smallest difference in values that can be detected reliably with a given measuring instrument.
sensitization. In austenitic stainless steels, the precipitation of chromium carbides, usually at grain boundaries, on exposure to temperatures of about 550 to 850 °C (about 1000 to 1550 °F), leaving the grain boundaries depleted of chromium and therefore susceptible to preferential attack by a corroding (oxidizing) medium.
sequence timer. In resistance welding, a device used for controlling the sequence and duration of any or all of the elements of a complete welding cycle except *weld time* or *heat time.*
sequence weld timer. Same as *sequence timer* except that either *weld time* or *heat time,* or both, are also controlled.
sequestering agent. A material that combines with metallic ions to form water-soluble complex compounds.
series welding. Resistance welding in which two or more spot, seam or projection welds are made simultaneously by a single welding transformer with three or more electrodes forming a series circuit.

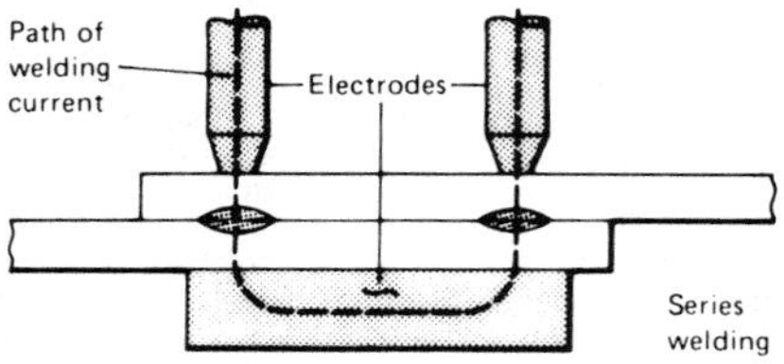

set. The shape of the solidifying surface of a metal, especially copper, with respect to concavity or convexity. May also be called pitch.
set copper. An intermediate copper product containing about 3.5% cuprous oxide, obtained at the end of the oxidizing portion of the fire-refining cycle.
settling. (1) Separation of solids from suspension in a fluid of lower density, solely by gravitational effects. (2) A process for removing iron from liquid magnesium alloys by holding the melt at a low temperature after manganese has been added to it.
shadowing. (1) Same as *shielding* in electroplating. (2) Directional deposition of carbon or a metallic film on a plastic replica so as to highlight features to be analyzed by transmission electron microscopy.
shakeout. Removal of castings from a sand mold.
shank. (1) The handle for carrying a small ladle or crucible. (2) The portion of a die, tool or forging by which it is held. (3) The main body of a lathe tool. If the tool is an inserted type, the shank is the portion that supports the insert. See sketch accompanying *single-point tool.*
shank-type cutter. A cutter having a straight or tapered shank to fit into a machine-tool spindle or adapter.
shaping. Producing flat surfaces using single-point tools. The work is held in a vise or fixture, or is clamped directly to the table. The ram supporting the tool is reciprocated in a linear motion past the work.
shatter cracks. See *flakes.*
shaving. (1) As a finishing operation, the accurate removal of a thin layer of a work surface by straight-line motion between a cutter and the surface. (2) Trimming parts such as stampings, forgings and tubes to remove uneven sheared edges or to improve accuracy.
shear. (1) That type of force that causes or tends to cause two contiguous parts of the same body to slide relative to each other in a direction parallel to their plane of contact. (2) A type of cutting tool with which a material in the form of wire, sheet, plate or rod is cut between two opposing blades. (3) The type of cutting action produced by *rake* so that the direction of chip flow is other than at right angles to the cutting edge.
shear angle. The angle that the *shear plane,* in metal cutting, makes with the work surface.
shear fracture. A ductile fracture in which a crystal (or a polycrystalline mass) has separated by sliding or tearing under the action of shear stresses.
shearing strain. See *strain.*
shear ledges. See *radial marks.*
shear lip. A narrow, slanting ridge along the edge of a fracture surface. The term sometimes also denotes a narrow, often crescent-shaped, fibrous region at the edge of a fracture that is otherwise of the cleavage type, even though this fibrous region is in the same plane as the rest of the fracture surface.
shear modulus. See *modulus of elasticity.*
shear plane. A confined zone along which shear takes place in metal cutting. It extends from the cutting edge to the work surface.
shear strain. Same as shearing strain; see *strain.*
shear strength. The stress required to produce fracture in the plane of cross section, the conditions of loading being such that the directions of force and of resistance are parallel and opposite although their paths are offset a specified minimum amount. The maximum load divided by the original cross-sectional area of a section separated by shear.
shear stress. See *stress.*
sheet. A flat-rolled metal product of some maximum thickness and minimum width arbitrarily dependent on the type of metal. It is thinner than plate, and has a width-to-thickness ratio greater than about 50.
sheet separation. In spot, seam or projection welding, the gap that exists between faying surfaces surrounding the weld, after the joint has been welded.
shelf roughness. Roughness on upward-facing surfaces where undissolved solids have settled on parts during a plating operation.
shell. (1) A hollow structure or vessel. (2) An article formed by deep drawing. (3) The metal sleeve remaining when a billet is extruded with a dummy block of somewhat smaller diameter. (4) In shell molding, a hard layer of sand and thermosetting plastic or resin formed over a pattern and used as the mold wall. (5) A tubular casting used in making seamless drawn tube. (6) A pierced forging.
shell core. A shell-molded sand core.
shell hardening. A surface-hardening process in which a suitable steel workpiece, when heated through and quench hardened, develops a martensitic layer or shell that closely follows the contour of the piece and surrounds a core of essentially pearlitic transformation product. This result is accomplished by a proper balance among section size, steel hardenability, and severity of quench.
shell molding. Forming a mold from thermosetting resin-bonded sand mixtures brought in contact with preheated (150 to 260 °C, or 300 to 500 °F) metal patterns, resulting in a firm shell with a cavity corresponding to the outline of the pattern. Also called *Croning process.*
shielded metal-arc welding. Arc welding in which metals are fused together by heating them with an arc between a *covered electrode* and the work. Decomposition of the covering on the consumable electrode provides shielding gas, and the electrode itself provides the filler metal. Pressure is not applied to the joint.
shielding. (1) A material barrier that prevents radiation or a flowing fluid from impinging on an object or a portion of an object. (2) Placing an object in an electrolytic bath so as to alter the current distribution on the cathode. A nonconductor is called a shield; a conductor is called a *robber,* a thief, or a guard.
shift. A casting imperfection caused by mismatch of cope and drag or of cores and mold.
shim. A thin piece of material used between two surfaces to obtain a proper fit, adjustment or alignment.
shimmy die. See *flat edge trimmer.*
shoe. (1) A metal block used in a variety of bending operations to form or support the part being processed. (2) An anvil cap or *sow block.*
Shore hardness test. Same as *Scleroscope test.*
short-circuiting transfer. In consumable-electrode arc welding, a type of metal transfer similar to globular transfer, but in which the drops are so large that the arc is short circuited momentarily during the transfer of each drop to the weld puddle. Compare with *spray transfer, globular transfer.*
shorts. The product that is retained on a specified screen in the screening of a crushed or ground material. See also *plus sieve.*
short transverse. See *transverse.*
shot. Small spherical particles of metal.
shot blasting. *Blasting* with metal *shot;* usually used to remove deposits or mill scale more rapidly or more effectively than can be done by sand blasting.
shot peening. Cold working the surface of a metal by metal-shot impingement.
shotting. The production of shot by pouring molten metal in finely divided streams. Solidified spherical particles are formed during the descent and are cooled in a tank of water.
shrinkage. See *casting shrinkage.*
shrinkage cavity. A void left in cast metal as a result of solidification shrinkage. See *casting shrinkage.*
shrinkage cracks. Hot tears associated with shrinkage cavities.
shrinkage rule. A measuring ruler with graduations expanded to compensate for the change in the dimensions of the solidified casting as it cools in the mold.
shrink fit. An interference fit produced by heating the outside member to a practical temperature for easy assembly. Usually the inside member is kept at or near room temperature. Sometimes the inside member is cooled to increase ease of assembly.
shrink forming. Forming of metal wherein the inner

fibers of a cross section undergo a reduction in a localized area by the application of heat, cold upset or mechanically induced pressures.

shut height. For a press, the distance from the top of the bed to the bottom of the slide with the stroke down and adjustment up. In general it is the maximum die height that can be accommodated for normal operation, taking the bolster plate into consideration.

side cutting-edge angle. Defined by sketch accompanying *single-point tool*.

side milling. Milling with cutters having peripheral and side teeth. They are usually profile sharpened but may be form relieved.

side rake. In a single-point turning tool, the angle between the tool face and a reference plane, corresponding to radial rake in milling. It lies in a plane perpendicular to the tool base and parallel to the rotational axis of the work. See sketch accompanying *single-point tool*.

single relief angle. Defined by sketch accompanying *single-point tool*.

sieve analysis. *Particle-size distribution;* usually expressed as the weight percentage retained on each of a series of standard sieves of decreasing size and the percentage passed by the sieve of finest size. Synonymous with sieve classification.

sieve classification. Same as *sieve analysis*.

sieve fraction. The portion of a powder sample that passes through a standard sieve of specified number and is retained by some finer sieve of specified number.

sigma phase. A hard, brittle, nonmagnetic intermediate phase with a tetragonal crystal structure, containing 30 atoms per unit cell, space group $P4_2/mnm$, occurring in many binary and ternary alloys of the transition elements. The composition of this phase in the various systems is not the same and the phase usually exhibits a wide range in homogeneity. Alloying with a third transition element usually enlarges the field of homogeneity and extends it deep into the ternary section.

silica flour. A sand additive, containing about 99.5% silica, commonly produced by pulverizing quartz sand in large ball mills to a mesh size of 80 to 325.

siliconizing. Diffusing silicon into solid metal, usually steel, at an elevated temperature.

silky fracture. A metal fracture in which the broken metal surface has a fine texture, usually dull in appearance. Characteristic of tough and strong metals. Contrast with *crystalline fracture, granular fracture*.

silver soldering. Nonpreferred term used to denote brazing with a silver-base filler metal. See preferred terms *furnace brazing, induction brazing, torch brazing*.

single-bevel groove weld. A groove weld in which the joint edge of one member is beveled from one side.

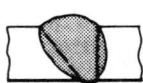 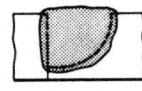

Single-bevel groove weld   Single-J groove weld

single-impulse welding. Spot, projection or upset welding by a single impulse of current. Where alternating current is used, an impulse may be any fraction or number of cycles.

single-J groove weld. A groove weld in which the joint edge of one member is prepared in the form of a J, from one side. See sketch.

single-point tool. See definition of nomenclature in accompanying sketch.

Nomenclature of single-point tool

End cutting-edge angle
Nose
Face | Shank
Cutting edge
Side cutting-edge angle
Side rake angle
Back rake angle
Flank
End relief | Lip | Shank
Side relief
End clearance angle
Base (heel)

single-stand mill. A rolling mill of such design that the product contacts only two rolls at a given moment. Contrast with *tandem mill*.

single-U groove weld. A groove weld in which each joint edge is prepared in the form of a J or half-U from one side. See sketch.

single-V groove weld. A groove weld in which each member is beveled from the same side. See sketch.

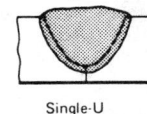 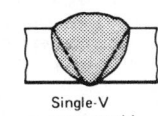

Single-U groove weld   Single-V groove weld

single welded joint. In arc and gas welding, any joint welded from one side only.

sinkhead. Same as *riser*.

sinking. See *tube sinking*.

sinter. To heat a mass of fine particles for a prolonged time below the melting point, usually to cause agglomeration.

sintering. The bonding of adjacent surfaces in a mass of particles by molecular or atomic attraction on heating at high temperatures below the melting temperature of any constituent in the material. Sintering strengthens a powder mass and normally produces densification and, in powdered metals, recrystallization. See also *liquid phase sintering*.

size effect. Effect of the dimensions of a piece of metal on its mechanical and other properties and on manufacturing variables such as forging reduction and heat treatment. In general, the mechanical properties are lower for a larger size.

size of weld. (1) The joint penetration in a groove weld. (2) The lengths of the nominal legs of a fillet weld. See illustrations.

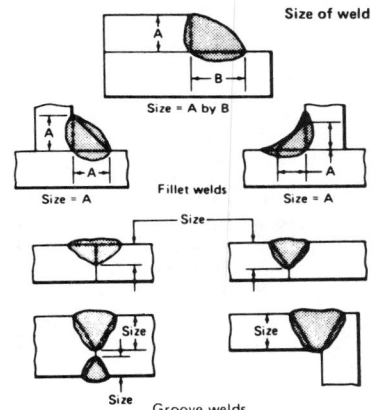

Size of weld

Size = A by B

Fillet welds   Size = A   Size = A

Groove welds

sizing. (1) Secondary forming or squeezing operations, required to square up, set down, flatten or otherwise correct surfaces to produce specified dimensions and tolerances. See *restriking*. (2) Some burnishing, broaching, drawing and shaving operations are also called sizing. (3) A finishing operation for correcting ovality in tubing. (4) Final pressing of a sintered powder metallurgy part.

skelp. The starting stock for making welded pipe or tubing; most often it is strip stock of suitable width, thickness and edge configuration.

skim gate. A gating arrangement designed to prevent the passage of slag and other undesirable materials into a casting.

skimmer. A tool for removing scum, slag and dross from the surface of molten metal.

skin. A thin outside metal layer, not formed by bonding as in cladding or electroplating, that differs in composition, structure or other characteristic from the main mass of metal.

skin lamination. In flat-rolled metals, a surface rupture resulting from the exposure of a subsurface lamination by rolling.

skin pass. See *temper rolling*.

skiving. (1) Removal of a material in thin layers or chips with a high degree of shear or slippage, or both, of the cutting tool. (2) A machining operation in which the cut is made with a form tool with its face so angled that the cutting edge progresses from one end of the work to the other as the tool feeds tangentially past the rotating workpiece.

skull. A layer of solidified metal or dross on the walls of a pouring vessel after the metal has been poured.

slab. A piece of metal, intermediate between ingot and plate, of which the width is at least twice the thickness.

slabbing mill. A primary mill that produces slabs.

slab milling. See preferred term, *peripheral milling*.

slack quenching. Incomplete hardening of steel due to quenching from the austenitizing temperature at a rate lower than the critical cooling rate for the particular steel, resulting in formation of one or more transformation products in addition to martensite.

slag. A nonmetallic product resulting from the mutual dissolution of flux and nonmetallic impurities in smelting, refining, and certain welding operations.

slag inclusion. Slag or dross entrapped in a metal.

slant fracture. A type of fracture appearance, typical of plane-stress fractures, in which the plane of metal separation is inclined at an angle (usually about 45°) to the axis of applied stress.

slide. Main reciprocating member of a mechanical press, guided in a press frame, to which the punch or upper die is fastened.

sliding fit. A loosely defined fit similar to a *slip fit*.

slime. (1) A material of extremely fine particle size encountered in ore treatment. (2) A mixture of metals and some insoluble compounds that forms on the anode in electrolysis.

slip. Plastic deformation by the irreversible shear displacement (translation) of one part of a crystal relative to another in a definite crystallographic direction and usually on a specific crystallographic plane. Sometimes called glide.

slip band. A group of parallel slip lines so closely spaced as to appear as a single line when observed under an optical microscope. See *slip line*.

slip direction. The crystallographic direction in which the translation of slip takes place.

slip fit. A loosely defined clearance fit between parts assembled by hand without force, but implying slipping contact.

slip flask. A tapered *flask* that depends on a movable strip of metal to hold the sand in position. After closing the mold, the strip is retracted and the flask can be removed and reused. Molds thus made are usually supported by a *mold jacket* during pouring.

slip-interference theory. Theory involving the resistance to deformation offered by a hard phase dispersed in a ductile matrix.

slip line. The trace of the slip plane on the viewing surface; the trace is (usually) observable only if the surface has been polished before deformation. The usual observation on metal crystals (under a light microscope) is of a cluster of slip lines known as a slip band.

slip plane. The crystallographic plane in which slip occurs in a crystal.

sliver. An imperfection consisting of a very thin elongated piece of metal attached by only one end to the parent metal into whose surface it has been worked.

slope control. Producing electronically a gradual increase or decrease in the welding current between definite limits and within a selected time interval.

slot furnace. A common batch furnace where stock is charged and removed through a slot or opening.

slotting. Cutting a narrow aperture or groove with a reciprocating tool in a vertical shaper or with a cutter, broach or grinding wheel.

slot weld. Similar to *plug weld,* the difference being that the hole is elongated and may extend to the edge of a member without closing.

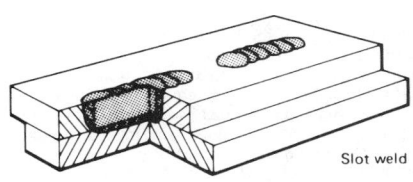

Slot weld

slug. (1) A short piece of metal to be placed in a die for forging or extrusion. (2) A small piece of material produced by piercing a hole in sheet material.

slugging. The unsound practice of adding a separate piece of material in a joint before or during welding, resulting in a welded joint in which the weld zone is not entirely built up by adding molten filler metal or by melting and recasting base metal, and which therefore does not comply with design, drawing or specification requirements.

slush casting. A hollow casting usually made of an alloy with a low but wide melting temperature range. After the desired thickness of metal has solidified in the mold, the remaining liquid is poured out.

smelting. Thermal processing wherein chemical reactions take place to produce liquid metal from a beneficiated ore.

smith forging. Manual forging with flat or simple-shaped dies that never completely confine the work.

smut. A reaction product sometimes left on the surface of a metal after pickling, electroplating or etching.

snagging. *Offhand grinding* on castings and forgings to remove surplus metal such as gate and riser pads, fins and parting lines.

snake. (1) The product formed by twisting and bending of hot rod prior to its next rolling process. (2) Any crooked surface imperfection in a plate, resembling a snake. (3) A flexible mandrel used in the inside of a shape to prevent flattening or collapse during a bending operation.

snaky edges. See *carbon edges.*

snap flask. A foundry flask hinged on one corner so that it can be opened and removed from the mold for reuse before the metal is poured.

snap temper. A precautionary interim stress-relieving treatment applied to high-hardenability steels immediately after quenching to prevent cracking because of delay in tempering them at the prescribed higher temperature.

S-N **diagram.** A plot showing the relationship of stress, *S*, and the number of cycles, *N*, before fracture in fatigue testing.

snowflakes. See *flakes.*

snug fit. A loosely defined fit implying the closest clearances that can be assembled manually for firm connection between parts.

soak cleaning. *Immersion cleaning* without electrolysis.

soaking. Prolonged holding at a selected temperature to effect homogenization of structure or composition.

soft soldering. See preferred term, *soldering.*

soft temper. Same as *dead soft temper.*

soil. Undesirable material on a surface that is not an integral part of the surface. Oil, grease, and dirt can be soils; a decarburized skin and excess *hard chromium* are not soils. Loose scale is soil; hard scale may be an integral part of the surface and, hence, not soil.

solderability. The ease with which a surface is wetted by solder.

solder embrittlement. Reduction in mechanical properties of a metal as a result of local penetration of solder along grain boundaries.

soldering. A group of processes that join metals by heating them to a suitable temperature below the solidus of the base metals and applying a filler metal having a liquidus not exceeding 450 °C (840 °F). Molten filler metal is distributed between the closely fitted surfaces of the joint by capillary action.

solder short. See *bridging* (5).

solid cutters. Cutters made of a single piece of material rather than a composite of two or more materials.

solidification. The change in state from liquid to solid on cooling through the melting temperature or melting range.

solidification shrinkage. See *casting shrinkage.*

solid shrinkage. See *casting shrinkage.*

solid solution. A single, solid, homogeneous crystalline phase containing two or more chemical species.

solid-state welding. A group of welding processes that join metals at temperatures essentially below the melting points of the base materials, without the addition of a brazing or soldering filler metal. Pressure may or may not be applied to the joint.

solidus. In a constitution or equilibrium diagram, the locus of points representing the temperatures at which various compositions finish freezing on cooling or begin to melt on heating. See also *liquidus.*

soluble oil. Specially prepared oil whose water emulsion is used as a cutting or grinding fluid.

solute. The component of either a liquid or solid solution that is present to a lesser or minor extent; the component that is dissolved in the *solvent.*

solution heat treatment. Heating an alloy to a suitable temperature, holding at that temperature long enough to cause one or more constituents to enter into solid solution, and then cooling rapidly enough to hold these constituents in solution.

solution potential. *Electrode potential* where half-cell reaction involves only the metal electrode and its ion.

solvent. The component of either a liquid or solid solution that is present to a greater or major extent; the component that dissolves the *solute.*

solvus. In a constitution or equilibrium diagram, the locus of points representing the temperatures at which the various compositions of the solid phases coexist with other solid phases—that is, the limits of solid solubility.

sonic testing. Any inspection method that uses sound waves (in the audible frequency range, about 20 to 20 000 Hz) to induce a response from a part or test specimen. Sometimes, but inadvisedly, used as a synonym for *ultrasonic testing.*

sorbite. (obsolete) A fine mixture of ferrite and cementite produced either by regulating the rate of cooling of steel or by tempering steel after hardening. The first type is very fine pearlite that is difficult to resolve under the microscope; the second type is tempered martensite.

sow block. In forging, a removable block of metal set into the hammer anvil to protect the anvil from shock and wear and occasionally to hold insert dies. Also called an anvil cap or a shoe.

space lattice. A regular, periodic array of points (lattice points) in space that represents the locations of atoms of the same kind in a perfect crystal. The concept may be extended, where appropriate, to crystalline compounds and other substances, in which case the lattice points often represent locations of groups of atoms of identical composition, arrangement and orientation.

spacer strip. A metal strip or bar inserted in the root of a joint prepared for groove welding, to serve as a backing and to maintain root opening throughout the course of the welding operation.

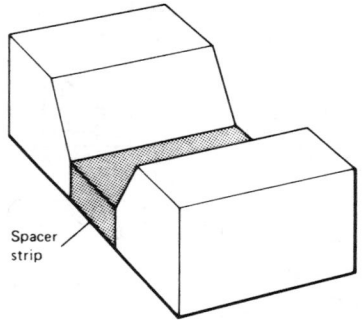

Spacer strip

spade drill. See preferred term, *flat drill.*

spalling. The cracking and flaking of particles out of a surface.

spangle. The characteristic crystalline form in which a hot dipped zinc coating solidifies on steel strip.

spatter. The metal particles expelled during arc or gas welding. They do not form part of the weld.

spatter loss. The metal lost due to *spatter.*

specific energy. In cutting or grinding, the energy expended or work done in removing a unit volume of material.

specific power. Same as *unit power.*

specimen. A test object, often of standard dimensions and/or configuration, that is used for destructive or nondestructive testing. One or more specimens may be cut from each unit of a *sample.*

speed of travel. In welding, the speed with which a weld is made along its longitudinal axis, usually measured in inches per minute or spots per minute.

speiss. Metallic arsenides and antimonides that result from smelting metal ores such as those of cobalt.

spelter. Crude zinc obtained in smelting zinc ores.

spelter solder. A brazing filler metal of approximately equal parts of copper and zinc.

spheroidite. An aggregate of iron or alloy carbides of essentially spherical shape dispersed throughout a matrix of ferrite.

spheroidizing. Heating and cooling to produce a spheroidal or globular form of carbide in steel. Spheroidizing methods frequently used are:

1 Prolonged holding at a temperature just below Ae$_1$.
2 Heating and cooling alternately between temperatures that are just above and just below Ae$_1$.
3 Heating to a temperature above Ae$_1$ or Ae$_3$ and then

cooling very slowly in the furnace or holding at a temperature just below Ae$_1$.

4 Cooling at a suitable rate from the minimum temperature at which all carbide is dissolved, to prevent the re-formation of a carbide network, and then reheating in accordance with method 1 or 2 above. (Applicable to hypereutectoid steel containing a carbide network.)

spherulitic graphite cast iron. Same as *ductile cast iron.*

spider die. Same as *porthole die.*

spiegeleisen (spiegel). A pig iron containing 15 to 30% Mn and 4.5 to 6.5% C.

spindle. (1) Shaft of a machine tool on which a cutter or grinding wheel may be mounted. (2) Metal shaft to which a mounted wheel is cemented.

spinning. Forming a seamless hollow metal part by forcing a rotating blank to conform to a shaped mandrel that rotates concentrically with the blank. In the usual application, a flat-rolled metal blank is forced against the mandrel by a blunt, rounded tool; however, other stock (notably, welded or seamless tubing) can be formed, and sometimes the working end of the tool is a roller.

spinodal structure. A fine homogeneous mixture of two phases that form by the growth of composition waves in a solid solution during suitable heat treatment. The phases of a spinodal structure differ in composition from each other and from the parent phase but have the same crystal structure as the parent phase.

spline. Any of a series of longitudinal, straight projections on a shaft that fit into slots on a mating part to transfer rotation to or from the shaft.

split die. Same as *segment die.*

sponge. A form of metal characterized by a porous condition that is the result of the decomposition or reduction of a compound without fusion. The term is applied to forms of iron, titanium, zirconium, uranium, plutonium and the platinum-group metals.

sponge iron. Either porous or powdered iron produced directly without fusion, such as by heating high-grade ore with charcoal, or an oxide with a reducing gas.

spot drilling. Making an initial indentation in a work surface, with a drill, to serve as a centering guide in a subsequent machining operation.

spot facing. Machining a flat seat for a bolt head, nut or other similar element at the end of and at right angles to the axis of a previously made hole.

spotting. Fitting one part of a die to another by applying an oil color to the surface of the finished part and bringing this against the surface of the intended mating part, the high spots being marked by the transferred color.

spotting out. Delayed, uneven staining of metal by entrapment of chemicals during the finishing operation.

spot welding. Welding of lapped parts in which fusion is confined to a relatively small circular area. It is generally resistance welding, but may also be gas tungsten-arc, gas metal-arc, or submerged-arc welding.

spray metallizing. See *metallizing.*

spray quenching. Quenching in a spray of liquid.

spray transfer. In consumable-electrode arc welding, a type of metal transfer in which the molten filler metal is propelled across the arc as fine droplets. Compare with *globular transfer, short-circuiting transfer.*

springback. (1) The elastic recovery of metal after cold forming. (2) The degree to which metal tends to return to its original shape or contour after undergoing a forming operation. (3) In flash, upset or pressure welding, the deflection in the welding machine caused by the upset pressure.

spring temper. A *temper* of nonferrous alloys and some ferrous alloys characterized by values of tensile strength and hardness about two-thirds of the way from those of *full hard* to those of *extra spring* temper.

sprue. (1) The mold channel that connects the *pouring basin* with the runner or, in the absence of a pouring basin, directly into which molten metal is poured. Sometimes referred to as downsprue or downgate. (2) Sometimes used to mean all gates, risers, runners and similar scrap that are removed from castings after shakeout.

square drilling. Making square holes by means of a specially constructed drill made to rotate and also to oscillate so as to follow accurately the periphery of a square guide bushing or template.

square groove weld. A groove weld in which the abutting surfaces are square.

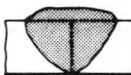 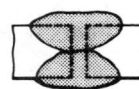

Square groove weld

squaring shear. A machine tool, used for cutting sheet metal or plate, consisting essentially of a fixed cutting knife (usually mounted on the rear of the bed) and another cutting knife mounted on the front of a reciprocally moving crosshead, which is guided vertically in side housings. Corner angles are usually 90°.

squeeze time. In resistance welding, the time between the initial applications of pressure and current.

stabilizing treatment. (1) Before finishing to final dimensions, repeatedly heating a ferrous or nonferrous part to or slightly above its normal operating temperature and then cooling to room temperature to ensure dimensional stability in service. (2) Transforming retained austenite in quenched hardenable steels, usually by *cold treatment*. (3) Heating a solution-treated stabilized grade of austenitic stainless steel to 870 to 900 °C (1600 to 1650 °F) to precipitate all carbon as TiC, NbC or TaC so that *sensitization* is avoided on subsequent exposure to elevated temperature.

stack cutting. *Oxyfuel gas cutting* of stacked metal plates arranged so that all are severed by a single cut.

stack molding. A molding method that makes use of both faces of a mold section, one face acting as the drag and the other as the cope. Sections, when assembled to other similar sections, form several tiers of mold cavities, all castings being poured together through a common sprue.

stack welding. Resistance *spot welding* of stacked plates, all being joined simultaneously.

staggered-intermittent fillet welding. Making a line of intermittent fillet welds on each side of a joint so that the increments on one side are not opposite those on the other. Contrast with *chain-intermittent fillet welding*.

staggered-tooth cutters. Milling cutters with alternate flutes of oppositely directed helixes.

stainless steel. Any of several steels containing 12 to 30% chromium as the principal alloying element; they usually exhibit *passivity* in aqueous environments.

staking. Fastening two parts together permanently by recessing one part within the other and then causing plastic flow at the joint.

stalagmometer. An apparatus for determining surface tension. The mass of a drop of liquid is measured by weighing a known number of drops or by counting the number of drops obtained from a given volume of the liquid.

stamping. A general term covering almost all press operations. It includes blanking, shearing, hot or cold forming, drawing, bending, coining.

stand. A piece of rolling mill equipment containing one set of working rolls. In the usual sense, any pass of a continuous, looping or cross-country hot rolling mill.

standard electrode potential. The reversible *electrode potential* where all reactants and products are at unit activity.

standard gold. A legally adopted alloy for coinage of gold. In the United States this alloy contains 10% Cu.

stardusting. An extremely fine form of roughness on the surface of a metal deposit.

starting sheet. A thin sheet of metal used as the cathode in electrolytic refining.

state of strain. A complete description of the deformation within a homogeneously deformed volume or at a point. The description requires, in general, the knowledge of six independent components of *strain*.

state of stress. A complete description of the stresses within a homogeneously stressed volume or at a point. The description requires, in general, the knowledge of six independent components of *stress*.

static fatigue. A term sometimes used to identify a form of hydrogen embrittlement in which a metal appears to fracture spontaneously under a steady stress less than the yield stress. There almost always is a delay between the application of stress (or exposure of the stressed metal to hydrogen) and the onset of cracking. More properly referred to as hydrogen-induced delayed cracking.

steadite. A hard structural constituent of cast iron that consists of a binary eutectic of ferrite (containing some phosphorus in solution) and iron phosphide (Fe_3P). The composition of the eutectic is 10.2% P, 89.8%

Fe, and the melting temperature is 1050 °C (1920 °F).

Stead's brittleness. A condition of brittleness that causes transcrystalline fracture in the coarse grain structure that results from prolonged annealing of thin sheets of low-carbon steel previously rolled at a temperature below about 705 °C (1300 °F). The fracture usually occurs at about 45° to the direction of rolling.

steadyrest. In cutting or grinding, a stationary support for a long workpiece.

Steckel mill. A cold reducing mill having two working rolls and two backup rolls, none of which is driven. The strip is drawn through the mill by a power reel in one direction as far as the strip will allow and then reversed by a second power reel, and so on until the desired thickness is attained.

steel. An iron-base alloy, malleable in some temperature ranges as initially cast, containing manganese, usually carbon, and often other alloying elements. In carbon steel and low-alloy steel, the maximum carbon is about 2.0%; in high-alloy steel, about 2.5%. The dividing line between low-alloy and high-alloy steels is generally regarded as being at about 5% metallic alloying elements.

Steel is to be differentiated from two general classes of "irons": the cast irons, on the high-carbon side, and the relatively pure irons such as ingot iron, carbonyl iron and electrolytic iron, on the low-carbon side. In some steels containing extremely low carbon, the manganese content is the principal differentiating factor, steel usually containing at least 0.25% and ingot iron considerably less.

step aging. Aging at two or more temperatures, by steps, without cooling to room temperature after each step. See *aging*, and compare with *interrupted aging* and *progressive aging*.

stepped extrusion. See *extrusion*.

stereoradiography. A technique for producing paired radiographs that may be viewed with a stereoscope to exhibit a shadowgraph in three dimensions with various sections in perspective and spatial relation.

sterling silver. A silver alloy containing at least 92.5% Ag, the remainder being unspecified but usually copper.

stick electrode. A shop term for *covered electrode*.

sticker breaks. Arc-shaped *coil breaks*, usually located near the center of sheet or strip.

stiffness. The ability of a metal or shape to resist elastic deflection. For identical shapes, the stiffness is proportional to the modulus of elasticity. For a given material, the stiffness increases with increasing moment of inertia, which is computed from cross-sectional dimensions.

stock. A general term for solid starting material that is formed, forged or machined to make parts.

stoking. (obsolete) Presintering, or sintering, in such a way that powder metallurgy compacts are advanced through the furnace at a fixed rate by manual or mechanical means; also called continuous sintering.

stop-off. See *resist*.

stopper rod. A device in a bottom-pour ladle for controlling the flow of metal through the nozzle into a mold. The stopper rod consists of a steel rod, protective refractory sleeves and a graphite stopper head.

stopping off. (1) Applying a *resist*. (2) Depositing a metal (copper, for example) in localized areas to prevent carburization, decarburization or nitriding in those areas. (3) Filling in a portion of a mold cavity to keep out molten metal.

stored-energy welding. Welding with electrical energy accumulated electrostatically, electromagnetically or electrochemically at a relatively low rate and made available at the higher rate required in welding.

straddle milling. Face milling a workpiece on both sides at once using two cutters spaced as required.

straight polarity. Direct-current arc welding circuit arrangement in which the electrode is connected to the negative terminal. Contrast with *reverse polarity*.

strain. A measure of the relative change in the size or shape of a body. Linear strain is the change per unit length of a linear dimension. True strain (or natural strain) is the natural logarithm of the ratio of the length at the moment of observation to the original gage length. Conventional strain is the linear strain over the original gage length. Shearing strain (or shear strain) is the change in angle (expressed in radians) between two lines originally at right angles. When the term "strain" is used alone it usually refers to the linear strain in the direction of applied stress. See also *state of strain*.

strain-age embrittlement. A loss in ductility accom-

panied by an increase in hardness and strength that occurs when low-carbon steel (especially rimmed or capped steel) is aged following plastic deformation. The degree of embrittlement is a function of aging time and temperature, occurring in a matter of minutes at about 200 °C (400 °F) but requiring a few hours to a year at room temperature.

strain aging. Aging induced by cold working. See *aging*.

strain energy. (1) The work done in deforming a body. (2) The work done in deforming a body within the elastic limit of the material. It is more properly termed elastic strain energy and can be recovered as work rather than heat.

strain hardening. An increase in hardness and strength caused by plastic deformation at temperatures below the recrystallization range.

strain-hardening exponent. A measure of rate of strain hardening. The constant n in the expression

$$\sigma = \sigma_0 \delta^n$$

where σ is true stress, σ_0 is true stress at unit strain, and δ is true strain.

strain rate. The time rate of straining for the usual tensile test. Strain as measured directly on the specimen gage length is used for determining strain rate. Because strain is dimensionless, the units of strain rate are reciprocal time.

strain-rate sensitivity. Qualitatively, the increase in stress (s) needed to cause a certain increase in plastic strain rate ($\dot{\epsilon}$) at a given level of plastic strain (ϵ) and a given temperature (T).

Strain-rate sensitivity =

$$m = \frac{\Delta \log s}{\Delta \log \dot{\epsilon}}\bigg|_{\epsilon, T}$$

strain rods. (1) Rods sometimes used on gapframe presses to lessen the frame deflection. (2) Rods used to measure elastic strains, and thus stresses, in frames of presses.

strain state. See *state of strain*.

strand casting. A generic term describing *continuous casting* of one or more elongated shapes such as billets, blooms or slabs; if two or more strands are cast simultaneously, they are often of identical cross section.

stray current. Current flowing in electrodeposition by way of an unplanned and undesired bipolar electrode that may be the tank itself or a poorly connected electrode.

stress. Force per unit area, often thought of as force acting through a small area within a plane. It can be divided into components, normal and parallel to the plane, called normal stress and shear stress, respectively. True stress denotes the stress where force and area are measured at the same time. Conventional stress, as applied to tension and compression tests, is force divided by original area. Nominal stress is the stress computed by simple elasticity formulas, ignoring stress raisers and disregarding plastic flow; in a notch bend test, for example, it is bending moment divided by minimum section modulus. See also *state of stress*.

stress amplitude. One-half the algebraic difference between the maximum and minimum stresses in one cycle of a repetitively varying stress.

stress-concentration factor (K_t). A multiplying factor for applied stress that allows for the presence of a structural discontinuity such as a notch or hole; K_t equals the ratio of the greatest stress in the region of the discontinuity to the nominal stress for the entire section.

stress-corrosion cracking. Failure by cracking under combined action of corrosion and stress, either external (applied) stress or internal (residual) stress. Cracking may be either intergranular or transgranular, depending on metal and corrosive medium. See also *season cracking*.

stress-intensity factor. A scaling factor, usually denoted by the symbol K, used in linear elastic fracture mechanics to describe the intensification of applied stress at the tip of a crack of known size and shape. At the onset of rapid crack propagation in any structure containing a crack, the factor is called the critical stress-intensity factor, or the *fracture toughness*. Various subscripts are used to denote different loading conditions or fracture toughnesses. The most common subscripts, and their meanings, are:

K_c.   Plane-stress fracture toughness. The value

of stress intensity at which crack propagation becomes rapid in sections thinner than those in which plane-strain conditions prevail.

K_I. Stress-intensity factor for a loading condition that displaces the crack faces in a direction normal to the crack plane (also known as the opening mode of deformation).

K_{Ic}. Plane-strain fracture toughness. The minimum value of K_c for any given material and condition, which is attained when rapid crack propagation in the opening mode is governed by plane-strain conditions.

K_{Id}. Dynamic fracture toughness. The fracture toughness determined under dynamic loading conditions; it is used as an approximation of K_{Ic} for very tough materials.

K_{Iscc}. Threshold stress intensity for stress-corrosion cracking. A value of stress intensity characteristic of a specific combination of material, material condition and corrosive environment above which stress-corrosion crack propagation occurs and below which the material is immune to stress-corrosion cracking.

stress raisers. Changes in contour or discontinuities in structure that cause local increases in stress.

stress range. The algebraic difference between the maximum and minimum stress in one cycle of a repetitively varying stress.

stress ratio. In fatigue, the ratio of the minimum stress to the maximum stress in one cycle, considering tensile stresses as positive, compressive stresses as negative.

stress relieving. Heating to a suitable temperature, holding long enough to reduce residual stresses and then cooling slowly enough to minimize the development of new residual stresses.

stress-rupture test. A method of evaluating elevated-temperature durability in which a tension-test specimen is stressed under constant load until it breaks. Data recorded commonly include: initial stress, time to rupture, initial extension, creep extension, reduction of area at fracture. Also known as creep-rupture test.

stress state. See *state of stress*.

stretcher leveling. Leveling a piece of metal (that is, removing warp and distortion) by gripping it at both ends and subjecting it to a stress higher than its yield strength. Sometimes called patent leveling.

stretcher straightening. Straightening rod, tubing or shapes by gripping the stock at both ends and applying tension. The products are elongated a definite amount to remove warpage.

stretcher strains. Elongated markings that appear on the surfaces of some materials when deformed just past the yield point. These markings lie approximately parallel to the direction of maximum shear stress and are the result of localized yielding. See also *Lüders lines*.

stretch former. (1) A machine used to perform *stretch forming* operations. (2) A device adaptable to a conventional press for accomplishing stretch forming.

stretch forming. Shaping of a sheet or part, usually of uniform cross section, by first applying suitable tension or stretch and then wrapping the sheet or part around a die of the desired shape.

stretch wipe forming. Same as *wiper forming*.

striation. A fatigue fracture feature, often observed in electron micrographs, that indicates the position of the crack front after each succeeding cycle of stress. The distance between striations indicates the advance of the crack front across that crystal during one stress cycle, and a line normal to the striations indicates the direction of local crack propagation.

strike. (1) A thin electrodeposited film of metal to be overlaid with other plated coatings. (2) A plating solution of high covering power and low efficiency designed to electroplate a thin, adherent film of metal.

striking. Electrodepositing, under special conditions, a very thin film of metal that will facilitate further plating with another metal or with the same metal under different conditions.

striking surface. Those areas on the faces of a set of dies that are designed to meet when the upper and lower dies are brought together. Striking surface helps protect impressions from impact shock and aids in maintaining longer die life. Also called beating area.

stringer. In wrought materials, an elongated configuration of microconstituents or foreign material aligned in the direction of working. Commonly, the term is associated with elongated oxide or sulfide inclusions in steel.

stringer bead. A continuous weld bead made without appreciable transverse oscillation. Contrast with *weave bead*.

strip. A flat-rolled metal product of some maximum thickness and width arbitrarily dependent on the type of metal. It is narrower than *sheet*.

stripper punch. A punch that serves as the top or bottom of the die cavity and later moves farther into the die to eject the part or compact. See also *ejector rod, knockout* (1).

stripping. Removing a coating from a metal surface.

structural shape. A piece of metal of any of several designs accepted as standard by the structural branch of the iron and steel industries.

stud arc welding. An arc welding process that produces coalescence of metals by heating them with an arc between a metal stud, or similar part, and another part. When the surfaces to be joined are properly heated, they are brought together under pressure. Partial shielding may be obtained by the use of a ceramic ferrule surrounding the stud. Shielding gas or flux may or may not be used.

subboundary structure. A network of low-angle boundaries (usually less than one degree) within the main crystals of a metallographic structure.

subcritical annealing. A process anneal performed on ferrous alloys at a temperature below Ac_1.

subgrain. A portion of a crystal or grain, with an orientation slightly different from the orientation of neighboring portions of the same crystal. Generally, neighboring subgrains are separated by low-angle boundaries such as *tilt boundaries* and *twist boundaries*.

submerged-arc welding. Arc welding in which the arc, between a bare metal electrode and the work, is shielded by a blanket of granular, fusible material overlying the joint. Pressure is not applied to the joint, and filler metal is obtained from the consumable electrode (and sometimes from a supplementary welding rod).

subsieve analysis. Size distribution of particles all of which will pass through a 44-μm (No. 325) standard sieve, as determined by specified methods.

subsieve fraction. That portion of a powdered sample which will pass through a 44-μm (No. 325) standard sieve.

substitutional solid solution. A solid solution in which the solute atoms are located at some of the lattice points of the solvent, the distribution being random. Contrast with *interstitial solid solution*.

substrate. Layer of metal underlying a coating, regardless of whether that layer is the basis metal.

substructure. Same as *subboundary structure*.

subsurface corrosion. Formation of isolated particles of corrosion products beneath a metal surface. This results from the preferential reactions of certain alloy constituents to inward diffusion of oxygen, nitrogen or sulfur.

sulfur dome. An inverted container, holding a high concentration of sulfur dioxide gas, used in die casting to cover a pot of molten magnesium to prevent burning.

sulfur print. A macrographic method of examining for distribution of sulfide inclusions by placing a sheet of wet acidified photographic paper in contact with the polished steel surface to be examined.

superalloy. See *heat-resisting alloy*.

superconductivity. The abrupt and large increase in electrical conductivity exhibited by some metals as the temperature approaches absolute zero.

supercooling. Cooling below the temperature at which an equilibrium phase transformation can take place, without actually obtaining the transformation.

superficial Rockwell hardness test. Form of Rockwell hardness test using relatively light loads that produce minimum penetration by the indenter. Used for determining surface hardness or hardness of thin sections or small parts, or where a large hardness impression might be harmful.

superfines. The portion of a metal powder that is composed of particles smaller than a specified size, usually 10 μm.

superfinishing. A form of *honing* in which the abrasive stones are spring supported.

superheating. (1) Heating above the temperature at which an equilibrium phase transformation should occur without actually obtaining the transformation. (2) Heating molten metal above the normal casting temperature so as to obtain more complete refining or greater fluidity.

superlattice. A lattice arrangement in which solute and solvent atoms of a solid solution occupy different preferred sites in the array. Contrast with *disordering*.

superplasticity. The ability of certain metals to undergo unusually large amounts of plastic deformation before local necking occurs.

supersonic. Pertains to phenomena in which the speed is higher than that of sound. Not synonymous with ultrasonic; see *ultrasonic frequency*.

support pins. Rods or pins of precise length used to support the overhang of irregularly shaped punches.

support plate. A plate that supports the draw ring or draw plate. It also serves as a spacer.

surface checking. Same as *checks*.

surface finish. (1) Condition of a surface as a result of a final treatment. (2) Measured surface profile characteristics, the preferred term being *roughness*.

surface grinding. Producing a plane surface by grinding.

surface hardening. A generic term covering several processes applicable to a suitable ferrous alloy that produces, by quench hardening only, a surface layer that is harder or more wear resistant than the core. There is no significant alteration of the chemical composition of the surface layer. The processes commonly used are *induction hardening, flame hardening* and *shell hardening*. Use of the applicable specific process name is preferred.

surface roughness. See *roughness*.

surface tension. Interfacial tension between two phases of which one is a gas.

surfacing. The deposition of filler metal on a metal surface by welding, spraying or braze welding, to obtain certain desired properties or dimensions. See also *hard facing*.

surfacing weld. A type of weld composed of one or more stringer or weave beads deposited on an unbroken surface to obtain desired properties or dimensions.

swaging. Tapering bar, rod, wire or tubing by forging, hammering or squeezing; reducing a section by progressively tapering lengthwise until the entire section attains the smaller dimension of the taper.

swarf. Intimate mixture of grinding chips and fine particles of abrasive and bond resulting from a grinding operation.

sweat. Exudation of a low-melting phase during solidification. Also known as sweatback. For tin bronzes, it is called tin sweat.

sweating. A soldering technique in which two or more parts are precoated (tinned), then reheated and joined without adding more solder. Also called sweat soldering.

sweating out. Bringing small globules of one of the low-melting constituents of an alloy to the surface during heat treatment, such as lead out of bronze.

sweep. A form or template used for shaping sand molds or cores by hand.

sweeps. Floor and table sweepings containing precious metal particles.

sweet roast. Same as *dead roast*.

swing forging machine. Equipment for continuously hot reducing ingots, blooms or billets to square flats, rounds or rectangles by the crank-driven oscillating action of paired dies.

swing-frame grinder. A grinding machine suspended by a chain at the center point so that it may be turned and swung in any direction for grinding of billets, large castings or other heavy work. Principal use is removing surface imperfections and roughness.

synchronous timing. In spot, seam or projection welding, a method of regulating the welding transformer primary current so that all the following conditions will prevail: (a) The first half-cycle is initiated at the proper time in relation to the voltage to ensure a balanced current wave; (b) each succeeding half-cycle is essentially identical to the first; and (c) the last half-cycle is of opposite polarity to the first.

syntectic. An isothermal reversible reaction in which a solid phase, on absorption of heat, is converted to two conjugate liquid phases.

synthetic cold rolled sheet. A hot rolled pickled sheet given a sufficient final temper pass to impart a surface approximating that of cold rolled steel.

T

tacking. Making *tack welds*.

tack welds. Small, scattered welds made to hold parts of a weldment in proper alignment while the final welds are being made.

taconite. A siliceous iron formation from which certain iron ores of the Lake Superior region are derived; consists chiefly of fine-grain silica mixed with magnetite and hematite.
tailings. The discarded portion of a crushed ore, separated during concentration.
tandem die. Same as *follow die*.
tandem mill. A rolling mill consisting of two or more stands arranged so that the metal being processed travels in a straight line from stand to stand. In continuous rolling, the various stands are synchronized so that the strip may be rolled in all stands simultaneously. Contrast with *single-stand mill*.
tandem welding. Arc welding in which two or more electrodes are in a plane parallel to the line of travel.
tangent bending. Forming one or more identical bends having parallel axes by wiping sheet metal around one or more radius dies in a single operation. The sheet, which may have side flanges, is clamped against the radius die, then made to conform to the radius die by pressure from a rocker-plate die that moves along the periphery of the radius die.
tangent modulus. See *modulus of elasticity*.
tank voltage. The total voltage between the anode and cathode of a plating bath or electrolytic cell during electrolysis. It is equal to the sum of: (*a*) the equilibrium reaction potential, (*b*) the *IR* drop and (*c*) the electrode potentials.
tap. A cylindrical or conical thread-cutting tool with one or more cutting elements having threads of a desired form on the periphery. By a combination of rotary and axial motions, the leading end cuts an internal thread, the tool deriving its principal support from the thread being produced.
tap density. The apparent density of a metal powder, obtained when the volume receptacle is tapped or vibrated during loading under specified conditions.
tapping. (1) Opening the outlet of a melting furnace to remove molten metal. (2) Removing molten metal from a furnace. (3) Cutting internal threads with a tap.
tarnish. Surface discoloration of a metal caused by formation of a thin film of corrosion product.
Taylor process. A process for making extremely fine wire by inserting a piece of larger-diameter wire into a glass tube and stretching the two together at high temperature.
technical cohesive strength. Fracture stress in a notch tensile test. Often used instead of merely "cohesive strength" to avoid confusion among the several definitions of cohesive strength.
tee joint. A joint in which the members are oriented in the form of a T.

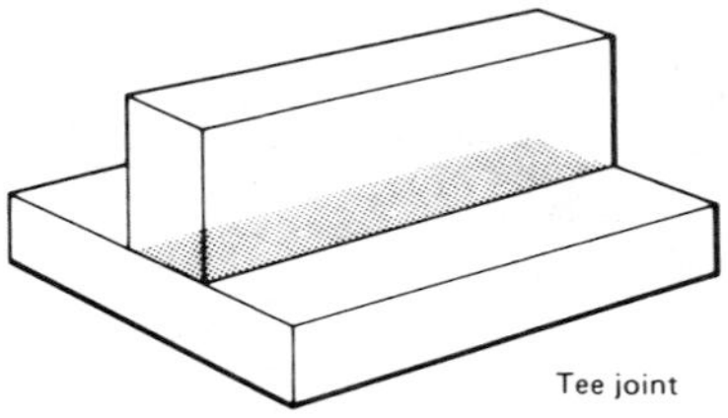

teeming. Pouring molten metal from a ladle into ingot molds. The term applies particularly to the specific operation of pouring either iron or steel into ingot molds.
temper. (1) In heat treatment, to reheat hardened steel or hardened cast iron to some temperature below the eutectoid temperature for the purpose of decreasing hardness and increasing toughness. The process also is sometimes applied to normalized steel. (2) In tool steels, temper is sometimes but inadvisedly, used to denote carbon content. (3) In nonferrous alloys and in some ferrous alloys (steels that cannot be hardened by heat treatment), the hardness and strength produced by mechanical or thermal treatment, or both, and characterized by a certain structure, mechanical properties or reduction in area during cold working. (4) To moisten sand for casting molds with water.
temper brittleness. Brittleness that results when certain steels are held within, or are cooled slowly through, a certain range of temperature below the transformation range. This brittleness is manifested as an upward shift in ductile-to-brittle transition temperature, but only rarely produces a low value of reduction in area in a smooth-bar tension test of the embrittled material.
temper carbon. Same as *annealing carbon*.
temper color. A thin, tightly adhering oxide skin (only a few molecules thick) that forms when steel is tempered at a low temperature, or for a short time, in air or a mildly oxidizing atmosphere. The color, which ranges from straw to blue depending on the thickness of the oxide skin, varies with both tempering time and temperature.
temper rolling. Light cold rolling of steel sheet. This operation is performed to improve flatness, to minimize the tendency toward formation of stretcher strains and flutes, and to obtain the desired texture and mechanical properties.
temper time. In resistance welding, that part of the postweld interval during which the current is suitable for tempering or heat treatment.
tensile strength. In tensile testing, the ratio of maximum load to original cross-sectional area. Also called *ultimate strength*. Compare with *yield strength*.
terminal phase. A solid solution having a restricted range of compositions, one end of the range being a pure component of an alloy system.
ternary alloy. An alloy that contains three principal elements.
terne. An alloy of lead containing 3 to 15% tin, used as a hot dip coating for steel sheet or plate. Terne coatings, which are smooth and dull in appearance, give the steel better corrosion resistance and enhance its ability to be formed, soldered or painted.
tertiary creep. See *creep*.
texture. In a polycrystalline aggregate, the state of distribution of crystal orientations. In the usual sense, it is synonymous with *preferred orientation*.
thermal analysis. A method for determining transformations in a metal by noting the temperatures at which thermal arrests occur. These arrests are manifested by changes in slope of the plotted or mechanically traced heating and cooling curves. When such data are secured under nearly equilibrium conditions of heating and cooling, the method is commonly used for determining certain critical temperatures required for the construction of equilibrium diagrams.
thermal electromotive force. The electromotive force generated in a circuit containing two dissimilar metals when one junction is at a temperature different from that of the other. See also *thermocouple*.
thermal fatigue. Fracture resulting from the presence of temperature gradients that vary with time in such a manner as to produce cyclic stresses in a structure.
thermal shock. The development of a steep temperature gradient and accompanying high stresses within a structure.
thermal spraying. A group of welding or allied processes in which finely divided metallic or nonmetallic materials are deposited in a molten or semimolten condition to form a coating. The coating material may be in the form of powder, ceramic rod, wire or molten materials. See also *flame spraying, plasma spraying*.
thermal stresses. Stresses in metal resulting from nonuniform temperature distribution.
thermit reactions. Strongly exothermic self-propagating reactions such as that where finely divided aluminum reacts with a metal oxide. A mixture of aluminum and iron oxide produces sufficient heat to weld steel, the filler metal being produced in the reaction.
thermit welding. Welding with heat produced by the reaction of aluminum with a metal oxide. Filler metal, if used, is obtained from reduction of an appropriate oxide.
thermocouple. A device for measuring temperatures, consisting of lengths of two dissimilar metals or alloys that are electrically joined at one end and connected to a voltage-measuring instrument at the other end. When one junction is hotter than the other, a *thermal electromotive force* is produced that is roughly proportional to the difference in temperature between the hot and cold junctions.
thermomechanical working. A general term covering a variety of processes combining controlled thermal and deformation treatments to obtain synergistic effects such as improvement in strength without loss of toughness. Same as thermal-mechanical treatment.
thief. In electroplating, same as *robber*.
Thomas converter. A Bessemer converter having a basic bottom and lining, usually dolomite, and employing a basic slag.
three-point bending. Bending of a piece of metal, or a structural member, in which the object is placed across two supports and force is applied between and in opposition to them. See *V-bend die*.
three-quarters hard. A *temper* of nonferrous alloys and some ferrous alloys characterized by values of tensile strength and hardness about midway between those of *half hard* and *full hard* tempers.
throat depth. On a resistance-welding machine, the distance from the centerline of the electrodes or platens to the nearest point of interference for flat work.
throat of a fillet weld. (theoretical) The distance from the beginning of the root of the joint perpendicular to the hypotenuse of the largest right triangle that can be inscribed within the fillet-weld cross section. (actual) The shortest distance from the root of a fillet to its face. (effective) The minimum distance from the root of the weld to its face, minus any reinforcement. See sketches accompanying *concave fillet weld, convex fillet weld*.
through weld. A nonpreferred term sometimes used to indicate a weld of substantial length made by melting through one member of a lap or tee joint and into the other member.
throwing power. The ability of a plating solution to produce a uniform metal distribution on an irregularly shaped cathode. Compare with *covering power*.
tiger stripes. Continuous bright lines on sheet or strip in the rolling direction.
tight fit. A loosely defined fit of slight negative allowance the assembly of which requires a light press or driving force.
TIG welding. Tungsten inert-gas welding; see preferred term, *gas tungsten-arc welding*.
tilt boundary. A subgrain boundary consisting of an array of edge *dislocations*.
tilt mold. A casting mold, usually a book mold, that rotates from a horizontal to a vertical position during pouring, which reduces agitation and thus the formation and entrapment of oxides.
tilt mold ingot. An ingot made in a *tilt mold*.
time quenching. Interrupted quenching in which the time in the quenching medium is controlled.
tinning. Coating metal with a very thin layer of molten solder or brazing filler metal.
tin pest. A polymorphic modification of tin that causes it to crumble into a powder known as gray tin. It is generally accepted that the maximum rate of transformation occurs at about −40 °C (−40 °F), but transformation can occur at as high as about 13 °C (55 °F).
tin sweat. See *sweat*.
tin tossing. Oxidizing impurities in molten tin by pouring it from one vessel to another in air, forming a dross that is mechanically separable.
TIR Abbreviation for *total indicator reading*.
TIV. Abbreviation for *total indicator variation*.
toe crack. A base-metal crack at the *toe of weld*.
toe of weld. The junction between the face of a weld and the base metal. See sketch accompanying *fillet weld*.
toggle press. A mechanical press in which the slide is actuated by one or more toggle links or mechanisms.
tolerance. The specified permissible deviation from a specified nominal dimension, or the permissible variation in size or other quality characteristic of a part.
tolerance limits. The boundaries that define the range of permissible variation in size or other *quality characteristic* of a part.
tong hold. The portion of a forging billet, usually on one end, that is gripped by the operator's tongs. It is removed from the part at the end of the forging operation. Common to drop-hammer and press-type forging.
tool steel. Any of a class of carbon and alloy steels commonly used to make tools. Tool steels are characterized by high hardness and resistance to abrasion, often accompanied by high toughness and resistance to softening at elevated temperature. These attributes are generally attained with high carbon and alloy contents.
tooth. (1) A projection on a multipoint tool (such as on a saw, milling cutter or file) designed to produce cutting. (2) A projection on the periphery of a wheel or segment thereof (as on a gear, spline or sprocket, for example), designed to engage another mechanism and thereby transmit force or motion, or both. A similar projection on a flat member such as a rack.
tooth point. On a face mill, the chamfered cutting edge of the blade, to which a flat is sometimes added to produce a shaving effect and to improve finish. See sketch accompanying *face mill*.
top-and-bottom process. A process for separating copper and nickel, in which their molten sulfides are separated into two liquid layers by the addition of

sodium sulfide. The lower layer holds most of the nickel.

torch. A gas burner used to solder, braze, weld or cut metals. For brazing or welding, it has two gas feed lines: one for fuel, such as acetylene or hydrogen, the other for oxygen. For cutting, there may be an additional feed line for oxygen. See *oxygen cutting*.

torch brazing. Brazing in which the heat is supplied by a fuel gas flame emanating from a *torch*.

torsion. A twisting action resulting in shear stresses and strains.

torsional moment. In a body being twisted, the algebraic sum of the couples or the moments of the external forces about the axis of twist, or both.

total carbon. The sum of the free and combined carbon (including carbon in solution) in a ferrous alloy.

total cyanide. Cyanide content of an electroplating bath (including both simple and complex ions).

total indicator reading. See preferred term, *total indicator variation*.

total indicator variation. The difference between the maximum and minimum indicator readings during a checking cycle.

toughness. Ability of a metal to absorb energy and deform plastically before fracturing. It is usually measured by the energy absorbed in a notch impact test, but the area under the stress-strain curve in tensile testing is also a measure of toughness.

tough pitch copper. Copper containing from 0.02 to 0.05% oxygen, obtained by refining copper in a reverberatory furnace.

tracer milling. Duplication of a three-dimensional form by means of a cutter controlled by a tracer that is directed by a master form.

traffic mark. See *abrasion*.

tramp alloys. Residual alloying elements that are introduced into steel when unidentified alloy steel is present in the scrap charge to a steelmaking furnace.

transcrystalline. Same as *intracrystalline*.

transference. The movement of ions through the electrolyte associated with the passage of the electric current. Also called transport or *migration*.

transference number. The proportion of total electroplating current carried by ions of a given kind. Also called transport number.

transformation-induced plasticity. A phenomenon, occurring chiefly in certain highly alloyed steels that have been heat treated to produce metastable austenite or metastable austenite plus martensite, whereby, on subsequent deformation, part of the austenite undergoes strain-induced transformation to martensite. Steels capable of transforming in this manner, commonly referred to as TRIP steels, are highly plastic after heat treatment, but exhibit a very high rate of strain hardening and thus have high tensile and yield strengths after plastic deformation at temperatures between about 20 and 500 °C (70 and 930 °F). Cooling to −195 °C (−320 °F) may or may not be required to complete the transformation to martensite. Tempering usually is done following transformation.

transformation ranges. Those ranges of temperature within which a phase forms during heating and transforms during cooling. The two ranges are distinct, sometimes overlapping but never coinciding. The limiting temperatures of the ranges depend on the composition of the alloy and on the rate of change of temperature, particularly during cooling. See *transformation temperature*.

transformation temperature. The temperature at which a change in phase occurs. This term is sometimes used to denote the limiting temperature of a transformation range. The following symbols are used for irons and steels:

Ac_{cm}. In hypereutectoid steel, the temperature at which solution of cementite in austenite is completed during heating.

Ac_1. The temperature at which austenite begins to form during heating.

Ac_3. The temperature at which transformation of ferrite to austenite is completed during heating.

Ac_4. The temperature at which austenite transforms to delta ferrite during heating.

Ae_{cm}, Ae_1, Ae_3, Ae_4. The temperatures of phase changes at equilibrium.

Ar_{cm}. In hypereutectoid steel, the temperature at which precipitation of cementite starts during cooling.

Ar_1. The temperature at which transformation of austenite to ferrite or to ferrite plus cementite is completed during cooling.

Ar_3. The temperature at which austenite begins to transform to ferrite during cooling.

Ar_4. The temperature at which delta ferrite transforms to austenite during cooling.

Ar'. The temperature at which transformation of austenite to pearlite starts during cooling.

M_f. The temperature at which transformation of austenite to martensite is completed during cooling.

M_s (or Ar''). The temperature at which transformation of austenite to martensite starts during cooling.

NOTE: All these changes, except formation of martensite, occur at lower temperatures during cooling than during heating, and depend on the rate of change of temperature.

transgranular. Same as *intracrystalline*.

transitional fit. A fit that may have either clearance or interference resulting from specified tolerances on hole and shaft.

transition lattice. An unstable crystallographic configuration that forms as an intermediate step in a solid-state reaction such as precipitation from solid solution or eutectoid decomposition.

transition metal. A metal in which the available electron energy levels are occupied in such a way that the *d*-band contains less than its maximum number of ten electrons per atom; for example, iron, cobalt, nickel and tungsten. The distinctive properties of the transition metals result from the incompletely filled *d*-levels.

transition point. At a stated pressure, the temperature (or at a stated temperature, the pressure) at which two solid phases exist in equilibrium—that is, an allotropic transformation temperature (or pressure).

transition temperature. (1) An arbitrarily defined temperature that lies within the temperature range in which metal fracture characteristics (as usually determined by tests of notched specimens) change rapidly, such as from primarily fibrous (shear) to primarily crystalline (cleavage) fracture. Commonly used definitions are "transition temperature for 50% cleavage fracture," "10 ft·lb transition temperature," and "transition temperature for half maximum energy." (2) Sometimes used to denote an arbitrarily defined temperature within a range in which the ductility changes rapidly with temperature.

transport. See *transference*.

transport number. Same as *transference number*.

transverse. Literally, "across," usually signifying a direction or plane perpendicular to the direction of working. In rolled plate or sheet, the direction across the width is often called long transverse, and the direction through the thickness, short transverse.

transverse rolling machine. Equipment for producing complex preforms or finished forgings from round billets inserted transversely between two or three rolls that rotate in the same direction and drive the billet. The rolls, carrying replaceable die segments with appropriate impressions, make several revolutions for each rotation of the workpiece.

trees. Visible projections of electrodeposited metal formed at sites of high current density.

trepanning. A type of *boring* where an annular cut is made into a solid material with the coincidental formation of a plug or solid cylinder.

triaxiality. In a *triaxial stress* state, the ratio of the smallest to the largest principal stress, all stresses being tensile.

triaxial stress. A state of stress in which none of the three *principal stresses* is zero.

tribology. The science and art concerned with the design, friction, lubrication and wear of contacting surfaces that move relative to each other (as in bearings, cams or gears, for example).

trimmer blades. The portion of *trimmers* through which a forging is pushed to shear off the flash.

trimmer punch. The upper portion of *trimmers*, which comes in contact with a forging and pushes it through the *trimmer blades*. The lower end of the trimmer punch is generally shaped to fit the surface of the forging against which it pushes.

trimmers. The combination of *trimmer punch, trimmer blades* and perhaps *trimming shoe* used to remove the flash from the forging.

trimming. (1) In drawing, shearing the irregular edge of the drawn part. (2) In forging or die casting, removing any parting-line flash and gates from the part by shearing. (3) In casting, the removal of gates, risers and fins.

trimming shoe. The holder used to support *trimmers*. Sometimes called trimming chair.

triple-action press. A mechanical or hydraulic press having three slides with three motions properly synchronized for triple-action drawing, redrawing and forming. Usually, two slides—the blankholder slide and the plunger—are located above and a lower slide is located within the bed of the press.

triple point. A point on a phase diagram where three phases of a substance coexist in equilibrium.

tripoli. Friable and dustlike silica used as an abrasive.

TRIP steel. A commercial steel product exhibiting *transformation-induced plasticity*.

trommel. A revolving cylindrical screen used in grading coarsely crushed ore.

troostite. (obsolete) A previously unresolvable, rapidly etching, fine aggregate of carbide and ferrite produced either by tempering martensite at low temperature or by quenching a steel at a rate lower than the critical cooling rate. Preferred terminology for the first product is tempered martensite; for the latter, fine pearlite.

true current density. See preferred term, *local current density*.

true rake. See preferred term, *effective rake*.

true strain. See *strain*.

true stress. See *stress*.

truing. Removal of an outside layer of abrasive grains on a grinding wheel to restore its face to running true or to alter the cutting face for grinding of special contours.

tube reducing. Reducing both the diameter and wall thickness of tubing with a mandrel and a pair of rolls with tapered grooves. The Rockrite process uses a fixed tapered mandrel, and the rolls reciprocate along the tubing with corresponding reversal in rotation. Roll reliefs at the initial and final diameters permit, respectively, advance and rotation of the tubing. The Pilger process uses a uniform rod (broach), which reciprocates with the tubing. The fixed rolls rotate continuously. During the gap in each revolution, the tubing is advanced and rotated and then, on roll contact, reduced and partially returned.

tube sinking. Drawing tubing through a die or passing it through rolls without the use of an interior tool (such as a mandrel or plug) to control inside diameter; sinking generally produces a tube of increased wall thickness and length.

tube stock. A semifinished tube suitable for subsequent reduction and finishing.

tumbling. Rotating workpieces, usually castings or forgings, in a barrel partly filled with metal slugs or abrasives, to remove sand, scale or fins. It may be done dry, or with an aqueous solution added to the contents of the barrel. Sometimes called rumbling or rattling.

tungsten inert-gas welding. See preferred term, *gas tungsten-arc welding*.

Turk's-head rolls. Four undriven working rolls, arranged in a square or rectangular pattern, through which strip, wire or tubing is drawn to form square or rectangular sections.

turning. Removing material by forcing a cutting tool (often a *single-point tool*) against the surface of a rotating workpiece. The tool may or may not be moved toward or along the axis of rotation while it cuts away material.

tuyere. An opening in the shell and refractory lining of a furnace, through which air is forced.

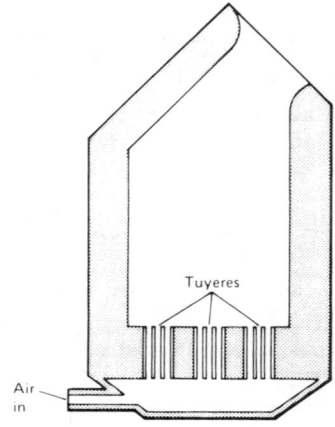

Tuyeres

Air in

twin. Two portions of a crystal having a definite crystallographic relationship; one may be regarded as the parent, the other as the twin. The orientation of the twin is either a mirror image of the orientation of the parent about a "twinning plane" or an orientation that can be derived by rotating the twin portion about a "twinning axis." See also *annealing twin, mechanical twin*.

twist boundary. A subgrain boundary consisting of an array of screw *dislocations*.

two-high mill. A type of rolling mill in which only two rolls, the working rolls, are contained in a single housing. Compare with *four-high mill, cluster mill*.

type metal. Any of a series of alloys containing 54 to 95% Pb, 2 to 28% Sb and 2 to 20% Sn, used to make printing type.

U

U-bend die. A die, commonly used in press-brake forming, machined horizontally with a square or rectangular cross-sectional opening that provides two edges over which metal is drawn into a channel shape.

Ugine-Sejournet process. A direct extrusion process for metals that uses molten glass to insulate the hot billet and to act as a lubricant.

ultimate strength. The maximum conventional stress (tensile, compressive or shear) that a material can withstand.

ultrasonic beam. A beam of acoustical radiation with a frequency higher than the frequency range for audible sound—that is, above about 20 kHz.

ultrasonic cleaning. Immersion cleaning aided by ultrasonic waves that cause microagitation.

ultrasonic frequency. A frequency, associated with elastic waves, that is greater than the highest audible frequency, generally regarded as being higher than 20 kHz.

ultrasonic machining. A form of abrasive machining in which a tool vibrating at ultrasonic frequency causes a grit-loaded slurry to impinge on the surface of a workpiece, and thereby remove material.

ultrasonic testing. A nondestructive test applied to sound-conductive materials having elastic properties for the purpose of locating inhomogeneities or structural discontinuities within a material by means of an *ultrasonic beam*.

ultrasonic welding. A solid-state process in which materials are welded by locally applying high-frequency vibratory energy to a joint held together under pressure.

underbead crack. A subsurface crack in the base metal near a weld.

undercooling. Same as *supercooling*.

undercut. (1) In weldments, a groove melted into the base metal adjacent to the toe of a weld and left unfilled. (2) For castings or forgings, same as *back draft*.

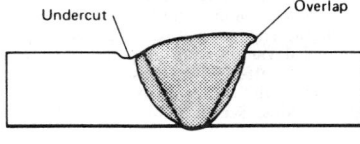

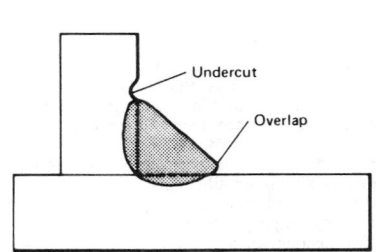

underdraft. A condition wherein a metal curves downward on leaving a set of rolls because of higher speed in the upper roll.

underfill. A portion of a forging that has insufficient metal to give it the true shape of the impression.

understressing. Applying a cyclic stress lower than the *endurance limit*. This may improve fatigue life if the member is later cyclically stressed at levels above the endurance limit.

uniaxial stress. A state of stress in which two of the three principal stresses are zero.

uniform strain. The strain occurring prior to the beginning of localization of strain (necking); the strain to maximum load in the tension test.

unit cell. In crystallography, the fundamental building block of a space lattice. Space lattices are constructed by stacking identical unit cells—that is, parallelepipeds of identical size, shape and orientation, each having a lattice point at every corner—face to face in perfect three-dimensional alignment.

unit die. A *die block* that contains several cavity inserts for making different kinds of castings.

unit power. The net amount of power required during machining to remove a unit volume of metal in unit time.

universal forging mill. A combination of four hydraulic presses arranged in one plane equipped with billet manipulators and automatic controls, used for radial or draw forging.

universal mill. A rolling mill in which rolls with a vertical axis roll the edges of the metal stock between some of the passes through the horizontal rolls.

upset. (1) The localized increase in cross-sectional area of a workpiece or weldment resulting from the application of pressure during mechanical fabrication or welding. (2) That portion of a welding cycle during which the cross-sectional area is increased by the application of pressure.

upset forging. A forging obtained by *upset* of a suitable length of bar, billet or bloom.

upsetter. A horizontal mechanical press used to make parts from bar stock or tubing by *upset forging*, piercing, bending or otherwise forming in dies. Also known as a header.

upsetting. Working metal so that the cross-sectional area of a portion or all of the stock is increased. See also *heading*.

upset welding. A resistance-welding process in which the weld is produced, simultaneously over the entire area of abutting surfaces or progressively along a joint, by applying mechanical force (pressure) to the joint, then causing electrical current to flow across the joint to heat the abutting surfaces. Pressure is maintained throughout the heating period. See also *open-gap upset welding*.

upslope time. In resistance welding, time associated with current increase using *slope control*.

V

vacancy. A type of lattice imperfection in which an individual atom site is temporarily unoccupied. Diffusion (of other than interstitial solutes) is generally visualized as the shifting of vacancies.

vacuum arc remelting. A *consumable-electrode remelting* process in which heat is generated by an electric arc between the electrode and the ingot. The process is performed inside a vacuum chamber. Exposure of the droplets of molten metal to the reduced pressure reduces the amount of dissolved gas in the metal. Sometimes abbreviated VAR.

vacuum deposition. Condensation of thin metal coatings on the cool surface of work in a vacuum.

vacuum fusion. An analytic technique for determining the amount of gases in metals; ordinarily used for hydrogen and oxygen, and sometimes for nitrogen. Applicable to many metals, but not to alkali or alkaline earth metals.

vacuum induction melting. A process for remelting and refining metals in which the metal is melted inside a vacuum chamber by induction heating. The metal may be melted in a crucible, then poured into a mold. This process may also be operated in a configuration similar to that used in *consumable-electrode remelting* except that the heat is supplied by an induction heating coil rather than from the passage of electric current through the electrode. Sometimes abbreviated VIM.

vacuum melting. Melting in a vacuum to prevent contamination from air, as well as to remove gases already dissolved in the metal; the solidification may also be carried out in a vacuum or at low pressure.

vapor blasting. Same as *liquid honing*.

vapor degreasing. Degreasing of work in the vapor over a boiling liquid solvent, the vapor being considerably heavier than air. At least one constituent of the soil must be soluble in the solvent. Modifications of this cleaning process include vapor–spray–vapor, warm liquid–vapor, boiling liquid–warm liquid–vapor, and ultrasonic degreasing.

vapor plating. Deposition of a metal or compound on a heated surface by reduction or decomposition of a volatile compound at a temperature below the melting points of the deposit and the base material. The reduction is usually accomplished by a gaseous reducing agent such as hydrogen. The decomposition process may involve thermal dissociation or reaction with the base material. Occasionally used to designate deposition on cold surfaces by vacuum evaporation—see *vacuum deposition*.

V-bend die. A die commonly used in press-brake forming, usually machined with a triangular cross-sectional opening to provide two edges as fulcrums for accomplishing three-point bending.

vector field. Same as *resultant field*.

Vegard's law. The relationship that states that the lattice parameters of substitutional solid solutions vary linearly between the values for the components, with composition expressed in atomic percentage.

veining. A type of subboundary structure that can be delineated because of the presence of a greater-than-average concentration of precipitate or possibly solute atoms.

vent. A small opening in a mold for the escape of gases.

vermicular iron. Same as *compacted graphite cast iron*.

vertical-position welding. Welding where the axis of the weld is essentially vertical.

vibratory finishing. A process for deburring and surface finishing in which the product and an abrasive mixture are placed in a container and vibrated.

Vickers hardness test. An indentation hardness test employing a 136° diamond pyramid indenter (Vickers) and variable loads enabling the use of one hardness scale for all ranges of hardness from very soft lead to tungsten carbide.

virgin metal. Same as *primary metal*.

voltage efficiency. The ratio, usually expressed as a percentage, of the equilibrium-reaction potential in a given electrochemical process to the bath voltage.

W

Wallner lines. A distinct pattern of intersecting sets of parallel lines, usually producing a set of V-shaped lines, sometimes observed in viewing brittle fracture surfaces at high magnifications in an electron microscope. Wallner lines are attributed to interaction between a shock wave and a brittle crack front propagating at high velocity. Sometimes Wallner lines are misinterpreted as *fatigue striations*.

wandering sequence. Same as *random sequence*.

warm working. Plastically deforming metal at a temperature above ambient (room) temperature but below the temperature at which the material undergoes recrystallization.

wash. (1) A coating applied to the face of a mold prior to casting. (2) An imperfection at a cast surface similar to a *cut*.

wash metal. Molten metal used to wash out a furnace, ladle or other container.

water break. The appearance of discontinuous film of water on a surface signifying nonuniform wetting and usually associated with a surface contamination.

waviness. A wavelike variation from a perfect surface, generally much larger and wider than the *roughness* caused by tool or grinding marks.

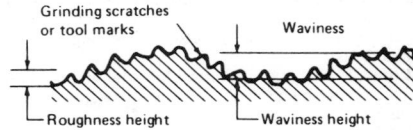

wear pad. In forming, an expendable pad of rubber or rubberlike material of nominal thickness that is placed against the diaphragm to lessen the wear on it. See *diaphragm* (2).

weave bead. A weld bead made with oscillations transverse to the axis of the weld. Contrast with *stringer bead*.

web. (1) For twist drills and reamers, the central portion of the tool body that joins the lands. (2) In forging, the thin section of metal remaining at the bottom of a cavity or depression or at the location of the top and bottom punches. The former type may be removed by piercing or machining; the latter, by the trim punch. (3) A plate or thin portion between stiffening ribs or flanges, as in an I-beam, H-beam or other similar section.

weight percent. Percentage composition by weight. Contrast with *atomic percent.*

weld. A union made by *welding.*

weldability. A specific or relative measure of the ability of a material to be welded under a given set of conditions. Implicit in this definition is the ability of the completed weldment to fulfill all functions for which the part was designed.

weld bead. A deposit of filler metal from a single welding *pass.*

weld crack. A crack in weld metal.

weld delay time. In spot, seam or projection welding, the time during which the current is delayed with respect to starting the forge delay timer in order to synchronize the forging pressure and the welding heat.

welder. A person who makes welds using manual or semiautomatic equipment. Formerly used as a synonym for *welding machine.*

weld gage. A device for checking the shapes and sizes of welds.

welding. (1) Joining two or more pieces of material by applying heat or pressure, or both, with or without filler material, to produce a localized union through fusion or recrystallization across the interface. The thickness of the filler material is much greater than the capillary dimensions encountered in *brazing.* (2) May also be extended to include brazing and soldering.

welding current. The current flowing through a welding circuit during the making of a weld. In resistance welding, the current used during preweld or postweld intervals is excluded.

welding cycle. The complete series of events involved in making a resistance weld. Also applies to semiautomatic mechanized fusion welds.

welding force. Same as *electrode force* in resistance welding.

welding generator. A generator used for supplying current for welding.

welding ground. Same as *work lead.*

welding leads. The electrical cables that serve as either *work lead* or *electrode lead* of an arc welding circuit.

welding machine. Equipment used to perform the welding operation—for example, spot welding machine, arc welding machine, seam welding machine.

welding procedure. The detailed methods and practices, including joint preparation and welding procedures, involved in the production of a *weldment.*

welding rod. Welding or brazing filler metal, usually in rod or wire form, but not a *consumable electrode.* Welding rod does not conduct the electric current to an arc, and may be either fed into the weld puddle or preplaced in the joint.

welding sequence. The order in which the various component parts of a weldment or structure are welded.

welding stress. *Residual stress* caused by localized heating and cooling during welding.

welding technique. The details of a welding operation that, within the limitations of a welding procedure, are performed by the *welder.*

welding tip. (1) A torch tip designed for welding. (2) The electrode tip that contacts the work in resistance spot welding.

weld interval. The total heat and cool times in making one multiple-impulse resistance weld.

weld-interval timer. A device used in resistance welding to control heat and cool times and weld interval when making multiple-impulse welds singly or simultaneously.

weld line. The junction of the weld metal and the base metal, or the junction of the base-metal parts when filler metal is not used.

weldment. An assembly whose component parts are joined by welding.

weld metal. That portion of a weld that has been melted during welding.

weld nugget. The weld metal in spot, seam or projection welding.

weldor. (obsolete) Formerly used to designate a person who makes welds. See preferred term, *welder.*

weld time. In single-impulse and flash welding, the time that the welding current is applied to the work.

weld timer. A device used in resistance welding to control the weld time only.

Wenstrom mill. A rolling mill similar to a universal mill but where the edges and sides of a rolled section are acted on simultaneously.

wet blasting. A process for cleaning or finishing by means of a slurry of abrasive in water directed at high velocity against the workpieces.

wetting. A condition in which the interfacial tension between a liquid and a solid is such that the contact angle is 0° to 90°.

wetting agent. A surface-active agent that produces *wetting* by decreasing the *cohesion* within the liquid.

whiskers. Metallic filamentary growths, often microscopic, sometimes formed during electrodeposition and sometimes spontaneously during storage or service, after finishing.

white cast iron. *Cast iron* that shows a white fracture because the carbon is in combined form.

whiteheart malleable. See *malleable cast iron.*

white metal. (1) A general term covering a group of white-colored metals of relatively low melting points (lead, antimony, bismuth, tin, cadmium and zinc) and the alloys based on these metals. (2) A copper matte of about 77% Cu obtained from smelting of sulfide copper ores.

white rust. Zinc oxide; the powdery product of corrosion of zinc or zinc-coated surfaces.

Widmanstätten structure. A structure characterized by a geometrical pattern resulting from the formation of a new phase along certain crystallographic planes of the parent solid solution. The orientation of the lattice in the new phase is related crystallographically to the orientation of the lattice in the parent phase. This structure was originally observed in meteorites, but is readily produced in many other alloys by appropriate heat treatment.

wildness. A condition that exists when molten metal, during cooling, evolves so much gas that it becomes violently agitated, forcibly ejecting metal from the mold or other container.

Williams riser. An *atmospheric riser.*

winning. Recovering a metal from an ore or chemical compound using any suitable hydrometallurgical, pyrometallurgical or electrometallurgical method.

wiped coat. A hot dipped galvanized coating from which virtually all free zinc is removed by wiping prior to solidification, leaving only a thin zinc-iron alloy layer.

wiped joint. A joint wherein filler metal is applied in liquid form and distributed by mechanical action.

wiper forming. A method of curving bars, tubes or rolled or extruded sections, in which the stock is bent so that it conforms to a fixed form block. Stock is clamped to the form block, then bent by applying force through a wiper block, shoe or roll that is moved along the periphery of the form block. Sometimes called compression forming. Contrast with *draw forming.*

wiping effect. Activation of a metal surface by mechanical rubbing or wiping to enhance the formation of conversion coatings, such as phosphate coatings.

wire. (1) A thin, flexible, continuous length of metal, usually of circular cross section, and usually produced by drawing through a die. See also *flat wire.* (2) A length of single metallic electrical conductor; it may be of solid, stranded or tinsel construction, and may be either bare or insulated.

wire bar. A cast shape, particularly of tough pitch copper, that has a cross section approximately square with tapered ends, designed for hot rolling to rod for subsequent drawing into wire.

wiredrawing. Reducing the cross section of wire by pulling it through a die. See *Taylor process.*

wire rod. Hot rolled coiled stock that is to be cold drawn into wire.

wiring. Formation of a curl along the edge of a shell, tube or sheet and insertion of a rod or wire within the curl for stiffening the edge. See *curling.*

wood flour. A pulverized wood product used in the foundry to furnish a reducing atmosphere in the mold, help overcome sand expansion, increase flowability, improve casting finish and provide easier shakeout.

woody structure. A macrostructure, found particularly in wrought iron and in extruded rods of aluminum alloys, that shows elongated surfaces of separation when fractured.

work angle. In arc welding, the angle between the electrode and one member of the joint, taken in a plane normal to the weld axis.

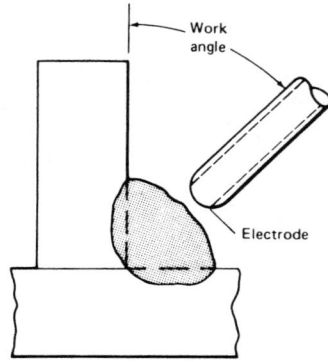

work hardening. Same as *strain hardening.*

work lead. The electrical conductor connecting the source of arc welding current to the work. Also called work connection, welding ground or ground lead.

worm. An exudation (sweat) of molten metal forced through the top crust of solidifying metal by gas evolution. See also *zinc worms.*

wrap forming. See *stretch forming.*

wringing fit. A fit of nominally zero allowance.

wrinkling. A wavy condition obtained in drawing, in the area of the metal that passes over the draw radius. Wrinkling may also occur in other forming operations when unbalanced compressive forces are set up.

wrought iron. A commercial iron consisting of slag (iron silicate) fibers entrained in a ferrite matrix.

X

x-ray. Electromagnetic radiation, of wavelength less than about 50 nm, emitted as the result of deceleration of fast-moving electrons (bremsstrahlung, continuous spectrum) or decay of atomic electrons from excited orbital states (*characteristic radiation*); specifically, the radiation produced when an electron beam of sufficient energy impinges on a target of suitable material.

Y

Y-block. A single *keel block.*

yield point. The first stress in a material, usually less than the maximum attainable stress, at which an increase in strain occurs without an increase in stress. Only certain metals exhibit a yield point. If there is a decrease in stress after yielding, a distinction may be made between upper and lower yield points.

yield strength. The stress at which a material exhibits a specified deviation from proportionality of stress and strain. An offset of 0.2% is used for many metals. Compare with *tensile strength.*

Young's modulus. See *modulus of elasticity.*

Z

zinc worms. Surface imperfections, characteristic of high-zinc brass castings, that occur when zinc vapor condenses at the mold/metal interface, where it is oxidized and then becomes entrapped in the solidifying metal.

zircon sand. A very refractory mineral, composed chiefly of zirconium silicate; it has low thermal expansion and high thermal conductivity.

zone melting. Highly localized melting, usually by induction heating, of a small volume of an otherwise solid piece, usually a rod. By moving the induction coil along the rod, the melted zone can be transferred from one end to the other. In a binary mixture where there is a large difference in composition on the liquidus and solidus lines, high purity can be attained by concentrating one of the constituents in the liquid as it moves along the rod.

Engineering Tables

Periodic table of the elements

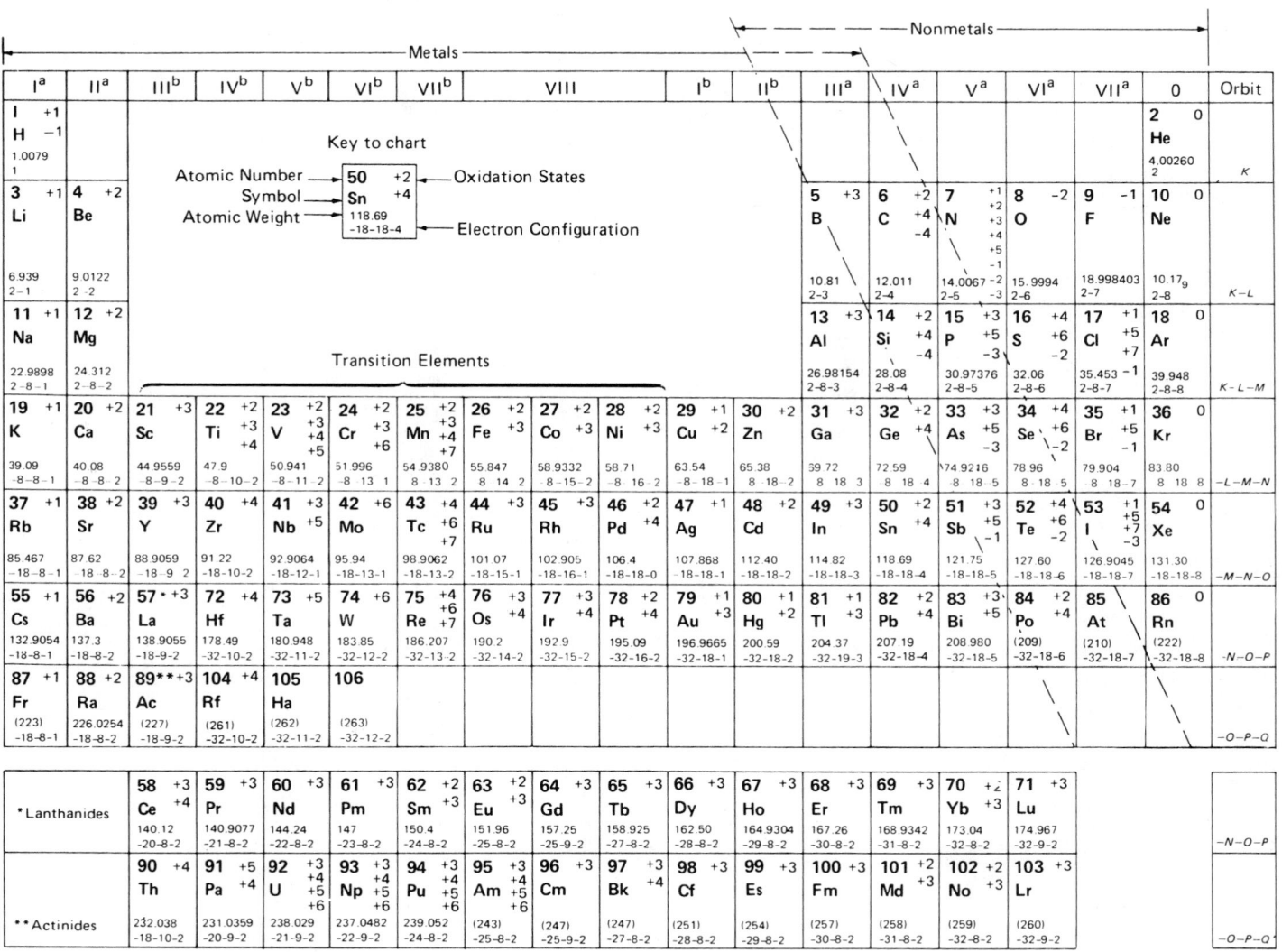

Numbers in parentheses are mass numbers of most stable isotope of that element

Physical properties of the elements

Element	Atomic No.	Atomic weight	Density(a), g/cm³ (lb/in.³)	Melting point, °C (°F)	Boiling point, °C (°F)	Specific heat(b), cal/g · °C (J/kg · K)	Heat of fusion, cal/g (Btu/lb)
Actinium (Ac)	89	227	... (...)	1050 ± 50 (1920 ± 90)	... (...)	... (...)	... (...)
Aluminum (Al)	13	26.98	2.70 (0.0974)	660 (1220)	2450 (4442)	0.215 (900)	94.5 (170)
Americium (Am)	95	243	11.87 (0.4285)	... (...)	... (...)	... (...)	... (...)
Antimony (Sb)	51	121.76	6.65 (0.240)	630.5 ± 0.1 (1166.9 ± 0.2)	1380 (2516)	0.049 (205)	38.3 (68.9)
Argon (A)	18	39.99	1.784(g) (0.06440)(g)	−189.4 ± 0.2 (−308.9 ± 0.4)	−185.8 (−302.4)	0.125 (523)	6.7 (12)
Arsenic (As)	33	74.91	5.72 (0.206)	817(j) (1503)(j)	613(k) (1135)(k)	0.082 (343)	88.5 (159.3)
Astatine (At)	85	211	... (...)	302(m) (576)(m)	... (...)	... (...)	... (...)
Barium (Ba)	56	137.36	3.6 (0.13)	714 (1317)	1640 (2980)	0.068 (285)	... (...)
Berkelium (Bk)	97	247	... (...)	... (...)	... (...)	... (...)	... (...)
Beryllium (Be)	4	9.01	1.85 (0.0668)	1277 (2332)	2770 (5020)	0.45 (190)	260 (470)
Bismuth (Bi)	83	209.00	9.80 (0.354)	271.3 (520.3)	1560 (2840)	0.0294 (123)	12.5 (22.5)
Boron (B)	5	10.82	2.45 (0.0884)	2030(q) (3690)(q)	... (...)	0.309 (1290)	... (...)
Bromine (Br)	35	79.92	3.12 (0.113)	−7.2 ± 0.2 (19.0 ± 0.4)	58 (136)	0.070 (290)	16.2 (29.2)
Cadmium (Cd)	48	112.41	8.65 (0.312)	320.9 (609.6)	765 (1409)	0.055 (230)	13.2 (23.8)
Calcium (Ca)	20	40.08	1.55 (0.0560)	838 (1540)	1440 (2625)	0.149(u) (624)(u)	52 (93.6)
Californium (Cf)	98	251	... (...)	... (...)	... (...)	... (...)	... (...)
Carbon, graphite (C) ...	6	12.01	2.25 (0.0812)	3727(k) (6740)(k)	4830 (8730)	0.165 (691)	... (...)
Cerium (Ce)	58	140.13	6.77 (0.244)	804 (1479)	3470 (6280)	0.045 (190)	8.5 (15.9)
Cesium (Cs)	55	132.91	1.87 (0.0675)	28.7 (83.6)	690 (1273)	0.04817 (201.7)	3.8 (6.8)
Chlorine (Cl)	17	35.46	3.214(g) (0.1160)(g)	−100.99 (−149.78)	−34.7 (−30.5)	0.116 (486)	21.6 (38.9)
Chromium (Cr)	24	52.01	7.19 (0.260)	1875 (3407)	2665 (4829)	0.11 (460)	96 (173)
Cobalt (Co)	27	58.94	8.85 (0.319)	1495 ± 1 (2723 ± 1.8)	2900 (5250)	0.099 (410)	58.4 (105)
Copper (Cu)	29	63.54	8.96 (0.323)	1083.0 ± 0.1 (1981.4 ± 0.18)	2595 (4703)	0.092 (380)	50.6 (91.1)
Curium (Cm)	96	247	7 (0.3)	... (...)	... (...)	... (...)	... (...)
Dysprosium (Dy)	66	162.51	8.55 (0.309)	1407 (2565)	2330 (4230)	0.041 (170)	25.2 (45.4)
Einsteinium (E)	99	254	... (...)	... (...)	... (...)	... (...)	... (...)
Erbium (Er)	68	167.27	9.15 (0.330)	1497 (2727)	2630 (4770)	0.040 (170)	24.5 (44.1)
Europium (Eu)	63	152.0	5.24 (0.189)	826 (1519)	1490 (2710)	0.039 (160)	16.5 (29.6)
Fermium (Fm)	100	253	... (...)	... (...)	... (...)	... (...)	... (...)
Fluorine (F)	9	19.00	1.696(g) (0.06123)(g)	−219.6 (−363.3)	−188.2 (−306.8)	0.18 (750)	10.1 (18.2)
Francium (Fr)	87	223	... (...)	27(m) (81)(m)	... (...)	... (...)	... (...)
Gadolinium (Gd)	64	157.26	7.86 (0.284)	1312 (2394)	2730 (4950)	0.071 (300)	23.5 (42.4)
Gallium (Ga)	31	69.72	5.91 (0.213)	29.78 (85.60)	2237 (4059)	0.079 (330)	19.16 (34.49)
Germanium (Ge)	32	72.60	5.32 (0.192)	937.4 ± 1.5 (1719.3 ± 2.7)	2830 (5125)	0.073 (310)	... (...)
Gold (Au)	79	197.0	19.3 (0.697)	1063.0 ± 0.0 (1945.4 ± 0.0)	2970 (5380)	0.0312(jj) (131)(jj)	16.1 (29.0)
Hafnium (Hf)	72	178.58	13.1 (0.473)	2222 ± 30 (4032 ± 54)	5400 (9750)	0.0351 (147)	... (...)
Helium (He)	2	4.00	0.1785(g) (0.006444)(g)	−269.7 (−453.5)	−268.9 (−452.0)	1.25 (5230)	... (...)
Holmium (Ho)	67	164.94	6.79 (0.245)	1461 (2662)	2330 (4230)	0.039 (160)	24.9 (44.7)
Hydrogen (H)	1	1.008	0.0899(g) (0.00325)(g)	−259.19 (−434.54)	−252.7 (−422.9)	3.45 (14 400)	15.0 (27.0)
Indium (In)	49	114.82	7.31 (0.264)	156.2 (313.1)	2000 (3632)	0.057 (240)	6.8 (12.2)
Iodine (I)	53	126.91	4.94 (0.178)	113.7 (236.7)	183 (361)	0.052 (220)	14.2 (25.6)
Iridium (Ir)	77	192.2	22.65 (0.8177)	2454 ± 3 (4449 ± 5)	5300 (9570)	0.0307 (129)	... (...)
Iron (Fe)	26	55.85	7.87 (0.284)	1536.5 ± 1 (2797.7 ± 1.8)	3000 ± 150 (5430 ± 270)	0.11 (460)	65.5 (117.9)
Krypton (Kr)	36	83.8	3.743(g) (0.1351)(g)	−157.3 (−251.1)	−152 (−242)	... (...)	... (...)
Lanthanum (La)	57	138.92	6.15 (0.222)	920 (1688)	3470 (6280)	0.048 (200)	17.3 (31.1)

(continued)

Physical properties of the elements (continued)

Symbol	Coefficient of linear thermal expansion(c), μin./in. °C (μin./in. °F)	Thermal conductivity(c), cal/cm²/cm/s/°C	Electrical resistivity, μΩ·cm	Modulus of elasticity in tension, 10^6 psi	Lattice constants(b), Å			Closest approach of atoms
					a	*b*	*c* (or axial angle)	
Ac.........	...	...	...	...	...	...	...	...
Al.........	23.6(d) (13.1)(d)	0.53	2.6548(b)	9	4.0491	...	...	2.862
Am.........	(...)	...	...	...	...	...	...	...
Sb.........	8.5 − 10.8(e) (4.7 − 6)(e)	0.045	39.0(f)	11.3	4.5065	...	57° 6.5′	2.904
A	(...)	0.406×10^{-4}	...	...	5.43(h)	...	...	3.84
As.........	4.7 (2.6)	...	33.3(b)	...	4.159	...	53° 49′	...
At.........	(...)	...	...	...	...	...	...	...
Ba.........	...	...	...	...	5.025	...	...	4.348
Bk.........	(...)	...	...	...	...	...	...	...
Be.........	11.6(n) (6.4)(n)	0.35	4(b)(p)	40-44	2.2858	...	3.5842	...
Bi.........	13.3 (7.4)	0.020	106.8(f)	4.6	4.7457	...	57° 14.2′	3.111
B..........	8.3(r) (4.6)(r)	...	1.8×10^{12}(f)	...	17.89	8.95	10.15	...
Br.........	(...)	...	...	...	4.49(s)	6.68(s)	8.74(s)	2.27
Cd	29.8 (16.55)	0.22	6.83(f)	8(t)	2.9787	...	5.617	...
Ca	22.3(v) (12.4)(v)	0.3	3.91(f)	3.2-3.8(w)	5.582	...	...	...
Cf.........	(...)	...	...	...	...	...	...	...
C..........	0.6-4.3(d) (0.3-2.4)(d)	0.057	1375(f)	0.7	2.4614	...	6.7041	1.42
Ce........	8 (4.44)	0.026(x)	75(y)	6(z)	5.16	...	...	...
Cs........	97(aa) (54)(aa)	...	20(b)	...	6.13(bb)	...	...	...
Cl.........	(...)	0.172×10^{-4}	...	...	8.58(cc)	...	6.13(cc)	1.81
Cr.........	6.2 (3.4)	0.16	12.9(f)	36	2.884	...	...	2.498
Co........	13.8 (7.66)	0.165	6.24(b)	30	2.5071	...	4.0686	2.4967
Cu	16.5 (9.2)	0.941 ± 0.005	1.6730(b)	16	3.6153	...	...	2.556
Cm........	(...)	...	...	...	...	...	...	...
Dy	9 (5)	0.024(x)	57(y)	10-14(z)	3.59	...	5.65	...
E..........	(...)	...	...	...	...	...	...	...
Er........	9 (5)	0.023(x)	107(y)	16(z)	3.65	...	5.58	...
Eu	26 (14.44)	...	90(y)	...	4.58	...	...	...
Fm	(...)	...	...	...	...	...	...	...
F..........	(...)	...	...	...	...	...	...	...
Fr.........	(...)	...	...	...	...	...	...	...
Gd	4(dd) (2.22)(dd)	0.021(x)	140.5(y)	8-14(z)	3.64	...	5.78	...
Ga	18(ee) (10)(ee)	0.07-0.09(ff)	17.4(gg)	...	4.524(y)	4.523(y)	7.661(y)	2.437
Ge	5.75 (3.19)	0.14	46(hh)	...	5.658	...	...	2.449
Au	14.2 (7.9)	0.71	2.35(b)	11.6	4.078	...	...	2.882
Hf.........	519(kk) (288)(kk)	0.223(mm)	35.1(y)	...	3.1883	...	5.0422	...
He	(...)	3.32×10^{-4}	...	...	3.58(nn)	...	5.84(nn)	3.58
Ho	(...)	...	87(y)	11(z)	3.58	...	5.62	...
H	(...)	4.06×10^{-4}	...	...	3.76(pp)	...	6.13(pp)	...
In	33 (18)	0.057	8.37(b)	1.57	4.594	...	4.951	3.25
I	93 (52)	10.4×10^{-4}	1.3×10^{15}(b)	...	4.787	7.266	9.793	2.71
Ir	6.8 (3.8)	0.14	5.3(b)	76	3.8389	...	...	2.714
Fe	11.76(qq) (6.53)(qq)	0.18(rr)	9.71(b)	28.5 ± 0.5	2.8664(y)	...	...	2.4824
Kr	(...)	0.21×10^{-4}	...	...	5.69(ss)	...	...	4.03
La.........	5 (2.77)	0.033(x)	57(y)	10-11(z)	3.77	...	12.16	...

(continued)

Physical properties of the elements (continued)

Element	Atomic No.	Atomic weight	Density(a), g/cm³ (lb/in.³)	Melting point, °C (°F)	Boiling point, °C (°F)	Specific heat(b), cal/g · °C (J/kg · K)	Heat of fusion, cal/g (Btu/lb)
Lawrencium (Lw)	103	257	... (...)	... (...)	... (...)	... (...)	... (...)
Lead (Pb)	82	207.21	11.34 (0.4094)	327.4258 (621.3664)	1725 (3137)	0.0309(f) (129)(f)	6.26 (11.27)
Lithium (Li)	3	6.94	0.534 (0.193)	180.54 (356.97)	1330 (2426)	0.79 (3300)	104.2 (187.6)
Lutetium (Lu)	71	174.99	9.85 (0.356)	1652(uu) (3006)(uu)	1930 (3510)	0.037 (150)	26.29 (47.32)
Magnesium (Mg)	12	24.32	1.74 (0.0628)	650 ± 2 (1202 ± 4)	1107 ± 10 (2025 ± 20)	0.245 (1030)	88 ± 2 (158 ± 4)
Manganese (Mn)	25	54.94	7.43 (0.268)	1245 (2273)	2150 (3900)	0.115(xx) (481)(xx)	63.7 (114.7)
Mendelevium (Mv)	101	256	... (...)	... (...)	... (...)	... (...)	... (...)
Mercury (Hg)	80	200.61	13.55 (0.4892)	−38.36 (−37.05)	357 (675)	0.033 (140)	2.8 (5.0)
Molybdenum (Mo)	42	95.95	10.2 (0.368)	2610 (4730)	5560 (10 040)	0.066 (280)	69.8(m) (125.6)(m)
Neodymium (Nd)	60	144.27	7.00 (0.253)	1019 (1866)	3180 (5756)	0.045 (190)	11.78 (21.20)
Neon (Ne)	10	20.18	0.8999(g) (0.03249)(g)	−248.6 ± 0.3 (−415.5 ± 0.5)	−246.0 (−410.8)	... (...)	... (...)
Neptunium (Np)	93	237	20.5 (0.740)	637 ± 2 (1179 ± 4)	... (...)	... (...)	... (...)
Nickel (Ni)	28	58.71	8.9 (0.32)	1453 (2647)	2730 (4950)	0.105 (440)	73.8 (132.8)
Niobium (Nb)	41	92.91	8.57 (0.309)	2468 ± 10 (4474 ± 18)	4927 (8901)	0.065(f) (270)(f)	69 (124.2)
Nitrogen (N)	7	14.01	1.250(g) (0.04513)(g)	−209.97 (−345.95)	−195.8 (−320.4)	0.247 (1030)	6.2 (11.2)
Nobelium (No)	102	247	... (...)	... (...)	... (...)	... (...)	... (...)
Osmium (Os)	76	190.2	22.61 (0.8162)	2700 ± 200(m) (4900 ± 350)(m)	5500 (9950)	0.031 (130)	... (...)
Oxygen (O)	8	16.00	1.429(g) (0.05159)(g)	−218.83 (−361.89)	−183.0 (−297.4)	0.218 (913)	3.3 (5.9)
Palladium (Pd)	46	106.4	12.02 (0.4339)	1552 (2826)	3980 (7200)	0.0584(f) (245)(f)	34.2 (61.6)
Phosphorus, white (P)	15	30.98	1.83 (0.0661)	44.25 (111.65)	280 (536)	0.177 (741)	5.0 (9.0)
Platinum (Pt)	78	195.09	21.45 (0.7743)	1769 (3217)	4530 (8185)	0.0314(f) (131)(f)	26.9 (48.4)
Plutonium (Pu)	94	242	19.4 (0.700)	640 (1184)	3235 (6000)	0.033(qqq) (140)(qqq)	... (...)
Polonium (Po)	84	210	9.40 (0.339)	254 ± 10 (489 ± 18)	... (...)	... (...)	... (...)
Potassium (K)	19	39.10	0.86 (0.031)	63.7 (146.7)	760 (1400)	0.177 (741)	14.6 (26.3)
Praseodymium (Pr)	59	140.92	6.77 (0.244)	919 (1686)	3020 (5468)	0.045 (188)	11.71 (21.08)
Promethium (Pm)	61	145	... (...)	1027(m) (1880)(m)	... (...)	... (...)	... (...)
Protactinium (Pa)	91	231.1	15.4 (0.556)	1230(m) (2246)(m)	... (...)	... (...)	... (...)
Radium (Ra)	88	226.05	5.0 (0.18)	700 (1292)	... (...)	... (...)	... (...)
Radon (Rn)	86	222	9.960(g) (0.3596)(g)	−71(m) (−96)(m)	−61.8 (−79.2)	... (...)	... (...)
Rhenium (Re)	75	186.22	21.0 (0.76)	3180 ± 20 (5755 ± 35)	5900 (10 650)	0.033 (140)	... (...)
Rhodium (Rh)	45	102.91	12.41 (0.4480)	1966 ± 3 (3571 ± 5)	4500 (8130)	0.059(f) (250)(f)	... (...)
Rubidium (Rb)	37	85.48	1.53 (0.0552)	38.9 (102)	688 (1270)	0.080 (330)	6.5 (11.79)
Ruthenium (Ru)	44	101.07	12.45 (0.4494)	2500 ± 100 (4530 ± 180)	4900 (8850)	0.057(f) (240)(f)	... (...)
Samarium (Sm)	62	150.35	7.49 (0.270)	1072 (1962)	1630 (2966)	0.042(xxx) (180)(xxx)	17.29 (31.12)
Scandium (Sc)	21	44.96	2.9 (0.10)	1539 (2802)	2730 (4946)	0.134 (561)	84.52 (152.14)
Selenium (Se)	34	78.96	4.8 (0.17)	217 (423)	685 ± 1 (1265 ± 2)	0.084(x) (350)(x)	16.4 (29.5)
Silicon (Si)	14	28.09	2.33 (0.0841)	1410 (2570)	2680 (4860)	0.162(f) (678)(f)	432 (778)
Silver (Ag)	47	107.88	10.49 (0.3787)	960.80 (1761.44)	2210 (4010)	0.0559(f) (234)(f)	25 (45)
Sodium (Na)	11	22.99	0.9712 (0.03506)	97.82 (208.08)	892 (1638)	0.295 (1240)	27.5 (49.5)
Strontium (Sr)	38	87.63	2.60 (0.0939)	768 (1414)	1380 (2520)	0.176 (737)	25 (45)
Sulfur, yellow (S)	16	32.07	2.07 (0.0747)	119.0 ± 0.5 (246.2 ± 0.9)	444.6 (832.3)	0.175 (733)	9.3 (16.7)
Tantalum (Ta)	73	180.95	16.6 (0.599)	2996 ± 50 (5425 ± 90)	5425 ± 100 (9800 ± 200)	0.034(y) (140)	38 (68)
Technetium (Tc)	43	98	11.5 (0.415)	2130(m) (3870)(m)	... (...)	... (...)	... (...)
Tellurium (Te)	52	127.61	6.24 (0.225)	449.5 ± 0.3 (841.1 ± 0.5)	989.8 ± 3.8 (1813.6 ± 6.8)	0.047 (200)	32 (58)
Terbium (Tb)	65	158.93	8.25 (0.298)	1356(uu) (2472)(uu)	2530 (4586)	0.044 (180)	24.54 (44.17)

(continued)

Physical properties of the elements (continued)

Symbol	Coefficient of linear thermal expansion(c), μin./in. °C (μin./in. °F)	Thermal conductivity(c), cal/cm²/cm/s/°C	Electrical resistivity, $\mu\Omega \cdot$ cm	Modulus of elasticity in tension, 10^6 psi	Lattice constants(b), Å a	b	c (or axial angle)	Closest approach of atoms
Lw	... (...)	...	...	...	...	...	...	...
Pb	29.3(tt) (16.3)(tt)	0.083(f)	20.648(b)	2	4.9489	...	...	3.499
Li	56 (31)	0.17	8.55(f)	...	3.5089	...	...	3.0387
Lu	... (...)	...	79(y)	...	3.50	...	5.50	...
Mg	27.1(vv) (15.05)(vv)	0.367	4.45(b)	6.35(ww)	3.2088(y)	...	5.2095(y)	3.196
Mn	22(yy) (12.22)(yy)	...	185(zz)	23	8.912	...	...	...
Mv	... (...)	...	...	...	...	...	...	...
Hg	... (...)	0.0196(f)	98.4(aaa)	...	3.005(bbb)	...	70° 31.7′(bbb)	3.005
Mo	4.9(d) (2.7)(d)	0.34	5.2(f)	47	3.1468(y)	...	...	2.725
Nd	6 (3.33)	0.031(ccc)	64(y)	...	3.66	...	11.80	...
Ne	... (...)	0.00011	...	...	4.53(ddd)	...	...	3.21
Np	... (...)	...	...	...	...	...	...	...
Ni	13.3(u) (7.39)(u)	0.22(y)	6.84(b)	30(eee)	3.5238	...	...	2.491
Nb	7.31 (4.06)	0.125(f)	12.5(f)	...	3.301	...	...	2.859
N	... (...)	0.000060	...	...	4.04(fff)	...	6.60(fff)	...
No	... (...)	...	...	...	...	...	...	...
Os	4.6(ggg) (2.6)(ggg)	...	9.5(b)	81	2.7341(hhh)	...	4.3197(hhh)	...
O	... (...)	0.000059	...	...	6.84(jjj)	...	...	...
Pd	11.76 (6.53)	1.68(jj)	10.8(b)	16.3	3.8902	...	...	2.750
P	125 (70)	...	1×10^{17}(kkk)	...	7.18(mmm)	...	...	...
Pt	8.9 (4.9)	0.165(nnn)	10.6(b)	21.3(ppp)	3.9310(y)	...	...	2.775
Pu	55(rrr) (30.55)(rrr)	0.020(y)	141.4(sss)	14(ttt)	6.182(y)	4.826(y)	10.956(y)	...
Po	... (...)	...	...	...	7.43	4.30	14.13	3.4
K	83 (46)	0.24	6.15(f)	...	5.334	...	...	4.624
Pr	4 (2.22)	0.028(ccc)	68(y)	7-14(z)	3.67	...	11.84	...
Pm	... (...)	...	...	...	...	...	...	...
Pa	... (...)	...	...	...	...	...	...	...
Ra	... (...)	...	...	...	...	...	...	...
Rn	... (...)	...	...	...	...	...	...	...
Re	6.7(uuu) (3.7)(uuu)	0.17	19.3(b)	66.7(b)	2.760	...	4.458	2.74
Rh	8.3 (4.6)	0.21(nnn)	4.51(b)	42.5(vvv)	3.804	...	...	2.689
Rb	90 (50)	...	12.5(b)	...	5.63(www)	...	...	4.88
Ru	9.1 (5.1)	...	7.6(f)	60(q)	2.7041	...	4.2814	...
Sm	... (...)	...	88(y)	8(z)	8.99	...	23° 13′	...
Sc	... (...)	...	61(yyy)	...	3.31	...	5.27	...
Se	37 (21)	$7\text{-}18.3 \times 10^{-4}$	12(f)	8.4	4.346	...	4.954	...
Si	2.8-7.3 (1.6-4.1)	0.20	10(f)	16.35(zzz)	5.428	...	...	2.351
Ag	19.68(u) (10.9)(u)	1.0(f)	1.59(b)	11	4.086 ± 0.0006(jj)	...	...	2.888
Na	71 (39)	0.32	4.2(f)	...	4.289	...	...	3.714
Sr	... (...)	...	23(b)	...	6.087	...	...	4.31
S	64 (36)	6.31×10^{-4}	2×10^{23}(b)	...	10.50	12.95	24.60	2.12
Ta	6.5 (3.6)	0.130	12.45(y)	27(b)	3.303	...	...	2.859
Tc	... (...)	...	...	...	...	...	...	...
Te	16.75 (9.3)	0.014	436 000(aaaa)	6	4.4570	...	5.9290	2.571
Tb	7 (3.88)	...	...	...	3.60	...	5.69	...

(continued)

Physical properties of the elements (continued)

Element	Atomic No.	Atomic weight	Density(a), g/cm³ (lb/in.³)	Melting point, °C (°F)	Boiling point, °C (°F)	Specific heat(b), cal/g · °C (J/kg · K)	Heat of fusion, cal/g (Btu/lb)
Thallium (Tl)	81	204.39	11.85 (0.4278)	303 (577)	1457 (2655)	0.031 (130)	5.04 (9.07)
Thorium (Th)	90	232.05	11.5 (0.415)	1750 (3182)	3850 ± 350 (7000 ± 600)	0.034 (140)	<19.82 (<35.68)
Thulium (Tm)	69	168.94	9.31 (0.336)	1545 (2813)	1720(www) (3130)(www)	0.038 (160)	26.04 (46.87)
Tin (Sn)	50	118.70	7.30 (0.264)	231.912 ± 0.000 (449.442 ± 0.000)	2270 (4120)	0.054 (230)	14.5 (26.1)
Titanium (Ti)	22	47.90	4.51 (0.163)	1668 ± 10 (3035 ± 18)	3260 (5900)	0.124 (519)	104(m) (188)(m)
Tungsten (W)..........	74	183.86	19.3 (0.697)	3410 (6170)	5930 (10 706)	0.033 (140)	44 (70)
Uranium (U)	92	238.07	19.07 (0.6884)	1132.3 ± 0.8 (2070.4 ± 1.5)	3818 (6904)	0.02709(jjjj) (113.4)(jjjj)	(...)
Vanadium (V)	23	50.95	6.11 (0.221)	1900 ± 25 (3450 ± 50)	3400 (6150)	0.119(t) (498)(t)	(...)
Xenon (Xe)	54	131.30	5.896(g) (0.2128)(g)	−111.9 (−169.4)	−108.0 (−162.4)	(...)	(...)
Ytterbium (Yb)	70	173.04	6.96 (0.251)	824 (1515)	1530 (2786)	0.035 (150)	12.71 (22.88)
Yttrium (Y)	39	88.92	4.47 (0.161)	1509(uu) (2748)(uu)	3030 (5490)	0.071 (300)	46 (83)
Zinc (Zn)	30	65.38	7.13 (0.257)	419.5050 (787.1090)	906 (1663)	0.0915 (383)	24.09 (43.36)
Zirconium (Zr)	40	91.22	6.49 (0.234)	1852 (3366)	3580 (6470)	0.067 ± 0.001 (280 ± 4)	60(m) (110)(m)

Symbol	Coefficient of linear thermal expansion(c), µin./in. °C (µin./in. °F)	Thermal conductivity(c), cal/cm²/cm/s/°C	Electrical resistivity, µΩ · cm	Modulus of elasticity in tension, 10⁶ psi	Lattice constants(b), Å — a	b	c (or axial angle)	Closest approach of atoms
Tl	28 (16)	0.093	18(f)	...	3.457	...	5.525	3.408
Th	12.5(bbbb) (6.9)(bbbb)	0.090(cccc)	13(f)	...	5.09	...	...	3.60
Tm	... (...)	...	79(y)	...	3.53	...	5.55	...
Sn	23(dddd) (13)(dddd)	1.50(e)	11(eeee)	6-6.5(ffff)	5.8314	...	3.1815	...
Ti	8.41 (4.67)	6.6(gggg)	42(b)	16.8	2.95030	...	4.68312	...
W	4.6 (2.55)	0.397(e)	5.65(hhhh)	50	3.158	...	...	2.734
U	6.8-14.1(kkkk) (3.8-7.8)(kkkk)	0.07(mmmm)	30(nnnn)	24	2.8545(y)	5.8681(y)	4.9566(y)	...
V	8.3(pppp) (4.6)(pppp)	0.074(cccc)	24.8-26.0(b)	18-20	3.039	...	...	2.632
Xe	... (...)	1.24 × 10⁻⁴	...	...	6.25(rrrr)	...	...	4.42
Yb	25 (13.9)	...	29(y)	...	5.49	...	...	...
Y	... (...)	0.035(ccc)	57(ssss)	17(z)	3.65	...	5.73	...
Zn	39.7(ssss) (22.0)(ssss)	0.27(y)	5.916(b)	(tttt)	2.6649	...	4.9470	2.6648
Zr	5.85(uuuu) (3.2)(uuuu)	0.211(vvvv)	40	13.7	3.2312(y)	...	5.1477(y)	3.17

(a) Density may depend considerably on previous treatment. (b) At 20 °C (68 °F). (c) Near 20 °C (68 °F). (d) From 20 to 100 °C (68 to 212 °F). (e) From 20 to 60 °C (68 to 140 °F). (f) At 0 °C (32 °F). (g) Gas, grams per litre at 20 °C (68 °F) and 760 mm (30 in.). (h) At −233 °C (−387 °F). (j) 28 atm. (k) Sublimes. (m) Estimated. (n) From 25 to 100 °C (77 to 212 °F). (p) Annealed, commercial purity. (q) Approximate. (r) From 20 to 750 °C (68 to 1380 °F). (s) At −150 °C (−238 °F). (t) Sand iron. (u) From 0 to 100 °C (32 to 212 °F). (v) For alpha at 0 to 400 °C (32 to 750 °F). (w) Annealed. (x) At 28 °C (82 °F). (y) At 25 °C (77 °F). (z) Measured from stress-strain relationship on as-cast metal. (aa) From 0 to 26 °C (32 to 70 °F). (bb) At −10 °C (14 °F). (cc) At −185 °C (−300 °F); the coefficient of expansion of gadolinium changes rapidly between −100 and +100 °C (−150 and +212 °F). (ee) From 0 to 30 °C (32 to 86 °F). (ff) At melting point. (gg) For a-axis; 8.1 for b-axis and 54.3 for c-axis. (hh) Ohm · cm of intrinsic germanium at 300 K. (jj) At 18 °C (64 °F). (kk) From 20 to 200 °C (68 to 390 °F). (mm) W/cm/°C at 50 °C (120 °F). (nn) At −271.5 °C (−456.7 °F). (pp) At −271 °C (−455.8 °F). (qq) At 25 °C (77 °F) for high-purity k iron. (rr) For ingot iron at 0 °C (32 °F). (ss) At −191 °C (−311.8 °F). (tt) From 17 to 100 °C (63 to 212 °F). (uu) Distilled metal. (vv) Along a-axis; 24.3 along c-axis. (ww) Dynamic; static, 5.77; both for 99.98% magnesium. (xx) For alpha; gamma is 0.120; both at 25.2 °C (77.3 °F). (yy) Alpha; gamma, 14; both from 0 to 100 °C (32 to 212 °F). (zz) Alpha at 20 °C (68 °F). (aaa) At 50 °C (122 °F). (bbb) At −50 °C (−58 °F). (ccc) At −2.22 °C (28 °F). (ddd) At −268 °C (−450.4 °F). (eee) At 0 °C (32 °F), unmagnetized. (fff) At −234 °C (−389 °F). (ggg) At 50 °C (122 °F), parallel to a-axis, mean value; parallel to c-axis at 50 °C (122 °F), 5.8. (hhh) At 26 °C (78.8 °F). (jjj) At −225 °C (−373 °F). (kkk) At 11 °C (51.8 °F). (mmm) At −35 °C (−31 °F). (nnn) At 17 °C (63 °F). (ppp) For small cyclic strains. (qqq) For alpha at 25 °C (77 °F). (rrr) From 21 to 104 °C (70 to 219 °F). (sss) At 107 °C (224.6 °F). (ttt) At 25 °C (77 °F), for cast metal. (uuu) From 20 to 500 °C (68 to 930 °F). (vvv) For hard wire. (www) At −173 °C (−279 °F). (xxx) Calculated. (yyy) Average value at 22 °C (72 °F), zone-refined bar. (zzz) Chill cast specimen 90.2 by 24.6 by 24.6 mm (3.55 by 0.97 by 0.97 in.). (aaaa) At 23 °C (73 °F). (bbbb) From 25 to 1000 °C (77 to 1830 °F). (cccc) At 100 °C (212 °F). (dddd) From 0 to 100 °C (32 to 212 °F), for iodide thorium. (eeee) At 0 °C (32 °F), for white tin. (ffff) Cast tin. (gggg) Btu · ft/h · ft² · °F at −400 °F. (hhhh) At 27 °C (80.6 °F). (jjjj) At 27 °C (80 °F). (kkkk) Rolled rods. (mmmm) At 70 °C (158 °F). (nnnn) Crystallographic average. (pppp) From 23 to 100 °C (73 to 212 °F). (qqqq) At −185 °C (−301 °F). (rrrr) Polycrystalline; c-axis, 135; basal plane, 72. (ssss) From 20 to 250 °C (68 to 480 °F), for polycrystalline metal. (tttt) Pure zinc has no clearly defined molulus of elasticity. (uuuu) Alpha, polycrystalline. (vvvv) W/cm/°C at 27 °C (80.6 °F).

Density of metals and alloys

Metal or alloy	Density g/cm³	lb/in.³
Aluminum and aluminum alloys		
Aluminum (99.996%)	2.6989	0.0975
Wrought alloys		
EC, 1060 alloys	2.70	0.098
1100	2.71	0.098
2011	2.82	0.102
2014	2.80	0.101
2024	2.77	0.100
2218	2.81	0.101
3003	2.73	0.099
4032	2.69	0.097
5005	2.70	0.098
5050	2.69	0.097
5052	2.68	0.097
5056	2.64	0.095
5083	2.66	0.096
5086	2.65	0.096
5154	2.66	0.096
5357	2.70	0.098
5456	2.66	0.096
6061, 6063	2.70	0.098
6101, 6151	2.70	0.098
7075	2.80	0.101
7079	2.74	0.099
7178	2.82	0.102
Casting alloys		
A13	2.66	0.096
43	2.69	0.097
108, A108	2.79	0.101
A132	2.72	0.098
D132	2.76	0.100
F132	2.74	0.099
138	2.95	0.107
142	2.81	0.101
195, B195	2.81	0.101
214	2.65	0.096
220	2.57	0.093
319	2.79	0.101
355	2.71	0.098
356	2.68	0.097
360	2.64	0.095
380	2.71	0.098
750	2.88	0.104
40E	2.81	0.101
Copper and copper alloys		
Wrought coppers		
Pure copper	8.96	0.324
Electrolytic tough pitch copper (ETP)	8.89	0.321
Deoxidized copper, high residual phosphorus (DHP)	8.94	0.323
Free-machining copper		
0.5% Te	8.94	0.323
1.0% Pb	8.94	0.323
Wrought alloys		
Gilding, 95%	8.86	0.320
Commercial bronze, 90%	8.80	0.318
Jewelry bronze, 87.5%	8.78	0.317
Red brass, 85%	8.75	0.316
Low brass, 80%	8.67	0.313
Cartridge brass, 70%	8.53	0.308
Yellow brass	8.47	0.306
Muntz metal	8.39	0.303
Leaded commercial bronze	8.83	0.319
Low-leaded brass (tube)	8.50	0.307
Medium-leaded brass	8.47	0.306
High-leaded brass (tube)	8.53	0.308
High-leaded brass	8.50	0.307
Extra-high-leaded brass	8.50	0.307
Free-cutting brass	8.50	0.307
Leaded Muntz metal	8.41	0.304
Forging brass	8.44	0.305
Architectural bronze	8.47	0.306
Inhibited admiralty	8.53	0.308
Naval brass	8.41	0.304
Leaded naval brass	8.44	0.305
Manganese bronze (A)	8.36	0.302
Phosphor bronze, 5% (A)	8.86	0.320
Phosphor bronze, 8% (C)	8.80	0.318

Metal or alloy	Density g/cm³	lb/in.³
Phosphor bronze, 10% (D)	8.78	0.317
Phosphor bronze, 1.25%	8.89	0.321
Free-cutting phosphor bronze	8.89	0.321
Cupro-nickel, 30%	8.94	0.323
Cupro-nickel, 10%	8.94	0.323
Nickel silver, 65-18	8.73	0.315
Nickel silver, 55-18	8.70	0.314
High-silicon bronze (A)	8.53	0.308
Low-silicon bronze (B)	8.75	0.316
Aluminum bronze, 5% Al	8.17	0.294
Aluminum bronze, (3)	7.78	0.281
Aluminum-silicon bronze	7.69	0.278
Aluminum bronze, (1)	7.58	0.274
Aluminum bronze, (2)	7.58	0.274
Beryllium copper	8.23	0.297
Casting alloys		
Chromium copper (1% Cr)	8.7	0.31
88Cu-10Sn-2Zn	8.7	0.31
88Cu-8Sn-4Zn	8.8	0.32
89Cu-11Sn	8.78	0.317
88Cu-6Sn-1.5Pb-4.5Zn	8.7	0.31
87Cu-8Sn-1Pb-4Zn	8.8	0.32
87Cu-10Sn-1Pb-2Zn	8.8	0.32
80Cu-10Sn-10Pb	8.95	0.323
83Cu-7Sn-7Pb-3Zn	8.93	0.322
85Cu-5Sn-9Pb-1Zn	8.87	0.320
78Cu-7Sn-15Pb	9.25	0.334
70Cu-5Sn-25Pb	9.30	0.336
85Cu-5Sn-5Pb-5Zn	8.80	0.318
83Cu-4Sn-6Pb-7Zn	8.6	0.31
81Cu-3Sn-7Pb-9Zn	8.7	0.31
76Cu-2.5Sn-6.5Pb-15Zn	8.77	0.317
72Cu-1Sn-3Pb-24Zn	8.50	0.307
67Cu-1Sn-3Pb-29Zn	8.45	0.305
61Cu-1Sn-1Pb-37Zn	8.40	0.304
Manganese bronze		
60 ksi	8.2	0.30
65 ksi	8.3	0.30
90 ksi	7.9	0.29
110 ksi	7.7	0.28
Aluminum bronze		
Alloy 9A	7.8	0.28
Alloy 9B	7.55	0.272
Alloy 9C	7.5	0.27
Alloy 9D	7.7	0.28
Nickel Silver		
12% Ni	8.95	0.323
16% Ni	8.95	0.323
20% Ni	8.85	0.319
25% Ni	8.8	0.32
Silicon bronze	8.30	0.300
Silicon brass	8.30	0.300
Iron and iron alloys		
Pure iron	7.874	0.2845
Ingot iron	7.866	0.2842
Wrought iron	7.7	0.28
Gray cast iron	7.15(a)	0.258(a)
Malleable iron	7.27(b)	0.262(b)
0.06% C steel	7.871	0.2844
0.23% C steel	7.859	0.2839
0.435% C steel	7.844	0.2834
1.22% C steel	7.830	0.2829
Low-carbon chromium-molybdenum steels		
0.5% Mo steel	7.86	0.283
1Cr-0.5Mo steel	7.86	0.283
1.25Cr-0.5Mo steel	7.86	0.283
2.25Cr-1.0Mo steel	7.86	0.283
5Cr-0.5Mo steel	7.78	0.278
7Cr-0.5Mo steel	7.78	0.278
9Cr-1Mo steel	7.67	0.276
Medium-carbon alloy steels		
1Cr-0.35Mo-0.25V steel	7.86	0.283
H11 die steel (5Cr-1.5Mo-0.4V)	7.79	0.281
Other iron-base alloys		
A-286	7.94	0.286
16-25-6 alloy	8.08	0.292
RA-330	8.03	0.290

(continued)

Metal or alloy	Density g/cm³	lb/in.³
Incoloy	8.02	0.290
Incoloy T	7.98	0.288
Incoloy 901	8.23	0.297
T1 tool steel	8.67	0.313
M2 tool steel	8.16	0.295
H41 tool steel	7.88	0.285
20W-4Cr-2V-12Co steel	8.89	0.321
Invar (36% Ni)	8.00	0.289
Hipernik (50% Ni)	8.25	0.298
4% Si	7.6	0.27
10.27% Si	6.97	0.252
Stainless steels and heat-resisting alloys		
Corrosion-resistant steel castings		
CA-15	7.612	0.2750
CA-40	7.612	0.2750
CB-30	7.53	0.272
CC-50	7.53	0.272
CE-30	7.67	0.277
CF-8	7.75	0.280
CF-20	7.75	0.280
CF-8M, CF-12M	7.75	0.280
CF-8C	7.75	0.280
CF-16F	7.75	0.280
CH-20	7.72	0.279
CK-20	7.75	0.280
CN-7M	8.00	0.289
Heat-resistant alloy castings		
HA	7.72	0.279
HC	7.53	0.272
HD	7.58	0.274
HE	7.67	0.277
HF	7.75	0.280
HH	7.72	0.279
HI	7.72	0.279
HK	7.75	0.280
HL	7.72	0.279
HN	7.83	0.283
HT	7.92	0.286
HU	8.04	0.290
HW	8.14	0.294
HX	8.14	0.294
Wrought stainless and heat-resisting steels		
Type 301	7.9	0.29
Type 302	7.9	0.29
Type 302B	8.0	0.29
Type 303	7.9	0.29
Type 304	7.9	0.29
Type 305	8.0	0.29
Type 308	8.0	0.29
Type 309	7.9	0.29
Type 310	7.9	0.29
Type 314	7.72	0.279
Type 316	8.0	0.29
Type 317	8.0	0.29
Type 321	7.9	0.29
Type 347	8.0	0.29
Type 403	7.7	0.28
Type 405	7.7	0.28
Type 410	7.7	0.28
Type 416	7.7	0.28
Type 420	7.7	0.28
Type 430	7.7	0.28
Type 430F	7.7	0.28
Type 431	7.7	0.28
Types 440A, 440B, 440C	7.7	0.28
Type 446	7.6	0.27
Type 501	7.7	0.28
Type 502	7.8	0.28
19-9DL	7.97	0.29
Precipitation-hardening stainless steels		
PH 15-7 Mo	7.804	0.2819
17-4 PH	7.8	0.28
17-7 PH	7.81	0.282
Nickel-base alloys		
D-979	8.27	0.299
Nimonic 80A	8.25	0.298
Nimonic 90	8.27	0.299
M-252	8.27	0.298
Inconel	8.51	0.307
Inconel "X" 550	8.30	0.300
Inconel 700	8.17	0.295
Inconel "713C"	7.913	0.2859

Density of metals and alloys (continued)

Metal or alloy	Density g/cm³	lb/in.³
Waspaloy	8.23	0.296
René 41	8.27	0.298
Hastelloy alloy B	9.24	0.334
Hastelloy alloy C	8.94	0.323
Hastelloy alloy X	8.23	0.297
Udimet 500	8.07	0.291
GMR-235	8.03	0.290

Cobalt-chromium-nickel-base alloys

Metal or alloy	g/cm³	lb/in.³
N-155 (HS-95)	8.23	0.296
S-590	8.36	0.301

Cobalt-base alloys

Metal or alloy	g/cm³	lb/in.³
S-816	8.68	0.314
V-36	8.60	0.311
HS-25	9.13	0.330
HS-36	9.04	0.327
HS-31	8.61	0.311
HS-21	8.30	0.300

Molybdenum-base alloy

Metal or alloy	g/cm³	lb/in.³
Mo-0.5Ti	10.2	0.368

Lead and lead alloys

Metal or alloy	g/cm³	lb/in.³
Chemical lead (99.90+% Pb)	11.34	0.4097
Corroding lead (99.73+% Pb)	11.36	0.4104
Arsenical lead	11.34	0.4097
Calcium lead	11.34	0.4097
5-95 solder	11.0	0.397
20-80 solder	10.2	0.368
50-50 solder	8.89	0.321

Antimonial lead alloys

Metal or alloy	g/cm³	lb/in.³
1% antimonial lead	11.27	0.407
Hard lead (96Pb-4Sb)	11.04	0.399
Hard lead (94Pb-6Sb)	10.88	0.393
8% antimonial lead	10.74	0.388
9% antimonial lead	10.66	0.385

Lead-base babbitt alloys

Metal or alloy	g/cm³	lb/in.³
Lead-base babbitt		
SAE 13	10.24	0.370
SAE 14	9.73	0.352
Alloy 8	10.04	0.363
Arsenical lead		
Babbitt (SAE 15)	10.1	0.365
"G" Babbitt	10.1	0.365

Magnesium and magnesium alloys

Metal or alloy	g/cm³	lb/in.³
Magnesium (99.8%)	1.738	0.06279

Casting alloys

Metal or alloy	g/cm³	lb/in.³
AM100A	1.81	0.065
AZ63A	1.84	0.066
AZ81A	1.80	0.065
AZ91A, B, C	1.81	0.065
AZ92A	1.82	0.066
HK31A	1.79	0.065
HZ32A	1.83	0.066
ZH42, ZH62A	1.86	0.067
ZK51A	1.81	0.065
ZE41A	1.82	0.066
EZ33A	1.83	0.066
EK30A	1.79	0.065
EK41A	1.81	0.065

Wrought alloys

Metal or alloy	g/cm³	lb/in.³
M1A	1.76	0.064
A3A	1.77	0.064
AZ31B	1.77	0.064
PE	1.76	0.064
AZ61A	1.80	0.065
AZ80A	1.80	0.065
ZK60A, B	1.83	0.066
ZE10A	1.76	0.064
HM21A	1.78	0.064
HM31A	1.81	0.065

Nickel and nickel alloys

Metal or alloy	g/cm³	lb/in.³
Nickel (99.95% Ni + Co)	8.902	0.322
"A" Nickel	8.885	0.321
"D" Nickel	8.78	0.317

Metal or alloy	Density g/cm³	lb/in.³
Duranickel	8.26	0.298
Cast nickel	8.34	0.301
Monel	8.84	0.319
"K" Monel	8.47	0.306
Monel (cast)	8.63	0.312
"H" Monel (cast)	8.5	0.31
"S" Monel (cast)	8.36	0.302
Inconel	8.51	0.307
Inconel (cast)	8.3	0.30
Ni-o-nel	7.86	0.294

Nickel-molybdenum-chromium-iron alloys

Metal or alloy	g/cm³	lb/in.³
Hastelloy B	9.24	0.334
Hastelloy C	8.94	0.323
Hastelloy D	7.8	0.282
Hastelloy F	8.17	0.295
Hastelloy N	8.79	0.317
Hastelloy W	9.03	0.326
Hastelloy X	8.23	0.297

Nickel-chromium-molybdenum-copper alloys

Metal or alloy	g/cm³	lb/in.³
Illium G	8.58	0.310
Illium R	8.58	0.310

Electrical resistance alloys

Metal or alloy	g/cm³	lb/in.³
80Ni-20Cr	8.4	0.30
60Ni-24Fe-16Cr	8.247	0.298
35Ni-45Fe-20Cr	7.95	0.287
Constantan	8.9	0.32

Tin and tin alloys

Metal or alloy	g/cm³	lb/in.³
Pure tin	7.3	0.264
Soft solder (30% Pb)	8.32	0.301
Soft solder (37% Pb)	8.42	0.304
Tin babbitt		
Alloy 1	7.34	0.265
Alloy 2	7.39	0.267
Alloy 3	7.46	0.269
Alloy 4	7.53	0.272
Alloy 5	7.75	0.280
White metal	7.28	0.263
Pewter	7.28	0.263

Titanium and titanium alloys

Metal or alloy	g/cm³	lb/in.³
99.9% Ti	4.507	0.1628
99.2% Ti	4.507	0.1628
99.0% Ti	4.52	0.163
Ti-6Al-4V	4.43	0.160
Ti-5Al-2.5Sn	4.46	0.161
Ti-2Fe-2Cr-2Mo	4.65	0.168
Ti-8Mn	4.71	0.171
Ti-7Al-4Mo	4.48	0.162
Ti-4Al-4Mn	4.52	0.163
Ti-4Al-3Mo-1V	4.507	0.1628
Ti-2.5Al-16V	4.65	0.168

Zinc and zinc alloys

Metal or alloy	g/cm³	lb/in.³
Pure zinc	7.133	0.2577
AG40A alloy	6.6	0.24
AC41A alloy	6.7	0.24
Commercial rolled zinc		
0.08% Pb	7.14	0.258
0.06 Pb, 0.06 Cd	7.14	0.258
0.3 Pb, 0.3 Cd	7.14	0.258
Copper-hardened, rolled zinc (1% Cu)	7.18	0.259
Rolled zinc alloy (1 Cu, 0.010 Mg)	7.18	0.259
Zn-Cu-Ti alloy (0.8 Cu, 0.15 Ti)	7.18	0.259

Precious metals

Metal or alloy	g/cm³	lb/in.³
Silver	10.49	0.379
Gold	19.32	0.698
70Au-30Pt	19.92	...
Platinum	21.45	0.775
Pt-3.5Rh	20.9	...
Pt-5Rh	20.65	...
Pt-10Rh	19.97	...
Pt-20Rh	18.74	...
Pt-30Rh	17.62	...
Pt-40Rh	16.63	...

Metal or alloy	Density g/cm³	lb/in.³
Pt-5Ir	21.49	...
Pt-10Ir	21.53	...
Pt-15Ir	21.57	...
Pt-20Ir	21.61	...
Pt-25Ir	21.66	...
Pt-30Ir	21.70	...
Pt-35Ir	21.79	...
Pt-5Ru	20.67	...
Pt-10Ru	19.94	...
Palladium	12.02	0.4343
60Pd-40Cu	10.6	0.383
95.5Pd-4.5Ru	12.07(a)	...
95.5Pd-4.5Ru	11.62(b)	...

Permanent magnet materials

Metal or alloy	g/cm³	lb/in.³
Cunico	8.30	0.300
Cunife	8.61	0.311
Comol	8.16	0.295
Alnico I	6.89	0.249
Alnico II	7.09	0.256
Alnico III	6.89	0.249
Alnico IV	7.00	0.253
Alnico V	7.31	0.264
Alnico VI	7.42	0.268
Barium ferrite	4.7	0.17
Vectolite	3.13	0.113

Pure metals

Metal or alloy	g/cm³	lb/in.³
Antimony	6.62	0.239
Beryllium	1.848	0.067
Bismuth	9.80	0.354
Cadmium	8.65	0.313
Calcium	1.55	0.056
Cesium	1.903	0.069
Chromium	7.19	0.260
Cobalt	8.85	0.322
Gallium	5.907	0.213
Germanium	5.323	0.192
Hafnium	13.1	0.473
Indium	7.31	0.264
Iridium	22.5	0.813
Lithium	0.534	0.019
Manganese	7.43	0.270
Mercury	13.546	0.489
Molybdenum	10.22	0.369
Niobium	8.57	0.310
Osmium	22.583	0.816
Plutonium	19.84	0.717
Potassium	0.86	0.031
Rhenium	21.04	0.756
Rhodium	12.44	0.447
Ruthenium	12.2	0.441
Selenium	4.79	0.174
Silicon	2.33	0.084
Silver	10.49	0.379
Sodium	0.97	0.035
Tantalum	16.6	0.600
Thallium	11.85	0.428
Thorium	11.72	0.423
Tungsten	19.3	0.697
Uranium	19.07	0.689
Vanadium	6.1	0.22
Zirconium	6.5	0.23

Rare earth metals

Metal or alloy	g/cm³	lb/in.³
Cerium	8.23(e)	...
Cerium	6.66(f)	...
Cerium	6.77(g)	...
Dysprosium	8.55(h)	...
Erbium	9.15(h)	...
Europium	5.245(g)	...
Gadolinium	7.86(h)	...
Holmium	6.79(h)	...
Lanthanum	6.19(f)	...
Lanthanum	6.18(e)	...
Lanthanum	5.97(g)	...
Lutetium	9.85(h)	...
Neodymium	7.00(f)	...
Neodymium	6.80(g)	...
Praseodymium	6.77(f)	...
Praseodymium	6.64(g)	...
Samarium	7.49(j)	...
Scandium	2.99(h)	...
Terbium	8.25(h)	...
Thulium	9.31(h)	...
Ytterbium	6.96(e)	...
Yttrium	4.47(h)	...

(a) 6.95 to 7.35 g/cm² (0.251 to 0.265 lb/in.³). (b) 7.20 to 7.34 g/cm³ (0.260 to 0.265 lb/in.³). (c) Annealed. (d) As cast. (e) Face-centered cubic. (f) Hexagonal. (g) Body-centered cubic. (h) Close-packed hexagonal. (j) Rhombohedral

Linear thermal expansion of metals and alloys

Metal or alloy	Temperature, °C	Coefficient of expansion, μ in./in. · °C
Aluminum and aluminum alloys		
Aluminum (99.996%)	20-100	23.6
Wrought alloys		
EC, 1060, 1100	20-100	23.6
2011, 2014	20-100	23.0
2024	20-100	22.8
2218	20-100	22.3
3003	20-100	23.2
4032	20-100	19.4
5005, 5050, 5052	20-100	23.8
5056	20-100	24.1
5083	20-100	23.4
5086	60-300	23.9
5154	20-100	23.9
5357	20-100	23.7
5456	20-100	23.9
6061, 6063	20-100	23.4
6101, 6151	20-100	23.0
7075	20-100	23.2
7079, 7178	20-100	23.4
Casting alloys		
A13	20-100	20.4
43 and 108	20-100	22.0
A108	20-100	21.5
A132	20-100	19.0
D132	20-100	20.5
F132	20-100	20.7
138	20-100	21.4
142	20-100	22.5
195	20-100	23.0
B195	20-100	22.0
214	20-100	24.0
220	20-100	25.0
319	20-100	21.5
355	20-100	22.0
356	20-100	21.5
360	20-100	21.0
750	20-100	23.1
40E	21-93	24.7
Copper and copper alloys		
Wrought coppers		
Pure copper	20	16.5
Electrolytic tough pitch copper (ETP)	20-100	16.8
Deoxidized copper, high residual phosphorus (DHP)	20-300	17.7
Oxygen-free copper	20-300	17.7
Free-machining copper, 0.5% Te or 1% Pb	20-300	17.7
Wrought alloys		
Gilding, 95%	20-300	18.1
Commercial bronze, 90%	20-300	18.4
Jewelry bronze, 87.5%	20-300	18.6
Red brass, 85%	20-300	18.7
Low brass, 80%	20-300	19.1
Cartridge brass, 70%	20-300	19.9
Yellow brass	20-300	20.3
Muntz metal	20-300	20.8
Leaded commercial bronze	20-300	18.4
Low-leaded brass	20-300	20.2
Medium-leaded brass	20-300	20.3
High-leaded brass	20-300	20.3
Extra-high-leaded brass	20-300	20.5
Free-cutting brass	20-300	20.5
Leaded Muntz metal	20-300	20.8
Forging brass	20-300	20.7
Architectural bronze	20-300	20.9
Inhibited admiralty	20-300	20.2
Naval brass	20-300	21.2
Leaded naval brass	20-300	21.2
Manganese bronze (A)	20-300	21.2
Phosphor bronze, 5% (A)	20-300	17.8
Phosphor bronze, 8% (C)	20-300	18.2

Metal or alloy	Temperature, °C	Coefficient of expansion, μ in./in. · °C
Phosphor bronze, 10% (D)	20-300	18.4
Phosphor bronze, 1.25%	20-300	17.8
Free-cutting phosphor bronze	20-300	17.3
Cupro-nickel, 30%	20-300	16.2
Cupro-nickel, 10%	20-300	17.1
Nickel silver, 65-18	20-300	16.2
Nickel silver, 55-18	20-300	16.7
Nickel silver, 65-12	20-300	16.2
High-silicon bronze (A)	20-300	18.0
Low-silicon bronze (B)	20-300	17.9
Aluminum bronze (3)	20-300	16.4
Aluminum-silicon bronze	20-300	18.0
Aluminum bronze (1)	20-300	16.8
Beryllium copper	20-300	17.8
Casting alloys		
88Cu-8Sn-4Zn	21-177	18.0
89Cu-11Sn	20-300	18.4
88Cu-6Sn-1.5Pb-4.5Zn	21-260	18.5
87Cu-8Sn-1Pb-4Zn	21-177	18.0
87Cu-10Sn-1Pb-2Zn	21-177	18.0
80Cu-10Sn-10Pb	21-204	18.5
78Cu-7Sn-15Pb	21-204	18.5
85Cu-5Sn-5Pb-5Zn	21-204	18.1
72Cu-1Sn-3Pb-24Zn	21-93	20.7
67Cu-1Sn-3Pb-29Zn	21-93	20.2
61Cu-1Sn-1Pb-37Zn	21-260	21.6
Manganese bronze		
60 ksi	21-204	20.5
65 ksi	21-93	21.6
110 ksi	21-260	19.8
Aluminum bronze		
Alloy 9A	...	17
Alloy 9B	20-250	17
Alloys 9C, 9D	...	16.2
Iron and iron alloys		
Pure iron	20	11.7
Fe-C alloys		
0.06% C	20-100	11.7
0.22% C	20-100	11.7
0.40% C	20-100	11.3
0.56% C	20-100	11.0
1.08% C	20-100	10.8
1.45% C	20-100	10.1
Invar (36% Ni)	20	0-2
13Mn-1.2C	20	18.0
13Cr-0.35C	20-100	10.0
12.3Cr-0.4Ni-0.09C	20-100	9.8
17.7Cr-9.6Ni-0.06C	20-100	16.5
18W-4Cr-1V	0-100	11.2
Gray cast iron	0-100	10.5
Malleable iron (pearlitic)	20-400	12
Lead and lead alloys		
Corroding lead (99.73+% Pb)	17-100	29.3
5-95 solder	15-110	28.7
20-80 solder	15-110	26.5
50-50 solder	15-110	23.4
1% antimonial lead	20-100	28.8
Hard lead (96Pb-4Sb)	20-100	27.8
Hard lead (94Pb-6Sb)	20-100	27.2
8% antimonial lead	20-100	26.7
9% antimonial lead	20-100	26.4
Lead-base babbitt		
SAE 14	20-100	19.6
Alloy 8	20-100	24.0
Magnesium and magnesium alloy		
Magnesium (99.8%)	20	25.2

Metal or alloy	Temperature, °C	Coefficient of expansion, μ in./in. · °C
Casting alloys		
AM100A	18-100	25.2
AZ63A	20-100	26.1
AZ91A, B, C	20-100	26
AZ92A	18-100	25.2
HZ32A	20-200	26.7
ZH42	20-200	27
ZH62A	20-200	27.1
ZK51A	20	26.1
EZ33A	20-100	26.1
EK30A, EK41A	20-100	26.1
Wrought alloys		
M1A, A3A	20-100	26
AZ31B, PE	20-100	26
AZ61A, AZ80A	20-100	26
ZK60A, B	20-100	26
HM31A	20-93	26.1
Nickel and nickel alloys		
Nickel (99.95% Ni + Co)	0-100	13.3
Duranickel	0-100	13.0
Monel	0-100	14.0
Monel (cast)	25-100	12.9
Inconel	20-100	11.5
Ni-o-nel	27-93	12.9
Hastelloy B	0-100	10.0
Hastelloy C	0-100	11.3
Hastelloy D	0-100	11.0
Hastelloy F	20-100	14.2
Hastelloy N	21-204	10.4
Hastelloy W	23-100	11.3
Hastelloy X	26-100	13.8
Illium G	0-100	12.19
Illium R	0-100	12.02
80Ni-20Cr	20-1000	17.3
60Ni-24Fe-16Cr	20-1000	17.0
35Ni-45Fe-20Cr	20-500	15.8
Constantan	20-1000	18.8
Tin and tin alloys		
Pure tin	0-100	23
Solder (70Sn-30Pb)	15-110	21.6
Solder (63Sn-37Pb)	15-110	24.7
Titanium and titanium alloys		
99.9% Ti	20	8.41
99.0% Ti	93	8.55
Ti-5Al-2.5Sn	93	9.36
Ti-8Mn	93	8.64
Zinc and zinc alloys		
Pure zinc	20-250	39.7
AG40A alloy	20-100	27.4
AC41A alloy	20-100	27.4
Commercial rolled zinc		
0.08 Pb	20-40	32.5
0.3 Pb, 0.3 Cd	20-98	33.9(a)
Rolled zinc alloy (1 Cu, 0.010 Mg)	20-100	34.8(b)
Zn-Cu-Ti alloy (0.8 Cu, 0.15 Ti)	20-100	24.9(c)
Pure metals		
Beryllium	25-100	11.6
Cadmium	20	29.8
Calcium	0-400	22.3
Chromium	20	6.2
Cobalt	20	13.8
Gold	20	14.2
Iridium	20	6.8
Lithium	20	56
Manganese	0-100	22
Palladium	20	11.76
Platinum	20	8.9
Rhenium	20-500	6.7
Rhodium	20-100	8.3
Ruthenium	20	9.1
Silicon	0-1400	5
Silver	0-100	19.68
Tungsten	27	4.6
Vanadium	23-100	8.3
Zirconium	...	5.85

(a) With the grain; 23.4 across the grain. (b) With the grain; 21.1 across the grain. (c) With the grain; 19.4 across the grain.

Thermal conductivity of metals and alloys

Metal or alloy	Thermal conductivity near room temperature, cal/cm² · cm · s · °C
Aluminum and aluminum alloys	
Wrought alloys	
EC (O)	0.57
1060 (O)	0.56
1100	0.53
2011 (T3)	0.34
2014 (O)	0.46
2024 (O)	0.45
2218 (T72)	0.37
3003 (O)	0.46
4032 (O)	0.37
5005	0.48
5050 (O)	0.46
5052 (O)	0.33
5056 (O)	0.28
5083	0.28
5086	0.30
5154	0.30
5357	0.40
5456	0.28
6061 (O)	0.41
6063 (O)	0.52
6101 (T6)	0.52
6151 (O)	0.49
7075 (T6)	0.29
7079 (T6)	0.29
7178	0.29
Casting alloys	
A13	0.29
43 (F)	0.34
108 (F)	0.29
A108	0.34
A132 (T551)	0.28
D132 (T5)	0.25
F132	0.25
138	0.24
142 (T21, sand)	0.40
195 (T4, T62)	0.33
B195 (T4, T6)	0.31
214	0.33
200 (T4)	0.21
319	0.26
355 (T51, sand)	0.40
356 (T51, sand)	0.40
360	0.35
380	0.23
750	0.44
40E	0.33
Copper and copper alloys	
Wrought coppers	
Pure copper	0.941
Electrolytic tough pitch copper (ETP)	0.934
Deoxidized copper, high residual phosphorus (DHP)	0.81
Free-machining copper (0.5% Te)	0.88
Free-machining copper (1% Pb)	0.92
Wrought alloys	
Gilding, 95%	0.56
Commercial bronze, 90%	0.45
Jewelry bronze, 87.5%	0.41
Red brass, 85%	0.38
Low brass, 80%	0.33
Cartridge brass, 70%	0.29
Yellow brass	0.28
Muntz metal	0.29
Leaded commercial bronze	0.43
Low-leaded brass (tube)	0.28
Medium-leaded brass	0.28
High-leaded brass (tube)	0.28
High-leaded brass	0.28
Extra-high-leaded brass	0.28
Leaded Muntz metal	0.29
Forging brass	0.28
Architectural bronze	0.29
Inhibited admiralty	0.26
Naval brass	0.28
Leaded naval brass	0.28
Manganese bronze (A)	0.26
Phosphor bronze, 5% (A)	0.17
Phosphor bronze, 8% (C)	0.15

Metal or alloy	Thermal conductivity near room temperature, cal/cm² · cm · s · °C
Phosphor bronze, 10% (D	0.12
Phosphor bronze, 1.25%	0.49
Free-cutting phosphor bronze	0.18
Cupro-nickel, 30%	0.07
Cupro-nickel, 10%	0.095
Nickel silver, 65-18	0.08
Nickel silver, 55-18	0.07
Nickel silver, 65-12	0.10
High-silicon bronze (A)	0.09
Low-silicon bronze (B)	0.14
Aluminum bronze, 5% Al	0.198
Aluminum bronze, (3)	0.18
Aluminum-silicon bronze	0.108
Aluminum bronze, (1)	0.144
Aluminum bronze, (2)	0.091
Beryllium copper	0.20(a)
Casting alloys	
Chromium copper (1% Cr)	0.4(a)
89Cu-11Sn	0.121
88Cu-6Sn-1.5Pb-4.5Zn	(b)
87Cu-8Sn-1Pb-4Zn	(c)
87Cu-10Sn-1Pb-2Zn	(c)
80Cu-10Sn-10Pb	(c)
Manganese bronze, 110 ksi	(d)
Aluminum bronze	
Alloy 9A	(e)
Alloy 9B	(f)
Alloy 9C	(b)
Alloy 9D	(c)
Propeller bronze	(g)
Nickel silver	
12% Ni	(h)
16% Ni	(h)
20% Ni	(j)
25% Ni	(k)
Silicon bronze	(h)
Iron and iron alloys	
Pure iron	0.178
Cast iron (3.16 C, 1.54 Si, 0.57 Mn)	0.112
Carbon steel (0.23 C, 0.64 Mn)	0.124
Carbon steel (1.22 C, 0.35 Mn)	0.108
Alloy steel (0.34 C, 0.55 Mn, 0.78 Cr, 3.53 Ni, 0.39 Mo, 0.05 Cu)	0.079
Type 410	0.057
Type 304	0.036
T1 tool steel	0.058
Lead and lead alloys	
Corroding lead (99.73+% Pb)	0.083
5-95 solder	0.085
20-80 solder	0.089
50-50 solder	0.111
1% antimonial lead	0.080
Hard lead (96Pb-4Sb)	0.073
Hard lead (94Pb-6Sb)	0.069
8% antimonial lead	0.065
9% antimonial lead	0.064
Lead-base babbitt (SAE 14)	0.057
Lead-base babbitt (alloy 8)	0.058
Magnesium and magnesium alloys	
Magnesium (99.8%)	0.367
Casting alloys	
AM100A	0.17
AZ63A	0.18
AZ81A (T4)	0.12
AZ91A, B, C	0.17
AZ92A	0.17
HK31A (T6, sand cast)	0.22
HZ32A	0.26
ZH42	0.27

Metal or alloy	Thermal conductivity near room temperature, cal/cm² · cm · s · °C
ZH62A	0.26
ZK51A	0.26
ZE41A (T5)	0.27
EZ33A	0.24
EK30A	0.26
EK41A (T5)	0.24
Wrought alloys	
M1A	0.33
AZ31B	0.23
AZ61A	0.19
AZ80A	0.18
ZK60A, B (F)	0.28
ZE10A (O)	0.33
HM21A (O)	0.33
HM31A	0.25
Nickel and nickel alloys	
Nickel (99.95% Ni + Co)	0.22
"A" nickel	0.145
"D" nickel	0.115
Monel	0.062
"K" Monel	0.045
Inconel	0.036
Hastelloy B	0.027
Hastelloy C	0.03
Hastelloy D	0.05
Illium G	0.029
Illium R	0.031
60Ni-24Fe-16Cr	0.032
35Ni-45Fe-20Cr	0.031
Constantan	0.051
Tin and tin alloys	
Pure tin	0.15
Soft solder (63Sn-37Pb)	0.12
Tin foil (92Sn-8Zn)	0.14
Titanium and titanium alloys	
Titanium (99.0%)	0.43
Ti-5Al-2.5Sn	0.19
Ti-2Fe-2Cr-2Mo	0.28
Ti-8Mn	0.26
Zinc and zinc alloys	
Pure zinc	0.27
AG40A alloy	0.27
AC41A alloy	0.26
Commercial rolled zinc 0.08 Pb	0.257
0.06 Pb, 0.06 Cd	0.257
Rolled zinc alloy (1 Cu, 0.010 Mg)	0.25
Zn-Cu-Ti alloy (0.8 Cu, 0.15 Ti)	0.25
Pure metals	
Beryllium	0.35
Cadmium	0.22
Chromium	0.16
Cobalt	0.165
Germanium	0.14
Gold	0.71
Indium	0.057
Iridium	0.14
Lithium	0.17
Molybdenum	0.34
Niobium	0.13
Palladium	0.168
Platinum	0.165
Plutonium	0.020
Rhenium	0.17
Rhodium	0.21
Silicon	0.20
Silver	1.0
Sodium	0.32
Tantalum	0.130
Thallium	0.093
Thorium	0.090
Tungsten	0.397
Uranium	0.071
Vanadium	0.074
Yttrium	0.035

(a) Depends on processing. (b) 18% of Cu. (c) 12% of Cu. (d) 9.05% of Cu. (e) 15% of Cu. (f) 16% of Cu. (g) 11% of Cu. (h) 7% of Cu. (j) 6% of Cu. (k) 6.5% of Cu.

Electrical conductivity and resistivity of metals and alloys

Metal or alloy	Conductivity, % IACS	Resistivity, μΩ · cm
Aluminum and aluminum alloys		
Aluminum (99.996%)	64.94	2.65
EC (O, H19)	62	2.8
5052 (O, H38)	35	4.93
5056 (H38)	27	6.4
6101 (T6)	56	3.1
Copper and copper alloys		
Wrought copper		
Pure copper	103.06	1.67
Electrolytic (ETP)	101	1.71
Oxygen-free copper (OF)	101	1.71
Free-machining copper		
0.5% Te	95	1.82
1.0% Pb	98	1.76
Wrought alloys		
Cartridge brass, 70%	28	6.2
Yellow brass	27	6.4
Leaded commercial bronze	42	4.1
Phosphor bronze, 1.25%	48	3.6
Nickel silver, 55-18	5.5	31
Low-silicon bronze (B)	12	14.3
Beryllium copper	22-30(a)	5.7-7.8(a)
Casting alloys		
Chromium copper (1% Cr)	80-90(a)	2.10
88Cu-8Sn-4Zn	11	15
87Cu-10Sn-1Pb-2Zn	11	15
Electrical contact materials		
Copper alloys		
0.04 oxide	100	1.72
1.25 Sn + P	48	3.6
5 Sn + P	18	11
8 Sn + P	13	13
15 Zn	37	4.7
20 Zn	32	5.4
35 Zn	27	6.4
2 Be + Ni or Co(b)	17-21	9.6-11.5
Silver and silver alloys		
Fine silver	106	1.59
92.5Ag-7.5Cu	85	2
90Ag-10Cu	85	2
72Ag-28Cu	87	2
72Ag-26Cu-2Ni	60	2.9
85Ag-15Cd	35	4.93
97Ag-3Pt	50	3.5
97Ag-3Pd	60	2.9
90Ag-10Pd	30	5.3
90Ag-10Au	40	4.2
60Ag-40Pd	8	23
70Ag-30Pd	12	14.3
Platinum and platinum alloys		
Platinum	16	10.6
95Pt-5Ir	9	19
90Pt-10Ir	7	25
85Pt-15Ir	6	28.5
80Pt-20Ir	5.6	31
75Pt-25Ir	5.5	33
70Pt-30Ir	5	35
65Pt-35Ir	5	36
95Pt-5Ru	5.5	31.5
90Pt-10Ru	4	43
89Pt-11Ru	4	43
86Pt-14Ru	3.5	46
96Pt-4W	5	36
Palladium and palladium alloys		
Palladium	16	10.8
95.5Pd-4.5Ru	7	24.2

Metal or alloy	Conductivity, % IACS	Resistivity, μΩ · cm
90Pd-10Ru	6.5	27
70Pd-30Ag	4.3	40
60Pd-40Ag	4.0	43
50Pd-50Ag	5.5	31.5
72Pd-26Ag-2Ni	4	43
60Pd-40Cu	5	35(c)
45Pd-30Ag-20Au-5Pt	4.5	39
35Pd-30Ag-14Cu-10Pt-10Au-1Zn	5	35
Gold and gold alloys		
Gold	75	2.35
90Au-10Cu	16	10.8
75Au-25Ag	16	10.8
72.5Au-14Cu-8.5Pt-4Ag-1Zn	10	17
69Au-25Ag-6Pt	11	15
41.7Au-32.5Cu-18.8Ni-7Zn	4.5	39
Electrical heating alloys		
Ni-Cr and Ni-Cr-Fe alloys		
78.5Ni-20Cr-1.5Si (80-20)	1.6	108.05
73.5Ni-20Cr-5Al-1.5Si	1.2	137.97
68Ni-20Cr-8.5Fe-2Si	1.5	116.36
60Ni-16Cr-22.5Fe-1.5Si	1.5	112.20
35Ni-20Cr-43.5Fe-1.5Si	1.7	101.4
Fe-Cr-Al alloys		
72Fe-23Cr-5Al	1.3	138.8
55Fe-37.5Cr-7.5Al	1.2	166.23
Pure metals		
Molybdenum	34	5.2
Platinum	16	10.64
Tantalum	13.9	12.45
Tungsten	30	5.65
Nonmetallic heating element materials		
Silicon carbide, SiC	1-1.7	100-200
Molybdenum disilicide, $MoSi_2$	4.5	37.24
Graphite	...	910.1
Instrument and control alloys		
Cu-Ni alloys		
98Cu-2Ni	35	4.99
94Cu-6Ni	17	9.93
89Cu-11Ni	11	14.96
78Cu-22Ni	5.7	29.92
55Cu-45Ni (constantan)	3.5	49.87
Cu-Mn-Ni alloys		
87Cu-13Mn (manganin)	3.5	48.21
83Cu-13Mn-4Ni (manganin)	3.5	48.21
85Cu-10Mn-4Ni (shunt manganin)	4.5	38.23
70Cu-20Ni-10Mn	3.6	48.88
67Cu-5Ni-27Mn	1.8	99.74
Ni-base alloys		
99.8 Ni	23	7.98
71Ni-29Fe	9	19.95
80Ni-20Cr	1.5	112.2
75Ni-20Cr-3Al + Cu or Fe	1.3	132.98
76Ni-17Cr-4Si-3Mn	1.3	132.98
60Ni-16Cr-24Fe	1.5	112.2
35Ni-20Cr-45Fe	1.7	101.4
Fe-Cr-Al alloy		
72Fe-23Cr-5Al-0.5Co	1.3	135.48

Metal or alloy	Conductivity, % IACS	Resistivity, μΩ · cm
Pure metals		
Iron (99.99%)	17.75	9.71
Thermostat metals		
75Fe-22Ni-3Cr	3	78.13
72Mn-18Cu-10Ni	1.5	112.2
67Ni-30Cu-1.4Fe-1Mn	3.5	56.52
75Fe-22Ni-3Cr	12	15.79
66.5Fe-22Ni-8.5Cr	3.3	58.18
Permanent magnet materials		
Carbon steel (0.65% C)	9.5	18
Carbon steel (1% C)	8	20
Chromium steel (3.5% Cr)	6.1	29
Tungsten steel (6% W)	6	30
Cobalt steel (17% Co)	6.3	28
Cobalt steel (36% Co)	6.5	27
Intermediate alloys		
Cunico	7.5	24
Cunife	9.5	18
Comol	3.6	45
Alnico alloys		
Alnico I	3.3	75
Alnico II	3.3	65
Alnico III	3.3	60
Alnico IV	3.3	75
Alnico V	3.5	47
Alnico VI	3.5	50
Magnetically soft materials		
Electrical steel sheet		
M-50	9.5	18
M-43	6-9	20-28
M-36	5.5-7.5	24-33
M-27	3.5-5.5	32-47
M-22	3.5-5	41-52
M-19	3.5-5	41-56
M-17	3-3.5	45-58
M-15	3-3.5	45-69
M-14	3-3.5	58-69
M-7	3-3.5	45-52
M-6	3.-3.5	45-52
M-5	3-3.5	45-52
Moderately high-permeability materials(d)		
Thermenol	0.5	162
16 Alfenol	0.7	153
Sinimax	2	90
Monimax	2.5	80
Supermalloy	3	65
4-79 Moly Permalloy, Hymu 80	3	58
Mumetal	3	60
1040 alloy	3	56
High Permalloy 49, A-L 4750, Armco 48	3.6	48
45 Permalloy	3.6	45
High-permeability materials(e)		
Supermendur	4.5	40
2V Permendur	4.5	40
35% Co, 1% Cr	9	20
Ingot iron	17.5	10
0.5% Si steel	6	28
1.75% Si steel	4.6	37
3.0% Si steel	3.6	47
Grain-oriented 3.0% Si steel	3.5	50
Grain-oriented 50% Ni iron	3.6	45
50% Ni iron	3.5	50

(continued)

Electrical conductivity and resistivity of metals and alloys (continued)

Metal or alloy	Conductivity, % IACS	Resistivity, μΩ · cm	Metal or alloy	Conductivity, % IACS	Resistivity, μΩ · cm	Metal or alloy	Conductivity, % IACS	Resistivity, μΩ · cm
Relay steels and alloys after annealing			**Stainless steels**			**Stainless and heat-resisting alloys**		
Low-carbon iron and steel			Type 410 3		57	Type 302 3		72
			Type 416 3		57	Type 309 .... 2.5		78
Low-carbon iron 17.5		10	Type 430 3		60	Type 316 .... 2.5		74
1010 steel 14.5		12	Type 443 3		68	Type 317 .... 2.5		74
			Type 446 3		61	Type 347 .... 2.5		73
Silicon steels			**Nickel irons**			Type 403 3		57
1% Si 7.5		23				Type 405 3		60
2.5% Si 4		41	50% Ni 3.5		48	Type 501 .... 4.5		40
3% Si 3.5		48	78% Ni 11		16	HH 2.5		80
3% Si, grain-oriented ... 3.5		48	77% Ni (Cu, Cr) 3		60	HK 2		90
4% Si 3		59	79% Ni (Mo) 3		58	HT 1.7		100

(a) Precipitation hardened; depends on processing. (b) A heat treatable alloy. (c) Annealed and quenched. (d) At low field strength and high electrical resistance. (e) At higher field strength; annealed for optimum magnetic properties.

Vapor pressures of the elements

Element	Pressure, atm											
	0.0001		0.001		0.01		0.1		0.50		1.0	
	°C	°F	°C	°F	°C	°F	°C	°F	°C	°F	°C	°F
Aluminum	1110	2030	1263	2305	1461	2662	1713	3115	1940	3524	2056	3733
Antimony	759	1398	872	1602	1013	1855	1196	2185	1359	2478	1440	2624
Arsenic	308	586	363	685	428	802	499	930	578	1072	610	1130
Bismuth	914	1677	1008	1846	1121	2050	1254	2289	1367	2493	1420	2588
Cadmium	307(a)	585(a)	384(b)	723(b)	471	880	594	1101	708	1306	765	1409
Calcium	688	1270	802(c)	1476(c)	958(b)	1756(b)	1175	2147	1380	2516	1487	2709
Carbon	3257	5895	3547	6417	3897	7047	4317	7803	4667	8433	4827	8721
Chromium	1420(a)	2588(a)	1594(b)	2901(b)	1813	3295	2097	3807	2351	4264	2482	4500
Copper	1412	2574	1602	2916	1844	3351	2162	3924	2450	4442	2595	4703
Gallium	1178	2152	1329	2424	1515	2759	1751	3184	1965	3569	2071	3760
Gold	1623	2953	1839	3342	2115	3839	2469	4476	2796	5065	2966	5371
Iron	1564	2847	1760	3200	2004	3639	2316	4201	2595	4703	2735	4955
Lead	815	1499	953	1747	1135	2075	1384	2523	1622	2952	1744	3171
Lithium	592	1098	707	1305	858	1576	1064	1947	1266	2311	1372	2502
Magnesium	516	961	608(a)	1126(a)	725(b)	1337(b)	886	1627	1030	1886	1107	2025
Manganese	1115(d)	2039(d)	1269(b)	2316(b)	1476	2889	1750	3182	2019	3666	2151	3904
Mercury	77.9(b)	172.2(b)	120.8	249.4	176.1	349.0	251.3	484.3	321.5	610.7	357	675
Molybdenum	2727	4941	3057	5535	3477	6291	4027	7281	4537	8199	4804	8679
Nickel	1586	2887	1782	3240	2025	3677	2321	4210	2593	4699	2732	4950
Platinum	2367	4293	2687	4869	3087	5589	3637	6579	4147	7497	4407	7965
Potassium	261	502	332	630	429	804	565	1051	704	1299	774	1425
Rubidium	223	433	288	550	377	711	497	927	617	1143	679	1254
Selenium	282	540	347	657	430	806	540	1004	634	1173	680	1256
Silicon	1572	2862	1707	3105	1867	3393	2057	3735	2217	4023	2287	4149
Silver	1169	2136	1334	2433	1543	2809	1825	3317	2081	3778	2212	4014
Sodium	349	660	429	804	534	993	679	1254	819	1506	892	1638
Strontium	....	....	(a)	(a)	877(b)	1629(b)	1081	1978	1279	2334	1384	2523
Tellurium	(a)	(a)	509(b)	948(b)	632	1170	810	1490	991	1816	1087	1989
Thallium	692	1277	809	1488	962	1764	1166	2131	1359	2478	1457	2655
Tin	....	....	....	....	....	....	1932(b)	3510(b)	2163	3925	2270	4118
Tungsten	3547	6417	3937	7119	4437	8019	5077	9171	5647	10197	5927	10701
Zinc	399(a)	750(a)	477(b)	891(b)	579	1074	717	1323	842	1548	907	1665

(a) In the solid state. (b) In the liquid state. (c) β. (d) γ.
Source: K. K. Kelley, Bureau of Mines Bulletin 383, 1935.

Metric conversion factors

To convert from	To	Multiply by	To convert from	To	Multiply by
angstrom	m	1.0000×10^{-10}(a)	hp(e)	W	7.4570×10^2
atm	Pa	1.0133×10^5	hp(f)	W	7.4600×10^2
Btu(b)	J	1.054×10^3	in.	m	2.5400×10^{-2}
Btu(b)/ft²·h	W/m²	3.1525	in.²	m²	6.4516×10^{-4}
Btu(b)/ft²·h·°F	W/m²·K	5.6745	in.³	m³	1.6387×10^{-5}
Btu(b)·ft/h·ft²·°F	W/m·K	1.7296	in. of Hg(g)	Pa	3.3864×10^3
Btu(b)/ft²·s	W/m²	1.135×10^4	in. of water(c)	Pa	2.4908×10^2
Btu(b)·in./ft²·h·°F	W/m·K	1.4413×10^{-1}	K	°C	$t_{°C} = t_K - 273.15$
Btu(b)·in./s·ft²·°F	W/m·K	5.1887×10^2	kgf	N	9.80665(a)
Btu(b)/lbm·°F	J/kg·K	4.1840×10^3	kgf/mm²	Pa	9.80665×10^6(a)
cal(b)	J	4.1840 (a)	ksi	MPa	6.8948
cal(b)/cm·s·°C	W/m·K	4.1840×10^2(a)	ksi	Pa	6.8948×10^6
cal(b)/g	J/kg	4.1840×10^3(a)	ksi√in.	MPa√m	1.089
cal(b)/g·°C	J/kg·K	4.1840×10^3(a)	lb(h)	kg	4.5359×10^{-1}
circ mil	m²	5.0671×10^{-10}	lb/in.³	kg/m³	2.7680×10^4
°C	K	$t_K = t_{°C} + 273.15$	lbf	N	4.4482
degree	rad	1.7453×10^{-2}	lbf·in.	N·m	1.1298×10^{-1}
dyne/cm²	Pa	1.0000×10^{-1}(a)	lbf·ft	N·m	1.3558
°F	°C	$t_{°C} = (t_F - 32)/1.8$	MPa√m	MNm$^{-3/2}$	1.0000(a)
°F	K	$t_K = (t_F + 459.67)/1.8$	µin.	m	2.5400×10^8(a)
ft	m	3.0480×10^{-1}	mil	m	2.5400×10^{-5}(a)
ft²	m²	9.2903×10^{-2}	N/m²	Pa	1.0000(a)
ft³	m³	2.8317×10^{-2}	oersted	A/m	79.578
ft of water(c)	Pa	2.9890×10^3	oz/ft²	kg/m²	3.0515×10^{-1}
ft²/h (thermal diffusivity)	m²/s	2.58064×10^{-5}(a)	psi	Pa	6.8948×10^3
ft·lbf	J	1.3558	°R	K	$t_K = t_{°R}/1.8$
ft·lbf/s	W	1.3558	ton(j)	kg	9.0718×10^2
ft/s	m/s	3.0480×10^{-1}	ton(k)	kg	1.0160×10^3
gauss	T	1.0000×10^{-4}(a)	ton/in.²	Pa	1.3786×10^4
gallon(d)	m³	3.7854×10^{-3}	tonne	kg	1.0000×10^3(a)
g/cm³	kg/m³	1.0000×10^3(a)	torr	Pa	1.3332×10^2
g/cm³	Mg/m³	1.0000(a)	Ω/circ mil·ft	Ω·m	1.6624×10^{-9}

(a) Exactly. (b) Thermochemical. (c) At 4 °C (39.2 °F). (d) U.S. liquid. (e) Mechanical (1 hp = 550 ft·lbf/s). (f) Electrical. (g) At 0 °C (32 °F). (g) Avoirdupois. (j) Short; equal to 2000 lbm. (k) Long; 2240 lbm.

Metric stress-intensity conversions

The middle column of figures (in bold-faced type) contains the reading (in MPa√m or ksi√in.) to be converted. If converting from ksi√in. to MPa√m, read the MPa√m equivalent in the column headed "MPa√m". If converting from MPa√m to ksi√in., read the ksi√in. equivalent in the column headed "ksi√in.". 1 ksi√in. = 1.098845 MPa√m.

ksi, √in.		MPa, √m	ksi, √in.		MPa, √m	ksi, √in.		MPa, √m	ksi, √in.		MPa, √m	ksi, √in.		MPa, √m
0.91005	1	1.0988	28.211	31	34.063	55.513	61	67.027	82.814	91	99.991	110.12	121	132.95
1.8201	2	2.1976	29.121	32	35.162	56.423	62	68.126	83.724	92	101.09	111.03	122	134.05
2.7301	3	3.2964	30.032	33	36.260	57.333	63	69.224	84.634	93	102.19	111.94	123	135.15
3.6402	4	4.3952	30.942	34	37.359	58.243	64	70.323	85.544	94	103.29	112.85	124	136.25
4.5502	5	5.4940	31.852	35	38.458	59.153	65	71.422	86.454	95	104.39	113.76	125	137.35
5.4603	6	6.5928	32.762	36	39.557	60.063	66	72.521	87.364	96	105.48	114.67	126	138.45
6.3703	7	7.6916	33.672	37	40.656	60.973	67	73.620	88.275	97	106.58	115.58	127	139.55
7.2804	8	8.7904	34.582	38	41.754	61.883	68	74.718	89.185	98	107.68	116.49	128	140.65
8.1904	9	9.8892	35.492	39	42.853	62.793	69	75.817	90.095	99	108.78	117.40	129	141.75
9.1005	10	10.988	36.402	40	43.952	63.703	70	76.916	91.005	100	109.88	118.31	130	142.84
10.011	11	12.087	37.312	41	45.051	64.613	71	78.015	91.915	101	110.98	119.22	131	143.94
10.921	12	13.186	38.222	42	46.150	65.523	72	79.114	92.825	102	112.08	120.13	132	145.04
11.831	13	14.284	39.132	43	47.248	66.433	73	80.212	93.735	103	113.18	121.04	133	146.14
12.741	14	15.383	40.042	44	48.347	67.343	74	81.311	94.645	104	114.28	121.95	134	147.24
13.651	15	16.482	40.952	45	49.446	68.253	75	82.410	95.555	105	115.37	122.86	135	148.34
14.561	16	17.581	41.862	46	50.545	69.164	76	83.509	96.465	106	116.47	123.77	136	149.44
15.471	17	18.680	42.772	47	51.644	70.074	77	84.608	97.375	107	117.57	124.68	137	150.54
16.381	18	19.778	43.682	48	52.742	70.984	78	85.706	98.285	108	118.67	125.59	138	151.63
17.291	19	20.877	44.592	49	53.841	71.893	79	86.805	99.195	109	119.77	126.50	139	152.73
18.201	20	21.976	45.502	50	54.940	72.804	80	87.904	100.11	110	120.87	127.41	140	153.83
19.111	21	23.075	46.412	51	56.039	73.714	81	89.003	101.02	111	121.97	128.32	141	154.93
20.021	22	24.174	47.322	52	57.138	74.624	82	90.102	101.93	112	123.07	129.23	142	156.03
20.931	23	25.272	48.232	53	58.236	75.534	83	91.200	102.84	113	124.16	130.14	143	157.13
21.841	24	26.371	49.143	54	59.335	76.444	84	92.300	103.75	114	125.26	131.05	144	158.23
22.751	25	27.470	50.053	55	60.434	77.354	85	93.398	104.66	115	126.36	131.96	145	159.33
23.661	26	28.569	50.963	56	61.533	78.264	86	94.497	105.57	116	127.46	132.87	146	160.42
24.571	27	29.668	51.873	57	62.632	79.174	87	95.596	106.48	117	128.56	133.78	147	161.52
25.481	28	30.766	52.783	58	63.730	80.084	88	96.694	107.39	118	129.66	134.69	148	162.62
26.391	29	31.865	53.693	59	64.829	80.994	89	97.793	108.30	119	130.76	135.60	149	163.72
27.301	30	32.964	54.603	60	65.928	81.904	90	98.892	109.21	120	131.86	136.51	150	164.82

(continued)

Metric stress-intensity conversions (continued)

The middle column of figures (in bold-faced type) contains the reading (in MPa√m or ksi√in.) to be converted. If converting from ksi√in. to MPa√m, read the MPa√m equivalent in the column headed "MPa√m". If converting from MPa√m to ksi√in., read the ksi√in. equivalent in the column headed "ksi√in.". 1 ksi√in. = 1.098845 MPa√m.

ksi√in.		MPa√m	ksi√in.		MPa√m	ksi√in.		MPa√m	ksi√in.		MPa√m	ksi√in.		MPa√m
137.42	**151**	165.92	146.52	**161**	176.91	155.62	**171**	187.90	164.72	**181**	198.88	173.82	**191**	209.87
138.33	**152**	167.02	147.43	**162**	178.01	156.53	**172**	189.00	165.63	**182**	199.98	174.73	**192**	210.97
139.24	**153**	168.12	148.34	**163**	179.10	157.44	**173**	190.10	166.54	**183**	201.08	175.64	**193**	212.07
140.15	**154**	169.22	149.25	**164**	180.20	158.35	**174**	191.19	167.45	**184**	202.18	176.55	**194**	213.17
141.06	**155**	170.31	150.16	**165**	181.30	159.26	**175**	192.29	168.36	**185**	203.28	177.46	**195**	214.27
141.97	**156**	171.41	151.07	**166**	182.40	160.17	**176**	193.39	169.27	**186**	204.38	178.37	**196**	215.36
142.88	**157**	172.51	151.98	**167**	183.50	161.08	**177**	194.49	170.18	**187**	205.48	179.28	**197**	216.46
143.79	**158**	173.61	152.89	**168**	184.60	161.99	**178**	195.59	171.09	**188**	206.57	180.19	**198**	217.56
144.70	**159**	174.71	153.80	**169**	185.70	162.90	**179**	196.69	172.00	**189**	207.67	181.10	**199**	218.66
145.61	**160**	175.81	154.71	**170**	186.80	163.81	**180**	197.78	172.91	**190**	208.77	182.01	**200**	219.76

Metric energy conversions

The middle column of figures (in bold-faced type) contains the reading (in J or ft·lb) to be converted. If converting from ft·lb to J, read the J equivalent in the column headed "J". If converting from J to ft·lb, read the equivalent in the column headed "ft·lb". 1 ft·lb = 1.355818 J.

ft·lb		J	ft·lb		J	ft·lb		J	ft·lb		J
0.7376	**1**	1.3558	28.7649	**39**	52.8769	56.7923	**77**	104.3980	129.0734	**175**	237.2681
1.4751	**2**	2.7116	29.5025	**40**	54.2327	57.5298	**78**	105.7538	132.7612	**180**	244.0472
2.2127	**3**	4.0675	30.2400	**41**	55.5885	58.2674	**79**	107.1096	136.4490	**185**	250.8263
2.9502	**4**	5.4233	30.9776	**42**	56.9444	59.0050	**80**	108.4654	140.1368	**190**	257.6054
3.6878	**5**	6.7791	31.7152	**43**	58.3002	59.7425	**81**	109.8212	143.8246	**195**	264.3845
4.4254	**6**	8.1349	32.4527	**44**	59.6560	60.4801	**82**	111.1771	147.5124	**200**	271.1636
5.1629	**7**	9.4907	33.1903	**45**	61.0118	61.2177	**83**	112.5329	154.8880	**210**	284.7218
5.9005	**8**	10.8465	33.9279	**46**	62.3676	61.9552	**84**	113.8887	162.2637	**220**	298.2799
6.6381	**9**	12.2024	34.6654	**47**	63.7234	62.6928	**85**	115.2445	169.6393	**230**	311.8381
7.3756	**10**	13.5582	35.4030	**48**	65.0793	63.4303	**86**	116.6003	177.0149	**240**	325.3963
8.1132	**11**	14.9140	36.1405	**49**	66.4351	64.1679	**87**	117.9562	184.3905	**250**	338.9545
8.8507	**12**	16.2698	36.8781	**50**	67.7909	64.9055	**88**	119.3120	191.7661	**260**	352.5126
9.5883	**13**	17.6256	37.6157	**51**	69.1467	65.6430	**89**	120.6678	199.1418	**270**	366.0708
10.3259	**14**	18.9815	38.3532	**52**	70.5025	66.3806	**90**	122.0236	206.5174	**280**	379.6290
11.0634	**15**	20.3373	39.0908	**53**	71.8583	67.1182	**91**	123.3794	213.8930	**290**	393.1872
11.8010	**16**	21.6931	39.8284	**54**	73.2142	67.8557	**92**	124.7452	221.2686	**300**	406.7454
12.5386	**17**	23.0489	40.5659	**55**	74.5700	68.5933	**93**	126.0911	228.6442	**310**	420.3036
13.2761	**18**	24.4047	41.3035	**56**	75.9258	69.3308	**94**	127.4469	236.0199	**320**	433.8617
14.0137	**19**	25.7605	42.0410	**57**	77.2816	70.0684	**95**	128.8027	243.3955	**330**	447.4199
14.7512	**20**	27.1164	42.7786	**58**	78.6374	70.8060	**96**	130.1585	250.7711	**340**	460.9781
15.4888	**21**	28.4722	43.5162	**59**	79.9933	71.5435	**97**	131.5143	258.1467	**350**	474.5363
16.2264	**22**	29.8280	44.2537	**60**	81.3491	72.2811	**98**	132.8702	265.5224	**360**	488.0944
16.9639	**23**	31.1838	44.9913	**61**	82.7049	73.0186	**99**	134.2260	272.8980	**370**	501.6526
17.7015	**24**	32.5396	45.7288	**62**	84.0607	73.7562	**100**	135.5818	280.2736	**380**	515.2108
18.4390	**25**	33.8954	46.4664	**63**	85.4165	77.4440	**105**	142.3609	287.6492	**390**	528.7690
19.1766	**26**	35.2513	47.2040	**64**	86.7723	81.1318	**110**	149.1400	295.0248	**400**	542.3272
19.9142	**27**	36.6071	47.9415	**65**	88.1282	84.8196	**115**	155.9191	302.4005	**410**	555.8854
20.6517	**28**	37.9629	48.6791	**66**	89.4840	88.5075	**120**	162.6982	309.7761	**420**	569.4435
21.3893	**29**	39.3187	49.4167	**67**	90.8398	92.1953	**125**	169.4772	317.1517	**430**	583.0017
22.1269	**30**	40.6745	50.1542	**68**	92.1956	95.8831	**130**	176.2563	324.5273	**440**	596.5599
22.8644	**31**	42.0304	50.8918	**69**	93.5514	99.5709	**135**	183.0354	331.9029	**450**	610.1181
23.6020	**32**	43.3862	51.6293	**70**	94.9073	103.2587	**140**	189.8145	339.2786	**460**	623.6762
24.3395	**33**	44.7420	52.3669	**71**	96.2631	106.9465	**145**	196.5936	346.6542	**470**	637.2344
25.0771	**34**	46.0978	53.1045	**72**	97.6189	110.6343	**150**	203.3727	354.0298	**480**	650.7926
25.8147	**35**	47.4536	53.8420	**73**	98.9747	114.3221	**155**	210.1518	361.4054	**490**	664.3508
26.5522	**36**	48.8094	54.5796	**74**	100.3305	118.0099	**160**	216.9308	368.7811	**500**	677.9090
27.2898	**37**	50.1653	55.3172	**75**	101.6863	121.6977	**165**	223.7099			
28.0274	**38**	51.5211	56.0547	**76**	103.0422	125.3856	**170**	230.4890			

Metric stress or pressure conversions

The middle column of figures (in bold-faced type) contains the reading (in MPa or ksi) to be converted. If converting from ksi to MPa, read the MPa equivalent in the column headed "MPa". If converting from MPa to ksi, read the ksi equivalent in the column headed "ksi". 1 ksi = 6.894757 MPa. 1 psi = 6.894757 kPa.

ksi		MPa	ksi		MPa	ksi		MPa	ksi		MPa
0.14504	1	6.895	9.8626	68	468.84	65.267	450	3102.6	179.85	1240	...
0.29008	2	13.790	10.008	69	475.74	66.717	460	3171.6	182.75	1260	...
0.43511	3	20.684	10.153	70	482.63	66.168	470	3240.5	185.65	1280	...
0.58015	4	27.579	10.298	71	489.53	69.618	480	3309.5	188.55	1300	...
0.72519	5	34.474	10.443	72	496.42	71.068	490	3378.4	191.45	1320	...
0.87023	6	41.369	10.588	73	503.32	72.519	500	3447.4	194.35	1340	...
1.0153	7	48.263	10.733	74	510.21	73.969	510	...	197.25	1360	...
1.1603	8	55.158	10.878	75	517.11	75.420	520	...	200.15	1380	...
1.3053	9	62.053	11.023	76	524.00	76.870	530	...	203.05	1400	...
1.4504	10	68.948	11.168	77	530.90	78.320	540	...	205.95	1420	...
1.5954	11	75.842	11.313	78	537.79	79.771	550	...	208.85	1440	...
1.7405	12	82.737	11.458	79	544.69	81.221	560	...	211.76	1460	...
1.8855	13	89.632	11.603	80	551.58	82.672	570	...	214.66	1480	...
2.0305	14	96.527	11.748	81	558.48	84.122	580	...	217.56	1500	...
2.1756	15	103.42	11.893	82	565.37	85.572	590	...	220.46	1520	...
2.3206	16	110.32	12.038	83	572.26	87.023	600	...	223.36	1540	...
2.4656	17	117.21	12.183	84	579.16	88.473	610	...	226.26	1560	...
2.6107	18	124.11	12.328	85	586.05	89.923	620	...	229.16	1580	...
2.7557	19	131.00	12.473	86	592.95	91.374	630	...	232.06	1600	...
2.9008	20	137.90	12.618	87	599.84	92.824	640	...	234.96	1620	...
3.0458	21	144.79	12.763	88	606.74	94.275	650	...	237.86	1640	...
3.1908	22	151.68	12.909	89	613.63	95.725	660	...	240.76	1660	...
3.3359	23	158.58	13.053	90	620.53	97.175	670	...	243.66	1680	...
3.4809	24	165.47	13.198	91	627.42	98.626	680	...	246.56	1700	...
3.6259	25	172.37	13.343	92	634.32	100.08	690	...	249.46	1720	...
3.7710	26	179.26	13.489	93	641.21	101.53	700	...	252.37	1740	...
3.9160	27	186.16	13.634	94	648.11	102.98	710	...	255.27	1760	...
4.0611	28	193.05	13.779	95	655.00	104.43	720	...	258.17	1780	...
4.2061	29	199.95	13.924	96	661.90	105.88	730	...	261.07	1800	...
4.3511	30	206.84	14.069	97	668.79	107.33	740	...	263.97	1820	...
4.4962	31	213.74	14.214	98	675.69	108.78	750	...	266.87	1840	...
4.6412	32	220.63	14.359	99	682.58	110.23	760	...	269.77	1860	...
4.7862	33	227.53	14.504	100	689.48	111.68	770	...	272.67	1880	...
4.9313	34	234.42	15.954	110	758.42	113.13	780	...	275.57	1900	...
5.0763	35	241.32	17.405	120	827.37	114.58	790	...	278.47	1920	...
5.2214	36	248.21	18.855	130	896.32	116.03	800	...	281.37	1940	...
5.3664	37	255.11	20.305	140	965.27	117.48	810	...	284.27	1960	...
5.5114	38	262.00	21.756	150	1034.2	118.93	820	...	287.17	1980	...
5.6565	39	268.90	23.206	160	1103.2	120.38	830	...	290.08	2000	...
5.8015	40	275.79	24.656	170	1172.1	121.83	840	...	292.98	2020	...
5.9465	41	282.69	26.107	180	1241.1	123.28	850	...	295.88	2040	...
6.0916	42	289.58	27.557	190	1310.0	124.73	860	...	298.78	2060	...
6.2366	43	296.47	29.008	200	1379.0	126.18	870	...	301.68	2080	...
6.3817	44	303.37	30.458	210	1447.9	127.63	880	...	304.58	2100	...
6.5267	45	310.26	31.908	220	1516.8	129.08	890	...	307.48	2120	...
6.6717	46	317.16	33.359	230	1585.8	130.53	900	...	310.38	2140	...
6.8168	47	324.05	34.809	240	1654.7	131.98	910	...	313.28	2160	...
6.9618	48	330.95	36.259	250	1723.7	133.43	920	...	316.18	2180	...
7.1068	49	337.84	37.710	260	1792.6	134.89	930	...	319.08	2200	...
7.2519	50	344.74	39.160	270	1861.6	136.34	940	...	321.98	2220	...
7.3969	51	351.63	40.611	280	1930.5	137.79	950	...	324.88	2240	...
7.5420	52	358.53	42.061	290	1999.5	139.24	960	...	327.79	2260	...
7.6870	53	365.42	43.511	300	2068.4	140.69	970	...	330.69	2280	...
7.8320	54	372.32	44.962	310	2137.4	142.14	980	...	333.59	2300	...
7.9771	55	379.21	46.412	320	2206.3	143.59	990	...	336.49	2320	...
8.1221	56	386.11	47.862	330	2275.3	145.04	1000	...	339.39	2340	...
8.2672	57	393.00	49.313	340	2344.2	147.94	1020	...	342.29	2360	...
8.4122	58	399.90	50.763	350	2413.2	150.84	1040	...	345.19	2380	...
8.5572	59	406.79	52.214	360	2482.1	153.74	1060	...	348.09	2400	...
8.7023	60	413.69	53.664	370	2551.1	156.64	1080	...	350.99	2420	...
8.8473	61	420.58	55.114	380	2620.0	159.54	1100	...	353.89	2440	...
8.9923	62	427.47	56.565	390	2689.0	162.44	1120	...	356.79	2460	...
9.1374	63	434.37	58.015	400	2757.9	165.34	1140	...	359.69	2480	...
9.2824	64	441.26	59.465	410	2826.9	168.24	1160	...	362.59	2500	...
9.4275	65	448.16	60.916	420	2895.8	171.14	1180	...			
9.5725	66	455.05	62.366	430	2964.7	174.05	1200	...			
9.7175	67	461.95	63.817	440	3033.7	176.95	1220	...			

Temperature Conversions

The general arrangement of this conversion table was devised by Sauveur and Boylston. The middle columns of numbers (in **boldface** type) contain the temperature readings (°F or °C) to be converted. When converting from degrees Fahrenheit to degrees Celsius, read the Celsius equivalent in the column headed "C". When converting from Celsius to Fahrenheit, read the Fahrenheit equivalent in the column headed "F".

F		C	F		C	F		C	F		C	F		C
.....	**−458**	−272.22	.....	**−308**	−188.89	−252.4	**−158**	−105.56	+17.6	**−8**	−22.22	287.6	**142**	61.11
.....	**−456**	−271.11	.....	**−306**	−187.78	−248.8	**−156**	−104.44	+21.2	**−6**	−21.11	291.2	**144**	62.22
.....	**−454**	−270.00	.....	**−304**	−186.67	−245.2	**−154**	−103.33	+24.8	**−4**	−20.00	294.8	**146**	63.33
.....	**−452**	−268.89	.....	**−302**	−185.56	−241.6	**−152**	−102.22	+28.4	**−2**	−18.89	298.4	**148**	64.44
.....	**−450**	−267.78	.....	**−300**	−184.44	−238.0	**−150**	−101.11	+32.0	**±0**	−17.78	302.0	**150**	65.56
.....	**−448**	−266.67	.....	**−298**	−183.33	−234.4	**−148**	−100.00	+35.6	**+2**	−16.67	305.6	**152**	66.67
.....	**−446**	−265.56	.....	**−296**	−182.22	−230.8	**−146**	−98.89	+39.2	**+4**	−15.56	309.2	**154**	67.78
.....	**−444**	−264.44	.....	**−294**	−181.11	−227.2	**−144**	−97.78	+42.8	**+6**	−14.44	312.8	**156**	68.89
.....	**−442**	−263.33	.....	**−292**	−180.00	−223.6	**−142**	−96.67	+46.4	**+8**	−13.33	316.4	**158**	70.00
.....	**−440**	−262.22	.....	**−290**	−178.89	−220.0	**−140**	−95.56	+50.0	**+10**	−12.22	320.0	**160**	71.11
.....	**−438**	−261.11	.....	**−288**	−177.78	−216.4	**−138**	−94.44	+53.6	**+12**	−11.11	323.6	**162**	72.22
.....	**−436**	−260.00	.....	**−286**	−176.67	−212.8	**−136**	−93.33	+57.2	**+14**	−10.00	327.2	**164**	73.33
.....	**−434**	−258.89	.....	**−284**	−175.56	−209.2	**−134**	−92.22	+60.8	**+16**	−8.89	330.8	**166**	74.44
.....	**−432**	−257.78	.....	**−282**	−174.44	−205.6	**−132**	−91.11	+64.4	**+18**	−7.78	334.4	**168**	75.56
.....	**−430**	−256.67	.....	**−280**	−173.33	−202.0	**−130**	−90.00	+68.0	**+20**	−6.67	338.0	**170**	76.67
.....	**−428**	−255.56	.....	**−278**	−172.22	−198.4	**−128**	−88.89	+71.6	**+22**	−5.56	341.6	**172**	77.78
.....	**−426**	−254.44	.....	**−276**	−171.11	−194.8	**−126**	−87.78	+75.2	**+24**	−4.44	345.2	**174**	78.89
.....	**−424**	−253.33	.....	**−274**	−170.00	−191.2	**−124**	−86.67	+78.8	**+26**	−3.33	348.8	**176**	80.00
.....	**−422**	−252.22	−457.6	**−272**	−168.89	−187.6	**−122**	−85.56	+82.4	**+28**	−2.22	352.4	**178**	81.11
.....	**−420**	−251.11	−454.0	**−270**	−167.78	−184.0	**−120**	−84.44	+86.0	**+30**	−1.11	356.0	**180**	82.22
.....	**−418**	−250.00	−450.4	**−268**	−166.67	−180.4	**−118**	−83.33	+89.6	**+32**	±0.00	359.6	**182**	83.33
.....	**−416**	−248.89	−446.8	**−266**	−165.56	−176.8	**−116**	−82.22	+93.2	**+34**	+1.11	363.2	**184**	84.44
.....	**−414**	−247.78	−443.2	**−264**	−164.44	−173.2	**−114**	−81.11	+96.8	**+36**	+2.22	366.8	**186**	85.56
.....	**−412**	−246.67	−439.6	**−262**	−163.33	−169.6	**−112**	−80.00	+100.4	**+38**	+3.33	370.4	**188**	86.67
.....	**−410**	−245.56	−436.0	**−260**	−162.22	−166.0	**−110**	−78.89	+104.0	**+40**	+4.44	374.0	**190**	87.78
.....	**−408**	−244.44	−432.4	**−258**	−161.11	−162.4	**−108**	−77.78	107.6	**42**	5.56	377.6	**192**	88.89
.....	**−406**	−243.33	−428.8	**−256**	−160.00	−158.8	**−106**	−76.67	111.2	**44**	6.67	381.2	**194**	90.00
.....	**−404**	−242.22	−425.2	**−254**	−158.89	−155.2	**−104**	−75.56	114.8	**46**	7.78	384.8	**196**	91.11
.....	**−402**	−241.11	−421.6	**−252**	−157.78	−151.6	**−102**	−74.44	118.4	**48**	8.89	388.4	**198**	92.22
.....	**−400**	−240.00	−418.0	**−250**	−156.67	−148.0	**−100**	−73.33	122.0	**50**	10.00	392.0	**200**	93.33
.....	**−398**	−238.89	−414.4	**−248**	−155.56	−144.4	**−98**	−72.22	125.6	**52**	11.11	395.6	**202**	94.44
.....	**−396**	−237.78	−410.8	**−246**	−154.44	−140.8	**−96**	−71.11	129.2	**54**	12.22	399.2	**204**	95.56
.....	**−394**	−236.67	−407.2	**−244**	−153.33	−137.2	**−94**	−70.00	132.8	**56**	13.33	402.8	**206**	96.67
.....	**−392**	−235.56	−403.6	**−242**	−152.22	−133.6	**−92**	−68.89	136.4	**58**	14.44	406.4	**208**	97.78
.....	**−390**	−234.44	−400.0	**−240**	−151.11	−130.0	**−90**	−67.78	140.0	**60**	15.56	410.0	**210**	98.89
.....	**−388**	−233.33	−396.4	**−238**	−150.00	−126.4	**−88**	−66.67	143.6	**62**	16.67	413.6	**212**	100.00
.....	**−386**	−232.22	−392.8	**−236**	−148.89	−122.8	**−86**	−65.56	147.2	**64**	17.78	417.2	**214**	101.11
.....	**−384**	−231.11	−389.2	**−234**	−147.78	−119.2	**−84**	−64.44	150.8	**66**	18.89	420.8	**216**	102.22
.....	**−382**	−230.00	−385.6	**−232**	−146.67	−115.6	**−82**	−63.33	154.4	**68**	20.00	424.4	**218**	103.33
.....	**−380**	−228.89	−382.0	**−230**	−145.56	−112.0	**−80**	−62.22	158.0	**70**	21.11	428.0	**220**	104.44
.....	**−378**	−227.78	−378.4	**−228**	−144.44	−108.4	**−78**	−61.11	161.6	**72**	22.22	431.6	**222**	105.56
.....	**−376**	−226.67	−374.8	**−226**	−143.33	−104.8	**−76**	−60.00	165.2	**74**	23.33	435.2	**224**	106.67
.....	**−374**	−225.56	−371.2	**−224**	−142.22	−101.2	**−74**	−58.89	168.8	**76**	24.44	438.8	**226**	107.78
.....	**−372**	−224.44	−367.6	**−222**	−141.11	−97.6	**−72**	−57.78	172.4	**78**	25.56	442.4	**228**	108.89
.....	**−370**	−223.33	−364.0	**−220**	−140.00	−94.0	**−70**	−56.67	176.0	**80**	26.67	446.0	**230**	110.00
.....	**−368**	−222.22	−360.4	**−218**	−138.89	−90.4	**−68**	−55.56	179.6	**82**	27.78	449.6	**232**	111.11
.....	**−366**	−221.11	−356.8	**−216**	−137.78	−86.8	**−66**	−54.44	183.2	**84**	28.89	453.2	**234**	112.22
.....	**−364**	−220.00	−353.2	**−214**	−136.67	−83.2	**−64**	−53.33	186.8	**86**	30.00	456.8	**236**	113.33
.....	**−362**	−218.89	−349.6	**−212**	−135.56	−79.6	**−62**	−52.22	190.4	**88**	31.11	460.4	**238**	114.44
.....	**−360**	−217.78	−346.0	**−210**	−134.44	−76.0	**−60**	−51.11	194.0	**90**	32.22	464.0	**240**	115.56
.....	**−358**	−216.67	−342.4	**−208**	−133.33	−72.4	**−58**	−50.00	197.6	**92**	33.33	467.6	**242**	116.67
.....	**−356**	−215.56	−338.8	**−206**	−132.22	−68.8	**−56**	−48.89	201.2	**94**	34.44	471.2	**244**	117.78
.....	**−354**	−214.44	−335.2	**−204**	−131.11	−65.2	**−54**	−47.78	204.8	**96**	35.56	474.8	**246**	118.89
.....	**−352**	−213.33	−331.6	**−202**	−130.00	−61.6	**−52**	−46.67	208.4	**98**	36.67	478.4	**248**	120.00
.....	**−350**	−212.22	−328.0	**−200**	−128.89	−58.0	**−50**	−45.56	212.0	**100**	37.78	482.0	**250**	121.11
.....	**−348**	−211.11	−324.4	**−198**	−127.78	−54.4	**−48**	−44.44	215.6	**102**	38.89	485.6	**252**	122.22
.....	**−346**	−210.00	−320.8	**−196**	−126.67	−50.8	**−46**	−43.33	219.2	**104**	40.00	489.2	**254**	123.33
.....	**−344**	−208.89	−317.2	**−194**	−125.56	−47.2	**−44**	−42.22	222.8	**106**	41.11	492.8	**256**	124.44
.....	**−342**	−207.78	−313.6	**−192**	−124.44	−43.6	**−42**	−41.11	226.4	**108**	42.22	496.4	**258**	125.56
.....	**−340**	−206.67	−310.0	**−190**	−123.33	−40.0	**−40**	−40.00	230.0	**110**	43.33	500.0	**260**	126.67
.....	**−338**	−205.56	−306.4	**−188**	−122.22	−36.4	**−38**	−38.89	233.6	**112**	44.44	503.6	**262**	127.78
.....	**−336**	−204.44	−302.8	**−186**	−121.11	−32.8	**−36**	−37.78	237.2	**114**	45.56	507.2	**264**	128.89
.....	**−334**	−203.33	−299.2	**−184**	−120.00	−29.2	**−34**	−36.67	240.8	**116**	46.67	510.8	**266**	130.00
.....	**−332**	−202.22	−295.6	**−182**	−118.89	−25.6	**−32**	−35.56	244.4	**118**	47.78	514.4	**268**	131.11
.....	**−330**	−201.11	−292.0	**−180**	−117.78	−22.0	**−30**	−34.44	248.0	**120**	48.89	518.0	**270**	132.22
.....	**−328**	−200.00	−288.4	**−178**	−116.67	−18.4	**−28**	−33.33	251.6	**122**	50.00	521.6	**272**	133.33
.....	**−326**	−198.89	−284.8	**−176**	−115.56	−14.8	**−26**	−32.22	255.2	**124**	51.11	525.2	**274**	134.44
.....	**−324**	−197.78	−281.2	**−174**	−114.44	−11.2	**−24**	−31.11	258.8	**126**	52.22	528.8	**276**	135.56
.....	**−322**	−196.67	−277.6	**−172**	−113.33	−7.6	**−22**	−30.00	262.4	**128**	53.33	532.4	**278**	136.67
.....	**−320**	−195.56	−274.0	**−170**	−112.22	−4.0	**−20**	−28.89	266.0	**130**	54.44	536.0	**280**	137.78
.....	**−318**	−194.44	−270.4	**−168**	−111.11	−0.4	**−18**	−27.78	269.6	**132**	55.56	539.6	**282**	138.89
.....	**−316**	−193.33	−266.8	**−166**	−110.00	+3.2	**−16**	−26.67	273.2	**134**	56.67	543.2	**284**	140.00
.....	**−314**	−192.22	−263.2	**−164**	−108.89	+6.8	**−14**	−25.56	276.8	**136**	57.78	546.8	**286**	141.11
.....	**−312**	−191.11	−259.6	**−162**	−107.78	+10.4	**−12**	−24.44	280.4	**138**	58.89	550.4	**288**	142.22
.....	**−310**	−190.00	−256.0	**−160**	−106.67	+14.0	**−10**	−23.33	284.0	**140**	60.00	554.0	**290**	143.33

F		C	F		C	F		C	F		C	F		C
557.6	292	144.44	870.8	466	241.11	1832.0	1000	537.78	3398.0	1870	1021.1	4964.0	2740	1504.4
561.2	294	145.56	874.4	468	242.22	1850.0	1010	543.33	3416.0	1880	1026.7	4982.0	2750	1510.0
564.8	296	146.67	878.0	470	243.33	1868.0	1020	548.89	3434.0	1890	1032.2	5000.0	2760	1515.6
568.4	298	147.78	881.6	472	244.44	1886.0	1030	554.44	3452.0	1900	1037.8	5018.0	2770	1521.1
572.0	300	148.89	885.2	474	245.56	1904.0	1040	560.00	3470.0	1910	1043.3	5036.0	2780	1526.7
575.6	302	150.00	888.8	476	246.67	1922.0	1050	565.56	3488.0	1920	1048.9	5054.0	2790	1532.2
579.2	304	151.11	892.4	478	247.78	1940.0	1060	571.11	3506.0	1930	1054.4	5072.0	2800	1537.8
582.8	306	152.22	896.0	480	248.89	1958.0	1070	576.67	3524.0	1940	1060.0	5090.0	2810	1543.3
586.4	308	153.33	899.6	482	250.00	1976.0	1080	582.22	3542.0	1950	1065.6	5108.0	2820	1548.9
590.0	310	154.44	903.2	484	251.11	1994.0	1090	587.78	3560.0	1960	1071.1	5126.0	2830	1554.4
593.6	312	155.56	906.8	486	252.22	2012.0	1100	593.33	3578.0	1970	1076.7	5144.0	2840	1560.0
597.2	314	156.67	910.4	488	253.33	2030.0	1110	598.89	3596.0	1980	1082.2	5162.0	2850	1565.6
600.8	316	157.78	914.0	490	254.44	2048.0	1120	604.44	3614.0	1990	1087.8	5180.0	2860	1571.1
604.4	318	158.89	917.6	492	255.56	2066.0	1130	610.00	3632.0	2000	1093.3	5198.0	2870	1576.7
608.0	320	160.00	921.2	494	256.67	2084.0	1140	615.56	3650.0	2010	1098.9	5216.0	2880	1582.2
611.6	322	161.11	924.8	496	257.78	2102.0	1150	621.11	3668.0	2020	1104.4	5234.0	2890	1587.8
615.2	324	162.22	928.4	498	258.89	2120.0	1160	626.67	3686.0	2030	1110.0	5252.0	2900	1593.3
618.8	326	163.33	932.0	500	260.00	2138.0	1170	632.22	3704.0	2040	1115.6	5270.0	2910	1598.9
622.4	328	164.44	935.6	502	261.11	2156.0	1180	637.78	3722.0	2050	1121.1	5288.0	2920	1604.4
626.0	330	165.56	939.2	504	262.22	2174.0	1190	643.33	3740.0	2060	1126.7	5306.0	2930	1610.0
629.6	332	166.67	942.8	506	263.33	2192.0	1200	648.89	3758.0	2070	1132.2	5324.0	2940	1615.6
633.2	334	167.78	946.4	508	264.44	2210.0	1210	654.44	3776.0	2080	1137.8	5342.0	2950	1621.1
636.8	336	168.89	950.0	510	265.56	2228.0	1220	660.00	3794.0	2090	1143.3	5360.0	2960	1626.7
640.4	338	170.00	953.6	512	266.67	2246.0	1230	665.56	3812.0	2100	1148.9	5378.0	2970	1632.2
644.0	340	171.11	957.2	514	267.78	2264.0	1240	671.11	3830.0	2110	1154.4	5396.0	2980	1637.8
647.6	342	172.22	960.8	516	268.89	2282.0	1250	676.67	3848.0	2120	1160.0	5414.0	2990	1643.3
651.2	344	173.33	964.4	518	270.00	2300.0	1260	682.22	3866.0	2130	1165.6	5432.0	3000	1648.9
654.8	346	174.44	968.0	520	271.11	2318.0	1270	687.78	3884.0	2140	1171.1	5450.0	3010	1654.4
658.4	348	175.56	971.6	522	272.22	2336.0	1280	693.33	3902.0	2150	1176.7	5468.0	3020	1660.0
662.0	350	176.67	975.2	524	273.33	2354.0	1290	698.89	3920.0	2160	1182.2	5486.0	3030	1665.6
665.6	352	177.78	978.8	526	274.44	2372.0	1300	704.44	3938.0	2170	1187.8	5504.0	3040	1671.1
669.2	354	178.89	982.4	528	275.56	2390.0	1310	710.00	3956.0	2180	1193.3	5522.0	3050	1676.7
672.8	356	180.00	986.0	530	276.67	2408.0	1320	715.56	3974.0	2190	1198.9	5540.0	3060	1682.2
676.4	358	181.11	989.6	532	277.78	2426.0	1330	721.11	3992.0	2200	1204.4	5558.0	3070	1687.8
680.0	360	182.22	993.2	534	278.89	2444.0	1340	726.67	4010.0	2210	1210.0	5576.0	3080	1693.3
683.6	362	183.33	996.8	536	280.00	2462.0	1350	732.22	4028.0	2220	1215.6	5594.0	3090	1698.9
687.2	364	184.44	1000.4	538	281.11	2480.0	1360	737.78	4046.0	2230	1221.1	5612.0	3100	1704.4
690.8	366	185.56	1004.0	540	282.22	2498.0	1370	743.33	4064.0	2240	1226.7	5702.0	3150	1732.2
694.4	368	186.67	1007.6	542	283.33	2516.0	1380	748.89	4082.0	2250	1232.2	5792.0	3200	1760.0
698.0	370	187.78	1011.2	544	284.44	2534.0	1390	754.44	4100.0	2260	1237.8	5882.0	3250	1787.7
701.6	372	188.89	1014.8	546	285.56	2552.0	1400	760.00	4118.0	2270	1243.3	5972.0	3300	1815.5
705.2	374	190.00	1018.4	548	286.67	2570.0	1410	765.56	4136.0	2280	1248.9	6062.0	3350	1843.3
708.8	376	191.11	1022.0	550	287.78	2588.0	1420	771.11	4154.0	2290	1254.4	6152.0	3400	1871.1
712.4	378	192.22	1040.0	560	293.33	2606.0	1430	776.67	4172.0	2300	1260.0	6242.0	3450	1898.8
716.0	380	193.33	1058.0	570	298.89	2624.0	1440	782.22	4190.0	2310	1265.6	6332.0	3500	1926.6
719.6	382	194.44	1076.0	580	304.44	2642.0	1450	787.78	4208.0	2320	1271.1	6422.0	3550	1954.4
723.2	384	195.56	1094.0	590	310.00	2660.0	1460	793.33	4226.0	2330	1276.7	6512.0	3600	1982.2
726.8	386	196.67	1112.0	600	315.56	2678.0	1470	798.89	4244.0	2340	1282.2	6602.0	3650	2010.0
730.4	388	197.78	1130.0	610	321.11	2696.0	1480	804.44	4262.0	2350	1287.8	6692.0	3700	2037.7
734.0	390	198.89	1148.0	620	326.67	2714.0	1490	810.00	4280.0	2360	1293.3	6782.0	3750	2065.5
737.6	392	200.00	1166.0	630	332.22	2732.0	1500	815.56	4298.0	2370	1298.9	6872.0	3800	2093.3
741.2	394	201.11	1184.0	640	337.78	2750.0	1510	821.11	4316.0	2380	1304.4	6962.0	3850	2121.1
744.8	396	202.22	1202.0	650	343.33	2768.0	1520	826.67	4334.0	2390	1310.0	7052.0	3900	2148.8
748.4	398	203.33	1220.0	660	348.89	2786.0	1530	832.22	4352.0	2400	1315.6	7142.0	3950	2176.6
752.0	400	204.44	1238.0	670	354.44	2804.0	1540	837.78	4370.0	2410	1321.1	7232.0	4000	2204.4
755.6	402	205.56	1256.0	680	360.00	2822.0	1550	843.33	4388.0	2420	1326.7	7322.0	4050	2232.2
759.2	404	206.67	1274.0	690	365.56	2840.0	1560	848.89	4406.0	2430	1332.2	7412.0	4100	2260.0
762.8	406	207.78	1292.0	700	371.11	2858.0	1570	854.44	4424.0	2440	1337.8	7502.0	4150	2287.7
766.4	408	208.89	1310.0	710	376.67	2876.0	1580	860.00	4442.0	2450	1343.3	7592.0	4200	2315.5
770.0	410	210.00	1328.0	720	382.22	2894.0	1590	865.56	4460.0	2460	1348.9	7682.0	4250	2343.3
773.6	412	211.11	1346.0	730	387.78	2912.0	1600	871.11	4478.0	2470	1354.4	7772.0	4300	2371.1
777.2	414	212.22	1364.0	740	393.33	2930.0	1610	876.67	4496.0	2480	1360.0	7862.0	4350	2398.8
780.8	416	213.33	1382.0	750	398.89	2948.0	1620	882.22	4514.0	2490	1365.6	7952.0	4400	2426.6
784.4	418	214.44	1400.0	760	404.44	2966.0	1630	887.78	4532.0	2500	1371.1	8042.0	4450	2454.4
788.0	420	215.56	1418.0	770	410.00	2984.0	1640	893.33	4550.0	2510	1376.7	8132.0	4500	2482.2
791.6	422	216.67	1436.0	780	415.56	3002.0	1650	898.89	4568.0	2520	1382.2	8222.0	4550	2510.0
795.2	424	217.78	1454.0	790	421.11	3020.0	1660	904.44	4586.0	2530	1387.8	8312.0	4600	2537.7
798.8	426	218.89	1472.0	800	426.67	3038.0	1670	910.00	4604.0	2540	1393.3	8402.0	4650	2565.5
802.4	428	220.00	1490.0	810	432.22	3056.0	1680	915.56	4622.0	2550	1398.9	8492.0	4700	2593.3
806.0	430	221.11	1508.0	820	437.78	3074.0	1690	921.11	4640.0	2560	1404.4	8582.0	4750	2621.1
809.6	432	222.22	1526.0	830	443.33	3092.0	1700	926.67	4658.0	2570	1410.0	8672.0	4800	2648.8
813.2	434	223.33	1544.0	840	448.89	3110.0	1710	932.22	4676.0	2580	1415.6	8762.0	4850	2676.6
816.8	436	224.44	1562.0	850	454.44	3128.0	1720	937.78	4694.0	2590	1421.1	8852.0	4900	2704.4
820.4	438	225.56	1580.0	860	460.00	3146.0	1730	943.33	4712.0	2600	1426.7	8942.0	4950	2732.2
824.0	440	226.67	1598.0	870	465.56	3164.0	1740	948.89	4730.0	2610	1432.2	9032.0	5000	2760.0
827.6	442	227.78	1616.0	880	471.11	3182.0	1750	954.44	4748.0	2620	1437.8	9122.0	5050	2787.7
831.2	444	228.89	1634.0	890	476.67	3200.0	1760	960.00	4766.0	2630	1443.3	9212.0	5100	2815.5
834.8	446	230.00	1652.0	900	482.22	3218.0	1770	965.56	4784.0	2640	1448.9	9302.0	5150	2843.3
838.4	448	231.11	1670.0	910	487.78	3236.0	1780	971.11	4802.0	2650	1454.4	9392.0	5200	2871.1
842.0	450	232.22	1688.0	920	493.33	3254.0	1790	976.67	4820.0	2660	1460.0	9482.0	5250	2898.8
845.6	452	233.33	1706.0	930	498.89	3272.0	1800	982.22	4838.0	2670	1465.6	9572.0	5300	2926.6
849.2	454	234.44	1724.0	940	504.44	3290.0	1810	987.78	4856.0	2680	1471.1	9662.0	5350	2954.4
852.8	456	235.56	1742.0	950	510.00	3308.0	1820	993.33	4874.0	2690	1476.7	9752.0	5400	2982.2
856.4	458	236.67	1760.0	960	515.56	3326.0	1830	998.89	4892.0	2700	1482.2	9842.0	5450	3010.0
860.0	460	237.78	1778.0	970	521.11	3344.0	1840	1004.4	4910.0	2710	1487.8	9932.0	5500	3037.7
863.6	462	238.89	1796.0	980	526.67	3362.0	1850	1010.0	4928.0	2720	1493.3	10022.0	5550	3065.5
867.2	464	240.00	1814.0	990	532.22	3380.0	1860	1015.6	4946.0	2730	1498.9	10112.0	5600	3093.3

Approximate Equivalent Hardness Numbers and Tensile Strengths for Vickers Hardness Numbers for Steel(a)

Vickers hardness No.	Brinell hardness No., 3000-kg load, 10-mm ball — Standard ball	Brinell — Tungsten carbide ball	Rockwell A scale, 60-kg load, Brale indenter	Rockwell B scale, 100-kg load, 1/16-in.-diam ball	Rockwell C scale, 150-kg load, Brale indenter	Rockwell D scale, 100-kg load, Brale indenter	Rockwell superficial 15N scale, 15-kg load	Rockwell superficial 30N scale, 30-kg load	Rockwell superficial 45N scale, 45-kg load	Knoop hardness No., 500-g load and greater	Shore Scleroscope hardness No.	Tensile strength (approx), 1000 psi	Vickers hardness No.
940	...	...	85.6	...	68.0	76.9	93.2	84.4	75.4	920	97	...	940
920	...	...	85.3	...	67.5	76.5	93.0	84.0	74.8	908	96	...	920
900	...	...	85.0	...	67.0	76.1	92.9	83.6	74.2	895	95	...	900
880	...	(767)	84.7	...	66.4	75.7	92.7	83.1	73.6	882	93	...	880
860	...	(757)	84.4	...	65.9	75.3	92.5	82.7	73.1	867	92	...	860
840	...	(745)	84.1	...	65.3	74.8	92.3	82.2	72.2	852	91	...	840
820	...	(733)	83.8	...	64.7	74.3	92.1	81.7	71.8	837	90	...	820
800	...	(722)	83.4	...	64.0	73.8	91.8	81.1	71.0	822	88	...	800
780	...	(710)	83.0	...	63.3	73.3	91.5	80.4	70.2	806	87	...	780
760	...	(698)	82.6	...	62.5	72.6	91.2	79.7	69.4	788	86	...	760
740	...	(684)	82.2	...	61.8	72.1	91.0	79.1	68.6	772	84	...	740
720	...	(670)	81.8	...	61.0	71.5	90.7	78.4	67.7	754	83	...	720
700	...	(656)	81.3	...	60.1	70.8	90.3	77.6	66.7	735	81	...	700
690	...	(647)	81.1	...	59.7	70.5	90.1	77.2	66.2	725	...	...	690
680	...	(638)	80.8	...	59.2	70.1	89.8	76.8	65.7	716	80	355	680
670	...	(630)	80.6	...	58.8	69.8	89.7	76.4	65.3	706	...	348	670
660	...	620	80.3	...	58.3	69.4	89.5	75.9	64.7	697	79	342	660
650	...	611	80.0	...	57.8	69.0	89.2	75.5	64.1	687	78	336	650
640	...	601	79.8	...	57.3	68.7	89.0	75.1	63.5	677	77	328	640
630	...	591	79.5	...	56.8	68.3	88.8	74.6	63.0	667	76	323	630
620	...	582	79.2	...	56.3	67.9	88.5	74.2	62.4	657	75	317	620
610	...	573	78.9	...	55.7	67.5	88.2	73.6	61.7	646	...	310	610
600	...	564	78.6	...	55.2	67.0	88.0	73.2	61.2	636	74	303	600
590	...	554	78.4	...	54.7	66.7	87.8	72.7	60.5	625	73	298	590
580	...	545	78.0	...	54.1	66.2	87.5	72.1	59.9	615	72	293	580
570	...	535	77.8	...	53.6	65.8	87.2	71.7	59.3	604	...	288	570
560	...	525	77.4	...	53.0	65.4	86.9	71.2	58.6	594	71	283	560
550	(505)	517	77.0	...	52.3	64.8	86.6	70.5	57.8	583	70	276	550
540	(496)	507	76.7	...	51.7	64.4	86.3	70.0	57.0	572	69	270	540
530	(488)	497	76.4	...	51.1	63.9	86.0	69.5	56.2	561	68	265	530
520	(480)	488	76.1	...	50.5	63.5	85.7	69.0	55.6	550	67	260	520
510	(473)	479	75.7	...	49.8	62.9	85.4	68.3	54.7	539	...	254	510
500	(465)	471	75.3	...	49.1	62.2	85.0	67.7	53.9	528	66	247	500
490	(456)	460	74.9	...	48.4	61.6	84.7	67.1	53.1	517	65	241	490
480	(448)	452	74.5	...	47.7	61.3	84.3	66.4	52.2	505	64	235	480
470	441	442	74.1	...	46.9	60.7	83.9	65.7	51.3	494	...	228	470
460	433	433	73.6	...	46.1	60.1	83.6	64.9	50.4	482	62	223	460
450	425	425	73.3	...	45.3	59.4	83.2	64.3	49.4	471	...	217	450
440	415	415	72.8	...	44.5	58.8	82.8	63.5	48.4	459	59	212	440
430	405	405	72.3	...	43.6	58.2	82.3	62.7	47.4	447	58	205	430
420	397	397	71.8	...	42.7	57.5	81.8	61.9	46.4	435	57	199	420
410	388	388	71.4	...	41.8	56.8	81.4	61.1	45.3	423	56	193	410
400	379	379	70.8	...	40.8	56.0	80.8	60.2	44.1	412	55	187	400
390	369	369	70.3	...	39.8	55.2	80.3	59.3	42.9	400	...	181	390
380	360	360	69.8	(110.0)	38.8	54.4	79.8	58.4	41.7	389	52	175	380
370	350	350	69.2	...	37.7	53.6	79.2	57.4	40.4	378	51	170	370
360	341	341	68.7	(109.0)	36.6	52.8	78.6	56.4	39.1	367	50	164	360
350	331	331	68.1	...	35.5	51.9	78.0	55.4	37.8	356	48	159	350
340	322	322	67.6	(108.0)	34.4	51.1	77.4	54.4	36.5	346	47	155	340
330	313	313	67.0	...	33.3	50.2	76.8	53.6	35.2	337	46	150	330
320	303	303	66.4	(107.0)	32.2	49.4	76.2	52.3	33.9	328	45	146	320
310	294	294	65.8	...	31.0	48.4	75.6	51.3	32.5	318	...	142	310
300	284	284	65.2	(105.5)	29.8	47.5	74.9	50.2	31.1	309	42	138	300
295	280	280	64.8	...	29.2	47.1	74.6	49.7	30.4	305	...	136	295
290	275	275	64.5	(104.5)	28.5	46.5	74.2	49.0	29.5	300	41	133	290
285	270	270	64.2	...	27.8	46.0	73.8	48.4	28.7	296	...	131	285
280	265	265	63.8	(103.5)	27.1	45.3	73.4	47.8	27.9	291	40	129	280
275	261	261	63.5	...	26.4	44.9	73.0	47.2	27.1	286	39	127	275
270	256	256	63.1	(102.0)	25.6	44.3	72.6	46.4	26.2	282	38	124	270
265	252	252	62.7	...	24.8	43.7	72.1	45.7	25.2	277	...	122	265
260	247	247	62.4	(101.0)	24.0	43.1	71.6	45.0	24.3	272	37	120	260
255	243	243	62.0	...	23.1	42.2	71.1	44.2	23.2	267	...	117	255
250	238	238	61.6	99.5	22.2	41.7	70.6	43.4	22.2	262	36	115	250
245	233	233	61.2	...	21.3	41.1	70.1	42.5	21.1	258	35	113	245
240	228	228	60.7	98.1	20.3	40.3	69.6	41.7	19.9	253	34	111	240
230	219	219	...	96.7	(18.0)	...	...	...	...	243	33	106	230
220	209	209	...	95.0	(15.7)	...	...	...	...	234	32	101	220
210	200	200	...	93.4	(13.4)	...	...	...	...	226	30	97	210
200	190	190	...	91.5	(11.0)	...	...	...	...	216	29	92	200
190	181	181	...	89.5	(8.5)	...	...	...	...	206	28	88	190
180	171	171	...	87.1	(6.0)	...	...	...	...	196	26	84	180
170	162	162	...	85.0	(3.0)	...	...	...	...	185	25	79	170
160	152	152	...	81.7	(0.0)	...	...	...	...	175	23	75	160
150	143	143	...	78.7	...	...	...	...	...	164	22	71	150
140	133	133	...	75.0	...	...	...	...	...	154	21	66	140
130	124	124	...	71.2	...	...	...	...	...	143	20	62	130
120	114	114	...	66.7	...	...	...	...	...	133	18	57	120
110	105	105	...	62.3	...	...	...	...	...	123	...	...	110
100	95	95	...	56.2	...	...	...	...	...	112	...	...	100
95	90	90	...	52.0	...	...	...	...	...	107	...	...	95
90	86	86	...	48.0	...	...	...	...	...	102	...	...	90
85	81	81	...	41.0	...	...	...	...	...	97	...	...	85

(a) For carbon and alloy steels in the annealed, normalized, and quenched-and-tempered conditions; less accurate for cold worked condition and for austenitic steels. The values in **boldface type** correspond to the values in the joint SAE-ASM-ASTM hardness conversions as printed in ASTM E140, Table 1. The values in parentheses are beyond normal range and are given for information only.

Approximate Equivalent Hardness Numbers and Tensile Strengths for Brinell Hardness Numbers for Steel(a)

Brinell indentation diam, mm	Brinell hardness No.(b), 3000 kg load, 10-mm ball — Standard ball	Tungsten carbide ball	Vickers hardness No.	Rockwell A scale, 60-kg load, Brale indenter	Rockwell B scale, 100-kg load, 1/16-in. diam ball	Rockwell C scale, 150-kg load, Brale indenter	Rockwell D scale, 100-kg load, Brale indenter	15N scale, 15-kg load	30N scale, 30-kg load	45N scale, 45-kg load	Knoop hardness No., 500-g load and greater	Shore Scleroscope hardness No.	Tensile strength (approx), 1000 psi	Brinell indentation diam, mm
2.25	...	(745)	840	84.1	...	65.3	74.8	92.3	82.2	72.2	852	91	...	2.25
2.30	...	(712)	783	83.1	...	63.4	73.4	91.6	80.5	70.4	808	...	...	2.30
2.35	...	(682)	737	82.2	...	61.7	72.0	91.0	79.0	68.5	768	84	...	2.35
2.40	...	(653)	697	81.2	...	60.0	70.7	90.2	77.5	66.5	732	81	...	2.40
2.45	...	627	667	80.5	...	58.7	69.7	89.6	76.3	65.1	703	79	347	2.45
2.50	...	601	640	79.8	...	57.3	68.7	89.0	75.1	63.5	677	77	328	2.50
2.55	...	578	615	79.1	...	56.0	67.7	88.4	73.9	62.1	652	75	313	2.55
2.60	...	555	591	78.4	...	54.7	66.7	87.8	72.7	60.6	626	73	298	2.60
2.65	...	534	569	77.8	...	53.5	65.8	87.2	71.6	59.2	604	71	288	2.65
2.70	...	514	547	76.9	...	52.1	64.7	86.5	70.3	57.6	579	70	273	2.70
2.75	(495)	...	539	76.7	...	51.6	64.3	86.3	69.9	56.9	571	...	269	2.75
2.75	...	495	528	76.3	...	51.0	63.8	85.9	69.4	56.1	558	68	263	
2.80	(477)	...	516	75.9	...	50.3	63.2	85.6	68.7	55.2	545	...	257	2.80
2.80	...	477	508	75.6	...	49.6	62.7	85.3	68.2	54.5	537	66	252	
2.85	(461)	...	495	75.1	...	48.8	61.9	84.9	67.4	53.5	523	...	244	2.85
2.85	...	461	491	74.9	...	48.5	61.7	84.7	67.2	53.2	518	65	242	
2.90	444	...	474	74.3	...	47.2	61.0	84.1	66.0	51.7	499	...	231	2.90
2.90	...	444	472	74.2	...	47.1	60.8	84.0	65.8	51.5	496	63	229	
2.95	429	429	455	73.4	...	45.7	59.7	83.4	64.6	49.9	476	61	220	2.95
3.00	415	415	440	72.8	...	44.5	58.8	82.8	63.5	48.4	459	59	212	3.00
3.05	401	401	425	72.0	...	43.1	57.8	82.0	62.3	46.9	441	58	202	3.05
3.10	388	388	410	71.4	...	41.8	56.8	81.4	61.1	45.3	423	56	193	3.10
3.15	375	375	396	70.6	...	40.4	55.7	80.6	59.9	43.6	407	54	184	3.15
3.20	363	363	383	70.0	...	39.1	54.6	80.0	58.7	42.0	392	52	177	3.20
3.25	352	352	372	69.3	(110.0)	37.9	53.8	79.3	57.6	40.5	379	51	172	3.25
3.30	341	341	360	68.7	(109.0)	36.6	52.8	78.6	56.4	39.1	367	50	164	3.30
3.35	331	331	350	68.1	(108.5)	35.5	51.9	78.0	55.4	37.8	356	48	159	3.35
3.40	321	321	339	67.5	(108.0)	34.3	51.0	77.3	54.3	36.4	345	47	154	3.40
3.45	311	311	328	66.9	(107.5)	33.1	50.0	76.7	53.3	34.4	336	46	149	3.45
3.50	302	302	319	66.3	(107.0)	32.1	49.3	76.1	52.2	33.8	327	45	146	3.50
3.55	293	293	309	65.7	(106.0)	30.9	48.3	75.5	51.2	32.4	318	43	142	3.55
3.60	285	285	301	65.3	(105.5)	29.9	47.6	75.0	50.3	31.2	310	42	138	3.60
3.65	277	277	292	64.6	(104.5)	28.8	46.7	74.4	49.3	29.9	302	41	134	3.65
3.70	269	269	284	64.1	(104.0)	27.6	45.9	73.7	48.3	28.5	294	40	131	3.70
3.75	262	262	276	63.6	(103.0)	26.6	45.0	73.1	47.3	27.3	286	39	127	3.75
3.80	255	255	269	63.0	(102.0)	25.4	44.2	72.5	46.2	26.0	279	38	123	3.80
3.85	248	248	261	62.5	(101.0)	24.2	43.2	71.7	45.1	24.5	272	37	120	3.85
3.90	241	241	253	61.8	100.0	22.8	42.0	70.9	43.9	22.8	265	36	116	3.90
3.95	235	235	247	61.4	99.0	21.7	41.4	70.3	42.9	21.5	259	35	114	3.95
4.00	229	229	241	60.8	98.2	20.5	40.5	69.7	41.9	20.1	253	34	111	4.00
4.05	223	223	234	...	97.3	(19.0)	...	...	...	...	247	...	107	4.05
4.10	217	217	228	...	96.4	(17.7)	...	...	...	...	242	33	105	4.10
4.15	212	212	222	...	95.5	(16.4)	...	...	...	...	237	32	102	4.15
4.20	207	207	218	...	94.6	(15.2)	...	...	...	...	232	31	100	4.20
4.25	201	201	212	...	93.7	(13.8)	...	...	...	...	227	...	98	4.25
4.30	197	197	207	...	92.8	(12.7)	...	...	...	...	222	30	95	4.30
4.35	192	192	202	...	91.9	(11.5)	...	...	...	...	217	29	93	4.35
4.40	187	187	196	...	90.9	(10.2)	...	...	...	...	212	...	90	4.40
4.45	183	183	192	...	90.0	(9.0)	...	...	...	...	207	28	89	4.45
4.50	179	179	188	...	89.0	(8.0)	...	...	...	...	202	27	87	4.50
4.55	174	174	182	...	88.0	(6.7)	...	...	...	...	198	...	85	4.55
4.60	170	170	178	...	87.0	(5.4)	...	...	...	...	194	26	83	4.60
4.65	167	167	175	...	86.0	(4.4)	...	...	...	...	190	...	81	4.65
4.70	163	163	171	...	85.0	(3.3)	...	...	...	...	186	25	79	4.70
4.75	159	159	167	...	83.9	(2.0)	...	...	...	...	182	...	78	4.75
4.80	156	156	163	...	82.9	(0.9)	...	...	...	...	178	24	76	4.80
4.85	152	152	159	...	81.9	...	...	...	...	...	174	...	75	4.85
4.90	149	149	156	...	80.8	...	...	...	...	...	170	23	73	4.90
4.95	146	146	153	...	79.7	...	...	...	...	...	166	...	72	4.95
5.00	143	143	150	...	78.6	...	...	...	...	...	163	22	71	5.00
5.10	137	137	143	...	76.4	...	...	...	...	...	157	21	67	5.10
5.20	131	131	137	...	74.2	...	...	...	...	...	151	...	65	5.20
5.30	126	126	132	...	72.0	...	...	...	...	...	145	20	63	5.30
5.40	121	121	127	...	69.8	...	...	...	...	...	140	19	60	5.40
5.50	116	116	122	...	67.6	...	...	...	...	...	135	18	58	5.50
5.60	111	111	117	...	65.4	...	...	...	...	...	131	17	56	5.60

(a) For carbon and alloy steels in the annealed, normalized, and quenched-and-tempered conditions; less accurate for cold worked condition and for austenitic steels. Values in **boldface type** correspond to the values in the joint SAE-ASM-ASTM hardness conversions as printed in ASTM E140, Table 3. Values in parentheses are beyond normal range and are given for information only.

(b) Brinell numbers are based on the diameter of impressed indentation. If the ball distorts (flattens) during test, Brinell numbers will vary in accordance with the degree of such distortion when related to hardnesses determined with a Vickers diamond pyramid, Rockwell Brale, or other indenter that does not sensibly distort. At high hardnesses, therefore, the relationship between Brinell and Vickers or Rockwell scales is affected by the type of ball used. Standard steel balls tend to flatten slightly more than tungsten carbide balls, resulting in a larger indentation and a lower Brinell number than shown by a tungsten carbide ball. Thus, on a specimen of about 539 to 547 HV, a standard ball will leave a 2.75-mm indentation (495 HB), and a tungsten carbide ball a 2.70-mm indentation (514 HB). Conversely, identical indentation diameters for both types of ball will correspond to different Vickers and Rockwell values. Thus, if indentations in two different specimens both are 2.75 mm in diameter (495 HB), the specimen tested with a standard ball has a Vickers hardness of 539 whereas the specimen tested with a tungsten carbide ball has a Vickers hardness of 528.

Approximate Equivalent Hardness Numbers and Tensile Strengths for Rockwell C and B Hardness Numbers for Steel(a)

Rockwell C-Scale Hardness Numbers

Rockwell C-scale hardness No.	Vickers hardness No.	Brinell 3000-kg load, Standard ball	Brinell 3000-kg load, Tungsten carbide ball	Rockwell A-scale 60-kg load Brale indenter	Rockwell B-scale 100-kg load 1/16-in. diam ball	Rockwell D-scale 100-kg load Brale indenter	15N 15-kg load	30N 30-kg load	45N 45-kg load	Knoop hardness No. 500-g load and greater	Shore Scleroscope hardness No.	Tensile strength (approx) 1000 psi	Rockwell C-scale hardness No.
68	940	…	…	85.6	…	76.9	93.2	84.4	75.4	920	97	…	68
67	900	…	…	85.0	…	76.1	92.9	83.6	74.2	895	95	…	67
66	865	…	…	84.5	…	75.4	92.5	82.8	73.3	870	92	…	66
65	832	…	(739)	83.9	…	74.5	92.2	81.9	72.0	846	91	…	65
64	800	…	(722)	83.4	…	73.8	91.8	81.1	71.0	822	88	…	64
63	772	…	(705)	82.8	…	73.0	91.4	80.1	69.9	799	87	…	63
62	746	…	(688)	82.3	…	72.2	91.1	79.3	68.8	776	85	…	62
61	720	…	(670)	81.8	…	71.5	90.7	78.4	67.7	754	83	…	61
60	697	…	(654)	81.2	…	70.7	90.2	77.5	66.6	732	81	…	60
59	674	…	(634)	80.7	…	69.9	89.8	76.6	65.5	710	80	351	59
58	653	…	615	80.1	…	69.2	89.3	75.7	64.3	690	78	338	58
57	633	…	595	79.6	…	68.5	88.9	74.8	63.2	670	76	325	57
56	613	…	577	79.0	…	67.7	88.3	73.9	62.0	650	75	313	56
55	595	…	560	78.5	…	66.9	87.9	73.0	60.9	630	74	301	55
54	577	…	543	78.0	…	66.1	87.4	72.0	59.8	612	72	292	54
53	560	…	525	77.4	…	65.4	86.9	71.2	58.6	594	71	283	53
52	544	(500)	512	76.8	…	64.6	86.4	70.2	57.4	576	69	273	52
51	528	(487)	496	76.3	…	63.8	85.9	69.4	56.1	558	68	264	51
50	513	(475)	481	75.9	…	63.1	85.5	68.5	55.0	542	67	255	50
49	498	(464)	469	75.2	…	62.1	85.0	67.6	53.8	526	66	246	49
48	484	(451)	455	74.7	…	61.4	84.5	66.7	52.5	510	64	238	48
47	471	442	443	74.1	…	60.8	83.9	65.8	51.4	495	63	229	47
46	458	432	432	73.6	…	60.0	83.5	64.8	50.3	480	62	221	46
45	446	421	421	73.1	…	59.2	83.0	64.0	49.0	466	60	215	45
44	434	409	409	72.5	…	58.5	82.5	63.1	47.8	452	58	208	44
43	423	400	400	72.0	…	57.7	82.0	62.2	46.7	438	57	201	43
42	412	390	390	71.5	…	56.9	81.5	61.3	45.5	426	56	194	42
41	402	381	381	70.9	…	56.2	80.9	60.4	44.3	414	55	188	41
40	392	371	371	70.4	…	55.4	80.4	59.5	43.1	402	54	182	40
39	382	362	362	69.9	…	54.6	79.9	58.6	41.9	391	52	177	39
38	372	353	353	69.4	…	53.8	79.4	57.7	40.8	380	51	171	38
37	363	344	344	68.9	…	53.1	78.8	56.8	39.6	370	50	166	37
36	354	336	336	68.4	(109.0)	52.3	78.3	55.9	38.4	360	49	161	36
35	345	327	327	67.9	(108.5)	51.5	77.7	55.0	37.2	351	48	157	35
34	336	319	319	67.4	(108.0)	50.8	77.2	54.2	36.1	342	47	153	34
33	327	311	311	66.8	(107.5)	50.0	76.6	53.3	34.9	334	46	149	33
32	318	301	301	66.3	(107.0)	49.2	76.1	52.1	33.7	326	44	145	32
31	310	294	294	65.8	(106.0)	48.4	75.6	51.3	32.5	318	43	141	31
30	302	286	286	65.3	(105.5)	47.7	75.0	50.4	31.3	311	42	138	30
29	294	279	279	64.7	(104.5)	47.0	74.5	49.5	30.1	304	41	135	29
28	286	271	271	64.3	(104.0)	46.1	73.9	48.6	28.9	297	40	131	28
27	279	264	264	63.8	(103.0)	45.2	73.3	47.7	27.8	290	39	128	27
26	272	258	258	63.3	(102.5)	44.6	72.8	46.8	26.7	284	38	125	26
25	266	253	253	62.8	(101.5)	43.8	72.2	45.9	25.5	278	38	122	25
24	260	247	247	62.4	(101.0)	43.1	71.6	45.0	24.3	272	37	119	24
23	254	243	243	62.0	(100.0)	42.1	71.0	44.0	23.1	266	36	117	23
22	248	237	237	61.5	99.0	41.6	70.5	43.2	22.0	261	35	114	22
21	243	231	231	61.0	98.5	40.9	69.9	42.3	20.7	256	35	112	21

Rockwell B-Scale Hardness Numbers

Rockwell B-scale hardness No.	Vickers hardness No.	Brinell 10-mm ball 500-kg load	Brinell 10-mm ball 3000-kg load	A-scale 60-kg load Brale indenter	C-scale 150-kg load Brale indenter	F-scale 60-kg load 1/16-in. diam ball	15T 15-kg load	30T 30-kg load	45T 45-kg load	Knoop hardness No. 500-g load and greater	Shore Scleroscope hardness No.	Tensile strength (approx) 1000 psi	Rockwell B-scale hardness No.
98	228	189	228	60.2	(19.9)	…	92.5	81.8	70.9	241	34	107	98
97	222	184	222	59.5	(18.6)	…	92.1	81.1	69.9	236	33	104	97
96	216	179	216	58.9	(17.2)	…	91.8	80.4	68.9	231	32	102	96
95	210	175	210	58.3	(15.7)	…	91.5	79.8	67.9	226	…	99	95
94	205	171	205	57.6	(14.3)	…	91.2	79.1	66.9	221	31	97	94
93	200	167	200	57.0	(13.0)	…	90.8	78.4	65.9	216	30	94	93
92	195	163	195	56.4	(11.7)	…	90.5	77.8	64.8	211	…	92	92
91	190	160	190	55.8	(10.4)	…	90.2	77.1	63.8	206	29	90	91
90	185	157	185	55.2	(9.2)	…	89.9	76.4	62.8	201	28	88	90
89	180	154	180	54.6	(8.0)	…	89.5	75.8	61.8	196	27	86	89
88	176	151	176	54.0	(6.9)	…	89.2	75.1	60.8	192	…	84	88
87	172	148	172	53.4	(5.8)	…	88.9	74.4	59.8	188	26	82	87
86	169	145	169	52.8	(4.7)	…	88.6	73.8	58.8	184	26	81	86
85	165	142	165	52.3	(3.6)	…	88.2	73.1	57.8	180	25	79	85
84	162	140	162	51.7	(2.5)	…	87.9	72.4	56.8	176	…	78	84
83	159	137	159	51.1	(1.4)	…	87.6	71.8	55.8	173	24	76	83
82	156	135	156	50.6	(0.3)	…	87.3	71.1	54.8	170	24	75	82
81	153	133	153	50.0		…	86.9	70.4	53.8	167	…	73	81
80	150	130	150	49.5		…	86.6	69.7	52.8	164	23	72	80
79	147	128	147	48.9		…	86.3	69.1	51.8	161	…	70	79
78	144	126	144	48.4		…	86.0	68.4	50.8	158	22	69	78
77	141	124	141	47.9		…	85.6	67.7	49.8	155	22	68	77
76	139	122	139	47.3		…	85.3	67.1	48.8	152	…	67	76
75	137	120	137	46.8		99.6	85.0	66.4	47.8	150	21	66	75
74	135	118	135	46.3		99.1	84.7	65.7	46.8	148	21	65	74
73	132	116	132	45.8		98.5	84.3	65.1	45.8	145	…	64	73
72	130	114	130	45.3		98.0	84.0	64.4	44.8	143	20	63	72
71	127	112	127	44.8		97.4	83.7	63.7	43.8	141	20	62	71
70	125	110	125	44.3		96.8	83.4	63.1	42.8	139	…	61	70
69	123	109	123	43.8		96.2	83.0	62.4	41.8	137	19	60	69
68	121	107	121	43.3		95.6	82.7	61.7	40.8	135	19	59	68
67	119	106	119	42.8		95.1	82.4	61.0	39.8	133	19	58	67
66	117	104	117	42.3		94.5	82.1	60.4	38.7	131	…	57	66
65	116	102	116	41.8		93.9	81.8	59.7	37.7	129	18	56	65
64	114	101	114	41.4		93.4	81.4	59.0	36.7	127	18		64
63	112	99	112	40.9		92.8	81.1	58.4	35.7	125	18		63
62	110	98	110	40.4		92.2	80.8	57.7	34.7	124	…		62
61	108	96	108	40.0		91.7	80.5	57.0	33.7	122	17		61
60	107	95	107	39.5		91.1	80.1	56.4	32.7	120	…		60
59	106	94	106	39.0		90.5	79.8	55.7	31.7	118	…		59
58	104	92	104	38.6		90.0	79.5	55.0	30.7	117	…		58
57	103	91	103	38.1		89.4	79.2	54.4	29.7	115	…		57
56	101	90	101	37.7		88.8	78.8	53.7	28.7	114	…		56
55	100	89	100	37.2		88.2	78.5	53.0	27.7	112	…		55

(a) For carbon and alloy steels in the annealed, normalized, and quenched-and-tempered conditions; less accurate for cold worked condition and for austenitic steels. The values in **boldface type** correspond to the values in the joint SAE-ASM-ASTM hardness conversions as printed in ASTM E140, Table 2. The values in parentheses are beyond normal range and are given for information only.

Physical Properties of Carbon, Low-Alloy, Stainless and Tool Steels

These tabular data have been compiled from previous editions of the Metals Handbook and other sources.

Densities of carbon and low-alloy steels

Nearest AISI-SAE grade	C	Mn	Si	Cr	Ni	Other	Treatment or condition	Density Mg/m³	lb/in.³
1008	0.06	0.38	0.01	...	...	...	Annealed	7.871	0.2844
1024	0.23	0.64	0.11	...	...	...	Annealed	7.859	0.2839
1042	0.44	0.69	0.20	...	...	...	Annealed	7.844	0.2834
(a)	1.22	0.35	0.16	...	...	...	Annealed	7.830	0.2829
5130	0.31	0.74	...	1.00	...	...	Hardened and tempered	7.84	0.283
52100	0.98	0.28	...	1.68	...	...	Annealed	7.81	0.282
(a)	0.51	0.22	...	1.72	3.52	...	Quenched in brine (BQ)	7.79	0.281
							BQ, tempered 190 °C (375 °F)	7.80	0.282
							BQ, tempered 365 °C (690 °F)	7.82	0.283
							BQ, tempered 600 °C (1110 °F)	7.835	0.2831
							Annealed	7.835	0.2831
18Ni250(b)	0.026	0.1	0.11	...	18.5	4.7 Mo; 7.0 Co; 0.22 Ti; 0.003 B;	...	8.0	0.289

(a) No AISI-SAE grade of similar composition. (b) Nominal composition.

Thermal conductivities of carbon and low-alloy steels

Nearest AISI-SAE grade	C	Mn	P	S	Si	Cr	Ni	Mo	Other	Treatment or condition
1008	0.08	0.31	...	...	...	0.045	0.07	0.02	...	Not known
1008	0.06	0.4	...	...	...	...	...	...	...	Annealed
1010	0.10	0.42	0.008	0.028	...	...	...	...	...	Not known
1025	0.23	0.64	...	...	...	Trace	0.074	...	0.13 Cu	Annealed
1042	0.42	0.64	...	...	...	Trace	0.063	...	0.12 Cu	Annealed
1078	0.80	0.32	...	...	...	0.11	0.13	0.01	0.07 Cu	Annealed
(d)	1.22	0.35	...	...	...	0.11	0.13	0.01	0.08 Cu	Annealed
1524	0.23	1.51	0.037	0.038	0.12	0.06	0.04	0.025	0.105 Cu; 0.033 Co; 0.015 Al	Annealed
4037	0.37	1.56	...	...	...	...	...	0.26	...	Hardened and tempered
4130(e)	0.3	0.5	...	...	0.3	0.95	...	0.5	...	Hardened and tempered
4140	0.41	0.67	...	...	...	1.01	...	0.23	...	Hardened and tempered
5132	0.32	0.69	...	...	...	1.09	0.073	0.012	0.07 Cu	Annealed
5140	0.39	0.79	...	...	...	1.03	...	...	...	Hardened and tempered
(d)	0.35	0.59	...	...	...	0.88	0.26	0.20	0.12 Cu	Not known
(d)	0.33	0.55	...	...	...	0.17	3.47	0.04	0.09 Cu	Not known
(d)	0.34	0.55	...	...	...	0.78	3.53	0.39	0.05 Cu	Hardened and tempered
(d)	1.22	13.0	...	...	0.22	0.03	0.07	...	0.07 Cu	Not known
18Ni250(e)	0.026	0.1	...	...	0.11	...	18.5	4.5	7.0 Co; 0.22 Ti; 0.003 B	Not known

Nearest AISI-SAE grade	°C: 0 / °F: 32	100 / 212	200 / 392	300 / 572	400 / 752	500 / 932	600 / 1112	700 / 1292	800 / 1472	1000 / 1832	1200 / 2192
1008	59.5	57.8	53.2	49.4	45.6	41.0	36.8	33.1	28.5	27.6	29.7
1008	65.3(b)	60.3	54.9	...	45.2	...	36.4	...	28.5	27.6	...
1010(c)	65.2	60.2	55.5	50.7	46.0	41.5	36.9	32.9	28.9	...	...
1025	51.9	51.1	49.0	46.1	42.7	39.4	35.6	31.8	26.0	27.2	29.7
1042	51.9	50.7	48.2	45.6	41.9	38.1	33.9	30.1	24.7	26.8	29.7
1078	47.8	48.2	45.2	41.4	38.1	35.2	32.7	30.1	24.3	26.8	30.1
(d)	45.2	44.8	43.5	41.0	38.5	36.0	33.5	31.0	23.9	26.0	28.5
1524	46.0	45.8	45.0	42.6	40.1	37.4	34.4	30.6	26.6	27.2	...
4037	...	48.2	45.6	...	39.4	...	33.9	...	...	...	...
4130(e)	...	42.7	...	40.6	...	37.3	...	31.0	...	28.1	30.1
4140	...	42.7	42.3	...	37.7	...	33.1	...	...	...	...
5132	48.6	46.5	44.4	42.3	38.5	35.6	31.8	28.9	26.0	28.1	30.1
5140	...	44.8	43.5	...	37.7	...	31.4	...	...	...	...
(d)	42.7	42.7	41.9	40.6	38.9	36.4	33.9	31.0	26.4	28.1	30.1
(d)	36.4	37.7	38.9	39.4	36.8	35.2	32.7	26.4	25.1	27.6	30.1
(d)	33.1	33.9	35.2	35.6	35.6	33.5	30.6	28.1	26.8	28.5	30.1
(d)	13.0	13.8	16.3	18.0	19.3	20.5	21.8	22.6	23.4	25.5	28.1
18Ni250(e)	19.7(b)	20.9	...	...	...	...	...	...	...	...	...

(a) To obtain conductivities in Btu/(ft · h · °F), multiply values in table by 0.5778; to obtain conductivities in cal/(cm · s · °C), multiply by 0.002388. (b) Thermal conductivity at 21 °C (70 °F). (c) 70.4 W/m · K at −100 °C (−148 °F). (d) No equivalent grade. (e) Nominal composition.

Specific heats of carbon and low-alloy steels

Nearest AISI-SAE grade	C	Mn	Cr	Ni	Mo	Other	Treatment or condition	Mean apparent specific heat, J/Kg·K, at temperatures (°C) of: 50 to 100	150 to 200	200 to 250	250 to 300	300 to 350	350 to 400	450 to 500	550 to 600	650 to 700	700 to 750	750 to 800	850 to 900
1008	0.06	0.38	...	...	...	...	Annealed	481	519	536	553	574	595	662	754	867	1105	875	846
1008	0.08	0.31	...	...	...	...	Annealed	481	523	544	557	569	595	662	741	858	1139	960	...
1010	0.10	0.42	...	...	...	0.008 P; 0.028 S	Not known	450	500	520	535	565	590	650	730	825	...	...	...
1025	0.23	0.64	...	...	...	...	Annealed	486	519	532	557	574	599	662	749	846	1432	950	...
1042	0.42	0.64	...	...	...	...	Annealed	486	515	528	548	569	586	649	708	770	1583	624	548
1078	0.80	0.32	...	...	...	...	Annealed	490	532	548	565	586	607	670	712	770	2081	615	...
(a)	1.22	0.35	...	...	...	...	Annealed	486	540	544	557	578	599	636	699	816	2089	649	...
1524	0.23	1.51	...	...	...	0.11 Cu	Annealed	477	511	528	544	565	590	649	741	837	1449	821	536
4130(b)	0.3	0.5	0.95	...	0.2	...	Hardened and tempered	477	515	...	544	...	595	657	737	825	...	833	
4140	0.41	0.67	1.01	...	0.23	...	Hardened and tempered	...	473(c)	...	...	...	519(c)	...	561(c)	...	...	...	...
5132	0.32	0.69	1.09	0.073	...	...	Annealed	494	523	536	553	574	595	657	741	837	1499	934	574
5140	0.39	0.79	1.03	...	...	...	Hardened and tempered	452(c)	473(c)	...	...	...	519(c)	...	561(c)	...	...	...	...
(a)	0.35	0.59	0.88	0.26	0.20	...	Annealed	477	515	528	544	569	595	657	737	825	1616	883	...
(a)	0.33	0.55	0.17	3.47	...	...	Not known	481	523	536	548	569	590	662	749	1637	955	603	640
(a)	0.34	0.55	0.78	3.53	0.39	...	Hardened and tempered	486	523	540	557	582	607	670	770	1051	1662	636	636
(a)	0.49	0.90	...	...	...	1.98 Si; 0.64 Cu	Not known	498	523	540	557	578	603	666	749	829	904	1365	...

(a) No equivalent grade. (b) Nominal composition. (c) Value presented is mean value for range of temperatures between room temperature and the higher of the cited temperatures.

Electrical resistivities of carbon and low-alloy steels

Nearest AISI-SAE grade	C	Mn	Si	Cr	Ni	Mo	Other	Treatment or condition
1008	0.06	0.38	...	...	...	...	...	Annealed
1008	0.08	0.31	...	...	...	...	...	Annealed
1025	0.23	0.64	...	...	...	...	...	Annealed
1042	0.42	0.64	...	...	...	...	...	Annealed
1078	0.80	0.32	...	...	...	...	...	Annealed
(a)	1.22	0.35	...	...	...	...	...	Annealed
1524	0.23	1.51	...	...	...	...	0.11 Cu	Not known
4130(b)	0.3	0.5	0.3	0.95	...	0.2	...	Hardened and tempered
4140	0.41	0.67	...	1.01	...	0.23	...	Hardened and tempered
4340	0.41	...	...	1.07	1.43	0.26	...	Hardened and tempered
5132	0.32	0.69	...	1.09	0.073	...	...	Annealed
5140	0.39	0.79	...	1.03	...	...	...	Hardened and tempered
(a)	0.35	0.59	...	0.88	0.26	0.20	...	Annealed
(a)	0.33	0.55	...	0.17	3.47	...	...	Not known
(a)	0.34	0.55	...	0.78	3.53	0.39	...	Hardened and tempered
(a)	0.49	0.90	1.98	...	...	...	0.64 Cu	Not known
18Ni250(b)	0.026	0.1	0.11	...	18.5	4.7	7.0 Co; 0.22 Ti; 0.003 B	Annealed

Nearest AISI-SAE grade	Resistivity, μΩ·m, at indicated temperature °C 20 / °F 68	100 / 212	200 / 392	400 / 752	600 / 1112	700 / 1292	800 / 1472	900 / 1652	1000 / 1832	1100 / 2012	1200 / 2192	1300 / 2372
1008	0.130	0.178	0.252	0.448	0.725	0.898	1.073	1.124	1.160	1.189	1.216	1.241
1008	0.142	0.190	0.263	0.458	0.734	0.905	1.081	1.130	1.165	1.193	1.220	1.244
1025	0.169	0.219	0.292	0.487	0.758	0.925	1.094	1.136	1.167	1.194	1.219	1.239
1042	0.171	0.221	0.296	0.493	0.766	0.932	1.111	1.149	1.179	1.207	1.230	...
1078	0.180	0.232	0.308	0.505	0.772	0.935	1.129	1.164	1.191	1.214	1.231	1.246
(a)	0.196	0.252	0.333	0.540	0.802	0.964	1.152	1.196	1.226	1.249	1.271	1.287
1524	0.208	0.259	0.333	0.523	0.786	0.946	1.103	1.143	1.174	1.202	1.227	1.250
4130	0.223	0.271	0.342	0.529	0.786	...	1.103	...	1.171	...	1.222	...
4140	0.222	0.263	0.326	0.475	0.646	...	...	...	...	...	...	...
4340	0.248	0.298	0.367	0.552	0.797	...	...	...	...	...	...	...
5132	0.210	0.259	0.330	0.517	0.778	0.934	1.106	1.145	1.177	1.205	1.230	1.251
5140	0.228	0.281	0.352	0.530	0.785	...	...	...	...	...	...	...
(a)	0.223	0.271	0.342	0.529	0.786	0.944	1.103	1.138	1.171	1.200	1.222	1.242
(a)	0.271	0.320	0.390	0.567	0.814	0.992	1.122	1.149	1.180	1.204	1.228	1.248
(a)	0.289	0.337	0.406	0.582	0.825	0.994	1.114	1.146	1.176	1.199	1.222	1.242
(a)	0.429	0.470	0.529	0.685	0.911	1.057	1.173	1.197	1.223	1.249	1.271	1.289
18Ni250: Annealed	0.6 to 0.7	...	...	...	...	...	...	...	...	...	...	...
Aged	0.36 to 0.6	...	...	...	...	...	...	...	...	...	...	...

(a) No AISI-SAE standard grade of similar composition. (b) Nominal composition.

Coefficients of linear thermal expansion of carbon and low-alloy steels

Nearest AISI-SAE grade	Treatment or condition	20 °C to: 68 °F to:	Average coefficients of expansion, µm/m · K(a)								
			100 212	200 392	300 572	400 752	500 932	600 1112	700 1292	800 1472	1000 1832
1008	Annealed		12.6(b)	13.1(b)	13.5(b)	13.8(b)	14.2(b)	14.6(b)	15.0(b)	14.7	13.8
1008	Annealed		11.6	12.5	13.0	13.6	14.2	14.6	15.0	. . .	. . .
1010	Annealed		12.2(b)	13.0(b)	13.5(b)	13.9(b)	14.3(b)	14.7(b)	15.0(b)	. . .	. . .
1010	Not known		11.9(c)	12.6	13.3	13.8	14.3	14.7	14.9	14.0	. . .
1015	Rolled		11.9(b)	12.5(b)	13.0(b)	13.6(b)	14.2(b)	. . .	. . .	. . .	. . .
1020	Annealed		11.7	12.1	12.8	13.4	13.9	14.4	14.8	. . .	. . .
1022	Not known		12.5	12.7	. . .	. . .	. . .	. . .	. . .	. . .	. . .
1022	Annealed		12.2(b)	12.7(b)	13.1(b)	13.5(b)	13.9(b)	14.4(b)	14.9(b)	12.6	13.4
1035	Annealed		. . .	12.6	13.3	13.8	14.3	14.8	15.2	. . .	. . .
1035	Annealed		11.1	11.9	12.7	13.4	14.0	14.4	14.8	. . .	. . .
1040	Annealed		11.3	12.0	12.5	13.3	13.9	14.4	14.8	. . .	. . .
1040	Annealed		11.2(b)	12.1(b)	13.0(b)	13.6(b)	14.0(b)	14.6(b)	14.8(b)	11.8	13.6
1045	Annealed		11.6(b)	12.3(b)	13.1(b)	13.7(b)	14.2(b)	14.7(b)	15.1(b)	. . .	. . .
1045	Annealed		11.2(d)	11.9(d)	12.7(d)	13.5(d)	14.1(d)	14.5(d)	14.8(d)	. . .	. . .
1052	Annealed		11.3(d)	11.8(d)	12.7(d)	13.7(d)	14.5(d)	14.7(d)	15.0(d)	. . .	. . .
1055	Annealed		11.0	11.8	12.6	13.4	14.0	14.5	14.8	. . .	. . .
1060	Annealed		11.1(d)	11.9(d)	12.9(d)	13.5(d)	14.1(d)	14.6(d)	14.9(d)	. . .	. . .
1070	Rolled		11.8(b)	12.6(b)	13.3(b)	14.0(b)	. . .	. . .	. . .	. . .	. . .
1078	Annealed		11.1	11.7	. . .	13.2	. . .	14.2	. . .	13.8	15.7
1080	Annealed		11.0	11.6	12.4	13.2	13.8	14.2	14.7	. . .	. . .
(e)	Not known		8.8(b)	9.8(b)	11.3(b)	12.3(b)	13.1(b)	13.6(b)	14.2(b)	. . .	. . .
1085	Annealed		11.1(b)	11.7(b)	12.5(b)	13.2(b)	13.6(b)	14.2(b)	14.7(b)	. . .	. . .
1095	Annealed		11.4(b)	. . .	. . .	. . .	. . .	. . .	. . .	. . .	. . .
	Hardened		13.0(b)	. . .	. . .	. . .	. . .	. . .	. . .	. . .	. . .
(f)	Annealed		10.6(b)	11.2(b)	12.1(b)	12.9(b)	13.5(b)	14.2(b)	14.7(b)	14.3	16.8
1145	Annealed		11.2(b)	12.1(b)	13.0(b)	13.6(b)	14.0(b)	14.6(b)	14.8(b)	. . .	. . .
1145	Annealed		11.6(b)	12.3(b)	13.1(b)	13.7(b)	14.2(b)	14.7(b)	15.1(b)	. . .	. . .
1524	Not known		11.9	12.7	. . .	13.9	. . .	14.7	. . .	12.1	13.8
(g)	Not known		8.8(b)	9.8(b)	11.3(b)	12.3(b)	13.1(b)	13.6(b)	14.2(b)	. . .	. . .
(h)	Oil hardened, tempered 600 °C (1110 °F)		11.9	12.6	. . .	13.8	. . .	14.5	. . .	. . .	. . .
2330	Annealed		10.9(d)	11.2(d)	12.1(d)	12.9(d)	13.4(d)	13.8(d)	. . .	. . .	. . .
3140	Hardened and tempered		11.8(b)	12.3(b)	12.9(b)	13.4(b)	14.0(b)	. . .	. . .	. . .	. . .
4137	Rolled		11.2(b)	11.8(b)	12.4(b)	13.0(b)	13.6(b)	. . .	. . .	. . .	. . .
4140	Oil hardened, tempered 600 °C (1110 °F)		12.3	12.7	. . .	13.7	. . .	14.5	. . .	. . .	. . .
4340	Oil hardened, tempered 630 °C (1170 °F)		(j)	12.4	. . .	13.6	. . .	14.3	. . .	. . .	. . .
(k)	Oil hardened, tempered 650 °C (1200 °F)		. . .	11.6	. . .	13.1	. . .	13.9	. . .	. . .	. . .
4615	Not known		11.5	12.1	12.7	13.2	13.7	14.1	. . .	. . .	. . .
4617	Carburized and hardened		12.5	13.1	. . .	. . .	. . .	. . .	. . .	. . .	. . .
5140	Annealed		. . .	12.6	13.4	13.9	14.3	14.6	15.0	. . .	. . .
(m)	Annealed		. . .	12.8	13.4	13.8	14.2	14.4	14.6	. . .	. . .
52100	Annealed		11.9(b)	. . .	. . .	. . .	. . .	. . .	. . .	. . .	. . .
	Hardened		12.6(b)	. . .	. . .	. . .	. . .	. . .	. . .	. . .	. . .
6150	Annealed		12.2	12.7	13.3	13.7	14.1	14.4	. . .	. . .	. . .
	Hardened, tempered 205 °C (400 °F)		12.0	12.5	12.9	13.0	13.3	13.7	. . .	. . .	. . .
6150	Annealed		12.4(d)	12.6(d)	13.3(d)	13.8(d)	14.2(d)	14.5(d)	14.7(d)	. . .	. . .
6150	Annealed		12.4	12.8	13.4	13.9	14.2	14.5	. . .	. . .	. . .
	Hardened, tempered 425 °C (800 °F)		11.8	12.4	13.1	13.6	13.9	14.1	. . .	. . .	. . .
	Hardened, tempered 650 °C (1200 °F)		12.3	12.7	13.4	13.9	14.3	14.7	. . .	. . .	. . .
18Ni250(n) . . .	Not known		. . .	. . .	10.1(p)	. . .	10.1	. . .	. . .	. . .	. . .

(a) To obtain coefficients in µin./in. · °F multiply values in table by 0.556. (b) Stated value represents average coefficient between 0 °C (32 °F) and indicated temperature. (c) 10.3 µm/m · K from −100 °C to 20 °C, 9.8 µm/m · K from −150 to 20 °C. (d) Stated value represents average coefficient between 25 °C (75 °F) and indicated temperature. (e) Nominal composition, 0.82 C, 1.65 Mn, 0.20 Si, 0.03 Cr. (f) Nominal composition, 1.22 C, 0.35 Mn, 0.009 P, 0.015 S, 0.16 Si, 0.11 Cr, 0.13 Ni, 0.01 Mo, 0.077 Cu, 0.006 Al, 0.025 As. (g) Nominal composition, 0.82 C, 1.65 Mn, 0.20 Si, 0.03 Cr. (h) Nominal composition, 0.40 C, 0.67 Mn, 0.80 Ni. (j) 11.2 µm/m · K from −100 to 20 °C; 10.4 µm/m · K from −150 to 20 °C. (k) Nominal composition, 0.32 C, 0.67 Cr, 2.60 Ni, 0.51 Mo. (m) Nominal composition, 0.37 C, 0.33 Mn, 0.21 si, 1.57 Cr. (n) Nominal composition, 0.026 C, 0.1 Mn, 0.11 Si, 18.5 Ni, 4.7 Mo, 7.0 Co, 0.22 Ti, 0.003 B. (p) Stated value represents average coefficient between 24 and 284 °C (74 and 540 °F).

Typical physical properties of wrought stainless steels, annealed condition

Type	UNS designation	Density Mg/m³	lb/in.³	Elastic modulus GPa	10⁶ psi	0 °C to: 32 °F to:	Mean coefficient of thermal expansion μm/m·°C 100 212	315 600	538 1000	μin./in.·°F 100 212	315 600	538 1000
201	S20100	7.8	0.28	197	28.6		15.7	17.5	18.4	8.7	9.7	10.2
202	S20200	7.8	0.28	...	...		17.5	18.4	19.2	9.7	10.2	10.7
205	S20500	7.8	0.28	197	28.6		...	17.9	19.1	...	9.9	10.6
301	S30100	8.0	0.29	193	28.0		17.0	17.2	18.2	9.4	9.6	10.1
302	S30200	8.0	0.29	193	28.0		17.2	17.8	18.4	9.6	9.9	10.2
302B	S30215	8.0	0.29	193	28.0		16.2	18.0	19.4	9.0	10.0	10.8
303	S30300	8.0	0.29	193	28.0		17.2	17.8	18.4	9.6	9.9	10.2
304	S30400	8.0	0.29	193	28.0		17.2	17.8	18.4	9.6	9.9	10.2
304L	S30403	8.0	0.29	...	...		...	...	...	...	...	...
S30430	S30430	8.0	0.29	193	28.0		17.2	17.8	...	9.6	9.9	...
304N	S30451	8.0	0.29	196	28.5		...	...	...	...	...	...
305	S30500	8.0	0.29	193	28.0		17.2	17.8	18.4	9.6	9.9	10.2
308	S30800	8.0	0.29	193	28.0		17.2	17.8	18.4	9.6	9.9	10.2
309	S30900	8.0	0.29	200	29.0		15.0	16.6	17.2	8.3	9.2	9.6
310	S31000	8.0	0.29	200	29.0		15.9	16.2	17.0	8.8	9.0	9.4
314	S31400	7.8	0.28	200	29.0		...	15.1	...	...	8.4	...
316	S31600	8.0	0.29	193	28.0		15.9	16.2	17.5	8.8	9.0	9.7
316L	S31603	8.0	0.29	...	...		...	...	...	...	...	...
316N	S31651	8.0	0.29	196	28.5		...	...	...	...	...	...
317	S31700	8.0	0.29	193	28.0		15.9	16.2	17.5	8.8	9.0	9.7
317L	S31703	8.0	0.29	200	29.0		16.5	...	18.1	9.2	...	10.1
321	S32100	8.0	0.29	193	28.0		16.6	17.2	18.6	9.2	9.6	10.3
329	S32900	7.8	0.28	...	...		...	...	...	...	...	...
330	N08330	8.0	0.29	196	28.5		14.4	16.0	16.7	8.0	8.9	9.3
347	S34700	8.0	0.29	193	28.0		16.6	17.2	18.6	9.2	9.6	10.3
384	S38400	8.0	0.29	193	28.0		17.2	17.8	18.4	9.6	9.9	10.2
405	S40500	7.8	0.28	200	29.0		10.8	11.6	12.1	6.0	6.4	6.7
409	S40900	7.8	0.28	...	...		11.7	...	...	6.5	...	...
410	S41000	7.8	0.28	200	29.0		9.9	11.4	11.6	5.5	6.3	6.4
414	S41400	7.8	0.28	200	29.0		10.4	11.0	12.1	5.8	6.1	6.7
416	S41600	7.8	0.28	200	29.0		9.9	11.0	11.6	5.5	6.1	6.4
420	S42000	7.8	0.28	200	29.0		10.3	10.8	11.7	5.7	6.0	6.5
422	S42200	7.8	0.28	...	...		11.2	11.4	11.9	6.2	6.3	6.6
429	S42900	7.8	0.28	200	29.0		10.3	...	...	5.7	...	...
430	S43000	7.8	0.28	200	29.0		10.4	11.0	11.4	5.8	6.1	6.3
430F	S43020	7.8	0.28	200	29.0		10.4	11.0	11.4	5.8	6.1	6.3
431	S43100	7.8	0.28	200	29.0		10.2	12.1	...	5.7	6.7	...
434	S43400	7.8	0.28	200	29.0		10.4	11.0	11.4	5.8	6.1	6.3
436	S43600	7.8	0.28	200	29.0		9.3	...	...	5.2	...	...
440A	S44002	7.8	0.28	200	29.0		10.2	...	...	5.7	...	...
440C	S44004	7.8	0.28	200	29.0		10.2	...	...	5.7	...	...
444	S44400	7.8	0.28	200	29.0		10.0	10.6	11.4	5.6	5.9	6.3
446	S44600	7.5	0.27	200	29.0		10.4	10.8	11.2	5.8	6.0	6.2
PH 13-8 Mo	S13800	7.8	0.28	203	29.4		10.6	11.2	11.9	5.9	6.2	6.6
15-5 PH	S15500	7.8	0.28	196	28.5		10.8	11.4	...	6.0	6.3	...
17-4 PH	S17400	7.8	0.28	196	28.5		10.8	11.6	...	6.0	6.4	...
17-7 PH	S17700	7.8	0.28	204	29.5		11.0	11.6	...	6.1	6.4	...

W/m·K		Btu/h·ft·°F		Specific heat(a)		Electrical resistivity,	Magnetic permea-	Melting range		
°C: 100 °F: 212	500 932	100 212	500 932	J/kg·K	Btu/lb·°F	nΩ·m	bility(b)	°C	°F	Type
16.2	21.5	9.4	12.4	500	0.12	69	1.02	1400-1450	2550-2650	201
16.2	21.6	9.4	12.5	500	0.12	69	1.02	1400-1450	2550-2650	202
...	...	...	...	500	0.12	...	...	...	...	205
16.2	21.5	9.4	12.4	500	0.12	72	1.02	1400-1420	2550-2590	301
16.2	21.5	9.4	12.4	500	0.12	72	1.02	1400-1420	2550-2590	302
15.9	21.6	9.2	12.5	500	0.12	72	1.02	1375-1400	2500-2550	.....302B
16.2	21.5	9.4	12.4	500	0.12	72	1.02	1400-1420	2550-2590	303
16.2	21.5	9.4	12.4	500	0.12	72	1.02	1400-1450	2550-2650	304
...	...	...	...	...	...	...	1.02	1400-1450	2550-2650	.....304L
11.2	21.5	6.5	12.4	500	0.12	72	1.02	1400-1450	2550-2650	.....S30430
...	...	...	...	500	0.12	72	1.02	1400-1450	2550-2650	.....304N
16.2	21.5	9.4	12.4	500	0.12	72	1.02	1400-1450	2550-2650	305
15.2	21.6	8.8	12.5	500	0.12	72	...	1400-1420	2550-2590	308
15.6	18.7	9.0	10.8	500	0.12	78	1.02	1400-1450	2550-2650	309
14.2	18.7	8.2	10.8	500	0.12	78	1.02	1400-1450	2550-2650	310
17.5	20.9	10.1	12.1	500	0.12	77	1.02	...	...	314
16.2	21.5	9.4	12.4	500	0.12	74	1.02	1375-1400	2500-2550	316
...	...	...	...	...	...	...	1.02	1375-1400	2500-2550	.....316L
...	...	...	...	500	0.12	74	1.02	1375-1400	2500-2550	.....316N
16.2	21.5	9.4	12.4	500	0.12	74	1.02	1375-1400	2500-2550	317
14.4	...	8.3	...	500	0.12	79	...	1375-1400	2500-2550	.....317L
16.1	22.2	9.3	12.8	500	0.12	72	1.02	1400-1425	2550-2600	321
...	...	...	...	460	0.11	75	...	...	...	329
...	...	...	...	460	0.11	102	1.02	1400-1425	2550-2600	330
16.1	22.2	9.3	12.8	500	0.12	73	1.02	1400-1425	2550-2600	347
16.2	21.5	9.4	12.4	500	0.12	79	1.02	1400-1450	2550-2650	384
27.0	...	15.6	...	460	0.11	60	...	1480-1530	2700-2790	405
...	...	...	...	...	...	...	...	1480-1530	2700-2790	409
24.9	28.7	14.4	16.6	460	0.11	57	700-1000	1480-1530	2700-2790	410
24.9	28.7	14.4	16.6	460	0.11	70	...	1425-1480	2600-2700	414
24.9	28.7	14.4	16.6	460	0.11	57	700-1000	1480-1530	2700-2790	416
24.9	...	14.4	...	460	0.11	55	...	1450-1510	2650-2750	420
23.9	27.3	13.8	15.8	460	0.11	...	...	1470-1480	2675-2700	422
25.6	...	14.8	...	460	0.11	59	...	1450-1510	2650-2750	429
26.1	26.3	15.1	15.2	460	0.11	60	600-1100	1425-1510	2600-2750	430
26.1	26.3	15.1	15.2	460	0.11	60	...	1425-1510	2600-2750	.....430F
20.2	...	11.7	...	460	0.11	72	...	...	...	431
...	26.3	...	15.2	460	0.11	60	600-1100	1425-1510	2600-2750	434
23.9	26.0	13.8	15.0	460	0.11	60	600-1100	1425-1510	2600-2750	436
24.2	...	14.0	...	460	0.11	60	...	1370-1480	2500-2700	.....440A
24.2	...	14.0	...	460	0.11	60	...	1370-1480	2500-2700	.....440C
26.8	...	15.5	...	420	0.10	62	...	...	...	444
20.9	24.4	12.1	14.1	500	0.12	67	400-700	1425-1510	2600-2750	446
14.0	22.0	8.1	12.7	460	0.11	102	...	1400-1440	2560-2625	.....PH 13-8 Mo
17.8	23.0	10.3	13.1	420	0.10	77	95	1400-1440	2560-2625	.....15-5 PH
18.3	23.0	10.6	13.1	460	0.11	80	95	1400-1440	2560-2625	.....17-4 PH
16.4	21.8	9.5	12.6	460	0.11	83	...	1400-1440	2560-2625	.....17-7 PH

(a) At 0 to 100 °C (32 to 212 °F). (b) Approximate values.

Density and thermal expansion of selected tool steels

Type	Density		μm/m · K from 20 °C to					μin./in. · °F from 68 °F to				
	Mg/m³	lb/in.³	100 °C	200 °C	425 °C	540 °C	650 °C	200 °F	400 °F	800 °F	1000 °F	1200 °F
W1	7.84	0.282	10.4	11.0	13.1	13.8(a)	14.2(b)	5.76	6.13	7.28	7.64(a)	7.90(b)
W2	7.85	0.283	...	...	14.4	14.8	14.9	...	...	8.0	8.2	8.3
S1	7.88	0.255	12.4	12.6	13.5	13.9	14.2	6.9	7.0	7.5	7.7	7.9
S2	7.79	0.281	10.9	11.9	13.5	14.0	14.2	6.0	6.6	7.5	7.8	7.9
S5	7.76	0.280	...	...	12.6	13.3	13.7	...	...	7.0	7.4	7.6
S6	7.75	0.279	...	...	12.6	13.3	...	...	...	7.0	7.4	...
S7	7.76	0.280	...	12.6	13.3	13.7(a)	13.3	...	7.0	7.4	7.6(a)	7.4
O1	7.85	0.283	...	10.6(c)	12.8	14.0(d)	14.4(d)	...	5.9(c)	7.1	7.8(d)	8.0(d)
O2	7.66	0.277	11.2	12.6	13.9	14.6	15.1	6.2	7.0	7.7	8.1	8.4
O7	7.8	0.282	...	...	...	...	...	...	...	...	...	...
A2	7.86	0.284	10.7	10.6(c)	12.9	14.0	14.2	5.96	5.91(c)	7.2	7.8	7.9
A6	7.84	0.283	11.5	12.4	13.5	13.9	14.2	6.4	6.9	7.5	7.7	7.9
A7	7.66	0.277	...	...	12.4	12.9	13.5	...	...	6.9	7.2	7.5
A8	7.87	0.284	...	...	12.0	12.4	12.6	...	...	6.7	6.9	7.0
A9	7.78	0.281	...	...	12.0	12.4	12.6	...	...	6.7	6.9	7.0
D2	7.70	0.278	10.4	10.3	11.9	12.2	12.2	5.8	5.7	6.6	6.8	6.8
D3	7.70	0.278	12.0	11.7	12.9	13.1	13.5	6.7	6.5	7.2	7.3	7.5
D4	7.70	0.278	...	...	12.4	...	...	...	...	6.9	...	...
D5	...	...	...	...	...	12.0	...	...	...	...	6.7	...
H10	7.81	0.281	...	...	12.2	13.3	13.7	...	...	6.8	7.4	7.6
H11	7.75	0.280	11.9	12.4	12.8	12.9	13.3	6.6	6.9	7.1	7.2	7.4
H13	7.76	0.280	10.4	11.5	12.2	12.4	13.1	5.8	6.4	6.8	6.9	7.3
H14	7.89	0.285	11.0	...	...	...	...	6.1	...	...	...	...
H19	7.98	0.288	11.0	11.0	12.0	12.4	12.9	6.1	6.1	6.7	6.9	7.2
H21	8.28	0.299	12.4	12.6	12.9	13.5	13.9	6.9	7.0	7.2	7.5	7.7
H22	8.36	0.302	11.0	...	11.5	12.0	12.4	6.1	...	6.4	6.7	6.9
H26	8.67	0.313	...	...	...	12.4	...	...	...	...	6.9	...
H42	8.15	0.295	...	...	...	11.9	...	...	...	...	6.6	...
T1	8.67	0.313	...	9.7	11.2	11.7	11.9	...	5.4	6.2	6.5	6.6
T2	8.67	0.313	...	...	...	...	...	...	...	...	...	...
T4	8.68	0.313	...	...	...	11.9	...	...	...	...	6.6	...
T5	8.75	0.316	11.2	...	...	11.5	...	6.2	...	...	6.4	...
T6	8.89	0.321	...	...	...	...	...	...	...	...	...	...
T8	8.43	0.305	...	...	...	...	...	...	...	...	...	...
T15	8.19	0.296	...	9.9	11.0	11.5	...	...	...	5.5(c)	6.1	6.4	...
M1	7.89	0.285	...	10.6(c)	11.3	12.0	12.4	...	5.9(c)	6.3	6.7	6.9
M2	8.16	0.295	10.1	9.4(c)	11.2	11.9	12.2	5.6	5.2(c)	6.2	6.6	6.8
M3, class 1	8.15	0.295	...	...	11.5	12.0	12.2	...	...	6.4	6.7	6.8
M3, class 2	8.16	0.295	...	...	11.5	12.0	12.8	...	...	6.4	6.7	7.1
M4	7.97	0.288	...	9.5(c)	11.2	12.0	12.2	...	5.3(c)	6.2	6.7	6.8
M7	7.95	0.287	...	9.5(c)	11.5	12.2	12.4	...	5.3(c)	6.4	6.8	6.9
M10	7.88	0.255	...	...	11.0	11.9	12.4	...	...	6.1	6.6	6.9
M30	8.01	0.289	...	...	11.2	11.7	12.2	...	...	6.2	6.5	6.8
M33	8.03	0.290	...	...	11.0	11.7	12.0	...	...	6.1	6.5	6.7
M36	8.18	0.296	...	...	...	...	...	...	...	...	...	...
M41	8.17	0.295	...	9.7	10.4	11.2	...	...	5.4	5.8	6.2	...
M42	7.98	0.288	...	...	...	...	...	...	...	...	...	...
M46	7.83	0.283	...	...	...	...	...	...	...	...	...	...
M47	7.96	0.288	10.6	11.0	11.9	...	12.6	5.9	6.1	6.6	...	7.0
L2	7.86	0.284	...	...	14.4	14.6	14.8	...	...	8.0	8.1	8.2
L6	7.86	0.284	11.3	12.6	12.6	13.5	13.7	6.3	7.0	7.0	7.5	7.6
P2	7.86	0.284	...	...	13.7	...	...	...	...	7.6	...	...
P5	7.80	0.282	...	...	...	...	...	...	...	...	...	...
P6	7.85	0.284	...	...	...	...	...	...	...	...	...	...
P20	7.85	0.284	...	...	12.8	13.7	14.2	...	...	7.1	7.6	7.9

(a) From 20 °C to 500 °C (68 °F to 930 °F). (b) From 20 °C to 600 °C (68 °F to 1110 F). (c) From 20 °C to 260 °C (68 °F to 500 °F). (d) From 38 °C (100 °F).

2 STRUCTURE AND PROPERTIES OF METALS

This section was condensed from Metals Handbook, Eighth Edition, Volume 8, Metallography, Structures and Phase Diagrams, pages 233 to 250, and from Principles of Physical and Chemical Metallurgy, by Giles F. Carter. For more detailed information on the topics covered in this section, the reader is referred to the larger works.

Crystal Structure

By C. S. Barrett, University of Denver
(Revisions in structure-type nomenclature by W. B. Pearson, University of Waterloo)

THE CRYSTAL structures presented in this article are those that are regarded as most important to metallurgists and that have been widely studied. More complete coverage is given in the references listed at the end of this article.

CRYSTALLOGRAPHIC TERMS AND CONCEPTS

The terms and concepts defined and explained in this section are basic to an understanding of the descriptions and illustrations of crystal structures presented in the next section, "Metallurgically Important Crystal Types."

Crystal Structure. The arrangement of atoms in the interior of a crystal is called its crystal structure. A fundamental unit of the arrangement repeats itself at regular intervals in three dimensions, throughout the interior of the crystal.

Unit Cell. A unit cell is a parallelepiped whose edges form the axes of a crystal. A unit cell is one unit of pattern of atomic arrangement. A crystal consists of unit cells stacked tightly together, each identical in size, shape and orientation with all others. The choice of the boundaries of a unit cell is somewhat arbitrary, being conditioned by symmetry considerations and by convenience.

Crystal Systems. Seven different systems of axes, each with a specified equality or inequality of edge lengths and angles, are used in crystallography. These are the basis of the seven crystal systems — triclinic (anorthic), monoclinic, orthorhombic, tetragonal, hexagonal, rhombohedral and cubic — employed in the classification of crystals.

The edge lengths a, b and c (along the corresponding crystal axes) are expressed in angstroms ($1 A = 10^{-8}$ cm). Faces of unit cells are identified by the capital letter A, B or C, when the faces contain axes b and c, c and a, or a and b, respectively.

Angles between the axes are expressed in degrees, with the angle in the A face denoted as α (alpha), the angle in the B face as β (beta), and the angle in the C face as γ (gamma). Table 1 shows the relationships of the edge lengths along the crystal axes, and of the interaxial angles, for each of the seven crystal systems. The edge lengths and angles are sometimes referred to as the lattice parameters, lattice spacings or lattice constants for a unit cell.

Lattices. A lattice (space lattice) is a regular, periodic array of points (lattice points) in space at each of which is located the same kind of atom or a group of atoms of identical composition, arrangement and orientation in a perfect crystal (at least, on a time-average basis).

There are five (actually, four plus rhombohedral) different basic arrangements for lattice points within a unit cell, and each is identified by a Hermann-Mauguin letter symbol in a space-lattice notation. These letter symbols, and the arrangements they identify, are: P, for primitive (simple), with lattice points only at cell corners; C, for base-face centered (end-centered), with lattice points centered on the C faces or ends of the crystal; F, for all-face centered, with lattice points centered on all faces; I, for innercentered, with lattice points at the center of volume of the unit cell (body-centered). The rhombohedral cell, also primitive, has R as its symbol. (The face having the base-face-centered lattice point may be designated as the C face, because the choice of axes is arbitrary and does not alter the atom positions in the space lattice. Rhombohedral crystals can be considered as having either a rhombohedral cell or a primitive hexagonal cell.)

The above letter symbols and definitions apply only to a basic arrangement of atoms, and do not limit the number of atoms in a unit cell. Atoms may be found at each corner of a base-centered, face-centered or innercentered cell and at other random positions on the cell faces or within the cell.

There are 14 kinds of space lattices, derived from all the combinations of equality and inequality of lengths of edges and interaxial angles. The 14 space lattices are listed in Table 2 with the Hermann-Mauguin and Pearson symbols for each. The Pearson symbols (Ref 1) consist of Hermann-Mauguin space-lattice letters preceded by a, m, o, t, h and c to denote, respectively, the six crystal systems: triclinic (anorthic), monoclinic, orthorhombic, tetragonal, hexagonal and cubic.

Structure symbols are arbitrary symbols that des-

Table 1. Relationships of edge lengths and of interaxial angles for the seven crystal systems

Crystal system	Edge lengths	Interaxial angles	Examples
Triclinic (anorthic)	$a \neq b \neq c$	$\alpha \neq \beta \neq \gamma \neq 90°$	HgK
Monoclinic	$a \neq b \neq c$	$\alpha = \gamma = 90° \neq \beta$	β-S; CoSb$_2$
Orthorhombic	$a \neq b \neq c$	$\alpha = \beta = \gamma = 90°$	α-S; Ga; Fe$_3$C (cementite)
Tetragonal	$a = b \neq c$	$\alpha = \beta = \gamma = 90°$	β-Sn (white); TiO$_2$
Hexagonal	$a = b \neq c$	$\alpha = \beta = 90°$; $\gamma = 120°$	Zn; Cd; NiAs
Rhombohedral(a)	$a = b = c$	$\alpha = \beta = \gamma \neq 90°$	As; Sb; Bi; calcite
Cubic	$a = b = c$	$\alpha = \beta = \gamma = 90°$	Cu; Ag; Au; Fe; NaCl

(a) Rhombohedral crystals (which are sometimes called trigonal) can also be described by using hexagonal axes (rhombohedral-hexagonal).

Table 2. The 14 space (Bravais) lattices and their Hermann-Mauguin and Pearson symbols

System	Space lattice	Hermann-Mauguin symbol	Pearson symbol
Triclinic (anorthic)	Simple	P	aP
Monoclinic	Simple	P	mP
	Base-centered(a)	C	mC
Orthorhombic	Simple	P	oP
	Base-centered(a)	C	oC
	Face-centered	F	oF
	Body-centered ...	I	oI
Tetragonal	Simple	P	tP
	Body-centered ...	I	tI
Hexagonal	Simple	P(b)	hP
Rhombohedral	Simple	R	hR
Cubic	Simple	P	cP
	Face-centered	F	cF
	Body-centered ...	I	cI

(a) The face that has a lattice point at its center may be chosen as the c face (the XY plane), denoted by the symbol C, or as the a or b face, denoted by A or B, because the choice of axes is arbitrary and does not alter the actual translations of the lattice. (b) The symbol C may be used for hexagonal crystals, because hexagonal crystals may be regarded as base-centered orthorhombic.

ignate the type of crystal structure. The Strukturbericht symbols (Ref 2) were widely used in the past, and are still used today, but this system of naming structure types has been overwhelmed by the number and complexity of types that we now recognize. Furthermore, the final publication of Strukturbericht was in 1939.

Today, the accepted system of naming the types of crystal structures that metals and alloys adopt is to arbitrarily select the formula of a phase with the structure type (i.e., a prototype) and to follow this by the Pearson symbol for the Bravais lattice of the structure and then the number of atoms in the conventionally chosen unit cell. Thus the nickel arsenide structure is referred to as the NiAs $hP4$ type, and rock salt as the NaCl $cF8$ type. The arbitrariness in the system appears not to be a problem, because norms become established by common usage. Thus the ordered AuCu structure should properly be described as AuCu $tP2$ according to the smallest primitive cell, but due to association of the structure with ordering from a face-centered-cubic solid solution ($cF4$), it is more usual to refer to it as AuCu $tP4$. The

advantage of this way of naming structure types is that it is "open-ended"—i.e., not limited in use by future discoveries of new crystal-structure types. Secondly, compared to using only a formula name, it is crystallographically informative due to the addition of the Pearson symbol, and therefore amenable to classification. Thus a person discovering a new intermetallic phase and establishing for it preliminary crystallographic information (the Bravais lattice, and the number of atoms in the unit cell) can consult a table of known structure types, classified by Pearson symbol, and readily see what already characterized types may be applicable to the newly discovered phase.

For convenience, Table 3 (Ref 3) lists Strukturbericht structure symbols, prototype names and corresponding Pearson symbols.

Space-group notation is a symbolic description of the space lattice and the symmetry of a crystal. The notation for a space group is comprised of the symbol for a space lattice followed by letters and numbers describing the symmetry of the crystal. These symmetry designations are not

Table 3. Conversion of Strukturbericht to Pearson symbol

Strukturbericht designation	Structure type	Pearson symbol	Strukturbericht designation	Structure type	Pearson symbol	Strukturbericht designation	Structure type	Pearson symbol
$A1$	Cu	$cF4$	$C11_b$	$MoSi_2$	$tI6$	$D5_1$	$\alpha\text{-}Al_2O_3$	$hR10$
$A2$	W	$cI2$	$C12$	$CaSi_2$	$hR6$	$D5_2$	La_2O_3	$hP5$
$A3$	Mg	$hP2$	$C14$	$MgZn_2$	$hP12$	$D5_3$	Mn_2O_3	$cI80$
$A4$	C	$cF8$	$C15$	Cu_2Mg	$cF24$	$D5_8$	S_3Sb_2	$oP20$
$A5$	Sn	$tF4$	$C15_b$	$AuBe_5$	$cF24$	$D5_9$	P_2Zn_3	$tP40$
$A6$	In	$tI2$	$C16$	Al_2Cu	$tI12$	$D5_{10}$	C_2Cr_3	$oP20$
$A7$	As	$hR2$	$C18$	FeS_2	$oP6$	$D5_{13}$	Al_3Ni_2	$hP5$
$A8$	Se	$hP3$	$C19$	$CdCl_2$	$hR3$	$D5_a$	Si_2U_3	$tP10$
$A10$	Hg	$hR1$	$C22$	Fe_2P	$hP9$	$D5_c$	C_3Pu_2	$cI40$
$A11$	Ga	$oC8$	$C23$	Cl_2Pb	$oP12$	$D7_1$	Al_4C_3	$hR7$
$A12$	α-Mn	$cI58$	$C32$	AlB_2	$hP3$	$D7_3$	P_4Th_3	$cI28$
$A13$	β-Mn	$cP20$	$C33$	Bi_2STe_2	$hR5$	$D7_b$	B_4Ta_3	$oI14$
$A15$	OW_3	$cP8$	$C34$	$AuTe_2$	$mC6$	$D8_1$	Fe_3Zn_{10}	$cI52$
$A20$	α-U	$oC4$	$C36$	$MgNi_2$	$hP24$	$D8_2$	Cu_5Zn_8	$cI52$
$B1$	ClNa	$cF8$	$C38$	Cu_2Sb	$tP6$	$D8_3$	Al_4Cu_9	$cP52$
$B2$	ClCs	$cP2$	$C40$	$CrSi_2$	$hP9$	$D8_4$	C_5Cr_{23}	$cF116$
$B3$	SZn	$cF8$	$C44$	GeS_2	$oF72$	$D8_5$	Fe_7W_6	$hR13$
$B4$	SZn	$hP4$	$C46$	$AuTe_2$	$oP24$	$D8_6$	$Cu_{15}Si_4$	$cI76$
$B8_1$	AsNi	$hP4$	$C49$	Si_2Zr	$oC12$	$D8_8$	Mn_5Si_3	$hP16$
$B8_2$	$InNi_2$	$hP6$	$C54$	Si_2Ti	$oF24$	$D8_9$	Co_9S_8	$cF68$
$B9$	HgS	$hP6$	C_c	Si_2Th	$tI12$	$D8_{10}$	Al_8Cr_5	$hR26$
$B10$	OPb	$tP4$	C_e	$CoGe_2$	$oC23$	$D8_{11}$	Al_5Co_2	$hP28$
$B11$	γ-CuTi	$tP4$	$D0_2$	As_3Co	$cI32$	$D8_a$	$Mn_{23}Th_6$	$cF116$
$B13$	NiS	$hR6$	$D0_3$	BiF_3	$cF16$	$D8_b$	σ-phase	$tP30$
$B16$	GeS	$oP8$	$D0_9$	O_3Re	$cP4$	$D8_f$	Ge_7Ir_3	$cI40$
$B17$	PtS	$tP4$	$D0_{11}$	CFe_3	$oP16$	$D8_i$	B_5Mo_2	$hR7$
$B18$	CuS	$hP12$	$D0_{18}$	$AsNa_3$	$hP8$	$D8_h$	B_5W_2	$hP14$
$B19$	AuCd	$oP4$	$D0_{19}$	Ni_3Sn	$hP8$	$D8_l$	B_3Cr_5	$tI32$
$B20$	FeSi	$cP8$	$D0_{20}$	Al_3Ni	$oP16$	$D8_m$	Si_3W_5	$tI32$
$B27$	BFe	$oP8$	$D0_{21}$	Cu_3P	$hP24$	$D10_1$	C_3Cr_7	$hP80$
$B31$	MnP	$oP8$	$D0_{22}$	Al_3Ti	$tI8$	$D10_2$	Fe_3Th_7	$hP20$
$B32$	NaTl	$cF16$	$D0_{23}$	Al_3Zr	$tI16$	$E0_1$	ClFPb	$tP6$
$B34$	PdS	$tP16$	$D0_{24}$	Ni_3Ti	$hP16$	$E1_1$	$CuFeS_2$	$tI16$
$B35$	CoSn	$hP6$	$D0_c$	SiU_3	$tI16$	$E2_1$	CaO_3Ti	$cP5$
$B37$	SeTl	$tI16$	$D0_e$	Ni_3P	$tI32$	$E3$	Al_2CdS_4	$tI14$
B_e	CdSb	$oP16$	$D1_3$	Al_4Ba	$tI10$	$E9_3$	CFe_3W_3	$cF112$
$B_f(B33)$	ζ-BCr	$oC8$	$D1_a$	$MoNi_4$	$tI10$	$E9_a$	Al_7Cu_2Fe	$tP40$
B_g	BMo	$tI16$	$D1_b$	Al_4U	$oI20$	$E9_b$	$AlLi_3N_2$	$cI96$
B_h	CW	$hP2$	$D1_c$	$PtSn_4$	$oC20$	$F0_1$	NiSSb	$cP12$
B_i	γ'-CMo (AsTi)	$hP8$	$D1_e$	B_4Th	$tP20$	$F5_1$	$CrNaS_2$	$hR4$
$C1$	CaF_2	$cF12$	$D1_f$	BMn_4	$oF40$	$H1_1$	Al_2MgO_4	$cF56$
$C1_b$	AgAsMg	$cF12$	$D2_1$	B_6Ca	$cP7$	$H2_4$	Cu_3S_4V	$cP8$
$C2$	FeS_2	$cP12$	$D2_3$	$NaZn_{13}$	$cF112$	$L1_0$	AuCu	$tP4$
$C3$	Cu_2O	$cP6$	$D2_b$	$Mn_{12}Th$	$tI26$	$L1_2$	$AuCu_3$	$cP4$
$C4$	O_2Ti	$tP6$	$D2_c$	MnU_6	$tI28$	$L2_1$	$AlCu_2Mn$	$cF16$
$C6$	CdI_2	$hP3$	$D2_d$	$CaCu_5$	$hP6$	$L'2_b$	H_2Th	$tI6$
$C7$	MoS_2	$hP6$	$D2_f$	$B_{12}U$	$cF52$	$L'3$	Fe_2N	$hP3$
$C11_a$	C_2Ca	$tI6$	$D2_h$	Al_6Mn	$oC28$	$L6_0$	$CuTi_3$	$tP4$

discussed here, but are described in various textbooks and are tabulated in the International Tables for X-ray Crystallography (Ref 4).

Each crystal belongs to at least one of the 230 possible space groups. Space-group notation follows structure type in the descriptions of crystal structures, which are given in Table 4.

Structure Prototype. As noted above, to aid in classification and identification, each structure type has been given the name of a representative substance (an element or phase) having that structure.

Unit cells with the same structure type generally do not have dimensions identical to the prototype or to each other, because different materials with the same type of atomic arrangement have atoms that differ in size, causing the lengths of the *a, b* and *c* edges to differ. Likewise, the atom-position coordinates *x, y* and *z* vary among different materials.

Atom Positions. The position of an atom, or the lattice point, in a unit cell is expressed by three coordinates (Wyckoff notation, Ref 5) — the three distances parallel to the *a, b* and *c* axes, respectively, from the origin at one corner of the cell to the atom in question. These distances are expressed in fractions of the edge lengths *a, b* and *c,* respectively, rather than in angstroms. Thus, $^1/_2,0,0$ is at the midpoint of the *a* edge, $^1/_2,^1/_2,0$ is at the center of the *C* face, and $^1/_2,^1/_2,^1/_2$ is at the center of the volume of the unit cell. The letters *x, y* and *z* are used for the coordinates that are not convenient fractions, or that differ among element or alloy phases.

A primitive (simple) unit cell has lattice points at its corners only — that is, at 0,0,0.

A body-centered unit cell has lattice points at the corners (at 0,0,0) and also at the center of volume (at $^1/_2,^1/_2,^1/_2$).

A face-centered unit cell has lattice points at the corners and at the centers of all six faces. The lattice points are at 0,0,0; $0,^1/_2,^1/_2$; $^1/_2,^1/_2,0$ and $^1/_2,0,^1/_2$.

A negative value for a coordinate is indicated by placing a bar over the letter — for example, $\bar{x}$.

Point Groups. Perhaps the most surprising fact to a person learning for the first time about symmetry operations and crystal structures is that a structure described by a specific space lattice (e.g., *cP*) may not have any atoms lying at the (space) lattice points; instead there may be groups of atoms with specific so-called *point symmetries,* located about each of the space-lattice points.

Table 4. Assorted structure types of metallurgical interest, arranged according to Pearson symbol

Pearson Symbol	Prototype	Space group	Unit-cell description	Pearson Symbol	Prototype	Space group	Unit-cell description
*cP*1	α-Po	*Pm3m*	One atom per cell, at 0,0,0. For α-Po, *a* = 3.34 A. See Fig. 1. **Examples:** Ag-Te (metastable), Au-Te (metastable), α-Po, Sb II (HP).	*cP*12	NiSSb	*P*2₁3	For NiSSb, *a* = 5.88 A. **Examples:** AsPdSe, BiPtTe, IrSbSe, RhSSb.
*cP*2	CsCl	*Pm3m*	One cesium atom at 0,0,0; and one chlorine atom at $^1/_2,^1/_2,^1/_2$. For CsCl, *a* = 4.11 A. See Fig. 1. **Examples:** AgCd, CoTi, CsCl, FeAl, FeCo, FeTi, FeV, β NiAl, β NiGa, δ NiIn, NiTi, β′ Cu-Zn.		FeS₂	*Pa*3	Pyrite. Four iron atoms at 0,0,0; $0,^1/_2,^1/_2$; $^1/_2,0,^1/_2$ and $^1/_2,^1/_2,0$; eight sulfur atoms at *x,x,x*; $^1/_2+x,^1/_2-x,\bar{x}$; $\bar{x},^1/_2+x,^1/_2-x$; $^1/_2-x,\bar{x},^1/_2+x$; $\bar{x},\bar{x},\bar{x}$; $^1/_2-x,^1/_2+x,x$; $x,^1/_2-x,^1/_2+x$ and $^1/_2+x,x,^1/_2-x$. For FeS₂ (pyrite), *x* = 0.386 and *a* = 5.42 A. **Examples:** CoPS, CoS₂, CoSe₂, FeS₂ (pyrite), MnS₂, MnTe₂, NiS₂₊ₓ, NiSe₂.
*cP*4	AuCu₃	*Pm3m*	One gold atom at 0,0,0; three copper atoms at $0,^1/_2,^1/_2$; $^1/_2,0,^1/_2$ and $^1/_2,^1/_2,0$. For AuCu₃, *a* = 3.74 A. See Fig. 1. **Examples:** α′ AlNi₃, AlZr₃, Au₃Cu, AuCu₃ I, CoPt₃, Cr₃Pt, Fe₃Ga, FePd₃, Ni₃Fe, Ni₃Mn, Sn₃U.	*cP*20	β-Mn	*P*4₁32	For β-Mn, *a* = 6.13 A. **Examples:** Ag₃Al, Au₉Nb₁₁, T C-Cr-Fe-W, γ Cu₅Si.
				*cP*36	BaHg₁₁	*Pm3m*	For BaHg₁₁, *a* = 9.59 A. **Examples:** Cd₁₁Ce, Hg₁₁K, Hg₁₁Sr.
*cP*5	CaTiO₃	*Pm3m*	Perovskite. One calcium atom at 0,0,0; one titanium atom at $^1/_2,^1/_2,^1/_2$; three oxygen atoms at $0,^1/_2,^1/_2$; $^1/_2,0,^1/_2$ and $^1/_2,^1/_2,0$. **Examples:** AlCFe₃, AlCMn₃, AlCTi₃, CaTiO₃, Fe₃C,In, Fe₃NNi, Fe₃NPd, Fe₃NSn.	*cP*52	Al₄Cu₉	*P*$\bar{4}$*3m*	Gamma brass. For Al₄Cu₉, *a* = 8.70 A. **Example:** Ga₄Cu₉.
				*cF*4	Cu	*Fm3m*	Four atoms at 0,0,0; $^1/_2,0,^1/_2$; $0,^1/_2,^1/_2$ and $^1/_2,^1/_2,0$. For Cu, *a* = 3.61 A. See Fig. 1. **Examples:** Ag, Al, Au, α-Ca, α-Ce, β-Co, Cu, γ-Fe, Ir, Ni, Pb, Pd, Pt, Rh, α-Sr, α-Th.
*cP*7	B₆Ca	*Pm3m*	One calcium atom at 0,0,0; six boron atoms at $x,^1/_2,^1/_2$; $^1/_2,x,^1/_2$; $^1/_2,^1/_2,x$; $\bar{x},^1/_2,^1/_2$; $^1/_2,\bar{x},^1/_2$; $^1/_2,^1/_2,\bar{x}$ with *x* = 0.207. For B₆Ca, *a* = 4.15 A. **Examples:** B₆Ba, B₆Gd, B₆Sc, B₆Th, B₆Yb.				
*cP*8	FeSi	*P*2₁3	Four iron atoms at *x,x,x*; $^1/_2+x,^1/_2-x,\bar{x}$; $\bar{x},^1/_2+x,^1/_2-x$ and $^1/_2-x,\bar{x},^1/_2+x$ (with *x* = 0.137); four silicon atoms at *x,x,x*; $^1/_2+x,^1/_2-x,\bar{x}$; $\bar{x},^1/_2+x,^1/_2-x$ and $^1/_2-x,\bar{x},^1/_2+x$ (with *x* = 0.842). For FeSi, *a* = 4.49 A. **Examples:** CoSi, FeSi, MnSi.	*cF*8	SZn	*F*$\bar{4}$*3m*	Sphalerite. Four zinc atoms at 0,0,0; $0,^1/_2,^1/_2$; $^1/_2,0,^1/_2$ and $^1/_2,^1/_2,0$; four sulfur atoms at $^1/_4,^1/_4,^1/_4$; $^1/_4,^3/_4,^3/_4$; $^3/_4,^1/_4,^3/_4$ and $^3/_4,^3/_4,^1/_4$. For SZn (sphalerite), *a* = 5.42 A. See Fig. 1. **Examples:** CdS, CdSe, CdTe, CuFeS₂ (HT), GaP, GaSb, InAs, InP, InSb, β MnS, β SiC, ZnO, ZnS (sphalerite), ZnSe.
	W₃O or Cr₃Si	*Pm3n*	Atom I, two at 0,0,0 and $^1/_2,^1/_2,^1/_2$; atom II, six at $^1/_4,0,^1/_2$; $^1/_2,^1/_4,0$; $0,^1/_2,^1/_4$; $^3/_4,0,^1/_2$; $^1/_2,^3/_4,0$ and $0,^1/_2,^3/_4$. For W₃O, *a* = 5.04 A. See Fig. 1. The prototype structure originally was attributed to β-W. This has since been shown to be the oxide, W₃O, having random distribution of atoms. **Examples:** AlV₃, AuTi₃, CoV₃, Cr₃O, Cr₃Si, Mo₃O, V₃Si, W₃O, W₃Si.		NaCl rocksalt	*Fm3m*	Four sodium atoms at 0,0,0; $0,^1/_2,^1/_2$; $^1/_2,0,^1/_2$ and $^1/_2,^1/_2,0$; four chlorine atoms at $^1/_2,^1/_2,^1/_2$; $^1/_2,0,0$; $0,^1/_2,0$ and $0,0,^1/_2$. For NaCl, *a* = 5.64 A. See Fig. 1. **Examples:** BaS, CdO, CdS, CrN, HfC, HfN, NaCl, NiO (HT), PbS, PbSe, TiO, UC, UO, UP, US, VO, ZrO.
					C	*Fd3m*	Diamond. Eight atoms per cell, at 0,0,0; $0,^1/_2,^1/_2$; $^1/_2,0,^1/_2$; $^1/_2,^1/_2,0$; $^1/_4,^1/_4,^1/_4$; $^1/_4,^3/_4,^3/_4$; $^3/_4,^1/_4,^3/_4$ and $^3/_4,^3/_4,^1/_4$. For C (diamond), *a* = 3.57 A. See Fig. 1. **Examples:** C (diamond), Ge, Si, α-Sn.

(continued)

Table 4. (continued)

Pearson symbol	Prototype	Space group	Unit-cell description
$cF12$	AgAsMg	$F\bar{4}3m$	Ternary version of fluorite. For AgAsMg, $a = 6.25$ A. **Examples:** AlBBe, CdCuSb, Cu-MnSb, NiSbV.
	CaF_2	$Fm3m$	Fluorite. Four calcium atoms at $0,0,0$; $0,^1/_2,^1/_2$; $^1/_2,0,^1/_2$ and $^1/_2,^1/_2,0$; eight fluorine atoms at $^1/_4,^1/_4,^1/_4$; $^1/_4,^3/_4,^3/_4$; $^3/_4,^1/_4,^3/_4$; $^3/_4,^3/_4,^1/_4$; $^3/_4,^3/_4,^3/_4$; $^3/_4,^1/_4,^1/_4$; $^1/_4,^3/_4,^1/_4$ and $^1/_4,^1/_4,^3/_4$. For CaF_2, $a = 5.46$ A. See Fig. 1. **Examples:** Be_2B, Be_2C, CaF_2, $CoSi_2$, rare-earth hydrides, K_2O, K_2S, Mg_2Pb, Mg_2Si, ζ $NiSi_2$, UN_2, UO_2.
$cF16$	$AlCu_2Mn$	$Fm3m$	Heusler alloy. For $AlCu_2Mn$, $a = 5.95$ A. **Examples:** $AgAuCd_2$, $AlNi_2Ta$, Co_2GaTi, CsK_2Sb.
	BiF_3 or $BiLi_3$	$Fm3m$	Four bismuth atoms at $0,0,0$; $0,^1/_2,^1/_2$; $^1/_2,0,^1/_2$ and $^1/_2,^1/_2,0$; 12 fluorine (or lithium) atoms at $^1/_2,^1/_2,^1/_2$; $^1/_2,0,0$; $0,^1/_2,0$; $0,0,^1/_2$; $^1/_4,^1/_4,^1/_4$; $^1/_4,^3/_4,^3/_4$; $^3/_4,^1/_4,^3/_4$; $^3/_4,^3/_4,^1/_4$; $^3/_4,^3/_4,^3/_4$; $^3/_4,^1/_4,^1/_4$; $^1/_4,^3/_4,^1/_4$ and $^1/_4,^1/_4,^3/_4$. For $BiLi_3$, $a = 6.71$ A. See Fig. 1. **Examples:** BiF_3, $BiLi_3$, Fe_3Al, γ Cu_3Sn (HT), α Fe_3Si, Mn_3Si, Ni_3Sn (HT).
	NaTl	$Fd3m$	For NaTl, $a = 7.49$ A. **Examples:** AlLi, CdLi, GaLi, InNa, LiZn.
$cF24$	$AuBe_5$	$F\bar{4}3m$ or $F23$	For $AuBe_5$, $a = 6.70$ A. **Examples:** Au_5Ca, Be_5Pd, $HfNi_5$, Ni_5U.
	Cu_2Mg	$Fd3m$	Eight magnesium atoms at $0,0,0$; $0,^1/_2,^1/_2$; $^1/_2,0,^1/_2$; $^1/_2,^1/_2,0$; $^1/_4,^1/_4,^1/_4$; $^1/_4,^3/_4,^3/_4$; $^3/_4,^1/_4,^3/_4$ and $^3/_4,^3/_4,^1/_4$; 16 copper atoms at $^5/_8,^5/_8,^5/_8$; $^5/_8,^1/_8,^1/_8$; $^1/_8,^5/_8,^1/_8$; $^1/_8,^1/_8,^5/_8$; $^5/_8,^7/_8,^7/_8$; $^5/_8,^3/_8,^3/_8$; $^1/_8,^7/_8,^3/_8$; $^1/_8,^3/_8,^7/_8$; $^7/_8,^5/_8,^7/_8$; $^7/_8,^1/_8,^3/_8$; $^3/_8,^5/_8,^3/_8$; $^3/_8,^1/_8,^7/_8$; $^7/_8,^7/_8,^5/_8$; $^7/_8,^3/_8,^1/_8$; $^3/_8,^7/_8,^1/_8$ and $^3/_8,^3/_8,^5/_8$. For Cu_2Mg, $a = 7.05$ A. **Examples:** Al_2Ca, Al_2U, CdCuZn, Co_2U, Co_2Zr, Cr_2Ti, Cu_2Mg, FeNiTa, Fe_2U, Fe_2Zr, MgNiZn, α $TiCo_2$, ZrW_2.
$cF52$	$B_{12}U$	$Fm3m$	For $B_{12}U$, $a = 7.48$ A. **Examples:** $B_{12}Dy$, $B_{12}Sc$, $B_{12}Zr$.
$cF56$	Al_2MgO_4	$Fd3m$	Spinel. **Examples:** Al_2CrS_4, Al_2MgO_4, Co_2NiS_4, Co_3O_4, Co_3S_4, CuS_4Ti_2, $FeNi_2S_4$, Fe_3O_4, Fe_3S_4 (greigite), Ni_3S_4 (LT).
$cF112$	$NaZn_{13}$	$Fm3c$	For $NaZn_{13}$, $a = 12.28$ A. **Examples:** $AmBe_{13}$, $Be_{13}Ho$, $Be_{13}Zr$, $Cd_{13}Cs$, KZn_{13}.
	CFe_3W_3	$Fd3m$	Eta-carbide. For CFe_3W_3, $a = 11.06$ A. **Examples:** $CoNb_2(C, N,O)_x$, Co_2Mo_4C, Cr_3Nb_3C, η Fe_2Nb_3, Fe_3Mo_3C, Fe_3Mo_3N, Fe_3W_3C, Mn_3Mo_3C, Mo_3Ni_3C, $NiTi_2$, Ni_3W_3C.
$cF116$	C_6Cr_{23}	$Fm3m$	For C_6Cr_{23}, $a = 10.66$ A. **Examples:** C_6Mn_{23}, B_2GaNi_7, $B_6Mg_3Ni_{20}$, $C_6Fe_{21}Mo_2$.
	$Mn_{23}Th_6$ ($Cu_{16}Mg_6Si_7$)	$Fm3m$	For $Mn_{23}Th_6$, $a = 12.52$ A. **Examples:** Dy_6Mn_{23}, $Mg_{23}Sr_6$, $Co_{16}Hf_6Si_7$, $Mn_6Ni_{16}Si_7$.
$cI2$	W	$Im3m$	Two atoms at $0,0,0$ and $^1/_2,^1/_2,^1/_2$. For W, $a = 3.16$ A. See Fig. 1. **Examples:** Ba, Nb, Cr, Cs, β Cu-Zn (HT), α-Fe, δ-Fe, K, β-Li, Mo, β-Na, Rb, Ta, V, W.
$cI26$	$Al_{12}W$	$Im3$	For $Al_{12}W$, $a = 7.58$ A. **Examples:** $Al_{12}Mo$, $Al_{12}Re$, $Al_{12}Tc$.
$cI28$	P_4Th_3	$I\bar{4}3d$	For P_4Th_3, $a = 8.62$ A. **Examples:** As_4Th_3, $BaLa_2S_4$, Bi_4U_3, Gd_2Se_4Sr, Se_4U_3.
$cI32$	As_3Co	$Im3$	For As_3Co, $a = 8.20$ A. **Examples:** As_3Ir, $CoSb_3$, IrP_3, $RhSb_3$.
$cI40$	C_3Pu_2	$I\bar{4}3d$	For C_3Pu_2, $a = 8.13$ A. **Examples:** C_3Ce_2, C_3Pr_2, C_3U_2.
$cI52$	Cu_5Zn_8	$I\bar{4}3m$	Gamma brass. For Cu_5Zn_8, $a = 8.86$ A. **Examples:** Ag_5Cd_8, Ag_5Zn_8, Al_8V_5, Cd_8Cu_5.
$cI58$	α-Mn	$I\bar{4}3m$	Alpha manganese appears to be an ordered array of either two or three physically distinguishable types of manganese atoms located on four crystallographically different sets of positions. These sets of positions have an ordered array of atoms in the chi-phase structure ($Fe_{36}Cr_{12}Mo_{12}$ and $Al_{12}Mg_{17}$). For α-Mn, $a = 8.91$ A. Other closely related structures are the mu, P, R and delta phases (see Ref 5, 6 and 7). **Examples:** γ $Al_{12}Mg_{17}$, χ $Co_5Cr_3Si_2$, $CrMn_2$, $Fe_{36}Cr_{12}Mo_{12}$, $Fe_5Si_2V_3$, α-Mn.
$cI80$	β Mn_2O_3	$Ia3$	Type C. For Mn_2O_3, $a = 9.41$ A. **Examples:** Am_2O_3, Be_3P_2, Cd_3N_2, N_2Zn_3, O_3Y_2.
$hP2$	WC	$P\bar{6}m2$	One tungsten atom at $0,0,0$; one carbon atom at $^1/_3,^2/_3,^1/_2$ or at $^2/_3,^1/_3,^1/_2$. For WC, $a = 2.91$ A and $c = 2.84$ A. **Examples:** γ MoC, TiS, WC, WN, Zr_3S_2.
	Mg	$P6_3/mmc$	Hexagonal close-packed. Atoms at $^1/_3,^2/_3,^1/_4$; $^2/_3,^1/_3,^3/_4$. These are the positions of the In atoms of the Ni_2In structure shown in Fig. 1. For Mg, $a = 3.21$ A, $c = 5.20$ A. **Examples:** α-Be, Cd, α-Co, Mg, α-Ti, Zn, α-Zr.
$hP3$	CdI_2	$P\bar{3}m1$	One cadmium atom at $0,0,0$; two iodine atoms at $^1/_3,^2/_3,z$; $^2/_3,^1/_3,\bar{z}$, with $z = ^1/_4$. For SnS_2, $a = 3.65$ A, $c = 5.88$ A. **Examples:** HfS_2, $PdTe_2$, S_2Ta, Se_2V, Te_2Zr.
	Fe_2N or CW_2	$P6_3/mmc$	Two iron atoms at $^1/_3,^2/_3,^1/_4$ and $^2/_3,^1/_3,^3/_4$; one nitrogen atom at either $0,0,0$ or $0,0,^1/_2$. See Fig. 1. **Examples:** Fe_2N, ζ Mn_2N, β Ta_2C, $Ta_{-2}N$, V_2C, W_2C (LT).
	Se	$P3_121$ or $P3_221$	Three atoms at $x,0,^1/_3$; $0,x,^2/_3$ and $\bar{x},\bar{x},0$ (or at $x,0,^2/_3$; $0,x,^1/_3$ and $\bar{x},\bar{x},0$). For Se, $a = 4.36$ A, $c = 4.96$ A and $x = 0.217$. **Example:** γ-Se.
	AlB_2	$P6/mmm$	One aluminum atom at $0,0,0$; two boron atoms at $^1/_3,^2/_3,^1/_2$; $^2/_3,^1/_3,^1/_2$. For AlB_2, $a = 3.01$ A, $c = 3.26$ A. **Examples:** Al_2Th, B_2Lu, B_2Mg, Ga_2Ho, Hg_2U, $PuZr_2$.

(continued)

Table 4. (continued)

Pearson symbol	Prototype	Space group	Unit-cell description
hP4	AsNi	$P6_3/mmc$	Two nickel atoms at $0,0,0$ and $0,0,^1/_2$; two arsenic atoms at $^1/_3,^2/_3,^1/_4$ and $^2/_3,^1/_3,^3/_4$. For AsNi, $a = 3.62$ A, $c = 5.03$ A. **Examples:** CoSb, CoSe, CoTe, CrH, CrSe, α'' FeS, MnSb, NiAs, NiSb, NiTe, TiS, VS, VSb.
	C	$P6_3/mmc$	Graphite. Two carbon(1) atoms at $0,0,^1/_4$; $0,0,^3/_4$. Two carbon(2) atoms at $^2/_3,^1/_3,^1/_4$; $^1/_3,^2/_3,^3/_4$. For hexagonal graphite, $a = 2.46$ A, $c = 6.71$ A. See Fig. 1.
	La	$P6_3/mmc$	Two lanthanum(1) atoms at $0,0,0$; $0,0,^1/_2$; two lanthanum(2) at $^1/_3,^2/_3,^1/_4$; $^2/_3,^1/_3,^3/_4$. For La, $a = 3.77$ A, $c = 12.16$ A. **Examples:** Am, α-Nd, α-Pr, α-Pm.
	SZn	$P6_3mc$	Wurtzite. Two zinc atoms at $^1/_3,^2/_3,z$ and $^2/_3,^1/_3,^1/_2+z$ (with $z = 0$); two sulfur atoms at $^1/_3,^2/_3,z$ and $^2/_3,^1/_3,^1/_2+z$ (with $z = 0.371$). For ZnS (wurtzite), $a = 3.82$ A and $c = 6.26$ A. See Fig. 1. **Examples:** AlN, BeO, CdS, CdSe, CuH, InN, InSb, γ MnS, ZnO, ZnS (wurtzite), ZnSe.
hP5	Al$_3$Ni$_2$	$P\bar{3}m1$	One aluminum(1) atom at $0,0,0$; two aluminum(2) at $^1/_3,^2/_3,z$; $^2/_3,^1/_3,\bar{z}$; with $z = 0.648$. Two nickel atoms as Al(2) with $z = 0.149$. For Al$_3$Ni$_2$, $a = 4.04$ A, $c = 4.90$ A. **Examples:** Al$_3$Pd$_2$, Al$_3$Ru$_2$, Ga$_3$Ni$_2$, In$_3$Pt$_2$.
	La$_2$O$_3$	$P\bar{3}m1$	Type A. One oxygen(1) atom at $0,0,0$; two oxygen(2) atoms at positions of Al(2) in Al$_3$Ni$_2$ hP5, with $z = 0.63$, La similarly with $z = 0.235$. For La$_2$O$_3$, $a = 3.94$ A, $c = 6.13$ A. **Examples:** Ce$_2$O$_3$, N$_3$Th$_2$, N$_3$U$_2$, O$_3$Pu$_2$.
hP6	CaCu$_5$	$P6/mmm$	One calcium atom at $0,0,0$; two copper(1) atoms at $^1/_3,^2/_3,0$; $^2/_3,^1/_3,0$; three copper(2) at $^1/_2,0,^1/_2$; $0,^1/_2,^1/_2$; $^1/_2,^1/_2,^1/_2$. For CaCu$_5$, $a = 5.09$ A, $c = 4.09$ A. **Examples:** Ag$_5$Ba, Be$_5$Sc, CeNi$_5$, Cu$_5$Y, LaNi$_5$, Pt$_5$Sr.
	CoSn	$P6/mmm$	For CoSn, $a = 5.28$ A, $c = 4.26$ A. **Examples:** FeGe, FeSn, InNi, PtTl.
	CaIn$_2$	$P6_3/mmc$	Two calcium atoms at $0,0,^1/_4$; $0,0,^3/_4$; four indium atoms at $^1/_3,^2/_3,z$; $^2/_3,^1/_3,^1/_2+z$; $^2/_3,^1/_3,\bar{z}$; $^1/_3,^2/_3,^1/_2-z$; with $z = 0.455$. For CaIn$_2$, $a = 4.90$ A, $c = 7.75$ A. **Examples:** EuIn$_2$, Ga$_2$Y, In$_2$Sr, SrTl$_2$.
	InNi$_2$	$P6_3/mmc$	Two nickel atoms at $0,0,0$ and $0,0,^1/_2$; two nickel atoms at $^1/_3,^2/_3,^3/_4$ and $^2/_3,^1/_3,^1/_4$; two indium atoms at $^1/_3,^2/_3,^1/_4$ and $^2/_3,^1/_3,^3/_4$. For InNi$_2$, $a = 4.18$ A and $c = 5.13$ A. See Fig. 1. **Examples:** AlZr$_2$, CoNiSn, Cu$_2$In, In$_2$Bi, Ni$_2$In, Ni$_{1.4}$Sn, Ti$_2$Sn.
hP8	H-AlCCr$_2$	$P6_3/mmc$	Two aluminum atoms at $^1/_3,^2/_3,^3/_4$; $^2/_3,^1/_3,^1/_4$; two carbon atoms at $0,0,0$; $0,0,^1/_2$; four chromium atoms at $^1/_3,^2/_3,z$; $^2/_3,^1/_3,\bar{z}$; $^2/_3,^1/_3,^1/_2+z$; $^1/_3,^2/_3,^1/_2-z$; with $z = 0.086$. For AlCCr$_2$, $a = 2.86$ A, $c = 12.82$ A. **Examples:**

Pearson symbol	Prototype	Space group	Unit-cell description
			AlCNb$_2$, CCdTi$_2$, CGeV$_2$, InNTi$_2$.
	AsNa$_3$	$P6_3/mmc$	For AsNa$_3$, $a = 5.10$ A, $c = 9.00$ A. **Examples:** AuMg$_3$, K$_3$P, Li$_3$Sb, Mg$_3$Pt.
hP8	Ni$_3$Sn	$P6_3/mmc$	Two tin atoms at $^1/_3,^2/_3,^1/_4$ and $^2/_3,^1/_3,^3/_4$; six nickel atoms at $x,2x,^1/_4$; $2\bar{x},\bar{x},^1/_4$; $x,\bar{x},^1/_4$; $\bar{x},2\bar{x},^3/_4$; $2x,x,^3/_4$ and $\bar{x},x,^3/_4$ (with $x = 0.833$). For Ni$_3$Sn, $a = 5.29$ A and $c = 4.24$ A. See Fig. 1. **Examples:** AlTi$_{2-3}$, Cd$_3$Mg, CdMg$_3$, Co$_3$Mo, Co$_3$W, β'' Fe$_3$Sn, γ Ni$_3$In, Ni$_3$Sn, Ti$_4$Pb.
hP9	Fe$_2$P	$P\bar{6}2m$	Three iron atoms at $x,0,0$; $0,x,0$ and $\bar{x},\bar{x},0$ (with $x = 0.256$); three iron atoms at $x,0,^1/_2$; $0,x,^1/_2$ and $\bar{x},\bar{x},^1/_2$ (with $x = 0.594$); three phosphorus atoms at $0,0,^1/_2$; $^1/_3,^2/_3,0$ and $^2/_3,^1/_3,0$. For Fe$_2$P, $a = 5.93$ A and $c = 3.45$ A. See Fig. 1. **Examples:** Fe$_2$P, Mn$_2$P, Ni$_2$P, Pt$_2$Si (HT).
hP12	MgZn$_2$	$P6_3/mmc$	Four magnesium atoms at $^1/_3,^2/_3,z$; $^2/_3,^1/_3,\bar{z}$; $^2/_3,^1/_3,^1/_2+z$ and $^1/_3,^2/_3,^1/_2-z$ (with $z = 0.062$); two zinc atoms at $0,0,0$ and $0,0,^1/_2$; six zinc atoms at $x,2x,^1/_4$; $2\bar{x},\bar{x},^1/_4$; $x,\bar{x},^1/_4$; $\bar{x},2\bar{x},^3/_4$; $2x,x,^3/_4$ and $\bar{x},x,^3/_4$ (with $x = 0.83$). For MgZn$_2$, $a = 5.18$ A and $c = 8.52$ A. See Fig. 1. **Examples:** Al$_2$Zr, Be$_2$Mo, CaCd$_2$, CaMg$_2$, CdCu$_2$, Fe$_2$Mo, FeSiW, Fe$_2$Ta, Fe$_2$Ti, Fe$_2$W, MgZn$_2$, TiZn$_2$.
hP16	Mn$_5$Si$_3$	$P6_3/mcm$	For Mn$_5$Si$_3$, $a = 6.91$ A, $c = 4.81$ A. **Examples:** Al$_3$Zr$_5$, Dy$_5$Ge$_3$, Ga$_3$Y$_5$, P$_3$Ti$_5$, Sn$_3$Zr$_5$.
	Ni$_3$Ti	$P6_3/mmc$	Four titanium atoms at $0,0,0$; $0,0,^1/_2$; $^1/_3,^2/_3,^1/_4$ and $^2/_3,^1/_3,^3/_4$; 12 nickel atoms at $^1/_2,0,0$; $0,^1/_2,0$; $^1/_2,^1/_2,0$; $^1/_2,0,^1/_2$; $0,^1/_2,^1/_2$; $^1/_2,^1/_2,^1/_2$; $x,2x,^1/_4$; $2\bar{x},\bar{x},^1/_4$; $x,\bar{x},^1/_4$; $\bar{x},2\bar{x},^3/_4$; $2x,x,^3/_4$ and $\bar{x},x,^3/_4$ (with $x = 0.833$). For Ni$_3$Ti, $a = 2.55$ A and $c = 8.31$ A. **Examples:** Co$_3$Ti, Ni$_3$Ti, Pd$_3$Zr.
hP20	Fe$_3$Th$_7$	$P6_3/mc$	For Fe$_3$Th$_7$, $a = 9.85$ A, $c = 6.15$ A. **Examples:** B$_3$Re$_7$, C$_3$Fe$_7$, Co$_3$Th$_7$, Ru$_3$Th$_7$.
hP24	HoH$_3$	$P\bar{3}c1$	For HoH$_3$, $a = 6.31$ A, $c = 6.56$ A. **Examples:** DyH$_3$, LuH$_3$, YH$_3$.
	MgNi$_2$	$P6_3/mmc$	For MgNi$_2$, $a = 4.82$ A, $c = 15.80$ A. **Examples:** CdCu$_2$, Cr$_2$Hf, Fe$_2$Sc, NbZn$_2$.
hR1	Hg	$R\bar{3}m$	One atom per cell, at $0,0,0$. For Hg, $a = 3.005$ A and $\alpha = 70°$ $32'$. A hexagonal cell, where $a = 3.47$ A and $c = 6.74$ A, has three atoms per cell, at $0,0,0$; $^1/_3,^2/_3,^2/_3$ and $^2/_3,^1/_3,^1/_3$. **Example:** Hg.
hR2	As	$R\bar{3}m$	In the cell based on hexagonal axes, there are six atoms, at $0,0,z$; $^1/_3,^2/_3,^2/_3+z$; $^2/_3,^1/_3,^1/_3+z$; $0,0,\bar{z}$; $^1/_3,^2/_3,^2/_3-z$ and $^2/_3,^1/_3,^1/_3-z$; where $z = 0.226$, $a = 3.76$ A and $c = 10.55$ A for α-As. The cell based on rhombohedral axes contains two atoms, at x,x,x and $\bar{x},\bar{x},\bar{x}$; where $x = 0.276$, $a = 4.13$ A and $\alpha = 54°$ $8'$ for α-As. **Examples:** α-As, Bi, Sb.

(continued)

Table 4. (continued)

Pearson symbol	Prototype	Space group	Unit-cell description	Pearson symbol	Prototype	Space group	Unit-cell description
$hR3$	Sm	$R\bar{3}m$	In a cell based on a hexagonal axis, there are nine atoms per cell, at $0,0,0$; $^1/_3,^2/_3,^2/_3$; $^2/_3,^1/_3,^1/_3$; $0,0,z$; $0,0,\bar{z}$; $^1/_3,^2/_3,^2/_3+z$; $^1/_3,^2/_3,^2/_3-z$; $^2/_3,^1/_3,^1/_3+z$ and $^2/_3,^1/_3,^1/_3-z$. For α-Sm, $a = 3.621$ A, $c = 26.25$ A and $z = {}^2/_9$. The cell based on rhombohedral axes contains three atoms per cell, at $0,0,0$; x,x,x and $\bar{x},\bar{x},\bar{x}$; with $a = 9.00$ A, $\alpha = 23° 19'$ and $x = {}^2/_9$. **Examples:** α-Sm, Ce-Y, δ Nd-Tm, δ Pr-Y.	$tP6$	Cu$_2$Sb (GeSeZr)	$P4/nmm$	Two copper atoms at $0,0,0$; $^1/_2,^1/_2,0$; two at $0,^1/_2,z$ and $^1/_2,0,\bar{z}$ (with $z = 0.27$); two antimony atoms at $0,^1/_2,z$ and $^1/_2,0,\bar{z}$ (with $z = 0.70$). Thus ordered ternary phases also take the structure. For Cu$_2$Sb, $a = 3.99$ A and $c = 6.09$ A. See Fig. 1. **Examples:** AlNaSi$_4$, AsCr$_2$, AsMn$_2$, Bi$_2$U, Cu$_2$Sb, Mn$_2$Sb, Pu$_2$U, Sb$_2$U.
$hR5$	Bi$_2$STe$_2$	$R\bar{3}m$	One sulfur atom at $0,0,0$; two bismuth atoms at x,x,x; $\bar{x},\bar{x},\bar{x}$ with $x = 0.392$. Tellurium atoms similarly with $x = 0.788$. For Bi$_2$STe$_2$, $a = 10.33$ A, $\alpha = 24° 10'$. **Examples:** Bi$_2$Se$_3$ Bi$_2$SeTe$_2$, Sb$_2$Te$_3$.		O$_2$Ti	$P4_2/mnm$	Rutile. Two titanium atoms at $0,0,0$ and $^1/_2,^1/_2,^1/_2$; four oxygen atoms at $x,x,0$; $\bar{x},\bar{x},0$; $^1/_2+x,^1/_2-x,^1/_2$ and $^1/_2-x,^1/_2+x,^1/_2$. For O$_2$Ti, $x = 0.3056$, $a = 4.59$ A and $c = 2.96$ A. See Fig. 1. **Examples:** CrO$_2$, β MnO$_2$, PbO$_2$, SnO$_2$, TaO$_2$, TeO$_2$, TiO$_2$ (rutile), VO$_2$ (HT), WO$_2$.
$hR10$	α Al$_2$O$_3$	$R\bar{3}c$	Ten atoms per unit rhombohedral cell or 30 atoms per unit hexagonal cell. (There are also other structures of alumina.) **Examples:** α Al$_2$O$_3$, α Fe$_2$O$_3$, Rh$_2$O$_3$, Ti$_2$O$_3$, V$_2$O$_3$ (HT).	$tP10$	Si$_2$U$_3$	$P4/mbm$	For Si$_2$U$_3$, $a = 7.33$ A, $c = 3.90$ A. **Examples:** Al$_2$Th$_3$, B$_2$V$_3$, Be$_2$Ta$_3$, Ga$_2$Ta$_3$, Si$_2$Zr$_3$.
$hR13$	Fe$_7$W$_6$	$R\bar{3}m$	Mu phase. For Fe$_7$W$_6$, $a = 9.04$ A, $\alpha = 30° 30'$. **Examples:** Co$_7$Mo$_6$, Fe$_7$Mo$_6$, (Fe,Si)$_7$Re$_6$.	$tP16$	CoGa$_3$	$P\bar{4}n2$	For CoGa$_3$, $a = 6.26$ A, $c = 6.48$ A. **Examples:** FeGa$_3$, Ga$_3$Os, In$_3$Ir, In$_3$Rh.
$hR19$	Th$_2$Zn$_{17}$	$R\bar{3}m$	For Th$_2$Zn$_{17}$ in the hexagonal cell containing 57 atoms, $a = 9.03$ A, $c = 13.20$ A. **Examples:** Ba$_2$Mg$_{17}$, Ce$_2$Fe$_{17}$, U$_2$Zn$_{17}$.	$tP20$	B$_4$Th	$P4/mbm$	For B$_4$Th, $a = 7.26$ A, $c = 4.11$ A. **Examples:** B$_4$Ce, B$_4$Ho, B$_4$Pu, B$_4$U, B$_4$Y.
$hR26$	Al$_8$Cr$_5$	$R3m$	Rhombohedrally distorted gamma brass. For Al$_8$Cr$_5$, $a = 7.80$ A, $\alpha = 109°$. **Examples:** Fe-Ga, Ga-Mn.	$tP30$	CrFe (sigma phase)	$P4_2/mnm$	This is a complex structure formed by transition-metal alloys in which the metal atoms occupy five independent site-sets. For CrFe, $a = 8.80$ A, $c = 4.54$ A. **Examples:** AlNb$_2$, CoCr, Co$_2$Mo$_3$, Cr$_2$Os, Cr$_2$Ru, FeMo, FeV, MoTc$_3$, Ni$_2$V$_3$, Re$_3$Ta$_2$.
$hR53$	R Co-Cr-Mo	$R\bar{3}$	R-phase. For R Co-Cr-Mo, $a = 9.01$ A, $\alpha = 74° 27'$. See α-Mn $cI58$ type, above; and Ref 6, 7 and 8. **Examples:** R-(Co-Cr-Mo), Co$_3$Cr$_3$Si$_2$, R-(Co-Mn-Mo), R-Fe$_{52}$Mn$_{16}$Mo$_{32}$, Fe$_2$SiV$_2$, Mn$_{78}$Mo$_3$Si$_{19}$, Mn$_6$Si, Ni$_3$SiV$_6$.	$tI2$	In	$I4/mmm$	Two atoms per cell, at $0,0,0$ and $^1/_2,^1/_2,^1/_2$. It is conventional also to use the cell that has four atoms, at $0,0,0$; $0,^1/_2,^1/_2$; $^1/_2,0,^1/_2$ and $^1/_2,^1/_2,0$. For In, $a = b = 4.60$ A and $c = 4.95$ A. At room temperature, the unit cell resembles that of Cu, which is shown in Fig. 1. **Examples:** δ GaNi$_2$, In, InPd$_3$.
$tP4$ ($tP2$)	AuCu	$P4/mmm$	Two gold atoms at $0,0,0$ and $^1/_2,^1/_2,0$; two copper atoms at $0,^1/_2,^1/_2$ and $^1/_2,0,^1/_2$. For AuCu, $a = 3.97$ A, $c = 3.67$ A. A smaller cell can also be taken with $a/\sqrt{2}$ and Au and Cu atoms at $0,0,0$ and $^1/_2,^1/_2,^1/_2$ respectively. See Fig. 1. **Examples:** AgTi, AlTi, AuCu I, θ CdPt, FePd, γ″ FePt, θ MnNi, NiPt.		Pa	$I4/mmm$	Atom positions as for In $tI2$, but $c/a < 1$. For Pa, $a = 3.93$ A, $c = 3.24$ A.
	CuTi$_3$	$P4/mmm$	One copper atom at $0,0,0$; one titanium(1) atom at $^1/_2,^1/_2,0$; two titanium(2) atoms at $0,^1/_2,^1/_2$; $^1/_2,0,^1/_2$. For CuTi$_3$, $a = 4.16$ A, $c = 3.59$ A. **Examples:** AgZr$_3$, BaBi$_3$, DyIn$_3$, Pd$_3$Tl.	$tI4$-x	Fe-C	$I4/mmm$	Martensite. In the unit cell there are iron atoms at $0,0,0$ and $^1/_2,^1/_2,^1/_2$; the carbon atoms are random, at $^1/_2,^1/_2,0$ and/or $0,0,^1/_2$, to provide two iron atoms and up to 0.12 carbon atoms per cell. **Examples:** Fe-C martensite, α' Fe-N martensite.
	γ CuTi	$P4/nmm$	Two copper atoms at $0,^1/_2,z$ and $^1/_2,0,\bar{z}$ (with $z = 0.10$); two titanium atoms at $0,^1/_2,z$ and $^1/_2,0,\bar{z}$ (with $z = 0.65$). For γ CuTi, $a = 3.12$ A and $c = 5.92$ A. See Fig. 1. **Examples:** AgZr, AuTi (LT), γ CuTi.	$tI4$	Sn	$I4_1/amd$	White Sn. Four atoms per cell, at $0,0,0$; $^1/_2,^1/_2,^1/_2$; $0,^1/_2,^1/_4$ and $^1/_2,0,^3/_4$. For β-Sn, $a = 5.83$ A and $c = 3.18$ A. See Fig. 1. **Examples:** AlSb II (HP), InSb II (HP), β-Sn (white).
	PbO	$P4/nmm$	Two oxygen atoms at $0,0,0$ and $^1/_2,^1/_2,0$; two lead atoms at $0,^1/_2,z$ and $^1/_2,0,\bar{z}$ (with $z = 0.237$). For PbO, $a = 3.97$ A, $c = 5.02$ A. See Fig. 1. **Examples:** FeS, β FeTe$_{0.9}$, PbO, SnO.	$tI6$	C$_2$Ca	$I4/mmm$	Two calcium atoms at $0,0,0$; $^1/_2,^1/_2,^1/_2$. Four carbon atoms at $0,0,z$; $0,0,\bar{z}$; $^1/_2,^1/_2,^1/_2+z$; $^1/_2,^1/_2,^1/_2-z$ with $z = 0.407$, form dumbbells along [001]. For C$_2$Ca, $a = 3.88$ A, $c = 6.37$ A. **Examples:** BaC$_2$, C$_2$Er, C$_2$Nd, C$_2$U, O$_2$Sr.

(continued)

Table 4. (continued)

Pearson symbol	Prototype	Space group	Unit-cell description	Pearson symbol	Prototype	Space group	Unit-cell description
$tI6$	$MoSi_2$	$I4/mmm$	Two molybdenum atoms at 0,0,0 and $^1/_2,^1/_2,^1/_2$; four silicon atoms at $0,0,z$; $0,0,\bar{z}$; $^1/_2,^1/_2,^1/_2+z$ and $^1/_2,^1/_2,^1/_2-z$ (with $z = 0.333$). For $MoSi_2$, $a = 3.20$ A and $c = 7.86$ A. See Fig. 1. **Examples:** $AgZr_2$, $AlCr_2$, Au_2Be, Au_2Mn, $CuTi_2$, Hg_2Mg, $MoSi_2$, Ni_2Ta, Si_2W.	$tI32$	B_3Cr_5	$I4/mcm$	For B_3Cr_5, $a = 5.46$ A, $c = 10.64$ A. **Examples:** Ag_3Ca_5, B_2Co_5P, Ba_5Pb_3, Si_3Ta_5.
					Si_3W_5	$I4/mcm$	For Si_3W_5, $a = 9.61$ A, $c = 4.96$ A. **Examples:** $CeCo_3Pu_4$, Cr_5Ge_3, Ga_3Ti_5, Re_5Si_3.
$tI8$	Al_3Ti	$I4/mmm$	Two titanium atoms at 0,0,0; $^1/_2,^1/_2,^1/_2$. Two aluminum(1) atoms at $0,0,^1/_2$; $^1/_2,^1/_2,0$. Four aluminum(2) atoms at $0,^1/_2,^1/_4$; $^1/_2,0,^1/_4$; $^1/_2,0,^3/_4$; $0,^1/_2,^3/_4$. For Al_3Ti, $a = 3.85$ A, $c = 8.60$ A. **Examples:** Al_3Nb, Ga_3Ti, Ni_3Ta, Pd_3V, Pt_3V.	$tI48$	$BaCd_{11}$	$I4_1/amd$	For $BaCd_{11}$, $a = 12.02$ A, $c = 7.74$ A. **Examples:** $Cd_{11}Eu$, $LaZn_{11}$, $PrZn_{11}$.
				$tI64$	$NaPb$	$I4_1/acd$	For $NaPb$, $a = 10.58$ A, $c = 17.75$ A. **Examples:** $CsPb$, KSn, $PbRb$, $RbSn$.
$tI10$	$MoNi_4$	$I4/m$	For $MoNi_4$, $a = 5.73$ A, $c = 3.57$ A. **Examples:** Au_4Cr, Au_4Ti, Ni_4W.	$oP4$	$AuCd$	$Pmma$	Two gold atoms at $^1/_4,^1/_2,z$ and $^3/_4,^1/_2,\bar{z}$ (with $z = 0.812$); two cadmium atoms at $^1/_4,0,z$ and $^3/_4,0,\bar{z}$ (with $z = 0.313$). For β' $AuCd$, $a = 4.76$ A, $b = 3.15$ A and $c = 4.86$ A. See Fig. 1. **Examples:** β'' $AgCd$, β' $AuCd$, $CdMg$, $IrMo$, IrW.
	Al_4Ba (Cu_2Si_2Th)	$I4/mmm$	Two barium atoms at 0,0,0; $^1/_2,^1/_2,^1/_2$. Four aluminum(1) atoms at $0,^1/_2,^1/_4$; $^1/_2,0,^1/_4$; $^1/_2,0,^3/_4$; $0,^1/_2,^3/_4$. Four aluminum(2) atoms at $0,0,z$; $0,0,\bar{z}$; $^1/_2,^1/_2,^1/_2+z$; $^1/_2,^1/_2,^1/_2-z$, with $z = 0.380$. The three atom-sites allow the formation of ordered ternary phases. For Al_4Ba, $a = 4.57$ A, $c = 11.25$ A. **Examples:** $BaGa_4$, $ThZn_4$, B_2Co_2Nd, Au_2GdSi_2. N.B. There are more recognized phases with this structure type than with any other structure of metallurgical interest.				
				$oP6$	FeS_2	$Pnnm$	Marcasite. Two iron atoms at 0,0,0 and $^1/_2,^1/_2,^1/_2$; four sulfur atoms at $x,y,0$; $\bar{x},\bar{y},0$; $^1/_2+x,^1/_2-y,^1/_2$ and $^1/_2-x,^1/_2+y,^1/_2$ (with $x = 0.200$ and $y = 0.378$). For FeS_2 (marcasite), $a = 4.44$ A, $b = 5.42$ A and $c = 3.39$ A. See Fig. 1. **Examples:** γ $CrSb_2$, FeP_2, FeS_2 (marcasite), $FeSe_2$, $FeTe_2$, $NiSb_2$.
$tI12$	Al_2Cu	$I4/mcm$	Four copper atoms at $0,0,^1/_4$; $^1/_2,^1/_2,^3/_4$; $0,0,^3/_4$ and $^1/_2,^1/_2,^1/_4$; eight aluminum atoms at $x,^1/_2+x,0$; $\bar{x},^1/_2-x,0$; $^1/_2+x,\bar{x},0$; $^1/_2-x,x,^1/_2$; $^1/_2-x,\bar{x},^1/_2$; $x,^1/_2-x,^1/_2$ and $\bar{x},^1/_2+x,^1/_2$ (with $x = 0.158$). For Al_2Cu, $a = 6.07$ A and $c = 4.87$ A. **Examples:** Co_2B, Cr_2B, θ $CuAl_2$, Fe_2B, $FeSn_2$, Mo_2B, Ni_2B, W_2B.				
				$oP8$	β Cu_3Ti	$Pmmn$	For Cu_3Ti, $a = 5.16$ A, $b = 4.35$ A, $c = 4.53$ A. **Examples:** Au_3Hf, $MoNi_3$, Ni_3Sb, Pt_3Ta.
	Si_2Th	$I4_1/amd$	For Si_2Th, $a = 4.14$ A, $c = 14.38$ A. **Examples:** $CeGe_2$, $DySi_{1.4}$, Ga_2Th, $LaSi_2$.		BFe	$Pnma$	For BFe, $a = 5.51$ A, $b = 2.95$ A, $c = 4.06$ A. **Examples:** BCo, BMn, $DyNi$, $LuPt$, $SiTi$.
$tI16$	$CuFeS_2$	$I\bar{4}2d$	Chalcopyrite. For $CuFeS_2$, $a = 5.25$ A, $c = 10.32$ A. **Examples:** $AgAlS_2$, $AlCuSe_2$, GeP_2Zn, CuS_2Tl.		GeS	$Pnma$	For GeS, $a = 10.44$ A, $b = 3.65$ A, $c = 4.30$ A. **Examples:** $GeSe$, $PbSe$ II, SnS, $SnTe$ II.
	Al_3Zr	$I4/mmm$	For Al_3Zr, $a = 4.01$ A, $c = 17.32$ A. **Examples:** Al_3Hf, Ga_3Zr.		MnP	$Pnma$	Four manganese atoms at $x,^1/_4,z$; $\bar{x},^3/_4,\bar{z}$; $^1/_2-x,^3/_4,^1/_2+z$ and $^1/_2+x,^1/_4,^1/_2-z$ (with $x = 0.20$ and $z = 0.005$); four phosphorus atoms at $x,^1/_4,z$; $\bar{x},^3/_4,\bar{z}$; $^1/_2-x,^3/_4,^1/_2+z$ and $^1/_2+x,^1/_4,^1/_2-z$ (with $x = 0.57$ and $z = 0.19$). For MnP, $a = 5.26$ A, $b = 3.17$ A, $c = 5.92$ A. **Examples:** CoP, CrP, FeP, MnP, WP.
	SiU_3	$I4/mcm$	For SiU_3, $a = 6.03$ A, $c = 8.70$ A. **Examples:** Ir_3Si, Pt_3Si.				
	$SeTl$	$I4/mcm$	For $SeTl$, $a = 8.04$ A, $c = 7.01$ A. **Examples:** $InTe$, STl.				
$tI18$	Te_4Ti_5	$I4/m$	For Te_4Ti_5, $a = 10.16$ A, $c = 3.77$ A. **Examples:** Nb_5Sb_4, Nb_5Te_4, Sb_4Ta_5.	$oP12$	Cl_2Pb ($NiSiTi$)	$Pnma$	The atoms occupy three independent site-sets, so that ordered ternary phases occur with this structure. For $AlPd_2$, $a = 5.41$ A, $b = 4.06$ A, $c = 7.77$ A. **Examples:** $AlPd_2$, As_2Zr, $CoMnSi$, $CoGeV$, $NiSiZr$.
$tI26$	$Mn_{12}Th$	$I4/mmm$	For $Mn_{12}Th$, $a = 8.74$ A, $c = 4.95$ A. **Examples:** $AgBe_{12}$, Al_8CeMn_4, $Mg_{12}Nd$, $Mn_{12}Y$.	$oP16$	$AlEr$	$Pmma$	For $AlEr$, $a = 5.57$ A, $b = 5.80$ A, $c = 11.27$ A. **Examples:** $AlDy$, $AlHo$, $AlTb$.
$tI28$	MnU_6	$I4/mcm$	For MnU_6, $a = 10.29$ A, $c = 5.24$ A. **Examples:** CoU_6, $FePu_6$, NiU_6.		Ge_3Rh_5	$Pbam$	For Ge_3Rh_5, $a = 5.42$ A, $b = 10.32$ A, $c = 3.96$ A. **Examples:** Al_3Pd_5, Ge_3Rh_5, In_3Pd_5, Rh_5Si_3.
					CFe_3	$Pnma$	Cementite. For CFe_3, $a = 5.09$ A, $b = 6.74$ A, $c = 4.52$ A. **Examples:** BCo_3, CCo_3, CNi_3, PPd_3.
$tI32$	Ni_3P	$I\bar{4}$	For Ni_3P, $a = 8.95$ A, $c = 4.39$ A. **Examples:** Cr_3P, Fe_3P, Mo_3P.	$oP20$	S_3Sb_2	$Pbnm$	For S_3Sb_2, $a = 11.25$ A, $b = 11.33$ A, $c = 3.84$ A. **Examples:** Bi_2S_3, Dy_2Se_3, Np_2S_3, Se_3U_2.

(continued)

Table 4. (continued)

Pearson symbol	Prototype	Space group	Unit-cell description	Pearson symbol	Prototype	Space group	Unit-cell description
$oP56$	P-$Cr_{18}Mo_{42}Ni_{40}$	$Pbnm$	P-phase. For $Cr_{18}Mo_{42}Ni_{40}$, $a = 9.07$ A, $b = 16.98$ A, $c = 4.75$ A. See α-Mn $cI58$ type, above, and Ref 6, 7 and 8. **Examples:** P-$Cr_{18}Mo_{42}Ni_{40}$, P-(Mo-Fe-Ni), P-(Mo-Mn-Co).	$oI12$	$GdSi_2$ ($GdSi_{1.4}$)	$Imma$	Atoms at $0,^1/_4,z$; $0,^3/_4,\bar{z}$; $^1/_2,^3/_4,^1/_2+z$; $^1/_2,^1/_4,^1/_2-z$ with Gd: $z = 0.375$; Si(1): $z = 0.786$; Si(2): $z = 0.964$. For $GdSi_{1.4}$, $a = 4.09$ A, $b = 4.01$ A, $c = 13.44$ A. **Examples:** $CeGe_2$, $DySi_{1.4}$, $HoSi_2$, $PrSi_{1.4}$.
$oC4$	α-U	$Cmcm$	Four atoms at $0,y,^1/_4$; $0,\bar{y},^3/_4$; $^1/_2,^1/_2+y,^1/_4$ and $^1/_2,^1/_2-y,^3/_4$. For α-U, $a = 2.85$ A, $b = 5.87$ A, $c = 4.95$ A and $y = 0.1024$.	$oI14$	B_4Ta_3	$Immm$	For B_4Ta_3, $a = 3.28$ A, $b = 13.38$ A, $c = 3.13$ A. **Examples:** B_4Cr_3, B_4Cr_2Ni, B_4Mn_3, B_4V_3.
$oC8$	BCr	$Cmcm$	Four chromium atoms at $0,y,^1/_4$; $0,\bar{y},^3/_4$; $^1/_2,^1/_2+y,^1/_4$; $^1/_2,^1/_2-y,^3/_4$; with $y = 0.146$. Four boron atoms at similar positions with $y = 0.440$. For BCr, $a = 2.97$ A, $b = 7.86$ A, $c = 2.93$ A. **Examples:** AgCa, BNi, CeNi, HfNi, NdRh, RuTh.	$oI20$	Al_4U	$Imma$	For Al_4U, $a = 4.41$ A, $b = 6.27$ A, $c = 13.71$ A. **Examples:** Al_4Np, Al_4Pu.
	Ga	$Cmca$	Eight atoms at $0,y,z$; $0,\bar{y},\bar{z}$; $^1/_2,y,^1/_2-z$; $^1/_2,\bar{y},^1/_2+z$; $^1/_2,^1/_2+y,z$; $^1/_2,^1/_2-y,\bar{z}$; $0,^1/_2+y,^1/_2-z$ and $0,^1/_2-y,^1/_2+z$. For Ga, $a = 2.90$ A, $b = 8.13$ A, $c = 3.17$ A, $y = 0.1549$ and $z = 0.081$.	$mP8$	Se_3Zr	$P2_1/m$	For Se_3Zr, $a = 5.42$ A, $b = 3.76$ A, $c = 19.0$ A, $\beta = 97.6°$. **Examples:** HfS_3, $HfSe_3$, TiS_3.
				$mP12$	$CoSb_2$	$P2_1/c$	For $CoSb_2$, $a = 6.52$ A, $b = 6.38$ A, $c = 6.55$ A. $\beta = 118.2°$. As the atoms occupy three different site-sets ordered ternary phases with the structure, occur. **Examples:** As_2Co, As_2Ir, IrP_2, $RhSb_2$, $As_{1.1}FeS_{0.9}$, AsOsS, AsOsTe, RuSbTe.
$oC12$	Si_2Zr	$Cmcm$	All atoms at the same sites as for BCr $oC8$ with Zr: $y = 0.102$; Si(1): $y = 0.752$; Si(2): $y = 0.447$. For Si_2Zr, $a = 3.72$ A, $b = 14.68$ A, $c = 3.68$ A. **Examples:** Ge_2Hf, Ge_2U, Si_2Ti.	$mP16$	AsLi	$P2_1/c$	For AsLi, $a = 5.79$ A, $b = 5.24$ A, $c = 10.70$ A, $\beta = 117.4°$. **Examples:** KSb, NaSb.
				$mC12$	As_2Nb	$C2$	For As_2Nb, $a = 9.36$ A, $b = 3.38$ A, $c = 7.79$ A, $\beta = 119.5°$. **Examples:** As_2Mo, $NbSb_2$, Sb_2Ta.
$oC28$	Al_6Mn	$Cmcm$	For Al_6Mn, $a = 7.55$ A, $b = 6.50$ A, $c = 8.87$ A. **Examples:** Al_6Fe, Al_6Re, Al_6Tc.	$mC14$	Cr_3S_4	$C2/m$	For Cr_3S_4, $a = 5.97$ A, $b = 3.43$ A, $c = 11.36$ A, $\beta = 91.2°$. **Examples:** Cr_3Se_4, Fe_3Se_4, $CoCr_2Se_4$.
$oF24$	Si_2Ti	$Fddd$	For Si_2Ti, $a = 8.25$ A, $b = 4.78$ A, $c = 8.54$ A. **Examples:** Al_2Ru, Ga_2Ru, Ge_2Ti, Sn_2Zr.	$mC28$	$CoZn_{13}$	$C2/m$	For $CoZn_{13}$, $a = 13.31$ A, $b = 7.54$ A, $c = 4.99$ A, $\beta = 126.8°$. **Examples:** $CrZn_{13}$, $FeZn_{13}$.
$oI6$	$MoPt_2$	$Immm$	Two molybdenum atoms at $0,0,0$; $^1/_2,^1/_2,^1/_2$. Four platinum atoms at $0,y,0$; $0,\bar{y},0$; $^1/_2,^1/_2+y,^1/_2$; $^1/_2,^1/_2-y,^1/_2$, with $y = 0.353$. For $MoPt_2$, $a = 2.75$ A, $b = 8.24$ A, $c = 3.92$ A. **Examples:** $NbPd_2$, Ni_2V, Pd_2V, Pt_2V.		C_2Mn_5	$C2/c$	For C_2Mn_5, $a = 5.09$ A, $b = 4.57$ A, $c = 11.66$ A, $\beta = 97.7°$. **Examples:** B_2Pd_5, C_2Fe_5.
				$aP8$	HgK	$P\bar{1}$	For HgK, $a = 6.59$ A, $b = 6.76$ A, $c = 7.06$ A, $\alpha = 106°$, $\beta = 102°$, $\gamma = 93°$.
$oI12$	$CeCu_2$	$Imma$	For $CeCu_2$, $a = 4.43$ A, $b = 7.06$ A, $c = 7.48$ A. **Examples:** Ag_2Ca, Cu_2Dy, Cu_2Pr, Hg_2K, $SrZn_2$.	$aP12$	$ReSe_2$	$P\bar{1}$	For $ReSe_2$, $a = 6.73$ A, $b = 6.61$ A, $c = 6.72$ A, $\alpha = 119°$, $\beta = 92°$, $\gamma = 105°$.

Nevertheless, the same space-lattice symmetry (cP) still pertains to the crystal structure. Thus, for example, in the close-packed-hexagonal structure Mg $hP2$, the primitive space-lattice points are vacant and the two Mg atoms are located within the unit cell. The structure type $hP1$, where only the primitive hexagonal space-lattice points are occupied, is unknown.

Alternatively, the space-lattice points may be occupied by atoms and, in addition, there may be groups of other atoms with various point-group symmetries surrounding the atoms on the space-lattice points. Such is the case in the CaF_2 $cF12$ structure (see Fig. 1) of Mg_2Sn. Sn occupies the F space-lattice sites, whereas the Mg atoms surround these sites on positions of point symmetry referred to as $\bar{4}3m$. Such point symmetry is in fact that of a cube, and there are cubes of Mg atoms about each Sn atom on its F space-lattice site in the structure.

Equivalent Positions. In each unit cell, there are positions that are equivalent because of crystal symmetry. This is often true of lattice points at special positions (such as $^1/_2,0,0$), and also of atoms at x, y or z, where the coordinates may have specific values. At each point of a set of equivalent positions, the same kind of atom will be found in each unit cell (of a perfect crystal). The coordinates listed for each kind of atom in the descriptions of crystal structure in Table 4 are thus coordinates of a set of equivalent positions.

The more complete tables, such as those in the International Tables for X-ray Crystallography (Ref 4), show clearly the number of equivalent points belonging to each set (the multiplicity) and thus the number of atoms that could be located at the equivalent points in a crystal. To save space in tabulating the equivalent center points in body-centered or face-centered lattices, the coordinates of the center points are given at the top of the list and, in turn, must be added to each of the coordinates listed below them.

As an example, for arsenic, in the cell based on hexagonal axes, the equivalent points are $(0,0,0; ^1/_3,^2/_3,^2/_3; ^2/_3,^1/_3,^1/_3) + 0,0,z; + 0,0,\bar{z}$. The full list for the six atoms in the unit cell, obtained by addition of coordinates, is $0,0,z$; $^1/_3,^2/_3,^2/_3+z$; $^2/_3,^1/_3,^1/_3+z$; $0,0,\bar{z}$; $^1/_3,^2/_3,^2/_3-z$; $^2/_3,^1/_3,^1/_3-z$.

Effect of Alloy Composition. In the descriptions of crystal structures given in Table 4, the structure prototypes and examples are listed, for simplicity, as if they were compounds with definite, unvarying composition—such as, for example, AuCd and Fe_3W_3C. In general, however, a composition can vary from the stoichiometric value without altering the type of structure, although the edge distances a, b and c generally change somewhat with composition, and angles other than 90° may change slightly.

The special atom-coordinate values such as 0, $^1/_8$, $^1/_4$, $^1/_2$ and $^1/_3$ do not change, but x, y and z values, which are variable, may change with composition across a single-phase region of a phase diagram.

Disordered vs Ordered Superstructures. Solid solutions may have more than one kind of atom at a

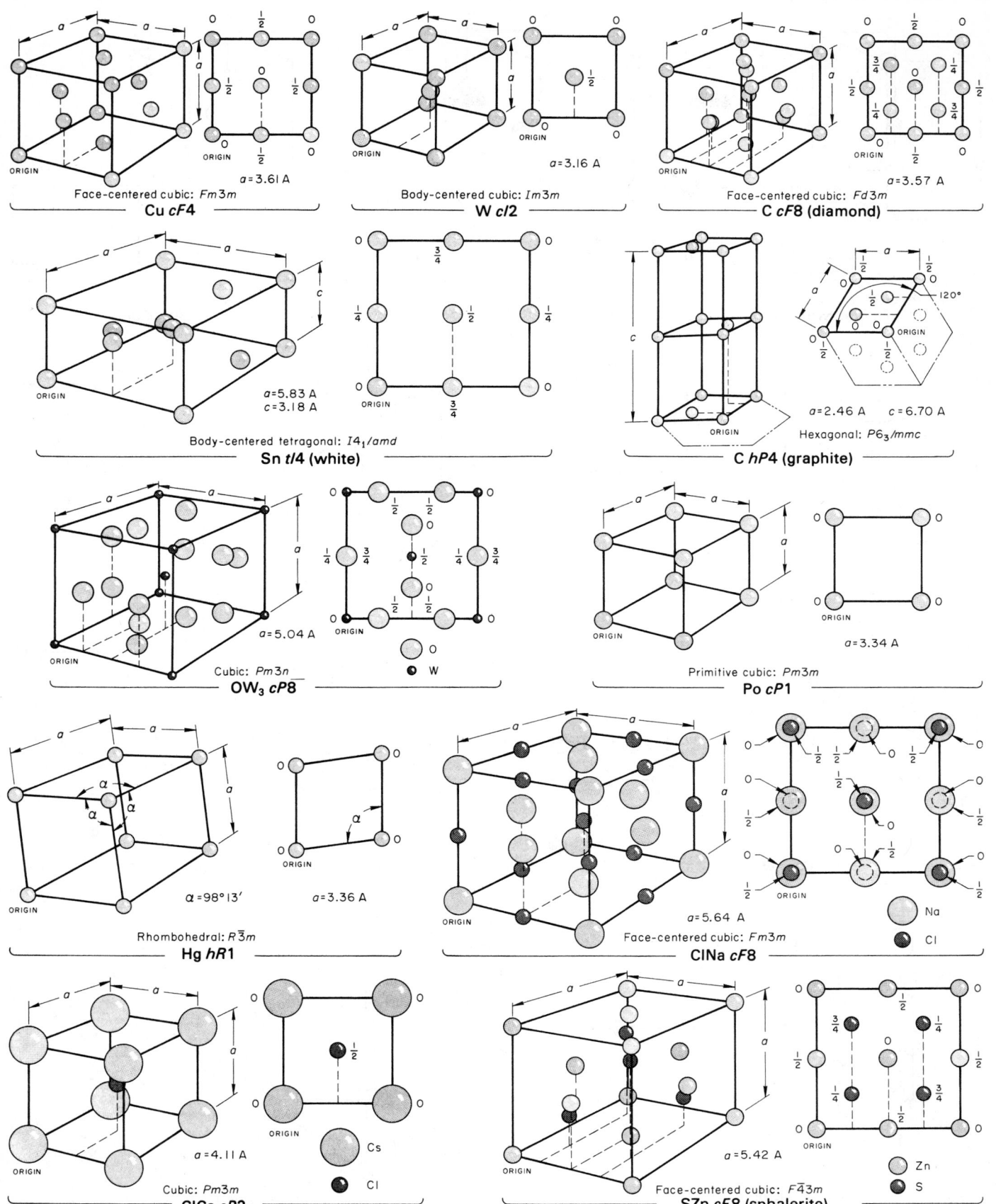

Fig. 1. Atom positions, prototypes, structure symbols, space-group notations and lattice parameters for some of the simple metallic crystals (continued on next two pages)

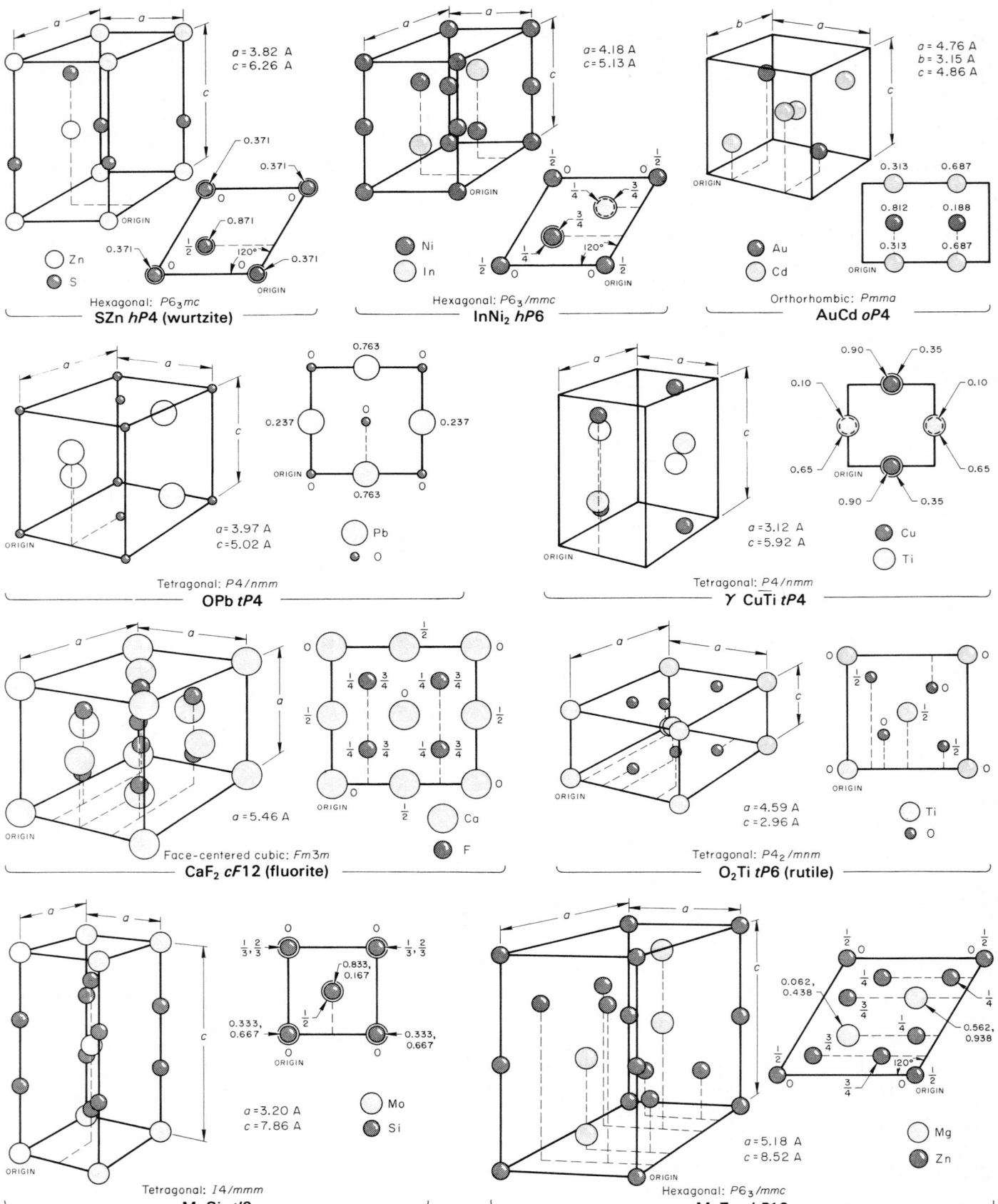

Fig. 1 (continued). Atom positions, prototypes, structure symbols, space-group notations and lattice parameters for some of the simple metallic crystals (continued on next page)

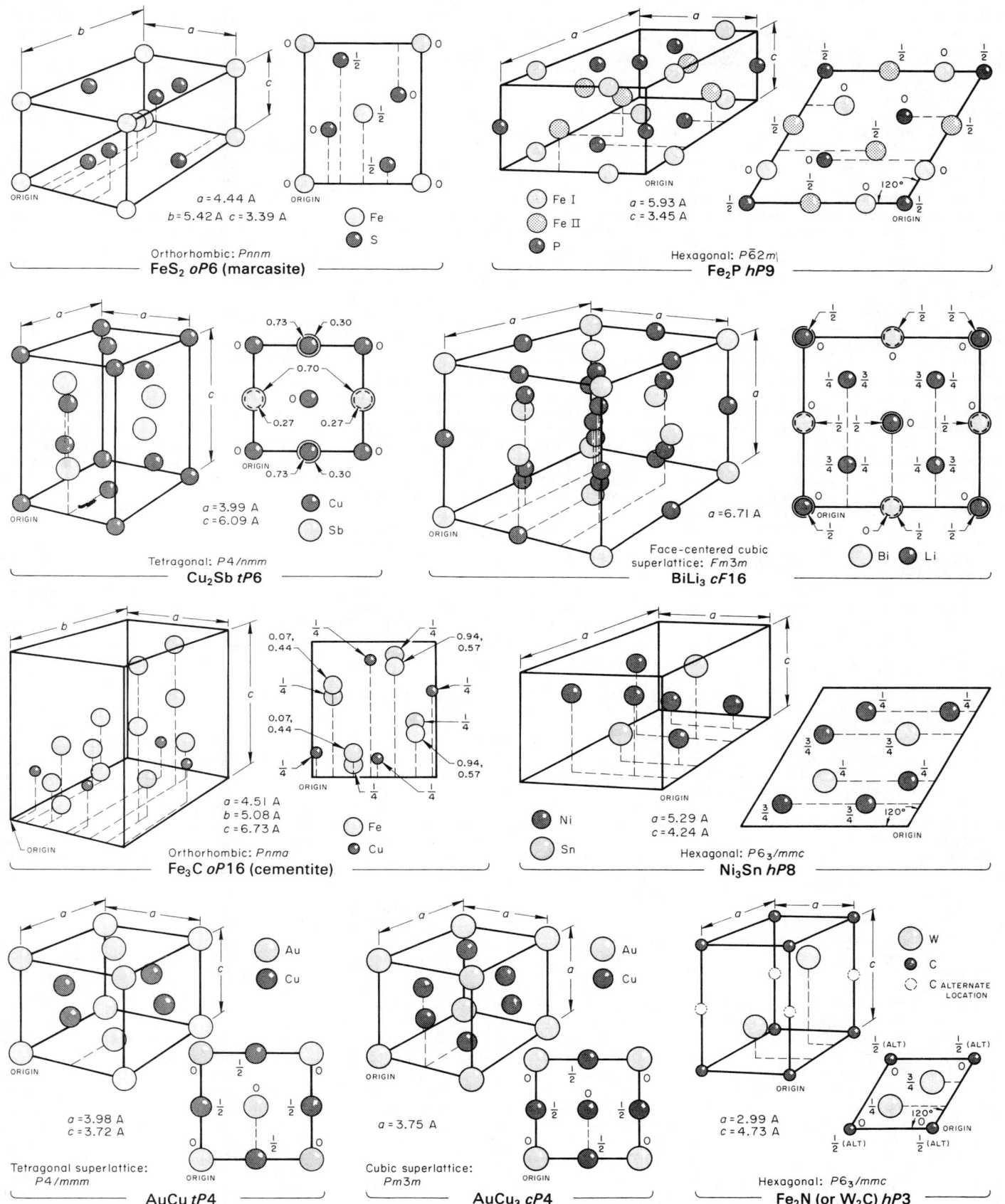

Fig. 1 (continued). Atom positions, prototypes, structure symbols, space-group notations and lattice parameters for some of the simple metallic crystals

set of equivalent positions. In an alloy of A and B atoms, the probability of finding an A atom (or a B atom) at a given atomic position is a function of the alloy composition, and, in simple alloys, is equal to the atomic percentage of A atoms (or B atoms) in the alloy. If occupation of the individual sites by A or B atoms is random, or nearly so, the solid solution is said to be disordered. The likelihood of a solid solution's being disordered is greatest at high temperatures. At lower temperatures, the likelihood for A atoms to locate at some positions and B atoms at others may overcome the randomizing action of thermal agitation and produce partial or complete order. When this occurs, the unit cell may be larger than that for the disordered state (usually with one, two or all three edges doubling in length), the number of atoms per unit cell is proportionately larger, and the crystal is said to be an ordered superstructure or superlattice. An example is Cd_3Mg with the Ni_3Sn $hP8$ structure (see Fig. 1), where the Mg $hP2$ structure of the Cd solid solution orders with doubling of the a edge of the unit cell.

Alternatively, the ordering of atoms randomly distributed in a solid solution may occur with a lowering of the crystal symmetry of the structure but without multiplication of the length of any of the unit-cell edges. An example of this is beta brass, which has a disordered body-centered-cubic structure (W $cI2$) at high temperatures and is designated as β Cu-Zn, but which upon cooling acquires an ordered (CsCl $cP2$) structure that is designated as β′ Cu-Zn, in which copper atoms are chiefly at positions 0,0,0 and zinc atoms are chiefly at positions $1/2, 1/2, 1/2$, or vice versa.

Number of Atoms per Cell. Calculation of the number of atoms in a unit cell requires inspection of the cell to determine (*a*) those atoms that are shared by other unit cells, (*b*) the number of cells that share each of these atoms, and (*c*) those atoms that are entirely within the cell.

The atoms at each of the eight corners of a unit cell, other than in hexagonal crystals, are shared by the eight cells that share that corner, and therefore count as only one atom per cell ($1/8 \times$ 8). In a face-centered cell (such as that of Cu $cF4$), there is, in addition, an atom in each of the six faces; each of these atoms is shared by two cells, and therefore an atom in a face counts as one-half an atom per cell. With six faces, there are three ($1/2 \times 6$) atoms per cell in the faces; thus, a face-centered cell has a total of four atoms. In a body-centered cell (such as that of W $cI2$), there is a total of one atom at the corners plus one atom in the center of the volume, for a total of two atoms per body-centered cell.

The atom at each corner of a primitive hexagonal unit cell is shared by the six cells that are in contact at that corner.

Miller Indices for Planes and Direction. Miller indices used for designating planes within a crystal are based on the intercepts of the plane with the crystal axes. If the unit-cell edges are of lengths a, b and c, and if a plane intersects these edges at a/h, b/k and c/l, the fractional intercepts are $1/h$, $1/k$ and $1/l$. The reciprocals of these fractional intercepts, when reduced to the smallest common denominator, are the Miller indices of the plane. They are written in parentheses—for example, (hkl). A negative intercept is indicated by the use of an overscore; for example, ($hk\bar{l}$) indicates a negative intercept along the c axis. Integers enclosed by braces designate all the equivalent nonparallel planes of a crystal; for ex-

ample, the entire set of cube faces of a cubic crystal is designated by {100}.

In Miller indices specifying directions in a crystal, the notation [uvw] is used to indicate the direction of a line from the origin to a point whose coordinates are u, v and w. It is customary to use square brackets, to avoid fractional coordinates, and to use the smallest integers that will locate a point on the line. Negative indices are indicated by the use of overscores. Because of symmetry, various directions in a crystal are equivalent. A full set of equivalent directions (directions of a form) is enclosed by carets, as in ⟨uvw⟩.

For additional information on Miller indices, see Ref 6, 7 and 8.

Miller-Bravais indices, instead of Miller indices, are used for hexagonal crystals. These crystals have three equal axes in the basal plane of the unit cell: a_1, a_2 and a_3. If the fractional intercepts on these three axes are $1/h$, $1/k$ and $1/i$, and the intercept on the c axis is $1/l$, the Miller-Bravais indices of the plane are ($hki\bar{l}$).

Crystal Zones and Zone Axes. A crystal zone is that set of nonparallel faces that intersect each other in a series of parallel straight lines. Faces belonging to the same zone lie parallel to a line that is called the zone axis and that defines a specific direction in the crystal. A line passing through the lattice point defined as the origin, and through a neighboring lattice point whose coordinates are u, v and w in terms of the axial lengths a, b and c, respectively, defines the direction of the zone axis or line, and is designated by the indices [uvw]. If the Miller indices of direction at the zone axis are low, the axis passes through a row of closely packed lattice points and therefore lies in one of a family of equidistant net planes of high atomic density, which is a possible crystal face. Two intersecting zone axes define a plane and thus a possible crystal face, or face of a unit cell.

METALLURGICALLY IMPORTANT CRYSTAL TYPES

In Table 4, which describes various common crystal-structure types, the structures are organized according to Pearson symbol, which designates crystal class, space lattice and number of atoms per cell. Thereafter follow the prototype formula name and the space group for the structure type. The atom positions are also specified for the simpler structures. Next follows the unit-cell description of the prototype, angles being given only when they differ from 90 or 120°. Finally, several examples of phases having the structure type in question are presented. Space limitations preclude full treatment of very complex structures. Because many readers wishing to find a particular structure type may still think of it in terms of its Strukturbericht symbol, they should refer to Table 3 to obtain the corresponding Pearson symbol.

Figure 1 illustrates many of the crystals described in this section; the illustrations include both perspective drawings of unit cells and projections onto cell faces. Small circles are used in Fig. 1 to indicate the positions of atom centers, without implying that the atoms are so small and so widely separated. Included in the illustrations in Fig. 1 are views in which the atoms are projected to the face of the unit cell containing the origin. The circles represent atom positions; the numbers adjacent to the circles indicate the distance the atom is from that face, in terms of fractions (or decimal equivalents of fractions) of the

length of the unit-cell edge extending from that face. The edge lengths and interaxial angles shown apply to the structure prototype. Cell dimensions for other element or alloy phases having the same structure prototype vary according to phase and composition, as discussed under "Effect of Alloy Composition," above. The crystal structures of the elements are summarized in Table 5.

Long-Period Superlattices. Some alloys in the ordered state have periodic structures with unit cells many times larger than in the disordered state, the long period being either along one axis, as in Fig. 2, or along two axes, as in Fig. 3. The superperiods, which have different values in different alloy systems, and which differ with composition, are known to depend on the average number of valence electrons per atom. Figure 2 shows an AuCu II structure with a one-dimensional long-range superlattice, with antiphase boundaries at intervals of five unit cells of the disordered state. It is orthorhombic when ordered, and face-centered cubic when disordered. Figure 3 shows a two-dimensional long-period superlattice as in certain AB_3 alloys (Cu-Pd, Au-Zn, Au-Mn), with antiphase boundaries spaced at intervals M_1 and M_2 and unit cell dimensions a, b and c, in the ordered state. This superlattice is orthorhombic when ordered, and face-centered cubic when disordered.

CRYSTAL DEFECTS

Crystal defects are important features in all real crystals. Some of the most significant defects are described in the following paragraphs. Additional information on defects, slip, twinning and cleavage can be found in Ref 6, 7 and 8.

Point defects include vacant atom positions that are occupied in perfect crystals. These vacancies increase in number as temperature is increased, and by jumping about from one lattice site to another they cause diffusion.

Interstitial B atoms are those located between the A atoms of the normal, perfect-crystal array; thus the carbon atoms in body-centered-cubic ferrite are interstitials in that they fit between the iron atoms of its bcc structure, which is similar to the W $cI2$ type illustrated in Fig. 1. Substitutional B atoms are those located at atom positions formerly occupied by A atoms in a normal, perfect-crystal array. A B atom in either substitutional or interstitial solid solution is another common form of point defect.

There are also many close pairs and clusters of point defects, such as divacancies, trivacancies and interstitial-vacancy pairs.

Line Defects. Dislocations are line defects that exist in all real crystals. An edge dislocation, which is the edge of an incomplete plane of atoms within a crystal, is represented in cross section in Fig. 4. In this illustration, the incomplete plane extends partway through the crystal from the top down, and the edge dislocation (which is indicated by the standard symbol T) is its lower edge.

If forces, as indicated by the arrows in Fig. 5, are applied to a crystal, such as the perfect crystal shown in Fig. 5(a), one part of the crystal will slip. The edge of the slipped region, shown as a dashed line in Fig. 5(b), is a dislocation. The portion of this line at the left near the front of the crystal and perpendicular to the arrows, in Fig. 5(b), is an edge dislocation, since the displacement involved is perpendicular to the dislocation.

Table 5. Crystal structures of the elements
Compiled by Donald T. Hawkins, Bell Laboratories.

Element	Phase(a)	Structure type	Source(b)	Element	Phase(a)	Structure type	Source(b)
Ac (actinium)	...	Cu cF4	1		β (HT)	W cI2	1
Ag (silver)	...	Cu cF4	1		Gd II (formed at 400 °C and 40 kbar, retained at normal T and P)	Sm hR3	1
Al (aluminum)	...	Cu cF4	1				
Am (americium)	α (RT)	La hP4	2	Ge (germanium)	RT	C cF8	1
	β (HT)	Cu cF4	1		Ge II (RT; 120 kbar)	Sn tI4	1
Ar (argon)	...	Cu cF4	2		Ge III (formed above 120 kbar, retained when pressure removed)	Ge tP12	1
As (arsenic)	α	As hR2	1				
	β	P oC8	1				
At (astatine)	...	...	...	H (hydrogen)	α	Mg hP2 (?)	2
Au (gold)	...	Cu cF4	1		β	fcc	2
B (boron)	α	B hR12	2	He (helium)	α	Mg hP2	2
	β	B hR105	9		β	Cu cF4	2
	γ	B tP190	2		γ	W cI2	2
Ba (barium)	...	W cI2	1	Hf (hafnium)	α (RT)	Mg hP2	1
	Ba II (62 kbar; RT)	Mg hP2	1		β (HT)	W cI2	1
Be (beryllium)	α	Mg hP2	1	Hg (mercury)	Below RT	Hg hR1	1
	β (HT)	W cI2	1		Hg II (formed at HP, retained at 77 K when pressure removed)	Pa tI2	1
Bi (bismuth)	α (RT)	As hR2	1				
	HP phases uncertain	...	9				
Bk (berkelium)	α (RT)	La hP4	3	Ho (holmium)	α (RT)	Mg hP2	1
	β (HT)	Cu cF4	3		Ho II (>75 kbar at RT)	Sm hR3	9
C (carbon)	Graphite	C hP4	2				
	Rhombohedral graphite	C hR2	9	In (indium)	...	In tF4	1
	Diamond	C cF8	1	Ir (iridium)	...	Cu cF4	1
	Hexagonal diamond	C hP4	9	K (potassium)	...	W cI2	1
Ca (calcium)	α (RT)	Cu cF4	1	Kr (krypton)	...	Cu cF4	2
	β (HT)	W cI2	1	La (lanthanum)	α (RT)	La hP4	1
Cd (cadmium)	...	Mg hP2	1		β (RT)	Cu cF4	1
	Cd II (HP: above about 100 kbar)	La hP4 (?)	7,8		γ (HT)	W cI2	1
Ce (cerium)	α (RT)	Cu cF4	2	Li (lithium)	α (RT)	W cI2	2
	β (<250 K)	La hP4	2		β (LT: 78 K)	Mg hP2	2
	γ (<110 K)	Cu cF4	2		γ (LT; strain induced at 20 K)	Cu cF4	1
	δ (HT)	W cI2	2				
Cm (curium)	α (LT)	La hP4	4	Lr (lawrencium)	...	...	...
	β (HT?)	Cu cF4	4	Lu (lutetium)	α (RT)	Mg hP2	1
Co (cobalt)	α	Mg hP2	1	Md (mendelevium)	...	...	...
	β	Cu cF4	1	Mg (magnesium)	...	Mg hP2	1
Cr (chromium)	...	W cI2	1	Mn (manganese)	α (RT)	Mn cI58	1
Cs (cesium)	...	W cI2	1		β (HT: >727 °C)	Mn cP20	1,9
Cu (copper)	...	Cu cF4	1		γ (HT: >1095 °C)	Cu cF4	1,9
Dy (dysprosium)	α (RT)	Mg hP2	1		δ (HT: >1133 °C)	W cI2	1,9
	β (HT)	W cI2	1	Mo (molybdenum)	...	W cI2	1
	γ (<86 K)	Tb oC4	1,9	N (nitrogen)	α	Cubic	2
	δ (at 75 kbar and RT)	Sm hR3	9		γ (HP)	Tetragonal	12
Er (erbium)	α (RT)	Mg hP2	1		β	Hexagonal	2
Es (einsteinium)	...	...	...	Na (sodium)	α (RT)	W cI2	1
Eu (europium)	...	W cI2	1		β (LT)	Mg hP2	1
F (fluorine)	α (<45.6 K)	Monoclinic	11	Nb (niobium)	...	W cI2	1
	β (>45.6 K)	Cubic	11	Nd (neodymium)	α (RT)	La hP4	1
Fe (iron)	α (RT)	W cI2	1		β (HT)	W cI2	1
	β (910-1390 °C)	Cu cF4	1,9		Nd II (RT; 50 kbar)	Cu cF4	9
	δ (>1390 °C)	W cI2	1,9	Ne (neon)	...	Cu cF4	2
	Fe II (≥130 kbar)	Mg hP2	1	Ni (nickel)	...	Cu cF4	1
Fm (fermium)	...	...	...	No (nobelium)	...	...	...
Fr (francium)	...	...	...	Np (neptunium)	α (RT)	Np oP8	1
Ga (gallium)	α (RT)	Ga oC8	1		β (HT: >280 °C)	Np tP4	1,9
	β (metastable, but stable above about 15 kbar at 0 °C)	Ga mC4	1		γ (HT: >577 °C)	W cI2	1,9
				O (oxygen)	α	Monoclinic	10
	γ (metastable or stable above 30 kbar at 50-70 °C)	Ga oC40	1		β	Hexagonal	2
					γ	Cubic	2
Gd (gadolinium)	α (RT)	Mg hP2	1	Os (osmium)	...	Mg hP2	1

(continued)

Table 5. (continued)

Element	Phase(a)	Structure type	Source(b)	Element	Phase(a)	Structure type	Source(b)
P (phosphorus)	White	Cubic	1,2	Sn (tin)	α (gray; LT)	C cF8	1
	Black	P oC8	1,2		β (white)	Sn tI4	1
	Red	P c-66	1,2		Sn II (314 °C; 39 kbar)	Pa tI2	1
	Hittorf's	P mP84	9		Sn III (RT; 110 kbar)	Cubic (?)	1
	P II (RT; 50-83 kbar)	As hR2	1	Sr (strontium)	α (RT)	Cu cF4	2
	P III (RT; 120 kbar)	Po cP1	1		β (HT)	W cI2	2
Pa (protactinium)	...	Pa tI2	1		Sr II (RT; 35 kbar)	W cI2	1
Pb (lead)	RT	Cu cF4	1	Ta (tantalum)	...	W cI2	1
	Pb II (RT; 130 kbar)	Mg hP2	9	Tb (terbium)	α (RT)	Mg hP2	1
Pd (palladium)	...	Cu cF4	1		β (HT)	W cI2	1
Pm (promethium)	α (RT)	La hP4	5		γ (<220 K)	Tb oC4	1,9
	β (HT)	W cI2	...		δ (60 kbar at RT)	Sm hR3	9
Po (polonium)	α (10 °C)	Po cP1	1	Tc (technetium)	...	Mg hP2	1
	β (75 °C)	Hg hR1	1	Te (tellurium)	...	Se hP3	1
Pr (praseodymium)	α (RT)	La hP4	1		Te II (>15 kbar)	As hR2 (?)	1
	β (HT)	W cI2	1		Te III (>70 kbar)	Hg hR1	9
	Pr II (RT; 40 kbar)	Cu cF4	1	Th (thorium)	α (RT)	Cu cF4	1
Pt (platinum)	...	Cu cF4	1		β (HT)	W cI2	1
Pu (plutonium)	α (RT)	Pu mP16	2,9	Ti (titanium)	α (RT)	Mg hP2	1
	β (>122 °C)	Pu mI32	2,9		β (HT)	W cI2	1
	γ (>206 °C)	Pu oF8	2,9		Ti II (HP; retained when pressure removed)	Ti hP3 (ω phase)	1
	δ (>319 °C)	Cu cF4	2,9				
	δ' (>451 °C)	In tF4	2,9				
	ε (>476 °C)	W cI2	2,9	Tl (thallium)	α (RT)	Mg hP2	1
Ra (radium)	...	W cI2	6		β (HT)	W cI2	1
Rb (rubidium)	...	W cI2	1		γ (HP: >40 kbar)	Cu cF4 (?)	9
Re (rhenium)	...	Mg hP2	1	Tm (thulium)	α (RT)	Mg hP2	1
Rh (rhodium)	...	Cu cF4	1	U (uranium)	α (RT)	U oC4	1
Rn (radon)	...	...	...		β (HT: 720 °C)	CrFe tP30 (σ phase)	1
Ru (ruthenium)	...	Mg hP2	1		γ (HT: 805 °C)	W cI2	1
S (sulfur)	α (RT)	S oF128	2	V (vanadium)	...	W cI2	1
	β (RT)	S mP48	2	W (tungsten)	...	W cI2	1
	γ (RT)	S hR6	2	Xe (xenon)	...	Cu cF4	2
Sb (antimony)	α (RT)	As hR2	1	Y (yttrium)	α (RT)	Mg hP2	2
	Sb II (RT; 50-70 kbar)	Po cP1	1		β (HT)	W cI2	2
	Sb III (RT; 90 kbar)	Mg hP2	1	Yb (ytterbium)	α (RT)	Cu cF4	2
Sc (scandium)	α (RT)	Mg hP2	1		β (HT; also at RT and 40 kbar)	W cI2	2
	β (HT)	W cI2	1				
Se (selenium)	α (RT)	Se(1) mP32	1		γ (LT: <270 K)	Mg hP2	9
	β (RT)	Se(2) mP32	1	Zn (zinc)	...	Mg hP2	1
	γ (RT)	Se hP3	1		Zn II (HP: above about 40 kbar)	La hP4 (?)	7,8
Si (silicon)	...	C cF8	1				
	Si II (RT; 195 kbar)	Sn tI4	1	Zr (zirconium)	α (RT)	Mg hP2	1
	Si III (110-160 kbar; retained when pressure removed)	Si cI16	1		β (HT)	W cI2	1
					Zr II (HP: retained when pressure removed)	Ti hP3	1
Sm (samarium)	α (RT)	Sm hR3	1				
	β (HT)	W cI2	2				
	Sm II (300 °C; 40 kbar)	La hP4	9				

(a) RT = room temperature; HT = high temperature; LT = low temperature; HP = high pressure; phases formed under high-pressure conditions are designated by roman numerals—e.g., Ge II, Ge III. (b) **Source 1:** *A Handbook of Lattice Spacings and Structures of Metals and Alloys,* Vol 1 and 2, by W. B. Pearson, Pergamon, 1958, 1967. **Source 2:** *Selected Values of the Thermodynamic Properties of the Elements,* by R. Hultgren *et al,* American Society for Metals, 1973. **Source 3:** J. R. Peterson, J. A. Fahey and R. D. Baybarz, *J Inorg Nucl Chem,* Vol 33, 1971, p 3345-3351. **Source 4:** P. K. Smith, W. H. Hale and M. C. Thompson, *J Chem Phys,* Vol 50, 1969, p 5066-5076. **Source 5:** P. G. Pallmer and T. D. Chikalla, *J Less-Common Metals,* Vol 24, 1971, p 233-236. **Source 6:** F. Weigel and A. Trinkl, *Radiochim Acta,* Vol 10, 1967, p 78-82. **Source 7:** R. W. Lynch and H. G. Drickamer, *J Phys Chem Solids,* Vol 26, 1965, p 63. **Source 8:** E. A. Perez-Albuerne *et al, Phys Rev,* Vol 142, 1966, p 392. **Source 9:** *The Structures of the Elements,* by J. Donohue, Wiley-Interscience, 1974. **Source 10:** C. S. Barrett and L. Meyer, *Phys Rev,* Vol 160, 1967, p 694. **Source 11:** L. Pauling *et al, J. Solid State Chem,* Vol 2, 1970, p 225. **Source 12:** A. F. Schach and R. L. Mills, *J Chem Phys,* Vol 1, No. 52, 1970, p 6000.

The slip deformation in Fig. 5(b) has also formed another type of dislocation. The part of the slipped region near the right side, where the displacement is parallel to the dislocation, is called a screw dislocation. In this part, the crystal no longer is made of parallel planes of atoms, but instead consists of a single plane in the form of a helical ramp (screw).

As the slipped region spread across the slip plane, the edge-type portion of the dislocation moved out of the crystal, leaving the screw-type portion still embedded, as shown in Fig. 5(c).

When all of the dislocation finally emerged from the crystal, the crystal was again perfect but with the upper part displaced one unit from the lower part, as shown in Fig. 5(d). Thus, Fig. 5 illustrates the mechanism of plastic flow by the slip process, which is actually flow by dislocation movement.

The displacement that occurs when a dislocation passes a point is described by a vector, known as the Burgers vector. The direction of the vector with respect to the dislocation line and the length of the vector with respect to the identity distance

in the direction of the vector are the fundamental characteristics of a dislocation. The perfection of a crystal lattice is restored after the passage of a dislocation, as indicated in Fig. 5(d), provided that no additional defects are generated in the process.

Each dislocation that remains in a crystal is the source of local stresses. The nature of these microstresses is indicated by the arrows in Fig. 6, which represents (qualitatively) the stresses acting on small volumes at different positions around the dislocation at the lower edge of the incom-

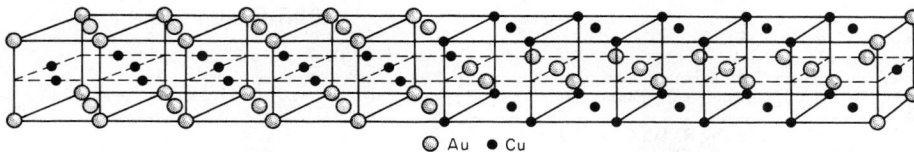

Fig. 2. AuCu II structure: a one-dimensional, long-period superlattice, with antiphase boundaries at intervals of five unit cells of the disordered state

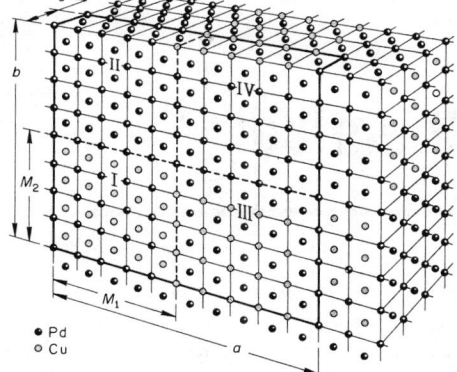

● Pd
○ Cu

Fig. 3. Two-dimensional, long-period superlattice, with antiphase boundaries spaced at intervals M_1 and M_2 and unit-cell dimensions a, b and c in the ordered state. The palladium atom has different positions in the small cubes in domains I, II, III and IV.

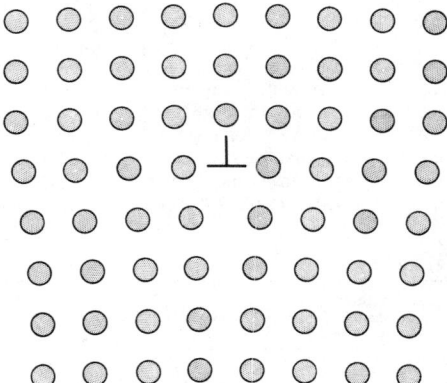

Fig. 4. Schematic representation of a section through an edge dislocation, which is perpendicular to the plane of the illustration and is indicated by the symbol ⊤

plete plane of atoms. Interstitial atoms usually cluster in regions where tensile stresses make more room for them, as in the lower central part of Fig. 6.

In addition to the large-angle boundaries that separate crystal grains, which have different lattice orientations, the individual grains are separated by small-angle boundaries (subboundaries) into subgrains that differ very little in orientation. These subboundaries may be considered as arrays of dislocations; tilt boundaries are arrays of edge dislocations, twist boundaries are arrays of screw dislocations. A tilt boundary is represented in Fig. 7 by the series of edge dislocations in a vertical row. Compared with large-angle boundaries, small-angle boundaries are less severe defects, obstruct plastic flow less, and are less effective as regions for chemical attack and segregation of alloying constituents. In general, mixed types of grain-boundary defects are common. All grain boundaries are sinks into which vacancies and dislocations can disappear, and may also serve as sources of these defects; they are important factors in creep deformation.

Stacking faults are two-dimensional defects that are planes where there is an error in the normal sequence of stacking of atom layers. Stacking faults may be formed during the growth of a crystal. They may also result from motion of partial dislocations. Contrary to a full dislocation, which produces a displacement of a full distance between the lattice points, a partial dislocation produces a movement that is less than a full distance.

Twins are portions of a crystal that have certain specific orientations with respect to each other. The twin relationship may be such that the lattice of one part is the mirror image of that of the other, or one part may be related to the other by a certain rotation about a certain crystallographic axis. Growth twins may occur frequently during crystallization from the liquid or the vapor state, by growth during annealing (by recrystallization or by grain-growth processes), or by the movement between different solid phases such as during phase transformation. Plastic deformation by shear may produce deformation (mechanical) twins. Twin boundaries generally are very flat, appear-

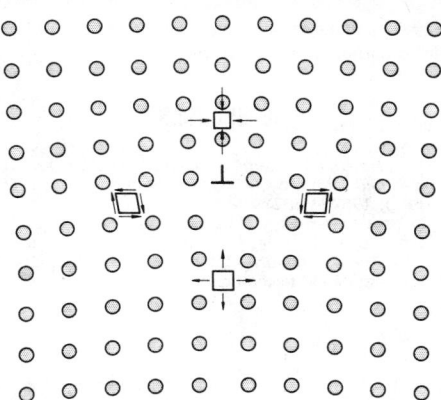

Fig. 6. Schematic representation of a crystal containing an edge dislocation, indicating qualitatively the stresses (shown by direction of arrows) at four positions around the dislocation

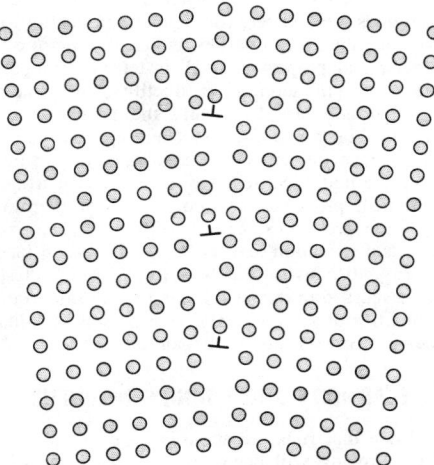

Fig. 7. Small-angle boundary (subboundary) of the tilt type, which consists of a vertical array of edge dislocations

ing as straight lines in micrographs, and are two-dimensional defects of lower energy than large-angle grain boundaries. Twin boundaries are therefore less effective as sources, and sinks, of other defects and are less active in deformation and corrosion than are ordinary grain boundaries. Textbooks and reference books (such as Ref 6, 7 and 8) list the indices of twinning planes (shear

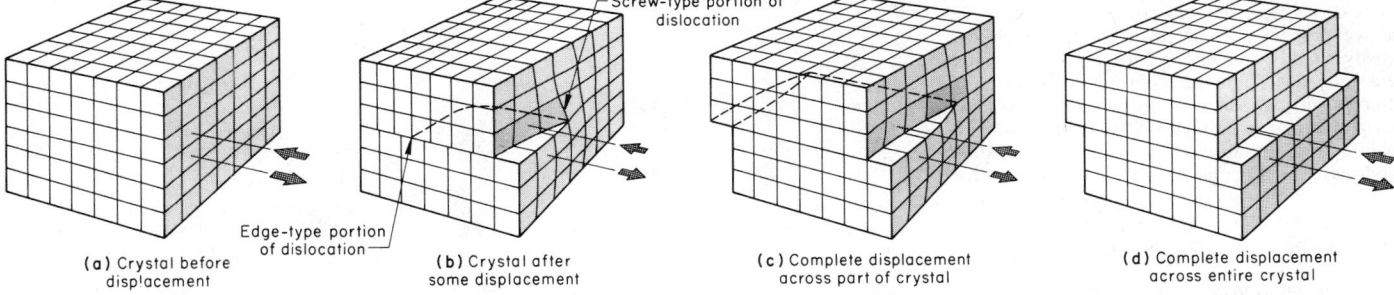

Screw-type portion of dislocation

Edge-type portion of dislocation

(a) Crystal before displacement   (b) Crystal after some displacement   (c) Complete displacement across part of crystal   (d) Complete displacement across entire crystal

Fig. 5. Schematic representation of four stages of slip deformation by formation and movement of a dislocation (dashed line) through a crystal

planes) and the directions of shear that occur when deformation twins are formed.

REFERENCES

1. *A Handbook of Lattice Spacings and Structures of Metals and Alloys*, by W. B. Pearson: Pergamon, 1958 (Vol 1), 1967 (Vol 2)
2. *Strukturbericht*, Akademische Verlagsgesellschaft m.b.H., Leipzig, Germany, 1913-1939; continued as *Structure Reports*, International Union of Crystallography, 1940 to the present
3. C. R. Hubbard and L. D. Calvert, *Bulletin of Alloy Phase Diagrams*, Vol 2, 1981, p 153-156
4. *International Tables for Crystallography*, Vol A, edited by T. Hahn: International Union of Crystallography, D. Reidel Publishing Co., Dordrecht, Holland, 1983
5. *Crystal Structures*, by R. W. G. Wyckoff: Interscience, 1963 (Vol 1), 1964 (Vol 2), 1965 (Vol 3)
6. *Structure of Metals*, by C. S. Barrett and T. B. Massalski: Pergamon, 1980
7. *Elements of Physical Metallurgy*, by A. G. Guy: Addison-Wesley, 1959
8. *Elements of Materials Science*, by L. H. Van Vlack: Addison-Wesley, 1964

ADDITIONAL READING

Metals Reference Book, 4th Ed., Vol 1, by C. J. Smithells: Plenum Press, 1967
Crystal Data, by J. D. H. Donnay: American Crystallographic Assn., 1963
Constitution of Binary Alloys, by M. Hansen (prepared in cooperation with K. Anderko): McGraw-Hill, 1958; First Supplement, R. P. Elliott, 1965; Second Supplement, F. A. Shunk, 1969
Powder Diffraction Data File: American Society for Testing and Materials
Intermetallic Compounds, edited by J. H. Westbrook: Wiley, 1967

Mechanical, Physical and Chemical Properties of Metals

By Giles F. Carter, Eastern Michigan University

THE PROPERTIES of materials determine their usefulness. However, for a given application, a combination of properties is usually required. Mechanical properties such as tensile strength, yield strength, elongation (ductility), toughness and hardness are frequently the properties of greatest concern in metals. There are many useful physical properties of metals, including optical, thermal, electrical and magnetic properties. Chemical properties are comprised of thermodynamic properties, chemical reactivity, corrosion and oxidation resistance, solubility, surface energy, diffusivity and basicity. Most of these are discussed in this article, and many important properties of various metals are presented in tables to serve as reference material.

ELASTICITY AND YOUNG'S MODULUS

When materials are stressed, one or more of the following will occur: (*a*) The material will undergo temporary deformation that lasts until the stress is removed, at which time the material returns to its original dimensions (for example, the stretching of a rubber band). This is elastic deformation. (*b*) The material may undergo permanent, or plastic, deformation, in which the material does not return to its original shape when the stress is removed (e.g., the bending of a metal coat hanger). (*c*) The material may fracture into two or more pieces (such as the shattering of a glass windowpane).

Obviously, gases do not have the bulk properties of elastic or plastic deformation, but in a given gas, each atom collides elastically with other atoms in the gas. Liquids that have high flow rates (low viscosities) do not normally show the above properties, whereas extremely high-viscosity liquids and crystalline solids do exhibit elasticity and sometimes plasticity. Glass is an amorphous material and may be considered to be a rigid, supercooled liquid. Over long periods, it exhibits the property of flow to a small extent and may crystallize to form a solid. On our time scale, glass appears to be a "solid" because it is elastic (glass fibers bend easily), it will fracture, and it exhibits minimal flow at room temperature.

Metals are almost always crystalline when cooled far beneath the melting point, so there is no question that they possess elasticity; many

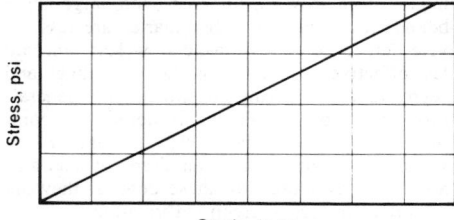

Fig. 1. Hooke's law

metals deform plastically, and all metals fracture when stressed beyond certain limits.

When a metal wire is loaded with a weight, it is subjected to a tensile stress (stress is force per unit area, such as g/cm^2 or lb/in.^{2}, psi), which is a stress applied in opposite directions tending to pull the material apart: it is represented by opposing forces $\uparrow$ $\downarrow$ or $\leftarrow\rightarrow$. The stretching of the wire is proportional to the load, provided that the load does not exceed a certain value. When the load is removed, the wire contracts to its original length: it has temporarily deformed elastically, provided that the wire has had no permanent deformation. The increase in the length of the wire is proportional to the tensile stress (see Fig. 1): $E = \sigma/\varepsilon$, where E is Young's modulus (the modulus of elasticity) in pounds per square inch; σ is the tensile stress in pounds per square inch; and ε is the strain in inches per inch.

Strain is fractional deformation (stretching or elongation): $\varepsilon = \Delta l/l$, which is unitless. Young's modulus is constant for a given metal at a given temperature. Table 1 lists E for a number of metals. Young's modulus may be visualized as being related to stiffness: the higher the value of E, the stiffer the material. For instance, metals in general are stiffer than plastics, which have Young's moduli ranging from about 0.1 to 1.5 × 10^6 psi, compared with 2 to 60 × 10^6 psi for metals. No-

Table 1. Elastic constants for polycrystalline metals at 20 °C

Metal	Young's modulus, E, 10^6 psi	Bulk modulus, K, 10^6 psi	Shear modulus, G, 10^6 psi	Poisson's ratio, ν
Aluminum	10.2	10.9	3.80	0.345
Brass, 30 Zn	14.6	16.2	5.41	0.350
Chromium	40.5	23.2	16.7	0.210
Copper	18.8	20.0	7.01	0.343
Iron (soft)	30.7	24.6	11.8	0.293
(cast)	22.1	15.9	8.7	0.27
Lead	2.34	6.64	0.811	0.44
Magnesium	6.48	5.16	2.51	0.291
Molybdenum	47.1	37.9	18.2	0.293
Nickel (soft)	28.9	25.7	11.0	0.312
(hard)	31.8	27.2	12.2	0.306
Nickel-silver, 55Cu-18Ni-27Zn	19.2	19.1	4.97	0.333
Niobium	15.2	24.7	5.44	0.397
Silver	12.0	15.0	4.39	0.367
Steel, mild	30.7	24.5	11.9	0.291
Steel, 0.75 C	30.5	24.5	11.8	0.293
Steel, 0.75 C, hardened	29.2	23.9	11.3	0.296
Steel, tool	30.7	24.0	11.9	0.287
Steel, tool, hardened	29.5	24.0	11.4	0.295
Steel, stainless, 2Ni-18Cr	31.2	24.1	12.2	0.283
Tantalum	26.9	28.5	10.0	0.342
Tin	7.24	8.44	2.67	0.357
Titanium	17.4	15.7	6.61	0.361
Tungsten	59.6	45.1	23.3	0.280
Vanadium	18.5	22.9	6.77	0.365
Zinc	15.2	10.1	6.08	0.249

tice that tungsten is stiffer than steel, which is stiffer than lead (see E values in Table 1).

Young's modulus is the slope of the straight line when stress is plotted versus elastic strain (see Fig. 1). In general, rubbers (elastomers) do not have a strictly linear stress-strain curve, but most solids have a straight-line relationship between stress and elastic strain.

Young's modulus is very roughly proportional to the boiling point of metals. Actually, both properties are related to the strength of the bond in the metal: the stronger the metal bonding, the higher the Young's modulus and the boiling point. For a given structural type—e.g., fcc, the Young's modulus is inversely proportional to the unit-cell dimension raised to the fourth power: for fcc crystals, $E = \text{constant}/a^4$.

Young's moduli are anisotropic: the value of E depends on crystallographic direction in a single crystal. Polycrystalline metals have values of E lying between the maximum and minimum values of E for single crystals.

Young's modulus also decreases with increasing temperature in a metal. This is not surprising because $E \propto a^{-4}$. As temperature increases, a increases and E decreases (see Fig. 2). Note the abrupt change in E when bcc iron transforms to fcc iron.

POISSON'S RATIO

When a metal is subjected to an axial tensile stress, it expands in the direction of the stress and contracts laterally. Figure 3 shows the top view of a cube of metal that has been pulled axially along the z-axis. If the material is isotropic, such as polycrystalline metals, ε_x and ε_y are equal. Poisson's ratio is defined as $\nu = -\varepsilon_x/\varepsilon_z$, where $-\varepsilon_x$ is the lateral contraction (the negative sign indicates contraction), and ε_z is the axial elongation. If a material maintains constant volume when it is elongated, it may be shown that $\nu = 0.5$. Generally, this value is the upper limit for Poisson's ratio: for polyisoprene, a type of rubber, $\nu = 0.49$. However, most metals have Poisson's ratios in the range from 0.25 to 0.40 (see Table 1).

BULK MODULUS OF ELASTICITY

Hydrostatic compression causes metals to contract in volume, $\Delta V/V$, and this relative volume contraction is initially proportional to the hydrostatic pressure: $\Delta V/V = \beta\sigma_{hyd}$, where $\Delta V/V$ is the relative volume contraction in in.³/in.³; β is a constant having units in.³/in.³ psi; and σ_{hyd} is the hydrostatic pressure in pounds per square inch. β is called the compressibility and is constant for a given metal.

The bulk modulus of elasticity is the constant that relates the ratio of hydrostatic pressure to the resulting volume change, or $K = 1/\beta$. Values of bulk moduli, K, are given in Table 1. Young's modulus, E, and the bulk modulus, K, are related through Poisson's ratio: $E = 3K(1 - 2\nu)$. Note that if $\nu = 0.50$, representing zero volume expansion on application of tensile stress, then $K = \infty$, or $\beta = 0$, representing no compressibility.

SHEAR MODULUS OF ELASTICITY

A shear stress is a slightly offset pair of stresses directed toward or away from each other: $\rightarrow \leftarrow$ or $\leftarrow \rightarrow$. The first pair of shear stresses represents a shear stress similar to the cutting or shear-

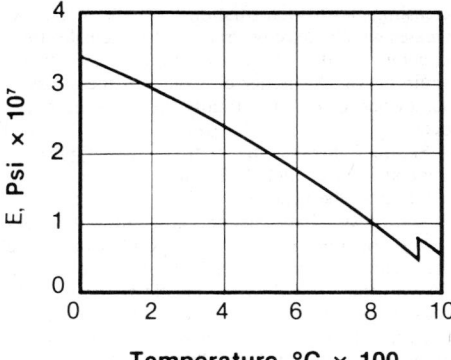

Fig. 2. Young's modulus vs temperature in iron

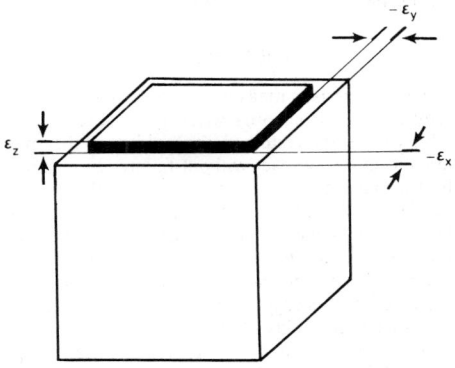

Fig. 3. Poisson's ratio and lateral contraction

ing action of scissors. When shear stresses are involved, the shear modulus, G, is defined as $G = \tau/\gamma$, where G is the shear modulus in pounds per square inch, τ is the shear stress in pounds per square inch, and γ is the shear strain in inches per inch. The shear modulus, G, is related to Young's modulus, E, as follows: $G = E/2(1 + \nu)$, or relating G to the bulk modulus, $G = 3K(1 - 2\nu)/2(1 + \nu)$. Because Poisson's ratio, ν, is usually about 0.3 for metals, G is generally about 40% of E. Refer to Table 1 for a tabulation of shear moduli, G, for various metals.

Shear strain may be measured as the amount of rotation of a hollow tube twisted by a radial torque. In fact, $\gamma = \tan \alpha = x/y$, as shown in Fig. 4.

The shear stress, τ, is found by determining the force at the tube wall and its cross-sectional area: force $= 1000$ in. · lb/1.25 in. $= 800$ lb. The cross-sectional area $= \pi \times$ diameter $\times$ thickness, or $\pi \times 2.50 \times 0.05 = 0.393$ in.². τ $=$ force/area $= 800$ lb/0.393 in.² $= 2036$ psi. Since $G = \tau/\gamma$, $G = 2036$ psi/2.94 $\times 10^{-4}$ in./ in. and $G = 6.9 \times 10^6$ psi. This is close to the shear modulus of copper (Table 1).

PLASTIC DEFORMATION

When a metal is stressed beyond a certain level, it deforms permanently and will not return to its original size and shape when the stress is removed. This is plastic deformation. The stress at which plastic deformation begins is called the yield stress or elastic limit. Brittle materials such as glass, chromium, other similar metals and most intermetallic compounds fracture before appre-

ciable plastic deformation occurs.

Plastic deformation is useful in shaping metal objects, but once the metal is in final form, plastic deformation is usually undesirable. Some metals such as iron have a sharp yield point, while others (e.g., aluminum) have gradual transitions from elastic to plastic deformation. Elongation is the relative permanent change in length of a plastically deformed metal. This is the same as strain: strain = elongation = $(l_f - l_o)/l_o$, where l_f is the final length and l_o is the original length. Notice that elongation, or strain, is unitless; but actually, strain is in inches per inch or centimetres per centimetre. Ductility is the ability of a metal to be elongated, usually by drawing through a die or stretching. Malleability refers to the ability of a metal to be deformed plastically by hammering. Although technically there is a difference between ductility and malleability (lead is highly malleable but not very ductile), the term ductility is commonly used to mean the ability to undergo plastic deformation. Ductility may also be measured by determining the reduction of area: reduction of area = $(A_o - A_f)/A_o$, where A_o indicates the original cross-sectional area, which is larger than A_f, the final cross-sectional area.

Tensile strength is defined as the maximum strength based on original dimensions: it is measured in pounds per square inch. When a cylindrical piece of metal is stressed in a tensile-testing machine, it stretches elastically until the yield stress is exceeded. Then the metal deforms plastically, usually by necking down, which is a moderate decrease in the cross-sectional area due to plastic deformation. Failure need not occur yet, and indeed in some materials it occurs long after necking down. Practically speaking, tensile strength is a very important property because it is a measure of the maximum load a material may withstand before failure. The true tensile strength equals the load divided by the reduced cross-sectional area due to necking down. Hence, the true tensile strength may be much greater than the tensile strength, but the concept of true tensile strength is not important from the practical standpoint: the maximum load that a given metal will sustain is important rather than the load divided by the "necked-down" cross-sectional area. The breaking strength is the stress at failure. Often, this is appreciably below the tensile strength, indicating that necking down and drawing can occur extensively prior to failure, causing strength to diminish from the maximum—i.e., the tensile strength.

STRESS-STRAIN CURVES

One of the most useful diagrams in metallurgy, particularly to design engineers, is the

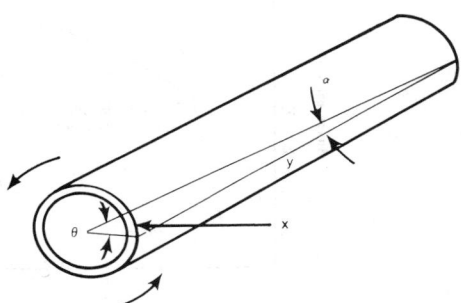

Fig. 4. Measurement of shear strain

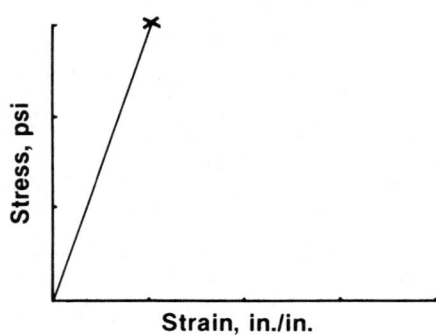

Fig. 5. Stress-strain curve for brittle metals

stress-strain curve, which is a plot of stress in pounds per square inch versus strain, usually measured in inches per inch along the x-axis. Stress-strain curves summarize quite a lot of useful information on mechanical properties: yield stress, tensile stress, breaking stress, elongation, Young's modulus of elasticity and toughness.

Figure 5 shows a stress-strain curve for a brittle material, such as chromium metal. Failure occurs before any appreciable plastic deformation. Thus, there is no real yield stress — only a breaking or fracture strength, which may also be called tensile strength. Figure 6 shows a stress-strain curve typical of iron; this element has a sharp yield stress, which is a definite stress at which plastic deformation begins. As soon as the yield stress is exceeded, the stress required for plastic deformation decreases appreciably — the minimum value is called the lower yield stress. Soon afterwards, the steel becomes stronger with in-

creasing plastic deformation, and the stress increases with increasing strain until the tensile stress is achieved; this is the maximum of the stress-strain curve. At higher elongations, the stress is somewhat lower, until finally the breaking stress occurs.

Figure 7 shows a stress-strain curve typical of a metal such as nickel. Because it has no sharp yield point, the yield stress is defined as the stress at which 0.2% permanent elongation has occurred. In order to locate this point, first locate 0.2% elongation on the x-axis; then from this point draw a line parallel to the elastic portion of the stress-strain curve until the newly constructed line intersects the stress-strain curve. The stress at this point is then called the yield stress. Note that at any point on the stress-strain curve the metal contracts elastically when stress is removed, and this contraction follows a line parallel with the elastic part of the stress-strain curve.

TOUGHNESS

Toughness is a measure of the amount of energy a material absorbs when it fractures. Usually, one is interested in the ability of the material to absorb impact without failure. In general, brittle materials are not tough at all: the ability of glass to absorb impact is relatively small unless the glass has been toughened by a thermal process to place the surface under compression compared with the interior. In order to absorb a fair amount of energy in impact, a material should be either highly elastic or ductile. High strength is also important if a very tough material is desired. Rubber is highly elastic and therefore may absorb a fair amount of energy without failing; however, it does not have a high tensile strength.

Armor plate is very strong yet fairly ductile. This combination makes an extremely tough material, and, of course, thickness increases the over-all toughness because the thicker the material, the more energy can be absorbed on impact without rupture.

Toughness is related to the area under the stress-strain curve: if this area is large, then the material is very tough, and if this area is relatively small, the material is brittle. Brittle materials have low toughness; however, they may be very strong but have poor ductility, or they may have adequate ductility but very low strength, or they may show other combinations of these two factors. Figure 5 illustrates stress-strain curves of brittle materials, while Fig. 6 shows the stress-strain curve of iron, a tough material.

Toughness is measured by any of a variety of impact tests. A common toughness tester holds a specimen securely and has a weighted arm or pendulum that falls freely from a preset height. The specimen must be broken to make the test, and the test is comprised of measuring the height of the swing of the weighted arm *after* fracture of the given specimen. In this way, the energy absorbed in impact is measured, and this is a measure of toughness. Often, the specimen has a notch cut in it because in certain instances materials are notch sensitive and fail much more easily when notches are present. The American Society for Testing and Materials (ASTM) may be consulted for standard impact tests. The units of toughness are often given as foot-pounds of energy absorbed for a specimen of a given size.

HARDNESS

Hardness is resistance to plastic indentation. This may involve a simple scratch test or indentation by loading a ball, diamond or other penetrator with a weight and measuring the length, width or depth of the indentation. The harder the metal, the smaller the indentation for a given load. The higher the load, the larger the size of indentation for a given metal. Of course, the indenter material must be much harder than the material tested to prevent deformation of the indenter material. That is why indenters are frequently made of diamond, tungsten carbide or hardened steel. The most common hardness tests are (a) Rockwell, in which a diamond cone or hardened steel sphere is used; (b) Brinell, in which a 10-mm steel or tungsten carbide sphere is used; (c) Vickers, using a square-base diamond pyramid, and (d) Knoop, using an elongated diamond pyramid. The hardness numbers are usually read directly from a table of values calculated for one parameter, such as the length, width or depth of the indentation.

Hardness testing may be either macro or micro; the latter involves use of a microscope and either a Vickers or Knoop diamond indenter. Because the indenter is extremely small, the hardness of coatings or of different phases may be measured rather than the over-all hardness.

Hardness is related to wear and abrasion resistance: it is usually desirable to have a very hard material in order to obtain good abrasion resistance. Soft materials scratch easily, and scratches generally are undesirable from the viewpoint of appearance. Therefore, a decorative metal should be hard enough to resist scratching.

Another useful concept is that, for a given metal, hardness is roughly proportional to strength. Thus, if one steel is harder than a second alloy,

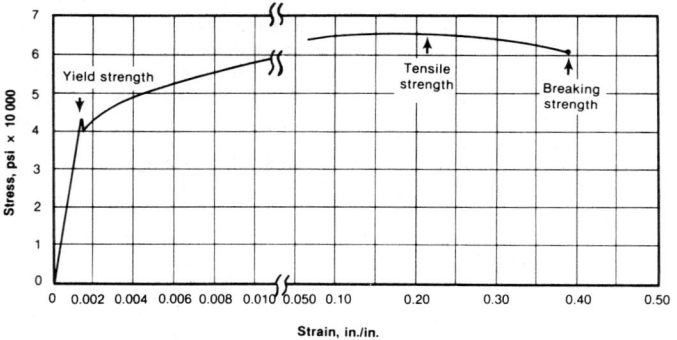

Fig. 6. Stress-strain curve for iron (containing 0.15% C)

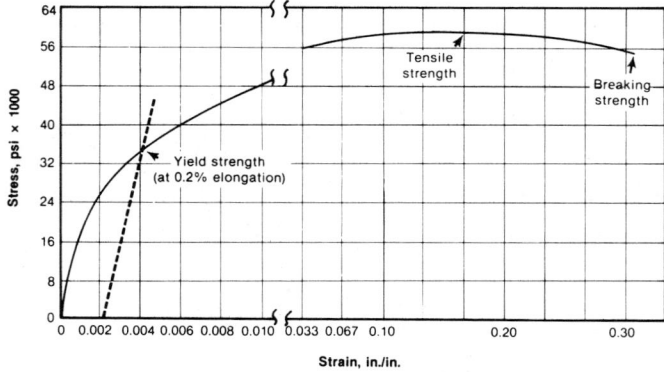

Fig. 7. Stress-strain curve for nickel

Table 2. Coefficients of friction of various metals

Metal	Static coefficient of friction on same metal, μ_s	Coefficient of friction on 0.13C-3.42Ni steel, μ_s
Aluminum	1.9	0.5
Calcium	0.5	...
Chromium	0.4	0.5
Copper	1.6	0.8
Gold	2.8	...
Indium	2	2
Iron	0.8	...
Lead	1.5	1.2
Magnesium	0.5	...
Molybdenum	0.8	0.5
Nickel	1.1	0.5
Platinum	3.0	...
Silver	1.5	0.5
Tin	...	0.9

then the first one is also stronger. Hardness may therefore be used as a rapid comparative strength test, and frequently, specifications call for minimum hardness. In applications such as machine tools, hardness is obviously a valuable property.

GENERAL PHYSICAL PROPERTIES OF METALS

Strictly speaking, mechanical properties are a subgroup of physical properties, but because they are so important in metals, they have been separated and placed in the foregoing sections. Physical properties include several important groups of properties: thermal, electric, magnetic, optical and general, the last of which will be discussed in this section. Each of the other groups will be discussed in later sections.

"General physical properties" include crystal structure, coefficient of friction, density, viscosity of liquid, and any other properties that do not readily fit into the other categories.

The coefficient of friction between two surfaces is the ratio of the force required to move one over the other to the total force pressing the two surfaces together: $\mu = F/W$, where μ is the coefficient of friction, F is the force required to move one surface over another, and W is the force pressing the surfaces together. Table 2 presents coefficients of friction of various metals. Of course, lubricants, which are used to prevent or reduce wear, markedly affect the coefficient of friction between metals, and normally they are used in moving metal parts. The subject of lubricants is extensive and of great importance to industry.

Several processes may occur in the wear of metals: (a) seizing, in which a particle or asperity, which is a projection from one of the surfaces, digs into the other metal and causes local welding, and finally the welded metal is broken, not necessarily at the weld, causing removal of some of the substrate metal; (b) deformation, in which the substrate may be locally deformed plastically followed by tearing loose of the deformed region; (c) corrosion, such as the removal of oxides, hydroxides or sulfides, which may reform rapidly in air or even in oil because true fresh metal surfaces are highly reactive chemically with nearly any environment (the repeated removal of oxides from the metal surface leads to wear); (d) heating of the metal surface due to friction (melting rarely occurs, but heating decreases hardness and increases corrosion, both of

which lead to wear). Hard surfaces are generally the most wear resistant, but the surface should also be corrosion resistant.

The densities of various metals are given in Table 3. Viscosity is a property of fluids (both liquids and gases) and is a resistance to change of form. It is the reciprocal of fluidity or flow rate. The formal definition of viscosity is $\eta = t(\tau/\gamma)$, where η is viscosity in poise (dyne·s/cm^2), τ is shear stress in dynes per square centimetre, γ is shear strain in centimetres per centimetre, and t is time in seconds. Viscosity is very temperature-sensitive; however, most liquid metals possess low viscosities because it is easy for metal atoms to flow past one another due to their small size and nondirectional bonding. Viscosity usually increases with decreasing temperature according to $\eta = \eta_o e^{Q/RT}$, where η_o is a constant, Q is an activation energy, measured in calories per mole, for viscous shear of atoms as they pass one another, R is the gas constant, 1.987 cal/Kmol, T is the temperature on the Kelvin scale, and e is the natural logarithm base, 2.71828. Viscosities of various metals at their melting points are presented in Table 3. Note that 1 cP = 0.01 P. The viscosity of water at 20 °C is 1 cP.

THERMAL PROPERTIES

Thermal properties include melting point, boiling point, coefficient of thermal expansion, thermal conductivity and thermodynamic properties. The following are brief descriptions of these terms:

Melting point: the temperature at which the solid and liquid phases of a pure material are in equilibrium

Boiling point: the temperature at which the vapor pressure of a liquid equals the pressure of the

surroundings; normally, this pressure is taken to be 1 at, or 760 torr

Coefficient of thermal expansion: the fractional increase in dimensions of a material when heated, resulting from increased thermal vibrations

Thermal conductivity: the rate of thermal energy transport in a material (see Table 3)

Thermodynamic properties: these include heat capacity, internal energy, enthalpy, entropy and free energy.

Thermal conductivity occurs in metals by two processes: (a) the rapid movement throughout the material of loosely bonded valence electrons; and (b) the motion of phonons, which are quantized elastic waves. Elastic waves—e.g., sound waves or heat waves—may move through solids by successive compression of planes of atoms as the wave front moves through the solid. A series of waves, or pulses, may pass through the solid, and these elastic waves are capable of transmitting heat energy. In metals, a large fraction of thermal energy is transferred by electrons, and thermal conductivity is therefore proportional to electrical conductivity. Thermal conductivity is measured in cal·cm/sK·cm^2, and the symbol is k. Table 3 contains thermal conductivities of various metals.

The ratio of thermal to electrical conductivity is called the Wiedemann-Franz ratio: WF = k/σ. This ratio is about 1.6×10^{-6} for metals at 20 °C (see Fig. 8). Because electrical resistivity, which is the reciprocal of electrical conductivity, is nearly proportional to the absolute temperature, the Lorentz number is approximately a constant: L = k/σT, where L, the Lorentz number, has a value of about 5.5×10^{-9} cal·ohm/sK^2 for many metals (see Table 4).

Thermal diffusivity may be defined as a constant similar to the diffusion coefficient: h = k/Cρ_d, or k = hCρ_d, where k is thermal conduc-

Table 3. Viscosities, surface tensions and other properties of metals at their melting points

Metal	Melting point, °C	Density, ρ_d, g/cm^3	Electrical resistivity, ρ, 10^{-6} ohm·cm	Thermal conductivity, k, cal·cm/sK·cm^2	Viscosity, η, cP	Surface tension, ergs/cm^2
Aluminum	660.1	2.39	20.0	0.22	4.5	860
Antimony	630.5	6.49	113.5	0.052	1.30	383
Beryllium	1284	...	(5)	...	...	1100
Bismuth	271	10.06	128.1	0.0262	1.68	393
Cadmium	320.9	8.02	33.7	0.105	1.4	666
Chromium	1875	6.46	36.6	0.06	0.684	1590
Copper	1083	7.96	21.1	0.118	3.36	1285
Gallium	29.8	6.20	2.8	0.08	2.04	735
Gold	1063	17.32	2.3	...	...	754
Indium	156.4	7.03	33.1	0.1	1.69	559
Iridium	2443	20.0	5.3	...	...	2250
Iron	1537	7.15	9.71	...	2.2	1675
Lead	327.4	10.68	95.0	0.039	2.634	470
Lithium	180	0.516	24.0	0.11	0.60	398
Magnesium	650	1.585	27.4	0.333	1.24	556
Mercury	Values at 20 °C	13.55	98.4	0.020	1.554	465
Molybdenum	2620	9.34	5.17	...	...	2250
Nickel	1453	7.90	6.84	...	...	1756
Palladium	1552	10.7	10.8	...	...	1500
Potassium	63.6	0.825	13.2	...	0.534	101
Platinum	1769	19.7	9.81	...	...	1740
Rhodium	1960	10.65	4.3	...	...	2000
Rubidium	38.8	1.475	11.3	0.1	0.673	76
Silver	960.8	9.33	17.2	...	3.9	930
Sodium	97.8	0.929	9.6	0.205	0.726	191
Thallium	303	11.29	73.1	0.06	...	490
Tantalum	2980	15.0	13.5	...	...	2150
Tin	231.9	6.97	48.0	0.08	1.97	579
Titanium	1670	4.13	55	...	...	1510
Tungsten	3380	17.6	5.5	...	...	2310
Zinc	419.5	6.64	37.4	0.144	3.93	824

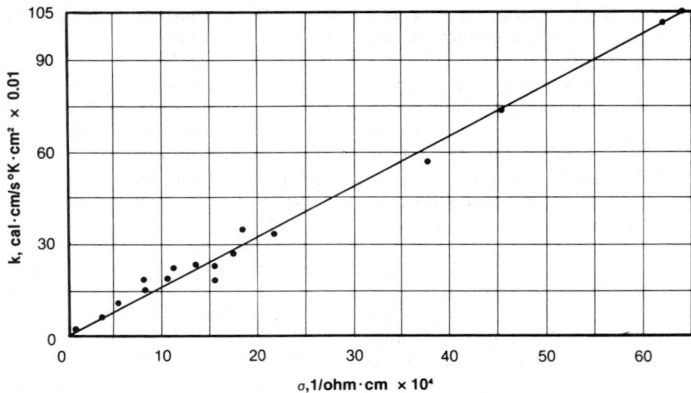

Fig. 8. Wiedemann-Franz ratio for metals

Table 4. Lorentz numbers of selected metals

Metal	Thermal conductivity, k, cal·cm/ sK·cm² at 20 °C	Electrical conductivity, σ, 10⁶ (ohm·cm)⁻¹ at 20 °C	Lorentz No., L, 10⁻⁹(cal· ohm/sK²)
Aluminum . . .	0.57	0.372	5.2
Copper	0.94	0.599	5.4
Gold	0.70	0.43	5.6
Iron	0.17	0.103	5.6
Lead	0.082	0.0485	5.8
Nickel	0.21	0.146	4.9
Platinum	0.17	0.094	6.2
Silver	1.00	0.625	5.5
Sodium	0.32	0.22	5.0
Tin	0.155	0.078	6.8
Titanium	0.041	0.0182	7.7
Zinc	0.265	0.169	5.4

Note: T = 20 °C = 293 K; L = k/σT.

tivity in cal·cm/sK·cm², C is the heat capacity in cal/gK, ρ_d is the density (the subscript "d" is placed here to remind the reader that this is density and not resistivity) in g/cm³, and h is a constant, the thermal diffusivity, in cm²/s.

ELECTRICAL PROPERTIES

Electrical properties of most interest in materials include conductivity, resistivity, polarization, dielectric constant and relative dielectric constant. In metals, the relatively free motion of valence electrons largely controls electrical properties. The following are brief descriptions of these terms:

Electrical conductivity: the reciprocal of resistivity, or the electrical charge flux per unit voltage gradient, V/d, V/cm: $\sigma = 1/\rho = $ flux/(V/d). The units of flux/(V/d) are (coulomb/cm²s)/(volt/cm) = (amp·s/cm²s)/(ohm·amp/cm) = 1/ohm·cm

Electrical resistivity: resistance per unit length times cross-sectional area, $\rho = R/1 \times A$, in units of ohm/cm × cm² = ohm·cm. Also, $\rho = 1/\sigma$, where ρ is electrical resistivity and σ is electrical conductivity.

Electrical conductivity results from the motion of electrons predominantly in one direction. In order for this to occur, valence electrons must gain velocity and momentum as they move toward a positive electrode. This requires more energy, and thus the electrons must occupy a previously unoccupied energy level. The criterion for electrical conductivity is the presence of vacant energy levels in the valence band, which is

a group of electron energy levels spaced very close to one another in solids. Metals in general have vacant energy levels available for conduction electrons, and hence metals are generally good conductors. Insulators or nonconductors do not have unoccupied energy levels in the valence electron bands.

The electrical conductivity of metals is limited by the mobility of electrons, not by the number of conducting electrons. As electrons move in metals, they are diffracted and scattered by irregularities in the structure; the average distance between scattering incidents is called the mean free path: the higher the mean free path, the greater the electrical conductivity, in general.

Decreased temperature increases the mean free path because of the greater structural order; therefore, electrical conductivity is inversely related to temperature and electrical resistivity is proportional to temperature (see Fig. 9). In general, the resistivity of metals varies with temperature as follows: $\rho_T = \rho_{273} [1 + 0.005(T - 273)]$. The temperature resistivity coefficient is 0.005 and is nearly the same for all pure metals.

When a metal dissolves a second metal, such as copper-nickel in U.S. nickels, the locale surrounding the solute atom is strained somewhat because the substituted atom is usually larger or smaller than the solvent atom and, also, the bonding is usually somewhat different between the atoms surrounding the solvent versus the solute atom. These structural variations reduce conductivity in solid solutions (see Fig. 10).

Likewise, ordered lattices have higher electrical conductivities than disordered lattices having the same chemical composition. Any imperfections in crystals, such as vacancies, interstitials, solute atoms, dislocations or grain boundaries, decrease conductivity. Annealing, which removes a high percentage of imperfections, increases electrical conductivity.

A dipole moment exists in a material when the centers of positive and negative charges do not coincide. Because metals usually have similar electronegativities (electron-drawing power), metals are unlikely to have very large dipole moments except in some intermetallic compounds. Polarization is defined as the dipole moment per unit volume; this is also small for most metals. The dielectric constant is the ratio of electric

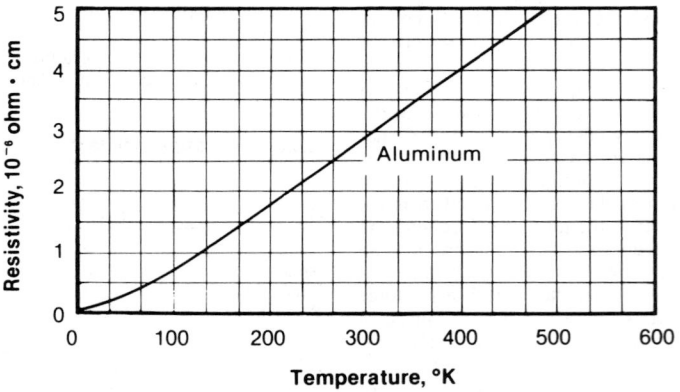

Fig. 9. Electrical resistivity as a function of temperature

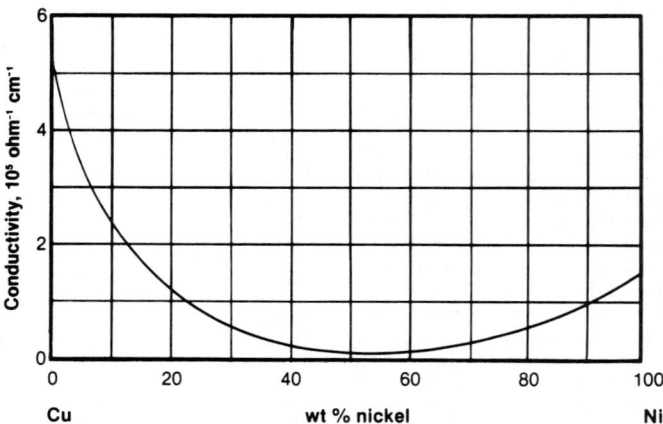

Fig. 10. Conductivity as a function of composition

charge density to field strength. Because of the high conductivities of metals, they are not used as dielectrics.

MAGNETIC PROPERTIES

In a vacuum, a magnetic field, H, produces a magnetic flux density of B, and the proportionality constant is μ_o, the magnetic permeability of a vacuum: $B = \mu_o H$. The units of this equation are $V \cdot s/m^2 = V \cdot s/amp \cdot m \times amp/m$. Because $1\ ohm \cdot s$ is defined as a henry and $1\ V \cdot s$ is a weber, $Wb/m^2 = henry/m \times amp/m$. $\mu_o = 4 \times 10^{-7}\ henry/m$.

When a material is placed in a magnetic field, the magnetic flux density is $B = \mu_o\mu_r H$, where μ_r is the relative permeability. For a vacuum, $\mu_r = 1$. μ_r is slightly more than one for paramagnetic materials, which are therefore slightly magnetic; paramagnetic metals usually have one or more unpaired electrons in the isolated atom. μ_r is slightly less than one in diamagnetic metals, which are slightly repelled by magnetic fields. Diamagnetic metals usually have all of their electron spins paired. Several metals have very high permeabilities. These are the ferromagnetic metals—i.e., the metals that are moderately or strongly magnetic. Iron, cobalt and nickel are the only elemental metals that are ferromagnetic at room temperature and above. Gadolinium and some other rare earth metals are ferromagnetic at very low temperatures.

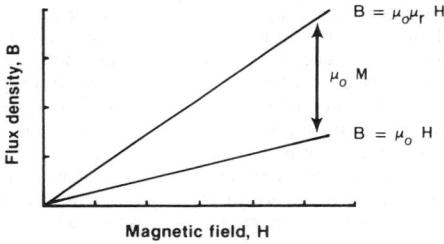

Fig. 11. Flux density as a function of magnetic field

Figure 11 shows a typical plot of flux density, B, as a function of magnetic field, H. The lower curve is the flux density of a vacuum, $B = \mu_o H$; the upper curve is the flux density of the material, $B = \mu_o\mu_r H$. If the flux density of the vacuum is subtracted from the flux density of the material, then we have a quantity that represents the added flux density over and above that of a vacuum due to the material. This difference is $\mu_o M$, where M is the magnetization of the material. For a given material, $B = \mu_o H + \mu_o M$, or $B = \mu_o(H + M)$; M has the same units as H, amp/m. Since $\mu_o M = B_{material} - B_{vac}$, $\mu_o M = \mu_o\mu_r H - \mu_o H = \mu_o(\mu_r - 1)H$, or $M = (\mu_r - 1)H$. The magnetic susceptibility, χ, is defined as $\chi = \mu_r - 1 = M/H$. Magnetic susceptibilities for metals are listed in Table 5.

If the magnetic susceptibility is slightly positive, ranging from 10^{-4} to 10^{-2}, as shown in Fig. 12, the material is paramagnetic (unpaired electrons usually cause this slight positive interaction with a magnetic field). If the magnetic susceptibility is slightly negative, ranging from 0 to -5×10^{-5} (Fig. 12), the metal is diamagnetic, usually having no unpaired electrons. Ferromagnetic materials have magnetic susceptibilities of 10^2 to 10^6.

Table 5. Work functions and magnetic susceptibilities of metals

Metal	Work function, $e\phi$, eV	Magnetic susceptibility, $10^6\ \chi$
Aluminum	4.2	+0.65
Antimony	4.1	−0.87
Arsenic	5.1	−0.31
Barium	2.5	+0.9
Beryllium	3.4	−1.00
Bismuth	4.4	−1.35
Cadmium	4.0	−0.18
Calcium	2.9	+1.10
Chromium	4.4	+3.08
Cobalt	4.0	Ferromagnetic
Cesium	1.9	−0.22
Copper	4.5	−0.086
Gallium	3.9	−0.24
Germanium	4.8	−0.12
Gold	4.8	−0.15
Hafnium	3.6	...
Iridium	4.6	+0.15
Iron (bcc)	4.48	Ferromagnetic
Iron (fcc)	4.21	...
Lead	4.0	−0.12
Lithium	2.4	+0.50
Magnesium	3.7	+0.55
Manganese	3.8	+11.8
Mercury	4.5	−0.168
Molybdenum	4.2	+0.04
Nickel	4.9	Ferromagnetic
Niobium	4.0	+1.5
Palladium	5.0	+5.4
Platinum	5.3	+1.10
Potassium	2.2	+0.52
Rhenium	5.0	...
Rhodium	4.6	+1.11
Rubidium	2.1	+0.21
Silicon	4.2	−0.13
Silver	4.7	−0.20
Sodium	2.25	+0.51
Strontium	2.7	−0.20
Tantalum	4.1	+0.93
Tellurium	4.8	−0.31
Thallium	3.8	−0.24
Thorium	3.5	+0.11
Tin	4.3	−0.25
Titanium	4.1	+1.25
Tungsten	4.5	+0.28
Uranium	3.6	...
Zinc	4.3	−0.157
Zirconium	3.8	−0.45

Note: Gadolinium and a few other rare earth metals are ferromagnetic at low temperatures.

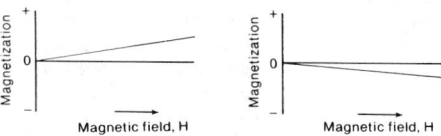

Fig. 12. Magnetization of metals: (left) paramagnetic; (right) diamagnetic

Additional information on the fundamentals of magnetism can be found in the article "Materials for Permanent Magnets," in Section 20 of this volume.

OPTICAL PROPERTIES

At first glance, one might think that optical properties of metals are unimportant because all metals are opaque except in very thin films. Of course, opaqueness is an optical property of metals and is due to the fact that light interacts with the valence electrons that can move easily throughout the lattice. The valence electrons can oscillate in the range of visible frequencies and therefore absorb and re-emit light. This accounts for the pleasing luster characteristic of metals. The fact that many finished objects have bright metal surfaces or finishes attests to the esthetic appeal of metallic luster.

Somewhat surprisingly, color is important in metals quite apart from painted or coated metals. If several different metals are placed side by side, a person with a trained eye may identify certain metals simply by the color. For instance, nickel is not as bright or "white" as chromium. Chromium is sometimes referred to as being blue-white. Stainless steel is somewhat yellow-white, depending on its composition. For persons trained in chromium electroplating, color is important. Thin oxides on the surfaces of metals, such as the "blueing" of steel, are quite attractive and of commercial value in certain instances.

Because the index of refraction depends on the transparency of the material involved, it will not be discussed further because of the opaqueness of metals. Likewise, many nonmetallic materials absorb certain wavelengths preferentially and hence are colored. Here again, absorption of light is not very important for metals; in general, they reflect most of the incident light. Metallic reflection of light is useful in mirrors and insulation: aluminum roofs reflect most of the incident sunlight, keeping the interior of a building fairly cool despite the high thermal conductivity of aluminum.

When light falls on a metal surface, it is either absorbed, reflected or transmitted. Absorptivity is the fraction of light absorbed, reflectivity is the fraction of light reflected, and transmissivity is the fraction of light transmitted. Because metals are opaque, transmissivity is very low except in extremely thin layers of metal (about 0.1 μm or less) and absorptivity and reflectivity are relatively high. Incident radiation of visible light over a wide range of frequencies excites electrons to unoccupied states of higher energy and thus is absorbed. The excited electrons soon decay back to lower energy levels, and light is re-emitted from the surface of the metal, giving rise to reflectance (the high reflectivity of polished metals was used in antiquity for producing mirrors made possible because the angle of reflection equals the angle of incidence).

Because silver is highly reflective over the entire visible range of light, its surface color is metallic white. Copper and gold exhibit red-orange and yellow colors, respectively, because incident light of short wavelengths excites electrons in filled d bands to empty levels in the s bands. The electrons decay by a different path, resulting in absorption of green, blue and violet light, whereas yellow, orange and red light are reflected. Many metals such as nickel and iron absorb light of various wavelengths and therefore have grayish or dull colors, or luster, due to relatively low reflectivity. Most metals highly reflect infrared light. Rough surfaces reflect light in a variety of directions, and the presence of oxides, hydroxides or other foreign materials greatly increases the absorption of light and decreases the reflectivity of metal surfaces.

When a metal is heated to a very high temperature, such as a tungsten filament in an incandescent light bulb, electrons are thermally excited to higher energy levels, and because many energy levels are involved, light of many wavelengths is emitted when the electrons decay back to lower energy levels. As the temperature of the

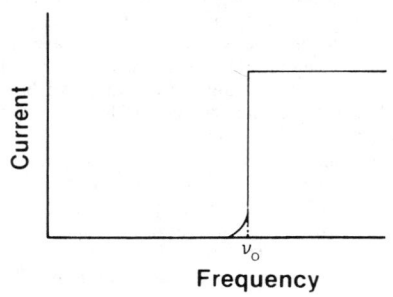

Fig. 13. Photoelectric current as a function of frequency (energy) of incident light, with light intensity being constant

metal is increased, the energies absorbed by the electrons increase, causing white light to be emitted, contrasted with red light emitted from metals at relatively low temperatures, about 600 to 1000 °C.

The photoelectric effect is the emission of electrons from a metal surface resulting from exposure to electromagnetic radiation, including visible light. Electrons are not emitted from a metal unless the frequency, or energy, of the light exceeds a specific value, ν_o, characteristic of the metal. If the intensity of light remains the same for various frequencies, the photoelectric current is constant for all frequencies above ν_o (see Fig. 13). The energy of emitted electrons increases with increased frequency (energy) of light as long as the frequency is greater than ν_o (see Fig. 14). All the energy of one photon of light is transferred to a single electron. If $\nu < \nu_o$, then the light is merely scattered, and the electron remains in the metal. If $\nu > \nu_o$, then the excess energy is retained by the electron after it escapes from the surface: $E_e = h\nu - h\nu_o$, where E_e is the energy of the electron; h is Planck's constant, 6.63 $\times 10^{-27}$ erg·s; ν is the frequency of the incident light; and ν_o is the critical frequency required to emit an electron.

Note that the slope of the curve (straight line) in Fig. 14 is equal to Planck's constant, h. The energy term, $h\nu_o$, is called the work function of the metal (see Table 5); it is the minimum energy needed to remove an electron from the surface of a metal. Photoelectric cells and devices are used in several applications: exposure meters in photography and sensors for doors, gates, signals and alarms. An even more important future application is solar cells, which convert sunlight to useful current. However, these are solid-state devices rather than pure metals.

When metals are heated to very high temperatures, such as several thousand degrees in an electric arc, transitions occur between electron states. The high temperature excites some electrons to higher energy levels, and when the electrons fall back to lower energy levels, light of characteristic wavelengths is emitted. This is the principle on which emission spectroscopy is based.

Emission spectroscopy is a valuable tool in the analysis of metals, particularly at trace-element levels (parts per million).

Bombardment of metals with high-energy electrons (electrons accelerated by voltages of about 30 000 V) results in the formation of characteristic x-rays, discussed in the section on x-ray fluorescence.

CHEMICAL PROPERTIES

Chemical properties in general are the properties that involve energy or energy changes, and, frequently, changes in electron structure. These include thermodynamic properties such as internal energy, enthalpy, entropy, free energy, and heat capacity, as well as the usual chemical properties, chemical reactivity, oxidation resistance in various environments, corrosion resistance in various environments, solubility, vapor pressure, surface energy, diffusivity and basicity.

Chemical reactivity is the ease with which metals react with the environment or with various chemicals. One aspect of chemical reactivity is oxidation resistance, which is the slowness of reaction of metals with oxygen or oxygen-containing compounds, particularly at elevated temperatures. Another aspect of chemical reactivity is corrosion resistance, which is the slowness of reaction of metals to form compounds through loss of electrons in electrochemical reactions with other elements or compounds.

Solubility is the ability to form a homogeneous, or single-phase, mixture of elements. In metals, solid as well as liquid solubility is important. Vapor pressure is the equilibrium gas pressure of a solid or liquid phase at a given temperature. Gaseous metals are not very important in most metallurgical processes, with the primary exception of extractive metallurgy. However, when metals are heated to elevated temperatures, the equilibrium vapor pressure may be of importance. By definition, at the normal boiling point of a metal, the vapor pressure in equilibrium with the liquid metal is 760 torr (millimetres of mercury).

Surface energy or surface tension is the work required to increase the surface area by 1 cm^2. Because surface atoms are attracted only toward the center of a liquid or solid, in order to increase the surface, work must be done to overcome the inward pull. Surface energies are usually expressed as ergs per square centimetre. Solid and liquid metals have high surface energies, from about 100 to 2000 ergs/cm^2 (see Table 3).

Diffusivity is the rate constant for the relative motion of atoms by processes other than convection or stirring. When solid metals are heated to relatively high temperatures, foreign atoms diffuse, or move, through the host lattice. The rate of diffusion is summarized by diffusivities.

Most metals are bases: they react with hydrogen ions, protons, and they are capable of supplying one or more valence electrons in a variety of chemical reactions. The greater the tendency

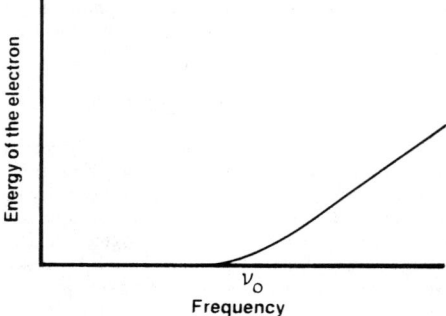

Fig. 14. Energy of photoelectrons as a function of frequency (energy) of incident light

to lose electrons in reactions, the stronger the basic character of metals. Hence, alkali metals, Group IA, are strong bases, and Group IIA metals are also fairly strong bases. The concept of metals acting as bases is generally more useful to chemists than to metallurgists.

Not only are most metals basic, but many metal oxides are bases. Group IA metal oxides, such as Na_2O, are very strong bases. They react with water to form hydroxides that are also strong bases. Although Group IIA metal oxides are moderately strong bases, the hydroxides are weak bases. Group IIIA metal oxides are amphoteric, meaning that the oxides, such as Al_2O_3, react with both strong acids and bases. The oxides of most transition metals are weak bases, and many of these are amphoteric, even iron oxide (Fe_2O_3), which may react with very strong bases to form ferrites [e.g., $Ca(FeO_2)_2$] or even ferrates having an oxidation number of +6 (e.g., K_2FeO_4). Generally, the metals farther to the right in the periodic table form oxides that are increasingly weaker bases. Because nearly all metals react with oxygen and moisture in the environment, clean metallic surfaces in reality are covered with extremely thin oxide and hydrated oxide layers. The properties of the surface oxides are extremely important in the corrosion resistance of many metals.

SELECTED REFERENCES

Materials Science for Engineers, by L. H. Van Vlack: Addison-Wesley, Reading, MA, 1970

Introduction to Mechanical Properties of Materials, by M. M. Eisenstadt: Macmillan, New York, 1971

Fundamentals of Physical Chemistry, by W. E. Wentworth and S. J. Ladner: Wadsworth, Belmont, CA, 1972

Handbook of Chemistry and Physics, 50th Ed., edited by R. C. Weast: CRC Press, Cleveland, 1969

Materials and Process Engineering Databook, 2nd Ed.: American Society for Metals, Metals Park, OH, 1970

Metals Reference Book, 4th Ed., Vol I, II and III, by C. J. Smithells: Plenum Press, New York, 1967

Thermophysical Properties of Matter—Metallic Elements and Alloys, Vol 4, by Y. S. Touloukian and E. H. Buyco: Plenum Press, New York, 1970

Phases and Phase Diagrams

By Giles F. Carter, Eastern Michigan University

A PHASE is a homogeneous part of a system. Although this definition may seem a little less than meaningful at first, with a little practice the reader will easily be able to identify the phases present in a system. First, most pure materials exist in three phases: solid, liquid and gas. At the melting point, two phases are in equilibrium: solid and liquid. At the boiling point, the phases in equilibrium are liquid and gas. Variables for a system comprising a single pure material are simply temperature and pressure. A beaker full of sand is one phase even though the particles of sand are physically separated from one another (the gaseous phase between sand particles could be counted as a second phase in this system). Sand placed in water comprises two phases: solid sand and liquid water. However, salt placed in water is one phase if the salt dissolves completely; if not, and solid salt remains in the saturated salt solution, there are two phases: NaCl(s) + H$_2$O(soln). At a unique pressure and temperature fixed by nature, the three phases of a pure substance, solid, liquid and gas, are in equilibrium with each other. This is called the triple point.

The phases present in a metal system and the way in which they are distributed (size, shape and number of precipitate particles) critically affect the properties of metals. Therefore, a knowledge of phases and phase diagrams is a prerequisite for understanding most metal systems and their properties.

PHASES AND COMPONENTS

In metallurgy, experiments frequently involve only solids and liquids; gas may be present as the atmosphere, but it may usually be ignored as an inert material, and gaseous metals usually are not involved in most metallurgy. In any system, there is at most only one gaseous phase, because all gases are soluble in one another, assuming that they do not react chemically. Because gaseous metals are relatively unimportant, this article will deal almost exclusively with solid and liquid metal systems.

A component is a pure metal or compound. A single element, such as mercury, is a single component whether it is present as a solid, liquid, gas, or some combination of these phases. An amalgam of silver and mercury, however, is comprised of two components. The metallurgist must be able to determine the number of phases and the number of components in a given system.

Binary alloys are defined as two-component systems, whereas ternary alloys comprise three components. Because ternary, quaternary, etc., systems are relatively complicated, these will not be dealt with. The reader should refer to texts written on phases and phase diagrams.

Binary alloys frequently form intermetallic compounds that have crystal structures different from either pure element and have no solid solubility; they are stoichiometric compounds. Intermediate phases are compounds different from either pure element and having appreciable solid-solubility ranges.

THE PHASE RULE

In metal systems, the primary variables are temperature and composition. Pressure is assumed to be constant at one atmosphere unless specifically stated otherwise. We have already defined phases and components and have seen how these are determined for a given system.

The phase rule is a useful equation for analyzing metal systems: it relates the number of phases and components to the degrees of freedom, F = C − P + 1, where F = degrees of freedom, C = number of components and P = number of phases. All three of these are integers. F, the degrees of freedom, gives the number of variables that may vary independently under given circumstances while maintaining the same phase or phases. For instance, nickel and copper dissolve completely in one another, even in the solid state. Therefore, there are F = C − P + 1 = 2 − 1 + 1 = 2 degrees of freedom in the solid state, because the number of components is two (copper, nickel) and there is one phase (solid copper-nickel solution). In solid copper-nickel alloy, it is possible to vary the weight percent copper and still retain the solid phase (one degree of freedom), or to change the temperature and retain the solid phase (the second degree of freedom). Because these two variables can be changed independently while maintaining the same phase, the number of degrees of freedom is two.

It has been assumed that the systems with which we have been dealing (copper-nickel, silver-copper, etc.) are at equilibrium. The phase rule holds true only for systems at equilibrium: it is possible to quench silver-copper alloys fast enough to prevent precipitation of copper. Thus, a one-phase alloy, a supersaturated solution of copper in silver, is obtainable at room temperature, but the phase rule does not apply. If the silver-copper supersaturated solution is annealed at an elevated temperature, copper precipitates and the alloy reaches equilibrium in time — then the phase rule will apply.

A supersaturated solution is metastable, which means it has temporary stability. A block of wood standing on its smaller end is metastable because it has a higher potential energy than when the block is lying down. The latter is the stable state. Metastable states may exist for indefinite periods if they are not disturbed: unless some force is applied to tip the wood over, it will not spontaneously tip over from the metastable to the stable state. Likewise, the silver-copper supersaturated solution (metastable) will remain at room temperature for a very long period. However, when its temperature is elevated several hundred degrees, copper precipitates, and the stable, or equilibrium, state is reached.

PHASE DIAGRAMS

One of the types of graphs most useful to metallurgists is the phase diagram. Metallurgical phase diagrams are usually plots of the stable phases for temperature versus composition for a metal system at equilibrium. A phase diagram for a pure metal, such as iron (see Fig. 1), re-

duces merely to a line because a single pure component cannot vary in concentration. However, if pressure is introduced as an additional variable, the phase diagram of a single component is two-dimensional, and the phase rule is F = C − P + 2. Variables are temperature and pressure. Binary phase diagrams (involving two components) contain a tremendous amount of useful information: melting points, phases present, compositions of phases present, relative amounts of phases present at a given set of conditions, and solubilities.

Copper and nickel are completely soluble in one another in both the solid and liquid states. This is indicated by the phase diagram of copper-nickel, shown in Fig. 2. Any point on the phase diagram indicates a unique combination of temperature and composition. The copper-nickel phase diagram shows that there are two phases in equilibrium for all points within the banana-shape area between the liquidus and solidus curves. The liquidus curve is the curve between the liquid phase and the two-phase region (liquid + solid), whereas the solidus curve is the curve between the solid phase and the two-phase region.

A tie-line is a horizontal line, representing a constant temperature, drawn across a two-phase region until the line intersects its boundaries. In Fig. 2, a tie-line has been drawn between points D and E. The line is horizontal because the temperature is 1400 °C all along the tie-line. The in-

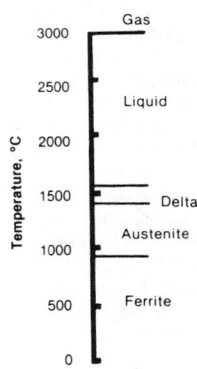

Fig. 1. Phase diagram for pure iron

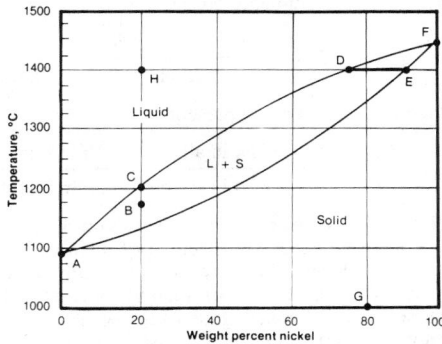

Fig. 2. Phase diagram for copper-nickel

tersections of the tie-line with the boundaries of a two-phase region give the chemical compositions of the two phases in equilibrium.

An over-all alloy content of 80Ni-20Cu can have exactly the same solid and liquid phase compositions at 1400 °C as an alloy of 85Ni-15Cu, because the proportions, or ratios, of the two phases are not the same in the above two cases. In the alloy containing 85% Ni, there is much more solid than liquid present, whereas in the 80% Ni alloy, there is slightly more liquid than solid.

At this point, the reader must make a clear distinction between two separate concepts: (a) that the chemical composition of a phase or phases is the weight percent of each component present; and (b) that the relative proportion or ratio of weights of the two phases is the weight in grams of one phase compared with the weight in grams of the second phase. This may be expressed as a ratio or fraction or given as weight in grams of the two phases. It is frequently convenient to assume a basis of 100 g of alloy and then to calculate the number of grams of phase 1 and the number of grams of phase 2.

The chemical compositions of two phases in equilibrium are read directly from the appropriate tie-line on the phase diagram. The relative amounts of two phases may be calculated by a principle, or equation, called the lever rule (see below).

Point D on the liquidus curve of Fig. 2 is 100% liquid and 0% solid, or, based on 100 g of alloy, there would be 100 g of liquid and 0 g of solid. Based on 100 g of alloy, point E in Fig. 2 corresponds to 100 g of solid and 0 g of liquid. At a point on the phase diagram very near to E, such as 86% Ni and 1400 °C, the two phases would have the following chemical compositions: 74Ni-26Cu (liquid) and 87Ni-13Cu (solid). However, the relative proportion of the two phases would consist of much solid and only a little liquid. Likewise, a point at 75% Ni and 1400 °C would correspond to an alloy comprised of two phases: mainly liquid with a small amount of solid.

To calculate quantitatively the relative amounts or ratio of the two phases present, it is necessary to know the chemical compositions of *three* points: (a) the liquid phase, (b) the solid phase in equilibrium with the liquid phase, and (c) the over-all composition of the alloy. Knowing these, the relative amount or fraction of liquid phase present is found by the following formula, which is known as the "lever rule" (Fig. 3):

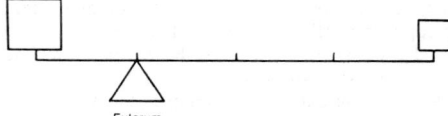

Fraction of liquid phase present

$$= \frac{\% \text{ Ni in solid} - \% \text{ Ni in alloy}}{\% \text{ Ni in solid} - \% \text{ Ni in liquid}}$$

Fraction of solid phase present

$$= \frac{\% \text{ Ni in alloy} - \% \text{ Ni in liquid}}{\% \text{ Ni in solid} - \% \text{ Ni in liquid}}$$

Notice that the fraction of the phase on the left, the liquid, is equal to the length of the *opposite* arm of the lever divided by the total length of the lever. The calculation of relative amounts of phases present is analogous to the weights of a father and small son on a see-saw. To balance the see-saw, the father, being heavier, must sit on the short arm of the see-saw, while the son, being lighter, sits on the long arm (Fig. 3). The

weight of the son is proportional to the opposite or short arm of the see-saw, and the weight of the father is proportional to the arm opposite him—namely, the long arm.

A material balance may be obtained by (a) determining the chemical compositions of the two phases present, (b) determining the relative amounts of the two phases present by the lever rule (usually on the basis of 100 g of alloy), and (c) calculating the weight of each component present in each phase by combining (a) and (b). The net result is the material balance, which is the number of grams of each component in each phase.

COOLING CURVES AND THE DETERMINATION OF PHASE DIAGRAMS

When phase changes occur, heat is generally evolved or absorbed. For instance, when water freezes, heat is given off (the heat of fusion), and when ice melts, heat is absorbed. Liquid metals have heats of fusion also, and heat is evolved when metals freeze. A pure metal freezes at a specific temperature: for instance, copper freezes (or melts) at 1083 °C, and silver freezes at 960.5 °C. If pure liquid silver is melted in a crucible and allowed to cool slowly (the crucible may be insulated, or it may contain a large amount of

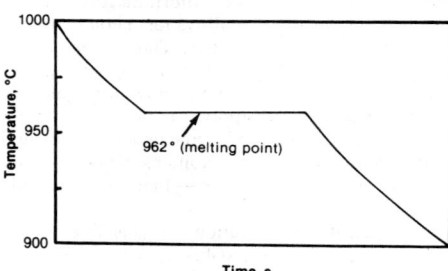

Balance is achieved if the large square on the left represents a weight three times that of the smaller square on the right.

Fig. 3. Lever rule

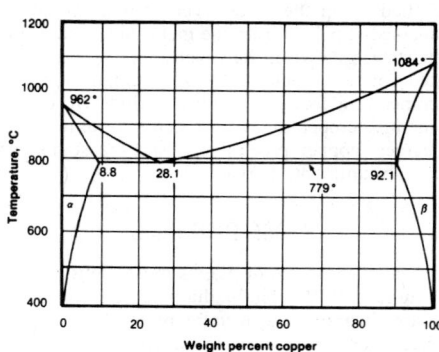

The temperature arrest at 962 °C is due to the heat of fusion being given off.

Fig. 4. Cooling curve for pure silver.

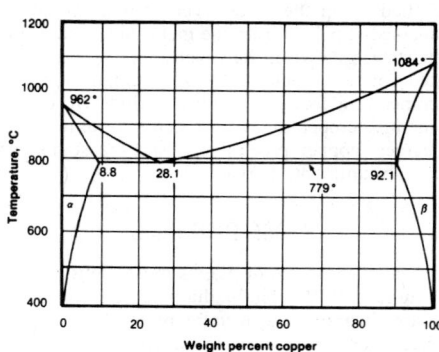

Fig. 5. Phase diagram for silver-copper

silver, such as 1 kg), the cooling rate will be nearly linear as heat is lost to the surroundings. The temperature may be obtained by a thermocouple connected to a recorder. As long as the silver is completely liquid, the temperature decrease is nearly linear with time; however, at 960.5 °C, silver solid begins to form and heat is given out. As soon as this occurs, the temperature remains at 960.5 °C until all silver liquid has frozen. This usually requires at least several minutes. Once the silver has completely frozen, then the temperature decreases nearly linearly with time until eventually the silver in the crucible attains the same temperature as the surroundings (often room temperature). The time-temperature curve is called a cooling curve (see Fig. 4) and is related to all phase diagrams based on silver as one of the components: for instance, refer to Fig. 5, the phase diagram of silver-copper.

In a like manner, the freezing temperature of pure copper may be obtained from a cooling curve. What would the cooling curve for a 50Ag-50Cu alloy look like? First, the cooling rate for the single liquid phase would be very nearly linear. At 875 °C, a copper-rich phase, called β, begins to solidify. When this happens, heat is given out, and the rate of cooling decreases (i.e., the slope of the cooling curve changes abruptly—see Fig. 6). Because we are now in a two-phase region, there is F = C − P + 1 = 2 − 2 + 1 = 1 degree of freedom. This means that the temperature can change while the two phases, liquid and solid, coexist. As the temperature decreases, several things happen. (a) The composition of the liquid changes: it takes on a composition higher in silver and lower in copper. (b) The composition of the solid changes: it too becomes slightly enriched in silver with a lower copper concentration. (c) Surprisingly, *both* phases shift the *same way* in concentration; this apparent contradiction is resolved when we realize that the ratio of solid to liquid phases also changes with cooling (different tie-lines are involved, and each tie-line at a lower concentration corresponds to a higher fraction of solid and a lower fraction of liquid). (d) The composition of the liquid eventually reaches a minimum of 28.1% Cu. This composition is called the eutectic composition because, in a binary phase diagram, three phases are in equilibrium (at a temperature of 779 °C): liquid, silver-rich α phase (solid), and copper-rich β phase (solid). A liquid 71.9Ag-28.1Cu alloy freezes at a specific temperature, 779 °C, to form a solid, but the solid consists of two phases: α and β. Likewise, if a solid 71.9Ag-28.1Cu alloy is heated, it has a sharp melting point at 779 °C.

Once the β phase begins to precipitate, the cooling curve will show an abrupt change of slope, as mentioned above. A plateau in the cooling curve occurs at the eutectic composition, because the eutectic liquid phase freezes at a constant temperature, 779 °C (Fig. 6). In experimental measurements of cooling curves, the alloy may not always be at equilibrium, because the cooling rate may be too rapid for complete equilibration, which requires diffusion over appreciable distances. Hence, the experimentally measured-cooling curve may deviate from the theoretical one.

When all of the liquid has frozen to α + β solids, the alloy will then cool at a uniform rate, and the cooling curve once again is approximately linear, but it has a different slope, depending on the amount of alloy present and the rate of cooling. Note that the important points in

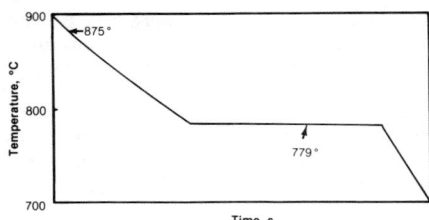

The slope of the cooling curve decreases at 875 °C due to the solidification of copper-rich solid. The temperature arrest at 779 °C is due to the heat of fusion given off by the eutectic.

Fig. 6. Cooling curve for 50Ag-50Cu alloy

the phase diagram are indicated by abrupt changes in slope of the cooling curve: 860 and 779 °C in the case of 50Ag-50Cu alloy. The cooling curve for a liquid having the eutectic composition is analogous to the cooling curve for pure silver or pure copper. Cooling curves may be used to obtain points all along the liquidus and solidus curves. The solvus curve is the curve between the single α solid phase and the two-solid-phase region, $\alpha + \beta$. Another solvus occurs between the β and $\alpha + \beta$ phase fields. Because of small heat effects involved in the solid-phase transformations, the solvus is difficult or impossible to obtain from cooling curves. However, the solvus curve actually represents the maximum solubility of copper in silver-rich α at a given temperature; likewise, the other solvus gives the maximum solid solubility of silver in copper at any temperature up to 779 °C.

THE USE OF PHASE DIAGRAMS IN METALLOGRAPHY

We have already seen how phase diagrams may be used to obtain useful information such as melting points, solubilities, compositions of phases and relative amounts of phases. One of the most useful applications of phase diagrams is in metallography: the interpretation of microstructures of metals. Other applications are covered elsewhere (e.g., age hardening, solidification, corrosion sensitization).

Because phase diagrams are equilibrium diagrams, they are very useful in interpreting the microstructures of slowly cooled metals. Figure 5, the phase diagram of silver-copper, is a good example of a common type of binary phase diagram, the eutectic system, so named because of the eutectic at 28.1% Cu and 779 °C. If liquid of eutectic composition is cooled slowly, it will freeze at 779 °C, and the resulting two solid phases are present in the ratio of 92.1 − 28.1/92.1 − 8.7 = 0.767 α and 0.233 β. One hundred grams of eutectic solid would consist of 76.7 g α (silver-rich) and 23.3 g β (copper-rich). Because the two phases form at the same time, at high magnification the microstructure of eutectic silver-copper shows a matrix of α (the continuous phase) with thin sheets of β spaced closely together. At medium magnification—e.g., 200×—the eutectic mixture of two phases simply looks gray because the two phases are not clearly resolved under the metallograph.

Compositions lying to the left of the eutectic are defined as hypoeutectic, while compositions to the right of the eutectic are designated hypereutectic. In the silver-copper phase diagram, compositions containing from 0 to 28.0% Cu are hypoeutectic, whereas compositions containing 28.2 to 100% Cu are hypereutectic. A hypoeutectic microstructure is shown in Fig. 7; the relatively large white areas are α (silver-rich) dendrites (treelike crystals) in a matrix of eutectic. During solidification upon cooling, the α dendrites crystallize first and form a skeletal solid in a matrix of liquid having the eutectic composition. At 779 °C, this eutectic liquid solidifies to the typical mixture of α and β phases typically found in eutectic solid. It is interesting that the silver dendrites are not separated from the silver matrix—the silver forms a continuous phase, and the copper particles in the eutectic are surrounded by silver.

In hypereutectic alloys (i.e., those containing more than 28.1% Cu), the copper-rich solid crystallizes first as β dendrites, a skeletal solid immersed in eutectic liquid. At 779 °C, the eutectic finally freezes (see Fig. 8). In this case also, the copper dendrites are completely surrounded by eutectic: silver is again the continuous phase even though more copper than silver may be present. Surface-energy relationships probably determine which is the continuous phase and which is non-continuous (it is possible to have two continuous phases).

Parts of microstructures that are clearly identifiable are called constituents. Thus, the eutectic solid mixture of two phases is a constituent. The copper dendrites that solidify before the eutectic are also a constituent. In this case, they are called the proeutectic constituent because the copper dendrites are the constituent formed *before* the eutectic freezes. In hypoeutectic compositions of silver-copper, the proeutectic constituent is α (silver-rich) phase.

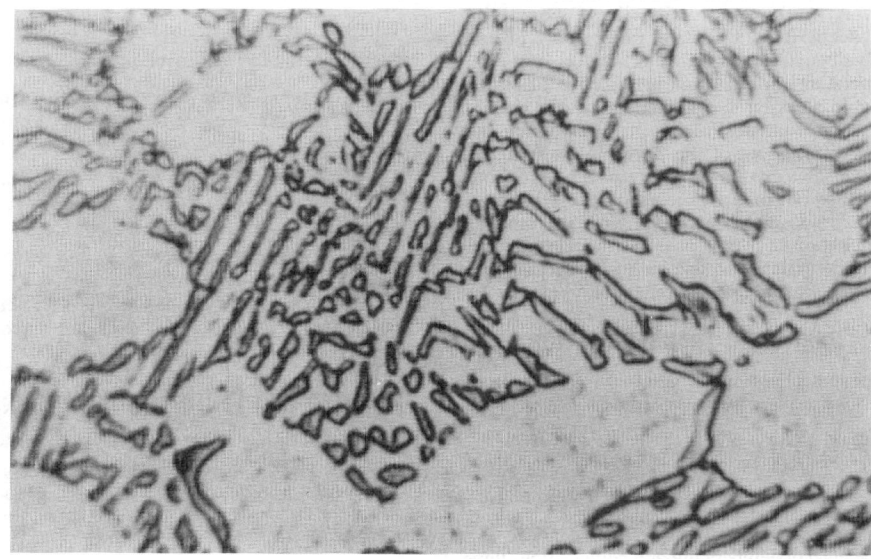

The eutectic is the fine white-in-gray structure. The hypoeutectoid silver is in white dendrites.

Fig. 7. Hypoeutectic microstructure in a silver-copper alloy

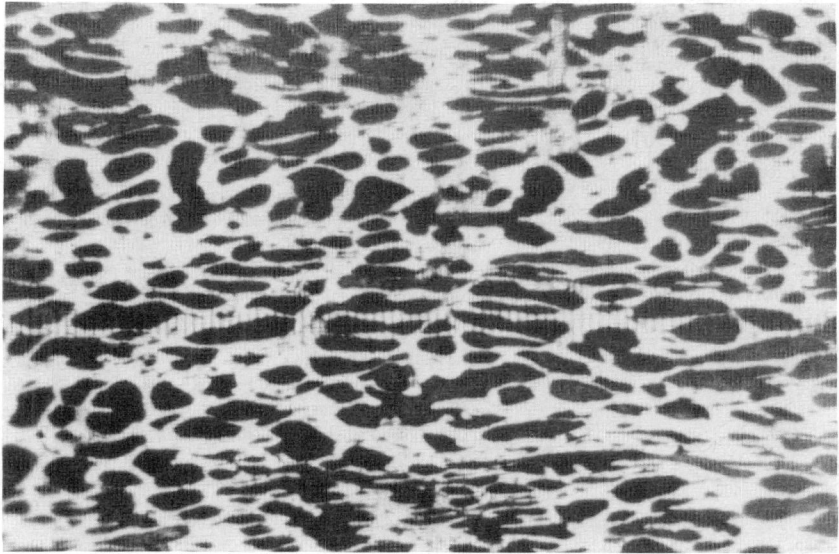

The large dark blobs are dendrites; the surrounding white regions are the eutectic, which is silver containing small particles of copper because copper is insoluble in silver at low temperatures. Scale: 12 mm = 25 μm.

Fig. 8. Hypereutectic silver-copper alloy

One may calculate the relative amounts of constituents present in an alloy by using the lever rule.

If there is a question in identifying a constituent in a microstructure, a knowledge of the approximate over-all composition of the alloy in combination with the phase diagram enables the question to be resolved. Very simply, if one knows whether the specimen is hypoeutectic or hypereutectic, then immediate identification of the proeutectic constituent is possible (e.g., for 40Ag-60Cu, the alloy is hypereutectic, and the dendritic proeutectic must therefore be copper-rich β).

If 91.3Ag-8.7Cu alloy is cooled to 779 °C, then there is only one phase present: α solid. However, if this alloy is cooled moderately slowly to room temperature, β phase (copper-rich solid) precipitates as very small particles in the α matrix. Also, hypereutectic alloys, containing proeutectic β, will contain a precipitate of very finely divided α in the proeutectic β phase at room temperature. Similar fine precipitates of α in β and β in α appear during moderately fast cooling of the eutectic constituent; on slow cooling, the silver diffuses out of the β and into silver, and vice versa. Very slow cooling will permit silver to diffuse out of the proeutectic β for hypereutectic compositions. Quenching usually prevents precipitation of α in β or β in α because there is insufficient time for precipitation to occur: the solid remains a supersaturated solution.

PERITECTIC TRANSFORMATION

The eutectic transformation in binary phase diagrams may be summarized as follows:

liquid phase
higher temperature

$\rightleftarrows$ solid phase 1 + solid phase 2
lower temperature

At the eutectic temperature and composition, three phases are in equilibrium, and F = 2 − 3 + 1 = 0, or there are no degrees of freedom: both the temperature and composition are fixed. Because three phases are in equilibrium, this is frequently called a three-phase reaction.

Another type of three-phase reaction is the peritectic reaction, in which a reaction occurs between a liquid and a solid phase to form another solid phase:

liquid phase + solid phase 1
higher temperature

$\rightleftarrows$ solid phase 2
lower temperature

An example of a peritectic reaction is shown in Fig. 9, which is a portion of the iron-nickel phase

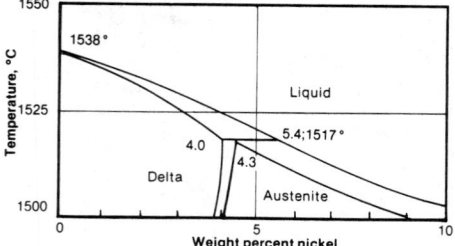

Fig. 9. Peritectic in the iron-nickel phase diagram

diagram. When iron containing 4.5% Ni is cooled slowly from 1600 °C, where the alloy is a single liquid phase, solid δ phase begins to precipitate at 1522 °C. The alloy then is comprised of two phases: liquid + solid δ. At 1517 °C, the liquid has a composition of 6.2% Ni and the solid contains 4.0% Ni. Upon cooling, these two phases transform to γ solid containing 4.3% Ni.

Peritectic reactions, or "peritectics," are common in binary phase diagrams. For instance, the copper-zinc phase diagram contains five peritectics.

MONOTECTIC TRANSFORMATION

The monotectic transformation is a three-phase transformation similar to the eutectic transformation except that *two* liquid phases and one solid phase are involved:

liquid phase 1
higher temperature $\rightleftarrows$

liquid phase 2 + solid
lower temperature

A good example is the copper-lead phase diagram. Monotectics are due to miscibility gaps in the liquid state. For instance, in the copper-lead phase diagram, at 60% Pb and 965 °C, two immiscible liquid phases are in equilibrium. The compositions of the liquid phases are 44% Pb in liquid phase 1 and 81% Pb in liquid phase 2. At the monotectic point, 36Pb-64Cu and 954 °C, three phases are in equilibrium, and F = 2 − 3 + 1 = 0, or both the temperature and the concentration are fixed. Monotectics are uncommon.

EUTECTOID AND PERITECTOID REACTIONS

Other types of three-phase reactions are the eutectoid and peritectoid reactions, in which three solid phases are involved. The eutectoid is analogous to the eutectic reaction except that all three phases are solid:

solid phase 1
higher temperature

$\rightleftarrows$ solid phase 2 + solid phase 3
lower temperature

The most famous eutectoid is in the iron-carbon system and has tremendous practical importance. Another eutectoid may be found in the copper-zinc phase diagram.

The last three-phase reaction to be described is the peritectoid transformation, which involves three solid phases:

solid phase 1 + solid phase 2
higher temperature

$\rightleftarrows$ solid phase 3
lower temperature

Peritectoid transformations are comparatively rare.

According to the phase rule, a maximum of three phases may be in equilibrium with each other for a binary alloy unless pressure is involved as an additional variable; then F = C − P + 2. If F = 0, then 0 = 2 − P + 2, or P = 4, or four phases may coexist in equilibrium when both temperature and pressure are variables in binary systems.

ANALYSIS OF BINARY PHASE DIAGRAMS

The simplest type of binary phase diagram is one in which both terminal phases, which are the

phases of the two pure components, have the same crystal structure and in which the atoms are of about the same size and are chemically similar. Then the solids form a single solution phase, as in the case of copper-nickel. In addition, there is a single liquid phase. Because single phases must be separated from one another by two-phase fields (although they may come together at a point), the liquid phase is separated from the solid phase by a two-phase field: liquid + solid.

The least complicated phase diagram after one such as copper-nickel is a diagram such as that exhibited by gold-copper. In this diagram, there are two two-phase (liquid + solid) lobes, and, at the minimum, the liquid has a sharp melting point: this is called a congruent point because the solid $\rightleftarrows$ liquid transformation occurs sharply at one composition. Congruent maximum points are common for phase diagrams involving several solid phases. At congruent points, freezing can occur with no change in composition.

Boundaries that surround two-phase regions can only meet at congruent points or at compositions of pure components.

Some binary phase diagrams contain solid miscibility gaps in which the metals form a single solid solution at higher temperatures but two immiscible (insoluble) solid phases at lower temperatures. An example is the gold-nickel phase diagram. Between 821 and 950 °C there is only one solid phase, while below 821 °C there are two solid phases, as indicated on the phase diagram (Fig. 10). According to Hume-Rothery, solid phases are miscible at room temperature when the atomic diameters of the two atoms differ by less than 15%, when the two metals are similar chemically (electrode potentials are reasonably close) and when the metals are near one another in the periodic table. Increasing temperature favors greater miscibility, and miscibility is greatly increased in liquid phases. Most liquid metals are miscible, particularly at high temperatures. Miscibility gaps are responsible for eutectics, peritectics and monotectics.

Terminal phases are defined as single solid phases at the extreme edges of the phase diagram, the phases ascribed to the pure solid components. Intermediate phases are solid phases of metallic compounds other than terminal phases. Many intermediate phases exist, such as in the copper-zinc phase diagram, which contains four intermediate phases. Intermediate phases may have broad ranges of solid solubility, such as the β, γ, and ε phases in copper-zinc, or they may have essentially no solid solubility, in which case they are stoichiometric compounds such as Fe_3C in the iron-carbon diagram. These stoichiometric com-

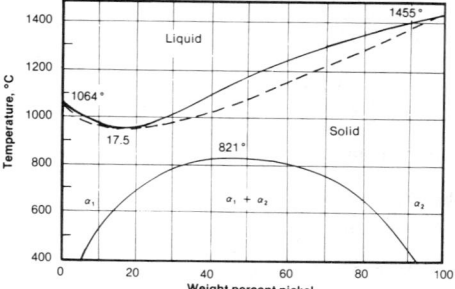

The dashed curve is only approximately known.

Fig. 10. Phase diagram for gold-nickel

pounds are often called intermetallic compounds (two examples are CaAl$_2$ and NbSi$_2$).

Intermediate phases either have a congruent maximum melting point or a peritectic point at their melting point. A few solid phases disappear on heating due to a peritectoid point—e.g., the β and δ phases in copper-silicon as well as some of the other intermediate phases. Some phases, particularly superlattices, have maximum "congruent" points, but a solid solution rather than a liquid is formed at the higher temperature.

If a horizontal line is drawn completely across any binary phase diagram at any given temperature, and if the line crosses two single-phase fields and one two-phase field, the sequence from left to right will be one-phase field, two-phase field, one-phase field. If intermediate phases are present, the sequence will be 1-2-1-2-1 for one intermediate phase and 1-2-1-2-1-2-1 for two intermediate phases, and so on. For instance, starting from the left of the copper-zinc phase diagram at 650 °C, the sequence of phase fields is α, α + β, β, β + γ, γ, γ + δ, δ, δ + liquid, liquid, or 1-2-1-2-1-2-1-2-1. Note that the two-phase region can always be identified by the single phases on either side of it. When the components of a phase diagram are pure elements, the terminal phases are always single (unless the terminal point is a melting point or a transformation point, such as 910 °C for iron, at which temperature both bcc and fcc iron are in equilibrium with each other). Phases are usually given names corresponding to letters of the Greek alphabet—α, β, γ, δ, ε, etc.—beginning at the left of the phase diagram.

Phase diagrams for the Pb-Sn and Al-Cb (Al-Nb) systems are presented in Fig. 11 and 12, respectively.

DIFFUSION IN BINARY ALLOYS

Diffusion couples involving binary alloys must comply with phase diagrams. Also, concentration gradients (in reality, activity gradients) in diffusion zones must consist of only single phases in binary diffusion couples. Two-phase regions in the phase diagram occur in binary diffusion couples as interfaces. In two-phase regions, according to the phase rule, F = C − P + 1, or F = 2 − 2 + 1 = 1, or one degree of freedom. The degree of freedom is "used up" when the temperature is set at a specific value. Therefore, in a diffusion zone of a binary alloy, a two-phase field cannot represent a concentration gradient (or activity gradient), and it must appear only as an interface separating two single phases.

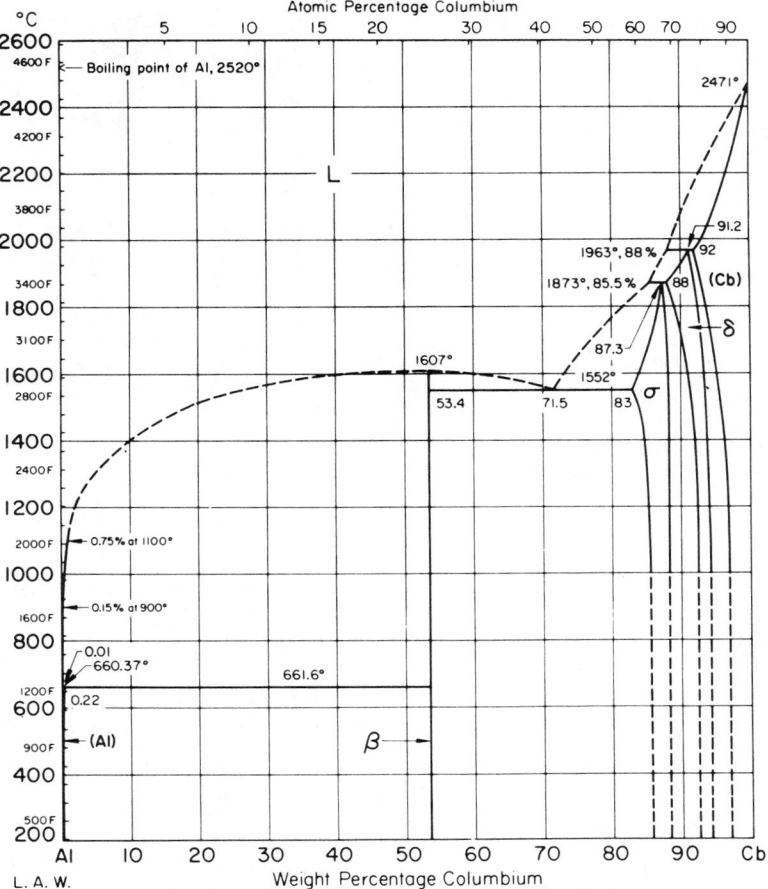

Fig. 12. Phase diagram for aluminum-columbium (niobium)

In a copper-nickel diffusion couple, the diffusion zone and the copper and nickel substrates all occur as the same phase. Because no intermediate phases exist, the concentration gradient will appear as shown in Fig. 13. In the silver-copper system, a diffusion couple prepared at 700 °C would appear as shown in Fig. 14. Note that the two-phase field occurs as an interface representing a discontinuity in copper concentration. However, the phase diagram will agree with the concentration gradient in the diffusion couple. As a matter of fact, diffusion couples may be used to obtain the portions of phase diagrams in which only solids are involved (or a liquid as one of the terminal phases).

The copper-zinc phase diagram at 650 °C contains the following phase fields: α Cu (0 to 37% Zn), α + β, β (42.7 to 52% Zn), β + γ, γ (57.7 to 70.3% Zn), γ + δ, δ (73.3 to 74.7% Zn), δ + liquid, liquid (85 to 100% Zn). If a diffusion couple were made by immersing copper in 15Cu-85Zn at 650 °C, then diffusion layers of α, β and δ should be observable. The limits for the concentrations of these diffusion layers would agree with the phase diagram as given above. The relative thicknesses of the various diffusion layers depends on several factors: (a) the crystal structure (diffusion is more rapid in bcc than in fcc, other things being equal), (b) the range of solid solubility (the greater the range, the thicker the

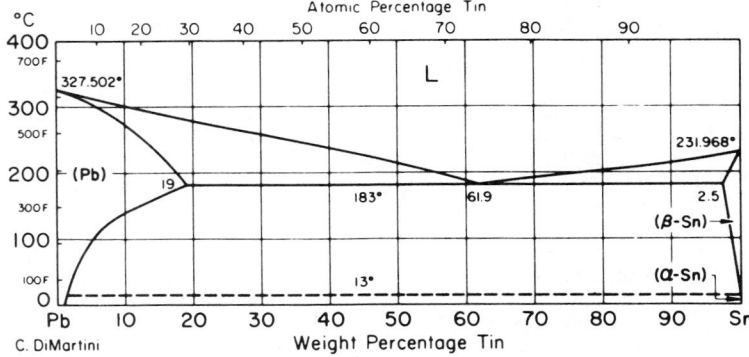

Fig. 11. Phase diagram for lead-tin

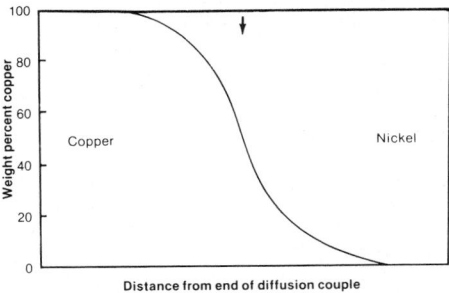

Arrow points to the Matano (original) interface.
Fig. 13. Concentration gradient in a diffusion couple of copper-nickel

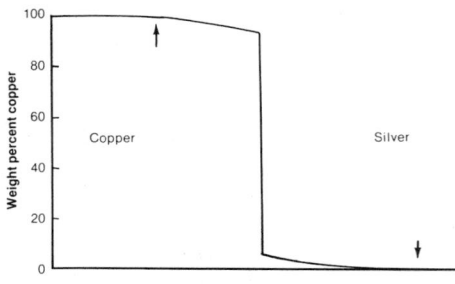

Arrow pointing up indicates the farthest diffusion of silver. Arrow pointing down indicates the farthest diffusion of copper. Large vertical line represents an interface where a discontinuity occurs in the weight percent copper because of the stability of two phases within this composition range (see Fig. 5).

Fig. 14. Concentration gradient in a diffusion couple of silver-copper at 700 °C

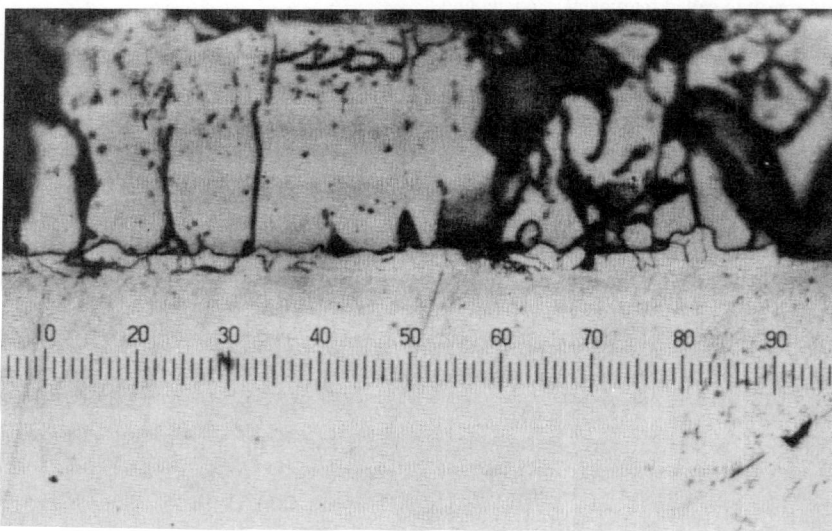

The thick layer at the top is NbAl$_3$; the grain boundaries are heavily etched, and the dark regions are where part of the brittle coating has accidentally been removed. The second layer from the top, about 6 to 7 μm thick, is Nb$_2$Al, and the very thin layer underneath is Nb$_3$Al. The white unetched material is Nb. Increments of 10 on the scale represent 25 μm.

Fig. 15. Aluminum alloy diffusion coating on niobium

diffusion layer, relatively speaking), and (c) the higher the relative value of D̄, the interdiffusion coefficient, the thicker the diffusion layer. Relative thickness depends on the relative rates of motion of the *two* boundaries of a given phase.

In some cases, the diffusion layers may be so thin relative to the other layers that their presence is difficult to determine experimentally (metallographically or by x-ray diffraction). However, the diffusion layers representing all single phases, including the terminal phases of the diffusion couple, must be present.

If essentially pure compounds occur as intermediate phases, the diffusion layers may occur as very thin layers. However, even slight solid solubility is enough to allow diffusion to occur and therefore to result in a measurably thick diffusion layer. If an intermediate phase is a stoichiometric compound having no measurable solid solubility, then the diffusion zone will be vanishingly thin: it would be an interface (no thickness) for zero solid solubility.

Relative thickness of diffusion layers is important in diffusion coating of metals for high-temperature service. For instance, if a system can be found in which a relatively thin diffusion layer exists next to the substrate alloy, then the diffusion coating may be highly protective in air at high temperatures, provided that the coating has excellent inherent oxidation resistance.

A practical example is coating of niobium or niobium-rich alloys with aluminum. If the coating process occurs at a temperature of 1200 °C, the diffusion layers, according to the phase diagram, should be comprised of Nb, Nb$_3$Al, Nb$_2$Al and NbAl$_3$. This has indeed been verified experimentally, as shown in Fig. 15; the layers have been identified by x-ray diffraction. Surprisingly, the NbAl$_3$ layer is relatively thick in spite of the fact that the phase diagram shows no solid solubility in NbAl$_3$. This must be due to the fact that D̄ is relatively high in NbAl$_3$ because of the presence of so much aluminum (also, the relatively low melting point of NbAl$_3$, 1607 °C, indicates that diffusion should be much greater than in alloys with higher melting points, such as Nb$_2$Al and Nb$_3$Al). In addition, the phase diagram is likely to be in error: there must be some solid solubility in NbAl$_3$. The Nb$_3$Al layer is very thin. Because it lies next to the niobium substrate and has a relatively flat concentration gradient, it

probably reduces the flux of aluminum into the niobium substrate.

An excellent oxidation-resistant diffusion coating should have the following properties: (a) excellent inherent resistance to oxidation, (b) resistance to spalling and cracking during thermal cycling, (c) impact resistance to particles in air colliding with the coating, and (d) low flux of diffusing atoms into the substrate metal (low diffusivity and/or flat concentration gradient). A shortcoming in any one of these areas may doom the coated metal to early failure in actual use at high temperatures. The first three are rather obvious requirements. If the diffusion coating dissipates itself by diffusion into the substrate or if it embrittles the substrate to relatively great depths, the diffusion coating may be useless even though the first three points are satisfactory.

Phase diagrams are useful in predicting structures of diffusion coatings and in interpreting the structures obtained. However, relative thicknesses of diffusion layers are not so readily predicted, although the general principles mentioned above help. Ultimately, diffusion-coated metals must be tested in use or under closely simulated conditions to prove the usefulness of a given system.

THE ASM/NBS PHASE DIAGRAM EVALUATION PROGRAM

The ASM/NBS Phase Diagram Evaluation Program has been in operation since 1978. Category editors are evaluating the existing phase diagram data for all binary and many higher order alloys with respect to the following features:

Phase Diagram. The principal line drawing of the phase diagram itself (with possible inserts) is being produced on a computer graphics display system. This allows each diagram to be displayed, if needed, with different scales, reference axes, etc.

Structures. Tabulated information about crystal structure and lattice parameters, with accompanying verbal comments and references.

Metastable. Information about metastable structures and/or metastable phase boundaries, including information about martensites, massive transformations, spinodal decomposition and metallic glass.

Thermodynamics. A brief analytical (thermodynamic) description of phase boundaries that can be calculated (if such information is available in the literature). Also included can be additional thermodynamic information, and thermodynamics of phase formation in specific important systems.

References. A complete list of references. The detailed individual evaluations are first published by ASM in the *Bulletin of Alloy Phase Diagrams,* and later collected into reference volumes.

The efforts providing the assessed data for the Program are world-wide, with alloy categories located in many different countries. The thermodynamic assessments (and often also the thermodynamic modeling) are frequently a key feature of the evaluation work, recognizing the fact that free energy considerations are the basis of all phase diagram features. Available information on metastable phases is being included for the first time, recognizing the importance of metastable situations in technological applications. Finally, the line phase diagrams are produced by computer graphics at the NBS, and a computerized database is being developed, with the ultimate goal of being able to "cross examine" the stored information and to obtain some important answers which might have required hours (or weeks) of work in the past. For example, which systems have eutectic temperatures between 900 and 1000 °C and at what compositions, or which systems include the well known superconducting compounds with the *A*15 structure, or which systems are prone to form metallic glass? The possibilities for obtaining the answers to such important questions from critically assessed data will greatly aid researchers in broadening their perspectives.

3 METALS ENGINEERING

Concepts and Criteria

By Charles O. Smith, University of Nebraska–Omaha, and Bruce E. Boardman, Deere & Co. Technical Center

ENGINEERING DESIGN of metal components is a complex task requiring consideration of many interrelated factors, not all of which are necessarily compatible. Thus, compromises and tradeoffs among various design factors are routinely made. A designer must know the relative importance of these factors and how they interact before intelligent choices between compatible requirements can be made. For convenience in discussion, design factors have been somewhat arbitrarily grouped into three categories: functional requirements, analysis of total life cycle and other major factors; the factors are listed in Tables 1, 2 and 3, respectively. These categories intersect and overlap, which constitutes a major problem in engineering design.

FUNCTIONAL REQUIREMENTS IN DESIGN

It should be obvious that any design must necessarily meet performance specifications; therefore, this item heads the list in Table 1. These specifications must reflect a full and complete analysis of the functions required of the product. An important distinction must be made between performance specifications, which enumerate the basic functional requirements of the product, and product specifications, which list requirements for configurations, tolerances, materials, manufacturing methods and the like. Performance specifications represent the basic parameters from which the design can be formulated; product specifications are codifications of designs, used for purchase or manufacture of the product. Excellence in design or product specification is not possible without complete and adequate performance specifications.

Performance specifications must reflect thorough consideration of the factors listed in Tables 2 and 3. Consequences and risks involved in possible product failures caused by predictable misuse or overload or by imperfections in workmanship or material must be considered in establishing performance specifications. Situations in which the consequences of product failure would be dire or in which only the very lowest risks of failure can be tolerated dictate the use of stringent performance specifications. When product failure does not involve a risk of personal injury and is not likely to result in great financial loss to the user, economic considerations usually imply that performance specifications be no more stringent than necessary to meet functional requirements. Realistic performance specifications result in design and manufacture of products that perform their required functions with little risk of failure, and that at the same time can be produced at the lowest possible cost.

As an example of setting performance specifications to suit the application, resistance to corrosion can be specified at any of three levels: (a) avoiding contamination by corrosion products, (b) preventing leaks into or out of closed containers, and (c) maintaining structural integrity and other mechanical and physical properties in spite of corrosive attack. For food-processing equipment, the first of these considerations has paramount importance. For a bridge, the third factor is critical; furthermore, a bridge must retain its structural integrity for many years. In a petrochemical plant, all three considerations are important; chemical process equipment may be designed for continuous operation for two or three years, and any breakdown between scheduled maintenance periods would be extremely expensive. Furthermore, leakage of dangerous chemicals from process equipment is unacceptable. In this case, the cost of a breakdown and the damage caused by leakage can justify use of expensive materials if their performance reduces the probability of leaks or breakdown to very low levels. On an automobile body, low-carbon steel with a corrosion-resistant surface treatment and coating provides corrosion resistance consistent with the anticipated lifetime of the vehicle. Thus, for these various applications, the appropriate criteria for corrosion resistance depend on one or more of the following factors: the degree of contamination permitted by the application, the intended lifetime of the object, the corrosion char-

Table 1. Functional requirements in design

Performance specifications
  Definition of need
  Risks and consequences of underspecification
  Consequences of overspecification
Design configuration
  Probabilistic or deterministic approach
  Stress or load considerations
  Restrictions on size, weight or volume
  Service hazards, such as cyclic loading or
    aggressive environment
  Failure anticipation
  Reliability, maintainability, availability and
    repairability
  Quantity to be produced
  Value analysis
  Candidate materials and manufacturing processes
Redesign
  Design review
  Simplification and standardization
  Functional substitution

Table 2. Total life cycle in design

Material selection
Producibility
Durability
Feasibility of recycling
Energy requirements
  For production
  During use
  For reclamation
Environmental compatibility
  Effect of product on environment
  Effect of environment on product
Inspection and quality-assurance testing
Handling
Packaging
Shipping and storage
Scrap value

Table 3. Other major factors in design

State of the art
  Prior knowledge
  Possible patent infringement
  Competitive products
Conformance to standards
  Codes for specific products, such as pressure vessels
  Safety requirements
    Products—Consumer Products
      Safety Commission
        Warnings
        Unintended uses
        Labels
    Manufacturing—Occupational Safety and Health
      Administration
    Environmental requirements—Environmental
      Protection Agency
  Industry standards
    ANSI
    ASTM
    SAE
    UL
    ISO
Human factors
  Ease of operation
  Ease of maintenance
Aesthetics
Cost

acteristics of the environment, and the consequences and risks associated with corrosion failure.

Considering the specific example of an automotive exhaust system, the performance specifications for that system must provide for the following basic functions:

• Conducting engine exhaust gases away from the engine
• Preventing noxious fumes from entering the automobile
• Cooling the exhaust gases
• Reducing engine noise
• Reducing exposure of automobile body parts to exhaust gases
• Affecting engine performance as little as possible
• Helping control undesirable exhaust emissions
• Having a service life that is acceptably long
• Having a reasonable cost, both as original equipment and as a replacement part.

In its simplest form, the exhaust system consists of a series of tubes that collect the gases at the engine and convey them to the rear of the automobile. The size of the tubing is determined by the volume of exhaust gases to be carried away and the extent to which the exhaust system can be allowed to impede the flow of these gases from the engine. An additional device, the muffler, is necessary to satisfy the requirement for noise reduction. Under current practice, the system must contain a catalyst to convert noxious gases to less-harmful emissions. The basic lifetime requirement is that the system must resist the attack of hot, moist exhaust gases for some specified period, whether that period be defined in years, miles or number of cold starts. The practical requirement that the exhaust system be placed under the automobile imposes additional corrosion hazards: the system must resist attack not only by the exhaust gases, but also by the atmosphere, water, mud and road salt. The location also requires that the exhaust system be designed as complex shapes that will not interfere with running gear of the car, road clearance or the passenger compartment. The number of automobiles produced each year requires that material used in exhaust systems be readily available at minimum cost. The choice of welded low-carbon steel tubing for most components of the system adequately meets every requirement.

The design process includes determination of the configuration of the product and its various component parts, and selection of the materials from which and processes by which it is to be made. In its early stages, the process consists of evaluating various combinations of preliminary configuration, candidate material and potential method of manufacture, and comparing these with the previously established performance specifications. The relationship between configuration and material (processed in some specific way) is that every configuration places certain demands on the material, which has certain capabilities to meet these demands. A common specific relationship is one between the stress imposed by the configuration and the strength of the material. Of course, there are other relationships as well. Changes in processing can change the properties of a material, and certain combinations of configuration and material cannot be made by some manufacturing processes.

Quantitative relationships between configura-

tional demands and material capabilities can be established by deterministic methods—the better-known approach—or probabilistic methods. In the former, nominal or average values of stress, dimensions and strength are used in design calculations; appropriate safety factors are used to compensate for expected variations in these parameters and for discontinuities in the material. In the probabilistic approach, described in greater detail in Ref 1, each design parameter is accorded a statistical distribution of values. From these distributions and from an allowable limit on probability of failure, minimum acceptable dimensions in critical areas (or minimum strength levels for critical components) can be calculated. Compared with deterministic methods, the probabilistic approach requires greater sophistication on the part of the designer and more elaborate calculations, but offers the potential for more compact parts that use less material. A significant handicap to use of probabilistic methods is the fact that statistical distributions of properties are not widely available and often must be determined before these methods can be applied to a specific design problem.

In either approach to design, the effects of notches and stress concentrations must be considered, because such features increase the vulnerability of all types of parts to failure. However, studies in fracture mechanics have shown that, under some circumstances, notches and discontinuities in the material may be benign, and therefore of little consequence in some design applications.

Cyclic loading, use at extreme temperatures and the presence of agents that cause either general corrosion or stress-corrosion cracking are special hazards that must be considered during the materials-selection process. Some common failure modes, and those mechanical properties most related to particular modes, are illustrated in Fig. 1. Cyclic loading, in particular, is a very common factor in the design of anything that has moving parts; it is also widely recognized that fatigue is responsible for a large portion of all service failures. In 1943, Almen (Ref 2) made an observation about fatigue that is still valid:

"Fully 90 percent of all fatigue failures occurring in service or during laboratory and road tests are traceable to design and production defects, and only the remaining 10 percent are primarily the responsibility of the metallurgist as defects in material, material specification, or heat treatment.

"Study of fatigue of materials is the joint duty of metallurgical, engineering, and production departments. There is no definite line between mechanical and metallurgical factors that contribute to fatigue. This overlapping of responsibility is not sufficiently understood.

"Hence, the engineers are constantly demanding new metallurgical miracles instead of correcting their own faults. Until metallurgists are less willing to look for metallurgical causes of fatigue and insist that equally competent examination for mechanical causes be made, we cannot hope to make full use of our engineering material."

Although Almen spoke specifically about fatigue, his comments can be applied to engineering design on a far broader basis. His comments must not be construed to excuse materials engineers from a proper share of the over-all design responsibility, nor to allow them to slacken their efforts to find better materials for specific appli-

cations. Almen's comments do imply that all aspects of design must be considered, because even apparently insignificant factors can have far-reaching effects. (For example, there is at least one known instance in which fatigue failure of an aircraft in flight was traced to an inspection stamp that was imprinted on a component using too heavy a hammer blow.)

Even with parts for which loading in service is known accurately and stress analysis is straightforward, gross design deficiencies may arise from reliance on static load-carrying capacity based solely on tensile and yield strengths. The possibility of failure in other modes, such as fatigue, stress-corrosion cracking and brittle fracture caused by impact loading, must be considered in the design process.

Any discussion of potential failures in service should include careful consideration of the possible consequences of failure. Those failure modes that might endanger life or limb, or destroy other components of the apparatus, are to be avoided if at all possible. Sometimes, a piece of equipment is designed so that one component will fail in a relatively harmless fashion and thus avoid the potentially more serious consequences of the failure of another. For example, a piece of earth-moving equipment might be designed so that the engine will stall if the operator attempts to lift a load so heavy that it might upset the equipment or damage any of its structural components. A blowout plug in a pressure vessel is another example. Conformance to codes and standards, such as those listed in Table 3, may preclude serious consequences of service failure, but designers still must exercise careful judgment in studying (and designing against) the consequences of possible modes of failure.

The size and weight of a part can affect the choice of both material and manufacturing process. Small parts often can be economically machined from solid bar stock, even in fairly large quantities. The material cost of a small part may be far less than the cost of manufacturing it, perhaps making relatively expensive materials feasible. Large parts may be difficult or impossible to heat treat to high strength levels. There are also limits on size for parts that can be formed by various manufacturing processes. Die castings, investment castings and powder metallurgy parts are generally limited to a few kilograms. When weight is a critical factor, parts often are made from materials having high strength-to-weight ratios.

The quantity of parts to be made can affect all aspects of the engineering design process. Low-quantity production runs can seldom justify the investment in tooling required by production processes such as forging or die casting, and may limit the choice of materials to those already in the designer's factory or those stocked by service centers. High-quantity production runs may be affected by the capability of materials producers to supply the required quantity. Mass-produced parts sometimes can be designed and redesigned, requiring a large expenditure for engineering and evaluation, but providing enough savings, considering the large quantity involved, to make the effort worthwhile. Design for small-quantity parts may be limited to finding the first design and material that serve the required purpose.

Products that may be manufactured in several locations can present additional problems for designers because the cost and availability of materials can vary from place to place. If the prod-

Material property

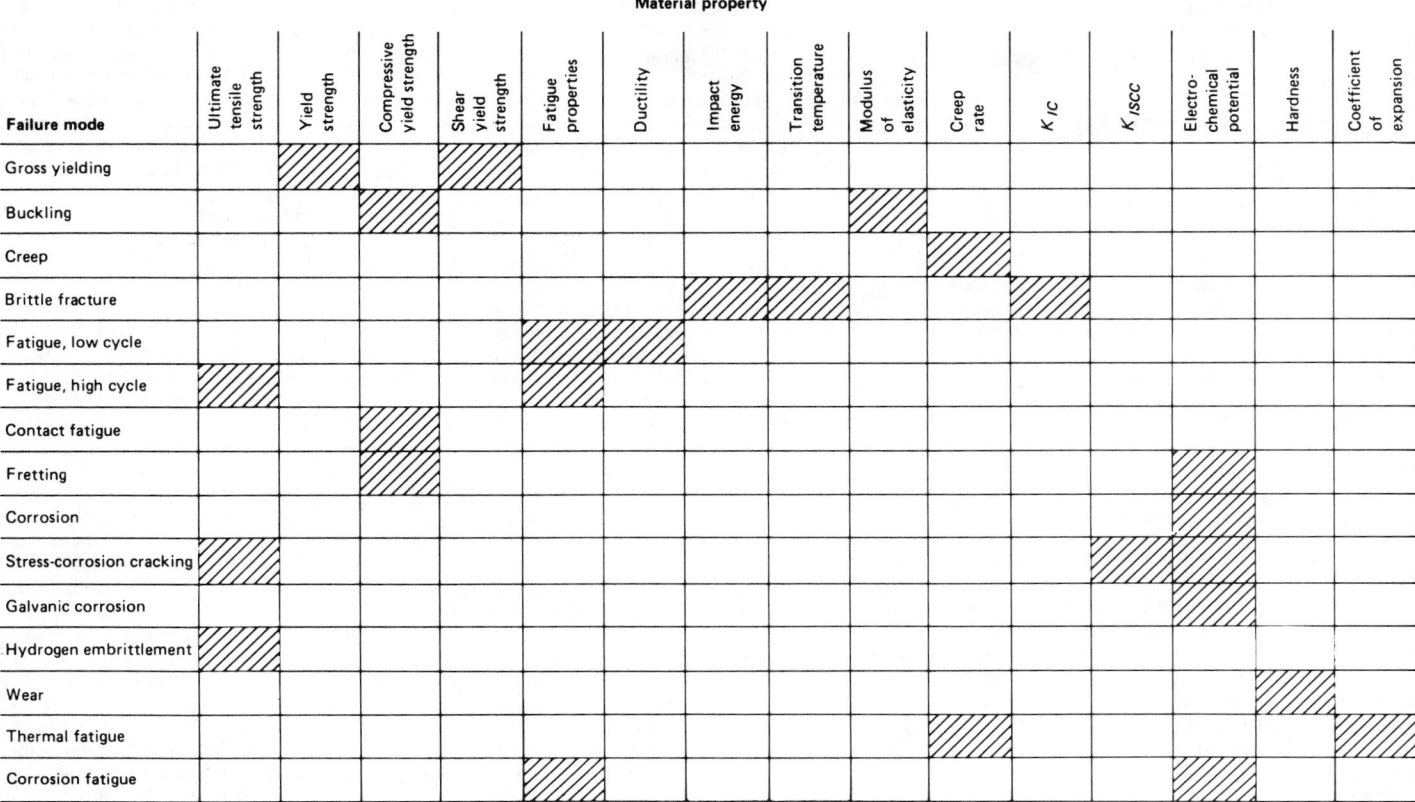

Failure mode	Ultimate tensile strength	Yield strength	Compressive yield strength	Shear yield strength	Fatigue properties	Ductility	Impact energy	Transition temperature	Modulus of elasticity	Creep rate	K_{IC}	K_{ISCC}	Electro-chemical potential	Hardness	Coefficient of expansion
Gross yielding		▨		▨											
Buckling			▨						▨						
Creep										▨					
Brittle fracture							▨	▨			▨				
Fatigue, low cycle					▨	▨									
Fatigue, high cycle	▨				▨										
Contact fatigue			▨												
Fretting			▨										▨		
Corrosion															
Stress-corrosion cracking	▨											▨			
Galvanic corrosion															
Hydrogen embrittlement	▨														
Wear														▨	
Thermal fatigue										▨					▨
Corrosion fatigue					▨								▨		

Shaded block at intersection of material property and failure mode indicates that a particular material property is influential in controlling a particular failure mode.

Fig. 1. Relationships between failure modes and material properties

uct is to be made in different countries, the nearest equivalent grades of steel, for example, might be different enough to affect service performance. In some areas that have low prevailing labor costs, it may be desirable to design a labor-intensive product; in a high-cost labor market, the designer often attempts to design the product to fit the capabilities of automated manufacturing equipment.

It is relatively late in the design process before designers, materials engineers and manufacturing engineers, by working together, can establish those factors listed in Table 4 on a detailed basis. Only then can the field of candidate materials and manufacturing processes be narrowed to a manageable number of alternatives. The implications of each of these alternative materials and manufacturing processes can then be evaluated, and any required changes in configuration can be made.

One of the complicating factors in materials selection is that virtually all materials properties, including fabricability, are interrelated. Substituting one material for another, or changing some aspect of processing in order to effect a change in one particular property, generally affects other properties simultaneously. Similar interrelations that are more difficult to characterize exist among the various mechanical and physical properties and variables associated with manufacturing processes. For example, cold drawing a wire to increase its strength also increases its electrical resistivity. Steels that have high carbon and alloy contents for high hardenability and strength generally are difficult to machine and weld. Addi-

Table 4. General factors in materials selection

Functional requirements and constraints
Mechanical properties
Design configuration
Available and alternative materials
Fabricability
Corrosion and degradation resistance
Stability
Properties of unique interest
Cost

tions of alloying elements such as lead to enhance machinability generally lower long-life fatigue strength and make welding and cold forming difficult. The list of these relationships is nearly limitless.

Value analysis using criteria such as those listed in Table 5 can provide both designers and managers with assurance that the final combination of configuration, material and manufacturing process is a good one.

Redesign often can improve performance and reduce cost, and may occur informally during the early stages of design. It may also result from formal design-review procedures. Failure-mode analysis is particularly useful in redesign to reduce the likelihood of further failures. These techniques often have resulted in greatly improved product performance.

Standardization and simplification of design can lead to substantial savings without loss of performance. Retaining only the most efficient sizes, types, grades and models of a product is an example; during World War II, the number of "standard" types of steel, brass and bronze valves

Table 5. Tests for value

Every material, every part, every operation should pass these tests—a negative response is passing:
- Can we do without it?
- Does it do more than is required?
- Does it cost more than it is worth?
- Is there something that does the job better?
- Can it be made by a less-costly method?
- Can a standard item be used?
- Considering the quantities used, could a less-costly tooling method be used?
- Does it cost more than the total of reasonable labor, overhead, material and profit?
- Can someone else provide it at less cost without affecting dependability?
- If it were your money, would you refuse to buy the item because it costs too much?

in the United States was reduced from 4080 to 2500. Utilization of standard off-the-shelf components can also lead to significant savings. It might be less costly, for example, to use a 1/4-20 bolt 2 in. long, which is a widely stocked size, even though design requirements could be satisfied by a bolt 1⅞ in. long, which would probably be a special-order item. Further savings could be realized if the assembly were to be redesigned to allow use of 1½-in. or 1¾-in. bolts.

Functional substitution offers great opportunity for improvement and cost reduction through redesign. The goal of functional substitution is to find a new and different way to meet a design requirement. For example, a bolted assembly might be redesigned for assembly by welding, by pressing mating parts together or by adhesive bonding. The scope of a functional redesign program might be rather modest, as in the example just mentioned, or it might entail complete redesign of the product.

TOTAL LIFE CYCLE IN DESIGN

It is an accepted principle of engineering to design a product for minimum cost, consistent with fulfilling the functional requirements of the product. It is tempting to define the cost of a product strictly on the basis of component and labor costs (plus allowances for overhead, selling costs and profit). However, such accounting neglects several important items in the total cost of the product to the user, such as cost of energy to operate the product, cost of maintaining the product, depreciation and cost of disposal. In addition to the cost of a product to its producer and user, there is the cost to society at large; some of the components of this cost are consumption of raw materials and energy and the impact of the product on the environment. The relative importance of considering the cost to the user and to society during design depends in part on the nature of the product. An automobile, for instance, has considerable user and societal costs; for a bolt, these costs are smaller, harder to identify and easier to neglect.

To a materials engineer, the concept of total life cycle must include the total life of the product, as discussed above and in Table 2, but should also include the total life cycle of the components and materials in the product. Whenever possible, a product should be designed so that components that do not wear out can be reused, or so that the materials in the product can be recycled.

Energy Requirements. Reuse of components and materials obviously conserves raw materials, such as the ore from which a metal is made, but it also conserves the energy required to extract the metal from the ore. Basic metal-production operations are highly energy-intensive. It has been estimated that 95% of the energy required to manufacture aluminum beverage cans from ore can be saved if metal in the cans is recycled. It may be possible to conserve energy by choosing materials that do not require heat treatment but still have adequate strength. Heavily drafted cold-drawn-and-stress-relieved steel bars can have yield strengths of 690 MPa (100 ksi); for many purposes, such bars can be used in place of hardened-and-tempered bars. The potential for conservation of scarce resources, particularly those that might be subject to manipulation for economic or political gain, is apparent in these two examples.

As suggested above, the energy required to operate a mechanical device can represent a significant portion of its total life-cycle cost. For example, the cost of electricity used to operate an air conditioner for the duration of the warranty period may be greater than its purchase price.

Producibility. The question of producibility affects engineering design at several levels. The most basic question is whether the technology required to produce economically the desired quantity of the item exists. A product should be designed in accordance with established technological practice whenever possible; working at the limit of technology leads to higher scrap levels and difficulties in production, and advancing the limit of technology requires a developmental program before production can begin. A second question concerns the capability of the factory (or entire industry) to produce the desired number of parts. The third question concerns utilization of existing equipment and personnel. Some of the answers to these questions are managerial prerogatives, but it is usually desirable for designers to analyze the various possibilities and present a set of alternatives, together with possible consequences, advantages and disadvantages of each, to management.

Durability, or intended service lifetime, is one of the parameters on which a design must be based. In general, basic goals for service lifetime are established for, rather than by, designers. As in the case of producibility, it may be desirable for a designer to develop alternative combinations of material and design that could significantly affect the anticipated service life of the product.

Quality assurance should be an integral part of the manufacturing process of any product, especially because the possible consequences of permitting deficient products to reach the marketplace can be dire. This article does not attempt to describe a quality assurance program, or how to develop one (see Ref 3); rather, it emphasizes the importance of such a program and points out how design and quality assurance are interrelated.

A designer must know the types and severities of discontinuities that can be detected in a quality assurance program and must have an appropriate level of confidence in the detection methods. By knowing what can be expected from the quality assurance program, the designer can then provide a margin of safety to compensate for the existence of discontinuities that cannot be detected. This is part of the basic philosophy of fracture mechanics. A similar approach is applicable to design of redundant systems in the product. In either case, the product should be designed with the capabilities and limitations of the quality assurance program in mind.

Handling requirements of various products are too frequently neglected in engineering design. Almost without exception, a product is made in one location and used in another. Thus, it is essential to provide means for transporting it from the manufacturer to the seller to the user; the means of transportation can affect the design of the product in several ways. The most obvious concern is to ensure that the product will not be degraded in handling, transit or storage. Whenever feasible, a product should be designed against the possibility of damage caused by normal handling. A product that may be easily damaged in handling must be carefully (and expensively) protected by packaging or special shipping procedures. It is also important to design a product to minimize shipping and storage space requirements. For example, wastebaskets are usually designed to be nested during handling; many wastebaskets have lugs or other design features that limit nesting in order to permit easy separation. Particularly for consumer products, the design of packages may not be considered a part of product design, but it is still appropriate for a product designer to consider packaging requirements that might result from various design alternatives.

OTHER MAJOR FACTORS IN DESIGN

Besides the functional factors listed in Table 1 and the total life-cycle considerations in Table 2, several other major factors affect design and materials selection; some of them are listed in Table 3 and discussed in the paragraphs that follow.

State of the Art. Most of the time, the state of the art in a particular field can be inferred from an engineering evaluation of products currently on the market. Existing products may lack capabilities that could be provided by a new product. The extent of improvement might range from purely cosmetic (which might be the case for a new product intended to capitalize on the market acceptance of an established product) to complete redesign resulting in an entirely new product unlike anything previously produced. The state of the art can be defined not only by existing products but also by industrial and societal standards, technical publications and patents. Patent infringement is often considered a matter for legal departments, but technical personnel are often able to analyze technological aspects of patents on devices and processes that might be applicable to new products.

Conformance to Standards. Many products are subject to either mandatory or voluntary standards. In general, the former are imposed through an act of law; fuel-economy requirements on automobiles and minimum pressure ratings for plumbing fittings are examples of legislated design standards. The requirement that small household appliances operate on 115-V, 60-Hz electric power is mandatory for products to be sold in the United States, not because of an act of law but because almost all electric utilities produce such electricity. Voluntary standards include those for standard nuts and bolts; grades of various product forms of aluminum, copper, steel and other metals; and methods of rating products.

Standards that pertain to the safety of either the builder or user of a product are particularly important. Several applicable safety standards for consumer products are acts of law. Other laws cover the safety of working conditions. Some standards, particularly those pertaining to consequences of the unintended use of products, have evolved from bodily injury litigation. All of these types of standards must be considered during the design process.

Conformance to standards is usually, but not necessarily, evidence of proper design. A designer must determine that a particular standard is appropriate to the product under consideration, and that the standard provides adequate assurance of satisfactory performance in service. Many standards, especially voluntary standards, have been derived from designs that have proven to be acceptable for most applications and that reflect the levels of quality and performance capability most manufacturers desire to produce.

Human factors in design describe the interactions between a device and anyone who uses or maintains it. Examples of human factors in design include designing the handle of a portable electric drill to fit comfortably into either hand of the user, arranging gages and controls on an automobile dashboard to be easily seen or reached, and designing an ejector to remove the beaters of an electric mixer. One branch of industrial engineering deals specifically with human factors; contact with persons involved in such research would be useful to designers, as would publications dealing with this subject, such as Ref 4.

Aesthetics is an important aspect of the design of almost any product, but one that should not be allowed to compromise functional requirements. Some consumer products are sold almost exclusively by aesthetic appeal; design of such products is usually assigned to artists. However, aesthetic appeal is important even for industrial products. A smooth or shiny finish, an artistically pleasing shape, or an appearance of ruggedness and durability may have little or no influence on the ability of the product to perform its function but may, nevertheless, have a considerable effect on the attitude of a potential buyer or on worker acceptance when the item is put into service.

Cost. Assuming that a product meets the basic functional requirements established for it, the most important single factor in its design and manufacture is cost. The cost of any product must be competitive with the costs of comparable products already on the market. Whether there is a directly comparable product or not, the cost must be low enough to convince a prospective purchaser that the benefits to be derived from the product exceed its cost. Cost-benefit studies almost always precede purchases of major pieces of manufacturing equipment.

Cost generally enters the design process as one of several criteria against which the merit of a design is judged. Whenever a choice exists between different materials, designs or manufacturing processes, the least costly alternative will be chosen, provided the basic functional requirements for the product can still be met. In some instances, the possibility of significant cost reduction will justify re-evaluation of basic functional requirements and performance specifications; it may be possible to modify the design goals slightly and thereby significantly reduce costs. However, it is important to remember that any modification of functional requirements and performance specifications may adversely affect utility and marketability.

FACTORS IN MATERIALS SELECTION

Two often-cited reasons for selecting a certain material for a particular application are (*a*) the material has always been used in that application and (*b*) the material has the right properties. Neither is evidence of original thinking or even careful analysis of the application. The collective experience gained from common usage of a material in a particular application is useful information, but not justification in itself for selecting a material. The time has passed when each application has its preferred material and a particular material its secure market. In the context in which it is frequently mentioned, the term "property" connotes something that a material inherently possesses. On the contrary, a property should be regarded as the response of the material to a given set of imposed conditions. It must also be recognized that this property should be that of the material in its final available and processed form. Tabulated properties data, such as those available from the sources listed in Table 6, are helpful, but such information must be used judiciously and must be relevant to a particular application.

Regardless of specific expertise, every engineer concerned with hardware of any description (and this includes essentially all engineers) must deal constantly with selecting an appropriate material (or combination of materials) for his design. Except in trivial applications, it certainly is not sufficient to indicate that the components should be "steel" or "aluminum" or "plastic." Rather, the engineer must focus his attention, knowledge and skill on the general factors in materials selection listed in Table 4. It is obvious that these are inseparable from and interwoven with the factors listed in Tables 1, 2 and 3.

In principle, one could write a mathematical expression describing the merit of an engineering design as a function of all these variables, differentiate it with respect to each of the criteria for evaluation and solve the resulting differential equations to obtain the ideal solution. No one has any illusion that we are in a position to do this. The principle is valid, however, and should provide a basis for action by the designer. In some instances, a standard, readily available component may be much less costly than, and yet nearly as effective as, a component of optimized, non-standard design.

Mechanical Properties. One question usually asked in selecting materials is whether strength is adequate to withstand the stresses imposed by service loading. Although the primary selection criterion is often strength, it also may be toughness, corrosion resistance, electrical conductivity, magnetic characteristics, thermal conductivity, specific gravity, strength-to-weight ratio or some other property. For example, in residential water service, where the water pressure is relatively low, weaker and more expensive copper tubing might be a better choice than stronger steel pipe. Steel pipe comes in sections and is joined by threaded connections with elbows at corners, whereas soft copper can be obtained in coils and can be easily bent around corners. Thus, the lower installation cost of copper would overcome its higher material cost. Also, because of the relatively low water pressure, the strength of copper would be adequate, and the greater strength of steel unnecessary. Furthermore, in the event of freezing, copper usually yields instead of bursting; in many regions of the country, it is more resistant to corrosion by the local water than is steel. In general, the usual criterion for selection is not just one property, such as strength, but some combination of properties, manufacturing characteristics and cost.

Table 6. Sources of reference data in the United States

American Society for Metals
American Iron and Steel Institute
American Society for Testing and Materials
Society of Automotive Engineers
Government-Industry Data Exchange Program (GIDEP), Corona, CA
Defense Documentation Center (DDC), Alexandria, VA
National Technical Information Service (NTIS), Springfield, VA
Machinability Data Center (MDC), Cincinnati, OH
Mechanical Properties Data Center (MPDC), Columbus, OH
Metals and Ceramics Information Center (MCIC), Columbus, OH
Nondestructive Testing Information Analysis Center (NTIAC), San Antonio, TX
Thermophysical and Electronic Properties Information Analysis Center (TEPIAC), West Lafayette, IN
Chemical Propulsion Information Agency (CPIA), Laurel, MD
Infrared Information and Analysis Center (IRIA), Ann Arbor, MI
Coordinating Agency for Supplier Evaluation (CASE), Sacramento, CA
National Bureau of Standards (NBS), Gaithersburg, MD
Smithsonian Science Information Exchange (SSIE), Washington, DC

Ultimate tensile strength is a commonly measured and widely reported indication of the ability of a material to withstand loads. However, the direct application of tensile-strength data to design problems is extremely difficult. First of all, the definition of "ultimate tensile strength" is the maximum stress (based on initial cross-sectional area) that a specimen can withstand before failure; it occurs at the onset of plastic instability. Thus, any component loaded to its ultimate tensile strength is likely to fracture immediately. Secondly, if the design is based on a fraction of the ultimate strength, there is the question of what fraction provides adequate strength and safety, together with efficient use of the material. Finally, there seems to be only a rough correlation between tensile strength and material properties such as hardness and fatigue strength at a specified number of cycles, and no correlation whatsoever between tensile strength and properties such as resistance to crack propagation, impact resistance or proportional limit.

Yield strength indicates the lowest stress at which measurable permanent deformation occurs. This information is necessary to estimate the forces required for forming operations. Yield strength is also useful in considering the effects of a single-application overload; most structures must be designed so that a foreseeable overload will not exceed the yield strength.

Hardness is another widely measured material property and is useful for estimating the wear resistance of materials and estimating approximate strength of steels. Its most widespread application is for quality assurance in heat treating. However, only rough correlation can be made between hardness and other mechanical properties or between hardness and behavior of materials in service.

Ductility of a metal, usually measured as the percent reduction in area or elongation that occurs during a tensile test, is often considered an important factor in material selection. It is assumed that, if a metal has a certain minimum elongation in tensile testing, it will not fail in service through brittle fracture. It is also assumed that if a little ductility is good, a lot is better. Neither of these assumptions is accurate. Metals normally considered very ductile can fail in a seemingly brittle manner, such as under fatigue or stress-corrosion conditions, or when the service temperature is below the ductile-to-brittle transition temperature. How much ductility is actually usable under service conditions and how best to measure it are very controversial. Several estimates of usable ductility (such as the ability of materials to absorb the movement within a large structure that is necessary to equalize the load among all of the members of the structure) fall in the range of 1 to 2% elongation in tensile testing. Larger amounts of ductility only indicate the possibility of more extensive permanent deformation of the structure. In many structures, any permanent deformation destroys the usefulness of the structure; for example, the aerodynamic efficiency of an aircraft wing can be substantially reduced by only 1 to 1½% deformation. The amount of ductility required for processing may far exceed that actually usable in service. Steel sheet, for example, often must have considerable ductility, far more than a part might need in service, but nevertheless enough to allow the part to be formed to its required shape.

Design Configuration. The shape of an object is partly responsible for service demands placed on the material from which it is made. Cross-sectional dimensions, for example, may determine the stresses imposed on the material by service loads. These apparent stresses may be increased through the effects of notches and changes in section size. Design configuration determines whether or not a component will be subjected to particular hazards, such as wear or exposure to a corrosive environment. It may also determine the product form from which the item is made. A long, slender part is often made from strip or bar simply because the raw material has the same general shape. A part with a high degree of rotational symmetry is often turned on a lathe. A hollow sphere may be made by joining two hemispheres stamped from flat sheet.

Available and Alternative Materials. Regardless of the merits of a material for a given application, if that material is not readily available in the desired form and quantity, it is a poor choice for that application. The question of material availability is compounded by shortages caused by market fluctuations, inability or reluctance of producers to supply the desired material in the desired form or quantity, and changes in the relative costs of various materials. A further complication may exist if a particular part is made in several locations, especially if these locations are in different countries. Thus, it is desirable to select more than one material for an application and, if possible, to give an order of preference.

Sometimes, an alternative material will be necessary as a substitute for a preferred but unavailable material. In other instances, the alternative material will be selected because of superiority in the specific situation. For example, a tool manufacturer had been purchasing AISI 1078 and 1086 carbon steel bars, which have overlapping composition ranges. This manufacturer decided to use only 1078 and thus reduce his inventory. Even more important, because it is a more widely used grade, 1078 is produced more frequently than 1086, and thus is more available and easier to schedule.

An alternative material for a pressure vessel may have slightly greater strength than a material it can replace. Even a small increase in strength, however, can mean a significant reduction in wall thickness. It also may permit fabrication of slightly larger vessels. The pressure-vessel designer, however, must be sure that the increase in strength does not concurrently reduce fracture resistance below acceptable values.

The desired form can also cause availability problems. A material that can be obtained only in castings obviously cannot readily be used in applications requiring drawn tubing, extrusions, wire cloth or other fabricated shapes.

Fabricability describes the capability of a metal to be fabricated by various manufacturing processes, such as machining, casting, forging, welding and stamping. Whenever possible, product design should incorporate those materials that can be readily fabricated using the desired process(es) without special precautions. Materials that can be fabricated by several methods offer convenient alternatives in the event that unforeseen conditions suddenly preclude use of the principal fabrication method.

Questions regarding production, such as equipment and manufacturing capacity available for fabricating components, are directly related to fabricability, as is the quantity required. If several thousand duplicate parts are necessary, the high cost of dies or other specialized equipment for quantity production may be economi-

cally justified. If few pieces are required, hand production from relatively expensive materials in stock may be less expensive than use of more elaborate methods and less expensive materials. Even so, an engineering estimate should be made of the relation between product cost and quantity for each proposed production method.

Design and fabricability are closely related. For example, a manufacturer of equipment for pleasure boats made lightweight anchors from 6061-T6 extrusions to replace steel anchors. The new aluminum alloy anchors had four major advantages over those made of steel: (1) more holding power for the same size, (2) half the weight, (3) freedom from rusting and staining and (4) interlocking construction rather than welded construction, which allowed shipping before assembly, thus saving transportation costs.

A material may not be commercially available in the desired mill form, but it may be possible, with relatively small-scale development, to produce it in the desired form. This entails considerable expense, but circumstances may justify the extra cost, as exemplified by the development of fabrication procedures for beryllium and zirconium components for nuclear reactors.

Fabricability and mechanical properties are interrelated, and they quite often act in opposite directions. As strength levels are increased, often by increasing carbon or alloy content, weldability and machinability frequently decrease. Although most materials can be welded, higher-strength materials generally require special techniques; the designer must be aware of the added cost built into the part in order to meet the welding requirements. In some instances, there may be no practical method of regaining the strength of the base metal lost through welding. As machinability is increased by additions of sulfur or lead, long-life fatigue resistance is decreased.

Corrosion and Degradation in Service. A material may be regarded as either resistant or susceptible to corrosion and degradation in service, depending on the nature of the application. Sometimes, it may be preferable to use a low-cost, frequently replaced component and assume that it will be corroded or degraded in a short time; in other situations, however, this approach may not be acceptable because of high potential danger to personnel or for other reasons.

There are other factors that should be considered when dealing with corrosive media, because neglect of these may lead to erroneous interpretations of corrosion tests and handbook data:

- A sample in a simple static test at a given temperature may corrode in a manner and at a rate significantly different than if the material were simultaneously transferring heat to the corrosive medium or if there were significant relative movement between the material and the corrosive medium.
- The test medium may become contaminated during the experiment, and its corrosive characteristics may change.
- Lack of correlation between laboratory tests and operating conditions may be caused by a limited ratio of solution volume to surface area of test material in the laboratory test.
- Pressure of gas or vapor above a corrosive liquid may appreciably affect the amounts of oxygen or other gases that may be dissolved in the solution.
- Alloys that owe their corrosion resistance to development of passive films are particularly

susceptible to development of concentration cells.
- The ability of the material to withstand stress-corrosion cracking or corrosion fatigue in service is not always accurately predicted by standard laboratory tests for these qualities. If the material is, in fact, susceptible to either failure process in the service environment, sudden and unpredictable service failures may occur because cracks propagate so rapidly under the combined action of stress and corrosion.

Stability of material in service can be affected by temperature, fluctuations in temperature, length of time at temperature and, in some applications, exposure to radiation. Elevated temperature not only reduces strength and induces creep but also can produce changes in microstructure, such as tempering of martensitic steels and overaging of precipitation-hardening alloys. Obviously, duration of exposure is important in determining the extent to which these phenomena occur and, consequently, in determining appropriate stability requirements. A rocket motor, for instance, may be required to operate only briefly, whereas a steam turbine is expected to operate for many years. In many applications, it may be essential to avoid any and all failures that would require the equipment to be shut down for repairs. In others, especially those involving mechanical wear, replacement at regular intervals is anticipated, and the affected part is designed to be readily replaced.

Other aspects of stability are the consequences of failure. For instance, a leak in a teakettle may be only a nuisance, but a leak in a vessel containing a flammable or radioactive fluid is critical. It should be noted that many designs for long-term operation involve extrapolation or educated guessing because the best available data typically represent lifetimes much shorter than those anticipated in actual service. The need for conservatism in design for such service is obvious.

Properties of Unique Interest. For most problems in materials selection, the choice represents a compromise among a large number of properties and fabrication characteristics of the material; each of these properties and fabrication characteristics has a significant effect on final choice. Often, however, one or two properties are far more important than all others. Some examples of applications where a single property of unique interest dominates the selection of a material are:

- *High density:* gyroscope rotors, and winding weights in self-winding watches
- *High stiffness-to-weight ratio:* aircraft frame components
- *Low melting point:* fusible links in sprinkler systems and fire doors
- *Expansion during freezing:* type metal
- *Special thermal-expansion characteristics:* metal-to-glass seals
- *Electrical conductivity (including superconductivity):* power-distribution systems
- *Wear resistance:* many applications, from plowshares and ore-crusher jaws to gear teeth and cam surfaces.

These are only a few examples of applications where a requirement for one or two properties of special interest severely limits the number of candidate materials.

A requirement for a particular combination of properties further limits the number of candidate materials. Beryllia (BeO) conducts heat as well as aluminum alloys but is an electrical insulator, the only known material with high thermal and low electrical conductivities. For applications demanding high toughness at low temperatures, the choice of materials is limited to metals with face-centered-cubic lattices and certain high-nickel steels. Turbine blades in the hot stages of gas turbine engines must resist deformation and corrosion at operating temperatures and must not fail in thermal fatigue.

Cost. In almost every situation, final selection of a material for a specific application depends on a compromise. In some applications, there are specialized requirements that restrict the choice to relatively few materials. Even then, there is a compromise among the contending factors. The compromise and final selection also involve economic considerations. Initial cost of a piece of equipment involves raw-material, fabrication, assembly and installation costs. The cost of the material is generally a relatively small part of the total life-cycle cost of the equipment. After the equipment has been placed in service, there may be additional costs, such as required rate of return on investment, depreciation, taxes, product-liability costs, replacements due to failure, shutdown expenses when equipment is undergoing repair or replacement, and production loss.

The tests for value listed in Table 5 are directly pertinent to economic considerations. If the answer to any of these questions (and perhaps other similar ones) is affirmative, the selection task is incomplete.

No industry is immune to savings through more effective application of materials. At least three major approaches may be taken to reduce cost through better use: (1) reconsider the material and mill form selected, (2) reconsider the shape of the part and its method of fabrication and (3) redesign to take full advantage of properties. Although these apply primarily to production-line parts rather than to tailor-made parts, the philosophy is applicable to all products.

In many situations, definite savings can be realized by the simple expedient of changing from one material to another without substantially changing the form or processing procedure. There are other applications for which two or more materials can be considered as alternates, with the choice at any given time dictated by current market prices.

Great savings can often be realized by changing the fabrication procedure or form in which material is used. One prominent example is the automobile-engine crankshaft. For many years, crankshafts were machined from forged steel that had to meet rather stringent specifications of strength and toughness. However, high toughness is an unnecessary requirement because a bent crankshaft is just as useless as a broken one, and impact loading on a crankshaft is not severe. Realization of these facts made it possible to effect a change to cast crankshafts. Today, there are millions of cast crankshafts in successful operation, all incapable of appreciable plastic bending, but having adequate rigidity and wear resistance. The cost of a cast and machined crankshaft is significantly less than that of a forged and machined one.

SUMMARY

Materials selection, which is a part of the overall design process, places several requirements on the designer. First, the designer must make a systematic study of the intended purpose of the product under design, noting all explicit and implied constraints on the design. The principal result of this study is characterization of the functional requirements of the product, including the relative priorities of potentially conflicting requirements. Another result of this study is a listing of the criteria to be used in evaluating various

possible designs; cost is almost always one of the most important. The second phase of the design process is to determine the required capabilities of each component of the product, then to compare each of these requirements with characteristics of the candidate materials. These characteristics, however, should not be considered merely as numbers in published tables of properties. They are better described as the response of that material *in the form of a specific component* to the imposed service conditions. (The effects of manufacturing processes on material properties are implicitly considered when the material is evaluated as a specific component.)

The third step in the design process is to consider various combinations of configuration, material and fabrication process, evaluating each against previously established criteria. Several methods of evaluation may be used, including, among others, design review, value analysis and simulated service testing of prototype and pilot-lot products. The final analysis in the material-selection process might include studies of field service reports and cost analysis of the production model.

REFERENCES

1. Probabilistic Design, by E. B. Haugen and P. H. Wirsching: *Machine Design,* Vol 47, No. 9, 17 Apr 1975, p 98-104; No. 10, 1 May 1975, p 80-85; No. 12, 15 May 1975, p 83-87; No. 13, 29 May 1975, p 54-58; No. 14, 12 June 1975, p 108-112. Also, *Probabilistic Approaches to Design,* by E. B. Haugen: Wiley, 1968
2. Probe Failures by Fatigue To Unmask Mechanical Causes, by J. O. Almen: *SAE Journal,* Vol 51, May 1943
3. *Quality Control Handbook,* 3rd Ed., edited by J. M. Juran: McGraw-Hill, 1974
4. *Human Factors in Engineering and Design,* 4th Ed., by E. J. McCormick: McGraw-Hill, 1976

Selection for Economy in Manufacture

By the ASM Committee on Selection for Economy in Manufacture*

FOR MAXIMUM ECONOMY IN MANUFACTURE, selection of the material for a production part must be done in conjunction with selection of the manufacturing method. Of course, it must be assumed that the material will be selected from among those candidate materials that meet all design and engineering requirements. Throughout this article, it is further assumed that the design characteristics of the components discussed meet all established functional performance standards, and that factors such as strength, durability and safety have been duly considered during design of the parts. When all of these preconditions have been met, selecting a material for economy in manufacture involves consideration of several, if not all, of the following:

Raw-material factors

- Chemical composition
- Form of mill product
- Size of mill product
- Material condition or temper

- Surface finish
- Quality characteristics

Processing factors

- Formability
- Machinability
- Weldability
- Response to heat treatment
- Coatability

General factors

- Quantity of material required
- Availability of grade and product form
- Plant standardization of grades and sizes
- Energy consumption
- Availability of required processing equipment.

Because these factors are interdependent and strongly influenced by variables in manufacture of the material and of the component made from it, none should be considered singly. Once all pertinent factors have been assembled, the different options should be evaluated for their ef-

fects on total manufacturing cost (including procurement, storage, handling and distribution), and final selection should be made on the basis of greatest economy in manufacture.

Standard specifications, such as ASTM, ASME or AMS, may establish unified requirements for any or all of the selection factors described below. For certain applications, such as pressure vessels, it is necessary to use a material that conforms to the applicable standard specification so that the finished product meets all requirements of the appropriate code.

COMPOSITION

In planning for economy in manufacture, choice of chemical composition must be correlated with mechanical-property requirements and performance requirements. For many applications, the specific composition itself is unimportant, except for the mechanical properties or heat treatment characteristics that it represents. More than one composition may meet the specified require-

*G. F. Bush, Ford Motor Co.; R. C. Barch, The Hoover Co.; A. M. Belavic, Eaton Corp.; R. E. Cook, Mills Alloy Steel Co.; S. Dinda, Chrysler Corp.; H. E. Fairman, Bristol Corp.; D. J. Hayes, United States Steel Corp.; V. A. Kortesoja, Ford Motor Co.; R. G. Mack, Abar Corp.; B. L. McMillan, International Harvester Co.; E. Ross, General Motors Corp.; R. E. Seward, General Electric Co.; K. E. Spray, Clark Equipment Co.; J. A. Sweet, Great Lakes Steel; V. J. Turk, General Motors Corp.; F. R. Varrese, Honeywell Corp.; G. Wood, IBM Corp.; D. S. Zamborsky, Warner & Swasey Corp.

ments. Maximum economy often can be achieved by specifying only the required mechanical and physical properties, and allowing purchasing and manufacturing personnel the greatest possible freedom in making decisions regarding factors such as form and size of purchased stock, quantity-price relationships, and substitution of thermally or mechanically treated rough stock so that an in-plant heat treating step can be eliminated. This freedom may allow, for example, a carbon steel with high residuals to be substituted for a more expensive alloy steel without compromising hardenability requirements. On the other hand, lack of such freedom may prevent savings if, for example, it prohibits use of alloys to which sulfur or lead has been added to improve machinability.

Economies may be realized by specifying nonstandard compositions. This generally requires purchase of relatively large quantities. If small quantities are required, it is economically unfeasible to use a nonstandard alloy; also, some standard grades are made only on order and cannot be purchased in quantities less than a full heat.

Consultation with various metal producers often can result in economies in specification of alloy composition. Many producers and users have developed alloys that contain greater quantities of low-cost alloying elements and smaller quantities of expensive alloying elements.

Because the pricing structure of alloys and commodities is constantly changing and because the relative costs of various fabricating operations are also subject to change, purchase specifications must be frequently reviewed to ensure that the grade being purchased remains the most economical.

FORM

For parts of simple shape, such as bolts or straight shafts, the most economical raw material and method of manufacture are readily apparent. For parts more complex in shape, applicability of two or more forms and fabrication methods adds complexity to the process of selection. A small gear, for example, may be completely machined from bar stock. On the other hand, depending largely on the total number of parts to be produced, it may be more economical to begin with a close-tolerance, forged gear blank. Gears for small mechanical devices may be most economically made by powder metallurgy.

Among such alternatives, final selection should be based on comparison of over-all costs. Cost studies should always be used to analyze the relative economic merits of all the various forms and methods of manufacture—machining from tube or bar, casting, hot forging, cold forging, powder metallurgy, extruding or welding.

Production quantity is the factor most likely to determine the form selected. For a small gear, the total cost of making 100 pieces might favor machining from bar. If the quantity were increased to 10 000 or more, use of forged gear blanks might constitute a considerable economic advantage. With a larger gear, if only 500 pieces were to be made, it might be more economical to leave the hub solid and drill the bore. If the required quantity were 5000 pieces, it would probably be advantageous to have the hub pierced in the forging operation. With even larger quantities, forging of minor contours might be justified. As quantity increases, it becomes progressively easier to amortize the initial cost of forging

dies, and savings in the costs of metal and machining increase accordingly.

Selection of product form may include consideration of bar or wire that is drawn to a special shape. Use of special shapes can sometimes reduce the amount of machining necessary to make a part from standard shapes and thereby reduce total cost.

SIZE

The design of the part determines the size of the raw material within a fairly narrow range. Nevertheless, there is some flexibility in selection, and savings depend on judicious choice of final size.

To illustrate the problem, consider a simple shaft with a constant diameter that must be held to close tolerances. Such a shaft may be made most economically from bar purchased in the turned-and-centerless-ground condition. This is particularly true if the shaft requires a surface finish that can be achieved only by a secondary operation such as grinding. Neither hot rolled nor cold drawn bars could satisfy this condition without additional machining.

However, for a more intricate shaft with several different diameters along its length, paying a premium for ground bar of special size would be uneconomical; either hot rolled or cold drawn bar would fit the need better. Because most of the surface will be machined off, there is no justification for a stringent surface requirement on the raw bar.

Ideally, the size of stock selected should be as close as possible to the size of the finished part. There are instances, however, where some additional removal of stock must be allowed to avoid an excessive scrap rate. When bar is slightly out-of-round or otherwise warped, additional stock serves as a safety factor against rejection. Even though extra stock removal is wasteful to some extent, it may be cheaper to purchase slightly larger stock than to pay for scrapped parts. In other instances, it may be more economical to select a bar that is oversize but that is lower in price because it carries a lower size extra.

The possibility of parts being rejected due to surface imperfections such as decarburization or seams is an additional reason for buying oversize stock. The excess stock should be adequate to permit complete removal of all decarburized metal and seams. It should also eliminate the need for magnetic-particle inspection to detect surface imperfections.

Conventional diametric machining allowances for hot rolled round bars are 3.2 mm ($^1/_8$ in.) for bars 38.1 to 76.2 mm ($1^1/_2$ to 3 in.) in diameter and 6.4 mm ($^1/_4$ in.) for bars over 76.2 mm (3 in.) in diameter. These allowances result in the following percentage increases in cross-sectional area:

Diameter before allowance		Diameter after allowance		Increase in area, %
mm	in.	mm	in.	
38.1	1.50	41.3	1.625	17.4
40.0	1.57	43.2	1.700	16.6
50.8	2.00	54.0	2.125	14.8
60.0	2.36	63.2	2.49	11.0
75.0	2.95	78.2	3.08	8.8
88.9	3.50	95.3	3.75	14.8
100.0	3.94	106.3	4.185	12.9
127.0	5.00	133.3	5.25	10.2

Thus, for some applications, it may be advantageous for the user to order bar of a smaller size and accept a scrap rate a few percentage points above normal in machining or some other operation, rather than to purchase 8 to 17% excess stock. In other applications, it might be preferable to use cold finished bars. Cold finished bars are more expensive, but may not have to be machined to size and often require little or no stock removal to eliminate surface imperfections.

CONDITION

Many metals are available in a wide variety of conditions. In almost all instances, a higher cost is associated with any condition other than that of the basic, commercial-quality product of the particular form being considered. It is the responsibility of the manufacturer of a part to determine if the additional material cost can be justified by savings in manufacturing. It is assumed that all functional requirements can be met by all materials under consideration.

Conditions for bar include hot drawn, cold drawn, and cold drawn and stress relieved. Hot rolled steel sheet is available in yield strengths as high as 690 MPa (100 ksi), whereas cold rolled steel sheet can be obtained with tensile strengths as high as 1380 MPa (200 ksi). Bars and sheets can be ordered to special internal-soundness requirements if normal inclusion content or internal cleanness is not acceptable. Carbon and alloy steel plate is available either quenched and tempered or normalized. Many other examples could be cited.

Choosing a different condition generally implies a change in the properties of the material and, consequently, in its reaction to manufacturing processes. For example, a quenched-and-tempered product may be specified to eliminate production-line heat treating and cleaning. The potential savings must be weighed against the higher material premium and the higher costs of machining the hardened stock. If it is possible to use heat treated stock, additional savings may be realized if distortion or quench cracking during heat treatment has been causing high rejection levels.

Sometimes savings are more difficult to evaluate. Two alternative designs for an automotive underbody subassembly involved a choice between HSLA (high-strength, low-alloy) 550-MPa (80-ksi) hot rolled steel and low-carbon cold rolled steel for the frame rails. Weight differences between the two proposals were minimal, but the design that incorporated HSLA steel gave a greater safety factor in crash testing. The higher cost of the HSLA material was offset by the fact that fewer parts were needed in the subassembly, thereby reducing the amount of capital equipment for fabrication and assembly and simplifying quality-control procedures. In many cases where strength-level changes in material are contemplated, the cost-benefit analysis must be carried beyond the individual part to consider potential benefits to the entire structure.

Often, consideration of condition can be fairly simple and easily overlooked. In many instances, developed blanks for stamped parts are laid out from a coil strictly on geometric considerations to minimize blanking scrap losses. Sometimes, a more judicious orientation of the blank relative to the rolling direction (such as rolling direction transverse to the critical bending axes) can allow utilization of a less expensive alloy, even though blanking scrap losses may be higher. In this instance, the manufacturer takes advantage of the fact that mechanical properties in sheet are generally better in the longitudinal direction.

Several steels are commercially available in a pretreated condition intended to enhance mechanical properties. Steels such as 1050, 1141 and 1144 are specially processed by cold drawing with heavy drafts. The bars are then stress relieved at about 370 to 510 °C (700 to 950 °F), depending on steel composition and mill practice.

As a result of cold drawing, tensile-strength values up to about 1035 MPa (150 ksi) can be obtained with hardnesses up to about 300 HB. The price of cold drawn stress-relieved steel bars is only slightly above that of conventionally cold drawn material. Certain steels (including those mentioned above) are also obtainable in the hot drawn condition. With hardnesses up to about 350 HB, these materials have even higher mechanical properties. (Hot drawn bars are passed through dies at about 370 °C, or 700 °F.)

At higher cost, carbon and alloy steels are also available in the quenched-and-tempered condition. In this condition, a common hardness range is 250 to 300 HB; hardnesses above or below this range can be special-ordered.

Whether or not the purchase of steel in the hardened-and-tempered condition is economical must be determined on the basis of direct costs. The premiums paid for steels in this condition, along with the added cost of machining at higher hardness, must be weighed against the cost of heat treating and cleaning parts in the production line. This is particularly true of those steels for which hardness is known to markedly decrease machining speeds or tool life, or both.

Part design is most likely to influence the decision to use hardened-and-tempered steel. Where little machining is required, such pretreatment may have almost no effect on final cost. When machining operations are extensive, the added time, effort and tool wear involved may rule out the use of pretreated steels.

If a part is likely to distort excessively in heat treatment, the use of steels that are hardened and tempered before machining is attractive. Savings realized through scrap reduction may more than offset the extra costs of raw material and fabrication.

SURFACE FINISH AND TEXTURE

Surface finish and texture are important features of metal products—not only for appearance, but also because they may affect performance. For example, excessive asperities in the surfaces of two parts in rolling or rubbing contact can increase friction, overheating and wear. As a result, operational costs are increased, and the product may fail prematurely. Evaluation and selection of optimum finishes and textures requires careful analysis of (*a*) the functional demands of the application on surface condition and (*b*) relative costs of producing or procuring specific finishes and textures, and of maintaining them throughout subsequent manufacturing steps. (Relative costs change because of fluctuations in price and availability, and thus should be frequently reviewed.)

A finish or texture is usually characteristic of a specific manufacturing process. For example, turned, milled and ground surfaces have different characteristic appearances (when viewed microscopically). Surface roughness is defined as the degree of deviation from an ideally smooth surface; specialized instrumentation and measuring techniques are necessary for control and measurement of surface roughness. Generally, the cost of producing various finishes and textures increases substantially as surface roughness is decreased.

The surface finish of a mill product can be affected by many variables in manufacture, including melting practice, hot rolling practice, employment of special techniques to remove injurious surface blemishes, technique of scale removal, and heat treatment and cold finishing practice (if either is used). The special measures required to produce mill products with low surface roughness represent additional costs. Because of the extra cost, the choice of cold finished rather than hot finished products must be justified by the functional requirements of the fabricated product (such as the appearance requirements for automobile body components) or by reduction in the number and cost of fabricating operations. Specifying restricted surface roughness, such as by requiring ground and polished bars or a reflective surface finish on sheet, requires further justification for the additional cost involved. To take full advantage of special surface finishes on mill products, the largest possible portion of the surface with the special finish should be retained in the finished part.

The surface finish of sheet can affect its formability. Lubricants do not adhere well to very smooth sheet. The appearance of rough sheet is objectionable for many applications, particularly if the sheet is finished with glossy paint or porcelain. Imperfections in the surface of the sheet, such as fluting, rolled-in dirt or scale, and skin lamination, make forming operations more difficult if not impossible.

For some applications, the surface of a product is deliberately textured. Hot rolled floor plates, for example, have a raised pattern to make them less slippery when wet or oily. Embossed sheet, which is made in a wide variety of patterns, is produced by passing the sheet through a set of rolls, of which one is smooth and the other is engraved with the reverse of the desired pattern. Rigidized sheet contains three-dimensional patterns, which are produced by passing the sheet through a set of synchronized rolls that have mating patterns cut into their surfaces. Both types of textured sheet are stronger and more rigid than smooth sheet of the same thickness; use of textured sheet results in increases in strength and stiffness and a significant decrease in elongation, and can reduce the weight of parts.

Some mill products are textured by grit blasting or shot peening. These processes, which can be readily applied to fabricated parts, are used both to induce residual compressive stresses at the surface and for cosmetic purposes.

QUALITY DESCRIPTORS

Quality descriptors are a convenient means of identifying mill products that have been produced so as to be especially well suited to certain applications or fabrication processes. The various quality descriptors imply that special manufacturing practices and more restrictive inspections have been utilized by the manufacturer to assure the customer that the product will perform as intended. These tests and manufacturing practices are not necessarily spelled out in a discussion of quality, and thus it might appear that the purchaser is forced to buy on faith. However, the system has been proved workable through many years of usage. The purchaser does not have to be familiar with mill processing, nor does the mill have to divulge proprietary information, but the purchaser is able to obtain material that is suited for a particular application (or fabrication process).

There are price extras associated with most quality descriptors, and the amount of extra cost increases as more restrictions are placed on the product. Thus, it is logical to purchase commercial quality materials unless the particular application requires the special attributes implied by some other quality descriptor.

SPECIFICATIONS

A specification is a statement of the attributes that a particular item must possess. By incorporating a specification into a purchase agreement for that item, the buyer and seller agree on what constitutes grounds for acceptability of that item. Specifications may be widely used published documents, such as those of ASTM or AMS, or they may be documents prepared for the benefit of a single user, such as government specifications or proprietary company specifications. The advantages of using a standard specification are that most producers are familiar with its requirements and that, furthermore, those requirements were probably written to be consistent with the capabilities of most producers. By comparison, before accepting orders that refer to proprietary specifications, the producers must analyze the requirements and compare them with their own capabilities. In some circumstances, a producer may accept an order to a standard specification, but decline an order (or charge a higher price) for the same item made to a proprietary specification.

A large proportion of steel bar stock is produced and marketed to AISI-SAE designations. These designations contain ranges and limits of chemical composition (and hardenability requirements for H-steels), but they are not specifications, for they contain no specific requirements for dimensional tolerances or mechanical properties. Incorporated in these designations, however, are implied specifications in the form of standard tolerances and practices that are published in various sections of the AISI Steel Products Manual. This common practice eliminates voluminous and repetitive specifications.

FORMABILITY

The formability of a material is the relative ease with which it can be shaped by various manufacturing processes into a component part. Formability depends on the mechanical properties of the material at various temperatures, the character of the forming process and the shape of the part. Significant savings in production costs may be achieved by proper matching of the manufacturing process, such as stamping, roll forming or extrusion, with the material's ability to be stretched, drawn, bent or extruded. Simplification of component shape can lessen formability requirements, thus permitting use of less costly forming processes and materials. For example, a requirement for drawing quality (DQ) steel sheet may be changed to one for commercial quality (CQ) sheet, which is less costly, by minor design modification. Application of a forming-analysis technique, such as the use of a forming-limit diagram (FLD) or strain analysis, in either the blueprint or the prototype stage, may save manufacturing costs by avoiding forming problems that increase scrap rates.

Forming-analysis techniques are useful to die shop personnel because they provide a visual image of metal flow in the stamping operation, and to die and tooling engineers because they provide an exact method of evaluating the variables involved in any stamping operation. Judicious use of FLD and strain analysis and concurrent base-material evaluation makes it possible to quantify the effects of blank development, lubrication, tooling modifications, press modifications and material substitutions. Strain analysis can be an invaluable aid for die development, quality-assurance programs and trouble-shooting.

Significant cost savings can sometimes be achieved by substituting thinner-gage high-strength materials for lower-strength materials. Although high-strength materials generally cost more, gage reduction sometimes is sufficient to effect cost savings. Formability of high-strength materials such as HSLA steel or nitrogenized steel is generally lower than that of lower-strength materials such as low-carbon steels. Through minor design modifications and better manufacturing techniques, these less-formable high-strength materials can be substituted for lower-strength materials, and savings in both cost and weight can be achieved. In applications that involve corrosion, however, it may not be desirable to make a part from a high-strength material; a given amount of corrosion will cause a proportionately greater loss of load-carrying capacity in a thin part made of high-strength material than in a thicker part made of low-strength material.

Generally, stamping from sheet metal is one of the least costly high-volume forming processes. Parts made from tubing can sometimes be changed to stampings made from cold rolled sheet to reduce the final piece cost. Forming cost decreases as volume of production increases. Cold heading processes are often found to be less expensive than machining or hot forging processes, even though a premium quality mill product may be required for cold heading.

Lubrication has a strong influence on formability. With better lubrication compounds and practice, less-formable material can be used in parts where costly, highly formable material would otherwise be required.

Many factors other than material selection also affect metal forming. In many instances, the manufacturer can realize significant savings by manipulating all factors involved in manufacturing a part (lubrication, blank design, tooling, pressing speed and others), rather than assuming that the process is a constant and that material must be found to fit the process.

MACHINABILITY

The cost of producing many machined parts is strongly dependent on the machinability of the materials selected for those parts. The machinability requirements of a particular part are determined by many factors, including the amount of metal that must be removed, type of machining operation, required surface finish, difficult machining steps such as deep-hole drilling, quantity to be produced and relative costs of ma-

terial and machining time. If extensive or difficult machining operations are required, selection of materials with high machinability ratings is justified. Ferrous and nonferrous alloys that contain small amounts of insoluble second phases such as lead, sulfides or selenides are widely preferred when the utmost in machinability is needed. Leaded steels, leaded brasses and free-machining stainless steels containing sulfur or selenium, for instance, are commonly specified for screw-machine products.

Generally, the harder the material, the more difficult it is to machine. However, machinability is directly affected more by microstructure than by hardness. The machinability of many classes of alloys can be improved if the microstructure is a two-phase structure consisting of either a brittle or an easily sheared second phase dispersed throughout a moderately ductile matrix. In turning 4140 steel, for example, far greater tool life and productive efficiency can be expected if the steel has a coarse lamellar pearlite-ferrite structure resulting from annealing than if it has been hardened and tempered to approximately 300 HB. Similarly, a high-carbon alloy steel having approximately 0.60% carbon will be more easily machined after spheroidize annealing than after a standard annealing treatment that produces a pearlitic structure. In alloys whose microstructures can be readily altered by heat treatment, such as many steels, it may prove more economical to purchase the material in a readily machinable condition (such as spheroidized) and to heat treat the material to another condition more compatible with service requirements after it has been machined to final configuration. In some instances, particularly with steel components that require close tolerances or fine finishes, parts are rough machined in a highly machinable condition, heat treated to final hardened condition and then finish machined (or, more often, ground) to final dimensional and surface-finish requirements.

The extent to which a metal has been cold worked can affect its machining behavior. Cold working a very soft steel can improve its machining characteristics, but extensive cold working, particularly of medium-carbon or alloy steels, can harden the steel sufficiently to greatly accelerate tool wear. Cold working causes residual stresses in the steel, which in turn can cause warping and loss of dimensional stability if a machining or thermal operation changes the distribution of the residual stresses. Residual stresses can be reduced by stress-relief annealing, but such a treatment may alter the mechanical properties of the part.

WELDABILITY

Weldability is the relative ease with which a metal or combination of metals can be joined by welding, with or without a filler metal. It is a complicated property, determined principally by (a) chemical composition of the metal(s) to be welded, (b) section shape and thickness, (c) cleanness of the surfaces to be joined and (d) mechanical properties of the metals.

Low-carbon steels are, in general, readily welded by all of the common welding techniques. For best weldability, the following ranges and limits of composition are preferred:

Carbon .... up to 0.20%
Manganese .... 0.30 to 0.60%
Silicon .... 0.10 to 0.20%
Sulfur .... 0.04% max
Phosphorus .... 0.04% max
Residual elements .... 0.10% max total

A composition that meets these ranges and limits should not be considered as a firm requirement for acceptable weldability, but as a preferred composition for which there are virtually no restrictions on welding method, type of joint, type of electrode, welding current, or position or speed of welding. Steels with compositions in these ranges are readily available in a variety of forms.

Most steels of other compositions are also weldable. However, special techniques, equipment and precautions may be required, all of which will probably raise welding costs and rejection rates.

Certain other metals are readily welded, but on the whole it cannot be assumed that materials other than low-carbon steel can be welded satisfactorily without special techniques or precautions.

RESPONSE TO HEAT TREATMENT

Satisfactory response to heat treatment is the capability of a material, fashioned into a particular part, to be heat treated to a desired microstructure, hardness and strength level without undergoing cracking, distortion or excessive size change. The most important factors affecting response to heat treatment are the recrystallization, phase-decomposition and/or transformation characteristics of the material, the configuration and mass of the part, and control of the heat treating operation.

Before determining what response to heat treatment is needed for a particular application, it is important to determine that heat treatment is absolutely necessary. The economy that may be realized by eliminating heat treatment can be considerable, but is often neglected due to the automatic assumption that a part—particularly a steel part—will be heat treated.

Because heat treatment can create substantial thermal stresses in the part (even if the part is not quenched), the part should be designed so that the deleterious effects of these thermal stresses are minimized; thermal stresses are greatest during quenching. The basic objective is to design the part so that the temperature distribution during heating and cooling is as uniform as possible. Sharp corners, particularly internal corners, must be avoided, because they concentrate thermal stresses, thus accentuating their effect on the part. The largest possible radii should be used in corners. Uniform section thickness facilitates uniform temperature distribution. Designing a part so as to minimize thermal stresses during heat treatment usually permits choice of the least costly material that can provide the required properties.

Perhaps the most common heat treatment is the austenitize, quench and temper sequence, which is the dominant process for hardening carbon and alloy steels. The critical aspect of this sequence is that the hardened steel part must contain a minimum percentage of martensite in order to meet the specified hardness. In other words, the hardenability of the steel must be sufficient to provide that percentage of martensite when the part is quenched into a medium that is appropriate to the size, shape and distortion tolerance of the part.

To achieve particular combinations of mechanical properties, it may be necessary to specify other heat treatments, such as austempering, induction hardening, flame hardening, carburizing or nitriding. Each of these processes places certain restrictions on the choice of steel; the restrictions may include carbon content, hardenability or the need for certain alloying elements.

Seemingly minor changes in the heat treating operation can often produce significant reductions in manufacturing costs, particularly if the change reduces the scrap rate, permits a change to a less costly alloy, or facilitates a subsequent fabricating operation.

COATINGS

Coating of materials, either by the supplier or by the user during manufacture of parts, must be carefully considered when assessing the cost of producing a given part or assembly. Obvious costs are those of the coating material and of the process for applying it. Less obvious are effects of the coating processes on the properties of the material being coated, which can adversely affect the quality and usefulness of the end product and produce high scrap costs. A few of the more important coatings used in manufacturing processes are:

Organic	Inorganic	Metallic
Paints	Phosphates	Metal
Lubricants	Chromates	spraying
		Dip coating
		Plating

Application of most of these coatings involves elevated temperatures ranging from 65 °C (150 °F) to as high as 455 °C (850 °F) for zinc coating or well over 650 °C (1200 °F) for aluminum coating. Metal spraying normally does not raise the substrate temperature much over 95 °C (200 °F).

Phosphate coatings and various lubricants are used to aid forming of sheet and cold heading of bar and rod. Low-carbon steel sheet can be adversely affected by the temperatures involved in phosphating and lubricant application. High scrap costs can result if these processes are not considered during initial specification of the sheet.

Paints are frequently used for corrosion protection and for enhancing the appearance of the finished part. Paint can be the least expensive way to achieve these desired features. However, the costs of applying paint must be carefully evaluated to avoid undue labor or process costs. Sometimes, use of an entirely different coating process may save money by eliminating costly hand operations.

Chromate dip coatings often are applied at the mill for corrosion protection of galvanized steel sheet. A chromate dip coated galvanized product is said to be "passivated" or "stabilized." In many manufacturing processes, the presence of a chromate coating causes no difficulties. However, chromate coating must not be used on galvanized steel parts that are to be resistance welded, or phosphated and painted, after forming. Once applied, a chromate coating on galvanized steel is almost impossible to remove without damaging the galvanized coating. Any chromate present will interfere with proper phosphating and thereby cause poor paint adhesion in the final product. Chromate also causes problems in resistance welding. Purchase agreements for galvanized steel sheet must be carefully written to avoid these problems.

A coating process such as metal spraying, which is most often done during the final stages of component production, normally does not adversely affect material selection unless the temperatures involved are capable of changing the mechanical properties of the part. The possibility of corrosion due to strong galvanic couples between the coating metal and the substrate metal or alloy also should be considered.

Higher-temperature dip coatings (galvanizing and aluminizing) have marked effects on the mechanical properties of the substrate material. The properties of commercially available coated sheet products are well defined, but the use of these processes on parts should be approached cautiously.

Mill-product form, size, condition and surface finish are not critical to coating processes, except where coating adhesion can be impaired by an improper condition of surface roughness (as in metal spraying), where irregularities can cause imperfections or objectionable appearance in the finish (as in plating), or where one surface coating can affect another coating (as in painting chromate-coated galvanized sheet).

Electroplating, electroless plating or mechanical plating of one metal on another is used to alter the appearance, wear resistance or corrosion resistance of both large and small parts. Before plating is specified, the designer must carefully consider the possibilities that changes in properties will be caused by plating and that the thickness of the plating may affect critical dimensions.

Electroplating of through-hardened or case hardened steel parts, or severely cold worked carbon steel parts that have not been stress relieved, can result in hydrogen embrittlement. However, preplating operations such as cathodic cleaning, acid pickling, and stripping of defective electroplate can cause as much or more hydrogen embrittlement as the electroplating process itself. To avoid delayed fracture of stressed parts due to such embrittlement, it is necessary to bake the plated parts at about 200 °C (400 °F) for at least 4 h. This must be done within a few hours of the plating operation to be truly effective. Baking is an additional operation that raises the cost of producing the parts, but it is an essential operation when there is a need to ensure freedom from hydrogen embrittlement.

Electroless plating is a chemical plating process rather than an electroplating process; it produces virtually no hydrogen embrittlement and is frequently used to plate case hardened screws and bolts that must not exhibit embrittlement in service. The chemical baths used for electroless plating require strict control of composition and temperature. Electroless plating generally is more expensive than electroplating. Not all metal plating systems are available as electroless processes; nickel and copper are most often plated by such processes. Imparting hardness and wear resistance to electroless nickel requires heat treatment at 370 to 425 °C (700 to 800 °F) under a protective atmosphere.

Mechanical plating is a proprietary process by which metal powders are mechanically welded to the surfaces of parts. The process requires activation of the surface during tumbling in a medium of glass beads and metal powder. Soft metals such as zinc, cadmium, tin and their alloys can be readily plated on steel by this process. Because electrical current is not required, hydrogen embrittlement does not occur.

Many organic paint systems are being converted from solvent-base to dry-powder or water-base systems to comply with restrictions on solvent emissions in environmental protection laws. Powdered paint, applied electrostatically, presents no pollution problems but is considerably more expensive than the older solvent-base systems. Use of electrophoretic paint is increasing rapidly—especially in the automotive industry. This type of paint system poses only minimal pollution problems and has the advantage of providing coverage that is uniform and complete, even in small openings and deep recesses. Some solvent-base paint systems use "exempt" solvents, which are subject to fewer restrictions. These solvents generally are more expensive than the hydrocarbon-base solvents used in earlier systems. In general, there appears to be no cost differential involved in the choice among electrostatic, electrophoretic and exempt-solvent paint systems.

QUANTITY AND AVAILABILITY

The criteria of quantity and availability are considered together because the quantity of material required for a particular product can limit the number of possible sources for that material and, indirectly, limit the number of product forms from which the designer can choose.

The number of parts to be made affects decisions regarding choice of fabrication process; but, because mill products are marketed by weight, it is the weight of material required that affects availability. For products that require large quantities of material, the critical factors in availability include (a) the ability of one or more mills to produce the required quantity, (b) production and shipping schedules and (c) whether or not the material can be purchased from multiple sources to ensure adequate supply and provide price competition. For products that require very small quantities of material, the most significant issue is where a product reasonably close to the preferred grade, shape and size can be obtained.

To establish what procurement options are available, it is necessary to determine the minimum order quantity for each purchased item and then to estimate the length of time such a quantity would last in normal production. If that time period is less than a few months, purchase from the mill should be considered a reasonable option; if not, the material should be selected from in-plant supplies or from items stocked in service centers.

The two principal reasons for ordering mill products directly from the producer are the possibilities of obtaining uncommon items (such as nonstandard sizes and shapes or seldom-produced grades) and of obtaining the lowest possible purchase price. An additional advantage is that large quantities can be purchased on a single order and delivered in partial shipments to fabrication facilities (either captive or vendor) at scattered locations. Disadvantages of ordering from the mill include the delay of several weeks or months between ordering and receiving the material, the requirement that relatively large quantities be ordered at one time, and costs of possessing the material from the time it is received to the time it is fabricated into a product. The costs of possession, which are all too easily neglected in estimating the relative costs of materials obtained from various sources, include the costs of storage facilities, wages and overhead for personnel who handle or guard the material, cost of money to buy the material (either as interest on borrowed money or interest lost by not investing capital), taxes on inventory, and losses from inventory due to material obsolescence, pilferage or atmospheric corrosion.

Certain types of material, such as prepainted sheet or coated wire, may be advantageously purchased from a converter or secondary processor. These firms specialize in buying mill products from manufacturers, applying coatings or performing some other work to the customer's specifications, and reselling the items.

Anticipated production quantities can affect the cost of manufacture by limiting the choice of production methods. Some operations, such as cold heading, stamping, welding and automatic screw machining, require extensive tooling and setup time, the costs of which must be spread over a large number of fabricated parts for economical production. Other operations, such as machining, casting and hand welding, require less tooling and setup time, and may be economical alternatives if only a few parts are to be made. However, these latter operations are relatively labor-intensive, and for production of large quantities of parts may be more costly than one of the more heavily automated processes.

When the parts to be produced are complex in configuration and either small or moderate in size, it is generally worthwhile to consider making the basic part from a casting or forging. Certain very large parts—such as casings for large pumps, and landing gear for jet aircraft—can be more economically produced from castings or forgings as well. Powder metallurgy parts are also attractive alternatives in many instances, especially for small parts made of materials of low to moderate strength. Powder metallurgy also makes it possible to use materials that are impossible or prohibitively expensive to make by conventional alloying techniques. Frequently, a detailed analysis of procurement options will be needed to establish the savings that can be achieved by using one of these "near net shape" product forms. Such an analysis must include amortization of the costs of dies, patterns, fixtures and other tooling that would not be needed if the part were made by another process. Depending on configuration, required strength and number of pieces, aluminum or iron castings often are the most economical.

SPECIAL REQUIREMENTS

The compositional and other related factors that determine suitability for carburizing, nitriding or similar case hardening processes may be classified as special requirements in selection. A carburized gear that is subject to high loads, for example, will require selection of a low-carbon medium-alloy steel with fairly high hardenability. Hardenability is particularly important in achieving a sufficiently high core hardness to support the compressive loads on the carburized case.

Plain carbon steels generally are not nitrided, because they build up an excessively brittle "white layer" that cracks and spalls readily. Obtaining a satisfactorily nitrided case requires selection of steels with particular ranges of alloy content.

There are many other situations that demand special requirements. Perhaps the most common of these are the surface conditions necessary for decorative electroplating or similar metal-finishing processes, and the compositions and material conditions necessary to ensure satisfactory welds in many materials.

ENERGY CONSIDERATIONS

The availability and cost of energy are problems that are assuming increasing importance in the determination of economy in manufacturing operations. Because of rapid and ongoing changes in this area, energy costs should be re-evaluated on a regular basis. Recent shortages of natural gas and oil illustrate the nature and severity of the problem. Also, substantial increases in the costs of all energy supplies have occurred for various and complex reasons. Thus, manufacturers of all classes of products are looking for new and better ways to use energy more efficiently. Most industries start with some basic raw materials; a finished product is then manufactured and marketed. What may be a finished product for one company becomes a raw material for another. Hence, energy consumption and attendant cost usually is cumulative during the total manufacturing process of an end product.

Refining and conversion of most metals into semifinished products (such as billets, castings, bar, sheet and strip) require high energy consumption. These products are then further pro-

cessed (machined, stamped, formed, welded or heat treated) into finished products by other industries. In each step of this complex process, energy is consumed. Because energy costs are usually considered part of indirect manufacturing cost (burden), it is difficult to directly relate energy savings to reduction of over-all product costs. Nevertheless, it is very desirable to determine the amount of energy consumed in each manufacturing step and to minimize total energy consumption. Whenever an operation in the manufacture of raw material or in subsequent processing can be eliminated or reduced, savings in energy and product cost can result. With the multitude of raw materials, equipment and processes used throughout the industry, specific guidelines for determining actual energy cost savings are most difficult to generate.

INTERACTIVE EFFECTS

Interactive effects among the selection factors mentioned above are quite common. A change in any of the factors concerning raw material may affect processing. For example, a change in chemical composition may affect surface finish, machinability or response to heat treatment. Likewise, a change in processing often requires a change in raw material. The general factors are also interactive with raw-material and processing factors. For example, a change in quality requirements may affect availability. Thus, in changing any of these factors affecting selection of a material for a particular part, the possibility of interactive effects on other factors must also be considered.

Assuming that both processing and operating requirements have been satisfied, the predominant factor in selection is cost. To some degree, all of the selection factors entail cost. Yet, individual cost factors (both higher and lower) may be balanced in such a way as to result in an over-all reduction in the cost of the finished part.

This balance of factors may lower final product cost by: (*a*) decreasing the cost of the raw material, (*b*) increasing or enhancing adaptability to specific manufacturing operations, (*c*) simplifying or eliminating one or more manufacturing operations, (*d*) simplifying or eliminating one or more inspection operations and (*e*) reducing or eliminating scrap losses.

Criteria for Selection

By Bernard Lement, Lement and Associates, and Eugene Rappaport, Rappaport Associates

CRITERIA FOR SELECTION OF LOW-EXPANSION ALLOYS

Definition

Low-expansion alloys are generally considered to be austenitic iron-nickel or austenitic iron-nickel-base alloys which possess relatively low thermal-expansion coefficients in the temperature range from absolute zero to their Curie points.

Desirable Characteristics

(a) A near-zero thermal-expansion coefficient in an application temperature range near room temperature.

(b) A selected value of thermal-expansion coefficient which matches that of another material in the application temperature range as achieved by a particular nickel content alone or by nickel in combination with cobalt or chromium.

(c) Low carbon content (under 0.05%) and a low-as-possible content of impurities and residual elements to achieve high dimensional stability (i.e., resistance to change in dimensions which occur with increase in time at constant temperature) with a near-zero thermal-expansion coefficient in an application temperature range near room temperature.

(d) Alloying with cobalt to achieve a near-zero thermal-expansion coefficient in an application temperature range near room temperature.

(e) Alloying with titanium to permit age hardening and achieve a combination of high strength with an acceptably low thermal-expansion coefficient in the application temperature range.

(f) Alloying with chromium (with or without tungsten, molybdenum, titanium or aluminum) and a corresponding nickel content to achieve a combination of a low thermoelastic coefficient (i.e., near-zero change of elastic modulus) and/or increased corrosion resistance with an acceptably low thermal-expansion coefficient in the application temperature range.

(g) Cold worked condition to achieve a combination of high strength with a near-zero thermal-expansion coefficient in an application temperature range near room temperature.

(h) Alloying with selenium or sulfur for improved machinability with an acceptably low thermal-expansion coefficient in the application temperature range.

CRITERIA FOR SELECTION OF WEAR-RESISTANT ALLOYS

Definition

The term "wear-resistant" signifies relatively high resistance to displacement and detachment of metallic particles from a metallic surface resulting from contact under lubricating, nonlubricating or corrosive conditions by either a metallic or nonmetallic abrasive (abrasive wear), contact with another metal (metallic wear) or contact with flowing liquids or gases (fluid erosion). Nonlubricated wear may involve coarse abrasion, grinding or high-stress abrasion, erosion or low-stress scratching abrasion. Lubricated wear may involve abrasion by airborne dirt or wear debris, galling or scuffing, spalling, indentation, pitting or fretting. The main groups of wear-resistant alloys are mentioned below.

Desirable Characteristics – General

(a) Low material and processing costs consistent with adequately long wear life under application conditions.

(b) Adequate strength, toughness and corrosion resistance to prevent premature failure.

(c) Fine surface finish.

(d) Low coefficient of friction.

Desirable Characteristics – Case Hardened Steels

(a) High case hardness consistent with adequate case toughness.

(b) Sufficient depth of case to ensure long wear life and attain high compressive residual stress at surface.

(c) Uniform distribution of fine excess carbides, nitrides or carbonitrides in microstructure of case as obtained by diffusion-type case hardening operations.

(d) Sufficient core hardenability to ensure adequate core strength.

(e) Combination of relatively low carbon content and moderate hardness of core to ensure adequate core toughness.

(f) Alloying with nickel to enhance impact fatigue resistance of case and toughness of core.

Desirable Characteristics – Steels

(a) High surface hardness consistent with adequate toughness.

(b) High carbon content consistent with adequate toughness.

(c) Uniform distribution of fine carbides in microstructure.

(d) Alloying with chromium and/or molybdenum and/or tungsten to form alloy carbides in microstructure.

(e) Combination of high corrosion resistance and high hardness as obtained in precipitation-hardened stainless steels.

(f) High toughness with long wear life as obtained by use of cast high-manganese austenitic steels.

Desirable Characteristics – Cast Irons

(a) Low silicon content to prevent graphite formation with chill casting.

(b) Microstructure consisting of pearlitic matrix and excess carbides.

(c) Alloying with chromium to form hard excess chromium carbides.

(d) Alloying with molybdenum to result in a martensitic matrix and excess carbides with chill casting.

Desirable Characteristics – Hard Facing Alloys

(a) Use of low-alloy iron-base alloys to achieve high toughness along with high wear resistance.

(b) Use of high-alloy iron-base alloys to achieve higher wear resistance along with somewhat lower toughness than low-alloy iron-base alloys.

(c) Use of nickel or cobalt-base alloys to achieve high heat resistance, toughness and corrosion resistance along with high wear resistance.

(d) Use of powder metallurgy alloys consisting of tungsten and/or titanium and/or tantalum carbides in various matrix compositions to achieve extremely high wear resistance along with other useful properties.

CRITERIA FOR SELECTION OF HIGH-TOUGHNESS ALLOYS

Definition

High-toughness alloys are those which possess relatively high resistance to crack initiation and propagation to complete fracture at the application temperature under static or dynamic loading and plane-stress or plane-strain conditions with or without a notch or other type of design stress concentration present.

Desirable Characteristics – Ferrous Alloys, General

(a) Low material and processing cost consistent with adequate fracture toughness under application conditions.

(b) Freedom from tool marks, corrosion or electrical arcing pits, deep scratches or machinery marks, bursts, flakes or various types of cracks.

(c) Low content of hydrogen and other impurities.

(d) Freedom from gross nonmetallic inclusions.

(e) Fine grain size.

(f) Compressive residual stress at surface.

(g) Uniform chemical composition throughout.

(h) Fracture-toughness transition-temperature parameter appropriate for application conditions and below minimum application temperature.

Desirable Characteristics – Cast, Malleable and Nodular Irons

(a) Sufficient silicon content to prevent chilled white iron surface layer.

(b) Low phosphorus and sulfur contents.

(c) Graphite present as a uniform distribution of spherulites, nodules or small flakes.

(d) Ferritic matrix.

(e) Alloyed with nickel.

Desirable Characteristics – Plain Carbon and Low-Alloy Steels

(a) Fully killed nickel-molybdenum or nickel-chromium-molybdenum steel of low carbon content consistent with required strength level.

(b) Highly tempered martensitic microstructure consistent with required strength level.

(c) Alloyed with molybdenum to prevent temper embrittlement.

(d) Low phosphorus and sulfur contents.

(e) Fine grain size.

Desirable Characteristics – High-Alloy Steels

(a) Tempered martensitic structure with moderately high nickel and low carbon contents for applications down to very low subzero temperatures.

(b) Maraged structure with high nickel and very low carbon content for applications down to very low temperatures.

(c) Austenitic structure with chromium-nickel or chromium-nickel-base composition with low carbon content for applications requiring both high corrosion resistance and high toughness.

(d) Austenitic structure with high manganese and high carbon contents for applications requiring high wear resistance and high toughness.

CRITERIA FOR SELECTION OF HEAT-RESISTANT ALLOYS

Definition

Heat-resistant alloys are generally considered to be those capable of being used at temperatures above 370 °C (700 °F) under load in particular gaseous environments as indicated by acceptable levels of stress-rupture and creep strengths which correspond to required times of service. The main groups of heat-resistant alloys are alloy steels, ferritic and martensitic stainless steels, austenitic stainless steels, higher-nickel austenitic alloys, precipitation-hardening stainless steels, Fe-Ni-Cr-Mo alloys, nickel-base alloys, Co-Cr-Ni-base alloys, molybdenum-titanium alloys, and refractory metals and alloys which are used with and without protective coatings.

Desirable Characteristics

(a) Low material and processing costs consistent with adequate high-temperature life under application conditions.

(b) Low oxygen, nitrogen and hydrogen contents.

(c) High ductility, fatigue strength and toughness at room temperature.

(d) High resistance to oxidation in applications requiring exposure to air or steam under load at elevated temperatures.

(e) Small decrease in elevated-temperature strength due to notch configurations as required in some applications.

(f) High resistance to products of combustion or gaseous chemical products at elevated temperatures under load as required in some applications.

(g) High resistance to thermal shock during heating or cooling.

(h) High fatigue strength at elevated temperatures as required in some applications.

(i) High creep strength under dynamic loading at elevated temperatures as required in some applications.

(j) High modulus of elasticity at application temperature and/or low thermal-expansion characteristic to result in small reversible dimensional changes as required in some applications.

(k) Adequate weldability to permit joining operations as required in some applications.

(l) Moderately large grain size to enhance stress-rupture strength.

CRITERIA FOR SELECTION OF STRESS-CORROSION-RESISTANT ALLOYS

Definition

Stress-corrosion-resistant alloys are those that possess relatively high resistance to failure by cracking under the combined action of applied or residual stress and corrosion in a particular corrosive environment.

Desirable Characteristics

(a) Low tensile residual stress at surface.

(b) Uniform chemical composition with respect to grain interiors and grain boundaries of microstructure.

(c) Fine grain size.

(d) Uniform microstructure throughout with mechanical properties independent of directionality.

(e) Absence of anodic precipitation effects resulting in localized differences in chemical composition at the application temperature.

(f) Relatively high immunity with respect to corrosive action by the application environment.

(g) High-purity metal or very low alloy content.

(h) Relatively low yield strength.

CRITERIA FOR SELECTION OF FATIGUE-RESISTANT ALLOYS

Definition

Fatigue resistance is the ability of a material to resist progressive localized permanent structural changes that occur as the material is subjected to repeated or fluctuating strains at stresses having a maximum value less than the tensile strength of the material. These structural changes may culminate in cracks or fracture after a sufficient number of fluctuations.

Desirable Alloy and Design Characteristics

(a) Resistance to crack nucleation by the presence of alloying-enhanced cross-slip, twinning or work-hardening properties.

(b) Increased tensile strength achieved by alloying or heat treatment to attain adequately high stress/life-cycle relationship.

(c) Increased fatigue strength by work hardening of certain alloys.

(d) Minimized workpiece section and surface discontinuities for increase fatigue strength; avoidance of notches, grooves, holes, fillets, threads, keyways and splines, especially in high-tensile-stress regions; use of smooth rather than sharp section changes, and polished rather than coarsely machined surfaces.

(e) Use of low stress levels, relative to the material's fatigue strength, for fluctuating loads.

(f) In some cases, reduced grain size increases fatigue life (high-cycle fatigue); in others, fatigue life is relatively independent of grain size (low-cycle fatigue).

(g) Absence of metallurgical discontinuities such as porosity, inclusions, internal bursts in forgings or roll-formed members, or localized alloy segregation to provide more difficult fatigue-crack-nucleation conditions.

(h) Presence of surface compressive residual stresses achieved by coining, surface rolling or shot peening to enhance surface resistance to fatigue-crack initiation and progression.

(i) Workpiece design accommodation and environmental conditioning to avoid the initiation of internal cracks from thermal fatigue (result of

temperature cycling) or from corrosion fatigue (fluctuating stresses in a corrosive environment).

(j) Avoidance of metallic coatings, such as chromium, nickel or cadmium platings, that can reduce fatigue life.

CRITERIA FOR SELECTION OF HIGH-STRENGTH ALLOYS

Definition

High-strength alloys are considered to be those having minimum values of yield strength of approximately 690 MPa (100 ksi). Groups of such alloys are as follows:

1. Hardenable high-, medium- and low-carbon steels.
2. High-strength low-alloy steels.
3. Medium- and high-alloy steels including tool steels.
4. Superalloys such as iron-, nickel- and cobalt-base alloys with additions of chromium and other alloying elements for elevated-temperature strength.
5. Nonferrous alloys based on copper, titanium, uranium and the refractory metals (tungsten, molybdenum, tantalum and niobium).

Desirable Characteristics

(a) Retention of ductility at low temperatures.

(b) Retention of strength properties at elevated temperatures.

(c) Capacity to sustain adequate ductile deformation at high stress levels. Cr-Ni-Mo steels have elongations of approximately 25 to 10% for corresponding yield strengths of 690 to 1550 MPa (100 to 225 ksi).

(d) Ability to withstand quenching with minimum distortion or cracking—favored by low carbon content in hardenable alloy steels.

(e) Low material and processing costs.

(f) Ability to sustain static or dynamic loading by deforming in a relatively ductile manner as indicated by high fracture toughness at the application temperature.

(g) Adequate fatigue strength in both smooth and notched conditions.

(h) Ability to meet special requirements dependent on particular service environmental conditions (e.g., resistance to corrosion, stress corrosion or corrosion-fatigue).

CRITERIA FOR SELECTION OF CORROSION-RESISTANT ALLOYS

Definition

Corrosion resistance is the ability to resist environmental deterioration by chemical or electrochemical reaction.

Desirable Characteristics

(a) High resistance to continuing uniform overall reactions with the specific environment of interest (general, or uniform, corrosion). Examples include oxidation and dissolution of metals in aggressive solutions.

(b) High resistance to local attack, with deep penetration as in localized pitting (pitting corrosion), or networks of local cracks as in stress-corrosion cracking or intergranular corrosion (often associated with local compositional inhomogeneity or local impurity agglomeration).

(c) Inertness, or nobility, in galvanic coupling with other metals in a conductive environment (galvanic corrosion).

(d) Resistance or immunity to enhanced corrosion due to the presence of applied or residual stress (stress-corrosion cracking), or application of fluctuating stress (corrosion fatigue).

(e) Resistance to damage effects due to collapse of cavities in the liquid at a solid/liquid interface under conditions of severe turbulent flow (cavitation), and to erosion due to exceeding critical flow velocities by surrounding fluids (erosion-corrosion).

(f) Resistance to enhanced corrosion or deterioration at the interface of two contacting, and slipping, surfaces under load in an aggressive environment (fretting corrosion).

(g) Resistance to the accelerated localized corrosion occurring at locations where easy access to the bulk environment is prevented, such as mating surfaces of assemblies (crevice corrosion).

(h) Resistance to selective dissolution of a more active constituent of an alloy leaving behind a weak deposit of the more noble metal (dealloying), and in particular dezincification of brass.

(i) Resistance to the combined action of corrosion (often pitting corrosion) and fluctuating stress to create conditions of corrosion fatigue.

(j) Resistance to cracking when metal is subjected to tensile stresses while in contact with liquid metals, such as mercury, cadmium, or silver-base brazing alloys (liquid-metal embrittlement).

(k) Resistance to corrosion degradation caused by stray electric currents flowing in the environment (stray-current corrosion).

CRITERIA FOR SELECTION OF MAGNETIC ALLOYS

Magnetically Soft Alloys

Definition. Magnetically soft alloys are generally iron and iron alloys that have little or no retentivity—that is, if they are magnetized in a magnetic field, and then are removed from that field, they lose most, if not all, of the magnetism they exhibited while in the field.

Desirable Characteristics.

(a) Low hysteresis loss (easy domain movement during magnetization) to minimize energy losses. This is favored by low interstitial impurity levels (particularly carbon, sulfur and nitrogen), large grain size, and suitable crystallographic texture.

(b) Low eddy-current loss from electric currents induced by flux changes to minimize energy losses.

(c) High magnetic permeability, and, in some cases, constant permeability at low field strengths.

(d) High magnetic saturation induction.

(e) Minimum or reproducible change in permeability with temperature in special applications.

(f) Low material and processing costs.

Alloys for Permanent Magnets

Definition. The term "permanent magnet" is used to describe materials that are normally used in a single magnetic state and that have sufficiently high resistance to demagnetizing fields and sufficiently high magnetic flux output to provide useful and stable magnetic fields. This implies insensitivity to temperature, mechanical shock, and demagnetizing field effects.

Desirable Alloy Characteristics.

(a) High magnetic saturation induction.

(b) High resistance to demagnetization, exemplified by a high value of coercive force. This property is determined less by composition than by specimen shape, crystal anisotropy, and microscopic mechanisms such as precipitation, lattice strains and fine particle consolidation.

(c) Maximum energy content as indicated by a large magnetic hysteresis loop.

(d) Low material and processing costs.

(e) Maintenance of magnetic properties at elevated temperatures.

CRITERIA FOR SELECTION OF ELECTRICAL ALLOYS

Electrical-Resistance Alloys

Definition. These alloys are high-electrical-resistance alloys generally utilized as follows:

1. *Resistance Alloys:* used for control or regulation of electrical properties of instrumentation.
2. *Heating-Element Alloys:* used for generation of heat by conduction of electrical current.
3. *Thermostat Metals:* used in applications where heat generated in a metal resistor is converted into mechanical energy; usually processed as bonded composite materials in the form of strip or sheet.

Desirable Characteristics – Resistance Alloys.

(a) Uniform electrical resistivity for known, predictable response.

(b) Stable electrical resistance with respect to time and temperature.

(c) Reproducible temperature coefficient of resistance to achieve desired known resistance

properties that will be maintained over lifetime usage.

(d) Low thermoelectric potential versus copper to minimize spurious voltages from temperature gradients.

(e) Adequate secondary properties, such as specific coefficient of thermal expansion, high mechanical strength, ductility, corrosion resistance, and joinability to other metals.

Desirable Characteristics – Heating-Element Alloys.

(a) High melting point to secure maximum temperature range of use.

(b) High electrical resistivity to achieve high temperatures with minimum electrical current.

(c) Reproducible temperature coefficient of resistance.

(d) Good oxidation resistance in furnace environments to enhance service life.

(e) Absence of volatile alloy components to avoid local contamination and to maintain stability of alloy electrical properties.

(f) Resistance to contamination at the application temperature.

(g) Adequate secondary properties, such as good elevated-temperature creep strength, high emissivity, low thermal expansion and elastic modulus (both of which help to minimize thermal fatigue), good resistance to thermal shock, and good strength and ductility at fabrication temperatures.

(h) Low material and processing costs.

Desirable Characteristics – Thermostat Metals.

(a) Reproducible coefficient of thermal expansion to maintain stability and constancy of mechanical response.

(b) Reproducible electrical resistivity.

(c) Good fabrication characteristics at forming temperatures.

(d) Long-time stability and reproducibility of temperature-motion properties.

Electrical-Contact Alloys

Definition. These alloys are low-electrical-resistance alloys used in electrical circuits where the flow of electrical current is interrupted.

Desirable Characteristics.

(a) High electrical conductivity to minimize the heat generated during passage of current.

(b) High thermal conductivity to dissipate both the resistive and arc heat developed.

(c) High reaction resistance to operating environments to minimize formation of insulating chemical compounds.

(d) Immunity to arcing damage by the making and breaking of electrical contact.

(e) Low force requirements to establish good electrical contact.

(f) Low electrical resistance between mating members of a contacting pair to minimize contact resistance and power loss.

(g) A melting temperature high enough to limit arc erosion, metal transfer, and metal welding or sticking, yet low enough to increase resistance to reignition in switching due to heating of local gas in the contact gap from the hot contacts.

(h) Low vapor pressure to minimize arc erosion and metal transfer.

(i) High hardness to provide wear resistance, with adequate ductility to permit fabrication and prevent chipping or flaking.

Processing Rankings

THE FOLLOWING rating system was developed as a general framework for ranking the alloys selected for inclusion in the MetalSelector computer program (see the software documentation at the end of this volume). It is included here for the benefit of those persons working with the MetalSelector.

HARDENABILITY

Definitions of the various degrees of hardenability indicated by letter designations for specific quench-hardenable iron-base alloys:

A – Extremely high hardenability to the degree that the end-quench curve is essentially a straight line for the length of the test bar — a condition that exists for many air-hardening steels.

B – High hardenability to the degree that hardness at the quenched end of the end-quench bar is fully retained for about 8/16 in.

C – Relatively high hardenability to the degree that hardness at the quenched end of the end-quench bar will be retained for approximately 4/16 in.

D – Moderate hardenability wherein the degree of hardness at the quenched end of the end-quench bar will be retained for approximately 2/16 in.

E – Low hardenability wherein the degree of hardness at the quenched end of the end-

quench bar extends for no more than about 1/16 in.

F – Very low hardenability wherein the hardness attained at the quenched end of the end-quench bar extends for an almost immeasurably short distance, then drops very sharply — a condition that characterizes many plain carbon, low-manganese steels.

FORMABILITY

Explanation of letter designations used to denote various degrees of formability for specific metals:

A – Highest rating; can be drawn into deep cups at room temperature without use of excessive amounts of power or special lubricants.

B – Good formability; can be drawn into deep cups at room temperature, but requires more power (compared with the A rating) and extra-pressure lubricants, and more attention must be given to generous die radii.

C – Can be drawn into relatively intricate shapes, but requires heating of the blanks (commonly up to 260 °C, or 500 °F), pigmented drawing compounds, powerful presses, and strict attention to generous radii in the dies.

D – Very difficult to cold form — in terms of power requirements and susceptibility to cracking — but can be used for shallow stampings where no metal thinning is involved. Radii must be

generous and pigmented drawing compounds are generally required.

E – Formability is poor, but some forming can be done by either heating the blank (up to approximately 425 °C, or 800 °F), or process annealing between forming operations.

F – Formability is very poor to the extent that any significant amount of *cold* forming is virtually impossible without danger of cracking. *Warm* forming (up to 700 °C, or 1300 °F, depending on metal composition) can be done.

X – Forming should not be considered.

WELDABILITY

Definitions of the various degrees of weldability indicated by letter designations for specific metals or alloys:

A – Readily weldable by virtually all methods without necessity for preheating or postheating.

B – Readily weldable by most of the common methods, but requires controlled cooling and/or postweld annealing to prevent formation of undesirable microstructures.

C – Can be welded by most of the common methods, but preheating should be used to ensure against cracking.

D – Can be welded by most of the available methods, but preheating, postweld stress relieving and close control of interpass temperature are all required. In addition, welding out-of-doors in cold weather is not recommended.

E – Can be welded, but preferably by the more sophisticated methods such as gas tungsten-arc and electron beam welding. Other precautionary procedures such as preheating, postheating, control of interpass temperature and control of environmental conditions should be used.

F – Should not normally be considered for welding, although, in case of extreme necessity, repair can be achieved by welding. As a rule, special filler metals are used, and in all instances preheating, postheating, close control of interpass temperature and close control of environmental conditions are mandatory.

X – Welding is never recommended.

MACHINABILITY

Definitions of the various degrees of machinability (with conventional cutting tools) indicated by letter designations for specific metals and alloys:

A – Easily machined by any or all types of operations at maximum feeds, speeds and depths of cut. Selection of cutting fluid is not critical, and cutting fluid often is not required.

B – Readily machinable, but containing no special additives, so that a cutting fluid is generally required. Also, to attain best finish, strict attention must be given to tool materials and tool geometry. Some reduction in speeds, feeds and depths of cut are usually required compared with machining of free-cutting materials.

C – Readily machinable, but only under well-controlled conditions in terms of requirements for rigid machine tools, tool materials, tool geometry, and specially prepared cutting fluids, and at reduced rates of speed, feed and depth of cut.

D – Machinable by most of the standard machining operations, but generally of moderate machinability. Thus, rigid equipment, chlorinated or sulfurized cutting fluids, special carbide tools and precise tool geometry are mandatory. Also, greatly reduced feeds, speeds and depths of cut are necessary.

E – Very difficult to machine to the degree that some machining operations are totally impractical. All of the normal procedures used for difficult machining are required — that is, rigid setups, special tools, special cutting fluids, low speeds and shallow feeds and depths of cut.

F – Conventional machining is not recommended, although in certain extreme demands some machining operations can be performed by use of all the approaches indicated in D and E above. It must be assumed that tool deterioration will be extremely high and cutting rates will be low.

X – Nonmachinable.

AVAILABILITY

Definitions of the various degrees of availability indicated by letter designations for specific metals and alloys:

A – Generally available as a stock item in virtually all service centers located in industrial areas.

B – Commonly exists as a stock item in some, but not all, service centers; more often in the very large centers.

C – May exist in some service centers, but more likely has to be ordered from a mill. Sometimes the mills can determine the location of the desired product, thus expediting delivery.

D – Must be ordered from the mill, but the product is one which is regularly produced so that delivery can be made in a reasonable length of time.

E – The desired product is produced by only a few suppliers and not on a regular basis by any. Thus, it may have to be "made to order," so that the delivery date could be intolerably long.

F – The product is in scarce supply and can be supplied by only one or two manufacturers. Thus, the delivery date might be unsatisfactory to the extent that a substitute material should be considered.

PROCESSING COST

Definitions of general processing cost indicated by letter designations for specific metals and alloys. This refers only to the relative cost of producing a specific item as it may be affected by metal selection. In no way is there an attempt made to compare costs of the various processes.

A – The material allows a direct processing sequence without the need for intermediate operations that may be required for some specific materials because of certain characteristics.

B – One secondary operation is required — for instance, an added machining or grinding operation to achieve surface-finish requirements.

C – Two intermediate operations are required — for example, process annealing and redrawing.

D – Three intermediate operations are needed to achieve the common objectives.

E – Four or more intermediate or secondary operations are needed to meet specifications.

F – A specific material not only requires several secondary or intermediate operations, but one (or more) of them is a hand operation, such as removal of pigmented drawing compounds.

Part II

PROPERTIES AND SELECTION

4 CARBON AND ALLOY STEELS

Edited by Conrad Mitchell, United States Steel Corp.

This section was condensed from Metals Handbook, Ninth Edition, Volume 1, Properties and Selection: Irons and Steels, pages 107 to 759. For more detailed information on the topics covered herein, the reader is referred to the larger work. Additional information on carbon and alloy steels can be found in Parts III and IV of this Desk Edition.

Classifications and Designations of Carbon and Alloy Steels

WROUGHT CARBON AND ALLOY STEELS are considered in this article. Ferrous materials that are cast or made by powder metallurgy methods are not included, but are described in other articles in this handbook. Tool steels, heat- and corrosion-resistant alloys and steels used primarily for their magnetic or electrical properties are described in Volume 3 of the 9th Edition of Metals Handbook, "Properties and Selection: Stainless Steels, Tool Materials and Special-Purpose Metals."

Classification is the systematic arrangement or division of steels into groups on the basis of some common characteristic. Steels can be classified on the basis of (a) composition, as carbon or alloy steel; (b) finishing methods, as hot rolled or cold rolled steel; or (c) product form, such as bar, plate, sheet, strip, tubing or structural shape. Classification by product form is very common within the steel industry because, by identifying the form of a product, the manufacturer can identify the mill equipment required for producing it and thereby schedule the utilization of these facilities.

Common usage has further subdivided these broad classifications. For example, carbon steels are often loosely and imprecisely classified according to carbon content as low-, medium- or high-carbon steels. They may be classified as rimmed, capped, semikilled or killed, depending on the deoxidation practice used in producing them. Alloy steels are often classified according to the principal alloying element (or elements)

present. Thus, there are nickel steels, chromium steels, chromium-vanadium steels and so on. Many other classification systems are in use, the names of which are usually self-explanatory. The classification systems most commonly used in the United States are those of the Society of Automotive Engineers (SAE) and the American Iron and Steel Institute (AISI).

"Grade," "type" and "class" are terms used to classify steel products. Within the steel industry, they have very specific uses: "grade" is used to denote chemical composition; "type" is used to indicate deoxidation practice; and "class" is used to describe some other attribute, such as strength level or surface smoothness.

In ASTM specifications, however, these terms are used somewhat interchangeably. In ASTM A533, for example, "type" denotes chemical composition, while "class" indicates strength level. In ASTM A515, "grade" identifies strength level; the maximum carbon content permitted by this specification depends on both plate thickness and strength level. In ASTM A302, "grade" connotes requirements for both chemical composition and mechanical properties. ASTM A514 and A517 are specifications for high-strength plate for structural and pressure-vessel applications, respectively; each contains several compositions that can provide the required mechanical properties. A514 type F has the identical composition limits as A517 grade F.

Designation is the specific identification of each grade, type or class of steel by a number, letter,

symbol, name or suitable combination thereof unique to a particular steel. Chemical composition is by far the most widely used basis for designation, followed by mechanical-property specifications. The most commonly used system of designation in the United States (described below) is that of the Society of Automotive Engineers (SAE) and the American Iron and Steel Institute (AISI).

Quality. The steel industry uses the term "quality" in a product description to imply special characteristics that make the mill product particularly well suited to specific applications or subsequent fabrication operations. The term does not necessarily imply that the mill product is better material, is made from better raw materials, or is more carefully produced than other mill products.

A specification is a written statement of attributes that a steel must possess in order to be suitable for a particular application, as determined by processing and fabrication needs and engineering and service requirements. It generally includes a list of the acceptable values for various attributes that the steel must possess and, possibly, restrictions on other characteristics of the steel that might be detrimental to its intended use.

A standard specification is a published document that describes a product acceptable for a wide range of applications and that can be produced by many manufacturers of such items. The most comprehensive and widely used standard specifications are those of the American Society for

4•1

Testing and Materials (ASTM). Even if there is no standard specification that completely describes the attributes required for a steel product to be used in a particular application, it may be preferable to cite the most nearly applicable standard specification and those exceptions necessitated by the particular application. By doing so, the familiarity of both producer and user with the standard specification is retained, while an individualized product can be obtained.

A specification can be advantageously used in purchasing steel (or any other product) by incorporating it into the purchase agreement. The specification clearly states what attributes the product must possess. The use of a designation alone as the basis for purchase indicates that the buyer is specifying only whatever attributes are described in the designation and permitting the supplier the latitude to produce the item according to his usual practice. The distinction between specifications and standard practices is discussed below.

QUALITY DESCRIPTORS

The need for communication among producers and between producers and users has resulted in the development of a group of terms known as "fundamental quality descriptors." These are names applied to various steel products to imply that the particular products possess certain characteristics that make them especially well suited for specific applications or fabrication processes. Some of the fundamental quality descriptors in common use are listed below.

Carbon Steels

Semifinished for forging
  Forging quality
    Special hardenability
    Special internal soundness
    Nonmetallic inclusion
      requirement
    Special surface
Carbon steel structural sections
  Structural quality
Carbon steel plates
  Regular quality
  Structural quality
  Cold drawing quality
  Cold pressing quality
  Cold flanging quality
  Forging quality
  Pressure vessel quality
  Marine quality
Hot rolled carbon steel bars
  Merchant quality
  Special quality
    Special hardenability
    Special internal soundness
    Nonmetallic inclusion
      requirement
    Special surface
  Scrapless nut quality
  Axle shaft quality
  Cold extrusion quality
  Cold heading and cold
    forging quality
Cold finished carbon steel bars
  Standard quality
    Special hardenability
    Special internal soundness
    Nonmetallic inclusion
      requirement
    Special surface

  Cold heading and cold
    forging quality
  Cold extrusion quality
Hot rolled sheets
  Commercial quality
  Drawing quality
  Drawing quality special killed
  Structural quality
Cold rolled sheets
  Commercial quality
  Drawing quality
  Drawing quality special killed
  Structural quality
Porcelain enameling sheets
  Commercial quality
  Drawing quality
Long terne sheets
  Commercial quality
  Drawing quality
  Drawing quality special killed
  Structural quality
Galvanized sheets
  Commercial quality
  Drawing quality
  Drawing quality special killed
  Structural quality
  Lock-forming quality
Electrolytic zinc-coated sheets
  Commercial quality
  Drawing quality
  Drawing quality special killed
  Structural quality
Hot rolled strip
  Commercial quality
  Drawing quality
  Drawing quality special killed
  Structural quality
Cold rolled strip
  Specific quality descriptors are not provided in
    cold rolled strip, since this product is largely
    produced for specific end use
Tin mill products
  Specific quality descriptors are not applicable
    to tin mill products
Carbon steel wire
  Industrial quality wire
  Cold extrusion wires
  Heading, forging and roll threading wires
  Mechanical spring wires
  Upholstery spring construction wires
  Welding wire
Carbon steel flat wire
  Stitching wire
  Stapling wire
Carbon steel pipe
Structural tubing
Line pipe
Oil country tubular goods
Steel specialty tubular products
  Pressure tubing
  Mechanical tubing
  Aircraft tubing
Hot rolled carbon steel wire rods
  Industrial quality
  Rods for manufacture of wire for
    electric welded chain
  Rods for heading, forging and
    roll threading wire
  Rods for lock washer wire
  Rods for scrapless nut wire
  Rods for upholstery spring wire
  Rods for welding wire

Alloy Steels

Alloy steel plates

Regular quality or structural
  quality
Drawing quality
Pressure vessel quality
Structural quality
Aircraft quality
Aircraft physical quality
Hot rolled alloy steel bars
  Regular quality
  Aircraft structural or steel subject
    to magnetic-particle inspection
  Axle shaft quality
  Bearing quality
  Cold heading quality
  Special cold heading quality
  Rifle barrel quality, gun quality,
    shell or AP shot quality
Alloy steel wire
  Aircraft quality
  Bearing quality
  Special surface quality
Cold finished alloy steel bars
  Regular quality
  Aircraft quality or steel subject
    to magnetic-particle inspection
  Axle shaft quality
  Bearing shaft quality
  Cold heading quality
  Special cold heading quality
  Rifle barrel quality, gun quality,
    shell or AP shot quality
Line pipe
Oil country tubular goods
Steel specialty tubular goods
  Pressure tubing
  Mechanical tubing
  Stainless and heat-resisting pipe,
    pressure tubing and mechanical
    tubing
  Aircraft tubing
  Pipe

Note: Detailed descriptions of many of the categories listed above appear in an appropriate section of the AISI Steel Products Manual.

Some of the quality descriptors listed in the above table, such as "forging quality" and "cold extrusion quality," are self-explanatory. The meanings of others are less obvious: for example, merchant-quality hot rolled carbon steel bars are made for noncritical applications requiring modest strength and mild bending or forming, but not forging or heat treating. The descriptor for one particular steel commodity is not necessarily carried over to subsequent products made from that commodity—for instance, standard-quality cold finished bars are made from special-quality hot rolled bars.

The various mechanical and physical attributes implied by a quality descriptor result from the combined effects of several factors, including:

• Degree of internal soundness
• Relative uniformity of chemical composition
• Relative freedom from surface imperfections
• Size of discard cropped from ingot
• Extensive testing during manufacture
• Number, size and distribution of nonmetallic inclusions
• Hardenability requirements.

Control of these factors during manufacture is necessary to achieve mill products having the desired characteristics. Extent of the control of these and other related factors is another piece of information conveyed by the quality descriptor.

Some, but not all, of the fundamental descriptors may be modified by one or more additional requirements, as may be appropriate: special discard, macroetch test, restricted chemical composition, maximum incidental (residual) alloy, special hardenability or austenitic grain size. These restrictions could be applied to forging-quality alloy steel bars, but not to merchant-quality carbon steel bars.

Understanding of the various quality descriptors is complicated by the fact that most of the requirements that qualify a steel for a particular descriptor are subjective. Only nonmetallic inclusion count, restrictions on chemical composition ranges and incidental alloying elements, austenitic grain size and special hardenability are quantified. Subjective evaluation of the other characteristics depends on the skill and experience of those who make the evaluation. Although the use of these rather subjective quality descriptors might seem imprecise and unworkable, the opposite is true: steel products made to meet the requirements of a particular quality descriptor can be relied upon to have those characteristics necessary for that product to be used in the indicated application or fabrication operation. Additional information about quality descriptors can be found in the AISI Steel Products Manual in the section pertaining to the appropriate product form.

SPECIFICATIONS

A specification is a written statement of the requirements, both technical and commercial, that a product must meet; it is a document that controls procurement. There are nearly as many formats for specifications as there are groups writing them, but any reasonably adequate specification will provide information about the items listed below.

- *Scope* may cover product classification, including size range when necessary, condition, and any comments on product processing deemed helpful to either the supplier or user. An informative title plus a statement of the required form may be used instead of a scope clause.
- *Chemical composition* may be detailed, or it may be indicated by a well-recognized designation based on chemical composition. The SAE-AISI designations are frequently used.
- The *quality* statement includes any appropriate quality descriptor and whatever additional requirements might be necessary. It may also include the type of steel and the steelmaking processes permitted.
- *Quantitative* requirements identify allowable ranges of composition and all physical and mechanical properties necessary to characterize the material. Test methods used to determine these properties should also be included, at least by reference to standard test methods. For reasons of economy, this section should be limited to those properties that are germane to the intended application.
- Any other requirement—such as special tolerances; surface preparation, edge and finish on flat rolled products; and identification, packaging and loading instructions—properly become part of a specification.

Engineering societies, associations and institutes whose members make, specify or purchase steel products publish standard specifications, many of which have become well known and highly respected. Some of the important specification-writing groups are listed in Table 1. It is obvious from the names of some of these that the specifications prepared by a particular group may be limited to its own specialized field. The most comprehensive and widely used specifications are those published by the American Society for Testing and Materials (ASTM). ASTM specifications pertaining to steel products exist at three distinct levels. ASTM A6 contains the general requirements for most carbon steel structural products. ASTM A588, for example, incorporates the general requirements of A6 and describes the more specific requirements of a family of HSLA steels. Other specifications, such as A231 for alloy steel spring wire, refer to a particular product intended for a specific application.

Table 1. Principal specification-writing groups in the United States

Name	Designation
Association of American Railroads ..	AAR
American Bureau of Shipping	ABS
American Petroleum Institute	API
American Railway Engineering Association	AREA
American Society of Mechanical Engineers	ASME
American Society for Testing and Materials	ASTM
United States government Department of Defense	MIL and JAN
General Services Administration	FED
Society of Automotive Engineers ...	SAE
Aerospace Material Specification (of SAE)	AMS

Other specifications for steel products have been prepared by various corporations and United States government agencies to serve their own special needs. They are used primarily for procurement by that corporation or agency, and they receive only limited distribution or usage beyond these channels.

There is an important difference between specifications and standard practices. As indicated above, a specification is a statement of the requirements that a product must meet. When it is cited by a purchaser and accepted by a supplier, it becomes part of the purchase agreement. Many manufacturers of steel mill products publish compilations of their standard manufacturing practices. These data represent the dimensions, tolerances and properties that might be expected in the absence of specific requirements that indicate otherwise. The AISI Steel Products Manuals are compilations of the AISI designations for carbon and alloy steels, the standard practices of many steelmakers, and related scientific and technical information that has been reported to the institute. AISI states that the Steel Products Manuals are not specifications; however, they are a good indication of what restrictions and tolerances many producers of steel mill products will accept. Commercial tolerances and practices described in these manuals should, whenever possible, be incorporated into a proprietary specification in order to minimize the additional cost incurred by ordering "nonstandard" steel products.

Table 2. Carbon steel heat composition ranges and limits — semifinished products for forging, hot rolled and cold finished bars, wire rod and seamless tubing

Element	Limit or max of specified range, %	Range, %
Carbon(a)	To 0.25 incl	0.05
	Over 0.25 to 0.40 incl	0.06
	Over 0.40 to 0.55 incl	0.07
	Over 0.55 to 0.80 incl	0.10
	Over 0.80	0.13
Manganese	To 0.40 incl	0.15
	Over 0.40 to 0.50 incl	0.20
	Over 0.50 to 1.65 incl	0.30
Phosphorus(b)	0.040(c) to 0.08 incl	0.03
	Over 0.08 to 0.13 incl	0.05
Sulfur(b)	0.050(c) to 0.09 incl	0.03
	Over 0.09 to 0.15	0.05
	Over 0.15 to 0.23 incl	0.07
	Over 0.23 to 0.35 incl	0.09
Silicon(d)	To 0.15 incl	0.08
	Over 0.15 to 0.20 incl	0.10
	Over 0.20 to 0.30 incl	0.15
	Over 0.30 to 0.60 incl	0.20

(a) Add 0.01 to specified carbon ranges for steels with manganese contents exceeding 1.10%. (b) Lower maximum limits on phosphorus and sulfur are required by certain quality descriptors. (c) Lowest permissible maximum for this element. (d) Silicon content not normally specified for acid bessemer steels.

Table 3. Carbon steel heat composition ranges and limits — structural shapes, plate, strip, sheet and welded tubing

Element	Limit or max of specified range, %	Range, %
Carbon(a)	0.08(b)(c) to 0.15 incl	0.05
	Over 0.15 to 0.30 incl	0.06
	Over 0.30 to 0.40 incl	0.07
	Over 0.40 to 0.60 incl	0.08
	Over 0.60 to 0.80 incl	0.11
	Over 0.80 to 1.35 incl	0.14
Manganese	0.40(b) to 0.50 incl	0.20
	Over 0.50 to 1.15 incl	0.30
	Over 1.15 to 1.65 incl	0.35
Phosphorus(d)	0.04(b) to 0.08 incl	0.03
	Over 0.08 to 0.15 incl	0.05
Sulfur(d)	0.05 to 0.08 incl	0.03
	Over 0.08 to 0.15 incl	0.05
	Over 0.15 to 0.23 incl	0.07
	Over 0.23 to 0.33 incl	0.10
Silicon	0.10 to 0.15 incl	0.08
	Over 0.15 to 0.30 incl	0.15
	Over 0.30 to 0.60 incl	0.30

(a) Add 0.01 to specified carbon range for steels with manganese contents exceeding 1.00%. (b) Lowest permissible maximum limit for this element. (c) 0.12% for structural shapes and plate. (d) Lower maximum limits on phosphorus and sulfur are required by certain quality descriptors.

CHEMICAL ANALYSIS

Chemical composition is often used as the basis for assigning standard designations to steels. Such designations (or other desired ranges of chemical composition) are often incorporated into specifications for steel products. Users and specifiers of steel products should be familiar with methods of sampling and analysis.

Chemical analyses of steels are usually performed by wet chemical methods (such as that of ASTM E350) or spectrochemical methods (such as those of ASTM E281 and E282). Wet analysis is most often used to determine the composition of small numbers of specimens or of specimens comprised of machine tool chips. Such chips

Table 4. Alloy steel heat composition ranges and limits—bars, blooms, billets and slabs

Element	Limit or max of specified range, %	Range, % Open hearth or basic oxygen steel	Electric furnace steel
Carbon	To 0.55 incl	0.05	0.05
	Over 0.55 to 0.70 incl	0.08	0.07
	Over 0.70 to 0.80 incl	0.10	0.09
	Over 0.80 to 0.95 incl	0.12	0.11
	Over 0.95 to 1.35 incl	0.13	0.12
Manganese	To 0.60 incl	0.20	0.15
	Over 0.60 to 0.90 incl	0.20	0.20
	Over 0.90 to 1.05 incl	0.25	0.25
	Over 1.05 to 1.90 incl	0.30	0.30
	Over 1.90 to 2.10 incl	0.40	0.35
Sulfur(a)	To 0.050 incl	0.015	0.015
	Over 0.050 to 0.07 incl	0.02	0.02
	Over 0.07 to 0.10 incl	0.04	0.04
	Over 0.10 to 0.14 incl	0.05	0.05
Silicon	To 0.15 incl	0.08	0.08
	Over 0.15 to 0.20 incl	0.10	0.10
	Over 0.20 to 0.40 incl	0.15	0.15
	Over 0.40 to 0.60 incl	0.20	0.20
	Over 0.60 to 1.00 incl	0.30	0.30
	Over 1.00 to 2.20 incl	0.40	0.35
Chromium	To 0.40 incl	0.15	0.15
	Over 0.40 to 0.90 incl	0.20	0.20
	Over 0.90 to 1.05 incl	0.25	0.25
	Over 1.05 to 1.60 incl	0.30	0.30
	Over 1.60 to 1.75 incl	(b)	0.35
	Over 1.75 to 2.10 incl	(b)	0.40
	Over 2.10 to 3.99 incl	(b)	0.50
Nickel	To 0.50 incl	0.20	0.20
	Over 0.50 to 1.50 incl	0.30	0.30
	Over 1.50 to 2.00 incl	0.35	0.35
	Over 2.00 to 3.00 incl	0.40	0.40
	Over 3.00 to 5.30 incl	0.50	0.50
	Over 5.30 to 10.00 incl	1.00	1.00
Molybdenum	To 0.10 incl	0.05	0.05
	Over 0.10 to 0.20 incl	0.07	0.07
	Over 0.20 to 0.50 incl	0.10	0.10
	Over 0.50 to 0.80 incl	0.15	0.15
	Over 0.80 to 1.15 incl	0.20	0.20
Tungsten	To 0.50 incl	0.20	0.20
	Over 0.50 to 1.00 incl	0.30	0.30
	Over 1.00 to 2.00 incl	0.50	0.50
	Over 2.00 to 4.00 incl	0.60	0.60
Copper	To 0.60 incl	0.20	0.20
	Over 0.60 to 1.50 incl	0.30	0.30
	Over 1.50 to 2.00 incl	0.35	0.35
Vanadium	To 0.25 incl	0.05	0.05
	Over 0.25 to 0.50 incl	0.10	0.10
Aluminum	Up to 0.10 incl	0.05	0.05
	Over 0.10 to 0.20 incl	0.10	0.10
	Over 0.20 to 0.30 incl	0.15	0.15
	Over 0.30 to 0.80 incl	0.25	0.25
	Over 0.80 to 1.30 incl	0.35	0.35
	Over 1.30 to 1.80 incl	0.45	0.45

Element	Steelmaking Process	Lowest max(c), %
Phosphorus	Basic open hearth, basic oxygen or basic electric furnace steels	0.035(d)
	Basic electric furnace "E" steels	0.025
	Acid open hearth or electric furnace steel	0.050
Sulfur	Basic open hearth, basic oxygen or basic electric furnace steels	0.040(d)
	Basic electric furnace "E" steels	0.025
	Acid open hearth or electric furnace steel	0.050

(a) A range of sulfur content normally indicates a resulfurized steel. (b) Not normally produced by open hearth process. (c) Not applicable to rephosphorized or resulfurized steels. (d) Lower maximum limits on phosphorus and sulfur are required by certain quality descriptors.

Table 5. Alloy steel heat composition ranges and limits—plate

Element	Limit or max of specified range, %	Range, % Open hearth or basic oxygen steels	Electric furnace steels
Carbon	To 0.25 incl	0.06	0.05
	Over 0.25 to 0.40 incl	0.07	0.06
	Over 0.40 to 0.55 incl	0.08	0.07
	Over 0.55 to 0.70 incl	0.11	0.10
	Over 0.70	0.14	0.13
Manganese	To 0.45 incl	0.20	0.15
	Over 0.45 to 0.80 incl	0.25	0.20
	Over 0.80 to 1.15 incl	0.30	0.25
	Over 1.15 to 1.70 incl	0.35	0.30
	Over 1.70 to 2.10 incl	0.40	0.35
Sulfur(a)	To 0.060 incl	0.02	0.02
	Over 0.060 to 0.100 incl	0.04	0.04
	Over 0.100 to 0.140 incl	0.05	0.05
Silicon	To 0.15 incl	0.08	0.08
	Over 0.15 to 0.20 incl	0.10	0.10
	Over 0.20 to 0.40 incl	0.15	0.15
	Over 0.40 to 0.60 incl	0.20	0.20
	Over 0.60 to 1.00 incl	0.30	0.30
	Over 1.00 to 2.20 incl	0.40	0.35
Chromium	To 0.40 incl	0.20	0.15
	Over 0.40 to 0.80 incl	0.25	0.20
	Over 0.80 to 1.05 incl	0.30	0.25
	Over 1.05 to 1.25 incl	0.35	0.30
	Over 1.25 to 1.75 incl	0.50	0.40
	Over 1.75 to 3.99 incl	0.60	0.50
Nickel	To 0.50 incl	0.20	0.20
	Over 0.50 to 1.50 incl	0.30	0.30
	Over 1.50 to 2.00 incl	0.35	0.35
	Over 2.00 to 3.00 incl	0.40	0.40
	Over 3.00 to 5.30 incl	0.50	0.50
	Over 5.30 to 10.00 incl	1.00	1.00
Molybdenum	To 0.10 incl	0.05	0.05
	Over 0.10 to 0.20 incl	0.07	0.07
	Over 0.20 to 0.50 incl	0.10	0.10
	Over 0.50 to 0.80 incl	0.15	0.15
	Over 0.80 to 1.15 incl	0.20	0.20
Tungsten	To 0.50 incl	0.20	0.20
	Over 0.50 to 1.00 incl	0.30	0.30
	Over 1.00 to 2.00 incl	0.50	0.50
	Over 2.00 to 4.00 incl	0.60	0.60
Copper	To 0.60 incl	0.20	0.20
	Over 0.60 to 1.50 incl	0.30	0.30
	Over 1.50 to 2.00 incl	0.35	0.35
Vanadium	To 0.25 incl	0.05	0.05
	Over 0.25 to 0.50 incl	0.10	0.10
Aluminum	Up to 0.10 incl	0.05	0.05
	Over 0.10 to 0.20 incl	0.10	0.10
	Over 0.20 to 0.30 incl	0.15	0.15
	Over 0.30 to 0.80 incl	0.25	0.25
	Over 0.80 to 1.30 incl	0.35	0.35
	Over 1.30 to 1.80 incl	0.45	0.45

Element	Steelmaking Process	Lowest max(b), %
Phosphorus	Basic open hearth or basic oxygen	0.035(c)
	Basic electric furnace	0.025
Sulfur	Basic open hearth or basic oxygen	0.040(c)
	Basic electric furnace	0.025

(a) A range of sulfur content normally indicates a resulfurized steel. (b) Not applicable to resulfurized or rephosphorized steels. (c) Lower maximum limits on phosphorus and sulfur are required by certain quality descriptors.

should be obtained in accordance with ASTM E59. Spectrochemical analysis is well-suited to routine determination of chemical composition of large numbers of specimens, such as are present in a steel mill.

Heat and Product Analyses. During the steelmaking process, a small sample of molten metal is removed from the ladle or steelmaking furnace, allowed to solidify and then analyzed for alloy content. In most steel mills, these heat analyses are performed using spectrochemical methods; as many as 14 different elements can be determined simultaneously. The heat analysis furnished to the customer, however, may include only those elements for which a range or a maximum or minimum limit exists in the appropriate designation or specification. A heat analysis is generally considered to be an accurate representation of the composition of the entire heat of metal. Producers of steel heave found that heat analyses for carbon and alloy steels can be consistently held within ranges that depend on the amount of the particular alloying element desired for the steel, the product form and the method of making the steel. These ranges have been published as commercial practice, then incorporated into standard specifications such as ASTM A29. Standard ranges and limits of heat analyses of carbon and alloy steels are given in Tables 2, 3, 4 and 5.

Because segregation of some alloying elements is inherent in the solidification of an ingot, different portions will have local chemical compositions that differ slightly from the average composition. Many lengths of bar stock can be made from a single ingot; thus, some variation in composition between individual bars must be expected. The compositions of individual bars might not conform to the applicable specification, even though the heat analysis does. The chemical composition of an individual bar (or other product) taken from a large heat of steel is called the "product analysis" or "check analysis." Ranges and limits for product analyses are generally broader and less restrictive than the corresponding ranges and limits for heat analyses. Such limits used in standard commercial practice are given in Tables 6, 7 and 8.

Residual elements usually enter steel products from raw materials used to produce pig iron or from scrap steel used in steelmaking. Through careful steelmaking practices, the amounts of these residual elements are generally held to acceptable levels. Sulfur and phosphorus are generally considered deleterious to the mechanical properties of steels; thus, there are restrictions on the allowable amounts of these elements for most grades. The amounts of sulfur and phosphorus are invariably reported in the analyses of both carbon and alloy steels. Other residual alloying elements generally exert lesser influences on the properties of steel than do sulfur and phosphorus. For many grades of steel, limitations on the amounts of these residual elements are either optional or omitted entirely. Amounts of residual alloying elements are generally not reported in either heat or product analyses, except for special reasons.

Silicon Content of Steels. The composition requirements for many steels, particularly plain carbon steels, contain no specific restriction on silicon content. The lack of a silicon requirement is not an omission, but rather a recognition that the amount of silicon in a steel can often be traced directly to the deoxidation practice employed in making it.

Table 6. Carbon steel product composition tolerances

Element	Limit or max of specified range, %	Tolerance over max or under min limits(a), %, for product with cross-sectional area of:			
		to 100 in.²	100-200 in.²	200-400 in.²	400-800 in.²
Carbon	To 0.25 incl	0.02	0.03	0.04	0.05
	Over 0.25 to 0.55 incl	0.03	0.04	0.05	0.06
	Over 0.55	0.04	0.05	0.06	0.07
Manganese	To 0.90 incl	0.03	0.04	0.06	0.07
	Over 0.90 to 1.65 incl	0.06	0.06	0.07	0.08
Phosphorus	Over maximum only, to 0.040 incl	0.008	0.008	0.010	0.015
Sulfur	Over maximum only, to 0.050 incl	0.008	0.010	0.010	0.015
Silicon	To 0.35 incl	0.02	0.02	0.03	0.04
	Over 0.35 to 0.60 incl	0.05	...	...	...
Copper	Under minimum only	0.02	0.03	...	...
Lead	0.15 to 0.35 incl	0.03	0.03	...	...

(a) Product composition requirements are not applicable to rimmed or capped steels, to boron content of boron steels, or to phosphorus and sulfur contents of rephosphorized and resulfurized steels. Product composition tolerances for alloying elements in HSLA steels are given in Table 7.

Table 7. Alloy steel product composition tolerances — bars, billets, blooms and slabs

Element	Limit or max of specified range, %	Tolerance over max or under min limits(a), %, for product with cross-sectional area of:			
		to 100 in.²	100-200 in.²	200-400 in.²	400-800 in.²
Carbon	To 0.30 incl	0.01	0.02	0.03	0.04
	Over 0.30 to 0.75 incl	0.02	0.03	0.04	0.05
	Over 0.75	0.03	0.04	0.05	0.06
Manganese	To 0.90 incl	0.03	0.04	0.05	0.06
	Over 0.90 to 2.10 incl	0.04	0.05	0.06	0.07
Phosphorus	Over max only	0.005	0.010	0.010	0.010
Sulfur	Over max only	0.005	0.010	0.010	0.010
Silicon	To 0.40 incl	0.02	0.02	0.03	0.04
	Over 0.40 to 2.20 incl	0.05	0.06	0.06	0.07
Chromium	To 0.90 incl	0.03	0.04	0.04	0.05
	Over 0.90 to 2.10 incl	0.05	0.06	0.06	0.07
	Over 2.10 to 3.99 incl	0.10	0.10	0.12	0.14
Nickel	To 1.00 incl	0.03	0.03	0.03	0.03
	Over 1.00 to 2.00 incl	0.05	0.05	0.05	0.05
	Over 2.00 to 5.30 incl	0.07	0.07	0.07	0.07
	Over 5.30 to 10.00 incl	0.10	0.10	0.10	0.10
Molybdenum	To 0.20 incl	0.01	0.01	0.02	0.03
	Over 0.20 to 0.40 incl	0.02	0.03	0.03	0.04
	Over 0.40 to 1.15 incl	0.03	0.04	0.05	0.06
Tungsten	To 1.00 incl	0.04	0.05	0.05	0.06
	Over 1.00 to 4.00 incl	0.08	0.09	0.10	0.12
Copper(b)	To 1.00 incl	0.03	...	...	...
	Over 1.00 to 2.00 incl	0.05	...	...	...
Vanadium	To 0.10 incl	0.01	0.01	0.01	0.01
	Over 0.10 to 0.25 incl	0.02	0.02	0.02	0.02
	Over 0.25 to 0.50 incl	0.03	0.03	0.03	0.03
	Min value specified, check under min limit(b)	0.01	0.01	0.01	0.01
Niobium(b)	To 0.10 incl	0.01(c)	...	...	...
Titanium(b)	To 0.10 incl	0.01(c)	...	...	...
Zirconium(b)	To 0.15 incl	0.03	...	...	...
Aluminum(b)	Up to 0.10 incl	0.03	...	...	...
	Over 0.10 to 0.20 incl	0.04	...	...	...
	Over 0.20 to 0.30 incl	0.05	...	...	...
	Over 0.30 to 0.80 incl	0.07	...	...	...
	Over 0.80 to 1.80 incl	0.10	...	...	...
Lead(b)	0.15 to 0.35 incl	0.03	...	...	...
Nitrogen(b)	To 0.030 incl	0.005	...	...	...

(a) Product composition requirements are not applicable to boron content of boron steels or sulfur content of resulfurized steels. (b) Tolerances shown apply only to cross-sectional areas of 100 in.² or less. (c) If the minimum of the range is 0.01%, the lower tolerance is 0.005%.

Rimmed and capped steels are not deoxidized; the only silicon present is the residual amount left from scrap or raw materials, typically less than 0.05% Si. Specifications and orders for these steels customarily indicate that the steel must be made rimmed or capped, as required by the purchaser; restrictions on silicon content are not customarily given.

The extent of rimming action during the solidification of semikilled steel ingots must be carefully controlled by matching the amount of deoxidizer with the oxygen content of the molten steel.

Table 8. Alloy steel product composition tolerances — plate

Element	Limit or max of specified range, %	Tolerance over max or under min limits, %
Carbon	To 0.30 incl	0.02
	Over 0.30 to 0.75 incl	0.03
	Over 0.75	0.04
Manganese	To 0.90 incl	0.04
	Over 0.90 to 2.10 incl	0.05
Phosphorus(a)	Over max only	0.01
Sulfur(a)(b)	... .	0.01
Silicon	To 0.40 incl	0.02
	Over 0.40 to 2.20 incl	0.06
Chromium	To 0.90 incl	0.04
	Over 0.90 to 2.10 incl	0.06
	Over 2.10 to 3.99 incl	0.10
Nickel	To 1.00 incl	0.03
	Over 1.00 to 2.00 incl	0.05
	Over 2.00 to 5.30 incl	0.07
	Over 5.30	0.10
Molybdenum	To 0.20 incl	0.01
	Over 0.20 to 0.40 incl	0.03
	Over 0.40 to 1.15 incl	0.04
Tungsten	To 1.00 incl	0.05
	Over 1.00 to 4.00 incl	0.09
Copper	To 1.00 incl	0.03
	Over 1.00 to 2.00 incl	0.05
Vanadium	To 0.10 incl	0.01
	Over 0.10 to 0.25 incl	0.02
	Over 0.25 to 0.50 incl	0.03
	Min value specified check under min limit	0.01
Aluminum	Up to 0.10 incl	0.03
	Over 0.10 to 0.20 incl	0.04
	Over 0.20 to 0.30 incl	0.05
	Over 0.30 to 0.80 incl	0.07
	Over 0.80 to 1.80 incl	0.10

(a) For pressure-vessel quality plate, the specified composition includes product composition tolerances for phosphorus and sulfur. (b) Product composition requirements not applicable to sulfur content of resulfurized steel.

The amount of silicon required for deoxidation may vary from heat to heat. Thus, the silicon content of the solid metal can vary slightly from heat to heat. A maximum silicon content of 0.10% is sometimes specified for semikilled steel, but this requirement is not very restrictive; for certain heats, a silicon addition sufficient to leave a residue of 0.10% may be enough of an addition to kill the steel.

Killed steels are fully deoxidized during their manufacture; deoxidation can be accomplished by additions of silicon, aluminum, or both, or by vacuum treatment of the molten steel. Because it is the least costly of these methods, silicon deoxidation is frequently used. For silicon-killed steels, a range of 0.15 to 0.35% Si is often specified, providing the manufacturer with adequate flexibility to compensate for variations in the steel-making process and ensuring a steel acceptable for most applications. Aluminum-killed or vacuum-deoxidized steels require no silicon; a requirement for minimum silicon content in such steel is unnecessary. A maximum permissible silicon content is appropriate for all killed plain carbon steels; a minimum silicon content implies a restriction that the steel must be silicon killed. Silicon is intentionally added to some alloy steels, in which it serves as both a deoxidizer and an alloying element to modify the properties of the steel. An acceptable range of silicon content would be appropriate for these steels.

Users and specifiers of steel mill products must realize that the silicon content of these items cannot be established independently of deoxidation practice. In ordering mill products, it is often desirable to cite a standard specification (such as an ASTM specification) where the various ramifications of restrictions on silicon content have already been considered in preparing the specification. In some instances, such as forming of low-carbon steel sheet, the choice of deoxidation practice can significantly affect the performance of the steel; in such cases, it is appropriate to specify the desired practice. Technical representatives of steel producers can be helpful in iden-

tifying a mill product well-suited to a particular application, selecting an appropriate standard specification or incorporating appropriate language into a proprietary specification.

AISI-SAE DESIGNATIONS

The most widely used system for designating carbon and alloy steels is that of the American Iron and Steel Institute (AISI) and the Society of Automotive Engineers (SAE). As a point of technicality, there are two separate systems, but they are nearly identical and are carefully coordinated by the two groups. The numerical designations summarized in Table 9 are used by both AISI and SAE. In this system, a particular designation implies the same limits and ranges of chemical composition for both an AISI steel and the corresponding SAE steel. The differences in listings occur as a result of differences in determining eligibility for listing. AISI uses production tonnage as the basis for including a steel. SAE includes a steel if it is used in significant quantity by two users or if it has unique engineering characteristics. The fact that a particular steel is listed by AISI or SAE implies only that it has been produced in appreciable quantity. It does not imply that other grades are unavailable, nor does it imply that any particular steel producer makes all of the listed grades.

As indicated in the preceding discussion on specifications, AISI designations and standard practices are not specifications. The SAE designations are published in the annual SAE handbook under various SAE standards. These standards are comprised entirely of listings of SAE designations and the limits and ranges of chemical composition defined by these designations. Either designation contains only a portion of the information necessary to properly describe a steel product for procurement purposes.

Carbon steels contain less than 1.65 Mn, 0.60 Si and 0.60 Cu; they comprise the 1xxx groups in the AISI-SAE system. Plain carbon steels in the 10xx group are listed in Tables 10 and 11;

Table 9. AISI-SAE system of designations

Numerals and digits(a)	Type of steel and/or nominal alloy content
Carbon steels	
10xx	Plain carbon (Mn 1.00% max)
11xx	Resulfurized
12xx	Resulfurized and rephosphorized
15xx	Plain carbon (max Mn range — 1.00 to 1.65%)
Manganese steels	
13xx	Mn 1.75
Nickel steels	
23xx	Ni 3.50
25xx	Ni 5.00
Nickel-chromium steels	
31xx	Ni 1.25; Cr 0.65 and 0.80
32xx	Ni 1.75; Cr 1.07
33xx	Ni 3.50; Cr 1.50 and 1.57
34xx	Ni 3.00; Cr 0.77
Molybdenum steels	
40xx	Mo 0.20 and 0.25
44xx	Mo 0.40 and 0.52
Chromium-molybdenum steels	
41xx	Cr 0.50, 0.80 and 0.95; Mo 0.12, 0.20, 0.25 and 0.30

Numerals and digits(a)	Type of steel and/or nominal alloy content
Nickel-chromium-molybdenum steels	
43xx	Ni 1.82; Cr 0.50 and 0.80; Mo 0.25
43BVxx	Ni 1.82; Cr 0.50; Mo 0.12 and 0.25; V 0.03 min
47xx	Ni 1.05; Cr 0.45; Mo 0.20 and 0.35
81xx	Ni 0.30; Cr 0.40; Mo 0.12
86xx	Ni 0.55; Cr 0.50; Mo 0.20
87xx	Ni 0.55; Cr 0.50; Mo 0.25
88xx	Ni 0.55; Cr 0.50; Mo 0.35
93xx	Ni 3.25; Cr 1.20; Mo 0.12
94xx	Ni 0.45; Cr 0.40; Mo 0.12
97xx	Ni 0.55; Cr 0.20; Mo 0.20
98xx	Ni 1.00; Cr 0.80; Mo 0.25
Nickel-molybdenum steels	
46xx	Ni 0.85 and 1.82; Mo 0.20 and 0.25
48xx	Ni 3.50; Mo 0.25
Chromium steels	
50xx	Cr 0.27, 0.40, 0.50 and 0.65
51xx	Cr 0.80, 0.87, 0.92, 0.95, 1.00 and 1.05

Numerals and digits(a)	Type of steel and/or nominal alloy content
Chromium steels	
50xxx	Cr 0.50 ⎫
51xxx	Cr 1.02 ⎬ C 1.00 min
52xxx	Cr 1.45 ⎭
Chromium-vanadium steels	
61xx	Cr 0.60, 0.80 and 0.95; V 0.10 and 0.15 min
Tungsten-chromium steel	
72xx	W 1.75; Cr 0.75
Silicon-manganese steels	
92xx	Si 1.40 and 2.00; Mn 0.65, 0.82 and 0.85; Cr 0.00 and 0.65
High-strength low-alloy steels	
9xx	Various SAE grades
Boron steels	
xxBxx	B denotes boron steel
Leaded steels	
xxLxx	L denotes leaded steel

(a) "xx" in the last two (or three) digits of these designations indicates that the carbon content (in hundredths of a percent) is to be inserted.

Table 10. Composition ranges and limits for AISI-SAE standard carbon steels containing less than 1.00% manganese—semifinished products for forging, hot rolled and cold finished bars, wire rod and seamless tubing

AISI-SAE designation	UNS designation	Heat composition ranges and limits(a), %		AISI-SAE designation	UNS designation	Heat composition ranges and limits(a), %		AISI-SAE designation	UNS designation	Heat composition ranges and limits(a), %	
		C	Mn			C	Mn			C	Mn
1005	G10050	0.06 max	0.35 max	1035	G10350	0.32-0.38	0.60-0.90	1074(b)	G10740	0.70-0.80	0.50-0.80
1006	G10060	0.08 max	0.25-0.40	1037	G10370	0.32-0.38	0.70-1.00	1075(b)	G10750	0.70-0.80	0.40-0.70
1008	G10080	0.10 max	0.30-0.50	1038	G10380	0.35-0.42	0.60-0.90	1078	G10780	0.72-0.85	0.30-0.60
1010	G10100	0.08-0.13	0.30-0.60	1039	G10390	0.37-0.44	0.70-1.00	1080	G10800	0.75-0.88	0.60-0.90
1011(b)	G10110	0.08-0.13	0.60-0.90	1040	G10400	0.37-0.44	0.60-0.90	1084	G10840	0.80-0.93	0.60-0.90
1012	G10120	0.10-0.15	0.30-0.60	1042	G10420	0.40-0.47	0.60-0.09	1085(b)	G10850	0.80-0.93	0.70-1.00
1013(b)	G10130	0.11-0.16	0.50-0.80	1043	G10430	0.40-0.47	0.70-1.00	1086	G10860	0.80-0.93	0.30-0.50
1015	G10150	0.13-0.18	0.30-0.60	1044	G10440	0.43-0.50	0.30-0.60	1090	G10900	0.85-0.98	0.60-0.90
1016	G10160	0.13-0.18	0.60-0.90	1045	G10450	0.43-0.50	0.60-0.90	1095	G10950	0.90-1.03	0.30-0.50
1017	G10170	0.15-0.20	0.30-0.60	1046	G10460	0.43-0.50	0.70-1.00				
1018	G10180	0.15-0.20	0.60-0.90	1049	G10490	0.46-0.53	0.60-0.90				
1019	G10190	0.15-0.20	0.70-1.00	1050	G10500	0.48-0.55	0.60-0.90				
1020	G10200	0.18-0.23	0.30-0.60	1053	G10530	0.48-0.55	0.70-1.00				
1021	G10210	0.18-0.23	0.60-0.90	1055	G10550	0.50-0.60	0.60-0.90				
1022	G10220	0.18-0.23	0.70-1.00	1059(c)	G10590	0.55-0.65	0.50-0.80				
1023	G10230	0.20-0.25	0.30-0.60	1060	G10600	0.55-0.65	0.60-0.90				
1025	G10250	0.22-0.28	0.30-0.60	1064	G10640	0.60-0.70	0.50-0.80				
1026	G10260	0.22-0.28	0.60-0.90	1065	G10650	0.60-0.70	0.60-0.90				
1029	G10290	0.25-0.31	0.60-0.90	1069(b)	G10690	0.65-0.75	0.40-0.70				
1030	G10300	0.28-0.34	0.60-0.90	1070	G10700	0.65-0.75	0.60-0.90				

(a) Limits on phosphorus and sulfur contents are given in Table 3; typical limits are 0.040% maximum phosphorus and 0.050% maximum sulfur. When silicon ranges or limits are required, the values in Table 3 apply. Steels listed in this table can be produced with additions of lead or boron. Leaded steels typically contain 0.15 to 0.35% lead and are identified by inserting the letter "L" in the designation—11L17; boron steels can be expected to contain 0.0005 to 0.003% boron and are identified by inserting the letter "B" in the designation—15B41. (b) SAE standard grade only. (c) AISI standard grade only.

Table 11. Composition ranges and limits for AISI-SAE standard carbon steels—structural shapes, plate, strip, sheet and welded tubing

AISI-SAE designation	UNS designation	Heat composition ranges and limits(a), %		AISI-SAE designation	UNS designation	Heat composition ranges and limits(a), %		AISI-SAE designation	UNS designation	Heat composition ranges and limits(a), %	
		C	Mn			C	Mn			C	Mn
1006	G10060	0.08 max	0.25-0.45	1038	G10380	0.34-0.42	0.60-0.90	1090	G10900	0.84-0.98	0.60-0.90
1008	G10080	0.10 max	0.25-0.50	1039	G10390	0.36-0.44	0.70-1.00	1095	G10950	0.90-1.04	0.30-0.50
1009	G10090	0.15 max	0.60 max	1040	G10400	0.36-0.44	0.60-0.90	1524(b)	G15240	0.18-0.25	1.30-1.65
1010	G10100	0.08-0.13	0.30-0.60	1042	G10420	0.39-0.47	0.60-0.90	1527(b)	G15270	0.22-0.29	1.20-1.55
1012	G10120	0.10-0.15	0.30-0.60	1043	G10430	0.39-0.47	0.70-1.00	1536(b)	G15360	0.30-0.38	1.20-1.55
1015	G10150	0.12-0.18	0.30-0.60	1045	G10450	0.42-0.50	0.60-0.90	1541(b)	G15410	0.36-0.45	1.30-1.65
1016	G10160	0.12-0.18	0.60-0.90	1046	G10460	0.42-0.50	0.70-1.00	1548(b)	G15480	0.43-0.52	1.05-1.40
1017	G10170	0.14-0.20	0.30-0.60	1049	G10490	0.45-0.53	0.60-0.90	1552(b)	G15520	0.46-0.55	1.20-1.55
1018	G10180	0.14-0.20	0.60-0.90	1050	G10500	0.47-0.55	0.60-0.90				
1019	G10190	0.14-0.20	0.70-1.00	1055	G10550	0.52-0.60	0.60-0.90				
1020	G10200	0.17-0.23	0.30-0.60	1060	G10600	0.55-0.66	0.60-0.90				
1021	G10210	0.17-0.23	0.60-0.90	1064	G10640	0.59-0.70	0.50-0 80				
1022	G10220	0.17-0.23	0.70-1.00	1065	G10650	0.59-0.70	0.60-0.90				
1023	G10230	0.19-0.25	0.30-0.60	1070	G10700	0.65-0.76	0.60-0.90				
1025	G10250	0.22-0.28	0.30-0.60	1074	G10740	0.69-0.80	0.50-0.80				
1026	G10260	0.22-0.28	0.60-0.90	1078	G10780	0.72-0.86	0.30-0.60				
1030	G10300	0.27-0.34	0.60-0.90	1080	G10800	0.74-0.88	0.60-0.90				
1033	G10330	0.29-0.36	0.70-1.00	1084	G10840	0.80-0.94	0.60-0.90				
1035	G10350	0.31-0.38	0.60-0.90	1085	G10850	0.80-0.94	0.70-1.00				
1037	G10370	0.31-0.38	0.70-1.00	1086	G10860	0.80-0.94	0.30-0.50				

(a) Limits on phosphorus and sulfur contents are given in Table 4; typical limits are 0.040% maximum phosphorus and 0.050% maximum sulfur. When silicon ranges or limits are required, the values in Table 4 apply. Steels listed in this table can be produced with additions of lead or boron. Leaded steels typically contain 0.15 to 0.35% lead and are identified by inserting the letter "L" in the designation—11L17; boron steels can be expected to contain 0.0005 to 0.003% boron and are identified by inserting the letter "B" in the designation—15B41. (b) Formerly designated 10xx grade.

Table 12. Composition ranges and limits for AISI-SAE merchant-quality steels

AISI-SAE designation	Heat composition ranges and limits, %			
	C	Mn	P max	S max
M1008	0.10 max	0.25-0.60	0.04	0.05
M1010	0.07-0.14	0.25-0.60	0.04	0.05
M1012	0.09-0.16	0.25-0.60	0.04	0.05
M1015	0.12-0.19	0.25-0.60	0.04	0.05
M1017	0.14-0.21	0.25-0.60	0.04	0.05
M1020	0.17-0.24	0.25-0.60	0.04	0.05
M1023	0.19-0.27	0.25-0.60	0.04	0.05
M1025	0.20-0.30	0.25-0.60	0.04	0.05
M1031	0.26-0.36	0.25-0.60	0.04	0.05
M1044	0.40-0.50	0.25-0.60	0.04	0.05

Table 13. Composition ranges and limits for AISI-SAE standard resulfurized carbon steels

AISI-SAE designation	UNS designation	Heat composition ranges and limits(a), %		
		C	Mn	S
1110	G11100	0.08-0.13	0.30-0.60	0.08-0.13
1117	G11170	0.14-0.20	1.00-1.30	0.08-0.13
1118	G11180	0.14-0.20	1.30-1.60	0.08-0.13
1137	G11370	0.32-0.39	1.35-1.65	0.08-0.13
1139	G11390	0.35-0.43	1.35-1.65	0.13-0.20
1140	G11400	0.37-0.44	0.70-1.00	0.08-0.13
1141	G11410	0.37-0.45	1.35-1.65	0.08-0.13
1144	G11440	0.40-0.48	1.35-1.65	0.24-0.33
1146	G11460	0.42-0.49	0.70-1.00	0.08-0.13
1151	G11510	0.48-0.55	0.70-1.00	0.08-0.13

(a) Limit on phosphorus content is given in Table 3; the typical value is 0.040% maximum phosphorus. Because of the adverse effect of silicon on machinability, steels listed in this table are generally not deoxidized with silicon. Steels listed in this table can be produced as leaded steels, typically containing 0.15 to 0.35% lead and identified by inserting the letter "L" in the designation—11L17.

Table 14. Composition ranges and limits for AISI-SAE standard resulfurized and rephosphorized carbon steels

| AISI-SAE designation | UNS designation | C max | Heat composition ranges and limits(a), % | | |
			Mn	P	S
1211	G12110	0.13	0.60-0.90	0.07-0.12	0.10-0.15
1212	G12120	0.13	0.70-1.00	0.07-0.12	0.16-0.23
1213	G12130	0.13	0.70-1.00	0.07-0.12	0.24-0.33
12L14(b)	G12144	0.15	0.85-1.15	0.04-0.09	0.26-0.35
1215	G12150	0.09	0.75-1.05	0.04-0.09	0.26-0.35

(a) Because of the adverse effect of silicon on machinability, steels listed in this table are generally not deoxidized with silicon. (b) Contains 0.15 to 0.35% lead; other steels listed in this table can be produced with the same lead content.

Table 15. Composition ranges and limits for AISI-SAE standard carbon steels with maximum manganese contents exceeding 1.10% — semifinished products for forging, hot rolled and cold finished bars, wire rod and seamless tubing

| AISI-SAE designation | UNS designation | Heat composition ranges and limits(a), % | | | | Former AISI-SAE designation |
		C	Mn	P max	S max	
1513	G15130	0.10-0.16	1.10-1.40	0.040	0.050	...
1518(b)	G15180	0.15-0.21	1.10-1.40	0.040	0.050	...
1522	G15220	0.18-0.24	1.10-1.40	0.040	0.050	...
1524	G15240	0.19-0.25	1.35-1.65	0.040	0.050	1024
1525(b)	G15250	0.23-0.29	0.80-1.10	0.040	0.050	...
1526	G15256	0.22-0.29	1.10-1.40	0.040	0.050	...
1527	G15270	0.22-0.29	1.20-1.50	0.040	0.050	1027
1536(b)	G15360	0.30-0.37	1.20-1.50	0.040	0.050	1036
1541	G15410	0.36-0.44	1.35-1.65	0.040	0.050	1041
1547(b)	G15470	0.43-0.51	1.35-1.65	0.040	0.050	1047
1548	G15480	0.44-0.52	1.10-1.40	0.040	0.050	1048
1551	G15510	0.45-0.56	0.85-1.15	0.040	0.050	1051
1552	G15520	0.47-0.55	1.20-1.50	0.040	0.050	1052
1561	G15610	0.55-0.65	0.75-1.05	0.040	0.050	1061
1566	G15660	0.60-0.71	0.85-1.15	0.040	0.050	1066
1572(b)	G15720	0.65-0.76	1.00-1.30	0.040	0.050	1072

(a) Limits on phosphorus and sulfur contents are given in Table 3; typical limits are 0.040% maximum phosphorus and 0.050% maximum sulfur. Killed steels commonly contain 0.15 to 0.30% silicon; other ranges are negotiable. Steels listed in this table can be produced with additions of lead or boron. Leaded steels typically contain 0.15 to 0.35% lead and are identified by inserting the letter "L" in the designation—11L17; boron can be expected to contain 0.0005 to 0.003% boron and are identified by inserting the letter "B" in the designation—15B41. (b) SAE standard grade only.

note that ranges and limits of chemical composition depend on the product form. Designations for merchant-quality steels, given in Table 12, include the prefix "M." Resulfurized carbon steels in the 11*xx* group are listed in Table 13, and resulfurized and rephosphorized carbon steels in the 12*xx* group are listed in Table 14. Both of these groups of steels are produced for applications requiring good machinability. Table 15 lists steels having nominal manganese contents of between 0.9 and 1.5%, but no other alloying additions; these steels now have 15*xx* designations in place of the 10*xx* designations formerly used. SAE Recommended Practice J776e includes hardenability bands for the steels given in Table 16. These steels have hardenability requirements in addition to the limits and ranges of chemical composition. They are distinguished from similar grades that have no hardenability requirement by the use of the suffix "H." Limits and ranges of chemical composition for all of these carbon steel products reflect the restrictions on heat and product analyses given in Tables 3, 4 and 5. Except where indicated, all of these designations for carbon steels are both AISI and SAE designations.

Alloy steels contain manganese, silicon or copper in quantities greater than those listed for the carbon steels; or they have specified ranges or minimums for one or more other alloying elements. The alloying additions enhance mechanical properties, fabricating characteristics or some other attribute of the steel. In the AISI-SAE system of designations, the major alloying elements in a steel are indicated by the first two digits of the designation. The amount of carbon, in hundredths of a percent, is indicated by the last two

Table 16. Composition ranges and limits for AISI-SAE standard carbon H-steels

| AISI-SAE designation | UNS designation | Heat composition ranges and limits(a), % | | | AISI-SAE designation | UNS designation | Heat composition ranges and limits(a), % | | |
		C	Mn	Si			C	Mn	Si
1038H	H10380	0.34-0.43	0.50-1.00	0.15-0.30	15B21H(b)	H15211	0.17-0.24	0.70-1.20	0.15-0.30
1045H	H10450	0.42-0.51	0.50-1.00	0.15-0.30	15B35H(b)	H15351	0.31-0.39	0.70-1.20	0.15-0.30
1522H	H15220	0.17-0.25	1.00-1.50	0.15-0.30	15B37H(b)	H15371	0.30-0.39	1.00-1.50	0.15-0.30
1524H	H15240	0.18-0.26	1.25-1.75	0.15-0.30	15B41H(b,c)	H15411	0.35-0.45	1.25-1.75	0.15-0.30
1526H	H15260	0.21-0.30	1.00-1.50	0.15-0.30	15B48H(b,c)	H15481	0.43-0.53	1.00-1.50	0.15-0.30
1541H	H15410	0.35-0.45	1.25-1.75	0.15-0.30	15B62H(b)	H15621	0.54-0.67	1.00-1.50	0.40-0.60

(a) Limits on phosphorus and sulfur content are given in Table 3; typical limits are 0.040% maximum phosphorus and 0.050% maximum sulfur. (b) Can be expected to contain 0.0005 to 0.003% boron. (c) AISI grade only.

Table 17. Composition ranges and limits for AISI-SAE standard alloy steels — bars, billets, blooms and slabs

| AISI-SAE designation | UNS designation | Heat composition ranges and limits, % | | | | | | | | |
		C	Mn	P max(a)	S max(a)	Si	Cr	Ni	Mo
1330	G13300	0.28-0.33	1.60-1.90	0.035	0.040	0.15-0.30	...	...	...
1335	G13350	0.33-0.38	1.60-1.90	0.035	0.040	0.15-0.30	...	...	...
1340	G13400	0.38-0.43	1.60-1.90	0.035	0.040	0.15-0.30	...	...	...
1345	G13450	0.43-0.48	1.60-1.90	0.035	0.040	0.15-0.30	...	...	...
4012	G40120	0.09-0.14	0.75-1.00	0.035	0.040	0.15-0.30	...	...	0.15-0.25
4023	G40230	0.20-0.25	0.70-0.90	0.035	0.040	0.15-0.30	...	...	0.20-0.30
4024	G40240	0.20-0.25	0.70-0.90	0.035	0.035-0.050(b)	0.15-0.30	...	...	0.20-0.30
4027	G40270	0.25-0.30	0.70-0.90	0.035	0.040	0.15-0.30	...	...	0.20-0.30
4028	G40280	0.25-0.30	0.70-0.90	0.035	0.035-0.050(b)	0.15-0.30	...	...	0.20-0.30
4032	G40320	0.30-0.35	0.70-0.90	0.035	0.040	0.15-0.30	...	...	0.20-0.30
4037	G40370	0.35-0.40	0.70-0.90	0.035	0.040	0.15-0.30	...	...	0.20-0.30
4042(c)	G40420	0.40-0.45	0.70-0.90	0.035	0.040	0.15-0.30	...	...	0.20-0.30
4047	G40470	0.45-0.50	0.70-0.90	0.035	0.040	0.15-0.30	...	...	0.20-0.30
4118	G41180	0.18-0.23	0.70-0.90	0.035	0.040	0.15-0.30	0.40-0.60	...	0.08-0.15
4130	G41300	0.28-0.33	0.40-0.60	0.035	0.040	0.15-0.30	0.80-1.10	...	0.15-0.25
4135(c)	G41350	0.33-0.38	0.70-0.90	0.035	0.040	0.15-0.30	0.30-1.10	...	0.15-0.25
4137	G41370	0.35-0.40	0.70-0.90	0.035	0.040	0.15-0.30	0.80-1.10	...	0.15-0.25
4140	G41400	0.38-0.43	0.75-1.00	0.035	0.040	0.15-0.30	0.80-1.10	...	0.15-0.25
4142	G41420	0.40-0.45	0.75-1.00	0.035	0.040	0.15-0.30	0.80-1.10	...	0.15-0.25
4145	G41450	0.43-0.48	0.75-1.00	0.035	0.040	0.15-0.30	0.80-1.10	...	0.15-0.25

(continued)

Table 17. (continued)

AISI-SAE designation	UNS designation	Heat composition ranges and limits, %							
		C	Mn	P max(a)	S max(a)	Si	Cr	Ni	Mo
4147	G41470	0.45-0.50	0.75-1.00	0.035		0.15-0.30	0.80-1.10	...	0.15-0.25
4150	G41500	0.48-0.53	0.75-1.00	0.035	0.040	0.15-0.30	0.80-1.10	...	0.15-0.25
4161	G41610	0.56-0.64	0.75-1.00	0.035	0.040	0.15-0.30	0.70-0.90	...	0.25-0.35
4320	G43200	0.17-0.22	0.45-0.65	0.035	0.040	0.15-0.30	0.40-0.60	1.65-2.00	0.20-0.30
4340	G43400	0.38-0.43	0.60-0.80	0.035	0.040	0.15-0.30	0.70-0.90	1.65-2.00	0.20-0.30
E4340(d)	G43406	0.38-0.43	0.65-0.85	0.025	0.025	0.15-0.30	0.70-0.90	1.65-2.00	0.20-0.30
4419(c)	G44190	0.18-0.23	0.45-0.65	0.035	0.040	0.15-0.30	...	...	0.45-0.60
4422(c)	G44220	0.20-0.25	0.70-0.90	0.035	0.040	0.15-0.30	...	...	0.35-0.45
4427(c)	G44270	0.24-0.29	0.70-0.90	0.035	0.040	0.15-0.30	...	...	0.35-0.45
4615	G46150	0.13-0.18	0.45-0.65	0.035	0.040	0.15-0.30	...	1.65-2.00	0.20-0.30
4617(c)	G46170	0.15-0.20	0.45-0.65	0.035	0.040	0.15-0.30	...	1.65-2.00	0.20-0.30
4620	G46200	0.17-0.22	0.45-0.65	0.035	0.040	0.15-0.30	...	1.65-2.00	0.20-0.30
4621(c)	G46210	0.18-0.23	0.70-0.90	0.035	0.040	0.15-0.30	...	1.65-2.00	0.20-0.30
4626	G46260	0.24-0.29	0.45-0.65	0.035	0.04	0.15-0.30	...	0.70-1.00	0.15-0.25
4718(c)	G47180	0.16-0.21	0.70-0.90	...	...	...	0.35-0.55	0.90-1.20	0.30-0.40
4720	G47200	0.17-0.22	0.50-0.70	0.035	0.040	0.15-0.30	0.35-0.55	0.90-1.20	0.15-0.25
4815	G48150	0.13-0.18	0.40-0.60	0.035	0.040	0.15-0.30	...	3.25-3.75	0.20-0.30
4817	G48170	0.15-0.20	0.40-0.60	0.035	0.040	0.15-0.30	...	3.25-3.75	0.20-0.30
4820	G48200	0.18-0.23	0.50-0.70	0.035	0.040	0.15-0.30	...	3.25-3.75	0.20-0.30
5015(e)	G50150	0.12-0.17	0.30-0.50	0.035	0.040	0.15-0.30	0.30-0.50	...	...
50B40(c)(e)	G50401	0.38-0.43	0.75-1.00	0.035	0.040	0.15-0.30	0.40-0.60	...	...
50B44(e)	G50441	0.43-0.48	0.75-1.00	0.035	0.040	0.15-0.30	0.40-0.60	...	...
5046(c)	G50460	0.43-0.48	0.75-1.00	0.035	0.040	0.15-0.30	0.20-0.35	...	...
50B46(e)	G50461	0.44-0.49	0.75-1.00	0.035	0.040	0.15-0.30	0.20-0.35	...	...
50B50(e)	G50501	0.48-0.53	0.75-1.00	0.035	0.040	0.15-0.30	0.40-0.60	...	...
5060(c)	G50600	0.56-0.64	0.75-1.00	0.035	0.040	0.15-0.30	0.40-0.60	...	...
50B60(e)	...	0.56-0.64	0.75-1.00	0.035	0.040	0.15-0.30	0.40-0.60	...	...
5115(c)	G51150	0.13-0.18	0.70-0.90	0.035	0.040	0.15-0.30	0.70-0.90	...	...
5117(f)	G51170	0.15-0.20	0.70-0.90	0.035	0.040	0.15-0.30	0.70-0.90	...	...
5120	G51200	0.17-0.22	0.70-0.90	0.035	0.040	0.15-0.30	0.70-0.90	...	...
5130	G51300	0.28-0.33	0.70-0.90	0.035	0.040	0.15-0.30	0.80-1.10	...	...
5132	G51320	0.30-0.35	0.60-0.80	0.035	0.040	0.15-0.30	0.75-1.00	...	...
5135	G51350	0.33-0.38	0.60-0.80	0.035	0.040	0.15-0.30	0.80-1.05	...	...
5140	G51400	0.38-0.43	0.70-0.90	0.035	0.040	0.15-0.30	0.70-0.90	...	...
5145(c)	G51450	0.43-0.48	0.70-0.90	0.035	0.040	0.15-0.30	0.70-0.90	...	...
5147(c)	G51470	0.46-0.51	0.70-0.95	0.035	0.040	0.15-0.30	0.85-1.15	...	...
5150	G51500	0.48-0.53	0.70-0.90	0.035	0.040	0.15-0.30	0.70-0.90	...	...
5155	G51550	0.51-0.59	0.70-0.90	0.035	0.040	0.15-0.30	0.70-0.90	...	...
5160	G51600	0.56-0.64	0.75-1.00	0.035	0.040	0.15-0.30	0.70-0.90	...	...
51B60(e)	G51601	0.56-0.64	0.75-1.00	0.035	0.040	0.15-0.30	0.70-0.90	...	...
50100	G50986	0.98-1.10	0.25-0.45	0.025	0.025	0.15-0.30	0.40-0.60	...	...
51100	G51986	0.98-1.10	0.25-0.45	0.025	0.025	0.15-0.30	0.90-1.15	...	...
52100	G52986	0.98-1.10	0.25-0.45	0.025	0.025	0.15-0.30	1.30-1.60	...	...
6118(g)	G61180	0.16-0.21	0.50-0.70	0.035	0.040	0.15-0.30	0.50-0.70	...	...
6150(h)	G61500	0.48-0.53	0.70-0.90	0.035	0.040	0.15-0.30	0.80-1.10	...	...
8115(c)	G81150	0.13-0.18	0.70-0.90	0.035	0.040	0.15-0.30	0.30-0.50	0.20-0.40	0.08-0.15
81B45(e)	G81451	0.43-0.48	0.75-1.00	0.035	0.040	0.15-0.30	0.35-0.55	0.20-0.40	0.08-0.15
8615	G86150	0.13-0.18	0.70-0.90	0.035	0.040	0.15-0.30	0.40-0.60	0.40-0.70	0.15-0.25
8617	G86170	0.15-0.20	0.70-0.90	0.035	0.040	0.15-0.30	0.40-0.60	0.40-0.70	0.15-0.25
8620	G86200	0.18-0.23	0.70-0.90	0.035	0.040	0.15-0.30	0.40-0.60	0.40-0.70	0.15-0.25
8622	G86220	0.20-0.25	0.70-0.90	0.035	0.040	0.15-0.30	0.40-0.60	0.40-0.70	0.15-0.25
8625	G86250	0.23-0.28	0.70-0.90	0.035	0.040	0.15-0.30	0.40-0.60	0.40-0.70	0.15-0.25
8627	G86270	0.25-0.30	0.70-0.90	0.035	0.040	0.15-0.30	0.40-0.60	0.40-0.70	0.15-0.25
8630	G86300	0.28-0.33	0.70-0.90	0.035	0.040	0.15-0.30	0.40-0.60	0.40-0.70	0.15-0.25
8637	G86370	0.35-0.40	0.75-1.00	0.035	0.040	0.15-0.30	0.40-0.60	0.40-0.70	0.15-0.25
8640	G86400	0.38-0.43	0.75-1.00	0.035	0.040	0.15-0.30	0.40-0.60	0.40-0.70	0.15-0.25
8642	G86420	0.40-0.45	0.75-1.00	0.035	0.040	0.15-0.30	0.40-0.60	0.40-0.70	0.15-0.25
8645	G86450	0.43-0.48	0.75-1.00	0.035	0.040	0.15-0.30	0.40-0.60	0.40-0.70	0.15-0.25
86B45(c)(e)	G86451	0.43-0.48	0.75-1.00	0.035	0.040	0.15-0.30	0.40-0.60	0.40-0.70	0.15-0.25
8650(c)	G86500	0.48-0.53	0.75-1.00	0.035	0.040	0.15-0.30	0.40-0.60	0.40-0.70	0.15-0.25
8655	G86550	0.51-0.59	0.75-1.00	0.035	0.040	0.15-0.30	0.40-0.60	0.40-0.70	0.15-0.25
8660(c)	G86600	0.56-0.64	0.75-1.00	0.035	0.040	0.15-0.30	0.40-0.60	0.40-0.70	0.15-0.25
8720	G87200	0.18-0.23	0.70-0.90	0.035	0.040	0.15-0.30	0.40-0.60	0.40-0.70	0.20-0.30
8740	G87400	0.38-0.43	0.75-1.00	0.035	0.040	0.15-0.30	0.40-0.60	0.40-0.70	0.20-0.30
8822	G88220	0.20-0.25	0.75-1.00	0.035	0.040	0.15-0.30	0.40-0.60	0.40-0.70	0.30-0.40
9254(c)	G92540	0.51-0.59	0.60-0.80	0.035	0.040	1.20-1.60	0.60-0.80	...	...
9255(c)	G92550	0.51-0.59	0.70-0.95	0.035	0.040	1.80-2.20	...	...	...
9260	G92600	0.56-0.64	0.75-1.00	0.035	0.040	1.80-2.20	...	...	...
9310(c)	G93106	0.08-0.13	0.45-0.65	0.025	0.025	0.15-0.30	1.00-1.40	3.00-3.50	0.08-0.15
94B15(c)(e)	G94151	0.13-0.18	0.75-1.00	0.035	0.040	0.15-0.30	0.30-0.50	0.30-0.60	0.08-0.15
94B17(c)	G94171	0.15-0.20	0.75-1.00	0.035	0.040	0.15-0.30	0.30-0.50	0.30-0.60	0.08-0.15
94B30(e)	G94301	0.28-0.33	0.75-1.00	0.035	0.040	0.15-0.30	0.30-0.50	0.30-0.60	0.08-0.15

(a) Limits for phosphorus and sulfur are for steel made by open hearth or basic oxygen processes; limits for steels made by other processes are given in Table 6. (b) A range of sulfur content normally indicates a resulfurized steel. (c) SAE standard grade only. (d) Prefix "E" indicates that the steel is made by electric furnace process. (e) Can be expected to contain 0.0005 to 0.003% boron. (f) AISI standard grade only. (g) Contains 0.10 to 0.15% vanadium. (h) Contains 0.15% min vanadium.

Table 18. Composition ranges and limits for AISI-SAE standard alloy steels—plates

AISI-SAE designation	UNS designation	C	Mn	Si(b)	Cr	Ni	Mo
				Heat composition ranges and limits(a), %			
1330	G13300	0.27-0.34	1.50-1.90	0.15-0.30	...	...	...
1335	G13350	0.32-0.39	1.50-1.90	0.15-0.30	...	...	...
1340	G13400	0.36-0.44	1.50-1.90	0.15-0.30	...	...	...
1345	G13450	0.41-0.49	1.50-1.90	0.15-0.30	...	...	...
4118	G41180	0.17-0.23	0.60-0.90	0.15-0.30	0.40-0.65	...	0.08-0.15
4130	G41300	0.27-0.34	0.35-0.60	0.15-0.30	0.80-1.15	...	0.15-0.25
4135	G41350	0.32-0.39	0.65-0.95	0.15-0.30	0.08-1.15	...	0.15-0.25
4137	G41370	0.33-0.40	0.65-0.95	0.15-0.30	0.80-1.15	...	0.15-0.25
4140	G41400	0.36-0.44	0.70-1.00	0.15-0.30	0.08-1.15	...	0.15-0.25
4142	G41420	0.38-0.46	0.70-1.00	0.15-0.30	0.80-1.15	...	0.15-0.25
4145	G41450	0.41-0.49	0.70-1.00	0.15-0.30	0.80-1.15	...	0.15-0.25
4340	G43400	0.36-0.44	0.55-0.80	0.15-0.30	0.60-0.90	1.65-2.00	0.20-0.30
E4340(c)	G43406	0.37-0.44	0.60-0.85	0.15-0.30	0.65-0.90	1.65-2.00	0.20-0.30
4615	G46150	0.12-0.18	0.40-0.65	0.15-0.30	...	1.65-2.00	0.20-0.30
4617	G46170	0.15-0.21	0.40-0.65	0.15-0.30	...	1.65-2.00	0.20-0.30
4620	G46200	0.16-0.22	0.40-0.65	0.15-0.30	...	1.65-2.00	0.20-0.30
5160	G51600	0.54-0.65	0.70-1.00	0.15-0.30	0.60-0.90	...	...
6150(d)	G61500	0.46-0.54	0.60-0.90	0.15-0.30	0.80-1.15	...	...
8615	G86150	0.12-0.18	0.60-0.90	0.15-0.30	0.35-0.60	0.40-0.70	0.15-0.25
8617	G86170	0.15-0.21	0.60-0.90	0.15-0.30	0.35-0.60	0.40-0.70	0.15-0.25
8620	G86200	0.17-0.23	0.60-0.90	0.15-0.30	0.35-0.60	0.40-0.70	0.15-0.25
8622	G86220	0.19-0.25	0.60-0.90	0.15-0.30	0.35-0.60	0.40-0.70	0.15-0.25
8625	G86250	0.22-0.29	0.60-0.90	0.15-0.30	0.35-0.60	0.40-0.70	0.15-0.25
8627	G86270	0.24-0.31	0.60-0.90	0.15-0.30	0.35-0.60	0.40-0.70	0.15-0.25
8630	G86300	0.27-0.34	0.60-0.90	0.15-0.30	0.35-0.60	0.40-0.70	0.15-0.25
8637	G86370	0.33-0.40	0.70-1.00	0.15-0.30	0.35-0.60	0.40-0.70	0.15-0.25
8640	G86400	0.36-0.44	0.70-1.00	0.15-0.30	0.35-0.60	0.40-0.70	0.15-0.25
8655	G86550	0.49-0.60	0.70-1.00	0.15-0.30	0.35-0.60	0.40-0.70	0.15-0.25
8742	G87420	0.38-0.46	0.70-1.00	0.15-0.30	0.35-0.60	0.40-0.70	0.20-0.30

(a) Indicated ranges and limits apply to steels made by open hearth or basic oxygen processes; maximum content for phosphorus is 0.035% and for sulfur 0.040%. For steels made by electric furnace process, the ranges and limits are reduced as follows: C—0.01%; Mn—0.05%; Cr—0.05% (under 1.25%), 0.10% (over 1.25%); maximum content for either phosphorus or sulfur is 0.025%. (b) Other silicon ranges may be negotiated. (c) Prefix "E" indicates that the steel is made by electric furnace process. (d) Contains 0.15% minimum vanadium.

Table 19. Composition ranges and limits for AISI-SAE standard alloy H-steels

AISI-SAE designation	UNS designation	C	Mn	Si	Cr	Ni	Mo
				Heat composition ranges and limits(a), %			
1330H	H13300	0.27-0.33	1.45-2.05	0.15-0.30	...	...	...
1335H	H13350	0.32-0.38	1.45-2.05	0.15-0.30	...	...	...
1340H	H13400	0.37-0.44	1.45-2.05	0.15-0.30	...	...	...
1345H	H13450	0.42-0.49	1.45-2.05	0.15-0.30	...	...	...
4027H	H40270	0.24-0.30	0.60-1.00	0.15-0.30	...	...	0.20-0.30
4028H(b)	H40280	0.24-0.30	0.60-1.00	0.15-0.30	...	...	0.20-0.30
4032H	H40320	0.29-0.35	0.60-1.00	0.15-0.30	...	...	0.20-0.30
4037H	H40370	0.34-0.41	0.60-1.00	0.15-0.30	...	...	0.20-0.30
4042H	H40420	0.39-0.46	0.60-1.00	0.15-0.30	...	...	0.20-0.30
4047H	H40470	0.44-0.51	0.60-1.00	0.15-0.30	...	...	0.20-0.30
4118H	H41180	0.17-0.23	0.60-1.00	0.15-0.30	0.30-0.70	...	0.08-0.15
4130H	H41300	0.27-0.33	0.30-0.70	0.15-0.30	0.75-1.20	...	0.15-0.25
4135H	H41350	0.32-0.38	0.60-1.00	0.15-0.30	0.75-1.20	...	0.15-0.25
4137H	H41370	0.34-0.41	0.60-1.00	0.15-0.30	0.75-1.20	...	0.15-0.25
4140H	H41400	0.37-0.44	0.65-1.10	0.15-0.30	0.75-1.20	...	0.15-0.25
4142H	H41420	0.39-0.46	0.65-1.10	0.15-0.30	0.75-1.20	...	0.15-0.25
4145H	H41450	0.42-0.49	0.65-1.10	0.15-0.30	0.75-1.20	...	0.15-0.25
4147H	H41470	0.44-0.51	0.65-1.10	0.15-0.30	0.75-1.20	...	0.15-0.25
4150H	H41500	0.47-0.54	0.65-1.10	0.15-0.30	0.75-1.20	...	0.15-0.25
4161H	H41610	0.55-0.65	0.65-1.10	0.15-0.30	0.65-0.95	...	0.25-0.35
4320H	H43200	0.17-0.23	0.40-0.70	0.15-0.30	0.35-0.65	1.55-2.00	0.20-0.30
4340H	H43400	0.37-0.44	0.55-0.90	0.15-0.30	0.65-0.95	1.55-2.00	0.20-0.30
E4340H(b)	H43406	0.37-0.44	0.60-0.95	0.15-0.30	0.65-0.95	1.55-2.00	0.20-0.30
4419H(c)	H44190	0.17-0.23	0.35-0.75	0.15-0.30	...	...	0.45-0.60
4620H	H46200	0.17-0.23	0.35-0.75	0.15-0.30	...	1.55-2.00	0.20-0.30
4621H(c)	H46210	0.17-0.23	0.60-1.00	0.15-0.30	...	1.55-2.00	0.20-0.30
4626H(d)	H46260	0.23-0.29	0.40-0.70	0.15-0.30	...	0.65-1.05	0.15-0.25
4718H(c)	H47180	0.15-0.21	0.60-0.95	0.15-0.30	0.30-0.60	0.85-1.25	0.30-0.40
4720H	H47200	0.17-0.23	0.45-0.75	0.15-0.30	0.30-0.60	0.85-1.25	0.15-0.25
4815H	H48150	0.12-0.18	0.30-0.70	0.15-0.30	...	3.20-3.80	0.20-0.30
4817H	H48170	0.14-0.20	0.30-0.70	0.15-0.30	...	3.20-3.80	0.20-0.30
4820H	H48200	0.17-0.23	0.40-0.80	0.15-0.30	...	3.20-3.80	0.20-0.30
50B40H(e)	H50401	0.37-0.44	0.65-1.10	0.15-0.30	0.30-0.70	...	...
50B44H(e)	H50441	0.42-0.49	0.65-1.10	0.15-0.30	0.30-0.70	...	...
5046H	H50460	0.43-0.50	0.65-1.10	0.15-0.30	0.13-0.43	...	...

(continued)

Table 19. (continued)

AISI-SAE designation	UNS designation	C	Mn	Heat composition ranges and limits(a), % Si	Cr	Ni	Mo
50B46H(e)	H50461	0.43-0.50	0.65-1.10	0.15-0.30	0.13-0.43	...	...
50B50H(e)	H50501	0.47-0.54	0.65-1.10	0.15-0.30	0.30-0.70	...	...
50B60H(e)	H50601	0.55-0.65	0.65-1.10	0.15-0.30	0.30-0.70	...	...
5120H	H51200	0.17-0.23	0.60-1.00	0.15-0.30	0.60-1.00	...	...
5130H	H51300	0.27-0.33	0.60-1.10	0.15-0.30	0.75-1.20	...	...
5132H	H51320	0.29-0.35	0.50-0.90	0.15-0.30	0.65-1.10	...	...
5135H	H51350	0.32-0.38	0.50-0.90	0.15-0.30	0.70-1.15	...	...
5140H	H51400	0.37-0.44	0.60-1.00	0.15-0.30	0.60-1.00	...	...
5145H(c)	H51450	0.42-0.49	0.60-1.00	0.15-0.30	0.60-1.00	...	...
5147H(c)	H51470	0.45-0.52	0.60-1.05	0.15-0.30	0.80-1.25	...	...
5150H	H51500	0.47-0.54	0.60-1.00	0.15-0.30	0.60-1.00	...	...
5155H	H51550	0.50-0.60	0.60-1.00	0.15-0.30	0.60-1.00	...	...
5160H	H51600	0.55-0.65	0.65-1.10	0.15-0.30	0.60-1.00	...	...
51B60H(e)	H51601	0.55-0.65	0.65-1.10	0.15-0.30	0.60-1.00	...	...
6118H(f)	H61180	0.15-0.21	0.40-0.80	0.15-0.30	0.40-0.80	...	...
6150H(g)	H61500	0.47-0.54	0.60-1.00	0.15-0.30	0.75-1.20	...	...
81B45H(e)	H81451	0.42-0.49	0.70-1.05	0.15-0.30	0.30-0.60	0.15-0.45	0.08-0.15
8617H	H86170	0.14-0.20	0.60-0.95	0.15-0.30	0.35-0.65	0.35-0.75	0.15-0.25
8620H	H86200	0.17-0.23	0.60-0.95	0.15-0.30	0.35-0.65	0.35-0.75	0.15-0.25
8622H	H86220	0.19-0.25	0.60-0.95	0.15-0.30	0.35-0.65	0.35-0.75	0.15-0.25
8625H	H86250	0.22-0.28	0.60-0.95	0.15-0.30	0.35-0.65	0.35-0.75	0.15-0.25
8627H	H86270	0.24-0.30	0.60-0.95	0.15-0.30	0.35-0.65	0.35-0.75	0.15-0.25
8630H	H86300	0.27-0.33	0.60-0.95	0.15-0.30	0.35-0.65	0.35-0.75	0.15-0.25
86B30H(e)	H86301	0.27-0.33	0.60-0.95	0.15-0.30	0.35-0.65	0.35-0.75	0.15-0.25
8637H	H86370	0.34-0.41	0.70-1.05	0.15-0.30	0.35-0.65	0.35-0.75	0.15-0.25
8640H	H86400	0.37-0.44	0.70-1.05	0.15-0.30	0.35-0.65	0.35-0.75	0.15-0.25
8642H	H86420	0.39-0.46	0.70-1.05	0.15-0.30	0.35-0.65	0.35-0.75	0.15-0.25
8645H	H86450	0.42-0.49	0.70-1.05	0.15-0.30	0.35-0.65	0.35-0.75	0.15-0.25
86B45H(e)	H86451	0.42-0.49	0.70-1.05	0.15-0.30	0.35-0.65	0.35-0.75	0.15-0.25
8650H	H86500	0.47-0.54	0.70-1.05	0.15-0.30	0.35-0.65	0.35-0.70	0.15-0.25
8655H	H86550	0.50-0.60	0.70-1.05	0.15-0.30	0.35-0.65	0.35-0.75	0.15-0.25
8660H	H86600	0.55-0.65	0.70-1.05	0.15-0.30	0.35-0.65	0.35-0.75	0.15-0.25
8720H	H87200	0.17-0.23	0.60-0.95	0.15-0.30	0.35-0.65	0.35-0.75	0.20-0.30
8740H	H87400	0.37-0.44	0.70-1.05	0.15-0.30	0.35-0.65	0.35-0.75	0.20-0.30
8822H	H88220	0.19-0.25	0.70-1.05	0.15-0.30	0.35-0.65	0.35-0.75	0.30-0.40
9260H	H92600	0.55-0.65	0.65-1.10	1.70-2.20	...	...	...
9310H(b)	H93100	0.07-0.13	0.40-0.70	0.15-0.30	1.00-1.45	2.95-3.55	0.08-0.15
94B15H(e)	H94151	0.12-0.18	0.70-1.05	0.15-0.30	0.25-0.55	0.25-0.65	0.08-0.15
94B17H(e)	H94171	0.14-0.20	0.70-1.05	0.15-0.30	0.25-0.55	0.25-0.65	0.08-0.15
94B30H(e)	H94301	0.27-0.33	0.70-1.05	0.15-0.30	0.25-0.55	0.25-0.65	0.08-0.15

(a) Limits on phosphorus and sulfur contents are given in Table 6; typical limits are 0.035% maximum phosphorus and 0.040% maximum sulfur. (b) Electric furnace steel. (c) SAE standard grade only. (d) AISI standard grade only. (e) Can be expected to contain 0.0005 to 0.003% boron. (f) Contains 0.10 to 0.15% vanadium. (g) Contains 0.15% minimum vanadium.

Table 20. Composition ranges and limits for SAE experimental steels

SAE designation	C	Mn	P max	Heat composition ranges and limits, % S max	Si	Cr	Ni	Mo
EX1	0.15-0.21	0.35-0.60	0.040	0.040	0.20-0.35	...	4.80-5.30	0.20-0.30
EX 9	0.19-0.24	0.95-1.25	0.035	0.040	0.050 max	0.25-0.40	0.20-0.40	0.05-0.10
EX 10	0.19-0.24	0.95-1.25	0.035	0.040	0.20-0.35	0.25-0.40	0.20-0.40	0.05-0.10
EX 11(a)	0.38-0.43	0.75-1.00	0.035	0.040	0.050 max	0.25-0.40	0.20-0.40	0.05-0.10
EX 12(a)	0.38-0.43	0.75-1.00	0.035	0.040	0.20-0.35	0.25-0.40	0.20-0.40	0.05-0.10
EX 13	0.66-0.75	0.80-1.05	0.025	0.025	0.050 max	0.25-0.40	0.20-0.40	0.05-0.10
EX 14	0.66-0.75	0.80-1.05	0.025	0.025	0.20-0.35	0.25-0.40	0.20-0.40	0.05-0.10
EX 15	0.18-0.23	0.90-1.20	0.035	0.040	0.20-0.35	0.40-0.60	...	0.13-0.20
EX 16	0.20-0.25	0.90-1.20	0.035	0.040	0.20-0.35	0.40-0.60	...	0.13-0.20
EX 17	0.23-0.28	0.90-1.20	0.035	0.040	0.20-0.35	0.40-0.60	...	0.13-0.20
EX 18	0.25-0.30	0.90-1.20	0.035	0.040	0.20-0.35	0.40-0.60	...	0.13-0.20
EX 19(a)	0.18-0.23	0.90-1.20	0.035	0.040	0.20-0.35	0.40-0.60	...	0.08-0.15
EX 20	0.13-0.18	0.90-1.20	0.035	0.040	0.20-0.35	0.40-0.60	...	0.13-0.20
EX 21	0.15-0.20	0.90-1.20	0.035	0.040	0.20-0.35	0.40-0.60	...	0.13-0.20
EX 24	0.18-0.23	0.75-1.00	0.035	0.040	0.20-0.35	0.45-0.65	...	0.20-0.30
EX 27	0.25-0.30	0.75-1.00	0.035	0.040	0.20-0.35	0.45-0.65	...	0.20-0.30
EX 29	0.18-0.23	0.75-1.00	0.035	0.040	0.20-0.35	0.45-0.65	0.40-0.70	0.30-0.40
EX 30	0.13-0.18	0.70-0.90	0.035	0.040	0.20-0.35	0.45-0.65	0.70-1.00	0.45-0.60
EX 31	0.15-0.20	0.70-0.90	0.035	0.040	0.20-0.35	0.45-0.65	0.70-1.00	0.45-0.60
EX 32	0.18-0.23	0.70-0.90	0.035	0.040	0.20-0.35	0.45-0.65	0.70-1.00	0.45-0.60
EX 33	0.17-0.24	0.85-1.25	0.035	0.040	0.20-0.35	0.20 min	0.20 min	0.05 min
EX 34	0.28-0.33	0.90-1.20	...	...	0.20-0.35	0.40-0.60	...	0.13-0.20
EX 35	0.35-0.40	0.90-1.20	...	...	0.20-0.35	0.45-0.65	...	0.13-0.20
EX 36	0.38-0.43	0.90-1.20	...	...	0.20-0.35	0.45-0.65	...	0.13-0.20
EX 37	0.40-0.45	0.90-1.20	...	...	0.20-0.35	0.45-0.65	...	0.13-0.20

(continued)

Table 20. (continued)

SAE designation	C	Mn	P max	S max	Si	Cr	Ni	Mo
EX 38	0.43-0.48	0.90-1.20	...	...	0.20-0.35	0.45-0.65	...	0.13-0.20
EX 39	0.48-0.53	0.90-1.20	...	...	0.20-0.35	0.45-0.65	...	0.13-0.20
EX 40	0.51-0.59	0.90-1.20	...	...	0.20-0.35	0.45-0.65	...	0.13-0.20
EX 41	0.56-0.64	0.90-1.20	...	...	0.20-0.35	0.45-0.65	...	0.13-0.20
EX 42	0.13-0.18	0.95-1.25	0.035	0.040	0.20-0.35	0.25-0.40	0.20-0.40	0.05-0.10
EX 43(a)	0.13-0.18	0.95-1.25	0.035	0.040	0.20-0.35	0.25-0.40	0.20-0.40	0.05-0.10
EX 44	0.15-0.20	0.95-1.25	0.035	0.040	0.20-0.35	0.25-0.40	0.20-0.40	0.05-0.10
EX 45(a)	0.15-0.20	0.95-1.25	0.035	0.040	0.20-0.35	0.25-0.40	0.20-0.40	0.05-0.10
EX 46	0.20-0.25	0.95-1.25	0.035	0.040	0.20-0.35	0.25-0.40	0.20-0.40	0.05-0.10
EX 47	0.23-0.28	0.95-1.25	0.035	0.040	0.20-0.35	0.25-0.40	0.20-0.40	0.05-0.10
EX 48	0.25-0.30	0.95-1.25	0.035	0.040	0.20-0.35	0.25-0.40	0.20-0.40	0.05-0.10
EX 49	0.28-0.33	0.95-1.25	0.035	0.040	0.20-0.35	0.25-0.40	0.20-0.40	0.05-0.10
EX 50	0.33-0.38	0.95-1.25	0.035	0.040	0.20-0.35	0.25-0.40	0.20-0.40	0.05-0.10
EX 51	0.35-0.40	0.95-1.25	0.035	0.040	0.20-0.35	0.25-0.40	0.20-0.40	0.05-0.10
EX 52	0.38-0.43	0.91-1.25	0.035	0.040	0.20-0.35	0.25-0.40	0.20-0.40	0.05-0.10
EX 53	0.40-0.45	0.95-1.25	0.035	0.040	0.20-0.35	0.25-0.40	0.20-0.40	0.05-0.10
EX 54	0.19-0.25	0.70-1.05	0.035	0.040	0.35 max	0.40-0.70	...	0.05 min
EX 55	0.15-0.20	0.70-1.00	0.035	0.040	0.20-0.35	0.45-0.65	1.65-2.00	0.65-0.80
EX 56	0.08-0.13	0.70-1.00	0.035	0.040	0.20-0.35	0.45-0.65	1.65-2.00	0.65-0.80

(a) Can be expected to contain 0.0005 to 0.003% boron.

Table 21. Composition ranges and limits for SAE HSLA steels

SAE designation(b)	Heat composition limits, %(a)		
	C max	Mn max	P max
942X	0.21	1.35	0.04
945A	0.15	1.00	0.04
945C	0.23	1.40	0.04
945X	0.22	1.35	0.04
950A	0.15	1.30	0.04
950B	0.22	1.30	0.04
950C	0.25	1.60	0.04
950D	0.15	1.00	0.15
950X	0.23	1.35	0.04
955X	0.25	1.35	0.04
960X	0.26	1.45	0.04
965X	0.26	1.45	0.04
970X	0.26	1.65	0.04
980X	0.26	1.65	0.04

(a) Maximum contents of sulfur and silicon for all grades: 0.050% S, 0.90% Si. (b) Second and third digits of designation indicate minimum yield strength in ksi. Suffix "X" indicates that the steel contains niobium, vanadium, nitrogen or other alloying elements. A second suffix "K" indicates that the steel is produced fully killed using fine grain practice; otherwise, the steel is produced semikilled.

Table 22. Composition ranges and limits for carbon steels formerly listed by SAE

SAE designation	C	Mn	P	S	Year last listed
1009	0.15 max	0.60 max	0.040 max	0.050 max	1965
1033	0.30-0.36	0.70-1.00	0.040 max	0.050 max	1965
1034	0.32-0.38	0.50-0.80	0.040 max	0.050 max	1968
1059	0.55-0.65	0.50-0.80	0.040 max	0.050 max	1968
1062	0.54-0.65	0.85-1.15	0.040 max	0.050 max	1953
1111	0.13 max	0.60-0.90	0.07-0.12	0.10-0.15	1969
1112	0.13 max	0.70-1.00	0.07-0.12	0.16-0.23	1969
1113	0.13 max	0.70-1.00	0.07-0.12	0.24-0.33	1969
1114	0.10-0.16	1.00-1.30	0.040 max	0.08-0.13	1952
1115	0.13-0.18	0.60-0.90	0.040 max	0.08-0.13	1965
1120	0.18-0.23	0.70-1.00	0.040 max	0.08-0.13	1965
1126	0.23-0.29	0.70-1.00	0.040 max	0.08-0.13	1965
1138	0.34-0.40	0.70-1.00	0.040 max	0.08-0.13	1965

The chemical ranges for these "H-Steels" are more generous. Slightly wider ranges of composition apply to plates than to bar products because the large rectangular ingots used for making plates are more susceptible to segregation during solidification than the smaller square or round ingots used in making bars. Table 19 lists the compositions of alloy steels that have specific hardenability requirements, described in SAE Standard J407d. The suffix "H" is used to distinguish these steels from corresponding grades that have no hardenability requirement. Limits and ranges of

(or three) digits. The chemical compositions of AISI-SAE standard grades of alloy steel bar and plate products are given in Tables 17 and 18, respectively. Most of the alloy steels listed in Table 17 also are available as "H-Steels" — that is, they comply with established hardenability bands.

chemical compositions for all of these alloy steel products reflect the restrictions on heat and product analyses given in Tables 6 through 9. The designations in Tables 18 and 19 are both AISI and SAE designations unless otherwise indicated.

"EX-" alloy steels are listed in SAE Information Report J1081 and Table 20. These are experimental grades to which no regular AISI-SAE designations have been assigned. Some were developed to minimize the nickel content; others were devised to improve a particular attribute of

Table 23. Composition ranges and limits for alloy steels formerly listed by SAE

SAE designation	C	Mn	P max	S max	Si	Cr	Ni	Mo	Year last listed
1320	0.18-0.23	1.60-1.90	0.040	0.040	0.20-0.35	...	...	...	1956
2317	0.15-0.20	0.40-0.60	0.040	0.040	0.20-0.35	...	3.25-3.75	...	1956
2330	0.28-0.33	0.60-0.80	0.040	0.040	0.20-0.35	...	3.25-3.75	...	1953
2340	0.38-0.43	0.70-0.90	0.040	0.040	0.20-0.35	...	3.25-3.75	...	1953
2345	043-0.48	0.70-0.90	0.040	0.040	0.20-0.35	...	3.25-3.75	...	1952
2512	0.09-0.14	0.45-0.60	0.025	0.025	0.20-0.35	...	4.75-5.25	...	1953
2515	0.12-0.17	0.40-0.60	0.040	0.040	0.20-0.35	...	4.75-5.25	...	1956
2517	0.15-0.20	0.45-0.60	0.025	0.025	0.20-0.35	...	4.75-5.25	...	1959
3115	0.13-0.18	0.40-0.60	0.040	0.040	0.20-0.35	0.55-0.75	1.10-1.40	...	1953
3120	0.17-0.22	0.60-0.80	0.040	0.040	0.20-0.35	0.55-0.75	1.10-1.40	...	1956
3130	0.28-0.33	0.60-0.80	0.040	0.040	0.20-0.35	0.55-0.75	1.10-1.40	...	1956
3135	0.33-0.38	0.60-0.80	0.040	0.040	0.20-0.35	0.55-0.75	1.10-1.40	...	1960
X3140	0.38-0.43	0.70-0.90	0.040	0.040	0.20-0.35	0.70-0.90	1.10-1.40	...	1947
3140	0.38-0.43	0.70-0.90	0.040	0.040	0.20-0.35	0.55-0.75	1.10-1.40	...	1964
3145	0.43-0.48	0.70-0.90	0.040	0.040	0.20-0.35	0.70-0.90	1.10-1.40	...	1952

(continued)

Table 23. (continued)

SAE designation	C	Mn	P max	S max	Si	Cr	Ni	Mo	Year last listed
3150	0.48-0.53	0.70-0.90	0.040	0.040	0.20-0.35	0.70-0.90	1.10-1.40	...	1952
3215	0.10-0.20	0.30-0.60	0.040	0.050	0.15-0.30	0.90-1.25	1.50-2.00	...	1941
3220	0.15-0.25	0.30-0.60	0.040	0.050	0.15-0.30	0.90-1.25	1.50-2.00	...	1941
3230	0.25-0.35	0.30-0.60	0.040	0.050	0.15-0.30	0.90-1.25	1.50-2.00	...	1941
3240	0.35-0.45	0.30-0.60	0.040	0.040	0.15-0.30	0.90-1.25	1.50-2.00	...	1941
3245	0.40-0.50	0.30-0.60	0.040	0.040	0.15-0.30	0.90-1.25	1.50-2.00	...	1941
3250	0.45-0.55	0.30-0.60	0.040	0.040	0.15-0.30	0.90-1.25	1.50-2.00	...	1941
3310	0.08-0.13	0.45-0.60	0.025	0.025	0.20-0.35	1.40-1.75	3.25-3.75	...	1964
3312	0.08-0.13	0.45-0.60	0.025	0.025	0.20-0.35	1.40-1.75	3.25-3.75	...	1948
3316	0.14-0.19	0.45-0.60	0.025	0.025	0.20-0.35	1.40-1.75	3.25-3.75	...	1956
3325	0.20-0.30	0.30-0.60	0.040	0.050	0.15-0.30	1.25-1.75	3.25-3.75	...	1936
3335	0.30-0.40	0.30-0.60	0.040	0.050	0.15-0.30	1.25-1.75	3.25-3.75	...	1936
3340	0.35-0.45	0.30-0.60	0.040	0.050	0.15-0.30	1.25-1.75	3.25-3.75	...	1936
3415	0.10-0.20	0.30-0.60	0.040	0.050	0.15-0.30	0.60-0.95	2.75-3.25	...	1941
3435	0.30-0.40	0.30-0.60	0.040	0.050	0.15-0.30	0.60-0.95	2.75-3.25	...	1936
3450	0.45-0.55	0.30-0.60	0.040	0.050	0.15-0.30	0.60-0.95	2.75-3.25	...	1936
4053	0.50-0.56	0.75-1.00	0.040	0.040	0.20-0.35	...	...	0.20-0.30	1956
4063	0.60-0.67	0.75-1.00	0.040	0.040	0.20-0.35	...	...	0.20-0.30	1964
4068	0.63-0.70	0.75-1.00	0.040	0.040	0.20-0.35	...	...	0.20-0.30	1957
4119	0.17-0.22	0.70-0.90	0.040	0.040	0.20-0.35	0.40-0.60	...	0.20-0.30	1956
4125	0.23-0.28	0.70-0.90	0.040	0.040	0.20-0.35	0.40-0.60	...	0.20-0.30	1950
4317	0.15-0.20	0.45-0.65	0.040	0.040	0.20-0.35	0.40-0.60	1.65-2.00	0.20-0.30	1953
4337	0.35-0.40	0.60-0.80	0.040	0.040	0.20-0.35	0.70-0.90	1.65-2.00	0.20-0.30	1964
4608	0.06-0.11	0.25-0.45	0.040	0.040	0.025 max	...	1.40-1.75	0.15-0.25	1956
46B12(a)	0.10-0.15	0.45-0.65	0.040	0.040	0.20-0.35	...	1.65-2.00	0.20-0.30	1957
X4620	0.18-0.23	0.50-0.70	0.040	0.040	0.20-0.35	...	1.65-2.00	0.20-0.30	1956
4640	0.38-0.43	0.60-0.80	0.040	0.040	0.20-0.35	...	1.65-2.00	0.20-0.30	1952
4812	0.10-0.15	0.40-0.60	0.040	0.040	0.20-0.35	...	3.25-3.75	0.20-0.30	1956
5045	0.43-0.48	0.70-0.90	0.040	0.040	0.20-0.35	0.55-0.75	...	...	1953
5117	0.15-0.20	0.70-0.90	0.040	0.040	0.20-0.35	0.70-0.90	...	...	1956
5152	0.48-0.55	0.70-0.90	0.040	0.040	0.20-0.35	0.90-1.20	...	...	1956
6115(b)	0.10-0.20	0.30-0.60	0.040	0.050	0.15-0.30	0.80.1.10	...	...	1936
6117(c)	0.15-0.20	0.70-0.90	0.040	0.040	0.20-0.35	0.70-0.90	...	...	1956
6120(c)	0.17-0.22	0.70-0.90	0.040	0.040	0.20-0.35	0.70-0.90	...	...	1961
6125(b)	0.20-0.30	0.60-0.90	0.040	0.050	0.15-0.30	0.80-1.10	...	...	1936
6130(b)	0.25-0.35	0.60-0.90	0.040	0.050	0.15-0.30	0.80-1.10	...	...	1936
6135(b)	0.30-0.40	0.60-0.90	0.040	0.050	0.15-0.30	0.80-1.10	...	...	1941
6140(b)	0.35-0.45	0.60-0.90	0.040	0.050	0.15-0.30	0.80-1.10	...	...	1936
6145(b)	0.43-0.48	0.70-0.90	0.040	0.050	0.20-0.35	0.80-1.10	...	...	1956
6195(b)	0.90-1.05	0.20-0.45	0.030	0.035	0.15-0.30	0.80-1.10	...	...	1936
71360(d)	0.50-0.70	0.30 max	0.035	0.040	0.15-0.30	3.00-4.00	...	...	1936
71660(e)	0.50-0.70	0.30 max	0.035	0.040	0.15-0.30	3.00-4.00	...	...	1936
7260(f)	0.50-0.70	0.30 max	0.035	0.040	0.15-0.30	0.50-1.00	...	...	1936
8632	0.30-0.35	0.70-0.90	0.040	0.040	0.20-0.35	0.40-0.60	0.40-0.70	0.15-0.25	1951
8635	0.33-0.38	0.75-1.00	0.040	0.040	0.20-0.35	0.40-0.60	0.40-0.70	0.15-0.25	1956
8641	0.38-0.43	0.75-1.00	0.040	0.040-0.060	0.20-0.35	0.40-0.60	0.40-0.70	0.15-0.25	1956
8653	0.50-0.56	0.75-1.00	0.040	0.040	0.20-0.35	0.50-0.80	0.40-0.70	0.15-0.25	1956
8647	0.45-0.50	0.75-1.00	0.040	0.040	0.20-0.35	0.40-0.60	0.40-0.70	0.15-0.25	1948
8715	0.13-0.18	0.70-0.90	0.040	0.040	0.20-0.35	0.40-0.60	0.40-0.70	0.20-0.30	1956
8717	0.15-0.20	0.70-0.90	0.040	0.040	0.20-0.35	0.40-0.60	0.40-0.70	0.20-0.30	1956
8719	0.18-0.23	0.60-0.80	0.040	0.040	0.20-0.35	0.40-0.60	0.40-0.70	0.20-0.30	1952
8735	0.33-0.38	0.75-1.00	0.040	0.040	0.20-0.35	0.40-0.60	0.40-0.70	0.20-0.30	1952
8742	0.40-0.45	0.75-1.00	0.040	0.040	0.20-0.35	0.40-0.60	0.40-0.70	0.20-0.30	1964
8745	0.43-0.48	0.75-1.00	0.040	0.040	0.20-0.35	0.40-0.60	0.40-0.70	0.20-0.30	1953
8750	0.48-0.53	0.75-1.00	0.040	0.040	0.20-0.35	0.40-0.60	0.40-0.70	0.20-0.30	1956
9250	0.45-0.55	0.60-0.90	0.040	0.040	1.80-2.20	...	...	...	1941
9261	0.55-0.65	0.75-1.00	0.040	0.040	1.80-2.20	0.10-0.25	...	...	1956
9262	0.55-0.65	0.75-1.00	0.040	0.040	1.80-2.20	0.25-0.40	...	...	1961
9315	0.13-0.18	0.45-0.65	0.025	0.025	0.20-0.35	1.00-1.40	3.00-3.50	0.08-0.15	1959
9317	0.15-0.20	0.45-0.65	0.025	0.025	0.20-0.35	1.00-1.40	3.00-3.50	0.08-0.15	1959
9437	0.35-0.40	0.90-1.20	0.040	0.040	0.20-0.35	0.30-0.50	0.30-0.60	0.08-0.15	1950
9440	0.38-0.43	0.90-1.20	0.040	0.040	0.20-0.35	0.30-0.50	0.30-0.60	0.08-0.15	1950
94B40(a)	0.38-0.43	0.75-1.00	0.040	0.040	0.20-0.35	0.30-0.50	0.30-0.60	0.08-0.15	1964
9442	0.40-0.45	0.90-1.20	0.040	0.040	0.20-0.35	0.30-0.50	0.30-0.60	0.08-0.15	1950
9445	0.43-0.48	0.90-1.20	0.040	0.040	0.20-0.35	0.30-0.50	0.30-0.60	0.08-0.15	1950
9447	0.45-0.50	0.90-1.20	0.040	0.040	0.20-0.35	0.30-0.50	0.30-0.60	0.08-0.15	1950
9747	0.45-0.50	0.50-0.80	0.040	0.040	0.20-0.35	0.10-0.25	0.40-0.70	0.15-0.25	1950
9763	0.60-0.67	0.50-0.80	0.040	0.040	0.20-0.35	0.10-0.25	0.40-0.70	0.15-0.25	1950
9840	0.38-0.43	0.70-0.90	0.040	0.040	0.20-0.35	0.70-0.90	0.85-1.15	0.20-0.30	1964
9845	0.43-0.48	0.70-0.90	0.040	0.040	0.20-0.35	0.70-0.90	0.85-1.15	0.20-0.30	1950
9850	0.48-0.53	0.70-0.90	0.040	0.040	0.20-0.35	0.70-0.90	0.85-1.15	0.20-0.30	1961
43BV12(a)(g)	0.08-0.13	0.75-1.00	...	...	0.20-0.35	0.40-0.60	1.65-2.00	0.20-0.30	...
43BV14(a)(g)	0.10-0.15	0.45-0.65	...	...	0.20-0.35	0.40-0.60	1.65-2.00	0.08-0.15	...

(a) Can be expected to contain 0.0005 to 0.003% boron. (b) Contains 0.15% minimum vanadium. (c) Contains 0.10% minimum vanadium. (d) Contains 12.00 to 15.00% tungsten. (e) Contains 15.00 to 18.00% tungsten. (f) Contains 1.50 to 2.00% tungsten. (g) Contains 0.03% minimum vanadium.

Table 24. ASTM specifications that incorporate AISI-SAE designations

A29 . . . . .Carbon and alloy steel bars, hot rolled and cold finished, generic
A108 . . . .Standard-quality cold finished carbon steel bars
A295 . . . .High carbon-chromium ball and roller bearing steel
A304 . . . .Alloy steel bars having hardenability requirements
A322 . . . .Hot rolled alloy steel bars
A331 . . . .Cold finished alloy steel bars
A434 . . . .Hot rolled or cold finished quenched-and-tempered alloy steel bars
A505 . . . .Hot rolled and cold rolled alloy steel sheet and strip, generic
A506 . . . .Regular-quality hot rolled and cold rolled alloy steel sheet and strip
A507 . . . .Drawing quality hot rolled and cold rolled alloy steel sheet and strip
A510 . . . .Carbon steel wire rods and coarse round wire, generic
A534 . . . .Carburizing steels for antifriction bearings
A535 . . . .Special-quality ball and roller bearing steel
A544 . . . .Scrapless nut quality carbon steel wire
A545 . . . .Cold heading quality carbon steel wire for machine screws
A546 . . . .Cold heading quality medium-high-carbon steel wire for hexagon-head bolts
A547 . . . .Cold heading quality alloy steel wire for hexagon head bolts
A548 . . . .Cold heading quality carbon steel wire for tapping or sheet metal screws
A549 . . . .Cold heading quality carbon steel wire for wood screws
A575 . . . .Merchant-quality hot rolled carbon steel bars
A576 . . . .Special-quality hot rolled carbon steel bars
A634 . . . .Aircraft-quality hot rolled and cold rolled alloy steel sheet and strip
A646 . . . .Premium-quality alloy steel blooms and billets for aircraft and aerospace forgings
A659 . . . .Commercial-quality hot rolled carbon steel sheet and strip
A680 . . . .Untempered spring quality cold rolled hard carbon steel strip
A682 . . . .Cold rolled spring quality carbon steel strip, generic
A684 . . . .Untempered spring quality cold rolled soft carbon steel strip
A689 . . . .Carbon and alloy steel bars for springs
A711 . . . .Carbon and alloy steel blooms, billets and slabs for forging
A713 . . . .High-carbon spring steel wire for heat treated components

Table 25. Generic ASTM specifications

A6 . . . . . .Rolled steel structural plate, shapes, sheet piling and bars, generic
A20 . . . . .Steel plate for pressure vessels, generic
A29 . . . . .Carbon and alloy steel bars, hot rolled and cold finished, generic
A505 . . . .Alloy steel sheet and strip, hot rolled and cold rolled, generic
A510 . . . .Carbon steel wire rod and coarse round wire, generic
A568 . . . .Carbon and HSLA, hot rolled and cold rolled steel sheet and hot rolled strip, generic
A646 . . . .Premium-quality alloy steel blooms and billets for aircraft and aerospace forgings
A711 . . . .Carbon and alloy steel blooms, billets and slabs for forging

Table 27. Composition ranges and limits for carbon steel structural shapes and plate (ASTM specifications)

ASTM speci-fication	Form, type or grade	UNS designation	C max	Mn	Si	Cu(b)
A36	Plate	. . . . . .	0.29	0.80-1.20	. . .	0.20
	Shapes	K02600	0.26	(c)	(d)	0.20
	Bars	. . . . . .	0.29	0.60-0.90	. . .	0.20
A283	Plate	. . . .	. . .	. . .	. . .	0.20
A284	Grade A	K01804 . .	0.24	0.90 max	0.10-0.30	. . .
	Grade B	K02001 . .	0.24	0.90 max	0.15-0.30	. . .
	Grade C	K02401 . .	0.36	0.90 max	0.15-0.30	. . .
	Grade D	K02702 . .	0.35	0.90 max	0.15-0.30	. . .
A529	Plate, bars and shapes	K02703 . .	0.27	1.20 max	. . .	0.20
A573	Grade 58	K02301 . .	0.23	0.60-0.90	0.10-0.35	. . .
	Grade 65	K02404 . .	0.26	0.85-1.20	0.15-0.30	. . .
	Grade 70	K02701 . .	0.28	0.85-1.20	0.15-0.30	. . .
A678	Grade A	K01600 . .	0.16	0.90-1.50	0.15-0.50	0.20
	Grade B	K02002 . .	0.20	0.70-1.60	0.15-0.50	0.20
	Grade C	K02204 . .	0.22	1.00-1.60	0.20-0.50	0.20

(a) Limits on phosphorus and sulfur contents are given in Table 4; typical limits are 0.040% maximum phosphorus and 0.050% maximum sulfur. (b) Minimum copper content applicable only if copper-bearing steel is specified. (c) 0.85-1.35% manganese required for shapes heavier than 634 kg/m (426 lb/ft). (d) 0.15-0.30% silicon required for shapes heavier than 634 kg/m (426 lb/ft).

ASTM (ASME) DESIGNATIONS

As noted previously, the most widely used standard specifications for steel products are those published by ASTM. These are complete specifications, generally adequate for procurement purposes. Many ASTM specifications apply to specific products, such as A574, for alloy steel socket head cap screws. These specifications are generally oriented toward performance of the fabricated end product, with considerable latitude in chemical composition of the steel used to make the end product.

ASTM specifications represent a consensus among producers, specifiers, fabricators and users of steel mill products. In many cases, the dimensions, tolerances, limits and restrictions in the ASTM specifications are the same as the corresponding items of the standard practices in the AISI Steel Products Manuals. Many of the ASTM specifications have been adopted by the Ameri-

a standard grade of alloy steel.

HSLA Steels. Several grades of HSLA steel are described in SAE Recommended Practice J410c; their chemical compositions are listed in Table 21. These steels have been developed as a compromise between the convenient fabrication characteristics and low cost of plain carbon steels and the high strength of heat treated alloy steels. These steels have excellent strength and ductility as-rolled.

Formerly Listed SAE Steels. A number of grades of carbon and alloy steels have been deleted from the list of SAE standard steels due to lack of use. For the convenience of those who might encounter an application for one of these grades, they are listed in Tables 22 and 23.

Table 26. Composition ranges and limits for sheet and strip, plain carbon and HSLA grades (ASTM specifications)

ASTM speci-fication	Description(a)	C max	Mn max	P max	S max	Other
A611	CRSQ:					
	Grades A, B, C	0.20	0.60	0.04	0.04	(b)
	Grade E	0.20	0.90	0.04	0.04	(b)
A366	CRCQ	0.15	0.60	0.035	0.04	(b)
A109	CR strip:					
	Tempers 1, 2, 3	0.25	0.60	0.035	0.04	(b)
	Tempers 4, 5	0.15	0.60	0.035	0.04	(b)
A619	CRDQ	0.10	0.50	0.025	0.035	(b)
A620	CR DQSK	0.10	0.50	0.025	0.035	(c)
A570	HR SQ:					
	Grades A, B, C	0.25	0.25-0.60	0.04	0.04	(b)
	Grades D, E	0.25	0.60-0.90	0.04	0.04	(b)
A569	HR CQ	0.15	0.60	0.035	0.04	(b)
A621	HR DQ	0.10	0.50	0.025	0.035	. . .
A622	HR DQSK	0.10	0.50	0.025	0.035	(c)
A414	Pressure vessel:					
	Grade A	0.15	0.90	0.035	0.04	(b)
	Grade B	0.22	0.90	0.035	0.04	(b)
	Grade C	0.25	0.90	0.035	0.04	(b)
	Grade D	0.25	1.20	0.035	0.04	(b)
	Grade E	0.27	1.20	0.035	0.04	(b)

ASTM speci-fication	Description	C max	Mn max	P max	S max	Other
A414 (continued):						
	Grade F	0.31	1.20	0.035	0.04	(b)
	Grade G	0.31	1.35	0.035	0.04	(b)
A606	HSLA	0.22	1.25	. . .	0.05	(d)
A607	Grade 45	0.22	1.35	0.04	0.05	(e)
	Grade 50	0.23	1.35	0.04	0.05	(e)
	Grade 55	0.25	1.35	0.04	0.05	(e)
	Grade 60	0.26	1.50	0.04	0.05	(e)
	Grade 65	0.26	1.50	0.04	0.05	(e)
	Grade 70	0.26	1.65	0.04	0.05	0.012 max N(e)
A715	Basic composition	0.15	1.65	0.025	0.035	0.012 max N

Type 1: 0.05 min Ti, 0.10 max Si(f)
Type 2: 0.02 min V, 0.60 max Si(g), 0.005 min N(f)(g)
Type 3: 0.005 min Nb, 0.08 max V(g), 0.60 max Si(g), 0.020 max N(f)(g)
Type 4: 0.05 min Zr, 0.90 max Si, 0.80 max Cr(g), 0.10 max Ti(g), 0.0025 max B(g), 0.005-0.06 Nb(f)(h)
Type 5: 0.03 min Nb(j), 0.20 min Mo(j), 0.30 max Si(f)
Type 6: 0.005-0.10 Nb, 0.90 max Si(f)
Type 7: 0.005 min Nb or V, or both, 0.60 max Si, 0.020 max N(f)

(a) CR cold rolled; SQ, structural quality; DQ, drawing quality; DQSK, drawing quality special killed. (b) Cu when specified as Cu-bearing steel: 0.20% min. (c) Aluminum as deoxidizer usually exceeds 0.010% in the product. (d) Other elements may be added if necessary to meet mechanical and corrosion requirements. (e) 0.005 min Nb or 0.01 min V for all grades. (f) These elements are added to basic composition. (g) Not added to grades 50 and 60. (h) Might not be added to grade 50. (j) Available as grade 80 only.

Table 28. Composition ranges and limits for HSLA and alloy steel plate (ASTM specifications)

ASTM specification	Type or grade	UNS designation	C	Mn	P max	S max	Si	Cr	Ni	Mo	V	Other
A242	Type 1	K11510	0.15 max	1.00 max	0.45	0.05	...	...	...	...	...	0.20 min Cu
	Type 2	K12010	0.20 max	1.35 max	0.04	0.05	...	...	...	...	...	0.20 min Cu if both 0.5 Si and 0.5 Cr not present
A440	...	K12810	0.28 max	1.10-1.60	0.04	0.05	0.30 max	...	...	...	...	0.20 min Cu
A441	...	K12211	0.22 max	0.85-1.25	0.04	0.05	0.30 max	...	...	...	...	0.20 min Cu; 0.02 min V
A514	Type A	K11856	0.15-0.21	0.80-1.10	0.035	0.04	0.40-0.80	0.50-0.80	...	0.18-0.28	...	0.05-0.15 Zr; 0.0025 max B
	Type B	K11630	0.12-0.21	0.70-1.00	0.035	0.04	0.20-0.35	0.40-0.65	...	0.15-0.25	0.03-0.08	0.01-0.03 Ti; 0.0005-0.005 B
	Type C	K11511	0.10-0.20	1.10-1.50	0.035	0.04	0.15-0.30	...	...	0.20-0.30	...	0.001-0.005 B
	Type D	K11662	0.13-0.20	0.40-0.70	0.035	0.04	0.20-0.35	0.85-1.20	...	0.15-0.25	...	0.04-0.10 Ti; 0.20-0.40 Cu; 0.0015-0.005 B
	Type E	K21604	0.12-0.20	0.40-0.70	0.035	0.04	0.20-0.35	1.40-2.00	...	0.40-0.60	...	0.04-0.10 Ti; 0.20-0.40 Cu; 0.0015-0.005 B
	Type F	K11576	0.10-0.20	0.60-1.00	0.035	0.04	0.15-0.35	0.40-0.65	0.70-1.00	0.40-0.60	0.03-0.08	0.15-0.50 Cu; 0.0005-0.006 B
	Type G	K11872	0.15-0.21	0.80-1.10	0.035	0.04	0.50-0.90	0.50-0.90	...	0.40-0.60	...	0.05-0.15 Zr; 0.0025 max B
	Type H	K11646	0.12-0.21	0.95-1.30	0.035	0.04	0.20-0.35	0.40-0.65	0.30-0.70	0.20-0.30	0.03-0.08	0.0005-0.005 B
	Type J	K11625	0.12-0.21	0.45-0.70	0.035	0.04	0.20-0.35	...	...	0.50-0.65	...	0.001-0.005 B
	Type K	K11523	0.10-0.20	1.10-1.50	0.035	0.04	0.15-0.30	...	...	0.45-0.55	...	0.001-0.005 B
	Type L	K11682	0.13-0.20	0.40-0.70	0.035	0.04	0.20-0.35	1.15-1.65	...	0.25-0.40	...	0.04-0.10 Ti; 0.20-0.40 Cu; 0.0015-0.005 B
	Type M	K11683	0.12-0.21	0.45-0.70	0.035	0.04	0.20-0.35	...	1.20-1.50	0.45-0.60	...	0.001-0.005 B
	Type N	K11847	0.15-0.21	0.80-1.10	0.035	0.04	0.40-0.90	0.50-0.80	...	0.25 max	...	0.05-0.15 Zr; 0.0005-0.0025 B
	Type P	K21650	0.12-0.21	0.45-0.70	0.035	0.04	0.20-0.35	...	1.20-1.50	...	...	
A572	Grade 42	...	0.21 max	1.35 max	0.04	0.05	0.30 max	...	...	...	...	0.20 min Cu(a)
	Grade 45	...	0.22 max	1.35 max	0.04	0.05	0.30 max	...	...	...	...	0.20 min Cu(a)
	Grade 50	...	0.23 max	1.35 max	0.04	0.05	0.30 max	...	...	...	...	0.20 min Cu(a)
	Grade 55	...	0.25 max	1.35 max	0.04	0.05	0.30 max	...	...	...	...	0.20 min Cu(a)
	Grade 60	...	0.25 max	1.35 max	0.04	0.05	0.30 max	...	...	...	...	0.20 min Cu(a)
	Grade 65	...	0.26 max	1.65 max	0.04	0.05	0.30 max	...	...	...	...	0.20 min Cu(a)
A588	Grade A	K11430	0.10-0.19	0.90-1.25	0.04	0.05	0.15-0.30	0.40-0.65	...	...	0.02-0.10	0.25-0.40 Cu
	Grade B	K12043	0.20 max	0.75-1.25	0.04	0.05	0.15-0.30	0.40-0.70	0.25-0.50	...	0.01-0.10	0.20-0.40 Cu
	Grade C	K11538	0.15 max	0.80-1.35	0.04	0.05	0.15-0.30	0.30-0.50	0.25-0.50	...	0.01-0.10	0.20-0.50 Cu
	Grade D	K11552	0.10-0.20	0.75-1.25	0.04	0.05	0.50-0.90	0.50-0.90	...	...	...	0.30 max Cu; 0.05-0.15 Zr; 0.04 max Nb
	Grade E	K11567	0.15 max	1.20 max	0.04	0.05	0.15-0.30	...	0.75-1.25	0.10-0.25	0.05 max	0.50-0.80 Cu
	Grade F	K11541	0.10-0.20	0.50-1.00	0.04	0.05	0.30 max	0.30 max	0.40-1.10	0.10-0.20	0.01-0.10	0.30-1.00 Cu
	Grade G	K12040	0.20 max	1.20 max	0.04	0.05	0.25-0.70	0.50-1.00	0.80 max	0.10 max	...	0.30-0.50 Cu; 0.07 max Ti
	Grade H	K12032	0.20 max	1.25 max	0.035	0.040	0.25-0.75	0.10-0.25	0.30-0.60	0.15 max	0.02-0.10	0.20-0.35 Cu; 0.005-0.030 Ti
	Grade J	K12044	0.20 max	0.60-1.00	0.04	0.05	0.30-0.50	...	0.50-0.70	...	...	0.30 min Cu; 0.03-0.05 Ti
A633	Grade A	K01802	0.18 max	1.00-1.35	0.04	0.05	0.15-0.50	...	...	...	...	0.05 max Nb
	Grade B	K01803	0.18 max	1.00-1.35	0.04	0.05	0.15-0.50	...	...	...	0.10 max	...
	Grade C	K12000	0.20 max	1.15-1.50	0.04	0.05	0.15-0.50	...	...	...	...	0.01-0.05 Nb
	Grade D	K02003	0.20 max	0.70-1.60	0.04	0.05	0.15-0.50	0.25 max	0.25 max	0.08 max	...	0.35 max Cu
	Grade E	K12202	0.22 max	1.15-1.50	0.04	0.05	0.15-0.50	...	...	...	0.04-0.11	0.01-0.03 N
A656	Grade 1	K11804	0.18 max	1.60 max	0.040	0.050	0.60 max	...	...	...	0.05-0.15	0.020 min Al; 0.005-0.030 N
	Grade 2	K11503	0.15 max	0.90 max	0.040	0.050	0.10 max	...	...	...	...	0.01 min Al; 0.05-0.50 Ti
A699	...	K10614	0.06 max	1.20-2.20	0.04	0.025	0.35 max	...	...	0.25-0.35	...	0.03-0.09 Nb; 0.20-0.35 Cu optional
A710	Grade A	K20747	0.07 max	0.40-0.70	0.025	0.025	0.35 max	0.60-0.90	0.70-1.00	0.15-0.25	...	1.00-1.30 Cu; 0.02 min Nb
	Grade B	K20622	0.06 max	0.40-0.65	0.025	0.025	0.20-0.35	...	1.20-1.50	...	...	1.00-1.30 Cu; 0.02 min Nb

(a) These grades may contain niobium, vanadium or nitrogen.

can Society of Mechanical Engineers (ASME) with little or no modification; ASME uses the prefix "S" and the ASTM designation for these specifications. For example, ASME SA-213 and ASTM A213 are identical.

Steel products can be identified by the number of the ASTM specification to which they are made. The number consists of the letter "A" (for ferrous materials) and an arbitrary, serially assigned number. Citing the specification number, however, is not always adequate to completely describe a steel product. For example, A434 is the specification for heat treated (hardened and tempered) alloy steel bars. To completely describe steel bars indicated by this specification, the grade (AISI-SAE designation in this case) and class (required strength level) must also be indicated. A434 also incorporates, by reference, two standards for test methods and A29, the general requirements for bar products.

AISI-SAE designations for the compositions of carbon and alloy steels are normally incorporated into the ASTM specifications for bars, wires and billets for forging. Some ASTM specifications for sheet products include AISI-SAE designations for composition. ASTM specifications for plates and structural shapes generally specify the limits and ranges of chemical composition directly, without the AISI-SAE designations. Table 24 includes a list of some of the ASTM specifications that incorporate AISI-SAE designations for compositions of the different grades of steel. These specifications are not included in Tables 26 through 30.

Generic Specifications. Several ASTM specifications, such as A29, contain the general requirements common to each member of a broad family of steel products. Table 25 lists several of these generic specifications, which generally must be supplemented by another specification describing a specific mill form or intermediate fabricated product.

Sheet Products. Limits and ranges of chemical compositions for sheet products with ASTM specifications that do not incorporate AISI-SAE

Table 29. Composition ranges and limits for carbon steel pressure-vessel plate (ASTM specifications)

Specification	Type or grade	UNS designation	C max	Mn	P max	S max	Si
A285	Grade A	K01700	0.17	...	0.035	0.045	...
	Grade B	K02200	0.22	0.90 max	0.035	0.045	...
	Grade C	K02801	0.28	0.90 max	0.035	0.045	...
A288	...	K02803	0.30	0.90-1.50	0.035	0.040	0.15-0.30
A442	Grade 55	K02202	0.24	0.60-1.10	0.04	0.05	0.15-0.30
	Grade 60	K02402	0.27	0.60-1.10	0.04	0.05	0.15-0.30
A455	Type I	K03300	0.33	0.85-1.20	0.040	0.050	0.10 max
	Type II	K02802	0.28	0.85-1.20	0.040	0.050	0.15-0.30
A515	Grade 55	K02001	0.28	0.90 max	0.035	0.040	0.15-0.30
	Grade 60	K02401	0.31	0.90 max	0.035	0.040	0.15-0.30
	Grade 65	K02800	0.33	0.90 max	0.035	0.040	0.15-0.30
	Grade 70	K03101	0.35	0.90 max	0.035	0.040	0.15-0.30
A516	Grade 55	K01800	0.26	0.60-1.20	0.035	0.04	0.15-0.30
	Grade 60	K02100	0.27	0.60-1.20	0.035	0.04	0.15-0.30
	Grade 65	K02403	0.29	0.85-1.20	0.035	0.04	0.15-0.30
	Grade 70	K02700	0.31	0.85-1.20	0.035	0.04	0.15-0.30
A537	...	K02400	0.24	0.70-1.60	0.035	0.040	0.15-0.30
A612(a)	...	K02900	0.27	1.00-1.50	0.035	0.040	0.15-0.30
A662	Grade A	K01701	0.17	0.90-1.35	0.035	0.040	0.15-0.30
	Grade B	K02203	0.19	0.85-1.50	0.035	0.040	0.15-0.30
A724(a)	Grade A	...	0.18	1.00-1.60	0.035	0.040	0.55 max

(a) Residual alloying elements restricted as follows: 0.35 max Cu; 0.25 max Ni; 0.25 max Cr; 0.08 max Mo; 0.08 max V.

Table 30. Composition ranges and limits for alloy steels for pressure-vessel plate (ASTM specifications)

ASTM specification	Type or grade	UNS designation	C	Mn	P max	S max	Si	Cr	Ni	Mo	V	Other
A202	Grade A	K11742	0.17 max	1.05-1.40	0.035	0.04	0.60-0.90	0.35-0.60	...	...	...	...
	Grade B	K12542	0.25 max	1.05-1.40	0.035	0.04	0.60-0.90	0.35-0.60	...	...	...	...
A203	Grade A	K21703	0.23 max	0.80 max	0.035	0.04	0.15-0.30	...	2.10-2.50	...	...	...
	Grade B	K22103	0.25 max	0.80 max	0.035	0.04	0.15-0.30	...	2.10-2.50	...	...	...
	Grade D	K31718	0.20 max	0.80 max	0.035	0.04	0.15-0.30	...	3.25-3.75	...	...	...
	Grade E	K32018	0.23 max	0.80 max	0.035	0.04	0.15-0.30	...	3.25-3.75	...	...	...
A204	Grade A	K11820	0.25 max	0.90 max	0.035	0.04	0.15-0.30	...	...	0.45-0.60	...	...
	Grade B	K12020	0.27 max	0.90 max	0.035	0.04	0.15-0.30	...	...	0.45-0.60	...	...
	Grade C	K12320	0.28 max	0.90 max	0.035	0.04	0.15-0.30	...	...	0.45-0.60	...	...
A225	Grade A	K11803	0.18 max	1.45 max	0.035	0.04	0.15-0.30	...	...	...	0.09-0.14	...
	Grade B	K12003	0.20 max	1.45 max	0.035	0.04	0.15-0.30	...	...	...	0.09-0.14	...
	Grade C	K12524	0.25 max	1.60 max	0.035	0.04	0.13-0.32	0.37-0.73	...	...	0.11-0.20	...
A302	Grade A	K12021	0.25 max	0.95-1.30	0.035	0.040	0.15-0.30	...	...	0.45-0.60	...	...
	Grade B	K12022	0.25 max	1.15-1.50	0.035	0.040	0.15-0.30	...	...	0.45-0.60	...	...
	Grade C	K12039	0.25 max	1.15-1.50	0.035	0.040	0.15-0.30	...	0.40-0.70	0.45-0.60	...	...
	Grade D	K12054	0.25 max	1.15-1.50	0.035	0.040	0.15-0.30	...	0.70-1.00	0.45-0.60	...	...
A353	...	K81340	0.13 max	0.90 max	0.035	0.040	0.15-0.30	...	8.50-9.50	...	...	...
A387	Grade 2	K12143	0.21 max	0.55-0.80	0.035	0.040	0.15-0.30	0.50-0.80	...	0.45-0.60	...	...
	Grade 12	K11757	0.17 max	0.40-0.65	0.035	0.040	0.15-0.30	0.80-1.15	...	0.45-0.60	...	...
	Grade 11	K11789	0.17 max	0.40-0.65	0.035	0.040	0.50-0.80	1.00-1.50	...	0.45-0.65	...	...
	Grade 22	K21590	0.15 max	0.30-0.60	0.035	0.035	0.50 max	2.00-2.50	...	0.90-1.10	...	...
	Grade 21	K31545	0.15 max	0.30-0.60	0.035	0.035	0.50 max	2.75-3.25	...	0.90-1.10	...	...
	Grade 5	K41545	0.15 max	0.30-0.60	0.040	0.030	0.50 max	4.00-6.00	...	0.45-0.65	...	...
	Grade 7	S50300	0.15 max	0.30-0.60	0.030	0.030	1.00 max	6.00-8.00	...	0.45-0.65	...	...
	Grade 9	S50400	0.15 max	0.30-0.60	0.030	0.030	1.00 max	8.00-10.00	...	0.90-1.10	...	...
A517	Grade A	K11856	0.15-0.21	0.80-1.10	0.035	0.04	0.40-0.80	0.50-0.80	...	0.18-0.28	...	0.05-0.15 Zr; 0.0025 max B
	Grade B	K11630	0.12-0.21	0.70-1.00	0.035	0.04	0.20-0.65	0.40-0.65	...	0.15-0.25	0.03-0.08	0.01-0.03 Ti; 0.0005-0.005 B
	Grade C	K11511	0.10-0.20	1.10-1.50	0.035	0.04	0.15-0.30	...	...	0.20-0.30	...	0.001-0.005 B
	Grade D	K11662	0.13-0.20	0.40-0.70	0.035	0.04	0.20-0.35	0.85-1.20	...	0.15-0.25	...	0.04-0.10 Ti; 0.20-0.40 Cu; 0.0015-0.005 B
	Grade E	K21604	0.12-0.20	0.40-0.70	0.035	0.04	0.20-0.35	1.40-2.00	...	0.40-0.60	...	0.04-0.10 Ti; 0.20-0.40 Cu; 0.0015-0.005 B
	Grade F	K11576	0.10-0.20	0.60-1.00	0.035	0.04	0.15-0.35	0.40-0.65	0.70-1.00	0.40-0.60	0.03-0.08	0.15-0.50 Cu; 0.002-0.006 B

(continued)

Table 30. (continued)

ASTM specification	Type or grade	UNS designation	C	Mn	P max	S max	Si	Cr	Ni	Mo	V	Other
A517	Grade G	K11872	0.15-0.21	0.80-1.10	0.035	0.04	0.50-0.90	0.50-0.90	...	0.40-0.60	...	0.05-0.15 Zr; 0.0025 max B
	Grade H	K11646	0.12-0.21	0.95-1.30	0.035	0.04	0.20-0.35	0.40-0.650	0.30-0.70	0.20-0.30	0.03-0.08	0.0005-0.005 B
	Grade J	K11625	0.12-0.21	0.45-0.70	0.035	0.04	0.20-0.35	...	...	0.50-0.65	...	0.001-0.005 B
	Grade K	K11523	0.10-0.20	1.10-1.50	0.035	0.04	0.15-0.30	...	...	0.45-0.55	...	0.001-0.005 B
	Grade L	K11682	0.13-0.20	0.40-0.70	0.035	0.04	0.20-0.35	1.15-1.65	...	0.25-0.40	...	0.04-0.10 Ti; 0.20-0.40 Cu; 0.0015-0.005 B
	Grade M	K11683	0.12-0.21	0.45-0.70	0.035	0.04	0.20-0.35	...	1.20-1.50	0.45-0.60	...	0.001-0.005 B
	Grade P	K21650	0.12-0.21	0.45-0.70	0.035	0.04	0.20-0.35	0.85-1.20	1.20-1.50	0.45-0.60	...	0.001-0.005 B
A533	Type A	K12521	0.25 max	1.15-1.50	0.035	0.040	0.15-0.30	...	...	0.45-0.60	...	...
	Type B	K12539	0.25 max	1.15-1.50	0.035	0.040	0.15-0.30	...	0.40-0.70	0.45-0.60	...	...
	Type C	K12554	0.25 max	1.15-1.50	0.035	0.040	0.15-0.30	...	0.70-1.00	0.45-0.60	...	...
	Type D	K12529	0.25 max	1.15-1.50	0.035	0.040	0.15-0.30	...	0.20-0.40	0.45-0.60	...	...
A538	Grade A	K92810	0.03 max	0.10 max	0.010	0.010	0.10 max	...	17.0-19.0	4.0-4.5	...	0.10-0.25 Ti; 7.0-8.5 Co; 0.05-0.15 Al; 0.003 B; 0.02 Zr; 0.05 Ca
	Grade B	K92890	0.03 max	0.10 max	0.010	0.010	0.10 max	...	17.0-19.0	4.6-5.1	...	0.30-0.50 Ti; 7.0-8.5 Co; 0.05-0.15 Al
	Grade C	K93120	0.03 max	0.10 max	0.010	0.010	0.10 max	...	18.0-19.0	4.6-5.2	...	0.55-0.80 Ti; 8.0-9.5 Co; 0.05-0.15 Al
A542	...	K21590	0.15 max	0.30-0.60	0.035	0.035	0.15-0.30	2.00-2.50	...	0.90-1.10	...	...
A543	Type A	K42338	0.23 max	0.40 max	0.035	0.040	0.20-0.35	1.50-2.00	2.60-4.00	0.45-0.60	0.03 max	...
	Type B	K42339	0.23 max	0.40 max	0.020	0.020	0.20-0.35	1.50-2.00	2.60-4.00	0.45-0.60	0.03 max	...
A553	Type I	K81340	0.13 max	0.90 max	0.035	0.040	0.15-0.30	...	8.50-9.50	...	...	...
	Type II	K71340	0.13 max	0.90 max	0.035	0.040	0.15-0.30	...	7.50-8.50	...	...	...
A562	...	K11224	0.12 max	1.20 max	0.04	0.05	0.15-0.50	...	...	...	...	(4 × %C) Ti; 0.15 max Cu
A590	...	K91890	0.03 max	0.10 max	0.010	0.010	0.10 max	4.50-5.50	11.5-12.5	2.75-3.25	...	0.20-0.35 Ti; 0.40 max Al
A605	...	K91401	0.13 max	0.20-0.40	0.010	0.010	0.10 max	0.65-0.85	8.5-9.5	0.90-1.10	0.06-0.12	4.25-5.00 Co
A645	...	K41583	0.30-0.60	0.30-0.60	0.025	0.025	0.20-0.35	...	4.75-5.25	0.20-0.35	...	0.020 max N; 0.02-0.12 Al
A734	Type A	....	0.17 max	0.45-0.75	0.035	0.015	0.35 max	0.90-1.20	0.90-1.20	0.25-0.40	...	0.06 max Al
	Type B	....	0.17 max	1.60 max	0.035	0.015	0.35 max	0.25 max	...	...	0.11 max	0.35 max Cu; 0.030 max N; 0.050 max Nb; 0.06 max Al
A735	...	....	0.06 max	1.20-2.20	0.04	0.025	0.35 max	...	...	0.23-0.47	...	0.20-0.35 Cu; 0.03-0.09 Nb
A736	...	....	0.07 max	0.40-0.70	0.025	0.025	0.35 max	0.60-0.90	0.70-1.00	0.15-0.25	...	1.00-1.30 Cu; 0.02 min Nb
A737	Grade A	....	0.20 max	1.00-1.35	0.035	0.030	0.15-0.50	...	...	...	0.10 max	...
	Grade B	....	0.20 max	1.15-1.50	0.035	0.030	0.15-0.50	...	...	...	...	0.05 max Nb
	Grade C	....	0.20 max	1.15-1.50	0.035	0.030	0.15-0.50	...	...	...	0.04-0.11	0.03 max N

Table 31. Product descriptions and carbon contents for wrought carbon steels (AMS designations)

AMS designation	Product form	Carbon content	Nearest AISI-SAE grade	AMS designation	Product form(a)	Carbon content	Nearest AISI-SAE grade
5010E	Bars—screw machine stock	...	1112	5062B	Bars, forgings, tubing, plate, sheet, strip	0.25 max	...
5020	Bars, forging, tubing	0.32-0.39	11L37	5069A	Bars, forgings, tubing	0.15-0.20	1018
5022G	Bars, forging, tubing	0.14-0.20(b)	1117	5070C	Bars, forgings (55 ksi TS)	0.18-0.23	1022
5024D	Bars, forging, tubing	0.32-0.39	1137	5075B	Tubing, seamless (55 ksi TS)	0.22-0.28	1025
5032B	Wire (annealed)	0.18-0.23	1020	5077B	Tubing, welded	0.22-0.28	1025
5040F	Sheet, strip (deep forming grade)	0.15 max	1010	5080D	Bars, forgings, tubing	0.31-0.38	1035
5041	Sheet, strip (cold rolled, extra deep drawing)	0.08 max	1006	5082A	Tubing, seamless (90 ksi TS)	0.31-0.38	1035
5042F	Sheet, strip (forming grade)	0.15 max	1010	5085A	Plate, sheet, strip (annealed)	0.47-0.55	1050
5044D	Sheet, strip (half hard temper)	Low	1010	5110B	Music wire—commercial	...	1080
5045C	Sheet, strip (hard temper)	Low	1020	5112E	Music spring wire—best quality	...	1090
5047A	Sheet, strip (aluminum killed)	Low	1010	5115C	Wire—spring	0.60-0.75	1070
5050F	Tubing, seamless (annealed)	0.15 max	1010	5120F	Strip	0.68-0.80	1074
5053C	Tubing, welded (annealed)	0.13 max	1010	5121C	Strip, spring	0.89-1.04	1095
5060C	Bars, forgings, tubing	0.13-0.18	1015	5122C	Strip (hard temper)	0.89-1.04	1095
5061B	Bars, wire	0.08-0.20	...	5132D	Bars	0.90-1.30	1095

(a) TS = tensile strength. (b) Contains 1.2% manganese.

Table 32. Product descriptions and nominal compositions for wrought alloy steels (AMS designations)

AMS designation	Product form(a)	C	Cr	Ni	Mo	Other	Nearest AISI-SAE grade
6242C	Bars, forgings	0.15-0.20	...	5	...	...	2517
6250F	Bars, forgings, tubing	0.07-0.13	1.5	3.5	...	...	3310
6260G	Bars, forgings, tubing	0.07-0.13	1.2	3.25	0.12	...	9310
6263D	Bars, forgings, tubing	0.11-0.17	1.2	3.25	0.12	...	9315
6264D	Bars, forgings, tubing	0.14-0.20	1.2	3.25	0.12	...	9317
6265C	Bars, forgings, tubing (P,VM)	0.07-0.13	1.2	3.25	0.12	...	9310
6266C	Bars, forgings, tubing	0.08-0.13	0.50	1.85	0.25	0.003 B	43BV12
6267A	Bars, forgings, tubing (P)	0.07-0.13	1.2	3.25	0.12	...	9310
6270G	Bars, forgings, tubing	0.11-0.17	0.50	0.55	0.20	...	8615
6272E	Bars, forgings, tubing	0.15-0.20	0.50	0.55	0.20	...	8617
6274G	Bars, forgings, tubing	0.18-0.23	0.50	0.55	0.20	...	8620
6275B	Bars, forgings, tubing	0.15-0.20	0.40	0.45	0.12	0.003 B	94B17
6276C	Bars, forgings, tubing (P,VM)	0.18-0.23	0.50	0.55	0.2	...	8620
6277A	Bars, forgings, tubing (P)	0.18-0.23	0.50	0.55	0.20	...	8620
6280E	Bars, forgings	0.28-0.33	0.50	0.55	0.20	...	8630
6281C	Tubing	0.28-0.33	0.5	0.55	0.2	...	8630
6282D	Tubing	0.33-0.38	0.50	0.55	0.25	...	8735
6290C	Bars, forgings	0.11-0.17	...	1.8	0.25	...	4615
6292C	Bars, forgings	0.15-0.20	...	1.8	0.25	...	4617
6294C	Bars, forgings	0.17-0.22	...	1.8	0.25	...	4620
6299A	Bars, forgings, tubing	0.17-0.23	0.50	1.8	0.25	...	4320H
6300	Bars, forgings	0.35-0.40	...	...	0.25	...	4037
6302B	Bars, forgings	0.28-0.33	1.25	...	0.50	0.65 Si; 0.25 V	...
6303A	Bars, forgings	0.25-0.30	1.25	...	0.5	0.65 Si; 0.85 V	...
6304C	Bars, forgings, tubing	0.40-0.50	0.95	...	0.55	0.30 V	...
6312A	Bars, forgings	0.38-0.43	...	1.8	0.25	...	4640
6317B	Bars, forgings (heat treated; 125 ksi TS)	0.38-0.43	...	1.8	0.25	...	4640
6320F	Bars, forgings	0.33-0.38	0.50	0.55	0.25	...	8735
6321A	Bars, forgings, tubing	0.38-0.43	0.43	0.30	0.12	0.003 B	81B40
6322F	Bars, forgings	0.38-0.43	0.50	0.55	0.25	...	8740
6323D	Tubing	0.38-0.43	0.50	0.55	0.25	...	8740
6324C	Bars, forgings	0.38-0.43	0.65	0.70	0.25	...	8740 mod
6325D	Bars, forgings (heat treated; 105 ksi TS)	0.38-0.43	0.50	0.55	0.25	...	8740
6327D	Bars, forgings (heat treated; 125 ksi TS)	0.38-0.43	0.50	0.55	0.25	...	8740
6328E	Bars, forgings, tubing	0.48-0.53	0.50	0.55	0.25	...	8750
6330A	Bars, forgings	0.33-0.38	0.6	1.25	...	...	3135
6342D	Bars, forgings, tubing	0.38-0.32	0.80	1.0	0.25	...	9840
6350D	Plate, sheet, strip (annealed)	0.28-0.33	0.95	...	0.20	...	4130
6351A	Plate, sheet, strip (spheroidized)	0.28-0.33	0.95	...	0.20	...	4130
6352B	Plate, sheet, strip (annealed)	0.32-0.39	0.95	...	0.2	...	4135H
6354	Plate, sheet, strip	0.10-0.17	0.6	...	0.2	0.75 Si; 0.1 Zr	NAX 9115-AC
6355G	Plate, sheet, strip (annealed)	0.28-0.33	0.50	0.55	0.20	...	8630
6356A	Plate, sheet, strip	0.30-0.35	0.95	...	0.20	...	...
6357D	Plate, sheet, strip	0.33-0.38	0.50	0.55	0.25	...	8735
6358B	Plate, sheet, strip	0.38-0.43	0.50	0.55	0.25	...	8740
6359B	Plate, sheet, strip	0.38-0.43	0.80	1.8	0.25	...	4340
6360F	Tubing, seamless	0.28-0.33	0.95	...	0.20	...	4130
6361	Tubing, seamless (125 ksi TS)	0.27-0.33	0.95	...	0.2	...	4130
6362	Tubing, seamless (150 ksi TS)	0.27-0.33	0.95	...	0.2	...	4130
6365E	Tubing, seamless	0.33-0.38	0.95	...	0.20	...	4135
6370F	Bars, forgings, rings	0.28-0.33	0.95	...	0.2	...	4130
6371D	Tubing	0.28-0.33	0.95	...	0.20	...	4130
6372D	Tubing	0.33-0.38	0.95	...	0.20	...	4135
6373A	Tubing, welded	0.28-0.33	0.95	...	0.20	...	4130
6378	Bars (die drawn and tempered; 130 ksi YS)	0.38-0.45	0.95	...	0.2	...	4140
6379	Bars (die drawn and tempered; 165 ksi YS)	0.40-0.53	0.95	...	0.2	...	4140
6381B	Tubing	0.38-0.43	0.95	...	0.20	...	4140
6382G	Bars, forgings	0.38-0.43	0.95	...	0.20	...	4140
6385B	Plate, sheet, strip	0.27-0.33	1.25	...	0.50	0.65 Si; 0.25 V	...
6386A	Plate, sheet (heat treated; 90 and 100 ksi YS)	...	...	...	...	...	...
6390A	Tubing (special quality)	0.38-0.43	0.95	...	0.20	...	4140
6395	Plate, sheet, strip	0.38-0.43	0.95	...	0.20	...	4140
6406A	Plate, sheet, strip	0.41-0.46	2.1	...	0.58	1.6 Si; 0.05 V	...
6407B	Bars, forgings, tubing	0.27-0.33	1.2	2.05	0.45	...	...
6411	Bars, forgings, tubing (P, CM)	0.28-0.33	0.85	1.8	0.40	...	...
6412F	Bars, forgings	0.35-0.40	0.80	1.8	0.25	...	4337
6413D	Tubing	0.35-0.40	0.8	1.8	0.25	...	4337
6414A	Bars, forgings, tubing (P)	0.38-0.43	0.80	1.8	0.25	...	4340
6415G	Bars, forgings, tubing	0.38-0.43	0.80	1.8	0.25	...	4340
6416	Bars, forgings, tubing	0.41-0.46	0.8	1.8	0.4	1.6 Si; 0.07 V	...

(continued)

Table 32. (continued)

AMS designation	Product form(a)	C	Cr	Ni	Mo	Other	Nearest AISI-SAE grade
						Nominal composition, %	
6417	Bars, forgings, tubing (P, CM)	0.38-0.43	0.82	1.8	0.40	1.6 Si; 0.07 V	4340 mod
6418C	Bars, forgings, tubing	0.23-0.28	0.30	1.8	0.40	1.3 Mn; 1.5 Si	Hy-Tuf
6419	Bars, forgings, tubing (P, CM)	0.41-0.46	0.82	1.8	0.40	1.6 Si; 0.07 V	300 M
6421A	Bars, forgings, tubing	0.35-0.40	0.80	0.85	0.20	0.003 B	Mod 98B37
6422C	Bars, forgings, tubing	0.38-0.43	0.80	0.85	0.20	0.003 B	Mod 98B40
6423A	Bars, forgings, tubing	0.40-0.46	0.92	0.75	0.52	0.003 B	...
6426A	Bars, forgings, tubing (VM)	0.80-0.90	1.0	...	0.58	0.75 Si	...
6427D	Bars, forgings, tubing	0.28-0.33	0.85	1.8	0.40	0.07 V	...
6428A	Bars, forgings, tubing	0.32-0.38	0.80	1.8	0.35	0.20 V	4335 mod
6429A	Bars, forgings, tubing (CM or VM)	0.33-0.38	0.80	1.8	0.35	0.20 V	4335 mod
6430A	Bars, forgings, tubing (special grade)	0.33-0.38	0.80	1.8	0.35	0.20 V	4335 mod
6431A	Bars, forgings, tubing (P, VM)	0.45-0.50	1.05	0.55	1.0	0.11 V	D6AC
6432	Bars, forgings, tubing	0.43-0.49	1.05	0.55	1.0	0.11 V	D6A
6433A	Plate, sheet, strip (special grade)	0.33-0.38	0.80	1.8	0.35	0.20 V	...
6434A	Plate, sheet, strip	0.31-0.38	0.80	1.8	0.35	0.20 V	4335 mod
6435A	Plate, sheet, strip (P, CM) (annealed)	0.33-0.38	0.80	1.8	0.35	0.20 V	4335 mod
6436A	Plate, sheet, strip	0.20-0.25	1.25	...	0.5	0.65 Si; 0.85 V	...
6437A	Plate, sheet, strip	0.38-0.43	5.0	...	1.3	0.5 V	H-11
6438A	Plate, sheet, strip (P, CM)	0.45-0.50	1.05	0.55	1.0	0.11 V	D6
6440E	Bars, wire, forgings	0.98-1.10	1.45	...	...	...	52100
6441D	Tubing (bearing quality)	0.98-1.10	1.45	...	...	...	52100
6442C	Bars, wire, forgings	0.98-1.10	0.50	...	...	...	50100
6443B	Bars, wire, forgings (P, VM)	0.95-1.1	1.05	...	...	...	51100
6444B	Bars, wire, forgings, tubing (P, VM)	0.95-1.10	1.45	...	...	...	52100
6445A	Bars, wire, forgings, tubing (P, VM)	0.92-1.02	1.05	...	...	1.1 Mn	51100 mod
6446	Bars, forgings (P)	0.95-1.10	1.05	...	...	...	51100
6447	Bars, forgings, tubing (P)	0.95-1.10	1.45	...	...	...	52100
6448C	Bars, forgings	0.48-0.53	0.95	...	...	0.22 V	6150
6450C	Wire (spring-annealed)	0.48-0.53	0.95	...	...	0.22 V	6150
6455C	Plate, sheet, strip (spring)	0.48-0.53	0.95	...	...	0.22 V	6150
6470F	Bars, forgings, tubing for nitriding	0.38-0.43	1.6	...	0.35	1.15 Al	...
6471	Bars, forgings, tubing for nitriding (P)	0.38-0.43	1.6	...	0.35	1.15 Al	...
6472	Bars, forgings for nitriding (heat treated; 112 ksi TS)	0.38-0.43	1.6	...	0.35	1.13 Al	...
6475C	Bars, forgings, tubing for nitriding	0.21-0.26	1.1	3.5	0.25	1.25 Al	...
6485B	Bars, forgings	0.38-0.43	5.0	...	1.3	0.50 V	H-11
6487C	Bars, forgings (P, VM)	0.38-0.43	5.0	...	1.3	0.50 V	H-11
6488	Bars, forgings (P)	0.38-0.43	5.0	...	1.3	0.50 V	H-11
6490B	Bars, forgings, tubing (P, VM)	0.77-0.85	4.0	...	4.25	1.0 V	M-50
6512	Bars, forgings, tubing, rings (CM) (annealed)	...	...	18	4.9	7.8 Co; 0.40 Ti; 0.10 Al	...
6514	Bars, forgings, tubing, rings (annealed) (CM)	...	...	18.5	4.9	9.0 Co; 0.65 Ti; 0.10 Al	...
6520	Plate, sheet, strip (solution heat treated) (CM)	...	...	18	4.9	7.8 Co; 0.40 Ti; 0.10 Al	...
6521	Plate, sheet, strip (solution heat treated) (CM)	...	...	18.5	4.9	9.0 Co; 0.65 Ti; 0.10 Al	...
6526	Bars, forgings, tubing (annealed) (P, CM)	0.29-0.34	1.0	7.5	1.0	4.5 Co; 0.09 V	...
6530E	Tubing, seamless	0.28-0.33	0.50	0.55	0.20	...	8630
6535D	Tubing, seamless	0.33-0.38	0.50	0.55	0.25	...	8735
6540A	Bars, forgings, rings, tubing (annealed)	0.24-0.30	0.48	8.0	0.48	4.0 Co; 0.09 V	...
6541A	Bars, forgings, rings, tubing (annealed) (P, CM)	0.24-0.30	0.48	8.0	0.48	4.0 Co; 0.09 V	...
6542A	Bars, forgings, tubing (annealed) (P, CM)	0.42-0.48	0.27	7.75	0.27	4.0 Co; 0.09 V	...
6545A	Plate, sheet, strip (annealed)	0.24-0.30	0.48	8.0	0.48	4.0 Co; 0.09 V	...
6546A	Plate, sheet, strip (annealed) (P, CM)	0.24-0.30	0.48	8.0	0.48	4.0 Co; 0.09 V	...
6550E	Tubing, welded	0.28-0.33	0.50	0.55	0.20	...	8630

(a) P, premium quality; VM, vacuum melted; CM, consumable electrode remelted. TS, tensile strength; YS, yield strength.

designations are listed in Table 26. ASTM specifications for sheet products that do include AISI-SAE designations are listed in Table 24.

Plate and Structural Shapes. ASTM specifications for plate and structural shapes generally do not incorporate AISI-SAE designations. Limits and ranges of chemical composition in ASTM specifications for carbon steel plate and structural shapes are listed in Table 27; compositions of HSLA and alloy plate and structural shapes are listed in Table 28.

Pressure-Vessel Plate. Steel plate intended for fabrication into pressure vessels must conform to different specifications than similar plate intended for structural applications. The major differences between the two groups of specifications are that pressure-vessel plate must meet requirements for notch toughness and has more stringent limits for allowable surface and edge imperfections. Limits and ranges of chemical composition for carbon steel plate are given in Table 29; those for HSLA and alloy pressure-vessel plate are given in Table 30.

AMS DESIGNATIONS

Aerospace Materials Specifications (AMS), published by SAE, are complete specifications that are generally adequate for procurement purposes. Most of the AMS designations pertain to materials intended for aerospace applications; the specifications may include mechanical-property requirements significantly more severe than those for grades of steel having similar compositions but intended for other applications. Processing requirements, such as consumable electrode remelting, are common in AMS steels. Chemical compositions for AMS grades of carbon and alloy steels are given in Tables 31 and 32, respectively.

UNS DESIGNATIONS

The Unified Numbering System (UNS) has been developed by ASTM and SAE and several other technical societies, trade associations and United States government agencies. A UNS number, which is a designation of chemical composition and not a specification, is assigned to each chemical composition of a metallic alloy. Available UNS designations are included in the tables of this article.

The UNS designation of an alloy consists of a letter and five numerals. The letters indicate the broad class of alloys; the numerals define specific alloys within that class. Existing systems of designation, such as the AISI-SAE system for steels, have been incorporated into UNS designations. The Unified Numbering System is described in greater detail in SAE Recommended Practice J1086 and ASTM E527.

Mechanical Properties of Carbon and Alloy Steels

This guide shows what to expect from a given grade of steel in the indicated condition. Data were obtained from specimens 0.505 in. in diameter which were machined from 1-in. rounds; gage lengths were 2 in. Average properties of hot rolled, normalized, and annealed material are listed, while properties of quenched and tempered grades are for single heats. Sources of the data are Bethlehem Steel Corp. and Republic Steel Corp.

Mechanical properties of selected carbon and alloy steels in the hot rolled, normalized and annealed condition

AISI No.(a)	Treatment	Austenitizing temperature °C	°F	Tensile strength MPa	ksi	Yield strength MPa	ksi	Elongation, %	Reduction in area, %	Hardness, HB	Izod impact strength J	ft·lb
1015	As-rolled	...	...	420.6	61.0	313.7	45.5	39.0	61.0	126	110.5	81.5
	Normalized	925	1700	424.0	61.5	324.1	47.0	37.0	69.6	121	115.5	85.2
	Annealed	870	1600	386.1	56.0	284.4	41.3	37.0	69.7	111	115.0	84.8
1020	As-rolled	...	...	448.2	65.0	330.9	48.0	36.0	59.0	143	86.8	64.0
	Normalized	870	1600	441.3	64.0	346.5	50.3	35.8	67.9	131	117.7	86.8
	Annealed	870	1600	394.7	57.3	294.8	42.8	36.5	66.0	111	123.4	91.0
1022	As-rolled	...	...	503.3	73.0	358.5	52.0	35.0	67.0	149	81.3	60.0
	Normalized	925	1700	482.6	70.0	358.5	52.0	34.0	67.5	143	117.3	86.5
	Annealed	870	1600	429.2	62.3	317.2	46.0	35.0	63.6	137	120.7	89.0
1030	As-rolled	...	...	551.6	80.0	344.7	50.0	32.0	57.0	179	74.6	55.0
	Normalized	925	1700	520.6	75.5	344.7	50.0	32.0	60.8	149	93.6	69.0
	Annealed	845	1550	463.7	67.3	341.3	49.5	31.2	57.9	126	69.4	51.2
1040	As-rolled	...	...	620.5	90.0	413.7	60.0	25.0	50.0	201	48.8	36.0
	Normalized	900	1650	589.5	85.5	374.0	54.3	28.0	54.9	170	65.1	48.0
	Annealed	790	1450	518.8	75.3	353.4	51.3	30.2	57.2	149	44.3	32.7
1050	As-rolled	...	...	723.9	105.0	413.7	60.0	20.0	40.0	229	31.2	23.0
	Normalized	900	1650	748.1	108.5	427.5	62.0	20.0	39.4	217	27.1	20.0
	Annealed	790	1450	636.0	92.3	365.4	53.0	23.7	39.9	187	16.9	12.5
1060	As-rolled	...	...	813.6	118.0	482.6	70.0	17.0	34.0	241	17.6	13.0
	Normalized	900	1650	775.7	112.5	420.6	61.0	18.0	37.2	229	13.2	9.7
	Annealed	790	1450	625.7	90.8	372.3	54.0	22.5	38.2	179	11.3	8.3
1080	As-rolled	...	...	965.3	140.0	586.1	85.0	12.0	17.0	293	6.8	5.0
	Normalized	900	1650	1010.1	146.5	524.0	76.0	11.0	20.6	293	6.8	5.0
	Annealed	790	1450	615.4	89.3	375.8	54.5	24.7	45.0	174	6.1	4.5
1095	As-rolled	...	...	965.3	140.0	572.3	83.0	9.0	18.0	293	4.1	3.0
	Normalized	900	1650	1013.5	147.0	499.9	72.5	9.5	13.5	293	5.4	4.0
	Annealed	790	1450	656.7	95.3	379.2	55.0	13.0	20.6	192	2.7	2.0
1117	As-rolled	...	...	486.8	70.6	305.4	44.3	33.0	63.0	143	81.3	60.0
	Normalized	900	1650	467.1	67.8	303.4	44.0	33.5	63.8	137	85.1	62.8
	Annealed	855	1575	429.5	62.3	279.2	40.5	32.8	58.0	121	93.6	69.0
1118	As-rolled	...	...	521.2	75.6	316.5	45.9	32.0	70.0	149	108.5	80.0
	Normalized	925	1700	477.8	69.3	319.2	46.3	33.5	65.9	143	103.4	76.3
	Annealed	790	1450	450.2	65.3	284.8	41.3	34.5	66.8	131	106.4	78.5
1137	As-rolled	...	...	627.4	91.0	379.2	55.0	28.0	61.0	192	82.7	61.0
	Normalized	900	1650	668.8	97.0	396.4	57.5	22.5	48.5	197	63.7	47.0
	Annealed	790	1450	584.7	84.8	344.7	50.0	26.8	53.9	174	49.9	36.8
1141	As-rolled	...	...	675.7	98.0	358.5	52.0	22.0	38.0	192	11.1	8.2
	Normalized	900	1650	706.7	102.5	405.4	58.8	22.7	55.5	201	52.6	38.8
	Annealed	815	1500	598.5	86.8	353.0	51.2	25.5	49.3	163	34.3	25.3
1144	As-rolled	...	...	703.3	102.0	420.6	61.0	21.0	41.0	212	52.9	39.0
	Normalized	900	1650	667.4	96.8	399.9	58.0	21.0	40.4	197	43.4	32.0
	Annealed	790	1450	584.7	84.8	346.8	50.3	24.8	41.3	167	65.1	48.0
1340	Normalized	870	1600	836.3	121.3	558.5	81.0	22.0	62.9	248	92.5	68.2
	Annealed	800	1475	703.3	102.0	436.4	63.3	25.5	57.3	207	70.5	52.0
3140	Normalized	870	1600	891.5	129.3	599.8	87.0	19.7	57.3	262	53.6	39.5
	Annealed	815	1500	689.5	100.0	422.6	61.3	24.5	50.8	197	46.4	34.2
4130	Normalized	870	1600	668.8	97.0	436.4	63.3	25.5	59.5	197	86.4	63.7
	Annealed	865	1585	560.5	81.3	360.6	52.3	28.2	55.6	156	61.7	45.5
4140	Normalized	870	1600	1020.4	148.0	655.0	95.0	17.7	46.8	302	22.6	16.7
	Annealed	815	1500	655.0	95.0	417.1	60.5	25.7	56.9	197	54.5	40.2
4150	Normalized	870	1600	1154.9	167.5	734.3	106.5	11.7	30.8	321	11.5	8.5
	Annealed	815	1500	729.5	105.8	379.2	55.0	20.2	40.2	197	24.7	18.2
4320	Normalized	895	1640	792.9	115.0	464.0	67.3	20.8	50.7	235	72.9	53.8
	Annealed	850	1560	579.2	84.0	609.5	61.6	29.0	58.4	163	109.8	81.0
4340	Normalized	870	1600	1279.0	185.5	861.8	125.0	12.2	36.3	363	15.9	11.7
	Annealed	810	1490	744.6	108.0	472.3	68.5	22.0	49.9	217	51.1	37.7
4620	Normalized	900	1650	574.3	83.3	366.1	53.1	29.0	66.7	174	132.9	98.0
	Annealed	855	1575	512.3	74.3	372.3	54.0	31.3	60.3	149	93.6	69.0
4820	Normalized	860	1580	75.0	109.5	484.7	70.3	24.0	59.2	229	109.8	81.0
	Annealed	815	1500	681.2	98.8	464.0	67.3	22.3	58.8	197	92.9	68.5
5140	Normalized	870	1600	792.9	115.0	472.3	68.5	22.7	59.2	229	38.0	28.0
	Annealed	830	1525	572.3	83.0	293.0	42.5	28.6	57.3	167	40.7	30.0
5150	Normalized	870	1600	870.8	126.3	529.5	76.8	20.7	58.7	255	31.5	23.2
	Annealed	825	1520	675.7	98.0	357.1	51.8	22.0	43.7	197	25.1	18.5

(continued)

Mechanical properties of selected carbon and alloy steels in the hot rolled, normalized and annealed condition (continued)

AISI No.(a)	Treat-ment	Austenitizing temperature		Tensile strength		Yield strength		Elongation, %	Reduction in area, %	Hardness, HB	Izod impact strength	
		°C	°F	MPa	ksi	MPa	ksi				J	ft·lb
5160	Normalized	855	1575	957.0	138.8	530.9	77.0	17.5	44.8	269	10.8	8.0
	Annealed	815	1495	722.6	104.8	275.8	40.0	17.2	30.6	197	10.0	7.4
6150	Normalized	870	1600	939.8	136.3	615.7	89.3	21.8	61.0	269	35.5	26.2
	Annealed	815	1500	667.4	96.8	412.3	59.8	23.0	48.4	197	27.4	20.2
8620	Normalized	915	1675	632.9	91.8	357.1	51.8	26.3	59.7	183	99.7	73.5
	Annealed	870	1600	536.4	77.8	385.4	55.9	31.3	62.1	149	112.2	82.8
8630	Normalized	870	1600	650.2	94.3	429.5	62.3	23.5	53.5	187	94.6	69.8
	Annealed	845	1550	564.0	81.8	372.3	54.0	29.0	58.9	156	95.2	70.2
8650	Normalized	870	1600	1023.9	148.5	688.1	99.8	14.0	40.4	302	13.6	10.0
	Annealed	795	1465	715.7	103.8	386.1	56.0	22.5	46.4	212	29.4	21.7
8740	Normalized	870	1600	929.4	134.8	606.7	88.0	16.0	47.9	269	17.6	13.0
	Annealed	815	1500	695.0	100.8	415.8	60.3	22.2	46.4	201	40.0	29.5
9255	Normalized	900	1650	932.9	135.3	579.2	84.0	19.7	43.4	269	13.6	10.0
	Annealed	845	1550	774.3	112.3	486.1	70.5	21.7	41.1	229	8.8	6.5
9310	Normalized	890	1630	906.7	131.5	570.9	82.8	18.8	58.1	269	119.3	88.0
	Annealed	845	1550	820.5	119.0	439.9	63.8	17.3	42.1	241	78.6	58.0

(a) All grades are fine grained except for those in the 1100 series which are coarse-grained. Heat treated specimens were oil quenched unless otherwise indicated.

Mechanical properties of selected carbon and alloy steels in the quenched-and-tempered condition

AISI No.(a)	Tempering temperature		Tensile strength		Yield strength		Elongation, %	Reduction in area, %	Hardness, HB
	°C	°F	MPa	ksi	MPa	ksi			
1030(b)	205	400	848	123	648	94	17	47	495
	315	600	800	116	621	90	19	53	401
	425	800	731	106	579	84	23	60	302
	540	1000	669	97	517	75	28	65	255
	650	1200	586	85	441	64	32	70	207
1040(b)	205	400	896	130	662	96	16	45	514
	315	600	889	129	648	94	18	52	444
	425	800	841	122	634	92	21	57	352
	540	1000	779	113	593	86	23	61	269
	650	1200	669	97	496	72	28	68	201
1040	205	400	779	113	593	86	19	48	262
	315	600	779	113	593	86	20	53	255
	425	800	758	110	552	80	21	54	241
	540	1000	717	104	490	71	26	57	212
	650	1200	634	92	434	63	29	65	192
1050(b)	205	400	1124	163	807	117	9	27	514
	315	600	1089	158	793	115	13	36	444
	425	800	1000	145	758	110	19	48	375
	540	1000	862	125	655	95	23	58	293
	650	1200	717	104	538	78	28	65	235
1050	205	400	. . .	. . .	. . .	. . .	. . .	. . .	. . .
	315	600	979	142	724	105	14	47	321
	425	800	938	136	655	95	20	50	277
	540	1000	876	127	579	84	23	53	262
	650	1200	738	107	469	68	29	60	223
1060	205	400	1103	160	779	113	13	40	321
	315	600	1103	160	779	113	13	40	321
	425	800	1076	156	765	111	14	41	311
	540	1000	965	140	669	97	17	45	277
	650	1200	800	116	524	76	23	54	229
1080	205	400	1310	190	979	142	12	35	388
	315	600	1303	189	979	142	12	35	388
	425	800	1289	187	951	138	13	36	375
	540	1000	1131	164	807	117	16	40	321
	650	1200	889	129	600	87	21	50	255
1095(b)	205	400	1489	216	1048	152	10	31	601
	315	600	1462	212	1034	150	11	33	534
	425	800	1372	199	958	139	13	35	388
	540	1000	1138	165	758	110	15	40	293
	650	1200	841	122	586	85	20	47	235
1095	205	400	1289	187	827	120	10	30	401
	315	600	1262	183	813	118	10	30	375
	425	800	1213	176	772	112	12	32	363
	540	1000	1089	158	676	98	15	37	321
	650	1200	896	130	552	80	21	47	269
1137	205	400	1082	157	938	136	5	22	352
	315	600	986	143	841	122	10	33	285
	425	800	876	127	731	106	15	48	262
	540	1000	758	110	607	88	24	62	229
	650	1200	655	95	483	70	28	69	197

(continued)

Mechanical properties of selected carbon and alloy steels in the quenched-and-tempered condition (continued)

AISI No.(a)	Tempering temperature °C	°F	Tensile strength MPa	ksi	Yield strength MPa	ksi	Elongation, %	Reduction in area, %	Hardness, HB
1137(b)	205	400	1496	217	1165	169	5	17	415
	315	600	1372	199	1124	163	9	25	375
	425	800	1103	160	986	143	14	40	311
	540	1000	827	120	724	105	19	60	262
	650	1200	648	94	531	77	25	69	187
1141	205	400	1634	237	1213	176	6	17	461
	315	600	1462	212	1282	186	9	32	415
	425	800	1165	169	1034	150	12	47	331
	540	1000	896	130	765	111	18	57	262
	650	1200	710	103	593	86	23	62	217
1144	205	400	876	127	627	91	17	36	277
	315	600	869	126	621	90	17	40	262
	425	800	848	123	607	88	18	42	248
	540	1000	807	117	572	83	20	46	235
	650	1200	724	105	503	73	23	55	217
1330(b)	205	400	1600	232	1455	211	9	39	459
	315	600	1427	207	1282	186	9	44	402
	425	800	1158	168	1034	150	15	53	335
	540	1000	876	127	772	112	18	60	263
	650	1200	731	106	572	83	23	63	216
1340	205	400	1806	262	1593	231	11	35	505
	315	600	1586	230	1420	206	12	43	453
	425	800	1262	183	1151	167	14	51	375
	540	1000	965	140	827	120	17	58	295
	650	1200	800	116	621	90	22	66	252
4037	205	400	1027	149	758	110	6	38	310
	315	600	951	138	765	111	14	53	295
	425	800	876	127	731	106	20	60	270
	540	1000	793	115	655	95	23	63	247
	650	1200	696	101	421	61	29	60	220
4042	205	400	1800	261	1662	241	12	37	516
	315	600	1613	234	1455	211	13	42	455
	425	800	1289	187	1172	170	15	51	380
	540	1000	986	143	883	128	20	59	300
	650	1200	793	115	689	100	28	66	238
4130(b)	205	400	1627	236	1462	212	10	41	467
	315	600	1496	217	1379	200	11	43	435
	425	800	1282	186	1193	173	13	49	380
	540	1000	1034	150	910	132	17	57	315
	650	1200	814	118	703	102	22	64	245
4140	205	400	1772	257	1641	238	8	38	510
	315	600	1551	225	1434	208	9	43	445
	425	800	1248	181	1138	165	13	49	370
	540	1000	951	138	834	121	18	58	285
	650	1200	758	110	655	95	22	63	230
4150	205	400	1931	280	1724	250	10	39	530
	315	600	1765	256	1593	231	10	40	495
	425	800	1517	220	1379	200	12	45	440
	540	1000	1207	175	1103	160	15	52	370
	650	1200	958	139	841	122	19	60	290
4340	205	400	1875	272	1675	243	10	38	520
	315	600	1724	250	1586	230	10	40	486
	425	800	1469	213	1365	198	10	44	430
	540	1000	1172	170	1076	156	13	51	360
	650	1200	965	140	855	124	19	60	280
5046	205	400	1744	253	1407	204	9	25	482
	315	600	1413	205	1158	168	10	37	401
	425	800	1138	165	931	135	13	50	336
	540	1000	938	136	765	111	18	61	282
	650	1200	786	114	655	95	24	66	235
50B46	205	400	...	...	...	...	...	...	560
	315	600	1779	258	1620	235	10	37	505
	425	800	1393	202	1248	181	13	47	405
	540	1000	1082	157	979	142	17	51	322
	650	1200	883	128	793	115	22	60	273
50B60	205	400	...	...	...	...	...	...	600
	315	600	1882	273	1772	257	8	32	525
	425	800	1510	219	1386	201	11	34	435
	540	1000	1124	163	1000	145	15	38	350
	650	1200	896	130	779	113	19	50	290
5130	205	400	1613	234	1517	220	10	40	475
	315	600	1496	217	1407	204	10	46	440
	425	800	1275	185	1207	175	12	51	379
	540	1000	1034	150	938	136	15	56	305
	650	1200	793	115	689	100	20	63	245

(continued)

Mechanical properties of selected carbon and alloy steels in the quenched-and-tempered condition (continued)

AISI No.(a)	Tempering temperature °C	°F	Tensile strength MPa	ksi	Yield strength MPa	ksi	Elongation, %	Reduction in area, %	Hardness, HB
5140	205	400	1793	260	1641	238	9	38	490
	315	600	1579	229	1448	210	10	43	450
	425	800	1310	190	1172	170	13	50	365
	540	1000	1000	145	862	125	17	58	280
	650	1200	758	110	662	96	25	66	235
5150	205	400	1944	282	1731	251	5	37	525
	315	600	1737	252	1586	230	6	40	475
	425	800	1448	210	1310	190	9	47	410
	540	1000	1124	163	1034	150	15	54	340
	650	1200	807	117	814	118	20	60	270
5160	205	400	2220	322	1793	260	4	10	627
	315	600	1999	290	1772	257	9	30	555
	425	800	1606	233	1462	212	10	37	461
	540	1000	1165	169	1041	151	12	47	341
	650	1200	896	130	800	116	20	56	269
51B60	205	400	...	...	...	...	...	...	600
	315	600	...	...	...	...	...	...	540
	425	800	1634	237	1489	216	11	36	460
	540	1000	1207	175	1103	160	15	44	355
	650	1200	965	140	869	126	20	47	290
6150	205	400	1931	280	1689	245	8	38	538
	315	600	1724	250	1572	228	8	39	483
	425	800	1434	208	1331	193	10	43	420
	540	1000	1158	168	1069	155	13	50	345
	650	1200	945	137	841	122	17	58	282
81B45	205	400	2034	295	1724	250	10	33	550
	315	600	1765	256	1572	228	8	42	475
	425	800	1407	204	1310	190	11	48	405
	540	1000	1103	160	1027	149	16	53	338
	650	1200	896	130	793	115	20	55	280
8630	205	400	1641	238	1503	218	9	38	465
	315	600	1482	215	1392	202	10	42	430
	425	800	1276	185	1172	170	13	47	375
	540	1000	1034	150	896	130	17	54	310
	650	1200	772	112	689	100	23	63	240
8640	205	400	1862	270	1669	242	10	40	505
	315	600	1655	240	1517	220	10	41	460
	425	800	1379	200	1296	188	12	45	400
	540	1000	1103	160	1034	150	16	54	340
	650	1200	896	130	800	116	20	62	280
86B45	205	400	1979	287	1641	238	9	31	525
	315	600	1696	246	1551	225	9	40	475
	425	800	1379	200	1317	191	11	41	395
	540	1000	1103	160	1034	150	15	49	335
	650	1200	903	131	876	127	19	58	280
8650	205	400	1937	281	1675	243	10	38	525
	315	600	1724	250	1551	225	10	40	490
	425	800	1448	210	1324	192	12	45	420
	540	1000	1172	170	1055	153	15	51	340
	650	1200	965	140	827	120	20	58	280
8660	205	400	...	...	...	...	...	...	580
	315	600	...	...	...	...	...	...	535
	425	800	1634	237	1551	225	13	37	460
	540	1000	1310	190	1213	176	17	46	370
	650	1200	1068	155	951	138	20	53	315
8740	205	400	1999	290	1655	240	10	41	578
	315	600	1717	249	1551	225	11	46	495
	425	800	1434	208	1358	197	13	50	415
	540	1000	1207	175	1138	165	15	55	363
	650	1200	986	143	903	131	20	60	302
9255	205	400	2103	305	2048	297	1	3	601
	315	600	1937	281	1793	260	4	10	578
	425	800	1606	233	1489	216	8	22	477
	540	1000	1255	182	1103	160	15	32	352
	650	1200	993	144	814	118	20	42	285
9260	205	400	...	...	...	...	...	...	600
	315	600	...	...	...	...	...	...	540
	425	800	1758	255	1503	218	8	24	470
	540	1000	1324	192	1131	164	12	30	390
	650	1200	979	142	814	118	20	43	295
94B30	205	400	1724	250	1551	225	12	46	475
	315	600	1600	232	1420	206	12	49	445
	425	800	1344	195	1207	175	13	57	382
	540	1000	1000	145	931	135	16	65	307
	650	1200	827	120	724	105	21	69	250

(a) All grades are fine-grained except for those in the 1100 series which are coarse-grained. Heat treated specimens were oil quenched unless otherwise indicated. (b) Water quenched.

Steel Sheet, Strip and Plate

Low-Carbon Steel Sheet and Strip

LOW-CARBON STEEL sheet and strip are used primarily in consumer goods. These applications require materials that are serviceable under a wide variety of conditions and that are especially adaptable to low-cost techniques of mass production into articles having good appearance. Therefore, these products must incorporate, in various degrees and combinations, ease of fabrication, adequate strength, excellent finishing characteristics to provide attractive appearance after fabrication, and compatibility with other materials and with various coatings and processes.

The steels used for these products are supplied over a wide range of chemical compositions; however, the vast majority are unalloyed, low-carbon steels selected for stamping applications, such as automobile bodies and appliances. For these major applications, typical compositions are 0.05 to 0.10% carbon, 0.25 to 0.50% manganese, 0.035% max phosphorus, and 0.04% max sulfur.

Generally, rimmed (or capped) steel is utilized because of its surface and formability characteristics. Rimmed or capped steel, however, will strain age, leading to associated changes in mechanical properties. Where strain aging is to be avoided and/or when exceptionally deep draws are employed, steel stabilized ("killed") with aluminum (drawing quality, special killed) usually is selected.

The width differentiation between sheet and strip made of low-carbon steel depends on the rolling process. In hot rolled products, widths up to 305 mm (12 in.) are classified as strip; widths in excess of 305 mm are classified as sheet. In cold rolled products, the strip category extends up to 608 mm ($23^{15}/_{16}$ in.). It should be noted that both sheet and strip can be purchased as either cut lengths or coils. The standard published dimensional tolerances for low-carbon steel strip are more restrictive than those for sheet. For standard size ranges of low-carbon steel sheet and strip, see Table 1. Typical characteristics of the various qualities of these products are listed in Table 2.

PRODUCTION OF SHEET AND STRIP

Most cold rolled low-carbon steel sheet is available in two classes (see Table 2). Class 1 (temper rolled) is intended for applications where surface appearance is important and where specified surface and flatness requirements must be met. Class 2 material may be annealed last or temper rolled. It is a product intended for applications where appearance is less important and where limitations on surface texture, imperfections, flatness and stretcher-strain tendency are not applicable.

Cold rolled low-carbon steel strip is available in five hardness tempers ranging from full hard to dead soft (see Table 4).

QUALITY DESCRIPTORS

The descriptors of "quality" that are used for hot rolled low-carbon steel sheet and strip and for cold rolled low-carbon steel sheet include commercial quality; drawing quality; drawing quality, special killed; and structural quality (see Table 2). Some of the as-rolled material made to these qualities is subject to surface disturbances known as coil breaks, fluting and stretcher strains, but it can be supplied so that fluting and stretcher strains will not be produced during subsequent forming if it is temper rolled and/or is roller leveled immediately prior to forming. It should be noted that any beneficial effects of roller leveling deteriorate rapidly. In addition to the require-

Table 2. Characteristics of low-carbon steel sheet and strip(a)

Quality designations	Applicable ASTM specification No.	AISI-SAE grade designation	Surface quality(b)	Strain aging(c)	Product form	Normally available thickness range mm	in.
Hot rolled							
Commercial	A569	1008-1012	U	Yes	Sheet	1.50-5.82	0.059-0.229
					Strip	0.86-5.82	0.034-0.229
	A635	1008-1012	U	Yes	Sheet	5.84-12.70	0.230-0.500
					Strip	0.86-5.82	0.034-0.229
	A659	1015-1023	U	Yes	Sheet	1.50-5.82	0.059-0.229
					Strip	0.86-5.82	0.034-0.229
Drawing	A621	1006-1008	U	Yes	Sheet	1.91-4.75	0.075-0.187
					Strip	. . .	. . .
Drawing, special killed	A622	1006-1008	U	No	Sheet	1.91-4.75	0.075-0.187
					Strip	. . .	. . .
Structural (physical)	A570	None(d)	U	Yes	Sheet	. . .	. . .
					Strip	. . .	. . .
Cold rolled							
Commercial							
Class 1 (temper rolled)	A366	1008-1012	E	Yes	Sheet	0.64-2.79	0.025-0.110
Class 2 (annealed last)	A366	1008-1012	U	No	Sheet	0.64-2.79	0.025-0.110
Drawing							
Class 1 (temper rolled)	A619	1006-1008	E	Yes	Sheet	0.64-2.79	0.025-0.110
Class 2 (annealed last)	A619	1006-1008	U	No	Sheet	0.64-2.79	0.025-0.110
Drawing, special killed							
Class 1 (temper rolled)	A620	1006-1008	E	No	Sheet	0.64-2.79	0.025-0.110
Class 2 (annealed last)	A620	1006-1008	U	No	Sheet	0.64-2.79	0.025-0.110
Structural (physical)	A611	None(d)	E	Yes	Sheet	. . .	. . .
Unspecified	A109	None(e)	E	Yes	Strip	. . .	. . .

(a) The general requirements for this class of products, except those covered by ASTM Specifications A109 and A635, are described in ASTM Specification A568. (b) E = suitable for exposed parts; U = suitable for unexposed parts. (c) Yes = has propensity to strain age; no = does not have propensity to strain age at room temperature. (d) Produced in five grades with specific mechanical-property limits; composition subordinate to mechanical properties. (e) Produced in five tempers with specific hardness and bend-test limits; composition subordinate to mechanical properties.

Table 1. Standard sizes of low-carbon steel sheet and strip

Product	Applicable ASTM specification No.	Thickness range mm	in.	Width range mm	in.	Delivery form
Hot rolled						
Sheet	A569, A570, A621, A622, A659	1.14-4.56	0.0449-0.1799	Over 300	Over 12.00	Cut lengths or coils
		4.57-5.84	0.1800-0.2299	301-1220	12.01-48.00	Cut lengths or coils
		4.57-5.84	0.1800-0.2299	Over 1220	Over 48.00	Coils
Sheet	A635	5.85-12.70	0.2300-0.5000	Over 300	Over 12.00	Coils
Strip	A569, A570, A621, A622, A659	0.65-0.87	0.0255-0.0343	12.7-90	0.50-3.50	Cut lengths or coils
		0.87-1.14	0.0344-0.0448	12.7-150	0.50-6.00	Cut lengths or coils
		1.14-5.16	0.0449-0.2030	12.7-300	0.50-12.00	Cut lengths or coils
		5.16-5.84	0.2031-0.2299	151-300	6.01-12.00	Cut lengths or coils
Strip	A635	5.85-12.70	0.2300-0.5000	201-300	8.01-12.00	Coils
Cold rolled						
Sheet	A366, A611, A619, A620	0.36 and over	0.0142 and over	Over 300	Over 12.00	Cut lengths or coils
Sheet, slit from wider coils	A366, A611, A619, A620	0.36-2.09	0.0142-0.0821	50-300	2.00-12.00	Cut lengths or coils
Strip	A109	Through 6.4	Through 0.2499	13-610	0.51-23.94	Cut lengths or coils

Table 3. Minimum mechanical properties of structural (physical) quality low-carbon steel sheet and strip

ASTM specification No.	Grade	Tensile strength MPa	ksi	Yield strength MPa	ksi	Elongation in 50 mm or 2 in., %	Bend diameter (a)
Hot rolled sheet and strip							
A570	A	310	45	170	25	23-27(b)	0
	B	340	49	205	30	21-25(b)	t
	C	360	52	230	33	18-23(b)	1.5t
	D	380	55	275	40	15-21(b)	2t
	E	400	58	290	42	13-19(b)	2.5t
Cold rolled sheet							
A611	A	290	42	170	25	26	0
	B	310	45	205	30	24	t
	C	330	48	230	33	22	1.5t
	D	360	52	275	40	20	2t
	E	570	82	550	80	...	(c)

(a) Value listed is the minimum diameter of a pin or rod, in multiples of the material thickness (t), around which the material can be bent 180° in any direction at room temperature without cracking. (b) Minimum elongation values depend on material thickness. (c) Bend test not applicable.

Table 4. Mechanical properties of cold rolled low-carbon steel strip (ASTM A109)

Temper	Rockwell B hardness requirements	Bend-test requirements	Approximate tensile strength MPa	ksi	Elongation in 50 mm or 2 in., %(d)
No. 1 (hard)	90 min(a), 84 min(b)	No bending in either direction	550-690	80-100	...
No. 2 (half-hard)	70-85	90° bend across rolling direction around a 1t radius(c)	380-520	55-75	4-16
No. 3 (quarter-hard) ...	60-75	180° bend across rolling direction and 90° bend along rolling direction, both around a 1t radius(c)	310-450	45-65	13-27
No. 4 (skin-rolled)	65 max	Bend flat on itself in any direction	290-370	42-54	24-40
No. 5 (dead-soft)	55 max	Bend flat on itself in any direction	260-340	38-50	31-47

(a) For strip less than 1.78 mm (0.070 in.) thick. (b) For strip 1.78 mm (0.070 in.) thick and greater. (c) t = thickness of strip. (d) For strip 1.27 mm (0.050 in.) thick.

ments listed below for the various qualities of low-carbon steel sheet and strip, internal soundness also may be specified.

Commercial quality (CQ) low-carbon steel sheet and strip are suitable for moderate forming; material of this quality has sufficient ductility to be bent flat on itself in any direction in a standard room-temperature bend test. Commercial quality material is not subject to any other mechanical test requirements and it is not expected to have exceptionally uniform chemical composition or mechanical properties. However, the hardness of cold rolled commercial quality sheet ordinarily is less than 60 HRB at the time of shipment.

Drawing Quality. When greater ductility or more uniform properties than those afforded by commercial quality are required, drawing quality (DQ) is specified. Drawing quality material is suitable for production of deep drawn parts and other parts requiring severe deformation. When the deformation is particularly severe or resistance to stretcher strains is required, drawing quality, special killed (DQSK) is specified.

Structural quality (SQ), formerly called *physical quality* (PQ), is applicable when specified strength and elongation values are required in addition to bend tests (see Table 3). Minimum values of tensile strength ranging up to 690 MPa (100 ksi) in hot rolled sheet and strip and up to 1030 MPa (150 ksi) in cold rolled sheet are available. Cold rolled strip, which does not have a "quality" descriptor, is available in five tempers that conform to specified Rockwell hardness ranges and bend-test requirements (see Table 4).

MECHANICAL PROPERTIES

The commonly measured tensile properties of low-carbon steel sheet and strip are not readily related to their performance in fabrication; the relationship between formability and values of n

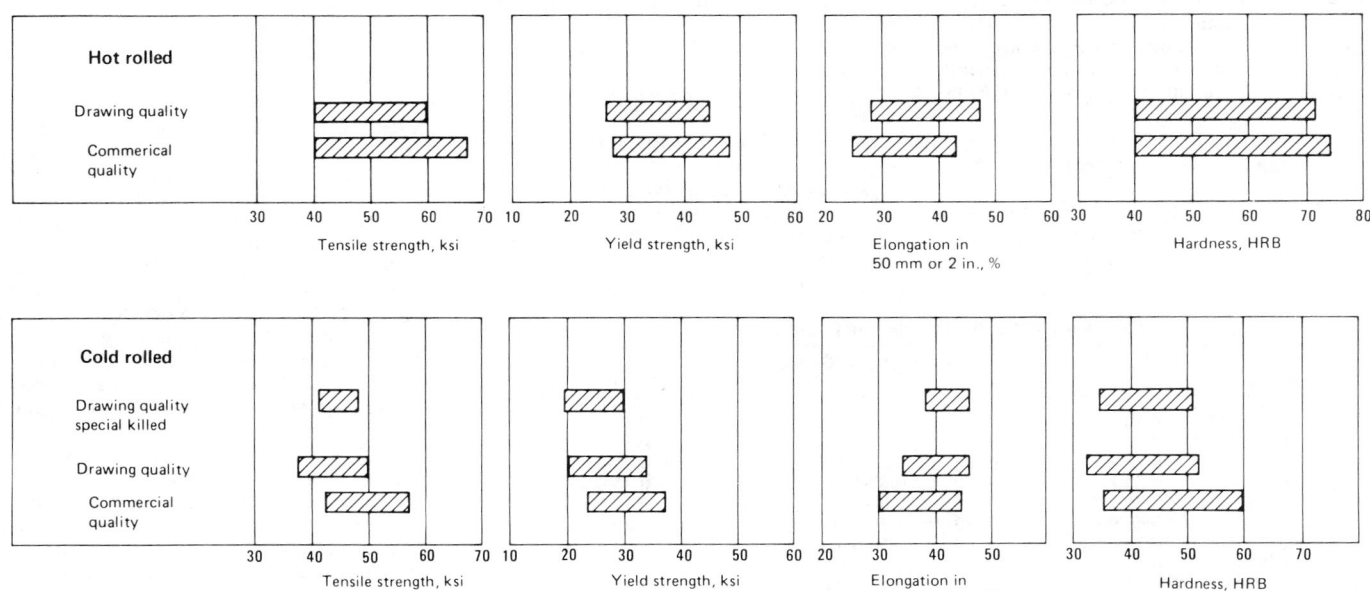

Range of properties of steel furnished by three mills. Hot rolled sheet thickness from 16 to 10 gage, inclusive (1.519 to 3.416 mm or 0.0598 to 0.1345 in.); cold rolled from 22 to 16 gage, inclusive (0.759 to 1.519 mm or 0.0299 to 0.598 in.). All cold rolled grades include a temper pass. All grades were rolled from rimmed steel except the one labeled "special killed." See Table 3 for mechanical properties of structural (physical) quality sheet.

Fig. 1. Typical mechanical properties of low-carbon steel sheet

and r (determined in tensile testing) is discussed in the article on formability. Mechanical properties are not applicable to commercial quality; drawing quality; and drawing quality, special killed sheet and strip. As soon as a mechanical property is specified, the material is identified as structural quality. As a matter of general interest, however, the ranges of mechanical properties typical of sheet produced by three mills in these qualities are given in Fig. 1. The bands would be wider if the product of the entire industry were represented (see also Tables 3 and 4).

It should be noted that the ranges are broader and the sheet harder for the hot rolled than for the cold rolled materials.

Alloy Steel Sheet and Strip

ALLOY STEEL sheet and strip are used primarily for those special applications that require the mechanical properties normally obtained by heat treatment. A sizeable selection of the standard alloy steels are available as sheet and strip, either hot rolled or cold rolled. The most commonly available alloys are listed in Table 5.

QUALITY DESCRIPTORS

As it is used for steel mill products, the term "quality" relates to the general suitability of the mill product to make a given class of parts. For alloy steel sheet and strip, the various quality descriptors imply certain inherent characteristics, such as degree of internal soundness and relative freedom from harmful surface imperfections.

The quality descriptors used for alloy steel sheet and plate include *regular quality, drawing quality* and *aircraft quality*, which are covered by ASTM specifications. The general requirements for these qualities are covered by ASTM A505. Additional qualities include *bearing quality, saw quality* and *aircraft structural quality*. Aircraft quality requirements are also covered in AMS specifications.

MILL HEAT TREATMENT

Hot rolled regular-quality alloy steel sheet and strip normally are available from the producer either as rolled or as heat treated. Standard mill

Table 6. Lowest maximum hardness for regular mill annealed and flattened hot rolled or cold rolled alloy steel sheet and strip(a)

AISI number	Maximum hardness	AISI number	Maximum hardness
4130	93 HRB	5140	98 HRB
4140	100 HRB	5150	100 HRB
4142	100 HRB	5160	27 HRC
4145	100 HRB	6150	100 HRB
4150	25 HRC	8630	93 HRB
4340	27 HRC	8640	100 HRB
		8645	100 HRB

(a) From AISI Steel Products Manual, Alloy Steel Sheets and Strip, April 1969. Specific microstructure requirements may result in different maximum hardness numbers. Lower hardness can be obtained with prolonged heating and special controlled cooling cycles.

heat treated conditions are: annealed, normalized, and normalized and tempered. Cold rolled regular-quality product normally is available only in the annealed condition.

The lowest maximum hardness that can be expected with regular mill annealed and flattened hot or cold rolled alloy steel sheet and strip is given in Table 6.

Precoated Steel Sheet

STEEL SHEET frequently is coated in coil form before fabrication either by the steel mills or by specialists called coil coaters.

The basic types of precoating include metallic, conversion, preprimed and prepainted finishing. Metallic coating may be done with zinc, aluminum, tin, terne metal, or combinations of metals such as aluminum and zinc. Conversion coatings are completed with phosphates, and preprimed finishes are done with zinc chromate and zinc-rich coatings. Prepainting consists of applying an organic paint system to steel sheet on a coil coating line either at a mill or at a coil coater.

ZINC COATINGS

Metallic zinc is applied to iron and steel by one of three processes: hot dip galvanizing, electrogalvanizing or zinc spraying. Most galvanized steel sheet is coated by the hot dip process.

Because service life of a zinc-coated part in a given atmosphere is directly proportional to the

amount of zinc in the coating, measurement of that amount is very important. The amount of zinc is most often measured by weight rather than thickness, usually by the method described in ASTM A90. Specimens are cut from one or three spots in samples of the sheet, as described in ASTM A525. These are weighed; the zinc is stripped (dissolved) in an acid solution; and the specimens are reweighed. The weight loss is reported in ounces per square foot of sheet. When specimens from three spots are checked (triple-spot test), the value of weight loss is the average of the three specimens (see Table 7).

Table 7. Designations and weights of zinc coating on hot dip galvanized sheet(a)

Coating designation(b)	Minimum coating weights(c)			
	Triple-spot test		Single-spot test	
	g/m²	oz/ft²	g/m²	oz/ft²
G 235	717	2.35	610	2.00
G 210	640	2.10	549	1.80
G 185	564	1.85	488	1.60
G 165	503	1.65	427	1.40
G 140	427	1.40	366	1.20
G 115	351	1.15	305	1.00
G 90	275	0.90	244	0.80
G 60	183	0.60	152	0.50
G 01	No minimum		No minimum	
A 60	183	0.60	152	0.50
A 40	122	0.40	91	0.30
A 01	No minimum		No minimum	

(a) From ASTM A525. (b) G, regular type of coating; A, zinc-iron alloy type of coating. (c) Total weight on both sides of sheet per unit area of sheet.

Hot dip galvanizing is a process in which an adherent, protective coating of zinc and iron-zinc alloys is developed on the surfaces of iron and steel products by immersing them in a bath of molten zinc. Most zinc coating of steel sheet is done by this process, usually on a continuous galvanizing line.

A typical hot dip galvanized coating consists of a series of layers. Starting from the basis steel at the bottom of the coating, each successive layer contains a higher proportion of zinc until the outer layer, which is relatively pure zinc, is reached. There is, therefore, no discrete line of demarcation between the iron and zinc, but a gradual transition through the series of iron-zinc alloys that provide a powerful bond between the basis metal and the coating.

Electrogalvanizing. Very thin formable zinc coatings ideally suited for deep drawing or painting can be obtained on steel products by electrogalvanizing. Zinc is electrodeposited on a variety of mill products, sheet, wire and, in some instances, pipe. Electrogalvanizing of sheet and wire in coil form produces a thin, uniform coating of pure zinc.

Electrodeposited zinc coatings are simpler in structure than hot dip galvanized coatings. They are composed of pure zinc, have a homogeneous structure, and are highly adherent. These coatings are not generally as thick as those produced by hot dip galvanizing. However, they do give good corrosion-free service. Electrogalvanized coatings in thicknesses up to 13.7 μm (0.54 mil, or 0.32 oz/ft²) have been applied to one or both sides of steel sheet. Common thicknesses are 1.65 and 3.56 μm (0.065 and 0.14 mil). To protect automobiles against the effects of road salts and entrapped moisture, heavier coatings 13.7 μm thick are now applied to the reverse sides of exposed body panels. Electrodeposited zinc is con-

Table 5. Standard alloy steels readily available as sheet and strip(a)

Grade	Chemical composition ranges and limits(b), %						
	C	Mn	Si	Cr	Ni	Mo	V, min
4130	0.28-0.33	0.40-0.60	0.15-0.30	0.80-1.10	...	0.15-0.25	...
4140	0.38-0.43	0.75-1.00	0.15-0.30	0.80-1.10	...	0.15-0.25	...
4142	0.40-0.45	0.75-1.00	0.15-0.30	0.80-1.10	...	0.15-0.25	...
4145	0.43-0.48	0.75-1.00	0.15-0.30	0.80-1.10	...	0.15-0.25	...
4150	0.48-0.53	0.75-1.00	0.15-0.30	0.80-1.10	...	0.15-0.25	...
4340	0.38-0.43	0.60-0.80	0.15-0.30	0.70-0.90	1.65-2.00	0.20-0.30	...
5140	0.38-0.43	0.70-0.90	0.15-0.30	0.70-0.90	...	...	...
5150	0.48-0.53	0.70-0.90	0.15-0.30	0.70-0.90	...	...	...
5160	0.55-0.65	0.75-1.00	0.15-0.30	0.70-0.90	...	...	...
6150	0.48-0.53	0.70-0.90	0.15-0.30	0.80-1.10	...	...	0.15
8615	0.13-0.18	0.70-0.90	0.15-0.30	0.40-0.60	0.40-0.70	0.15-0.25	...
8617	0.15-0.20	0.70-0.90	0.15-0.30	0.40-0.60	0.40-0.70	0.15-0.25	...
8620	0.18-0.23	0.70-0.90	0.15-0.30	0.40-0.60	0.40-0.70	0.15-0.25	...
8630	0.28-0.33	0.70-0.90	0.15-0.30	0.40-0.60	0.40-0.70	0.15-0.25	...
8640	0.38-0.43	0.75-1.00	0.15-0.30	0.40-0.60	0.40-0.70	0.15-0.25	...
8645	0.43-0.48	0.75-1.00	0.15-0.30	0.40-0.60	0.40-0.70	0.15-0.25	...

(a) Open hearth, basic oxygen and electric furnace steels. (b) For open hearth and basic oxygen processes, these steels may have a maximum of 0.035% P and 0.040% S; for electric furnace process, the maximums are 0.025% P and 0.025% S.

sidered to adhere to steel as well as any other metallic coating. Because of its excellent adhesion, electrogalvanized coils of steel sheet and wire have good working properties, and the coating remains intact after severe deformation. Good adhesion depends on very close physical conformity of the coating with the basis metal. Therefore, particular care must be taken during initial cleaning. Electrodeposition also affords a means of applying zinc coatings to finished parts that cannot be predipped. It is especially useful where a high processing temperature could injure the part. One advantage of electrodeposition is that it can be done cold and thus does not change the mechanical properties of the steel.

Zinc spraying consists of projecting atomized particles of molten zinc onto a prepared surface. Three types of spraying pistols are in commercial use today: the molten metal pistol, the powder pistol and the wire pistol. The sprayed coating is slightly rough and slightly porous; the specific gravity of a typical coating is approximately 6.35, compared with 7.1 for cast zinc. This slight porosity does not affect the protective value of the coating, because zinc is anodic to steel. The zinc corrosion products that form during service fill the pores of the coating, giving a solid appearance. The slight roughness of the surface makes it an ideal basis for paint, when properly pretreated.

Chromate Passivation. Several types of finishes can be applied to zinc-coated surfaces to provide extra corrosion resistance. The simplest type of finish applicable to fresh zinc surfaces is a chromate passivation treatment. It is equally suitable for use on hot dip galvanized, zinc-sprayed and zinc-plated articles. Usually, the treatment consists of simply cleaning the articles and then dipping them in a chromic acid or sodium dichromate solution at about 20 to 30 °C (68 to 86 °F), followed by rinsing in cold fresh water and drying in warm air.

Anodizing of galvanized steel produces a surface that exhibits exceptional resistance to corrosion, wear, heat and abrasion. The anodized film is an electrical insulator. The film is formed by immersing finished parts in a simple electrolyte and applying an increasing voltage for up to 10 min. A nonreflective functional coating, anodized zinc demonstrates superior properties when compared with chromate conversion coatings. The anodized zinc coatings range from 25 to 33 μm (1.0 to 1.3 mils) in thickness and will not deteriorate at temperatures up to the melting point of zinc.

Painting. The selection of galvanized steel as a material for barns, buildings, roofs, sidings, appliances and many hardware items is based on the sacrificial protection plus the barrier coating afforded the basis metal by zinc coating. For additional protection and pleasing appearance, paint coatings are often applied to the galvanized steel.

Weldability. Direct spot welding is recommended for zinc-coated steel, which may be either hot dip galvanized or electroplated, because the shunting current associated with series welding, when added to the higher-than-normal current needed to weld zinc-coated steel, results in excessive electrode heating and short electrode life. Weldability of thin sheets electroplated with zinc decreases as the coating thickness increases, in the range from 0.005 to 0.025 mm (0.0002 to 0.001 in.). However, as sheet thickness increases above 1.5 mm (0.060 in.), weldability increases regardless of coating thickness. The welding behavior of hot dip galvanized steel is affected by the thickness and uniformity of the zinc-iron alloy layer, as well as the thickness of unalloyed zinc.

ALUMINUM COATINGS

Aluminized (aluminum-coated) steel sheet is used for applications where heat resistance, heat reflectivity or resistance to corrosion are required in an esthetically pleasing, economical sheet. Aluminum coating is done on continuous lines similar to those used for hot dip galvanizing of steel sheet. Cold rolled steel sheet is hot dipped into molten aluminum or an aluminum alloy containing 5 to 10% silicon. The coating consists of two interfacial layers. Between the exterior layer of aluminum-silicon alloy and the steel basis metal, an aluminum-iron-silicon alloy layer is formed. This alloy can significantly affect the ductility, adhesion, uniformity, smoothness and appearance of the surface and is controlled for optimum properties.

Coating Weight. Aluminum coatings on steel sheet are designated according to total coating weight on both surfaces in ounces per square foot of sheet, as shown in Table 8. Type 1 is produced to ASTM A463. Type 1, light coating, is recommended for drawing applications and also when welding is a significant portion of the fabrication sequence. Type 1, regular or commercial, has approximately a 1-mil-thick coating on each surface. It is designated for applications requiring excellent heat resistance. Type 2 has a coating approximately 2 mils thick on each side. It is frequently used for atmospheric corrosion resistance.

Corrosion Resistance. Aluminum's value as a protective coating for steel sheet lies principally in its inherent corrosion resistance. In most environments, the long-term corrosion rate of aluminum is only about 15 to 25% of that of zinc. Generally, the protective value of an aluminum coating on steel is a function of coating thickness. The coating tends to remain intact and thus provides long-term protection.

Aluminum coatings do not provide sacrificial protection in most environments, particularly in atmospheric exposure.

Heat Resistance. Aluminum-coated sheet steel has excellent resistance to high-temperature oxidation. At surface temperatures above about 510 °C (950 °F), the aluminum coating protects the steel basis metal against oxidation without discoloration. Between 510 and 675 °C (950 and 1250 °F), the coating provides protection to the steel, but some darkening may result from the formation of aluminum-iron-silicon alloy. The alloy is extremely heat resistant, but on long exposure, the coating may become brittle and spall at temperatures above 675 °C (1250 °F) due to a different coefficient of expansion from that of the steel.

Weldability. Aluminum-coated steel sheet can be joined by electric resistance welding (spot welding or seam welding). It also can be metal-arc welded, flash welded or oxyacetylene welded. Thorough removal of grease, oil, paint and dirt followed by wire brushing is recommended before joining. Special fluxes are required for metal arc and oxyacetylene welding. During spot welding, electrodes tend to pick up aluminum, and the tips must be dressed more frequently than during spot welding of uncoated steel. Also, current density should be higher.

TIN COATINGS

Tin coatings are applied to steel sheet either by electrolytic deposition or by immersion in a molten bath of tin (hot dip process). Hot dip tin coatings are applied to provide nontoxic, protective and decorative coating of food-handling, packaging and dairy equipment, and to facilitate soldering of components used in electronic and electrical equipment. In the United States, hot dip tin coating has been replaced by electrolytic tin coating.

Electrolytic tin-coated steel sheet is used where solderability, appearance or corrosion resistance under certain conditions is important, such as in electronic equipment, food-handling and processing equipment, and laboratory clamps. It is generally produced with a matte finish formed by applying the coating to basis metal sheet called "black plate," which has a dull surface texture, and leaving the coating unmelted. It can also be produced with a bright finish formed by applying the coating to basis metal having a smooth surface texture and then melting the coating. Electrolytic tin-coated sheet is usually produced in nominal thicknesses from 0.38 to 0.84 mm (0.015 to 0.033 in.) and in widths from 300 to 910 mm (12 to 36 in.).

Electrolytic tin-coated steel sheet may be specified to one of the five coating-weight designations listed in Table 9. The coating weight is the total amount of tin on both surfaces, expressed in ounces per square foot of sheet.

Table 9. Designations and weights of tin coating on electrolytic tin-coated steel sheet(a)

| Coating designation | Minimum coating weight(b) | | | |
| | Triple-spot test | | Single-spot test | |
	g/m²	oz/ft²	g/m²	oz/ft²
25	3.7	0.012	2.8	0.009
50	7.3	0.024	5.6	0.018
75	11.0	0.036	8.2	0.027
100	14.6	0.048	11.0	0.036
125	18.3	0.060	13.8	0.045

(a) From ASTM A599. (b) Total weight on both sides of sheet per unit area of sheet.

TERNE COATINGS

Long terne steel sheet is carbon steel sheet continuously coated by various hot dip processes with terne metal (lead with 3 to 15% tin). This coated sheet is duller in appearance than conventional tin-coated sheet, hence the name (terne) from the French, which means "dull" or "tarnished." The smooth, dull coating gives the sheet corrosion resistance, formability, excellent solderability and paintability. The term "long terne" is used to describe terne-coated sheet, while "short terne" is used for terne-coated plate.

Because of its unusual properties, long terne sheet has been adapted to a wide variety of applications. Its greatest use is in automotive gasoline tanks. Its excellent solderability and special

Table 8. Designations and weights of aluminum coating on aluminized steel sheet(a)

| Coating designation | Minimum coating weight(b) | | | |
| | Triple-spot test | | Single-spot test | |
	g/m²	oz/ft²	g/m²	oz/ft²
T1 25 (light)	80	0.25	60	0.20
T1 40 (regular)	120	0.40	90	0.30
T2 (regular)	230	0.75	200	0.65

(a) From ASTM A463 and A428. (b) Total weight on both sides of sheet per unit area of sheet.

Table 10. Designations and weights of lead-tin coating on terne-coated steel sheet(a)

Coating designation	Minimum coating weight(b)			
	Triple-spot test		Single-spot test	
	g/m²	oz/ft²	g/m²	oz/ft²
LT01	No minimum		No minimum	
LT25	76	0.25	61	0.20
LT35	107	0.35	76	0.25
LT40	122	0.40	92	0.30
LT55	168	0.55	122	0.40
LT85	259	0.85	214	0.70
LT110	336	1.10	275	0.90

(a) From ASTM A308. (b) Total weight on both sides of sheet per unit area of sheet.

corrosion resistance makes the product well-suited for this application.

Long terne sheet is often produced to ASTM A308. The coatings are designated according to total coating weight on both surfaces in ounces per square foot of sheet, as shown in Table 10. For applications requiring good formability, the coating is applied over low-carbon steel sheet of commercial quality, drawing quality, or drawing quality special killed. The terne coating acts as a lubricant and facilitates forming, and the strong bond of the terne metal allows it to be formed along with the basis metal. When higher strength is required, the coating can be applied over low-carbon steel sheet of structural (physical) quality, although at some sacrifice in ductility.

PHOSPHATE COATINGS

Phosphate coating of iron and steel consists of treatment with a dilute solution of phosphoric acid and other chemicals whereby the surface of the metal, reacting chemically with the phosphoric acid, is converted to an integral layer of insoluble crystalline phosphate compound. This layer is less reactive than the metal surface and at the same time is more absorbent of lubricants or paints. Because the coating is an integral part of the surface, it adheres to the basis metal tenaciously.

The chief application for iron phosphate coatings is as a paint base for nongalvanized carbon steel sheet; such a coating is usually applied on coil-coating lines. Manganese phosphate coatings are used chiefly as an oil base on engine parts for break-in and to prevent galling.

Steel Sheet for ——Porcelain Enameling——

PORCELAIN ENAMELS are inorganic coatings fired at temperatures around 730 to 870 °C (1350 to 1600 °F). They are distinct from ceramic coatings used to protect metal at temperatures of 1000 °C (1800 °F) or higher because of their different compositions.

The base metal on which porcelain enamel is applied may be gold, iron, aluminum, copper or stainless steel, but most commercial porcelain enamel is applied to low-carbon steel. Properties of the particular steel sheet should be evaluated before its selection for enameling; some steel sheet is not recommended because of problems that complicate the process of coating the sheet with porcelain enamel. The four major problems are:

1. Distortion caused by sag and warpage occur-

ring at the temperatures reached during firing of the enamel
2. Improper surface preparation that causes poor adherence
3. Fishscale imperfections, particularly of the delayed variety, caused by hydrogen evolution
4. Carbon boiling in the enamel coating caused by surface carbides in the steel.

Steels for porcelain enameling with a single cover coat (white or colored) have since been developed. These steels, in addition to possessing single-coat coverage, have good sag resistance and freedom from carbon boiling, fishscale formation and the other problems associated with carbon and internal imperfections in the steel. These steels are highly formable and may be drawn into more complex shapes than enameling iron.

TYPES OF SHEET

Six types of steel sheet are available for porcelain enameling. These are described below. In most instances, a detailed review should be made with the steel supplier before ordering a particular grade or beginning production. Typical chemical analysis of the steels recommended for enameling are listed in Table 11.

Cold rolled aluminum-killed steel is a deep drawing grade of low-carbon steel that can be formed into complex shaped parts. However, it is inherently subject to blemishes and certain other surface imperfections.

Cold rolled rimmed steel is less expensive than aluminum-killed steel and is used in moderately formed, unexposed parts, where warpage and distortion are not critical. Normally, only a ground coat is applied and enamel firing temperatures of less than 790 °C (1450 °F) are used to minimize sag and warpage.

Enameling iron is a premium grade of commercially pure iron. For applications requiring improved formability to produce a given shaped part, as much as 0.20% manganese may be added. Enameling iron requires a ground coat containing adherence-promoting oxides prior to application of the white (or colored) cover coat. Thus, the material receives a costly double firing cycle at 815 to 830 °C (1500 to 1525 °F).

There are two main advantages to using enameling iron instead of cold rolled steel for porcelain enameling. First, enameling iron, due to its low carbon and manganese levels, retains formed shape better during fusion of the porcelain enamel. This is often described as improved sag resistance. Second, due to its low carbon content, enameling iron is less subject to carbon boiling during firing.

Decarburized enameling steels (also called direct-on cover coat steels) receive a decarburizing anneal to prevent carbon boiling. They were developed primarily for use with a direct-on cover coat enamel (white or colored). Resistance to fishscale is achieved mainly by controlling ladle carbon content, hot coiling off the strip mill, and the percentage of cold reduction. These steels may be either rimmed or aluminum-killed, the latter being used only if aging due to prolonged storage prior to forming is necessary.

Interstitial-free enameling steel is not subject to critical grain growth at the enamel firing temperature used for single cover coat enamel and will, therefore, not sag or distort excessively. It is essentially a steel with very low carbon and is not subject to fishscale. The surface quality is not always as good as that of rimmed steel. Consequently, some enamel problems are possible, such as black specks that appear when single-coat white enamels are fired. It has a highly developed texture, and because it has a plastic strain ratio of around 1.7 r_m, it can be deep drawn into more complex shapes than other enameling grades. Lack of distortion and retention of shape are superior in enameled products made with interstitial-free enameling steel.

Carbon and Low-Alloy ——Steel Plate——

STEEL PLATE is any hot finished flat-rolled steel product over 203 mm (8 in.) wide and over 5.84 mm (0.230 in.) thick or over 1220 mm (48 in.) wide and over 4.57 mm (0.180 in.) thick. The majority of plate mills for rolling steel have a working-roll width between 2030 and 5590 mm (80 and 220 in.). Thus, the width of product normally available ranges from 1520 to 5080 mm (60 to 200 in.). Most steel plate consumed in the United States ranges in width from 2030 to 3050 mm (80 to 120 in.) and ranges in thickness from 5 to 200 mm (³/₁₆ to 8 in.). Some plate mills, however, have the capability to roll steel more than 640 mm (25 in.) thick.

Steel plate usually is used in the hot finished condition, but control of the rolling temperatures may be used to improve both strength and toughness. Heat treatment is also used to improve the mechanical properties of some plate.

BASIC REQUIREMENTS

Steel plate is used mainly in the construction of buildings, bridges, ships, railroad cars, storage tanks, pressure vessels, pipe, large machines and other heavy structures, where good form-

Table 11. Chemical compositions of steel sheet for porcelain enameling

Steel grade	Composition, %					
	C max	Mn avg	P max	S max	Al avg	Nb and Ti avg
Cold rolled steel, rimmed	0.08	0.35	0.025	0.025	...	...
Cold rolled steel, aluminum killed	0.08	0.35	0.025	0.025	0.05	...
Enameling iron	0.03	0.05	0.025	0.025	...	...
Decarburized rimmed steel	0.008	0.20	0.025	0.025	...	...
Interstitial-free steel (heat analysis)	0.010	0.30	0.015	0.025	0.05	0.060

ability, weldability and machinability are required. The impairment of these desirable characteristics with increasing carbon content usually limits the steel to the low-carbon and medium-carbon constructional grades, with the low-carbon grades predominating. Many alloy steels are also produced as plate. In the final structure, however, alloy steel plate is usually heat treated to achieve mechanical properties superior to those typical of the hot finished product.

STEELMAKING PRACTICES

Killed steel is fully deoxidized, and from the viewpoint of minimum chemical segregation and uniform mechanical properties, killed steel is utilized for higher quality requirements. Therefore, killed steel is generally specified when homogeneous structure and internal soundness of the plate are required.

Vacuum degassing is used to remove dissolved oxygen and hydrogen from steel, thus reducing the number and size of indigenous nonmetallic inclusions. It also reduces the likelihood of internal fissures or "flakes" caused when hydrogen content is higher than desired.

Desulfurization. Using either hot metal desulfurization or ladle desulfurizing agents immediately before teeming (e.g., calcium or rare earth additions), final plate sulfur content can be reduced to less than 0.010%. Lower sulfur content improves plate ductility and impact properties, but adds to the cost of the steel.

Electroslag remelting is a consumable remelting process, used to produce the highest levels of quality in plate steels, particularly in thick plates. First, an ingot is melted by conventional methods, allowed to solidify in the form of an electrode and prepared for remelting. It then is remelted through a reactive flux using a powerful electric current and again allowed to solidify. The resultant ingot is very low in sulfur and oxygen contents and free from internal blowholes and segregation.

QUALITY DESCRIPTORS

Steel quality, as the term applies to steel plate, is indicative of many conditions such as degree of internal soundness, relative uniformity of mechanical properties and chemical composition, and relative freedom from injurious surface imperfections. The three main quality descriptors used to describe steel plate are regular quality, structural quality and pressure-vessel quality.

Regular quality carbon steel is produced with a maximum carbon content of 0.33%. Plates of this quality are not expected to have the same degree of chemical uniformity, internal soundness or freedom from surface imperfections that is associated with structural quality or pressure-vessel quality plate.

Structural quality and pressure-vessel quality steel plates usually are produced to meet specific standard specifications prepared by one of several specification-writing bodies. ASTM A6 covers the general requirements for the structural

Table 12. ASTM specifications for structural quality steel plate(a)

Specification	Steel type and condition	Specification	Steel type and condition
Carbon steels		**Alloy steels**	
A36(b)(c)	Carbon steel plates, bars and shapes	A514	Quenched-and-tempered alloy steel plates with high yield strength, for welded bridges and other structures
A113(d)	Carbon steel plates, bars and shapes for locomotives and railroad cars	A699	Low-carbon Mn-Mo-Nb alloy steel plates, bars and shapes
A131(e)	Carbon and HSLA steel plates, bars, shapes and rivets for ships	A709	(See above under "Carbon steels")
A283(b)	Carbon steel plates of low or intermediate tensile strength	A710	Age-hardening low-carbon Ni-Cu-Cr-Mo-Nb and Ni-Cu-Nb alloy steel plates, bars and shapes
A284	Carbon-silicon steel plates of low or intermediate tensile strength for machine parts and general construction	**High-strength low-alloy steels**	
A440	Carbon steel plates, bars and shapes of high tensile strength	A131(e)	(See above under "Carbon steels")
		A242	HSLA steel plates, bars and shapes
A529	Carbon steel plates, bars, shapes and sheet piling with 290-MPa (42-ksi) minimum yield point(f)	A441	Mn-V HSLA steel plates, bars and shapes
A573	Carbon steel plates for applications requiring toughness at atmospheric temperatures	A572	Nb-V HSLA steel plates, bars, shapes, and sheet piling
		A588	HSLA steel plates, bars and shapes with 345-MPa (50-ksi) minimum yield point(g)
A678	Quenched-and-tempered carbon steel plates	A633	Normalized HSLA steel plates, bars and shapes
A709	Carbon, alloy and HSLA steel plates, bars and shapes for bridges	A656	V-Al-N and Ti-Al HSLA steel plates
		A709	(See above under "Carbon steels")

(a) Covered in A6. (b) This ASTM specification is also published by ASME, which adds an "S" in front of the "A" (for example, SA36). (c) See also GSA specification G40.8. (d) See also AAR specification M113. (e) See also Section 39 of the ABS specifications. (f) For thicknesses of 12.7 mm (1/2 in.) and under. (g) For thickness of 101.6 mm (4 in.). Lower minimum yield strengths apply for sections from 101.6 to 203.2 mm (4 to 8 in.) thick.

quality steel plate products described in the individual ASTM specifications, while ASTM A20 covers general requirements for pressure-vessel quality products.

TYPES OF STEEL

Carbon steel plate is available in all qualities except aircraft quality; it is available in many grades. The chemical compositions of the standard AISI-SAE grades of carbon steel for plate are listed in the article "Classifications and Designations of Carbon and Alloy Steels," in this volume. Individual ASTM specifications are listed and described in Table 12. The compositions of some ASTM grades and types of carbon steel plate are listed in Table 27 on page 4•14 and Table 29 on page 4•16.

Alloy Steel. Steel is considered to be alloy steel when either (a) the maximum of the range given for the content of alloying elements exceeds one or more of the following limits: 1.65 Mn; 0.60 Si; 0.60 Cu or (b) any definite range or definite minimum quantity of any of the following elements is specified or required within the limits of the recognized field of constructional alloy steels: aluminum, boron, chromium up to 3.99%, cobalt, niobium, molybdenum, nickel, titanium, tungsten, vanadium, zirconium, or any other alloying element added to obtain the desired alloying effect.

Representative compositions of ASTM grades and types of alloy steel plate are listed in Table 28 on page 4•15 and Table 30 on pages 4•16 and 4•17.

High-Strength Low-Alloy Steel. A special category of steel has been developed that offers higher

mechanical properties and, in certain of these steels, materially greater resistance to atmospheric corrosion than is obtainable from conventional carbon structural steels. This category has been given the name high-strength low-alloy (HSLA) steel, which is generally produced with emphasis on mechanical-property requirements rather than chemical-composition limits. It is not considered to be an alloy steel as described in the AISI steel products manuals, even though utilization of any intentionally added alloy content would technically qualify as such.

There are two groups of compositions in this category:

1. Vanadium and/or niobium steels, with a manganese content generally not exceeding 1.35% maximum, and with the addition of 0.2% minimum copper when specified
2. High-strength intermediate-manganese steels, with a manganese content in the range of 1.10 to 1.65%, and with the addition of 0.2% minimum copper when specified.

Other elements commonly added to these HSLA steels to give the desired properties include silicon, chromium, nickel, molybdenum, titanium, zirconium, boron, aluminum and nitrogen. The chemical compositions of ASTM grades and types of HSLA steels are listed in Table 28 on page 4•15 and Table 30 on pages 4•16 and 4•17.

Ultrahigh-strength steel used in pressure-vessel quality plate contains up to 19% nickel and up to about 5% molybdenum, and may contain substantial amounts of chromium and cobalt. The chemical compositions of ASTM grades of ultrahigh-strength steel plate are listed in Table 30 on pages 4•16 and 4•17.

Steel Bar, Rod and Wire

Hot Rolled Steel Bar and Shapes

HOT ROLLED STEEL BARS and other hot rolled steel shapes are produced from ingots, blooms, or billets converted from ingots, or from strand cast blooms and billets, and comprise a variety of sizes and cross sections. These bars and shapes are most often produced in straight lengths, but bars in some sizes and cross sections are also produced in coils.

The term "bar" includes: rounds, squares, hexagons and similar cross sections 9.52 mm ($^3/_8$ in.) and greater across; flats greater than 5.16 mm (0.203 in.) in thickness and 152.4 mm (6 in.) and less in width, or 5.84 mm (0.230 in.) and greater in thickness and 203.2 mm (8 in.) and less in width; small angles, channels, tees and other standard shapes less than 76 mm (3 in.) across; and concrete-reinforcing bars. The term "shapes" includes structural shapes and special shapes. Structural shapes are flanged, are 76 mm (3 in.) or greater in at least one cross-sectional dimension, and are used in structures such as bridges, buildings, ships and railroad cars. Special shapes are those designed by users for specific applications.

DIMENSIONS AND TOLERANCES

The nominal dimensions of hot rolled steel bars and shapes are designated in inches or millimetres with applicable tolerances, as shown in ASTM specifications A6 and A29.

SURFACE IMPERFECTIONS

Surface seams or laps are the most common imperfections in hot rolled bars and shapes; they appear as longitudinal fissures. Seams are crevices in the steel that have been closed, but not welded. Laps are somewhat similar imperfections that result from rolling fins or protrusions into the surface of the bar.

Allowances for Surface Imperfections. Experience has shown that when purchasers order hot rolled or heat treated bars that are to be machined, it is advisable for the purchaser to make adequate allowances for removal of surface imperfections and to specify the sizes accordingly. These allowances depend on the way the surface metal is removed, the length and size of bars, the straightness, the size tolerance and the out-of-round tolerance. Bars are generally straightened before machining. For special quality carbon steel bars and regular quality alloy steel bars, either resulfurized or nonresulfurized, it is advisable that allowances be adequate to permit stock removal of not less than the amount shown below:

Type of machining	Minimum machining allowance per side, % of specific size	
	Nonresulfurized	Resulfurized
Turning on centers	3.0	3.8
Centerless turning or grinding	2.6	3.4

Note that these allowances are based on bars within straightness tolerance. Also, because

Table 1. Minimum stock removal for steel bars subject to magnetic-particle inspection

Hot rolled size		Minimum stock removal from the surface(a)	
mm	in.	mm	in.
Up to 12.7	Up to $^1/_2$	0.76	0.030
Over 12.7 to 19	Over $^1/_2$ to $^3/_4$ ...	1.14	0.045
Over 19 to 25	Over $^3/_4$ to 1	1.52	0.060
Over 25 to 38	Over 1 to $1^1/_2$ ...	1.90	0.075
Over 38 to 51	Over $1^1/_2$ to 2 ...	2.29	0.090
Over 51 to 64	Over 2 to $2^1/_2$ ...	3.18	0.125
Over 64 to 89	Over $2^1/_2$ to $3^1/_2$ ...	3.96	0.156
Over 89 to 114	Over $3^1/_2$ to $4^1/_2$ ...	4.75	0.187
Over 114 to 152	Over $4^1/_2$ to 6 ...	6.35	0.250
Over 152 to 191	Over 6 to $7^1/_2$ ...	7.92	0.312
Over 191 to 229	Over $7^1/_2$ to 9 ...	9.52	0.375
Over 229 to 254	Over 9 to 10 ....11.10		0.437

(a) The minimum reduction in diameter of rounds is twice the minimum stock removal from the surface.

straightness is a function of length, additional machining allowance may be required for turning long bars on centers. For steel bars subject to magnetic-particle inspection, additional stock removal is recommended, as shown in Table 1.

PRODUCT REQUIREMENTS

Hot rolled steel bars and shapes can be produced to chemical-composition ranges or limits only. These products also can be produced to meet mechanical-property requirements as well as limited composition requirements.

Mechanical testing of hot rolled steel bars and shapes can include tensile, Brinell or Rockwell hardness, bend, Charpy impact, fracture toughness, and short-time elevated-temperature tests, as well as tests for elastic limit, proportional limit and offset yield strength, which require use of an extensometer or plotting of a stress-strain curve. These tests are covered by ASTM A370 and other ASTM standards.

Other tests sometimes required include measurement of grain size and hardenability. Soundness and homogeneity may be evaluated by macroetching or fracturing. The fracture test is commonly applied only to high-carbon bearing quality steel. Location of samples, number of tests, details of testing technique, and acceptance limits based on the test should be established in each instance.

Testing for nonmetallic inclusions consists of careful microscopic examination (at 100×) of prepared and polished specimens. The specimens should be taken on a longitudinal plane midway between the center and surface of the product. Location of specimens, number of tests, and interpretation of results should be established in each instance. Typical testing procedures are described in ASTM E45.

Tensile and hardness tests are the most common mechanical-property tests performed on hot rolled steel bars and shapes. Hardness is a relatively simple property to measure, and it is closely related to tensile strength, as shown in Fig. 1. When Fig. 2 is used together with Fig. 1, a simple hardness test can give an estimate of yield strength and elongation, as well as of tensile strength.

It is not practicable to set definite limitations

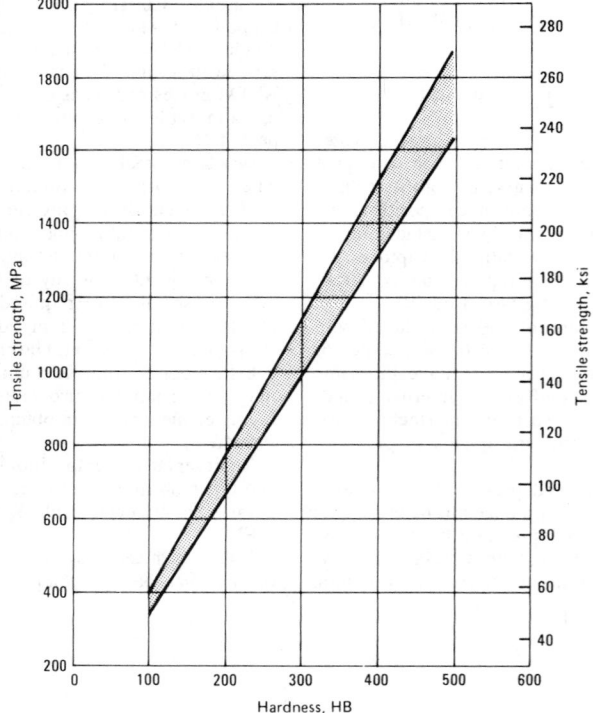

Range up to 300 HB is applicable to hot finished steel discussed here. (SAE Handbook).
Fig. 1. Relation between hardness and tensile strength of steel

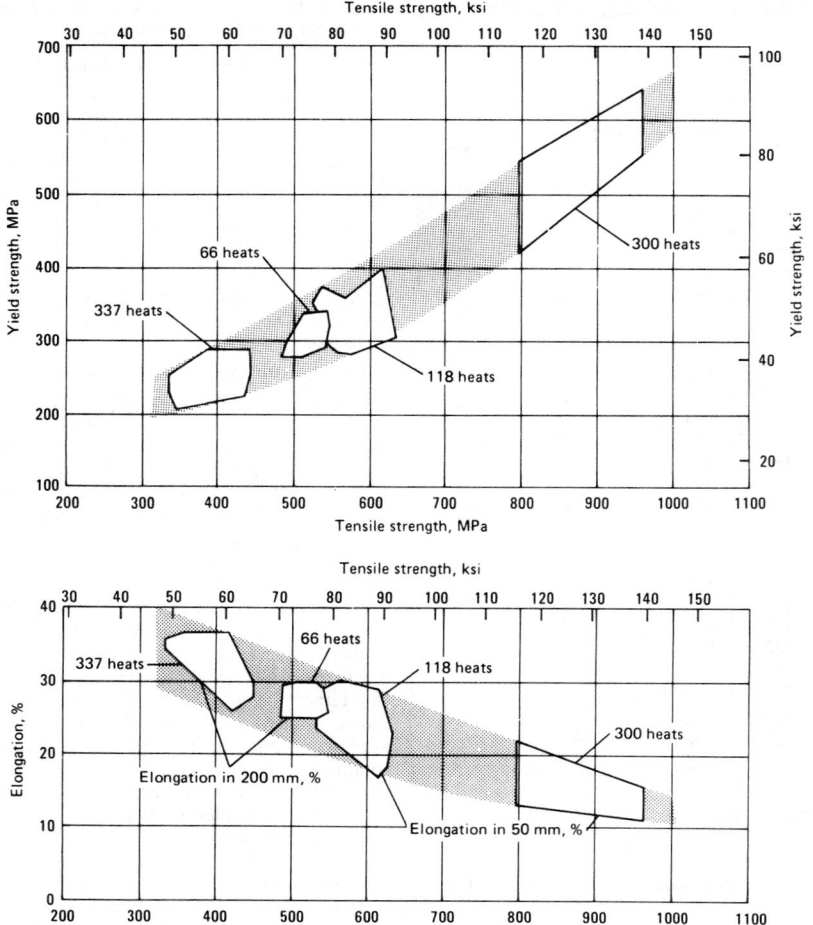

Fig. 2. Relation of tensile properties for hot rolled carbon steel

on tensile strength or hardness for carbon or alloy steel bars in the as-rolled condition. For mill-annealed steel bars, there is a maximum tensile strength or a maximum hardness (Table 2) that can be expected for each grade of steel. For steel bars in the normalized condition, maximum hardness, maximum tensile strength, minimum hardness or minimum tensile strength can be specified.

PRODUCT CATEGORIES

Hot rolled carbon steel bars are produced in two fundamental categories identified by the quality descriptors "merchant quality" and "special quality." Differentiation between the two qualities is based on conditions such as internal cleanness, degree of segregation, relative freedom from surface imperfections, and other individual characteristics for particular applications.

MERCHANT QUALITY BARS

Merchant quality is the least restrictive descriptor for hot rolled carbon content, uniformity of chemical composition, and freedom from surface imperfections. The extent to which these characteristics can be obtained is limited by the raw materials available, by teeming, rolling and other steelmaking practices, and by inspection techniques.

Carbon steel structural shapes and special shapes do not have specific quality descriptors but are covered by several ASTM specifications.

Grades. Merchant quality bars are commonly produced to chemical compositions (heat analysis only) including 0.50 max C, 0.60 max Mn, 0.04 max P and 0.05 max S. Merchant quality bars are not produced to any specified silicon content, grain size or other requirement that would dictate the type of steel produced.

Merchant quality steel bars do not require the chemical ranges typical of standard steels. Consequently, they are produced to wider carbon and manganese ranges and are designated by the prefix "M" (see Table 3).

Mechanical Properties. Some merchant quality steel bars are produced to specified mechanical properties. ASTM A663 gives the tensile properties of such bars (see Table 4).

SPECIAL QUALITY BARS

Special quality steel bars are employed when end use, method of fabrication, or subsequent processing treatment requires characteristics not available in merchant quality bars. Typical applications, including many structural uses, require hot forging, heat treating, cold drawing, cold forming and machining.

Special quality bars are required to be free from visible pipe and excessive chemical segregation. Also, they are rolled from blooms and billets that have been inspected and conditioned, as necessary, to minimize surface imperfections.

Mechanical Properties. Some special quality steel bars are produced to specified mechanical properties. Tensile properties from ASTM A675 are given in Table 5.

Cold Finished Steel Bar

CARBON AND ALLOY STEEL bar products (round, square, hexagonal, flat or special shapes) that are cold finished either by turning or grinding or by drawing through a die are discussed in this article. Not covered here are flat-rolled products such as sheet, strip or plate, which normally are cold finished by cold rolling, or cold drawn tubular products.

Cold finished bars fall into four classifications: cold drawn bars; turned and polished bars; cold drawn, ground and polished bars; and turned, ground and polished bars.

Cold drawn bars represent the largest-tonnage production and are widely used in mass production of machined and other parts. They have attractive combinations of mechanical and dimensional properties.

Turned and polished bars have the mechanical properties of hot rolled products but greatly improved surface finish and dimensional accuracy. These bars are available in sizes larger than those that can be cold drawn.

Cold drawn, ground and polished bars have the increased machinability, tensile strength and yield strength of cold drawn bars together with very close size tolerances.

Turned, ground and polished bars have superior surface finish, dimensional accuracy and straightness and find application in precision shafting and in plating, where such factors are of prime importance.

Cold finished steel bars are available in a wide variety of sizes and cross-sectional shapes. Normally, they are furnished in straight lengths, but in some sizes and cross sections they are furnished in coils. Cold finished steel bars are available with nominal dimensions designated in either inches or millimetres. The sizes in which they are commonly available are as follows:

Rounds ....Through 229 mm (9 in.)
Squares ....Through 127 mm (5 in.)
Hexagons ......Through 102 mm (4 in.)
Flats ....3.18 mm ($^1/_8$ in.) and over in thickness; up to 305 mm (12 in.) in width

Rounds, squares and hexagons from $^1/_8$ through 1 in. commonly are available in size increments of $^1/_{64}$ in.; over 1 through 2 in., increments of $^1/_{32}$ in.; and over 2 in., increments of $^1/_{16}$ in. The preferred millimetre sizes published in ANSI B32.4 do not correspond to the same increments, and for the larger bar sizes, the millimetre increments are much larger.

PRODUCT TYPES

In the manufacture of cold finished bars, the steel is first hot rolled oversize to appropriate shape

Table 2. Lowest maximum hardness that can be expected for hot rolled steel bars, billets and slabs with ordinary mill annealing

Steel grade	Maximum hardness, HB(a)		Steel grade	Maximum hardness, HB(a)		Steel grade	Maximum hardness, HB(a)	
	Straight-ened	Non-straight-ened		Straight-ened	Non-straight-ened		Straight-ened	Non-straight-ened
Carbon steels			**Alloy steels**			**Alloy steels**		
1141	201	192	4150	235	223	5160	235	223
1144	207	197	4161	241	229	51B60	235	223
1151	207	201	4320	207	197	6118	163	156
1541	207	197	4340	235	223	6150	217	207
1548	212	207	4419	170	163	81B45	201	192
1552	212	207	4615	174	167	8615	163	156
15B41	207	197	4620	179	170	8617	163	156
15B48	212	207	4621	179	170	8620	170	163
			4626	187	179	8622	179	170
Alloy steels			4718	179	170	8625	179	170
1330	187	179	4720	170	163	8627	183	174
1335	197	187	4815	223	192	8630	187	179
1340	201	192	4817	229	197	8637	201	192
1345	212	201	4820	229	197	8640	207	197
4012	149	143	5015	156	149	8642	212	201
4023	156	149	50B44	207	197	8645	217	207
4024	156	149	50B46	217	201	8655	235	223
4027	170	163	50B50	217	201	8720	170	163
4028	170	163	50B60	229	217	8740	212	201
4037	192	183	5120	170	163	8822	187	179
4047	212	201	5130	183	174	9254	241	229
4118	170	163	5132	187	179	9255	241	229
4130	183	174	5135	192	183	9260	248	235
4137	201	192	5140	197	187	94B17	156	149
4140	207	197	5145	229	197	94B30	183	174
4142	212	201	5147	217	207			
4145	217	207	5150	212	201	(a) Specific microstructure requirements may necessitate modification of these hardness numbers.		
4147	223	212	5155	229	217			

Table 3. Chemical limits (heat analysis) for M series carbon steels for merchant quality bars

Grade	Chemical composition, %			
	C	Mn	P, max	S, max
M1008	0.10 max	0.25-0.60	0.04	0.05
M1010	0.07-0.14	0.25-0.60	0.04	0.05
M1012	0.09-0.16	0.25-0.60	0.04	0.05
M1015	0.12-0.19	0.25-0.60	0.04	0.05
M1017	0.14-0.21	0.25-0.60	0.04	0.05
M1020	0.17-0.24	0.25-0.60	0.04	0.05
M1023	0.19-0.27	0.25-0.60	0.04	0.05
M1025	0.20-0.30	0.25-0.60	0.04	0.05
M1031	0.26-0.36	0.25-0.60	0.04	0.05
M1044	0.40-0.50	0.25-0.60	0.04	0.05

Cold drawn carbon steels are most common in grades with less than 0.55% carbon. Grades above this carbon level must be annealed before cold drawing, which increases cost, and their ultimate mechanical properties after annealing and cold drawing may be lower than those of lower-carbon grades cold drawn directly from hot rolled stock. Some improvement in machining characteristics of the higher-carbon grades can be obtained by annealing.

Alloy steels containing more than 0.38% carbon must be annealed before cold drawing.

Machined Bars. Bar products cold finished by stock removal may be (*a*) turned and polished; (*b*) turned, ground and polished; or (*c*) cold drawn, ground and polished. Turning is done in special machines with cutting tools mounted in rotating heads, thus eliminating the problem of having to support long bars as in a lathe. Grinding is done in centerless machines. Polishing is usually done in a roll straightener of the crossed-axis (Medart) type with polished rolls to provide a smooth finish.

The published range of diameters both for turned and for turned and ground bars is 12.7 to 229 mm ($^1/_2$ to 9 in.), inclusive; for cold drawn and ground bars, it is 3.18 to 102 mm ($^1/_8$ to 4 in.), inclusive. These are composites of size ranges throughout the industry; an individual producer may be unable to furnish a full range of sizes. For example, one well-known producer supplies turned rounds from 12.7 to 229 mm ($^1/_2$ to 9 in.), and another from 29 to 203 mm ($1^1/_8$ to 8 in.) — all finished sizes.

Stock removal in turning, or in turning and grinding, measured on the diameter, is normally 1.59 mm ($^1/_{16}$ in.) for sizes up to 38 mm ($1^1/_2$ in.), 3.17 mm ($^1/_8$ in.) for the 38-to-76-mm ($1^1/_2$-to-3-in.) range, 4.75 mm ($^3/_{16}$ in.) for the 76-to-127 mm (3-to-5-in.) range, and 6.35 mm ($^1/_4$ in.) for diameters of 127 mm (5 in.) and greater.

Cold drawn round bars are available in a range of diameters from 3.17 to 152 mm ($^1/_8$ to 6 in.). Maximum diameters available from individual producers, however, may vary from 76 to 152 mm (3 to 6 in.). The reduction in diameter in

Table 4. Tensile properties of merchant quality hot rolled steel bars(a)

Grade	Tensile strength		Yield strength		Min elongation, %	
	MPa	ksi	MPa	ksi	200-mm, or 8-in., gage length	50-mm, or 2-in., gage length
45	310-380	45-55	155	22.5	27	33
50	345-415	50-60	175	25.0	25	30
55	380-450	55-65	190	27.5	23	26
60	415-495	60-72	210	30.0	21	22
65	450-530	65-77	225	32.5	17	20
70	485-585	70-85	240	35.0	14	18
75	515-620	75-90	260	37.5	14	18
80	550 min	80 min	275	40.0	13	17
(a) Values from ASTM A663.						

Table 5. Tensile properties of special quality hot rolled steel bars(a)

Grade(b)	Tensile strength		Yield point, min		Min elongation, %	
	MPa	ksi	MPa	ksi	200-mm, or 8-in., gage length(c)	50-mm, or 2-in., gage length
45	310-380	45-55	155	22.5	27	33
50	345-415	50-60	170	25.0	25	30
55	380-450	55-65	190	27.5	23	26
60	415-495	60-72	205	30.0	21	22
65	450-530	65-77	225	32.5	17	20
70	485-585	70-85	240	35.0	14	15(d)
75	515-620	75-90	260	37.5	14	15(d)
80	550 min	80 min	275	40.0	13	14(d)
90	620 min	90 min	380	55.0	10	11(d)
(a) Values from ASTM A675. (b) When lead is required, add the letter "L" after the grade designation. (c) For bars over 19 mm ($^3/_4$ in.) in thickness or diameter, the minimum elongation listed is reduced by 0.25% for each additional 0.79 mm ($^1/_{32}$ in.). (d) Minimum value depends on bar thickness or diameter.						

and is then subjected to mechanical operations (other than those intended primarily for scale removal) that affect its finish, size and mechanical properties. As indicated above, the two common methods of cold finishing bars are (*a*) removal of surface material by turning or grinding, employed either singly or in combination, and (*b*) drawing the material through a die of suitable configuration.

Commercial Grades. Any grade of carbon or alloy steel that can be hot rolled also can be cold finished. The grades of carbon and alloy steels most commonly available as cold finished bars are noted in Table 6, although other grades are also available from producers. The choice of grade is based on the hardenability and tempering characteristics necessary to obtain the required mechanical properties.

Table 6. Grades of steel most commonly available as cold finished bars(a)

Carbon steels		Alloy steels	
1018	1141	1335	4615
1020	1144	4037	5120
1045	1212	4130	5140
1050	1213	4140	8620
1117	12L14	4340	8640
1137	1215		

(a) Leaded steels (0.15 to 0.35% Pb) can be produced in most grades and are indicated by "L" after the second digit, as in 12L14 above.

cold drawing, called draft, commonly ranges from 0.79 mm ($^1/_{32}$ in.) to 1.58 mm ($^1/_{16}$ in.). The draft selected depends on the material being drawn, physical properties desired and the equipment being used. Some special processes use heavier drafts followed by stress relieving. One producer employs heavy drawing at elevated temperature. With this exception, drawing operations are begun with the material at room temperature to start, and the only elevated temperature involved is that developed in the bar as a result of drawing; this temperature rise is small and of little significance.

Originally, cold finishing, whether by turning or by cold rolling, was employed only for sizing to produce a bar with closer dimensional tolerances and a smoother surface. As cold finished bar products were developed and improved, increased attention was paid to the substantial enhancement of mechanical properties that could be obtained by cold working. This additional advantage is now more fully appreciated, as attested to by the fact that in about 40% of the applications increased mechanical properties are an important consideration. In approximately half of these applications, or 20% of the total, cold drawing is used only to increase strength; in the other 20%, close tolerances and better surface finish are desired in addition to increased strength.

Machining Versus Cold Drawing. Basic differences exist between bars finished by machining and those finished by cold drawing. First, it is obvious that turning and centerless grinding are applicable only to round bars, whereas drawing can be applied to a variety of shapes. Drawing, therefore, is more versatile than machining.

Different size tolerances are applicable to cold finished products, depending on shape, carbon content and heat treatment. Shown in Tables 7 through 13 are the tolerances for cold finished carbon and alloy steel bars published in ASTM A29. These tables include cold drawn bars; turned and polished rounds; cold drawn, ground and polished rounds; and turned, ground and polished rounds. From the data in these tables, certain generalizations can be drawn. The tolerances for cold drawn and for turned and polished rounds, for example, are the same for sizes up to and including 102 mm (4 in.). There are differences, however, between the tolerances that apply to carbon steels and those that apply to alloy steels.

CARBON STEEL QUALITY DESCRIPTORS

Standard quality is the descriptor applied to the basic quality level to which cold finished carbon steel bars are produced. Standard quality cold finished bars are produced from hot rolled carbon steel of "special" quality (the standard quality for hot rolled bars for cold finishing). Steel bars of standard quality must be free from visible pipe and excessive chemical segregation. They may contain surface imperfections. In general, the size of surface imperfections increases with bar size; therefore, where freedom from surface imperfections is required, surface removal is recommended.

Restrictive requirement quality A (RRA) incorporates all the features of standard quality carbon steel bars described above, plus any one of the following restrictive requirements.

Special surface bars are produced with special surface preparation to minimize the frequency and size of seams and other surface imperfections.

Table 7. Size tolerances for cold drawn carbon steel bars(a)

Size range(b), mm (in.)	Minus size tolerances for bars									
	Annealed or spheroidize annealed before drawing; for grades with max carbon content of:						Stress relieved or annealed after drawing; for grades with max carbon through 0.55%		Quenched, or normalized, and tempered before drawing; for all grades	
	Through 0.28%		Over 0.28% through 0.55%		Over 0.55%					
	mm	in.	mm	in.	mm	in.	mm	in.	mm	in.
Rounds										
To 38.1 ($1^1/_2$) ..	0.051	0.002	0.076	0.003	0.127	0.005	0.102	0.004	0.127	0.005
Over 38.1 ($1^1/_2$) to 63.5 ($2^1/_2$) ...	0.076	0.003	0.102	0.004	0.152	0.006	0.127	0.005	0.152	0.006
Over 63.5 ($2^1/_2$) to 101.6 (4) ...	0.102	0.004	0.127	0.005	0.178	0.007	0.152	0.006	0.178	0.007
Over 101.6 (4) to 152.4 (6) ...	0.127	0.005	0.152	0.006	0.203	0.008	0.178	0.007	0.203	0.008
Hexagons(c)										
To 19.05 ($^3/_4$) ..	0.051	0.002	0.076	0.003	0.152	0.006	0.102	0.004	0.152	0.006
Over 19.05 ($^3/_4$) to 38.1 ($1^1/_2$) ...	0.076	0.003	0.102	0.004	0.178	0.007	0.127	0.005	0.178	0.007
Over 38.1 ($1^1/_2$) to 63.5 ($2^1/_2$) ...	0.102	0.004	0.127	0.005	0.203	0.008	0.152	0.006	0.203	0.008
Over 63.5 ($2^1/_2$) to 79.4 ($3^1/_8$) ...	0.127	0.005	0.152	0.006	0.229	0.009	0.178	0.007	0.229	0.009
Squares(c)										
To 19.05 ($^3/_4$) ..	0.051	0.002	0.102	0.004	0.178	0.007	0.127	0.005	0.178	0.007
Over 19.05 ($^3/_4$) to 38.1 ($1^1/_2$) ...	0.076	0.003	0.127	0.005	0.203	0.008	0.152	0.006	0.203	0.008
Over 38.1 ($1^1/_2$) to 63.5 ($2^1/_2$) ...	0.102	0.004	0.152	0.006	0.229	0.009	0.178	0.007	0.229	0.009
Over 63.5 ($2^1/_2$) to 101.6 (4) ...	0.127	0.005	0.203	0.008	0.279	0.011	0.229	0.009	0.279	0.011
Flats										
Width range(d), mm (in.)										
To 19.05 ($^3/_4$) ..	0.076	0.003	0.102	0.004	0.203	0.008	0.152	0.006	0.203	0.008
Over 19.05 ($^3/_4$) to 38.1 ($1^1/_2$) ...	0.102	0.004	0.127	0.005	0.254	0.010	0.203	0.008	0.254	0.010
Over 38.1 ($1^1/_2$) to 76.2 (3)	0.127	0.005	0.152	0.006	0.305	0.012	0.254	0.010	0.305	0.012
Over 76.2 (3) to 101.6 (4) ...	0.152	0.006	0.203	0.008	0.406	0.016	0.279	0.011	0.406	0.016
Over 101.6 (4) to 152.4 (6) ...	0.203	0.008	0.254	0.010	0.508	0.020	0.305	0.012	0.508	0.020
Over 152.4 (6)	0.330	0.013	0.381	0.015	...	...	...	...	...	...

(a) Data from ASTM A29. Table includes tolerances for bars that have been annealed, spheroidize annealed, normalized and tempered, or quenched and tempered before cold drawing and for bars that have been stress relieved or annealed after cold drawing. Table does not include tolerances for bars that are spheroidize annealed, normalized and tempered or quenched and tempered after cold drawing. (b) For sizes larger than those shown, the producer should be consulted regarding size tolerances. (c) Size of a hexagon or square is the distance between opposite sides. (d) Width governs tolerances for both width and thickness of flats. For example, when maximum of carbon range is 0.28% or less, for a flat 50.8 mm (2 in.) wide and 25.4 mm (1 in.) thick, width tolerance is 0.127 mm (0.005 in.) and thickness tolerance is the same, namely, 0.127 mm (0.005 in.).

Special internal soundness bars have greater freedom from chemical segregation and porosity than standard quality bars.

Special hardenability bars are produced to hardenability requirements other than those of standard H-steels.

Cold finished carbon steel bars are also produced to inclusion ratings as determined by standard nonmetallic inclusion testing.

Special discard is specified when minimized chemical segregation, special steel cleanliness, or internal soundness requirements dictate that the product be selected from certain positions in the ingot.

Minimized decarburization is specified whenever decarburization is important, such as in heat treating for surface hardness requirements.

Restrictions other than those noted above, such as special chemical limitations, special processing techniques, and other special characteristics not previously anticipated, are also covered by this quality level.

Multiple restrictive requirement quality (MRR) applies when two or more of the above described restrictive requirements are involved.

Cold forging quality A and cold extrusion quality A apply to cold finished carbon steel bars used in production of solid or hollow shapes by means of

Table 8. Diameter tolerances for turned and polished carbon steel rounds(a)

| Diameter range, mm (in.) | Annealed or spheroidize annealed before finishing; for grades with max carbon content of: | | | | | | Stress relieved or annealed after finishing; for grades with max carbon through 0.55% | | Quenched, or normalized, and tempered before finishing; for all grades | |
| | Through 0.28% | | Over 0.28% through 0.55% | | Over 0.55% | | | | | |
	mm	in.	mm	in.	mm	in.	mm	in.	mm	in.
To 38.1 (1½)	0.051	0.002	0.076	0.003	0.127	0.005	0.102	0.004	0.127	0.005
Over 38.1 (1½) to 63.5 (2½)	0.076	0.003	0.102	0.004	0.152	0.006	0.127	0.005	0.152	0.006
Over 63.5 (2½) to 101.6 (4)	0.102	0.004	0.127	0.005	0.178	0.007	0.152	0.006	0.178	0.007
Over 101.6 (4) to 152.4 (6)	0.127	0.005	0.152	0.006	0.203	0.008	0.178	0.007	0.203	0.008
Over 152.4 (6) to 203.2 (8)	0.152	0.006	0.178	0.007	0.229	0.009	0.203	0.008	0.229	0.009
Over 203.2 (8) to 228.6 (9)	0.178	0.007	0.203	0.008	0.254	0.010	0.229	0.009	0.254	0.010
Over 228.6 (9)	0.203	0.008	0.229	0.009	0.279	0.011	0.254	0.010	0.279	0.011

(a) Data from ASTM A29. Table includes tolerances for bars that have been annealed, spheroidize annealed, normalized, normalized and tempered, or quenched and tempered before cold finishing and for bars that have been stress relieved or annealed after cold finishing. Table does not include tolerances for bars that are spheroidize annealed, normalized, normalized and tempered, or quenched and tempered after cold finishing.

severe cold plastic deformation involving movement of metal by expansion and/or compression with no expansion of the surface and not requiring special inspection standards.

Cold working quality applies to steel used in the production of solid or hollow shapes by means of very severe cold plastic deformation involving cold working by expansion and/or compression. Such steels are produced by closely controlled steelmaking practices and are subject to special inspection standards for internal soundness, surface quality and uniform chemical composition. The severe cold forming operations involved for this quality level normally require a suitable ther-

mal treatment to obtain proper hardness and microstructure for cold working.

Other Carbon Steel Qualities. The quality descriptors listed below are some of those that apply to cold finished carbon steel bars intended for specific requirements and applications. They may have requirements for surface quality, amount of discard, macroetch tests, mechanical properties

or chemical uniformity as indicated in product specifications.

- Axle shaft quality
- Shell steel quality A
- Shell steel quality C
- Rifle barrel quality
- Spark plug quality.

ALLOY STEEL QUALITY DESCRIPTORS

Regular quality is the descriptor applied to the basic, or standard, quality level to which cold finished alloy steel bars are produced. Steels for this quality are killed and usually are produced to a fine grain size. They are melted to chemical ranges and limits and are inspected and tested to meet normal requirements for regular constructional alloy steel applications. Regular quality cold finished alloy steel bars may contain surface imperfections. In general, the size of detrimental surface imperfections increases with bar size.

Cold working quality applies to steel used in the production of solid or hollow shapes by means of very severe cold plastic deformation involving cold working by expansion and/or compression. Such steels are produced by closely controlled steelmaking practices and are subject to special inspection standards for internal soundness, surface quality and uniform chemical composition. The severe cold forming operations involved for this quality level normally require a suitable thermal treatment to obtain proper hardness and microstructure for cold working.

Aircraft quality and magnaflux quality apply to cold finished alloy steel bars for important or highly stressed parts of aircraft and for other similar or

Table 9. Diameter tolerances for cold drawn, ground and polished carbon and alloy steel rounds(a)

| Diameter range, mm (in.) | Minus diameter tolerances | |
	mm	in.
To 38.1 (1½)	0.025	0.001
Over 38.1 (1½) to 63.5 (2½)	0.038	0.0015
Over 63.5 (2½) to 76.2 (3)	0.051	0.002
Over 76.2 (3) to 101.6 (4)	0.076	0.003

(a) Data from ASTM A29.

Table 10. Diameter tolerances for turned, ground and polished carbon steel rounds(a)

| Diameter range, mm (in.) | Minus diameter tolerance | |
	mm	in.
To 38.1 (1½)	0.025	0.001
Over 38.1 (1½) to 63.5 (2½)	0.038	0.0015
Over 63.5 (2½) to 76.2 (3)	0.051	0.002
Over 76.2 (3) to 101.6 (4)	0.076	0.003
Over 101.6 (4) to 152.4 (6)	0.102(b)	0.004(b)
Over 152.4 (6)	0.127(b)	0.005(b)

(a) Data from ASTM A29. (b) For nonresulfurized steels (steels specified to maximum sulfur limits under 0.08%) and for heat treated steels, the tolerance is increased by 0.025 mm (0.001 in.).

Table 11. Size tolerances for cold drawn alloy steel bars(a)

| Size range, mm (in.) | Annealed, spheroidize annealed, or normalized before drawing; for grades with max carbon content of: | | | | Stress relieved or annealed after drawing; for grades with max carbon content through 0.55% | | With or without stress relief or anneal after drawing; for grades with max carbon content over 0.55% | | Quenched, or normalized, and tempered before drawing; for all levels of carbon | |
| | Through 0.28% | | Over 0.28% through 0.55% | | | | | | | |
	mm	in.	mm	in.	mm	in.	mm	in.	mm	in.
Rounds										
To 25.4 (1), in coils	0.051	0.002	0.076	0.003	0.102	0.004	0.127	0.005	0.127	0.005
To 38.1 (1½)	0.076	0.003	0.102	0.004	0.127	0.005	0.152	0.006	0.152	0.006
Over 38.1 (1½) to 63.5 (2½)	0.102	0.004	0.127	0.005	0.152	0.006	0.178	0.007	0.178	0.007
Over 63.5 (2½) to 101.6 (4)	0.127	0.005	0.152	0.006	0.178	0.007	0.203	0.008	0.203	0.008
Over 101.6 (4) to 152.4 (6)	0.152	0.006	0.178	0.007	0.203	0.008	0.229	0.009	0.229	0.009
Hexagons(b)										
To 19.05 (¾)	0.076	0.003	0.102	0.004	0.127	0.005	0.178	0.007	0.178	0.007
Over 19.05 (¾) to 38.1 (1½)	0.102	0.004	0.127	0.005	0.152	0.006	0.203	0.008	0.203	0.008
Over 38.1 (1½) to 63.5 (2½)	0.127	0.005	0.152	0.006	0.178	0.007	0.229	0.009	0.229	0.009
Over 63.5 (2½) to 79.4 (3⅛)	0.152	0.006	0.178	0.007	0.203	0.008	0.254	0.010	0.254	0.010
Over 79.4 (3⅛) to 101.6 (4)	0.152	0.006	...	...	...	...	...	...	...	...

(continued)

Table 11. Size tolerances for cold drawn alloy steel bars(a)

Size range, mm (in.)	Annealed, spheroidize annealed, or normalized before drawing; for grades with max carbon content of: Through 0.28% mm	in.	Over 0.28% through 0.55% mm	in.	Stress relieved or annealed after drawing; for grades with max carbon content through 0.55% mm	in.	With or without stress relief or anneal after drawing; for grades with max carbon content over 0.55% mm	in.	Quenched, or normalized, and tempered before drawing; for all levels of carbon mm	in.
Squares(b)										
To 19.05 (³/₄)	0.076	0.003	0.127	0.005	0.152	0.006	0.203	0.008	0.203	0.008
Over 19.05 (³/₄) to 38.1 (1¹/₂)	0.102	0.004	0.152	0.006	0.178	0.007	0.229	0.009	0.229	0.009
Over 38.1 (1¹/₂) to 63.5 (2¹/₂)	0.127	0.005	0.178	0.007	0.203	0.008	0.254	0.010	0.254	0.010
Over 63.5 (2¹/₂) to 79.4 (3¹/₈)	0.178	0.007	0.229	0.009	0.254	0.010	0.305	0.012	0.305	0.012
Over 101.6 (4) to 127.0 (5)	0.279	0.011	...	...	...	...	...	...	...	...
Flats(c)										
To 19.05 (³/₄)	0.102	0.004	0.127	0.005	0.178	0.007	0.229	0.009	0.229	0.009
Over 19.05 (³/₄) to 38.1 (1¹/₂)	0.127	0.005	0.152	0.006	0.229	0.009	0.279	0.011	0.279	0.011
Over 38.1 (1¹/₂) to 76.2 (3)	0.152	0.006	0.178	0.007	0.279	0.011	0.330	0.013	0.330	0.013
Over 76.2 (3) to 101.6 (4)	0.178	0.007	0.229	0.009	0.305	0.012	0.432	0.017	0.432	0.017
Over 101.6 (4) to 152.4 (6)	0.229	0.009	0.279	0.011	0.330	0.013	0.533	0.021	0.533	0.021
Over 152.4 (6)	0.356	0.014	...	...	...	...	...	...	...	...

(a) Data from ASTM A29. Table includes tolerances for bars that have been annealed, spheroidize annealed, normalized, quenched and tempered, or normalized and tempered before cold drawing and for bars with or without stress relief or anneal after cold drawing. Table does not include tolerances for bars that are spheroidize annealed, normalized, quenched and tempered, or normalized and tempered after cold drawing. (b) The size of a hexagon or square is the distance between opposite sides. (c) Width governs tolerances for both width and thickness of flats. For example: when maximum of carbon range is 0.28% or less, for a flat 50.8 mm (2 in.) wide and 25.4 mm (1 in.) thick, width tolerance is 0.152 mm (0.006 in.) and thickness tolerance is the same, namely 0.152 mm (0.006 in.).

Table 12. Diameter tolerances for turned and polished alloy steel rounds(a)

Size range, mm (in.)	Annealed, spheroidize annealed, or normalized before finishing; for grades with max carbon content of: Through 0.28% mm	in.	Over 0.28% through 0.55% mm	in.	Stress relieved or annealed after finishing; for grades with max carbon content through 0.55% mm	in.	With or without stress relief or anneal after finishing; for grades with max carbon content over 0.55% mm	in.	Quenched, or normalized, and tempered before drawing; for all levels of carbon mm	in.
To 38.1 (1¹/₂)	0.076	0.003	0.102	0.004	0.127	0.005	0.152	0.006	0.152	0.006
Over 38.1 (1¹/₂) to 63.5 (2¹/₂)	0.102	0.004	0.127	0.005	0.152	0.006	0.178	0.007	0.178	0.007
Over 63.5 (2¹/₂) to 101.6 (4)	0.127	0.005	0.152	0.006	0.178	0.007	0.203	0.008	0.203	0.008
Over 101.6 (4) to 152.4 (6)	0.152	0.006	0.178	0.007	0.203	0.008	0.229	0.009	0.229	0.009
Over 152.4 (6) to 203.2 (8)	0.178	0.007	0.203	0.008	0.229	0.009	0.254	0.010	0.254	0.010
Over 203.2 (8) to 228.6 (9)	0.203	0.008	0.229	0.009	0.254	0.010	0.279	0.011	0.279	0.011
Over 228.6 (9)	0.229	0.009	0.254	0.010	0.279	0.011	0.305	0.012	0.305	0.012

(a) Data from ASTM A29. Table includes tolerances for bars that have been annealed, spheroidize annealed, normalized, quenched and tempered or normalized and tempered before cold finishing and for bars with or without stress relief after cold finishing. Table does not include tolerances for bars that are spheroidize annealed, normalized, normalized and tempered or quenched and tempered after cold finishing.

Table 13. Diameter tolerances for turned, ground and polished alloy steel rounds(a)

Diameter range, mm (in.)	Minus size tolerances for bars with all levels of carbon content Not heat treated mm	in.	Heat treated(b) mm	in.
To 38.1 (1¹/₂)	0.025	0.001	0.025	0.001
Over 38.1 (1¹/₂) to 63.5 (2¹/₂)	0.038	0.0015	0.038	0.0015
Over 63.5 (2¹/₂) to 76.2 (3)	0.051	0.002	0.051	0.002
Over 76.2 (3) to 101.6 (4)	0.076	0.003	0.076	0.003
Over 101.6 (4) to 152.4 (6)	0.102	0.004	0.127	0.005
Over 152.4 (6)	0.127	0.005	0.152	0.006

(a) Data from ASTM A29. (b) Applies to quenched and tempered (heat treated), normalized and tempered, or any similar double treatment prior to turning.

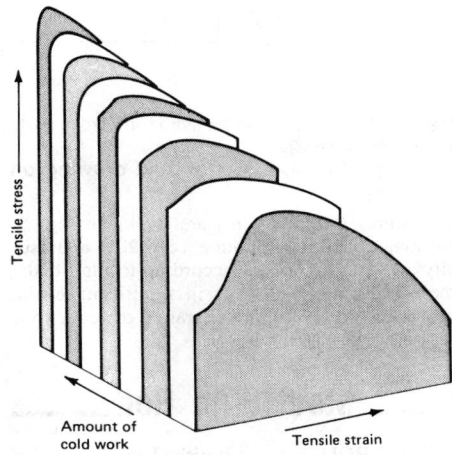

Fig. 3. Effect of cold work on tensile stress-strain curve for low-carbon steel bars

corresponding purposes involving additional stringent requirements, such as magnetic-particle inspection.

Other Alloy Steel Qualities. The quality descriptors listed below apply to cold finished alloy steel bars intended for rifles, guns, shell, shot and similar applications. They may have requirements for amount of discard, macroetch testing, surface requirements or magnetic-particle testing as indicated in the product specifications.

- Armor piercing (AP) shot quality
- Armor piercing (AP) shot magnaflux quality
- Gun quality
- Rifle barrel quality
- Shell quality
- Shell magnaflux quality.

MECHANICAL PROPERTIES

A major difference between machined and cold drawn round bars is the improvement in tensile and yield strengths that results from the cold work of drawing. Cold work also changes the shape of the stress-strain diagram, as shown in Fig. 3. Within the range of commercial drafts, cold work markedly affects certain mechanical properties (see Fig. 4). The variations in percentage of reduction of cross section for bars drawn with normal commercial drafts of 0.79 and 1.59 mm (¹/₃₂ and ¹/₁₆ in.) and with heavy drafts of 3.18 and

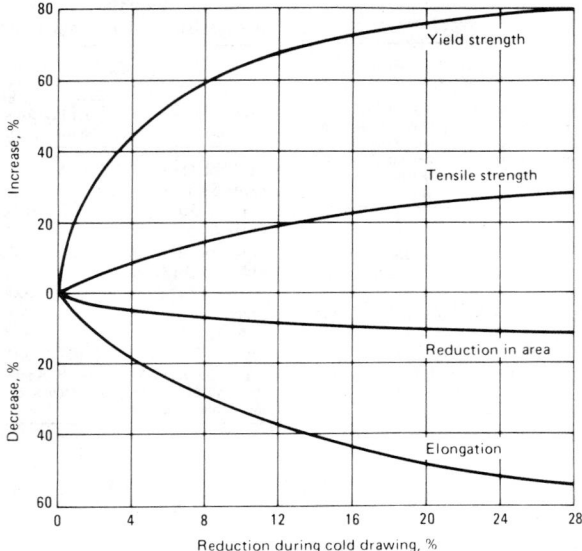

Data are for bars of up to 25 mm (1 in.) cross section having a tensile strength of 690 MPa (100 ksi) or less before cold drawing.

Fig. 4. Effect of cold drawing on the tensile properties of steel bars

4.76 mm ($^1/_8$ and $^3/_{16}$ in.) are shown in Fig. 5. Normal reductions seldom exceed 20% and usually are less than 12%. According to Fig. 4, the more pronounced changes in significant tensile properties occur within this range of reductions (up to about 15%).

____Steel Wire Rod____

WIRE ROD is a semifinished product rolled from billet on a rod mill and is used primarily for the manufacture of wire. The steel for wire rod is produced by all the modern processes, including the basic oxygen, basic open hearth, and electric furnace processes. Steel wire rod usually is cold drawn into wire suitable for further drawing; for cold rolling, cold heading, cold upsetting, cold extrusion, or cold forging; for hot forging; or for carburizing.

Although wire rod may be produced in several regular shapes, most is round in cross section. Round rod usually is produced in nominal diameters of $^7/_{32}$ to $^{47}/_{64}$ in., advancing in increments of $^1/_{64}$ in. The $^7/_{32}$-in.-diam rod, which is the smallest round that is practical to produce on a rod mill, is considered a standard because most wire 4.11 mm (0.162 in.) in diameter and smaller is drawn from rod of this size. (Wire coarser than 4.11 mm, or 0.162 in., is usually drawn from larger-diameter rod.)

As the rod comes off the rolling mill, it is wound into coils. These coils are usually about 760 mm (30 in.) in inside diameter and weigh from 135 to 1810 kg (300 to 4000 lb).

Producers of wire rod may market their product as rolled, as cleaned and coated, or as heat treated. These operations are explained in the following sections, along with the several recognized quality and commodity classifications applicable to steel wire rods.

Cleaning and Coating. Mill scale is cleaned from steel wire rods by pickling or caustic cleaning followed by water rinsing, or by mechanical means such as shot blasting with abrasive particles or reverse bending over sheaves.

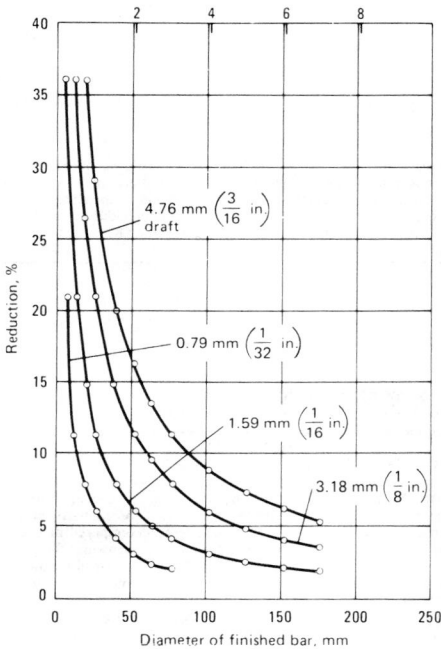

Fig. 5. Effect of draft on reduction of cross section of steel bars

Cleaning of steel wire rods is always followed by a supplementary coating operation. Lime, borax or phosphate coating is performed to provide a carrier for the lubricant necessary for subsequent processing into wire. In lime coating, practices may be varied to apply differing amounts of lime on the rods depending on customer requirements. Phosphate-coated rods may have a supplementary coating of lime, borax or water-soluble soap.

Heat treatments commonly applied to steel wire rod, either before or during processing into wire, include annealing, spheroidize annealing, patenting, and controlled cooling.

Annealing commonly involves heating to a temperature near or below the lower critical temperature and holding at that temperature for a sufficient period of time, followed by slow cooling. This process softens the steel for further processing, but not to the same degree as spheroidize annealing.

Spheroidize annealing involves prolonged heating at a temperature near or slightly below the lower critical temperature (or thermal cycling about the lower critical temperature), followed by slow cooling, with the object of changing the shape of carbides in the microstructure to globular (spheroidal), which produces maximum softness.

Patenting is a heat treatment usually confined to medium-high-carbon and high-carbon steels. In this process, individual strands of rod or wire are heated well above the upper critical temperature, and then are cooled comparatively rapidly in air, molten salt or molten lead. This treatment generally is employed to prepare the material for subsequent wiredrawing.

Controlled cooling is a heat treatment performed in modern rod mills in which the rate of cooling after hot rolling is carefully controlled. The process imparts uniformity of properties and some degree of control over scale, grain size and microstructure.

QUALITIES AND COMMODITIES OF CARBON STEEL ROD

Rod for the manufacture of carbon steel wire is produced with manufacturing controls and inspection procedures intended to ensure the degree of soundness and freedom from injurious surface imperfections necessary for specific applications. The quality descriptors and commodities applicable to carbon steel wire rod are:

- Industrial quality rod
- Chain quality rod
- Fine wire quality rod
- Cold finishing quality rod
- Heading, cold extrusion or cold rolling quality rod
- Wood screw quality rod
- Scrapless nut quality rod.

MECHANICAL PROPERTIES OF CARBON STEEL ROD

Hot rolled rod seldom is sold to specific mechanical properties because of the inherent variations of such properties. These properties for a given grade of steel vary from mill to mill and are influenced by both the type of mill and the source of steel being rolled. In older mills, where rod is coiled hot, there is considerable variation within each coil because of the effect of varying cooling rates from the center to the periphery of the coil.

Table 14 lists typical values of tensile strength for $^7/_{32}$-in. low-carbon steel rod rolled on a modern rod mill equipped with controlled cooling facilities. The values shown are for rods rolled with full air cooling. Tensile-strength values for larger-diameter rod are lower, decreasing by approximately 1.9 MPa (270 psi) for each $^1/_{64}$-in. increment by which rod diameter exceeds $^7/_{32}$ in. Similar analyses of rods rolled without full air cooling or rods rolled on an older mill, where the steel is coiled hot, would be expected to reveal lower tensile strength.

Table 14. Tensile strengths of ⁷/₃₂-in.-diam hot rolled low-carbon steel rod(a)

| Steel grade | Tensile strength for steel of three deoxidation types | | | | | |
| | Rimmed | | Capped | | Killed or semikilled | |
	MPa	ksi	MPa	ksi	MPa	ksi
1005	352	51	. . .	. . .	393	57
1006	360	52	365	53	400	58
1008	372	54	386	56	421	61
1010	386	56	400	58	434	63
1012	407	59	421	61	455	66
1015	434	63	448	65	483	70
1017	455	66	469	68	510	74
1018	. . .	. . .	476	69	517	75
1020	490	71	503	73	538	78
1022	. . .	. . .	. . .	. . .	545	79

(a) Data obtained from rod produced with controlled cooling.

Table 15. Tensile strengths of ⁷/₃₂-in.-diam hot rolled medium-high-carbon and high-carbon steel rod(a)

| Carbon content of steel, % | Tensile strength for steel with manganese content of: | | | | | |
| | 0.60% | | 0.80% | | 1.00% | |
	MPa	ksi	MPa	ksi	MPa	ksi
0.30	641	93	676	98	717	104
0.35	689	100	731	106	793	115
0.40	745	108	779	113	820	119
0.45	793	115	834	121	869	126
0.50	848	123	883	128	931	135
0.55	896	130	938	136	972	141
0.60	951	138	986	143	1020	148
0.65	1000	145	1041	151	1076	156
0.70	1055	153	1089	158	1124	163
0.75	1103	160	1138	165	1179	171
0.80	1151	167	1193	173	1227	178
0.85	1207	175	1241	180	1282	186

(a) Data obtained from rod produced with controlled cooling.

Table 15 presents typical expected tensile-strength values for ⁷/₃₂-in. medium-high-carbon and high-carbon steel rods rolled on a mill utilizing controlled cooling.

ALLOY STEEL ROD

Alloy steels are those steels for which maximum specified manganese content exceeds 1.65% or maximum specified silicon or copper content exceeds 0.60%; or for which a definite range or definite minimum quantity of any other element is specified in order to obtain desired effects on properties. For detailed information on composition ranges and limits of alloy steels, see the article "Classifications and Designations of Carbon and Alloy Steels," at the beginning of this section.

Steel Wire

WIRE may be cold drawn from any of the types of carbon steel or alloy steel rod described in the preceding article on steel wire rod. For convenience in describing the various grades of carbon steel wire, they may be divided into the same four classes used for carbon steel wire rod. Based on carbon content, these classes are: low-carbon steel wire (0.15% carbon max), medium-low-carbon steel wire (>0.15 to 0.23% carbon), medium-high-carbon steel wire (>0.23 to 0.44% carbon), and high-carbon steel wire (>0.44% carbon). The conventional four-digit or five-digit AISI-SAE designation is used to specify the car-

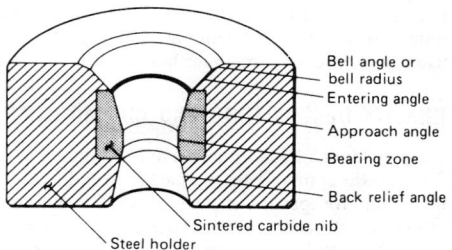

Schematic cross section (not to scale) of a single-hole wiredrawing die employing a nib of sintered carbide mounted in a circular steel holder.

Fig. 6. Typical wiredrawing die

bon or alloy steel used to make the wire. Carbon and alloy steel wire can be produced in qualities suitable for cold rolling, cold drawing, cold heading, cold upsetting, cold extrusion, cold forging, hot forging, cold coiling, heat treatment or carburizing, and for a wide variety of fabricated products.

Shapes of Wires. Although wire ordinarily is thought of as being only round, it may have any one of an infinite number of sectional shapes, as required by end use. After ordinary round wire, the most common shapes are square, hexagonal, octagonal, oval, half-oval, half-round, triangular and flat. Besides these regular (symmetrical) shapes, wire also is made in various odd and irregular shapes for specific purposes.

Sizes of Wire. The size limits for the product commonly known as wire range from approximately 0.13 mm (0.005 in.) to (but not including) 25.4 mm (1 in.) for round sections, and from a few tenths of a millimetre (a few thousandths of an inch) to approximately 13 mm (¹/₂ in.) for square sections. Larger rounds and squares, and all sizes of hexagonal and octagonal sections, are commonly known as bars.

WIREMAKING PRACTICES

Wiredrawing. Steel wire is produced from coils of wire rod, after removal of scale, by one or more cold reduction processes intended primarily for the purpose of obtaining the size desired. Wiredrawing, which improves surface finish and dimensional accuracy, is the most common cold reduction process. A natural result of cold drawing is mechanical and physical properties different from those of hot rolled steel of like composition. By varying the amount of cold reduction and other wire mill practices, including heat treatment, a wide diversity of properties and finishes can be obtained.

Mechanical characteristics of steel wire (strength, stiffness, hardness, etc.) result from wire mill treatment as well as from chemical composition, and thus mechanical properties of wire are less dependent on chemical composition than those of other steel mill forms. In most instances, the purchaser of wire is interested in suitability for a given application rather than in chemical composition.

Before drawing, the scale is removed from the material by acid pickling or mechanical descaling. If acid pickled, the coil is then rinsed with water, dipped in a vat containing hydrated lime in suspension or other material in solution, and baked to dry the coating and to liberate any hydrogen that may have been absorbed by the steel during pickling.

In cold drawing of wire, coiled rod is drawn through the tapered hole of a die or through a series of dies, the number of dies employed depending on the finished diameter required. To begin drawing, one end of the rod is pointed, inserted through the die and attached to a power-driven reel (block); the block then pulls the material through the die and coils the drawn wire.

The common design of wiredrawing die (see Fig. 6) consists of a supporting ring of steel encasing a hard, wear-resistant nib. The nib consists of one or more carbides (such as tungsten, tantalum or titanium) mixed with a bonding agent, such as cobalt, pressed into the desired shape and sintered into its hardened rough form, after which it is mounted, sized and polished. For certain types of fine wire, diamond dies sometimes are used instead of carbide dies.

The amount of reduction during drawing is expressed as a percent of the original cross-sectional area and is known as draft. As soon as a wire rod has been reduced by drawing, it is called a wire, even though many more reductions (drafts) may be necessary to reduce it to final size.

Cleaning and Coating. Whenever necessary, scale and other surface contaminants are cleaned from steel wire by acid pickling followed by water rinsing. Cleaning is always followed by a supplementary coating operation.

Lime, borax or phosphate coating is performed to provide a carrier for the lubricant necessary for subsequent processing. In lime coating, practices may be varied to apply differing amounts of lime, depending on the end use. Phosphate-coated wire may have a supplementary coating of lime, borax or a water-soluble soap.

Heat treatments for steel wire include annealing, patenting and oil tempering.

Annealing-in-process is performed on dry drawn low-carbon or medium-low-carbon steel wire. The product is sometimes called "processed wire" or "bright soft wire." In producing annealed-in-process wire, an annealing heat treatment (followed by a separate cleaning and coating operation) is performed at an intermediate stage of wiredrawing to produce a softer wire for applications in which direct drawn wire would be too hard or too stiff.

Patenting is a thermal treatment usually confined to medium-high-carbon and high-carbon steels. In this process, individual strands of rod or wire are heated well above the upper critical temperature and then cooled comparatively rapidly in air, molten salt or molten lead. This treatment is generally employed to prepare the material for subsequent wiredrawing.

Oil tempering is a heat treatment for high-carbon steel wire in which strands of the wire at finish size are continuously heated to an appropriate temperature above the critical temperature range, oil quenched, and finally passed through a tempering bath. Oil tempering is used in the production of such commodities as oil-tempered spring wire, which is used in certain types of mechanical springs that are not subjected to a final heat treatment after forming.

SPECIFICATION WIRE

There are some applications for low-carbon and medium-low-carbon steel wire that involve special requirements, such as specific tensile-strength ranges or hardness limitations, the attainment of which involves special selection of steel and modification of conventional wire mill practices

Table 16. Ranges of tensile strength for coarse, round specification wire

Wire diameter		Tensile-strength range(a)	
mm	in.	MPa	ksi
0.89-2.51	0.035-0.099	205	30
2.69-4.50	0.106-0.177	170	25
Over 4.50	Over 0.177	140	20

(a) Difference between specified minimum and maximum values.

and/or heat treatment (for example, annealed-in-process wire). Such wire commonly is designated as specification wire.

The highest useful tensile strength obtainable by cold drawing coarse, low-carbon steel round wire is approximately 830 MPa (120 ksi). Table 16 gives tensile-strength ranges commonly obtained by modifying conventional wire mill procedure through introduction of intermediate heat treatments (process annealing, for example) or other special practices to produce equivalent properties during wiredrawing.

METAL-COATED WIRE

Metallic coatings can be applied to wire by various methods, including both hot dip processes and electrolytic processes.

Aluminized wire (aluminum-coated wire) is produced by passing strands of wire through a bath of molten aluminum or aluminum alloy.

Brass-plated wire is produced by passing strands of wire through an electrolytic cell containing a solution of both copper and zinc salts. Generally, such wire is used when rubber adhesion is required or when pleasing appearance is important. Brass-plated wire is not intended for applications requiring corrosion resistance.

Galvanized wire (zinc-coated wire) is produced by passing strands of wire through a bath of molten zinc (hot dip galvanized) or through an electrolytic cell containing a solution of a zinc salt (electrogalvanized). The wire usually is annealed in the same operation by being passed through molten lead, molten salt or a furnace, followed by cleaning or pickling, prior to galvanizing. The general requirements for galvanized carbon steel wire are given in ASTM A641.

The term "temper" as applied to galvanized wire is a reference to stiffness or resistance to bending, not to heat treatment. It customarily is expressed as soft, medium or hard. Tensile strengths corresponding to these three tempers are given in Table 17.

Tinned wire is produced by passing strands of wire continuously through a molten tin bath and then through tightly compressed wipes as the strands emerge from the bath. Tinned wire is commonly manufactured in three temperatures: soft, medium hard, and hard. Soft tinned wire is tinned after being annealed at or near finish size. Medium hard tinned wire is produced from heat

treated wire. Hard tinned wire is obtained by tinning wire that has been cold drawn to final size, usually without intermediate heat treatment.

QUALITY DESCRIPTORS AND COMMODITIES

Many kinds of steel wire have been developed for specific components of machines and equipment and for particular end uses. The unique properties of each of these types of wire are obtained by employing a specific combination of steel composition, steel quality, process heat treatment and cold drawing practice.

These wires normally are grouped into broad usage categories; these categories include:

- Bailing wire
- Strapping wire
- Wire for structural applications
- Strand wire
- Concrete reinforcing wire
- High-carbon wire (see Table 18)
- Wire for conductor applications
- Telephone and telegraph wire.

Table 18. Composition of uncoated round high-carbon steel wire for tensioning by mechanical methods

Element	Content, %		
	Class I	Class II	Class III
Carbon(a)	0.45-0.75	0.50-0.85	0.55-0.88
Manganese(b) ..	0.60-1.10	0.60-1.10	0.60-1.10
Phosphorus	0.040 max	0.040 max	0.040 max
Sulfur	0.050 max	0.050 max	0.050 max
Silicon	0.10-0.35	0.10-0.35	0.10-0.35

(a) Not varying more than 0.13% in any one lot. (b) Not varying more than 0.30% in any one lot.

WIRE FOR FASTENERS

Included in wire for fasteners is wire intended for such applications as bolts and cap screws, rivets, wood and self-tapping screws, and scrapless nuts. Depending on the application, such wire must be able to be forged, extruded, cold upset, roll or cut threaded, drilled, and hardened by suitable heat treatment.

The type and grade of steel used in wire for fasteners depend on the requirements of the finished product and the nature of the required forming operations. Compositions range from that of 1006 steel, which is used for such items as common rivets, to that of a 0.55 to 0.65% carbon steel intended for lockwashers or screwdrivers. The wire may be drawn from hot rolled or annealed wire rod, and it may be either annealed or spheroidized in process.

The coating on the wire must provide sufficient lubrication in the header dies and must have the necessary lubricating qualities to prevent galling or undesirable die wear. Although lime-soap coatings are common, phosphate coatings are frequently used for more demanding applications. Producing phosphated wire may involve

coating the cleaned rod or process wire with zinc phosphate and then coating with lime or borax to carry the lubricant during subsequent drawing.

It is important that the wire be internally sound and free from seams and other surface imperfections. Decarburization must be held to a minimum for those products that are to be hardened by heat treatment.

MECHANICAL SPRING WIRE FOR GENERAL USE

The several types of steel wire used for mechanical springs are produced in a variety of chemical compositions, but the primary consideration is that the wire have the specific properties necessary for the application. The required properties vary with the intended use of the spring and with problems involved in its fabrication. Among the factors governing the selection of spring wire are: (a) the load range through which the spring must operate, (b) the corresponding stress range for the wire, (c) weight and space limitations, (d) expected life of the spring, (e) temperatures and other environmental conditions to be encountered in service, and (f) severity of deformation to be encountered in fabrication.

As stress on the wire is increased, wire with higher strength is required. Because the surface of the wire is the most highly stressed part of a spring, freedom from surface imperfections becomes increasingly important as maximum stress or required service life is increased. Surface condition is very important in music spring steel wire and is even more important in valve spring quality wire.

There are three types of spring wire for general use. They are *hard drawn spring wire* (covered in ASTM A227 and A679), *oil tempered spring wire* (ASTM A229) and *spring steel wire for heat treated components* (ASTM A713).

Music spring steel wire is used in springs subject to high stresses and requiring good fatigue properties. Final cold drawing commonly is performed by the wet white liquor method to develop a characteristic smooth bright luster. Manufacturers employ specialized coiling tests, twist tests, torsion tests and bend tests to verify that the exacting requirements of this type of wire are met. ASTM A228 describes music spring steel wire in detail.

FINE WIRE

Fine wire is considered to include all wire less than 0.89 mm (0.035 in.) in diameter, as well as some coarser wire up to 1.57 mm (0.062 in.) in diameter when so designated. Fine wire commonly is produced with bright, liquor, coppered or phosphate finishes; with galvanized, tin or cadmium coatings; and in the hard drawn, annealed or oil-tempered conditions. Aircraft cords, brooms, brushes, fishhooks, florist wire, hose reinforcement, paper clips, insect screens and safety pins are examples of products produced entirely or in part from various kinds of fine wire.

Aircraft cord wire is a hard drawn, high-tensile-strength, high-carbon steel wire designed for manufacture of flexible cords and multiple-wire strands for aircraft controls.

ALLOY WIRE

Chemical compositions and quality descriptions, as well as requirements and tests, appli-

Table 17. Tensile strengths of galvanized wire

Wire diameter		Tensile strength					
		Soft temper wire		Medium temper wire		Hard temper wire	
mm	in.	MPa	ksi	MPa	ksi	MPa	ksi
0.89-1.93	0.035-0.076	515 max	75 max	485-690	70-100	620-825	90-120
2.03-2.51	0.080-0.099	515 max	75 max	485-655	70-95	585-795	85-115
2.69-4.50	0.106-0.177	485 max	70 max	450-620	65-90	550-760	80-110
Over 4.50	Over 0.177	485 max	70 max	415-585	60-85	515-725	75-105

cable to alloy wire are described in the preceding article on steel wire rod. Many alloy steel wires have been developed for specific applications, which include wires for bearings, chains and springs, and for cold heading and cold forging applications.

Alloy steel spring wire is used for manufacture of springs intended for operation at moderately elevated temperatures. Two grades in common use are 6150 chromium-vanadium steel and 9254 chromium-silicon steel, which are covered in ASTM A231, A232 and A401. The wire is commonly produced in one of two conditions: oil tempered or spheroidize annealed. Oil-tempered alloy spring wire, commonly produced in diameters up to 12.7 mm (0.500 in.) to tensile-strength or hardness requirements, is intended for very light forming and generally is used for coiling into common types of springs. Table 19 shows tensile-strength ranges for several oil-tempered wires.

Table 19. Tensile-strength ranges for oil-tempered alloy steel spring wires

Wire diameter(a)		Chromium-vanadium spring wire, aircraft quality spring wire and valve-spring quality wire		Chromium-silicon spring wire	
mm	in.	MPa	ksi	MPa	ksi
0.51	0.020	2070-2240	300-325	...	...
0.89	0.035	2000-2170	290-315	2070-2240	300-325
1.22	0.048	1930-2100	280-305	2030-2200	295-320
1.57	0.062	1830-2000	265-290	2000-2170	290-315
1.83	0.072	1790-1930	260-280	1990-2160	288-313
2.03	0.080	1760-1900	255-275	1960-2140	285-310
2.34	0.092	1720-1860	250-270	1930-2100	280-305
2.68	0.105	1690-1830	245-265	1900-2070	275-300
3.43	0.135	1620-1760	235-255	1860-2030	270-295
4.11	0.162	1550-1690	225-245	1830-2000	265-290
4.88	0.192	1520-1660	220-240	1790-1950	260-283
6.35	0.250	1450-1590	210-230	1720-1900	250-275
7.19	0.283	1410-1550	205-225	1710-1880	248-273
7.94	0.312	1400-1540	203-223	1690-1860	245-270
9.52	0.375	1380-1520	200-220	1660-1830	240-265
11.12	0.438	1340-1480	195-215	1620-1790	235-260
12.70	0.500	1310-1450	190-210	...	...

(a) For diameters other than those shown in the above table, tensile strengths may be determined by interpolation.

Steel Tubular Products

STEEL TUBULAR PRODUCTS is the term used to cover all hollow steel products. Although these products usually are produced in cylindrical form, often they are subsequently altered by various processing methods to produce square, oval, rectangular and other symmetrical shapes. Such products have applications that are almost innumerable, but they are most commonly used as conveyors of fluids and as structural members.

PRODUCT CLASSIFICATION

The two simplest and broadest commercial classifications of steel tubular products are tube and pipe. (Although application of the terms "pipe" and "tube" is not always consistent, "pipe" commonly is used to describe cylindrical tubular products made to standard combinations of outside diameter and wall thickness.) These two broad classifications are subdivided into several named use groups. For example, the term "tube" covers three such groups: pressure tubes, structural tubing and mechanical tubing.

The term "pipe" covers five such groups: standard pipe, line pipe, oil country tubular goods, water-well pipe and pressure pipe. There is also pipe for some special applications, such as conduit pipe and tubular piling, that do not fit any of these classifications. Each of these use groups has, in turn, a number of uses or named use subdivisions. These are shown in Table 1.

The named uses in Table 1 have been developed without regard to method of manufacture, size range or wall thickness of product, or degree of finish. For example, the names do not distinguish between those products commonly called pipe, with size and wall thickness the same as standard pipe, and those having different basic dimensions; all are termed "pipe."

On a use basis, pressure pipe is distinguished from pressure tubes in that the latter are suitable for those applications in which heat is applied externally. The principal use groups and types of pressure tubes are shown in Table 2.

Structural and mechanical tubing do not follow this system of nomenclature. Instead, their names are derived from method of fabrication and degree of finish, such as "cold formed welded" or "seamless hot finished."

Steel tubular products can be made by forming a flat skelp, sheet, strip or plate into a hollow cylinder and welding the resulting longitudinal seam together or by generating an opening in a solid cylinder by piercing and elongating the resultant hollow cylinder.

WELDING PROCESSES

In producing welded steel tubular products, flat rolled skelp, strip, sheet or plate is formed into cylinders that are then joined at the longitudinal seam by one of the following processes.

Electric resistance welding employs a series of operations, in the first of which the flat rolled steel is cold shaped into tubular form. Welding is effected by the application pressure and heat generated by induction or by an electric current through the seam. The welding pressure is generated by constricting rolls and the electromagnetic effects of the high welding current. Electric resistance welded tubular products having longitudinal seams are usually made in sizes from 3.2 mm ($^1/_8$ in.) nominal to 0.6 m (24 in.) actual outside diameter, but larger sizes are also available.

Furnace Butt Welding. In furnace butt welding, skelp with square or slightly beveled edges is furnace heated to the welding temperature. The heated stock is roll formed into cylindrical shape as it emerges from the furnace, additional heat is usually provided by an oxygen or air jet impinging on the seam edges, and the tube passes through constricting rolls where the seam edges are welded by the pressure of the rolls. Furnace butt welded products are available in nominal diameters from 3.2 to 100 mm ($^1/_8$ to 4 in.).

Fusion Welding. In fusion welding, the flat rolled steel, with edges suitably prepared, is formed into tubular shape by either hot or cold shaping. The flat rolled steel may be shaped longitudinally (straight seam) or bent into helical form (spiral welded). The edges are welded with or without

Table 1. Major types and uses of pipe

Type of pipe	Uses
Standard	Industrial or residential water, steam, oil or gas transmission, distribution or service lines, structural uses
Special	Conduit, piling, pipe for nipples, etc.
Line	Oil- or gas-transmission pipe, water-main pipe, slurry pipe
Oil country tubular goods	Drill pipe, casing, tubing
Water well	Drive pipe, driven well pipe, casing, reamed and drifted pipe
Pressure	Pipe for elevated-temperature or pressure service

Table 2. Principal uses and types of pressure tubes

Use groups	Types
Water-tube boilers	Generating tubes, superheater tubes, economizer tubes, circulator tubes, furnace-wall tubes
Fire-tube boilers	Boiler flues, superheater flues, superheater tubes, arch tubes, stay tubes, safe ends
Others	Feed-water heater tubes, oil-still tubes

simultaneously depositing filler metal in a molten or molten-and-vapor state. Mechanical pressure is not required to effect welding. Fusion may be accomplished by either electric arc or gas heating, or by a combination of both.

SEAMLESS PROCESSES

Steel tubular products produced by seamless processes are made in diameters up to 0.66 m (26 in.) by the rotary piercing method and up to 1.22 m (48 in.) by hot extrusion.

Rotary Piercing. In rotary piercing, rounds of the necessary diameter and length are first heated to rolling temperature. Each hot round is fed into a set of rolls having crossed axes and surface contours that pull it through the rolls, thus rupturing it longitudinally. The force of the rolls then causes the metal to flow around a piercing point, enlarging the axial hole, smoothing the inside surface and forming a tube. After being pierced, the rough tube is usually hot rolled to final dimensions.

Press Piercing. A press piercing mill is composed of three basic elements: a roll stand with a round pass between a pair of driven rolls; a billet pusher; and a fixed plug located between the two rolls. The billet, enveloped in a four-sided guide, is forced against the plug by the combined action of the pusher and the driven rolls. The material deformation inherent in this process is mainly compressive, with low elongation (1.2% maximum), and thus the billet material (wrought or continuously cast) is not subjected to high tensile stresses. After being pierced, the rough tube is hot rolled to final dimension.

Hot extrusion is a hot working process for making hollows, suitable for processing into finished tubing of regular and irregular form, by forcing hot, prepierced billets through a suitably shaped orifice formed by an external die and internal mandrel.

COLD FINISHING

Pipe in suitable sizes and most products classified as tubing, both seamless and welded, may be cold finished. The process may be used to increase or decrease the diameter, to produce shapes other than round, to produce a smoother surface or closer dimensional tolerances, or to modify mechanical properties. The process most commonly used is cold drawing, in which the descaled hot worked tube is plastically deformed by drawing it through a die and over a mandrel (mandrel drawing) to work both exterior and interior surfaces. Cold drawing through the die only (without a mandrel) is called "sink drawing" or "sinking."

Tube Reducing and Swaging. In tube reducing by rotorolling or pilgering, and in swaging, a reducing die works the tube hollow over a mandrel; swaging may, however, be done without a mandrel. The commercial importance of tube reducing is, first, that very heavy reductions (up to 85%) can be applied to mill-length tubes, and second, that the process can be applied to refractory alloys that are difficult to cold draw because of high power requirements.

Cold Finishing. Tubular products of circular cross section may be cold finished on the outside by turning, grinding or polishing, or by any combination of these processes. They may be bored on the inside. Because these operations involve only stock removal, with negligible plastic deformation, there is no enhancement of mechanical properties.

Cold drawing may be employed to improve the surface finish and dimensional accuracy, and to increase the strength, of tubular products. Some customer specifications prescribe strength levels that can best be attained by cold working.

PIPE SIZES AND SPECIFICATIONS

Pipe is distinguished from tubing by the fact that it is produced in relatively few sizes and, therefore, in comparatively large quantities of each size.

For a reasonably complete list of the standardized sizes and weights of pipe for the major named uses, the AISI Steel Products Manual should be consulted. For oil country tubular goods, the specifications of the American Petroleum Institute (API) govern. Table 3 lists the current ASTM and API specifications covering pipe. Some of these involve several grades. The specifications listed cover carbon and alloy steels other than stainless, all methods of manufacture, and a wide range of service temperatures.

COMMON TYPES OF PIPE

The following brief descriptions concern the end uses of some of the more common types of pipe.

Standard pipe is standard-weight, extra-strong and double-extra-strong welded or seamless pipe of ordinary finish and dimensional tolerances, produced in sizes up to 26 in. in nominal diameter, inclusive.

Conduit pipe is welded or seamless pipe intended especially for fabrication into rigid conduit, a product used for protection of electrical wiring systems.

Piling pipe is welded or seamless pipe for use as piles, where the cylinder section acts as a permanent load-carrying member or where it acts as a shell to form cast-in-place concrete piles.

Transmission or line pipe is welded or seamless pipe currently produced in sizes from $1/8$ in. nominal to 56 in. actual outside diameter and is used principally for conveying gas or oil.

Water-main pipe is welded or seamless steel pipe used for conveying water for municipal and industrial purposes. Pipe lines for such purposes are commonly designated as flow mains, transmission mains, force mains, water mains or distribution mains. The mains are generally laid underground. Sizes range from $1^1/2$ to 96 in. in nominal diameter in a variety of wall thicknesses.

Oil country tubular goods is a collective term applied in the oil and gas industries to three kinds of pipe used in oil wells: drill pipe, casing and tubing. These products conform to API Specifications 5A, 5AC, and 5AX.

Drill pipe is used to transmit power by rotary motion from ground level to a rotary drilling tool below the surface, and also to convey flushing media to the cutting face of the tool. Drill pipe is produced in sizes from $2^3/8$ to $6^5/8$ in. in outside diameter. Size designations refer to actual outside diameter and weight per foot. Drill pipe is usually upset, either internally or externally, or both, and is prepared to accommodate welded-on types of joints.

Casing is used as a structural retainer for the walls of oil or gas wells, to exclude undesirable fluids, and to confine and conduct oil or gas from productive subsurface strata to ground level. Casing is produced in sizes from $4^1/2$ to 20 in.

Table 3. Specifications for carbon and alloy steel pipe (ASTM and API)

Specification	Product
ASTM specifications	
A53(a)	Welded or seamless steel pipe, black or hot dip galvanized
A106(a)	Seamless carbon steel pipe for high-temperature service
A120	Welded or seamless steel pipe for ordinary uses, black or hot dip galvanized
A134(a)	Arc-welded steel-plate pipe (sizes 16 in. and over)
A135(a)(b)	Resistance-welded steel pipe
A139	Arc-welded steel pipe (sizes 4 in. and over)
A155(a)(c)	Arc-welded steel pipe for high-pressure service
A211	Spiral-welded steel or iron pipe
A252	Welded or seamless steel-pipe piles
A333(a)	Welded or seamless steel pipe for low-temperature service
A335(a)	Seamless ferritic alloy steel pipe for high-temperature service
A381	Double submerged-arc welded steel pipe for high-pressure transmission systems
A405	Seamless ferritic alloy steel pipe, specially heat treated for high-temperature service
A523	Resistance-welded or seamless steel pipe (plain end) for high-pressure electric cable conduit
A524(a)	Seamless carbon-steel pipe for atmospheric and lower temperatures
A587(a)	Resistance-welded low-carbon steel pipe for the chemical industry
A589	Welded or seamless carbon steel water-well pipe
A671(a)	Arc-welded steel pipe for atmospheric and lower temperatures
A672(a)	Arc-welded steel pipe for high-pressure service at moderate temperatures
A691	Arc-welded carbon or alloy steel pipe for high-pressure service at high temperatures
A714	Welded or seamless HSLA steel pipe
API specifications	
2B	Welded steel-plate pipe for construction of offshore drilling platforms
5A	Welded or seamless steel pipe for oil or gas well casing, tubing or drill pipe
5AC	Welded or seamless steel pipe with restricted yield-strength range for oil or gas well casing or tubing
5AX	High-strength seamless steel pipe for oil or gas well casing, tubing or drill pipe
5L	All welded or seamless steel line pipe for oil or gas transmission

(a) This ASTM specification is also published by ASME, which adds an "S" in front of the "A" (for example, SA53). (b) This specification has been withdrawn by ASTM. (c) This specification has been replaced by ASTM A671, A672 and A691.

in outside diameter.

Tubing is used within the casings of oil wells to conduct oil and gas to ground level. It is produced in sizes from 1.05 to 4.50 in. in outside diameter, in several weights per foot. Ends are threaded for special integral-type joints or fitted with couplings, and may or may not be upset externally.

Water-well pipe is a collective term applied to four types of pipe that are used in water wells and that conform to ASTM A589: Type I, drive pipe; Type II, reamed and drifted pipe; Type III, driven well pipe; and Type IV, casing pipe.

Drive pipe is used to transmit power from ground level to a rotary drill tool below the surface, and to convey flushing media to the cutting face of the tool. The lengths of pipe have specially threaded ends that permit the lengths to butt inside the coupling. Drive pipe is produced in nominal sizes of 6, 8, 12, 14 and 16 in. in outside diameter.

Driven well pipe is threaded pipe in short lengths used for manual driving of a drill tool or for use with short rigs. It may be furnished in random lengths ranging from 0.9 to 1.8 m (3 to 6 ft) or in random lengths ranging from 1.8 to 3.0 m (6 to 10 ft).

Casing is used to confine and conduct water to ground level and as a structural retainer for the walls of water wells. It is produced as threaded pipe in random lengths from 4.9 to 6.7 m (16 to 22 ft) and in sizes from $3^1/_2$ to $8^5/_8$ in. in outside diameter. In western water-well practice, welded strings are sometimes used.

Pressure pipe, as distinguished from pressure tubes, is a commercial term for pipe used for conveying fluids at elevated temperature or pressure, or both, but not subjected to external application of heat. This commodity is not differentiated from other types of pipe by ASTM, and the applicable specifications are listed with the other types in Table 3. Pressure pipe ranges in size from $^1/_8$ in. nominal to 26 in. actual outside diameter in various wall thicknesses.

PRESSURE TUBES

Pressure tubes are given a separate classification by both ASTM and the producers. Pressure tubes are distinguished from pressure pipe in that they are suitable for application of external heat while conveying pressurized fluids.

The principal named uses of pressure tubing are given in Table 2. They are produced to actual outside diameter and minimum or average wall thickness (as specified by the purchaser) and may be hot finished or cold finished, as specified.

Double-wall brazed tubing is a specialty tubing confined to small sizes (see ASTM A254). It is used in large quantities by the automotive industry for brake lines and fuel lines, and by the refrigeration industry for refrigerant lines. It is made by forming copper-coated strip into a tubular section with double walls using either single-strip or double-strip construction. The tubing is then heated in a reducing atmosphere to join all mating surfaces completely. The resulting product is thus copper coated both inside and outside. When required by the intended service, a tin coating may be supplied. Available sizes range from 3.2 to 16 mm ($^1/_8$ to $^5/_8$ in.) in outside diameter with wall thickness from 0.64 mm (0.025 in.) for 3.2-mm ($^1/_8$-in.) OD to 0.89 mm (0.035 in.) for 16-mm ($^5/_8$ in.) OD.

STRUCTURAL TUBING

Structural tubing is used for welded, riveted or bolted construction of bridges and buildings, and for general structural purposes. It is available as round, square, rectangular or special-shape tubing, as well as tapered tubing. These products are covered by ASTM specifications.

MECHANICAL TUBING

Mechanical tubing includes welded and seamless tubing used for a wide variety of mechanical purposes. It usually is produced to meet specific end-use requirements and therefore is produced in many shapes, to a variety of chemical compositions and mechanical properties, and with hot rolled or cold finished surfaces. However, some mechanical tubing is covered by ASTM specifications.

Mechanical tubing is not produced to specified standard sizes; instead, it is produced to specified dimensions, which may be anything the customer requires within the limits of production equipment or processes.

Welded mechanical tubing usually is made by electric resistance welding, but some is made by the various fusion welding processes. In all instances, the exterior welding flash may be removed (if necessary) by cutting, grinding or hammering.

Electric resistance welded mechanical tubing is made from either hot rolled or cold rolled carbon steel or alloy steel and is produced either as welded or as cold finished. Sizes produced range in outside diameter from 6.4 to 195 mm ($^1/_4$ to $7^5/_8$ in.) and in wall thickness from 1.65 to 9.52 mm (0.065 to 0.375 in.) for hot rolled steel and from 0.64 to 4.19 mm (0.025 to 0.165 in.) for cold rolled steel.

Butt welded cold finished mechanical tubing, as its name implies, is tubing that has been hot formed by furnace butt welding and cold finished. It is furnished sink drawn or mandrel drawn and is available in outside diameters up to 90 mm ($3^1/_2$ in.) and wall thicknesses from 0.89 to 12.7 mm (0.035 to 0.500 in.). The material is low-carbon steel and the product is, in effect, a form of cold drawn pipe. Although furnished in a narrower size range than electric resistance welded tubing, it has two advantages: (*a*) within the available size range, heavier walls are available and (*b*) there is no problem with flash.

Seamless mechanical tubing, both hot and cold finished, is available in a wide variety of finishes and mechanical properties. It is made from carbon and alloy steels in sizes up to and including 325 mm ($12^3/_4$ in.) in outside diameter.

Closed-Die Steel Forgings

FORGING is the process of working hot metal between dies, usually under successive blows and sometimes by continuous squeezing. Closed-die forgings, hot upset parts and extrusions are shaped within a cavity formed by the closed dies.

Justification for selecting forging in preference to other and sometimes more economical methods of producing useful shapes is based on several considerations. Mechanical properties in wrought materials are maximized in the direction of major metal flow during working.

Type of Forging. Forgings are classified in several ways, beginning with the general classifications "open-die" and "closed-die." They are also classified in terms of the "close-to-finish" factor, or the amount of stock (cover) that must be removed from the forging by machining to satisfy the dimensional and detail requirements of the finished part (see Fig. 1). Finally, forgings are further classified in terms of the forging equipment required for their manufacture, as, for example, hammer upset forgings, ring-rolled forgings and multiple-ram press forgings.

Of the various classifications, those based on the close-to-finish factor are most closely related to the inherent properties of the forging, such as strength and resistance to stress corrosion. In general, the type of forging that requires the least machining to satisfy finished-part requirements has the best properties. For this reason, a finished part that is machined from a blocker-type forging usually exhibits mechanical properties and corrosion characteristics that are inferior to those of a part produced from a close-tolerance, no-draft forging.

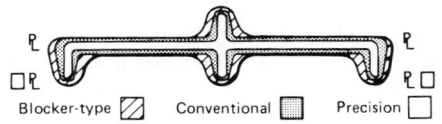

Schematic composite of cross sections of blocker-type, conventional and precision forgings.

Fig. 1. Types of forgings

SELECTION OF STEEL

Selection of a steel for a forged component is an integral part of the design process, and acceptable performance is dependent on this choice. A thorough understanding of the end use of the finished part will serve to define the required mechanical properties, surface-finish requirements, tolerance to nonmetallic inclusions and the attendant inspection methods and criteria.

Forging-quality steels are produced to a wide range of chemical compositions by electric furnace, open hearth or pneumatic steelmaking processes.

"Forgeability" describes the relative ability of a steel to flow under compressive loading without fracturing. Except for resulfurized and rephosphorized grades, most carbon and low-alloy steels are usually considered to have good forgeability. Differences in forging behavior among the various grades of steel are small enough that selection of the steel is seldom affected by forging behavior. However, the choice of a resulfurized or rephosphorized steel for a forging is usually justified only if the forging must be extensively machined; because one of the principal reasons for considering manufacture by forging is the avoidance of subsequent machining operations, this situation is uncommon.

Design Requirements. Selection of a steel for a forged part usually requires some compromise between opposing factors—for instance, strength versus toughness, stress-corrosion resistance versus weight, manufacturing cost versus useful load-carrying ability, production cost versus maintenance cost, and the cost of the steel.

Material selection also involves consideration of melting practices, forming methods, machining operations, heat treating procedures, and deterioration of properties with time in service, as well as the conventional mechanical and chemical properties of the steel to be forged.

An efficient forging design obtains maximum performance from the minimum amount of material consistent with the loads to be applied, producibility, and desired life expectancy. To match a steel to its design component, the steel is first appraised for strength and toughness and then qualified for stability to temperature and environment. Optimum steels are then analyzed for producibility and finally for economy.

Cost. The cost of steel as a percentage of the total manufacturing cost of forgings is shown in Fig. 2. These curves are based on an average of many actual forgings that are different in number of forging and heat treating operations required, cost of steel, quantity and setup cost. It should not be inferred from these data that an average 14-kg (30-lb) stainless steel forging will cost 34% more than an average carbon steel forging of the same weight.

MATERIAL CONTROL

After completion of a forging design, there remains the responsibility of ensuring and verifying that the delivered forging will have all of the properties and characteristics specified on the forging drawing.

Responsibility for material control is subject to agreement between the purchaser and forging supplier. In many such agreements, the purchaser is responsible for design, material selection and controls during manufacture; the forging supplier is responsible for performing raw-material inspection as well as maintaining adequate process control and product inspection.

Tests and Test Coupons. Tests contained in the material specifications are intended to provide correlation with, and interpretation of, the behavior of material in actual use. The dynamic behavior of a full-size structural component seldom can be accurately predicted from simple room-temperature tests on small specimens. Analytical studies coupled with model or full-scale testing can augment simple tests in interpreting the complex behavior of materials.

The kinds of test specimens and tests specified for quality assurance depend on the conditions imposed on the final component in service. If, for example, a critical forging is to be subjected to large tensile loads, the designer would specify tests to measure fracture toughness and tensile yield strength. For components for elevated-temperature service, tests measuring strength, ductility and creep at appropriate temperatures would be specified.

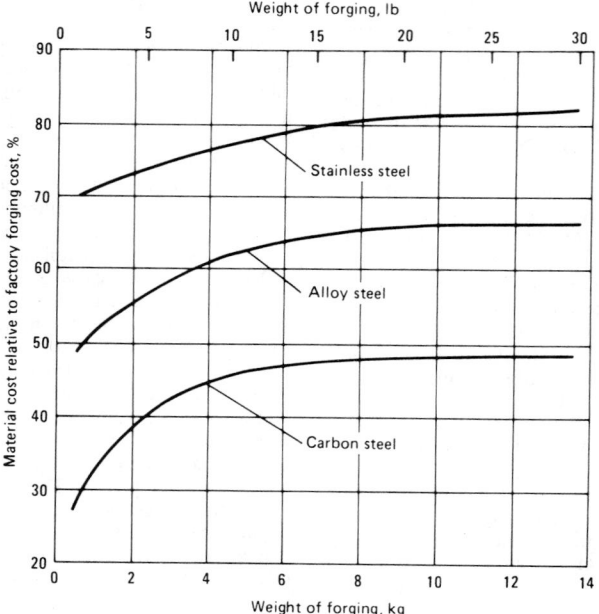

Cost of steel as a percentage of total cost of forgings.

Fig. 2. Cost of steel forgings

Table 1. Testing plan for determining mechanical properties of forging material

	Number of tests(a) for:							
	9Ni-4Co-0.30C Steel (1515 to 1660 MPa, or 220 to 240 ksi)(b)				9Ni-4Co-0.45C Steel (1790 to 1930 MPa, or 260 to 280 ksi)(c)			
Temperature and test	L	LT	ST	Total	L	LT	ST	Total
At −80 °C (−110 °F):								
Tension	2	3	2	7	2	4	2	8
Compression	1	3	1	5	1	4	...	5
Shear	1	3	1	5	1	3	1	5
Bearing e/D = 1.5(d)	1	2	1	4	1	1	1	3
Bearing e/D = 2.0(d)	1	2	1	4	1	4	1	6
At room temperature:								
Tension	12	12	12	36	18	17	18	53
Compression	3	3	3	9	2	4	3	9
Shear	3	3	3	9	3	4	2	9
Bearing e/D = 1.5(d)	3	3	3	9	3	3	3	9
Bearing e/D = 2.0(d)	3	3	3	9	3	3	3	9
Modulus of elasticity	1	1	...	2	1	1	...	2
At 150 °C (300 °F):								
Tension	1	3	1	5	2	4	2	8
Compression	1	1	1	3	1	4	...	5
Shear	1	1	1	3	1	3	1	5
Bearing e/D = 1.5(d)	1	1	1	3	1	2	1	4
Bearing e/D = 2.0(d)	...	...	...	0	1	4	1	6
At 260 °C (500 °F):								
Tension	2	3	2	7	...	...	...	0
Compression	1	3	1	5	...	...	...	0
Shear	1	3	1	5	...	...	...	0
Bearing e/D = 1.5(d)	1	2	1	4	...	...	...	0
Bearing e/D = 2.0(d)	2	2	1	5	...	...	...	0
Total number of tests	42	57	40	139	42	65	39	146

(a) L, longitudinal; LT, long transverse; ST, short transverse. (b) Three heats. (c) Four heats. (d) D, hole diameter; e, edge distance measured from the hole center to the edge of the material in direction of applied stress.

Test Plans. Frequently, specifications are prepared from the results of tests on laboratory specimens, because the cost and time required for full-scale testing are usually prohibitive. Test plans for evaluation of the mechanical properties of two high-strength 9Ni-4Co steels used in aircraft service at temperatures ranging from −45 to greater than 205 °C (−50 to greater than 400 °F) are shown in Table 1. This table illustrates the range and number of tests required for a very extensive type of evaluation.

As shown in Table 1, test plans for mechanical properties include tension, compression, shear and bearing strength tests; the effect of grain orientation is evaluated by testing specimens representative of the longitudinal, long-transverse and short-transverse directions, as required. In addition to room-temperature tests, specimens are tested at −80, 150 and 260 °C (−110, 300 and 500 °F). The plan encompasses a total of 285 individual tests.

Ductility and the Amount of Forging Reduction. A

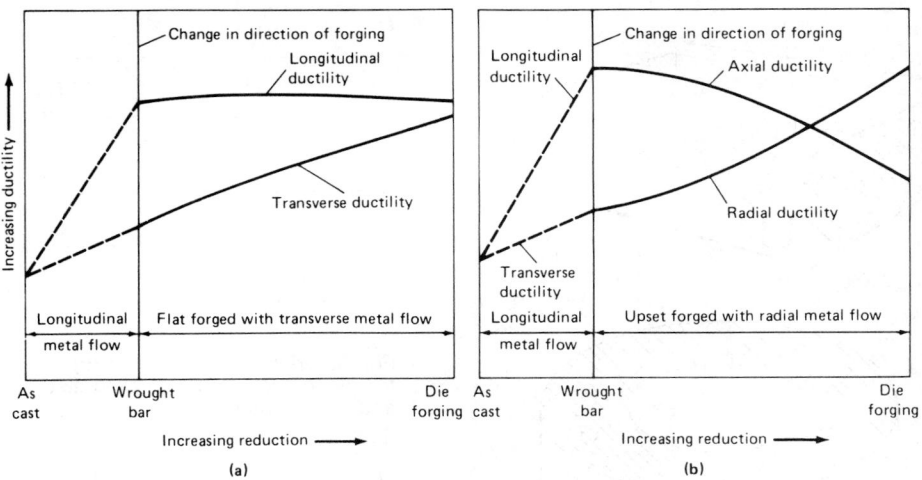

Effect of extent and direction of metal flow during forging on ductility. (a) Longitudinal and transverse ductility in flat forged bars. (b) Axial and radial ductility in upset forged bars.

Fig. 3. Metal flow in forging

principal objective of material control is to ensure that optimum mechanical properties will be obtained in the finished forging. The amount of reduction achieved in forging has a marked effect on ductility, as shown in Fig. 3, which compares ductility in the cast ingot, the wrought (rolled) bar or billet and the forging. The curves in Fig. 3(a) indicate that when a wrought bar or billet is flat forged in a die, an increase in forging reduction does not affect longitudinal ductility, but does result in a gradual increase in transverse ductility. When a similar bar or billet is upset forged in a die, an increase in forging reduction results in a gradual decrease in axial ductility and a gradual increase in radial ductility.

Grain Flow. Macroetching permits direct observation of grain direction and contour and also serves to detect folds, laps and re-entrant flow. By macroetching of suitable specimens, grain flow can be examined in the longitudinal, long-transverse and short-transverse directions. Macroetching also permits evaluation of complete sections, end to end and side to side, and a review of uniformity of macro grain size. Figure 4 illustrates grain flow in a representative forged part.

End-Grain Exposure. Lowered resistance to stress-corrosion cracking in the long-transverse and short-transverse directions is related to end-grain exposure. A long, narrow test specimen sectioned so that the grain is parallel to the longitudinal axis of the specimen has no exposed end grain, except at the extreme ends, which are not subjected to loading. In contrast, a corresponding specimen cut in the transverse direction has end-grain exposure at all points along its length. End grain is especially pronounced in the short-transverse direction on die forgings designed with a flash line. Consequently, forged components designed to reduce or eliminate end grain have better resistance to stress-corrosion cracking.

MECHANICAL PROPERTIES

A major advantage of shaping metal parts by rolling, forging or extrusion stems from the opportunities such processes offer the designer with respect to control of grain flow. The strength of

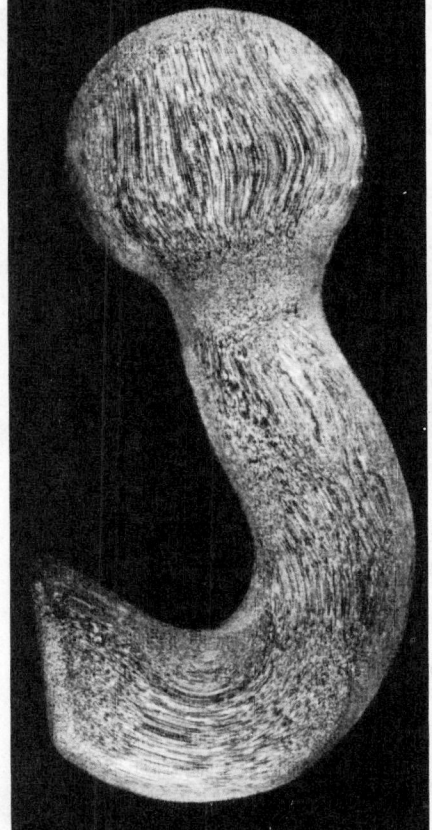

Flow lines, as shown in a longitudinal section taken through a hook that was forged from 4140 steel.

Fig. 4. Metal flow in a forged hook

these and similar wrought products is almost always greatest in the longitudinal direction (or equivalent) of grain flow, and the maximum load-carrying ability in the finished part is attained by providing a grain-flow pattern parallel to the direction of the major applied service loads when,

in addition, sound, dense, good-quality metal of sufficiently fine grain size has been produced throughout.

Grain Flow and Anisotropy. Metal that is rolled, forged or extruded develops and retains a fiber-like grain structure aligned in the principal direction of working. This characteristic becomes visible on external and sectional surfaces of wrought products when the surfaces are suitably prepared and etched. The "fibers" are the result of elongation of the microstructural constituents of the metal in the direction of working. Thus, the phrase "direction of grain flow" is commonly used to describe the dominant direction of these fibers within wrought metal products.

In wrought metal, the direction of grain flow is also evidenced by measurements of mechanical properties. Strength and ductility are almost always greater in the direction parallel to that of working. The characteristic of exhibiting different strength and ductility values with respect to the direction of working is referred to as "mechanical anisotropy" and is exploited in the design of wrought products.

Although best properties in wrought metals are most frequently the longitudinal (or equivalent), properties in other directions may yet be superior to those in products not wrought—that is, in cast ingots or in forging stock taken from only lightly worked ingot.

The square rolled section shown schematically in Fig. 5(a) is anisotropic with respect to average mechanical properties of test bars such as those shown in phantom. Average mechanical properties of the longitudinal bar 1 are superior to the average properties of the transverse bars 2 and 3. Mechanical properties are equivalent for bars 2 and 3 because the section is square, which implies equal reduction in section in both transverse directions.

Mechanical anisotropy also is found in rectangular sections such as that shown in Fig. 5(b), in cylinders as in Fig. 5(c), and in rolled rings as in Fig. 5(d). Again, best strength properties are, on the average, those of the longitudinal, as in test bar 1. Flat rolling of a section such as that shown in Fig. 5(a) to a rectangular section (Fig. 5b) enhances the average "long transverse" properties of test bar 4 when compared with "short transverse" properties of test bar 5. Thus, such rectangular sections exhibit anisotropy among all three principal directions—longitudinal, long transverse and short transverse. A design that employs a rectangular section such as that shown in Fig. 5(b) involves the properties in all these directions, not just the longitudinal. Thus, the longitudinal, long transverse and short transverse service loads of rectangular sections are analyzed separately. The same concept can be applied to cylinders, whether extruded or rolled; longitudinal direction changes with the forging process used, as indicated in Fig. 5(c) and (d).

FUNDAMENTALS OF HAMMER AND PRESS FORGINGS

Many small forgings are made in a die that has successive cavities to preshape the stock progressively into its final shape in the last, or "finish," cavity. Dies for large forgings are usually made to perform one operation at a time. The upper half of the die, having the deeper and more intricate cavity, is keyed or dovetailed into the

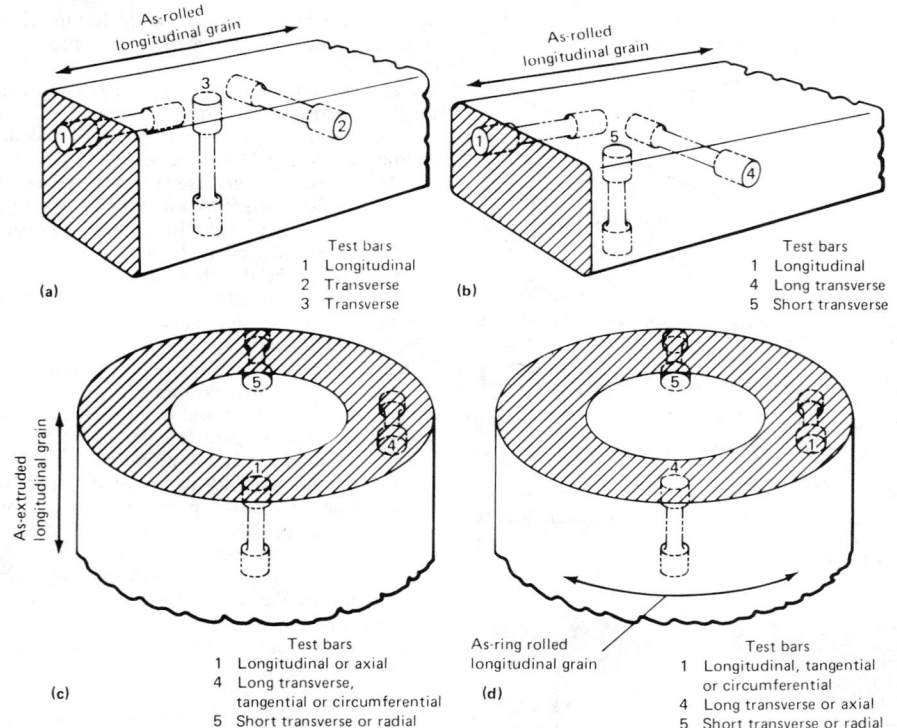

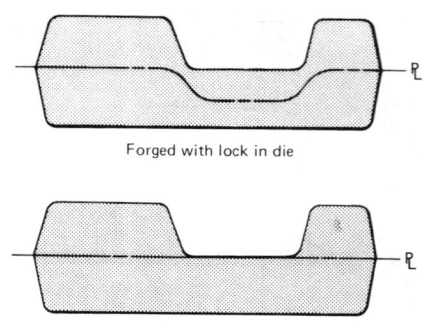

Fig. 8. Orientation of a forging in the die to avoid counterlocked dies and to eliminate draft

Schematic views of sections from (a) square rolled stock, (b) rectangular rolled stock, (c) a cylindrical extruded section, and (d) a ring-rolled section, illustrating the effect of section configuration or forging process, or both, on the longitudinal direction in a forging.

Fig. 5. Anisotropy and mechanical properties in forgings

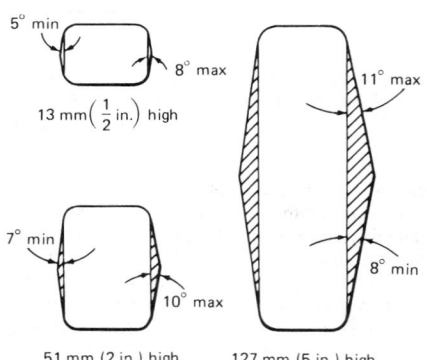

Fig. 9. Effect of part size on the amount of metal needed for draft

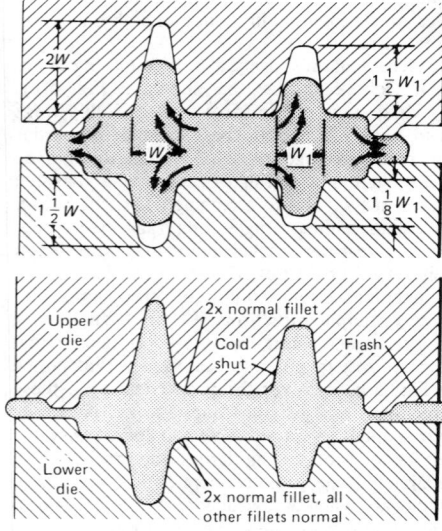

Two stages in completion of a forging. Top diagram shows limitations on height of ribs above and below the parting line.

Fig. 6. Two stages of metal flow in forging

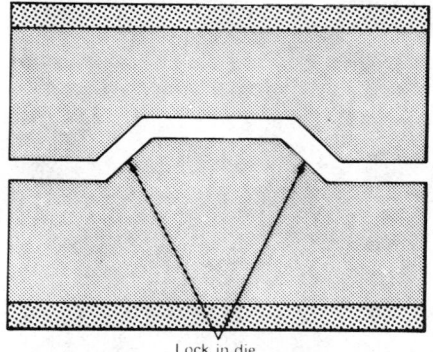

Fig. 7. Locked dies

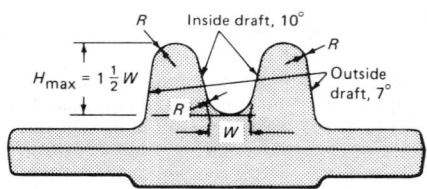

Fig. 10. Definition of inside and outside draft and limitations on the depth of cavities, as between ribs

hammer or press ram. The lower half is keyed to the "sow block" or bed of the hammer or press in precise alignment with the upper die. After being heated, the forging stock is placed in one cavity after another and is thus forged progressively to final shape.

The parting line is the plane along the periphery of the forging where the striking faces of the up-

per and lower dies come together. Usually, the die has a gutter or recess just outside of the parting line to receive overflow metal or flash forced out between the two dies in the finish cavity (Fig. 6). More complex forgings may have other parting lines around holes and other contours within the forging that may or may not be in the same plane as the outer parting line.

For greatest economy, the outer parting line should be in a single plane. When it must be along a contour, either step or locked dies may be necessary to equalize thrust, as shown in Fig. 7. This may increase costs as much as 20%, because of the increased cost of dies and cost increases from processing difficulties in forging and trimming. Sharp steps or drops in the parting line should be limited to about 15° from the vertical in small parts and 25° in large forgings, to prevent a tearing instead of a cutting action in trimming off the flash. Locked dies may sometimes be avoided

by locating the parting line as shown in the lower sketch of Fig. 8.

Specification of optional parting lines on forgings to be made in different shops allows these lines to vary from shop to shop. Unless the draft has been removed, this variation may cause difficulties in locating forgings when they are being chucked for subsequent machining. However, shearing the draft is not always an adequate remedy if trimming angles vary. Forgings made in different shops are likely to be more consistent in quality and to have less variation in shape when a definite parting line is specified.

Draft on the sides of a forging is an angle or taper necessary for releasing the forging from the die and is desirable for long die life and economical production. Draft requirements vary with the shape and size of the forging. The effect of part size on the amount of metal needed for draft is illustrated by Fig. 9.

"Inside draft" is draft on surfaces that tightens on the die as the forging shrinks during cooling; examples are cavities such as narrow grooves or pockets. "Outside draft" is draft on surfaces such

Table 2. Draft angles and tolerances for steel forgings

| Height or depth of draft | | Commercial standard | | Special standard | |
mm	in.	Draft, degrees	Plus tolerance(a), degrees	Draft, degrees	Plus tolerance(a), degrees
Outside draft					
6.35 to 12.7	1/4 to 1/2	...	...	3	2
19 to 25	3/4 to 1	5	3	...	...
Over 12.7, up to 25	Over 1/2, up to 1	...	...	5	2
Over 25, up to 76	Over 1, up to 3	7	3	5	3
Over 76	Over 3	7	4	7	3
Inside draft					
6.35 to 25.4	1/4 to 1	7	3	5	3
Over 25.4	Over 1	10	3	10	3

(a) The minus tolerance is zero.

Table 3. Recommended commercial tolerances on length and location

| Max length of forging | | Tolerance on length or location Plus | | Minus | |
mm	in.	mm	in.	mm	in.
150	6	1.19	0.047	0.79	0.031
380	15	1.57	0.062	1.19	0.047
610	24	3.18	0.125	1.57	0.062
910	36	3.18	0.125	1.57	0.062
1220	48	3.18	0.125	3.18	0.125
1520	60	4.75	0.187	3.18	0.125
1830	72	5.56	0.219	3.18	0.125

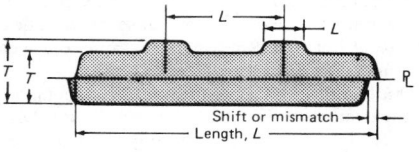

Shift or mismatch
Length, L

Table 5. Tolerances on burr for steel forgings

| Weight | | Trim size(a) | | Tolerance, plus(b) | |
kg	lb	mm	in.	mm	in.
0.45	1	50	2	0.79	0.031
4.5	10	150	6	1.57	0.062
11	25	200	8	3.18	0.125
45	100	625	25	6.35	0.250

(a) The trim size refers to the greatest distance across the forging at the trim line. (b) The minus tolerance is zero.

TOLERANCES

Forging tolerances, based on area and weight, that represent good commercial practice are listed in Tables 3 and 4. These tolerances apply to the dimensions shown in the illustration accompanying Table 3. In using these tables to determine the size of the forging, the related tolerances, such as mismatch, die wear and length, should be added to allowance for machining plus machined dimensions. On the average, tolerances listed in Tables 3 and 4 conform to the full process tolerances of actual production parts and yield more than 99% acceptance of any dimension specified from this table. In particular, instances may be found of precise accuracy or rarely as much as

as ribs or bosses that shrink *away from* the die during cooling. Both are illustrated in Fig. 10, which shows inside draft to be greater than outside draft (the usual relation). Recommended draft angles and tolerances for steel forgings are given in Table 2.

Table 4. Recommended commercial tolerances for steel forgings

Forging size Area 10³ mm²	in.²	Weight kg	lb	Thickness(a) Plus mm	in.	Minus mm	in.	Tolerance Mismatch(a), plus mm	in.	Die wear, plus mm	in.
3.2	5.0	0.45	1	0.79	0.031	0.41	0.016	0.41 to 0.79	0.016 to 0.031	0.79	0.031
4.5	7.0	3.2	7	1.57	0.062	0.79	0.031	0.41 to 0.79	0.016 to 0.031	1.57	0.062
6.5	10.0	0.7	1.5	0.79	0.031	0.79	0.031	0.41 to 0.79	0.016 to 0.031	0.79	0.031
7.7	12.0	5.5	12	1.57	0.062	0.79	0.031	0.41 to 0.79	0.016 to 0.031	1.57	0.062
12.9	20.0	0.9	2	1.57	0.062	0.79	0.031	0.41 to 0.79	0.016 to 0.031	1.57	0.062
12.9	20.0	14	30	1.57	0.062	0.79	0.031	0.51 to 1.02	0.020 to 0.040	1.57	0.062
24.5	38.0	2	4.5	1.57	0.062	0.79	0.031	0.41 to 0.79	0.016 to 0.031	1.57	0.062
24.5	38.0	36	80	1.57	0.062	0.79	0.031	0.64 to 1.27	0.025 to 0.050	1.57	0.062
32.3	50.0	3	8	1.57	0.062	0.79	0.031	0.51 to 1.02	0.020 to 0.040	1.57	0.062
32.3	50.0	27	60	1.57	0.062	0.79	0.031	0.51 to 1.02	0.020 to 0.040	1.57	0.062
32.3	50.0	45	100	1.57	0.062	0.79	0.031	0.64 to 1.27	0.025 to 0.050	1.57	0.062
61.3	95.0	5	11	1.57	0.062	0.79	0.031	0.51 to 1.02	0.020 to 0.040	1.57	0.062
85.2	132.0	8	17	1.57	0.062	0.79	0.031	0.64 to 1.27	0.025 to 0.050	1.57	0.062
107	166.0	33	73	2.39	0.094	0.79	0.031	0.76 to 1.52	0.030 to 0.060	2.39	0.094
113	175.0	68	150	2.39	0.094	0.79	0.031	0.76 to 1.52	0.030 to 0.060	2.39	0.094
130	201.0	18	40	1.57	0.062	0.79	0.031	0.64 to 1.27	0.025 to 0.050	1.57	0.062
155	240.0	23	51.5	2.39	0.094	0.79	0.031	0.76 to 1.52	0.030 to 0.060	2.39	0.094
161	250.0	114	250	2.39	0.094	0.79	0.031	0.76 to 1.52	0.030 to 0.060	2.39	0.094
171	265.0	27	60	2.39	0.094	0.79	0.031	0.76 to 1.52	0.030 to 0.060	2.39	0.094
177	275.0	30	65	3.18	0.125	0.79	0.031	1.19 to 2.39	0.047 to 0.094	3.18	0.125
194	300.0	34	75	3.18	0.125	1.57	0.062	1.19 to 2.39	0.047 to 0.094	3.18	0.125
194	300.0	159	350	2.39	0.094	0.79	0.031	0.76 to 1.52	0.030 to 0.060	2.39	0.094
242	375.0	205	450	3.18	0.125	0.79	0.031	1.19 to 2.39	0.047 to 0.094	3.18	0.125
268	415.0	139	306	3.18	0.125	1.57	0.062	1.19 to 2.39	0.047 to 0.094	3.18	0.125
339	525.0	340	750	3.18	0.125	1.57	0.062	1.19 to 2.39	0.047 to 0.094	3.18	0.125
580	900.0	455	1000	3.18	0.125	1.57	0.062	1.19 to 2.39	0.047 to 0.094	3.18	0.125

(a) The sketch in Table 3 shows locations of thickness and mismatch.

±50% error in the tolerances recommended in Table 4.

The characteristics of die wear are shown graphically in Fig. 11. The part represented was made of 4140 steel, using ten blows in a 1150-kg (2500-lb) board hammer. Tolerances were commercial standard, and the part was later coined to a thickness tolerance of +0.25 mm, −0.000 (+0.010 in., −0.000). The die block, 250 by 455 by 455 mm (10 by 18 by 18 in.), was hardened to 42 HRC. After 30 000 forgings had been produced, the die wore as indicated and the dies were resunk.

Ranges of mismatch tolerance are given in Table 4. The higher values are to be added to tolerances for forgings that need locked dies or involve side thrust on the dies during forging. On forgings heavier than 23 kg (50 lb), it is sometimes necessary to grind out mismatch defects up to 3.18 mm (1/8 in.) maximum.

Flash is trimmed in a press with a trimming die shaped to suit the plan view, outline and side view contour of the parting line. The forging may be trimmed with a stated amount of burr or flash left around the periphery at the parting line. It may also be trimmed flush to the side face of the forging or some of the draft may be trimmed off, provided that the serrations or score marks left by the shearing operation are not an objectionable feature. In most commercial forgings, some draft is sheared away.

Trimming Tolerance. When the trim must cut through the flash only and leave the side of the forging untouched, it is necessary to use a trim dimension that includes burr tolerance, mismatch, draft tolerance, and die-wear-plus-shrink tolerance. When it is satisfactory to trim draft partially, a closer trim tolerance may be held. Burr tolerance, listed in Table 5, applies to the amount of flash that should remain between the side of the forging at the parting line and the outside edge of the trim cut.

DESIGN OF HOT UPSET FORGINGS

Hot heading, upset forging or, more broadly, machine forging consists primarily of holding a bar of uniform cross section, usually round, between grooved dies and applying pressure on the end in the direction of the axis of the bar by using a heading tool so as to upset or enlarge the end into an impression of the die. The shapes generally produced include a variety of enlargements of the shank, or multiple enlargements of the shank and "re-entrant angle" configurations. Transmission cluster gears, pinion blanks, shell bodies and many other shaped parts are adapted to production by the upset machine forging process. This process produces a "looped" grain flow of major importance for gear teeth. Simple, headed forgings may be completed in one step, while some that have large, configured heads or multiple upsets may require as many as six steps. Upset forgings are produced weighing from less than 0.45 kg (1 lb) to about 225 kg (500 lb).

Machining Stock Allowances. The standard for machining stock allowance on any upset portion of the forging is 2.39 mm (0.094 in.), although allowances vary from 1.58 to 3.18 mm (0.062 to 0.125 in.), depending on size of upset, material and shape of the part (Fig. 12a).

Mismatch and shift of dies are each limited to 0.406 mm (0.016 in.) maximum. Mismatch is the location of the gripper dies with respect to each other.

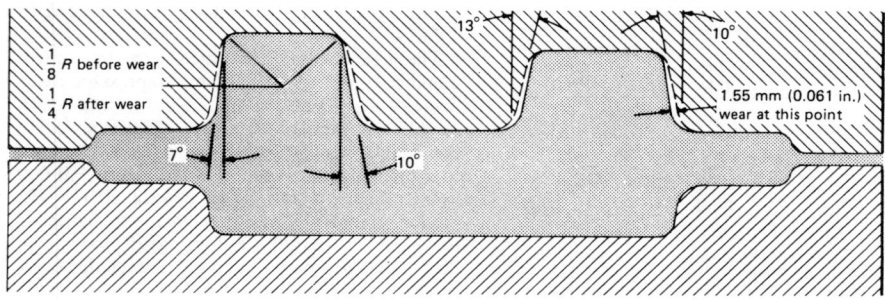

Fig. 11. An example of die wear

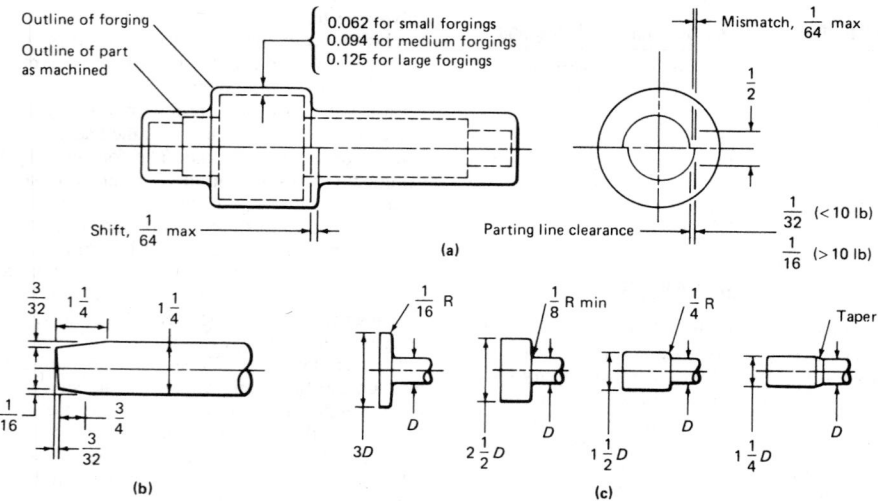

(a) Hot upset forging terminology and standards. (b) Probable shape of shear-cut ends. (c) Variation of corner radius with thickness of upset. These parts are the simplest forms of upset forgings. All linear dimensions are in inches.

Fig. 12. Allowances for hot upset forging

Parting-line clearance is required in gripper dies for tangential clearance in order to avoid undercut and difficulty in removal of the forging from the dies (Fig. 12a).

Tolerances for shear-cut ends have not been established. Figure 12(b) shows a shear-cut end on a 31.8-mm ($1^1/_4$-in.) diameter shank. Straight ends may be produced by torch cutting, hacksawing or abrasive wheel cutoff, at a higher cost than that of shearing.

Corner radii should follow the contours of the finished part, with a minimum radius of 1.59 mm ($^1/_{16}$ in.). Radii at the outer diameter of the upset face are not required, but may be specified as desired. Variations in thickness of the upset require variations in radii, as shown in Fig. 12(c), because the source of the force is farther removed and the die cavity is more difficult to fill. When a long upset is only slightly larger than the original bar size, a taper is advisable instead of a radius. Fillets can conform to the finished contour in most instances. The absolute minimum should be 3.18 mm ($^1/_8$ in.) on simple upsets.

Tolerances for all upset forged diameters are generally +1.59 mm, −0 (+$^1/_{16}$ in., −0) except for thin sections of flanges and upsets relatively large in ratio to the stock sizes used, where they are +2.38 mm, −0 (+$^3/_{32}$ in., −0). The increase of tolerances over the standard +1.59 mm, −0 is sometimes a necessity, because of variations in size of hot rolled mill bars, extreme die wear or complexity of the part.

Draft angles may vary from 1 to 7°, depending on the characteristics of the forging design. Draft is needed to release the forging from the split dies; it also reduces the shearing of face surfaces in transfer from impression to impression.

For an upset forged part that requires several operations or passes, the dimensioning of lengths is determined on the basis of the design of each individual pass or operation.

Steel Castings

STEEL CASTINGS can be made from any of the many types of carbon and alloy steel produced in wrought form. Such castings are produced by pouring molten steel of the desired composition into a mold of the desired configuration and allowing the steel to solidify. The mold material may be silica, zircon, chromite or olivine sand, graphite, metal or ceramic. Choice of mold material depends on the size, intricacy and dimensional accuracy of the casting and on cost. While the producible size, surface finish and dimensional accuracy of castings vary widely with the type of mold, the properties of the cast steel are not affected significantly. Steel castings produced in any of the various types of molds and wrought steel of equivalent chemical composition respond similarly to heat treatment, have the same weldability, and have similar physical and mechanical properties. Cast steels do not exhibit the effects of directionality on mechanical properties that are typical of wrought steels.

The steel castings discussed in this article are classified into four general groups according to their carbon or alloy contents. Carbon steel castings account for three of these groups: (a) low-carbon steel castings with less than 0.20% carbon, (b) medium-carbon castings with 0.20 to 0.50% carbon, and (c) high-carbon castings with more than 0.50% carbon. The fourth group, low-alloy steel castings, is limited to grades with a total alloy content of less than 8%.

SPECIFICATIONS

Steel castings are usually purchased to meet specified mechanical properties, with some restrictions on chemical composition.

Table 1 lists the requirements given in various standard ASTM, SAE and government specifications. In the low-strength ranges, some specifications limit carbon and manganese contents, usually to ensure satisfactory weldability. In SAE J435c, carbon and manganese are specified to ensure that the minimum desired hardness and strength are obtained after heat treatment. For special applications, other elements may be specified either as maximum or minimum percentages, depending on the characteristics desired.

Other ASTM specifications that include carbon and low-alloy grades of steel castings are A216, A217, A352, A356, A389, A426, A486, A487, A643 and A757.

Where only mechanical properties are specified, the chemical compositions of castings for general engineering applications usually are left to the discretion of the casting supplier. For specific applications, however, certain chemical-composition limits have been established to ensure development of specified mechanical properties after proper heat treatment as well as to facilitate welding, uniform response to heat treatment, or other requirements. Hardness is specified for most grades of SAE J435c to ensure machinability, ease of inspection for high-production-rate items, or certain characteristics pertaining to wear.

SAE J435c includes three grades, HA, HB and HC, with specified hardenability requirements. Figure 1 plots hardenability requirements, both minimum and maximum, for these steels. Hardenability is determined by the end-quench hardenability test described in the article on hardenability, in this section. Other specifications require minimum hardness at one or two locations on the end-quench specimen. In general, hardenability

Table 1. Summary of specification requirements for steel castings

Class or grade	Minimum tensile strength MPa	ksi	Minimum yield strength MPa	ksi	Minimum elongation in 50 mm or 2 in., %	Minimum reduction in area, %	Hardness(a), HB	Chemical composition(b), % C	Mn
Federal QQ-S-681E									
65-35	448	65	241	35	24	35	...	0.30(c)	0.70(c)
70-36	483	70	248	36	22	30	...	0.35(c)	0.70(c)
80-40	552	80	276	40	18	30	...	...	...
0050A	586	85	310	45	16	24	...	0.50	0.90
0050B	689	100	483	70	10	15	...	0.50	0.90
80-50	552	80	345	50	22	35	...	...	...
90-60	621	90	414	60	20	40	...	...	...
105-85	724	105	586	85	17	35	...	...	...
120-95	827	120	655	95	14	30	...	...	...
150-125	1034	150	862	125	9	22	...	...	...
175-145	1207	175	1000	145	6	12	...	...	...
ASTM A27-77									
N-1	...	...	...	...	...	...	...	0.25(c)	0.75(c)
N-2	...	...	...	...	...	...	...	0.35(c)	0.60(c)
U60-30	415	60	205	30	22	30	...	0.25(c)	0.75(c)
60-30	415	60	205	30	24	35	...	0.30(c)	0.60(c)
65-35	450	65	240	35	24	35	...	0.30(c)	0.70(c)
70-36	485	70	250	36	22	30	...	0.35(c)	0.70(c)
70-40	485	70	275	40	22	30	...	0.25(c)	1.20(c)
ASTM A148-73									
80-40	552	80	276	40	18	30	...	...	...
80-50	552	80	345	50	22	35	...	...	...
90-60	621	90	414	60	20	40	...	...	...
105-85	724	105	586	85	17	35	...	...	...
120-95	827	120	655	95	14	30	...	...	...
150-125	1034	150	862	125	9	22	...	...	...
175-145	1207	175	1000	145	6	12	...	...	...
SAE J435c									
0022	...	...	...	...	...	...	187 max	0.12-0.22	0.50-0.90
0025	414	60	207	30	22	30	187 max	0.25(c)	0.75(c)
0030	448	65	241	35	24	35	131-187	0.30(c)	0.70(c)
0050A	586	85	310	45	16	24	170-229	0.40-0.50	0.50-0.90
0050B	690	100	483	70	10	15	207-255	0.40-0.05	0.50-0.90
080	552	80	345	50	22	35	163-207	...	...
090	621	90	414	60	20	40	187-241	...	...
0105	724	105	586	85	17	35	217-248	...	...
0120	827	120	655	95	14	30	248-311	...	...
0150	1034	150	862	125	9	22	311-363	...	...
0175	1207	175	1000	145	6	12	363-415	...	...
HA, HB, HC(d)	...	...	...	...	...	...	...	0.25-0.34	...

(a) Hardness values apply to cast parts in locations not over 76 mm (3 in.) thick. (b) Carbon and manganese are maximum limits unless a range is given. Federal QQ-S-681E, ASTM A27-73 and ASTM A148-73 restrict phosphorus to 0.050% and sulfur to 0.060% max. SAE J435c restricts phosphorus to 0.040% and sulfur to 0.045% max. Silicon and alloying elements are restricted on some grades. (c) For each reduction of 0.01% carbon below the maximum specified, an increase of 0.04% manganese above the maximum specified is permitted to a maximum of 1.00%, except for A27, grade 70-40, where the maximum permitted is 1.40%. (d) Purchased on the basis of hardenability. Manganese and other elements added as required (see Fig. 1).

is specified to ensure a predetermined degree of transformation from austenite to martensite during quenching, in the thickness required. This is important in critical parts requiring toughness and optimum resistance to fatigue.

Particularly where the purchaser heat treats the part after other processing, a casting will be ordered to composition limits closely equivalent to the AISI-SAE wrought steel compositions, with somewhat higher silicon permitted. As in other steel castings, it is best not to specify a range of silicon, but to permit the foundry to utilize the silicon and manganese combination needed to achieve required soundness in the shape being cast. The silicon content is frequently higher in cast steels than for the same nominal composition in wrought steel. Silicon above 0.80% is considered an alloy addition because it contributes significantly to resistance to tempering.

MECHANICAL PROPERTIES

If ferritic steels are compared at a given level of hardness and hardenability, the tensile and yield strengths of cast, rolled, forged and welded metal are virtually identical, regardless of alloy content. Consequently, where tensile and yield properties are controlling criteria, the designer can interchange rolled, forged, welded and cast steel.

Ductility. The ductility of cast steels is nearly the same as that of forged, rolled or welded steels of the same hardness. The longitudinal properties of rolled or forged steel are somewhat higher than the properties of cast steel or weld metal. However, the transverse properties are lower by an amount that depends on the amount of working. When service conditions involve multidirectional loading, the nondirectional characteristic of cast steels may be advantageous.

Toughness. The notched-bar impact test is often used as a measure of the toughness of materials and is particularly useful in determining the transition temperature from ductile to brittle fracture. Nil Ductility Transition Temperature, NDTT (determined as per Method ASTM E208), lateral expansion values, and the energy absorbed values at specific temperatures are some of the different criteria for evaluating impact properties. The impact properties of wrought steels are usually listed for the longitudinal direction; the values shown are higher than those for cast steels of equivalent composition and thermal treatment. The transverse impact properties of wrought steels are usually 50 to 75% of those in the longitudinal direction above the transition temperature and, in some conditions of composition and degree of working, even lower. Because cast steels are nondirectional, their impact properties usually fall somewhere between the longitudinal and transverse properties of wrought steel of similar composition.

Impact properties are controlled by microstructure and, in general, are not significantly affected by microshrinkage or hydrogen. The effect of microstructure, as controlled by chemical composition and heat treatment, is discussed in the article on notch toughness in this section. Curves of impact energy versus temperature for casting steels designed specifically for pressure-containing parts for low-temperature service are presented in Fig. 2. These curves illustrate the significant changes in impact properties that can be effected by changes in steel grade and/or heat treatment.

Section size also affects the impact properties that are obtained. Figure 3 illustrates this effect for one of the grades of steel castings described in Fig. 2 (grade LCB). When the section size is increased from 25 to 127 mm (1 to 5 in.), the temperature at which the impact energy is reduced to an average of 18 J (13 ft·lb) is increased by 28 °C (50 °F).

LOW-CARBON CAST STEELS

Low-carbon cast steels are those with carbon contents less than 0.20%. Most of the tonnage produced in the low-carbon classification contains between 0.16 and 0.19 C, with 0.50 to 0.80 Mn, 0.05 max P and 0.06 max S and 0.35 to 0.70 Si. In order to obtain high magnetic properties in electrical equipment, the manganese content is usually held between 0.10 and 0.20%.

Low-carbon steel castings are made in two important classes. One may be termed "railroad castings" and the other "miscellaneous jobbing castings." The railroad castings consist mainly of comparatively symmetrical and well-designed castings where possible adverse stress conditions have been carefully studied and avoided. Miscellaneous jobbing castings present a wide variation in design and frequently involve joining of light and heavy sections. Varying sections make it more difficult to avoid high residual stress in the as-cast shape. Because residual stresses of large magnitude cannot be tolerated in many service applications, stress relieving becomes necessary. Therefore annealing of those castings is decidedly beneficial even though it may cause little improvement of mechanical properties.

The composition and properties of low-carbon cast steel in the as-cast condition, averaged from a study of over 2000 consecutive heats, are

Carbon	0.189%
Manganese	0.740%
Silicon	0.370%
Phosphorus	0.013%
Sulfur	0.026%
Tensile strength	444.473 MPa (64.465 ksi)
Yield strength	239.180 MPa (34.690 ksi)
Elongation in 50 mm or 2 in.	32.9%
Reduction in area	53.0%

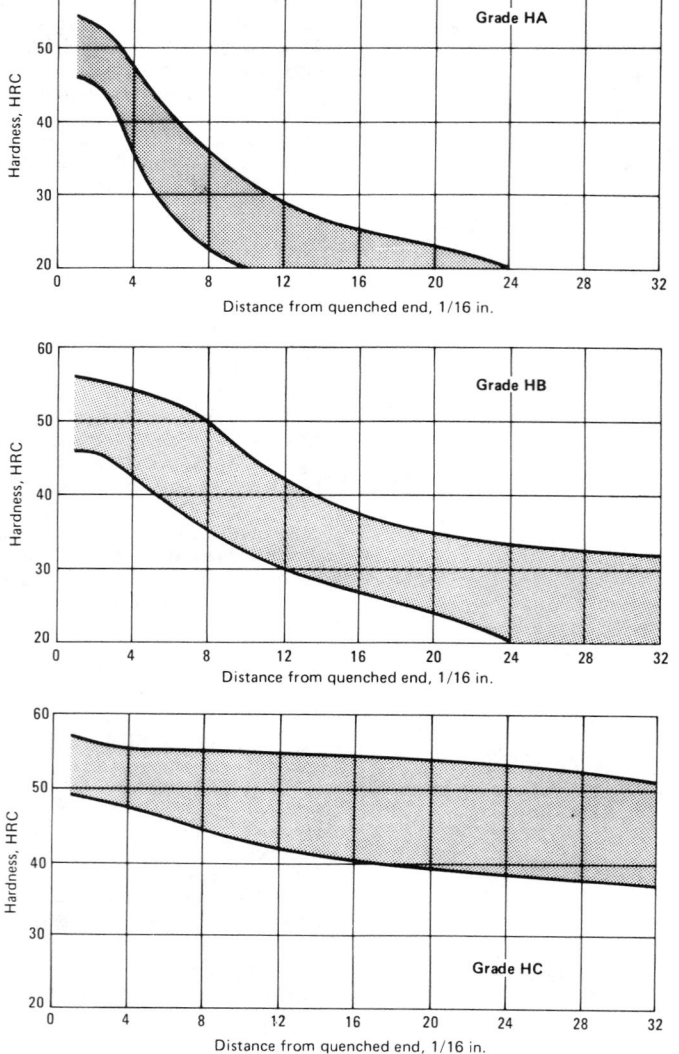

Nominal carbon content of these steels is 0.30% carbon (see Table 1). Manganese and other alloying elements are added as required to produce castings that meet these limits.

Fig. 1. End-quench hardenability limits for three SAE grades of cast steel

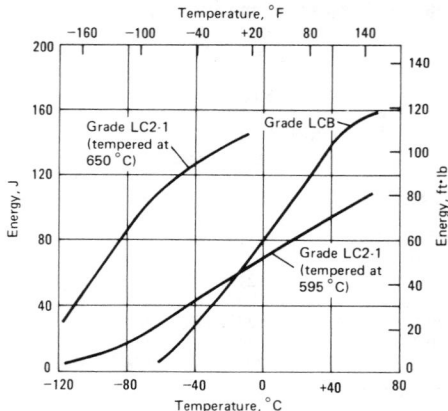

Steel grades conformed to ASTM A352. Heat treatments were as follows: grade LCB, water quenched from 890 °C (1650 °F), tempered at 650 °C (1200 °F) and water quenched, aged 40 h at 425 °C (800 °F), and stress relieved 40 h at 595 °C (1100 °F); grade LC2-1, normalized at 955 °C (1750 °F) and air cooled, reheated to 890 °C (1650 °F) and water quenched, either tempered at 595 °C (1100 °F) and aged 40 h at 425 °C (800 °F) or tempered at 650 °C (1200 °F) and aged 64 h at 425 °C (800 °F). All specimens were taken at locations greater than one-fourth the thickness in from the surface of test blocks 51 by 210 by 229 mm (2 by 8¼ by 9 in.) having an ASTM grain size of 6 to 8. The curves represent average values for several tests at each test temperature.

Fig. 2. Effect of temperature on Charpy V-notch impact energy of cast steels for low-temperature service

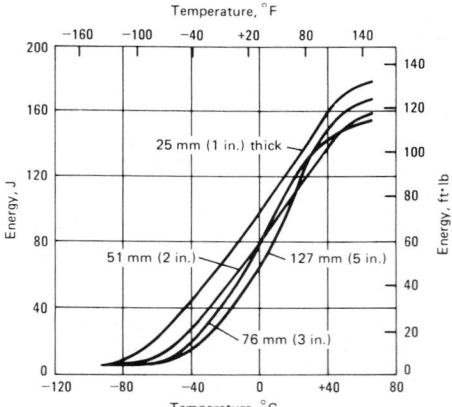

Steel grade conformed to ASTM A352. Heat treatment was the same as for Fig. 2. All specimens were taken at locations greater than one-fourth the thickness in from the surface of test blocks of four sizes: 25 by 25 by 279 mm (1 by 1 by 11 in.), 51 by 210 by 229 mm (2 by 8¼ by 9 in.), 76 by 229 by 283 mm (3 by 9 by 11⅛ in.) and 127 by 381 by 381 mm (5 by 15 by 15 in.). The ASTM grain size of the blocks was 6 to 8. The curves represent average values for several tests at each test temperature.

Fig. 3. Effect of section thickness on Charpy V-notch impact curves of grade LCB steel castings

Fatigue Strength. The ratio of fatigue limit to tensile strength for the low-carbon cast steels varies somewhat, but is approximately 45%. This approximate ratio is also maintained at low and high temperatures and is not much affected by the various types of heat treatment that the steel may receive.

In designing cast steel structures based on the fatigue ratio, it is advisable to use 40% of the tensile strength for a smooth bar when actual fatigue test values cannot be obtained. A factor of safety is added to this approximate figure.

MEDIUM-CARBON CAST STEELS

The medium-carbon grades of cast steel contain 0.20 to 0.50% carbon and represent the bulk of total steel casting production. In addition to carbon, they contain 0.50 to 1.50 Mn, 0.05 max P, 0.06 max S, and 0.35 to 0.80 Si. Room-temperature tensile strengths of as-cast steels containing from 0.20 to 0.50% carbon are plotted in

Fig. 4. Steels in this carbon range are always heat treated, which relieves casting strains, refines the as-cast structure, and improves the ductility of the steel.

The tensile strength (as well as the hardness) of the as-cast steel falls off slightly following full annealing (see Fig. 4). A very large proportion of steel castings of this grade are given a normalizing treatment, followed by a tempering treatment (Fig. 4).

Effect of Mass. The effect of increasing mass on the tensile strength of a medium-carbon cast steel in the as-cast and annealed conditions is illustrated in Fig. 5. In cross sections up to 200 mm (8 in.) square, tensile strength decreases as size increases. Most of the total effect occurs as the cross section is increased to 100 mm (4 in.) square. The difference between the as-cast and annealed tensile strength is minor.

If the design of a casting is suitable for liquid quenching, further improvements are possible in the mechanical properties. In fact, to develop

mechanical properties to the fullest degree, steel castings should be heat treated by liquid quenching and tempering. Commercial procedure calls for tempering to obtain the desired strength level. Tempering temperatures of 650 to 705 °C (1200 to 1300 °F) are usually used to obtain higher ductility and impact properties.

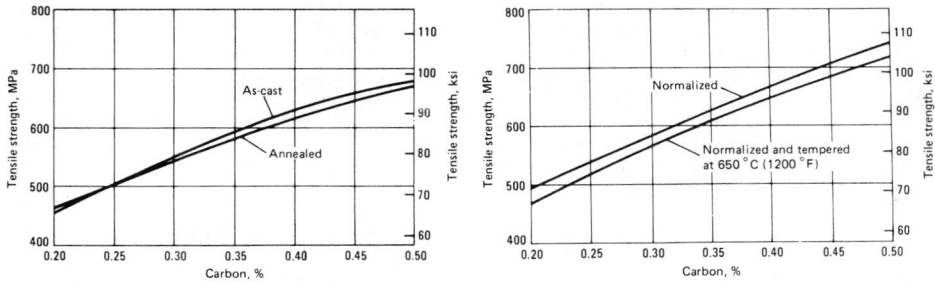

Fig. 4. Effect of carbon content on tensile strength of medium-carbon steel castings

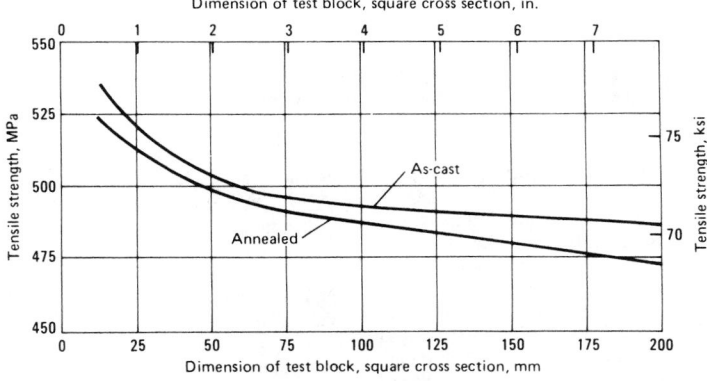

Fig. 5. Decrease in tensile strength with increase in mass of medium-carbon steel castings

HIGH-CARBON CAST STEELS

Cast steels containing more than 0.50% carbon are classified as high-carbon steels. This grade also contains 0.50 to 1.50 Mn, 0.05 max P and S, and 0.35 to 0.70 Si. The mechanical properties of high-carbon steels at room temperature are shown in Fig. 6. These test results were obtained on cast steel in the fully annealed condition. Occasionally, a normalizing and tempering treatment is given, and for certain applications, an oil quenching and tempering treatment may be used.

The microstructure of high-carbon steel is controlled by heat treatment. Carbon also has a marked influence—for example, giving 100% pearlitic structure at eutectoid composition (approximately 0.83% carbon). Higher proportions of carbon than eutectoid composition will increase the proeutectoid cementite, which is detrimental to the casting if it forms a network at the grain boundaries due to improper heat treatment (for example, slow cooling from above the A_{cm} temperature). Faster cooling will prevent the formation of this network and, hence, improve the properties.

LOW-ALLOY CAST STEELS

Low-alloy cast steels contain a total alloy content of less than 8%. These steels have been developed and extensively used for meeting special requirements that cannot be met by ordinary plain carbon steels with low hardenability. The addition of alloying elements to plain carbon steel castings may be made for any of several reasons, such as to provide higher hardenability, increased wear resistance, higher impact resistance at increased strength, good machinability even at higher hardness, higher strength at elevated and low temperatures, or better resistance to corrosion and oxidation. These materials are produced to meet tensile-strength requirements of 485 to 1380 MPa (70 to 200 ksi), together with some of the above special requirements.

PHYSICAL PROPERTIES

The physical properties of cast steel generally are similar to those of wrought steel.

Elastic constants of carbon and low-alloy cast steels as determined at room temperature are only slightly affected by changes in composition and structure. The modulus of elasticity, E, is about 207 GPa (30 million psi), Poisson's ratio is 0.3, and the modulus of rigidity is 77.2 GPa (11.2 million psi). Increasing temperature has a marked effect on the modulus of elasticity and the modulus of rigidity. Moduli of elasticity at some elevated temperatures are as follows: 193 GPa (28 million psi) at 200 °C (400 °F); 179 GPa (26 million psi) at 360 °C (680 °F); 165 GPa (24 million psi) at 445 °C (830 °F); and 152 GPa (22 million psi) at 490 °C (910 °F). Above 480 °C (900 °F), the value of the modulus of elasticity is rapidly reduced.

Density of cast steel is sensitive to changes in composition, structure and temperature. The density of medium-carbon cast steel is in the range 7.825 to 7.830 Mg/m³. Steel castings have a weight of 490 lb/ft³ or 0.283 lb/in.³. The density of cast steel is also affected somewhat by mass or size of section.

Electrical properties of carbon and low-alloy steel castings do not account for any significant usage. The only electrical property that may be regarded to be of any importance is resistivity, which, for various annealed carbon steel castings with 0.07 to 0.20% carbon, is 0.13 to 0.14 µΩ·m. Resistivity increases with carbon content and is about 0.20 µΩ·m at 1.0% carbon.

Magnetic Properties. Steel castings form the housings for electrical machinery and magnetic equipment and carry only stray fluxes around the machines; for this reason, the magnetic properties of steel castings are less important than they were formerly when core material was manufactured from commercial cast iron and steel. Low-carbon cast dynamo steel has now supplanted other cast metals for housings and frames for magnetic circuits.

The carbon content of the steel is very important in determining its magnetic properties. The maximum permeability and the saturation magnetization decrease and the coercive force increases as the carbon content increases. Manganese, phosphorus, sulfur and silicon also increase the magnetic hysteresis loss in cast steels. This loss is equal to about 10 J/m³ per cycle for $B = 1$ T for each 0.10 Mn, 0.01 S and 0.01 P. With other factors unchanged, the magnetic hysteresis loss is unaffected by more than 0.02% phosphorus. Magnetic properties change considerably, depending on the mechanical treatment and heat treatment of the steel.

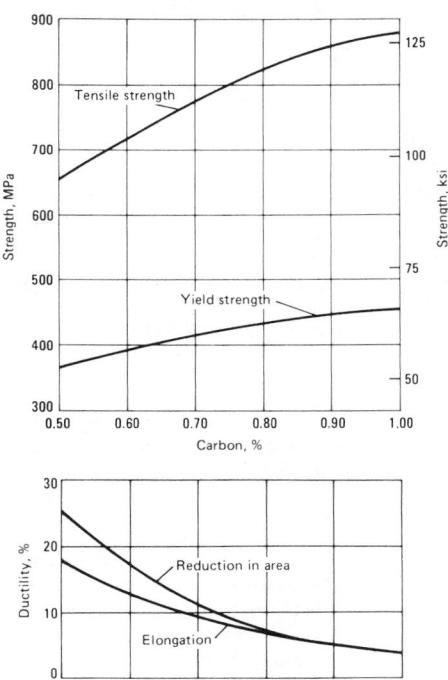

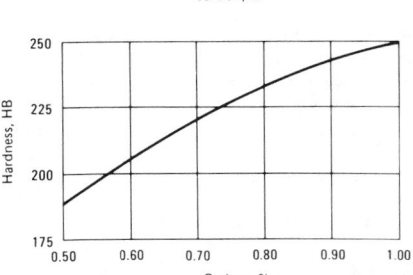

Fig. 6. Effect of carbon content on mechanical properties of fully annealed high-carbon cast steel

Cast dynamo steels contain about 0.10% carbon, with other alloying elements held to a minimum; the castings are furnished in the annealed condition. Specifications require 0.05 to 0.15 C, 0.20 Mn, and 0.35 to 0.60 or 1.50 to 2.00 Si.

The magnetic properties of annealed cast dynamo steel that may normally be expected are as follows:

Maximum permeability, mH/m 18.6
Hysteresis loss (induction for H
= 11.9 kA/m), T 1.91
Saturation magnetization, T 2.14
Residual induction, T 1.10
Coercive force, A/m 29

As the carbon content is increased, maximum permeability and saturation magnetization decrease and coercive force increases. Also, an increase in manganese and sulfur content increases the magnetic hysteresis loss.

Silicon and aluminum eliminate the allotropic transformations in iron and permit annealing at high temperature without recrystallization during cooling; thus, large grains can be obtained.

NONDESTRUCTIVE INSPECTION

Highly stressed steel castings for aircraft, and for high-pressure or high-temperature service, must pass rigid nondestructive inspection. ASTM specifications E186, E280 and E446 cover radiographic standards for steel castings, E94 covers ASTM Recommended Practice for Radiographic Testing, and E142 covers the quality control of radiographic testing. Radiographic acceptance standards must be agreed upon by the user and producer before production begins. Critical areas to be radiographed may be identified on the casting drawing.

Magnetic-particle inspection is used on highly stressed steel castings to detect surface discontinuities or imperfections at or just below the surface.

High-Strength Steels

High-Strength Structural and High-Strength Low-Alloy Steels

THE STEELS discussed in this article are characterized by higher yield strengths than those of plain carbon structural steels. There are four categories: (*a*) proprietary as-rolled pearlitic structural steels with minimum yield strengths of 275 to 345 MPa (40 to 50 ksi), known commercially as either high-strength steels or high-strength low-alloy steels — the latter fulfilling requirements of specifications such as ASTM A242 or A568 or of SAE J410c; (*b*) proprietary microalloyed high-strength low-alloy steels, with properties that result from a combination of alloy additions and controlled hot rolling; (*c*) carbon steel grades — either normalized or quenched and tempered — having minimum yield strengths of 290 to 690 MPa (42 to 100 ksi) and, if specified, a minimum toughness or impact strength; and (*d*) quenched-and-tempered low-alloy steel grades having minimum yield strengths of 552 to 758 MPa (80 to 110 ksi) that meet the requirements of specifications such as ASTM A514.

Availability. High-strength and high-strength low-alloy grades are generally available in all standard wrought forms: sheet, strip, plate, structural shapes, bars, bar-size shapes and special sections. These steels are also furnished as cold rolled sheet or strip in gages up to about 1.6 mm ($^1/_{16}$ in.) for greater control of thickness in structures such as trailer bodies or for improved surface finish in instances where parts are to be plated.

The heat treated grades are available as plate, bar and, occasionally, sheet and structural shapes. When intended for applications involving hot forming, they are generally purchased in the non-heat-treated condition. Also, semifinished mill products are available for forging. After forming or forging, such products may be heat treated to develop high strength.

High-strength steels are produced to specified mechanical properties, which may differ slightly for different thickness ranges. Maximum limits of chemical composition are generally published because the carbon and alloy contents are varied as necessary to maintain mechanical properties in products of various thicknesses.

COMPOSITIONS AND SPECIFICATIONS

Chemical compositions of typical high-strength structural and high-strength low-alloy (HSLA) steels are given in Table 1. Mechanical properties of as-rolled pearlitic structural steels and microalloyed HSLA steels are generally intermediate between those of plain carbon structural steels and those of heat treated high-strength carbon and low-alloy constructional steels. Other steels of similar carbon, manganese and copper contents (steels 1 and 2 in Table 1, for example) have mechanical properties similar to those of HSLA steels. Such manganese-copper steels, commercially designated simply high-strength steels, are often used interchangeably with HSLA steels.

Tables 2, 3 and 4 list alloying elements, forms, intended uses, grades and mechanical properties as designated in the ASTM specifications.

High-strength low-alloy steels are not normally considered to be alloy steels. This is a commercial fact of metallurgical importance because these steels are not subject to complete chemical check analysis and certain other tolerances normally imposed on alloy steels. They are marketed as a separate commodity even though their alloy contents would qualify many of them as alloy steels in the metallurgical sense. Separate product recognition has resulted because of the extensive application of high-strength low-alloy steels for reasons of weight savings and/or atmospheric corrosion resistance. HSLA steels generally are priced from the base price for carbon steels, not from the base price for alloy steels.

As-rolled pearlitic structural steels are a specific group of steels in which enhanced mechanical properties (and, in some cases, resistance to atmospheric corrosion) are obtained by addition of moderate amounts of one or more alloying elements other than carbon.

As-rolled pearlitic structural steels are characterized by as-rolled yield strengths in the range from 290 to 345 MPa (42 to 50 ksi). They are not intended for quenching and tempering and should not be subjected to such treatment. For certain applications, they may be annealed, normalized or stress relieved, which may alter their mechanical properties.

When these steels are used in welded structures, care must be exercised in selecting the grade and specifying details of the welding process.

Certain grades may be welded without preheating or postheating.

Microalloyed high-strength low-alloy steels contain alloying elements such as niobium and vanadium which are added to increase strength (and thereby increase load-carrying ability) of hot rolled steel without increasing carbon and/or manganese contents. Increasing the level of carbon and/or manganese is detrimental to both weldability and notch toughness. At carbon and manganese contents beyond about 0.30 and 1.6%, respectively, cracking during welding may occur and microstructures characterized by low notch toughness are readily formed in as-rolled plate.

High-strength structural carbon steels are basically carbon-manganese-silicon steels, with some grades containing small alloying additions of other elements. Such steels are either normalized or quenched and tempered to produce minimum yield strengths of 290 to 690 MPa (42 to 100 ksi) and, when specified, minimum toughness or impact values.

The normalized carbon grades meet the requirements of ASTM A633 and are available in minimum yield strengths up to 415 MPa (60 ksi), with minimum Charpy V-notch impact strengths as high as 27 J (20 ft·lb) at temperatures as low as −68 °C (−90 °F), depending on grade and producer.

Heat treated structural low-alloy grades have better hardenability than carbon grades and thus can provide both high strength and good toughness in thicker sections. Their alloy content provides improved heat and corrosion resistance. However, these grades are somewhat more difficult to weld and generally are more expensive than carbon or HSLA grades. Quenched-and-tempered low-alloy grades may be ordered to fulfill the requirements of ASTM A514 or A517. The properties required of these steels are produced by heating to at least 900 °C (1650 °F), quenching in water or oil, and tempering at not less than 600 °C (1100 °F) to provide microstructures of tempered martensite or tempered martensite plus bainite, depending on section thickness.

Quenched-and-tempered structural high-strength low-alloy grades provide minimum yield strengths of 550 to 760 MPa (80 to 110 ksi) and minimum Charpy V-notch impact strengths of 20 to 27 J (15 to 20 ft·lb) at −45 °C (−50 °F). Mill products are available in thicknesses up to 203 mm (8 in.).

Table 1. Compositions of typical high-strength structural and HSLA steels

Steel	C	Mn	P	S	Si	Cr	Ni	Mo	Cu	V	Others
As-rolled high-strength structural steels											
1	0.20	1.25	0.025	...	0.25	...	...	...	0.25	...	...
2	0.22	1.35	0.025	...	0.25	...	...	...	0.25	...	...
3	0.17	1.10	0.12	...	0.08	...	...	...	0.30	0.05	...
4	0.15	0.75	0.060	...	0.25	0.30	0.75	...	0.40	...	...
5	0.10	0.75	0.090	...	0.10	...	0.60	0.13	1.15	...	0.20 Al
6	0.15	0.75	0.080	...	0.25	0.25	0.75	0.30	0.35	...	...
7	0.08	0.65	0.025	...	0.08	...	...	...	0.15	0.04	...
8	0.20	1.10	0.025	...	0.25	...	...	...	0.25	0.04	0.01 Ti
9	0.14	1.05	0.025	...	0.20	...	...	...	0.25	0.04	...
10	0.15	0.75	0.015	...	0.65	...	...	...	0.20	...	0.04 Ti; 0.04 Zr
11	0.15	0.70	0.040	...	0.75	0.55	...	0.15	...	...	0.04 Zr
12	0.15	0.75	0.055	...	0.10	...	0.75	...	0.65	...	...
13	0.08	0.50	0.025	...	0.25	0.30	0.50	...	0.40	...	...
14	0.11	0.65	0.015	...	0.50	0.21	0.15	0.06	0.28	0.08	0.01 Ti
15	0.10	0.75	0.100	...	0.60	0.75	0.85	...	0.85	...	0.06 Zr
16	0.10	0.35	0.120	...	0.50	0.80	0.50	...	0.50	...	...
17	0.12	0.75	0.080	...	0.25	...	0.55	0.10	0.55	...	...
18	0.12	0.75	0.025	...	0.25	0.25	0.75	0.13	0.75	...	...
Heat treated high-strength low-alloy steels											
19	0.28	1.50	0.025	...	0.25	...	...	0.15	...	...	...
20	0.28	0.75	0.025	...	0.75	0.50	...	0.16	...	...	0.12 Zr
21	0.12	0.50	0.025	...	0.25	...	1.85	...	1.00	...	...
22	0.18	1.00	0.025	...	0.25	...	1.35	0.25	0.65	...	...
23	0.12	0.75	0.025	...	0.10	...	1.25	0.25	1.15	...	...
24	0.18	0.75	0.025	...	0.10	...	1.50	0.25	1.25	...	...
25	0.15	0.80	0.025	...	0.25	0.60	0.85	0.50	0.35	0.05	0.004 B
26	0.12-0.21	0.45-0.70	0.035	...	0.20-0.35	...	...	0.50-0.65	...	...	0.001-0.005 B
27	0.12-0.21	0.45-0.70	0.035	...	0.20-0.35	...	1.20-1.50	0.45-0.60	...	...	0.001-0.005 B
28	0.12-0.21	0.45-0.70	0.035	...	0.20-0.35	0.85-1.20	1.20-1.50	0.45-0.60	...	...	0.001-0.005 B
29	0.10-0.20	0.60-1.00	0.035	...	0.15-0.35	0.40-0.65	0.70-1.00	0.40-0.60	0.15-0.50	0.03-0.08	0.0005-0.006 B
30	0.12-0.21	0.70-1.00	0.035	...	0.20-0.35	0.40-0.65	...	0.15-0.25	...	0.03-0.08	0.01-0.03 Ti; 0.0005-0.005 B
31	0.12-0.21	0.95-1.30	0.035	...	0.20-0.35	0.40-0.65	0.30-0.70	0.20-0.30	...	0.03-0.08	0.0005 B
32	0.21	0.60-1.10	0.04	...	0.40-0.90	0.40-0.90	...	0.28	...	...	0.05 Zr; 0.0025 max B
33	0.18	0.10-0.40	0.025	...	0.15-0.35	1.00-1.80	2.00-3.25	0.20-0.60	...	...	...
34	0.20	0.10-0.40	0.025	...	0.15-0.35	1.00-1.80	2.20-3.50	0.20-0.60	...	...	...
35	0.12	0.60-0.90	0.015	...	0.15-0.35	0.40-0.70	4.75-5.25	0.30-0.65	...	0.05-0.10	...
36	0.16	0.60-0.90	0.015	...	0.15-0.35	0.40-0.70	4.75-5.25	0.30-0.65	...	0.05-0.10	...
Microalloyed high-strength low-alloy steels											
37	0.09	0.35	0.010	0.012	0.01	...	...	...	...	<0.01	0.05 Al; 0.005 N; 0.015 Nb
38	0.09	1.25	0.010	0.012	0.30	...	...	...	...	0.09	0.05 Al; 0.005 N; 0.09 Nb
39	0.14	1.25	0.03	0.03	0.30	...	...	...	...	0.02	0.01 Nb
40	0.16	1.40	0.03	0.03	0.30	...	...	...	...	0.02	0.01 Nb
41	0.18	1.50	0.03	0.03	0.30	...	...	...	...	0.02	0.01 Nb
42	0.18	1.60	0.03	0.03	0.60	...	...	...	...	0.05	...
43	0.06	0.45	0.01	0.02	0.10	...	...	...	...	...	0.02 Nb
44	0.06	0.75	0.01	0.02	0.10	...	...	...	...	...	0.04 Nb
45	0.06	0.95	0.01	0.02	0.10	...	...	...	...	...	0.10 Nb
46	0.10	0.40	...	...	...	...	...	...	...	...	0.05 Ti; 0.01 Al
47	0.10	0.40	...	...	...	...	...	...	...	...	0.05 Ti; 0.01 Al
48	0.10	0.40	...	...	...	...	...	...	...	...	0.05 Ti; 0.01 Al

(a) Single values are typical values unless otherwise noted.

Table 2. Summary of characteristics and intended uses of HSLA steels described in ASTM specifications(a)

ASTM specification	Title	Alloying elements(b)	Available mill forms	Special characteristics	Intended uses
A242	High-strength low-alloy structural steel	Cr, Cu, N, Ni, Si, Ti, V, Zr	Plate, bar and shapes up to 102 mm (4 in.) thick	Atmospheric corrosion resistance four times that of carbon steel	Structural members, in welded, bolted or riveted constructions
A440	High-strength structural steel	Cu, Si	Plate, bar and shapes up to 102 mm (4 in.) thick	Atmospheric corrosion resistance twice that of carbon steel	Structural members, primarily in bolted or riveted constructions
A441	High-strength low-alloy structural manganese-vanadium steel	V, Cu, Si	Plate, bar and shapes up to 203 mm (8 in.) thick	Atmospheric corrosion resistance twice that of carbon steel	Welded, bolted or riveted structures, but primarily welded bridges and buildings
A572	High-strength low-alloy niobium-vanadium steels of structural quality	Nb, V, N	Plate, bar, shapes and sheet piling up to 152 mm (6 in.) thick	Yield strengths of 290 to 450 MPa (42 to 65 ksi), in six grades	Welded, bolted or riveted structures, but mainly bolted or riveted bridges and buildings
A588	High-strength low-alloy structural steel with 50 ksi minimum yield point to 4 in. thick	Nb, V, Cr, Ni, Mo, Cu, Si, Ti, Zr	Plate, bar and shapes up to 203 mm (8 in.) thick	Atmospheric corrosion resistance four times that of carbon steel; nine grades of similar strength	Welded, bolted or riveted structures, but primarily welded bridges and buildings where weight savings or added durability is important

(continued)

Table 2. (continued)

ASTM specification	Title	Alloying elements(b)	Available mill forms	Special characteristics	Intended uses
A606	Steel sheet and strip, hot rolled and cold rolled, high-strength low-alloy with improved corrosion resistance	Not specified	Hot rolled and cold rolled sheet and strip	Atmospheric corrosion resistance twice that of carbon steel (type 2) or four times that of carbon steel (type 4)	Structural and miscellaneous purposes where weight savings or added durability is important
A607	Steel sheet and strip, hot rolled and cold rolled, high-strength low-alloy niobium and/or vanadium	Nb, V, N, Cu	Hot rolled and cold rolled sheet and strip	Atmospheric corrosion resistance twice that of carbon steel, but only when copper content is specified; yield strengths of 310 to 485 MPa (45 to 70 ksi) in six grades	Structural and miscellaneous purposes where greater strength or weight savings is important
A618	Hot-formed welded and seamless high-strength low-alloy structural tubing	Nb, V, Si, Cu	Square, rectangular, round and special-shape structural tubing, welded or seamless	Three grades of similar yield strength; may be purchased with atmospheric corrosion resistance twice that of carbon steel	General structural purposes, including welded, bolted or riveted bridges and buildings
A633	Normalized high-strength low-alloy structural steel	Nb, V, Cr, Ni, Mo, Cu, N, Si	Plate, bar and shapes up to 152 mm (6 in.) thick	Enhanced notch toughness; yield strengths of 290 to 415 MPa (42 to 60 ksi) in five grades	Welded, bolted or riveted structures for service at temperatures down to −45 °C (−50 °F)
A656	High-strength low-alloy hot rolled structural vanadium-aluminum-nitrogen and titanium-aluminum steels	V, Al, N, Ti, Si	Plate, normally up to 15.9 mm (⁵/₈ in.) thick	Yield strength of 552 MPa (80 ksi)	Truck frames, brackets, crane booms, rail cars and other applications where weight savings is important
A690	High-strength low-alloy steel H-piles and sheet piling for use in marine environments	Ni, Cu, Si	Structural quality H-piles and sheet piling	Corrosion resistance two to three times greater than carbon steel in splash zone of marine structures	Dock walls, sea walls, bulkheads, excavations and similar structures exposed to seawater
A715	Steel sheet and strip, hot rolled, high-strength low-alloy with improved formability	Nb, V, Cr, Mo, N, Si, Ti, Zr, B	Hot rolled sheet and strip	Improved formability(c) compared to A606 and A607; yield strengths of 345 to 550 MPa (50 to 80 ksi) in four grades	Structural and miscellaneous applications where high strength, weight savings, improved formability and good weldability are important

(a) For grades and mechanical properties, see Table 3. (b) In addition to C, Mn, P and S, a given grade may contain one or more of the listed elements, but not necessarily all of them; for specified composition limits, see Table 4. (c) Obtained by producing killed steel, made to fine grain practice, and with microalloying elements such as Nb, V, Ti and Zr in the composition.

Table 3. Mechanical properties of HSLA steel grades described in ASTM specifications(a)

ASTM specification	Type, grade or condition	UNS designation	Min tensile strength(b) MPa	ksi	Min yield strength(b) MPa	ksi	Min elongation(b), % In 200 mm or 8 in.	In 50 mm or 2 in.	Impact energy(c) J	ft · lb	Bend radius(b) Longitudinal	Transverse
A242	Type 1	K11510	435 to 480	63 to 70	290 to 345	42 to 50	18	21	...	...	(d)	...
	Type 2	K12010	435 to 480	63 to 70	290 to 345	42 to 50	18	21	...	...	(d)	...
A440	...	K12810	435 to 485	63 to 70	290 to 345	42 to 50	18	21	...	...	(d)	...
A441	...	K12211	415 to 485	60 to 70	275 to 345	40 to 50	18	21	...	...	(d)	...
A572	Grade 42	...	415	60	290	42	20	24	...	...	(d)	...
	Grade 45	...	415	60	310	45	19	22	...	...	(d)	...
	Grade 50	...	450	65	345	50	18	21	...	...	(d)	...
	Grade 55	...	485	70	380	55	17	20	...	...	(d)	...
	Grade 60	...	520	75	415	60	16	18	...	...	(d)	...
	Grade 65	...	550	80	450	65	15	17	...	...	(d)	...
A588	Grades A to J	(e)	435 to 485	63 to 70	290 to 345	42 to 50	18	21	...	...	(d)	...
A606	Hot rolled	...	480	70	345	50	...	22	...	...	t	2t to 3t
	Hot rolled and annealed or normalized	...	450	65	310	45	...	22	...	...	t	2t to 3t
	Cold rolled	...	450	65	310	45	...	22	...	...	t	2t to 3t
A607	Grade 45	...	410	60	310	45	...	22 to 25	...	...	t	1.5t
	Grade 50	...	450	65	345	50	...	20 to 22	...	...	t	1.5t
	Grade 55	...	480	70	380	55	...	18 to 20	...	...	1.5t	2t
	Grade 60	...	520	75	415	60	...	16 to 18	...	...	2t	3t
	Grade 65	...	550	80	450	65	...	15 to 16	...	...	2.5t	3.5t
	Grade 70	...	590	85	485	70	...	14	...	...	3t	4t
A618	Grade I	K02601	483	70	345	50	19	22	...	...	t to 2t	...
	Grade II	K12609	483	70	345	50	18	22	...	...	t to 2t	...
	Grade III	K12700	448	65	345	50	18	20	...	...	t to 2t	...
A633	Grades A and B	(e)	430 to 570	63 to 83	290	42	18	23	(f)	(f)	(d)	...
	Grades C and D	(e)	450 to 620	65 to 90	315 to 345	46 to 50	18	23	(f)	(f)	(d)	...
	Grade E	K12202	520 to 690	75 to 100	380 to 415	55 to 60	18	23	(f)	(f)	(d)	...
A656	Grades 1 and 2	(e)	655 to 793	95 to 115	552	80	12	...	...	...	(d)	...
A690	...	K12249	485	70	345	50	18	...	...	...	2t	...
A715	Grade 50	...	415	60	345	50	...	22 to 24	...	...	0	t
	Grade 60	...	485	70	415	60	...	20 to 22	...	...	0	t
	Grade 70	...	550	80	485	70	...	18 to 20	...	...	t	1.5t
	Grade 80	...	620	90	550	80	...	16 to 18	...	...	t	1.5t

(a) For characteristics and intended uses, see Table 2; for specified composition limits, see Table 4. (b) May vary with product size and mill form; for specific limits, see the article in this Handbook on the appropriate mill form. (c) Charpy V-notch. (d) Optional supplementary requirement given in ASTM A6, section S14. (e) See Table 4. (f) Optional supplementary requirement given in ASTM A6, section S1.

Table 4. Composition limits for HSLA steel grades described in ASTM specifications(a)

| ASTM specification | Type or grade | UNS designation | Heat composition limits(b), % ||||||||||
			C	Mn	P	S	Si	Cr	Ni	Cu	V	Other
A242	Type 1	K11510	... 0.15	1.00	0.45	0.05	...	...	...	0.20 min	...	...
	Type 2	K12010	... 0.20	1.35	0.04	0.05	...	...	...	0.20 min(c)	...	...
A440	...	K12810	... 0.28	1.10-1.60	0.04	0.05	0.30	...	...	0.20 min	...	...
A441	...	K12211	... 0.22	0.85-1.25	0.04	0.05	0.30	...	...	0.20 min	0.02 min	...
A572	Grade 42	... .	0.21	1.35(d)	0.04	0.05	0.30(d)	...	...	0.20 min(e)	...	(f)
	Grade 45		0.22	1.35(d)	0.04	0.05	0.30(d)	...	...	0.20 min(e)	...	(f)
	Grade 50		0.23	1.35(d)	0.04	0.05	0.30(d)	...	...	0.20 min(e)	...	(f)
	Grade 55		0.25	1.35(d)	0.04	0.05	0.30	...	...	0.20 min(e)	...	(f)
	Grade 60		0.26	1.35(d)	0.04	0.05	0.30	...	...	0.20 min(e)	...	(f)
	Grade 65		0.23(d)	1.65(d)	0.04	0.05	0.30	...	...	0.20 min(e)	...	(f)
A588	Grade A	K11430	...0.10-0.19	0.90-1.25	0.04	0.05	0.15-0.30	0.40-0.65	...	0.25-0.40	0.02-0.10	...
	Grade B	K12043	... 0.20	0.75-1.25	0.04	0.05	0.15-0.30	0.40-0.70	0.25-0.50	0.20-0.40	0.01-0.10	...
	Grade C	K11538	... 0.15	0.80-1.35	0.04	0.05	0.15-0.30	0.30-0.50	0.25-0.50	0.20-0.50	0.01-0.10	...
	Grade D	K11552	...0.10-0.20	0.75-1.25	0.04	0.05	0.50-0.90	0.50-0.90	...	0.30	...	0.04 Nb; 0.05-0.15 Zr
	Grade E	K11567	... 0.15	1.20	0.04	0.05	0.15-0.30	...	0.75-1.25	...	0.05	0.10-0.25 Mo
	Grade F	K11541	...0.10-0.20	0.50-1.00	0.04	0.05	0.30	0.30	0.40-1.10	0.30-1.00	0.01-0.10	0.10-0.20 Mo
	Grade G	K12040	... 0.20	1.20	0.04	0.05	0.25-0.70	0.50-1.00	0.80	0.30-0.50	...	0.10 Mo; 0.07 Ti
	Grade H	K12032	... 0.20	1.25	0.035	0.040	0.25-0.75	0.10-0.25	0.30-0.60	0.20-0.35	0.20-0.10	0.15 Mo; 0.005-0.030 Ti
	Grade J	K12044	... 0.20	0.60-1.00	0.04	0.05	0.30-0.50	...	0.50-0.70	0.30 min	...	0.03-0.05 Ti
A606	...	... .	0.22	1.25	...	0.05	...	...	...	...	...	...
A607	Grade 45	... .	0.22	1.35	0.04	0.05	...	...	...	0.20 min(e)	...	(f)
	Grade 50		0.23	1.35	0.04	0.05	...	...	...	0.20 min(e)	...	(f)
	Grade 55		0.25	1.35	0.04	0.05	...	...	...	0.20 min(e)	...	(f)
	Grade 60		0.26	1.50	0.04	0.05	...	...	...	0.20 min(e)	...	(f)
	Grade 65		0.26	1.50	0.04	0.05	...	...	...	0.20 min(e)	...	(f)
	Grade 70	... .	0.26	1.65	0.04	0.05	...	...	...	0.20 min(e)	...	(f)
A618	Grade I	K02601	... 0.22	1.25	...	0.05	...	...	...	(g)	...	(f)
	Grade II	K12609	... 0.22	0.85-1.25	0.04	0.05	0.30	...	...	0.20 min	0.02 min	...
	Grade III	K12700	... 0.23	1.35	0.04	0.05	0.30	...	...	...	0.02 min	0.005 min Nb (h)
A633	Grade A	K01802	... 0.18	1.00-1.35	0.04	0.05	0.15-0.30	...	...	...	...	0.05 Nb
	Grade B	K01803	... 0.18	1.00-1.35	0.04	0.05	0.15-0.50	...	...	...	0.10	...
	Grade C	K12000	... 0.20	1.15-1.50	0.04	0.05	0.15-0.50	...	...	...	...	0.01-0.05 Nb
	Grade D	K02003	... 0.20	0.70-1.60(d)	0.04	0.05	0.15-0.50	0.25	0.25	0.35	...	0.08 Mo
	Grade E	K12202	... 0.22	1.15-1.50	0.04	0.05	0.15-0.50	...	...	...	0.04-0.11	0.01-0.05 Nb(e); 0.01-0.03 N
A656	Grade 1	K11804	... 0.18	1.60	0.040	0.050	0.60	...	...	...	0.05-0.15	0.020 min Al; 0.005-0.030 N
	Grade 2	K11503	... 0.15	0.90	0.040	0.050	0.10	...	...	...	...	0.05-0.50 Ti; 0.01 min Al
A690	...	K12249	... 0.22	0.60-0.90	0.08-0.15	0.05	0.10	...	0.40-0.75	0.50 min	...	...
A715	Type 1	... .	0.15	1.65	0.025	0.035	0.10	...	...	...	...	0.05 min Ti
	Type 2	... .	0.15	1.65	0.025	0.035	0.60(j)	...	...	...	...	0.005 min N(j)
	Type 3	... .	0.15	1.65	0.025	0.035	0.60(j)	...	...	...	0.08(j)	0.005 min Nb; 0.020 N(j)
	Type 4	... .	0.15	1.65	0.025	0.035	0.90	0.80(j)	...	...	...	0.005-0.06 Nb(k); 0.10 Ti(j); 0.05 min Zr; 0.0025B(j)
	Type 5(m)	... .	0.15	1.65	0.025	0.035	0.30	...	...	...	...	0.20 min Mo; 0.03 min Nb
	Type 6	... .	0.15	1.65	0.025	0.035	0.90	...	...	...	...	0.005-0.10 Nb
	Type 7	... .	0.15	1.65	0.025	0.035	...	...	...	...	0.005 min(n)	0.020 N

(a) For characteristics and intended uses, see Table 2; for mechanical properties, see Table 3. (b) Where a single value is shown, it is a maximum unless otherwise stated. (c) Applicable only if Si and/or Cr content is below 0.5%. (d) Values may vary, or minimum value may exist, depending on product size and mill form. (e) Optional or when specified. (f) May be purchased as type 1 (0.005-0.05 Nb), type 2 (0.01-0.15 V), type 3 (0.05 max Nb plus 0.02-0.15 V) or type 4 (0.015 max N plus V ≥ 4 N). (g) As agreed between producer and purchaser. (h) May be substituted for all or part of the V. (j) Not added to grades 50 and 60. (k) Might not be added to grade 50. (m) Available as grade 80 only. (n) 0.005 min Nb may be added in place of or in addition to the V.

Ultrahigh-Strength Steels

STRUCTURAL STEELS with very high strength levels are often referred to as ultrahigh-strength steels. The designation "ultrahigh-strength" is arbitrary because no universally accepted strength level for use of the term has been set. Also, as structural steels with greater and greater strength have been developed, the strength range for which the term may be applied has gradually increased. This article describes those commercial structural steels capable of a minimum yield strength of 1380 MPa (200 ksi).

Besides the steels discussed in this article, many other proprietary and standard steels are used for essentially the same types of applications, but at strength levels slightly below the arbitrary lower limit of 1380 MPa (200 ksi) established above for the ultrahigh-strength class of constructional steels. Medium-alloy steels such as 4330V and 4335V (vanadium-modified versions of the corresponding AISI standard steels) are among the more widely used steels for yield strengths of 1240 to 1380 MPa (180 to 200 ksi). Certain proprietary steels such as Hy Tuf (a silicon-modified steel similar to 300M) exhibit excellent toughness at strengths up to or slightly above 1380 MPa — the toughness of Hy Tuf is about the same as that of a maraging steel in this strength range.

MEDIUM-CARBON LOW-ALLOY STEELS

The medium-carbon low-alloy family of ultrahigh-strength steels includes AISI/SAE 4130, the higher-strength 4140, and the deeper-hardening, higher-strength 4340. Several modifications of the basic 4340 steel have been developed. In one modification (300M), silicon content is increased to prevent embrittlement when the steel is tempered at the low temperatures required for very high strength. In AMS 6434, vanadium is added as a grain refiner to increase toughness, and carbon is slightly reduced to promote weldability. Ladish D-6ac contains the grain refiner vanadium, slightly higher carbon, chromium and molybdenum than 4340, and slightly lower nickel.

Other less widely used steels that may be included in this family are 6150 and 8640. Chemical compositions are given in Table 5.

No distinctly different or new commercial steels have been added to this class of steels in recent years. Rather, developmental efforts have been aimed mostly at increasing ductility and toughness by improving melting and processing techniques as well as by using stricter process control and inspection.

4130

AISI/SAE 4130 is a water-hardening alloy steel of low to intermediate hardenability. It retains good tensile, fatigue and impact properties up to about 370 °C (700 °F); however, it has poor impact properties at cryogenic temperatures. This steel is not subject to temper embrittlement and can be nitrided. It usually is forged at 1100 to 1200 °C (2000 to 2200 °F); finishing temperature should never fall below 980 °C (1800 °F).

Properties. Table 6 summarizes the typical mechanical properties obtained by tempering water-quenched and oil-quenched 4130 steel bars at various temperatures.

4140

AISI/SAE 4140 is similar in composition to 4130 except for a higher carbon content. It is used in applications requiring a combination of moderate hardenability and good strength and toughness, but in which service conditions are only moderately severe.

Forging of 4140 steel can be done readily, usually at 1100 to 1200 °C (2000 to 2200 °F); the finishing temperature should not be below 980 °C (1800 °F). Parts should be cooled slowly after hot forming. This steel can be welded by any of the standard welding methods provided that proper procedures are employed. For welding, preheating at 150 to 260 °C (300 to 500 °F) and post heating at 600 to 675 °C (1100 to 1250 °F) followed by slow cooling are recommended.

Properties. Table 7 summarizes the typical mechanical properties obtained by tempering oil-quenched 4140 steel at various temperatures.

4340

AISI/SAE 4340 steel is considered the standard to which other ultrahigh-strength steels are compared. It combines deep hardenability with high ductility, toughness and strength. It has high fatigue resistance and is often used where severe service conditions exist and where high strength in heavy sections is required.

4340 steel usually is forged at 1065 to 1230 °C (1950 to 2250 °F); after forging, parts may be air cooled in a dry place or, preferably, furnace cooled. The machinability rating of 4340 is 55% for cold drawn material and 45% for annealed material (cold rolled B1112, 100%). A partly spheroidized structure obtained by normalizing and then tempering at 650 °C (1200 °F) is best for optimum machinability. 4340 steel has good welding characteristics. It can be readily gas or arc welded, but welding rods of the same composition should be used. Because 4340 is air hardening, welded parts should be either (a) annealed or (b) normalized and tempered shortly after welding.

Properties. Through hardening of 4340 steel can be done by oil quenching for round sections up to 75 mm (3 in.) in diameter, and by water quenching for larger sections (up to the limit of hardenability). Typical mechanical properties of oil-quenched 4340 are given in Table 8.

300M

Alloy 300M is basically a silicon-modified (1.6% Si) 4340 steel, but has slightly higher carbon and molybdenum contents and also contains vanadium. This steel exhibits deep hardenability and has ductility and toughness at tensile strengths of 1860 to 2070 MPa (270 to 300 ksi). Many of the properties of this steel are similar to those of 4340 steel, except that the increased silicon content provides deeper hardenability, increased solid-solution strengthening and better resistance to softening at elevated temperatures. Compared with 4340 of similar strength, 300M can be tempered at a higher tempering temperature, which provides greater relief of quenching stresses. "500 °F" embrittlement is displaced to higher temperatures.

300M is forged at 1065 to 1095 °C (1950 to 2000 °F). Forging should not be continued below 925 °C (1700 °F). After forging, parts preferably

Table 5. Compositions of the ultrahigh-strength steels described in this article

Designation or trade name	C	Mn	Si	Cr	Ni	Mo	V	Others
Medium-carbon low-alloy steels								
4130	0.28-0.33	0.40-0.60	0.20-0.35	0.80-1.10	...	0.15-0.25	...	...
4140	0.38-0.43	0.75-1.00	0.20-0.35	0.80-1.10	...	0.15-0.25	...	...
4340	0.38-0.43	0.60-0.80	0.20-0.35	0.70-0.90	1.65-2.00	0.20-0.30	...	...
AMS 6434	0.31-0.38	0.60-0.80	0.20-0.35	0.65-0.90	1.65-2.00	0.30-0.40	0.17-0.23	...
300M	0.40-0.46	0.65-0.90	1.45-1.80	0.70-0.95	1.65-2.00	0.30-0.45	0.05 min	...
D-6a	0.42-0.48	0.60-0.90	0.15-0.30	0.90-1.20	0.40-0.70	0.90-1.10	0.05-0.10	...
6150	0.48-0.53	0.70-0.90	0.20-0.35	0.80-1.10	...	...	0.15-0.25	...
8640	0.38-0.43	0.75-1.00	0.20-0.35	0.40-0.60	0.40-0.70	0.15-0.25	...	...
Medium-alloy air-hardening steels								
H11 Mod ..	0.37-0.43	0.20-0.40	0.80-1.00	4.75-5.25	...	1.20-1.40	0.40-0.60	...
H13	0.32-0.45	0.20-0.50	0.80-1.20	4.75-5.50	...	1.10-1.75	0.80-1.20	...
9Ni-4Co steels								
HP 9-4-20	0.16-0.23	0.20-0.40	0.20 max	0.65-0.85	8.50-9.50	0.90-1.10	0.06-0.12	4.25-4.75 Co
HP 9-4-30	0.29-0.34	0.10-0.35	0.20 max	0.90-1.10	7.0-8.0	0.90-1.10	0.06-0.12	4.25-4.75 Co

(a) P and S contents may vary with steelmaking practice. Usually, these steels contain no more than 0.035 P and 0.040 S; 9Ni-4Co steels are specified to have 0.10 max P and 0.10 max S. (b) ASTM A681; composition ranges utilized by some producers are narrower.

Table 6. Typical mechanical properties of heat treated 4130 steel

Tempering temperature		Tensile strength		Yield strength		Elongation in 50 mm or 2 in., %	Reduction in area, %	Hardness, HB	Izod impact energy	
°C	°F	MPa	ksi	MPa	ksi				J	ft·lb
Water quenched and tempered(a)										
205	400	1765	256	1520	220	10.0	33.0	475	18	13
260	500	1670	242	1430	208	11.5	37.0	455	14	10
315	600	1570	228	1340	195	13.0	41.0	425	14	10
370	700	1475	214	1250	182	15.0	45.0	400	20	15
425	800	1380	200	1170	170	16.5	49.0	375	34	25
540	1000	1170	170	1000	145	20.0	56.0	325	81	60
650	1200	965	140	830	120	22.0	63.0	270	135	100
Oil quenched and tempered(b)										
205	400	1550	225	1340	195	11.0	38.0	450	...	...
260	500	1500	218	1275	185	11.5	40.0	440	...	...
315	600	1420	206	1210	175	12.5	43.0	418	...	...
370	700	1320	192	1120	162	14.5	48.0	385	...	...
425	800	1230	178	1030	150	16.5	54.0	360	...	...
540	1000	1030	150	840	122	20.0	60.0	305	...	...
650	1200	830	120	670	97	24.0	67.0	250	...	...

(a) 25-mm (1-in.) diam round bars quenched from 845 to 870 °C (1550 to 1600 °F). (b) 25-mm (1-in.) diam round bars quenched from 860 °C (1575 °F).

Table 7. Typical mechanical properties of heat treated 4140 steel(a)

Tempering temperature		Tensile strength		Yield strength		Elongation in 50 mm or 2 in., %	Reduction in area, %	Hardness, HB	Izod impact energy	
°C	°F	MPa	ksi	MPa	ksi				J	ft·lb
205	400	1965	285	1740	252	11.0	42	578	15	11
260	500	1860	270	1650	240	11.0	44	534	11	8
315	600	1720	250	1570	228	11.5	46	495	9	7
370	700	1590	231	1460	212	12.5	48	461	15	11
425	800	1450	210	1340	195	15.0	50	429	28	21
480	900	1300	188	1210	175	16.0	52	388	46	34
540	1000	1150	167	1050	152	17.5	55	341	65	48
595	1100	1020	148	910	132	19.0	58	311	93	69
650	1200	900	130	790	114	21.0	61	277	112	83
705	1300	810	117	690	100	23.0	65	235	136	100

(a) 12.7-mm (1/2-in.) diam round bars, oil quenched from 845 °C (1550 °F).

Table 8. Typical mechanical properties of 4340 steel(a)

Tempering temperature		Tensile strength		Yield strength		Elongation in 50 mm or 2 in., %	Reduction in area, %	Hardness		Izod impact energy	
°C	°F	MPa	ksi	MPa	ksi			HB	HRC	J	ft·lb
205	400	1980	287	1860	270	11	39	520	53	20	15
315	600	1760	255	1620	235	12	44	490	49.5	14	10
425	800	1500	217	1365	198	14	48	440	46	16	12
540	1000	1240	180	1160	168	17	53	360	39	47	35
650	1200	1020	148	860	125	20	60	290	31	100	74
705	1300	860	125	740	108	23	63	250	24	102	75

(a) Oil quenched from 845 °C (1550 °F) and tempered at various temperatures.

should be slowly cooled in a furnace, but may be allowed to cool in air in a dry place. Although 300M can be readily gas or arc welded, welding is generally not recommended.

Properties. Variations in hardness and mechanical properties of 300M with tempering temperature are presented in Table 9.

D-6a and D-6ac

Ladish D-6a is a low-alloy ultrahigh-strength steel developed for aircraft and missile structural applications. It is designed primarily for use at room-temperature tensile strengths of 1800 to 2000 MPa (260 to 290 ksi). D-6a maintains a very high ratio of yield strength to tensile strength up to a tensile strength of 1930 MPa (280 ksi), combined with good ductility. It has good notch toughness, which results in high resistance to impact loading. It is deeper hardening than 4340 and does not exhibit temper embrittlement. It retains high strength at elevated temperature. Susceptibility of D-6a to stress-corrosion cracking and corrosion fatigue in moist and aqueous environments is comparable to that of 300M steel at the same strength level. The alloy is called D-6a when produced by air melting in an electric furnace, and D-6ac when produced by air melting followed by vacuum-arc melting.

Processing. To forge D-6a, heat to a maximum temperature of 1230 °C (2250 °F); forging should be finished above 980 °C (1800 °F). Finished forgings should be cooled slowly, either in a furnace or embedded in a suitable insulating medium such as ashes or lime. For maximum machinability, charge the parts into a 650 °C (1200 °F) furnace immediately after forging and hold 12 h, increase temperature to 900 °C (1650 °F) and hold for a time period that depends on section size, cool to 650 °C and hold 10 h, and finally air cool to room temperature.

D-6a can be welded (even heavy sections) provided that the techniques and controls normally employed for welding medium-carbon, high-hardenability alloy steels are used. Welding rod of the same composition should be used.

Properties. Typical mechanical properties of heat treated D-6a bar are given in Table 10.

6150

AISI/SAE 6150 is a tough, shock-resisting, shallow-hardening chromium-vanadium steel. It has high fatigue and impact resistance in the heat treated condition. It can be nitrided for maximum surface hardness and abrasion resistance; nitriding characteristics are similar to those of 4140 and 4340 steels. 6150 may be forged from temperatures up to 1200 °C (2200 °F), but the usual temperature range is 1175 to 950 °C (2150 to 1750 °F). Parts made of 6150 steel can be readily welded using any of the standard welding methods. After welding, parts should be normalized, then hardened and tempered to the desired hardness.

Properties. Typical room-temperature mechanical properties of small-diameter round sections of 6150 tempered at various temperatures are given in Table 11.

8640

AISI/SAE 8640 was especially designed to provide the maximum hardenability and best combination of properties possible with minimum alloying additions. 8640 is normally an oil-hardening steel.

Processing. 8640 steel may be forged from temperatures up to 1200 °C (2200 °F), but is usually forged in the range 1175 to 950 °C (2150 to 1750 °F). Forged parts are cooled slowly from the forging temperature, then annealed prior to machining.

8640 can be welded by any of the standard welding methods. Because 8640 has some air-hardening tendencies, preheating to 150 to 260 °C (300 to 500 °F) before welding and postheating after welding are recommended. Stress relieving at 600 to 650 °C (1100 to 1200 °F) is quite satisfactory for most welded parts.

Properties. Variations in properties of heat treated round sections of 8640 with tempering temperature are given in Table 12.

MEDIUM-ALLOY AIR-HARDENING STEELS

The ultrahigh-strength steels H11 Modified (H11 Mod) and H13, which are popularly known as 5% Cr hot work die steels, are discussed in this section. Besides being extensively used in dies, these steels are widely used for structural applications, but not as widely as they once

were—primarily because of the development of several other steels of essentially the same cost but substantially greater fracture toughness at equivalent strength. Nonetheless, H11 Mod and H13 possess some attractive features. Both can be hardened through in large sections by air cooling. Chemical compositions of these steels are given in Table 5.

H11 Modified

This steel is a modification of the martensitic hot work die steel AISI H11, the significant difference being a slightly higher carbon content. H11 Mod can be heat treated to strengths exceeding 2070 MPa (300 ksi). It is air hardening, which results in minimal residual stress after hardening. H11 Mod is a secondary-hardening steel, and thus develops optimum properties when tempered at temperatures above 510 °C (950 °F).

Processing. H11 Mod is readily forged from 1120 to 1150 °C (2050 to 2100 °F). Preferably, stock should be preheated at 790 to 815 °C (1450 to 1500 °F), then heated uniformly to the forging temperature. Forging should not be continued below 925 °C (1700 °F); stock may be reheated

Table 9. Typical mechanical properties of 300M steel(a)

| Tempering temperature | | Tensile strength | | Yield strength | | Elongation in 50 mm or 2 in., % | Reduction in area, % | Charpy V-notch impact energy | | Hardness, HRC |
°C	°F	MPa	ksi	MPa	ksi			J	ft·lb	
90	200	2340	340	1240	180	6.0	10.0	17.6	13.0	56.0
205	400	2140	310	1650	240	7.0	27.0	21.7	16.0	54.5
260	500	2050	297	1670	242	8.0	32.0	24.4	18.0	54.0
315	600	1990	289	1690	245	9.5	34.0	29.8	22.0	53.0
370	700	1930	280	1620	235	9.0	32.0	23.7	17.5	51.0
425	800	1790	260	1480	215	8.5	23.0	13.6	10.0	45.5

(a) Round bars, 25 mm (1 in.) in diameter, oil quenched from 860 °C (1575 °F) and tempered at various temperatures.

Table 10. Typical mechanical properties of D-6a steel bar(a)

| Tempering temperature | | Tensile strength | | Yield strength | | Elongation in 50 mm or 2 in., % | Reduction in area, % | Charpy V-notch impact energy | |
°C	°F	MPa	ksi	MPa	ksi			J	ft·lb
150	300	2060	299	1450	211	8.5	19.0	14	10
205	400	2000	290	1620	235	8.9	25.7	15	11
315	600	1840	267	1700	247	8.1	30.0	16	12
425	800	1630	236	1570	228	9.6	36.8	16	12
540	1000	1450	210	1410	204	13.0	45.5	26	19
650	1200	1030	150	970	141	18.4	60.8	41	30

(a) Normalized at 900 °C (1650 °F), oil quenched from 845 °C (1550 °F) and tempered at various temperatures.

Table 11. Room-temperature mechanical properties of heat treated 6150 steel

| Tempering temperature | | Tensile strength | | Yield strength | | Elongation in 50 mm or 2 in., % | Reduction in area, % | Hardness, HB | Izod impact energy | |
°C	°F	MPa	ksi	MPa	ksi				J	ft·lb
Round bars, 14 mm (0.55 in.) in diameter(a)										
205	400	2050	298	1810	263	1	5	610	...	...
260	500	2070	300	1810	263	4	12	570	...	...
315	600	1950	283	1720	250	7	27	540	...	...
370	700	1770	257	1620	235	10	37	505	9	7
425	800	1585	230	1490	216	11	42	470	14	10
480	900	1410	204	1340	195	12	44	420	16	12
540	1000	1250	182	1210	175	13	46	380	20	15
595	1100	1150	167	1080	157	16	47	350	28	21
Round bars, 25 mm (1 in.) in diameter(b)										
425	800	1570	228	1450	210	10	37	461	...	...
480	900	1360	197	1210	175	11	41	401	...	...
540	1000	1180	171	1030	150	12	45	341	...	...
595	1100	1030	150	875	127	15	50	302	...	...
650	1200	920	133	760	110	19	55	262	...	...
705	1300	810	118	660	96	23	61	235	...	...

(a) Normalized at 870 °C (1600 °F), oil quenched from 860 °C (1575 °F) and tempered at various temperatures. (b) Oil quenched from 860 °C and tempered at various temperatures.

Table 12. Typical room-temperature mechanical properties of 8640 steel

Tempering temperature °C	°F	Tensile strength MPa	ksi	Yield strength MPa	ksi	Elongation in 50 mm or 2 in., %	Reduction in area, %	Impact energy J	ft·lb	Hardness HB	HRC
Round bars, 13.5 mm (0.53 in.) in diameter(a)											
205	400	...1810	263	1670	242	8.0	25.8	11.5(b)	8.5(b)	555	55
315	600	...1585	230	1430	208	9.0	37.3	15.6(b)	11.5(b)	461	48
425	800	...1380	200	1230	179	10.5	46.3	27.8(b)	20.5(b)	415	44
540	1000	...1170	170	1050	152	14.0	53.3	56.3(b)	41.5(b)	341	37
650	1200	... 870	126	760	110	20.5	61.0	96.9(b)	71.5(b)	269	28
Round bars, 25 mm (1 in.) in diameter(a)											
425	800	...1382	200.5	1230	179	10	46	27(c)	20(c)	415	44
480	900	...1250	181	1120	162	13	51	41(c)	30(c)	388	42
540	1000	...1070	155	940	137	17	56	54(c)	40(c)	331	36
595	1100	...1020	148	910	132	16	57	73(c)	54(c)	302	32
650	1200	... 865	125.5	760	110.5	20	61	83(c)	61(c)	269	28

(a) Oil quenched from 830 °C (1525 °F) and tempered at indicated temperature. (b) Izod. (c) Charpy V-notch.

Table 13. Typical longitudinal mechanical properties of H11 Mod steel(a)

Tempering temperature °C	°F	Tensile strength MPa	ksi	Yield strength MPa	ksi	Elongation in 50 mm or 2 in., %	Reduction in area, %	Charpy V-notch impact energy J	ft·lb	Hardness, HRC
510	950	2120	308	1710	248	5.9	29.5	13.6	10.0	56.5
540	1000	2010	291	1675	243	9.6	30.6	21.0	15.5	56.0
565	1050	1850	269	1565	227	11.0	34.5	26.4	19.5	52.0
595	1100	1540	223	1320	192	13.1	39.3	31.2	23.0	45.0
650	1200	1060	154	850	124	14.1	41.2	40.0	29.5	33.0
705	1300	 940	136	700	101	16.4	42.2	90.6	66.8	29.0

(a) Air cooled from 1010 °C (1850 °F); double tempered, 2 + 2 h at indicated temperature.

Table 14. Typical longitudinal room-temperature mechanical properties of H13 steel(a)

Tempering temperature °C	°F	Tensile strength MPa	ksi	Yield strength MPa	ksi	Elongation in 4 D, %	Reduction in area, %	Charpy V-notch impact energy J	ft·lb	Hardness, HRC
527	980	1960	284	1570	228	13.0	46.2	16	12	52
555	1030	1835	266	1530	222	13.1	50.1	24	18	50
575	1065	1730	251	1470	213	13.5	52.4	27	20	48
593	1100	1580	229	1365	198	14.4	53.7	28.5	21	46
605	1120	1495	217	1290	187	15.4	54.0	30	22	44

(a) Round bars, oil quenched from 1010 °C (1850 °F) and double tempered, 2 + 2 h at indicated temperature.

Table 15. Room-temperature mechanical properties of HP 9-4-30 steel

Property	Typical value for hardness of: 49-53 HRC(a)	44-48 HRC(b)	Minimum value(c)
Tensile strength	1650-1790 MPa (240-260 ksi)	1520-1650 MPa (220-240 ksi)	1520 MPa (220 ksi)
Yield strength(d)	1380-1450 MPa (200-210 ksi)	1310-1380 MPa (190-200 ksi)	1310 MPa (190 ksi)
Elongation in 4D	8-12%	12-16%	10%
Reduction in area	25-35%	35-50%	35%
Charpy V-notch impact energy	20-27 J (15-20 ft·lb)	24-34 J (18-25 ft·lb)	24 J (18 ft·lb)
Fracture toughness (K_{Ic})	66-99 MPa $\sqrt{m}$ (60-90 ksi $\sqrt{in.}$)	99-115 MPa $\sqrt{m}$ (90-105 ksi $\sqrt{in.}$)	...

(a) Oil quenched from 845 °C (1550 °F), refrigerated to −73 °C (−100 °F) and double tempered at 205 °C (400 °F). (b) Same heat treatment as (a) except double tempered at 550 °C (1025 °F). (c) For sections forged to 75 mm (3 in.) or less in thickness (or to less than 0.016 m², or 25 in.², in total cross-sectional area), quenched to martensite and double tempered at 540 °C (1000 °F).

as often as necessary. Because H11 Mod is air hardening, it must be cooled slowly after forging to prevent stress cracks. After forging, the part should be charged into a furnace at about 790 °C (1450 °F), soaked until the temperature is uniform, and then slowly cooled, either in the furnace or buried in an insulating medium such as lime, mica or silocel. When the forging has cooled, it should be annealed.

Properties. Typical longitudinal mechanical properties of H11 Mod after heat treatment are presented in Table 13.

H13

AISI H13 is a 5% Cr ultrahigh-strength steel similar to H11 Mod in composition, heat treat-

ment and many properties. The main difference in composition is the higher vanadium content of H13 (see Table 5); this leads to a greater dispersion of hard vanadium carbides, which results in higher wear resistance.

Like H11 Mod, H13 is a secondary-hardening steel. It has good temper resistance and maintains high hardness and strength at elevated temperatures. It is deep hardening, which allows large sections to be hardened by air cooling. H13 steel can be heat treated to strengths exceeding 2070 MPa (300 ksi); like H11 Mod, it has good ductility and impact strength.

Processing procedures for H13, including heat treatment, are almost identical to those outlined above for H11 Mod. Typical longitudinal room-temperature mechanical properties of H13 steel round bars after heat treatment are presented in Table 14.

9Ni-4Co STEELS

During the 1960's, Republic Steel Corporation introduced a family of four weldable steels which have high fracture toughness when heat treated to very high strength levels. These steels nominally contain about 9% Ni and 4% Co and basically differ only in carbon content. The four steels, designated HP 9-4-20, HP 9-4-25, HP 9-4-30 and HP 9-4-45, nominally containing 0.20, 0.25, 0.30 and 0.45% C, respectively, were introduced. Only HP 9-4-20 and HP 9-4-30 are produced in significant quantities; their chemical compositions are given in Table 5. HP 9-4-20 is not described here because, even though it has good weldability and high fracture toughness, it cannot develop a yield strength of 1380 MPa (200 ksi), which was selected as a criterion for the ultrahigh-strength steels discussed in this article. HP 9-4-30 steel, which is described below, can be considered representative of the 9Ni-4Co family of steels.

HP 9-4-30

HP 9-4-30 steel is capable of developing a tensile strength of 1520 to 1650 MPa (220 to 240 ksi) with a minimum plane-strain fracture toughness of 99 MPa $\sqrt{m}$ (90 ksi $\sqrt{in.}$). This steel has deep hardenability and can be fully hardened to martensite in sections up to 150 mm (6 in.) thick. HP 9-4-30 (in the heat treated condition) can be formed by bending, rolling or shear spinning. Heat treated parts can be readily welded; tungsten-arc welding under inert-gas shielding is the preferred welding process. Neither postheating nor postweld heat treatment is required. After welding, parts may be stress relieved at about 540 °C (1000 °F) for 24 h; this stress-relief treatment results in no adverse effect on the strength or the toughness of either the weld metal or the base metal.

Heat Treatments. The following heat treatments apply to HP 9-4-30 steel:

• *Normalize:* Heat to 870 to 925 °C (1600 to 1700 °F) and hold 1 h for each 25 mm (1 in.) of thickness (1 h minimum); air cool.

• *Anneal:* Heat to 620 °C (1150 °F) and hold 24 h; air cool.

• *Harden:* Austenitize at 830 to 860 °C (1525 to 1575 °F) and hold 1 h for each 25 mm (1 in.) of thickness (1 h minimum); water or oil quench. Complete the martensitic transformation by refrigerating at least 1 h at −87 to −60 °C (−125 to −75 °F); allow to warm to room temperature.

• *Temper:* At 200 to 600 °C (400 to 1100 °F), depending on desired final strength; double tempering is preferred. The most widely used tempering treatment is double tempering—2 h at room temperature, air cool, then 2 h more at temperature—at a temperature from 540 to 575 °C (1000 to 1075 °F).

Stress relieve: Usually required only after welding of restrained sections. Heat to 540 °C (1000 °F) and hold 24 h; air cool to room temperature.

Properties. Table 15 presents room-temperature mechanical properties of HP 9-4-30 double tempered at three different temperatures.

Maraging Steels

MARAGING STEELS comprise a special class of high-strength steels that differ from conventional steels in that they are hardened by a metallurgical reaction that does not involve carbon. Instead, these steels are strengthened by the precipitation of intermetallic compounds at temperatures of about 480 °C (900 °F). The term "maraging" is derived from "martensite age hardening" and denotes age hardening of a low-carbon martensite matrix.

Commercial maraging steels are designed to provide specific levels of yield strength from 1030 to 2420 MPa (150 to 350 ksi). Some experimental maraging steels have yield strengths as high as 3450 MPa (500 ksi). These steels typically have very high nickel, cobalt and molybdenum contents and very low carbon contents.

PHYSICAL METALLURGY

Maraging steels can be considered highly alloyed low-carbon martensites. The phase transformations in these steels can be explained with

Table 16. Nominal compositions of commercial maraging steels

Grade	Composition(a), %				
	Ni	Mo	Co	Ti	Al
18Ni(200)	18	3.3	8.5	0.2	0.1
18Ni(250)	18	5.0	8.5	0.4	0.1
18Ni(300)	18	5.0	9.0	0.7	0.1
18Ni(350)	18	4.2(b)	12.5	1.6(b)	0.1
18Ni(Cast)	17	4.6	10.0	0.3	0.1
12-5-3(180)(c)	12	3	...	0.2	0.3

(a) All grades contain no more than 0.03% C. (b) Some producers use a combination of 4.8% Mo and 1.4% Ti, nominal. (c) Also contains 5% Cr.

the help of the two phase diagrams shown in Fig. 1, which depict the iron-rich end of the Fe-Ni binary system. On the left is the metastable diagram plotting the austenite-to-martensite transformation on cooling and the martensite-to-austenite reversion on heating. On the right is the equilibrium diagram showing that at higher nickel contents the equilibrium phases at low temperatures are austenite and ferrite.

Age hardening of maraging steels is produced by heat treating for several hours at temperatures on the order of 480 °C (900 °F). The metallurgical reactions that take place during such treatment can be explained using the equilibrium diagram at the right in Fig. 1.

COMMERCIAL ALLOYS

Table 16 lists the chemical compositions of the more common grades of maraging steel. The nomenclature that has become established for these steels is nominal nickel content followed by nominal yield strength (ksi units) in parentheses. Thus, for example, 18Ni(200) steel normally is age hardened to a yield strength of 200 ksi (1380 MPa). The first three steels in Table 16—18Ni(200), 18Ni(250) and 18Ni(300)—are the most widely used and most commonly available

grades. The 18Ni(350) grade is an ultrahigh-strength variety made in limited quantities for special applications. Two 18Ni(350) compositions have been produced (see footnote(b) in Table 16). The 18Ni(Cast) grade was developed specifically as a cast composition.

PROCESSING

Hot Working. Maraging steels can be hot worked by conventional steel mill techniques, even though allowances must be made for several unique characteristics. Steels with high titanium contents have greater hot strength than conventional steels and require higher hot working loads or higher working temperatures. Working above about 1260 °C (2300 °F) should be avoided.

Cold Working. Maraging steels can be cold worked by any conventional technique when in the solution annealed (unaged or as-transformed) condition. They have very low work-hardening rates and can be subjected to very heavy reductions with only slight accompanying gains in hardness.

Machining. Maraging steels can be machined by any conventional technique when in the solution annealed or age hardened condition. Machinability generally is as good as or better than that of conventional steels of the same hardness.

Table 17. Heat treatments and typical mechanical properties of standard 18Ni maraging steels

Grade	Heat treatment (a)	Tensile strength		Yield strength		Elongation in 50 mm or 2 in., %	Reduction in area, %	Fracture toughness	
		MPa	ksi	MPa	ksi			MPa $\sqrt{m}$	ksi $\sqrt{in.}$
18Ni(200)	A	1500	218	1400	203	10	60	155-200	140-220
18Ni(250)	A	1800	260	1700	247	8	55	120	110
18Ni(300)	A	2050	297	2000	290	7	40	80	73
18Ni(350)	B	2450	355	2400	348	6	25	35-50	32-45
18Ni(Cast)	C	1750	255	1650	240	8	35	105	95

(a) Treatment A: solution treat 1 h at 820 °C (1500 °F), then age 3 h at 480 °C (900 °F). Treatment B: solution treat 1 h at 820 °C (1500 °F), then age 12 h at 480 °C (900 °F). Treatment C: anneal 1 h at 1150 °C (2100 °F), age 1 h at 595 °C (1100 °F), solution treat 1 h at 820 °C (1500 °F) and age 3 h at 480 °C (900 °F).

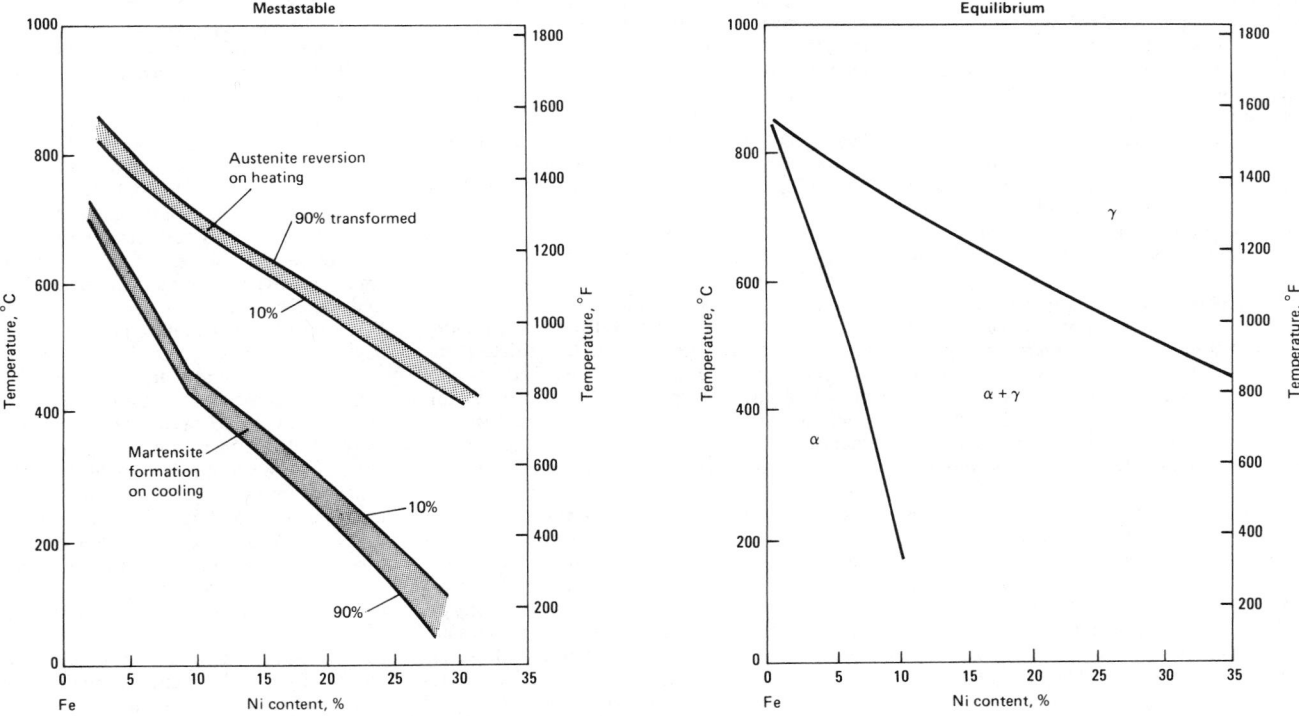

Fig. 1. Metastable and equilibrium phase relationships in the Fe-Ni system

Heat Treating. Properties of conventionally heat treated maraging steels are given in Table 17. Maraging steels normally are solution annealed (austenitized) one hour for each 25 mm (1 in.) of section size. It may be necessary to heat treat sheet products in dry hydrogen or dissociated ammonia to minimize surface damage.

MECHANICAL PROPERTIES

Typical tensile properties and fracture toughness values of the conventional grades of maraging steel are listed in Table 17. One of the distinguishing features of maraging steels is superior toughness compared to conventional steels. Figure 2 compares K_{Ic} values for several maraging steels with those of quenched and tempered alloy steels. Toughness of maraging steels is sensitive to purity level, and carbon and sulfur in particular should be kept low to obtain optimum fracture toughness.

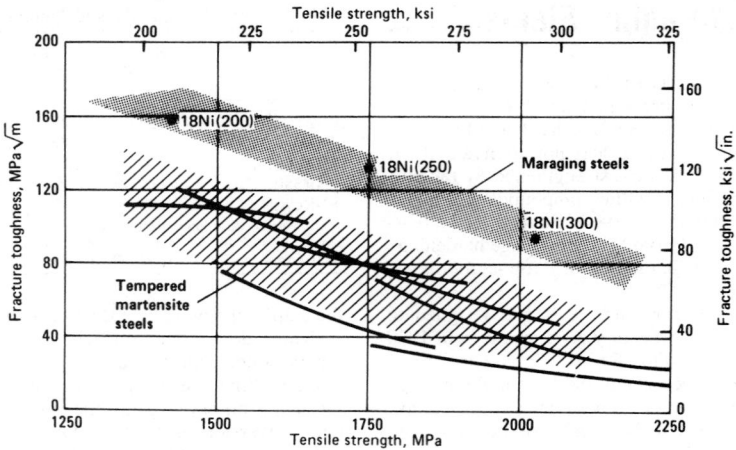

Fig. 2. Plane-strain fracture toughness of maraging steels compared with fracture toughness of several ultrahigh-strength steels

Hardenability of Carbon and Alloy Steels

HARDENABILITY OF STEEL is the property that determines the depth and distribution of hardness induced by quenching. Steels that exhibit deep hardness penetration are considered to have high hardenability, while those that exhibit shallow hardness penetration are of low hardenability. Because the primary objective in quenching is to obtain satisfactory hardening to some desired depth, it follows that hardenability is usually the most important single factor in the selection of steel for heat treated parts.

Hardenability should not be confused with hardness as such or with maximum hardness. The maximum attainable hardness of any steel depends solely on carbon content. Also, the maximum hardness values that can be obtained with small test specimens under the fastest cooling rates of water quenching are nearly always higher than those developed under production heat treating conditions, because hardenability limitations in quenching larger sizes may result in less than 100% martensite formation. Effects of carbon and martensite content on hardness are shown in Fig. 1.

The hardenability of steel is governed almost entirely by the chemical composition (carbon and alloy content) at the austenitizing temperature and the austenite grain size at the moment of quenching. In some instances, the chemical composition of the austenite may not be the same as that determined by chemical analysis, because some carbide may be undissolved at the austenitizing temperature. Such carbides would be reflected in the chemical analysis, but, being undissolved in the austenite, neither their carbon nor alloy content can contribute to hardenability; also, by nucleating transformation products, undissolved carbides can actively decrease hardenability. This is especially important in high-carbon (0.50 to 1.10%) and alloy carburizing steels, which may contain excess carbides at the austenitizing temperature. Consequently, such factors as austenitizing temperature, time at temperature and prior microstructure are sometimes very important

variables when determining the basic hardenability of a specific steel composition.

HARDENABILITY TESTING

Hardenability of a steel is best assessed by studying the hardening response of the steel to cooling in a standardized configuration in which a variety of cooling rates can be easily and consistently reproduced from one test to another.

The end-quench, or Jominy, test fulfills the cooling rate requirements of hardenability testing most conveniently. The test specimen, a 1-in. (25.4-mm) diam bar 4 in. (102 mm) in length, is water quenched on one end face. The bar from which the specimen is made must be normalized before the test specimen is machined. The test involves heating the test specimen to the proper austenitizing temperature and then transferring it to a quenching fixture so designed that the specimen is held vertically 12.7 mm ($^{1}/_{2}$ in.) above an opening through which a column of water may be directed against the bottom face of the specimen. While the bottom end is being quenched by the column of water, the opposite end is cooling slowly in air, and intermediate positions along the specimen are cooling at intermediate rates. After the specimen has been quenched, parallel flats 180° apart are ground 0.015 in. (0.38 mm) deep on the cylindrical surface. Rockwell C hardness is measured at intervals of $^{1}/_{16}$ in. (1.59 mm) for alloy steels and $^{1}/_{32}$ in. (0.79 mm) for carbon steels, starting from the water-quenched end. A plot of these hardness values and their positions on the test bar, as shown in Fig. 2, indicates the relation between hardness and cooling rate, which, in effect, is the hardenability of the steel. Figure 2 also shows the cooling rate for the designated test positions. Details of the standard test method are contained in specifications of the American Society for Testing and Materials (ASTM Method A255) and the Society of Automotive Engineers (Standard J406); in these specifications, dimensions are given in inches.

The measured hardenability curve shown in Fig. 2 for one specimen of 8650 steel falls about midway between the minimum and maximum hardenability curves that would be obtained from many heats of 8650 steel, and that define the limits of the so-called H-band for the 8650 steel. (Each grade of steel purchased to hardenability limits is represented by such an H-band that defines these limits.)

The Carburized Hardenability Test. It is often necessary to determine the hardenability of the high-carbon case regions of carburized steels. Such information is important in controlling carburizing and quenching practice and in determining the ability of a specific steel to meet the hardness and case depth requirements of the carburized component manufactured from the steel.

Measurements of case hardenability are performed as follows. A standard end-quench bar is pack carburized for 8 h at 925 °C (1700 °F) and end quenched in the usual manner. A comparison bar is simultaneously carburized in the same pack to determine carbon penetration. Successive layers are removed from it and analyzed chemically to determine the carbon content at various depths. When a carbon-penetration curve is established, depths to various carbon levels can be determined in the Jominy bar, assuming that distribution of carbon in the end-quench specimen is the same as in the carbon gradient bar. Longitudinal flats are then carefully ground to various depths on the end-quench bar (usually to carbon concentrations of 1.1, 1.0, 0.9, 0.8, and in some instances to as low as 0.6%), and hardenability is determined at these carbon levels by hardness traverses. In grinding, care must be exercised to avoid overheating and tempering, and in conducting hardness surveys, similar concern must be shown to ensure relating hardness to a single carbon level by remaining in the exact center of the flat. Rockwell A hardness readings are preferable to Rockwell C readings because they minimize the depth of indenter penetration into softer subsurface layers. Rockwell A values are con-

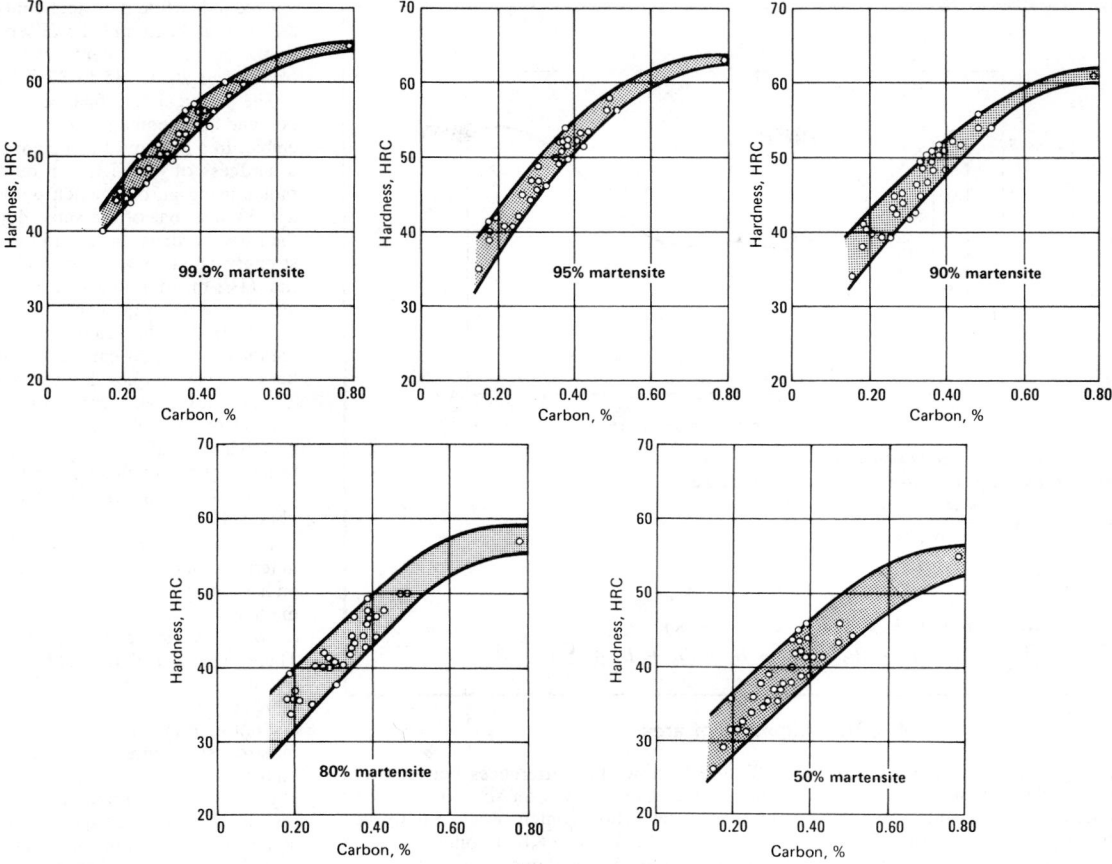

Fig. 1. Effect of carbon on hardness of martensite structures

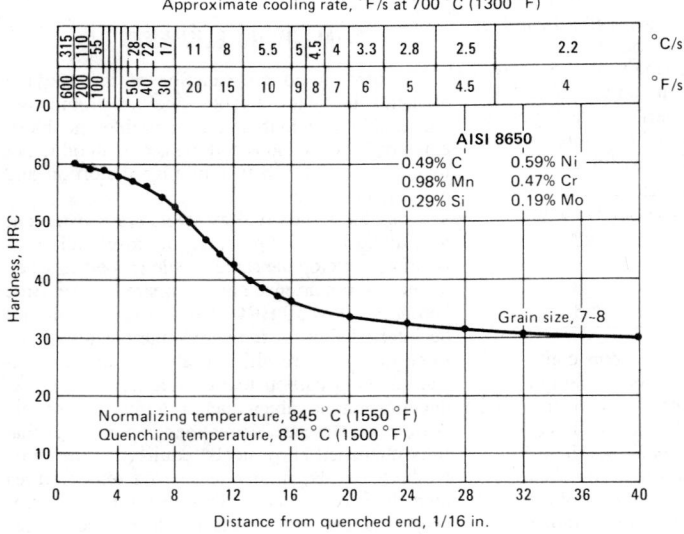

Fig. 2. Method for presenting end-quench hardenability data

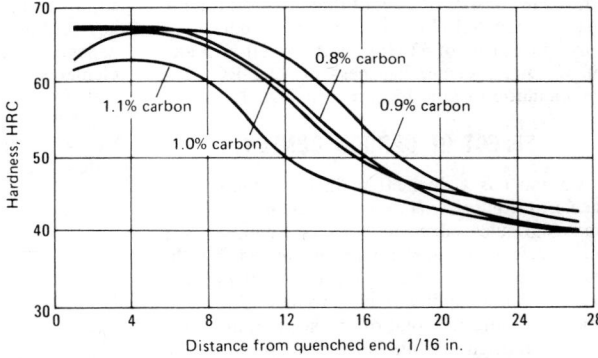

Composition: 0.18 to 0.23 C, 0.90 to 1.20 Mn, 0.40 to 0.60 Cr, 0.08 to 0.15 Mo, 0.0005 min B.

Fig. 3. Carburized hardenability, EX19 steel

verted to Rockwell C values for plotting, as shown in Fig. 3, which shows the curves of carburized hardenability of an EX19 steel.

SAC Hardenability Test. This test is applicable to all plain carbon and low-alloy steels other than carbon tool steels and is suitable only for shallow-hardening steels that will not through harden in sizes larger than 25 mm (1 in.) in diameter. Tha acronym "SAC" denotes surface-area-cen-ter. Some have found the SAC test to be more discriminating than the end-quench test for determining hardenability of shallow-hardening steels because of the sharp gradient on the end-quench curve.

The SAC test surveys hardnesses on an austenitized and quenched cross section. The specimen is 140 mm (5½ in.) long by 25.4 mm (1 in.) in diameter. After normalizing at the spec-ified temperature for 1 h and cooling in air, it is austenitized by holding at temperature for 30 min and quenched in water at 24 ± 5.5 °C (75 ± 10 °F), where it is allowed to remain until the temperature is uniform throughout the specimen.

After quenching, a cylinder 25.4 mm (1 in.) in length is cut from the middle of the hardened specimen. The cut faces of the cylinder are carefully ground parallel to remove any burning or tempering that might result from cutting and to assure that a flat face will be presented to the anvil or fixture of the hardness-testing machine. Rockwell C hardness is measured on the cylindrical surface of the specimen at a minimum of

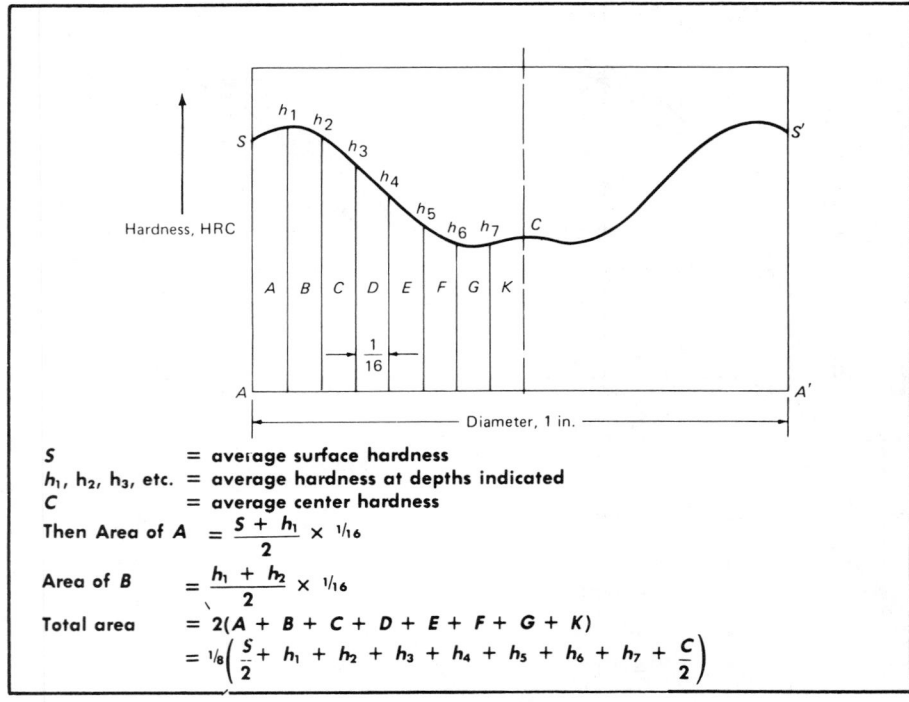

S = average surface hardness
h_1, h_2, h_3, etc. = average hardness at depths indicated
C = average center hardness

Then Area of A = $\dfrac{S + h_1}{2} \times \text{1/16}$

Area of B = $\dfrac{h_1 + h_2}{2} \times \text{1/16}$

Total area = $2(A + B + C + D + E + F + G + K)$

$= \text{1/8}\left(\dfrac{S}{2} + h_1 + h_2 + h_3 + h_4 + h_5 + h_6 + h_7 + \dfrac{C}{2}\right)$

Fig. 4. SAC estimation of area

four points at 90° angles to each other. The average of these readings then becomes the surface reading. Next, a series of Rockwell C readings is taken on the cross section in steps of 1.59 mm (1/16 in.) from the surface to center. From these, a quantitative value can be computed and designated by a code known as the SAC number. The code consists of a set of three two-digit numbers, e.g., SAC No. 63-52-42. This indicates a surface hardness of 63 HRC, a Rockwell-inch area of 52, and a center hardness of 42 HRC. The computation of area is shown in Fig. 4.

EFFECT OF CARBON CONTENT

Carbon has a dual effect in hardenable alloy steels: it controls maximum attainable hardness and contributes substantially to hardenability. The latter effect is substantially enhanced by the quantity and type of alloying elements present, because hardenability is the product of the carbon factor and the total of the alloy multiplying factors. It might be concluded, therefore, that increasing carbon content is the least expensive approach to improving hardenability. This is true to a degree, but several factors rule against the use of large amounts of carbon:

1. High carbon content generally decreases toughness at room and subzero temperatures.
2. It produces harder and more abrasive microstructures in the annealed conditions, which makes cold shearing, sawing, machining and other forms of cold processing more difficult.
3. It makes the steel more susceptible to hot shortness in hot working.
4. It makes the steel more prone to cracking and distortion in heat treatment. Because of these disadvantages, more than 0.60% carbon is seldom used in steels for machine parts, except for springs and bearings, and steels with 0.50 to 0.60% carbon are used less frequently than those containing less than 0.50%.

Figure 5 shows the differences between minimum hardenability curves for six series of steels. In each series, alloy content is essentially constant, and the effect of carbon content on hardenability can be observed over a range from 0.15 to 0.60%. The hardness effect is shown by the vertical distance between the curves at any position on the end-quench specimen—that is, for any cooling rate. This effect varies significantly, depending on the kind and amounts of alloying elements. For instance, referring to Fig. 5(d), (e) and (f), an increase in carbon content from 0.35 to 0.50% in each of the three series of steels causes hardness increases (in Rockwell C points) at four different end-quench positions as follows:

Series	Distance from quenched surface, in.			
	1/16	4/16	8/16	12/16
41xxH	8	10	17	20
51xxH	8	13	9	8
86xxH	8	12	18	12

The hardenability effect of carbon content is read on the horizontal axis. For example, in comparing the effect of carbon content on hardenability in 8650 versus 8630 at 45 HRC, the effect can be expressed as 6/16 (10/16 minus 4/16).

Similarly, at 45 HRC and with nominal carbon contents of 0.35 and 0.50%, the hardenability effect of carbon content is found to be much less in 51xx series steels and much greater in 41xx steels.

Considering the hardenability effect in terms of quenching speed, the cooling rate (or quenching speed) required to produce 45 HRC is affected more by 0.15% carbon with certain combinations of alloying elements than it is by other combinations. For instance, in a steel containing 0.75 Cr and 0.15 Mo (a 41xxH series steel, for example), increasing the carbon content by 0.15% lowers the required or critical cooling rate to obtain 45 HRC from 25 to 4.6 °C (45 to 8.3 °F)

per second, while in a steel containing 0.75% Cr and no molybdenum (51xxH series), the same increase in carbon content lowers the cooling rate from 47 to 26 °C (85 to 37 °F) per second.

The practical significance of the effect of carbon and alloy contents on cooling rate is considerable. In a 51-mm (2-in.) diam bar of 4150 steel, a hardness of 45 HRC can be obtained at half-radius using an oil quench without agitation. In a 4135 steel bar of the same diameter, to obtain the same hardness at half-radius would require a strongly agitated water quench. Comparing 32-mm (11/4-in.) diam bars of 5135 and 5150 steel, an agitated water quench will produce a hardness of 45 HRC at half-radius in the 5135 bar; the identical condition can be obtained in the 5150 bar using an oil quench with moderate agitation. To summarize, an increase or decrease in carbon content or an alloying addition, such as 0.15% molybdenum, affects the results obtained both in terms of the quenching severity required and the section size in which the desired results can be obtained.

The above method of rating differs from rating in terms of ideal critical diameter, which is based on hardening not to a given hardness, but to 50% martensite at the center of a section, the hardness of which varies according to carbon content. Therefore, a part that must be hardened to a minimum hardness regardless of carbon content requires a steel rated in terms of that one hardness and not in terms of 50% martensite, which varies in hardness. Figure 6 shows how steels are rated on the basis of ideal critical diameter by expressing the effect of carbon and alloy content on the section size that will harden to 50% martensite at the center, assuming an ideal quench. An ideal quench is defined as one that removes heat from the surface of the steel as fast as it is delivered to the surface.

ALLOYING ELEMENTS

The most important function of the alloying elements in heat treatable steel is to increase hardenability, which makes possible the hardening of larger sections and the use of an oil rather than a water quench to minimize distortion and avoid quench cracking.

When the standard alloy steels are considered, it is found that, for practical purposes, all compositions develop the same tensile properties when quenched to martensite and tempered to the same hardness below 50 HRC. However, it should not be inferred that all tempered martensites of the same hardness are alike in all respects. For instance, plain carbon martensites have lower reduction-in-area values than do alloy martensites. A further difference, sometimes important, is that fully quenched alloy steels require, for the same hardness levels, higher tempering temperatures than do carbon steels. This difference in tempering temperature may serve to reduce the residual stress level in finished parts. The stress reduction could be an advantage or a disadvantage, depending on whether a controlled compressive stress is desired in the part. Although tensile properties may not differ significantly from one alloy steel to another, considerable differences may exist in fracture toughness and low-temperature impact properties. In general, steels with a higher nickel content, such as 4320, 3310 and 4340, offer much greater toughness at a given hardness or hardenability level. In some applications, the toughness factor rather than hard-

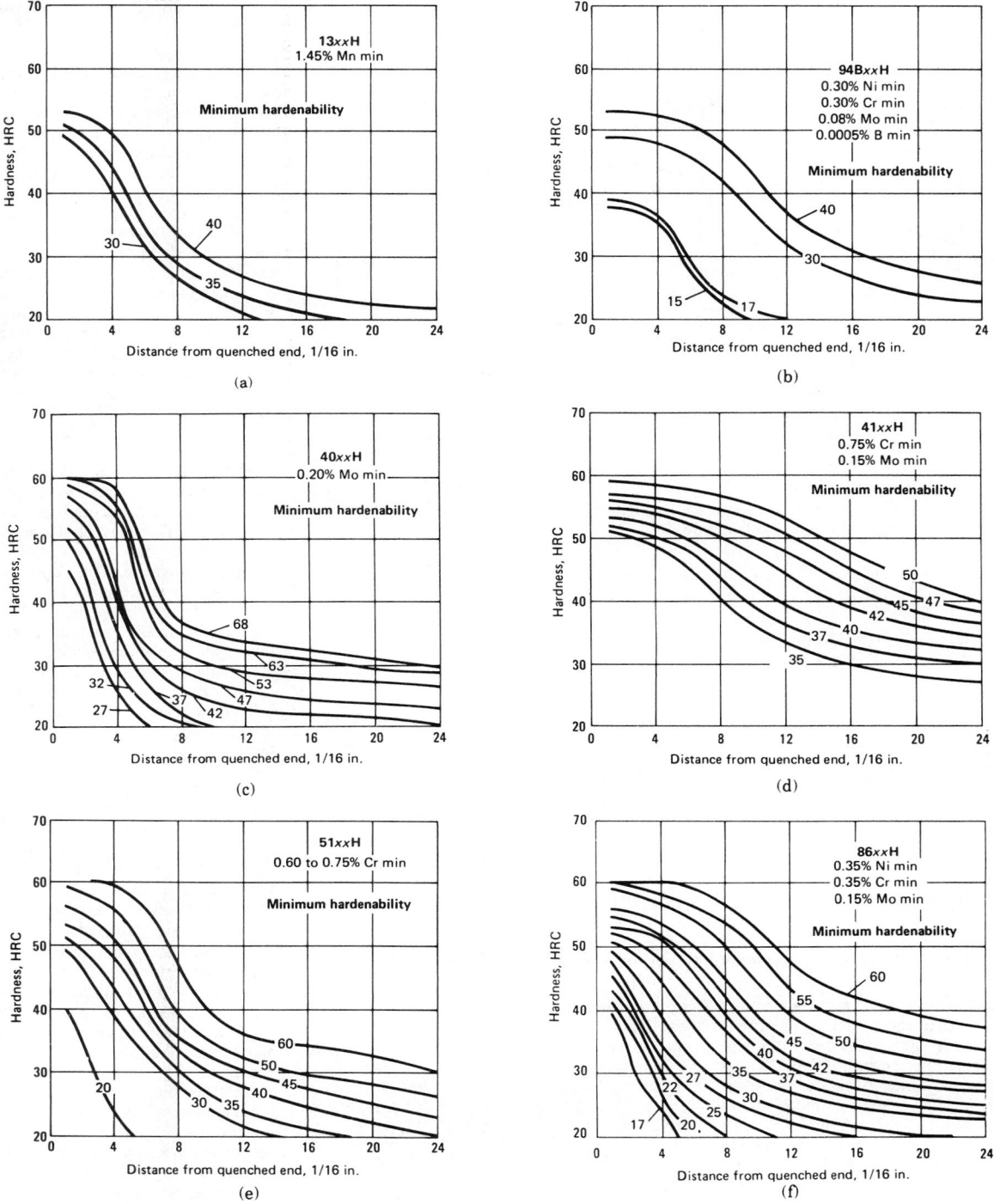

Number adjacent to each curve indicates carbon content of steel, to be inserted in place of *xx* in alloy designation.

Fig. 5. Effect of carbon content on the minimum end-quench hardenability of six series of alloy H-steels

enability may dictate steel selection, but hardenability is still important, because steels that can be fully quenched to 100% martensite are much tougher than those that cannot.

Usually, the least expensive means of increasing hardenability at a given carbon content is by increasing the manganese content. Chromium and molybdenum, already referred to as increasing hardenability, are also among the most economical elements per unit of increased hardenability. Nickel is the most expensive per unit, but is warranted when toughness is a prime consideration.

Another potent and economical alloying element is boron, which markedly increased hardenability when added to a fully deoxidized steel. The effects of boron on hardenability are unique in several respects: (*a*) a very small amount of boron (about 0.001%) has a powerful effect on hardenability; (*b*) the effect of boron on hardenability is much less in high-carbon than in low-carbon steels; (*c*) nitrogen and deoxidizers influence the effectiveness of boron; and (*d*) high-temperature treatment reduces the hardenability effect of boron. (Recommended austenitizing temperatures for boron H-steels are given with the H-bands.)

EFFECT OF GRAIN SIZE

The hardenability of a carbon steel may increase as much as 50% with an increase in austenite grain size from ASTM 8 (6 to 10) to ASTM 3 (1 to 4). The effect becomes more pronounced

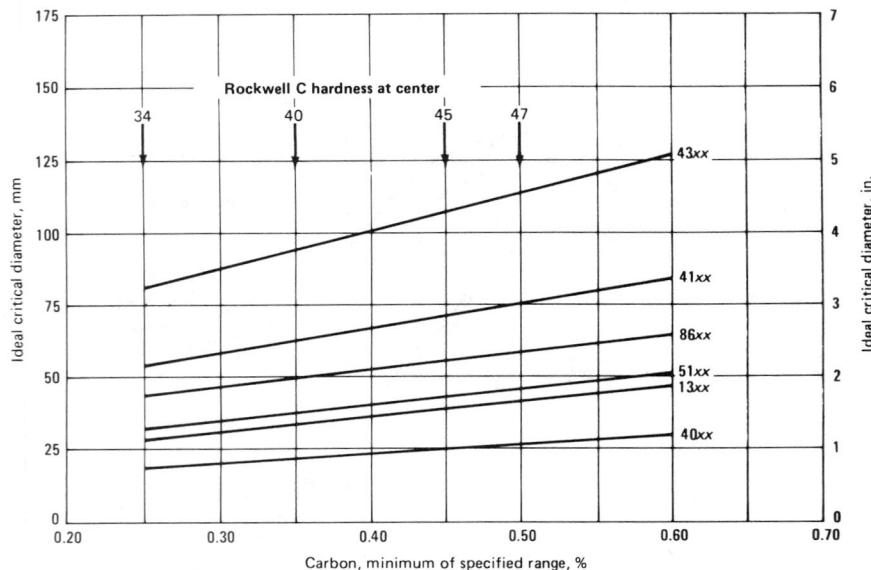

Fig. 6. Effect of carbon content on ideal critical diameter, calculated for the minimum chemical composition of each grade

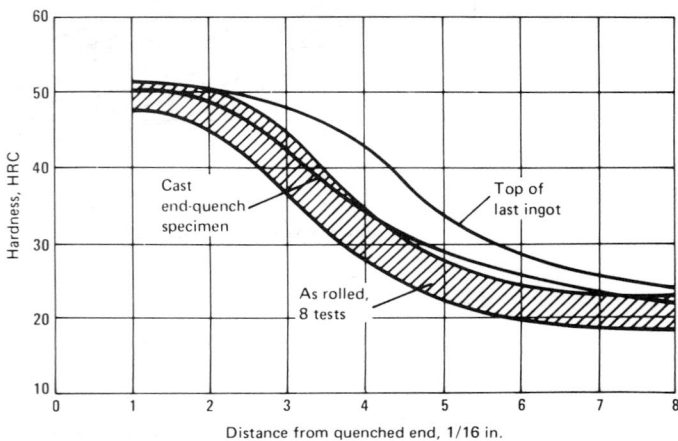

Fig. 7. Variation of hardenability within a heat of 4028 steel

DETERMINING HARDENABILITY REQUIREMENTS

The basic information needed before a steel with adequate hardenability can be specified includes: (*a*) the as-quenched hardness required prior to tempering to final hardness that will produce the best stress-resisting microstructure; (*b*) the depth below the surface to which this hardness must extend, and (*c*) the quenching medium that should be used in hardening.

As-Quenched Hardness. A common and economical practice is to select the steel with the lowest carbon content that will produce the indicated as-quenched hardness using the quenching medium available (or one that can be made available). Following this procedure, the structures possessing the indicated hardnesses would be fully hardened — that is, they would contain more than 90% martensite, which is a common and practical definition of full hardening.

Percentage Martensite. Because of the assumption made above regarding hardening to 90% martensite, it is of interest to compare the percentages of martensite represented by the hardnesses specified and the steels recommended in any specific application. This may be done using the Hodge and Orehoski data.

Depth of Hardening. The depth and percentage martensite to which parts are hardened may affect their serviceability, but it always affects the hardenability required and, therefore, cost. In parts less highly stressed in bending, hardening to 80% martensite at three-quarter-radius of the part as finished may be sufficient; in other parts, even less depth may be required. The latter include principally those parts designed for low deflection under load, in which even the exterior regions are only moderately stressed. In contrast, some parts loaded principally in tension and others operating at high hardness levels, such as springs of all types, are usually hardened more nearly through the section. In automobile leaf springs, the leaves are designed with a low section modulus in the direction of loading. The allowable deflection is large, and most of the cross section is highly stressed.

In general, hardening need be no deeper than is required to provide the strength to sustain the load at any given depth below the surface. When these requirements are not wholly justified, the results are overspecification of steel at higher cost, and greater likelihood of distortion and quench cracking.

Quenching Media. The cooling potential of quenching media is a critical factor in heat treating processes because of its contribution to attaining the minimum hardenability requirement of the part or section being heat treated. The cooling potential, a measure of quenching severity, can be varied over a rather wide range by (*a*) selection of a particular quenching medium, (*b*) control of agitation, and (*c*) additives that improve the cooling capability of the quenchant.

In general, the more severe the quenchant and the less symmetrical the part being quenched, the greater are the size and shape changes that result from quenching and the greater is the risk of quench cracking. Consequently, although water quenching is less costly than oil quenching and water-quenching steels are less expensive than those requiring oil quenching, it is important that parts to be hardened be carefully reviewed to determine whether the amount of distortion and the possibility of cracking as a result of water quenching will permit taking advantage of the

if the carbon content is increased at the same time. When the danger of quench cracking is remote (no abrupt changes in section thickness) and engineering considerations permit, it may sometimes appear to be more practical to use a coarser-grained steel than a fine-grained or more expensive alloy steel to obtain hardenability. However, this is not recommended, because the use of coarser-grained steels usually involves a serious sacrifice in notch toughness and may lead to other difficulties.

VARIATIONS WITHIN HEATS

Segregation of carbon, manganese and other elements always occurs during ingot pouring and solidification. As a result, hardenability of the steel in these segregated portions will differ from that in the remainder of the ingot. In general, specimens taken from the top of the ingot have higher hardenability than steel from the middle, and specimens from the bottom of the ingot will have lower hardenability than steel from the middle.

The same effect is observed in alloy steels. End-quench hardenability test results for one heat of 4028 steel (Fig. 7), show higher hardenability for a cast bar taken from the top of the last ingot of the heat than for a specimen from the melting floor and labeled "cast end-quench specimen." The latter was taken from about the middle of the heat.

Table 1. Quenching severities for various media and quenching conditions

| Quenchant agitation | Typical flow rates | | Typical "H" values | | | |
	m/min	ft/min	Air	Mineral oil	Water	Brine
None	0	0	0.02	0.02/0.30	0.9/1.0	2.0
Mild	15	50	...	0.20/0.35	1.0/1.1	2.1
Moderate	30	100	...	0.35/0.40	1.2/1.3	...
Good	61	200	0.05	0.40/0.60	1.4/2.0	...
Strong	230	750	...	0.60/0.80	1.6/2.0	4.0

Jominy equivalent cooling (J_{eh}) rates are determined by comparing hardnesses of cross sections of parts receiving the established production heat treatment to hardnesses obtained on end-quenched bars of the same steel. A typical procedure is as follows:

1 Select hardening and quenching conditions which your production hardening equipment can fulfill easily.
2 Select a low hardenability steel, such as 8620, 4023 or 1040, and manufacture a quantity of finished components: gears, bearings, shafts.
3 Quench a number of these components (in the uncarburized condition) in the production facility.
4 Measure hardnesses obtained at all critical locations from the surface to the core.
5 Compare the measured hardness values at these locations with equivalent hardness values produced at some end quench (J_{eh}) location on a Jominy bar made from the same heat and end quenched from the same thermal conditions.
6 The J_{eh} values obtained in this fashion define the equal hardness cooling conditions for each location in

the production quenched component.
7 Finally, select from available end quench data a steel which will produce the hardnesses required at each critical J_{eh} location in the finished production part. If end quench data are not available, calculate a suitable composition by one of the standard methods.

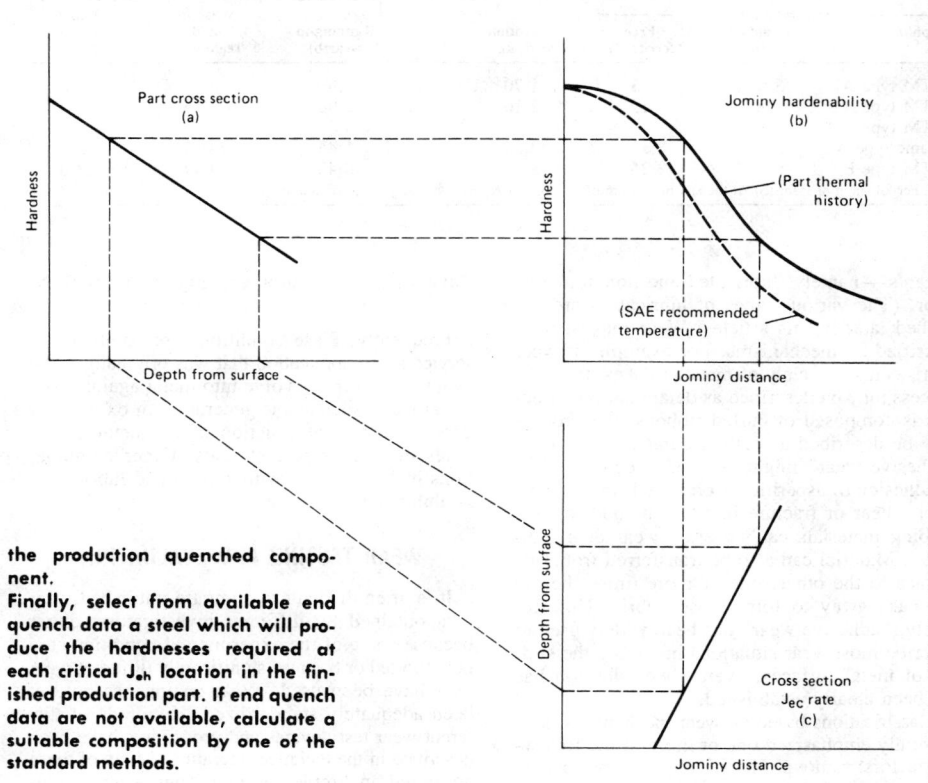

Fig. 8. Determination of Jominy equivalent cooling (J_{eh}) rates

lower cost of water quenching. Oil, salt and synthetic water-polymer quenchants are alternatives, but their use often requires steels of higher alloy content to satisfy hardenability requirements.

A rule regarding selection of a steel and quenching medium for a given part is that the steel should have a minimum hardenability not exceeding that required by the quenching severity of the medium selected. The steel should also contain the lowest carbon content compatible with required hardness and strength properties. This rule is based on the fact that the quench cracking susceptibility of steels increases with a decrease

in M_s temperature and/or an increase in carbon content.

Table 1 lists typical quenching severities, or "H" values, for the common quenching media and conditions.

Hardenability Versus Size and Shape. When end-quench data such as those shown in Fig. 2 are available, either of two methods can be used to estimate the hardenability that a steel part of given size and configuration must have in order for it to achieve desired hardness, strength and microstructure at critical locations when quenched in various production media. These methods comprise the following:

1. The correlation of end-quench hardness data (J_{eh}) with equivalent hardness locations in variously quenched shapes.
2. The correlation of end-quench cooling-rate data (J_{ec}) with equivalent cooling-rate locations in variously quenched production shapes.

Method 1, which is described in Fig. 8, is the more accurate and preferred method, because in practice it has been found that, when cooling at the same rates, large sections produce somewhat lower hardnesses than do smaller sections.

Service Characteristics of Carbon and Alloy Steels

Wear Resistance

WEAR of metals occurs by the plastic displacement of surface and near-surface material and by the detachment of particles that form wear debris. In metals, this process may occur by contact with other metals, nonmetallic solids, flowing liquids, or solid particles or liquid droplets entrained in flowing gases.

Wear involves damage to a solid surface due to relative motion between that surface and one

or more contacting substances and generally consists of a progressive loss of material from the wearing surface. It may include oxidation, corrosion, creep, fatigue, frictional effects, battering from impact, pseudomachining due to rough surfaces, and the cutting and deformation action of abrasive particles. Rigorous analysis of such a complex phenomenon is difficult; consequently, investigations are usually limited to only certain facets of wear. Mechanical effects are those usually considered, whereby local deformation or metal removal results from stresses applied

either directly or indirectly to a surface.

Wear can result in alteration of critical dimensions, increased vibration, fatigue damage, clogging of critical passages and contacts by debris, generation of abrasive particles, leakage of fluid from enclosures, inefficient operation, and system failure.

CLASSIFICATION OF WEAR

One of the simplest classifications of wear is based on the presence or absence of effective lu-

Table 1. Wear of gray iron pins in pin-on-disk tests

Graphite form	Structure of the iron Graphite size	Free ferrite, %	Transition load, kg	Running-in wear(b)	Weight loss(a), g Wear in mild regions(c)	Wear in severe regions(c)
ASTM type A	3 to 4	5	1.70	1.90	0.0	5.0
ASTM type A	4	10	2.10	2.90	0.016	5.0
ASTM type D, some type A	5 to 6	15	3.55	1.45	0.083	13.2
ASTM type E	6 to 7	25	3.20	0.40	0.049	22.0

(a) Per kilogram of load. (b) At the end of 6700 m of sliding. (c) At the end of 10 000 m of sliding.

bricants—namely, lubricated and nonlubricated wear. (The various types of lubrication are described later in this article.) Wear may also be classified by mechanisms. For example, if wear debris consists mainly of metallic flakes, the wear process may be described as delamination. If debris is composed of curled ribbons, the process may be described as cutting abrasion. The term "adhesive wear" might be used if cold welding or adhesion of asperities is observed. In adhesive wear, shear or fracture in either or both of two rubbing materials can generate wear debris directly. Material can also be transferred from one surface to the other, one or more times, before it breaks away to form loose debris. Unfortunately, "adhesive wear" has been widely used to describe most wear situations involving the sliding of metal surfaces, even when adhesion has not been clearly established.

Classifications based on wear mechanisms have generally emphasized one or more of the following factors: removal of individual atoms, plastic deformation, ductile or brittle fracture, cutting, fatigue, adhesion and chemical-electrochemical effects.

An inherent problem exists in any classification scheme that relies mainly on labels implying mechanisms. Overuse of any such label (i.e., use without adequate experimental evidence) could lead to the misconception that a particular wear situation is well understood. This, in turn, may discourage innovative efforts in wear control.

A third approach to wear classification emphasizes the nature of the contacting materials and the experimental conditions, using descriptive terms that are widely understood and accepted. The following is an example of this type of classification:

Metal against nonmetallic abrasive
High-stress gouging or grinding:
    Wet, as in ball and rod mills
    Dry, as in jaw-type or roll-type ore crushers
Low-stress scratching or sliding:
    Wet, such as conveyor screws for wet sand
    Dry, as against plows or earthmoving devices operating in dry soil
Impact of loose abrasive (erosion):
    Wet, as against impellers in slurry pumps
    Dry, as in sand blasting.
Metal against metal
Sliding:
    Lubricated, such as engine crosshead or shaft in a bearing
    Nonlubricated, such as fasteners, nuts and bolts
Rolling:
    Lubricated, such as roller bearings and gears
    Nonlubricated, such as wheels on tracks
Liquid or vapor impingement on metals
Wet steam, such as turbines
Combustion gases, such as gas turbines
Water, such as pump impellers

Cavitations, as in turbulent, high-velocity flowing liquids.

Frequently, these conditions are combined in service so an application that was originally metal against metal may evolve into metal against nonmetal wear, such as the generation of oxide wear debris and the introduction of nonmetallic particles through imperfect seals. Other combinations include rolling with sliding and lubricated-nonlubricated situations.

WEAR TESTING AND EVALUATION

It is often difficult to compare published wear data obtained by different investigators, either because some of the experimental conditions are not reported or because significantly different wear tests have been used. Wear testing has not yet been adequately standardized. Hundreds of different wear test devices and procedures have been described in the technical literature. Many of these are based on simple configurations, such as pin-on-disk, block-on-ring, ring-on-ring or cylinder-on-cylinder. Many other devices have been designed to simulate as closely as possible actual operating conditions for actual components. This is frequently a desirable approach, particularly if the designer has difficulty adapting published wear data to his own situation, although it contributes to the proliferation of wear tests.

Despite the great variety of wear tests, the main variables in sliding mechanical wear may be identified by asking three simple questions:

1. Is the system lubricated or unlubricated?
2. Is most of the wear debris removed?
3. Is the wearing surface in steady contact with another surface?

Reporting wear data is not well standardized; however, some of the more common reporting methods are discussed below. The simplest and most widely used way to summarize wear data is to note changes in length, volume or mass (or these same measures per unit time) for specified conditions. Wear resistance is sometimes given as the inverse of one of these quantities. Of course, when expressed in this manner, the results are not materials constants, because they are based on experimental conditions.

Several models of sliding wear lead to the prediction for volume change:

$$\Delta V = kLS/3H$$

where L is load, S is sliding distance, H is Vickers hardness, and k is a wear coefficient. Therefore, wear rate is commonly reported as $\Delta V/L$. For mild wear, k would be approximately 10^{-8} to 10^{-7}; for severe wear, it would be about 10^{-4} to 10^{-3}. Sometimes a wear factor defined as $\Delta V/SL$ or $k/3H$ is used.

Wear is sometimes reported as "specific wear," or d/PS, where d is depth of wear, P is bearing pressure, and S is sliding distance. Wear data may also be expressed as lifetime-in-service when wear is the main determinant of lifetime. Wear rates are commonly reported as values that are relative to the wear of a standard material subjected to the same conditions as the test specimen. The use of a good reference material is almost imperative. It is the only reasonable way of ranking materials for comparison with other wear situations, because values of metal loss generally do not agree even when different investigators make every effort to subject specimens to identical test conditions. In fact, the amount of metal loss from different specimens of a standard reference material tested under standard conditions in a single test apparatus can vary by as much as 25% over an extended period, with no apparent trend implying that test conditions are changing systematically.

Wear-Rate Transitions. Wear testing of many materials results in a linear variation of wear with load. It is therefore tempting to extrapolate such data to higher load situations, especially because simple wear models generally predict a linear load dependence. However, this is not a reliable practice, because one or more wear-rate transitions associated with changes in microstructure, temperature or other test conditions may occur.

Several examples of such transitions are summarized in Table 1 for wear of gray iron pins in a series of pin-on-disk tests. At low loads, the wear debris was mainly oxide, but at higher loads, the debris particles were metal flakes approximately five times larger than the oxide debris particles. Lowest wear rates in both regimes were found for ASTM type A flake graphite of size 3 to 4, randomly distributed in a pearlite matrix. Similar transitions with load are observed for many ferrous materials.

White-Etching Layers. In the wear literature, reference is frequently made to a "white layer" or "white-etching layer." This white layer is usually much harder than the adjacent material, but seldom are any other characteristics of the layer clearly defined. Some confusion arises because there are many distinct types of "white layer." For example, the white layer may be a region of martensite developed during frictional heating and rapid cooling. In other wear situations, it may be a nitride layer formed by special surface treatment or it may be a highly deformed region with a fine cell microstructure.

Wear debris is pieces of material (varied in size and shape) that become loose during wear. The structure and properties of the debris can provide clues to the type of wear involved and the prevailing conditions that existed during their formation.

Wear debris is a sensitive indicator of wear mechanisms and operating conditions. When the debris from two different test devices is similar, comparable wear situations can be expected. This principle can be extended to comparisons of published wear data (when wear debris information is given) and even to actual operating equipment. In the latter case, wear-debris analysis facilitates monitoring of wear conditions, thus providing advance warning of wear failure.

ABRASIVE WEAR

Abrasive wear involves plowing of localized surface contacts through a softer mating mate-

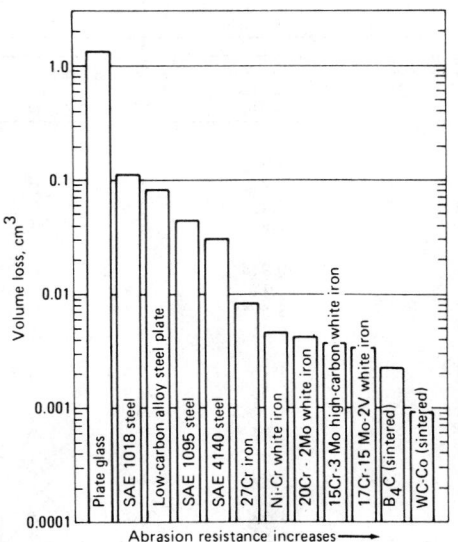

Abrasion resistance of various materials as determined by the rubber wheel abrasion test in terms of volume loss, normalized to 55 Durometer hardness of the rubber wheel. Note the logarithmic scale for wear loss.

Fig. 1. Rubber wheel abrasion test results

rial. Such wear is most frequently caused by nonmetallic materials, but metallic particles can also cause abrasion. Generally, a substance is seriously abraded or scratched only by materials that are harder than itself. Most protection against abrasion is based on this generalization, but there are exceptions.

When abrasion takes place under high surface loads, such as in ball or rod mills or in ore-crushing devices, the action is more complex. The abrasive is continually broken into smaller and smaller fragments. As a result, action that starts as gouging and surface deformation under the forces that drive large particles against the surface becomes transformed into cutting and scratching as the particles become finer and more angular.

Types of Abrasive Wear. Abrasive wear is commonly divided into the following three types:

1. *Gouging abrasion* involves removal of relatively coarse particles from the steel wearing surface and is similar to the removal of metal by machining or grinding with a coarse-grit grinding wheel. In service, this occurs on parts such as dipper teeth handling large sharp rocks, impact pulverizer hammers, and some chute liners. Gouging abrasion is at least partially responsible for wear in crusher liners.

2. *High-stress grinding abrasion* involves removal of relatively fine (microscopic) particles from the wearing surface. The pinching action of two metal surfaces causes the abrasive to fragment. Unit compressive or shear stresses are very high. Consequently, the harder abrasives, such as quartz, are capable of indenting or scratching steel with a hardness of 65 to 70 HRC. Metal may be removed from the wearing surface by microscopic gouging or by a combination of local plastic flow and microcracking. Ball milling is a good example of grinding abrasion. In machinery, such abrasion occurs where parts rub against each other in a gritty environment.

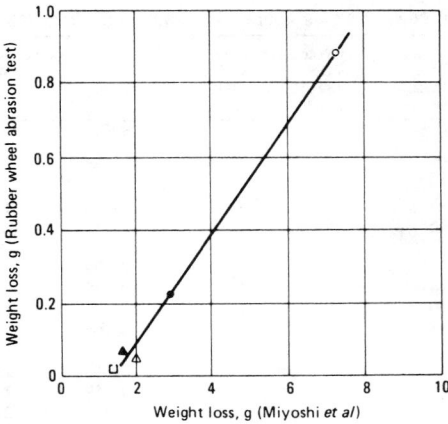

	Rubber Wheel Abrasion Test SAE 1018 steel (normalized)	Miyoshi et al Mild steel (0.18% C)
●	1% Cr cast iron (3.25% C), as-cast	2% Cr white iron (3.21% C), as-cast
△	2% Cr-4% Ni iron (3.21% C), as-cast	2.5% Cr-4% Ni iron (3.32% C), as-cast
▲	27Cr iron (2.83% C), as-cast	25% Cr iron (2.68% C), heat treated
□	15Cr-3Mo alloy (3.40% C), heat treated	15Cr-3Mo alloy (3.07% C), heat treated

Correlation between results of the rubber wheel abrasion test and Sumitomo Metal Industries abrasion testing machine used by Miyoshi *et al*.

Fig. 2. Correlation between dissimilar abrasion tests

3. *Erosion and low-stress scratching abrasion* occur by very light rubbing contact from sharp abrasive particles. The stresses are due mainly to velocity and are normally insufficient to cause much fragmentation of the abrasive. Examples of this exist on screens, chute liners handling sand, parts handling sand slurries, or parts exposed to airborne abrasives. Erosion also may occur when a flowing liquid containing suspended abrasive particles flows over a surface. Liquids themselves may cause erosion when a stream of liquid impinges directly on a surface, when fluid-flow conditions induce cavitation in the liquid adjacent to a surface, or when droplets carried by a flowing gas impact against a surface. Corrosion may be involved in the erosion process, especially in liquids or when the temperature is above room ambient.

Abrasion Testing. When abrasion testing of materials under actual service conditions is not practical, a laboratory abrasion test that closely simulates service conditions should be selected. However, in laboratory testing, the following factors must be considered:

1. The position of the abrasive — whether fixed, as in abrasive paper (two-body abrasion), or free to rotate, as in loose sand (three-body abrasion)
2. The size, shape and hardness values of the dominant abrasive (usually the hardest component in a mixture)
3. The direction and speed of relative motion during abrasion
4. The contact pressures or loads in the system.

To simulate low-stress or scratching abrasion with loose abrasive, a rubber wheel test has been found useful. Figure 1 shows test results for a wide range of materials. The least wear-resistant material in Fig. 1 wears 1500 times faster than the most wear-resistant material.

Correlations Between Dissimilar Abrasion Tests. In some applications, data from one type of abrasion test can be used to predict relative wear rates under quite different conditions. For instance, several steels and cast irons were tested in a rubber wheel device and also by movement through an abrasive slurry. Despite the obvious differences in test conditions, correlation between results from both types of tests is quite good (see Fig. 2). In both cases, the abrasive was loose and the stresses were low. If test conditions were more dissimilar, the simple correlation could not be expected to continue.

Abrasive Particle Size. Airborne dirt is a chronic cause of excessive ring wear in internal-combustion engines. Proper maintenance and design of filter equipment are essential controls. Prevention of air leaks between filters and the engine is also imperative. Contamination of lubricating oil by dirt is a less-frequent problem, but proper design and maintenance of oil filters and seals are required if such contamination is to be controlled. When the particle size of dirt in an engine reaches a certain size, wear subsides and the dirt particles often allay rather than increase friction. Test data in Fig. 3 for piston-ring wear indicate that abrasion is most severe at a particle size of about 20 μm. Apparently, the lubricating effectiveness of the oil over a wide range of viscosities has little influence on the wear peak established by particle size.

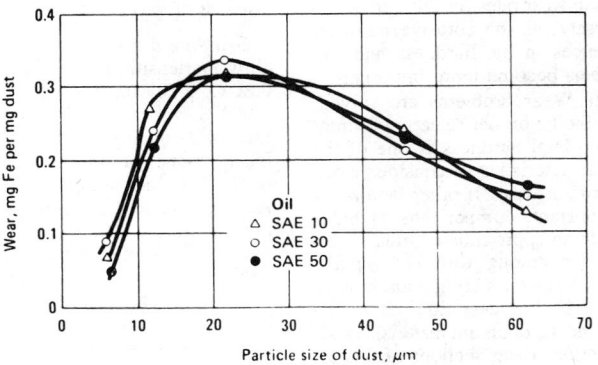

Effect of dust particle size on wear of engine piston rings lubricated by oils of three different viscosities.

Fig. 3. Wear versus abrasive particle size

Table 2. Relative hardness and abrasion resistance of several steels against hard and soft abrasives

Material	C, %	Hardness, HV	Wear resistance(a) SiC, 180 grit	Wear resistance(a) Al₂O₃, 180 grit	Flint abrasive(b) Relative hardness (d)	Flint abrasive(b) Wear resistance(a) 180 grit	Flint abrasive(b) Wear resistance(a) 36 grit	Glass abrasive(c) Relative hardness (d)	Glass abrasive(c) Wear resistance(a) 180-200 grit	Glass abrasive(c) Wear resistance(a) 30 grit
1.6Ni-1.1Cr-0.26Mo steel, hardened and tempered	0.37	352	...	1.47	0.735	1.40	...	1.32	1.95	...
		433	...	1.60	0.730	...	...	1.31	3.04	...
		498	1.62	1.55	0.828	1.49	1.69	1.49	2.81	1.86
		626	1.96	1.87	1.03	2.01	...	1.85	200	26.0
Carbon spring steel, hardened and tempered	0.43	174	...	1.31	0.509	...	...	0.914	1.57	...
		344	1.47	1.44	0.626	...	...	1.12	1.87	...
		590	1.84	1.74	0.875	1.87	2.01	1.57	...	...
		688	1.75	1.85	0.968	1.87	1.87	1.74	∞	113
Carbon spring steel, hardened and tempered	0.74	221	...	1.42	0.645	1.45	...	1.16	1.96	1.56
		503	1.75	1.70	0.849	1.71	...	1.53	4.98	2.24
		650	2.03	1.86	0.950	2.35	2.19	1.70	∞	69.4
		813	2.51	2.32	1.18	3.07	2.93	2.12	∞	∞
Austenitic 18Cr-10Ni stainless steel	0.06	158	1.80	1.65	0.610	1.65	...	1.10	1.63	...
Austenitic 12% manganese steel	1.20	220	2.17	2.11	0.850	2.39	...	1.53	41.2	5.78
Armco ingot iron	<0.02	100	1.00	1.00	0.378	1.00	1.00	0.680	1.00	1.00

(a) Volumetric wear of reference material divided by volumetric wear of test material; reference material, Armco ingot iron. (b) 1060 HV. (c) 590 HV. (d) HV_μ/HV_a, where HV_μ is maximum Vickers hardness on abraded face and HV_a is Vickers hardness of abrasive.

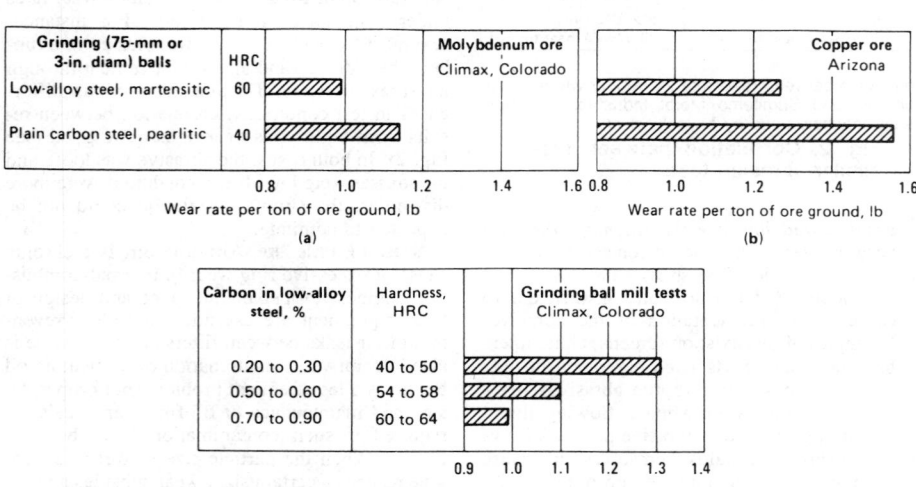

Effect of steel composition, hardness and type of material being ground on the wear rate of 75-mm (3-in.) diameter steel grinding balls.

Fig. 4. Wear of grinding balls

Hardness of the abrasive significantly influences abrasion. As the hardness of the abrasive increases, variations in wear rates for different steels diminish. Conversely, as the abrasive hardness decreases, differences in the hardness and microstructure of steels become more important.

Hardness of Metal. Wear problems are solved, wholly or in part, by the proper selection of metals (see Table 2). Metal hardness is one of the most important characteristics to consider when selecting for maximum wear. If other factors remain relatively constant, comparisons of hardness values provide an approximate guide to relative wear behavior among different metals, especially in metal-to-metal sliding and abrasive applications. However, hardness alone cannot be expected to determine the optimum material choice for a given application. Later sections of this article describe other factors to be considered when selecting a material for wear resistance, one of the most important of which is microstructure.

In the example that follows, hardness appears to be a dominant factor governing the wear resistance of various steels.

Grinding Balls (Fig. 4). Performances of low-alloy martensitic steel grinding balls containing 0.80 C, 0.70 Mn, 0.30 Si, 0.20 Cr, 0.15 Mo and 0.002 B, water quenched to 60 HRC, and pearlitic steel containing 0.80 C, 0.70 Mn and 0.30 Si, oil quenched to give a fine pearlitic structure at 40 HRC, are compared in Fig. 4(a) and (b). Adding balls to a ball mill requires no shutdown time; therefore, selection is based on cost per pound compared with wear rate, as well as availability. The low-alloy martensitic balls cost more per pound, but their high wear resistance justifies the increased cost in these and numerous other grinding applications.

In another series of tests, summarized in Fig. 4(c), three types of balls of fully hardened, low-alloy steel with varying carbon contents were

compared. Their compositions and hardnesses were as follows:

1. Low-carbon steel: 0.25 C, 0.90 Mn, 0.30 Si, 0.60 Cr, 0.20 Mo, 0.002 B at 40 to 50 HRC
2. Medium-carbon steel: 0.55 C, 0.90 Mn, 0.25 Si, 0.30 Cr, 0.05 Mo, 0.001 B at 54 to 58 HRC
3. High-carbon steel: 0.80 C, 0.70 Mn, 0.30 Si, 0.20 Cr, 0.15 Mo, 0.002 B at 60 to 64 HRC.

LUBRICATION AND LUBRICATED WEAR

Use of an effective lubricant can reduce wear in several ways. The lubricant serves to separate the rubbing surfaces, preventing or greatly reducing metal-to-metal contact. Complete separation is more readily achieved if the mating surfaces are smooth (except for microscopic asperities). The lubricant can also serve as a vehicle to remove loose wear debris from the rubbing areas before abrasive damage occurs. The lubricant, especially in a flowing system, can control increases in temperature due to frictional heating.

Hydrodynamic lubrication occurs when the viscous property of a fluid during shearing between two surfaces in relative motion produces sufficient fluid pressure to separate the surfaces and support a load. The hydrodynamic effect occurs when the fluid is forced to flow into a converging wedge configuration. In a journal bearing, this occurs when the rotating shaft surface drags a film of oil into a converging wedge formed by the displacement of the shaft center from the bearing center under applied load. Surface separation during hydrodynamic lubrication is sufficient to prevent any contact between asperities. Bearing friction is proportional to fluid viscosity and surface velocity. Minimum film thickness is proportional to viscosity and surface velocity. Hydrodynamic lubrication is also called "thick-film" and "fluid-film" lubrication.

Hydrostatic lubrication refers to a condition in which lubricant is supplied under sufficient pressure to separate the surfaces.

Elastohydrodynamic lubrication refers to the fluid-film lubrication of high-speed rolling contacts.

The contact area between the rolling elements is separated by a thin lubricant film (as small as 0.25 μm, or 10^{-5} in.). Fluid dynamic conditions exist in the thin film. Under concentrated contact, such as a ball in contact with a race, a Hertzian pressure distribution is assumed to exist between the contact areas and elastic deformation is assumed to cause a flattening in the contact zone. The fluid film is separated by approximately parallel surfaces. The entry condition is the hydrodynamic wedge and the elastically deformed region is one in which the fluid is subjected to high pressure and high shear rates. These conditions alter the apparent viscosity of the fluid (viscosity increases with pressure). Elastohydrodynamic lubrication requires very smooth surface finishes to prevent asperity contact.

Boundary lubrication denotes a range of contact conditions that occur when the hydrodynamic film collapses under high load or with too low a shear rate. Solid surface contact occurs between adsorbed soft surface films. The surface film can be an array of oriented long-chain fatty-acid molecules that have sufficient penetration resistance to prevent metal-to-metal contact. They also have a low shear resistance parallel to the surface, which allows sliding contact with reasonable friction levels (0.05 to 0.1 friction coefficient) and mild wear. Friction no longer becomes a function of viscosity or surface velocity. Boundary lubrication usually involves some metal-to-metal contact. A boundary film can be physically adsorbed or chemisorbed. The film is formed and replenished from the fluid lubricant environment. The success of boundary lubrication depends greatly on the surface chemistry of the lubricated surfaces and the operating conditions.

Solid lubrication describes a condition in which a soft solid is used instead of a fluid lubricant. The solid lubricant adheres to the rubbing surfaces and forms a low-shear-strength solid film. Natural solid lubricants include graphite and MoS_2. Once a solid film of lubricant becomes established, it tends to wear away unless continually replenished. The soft solid is suspended as a pigment in a resin or inorganic binder in solid lubricant coatings. Many of the effective solid lubricants have layer structures; however, this is not a prerequisite for classification as a solid lubricant.

Dry wear generally implies wear between surfaces without benefit of lubrication. Under certain conditions of load, speed, surface finish and temperature, dry wear can occur even when a lubricant is present. Dry friction can lead to temperature increases that are sufficient to change microstructure and mechanical properties. Materials that achieve their hardness and strength by heat treatment are most likely to be affected by frictional heat. The actual surface temperatures achieved during dry wear are virtually impossible to measure, and both the temperatures themselves and the methods used to obtain them are matters of continuous controversy.

SELECTION OF STEELS FOR WEAR RESISTANCE

Wear problems are solved, wholly or in part, by the proper selection of metals. Stress analysis and laboratory investigation usually provide only a partial solution and seldom completely resolve wear problems. Frequently, a metal is selected for trial, shaped into an experimental or service part, incorporated in an operating mechanism, and

observed for wear rate. If superior performance is proved in such limited service, the part may be incorporated in production models and run in extended service under more widely varying conditions, so that further observations can be recorded on its life and suitability. Lubrication and design of contacting components also play important roles in wear control.

Mild steel demonstrates poor wear resistance and resistance to surface damage during dry sliding. The use of mild steel in sliding surface contact requires surface treatment, such as hardening or coating, and/or selection of a "compatible" mating material such as bronze or babbitt. Where hard minerals come in contact with steel, wear is very rapid unless the steel surface is hardened or coated with a very hard material.

Steel is subject to accelerated wear in corrosive environments. Unprotected steel is also susceptible to fretting damage or formation of oxidized wear debris between two contacting surfaces in low amplitude oscillating motion. A wide variety of microstructures is possible in the heat treatment of steel or cast iron. Wear properties can be related to specific microstructures.

Relative Costs. In selection of steel for wear resistance, the total cost of the steel and its heat treatment must be considered. The following steels, which may have suitable wear-resistance properties in specific applications, are listed in order of increasing total cost.

1. Low-carbon steels, such as 1020, not heat treated
2. Simple high-carbon steels, such as 1095, not heat treated
3. Directly hardened carbon or low-alloy steels, either through hardened or surface hardened by induction or flame
4. Low-carbon or low-alloy steels that are surface hardened by carburizing, cyaniding or carbonitriding
5. Medium-carbon chromium or chromium-aluminum steels that are hardened by nitriding
6. Directly hardened high-alloy steels, such as D2 high-carbon, high-chromium tool steel (1.50 C, 12 Cr), that contain particles of free carbide
7. Precipitation-hardening stainless steels (mainly for applications involving heat or corrosion, or both, as well as wear)
8. Specialty steels produced by powder metallurgical or mechanical alloying techniques
9. Alloy carbides bonded by steel matrices.

Other ferrous materials, such as high manganese austenitic steels and various classes of cast irons, are also widely used for wear-resistance applications.

Depth of Hardened Regions. Skids, grinding rods, chute liners and similar parts may be considerably reduced in section before being replaced. In such parts, a more expensive deep-hardening steel may be more economical than a shallow-hardening steel. For example, a round bar 64 mm (2^1/$_2$ in.) in diameter with a surface hardness of 50 HRC may be made of either a water-quenched 1040 or an oil-quenched 5160 steel. However, by the time the bar has been worn to three-fourths of its original diameter (about 48 mm, or 1^7/$_8$ in.), the 1040 steel will have a surface hardness of about 25 HRC and thus would wear at a much faster rate than the 5160 steel with its hardness of about 37 HRC at the same location.

Toughness. Wear resistance tends to increase with hardness, but it decreases as toughness increases.

This is an important relationship in applications that require both wear resistance and impact resistance.

A correlation of wear resistance and toughness for a variety of ferrous alloys is shown in Fig. 5. The scatter arises, at least in part, from microstructural effects. For example, point 22B refers to AISI 4340 steel, quenched and tempered at a high temperature of about 650 °C (1200 °F) to produce fine carbides in a ferrite matrix. Point 22A represents the same steel, normalized to produce fine pearlite; point 22C represents a quenched sample tempered at 205 °C (400 °F), a relatively low tempering temperature. Steels in the lower band of Fig. 5 combine toughness with wear resistance; these are mainly the austenitic manganese steels. Figure 5 indicates that for most ferrous alloys there is a trade-off between wear resistance and toughness. In some alloys, altering carbon content is a simple way to adjust these properties.

Carbon Content. The wear resistance of ferritic steel is improved by hardening, either throughout the section or superficially. The maximum hardness depends on the carbon content of the steel and the amount of martensite (efficiency of quenching), as shown in Fig. 6.

Standard hardness measurements may indicate that a martensitic steel is largely transformed, although it may retain some austenite. Exposure to ultra-low temperatures (followed by tempering) can help complete the transformation to martensite and improve wear resistance. Because martensite is a metastable structure, it begins to transform to more stable structures as the temperature is raised. Consequently, martensitic steels are not suitable for wear resistance at elevated temperatures or for applications in which the heat of friction can raise the temperature significantly. They should not be used at temperatures above 200 °C. Special alloy steels, such as tool steels or martensitic stainless steels, are appropriate for service at higher temperatures. The thermal instability of martensite should also be considered during finishing operations (such as grinding), when a heat-affected zone could be produced at the surface. The resultant tempering effects could be localized or general; in either case, wear resistance is likely to be reduced.

Carbon content also affects hardness and wear resistance through formation of various simple and complex carbides. Wear properties depend on the type, amount, shape, size and distribution of carbides present, as well as the properties of the matrix (hardness, toughness and stability). Despite this complexity, correlation of relative wear rates with carbon content is possible. An example of gouging wear is shown in Fig. 7. Because most carbides are relatively stable compared with martensite, wear resistance achieved with the aid of these hard microconstituents is retained at higher temperatures.

Hardness and Microstructure. The frequent use of bulk hardness as a guide to abrasive wear resistance is supported by the data shown in Fig. 8 for annealed unalloyed metals. These data were obtained using abrasive cloth (two-body abrasion) with an abrasive hardness much greater than that of the metal samples. The data points are approximate; the experimental scatter of the measurements is not shown.

Corresponding correlations with other properties related to hardness (such as elastic modulus) have also been presented. In all cases, if the metals are unalloyed, a simple correlation is ob-

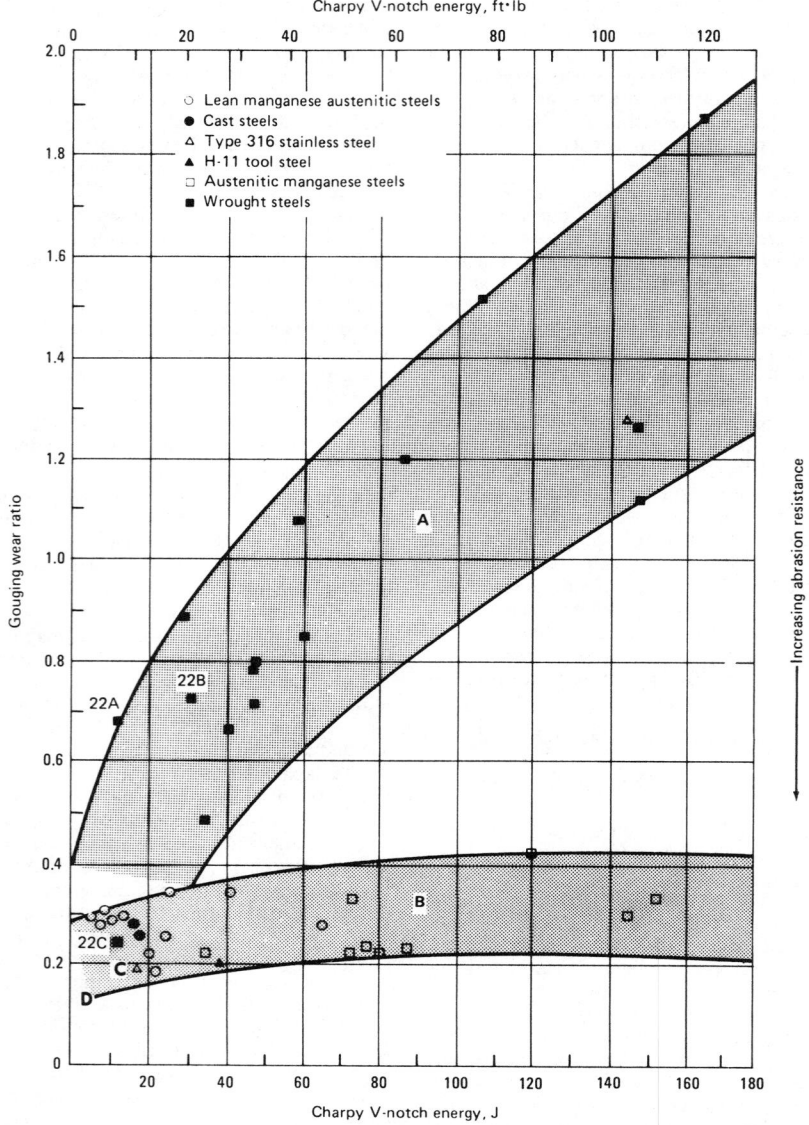

Relationship between resistance to gouging abrasion and toughness of various materials. Area A, wrought and cast low-alloy steels; area B, austenitic manganese steels; area C, variety of heat treated steels; area D, high-chromium white cast irons.

Fig. 5. Wear resistance versus toughness

tained for controlled tests of two-body abrasion. Different crystal structures would be expected to yield different correlations, but the data in Fig. 8 do not show such an effect.

Care must be exercised in extending the simple hardness correlation to metals containing impurities or solutes, or to more complicated alloys. Figure 9 shows how wear resistance and hardness correlate for various classes of materials. The linear plot shown for pure metals in Fig. 8 is repeated as the steep line in Fig. 9. Another straight line describes brittle ceramics reasonably well. The differences in bonding type may account for these two distinct lines.

WEAR AND MICROSTRUCTURE

The preceding discussion established the importance of hardness and microstructure as factors in resistance to wear. Metallurgically, hardness and microstructure are commonly interrelated. Increasing the carbon content of a carbon steel, for example, results in microstructural alteration that increases as-quenched hardness and decreases ductility or toughness.

Wear Resistance at Constant Hardness. Very few tests of the comparative wear resistance of different steels at a constant hardness have been reported. However, in one series of tests for seizure resistance under heavy loads, various steels were selected to provide a wide range of mechanical properties and microstructures when heat treated to 40 HRC. The steels, which are listed in Table 3, included a plain carbon steel, a tool steel, an alloy steel, a martensitic stainless steel, two precipitation-hardening steels and a maraging steel. Details of heat treatment and descriptions of the various microstructures are also given in Table 3.

Effect of Carbides on Wear. The amount, size and distribution of carbides in a steel microstructure have a distinct influence on wear resistance. For the most part, wear resistance increases as the amount or size of carbide particles at the wear surface increases. This effect is most often attributed to the proportion of the surface area occupied by the carbides, and is more pronounced when gouging abrasion is not a factor (see the specific example that follows).

Wear Plate in a Textile Machine (Fig. 10). Experience with a wear plate over which textile fibers were drawn proved that an increase in the size and amount of carbides can decrease wear.

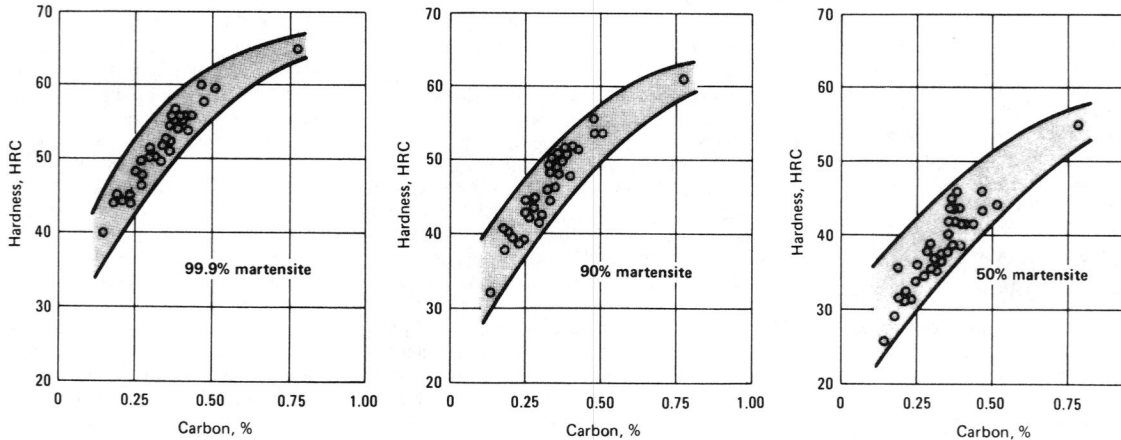

Fig. 6. Effect of carbon content and percentage martensite on the hardness of as-quenched steel

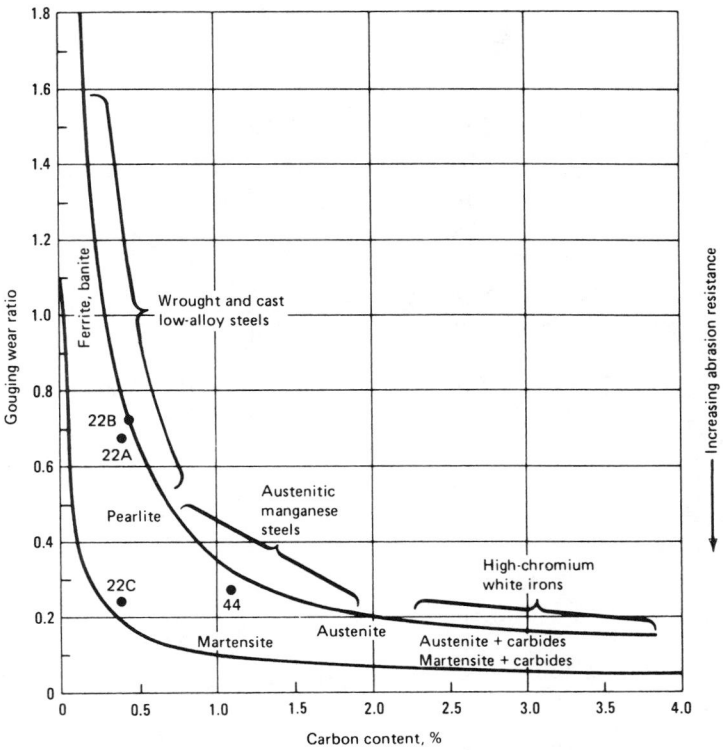

Fig. 7. Relation between gouging wear and carbon content for various types of steel and cast iron

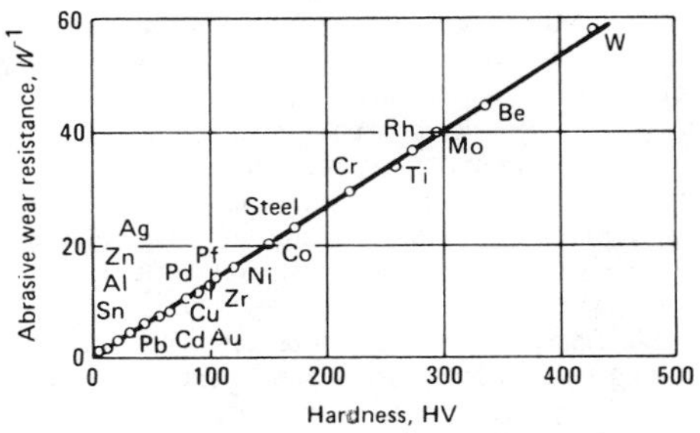

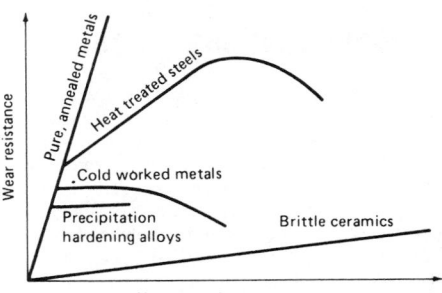

Fig. 8. Abrasive wear resistance versus hardness for annealed unalloyed metals and steel

Fig. 9. Wear versus hardness for several classes of material

A wear plate made of 52100 steel hardened to a minimum of 63 HRC had given satisfactory service. When new fibers with more abrasive qualities were processed, the plates wore out in three months. Plates were then made from D2 tool steel (1.50 C, 12.00 Cr, 1.00 Mo, 0.75 V). These were still serviceable after one year of use under the same operating conditions.

The difference in wear resistance was attributed to the presence of hard carbide particles, their hardness not being registered by the Rockwell test. The difference in hardness and microstructure of these two steels is shown in Fig. 10. The Knoop indentation test proved that the large carbide particle is much harder (1960) than the matrix (850). The latter is about the same as the surface hardness of 52100 steel.

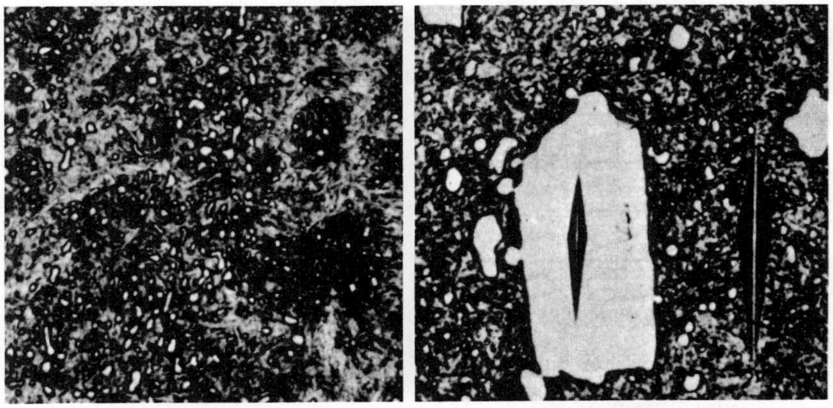

HARDNESS EVALUATION

With some exceptions, the data in this article have indicated that hardness is the most significant factor in controlling wear of steel. Conven-

SAE 52100 steel

Type D2 tool steel

The 52100 steel was hardened to 63 HRC (850 HK). Type D2 high-carbon high-chromium tool steel also was hardened to 63 HRC (850 HK on matrix); hardness of carbide particle shown is 1960 HK. Both micrographs at magnification 1000×.

Fig. 10. Carbide size and distribution in 52100 steel and D2 tool steel

Table 3. Steels used in seizure-resistance tests

Type of steel	Type of heat treatment	Details of heat treatment(a)	Description of structure
1040	Quenched and tempered	830 °C, 20 min, WQ; 415 °C, 4 h, AC	Tempered martensite with very fine carbide distribution
W1 (I)	Quenched and tempered	790 °C, 20 min, WQ; 440 °C 4 h, AC	Tempered martensite with very fine carbide distribution
W1 (II)	Quenched and tempered	900 °C, 20 min, OQ; 470 °C, 4 h, AC	Primary carbides (4 to 8 μm) in tempered martensite matrix
4340	Quenched and tempered	830 °C, 20 min, OQ, 485 °C, 4 h, AC	Tempered martensite with very fine carbide distribution
440C	Quenched and tempered	1065 °C, 20 min, OQ, 580 °C, 2 h, AC	Large primary Cr and Fe carbides (20 μm) in tempered martensitic matrix
15-5 PH	Solution treated and aged	1040 °C, 30 min, OQ; 530 °C, 4 h, AC	Martensite
17-4 PH	Solution treated and aged	1040 °C, 30 min, OQ; 530 °C, 4 h, AC	δ-ferrite stringers in martensitic matrix
18Ni(250) (I)	Solution treated and underaged	815 °C, 60 min (in vacuum), AC; 320 °C, 2 h, AC	Lath martensite
18Ni(250) (II)	Solution treated and overaged	815 °C, 60 min (in vacuum), AC; 630 °C, 2 h, AC	Lath martensite and very fine reverted austenite

(a) WQ, water quenched; OQ, oil quenched; AC, air cooled.

tional methods of hardness evaluation such as Rockwell and Brinell tests may not always provide sufficient information or may be insensitive to minor, but significant, aberrations. Small amounts of retained austenite, decarburization or variation in microconstituents, and other factors that are not always evaluated in conventional hardness tests, may have a marked influence on wear resistance.

Microhardness testing is more effective for detecting local differences. Hardness traverses taken on polished cross sections may reveal conditions not registered by other means.

Elevated-Temperature Properties of ─ Constructional Steels ─

CARBON AND LOW-ALLOY constructional steels in several forms are used extensively in a variety of applications that involve exposure to elevated temperatures and, possibly, reactive environments. Typical applications include equipment for generation of electric power, chemical-process equipment and aircraft power plants.

Within the context of this article, low-alloy steels are those that contain no more than 10% total alloy content.

The term "elevated temperature" can be used to mean any temperature above room temperature. As applied to the carbon and low-alloy steels considered in this article, elevated temperature relates primarily to the temperature range of almost 370 to 650 °C (700 to 1200 °F). The properties of carbon and low-alloy steels vary considerably over this temperature range. Properties of carbon and low-alloy steels at temperatures from 800 to 1200 °C (1500 to 2200 °F) are important in mill processing (rolling, drawing, forging and heat treating).

Some devices, such as steam power plants and gas turbines, operate with increasing efficiency as the operating temperature is raised. The maximum temperature at which such equipment can operate is limited by the capabilities of the materials from which it is made. Resistance of steels to degradation of properties at elevated temperatures usually increases with alloy content; the cost of these steels also increases with alloy content. Thus, selection of steels to be used at elevated temperatures generally involves a compromise between the higher efficiencies obtained at higher operating temperatures and the cost of the equipment, including materials, fabrication, replacement and down-time costs.

To illustrate the tonnage requirements for carbon and low-alloy steels in industrial construction, 1500 tons of pressure tubing were required for the construction of a single 500-MW coal-fired generating plant. The quantities of the various carbon and low-alloy steels used in this pressure tubing were as follows:

Steel type	Tons	% of total tonnage
Carbon	540	36
C-$\frac{1}{2}$Mo	150	10
1$\frac{1}{4}$Cr-1Mo	495	33
2$\frac{1}{4}$Cr-1Mo	150	10
9Cr-1Mo	165	11

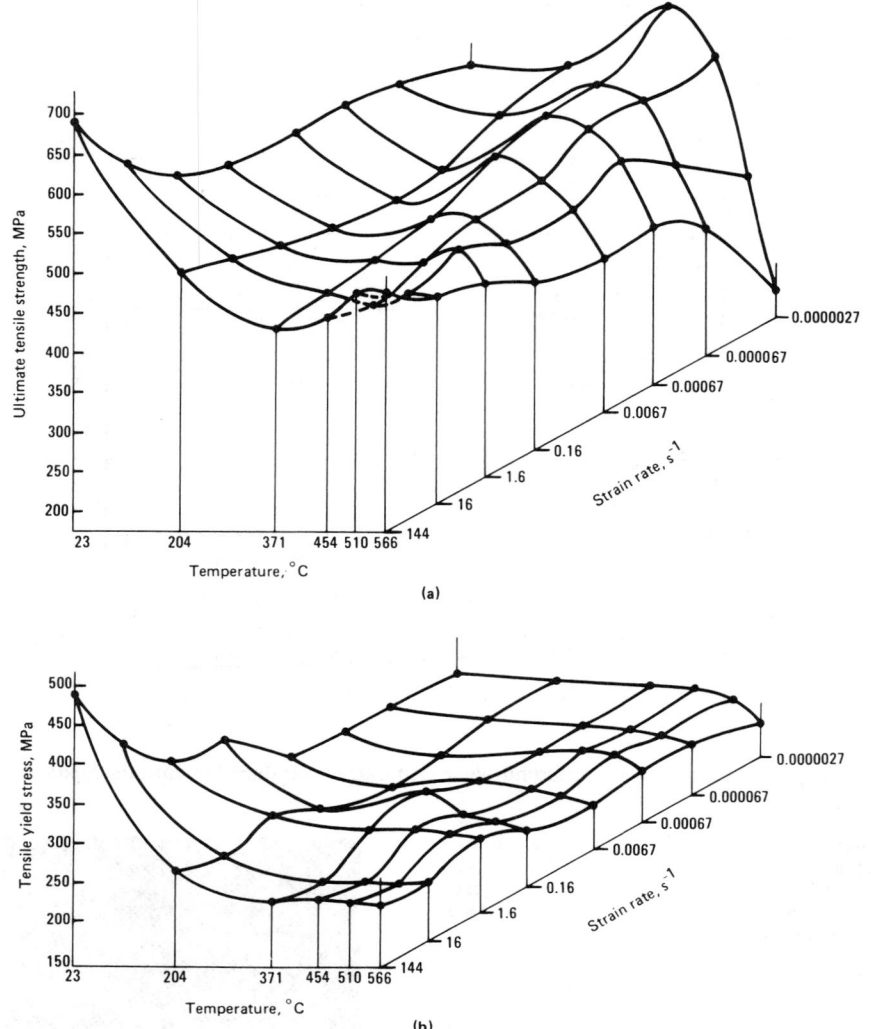

Pictorial representation of (a) tensile strength and (b) yield strength of 2$\frac{1}{4}$Cr-1Mo steel tested at various temperatures and strain rates.

Fig. 11. Effect of test temperature and strain rate on strength of 2$\frac{1}{4}$Cr-1Mo steel

Measurement of Elevated-Temperature Mechanical Properties. The types of tests used to evaluate the mechanical properties of steels at elevated temperatures can be categorized into six groups: (*a*) short-term elevated-temperature tests; (*b*) long-term elevated-temperature tests; (*c*) short-term and long-term tests following long-term exposure to elevated temperature; (*d*) thermal shock tests, including thermal-fatigue tests; (*e*) time-dependent fatigue tests; and (*f*) qualification tests under simulated service conditions.

Short-term elevated-temperature tests include the elevated-temperature tensile test (described in ASTM Recommended Practice E21), a test for elastic modulus (ASTM E231), compression tests, pin bearing load tests and the hot hardness test. The mechanical properties determined by means of the tensile test include ultimate tensile strength, yield strength, percent elongation and percent reduction in area. Because elevated-temperature tensile properties are sensitive to strain rate, these tests are conducted at carefully controlled strain rates. Tensile-strength data obtained on annealed specimens of 2$\frac{1}{4}$Cr-1Mo steel at various temperatures and at strain rates ranging from 2.7 μm/

m · s to 144 m/m · s are shown in three-dimensional representation in Fig. 11.

In designing components that are to be produced from low-alloy steels and to be exposed to temperatures up to 370 °C (700 °F), the yield and ultimate strengths at the maximum service temperature may be used much as they would be used in the design of components for service at room temperature. Certain codes require that appropriate factors be applied in calculating allowable stresses.

Elevated-temperature values of elastic modulus may be determined during tensile testing or dynamic testing by measuring the natural frequency of a test bar at the designated test temperature. Figure 12 shows values of elastic modulus at temperatures between room temperature and 650 °C (1200 °F) for several low-alloy steels, determined during "static" tensile loading and during dynamic loading.

Compression tests and pin bearing load tests (ASTM E209 and E238) may be used to evaluate materials for applications in which the components will be subjected to these types of loading at elevated temperatures. Hot hardness tests may

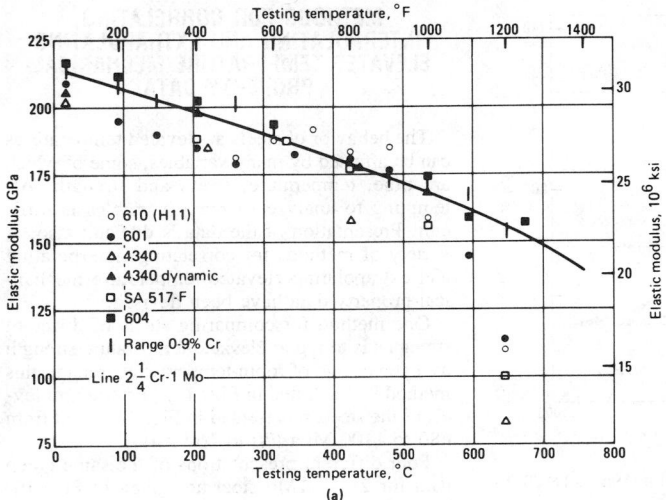

(a)

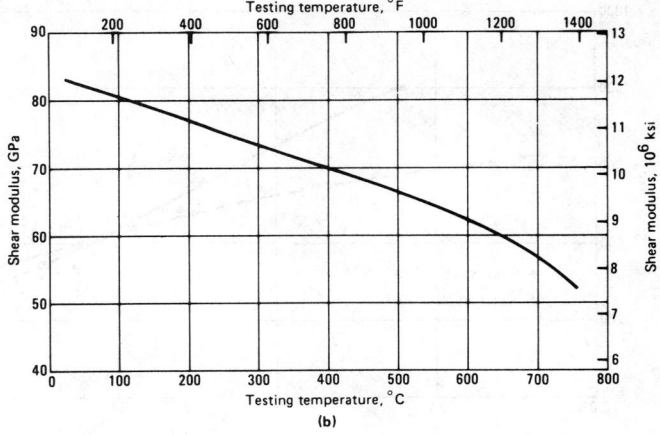

(b)

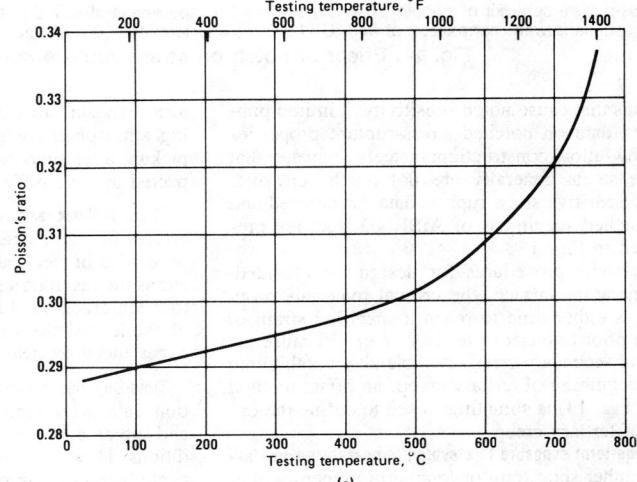

(c)

(a) Effect of test temperature on elastic modulus for several steels commonly used at elevated temperatures. Dynamic measurements of elastic modulus were made by determining the natural frequencies of test specimens; static measurements were made during tensile testing. (b) Effect of test temperature on shear modulus of 2¼Cr-1Mo steel. (c) Effect of test temperature on Poisson's ratio of 2¼Cr-1Mo steel.

Fig. 12. Effect of test temperature on elastic modulus, shear modulus and Poisson's ratio

be used to evaluate materials for elevated-temperature service and may be applied to qualification of materials in the same way in which room-temperature hardness tests are applied.

Components for many elevated-temperature applications are joined by welding. Elevated-temperature properties of both the weld metal and the heat-affected zones may be determined by the same methods used in evaluating the properties of the base metal.

Long-term elevated-temperature tests include creep, creep-rupture and stress-rupture tests (ASTM E139); notched-bar rupture tests (ASTM E292); and relaxation tests. Any of these tests may be combined with exposure to an aggressive environment to evaluate the effect of the environment on creep behavior.

Creep is defined as the time-dependent strain that occurs under constant load at constant temperature. Typical creep behavior is illustrated in Fig. 13. The initial creep strain which occurs at a diminishing rate is designated as primary (first-stage) creep, that which occurs at a minimum and almost constant rate as secondary (second-stage) creep, and that which occurs at an accelerating rate as tertiary (third-stage) creep. If only time to rupture is measured, the test is a stress-rupture test. If the test is conducted to measure only the deformation behavior at the elevated temperature, it is a creep test. If both deformation and time to rupture are measured, it is a creep-rupture test.

An increase in either stress or temperature accelerates the creep process; thus, the minimum creep rate is increased. If either stress or temperature is increased beyond the design levels for boiler tubes or heat exchangers, for example, the increased deformation can alter the temperature distribution in the system, which may cause local zones to be overheated, eventually resulting in failure.

For those alloys in which failure occurs before a well-defined start of tertiary creep, it is useful to use notched specimens or specimens with both smooth and notched test sections (with the cross-sectional area of the notch equal to that of the smooth test section). If the material is notch sensitive, the specimen will fail in the notch before failure occurs in the smooth section. The purpose is to identify notch-sensitive materials and con-

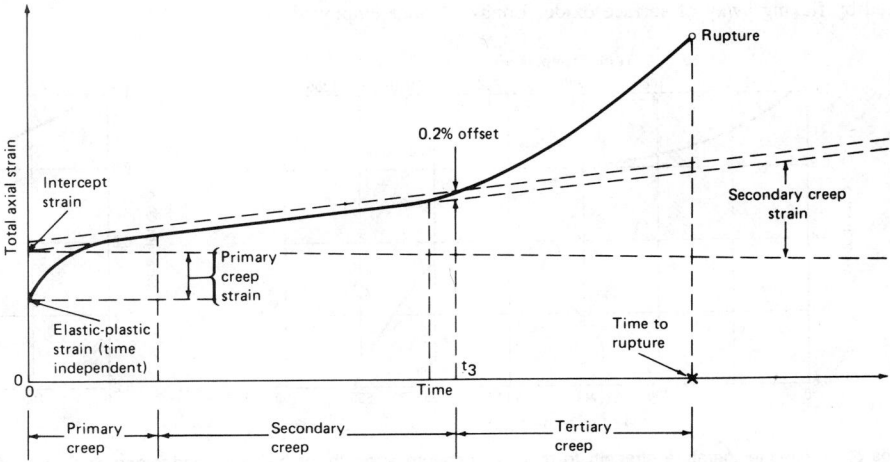

Fig. 13. Schematic representation of creep behavior

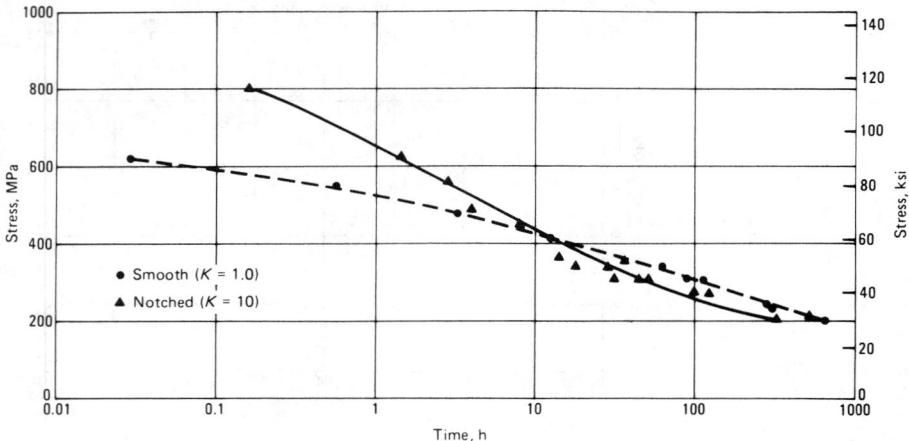

Stress-rupture behavior of smooth ($K = 1.0$) and notched specimens of AISI 603 steel tested at 595 °C (1100 °F). All specimens were normalized at 980 °C (1800 °F) and tempered 6 h at 650 °C (1250 °F).

Fig. 14. Effect of notch on stress-rupture behavior

ditions that cause notch sensitivity. Limited published data on notched stress-rupture properties of low-alloy constructional steels indicate that these steels generally are not notch sensitive. Representative stress-rupture data for notched and unnotched specimens of AISI 603 steel are presented in Fig. 14.

In some procedures for design for elevated-temperature service, the critical materials property is either time to reach a specified strain or time prior to onset of tertiary creep. Because the end of secondary creep often blends smoothly into the beginning of tertiary creep, an offset method (see Fig. 13) is sometimes used to define the onset of tertiary creep.

Long-term exposure to elevated temperature may affect either short-term or long-term properties. For example, exposure of plain carbon steels to elevated temperatures (such as 650 °C, or 1200 °F) can cause overtempering or spheroidization. Such a change in microstructure can logically be expected to change the mechanical properties of the steel. Other metallurgical instabilities may affect the mechanical properties of steels.

Long-term exposure to elevated temperature may also affect the properties of steels through alteration of the steel by the environment. Carbon steels are subject to oxidation in air, and the cross-sectional area of a steel part may be reduced by flaking away of surface oxide. Long-

term exposure to carbon-bearing or sulfur-bearing atmospheres will result in carbon or sulfur pickup; after such exposure, the steel can be expected to have different properties.

Thermal shock tests may be used to determine the effects of rapid changes in temperature on the properties of steel parts. Thermal shock may occur as the result of a single change in temperature (quench cracking, for example), or as the result of cyclic variations in temperature such as those experienced by heat treating fixtures.

Time-Dependent Fatigue Tests. Creep and relaxation data are obtained at constant temperatures and either constant-load or constant-strain conditions. However, in many service applications, cyclic variations in either temperature or applied stress (or strain) may occur. These variations may occur randomly or in regular, uniform cycles. In order to determine the effect of cyclic loading superimposed on a constant load at elevated temperatures, several types of fatigue testing may be employed: constant alternating stress, constant alternating strain, tension-tension loading with the stress ratios greater than zero, and special waveforms that provide specific holding times at maximum load. Results of these tests show what factors are most contributory to deformation and fracture of the specimens for the testing conditions employed.

METHODS FOR CORRELATING, INTERPOLATING AND EXTRAPOLATING ELEVATED-TEMPERATURE MECHANICAL-PROPERTY DATA

The behavior of steels at elevated temperatures can be affected by many variables, some of which are time, temperature, stress and strength. Attempting to analyze so many variables is difficult. Presentation of the data is difficult also. A variety of methods for correlating, interpolating and extrapolating elevated-temperature mechanical-property data have been devised.

One method for comparing steels of different strengths is to report elevated-temperature strength as a percentage of room-temperature strength; this method is illustrated in Fig. 15. The strength levels of the steels represented in Fig. 15 varied from 480 to 1100 MPa (70 to 160 ksi).

Four different presentations of the same creep data for 2^1/$_4$Cr-1Mo steel are given in Fig. 16. In parts (a), (b) and (c), only the creep strain is plotted. In the isochronous stress-strain diagram (part d), total strain is used. The over-all format of Fig. 16(d) is particularly useful in design problems where total strain is a major consideration; the linear scales make unjustified extrapolation a little more difficult.

EFFECTS OF COMPOSITION ON ELEVATED-TEMPERATURE MECHANICAL PROPERTIES OF CARBON AND LOW-ALLOY STEELS

The mechanical properties of carbon and low-alloy steels are determined primarily by composition and heat treatment. The effects of alloying elements in annealed, normalized-and-tempered, and quenched-and-tempered steels are as follows:

Carbon increases both strength and hardenability of steel at room temperature. In plain carbon and carbon-molybdenum steels intended for elevated-temperature service, carbon content usually is limited to about 0.20%; in some classes of tubing for boilers, however, carbon may be as high as 0.35%. For chromium-containing steels, carbon content usually is limited to 0.15%. Carbon increases short-term tensile strength, but does not add appreciably to creep resistance at temperatures above 540 °C (1000 °F) because carbides eventually become spheroidized at such temperatures.

Manganese, in addition to its normal function of

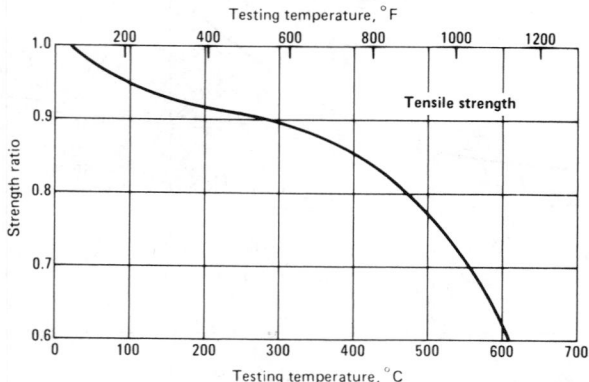

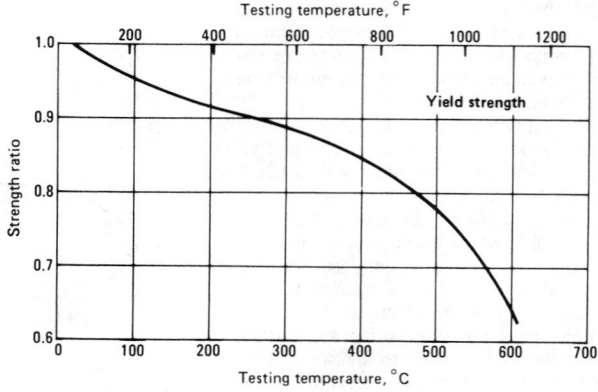

Ratios of elevated-temperature strength to room-temperature strength for hardened-and-tempered 2^1/$_4$Cr-1Mo steel tempered to room-temperature tensile strengths ranging from 480 to 1100 MPa (70 to 160 ksi).

Fig. 15. Ratios of elevated-temperature strength to room-temperature strength

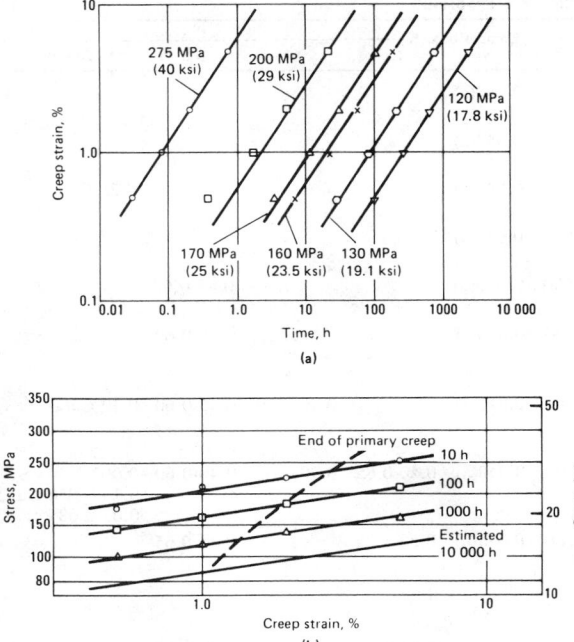

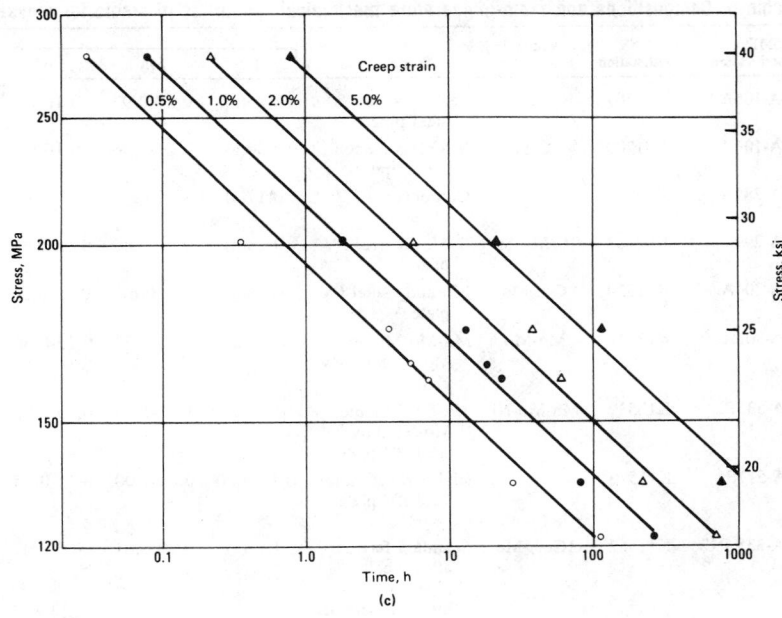

Creep behavior of 2¼Cr-1Mo steel tested at 540 °C (1000 °F). (a) Creep strain–time plot; constant-stress lines have been drawn parallel. (b) Stress–creep strain plot. (c) Stress-time plot; constant-strain lines have been drawn parallel. (d) Isochronous stress-strain curves.

Fig. 16. Analysis of creep data

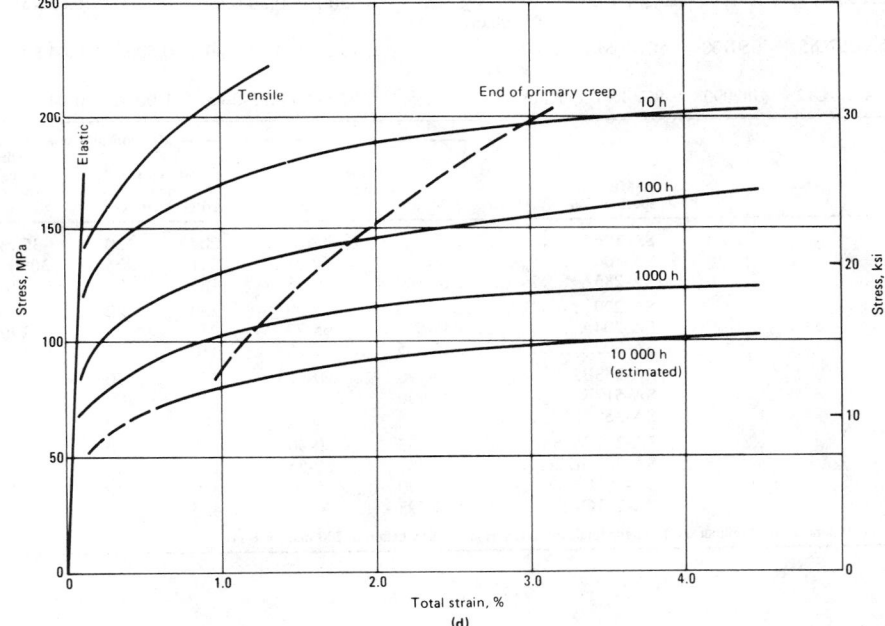

preventing hot shortness by forming dispersed manganese sulfide inclusions, appears to enhance the effectiveness of nitrogen in increasing the strength of plain carbon steels at elevated temperatures.

Silicon increases the elevated-temperature strength of steel; it also increases the resistance to scaling of the low-chromium steels in air at elevated temperatures.

Chromium, in small amounts (about 0.5%), is a carbide former and stabilizer; in larger amounts (up to 9% or more), it increases the resistance of steels to scaling in air. In the lower-carbon grades, increasing chromium content does not increase resistance to deformation at elevated temperatures.

Molybdenum, even in small amounts (0.1 to 0.5%), increases the resistance of these steels to deformation at elevated temperatures. It is a carbide stabilizer and prevents graphitization. For certain ranges of stress and temperature, the dissolving of iron carbide and the concurrent precipitation of molybdenum carbide cause strain hardening in these steels.

Vanadium is added to some of the higher-carbon constructional steels to provide additional resistance to tempering.

Niobium sometimes is added to these steels to increase their strength through formation of carbides.

CARBON AND LOW-ALLOY STEELS FOR ELEVATED-TEMPERATURE SERVICE

Carbon and low-alloy steels used for elevated-temperature service usually are identified by either AISI designation; AMS, ASME or ASTM specification number; nominal composition; or trade name. These steels also have been assigned numbers in the Unified Numbering System. In addition, there is a series of Military and Federal specifications covering many of these steels.

Steel products manufactured for use under the ASME Boiler and Pressure Vessel Code must comply with provisions of the appropriate ASME specification. Each specification includes information on ranges and limits of composition, dimensions and tolerances, minimum mechanical properties and other functional requirements. The designations applied to these products include the letters "SA," the number of the specification and possibly other letters or numbers to distinguish among the various types, grades and classes within a single specification. Most ASTM specifications are identical to the ASME specification of

the same number except that the ASTM designations begin with the letter "A." Some examples of ASME specifications for elevated-temperature steels, and their compositions and typical room-temperature mechanical properties, are given in Table 4.

Aerospace Material Specifications, as the name suggests, are specifications for products intended for the aerospace industry. The nominal compositions, typical applications and typical mechanical properties of steels often identified by AMS numbers are given in Table 5.

The AISI designation for steels intended for elevated-temperature service is a three-digit number beginning with a 6, such as 601. The AISI designations for comparable AMS designations are also included in Table 5.

Table 4. Compositions and room-temperature mechanical properties of steels for elevated-temperature service

ASME specification	UNS designation	Nominal composition	Product form	C	Mn	Si	P	S	Cr	Ni	Mo	Others
SA-106A	K02501	C	Seamless carbon steel pipe	0.25(a)	0.27-0.93	0.10(b)	0.048(a)	0.058(a)	...	...	...	...
SA-106B	K01700	C-Si	Seamless carbon steel pipe	0.30(a)	0.29-1.06	0.10(b)	0.048(a)	0.058(a)	...	...	...	...
SA-285A	K03006	C	Carbon steel PV plate	0.17(a)	0.90(a)	...	0.035(a)	0.045(a)	...	...	...	0.25 Cu(a)
SA-299	K02803	C-Mn-Si	C-Mn-Si steel PV plate	0.28(a)	0.90-1.40	0.15-0.30	0.035(a)	0.040(a)	...	...	...	...
SA-204A	K11820	C-$\frac{1}{2}$Mo	Mo alloy steel PV plate	0.18(a)	0.90(a)	0.15-0.30	0.035(a)	0.040(a)	...	...	0.45-0.60	...
SA-302A	K12021	Mn-Mo	Mn-Mo-Mn and Mo-Ni-alloy PV plate	0.20(a)	0.95-1.30	0.15-0.30	0.035(a)	0.040(a)	...	...	0.45-0.60	...
SA-533B2	K12539	Mn-Mo-Ni	Mn-Mo-Mn and Mo-Ni alloy steel PV plate	0.25(a)	1.15-1.50	0.15-0.30	0.035(a)	0.040(a)	...	0.40-0.70	0.45-0.60	0.10 Cu(a)
SA-517F	K11576	...	High-strength alloy steel PV plate	0.10-0.20	0.60-1.00	0.15-0.35	0.035(a)	0.040(a)	0.40-0.65	0.70-1.00	0.40-0.60	0.002-0.006 B, 0.15-0.050 Cu, 0.03-0.08V
SA-335 P12	K11562	1Cr-$\frac{1}{2}$Mo	Seamless ferritic alloy steel pipe for high-temperature service	0.15(a)	0.30-0.61	0.50(a)	0.045(a)	0.045(a)	0.50-1.25	...	0.44-0.65	...
SA-217WC6	J12072	1$\frac{1}{4}$Cr-$\frac{1}{2}$Mo	Alloy steel castings	0.20(a)	0.50-0.80	0.60(a)	0.04(a)	0.045(a)	1.00-1.50	...	0.45-0.65	...
SA-387Gr22	K21590	2$\frac{1}{4}$Cr-1Mo	Cr-Mo alloy steel PV plate	0.15(a)	0.30-0.60	0.50(a)	0.035(a)	0.035(a)	2.0-2.5	...	0.90-1.10	...
SA-387Gr5	S50100	5Cr-$\frac{1}{2}$Mo	Cr-Mo alloy steel PV plate	0.15(a)	0.30-0.60	0.50(a)	0.040(a)	0.030(a)	4.0-6.0	...	0.45-0.65	...
SA-217C12	J82090	9Cr-1Mo	Alloy steel castings	0.02(a)	0.35-0.65	1.00(a)	0.04(a)	0.045(a)	8.0-10.0	...	0.90-1.20	...

ASME specification	Tensile strength MPa	ksi	Yield strength, min MPa	ksi	Minimum elongation in 50 mm or 2 in., %	Minimum reduction in area, %
SA-106A	390	48(b)	207	30	35(c), 25(d)	...
SA-106B	415	60(b)	241	35	30(c), 16.5(d)	...
SA-285A	310-380	45-55	165	24	27(e), 30	...
SA-299	515-620	75-90	290	42	16(e)	...
SA-204A	445-530	65-77	255	37	19(e), 23	...
SA-302A	515-655	75-95	310	45	15(e), 19	...
SA-533B2	620-790	90-115	475	70	16	...
SA-517F	795-930	115-135	689	100	16	35-45
SA-335P12	415	60(b)	207	30	30(c), 20(d)	...
SA-217WC6	485-655	70-90	275	40	20	35
SA-387Gr22-1	415-585	60-85	207	30	18(e), 45	40
SA-387Gr5-2	515-690	75-100	310	45	18(e), 22	45
SA-217C12	620-795	90-115	415	60	18	35

(a) Maximum. (b) Minimum. (c) Longitudinal. (d) Transverse. (e) Elongation in 200 mm, or 8 in.

Table 5. Compositions and mechanical properties of AISI steels for elevated-temperature service

AISI designation	AMS designations	Commercial designation	UNS designations	Typical applications	C	Mn	Si	Cr	Mo	V
601	6304	...	K14675	Bolting and structural parts	0.46	0.60	0.26	1.00	0.50	0.30
602	6302, 6385, 6458	17-22 AS	K23015	Bolting and structural parts	0.30	0.55	0.65	1.25	0.50	0.25
603	6303, 6436	17-22 AV	K22770	Turbine rotors and aircraft parts	0.27	0.75	0.65	1.25	0.50	0.85
610	6437, 6485	H11 mod	T20811, K74015	Ultrahigh-strength components	0.40	0.30	0.90	5.00	1.30	0.50

AISI designation	Yield strength MPa	ksi	Tensile strength MPa	ksi	Elongation in 50 mm or 2 in., %	Reduction in area, %	1000 h °C	1000 h °F	10 000 h °C	10 000 h °F	1 µm/m·h °C	1 µm/m·h °F	0.1 µm/m·h °C	0.1 µm/m·h °F
601	710	103	855	124	29	61	620	1150	595	1100	...	...	...	...
602	745-930	108-135	880-1060	128-154	16-21	53-63	625	1160	590	1090	...	...	554	1030
603	1000	145	1100	160	17	52	650	1200	613	1135	...	...	566	1050
610	1480	215	1805	262	10	36	630	1170	595	1100	560	1040	538	1000

The columns "Temperature at which 70 MPa (10 ksi) will cause rupture in" span the 1000 h and 10 000 h columns. The columns "Temperature to produce min creep rate at 70 MPa (10 ksi)" span the 1 µm/m·h and 0.1 µm/m·h columns.

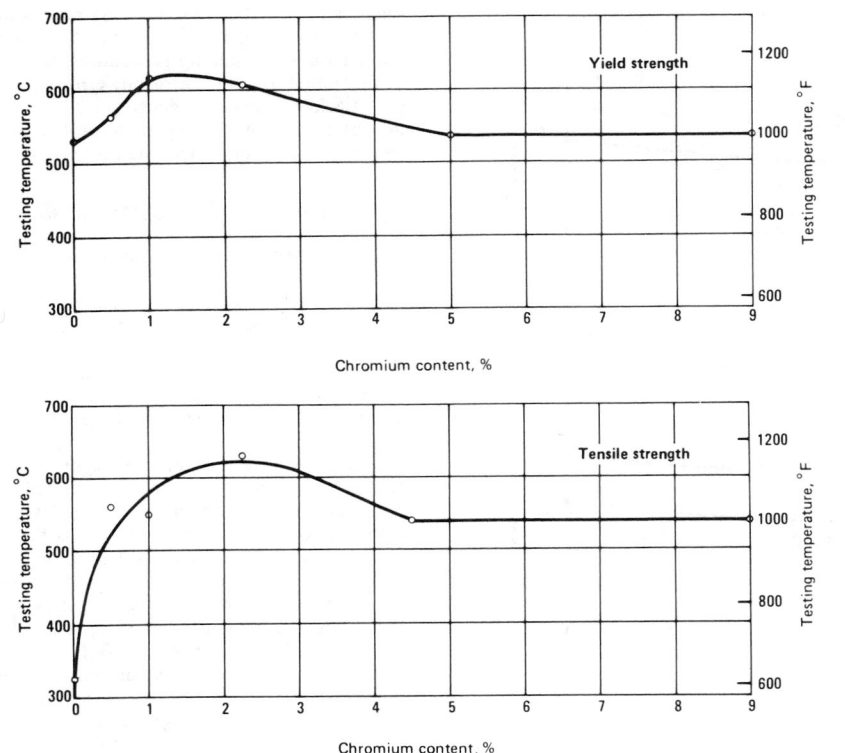

Test temperature required to reduce tensile strength and yield strength to 60% of their room-temperature values for Cr-Mo steels containing 0.5 to 1.0% Mo and the indicated amount of chromium.

Fig. 17. Effect of chromium content on strength

ELEVATED-TEMPERATURE PROPERTIES OF CARBON AND LOW-ALLOY STEELS

Typical elevated-temperature mechanical properties of the carbon and low-alloy steels identified in the preceding tables are presented in this section. Although the values given for elevated-temperature properties usually are based on results obtained from specimens from more than one heat, for only a very few of these steels has the number of tests and heats been sufficient to permit the determination of upper and lower confidence limits for design.

Short-term elevated-temperature mechanical properties of primary interest in the design of structures for elevated-temperature service are ultimate tensile strength, yield strength and elastic modulus. Reduction in area during tensile testing and compressive yield strength sometimes are useful, particularly in analysis of hot forming operations.

Figure 17 summarizes the effects of chromium content on the tensile and yield strengths of Cr-Mo steels containing 0.5 to 1.0% Mo and various amounts of chromium. The effect of temperature is reported as the test temperature at which strength is reduced to 60% of its room-temperature value. The elevated-temperature strength data in Table 5 are reported in the same manner. Chromium is most effective in strengthening these Cr-Mo steels when it is used in amounts of 1 to $2\frac{1}{2}$%.

The effect of temperature on modulus of elasticity for several steels is illustrated in Fig. 12 and Table 6.

Long-term elevated-temperature mechanical properties of primary interest in the design of structures for elevated-temperature service are creep rate, rupture life and relaxation behavior.

Figure 18 summarizes the effects of stress level and chromium content on creep rate and rupture life of Cr-Mo steels containing 0.5 to 1.0% Mo and various amounts of chromium. The effect of temperature is reported as the temperature required to produce the indicated creep rate or rupture life. The most effective chromium content

for resisting creep is about $2\frac{1}{4}$%. Creep-rate and rupture-life data for four steels that have AISI designations are presented in Table 5; these data are given as temperature required to produce the indicated creep rate or rupture life.

Relaxation behavior of AISI 602 and 603 steels is illustrated in Fig. 19. Relaxation data are applicable to threaded fasteners for elevated-temperature service; fasteners must be capable of maintaining a specified force on the contact area between the parts that they hold together.

Fatigue Resistance of Steels

FATIGUE is the progressive localized permanent structural change that occurs in a material subjected to repeated or fluctuating strains at nominal stresses having maximum values often much less than the tensile strength of the material. Fatigue may culminate in cracks or fracture after a sufficient number of fluctuations.

Fatigue fractures are caused by the simultaneous action of cyclic stress, tensile stress and plastic strain. If any one of these three is not present, a fatigue crack will not initiate and propagate. The plastic strain resulting from cyclic stress initiates the crack; the tensile stress promotes crack growth (propagation). Careful measurement of strain shows that plastic strain can be present at low levels of stress where the strain might otherwise appear to be totally elastic. Although compressive stresses will not cause fatigue, compression loads may result in local tensile stresses.

In early literature, fatigue fractures were attributed to "crystallization," because of their crystalline appearance. Because metals are crystalline solids, the use of the term "crystallization" in connection with fatigue is confusing and should be avoided.

The process of fatigue consists of three stages:

1. Initial fatigue damage leading to crack initiation
2. Crack propagation to critical size (size at which the remaining uncracked cross section of the part becomes too weak to carry the imposed loads)
3. Final, sudden fracture of the remaining cross section.

FATIGUE RESISTANCE

Fatigue strength is usually proportional to hardness and tensile strength, although this generalization is not true for high tensile-strength values. Processing, fabrication and heat treatment techniques, surface treatment, finishing, and

Table 6. Effect of temperature on modulus of elasticity for several steels

Material	20 °C	70 °F	95 °C	200 °F	150 °C	300 °F	200 °C	400 °F	260 °C	500 °F	315 °C	600 °F	370 °C	700 °F
Carbon steels with carbon content of 0.30% or less, $3\frac{1}{2}$ Ni	192	27.9	191	27.7	189	27.4	186	27.0	182	26.4	177	25.7	171	24.8
Carbon steels with carbon content above 0.30%	206	29.9	203	29.5	200	29.0	195	28.3	189	27.4	184	26.7	175	25.4
Carbon-molybdenum steels, low-chromium steels through 3 Cr	206	29.9	203	29.5	200	29.0	197	28.6	193	28.0	189	27.4	183	26.6
Intermediate-chromium steels (5 Cr through 9 Cr)	189	27.4	187	27.1	185	26.8	182	26.4	179	26.0	175	25.4	172	24.9

(a) Values listed under Celsius temperatures are in GPa; values listed under Fahrenheit temperatures are in 10^6 psi.

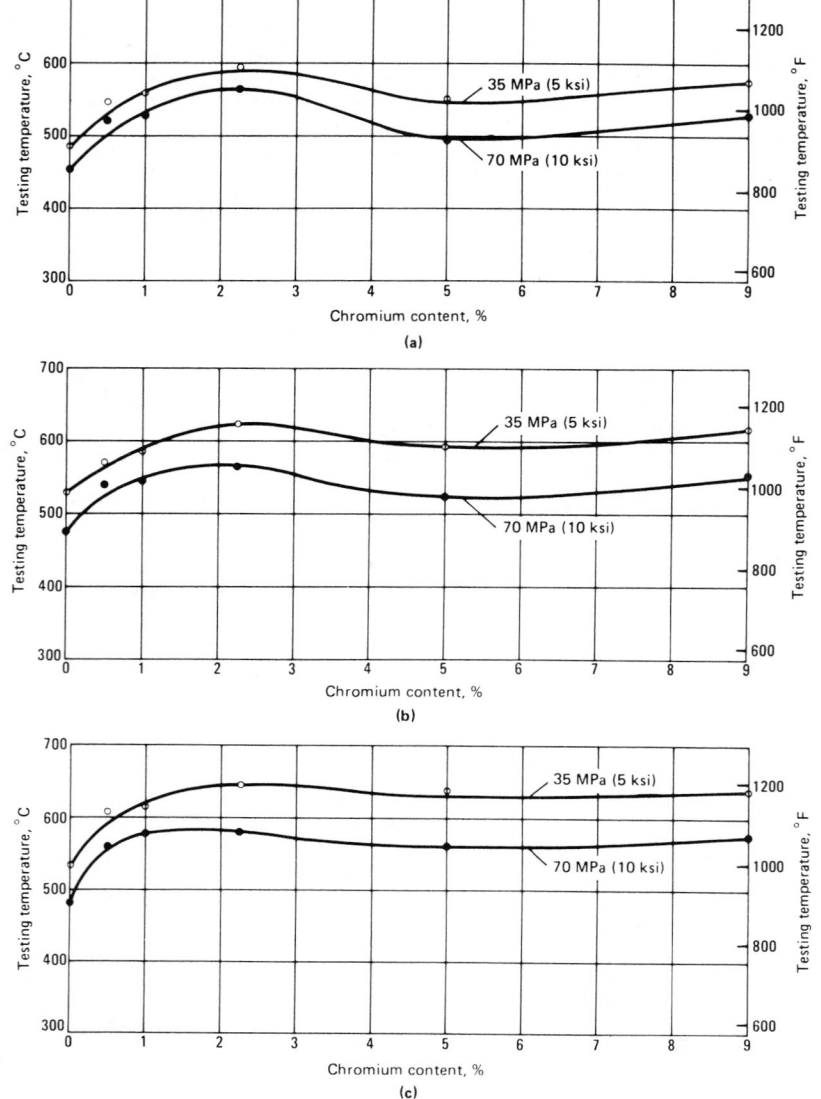

Test temperature required for (a) minimum creep rate of 0.1 μm/m·h, (b) minimum creep rate of 1 μm/m·h and (c) rupture in 10 000 hours for Cr-Mo steels containing 0.5 to 1.0% Mo and the indicated amount of chromium. Data are given for stresses of 35 and 70 MPa (5 and 10 ksi).

Fig. 18. Effect of chromium content on minimum creep rate and rupture life

service environment significantly influence the ultimate behavior of a metal subjected to cyclic stressing.

Prediction of the fatigue life of a metal part is complicated because materials are sensitive to small changes in loading conditions, stress concentrations, and local characteristics. The significant structural resistance of a metal to fatigue is also affected by manufacturing procedures such as cold forming, welding, brazing and plating, and by surface conditions such as surface roughness and residual stresses. Fatigue tests performed on small specimens are not sufficient for precisely establishing the fatigue life of a part. These tests are useful for rating the relative resistance of a material to cyclic stressing.

In addition to material properties, the fatigue strength of a part is affected by type of loading (uniaxial, bending or torsional), loading pattern (periodic loading at constant or variable amplitude, or random loading), magnitude of peak stresses, over-all size of the part, fabrication method, surface roughness, presence of fretting surface, operating temperature and environment, and occurrence of service-induced imperfections.

PREVENTION OF FATIGUE FAILURE

The incidence of fatigue failure can be considerably reduced by careful attention to design details and manufacturing processes. As long as the metal is sound and free from major flaws, a change in material composition is not as effective in achieving satisfactory fatigue life as are care in design, fabrication, and maintenance during service. The most effective and economical method of improving fatigue strength is improvement in design to: (a) eliminate stress raisers by streamlining the part, (b) avoid sharp surface tears resulting from punching, stamping, shearing, etc., (c) prevent the development of surface discontinuities or decarburization during processing or heat treatment, and (d) improve the details of fabrication and fastening procedures. Control of or protection against corrosion, erosion, chemical attack or service-induced nicks and other gouges is an important part of proper maintenance of fatigue strength during active service life.

SYMBOLS AND DEFINITIONS

Applied Stresses. The mean stress, S_m, is the algebraic average of the maximum and minimum stresses in one cycle, $S_m = (S_{max} + S_{min})/2$. In the completely reversed test, the mean stress is zero. The range of stress, S_r, is the algebraic difference between the maximum and minimum stresses in one cycle, $S_r = S_{max} - S_{min}$. The stress amplitude, S_a, is one-half the range of stress, $S_a = S_r/2 = (S_{max} - S_{min})/2$.

During a fatigue test, the stress cycle is usually maintained constant so that the applied stress conditions can be written as $S_m \pm S_a$, where S_m is the static or mean stress, and S_a is the alternating stress equal to half the stress range. The positive sign is used to denote a tensile stress; and the negative sign, a compressive stress. Some of the possible combinations of S_m and S_a are illustrated in Fig. 20. When $S_m = 0$ (Fig. 20a), the maximum tensile stress is equal to the maximum compressive stress; this is called an alternating stress or a completely reversed stress. When $S_m = S_a$ (Fig. 20b), the minimum stress of the cycle is zero; this is called a pulsating or repeated tensile (or compressive) stress. Any other combination is known as a fluctuating stress, which may be a fluctuating tensile stress (Fig. 20c), a fluctuating compressive stress, or a stress that fluctuates between a tensile and a compressive value (Fig. 20d).

Nominal stresses can be calculated on the net section of a part ($S = P/A$) without consideration of variations in stress conditions caused by holes, grooves, fillets, etc. Nominal stresses are often used in these calculations, although a closer estimate of actual stresses through the use of a stress-concentration factor might be preferred.

Stress ratio is the algebraic ratio of two specified stress values in a stress cycle. Two commonly used stress ratios are the ratio, A, of the alternating stress amplitude to the mean stress ($A = S_a/S_m$) and the ratio, R, of the minimum stress to the maximum stress ($R = S_{min}/S_{max}$). If the stresses are fully reversed, the stress ratio R becomes -1; if the stresses are partially reversed, R becomes a negative number less than 1. If the stress is cycled between a maximum stress and no load, the stress ratio R becomes zero. If the stress is cycled between two tensile stresses, the stress ratio R becomes a positive number less than 1. A stress ratio R of 1 indicates no variation in stress, and the test would become a sustained-load creep test rather than a fatigue test.

S-N Curves. The results of fatigue tests are usually plotted as maximum stress or stress amplitude to number of cycles, N, to fracture using a logarithmic scale for the number of cycles. Stress is plotted on either a linear or a logarithmic scale. The resulting curve of data points is called an S-N curve. A family of S-N curves for a material tested at various stress ratios is shown schematically in Fig. 21. For carbon and low-alloy steels, S-N curves typically have a fairly straight slanting portion at low cycles changing into a straight, horizontal line at higher cycles, with a sharp transition between the two.

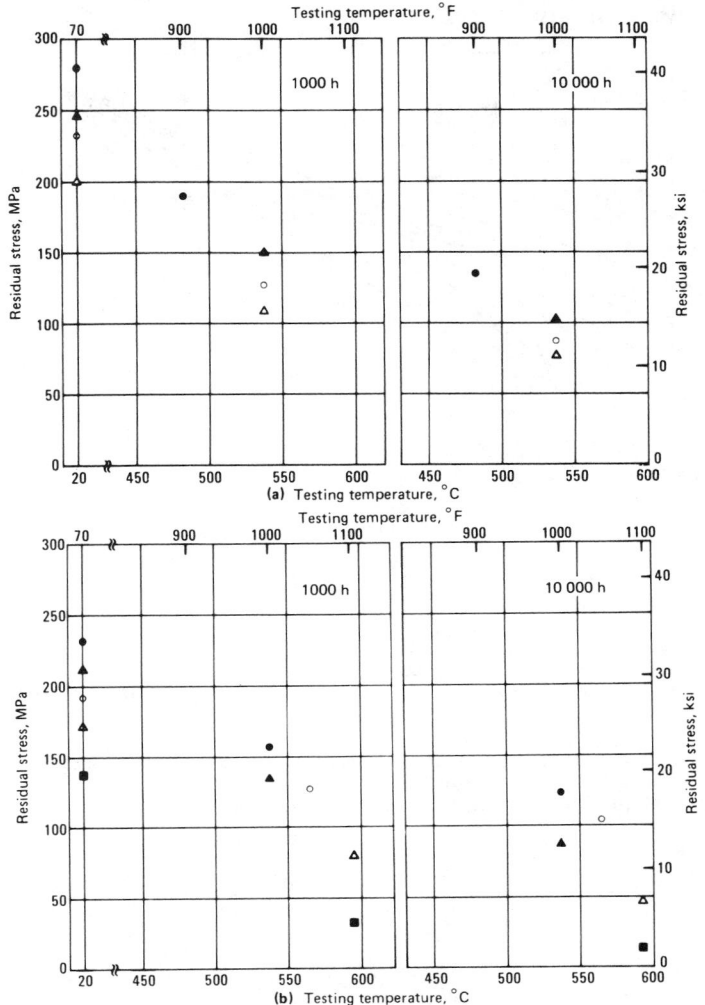

Relaxation behavior of (a) four specimens of AISI 602 and (b) five specimens of AISI 603. All specimens were strained at room temperature to produce the residual stress levels indicated at far left in the graphs. The specimens were then exposed to elevated temperature, as shown for 1000 and 10 000 h, which reduced the residual stress levels, as shown.

Fig. 19. Relaxation behavior of AISI 602 and 603

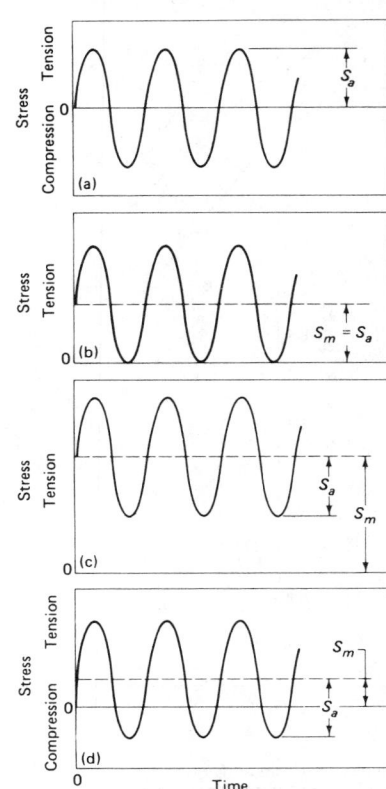

(a) Alternating stress in which $S_m = 0$ and $R = -1$.
(b) Pulsating tensile stress in which $S_m = S_a$ the minimum stress is zero, and $R = 0$. (c) Fluctuating tensile stress in which both the minimum and maximum stresses are tensile stresses, $R = \frac{1}{3}$. (d) Fluctuating tensile-to-compressive stress in which the minimum stress is a compressive stress and the maximum stress is a tensile stress, $R = -\frac{1}{3}$.

Fig. 20. Types of fatigue-test stress

An *S-N* curve usually represents the median life for a given stress—the life that half the specimens attain or surpass and half fail to attain. Scatter of fatigue lives can cover a very wide range.

Fatigue limit (or endurance limit) is the value of the stress below which a material can presumably endure an infinite number of stress cycles—that is, the stress at which the *S-N* diagram becomes and appears to remain horizontal. The existence of a fatigue limit is typical for carbon and low-alloy steels.

Fatigue strength, which should not be confused with fatigue limit, is the stress to which the material can be subjected for a specified number of cycles. Fatigue strength is used for materials, such as most nonferrous metals, that do not exhibit well-defined fatigue limits. It is also used to describe the fatigue behavior of carbon and low-alloy steels at stresses higher than the fatigue limit.

Stress-Concentration Factor. Concentrated stress in a metal is evidenced by discontinuities such as notches, holes and scratches, and by changes in microstructure. The stress-concentration factor,

K_t, is the ratio of the greatest stress in the region of the notch (or other stress concentrators) to the corresponding nominal stress. For determination of K_t, the greatest stress in the region of the notch is calculated from the theory of elasticity, or equivalent values may be derived experimentally. An experimental stress-concentration factor is a ratio of stress in a notched specimen to the stress in a smooth (unnotched) specimen.

Fatigue notch factor, K_f, is the ratio of the fatigue strength of a smooth (unnotched) specimen to the fatigue strength of a notched specimen at the same number of cycles. The fatigue notch factor will vary with the position on the *S-N* curve and with the mean shear stress. At high stress levels and short cycles, the factor is usually less than at lower stress levels and longer cycles due to blunting of the notch by plastic deformation.

Fatigue notch sensitivity, q, is determined by comparing the fatigue notch factor, K_f, and the stress-concentration factor, K_t, for a specimen of a given size containing a stress concentrator of a given shape and size. A common definition of fatigue notch sensitivity is $q = (K_f - 1)/(K_t - 1)$, in

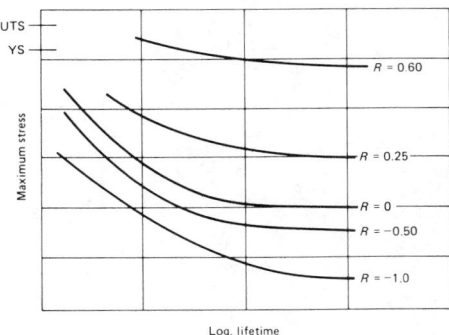

Schematic *S-N* curves for a material at various stress ratios. UTS and YS indicate ultimate tensile strength and yield strength, respectively, in uniaxial tensile testing.

Fig. 21. Typical *S-N* curves

which q may vary between zero (where $K_f = 1$) and 1 (where $K_f = K_t$). This value may be stated as a percentage. As the fatigue notch factor varies with the position on the *S-N* curve, so will the notch sensitivity. Most metals are fully notch sensitive at low stresses and long cycles. If they are not, it may be that the fatigue strengths for the smooth (unnotched) specimens are lower than they could be because of surface imperfections.

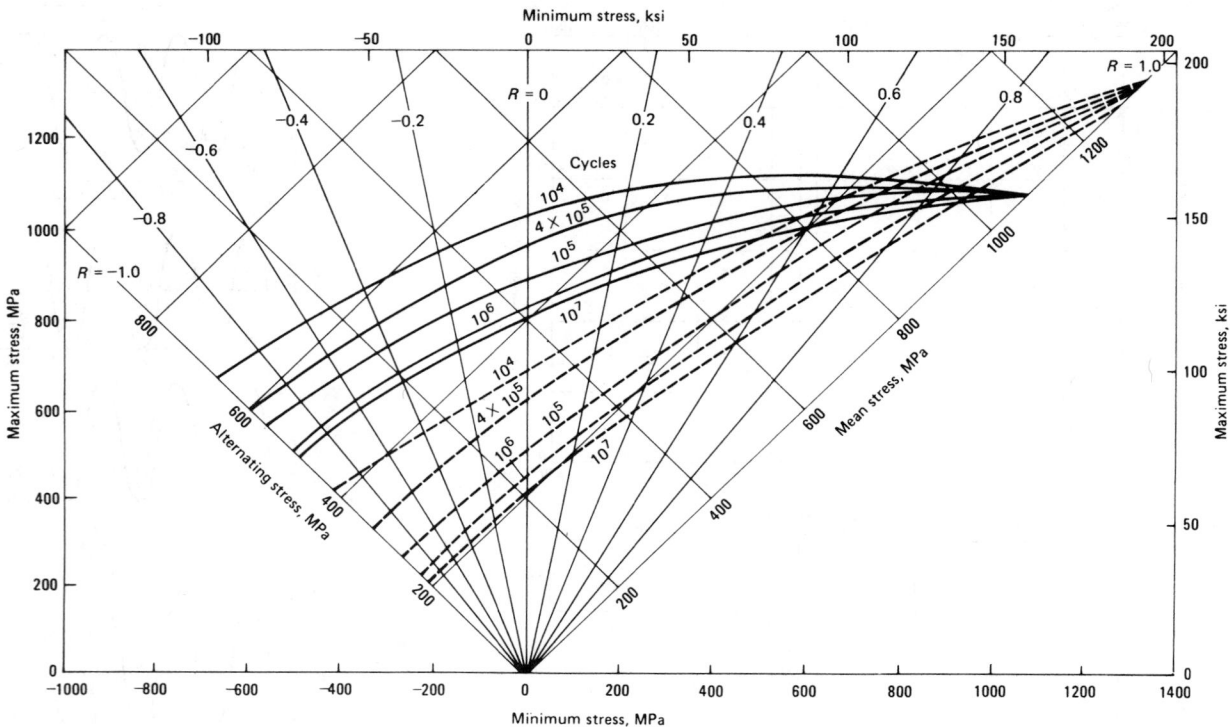

Constant-lifetime fatigue diagram for AISI-SAE 4340 alloy steel (bar), hardened and tempered to a tensile strength of 1035 MPa (150 ksi). Solid lines represent data obtained from unnotched specimens; dashed lines represent data from specimens having notches with K_t = 3.3.

Fig. 22. Comparison of constant-lifetime fatigue behavior of notched and unnotched specimens

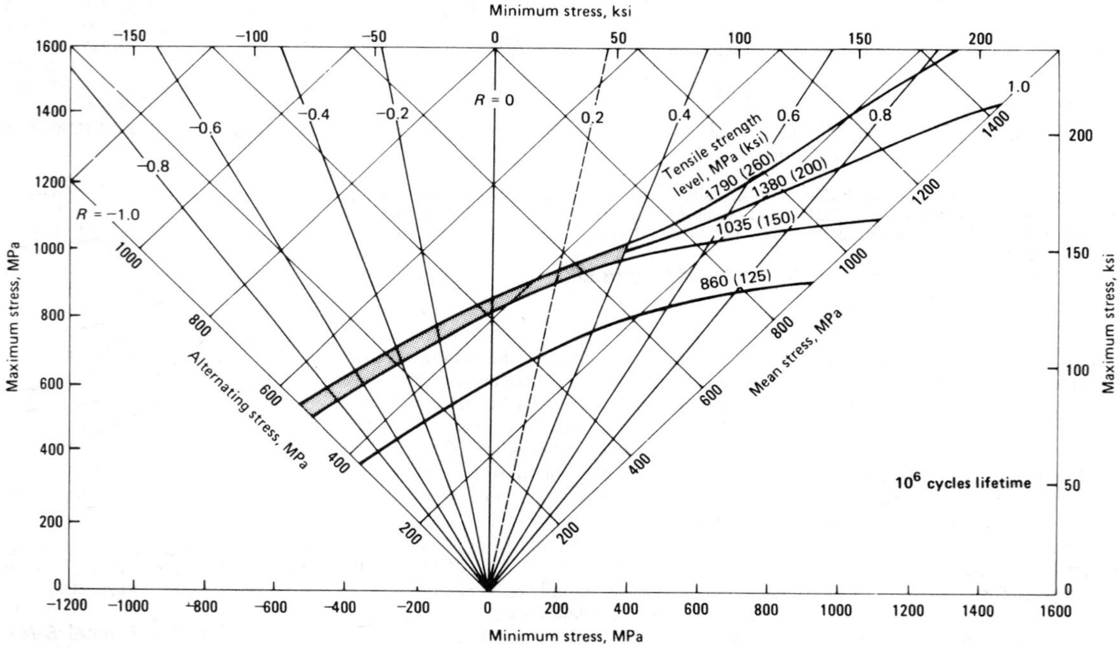

Constant-lifetime fatigue diagram for AISI-SAE 4340 alloy steel (bar), hardened and tempered to tensile-strength levels of 860 MPa (125 ksi), 1035 MPa (150 ksi), 1380 MPa (200 ksi) and 1790 MPa (260 ksi). All lines represent fatigue lifetimes of one million cycles.

Fig. 23. Effect of strength level on constant-lifetime fatigue behavior

STRESS-BASED APPROACH TO FATIGUE

The design of a machine element that will be subjected to cyclic loading can be approached by adjusting the configuration of the part so that the calculated stresses fall safely within the required life line on a constant-life diagram. In stress-based analysis, the material is assumed to deform in a nominally elastic manner; local plastic strains are neglected. To the extent that these approximations are valid, the stress-based approach is useful. These assumptions imply that the stresses will be essentially elastic.

The constant-life fatigue diagram shown in Fig. 22 presents data for AISI-SAE 4340 steel, heat treated to a tensile strength of 1035 MPa (150 ksi). Figure 23 shows the combinations of cyclic stresses that can be tolerated for a million cycles by the same steel; the specimens were heat treated to tensile strengths of 860 to 1790 MPa (125 to 260 ksi). Note that for stress ratios below about

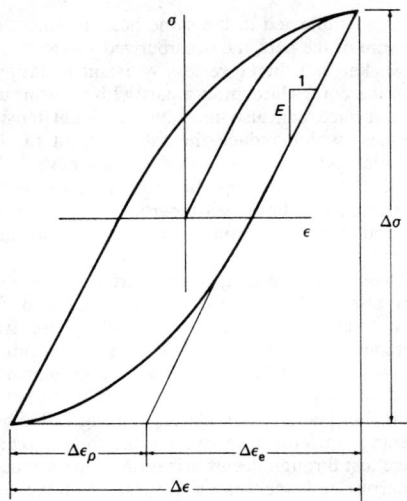

Fig. 24. Stress-strain hysteresis loop

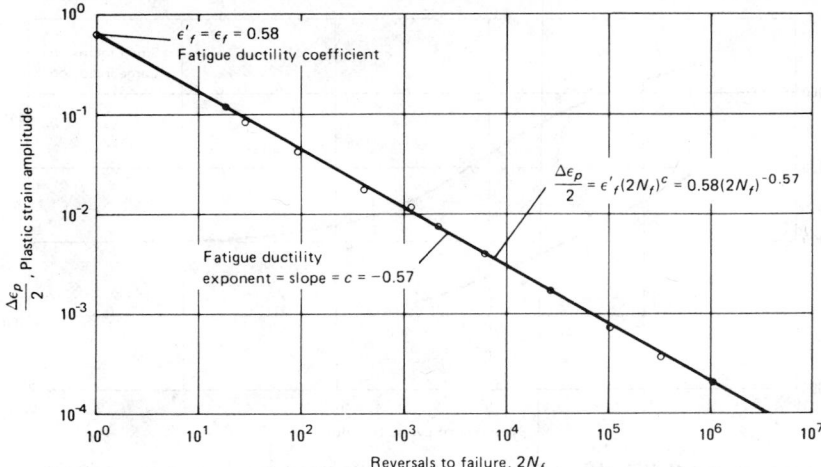

Ductility versus fatigue life for annealed AISI-SAE 4340 steel.
Fig. 25. Ductility versus fatigue life

0.2 the fatigue lives of the specimens at the three higher strength levels are about the same; the scatter in the experimental data is at least as great as any real differences in fatigue life among these specimens.

STRAIN-BASED APPROACH TO FATIGUE

A strain-based approach to fatigue, developed for analysis of low-cycle fatigue data, has proven useful for analyzing long-life fatigue data as well. This approach can account for both elastic and plastic responses to applied loadings, and the data are presented on a log-log plot similar in shape to an S-N curve; the value plotted in the x-direction is the number of strain reversals (twice the number of cycles) to failure, and the y-direction value is the strain amplitude (half of the strain range).

During cyclic loading, the stress-strain relationship can be generally described by a loop, such as that shown in Fig. 24. For purely elastic loading, the loop becomes a straight line whose slope is the elastic modulus, E, of the material. The occurrence of a hysteresis loop is most common. The definitions of the plastic strain range, $\Delta\varepsilon_p$, the elastic strain range, $\Delta\varepsilon_e$, the total strain range, $\Delta\varepsilon$, and the stress range, $\Delta\sigma$, are indicated in Fig. 24. A series of fatigue tests, each having a different total strain range, will generate a series of hysteresis loops. For each set of conditions, a characteristic number of strain reversals is necessary to cause failure.

As illustrated in Fig. 25, a plot on logarithmic coordinates of the plastic portion of the strain amplitude (half of the plastic strain range) versus the fatigue life yields a straight line described by the equation

$$\frac{\Delta\varepsilon_p}{2} = \varepsilon'_f (2N_f)^c$$

where ε'_f is the fatigue ductility coefficient; c is the fatigue ductility exponent; and N_f is the number of cycles to failure.

METALLURGICAL VARIABLES IN FATIGUE

The metallurgical variables having the most pronounced effects on the fatigue behavior of carbon and low-alloy steels are strength level,

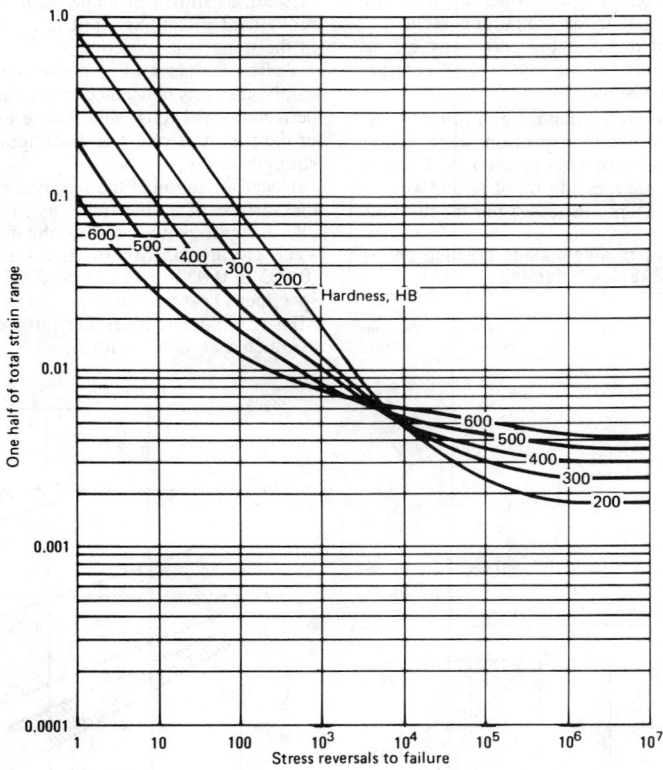

Predicted plots of total strain versus fatigue life for typical medium-carbon steel at the indicated hardness levels.
Fig. 26. Effect of hardness level on plot of total strain versus fatigue life

ductility, cleanness of the steel, residual stresses, surface conditions and aggressive environments. At least partly because of the characteristic scatter of fatigue testing data, it is difficult to distinguish the direct effects of other variables such as composition on fatigue from their effects on the strength level of steel.

Strength Level. For most steels whose hardness is below 400 HB (not including precipitation-hardening steels), the fatigue limit is about half of the ultimate tensile strength. Thus, any heat treatment or alloying addition that increases the strength of the steel can be expected to increase its fatigue limit. The relationship between strength

and fatigue limit is illustrated in Fig. 23. The relationship between hardness and low-cycle fatigue behavior is illustrated in Fig. 26.

Cleanness of a steel refers to its relative freedom from nonmetallic inclusions. These inclusions generally have a deleterious effect on the fatigue behavior of steels (see Fig. 27), particularly for long-life applications. The type, number, size and distribution of nonmetallic inclusions may have a greater effect on fatigue life of carbon and alloy steels than differences in chemical composition, microstructure or stress gradients. Nonmetallic inclusions, however, are rarely the prime cause of fatigue failure of production

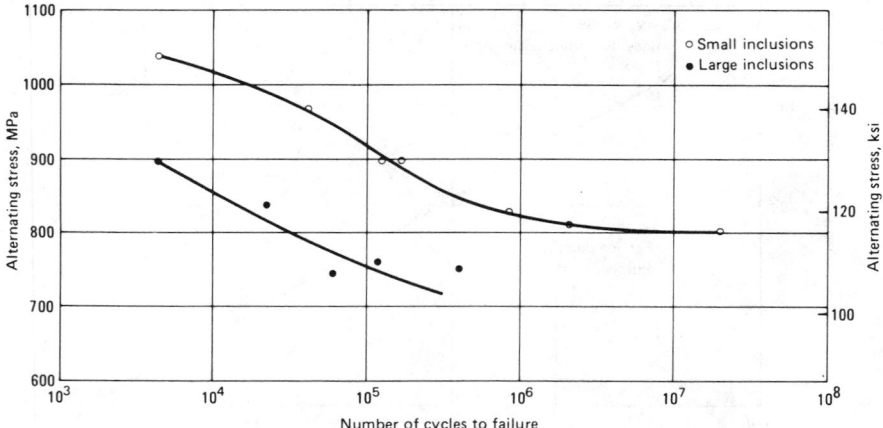

Fatigue life of two lots of AISI-SAE 4340H steel; one lot (lower curve) contained abnormally large inclusions; the other lot (upper curve) contained small inclusions.

Fig. 27. Effect of nonmetallic inclusion size on fatigue

parts; if the design fatigue properties were determined using specimens containing inclusions representative of those in the parts, any effects of these inclusions would already be incorporated in the test results.

Large nonmetallic inclusions can often be detected by nondestructive inspection; steels can be selected on the basis of such inspection. Vacuum melting, which reduces the number and size of nonmetallic inclusions, increases the fatigue limit

Table 7. Influence of steelmaking practice on fatigue limit of SAE 4340 steel(a)

	Electric furnace-melted	Vacuum-melted
Longitudinal fatigue limit		
MPa	800	960
ksi	116	139
Transverse fatigue limit		
MPa	545	825
ksi	79	120
Ratio transverse to longitudinal	0.68	0.86
Hardness, HRC	27	29

(a) Determined in repeated-bending fatigue test ($R = 0$).

of steel, as shown in Table 7 for 4340 steel. Improvement in fatigue limit is especially evident in the transverse direction.

Surface conditions of a metal part can significantly affect its resistance to fatigue. Surface imperfections and roughness reduce the fatigue limit of the part; this effect is most apparent for high-strength steels.

Decarburization is the removal of carbon from the surface of a steel part; as indicated in Fig. 28, it significantly reduces the fatigue limit of steel. Decarburization of from 0.08 to 0.76 mm (0.003 to 0.030 in.) on AISI-SAE 4340 notched specimens heat treated to a strength level of 1860 MPa (270 ksi) reduces the fatigue limit almost as much as a notch with $K_t = 3$.

When subjected to the same heat treatment as the core of the part, the decarburized surface layer is weaker and therefore less resistant to fatigue than the core. Hardening a part with a decarburized surface can also introduce residual tensile stresses, which reduce the fatigue limit of the material. Results of research studies have indicated that fatigue properties lost through decarburization can be at least partially regained by recarburization (carbon restoration in the surfaces).

Mechanical working of the surface of a steel part effectively increases the resistance to fatigue. Shot peening and skin rolling are two methods for developing compressive residual stresses at the surface of the part. The improvement in fatigue life of a crankshaft that results from shot peening is illustrated in Fig. 29. Shot peening is useful in recovering the fatigue resistance lost through decarburization of the surface; decarburized specimens similar to those described in Fig. 28 were shot peened, raising the fatigue limit from 275 MPa (40 ksi) after decarburizing to 655 MPa (95 ksi) after shot peening.

Embrittlement of Steels

EMBRITTLEMENT, in many forms, can lead to brittle fracture of steel parts. At least nine forms can occur during thermal treatment or elevated-temperature service. These forms of embrittlement (and the types of steel that some forms specifically affect) are:

- Strain-age embrittlement (low-carbon steel)
- Quench-age embrittlement (low-carbon steel)
- Blue brittleness
- Temper embrittlement (alloy steels)

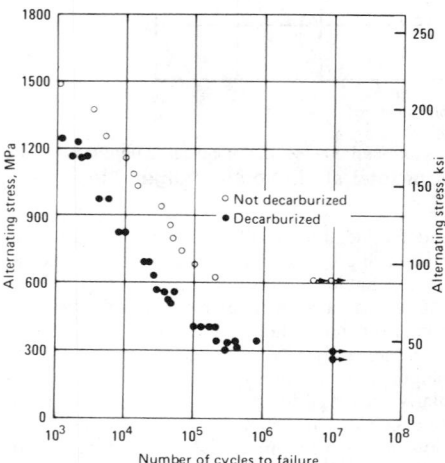

Fig. 28. Effect of decarburization on fatigue behavior of a steel

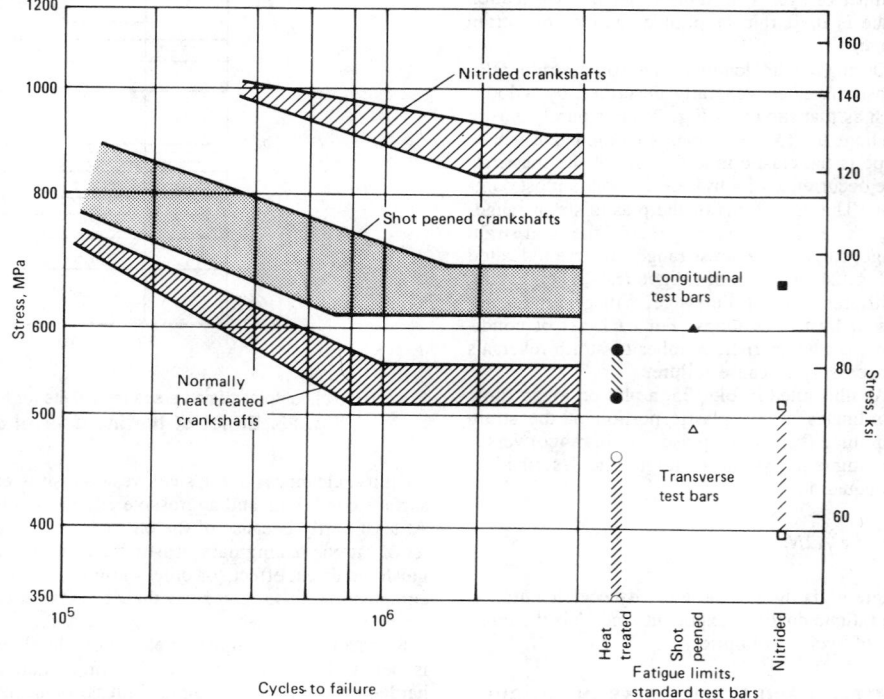

Comparison between fatigue limits of crankshafts (*S-N* bands) and fatigue limits for separate test bars, which are indicated by plotted points at right. Steel was 4340.

Fig. 29. Effect of nitriding and shot peening on fatigue behavior

- 500 °F (also called "350 °C") embrittlement (high-strength low-alloy steels)
- 400 to 500 °C (750 to 930 °F) embrittlement (ferritic stainless steels)
- Sigma-phase embrittlement (ferritic and austenitic stainless steels)
- Graphitization (carbon and low-alloy steels)
- Embrittlement by intermetallic compounds (galvanized steel).

In addition, steels (and other metals and alloys) can be embrittled by environmental conditions. The four forms of environmental embrittlement are:

- Neutron embrittlement
- Hydrogen embrittlement
- Stress-corrosion cracking
- Liquid-metal embrittlement.

The various forms of embrittlement in the two lists above are discussed in this section.

STRAIN-AGE EMBRITTLEMENT

If low-carbon steel is deformed, its hardness and strength will increase on aging at room or slightly elevated temperature, but with a concurrent loss of ductility. Rimmed or capped sheet steels are particularly susceptible to strain-age embrittlement, although strain-aging has also been encountered in plate steels and weld heat-affected zones. These steels often are temper rolled to suppress the yield point. The return of the yield point or the presence of Lüders strain in the stress-strain curve is evidence that strain-age embrittlement has occurred. The degree of embrittlement is a function of the amount of cold work, the aging temperature and the time at temperature. Room-temperature aging may require from a few hours to a year. However, as the aging temperature is increased, the required time decreases, with embrittlement occurring in a matter of minutes at about 200 °C (400 °F).

QUENCH-AGE EMBRITTLEMENT

If low-carbon steels are rapidly cooled from temperatures slightly below the lower critical temperature (Ac_1) of the steel, the hardness of the steel increases, with a resultant loss of ductility upon aging at room temperature. As with strain-aging, quench-age embrittlement is a function of time at the aging temperature until the maximum degree of embrittlement is reached. An aging period of several weeks at room temperature is required for maximum embrittlement.

A decrease in the quenching temperature decreases the extent of the embrittlement. Quenching from temperatures of 560 °C (1040 °F) and below does not produce quench-age embrittlement. Steels with carbon contents of 0.04 to 0.12% appear most susceptible to quench-age embrittlement; increasing the carbon content above 0.12% reduces the effect. Quench-age embrittlement results from (a) precipitation of solute carbon at existing dislocations and (b) precipitation hardening because of differences in the solid solubility of carbon in ferrite at different temperatures.

BLUE BRITTLENESS

When plain carbon steels and some alloy steels are heated between 230 and 370 °C (450 and 700 °F), there is an increase in strength and a marked decrease in ductility and impact strength. This embrittling phenomenon is known as "blue brittleness," because it occurs in the blue-heat range. Blue brittleness is an accelerated form of strain-age embrittlement. The increase in strength and decrease in ductility are caused by precipitation hardening within the critical-temperature range. Deformation while the steel is heated in the blue-heat range results in even higher hardness and tensile strength after cooling to room temperature. If the strain rate is increased, the blue-brittle temperature range increases.

Use of susceptible steels that have been heated in the blue-brittleness range should be avoided, especially if the steels are subjected to impact loads, because the toughness of these materials will be considerably less than optimum.

TEMPER EMBRITTLEMENT

The technical importance of temper brittleness arises from the fact that many of the common low-alloy steels exhibit an increase in their ductile-to-brittle transition temperatures after being heated in the range of 375 to 575 °C (700 to 1070 °F) or slowly cooled through this temperature range. Plain carbon steels containing less than 0.30% manganese are not susceptible to temper brittleness, although any steel containing appreciable amounts of manganese, nickel or chromium will be susceptible provided that the steel also contains one or more of the impurities antimony, phosphorus, tin and arsenic. Thus, a combination of an alloying element and an impurity is required to cause temper embrittlement.

Detection. Temper embrittlement usually is detected by an upward shift in the ductile-to-brittle transition temperature in a notched-bar test such as the Charpy V-notch impact test. For common commercial levels of phosphorus and tin in nickel-chromium steels, the transition temperature may increase as much as 200 to 300 °C (360 to 540 °F) for steel held at the temperature of most rapid embrittlement for 1000 h. Where large amounts of antimony (about 0.05%) have been deliberately added in experimental materials, shifts of as much as 500 to 600 °C (900 to 1080 °F) have been found. Generally, there is no detectable change in smooth-bar tension-test behavior at room temperature (either in the yield strength or the flow curve, or in the elongation or reduction in area) accompanying such large-scale changes in notched-bar behavior. In cases of extreme embrittlement, there may be a detectable drop in reduction of area, which becomes more marked at cryogenic temperatures.

An embrittling treatment called "step cooling" is sometimes used to estimate the effects of long-time exposures. Step cooling consists of cooling the material through the embrittling range in a series of steps from about 600 to 300 °C (1100 to 570 °F) with the time at each step increasing as the temperature is lowered.

Embrittlement Mechanism. The increase in transition temperature due to temper embrittlement is accompanied by a gradual change in brittle fracture (below the transition temperature) from completely transcrystalline to completely intercrystalline fracture. The fracture path generally follows the prior austenite grain boundaries in embrittled material. Upon de-embrittlement, the brittle fracture mode reverts to transcrystalline. There is no detectable phase change accompanying this change in fracture mode; rather, the increase in transition temperature and change in fracture mode result from segregation of the responsible impurity along prior austenite grain boundaries.

Control of Temper Embrittlement. Steels that have become embrittled can be restored to their original toughness by heating to about 600 °C (1100 °F) or above, then cooling rapidly to below about 300 °C (570 °F). Alternatively, embrittlement can be forestalled by reducing susceptibility. The principal means of reducing susceptibility is to reduce embrittling impurities as much as possible by control of raw materials and melting practice.

500 °F EMBRITTLEMENT

500 °F ("350 °C") embrittlement of quenched-and-tempered high-strength low-alloy steels occurs over a temperature range of approximately 400 to 700 °F (200 to 370 °C). It occurs mainly in steels that have been heat treated to a microstructure of tempered martensite; thus, the more descriptive term "tempered-martensite embrittlement" is sometimes used. Steels with microstructures of tempered lower bainite also are susceptible to 500 °F embrittlement, but steels with pearlitic microstructures and other bainitic steels are not.

500 °F embrittlement is evaluated by measuring the effect of tempering temperature on room-temperature impact energy; this is in contrast to temper embrittlement, which is evaluated by measuring the effect of tempering temperature on the ductile-to-brittle transition temperature.

500 °F embrittlement is believed to be caused by ferrite networks resulting from precipitation of cementite platelets along prior austenite grain boundaries. However, some investigators believe that the precipitation of grain-boundary cementite platelets as such is responsible for 500 °F embrittlement.

Steels containing substantial amounts of chromium or manganese are highly susceptible to 500 °F embrittlement. Aluminum contents above 0.04% reduce embrittlement, and additions of 0.1% aluminum usually eliminate the problem. Some degree of embrittlement has been observed when phosphorus, antimony, arsenic, tin, silicon, manganese or nitrogen additions were made to high-purity steels. Additions of nitrogen produced intergranular fractures, but the other embrittling agents did not. (Commercial grades of steel that are subjected to 500 °F embrittlement treatments fracture intergranularly.)

Embrittlement in low-alloy steels heat treated to high strength levels can be minimized by:

- Developing special steels with retarded martensite-tempering characteristics
- Developing steels with faster rates of martensite tempering
- Using steels capable of transformation to 100% upper bainite at the desired strength level and section size
- Avoiding tempering in the region of susceptibility
- Using the lowest possible carbon content consistent with the desired strength level.

400 TO 500 °C EMBRITTLEMENT

Fine-grained, high-chromium stainless steels normally possess good ductility. However, if they are held for long periods of time at temperatures in the range of 400 to 500 °C (750 to 930 °F), they will become harder, but embrittled. Embrittled high-chromium ferritic stainless steels contain two ferrites, one rich in iron and one rich in chromium.

Susceptibility to 400 to 500 °C embrittlement increases with increasing chromium content, with the highest degree of embrittlement occurring at chromium contents greater than 19%. At least 15% chromium is necessary for embrittlement to occur. The effect of carbon content on embrittlement is minimal. High-chromium steels that contain at least 1% titanium are more susceptible to embrittlement than are similar steels with lower titanium contents.

The embrittlement caused by prolonged soaking within the 400 to 500 °C range can be removed by soaking at somewhat higher temperatures for several hours.

SIGMA-PHASE EMBRITTLEMENT

The formation of sigma phase in ferritic and austenitic stainless steels during long periods of exposure to temperatures between approximately 560 and 980 °C (1050 and 1800 °F) results in considerable embrittlement after cooling to room temperature. Sigma phase, an iron-chromium compound approximately equivalent to FeCr, can be formed by either (a) slow cooling from temperatures of 1040 to 1150 °C (1900 to 2100 °F) or (b) water quenching from 1040 to 1150 °C followed by heating at 560 to 980 °C, with heating at 850 °C (1560 °F) producing the greatest effect. The embrittlement is most detrimental after the steel has cooled to temperatures below 260 °C (500 °F). At higher temperatures, stainless steels containing sigma phase usually can withstand normal design stresses. However, cooling to 260 °C or below results in essentially complete loss of toughness.

The presence of sigma phase greatly increases notch sensitivity, particularly in ferritic stainless steels and austenitic alloys that also contain some ferrite. The hardness and tensile strengths are usually not significantly affected by the presence of sigma phase, but the impact strength is greatly affected. Sigma phase exerts a strengthening effect at high temperatures; however, the impact strength at high temperatures of an alloy containing sigma phase is lower than the impact strength at room temperature of an alloy without sigma phase.

GRAPHITIZATION

Graphitization of carbon and carbon-molybdenum steel piping during service at elevated temperatures above 425 °C (800 °F) has caused numerous failures in steam power plants and refineries. Graphite formation generally occurs in a narrow region in the heat-affected zone of a weld where the metal has been briefly heated above the lower critical temperature. The graphitization tendency of carbon and carbon-molybdenum steels is increased when the aluminum content exceeds about 0.025%. Steels deoxidized with silicon may also be susceptible to graphitization. Deoxidation with titanium usually will produce good resistance to graphitization. Carbon-molybdenum steels exhibit greater resistance to graphitization than do carbon steels.

The degree of embrittlement depends on the distribution, size and shape of the graphite. The severity of graphitization is frequently evaluated by bend testing. If graphitization is detected in

its early stages, the material often can be rehabilitated by normalizing and tempering just below the lower critical temperature. Steel that has undergone more severe graphitization cannot be salvaged in this manner; the defective region must be cut out and rewelded, or the section must be replaced. Carbon and carbon-molybdenum steels can be rendered less susceptible to graphitization by tempering just below the lower critical temperature.

INTERMETALLIC-COMPOUND EMBRITTLEMENT

Embrittlement of galvanized steel can result from long periods of exposure at elevated temperatures below the melting point of the zinc in the coating. In this type of embrittlement, zinc diffuses from the galvanized coating to grain boundaries in the steel, resulting in the formation of a brittle intergranular network of iron-zinc intermetallic compound. The presence of this compound can lead to brittle fracture.

ENVIRONMENTALLY ASSISTED EMBRITTLEMENT

Neutron Embrittlement. Neutron irradiation of steel components in nuclear reactors usually results in a significant rise in the ductile-to-brittle transition temperature of the steel. The transition temperature, as determined by Charpy V-notch impact tests, may be raised substantially, depending on such factors as the neutron dose, neutron spectrum, irradiation temperature and steel composition. The amount by which the transition temperature increases because of neutron irradiation is usually at least 17 °C (30 °F) but less than 200 °C (360 °F).

The increase in the ductile-to-brittle transition temperature, or in the nil-ductility transition temperature, has been determined for many steels used for these applications under various conditions of irradiation. High-strength steels, which have lower initial nil-ductility transition temperatures than low-strength steels, generally are less susceptible to radiation embrittlement. Steels with low initial nil-ductility transition temperatures, fine-grained microstructures and high dislocation densities generally offer greater resistance to neutron embrittlement. Neutron embrittlement renders the steel more susceptible to intergranular fracture.

Heat treatment practice greatly affects the susceptibility of a steel to neutron embrittlement. Steels with tempered-martensite microstructures are less susceptible than those with tempered-upper-bainite or ferritic microstructures. Vacuum degassing and control of elements such as copper and phosphorus help reduce susceptibility to neutron embrittlement.

Hydrogen embrittlement has been a long-time problem. Quantitative knowledge regarding the influence of hydrogen on metals is difficult to obtain; however, from a qualitative standpoint, the effects of hydrogen have been well documented in the literature.

Absorption of hydrogen results in a general loss of ductility. Historically, hydrogen-embrittlement effects have been evaluated by reversed-bend tests, single-bend tests and fatigue tests. Reduction-in-area and elongation values determined by standard tensile tests also show the effect of hydrogen embrittlement. Impact tests generally are not a good method for detecting hydrogen embrittlement.

The degree of hydrogen embrittlement depends greatly on the strength level of the steel; resistance to hydrogen embrittlement decreases as the strength level is increased.

Hydrogen embrittlement of steel may occur during pickling to remove scale and rust. The rate of hydrogen pickup depends on the type and concentration of acid, the temperature of the solution, the pickling time, and the presence and concentration of inhibitors. Strongly ionized acids, such as hydrochloric, sulfuric and hydrofluoric, cause severe embrittlement.

Hydrogen embrittlement of steel may occur during electroplating. The coating itself may become embrittled if hydrogen becomes trapped in the coating during plating. Embrittled coatings frequently develop blisters; ruptures of the plating may result if sufficient hydrogen pressure develops in a blister.

The effects of hydrogen embrittlement decrease as the strain rate increases. These effects are most pronounced at intermediate service temperatures and disappear at high and low temperatures. Because the solubility of hydrogen is greater at high temperatures than at low temperatures, the concentration of hydrogen can exceed solubility limits at room temperature on rapid cooling from high temperatures, such as after forging and heat treatment. Any excess hydrogen present, therefore, must diffuse to the surface of the steel or form discontinuities in the interior of the part. Pockets of hydrogen in these discontinuities may develop enough pressure to form hairline cracks or to shatter the steel.

A form of hydrogen embrittlement known as sulfide stress cracking is encountered in "sour" environments typical of deep oil and gas wells. Corrosion of steel in aqueous environments containing hydrogen sulfide or in moist atmospheres containing gaseous hydrogen sulfide releases atomic hydrogen that is then absorbed into the steel, where it may cause embrittlement. This problem seriously limits the use of hardenable high-strength steels for well drilling and oil and gas production equipment. For instance, alloy steels heat treated to yield strengths above 560 MPa (80 ksi) have been known to fracture through in a very short time.

The best countermeasures developed to date have been to increase well temperatures to at least 150 °C (300 °F) or to increase the pH of the aqueous environment to 6.0 or above. Austenitic steels, nickel alloys and cobalt alloys perform better than alloy steels in sour environments, but even these alloys are not totally resistant to cracking. Cracking of alloy steels appears to be greater when they have been cold worked, or when the operating conditions include high-pressure hydrogen sulfide, low pH, or temperatures between −45 and +65 °C (−50 and +150 °F).

Stress-corrosion cracking is a mechanical-environmental failure process in which mechanical stress and chemical attack combine to initiate and propagate fracture in a metal part. Stress-corrosion cracking is produced by the synergistic action of sustained tensile stress and a specific corrosive environment; this action causes failure to occur more rapidly than it would if the separate effects of the stress and the corrosive environment were simply added together.

Failure by stress-corrosion cracking frequently is caused by simultaneous exposure to a seemingly mild chemical environment and to a tensile stress well below the yield strength of the metal. Under such conditions, fine cracks can penetrate deeply into the part while the surface exhibits only insignificant amounts of corrosion. Hence, there may be no macroscopic indications of an impending failure.

Liquid-metal embrittlement can cause cracking and fracture in stressed parts of many metals. Not all combinations of solid and liquid metals produce embrittlement. For example, aluminum is embrittled by liquid gallium, sodium and tin, and steel is embrittled by liquid cadmium and lithium. (Liquid-metal embrittlement of steel also has been reported to be caused by liquid copper, brass, aluminum bronze, antimony and tellurium.) Both aluminum and steel are embrittled by liquid indium, zinc and mercury.

The melting temperature and chemical reactivity of a liquid metal are not deciding factors as to whether it will cause embrittlement or not. Instead, most instances of embrittlement are accompanied by low intersolubilities and absence of intermetallic-compound formation.

Notch Toughness
of Steels

TOUGHNESS is the ability of a metal to absorb energy and deform plastically before fracturing; the amount of energy absorbed during deformation and fracture is a measure of the toughness of the metal. By contrast, the amount of deformation that occurs prior to fracture is a measure of the metal's ductility, and the force necessary to cause fracture is a measure of its strength. For an application in which a metal object must withstand some specified load, the strength of the metal is the controlling property. Ductility may be the governing property if the metal must be formed to a specified shape. But if the metal must be able to absorb a certain quantity of mechanical energy without fracturing, its toughness is the limiting property.

Systematic investigations into the failures of various types of steel structures, such as bridges, storage tanks, pressure vessels and gas pipelines, have firmly established notch toughness as an important parameter for selecting the material to be used in any structure that might be subjected to impulsive loading at low temperatures.

Notch toughness usually is evaluated by testing specimens of prescribed size and shape at a known temperature in a single-blow pendulum-type impact machine. Two commonly used methods of impact testing are the Charpy and Izod tests, which are discussed in detail in ASTM Standard E23. Notched specimens, such as those illustrated in Fig. 30, usually are used to evaluate the toughness of most metals. Although somewhat redundant, the term notch toughness is very descriptive of the results of these tests and it reduces the likelihood of data being misinterpreted. Unnotched test specimens usually are used to evaluate the toughness of less ductile metals, such as gray cast irons.

Notch-toughness values derived from test results cannot be used directly in engineering design calculations. The values become significant for design only when correlated with a particular type of structure in a particular kind of service, or when these values are used to compare different materials.

Notch toughness of a metal is influenced by chemical composition and many physical factors. For steels, carbon content, alloying ele-

ments, gas content and impurities are chemical factors that affect this property. The physical factors include microstructure, grain size, section size, hot and cold working temperature, method of fabrication, and specimen orientation in relation to working direction. Surface conditions, such as carburization and decarburization, are important also.

Reproducibility of Test Results. The Army Materials and Mechanics Research Center (formerly Watertown Arsenal Laboratory) conducted a closely controlled experiment that established the Charpy V-notch impact test as both reliable and reproducible. A total of 1200 specimens from a single heat of aircraft-quality 4340 steel were divided into three groups and heat treated to three different ranges of hardness: 43 to 46, 32.5 to 36.5 and 26 to 29 HRC. Two hundred specimens at each hardness level were impact tested in each of two Charpy machines manufactured by two companies; the average impact-energy values and distribution of results are shown in Fig. 31.

The test program just described clearly demonstrated the narrow spread of results that can be obtained under carefully controlled testing conditions. However, the experience of other laboratories indicates that even when preparation and testing of impact specimens are closely controlled, a considerable spread of test results can still occur.

DUCTILE-TO-BRITTLE FRACTURE TRANSITION

Plain carbon and low-alloy steels, like many metals with body-centered-cubic lattice structures, are susceptible to a lowering of absorbed impact energy with decreasing temperature, either in service or during testing. This change is accompanied by a transition from a fibrous to a crystalline-appearing fracture surface. A somewhat arbitrarily defined temperature in the transition range is called the ductile-to-brittle fracture transition temperature, or T_c. Because there are several ways of defining T_c, the criterion used to establish the value should be stated whenever a value of T_c is given. Because brittleness is seldom desirable, it usually is necessary to prevent

brittle behavior under impact loading by keeping T_c below the expected service temperature.

Charpy notched-bar impact tests are especially effective in determining T_c for plain carbon and low-alloy steels. Figure 32 shows that, in a plot of energy absorbed during fracture versus testing temperature, there is a sharp drop in absorbed energy as the testing temperature decreases. This drop is called the energy transition. The value of T_c associated with energy transition is called the energy transition temperature. As shown in Fig. 32, in this series of tests on low-carbon steel plate the drop in energy for the Charpy keyhole specimens was steeper and more sharply defined than for the Charpy V-notch specimens; thus, the value of T_c for keyhole specimens was more precise. However, the value of T_c for keyhole specimens was lower than the temperatures at which brittle fractures occured in service and lower than T_c for V-notch specimens. On the other hand, T_c for V-notch specimens correlated well with the temperatures at which service failures occurred in components made of this steel.

The wide range of temperature over which V-notch energy transition occurs (see Fig. 32) emphasizes the importance of the criterion for defining T_c. One criterion is the average-energy criterion (the temperature corresponding to the median between the maximum energy, or upper-shelf energy, and the minimum energy, or lower-shelf energy). In many instances, the temperature at which the transition curve crosses an arbitrary value of absorbed energy (such as 20 or 40 J, or 15 or 30 ft · lb) is chosen. An alternative criterion is the average-temperature criterion (the median temperature of the transition range).

Transition temperature also can be defined by observing the change in fracture appearance from fibrous to crystalline as the test temperature is reduced. The fracture appearance changes because above T_c fracture occurs predominantly by microvoid coalescence whereas below T_c fracture occurs predominantly by cleavage. Some specifications (particularly those of ASTM) define the temperature at which specimens fracture 50% in shear and 50% in cleavage as the transition temperature; this value of T_c is usually referred to as the 50% FATT. Other specifications use 100%

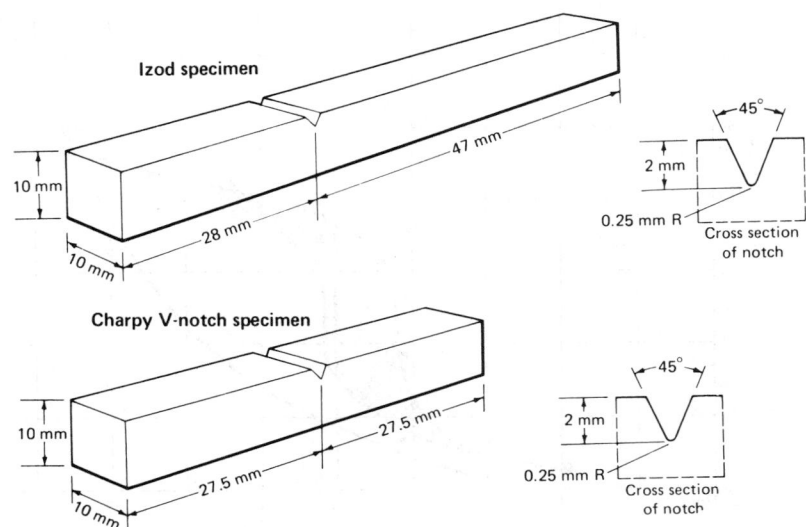

Izod and Charpy V-notch test specimens used for evaluation of notch toughness (ASTM E23)

Fig. 30. Impact test specimens

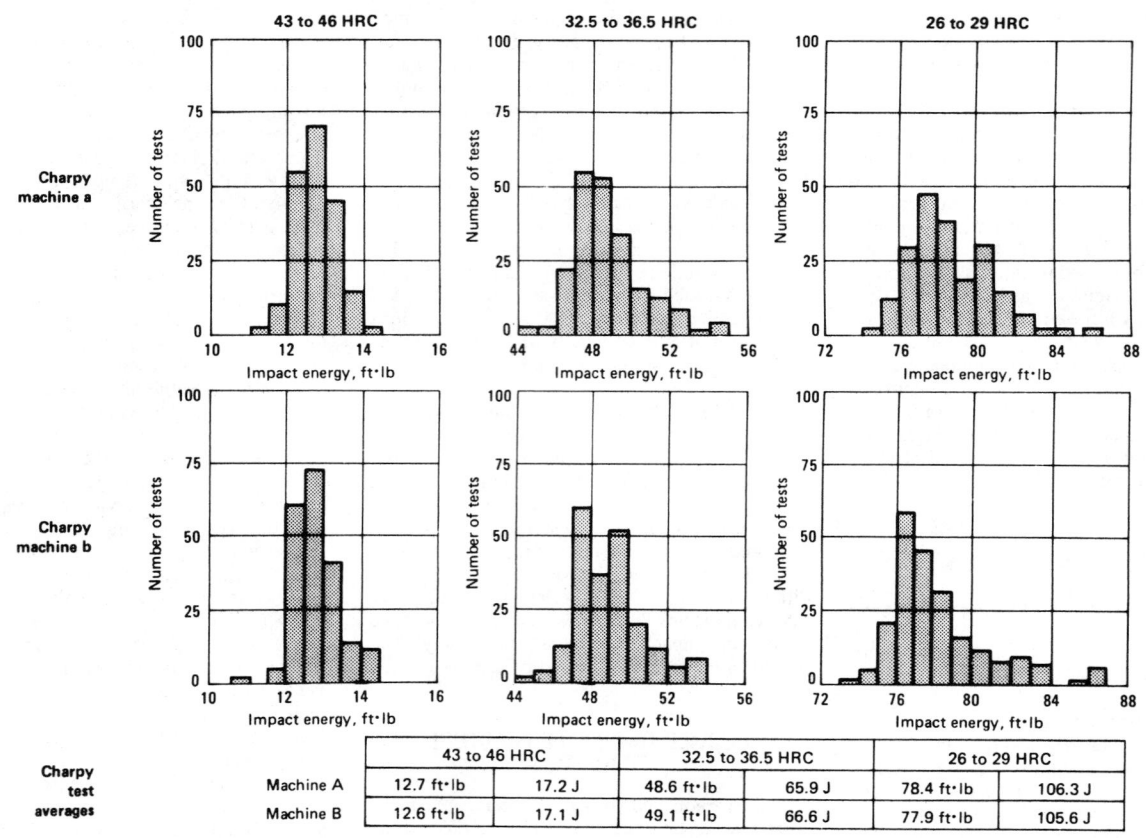

		43 to 46 HRC		32.5 to 36.5 HRC		26 to 29 HRC	
Charpy test averages	Machine A	12.7 ft·lb	17.2 J	48.6 ft·lb	65.9 J	78.4 ft·lb	106.3 J
	Machine B	12.6 ft·lb	17.1 J	49.1 ft·lb	66.6 J	77.9 ft·lb	105.6 J

Comparison of test results from two Charpy impact machines manufactured by two companies. 1200 specimens were made from a single heat of aircraft-quality 4340 steel. Specimens were hardened and tempered to three hardness levels: 43 to 46 HRC, 32.5 to 36.5 HRC and 26 to 29 HRC. 200 specimens at each of the three hardness levels were tested at 21 °C (70 °F) on each of the impact machines.

Fig. 31. Reproducibility of impact testing

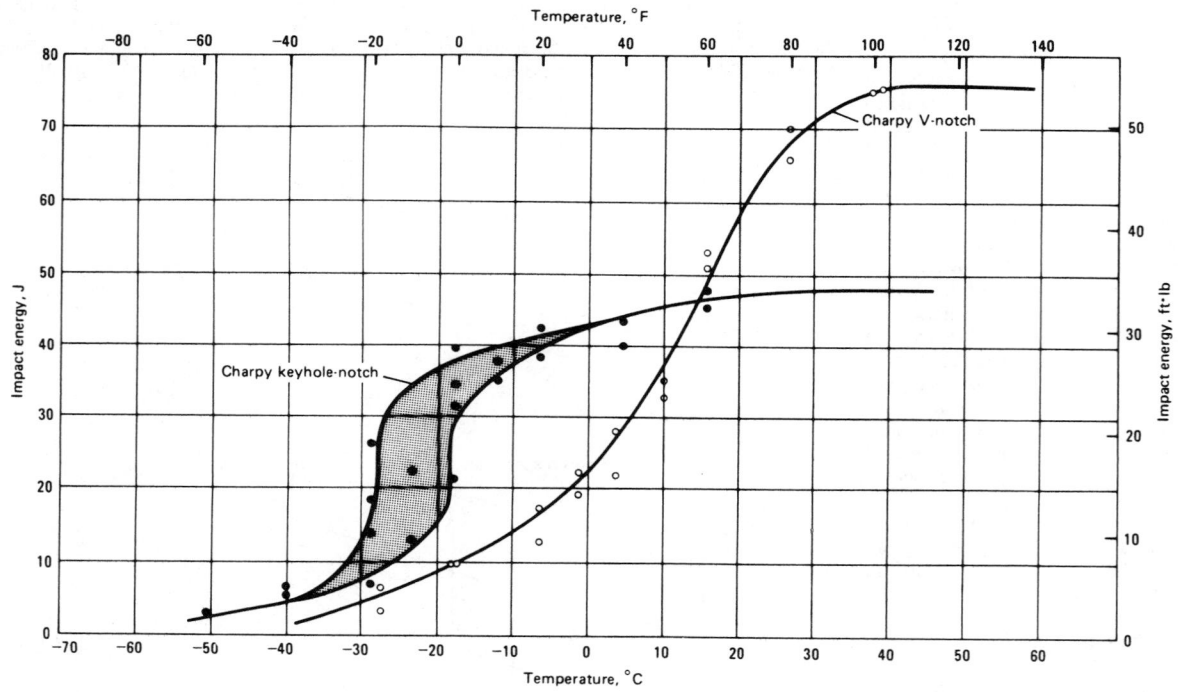

Charpy V-notch and Charpy keyhole-notch impact energy for semikilled low-carbon steel (0.18 C, 0.54 Mn, 0.07 Si) tested over the ductile-to-brittle transition temperature range.

Fig. 32. Transition temperature behavior of low-carbon steel

shear as the criterion, which defines the 100% fracture appearance transition temperature (100% FATT). Still other specifications, such as those for ship plate, require a minimum impact energy at a given testing temperature — 20 J at 5 °C (15 ft·lb at 40 °C), for example.

Selection of the method most appropriate for a given application is difficult and requires an understanding of both toughness testing and service behavior. Often, additional tests are needed to establish a correlation between the transition temperature determined using a certain method and service behavior of a specific structure made of the same material.

EFFECTS OF COMPOSITION

The composition of a steel, as well as its microstructure and processing history, significantly affects both T_c and the energy absorbed during fracture at any particular temperature. The effects of the various alloying elements and those of microstructural and processing variables are intimately interrelated.

Carbon increases transition temperature and decreases upper-shelf fracture energy. These effects, measured by Charpy V-notch impact tests, are shown in Fig. 33. Carbon is one of the more potent alloying elements in its effect on notch toughness. Consequently, for maximum toughness, the carbon content should be kept as low as possible, consistent with strength requirements.

Manganese can substantially reduce the transition temperature of low-carbon steels, as shown in Fig. 34.

In a hardened-and-tempered steel, manganese can have the opposite effect, as illustrated in Fig. 35. Manganese can make the steel susceptible to temper embrittlement and it may cause the formation of brittle upper bainite (rather than fine pearlite) during normalizing.

Sulfur. The effect of sulfur on notch toughness of steels is directly related to deoxidation practice. For rimmed, semikilled and Si-killed steels, sulfur in amounts up to about 0.04% has a negligible effect on notch toughness. However, for Si-Al-killed steels, a reduction in sulfur content can substantially increase upper-shelf energy.

Phosphorus has a strongly deleterious effect on the notch toughness of steel; it raises T_c about 7 °C (13 °F) for each 0.01% P and also reduces uppershelf energy. Phosphorus also increases the susceptibility of some steels to temper embrittlement.

Silicon, used in amounts of 0.15 to 0.30% to deoxidize steels, lowers T_c and raises upper-shelf energy. Compared with rimmed or semikilled steels, Si-killed steels are cleaner and have more uniform ferrite grains. These effects probably are caused by variations in steelmaking practice characteristic of the deoxidation methods used rather than by the silicon content. T_c is appreciably raised by silicon in excess of 0.06%.

Nitrogen, by itself, lowers upper-shelf energy and raises transition temperature. However, most nitrogenized steels are deoxidized with silicon and aluminum, both of which combine with nitrogen. Aluminum nitrides precipitated during cooling serve to stabilize grain size and thus improve the notch toughness of these steels.

Nickel, like manganese, is useful for improving notch toughness of steels at low temperatures. The effects of nickel on the notch toughness (and the fracture toughness) of steels are illustrated in Fig. 36. Nickel is less effective in improving the

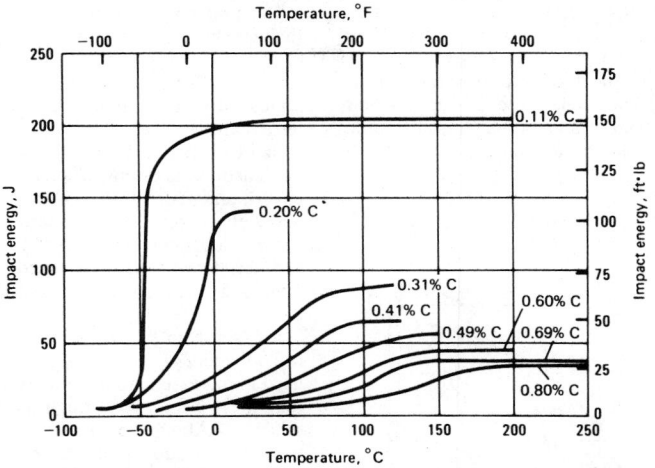

Variation in Charpy V-notch impact energy with temperature for normalized plain carbon steels of various carbon contents.

Fig. 33. Effect of carbon content on notch toughness

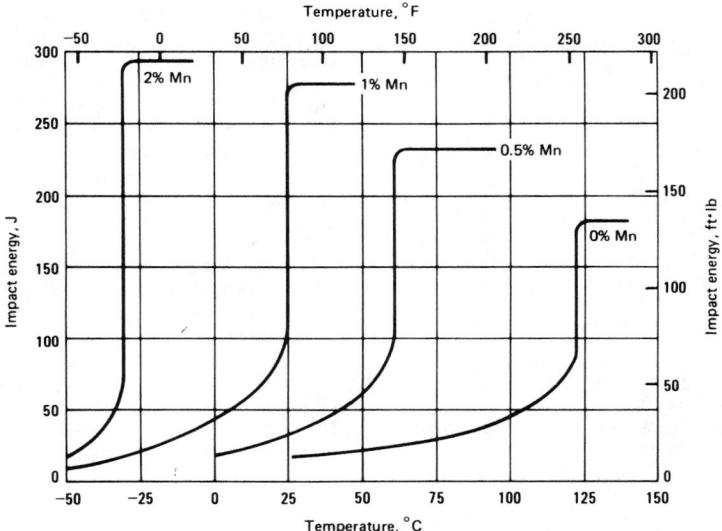

Variation in Charpy V-notch impact energy with temperature for furnace-cooled Fe-Mn-0.05C alloys containing various amounts of manganese.

Fig. 34. Effect of manganese content on notch toughness

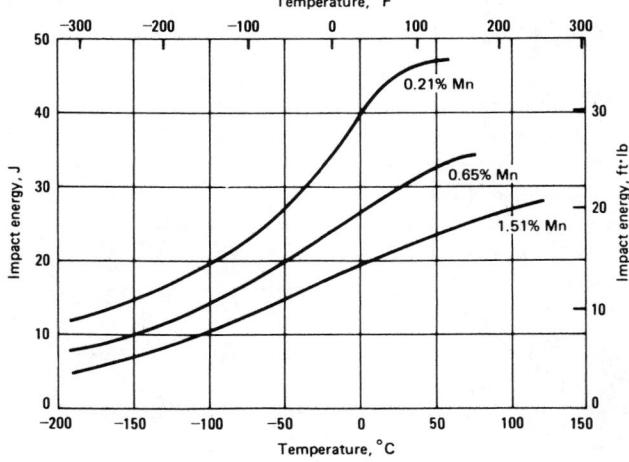

Variation in Charpy V-notch impact energy with temperature for alloy steels containing 0.35 C, 0.35 Si, 0.80 Cr, 3.00 Ni, 0.30 Mo, 0.10 V and the indicated amounts of manganese; the steels were hardened and tempered to a yield strength of approximately 1175 MPa (170 ksi). The microstructures of these steels contained tempered martensite.

Fig. 35. Effect of manganese content on notch toughness

toughness of medium-carbon than of low-carbon steels. Some high-nickel alloy steels, such as maraging steels and austenitic stainless steels, do not exhibit the typical ductile-to-brittle transition; the high nickel content reduces upper-shelf fracture energy, but to a level which is still quite acceptable for most applications.

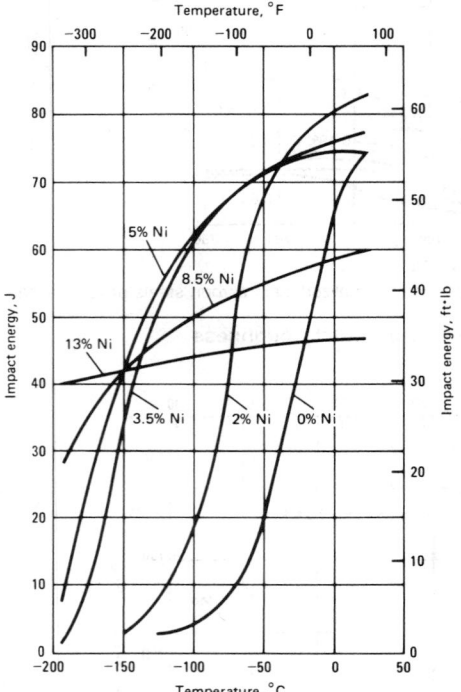

Variation in Charpy keyhole-notch impact energy with temperature for pearlitic low-carbon steels containing various amounts of nickel.

Fig. 36. Effect of nickel content on notch toughness

Chromium raises transition temperature slightly. In steels having chromium contents in excess of 0.90%, it is very difficult to develop those microstructures and mechanical properties that are typical of plain carbon steels; hence impact test results are not comparable. Chromium is usually added to increase hardenability. The increase in hardenability is often sufficient to develop a martensitic microstructure, which provides high upper-shelf energy. Medium-carbon, straight chromium alloy steels, such as 5140, are susceptible to embrittlement when quenched to martensite and tempered between 375 and 575 °C (700 and 1070 °F).

Molybdenum in the quantities usual in alloy steels (up to about 0.40%) raises T_c. It is frequently used to increase hardenability; it influences notch toughness primarily through its effect on microstructure. About 0.5 to 1.0% molybdenum may be added to alloy steels to reduce susceptibility to temper embrittlement, but it is effective only for relatively short times of heating at embrittling temperatures. Molybdenum appears to delay rather than to eliminate temper embrittlement, because steels containing small amounts of this element have become embrittled upon prolonged exposure within the embrittling temperature range.

Copper in steels that have not been subjected to precipitation hardening appears to be moderately beneficial to low-temperature notch toughness. Copper promotes precipitation hardening in steel and, as a result, may adversely affect notch toughness, particularly if the tempering temperature is between 400 and 565 °C (750 and 1050 °F).

Vanadium, niobium and titanium are most often used in steels that receive controlled thermomechanical treatment. Consequently, the toughness of steels containing these elements is largely a function of mill processing. When the steel is finished at temperatures below about 925 °C (1700 °F), such as is characteristic of certain HSLA steels, vanadium, niobium and titanium improve toughness primarily by refining the fer-

rite grain size. At higher finishing temperatures, these elements may be detrimental to toughness.

EFFECTS OF MANUFACTURING PRACTICE

Deoxidation Practice. The effect of deoxidation practice on notch toughness of steels is directly traceable to the presence of those alloying elements and impurities characteristic of the deoxidation practice. Rimmed steels typically contain appreciable quantities of oxygen and nitrogen; soundness and homogeneity of rimmed steel ingots is often poor. These characteristics of rimmed steels account for the poor notch toughness of rimmed steels. Killed steels, particularly Si-Al-killed steels, have lower transition temperatures and higher upper-shelf energy values than rimmed steels. Semikilled steels have toughness properties between those of rimmed and killed steels. A comparison of the notch toughness of these three types of steel is shown in Fig. 37.

Hot Rolling Temperature. The effect of hot rolling practice on notch toughness of steel can be directly related to the microstructure produced during hot rolling. Steels rolled at temperatures above about 980 °C (1800 °F) undergo considerable recrystallization and grain growth during rolling; the structure thus obtained is only slightly affected by the rolling process. Notch toughness of steels rolled at such high temperatures is determined largely by the size to which the austenite grains grow after recrystallization. Petch has shown the influence of grain size on transition temperature (see Fig. 38). When steels are hot rolled at lower temperatures, recrystallization and growth of austenite grains cannot proceed to the extent possible at higher rolling temperatures. Thus the transition temperature can be significantly lowered by rolling in the lower portion of the austenite temperature range, as shown in Fig. 39.

Toughness Anisotropy. Steels can acquire strongly anisotropic microstructures as a result of working. Anisotropic microstructures often are indic-

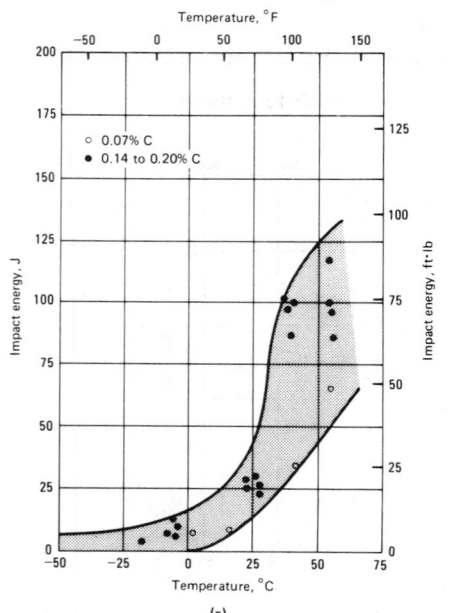

(a)

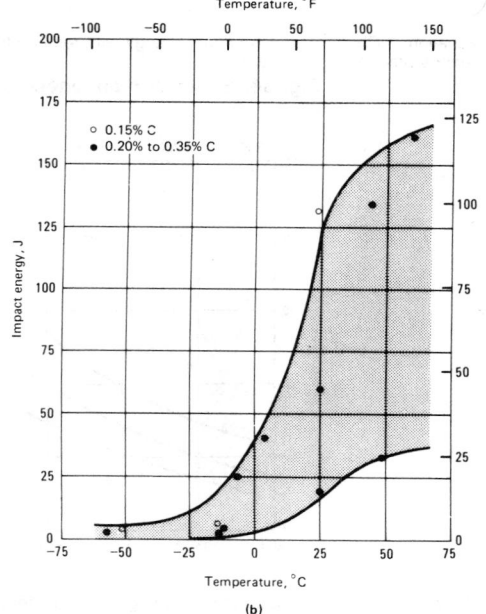

(b)

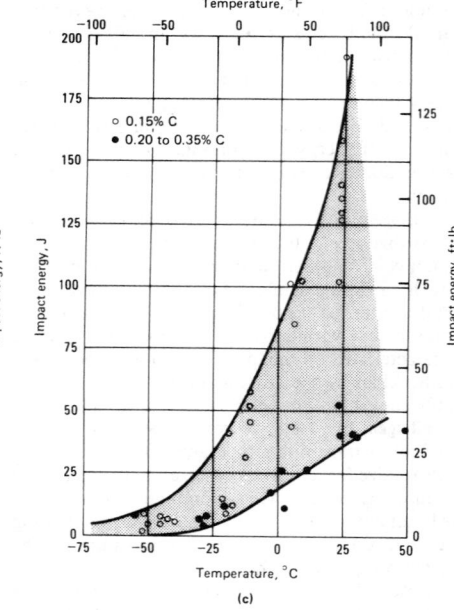

(c)

Variation in Charpy V-notch impact energy with temperature for (a) rimmed, (b) semikilled and (c) killed plain carbon steels.
Fig. 37. Effect of deoxidation practice on notch toughness

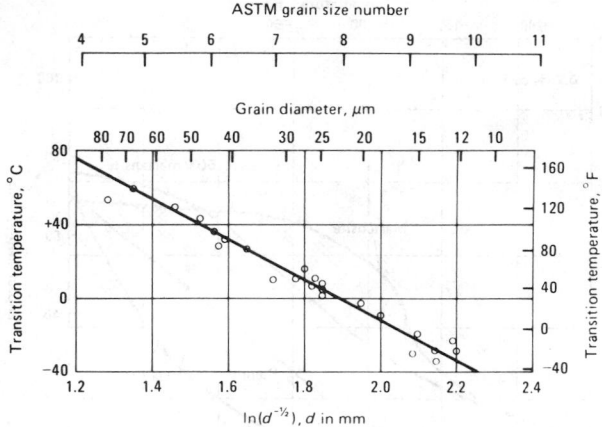

Variation in fracture appearance transition temperature with ferritic grain size for 0.11% C mild steel. Transition temperature varies linearly with $\ln(d^{-1/2})$ and is lower for fine-grained steel.

Fig. 38. Effect of grain size on transition temperature

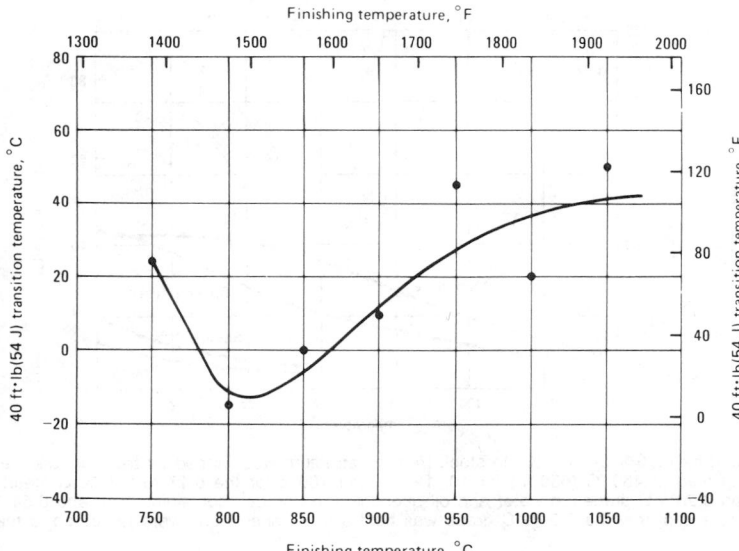

Variation in 40 ft·lb (54 J) Charpy V-notch transition temperature with hot rolling finishing temperature for Si-killed 0.24%C-1.69%Mn steel.

Fig. 39. Effect of finishing temperature on notch toughness

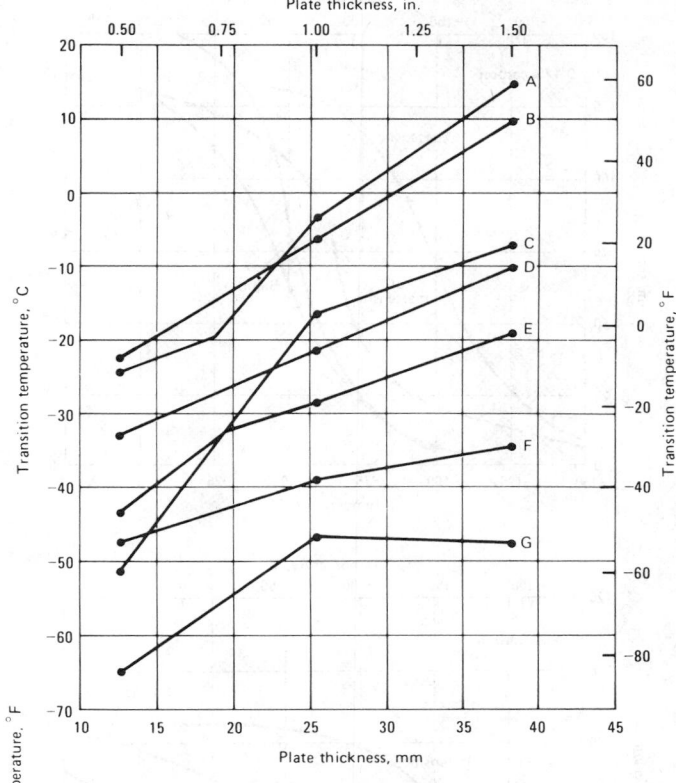

Steel	Composition, %	
	C	Mn
A (semikilled)	0.23	0.45
B (aluminum killed)	0.23	0.45
C (silicon-aluminum killed)	0.16	0.45
D (silicon killed)	0.23	0.45
E (semikilled)	0.16	0.75
F (semikilled)	0.16	0.95
G (silicon-aluminum killed)	0.16	0.75

Variation in Charpy keyhole-notch transition temperature with plate thickness for seven low-carbon steels.

Fig. 41. Effect of plate thickness on notch toughness

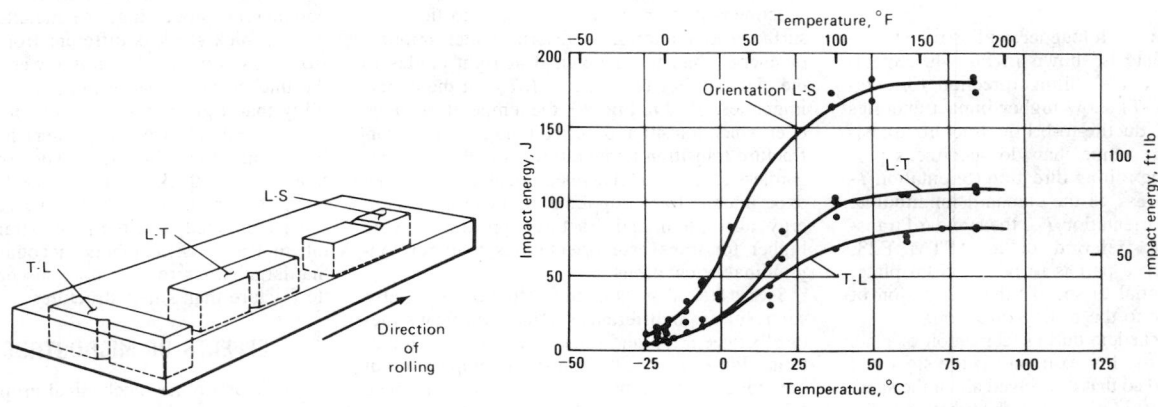

Variation of Charpy V-notch impact energy with notch orientation and temperature for steel plate containing 0.012% C.

Fig. 40. Toughness anisotropy in as-rolled low-carbon steel plate

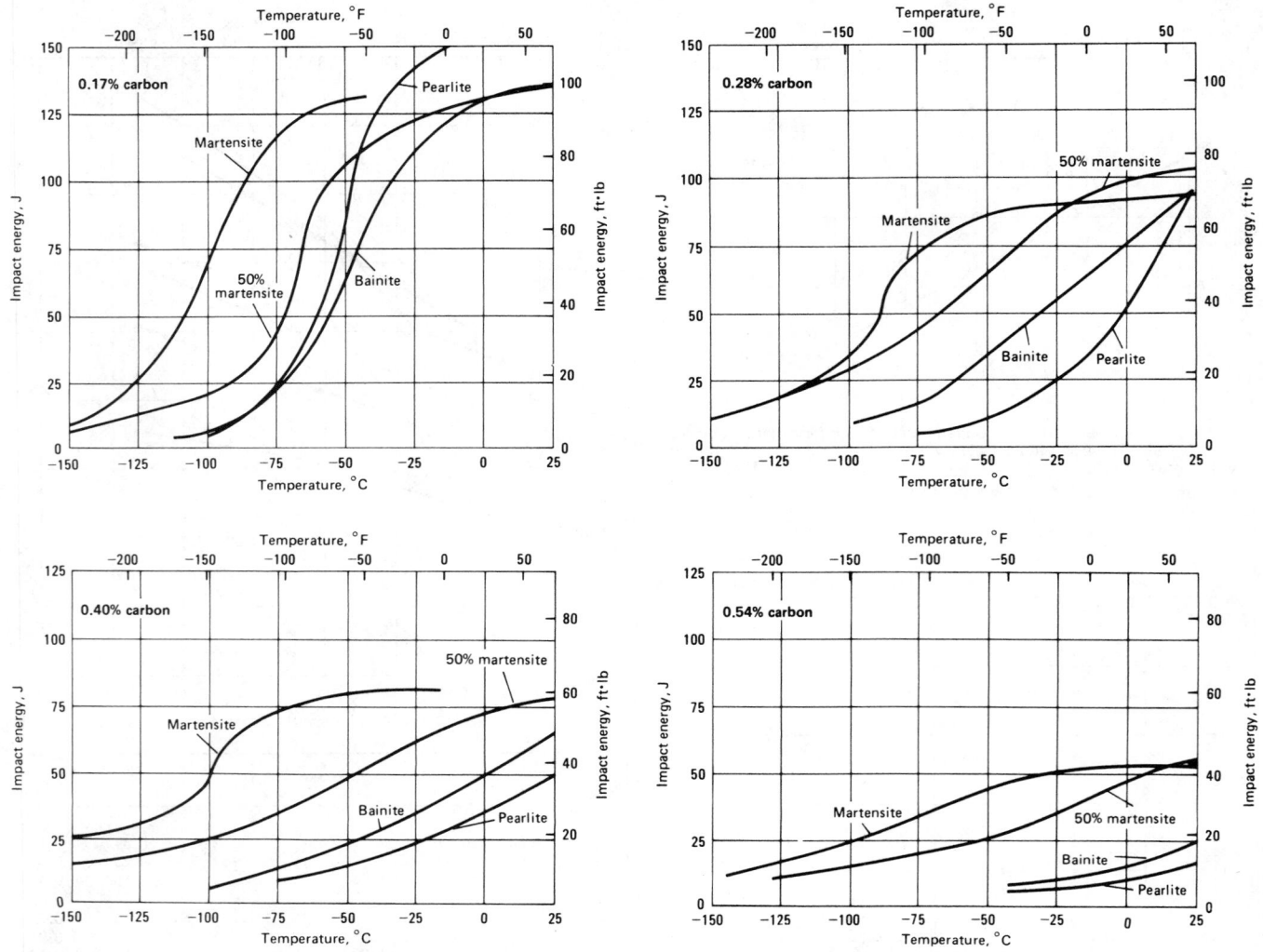

Variation in Charpy V-notch impact energy with microstructure and carbon content for 0.70% Cr, 0.32% Mo steel. Pearlitic structure was formed by transformation at 650 °C (1200 °F). A structure with 50% martensite was formed by quenching in lead at 450 °C (850 °F) for 10, 19, 35 and 100 s for the 0.17 to 0.54% C steels, respectively. Fully martensitic structures were formed by quenching the 0.17 and 0.28% C grades in water and oil quenching the grades containing 0.40 and 0.54% C. Bainite was formed by quenching in lead at 450 °C (850 °F) and holding 1 h, except that the 0.54% C grade was held 3 h. All specimens were tempered to the same hardness level.

Fig. 42. Effect of microstructure on notch toughness

ative of anisotropy of mechanical properties, particularly notch toughness. Anisotropy, therefore, is an important consideration in design and fabrication of rolled, forged, drawn or extruded steel products.

Anisotropy in notch toughness of as-rolled low-carbon steel plate is shown in Fig. 40. Specimens parallel to the rolling direction (orientations *L-S* and *L-T*) show higher impact energies throughout the ductile-to-brittle fracture transition temperature range than do specimens perpendicular to the rolling direction (orientation *T-L*). Orientation *L-T* is the standard longitudinal specimen, and orientation *T-L* the standard transverse specimen, referred to in ASTM E23. Therefore, when a part is to be cut from plate, it may be essential to specify the orientation of the part relative to the rolling direction.

The rolling schedule during fabrication of plate affects anisotropy. For example, if the steel had been cross rolled so that it received about the same amount of hot reduction in both directions, the curves for orientations *L-T* and *T-L* would nearly

coincide at a position between the *L-T* and *T-L* curves for material reduced by conventional rolling.

Regardless of the amount of cross rolling, specimens that are notched parallel to the plate surface (orientation *L-S*) absorb greater amounts of energy than those notched at right angles to the plate surface (orientation *L-T*). In the experiment described in Fig. 40, the temperature range over which transition occurred (and also the shear-fracture transition temperature) was the same regardless of notch or specimen orientation. In other experiments, transition temperatures for both energy absorption and fracture appearance were higher for transverse orientations than for longitudinal orientations.

Section Size. Variations in fracture behavior also can result from differences in metallurgical structure between thin and thick stock of a given material. For example, the transition temperature of hot rolled low-carbon steel varies with plate thickness, as shown in Fig. 41. In these tests, specimen size was constant, yet T_c increased with

increasing plate thickness. Reduction of toughness with increasing plate thickness is not limited to low-carbon steel, but apparently applies to all steels. Because of the characteristics of normal commercial processing, the metallurgical structure of thick stock is different from that of thin stock, resulting in inherently lower toughness for the thicker stock. More importantly, the probability that a given part will contain a crack or a flaw of critical size (or greater) increases with increasing stock thickness. The lower inherent toughness of thick stock and the higher probability that thick stock contains a large crack or flaw, combined with the plane-strain conditions inherent in thick members, account for the fact that large structures are more susceptible to brittle fracture than small structures.

EFFECTS OF MICROSTRUCTURE

Like many other mechanical properties, notch toughness of steel usually can be related directly to microstructure. Because microstructures of steels are readily observed and classified, it is convenient to attribute the various mechanical properties to the microstructure, even though the properties might be more properly attributed to the composition and manufacturing history of the steel.

Microstructural Constituents. Of the major microstructural constituents found in steels, ferrite has the highest transition temperature, followed by pearlite, upper bainite, and finally tempered martensite or lower bainite. Values of notch toughness for similar steels having different carbon contents and microstructural constituents are shown in Fig. 42; to facilitate comparison of the different microstructures, these steels were tempered to uniform strength levels before testing. In practice, the cooling or quenching rate and the time-temperature transformation characteristics of a steel (including its conventional hardenability) determine the resulting microstructure or mixture of structures. Transformation characteristics, in turn, are controlled by alloy content, austenitizing temperature and austenite grain size.

Generally, treatments that produce microstructures with inferior room-temperature toughness also raise T_c. Precipitated second-phase particles are detrimental to toughness, especially if located at grain boundaries. However, spheroidization treatment can improve toughness by reducing strength and eliminating ferrite plates, which are paths of easy cleavage fracture in pearlite. Steels also can lose toughness because of various embrittlement phenomena.

Isothermally transformed lower bainite has superior toughness, and a slightly lower transition temperature, than tempered martensite of the same strength (see Fig. 43). However, mixed structures, which result from incomplete bainitic treatments causing partial transformation to martensite, have lower toughness and much higher transition temperatures than either 100% tempered martensite or 100% lower bainite. Thus, it is important that bainitic treatments be carried to completion to avoid the adverse effects of mixed structure.

Finally, the presence of austenite inhibits the fast propagation of cleavage fracture in some ferritic and martensitic steels.

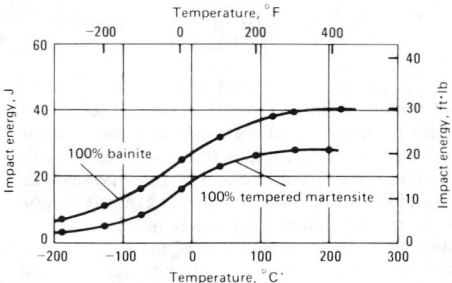

Variation in Charpy V-notch impact energy with temperature for specimens of 4340 steel having martensite and 100% bainite microstructures. All specimens were austenitized for 30 min at 843 °C (1550 °F) in neutral salt. 100% bainite was produced by isothermal transformation for 1 h at 315 °C (600 °F) in agitated salt. 100% tempered martensite was produced by quenching in agitated oil at 49 to 57 °C (120 to 135 °F) and tempering at 315 °C (600 °F). All specimens had the same tensile strength.

Fig. 43. Comparison of notch toughness for different microstructures

Corrosion Characteristics of Carbon and Alloy Steels

Types of Corrosive Environments

CORROSION OF METALS is defined as deterioration caused by chemical or electrochemical reaction of the metal with its environment. Usually, the corroding medium is a liquid, but gases and solids can also act as corroding media.

Corroding media are generally classified as aqueous or nonaqueous. Aqueous media include all types of water—natural, purified and contaminated—ranging from deionized and distilled waters, which contain relatively few dissolved substances, through natural fresh waters, which may contain up to 10 000 ppm or more of various minerals or pollutants, to mine waters, deep-well brines, seawater and certain process waters, which may contain several percent of one or more dissolved chemical substances such as acids, bases, salts and gases.

Nonaqueous corrodents include dry gases, organic liquids and liquid metals.

CORROSION BY AQUEOUS MEDIA

Wherever water is involved, corrosion of iron and steel can be presumed to proceed by electrochemical action—that is, local anodic (active) and cathodic (noble) sites develop on the surface of the metal, and corrosion (oxidation) occurs at the anodic sites: $Fe^0 \rightarrow Fe^{++} + 2e^-$. At the cathodic sites, a reduction reaction occurs which, depending on the pH of the environment, can be hydrogen evolution ($2H^+ + 2e^- \rightarrow H_2\uparrow$) or oxygen reduction [$O_2 + 2H_2O + 4e^- \rightarrow 4(OH)^-$]. The electrodes liberated at the anode flow through the metal to the cathodic sites, where they are consumed by the cathodic reaction. In near-neutral or alkaline pH solutions, some dissolved oxygen or other reducible species must be present to support the cathodic reaction in order for appreciable corrosion to occur. The presence of dissolved salts generally increases corrosion rate, because it increases the conductivity of the solution and thus increases corrosion currents.

Galvanic cells can be set up by placing iron or steel in an aqueous solution while the ferrous metal is in electrical contact with another metal whose galvanic potential is more noble (more cathodic) than that of the ferrous metal. In this case a larger portion of the surface of the ferrous metal becomes anodic, and corrosion rate increases. Cells also can be set up when a local area of a ferrous metal is shielded from contact with water containing oxygen; this is called oxygen concentration-cell corrosion or crevice corrosion, because it is often seen in crevices, where oxygen is available at the edge of the crevice but is effectively absent within the crevice.

Water is the active corrodent even for some media that are not considered aqueous. Corrosion of ferrous metals in air increases when the surface is moistened regularly, as occurs in locations where dew regularly forms at night.

CORROSION BY NONAQUEOUS MEDIA

Corrosion by most nonaqueous media except liquid metals occurs most often by direct chemical action resulting in loss of metal (e.g., high-temperature oxidation in air) or formation of a converted surface layer (e.g., carburization, nitriding). Most dry gases and organic liquids are noncorrosive to steel at ordinary temperatures, but can become corrosive at elevated temperatures.

Corrosion in liquid-metal systems almost always involves some form of dissolution or leaching, in which atoms leave the solid surface and go into solution in the liquid metal. In a flowing system, particularly a recirculating system, the corrosion process is often controlled by the difference in temperature between the hottest and coldest points in the system, because solubility is greater as temperature increases.

Carbon and low-alloy steels can be embrittled by contact with liquid brass, aluminum bronze, copper, zinc, lead-tin solders, lithium, cadmium and indium. The degree of embrittlement is greater for steels that have been heat treated or alloyed to produce higher strength.

Atmospheric Corrosion

ALL UNPROTECTED carbon and alloy steels corrode to some extent when exposed to the atmosphere. The corrosion process can occur relatively rapidly for unalloyed iron, especially in moist air containing specific airborne contaminants. The addition of certain alloying elements—notably silicon, copper, chromium and nickel—slows the rate of corrosion. The degree of effectiveness of each alloying element on slowing corrosion depends on both the climatic conditions of exposure and the amount of the alloying element present.

Atmospheric corrosion of ferrous metals proceeds by an electrochemical reaction in which the electrolyte is based on moisture from the air—either moisture that falls as precipitation or moisture that condenses as dew on exposed surfaces when the temperature of relatively humid air decreases.

GEOGRAPHIC AND METEOROLOGICAL FACTORS

It is generally accepted that the principal environmental contaminants responsible for increased ferrous-metal corrosion in urban areas are the gaseous sulfur oxides that originate from the combustion of fossil fuels. The airborne seawater particles that originate in the ocean are responsible for accelerated corrosion in coastal areas. Airborne salts from arid soils of the western states and windborne fertilizer plus silicious and calcareous dust from midwestern rural areas likewise accelerate the atmospheric corrosion of ferrous metals. Complementing these ionic materials are nonionic dust particles that serve as nuclei for condensing dew.

ROLE OF DEW

Dew condensate can be a major factor in promoting corrosion. Condensed dew from a semi-industrial location was reported to have a pH of 4.5; rainwater plus dew collected from a gutter in an eastern industrial location had a pH of 3.5. Such values are sufficiently acidic to stimulate the corrosion of carbon steel through acid attack. Dew cleans the atmosphere by nucleating on particulate matter, dissolving gaseous contaminants, and subsequently clearing both from the air as it condenses on surfaces. Unfortunately, such condensed dew is damaging to both metals and concrete.

CARBON STEELS

For many years, Committee A-5 of the American Society for Testing and Materials has exposed carbon steel specimens for the calibration of various test sites. The standard rural test site, State College, PA, is very consistent in performance, as shown by corrosion rates of 1- and 2-year calibrating specimens for two different time periods about a decade apart (see Table 1).

Table 1. Corrosion rates of carbon steel calibrating specimens at State College, PA

Exposure period	Length of exposure, yr	Average corrosion rate µm/yr	mils/yr
1948-1955	1	26.9	1.06
	2	23.1	0.91
	4	23.4	0.92
1960-1962	1	25.1	0.99
	2	23.1	0.91

From 1960 to 1964, specimens of carbon steel were exposed for periods of 1 and 2 years in 46 locations, 14 of which were in foreign countries. The locations ranged from tropical to polar, industrial to rural, and marine to arid. The average corrosion rates for 2-year exposures are given in Table 2 for 34 of the 46 locations. The greatest difference in corrosivity (Table 2) is the 1400-fold span between polar Norman Wells and near-tropical Cape Kennedy (beach site). The span between arid, rural Phoenix and rural State College shows a fivefold difference in corrosivity. A twofold difference exists between rural State College and industrial Newark. These data offer a preliminary appraisal of the atmospheric corrosion that might be expected when unprotected carbon steel structures are exposed in a particular location.

HIGH-STRENGTH LOW-ALLOY STEELS

One of the significant observations that led to the development of high-strength low-alloy steels was the improved atmospheric corrosion resistance exhibited by carbon steel containing 0.05 to 0.2% copper. Buck performed one of the earliest systematic studies, in which two levels of copper (0.15 and 0.25%) were added to corrugated carbon steel sheets used in 18° inclined roofs for outdoor structures. Weight-loss specimens were exposed simultaneously. The results from three environments — severe industrial, marine and rural — indicated that copper reduced corrosion. In the industrial location, copper-bearing steel lost 40 to 48% less weight than did copper-free steel. In the marine environment, copper-bearing steel lost 51 to 61% less weight than copper-free steel.

Effect of Composition. Corrosion test results from a systematic study, involving 15.5 years of atmospheric exposure of 270 steels in three environments, were reported by Larrabee and Coburn. The alloying elements studied were copper, nickel, chromium, silicon and phosphorus, all at levels of less than 1.3%. All these alloying elements were found to be beneficial, with the greatest change in corrosion resistance being caused by increasing copper content. The industrial site was at Kearny, NJ, the semirural site at South Bend, PA, and the marine site at Kure Beach, NC, about 240 m (800 ft) from the ocean. Some of the results of the study at the industrial site are illustrated in Fig. 1. The "HSLA steel" (bottom curve) was comparable to ASTM A242.

Corrosion in Selected Environments. Table 3 compares carbon steel with copper-bearing steel and proprietary high-strength low-alloy steels of the

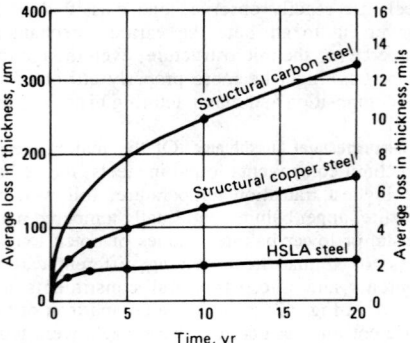

Fig. 1. Atmospheric corrosion versus time in a semi-industrial or industrial environment

ASTM types A242, A588 and A514 in various rural, industrial and marine environments. A significant improvement in performance has been achieved by all of the high-strength low-alloy steels in comparison with structural copper steel and structural carbon steel.

Soil Corrosion

SOIL CORROSION may be considered to encompass all corrosion taking place on buried structures. Soil corrosion has been ascribed to low pH, stray currents, reactive chemicals, low resistivity, and bacterial action; however, oxygen and water are considered to be the key factors necessary for corrosion.

Table 2. Corrosion rates of carbon steel calibrating specimens at various locations

Location	Type of environment	Corrosion rate(a) µm/yr	mils/yr
Norman Wells, NWT, Canada	Polar	0.76	0.03
Phoenix, AZ	Rural arid	4.6	0.18
Esquimalt, Vancouver Island BC, Canada	Rural marine	13	0.5
Detroit, MI	Industrial	14.5	0.57
Fort Amidor Pier, CZ	Marine	14.5	0.57
Morenci, MI	Urban	19.5	0.77
Potter County, PA	Rural	20	0.8
Waterbury, CT	Industrial	22.8	0.89
State College, PA	Rural	23	0.9
Montreal, Que, Canada	Urban	23	0.9
Durham, NH	Rural	28	1.1
Middletown, OH	Semi-industrial	28	1.1
Pittsburgh, PA	Industrial	30	1.2
Columbus, OH	Industrial	33	1.3
Trail, BC, Canada	Industrial	33	1.3
Cleveland, OH	Industrial	38	1.5
Bethlehem, PA	Industrial	38	1.5
London, Battersea, England	Industrial	46	1.8
Monroeville, PA	Semi-industrial	48	1.9
Newark, NJ	Industrial	51	2.0
Manila, Philippine Islands	Tropical marine	51	2.0
Limon Bay, Panama, CZ	Tropical marine	61	2.4
Bayonne, NJ	Industrial	79	3.1
East Chicago, IN	Industrial	84	3.3
Brazos River, TX	Industrial marine	94	3.7
Cape Kennedy, FL(b) (60 ft elev., 60 yd from ocean)	Marine	132	5.2
Kure Beach, NC (800 ft from ocean)	Marine	147	5.8
Cape Kennedy, FL(b) (30 ft elev., 60 yd from ocean)	Marine	165	6.5
Daytona Beach, FL	Marine	295	11.6
Cape Kennedy, FL(b) (ground level, 60 yd from ocean)	Marine	442	17.4
Point Reyes, CA	Marine	500	19.7
Kure Beach, NC (80 ft from ocean)	Marine	533	21.0
Galeta Point Beach, Panama, CZ	Marine	686	27.0
Cape Kennedy, FL(b) (beach)	Marine	1070	42.0

(a) Two-year average. (b) Now known as Cape Canaveral.

Table 3. Corrosion of structural steels in various environments

Type of atmosphere	Time, yr	Structural carbon steel	Structural copper steel	UNS K11510(b)	UNS K11430(c)	UNS K11630(d)	UNS K11576(e)
Industrial	3.5	3.3	2.6	1.3	1.8	1.4	2.2
(Newark, NJ)	7.5	4.1	3.2	1.5	2.1	1.7	...
	15.5	5.3	4.0	1.8	...	2.1	...
Semi-industrial	1.5	2.2	1.7	1.1	1.4	1.2	1.6
(Monroeville, PA)	3.5	3.7	2.5	1.2	2.1	1.4	2.4
	7.5	5.1	3.2	1.4	2.4	1.7	...
	15.5	7.3	4.7	1.8	...	1.8	...
Semi-industrial	1.5	1.8	1.4	1.0	1.3	1.0	1.5
(South Bend, PA)	3.5	2.9	2.2	1.3	1.9	1.5	2.4
	7.5	4.6	3.2	1.8	2.7	1.9	...
	15.5	7.0	4.8	2.2	...	2.5	...
Rural	2.5	...	1.3	0.8	1.2	...	...
(Potter County, PA)	3.5	2.0	1.7	1.1	1.4	1.2	1.8
	7.5	3.0	2.5	1.3	1.5	1.5	...
	15.5	4.7	3.8	1.4	...	2.0	...
Moderate marine	0.5	0.9	0.8	0.6	0.8	0.7	1.0
(Kure Beach, NC,	1.5	2.3	1.9	1.1	1.7	1.2	1.7
800 ft from ocean)	3.5	4.9	3.3	1.8	2.5	1.9	2.2
	7.5	5.6	4.5	2.5	3.7	2.9	...
Severe marine	0.5	7.2	4.3	2.2	3.8	1.1	0.7
(Kure Beach, NC,	2.0	36.0	19.0	3.3	12.2	...	2.1
80 ft from ocean)	3.5	57.0	38.0	...	28.7	3.9	3.9
	5.0	(f)	(f)	19.4	38.8	5.0	...

(a) To obtain equivalent values in μm, multiply listed value by 25. (b) ASTM A242 (type 1). (c) ASTM A588 (grade A). (d) ASTM A514 (type B) and A517 (grade B). (e) ASTM A514 (type F) and A517 (grade F). (f) Specimen corroded completely away.

SOIL CHARACTERISTICS

Soil is a heterogeneous, porous material that ranges in character from soft spongy peats through soft clays, loams, and silts to coarse-grained sands and gravels. There are two large soil classes in the United States; west of the Mississippi River, soils generally accumulate lime and tend to be alkaline, whereas east of the Mississippi they do not and, therefore, tend to be acidic. Soil classes may be subdivided into texture groups containing varying amounts of clay, silt and sand. Well-aerated soils are red, yellow or brown due to the presence of ferric iron compounds; gray soils contain ferrous iron compounds. Dark soils contain decaying organic matter. Soil analysis in corrosion studies involves only substances present as water-soluble compounds, including base formers such as sodium, potassium, calcium and magnesium, and acid formers such as carbonate, bicarbonate, chloride, nitrate and sulfate.

Aeration and water-retention characteristics are the chief physical attributes of soil. Coarse soils (sands and gravels) that have good drainage and ample aeration corrode steel at a rate approaching that of the local atmosphere. Clay and silt soils are characterized by fine texture, high water retention, poor aeration and poor drainage, all of which increase corrosion rates for steel. The most severe corrosion usually takes place at low elevations in poorly drained soils where there is minimal aeration.

Differential Aeration. Corrosion by differential aeration can result from various conditions. For example, when a pipe passes through two different soils that differ in oxygen permeability, a galvanic current will flow from the more poorly aerated (anodic) surface of the pipe through the soil to the better aerated (cathodic) surface. If the two surface conditions are widely separated, these galvanic currents, known as long-line currents, can be detected and used to locate the anodic areas.

Dissolved Salts. The most corrosive soils are those

that contain large concentrations of soluble salts. Because of the presence of salts, such soils have relatively high electrical conductivities (or low electrical resistivities). The least-corrosive soils have low concentrations of soluble salts and high resistivities. Resistivity measurements can be obtained readily and provide as much information on corrosion characteristics as do measurements of any other single soil property. Consequently, resistivity is the property most often used to approximate the aggressiveness of a soil. Observations of soil drainage, and/or measurements of pH, supplement resistivity measurements. The following table lists the general relationship that exists between soil resistivity and corrosion of ferrous metals; however, because of other factors this relationship may not always be valid.

Soil resistivity, Ω·m	Classification
<7	Very corrosive
7-20	Corrosive
20-50	Moderately corrosive
>50	Mildly corrosive to noncorrosive

BACTERIAL ACTION

Bacterial action is another factor that influences underground corrosion. Of the various bacteria in soils, the anaerobic sulfate-reducing bacteria are of greatest concern. These bacteria can act as depolarizers for the cathodic reaction $H_2O + e^- \rightarrow H^0 + OH^-$, and thereby allow corrosion to proceed in oxygen-free soils. It should be noted that the conditions favorable to the activity of these bacteria (soluble sulfate) would be detected by the resistivity measurements mentioned above.

PREVENTIVE MEASURES

Although other methods of corrosion control, such as the use of protective coatings, are often

used to combat soil corrosion of steel, the cathodic protection method is regarded as the most economical and effective method available. Cathodic protection offers two important advantages. First, it can be retrofitted to existing structures, usually at a small fraction of the cost of replacing the structure and without major disruption of the facility or service; and second, the effectiveness of the cathodic protection in arresting corrosion can be measured by a simple, nondestructive electrical measurement.

Corrosion in Fresh Water

FRESH WATER includes all nonsaline natural waters, polluted or unpolluted, found in inland bodies such as streams, rivers, ponds and lakes. In addition to nonsaline natural waters, which originate as precipitation such as rain or snow, fresh water includes well water, treated potable water and various process waters that may or may not be treated and may or may not contain higher levels of contaminants than the natural waters from which they were derived. Data on the useful life of mild steel in contact with fresh water are incomplete without information on: (a) mineral quality of the water, (b) acidity, (c) presence or absence of dissolved oxygen, (d) velocity of flow, (e) temperature and (f) environmental conditions such as nonuniformity of a corrosion-product deposit, contact with copper-bearing metals, and the existence of crevices, all of which can lead to severe localized corrosion.

EFFECT OF DISSOLVED GASES

Of all the characteristics of water that influence the rate of corrosion, dissolved gases are probably the most important.

Oxygen. The rate at which oxygen reaches a metal surface controls the rate of corrosion and can determine whether aerobic or anaerobic corrosion results. In addition, oxygen accelerates the action of carbon dioxide, hydrogen sulfide, and other dissolved gases and solids.

Carbon dioxide is only about one-tenth as corrosive as oxygen, and in natural waters the carbon dioxide concentration is important only as it affects the solubility and precipitation of calcium carbonate.

Hydrogen sulfide can cause rapid attack of steel, even in the absence of oxygen; it also produces severe graphitic corrosion of cast iron. The anaerobic corrosion products are sulfur and ferrous sulfide, both of which are cathodic to iron and tend to cause severe pitting. Most of the corrosion problems due to hydrogen sulfide arise when this chemical is produced by bacteria.

EFFECT OF pH

There is little difference in the corrosion rate of steel in natural waters having pH values between 4.5 and 9.5. This has been confirmed in both tap water and distilled water. In this range of pH, the corrosion products maintain a pH of 9.5 next to the steel surface, regardless of the pH of the solution. At a pH of 4 or below, hydrogen evolution begins and corrosion increases rapidly (see Fig. 2).

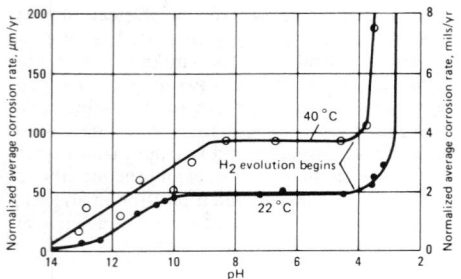

Corrosion rates are normalized to a solution containing 1 mL O_2 per litre of water. To estimate corrosion rates at other concentrations, multiply values derived from this graph by the oxygen concentration in mL/L.

Fig. 2. Effect of pH on corrosion of steel in aerated water

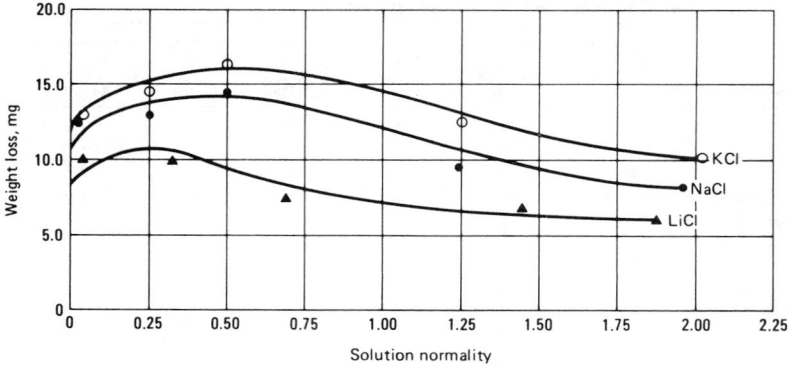

Fig. 3. Effect of alkali metal salt concentration on corrosion of steel at 35 °C (95 °F)

EFFECT OF DISSOLVED SALTS

Most salts dissolved in water tend to decrease the solubility of both oxygen and ferrous hydroxide. Therefore, rates of corrosion in concentrated salt solutions are usually less than those in dilute salt solutions. Some salts also tend to buffer the pH value, which can overcome a tendency toward acidic corrosion. On the other hand, most dissolved salts become ionized and increase the conductivity of water which, in turn, tends to localize corrosion, thus leading to severe pitting. The increased conductivity permits cathodic areas to become larger and causes anode and cathode reaction products to combine in the water instead of at the corroding surface. Cations stimulate corrosion in the following order (magnesium ions having the least effect and ferric ions the greatest): magnesium, cadmium, manganese, calcium, strontium, barium, lithium, sodium, potassium, ammonium, trivalent chromium and trivalent iron (ferric).

Ammonium salts and salts of trivalent elements such as chromium and iron greatly stimulate corrosion because they hydrolyze, creating acidic conditions. Corrosion then occurs by hydrogen evolution as well as by oxygen depolarization. Ammonium chloride is particularly active, even at concentrations as low as 10 ppm.

Copper and Mercury Ions. Certain soluble ions, especially copper and mercury, are capable of undergoing a reaction termed mutual replacement, in which the dissolved ions (which are more noble than iron) deposit on an iron or steel surface and an equal number of iron atoms go into solution. Mutual replacement can take place in the absence of oxygen, and there is no known inhibitor that can prevent this reaction.

Alkali Metal Salts. When salts of lithium, sodium or potassium are added to aerated water, the corrosion rate varies with concentration. At low concentrations, addition of an alkali metal salt increases the corrosion rate; at higher concentrations, the corrosion rate reaches a maximum and then decreases with increasing concentration, as shown in Fig. 3.

CORROSION IN POTABLE WATER SYSTEMS

Steel pipe is not used for domestic potable water. Other steel containers seldom are used for household or industrial water or for water distribution systems without protective coatings of zinc,

cement, enamel, paint, coal tar or a coal-tar-base product. Such coatings are necessary not only to reduce corrosion but also to avoid red water. Steels exposed to water at gaps (holidays) in coatings or at joints will not resist aggressive water, but will remain unattacked in nonaggressive water, in water that has been treated with the proper inhibitor, or when placed under cathodic protection.

CORROSION IN NATURAL WATERS

Steel is used for structures, pilings, and water-intake piping that are completely or partly submerged in various natural waters. In virtually all applications, the corrosion rate is influenced not only by the prevailing oxygen content of the natural water, but also by the presence of minerals or industrial pollutants, or both. The prevailing pH and degree of movement around an exposed steel structure (producing turbulent flow, laminar flow or stagnation) also affect the corrosion rate. Frequently, it is impossible to predict useful life with any reasonable certainty without conducting corrosion tests at the site.

Corrosion
in Seawater

SEAWATER is a relatively uniform saline solution consisting predominantly of sodium and magnesium chlorides dissolved in water. Although many other soluble minerals are present in very small quantities, both the individual and cumulative effects of these minerals are so greatly overshadowed by the effect of the dominant chlorides that seawater can be considered equivalent to a $0.5N$ sodium chloride solution. At this concentration, a sodium chloride solution is at a peak in corrosivity, acting more aggressively toward steel than at either higher or lower concentrations. Factors other than chloride concentration that affect corrosion in seawater include

oxygen concentration, biofouling, water velocity and water temperature. These factors are discussed later in this article.

The corrosion rate most commonly cited for carbon steel immersed in seawater is 5 mils/yr (130 μm/yr). This rate may be considered linear with time up to about 8 years. After about 8 years, the corrosion rate decreases to a steady, but slower, rate. In one instance, examination of steel H-piles exposed to unpolluted seawater in the Santa Barbara Channel indicated that the loss due to corrosion is about 1 mm (40 mils) over a 20-year interval, or a rate of about 50 μm/yr (2 mils/yr), which can be reduced to about 25 μm/yr (1 mil/yr) thereafter.

Corrosion in brackish waters, such as is found in bays and estuaries, parallels corrosion in seawater to a certain extent. However, because of the periodic reversal in flow typical of many brackish waters and because of certain other characteristics, the pattern of corrosion in brackish waters differs from the pattern in ordinary seawater.

Effect of Mill Scale. Corrosion rates around the world are somewhat similar, regardless of whether they are measured in terms of weight loss or of maximum pit depth. However, in almost all instances, maximum pit depth is much greater in carbon steel specimens covered with mill scale than in specimens that are free of mill scale, as shown in Table 4. Mill scale is very cathodic to bare steel.

Effect of Oxygen Concentration. It is the cathodic depolarizing effect of oxygen that permits most corrosion reactions to proceed. In practice, the various natural forms of inhibition that prevent access of oxygen to a metal surface reduce the corrosion rate and often stifle corrosion completely.

In one instance, the effects of such parameters of the deep-sea environment as oxygen concentration, salinity, temperature and pH were investigated. Some of the data from that study are given in Table 5. The corrosion rate for steel was de-

Table 4. Corrosion of mill-scaled surface of carbon steel versus corrosion of scale-free surface

| | Pickled surface | | | | Mill-scaled surface | | | |
| | Corrosion loss | | Max pit depth | | Corrosion loss | | Max pit depth | |
Location	μm/yr	mils/yr	mm	mils	μm/yr	mils/yr	mm	mils
Halifax, Nova Scotia	120	4.8	1.9	75	110	4.3	1.7	68
Auckland, New Zealand	75	3.0	1.1	43	85	3.3	3.7	147
Plymouth, England	60	2.4	1.65	65	60	2.4	4.0	156
Colombo, Ceylon	90	3.6	1.6	64	100	3.9	6.1	240

Table 5. Corrosion rates of carbon steel for near-surface and deep-sea immersion

Immersion depth		Temperature, °C	Oxygen concentration, ppm	Immersion time, days	Corrosion rate	
m	ft				μm/yr	mils/yr
Near-surface site(a)						
...	...	5 to 30	5 to 10	365	130	5.0
Deep-sea site(b)						
715	2340	7.2	0.6	197	43	1.7
1615	5300	2.5	1.8	1604	23	0.9
1720	5640	2.8	1.2	123	50	2.0
		2.3	2.1	751	20	0.8
2065	6780	2.7	1.7	403	58	2.3

(a) Wrightsville Beach, NC. (b) Pacific Ocean.

Table 6. Galvanic series in seawater

1 Magnesium (most active)	18 Inconel (active)
2 Magnesium alloys	19 Hastelloy B
3 Zinc	20 Brasses
4 Aluminum 1100	21 Copper
5 Cadmium	22 Bronzes
6 Aluminum 2017	23 Copper-nickel alloys
7 Steel (plain)	24 Titanium
8 Cast iron	25 Monel
9 Chromium iron (active)	26 Silver solder
10 Nickel cast iron	27 Nickel (passive)
11 304 stainless (active)	28 Inconel (passive)
12 316 stainless (active)	29 Chromium iron (passive)
13 Hastelloy C	30 304 stainless (passive)
14 Lead-tin solders	31 316 stainless (passive)
15 Lead	32 Silver
16 Tin	33 Graphite (least active)
17 Nickel (active)	

termined to be a linear function of dissolved oxygen content.

GALVANIC CORROSION

Galvanic corrosion in seawater is a matter of concern because the corroding medium has a fairly high conductivity. A practical version of the galvanic relationship of metals in seawater is shown in Table 6. Service conditions can differ considerably because of solution composition, solute concentration, agitation, aeration, temperature, and purity of the metals, as well as corrosion-product formation and biological growth, each of which can result in a different galvanic series.

BASIC CORROSION RATES IN SEAWATER

Carbon steel and many of the low-alloy steels are frequently cited as having short-term corrosion rates of about 130 μm/yr (5 mils/yr) when completely immersed in seawater. The long-term corrosion rates of these steels (after immersion for 5 to 10 years) are somewhat lower.

It is generally conceded that steel making practice has no effect on corrosion in seawater; there is no substantial difference in behavior between rimmed and killed steel. Likewise, alloying with small amounts of either copper or chromium has no effect on corrosion. In one investigation, for example, seven different open-hearth steels, six copper-bearing steels and several low-chromium copper-bearing steels were immersed in seawater for 203 days; the corrosion rates for all these steels were in the relatively narrow range of 135 to 150 μm/yr (5.4 to 5.8 mils/yr).

Effect of Velocity. Increasing the velocity of seawater generally increases the rate of corrosion.

For instance, one study showed that the corrosion rate in stagnant seawater was the same for both carbon steel and copper-bearing steel — about 70 μm/yr (2.8 mils/yr) — for immersion periods of 5 to 10 years. In seawater flowing at low velocity, the corrosion rate for carbon steel was about 95 μm/yr (3.8 mils/yr), but at higher velocity the corrosion rate was 380 μm/yr (15 mils/yr). The corrosion rate for copper-bearing steel in high-velocity seawater was only about 120 μm/yr (5.5 mils/yr), which indicates a possible benefit from alloying with small amounts of copper. Most investigators consider the effect of velocity on corrosion rates to be an erosion effect, whereby corrosion products are continually removed from the surface as they form, which reduces their effectiveness in stifling further corrosion.

Pitting. Where pitting is a serious consideration, steels containing up to 3% Ni or 3% Cr may be used. Both nickel steels and chromium steels pit at about half the rate of carbon steel. For instance, in one 6-year immersion test, a 2%-Ni steel pitted to an average depth of about 2 mm (80 mils), compared to a depth of 5 mm (200 mils) for carbon steel. Increasing the amount of chromium above 3% changes the mode of corrosion to an increasingly more localized attack with deeper pits. For instance, a 3%-Cr steel typically exhibits pitting rates of about 0.2 mm/yr (8 mils/yr), compared to about 0.9 mm/yr (37 mils/yr) for a 13%-Cr steel and 1.75 mm/yr (69 mils/yr) for a 17%-Cr steel.

Corrosion in Tropical Waters. The higher temperature of seawater in tropical climates than in temperate or arctic climates results in higher rates of corrosion. Results of a systematic study conducted in both fresh water and seawater in a tropical climate are given in Table 7; in this climate, as in more temperate climates, the average corrosion rates for both carbon and HSLA steels decrease with time in both fresh and salt water. Furthermore, the average depth of the 20 deepest pits is about the same for both carbon and copper-bearing steels, higher after long-term expo-

sure for Ni steels, and slightly lower for long-term exposure only for 5%-Cr steel. Besides the results given in Table 7, a 3%-Cr steel and several HSLA steels containing various amounts of Ni, Cr, P, Cu and Si were subjected to the same conditions, and gave about the same results, as those reported for similar steels in Table 7.

PREVENTIVE MEASURES

The use of organic coatings in conjunction with cathodic protection offers the most economical and effective method of corrosion prevention in seawater. Offshore oil rigs, pipelines and other underwater equipment make use of this combination. The less porous the coating, the less current needed for cathodic protection and the more economical the system. Aluminum anodes, with their high current efficiency, are most often used in seawater. However, zinc anodes also have been used extensively in seawater applications, while magnesium anodes have been used less extensively. The slight advantage of low-alloy steels over carbon steels usually is lost in the final economic analysis of corrosion prevention.

Inhibitors are sometimes used in seawater cooling equipment. Wherever possible, all equipment in contact with seawater should be designed without galvanic couples.

Protection of Steel from Corrosion

CORROSION PROTECTION is often an essential consideration in selection of carbon or alloy steel for a given structural application. Corrosion can reduce the load-carrying capacity of a component either by generally reducing its size (cross section) or by pitting, which not only reduces the effective cross section in the pitted region, but also introduces stress raisers that may initiate cracks. Obviously, any measure that reduces or eliminates corrosion will extend the life of a component and increase its reliability (see Table 8, which is intended to be a general guide).

Over-all economics, environmental conditions, degree of protection needed for the projected life of the part, consequences of unexpected service failure, and importance of appearance are the chief factors that determine not only whether a steel part needs to be protected against corrosion, but the most effective and economic method of achieving that protection as well. Protection against corrosion can be done by (a) providing an impervious barrier between the steel and the corroding medium, (b) introducing a substance that inhibits the chemical

Table 7. Typical corrosion rates in tropical waters

Material	Average penetration rate						Average depth of 20 deepest pits					
	1 yr		8 yr		16 yr		1 yr		8 yr		16 yr	
	μm/yr	mils/yr	μm/yr	mils/yr	μm/yr	mils/yr	μm/yr	mils/yr	μm/yr	mils/yr	μm/yr	mils/yr
Fresh water(a)												
Carbon steel(b) ...	195	7.7	65	2.58	45	1.75	510	20	1470	58	1830	72
0.3 Cu steel	200	7.9	75	2.9	45	1.8	560	22	1630	64	1630	64
Seawater(c)												
Carbon steel(b) ...	150	5.9	80	3.2	75	2.9	1040	41	1680	66	2290	90
0.3 Cu steel	150	5.9	90	3.5	80	3.1	915	36	1600	63	2160	85
2 Ni steel	190	7.4	100	4	85	3.3	840	33	2390	94	(d)	(d)
5 Ni steel	160	6.3	100	4	85	3.3	735	29	2970	117	(d)	(d)
5 Cr steel	70	2.7	100	4	85	3.5	685	27	1600	63	1750	69

(a) Gatun Lake, CZ. (b) Pickled. (c) Pacific Ocean, near Panama Canal Zone. (d) Specimen corroded through.

Table 8. Guide to corrosion prevention for carbon steels in various environments

Preventive method	Atmosphere	Soil	Fresh water	Seawater	Steam systems	Acids and pickling baths
Metal coatings: electroplating, galvanizing	Galvanizing very effective; plating with other metals used for both decorative appearance and corrosion protection	Not recommended	Galvanizing used in potable water	Not recommended	Not recommended	Not recommended
Painting: chemical treatment, priming and painting	Most economical and effective corrosion prevention	Seldom used	Fairly effective	Special paint systems used	Not recommended	Not recommended
Cathodic protection	Not recommended	Most economical and effective method, especially with organic coatings other than paint	Fairly effective with organic coatings	Very effective	Not recommended	Effective under special conditions
Inhibitors: liquid and vapor	Not recommended	Not recommended	Effective in some applications, especially cooling waters	Fairly effective in some applications	Very effective	Very effective
Alloying additions to steel	Very effective, especially copper-bearing and HSLA steels	Not effective	Not effective	Only effective with much alloying	Chromium-molybdenum steels are very effective	Only effective with much alloying
Removal of oxygen from environment	Not recommended	Not recommended	Seldom used	Very effective, especially in desalination and hot seawater	Very effective	Not recommended
Removal of more noble metals; elimination of galvanic couples	Usually not necessary	Fairly effective	Effective	Necessary	Advisable	Not effective
Organic coatings other than paint	Seldom used to replace painting	Used to advantage with cathodic protection	Fairly effective with cathodic protection	Used to advantage with cathodic protection	Not recommended	Have been used

action, or (c) inducing galvanic activity that counteracts the galvanic activity involved in corrosion.

COATINGS

The types of coatings available seem endless, varying from simply oiling the surface for low-cost temporary protection to electroplating with a multilayer copper-nickel-chromium coating to obtain a high-gloss metallic appearance and superior corrosion protection. For economic reasons, the desired degree of protection must be determined before a coating is selected.

Temporary protection during storage or shipment usually can be done effectively by coating the steel with mineral oil, solvents combined with inhibitors and film formers, emulsions of petroleum-based coatings, or water plus waxes. These coatings are applied (a) after acid pickling to remove mill scale or (b) between coating sequences, and are not expected to give long-term corrosion protection.

Cleaning is the most important prerequisite for any coating process. Any oxides on a steel surface must be removed by acid pickling or sand blasting. Degreasing is necessary after oxide removal or when the steel has been given a temporary coating. Degreasing can be accomplished by alkaline cleaning, vapor degreasing or emulsion cleaning.

Hot dip coating processes are particularly well suited for applying coatings of aluminum, lead, tin, zinc and some of their alloys. Hot dipping consists of immersing the steel in a bath of mol-

ten metal that usually is covered with a layer of molten flux.

Electroplating is perhaps the most expensive of all processes for coating steel. Electrodeposits, therefore, are used when appearance, solderability or some other requirement demands a metal coating. If only corrosion protection is needed, many nonmetallic coatings will give better protection at lower cost. All conventional electroplates can be applied to steel.

Conversion coatings offer only mild corrosion protection to steel if not covered with another system but become very protective when used with paints. Phosphate coatings and chromate conversion coatings are widely used as paint bases on both uncoated and galvanized steel.

Thin organic coatings (paints) are used more often for corrosion protection than any of the other coatings because the cost is low for the degree of protection afforded. A good paint system includes proper cleaning, a conversion coating, a primer and a compatible top coat. Application of the various coats can be accomplished by spraying, dipping, flow-coating, roller coating or electrophoretic deposition.

INHIBITORS

Inhibitors find their major uses in acid pickling solutions, acidic service environments, steam systems, and neutral and near-neutral aqueous solutions. Inhibitors may be organic or inorganic compounds and usually are dissolved in aqueous environments. Inhibitors have been added to chemical conversion treatment baths and to paint

primers. A few vapor-phase inhibitors are used in confined atmospheres.

Corrosion Inhibitors in Acid Pickling. Mill scale is most often removed from steel by pickling in either sulfuric or hydrochloric acid solutions. Because pickling is essentially a chemical dissolution (corrosion) process, some of the underlying metal is inevitably removed from the surface along with the mill scale. Inhibitors are added to pickling baths to minimize metal loss, minimize the extent of hydrogen pickup (which can lead to embrittlement), protect the metal against pitting and poor surface quality, reduce acid fumes and reduce acid consumption. Inhibitors prolong pickling time, but the benefits outweigh the increase in time.

Packaging Applications. The major application utilizing vapor-phase inhibition has been the packaging of steel articles ranging from small hand tools to large coils of steel strip. Paper can be impregnated with varying quantities of inhibitor (0.2 to 2 g/m^2) depending on the prevailing temperature and on the period of time over which a nonaggressive environment must be maintained. Packaging for long-term storage imposes different volatility requirements compared to packaging only for protection in transit. Paper impregnated with 0.2 g/m^2 of dicyclohexylamine nitrite is capable of conferring protection up to 10 years if the temperature does not exceed 23 °C (73 °F). However, protection is effective for only about 100 days at 75 °C (167 °F). It has been found that the logarithm of the effective service time of paper impregnated with inhibitor is inversely proportional to the temperature.

5 CAST IRONS

Edited by R. K. Buhr, Physical Metallurgy Research Laboratories, Canada Centre for Mineral and Energy Technology

This section was condensed from Metals Handbook, Ninth Edition, Volume 1, Properties and Selection: Irons and Steels, pages 3 to 106. For more detailed information on the topics covered herein, the reader is referred to the larger work. Additional articles on cast irons can be found within this volume in the following sections: Casting (Section 23), Machining (Section 27), Heat Treating (Section 28), Joining (Section 30) and Metallography (Section 35). The reader should also consult the index to locate information not otherwise categorized.

Basic Metallurgy of Cast Iron

The term "CAST IRON," like the term "steel," identifies a large family of ferrous alloys. Cast irons primarily are alloys of iron that contain more than 2% carbon and from 1 to 3% silicon. Wide variations in properties can be achieved by varying the balance between carbon and silicon, by alloying with various metallic or nonmetallic elements, and by varying melting, casting and heat treating practices.

The four basic types of cast iron are white iron, gray iron, ductile iron and malleable iron. White iron and gray iron derive their names from the appearances of their respective fracture surfaces: white iron exhibits a white, crystalline fracture surface, and gray iron exhibits a gray fracture surface with exceedingly tiny facets. Ductile iron derives its name from the fact that, in the as-cast form, it exhibits measurable ductility. By contrast, neither white nor gray iron exhibits significant ductility in a standard tensile test. Malleable iron is initially cast as white iron, then "malleablized" — that is, heat treated to impart ductility to an otherwise exceedingly brittle material.

THE IRON–IRON CARBIDE–SILICON SYSTEM

A section through the ternary Fe-Fe$_3$C-Si diagram at 2% Si (which approximates the silicon contents of many cast irons) provides a convenient reference for discussing the metallurgy of cast iron. The diagram in Fig. 1 resembles the binary Fe-Fe$_3$C diagram but exhibits important differences characteristic of ternary systems. Eutectic and eutectoid temperatures change from single values in the Fe-Fe$_3$C system to temperature ranges in the Fe-Fe$_3$C-Si system; the eutectic and eutectoid points shift to lower carbon contents.

Figure 1 represents the metastable equilibrium between iron and iron carbide (cementite), a metastable system. The silicon that is present remains in solid solution in the iron, in both ferrite and austenite, and so does not affect the composition of the carbide phase but only the conditions and the kinetics of carbide formation on cooling. The designations α, γ and Fe$_3$C, therefore, are used in the ternary system to identify the same phases that occur in the Fe-Fe$_3$C binary system. Some of the silicon may precipitate along with the carbide, but it cannot be distinguished as a different phase. The solidification of certain compositions does not occur in the metastable system, but rather in the stable system, where the products are iron and graphite rather than iron and carbide. These compositions encompass the gray, ductile and compacted graphite cast irons.

If the section through the ternary diagram at 2% Si is to be used in tracing out the phase changes that occur, its use can be justified only on the assumption that the silicon concentration remains at 2% in all parts of the alloy under all conditions. This obviously is not strictly true, but there is little evidence that silicon segregates to any marked degree in cast iron.

CARBON EQUIVALENCE

Carbon equivalence (CE) is a method that often is used to simplify evaluation of the effect of composition in unalloyed cast irons. CE equals total carbon content (TC) plus about one-third the sum of the silicon and phosphorus contents:

$$CE = TC + (\%Si + \%P)/3$$

Comparison of CE with the eutectic composition in the Fe-C system (4.3% C) will indicate whether a cast iron will behave as a hypoeutectic or hypereutectic alloy during solidification. When CE is near the eutectic value, the liquid state persists to a relatively low temperature and solidification takes place over a small temperature range. The latter characteristic can be important in promoting uniformity of properties within a given casting.

In hypereutectic irons (CE greater than about 4.3%), there is a tendency for kish graphite — proeutectic graphite that forms and floats free in the molten iron — to precipitate on solidification under normal cooling conditions. In hypoeutectic irons, the lower the CE, the greater the tendency for white or mottled iron to form on solidification.

WHITE CAST IRON

White iron is formed when the carbon in solution in the molten iron does not form graphite on solidification but remains combined with the iron, often in the form of massive carbides. White irons are hard and brittle and produce white, crystalline fracture surfaces.

White cast irons have high compressive strength and good retention of strength and hardness at elevated temperature, but they are most often used for their excellent resistance to wear and abrasion. The massive carbides in the microstructure are chiefly responsible for these properties.

GRAY CAST IRON

When the composition of the iron and the cooling rate at solidification are suitable, a substantial portion of the carbon content separates out of the liquid to form flakes of graphite. When a piece of the solidified alloy is broken, the fracture path follows the graphite flakes, and the fracture surfaces appear gray because of the predominance of exposed graphite.

Gray cast iron has several unique properties that are derived from the existence of flake graphite in the microstructure. Gray iron can be machined easily at hardnesses conducive to good wear resistance. It resists galling under boundary-lubrication conditions (conditions wherein the flow of lubricant is insufficient to maintain a full fluid film). It has outstanding properties for applications involving vibrational damping or moderate thermal shock.

DUCTILE CAST IRON

Ductile iron, which is also known as nodular iron or spherulitic-graphite cast iron, is very sim-

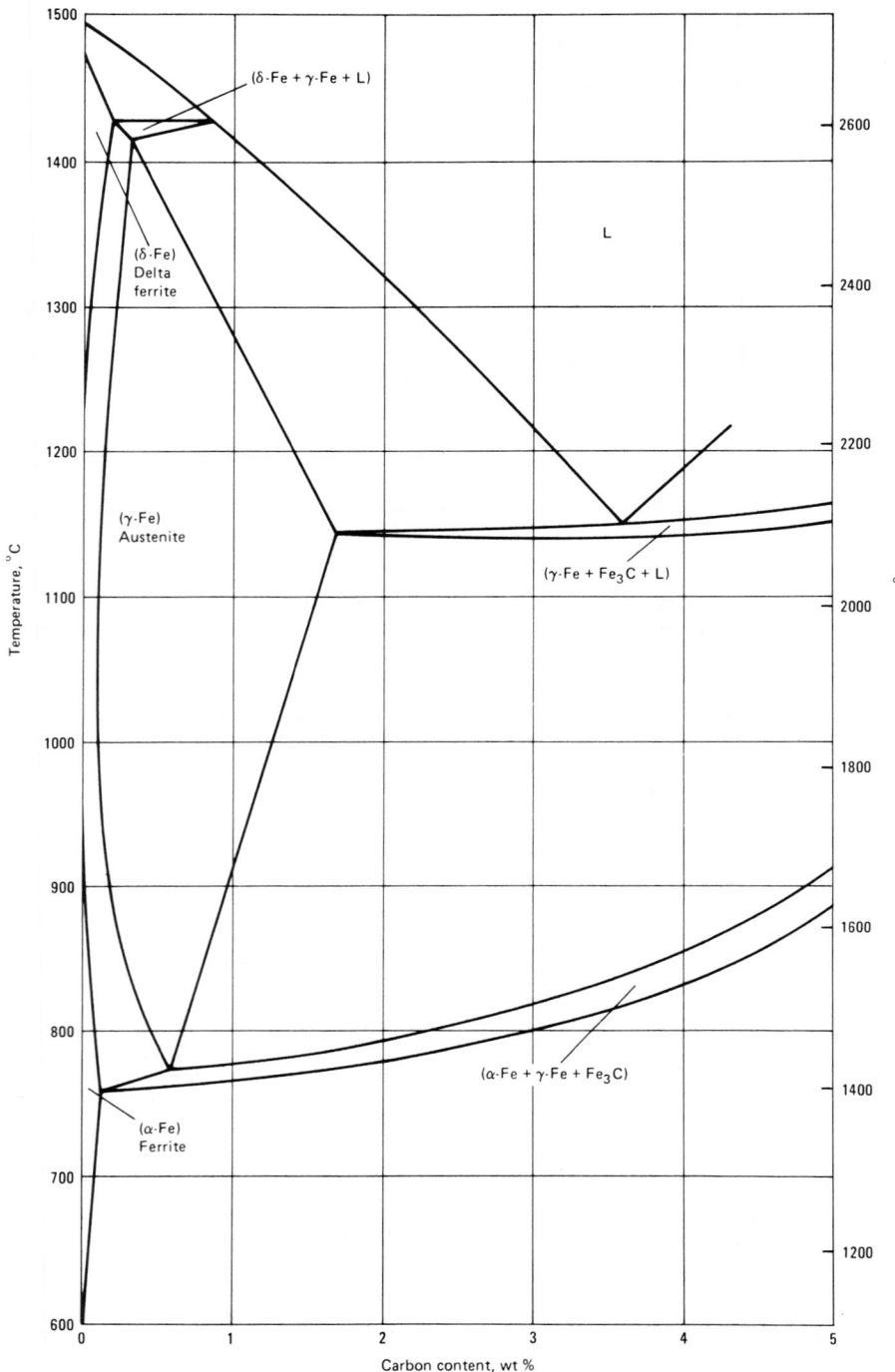

Fig. 1. Section through the iron–iron carbide–silicon ternary equilibrium diagram at 2% silicon

Table 1. Summary and description of ASTM and equivalent ISO classifications of graphite shapes in cast iron

ASTM type(a)	Equivalent ISO form(b)	Description
I	VI	Nodular (spheroidal) graphite
II	VI	Nodular (spheroidal) graphite, imperfectly formed
III	IV	Aggregate, or temper carbon
IV	III	Quasi flake graphite
V	II	Crab-form graphite
VI	V	Irregular or "open" type nodules
VII(c)	I	Flake graphite

(a) As defined in ASTM A247. (b) As defined in ISO/R 945-1969 (E). (c) Divided into five subtypes: uniform flakes; rosette grouping; superimposed flake size; interdendritic, random orientation; and interdendritic, preferred orientation.

higher tensile strengths (1000 MPa, or 145 ksi).

Graphite Shape. For most applications, some deviation from true spheroidal shape can be tolerated without unacceptable loss of properties. However, the quasi flake and crab forms (see Table 1) are unacceptable for most applications.

COMPACTED GRAPHITE CAST IRON

Compacted graphite (CG) cast iron (CG is now the officially recognized category) is characterized by graphite that is interconnected within eutectic cells as is the flake graphite in gray iron. Compared with the graphite in gray iron, however, the graphite in CG iron is coarser and more rounded, similar in metallographic appearance to ASTM type IV (ISO form III), quasi flake graphite. The structure can be considered intermediate between those of gray iron and ductile iron. Individual properties also can be considered intermediate between those of gray and ductile irons, but the unique combinations of properties obtainable in CG irons make them superior to either gray or ductile iron in applications such as disc-brake rotors and diesel-engine heads.

Compacted graphite cast iron can be obtained by very carefully controlling the amount of magnesium added as an inoculant in a process very similar to the process used to make ductile iron. Unfortunately, either undertreatment, resulting in a gray iron structure, or overtreatment, resulting in a ductile iron structure, can occur if the ideal quantity of magnesium is missed by as little as 50 ppm. Current commercial production of CG iron is accomplished by inoculation with magnesium to give a residual content of 50 to 600 ppm in the presence of 0.15 to 0.5% titanium and 10 to 150 ppm of rare earths, such as cerium. In effect, the process is one in which the nodulizing reaction due to the addition of magnesium is poisoned by the presence of a controlled amount of titanium, and in which cerium is added to eliminate a need to control sulfur at a low concentration.

Foundry practices used in casting gray iron can be used for CG iron with only slight modifications. The yield of finished castings for a given amount of metal poured is comparable to that for gray iron and substantially greater than that for ductile iron.

Effect of Graphite Type on Properties. Depending on composition and on amount of pearlite in the microstructure, the tensile strength of CG iron can

ilar to gray iron in composition, but during casting of ductile iron the graphite is caused to nucleate as spherical particles, or spherulites, rather than as flakes. This is accomplished through the addition of a very small but definite amount of magnesium and/or cerium to the molten iron in a process step called nodulizing.

Ductile iron is produced from the same types of raw materials as gray iron, but usually requires slightly higher purity, especially in regard to sulfur. Casting properties of ductile iron, such as fluidity, are comparable to those of gray iron.

The chief advantage of ductile iron over gray iron is its combination of high strength and ductility—up to 18% minimum elongation for ferritic ductile iron with a tensile strength of 415 MPa (60 ksi) as opposed to only about 0.6% elongation for a gray iron of comparable strength. Martensitic ductile irons with tensile strengths of about 830 MPa (120 ksi) exhibit at least 2% elongation, and the newer austempered ductile irons exhibit in excess of 5% elongation at even

be as high as 520 MPa (75 ksi); ductility usually is about 1 to 6%. Tensile and yield strengths are about the same as those of high-strength gray cast irons with carbon-equivalence values in the lower portion of the gray iron range. Thermal conductivity and damping capacity of CG irons also are about the same as those of high-strength gray irons. Impact and fatigue properties, although not as good as those of ductile iron, are substantially better than those of gray cast iron. It is this combination of high strength and good impact resistance, coupled with a good capacity for heat dissipation (derived from relatively high thermal conductivity), that makes CG irons well suited for applications where neither gray iron nor ductile iron is entirely satisfactory.

It is believed that the superiority in mechanical properties of CG iron over gray iron occurs because the rounded graphite flakes in CG iron constitute less-severe internal notches than the sharper and more angular flakes in gray iron. Furthermore, the interconnected nature of the eutectic graphite in CG iron makes its thermal conductivity definitely better than that of ductile iron.

Gray Iron

CAST IRONS are alloys of iron, carbon and silicon in which more carbon is present than can be retained in solid solution in austenite at the eutectic temperature. In gray cast iron, the carbon that exceeds the solubility in austenite precipitates as flake graphite. Gray irons usually contain 2 to 4% C and 1 to 3% Si.

Classes of Gray Iron. A simple and convenient classification of gray irons is found in ASTM specification A48, which classifies the various types in terms of tensile strength expressed in ksi. The ASTM classification by no means connotes a scale of ascending superiority from class 20 (minimum tensile strength, 140 MPa) to class 60 (minimum tensile strength, 410 MPa). In many applications strength is not the major criterion for the choice of grade.

Generally, it can be assumed that the following properties of gray cast irons increase with increasing tensile strength from class 20 to class 60:

- All strengths, including strength at elevated temperature.
- Ability to produce a fine machined finish
- Modulus of elasticity
- Wear resistance.

On the other hand, the following properties decrease with increasing tensile strength, so that low-strength irons often perform better than high-strength irons when these properties are important:

- Machinability
- Resistance to thermal shock
- Damping capacity
- Ability to be cast in thin sections.

CASTABILITY

Successful production of a gray iron casting depends on fluidity of the molten metal and on casting design, especially minimum section thickness and section-thickness variations. Casting design often is described in terms of "section sensitivity," which attempts to correlate properties in critical sections of the casting with the combined effects of composition and cooling rate. All these factors are interrelated and may be condensed into a single term, "castability," which for gray iron may be defined as the minimum section thickness that can be produced in a mold cavity with given volume/area ratio and mechanical properties consistent with the type of iron being poured.

Fluidity. Scrap losses resulting from misruns, cold shuts and round corners often are attributed to lack of fluidity of the metal poured.

Mold conditions, pouring rate and other pro-

cess variables being equal, the fluidity of commercial gray irons depends primarily on the amount of superheat above the freezing temperature (liquidus). As the total carbon content decreases, the liquidus temperature increases, and the fluidity at a given pouring temperature therefore decreases. Fluidity commonly is measured as the length of flow into a spiral-type fluidity test mold. The relation between fluidity and superheat is shown in Fig. 1 and Table 1 for unalloyed gray irons of different carbon contents.

MICROSTRUCTURE

The usual microstructure of gray iron is a matrix predominantly of pearlite, with graphite flakes dispersed throughout. Foundry practice can be varied so that nucleation and growth of graphite flakes occur in a pattern that enhances the desired properties. The amount, size and distribution of graphite are important. Cooling that is too rapid may produce "mottled iron," in which carbon is present in the form of both primary cementite (iron carbide) and graphite. Very slow cooling of irons that contain large percentages of silicon and carbon is likely to produce a matrix predominantly of ferrite, together with coarse graphite flakes.

Flake graphite is one of seven types (shapes or forms) of graphite established in ASTM A247. Flake graphite is subdivided into five types (pat-

Table 1. Superheat above liquidus for 2% Si irons of various carbon contents poured at 1455 °C (2650 °F)

Carbon, %	Liquidus temperature °C	°F	Superheat above liquidus °C	°F
2.52	1295	2360	160	290
3.04	1245	2270	210	380
3.60	1175	2150	280	500

terns), which are designated by the letters A through E (see Fig. 2). Graphite size is established by comparison with an ASTM size chart, which shows the typical appearances of flakes of eight different sizes at a magnification of 100×.

Type A flake graphite (random orientation) is preferred for most applications. In the intermediate flake sizes, type A flake graphite is superior to other types in certain wear applications such as the cylinders of internal-combustion engines. Type B flake graphite (rosette pattern) is typical of fairly rapid cooling, such as is common in moderately thin sections (about 10 mm, or 3/8 in.) and along the surfaces of thicker sections, and sometimes results from poor inoculation. The large flakes of type C flake graphite are typical of kish graphite that is formed in hypereutectic irons. These large flakes enhance resistance to thermal shock by increasing thermal

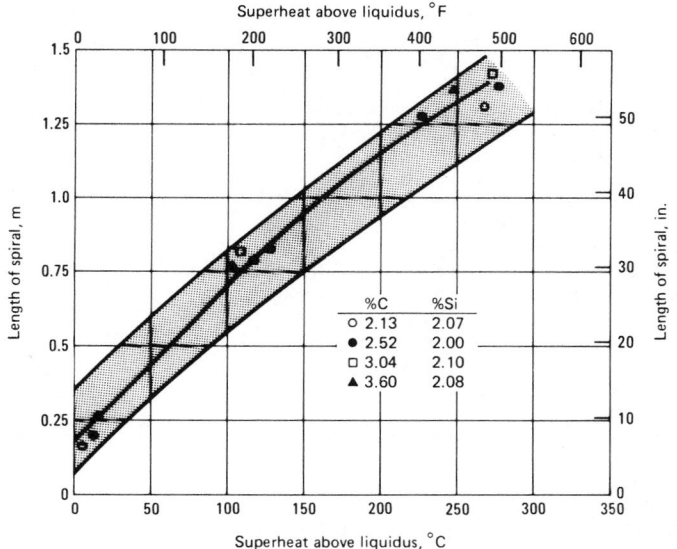

Fig. 1. Fluidity versus degree of superheat for four gray irons of different carbon contents

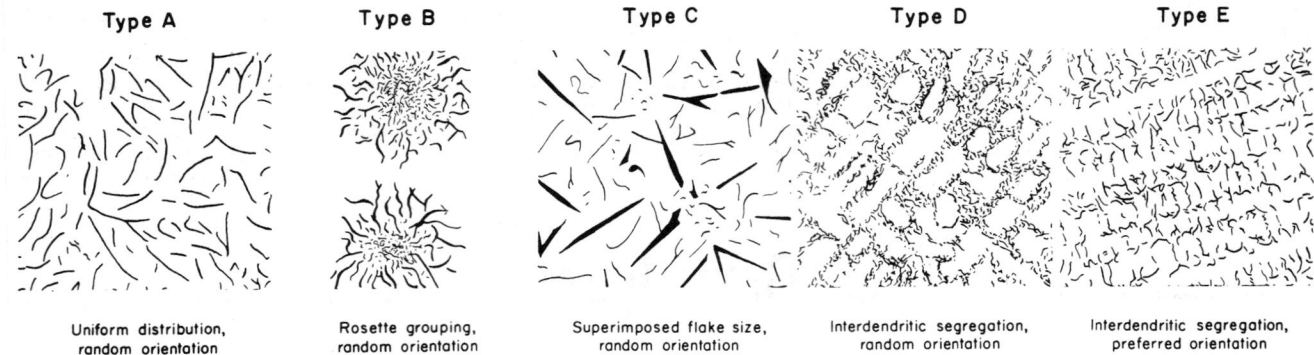

Type A — Uniform distribution, random orientation

Type B — Rosette grouping, random orientation

Type C — Superimposed flake size, random orientation

Type D — Interdendritic segregation, random orientation

Type E — Interdendritic segregation, preferred orientation

In the recommended practice (ASTM A247), these charts are shown at a magnification of 100×. They have been reduced to one-third size for reproduction here. This figure may be purchased from ASTM at a nominal fee.

Fig. 2. Types of graphite flakes in gray iron (AFS-ASTM)

conductivity and decreasing elastic modulus, both of which reduce thermal stress. On the other hand, large flakes are not conducive to good surface finishes on machined parts or to high strength or good impact resistance. The small, randomly oriented interdendritic flakes in type D flake graphite promote a fine machined finish by minimizing surface pitting, but it is difficult to obtain a pearlitic matrix with this type of graphite. Type D flake graphite may be formed near rapidly cooled surfaces or in thin sections. Frequently, such graphite is surrounded by a ferrite matrix, resulting in soft spots in the casting. Type E flake graphite is an interdendritic form that has a preferred rather than a random orientation. Unlike type D graphite, type E can be associated with a pearlite matrix, thus producing a casting whose wear properties are as good as those of a casting containing only type A graphite in a pearlitic matrix.

SECTION SENSITIVITY

In practice, the minimum thickness of section in which any given class of gray iron may be poured is more likely to depend on the cooling rate of the section than on the fluidity of the metal. For example, although a plate 300 mm square by 6 mm thick can be poured in class 50 as well as in class 25 iron, the former casting would not be gray iron, because the cooling rate would be so rapid that massive carbides would be formed. Yet it is entirely feasible to use class 50 iron for a diesel-engine cylinder head that has predominantly 6-mm wall sections in the water jackets above the firing deck. This is simply because the cooling rate of the cylinder head has been reduced by the "mass effect" resulting from enclosed cores and the proximity (often less than 12 mm) of one 6-mm wall to the other.

It should be recognized that the smallest section that can be cast gray, without massive carbides, depends not only on metal composition but also on foundry practices. For example, by adjusting silicon content or by adding inoculants (graphitizing agents) to the ladle (or, more effectively, to the stream as the mold is poured), the foundryman may decrease the minimum section size for freedom from carbides for a given basic composition of gray iron.

Typical Effects of Section Size. When a wedge-shape bar with about 10° taper is cast in a sand mold and sectioned near the center of its length, and Rockwell hardness determinations are made on the cut surface from the point of the wedge progressively into the thicker sections, the curves so

determined show to what extent continually increasing section size will affect hardness, as shown in Fig. 3.

Progressing along the curve from the left in Fig. 3, the following metallographic constituents occur. The tip of the wedge is white iron (a mixture of carbide and pearlite) with hardness greater than 50 HRC. As the iron becomes mottled (a mixture of white iron and gray iron), the hardness decreases sharply. A minimum is reached because of the occurrence of fine type D flake graphite, which usually has associated with it large amounts of ferrite. With a slightly lower cooling rate, the structure becomes fine type A flake graphite in a pearlite matrix, and the hardness rises to another maximum on the curve. This structure usually is the most desirable for wear resistance and strength. With increasing section thickness beyond this point, the graphite flakes become coarser and the pearlite lamellae more widely spaced, resulting in slightly lower hardness. With further increase in wedge thickness and decrease in cooling rate, pearlite decomposes progressively to a mixture of ferrite and graphite, resulting in softer and weaker iron.

The structures of most commercial gray iron castings are represented by the right-hand down-ward-sloping portion of the curve in Fig. 3, beyond 5-mm wedge thickness, and normally increasing section size is reflected by gradual lowering of hardness and strength. However, thin sections may be represented by the left-hand downward-sloping portion.

The graph at left in Fig. 4 shows average tensile strength (up to 10 tests per point) of two irons, for each of which six sizes of cylindrical round bars were cast and appropriate tensile specimens machined. With the class 20 iron, strength increases as the as-cast section decreases down to the 6-mm cast bar. However, for the class 30 iron, a section 6 mm in diameter is so small that the strength falls off sharply, because of the occurrence of type D flake graphite or mottled iron, or both. The graph at right in Fig. 4 shows similar data for the same two classes of iron and for three higher classes.

PREVAILING SECTIONS

Although the ASTM size B test bar (30.5-mm diam, or 1.2-in. diam) is the bar most commonly used for all gray irons from class 20 to class 60, ASTM specification A48 provides a selection of bar sizes to approximate the cooling rate of the

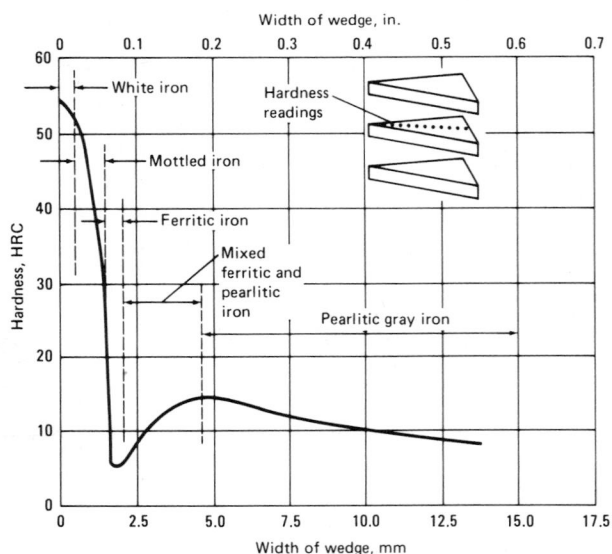

Hardness readings were taken at increasing distance from the tip of a cast wedge section, as shown by inset. Composition of iron: 3.52 C, 2.55 Si, 1.01 Mn, 0.215 P and 0.086 S.

Fig. 3. Effect of section thickness on hardness and structure

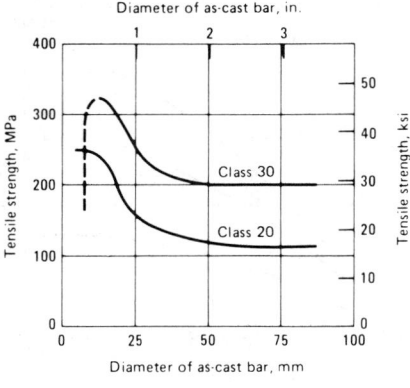

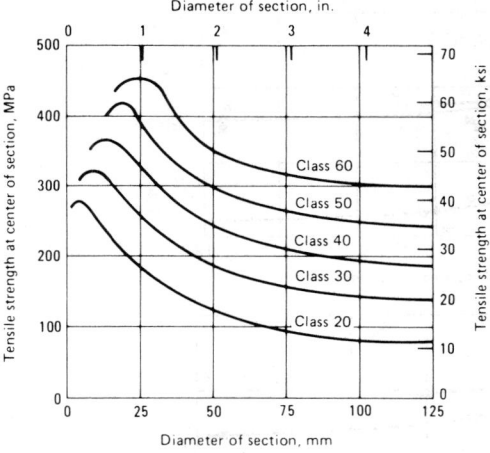

Fig. 4. Effect of section diameter on tensile strength at center of cast specimen for five classes of gray iron

critical section of the casting. In practice, it is customary to be somewhat more definite regarding the prevailing values of minimum casting section considered feasible for the various ASTM classes of cast iron. As summarized in Table 2, these minimum prevailing sections include the requirement for freedom from carbidic areas. In a platelike section, occasional thinner walls (such as ribs) are of no importance unless they are ridiculously thin or appended to the outer edges of the casting.

Mechanical properties of class 30 and class 50 gray irons in various sections are shown in Fig. 5. For class 30 iron, the combined carbon content and hardness are still at a safe level in sections equivalent to a 10-mm (0.4-in.) plate, which has a V/A ratio of about 5 mm (0.20 in.). For class 50 iron, however, both combined carbon and Brinell hardness show marked increases when the thickness of the equivalent plate section is decreased to about 15 mm (0.6 in.), with a V/A ratio of around 7 mm (0.27 in.). These results are consistent with the minimum prevailing casting sections recommended in Table 2.

The hazards involved in pouring a given class of gray iron in a plate section thinner than recommended are discovered when the casting is machined. Typical losses as a result of specifying too high a strength for a prevailing section of 9.5 mm ($^3/_8$ in.) are given below (rejections were for "hard spots" that made it impossible to machine the castings by normal methods):

Class 35 . . . . Rejections negligible
Class 45 . . . . 15 to 25% rejections
Class 55 . . . . 80 to 100% rejections

In marginal applications a higher class of iron sometimes may be used if the casting is cooled slowly (thus, in effect, increasing the section thickness) by judicious placement of flow-offs and risers.

TEST-BAR PROPERTIES

Mechanical-property values obtained from test bars are sometimes the only available guides to the actual mechanical properties of the metal in production castings. When test bars and castings are both poured from the same metal and have subsequent thermal histories that are nearly identical, the strength of the test bars gives a general indication of the strength of the metal in the cast-

Table 2. Minimum prevailing casting sections recommended for gray irons

ASTM class	Minimum thickness in.	mm	V/A ratio(a) in.	mm
20	$^1/_8$	3.2	0.06	1.5
25	$^1/_4$	6.4	0.12	3.0
30	$^3/_8$	9.5	0.17	4.3
35	$^3/_8$	9.5	0.17	4.3
40	$^5/_8$	15.9	0.28	7.1
50	$^3/_4$	19.0	0.33	8.4
60	1	25.4	0.42	10.7

(a) V/A ratios are for square plates.

ings. However, test-bar results cannot be assumed to represent accurately the properties of the metal in every section of every casting, because of differences in section size and cooling rate.

Usual Tests. Tension and transverse tests on bars cast specifically for such tests are the most common for evaluating the strength of gray iron.

Yield strength, elongation and reduction of area are seldom determined for gray iron in standard tension tests. The transverse test measures strength in bending and has the additional advantage that a deflection value may be obtained readily. Minimum specification values are given in Table 3. Data usually can be obtained faster from the transverse test than from the tension test because machining of the specimen is unnecessary in transverse testing. The surface condition of the bar will affect the transverse test but not the tension test made on a machined specimen. Conversely, the presence of coarse graphite in the

Table 4. Test bars designed to match controlling sections of castings (ASTM A48)

Controlling section in.	mm	Test bar	Diameter of as-cast test bar mm	in.
Less than 0.25	Less than 6	S	(a)	(a)
0.25 to 0.50	6 to 12	A	22.4	0.88
0.51 to 1.00	13 to 25	B	30.5	1.20
1.01 to 2.00	26 to 50	C	50.8	2.00
More than 2.00	More than 50	S	(a)	(a)

(a) All dimensions of test bar established by agreement between manufacturer and purchaser.

center of the bar, which can occur in an iron that is very section sensitive, will affect the tension test but not the transverse test.

Typical Specifications. ASTM A48 is typical of specifications based on test bars. In practice, separately cast test bars in one of three different standard sizes are used to evaluate the properties in the controlling section of the castings. After manufacturer and purchaser agree on a controlling section of the casting, the size of test bar that corresponds, approximately, to the cooling rate expected in that section is specified by letter (see Table 4).

Most gray iron castings for general engineering use are specified as class 25, 30 or 35. Specification A48 is based entirely on mechanical properties, and the composition that provides the required properties can be selected by the individual producer. A manufacturer whose major production is medium-section castings of class 35 iron will find, for heavy-section castings where the 51-mm (2-in.) test bar is required for qualifying, that the same composition will not meet the requirements for class 35. It will qualify only for some lower class, such as 25 or 30. As thickness of the controlling section increases, the composition must be adjusted to maintain the same tensile strength.

Compressive Strength. When gray iron is used for structural applications such as machinery foundations or supports, the engineer is usually designing to support weight only, and he bases his calculations on the compressive strength of the material. Table 5, which summarizes typical values for mechanical properties of the various grades, shows the high compressive strength of gray irons. If loads other than dead weights are involved (unless these loads are constant), the problem is one of dynamic stresses, which are discussed later in this article.

Tensile strength is considered in selecting gray iron for parts intended for static loads in direct tension or bending. Such parts include pressure

Table 3. Transverse breaking loads (ASTM A438)(a)

ASTM class	Approximate tensile strength MPa	ksi	A bar(b) kg	lb	Corrected transverse breaking load B bar(c) kg	lb	C bar(d) kg	lb
20	138	20	408	900	816	1800	2720	6 000
25	172	25	465	1025	907	2000	3080	6 800
30	207	30	522	1150	998	2200	3450	7 600
35	241	35	578	1275	1089	2400	3760	8 300
40	276	40	635	1400	1179	2600	4130	9 100
45	310	45	699	1540	1270	2800	4400	9 700
50	345	50	760	1675	1361	3000	4900	10 300
60	414	60	873	1925	1542	3400	5670	12 500

(a) For separately cast test specimens produced in accordance with ASTM A48 or A278, with ASME SA278, with FED QQ-I-652, or with any other specification that designates ASME A438 as the test method. Included in specifications only by agreement between producer and purchaser. (b) 0.88 in. (22.4 mm) diam; 12 in. (305 mm) between supports. (c) 1.20 in. (30.5 mm) diam; 18 in. (457 mm) between supports. (d) 2.00 in. (50.8 mm) diam; 24 in. (610 mm) between supports.

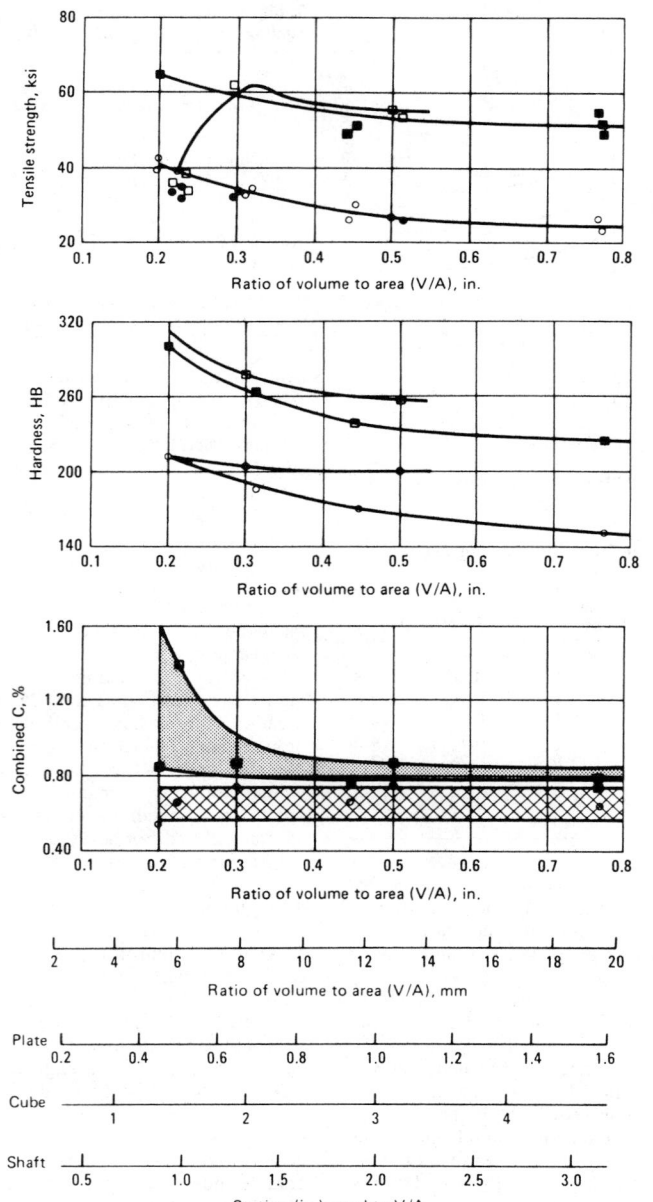

Composition of the class 30 iron: 3.40 C, 0.71 Mn, 2.38 Si, 0.423 P and 0.152 S; for the class 50 iron: 2.96 C, 1.05 Mn, 1.63 Si, 0.67 Mo, 0.114 P and 0.072 S. Note: ○, Class 30 Y blocks; ●, class 30 arbitration bars; □, class 50 arbitration bars; ■, class 50 Y blocks.

Fig. 5. Mechanical properties of class 30 and class 50 gray irons

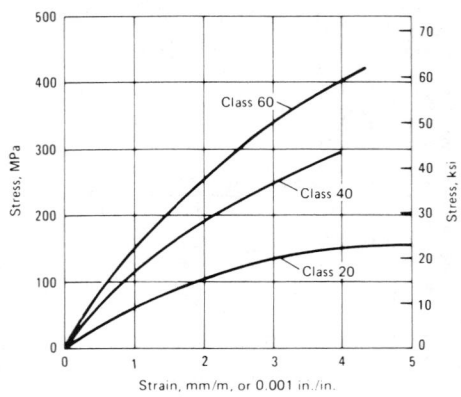

Fig. 6. Typical stress-strain curves for three classes of gray iron in tension

beam formula. The value so determined is arbitrarily called the "modulus of rupture." The values for modulus of rupture are useful for production control, but cannot be used in the design of castings without further analysis and interpretation.

Elongation of gray iron at fracture is very small (on the order of 0.6%) and hence is seldom reported. The designer cannot use the numerical value of permanent elongation in any quantitative manner.

Torsional Shear Strength. As shown in Table 5, most gray irons have high torsional shear strength. Many grades have torsional strength greater than some grades of steel. This characteristic, along with low notch sensitivity, makes gray iron a suitable material for shafting of various types, particularly in the grades of higher tensile strength.

Modulus of Elasticity. Typical stress-strain curves for gray iron are shown in Fig. 6. Gray iron does not obey Hooke's law, and the modulus in tension is usually determined arbitrarily as the slope of the line connecting the origin of the stress-strain curve with the point corresponding to $1/4$ of the tensile strength (secant modulus). Some engineers use the slope of the stress-strain curve near the origin (tangent modulus).

Hardness of gray iron, as measured by Brinell or Rockwell testers, is an intermediate value between the hardness of the soft graphite in the iron and that of the harder metallic matrix. Variations in graphite size and distribution will cause wide variations in hardness (particularly Rockwell hardness) even though the hardness of the metallic matrix is constant. To illustrate this effect, matrix microhardness measurements for five types of hardened iron, as compared with Rockwell C measurements on the same iron, are shown in Table 6.

Table 5. Typical mechanical properties of standard gray iron test bars, as cast

ASTM class	Tensile strength		Torsional shear strength		Compressive strength		Reversed bending fatigue limit		Transverse load on test bar B		Hardness, HB
	MPa	ksi	MPa	ksi	MPa	ksi	MPa	ksi	kg	lb	
20	152	22	179	26	572	83	69	10	839	1850	156
25	179	26	220	32	669	97	79	11.5	987	2175	174
30	214	31	276	40	752	109	97	14	1145	2525	210
35	252	36.5	334	48.5	855	124	110	16	1293	2850	212
40	293	42.5	393	57	965	140	128	18.5	1440	3175	235
50	362	52.5	503	73	1130	164	148	21.5	1638	3600	262
60	431	62.5	610	88.5	1293	187.5	169	24.5	1678	3700	302

Table 6. Influence of graphite type and distribution on hardness of hardened gray irons

Type of graphite	Total carbon, %	Conventional hardness, HRC(a)	Matrix hardness, HRC(b)
A	3.06	45.2(c)	61.5
A	3.53	43.1	61.0
A	4.00	32.0	62.0
D	3.30	54.0	62.5
D	3.60	48.7	60.5

(a) Measured by conventional Rockwell C test. (b) Hardness of matrix, measured with superficial hardness tester and converted to Rockwell C. (c) Although this value was obtained in the specific test cited, it is not typical of gray iron containing 3.06% C. Ordinarily the hardness of such iron is 48 to 50 HRC.

vessels, autoclaves, housings and other enclosures, valves, fittings and levers. Depending on the uncertainty of loading, safety factors of 2 to 12 have been used in figuring allowable design stresses.

Transverse Strength and Deflection. When an arbitration bar is loaded as a simple beam and the load and deflection required to break it are determined, the resulting value is converted into a nominal index of strength by using the standard

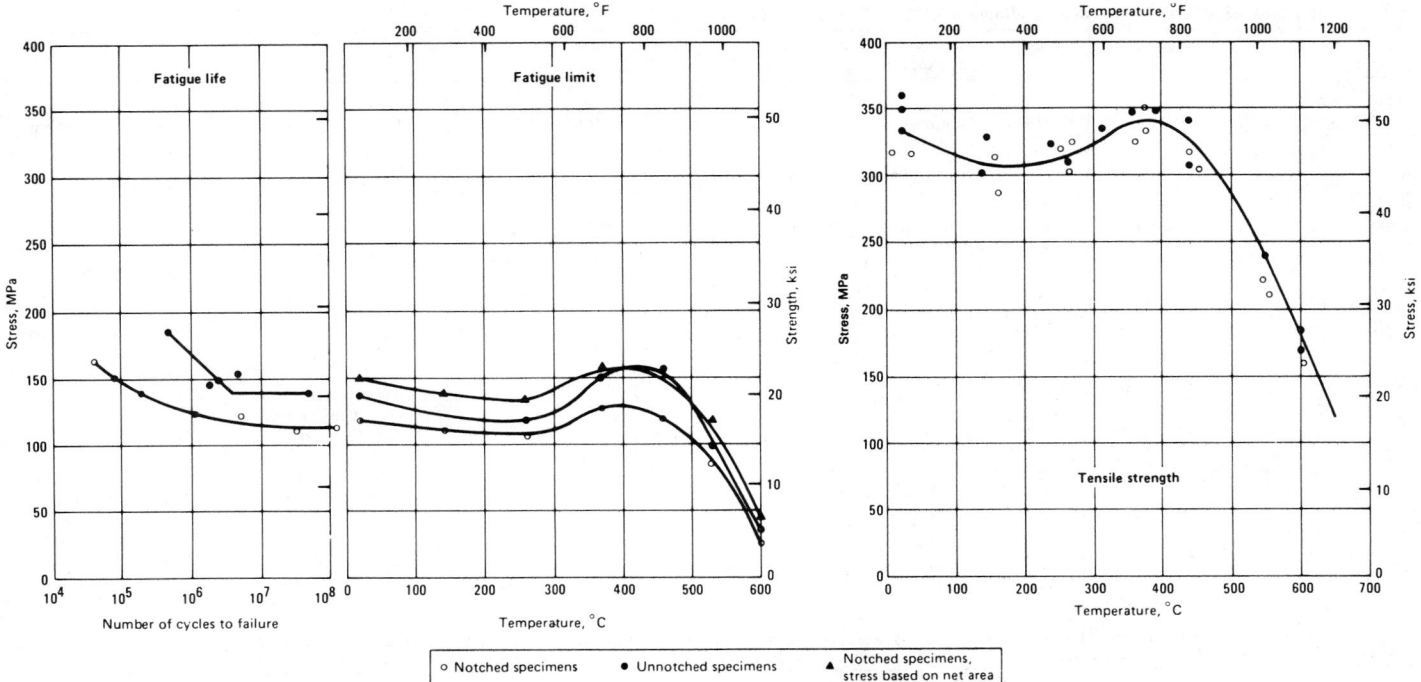

Composition: 2.84 C, 1.52 Si, 1.05 Mn, 0.07 P, 0.12 S, 0.31 Cr, 0.20 Ni, 0.37 Cu.

Fig. 7. *S-N* **curves and effects of temperature on reverse-bending fatigue limit (each point represents one fatigue limit) of gray iron of indicated tensile strength (each point represents one test)**

If any hardness correlation is to be attempted, the type and amount of graphite must be constant in the irons being compared. Rockwell hardness tests are considered appropriate only for hardened castings (camshafts, for example), and even on hardened castings, Brinell tests are preferred. Brinell tests must be used when attempting any strength correlations for unhardened castings.

FATIGUE LIMIT IN REVERSED BENDING

Because fatigue limits are expensive to determine, the designer usually has incomplete infor-mation on this property. Typical *S-N* curves for notched and unnotched specimens of gray iron under completely reversed cycles of bending stress are shown in the graph at left in Fig. 7; each point represents the data from one specimen.

The effects of temperature on fatigue limit and tensile strength of notched and unnotched specimens of gray iron are shown in the middle and right graphs in Fig. 7.

Axial loading or torsional loading cycles are frequently encountered in designing parts of cast iron, and in many instances these are not completely reversed loads. Types of regularly repeated stress variation usually can be expressed as a function of a mean stress and a stress range. Wherever possible, the designer should use actual data from the limited information available. In the absence of precisely applicable test data, it is possible to make an estimate of the reverse-bending fatigue limit of machined parts by using about 35% of the minimum specified tensile strength of the particular grade of gray iron being considered for the application. This percentage is probably a safe value rather than an average of the few data available concerning the fatigue limit for gray iron.

Ductile Iron

DUCTILE CAST IRON, also known as nodular iron or spheroidal-graphite cast iron, is cast iron in which the graphite is present as tiny spheres. In ductile iron, eutectic graphite separates from the molten iron during solidification in a manner similar to that in which eutectic graphite separates in gray iron. However, because of additives that are introduced in the molten iron just before casting, the graphite grows as spheres, rather than as flakes of any of the forms that are characteristic of gray iron. Cast iron containing nodular graphite has markedly greater strength and greater ductility than gray iron of similar composition.

The relatively high strength and toughness of ductile iron give it an advantage over gray iron in many structural applications. Also, because ductile iron does not require prolonged heat treatment to produce graphite nodules (as does malleable iron), it can compete with malleable iron even though it is more expensive to melt and cast.

Typically, the composition of unalloyed ductile iron is similar to that of gray iron. Ductile iron and gray iron are produced from the same types of raw materials, but those materials used for ductile iron usually are of higher purity. Like gray iron, ductile iron can be melted in cupola, electric-arc or induction furnaces. As liquid iron, it has high fluidity and excellent castability. The sands and molding equipment used for ductile iron castings are similar to those used for gray iron castings.

The formation of graphite during solidification causes an attendant increase in volume that in part counteracts the loss in volume due to the liquid-to-solid phase change in the metallic constituent. Ductile iron castings made in rigid (chemically bonded) sand molds require only minimal use of risers (reservoirs in the mold that feed molten metal into the mold cavity to compensate for liquid contraction during cooling and solidification). Gray irons often do not require risers to ensure shrinkage-free castings, but on the other hand, steels and malleable iron generally require heavy risering. This means that the yield of ductile iron castings—that is, the ratio of the weight of usable castings to the weight of metal poured—is much higher than that of either steel castings or malleable iron castings, and almost as high as that of gray iron castings.

Shrinkage of cast iron during both solidification and subsequent cooling to room temperature often requires designers to compensate by making patterns with dimensions larger than those desired in the finished castings. Typically, ductile iron requires less compensation than any other cast ferrous metal. Patternmakers' rules (shrink

Table 1. Compositions and general uses of standard grades of ductile iron(a)

Specification No.	Grade or class	UNS No.	Typical composition, %					Description	General uses
			TC(b)	Si	Mn	P	S		
ASTM A395; ASME SA395	60-40-18	F32800	3.00 min	2.50 max(c)	...	0.08 max	...	Ferritic; annealed	Pressure-containing parts for use at elevated temperatures
ASTM A476; SAE AMS5316	80-60-03	F34100	3.00 min(d)	3.0 max	...	0.08 max	0.05 max	As cast	Paper-mill dryer rolls, at temperatures up to 230 °C (450 °F)
ASTM A536; MIL-I-11466B(MR)	60-40-18(e)	F32800						Ferritic; may be annealed	Shock-resistant parts; low-temperature service
	65-45-12(e)	F33100						Mostly ferritic; as cast or annealed	General service
	80-55-06(e)	F33800						Ferritic/pearlitic; as cast	General service
	100-70-03(e)	F34800						Mostly pearlitic; may be normalized	Best combination of strength, wear resistance and response to surface hardening
	120-90-02(e)	F36200						Martensitic; oil quenched and tempered	Highest strength and wear resistance

(a) For mechanical properties and typical applications, see Table 2. (b) Total carbon. (c) The silicon limit may be increased by 0.08% for each 0.01% reduction in phosphorus content, up to 2.75 Si. (d) Carbon equivalent (CE), 3.8-4.5; CE = TC + 0.3 (Si + P). (e) Composition subordinate to mechanical properties; composition range for any element may be specified by agreement between supplier and purchaser.

rules) usually incorporate the following allowances:

Type of cast metal	Shrinkage allowance, %
Ductile iron	0-0.7
Gray iron	1.0
Malleable iron	1.0
Austenitic alloy iron	1.3-1.5
White iron	2.0
Carbon steel	2.0
Alloy steel	2.5

Shrinkage allowance can vary somewhat around the percentages given above, and often different percentages must be used for different directions in the casting because of the influence of the solidification pattern on the amount of contraction that takes place in different directions.

Many ductile iron castings are used as cast, but in some foundries, 50% or more are heat treated before being shipped. Heat treatment varies according to the desired effect on properties. Many of the heat treated castings are given a ferritizing anneal or are normalized to produce a uniform pearlitic microstructure. Other castings are given hardening treatments that produce bainitic or martensitic matrixes. The relative effect of the matrix structure on properties can be considered analogous to the effect of microstructure on properties of steel. As the matrix structure is varied progressively from ferrite to ferrite plus pearlite to pearlite to bainite and finally to martensite, hardness, strength and wear resistance increase, but impact resistance, ductility and machinability decrease. An exception to this is austempered ductile iron, in which considerable elongation (as high as 10%) can be obtained even at high strengths (1000 MPa, or 145 ksi).

Ductile iron can be alloyed with small amounts of nickel, molybdenum or copper to improve its strength and hardenability. Larger amounts of silicon, chromium, nickel or copper can be added for improved resistance to corrosion or for high-temperature applications.

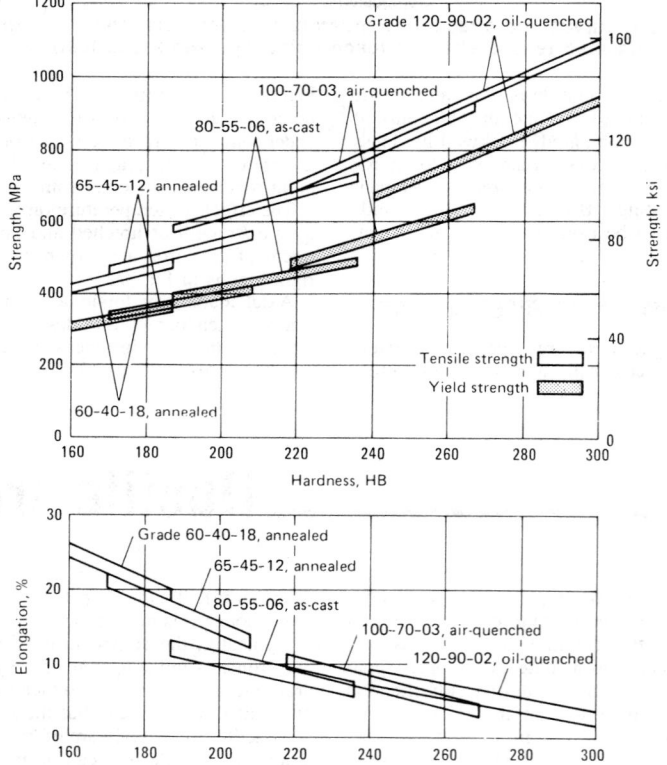

Mechanical properties determined on specimens taken from a 25-mm (1-in.) keel block.

Fig. 1. Tensile properties of ductile iron versus hardness

SPECIFICATIONS

Most of the specifications for standard grades of ductile iron are based on properties—that is, strength and/or hardness is specified for each grade of ductile iron, and composition is either loosely specified or made subordinate to the mechanical properties. Tables 1 and 2 list compositions, properties, uses and typical applications for most of the ductile irons that are defined by current standard specifications (except for the high-nickel corrosion-resistant and heat-resistant irons defined in ASTM A439). As shown in Table 2, the ASTM system for designating the grade of

Table 2. Mechanical properties and typical applications of standard grades of ductile iron(a)

Specification No.	Grade or class	Hardness HB(b)	Tensile strength, min(c) MPa	ksi	Yield strength, min(c) MPa	ksi	Elongation in 50 mm or 2 in., min, %(c)	Typical applications
ASTM A395-76; ASME SA395	60-40-18	143-187	414	60	276	40	18	Valves and fittings for steam and chemical-plant equipment
ASTM A476-70(d); SAE AMS5316	80-60-03	201 min	552	80	414	60	3	Paper-mill dryer rolls
ASTM A536-72, MIL-I-11466B(MR) ..	60-40-18	...	414	60	276	40	18	Pressure-containing parts such as valve and pump bodies
	65-45-12	...	448	65	310	45	12	Machine components subject to shock and fatigue loads
	80-55-06	...	552	80	379	55	6	Crankshafts, gears and rollers
	100-70-03	...	689	100	483	70	3	High-strength gears and machine components
	120-90-02	...	827	120	621	90	2	Pinions, gears, rollers and slides
SAE J434c	D4018	170 max	414	60	276	40	18	Steering knuckles
	D4512	156-217	448	65	310	45	12	Disc-brake calipers
	D5506	187-255	552	80	379	55	6	Crankshafts
	D7003	241-302	689	100	483	70	3	Gears
	DQ & T	(e)	(f)	(f)	(f)	(f)	(f)	Rocker arms
MIL-I-24137(Ships) ...	Class A	190 max	414	60	310	45	15	Electric equipment, engine blocks, pumps, housings, gears, valve bodies, clamps and cylinders
	Class B	190 max	379	55	207	30	7	Pressure parts, machine components and propellers
	Class C	175 max	345	50	172	25	20	Pressure parts, machine components and propellers

(a) For compositions, descriptions and general uses, see Table 1. (b) Measured at a predetermined location on the casting. (c) Determined using a standard specimen taken from a separately cast test block, as set forth in the applicable specification. (d) Reapproved in 1976. (e) Range specified by mutual agreement between producer and purchaser. (f) Value must be compatible with minimum hardness specified for production castings.

ductile iron incorporates the numbers indicating tensile strength in ksi, yield strength in ksi and elongation in percent. This system makes it easy to specify nonstandard grades that meet the general requirements of ASTM A536. For example, grade 80-60-03 (552 MPa min tensile strength, 414 MPa min yield strength and 3% min elongation) is widely used in applications where relatively high ductility is not important. Grades 60-42-10 and 60-45-10 are annealed grades that have been widely used for pipe. Grade 70-50-05 has been used extensively to make as-cast pipe fittings for water and other liquids. Grades other than those listed in ASTM A536 or mentioned above can be made to the general requirements of A536, but with the mechanical properties specified by mutual agreement between purchaser and producer.

MANUFACTURE AND METALLURGICAL CONTROL

Greater metallurgical and process control is required in production of ductile iron than in production of other cast irons. Frequent chemical, mechanical and metallurgical testing is needed to ensure that the required quality is maintained and that specifications are met.

Manufacture of high-quality ductile iron begins with careful selection of charge materials that will give a relatively pure cast iron free of the undesirable residual elements sometimes found in other cast irons. Carbon, manganese, silicon, phosphorus and sulfur must be held at specified levels. Magnesium, cerium and certain other elements must be controlled in order to attain the desired graphite shape and to offset the deleterious effects of subversive elements; elements such as antimony, lead, titanium, tellurium, bismuth and zirconium interfere with the nodulizing process, and must be either eliminated or restricted to very low concentrations.

Reduction of the sulfur content to less than 0.02% is necessary prior to the nodulizing process; this can be accomplished through basic melting alone, by use of low-sulfur charge material or by desulfurization of the base metal before the magnesium-nodulizing alloy is added. If base sulfur is not so reduced, excessive amounts of costly nodulizing alloys will be required and melting efficiency will be impaired.

Graphite Shape and Distribution. There are three major types of nodulizing agents, all of which contain magnesium: unalloyed magnesium, nickel-base nodulizers and magnesium-ferrosilicon nodulizers. Unalloyed magnesium has been added to molten iron as wire, ingots or pellets; as briquets, in combination with sponge iron; or in the cellular pores of metallurgical coke. The method of introducing the alloy has varied from an open-ladle method (in which the alloy is placed at the bottom of the ladle and iron is poured rapidly over the alloy) to a pressure-container method (in which unalloyed magnesium is placed inside a

container holding molten iron and the container is rotated so that the iron flows over the magnesium). In all cases, magnesium is vaporized and the vapors travel through the molten iron, lowering the sulfur content and promoting formation of spheroidal graphite.

Testing and Inspection. Various tests are used to control the processing of ductile iron, starting with analyses of raw materials and of the molten metal both before and after the nodulizing treatment. Rapid thermal-arrest methods are used to determine carbon, silicon, and carbon equivalence in the molten iron. Silicon content is also determined by thermoelectric and spectrometric techniques. Chill tests are used for production-line testing for silicon.

After the nodulizing step, a standard test coupon for microscopic examination should be poured from each batch of metal, as recommended by AFS and as specified in ASTM A395. One ear of the test coupon is broken off and polished to reveal graphite shape and distribution, plus matrix structure. These characteristics are evaluated by comparison with standard ASTM/AFS photomicrographs, and acceptance or rejection of castings is based on this comparison.

Tensile-test specimens are machined from separately cast keel blocks, Y blocks or modified keel blocks, as described in ASTM A395. If the terms of purchase require tensile specimens to be taken from castings, the part drawing must identify the area of the casting and the size of the test specimen. These terms also must be mutually acceptable to both producer and purchaser.

Hardness testing of production castings also is used to evaluate conformance to specified properties. Some standard specifications, such as SAE J434b, relate strength and hardness as shown in Fig. 1.

Heat Treatment. When the properties desired are difficult to obtain in the as-cast metal, ductile iron may be heat treated. Heat treated ductile iron usually has more uniform mechanical properties than as-cast ductile iron, particularly in castings with wide variations in section thickness.

Ductile iron castings of large or nonuniform cross section occasionally are stress relieved at 540 to 660 °C (1000 to 1100 °F), which reduces warping and distortion during subsequent machining. Mechanical properties of castings are essentially unchanged by the stress-relieving process.

Full ferritizing annealing produces grade 60-40-18 for applications requiring maximum impact resistance and ductility. This heat treatment usually involves heating to 900 °C (1650 °F) and holding, then cooling to about 700 °C (1300 °F) and holding, followed by controlled cooling to near room temperature.

Subcritical annealing produces either grade 60-40-18 or grade 65-45-12 for applications requiring high toughness and ductility. Subcritical annealing is usually done by heating to 730 °C (1350 °F) and holding until a ferritic structure is obtained, followed by controlled cooling to near room temperature.

Normalizing and tempering produces pearlitic grade 100-70-03, which is widely used for applications requiring good strength and wear resistance. Castings are heated to about 900 °C (1650 °F), held there long enough to stabilize the structure, and rapidly cooled with a fan or air blast. Then the castings are reheated to 540 to 675 °C (1000 to 1250 °F), which provides both stress relief and control of final hardness.

Martensitic ductile iron (grade 120-90-02) is produced by heating to about 900 °C (1650 °F) and holding, then quenching in agitated oil. This treatment produces castings of the highest strength and best wear resistance. Stress relief and control of final hardness are accomplished by tempering at 510 to 565 °C (950 to 1050 °F). Austempered ductile iron is being used commercially but has, as yet, no ASTM specification. Following austenitizing at about 900 °C (1650 °F), it is cooled, at a rate high enough to prevent pearlite formation, to about 350 °C (660 °F) and held for about 2 h to produce a bainitic microstructure, and then is cooled to room temperature. In heavier sections, some additional alloying is needed to ensure that no pearlite is formed when the metal is cooled to the austempering temperature. For further information on heat treating procedures, see the heat treating section of this Handbook.

MECHANICAL PROPERTIES

Most of the standard specifications for ductile iron require minimum strength and ductility, as determined using separately cast standard ASTM test bars. The various specification limits have been established by evaluating the results from thousands of these test bars. Properties of test bars are useful approximations of the properties of finished castings. Test-bar properties also make it possible to compare many different batches of metal without having to account for the variations due to differences in the shapes being cast or differences in the production practices used in different foundries.

Test bars are machined from keel blocks, Y blocks or modified keel blocks (see ASTM A395 for details and dimensions). These test blocks are designed for ideal feeding from heavy molten metal heads over the mold and for controlled cooling at optimum rates. In practice, these characteristics may not be economically feasible, or may be impossible because of the configuration of the casting. As a result, actual properties of production castings may differ from those of test bars cast from the same heat of molten metal, a fact that sometimes is overlooked.

Effect of Composition. The properties of ductile iron depend first on composition. Composition should be uniform within each casting and among all castings poured from the same melt. Many elements influence casting properties, but those of greatest importance are the elements that exert powerful influences on matrix structure or on shape and distribution of graphite nodules.

Carbon influences the fluidity of the molten iron and the shrinkage characteristics of the cast metal, both of which affect casting design. Carbon also influences the size and number of graphite particles that are formed on solidification. The size and number of graphite particles is also influenced by inoculation procedures.

Silicon is a powerful graphitizing agent. Within the normal composition limits, increasing amounts of silicon promote structures that have progressively greater amounts of ferrite; furthermore, silicon contributes to the solution strengthening of ferrite. Increasing the amount of ferrite increases ductility and slightly increases yield strength, but concurrently reduces tensile strength and Brinell hardness.

Among the alloying elements commonly used to improve the mechanical properties of ductile iron, manganese acts as a pearlite stabilizer, and increases strength but reduces ductility. Nickel is

frequently used to increase strength by promoting formation of fine pearlite. Nickel is also used to increase hardenability, especially for surface hardening applications. Copper has been used as a pearlite stabilizer and, as such, increases strength. Molybdenum can be added to stabilize the structure at elevated temperature, thus promoting better retention of strength at temperatures up to about 650 °C (1200 °F) in unalloyed or low-alloy ductile irons.

Effect of Graphite Shape. Conversion of graphite from flakes to nodules, which is caused by addition of magnesium (or magnesium and cerium) to the molten iron, results in a fivefold to sevenfold increase in the strength of the cast metal. Shapes that are intermediate between a true nodular form and a flake form (such as, respectively, ASTM types I and VII, as established by ASTM A247) yield mechanical properties that are inferior to those of ductile iron with a true nodular graphite but that are still better than the properties of gray iron of similar composition.

Effect of Section Size. Cooling rate is the variable chiefly affected by section size. The cooling rate, in turn, affects both the size of the graphite nodules and the microstructure of the matrix. The heavier the section, the more slowly it cools, and therefore the larger the graphite nodules that can form on solidification. When ductile iron is cast in sections greater than about 65 mm (2$\frac{1}{2}$ in.), there is the possibility that degenerate graphite shapes (vermicular, crab, etc.) will be produced. Careful control of residuals and/or the presence of small amounts of cerium are usually effective in combating this problem.

The structure of the matrix is essentially determined by the cooling rate through the eutectoid temperature range, although the specific effects of cooling rate are modified by the presence of alloying elements, as discussed above in the section on effect of composition. Slow cooling rates prevalent in heavy sections promote transformation to ferrite. For a given silicon content, a decrease in section size and an increase in cooling rate tends to promote pearlite formation, along with an increase in strength and hardness and a decrease in ductility. Bainitic or martensitic structures are not often found in as-cast ductile iron, although it is possible for such structures

to occur in very thin sections; these structures are normally obtained by heat treatment. Bainite and martensite are the main constituents in high-strength, heat treated ductile irons such as grade 120-90-02.

Tensile Properties of one heat of ductile iron heat treated to strength levels approximately equivalent to four standard ductile irons are given in Table 3. These values are not necessarily the average property values that can be expected for metal produced to the indicated grades. Within each grade, strength and ductility vary somewhat with hardness, as shown in Fig. 1. In some instances, the ranges of expected strength and ductility overlap those for the next higher or lower grade.

As shown in Table 3, the modulus of elasticity in tension ranges from 164 to 169 GPa (23.8 to 24.5 × 10⁶ psi) and does not vary greatly with grade. Similarly, Poisson's ratio lies in a narrow range near 0.30 and does not vary greatly with grade. The values of tensile modulus shown in Table 3 were determined using standard 12.83-mm-diam (0.505-in.-diam) tensile bars equipped with strain gages affixed to the reduced section.

Compressive Properties. The 0.2% offset yield strength of ductile iron in compression generally is reported as 1.0 to 1.2 times the 0.2% offset yield strength in tension. The compressive properties shown in Table 3 were determined using specimens from the same single heat of ductile iron described above under "Tensile properties."

Torsional Properties. Very few data are available on the ultimate shear strength of ductile iron because it is very difficult to obtain accurate shear data on materials that exhibit some ductility. It is generally agreed that the ultimate shear strength of ductile cast iron is about 0.9 to 1.0 times the ultimate tensile strength. Table 3 gives data for shear strength and 0.0375% offset yield strength in torsion for a single heat of ductile iron heat treated to strength levels approximately equivalent to four standard ductile irons.

Damping Capacity. The average damping capacity of ductile iron in the hardness range of 156 to 241 HB is about 6.6 times that of 1018 steel and about 0.12 times that of class 30 gray iron.

Impact Properties. Data from a comprehensive study of impact properties of ductile iron are

Table 3. Average mechanical properties of ductile irons heat treated to various strength levels(a)

Nearest standard grade	Hardness, HB	Ultimate strength MPa	Ultimate strength ksi	Yield strength MPa	Yield strength ksi	Elongation, %(b)	Modulus GPa	Modulus 10⁶ psi	Poisson's ratio
Tension									
60-40-18	167	461	66.9	329(c)	47.7(c)	15.0	169	24.5	0.29
65-45-12	167	464	67.3	332(c)	48.2(c)	15.0	168	24.4	0.29
80-55-06	192	559	81.1	362(c)	52.5(c)	11.2	168	24.4	0.31
120-90-02	331	974	141.3	864(c)	125.3(c)	1.5	164	23.8	0.28
Compression									
60-40-18	167	...	...	359(c)	52.0(c)	...	164	23.8	0.26
65-45-12	167	...	...	362(c)	52.5(c)	...	163	23.6	0.31
80-55-06	192	...	...	386(c)	56.0(c)	...	165	23.9	0.31
120-90-02	331	...	...	920(c)	133.5(c)	...	164	23.8	0.27
Torsion									
60-40-18	167	472	68.5	195(d)	28.3(d)	...	63 / 65.5(e)	9.1 / 9.5(e)	...
65-45-12	167	475	68.9	297(d)	30.0(d)	...	64 / 65(e)	9.3 / 9.4(e)	...
80-55-06	192	504	73.1	193(d)	28.0(d)	...	62 / 64(e)	9.0 / 9.3(e)	...
120-90-02	331	875	126.9	492(d)	71.3(d)	...	63.4 / 64(e)	9.2 / 9.3(e)	...

(a) Determined for a single heat of ductile iron, heat treated to approximate various standard grades. Properties were obtained using test bars machined from 25-mm (1-in.) keel blocks. (b) In 50 mm or 2 in. (c) 0.2% offset. (d) 0.0375% offset. (e) Calculated from tensile modulus and Poisson's ratio in tension.

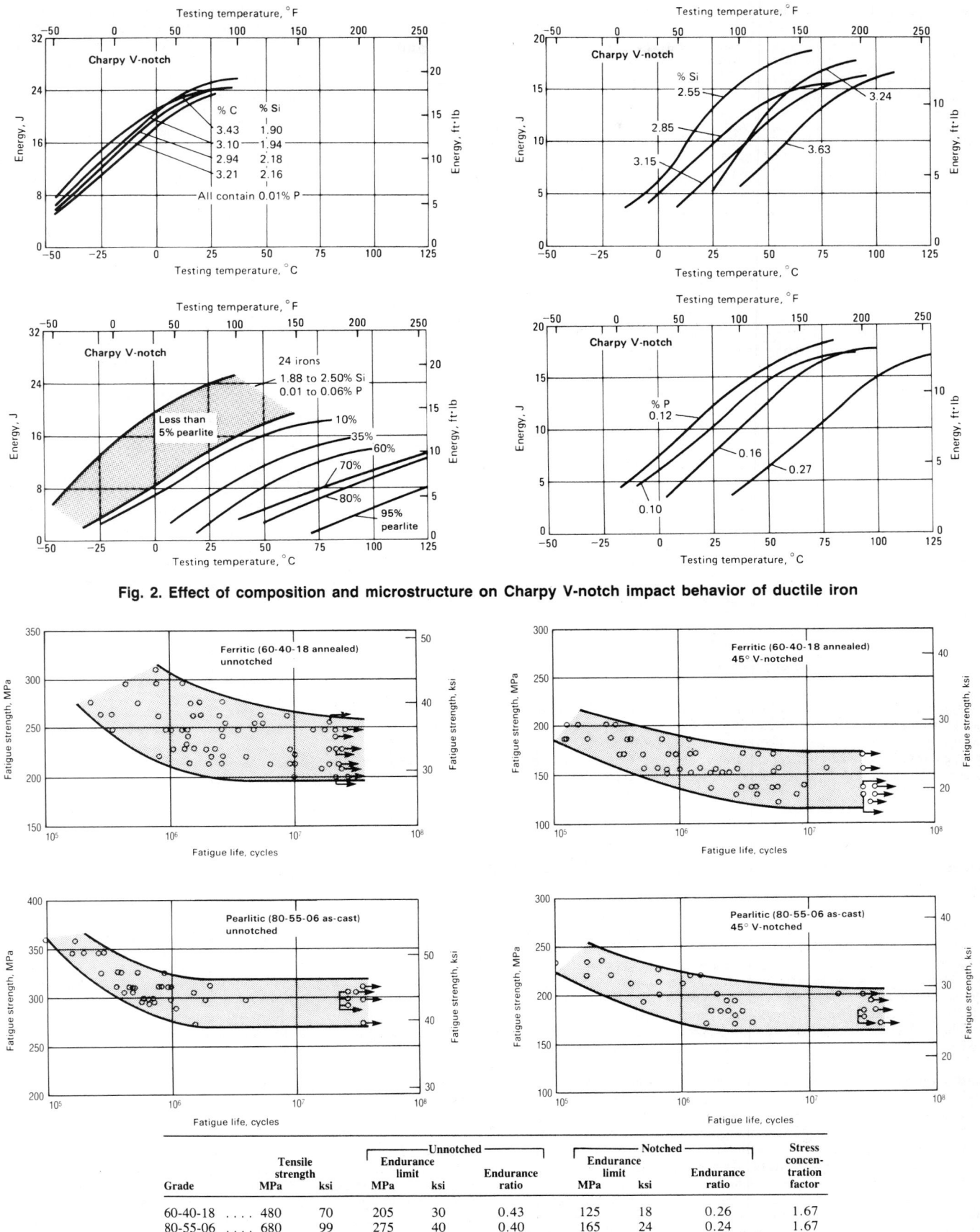

Fig. 2. Effect of composition and microstructure on Charpy V-notch impact behavior of ductile iron

Fig. 3. Fatigue properties of ductile iron

Grade	Tensile strength MPa	ksi	Unnotched Endurance limit MPa	ksi	Endurance ratio	Notched Endurance limit MPa	ksi	Endurance ratio	Stress concentration factor
60-40-18	480	70	205	30	0.43	125	18	0.26	1.67
80-55-06	680	99	275	40	0.40	165	24	0.24	1.67

Table 4. Fracture toughness of ductile iron

Type of iron	Condition	Ultimate tensile strength, ksi	Yield strength, ksi	Elongation, %	K_{Ic}, ksi·in.$^{1/2}$, at: 20 °C (70 °F)	−40 °C (−40 °F)	−110 °C (−160 °F)
Ferritic							
3.0% Si	As cast	75.6	62.0	11.0	...	32.0	27.5
3.5% Si	As cast	79.4	68.3	9.0	...	24.6	...
Pearlitic							
2.5% Si	As cast	102.0	54.2	7.5	...	33.8	...
	Normalized ...	133.2	80.0	3.6	41.3	...	...
	Austempered		90.0(a)	...	33.3	...	...

(a) Estimated.

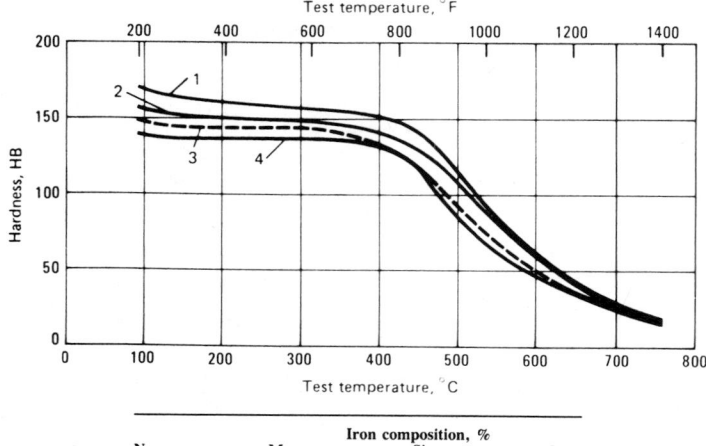

No.	Mn	Iron composition, % Si	Ni
1	0.59	2.63	1.45
2	0.42	2.41	0.72
3	0.26	2.30	0.96
4	0.57	1.85	...

Fig. 4. Hot hardness of four annealed (ferritic) ductile irons

shown in Fig. 2. These data show that increasing pearlite decreases impact energy and that increasing phosphorus and/or silicon decreases impact energy. The transition temperature is significantly affected by phosphorus and/or silicon

content, but is little affected by other elements present within the normal variations in composition.

Fracture Toughness. Certain lower-strength grades of ductile iron do not fracture in a brittle manner

when tested under nominal plane-strain conditions in a standard fracture-toughness test. This behavior is contrary to the basic tenets of fracture mechanics and has been attributed to localized deformation in the ferrite envelope surrounding each graphite nodule. In the low-strength ductile irons, plane-strain conditions are established only at temperatures low enough to embrittle the ferrite. Otherwise, increasing the size of the fracture-toughness test specimens does not provide the degree of mechanical constraint necessary to obtain a valid measurement of K_{Ic}.

Selected values of fracture toughness are given in Table 4. These values were determined using compact tension (CT) specimens 21 mm (0.83 in.) in width. All tests were made in accordance with ASTM E399-72. The fracture-toughness data given in Table 4 are for ductile iron with a nodule shape approximately corresponding to 50% ASTM type I.

Fatigue Strength. Figure 3 shows fatigue-strength curves for ferritic and pearlitic ductile iron in both the notched and unnotched conditions. The tests were made on Wohler-type fatigue machines with polished specimens 10.6 mm (0.417 in.) in diameter.

The endurance limit for a given grade of ductile iron depends on surface condition. Endurance ratio is defined as endurance limit divided by tensile strength. Because the endurance ratio of ductile iron declines as tensile strength increases, regardless of matrix structure, there may be little value in specifying a higher-strength ductile iron for a structure that is prone to fatigue failure; redesigning the structure to reduce stresses may prove to be a better solution.

MECHANICAL PROPERTIES AT ELEVATED TEMPERATURES

The hardness and strength of all standard grades of ductile iron are relatively constant up to about 425 °C (800 °F). Hot hardness data for four annealed (ferritic) ductile irons are shown in Fig. 4.

Dimensional growth of ductile iron is much less than that of gray iron at elevated temperatures.

Malleable Iron

MALLEABLE IRON is a cast ferrous metal that is initially produced as white cast iron, then is heat treated to convert the carbon-containing phase from iron carbide to a nodular form of graphite called temper carbon. There are two types of ferritic malleable iron, "blackheart" and "whiteheart." Only the blackheart type is produced in the United States. This material has a matrix of ferrite with interspersed nodules of temper carbon. "Cupola malleable iron" is a blackheart malleable iron produced by cupola melting and used for pipe fittings and similar thin-section castings. Because of its low strength and ductility, cupola malleable iron usually is not specified for structural applications. Pearlitic malleable iron is designed to have combined carbon in the matrix, resulting in higher strength and hardness than are available in ferritic malleable iron. Martensitic malleable iron is produced by quenching and tempering pearlitic malleable iron.

Malleable iron, like ductile iron, possesses considerable ductility and toughness because of its combination of nodule-shape graphite in a low-carbon metallic matrix. Because of the way in which graphite is formed in malleable iron, however, the nodules are not truly spherical as they are in ductile iron but rather than somewhat irregular aggregates.

Malleable iron and ductile iron are used for some of the same applications where ductility and toughness are important. In many instances, the choice between malleable and ductile iron is based on economy or availability rather than on properties. In certain applications, however, malleable iron has a clear-cut advantage. It is preferred for thin-section castings; for parts that are to be pierced, coined or cold formed; for parts requiring maximum machinability; for parts that must retain good impact resistance at low temperatures; and for some parts requiring wear resis-

tance (martensitic malleable iron only). It is also preferred for certain applications requiring high modulus of elasticity, because its modulus is 10 to 15% greater than that of ductile iron. Ductile iron has a clear advantage where low solidification shrinkage is needed to avoid hot tears, or where the section is too thick to permit solidification as white iron.

Malleable iron castings are produced in section thicknesses ranging from about 1.5 to 100 mm ($^1/_{16}$ to 4 in.), and in weights from less than 30 g (1 oz) to 180 kg (400 lb) or more. In one instance, ingot molds weighing about 1800 kg (3970 lb) and having sections as thick as 180 mm (7 in.) were made in malleable iron.

Melting may be accomplished by batch "cold melting" or by "duplexing."

Composition ranges typical of ferritic and pearlitic malleable irons are shown in Table 1. Chemical composition normally is not specified by the

Table 1. Typical composition ranges for ferritic and pearlitic malleable irons, analyzed in the white iron condition

Type of iron	Composition, %				
	TC	Mn	Si	P	S
Ferritic:					
Grade 32510	2.30-2.70	0.25-0.55	1.00-1.75	0.05 max	0.03-0.18
Grade 35018	2.00-2.45	0.25-0.55	1.00-1.35	0.05 max	0.03-0.18
Pearlitic	2.00-2.70	0.25-1.25	1.00-1.75	0.05 max	0.03-0.18

user; the producer determines the composition after consideration of section size, melting process, availability of raw materials, and heat treatment practices. Occasionally, castings are made to composition limits established by mutual agreement between producer and user.

Control of Melting. Metallurgical control of the melting operation is designed to ensure that the molten iron will have a certain composition and will:

- Solidify white in the castings to be produced
- Anneal on an established time-temperature cycle set to minimum values in the interest of economy
- Produce the desired graphite distribution (nodule count) on annealing.

Changes in melting practice or composition that would satisfy the first requirement above generally are opposed to satisfaction of the second and third, while attempts to improve annealability beyond a certain point may result in an unacceptable tendency for the as-cast iron to be mottled instead of white.

The common elements in malleable iron are generally controlled within about ±0.05 to ±0.15%. For example, one high-production malleable foundry operates to a composition standard of:

Carbon ....2.40 to 2.70
Manganese ....0.45 to 0.55
Silicon ....1.25 to 1.55
Phosphorus ....0.05 max
Sulfur ....0.15 max
Chromium ....0.08 max
Boron ....0.002 to 0.003

A limiting minimum carbon content is required in the interest of mechanical quality and annealability, because decreasing carbon content reduces fluidity of the molten iron, increases shrinkage during solidification, and reduces annealability. A limiting maximum carbon content is imposed by the requirement that the casting be white as cast.

MICROSTRUCTURE

Malleable iron is characterized by microstructures consisting of uniformly dispersed fine particles of free carbon in a matrix of ferrite or tempered martensite. These microstructures can be produced in base metal of essentially the same composition. Structural differences between ferritic malleable iron and the various grades of pearlitic or martensitic malleable iron are achieved through variations in heat treatment. The microstructure of a casting of any type of malleable iron is derived by controlled annealing of white iron of suitable composition. During the annealing cycle, carbon that exists in combined form, either as massive carbides or as a microconstituent in pearlite, is converted to a form of free graphite known as temper carbon.

Ferritic malleable iron requires a two-stage annealing cycle: the first stage converts primary carbides to temper carbon; the second stage converts the carbon dissolved in austenite at the first-stage annealing temperature to temper carbon and ferrite.

The microstructure of ferritic malleable iron is shown in Fig. 1; satisfactory structure consists of temper carbon in a matrix of ferrite. There should be no flake graphite and essentially no combined carbon in ferritic malleable iron.

Pearlitic and martensitic malleable irons contain a controlled quantity of combined carbon, which, depending on heat treatment, may appear in the metallic matrix as lamellar pearlite, tempered martensite, or spheroidite.

Molten iron produced under properly controlled melting conditions solidifies with all carbon in the combined form, producing the white iron structure fundamental to the manufacture of either ferritic or pearlitic malleable iron (see Fig. 2). Because the base iron must contain balanced quantities of carbon and silicon to simultaneously provide (*a*) castability, (*b*) white iron in even the thickest sections of the castings, and (*c*) annealability, precise metallurgical control is necessary for quality production.

Control of Nodule Count. Proper annealing in short time cycles and attainment of high levels of casting quality require that controlled distribution of graphite particles be obtained during first-stage heat treatment. With low nodule count (few graphite particles per unit area or volume), me-

chanical properties are reduced from optimum, and second-stage annealing time is unnecessarily long because of long diffusion distances. Excessive nodule count is also undesirable, because graphite particles may become aligned in a configuration corresponding to the boundaries of the original primary cementite. In martensitic malleable iron, very high nodule counts are sometimes associated with low hardenability and nonuniform tempering. Generally, a nodule count of 80 to 150 discrete graphite particles per square millimetre (80 to 150 in 15.5 in.^{2} of a photomicrograph at 100×) appears to be optimum. This produces random particle distribution, with short distances between particles.

Control of Annealing. The rate of annealing of a hard iron casting depends on chemical composition, nucleation tendency as discussed above, and annealing temperature. With proper balance of boron content and graphitic materials in the charge, optimum number and distribution of graphite nuclei are developed in the early part of first-stage annealing, and growth of the temper carbon particles proceeds rapidly at any annealing temperature. An optimum iron will anneal completely through the first-stage reaction in approximately 3^1/$_2$ h at 940 °C (1720 °F). Irons with lower silicon contents or less-than-optimum nodule counts may require as much as 20 h for completion of first-stage annealing.

Castings are generally heated at a rate no greater than that at which they would reach first-stage temperature in 4 h, to minimize cracking of the brittle white iron due to uneven heating.

First-stage annealing time and temperature are adjusted to accommodate the composition and melting practice used in the foundry. Normally, soaking time is about 50% longer than experimentation shows would be necessary for complete first-stage annealing. Depending on the composition of the iron and the temperature of first-stage annealing, soaking time varies from about 2 to 36 h; the longer soaking times are re-

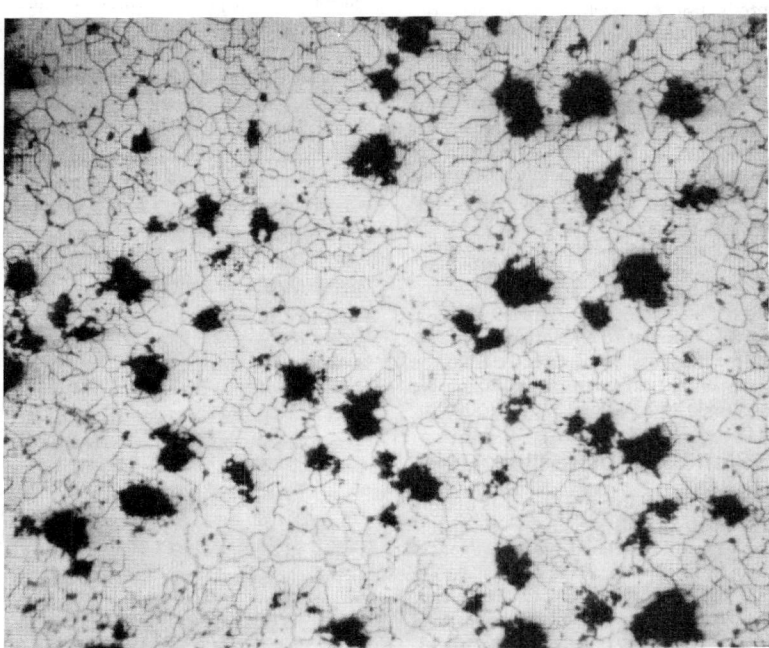

Temper carbon in ferrite. Etched; magnification, 100×.
Fig. 1. Structure of annealed ferritic malleable iron

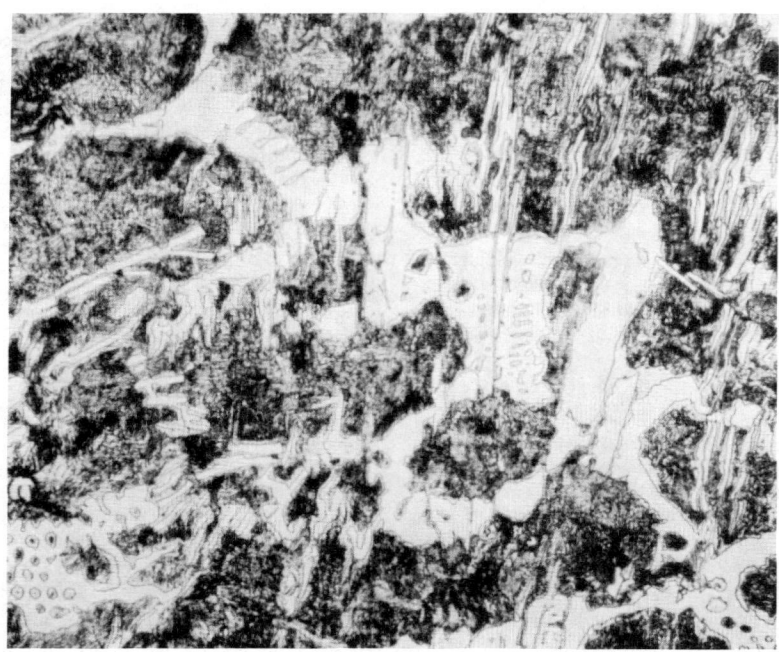

Mixture of pearlite and eutectic carbides. Etched; magnification, 400×.
Fig. 2. Structure of as-cast malleable white iron

quired for low-silicon compositions.

After first-stage annealing, the castings are cooled as rapidly as practical to 740 to 760 °C (1360 to 1400 °F) in preparation for second-stage annealing. The fast cooling step requires 1 to 6 h, depending on the equipment employed. Castings then are cooled slowly at a rate of about 3 to 11 °C (5 to 20 °F) per hour. During cooling, the carbon dissolved in the austenite is converted to graphite and deposited on the existing particles of temper carbon. This results in a fully ferritic matrix.

In the production of pearlitic malleable iron, the first-stage heat treatment is identical to that employed for ferritic malleable iron. However, some foundries then slowly cool the castings to about 870 °C (1600 °F). During cooling, the combined carbon content of the austenite is reduced to about 0.75%, and the castings then are air cooled. Air cooling is accelerated by an air blast to avoid the formation of ferrite envelopes around the temper carbon particles (bull's-eye structure) and to produce a fine pearlitic matrix. The castings then are tempered to specification, or are reheated to reaustenitize at about 870 °C, oil quenched, and tempered to specification. Large foundries usually eliminate the reaustenitizing step and quench the castings in oil directly from the first-stage annealing furnace after stabilizing the temperature at 845 to 870 °C (1550 to 1600 °F).

FERRITIC MALLEABLE IRON

Because ferritic malleable iron consists of only ferrite and temper carbon, the properties of ferritic malleable castings depend on the quantity, size, shape and distribution of the temper carbon particles and on the composition of the ferrite. Fully annealed ferritic malleable iron castings contain 2.00 to 2.70% graphitic carbon by weight, which is equivalent to about 6 to 8% by volume. Because the graphitic carbon contributes nothing to the strength of the castings, those with lesser

amounts of graphite are somewhat stronger and more ductile than those containing greater amounts (assuming equal size and distribution of graphite particles). Elements such as silicon and manganese in solid solution in the ferritic matrix contribute to the strength and reduce the elongation of the ferrite. Therefore, by varying base-metal composition, slightly different strength levels can be obtained in the fully annealed ferritic product.

Specifications. Ferritic malleable iron is produced to one of four existing grades, depending on the melting practice employed and the applicable ASTM specification for the casting (see listings under ASTM A47, A197, and A602, grade M3210, in Table 2).

Mechanical properties that are most important for design purposes are tensile strength, yield strength, modulus of elasticity, fatigue strength and impact strength. The hardness can be considered no more than an approximate indicator that the ferritizing anneal was complete, and is seldom used for any other purpose. The hardness of ferritic malleable iron is almost always within the range from 110 to 156 HB.

Tensile properties of ferritic malleable iron usually are measured on unmachined test bars. (Machined test bars normally are used for pearlitic malleable.)

The fatigue limit of unnotched ferritic malleable iron is about 50% of the tensile strength, or from 170 to 205 MPa (25 to 30 ksi). Figure 3 summarizes the effects of notches on fatigue strength. Generally, notch radius has little effect on fatigue strength, but fatigue strength decreases with increasing notch depth.

PEARLITIC AND MARTENSITIC MALLEABLE IRONS

Pearlitic malleable iron is produced either by controlled heat treatment of the same base white iron used to produce ferritic malleable iron or by alloying to prevent decomposition of carbides dissolved in austenite during cooling from the first-stage annealing temperature. It may be produced by air cooling after first-stage annealing and subsequent tempering to develop specified properties. Martensitic malleable iron is produced by liquid quenching and tempering to develop specified properties. (See earlier subsection on "Control of Annealing," under "Microstructure.") Variations in heat treatment, coupled with variations in base composition and melting practice, make it possible to obtain a wide range of properties in pearlitic or martensitic malleable iron.

Specifications. Some specifications for pearlitic and martensitic malleable iron are based on grade designations that require certain minimum tensile, yield and elongation values, with advisory

Table 2. Properties of malleable iron castings

Specification No.	Class or grade	Tensile strength MPa	ksi	Yield strength MPa	ksi	Hardness HB	Elongation(a), %
Ferritic							
ASTM A47, A338, ANSI G48.1; FED QQ-I-666c	32510	345	50	224	32	156 max	10
	35018	365	53	241	35	156 max	18
ASTM A197	....	276	40	207	30	156 max	5
Pearlitic and martensitic							
ASTM A220; ANSI G48.2; MIL-I-11444B	40010	414	60	276	40	149-197	10
	45008	448	65	310	45	156-197	8
	45006	448	65	310	45	156-207	6
	50005	483	70	345	50	179-229	5
	60004	552	80	414	60	197-241	4
	70003	586	85	483	70	217-269	3
	80002	655	95	552	80	241-285	2
	90001	724	105	621	90	269-321	1
Automotive							
ASTM A602; SAE J158	M3210(b)	345	50	224	32	156 max	10
	M4504(c)	448	65	310	45	163-217	4
	M5003(c)	517	75	345	50	187-241	3
	M5503(d)	517	75	379	55	187-241	3
	M7002(d)	621	90	483	70	229-269	2
	M8501(d)	724	105	586	85	269-302	1

(a) Minimum in 50 mm or 2 in. (b) Annealed. (c) Air quenched and tempered. (d) Liquid quenched and tempered.

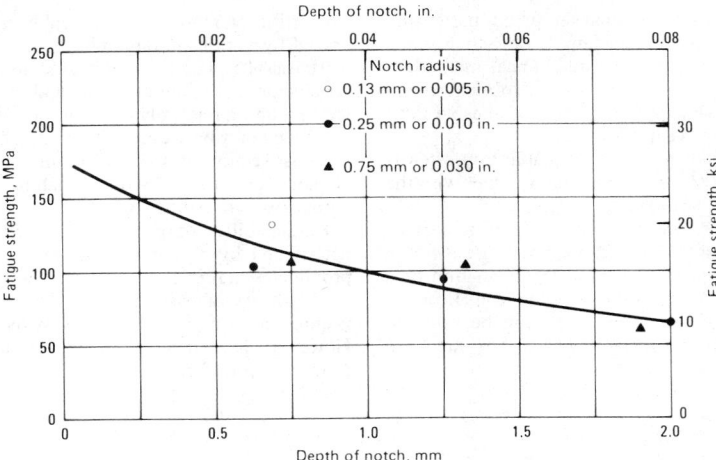

Fig. 3. Effects of notch radius and notch depth on fatigue strength of ferritic malleable iron

Brinell hardnesses (Table 2). (The hardnesses are termed "advisory" because the hardness ranges corresponding to the various grades overlap so much that hardness alone cannot ensure that a given casting meets the specification for a specific grade.) Other specifications (ASTM A602 and SAE J158, for example) are based on re-

quired hardness ranges and microstructures, and tensile-test data are considered advisory.

Mechanical properties of pearlitic and martensitic malleable irons vary in a substantially linear relationship with Brinell hardness (see Fig. 4). In the low hardness ranges, below about 207 HB, the properties of air-quenched-and-tempered

pearlitic malleable are essentially the same as those of oil-quenched-and-tempered martensitic malleable. This results from the fact that attainment of the low hardnesses requires considerable coarsening of the matrix carbides and partial second-stage graphitization. Either an air-quenched pearlitic structure or an oil-quenched martensitic structure can be coarsened and decarburized to meet this hardness requirement.

At higher hardnesses, oil-quenched-and-tempered malleable has higher yield strength and elongation than air-quenched-and-tempered malleable, because of greater uniformity of matrix structure and finer distribution of carbide particles. Oil-quenched-and-tempered pearlitic malleable is produced commercially to hardnesses as high as 321 HB, whereas the maximum hardness for high-production air-quenched-and-tempered pearlitic malleable is about 255 HB. The lower maximum hardness is applied to the air-quenched material for several reasons:

- Because hardness on air quenching normally does not exceed 321 HB and may be as low as 269 HB, attempts to temper to a hardness range above 255 HB produce nonuniform hardness and make the process control limits excessive.
- Very little structural alteration occurs during the tempering heat treatment to a higher hardness, and the resulting structure is more difficult to machine than an oil-quenched-and-tempered structure at the same hardness.

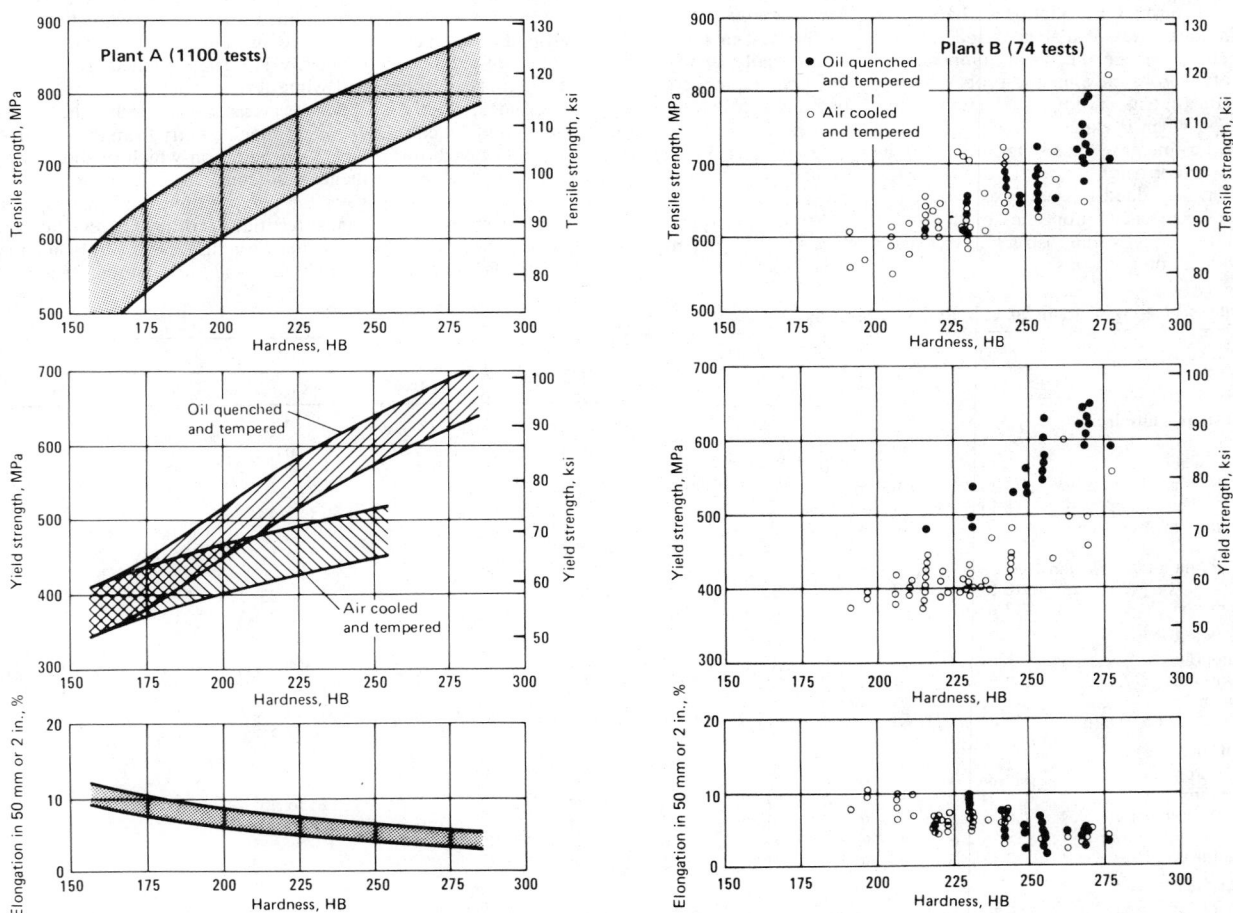

The mechanical properties of these irons vary in substantially a linear relationship with Brinell hardness and, in the low hardness ranges (below about 207 HB), the properties of air-quenched-and-tempered material are essentially the same as those produced by oil quenching and tempering.

Fig. 4. Relation of tensile properties to Brinell hardness for pearlitic malleable irons from two foundries

• There is only a slight improvement in other mechanical properties with increased hardness above 255 HB.

Because of these considerations, applications for air-quenched-and-tempered pearlitic malleable are usually those requiring moderate strength levels, while the higher-strength applications need the oil-quenched-and-tempered material.

Modulus of elasticity in tension of pearlitic malleable iron is 175 to 195 GPa (25.5 to 28.0 × 10⁶ psi). For automobile crankshafts, the modulus is important, and so must be determined with greater precision.

Fatigue limits of pearlitic and martensitic malleable irons are about 40 to 50% of tensile strength. Oil-quenched-and-tempered martensitic iron usually has a higher fatigue ratio than pearlitic iron made by the arrested anneal method.

Wear Resistance. Because of their structure and hardness, pearlitic and martensitic malleable irons have excellent wear resistance.

Machinability of pearlitic malleable iron often is from 10 to 30% better than that of steel with the same Brinell hardness.

Welding and Brazing. Welding of pearlitic or martensitic malleable iron is difficult because the high temperatures used can cause the formation of a brittle layer of graphite-free white iron. Pearlitic and martensitic malleable iron can be successfully welded if the surface to be welded has been heavily decarburized.

Pearlitic or martensitic malleable iron can be brazed by various commercial processes.

Hardenability. Pearlitic malleable iron contains considerable alloying elements and, at hardening temperature, a relatively high carbon content, both of which contribute to relatively high hardenability. Cast sections as thick as 30 mm (1¼ in.) can be hardened through by oil quenching to develop hardnesses of about 555 HB.

Hardenability of pearlitic malleable iron, as determined by a standard end-quench test, approximates that of SAE 4140 or 8640 steel. The hardenability of castings from any one foundry is quite uniform, within a relatively narrow band. However, hardenability may vary in castings produced in different foundries.

Alloy Cast Irons

ALLOY CAST IRONS are considered to be those casting alloys based on the iron-carbon-silicon system that contain one or more alloying elements intentionally added to enhance one or more useful properties. Addition to the ladle of small amounts of substances (such as ferrosilicon, cerium or magnesium) that are used to control the size, shape and/or distribution of graphite particles is termed inoculation rather than alloying. The quantities of material used for inoculation neither change the basic composition of the solidified iron nor alter the properties of individual constituents. Alloying elements, including silicon when it exceeds about 3%, usually are added to increase the strength, hardness, hardenability or corrosion resistance of the basic iron and often are added in quantities sufficient to affect the occurrence, properties or distribution of constituents in the microstructure (see Table 1 for typical compositions).

In gray and ductile irons, small amounts of alloying elements such as chromium, molybdenum and nickel are used primarily to achieve high strength or to ensure the attainment of a specified minimum strength in heavy sections. Otherwise, alloying elements are used almost exclusively to enhance resistance to abrasive wear or chemical corrosion, or to extend service life at elevated temperatures.

White cast irons, so named because of their characteristically white fracture surfaces, do not have any graphite in their microstructures. Instead, the carbon is present in the form of carbides, chiefly of the types Fe_3C and Cr_7C_3. Often, complex carbides such as $(Fe,Cr)_3C$ and $(Cr,Fe)_7C_3$, or those containing other carbide-forming elements, are present also.

White cast irons usually are very hard, which is the single property most responsible for their excellent resistance to abrasive wear. White iron can be produced either throughout the section (chiefly by adjusting the composition) or only partly inward from the surface (chiefly by casting against a chill). The latter iron is sometimes referred to as chilled iron to distinguish it from iron that is white throughout (see Fig. 1).

The main difference in microstructure between chilled iron and white iron is that chilled iron is fine grained and exhibits directionality perpendicular to the chilled face, whereas white iron ordinarily is coarse grained, randomly oriented and white throughout.

Corrosion-resistant irons derive their resistance to chemical attack chiefly from their high alloy content. Depending on which of three alloying elements—silicon, chromium or nickel—dominates the composition, a corrosion-resistant iron can be ferritic, pearlitic, martensitic or austenitic in its microstructure. Depending on composition, cooling rate and inoculation practice, a corro-

Table 1. Ranges of alloy content for various types of alloy cast irons

Description	TC(b)	Mn	P	S	Composition(a), wt % Si	Ni	Cr	Mo	Cu	Matrix structure, as-cast(c)
Abrasion-resistant white irons										
Low-C white iron(d)	2.2 to 2.8	0.2 to 0.6	0.15	0.15	1.0 to 1.6	1.5	1.0	0.5	(e)	CP
High-C, low-Si white iron	2.8 to 3.6	0.3 to 2.0	0.30	0.15	0.3 to 1.0	2.5	3.0	1.0	(e)	CP
Malleable white iron	2.2 to 2.5	0.3 to 0.5	0.15	0.15	1.0 to 1.6	...	...	...	...	CP
Martensitic nickel-chromium iron	2.5 to 3.7	1.3	0.30	0.15	0.8	2.7 to 5.0	1.1 to 4.0	1.0	...	M, A
Martensitic nickel, high-chromium iron	2.5 to 3.6	1.3	0.10	0.15	1.0 to 2.2	5 to 7	7 to 11	1.0	...	M, A
Martensitic chromium-molybdenum iron	2.0 to 3.6	0.5 to 1.5	0.10	0.06	1.0	1.5	11 to 23	0.5 to 3.5	1.2	M, A
High-chromium iron	2.3 to 3.0	0.5 to 1.5	0.10	0.06	1.0	1.5	23 to 28	1.5	1.2	M
Corrosion-resistant irons										
High-silicon iron(f)	0.4 to 1.1	1.5	0.15	0.15	14 to 17	...	5.0	1.0	0.5	F
High-chromium iron	1.2 to 4.0	0.3 to 1.5	0.15	0.15	0.5 to 3.0	5.0	12 to 35	4.0	3.0	M, A
Nickel-chromium gray iron(g)	3.0	0.5 to 1.5	0.08	0.12	1.0 to 2.8	13.5 to 36	1.5 to 6.0	1.0	7.0	A
Nickel-chromium ductile iron(h)	3.0	0.7 to 4.5	0.08	0.12	1.0 to 3.0	18 to 36	1.0 to 5.5	1.0	...	A
Heat-resistant gray irons										
Medium-silicon iron(j)	1.6 to 2.5	0.4 to 0.8	0.30	0.10	4.0 to 7.0	...	...	...	...	F
High-chromium iron	1.8 to 3.0	0.3 to 1.5	0.15	0.15	0.5 to 2.5	5.0	15 to 35	...	...	F, CP
Nickel-chromium iron (g)	1.8 to 3.0	0.4 to 1.5	0.15	0.15	1.0 to 2.75	13.5 to 36	1.8 to 6.0	1.0	7.0	A
Nickel-chromium-silicon iron(k)	1.8 to 2.6	0.4 to 1.0	0.10	0.10	5.0 to 6.0	13 to 43	1.8 to 5.5	1.0	10.0	A
High-aluminum iron	1.3 to 2.0	0.4 to 1.0	0.15	0.15	1.3 to 6.0	...	20 to 25 Al	...	...	F
Heat-resistant ductile irons										
Medium-silicon ductile iron	2.8 to 3.8	0.2 to 0.6	0.08	0.12	2.5 to 6.0	1.5	...	...	...	F
Nickel-chromium ductile iron(h)	3.0	0.7 to 2.4	0.08	0.12	1.75 to 5.5	18 to 36	1.75 to 3.5	1.0	...	A

(a) Where a single value is given rather than a range, that value is a maximum limit. (b) Total carbon. (c) CP, coarse pearlite; M, martensite; A, austenite; F, ferrite. (d) May be produced from a malleable-iron base composition. (e) Cu may replace all or part of the Ni. (f) Such as Duriron, Durichlor 51, Superchlor. (g) Such as Ni-Resist austenitic iron (ASTM A436). (h) Such as Ni-Resist austenitic ductile iron (ASTM A439). (j) Such as Silal. (k) Such as Nicrosilal.

White, mottled and gray portions are shown at full size, top to bottom.

Fig. 1. Fracture surface of as-cast chilled iron

sion-resistant iron can be white, gray or nodular in both form and distribution of carbon.

Heat-resistant irons combine resistance to high-temperature oxidation and scaling with resistance to softening or microstructural degradation. Resistance to scaling depends chiefly on high alloy content, and resistance to softening on initial microstructure plus stability of the carbon-containing phase. Heat-resistant irons usually are ferritic or austenitic as cast; carbon exists predominantly as graphite, either in flake or spherulitic form, which subdivides heat-resistant irons into either gray or ductile irons.

EFFECTS OF ALLOYING ELEMENTS

In most cast irons, it is the interaction among alloying elements (including carbon and silicon) that has the greatest effect of properties. This influence is exerted largely by effects on the amount and shape of graphitic carbon present in the casting. For example, depth of chill or the tendency of the iron to be white as cast depends greatly on the carbon equivalent and the balance between carbon and silicon in the composition; addition of other elements can only modify the basic tendency established by the carbon-silicon relationship.

In general, only small amounts of alloying elements are needed to improve depth of chill, hardness and strength. Typical effects on depth of chill are given in Fig. 2 for the alloying elements commonly used in low to moderately alloyed cast irons. High alloy contents are needed for the most significant improvements in corrosion resistance or elevated-temperature properties.

Carbon. In chilled irons, the depth of chill decreases (Fig. 2), and the hardness of the chilled zone increases, with increasing carbon content. Carbon also increases the hardness of white irons.

Silicon is present in all cast irons. In alloy cast irons, as in other types, silicon is the chief factor that determines the carbon content of the eutectic. Increasing the silicon content lowers the carbon content of the eutectic and also promotes the formation of graphite on solidification. Thus, the silicon content is the principal factor controlling

the depth of chill in unalloyed or low-chromium chilled and white irons. This effect for relatively high-carbon irons is summarized in the graph at top left in Fig. 2.

Silicon additions of 4.5 to 8% improve high-temperature properties by raising the eutectoid transformation and by reducing the rates of scaling and growth. Additions of 14 to 17% Si (often accompanied by addition of about 5% Cr and 1% Mo) yield cast iron that is very resistant to corrosive acids, although resistance varies somewhat with acid concentration.

High-silicon irons are difficult to cast and are virtually unmachinable. High-silicon irons have particularly low resistance to mechanical and thermal shock at room temperature or moderately elevated temperature. However, above about 250 °C (500 °F), the shock resistance exceeds that of ordinary gray iron.

Manganese and sulfur should be considered together in their effects on gray or white iron. Alone, either manganese or sulfur increases the depth of chill, but when one is present, addition of the other decreases the depth of chill until the residual concentration has been neutralized by the formation of manganese sulfide. Generally, sulfur is the residual element, and excess manganese can be used to increase chill depth and hardness,

as shown in Fig. 2. Because it promotes formation of finer and harder pearlite, manganese often is used to decrease or prevent mottling in heavy-section castings.

Manganese, in excess of the amount needed to scavenge sulfur, mildly suppresses pearlite formation. It also is a relatively strong austenite stabilizer and normally is kept below about 0.7% in martensitic white irons. In some pearlitic or ferritic alloy cast irons, up to about 1.5% Mn may be used to help ensure that specified strength levels are obtained.

Phosphorus is a mild graphitizer in unalloyed irons; it mildly reduces chill depth in chilled irons (Fig. 2). In alloyed irons, the effects of phosphorus are somewhat obscure. There is some evidence that it reduces the toughness of martensitic white irons.

Chromium has three major uses in cast irons: to form carbides, to impart corrosion resistance and to stabilize the structure for high-temperature applications. Small amounts of chromium are added routinely to stabilize pearlite in gray iron, to control chill depth in chilled iron or to ensure a graphite-free structure in white iron containing less than 1% silicon. At such low percentages, usually no greater than 2 to 3%, chromium has little or no effect on hardenability, chiefly be-

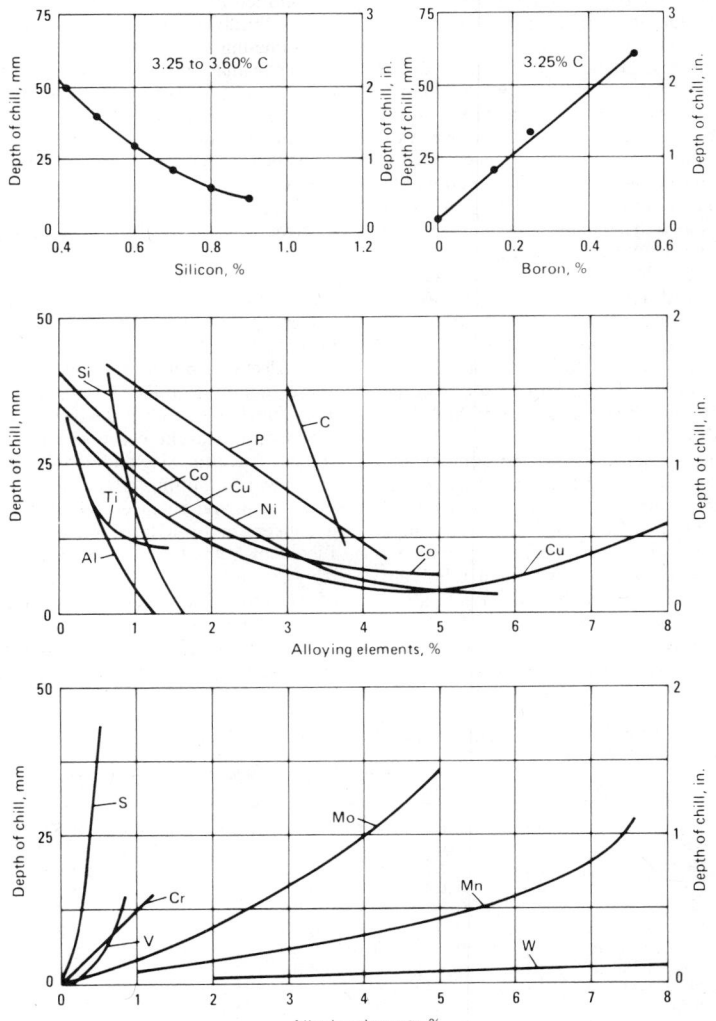

Fig. 2. Typical effects of alloying elements on depth of chill

cause most of the chromium is tied up in carbides. However, chromium does influence the fineness and hardness of pearlite and also tends to increase the amount and hardness of the eutectic carbides.

When the chromium content of cast iron is greater than about 10%, eutectic carbides of the M_7C_3 type are formed, rather than the M_3C type that predominates at lower chromium contents. More significantly, however, the higher chromium content causes a change in solidification pattern to a structure in which the M_7C_3 carbides are surrounded by a matrix of austenite or its transformation products. At lower chromium contents, the M_3C carbide forms the matrix. Because of the solidification characteristics, hypoeutectic irons containing M_7C_3 carbides normally are stronger and tougher than irons containing M_3C carbides.

The relatively good abrasion resistance, toughness and corrosion resistance found in high-chromium white irons have led to the development of a series of commercial martensitic or austenitic white irons containing 12 to 28% chromium. Because much of the chromium in these irons is present in combined form as carbides, chromium is much less effective than molybdenum, nickel, manganese or copper in suppressing the eutectoid transformation to pearlite and thus has a lesser effect on hardenability than it has in steels. Martensitic white irons usually contain one or more of the elements molybdenum, nickel, manganese and copper to give the required hardenability. These elements ensure that martensite will form on cooling from above the upper transformation temperature either while the casting is cooling in the mold or during subsequent heat treatment.

Nickel is almost entirely distributed in the austenite phase or its transformation products. Nickel tends to promote graphite formation, and in white and chilled irons, this effect usually is balanced by the addition of about one part chromium for every three parts nickel in the composition.

One of the Ni-Hard family of commercial alloy white irons (Type IV Ni Hard) contains 1.0 to 2.2% Si, 5 to 7% Ni and 7 to 11% Cr. In the as-cast condition, it has a structure of M_7C_3 eutectic carbides in a martensitic matrix. If retained austenite is present, the martensite content and hardness of the alloy can be increased by refrigeration treatment or by reaustenitizing and air cooling. Ni-Hard IV often is specified for pumps and other equipment used for handling abrasive slurries because of its combination of relatively good strength, toughness and resistance to abrasion.

Nickel additions of more than 12% are needed for optimum resistance to corrosion or heat. High-nickel gray or ductile irons usually contain 1 to 6% chromium, and may contain as much as 10% copper. These elements act in conjunction with the nickel to promote resistance to corrosion and scaling, especially at elevated temperatures. All types of cast iron with nickel contents above 18% are fully austenitic.

Copper in moderate amounts can be used to suppress pearlite formation in both low- and high-chromium martensitic white irons. It has a relatively mild effect compared with that of nickel, and, because of the limited solubility of copper in austenite, copper additions probably should be limited to about 2.5% This limitation means that copper cannot completely replace nickel in Ni-Hard-type irons.

Copper is used in amounts of about 3 to 10% in some high-nickel gray and ductile irons that are specified for corrosion or high-temperature service. Here, copper enhances corrosion resistance, particularly resistance to oxidation or scaling.

Molybdenum in chilled and white iron compositions is distributed between the eutectic carbides and the matrix. In graphitic irons, its main functions are to promote deep hardening and to improve corrosion resistance and high-temperature strength.

In chilled iron compositions, molybdenum additions mildly increase depth of chill (they are about one-third as effective as chromium; see Fig. 2). The primary purpose of small additions (0.25 to 0.75%) of molybdenum to chilled iron is to improve the resistance of the chilled face to spalling, pitting, chipping and heat checking. Molybdenum hardens and toughens the pearlitic matrix.

Where a martensitic white iron is desired for superior abrasion resistance, additions of 0.5 to 3.0% molybdenum effectively suppress formation of pearlite and other high-temperature transformation products when used in combination with copper, chromium, nickel, or both chromium and nickel.

Vanadium is a potent carbide stabilizer and increases depth of chill. The magnitude of the increase of depth of chill depends on the amount of vanadium and the composition of the iron, as well as on section size and conditions of casting. The powerful chilling effect of vanadium in thin sections may be balanced by additions of nickel or copper, or by a large increase in carbon or silicon, or both. In addition to its carbide-stabilizing influence, vanadium in amounts of 0.10 to 0.50% refines the structure of the chill and minimizes coarse columnar grain structure.

Because of its strong carbide-forming tendency, vanadium is rarely used in gray or ductile irons for corrosion or elevated-temperature service.

ABRASION-RESISTANT CAST IRONS

The assessment of resistance of cast irons — or, for that matter, of any substance — to the ravages of abrasive wear is, at best, a difficult task. Wear resistance is so intimately dependent on the properties of the abrasive that it is impossible to obtain valid quantitative data that can be translated into a prediction of service life.

Typical Compositions. The first three lines of Table 1 list the composition ranges for the typical commercial unalloyed and low-alloy grades of white and chilled irons used for abrasion-resistant castings. These are nominally classed as pearlitic white irons. Historically, most of the early white iron castings produced for abrasion resistance were cast from low-carbon, 1.0 to 1.6% Si unalloyed compositions, which were also used for malleable iron castings. As changes in demand and specific uses have occurred, the trend has been to produce a more abrasion-resistant 2.8 to 3.6% C, low-silicon grade, which usually is alloyed with chromium to suppress graphite and to increase the fineness and hardness of the pearlite. Other alloying elements such as nickel, molybdenum, copper and manganese are used for specific applications.

Martensitic white irons have largely displaced pearlitic white irons for making many types of abrasion-resistant castings, with the possible exception of chilled iron rolls and grinding balls. Although martensitic white irons cost more than pearlitic irons, their much superior abrasion resistance, combined with increasing costs of all castings, makes martensitic alloy white irons economically attractive. The better strength and toughness of martensitic irons favor their use, and the trend toward replacing cupola melting with electric-furnace melting makes martensitic white irons relatively easy to produce.

Table 2 lists the composition ranges of commercial martensitic white cast irons, as shown in ASTM A532.

CORROSION-RESISTANT CAST IRONS

The corrosion resistance of gray cast iron is enhanced by the addition of appreciable amounts of nickel, chromium and copper, singly or in combination, or silicon in excess of about 3%. Chemical composition ranges bracketing some of the more widely used corrosion-resistant cast irons are given in Table 1.

Up to 3% silicon is normally present in all cast irons; in larger percentages, silicon is considered an alloying element. It promotes the formation of a strongly protective surface film under oxidizing conditions such as exposure to oxidizing acids. Relatively small amounts of molybdenum and/or chromium may be added in combination with high silicon. The addition of nickel to gray iron improves resistance to reducing acids and provides a high resistance to caustic alkalis. Chromium assists in forming a protective oxide that resists oxidizing acids, although it is of little benefit under reducing conditions. Copper has a smaller beneficial effect on resistance to sulfuric acid.

High-silicon irons are the most universally corrosion-resistant alloys available at moderate cost. They are widely used for handling corrosive media common in chemical plants, even when

Table 2. Chemical compositions of standard martensitic white cast irons(a)

Class	Type	Designation	TC(c)	Mn	P	S	Si	Cr	Ni	Mo	Cu
I	A	Ni-Cr-HC	3.0-3.6	1.3	0.30	0.15	0.8	1.4-4.0	3.3-5.0	1.0	...
I	B	Ni-Cr-LC	2.5-3.0	1.3	0.30	0.15	0.8	1.4-4.0	3.3-5.0	1.0	...
I	C	Ni-Cr-GB	2.9-3.7	1.3	0.30	0.15	0.8	1.1-1.5	2.7-4.0	1.0	...
I	D	Ni-Hi Cr	2.5-3.6	1.3	0.10	0.15	1.0-2.2	7-11	5-7	1.0	...
II	A	12% Cr	2.4-2.8	0.5-1.5	0.10	0.06	1.0	11-14	0.5	0.5-1.0	1.2
II	B	15% Cr-Mo-LC	2.4-2.8	0.5-1.5	0.10	0.06	1.0	14-18	0.5	1.0-3.0	1.2
II	C	15% Cr-Mo-HC	2.8-3.6	0.5-1.5	0.10	0.06	1.0	14-18	0.5	2.3-3.5	1.2
II	D	20% Cr-Mo-LC	2.0-2.6	0.5-1.5	0.10	0.06	1.0	18-23	1.5	1.5	1.2
II	E	20% Cr-Mo-HC	2.6-3.2	0.5-1.5	0.10	0.06	1.0	18-23	1.5	1.0-2.0	1.2
III	A	25% Cr	2.3-3.0	0.5-1.5	0.10	0.06	1.0	23-28	1.5	1.5	1.2

(a) From ASTM A532-75a. Certain specific compositions of alloys II-B, II-C, II-D and II-E are covered by U.S. Patent No. 3,410,682. (b) Where a single value is given rather than a range, that value is a maximum limit. (c) Total carbon.

Table 3. Typical mechanical properties of heat-resistant alloy cast irons

Type of iron(a)	Hardness, HB	Tensile strength MPa	ksi	Compressive strength MPa	ksi	Impact energy J	ft·lb	Transverse breaking load(b) kg	lb	Transverse deflection(b) mm	in.
Medium-silicon gray iron	170 to 250	170 to 310	25 to 45	620 to 1040	90 to 150	20 to 31(c)	15 to 23(c)	455 to 1090	1000 to 2400	4.6 to 8.9	0.18 to 0.35
High-chromium gray iron	250 to 500	210 to 620	30 to 90	690	100	27 to 47(c)	20 to 35(c)	910 to 1590	2000 to 3500	1.5 to 3.8	0.06 to 0.15
High-nickel gray iron	130 to 250	170 to 310	25 to 45	690 to 1100	100 to 160	80 to 200(c)	60 to 150(c)	820 to 1360	1800 to 3000	5 to 25	0.2 to 1.0
Ni-Cr-Si gray iron	110 to 210	140 to 310	20 to 45	480 to 690	70 to 100	110 to 200(c)	80 to 150(c)	820 to 1130	1800 to 2500	7 to 35	0.3 to 1.4
High-aluminum gray iron	180 to 350	235 to 620	34 to 90	...	...	...	...	...	...	...	...
Medium-silicon ductile iron	140 to 300	415 to 690	60 to 100(c)	...	...	7 to 155(d)	5 to 115(d)	...	...	...	...
High-nickel ductile iron (20 Ni)	140 to 200	380 to 415	55 to 60(e)	1240 to 1380	180 to 200	16(f)	12(f)	...	...	...	...
High-nickel ductile iron (23 Ni)	130 to 170	400 to 450	58 to 65(g)	...	...	38(f)	28(f)	...	...	...	...

(a) For composition ranges, see Table 1. (b) Unnotched 30.5-mm (1.2-in.) diam test bar broken on 152-mm (6-in.) supports in a Charpy testing machine. (c) Yield strength, 310 to 520 MPa (45 to 75 ksi); elongation, 0.2%. (d) Standard Charpy test on 10-mm unnotched specimen. (e) Yield strength, 210 to 240 MPa (30 to 35 ksi); elongation, 8 to 20%. (f) Standard Charpy test on 10-mm notched specimen. (g) Yield strength, 195 to 240 MPa (28 to 35 ksi); elongation, 20 to 40%.

abrasive conditions also are encountered. When the silicon content is 14.2% or higher, these irons exhibit a very high resistance to boiling sulfuric acid.

The 14.5% Si iron is less resistant to the corrosive action of hydrochloric acid, but this resistance can be improved by additions of chromium and molybdenum and can be further enhanced by increasing the silicon content to 17%. The chromium-bearing silicon irons are very useful in contact with solutions containing copper salts, "free" wet chlorine or other strongly oxidizing contaminants.

High-chromium irons containing from 20 to 35% Cr give good service with oxidizing acids, particularly nitric, but are not resistant to reducing acids. These irons are also reliable for use in weak acids under oxidizing conditions, in numerous salt solutions, in organic acid solutions and in marine or industrial atmospheres.

HEAT-RESISTANT CAST IRONS

Heat-resistant cast irons are basically alloys of iron, carbon and silicon having high-temperature properties markedly improved by the addition of certain alloying elements, singly or in combination, principally chromium, nickel, molybdenum, aluminum, and silicon in excess of 3%. Silicon and chromium increase resistance to heavy scaling by forming a light surface oxide that is impervious to oxidizing atmospheres. Both elements reduce the toughness and thermal shock resistance of the metal. Although nickel does not appreciably affect oxidation resistance, it increases strength and toughness at elevated temperatures. Molybdenum also increases high-temperature strength. Aluminum additions reduce both growth and scaling but adversely affect mechanical properties at room temperature.

The chemical composition ranges of some of the more widely used heat-resistant irons (both gray and ductile types) suitable for elevated-temperature service are given in Table 1; mechanical properties are presented in Table 3.

Growth is the permanent increase in volume that

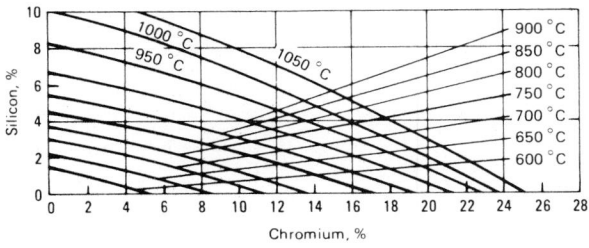

Indicated are the temperatures at which various irons may be used with very little or insignificant scaling in sulfur-free oxidizing atmospheres.

Fig. 3. Relation of silicon and chromium contents to the scaling resistance of silicon-chromium irons

occurs in some cast irons after prolonged exposure to elevated temperature or after repeated cyclic heating and cooling. It is produced by the expansion that accompanies graphitization, expansion and contraction at the transformation temperature, and internal oxidation of the iron. Gases can penetrate the surface of hot cast iron at the graphite flakes and oxidize the graphite as well as the iron and silicon. The occurrence of fine cracks, or "crazing," may accompany repeated heating and cooling through the transformation temperature range because of thermal and transformational stresses.

Silicon contents of less than about 3.5% increase the rate of growth by promoting graphitization, but silicon contents of 4% or more retard growth.

The carbide-stabilizing alloying elements, particularly chromium, effectively reduce growth in gray irons at 450 °C (850 °F) or above. Growth is not a problem below 400 °C (750 °F), except in the presence of superheated steam, where it can occur in coarse-grain irons at about 310 °C (600 °F). Even small amounts of chromium, molybdenum and vanadium produce marked reductions in growth at the higher temperatures.

Scaling. In addition to the internal oxidation that contributes to growth, a surface scale forms on unalloyed gray iron after exposure at a sufficiently high temperature.

Silicon, chromium and aluminum increase the scaling resistance of cast iron by forming a light surface oxide that is impervious to oxidizing atmospheres. Unfortunately, these elements tend to reduce toughness and thermal shock resistance. The presence of nickel improves the scaling resistance of most alloys containing chromium and, more importantly, increases their toughness and strength at elevated temperatures. Carbon has a somewhat damaging effect above 700 °C (1300 °F) as a result of the mechanism of decarburization and the evolution of carbon monoxide and carbon dioxide. When these gases are evolved at the metal surface, the formation of protective oxide layers is hindered, and cracks and blisters may develop in the scale.

Figure 3 indicates the temperatures at which various silicon-chromium irons may be used with only slight or insignificant scaling in sulfur-free oxidizing atmospheres. Greater scaling rates may be tolerated in some applications, so that higher useful temperatures are often possible.

High-Nickel Irons. The austenitic cast irons containing 18 to 36% Ni, up to 7% Cu, and 1.75 to 4% Cr are used for both heat-resistant and corrosion-resistant applications. Known as Ni-Resist, this type of iron exhibits good resistance to high-temperature scaling and growth up to 815 °C (1500 °F) in most oxidizing atmospheres and good performance in steam service up to 530 °C

(990 °F), and can handle sour gases and liquids up to 400 °C (750 °F). The maximum temperature of use is 540 °C (1000 °F) if appreciable sulfur is present in the atmosphere. Austenitic cast irons can be employed at temperatures as high as 950 °C (1740 °F). Austenitic irons have the advantage of considerably greater toughness and thermal shock resistance than the other heat-resistant alloy irons, although their strength is rather low.

High-nickel ductile irons are considerably stronger and tougher than the comparable gray irons. Tensile strengths of 400 to 470 MPa (58 to 68 ksi), yield strengths of 200 to 275 MPa (30 to 40 ksi) and elongations of 10 to 40% may be realized in high-nickel ductile irons.

High-Aluminum Irons. Alloy cast irons containing 6 to 7% Al, 18 to 25% Al, or 12 to 25% Cr plus 4 to 16% Al are reported to have considerably better resistance to scaling than several other alloy irons, including the high-silicon type. These high-aluminum irons have been little used commercially because of their brittleness and poor castability.

Selected References on Cast Irons

The Physical Metallurgy of Cast Iron, by I. Minkoff: John Wiley, New York, 1983

Cast Irons: Physical and Engineering Properties, 2nd Ed., by H. T. Angus: Butterworths, London, 1976

Malleable Iron Castings: Malleable Founders' Society, Cleveland, 1960

Cupola Handbook, 4th Ed.: American Foundryman's Society, 1975

The Structure of Cast Iron, by A. Boyles: American Society for Metals, 1949

Solidification and Casting, by G. J. Davies: Applied Science Publishers, London, 1973

Ductile Iron, Vol I and II, by S. Karsay: Quebec Iron and Titanium Corp., 1976

Iron Castings Handbook, edited by C. F. Walton and T. J. Opar: Iron Castings Society, Inc., 1981

ADDITIONAL READING

Cast Irons in High-Temperature Service, by R. J. Greene and F. G. Sefing: National Association of Corrosion Engineers, Mar 1954

Composition and Properties of Gray Iron, Parts I and II, by R. Schneidewind and R. G. McElwee: *Trans AFS,* Vol 58, 1950, p 312–330

Development of Low-Alloy Ductile Irons for Service at 1200–1500 F, by D. L. Sponseller, W. G. Scholz and D. F. Rundle: *Trans AFS,* Vol 76, 1968, p 353–368

Elevated Temperature Properties of Ductile Cast Irons, by C. R. Wilks, N. A. Matthews and R. W. Kraft, Jr.: *Trans ASM,* Vol 47, 1954

Gray Iron—A Unique Engineering Material, by D. E. Krause: STP 455, American Society for Testing and Materials, Philadelphia, 1969, p 3–28

Gray Iron Castings Section Sensitivity, by H. C. Winte: *Trans AFS,* Vol 54, 1946, p 436–443

Hardenability of Ductile Iron, by C. C. Reynolds, W. T. Whittington and H. F. Taylor: *Trans AFS,* Vol 63, 1955, p 116–122

Hot Hardness and Structure of Cast Irons, by H. D. Merchant and M. H. Moulton: *British Foundryman,* Vol 57, Part 2, Feb 1964, p 62–73

High-Strength Cast Iron at Elevated Temperatures: *Proc ASTM,* Vol 41, 1941, p 797–807

Improved Test Bars for Standard and Ductile Grades

of Cast Iron, by R. A. Flinn and R. W. Kraft: *Trans AFS,* Vol 58, 1950, p 153–167

Influence of Silicon on Ductile Cast Iron, by J. Pelleg: *Trans AFS,* Vol 71, 1963

Machinable 1.5 Per Cent and 2.0 Per Cent Chromium Cast Irons to Resist Deterioration at High Temperatures, by C. O. Burgess and A. E. Shrubsall: *Trans AFS,* Vol 50, Dec 1942

Mechanical Properties at Elevated Temperatures of Ductile Cast Iron, by F. B. Foley: *Trans ASME,* Vol 78, 1956, p 1435–1438

Notch Ductility of Nodular Irons, by W. S. Pellini, G. Sandoz and H. F. Bishop: *Trans ASM,* Vol 46, 1954, p 418–445

Static and Dynamic Toughness of Ductile Cast Iron, by R. K. Nanstad, F. J. Worzala and C. R. Loper, Jr.: *Trans AFS,* Vol 83, 1975

Stress Relief Heat Treatment of Gray Cast Iron, by J. H. Schaum: *Trans AFS,* Vol 61, 1953, p 646–650

Tensile and Fatigue Tests on Normalized Pearlitic Nodular Irons, by G. N. J. Gilbert: *Journal of Research,* British Cast Iron Research Assn., Vol 6, No. 10, Feb 1957, p 498–504

6 ALUMINUM

By Harold Hunsicker (retired), Alcoa Laboratories

Additional and more detailed information on aluminum and aluminum alloys can be found in Metals Handbook, Ninth Edition, Volume 2, Properties and Selection: Nonferrous Alloys and Pure Metals, pages 1 to 236. For additional information within this volume, the reader should consult the index.

Introduction to Aluminum and Aluminum Alloys

THE UNIQUE COMBINATIONS of properties provided by aluminum and its alloys make aluminum one of the most versatile, economical and attractive metallic materials for a broad range of uses from soft, highly ductile wrapping foil to the most demanding engineering applications. Its low density and high specific strength (strength-to-weight ratio) are most useful characteristics. It weighs only about 2.7 Mg/m^3 (0.1 $lb/in.^3$), approximately one-third as much as the same volume of steel, copper or brass. Such light weight, coupled with the high strength of some aluminum alloys (exceeding that of structural steel), permits design and construction of strong, lightweight structures which are particularly advantageous for anything that moves—space vehicles and aircraft as well as all types of land- and water-borne vehicles. Aluminum has high resistance to corrosion in atmospheric environments, in fresh and salt waters, and in many chemicals and their solutions. No colored salts are formed to discolor or stain adjacent materials or products with which the aluminum comes into contact. It has no toxic reactions, and thus is highly suitable for processing, handling, storing and packaging of foods and beverages.

The high electrical and thermal conductivities of aluminum account for its use in many applications. Aluminum is highly reflective to radiant energy-visible light, radiant heat and electromagnetic waves. It is nonferromagnetic, a property of importance in the electrical and electronic industries. It has nonsparking characteristics, which permit its use in tools to be employed near flammable or explosive materials.

Aluminum and many of its alloys can be worked readily into any form needed and can be cast by all foundry processes. Aluminum accepts a wide variety of attractive, durable and functional surface finishes. Natural-color finishes can range from soft and lustrous to bright and specular. Aluminum can be finished to almost any color and surface texture, many combinations of which are unique to this metal.

Production and Markets. Among the commercial metals, aluminum is second only to iron in production and consumption on a weight as well as a volume basis. Total annual U.S. shipments of aluminum products, including ingot for castings, ranged from 4500 to 6700 million kilograms (9900 to 14 800 million pounds) during the period from 1972 to 1982. Annual shipments of castings during the same period ranged from 700 to 940 million kilograms (1540 to 2080 million pounds). For the last 6 or 7 years, the relative proportions of primary, secondary and imported metal used in the United States in producing these products remained relatively stable, averaging 66% from primary metal produced domestically, 23% recovered from all secondary sources and 11% imported.

Mill products constitute the major share of total aluminum product shipments, followed by castings and ingot other than for castings (see Table 1). In decreasing order of current market size, the major application categories are containers and packaging (29.3%), building and construction (21.2%), transportation (17.9%), electrical (11.1%), consumer durables (8.2%), machinery and equipment (7.0%), and others (5.3%). Typical applications in each of these market categories are discussed later.

Primary aluminum is produced by direct-current electrolysis of aluminum oxide (alumina) dissolved in a molten sodium fluoride – aluminum fluoride bath at temperatures of 940 to 980 °C, or 1725 to 1800 °F (Hall/Heroult Process). Although there are many minerals having substantial aluminum contents (aluminum comprises 8.8 wt %, or 6.6 at. %, of the earth's crustal mineral content), the principal current source is the mineral bauxite, from which aluminum oxide is extracted and prepared for the smelter by crushing, grinding, chemical processing and calcination.

Primary aluminum contains iron and silicon as major impurities. Iron contents may vary from 0.05 to 0.6% and silicon from 0.04 to 0.3%. Additionally, very small amounts of many other elements are present as impurities. The principal trace impurities are Cu, Mn, Ni, Zn, Ti, V, Na and Ga, most of which are present in amounts below 100 ppm (0.01%). These impurities derive from residual impurities in the smelter-grade alumina and in the petroleum coke used in producing anodes and linings for the electrolytic cells. Because there are variations in these materials from various sources and at different times, and because their transfer to the metal is influenced by the age of the cell and by operating conditions, the metal from the smelting cells is normally analyzed and graded on regular schedules. High-purity smelter grades, which have preferred characteristics for certain uses, may be priced at a slight premium over the base grade and are generally available in smaller quantities as purity increases.

Most of the applications for aluminum products require properties or characteristics that cannot be obtained using the metal that comes directly from the smelting cells, unaltered except for removal of stray nonmetallics. The changes in composition that are required to produce the needed properties can be effected by (a) increasing the purity of the metal (refining) or (b) alloying. In some cases, detrimental elements are removed and beneficial elements are added.

Impurity Removal. Aluminum of smelter grade is refined to remove impurity elements that degrade electrical conductivity, bright finishing capability, corrosion resistance, fabricability or electrochemical characteristics. For high-magnesium 5xxx-type alloys, fluxing, filtering and metal-transfer practices are designed to ensure very low contents of sodium and calcium, which are extremely detrimental to hot-line recovery (causing excessive ingot cracking during hot rolling). For electrical conductor grades, such as 1350, smelter metal is commonly selected for low silicon con-

Table 1. Average annual shipments of aluminum products for the ten-year period from 1972 to 1981 (inclusive)

Type of product	Average annual shipments		
	10^6 kg	10^6 lb	% of total
Mill products	4608	10 158	77.7
Castings	824	1 816	13.9
Ingot (other than for castings)	498	1 098	8.4
Total	5930	13 072	100.0

tent in particular, and is treated in the molten state with small additions of boron, which combine with the conductivity-decreasing impurity elements titanium, vanadium and zirconium to form intermetallic boride compounds of extremely low solubility. This may increase the electrical conductivity of the final product, largely cable stranded from hard-drawn wire, by as much as 1% IACS, depending on the amounts of titanium, vanadium and zirconium that are initially present.

Super-Purity Aluminum. Several million pounds of super-purity aluminum (99.99% Al, minimum) are produced and used annually, principally for manufacture of electrolytic capacitor foil and for increasing the purity of alloys used in bright-finished applications. In the latter case, certain alloy additions that do not impair finishing are made to obtain the desired higher mechanical properties, while the impurities are decreased by blending the super-purity metal with smelter-grade metal to achieve the desired finishing characteristics. In electrolytic capacitor foil, use of the highest possible purity minimizes current leakage caused by minute second-phase particles formed by many impurity elements. For this application, strength is a relatively minor consideration. The individual impurity elements in super-purity aluminum range from 0.0001% or less to a maximum of about 0.0004%, with a typical total for all impurities of about 0.007%.

Production of super-purity aluminum requires relatively expensive further processing of selected-grade smelter metal, usually either by a second electrolysis in a molten salt bath (electrolytic refining) or by a combination process wherein the peritectic-forming elements (Ti, V, Zr, Cr) are removed by forming borides, which precipitate from the molten metal because of their higher density and melting temperature, followed by fractional crystallization to remove the eutectic-forming elements (Cu, Si, Mn, Zn). Because of the additional processing costs involved in producing this metal, it is priced at a premium over smelter grades.

Alloying. The other direction of change in composition from smelter-grade metal—i.e., alloying—is more obvious and far more prevalent, in terms of tonnage, than refining. Although the preeminent reason for alloying is to increase mechanical properties (strength, hardness, and resistance to fatigue, creep or wear), in many alloys other characteristics are either primary or important supplementary reasons for the amounts and types of alloying elements used. Specific additions improve casting characteristics, whereas others improve fabricating, machining or finishing characteristics; corrosion resistance in certain environments is improved by some additions. The alloys used for joining by welding or brazing processes have compositions tailored to obtain specific melting-temperature ranges, with optimized surface tension in the molten state, and may contain special additions to nucleate grains during solidification of the weld deposit, thereby refining the grain size and minimizing cracking during solidification and cooling. For parts used at elevated temperatures and fitted to operate at close clearances, such as pistons of internal-combustion engines as well as engine blocks, cylinder heads, etc., the coefficient of thermal expansion is an important property. Reduced expansion coefficient is provided by increasing silicon and nickel contents, so that these elements are prominent in the alloy compositions preferred for such parts.

The maximum alloy content (total of all alloying elements) included in the alloys for producing wrought products (wrought alloys) is about 11%. Casting alloys may have up to about 24% total alloy content. In terms of annual usage as an alloying addition to aluminum, magnesium stands in first place, and a greater volume of this other low-density metal is used for this purpose than for any other. Silicon is next at only 40% (by weight) of the magnesium usage, followed by copper at 23%, manganese at 17% and zinc at 15%. These relationships obviously depend on the relative tonnages of the different alloy types that are produced and sold, and may shift as the different market categories vary in relative importance.

Alloying is almost always accomplished by adding either pure alloying metals or previously prepared master alloys (concentrated alloys with an aluminum base) to the base metal in the molten state. The molten base metal may have come directly from smelting cells or may have been solidified and later remelted. Some production of alloys can be performed by making additions in the smelting cells. Powder metallurgy processes may include other means of producing alloys. In the case of remelted secondary metal, some unwanted alloying elements may be removed and the contents of others adjusted by additions.

EFFECTS OF ALLOYING

Microstructural Effects. Many effects of alloying elements on microstructures can be predicted from and related directly to the equilibrium constitutional relationships. Pure unalloyed aluminum is a single-phase material, and its optical microstructure is composed of only grains and grain boundaries. This is a complete description also of the microstructures of the solid solutions formed with pure aluminum by any of the alloying elements, the phase diagrams of which show a solid-solution field at the aluminum end of the diagram. If the pure aluminum or purely solid-solution alloy has been strained plastically (mechanically deformed), there will also be a substructure of tangled dislocations, which with increased strain develop into cells or subgrains. These extremely fine-scale features can be resolved only by transmission electron microscopy.

Alloying-element contents that exceed the solid-solubility limit produce "second-phase" microstructural constituents that may consist of either the pure alloying ingredient or an intermetallic-compound phase. In the first group are silicon, tin and beryllium. If the alloy is a ternary or higher-order alloy, however, silicon or tin may form intermetallic-compound phases. Most of the other alloying elements form such compounds with aluminum in binary alloys and more complex phases in ternary or higher-order alloys. Details depend on ratios and total amounts of alloying elements present and require reference to the phase diagrams for prediction. It must be kept in mind, however, that metastable conditions frequently prevail that are characterized by the presence of phases that are not shown on the equilibrium diagrams.

Smelter-grade primary metal, whether in ingot or wrought-product form, contains a small volume fraction of second-phase particles, chiefly iron-bearing phases—the metastable Al_6Fe, the stable Al_3Fe, which forms from Al_6Fe on solid-state heating, and $Al_{12}Fe_3Si$. Proportions of the binary and ternary phases depend on relative iron and silicon contents.

Physical Properties. Most of the physical properties—density, melting-temperature range, heat content, coefficient of thermal expansion, and electrical and thermal conductivities—are changed by addition of one or more alloying elements. The rates of change in these properties with each incremental addition are specific for each element and depend, in many cases drastically, on whether a solid solution or a second phase is formed. In those cases in which the element or elements may be either dissolved or precipitated by heat treatment, certain of these properties, particularly density and conductivity, can be altered substantially by heat treatment. Density and conductivity of such alloys show relatively large differences from one temper to another.

Electrochemical properties and corrosion resistance are strongly affected by alloying elements that form either solid solutions or additional phases, or both. For those systems exhibiting substantial changes in solid solubility with temperature, these properties may change markedly with heat treated tempers and, although infrequently, even with room-temperature aging (for example, the stress-corrosion resistance of high-Mg 5xxx alloys in strain-hardened tempers). The strongest electrochemical effects are from copper or zinc in solid solution. Additions of copper in solid solution change the electrochemical solution potential in the cathodic direction at the rate of 0.047 V/wt % (0.112 V/at. %), and additions of zinc change it in the anodic direction at the rate of 0.063 V/wt % (0.155 V/at. %). These potentials are those measured in an aqueous solution of 53 g NaCl + 3 g H_2O_2 per litre. Magnesium and silicon, which are the basis for the 4xxx, 5xxx and 6xxx series wrought alloys and the 3xx.x, 4xx.x and 5xx.x series casting alloys, and which are prominent in the compositions of many other alloys, have relatively mild effects on solution potential and are not detrimental to corrosion resistance.

Although aluminum is a thermodynamically reactive metal, it has excellent resistance to corrosion in most environments, which may be attributed to the passivity afforded by a protective film of aluminum oxide. This film is strongly bonded to the surface of the metal and, if damaged, re-forms almost immediately. The continuity of the film is affected by the microstructure of the metal—in particular, by the presence and volume fraction of second-phase particles. Corrosion resistance is affected by this factor and by the solution-potential relationships between the second-phase particles or constituents and the solid-solution matrix in which they occur. In most environments, resistance to corrosion of unalloyed aluminum increases with increasing purity. The resistance of an alloy depends not only on the microstructural relationships involving the specific types, amounts and distributions of the second-phase constituents but even more strongly on the nature of the solid solutions in which they are present. Copper reduces corrosion resistance despite the fact that when in solid solution it makes the alloy more cathodic (less active thermodynamically). This is explained by the fact that copper ions taken into solution in aqueous, corroding media replate on the aluminum alloy surface as minute particles of metallic copper, forming even more active corrosion couples, because metallic copper is highly cathodic to the alloy. Manganese, which in solid solution changes solution potential in the cathodic direction as strongly as does copper, does not impair corrosion resistance of commercial alloys that contain it, because the amounts left in solid solution in commercial products, which undergo extensive solid-state heating in process, are very small, and the manganese does not replate from solution as does copper.

The differences in solution potential among alloys of different compositions are used to great advantage in the composite Alclad products. In these products, the structural component of the composite, usually a strong or heat treatable alloy, is made the core of the product and is covered by a cladding alloy of a composition which not only is highly corrosion resistant but also has a solution potential that is anodic to that of the core. Analogous to the protection of the underlying steel afforded by zinc on the surfaces of galvanized steel products, the aluminum alloy core is protected electrolytically by the more-anodic cladding. The composition of the cladding material is designed specifically to protect the core alloy, so that, for those containing copper as the principal alloying ingredient (2xxx type, the more anodic unalloyed aluminum (1xxx type) serves to protect the core electrolytically. In the case of the strong alloys containing zinc along with magnesium and copper (such as 7049, 7050, 7075 and 7178), an aluminum-zinc alloy (7072) or an aluminum-zinc-magnesium alloy (7008 or 7011) provides protection. The latter provides higher strength.

Impurity Effects. Although major differences in properties and characteristics are usually associated with alloying additions of one to several percent, many alloying elements produce highly significant effects when added in small fractions of one percent or when increased by such small amounts. With respect to mechanical properties, this is particularly true for combinations of certain elements. The interactions are quite complex, and a given element may be either highly beneficial or highly detrimental depending on what other elements are involved and on which property or combination of properties is needed.

The presence or absence of amounts on the order of one thousandth of one percent of certain impurities — sodium and calcium, for example — may make the difference between success or complete failure in fabricating high-Mg 5xxx alloy ingots into useful wrought products. There are many other examples of equal practical importance. Impurity limits specified for commercial alloys reflect some of these effects, but producers of mill products must adhere to even more restrictive limits in many cases to ensure good product recovery.

Silicon-Modifying Additions. Additions of similarly small percentages of both metallic and nonmetallic substances — sodium and phosphorus, for example — are used to enhance the mechanical and machining properties of silicon-containing casting alloys.

Grain-Refining Additions. Most alloys produced as "fabricating ingots" for fabricating wrought products, as well as those in the form of foundry ingot, have small additions of titanium or boron, or combinations of these two elements, in controlled proportions. The purpose of these additions is to control grain size and shape in the as-cast fabricating ingot or in castings produced from the foundry ingot. These grain-refining additions have little effect on changes in grain size that occur during or as a result of working or recrystallization. Welding filler alloys and casting alloys generally have higher contents of the grain-refining elements to ensure highest resistance to cracking during solidification of welds and castings.

The elements that have relatively great and controlling effects on grain sizes and shapes produced by the mechanical working required to produce wrought products (their thermomechanical history) are manganese, chromium and zirconium. Small amounts (fractional percentages) of these elements, singly or in combination, are included in the compositions of many alloys to control grain size and recrystallization behavior through fabrication and heat treatment. Such grain control has many purposes, which include ensuring good resistance to stress-corrosion cracking (SCC), high fracture toughness and good forming characteristics. In specific alloys, these elements have highly significant supplementary beneficial effects on strength, resistance to fatigue, or strength at elevated temperatures. In order to fulfill their grain-control functions, these elements must be precipitated as finely distributed particles termed "dispersoids." Their precipitation is accomplished primarily by the high-temperature, solid-state heating involved in ingot preheating.

Secondary Aluminum. Aluminum recovered from scrap (secondary aluminum) has been an important contributor to the total metal supply for many years. New scrap is defined as that generated by plants making end products, whereas old scrap is that recovered from metal that has been previously used by consumers. In addition, considerable amounts of scrap generated at various stages of mill processing are recycled. The increased concern with and economic implications of energy supply in recent years have focused even more attention on recycling of aluminum because of its energy-intensive nature. The energy required to remelt secondary aluminum preparatory to fabrication for reuse is only 5% of that required to produce new aluminum. Consequently, recycling is increasing. In particular, recovery of used aluminum beverage containers has multiplied many times in the last 10 years, so that in 1981 over one billion pounds of metal was reclaimed from this source — enough aluminum to equal the output of two large smelters each having a capacity of 250 000 metric tons per year.

For some uses, secondary aluminum alloys may be treated to remove certain impurities or alloying elements. Chief among the alloying elements removed is magnesium, which is frequently present in greater amounts in secondary metal than in the alloys to be produced from it. Magnesium is usually removed by fluxing with chlorine gas or halide salts.

Mechanical Properties. The predominant reason for alloying is to increase strength, hardness, and resistance to wear, creep, stress relaxation or fatigue. Effects on these properties are specific to the different alloying elements and combinations of them, and are related to their alloy phase diagrams and to the microstructures and substructures which they form as a result of solidification, thermomechanical history, heat treatment and/or cold working.

The tensile yield strength of super-purity aluminum in its annealed (softest) state is about 10 MPa (1.5 ksi), whereas those of some heat treated commercial high-strength alloys exceed 550 MPa (80 ksi). When one considers the magnitude of this difference (an increase of over 5000%), this practical, everyday accomplishment, which is just one aspect of the physical metallurgy of aluminum, is truly remarkable. Higher strengths, up to a yield strength of 690 MPa (100 ksi) and over, may be readily produced, but the fracture toughness of such alloys does not meet levels consid-

ered essential for aircraft or other critical-structure applications.

The elements that are most commonly present in commercial alloys to provide increased strength — particularly when coupled with strain hardening by cold working or with heat treatment, or both — are copper, magnesium, manganese, silicon and zinc. These elements all have significant solid solubility in aluminum, and in all cases the solubility increases with increasing temperature (see Fig. 1).

For those elements that form solid solutions, the strengthening effect when the element is in solution tends to increase with increasing difference in the atomic radii of the solvent (Al) and solute (alloying element) atoms. This factor is evident in data obtained from super-purity binary solid-solution alloys in the annealed state, presented in Table 2, but it is evident that other effects are involved, chief among which is an electronic bonding factor. The effects of multiple solutes in solid solution are somewhat less than additive and are nearly the same when one solute has a larger and the other a smaller atomic radius than that of aluminum as when both are either smaller or larger. Manganese in solid solution is highly effective in strengthening binary alloys. Its contribution to the strength of commercial alloys is less, because in these compositions, as a result of commercial mill fabricating operations, the manganese is largely precipitated.

Among the commercial alloys, only the bright-finishing alloys such as 5657 and 5252, which contain 0.8 and 2.5% Mg (nominal), respectively, with very low limits on all impurities, may be regarded as nearly pure solid solutions.

Elements and combinations that form predominantly second-phase constituents with relatively low solid solubility include iron, nickel, titanium, manganese and chromium, and combinations thereof. The presence of increasing volume fractions of the intermetallic-compound phases formed by these elements and the elemental silicon constituent formed by silicon during solidification or by precipitation in the solid state during postsolidification heating also increases strength and hardness. The rates of increase per unit weight of alloying element added are frequently similar to but usually lower than those resulting from solid solution. This "second-phase" hardening occurs even though the constituent particles are of sizes readily resolved by optical microscopy. These irregularly shaped particles form during solidification and occur mostly along grain boundaries and between dendrite arms.

Manganese and chromium are included in the group of elements that form predominantly second-phase constituents, because in commercial alloys they have very low equilibrium solid solubilities. In the case of many compositions containing manganese, this is because iron and silicon are also present and form the quaternary phase $Al_{12}(Fe,Mn)_3Si$. In alloys containing copper and manganese, the ternary phase $Al_{20}Cu_2Mn_3$ is formed. Most of the alloys in which chromium is present also contain magnesium, so that during solid-state heating they form $Al_{12}Mg_2Cr$, which also has very low equilibrium solid solubility. The concentrations of manganese and/or chromium held in solid solution in as-cast ingot that has been rapidly solidified and cooled from the molten state greatly exceed the equilibrium solubility. The solid solution is thus supersaturated and metastable. Ingot preheating for wrought commercial alloys containing these elements is designed to cause solid-state precipitation of the complex phase

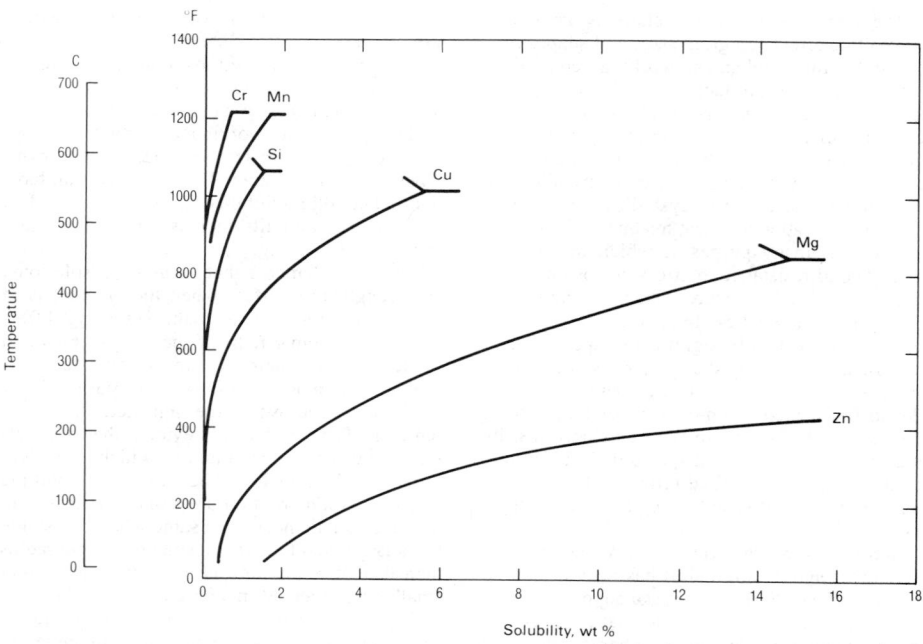

Fig. 1. Equilibrium binary solid solubility as a function of temperature for alloying elements most frequently added to aluminum

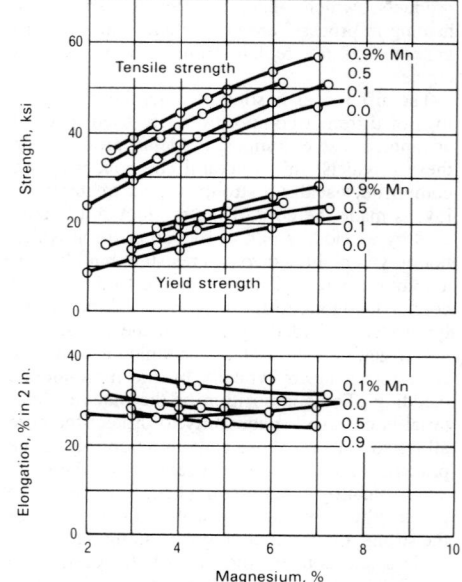

Fig. 2. Tensile properties in Al-Mg-Mn alloys in the form of annealed (O temper) plate 0.5 in. thick

containing one or the other of these elements that is appropriate to the alloy composition. This precipitation does not cause appreciable hardening, nor is it intended that it should. Its purpose is to produce finely divided and dispersed particles that retard or inhibit recrystallization and grain growth in the alloy during subsequent heatings. The precipitate particles of $Al_{12}(Fe,Mn)_3Si$, $Al_{20}Cu_2Mn_3$ or $Al_{12}Mg_2Cr$ are incoherent with the matrix, and concurrent with their precipitation the original solid solution becomes less concentrated. These conditions do not provide appreciable precipitation hardening. Changes in electrical conductivity constitute an effective measure of the completeness of these precipitation reactions that occur in preheating.

The newer "in-line" or integrated processes that shorten the path from molten metal to wrought product, avoiding ingot preheating and reducing the over-all time-temperature history, are changing this conventional or traditional picture. It seems very probable that in order to obtain the best results from such processes, traditional alloy compositions should be adjusted taking into account the fact that larger proportions of these elements would be expected to remain in solid solution through such abbreviated and truncated thermomechanical operations. New capabilities may be obtained with currently standard alloys

in some instances, but it would not be expected that a particular alloy would exhibit the same properties when produced by the two types of processes.

For alloys that are composed of both solid-solution and second-phase constituents and/or dispersoid precipitates, all of these components of microstructure contribute to strength, in a roughly additive manner. This is shown in Fig. 2 for Al-Mg-Mn alloys in the annealed condition.

Non-Heat-Treatable Alloys. By definition, the group of commercial alloys that are classed as non-heat-treatable are those that are not appreciably strengthened by heat treatment—that is, show no effective precipitation hardening. The strengthening mechanisms discussed so far (solid-solution formation, second-phase microstructural constituents and dispersoid precipitates) are those that provide the basis for the non-heat-treatable alloys. Wrought alloys of this type are mainly those of the 3xxx and 5xxx groups containing magnesium, manganese and/or chromium as well as the 1xxx aluminums and some alloys of the 4xxx group that contain only silicon. Non-heat-treatable casting alloys are of the 4xx.x or 5xx.x groups, containing silicon or magnesium, respectively, and the 1xx.x aluminums.

Strain Hardening. Strain hardening by cold rolling, drawing or stretching is a highly effective

means of increasing the strength of non-heat-treatable alloys. Work- or strain-hardening curves for several typical non-heat-treatable commercial alloys (Fig. 3) illustrate the increases in strength that accompany increasing reduction by cold rolling of initially annealed temper sheet. This increase is obtained at the expense of ductility as measured by percent elongation in a tensile test and by reduced formability in operations such as bending and drawing. It is frequently advantageous to employ material in a partially annealed (H2x) or stabilized (H3x) temper when bending, forming or drawing is required, since material in

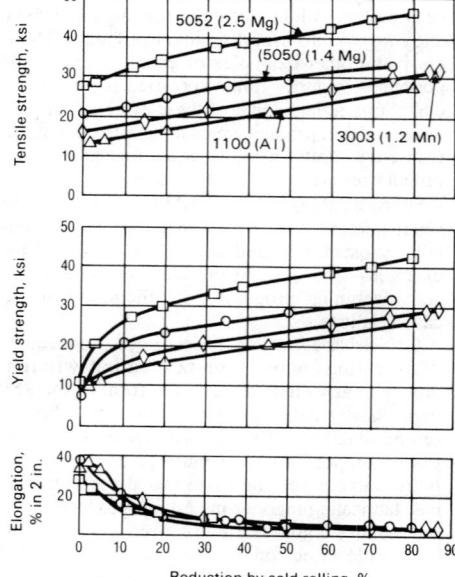

Fig. 3. Strain-hardening curves for aluminum (1100) and for Al-Mn (3003) and Al-Mg (5050 and 5052) alloys

Table 2. Solid-solution effects on strength of principal solute elements in super-purity aluminum(a)

Element	Difference in atomic radii, $r_x - r_{Al}$, %(b)	Yield strength/% addition(c) MPa/at. %	ksi/at. %	MPa/wt %	ksi/wt %	Tensile strength/% addition(d) MPa/at. %	ksi/at. %	MPa/wt %	ksi/wt %
Si	−3.8	9.3	1.35	9.2	1.33	40.0	5.8	39.6	5.75
Zn	−6.0	6.6	0.95	2.9	0.42	20.7	3.0	15.2	2.2
Cu	−10.7	16.2	2.35	13.8	2.0	88.3	12.8	43.1	6.25
Mn	−11.3	(e)	(e)	30.3	4.4	(e)	(e)	53.8	7.8
Mg	+11.8	17.2	2.5	18.6	2.7	51.0	7.4	50.3	7.3

(a) Some property–percent addition relationships are nonlinear. Generally, the unit effects of smaller additions are greater. (b) Listed in order of increasing percent difference in atomic radii. (c) Increase in yield strength (0.2% offset) for 1% (atomic or weight basis) alloy addition. (d) Increase in ultimate tensile strength for 1% (atomic or weight basis) alloy addition. (e) 1 at. % of manganese is not soluble.

Table 3. Tensile-property data illustrating typical relationships between strength and elongation for non-heat-treatable alloys in H1*x* vs H2*x* or H3*x* tempers

Alloy and temper	Tensile strength MPa	ksi	Yield strength MPa	ksi	Elongation, %
3105-H14	172	25	152	22	5
3105-H25	179	26	159	23	8
3105-H16	193	28	172	25	4

these tempers has greater forming capability for the same strength levels than does strain-hardened-only (H1*x*) material (see Table 3, for example).

All mill products can be supplied in the strain-hardened condition although there are limitations on the amounts of strain that can be applied to products such as die forgings and impacts. Even aluminum castings have been strengthened by cold pressing for certain applications.

Heat Treatable Alloys and Precipitation Hardening. Again by definition, heat treatable alloys are those that can be strengthened by suitable thermal treatment and include compositions used for wrought products as well as alloys for producing castings. Temperature-dependent solid solubility of the type shown for individual solutes in Fig. 1, the solubility increasing with increasing temperature, is a prerequisite. However, this feature alone does not make an alloy capable of precipitation hardening (or heat treatable). The strengths of most binary alloys containing Mg, Si, Zn, Cr or Mn alone exhibit little change from thermal treatments regardless of whether the solute is completely in solid solution, partially precipitated or substantially completely precipitated.

In contrast, alloys of the binary Al-Cu system having 3% Cu or more exhibit natural aging (hardening with time at ambient temperatures) after being solution heat treated and quenched. The amounts by which strength and hardness increase become larger with time of natural aging and with the copper content of the alloy from about 3% to the limit of solid solubility (5.65%). Natural aging curves for slowly quenched, high-purity Al-Cu alloys with 1 to 4.5% Cu are shown in Fig. 4. The rates and amounts of the changes in strength and hardness can be increased by holding the alloys at moderately elevated temperatures (for alloys of all types, the useful range is about 120 to 230 °C, or 250 to 450 °F). This treatment is called precipitation heat treating or artificial aging. In the Al-Cu system, alloys with as little as 1% Cu, again slowly quenched, start to harden after about 20 days at a temperature of 150 °C or 300 °F (see Fig. 5). The alloys of this system having less than about 3% Cu show little or no natural aging after low-cooling-rate quenching, which introduces little stress.

The characteristic that distinguishes between the systems having the required temperature–solid solubility relationship that do or do not exhibit precipitation hardening is the type or types of precipitate structures formed. Precipitation hardening is caused by a sequence of submicroscopic structure changes resulting from precipitation reactions that are responsible for the strength changes and can be revealed and analyzed only by such methods as x-ray diffraction and transmission electron microscopy. Room-temperature age hardening (natural aging) is a result of spontaneous formation of G-P zone structure, named for the co-discoverers, Guinier and Preston. Solute atoms either cluster or segregate to selected

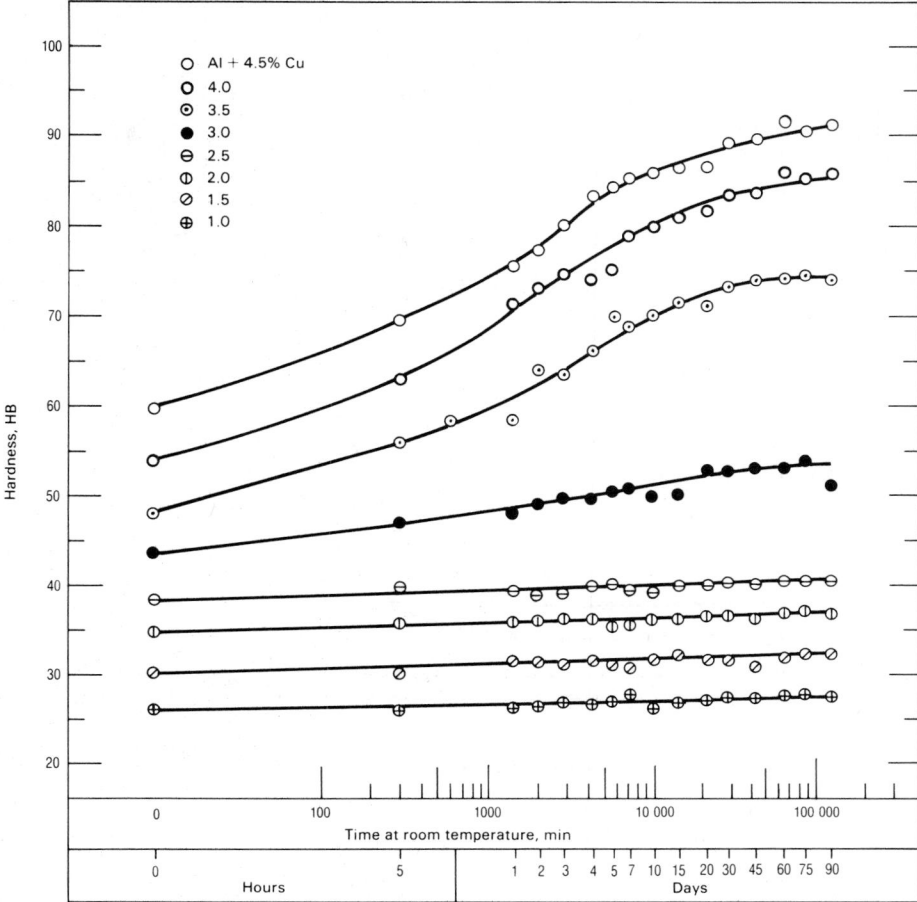

Fig. 4. Natural aging curves for binary Al-Cu alloys quenched in water at 100 °C (212 °F)

atomic lattice planes, depending on the alloy system, to form the G-P zones, and this structure is more resistant to movement of dislocations through the lattice, and hence is stronger.

Curves showing the changes in tensile yield strength with time at room temperature (natural aging curves) for three wrought commercial heat treatable alloys of different alloy systems are shown in Fig. 6. The magnitudes of increase in this property are considerably different for the three alloys, and the differences in rate of change with time are of practical importance. Because 7075 and similar alloys never become completely stable under these conditions, they are rarely used in the naturally aged temper. On the other hand, 2024 is widely used in this condition.

Precipitation heat treating (or artificial aging) at higher temperatures produces transition, metastable forms of the equilibrium precipitate of the particular alloy system. These transition precipitates are still coherent with the solid-solution matrix. The characteristic that determines whether a precipitate phase is coherent or noncoherent is the closeness of match or degree of disregistry between the atomic spacings on the lattice of the matrix and on that of the precipitate. The presence of the precipitate particles, and, even more importantly in most cases, the strain fields in the matrix surrounding the coherent particles, obstruct and retard the movement of dislocations, thus providing increased resistance to deforma-

tion—in other words, higher strength. These particles, at the maximum-strength stage, are extremely fine, are resolvable only by TEM, and constitute a relatively large volume fraction, particularly when the strain fields are considered.

With further heating at temperatures that cause strengthening or at higher temperatures, the precipitate particles grow, but even more importantly convert to the equilibrium phases, which generally are not coherent. These changes soften the material and, carried further, produce the softest or annealed condition. Even at this stage, the precipitate particles are still too small to be clearly resolved by optical microscopy, although etching effects are readily observed—particularly in alloys containing copper.

Precipitation heat treatment or artificial aging curves for the Al-Mg-Si wrought alloy 6061 are shown in Fig. 7. This is a typical family of curves showing the changes in tensile yield strength that accrue with increasing time at each of a series of temperatures. In all cases, the material had been given a solution heat treatment followed by a quench just prior to the start of the precipitation heat treatment. For detailed presentation of heat treating operations, parameters and practices see the section on Heat Treating in this volume.

The above description of the precipitation process and its effects on strength and metallurgical structures applies not only to heat treatable binary compositions, none of which is used commercially, but also to the commercial alloys,

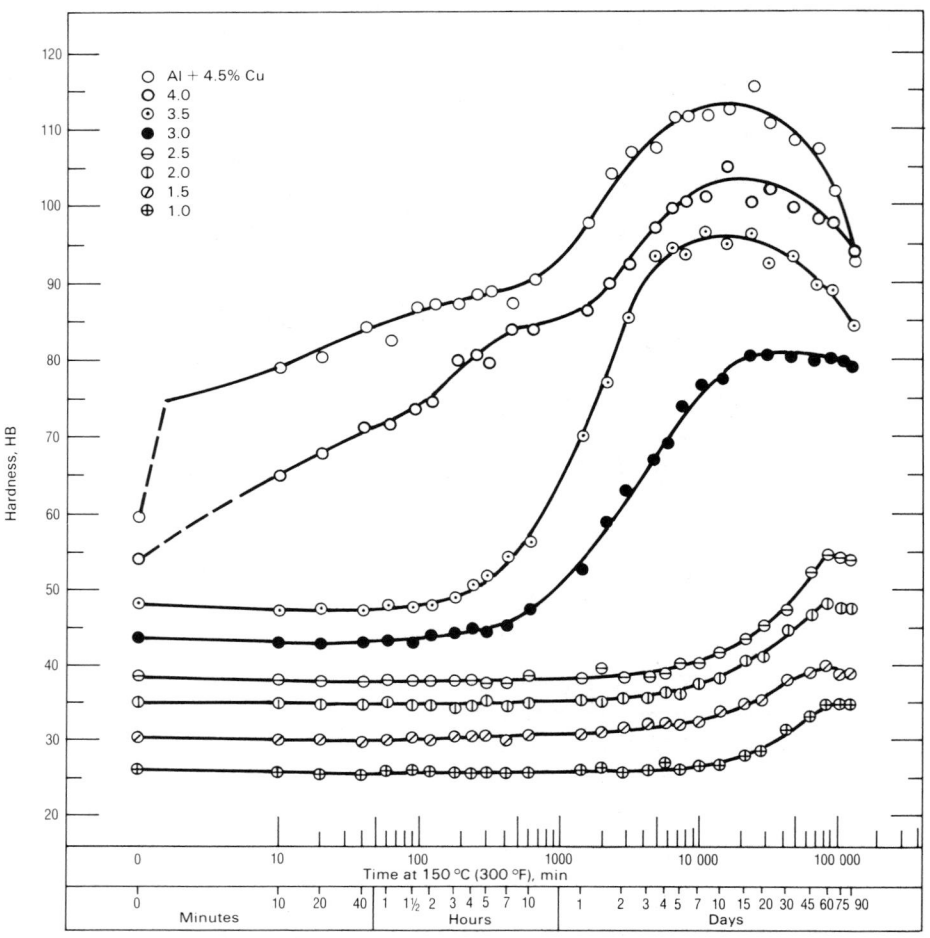

Fig. 5. Precipitation hardening curves for binary Al-Cu alloys quenched in water at 100 °C (212 °F) and aged at 150 °C (300 °F)

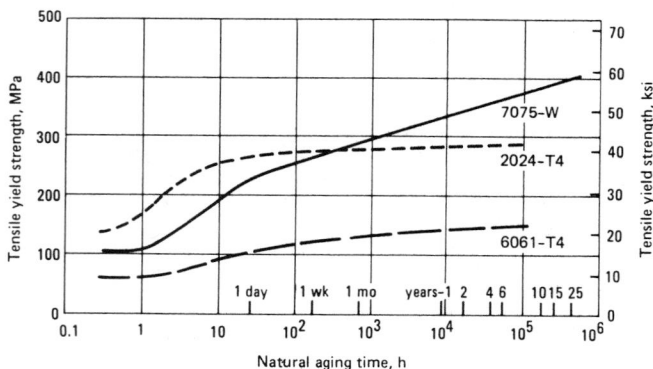

Fig. 6. Natural aging curves for three solution heat treated wrought aluminum alloys

conductivity than the same elements combined with others as intermetallic compounds, thermal treatments are applied to ingots used for fabrication of electrical conductor products. These thermal treatments are intended to precipitate as much as possible of the dissolved impurities. Iron is the principal element involved, and, although the amount precipitated is only a few hundredths of a percent, the effect on electrical conductivity of the wire, cable or other product made from the ingot is of considerable practical importance. These alloys may or may not be heat treatable with respect to mechanical properties. Electrical conductor alloys 6101 and 6201 are heat treatable. These alloys are used in tempers in which their strengthening precipitate, the transition form of Mg_2Si, is largely out of solid solution to optimize both strength and conductivity.

The commercial heat treatable alloys are, with few exceptions, based on ternary or quaternary systems with respect to the solutes involved in developing strength by precipitation. The most prominent systems are: Al-Cu-Mg, Al-Cu-Si and Al-Cu-Mg-Si, alloys of which are in the 2*xxx* and 2*xx.x* groups (wrought and casting alloys, respectively); Al-Mg-Si (6*xxx* wrought alloys); Al-Si-Mg, Al-Si-Cu and Al-Si-Mg-Cu (3*xx.x* casting alloys); and Al-Zn-Mg and Al-Zn-Mg-Cu (7*xxx* wrought and 7*xx.x* casting alloys). In each case the solubility of the multiple solute elements decreases with decreasing temperature, as can be seen from solvus diagrams included in Vol 8 of Metals Handbook, 8th Edition. Reference to these diagrams also shows the equilibrium phase (or phases) that precipitates in a particular system but does not show whether a transition phase may occur or its composition.

These multiple alloying additions of both major solute elements and supplementary elements employed in commercial alloys are strictly functional and serve with different heat treatments to provide the many different combinations of properties — physical, mechanical and electrochemical — that are required for different applications. Some alloys, particularly those for foundry production of castings, contain amounts of silicon far in excess of the amount which is soluble or needed for strengthening alone. The function here is chiefly to improve casting soundness and freedom from cracking, but the excess silicon also serves to increase wear resistance, as do other microstructural constituents formed by manganese, nickel and iron. Parts made of such alloys are commonly used in gasoline and diesel engines (pistons, cylinder blocks, etc.). The system of numerical nomenclature used to designate the alloys, and that for the strain-hardened and heat treated tempers, will be described later in this section.

Alloys containing the elements silver, lithium and germanium are also capable of providing high strength with heat treatment and, in the case of lithium, both increased elastic modulus and lower density, which are highly advantageous — particularly for aerospace applications. Commercial use of alloys containing these elements has been restricted either by cost or by difficulties encountered in producing them. Such alloys are used to some extent, however, and research is being directed toward overcoming their disadvantages.

In the case of alloys having copper as the principal alloying ingredient and no magnesium, strengthening by precipitation can be greatly increased by adding small fractional percentages of tin, cadmium or indium, or combinations of these

having generally much greater complexity of composition. As noted before, not only do the mechanical properties change with these heat treatments, and with natural aging, but also physical properties (density and electrical and thermal conductivities) and electrochemical properties (solution potential). On the microstructural and submicroscopic scales, the electrochemical properties develop point-to-point nonuniformities that account for changes in corrosion resistance.

Measurements of changes in physical and electrochemical properties have played an important role in completely describing precipitation reactions and are very useful in analyzing or diagnosing whether heat treatable products have been properly or improperly heat treated. Although they may be indicative of the strength levels of products, they cannot be relied upon to determine whether or not the product meets specified mechanical-property limits. Since elements in solid solution are always more harmful to electrical

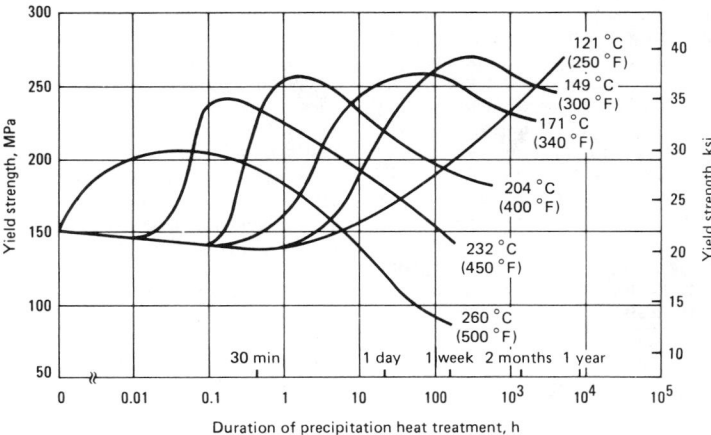

Fig. 7. Precipitation heat treatment or artificial aging curves for solution heat treated aluminum alloy 6061

elements. Alloys based on these effects have been produced commercially but not in large volumes because of costly special practices and limitations required in processing and, in the case of cadmium, the need for special facilities to avoid health hazards from formation and release of cadmium vapor during alloying. Such alloys, as well as those containing silver, lithium or germanium, may be used on a selective basis in the future.

Strength at elevated temperatures is improved mainly by solid-solution and second-phase hardening, because, at least for temperatures exceeding those of the precipitation-hardening range— 230 °C (450 °F) and over—the precipitation reactions continue into the softening regime. For supersonic aircraft and space vehicle applications subject to aerodynamic heating, the heat treatable alloys of the 2*xxx* group can be used for temperatures up to about 150 °C (300 °F).

Effects of Strengthening on Other Mechanical Properties. Resistance to fatigue (from application of dynamic stresses into the tensile range) increases generally with increasing strength whether from alloying effects alone, from strain hardening or from heat treatment. This improvement, in terms of either time to initiate cracks or cycles to failure, is generally less than the improvement in static strength, and the highest resistance to fatigue for a given alloy is sometimes provided by a temper having static strength levels lower than the highest strength possible—e.g., 2024-T3*x* or -T4*x* versus 2024-T6*x* or -T8*x*. This occurs mainly because the resistance to fatigue-crack growth at high levels of stress-intensity factor requires good fracture toughness, and this property decreases generally with increasing strength at high levels.

As strength, defined as resistance to deformation, increases, the properties of ductility, malleability, ease of forming, and fracture toughness tend to decrease. This inverse relationship between these properties and strength, however, is not universal. As indicated previously, for the non-heat-treatable alloys, the H2*x* or H3*x* tempers have advantages over those of the H1*x* series in the ductility/strength relationship. It is also true that, in the relationship of fracture toughness to yield strength, alloys of the 7*xxx* group (Al-Zn-Mg-Cu types) are superior to those of the 2*xxx* and 6*xxx* groups.

All of these properties must be carefully considered in application of alloys to critical engineering structures such as aircraft as well as to products such as truck wheels and other automotive and truck components. Good design, avoiding stress-concentrating features and any features that promote localized corrosion, is in most cases at least as important as good alloy/temper selection.

Other Considerations Involved in Alloy Development. The engineering properties and characteristics that have immediate effects on the functional behavior of end products is only a partial list of the features that must be considered in alloy design. One of the foremost considerations is that, to be economically viable, an alloy must be capable of being cast and fabricated to the form desired with reasonable freedom from scrap losses resulting from cracking or other in-process damage. Many alloys and specialty products, including those for extremely large-tonnage items such as beverage cans as well as Alclad products and brazing composites, are tailored to very specific uses, and cost-effective principles apply to the commercial viability of these also. All of these products must compete for markets with other metallic materials as well as polymeric materials, glasses, ceramics, etc.

MILL FABRICATION OF WROUGHT PRODUCTS

Aluminum and many of its alloys can be fabricated by virtually all of the known metalworking processes—hot and cold rolling, direct and indirect extrusion, forging, swaging, upsetting, impacting, reverse extrusion, cupping, deep drawing, wire drawing, etc.—to form a wide variety of wrought product shapes and sizes. Combinations of working processes are commonly used in producing mill products; in producing wire, for example, a 380-mm (15-in.) diam ingot may be extruded to a bar 150 by 150 mm (6 by 6 in.) in cross section, hot rolled from that size to 9.5-mm ($^3/_8$-in.) diam drawing stock, and cold drawn from that size to the required wire size. Stock for production of forgings is normally produced by extrusion or hot rolling. Preforging of ingots by means of hydraulic presses prior to hot rolling has been employed at various times in producing thick plate to improve soundness and mechanical properties in the short-transverse direction.

Ingots used for fabrication of wrought products (fabricating ingots) are generally of square, rectangular or round cross section with thicknesses up to about 600 mm (about 24 in.). They are generally produced by semicontinuous casting processes in which very rapid solidification and postsolidification cooling are provided by direct spraying or flooding of the ingot surfaces with water. This rapid solidification produces fine grain, dendrite and constituent sizes, which favor good mechanical properties in the resulting wrought products.

Before fabrication, most ingots are preheated at relatively high temperatures, an operation generically termed "ingot preheating" but sometimes referred to as "homogenization." The latter term is only partially appropriate for ingot preheating since the operation may have one or more of several functions: to improve workability; to establish dispersoid precipitate, the functions of which have been stated earlier; and to dissolve soluble constituents that were rejected during solidification and redistribute solute within cored solid solutions (the homogenization function).

The amounts of effective deformation involved in producing wrought products may range from a minimum of about 60% reduction in cross section or thickness (for very large products) to well over 99.99%. For the 0.11-mm (0.0045-in.) thick sidewalls of beverage cans, produced literally by the billions annually, the thickness reduction from the starting ingot is 99.976%. For foil 0.0043 mm (0.00017 in.) thick, produced from ingot 470 mm (18.5 in.) thick, the reduction is 99.999%, while for wire 0.15 mm (0.006 in.) in diameter, which started as a 380-mm (15-in.) diam ingot, the reduction in cross-sectional area is a remarkable 99.99998%.

In current practice, quite different processes are in use for fabricating some products, particularly sheet and rod. In-line casting and rolling facilities process these products directly from molten metal. In these cases, the amounts of deformation (reduction in thickness or cross-sectional area from cast section to final product) are much less than those encountered when ingots are cast, cooled, and transferred to mills for fabrication. Not all of the traditional alloys are amenable to semicontinuous in-line operations.

Hot working temperatures may vary from 300 °C (600 °F) to a temperature at which an alloy may become "hot short," usually well below the temperature at which melting starts. Cold rolling develops adiabatic heat so that temperatures may rise during such operations into the range from 150 to 175 °C (300 to 350 °F). Annealing may be required between hot and cold working and sometimes at intermediate stages of cold working. The sequences of cold working and annealing operations affect directionality in the surface plane of sheet products, which in turn influences earing characteristics in deep drawing. Particular sequences are developed to produce "non-earing" sheet products.

WROUGHT ALUMINUM P/M ALLOYS

A very different approach to production of high-strength wrought aluminum products has been progressing through research and development for a number of years and has recently advanced to initial limited-scale production. These products are produced by prealloying in the molten state, atomizing to powder, consolidating under pressure followed by encapsulation, vacuum preheating and hot consolidation, scalping, hot working by extrusion, forging or rolling, and heat treating. Two experimental alloys of the 7*xxx* group (containing the unusual element cobalt) offer mechanical and corrosion-resisting properties superior to those of conventional 7xxx alloy products produced from cast ingot. Evaluation in aircraft applications, now underway, will determine whether these products are cost effective.

"Mechanically alloyed" aluminum alloy products are a variant of those discussed above. To produce these materials, fine powders of elemental metals are alloyed by high-energy milling, which also introduces strengthening oxide and carbide dispersoids. Billets are consolidated by pressure and hot working and, after being shaped, have property combinations superior to those of conventionally produced material. These are also under evaluation.

CASTINGS

Aluminum alloy engineered castings, designed for specific purposes, are produced in a great variety of shapes and sizes by all casting processes. Die castings, permanent and semipermanent mold castings, and sand castings constitute most of this production, with smaller quantities cast by other processes (see Table 4). Based on average annual shipments over the last ten years (1972 to 1981, inclusive), castings of all types amounted to 13.9% of all aluminum product shipments. Transportation is the leading market, and the trend toward increasing use in automobile applications is increasing the importance of castings in the total industry picture. Castings share most of the other markets listed for wrought products with the exception of containers and packaging.

Die castings generally are not heat treated. Those produced by the other processes may or may not be heat treated, depending on their intended applications, and alloy selection is affected by both the casting process and whether or not heat treatment is to be applied. The most popular alloys differ from those used in producing wrought products chiefly in their higher silicon contents. These compositions greatly favor freedom from unsoundness and cracking.

Table 4. Average annual shipments of aluminum castings by type for the ten-year period from 1972 to 1981 (inclusive)

Type of castings	Average annual shipments		
	10^6 kg	10^6 lb	% of total
Die castings	517.6	1141.1	63
Permanent mold and semipermanent mold castings	182.1	401.5	22
Sand castings	109.9	242.2	13
Others	14.2	31.2	2
Total	823.8	1816.0	100

POWDER METALLURGY PARTS

Production of powder metallurgy (P/M) parts consists of compressing metal powder in a shaped die to produce green compacts, and then sintering (diffusion bonding) the compacts at elevated temperature in a furnace with a protective atmosphere. During sintering, the compacts are consolidated and strengthened. Re-pressing to improve dimensional accuracy (sizing) or to improve configuration (coining) also increases density and strength. Powder metallurgy parts frequently are competitive with castings, machined components, stampings and fabricated assemblies. Products with controlled intentional porosity, such as filter elements and self-lubricating bearings, are produced by powder metallurgy.

Particles, Powder and Flake. There are many commercial forms of particulate aluminum ranging from globules as large as 13 mm ($1/2$ in.) in diameter to flakes only a few microns in thickness. The coarse particles, called "shot," are used mainly for deoxidizing steel. Powder produced by atomization may be granular or spherical and ranges in particle size from -12 to -325 mesh. The particles in paste and flake powders are produced in many grades and sizes for uses ranging from printing and other decorative and reflecting applications to the metallic finishes used in the automobile industry.

One of the substantial current applications of particulate aluminum is its use as one of the fuel components in solid propellants. The two solid propellant boosters of the space shuttle oxidize 180 000 kg (400 000 lb) of aluminum during one launch before breaking away from the main craft. Powder is also widely used in explosives and fireworks. Production and shipment of powder during the last five years (1977 through 1981) was at an average annual rate of 51 million kg (113 million lb), constituting 0.7% of all aluminum shipments. This product is included in mill products in the Aluminum Association "Aluminum Statistical Review," which is the source of the production statistics cited in this article.

Alloy and Temper Designation Systems for Aluminum

THE SYSTEMS for designating aluminums and aluminum alloys in wrought form, aluminums and aluminum alloys in the form of castings and foundry ingot, and the tempers in which these wrought or cast products are produced, are covered by national standard ANSI H35.1-1982. The rules by which the distinction is made between "aluminums" and "aluminum alloys" and by which the numerical designations are assigned are a part of this document.

WROUGHT ALUMINUM AND ALUMINUM ALLOY DESIGNATION SYSTEM

A system of four-digit numerical designations is used to identify wrought aluminums and aluminum alloys. The first digit indicates the group, as follows:

Aluminum, 99.00% minimum and
    greater 1xxx
Aluminum alloys grouped by major alloying
    element(s):
    Copper 2xxx
    Manganese 3xxx
    Silicon 4xxx
    Magnesium 5xxx
    Magnesium and silicon 6xxx
    Zinc 7xxx
    Other element 8xxx
    Unused series 9xxx

For the 2xxx through 7xxx series, the alloy group is determined by the alloying element present in the greatest mean percentage. An exception is the 6xxx series alloys in which the proportions of magnesium and silicon available to form Mg_2Si are predominant. Another exception is made in those cases in which the alloy qualifies as a modification of a previously registered alloy. If the greatest mean percentage is the same for more than one element, the choice of group is in order of group sequence: Cu, Mn, Si, Mg, Mg_2Si, Zn, or others.

The second digit indicates a modification of impurity limits for aluminums (group 1xxx) and a modification of the original alloy for groups 2xxx through 8xxx. The last two digits indicate purity for aluminums and indicate the specific alloy for groups 2xxx through 8xxx.

Aluminums. In the 1xxx group, the series 10xx is used to designate "unalloyed" compositions— i.e., metal having only "natural" impurities. The last two of the four digits in the designation indicate the minimum aluminum percentage. These digits are the same as the two digits to the right of the decimal point in the minimum aluminum percentage when expressed to the nearest 0.01%.

Designations having second digits other than zero (integers 1 through 9, assigned consecutively as needed) indicate special control of one or more individual impurities.

Aluminum Alloys. In the 2xxx through 8xxx alloy groups, the second digit in the designation indicates alloy modification. If the second digit is zero, it indicates the original alloy; integers 1 through 9, assigned consecutively, indicate modifications of the original alloy. Explicit rules are established for determining whether a proposed composition is a modification of a previously registered alloy or is a new alloy.

The last two of the four digits in the 2xxx through 8xxx groups have no special significance but serve only to identify the different aluminum alloys in the group.

CAST ALUMINUM AND ALUMINUM ALLOY DESIGNATION SYSTEM

A system of four-digit numerical designations incorporating a decimal point is used to identify aluminums and aluminum alloys in the form of castings and foundry ingot. The first digit indicates the alloy group, as follows:

Aluminum, 99.00% minimum and
greater 1*xx.x*
Aluminum alloys grouped by major alloying
element(s):
Copper 2*xx.x*
Silicon, with added copper and/or
magnesium 3*xx.x*
Silicon 4*xx.x*
Magnesium 5*xx.x*
Zinc 7*xx.x*
Tin 8*xx.x*
Other element 9*xx.x*
Unused series 6*xx.x*

For 2*xx.x* through 8*xx.x* alloys, the alloy group
is determined by the alloying element present in
the greatest mean percentage except in cases in
which the composition being registered qualifies
as a modification of a previously registered alloy. If the greatest mean percentage is common
to more than one alloying element, the alloy group
is determined by the element that comes first in
the sequence shown above.

The second two digits identify the specific aluminum alloy or, for the aluminums (1*xx.x* series), indicate purity. The last digit, which is
separated from the others by a decimal point, indicates the product form, whether castings or ingot. A modification of an original alloy, or of
the impurity limits for an aluminum, is indicated
by a serial letter preceding the numerical designation. The serial letters are assigned in alphabetical sequence starting with A but omitting I,
O, Q and X, the X being reserved for experimental alloys. Explicit rules are established for
determining whether a proposed composition is
a modification of an existing alloy or is a new
alloy.

Aluminum Castings and Ingot. For the 1*xx.x* group,
the second two of the four digits in the designation indicate the minimum aluminum percentage. These digits are the same as the two digits
to the right of the decimal point in the minimum
aluminum percentage when expressed to the
nearest 0.01%. The last digit, which is to the right
of the decimal point, indicates the product form:
1*xx.*0 indicates castings, and 1*xx.*1 indicates ingot.

Aluminum Alloy Castings and Ingot. For the 2*xx.x*
through 9*xx.x* alloy groups, the second two of
the four digits in the designation have no special
significance but serve only to identify the different alloys in the group. The last digit, which is
to the right of the decimal point, indicates the
product form: *xxx.*0 indicates castings, and *xxx.*1
indicates ingot having limits for alloying elements the same as those for the alloy in the form
of castings, except for the following:

Maximum iron percentage:
For sand and permanent mold castings	**For ingot**
Up through 0.15	0.03 less than for castings
Over 0.15 through 0.25	0.05 less than for castings
Over 0.6 through 1.0	0.2 less than for castings
Over 1.0	0.3 less than for castings

For die castings | **For ingot**
| Up through 1.3 | 0.3 less than for castings |
| Over 1.3 | 1.1 maximum |

Minimum magnesium percentage:
For all castings	**For ingot**
Less than 0.50	0.05 more than for castings(a)
0.50 and greater	0.1 more than for castings(a)

Maximum zinc percentage:
For die castings	**For ingot**
Over 0.25 through 0.6	0.10 less than for castings
Over 0.6	0.1 less than for castings

(a) Applicable only when the specified magnesium range for castings is greater than 0.15%.

DESIGNATIONS FOR EXPERIMENTAL ALLOYS

Experimental alloys also are designated in accordance with the systems for wrought and cast
alloys, but they are indicated by the prefix X.
The prefix is dropped when the alloy is no longer
experimental. During development and before they
are designated as experimental, new alloys may
be identified by serial numbers assigned by their
originators. Use of the serial number is discontinued when the X number is assigned.

UNIFIED NUMBERING SYSTEM

UNS numbers correlate many nationally used
numbering systems currently administered by societies, trade associations, and individual users
and producers of metals and alloys. Table 1 correlates AA (Aluminum Association) numbers to
UNS numbers.

TEMPER DESIGNATION SYSTEM FOR ALUMINUM AND ALUMINUM ALLOYS

Basic Temper Designations

F As fabricated. Applies to products shaped by
cold working, hot working or casting processes in which no special control over thermal conditions or strain hardening is employed. For wrought products, there are no
mechanical-property limits.

O Annealed. Applies to wrought products that are
annealed to obtain lowest strength temper, and
to cast products that are annealed to improve
ductility and dimensional stability. The O may
be followed by a digit other than zero.

H Strain hardened (wrought products only). Applies
to products that have been strengthened by
strain hardening, with or without supplementary heat treatment to produce some reduction in strength. The H is always followed by
two or more digits, as discussed in the following section.

W Solution heat treated. An unstable temper applicable only to alloys that naturally age
(spontaneously age at room temperature) after
solution heat treatment. This designation is
specific only when the period of natural aging is indicated—for example, W ¹/₂ hr. (See
also the discussion of the T*x*51, T*x*52 and T*x*54

Table 1. UNS (Unified Numbering System) numbers corresponding to AA (Aluminum Association) numbers for aluminums and aluminum alloys

AA No.	UNS No.	AA No.	UNS No.	AA No.	UNS No.	AA No.	UNS No.
100.1	A01001	324.2	A03242	413.0	A04130	1145	A91145
130.1	A01301	328.0	A03280	413.2	A04132	1170	A91170
150.1	A01501	328.1	A03281	443.0	A04430	1175	A91175
160.1	A01601	333.0	A03330	443.1	A04431	1180	A91180
170.1	A01701	333.1	A03331	443.2	A04432	1185	A91185
201.0	A02010	343.0	A03430	444.0	A04440	1188	A91188
201.2	A02012	343.1	A03431	444.2	A04442	1193	A91193
202.0	A02020	354.0	A03540	514.0	A05140	1199	A91199
202.2	A02022	354.1	A03541	514.1	A05141	1200	A91200
203.0	A02030	355.0	A03550	514.2	A05142	1230	A91230
203.2	A02032	355.1	A03551	518.0	A05180	1235	A91235
204.0	A02040	355.2	A03552	518.1	A05181	1250	A91250
204.2	A02042	356.0	A03560	518.2	A05182	1260	A91260
208.0	A02080	356.1	A03561	520.0	A05200	1285	A91285
208.1	A02081	356.2	A03562	520.2	A05202	1345	A91345
208.2	A02082	357.0	A03570	535.0	A05350	1350	A91350
213.0	A02130	357.1	A03571	535.2	A05352	1435	A91435
213.1	A02131	359.0	A03590	705.0	A07050	2011	A92011
222.0	A02220	359.2	A03592	705.1	A07051	2014	A92014
222.1	A02221	360.0	A03600	707.0	A07070	2017	A92017
224.0	A02240	360.2	A03602	707.1	A07071	2018	A92018
224.2	A02242	363.0	A03630	713.0	A07130	2020	A92020
238.0	A02380	363.1	A03631	713.1	A07131	2021	A92021
238.1	A02381	364.0	A03640	771.0	A07710	2024	A92024
238.2	A02382	364.2	A03642	771.2	A07712	2025	A92025
242.0	A02420	380.0	A03800	850.0	A08500	2036	A92036
242.1	A02421	380.2	A03802	850.1	A08501	2048	A92048
242.2	A02422	383.0	A03830	1030	A91030	2117	A92117
249.0	A02490	383.1	A03831	1035	A91035	2124	A92124
249.2	A02492	383.2	A03832	1040	A91040	2214	A92214
295.0	A02950	384.0	A03840	1045	A91045	2218	A92218
295.1	A02951	384.1	A03841	1050	A91050	2219	A92219
295.2	A02952	384.2	A03842	1055	A91055	2319	A92319
305.0	A03050	390.0	A03900	1060	A91060	2419	A92419
305.2	A03052	390.2	A03902	1065	A91065	2618	A92618
308.0	A03080	392.0	A03920	1070	A91070	3002	A93002
308.1	A03081	392.1	A03921	1075	A91075	3003	A93003
308.2	A03082	393.0	A03930	1080	A91080	3004	A93004
319.0	A03190	393.1	A03931	1085	A91085	3005	A93005
319.1	A03191	393.2	A03932	1090	A91090	3006	A93006
319.2	A03192	408.2	A04082	1095	A91095	3007	A93007
324.0	A03240	409.2	A04092	1100	A91100	3102	A93102
324.1	A03241	411.2	A04112	1135	A91135	3105	A93105

(continued)

Table 1 (continued)

AA No.	UNS No.	AA No.	UNS No.	AA No.	UNS No.	AA No.	UNS No.
3303	A93303	6007	A96007	8040	A98040	A514.0	A15140
4002	A94002	6011	A96011	8076	A98076	A514.2	A15142
4004	A94004	6053	A96053	8079	A98079	A535.0	A15350
4032	A94032	6061	A96061	8081	A98081	A535.1	A15351
4043	A94043	6063	A96063	8112	A98112	A712.0	A17120
4044	A94044	6066	A96066	8280	A98280	A712.1	A17121
4045	A94045	6070	A96070	A201.0	A12010	A850.0	A18500
4047	A94047	6101	A96101	A201.1	A12012	A850.1	A18501
4145	A94145	6105	A96105	A240.0	A12400	B295.0	A22950
4343	A94343	6151	A96151	A240.1	A12401	B295.1	A22951
4543	A94543	6162	A96162	A242.0	A12420	B295.2	A22952
4643	A94643	6201	A96201	A242.1	A12421	B358.0	A23580
5005	A95005	6205	A96205	A242.2	A12422	B358.2	A23582
5010	A95010	6253	A96253	A305.0	A13050	B380.0	A23800
5034	A95034	6261	A96261	A305.1	A13051	B380.1	A23801
5039	A95039	6262	A96262	A305.2	A13052	B384.0	A23840
5040	A95040	6301	A96301	A319.0	A13190	B384.1	A23841
5050	A95050	6351	A96351	A319.1	A13191	B443.0	A24430
5051	A95051	6463	A96463	A332.0	A13320	B443.1	A24431
5052	A95052	6763	A96763	A332.1	A13321	B444.2	A24442
5056	A95056	6951	A96951	A332.2	A13322	B514.0	A25140
5082	A95082	7001	A97001	A333.0	A13330	B514.2	A25142
5083	A95083	7004	A97004	A333.1	A13331	B535.0	A25350
5086	A95086	7005	A97005	A356.0	A13560	B535.2	A25352
5151	A95151	7008	A97008	A356.1	A13561	B771.0	A27710
5154	A95154	7011	A97011	A356.2	A13562	B771.2	A27712
5182	A95182	7013	A97013	A357.0	A13570	B850.0	A28500
5183	A95183	7039	A97039	A357.2	A13572	B850.1	A28501
5205	A95205	7049	A97049	A360.0	A13600	C355.0	A33550
5252	A95252	7050	A97050	A360.1	A13601	C355.1	A33551
5254	A95254	7070	A97070	A360.2	A13602	C355.2	A33552
5352	A95352	7072	A97072	A380.0	A13800	C443.0	A34430
5356	A95356	7075	A97075	A380.1	A13801	C443.1	A34431
5357	A95357	7076	A97076	A380.2	A13802	C443.2	A34432
5454	A95454	7079	A97079	A384.0	A13840	C712.0	A37120
5456	A95456	7104	A97104	A384.1	A13841	C712.1	A37121
5457	A95457	7149	A97149	A390.0	A13900	D712.0	A47120
5554	A95554	7175	A97175	A390.1	A13901	D712.2	A47122
5556	A95556	7178	A97178	A413.0	A14130	F332.0	A63320
5652	A95652	7179	A97179	A413.1	A14131	F332.1	A63321
5654	A95654	7277	A97277	A413.2	A14132	F332.2	A63322
5657	A95657	7472	A97472	A443.0	A14430	F356.0	A63560
6003	A96003	7475	A97475	A443.1	A14431	F356.2	A63562
6004	A96004	8001	A98001	A444.0	A14440	F514.0	A65140
6005	A96005	8013	A98013	A444.1	A14441	F514.1	A65141
6006	A96006	8020	A98020	A444.2	A14442	F514.2	A65142

tempers, in the section on heat treatable alloys.)

T **Heat treated to produce stable tempers other than F, O or H.** Applies to products that are thermally treated, with or without supplementary strain hardening, to produce stable tempers. The T is always followed by one or more digits, as discussed in a later section.

System for Strain-Hardened Products

Temper designations for wrought products that are strengthened by strain hardening consist of an H followed by two or more digits. The first digit following the H indicates the specific sequence of basic operations, as follows:

H1 **Strain hardened only.** Applies to products that are strain hardened to obtain the desired strength without supplementary thermal treatment. The digit following the H1 indicates the degree of strain hardening.

H2 **Strain hardened and partially annealed.** Applies to products that are strain hardened more than the desired final amount and then reduced in strength to the desired level by partial annealing. The digit following the H2 indicates the degree of strain hardening remaining after the product has been partially annealed.

H3 **Strain hardened and stabilized.** Applies to products that are strain hardened and whose mechanical properties are stabilized by a low-temperature thermal treatment that slightly decreases tensile strength and improves ductility. This designation is applicable only to those alloys that, unless stabilized, gradually age soften at room temperature. The digit following the H3 indicates the degree of strain hardening after stabilization.

For alloys that age soften at room temperature, each H2x temper has the same minimum ultimate tensile strength as the H3x temper with the same second digit. For other alloys, each H2x temper has the same minimum ultimate tensile strength as the H1x with the same second digit, and slightly higher elongation.

The digit following the designations H1, H2 and H3, which indicates the degree of strain hardening, is a numeral from 1 through 8. Numeral 8 indicates tempers with ultimate tensile strength equivalent to that achieved by about 75% cold reduction (temperature during reduction not to exceed 50 °C, or 120 °F) following full annealing. Tempers between 0 (annealed) and 8 are designated by numerals 1 through 7. Material having an ultimate tensile strength approximately

midway between that of the 0 temper and that of the 8 temper is designated by the numeral 4; approximately midway between the 0 and 4 tempers by the numeral 2; and approximately midway between the 4 and 8 tempers by the numeral 6. Numeral 9 designates tempers whose minimum ultimate tensile strength exceeds that of the 8 temper by 10 MPa or more (or 2.0 ksi or more when English unit strengths are used). For two-digit H tempers whose second digits are odd, the standard limits for strength are the arithmetic mean, rounded to the nearest multiple of 5 MPa or 0.5 ksi (in conformance with ASTM Recommended Practice E29), of the standard limits for the adjacent two-digit H tempers whose second digits are even.

For alloys that cannot be sufficiently cold reduced to establish an ultimate tensile strength applicable to the 8 temper (75% cold reduction after full annealing), the 6-temper tensile strength may be established by cold reduction of approximately 55% following full annealing, or the 4-temper tensile strength may be established by cold reduction of approximately 35% after full annealing.

When it is desirable to identify a variation of a two-digit H temper, a third digit (from 1 to 9) may be assigned. Zero has been assigned to indicate variations negotiated between the manufacturer and purchaser which are not used widely enough to justify registration. The third digit is used when the degree of control of temper or the mechanical properties are different from but close to those for the two-digit H temper designation to which it is added, or when some other characteristic is significantly affected. The minimum ultimate tensile strength of a three-digit H temper is at least as close to that of the corresponding two-digit H temper as it is to either of the adjacent two-digit H tempers. Products in H tempers whose mechanical properties are below those of Hx1 tempers are assigned variations of Hx1. Some three-digit H temper designations have already been assigned; these are described below. The following designations have been assigned for wrought products in all alloys:

Hx11 Applies to products that incur sufficient strain hardening after final annealing that they fail to qualify as O temper, but not so much or consistent amount of strain hardening that they qualify as Hx1 temper.

H112 Applies to products that may acquire some strain hardening during working at elevated temperature and for which there are mechanical-property limits.

The following designations have been assigned for wrought products in alloys with nominal magnesium contents greater than 4%.

H311 Applies to products that are strain hardened less than the amount required for a controlled H31 temper.

H321 Applies to products that are strain hardened less than the amount required for a controlled H32 temper.

H323,
H343 These designations apply to products that are specially fabricated to have acceptable resistance to stress-corrosion cracking.

The following three-digit H-temper designations have been assigned for patterned or embossed sheet.

Patterned or embossed sheet	Temper fabricated from (respectively)
H114	O
H124, H224, H324	H11, H21, H31
H134, H234, H334	H12, H22, H32
H144, H244, H344	H13, H23, H33
H154, H254, H354	H14, H24, H34
H164, H264, H364	H15, H25, H35
H174, H274, H374	H16, H26, H36
H184, H284, H384	H17, H27, H37
H194, H294, H394	H18, H28, H38
H195, H295, H395	H19, H29, H39

System for Heat Treatable Alloys

The temper designation system for wrought and cast products that are strengthened by heat treatment employs the W and T designations described in the section on basic temper designations. The W designation denotes an unstable temper, whereas the T designation denotes a stable temper other than F, O or H. The T is followed by a number from 1 to 10; each number indicates a specific sequence of basic treatments, as follows:

T1 **Cooled from an elevated temperature shaping process and naturally aged to a substantially stable condition.** Applies to products that are not cold worked after an elevated temperature shaping process such as casting or extrusion, and for which mechanical properties have been stabilized by room-temperature aging. If the products are flattened or straightened after cooling from the shaping process, the effects of the cold work imparted by flattening or straightening are not accounted for in specified property limits.

T2 **Cooled from an elevated temperature shaping process, cold worked, and naturally aged to a substantially stable condition.** Applies to products that are cold worked specifically to improve strength after cooling from a hot working process such as rolling or extrusion, and for which mechanical properties have been stabilized by room-temperature aging. The effects of cold work, including any cold work imparted by flattening or straightening, are accounted for in specified property limits.

T3 **Solution heat treated, cold worked, and naturally aged to a substantially stable condition.** Applies to products that are cold worked specifically to improve strength after solution heat treatment, and for which mechanical properties have been stabilized by room-temperature aging. The effects of cold work, including any cold work imparted by flattening or straightening, are accounted for in specified property limits.

T4 **Solution heat treated and naturally aged to a substantially stable condition.** Applies to products that are not cold worked after solution heat treatment, and for which mechanical properties have been stabilized by room-temperature aging. If the products are flattened or straightened, the effects of the cold work

imparted by flattening or straightening are not accounted for in specified property limits.

T5 **Cooled from an elevated temperature shaping process and artificially aged.** Applies to products that are not cold worked after an elevated temperature shaping process such as casting or extrusion, and for which mechanical properties or dimensional stability, or both, have been substantially improved by precipitation heat treatment. If the products are flattened or straightened after cooling from the shaping process, the effects of the cold work imparted by flattening or straightening are not accounted for in specified property limits.

T6 **Solution heat treated and artificially aged.** Applies to products that are not cold worked after solution heat treatment, and for which mechanical properties or dimensional stability, or both, have been substantially improved by precipitation heat treatment. If the products are flattened or straightened, the effects of the cold work imparted by flattening or straightening are not accounted for in specified property limits.

T7 **Solution heat treated and stabilized.** Applies to products that have been precipitation heat treated to the extent that they are overaged. Stabilization heat treatment carries the mechanical properties beyond the point of maximum strength to provide some special characteristic, such as enhanced resistance to stress corrosion cracking or exfoliation corrosion.

T8 **Solution heat treated, cold worked, and artificially aged.** Applies to products that are cold worked specifically to improve strength after solution heat treatment, and for which mechanical properties or dimensional stability, or both, have been substantially improved by precipitation heat treatment. The effects of cold work, including any cold work imparted by flattening or straightening, are accounted for in specified property limits.

T9 **Solution heat treated, artificially aged, and cold worked.** Applies to products that are cold worked specifically to improve strength after they have been precipitation heat treated.

T10 **Cooled from an elevated temperature shaping process, cold worked, and artificially aged.** Applies to products that are cold worked specifically to improve strength after cooling from a hot working process such as rolling or extrusion, and for which mechanical properties or dimensional stability, or both, have been substantially improved by precipitation heat treatment. The effects of cold work, including any cold work imparted by flattening or straightening, are accounted for in specified property limits.

When it is desirable to identify a variation of one of the ten major T tempers described above, additional digits, the first of which cannot be zero, may be added to the designation.

The following specific sets of additional digits have been assigned to stress-relieved wrought products:

Tx51 **Stress relieved by stretching.** Applies to the following products when stretched to the indicated amounts after solution heat treatment or after cooling from an elevated-temperature shaping process:

Product form	Permanent set, %
Plate	1½ to 3
Rod, bar, shapes, extruded tube	1 to 3
Drawn tube	½ to 3

Applies directly to plate and to rolled or cold finished rod and bar. These products receive no further straightening after stretching. Applies to extruded rod, bar, shapes and tubing, and to drawn tubing, when designated as follows:

Tx510 Products that receive no further straightening after stretching

Tx511 Products that may receive minor straightening after stretching to comply with standard tolerances

Tx52 Stress relieved by compressing. Applies to products that are stress relieved by compressing after solution heat treatment, or after cooling from a hot working process to produce a permanent set of 1 to 5%

Tx54 Stress relieved by combining stretching and compressing. Applies to die forgings that are stress relieved by restriking cold in the finish die. (These same digits — and 51, 52, and 54 — may be added to the designation W to indicate unstable solution heat treated and stress-relieved tempers.)

The following temper designations have been assigned to wrought products heat treated from the O or the F temper to demonstrate response to heat treatment:

T42 **Solution heat treated from the O or the F temper to demonstrate response to heat treatment, and naturally aged to a substantially stable condition**

T62 **Solution heat treated from the O or the F temper to demonstrate response to heat treatment, and artificially aged**

Temper designations T42 and T62 also may be applied to wrought products heat treated from any temper by the user when such heat treatment results in the mechanical properties applicable to these tempers.

System for Annealed Products

A digit following the "O", when used, indicates a product in annealed condition having special characteristics. For example, for heat treatable alloys, O1 indicates a product that has been heat treated at approximately the same time and temperature required for solution heat treatment and then air cooled to room temperature; this designation applies to products that are to be machined prior to solution heat treatment by the user.

Aluminum Mill and Engineered Wrought Products

COMMERCIAL WROUGHT ALUMINUM PRODUCTS are divided into five major categories based on production method as well as geometric configuration. These categories are (*a*) flat rolled products (sheet, plate and foil); (*b*) rod, bar and wire; (*c*) tubular products; (*d*) shapes; and (*e*) forgings. In the aluminum industry, rod, bar, wire, tubular products and shapes are termed "mill" products, as they are in the steel industry, even though they often are produced by extrusion rather than by rolling. Aluminum forgings usually are not classified as "mill products" but as "engineered products." Engineered products are those designed for one specific application in contrast to "off-the-shelf" products such as standard sizes of sheet, plate, rod, bar, wire, tube, pipe and standard structural shapes. These standard items may be available from distributors. Engineered wrought products include special extruded shapes, die forgings, hand forgings, impacts, and special sizes of standard products.

GENERAL CHARACTERISTICS OF WROUGHT PRODUCTS

Wrought products are those that have been shaped by plastic deformation. This deformation, which is done by hot and cold working processes such as rolling, extruding, forging and drawing, either singly or in combination, transforms the cast ingot into the desired product form. As the deformation proceeds, the metallurgical structure also changes from a cast structure to a fully wrought structure. In this process, grain size and shape may be radically changed, the final configuration depending on the entire thermomechanical history (including any final annealing stages or heat treatments). During deformation, the second-phase microconstituents present in irregular forms in the ingot are fragmented into more equiaxed particles which tend to align in the direction of greatest extension. The grains are usually also elongated in this direction and thinned or flattened in flat rolled products, in thin extruded products and in the flash-plane areas of die forgings. Thus, the wrought metallurgical structure has directionality, the degree of which depends on the directionality of the deformation imposed during shaping of the product.

These changes in metallurgical structure are accompanied by changes in properties, particularly mechanical properties, which are generally higher in the wrought products than in the cast ingot from which they originate. The wrought products generally exhibit some pronounced directionality of mechanical properties (i.e., anisotropy), which is negligible in the ingot from which they are produced. The directionality in sheet is not pronounced with respect to tensile properties but can be quite significant with respect to performance in deep drawing or cupping operations. Nonuniformity in different directions in the plane of the sheet causes formation of protuberances called "ears" in circular cups. These are primarily the result of crystallographic texture in the sheet, a nonrandom or preferred orientation of the grains. For deep drawing and cupping operations such as those employed in making beverage and food cans, special "non-earing quality" sheet is supplied.

Thicker products that can be stressed or tested in three orthogonal directions generally exhibit differences in tensile and compressive properties as well as in resistance to fatigue stresses and stress-corrosion cracking in the three directions. In respect to these characteristics, the longitudinal direction (that in which the product was lengthened or extended most in working) generally is superior. In the long-transverse direction (which may be either an extension or compression direction, but with less extension or less compression than either of the other two directions), the properties and resistance are intermediate to those of the other directions. In the short-transverse direction (direction of greatest compression in working), the properties and resistance are generally lower than in the other directions. In most cases the longitudinal, long-transverse and short-transverse directions correspond to the greatest, intermediate and smallest dimensions of product having a rectangular cross section. For products of axisymmetric cross section (round, square, hexagonal, etc.) there is no long- and short-transverse distinction, any direction in the cross-sectional plane being regarded as transverse.

Directionality is considerably affected by whether the grain structure is recrystallized or unrecrystallized, the latter condition generally being more highly directional. In the case of heat treated extruded shapes there is usually a very thin surface or peripheral layer of recrystallized grain structure surrounding an unrecrystallized core. Since the recrystallized grain material is somewhat less resistant to fatigue than the unrecrystallized structure, its thickness should be minimized. For critical applications, a considerable percentage of the length from the rear of the extrusion is discarded to accomplish this, because the recrystallized layer thickness increases from front to rear of the extrusion. In the most critical applications, complete removal of the recrystallized material by machining or chem-milling may be required.

Extrusions, which generally have unrecrystallized structures, exhibit somewhat higher directionality of tensile properties than rolled products of equivalent cross section that also are unrecrystallized. This difference, primarily evidenced by higher longitudinal strength values in the extruded product, which may be as much as 10% higher, is attributed to more pronounced preferred grain orientation generated by the deformation of extrusion.

PRODUCT TYPES

Flat rolled products include sheet, plate and foil. They are manufactured by either hot rolling or hot and cold rolling, are rectangular in cross section and form, and have uniform thickness. This category comprises about two-thirds of the total aluminum mill products purchased by U.S. fabricators.

"Plate" refers to a product having a thickness greater than 6.3 mm (0.250 in.). Plate up to 200 mm (8 in.) thick is available in some alloys. Extra-large plates—e.g., 22 mm ($^7/_8$ in.) thick by 2.25 m (89 in.) wide by 32 m (105 ft) long—are supplied for construction of the wings of widebody aircraft. Plate usually has either sheared or sawed edges and can be cut into circles, rectangles or odd-shape blanks. Plate of certain alloys—notably the high-strength 2*xxx* and 7*xxx* series alloys—also is available in clad form, which comprises an aluminum alloy core having on one or both sides a metallurgically bonded aluminum or aluminum alloy coating that is anodic to the core, thus electrolytically protecting the core against corrosion. Most often, the coating consists of a high-purity aluminum, a low Mg-Si alloy, or an alloy containing 1% Zn or Zn plus Mg. Usually, coating thickness (one side) is from $1^1/_2$ to $2^1/_2$% of total thickness. The most commonly used plate alloys are 2024, 2124, 2219, 7050, 7178 and 7475 for aircraft structures; 5083, 5086 and 5456 for marine, cryogenic and pressure-vessel applications; and 1100, 3003, 5052 and 6061 for general applications.

When a flat rolled product is over 0.15 through 6.3 mm (0.006 through 0.249 in.) in thickness, it is classified as "sheet." Sheet edges can be sheared, slit or sawed. Sheet is supplied in flat form, in coils, or in pieces cut to length from coils. Current facilities permit production of a limited amount of extra-large sheet, for example, up to 5 m (200 in.) wide by 25 m (1000 in.) long. Aluminum sheet is available in several surface finishes that range from "mill finishes," which have uncontrolled surface appearance that may vary from sheet to sheet, to bright finishes on one or two sides, to "aircraft skin quality." It may also be supplied embossed, patterned, painted or otherwise surface treated and with combinations of such treatments. Special products include corrugated, V-beam and ribbed roofing and siding, duct sheet, fin stock, recording circles and computer memory disks.

Among standard products are the alclad composites, which consist of heat treated 2*xxx* or 7*xxx* alloys clad on either or both sides with an appropriate anodic alloy or with aluminum. For sheet thicknesses, cladding thickness may range from $1^1/_2$ to as much as 10% of sheet thickness, the greater percentages applying to thinner products. A series of products termed "brazing sheet" is available. These products are also composites clad one or both sides with brazing alloy. For architectural uses, clad non-heat-treatable alloys may be supplied. These provide a variety of special finishing characteristics, integral color finishing capability, greater uniformity in appearance and improved corrosion resistance.

With a few exceptions, most alloys in the 1*xxx*, 2*xxx*, 3*xxx*, 5*xxx* and 7*xxx* series are available in sheet form. Along with alloy 6061, they cover a wide range of applications from builders' hardware to transportation equipment and from appliances to aircraft structures. Alloys 2036, 6009 and 6010 are widely used for automobile body panels and hood and deck stampings.

Foil is a product up through 0.15 mm (0.006 in.) thick. Most foil is supplied in coils, although it is also available in rectangular form (sheets).

One of the largest end uses of foil is household wrap. There is a wider variety of surface finishes for foil than for sheet. Foil often is treated chemically or mechanically to meet the needs of specific applications. Common foil alloys are 1100, 1145, 1235, 3003, 5052 and 5056. Higher-strength foil of alloy 2024, 5052 or 5056 is used to produce the honeycomb cores used in bonded honeycomb sandwich panels.

Bar, rod and wire are defined as solid products that are extremely long in relation to their cross section. They differ from each other only in cross-sectional shape and in thickness or diameter. When the cross section is round or nearly round and is over 10 mm (3/8 in.) in diameter, it is called "rod." It is called "bar" when the cross section is square, rectangular or in the shape of a regular polygon and when at least one perpendicular distance between parallel faces (thickness) is over 10 mm. "Wire" refers to a product, regardless of its cross-sectional shape, whose diameter or greatest perpendicular distance between parallel faces is 10 mm or less.

Rod and bar can be produced by either hot rolling or hot extruding and brought to final dimensions with or without additional cold working. Wire usually is produced and sized by drawing through one or more dies, although roll flattening also is used. Alclad rod or wire for additional corrosion resistance is available only in certain alloys. Many aluminum alloys are available as bar, rod and wire; among these alloys, 2011 and 6262 are specially designed for screw-machine products, and 2117 and 6053 for rivets and fittings. Alloy 2024-T4 is a standard material for bolts and screws. Alloys 1350, 6101 and 6201 are extensively used as electrical conductors. Alloy 5056 is used for zippers, and alclad 5056 for insect-screen wire.

Tubular products include tube and pipe. They are hollow wrought products that are long in relation to their cross section and have uniform wall thickness except as affected by corner radii. Tube is round, elliptical, square, rectangular or regular polygonal in cross section. When round tubular products are in standardized combinations of outside diameter and wall thickness, commonly designated by "nominal pipe sizes" and "ANSI schedule numbers," they are classified as pipe.

Tube and pipe may be produced from a hollow extrusion ingot, by piercing a solid extrusion ingot or by extruding through a porthole die or bridge die. They also may be made by forming and welding sheet. Tube may be brought to final dimensions by drawing through dies. Tube (both extruded and drawn) for general applications is available in such alloys as 1100, 2014, 2024, 3003, 5050, 5086, 6061, 6063 and 7075. For heat-exchanger tube, alloys 1060, 3003, alclad 3003, 5052, 5454 and 6061 are most widely used. Clad tube is available only in certain alloys and is clad only on one side (either inside or outside). Pipe is available only in alloy 3003, 6061 and 6063.

Shapes are products that are long in relation to their cross-sectional dimensions and that have cross-sectional shapes other than those of sheet, plate, rod, bar, wire or tube. Most shapes are produced by extruding or by extruding plus cold finishing; shapes are now rarely produced by rolling because of economic advantages of the extrusion process. Shapes may be solid, hollow (with one or more voids) or semihollow. The 6xxx series (Al-Mg-Si) alloys, because of their easy extrudability, are the most popular alloys for pro-

ducing shapes. Alloys of the 2xxx and 7xxx series are used in applications requiring higher strength.

Standard structural shapes such as I-beams, channels and angles produced in alloy 6061 are made in different and fewer configurations than similar shapes made of steel; the patterns especially designed for aluminum offer better section properties and greater structural stability than those designed for steel, as a result of more efficient metal usage. The dimensions, weights and properties of the alloy 6061 standard structural shapes, along with other information needed by structural engineers and designers, are contained in the Aluminum Construction Manual, published by the Aluminum Association, Inc.

Most aluminum alloys can be obtained as precision extrusions with good as-extruded surfaces; major dimensions usually do not need to be machined, because tolerances of the as-extruded product often permit manufacturers to complete the part by simple cutoff, drilling or other minor operations.

In many instances, long aircraft structural elements incorporate large attachment fittings at one end. Such elements often are more economical to machine from stepped aluminum extrusions, with two or more cross sections in one piece, rather than from extrusions of uniform cross section that are large enough for the attachment fittings.

Aluminum shapes are produced in a great variety of cross-sectional designs that place the metal where it is needed to meet functional and appearance requirements. Full utilization of this capability of the extrusion process depends on the ingenuity of designers in creating new and useful configurations. However, the alloy extruded and the cross-sectional design greatly influence tooling cost, production rate, surface finish and production cost. Therefore, the extruder should be consulted during the design process to ensure producibility, dimensional control and finish capabilities that are required for the application.

Producibility is limited by metal-flow characteristics, and is a function of alloy composition, extrusion temperature, press size and shape complexity. Shapes are classified with respect to producibility into solid, hollow and semihollow types and are further classified by rules based on the dimensions of the features. The difficulty of extrusion can be estimated from the dimensions, taking into account the alloy to be extruded.

The over-all size of the shape affects ease of extrusion and dimensional tolerances. As the circumscribing circle size (smallest diameter that completely encloses the shape; see Fig. 1) increases, extrusion becomes more difficult. Metal flow is most rapid at the center of the die face. As circle size increases, differences in flow rate from the center to the outside of the shape increase, and die design and construction to counteract this effect are more difficult.

Complexity and production difficulty also increase with increasing "shape factor," which is the ratio of the perimeter of a shape to its weight per unit length. Increasing thickness aids extrusion, and shapes having uniform thickness are most easily extruded. Although weight and metal cost decrease with decreasing thickness, the increasing extrusion cost may offset the savings in metal cost. Limits on minimum practical thickness, which depend on circle size, classification and alloy, are given in Table 1.

Alloy selection for extruded shapes has an important effect on producibility and cost as well as on minimum thickness. The extrusion speed possible for a given shape is strongly affected by the composition being extruded and may vary by as much as a factor of 20 (see Table 2).

More detailed information on design of extruded shapes is contained in the ASM Metals Handbook, 9th Ed., Vol 2, pages 53 to 58. Publications of the Aluminum Association, Inc., and the Aluminum Extruders Council — "A Guide to Aluminum Extrusions," 1st Ed., May 1979, and "Aluminum Extrusion Application Guide," 2nd Ed., Aug 1980 — will be helpful to the designer.

Forgings. The term "hand forgings" is applied to most of the open-die forgings produced on flat or contoured dies generally in hydraulic presses with capacities up to 50 000 tons. The most usual are of rectangular or cylindrical cross section and may be produced in economical lengths (multiple length) and later cut into shorter pieces. The general category includes disk-shape parts sometimes referred to as "biscuits" as well as more complicated pieces which may vary in cross section throughout the length or be bent, curved or contoured. These forgings fill a frequent need in which the number of pieces required does not justify the time and expense of impression dies. Mandrel-forged rings are another type of open-die-forged product.

Forgings of all types, produced on hammers, mechanical presses or hydraulic presses, range in size from 45 g (0.1 lb) to 1360 kg (3000 lb) or over. Those weighing up to 450 kg (1000 lb) are produced regularly, those weighing between 450 and 900 kg (1000 and 2000 lb) are less common, and pieces weighing more than 900 kg are special items.

Most aluminum forgings are produced in closed dies and can vary widely in detail and closeness of approach to the final dimensions desired. The ultimate in shaping parts by forging is represented by "precision forgings," which are essentially net-shape parts requiring little or no machining. The advantage of closed-die forgings is that the metallurgical-structure alignment follows the part contours. This is sometimes called the "grain-flow pattern." This alignment is highly

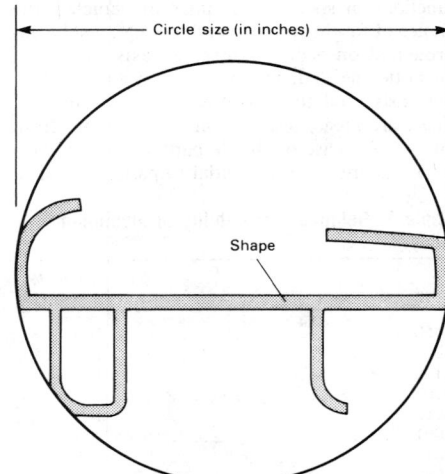

Fig. 1. Illustration of "circumscribing circle" method of characterizing the size of an extruded shape

Table 1. Standard manufacturing limits (in inches) for aluminum extrusions

Diameter of circumscribing circle, in.	Minimum wall thickness, in.				
	1060, 1100, 3003	6063	6061	2014, 5086, 5454	2024, 2219, 5083, 7001, 7075, 7079, 7178
Solid and semihollow shapes, rod and bar					
0.5 to 2	0.040	0.040	0.040	0.040	0.040
2 to 3	0.045	0.045	0.045	0.050	0.050
3 to 4	0.050	0.050	0.050	0.050	0.062
4 to 5	0.062	0.062	0.062	0.062	0.078
5 to 6	0.062	0.062	0.062	0.078	0.094
6 to 7	0.078	0.078	0.078	0.094	0.109
7 to 8	0.094	0.094	0.094	0.109	0.125
8 to 10	0.109	0.109	0.109	0.125	0.156
10 to 11	0.125	0.125	0.125	0.125	0.156
11 to 12	0.156	0.156	0.156	0.156	0.156
12 to 17	0.188	0.188	0.188	0.188	0.188
17 to 20	0.188	0.188	0.188	0.188	0.250
20 to 24	0.188	0.188	0.188	0.250	0.500
Class 1 hollow shapes(a)					
1.25 to 3	0.062	0.050	0.062	...	...
3 to 4	0.094	0.050	0.062	...	...
4 to 5	0.109	0.062	0.062	0.156	0.250
5 to 6	0.125	0.062	0.078	0.188	0.281
6 to 7	0.156	0.078	0.094	0.219	0.312
7 to 8	0.188	0.094	0.125	0.250	0.375
8 to 9	0.219	0.125	0.156	0.281	0.438
9 to 10	0.250	0.156	0.188	0.312	0.500
10 to 12.75	0.312	0.188	0.219	0.375	0.500
12.75 to 14	0.375	0.219	0.250	0.438	0.500
14 to 16	0.438	0.250	0.375	0.438	0.500
16 to 20.25	0.500	0.375	0.438	0.500	0.625
Class 2 and 3 hollow shapes(b)					
0.5 to 1	0.062	0.050	0.062	...	...
1 to 2	0.062	0.055	0.062	...	...
2 to 3	0.078	0.062	0.078	...	...
3 to 4	0.094	0.078	0.094	...	...
4 to 5	0.109	0.094	0.109	...	...
5 to 6	0.125	0.109	0.125	...	...
6 to 7	0.156	0.125	0.156	...	...
7 to 8	0.188	0.156	0.188	...	...
8 to 10	0.250	0.188	0.250	...	...

(a) Minimum inside diameter is one-half the circumscribing diameter, but never under 1 in. for alloys in first three columns or under 2 in. for alloys in last two columns. (b) Minimum hole size for all alloys is 0.110 sq in. in area or 0.375 in. in diam.

favorable for static strength and resistance to dynamic stresses (fatigue conditions). In working operations preliminary to final shaping, the metal may be upset and drawn in ways that further improve final mechanical properties.

Particularly for die forgings, which represent engineered products designed to perform specific functions in specific machines or vehicles, the choice of forging process and tooling must be approached on a cost-effective basis involving quantities needed, tooling, forging and machining costs, and the net effects of these factors. There are great variations in this problem, from the extreme case of simple parts for which tooling costs are inconsequential to parts ordered in

Table 2. Relative extrudability of aluminum alloys

Alloy	Extrudability, % of rate for alloy 6063
1350	160
1060	135
1100	135
3003	120
6063	100
6061	60
2011	35
5086	25
2014	20
5083	20
2024	15
7075	9
7178	8

large quantities which usually justify the tooling cost of dies to produce a part with conventional machining allowances or even the higher costs of precision forging (see Fig. 2).

The alloy used affects cost, reflecting the relative alloy forgeability (see Fig. 3 and 4). This factor, combined with those discussed in the pre-

ceding paragraph, may account for a price ratio of up to 10:1 for parts of the same weight. Forging dies are more expensive than extrusion dies by a factor varying from 4:1 to 3:1, for small and large sections. Factors in formulating decisions as to whether parts should be shaped by machining from bar or plate stock or be produced as hand or die forgings are considered in the next section of this article (Product Economics, Selection and Design). In many cases, a die forging may serve to replace an assembly of parts produced from mill products and stampings or shapes.

The alloys most prominent in forgings are of the 2*xxx*, 6*xxx* and 7*xxx* types; 1*xxx*, 3*xxx*, 4*xxx* and 5*xxx* alloys are producible but are used with less frequency. Limitations on controlled strain-hardening capability and on use of strain to produce T8-type tempers influence the use of some of these alloys. The largest forging tonnages in markets other than aerospace are accounted for by automobile and truck wheels of alloy 6061, whereas the aerospace industry uses forgings principally of 2*xxx* and 7*xxx* alloys (2014, 2219, 7049, 7050, 7075 and 7175).

Mandrel-forged or ring-rolled rings are produced in some forging plants. More information on a variety of representative die forgings, including the alloys and tempers used, the types of forging equipment used, and comments on selection factors, is given in the Aluminum Association publication "Aluminum Forgings Application Guide," 1st Ed., Nov. 1975.

Impacts are formed in a confining die from a lubricated slug, usually cold, by a single-stroke application of force through a metal punch, causing the metal to flow around the punch or die. The process lends itself to high production rates, with precision parts being produced to exacting quality standards. Impacts involve a combination of cold extrusion and cold forging and, as such, combine most of the advantages of both the forging and extrusion processes.

There are three basic types of impacting, all of which are used on aluminum. Reverse impacting is used to make shells with forged bases and extruded sidewalls. The slug is placed in a die cavity and struck by a punch, which forces the metal to flow back (upward) around the punch, and then through the opening between the punch

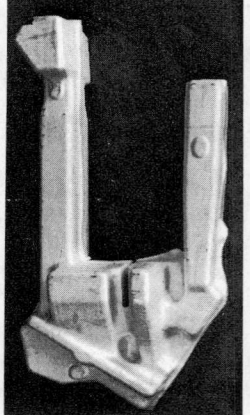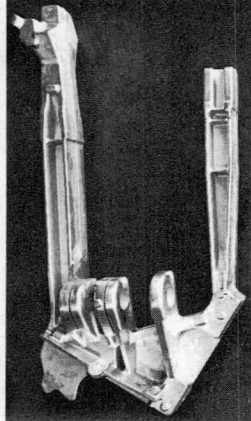

(Left) Special or contoured hand forging, weighing 635 kg (1400 lb), for fast, low-cost prototype development. (Center) Blocker forging, weighing 397 kg (875 lb), produced at minimum die expense. (Right) Precision forging, 1.65 m (65 in.) long and weighing 139 kg (306 lb), for quantity production with minimum finish machining. SOURCE: *Aluminum*, American Society for Metals, 1967, Vol 2, p 163 (Fig. 2), or Vol 3, p 140 (Fig. 4).

Fig. 2. Three parts for the same end use, produced by three different forging methods

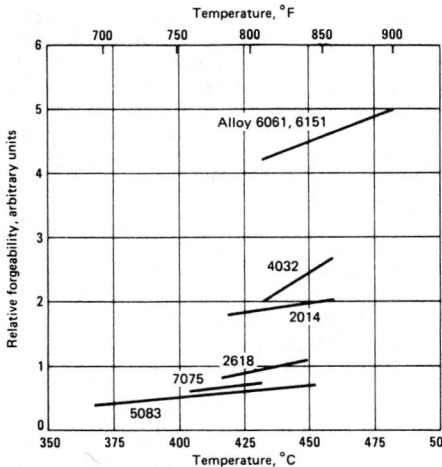

Forgeability increases as the arbitrary unit increases.

Fig. 3. Forgeability vs forging temperature for seven aluminum alloys

and die, to form a simple shell. Forward impacting somewhat resembles conventional extrusion in that the metal is forced through an orifice in the die by the action of a punch, causing the metal to flow in the direction of punch travel. The punch fits the walls of the die so closely that no metal escapes backwards. Forward impacting with a flat-face punch is used to form round, nonround, straight and ribbed rods, and forward impacting with a stop-face punch is used to form thin-wall tubes with one or both ends open, and with parallel or tapered sidewalls. If the punch is smaller than the die and the die contains an orifice, reverse and forward impacting can be combined to produce a combination impact.

A major consideration in designing aluminum impacts is selection of the appropriate alloy. Alloys 1100, 2014, 3003, 6061, 6351 and 7075 are most often utilized in aluminum impacts. These alloys offer a range of mechanical properties that fit most applications. Generally, the stronger the alloy impacted, the shorter the tool life and the higher the production costs. Although each part must be considered individually, the stronger alloys generally require greater minimum wall thicknesses (see Table 3).

Alloy 1100, which has excellent corrosion resistance in rural, industrial and marine atmospheres, is commonly impacted to form containers for liquid and semiliquid materials such as food preserves and products sprayed by aerosols. Alloy 3003 is used for many of the same applications as alloy 1100, but is selected when higher strength than that of 1100 is required. Alloy 6061, which is heat treatable and has excellent corrosion resistance, is widely used in the manufacture of parts for automotive, aircraft and marine applications, especially where welding is involved or high strength is required. Alloy 6351 is a medium-to-high-strength heat treatable alloy with good corrosion resistance. Alloy 2014 is a heat treatable alloy used for general applications where high tensile and yield strengths, combined with good ductility and good fatigue resistance, are essential. It is widely used in structural applications and in aircraft, automobile and ordnance parts. Alloy 7075 has the highest strength and hardness of these alloys. This heat treatable alloy is used for many of the same applications as those of alloy 2014, but is selected where highest stresses are expected or for maximum weight savings.

Impacts usually have properties in the longitudinal direction equal to those specified for other product forms of similar composition. Figure 5 presents properties of a bomb ejector foot. Here, the properties of both the impact made at room temperature and the hot impact are equivalent and exceed the specified die forging properties for the same alloy.

PRODUCT ECONOMICS, SELECTION AND DESIGN

In the "cost-effective" approach to the problem of material selection, the only valid basis for choosing a particular material is that it will perform all required functions at the lowest over-all cost. The material chosen may be the most cost-effective (*a*) because it is lowest in first cost and provides service and durability at least equal to those offered by any alternative material; (*b*) because it is most economical in the long run due to lowest operating or maintenance costs; or (*c*) because it has special characteristics not matched by any alternative material. These considerations at times are "warped" by artificial factors arising from such sources as legislation or the "energy crisis." Also, they may at times be greatly influenced, or even outweighed, by factors of availability or delivery time.

Table 3. Practical minimum wall thicknesses for reverse aluminum impacts

Nominal outside diameter		1100		6061		2014		7075	
mm	in.	mm	in.	mm	in.	mm	in.	mm	in.
25	1	0.25	0.010	0.40	0.015	0.90	0.035	1.00	0.040
50	2	0.50	0.020	0.75	0.030	1.80	0.070	2.05	0.080
75	3	0.75	0.030	1.15	0.045	2.65	0.105	3.05	0.120
100	4	1.00	0.040	1.50	0.060	3.55	0.140	4.05	0.160
125	5	1.25	0.050	1.90	0.075	4.45	0.175	5.10	0.200
150	6	1.50	0.060	2.30	0.090	5.35	0.210	6.10	0.240
180	7	1.90	0.075	2.80	0.110	6.20	0.245	7.10	0.280
205	8	2.55	0.100	3.30	0.130	7.10	0.280	8.15	0.320
230	9	2.80	0.110	3.70	0.145	8.00	0.315	9.15	0.360
255	10	3.15	0.125	4.20	0.165	8.90	0.350	10.15	0.400

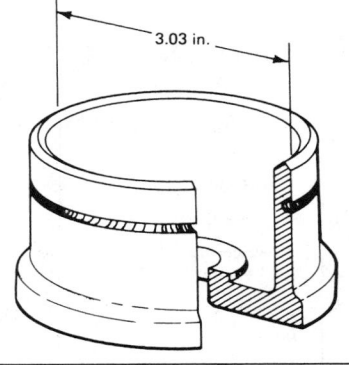

Method	Tensile strength MPa	ksi	Yield strength MPa	ksi	Elongation, %
Die forging(a)	517	75.0	448	65.0	10.0
Impacting at room temperature	563	81.7	503	73.0	10.3
Impacting at 230 °C (450 °F)	561	81.3	492	71.3	10.0
(a) Specified properties for 7075-T6 in the required section size.					

Impacts were stress relieved at 340 °C (645 °F) after impacting, then solution treated and aged to T6 temper. Each value in the tabulation above is the average of five test specimens taken from the cylinder wall, parallel to the axis of impacting.

Fig. 5. Comparison of mechanical properties for bomb ejector feet made by forging and by impacting from alloy 7075

The choice between aluminum and some other material on the cost-effective basis is sometimes simple and at other times quite complex; in these cases the choice may shift from one time to another as relative costs change. Machining, joining or finishing capabilities, as well as physical properties (predominantly density, conductivity or reflectivity), are foremost considerations in

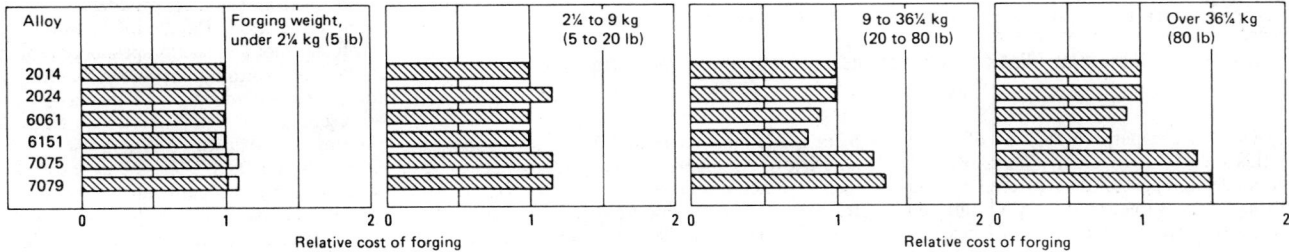

Although included in this comparison, alloy 2024 is not a standard forging alloy.

Fig. 4. Comparative die forging costs of different aluminum alloys as a function of forging weight

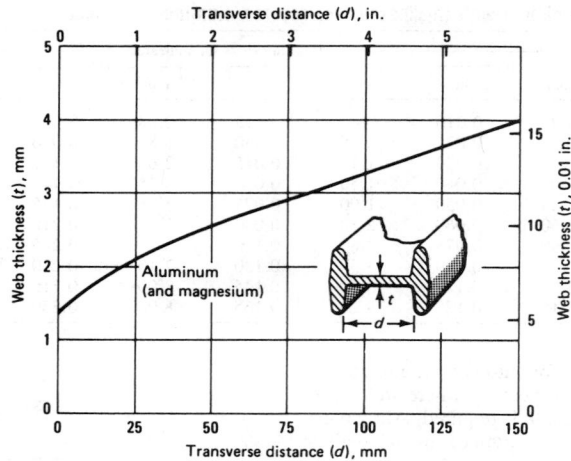

Fig. 6. Limiting web proportions for aluminum die forgings

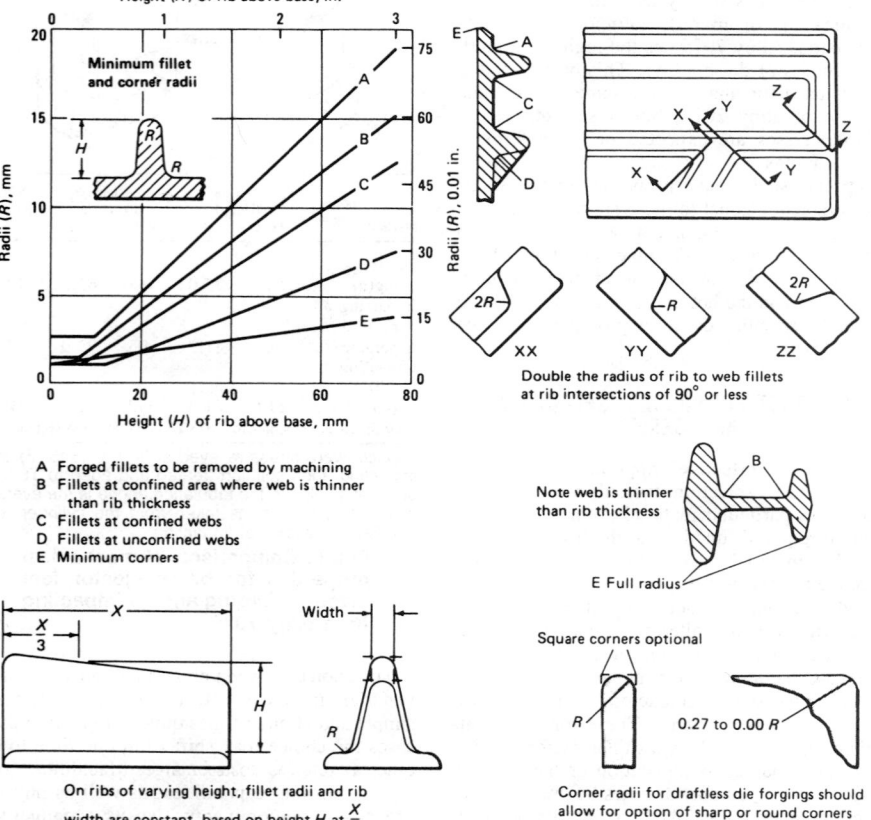

A Forged fillets to be removed by machining
B Fillets at confined area where web is thinner
    than rib thickness
C Fillets at confined webs
D Fillets at unconfined webs
E Minimum corners

Double the radius of rib to web fillets at rib intersections of 90° or less

Note web is thinner than rib thickness

E Full radius

On ribs of varying height, fillet radii and rib width are constant, based on height H at $\frac{X}{3}$

Square corners optional

0.27 to 0.00 R

Corner radii for draftless die forgings should allow for option of sharp or round corners

Fillet radii occur at internal junctures of intersecting surfaces; corner radii, at external junctures. Large radii improve the forgeability of a part and should be used when practical. Radii less than 3.3 mm (0.13 in.) are special. Economy in die sinking will result from the use of one standard fillet radius as well as one standard corner radius on a part whenever possible.

Fig. 7. Minimum fillet and corner radii for aluminum die forgings

many cases. The competitive position of aluminum in this race is often greatly augmented by the myriad design possibilities offered by aluminum shapes and forged parts. (Aluminum engineered castings also offer many of these advantages.)

Product selection is based primarily on shape, dimensional and mechanical requirements. The piece needed may be required to cover or enclose an area, to fill a certain space, to connect or attach to other pieces or to conduct or contain a fluid or gas, or may be limited by weight or by other factors. There is seldom any question concerning the best choice in the case of products like foil, wire, tube or large-area applications for sheet. In other cases, an intelligent choice from among the various products is a complex engineering problem involving many factors in addition to first cost of the product itself. Shapes or die forgings frequently offer the advantage of a single piece having such features as ribs for stiffening, fins for heat dissipation, and bosses or pads for attachment, replacing several pieces which must be cut, formed and joined. Mechanical-property considerations may be important in some cases—i.e., such matters as the availability or unavailability of an alloy and temper in a particular product type, the higher longitudinal static strength of extruded versus rolled or forged products, or the matching of metallurgical-structure alignment (or grain-flow pattern) with surface contours that is characteristic of die forgings.

For the most stringent structural applications, the greatest hazards are those posed by dynamic stresses and by combinations of static stresses from assembly interferences or misfits with dynamic applied stresses and corrosive environments. To avoid premature failure from fatigue cracking or stress-corrosion cracking, good mechanical design to minimize such stresses, stress concentrations and exposure to and entrapment of corrosive media is of paramount importance. The type of product as well as the alloy and temper selected also play a strong role in the design process.

Product Dimensions and Dimensional Tolerances

The ranges of thickness in which foil, sheet and plate are available have been stated previously. By definition, foil is no greater than 0.15 mm (0.006 in.) thick and for some purposes is as thin as 0.0043 mm (0.00017 in.), sheet ranges from over 0.15 to 6.3 mm (over 0.006 to 0.249 in.) thick, and plate ranges from 6.3 to about 200 mm (0.250 to about 8 in.) thick. Corresponding width, length and coil-size limitations are matters for inquiry with producers.

Standard dimensional tolerances for all mill products are contained in the American National Standard document ANSI H35.2-1982, "Standard Dimensional Tolerances for Aluminum Mill Products," and are also shown in the Aluminum Association publication "Aluminum Standards and Data." These dimensional tolerances on thickness, width, length, diameter, squareness, flatness, straightness, lateral bow and twist, as applicable, vary with product type and are not tabulated in this handbook.

The limiting dimensions for extruded products, including rod and bar, and for other solid, hollow and semihollow shapes, with respect to maximum circumscribing circle size and minimum wall thickness, are alloy-dependent, as indicated in Table 1. Various sizes of structural shapes (angles, channels, I-beams, H-beams, Tees and Zees) produced from alloy 6061-T6 have been established as standard products by the Aluminum Association. Detailed dimensions, weights per lineal foot and section properties (moment of inertia, section modulus and radius of gyration) for these standard structural shapes—data which are useful for design—are tabulated in the Aluminum Association publication "Aluminum Standards and Data." These data, as well as those given in the same publication for standard pipe sizes, are not repeated in this handbook.

Limiting dimensions in design of die forgings include total plan area, which may be as great as about 0.3 m² (about 500 in.²) for forgings produced in mechanical presses, about 1.9 m² (about

3000 in.²) for hammer forgings, and about 3.2 m² (about 5000 in.²) for parts made in the largest hydraulic presses. Other limiting design features are web thickness, rib thickness and height, draft angles, and minimum radii for fillets and corners. Limits for conventional forgings, and interactions among some of these factors, are illustrated in Fig. 6 and 7.

Cylindrical and other axisymmetric shapes are easiest to produce as impacts, but nonsymmetrical shapes may also be produced. Longitudinal ribs can be incorporated in sidewalls, either inside or outside. Lugs, bosses, grooves, depressions and even ribs can be incorporated in the end or base configuration. Dimensional limitations apply to maximum diameter (dependent on press size), length-to-diameter ratio (a maximum of 18 to 1, with 12 to 1 more normal), and minimum wall thickness. All of these shape factors are alloy dependent. The dependence of minimum wall thickness on alloy and diameter is shown in Table 3.

Effective Design and Use of Wrought Product Capabilities

Good design for aluminum can be defined as making the most effective use of its capabilities. An outstanding example of this is the two-piece all-aluminum beverage container. Thin, coiled sheet of relatively high-strength alloys, rolled at extremely high speeds, is drawn and ironed to produce the body and formed to produce easy-open ends, both of these manufacturing operations producing the parts at extremely high rates. The favorable economics of lowest first cost, lower shipping costs (because of low weight), and recyclability combine with functional advantages to make these containers a viable and successful product in a highly competitive market. Such containers are pressure vessels in addition to being subject to considerable abuse from handling, and these factors require the best possible balance among alloy strength, fabricability and design. Tool design for production of these containers played a key role in making them successful.

Roofing and siding for highway freight trailers represent another highly effective, high-volume use of aluminum alloy sheet with advantages in both first and lifetime costs. These applications employ semi-monocoque (stressed skin) construction and moderately high-strength, non-heat-treatable alloys. Skins for aircraft wings, control surfaces, and fuselages (pressure vessels) are likewise effective in both cost and function. These applications employ high-strength, heat treatable alloys and forming operations ranging from mild to severe. In some cases the good formability of freshly quenched very long wing-skin panels is preserved by refrigeration during transcontinental shipment by rail in special cars from mill to aircraft factory. The fact that sheet accounts for such a high percentage of all aluminum mill products produced and shipped annually (55% in 1981) attests to the versatility and high efficiency of this form in a myriad of applications.

Plate is an obvious selection for many purposes. One example is large welded storage tanks. The Saturn V space vehicle was "an aluminum bird" with formed and welded plate members serving the dual function of containing the fuel and oxidizers (cryogenic liquids) as well as forming the structure of the vehicle. For such applications, as well as for aircraft wings, plate may be extensively machined or chem-milled to form integral stiffening ribs or waffle patterns. When

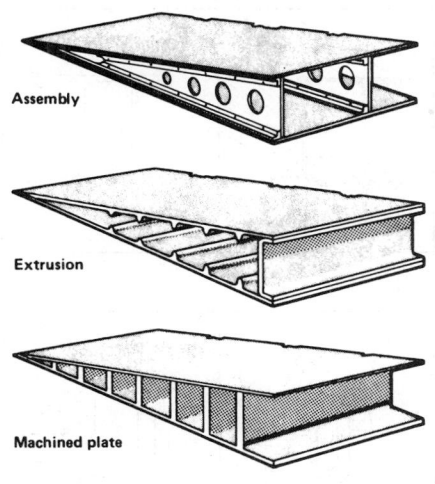

Assembly

Extrusion

Machined plate

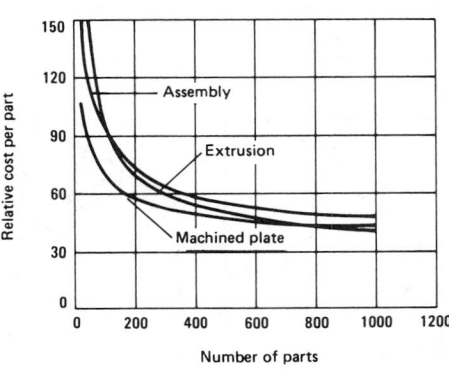

Fig. 8. Cost comparison for producing an alloy 2024-T4 part by three methods

only longitudinal stiffening ribs are required, an alternative product is wide extruded panels with integral ribs.

For some designs, machining may be the most economical production technique, because the setup time may account for a large part of the cost of machining. In Fig. 8, for instance, the machined plate was the least expensive method of producing the aluminum part until production reached 600 pieces. At that point, the quantity was sufficient to compensate for the cost of the extrusion die. The built-up design, with skin on both surfaces, was heavier and costlier.

Chemical milling (removal of metal by dissolution in an alkaline or acid solution) is routine for specialized operations on aluminum. For flat parts on which large areas having complex or wavy peripheral outlines are to be reduced only slightly in thickness, chemical milling is usually the most economical method. A typical curve used by one company for choosing the most economical method of removing metal from such parts is shown in Fig. 9.

For sheet metal parts that cannot be formed after machining, chemical milling is the only practical method by which metal can be removed to obtain a waffle-type grid with uniform skin thickness. Even then, allowance must often be made for some springback resulting from metal removal and the consequent redistribution of residual forming stresses.

Chemical milling can normally produce a stiffened skin to a thickness tolerance of ±0.13 mm

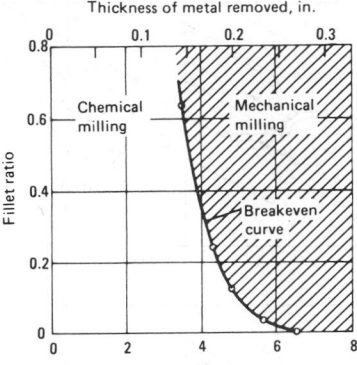

Fig. 9. A parameter for choice of method for milling aluminum

The parameter applies to the removal of metal from large areas having complex or wavy peripheral outlines. The curve shows that the chemical process should always be used for metal removal of 3 mm (0.125 in.) or less. The choice of method for thicknesses of 3 to 6 mm (0.125 to 0.250 in.) depends on fillet ratio or the amount of weight penalty. Metal thicknesses above 6 mm (0.250 in.) should be mechanically milled.

(±0.005 in.), with the cladding left intact on the unmilled surface. Mechanical milling of skins from sheet thicker than 3 mm (¹/₈ in.) requires a cleanup "skim" cut for flatness on the holddown surface, because of holddown limitations.

Extruded Shapes. Extrusions, with their great design versatility, good surface quality and precise dimensions, frequently do not have to be machined extensively; the configuration and dimensional precision of the as-extruded product often permits a manufacturer to complete the part by simple cutoff, drilling, broaching, or other minor machining operations. For any part that can be produced as an extrusion, the cost of the extrusion die is usually written off after a few parts have been produced.

Cost of machining may be the only selection consideration. This is illustrated in Fig. 10 by the cost figures for a fuel-tank attachment fitting. The

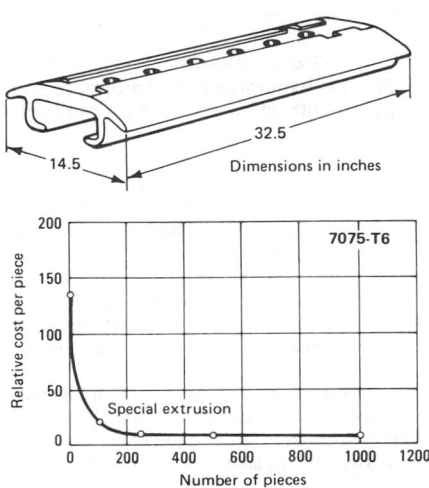

Dimensions in inches

A part completely machined from bar stock is rated 100.

Fig. 10. Cost of an extruded fuel-tank attachment fitting as a function of quantity

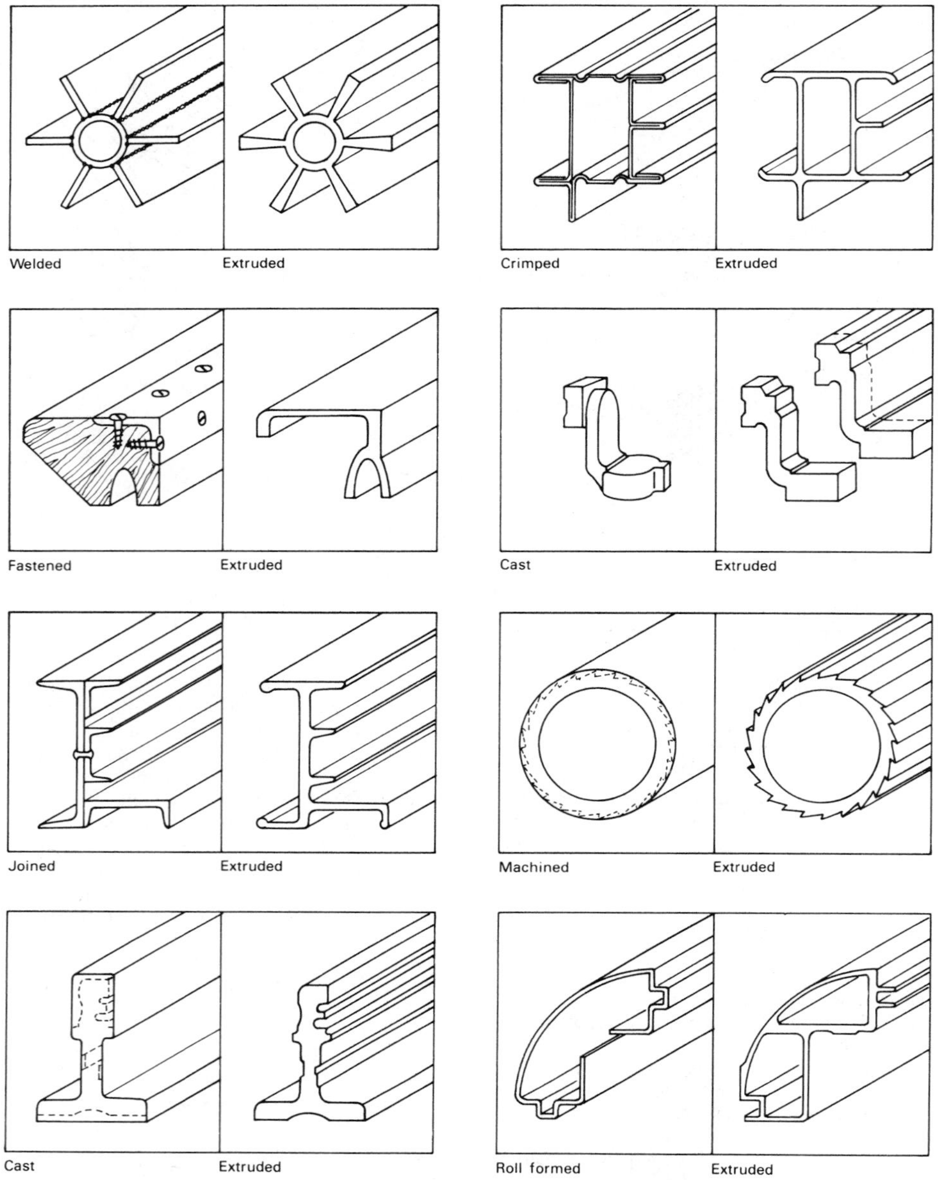

Welded	Extruded

Fastened	Extruded

Joined	Extruded

Cast	Extruded

Crimped	Extruded

Cast	Extruded

Machined	Extruded

Roll formed	Extruded

SOURCE: *A Guide to Aluminum Extrusion,* Aluminum Association, Inc., p 5 and 6.

Fig. 11. Examples of functional shapes, illustrating advantages of extrusion compared with other production methods

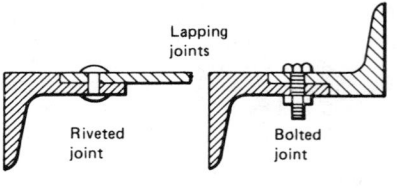

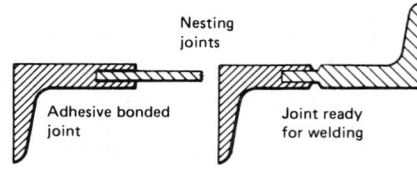

Special angle and sheet   Two special angles

Fig. 12. Four examples of interconnecting extrusions that fit together or fit other products, and four examples of joining methods

Free-moving hinge

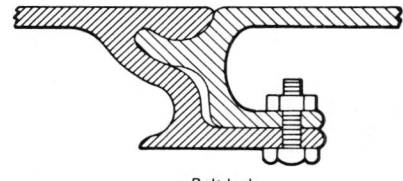

Bolt-lock

Fig. 13. Two examples of extrusions with nonpermanent interconnections

design of this part permitted the use of an extrusion, which required very little machining compared with the same part fabricated from solid bar. After about 100 pieces, the cost per piece decreased substantially.

Examples illustrating advantageous use of extrusion in place of other methods of producing parts of various configurations are shown in Fig. 11.

Interconnecting Shapes. It is becoming increasingly common to include an interconnecting feature in the design of an extruded shape to facilitate its assembly to a similar shape or to another product. It can be a simple step to provide a smooth lapping joint, or a tongue and groove for a nesting joint (see Fig. 12). Such connections can be secured by any of the common joining methods. Of special interest when the joint is to be arc welded is the fact that lapping and nesting

types of interconnections can be designed to provide edge preparation and/or integral backing for the weld (see sketch at bottom right in Fig. 12).

Interlocking joints can be designed to incorporate a free-moving hinge (see top sketch in Fig. 13) when one part is slid lengthwise into the mating portion of the next. Panel-type extrusions with hinge joints have found application in conveyor belts and roll-up doors.

A more common type of interlocking feature used in interconnecting extrusions is the nesting type that requires rotation of one part relative to the mating part for assembly (see bottom sketch in Fig. 13). Such joints can be held together by gravity or by mechanical devices. If a nonpermanent joint is desired, a bolt or other fastener can be used, as illustrated in the bottom sketch in Fig. 13.

When a permanent joint is desired, a snapping

or crimping feature can be added to interlocking extrusions (see Fig. 14). Crimping also can be used to make a permanent joint between an interlocking extrusion and sheet (Fig. 14). Extrusions also can be provided with longitudinal teeth or serrations, which will permanently grip smooth surfaces as well as surfaces provided with mating teeth or serrations; this is illustrated in the sketch at the bottom of Fig. 14.

Applications for interconnecting extrusions include doors; wall, ceiling and floor panels; pallets; aircraft landing mats; highway signs; window frames; and large cylinders.

Die Forgings. The diversity of geometrical possibilities offered by aluminum die forgings frequently makes them highly advantageous choices in that they can be produced without costly machining, joining or assembly operations. When compared with alternative methods of achieving the same functions, the integral product is often ahead in a value analysis. Tool (die) costs play a major role in the decision when the choice is between machining parts from bar or plate stock and purchasing die forgings. In other cases, the differences in directionality of wrought structure of the different products, which affect expected resistance to directional stresses and service environments, may be the principal deciding factor.

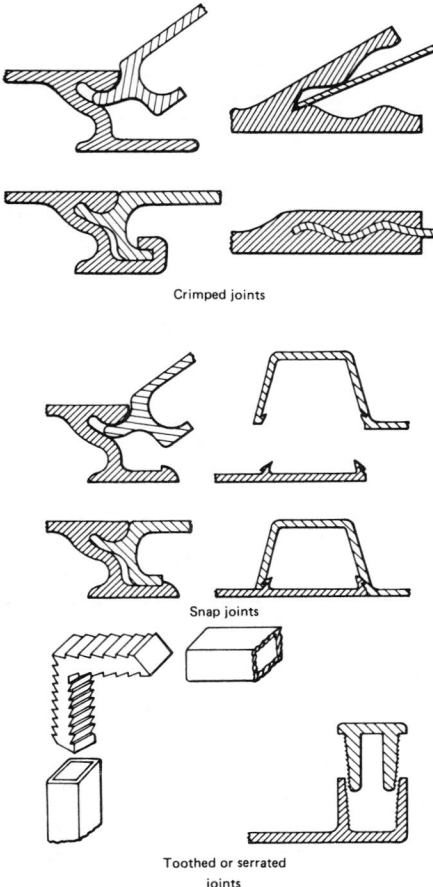

Fig. 14. Six examples of interconnecting extrusions that lock together or lock to other products

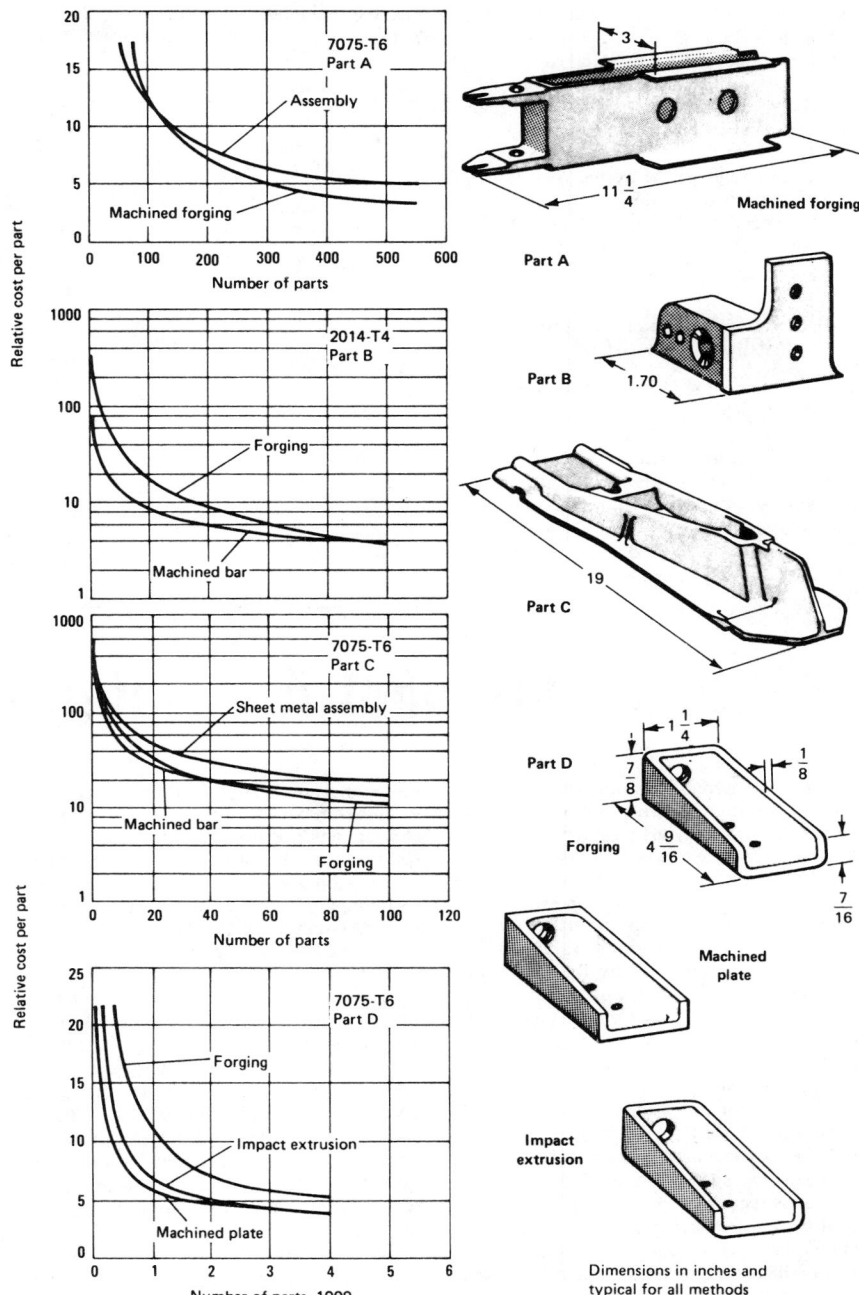

The comparison for Part A is between a built-up design and a die forging. Although the rough forging was machined on all surfaces, a saving in fabrication cost was evident after about 125 forged fittings had been made. Part B is a simpler part, and the costs of forging compared with machining from bar were the same at about the 100th piece. For a more complicated forging (part C), the crossover point where the machined bar became more expensive than the forging occurred at 40 pieces. Part D, a relatively simple fitting, was made as a die forging, an impact extrusion, and a hollowed-out fitting. Forging was the most expensive approach. The cost of extrusion and of machining from plate were about the same for 3000 fittings.

Fig. 15. Relative costs of aluminum die forgings and similar components fabricated by other methods

The cost of machining a few parts from a bar or slab is usually less than the cost of making a die and producing the parts by forging. When greater numbers of parts are to be produced, forging usually is the less-expensive method. In borderline cases, a detailed study of machining and die costs is necessary to determine the crossover point. In determining this point, it is necessary to calculate only the original cost of the die, because the supplier of the forging is responsible for die replacements caused by breakage or wear. This replacement cost is included in the price of the forgings. Die cost varies with the size and intricacy of the part.

In Fig. 15, a die forging (part A) is compared with a built-up design. Although 75% of the metal was machined away from the rough forging, the machined forging was more economical than the assembly for quantities greater than 125.

In some large, complicated forgings, the breakeven point may be at the first or second forging. It also may be desirable to rough forge the part in relatively inexpensive roughing dies and complete the part by machining if only a few pieces are desired. When this technique is used, the desirable flow of metal induced by forging and the consequent improvement in properties can be obtained at a lower cost than would have to paid for a part forged to final dimensions.

The curves in Fig. 15 compare costs of parts of different size and shape, produced by competitive methods. For the simpler part B, the crossover point occurs at about 100 pieces. For the complicated forging (part C), the crossover point occurs at 40 pieces. The items considered in determining the costs of these two parts include: fabrication-shop learning curves, unit-run labor, amortized setup, labor, tooling costs (including dies and fixtures), raw materials and overhead charges.

Die forging is not always the cheapest method of producing a large quantity of parts. A relatively simple fitting (part D in Fig. 15) was analyzed for production costs as a die forging, as an impact extrusion and as a part machined from plate. The machined fitting was more economical for all quantities, because the cost of the finishing operations required for the die forging closely approached the cost of producing the part by machining only. The thin walls and deep crevices of this fitting should have made it ideally suited to impact extrusion. Analysis showed that this method of manufacture was only slightly more expensive than machining for small quantities and identical in cost for quantities greater than about 3000 parts.

The examples represented by the parts shown in Fig. 15 serve to relate design and cost, and emphasize the necessity for conducting a detailed cost analysis of each of the several methods of fabrication.

WROUGHT MILL AND ENGINEERED PRODUCT STATISTICS

Statistics on annual shipments of wrought mill and engineered products compiled by the Aluminum Association for 1981 are shown in Table 4. These data show the distribution of production and shipments of the various product categories and also compare tonnages of non-heat-treatable or "soft" aluminum alloys with those of heat treatable or "hard" alloys. Production of non-heat-treatable alloys greatly exceeds that of heat treatable types.

Table 4. Shipments of mill products in 1981
SOURCE: "Aluminum Statistical Review for 1981," Aluminum Association, Inc.

Type of mill product	1981 shipments 10^6 kg	10^6 lb	% of total
Sheet:			
Non-heat-treatable	2504	5 520	53.4
Heat treatable	93	204	2.0
Plate:			
Non-heat-treatable	48	106	1.0
Heat treatable	76	167	1.6
Foil	386	850	8.2
Extrusions:			
Rod and bar	47	103	1.0
Shapes:			
"Soft" alloys	786	1 733	16.7
"Hard" alloys	32	70	0.7
Pipe and tube	51	112	1.1
Drawn tube	34	75	0.7
Welded tube	52	114	1.1
Bar and rod	122	269	2.6
Wire, bare	31	68	0.7
ACSR and bare cable	172	379	3.7
Insulated wire and cable	148	327	3.2
Forgings	49	109	1.0
Impacts	14	30	0.3
Powder	48	105	1.0
Total	4693	10 341	100.0

Wrought Aluminum Alloys

THE DATA on wrought aluminum alloys presented in this section, which are mostly in tabular form, are of two types: (a) typical values and (b) limits.

Typical Values. One type of table lists values considered as nominal, typical or representative. Those for the chemical compositions of the alloys (Table 1) show the midrange values of the alloying elements established by the chemical-composition limits. Physical properties (Tables 6 and 7) are median values determined in laboratory tests of representative commercial products. Mechanical properties (Tables 8A, 8B, 8C, 8D, 9, 10 and 15) are average or median values, near the peaks of distribution curves derived from routine quality-control tests of commercial products processed by standard mill procedures. The values listed are representative of products of moderate cross section or thickness, and are most useful for demonstrating relationships among alloys and tempers. These data are not intended to be used for critical design purposes. Static-strength values from tensile tests listed as typical do not represent the somewhat higher values (5 to 10% higher) obtained in tests (longitudinal direction) of extruded products of moderate section thickness nor do they represent the lower values expected in tests of very thick, heat treated products.

Representative data, on a selective basis, are tabulated for static tensile properties at low and elevated temperatures (Table 9) and for the effects of holding time at elevated temperatures (Table 10). Fracture-toughness data for the most popular high-strength alloys are given in Table 15.

Limits. Tables of the second type list limiting values (minimum or maximum). Of these limiting values, *only those given in English units* are American industry standards established by use of accepted and standardized methods of sam-

pling, analysis, testing and data processing. Mechanical-property limits are established at levels at which 99% of the material is expected to conform at a confidence level of 0.95. Additional information concerning methods is contained in "Aluminum Standards and Data" (a publication of the Aluminum Association, Inc.) and in methods publications referenced therein.

Since 1978, standards for aluminum products have been established in both English and SI (metric) units. Because the subdivisions or groupings by dimensional ranges, and the corresponding properties, in many of the mechanical-property-limit tables of the English and SI documents are not identical, it is impractical to list standard values in both types of units in the same tables. Equivalent values for dimensions and strengths in SI units included in the tables presented in this handbook were obtained by direct conversion from values listed in English units and should not be used for specification purposes. For such uses, values listed in "Aluminum Standards and Data—Metric SI" are applicable.

Chemical-composition limits are presented in Table 2. Minimum tensile-property values, which may be used as a basis for design, are given for only two product categories used in the greatest volumes—sheet and plate (Tables 11A and 11B) and extruded wire, rod, bar and shapes (Tables 12A and 12B)—and for alloy 6061-T6 structural shapes (Table 13) and die forgings (Table 14). Data of this type for many other mill, engineered and specialty products (fin stock; rolled or cold finished wire, rod and bar; rivet and cold heading wire and rod; extruded tube; extruded coiled tube; drawn tube; heat-exchanger tube; welded tube; pipe; hand forgings; rolled rings; impacts; and electrical-conductor products) may be found in "Aluminum Standards and Data," which is updated on a biennial basis.

In addition to these tables, this section in-

cludes several figures illustrating tensile-property experience with some typical die forgings (Fig. 1 and 2) and fatigue data from tests of different products illustrating similarities and differences (Fig. 3 to 6). Fatigue-crack-growth rates as affected by product thickness, direction and environment are shown in Fig. 7.

Chemical Compositions. Nominal (Table 1); limits (Table 2). Conformance with the limiting values of Table 2 is normally determined by the producer by analysis of samples taken from the molten metal at the time the ingots are poured.

Components of Clad Products. Component alloys and cladding thicknesses, both nominal values and limits (Table 3).

Specialty Mill Products. Specialty products, available alloys and tempers, and product forms (Table 4).

Comparative Characteristics and Typical Applications. Ratings for aluminum alloys with respect to corrosion, workability, machinability and joining characteristics are listed in Table 5 along with a few applications of each alloy.

Physical Properties. Listed typical physical-property values (Table 6) are given only as a basis for comparing alloys and tempers and should not be specified as engineering requirements or used for design purposes. They are not guaranteed values since in most cases they are averages for various sizes, product forms and methods of manufacture and may not be exactly representative of any particular size or product. Density values for the annealed (O) temper are listed in Table 7.

Typical Mechanical Properties. Data from tension, hardness, shear and simple, basic fatigue tests of the various alloys and tempers are tabulated in Tables 8A to D. These values are intended to serve only as a basis for comparing alloys and tempers and should not be specified as engineering requirements or used for design pur-

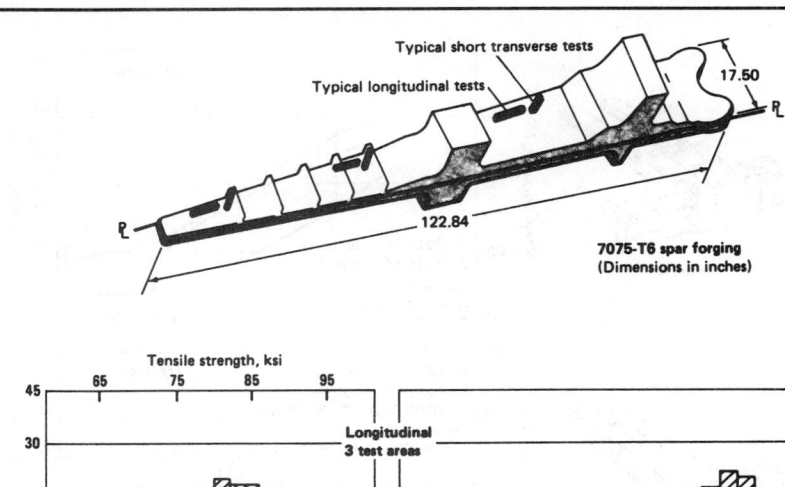

7075-T6 spar forging
(Dimensions in inches)

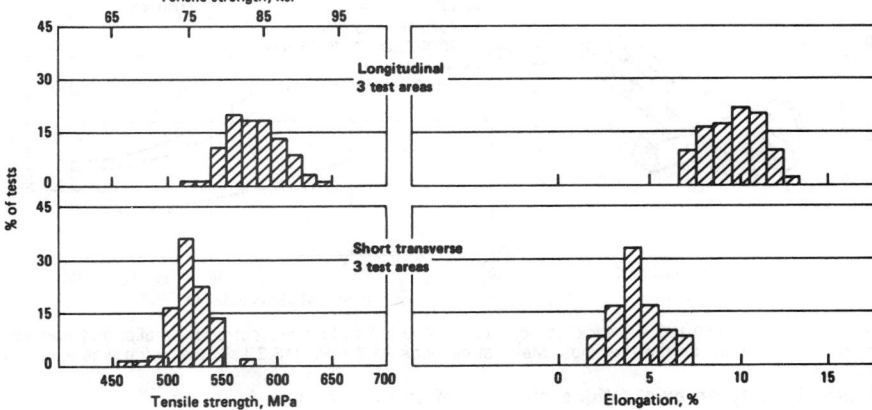

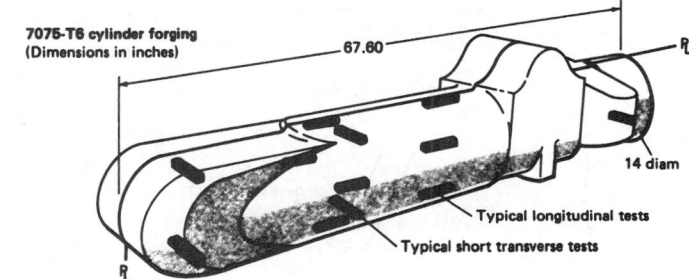

7075-T6 cylinder forging
(Dimensions in inches)

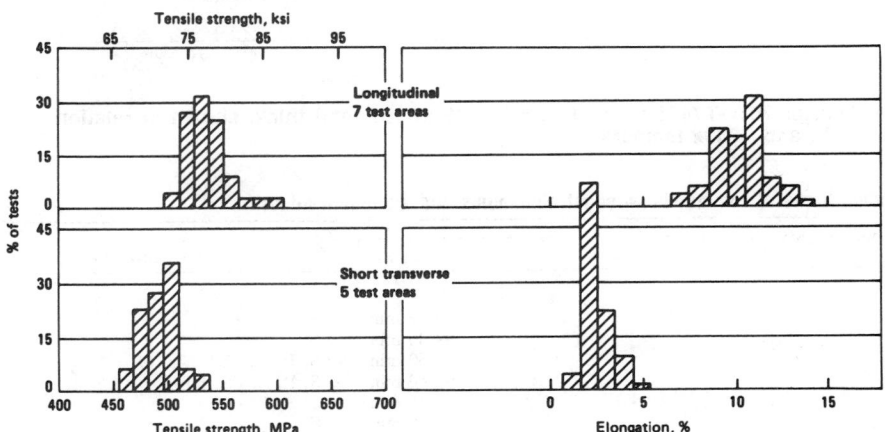

Both forgings were forged on a 31.71-kt (35 000-ton) press, solution treated at 470 °C (880 °F) for 3 h, quenched in water at 60 °C (140 °F) and aged at 120 °C (250 °F) for 24 h. **Spar forging.** Cast ingots were received in 17 lots from three sources. Two lots were in 19-in. rounds and 15 lots in 10 to 12-in. squares. Twenty-eight forgings were tested at the locations indicated. **Cylinder forging.** All cylinders were forged from 19-in. rounds. Six of the forgings were tested at locations shown.

Fig. 1. Variation in tensile properties of two different alloy 7075-T6 forgings

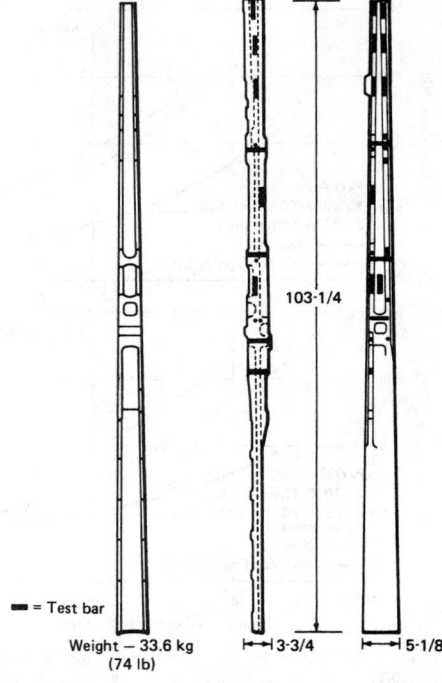

= Test bar

Weight — 33.6 kg (74 lb)

Test-bar direction	Tensile strength MPa	ksi	Yield strength MPa	ksi	Elonga-tion, %
Longitudinal:					
Average . . .	549	79.6	502	72.8	13.8
Minimum . .	524	76.0	455	66.0	7.0
Transverse:					
Average . . .	519	75.3	456	66.1	9.9
Minimum . .	490	71.0	427	62.0	4.0

Fig. 2. Variation in tensile properties of alloy 7175-T736 premium-strength spar forging for F-5 aircraft (based on 200 production test specimens)

poses. Typical tensile properties at various temperatures after long-time heating, to which the same limitations on use as stated above apply, are presented in Table 9.

Tensile-Property Limits. Sheet and plate in non-heat-treatable alloys (Table 11A); sheet and plate in heat treatable alloys (Table 11B); extruded wire, rod, bar and shapes in non-heat-treatable alloys (Table 12A); extruded wire, rod, bar and shapes in heat treatable alloys (Table 12B); structural shapes (Table 13); die forgings (Table 14).

Other Mechanical Properties. Fracture-toughness values, both minimum and typical (Table 15); tensile-property histograms (Fig. 1) and statistics (Fig 2); fatigue data (Fig. 3 to 6); fatigue-crack-growth rates (Fig. 7).

Table 13. Mechanical-property limits for aluminum alloy 6061-T6 structural shapes(a)

Alloy and temper	Tensile strength (minimum) MPa	ksi	Yield strength (minimum) MPa	ksi	Elonga-tion (min), %(b)
6061-T6 . . .	262	38.0	241	35.0	10(c)

(a) Converted SI (metric) values are for information only and are not to be used for purposes of specification, acceptance or rejection. (b) In 50 mm, 2 in. or 4*d*. (c) For thicknesses up through 0.249 in., elongation is 8%.

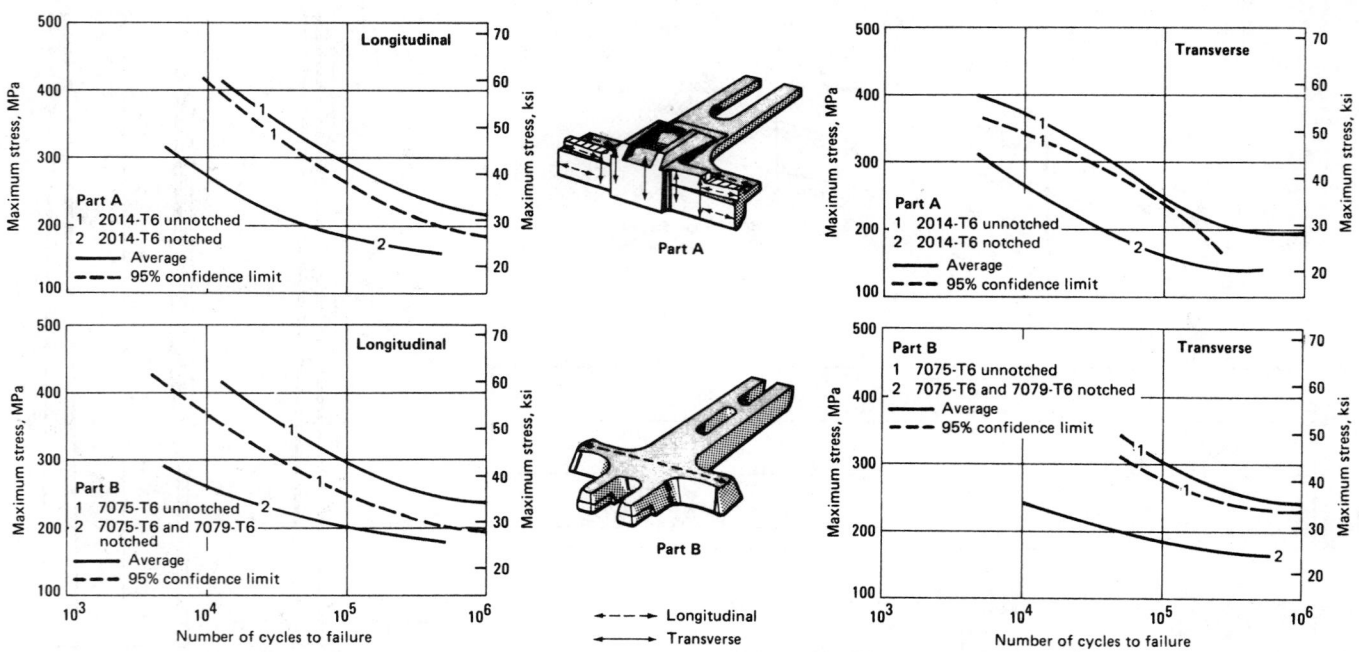

Data apply to Parts A and B, as shown. Sheet-type fatigue specimens, 3.2 mm (0.125 in.) thick and 6.4 mm (0.250 in.) wide, were cut both parallel and transverse to the forging flow lines. Locations from which specimens were taken are shown on the drawings. Mean stress was 91.7 MPa (13.3 ksi); the notch was a hole 1.2 mm (0.047 in.) in diameter in center of specimen. K_t = 2.5.

Fig. 3. Effect of alloy, design and directionality on axial fatigue strength of aluminum alloy forgings

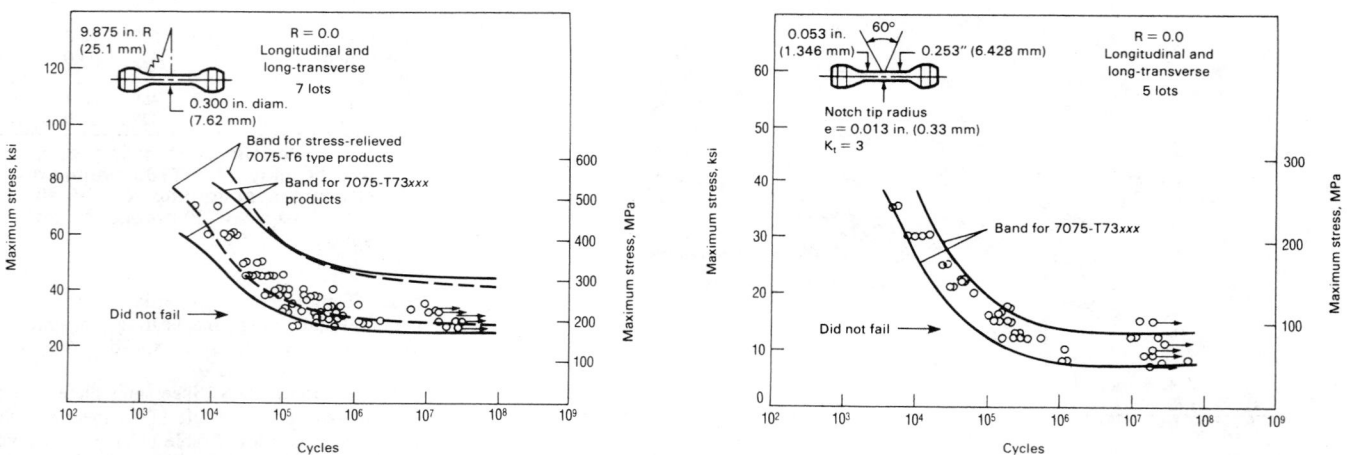

Fig. 4. Smooth and notched axial stress fatigue data for 7050-T7451 plate, 1 to 6 in. (25 to 152 mm) thick, shown in relation to bands established for 7075 wrought products in T6 and T73xx tempers

Table 1. Nominal chemical compositions and standard product forms of commercial wrought aluminums and aluminum alloys

AA No.	Si	Cu	Mn	Mg	Cr	Zn	Others	Al	Product(s)(a)
Non-heat-treatable									
1199	...	...	...	...	...	...	...	99.99 min	F, I
1180	...	...	...	...	...	...	...	99.80 min	F
1060	...	...	...	...	...	...	...	99.60 min	S, P, ET, DT
1350	...	...	...	...	...	...	...	99.50 min	S, P, E, ES, ET, C
1145	...	...	...	...	...	...	...	99.45 min	S, P, F
1235	...	...	...	...	...	...	...	99.35 min	F
1100	0.12	...	...	...	...	...	...	99.00 min	S, P, F, E, ES, ET, C, DT
3102	...	...	0.22	...	...	...	...	Rem	S, DT
3003	0.12	1.2	...	...	...	...	...	Rem	S, P, F, E, ES, ET, C, DT
3004	...	1.2	1.0	...	...	...	...	Rem	S, P, ET, DT, I
3104	0.15	1.1	1.0	...	...	...	...	Rem	S
3005	...	1.2	0.40	...	...	...	...	Rem	S, P
3105	...	0.6	0.50	...	...	...	...	Rem	S

(continued)

Table 1. (continued)

AA No.	Si	Cu	Mn	Mg	Cr	Zn	Others	Al	Product(s)(a)
4043	5.2	...	...	...	...	...	...	Rem	C
4343	7.5	...	...	...	...	...	...	Rem	S, C
4643	4.1	...	...	0.20	...	...	...	Rem	C
4045	10.0	...	...	...	...	...	...	Rem	C
4145	10.0	4.0	...	...	...	...	...	Rem	C
4047	12.0	...	...	...	...	...	...	Rem	C
5005	...	...	...	0.8	...	...	...	Rem	S, P, C
5042	...	...	0.35	3.5	...	...	...	Rem	S
5050	...	...	...	1.4	...	...	...	Rem	S, P, C, DT
5052	...	...	...	2.5	0.025	...	...	Rem	S, P, F, C, DT, R
5252	...	...	...	2.5	...	...	...	Rem	S
5154	...	...	...	3.5	0.25	...	...	Rem	S, P, E, ES, ET, C, DT
5454	...	...	0.8	2.7	0.12	...	...	Rem	S, P, E, ES, ET
5654	...	...	...	3.5	0.025	...	0.10 Ti	Rem	C
5056	...	...	0.12	5.0	0.12	...	...	Rem	F, C, R
5456	...	...	0.8	5.1	0.12	...	...	Rem	S, P, E, ES, ET, DT, FG
5457	...	...	0.30	1.0	...	...	...	Rem	S
5657	...	...	...	0.8	...	...	...	Rem	S
5082	...	...	...	4.5	...	...	...	Rem	S
5182	...	...	0.35	4.5	...	...	...	Rem	S
5083	...	...	0.7	4.4	0.15	...	...	Rem	S, P, E, ES, ET, FG
5086	...	...	0.45	4.0	0.12	...	...	Rem	S, P, E, ES, ET, DT
7072	...	...	...	...	...	1.0	...	Rem	S, F
8001	...	...	...	...	...	...	0.6 Fe, 1.1 Ni	Rem	S, P, E, ES, ET, I
8280	1.5	1.0	...	...	...	...	0.45 Ni, 6.2 Sn	Rem	S, P
8081	...	1.0	...	...	...	...	20 Sn	Rem	S

Heat treatable

AA No.	Si	Cu	Mn	Mg	Cr	Zn	Others	Al	Product(s)(a)
2011	...	5.5	...	...	...	...	0.40 Bi, 0.40 Pb	Rem	E, ES, ET, C, DT, R
2014	0.8	4.4	0.8	0.5	...	...	...	Rem	S, P, E, ES, ET, C, DT, FG, I, RS
2017	0.5	4.0	0.7	0.6	...	...	...	Rem	E, C, FG, R
2117	...	2.6	...	0.35	...	...	...	Rem	C, R
2218	...	4.0	...	1.5	...	...	2.0 Ni	Rem	FG
2618	0.18	2.3	...	1.6	...	...	1.1 Fe, 1.0 Ni, 0.07 Ti	Rem	FG, I
2219, 2419	...	6.3	0.30	...	...	...	0.10 V, 0.18 Zr, 0.06 Ti	Rem	S, P, E, ES, ET, C, FG, R
2024	...	4.4	0.6	1.5	...	...	...	Rem	S, P, E, ES, ET, C, DT, FG, I, R
2124, 2224	...	4.4	0.6	1.5	...	...	...	Rem	P, ES, FG
2025	0.8	4.4	0.8	...	...	...	...	Rem	FG
2036	...	2.6	0.25	0.45	...	...	...	Rem	S
4032	12.2	0.9	...	1.0	...	...	0.9 Ni	Rem	FG
6101	0.50	...	...	0.6	...	...	...	Rem	E, ES, ET, DT, FG, RS
6201	0.7	...	...	0.8	...	...	...	Rem	C
6009	0.8	0.37	0.50	0.6	...	...	...	Rem	S
6010	1.0	0.37	0.50	0.8	0.09	...	...	Rem	S
6151	0.9	...	...	0.6	0.25	...	...	Rem	C, FG, I, R
6351	1.0	...	0.6	0.6	...	...	...	Rem	ES
6951	0.35	0.28	...	0.6	...	...	...	Rem	S, P, C, DT
6053	0.7	...	...	1.2	0.25	...	...	Rem	ES, ET, DT, C, FG, I, R
6061	0.6	...	0.28	1.0	0.20	...	...	Rem	S, P, E, ES, ET, C, DT, FG, I, R, RS
6262	0.2	0.28	...	1.0	...	...	0.6 Bi, 0.6 Pb	Rem	E, ES, ET, DT, C
6063	0.4	...	...	0.7	...	...	...	Rem	E, ES, ET, DT, I
6066	1.3	1.0	0.8	1.1	...	...	...	Rem	E, ES, ET, DT, FG
6070	1.3	0.28	0.7	0.8	...	...	...	Rem	E, ES, ET
7001	...	2.1	...	3.0	0.26	7.4	...	Rem	E, ES
7005	...	...	0.45	1.4	0.13	4.5	0.14 Zr, 0.03 Ti	Rem	S, E, ES, DT
7016	...	0.8	...	1.1	...	4.7	...	Rem	S
7021	...	...	...	1.5	...	5.5	0.13 Zr	Rem	S, P
7029	...	0.7	...	1.2	...	4.7	...	Rem	S, ES
7049	...	1.6	...	2.4	0.16	7.7	...	Rem	FG
7050	...	2.3	...	2.2	...	6.2	0.12 Zr	Rem	S, P, ES, FG, R
7150	...	2.2	...	2.4	...	6.4	0.12 Zr	Rem	ES
7075, 7175(b)	...	1.6	...	2.5	0.23	5.6	...	Rem	S, P, E, ES, ET, C, DT, FG, I, R
7475	...	1.6	...	2.2	0.22	5.7	...	Rem	S, P
7076	...	0.6	0.50	1.6	...	7.5	...	Rem	FG
7178	...	2.0	...	2.7	0.26	6.8	...	Rem	S, P, E, ES, ET, DT, C, FG, I, R

(a) S = sheet; P = plate; F = foil; E = extruded rod, bar and wire; ES = extruded shapes; ET = extruded tube; C = cold finished rod, bar and wire; DT = drawn tube; FG = forgings; R = rivets; RS = rolled shapes; I = impacts. (b) 7175: FG only.

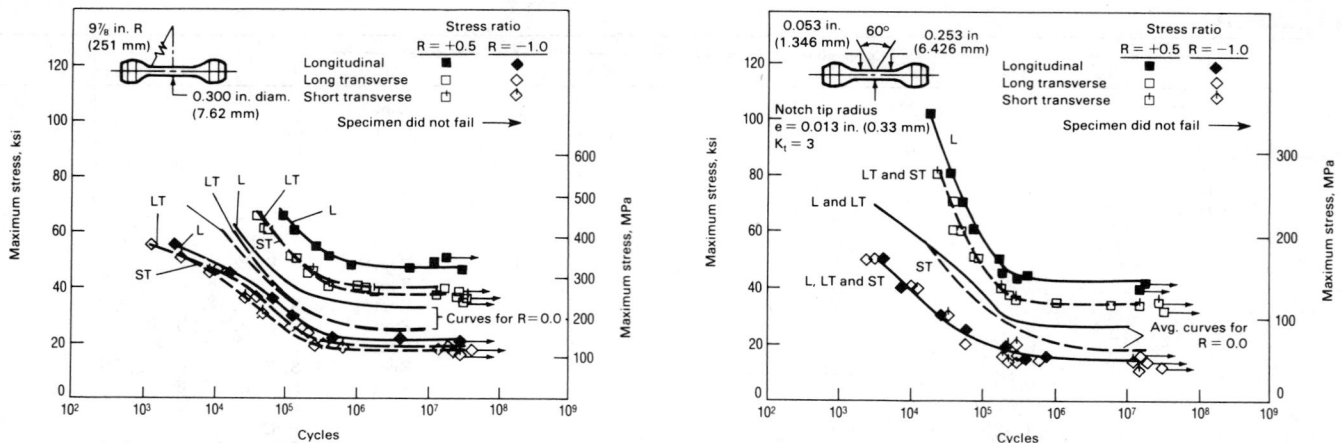

Fig. 5. Smooth and notched axial stress fatigue data for 7050-T7452 hand forgings, 4¹/₂ by 22 by 84 in. (144 by 559 by 2133 mm)

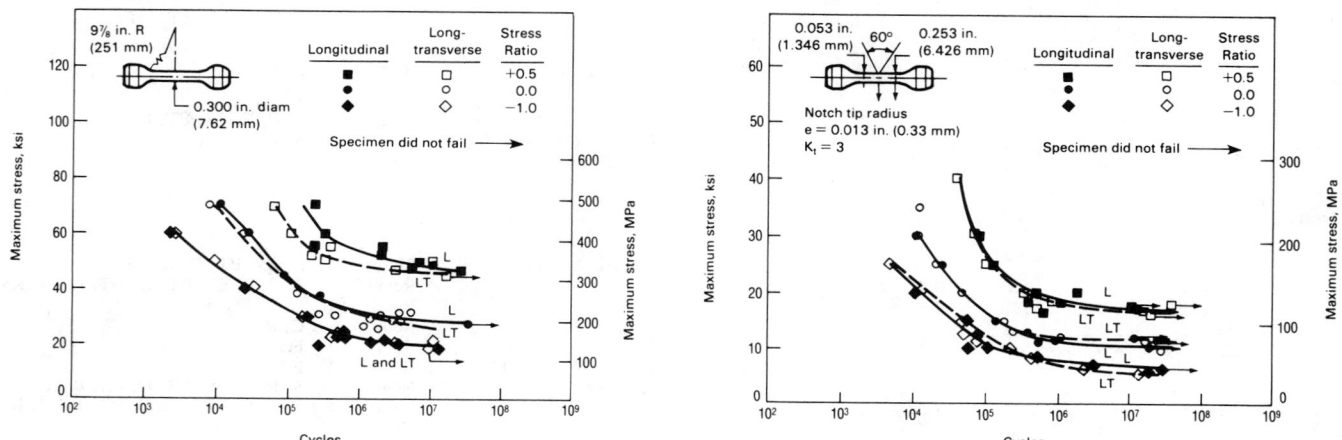

Fig. 6. Smooth and notched axial stress fatigue data for 7050-T7651x extruded shapes, 1.161 in. (29.5 mm) thick

Table 2. Chemical-composition limits of wrought aluminum alloys

AA No.	Si	Fe	Cu	Mn	Mg	Cr	Ni	Zi	Ti	Others(c) Each(d)	Others(c) Total	Al(e)
1050	0.25	0.40	0.05	0.05	0.05	...	...	0.05	0.03	0.03(f)	...	99.50 min
1060	0.25	0.35	0.05	0.03	0.03	...	...	0.05	0.03	0.03(f)	...	99.60 min
1100	1.0 Si + Fe		0.05-0.20	0.05	...	...	...	0.10	...	0.05(g)	0.15	99.00 min
1145(h)	0.55 Si + Fe		0.05	0.05	0.05	...	...	0.05	0.03	0.03(f)	...	99.45 min
1175(j)	0.15 Si + Fe		0.10	0.02	0.02	...	...	0.04	0.02	0.02(k)	...	99.75 min
1200	1.0 Si + Fe		0.05	0.05	...	...	...	0.10	0.05	0.05	0.15	99.00 min
1230(j)	0.7 Si + Fe		0.10	0.05	0.05	...	...	0.10	0.03	0.03(f)	...	99.30 min
1235	0.65 Si + Fe		0.05	0.05	0.05	...	...	0.10	0.06	0.03(f)	...	99.35 min
1345	0.30	0.40	0.10	0.05	0.05	...	...	0.05	0.03	0.03(f)	...	99.45 min
1350(m)	0.10	0.40	0.05	0.01	...	0.01	...	0.05	...	0.03(n)	0.10	99.50 min
2011	0.40	0.7	5.0-6.0	...	...	...	...	0.30	...	0.05(p)	0.15	Rem
2014	0.50-1.2	0.7	3.9-5.0	0.40-1.2	0.20-0.8	0.10	...	0.25	0.15	0.05	0.15	Rem
2017	0.20-0.8	0.7	3.5-4.5	0.40-1.0	0.40-0.8	0.10	...	0.25	0.15	0.05	0.15	Rem
2018	0.9	1.0	3.5-4.5	0.20	0.45-0.9	0.10	1.7-2.3	0.25	...	0.05	0.15	Rem
2024	0.50	0.50	3.8-4.9	0.30-0.9	1.2-1.8	0.10	...	0.25	0.15	0.05	0.15	Rem
2025	0.50-1.2	1.0	3.9-5.0	0.40-1.2	0.05	0.10	...	0.25	0.15	0.05	0.15	Rem
2036	0.50	0.50	2.2-3.0	0.10-0.40	0.30-0.6	0.10	...	0.25	0.15	0.05	0.15	Rem
2117	0.8	0.7	2.2-3.0	0.20	0.20-0.50	0.10	...	0.25	...	0.05	0.15	Rem
2124	0.20	0.30	3.8-4.9	0.30-0.9	1.2-1.8	0.10	...	0.25	0.15	0.05	0.15	Rem
2218	0.9	1.0	3.5-4.5	0.20	1.2-1.8	0.10	1.7-2.3	0.25	...	0.05	0.15	Rem
2219	0.20	0.30	5.8-6.8	0.20-0.40	0.02	...	...	0.10	0.02-0.10	0.05(q)	0.15	Rem
2319	0.20	0.30	5.8-6.8	0.20-0.40	0.02	...	...	0.10	0.10-0.20	0.05(q)	0.15	Rem
2618	0.10-0.25	0.9-1.3	1.9-2.7	...	1.3-1.8	...	0.9-1.2	0.10	0.04-0.10	0.05	0.15	Rem

(continued)

Table 2. (continued)

AA No.	Si	Fe	Cu	Mn	Mg	Cr	Ni	Zi	Ti	Others(c) Each(d)	Total	Al(e)
3003	0.6	0.7	0.05-0.20	1.0-1.5	...	...	...	0.10	...	0.05	0.15	Rem
3004	0.30	0.7	0.25	1.0-1.5	0.8-1.3	...	...	0.25	...	0.05	0.15	Rem
3005	0.6	0.7	0.30	1.0-1.5	0.20-0.6	0.10	...	0.25	0.10	0.05	0.15	Rem
3105	0.6	0.7	0.30	0.30-0.8	0.20-0.8	0.20	...	0.40	0.10	0.05	0.15	Rem
4032	11.0-13.5	1.0	0.50-1.3	...	0.8-1.3	0.10	0.50-1.3	0.25	...	0.05	0.15	Rem
4043	4.5-6.0	0.8	0.30	0.05	0.05	...	...	0.10	0.20	0.05(g)	0.15	Rem
4045(r)	9.0-11.0	0.8	0.30	0.05	0.05	...	...	0.10	0.20	0.05	0.15	Rem
4047(r)	11.0-13.0	0.8	0.30	0.15	0.10	...	...	0.20	...	0.05(g)	0.15	Rem
4145(r)	9.3-10.7	0.8	3.3-4.7	0.15	0.15	0.15	...	0.20	...	0.05(g)	0.15	Rem
4343(r)	6.8-8.2	0.8	0.25	0.10	...	...	...	0.20	...	0.05	0.15	Rem
4643	3.6-4.6	0.8	0.10	0.05	0.10-0.30	...	...	0.10	0.15	0.05(g)	0.15	Rem
5005	0.30	0.7	0.20	0.20	0.50-1.1	0.10	...	0.25	...	0.05	0.15	Rem
5050	0.40	0.7	0.20	0.10	1.1-1.8	0.10	...	0.25	...	0.05	0.15	Rem
5052	0.25	0.40	0.10	0.10	2.2-2.8	0.15-0.35	...	0.10	...	0.05	0.15	Rem
5056	0.30	0.40	0.10	0.05-0.20	4.5-5.6	0.05-0.20	...	0.10	...	0.05	0.15	Rem
5083	0.40	0.40	0.10	0.40-1.0	4.0-4.9	0.05-0.25	...	0.25	0.15	0.05	0.15	Rem
5086	0.40	0.50	0.10	0.20-0.7	3.5-4.5	0.05-0.25	...	0.25	0.15	0.05	0.15	Rem
5154	0.25	0.40	0.10	0.10	3.1-3.9	0.15-0.35	...	0.20	0.20	0.05	0.15	Rem
5183	0.40	0.40	0.10	0.50-1.0	4.3-5.2	0.05-0.25	...	0.25	0.15	0.05(g)	0.15	Rem
5252	0.08	0.10	0.10	0.10	2.2-2.8	...	...	0.05	...	0.03(f)	0.10	Rem
5254	0.45 Si + Fe		0.05	0.01	3.1-3.9	0.15-0.35	...	0.20	0.05	0.05	0.15	Rem
5356	0.25	0.40	0.10	0.05-0.20	4.5-5.5	0.05-0.20	...	0.10	0.06-0.20	0.05(g)	0.15	Rem
5454	0.25	0.40	0.10	0.50-1.0	2.4-3.0	0.05-0.20	...	0.25	0.20	0.05	0.15	Rem
5456	0.25	0.40	0.10	0.50-1.0	4.7-5.5	0.05-0.20	...	0.25	0.20	0.05	0.15	Rem
5457	0.08	0.10	0.20	0.15-0.45	0.8-1.2	...	...	0.05	...	0.03(f)	0.10	Rem
5554	0.25	0.40	0.10	0.50-1.0	2.4-3.0	0.05-0.20	...	0.25	0.05-0.20	0.05(g)	0.15	Rem
5556	0.25	0.40	0.10	0.50-1.0	4.7-5.5	0.05-0.20	...	0.25	0.05-0.20	0.05(g)	0.15	Rem
5652	0.40 Si + Fe		0.04	0.01	2.2-2.8	0.15-0.35	...	0.10	...	0.05	0.15	Rem
5654	0.45 Si + Fe		0.05	0.01	3.1-3.9	0.15-0.35	...	0.20	0.05-0.15	0.05(g)	0.15	Rem
5657	0.08	0.10	0.10	0.03	0.6-1.0	...	...	0.05	...	0.02(k)	0.05	Rem
6003(j)	0.35-1.0	0.6	0.10	0.8	0.8-1.5	0.35	...	0.20	0.10	0.05	0.15	Rem
6005	0.6-0.9	0.35	0.10	0.10	0.40-0.6	0.01	...	0.10	0.10	0.05	0.15	Rem
6009(s)	0.6-1.0	0.50	0.15-0.6	0.20-0.8	0.40-0.8	0.10	...	0.25	0.10	0.05	0.15	Rem
6010(s)	0.8-1.2	0.50	0.15-0.6	0.20-0.8	0.60-1.0	0.10	...	0.25	0.10	0.05	0.15	Rem
6053	(t)	0.35	0.10	...	1.1-1.4	0.15-0.35	...	0.10	...	0.05	0.15	Rem
6061	0.40-0.8	0.7	0.15-0.40	0.15	0.8-1.2	0.04-0.35	...	0.25	0.15	0.05	0.15	Rem
6063	0.20-0.6	0.35	0.10	0.10	0.45-0.9	0.10	...	0.10	0.10	0.05	0.15	Rem
6066	0.9-1.8	0.50	0.7-1.2	0.6-1.1	0.8-1.4	0.40	...	0.25	0.20	0.05	0.15	Rem
6070	1.0-1.7	0.50	0.15-0.40	0.40-1.0	0.50-1.2	0.10	...	0.25	0.15	0.05	0.15	Rem
6101(u)	0.30-0.7	0.50	0.10	0.03	0.35-0.8	0.03	...	0.10	...	0.03(v)	0.10	Rem
6105	0.6-1.0	0.35	0.10	0.10	0.45-0.9	0.10	...	0.10	0.10	0.05	0.15	Rem
6151	0.6-1.2	1.0	0.35	0.20	0.45-0.8	0.15-0.35	...	0.25	0.15	0.05	0.15	Rem
6162	0.40-0.8	0.50	0.20	0.10	0.7-1.1	0.10	...	0.25	0.10	0.05	0.15	Rem
6201	0.50-0.9	0.50	0.10	0.03	0.6-0.9	0.03	...	0.10	...	0.03(v)	0.10	Rem
6253(j)	(t)	0.50	0.10	...	1.0-1.5	0.04-0.35	...	1.6-2.4	...	0.05	0.15	Rem
6262	0.40-0.8	0.7	0.15-0.40	0.15	0.8-1.2	0.04-0.14	...	0.25	0.15	0.05(w)	0.15	Rem
6351	0.7-1.3	0.50	0.10	0.40-0.8	0.40-0.8	...	...	0.20	0.20	0.05	0.15	Rem
6463	0.20-0.6	0.15	0.20	0.05	0.45-0.9	...	...	0.05	...	0.05	0.15	Rem
6951	0.20-0.50	0.8	0.15-0.40	0.10	0.40-0.8	...	...	0.20	...	0.05	0.15	Rem
7001	0.35	0.40	1.6-2.6	0.20	2.6-3.4	0.18-0.35	...	6.8-8.0	0.20	0.05	0.15	Rem
7005	0.35	0.40	0.10	0.20-0.7	1.0-1.8	0.06-0.20	...	4.0-5.0	0.01-0.06	0.05(x)	0.15	Rem
7008(j)	0.10	0.10	0.05	0.05	0.7-1.4	0.12-0.25	...	4.5-5.5	0.05	0.05	0.10	Rem
7016(s)	.010	0.12	0.45-1.0	0.03	0.8-1.4	...	...	4.0-5.0	0.03	0.03(y)	0.10	Rem
7021(s)	0.25	0.40	0.25	0.10	1.2-1.8	0.05	...	5.0-6.0	0.10	0.05	0.15	Rem
7029(s)	0.10	0.12	0.50-0.9	0.03	1.3-2.0	...	...	4.2-5.2	0.05	0.03(y)	0.10	Rem
7049(s)	0.25	0.35	1.2-1.9	0.20	2.0-2.9	0.10-0.22	...	7.2-8.2	0.10	0.05	0.15	Rem
7050(s)	0.12	0.15	2.0-2.6	0.10	1.9-2.6	0.04	...	5.7-6.7	0.06	0.05	0.15	Rem
7072(j)	0.7 Si + Fe		0.10	0.10	0.10	...	...	0.8-1.3	...	0.05	0.15	Rem
7075	0.40	0.50	1.2-2.0	0.30	2.1-2.9	0.18-0.28	...	5.1-6.1	0.20	0.05	0.15	Rem
7175(s)	0.15	0.20	1.2-2.0	0.10	2.1-2.9	0.18-0.28	...	5.1-6.1	0.10	0.05	0.15	Rem
7178	0.40	0.50	1.6-2.4	0.30	2.4-3.1	0.18-0.28	...	6.3-7.3	0.20	0.05	0.15	Rem
7475(s)	0.10	0.12	1.2-1.9	0.06	1.9-2.6	0.18-0.25	...	5.2-6.2	0.06	0.05	0.15	Rem

(a) Composition in percent maximum unless shown as a range or a minimum. (b) For purposes of determining conformance to these limits, an observed value or a calculated value obtained from analysis is rounded off to the nearest unit in the last right-hand place of figures used in expressing the specified limit, in accordance with ASTM Recommended Practice E29. (c) Analysis is required for elements other than aluminum for which specified limits are shown. Analysis for other elements is made when their presence is suspected to be, or in the course of routine analysis is indicated to be, in excess of the specified limit. (d) In addition to those alloys referencing footnote (g), any alloy to be used as welding electrodes or welding rod shall have 0.0008% max beryllium. (e) The aluminum content for unalloyed aluminum not made by a refining process is the difference between 100.0% and the sum of all other metallic elements present in amounts of 0.010% or more each, expressed to the second decimal before determining the sum. (f) Vanadium, 0.05% max. (g) Beryllium, 0.0008% max for welding electrodes and welding rod only. (h) Foil. (j) Cladding alloy: see Table 3. (k) Gallium, 0.03% max; vanadium, 0.05% max. (m) Electrical conductor. Formerly designated EC. (n) Vanadium plus titanium, 0.02% max; boron, 0.05% max; gallium, 0.03% max. (p) Also contains 0.20 to 0.6% each of lead and bismuth. (q) Vanadium, 0.05 to 0.15%; zirconium, 0.10 to 0.25%. (r) Brazing alloy. (s) Not included in "Aluminum Standards and Data—1982." (t) Silicon, 45 to 65% of actual magnesium content. (u) Bus conductor. (v) Boron, 0.06% max. (w) Also contains 0.40 to 0.7% each of lead and bismuth. (x) Zirconium, 0.08 to 0.20%. (y) Vanadium, 0.05% max.

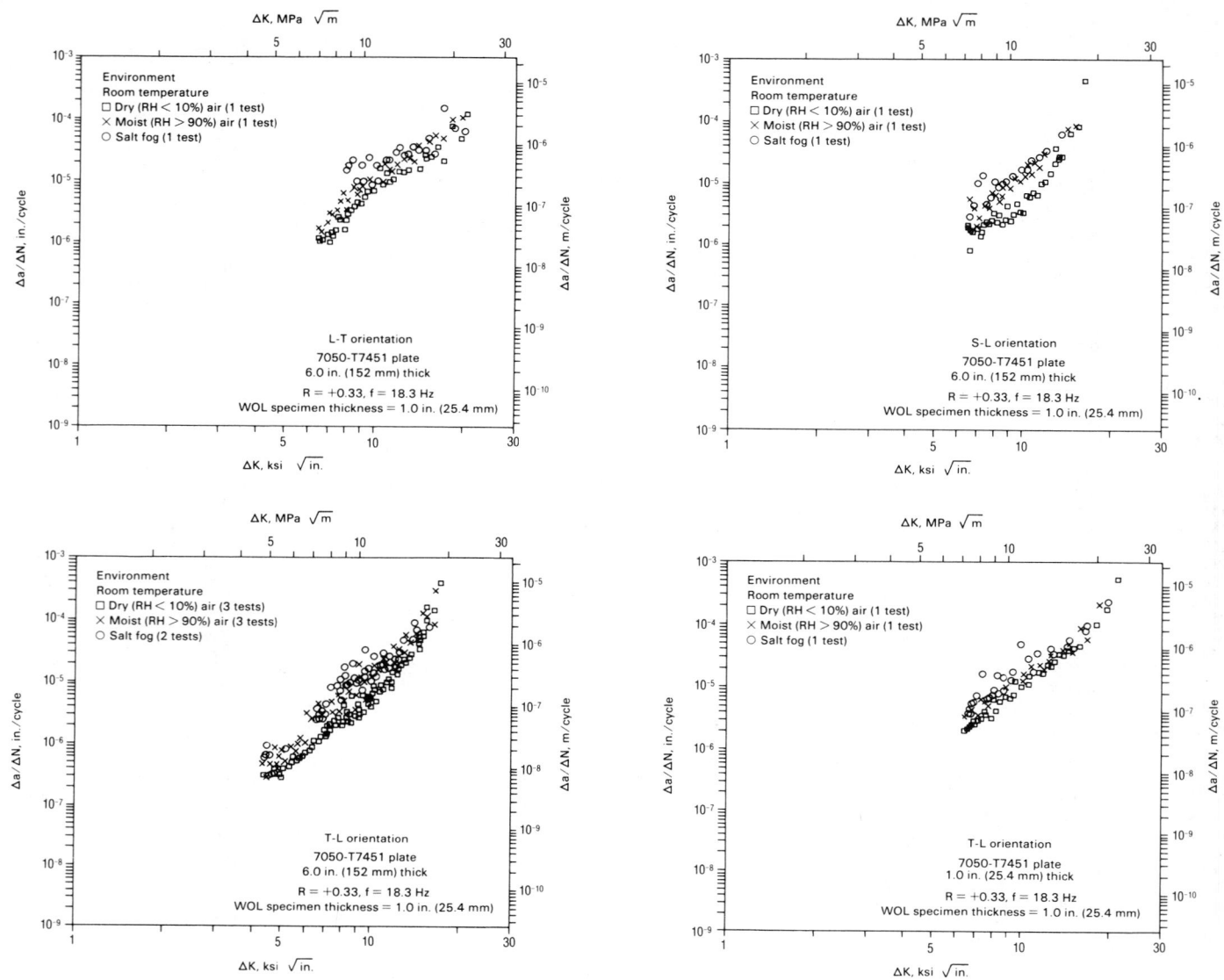

Fig. 7. Fatigue-crack-growth rates as functions of stress-intensity factor for two thicknesses of 7050-T7451 plate tested in three directions and in three environments

Table 3. Components of clad products

Designation	Component alloys(a) Core	Cladding	Total specified thickness of composite product mm Over	Thru	in.	Sides clad	Cladding thickness per side, % of composite thickness Nominal	Average(b) min	max
Alclad 2014 sheet and plate 2014		6003	...	0.63	Up thru 0.024	Both	10	8	...
			0.63	1.00	0.025-0.039	Both	7 1/2	6	...
			1.00	2.50	0.040-0.099	Both	5	4	...
			2.50	...	0.100 and over	Both	2 1/2	2	3(c)
Alclad 2024 sheet and plate 2024		1230	...	1.60	Up thru 0.062	Both	5	4	...
			1.60	...	0.063 and over	Both	2 1/2	2	3(c)
1 1/2% Alclad 2024 sheet and plate 2024		1230	4.00	...	0.188 and over	Both	1 1/2	1.2	3(c)
Alclad one side 2024 sheet and plate 2024		1230	...	1.60	Up thru 0.062	One	5	4	...
			1.60	...	0.063 and over	One	2 1/2	2	3(c)
1 1/2% Alclad one side 2024 sheet and plate 2024		1230	4.00	...	0.188 and over	One	1 1/2	1.2	3(c)
Alclad 2219 sheet and plate 2219		7072	...	1.00	Up thru 0.039	Both	10	8	...
			1.00	2.50	0.040-0.099	Both	5	4	...
			2.50	...	0.100 and over	Both	2 1/2	2	3(c)
Alclad 3003 sheet and plate 3003		7072	All		All	Both	5	4	6(c)
Alclad 3003 tube 3003		7072	All		All	Inside	10	...	...
			All		All	Outside	7	...	...
Alclad 3004 sheet and plate 3004		7072	All		All	Both	5	4	6(c)

(continued)

Table 3. (continued)

Designation	Core	Cladding	mm Over	mm Thru	in.	Sides clad	Nominal	min	max
Alclad 5056 rod and wire	5056	6253	All		All	Outside	20 (of total cross-sectional area)	16	...
Alclad 6061 sheet and plate	6061	7072	All		All	Both	5	4	6(c)
Alclad 7075 sheet and plate	7075	7072	...	1.60	Up thru 0.062	Both	4	3.2	...
			1.60	4.00	0.063-0.187	Both	2½	2	...
			4.00	...	0.188 and over	Both	1½	1.2	3(c)
2½% Alclad 7075 sheet and plate	7075	7072	...	...	0.188 and over	Both	2½	2	4(c)
Alclad one side 7075 sheet and plate	7075	7072	...	1.60	Up thru 0.062	One	4	3.2	...
			1.60	4.00	0.063-0.187	One	2½	2	...
			4.00	...	0.188 and over	One	1½	1.2	3(c)
2½% Alclad one side 7075 sheet and plate	7075	7072	...	...	0.188 and over	One	2½	2	4(c)
7008 Alclad 7075 sheet and plate	7075	7008	...	1.60	Up thru 0.062	Both	4	3.2	...
			1.60	4.00	0.063-0.187	Both	2½	2	...
			4.00	...	0.188 and over	Both	1½	1.2	3(c)
Alclad 7178 sheet and plate	7178	7072	...	1.60	Up thru 0.062	Both	4	3.2	...
			1.60	4.00	0.063-0.187	Both	2½	2	...
			4.00	...	0.188 and over	Both	1½	1.2	3(c)
No. 11 brazing sheet	3003	4343(d)	...	1.60	Up thru 0.063	One	10	8	12
			1.60	...	0.064 and over	One	5	4	6
No. 12 brazing sheet	3003	4343(d)	...	1.60	Up thru 0.063	Both	10	8	12
			1.60	...	0.064 and over	Both	5	4	6
No. 21 brazing sheet	6951	4343	...	2.50	Up thru 0.090	One	10	8	12
			2.50	...	0.091 and over	One	5	4	6
No. 22 brazing sheet	6951	4343	...	2.50	Up thru 0.090	Both	10	8	12
			2.50	...	0.091 and over	Both	5	4	6
No. 23 brazing sheet	6951	4045	...	2.50	Up thru 0.090	One	10	8	12
			2.50	...	0.091 and over	One	5	4	6
No. 24 brazing sheet	6951	4045	...	2.50	Up thru 0.090	Both	10	8	12
			2.50	...	0.091 and over	Both	5	4	6
Clad 1100 reflector sheet	1100	1175	...	1.60	Up thru 0.064	Both	15	12	18
			1.60	...	0.065 and over	Both	7½	6	9
Clad 3003 reflector sheet	3003	1175	...	1.60	Up thru 0.064	Both	15	12	18
			1.60	...	0.065 and over	Both	7½	6	9

(a) Cladding composition is applicable only to the aluminum or aluminum alloy bonded to the alloy ingot or slab preparatory to processing to the specified composite product. The composition of the cladding may be subsequently altered by diffusion between the core and cladding due to thermal treatment. (b) Average thickness per side as determined by averaging cladding-thickness measurements taken at a magnification of 100 diameters on the cross section of a transverse specimen polished and etched for microscopic examination. (c) Applicable for thicknesses of 0.500 in. and greater. (d) The cladding component, in lieu of alloy 4343, may be 5% 1xxx clad 4343.

Table 4. Specialty mill products

Specialty-product designation	Alloy(a)	Specialty-product description Temper(a)	Product form
Brazing sheet:			
No. 11 and 12	3003 clad with 4343 on one side (No. 11) or both (No. 12)	O, H12, H14	Sheet
No. 21 and 22	6951 clad with 4343 on one side (No. 21) or both (No. 22)	O	Sheet
No. 23 and 24	6951 clad with 4045 on one side (No. 23) or both (No. 24)	O	Sheet
Reflector sheet:			
Clad 1100	1100 clad with 1175 on one or both sides	...	Sheet
Clad 3003	3003 clad with 1175 on one or both sides	...	Sheet
Painted sheet	1100, 3003	O, H12, H14, H16, H18	Coiled sheet
	3105	O, H12, H14, H16, H18, H25	Coiled sheet
	5005, 5050, 5052	O, H32, H34, H36, H38	Coiled sheet
Commercial roofing and siding:			
Corrugated roofing and siding	3004, Alclad 3004	...	Sheet
V-beam roofing and siding	3004, Alclad 3004	...	Sheet
Ribbed roofing	Alclad 3004	...	Sheet
Ribbed siding	3004, Alclad 3004	...	Sheet
Duct sheet	Alloy and temper with minimum tensile strength of 16.0 ksi		Coiled or flat sheet
Tread plate	6061	O, T4, T6	Sheet and plate with raised pattern on one surface
Heat-exchanger tube	1060	H14	Tube
	3003	H14, H25	Tube
	Alclad 3003	H14, H25	Tube
	5052	H32, H34	Tube
	5454	H32, H34	Tube
	6061	T4, T6	Tube
Rigid electrical conduit	3003, 6063	H12, T1	Tube

(a) Other alloys and tempers may be available from individual producers for some of these products.

Table 5. Comparative corrosion and fabrication characteristics and typical applications of wrought aluminum alloys

Alloy temper	Resistance to corrosion General(a)	Stress-corrosion cracking(b)	Workability (cold)(e)	Machinability(e)	Weldability(f) Gas	Arc	Resistance spot and seam	Brazeability(f)	Solderability(g)	Typical applications
1050 O	A	A	A	E	A	A	B	A	A	Chemical equipment, railroad tank cars
H12	A	A	A	E	A	A	A	A	A	
H14	A	A	A	D	A	A	A	A	A	
H16	A	A	B	D	A	A	A	A	A	
H18	A	A	B	D	A	A	A	A	A	
1060 O	A	A	A	E	A	A	B	A	A	Chemical equipment, railroad tank cars
H12	A	A	A	E	A	A	A	A	A	
H14	A	A	A	D	A	A	A	A	A	
H16	A	A	B	D	A	A	A	A	A	
H18	A	A	B	D	A	A	A	A	A	
1100 O	A	A	A	E	A	A	B	A	A	Sheet-metal work, spun hollowware, fin stock
H12	A	A	A	E	A	A	A	A	A	
H14	A	A	A	D	A	A	A	A	A	
H16	A	A	B	D	A	A	A	A	A	
H18	A	A	C	D	A	A	A	A	A	
1145 O	A	A	A	E	A	A	B	A	A	Foil, fin stock
H12	A	A	A	E	A	A	A	A	A	
H14	A	A	A	D	A	A	A	A	A	
H16	A	A	B	D	A	A	A	A	A	
H18	A	A	B	D	A	A	A	A	A	
1199 O	A	A	A	E	A	A	B	A	A	Electrolytic capacitor foil, chemical equipment, railroad tank cars
H12	A	A	A	E	A	A	A	A	A	
H14	A	A	A	D	A	A	A	A	A	
H16	A	A	B	D	A	A	A	A	A	
H18	A	A	B	D	A	A	A	A	A	
1350 O	A	A	A	E	A	A	B	A	A	Electrical conductors
H12, H111	A	A	A	E	A	A	A	A	A	
H14, H24	A	A	A	D	A	A	A	A	A	
H16, H26	A	A	B	D	A	A	A	A	A	
H18	A	A	B	D	A	A	A	A	A	
2011 T3	D(c)	D	C	A	D	D	D	D	C	Screw-machine products
T4, T451	D(c)	D	B	A	D	D	D	D	C	
T8	D	B	D	A	D	D	D	D	C	
2014 O	...	...	...	D	D	D	B	D	C	Truck frames, aircraft structures
T3, T4, T451	D(c)	C	C	B	D	B	B	D	C	
T6, T651, T6510, T6511	D	C	D	B	D	B	B	D	C	
2024 O	...	...	...	D	D	D	D	D	C	Truck wheels, screw-machine products, aircraft structures
T4, T3, T351, T3510, T3511	D(c)	C	C	B	C	B	B	D	C	
T361	D(c)	C	D	B	D	C	B	D	C	
T6	D	B	C	B	D	C	B	D	C	
T861, T81, T851, T8510, T8511	D	B	D	B	D	C	B	D	C	
T72	...	...	...	B	...	...	...	...	...	
2036 T4	C	...	B	C	...	B	B	D	...	Auto-body panel sheet
2124 T851	D	B	D	B	D	C	B	D	C	Military supersonic aircraft
2218 T61	D	C	...	...	...	...	C	...	C	Jet engine impellers and rings
T72	D	C	...	B	D	C	B	D	C	
2219 O	...	...	...	...	D	A	B	D		Structural uses at high temperatures (to 315 °C or 600 °F) high-strength weldments
T31, T351, T3510, T3511	D(c)	C	C	B	A	A	A	D		
T37	D(c)	C	D	B	A	A	A	D	NA	
T81, T851, T8510, T8511	D	B	D	B	A	A	A	D		
T87	D	B	D	B	A	A	A	D		
2618 T61	D	C	...	B	D	C	B	D	NA	Aircraft engines
3003 O	A	A	A	E	A	A	B	A	A	Cooking utensils, chemical equipment, pressure vessels, sheet-metal work, builder's hardware, storage tanks, trailer roof and panel sheet, heat-exchanger tube
H12	A	A	A	E	A	A	A	A	A	
H14	A	A	B	D	A	A	A	A	A	
H16	A	A	C	D	A	A	A	A	A	
H18	A	A	C	D	A	A	A	A	A	
H25	A	A	B	D	A	A	A	A	A	
3004 O	A	A	A	D	B	A	B	B	B	Sheet-metal work, storage tanks, drawn and ironed beverage cans, trailer panel sheet, industrial siding
H32	A	A	B	D	B	A	A	B	B	
H34	A	A	B	C	B	A	A	B	B	
H36	A	A	C	C	B	A	A	B	B	
H38	A	A	C	C	B	A	A	B	B	
3105 O	A	A	A	E	B	A	B	B	B	Residential siding, mobile homes, rain-carrying goods, sheet-metal work
H12	A	A	B	E	B	A	A	B	B	
H14	A	A	B	D	B	A	A	B	B	
H16	A	A	C	D	B	A	A	B	B	
H18	A	A	C	D	B	A	A	B	B	
H25	A	A	B	D	B	A	A	B	B	
4032 T6	C	B	...	B	D	B	C	D	NA	Pistons
4043	B	A	NA	C	NA	NA	NA	NA	NA	Welding electrode
5005 O	A	A	A	E	A	A	B	B	B	Appliances, utensils, architectural, electrical conductors, boat sheet
H12	A	A	A	E	A	A	A	B	B	
H14	A	A	B	D	A	A	A	B	B	
H16	A	A	C	D	A	A	A	B	B	

(continued)

Table 5. (continued)

Alloy temper	Resistance to corrosion — General(a)	Stress-corrosion cracking(b)	Workability (cold)(e)	Machinability(e)	Weldability(f) Gas	Arc	Resistance spot and seam	Brazeability(f)	Solderability(g)	Typical applications
H18	A	A	C	D	A	A	A	B	B	
H32	A	A	B	E	A	A	A	B	B	
H34	A	A	C	D	A	A	A	B	B	
H36	A	A	C	D	A	A	A	B	B	
H38	A	A		D	A	A	A	B	B	
5050 O	A	A	A	E	A	A	B	B	C	Builders' hardware, refrigerator trim, coiled tubes
H32	A	A	A	D	A	A	A	B	C	
H34	A	A	B	D	A	A	A	B	C	
H36	A	A	C	C	A	A	A	B	C	
H38	A	A	C	C	A	A	A	B	C	
5052 O	A	A	A	D	A	A	B	C	D	Sheet-metal work, hydraulic tube, appliances, bus body sheet, welded structures, boat sheet
H32	A	A	B	D	A	A	A	C	D	
H34	A	A	B	C	A	A	A	C	D	
H36	A	A	C	C	A	A	A	C	D	
H38	A	A	C	C	A	A	A	C	D	
5056 O	A(d)	B(d)	A	D	C	A	B	D	D	Cable sheathing, rivets for magnesium, screen wire, zippers
H111	A(d)	B(d)	A	D	C	A	A	D	D	
H12, H32	A(d)	B(d)	B	D	C	A	A	D	D	
H14, H34	A(d)	B(d)	B	C	C	A	A	D	D	
H18, H38	A(d)	C(d)	C	C	C	A	A	D	D	
H192	B(d)	D(d)	D	B	C	A	A	D	D	
H392	B(d)	D(d)	D	B	C	A	A	D	D	
5083 O	A(d)	B(d)	B	D	C	A	B	D	D	Unfired, welded pressure vessels, marine, cryogenics, TV towers, drilling rigs, transportation equipment, missile components
H321, H116	A(d)	B(d)	C	D	C	A	A	D	D	
H323	A(d)	B(d)	C	D	C	A	A	D	D	
H343	A(d)	B(d)	C	C	C	A	A	D	D	
H111	A(d)	B(d)	C	D	C	A	A	D	D	
5086 O	A(d)	A(d)	A	D	C	A	B	D	D	Welded structures; pressure vessels; marine, cryogenic and transportation equipment; computer memory disks
H32, H116	A(d)	A(d)	B	D	C	A	A	D	D	
H34	A(d)	B(d)	B	C	C	A	A	D	D	
H36	A(d)	B(d)	C	C	C	A	A	D	D	
H38	A(d)	B(d)	C	C	C	A	A	D	D	
H111	A(d)	A(d)	B	D	C	A	A	D	D	
5154 O	A(d)	A(d)	A	D	C	A	B	D	D	Welded structures, storage tanks, pressure vessels, salt-water service
H32	A(d)	A(d)	B	D	C	A	A	D	D	
H34	A(d)	A(d)	B	C	C	A	A	D	D	
H36	A(d)	A(d)	C	C	C	A	A	D	D	
H38	A(d)	A(d)	C	C	C	A	A	D	D	
5182 O	A	A(d)	A	D	C	A	B	D	D	Automobile body sheet, can ends
H19	A	A(d)	D	B	C	A	A	D	D	
5252 H24	A	A	B	D	A	A	A	C	D	Automotive and appliance trim
H25	A	A	B	C	A	A	A	C	D	
H28	A	A	C	C	A	A	A	C	D	
5254 O	A(d)	A(d)	A	D	C	A	B	D	D	Hydrogen peroxide and chemical storage vessels
H32	A(d)	A(d)	B	D	C	A	A	D	D	
H34	A(d)	A(d)	B	C	C	A	A	D	D	
H36	A(d)	A(d)	C	C	C	A	A	D	D	
H38	A(d)	A(d)	C	C	C	A	A	D	D	
5356	A	A	NA	B	NA	NA	NA	NA	NA	Welding electrode
5454 O	A	A	A	D	C	A	B	D		Welded structures, pressure vessels, marine service, tank trailer bodies, dump truck bodies
H32	A	A	B	D	C	A	A	D	NA	
H34	A	A	B	C	C	A	A	D		
H111	A	A	B	D	C	A	A	D		
5456 O	A(d)	B(d)	B	D	C	A	B	D		High-strength welded structures, storage tanks, pressure vessels, marine applications
H111	A(d)	B(d)	C	D	C	A	A	D		
H321, H115	A(d)	B(d)	C	D	C	A	A	D	NA	
H323	A(d)	B(d)	C	D	C	A	A	D		
H343	A(d)	B(d)	C	C	C	A	A	D		
5457 O	A	A	A	E	A	A	B	B	B	
5652 O	A	A	A	D	A	A	B	C	D	Hydrogen peroxide and chemical storage vessels
H32	A	A	B	D	A	A	A	C	D	
H34	A	A	B	C	A	A	A	C	D	
H36	A	A	C	C	A	A	A	C	D	
H38	A	A	C	C	A	A	A	C	D	
5657 H241	A	A	A	D	A	A	A	B		Anodized auto and appliance trim
H25	A	A	B	D	A	A	A	B	NA	
H26	A	A	B	D	A	A	A	B		
H28	A	A	C	D	A	A	A	B		
6005 T5	B	A	C	C	A	A	A	A	NA	Heavy-duty structures requiring good corrosion resistance applications, truck and marine, railroad cars, furniture, pipelines
6009 T4	A	A	A	C	A	A	A	A	B	Automobile body sheet
6010 T4	A	A	B	C	A	A	A	A	B	Automobile body sheet, canoes

(continued)

Table 5. (continued)

Alloy temper	Resistance to corrosion General(a)	Stress-corrosion cracking(b)	Work-ability (cold)(e)	Machin-ability(e)	Weldability(f) Gas	Arc	Resistance spot and seam	Braze-ability(f)	Solder-ability(g)	Typical applications
6061 O	B	A	A	D	A	A	B	A	B	Heavy-duty structures requiring good corrosion resistance, truck and marine, railroad cars, furniture, pipelines
T4, T451, T4510, T4511	B	B	B	C	A	A	A	A	B	
T6, T651, T652, T6510, T6511	B	A	C	C	A	A	A	A	B	
6063 T1	A	A	B	D	A	A	A	A	B	Pipe railing, furniture, architectural extrusions
T4	A	A	B	D	A	A	A	A	B	
T5, T52	A	A	B	C	A	A	A	A	B	
T6	A	A	C	C	A	A	A	A	B	
T83, T831, T832	A	A	C	C	A	A	A	A	B	
6066 O	C	A	B	D	D	B	B	D		Forgings and extrusions for welded structures
T4, T4510, T4511	C	B	C	C	D	B	B	D	NA	
T6, T6510, T6511	C	B	C	B	D	B	B	D		
6070 T4, T4511	B	B	B	C	A	A	A	B		Heavy-duty welded structures, pipelines
T6	B	B	C	C	A	A	A	B	NA	
6101 T6, T63	A	A	C	C	A	A	A	A		High-strength bus conductors
T61, T64	A	A	B	D	A	A	A	A		
6151 T6, T652	...	...	...	...	...	...	...	...	B	Moderate-strength, intricate forgings for machine and auto parts
6201 T81	A	A	...	C	A	A	A	A	NA	High-strength electric conductor wire
6262 T6, T651, T6510, T6511	B	A	C	B	A	A	A	A		Screw-machine products
T9	B	A	D	B	A	A	A	A	NA	
6351 T5, T6	B	A	C	C	A	A	A	A	B	Heavy-duty structures requiring good corrosion resistance, truck and tractor extrusions
6463 T1	A	A	B	D	A	A	A	A		Extruded architectural and trim sections
T5	A	A	B	C	A	A	A	A	NA	
T6	A	A	C	C	A	A	A	A		
7005 T53, T63	B	B	C	A	B	B	B	B	B	Heavy-duty structures requiring good corrosion resistance, trucks, trailers, dump bodies
7049 T73, T7351, T7352	C	B	D	B	D	C	B	D	D	Aircraft and other structures
T76, T7651	C	B	D	B	D	C	B	D	D	
7050 T74, T7451, T7452	C	B	D	B	D	C	B	D	D	Aircraft and other structures
T76, T761	C	B	D	B	D	C	B	D	D	
7072	A	A	A	D	A	A	A	A	A	Fin stock, cladding alloy
7075 O	...	...	...	D	D	C	B	D	D	Aircraft and other structures
T6, T651, T652, T6510, T6511	C(c)	C	D	B	D	C	B	D	D	
T73, T7351	C	B	D	B	D	C	B	D	D	
7175 T736, T73652	C	B	D	B	D	C	B	D	D	Aircraft and other structures, forgings
7178 O	...	...	...	...	D	C	B	D	D	Aircraft and other structures
T6, T651, T6510, T6511	C(c)	C	D	B	D	C	B	D	D	
7475 T6, T651	C	C	D	B	D	C	B	D	D	Aircraft and other structures
T73, T7351, T7352	C	B	D	B	D	C	B	D	D	
T76, T7651	C	B	D	B	D	C	B	D	D	

(a) Ratings A through E are relative ratings in decreasing order of merit, based on exposures to sodium chloride solution by intermittent spraying or immersion. Alloys with A and B ratings can be used in industrial and seacoast atmospheres without protection. Alloys with C, D and E ratings generally should be protected at least on faying surfaces. (b) Stress-corrosion cracking ratings are based on service experience and on laboratory tests of specimens exposed to the 3.5% sodium chloride alternate immersion test. A = No known instance of failure in service or in laboratory tests. B = No known instance of failure in service; limited failures in laboratory tests of short transverse specimens. C = Service failures with sustained tension stress acting in short transverse direction relative to grain structure; limited failures in laboratory tests of long transverse specimens. D = Limited service failures with sustained longitudinal or long transverse stress. (c) In relatively thick sections the rating would be E. (d) This rating may be different for material held at elevated temperature for long periods. (e) Ratings A through D for workability (cold), and A through E for machinability, are relative ratings in decreasing order of merit. (f) Ratings A through D for weldability and brazeability are relative ratings defined as follows: A = Generally weldable by all commercial procedures and methods. B = Weldable with special techniques or for specific applications which justify preliminary trials or testing to develop welding procedure and weld performance. C = Limited weldability because of crack sensitivity or loss in resistance to corrosion and mechanical properties. D = No commonly used welding methods have been developed. (g) Ratings A through D and NA for solderability are relative ratings defined as follows: A = Excellent. B = Good. C = Fair. D = Poor. NA = Not Applicable.

Table 6. Typical physical properties of wrought aluminum alloys

Alloy	Temper	Melting range (approx) °F	°C	Electrical conductivity Vol	Wt	Electrical resistivity nΩ·m	ohms	Thermal conductivity W/m·K	Btu/ft·h·°F	Average coefficient of thermal expansion over temperature range 20 to 100 °C or 68 to 212 °F(a) Per °C	Per °F
1050	O	...	...	61	190	28	17	231	133	23.6	13.1
1060	O	1195-1215	646-657	62	204	28	17	234	135	23.6	13.1
	H18			61	201	28	17	230	135		
1100	O	1190-1215	643-657	59	194	29	18	222	128	23.6	13.1
	H18			57	187	30	18	218	126		
1145	O	...	...	61	202	28	17	230	133	23.6	13.1
	H18			60	198	29	18	227	131		
1199	O	...	...	65	215	27	16	243	140	23.6	13.1
1350	O	1195-1215	646-657	62	204	28	17	234	135	23.8	13.2
	H1x			61	201	28	17	230	133		
2011	T3, T4	1005-1190	541-643	39	123	44	27	152	88	22.9	12.7
	T8			45	142	38	23	173	100		

(continued)

Table 6. (continued)

Alloy	Temper	Melting range (approx) °F	°C	Electrical conductivity Vol	Wt	Electrical resistivity nΩ·m	ohms	Thermal conductivity W/m·K	Btu/ft·h·°F	Average coefficient of thermal expansion over temperature range 20 to 100 °C or 68 to 212 °F(a) Per °C	Per °F
2014	O	945-1180	507-638	50	159	34	21	192	111	23.0	12.8
	T3, T4			34	108	51	31	134	77		
	T6			40	127	43	26	155	89		
2017	O	955-1185	512-641	50	159	34	21	193	111	23.6	13.1
	T4			34	108	51	31	134	77		
2018	T61	945-1180	507-638	40	127	43	26	155	90	22.3	12.4
2024	O	935-1180	502-638	50	160	34	21	193	111	23.2	12.9
	T3, T4, T361			30	96	57	35	120	69		
	T6, T81, T861			38	122	45	27	151	87		
2025	T6	970-1185	521-641	40	128	43	26	155	89	22.7	12.6
2036	O	1030-1200	554-649	52	169	33	20	198	114	23.4	13.0
	T4			41	135	42	25	159	92		
2048	T851	...	...	42	137	41	24	162	94		
2117	T4	1030-1200	554-649	40	130	43	26	155	89	23.8	13.2
2124	O			50	161	34	21	193	111	23.2	12.9
	T851			39	126	44	27	152	88		
2218	T61	940-1175	504-635	38	121	45	27	148	86	22.3	12.4
	T72			40	128	43	26	155	90		
2219	O	1010-1190	543-643	44	138	39	24	170	98	22.3	12.4
	T31, T37, T351			28	88	62	37	116	67		
	T62, T81, T87, T851			30	95	57	35	120	69		
2319	O			44	139	39	24	170	98	22.3	12.4
2618	T61	1020-1180	549-638	37	120	47	28	146	84	22.3	12.4
3003	O	1190-1210	643-654	50	163	34	22	193	111	23.2	12.9
	H12			42	138	41	25	162	94		
	H14			41	134	42	25	159	92		
	H18			40	130	43	26	155	90		
3004	O (all)	1165-1210	629-654	42	137	41	25	162	94	23.9	13.3
3105	O (all)	1175-1210	635-654	45	148	38	23	173	100	23.6	13.1
4032	O	990-1060	532-571	40	132	43	26	155	90	19.4	10.8
	T6			35	116	48	29	138	80		
4043	O	1065-1170	574-632	42	140	41	25	163	94	22.0	12.3
4045	All	1065-1110	574-599	45	151	38	23	171	99	21.1	11.7
4343	All	1070-1135	577-613	47	158	37	22	180	104	21.6	12.0
5005	All	1170-1210	632-654	52	172	33	20	200	114	23.8	13.2
5050	All	1155-1205	624-652	50	165	34	21	191	110	23.8	13.2
5052	All	1125-1200	607-649	35	116	49	30	137	79	23.8	13.2
5056	O	1055-1180	568-638	29	98	59	36	117	68	24.1	13.4
	H38			27	91	64	38	112	64		
5083	O	1095-1180	591-638	29	98	59	36	117	68	23.8	13.2
5086	All	1085-1185	585-641	31	104	56	33	126	71	23.8	13.2
5154	All	1100-1190	593-643	32	107	54	32	127	73	23.9	13.3
5182	All	...	...	31	104	56	33	126	71	23.9	13.3
5252	All	1125-1200	607-649	35	117	49	30	138	80	23.8	13.2
5254	All	1100-1190	593-643	32	107	54	32	127	73	23.9	13.3
5356	O	1060-1175	571-635	29	98	59	36	116	67	24.1	13.4
5454	All	1115-1195	602-646	34	113	51	31	134	77	23.6	13.1
5456	O	1055-1180	568-638	29	98	59	36	116	67	23.9	13.3
5457	All	1165-1210	629-654	46	153	37	22	177	102	23.8	13.2
5652	All	1125-1200	607-649	35	116	49	30	137	79	23.8	13.2
5657	All	1180-1215	638-657	54	180	32	19	205	118	23.8	13.2
6005	T1	1125-1210	607-654	47	155	36	22	180	105	23.8	13.2
	T5			49	161	35	21	188	108	23.6	13.0
6009	O	1090-1202	588-650	54	180	32	19	205	118	23.6	13.0
	T4			44	145	39	24	172	99		
	T6			47	155	37	22	180	104		
6010	O	1085-1200	585-649	53	175	33	20	202	117	23.2	12.9
	T4			39	123	44	27	151	87		
	T6			44	145	39	24	180	99		
6053	O	1070-1205	577-652	45	148	38	23	173	100	23.0	12.8
	T4			40	132	43	26	155	90		
	T6			42	139	41	25	162	94		
6061	O	1080-1205	582-652	47	155	37	22	180	104	23.6	13.1
	T4			40	132	43	26	154	89		
	T6			43	142	40	24	167	96		
6063	O	1140-1210	616-654	58	191	30	18	218	126	23.4	13.0
	T1			50	165	34	21	193	112		
	T5			55	181	31	19	209	121		
	T6, T83			53	175	33	20	201	116		
6066	O	1045-1195	563-646	40	132	43	26	155	90	23.2	12.9
	T6			37	122	47	28	147	85		
6070	T6	1050-1200	566-649	44	145	39	24	172	99	23.2	12.9

(continued)

Table 6. (continued)

Alloy	Temper	Melting range (approx) °F	Melting range (approx) °C	Electrical conductivity Vol	Electrical conductivity Wt	Electrical resistivity $n\Omega \cdot m$	Electrical resistivity ohms	Thermal conductivity $W/m \cdot K$	Thermal conductivity $Btu/ft \cdot h \cdot °F$	Average coefficient of thermal expansion over temperature range 20 to 100 °C or 68 to 212 °F(a) Per °C	Average coefficient of thermal expansion over temperature range 20 to 100 °C or 68 to 212 °F(a) Per °F
6101	T6	1150-1210	621-654	57	188	30	18	218	126	23.4	13.0
	T8			54	178	32	19	205	118		
	T61			59	194	29	18	222	128		
	T63			58	191	30	18	218	126		
	T64			60	198	29	18	226	131		
	T65			58	191	30	18	218	126		
6105	T1	1110-1200	599-649	46	151	37	22	177	102	23.4	13.0
	T5			50	165	34	21	193	112		
6151	O	1090-1200	588-649	54	178	32	19	205	118	23.2	12.9
	T4			42	138	41	25	163	94		
	T6			45	148	38	23	175	101		
6201	T81	1125-1210	607-654	54	180	32	19	205	118	23.4	13.0
6205	T1	...	...	45	149	37	22	172	99	23.4	13.0
	T5			49	162	35	21	188	109		
6253	...	1100-1205	593-652	...	...	...	...	...	...	...	...
6262	T9	1080-1205	582-652	44	145	39	24	170	99	23.4	13.0
6351	T6	...	...	46	152	38	23	176	102	...	...
6463	T1	1140-1210	616-654	50	165	34	21	192	111	23.4	13.0
	T5			55	181	31	19	209	121		
	T6			53	175	33	20	201	116		
6951	O	1140-1210	616-654	56	186	31	19	214	124	23.4	13.0
	T6			52	172	33	20	199	115		
7001	T6	890-1160	477-627	31	98	56	33	126	72	23.4	13.0
7005	O	...	...	43	138	40	24	166	96	23.6	13.1
	T53			38	122	45	27	148	86		
	T6			35	113	49	30	137	79		
	T63			38	122	45	27	148	86		
7049	T73	890-1160	477-627	38	120	43	27	148	86	23.4	13.0
7050	O	870-1175	466-635	47	155	37	22	180	104	23.6	13.1
	T76			39.5	123	44	26	153	89		
	T74			40.5	130	43	27	157	91		
7072	O	...	...	59	193	29	17	222	129	23.6	13.1
7075	O	890-1175	477-635	45	148	38	23	173	100	23.6	13.1
	T6			33	105	52	31	130	75		
	T73			40	127	43	26	155	90		
	T76			38.5	123	45	27	150	87		
7175	O	...	...	46	147	38	23	177	102	23.6	13.1
	T66			36	115	48	29	142	82		
	T73			40	128	43	26	155	90		
7178	T6	890-1165	477-629	31	98	56	33	124	71	23.4	13.0
7475	O	890-1175	477-635	46	153	38	23	177	102	23.6	13.1
	T6			36	118	48	29	142	82		
	T73			42	140	41	25	163	94		
	T76			40	132	43	26	155	90		
	T735			42	140	41	25	163	94		

(a) Coefficient to be multiplied by 10^{-6}. Example: $23.6 \times 10^{-6} = 0.0000236$.

Table 7. Nominal densities and specific gravities of wrought aluminums and aluminum alloys

Density and specific gravity are dependent on composition, and variations are discernible from one cast to another for most alloys. The nominal values given here should not be specified as engineering requirements but are used in calculating typical values for weight per unit length, weight per unit area, covering area, etc. The density values were derived from the metric and subsequently rounded; these values are not to be back-converted to the metric.

Alloy	Density, lb/in.^{3}	Specific gravity	Alloy	Density, lb/in.^{3}	Specific gravity	Alloy	Density, lb/in.^{3}	Specific gravity
1050	0.0975(a)	2.705(a)	2124	0.100	2.78	5056	0.095	2.64
1060	0.0975(a)	2.705(a)	2218	0.101	2.81	5082	0.096	2.65
1100	0.098	2.71	2219	0.103	2.84	5083	0.096	2.66
1145	0.0975(a)	2.700(a)	2618	0.100	2.76	5086	0.096	2.66
1175	0.0975(a)	2.700(a)	3003	0.099	2.73	5154	0.096	2.66
1200	0.098	2.70	3004	0.098	2.72	5182	0.096	2.65
1230	0.098	2.70	3005	0.098	2.73	5183	0.096	2.66
1235	0.0975(a)	2.705(a)	3105	0.098	2.72	5252	0.096	2.67
1345	0.0975(a)	2.705(a)	4032	0.097	2.68	5254	0.096	2.66
1350	0.0975(a)	2.705(a)	4043	0.097	2.69	5352	0.097	2.67
2011	0.102	2.83	4045	0.096	2.67	5356	0.096	2.64
2014	0.101	2.80	4047	0.096	2.66	5454	0.097	2.69
2017	0.101	2.79	4145	0.099	2.74	5456	0.096	2.66
2018	0.102	2.82	4343	0.097	2.68	5457	0.097	2.69
2024	0.101	2.78	4643	0.097	2.69	5554	0.097	2.69
2025	0.101	2.81	5005	0.098	2.70	5556	0.096	2.66
2036	0.100	2.75	5050	0.097	2.69	5652	0.097	2.67
2117	0.099	2.75	5052	0.097	2.68	5654	0.096	

(continued)

Table 7. (continued)

Alloy	Density, lb/in.³	Specific gravity	Alloy	Density, lb/in.³	Specific gravity	Alloy	Density, lb/in.³	Specific gravity
5657	0.097	2.69	6105	0.097	2.69	7016	0.100	2.78
6003	0.097	2.70	6151	0.098	2.71	7021	0.101	2.78
6005	0.097	2.70	6162	0.097	2.70	7029	0.100	2.77
6009	0.098	2.71	6201	0.097	2.69	7049	0.102	2.82
6010	0.098	2.71	6262	0.098	2.72	7050	0.102	2.83
6053	0.097	2.69	6351	0.098	2.71	7072	0.098	2.72
6061	0.098	2.70	6463	0.097	2.69	7075	0.101	2.81
6063	0.097	2.70	6951	0.098	2.70	7178	0.102	2.83
6066	0.098	2.72	7001	0.103	2.84	7475	0.101	2.80
6070	0.098	2.71	7005	0.100	2.78			
6101	0.097	2.70	7008	0.100	2.78			

(a) Limited to aluminum with a purity of 99.35% or higher.

Table 8A. Typical mechanical properties of unalloyed aluminum

Alloy	Temper	Tensile strength MPa	ksi	Yield strength(a) MPa	ksi	Elongation, %(b)	Hardness, HB(c)	Shear strength MPa	ksi	Fatigue limit(d) MPa	ksi
1199	O	45	6.5	10	1.5	50	...	...	...	...	...
	H18	115	17	110	16	5	...	...	...	...	...
1180	O	60	9	20	3	45	...	...	...	...	...
	H18	125	18	115	17	5	...	...	...	...	...
1060	O	70	10	30	4	43	19	50	7	20	3
	H14	100	14	90	13	12	26	60	9	35	5
	H18	130	19	125	18	6	35	75	11	45	6.5
1350	O	85	12	30	4	23(e)	...	55	8	...	...
	H14	110	16	95	14	...	...	70	10	...	...
	H19	185	27	165	24	2.5(e)	...	105	15	...	...
1145	O	75	11	35	5	40	...	55	8	...	...
	H18	145	21	115	17	5	...	85	12	...	...

(a) At 0.2% offset. (b) In 50 mm or 2 in. (c) 500-kg load, 10-mm ball, 30 s. (d) Based on 500 million cycles using an R. R. Moore–type rotating-beam machine. (e) Elongation, % in 250 mm.

Table 8B. Typical mechanical properties of non-heat-treatable aluminum alloys

Alloy	Temper	Tensile strength MPa	ksi	Yield strength(a) MPa	ksi	Elongation, %(b)	Hardness, HB(c)	Shear strength MPa	ksi	Fatigue limit(d) MPa	ksi
1100	O	90	13	35	5	35	23	60	9	35	5
	H14	125	18	115	17	9	32	75	11	50	7
	H18	165	24	150	22	5	44	90	13	60	9
3003	O	110	16	40	6	30	28	75	11	50	7
	H14	150	22	145	21	8	40	95	14	60	9
	H18	200	29	185	27	4	55	110	16	70	10
Alclad 3003	O	110	16	40	6	30	...	75	11	...	...
	H14	150	22	145	21	8	...	95	14	...	...
	H18	200	29	185	27	4	...	110	16	...	...
3004	O	180	26	70	10	20	45	110	16	95	14
	H34	240	35	200	29	9	63	125	18	105	15
	H38	285	41	250	36	5	77	145	21	110	16
	H19	295	43	285	41	2	...	...	...	...	...
Alclad 3004	O	180	26	70	10	20	...	110	16	...	...
	H34	240	35	200	29	9	...	125	18	...	...
	H38	285	41	250	36	5	...	145	21	...	...
	H19	295	43	285	41	2	...	...	...	...	...
3104	H19	290	42	260	38	4	...	...	...	...	...
3005	O	130	19	55	8	25	...	...	...	...	...
	H14	180	26	165	24	7	...	...	...	...	...
	H18	240	35	225	32	4	...	...	...	...	...
3105	O	115	17	55	8	24	...	85	12	...	...
	H25	180	26	160	23	8	...	105	15	...	...
	H18	215	31	195	28	3	...	115	17	...	...
5005	O	125	18	40	6	25	28	75	11	...	...
	H34	160	23	140	20	8	41	95	14	...	...
	H38	200	29	185	27	5	55	110	16	...	...
5042	H19	360	52	345	50	4.5	...	...	...	...	...
5050	O	145	21	55	8	24	36	105	15	85	12
	H34	190	28	165	24	8	53	125	18	90	13
	H38	220	32	200	29	6	63	140	20	95	14
5052	O	195	28	90	13	25	47	125	18	110	16
	H34	260	38	215	31	10	68	145	21	125	18
	H38	290	42	255	37	7	77	165	24	140	20
5252	O	180	26	85	12	23	46	115	17	...	...
	H25	235	34	170	25	11	68	145	21	...	...
	H28	285	41	240	35	5	75	160	23	...	...

(continued)

Table 8B. (continued)

Alloy	Temper	Tensile strength MPa	ksi	Yield strength(a) MPa	ksi	Elongation, %(b)	Hardness, HB(c)	Shear strength MPa	ksi	Fatigue limit(d) MPa	ksi
5154	O	240	35	115	17	27	58	150	22	115	17
	H34	290	42	230	33	13	73	165	24	130	19
	H38	330	48	270	39	10	80	195	28	145	21
	H112	240	35	115	17	25	63	...	...	115	17
5454	O	250	36	115	17	22	62	160	23	...	...
	H34	305	44	240	35	10	81	180	26	...	...
	H111	260	38	180	26	14	70	160	23	...	...
	H112	250	36	125	18	18	62	160	23	...	...
5056	O	290	42	150	22	35	65	180	26	140	20
	H18	435	63	405	59	10	105	235	34	150	22
	H38	310	60	345	50	15	100	220	32	150	22
5456	O	310	45	160	23	24	...	...	...	...	...
	H112	310	45	165	24	22	...	...	...	...	...
	H116	350	51	255	37	16	90	205	30	...	...
5457	O	130	19	50	7	22	32	85	12	...	...
	H25	180	26	160	23	12	48	110	16	...	...
	H28	205	30	185	27	6	55	125	18	...	...
5657	O	110	16	40	6	25	28	75	11	...	...
	H25	160	23	140	20	12	40	95	14	...	...
	H28	195	28	165	24	7	50	105	15	...	...
5082	H19	395	57	370	54	4	...	...	...	...	...
5182	O	275	40	130	19	21	...	...	...	...	...
	H19	420	61	395	57	4	...	...	...	...	...
5083	O	290	42	145	21	22	...	170	25	...	...
	H116	315	46	230	33	16	...	...	...	160	23
5086	O	260	38	115	17	22	...	160	23	...	...
	H34	325	47	255	37	10	...	185	27	...	...
	H112	270	39	130	19	14	...	...	...	...	...
	H116	290	42	205	30	12	...	...	...	...	...
7072	O	70	10	...	...	15	...	...	...	...	...
	H113	75	11	...	...	15	...	...	...	...	...
8001	O	110	16	40	6	30	...	...	...	...	...
	H18	200	29	185	27	4	...	...	...	...	...
8081	H25	165	24	145	21	13	...	...	...	...	...
	H112	195	28	170	25	10	...	...	...	...	...
8280	O	115	17	50	7	28	...	...	...	...	...
	H18	220	32	205	30	4	...	...	...	...	...

(a) At 0.2% offset. (b) In 50 mm or 2 in. (c) 500-kg load, 10-mm ball, 30 s. (d) Based on 500 million cycles using an R. R. Moore–type rotating-beam machine.

Table 8C. Typical mechanical properties of heat treatable aluminum alloys

Alloy	Temper	Tensile strength MPa	ksi	Yield strength(a) MPa	ksi	Elongation, %(b)	Hardness, HB(c)	Shear strength MPa	ksi	Fatigue limit(d) MPa	ksi
2011	T3	380	55	295	43	15	95	220	32	125	18
	T6	395	57	270	39	17	97	235	34	125	18
	T8	405	59	310	45	12	100	240	35	125	18
2014	O	185	27	95	14	18	45	125	18	90	13
	T4, T451	425	62	290	42	20	105	260	38	140	20
	T6, T651	485	70	415	60	13	135	290	42	125	18
Alclad 2014	O	170	25	70	10	21	...	125	18	...	...
	T3	435	63	275	40	20	...	255	37	...	...
	T4, T451	420	61	255	37	22	...	255	37	...	...
	T6, T651	470	68	415	60	10	...	285	41	...	...
2017	O	180	26	70	10	22	45	125	18	90	13
	T4, T451	425	62	275	40	22	105	260	38	125	18
2117	T4	300	43	165	24	27	70	195	28	95	14
2218	T72	330	48	255	37	11	95	205	30	...	...
2618	T61	435	63	370	54	10	...	...	...	130	19
2219	O	170	25	70	11	18	...	...	...	...	...
	T42	360	52	185	27	20	...	...	...	...	...
	T31, T351	360	52	250	36	17	100	230	33	...	...
	T37	395	57	315	46	11	117	255	37	...	...
	T62	415	60	290	42	10	115	255	37	105	15
	T81, T85x	455	66	350	51	10	130	285	41	105	15
	T87	475	69	395	57	10	130	280	40	105	15
2024	O	185	27	75	11	20	47	125	18	90	13
	T3	485	70	345	50	18	120	280	40	140	20
	T361	495	72	395	57	13	130	290	42	125	18
	T4, T351	470	68	325	47	20	120	285	41	140	20
	T81, T851	485	70	450	65	6	128	295	43	125	18
	T86	515	75	490	71	6	135	310	45	125	18
Alclad 2024	O	180	26	76	11	20	...	125	18	...	...
	T3	450	65	310	45	18	...	275	40	...	...
	T36	460	67	365	53	11	...	285	41	...	...
	T4, T351	440	64	290	42	19	...	275	40	...	...

(continued)

Table 8C. (continued)

Alloy	Temper	Tensile strength MPa	ksi	Yield strength (a) MPa	ksi	Elongation, %(b)	Hardness, HB(c)	Shear strength MPa	ksi	Fatigue limit(d) MPa	ksi
	T8, T851	450	65	415	60	6	...	275	40	...	...
	T86	485	70	455	66	6	...	290	42	...	...
2124	T351	470	68	325	47	20	120	285	41	140	20
	T851	485	70	450	65	6	128	295	43	125	18
2025	T6	400	58	255	37	19	110	240	35	125	18
2036	T4	340	49	195	28	24	...	...	...	125	18
4032	T6	380	55	315	46	9	120	260	38	110	16
6101	T6	220	32	195	28	15	71	140	20	...	...
6201	T81	330	48	...	...	6(e)	...	...	...	105	15
6009	T4	230	33	125	18	25	62	150	22	115	17
6010	T4	290	42	170	25	24	78	195	28	125	18
6151	T6	330	48	295	43	17	100	220	32	85	12
6351	T4, T451	290	42	185	27	20	60	150	22	90	13
	T6, T651	340	49	295	43	13	95	200	29	90	13
6951	O	110	16	40	6	30	28	75	11	...	...
	T6	270	39	230	33	13	82	180	26	...	...
6053	O	110	16	55	8	35	26	75	11	55	8
	T6	255	37	220	32	13	80	160	23	90	13
6063	O	90	13	50	7	...	25	70	10	55	8
	T1	150	22	90	13	20	42	95	14	70	10
	T4	170	25	90	13	22	...	110	16	...	...
	T5	185	27	145	21	12	60	115	17	70	10
	T6	240	35	215	31	12	73	150	22	70	10
	T83	255	37	240	35	9	82	150	22	...	...
	T831	205	30	185	27	10	70	125	18	...	...
	T832	290	42	270	39	12	95	185	27	...	...
	T835	330	48	295	43	8	105	205	30	...	...
6463	O	90	13	50	7	...	25	70	10	55	8
	T1	150	22	90	13	20	42	95	14	70	10
	T4	170	25	90	13	22	...	110	16	...	...
	T5	185	27	145	21	12	60	115	17	70	10
	T6	240	35	215	31	12	74	150	22	70	10
6061	O	125	18	55	8	25	30	80	12	60	9
	T4, T451	240	35	145	21	22	65	165	24	90	13
	T6, T651	310	45	275	40	12	95	205	30	95	14
	T91	405	59	395	57	12	...	230	33	95	14
	T913	460	67	455	66	10	...	240	35	...	...
Alclad 6061	O	115	17	50	7	25	...	75	11	...	...
	T4, T451	230	33	130	19	22	...	150	22	...	...
	T6, T651	290	42	255	37	12	...	185	27	...	...
6262	T9	400	58	380	55	10	120	240	35	90	13
6066	O	150	22	85	12	18	43	95	14	...	...
	T4, T451	360	52	205	30	18	90	200	29	...	...
	T6, T651	395	57	360	53	12	120	235	34	110	16
6070	O	145	21	70	10	20	35	95	14	60	9
	T6	380	55	350	51	10	120	235	34	95	14
7001	O	255	37	150	22	14	60	...	...	...	...
	T6, T651	675	98	625	91	9	160	...	...	150	22
	T75	580	84	495	72	12	...	...	...	...	...
7005	O	195	28	80	12	20	...	...	...	...	...
	W	345	50	205	30	20	...	...	...	...	...
	T6	350	51	290	42	13	...	215	31	150	22
7016	T5	360	52	315	46	15	96	...	...	...	...
7021	T62	420	61	380	55	13	...	...	...	138	20
7029	T5	430	62	380	55	15	...	...	...	...	...
7049	T73	540	78	475	69	10	146	...	...	...	...
7050	T74, T7451, T7452	510	74	450	65	13	142	...	...	...	...
7075	O	230	33	105	15	17	60	150	22	115	17
	T6, T651	570	83	505	73	11	150	330	48	160	23
	T73, T735x	505	73	435	63	13	...	...	...	150	22
Alclad 7075	O	220	32	95	14	17	...	150	22	...	...
	T6	525	76	460	67	11	...	315	46	...	...
	T651	525	76	460	67	11	...	315	46	...	...
7175	T736, T7365x	550	80	485	70	12	145	...	...	...	...
	T7351	505	73	435	63	13	...	...	...	...	...
7475	T7351	505	73	435	63	14	...	...	...	...	...
7076	T61	510	74	470	68	14	150	...	...	...	...
7178	O	230	33	105	15	15	60	150	22	...	...
	T6, T651	605	38	540	78	10	160	360	52	150	22
	T76, T7651	570	83	505	73	9	...	...	...	...	...
Alclad 7178	O	220	32	95	14	16	...	150	22	...	...
	T6	560	81	490	71	10	...	340	49	...	...
	T651	560	81	490	71	10	...	340	49	...	...

(a) At 0.2% offset. (b) In 50 mm or 2 in. (c) 500-kg load, 10-mm ball. (d) Based on 500 million cycles using an R. R. Moore–type rotating-beam machine. (e) Elongation, % in 250 mm.

Table 8D. Typical mechanical properties of aluminum alloy foil(a)

Alloy	Temper	Tensile strength MPa	ksi	Yield strength(b) MPa	ksi	Elongation, %(c)
Non-heat-treatable foil						
1145, 1235	O	75	11	30	4.5	2.4
	H19	165	24	145	21	2.5
1100	H19	205	30	165	24	3.0
3003	H19	250	36	220	32	3.5
5052	H19	330	48	325	47	4.0
5056	H191	450	65	435	63	3.5
Heat treatable foil						
2024	T81	450	65	415	60	2.0

(a) Properties of non-heat-treatable alloy foils were determined in the longitudinal direction. Properties of 2024 heat treatable foil are for the long transverse direction. (b) At 0.2% offset. (c) In 125 mm or 5 in.

Table 9. Typical tensile properties of selected wrought aluminum alloys at various temperatures(a)

The following typical properties are not guaranteed, because in most cases they are averages for various sizes, product forms and methods of manufacture and may not be exactly representative of any particular product or size. These data are intended for use only as a basis for comparing alloys and tempers and should not be specified as engineering requirements or used for design purposes.

Alloy and temper	Temperature °C	°F	Tensile strength MPa	ksi	Yield strength(b) MPa	ksi	Elongation, %(c)	Alloy and temper	Temperature °C	°F	Tensile strength MPa	ksi	Yield strength(b) MPa	ksi	Elongation, %(c)
2014-T6, -T651	−195	−320	580	84	495	72	14		260	500	75	11	62	9	55
	−80	−112	510	74	450	65	13		315	600	52	7.5	41	6	75
	−30	−18	495	72	425	62	13		370	700	34	5	28	4	100
	25	75	485	70	415	60	13	2219-T62	−195	−320	505	73	340	49	16
	100	212	435	63	395	57	15		−80	−112	435	63	305	44	13
	150	300	275	40	240	35	20		−30	−18	415	60	290	42	12
	205	400	110	16	90	13	38		25	75	400	58	275	40	12
	260	500	65	9.5	52	7.5	52		100	212	370	54	255	37	14
	315	600	45	6.5	34	5	65		150	300	310	45	230	33	17
	370	700	30	4.3	24	3.5	72		205	400	235	34	170	25	20
2024-T3 (sheet)	−195	−320	585	85	425	62	18		260	500	185	27	140	20	21
	−80	−112	505	73	360	52	17		315	600	69	10	55	8	40
	−30	−18	495	72	350	51	17		370	700	30	4.4	26	3.7	75
	25	75	485	70	345	50	17	2219-T81, -T851	−195	−320	570	83	420	61	15
	100	212	455	66	330	48	16		−80	−112	490	71	370	54	13
	150	300	380	55	310	45	11		−30	−18	475	69	360	52	12
	205	400	185	27	140	20	23		25	75	455	66	345	50	12
	260	500	75	11	62	9	55		100	212	415	60	325	47	15
	315	600	52	7.5	41	6	75		150	300	340	49	275	40	17
	370	700	34	5	28	4	100		205	400	250	36	200	29	20
2024-T4, -T351 (plate)	−195	−320	580	84	420	61	19		260	500	200	29	160	23	21
	−80	−112	490	71	340	49	19		315	600	48	7	41	6	55
	−30	−18	475	69	325	47	19		370	700	30	4.4	26	3.7	75
	25	75	470	68	325	47	19	2618-T61	−195	−320	540	78	420	61	12
	100	212	435	63	310	45	19		−80	−112	460	67	380	55	11
	150	300	310	45	250	36	17		−30	−18	440	64	370	54	10
	205	400	180	26	130	19	27		25	75	440	64	370	54	10
	260	500	75	11	62	9	55		100	212	425	62	370	54	10
	315	600	52	7.5	41	6	75		150	300	345	50	305	44	14
	370	700	34	5	28	4	100		205	400	220	32	180	26	24
2024-T6, -T651	−195	−320	580	84	470	68	11		260	500	90	13	62	9	50
	−80	−112	495	72	405	59	10		315	600	52	7.5	31	4.5	80
	−30	−18	485	70	400	58	10		370	700	34	5	24	3.5	120
	25	75	475	69	395	57	10	4032-T6	−195	−320	455	66	330	48	11
	100	212	450	65	370	54	10		−80	−112	400	58	315	46	10
	150	300	310	45	250	36	17		−30	−18	385	56	315	46	9
	205	400	180	26	130	19	27		25	75	380	55	315	46	9
	260	500	75	11	62	9	55		100	212	345	50	305	44	9
	315	600	52	7.5	41	6	75		150	300	255	37	230	33	9
	370	700	34	5	28	4	100		205	400	90	13	62	9	30
2024-T81, -T851	−195	−320	585	85	540	78	8		260	500	55	8	38	5.5	50
	−80	−112	510	74	475	69	7		315	600	34	5	22	3.2	70
	−30	−18	505	73	470	68	7		370	700	23	3.4	14	2	90
	25	75	485	70	450	65	7	5052-O	−195	−320	305	44	110	16	46
	100	212	455	66	425	62	8		−80	−112	200	29	90	13	35
	150	300	380	55	340	49	11		−30	−18	195	28	90	13	32
	205	400	185	27	140	20	23		25	75	195	28	90	13	30
	260	500	75	11	62	9	55		100	212	195	28	90	13	36
	315	600	52	7.5	41	6	75		150	300	160	23	90	13	50
	370	700	34	5	28	4	100		205	400	115	17	75	11	60
2024-T861	−195	−320	635	92	585	85	5		260	500	85	12	52	7.5	80
	−80	−112	560	81	530	77	5		315	600	52	7.5	38	5.5	110
	−30	−18	540	78	510	74	5		370	700	34	5	21	3.1	130
	25	75	515	75	490	71	5	5052-H34	−195	−320	380	55	250	36	28
	100	212	485	70	460	67	6		−80	−112	275	40	220	32	21
	150	300	370	54	330	48	11		−30	−18	260	38	215	31	18
	205	400	145	21	115	17	28								

(continued)

Table 9. (continued)

Alloy and temper	°C	°F	Tensile strength MPa	ksi	Yield strength(b) MPa	ksi	Elongation, %(c)
	25	75	260	38	215	31	16
	100	212	260	38	215	31	18
	150	300	205	30	185	27	27
	205	400	165	24	105	15	45
	260	500	85	12	52	7.5	80
	315	600	52	7.5	38	5.5	110
	370	700	34	5	21	3.1	130
5052-H38	−195	−320	415	60	305	44	25
	−80	−112	305	44	260	38	18
	−30	−18	290	42	255	37	15
	25	75	290	42	255	37	14
	100	212	275	40	250	36	16
	150	300	235	34	195	28	24
	205	400	170	25	105	15	45
	260	500	85	12	52	7.5	80
	315	600	52	7.5	38	5.5	110
	370	700	34	5	21	3.1	130
5083-O	−195	−320	405	59	165	24	36
	−80	−112	295	43	145	21	30
	−30	−18	290	42	145	21	27
	25	75	290	42	145	21	25
	100	212	275	40	145	21	36
	150	300	215	31	130	19	50
	205	400	150	22	115	17	60
	260	500	115	17	75	11	80
	315	600	75	11	52	7.5	110
	370	700	41	6	29	4.2	130
5086-O	−195	−320	380	55	130	19	46
	−80	−112	270	39	115	17	35
	−30	−18	260	38	115	17	32
	25	75	260	38	115	17	30
	100	212	260	38	115	17	36
	150	300	200	29	110	16	50
	205	400	150	22	105	15	60
	260	500	115	17	75	11	80
	315	600	75	11	52	7.5	110
	370	700	41	6	29	4.2	130
5454-O	−195	−320	370	54	130	19	39
	−80	−112	255	37	115	17	30
	−30	−18	250	36	115	17	27
	25	75	250	36	115	17	25
	100	212	250	36	115	17	31
	150	300	200	29	110	16	50
	205	400	150	22	105	15	60
	260	500	115	17	75	11	80
	315	600	75	11	52	7.5	110
	370	700	41	6	29	4.2	130
5454-H32	−195	−320	405	59	250	36	32
	−80	−112	290	42	215	31	23
	−30	−18	285	41	205	30	20
	25	75	275	40	205	30	18
	100	212	270	39	200	29	20
	150	300	220	32	180	26	37
	205	400	170	25	130	19	45
	260	500	115	17	75	11	80
	315	600	75	11	52	7.5	110
	370	700	41	6	29	4.2	130
5454-H34	−195	−320	435	63	285	41	30
	−80	−112	315	46	250	36	21
	−30	−18	305	44	240	35	18
	25	75	305	44	240	35	16
	100	212	295	43	235	34	18
	150	300	235	34	195	28	32
	205	400	180	26	130	19	45
	260	500	115	17	75	11	80
	315	600	75	11	52	7.5	110
	370	700	41	6	29	4.2	130
5456-O	−195	−320	425	62	180	26	32
	−80	−112	315	46	160	23	25
	−30	−18	310	45	160	23	22
	25	75	310	45	160	23	20
	100	212	290	42	150	22	31
	150	300	215	31	140	20	50
	205	400	150	22	115	17	60
	260	500	115	17	75	11	80
	315	600	75	11	52	7.5	110
	370	700	41	6	29	4.2	130
6061-T6, -T651	−195	−320	415	60	325	47	22
	−80	−112	340	49	290	42	18
	−30	−18	325	47	285	41	17
	25	75	310	45	275	40	17
	100	212	290	42	260	38	18
	150	300	235	34	215	31	20
	205	400	130	19	105	15	28
	260	500	52	7.5	34	5	60
	315	600	32	4.6	19	2.7	85
	370	700	21	3	12	1.8	95
6063-T1	−195	−320	235	34	110	16	44
	−80	−112	180	26	105	15	36
	−30	−18	165	24	95	14	34
	25	75	150	22	90	13	33
	100	212	150	22	95	14	18
	150	300	145	21	105	15	20
	205	400	62	9	45	6.5	40
	260	500	31	4.5	24	3.5	75
	315	600	22	3.2	17	2.5	80
	370	700	16	2.3	14	2	105
6063-T5	−195	−320	255	37	165	24	28
	−80	−112	200	29	150	22	24
	−30	−18	195	28	150	22	23
	25	75	185	27	145	21	22
	100	212	165	24	140	20	18
	150	300	140	20	125	18	20
	205	400	62	9	45	6.5	40
	260	500	31	4.5	24	3.5	75
	315	600	22	3.2	17	2.5	80
	370	700	16	2.3	14	2	105
6063-T6	−195	−320	325	47	250	36	24
	−80	−112	260	38	230	33	20
	−30	−18	250	36	220	32	19
	25	75	240	35	215	31	18
	100	212	215	31	195	28	15
	150	300	145	21	140	20	20
	205	400	62	9	45	6.5	40
	260	500	31	4.5	24	3.5	75
	315	600	23	3.3	17	2.5	80
	370	700	16	2.3	14	2	105
7050-T7451 (plate)	25	75	510	74	455	66	11
	100	212	440	64	420	61	15
	150	300	220	32	195	28	29
	175	350	160	23	125	18	40
	205	400	115	17	90	13	54
7050-T7452 (forgings)	−195	−320	660	96	570	83	13
	−80	−112	585	85	505	73	14
	−30	−18	550	80	475	69	15
	25	75	525	76	455	66	15
	100	212	460	67	420	61	17
	150	300	220	32	195	28	29
	175	350	160	23	125	18	40
	205	400	115	17	90	13	54
7075-T6, -T651	−195	−320	705	102	635	92	9
	−80	−112	620	90	545	79	11
	−30	−18	595	86	515	75	11
	25	75	570	83	505	73	11
	100	212	485	70	450	65	14
	150	300	215	31	185	27	30
	205	400	110	16	90	13	55
	260	500	75	11	62	9	65
	315	600	55	8	45	6.5	70
	370	700	41	6	32	4.6	70
7075-T73, -T7351	−195	−320	635	92	495	72	14
	−80	−112	545	79	460	67	14
	−30	−18	525	76	450	65	13
	25	75	505	73	435	63	13
	100	212	435	63	400	58	15
	150	300	215	31	185	27	30
	205	400	110	16	90	13	55
	260	500	75	11	62	9	65
	315	600	55	8	45	6.5	70
	370	700	41	6	32	4.6	70

(a) These data are based on a limited amount of testing and represent the lowest strength exhibited during 10 000 h of exposure at test temperature under no load; stress applied at 5 ksi/min to yield strength and then at a strain rate of 0.05 in./in. · min to failure. Under some conditions of temperature and time, application of heat will adversely affect certain other properties of some alloys. (b) At 0.2% offset. (c) In 50 mm, 2 in. or 4d.

Table 10. Effects of temperature and of time at elevated temperature on tensile properties of three wrought aluminum alloys

Temperature °C	°F	Time at temperature, h	2014-T6 Tensile strength MPa	ksi	Yield strength MPa	ksi	Elongation, %(a)	2024-T81 Tensile strength MPa	ksi	Yield strength MPa	ksi	Elongation, %(a)	Temperature °C	°F	Time at temperature, h	2219-T81 Tensile strength MPa	ksi	Yield strength MPa	ksi	Elongation, %(a)
−260	−435	...	710	103	560	81	12	...	...	...	...	...	−260	−435	...	685	99	470	68	15
−200	−330	...	580	84	495	72	14	585	85	540	78	8	−200	−330	...	560	81	435	63	13
−80	−110	...	510	74	450	65	13	510	74	475	69	7	−80	−110	...	490	71	375	54	10
−28	−18	...	495	72	430	62	13	505	73	470	68	7	−28	−18	...	475	69	360	52	10
24	75	...	485	70	415	60	13	485	70	450	65	7	24	75	...	455	66	345	50	10
100	212	0.5	435	63	395	57	14	455	66	430	62	8	100	212	0.5	415	60	325	47	13
		10	435	63	395	57	14	455	66	430	62	8			10	415	60	325	47	13
		100	435	63	400	58	14	455	66	430	62	8			100	415	60	325	47	13
		1 000	440	64	405	59	15	455	66	430	62	8			1 000	415	60	325	47	13
		10 000	440	64	400	58	15	455	66	430	62	8			10 000	415	60	325	47	13
150	300	0.5	380	55	350	51	15	420	61	400	58	9	150	300	0.5	375	54	305	44	15
		10	385	56	350	51	16	415	60	395	57	9			10	375	54	305	44	15
		100	385	56	350	51	16	415	60	395	57	10			100	375	54	305	44	15
		1 000	350	51	315	46	17	405	59	380	55	10			1 000	360	52	295	43	15
		10 000	275	40	240	35	20	380	55	340	49	11			10 000	340	49	275	40	15
175	345	0.5	350	51	325	47	14	385	56	360	52	10	200	390	0.5	295	43	250	36	16
		10	345	50	315	46	17	380	55	350	51	10			10	275	40	235	34	16
		100	310	45	275	40	18	365	53	330	48	11			100	260	38	220	32	16
		1 000	235	34	205	30	20	330	48	305	44	13			1 000	250	36	205	30	16
		10 000	170	25	140	20	28	305	44	255	37	15			10 000	250	36	200	29	16
200	390	0.5	310	45	285	41	14	350	51	330	48	11	230	445	0.5	240	35	205	30	16
		10	285	41	255	37	18	340	49	310	45	11			10	235	34	195	28	16
		100	205	30	185	27	22	305	44	270	39	13			100	230	33	185	27	16
		1 000	145	21	125	18	29	260	38	220	32	15			1 000	220	32	185	27	16
		10 000	110	16	90	13	38	185	27	140	20	23			10 000	220	32	180	26	16
260	500	0.5	170	25	160	23	18	250	36	220	32	14	260	500	0.5	200	29	170	25	16
		10	110	16	105	15	27	215	31	185	27	16			10	200	29	165	24	16
		100	90	13	75	11	34	150	22	110	16	25			100	200	29	165	24	16
		1 000	75	11	65	9	43	105	15	75	11	40			1 000	200	29	165	24	16
		10 000	65	9	50	7	52	80	12	60	9	55			10 000	200	29	165	24	16
315	600	0.5	75	11	65	9	28	140	20	115	17	20	290	555	0.5	170	25	150	22	16
		10	60	9	50	7	39	80	12	70	10	45			10	165	24	145	21	16
		100	50	7	40	6	48	70	10	55	8	55			100	165	24	140	20	16
		1 000	50	7	40	6	55	60	9	45	7	65			1 000	160	23	130	19	16
		10 000	45	7	35	5	65	50	7	40	6	75			10 000	115	17	75	11	19
370	700	0.5	40	6	35	5	50	60	9	45	7	55	315	600	0.5	140	20	125	18	18
		10	35	5	30	4	56	50	7	35	5	70			10	130	19	115	17	18
		100	35	5	26	4	62	40	6	30	4	85			100	125	18	105	15	18
		1 000	30	4	26	4	68	40	6	25	4	95			1 000	95	14	85	12	23
		10 000	30	4	24	3	72	35	5	25	4	100			10 000	45	7	40	6	55

(a) In 50 mm, 2 in. or 4d.

Table 11A. Mechanical-property limits for non-heat-treatable aluminum alloy sheet and plate(a)

Alloy and temper	Specified thickness(b) mm	in.	Tensile strength Minimum MPa	ksi	Maximum MPa	ksi	Yield strength Minimum(c) MPa	ksi	Maximum MPa	ksi	Elongation (min), %(d)
1100-O	0.15-0.48	0.006-0.019	76	11.0	107	15.5	24	3.5	...	...	15
	0.51-0.79	0.020-0.031	76	11.0	107	15.5	24	3.5	...	...	20
	0.81-1.27	0.032-0.050	76	11.0	107	15.5	24	3.5	...	...	25
	1.30-6.32	0.051-0.249	76	11.0	107	15.5	24	3.5	...	...	30
	6.35-76.20	0.250-3.000	76	11.0	107	15.5	24	3.5	...	...	28
1100-H12(e)	0.43-0.48	0.017-0.019	97	14.0	131	19.0	76	11.0	...	...	3
	0.51-0.79	0.020-0.031	97	14.0	131	19.0	76	11.0	...	...	4
	0.81-1.27	0.032-0.050	97	14.0	131	19.0	76	11.0	...	...	6
	1.30-2.87	0.051-0.113	97	14.0	131	19.0	76	11.0	...	...	8
	2.90-12.67	0.114-0.499	97	14.0	131	19.0	76	11.0	...	...	9
	12.70-50.80	0.500-2.000	97	14.0	131	19.0	76	11.0	...	...	12
1100-H14(e)	0.23-0.30	0.009-0.012	110	16.0	145	21.0	97	14.0	...	...	1
	0.33-0.48	0.013-0.019	110	16.0	145	21.0	97	14.0	...	...	2
	0.51-0.79	0.020-0.031	110	16.0	145	21.0	97	14.0	...	...	3
	0.81-1.27	0.032-0.050	110	16.0	145	21.0	97	14.0	...	...	4
	1.30-2.87	0.051-0.113	110	16.0	145	21.0	97	14.0	...	...	5
	2.90-12.67	0.114-0.499	110	16.0	145	21.0	97	14.0	...	...	6
	12.70-25.40	0.500-1.000	110	16.0	145	21.0	97	14.0	...	...	10
1100-H16(e)	0.15-0.48	0.006-0.019	131	19.0	165	24.0	117	17.0	...	...	1
	0.51-0.79	0.020-0.031	131	19.0	165	24.0	117	17.0	...	...	2
	0.81-1.27	0.032-0.050	131	19.0	165	24.0	117	17.0	...	...	3
	1.30-4.11	0.051-0.162	131	19.0	165	24.0	117	17.0	...	...	4
1100-H18	0.15-0.48	0.006-0.019	152	22.0	...	...	...	...	...	...	1
	0.51-0.79	0.020-0.031	152	22.0	...	...	...	...	...	...	2
	0.81-1.27	0.032-0.050	152	22.0	...	...	...	...	...	...	3
	1.30-3.25	0.051-0.128	152	22.0	...	...	...	...	...	...	4

(continued)

Table 11A. (continued)

Alloy and temper	Specified thickness(b) mm	in.	Tensile strength Minimum MPa	ksi	Maximum MPa	ksi	Yield strength Minimum(c) MPa	ksi	Maximum MPa	ksi	Elonga-tion (min), %(d)
1100-H112	6.35-12.67	0.250-0.499	90	13.0	...	...	48	7.0	...	...	9
	12.70-50.80	0.500-2.000	83	12.0	...	...	34	5.0	...	...	14
	50.83-76.20	2.001-3.000	79	11.5	...	...	28	4.0	...	...	20
3003-O	0.15-0.18	0.006-0.007	97	14.0	131	19.0	34	5.0	...	...	14
	0.20-0.30	0.008-0.012	97	14.0	131	19.0	34	5.0	...	...	18
	0.33-0.79	0.013-0.031	97	14.0	131	19.0	34	5.0	...	...	20
	0.81-1.27	0.032-0.050	97	14.0	131	19.0	34	5.0	...	...	23
	1.30-6.32	0.051-0.249	97	14.0	131	19.0	34	5.0	...	...	25
	6.35-76.20	0.250-3.000	97	14.0	131	19.0	34	5.0	...	...	23
3003-H12(e)	0.43-0.48	0.017-0.019	117	17.0	159	23.0	83	12.0	...	...	3
	0.51-0.79	0.020-0.031	117	17.0	159	23.0	83	12.0	...	...	4
	0.81-1.27	0.032-0.050	117	17.0	159	23.0	83	12.0	...	...	5
	1.30-2.87	0.051-0.113	117	17.0	159	23.0	83	12.0	...	...	6
	2.90-4.09	0.114-0.161	117	17.0	159	23.0	83	12.0	...	...	7
	4.11-6.32	0.162-0.249	117	17.0	159	23.0	83	12.0	...	...	8
	6.35-12.67	0.250-0.499	117	17.0	159	23.0	83	12.0	...	...	9
	12.70-50.80	0.500-2.000	117	17.0	159	23.0	83	12.0	...	...	10
3003-H14(e)	0.23-0.30	0.009-0.012	138	20.0	179	26.0	117	17.0	...	...	1
	0.33-0.48	0.013-0.019	138	20.0	179	26.0	117	17.0	...	...	2
	0.51-0.79	0.020-0.031	138	20.0	179	26.0	117	17.0	...	...	3
	0.81-1.27	0.032-0.050	138	20.0	179	26.0	117	17.0	...	...	4
	1.30-2.87	0.051-0.113	138	20.0	179	26.0	117	17.0	...	...	5
	2.90-4.09	0.114-0.161	138	20.0	179	26.0	117	17.0	...	...	6
	4.11-6.32	0.162-0.249	138	20.0	179	26.0	117	17.0	...	...	7
	6.35-12.67	0.250-0.499	138	20.0	179	26.0	117	17.0	...	...	8
	12.70-25.40	0.500-1.000	138	20.0	179	26.0	117	17.0	...	...	10
3003-H16(e)	0.15-0.48	0.006-0.019	165	24.0	207	30.0	145	21.0	...	...	1
	0.51-0.79	0.020-0.031	165	24.0	207	30.0	145	21.0	...	...	2
	0.81-1.27	0.032-0.050	165	24.0	207	30.0	145	21.0	...	...	3
	1.30-4.11	0.051-0.162	165	24.0	207	30.0	145	21.0	...	...	4
3003-H18(e)	0.15-0.48	0.006-0.019	186	27.0	...	...	165	24.0	...	...	1
	0.51-0.79	0.020-0.031	186	27.0	...	...	165	24.0	...	...	2
	0.81-1.27	0.032-0.050	186	27.0	...	...	165	24.0	...	...	3
	1.30-3.25	0.051-0.128	186	27.0	...	...	165	24.0	...	...	4
3003-H112	6.35-12.67	0.250-0.499	117	17.0	...	...	69	10.0	...	...	8
	12.70-50.80	0.500-2.000	103	15.0	...	...	41	6.0	...	...	12
	50.83-76.20	2.001-3.000	100	14.5	...	...	41	6.0	...	...	18
3004-O	0.15-0.18	0.006-0.007	152	22.0	200	29.0	59	8.5	...	...	...
	0.20-0.48	0.008-0.019	152	22.0	200	29.0	59	8.5	...	...	10
	0.51-0.79	0.020-0.031	152	22.0	200	29.0	59	8.5	...	...	14
	0.81-1.27	0.032-0.050	152	22.0	200	29.0	59	8.5	...	...	16
	1.30-6.32	0.051-0.249	152	22.0	200	29.0	59	8.5	...	...	18
	6.35-76.20	0.250-3.000	152	22.0	200	29.0	59	8.5	...	...	16
3004-H32(e)	0.43-0.48	0.017-0.019	193	28.0	241	35.0	145	21.0	...	...	1
	0.51-0.79	0.020-0.031	193	28.0	241	35.0	145	21.0	...	...	3
	0.81-1.27	0.032-0.050	193	28.0	241	35.0	145	21.0	...	...	4
	1.30-2.87	0.051-0.113	193	28.0	241	35.0	145	21.0	...	...	5
	2.90-50.80	0.114-2.000	193	28.0	241	35.0	145	21.0	...	...	6
3004-H34(e)	0.23-0.48	0.009-0.019	221	32.0	262	38.0	172	25.0	...	...	1
	0.51-1.27	0.020-0.050	221	32.0	262	38.0	172	25.0	...	...	3
	1.30-2.87	0.051-0.113	221	32.0	262	38.0	172	25.0	...	...	4
	2.90-25.40	0.114-1.000	221	32.0	262	38.0	172	25.0	...	...	5
3004-H36(e)	0.15-0.18	0.006-0.007	241	35.0	283	41.0	193	28.0	...	...	...
	0.20-0.48	0.008-0.019	241	35.0	283	41.0	193	28.0	...	...	1
	0.51-0.79	0.020-0.031	241	35.0	283	41.0	193	28.0	...	...	2
	0.81-1.27	0.032-0.050	241	35.0	283	41.0	193	28.0	...	...	3
	1.30-4.11	0.051-0.162	241	35.0	283	41.0	193	28.0	...	...	4
3004-H38(e)	0.15-0.18	0.006-0.007	262	38.0	...	...	214	31.0	...	...	...
	0.20-0.48	0.008-0.019	262	38.0	...	...	214	31.0	...	...	1
	0.51-0.79	0.020-0.031	262	38.0	...	...	214	31.0	...	...	2
	0.81-1.27	0.032-0.050	262	38.0	...	...	214	31.0	...	...	3
	1.30-3.25	0.051-0.128	262	38.0	...	...	214	31.0	...	...	4
3004-H112	6.35-76.20	0.250-3.000	159	23.0	...	...	62	9.0	...	...	7
5052-O	0.15-0.18	0.006-0.007	172	25.0	214	31.0	65	9.5	...	...	...
	0.20-0.30	0.008-0.012	172	25.0	214	31.0	65	9.5	...	...	14
	0.33-0.48	0.013-0.019	172	25.0	214	31.0	65	9.5	...	...	15
	0.51-0.79	0.020-0.031	172	25.0	214	31.0	65	9.5	...	...	16
	0.81-1.27	0.032-0.050	172	25.0	214	31.0	65	9.5	...	...	18
	1.30-2.87	0.051-0.113	172	25.0	214	31.0	65	9.5	...	...	19
	2.90-6.32	0.114-0.249	172	25.0	214	31.0	65	9.5	...	...	20
	6.35-76.20	0.250-3.000	172	25.0	214	31.0	65	9.5	...	...	18
5052-H32(e)	0.43-0.48	0.017-0.019	214	31.0	262	38.0	159	23.0	...	...	4
	0.51-1.27	0.020-0.050	214	31.0	262	38.0	159	23.0	...	...	5
	1.30-2.87	0.051-0.113	214	31.0	262	38.0	159	23.0	...	...	7
	2.90-6.32	0.114-0.249	214	31.0	262	38.0	159	23.0	...	...	9
	6.35-12.67	0.250-0.499	214	31.0	262	38.0	159	23.0	...	...	11
	12.70-50.80	0.500-2.000	214	31.0	262	38.0	159	23.0	...	...	12

(continued)

Table 11A. (continued)

Alloy and temper	Specified thickness(b) mm	in.	Tensile strength Minimum MPa	ksi	Maximum MPa	ksi	Yield strength Minimum(c) MPa	ksi	Maximum MPa	ksi	Elongation (min), %(d)
5052-H34(e)	0.23-0.48	0.009-0.019	234	34.0	283	41.0	179	26.0	...	...	3
	0.51-1.27	0.020-0.050	234	34.0	283	41.0	179	26.0	...	...	4
	1.30-2.87	0.051-0.113	234	34.0	283	41.0	179	26.0	...	...	6
	2.90-6.32	0.114-0.249	234	34.0	283	41.0	179	26.0	...	...	7
	6.35-25.40	0.250-1.000	234	34.0	283	41.0	179	26.0	...	...	10
5052-H36(e)	0.15-0.18	0.006-0.007	255	37.0	303	44.0	200	29.0	...	...	2
	0.20-0.79	0.008-0.031	255	37.0	303	44.0	200	29.0	...	...	3
	0.81-4.11	0.032-0.162	255	37.0	303	44.0	200	29.0	...	...	4
5052-H38(e)	0.15-0.18	0.006-0.007	269	39.0	...	...	221	32.0	...	...	2
	0.20-0.79	0.008-0.031	269	39.0	...	...	221	32.0	...	...	3
	0.81-3.25	0.032-0.128	269	39.0	...	...	221	32.0	...	...	4
5052-H112	6.35-12.67	0.250-0.499	193	28.0	...	...	110	16.0	...	...	7
	12.70-50.80	0.500-2.000	172	25.0	...	...	65	9.5	...	...	12
	50.83-76.20	2.001-3.000	172	25.0	...	...	65	9.5	...	...	16
5083-O	1.30-38.10	0.051-1.500	276	40.0	352	51.0	124	18.0	200	29.0	16
	38.13-76.20	1.501-3.000	269	39.0	345	50.0	117	17.0	200	29.0	16
	76.23-101.60	3.001-4.000	262	38.0	...	...	110	16.0	...	...	16
	101.63-127.00	4.001-5.000	262	38.0	...	...	110	16.0	...	...	14
	127.03-177.80	5.001-7.000	255	37.0	...	...	103	15.0	...	...	14
	177.83-203.20	7.001-8.000	248	36.0	...	...	97	14.0	...	...	12
5083-H112	6.35-38.10	0.250-1.500	276	40.0	...	...	124	18.0	...	...	12
	38.13-76.20	1.501-3.000	269	39.0	...	...	117	17.0	...	...	12
5083-H116(f)(g)	0.10-12.67	0.063-0.499	303	44.0	...	...	214	31.0	...	...	10
	12.70-31.75	0.500-1.250	303	44.0	...	...	214	31.0	...	...	12
	31.78-38.10	1.251-1.500	303	44.0	...	...	214	31.0	...	...	12
	38.13-76.20	1.501-3.000	283	41.0	...	...	200	29.0	...	...	12
5083-H321	4.78-38.10	0.188-1.500	303	44.0	386	56.0	214	31.0	296	43.0	12
	38.13-76.20	1.501-3.000	283	41.0	386	56.0	200	29.0	296	43.0	12
5086-O	0.51-1.27	0.020-0.50	241	35.0	303	44.0	97	14.0	...	...	15
	1.30-6.32	0.051-0.249	241	35.0	303	44.0	97	14.0	...	...	18
	6.35-50.80	0.250-2.000	241	35.0	303	44.0	97	14.0	...	...	16
5086-H32(e)	0.51-1.27	0.020-0.050	276	40.0	324	47.0	193	28.0	...	...	6
	1.30-6.32	0.051-0.249	276	40.0	324	47.0	193	28.0	...	...	8
	6.35-50.80	0.250-2.000	276	40.0	324	47.0	193	28.0	...	...	12
5086-H34(e)	0.23-0.48	0.009-0.019	303	44.0	352	51.0	234	34.0	...	...	4
	0.51-1.27	0.020-0.050	303	44.0	352	51.0	234	34.0	...	...	5
	1.30-6.32	0.051-0.249	303	44.0	352	51.0	234	34.0	...	...	6
	6.35-25.40	0.250-1.000	303	44.0	352	51.0	234	34.0	...	...	10
5086-H36(e)	0.15-0.48	0.006-0.019	324	47.0	372	54.0	262	38.0	...	...	3
	0.51-1.27	0.020-0.050	324	47.0	372	54.0	262	38.0	...	...	4
	1.30-4.11	0.051-0.162	324	47.0	372	54.0	262	38.0	...	...	6
5086-H38(e)	0.15-0.51	0.006-0.020	345	50.0	...	...	283	41.0	...	...	3
5086-H112	4.78-12.67	0.188-0.499	248	36.0	...	...	124	18.0	...	...	8
	12.70-25.40	0.500-1.000	241	35.0	...	...	110	16.0	...	...	10
	25.43-50.80	1.001-2.000	241	35.0	...	...	97	14.0	...	...	14
	50.83-76.20	2.001-3.000	234	34.0	...	...	97	14.0	...	...	14
5086-H116(f)(g)	0.10-6.32	0.063-0.249	276	40.0	...	...	193	28.0	...	...	8
	6.35-12.67	0.250-0.499	276	40.0	...	...	193	28.0	...	...	10
	12.70-31.75	0.500-1.250	276	40.0	...	...	193	28.0	...	...	10
	31.78-50.80	1.251-2.000	276	40.0	...	...	193	28.0	...	...	10
5154-O	0.51-0.79	0.020-0.031	207	30.0	283	41.0	76	11.0	...	...	12
	0.81-1.27	0.032-0.050	207	30.0	283	41.0	76	11.0	...	...	14
	1.30-2.87	0.051-0.113	207	30.0	283	41.0	76	11.0	...	...	16
	2.90-76.20	0.114-3.000	207	30.0	283	41.0	76	11.0	...	...	18
5154-H32(e)	0.51-1.27	0.020-0.050	248	36.0	296	43.0	179	26.0	...	...	5
	1.30-6.32	0.051-0.249	248	36.0	296	43.0	179	26.0	...	...	8
	6.35-50.80	0.250-2.000	248	36.0	296	43.0	179	26.0	...	...	12
5154-H34(e)	0.23-1.27	0.009-0.050	269	39.0	317	46.0	200	29.0	...	...	4
	1.30-4.09	0.051-0.161	269	39.0	317	46.0	200	29.0	...	...	6
	4.11-6.32	0.162-0.249	269	39.0	317	46.0	200	29.0	...	...	7
	6.35-25.40	0.250-1.000	269	39.0	317	46.0	200	29.0	...	...	10
5154-H36(e)	0.15-1.27	0.006-0.050	290	42.0	338	49.0	221	32.0	...	...	3
	1.30-2.87	0.051-0.113	290	42.0	338	49.0	221	32.0	...	...	4
	2.90-4.11	0.114-0.162	290	42.0	338	49.0	221	32.0	...	...	5
5154-H38(e)	0.15-1.27	0.006-0.050	310	45.0	...	...	241	35.0	...	...	3
	1.30-2.87	0.051-0.113	310	45.0	...	...	241	35.0	...	...	4
	2.90-3.25	0.114-0.128	310	45.0	...	...	241	35.0	...	...	5
5154-H112	6.35-12.67	0.250-0.499	221	32.0	...	...	124	18.0	...	...	8
	12.70-50.80	0.500-2.000	207	30.0	...	...	76	11.0	...	...	11
	50.83-76.20	2.001-3.000	207	30.0	...	...	76	11.0	...	...	15
5454-O	0.51-0.79	0.020-0.031	214	31.0	283	41.0	83	12.0	...	...	12
	0.81-1.27	0.032-0.050	214	31.0	283	41.0	83	12.0	...	...	14
	1.30-2.87	0.051-0.113	214	31.0	283	41.0	83	12.0	...	...	16
	2.90-76.20	0.114-3.000	214	31.0	283	41.0	83	12.0	...	...	18
5454-H32(e)	0.51-1.27	0.020-0.050	248	36.0	303	44.0	179	26.0	...	...	5
	1.30-6.32	0.051-0.249	248	36.0	303	44.0	179	26.0	...	...	8
	6.35-50.80	0.250-2.000	248	36.0	303	44.0	179	26.0	...	...	12

(continued)

Table 11A. (continued)

Alloy and temper	Specified thickness(b) mm	in.	Tensile strength Minimum MPa	ksi	Maximum MPa	ksi	Yield strength Minimum(c) MPa	ksi	Maximum MPa	ksi	Elonga-tion (min), %(d)
5454-H34(e)	0.51-1.27	0.020-0.050	269	39.0	324	47.0	200	29.0	...	...	4
	1.30-4.09	0.051-0.161	269	39.0	324	47.0	200	29.0	...	...	6
	4.11-6.32	0.162-0.249	269	39.0	324	47.0	200	29.0	...	...	7
	6.35-25.40	0.250-1.000	269	39.0	324	47.0	200	29.0	...	...	10
5454-H112	6.35-12.67	0.250-0.499	221	32.0	...	...	124	18.0	...	...	8
	12.70-50.80	0.500-2.000	214	31.0	...	...	83	12.0	...	...	11
	50.83-76.20	2.001-3.000	214	31.0	...	...	83	12.0	...	...	15
5456-O	1.30-38.10	0.051-1.500	290	42.0	365	53.0	131	19.0	207	30.0	16
	38.13-76.20	1.501-3.000	283	41.0	358	52.0	124	18.0	207	30.0	16
	76.23-127.00	3.001-5.000	276	40.0	...	...	117	17.0	...	...	14
	127.03-177.80	5.001-7.000	269	39.0	...	...	110	16.0	...	...	14
	177.83-203.20	7.001-8.000	262	38.0	...	...	103	15.0	...	...	12
5456-H112	6.35-38.10	0.250-1.500	290	42.0	...	...	131	19.0	...	...	12
	38.13-76.20	1.501-3.000	283	41.0	...	...	124	18.0	...	...	12
5456-H116(f)(g)	1.60-12.67	0.063-0.499	317	46.0	...	...	228	33.0	...	...	10
	12.70-31.75	0.500-1.250	317	46.0	...	...	228	33.0	...	...	12
	31.78-38.10	1.251-1.500	303	44.0	...	...	214	31.0	...	...	12
	38.13-76.20	1.501-3.000	283	41.0	...	...	200	29.0	...	...	12
	76.23-101.60	3.001-4.000	276	40.0	...	...	172	25.0	...	...	12
5456-H321	4.78-12.67	0.188-0.499	317	46.0	407	59.0	228	33.0	317	46.0	12
	12.70-38.10	0.500-1.500	303	44.0	386	56.0	214	31.0	303	44.0	12
	38.13-76.20	1.501-3.000	283	41.0	372	54.0	200	29.0	296	43.0	12

(a) Converted SI (metric) values are for information only and are not to be used for purposes of specification, acceptance or rejection. (b) Type of test specimen used depends on thickness of material. (c) Minimum yield strengths are not determined unless specifically requested. (d) In 50 mm, 2 in. or 4d. (e) For the corresponding H2 temper, limits for maximum tensile strength and minimum yield strength do not apply. (f) When tested upon receipt by the purchaser, material in this temper is required to pass the exfoliation corrosion resistance test (ASSET method). The improved resistance to exfoliation corrosion of individual lots is determined by microscopic examination to ensure a microstructure that is predominantly free of a continuous grain-boundary network of aluminum-magnesium precipitate. The microstructure is compared with that in a previously established acceptable reference photomicrograph. (g) Also applies to material previously designated H117.

Table 11B. Mechanical-property limits for heat treatable aluminum alloy sheet and plate(a)

Alloy and temper	Specified thickness mm	in.	Tensile strength Minimum MPa	ksi	Maximum MPa	ksi	Yield strength Minimum MPa	ksi	Maximum MPa	ksi	Elongation (min), %(b)
2014-O sheet and plate	0.51-12.67	0.020-0.499	...	...	221	32.0	...	...	110	16.0	16
	12.70-25.40	0.500-1.000	...	...	221	32.0	...	...	...	...	10
2014-T3 flat sheet	0.51-0.99	0.020-0.039	407	59.0	...	...	241	35.0	...	...	14
	1.02-6.32	0.040-0.249	407	59.0	...	...	248	36.0	...	...	14
2014-T4 coiled sheet	0.51-6.32	0.020-0.249	407	59.0	...	...	241	35.0	...	...	14
2014-T451(c)(d) plate	6.35-12.67	0.250-0.499	400	58.0	...	...	248	36.0	...	...	14
	12.70-25.40	0.500-1.000	400	58.0	...	...	248	36.0	...	...	14
	25.43-50.80	1.001-2.000	400	58.0	...	...	248	36.0	...	...	12
	50.83-76.20	2.001-3.000	393	57.0	...	...	248	36.0	...	...	8
2014-T42(e)(f) sheet and plate	0.51-25.40	0.020-1.000	400	58.0	...	...	234	34.0	...	...	14
2014-T6 and -T62(e)(f) sheet	0.51-0.99	0.020-0.039	441	64.0	...	...	393	57.0	...	...	6
	1.02-6.32	0.040-0.249	441	66.0	...	...	400	58.0	...	...	7
2014-T62(e)(f) and -T651(c) plate	6.35-12.67	0.250-0.499	462	67.0	...	...	407	59.0	...	...	7
	12.70-25.40	0.500-1.000	462	67.0	...	...	407	59.0	...	...	6
	25.43-50.80	1.001-2.000	462	67.0	...	...	407	59.0	...	...	4
	50.83-63.50	2.001-2.500	448	65.0	...	...	400	58.0	...	...	2
	63.53-76.20	2.501-3.000	434	63.0	...	...	393	57.0	...	...	2
	76.23-101.60	3.001-4.000	407	59.0	...	...	379	55.0	...	...	1
2024-O sheet and plate	0.25-12.67	0.010-0.499	...	...	221	32.0	...	...	97	14.0	12
	12.70-44.45	0.500-1.750	...	...	221	32.0	...	...	...	...	12
2024-T3(d) flat sheet	0.20-0.23	0.008-0.009	434	63.0	...	...	290	42.0	...	...	10
	0.25-0.51	0.010-0.020	434	63.0	...	...	290	42.0	...	...	12
	0.53-3.25	0.021-0.128	434	63.0	...	...	290	42.0	...	...	15
	3.28-6.32	0.129-0.249	441	64.0	...	...	290	42.0	...	...	15
2024-T361(d)(g) flat sheet and plate	0.51-1.57	0.020-0.062	462	67.0	...	...	345	50.0	...	...	8
	1.60-6.32	0.063-0.249	469	68.0	...	...	352	51.0	...	...	9
	6.35-12.67	0.250-0.499	455	66.0	...	...	338	49.0	...	...	9
	12.70	0.500	455	66.0	...	...	338	49.0	...	...	10
2024-T4 coiled sheet	0.25-0.51	0.010-0.020	427	62.0	...	...	276	40.0	...	...	12
	0.53-6.32	0.021-0.249	427	62.0	...	...	276	40.0	...	...	15
2024-T351(c)(d) plate	6.35-12.67	0.250-0.499	441	64.0	...	...	290	42.0	...	...	12
	12.70-25.40	0.500-1.000	434	63.0	...	...	290	42.0	...	...	8
	25.43-38.10	1.001-1.500	427	62.0	...	...	290	42.0	...	...	7
	38.13-50.80	1.501-2.000	427	62.0	...	...	290	42.0	...	...	6
	50.83-76.20	2.001-3.000	414	60.0	...	...	290	42.0	...	...	4
	76.23-101.60	3.001-4.000	393	57.0	...	...	283	41.0	...	...	4
2024-T42(e)(f) sheet and plate	0.25-0.51	0.010-0.020	427	62.0	...	...	262	38.0	...	...	12
	0.53-6.32	0.021-0.249	427	62.0	...	...	262	38.0	...	...	15
	6.35-12.67	0.250-0.499	427	62.0	...	...	262	38.0	...	...	12
	12.70-25.40	0.500-1.000	421	61.0	...	...	262	38.0	...	...	8
	25.43-38.10	1.001-1.500	414	60.0	...	...	262	38.0	...	...	7
	38.13-50.80	1.501-2.000	414	60.0	...	...	262	38.0	...	...	6
	50.83-76.20	2.001-3.000	400	58.0	...	...	262	38.0	...	...	4

(continued)

Table 11B. (continued)

Alloy and temper	Specified thickness mm	in.	Tensile strength Minimum MPa	ksi	Maximum MPa	ksi	Yield strength Minimum MPa	ksi	Maximum MPa	ksi	Elongation (min), %(b)
2024-T62(e)(f) sheet and plate	0.25-12.67	0.010-0.499	441	64.0	...	...	345	50.0	...	...	5
	12.70-76.20	0.500-3.000	434	63.0	...	...	345	50.0	...	...	5
2024-T72(e)(f) sheet	0.25-6.32	0.010-0.249	414	60.0	...	...	317	46.0	...	...	5
2024-T81 flat sheet	0.25-6.32	0.010-0.249	462	67.0	...	...	400	58.0	...	...	5
2024-T851(c) plate	6.35-12.67	0.250-0.499	462	67.0	...	...	400	58.0	...	...	5
	12.70-25.40	0.500-1.000	455	66.0	...	...	400	58.0	...	...	5
	25.43-38.07	1.001-1.499	455	66.0	...	...	393	57.0	...	...	5
2024-T861(g) flat sheet and plate	0.51-1.57	0.020-0.062	483	70.0	...	...	427	62.0	...	...	3
	1.60-6.32	0.063-0.249	489	71.0	...	...	455	66.0	...	...	4
	6.35-12.67	0.250-0.499	483	70.0	...	...	441	64.0	...	...	4
	12.70	0.500	483	70.0	...	...	441	64.0	...	...	4
Alclad 2024-O sheet and plate	0.20-0.23	0.008-0.009	...	...	207	30.0	...	...	97	14.0	10
	0.25-0.81	0.010-0.032	...	...	207	30.0	...	...	97	14.0	12
	0.84-1.57	0.033-0.062	...	...	207	30.0	...	...	97	14.0	12
	1.60-4.75	0.063-0.187	...	...	221	32.0	...	...	97	14.0	12
	4.78-12.67	0.188-0.499	...	...	221	32.0	...	...	97	14.0	12
	12.70-44.45	0.500-1.750	...	...	221	32.0(h)	...	...	...	...	12
Alclad 2024-T3(d) flat sheet	0.20-0.23	0.008-0.009	400	58.0	...	...	269	39.0	...	...	10
	0.25-0.51	0.010-0.020	407	59.0	...	...	269	39.0	...	...	12
	0.53-1.57	0.021-0.062	407	59.0	...	...	269	39.0	...	...	15
	1.60-3.25	0.063-0.128	421	61.0	...	...	276	40.0	...	...	15
	3.28-6.32	0.129-0.249	427	62.0	...	...	276	40.0	...	...	15
Alclad 2024-T361(d)(g) flat sheet and plate	0.51-1.57	0.020-0.062	421	61.0	...	...	324	47.0	...	...	8
	1.60-4.75	0.063-0.187	441	64.0	...	...	331	48.0	...	...	9
	4.78-6.32	0.188-0.249	441	64.0	...	...	331	48.0	...	...	9
	6.35-12.67	0.250-0.499	441	64.0	...	...	331	48.0	...	...	9
	12.70	0.500	455	66.0(h)	...	...	338	49.0(h)	...	...	10
Alclad 2024-T4 coiled sheet	0.25-0.51	0.010-0.020	400	58.0	...	...	248	36.0	...	...	12
	0.53-1.57	0.021-0.062	400	58.0	...	...	248	36.0	...	...	15
	1.60-3.25	0.063-0.128	421	61.0	...	...	262	38.0	...	...	15
Alclad 2024-T351(c)(d) plate	6.35-12.67	0.250-0.499	427	62.0	...	...	276	40.0	...	...	12
	12.70-25.40	0.500-1.000	434	63.0(h)	...	...	290	42.0(h)	...	...	8
	25.43-38.10	1.001-1.500	427	62.0(h)	...	...	290	42.0(h)	...	...	7
	38.13-50.80	1.501-2.000	427	62.0(h)	...	...	290	42.0(h)	...	...	6
	50.83-76.20	2.001-3.000	414	60.0(h)	...	...	290	42.0(h)	...	...	4
	76.23-101.60	3.001-4.000	393	57.0(h)	...	...	283	41.0(h)	...	...	4
Alclad 2024-T42(e)(f) sheet and plate	0.20-0.23	0.008-0.009	379	55.0	...	...	234	34.0	...	...	10
	0.25-0.51	0.010-0.020	393	57.0	...	...	234	34.0	...	...	12
	0.53-1.57	0.021-0.062	393	57.0	...	...	234	34.0	...	...	15
	1.60-4.75	0.063-0.187	414	60.0	...	...	248	36.0	...	...	15
	4.78-6.32	0.188-0.249	414	60.0	...	...	248	36.0	...	...	15
	6.35-12.67	0.250-0.499	414	60.0	...	...	248	36.0	...	...	12
	12.70-25.40	0.500-1.000	421	61.0(h)	...	...	262	38.0(h)	...	...	8
	25.43-38.10	1.001-1.500	414	60.0(h)	...	...	262	38.0(h)	...	...	7
	38.13-50.80	1.501-2.000	414	60.0(h)	...	...	262	38.0(h)	...	...	6
	50.83-76.20	2.001-3.000	400	58.0(h)	...	...	262	38.0(h)	...	...	4
Alclad 2024-T62(e)(f) sheet and plate	0.25-1.57	0.010-0.062	414	60.0	...	...	324	47.0	...	...	5
	1.60-4.75	0.063-0.187	427	62.0	...	...	338	49.0	...	...	5
	4.78-12.67	0.188-0.499	427	62.0	...	...	338	49.0	...	...	5
Alclad 2024-T72(e)(f) sheet	0.25-1.57	0.010-0.062	386	56.0	...	...	296	43.0	...	...	5
	1.60-4.75	0.063-0.187	400	58.0	...	...	310	45.0	...	...	5
	4.78-6.32	0.188-0.249	400	58.0	...	...	310	45.0	...	...	5
Alclad 2024-T81 flat sheet	0.25-1.57	0.010-0.062	427	62.0	...	...	372	54.0	...	...	5
	1.60-4.75	0.063-0.187	448	65.0	...	...	386	56.0	...	...	5
	4.78-6.32	0.188-0.249	448	65.0	...	...	386	56.0	...	...	5
Alclad 2024-T851(c) plate	6.35-12.67	0.250-0.499	448	65.0	...	...	386	56.0	...	...	5
	12.70-25.40	0.500-1.000	455	66.0(h)	...	...	400	58.0(h)	...	...	5
Alclad 2024-T861(g) flat sheet and plate	0.51-1.57	0.020-0.062	441	64.0	...	...	400	58.0	...	...	3
	1.60-4.75	0.063-0.187	476	69.0	...	...	441	64.0	...	...	4
	4.78-6.32	0.188-0.249	476	69.0	...	...	441	64.0	...	...	4
	6.35-12.67	0.250-0.499	469	68.0	...	...	427	62.0	...	...	4
	12.70	0.500	483	70.0(h)	...	...	441	64.0	...	...	4
2036-T4 flat sheet	0.64-3.18	0.025-0.125	290	42.0	...	...	.59	23.0	...	...	20
2219-O sheet and plate	0.51-50.83	0.020-2.000	...	...	221	32.0	...	...	110	16.0	12
2219-T31(d) flat sheet	0.51-0.99	0.020-0.039	317	46.0	...	...	200	29.0	...	...	8
	1.02-6.32	0.040-0.249	317	46.0	...	...	193	28.0	...	...	10
2219-T351(c)(d) plate	6.35-50.80	0.250-2.000	317	46.0	...	...	193	28.0	...	...	10
	50.83-76.20	2.001-3.000	303	44.0	...	...	193	28.0	...	...	10
	76.23-101.60	3.001-4.000	290	42.0	...	...	186	27.0	...	...	9
	101.63-127.00	4.001-5.000	276	40.0	...	...	179	26.0	...	...	9
	127.03-152.40	5.001-6.000	269	39.0	...	...	172	25.0	...	...	8
2219-T37(d) flat sheet and plate	0.51-0.99	0.020-0.039	338	49.0	...	...	262	38.0	...	...	6
	1.02-50.80	0.040-2.000	338	49.0	...	...	255	37.0	...	...	6
	50.83-63.50	2.001-2.500	338	49.0	...	...	255	37.0	...	...	6
	63.53-76.20	2.501-3.000	324	47.0	...	...	248	36.0	...	...	6
	76.23-101.60	3.001-4.000	310	45.0	...	...	241	35.0	...	...	5
	101.63-127.00	4.001-5.000	296	43.0	...	...	234	34.0	...	...	4

(continued)

Table 11B. (continued)

| Alloy and temper | Specified thickness | | Tensile strength | | | | Yield strength | | | | Elongation |
| | mm | in. | Minimum | | Maximum | | Minimum | | Maximum | | (min), %(b) |
			MPa	ksi	MPa	ksi	MPa	ksi	MPa	ksi	
2219-T62(e)(f) sheet and plate	0.51-0.99	0.020-0.039	372	54.0	. . .	. . .	248	36.0	. . .	. . .	6
	1.02-6.32	0.040-0.249	372	54.0	. . .	. . .	248	36.0	. . .	. . .	7
	6.35-25.40	0.250-1.000	372	54.0	. . .	. . .	248	36.0	. . .	. . .	8
	25.43-50.80	1.001-2.000	372	54.0	. . .	. . .	248	36.0	. . .	. . .	7
2219-T81 flat sheet	0.51-0.99	0.020-0.039	427	62.0	. . .	. . .	317	46.0	. . .	. . .	6
	1.02-6.32	0.040-0.249	427	62.0	. . .	. . .	317	46.0	. . .	. . .	7
2219-T851(c) plate	6.35-25.40	0.250-1.000	427	62.0	. . .	. . .	317	46.0	. . .	. . .	8
	25.43-50.80	1.001-2.000	427	62.0	. . .	. . .	317	46.0	. . .	. . .	7
	50.83-76.20	2.001-3.000	427	62.0	. . .	. . .	310	45.0	. . .	. . .	6
	76.23-101.60	3.001-4.000	414	60.0	. . .	. . .	303	44.0	. . .	. . .	5
	101.63-127.00	4.001-5.000	407	59.0	. . .	. . .	296	43.0	. . .	. . .	5
	127.03-152.40	5.001-6.000	393	57.0	. . .	. . .	290	42.0	. . .	. . .	4
2219-T87 flat sheet and plate	0.51-0.99	0.020-0.039	441	64.0	. . .	. . .	358	52.0	. . .	. . .	5
	1.02-6.32	0.040-0.249	441	64.0	. . .	. . .	358	52.0	. . .	. . .	6
	6.35-25.40	0.250-1.000	441	64.0	. . .	. . .	352	51.0	. . .	. . .	7
	25.43-50.80	1.001-2.000	441	64.0	. . .	. . .	352	51.0	. . .	. . .	6
	50.83-76.20	2.001-3.000	441	64.0	. . .	. . .	352	51.0	. . .	. . .	6
	76.23-101.60	3.001-4.000	427	62.0	. . .	. . .	345	50.0	. . .	. . .	4
	101.63-127.00	4.001-5.000	421	61.0	. . .	. . .	338	49.0	. . .	. . .	3
6061-O sheet and plate	0.15-0.18	0.006-0.007		. . .	152	22.0	. . .	. . .	83	12.0	10
	0.20-0.23	0.008-0.009		. . .	152	22.0	. . .	. . .	83	12.0	12
	0.25-0.51	0.010-0.020		. . .	152	22.0	. . .	. . .	83	12.0	14
	0.53-3.25	0.021-0.128		. . .	152	22.0	. . .	. . .	83	12.0	16
	3.28-12.67	0.129-0.499		. . .	152	22.0	. . .	. . .	83	12.0	18
	12.70-25.40	0.500-1.000		. . .	152	22.0	. . .	. . .	. . .	. . .	18
	25.43-76.20	1.001-3.000		. . .	152	22.0	. . .	. . .	. . .	. . .	16
6061-T4 sheet	0.15-0.18	0.006-0.007	207	30.0	. . .	. . .	110	16.0	. . .	. . .	10
	0.20-0.23	0.008-0.009	207	30.0	. . .	. . .	110	16.0	. . .	. . .	12
	0.25-0.51	0.010-0.020	207	30.0	. . .	. . .	110	16.0	. . .	. . .	14
	0.53-6.32	0.021-0.249	207	30.0	. . .	. . .	110	16.0	. . .	. . .	16
6061-T451(c)(d) plate	6.35-25.40	0.250-1.000	207	30.0	. . .	. . .	110	16.0	. . .	. . .	18
	25.43-76.20	1.001-3.000	207	30.0	. . .	. . .	110	16.0	. . .	. . .	16
6061-T42(e)(f) sheet and plate	0.15-0.18	0.006-0.007	207	30.0	. . .	. . .	97	14.0	. . .	. . .	10
	0.20-0.23	0.008-0.009	207	30.0	. . .	. . .	97	14.0	. . .	. . .	12
	0.25-0.51	0.010-0.020	207	30.0	. . .	. . .	97	14.0	. . .	. . .	14
	0.53-6.32	0.021-0.249	207	30.0	. . .	. . .	97	14.0	. . .	. . .	16
	6.35-25.40	0.250-1.000	207	30.0	. . .	. . .	97	14.0	. . .	. . .	18
	25.43-76.20	1.001-3.000	207	30.0	. . .	. . .	97	14.0	. . .	. . .	16
6061-T6 and -T62(e)(f) sheet	0.15-0.18	0.006-0.007	290	42.0	. . .	. . .	241	35.0	. . .	. . .	4
	0.20-0.23	0.008-0.009	290	42.0	. . .	. . .	241	35.0	. . .	. . .	6
	0.25-0.51	0.010-0.020	290	42.0	. . .	. . .	241	35.0	. . .	. . .	8
	0.53-6.32	0.021-0.249	290	42.0	. . .	. . .	241	35.0	. . .	. . .	10
6061-T62(e)(f) and -T651(c) plate	6.35-12.67	0.250-0.499	290	42.0	. . .	. . .	241	35.0	. . .	. . .	10
	12.70-25.40	0.500-1.000	290	42.0	. . .	. . .	241	35.0	. . .	. . .	9
	25.43-50.80	1.001-2.000	290	42.0	. . .	. . .	241	35.0	. . .	. . .	8
	50.83-101.60	2.001-4.000	290	42.0	. . .	. . .	241	35.0	. . .	. . .	6
	101.63-152.40	4.001-6.000(j)	276	40.0	. . .	. . .	241	35.0	. . .	. . .	6
7075-O sheet and plate	0.38-12.67	0.015-0.499		. . .	276	40.0	. . .	. . .	145	21.0	10
	12.70-50.80	0.500-2.000		. . .	276	40.0	. . .	. . .	. . .	. . .	10
7075-T6 and -T62(e)(f) sheet	0.20-0.28	0.008-0.011	510	74.0	. . .	. . .	434	63.0	. . .	. . .	5
	0.30-0.99	0.012-0.039	524	76.0	. . .	. . .	462	67.0	. . .	. . .	7
	1.02-3.18	0.040-0.125	538	78.0	. . .	. . .	469	68.0	. . .	. . .	8
	3.21-6.32	0.126-0.249	538	78.0	. . .	. . .	476	69.0	. . .	. . .	8
7075-T62(e)(f) and -T651(c) plate	6.35-12.67	0.250-0.499	538	78.0	. . .	. . .	462	67.0	. . .	. . .	9
	12.70-25.40	0.500-1.000	538	78.0	. . .	. . .	469	68.0	. . .	. . .	7
	25.43-50.80	1.001-2.000	531	77.0	. . .	. . .	462	67.0	. . .	. . .	6
	50.83-63.50	2.001-2.500	524	76.0	. . .	. . .	441	64.0	. . .	. . .	5
	63.53-76.20	2.501-3.000	496	72.0	. . .	. . .	421	61.0	. . .	. . .	5
	76.23-88.90	3.001-3.500	489	71.0	. . .	. . .	400	58.0	. . .	. . .	5
	88.93-101.60	3.501-4.000	462	67.0	. . .	. . .	372	54.0	. . .	. . .	3
7075-T73(k) sheet	1.02-6.32	0.040-0.249	462	67.0	. . .	. . .	386	56.0	. . .	. . .	8
7075-T7351(c)(k) plate	6.35-25.40	0.250-1.000	476	69.0	. . .	. . .	393	57.0	. . .	. . .	7
	25.43-50.80	1.001-2.000	476	69.0	. . .	. . .	393	57.0	. . .	. . .	6
	50.83-63.50	2.001-2.500	455	66.0	. . .	. . .	358	52.0	. . .	. . .	6
	63.53-76.20	2.501-3.000	441	64.0	. . .	. . .	338	49.0	. . .	. . .	6
7075-T76(m) sheet	3.18-6.32	0.125-0.249	503	73.0	. . .	. . .	427	62.0	. . .	. . .	8
7075-T7651(e)(m) plate	6.35-12.67	0.250-0.499	496	72.0	. . .	. . .	421	61.0	. . .	. . .	8
	12.70-25.40	0.500-1.000	489	71.0	. . .	. . .	414	60.0	. . .	. . .	6
7178-O sheet and plate	0.38-12.67	0.015-0.499		. . .	276	40.0	. . .	. . .	145	21.0	10
	12.70	0.500		. . .	276	40.0	. . .	. . .	. . .	. . .	10
7178-T6 and T62(e)(f) sheet	0.38-1.12	0.015-0.044	572	83.0	. . .	. . .	496	72.0	. . .	. . .	7
	1.14-6.32	0.045-0.249	579	84.0	. . .	. . .	503	73.0	. . .	. . .	8
7178-T62(e)(f) and -T651(c) plate	6.35-12.67	0.250-0.499	579	84.0	. . .	. . .	503	73.0	. . .	. . .	8
	12.70-25.40	0.500-1.000	579	84.0	. . .	. . .	503	73.0	. . .	. . .	6
	25.43-38.10	1.001-1.500	579	84.0	. . .	. . .	503	73.0	. . .	. . .	4
	38.13-50.80	1.501-2.000	552	80.0	. . .	. . .	483	70.0	. . .	. . .	3

(continued)

Table 11B. (continued)

Alloy and temper	Specified thickness mm	Specified thickness in.	Tensile strength Minimum MPa	ksi	Maximum MPa	ksi	Yield strength Minimum MPa	ksi	Maximum MPa	ksi	Elongation (min), %(b)
7178-T76(m) sheet	1.14-6.32	0.045-0.249	517	75.0	...	...	441	64.0	...	...	8
7178-T7651(c)(m) plate	6.35-12.67	0.250-0.499	510	74.0	...	...	434	63.0	...	...	8
	12.70-25.40	0.500-1.000	503	73.0	...	...	427	62.0	...	...	6

Alloy and temper	Specified thickness mm	Specified thickness in.	Axis of test specimen(n)	Tensile strength Minimum MPa	ksi	Maximum MPa	ksi	Yield strength Minimum MPa	ksi	Maximum MPa	ksi	Elongation (min), %(b)
2124-T851(c) plate	38.10-50.80	1.500-2.000	L	455	66.0	...	...	393	57.0	...	...	6
			LT	455	66.0	...	...	393	57.0	...	...	5
			ST	441	64.0	...	...	379	55.0	...	...	1.5
	50.83-76.20	2.001-3.000	L	448	65.0	...	...	393	57.0	...	...	6
			LT	448	65.0	...	...	393	57.0	...	...	4
			ST	434	63.0	...	...	379	55.0	...	...	1.5
	76.23-101.60	3.001-4.000	L ....'.	448	65.0	...	...	386	56.0	...	...	5
			LT	448	65.0	...	...	386	56.0	...	...	4
			ST	427	62.0	...	...	372	54.0	...	...	1.5
	101.63-127.00	4.001-5.000	L	441	64.0	...	...	379	55.0	...	...	5
			LT	441	64.0	...	...	379	55.0	...	...	4
			ST	421	61.0	...	...	365	53.0	...	...	1.5
	127.03-152.40	5.001-6.000	L	434	63.0	...	...	372	54.0	...	...	5
			LT	434	63.0	...	...	372	54.0	...	...	4
			ST	400	58.0	...	...	352	51.0	...	...	1.5
7050-T7451(c)(k) (formerly 7050-T73651) plate	6.35-50.80	0.250-2.000	L	510	74.0	...	...	441	64.0	...	...	10
			LT	510	74.0	...	...	441	64.0	...	...	9
			ST	...	...	...	...	...	...	...	...	...
	50.83-76.20	2.001-3.000	L	503	73.0	...	...	434	63.0	...	...	9
			LT	503	73.0	...	...	434	63.0	...	...	8
			ST	469	68.0	...	...	407	59.0	...	...	2
	76.23-101.60	3.001-4.000	L	496	72.0	...	...	427	62.0	...	...	9
			LT	496	72.0	...	...	427	62.0	...	...	6
			ST	469	68.0	...	...	400	58.0	...	...	2
	101.63-127.00	4.001-5.000	L	489	71.0	...	...	421	61.0	...	...	9
			LT	489	71.0	...	...	421	61.0	...	...	5
			ST	462	67.0	...	...	393	57.0	...	...	2
	127.03-152.40	5.001-6.000	L	483	70.0	...	...	414	60.0	...	...	8
			LT	483	70.0	...	...	414	60.0	...	...	4
			ST	462	67.0	...	...	393	57.0	...	...	2
7050-T7651(c)(m) plate	6.35-25.40	0.250-1.000	L	524	76.0	...	...	455	66.0	...	...	9
			LT	524	76.0	...	...	455	66.0	...	...	8
			ST	...	...	...	...	...	...	...	...	...
	25.43-38.10	1.001-1.500	L	531	77.0	...	...	462	67.0	...	...	9
			LT	531	77.0	...	...	462	67.0	...	...	8
			ST	...	...	...	...	...	...	...	...	...
	38.13-50.80	1.501-2.000	L	524	76.0	...	...	455	66.0	...	...	9
			LT	524	76.0	...	...	455	66.0	...	...	8
			ST	...	...	...	...	...	...	...	...	...
	50.83-76.20	2.001-3.000	L	524	76.0	...	...	455	66.0	...	...	8
			LT	524	76.0	...	...	455	66.0	...	...	7
			ST	...	...	...	...	...	...	...	...	1.5

(a) Converted SI (metric) values are for information only and are not to be used for purposes of specification, acceptance or rejection. (b) In 50 mm, 2 in. or 4d. (c) For stress-relieved tempers, the characteristics and properties other than those specified may differ somewhat from the corresponding characteristics and properties of material in the basic temper. (d) Upon artificial aging, material in the T3/T31, T37, T351, T361 and T451 tempers shall be capable of developing the mechanical properties applicable to material in the T81, T87, T851, T861 and T651 tempers, respectively. (e) These properties usually can be obtained by the user when the material is properly solution heat treated or solution and precipitation heat treated from the O (annealed) or F (as fabricated) temper. These properties also apply to samples of material in the O and F tempers which are solution heat treated or solution and precipitation heat treated by the producer to determine that the material will respond to proper heat treatment. Properties attained by the user, however, may be lower than those listed if the material has been formed or otherwise cold or hot worked, particularly in the annealed temper, prior to solution heat treatment. (f) This temper is not available from the material producer. (g) Tempers T361 and T861 were formerly designated T36 and T86, respectively. (h) This table specifies properties applicable to test specimens, and, because for plate in thicknesses of 0.500 in. or greater the cladding material is removed during preparation of specimens, the listed properties are applicable to the core material only. Tensile and yield strengths of the composite plate are slightly lower depending on cladding thickness. (j) The properties given for this thickness apply only to the T651 temper. (k) When subjected to stress-corrosion testing, material in this temper is capable of exhibiting no evidence of stress-corrosion cracking when exposed for a period of 30 days in the short-transverse direction at a stress level of 75% of the specified yield strength. The stress-corrosion resistance capabilities of individual lots are determined by testing the previously selected tensile-test specimens in accordance with the applicable electrical conductivity acceptance criteria. (m) Material in this temper, when tested upon receipt by the purchaser, is capable of passing an exfoliation corrosion resistance test and the stress-corrosion criteria of some tests except that the stress level is to be 25.0 ksi. The improved resistance to exfoliation corrosion and stress-corrosion cracking of individual lots is determined by testing the previously selected tensile-test specimens in accordance with the applicable electrical conductivity acceptance criteria. (n) L = longitudinal; LT = long transverse; ST = short transverse.

Table 12A. Mechanical-property limits for non-heat-treatable aluminum alloy extruded wire, rod, bar and shapes(a)

Alloy and temper	Specified diameter or thickness(b) mm	Specified diameter or thickness(b) in.	Area cm²	Area in.²	Tensile strength Minimum MPa	ksi	Maximum MPa	ksi	Yield strength (minimum) MPa	ksi	Elongation (min), %(c)
1100-O	All	All	All	All	76	11.0	107	15.5	21	3.0	25
1100-H112	All	All	All	All	76	11.0	...	...	21	3.0	...
3003-O	All	All	All	All	97	14.0	131	19.0	34	5.0	25
3003-H112	All	All	All	All	97	14.0	...	...	34	5.0	...
5083-O	Up thru 127	Up thru 5.000	Up thru 206	Up thru 32	269	39.0	352	51.0	110	16.0	14
5083-H111	Up thru 127	Up thru 5.000	Up thru 206	Up thru 32	276	40.0	...	...	165	24.0	12
5083-H112	Up thru 127	Up thru 5.000	Up thru 206	Up thru 32	269	39.0	...	...	110	16.0	12

(continued)

Table 12A. (continued)

Alloy and temper	Specified diameter or thickness(b) mm	in.	Area cm²	in.²		Tensile strength Minimum MPa	ksi	Maximum MPa	ksi	Yield strength (minimum) MPa	ksi	Elongation (min), %(c)
5086-O	Up thru 127	Up thru 5.000	Up thru 206	Up thru 32	....	241	35.0	317	46.0	97	14.0	14
5086-H111	Up thru 127	Up thru 5.000	Up thru 206	Up thru 32		248	36.0	...	...	145	21.0	12
5086-H112	Up thru 127	Up thru 5.000	Up thru 206	Up thru 32		241	35.0	...	...	97	14.0	12
5154-O	All	All	All	All	....	207	30.0	283	41.0	76	11.0	...
5154-H112	All	All	All	All		207	30.0	...	...	76	11.0	...
5454-O	Up thru 127	Up thru 5.000	Up thru 206	Up thru 32	....	214	31.0	283	41.0	83	12.0	14
5454-H111	Up thru 127	Up thru 5.000	Up thru 206	Up thru 32		228	33.0	...	...	131	19.0	12
5454-H112	Up thru 127	Up thru 5.000	Up thru 206	Up thru 32		214	31.0	...	...	83	12.0	12

(a) Converted SI (metric) values are for information only and are not to be used for purposes of specification, acceptance or rejection. (b) The thickness of the cross section from which the tensile-test specimen is taken determines the applicable mechanical properties. (c) In 50 mm, 2 in. or 4d.

Table 12B. Mechanical-property limits for heat treatable aluminum alloy extruded wire, rod, bar and shapes(a)

Alloy and temper	Specified diameter or thickness(b) mm	in.	Area cm²	in.²		Tensile strength Minimum MPa	ksi	Maximum MPa	ksi	Yield strength Minimum MPa	ksi	Maximum MPa	ksi	Elongation (min), %(c)
2014-O	All	All	All	All	....	...	...	207	30.0	...	...	124	18.0	12
2014-T4, T4510(d)(e) and T4511(d)(e)	All	All	All	All	....	345	50.0	...	...	241	35.0	...	...	12
2014-T42(f)(g)	All	All	All	All	....	345	50.0	...	...	200	29.0	...	...	12
2014-T6, T6510(d) and T6511(d)	Up thru 12.67	Up thru 0.499	All	All	....	414	60.0	...	...	365	53.0	...	...	7
	12.70-19.02	0.500-0.749	All	All	....	441	64.0	...	...	400	58.0	...	...	7
	19.05 and over	0.750 and over	Up thru 161	Up thru 25	....	469	68.0	...	...	414	60.0	...	...	7
	19.05 and over	0.750 and over	Over 161 thru 206	Over 25 thru 32	....	469	68.0	...	...	400	58.0	...	...	6
2014-T62(f)(g)	Up thru 19.02	Up thru 0.749	All	All	....	414	60.0	...	...	365	53.0	...	...	7
	19.05 and over	0.750 and over	Up thru 161	Up thru 25	....	414	60.0	...	...	365	53.0	...	...	7
	19.05 and over	0.750 and over	Over 161 thru 206	Over 25 thru 32	....	414	60.0	...	...	365	53.0	...	...	6
2024-O	All	All	All	All	....	...	...	241	35.0	...	...	131	19.0	12
2024-T3, T3510(d)(e) and T3511(d)(e)	Up thru 6.32	Up thru 0.249	All	All	....	393	57.0	...	...	290	42.0	...	...	12
	6.35-19.02	0.250-0.749	All	All	....	414	60.0	...	...	303	44.0	...	...	12
	19.05-38.07	0.750-1.499	All	All	....	448	65.0	...	...	317	46.0	...	...	10
	38.10 and over	1.500 and over	Up thru 161	Up thru 25	....	483	70.0	...	...	358	52.0	...	...	10
	38.10 and over	1.500 and over	Over 161 thru 206	Over 25 thru 32	....	469	68.0	...	...	331	48.0	...	...	8
2024-T42(f)(g)	Up thru 19.02	Up thru 0.749	All	All	....	393	57.0	...	...	262	38.0	...	...	12
	19.05-38.07	0.750-1.499	All	All	....	393	57.0	...	...	262	38.0	...	...	10
	38.10 and over	1.500 and over	Up thru 161	Up thru 25	....	393	57.0	...	...	262	38.0	...	...	10
	38.10 and over	1.500 and over	Over 161 thru 206	Over 25 thru 32	....	393	57.0	...	...	262	38.0	...	...	8
2024-T81, T8510(d) and T8511(d)	1.27-6.32	0.050-0.249	All	All	....	441	64.0	...	...	386	56.0	...	...	4
	6.35-38.07	0.250-1.499	All	All	....	455	66.0	...	...	400	58.0	...	...	5
	38.10 and over	1.500 and over	Up thru 206	Up thru 32	....	455	66.0	...	...	400	58.0	...	...	5
2219-O		All	All	All	....	...	...	221	32.0		...	124	18.0	12
2219-T31, T3510(d)(e) and T3511(d)(e)	Up thru 12.67	Up thru 0.499	Up thru 161	Up thru 25	....	290	42.0	...	...	179	26.0	...	...	14
	12.70-76.17	0.500-2.999	Up thru 161	Up thru 25	....	310	45.0	...	...	186	27.0	...	...	14
2219-T62(f)(g)	Up thru 25.37	Up thru 0.999	Up thru 161	Up thru 25	....	372	54.0	...	...	248	36.0	...	...	6
	25.40 and over	1.000 and over	Up thru 206	Up thru 32	....	372	54.0	...	...	248	36.0	...	...	6
2219-T81, T8510(d) and T8511(d)	Up thru 76.17	Up thru 2.999	Up thru 161	Up thru 25	....	400	58.0	...	...	290	42.0	...	...	6
6005-T1	Up thru 12.70	Up thru 0.500	All	All	....	172	25.0	...	...	103	15.0	...	...	16
6005-T5	Up thru 3.15	Up thru 0.124	All	All	....	262	38.0	...	...	241	35.0	...	...	8
	3.18-25.40	0.125-1.000	All	All	....	262	38.0	...	...	241	35.0	...	...	10
6061-O	All	All	All	All	....		...	152	22.0		...	110	16.0	16
6061-T1	Up thru 15.88	Up thru 0.625	All	All	....	179	26.0	...	...	97	14.0	...	...	16
6061-T4, T4510(d)(e) and T4511(d)(e)	All	All	All	All	....	179	26.0	...	...	110	16.0	...	...	16
6061-T42(f)(g)	All	All	All	All	....	179	26.0	...	...	83	12.0	...	...	16
6061-T51	Up thru 15.88	Up thru 0.625	All	All	....	241	35.0	...	...	207	30.0	...	...	8
6061-T6, T62(f)(g) T6510(d) and T6511(d)	Up thru 6.32	Up thru 0.249	All	All	....	262	38.0	...	...	241	35.0	...	...	8
	6.35 and over	0.250 and over	All	All	....	262	38.0	...	...	241	35.0	...	...	10
6063-O	All	All	All	All	....		...	131	19.0	...	...	...	...	18
6063-T1	Up thru 12.70	Up thru 0.500	All	All	....	117	17.0	...	...	62	9.0	...	...	12
	12.73-25.40	0.501-1.000	All	All	....	110	16.0	...	...	55	8.0	...	...	12
6063-T4 and T42(f)(g)	Up thru 12.70	Up thru 0.500	All	All	....	131	19.0	...	...	69	10.0	...	...	14
	12.73-25.40	0.501-1.000	All	All	....	124	18.0	...	...	62	9.0	...	...	14
6063-T5	Up thru 12.70	Up thru 0.500	All	All	....	152	22.0	...	...	110	16.0	...	...	8
	12.73-25.40	0.501-1.000	All	All	....	145	21.0	...	...	103	15.0	...	...	8
6063-T52	Up thru 25.40	Up thru 1.000	All	All	....	152	22.0	207	30.0	110	16.0	172	25.0	8
6063-T6 and T62(f)(g)	Up thru 3.15	Up thru 0.124	All	All	....	207	30.0	...	...	172	25.0	...	...	8
	3.18-25.40	0.125-1.000	All	All	....	207	30.0	...	...	172	25.0	...	...	10

(continued)

Table 12B. (continued)

Alloy and temper	Specified diameter or thickness(b) mm	in.	Area cm²	in.²	Tensile strength Minimum MPa	ksi	Maximum MPa	ksi	Yield strength Minimum MPa	ksi	Maximum MPa	ksi	Elongation (min), %(c)
6066-O	All	All	All	All			200	29.0			124	18.0	16
6066-T4, T4510(d)(e) and T4511(d)(e)	All	All	All	All	276	40.0			172	25.0			14
6066-T42(f)(g)	All	All	All	All	276	40.0			165	24.0			14
6066-T6, T6510(d) and T6511(d)	All	All	All	All	345	50.0			310	45.0			8
6066-T62(f)(g)	All	All	All	All	345	50.0			290	42.0			8
6070-T6 and T62(f)(g)	Up thru 76.17	Up thru 2.999	Up thru 206	Up thru 32	331	48.0			310	45.0			6
6105-T1	Up thru 12.70	Up thru 0.500	All	All	172	25.0			103	15.0			16
6105-T5	Up thru 12.70	Up thru 0.500	All	All	262	38.0			241	35.0			8
6162-T5, T5510(d) and T5511(d)	Up thru 25.40	Up thru 1.000	All	All	255	37.0			234	34.0			.
6162-T6,	Up thru 6.32	Up thru 0.249	All	All	262	38.0			241	35.0			8
T6510(d) and T6511(d)	6.35-12.67	0.250-0.499	All	All	262	38.0			241	35.0			10
6262-T6, T62(f)(g), T6510(d) and T6511(d)	All	All	All	All	262	38.0			241	35.0			10
6351-T54	Up thru 12.70	Up thru 0.500	Up thru 129	Up thru 20	207	30.0			138	20.0			10
6463-T1	Up thru 12.70	Up thru 0.500	Up thru 129	Up thru 20	117	17.0			62	9.0			12
6463-T5	Up thru 12.70	Up thru 0.500	Up thru 129	Up thru 20	152	22.0			110	16.0			8
6463-T6 and T62(f)(g)	Up thru 3.15	Up thru 0.124	Up thru 129	Up thru 20	207	30.0			172	25.0			8
	3.18-12.70	0.125-0.500	Up thru 129	Up thru 20	207				172	25.0			10
7001-O	All	All	All	All			290	42.0			179	26.0	10
7001-T6, T62(f)(g), T6510(d) and T6511(d)	Up thru 6.32	Up thru 0.249	All	All	614	89.0			565	82.0			5
	6.35-12.67	0.250-0.499	All	All	634	92.0			579	84.0			5
	12.70-50.77	0.500-1.999	All	All	648	94.0			607	88.0			5
	50.80-76.17	2.000-2.999	All	All	620	90.0			579	84.0			5
7005-T53	Up thru 19.05	Up thru 0.750	All	All	345	50.0			303	44.0			10
7075-O	All	All	All	All			276	40.0			165	24.0	10
7075-T6, T62(f)(g), T6510(d) and T6511(d)	Up thru 6.32	Up thru 0.249	All	All	538	78.0			483	70.0			7
	6.35-12.67	0.250-0.499	All	All	558	81.0			503	73.0			7
	12.70-38.07	0.500-1.499	All	All	558	81.0			496	72.0			7
	38.10-76.17	1.500-2.999	All	All	558	81.0			496	72.0			7
	76.20-114.27	3.000-4.499	Up thru 129	Up thru 20	558	81.0			489	71.0			7
	76.20-114.27	3.000-4.499	Over 129 thru 206	Over 20 thru 32	538	78.0			483	70.0			6
	114.30-127	4.500-5.000	Up thru 206	Up thru 32	538	78.0			469	68.0			6
7075-T73(h), T73510(d)(h) and T73511(d)(h)	1.57-6.32	0.062-0.249	Up thru 129	Up thru 20	469	68.0			400	58.0			7
	6.35-38.07	0.250-1.499	Up thru 161	Up thru 25	483	70.0			421	61.0			8
	38.10-76.17	1.500-2.999	Up thru 161	Up thru 25	476	69.0			407	59.0			8
	76.20-114.27	3.000-4.499	Up thru 129	Up thru 20	469	68.0			393	57.0			7
	76.20-114.27	3.000-4.499	Over 129 thru 206	Over 20 thru 32	448	65.0			379	55.0			7
7075-T76(j), T76510(d)(j) and T76511(d)(j)	Up thru 3.15	Up thru 0.124	All	All	496	72.0			427	62.0			7
	3.18-6.32	0.125-0.249	Up thru 129	Up thru 20	510	74.0			441	64.0			7
	6.35-12.67	0.250-0.499	Up thru 129	Up thru 20	517	75.0			448	65.0			7
	12.70-25.40	0.500-1.000	Up thru 129	Up thru 20	517	75.0			448	65.0			7
7178-O	All	All	Up thru 206	Up thru 32			276	40.0			165	24.0	10
7178-T6, T6510(d) and T6511(d)	Up thru 1.55	Up thru 0.061	All	All	565	82.0			524	76.0			...
	1.58-6.32	0.062-0.249	Up thru 129	Up thru 20	579	84.0			524	76.0			5
	6.35-38.07	0.250-1.499	Up thru 161	Up thru 25	600	87.0			538	78.0			5
	38.10-63.47	1.500-2.499	Up thru 161	Up thru 25	593	86.0			531	77.0			5
	38.10-63.47	1.500-2.499	Over 161 thru 206	Over 25 thru 32	579	84.0			517	75.0			5
	63.50-76.17	2.500-2.999	Up thru 206	Up thru 32	565	82.0			489	71.0			5
7178-T62(f)(g)	Up thru 1.55	Up thru 0.061	All	All	545	79.0			503	73.0			...
	1.58-6.32	0.062-0.249	Up thru 129	Up thru 20	565	82.0			510	74.0			5
	6.35-38.07	0.250-1.499	Up thru 161	Up thru 25	593	86.0			531	77.0			5
	38.10-63.47	1.500-2.499	Up thru 161	Up thru 25	593	86.0			531	77.0			5
	38.10-63.47	1.500-2.499	Over 161 thru 206	Over 25 thru 32	579	84.0			517	75.0			5
	63.50-76.17	2.500-2.999	Up thru 206	Up thru 32	565	82.0			489	71.0			5
7178-T76(j), T76510(d)(j) and T76511(d)(j)	3.18-6.32	0.125-0.249	Up thru 129	Up thru 20	524	76.0			455	66.0			7
	6.35-12.67	0.250-0.499	Up thru 129	Up thru 20	531	77.0			462	67.0			7
	12.70-25.40	0.500-1.000	Up thru 129	Up thru 20	531	77.0			462	67.0			7

(a) Converted SI (metric) values are for information only and are not to be used for purposes of specification, acceptance or rejection. (b) The thickness of the cross section from which the tensile-test specimen is taken determines the applicable mechanical properties. (c) In 50 mm, 2 in. or 4d. For material of such dimensions that a standard test specimen cannot be taken, or for shapes thinner than 0.062 in., the test for elongation is not required. (d) For stress-relieved tempers, the characteristics and properties other than those specified may differ somewhat from the corresponding characteristics and properties of material in the basic temper. (e) Upon artificial aging, material in the T3/T31, T3510, T3511, T4, T4510 and T4511 tempers shall be capable of developing the mechanical properties applicable to material in the T81, T8510, T8511, T6, T6510 and T6511 tempers, respectively. (f) These properties usually can be obtained by the user when the material is properly solution heat treated or solution and precipitation heat treated from the O (annealed) or F (as fabricated) temper. These properties also apply to samples of material in the O and F tempers which are solution heat treated or solution and precipitation heat treated by the producer to determine that the material will respond to proper heat treatment. Properties attained by the user, however, may be lower than those listed if the material has been formed or otherwise cold or hot worked, particularly in the annealed temper, prior to solution heat treatment. (g) This temper is not available from the material producer. (h) When subjected to stress-corrosion testing, material in this temper is capable of exhibiting no evidence of stress-corrosion cracking when exposed for a period of 30 days in the short-transverse direction at a stress level of 75% of the specified yield strength. The stress-corrosion resistance capabilities of individual lots are determined by testing the previously selected tensile-test specimens in accordance with the applicable electrical conductivity acceptance criteria. (j) Material in this temper, when tested upon receipt by the purchaser, is capable of passing an exfoliation corrosion resistance test and the stress-corrosion resistance criteria of note (h) above except that the stress level is to be 25.0 ksi. The improved resistance to exfoliation corrosion and stress-corrosion cracking of individual lots is determined by testing the previously selected tensile-test specimens in accordance with the applicable electrical conductivity acceptance criteria.

NOTE: Table 13 (Mechanical property limits for aluminum alloy 6061-T6 structural shapes) is on page 6•21.

Table 14. Mechanical-property limits for aluminum alloy die forgings(a)

Alloy and temper	Specified thickness(b) mm	in.	Specimen axis parallel to direction of grain flow Tensile strength MPa	ksi	Yield strength MPa	ksi	Elongation (min),%(c) Coupon	Forging	Specimen axis not parallel to direction of grain flow Tensile strength MPa	ksi	Yield strength MPa	ksi	Elongation (min),%(c) (Forging)	Hardness, HB(d)
1100-H112(e)	Up thru 100	Up thru 4	76	11.0	28	4.0	25	18	. . .	. . .	. . .	. . .	. . .	20
2014-T4	Up thru 100	Up thru 4	379	55.0	207	30.0	16	11	. . .	. . .	. . .	. . .	. . .	100
2014-T6	Up thru 25	Up thru 1	448	65.0	386	56.0	8	6	441	64.0	379	55.0	3	125
	Over 25 thru 50	Over 1 thru 2	448	65.0	386	56.0	(f)	6	441	64.0	379	55.0	2	125
	Over 50 thru 75	Over 2 thru 3	448	65.0	379	55.0	(f)	6	434	63.0	372	54.0	2	125
	Over 75 thru 100	Over 3 thru 4	434	63.0	379	55.0	(f)	6	434	63.0	372	54.0	2	125
2018-T61	Up thru 100	Up thru 4	379	55.0ʹ	276	40.0	10	7	. . .	. . .	. . .	. . .	. . .	100
2025-T6	Up thru 100	Up thru 4	359	52.0	228	33.0	16	11	. . .	. . .	. . .	. . .	. . .	100
2218-T61	Up thru 100	Up thru 4	379	55.0	276	40.0	10	7	. . .	. . .	. . .	. . .	. . .	100
2218-T72	Up thru 100	Up thru 4	262	38.0	200	29.0	8	5	. . .	. . .	. . .	. . .	. . .	85
2219-T6	Up thru 100	Up thru 4	400	58.0	262	38.0	10	8	386	56.0	248	36.0	4	100
2618-T61	Up thru 100	Up thru 4	400	58.0	310	45.0	6	4	379	55.0	290	42.0	4	115
3003-H112(e)	Up thru 100	Up thru 4	97	14.0	34	5.0	25	18	. . .	. . .	. . .	. . .	. . .	25
4032-T6	Up thru 100	Up thru 4	359	52.0	290	42.0	5	3	. . .	. . .	. . .	. . .	. . .	115
5083-H111(e)	Up thru 100	Up thru 4	290	42.0	152	22.0	. . .	14	269	39.0	138	20.0	12	. . .
5083-H112(e)	Up thru 100	Up thru 4	276	40.0	124	18.0	. . .	16	269	39.0	110	16.0	14	. . .
5456-H112(e)	Up thru 100	Up thru 4	303	44.0	138	20.0	. . .	16	. . .	. . .	. . .	. . .	. . .	. . .
6053-T6	Up thru 100	Up thru 4	248	36.0	207	30.0	16	11	. . .	. . .	. . .	. . .	. . .	75
6061-T6	Up thru 100	Up thru 4	262	38.0	241	35.0	10	7	262	38.0	241	35.0	5	80
6066-T6	Up thru 100	Up thru 4	345	50.0	310	45.0	12	8	. . .	. . .	. . .	. . .	. . .	100
6151-T6	Up thru 100	Up thru 4	303	44.0	255	37.0	14	10	303	44.0	255	37.0	6	90
7075-T6	Up thru 25	Up thru 1	517	75.0	441	64.0	10	7	490	71.0	421	61.0	3	135
	Over 25 thru 50	Over 1 thru 2	510	74.0	434	63.0	(f)	7	490	71.0	421	61.0	3	135
	Over 50 thru 75	Over 2 thru 3	510	74.0	434	63.0	(f)	7	483	70.0	414	60.0	3	135
	Over 75 thru 100	Over 3 thru 4	503	73.0	427	62.0	(f)	7	483	70.0	414	60.0	2	135
7075-T73	Up thru 75	Up thru 3	455	66.0	386	56.0	. . .	7	427	62.0	365	53.0	3	125
	Over 75 thru 100	Over 3 thru 4	441	64.0	379	55.0	. . .	7	421	61.0	359	52.0	2	125
7075-T7352	Up thru 75	Up thru 3	455	66.0	386	56.0	. . .	7	427	62.0	352	51.0	3	125
	Over 75 thru 100	Over 3 thru 4	441	64.0	365	53.0	. . .	7	421	61.0	338	49.0	2	125

(a) Converted SI (metric) values are for information only and are not to be used for purposes of specification, acceptance or rejection. (b) As-forged thickness. When forgings are machined prior to heat treatment, the properties given here also will apply to the machined thickness provided that the machined thickness is not less than one-half the original (as-forged) thickness. (c) In 50 mm, 2 in. or 4d. (d) For information only. Brinell hardness usually is measured on the surface of the heat treated forging using a 500-kg load and a 10-mm penetrator ball. (e) Properties of forgings in H111 and H112 tempers depend on the equivalent cold work in the forgings. The properties listed should be attainable in any forging within the prescribed thickness range and may be considerably exceeded in some instances. (f) When separately forged coupons are used to verify acceptability of forgings in the indicated thicknesses, the properties shown for thicknesses up through 1 in., including test-coupon elongation, apply.

Table 15. Minimum and typical room-temperature plane-strain fracture-toughness values for several high-strength aluminum alloys

Product form	Alloy and temper	Thickness mm	in.	L-T direction Minimum MPa√m	ksi√in.	Typical MPa√m	ksi√in.	T-L direction Minimum MPa√m	ksi√in.	Typical MPa√m	ksi√in.	S-L direction Minimum MPa√m	ksi√in.	Typical MPa√m	ksi√in.
Plate	7050-T7451	25.40-50.80	1.000-2.000 . . .	31.9	29.0	37	34	27.5	25.0	33	30	. . .	. . .	. . .	. . .
		50.83-76.20	2.001-3.000 . . .	29.7	27.0	36	33	26.4	24.0	32	29	23.1	21.0	28	25
		76.23-101.60	3.001-4.000 . . .	28.6	26.0	35	32	25.3	23.0	31	28	23.1	21.0	28	25
		101.63-127.00	4.001-5.000 . . .	27.5	25.0	32	29	24.2	22.0	29	26	23.1	21.0	28	25
		127.03-152.40	5.001-6.000 . . .	26.4	24.0	31	28	24.2	22.0	28	25	23.1	21.0	28	25
	7050-T7651	25.40-50.80	1.000-2.000 . . .	28.6	26.0	34	31	26.4	24.0	31	28	. . .	. . .	. . .	. . .
		50.83-76.20	2.001-3.000 . . .	26.4	24.0	. . .	. . .	25.3	23.0	. . .	. . .	22.0	20.0	26	24
	7475-T651	. . . .		33.0	30.0	46	42	30.8	28.0	41	37	. . .	. . .	. . .	. . .
	7475-T7651	. . . .		36.3	33.0	47	43	33.0	30.0	41	37	. . .	. . .	. . .	. . .
	7475-T7351	. . . .		41.8	38.0	55	50	35.2	32.0	45	41	27.5	25.0	36	33
	7075-T651	. . . .		. . .	. . .	29	26	. . .	. . .	25	23	. . .	. . .	20	18
	7075-T7651	. . . .		. . .	. . .	30	27	. . .	. . .	24	22	. . .	. . .	20	18
	7075-T7351	. . . .		. . .	. . .	32	30	. . .	. . .	29	26	. . .	. . .	20	18
	7079-T651	. . . .		. . .	. . .	30	27	. . .	. . .	25	23	. . .	. . .	18	16
	2124-T851	. . . .		26.4	24.0	32	29	22.0	20.0	26	24	19.8	18.0	26	24
	2024-T351	. . . .		. . .	. . .	37	34	. . .	. . .	32	29	. . .	. . .	26	24
Die forgings	7050-T74, -T7452	. . . .		27.5	25.0	38	35	20.9	19.0	32	29	20.9	19.0	29	26
	7175-T736, -T73652	. . . .		29.7	27.0	38	35	23.1	21.0	34	31	23.1	21.0	31	28
	7075-T7352	. . . .		. . .	. . .	32	29	. . .	. . .	30	27	. . .	. . .	29	26

(continued)

Table 15. (continued)

Product form	Alloy and temper	Thickness mm	Thickness in.	L-T direction Minimum MPa√m	ksi√in.	Typical MPa√m	ksi√in.	T-L direction Minimum MPa√m	ksi√in.	Typical MPa√m	ksi√in.	S-L direction Minimum MPa√m	ksi√in.	Typical MPa√m	ksi√in.
Hand forgings	7050-T7452			29.7	27.0	36	33	18.7	17.0	28	25	17.6	16.0	29	26
	7075-T73, -T7352	...	...			42	38	...	...	28	25	...	...	28	25
	7175-T73652			33.0	30.0	40	36	27.5	25.0	30	27	23.1	21.0	28	25
	2024-T852	...	...			26	24	...	...	22	20	...	...	20	18
Extrusions	7050-T7651x	...	...			44	40	...	...	31	28	...	...	28	25
	7050-T7351x	...	...	...	...	...	...	...	...	...	...	...	...	...	...
	7075-T651x	...	...			34	31	...	...	22	20	...	...	20	18
	7075-T7351x	...	...			33	30	...	...	26	24	...	...	22	20
	7150-T7351x	24.2	22.0			31	28	...	...	...	...	...	...	...	...
	7175-T7351x			33.0	30.0	40	36	30.8	28.0	34	31	...	...	...	...

Aluminum Foundry Products

ALUMINUM ALLOY ENGINEERING CASTINGS were produced and marketed at an average annual rate of 824 000 metric tons (1816 million pounds) during the period from 1972 to 1981, and constituted 13.9% of total U.S. shipments of aluminum products during that period (see Fig. 1). No long-term trends should be inferred from the changes that took place during this ten-year period. The sharp decline for the years 1980 and 1981 reflects recession years with sharp drops in automobile and truck production, which affected production of aluminum castings more strongly than that of wrought aluminum products. Of the total weight of all castings shipped during this ten-year period, 63% was accounted for by die castings, 22% by permanent mold and semipermanent mold castings, 13% by sand castings and 2% by castings of other types. Over this period, 60% of all die castings, 45% of all permanent mold castings and 22% of all sand castings were produced for use in equipment or consumer products marketed by castings producers.

Aluminum alloy castings are the cost-effective answer to many needs and problems in construction of machines, equipment, appliances, vehicles and structures, usually serving a primarily mechanical function, but often combining this with an appearance or decorative function. This requires cast parts in a great variety of geometric configurations, frequently combining several different basic forms in an integral or monolithic piece. Such parts include covers and housings, which may be ribbed for reinforcement or finned for heat conduction or dissipation; frames and boxlike parts; cylindrical or spherical tanks for containment of gases or fluids; brackets; pistons; wheels; disks; impellers; bulkheads; and clamps. The list of forms produced is nearly endless, and many such parts have a multitude of cored holes in bosses for fastening or as passages for fluids.

CASTING PROCESSES

Almost all known casting processes are used in producing aluminum alloy castings.

Die Casting. In die casting, which is the process used for the highest volume of production, mol-ten metal is injected into cavities formed by heat treated steel dies and cores under pressures up to 140 MPa (20 ksi). Part size is limited only by machine capacity, and some machines are capable of producing castings weighing over 50 kg (100 lb), although casting weights up to about 5 kg (10 lb) are more common. As-cast surfaces are very smooth and detailed; machining is generally required only to provide fits or seals with other parts. Die casting permits metal sections thinner than those obtainable by sand or permanent mold casting. High mechanical properties and resistance to fatigue can be developed in die castings without heat treatment because of the fine microstructures produced by the rapid solidification.

Alloys most frequently used in die casting differ in composition from those employed in other casting processes, because the fine die cast microstructure allows higher volume fractions of second-phase constituents, silicon and intermetallic compounds without damaging effects on ductility or impact resistance. Iron contents up to about 1%, which would impair both strength and ductility of other more slowly solidified castings, are quite beneficial in die casting compositions, minimizing the tendency of the aluminum to "solder" or adhere to the steel dies.

Fig. 1. Annual U.S. shipments of aluminum castings, and shipments of castings as a percentage of all aluminum shipments (weight basis)

Permanent mold casting employs metal molds and cores with either gravity or low-pressure introduction of molten metal. Most permanent mold castings weigh less than 10 kg (20 lb), but castings weighing up to 25 kg (50 lb), and sometimes even up to 100 kg (200 lb), are not uncommon. Permanent mold castings have smoother surfaces than those of sand castings and exhibit superior pressure tightness. Tapered metal cores are used to form straight-wall cavities, and collapsible metal cores are used to form internal ribs and undercuts, which also can be formed by expendable (dry sand or plastic) cores. When the latter are used, the process is referred to as semipermanent mold casting. Mechanical properties (including fatigue resistance) of permanent mold castings are very high because of the fine microstructure and the capability for heat treatment.

Sand casting is in some ways the most versatile foundry method, with few limitations on the type of alloy that may be used or on part size or shape and extent of coring to form internal cavities and passages. This process is employed for relatively large parts, when required quantities are small, or when the design or the alloy dictates use of an expendable mold material. Size or weight limitations are generally established by melting, metal or mold-handling capabilities. Sand castings weighing over 9000 kg (18 000 lb) and having dimensions of 5.5 m (18 ft) and over have been produced. As-cast surfaces are rougher than for other processes, and required dimensional tolerances are greater. A full range of mechanical properties is available since both non-heat-treatable alloys and heat treatable alloys of high strength capability can be used.

Shell Mold Casting. In this process the mold cavity is formed by a shell of resin-bonded sand only 10 to 20 mm (0.4 to 0.8 in.) thick—much thinner and lighter than the massive molds commonly used in sand foundries. Shell mold castings surpass ordinary sand castings in surface finish and dimensional accuracy and cool at slightly higher rates. Equipment costs are higher, and the size and complexity of castings that can be produced are more limited.

Plaster and Investment Casting. These processes are capable of producing intricate, thin-section parts with great dimensional accuracy, sharp reproduction of fine detail and very smooth surfaces. Most castings produced by these methods are small and of thin section. Other mold materials or composite molds are generally required to provide adequate chilling capacity for castings having thick sections.

Composite Mold Casting. Use of more than one mold material having different heat-extraction (chilling) capabilities is the basis for composite mold casting. The mold materials are arranged to optimize the sequence of solidification from one part of the casting to another, providing increased soundness and strength with good dimensional control.

Premium-Quality Castings. The ultimate in mold and process engineering and control, usually employing composite mold materials, is used in producing premium-quality castings. These techniques provide exceptional combinations of dimensional accuracy, finish, soundness and mechanical properties not usually produced by more conventional casting processes. Unit costs are usually higher, but may be partially offset by reduced machining costs. Higher and less-variable mechanical properties are usually guaranteed.

Evaporative Casting Process. One of the newest casting processes—the evaporative casting process—uses patterns made from foamed or expanded polystyrene. The consumable pattern is not removed from the mold but is vaporized as the molten metal fills the mold.

Centrifugal Casting. Axisymmetric parts, such as those of tubular shape, are sometimes produced by true centrifugal casting in a metal mold. For nonsymmetrical shapes, mold filling may be assisted by centrifuging the molten metal into the mold cavity. In this process, multiple mold cav-

ities are usually arranged around a central pouring sprue.

Continuous Casting. Long shapes of simple cross section, such as round, square or hexagonal rod, can be produced by continuous casting. Use of this process is limited, however, because the same shapes can be produced by extrusion at approximately the same cost but with better mechanical properties. Occasionally, semicontinuous cast (D.C. type) ingots of large rectangular, square or round cross section have been used to produce very large components by extensive machining.

SELECTION OF CASTING PROCESS

Many factors affect the selection of a casting process for producing a specific part. In many cases the decision is influenced strongly by geometric configuration and design features, which determine the feasibilities of the different processes. When the casting can be produced by several of the available methods, relative weights, quantity requirements and unit costs are frequently determinative. Selection can also depend on quality factors, such as soundness, pressure tightness, surface finish, machining requirements, mechanical properties, environmental service conditions or special conditions of installation, assembly or joining (such as by weld-

ing or brazing). Close cooperation between designers and potential producers is essential in coupling design and process to achieve the most cost-effective results. As with all procurement, availability and delivery may sometimes be prevailing factors.

The principal technical factors of comparison among the basic sand, permanent mold and die casting processes are listed in Table 1. Comparative microstructures of an Al-5%Si alloy (443.0 type) cast by these three processes, and the corresponding typical tensile properties, are presented in Fig. 2. These differences may be important factors in process selection.

Unit costs are highly quantity-dependent because of tooling amortization costs. In general, the lower the tooling investment (patterns, molds, dies and auxiliary equipment), the greater the cost of producing each piece. When only a few pieces are required, the method involving the least expensive tooling will result in the lowest unit cost, even though the cost of casting each piece is high. For very large production runs, where the tooling cost is shared by a large number of castings, more elaborate tooling decreases unit casting cost sufficiently to outweigh the initial investment. A typical analysis for production of a small part by various casting processes (sand, permanent mold, die and investment casting) and by die forging,

Table 1. Factors affecting selection of casting process for aluminum alloys

Factor	Sand casting	Permanent mold casting	Die casting
Cost of equipment	Lowest cost if only a few items required	Less than die casting	Highest
Casting rate	Lowest rate	11 kg/h (25 lb/h) common; higher rates possible	4.5 kg/h (10 lb/h) common; 45 kg/h (100 lb/h) possible
Maximum casting size	Largest of any casting method	Limited by size of machine	Limited by size of machine
External and internal shape	Best suited for complex shapes where coring required	Cavities formed by metal cores with draft or collapsible to form undercuts or ribs	Cavities formed by metal cores with draft
Minimum wall thickness	3.0-5.0 mm (0.125-0.200 in.) required; 4.0 mm (0.150 in.) normal	3.0-5.0 mm (0.125-0.200 in.) required; 3.5 mm (0.140 in.) normal	1.0-2.5 mm (0.100-0.040 in.); depends on casting size
Type of cores	Complex baked sand cores can be used	Metal cores, including collapsible; semipermanent mold casting uses expendable cores	Steel cores; must be simple and straight so they can be pulled
Tolerance obtainable	Poorest; best linear tolerance is 300 mm/m (300 mils/in.)	Best linear tolerance is 10 mm/m (10 mils/in.)	Best linear tolerance is 4 mm/m (4 mils/in.)
Surface finish	6.5-12.5 μm (250-500 μin.)	4.0-10 μm (150-400 μin.)	1.5 μm (50 μin.); best finish of the three casting processes
Gas porosity	Lowest porosity possible with good technique	Best pressure tightness; low porosity possible with good technique	Porosity may be present
Cooling rate	0.1-0.5 °C/s (0.2-0.9 °F/s)	0.3-1.0 °C/s (0.5-1.8 °F/s)	50-500 °C/s (90-900 °F/s)
Grain size	Coarse	Fine	Very fine on surface
Dendrite-arm spacing	0.05-0.5 mm (0.002-0.02 in.)	0.03-0.07 mm (0.001-0.003 in.)	0.005-0.015 mm (0.0002-0.0006 in.)
Strength	Lowest	Excellent	Highest, usually used in the as-cast condition
Fatigue properties	Good	Good	Excellent
Wear resistance	Good	Good	Excellent
Over-all quality	Depends on foundry technique	Highest quality	Tolerance and repeatability very good
Remarks	Very versatile as to size, shape, internal configurations	. . .	Excellent for high production rates

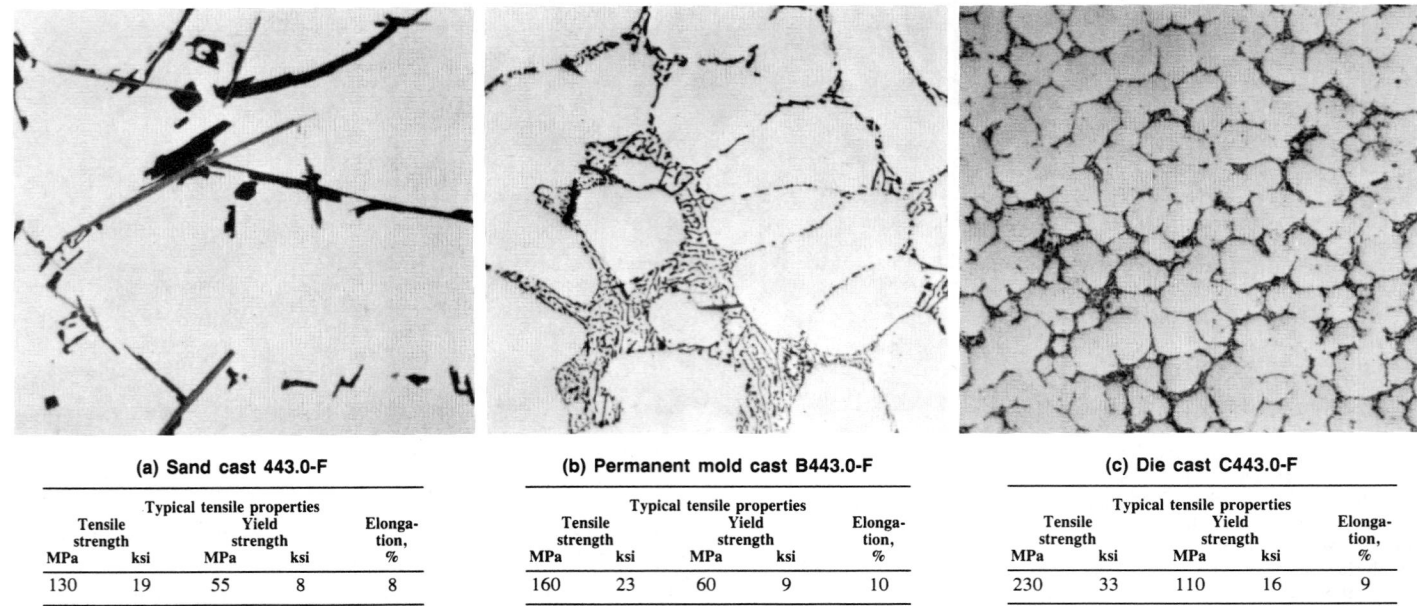

(a) Sand cast 443.0-F

Typical tensile properties				
Tensile strength		Yield strength		Elonga-tion,
MPa	ksi	MPa	ksi	%
130	19	55	8	8

(b) Permanent mold cast B443.0-F

Typical tensile properties				
Tensile strength		Yield strength		Elonga-tion,
MPa	ksi	MPa	ksi	%
160	23	60	9	10

(c) Die cast C443.0-F

Typical tensile properties				
Tensile strength		Yield strength		Elonga-tion,
MPa	ksi	MPa	ksi	%
230	33	110	16	9

Fig. 2. Aluminum–5% silicon alloy microstructures resulting from different solidification rates characteristic of different casting processes. Dendrite cell size and constituent particle size decrease with increasing cooling rate, from sand cast to permanent mold cast to die cast. Etchant, 0.5% hydrofluoric acid; magnification, 500×.

on the basis of relative unit cost versus quantity required, is illustrated graphically in Fig. 3.

Design and Function. The objective of casting design for mechanical function is to prevent failure due to the loads and stresses of normal operation plus those resulting from abuse and accidents. Basic principles applicable to any mechanical metal part apply, such as avoidance of sharp corners and other stress-concentrating features. Sections varying in thickness must be carefully blended, and wide thickness variations must be avoided. Thick sections isolated by thin sections cannot be "fed," and internal unsoundness or surface shrinkage may result. Each process has generally applicable minimum-section-thickness requirements, but these vary considerably depending on the area of the section and the character of surrounding features. As indicated previously, there is no substitute for experience in linking design, producibility and quality, and early consultation with foundry personnel during casting design will save time and avoid difficulties.

Casting Quality. When applied to castings, the term "quality" embraces many features including fidelity of reproduction of intended shape and dimensions, surface continuity and smoothness, as well as degree of soundness (freedom from cracks, porosity and surface imperfections). These quality features are affected by design, casting process, alloy, and foundry procedures and control. Metallurgical microstructures are highly dependent on local rates of solidification and may vary substantially from one part of the casting to another. Solidification rate depends on section thickness and on the heat-extraction (chilling) capability of the mold or die, which in turn depends on two factors — the mold or die material, and the degree of persistence of thermal contact between the solidifying metal and the cavity surfaces. This obviously is process-dependent, because not only the mold materials but also the pressure of the metal in the cavity vary with the process. Thus, solidification rates increase in the order: plaster, sand, permanent mold, centrifugal

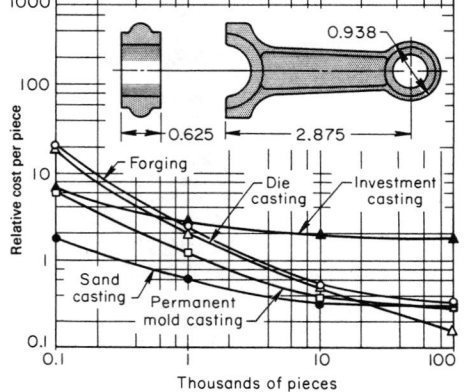

Fig. 3. Relative unit costs of producing a small connecting rod in various quantities by four casting processes and by forging. Tooling costs appropriate to process and quantity requirements are assumed. Dimensions are in inches.

and die casting. Some nominal cooling rates are listed in Table 1.

Grain size is refined by increasing solidification rate but is also highly dependent on the presence of grain-refining elements (principally titanium and boron) in the alloy. Even more importantly, solidification rate establishes the fineness of the microstructure — that is, the size and spacing of dendrites and the size and distribution of second-phase constituent particles. Relative values of dendrite-arm spacing (DAS), the most convenient feature for metallographic measurement and quantification, are closely correlated with constituent size and the effects on strength and ductility. Nominal DAS values are listed for three casting processes (sand, permanent mold and die casting) in Table 1, and the correlation between mechanical properties and

microstructure for alloy 443.0-F is apparent in Fig. 2.

In general, two types of porosity may occur in cast aluminum: gas porosity and shrinkage porosity. Gas porosity, which generally is fairly spherical in shape, results either from precipitation of hydrogen during solidification (because the solubility of this gas is much higher in the molten metal than in the solid metal) or from occlusion of gas bubbles during the high-velocity injection of molten metal in die casting. The other source of porosity is the liquid-to-solid shrinkage which frequently takes the form of interdendritically distributed voids. These voids may be enlarged by hydrogen, and, because larger dendrites result from slower solidification, the size of such porosity also increases as solidification rate decreases. It is not fair to the various processes to establish inherent ratings with respect to anticipated porosity, because castings made by any process can vary substantially in soundness — from nearly completely sound to very unsound — depending on casting size and design as well as on foundry techniques.

All aspects of casting quality are affected significantly by alloy composition, and each process tends to operate most effectively, providing the highest productivity and quality, with a specific alloy or with alloys of a specific type. It is on the basis of experience with this factor that certain alloys are indicated as being intended for use primarily in sand, permanent mold or die casting. Some alloys are used for several (or all) processes.

ALLOY SYSTEMS

Aluminum casting alloys are based on the same alloy systems as those of wrought aluminum alloys, are strengthened by the same mechanisms (with the general exception of strain hardening), and are similarly classified into non-heat-treatable and heat treatable types. The major difference is that the casting alloys used in the greatest volumes contain alloying additions of silicon far

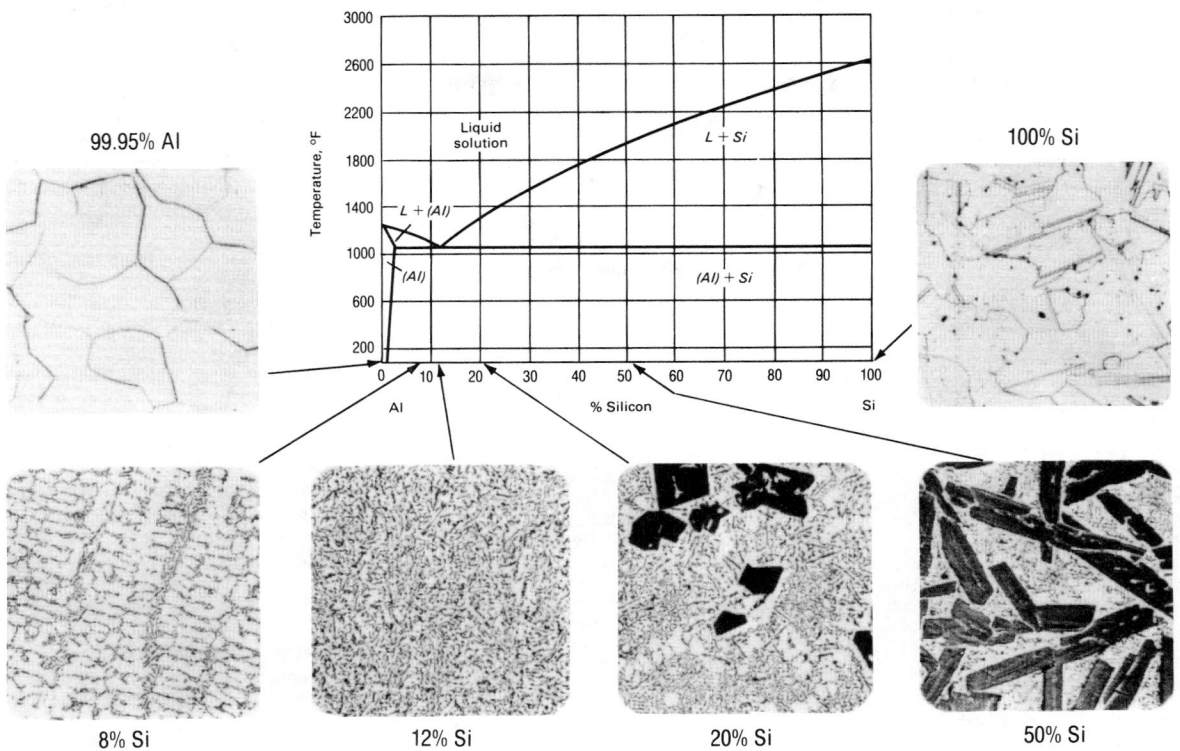

Fig. 4. Aluminum-silicon phase diagram, and cast microstructures of pure components and of alloys of various compositions

in excess of the amounts in most wrought alloys.

Alloy selection for some types of applications is purely a matter of meeting functional requirements. For example, where very high electrical conductivity is essential, as in the pressure-cast integral conductor bars and collector rings of electric-motor rotors, only the unalloyed 1xx.x aluminum compositions can be used. For marine and salt water exposures, highest resistance to corrosion requires use of 5xx.x aluminum-magnesium alloys, as do many decorative, bright-finish and food-processing-machinery applications.

The 2xx.x aluminum-copper group includes compositions capable of developing the highest strengths among all casting alloys, and these alloys are used where this is a predominant requirement. These alloys (A201.0, 202.0, 204.0 and A206.0) contain 4 to 6% Cu and 0.25 to 0.35% Mg, with highly restrictive impurity (Fe and Si) limits, and in some cases also contain 0.25 to 0.35% Mn or Cr and (in alloys 201.0, A201.0 and 202.0) 0.7% Ag. Good casting design and foundry techniques must be employed to realize the full mechanical-property capabilities of these alloys and consistently high quality.

The 2xx.x alloys also have the highest strengths and hardnesses of all casting alloys at elevated temperatures (to 300 °C, or 600 °F), and this factor accounts for their use in some applications. Alloys 222.0, 224.0, 238.0, 240.0, 242.0 and 243.0, some with higher copper contents, up to 2% Mg (6% in alloy 240.0), and additions of Mn, Ni, V and/or Zr, are used primarily at elevated temperatures.

Heat treatment is required with the 2xx.x alloys for development of highest strength and ductility and must be properly applied to ensure high resistance to stress-corrosion cracking. General corrosion resistance of these alloys is

lower than those of other types of casting alloys, and protection by surface coatings is required in critical applications.

The 7xx.x aluminum-zinc-magnesium alloys are notable for their combinations of good finishing characteristics, good general corrosion resistance and the capability of developing high strength through natural aging without heat treatment. The 8xx.x aluminum-tin alloys are special-purpose compositions used for sleeve bearings and bushings.

Almost all of the alloys in the 2xx.x, 5xx.x, 7xx.x and 8xx.x groups are limited with respect to the casting configurations that can be produced with good recovery and quality. Greater care and adaptations in mold design are required to produce highest-quality castings with these alloys than with those of the 3xx.x and 4xx.x groups that have higher silicon contents.

Silicon is the alloying element that literally makes possible the commercial viability of the high-volume aluminum casting industry. Silicon contents from about 4% to the eutectic level of about 12% reduce scrap losses, permit production of much more intricate designs with greater variations in section thickness, and yield castings with higher surface and internal quality. These benefits derive from the effects of silicon in increasing fluidity, reducing cracking and improving feeding to minimize shrinkage porosity.

The complete phase diagram of the binary aluminum-silicon system is shown in Fig. 4. This is a simple eutectic system with limited terminal solubility and is the basis for the 4xx.x alloys. Metallographic structures of the pure components and of several intermediate compositions show typical morphologies. The intermediate compositions are mixtures of aluminum containing about 1% Si in solid solution as the contin-

uous phase, with particles of essentially pure silicon. Alloys with less than 12% Si are referred to as "hypoeutectic," those with close to 12% Si as "eutectic" and those with over 12% Si as "hypereutectic."

Resistance to cracking during casting is favored by a small range of solidification temperature, which drops from about 78 °C (140 °F) at 1% Si to zero at about 12% Si. Good feeding characteristics to minimize shrinkage porosity are benefited by a profile of volume fraction solidified versus temperature which is weighted toward the lower portion of the temperature range — that is, toward increased eutectic. In the binary system, under the nonequilibrium conditions of casting, the volume fraction of eutectic increases linearly from about 0 to 1 as silicon content increases from 1 to 12%.

The highest-volume-usage alloys are those in the 3xx.x group which, in addition to silicon, contain magnesium or copper, or both, and in specific cases supplementary additions of nickel or beryllium. In general, they fall into one of three types: Al-Si-Mg, Al-Si-Cu or Al-Si-Cu-Mg. Silicon contents range from 5 to 22%. Copper contents range from 0% (alloys 356.0 through 361.0) to a maximum of 4.5%. Most of these alloys have nominal magnesium contents ranging from as low as 0.3% to about 0.6% for the high-strength compositions and 1.0% for the piston alloys 332.0 and 336.0. The principal alloys of this group requiring low magnesium contents (0.10% maximum) are the die casting compositions 380.0 through 384.0.

Both copper and magnesium increase strength and hardness in the as-cast (F) temper through increased solid-solution hardening. Much greater increases are afforded by artificial aging only (T5-type tempers) or by complete solution plus arti-

Table 2. Typical applications of aluminum alloy die, permanent mold, sand and premium-quality castings

Alloy(s)	Applications	Alloys(s)	Applications
Die castings		356.0, A356.0	Machine-tool parts, aircraft wheels, pump parts, marine hardware, valve bodies
380.0	Lawnmower housings, automotive transmission and gear cases, cylinder heads for air-cooled engines	B443.0	Carburetor bodies, waffle irons
A380.0	Streetlamp housings, typewriter frames, dental equipment	513.0	Ornamental hardware and architectural fittings
360.0	Frying skillets, cover plates, instrument cases, parts requiring corrosion resistance	**Sand and premium-quality castings**	
		A201.0, A206.0	Aircraft structural components
		C355.0	Air-compressor fittings, crankcases, gear housings
413.0	Outboard-motor parts such as pistons, connecting rods and housings	A356.0	Automobile and truck transmission cases, oil pans and rear-axle housings
518.0	Escalator parts, conveyor components, aircraft and marine hardware and fittings	357.0	Pump bodies, cylinder blocks for water-cooled engines
Permanent mold castings		443.0	Pipe fittings, cooking utensils, ornamental fittings, marine fittings
336.0	Automotive pistons	535.0	Aircraft fittings, components, levers, brackets
355.0, C355.0, A357.0	Timing gears, impellers, compressors, and aircraft and missile components requiring high strength	713.0	General-purpose casting alloy for applications that require strength without heat treatment or that involve brazing

ficial aging treatments (T6- or T7-type tempers). Depending on composition, the precipitation hardening is the result of precipitate structures based on Mg_2Si, Al_2Cu, Al_2CuMg or combinations of these phases. The alloys containing both copper and magnesium have higher strengths at elevated temperatures.

Higher-silicon-content alloys are preferred for casting by the permanent mold and die casting processes. The thermal expansion coefficient decreases with increasing silicon and nickel contents. A low expansion coefficient is beneficial for engine applications such as pistons and cylinder blocks. When the silicon content exceeds 12%, as in alloys 390.0 through 393.0, primary silicon crystals are present and, if fine and well distributed, enhance wear resistance.

Alloys of the 4xx.x group, based on the binary aluminum-silicon system and containing from 5 to 12% Si, find many applications where combinations of moderate strength and high ductility and impact resistance are required. Bridge railing support castings are a representative example.

MECHANICAL PROPERTIES

The mechanical properties listed for aluminum casting alloys in the compilations of the following article are of two types. One type comprises nominal values obtained in tests of separately cast tensile specimens. Such specimens have a relatively high degree of soundness and good metallurgical structure so that they provide representative values for comparing alloys and tempers. Minimum properties based on separately cast specimens, however, cannot be used as design limits for production castings, because in many cases the soundness and microstructure of such castings differ from those of separately cast specimens.

The mechanical properties in various locations of engineering castings may vary from values higher to values considerably lower than those determined using separately cast specimens poured from the same metal and heat treated in the same heat treat lot as the castings. To increase the correspondence of test-specimen properties to casting properties, test specimens cast integrally with the casting are sometimes used to monitor and determine acceptability. For the most critical castings, minimum mechanical-property values are specified for tensile specimens machined from various locations in the castings on a sampling-plan basis.

APPLICATIONS

The applications of aluminum alloy foundry products are numerous and quite varied. Competition with other materials, the most prominent of which are polymeric materials, is keen and increasing. For many years die cast aluminum housings were the standard for hand tools such as electric drills, saws, routers and hedge trimmers. Many of these are now made of polymeric materials. In the automotive field, aluminum castings have made considerable inroads into applications formerly monopolized by cast iron, such as transmission cases, intake manifolds, and some engine blocks and cylinder heads. Relationships and viabilities for different applications change greatly with time so that the aluminum castings industry may be characterized as continually dynamic. A brief listing of typical applications is given in Table 2.

Commercial Aluminum Casting Alloys

THIS ARTICLE presents data for aluminum foundry alloys and products. Included are a cross-reference chart of frequently used designations and specifications (Table 1), nominal compositions (Table 2), typical properties (Tables 7, 8A, 8B and 8C) and characteristics (Tables 6A, B and C). The bases for the nominal compositions and typical properties are the same as those for the corresponding attributes of commercial wrought aluminum alloys and products.

Composition limits (Tables 3, 4 and 5) are current industry standards. Mechanical-property limits (Tables 9A, 9B, 10 and 11) are those listed in various applicable specifications.

Curves showing the effects of temperature on tensile properties of several alloys are presented (Fig. 1 to 4). Creep-rupture data for three representative alloys are listed (Tables 12A, B and C). Fatigue data are given for separate cast specimens of two alloys (Tables 13A and B), and comparisons of fatigue data are shown for various alloys, casting configurations and processes, and stress conditions (Fig. 5, 6 and 7).

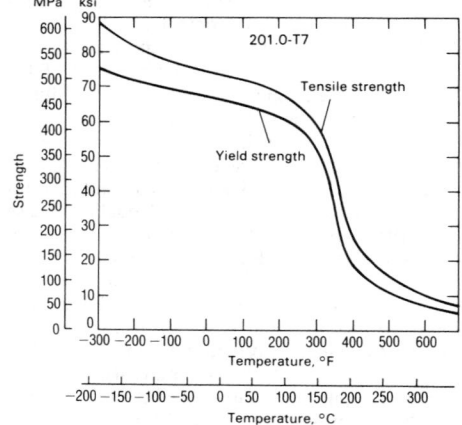

Fig. 1. Tensile and yield strengths of alloy 201.0-T7 as functions of temperature

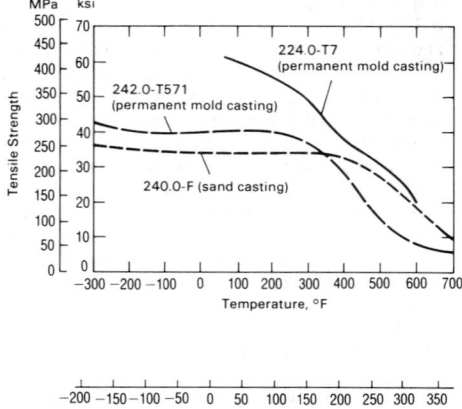

Fig. 2. Tensile strengths of alloys 240.0-F, 224.0-T7 and 242.0-T571 as functions of temperature

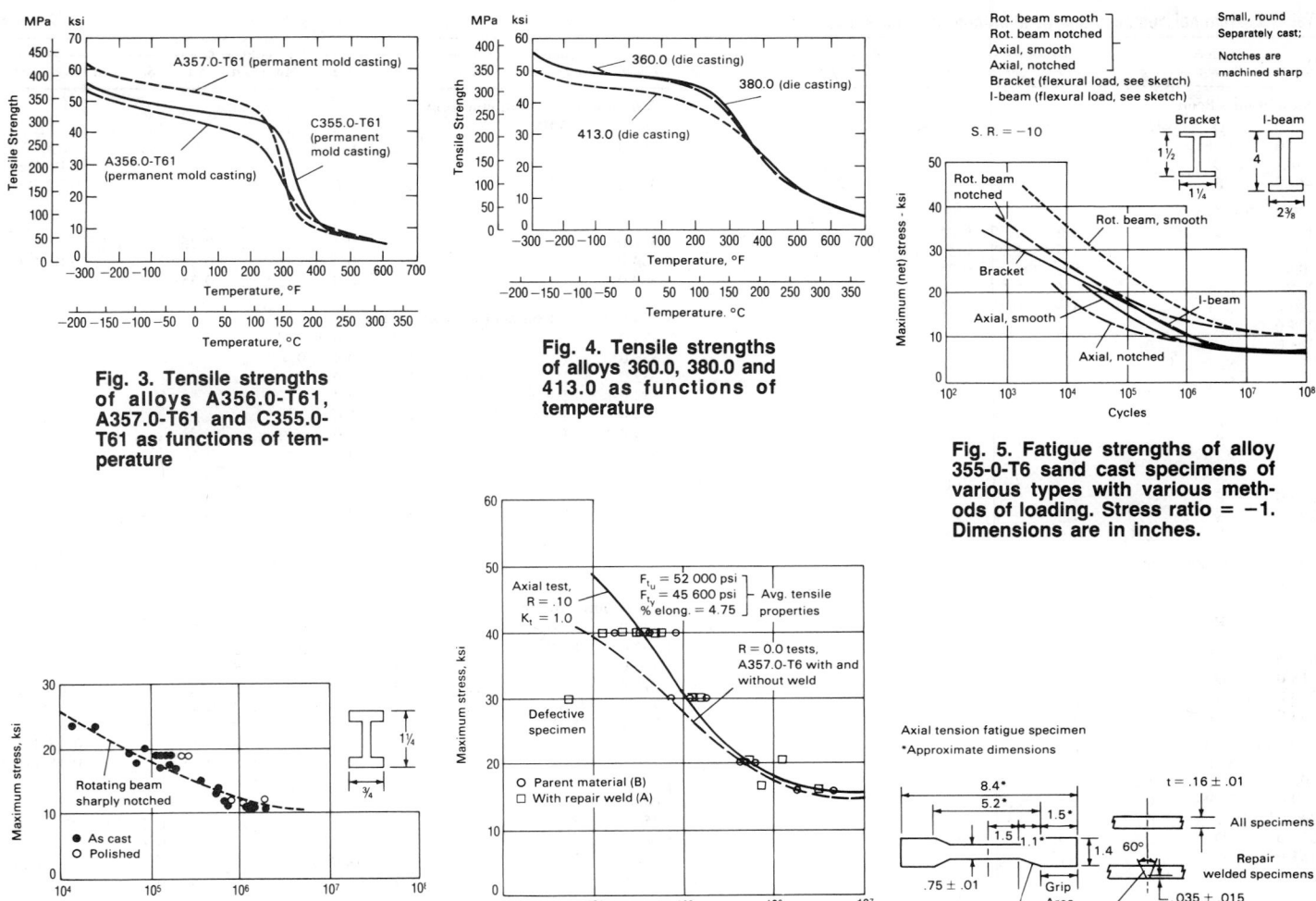

Fig. 3. Tensile strengths of alloys A356.0-T61, A357.0-T61 and C355.0-T61 as functions of temperature

Fig. 4. Tensile strengths of alloys 360.0, 380.0 and 413.0 as functions of temperature

Fig. 5. Fatigue strengths of alloy 355-0-T6 sand cast specimens of various types with various methods of loading. Stress ratio = −1. Dimensions are in inches.

Fig. 6. Flexural fatigue strength of alloy 356.0-T6 permanent mold production castings. Stress ratio = −1. Dimensions are in inches.

Fig. 7. Axial-stress fatigue curves for specimens cut from alloy A357.0-T6 premium-quality pylon castings for F5 military aircraft. Dimensions are in inches.

Table 1. Cross-reference chart of frequently used specifications for aluminum alloy sand and permanent mold castings

AA No.	Former designation	QQ-A-601E (sand)	QQ-A-596d (P.M.)	B26 (sand)	B108 (P.M.)	SAE	Military (MIL-21180c)	AA No.	Former designation	QQ-A-601E (sand)	QQ-A-596d (P.M.)	B26 (sand)	B108 (P.M.)	SAE	Military (MIL-21180c)
208.0	108	108	...	CS43A	CS43A	...	...	512.0	B514.0	B214	...	GS42A	GS42A	...	...
213.0	C113	...	113	CS74A	CS74A	33	...	513.0	A514.0	...	A214	...	GZ42A	...	...
222.0	122	122	122	CG100A	CG100A	34	...	514.0	214	214	...	G4A	...	320	...
242.0	142	142	142	CN42A	CN42A	39	...	520.0	220	220	...	G10A	...	324	...
295.0	195	195	195	C4A	...	38	...	535.0	Almag 35	Almag 35	...	GM70B	GM70B	...	...
296.0	B295.0	...	B195	...	...	380	...	705.0	603, Tern-alloy 5	Tern-alloy 5	Tern-alloy 5	ZG32A	ZG32A	311	...
308.0	A108	...	A108	...	...	...	...								
319.0	319, Allcast	319	319	SC64D	SC64D	326	...	707.0	607, Tern-alloy 7	Tern-alloy 7	Tern-alloy 7	ZG42A	ZG42A	312	...
328.0	Red X-8	Red X-8	...	SC82A	...	327	...								
332.0	F332.0	...	F132	...	SC103A	332	...	710.0	A712.0	A612	...	ZG61B	...	313	...
333.0	333	...	333	...	...	...	...	712.0	D712.0	40E	...	ZG61A	...	310	...
336.0	A332.0	...	A132	...	SN122A	321	...	713.0	613, Tenz-aloy	Tenzaloy	...	ZC81A	...	315	...
354.0	354	...	...	...	...	...	C354								
355.0	355	355	355	SC51A	SC51A	322	...	771.0	Prece-dent 71A	Prece-dent 71A	...	...	...	...	...
C355.0	C355	...	C355	...	SC51B	355	C355								
356.0	356	356	356	SG70A	SG70A	323	...	850.0	750	750	750	...	...	...	...
A356.0	A356	...	A356	...	SG70B	336	A356	851.0	A850.0	A750	A750	...	...	...	...
357.0	357	...	357	...	...	...	...	852.0	B850.0	B750	B750	...	...	...	...
A357.0	A357	...	...	...	...	...	A357								
359.0	359	...	...	...	...	...	359								
B443.0	43	43	43	S5A	S5A	...	...								

(a) Former designations. ASTM adopted the Aluminum Association designation system in 1974.

Table 2. Nominal compositions of aluminum casting alloys

AA No.	Product form(a)	Si	Cu	Mg	Mn	Zn	Cr	Sn	Ni	Other
Aluminum-silicon										
413.0	D	12	...	...	...	...	...	...	...	...
443.0	S, P	5	...	...	...	...	...	...	...	...
C443.0	D	5	...	...	...	...	...	...	...	...
444.0	P	7	...	...	...	...	...	...	...	...
Aluminum-copper										
201.0	S	...	4.6	0.37	0.35	...	...	...	...	0.7 Ag
202.0	S	...	4.6	0.37	0.5	...	0.4	...	...	0.7 Ag
203.0	S	...	5.0	...	0.25	...	...	...	1.5	0.25 Sb
204.0	S, P	...	4.6	0.27	...	...	...	...	...	...
206.0	S, P	...	4.6	0.27	0.35	...	...	...	...	...
208.0	S	3	4	...	...	...	...	...	...	...
213.0	S, P	2	7	...	...	...	...	...	...	...
222.0	S, P	...	10	0.27	...	...	...	...	...	...
224.0	S, P	...	5	...	0.35	...	...	...	...	0.17 Zr, 0.1 V
238.0	P	4	10	0.27	...	...	...	...	...	...
240.0	S	...	8	6	0.5	...	...	...	0.5	...
242.0	S, P	...	4	1.5	...	...	...	...	2	...
243.0	S	...	4	2.0	0.3	...	...	...	2.1	0.13 V
249.0	P	...	4.2	0.4	0.37	3	...	...	...	...
295.0	S	...	4.5	...	...	...	...	...	...	...
296.0	P	2.5	4.5	...	...	...	...	...	...	...
Aluminum-magnesium										
511.0	S	0.5	...	4	...	...	...	...	...	...
512.0	S	1.8	...	4	...	...	...	...	...	...
513.0	P	...	...	4	...	1.8	...	...	...	...
514.0	S	...	...	4	...	...	...	...	...	...
515.0	D	0.75	...	3.2	...	...	...	...	...	...
518.0	D	...	...	8	...	...	...	...	...	...
520.0	S	...	...	10	...	...	...	...	...	...
535.0	S, P	...	...	7	0.2	...	...	...	...	...
Aluminum-silicon-copper										
305.0	S, P	5	1.25	...	...	...	...	...	...	...
308.0	S, P	5.5	4.5	...	...	...	...	...	...	...
319.0	S, P	6	3.5	...	...	...	...	...	...	...
380.0	D	8.5	3.5	...	...	...	...	...	...	...
383.0	D	10.5	2.5	...	...	...	...	...	...	...
384.0	D	11.2	3.75	...	...	...	...	...	...	...
385.0	D	12	3	...	...	...	...	...	...	...
Aluminum-silicon-magnesium										
356.0	S, P	7	...	0.32	...	...	...	...	...	...
357.0	S, P	7	...	0.52	...	...	...	...	...	...
A357.0	S, P	7	...	0.55	...	...	...	...	...	0.05 Be
358.0	S, P	8	...	0.5	...	...	...	...	...	0.22 Be
359.0	S, P	9	...	0.6	...	...	...	...	...	...
360.0	D	9.5	...	0.5	...	...	...	...	...	...
361.0	D	10	...	0.5	...	0.25	...	0.25	...	...
364.0	D	8.5	...	0.3	...	0.3	...	...	...	0.03 Be
369.0	D	11.5	...	0.35	...	0.35	...	...	...	...
Aluminum-silicon-copper-magnesium										
324.0	P	7.5	0.5	0.6	...	...	...	...	...	...
328.0	S	8	1.5	0.4	0.4	...	...	...	...	...
332.0	P	9.5	3	1.0	...	...	...	...	...	...
333.0	P	9	3.5	0.3	...	...	...	...	...	...
336.0	P	12	1.0	1.0	...	...	...	...	2.5	...
339.0	P	12	2.2	1.0	...	...	...	...	1.0	...
354.0	P	9	1.8	0.5	...	...	...	...	...	...
355.0	S, P	5	1.2	0.5	...	...	...	...	...	...
363.0	S, P	5	3	0.3	...	3.7	...	...	...	...
390.0	D	17	4.5	0.6	...	...	...	...	...	...
392.0	D	19	0.6	1.0	0.4	...	...	...	...	...
393.0	S, P, D	22	0.9	1.0	...	...	...	...	2.2	0.1 V, 0.15 Ti
Aluminum-zinc-magnesium										
705.0	S, P	...	...	1.6	0.5	3	0.3	...	...	...
707.0	S, P	...	...	2.1	0.5	4.2	0.3	...	...	...
710.0	S	...	0.5	0.7	...	6.5	...	...	...	...
711.0	P	...	0.5	0.35	...	6.5	...	...	...	1.0 Fe
712.0	S	...	...	0.57	...	5.7	0.5	...	...	0.2 Ti
713.0	S, P	...	0.7	0.35	...	7.0	...	...	...	...
771.0	S	...	...	0.9	...	7.0	0.13	...	...	0.15 Ti
772.0	S	...	...	0.7	...	6.5	0.13	...	...	0.15 Ti
Aluminum-tin										
850.0	S, P	...	1.0	...	...	...	...	6.2	1.0	...
851.0	S, P	2.5	1.0	...	...	...	...	6.2	0.5	...
852.0	S, P	...	2.0	0.75	...	...	...	6.2	1.2	...
853.0	S, P	6.0	3.5	...	...	...	...	6.2	...	...

(a) D = die castings; S = sand castings; P = permanent mold castings.

Table 3. Chemical-composition limits for aluminum casting alloys

AA No.	Si	Fe	Cu	Mn	Mg	Cr	Ni	Zn	Sn	Ti	Others Each	Others Total
201.0	0.10	0.15	4.0-5.2	0.20-0.50	0.15-0.55	...	...	...	...	0.15-0.35	0.05(b)	0.10
A201.0	0.05	0.10	4.0-5.0	0.20-0.40	0.15-0.35	...	...	...	...	0.15-0.35	0.03(b)	0.10
202.0	0.10	0.15	4.0-5.2	0.20-0.8	0.15-0.55	0.20-0.6	...	...	...	0.15-0.35	0.05(b)	0.10
203.0	0.30	0.50	4.5-5.5	0.20-0.30	0.10	...	1.3-1.7	0.10	...	0.15-0.25(c)	0.05(d)	0.20
204.0	0.20	0.35	4.2-5.0	0.10	0.15-0.35	...	0.05	0.10	0.05	0.15-0.30	0.05	0.15
206.0	0.10	0.15	4.2-5.0	0.20-0.50	0.15-0.35	...	0.05	0.10	0.05	0.15-0.30	0.05	0.15
A206.0	0.05	0.10	4.2-5.0	0.20-0.50	0.15-0.35	...	0.05	0.10	0.05	0.15-0.30	0.05	0.15
208.0	2.5-3.5	1.2	3.5-4.5	0.50	0.10	...	0.35	1.0	...	0.25	...	0.50
213.0	1.0-3.0	1.2	6.0-8.0	0.6	0.10	...	0.35	2.5	...	0.25	...	0.50
222.0	2.0	1.5	9.2-10.7	0.50	0.15-0.35	...	0.50	0.8	...	0.25	...	0.35
224.0	0.06	0.10	4.5-5.5	0.20-0.50	...	...	...	...	...	0.35	0.03(e)	0.10
238.0	3.5-4.5	1.5	9.0-11.0	0.6	0.15-0.35	...	1.0	1.5	...	0.25	...	0.50
240.0	0.50	0.50	7.0-9.0	0.30-0.7	5.5-6.5	...	0.30-0.7	0.10	...	0.20	0.05	0.15
242.0	0.7	1.0	3.5-4.5	0.35	1.2-1.8	0.25	1.7-2.3	0.35	...	0.25	0.05	0.15
A242.0	0.6	0.8	3.7-4.5	0.10	1.2-1.7	0.15-0.25	1.8-2.3	0.10	...	0.07-0.20	0.05	0.15
243.0	0.35	0.40	3.5-4.5	0.15-0.45	1.8-2.3	0.20-0.40	1.9-2.3	0.05	...	0.06-0.20	0.05(f)	0.15
249.0	0.05	0.10	3.8-4.6	0.25-0.50	0.25-0.50	...	...	2.5-3.5	...	0.02-0.35	0.03	0.10
295.0	0.7-1.5	1.0	4.0-5.0	0.35	0.03	...	...	0.35	...	0.25	0.05	0.15
296.0	2.0-3.0	1.2	4.0-5.0	0.35	0.05	...	0.35	0.50	...	0.25	...	0.35
305.0	4.5-5.5	0.6	1.0-1.5	0.50	0.10	0.25	...	0.35	...	0.25	0.05	0.15
A305.0	4.5-5.5	0.20	1.0-1.5	0.10	0.10	...	...	0.10	...	0.20	0.05	0.15
308.0	5.0-6.0	1.0	4.0-5.0	0.50	0.10	...	...	1.0	...	0.25	...	0.50
319.0	5.5-6.5	1.0	3.0-4.0	0.50	0.10	...	0.35	1.0	...	0.25	...	0.50
A319.0	5.5-6.5	1.0	3.0-4.0	0.50	0.10	...	0.35	3.0	...	0.25	...	0.50
324.0	7.0-8.0	1.2	0.40-0.6	0.50	0.40-0.7	...	0.30	1.0	...	0.20	0.15	0.20

(continued)

Table 3. (continued)

AA No.	Si	Fe	Cu	Mn	Mg	Cr	Ni	Zn	Sn	Ti	Others Each	Others Total
						Composition(a), wt%						
328.0	7.5-8.5	1.0	1.0-2.0	0.20-0.6	0.20-0.6	0.35	0.25	1.5	...	0.25	...	0.50
332.0	8.5-10.5	1.2	2.0-4.0	0.50	0.50-1.5	...	0.50	1.0	...	0.25	...	0.50
333.0	8.0-10.0	1.0	3.0-4.0	0.50	0.05-0.50	...	0.50	1.0	...	0.25	...	0.50
A333.0	8.0-10.0	1.0	3.0-4.0	0.50	0.05-0.50	...	0.50	3.0	...	0.25	...	0.50
336.0	11.0-13.0	1.2	0.50-1.5	0.35	0.7-1.3	...	2.0-3.0	0.35	...	0.25	0.05	...
339.0	11.0-13.0	1.2	1.5-3.0	0.50	0.50-1.5	...	0.50-1.5	1.0	...	0.25	...	0.50
343.0	6.7-7.7	1.2	0.50-0.9	0.50	0.10	0.10	...	1.2-2.0	0.50	...	0.10	0.35
354.0	8.6-9.4	0.20	1.6-2.0	0.10	0.40-0.6	...	...	0.10	...	0.20	0.05	0.15
355.0	4.5-5.5	0.6(g)	1.0-1.5	0.50(g)	0.40-0.6	0.25	...	0.35	...	0.25	0.05	0.15
C355.0	4.5-5.5	0.20	1.0-1.5	0.10	0.40-0.6	...	...	0.10	...	0.20	0.05	0.15
356.0	6.5-7.5	0.6	0.25	0.35	0.20-0.45	...	...	0.35	...	0.25	0.05	0.15
A356.0	6.5-7.5	0.20	0.20	0.10	0.25-0.45	...	...	0.10	...	0.20	0.05	0.15
F356.0	6.5-7.5	0.20	0.20	0.10	0.17-0.25	...	...	0.10	...	0.20	0.05	0.15
357.0	6.5-7.5	0.15	0.05	0.03	0.45-0.6	...	...	0.05	...	0.20	0.05	0.15
A357.0	6.5-7.5	0.20	0.20	0.10	0.40-0.7	...	...	0.10	...	0.10-0.20	0.05(h)	0.15
358.0	7.6-8.6	0.30	0.20	0.20	0.40-0.6	0.20	...	0.20	...	0.10-0.20	0.05(h)	0.15
359.0	8.5-9.5	0.20	0.20	0.10	0.50-0.7	...	...	0.10	...	0.20	0.05	0.15
360.0	9.0-10.0	2.0	0.6	0.35	0.40-0.6	...	0.50	0.50	0.15	...	...	0.25
A360.0	9.0-10.0	1.3	0.6	0.35	0.40-0.6	...	0.50	0.50	0.15	...	...	0.25
361.0	9.5-10.5	1.1	0.50	0.25	0.40-0.6	0.20-0.30	0.20-0.30	0.50	0.10	0.20	0.05	0.15
363.0	4.5-6.0	1.1	2.5-3.5	(j)	0.15-0.40	(j)	0.25	3.0-4.5	0.25	0.20	(k)	0.30
364.0	7.5-9.5	1.5	0.20	0.10	0.20-0.40	0.25-0.50	0.15	0.15	0.15	...	0.05(m)	0.15
369.0	11.0-12.0	1.3	0.50	0.35	0.25-0.45	0.30-0.40	0.05	1.0	0.10	...	0.05	0.15
380.0	7.5-9.5	2.0	3.0-4.0	0.50	0.10	...	0.50	3.0	0.35	...	...	0.50
A380.0	7.5-9.5	1.3	3.0-4.0	0.50	0.10	...	0.50	3.0	0.35	...	...	0.50
B380.0	7.5-9.5	1.3	3.0-4.0	0.50	0.10	...	0.50	1.0	0.35	...	...	0.50
383.0	9.5-11.5	1.3	2.0-3.0	0.50	0.10	...	0.30	3.0	0.15	...	...	0.50
384.0	10.5-12.0	1.3	3.0-4.5	0.50	0.10	...	0.50	3.0	0.35	...	...	0.50
A384.0	10.5-12.0	1.3	3.0-4.5	0.50	0.10	...	0.50	1.0	0.35	...	...	0.50
385.0	11.0-13.0	2.0	2.0-4.0	0.50	0.30	...	0.50	3.0	0.30	...	...	0.50
390.0	16.0-18.0	1.3	4.0-5.0	0.10	0.45-0.65	...	...	0.10	...	0.20	0.10	0.20
A390.0	16.0-18.0	0.50	4.0-5.0	0.10	0.45-0.65	...	...	0.10	...	0.20	0.10	0.20
B390.0	16.0-18.0	1.3	4.0-5.0	0.50	0.45-0.65	...	0.10	1.5	...	0.20	0.10	0.20
392.0	18.0-20.0	1.5	0.40-0.8	0.20-0.6	0.8-1.2	...	0.50	0.50	0.30	0.20	0.15	0.50
393.0	21.0-23.0	1.3	0.7-1.1	0.10	0.7-1.3	...	2.0-2.5	0.10	...	0.10-0.20	0.05(n)	0.15
413.0	11.0-13.0	2.0	1.0	0.35	0.10	...	0.50	0.50	0.15	...	...	0.25
A413.0	11.0-13.0	1.3	1.0	0.35	0.10	...	0.50	0.50	0.15	...	...	0.25
443.0	4.5-6.0	0.8	0.6	0.50	0.05	0.25	...	0.50	...	0.25	...	0.35
A443.0	4.5-6.0	0.8	0.30	0.50	0.05	0.25	...	0.50	...	0.25	...	0.35
B443.0	4.5-6.0	0.8	0.15	0.35	0.05	...	...	0.35	...	0.25	0.05	0.15
C443.0	4.5-6.0	2.0	0.6	0.35	0.10	...	0.50	0.50	0.15	...	...	0.25
444.0	6.5-7.5	0.6	0.25	0.35	0.10	...	...	0.35	...	0.25	0.05	0.15
A444.0	6.5-7.5	0.20	0.10	0.10	0.05	...	...	0.10	...	0.20	0.05	0.15
511.0	0.30-0.7	0.50	0.15	0.35	3.5-4.5	...	...	0.15	...	0.25	0.05	0.15
512.0	1.4-2.2	0.6	0.35	0.8	3.5-4.5	0.25	...	0.35	...	0.25	0.05	0.15
513.0	0.30	0.40	0.10	0.30	3.5-4.5	...	...	1.4-2.2	...	0.20	0.05	0.15
514.0	0.35	0.50	0.15	0.35	3.5-4.5	...	...	0.15	...	0.25	0.05	0.15
515.0	0.50-1.0	1.3	0.20	0.40-0.6	2.5-4.0	...	...	0.10	...	...	0.05	0.15
518.0	0.35	1.8	0.25	0.35	7.5-8.5	...	0.15	0.15	0.15	...	...	0.25
520.0	0.25	0.30	0.25	0.15	9.5-10.6	...	...	0.15	...	0.25	0.05	0.15
535.0	0.15	0.15	0.05	0.10-0.25	6.2-7.5	...	...	...	...	0.10-0.25	0.05(p)	0.15
A535.0	0.20	0.20	0.10	0.10-0.25	6.5-7.5	...	...	...	...	0.25	0.05	0.15
B535.0	0.15	0.15	0.10	0.05	6.5-7.5	...	...	...	...	0.10-0.25	0.05	0.15
705.0	0.20	0.8	0.20	0.40-0.6	1.4-1.8	0.20-0.40	...	2.7-3.3	...	0.25	0.05	0.15
707.0	0.20	0.8	0.20	0.40-0.6	1.8-2.4	0.20-0.40	...	4.0-4.5	...	0.25	0.05	0.15
710.0	0.15	0.50	0.35-0.65	0.05	0.6-0.8	...	...	6.0-7.0	...	0.25	0.05	0.15
711.0	0.30	0.7-1.4	0.35-0.65	0.05	0.25-0.45	...	...	6.0-7.0	...	0.20	0.05	0.15
712.0	0.30	0.50	0.25	0.10	0.50-0.65	0.40-0.6	...	5.0-6.5	...	0.15-0.25	0.05	0.20
713.0	0.25	1.1	0.40-1.0	0.6	0.20-0.50	0.35	0.15	7.0-8.0	...	0.25	0.10	0.25
771.0	0.15	0.15	0.10	0.10	0.8-1.0	0.06-0.20	...	6.5-7.5	...	0.10-0.20	0.05	0.15
772.0	0.15	0.15	0.10	0.10	0.6-0.8	0.06-0.20	...	6.0-7.0	...	0.10-0.20	0.05	0.15
850.0	0.7	0.7	0.7-1.3	0.10	0.10	...	0.7-1.3	...	5.5-7.0	0.20	...	0.30
851.0	2.0-3.0	0.7	0.7-1.3	0.10	0.10	...	0.30-0.7	...	5.5-7.0	0.20	...	0.30
852.0	0.40	0.7	1.7-2.3	0.10	0.6-0.9	...	0.9-1.5	...	5.5-7.0	0.20	...	0.30
853.0	5.5-6.5	0.7	3.0-4.0	0.50	...	...	...	...	5.5-6.0	0.20	...	0.30

(a) All values except those shown as ranges are maximum values. Remainder aluminum. (b) 0.40 to 1.0% Ag. (c) 0.50% max Ti + Zr. (d) 0.20 to 0.30% Sb; 0.20 to 0.30% Co; 0.10 to 0.30% Zr. (e) 0.05 to 0.15% V; 0.10 to 0.25% Zr. (f) 0.06 to 0.20 V. (g) If Fe content exceeds 0.45%, Mn content must be no less than one-half Fe content. (h) 0.04 to 1.0 Be. (j) 0.8% max Mn + Cr. (k) 0.25% max Pb. (m) 0.02 to 0.04% Be. (n) 0.08 to 0.15% V. (p) 0.003 to 0.007% Be; 0.002% max B.

Table 4. Examples of relation between chemical-composition limits for aluminum alloy ingot and castings

| AA No. | Product form(a) | Composition, % | | | | | | | | | Other | |
		Si	Fe	Cu	Mn	Mg	Ni	Zn	Sn	Ti	Each	Total
356.0	S, P	6.5-7.5	0.6	0.25	0.35	0.20-0.45	...	0.35	...	0.25	0.05	0.15
356.1	I	6.5-7.5	0.50	0.25	0.35	0.20-0.45	...	0.35	...	0.35	0.05	0.15
356.2	I	6.5-7.5	0.13-0.25	0.10	0.05	0.30-0.45	...	0.05	...	0.20	0.05	0.15
380.0	D	7.5-9.5	2.0	3.0-4.0	0.50	0.10	0.50	3.0	0.35	...	...	0.50
380.2	I	7.5-9.5	0.7-1.1	3.0-4.0	0.10	0.10	0.10	0.10	0.10	...	...	0.2

(a) S = sand castings; P = permanent mold castings; I = ingot; D = die castings.

Table 5. Chemical-composition limits for unalloyed aluminum ingot for casting electric-motor rotors

| AA No. | Electrical conductivity (min), % IACS | Composition, % | | | | | | | Others | | Al (min) |
		Si	Fe	Cu	Mn	Cr	Zn	Ti	Each	Total	
100.1	54	0.15	0.6-0.8	0.10	(a)	(a)	0.05	(a)	0.03(a)	0.10	99.00
130.1	55	(b)	(b)	0.10	(a)	(a)	0.05	(a)	0.03(a)	0.10	99.30
150.1	57	(c)	(c)	0.05	(a)	(a)	0.05	(a)	0.03(a)	0.10	99.50
160.1	... .	0.10(c)	0.25(c)	...	(a)	(a)	0.05	(a)	0.03(a)	0.10	99.60
170.1	59	(d)	(d)	...	(a)	(a)	0.05	(a)	0.03(a)	0.10	99.70

(a) Mn + Cr + Ti + V ≤ 0.025. (b) Fe/Si ≥ 2.5. (c) Fe/Si ≥ 2.0. (d) Fe/Si ≥ 1.5.

Table 6A. Ratings of aluminum sand casting alloys: castability, corrosion resistance, machinability and weldability(a)

AA No.	Resistance to hot cracking(b)	Pressure tightness	Fluidity(c)	Shrinkage tendency(d)	Resistance to corrosion(e)	Machinability(f)	Weldability(g)	AA No.	Resistance to hot cracking(b)	Pressure tightness	Fluidity(c)	Shrinkage tendency(d)	Resistance to corrosion(e)	Machinability(f)	Weldability(g)
201.0 ...	4	3	3	4	4	1	2	511.0 ...	4	5	4	5	1	1	4
208.0 ...	2	2	2	2	4	3	3	512.0 ...	3	4	4	4	1	2	4
213.0 ...	3	3	2	3	4	2	2	514.0 ...	4	5	4	5	1	1	4
222.0 ...	4	4	3	4	4	1	3	520.0 ...	2	5	4	5	1	1	5
240.0 ...	4	4	3	4	4	3	4	535.0 ...	4	5	4	5	1	1	3
242.0 ...	4	3	4	4	4	2	3	A535.0 ..	4	5	4	4	1	1	4
A242.0 ..	4	4	3	4	4	2	3	B535.0 ..	4	5	4	4	1	1	4
295.0 ...	4	4	4	3	3	2	2								
319.0 ...	2	2	2	2	3	3	2	705.0 ...	5	4	4	4	2	1	4
354.0 ...	1	1	1	1	3	3	2	707.0 ...	5	4	4	4	2	1	4
355.0 ...	1	1	1	1	3	3	2	710.0 ...	5	3	4	4	2	1	4
A356.0 ..	1	1	1	1	2	3	2	711.0 ...	5	4	5	4	3	1	3
357.0 ...	1	1	1	1	2	3	2	712.0 ...	4	4	3	3	3	1	4
359.0 ...	1	1	1	1	2	3	1	713.0 ...	4	4	3	4	2	1	3
A390.0 ..	3	3	3	3	2	4	2	771.0 ...	4	4	3	3	2	1	...
								772.0 ...	4	4	3	3	2	1	...
A443.0 ..	1	1	1	1	2	4	4	850.0 ...	4	4	4	4	3	1	4
444.0 ...	1	1	1	1	2	4	1	851.0 ...	4	4	4	4	3	1	4
								852.0 ...	4	4	4	4	3	1	4

(a) For ratings of characteristics, 1 is the best and 5 is the poorest of the alloys listed. Individual alloys may have different ratings for other casting processes. (b) Ability of alloy to withstand stresses from contraction while cooling through hot-short or brittle temperature range. (c) Ability of molten alloy to flow readily in mold and fill thin sections. (d) Decrease in volume accompanying freezing of alloy and measure of amount of compensating feed metal required in form of risers. (e) Based on resistance of alloy in standard salt spray test. (f) Composite rating, based on ease of cutting, chip characteristics, quality of finish, and tool life. In the case of heat treatable alloys, rating is based on T6 temper. Other tempers, particularly the annealed temper, may have lower ratings. (g) Based on ability of material to be fusion welded with filler rod of same alloy.

Table 6B. Ratings of aluminum permanent mold casting alloys: castability, corrosion resistance, machinability and weldability(a)

AA No.	Resistance to hot cracking(b)	Pressure tightness	Fluidity(c)	Shrinkage tendency(d)	Resistance to corrosion(e)	Machinability(f)	Weldability(g)	AA No.	Resistance to hot cracking(b)	Pressure tightness	Fluidity(c)	Shrinkage tendency(d)	Resistance to corrosion(e)	Machinability(f)	Weldability(g)
201.0 ...	4	3	3	4	4	1	2	357.0 ...	1	1	1	1	2	3	2
213.0 ...	3	3	2	3	4	2	2	A357.0 ..	1	1	1	1	2	3	2
222.0 ...	4	4	3	4	4	1	3	359.0 ...	1	1	1	1	2	3	1
238.0 ...	2	3	2	2	4	2	3	A390.0 ..	2	2	2	3	2	4	2
240.0 ...	4	4	3	4	4	3	4	443.0 ...	1	1	2	1	2	5	1
296.0 ...	4	3	4	3	4	3	4	A444.0 ..	1	1	1	1	2	3	1
308.0 ...	2	2	2	2	4	3	3	512.0 ...	3	4	4	4	1	2	4
319.0 ...	2	2	2	2	3	3	2	513.0 ...	4	5	4	4	1	1	5
332.0 ...	1	2	1	2	3	4	2	711.0 ...	5	4	5	4	3	1	3
333.0 ...	1	1	2	2	3	3	3	771.0 ...	4	4	3	3	2	1	...
336.0 ...	1	2	2	3	3	4	2	772.0 ...	4	4	3	3	2	1	...
354.0 ...	1	1	1	1	3	3	2	850.0 ...	4	4	4	4	3	1	4
355.0 ...	1	1	1	2	3	3	2	851.0 ...	4	4	4	4	3	1	4
C355.0 ..	1	1	1	2	3	3	2	852.0 ...	4	4	4	4	3	1	4
356.0 ...	1	1	1	1	2	3	2								
A356.0 ..	1	1	1	1	2	3	2								

(a) For ratings of characteristics, 1 is the best and 5 is the poorest of the alloys listed. Individual alloys may have different ratings for other casting processes. (b) Ability of alloy to withstand stresses from contraction while cooling through hot-short or brittle temperature range. (c) Ability of molten alloy to flow readily in mold and fill thin sections. (d) Decrease in volume accompanying freezing of alloy and measure of amount of compensating feed metal required in form of risers. (c) Based on resistance of alloy in standard salt spray test. (f) Composite rating, based on ease of cutting, chip characteristics, quality of finish, and tool life. In the case of heat treatable alloys, rating is based on T6 temper. Other tempers, particularly the annealed temper, may have lower ratings. (g) Based on ability of material to be fusion welded with filler rod of same alloy.

Table 6C. Ratings of aluminum die casting alloys: castability, corrosion resistance and machinability(a)

AA No.	Resistance to hot cracking(b)	Pressure tightness	Fluidity(c)	Resistance to corrosion(d)	Machinability(e)	Weldability(f)	AA No.	Resistance to hot cracking(b)	Pressure tightness	Fluidity(c)	Resistance to corrosion(d)	Machinability(e)	Weldability(f)
360.0	1	1	2	2	3	4	390.0	2	2	2	2	4	2
A360.0	1	1	2	2	3	4	413.0	1	2	1	2	4	4
364.0	2	2	1	3	4	3	C443.0	2	3	3	2	5	4
380.0	2	1	2	5	3	4	515.0	4	5	5	1	2	4
A380.0	2	2	2	4	3	4	518.0	5	5	5	1	1	4
384.0	2	2	2	3	3	4							

(a) For ratings, 1 is the best and 5 is the poorest of the alloys listed. Individual alloys may have different ratings for other casting processes. (b) Ability of alloy to withstand stresses from contraction while cooling through hot-short or brittle temperature range. (c) Ability of liquid alloy to flow readily in mold and fill thin sections. (d) Based on resistance of alloy in standard salt spray test. (e) Composite rating based on ease of cutting, chip characteristics, quality of finish, and tool life. (f) Based on ability of material to be fusion welded with filler rod of same alloy.

Table 7. Physical properties of aluminum casting alloys

AA No.	Temper(a)	Specific gravity(b)	Density(b) kg/m³	Density(b) lb/in.³	Approximate melting range °C	Approximate melting range °F	Electrical conductivity, % IACS	Thermal conductivity at 25 °C (77 °F), cgs units	Coefficient of thermal expansion per °C × 10⁻⁶ 20 to 100 °C	(per °F × 10⁻⁶) 68 to 212 °F	per °C × 10⁻⁶ 20 to 300 °C	(per °F × 10⁻⁶) 68 to 570 °F
201.0	T6 (S)	2.80	2796	0.101	570-650	1060-1200	27-32	0.29	34.7	19.3	44.5	24.7
	T7 (P)	2.80	2796	0.101	570-650	1060-1200	32-34	0.29	34.7	19.3	44.5	24.7
206.0	. . .	2.8	2796	0.101	570-650	1060-1200	. . .	0.29	. . .	. . .	. . .	. . .
A206.0	. . .	2.8	2796	0.101	570-650	1060-1200	. . .	0.29	. . .	. . .	. . .	. . .
208.0	F (S)	2.79	2796	0.101	520-630	970-1170	31	0.29	22.0	12.2	23.9	13.3
	O (S)	2.79	2796	0.101	520-630	970-1170	38	0.35	. . .	. . .	. . .	. . .
222.0	F (P)	2.95	2962	0.107	520-625	970-1160	34	0.32	22.1	12.3	23.6	13.1
	O (S)	2.95	2962	0.107	520-625	970-1160	41	0.38	. . .	. . .	. . .	. . .
	T61 (S)	2.95	2962	0.107	520-625	970-1160	33	0.31	22.1	12.3	23.6	13.1
224.0	T62 (S)	2.81	2824	0.102	550-645	1020-1190	30	0.28	. . .	. . .	. . .	. . .
238.0	F (P)	2.95	1938	0.107	510-600	950-1110	25	0.25	21.4	11.9	22.9	12.7
240.0	F (S)	2.78	2768	0.100	515-605	960-1120	23	0.23	22.1	12.3	24.3	13.5
242.0	O (S)	2.81	2823	0.102	530-635	990-1180	44	0.40	. . .	. . .	. . .	. . .
	T77	2.81	2823	0.102	525-635	980-1180	38	0.36	22.1	12.3	23.6	13.1
	T571 (P)	2.81	2823	0.102	525-635	980-1180	34	0.32	22.5	12.5	24.5	13.6
	T61 (P)	2.81	2823	0.102	525-635	980-1180	33	0.32	22.5	12.5	24.5	13.6
295.0	T4 (S)	2.81	2823	0.102	520-645	970-1190	35	0.33	22.9	12.7	24.8	13.8
	T62 (S)	2.81	2823	0.102	520-645	970-1190	35	0.34	22.9	12.7	24.8	13.8
296.0	T4 (P)	2.80	2796	0.101	520-630	970-1170	33	0.32	22.0	12.2	23.9	13.3
	T6 (P)	2.80	2796	0.101	520-630	970-1170	33	0.32	22.0	12.2	23.9	13.3
	T62 (S)	2.80	2796	0.101	520-630	970-1170	33	0.32	. . .	. . .	. . .	. . .
308.0	F (P)	2.79	2796	0.101	520-615	970-1140	37	0.34	21.4	11.9	22.9	12.7
319.0	F (S)	2.79	2796	0.101	520-605	970-1120	27	0.27	21.6	12.0	24.1	13.4
	F (P)	2.79	2796	0.101	520-605	970-1120	28	0.28	21.6	12.0	24.1	13.4
324.0	F (P)	2.67	2658	0.096	545-605	1010-1120	34	0.37	21.4	11.9	23.2	12.9
332.0	T5 (P)	2.76	2768	0.100	520-580	970-1080	26	0.25	20.7	11.5	22.3	12.4
333.0	F (P)	2.77	2768	0.100	520-585	970-1090	26	0.25	20.7	11.5	22.7	12.6
	T5 (P)	2.77	2768	0.100	520-585	970-1090	29	0.29	20.7	11.5	22.7	12.6
	T6 (P)	2.77	2768	0.100	520-585	970-1090	29	0.28	20.7	11.5	22.7	12.6
	T7 (P)	2.77	2768	0.100	520-585	970-1090	35	0.34	20.7	11.5	22.7	12.6
336.0	T551 (P)	2.72	2713	0.098	540-570	1000-1060	29	0.28	18.9	10.5	20.9	11.6
354.0	F (P)	2.71	2713	0.098	540-600	1000-1110	32	0.30	20.9	11.6	22.9	12.7
355.0	T51 (S)	2.71	2713	0.098	550-620	1020-1150	43	0.40	22.3	12.4	24.7	13.7
	T6 (S)	2.71	2713	0.098	550-620	1020-1150	36	0.34	22.3	12.4	24.7	13.7
	T61 (S)	2.71	2713	0.098	550-620	1020-1150	37	0.35	22.3	12.4	24.7	13.7
	T7 (S)	2.71	2713	0.098	550-620	1020-1150	42	0.39	22.3	12.4	24.7	13.7
	T6 (P)	2.71	2713	0.098	550-620	1020-1150	39	0.36	22.3	12.4	24.7	13.7
C355.0	T61 (S)	2.71	2713	0.098	550-620	1020-1150	39	0.35	22.3	12.4	24.7	13.7
356.0	T51 (S)	2.68	2685	0.097	560-615	1040-1140	43	0.40	21.4	11.9	23.4	13.0
	T6 (S)	2.68	2685	0.097	560-615	1040-1140	39	0.36	21.4	11.9	23.4	13.0
	T7 (S)	2.68	2685	0.097	560-615	1040-1140	40	0.37	21.4	11.9	23.4	13.0
	T6 (P)	2.68	2685	0.097	560-615	1040-1140	41	0.37	21.4	11.9	23.4	13.0
A356.0	T6 (S)	2.69	2713	0.098	560-610	1040-1130	40	0.36	21.4	11.9	23.4	13.0
357.0	T6 (S)	2.68	2713	0.098	560-615	1040-1140	39	0.36	21.4	11.9	23.4	13.0
A357.0	T6 (S)	2.69	2713	0.098	555-610	1030-1130	40	0.38	21.4	11.9	23.6	13.1
358.0	T6 (S)	2.68	2658	0.096	560-600	1040-1110	39	0.36	21.4	11.9	23.4	13.0
359.0	T6 (S)	2.67	2685	0.097	565-600	1050-1110	35	0.33	20.9	11.6	22.9	12.7
360.0	F (D)	2.68	2685	0.097	570-590	1060-1090	37	0.35	20.9	11.6	22.9	12.7
A360.0	F (D)	2.68	2685	0.097	570-590	1060-1090	37	0.35	21.1	11.7	22.9	12.7
364.0	F (D)	2.63	2630	0.095	560-600	1040-1110	30	0.29	20.9	11.6	22.9	12.7
380.0	F (D)	2.76	2740	0.099	520-590	970-1090	27	0.26	21.2	11.8	22.5	12.5
A380.0	F (D)	2.76	2740	0.099	520-590	970-1090	27	0.26	21.1	11.7	22.7	12.6
384.0	F (D)	2.70	2713	0.098	480-580	900-1080	23	0.23	20.3	11.3	22.1	12.3
390.0	F (D)	2.73	2740	0.099	510-650	950-1200	25	0.32	18.5	10.3	. . .	. . .
	T5 (D)	2.73	2740	0.099	510-650	950-1200	24	0.32	18.0	10.0	. . .	. . .
392.0	F (P)	2.64	2630	0.095	550-670	1020-1240	22	0.22	18.5	10.3	20.2	11.2
413.0	F (D)	2.66	2657	0.096	575-585	1070-1090	39	0.37	20.5	11.4	22.5	12.5
A413.0	F (D)	2.66	2657	0.096	575-585	1070-1090	39	0.37	. . .	. . .	. . .	. . .
443.0	F (S)	2.69	2685	0.097	575-630	1070-1170	37	0.35	22.1	12.3	24.1	13.4
	O (S)	2.69	2685	0.097	575-630	1070-1170	42	0.39	. . .	. . .	. . .	. . .

(continued)

Table 7. (continued)

AA No.	Temper(a)	Specific gravity(b)	Density(b) kg/m³	lb/in.³	Approximate melting range °C	°F	Electrical conductivity, ; IACS	Thermal conductivity at 25 °C (77 °F), cgs units	Coefficient of thermal expansion per °C × 10⁻⁶ (per °F × 10⁻⁶) 20 to 100 °C	68 to 212 °F	20 to 300 °C	68 to 570 °F
	F (D)	2.69	2685	0.097	575-630	1070-1170	37	0.34	...	...	...	...
A444.0	F (P)	2.68	2685	0.097	575-630	1070-1170	41	0.38	21.8	12.1	23.8	13.2
511.0	F (S)	2.66	2657	0.096	590-640	1090-1180	36	0.34	23.6	13.1	25.7	14.3
512.0	F (S)	2.65	2657	0.096	590-630	1090-1170	38	0.35	22.9	12.7	24.8	13.8
513.0	F (P)	2.68	2685	0.097	580-640	1080-1180	34	0.32	23.9	13.3	25.9	14.4
514.0	F (S)	2.65	2657	0.096	600-640	1110-1180	35	0.33	23.9	13.3	25.9	14.4
518.0	F (D)	2.53	2519	0.091	540-620	1000-1150	24	0.24	24.1	13.4	26.1	14.5
520.0	T4 (S)	2.57	2574	0.093	450-600	840-1110	21	0.21	25.2	14.0	27.0	15.0
535.0	F (S)	2.62	2519	0.091	550-630	1020-1170	23	0.24	23.6	13.1	26.5	14.7
A535.0	F (D)	2.54	2547	0.092	550-620	1020-1150	23	0.24	24.1	13.4	26.1	14.5
B535.0	F (S)	2.62	2630	0.095	550-630	1020-1170	24	0.23	24.5	13.6	26.5	14.7
705.0	F (S)	2.76	2768	0.100	600-640	1110-1180	25	0.25	23.6	13.1	25.7	14.3
707.0	F (S)	2.77	2768	0.100	585-630	1090-1170	25	0.25	23.8	13.2	25.9	14.4
710.0	F (S)	2.81	2823	0.102	600-650	1110-1200	35	0.33	24.1	13.4	26.3	14.6
711.0	F (P)	2.84	2851	0.103	600-645	1110-1190	40	0.38	23.6	13.1	25.6	14.2
712.0	F (S)	2.82	2823	0.102	600-640	1110-1180	40	0.38	23.6	13.1	25.6	14.2
713.0	F (S)	2.84	2879	0.104	595-630	1100-1170	37	0.37	23.9	13.3	25.9	14.4
850.0	T5 (S)	2.87	2851	0.103	225-650	440-1200	47	0.44	...	...	...	...
851.0	T5 (S)	2.83	2823	0.102	230-630	450-1170	43	0.40	22.7	12.6	...	...
852.0	T5 (S)	2.88	2879	0.104	210-635	410-1170	45	0.42	23.2	12.9	...	...

(a) S = sand cast; P = permanent mold; D = die cast. (b) The specific gravity and density data in this table assume solid (void-free) metal. Because some porosity cannot be avoided in commercial castings, their specific gravities and densities are slightly less than the theoretical values.

Table 8A. Typical mechanical properties of aluminum sand casting alloys based on tests of separately cast specimens(a)

AA No.	Temper	Tensile strength MPa	ksi	Tensile yield strength(b) MPa	ksi	Elongation in 50 mm or 2 in., %	Hardness(c), HB	Compressive yield strength(b) MPa	ksi	Shear strength MPa	ksi	Endurance limit(d) MPa	ksi	Modulus of elasticity(e) kPa × 10⁶	psi × 10⁶
201.0	T43	414	60	255	37	17.0	...	...	...	...	...	...	...	...	...
	T6	448	65	379	55	8.0	130	386	56	290	42	...	...	...	...
	T7	467	68	414	60	5.5	...	...	...	...	...	97	14	...	...
A206.0	T4	354	51	250	36	7.0	...	264	38	278	40	...	...	...	...
208.0	F	145	21	97	14	2.5	55	103	15	117	17	76	11	...	...
213.0	F	165	24	103	15	1.5	70	110	16	138	20	62	9	...	...
222.0	O	186	27	138	20	1.0	80	138	20	145	21	65	9.5	...	...
	T61	283	41	276	40	<0.5	115	296	43	221	32	59	8.5	74	10.7
	T62	421	61	331	48	4.0	...	338	49	262	38	...	...	...	...
224.0	T72	380	55	276	40	10.0	123	283	41	245	36	62	9.0	72	10.5
240.0	F	235	34	200	29	1.0	90	207	30	...	...	...	...	...	...
242.0	F	214	31	217	31	0.5	...	...	...	...	...	...	...	...	...
	O	186	27	124	18	1.0	70	124	18	145	21	55	8	71	10.3
	T571	221	32	207	30	0.5	85	234	34	179	26	76	11	71	10.3
	T77	207	30	159	23	2.0	75	165	24	165	24	72	10.5	71	10.3
A242.0	T75	214	31	...	...	2.0	...	483	70	...	...	...	...	...	...
295.0	T4(f)	221	32	110	16	8.5	60	117	17	179	26	48	7	69	10.0
	T6	250	36	165	24	5.0	75	172	25	217	31	52	7.5	69	10.0
	T62	283	41	221	32	2.0	90	234	34	228	33	55	8	69	10.0
319.0	F	186	27	124	18	2.0	70	131	19	152	22	69	10	74	10.7
	T5	207	30	179	26	1.5	80	186	27	164	24	76	11	75	10.7
	T6	250	36	164	24	2.0	80	172	25	200	29	76	11	74	10.7
355.0	F	159	23	83	12	3.0	...	...	...	...	...	...	...	...	...
	T51	193	28	159	23	1.5	65	164	24	152	22	55	8	70	10.2
	T6	241	35	172	25	3.0	80	179	26	193	28	62	9	70	10.2
	T61	269	39	241	35	1.0	90	255	37	214	31	65	9.5	...	...
	T7	264	38	250	36	0.5	85	264	38	193	28	69	10	70	10.2
	T71	241	35	200	29	1.5	75	207	30	179	26	69	10	70	10.2
C355.0	T6	269	39	200	29	5.0	85	...	...	...	...	...	...	...	...
356.0	F	164	24	124	18	6.0	...	...	...	...	...	...	...	...	...
	T51	172	25	138	20	2.0	60	145	21	138	20	55	8	72	10.5
	T6	228	33	164	24	3.5	70	172	25	179	26	59	8.5	72	10.5
	T7	234	34	207	30	2.0	75	214	31	164	24	62	9	72	10.5
	T71	193	28	145	21	3.5	60	152	22	138	20	59	8.5	72	10.5
A356.0	F	159	23	83	12	6.0	...	...	...	...	...	...	...	...	...
	T51	179	26	124	18	3.0	...	...	...	...	...	...	...	...	...
	T6	278	40	207	30	6.0	75	...	...	...	...	...	...	...	...
	T71	207	30	138	20	3.0	...	...	...	...	...	...	...	...	...
357.0	F	172	25	90	13	5.0	...	...	...	...	...	...	...	...	...
	T51	179	26	117	17	3.0	...	...	...	...	...	...	...	...	...
	T6	345	50	296	43	2.0	90	...	...	...	...	...	...	...	...
	T7	278	40	234	34	3.0	60	...	...	...	...	...	...	...	...
A357.0	T6	317	46	248	36	3.0	85	241	35	278	40	83	12	...	...
A390.0	F	179	26	179	26	<1.0	100	...	...	...	...	...	...	82	11.9
	T5	179	26	179	26	<1.0	100	...	...	...	...	...	...	...	...
	T6	278	40	278	40	<1.0	140	...	...	...	...	90	13	...	...
	T7	250	36	250	36	<1.0	115	...	...	...	...	...	...	...	...

(continued)

Table 8A. (continued)

AA No.	Temper	Tensile strength MPa	ksi	Tensile yield strength(b) MPa	ksi	Elongation in 50 mm or 2 in., %	Hardness(c), HB	Compressive yield strength(b) MPa	ksi	Shear strength MPa	ksi	Endurance limit(d) MPa	ksi	Modulus of elasticity(e) kPa × 10⁶	psi × 10⁶
443.0	F	131	19	55	8	8.0	40	62	9	97	14	55	8	71	10.2
A444.0	F	145	21	62	9	9.0	...	...	...	...	...	...	...	...	...
	T4	159	23	62	9	12.0	...	...	...	...	...	...	...	...	...
511.0	F	145	21	83	12	3.0	50	90	13	117	17	55	8	...	...
512.0	F	138	20	90	13	2.0	50	97	14	117	17	59	8.5	...	...
514.0	F	172	25	83	12	9.0	50	83	12	138	20	48	7	71	10.2
520.0	T4	331	48	179	26	16.0	75	186	27	234	34	55	8	65	9.5
A535.0	F	250	36	124	18	9.0	65	...	...	...	...	...	...	...	...
710.0	F	241	35(g)	172	25(g)	5.0(g)	75(g)	172	25	179	26	55	8	67	9.7
712.0(h)	F	241	35	172	25	5.0	75	172	25	179	26	62	9	...	...
713.0(h)	F	241	35	172	25	5.0	74	...	...	...	...	...	...	67	9.7
850.0	T5	138	20	76	11	8.0	45	76	11	97	14	...	...	71	10.2
851.0	T5	138	20	76	11	5.0	45	76	11	97	14	...	...	71	10.2
852.0	T5	186	27	152	22	2.0	65	152	22	124	18	69	10	71	10.2

(a) Tensile and hardness values determined by tests on standard 13-mm (¹/₂-in.) diam test specimens, without surface machining, each cast in a green sand mold. (b) Offset: 0.2%. (c) 500 kg (1102-lb) load on 10-mm (0.4-in.) ball. (d) Endurance limits based on 500 million cycles of completely reversed stresses using rotating beam–type machine and specimen. (e) Average tension and compression moduli, compression modulus is about 2% greater than tension modulus. (f) Properties of T4 approach those of T6 after standing for several weeks at room temperature. (g) Tests made approximately 30 days after casting. (h) Tests made 10 days after casting.

Table 8B. Typical mechanical properties of aluminum permanent mold casting alloys based on tests of separately cast specimens(a)

AA No.	Temper	Tensile strength MPa	ksi	Tensile yield strength(b) MPa	ksi	Elongation in 50 mm or 2 in., %	Hardness(c), HB	Compressive yield strength(b) MPa	ksi	Shear strength MPa	ksi	Endurance limit(d) MPa	ksi	Modulus of elasticity(e) kPa × 10⁶	psi × 10⁶
201.0	T43	414	60	255	37	17.0	...	...	...	...	...	...	...	...	...
	T6	448	65	379	55	8.0	130	386	56	290	42	...	...	...	...
	T7	469	68	414	60	5.0	...	...	...	...	...	97	14	...	...
A206.0	T4	431	62	264	38	17.0	...	285	41	292	42	...	...	...	...
	T7	436	63	347	50	11.7	...	372	54	257	37	...	...	...	...
213.0	F	207	30	165	24	1.5	85	172	25	165	24	66	9.5	...	...
222.0	T52	241	35	214	31	1.0	100	214	31	172	25	...	...	...	...
	T551	255	37	241	35	<0.5	115	276	40	207	30	59	8.5	74	10.7
	T65	331	48	248	36	<0.5	140	248	36	248	26	62	9	74	10.7
238.0	F	207	30	165	24	1.5	100	207	30	165	24	...	...	...	...
242.0	T571	276	40	234	34	1.0	105	234	34	207	30	72	10.5	71	10.3
	T61	324	47	290	42	0.5	110	303	44	241	35	66	9.5	71	10.3
249.0	T63	476	69	414	60	6.0	...	...	...	...	...	...	...	...	...
	T7	278	62	359	52	9.0	...	414	60	276	40	55	8.0	72	10.5
296.0	T4(f)	255	37	131	19	9.0	75	138	20	207	30	66	9.5	70	10.1
	T6	276	40	179	26	5.0	90	179	26	221	32	69	10	70	10.1
	T7	270	39	138	20	4.5	80	138	20	207	30	63	9	70	10.1
308.0	F	193	28	110	16	2.0	70	117	17	152	22	90	13	...	...
319.0	F	234	34	131	19	2.5	85	131	19	165	24	...	...	...	...
	T6	276	40	186	27	3.0	95	186	27	...	...	...	...	...	...
324.0	F	207	30	110	16	4.0	70	...	...	...	...	...	...	...	...
	T5	248	36	179	26	3.0	90	...	...	...	...	...	...	...	...
	T62	310	45	269	39	3.0	105	...	...	...	...	...	...	...	...
332.0	T5	248	36	193	28	1.0	105	...	...	...	...	...	...	77	11.2
333.0	F	234	34	131	19	2.0	90	131	19	186	27	100	14.5	...	...
	T5	234	34	172	25	1.0	100	172	25	186	27	83	12	...	...
	T6	290	42	207	30	1.5	105	207	30	228	33	103	15	...	...
	T7	255	37	193	28	2.0	90	193	28	193	28	83	12	...	...
336.0	T551	248	36	193	28	...	105	193	28	193	28	93	13.5	...	...
	T65	324	47	296	43	0.5	125	296	43	248	36	...	...	...	...
355.0	T51	205	30	165	24	2.0	75	165	24	165	24	...	...	...	...
	T6	290	42	185	27	4.0	90	185	27	235	34	69	10	...	...
	T61	310	45	275	40	1.5	105	275	40	250	36	69	10	...	...
	T7	275	40	205	30	2.0	85	205	30	205	30	69	10	...	...
	T71	250	36	215	31	3.0	85	215	31	185	27	69	10	...	...
C355.0	T61	303	44	234	34	3.0	90	248	36	221	32	97	14	...	...
356.0	F	179	26	124	18	5.0	...	...	...	...	...	...	...	...	...
	T51	186	27	138	20	2.0	...	...	...	...	...	...	...	...	...
	T6	262	38	186	27	5.0	80	186	27	207	30	90	13	72	10.5
	T7	221	32	165	24	6.0	70	165	24	172	25	76	11	72	10.5
A356.0	T61	283	41	207	30	10.0	90	221	32	193	28	90	13	72	10.5
357.0	F	193	28	103	15	6.0	...	...	...	...	...	...	...	...	...
	T51	200	29	145	21	4.0	...	...	...	...	...	...	...	...	...
	T6	359	52	296	43	5.0	100	303	44	241	35	90	13	...	...
A357.0	T6	359	52	290	42	5.0	100	296	43	241	35	103	15	...	...
359.0	T62	345	50	290	42	5.5	...	...	...	...	...	110	16	...	...
A390.0	F	200	29	200	29	<1.0	110	...	...	...	...	...	...	82	11.9
	T5	200	29	200	29	<1.0	110	...	...	...	...	...	...	...	...
	T6	310	45	310	45	<1.0	145	414	60	...	...	117	17	...	...
	T7	262	38	262	38	<1.0	120	359	52	...	...	100	14.5	...	...

(continued)

Table 8B. (continued)

AA No.	Temper	Tensile strength MPa	ksi	Tensile yield strength(b) MPa	ksi	Elongation in 50 mm or 2 in., %	Hardness(c), HB	Compressive yield strength(b) MPa	ksi	Shear strength MPa	ksi	Endurance limit(d) MPa	ksi	Modulus of elasticity(e) kPa × 10⁶	psi × 10⁶
443.0	F	159	23	62	9	10.0	45	62	9	110	16	55	8	71	10.3
A444.0	F	165	24	76	11	13.0	44	...	...	...	...	...	...	...	...
	T4	159	23	69	10	21.0	45	76	11	110	16	55	8	...	...
513.0	F	186	27	110	16	7.0	60	117	17	152	22	69	10	...	...
711.0	F	241	35(g)	124	18(g)	8.0(g)	70(g)	...	...	...	...	76	11	76	11.0
850.0	T5	159	23	76	11	12.0	45	76	11	103	15	62	9	71	10.3
851.0	T5	138	20	76	11	5.0	45	76	11	97	14	62	9	71	10.3
852.0	T5	221	32	159	23	5.0	70	159	23	148	21	76	11	71	10.3

(a) Tension and hardness values determined by tests on standard 13-mm (1/2-in.) diam test specimens, without surface machining, each cast in permanent mold. (b) At 0.2% offset. (c) 500-kg (1102-lb) load on 10-mm (0.4-in.) ball. (d) Endurance limits based on 500 million cycles of completely reversed stresses using rotating beam–type machine and specimen. (e) Average of tension and compression moduli; compression modulus is about 2% greater than tension modulus. (f) Properties of T4 approach those of T6 after standing for several weeks at room temperature. (g) Tests made approximately 30 days after casting.

Table 8C. Typical mechanical properties of aluminum die casting alloys based on tests of separately cast specimens(a)

AA No.	Temper	Tensile strength MPa	ksi	Yield strength(b) MPa	ksi	Elongation in 50 mm or 2 in., %	Hardness(c), HB	Shear strength MPa	ksi	Endurance limit(d) MPa	ksi	Modulus of elasticity(e) kPa × 10⁶	psi × 10⁶
360.0	F	324	47	172	25	3.0	75	207	30	131	19	71	10.3
A360.0	F	317	46	165	24	5.0	75	200	29	124	18	...	...
364.0	F	296	43	159	23	7.5	...	179	26	124	18	...	...
380.0	F	331	48	165	24	3.0	80	214	31	145	21	71	10.3
A380.0	F	324	47	159	23	4.0	80	207	30	138	20	...	...
384.0	F	324	47	172	25	1.0	...	207	30	145	21	71	10.3
390.0	F	279	40.5	241	35	1.0	120	...	...	138	20	82	11.9
	T5	296	43	265	38.5	1.0	...	...	...	...	...	...	...
392.0	F	290	42	262	38	<0.5	...	234(f)	34(f)	103(f)	15(f)	...	...
413.0	F	296	43	145	21	2.5	80	193	28	131	19	71	10.3
A413.0	F	241	35	110	16	3.5	80	172	25	131	19	...	...
443.0	F	228	33	110	16	9.0	50	145	21	117	17	71	10.3
513.0	F	276	40	152	22	10.0	...	179	26	124	18	...	...
515.0	F	283	41	...	...	10.0	...	...	...	...	...	...	...
518.0	F	310	45	186	27	8.0	80	200	29	138	20	...	...

(a) Tension properties are average values determined from ASTM standard 6-mm (1/4-in.) diam test specimens cast on a cold chamber (high-pressure) die casting machine. (b) At 0.2% offset. (c) 500-kg (1102-lb) load on 10-mm (0.4-in.) ball. (d) Endurance limits based on 500 million cycles of completely reversed stresses using rotating beam–type machine and specimen. (e) Average of tension and compression moduli; compression modulus is about 2% greater than tension modulus. (f) Estimated.

Table 9A. Mechanical-property limits for aluminum sand casting alloys

AA No.	Temper	Tensile strength MPa	ksi	Yield strength MPa	ksi	Elongation in 50 mm or 2 in., %	AA No.	Temper	Tensile strength MPa	ksi	Yield strength MPa	ksi	Elongation in 50 mm or 2 in., %
201.0	T6	414	60.0	345	50.0	5.0		T6	207	30.0	138	20.0	3.0
	T7	414	60.0	345	50.0	3.0		T7	214	31.0	200	29.0	...
204.0	T4	310	45.0	193	28.0	6.0		T71	172	25.0	124	18.0	3.0
208.0	F	131	19.0	83	12.0	1.5	A356.0	T6	234	34.0	165	24.0	3.5
	T55	145	21.0	...	...	...	A357.0	(b)	...	...	...	...	...
222.0	O	159	23.0	...	...	...	359.0	(b)	...	...	...	...	...
	T61	207	30.0	...	...	...	443.0	F	117	17.0	48	7.0	3.0
242.0	O	159	23.0	...	...	...	B443.0	F	117	17.0	41	6.0	3.0
	T571	200	29.0	...	...	...	514.0	F	152	22.0	62	9.0	6.0
	T61	221	32.0	138	20.0	...	520.0	T4	290	42.0	152	22.0	12.0
	T77	165	24.0	90	13.0	1.0	535.0	F or T5	241	35.0	124	18.0	9.0
295.0	T4	200	29.0	90	13.0	6.0	705.0	F or T5	207	30.0	117	17.0	5.0
	T6	221	32.0	138	20.0	3.0	707.0	F or T5	227	33.0	152	22.0	2.0
	T62	248	36.0	193	28.0	...		T7	255	37.0	207	30.0	1.0
	T7	200	29.0	110	16.0	3.0	710.0	F or T5	221	32.0	138	20.0	2.0
319.0	F	159	23.0	90	13.0	1.5	712.0	F or T5	234	34.0	172	25.0	4.0
	T5	172	25.0	...	...	...	713.0	F or T5	221	32.0	152	22.0	3.0
	T6	214	31.0	138	20.0	1.5	771.0	T5	290	42.0	262	38.0	1.5
328.0	F	172	25.0	97	14.0	1.0		T51	221	32.0	186	27.0	3.0
	T6	234	34.0	145	21.0	1.0		T52	248	36.0	207	30.0	1.5
354.0	(b)	...	...	...	...	...		T53	248	36.0	186	27.0	1.5
355.0	T51	172	25.0	124	18.0	...		T6	290	42.0	241	35.0	5.0
	T6	221	32.0	138	20.0	2.0		T71	331	48.0	310	45.0	2.0
	T7	241	35.0	...	...	...	850.0	T5	110	16.0	...	...	5.0
	T71	207	30.0	152	22.0	...	851.0	T5	117	17.0	...	...	3.0
C355.0	T6	248	36.0	172	25.0	2.5	852.0	T5	165	24.0	124	18.0	...
356.0	F	131	19.0	...	...	2.0							
	T51	159	23.0	110	16.0	...							

(a) These values represent properties obtained from separately cast test bars and are derived from ASTM B26, Standard Specification for Aluminum Alloy Sand Castings; Federal Specification QQ-A-601e, Aluminum Alloy Sand Castings; and Military Specification MIL-A-21180c, Aluminum Alloy Castings, High Strength. Unless otherwise specified, the average values of tensile strength, yield strength and elongation for specimens cut from castings shall be not less than 75% of the tensile and yield strength values, and not less than 25% of the elongation values, given above. The customer should keep in mind that some foundries may offer additional tempers for the above alloys and that foundries are constantly improving casting techniques and, as a result, may offer minimum properties in excess of those given here. (b) Mechanical properties of this alloy depend on the casting process. For further information, consult the individual foundries.

Table 9B. Mechanical-property limits for aluminum permanent mold casting alloys

AA No.	Temper	Tensile strength MPa	ksi	Yield strength MPa	ksi	Elongation in 50 mm or 2 in., %
204.0	T4	331	48.0	200	29.0	8.0
208.0	T4	227	33.0	103	15.0	4.5
	T6	241	35.0	152	22.0	2.0
	T7	227	33.0	110	16.0	3.0
213.0	F	159	23.0	...	...	...
222.0	T551	207	30.0	...	...	...
	T65	276	40.0	...	...	...
242.0	T571	234	34.0	...	...	...
	T61	276	40.0	...	...	...
296.0	T4	227	33.0	...	...	4.5
	T6	241	35.0	...	...	2.0
	T7	227	33.0	...	...	3.0
308.0	F	165	24.0	...	...	...
319.0	F	193	28.0	97	14.0	1.5
	T6	234	34.0	...	...	2.0
332.0	T5	214	31.0	...	...	...
333.0	F	193	28.0	...	...	...
	T5	207	30.0	...	...	...
	T6	241	35.0	...	...	...
	T7	214	31.0	...	...	...
336.0	T551	214	31.0	...	...	...
	T65	276	40.0	...	...	...
354.0	(b)	...	...	...	...	...
355.0	T51	186	27.0	...	...	...
	T6	255	37.0	...	...	1.5
	T62	290	42.0	...	...	...
	T7	248	36.0	...	...	3.0
	T71	234	34.0	186	27.0	...
C355.0	T61	276	40.0	207	30.0	3.0
356.0	F	145	21.0	...	...	3.0
	T51	172	25.0	...	...	...
	T6	227	33.0	152	22.0	3.0
	T7	172	25.0	...	...	3.0
	T71	172	25.0	...	...	3.0
A356.0	T61	255	37.0	179	26.0	5.0
357.0	T6	310	45.0	...	...	3.0
A357.0	(b)	...	...	...	...	...
359.0	(b)	...	...	...	...	...
443.0	F	145	21.0	48	7.0	2.0
B443.0	F	145	21.0	41	6.0	2.5
A444.0	T4	138	20.0	...	...	20.0
513.0	F	152	22.0	83	12.0	2.5
535.0	F	241	35.0	124	18.0	8.0
705.0	T5	255	37.0	117	17.0	10.0
707.0	T5	290	42.0	172	25.0	4.0
	T7	310	45.0	241	35.0	3.0
711.0	T1	193	28.0	124	18.0	7.0
713.0	T5	221	32.0	152	22.0	4.0
850.0	T5	124	18.0	...	...	8.0
851.0	T5	117	17.0	...	...	3.0
	T6	124	18.0	...	...	8.0
852.0	T5	186	27.0	...	...	3.0

(a) These values represent properties obtained from separately cast test bars and are derived from ASTM B108, Standard Specification for Aluminum Alloy Permanent Mold Castings; Federal Specification QQ-A-596d, Aluminum Alloy Permanent and Semipermanent Mold Castings; and Military Specification MIL-A-21180c, Aluminum Alloy Castings, High Strength. Unless otherwise specified, the average values of tensile strength, yield strength and elongation for specimens cut from castings shall be not less than 75% of the tensile and yield strength values, and not less than 25% of the elongation values, given above. The customer should keep in mind that some foundries may offer additional tempers for the above alloys and that foundries are constantly improving casting techniques and, as a result, may offer minimum properties in excess of those given here. (b) Mechanical properties of this alloy depend on the casting process. For further information, consult the individual foundries

Table 10. Mechanical-property limits (when specified) for specimens cut from aluminum alloy permanent mold castings (from ASTM B108)

Alloy and temper	Tensile strength MPa	ksi	Yield strength MPa	ksi	Elongation, %
Specimens cut from designated areas of castings					
354.0-T61	324	47	248	36	3.0
354.0-T62	344	50	290	42	2.0
C355.0-T61	276	40	207	30	3.0
A356.0-T61	228	33	179	26	5.0
A357.0-T61	317	46	248	36	3.0
359.0-T61	310	45	234	34	4.0
359.0-T62	324	47	262	38	3.0
A444.0-T4	138	20	...	...	20.0
Specimens cut from any location in castings					
354.0-T61, -T62	297	43	228	33	2.0
C355.0-T61	255	37	207	30	3.0
A356.0-T61	193	28	179	26	3.0
A357.0-T61	283	41	214	31	3.0
359.0-T61, -T62	276	40	207	30	3.0
A444.0-T4	138	20	...	...	20.0

Table 11. Mechanical-property limits for specimens cut from aluminum alloy premium-quality castings (from MIL-A-21180c)

Alloy and temper	Class No.	Tensile strength MPa	ksi	Yield strength MPa	ksi	Elongation, %
Specimens cut from designated areas of castings						
A201.0-T7	1	414	60	345	50	5.0
	2	414	60	345	50	3.0
224.0-T72	1	345	50	255	37	3.0
	2	379	55	255	37	5.0
249.0-T7	1	345	50	276	40	2.0
	2	379	55	310	45	3.0
	3	414	60	345	50	5.0
354.0-T61	1	324	47	248	36	3.0
	2	345	50	290	42	2.0
C355.0-T61	1	283	41	214	31	3.0
	2	303	44	228	33	3.0
	3	345	50	276	40	2.0
A356.0-T61	1	262	38	193	28	5.0
	2	276	40	207	30	3.0
	3	310	45	234	34	3.0
A357.0-T6	1	310	45	241	35	3.0
	2	345	50	276	40	5.0
Specimens cut from any location in castings						
A201.0-T7	10	386	56	331	48	3.0
	11	386	56	331	48	1.5
224.0-T72	10	310	45	241	35	2.0
	11	345	50	255	37	3.0
249.0-T7	10	379	55	310	45	3.0
	11	345	50	276	40	2.0
354.0-T61	10	324	47	248	36	3.0
	11	296	43	228	33	2.0
C355.0-T61	10	283	41	214	31	3.0
	11	255	37	207	30	1.0
	12	241	35	193	28	1.0
A356.0-T61	10	262	38	193	28	5.0
	11	228	33	186	27	3.0
	12	221	32	152	22	2.0
A357.0-T6	10	262	38	193	28	5.0
	11	283	41	214	31	3.0

Table 12A. Creep-rupture properties of separately sand cast test bars of alloy 201.0

Temperature °C	°F	Time under stress(a), h	Rupture stress MPa	ksi	Minimum creep rate at rupture stress, % per h	Stress for creep of: 1.0% MPa	ksi	0.5% MPa	ksi	0.25% MPa	ksi
150	300	10 .... Above yield			...	...	...	...	...	...	...
		100 .... Above yield			...	...	...	...	...	...	...
		1000 270		39	0.00013	260	38	250	36	250	36
		10 000 195		28	0.000023	195	28	185	27	180	76
205	400	10 250		36	0.0145	240	35	230	33	220	32
		100 180		26	0.0024	180	26	170	24	170	24
		1000 130		19	0.00046	130	19	125	18	110	16
		10 000 95		14	0.000088	95	14	90	13	85	12
260	500	10 140		20	0.047	140	20	125	18	110	16
		100 95		14	0.0080	95	14	95	14	85	12
		1000 70		10	0.00130	70	10	70	9.8	60	8.6
		10 000 50		7.5	0.00028	50	7.5	50	7.2	45	6.3

(a) 10 000-h data are extrapolated.

Table 12B. Creep-rupture properties of separately cast test bars of alloy C355.0-T61

Temperature °C	°F	Time under stress, h	Rupture stress MPa	ksi	1% MPa	ksi	0.5% MPa	ksi	0.2% MPa	ksi	0.1% MPa	ksi
150	300	0.1 285		41	275	40	270	39	240	35	230	33
		1 285		41	270	39	260	38	235	34	220	32
		10 275		40	260	38	250	36	230	33	205	30
		100 260		38	250	36	235	34	215	31	170	25
		1000 220		32	215	31	206	30	185	27	140	20
205	400	0.1 250		36	250	36	240	35	230	33	170	25
		1 230		33	220	32	205	30	170	25	140	20
		10 180		26	120	25	160	23	130	19	110	16
		100 130		19	130	19	125	18	97	14	...	...
		1000 97		14	90	13	83	12	...	...	...	...
260	500	0.1 165		24	145	21	130	19	105	15	83	12
		1 125		18	110	16	97	14	83	12	59	8.5
		10 90		13	83	12	76	11	59	8.5	41	6
		100 62		9	62	9	55	8	41	6	...	...
		1000 45		6.5	45	6.5	41	6	...	...	...	...

Table 12C. Creep-rupture properties of separately cast test bars of alloy A356.0-T61 at 150 °C (300 °F)

Time under stress, h	Rupture stress MPa	ksi	1% MPa	ksi	0.5% MPa	ksi	0.2% MPa	ksi	0.1% MPa	ksi
0.1 235		34	215	31	205	30	195	28	185	27
1 235		34	215	31	200	29	185	27	180	26
10 230		33	205	30	195	28	180	26	170	25
100 200		29	195	28	185	27	170	25	165	24
1000 165		24	165	24	160	23	...	...	...	...

Table 13A. Fatigue strengths of separately cast test bars of alloy 354.0-T61(a)

Temperature °C	°F	10^4 cycles MPa	ksi	10^5 cycles MPa	ksi	10^6 cycles MPa	ksi	10^7 cycles MPa	ksi	10^8 cycles MPa	ksi	5×10^8 cycles MPa	ksi
24	75	 345	50	275	40	215	31	175	25.5	145	21	135	19.5
150	300	....	...	255	37	200	29	150	21.5	115	17	110	16
205	400	....	...	215	31	150	22	105	15	70	10	60	9
260	500	 195	28	140	20.5	96	14	60	9	40	6	40	6
315	600	....	...	75	11	55	8	40	6	30	4	30	4

(a) R. R. Moore–type test.

Table 13B. Fatigue properties of separately cast test bars of alloy C355.0-T61

Temperature °C	°F	No. of cycles	Fatigue strength(a) MPa	ksi	Temperature °C	°F	No. of cycles	Fatigue strength(a) MPa	ksi
24	75	10^5 195		28.0	260	500	10^5 125		18.0
		10^6 130		19.0			10^6 80		11.5
		10^7 110		16.0			10^7 50		7.5
		10^8 100		14.5			10^8 40		5.5
		5×10^8 95		14.0			5×10^8 35		5.0

(a) Based on rotating-beam tests at room temperature and cantilever-beam (rotating-load) tests at elevated temperature.

Aluminum Powder Metallurgy Products

CONVENTIONAL aluminum powder metallurgy (P/M) consists of compressing metal powder in a shaped die to produce green compacts, and then sintering (diffusion bonding) the compacts at elevated temperature in a furnace with a protective atmosphere. During sintering, the compacts become consolidated and strengthened.

The density of sintered compacts may be increased by re-pressing. When re-pressing is performed primarily to improve the dimensional accuracy of a compact, it usually is termed "sizing"; when performed to improve configuration, it is termed "coining." Re-pressing may be followed by resintering, which relieves stress due to cold work in re-pressing and may further consolidate the compact. By pressing and sintering only, parts of over 80% theoretical density can be produced. By re-pressing, with or without resintering, parts of 90% theoretical density or more can be produced. The density attainable is limited by the size and shape of the compact.

Powder metallurgy parts frequently are competitive with forgings, castings, stampings, machined components and fabricated assemblies. Within the limitations inherent in the P/M processes, parts can be fabricated to final or nearly final shape, thus eliminating or reducing scrap metal and secondary machining and assembly operations. Although the unit cost of metal powder usually is higher than that of solid metal in bars, forgings or castings, the savings achieved by eliminating fabricating operations and minimizing scrap losses often result in lower total cost for P/M parts than for parts made by other processes.

Certain metal products can be produced only by powder metallurgy; among these products are materials whose porosity (number, distribution and size of pores) must be controlled; two examples of controlled-porosity materials are filter elements and self-lubricating bearings.

There is no industry standard alloy designation system for P/M products; powder producers assign numbers to the products they offer. Principal alloying elements are copper, magnesium and silicon, employed individually or in combination in amounts ranging from those retained in solid solution to those of the moderate- and high-strength precipitation-hardening wrought alloys. The latter may be heat treated to increase hardness and strength.

Compositions, properties, and characteristics pertinent to applications for P/M materials are listed in Table 1. Effects of density (as pressed) and temper on mechanical properties of two of these materials (alloys 601AB and 201AB) are shown in Table 2.

Table 1. Compositions, characteristics and properties of aluminum powder metallurgy (P/M) materials

Designation	Composition	Characteristics affecting application	Density, g/cm³	Treatment(a)	Tensile strength MPa	ksi	Yield strength MPa	ksi	Hardness	Elongation, %(b)
601AB (Alcoa)	0.25 Cu, 0.6 Si, 1.0 Mg, 1.5 lubricant, rem 1202(c)	Similar to wrought 6061 in strength, ductility and corrosion resistance	2.55	S (T1)	145	21.0	94	13.7	65-70 HRH	6.0
				H(d)	238	34.5	230	33.4	80-85 HRE	2.0
201AB (Alcoa)	4.4 Cu, 0.8 Si, 0.5 Mg, 1.5 lubricant, rem 1202(c)	Similar to wrought 2014 but without manganese. Good strength properties	2.64	S (Ti)	209	30.3	181	26.2	70-75 HRE	3.0
				H(d)	332	48.1	328	47.5	85-90 HRE	2.0
202AB (Alcoa)	4.0 Cu, 1.5 lubricant, rem 1202(c)	Good ductility. Suitable for cold formed parts	2.56	S (T1)	160	23.2	75	10.9	55-60 HRH	10.0
				H(e)	228	33.0	147	21.3	45-50 HRE	7.3
602AB (Alcoa)	0.4 Si, 0.6 Mg, 1.5 lubricant, rem 1202(c)	Good electrical conductivity (from 42.0 to 48.5% IACS, depending on treatment), ductility and finishability	2.55	S (T1)	121	17.5	59	8.5	55-60 HRH	9.0
				H(d)	179	26.0	169	24.5	65-70 HRE	2.0
22 (Alcan)(f)	2.0 Cu, 1.0 Mg, 0.3 Si, rem Al	Good mechanical properties in sintered and heat treated forms	2.53	S	165	24.0	110	16.0	83 HRH	6.0
				H(g)	262	38.0	200	29.0	74 HRE	3.0
24 (Alcan)(f) (2014)	4.4 Cu, 0.5 Mg, 0.9 Si, 0.4 Mn, rem Al	Properties similar to those of wrought counterpart 2014. Good mechanical properties	2.54	S	165	24.0	97	14.0	80 HRH	5.0
				H(h)	241	35.0	193	28.0	72 HRE	3.0
67 (Alcan)(f)	0.5 Cu, rem Al	High electrical conductivity (48% IACS) and ductility. Similar to wrought 1100	2.52	S (T1)	103	15.0	55	8.0	60 HRH	12.0
68 (Alcan)(f)	0.6 Mg, 0.4 Si, rem Al	Good surface finish; high ductility and conductivity (42% IACS). Similar to wrought 6101	2.52	S (T1)	117	17.0	62	9.0	64 HRH	9.0
69 (Alcan)(f) (6061)	0.25 Cu, 1.0 Mg, 0.6 Si, 0.10 Cr	Properties similar to those of 6061. Good strength, corrosion resistance, ductility and conductivity (40% IACS)	2.50	S	128	18.5	69	10.0	66 HRH	10.0
				H(g)	207	30.0	193	28.0	71 HRE	2.0
76 (Alcan)(f) (7075)	1.6 Cu, 2.5 Mg, 0.20 Cr, 5.6 Zn	Properties similar to those of 7075. High strength and hardness	2.51	S	207	30.0	152	22.0	90 HRH	3.0
				H(j)	310	45.0	276	40.0	80 HRE	2.0
91 (Alcan)(f)	26.3 Tribaloy	Excellent wear resistance	3.05	S (T1)	94	13.6	...	...	...	2.0
				H	106	15.4	...	...	...	1.0
4040 (Alcan)(k)	1.0 Cu, 1.0 Si, rem Al	High-porosity parts for controlling contamination, pressure, sound, catalytic reactions, etc.	1.40	LS (T1)	(m)	(m)	...	...	...	...
4090 (Alcan)(n)	1.0 Cu, 1.0 Si, rem Al	Same as above	1.35	LS (T1)	(m)	(m)	...	...	...	...
4160 (Alcan)(p)	1.0 Cu, 1.0 Si, rem Al	Same as above	1.30	LS (T1)	(m)	(m)	...	...	...	...

(a) S = sintered; H = heat treated; LS = loose sintered. (b) In 25 mm or 1 in. (c) Composition of 1202: 99.4 Al, 0.3 Al₂O₃, 0.15 Fe, 0.07 Si, rem other metallics. (d) Solutioned for 30 min at 520 °C or 970 °F (505 °C or 940 °F for grade 201AB), quenched in cold water, and aged for 18 h at 160 °C (320 °F) to the T6 condition. (e) Solutioned for 30 min at 540 °C (1000 °F), quenched in cold water, and aged for 20 h at 150 °C (300 °F) to the T6 condition. (f) Grade designation includes a suffix: FF = premix with 1.5% lubricant; NL = premix without lubricant. (g) Solutioned for 30 min at 520 °C (970 °F), quenched in water, and aged for 18 h at 150 °C (300 °F). (h) Solutioned for 60 min at 500 °C (930 °F), quenched in water, and aged for 18 h at 150 °C (300 °F). (j) Solutioned for 60 min at 475 °C (890 °F), quenched in water, and aged for 18 h at 125 °C (260 °F). (k) −150, +325 mesh. (m) Breaking strength, 100 lb. (n) −60, +150 mesh. (p) −30, +60 mesh.

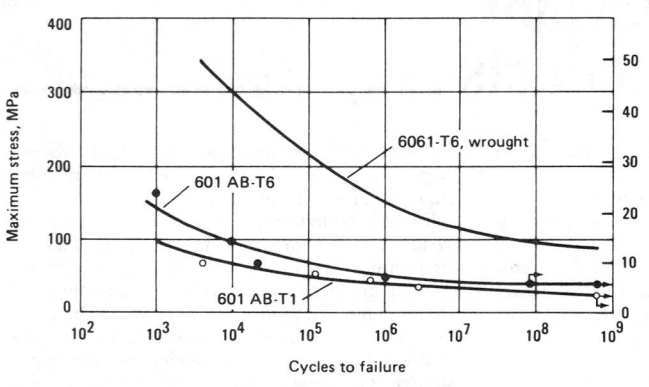

Fig. 1. Fatigue curves for P/M 601AB and wrought 6061

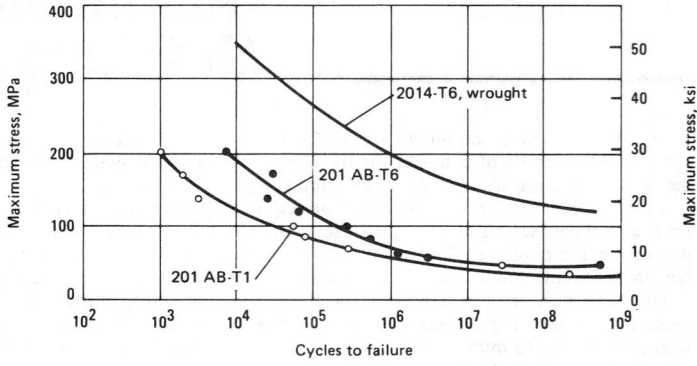

Fig. 2. Fatigue curves for P/M 201AB and wrought 2014

Impact tests are used to provide a measure of toughness of powder metal materials, which are somewhat less ductile than similar wrought compositions. Annealed specimens develop the highest impact strength, whereas fully heat treated parts have the lowest impact values. Alloy 201AB generally exhibits higher impact resistance than alloy 601AB at the same percent density, and impact strength of 201AB increases with increasing density. A desirable combination of strength and impact resistance is attained in the T4 temper for both alloys. In the T4 temper, 95% density 201AB develops strength and impact properties exceeding those for as-sintered 99Fe-1C alloy, a P/M material frequently employed in applications requiring tensile strengths under 340 MPa (50 ksi).

Fatigue is an important design consideration for P/M parts subjected to dynamic stresses. Fatigue strengths of powder metallurgy parts may be expected to be about half those of the wrought alloys of corresponding compositions (see comparisons of two P/M alloys with two wrought alloys in Fig. 1 and 2). These fatigue-strength levels are suitable for many applications.

Table 2. Sintered properties of aluminum P/M alloys 601AB and 201AB

%	Green density g/cm³	Temper(a)	Tensile strength MPa	ksi	Yield strength MPa	ksi	Elonga- tion, %(b)	Apparent hardness
Alloy 601AB(c)								
85	2.29	T1	110	16.0	50	7.0	6.0	55-60 HRH
		T4	140	20.5	95	14.0	5.0	80-85 HRH
		T6	185	26.5	. . .	. . .	1.0	70-75 HRE
90	2.42	T1	120	17.5	55	8.0	7.0	60-65 HRH
		T4	. . .	. . .	. . .	. . .	. . .	80-85 HRH
		T6	225	32.5	215	31.0	2.0	75-80 HRE
95	2.55	T1	125	18.0	60	8.5	8.0	65-70 HRH
		T4	150	22.0	105	15.0	5.0	85-90 HRH
		T6	250	36.5	240	35.0	2.0	80-85 HRE
Alloy 201AB(d)								
85	2.36	T1	170	24.5	145	21.0	2.0	60-65 HRE
		T4	210	30.5	180	26.0	3.0	70-75 HRE
		T6	250	36.0	. . .	. . .	. . .	75-80 HRE
90	2.50	T1	200	29.2	170	24.6	3.0	70-75 HRE
		T4	245	35.6	205	29.8	3.5	75-80 HRE
		T6	325	46.8	. . .	. . .	. . .	85-90 HRE
95	2.64	T1	210	30.3	180	26.2	3.0	70-75 HRE
		T4	260	38.0	215	31.0	5.0	80-85 HRE
		T6	330	48.1	325	47.5	2.0	85-90 HRE

(a) T1 = as sintered; T4 = solution heat treated and naturally aged; T6 = solution heat treated and artificially aged. (b) In 25 mm or 1 in. (c) Sintered for 30 min at 620 °C (1150 °F) in N_2 with average dewpoint of 7 °C (45 °F). (d) Sintered for 30 min at 590 °C (1100 °F) in N_2 with average dewpoint of 7 °C (45 °F).

Corrosion Resistance of
Aluminum and Aluminum Alloys

ALUMINUM and most of its alloys have good corrosion resistance in natural atmospheres, fresh waters, seawater, many soils, many chemicals and their solutions, and most foods. This resistance to corrosion is the result of the presence of a very thin, compact and adherent film of aluminum oxide on the metal surface. Whenever a fresh surface is created by cutting or abrasion and is exposed to either air or water, a new film forms rapidly, growing to a stable thickness. The film formed in air at ambient temperatures is about 5 nm (50 Å) thick. The thickness increases with increasing temperature and in the presence of water. The oxide film is soluble in alkaline solutions and in strong acids, with some exceptions, but is stable over a pH range of about 4.0 to 9.0.

There are different types of corrosion and various interactions with induced or imposed stresses so that the effects can range from unimportant to highly damaging. For some types of applications, a distinction should be made between appearance and durability. The surface may become unattractive because of roughening by shallow pitting and many darken with dirt retention, but these conditions may have no effect on durability or function. On the other hand, stress-corrosion cracking or highly localized, severe corrosion due to heavy metal ions in solutions, stray electrical currents or galvanic couples with more-anodic metals can be quite damaging. Good design and application practices must be observed to avoid these conditions. This includes selection of alloys appropriate for the conditions of the application.

EFFECTS OF ALLOY COMPOSITION

Among the principal alloying elements (Mg, Si, Cu, Mn and Zn), the one that has the greatest effect on general corrosion resistance is copper. Copper reduces resistance because it replates from solution as minute metallic particles forming highly active corrosion couples. The effects are apparent at copper contents exceeding a few tenths of one percent. In the heat treatable 2*xxx* or 2*xx.x* alloys with several percent copper, the state of solution or precipitation affects the type of corrosion attack as well as susceptibility to stress-corrosion cracking. These effects are recognized in the alloy/temper ratings for general corrosion and stress-corrosion cracking that appear in Table 1 for wrought alloys and Table 2 for casting alloys.

Among the common impurity elements, iron is probably the most important, degrading general corrosion resistance by increasing the volume fraction of cathodic and surface-film-weakening intermetallic-phase microconstituents.

The 7*xxx* Al-Zn-Mg alloys without copper have

high general corrosion resistance. The alloys of this group that contain more than 1% Cu are less resistant to general corrosion. Appropriate tempers should be used to avoid stress-corrosion problems.

The 3xxx alloys are generally among those having the highest general corrosion resistance, as are those of the 5xxx group, which outperform any of the others in marine exposures. The 6xxx alloys also have high resistance.

ATMOSPHERIC EXPOSURES—WEATHERING

Alloys other than those with the higher copper contents have excellent resistance to atmospheric exposures and in many outdoor applications require no protection or maintenance. Products widely used under such conditions include electrical conductors, outdoor lighting poles, bridge railings and ladders. These often retain a bright metallic appearance for many years but may darken with mild surface roughening caused by shallow pitting and with accumulation of dirt.

An important characteristic of weathering, as well as corrosion of aluminum under many other environmental conditions, is that the rate of corrosion decreases with time. Typical curves of the average changes that occurred in exposures of two generic types, seacoast and industrial, over a 30-year period with sheet specimens of 1100, 3003 and 3004 alloys, all in H14 temper, are shown in Fig. 1. The curve shapes are similar, whether the amount of corrosion is measured by weight loss, depth of pitting or loss in strength. Leveling-off usually occurs in 6 months to 2 years, after which the rate becomes approximately linear at a low rate. In rural atmospheres the weight-loss rate may be less than 0.025 μm/yr (0.001 mil/yr). In industrial locations, rates vary from about 0.75 to 2.75 μm/yr (0.03 to 0.11 mil/yr) with a few particularly aggressive sites at which rates up to 12.5 μm/yr (0.5 mil/yr), based on maximum pit depth, were observed.

Castings of 4xx.x and low-Cu 3xx.x alloys have been used for many years in such applications as bridge guard-rail supports and lighting-pole bases with little adverse effect from corrosion. The 5xx.x casting alloys with 4 to 8% Mg are particularly suitable for cast parts that are used in marine and seawater exposure sites. Copper-containing alloys such as 295.0, 333.0, 380.0 and even 355.0 require surface protection for satisfactory use in corrosive marine or industrial atmospheres. Stress-corrosion cracking has been encountered occasionally with some of the 7xx.x alloys and alloy 520.0-T4 in atmospheric exposures of stressed parts. Applications for these alloys must be carefully engineered to avoid such problems.

EXPOSURE IN WATERS

Corrosion resistance of aluminum in high-purity water, distilled or deionized, and in steam condensate is so high that these fluids are regularly contained and handled in aluminum equipment. Resistance is also high in most natural fresh waters. Soft waters have the least pitting tendency. Components of natural waters that increase pitting are copper ions, bicarbonate, chloride, sulfate and oxygen. Thus, harder waters with more bicarbonate have a higher pitting tendency.

Service experience with aluminum alloys in marine and coastal applications, including structures, buoys, pipelines, lifeboats, motor launches,

Table 1. Relative ratings of resistance to general corrosion and to stress-corrosion cracking of wrought aluminum alloys

Alloy	Temper	General(a)	Stress-corrosion cracking(b)	Alloy	Temper	General(a)	Stress-corrosion cracking(b)
		Resistance to corrosion				Resistance to corrosion	
1060	All	A	A	5154	All	A(d)	A(d)
1100	All	A	A	5252	All	A	A
1350	All	A	A	5254	All	A(d)	A(d)
2011	T3, T4, T451	D(c)	D	5454	All	A	A
	T8	D	B	5456	All	A(d)	B(d)
2014	T6, T651, T6510, T6511	D	C	5457	O	A	A
				5652	All	A	A
2017	T4, T451	D(c)	C	5657	All	A	A
2024	T4, T3, T351, T3510, T3511, T361	D(c)	C	6053	T6, T61	A	A
				6061	O	B	A
	T6, T861, T81, T851, T8510, T8511	D	B		T4, T451, T4510, T4511	B	B
2025	T6	D	C		T6, T651, T652, T6510, T6511	B	A
2036	T4	C	...	6063	All	A	A
2117	T4	C	A	6066	O	C	A
2218	T61, T72	D	C		T4, T4510, T4511, T6, T6510, T6511	C	B
2219	T31, T351, T3510, T3511, T37	D(c)	C	6070	T4, T4511, T6	B	B
				6101	T6, T63, T61, T64	A	A
	T81, T851, T8510, T8511, T87	D	B	6151	T6, T652	B	B
2618	T61	D	C	6201	T81	A	A
3003	All	A	A	6262	T6, T651, T6510, T6511, T9	B	A
3004	All	A	A				
3105	All	A	A	6463	All	A	A
4032	T6	C	B	7001	T6, T651, T6510, T6511	C(c)	C
5005	All	A	A				
5050	All	A	A	7075	T6, T651, T652, T6510, T6511	C(c)	C
5052	All	A	A		T73, T7351	C	B
5056	O, H11, H12, H32, H14, H34	A(d)	B(d)	7178	T6, T651, T6510, T6511	C(c)	C
	H18, H38	A(d)	C(d)				
	H192, H392	B(d)	D(d)				
5083	All	A(d)	B(d)				
5086	O, H32, H116, H34, H36, H38, H111	A(d)	A(d)				

(a) Ratings A through E are relative ratings in decreasing order of merit, based on exposures to NaCl solution by intermittent spraying or immersion. Alloys with A and B ratings can be used in industrial and seacoast atmospheres without protection. Alloys with C, D and E ratings generally should be protected, at least on faying surfaces. (b) Stress-corrosion-cracking ratings are based on service experience and on laboratory tests of specimens exposed to alternate immersion in 3.5% NaCl solution. A—no known instance of failure in service or in laboratory tests. B—no known instance of failure in service; limited failures in laboratory tests of short-transverse specimens. C—service failures when sustained tensile stress acts in short-transverse direction relative to grain structure; limited failures in laboratory tests of long-transverse specimens. D—limited service failures when sustained stress acts in longitudinal or long-transverse direction relative to grain structure. (c) In relatively thick sections the rating would be E. (d) This rating may be different for material held at elevated temperature for long periods.

Table 2. Relative ratings of resistance to general corrosion and to stress-corrosion cracking of aluminum casting alloys

Alloy	Temper	General(a)	Stress-corrosion cracking(b)
		Resistance to corrosion	
Sand castings			
208.0	F	B	B
224.0	T7	C	B
240.0	F	D	C
242.0	All	D	C
A242.0	T75	D	C
249.0	T7	C	B
295.0	All	C	C
319.0	F, T5	C	B
	T6	C	C
355.0	All	C	A
C355.0	T6	C	A
356.0	T6, T7, T71, T51	B	A
A356.0	T6	B	A
443.0	F	B	A
512.0	F	A	A
513.0	F	A	A
514.0	F	A	A
520.0	T4	A	C

(continued)

Table 2. (continued)

Alloy	Temper	Resistance to corrosion General(a)	Stress-corrosion cracking(b)
Sand castings (continued)			
535.0	F	A	A
B535.0	F	A	A
705.0	T5	B	B
707.0	T5	B	C
710.0	T5	B	B
712.0	T5	B	C
713.0	T5	B	B
771.0	T6	C	C
850.0	T5	C	B
851.0	T5	C	B
852.0	T5	C	B
Permanent mold castings			
242.0	T571, T61	D	C
308.0	F	C	B
319.0	F	C	B
	T6	C	C
332.0	T5	C	B
336.0	T551, T65	C	B
354.0	T61, T62	C	A
355.0	All	C	A
C355.0	T61	C	A
356.0	All	B	A
A356.0	T61	B	A
F356.0	All	B	A
A357.0	T61	B	A
358.0	T6	B	A
359.0	All	B	A
B443.0	F	B	A
A444.0	T4	B	A
513.0	F	A	A
705.0	T5	B	B
707.0	T5	B	C
711.0	T5	B	A
713.0	T5	B	B
850.0	T5	C	B
851.0	T5	C	B
852.0	T5	C	B
Die castings			
360.0	F	C	A
A360.0	F	C	A
364.0	F	C	A
380.0	F	E	A
A380.0	F	E	A
383.0	F	E	A
384.0	F	E	A
390.0	F	E	A
392.0	F	E	A
413.0	F	C	A
A413.0	F	C	A
C443.0	F	B	A
518.0	F	A	A
Rotor metal(c)			
100.1		A	A
150.1		A	A
170.1		A	A

(a) Relative ratings of general corrosion resistance A through E are in decreasing order of merit, based on exposures to NaCl solution by intermittent spray or immersion. (b) Relative ratings of resistance to stress-corrosion cracking are based on service experience and on laboratory tests of specimens exposed to alternate immersion in 3.5% NaCl solution. A—no known instance of failure in service when properly manufactured. B—failure not anticipated in service from residual stresses or from design and assembly stresses below about 45% of the minimum guaranteed yield strength given in applicable specifications. C—failures have occurred in service with either this specific alloy/temper combination or with alloy/temper combinations of this type; designers should be aware of the potential stress-corrosion-cracking problem that exists when these alloys and tempers are used under adverse conditions. (c) For electric motor rotors.

cabin cruisers, patrol boats, barges and larger vessels, has demonstrated their good resistance and long life under conditions of partial, intermittent and total immersion. Wrought alloys of the 3xxx, 5xxx and 6xxx groups are used. Those of the 5xxx Al-Mg group are most resistant and most widely used because of their favorable strength and good weldability. The rate of corrosion based on weight loss does not exceed about 5 μm/yr (0.2 mil/yr), which is less than 5% of the rate for unprotected low-carbon steel in seawater. Corrosion is mainly pitting, decelerating with time from rates of 2.5 to 5 μm/yr (0.1 to 0.2 mil/yr) in the first year to average rates over a 10-year period of 0.75 to 1.5 μm/yr (0.03 to 0.06 mil/yr). The curve of maximum depth of pitting versus time follows an approximate cube-root law, from which it follows that doubling material thickness increases time to perforation by a factor of 8.

Casting alloys of the 5xx.x group with up to 8% Mg have high resistance to seawater corrosion and are used for fittings. Al-Si and Al-Si-Mg alloys are less resistant although 443.0 and 356.0T6 are sometimes used. Resistance of 2xxx, 2xx.x, 7xxx and 7xx.x alloys is distinctly inferior, and their use without cladding or metallizing is not recommended.

EXPOSURE IN SOILS

Soils differ widely in mineral content, texture, permeability, moisture, pH, and electrical conductivity as well as aeration, organic matter and micro-organisms. With this variability, corrosion performance of unprotected buried aluminum, like that of other metals, varies considerably. However, in many cases where carbon steel requires protective coatings, unprotected aluminum alloys have performed well. In most cases protection is recommended for buried applications. Use of 2xxx or 7xxx alloys or their cast counterparts is generally not recommended. Stray-current effects and contact with more-cathodic metals should be avoided. Successful applications include pipelines employing alloys 3003 and 6063 and culverts made of alclad 3004.

EXPOSURE TO FOODS

Aluminum alloys of the 3xxx and 5xxx groups are resistant to most foods and beverages. Aluminum products constitute a substantial share of the domestic cooking-utensil market and are used extensively for commercial handling and processing of foods. Large quantities of foil, foil laminated to plastics or paper, and cans are used for packaging and marketing of foods and beverages. Beverage-can bodies are generally produced from alloy 3004, food-can bodies from 5352 or 5050, and can ends from 5182. These cans generally have internal and external organic coatings, not for corrosion protection but for decoration and prevention of effects on product taste.

EXPOSURE TO CHEMICALS

Aluminum alloys are used in storing, processing, handling and packaging of a variety of chemical products. They are compatible with most dry inorganic salts. Within the passive pH range, about 4 to 9, they resist corrosion in solutions of most inorganic chemicals but are subject to pitting in aerated solutions, particularly of halides. Aluminum alloys are not suitable for containing or handling mineral acids with the exceptions of nitric acid in concentrations over 82 wt % and sulfuric acid in concentrations from 98 to 100 wt %.

Aluminum alloys resist most alcohols; however, some may cause corrosion when extremely dry and at elevated temperatures. Similar characteristics are associated with phenol. Aldehydes have little or no corrosive effects. Care must be taken in using aluminum with halogenated organic compounds because under some conditions when moisture is present they may hydrolyze and react violently. A list of bulk raw materials, chemicals and products that have been stored and handled in aluminum equipment is presented in Table 3.

EXPOSURE TO BUILDING MATERIALS

Aluminum alloys are not corroded seriously by long-time embedment in portland cement concrete, standard or lime brick mortar, hardwall plaster or stucco. Superficial etching of the alu-

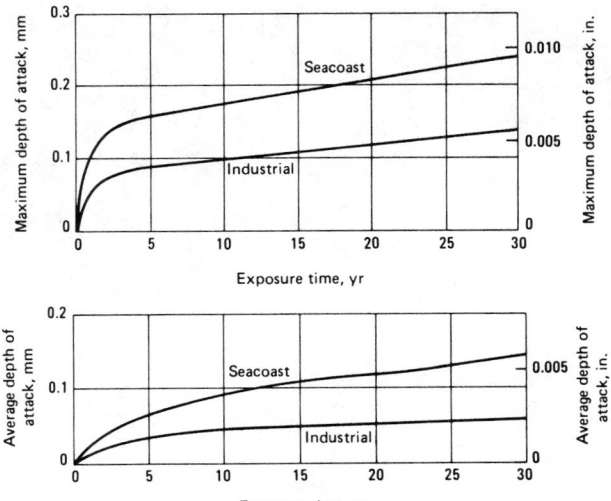

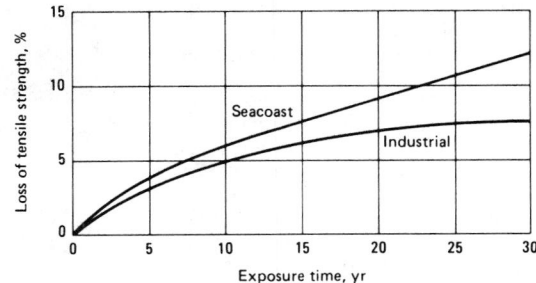

Shown is the average performance of the three alloys, all in H14 temper. Seacoast exposure was at a severe location (Pt. Judith, RI); industrial exposure was at New Kensington, PA. Tensile strengths were computed using original cross-sectional area, and loss in strength is expressed as percentage of original tensile strength.

Fig. 1. Effects of weathering on depth of corrosion and loss of tensile strength for alloys 1100, 3003 and 3004

Table 3. Typical bulk chemicals handled in aluminum equipment

Acetaldehyde	Cod liver oil	Methyl ethyl ketone	Propylene glycol
Acetic acid	Corn syrup	Methyl salicylate	Pyridine(b)
Acetic anhydride	Creosote	Mono-chloro-difluoro methane	Rice(b)
Acetone	Cresol	Molasses(e)	Ricinoleic acid
Acrolein	Crotonaldehyde	Naphthalene	Rubber and rubber products
Acrylonitrile	Cyclohexane	Naphthenic acid	Rye(b)
Adipic acid	Cyclopentane	Naval stores	Safflower
Alcohols (except dry and boiling)	Cyclopropane	Nitric acid (concentrate)	Salicylic acid (crystalline)
Aldol	Dairy products	Nitrocellulose	Shellac
Alumina and its hydrates	Dichlorobenzene	Nitrogen fertilizers	Soap(c)
Aluminum chloride(a)	Ebonite	Nitroglycerine	Sodium bicarbonate
Aluminum potassium sulfate	Essential oils	Nitrous oxide(b)	Sodium carbonate
Aluminum silicate	Ester gum	Nylon and nylon salts	Sodium chloride
Aluminum sulfate	Ethyl acetate	Oils, edible	Sodium nitrate
Ammonia	Ethyl acetoacetate	Oleic acid	Sodium phosphate, dibasic
Ammoniated ammonium nitrate solutions	Ethyl alcohol	Oxalic acid	Sodium phosphate, monobasic
Ammonium nitrate	Ethylene glycol	Oxygen	Sodium sulfate
Ammonium sulfate	Fatty acids	Paints, varnishes and paint materials	Sorbitol
Ammonium thiocyanate	Feeds(b)	Paraffins	Soybeans and soybean products
Aniline	Ferrous sulfate	Paraformaldehyde	Starch
Anthracene	Flour	Paraldehyde	Stearic acid
Baking powder(b)	Formaldehyde	Peanuts and peanut products	Styrene monomer and polymer
Barium carbonate(b)	Furfural	Pentaerythritol	Sugars
Benzene	Gasoline	Pentaerythritol tetranitrate	Sulfur
Benzoic acid	Glucose	Pentane	Sulfur dioxide
Bone black	Glue(c)	Perchloroethylene	Tall oil
Boric acid	Glycerin	Petroleum products, refined	Tar
Butyl acetate	Grains(b)	Phthalic acid	Tobacco stems
Calcium carbide(b)	Grits, hominy(b)	Phthalic anhydride	Toluene
Calcium chromate	Helium	Pitch	Triacetin
Caprylic	Hexamine	Polyethylene	Trichlorobenzene
Carbon(b)	Hydrocyanic acid	Polystyrene	Trichloroethylene
Carbon dioxide	Hydrogen	Potassium carbonate	Urea
Carbon disulfide	Hydrogen peroxide	Potassium chloride	Vegetable oils
Carbonic acid	Isobutyric acid	Potassium iodide	Vinyl acetate
Castor oil	Lacquer and its solvents	Potassium nitrate	Vinyl resins
Cellulose acetate	Linseed oil	Potassium sulfate	Water, high purity
Cement	Malt	Propane	Wood chips(b)
Cereals	Manganese dioxide	Propionic acid	Xylene
Coal	Maple syrup	Propionic anhydride	Zinc sulfide

(a) Anhydrous. (b) Dry. (c) Neutral. (d) Extremely alloy-sensitive. (e) Little or no copper. (f) At normal temperature.

minum occurs while these products are setting, but after curing further attack is minimal. Special consideration should be given to protection where crevices between the concrete and the metal may entrap environmental contaminants. For example, highway railings and streetlight standards or stanchions usually are coated with a sealing compound where they are fastened to concrete, to prevent entry of salt-laden road splash into crevices.

Absorbent materials such as paper, wood ref-

use and wallboard in contact with aluminum under conditions where it may become wet will cause corrosion. Composite-bonded insulated aluminum panels employ a moisture barrier on the inside to prevent condensation and wetting of the insulation. Some insulating materials, such as magnesia, are alkaline and quite aggressive to aluminum. Magnesium oxychloride, a flooring compound sometimes used in subway and railway passenger cars and ship decks, is very corrosive and should not be used with aluminum.

Although wood usually is not corrosive, it can become so if its moisture content exceeds 18 to 20%. Wood preservatives containing copper are detrimental, and those containing mercury should not be used where aluminum is involved.

FORM AND TYPE OF CORROSION

Form and type of corrosion may be categorized by the morphology, which may or may not be related to the metal's microstructure, or by the conditions causing the corrosion. Uniform attack or dissolution, which may occur if the surface oxide film is soluble in the corroding medium, is infrequent in service. Most corrosion in service is localized in one way or another. When the oxide film is insoluble in the corroding medium, corrosion is localized at weak spots in the film, which may result from microstructural features such as the presence of microconstituents. Local cells are formed by such nonuniformities in the metal as well as environmental nonuniformities such as those created by differential aeration cells or by heavy metals plated out on the surface. Localized corrosion in a microscopic sense results from galvanic coupling and stray-current effects.

MORPHOLOGICAL FORMS OF CORROSION

Pitting is the most common form of localized corrosion and frequently is difficult to associate with specific metallographic features. Pit shape may vary from shallow depressions to cylindrical or roughly hemispherical cavities. These shapes distinguish pitting from intergranular or exfoliation corrosion. Superpurity aluminum has the highest resistance to pitting, and, among the 1xxx and 1xx.x aluminums, resistance improves with purity. Among commercial alloys, those of the 5xxx group have the lowest pitting probability and penetration rates, followed by those of the 3xxx group.

Intergranular corrosion is selective attack of grain boundaries. The mechanism is electrochemical, resulting from local cell action in the boundaries. Microconstituents precipitated in grain boundaries have a corrosion potential differing from that of adjacent solid-solution and transition precipitate structure and form cells with it. In alloys of the 5xxx and 7xxx groups, the precipitates (Al_8Mg_5, $MgZn_2$ and $Al_2Mg_3Zn_3$) are anodic to the matrix. In 2xxx alloys, the precipitates (Al_2Cu and Al_2CuMg) are cathodic. Intergranular corrosion may occur with either type. Susceptibility depends on the extent of intergranular precipitation, which is controlled by fabricating or heat treating parameters.

In 2xxx alloys, grain-boundary precipitation is caused by an inadequate cooling rate during the quenching operation of heat treatment. Thick-section products cannot be cooled sufficiently rapidly to completely avoid susceptibility to intergranular corrosion in T3- and T4-type tempers. Resistance to this type of attack is much higher in T6- and T8-type tempers. Alloys of the 6xxx group with a balanced Mg/Si ratio (Mg_2Si proportions) show little tendency toward intergranular corrosion; susceptibility is higher in those with excess silicon over the Mg_2Si ratio. Alloys of the 7xxx group may corrode intergranularly; overaging to T7-type tempers provides high resistance.

Exfoliation corrosion is selective attack that proceeds along multiple subsurface paths roughly parallel to the surface. It can be intergranular but also is associated with striated insoluble microconstituents and dispersoid bands aligned parallel to the product surface. It is most common in thin-section products with highly worked, flattened and elongated metallurgical structures. Leafing or delamination accompanied by swelling caused by expansion of corrosion products is characteristic and is apparent in metallographic section (see Fig. 2). Exfoliation frequently proceeds from sheared edges and may be initiated

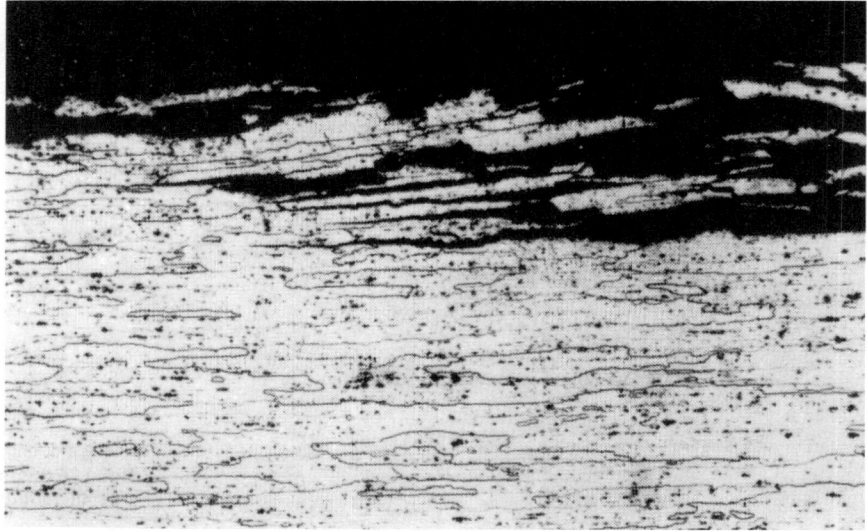

Cross section of plate, showing how exfoliation develops by corrosion along boundaries of thin, elongated grains.

Fig. 2. Exfoliation corrosion in alloy 7178-T651 plate exposed in a seacoast environment

at pit surfaces. It is not accelerated by applied stress but is intensified by slightly acidic solutions and by galvanic coupling.

In 2xxx and copper-containing 7xxx alloys, exfoliation is considerably affected by section thickness and corresponding microstructure and by temper. The 2xxx alloys are susceptible only in T3- and T4-type tempers, resistant in T6 or T8 type tempers. In the copper-containing 7xxx group alloys, resistance is greatly improved by overaging beyond peak strength (T6 type tempers), so that these materials are resistant in T7-type tempers. In extrusions of these heat treated alloys, the recrystallized peripheral zone near surfaces is often highly resistant to exfoliation while the underlying unrecrystallized portion may be vulnerable to this type of attack.

The 1xxx aluminums and 3xxx alloys are highly resistant to exfoliation in all tempers. Exfoliation has been encountered in highly cold worked, high-magnesium 5xxx alloys such as 5456-H321 boat hull plate. An improved temper (H116) with high resistance has been established.

GALVANIC CORROSION

Most forms of corrosion discussed above are electrochemical in nature and involve cells formed by microstructural or environmental features. A number of other conditions establish potential differences that intensify and localize corrosion. The accelerated corrosion resulting from electrical contact with a more-noble metal or with a nonmetallic conductor such as graphite is termed "galvanic corrosion." The most common examples occur when aluminum alloys are joined to steel or copper and are exposed to wet saline environments. In such situations the aluminum is more rapidly corroded than it would be in the absence of the dissimilar metal.

For each environment, metals can be arranged in a galvanic series from most to least active. Potentials based on measurements in sodium chloride solution are listed in Table 4. The rate of corrosive attack when two metals are coupled depends on several factors: (*a*) the potential dif-

ference, (*b*) the electrical resistance between the metals, (*c*) the conductivity of the electrolyte, (*d*) the cathode/anode area ratio, and (*e*) the polarization characteristics of the metals. Although the corrosion potential can be used to predict which metal will be attacked galvanically, the extent of attack cannot be predicted because of polarization. For example, the potential difference between aluminum and stainless steel exceeds that between aluminum and copper; however, the

Table 4. Electrode potentials of representative aluminum alloys and other metals

Aluminum alloy(a) or other metal	Potential(b), V
Chromium	+0.18 to −0.40
Nickel	−0.07
Silver	−0.08
Stainless steel (300 series)	−0.09
Copper	−0.20
Tin	−0.49
Lead	−0.55
Mild carbon steel	−0.58
2219-T3, -T4	−0.64(c)
2024-T3, -T4	−0.69(c)
295.0-T4 (S or PM)	−0.70
295.0-T6 (S or PM)	−0.71
2014-T6, 355.0-T4 (S or PM)	−0.78
355.0-T6 (S or PM)	−0.79
2219-T6, 6061-T4	−0.80
2024-T6	−0.81
2219-T8, 2024-T8, 356.0-T6 (S or PM), 443.0-F (PM), cadmium	−0.82
1100, 3003, 6061-T6, 6063-T6, 7075-T6(c), 443.0-F (S)	−0.83
1060, 1350, 3004, 7050-T73(c), 7075-T73(c)	−0.84
5052, 5086	−0.85
5454	−0.86
5456, 5083	−0.87
7072	−0.96
Zinc	−1.10
Magnesium	−1.73

(a) The potential of an aluminum alloy is the same in all tempers wherever the temper is not designated. S = sand cast; PM = permanent mold cast. (b) Measured in an aqueous solution of 53 g of NaCl and 3 g of H_2O_2 per litre at 25 °C; 0.1N calomel reference electrode. (c) Potential varies ±0.01 to 0.02 V with quenching rate.

galvanic effect of stainless steel on aluminum, because of polarization, is much less than that of copper, which shows little polarization.

In natural environments, including saline solutions, zinc is anodic to aluminum and corrodes preferentially, giving protection to the aluminum. Magnesium is also protective, although in severe marine environments it may cause corrosion of aluminum because of an alkaline reaction. Cadmium is neutral to aluminum and can be used safely in contact with it. Copper and its alloys, brass, bronze and monel, are the most harmful, followed closely by carbon steel in saline environments. Nickel is less aggressive than copper, approaching stainless steel in effect, as does chromium electroplate. Lead may be used with aluminum except in marine environments.

Stray-Current Corrosion. Whenever an electric current is conducted from aluminum to an environment such as water, soil or concrete, the aluminum is corroded in the area of anodic reaction in proportion to the current. At low current densities the corrosion may be in the form of pitting, while at higher current densities considerable metal destruction may occur at rates that do not diminish with time.

In soils, stray-current corrosion can be caused by close proximity to other buried metal systems which are protected by an impressed-current cathodic-protection system. The ground current may leak onto a buried aluminum structure at one point, then off at another where the corrosion occurs, taking a lower-resistance path between the driven buried anode and the nearby structure being protected. Common bonding of all buried metal systems in close proximity is the usual way to avoid such attack.

Deposition corrosion is a special form of galvanic corrosion that causes pitting. It occurs when particles of a more-cathodic metal plate out of solution on the aluminum surface, setting up local galvanic cells. The ions aggressive to aluminum are Cu, Pb, Hg, Ni and Sn, often referred to as heavy metals. Their effects are greater in acidic solutions and much less severe in alkaline solutions in which their solubility is low.

Copper ions most commonly cause this type of corrosion in applications of aluminum. For example, rain runoff from copper roof flashing may cause corrosion of aluminum gutters with no electrical contact between the two metals. Very small amounts of copper in solution (as low as 0.05 ppm) can be detrimental. The inferior general corrosion resistance of the alloys containing copper is attributed to deposition corrosion by copper replated from the dissolved corrosion products.

Mercury is the ion most aggressive to aluminum, and even traces can cause serious problems. Liquid mercury does not wet aluminum, but if the natural oxide film on the aluminum surface is broken, aluminum dissolves in the mercury, forming amalgam, and the corrosion reaction becomes catastrophic. In corrosive solutions, any concentration of mercury greater than a few ppb should be cause for concern.

Crevice Corrosion. If an electrolyte is present in a crevice formed between two facing aluminum surfaces or between an aluminum surface and a nonmetallic material such as a gasket, localized corrosion in the form of pits or etch patches may occur. This is the result of formation of a concentration cell or differential aeration cell. Staining that occurs on inter-wrap surfaces of coiled

sheet or foil or in packages of flat sheet or circles is a result of the same mechanism and may be preliminary to more severe corrosion that will make separation difficult. Such damage can be prevented by ensuring that the product is initially dry and by avoiding ingress of moisture by protecting it against condensation, rain and other sources of contamination. Filiform corrosion is a special case of crevice corrosion that may occur particularly under organic coatings. It takes the form of randomly distributed threadlike filaments and is sometimes called "vermiform" or "worm track" corrosion. Its occurrence on lacquered surfaces in aircraft exposed to marine and other high-humidity environments has been controlled by use of chemical conversion coatings, anodizing, or application of chromate-inhibited primers prior to final coating.

STRESS-CORROSION CRACKING

Time-dependent cracking under the combined influence of sustained tensile stress and a corrosive environment is labeled stress-corrosion cracking (SCC). In aluminum products, SCC has been experienced only in higher-strength alloys and tempers of the 2xxx, 7xxx and 5xxx types (with more than 3% Mg) and of the 6xxx type (with excess Si). The relative resistances of aluminum products made of such alloys in various tempers, and with respect to the direction of tensile stress, is indicated by the ratings listed in Table 5. No SCC problems have been encountered in service with 1xxx aluminum or with 3xxx, 6xxx (Mg_2Si ratio) or 5xxx (containing 3% Mg or less) alloys. 6061-T6, which is included in Table 5, is a balanced-ratio alloy.

Table 5. Relative stress-corrosion-cracking ratings for wrought products of high-strength aluminum alloys(a)

Alloy and temper(b)	Test direction(c)	Rolled plate	Rod and bar(d)	Extruded shapes	Forgings
2011-T3, -T4	L	(e)	B	(e)	(e)
	LT	(e)	D	(e)	(e)
	ST	(e)	D	(e)	(e)
2011-T8	L	(e)	A	(e)	(e)
	LT	(e)	A	(e)	(e)
	ST	(e)	A	(e)	(e)
2014-T6	L	A	A	A	B
	LT	B(f)	D	B(f)	B(f)
	ST	D	D	D	D
2024-T3, -T4	L	A	A	A	(e)
	LT	B(f)	D	B(f)	(e)
	ST	D	D	D	(e)
2024-T6	L	(e)	A	(e)	A
	LT	(e)	B	(e)	A(f)
	ST	(e)	B	(e)	D
2024-T8	L	A	A	A	A
	LT	A	A	A	A
	ST	B	A	B	C
2124-T851	L	A	(e)	(e)	(e)
	LT	A	(e)	(e)	(e)
	ST	B	(e)	(e)	(e)
2219-T3, -T37	L	A	(e)	A	(e)
	LT	B	(e)	B	(e)
	ST	D	(e)	D	(e)
2219-T6, -T8	L	A	A	A	A
	LT	A	A	A	A
	ST	A	A	A	A
6061-T6	L	A	A	A	A
	LT	A	A	A	A
	ST	A	A	A	A
7005 -T53, -T63	L	(e)	(e)	A	A
	LT	(e)	(e)	A(f)	A(f)
	ST	(e)	(e)	D	D
7039 -T63, -T64	L	A	(e)	A	(e)
	LT	A(f)	(e)	A(f)	(e)
	ST	D	(e)	D	(e)
7049-T73	L	A	(e)	A	A
	LT	A	(e)	A	A
	ST	A	(e)	B	A
7149-T73	L	(e)	(e)	A	A
	LT	(e)	(e)	A	A
	ST	(e)	(e)	B	A
7050-T74	L	A	(e)	A	A
	LT	A	(e)	A	A
	ST	B	(e)	B	B
7050-T76	L	A	A	A	(e)
	LT	A	B	A	(e)
	ST	C	B	C	(e)
7075-T6	L	A	A	A	A
	LT	B(f)	D	B(f)	B(f)
	ST	D	D	D	D
7075-T73	L	A	A	A	A
	LT	A	A	A	A
	ST	A	A	A	A
7075-T736	L	(e)	(e)	(e)	A
	LT	(e)	(e)	(e)	A
	ST	(e)	(e)	(e)	B
7075-T76	L	A	(e)	A	(e)
	LT	A	(e)	A	(e)
	ST	C	(e)	C	(e)
7175-T736	L	(e)	(e)	(e)	A
	LT	(e)	(e)	(e)	A
	ST	(e)	(e)	(e)	B
7475-T6	L	A	(e)	(e)	(e)
	LT	B(f)	(e)	(e)	(e)
	ST	D	(e)	(e)	(e)
7475-T73	L	A	(e)	(e)	(e)
	LT	A	(e)	(e)	(e)
	ST	A	(e)	(e)	(e)
7475-T76	L	A	(e)	(e)	(e)
	LT	A	(e)	(e)	(e)
	ST	C	(e)	(e)	(e)
7178-T6	L	A	(e)	A	(e)
	LT	B(f)	(e)	B(f)	(e)
	ST	D	(e)	D	(e)
7178-T76	L	A	(e)	A	(e)
	LT	A	(e)	A	(e)
	ST	C	(e)	C	(e)
7079-T6	L	A	(e)	A	A
	LT	B(f)	(e)	B(f)	B(f)
	ST	D	(e)	D	D

(a) Resistance ratings are as follows: A—very high. No record of service problems; stress-corrosion cracking not anticipated in general applications. B—high. No record of service problems; stress-corrosion cracking not anticipated at stresses of the magnitude caused by solution heat treatment. Precautions must be taken to avoid high sustained tensile stresses (exceeding 50% of the minimum specified yield strength) produced by any combination of sources, including heat treatment, straightening, forming, fit-up and sustained service loading. C—intermediate. Stress-corrosion cracking not anticipated if total sustained tensile stress is maintained below 25% of minimum specified yield strength. This rating is designated for the short-transverse direction in products used primarily for high resistance to exfoliation corrosion in relatively thin sections, where appreciable stresses in the short-transverse direction are unlikely. D—low. Failure from stress-corrosion cracking is anticipated in any application involving sustained tensile stress in the designated test direction. This rating is currently designated only for the short-transverse direction in certain products. (b) Ratings apply to standard mill products in the types of tempers indicated and also in Tx5x and Tx5xx (stress-relieved) tempers, and may be invalidated in some instances by use of nonstandard thermal treatments, or mechanical deformation at room temperature, by the user. (c) Test direction refers to orientation of direction in which stress is applied relative to the directional grain structure typical of wrought alloys, which for extrusions and forgings may not be predictable on the basis of the cross-sectional shape of the product: L—longitudinal; LT—long transverse; ST—short transverse. (d) Sections with width-to-thickness ratios equal to or less than two, for which there is no distinction between LT and ST properties. (e) Rating not established because product not offered commercially. (f) Rating is one class lower for thicker sections: extrusions, 25 mm (1 in.) and thicker; plate and forgings, 40 mm (1.5 in.) and thicker.

In general, high tensile stress is a prerequisite to cracking when the stress direction is parallel to either the longitudinal or the long-transverse direction. When the tensile stress is in the short-transverse direction (perpendicular to the surfaces of plate or across the flash plane of die forgings), SCC can occur in susceptible alloy/temper combinations at relatively low stresses. Cracking is accelerated by aggressive, chloride-containing environments, but can occur in humid air.

Because of the orientation-dependence of SCC, it is important to minimize stresses in the most susceptible direction. In addition to the stresses imposed by service loading, the residual stresses from quenching or forming, and any resulting from interference fits or assembly misfits, must be taken into account. Minimizing these stresses in the short-transverse direction greatly reduces the probability of SCC failure of susceptible alloy/temper combinations. Use of resistant tempers is recommended.

For thin-section products used under conditions that induce little or no stress in the short-transverse direction, resistance of 2xxx alloys in T3- or T4-type tempers or of 7xxx alloys in T6-type tempers is often satisfactory. For rolled, extruded or forged thick-section products, resistance in the short-transverse direction or across the flash plane of die forgings usually controls their use. Overaging stress-relief treatments that sacrifice some strength are very effective in providing high resistance to SCC in the copper-containing 7xxx alloys — 7075, 7175, 7475, 7049 and 7050 — in the T73, T736 and T74 tempers. Overaging of the premium-strength 2xx.x casting alloys to T63 to T7 tempers provides good resistance to SCC, and 2219 in the T6 temper is similarly overaged with respect to strength to achieve high SCC resistance.

Medium-strength Cu-free and low-Cu alloys of the 7xxx group tend to be susceptible to SCC. Successful use of 7039, which has poor resistance, in armor-plate applications requires control of short-transverse stresses and weld overlays. Alloys 7016, 7021 and 7029 with copper contents up to 1% have good formability and finishing properties for automotive applications such as bumpers. Maximum SCC resistance is obtained by forming in the freshly quenched (W) temper followed by two-step aging. Casting alloys of the 7xx.x group have compositions similar to those of the aforementioned wrought alloys, and some SCC problems have occasionally been encountered in their applications.

Other Interactive Effects. Other forms of damage involving interactions of corrosive environments and mechanical actions are identified as corrosive fatigue, cavitation, impingement and erosion corrosion, and fretting corrosion.

In the presence of a corrosive environment, the stress-life curve (S/N curve) is lowered and tends to have increased slope at high numbers of stress

Table 6. Combinations of aluminum alloys used in some alclad products

Core alloy	Cladding alloy
2014	6003 or 6053
2024	1230
2219	7072
3003	7072
3004	7072 or 7013
5056	6253
6061	7072
7075	7072, 7008 or 7011
7178	7072

cycles. These corrosion-fatigue effects are substantial in aggressive environments and must be accounted for in designing for such conditions.

Cavitation damage results from exposure to highly turbulent liquids causing pitting. Impingement, caused by a liquid stream striking the metal surface at an angle, is similar to cavitation, involving both mechanical and chemical actions, as does erosion corrosion. Damage of these types can be minimized by designing to eliminate or minimize the high velocity, turbulence and impingement conditions. Cavitation in automotive coolant systems is reduced by use of appropriate inhibitors.

Fretting corrosion is a process of attrition and oxidation of metal surfaces that are in pressure contact but that undergo minute cyclic slip under vibratory conditions. It occurs in assembled machine and engine parts, at bolted or riveted joints in structures such as truck bodies, and even in packages of aluminum products during shipment. Design of fastened joints to minimize motion is important, and use of nonmetallic gasket-type materials is beneficial in some cases.

Corrosion at Joints. Joints fastened by mechanical fasteners — bolts, screws, cold headed rivets and threaded connections — or by pressure or adhesive bonding require special attention to avoid galvanic effects, crevice corrosion and assembly stresses in products susceptible to SCC. Methods of joining which involve heat — welding, brazing and soldering — alter metallurgical structures of the parent material adjacent to the fusion zones and introduce composition differences between the joint proper and the parent metal. These can be a source of galvanic effects, particularly in heat treatable alloys, and are a factor of importance in selection of appropriate joining materials.

Alclad Products. In alclad aluminum products, the difference in solution potential between the core alloy and the cladding alloy is used to provide cathodic protection to the core. These products, primarily sheet and tube, consist of a core coated on one or both surfaces with a metallurgically bonded layer of an alloy that is anodic to the core alloy. The thickness of the cladding layer usually is less than 10% of the over-all thickness of the product. Cladding alloys generally are of the non-

heat-treatable type, although, for higher strength, heat treatable alloys sometimes are used. Composition relationships of core and cladding alloys generally are designed so that the cladding is 80 to 100 mV more anodic than the core. Several core alloy/cladding alloy combinations for common alclad products are listed in Table 6. Because of the cathodic protection provided by the cladding, corrosion progresses only to the core/cladding interface, and then spreads laterally. This is highly effective in eliminating perforation of thin products.

COATINGS

Paints and other coatings are applied to aluminum alloy products for decorative as well as protective purposes. Almost any type of paint for metals (acrylic, alkyl, polyester, vinyl, etc.) is suitable; the performance of a particular paint on aluminum, when applied properly, is better than its performance on steel. As with any metal, surface preparation is important. Conversion coatings, of either the chromate or the phosphate type, are recommended for preparation of aluminum alloys. For milder environments, the paint may be applied to the conversion coating; for more aggressive environments such as those containing chlorides, a chromated primer should be applied first. Aluminum alloys are especially amenable to waterborne paints, which are being used increasingly because of environmental considerations. Many products precoated in a variety of colors for agricultural, industrial and residential applications are available commercially.

Although more expensive, and restricted to in-plant application, anodized coatings provide excellent protection for aluminum alloys. They are also sometimes used as bases for paints. The many monumental buildings with outer walls of anodized aluminum alloys attest to the durability of these materials under conditions of weathering. Anodized coatings also provide a variety of colorations, most commonly shades of gray and bronze, produced by selection of both alloy and anodizing process.

Anodized coatings are produced by an electrolytic process in which the surface of an alloy that is made the anode is converted to aluminum oxide; this oxide is bound to the alloy as tenaciously as the natural oxide film but is much thicker. Coatings used to provide corrosion resistance range in thickness from 5 to 30 μm (0.2 to 1.2 mils); little or no additional protection is provided by thicker coatings. As with the alloys themselves, anodized coatings are not resistant to most environments with pH values outside the range from 4 to 9. Within this range, resistance to corrosion can be improved by an order of magnitude or more; in atmospheric weathering tests, the number of pits that developed in the base metal was found to decrease exponentially with coating thickness.

7 COPPER

Edited by Derek E. Tyler, Olin Corp.

Review Committee: Eugene Shapiro, Olin Corp.; Ned W. Polan, Olin Corp.; Raymond Cribb, Brush Wellman Inc.; J. C. Harkness, Brush Wellman Inc.; William Black, Copper Development Association Inc.

This section was condensed from Metals Handbook, Ninth Edition, Volume 2, Properties and Selection: Nonferrous Alloys and Pure Metals, pages 237 to 490. Additional articles on copper and copper alloys can be found in the following sections: 23, Casting; 24, Forging; 26, Forming; 27, Machining; 28, Heat Treating; 29, Surface Technology; 30, Joining; and 35, Metallography.

Copper and Copper Alloys

COPPER and its alloys constitute one of the major groups of commercial metals. They are widely used because of their excellent electrical and thermal conductivities, outstanding resistance to corrosion, and ease of fabrication, together with good strength and fatigue resistance. They are generally nonmagnetic. They can be readily soldered and brazed, and many coppers and copper alloys can be welded by various gas, arc and resistance methods. For decorative parts, standard alloys having specific colors are readily available. Copper alloys can be polished and buffed to almost any desired texture and luster. They can be plated, coated with organic substances or chemically colored to further extend the variety of available finishes.

Pure copper is used extensively for cables and wires, electrical contacts, and a wide variety of other parts that are required to pass electrical current. Coppers and certain brasses, bronzes and cupronickels are used extensively for automobile radiators, heat exchangers, home heating systems, panels for absorbing solar energy, and various other applications requiring rapid conduction of heat across or along a metal section. Because of their outstanding ability to resist corrosion, coppers, brasses, some bronzes, and cupronickels are used for pipes, valves and fittings in systems carrying potable water, process water or other aqueous fluids.

In all classes of copper alloys, certain alloy compositions for wrought products have counterparts among the cast alloys, which enables the designer to make an initial alloy selection before deciding on the manufacturing process. Most wrought alloys are available in various cold worked conditions, which have room-temperature strengths and fatigue resistances that depend on the amount of cold work as well as the alloy content. Typical applications of cold worked conditions (cold worked tempers) include springs, fasteners, hardware, small gears, cams, electrical contacts, and components.

Certain types of parts—most notably plumbing fittings and valves—are produced by hot forging simply because no other fabrication process can produce the required shapes and properties as economically.

Copper alloys containing 1 to 6% Pb are free-machining grades, and are used widely for machined parts—especially those produced in screw machines.

Although fewer alloys are produced now than in the 1930's, new alloys continue to be developed and introduced, in particular to meet the challenging requirements of the electronics industry.

Properties and applications of wrought copper alloys are presented in Tables A and A-1. Similar data for cast copper alloys are given in Table B.

PROPERTIES OF IMPORTANCE

Good resistance to corrosion, good electrical conductivity, good thermal conductivity, color and ease of fabrication coupled with strength, resistance to fatigue and ability to take a good finish are criteria by which copper or one of its alloys is selected.

Corrosion Resistance. Copper is a noble metal but, unlike gold and other precious metals, can be attacked by common reagents and environments.

Pure copper resists attack quite well under most corrosive conditions. Some copper alloys, however, sometimes have limited usefulness in certain environments because of hydrogen embrittlement or stress-corrosion cracking.

Hydrogen embrittlement is observed when tough pitch coppers, which are alloys containing cuprous oxide, are exposed to a reducing atmosphere. Most of the copper alloys are deoxidized, and thus are not subject to hydrogen embrittlement.

Stress-corrosion cracking (season cracking) most commonly occurs in brass that is exposed to ammonia or amines. Brasses containing more than 15% zinc are the most susceptible. Copper and most copper alloys that either do not contain zinc or are low in zinc content are generally not susceptible to stress-corrosion cracking. Because stress-corrosion cracking requires both tensile stress and a specific chemical species to be present at the same time, removal of either the stress or the chemical species can prevent cracking. Annealing or stress relieving after forming alleviates stress-corrosion cracking by relieving residual stresses. Stress relieving is effective only if the parts are not subsequently bent or strained in service; such operations reintroduce stresses and resensitize the parts to stress-corrosion cracking.

Electrical and Thermal Conductivity. Copper and its alloys are relatively good conductors of electricity and heat. In fact, copper is used for these purposes more often than any other metal. Alloying invariably decreases electrical conductivity and, to a lesser extent, thermal conductivity. The amount of reduction due to alloying does not depend on the conductivity or any other bulk property of the alloying element, but only on the effect that the particular solute atoms have on the copper lattice. For this reason, coppers and high-copper alloys are preferred over copper alloys containing more than a few percent total alloy content when high electrical or thermal conductivity is required for the application.

Color. Copper and certain copper alloys are used for decorative purposes alone, or when a particular color and finish is combined with a desirable mechanical or physical property of the alloy. Table 1 lists the range of colors that can be obtained with standard copper alloys.

Ease of Fabrication. Copper and its alloys are generally capable of being shaped to the required form and dimensions by any of the common fabricating processes. They are routinely rolled, stamped, drawn and headed cold; they are rolled, extruded, forged and formed at elevated temperature; there are casting alloys for all of the generic families of coppers and copper alloys.

Copper metals can be polished, textured, plated or coated to provide a wide variety of functional or decorative surfaces.

(Text is continued on page 7·14.)

Table A. Properties of wrought coppers and copper alloys

Alloy number (and name)	Nominal composition, %	Commercial forms(a)	Tensile strength MPa	ksi	Yield strength MPa	ksi	Elongation in 2 in., %	Corrosion resistance (c)	Machinability rating (d)
C10100 (oxygen-free electronic)	99.99 Cu	F, R, W, T, P, S	221-455	32-66	69-365	10-53	55-4	G-E	20
C10200 (oxygen-free copper)	99.95 Cu	F, R, W, T, P, S	221-455	32-66	69-365	10-53	55-4	G-E	20
C10300 (oxygen-free, extra-low phosphorus)	99.95 Cu, 0.003 P	F, R, T, P, S	221-379	32-55	69-345	10-50	50-6	G-E	20
C10400, C10500, C10700 (oxygen-free, silver-bearing)	99.95 Cu(e)	F, R, W, S	221-455	32-66	69-365	10-53	55-4	G-E	20
C10800 (oxygen-free, low-phosphorus)	99.95 Cu, 0.009 P	F, R, T, P	221-379	32-55	69-345	10-50	50-4	G-E	20
C11000 (electrolytic tough pitch copper)	99.90 Cu, 0.04 O	F, R, W, T, P, S	221-455	32-66	69-365	10-53	55-4	G-E	20
C11100 (electrolytic tough pitch, anneal resistant)	99.90 Cu, 0.04 O, 0.01 Cd	W	455	66	...	...	1.5 in 60 in.	G-E	20
C11300, C11400, C11500, C11600 (silver-bearing tough pitch copper)	99.90 Cu, 0.04 O, Ag(f)	F, R, W, T, S	221-455	32-66	69-365	10-53	55-4	G-E	20
C12000, C12100	99.9 Cu(g)	F, T, P	221-393	32-57	69-365	10-53	55-4	G-E	20
C12200 (phosphorus-deoxidized copper, high residual phosphorus)	99.90 Cu, 0.02 P	F, R, T, P	221-379	32-55	69-345	10-50	45-8	G-E	20
C12500, C12700, C12800, C12900, C13000 (fire-refined tough pitch with silver)	99.88 Cu(h)	F, R, W, S	221-462	32-67	69-365	10-53	55-4	G-E	20
C14200 (phosphorus-deoxidized, arsenical)	99.68 Cu, 0.3 As, 0.02 P	F, R, T	221-379	32-55	69-345	10-50	45-8	G-E	20
C19200	98.97 Cu, 1.0 Fe, 0.03 P	F, T	255-531	37-77	76-510	11-74	40-2	G-E	20
C14300	99.9 Cu, 0.1 Cd	F	221-400	32-58	76-386	11-56	42-1	G-E	20
C14310	99.8 Cu, 0.2 Cd	F	221-400	32-58	76-386	11-56	42-1	G-E	20
C14500 (phosphorus-deoxidized, tellurium-bearing)	99.5 Cu, 0.50 Te, 0.008 P	F, R, W, T	221-386	32-56	69-352	10-51	50-3	G-E	85
C14700 (sulfur-bearing)	99.6 Cu, 0.40 S	R, W	221-393	32-57	69-379	10-55	52-8	G-E	85
C15000 (zirconium copper)	99.8 Cu, 0.15 Zr	R, W	200-524	29-76	41-496	6-72	54-1.5	G-E	20
C15500	99.75 Cu, 0.06 P, 0.11 Mg, Ag(i)	F	276-552	40-80	124-496	18-72	40-3	G-E	20
C15710	99.8 Cu, 0.2 Al$_2$O$_3$	R, W	324-724	47-105	268-689	39-100	20-10	...	...
C15720	99.6 Cu, 0.4 Al$_2$O$_3$	F, R	462-614	67-89	365-586	53-85	20-3.5	...	...
C15735	99.3 Cu, 0.7 Al$_2$O$_3$	R	483-586	70-85	414-565	60-82	16-10	...	...
C15760	98.9 Cu, 1.1 Al$_2$O$_3$	F, R	483-648	70-94	386-552	56-80	20-8	...	...
C16200 (cadmium copper)	99.0 Cu, 1.0 Cd	F, R, W	241-689	35-100	48-476	7-69	57-1	G-E	20
C16500	98.6 Cu, 0.8 Cd, 0.6 Sn	F, R, W	276-655	40-95	97-490	14-71	53-1.5	G-E	20
C17000 (beryllium copper)	99.5 Cu, 1.7 Be, 0.20 Co	F, R	483-1310	70-190	221-1172	32-170	45-3	G-E	20
C17200 (beryllium copper)	99.5 Cu, 1.9 Be, 0.20 Co	F, R, W, T, P, S	469-1462	68-212	172-1344	25-195	48-1	G-E	20
C17300 (beryllium copper)	99.5 Cu, 1.9 Be, 0.40 Pb	R	469-1479	68-200	172-1255	25-182	48-3	G-E	50
C17500 (copper-cobalt-beryllium alloy)	99.5 Cu, 2.5 Co, 0.6 Be	F, R	310-793	45-115	172-758	25-110	28-5	G-E	...
C18200, C18400, C18500 (chromium copper)	99.5 Cu(j)	F, W, R, S, T	234-593	34-86	97-531	14-77	40-5	G-E	20
C18700 (leaded copper)	99.0 Cu, 1.0 Pb	R	221-379	32-55	69-345	10-50	45-8	G-E	85
C18900	98.75 Cu, 0.75 Sn, 0.3 Si, 0.20 Mn	R, W	262-655	38-95	62-359	9-52	48-14	G-E	20
C19000 (copper-nickel-phosphorus alloy)	98.7 Cu, 1.1 Ni, 0.25 P	F, R, W	262-793	38-115	138-552	20-80	50-2	G-E	30
C19100 (copper-nickel-phosphorus-tellurium alloy)	98.15 Cu, 1.1 Ni, 0.50 Te, 0.25 P	R, F	248-717	36-104	69-634	10-92	27-6	G-E	75
C19400	97.5 Cu, 2.4 Fe, 0.13 Zn, 0.03 P	F	310-524	45-76	165-503	24-73	32-2	G-E	20
C19500	97.0 Cu, 1.5 Fe, 0.6 Sn, 0.10 P, 0.80 Co	F	552-669	80-97	448-655	65-95	15-2	G-E	20
C21000 (gilding, 95%)	95.0 Cu, 5.0 Zn	F, W	234-441	34-64	69-400	10-58	45-4	G-E	20
C22000 (commercial bronze, 90%)	90.0 Cu, 10.0 Zn	F, R, W, T	255-496	37-72	69-427	10-62	50-3	G-E	20

(continued)

(a) F, flat products; R, rod; W, wire; T, tube; P, pipe; S, shapes. (b) Softest to hardest commercial forms. The strength of the standard copper alloys depends on the temper (annealed grain size or degree of cold work) and the section thickness of the mill product. Ranges cover standard tempers for each alloy. (c) E, excellent; G, good; F, fair. (d) Based on 100% for C36000. (e) C10400, 8 oz/ton Ag; C10500, 10 oz/ton; C10700, 25 oz/ton. (f) C11300, 8 oz/ton Ag; C11400, 10 oz/ton; C11500, 16 oz/ton; C11600, 25 oz/ton. (g) C12000, 0.008 P; C12100, 0.008 P and 4 oz/ton Ag. (h) C12700, 8 oz/ton Ag; C12800, 10 oz/ton; C12900, 16 oz/ton; C13000, 25 oz/ton. (i) 8.30 oz/ton Ag. (j) C18200, 0.9 Cr; C18400, 0.8 Cr; C18500, 0.7 Cr. (k) Rod, 61.0 Cu min. Compiled by Copper Development Assn. Inc., New York.

Table A. (continued)

Alloy number (and name)	Nominal composition, %	Commercial forms(a)	Tensile strength MPa	Tensile strength ksi	Yield strength MPa	Yield strength ksi	Elongation in 2 in., %	Corrosion resistance (c)	Machinability rating (d)
C22600 (jewelry bronze, 87.5%)	87.5 Cu, 12.5 Zn	F, W	269-669	39-97	76-427	11-62	46-3	G-E	30
C23000 (red brass, 85%)	85.0 Cu, 15.0 Zn	F, W, T, P	269-724	39-105	69-434	10-63	55-3	G-E	30
C24000 (low brass, 80%)	80.0 Cu, 20.0 Zn	F, W	290-862	42-125	83-448	12-65	55-3	F-E	30
C26000 (cartridge brass, 70%)	70.0 Cu, 30.0 Zn	F, R, W, T	303-896	44-130	76-448	11-65	66-3	F-E	30
C26800, C27000 (yellow brass)	65.0 Cu, 35.0 Zn	F, R, W	317-883	46-128	97-427	14-62	65-3	F-E	30
C28000 (Muntz metal)	60.0 Cu, 40.0 Zn	F, R, T	372-510	54-74	145-379	21-55	52-10	F-E	40
C31400 (leaded commercial bronze)	89.0 Cu, 1.75 Pb, 9.25 Zn	F, R	255-414	37-60	83-379	12-55	45-10	G-E	80
C31600 (leaded commercial bronze, nickel-bearing)	89.0 Cu, 1.9 Pb, 1.0 Ni, 8.1 Zn	F, R	255-462	37-67	83-407	12-59	45-12	G-E	80
C33000 (low-leaded brass tube)	66.0 Cu, 0.5 Pb, 33.5 Zn	T	324-517	47-75	103-414	15-60	60-7	F-E	60
C33200 (high-leaded brass tube)	66.0 Cu, 1.6 Pb, 32.4 Zn	T	359-517	52-75	138-414	20-60	50-7	F-E	80
C33500 (low-leaded brass)	65.0 Cu, 0.5 Pb, 34.5 Zn	F	317-510	46-74	97-414	14-60	65-8	F-E	60
C34000 (medium-leaded brass)	65.0 Cu, 1.0 Pb, 34.0 Zn	F, R, W, S	324-607	47-88	103-414	15-60	60-7	F-E	70
C34200 (high-leaded brass)	64.5 Cu, 2.0 Pb, 33.5 Zn	F, R	338-586	49-85	117-427	17-62	52-5	F-E	90
C34900	62.2 Cu, 0.35 Pb, 37.45 Zn	R, W	365-469	53-68	110-379	16-55	72-18	F-E	50
C35000 (medium-leaded brass)	62.5 Cu, 1.1 Pb, 36.4 Zn	F, R	310-655	45-95	90-483	13-70	66-1	F-E	70
C35300 (high-leaded brass)	62.0 Cu, 1.8 Pb, 36.2 Zn	F, R	338-586	49-85	117-427	17-62	52-5	F-E	90
C35600 (extra-high-leaded brass)	63.0 Cu, 2.5 Pb, 34.5 Zn	F	338-510	49-74	117-414	17-60	50-7	F-E	100
C36000 (free-cutting brass)	61.5 Cu, 3.0 Pb, 35.5 Zn	F, R, S	338-469	49-68	124-310	18-45	53-18	F-E	100
C36500 to C36800 (leaded Muntz metal)	60.0 Cu(k), 0.6 Pb, 39.4 Zn	F	372	54	138 (As hot rolled)	20	45	F-E	60
C37000 (free-cutting Muntz metal)	60.0 Cu, 1.0 Pb, 39.0 Zn	T	372-552	54-80	138-414	20-60	40-6	F-E	70
C37700 (forging brass)	59.0 Cu, 2.0 Pb, 39.0 Zn	R, S	359	52	138 (As extruded)	20	45	F-E	80
C38500 (architectural bronze)	57.0 Cu, 3.0 Pb, 40.0 Zn	R, S	414	60	138 (As extruded)	20	30	F-E	90
C40500	95 Cu, 1 Sn, 4 Zn	F	269-538	39-78	83-483	12-70	49-3	G-E	20
C40800	95 Cu, 2 Sn, 3 Zn	F	290-545	42-79	90-517	13-75	43-3	G-E	20
C41100	91 Cu, 0.5 Sn, 8.5 Zn	F, W	269-731	39-106	76-496	11-72	13-2	G-E	20
C41300	90.0 Cu, 1.0 Sn, 9.0 Zn	F, R, W	283-724	41-105	83-565	12-82	45-2	G-E	20
C41500	91 Cu, 1.8 Sn, 7.2 Zn	F	317-558	46-81	117-517	17-75	44-2	G-E	30
C42200	87.5 Cu, 1.1 Sn, 11.4 Zn	F	296-607	43-88	103-517	15-75	46-2	G-E	30
C42500	88.5 Cu, 2.0 Sn, 9.5 Zn	F	310-634	45-92	124-524	18-76	49-2	G-E	30
C43000	87.0 Cu, 2.2 Sn, 10.8 Zn	F	317-648	46-94	124-503	18-73	55-3	G-E	30
C43400	85.0 Cu, 0.7 Sn, 14.3 Zn	F	310-607	45-88	103-517	15-75	49-3	G-E	30
C43500	81.0 Cu, 0.9 Sn, 18.1 Zn	F, T	317-552	46-80	110-469	16-68	46-7	G-E	30
C44300, C44400, C44500 (inhibited admiralty)	71.0 Cu, 28.0 Zn, 1.0 Sn	F, W, T	331-379	48-55	124-152	18-22	65-60	G-E	30
C46400 to C46700 (naval brass)	60.0 Cu, 39.25 Zn, 0.75 Sn	F, R, T, S	379-607	55-88	172-455	25-66	50-17	F-E	30
C48200 (naval brass, medium-leaded)	60.5 Cu, 0.7 Pb, 0.8 Sn, 38.0 Zn	F, R, S	386-517	56-75	172-365	25-53	43-15	F-E	50
C48500 (leaded naval brass)	60.0 Cu, 1.75 Pb, 37.5 Zn, 0.75 Sn	F, R, S	379-531	55-77	172-365	25-53	40-15	F-E	70
C50500 (phosphor bronze, 1.25% E)	98.75 Cu, 1.25 Sn, trace P	F, W	276-545	40-79	97-345	14-50	48-4	G-E	20
C51000 (phosphor bronze, 5% A)	95.0 Cu, 5.0 Sn, trace P	F, R, W, T	324-965	47-140	131-552	19-80	64-2	G-E	20
C51100	95.6 Cu, 4.2 Sn, 0.2 P	F	317-710	46-103	345-552	50-80	48-2	G-E	20
C52100 (phosphor bronze, 8% C)	92.0 Cu, 8.0 Sn, trace P	F, R, W	379-965	55-140	165-552	24-80	70-2	G-E	20
C52400 (phosphor bronze, 10% D)	90.0 Cu, 10.0 Sn, trace P	F, R, W	455-1014	66-147	193 (Annealed)	28	70-3	G-E	20
C54400 (free-cutting phosphor bronze)	88.0 Cu, 4.0 Pb, 4.0 Zn, 4.0 Sn	F, R	303-517	44-75	131-434	19-63	50-16	G-E	80
C60800 (aluminum bronze, 5%)	95.0 Cu, 5.0 Al	T	414	60	186	27	55	G-E	20
C61000	92.0 Cu, 8.0 Al	R, W	483-552	70-80	207-379	30-55	65-25	G-E	20

(continued)

(a) F, flat products; R, rod; W, wire; T, tube; P, pipe; S, shapes. (b) Softest to hardest commercial forms. The strength of the standard copper alloys depends on the temper (annealed grain size or degree of cold work) and the section thickness of the mill product. Ranges cover standard tempers for each alloy. (c) E, excellent; G, good; F, fair. (d) Based on 100% for C36000. (e) C10400, 8 oz/ton Ag; C10500, 10 oz/ton; C10700, 25 oz/ton. (f) C11300, 8 oz/ton Ag; C11400, 10 oz/ton; C11500, 16 oz/ton; C11600, 25 oz/ton. (g) C12000, 0.008 P; C12100, 0.008 P and 4 oz/ton Ag. (h) C12700, 8 oz/ton Ag; C12800, 10 oz/ton; C12900, 16 oz/ton; C13000, 25 oz/ton. (i) 8.30 oz/ton Ag. (j) C18200, 0.9 Cr; C18400, 0.8 Cr; C18500, 0.7 Cr. (k) Rod, 61.0 Cu min. Compiled by Copper Development Assn. Inc., New York.

Table A. (continued)

Alloy number (and name)	Nominal composition, %	Commercial forms(a)	Mechanical properties(b) Tensile strength MPa	ksi	Yield strength MPa	ksi	Elongation in 2 in., %	Corrosion resistance (c)	Machinability rating (d)
C61300	92.65 Cu, 0.35 Sn, 7.0 Al	F, R, T, P, S	483-586	70-85	207-400	30-58	42-35	G-E	30
C61400 (aluminum bronze, D)	91.0 Cu, 7.0 Al, 2.0 Fe	F, R, W, T, P, S	524-614	76-89	228-414	33-60	45-32	G-E	20
C61500	90.0 Cu, 8.0 Al, 2.0 Ni	F	483-1000	70-145	152-965	22-140	55-1	G-E	30
C61800	89.0 Cu, 1.0 Fe, 10.0 Al	R	552-586	80-85	269-293	39-42.5	28-23	G-E	40
C61900	86.5 Cu, 4.0 Fe, 9.5 Al	F	634-1048	92-152	338-1000	49-145	30-1	G-E	...
C62300	87.0 Cu, 3.0 Fe, 10.0 Al	F, R	517-676	75-98	241-359	35-52	35-22	G-E	50
C62400	86.0 Cu, 3.0 Fe, 11.0 Al	F, R	621-724	90-105	276-359	40-52	18-14	G-E	50
C62500	82.7 Cu, 4.3 Fe, 13.0 Al	F, R	689 (As extruded)	100	379	55	1	G-E	20
C63000	82.0 Cu, 3.0 Fe, 10.0 Al, 5.0 Ni	F, R	621-814	90-118	345-517	50-75	20-15	G-E	30
C63200	82.0 Cu, 4.0 Fe, 9.0 Al, 5.0 Ni	F, R	621-724	90-105	310-365	45-53	25-20	G-E	30
C63600	95.5 Cu, 3.5 Al, 1.0 Si	R, W	414-579	60-84	...	...	64-29	G-E	40
C63800	95.0 Cu, 2.8 Al, 1.8 Si, 0.40 Co	F	565-896	82-130	372-786	54-114	36-4	G-E	...
C64200	91.2 Cu, 7.0 Al	F, R	517-703	75-102	241-469	35-68	32-22	G-E	60
C65100 (low-silicon bronze, B)	98.5 Cu, 1.5 Si	R, W, T	276-655	40-95	103-476	15-69	55-11	G-E	30
C65500 (high-silicon bronze, A)	97.0 Cu, 3.0 Si	F, R, W, T	386-1000	56-145	145-483	21-70	63-3	G-E	30
C66700 (manganese brass)	70.0 Cu, 28.8 Zn, 1.2 Mn	F, W	315-689	45.8-100	83-638	12-92.5	60-2	G-E	30
C67400	58.5 Cu, 36.5 Zn, 1.2 Al, 2.8 Mn, 1.0 Sn	F, R	483-634	70-92	234-379	34-55	28-20	F-E	25
C67500 (manganese bronze, A)	58.5 Cu, 1.4 Fe, 39.0 Zn, 1.0 Sn, 0.1 Mn	R, S	448-579	65-84	207-414	30-60	33-19	F-E	30
C68700 (aluminum brass, arsenical)	77.5 Cu, 20.5 Zn, 2.0 Al, 0.1 As	T	414	60	186	27	55	G-E	30
C68800	73.5 Cu, 22.7 Zn, 3.4 Al, 0.40 Co	F	565-889	82-129	379-786	55-114	36-2	G-E	...
C69000	73.3 Cu, 3.4 Al, 0.6 Ni, 22.7 Zn	F	496-896	72-130	345-807	50-117	40-2	G-E	...
C69400 (silicon red brass)	81.5 Cu, 14.5 Zn, 4.0 Si	R	552-689	80-100	276-393	40-57	25-20	G-E	30
C70400	92.4 Cu, 1.5 Fe, 5.5 Ni, 0.6 Mn	F, T	262-531	38-77	276-524	40-76	46-2	G-E	20
C70600 (copper nickel, 10%)	88.7 Cu, 1.3 Fe, 10.0 Ni	F, T	303-414	44-60	110-393	16-57	42-10	E	20
C71000 (copper nickel, 20%)	79.0 Cu, 21.0 Ni	F, W, T	338-655	49-95	90-586	13-85	40-3	E	20
C71500 (copper nickel, 30%)	70.0 Cu, 30.0 Ni	F, R, T	372-517	54-75	138-483	20-70	45-15	E	20
C71700	67.8 Cu, 0.7 Fe, 31.0 Ni, 0.5 Be	F, R, W	483-1379	70-200	207-1241	30-180	40-4	G-E	20
C72500	88.2 Cu, 9.5 Ni, 2.3 Sn	F, R, W, T	379-827	55-120	152-745	22-108	35-1	E	20
C73500	72.0 Cu, 10.0 Zn, 18.0 Ni	F, R, W, T	345-758	50-110	103-579	15-84	37-1	E	20
C74500 (nickel silver, 65-10)	65.0 Cu, 25.0 Zn, 10.0 Ni	F, W	338-896	49-130	124-524	18-76	50-1	E	20
C75200 (nickel silver, 65-18)	65.0 Cu, 17.0 Zn, 18.0 Ni	F, R, W	386-710	56-103	172-621	25-90	45-3	E	20
C75400 (nickel silver, 65-15)	65.0 Cu, 20.0 Zn, 15.0 Ni	F	365-634	53-92	124-545	18-79	43-2	E	20
C75700 (nickel silver, 65-12)	65.0 Cu, 23.0 Zn, 12.0 Ni	F, W	359-641	52-93	124-545	18-79	48-2	E	20
C76200	59.0 Cu, 29.0 Zn, 12.0 Ni	F, T	393-841	57-122	145-758	21-110	50-1	G-E	...
C77000 (nickel silver, 55-18)	55.0 Cu, 27.0 Zn, 18.0 Ni	F, R, W	414-1000	60-145	186-621	27-90	40-2	E	30
C72200	82.0 Cu, 16.0 Ni, 0.5 Cr, 0.8 Fe, 0.5 Mn	F, T	317-483	46-70	124-455	18-66	46-6	G-E	...
C78200 (leaded nickel silver, 65-8-2)	65.0 Cu, 2.0 Pb, 25.0 Zn, 8.0 Ni	F	365-627	53-91	159-524	23-76	40-3	E	60

(a) F, flat products; R, rod; W, wire; T, tube; P, pipe; S, shapes. (b) Softest to hardest commercial forms. The strength of the standard copper alloys depends on the temper (annealed grain size or degree of cold work) and the section thickness of the mill product. Ranges cover standard tempers for each alloy. (c) E, excellent; G, good; F, fair. (d) Based on 100% for C36000. (e) C10400, 8 oz/ton Ag; C10500, 10 oz/ton; C10700, 25 oz/ton. (f) C11300, 8 oz/ton Ag; C11400, 10 oz/ton; C11500, 16 oz/ton; C11600, 25 oz/ton. (g) C12000, 0.008 P; C12100, 0.008 P and 4 oz/ton Ag. (h) C12700, 8 oz/ton Ag; C12800, 10 oz/ton; C12900, 16 oz/ton; C13000, 25 oz/ton. (i) 8.30 oz/ton Ag. (j) C18200, 0.9 Cr; C18400, 0.8 Cr; C18500, 0.7 Cr. (k) Rod, 61.0 Cu min. Compiled by Copper Development Assn. Inc., New York.

Table A-1. Fabricating characteristics and typical applications of wrought coppers and copper alloys

Alloy number (and name)	Fabricating characteristics and typical applications
C10100 (oxygen-free electronic)	Excellent hot and cold workability; good forgeability. Fabricated by coining, coppersmithing, drawing and upsetting, hot forging and pressing, spinning, swaging, stamping. Uses: busbars, bus conductors, waveguides, hollow conductors, lead-in wires and anodes for vacuum tubes, vacuum seals, transistor components, glass to metal seals, coaxial cables and tubes, klystrons, microwave tubes, rectifiers.
C10200 (oxygen-free copper)	Fabricating characteristics same as C10100. Uses: busbars, waveguides.
C10300 (oxygen-free, extra-low phosphorus)	Fabricating characteristics same as C10100. Uses: busbars, electrical conductors, tubular bus and applications requiring good conductivity and welding or brazing properties.
C10400, C10500, C10700 (oxygen-free, silver-bearing)	Fabricating characteristics same as C10100. Uses: auto gaskets, radiators, busbars, conductivity wire, contacts, radio parts, winding, switches, terminals, commutator segments; chemical process equipment, printing rolls, clad metals, printed circuit foil.
C10800 (oxygen-free, low-phosphorus)	Fabricating characteristics same as C10100. Uses: refrigerators, air conditioners, gas and heater lines, oil burner tubes, plumbing pipe and tube, brewery tubes, condenser and heat exchanger tubes, dairy and distiller tubes, pulp and paper lines, tanks, air, gasoline, hydraulic and oil lines.
C11000 (electrolytic tough pitch copper)	Fabricating characteristics same as C10100. Uses: downspouts, gutters, roofing, gaskets, auto radiators, busbars, nails, printing rolls, rivets, radio parts.
C11100 (electrolytic tough pitch, anneal resistant)	Fabricating characteristics same as C10100. Uses: electrical power transmission where resistance to softening under overloads is desired.
C11300, C11400, C11500, C11600 (silver-bearing tough pitch copper)	Fabricating characteristics same as C10100. Uses: gaskets, radiators, busbars, windings, switches, chemical process equipment, clad metals, printed circuit foil.
C12000, C12100	Fabricating characteristics same as C10100. Uses: busbars, electrical conductors, tubular bus, and applications requiring welding or brazing.
C12200 (phosphorus-deoxidized copper, high residual phosphorus)	Fabricating characteristics same as C10100. Uses: gas and heater lines; oil burner tubing; plumbing pipe and tubing; condenser, evaporator, heat exchanger, dairy, and distiller tubing; steam and water lines; air, gasoline, and hydraulic lines.
C12500, C12700, C12800, C12900, C13000 (fire-refined tough pitch with silver)	Fabricating characteristics same as C10100. Uses: same as C11000.
C14200 (phosphorus-deoxidized, arsenical)	Fabricating characteristics same as C10100. Uses: plates for locomotive fireboxes, staybolts, heat exchanger and condenser tubes.
C19200	Excellent hot and cold workability. Uses: automotive hydraulic brake lines, flexible hose, electrical terminals, fuse clips, gaskets, gift hollow ware, applications requiring resistance to softening and stress corrosion, air conditioning and heat exchanger tubing.
C14300	Fabricating characteristics same as C10100. Uses: anneal resistant electrical applications requiring thermal softening and embrittlement, resistance, lead frames, contacts, terminals, solder-coated and solder-fabricated parts, furnace-brazed assemblies and welded components, cable wrap.
C14310	Same as C14300.
C14500 (phosphorus-deoxidized, tellurium-bearing)	Fabricating characteristics same as C10100. Uses: forgings and screw machine products, and parts requiring high conductivity, extensive machining, corrosion resistance, copper color, or a combination of these; electrical connectors, motor and switch parts, plumbing fittings, soldering coppers, welding torch tips, transistor bases and furnace-brazed articles.
C14700 (sulfur-bearing)	Fabricating characteristics same as C10100. Uses: screw machine products and parts requiring high conductivity, extensive machining, corrosion resistance, copper color, or a combination of these; electrical connectors, motor and switch components, plumbing fittings, cold headed and machined parts, cold forgings, furnace brazed articles, screws, soldering coppers, rivets and welding torch tips.
C15000 (zirconium copper)	Fabricating characteristics same as C10100. Uses: switches, high-temperature circuit breakers; commutators, stud bases for power transmitters, rectifiers, soldering welding tips.
C15500	Fabricating characteristics same as C10100. Uses: high-conductivity light-duty springs, electrical contacts, fittings, clamps, connectors, diaphragms, electronic components, resistance welding electrodes.
C15710	Excellent cold workability. Fabricated by extrusion, drawing, rolling, impacting, heading, swaging, bending, machining, blanking, roll threading. Uses: electrical connectors, light-duty current-carrying springs, inorganic insulated wire, thermocouple wire, lead wire, resistance welding electrodes for aluminum, heat sinks.
C15720	Excellent cold workability. Fabricated by extrusion, drawing, rolling, impacting, heading, swaging, machining, blanking. Uses: relay and switch springs, lead frames, contact supports, heat sinks, circuit breaker parts, rotor bars, resistance welding electrodes and wheels, connectors, high-strength high-temperature parts.
C15735	Excellent cold workability. Fabricated by extrusion, drawing, heading, impacting, machining. Uses: resistance welding electrodes, circuit breakers, feed-through conductors, heat sinks, motor parts, high-strength high-temperature parts.
C15760	Excellent cold workability. Fabricated by extrusion and drawing. Uses: resistance welding electrodes, circuit breakers, electrical connectors, wire feed contact tips, plasma spray nozzles, high-strength high-temperature parts.
C16200 (cadmium copper)	Excellent cold workability; good hot formability. Uses: trolley wire, heating pad, electric-blanket elements, spring contacts, railbands, high-strength transmission lines, connectors, cable wrap, switch gear components and waveguide cavities.
C16500	Fabricating characteristics same as C16200. Uses: electrical springs and contacts, trolley wire, clips, flat cable, resistance welding electrodes.
C17000 (beryllium copper)	Fabricating characteristics same as C16200. Commonly fabricated by blanking, forming and bending, turning, drilling, tapping. Uses: bellows, bourdon tubing, diaphragms, fuse clips, fasteners, lock-washers, springs, switch parts, roll pins, valves, welding equipment.
C17200 (beryllium copper)	Similar to C17000, particularly for its nonsparking characteristics.
C17300 (beryllium copper)	Combines superior machinability with good fabricating characteristics of C17200.

(continued)

Table A-1. (continued)

Alloy number (and name)	Fabricating characteristics and typical applications
C17500 (copper-cobalt-beryllium alloy)	Fabricating characteristics same as C16200. Uses: fuse clips, fasteners, springs, switch and relay parts, electrical conductors, welding equipment.
C18200, C18400, C18500 (chromium copper)	Excellent cold workability, good hot workability. Uses: resistance welding electrodes, seam welding wheels, switch gear, electrode holder jaws, cable connectors, current carrying arms and shafts, circuit breaker parts, molds, spot welding tips, flash welding electrodes, electrical and thermal conductors requiring strength, switch contacts.
C18700 (leaded copper)	Good cold workability; poor hot formability. Uses: connectors, motor and switch parts, screw machine parts requiring high conductivity.
C18900	Fabricating characteristics same as C10100. Uses: welding rod and wire for inert gas tungsten arc and metal arc welding and oxyacetylene welding of copper.
C19000 (copper-nickel-phosphorus alloy)	Fabricating characteristics same as C10100. Uses: springs, clips, electrical connectors, power tube and electron tube components, high-strength electrical conductors, bolts, nails, screws, cotter pins, and parts requiring some combination of high-strength, high-electrical or thermal conductivity, high resistance to fatigue and creep, and good workability.
C19100 (copper-nickel-phosphorus-tellurium alloy)	Good hot and cold workability. Uses: forgings and screw machine parts requiring high strength, hardenability, extensive machining, corrosion resistance, copper color, good conductivity, or a combination of these; bolts, bushings, electrical connectors, gears, marine hardware, nuts, pinions, tie rods, turnbuckle barrels, welding torch tips.
C19400	Excellent hot and cold workability. Uses: circuit breaker components, contact springs, electrical clamps, electrical springs, electrical terminals, flexible hose, fuse clips, gaskets, gift hollow ware, plug contacts, rivets, and welded condenser tubes.
C19500	Excellent hot and cold workability. Uses: electrical springs, sockets, terminals, connectors, clips and other current carrying parts having strength.
C21000 (gilding, 95%)	Excellent cold workability, good hot workability for blanking, coining, drawing, piercing and punching, shearing, spinning, squeezing and swaging, stamping. Uses: coins, medals, bullet jackets, fuse caps, primers, plaques, jewelry base for gold plate.
C22000 (commercial bronze, 90%)	Fabricating characteristics same as C21000, plus heading and upsetting, roll threading and knurling, hot forging and pressing. Uses: etching bronze, grillwork, screen cloth, weatherstripping, lipstick cases, compacts, marine hardware, screws, rivets.
C22600 (jewelry bronze, 87.5%)	Fabricating characteristics same as C21000, plus heading and upsetting, roll threading and knurling. Uses: angles, channels, chain, fasteners, costume jewelry, lipstick cases, compacts, base for gold plate.
C23000 (red brass, 85%)	Excellent cold workability, good hot formability. Uses: weatherstripping, conduit, sockets, fasteners, fire extinguishers, condenser and heat exchanger tubing, plumbing pipe, radiator cores.
C24000 (low brass, 80%)	Excellent cold workability. Fabricating characteristics same as C23000. Uses: battery caps, bellows, musical instruments, clock dials, pump lines, flexible hose.
C26000 (cartridge brass, 70%)	Excellent cold workability. Fabricating characteristics same as C23000, except for coining, roll threading, and knurling. Uses: radiator cores and tanks, flashlight shells, lamp fixtures, fasteners, locks, hinges, ammunition components, plumbing accessories, pins, rivets.
C26800, C27000 (yellow brass)	Excellent cold workability. Fabricating characteristics same as C23000. Uses: same as C26000 except not used for ammunition.
C28000 (Muntz metal)	Excellent hot formability and forgeability for blanking, forming and bending, hot forging and pressing, hot heading and upsetting, shearing. Uses: architectural, large nuts and bolts, brazing rod, condenser plates, heat exchanger and condenser tubing, hot forgings.
C31400 (leaded commercial bronze)	Excellent machinability. Uses: screws, machine parts, pickling crates.
C31600 (leaded commercial bronze, nickel-bearing)	Good cold workability; poor hot formability. Uses: electrical connectors, fasteners, hardware, nuts, screws, screw machine parts.
C33000 (low-leaded brass tube)	Combines good machinability and excellent cold workability. Fabricated by forming and bending, machining, piercing and punching. Uses: pump and power cylinders and liners, ammunition primers, plumbing accessories.
C33200 (high-leaded brass tube)	Excellent machinability. Fabricated by piercing, punching, and machining. Uses: general-purpose screw machine parts.
C33500 (low-leaded brass)	Similar to C33200. Commonly fabricated by blanking, drawing, machining, piercing and punching, stamping. Uses: butts, hinges, watch backs.
C34000 (medium-leaded brass)	Similar to C33200. Fabricated by blanking, heading and upsetting, machining, piercing and punching, roll threading and knurling, stamping. Uses: butts, gears, nuts, rivets, screws, dials, engravings, instrument plates.
C34200 (high-leaded brass)	Combines excellent machinability with moderate cold workability. Uses: clock plates and nuts, clock and watch backs, gears, wheels and channel plate.
C34900	Good cold workability, fair hot workability for bending and forming, heading and upsetting, machining, roll threading and knurling. Uses: building hardware, rivets and nuts, plumbing goods, and parts requiring moderate cold working combined with some machining.
C35000 (medium-leaded brass)	Fair cold workability; poor hot formability. Uses: bearing cages, book dies, clock plates, engraving plates, gears, hinges, hose couplings, keys, lock parts, lock tumblers, meter parts, nuts, sink strainers, strike plates, templates, type characters, washers, wear plates.
C35300 (high-leaded brass)	Similar to C34200.
C35600 (extra-high-leaded brass)	Excellent machinability. Fabricated by blanking, machining, piercing and punching, stamping. Uses: same as C34200 and C35300.
C36000 (free-cutting brass)	Excellent machinability. Fabricated by machining, roll threading and knurling. Uses: gears, pinions, automatic high-speed screw machine parts.
C36500 to C36800 (leaded Muntz metal)	Combines good machinability with excellent hot formability. Uses: condenser tube plates.
C37000 (free-cutting Muntz metal)	Fabricating characteristics similar to C36500 to C36800. Uses: automatic screw machine parts.

(continued)

Table A-1. (continued)

Alloy number (and name)	Fabricating characteristics and typical applications
C37700 Forging brass	Excellent hot workability. Fabricated by heading and upsetting, hot forging and pressing, hot heading and upsetting, machining. Uses: forgings and pressings of all kinds.
C38500 (architectural bronze)	Excellent machinability and hot workability. Fabricated by hot forging and pressing, forming, bending and machining. Uses: architectural extrusions, store fronts, thresholds, trim, butts, hinges, lock bodies and forgings.
C40500	Excellent cold workability. Fabricated by blanking, forming and drawing. Uses: meter clips, terminals, fuse clips, contact and relay springs, washers.
C40800	Excellent cold workability. Fabricated by blanking, stamping and shearing. Uses: electrical connectors.
C41100	Excellent cold workability, good hot formability. Fabricated by blanking, forming and drawing. Uses: bushings, bearing sleeves, thrust washers, terminals, connectors, flexible metal hose, electrical conductors.
C41300	Excellent cold workability; good hot formability. Uses: plater bar for jewelry products, flat springs for electrical switchgear.
C41500	Excellent cold workability. Fabricated by blanking, drawing, bending, forming, shearing and stamping. Uses: spring applications for electrical switches.
C42200	Excellent cold workability; good hot formability. Fabricated by blanking, piercing, forming and drawing. Uses: sash chains, fuse clips, terminals, spring washers, contact springs, electrical connectors.
C42500	Excellent cold workability. Fabricated by blanking, piercing, forming and drawing. Uses: electrical switches, springs, terminals, connectors, fuse clips, pen clips, weather stripping.
C43000	Excellent cold workability; good hot formability. Fabricated by blanking, coining, drawing, forming, bending, heading, and upsetting. Uses: same as C42500.
C43400	Excellent cold workability. Fabricated by blanking, drawing, bonding, forming, stamping and shearing. Uses: electrical switch parts, blades, relay springs, contacts.
C43500	Excellent cold workability for fabrication by forming and bending. Uses: bourdon tubing and musical instruments.
C44300, C44400, C44500 (inhibited admiralty)	Excellent cold workability for forming and bending. Uses: condenser, evaporator and heat exchanger tubing, condenser tubing plates, distiller tubing, ferrules.
C46400 to C46700 (naval brass)	Excellent hot workability and hot forgeability. Fabricated by blanking, drawing, bending, heading and upsetting, hot forging, pressing. Uses: aircraft turnbuckle barrels, balls, bolts, marine hardware, nuts, propeller shafts, rivets, valve stems, condenser plates, welding rod.
C48200 (naval brass, medium-leaded)	Good hot workability for hot forging, pressing, and machining operations. Uses: marine hardware, screw machine products, valve stems.
C48500 (leaded naval brass)	Combines excellent hot forgeability and machinability. Fabricated by hot forging and pressing, machining. Uses: marine hardware, screw machine parts, valve stems.
C50500 (phosphor bronze, 1.25% E)	Excellent cold workability; good hot formability. Fabricated by blanking, bending, heading and upsetting, shearing and swaging. Uses: electrical contacts, flexible hose, pole-line hardware.
C51000 (phosphor bronze, 5% A)	Excellent cold workability. Fabricated by blanking, drawing, bending, heading and upsetting, roll threading and knurling, shearing, stamping. Uses: bellows, bourdon tubing, clutch discs, cotter pins, diaphragms, fasteners, lock washers, wire brushes, chemical hardware, textile machinery, welding rod.
C51100	Excellent cold workability. Uses: bridge bearing plates, locator bars, fuse clips, sleeve bushings, springs, switch parts, truss wire, wire brushes, chemical hardware, perforated sheets, textile machinery, welding rod.
C52100 (phosphor bronze, 8% C)	Good cold workability for blanking, drawing, forming and bending, shearing, stamping. Uses: generally for more severe service conditions than C51000.
C52400 (phosphor bronze, 10% D)	Good cold workability for blanking, forming and bending, shearing. Uses: heavy bars and plates for severe compression, bridge and expansion plates and fittings, articles requiring good spring qualities, resiliency, fatigue resistance, good wear and corrosion resistance.
C54400 (free-cutting phosphor bronze)	Excellent machinability; good cold workability. Fabricated by blanking, drawing, bending, machining, shearing, stamping. Uses: bearings, bushings, gears, pinions, shafts, thrust washers, valve parts.
C60800 (aluminum bronze, 5%)	Good cold workability; fair hot formability. Uses: condenser, evaporator and heat exchanger tubes, distiller tubes, ferrules.
C61000	Good hot and cold workability. Uses: bolts, pump parts, shafts, tie rods, overlay on steel for wearing surfaces.
C61300	Good hot and cold formability. Uses: nuts, bolts, stringers and threaded members, corrosion-resistant vessels and tanks, structural components, machine parts, condenser tube and piping systems, marine protective sheathing and fastening, munitions mixing troughs and blending chambers.
C61400 (aluminum bronze, D)	Similar to C61300.
C61500	Good hot and cold workability. Fabricating characteristics similar to C52100. Uses: hardware, decorative metal trim, interior furnishings and other articles requiring high tarnish resistance.
C61800	Fabricated by hot forging and hot pressing. Uses: bushings, bearings, corrosion-resistant applications, welding rods.
C61900	Excellent hot formability for fabricating by blanking, forming, bending, shearing, and stamping. Uses: springs, contacts, and switch components.
C62300	Good hot and cold formability. Fabricated by bending, hot forging, hot pressing, forming, and welding. Uses: bearings, bushings, valve guides, gears, valve seats, nuts, bolts, pump rods, worm gears, and cams.
C62400	Excellent hot formability for fabrication by hot forging and hot bending. Uses: bushings, gears, cams, wear strips, nuts, drift pins, tie rods.
C62500	Excellent hot formability for fabrication by hot forging and machining. Uses: guide bushings, wear strips, cams, dies, forming rolls.
C63000	Good hot formability. Fabricated by hot forming and forging. Uses: nuts, bolts, valve seats, plunger tips, marine shafts, valve guides, aircraft parts, pump shafts, structural members.
C63200	Good hot formability. Fabricated by hot forming and welding. Uses: nuts, bolts, structural pump parts, shafting requiring corrosion resistance.

(continued)

Table A-1. (continued)

Alloy number (and name)	Fabricating characteristics and typical applications
C63600	Excellent cold workability; fair hot formability. Fabricated by cold heading. Uses: components for pole line hardware, cold-headed nuts for wire and cable connectors, bolts and screw products.
C63800	Excellent cold workability and hot formability. Uses: springs, switch parts, contacts, relay springs, glass sealing and porcelain enameling.
C64200	Excellent hot formability. Fabricated by hot forming, forging, machining. Uses: valve stems, gears, marine hardware, pole-line hardware, bolts, nuts, valve bodies and components.
C65100 (low-silicon bronze, B)	Excellent hot and cold workability. Fabricated by forming and bending, heading and upsetting, hot forging and pressing, roll threading and knurling, squeezing and swaging. Uses: hydraulic pressure lines, anchor screws, bolts, cable clamps, cap screws, machine screws, marine hardware, nuts, pole-line hardware, rivets, U-bolts, electrical conduits, heat exchanger tubing, welding rod.
C65500 (high-silicon bronze, A)	Excellent hot and cold workability. Fabricated by blanking, drawing, forming and bending, heading and upsetting, hot forging and pressing, roll threading and knurling, shearing, squeezing and swaging. Uses: similar to C65100 including propeller shafts.
C66700 (manganese brass)	Excellent cold formability. Fabricated by blanking, bending, forming, stamping, welding. Uses: brass products resistance welded by spot, seam, and butt welding.
C67400	Excellent hot formability. Fabricated by hot forging and pressing, machining. Uses: bushings, gears, connecting rods, shafts, wear plates.
C67500 (manganese bronze, A)	Excellent hot workability. Fabricated by hot forging and pressing, hot heading and upsetting. Uses: clutch discs, pump rods, shafting, balls, valve stems and bodies.
C68700 (aluminum brass, arsenical)	Excellent cold workability for forming and bending. Uses: condenser, evaporator and heat exchanger tubing, condenser tubing plates, distiller tubing, ferrules.
C68800	Excellent hot and cold formability. Fabricated by blanking, drawing, forming and bending, shearing and stamping. Uses: springs, switches, contacts, relays, drawn parts.
C69000	Fabricating characteristics same as C68800. Uses: wiring devices, relays, switches, springs, high-strength shells.
C69400 (silicon red brass)	Excellent hot formability for fabrication by forging, screw machine operations. Uses: valve stems where corrosion resistance and high strength are critical.
C70400	Excellent cold workability; good hot formability. Fabricated by forming, bending and welding. Uses: condensers, evaporators, heat exchangers, ferrules, salt water piping, lithium bromide absorption tubing, shipboard condenser intake systems.
C70600 (copper nickel, 10%)	Good hot and cold workability. Fabricated by forming and bending, welding. Uses: condensers, condenser plates, distiller tubing, evaporator and heat exchanger tubing, ferrules, salt water piping.
C71000 (copper nickel, 20%)	Good hot and cold formability. Fabricated by blanking, forming and bending, welding. Uses: communication relays, condensers, condenser plates, electrical springs, evaporator and heat exchanger tubes, ferrules, resistors.
C71500 (copper nickel, 30%)	Similar to C70600.
C71700	Good hot and cold formability. Uses: high-strength constructional parts for sea water corrosion resistance, hydrophone cases, mooring cable wire, springs, retainer rings, bolts, screws, pins for ocean telephone cable applications.
C72500	Excellent cold and hot formability. Fabricated by blanking, brazing, coining, drawing, etching, forming and bending, heading and upsetting, roll threading and knurling, shearing, spinning, squeezing, stamping and swaging. Uses: relay and switch springs, connectors, brazing alloy, lead frames, control and sensing bellows.
C73500	Fabricating characteristics same as C74500. Uses: holloware, medallions, jewelry, base for silver plate, cosmetic cases, musical instruments, name plates, contacts.
C74500 (nickel silver, 65-10)	Excellent cold workability. Fabricated by blanking, drawing, etching, forming and bending, heading and upsetting, roll threading and knurling, shearing, spinning, squeezing and swaging. Uses: rivets, screws, slide fasteners, optical parts, etching stock, hollow ware, nameplates, platers' bars.
C75200 (nickel silver, 65-18)	Fabricating characteristics similar to C74500. Uses: rivets, screws, table flatware, truss wire, zippers, bows, camera parts, core bars, temples, base for silver plate, costume jewelry, etching stock, hollow ware, nameplates, radio dials.
C75400 (nickel silver, 65-15)	Fabricating characteristics similar to C74500. Uses: camera parts, optical equipment, etching stock, jewelry.
C75700 (nickel silver, 65-12)	Fabricating characteristics similar to C74500. Uses: slide fasteners, camera parts, optical parts, etching stock, nameplates.
C76200	Fabricating characteristics same as C77000. Uses: electrical terminals, contact springs, release brackets, ornamental bits and spurs, optical parts, surgical instruments, electrical contacts.
C77000 (nickel silver, 55-18)	Good cold workability. Fabricated by blanking, forming and bending, and shearing. Uses: optical goods, springs and resistance wire.
C72200	Good hot and cold workability. Fabricated by forming, bending and welding. Uses: condenser and heat exchanger tubing, salt water piping.
C78200 (leaded nickel silver, 65-8-2)	Good cold formability. Fabricated by blanking, milling and drilling. Uses: key blanks, watch plates, watch parts.

Table B. Properties and applications of cast coppers and copper alloys

UNS designation(a)	Nominal composition, %(a)	Tensile strength		Yield strength		Elongation in 2 in., %	Hardness			Machinability rating(c)	Casting types(d)	Typical applications
		MPa	ksi	MPa	ksi		Rockwell	Brinell 500 kg	Brinell 3 000 kg			
C80100	99.95 Cu + Ag min, 0.05 others max	172	25	62	9	40	...	44	...	10	C, T, I M, P, S	Electrical and thermal conductors; corrosion and oxidation resistant applications.
C80300	99.95 Cu + Ag min, 0.034 Ag min, 0.05 others max	172	25	62	9	40	...	44	...	10	C, T, I M, P, S	Electrical and thermal conductors; corrosion and oxidation resistant applications.
C80500	99.75 Cu + Ag min, 0.034 Ag min, 0.02 B max, 0.23 others max	172	25	62	9	40	...	44	...	10	C, T, I M, P, S	Electrical and thermal conductors; corrosion and oxidation resistant applications.
C80700	99.75 Cu + Ag min, 0.02 B max, 0.23 others max	172	25	62	9	40	...	44	...	10	C, T, I M, P, S	Electrical and thermal conductors; corrosion and oxidation resistant applications.
C80900	99.70 Cu + Ag min, 0.034 Ag min, 0.30 others max	172	25	62	9	40	...	44	...	10	C, T, I M, P, S	Electrical and thermal conductors; corrosion and oxidation resistant applications.
C81100	99.70 Cu + Ag min, 0.30 others max	172	25	62	9	40	...	44	...	10	C, T, I M, P, S	Electrical and thermal conductors; corrosion and oxidation resistant applications.
High-copper alloys												
C81300	98.5 Cu min, 0.06 Be, 0.80 Co, 0.40 others max	(365)	(53)	(248)	(36)	(11)	...	(39)	...	20	C, T, I, M, P, S	Higher hardness electrical and thermal conductors.
C81400	98.5 Cu min, 0.06 Be, 0.80 Cr, 0.40 others max	(365)	(53)	(248)	(36)	(11)	(B 69)	...	...	20	C, T, I, M, P, S	Higher hardness electrical and thermal conductors.
C81500	98.0 Cu min, 1.0 Cr, 0.50 others max	(352)	(51)	(276)	(40)	(17)	...	(105)	...	20	C, T, I, M, P, S	Electrical and/or thermal conductors used as structural members where strength and hardness greater than that of C80100-81100 are required.
C81700	94.25 Cu min, 1.0 Ag, 0.4 Be, 0.9 Co, 0.9 Ni	(634)	(92)	(469)	(68)	(8)	...	...	(217)	30	C, T, I, M, P, S	Electrical and/or thermal conductors used as structural members where strength and hardness greater than that of C80100-81100 are required. Also used in place of C81500 where electrical and/or thermal conductivities can be sacrificed for hardness and strength.
C81800	95.6 Cu min, 1.0 Ag, 0.4 Be, 1.6 Co	345 (703)	50 (102)	172 (517)	25 (75)	20 (8)	B 55 (B 96)	...	...	20	C, T, I, M, P, S	Resistance welding electrodes, dies.
C82000	96.8 Cu, 0.6 Be, 2.6 Co	345 (689)	50 (100)	138 (517)	20 (75)	20 (8)	B 55 (B 95)	...	(195)	20	C, T, I, M, P, S	Current carrying parts, contact and switch blades, bushings and bearings, soldering iron and resistance welding tips.
C82100	97.7 Cu, 0.5 Be, 0.9 Co, 0.9 Ni	(634)	(92)	(469)	(68)	(8)	...	...	(217)	30	C, T, I, M, P, S	Electrical and/or thermal conductors used as structural members where strength and hardness greater than that of C80100-81100 are required. Also used in place of C81500 where electrical and/or thermal conductivities can be sacrificed for hardness and strength.

(continued)

(a) Nominal composition, unless otherwise noted. For seldom-used alloys, only compositions are available. (b) Values for C82700, 84200, 96200, 96300 are minimum, not typical. As-cast values are for sand casting except C93900, continuous cast; and C85800, 87800, 87900, die cast. Heat treated values, in parentheses, indicate that the alloy responds to heat treatment. If heat treated values are not shown, the copper or copper alloy does not respond. (c) Free cutting brass = 100. (d) C, centrifugal; T, continuous; D, die; I, investment; M, permanent mold; P, plaster; S, sand. (e) As-heat treated value for C94700, 20; for C94800, 40. (Note: C82000, 82400, 82500, 82600, 82800 are also pressure cast.) **Source**: Copper Development Assn. Inc., New York. (Revised March 1980)

Table B. (continued)

UNS designation(a)	Nominal composition, %(a)	Tensile strength MPa	ksi	Yield strength MPa	ksi	Elongation in 2 in., %	Hardness Rockwell	Brinell 500 kg	3 000 kg	Machinability rating(c)	Casting types(d)	Typical applications
High-copper alloys (continued)												
C82200	96.5 Cu min, 0.6 Be, 1.5 Ni	393 (655)	57 (95)	207 (517)	30 (75)	20 (8)	B 60 (B 96)	...	...	20	C, T, I, M, P, S	Clutch rings, brake drums, seam welder electrodes, projection welding dies, spot welding tips, beam welder shapes, bushings, water-cooled holders.
C82400	96.4 Cu min, 1.70 Be, 0.25 Co	496 (1034)	72 (150)	255 (965)	37 (140)	20 (1)	B 78 (C 38)	...	...	20	C, I, M, P, S	Safety tools, molds for plastic parts, cams, bushings, bearings, valves, pump parts, gears.
C82500	97.2 Cu, 2.0 Be, 0.5 Co, 0.25 Si	552 (1103)	80 (160)	310	45	20 (1)	B 82 (C 40)	...	...	20	C, I, M, P, S	Safety tools, molds for plastic parts, cams, bushings, bearings, valves, pump parts.
C82600	95.2 Cu min, 2.3 Be, 0.5 Co, 0.25 Si	565 (1138)	82 (165)	324 (1069)	47 (155)	20 (1)	B 83 (C 43)	...	...	20	C, I, M, P, S	Bearings and molds for plastic parts.
C82700	96.3 Cu, 2.45 Be, 1.25 Ni	(1069)	155	(896)	(130)	(0)	(C 39)	...	...	20	C, I, M, P, S	Bearings and molds for plastic parts.
C82800	96.6 Cu, 2.6 Be, 0.5 Co, 0.25 Si	669 (1138)	97 (165)	379 (1000)	55 (145)	20 (1)	B 85 (C 45)	...	...	10	C, I, M, P, S	Molds for plastic parts, cams, bushings, bearings, valves, pump parts, sleeves.
Red brasses and leaded red brasses												
C83300	93 Cu, 1.5 Sn, 1.5 Pb, 4 Zn	221	32	69	10	35	...	35	...	35	S	Terminal ends for electrical cables.
C83400	90 Cu, 10 Zn	241	35	69	10	30	F 50	...	...	60	C, S	Moderate strength, moderate conductivity castings; rotating bands.
C83600	85 Cu, 5 Sn, 5 Pb, 5 Zn	255	37	117	17	30	...	60	...	84	C, T, I, S	Valves, flanges, pipe fittings, plumbing goods, pump castings, water pump impellers and housings, ornamental fixtures, small gears.
C83800	83 Cu, 4 Sn, 6 Pb, 7 Zn	241	35	110	16	25	...	60	...	90	C, T, S	Low-pressure valves and fittings, plumbing supplies and fittings, general hardware, air-gas-water fittings, pump components, railroad catenary fittings.
Semi-red brasses and leaded semi-red brasses												
C84200	80 Cu, 5 Sn, 2.5 Pb, 12.5 Zn	193	28	103	15	27	...	60	...	80	C, T, S	Pipe fittings, elbows, T's, couplings, bushings, locknuts, plugs, unions.
C84400	81 Cu, 3 Sn, 7 Pb, 9 Zn	234	34	103	15	26	...	55	...	90	C, T, S	General hardware, ornamental castings, plumbing supplies and fixtures, low-pressure valves and fittings.
C84500	78 Cu, 3 Sn, 7 Pb, 12 Zn	241	35	97	14	28	...	55	...	90	C, T, S	Plumbing fixtures, cocks, faucets, stops, waste, air and gas fittings, low-pressure valve fittings.
C84800	76 Cu, 3 Sn, 6 Pb, 15 Zn	248	36	97	14	30	...	55	...	90	C, S	Plumbing fixtures, cocks, faucets, stops, waste, air and gas fittings, general hardware, and low-pressure valve fittings.
Yellow brasses and leaded yellow brasses												
C85200	72 Cu, 1 Sn, 3 Pb, 24 Zn	262	38	90	13	35	...	45	...	80	C, T	Plumbing fittings and fixtures, ferrules, valves, hardware, ornamental brass, chandeliers, and irons.
C85400	67 Cu, 1 Sn, 3 Pb, 29 Zn	234	34	83	12	35	...	50	...	80	C, T, M, P, S	General purpose yellow casting alloy not subject to high internal pressure. Furniture hardware, ornamental castings, radiator fittings, ship trimmings, cocks, battery clamps, valves and fittings.

Table B. (continued)

UNS designation(a)	Nominal composition, %(a)	Typical mechanical properties, as cast (heat treated)(b)								Machin- ability rating(c)	Casting types(d)	Typical applications
		Tensile strength		Yield strength		Elon- gation in 2 in., %	Hardness					
		MPa	ksi	MPa	ksi		Rockwell	Brinell 500 kg	3 000 kg			

Yellow brasses and leaded yellow brasses (continued)

UNS designation(a)	Nominal composition, %(a)	MPa	ksi	MPa	ksi	Elong %	Rockwell	500 kg	3 000 kg	Mach rating	Casting types	Typical applications
C85500	61 Cu, 0.8 Al, bal Zn	414	60	159	23	40	B 55	85	...	80	C, S	Ornamental castings.
C85700	63 Cu, 1 Sn, 1 Pb, 34.7 Zn, 0.3 Al	345	50	124	18	40	...	75	...	80	C, M, P, S	Bushings, hardware fittings, ornamental castings.
C85800	58 Cu, 1 Sn, 1 Pb, 40 Zn	379	55	207	30	15	B 55	...	...	80	D	General purpose die casting alloy having moderate strength.
Manganese and leaded manganese bronze alloys												
C86100	67 Cu, 21 Zn, 3 Fe, 5 Al, 4 Mn	655	95	345	50	20	...	...	180	30	C, I, P, S	Marine castings, gears, gun mounts, bushings and bearings, marine racing propellers.
C86200	64 Cu, 26 Zn, 3 Fe, 4 Al, 3 Mn	655	95	331	48	20	...	...	180	30	C, T, D, I, P, S	Marine castings, gears, gun mounts, bushings and bearings.
C86300	63 Cu, 25 Zn, 3 Fe, 6 Al, 3 Mn	793	115	572	83	15	...	...	225	8	C, I, P, S	Extra-heavy duty, high-strength alloy. Large valve stems, gears, cams, slow-speed heavy-load bearings, screwdown nuts, hydraulic cylinder parts.
C86400	59 Cu, 1 Pb, 40 Zn	448	65	172	25	20	...	90	105	65	C, D, M, P, S	Free-machining manganese bronze. Valve stems, marine fittings, lever arms, brackets, light-duty gears.
C86500	58 Cu, 0.5 Sn, 39.5 Zn, 1 Fe, 1 Al	490	71	193	28	30	...	100	130	26	C, I, P, S	Machinery parts requiring strength and toughness, lever arms, valve stems, gears.
C86700	58 Cu, 1 Pb, 41 Zn	586	85	290	42	20	B 80	...	155	55	C, S	High strength, free-machining manganese bronze. Valve stems.
C86800	55 Cu, 37 Zn, 3 Ni, 2 Fe, 3 Mn	565	82	262	38	22	...	...	80	30	S	Marine fittings, marine propellers.
Silicon bronzes and silicon brasses												
C87200	89 Cu min, 4 Si	379	55	172	25	30	...	85	...	40	C, I, M, P, S	Bearings, bells, impellers, pump and valve components, marine fittings, corrosion-resistant castings.
C87400	83 Cu, 14 Zn, 3 Si	379	55	165	24	30	...	70	100	50	C, D, I, M, P, S	Bearings, gears, impellers, rocker arms, valve stems, clamps.
C87500	82 Cu, 14 Zn, 4 Si	462	67	207	30	21	...	115	134	50	C, D, I, M, P, S	Bearings, gears, impellers, rocker arms, valve stems, small boat propellers.
C87600	90 Cu, 5.5 Zn, 4.5 Si	455	66	221	32	20	B 76	110	135	40	S	Valve stems.
C87800	82 Cu, 14 Zn, 4 Si	586	85	345	50	25	B 85	...	...	40	D	High-strength, thin-wall die castings; brush holders, lever arms, brackets, clamps, hexagonal nuts.
C87900	65 Cu, 34 Zn, 1 Si	483	70	241	35	25	B 70	...	...	80	D	General purpose die casting alloy having moderate strength.
Tin bronzes												
C90200	93 Cu, 7 Sn	262	38	110	16	30	...	70	...	20	C, S	Bearings and bushings.
C90300	88 Cu, 8 Sn, 4 Zn	310	45	145	21	30	...	70	...	30	C, T, I, P, S	Bearings, bushings, pump impellers, piston rings, valve components, seal rings, steam fittings, gears.
C90500	88 Cu, 10 Sn, 2 Zn	310	45	152	22	25	...	75	...	30	C, T, I, S	Bearings, bushings, pump impellers, piston rings, valve components, steam fittings, gears.
C90700	89 Cu, 11 Sn	303 (379)	44 (55)	152 (207)	22 (30)	20 (16)	...	80 (102)	...	20	C, T, I M, S	Gears, bearings, bushings.
C90800	87 Cu, 12 Sn											
C90900	87 Cu, 13 Sn	276	40	138	20	15	...	90	...	20	C, S	Bearings and bushings.

(continued)

(a) Nominal composition, unless otherwise noted. For seldom-used alloys, only compositions are available. (b) Values for C82700, 84200, 96200, 96300 are minimum, not typical. As-cast values are for sand casting except C93900, continuous cast; and C85800, 87800, 87900, die cast. Heat treated values, in parentheses, indicate that the alloy responds to heat treatment. If heat treated values are not shown, the copper or copper alloy does not respond. (c) Free cutting brass = 100. (d) C, centrifugal; T, continuous; D, die; I, investment; M, permanent mold; P, plaster; S, sand. (e) As-heat treated value for C94700, 20; for C94800, 40. (Note: C82000, 82400, 82500, 82600, 82800 are also pressure cast.) **Source:** Copper Development Assn. Inc., New York. (Revised March 1980)

Table B. (continued)

UNS designation(a)	Nominal composition, %(a)	Typical mechanical properties, as cast (heat treated)(b)									Machin-ability rating(c)	Casting types(d)	Typical applications
		Tensile strength		Yield strength		Elon-gation in 2 in., %	Hardness						
		MPa	ksi	MPa	ksi		Rockwell	Brinell 500 kg	3 000 kg				

Tin bronzes (continued)

C91000	85 Cu, 14 Sn, 1 Zn	221	32	172	25	2	...	105	...	20	C, T, I, S	Piston rings and bearings.
C91100	84 Cu, 16 Sn	241	35	172	25	2	...	...	135	10	S	Piston rings, bearings, bushings, bridge plates.
C91300	81 Cu, 19 Sn	241	35	207	30	0.5	...	...	170	10	C, T, M, S	Piston rings, bearings, bushings, bridge plates, bells.
C91600	88 Cu, 10.5 Sn, 1.5 Ni	303 (414)	44 (60)	152 (221)	22 (32)	16 (16)	...	85 (106)	...	20	C, T, M, S	Gears.
C91700	86.5 Cu, 12 Sn, 1.5 Ni	303 (414)	44 (60)	152 (221)	22 (32)	16 (16)	...	85 (106)	...	20	C, T, I, M, S	Gears.

Leaded tin bronzes

C92200	88 Cu, 6 Sn, 1.5 Pb, 4.5 Zn	276	40	138	20	30	...	65	...	42	C, T, I, M, P, S	Valves, fittings, and pressure-containing parts for use up to 550 °F.
C92300	87 Cu, 8 Sn, 4 Zn	276	40	138	20	25	...	70	...	42	C, T, S	Valves, pipe fittings, and high-pressure steam castings. Superior machinability to C90300.
C92400	88 Cu, 10 Sn, 2 Pb, 2 Zn											
C92500	87 Cu, 11 Sn, 1 Pb, 1 Ni	303	44	138	20	20	...	80	...	30	C, T, M, S	Gears, automotive synchronizer rings.
C92600	87 Cu, 10 Sn, 1 Pb, 2 Zn	303	44	138	20	30	F 78	70	...	40	C, T, S	Bearings, bushings, pump impellers, piston rings, valve components, steam fittings, and gears. Superior machinability to C90500.
C92700	88 Cu, 10 Sn, 2 Pb	290	42	145	21	20	...	77	...	45	C, T, S	Bearings, bushings, pump impellers, piston rings, valve components, steam fittings, and gears. Superior machinability to C90500.
C92800	79 Cu, 16 Sn, 5 Pb	276	40	207	30	1	B 80	...	...	70	C, S	Piston rings.
C92900	84 Cu, 10 Sn, 2.5 Pb, 3.5 Ni	324 (324)	47 (47)	179 (179)	26 (26)	20 (20)	...	80 (80)	...	40	C, T, M, S	Gears, wear plates, guides, cams, parts requiring machinability superior to that of C91600 or 91700.

High-leaded tin bronzes

C93200	83 Cu, 7 Sn, 7 Pb, 3 Zn	241	35	124	18	20	...	65	...	70	C, T, M, S	General-utility bearings and bushings.
C93400	84 Cu, 8 Sn, 8 Pb	221	32	110	16	20	...	60	...	70	C, T, S	Bearings and bushings.
C93500	85 Cu, 5 Sn, 9 Pb	221	32	110	16	20	...	60	...	70	C, T, S	Small bearings and bushings, bronze backing for babbit-lined automotive bearings.
C93700	80 Cu, 10 Sn, 10 Pb	241	35	124	18	20	...	60	...	80	C, T, M, S	Bearings for high speed and heavy pressures, pumps, impellers, corrosion-resistant applications, pressure tight castings.
C93800	78 Cu, 7 Sn, 15 Pb	207	30	110	16	18	...	55	...	80	C, T, M, S	Bearings for general service and moderate pressures, pump impellers and bodies for use in acid mine water.
C93900	79 Cu, 6 Sn, 15 Pb	221	32	152	22	7	...	63	...	80	T	Continuous castings only. Bearings for general service, pump bodies and impellers for mine waters.
C94000	70.5 Cu, 13.0 Sn, 15.0 Pb, 0.50 Zn, 0.75 Ni, 0.25 Fe, 0.05 P, 0.35 Sb(h)											
C94100	70.0 Cu, 5.5 Sn, 18.5 Pb, 3.0 Zn, 1.0 others max											
C94300	70 Cu, 5 Sn, 25 Pb	186	27	90	13	15	...	48	...	80	C, S	High-speed bearings for light loads.
C94400	81 Cu, 8 Sn, 11 Pb	221	32	110	16	18	...	55	...	80	C, T, S	General-utility alloy for bushings and bearings.

(continued)

Table B. (continued)

UNS designa-tion(a)	Nominal composition, %(a)	Tensile strength MPa	ksi	Yield strength MPa	ksi	Elongation in 2 in., %	Hardness Rockwell	Brinell 500 kg	Brinell 3 000 kg	Machin-ability rating(c)	Casting types(d)	Typical applications
High-leaded tin bronzes (continued)												
C94500	73 Cu, 7 Sn, 20 Pb	172	25	83	12	12	...	50	...	80	C, S	Locomotive wearing parts, high-speed low-load bearings.
Nickel-tin bronzes												
C94700	88 Cu, 5 Sn, 2 Zn, 5 Ni	345 (586)	50 (85)	159 (414)	23 (60)	35 (10)	...	85	(180)	30 (d)	C, T, I, M, S	Valve stems and bodies, bearings, wear guides, shift forks, feeding mechanisms, circuit breaker parts, gears, piston cylinders, nozzles.
C94800	87 Cu, 5 Sn, 5 Ni	310 (414)	45 (60)	159 (207)	23 (30)	35 (8)	...	80 (120)	...	50 (d)	M, S	Structural castings, gear components, motion translation devices, machinery parts, bearings.
C94900	80 Cu, 5 Sn, 5 Pb, 5 Zn, 5 Ni											
Aluminum bronzes												
C95200	88 Cu, 3 Fe, 9 Al	552	80	186	27	35	...	...	125	50	C, T, M, P, S	Acid-resisting pumps, bearings, gears, valve seats, guides, plungers, pump rods, bushings.
C95300	89 Cu, 1 Fe, 10 Al	517 (586)	75 (85)	186 (290)	27 (42)	25 (15)	...	...	140 (174)	55	C, T, M, P, S	Pickling baskets, nuts, gears, steel mill slippers, marine equipment, welding jaws.
C95400	85 Cu, 4 Fe, 11 Al	586 (724)	85 (105)	241 (372)	35 (54)	18 (8)	...	...	170 (195)	60	C, T, M, P, S	Bearings, gears, worms, bushings, valve seats and guides, pickling hooks.
C95410	85 Cu, 4 Fe, 11 Al, 2 Ni											
C95500	81 Cu, 4 Ni, 4 Fe, 11 Al	689 (827)	100 (120)	303 (469)	44 (68)	12 (10)	...	...	192 (230)	50	C, T, M, P, S	Valve guides and seats in aircraft engines, corrosion-resistant parts, bushings, gears, worms, pickling hooks and baskets, agitators.
C95600	91 Cu, 7 Al, 2 Si	517	75	234	34	18	...	...	140	60	C, T, M, P, S	Cable connectors, terminals, valve stems, marine hardware, gears, worms, pole-line hardware.
C95700	75 Cu, 2 Ni, 3 Fe, 8 Al, 12 Mn	655	95	310	45	26	...	...	180	50	C, T, M, P, S	Propellers, impellers, stator clamp segments, safety tools, welding rods, valves, pump casings.
C95800	81 Cu, 5 Ni, 4 Fe, 9 Al, 1 Mn	655	95	262	38	25	...	...	159	50	C, T, M, P, S	Propeller hubs, blades, and other parts in contact with salt water.
Copper-nickels												
C96200	88.6 Cu, 10 Ni, 1.4 Fe	310	45	172	25	20	...	...	...	10	C, S	Components of items being used for sea water corrosion resistance.
Copper-nickels (continued)												
C96300	79.3 Cu, 20 Ni, 0.7 Fe	517	75	379	55	10	...	150	...	15	C, S	Centrifugally cast tailshaft sleeves.
C96400	69.1 Cu, 30 Ni, 0.9 Fe	469	68	255	37	28	...	...	140	20	C, T, S	Valves, pump bodies, flanges, elbows used for sea-water corrosion resistance.
C96600	68.5 Cu, 30 Ni, 1 Fe, 0.5 Be	(758)	(110)	(482)	(70)	(7)	...	...	(230)	20	C, T, I, M, S	High-strength constructional parts for sea-water corrosion resistance.
C96700	67.6 Cu, 30 Ni, 0.9 Fe, 1.15 Be, 0.15 Zr, 0.15 Ti	(1207)	(175)	(552)	(80)	(10)	C26	...	...	40	I, M, S	Corrosion-resistant molds for plastics, high-strength constructional parts for sea-water use.
Nickel silvers												
C97300	56 Cu, 2 Sn, 10 Pb, 12 Ni, 20 Zn	241	35	117	17	20	...	55	...	70	I, M, S	Hardware fittings, valves and valve trim, statuary, ornamental castings.
C97400	59 Cu, 3 Sn, 5 Pb, 17 Ni, 16 Zn	262	38	117	17	20	...	70	...	60	C, I, S	Valves, hardware, fittings, ornamental castings.

(continued)

(a) Nominal composition, unless otherwise noted. For seldom-used alloys, only compositions are available. (b) Values for C82700, 84200, 96200, 96300 are minimum, not typical. As-cast values are for sand casting except C93900, continuous cast; and C85800, 87800, 87900, die cast. Heat treated values, in parentheses, indicate that the alloy responds to heat treatment. If heat treated values are not shown, the copper or copper alloy does not respond. (c) Free cutting brass = 100. (d) C, centrifugal; T, continuous; D, die; I, investment; M, permanent mold; P, plaster; S, sand. (e) As-heat treated value for C94700, 20; for C94800, 40. (Note: C82000, 82400, 82500, 82600, 82800 are also pressure cast.) **Source:** Copper Development Assn. Inc., New York. (Revised March 1980)

Table B. (continued)

UNS designation(a)	Nominal composition, %(a)	Tensile strength		Yield strength		Elongation in 2 in., %	Hardness			Machinability rating(c)	Casting types(d)	Typical applications
								Brinell				
		MPa	ksi	MPa	ksi		Rockwell	500 kg	3 000 kg			
Nickel silvers (continued)												
C97600	64 Cu, 4 Sn, 4 Pb, 20 Ni, 8 Zn	310	45	165	24	20	...	80	...	70	C, I, S	Marine castings, sanitary fittings, ornamental hardware, valves, pumps.
C97800	66 Cu, 5 Sn, 2 Pb, 25 Ni, 2 Zn	379	55	207	30	15	...	...	130	60	I, M, S	Ornamental and sanitary castings, valves and valve seats, musical instrument components.
Leaded coppers												
C98200	76.0 Cu, 24.0 Pb											
C98400	70.5 Cu, 28.5 Pb, 1.5 Ag											
C98600	65.0 Cu, 35.0 Pb, 1.5 Ag											
C98800	59.5 Cu, 40.0 Pb, 5.5 Ag											
Special alloys												
C99300	71.8 Cu, 15 Ni, 0.7 Fe, 11 Al, 1.5 Co	655	95	379	55	2	...	200	20	20	T, S	Glass making molds, plate glass rolls, marine hardware.
C99400	90.4 Cu, 2.2 Ni, 2.0 Fe, 1.2 Al, 1.2 Si, 3.0 Zn	455 (545)	66 (79)	234 (372)	34 (54)	25	...	...	125 (170)	50	C, T, I, S	Valve stems, marine and other uses requiring resistance to dezincification and dealuminification, propeller wheels, electrical parts, mining equipment gears.
C99500	87.9 Cu, 4.5 Ni, 4.0 Fe, 1.2 Al, 1.2 Si, 1.2 Zn	483	70	276	40	12	...	145	50	50	C, T, S	Same as C99400, but where higher yield strength is required.
C99600	58 Cu, 2 Al, 40 Mn	558 (558)	81 (81)	248 (303)	36 (44)	34 (27)	B 72	...	130	...	C, T, M, S	Damping alloys to reduce noise and vibration.
C99700	56.5 Cu, 1 Al, 1.5 Pb, 12 Mn, 5 Ni, 24 Zn	379	55	172	25	25	...	...	110	80	C, D, I, M, P, S	
C99750	58 Cu, 1 Al, 1 Pb, 20 Mn, 20 Zn	448 (517)	65 (75)	221 (276)	32 (40)	30 (20)	B 77 (B 82)	110	...	...	D, I, M, P, S	

(a) Nominal composition, unless otherwise noted. For seldom-used alloys, only compositions are available. (b) Values for C82700, 84200, 96200, 96300 are minimum, not typical. As-cast values are for sand casting except C93900, continuous cast; and C85800, 87800, 87900, die cast. Heat treated values, in parentheses, indicate that the alloy responds to heat treatment. If heat treated values are not shown, the copper or copper alloy does not respond. (c) Free cutting brass = 100. (d) C, centrifugal; T, continuous; D, die; I, investment; M, permanent mold; P, plaster; S, sand. (e) As-heat treated value for C94700, 20; for C94800, 40. (Note: C82000, 82400, 82500, 82600, 82800 are also pressure cast.) **Source:** Copper Development Assn. Inc., New York. (Revised March 1980)

Table 1. Standard color-controlled wrought copper alloys

UNS number	Common name	Color description
C11000	Electrolytic tough pitch copper	Soft pink
C21000	Gilding, 95%	Red brown
C22000	Commercial bronze, 90%	Bronze gold
C23000	Red brass, 85%	Tan gold
C26000	Cartridge brass, 70%	Green gold
C28000	Muntz metal, 60%	Light brown gold
C63800	Aluminum bronze	Gold
C65500	High-silicon bronze, A	Lavender-brown
C70600	Copper-nickel, 10%	Soft lavender
C74500	Nickel silver, 65-10	Gray white
C75200	Nickel silver, 65-18	Silver

Copper and copper alloys are readily assembled by any of the various mechanical or bonding processes commonly used to join metal components. Crimping, staking, riveting and bolting are mechanical means of maintaining joint integrity. Soldering, brazing and welding are the most widely used processes for bonding copper metals. Selection of the best joining process is governed by service requirements, joint configuration, thickness of the components, and alloy composition(s). These factors are reviewed in Section 30 ("Joining") of this Handbook, and discussed in detail in Metals Handbook, 9th Edition, Volume 6.

MECHANICAL WORKING

High-purity copper is a very soft metal. It is softest in its undeformed, single-crystal form, requiring a shear stress of only 3.9 MPa (570 psi) on {111} crystal planes for slip. Annealed tough pitch copper is almost as soft as high-purity copper, but many of the copper alloys are much harder and stiffer, even in annealed tempers.

Copper is easily deformed cold. Once flow has been started it takes little energy to continue, and thus extremely large changes in shape or reductions in section are possible in a single pass. The only limitation appears to be the ability to design and build the necessary tools. Very heavy reductions are possible, especially with continuous flow. Rolling reductions of more than 90% in one pass are used for rolling strip.

Copper and many of its alloys also respond well to sequential cold working. Tandem rolling and gang-die drawing are common. Some copper alloys work harden rapidly, so there is a limit to the number of operations that can be performed before annealing to resoften the metal.

Copper can be cold reduced almost limitlessly without annealing but heavy deformation (more than about 80 to 90%) may induce preferred crystal orientation, or texturing. Textured metal has different properties in different directions, which is undesirable for some applications.

Cold working increases both tensile strength and yield strength, but the effect is more pronounced on the latter. For most coppers and copper alloys, the tensile strength of the hardest cold worked temper is approximately twice the tensile strength of the annealed temper. For the same alloys, the yield strength of the hardest cold worked temper may be as much as five to six times that of the annealed temper.

Hardness as a measure of temper is inaccurate—the relation between hardness and strength is different for different alloys. Usually, hardness and strength for a given alloy can be correlated only over a rather narrow range of conditions. Also, the range of correlation is often different for different methods of hardness determination.

Hot Working. Not all shaping is confined to cold deformation. Hot working is commonly used for alloys that remain ductile above the recrystallization temperature. Hot working permits more extensive changes in shape than cold working, so that a single operation can replace a sequence of forming and annealing operations. To avoid preferred orientation and textures, as well as to achieve processing economy, copper and many of its alloys are hot worked to nearly finished size. Hot working reduces as-cast grain size from about 1 to 10 mm to about 0.1 mm or less, and yields a soft, texture-free structure suitable for cold finishing.

Some hot working operations may produce strengths that exceed that of the annealed temper. However, property control by hot working is very difficult, and is rarely attempted.

HEAT TREATING

Work hardened metal can be returned to a soft state by heating, or annealing. During the annealing of simple, single-phase alloys, deformed and highly stressed crystals are transformed into unstressed crystals by recovery, recrystallization and grain growth. In severely deformed metal, recrystallization occurs at lower temperatures than in lightly deformed metal. Also, the grains are smaller and more uniform in size when severely deformed metal is recrystallized.

Grain size can be controlled by proper selection of cold working and annealing practices. Large amounts of prior cold work, fast heating to annealing temperature and short annealing times favor fine grain sizes. Larger grain sizes are normally produced by a combination of limited deformation and long annealing times. In normal commercial practice, annealed grain sizes are controlled about a median value in the range 0.01 to 0.10 mm.

Variations in annealed grain size produce variations in hardness and other mechanical properties that are smaller than the variations that occur in cold worked material, but by no means negligible. Fine grain sizes often are required to enhance end-product characteristics such as load-carrying capacity, fatigue resistance, resistance to stress-corrosion cracking, and surface quality for polishing or buffing of either annealed or cold formed parts.

Heat treating processes may also be applied to copper and copper alloys to achieve homogenization, stress relieving, solutionizing, precipitation hardening, and quench hardening and tempering. These aspects are referred to throughout this section, and are reviewed in more detail in Section 28 ("Heat Treating") in this Handbook.

TEMPER DESIGNATIONS

The temper designations for wrought copper and copper alloys were traditionally specified on the basis of cold reduction imparted by rolling or drawing. This scheme related the nominal temper designations to the amount of reduction stated in Browne & Sharpe (B & S) gage numbers for rolled sheet and drawn wire. Heat-treatable alloys and product forms such as rod, tube, extrusions and castings, were not readily described by this system. For simple, single-phase alloys the

annealed temper is generally specified by the recrystallized grain size. Grain-stabilized alloys which resist recrystallization at high annealing temperatures are, however, frequently supplied in terms "light annealed," "soft annealed," or "annealed to temper." The Standard Recommended Practice ASTM B60, "Temper Designations for Copper and Copper Alloys—Wrought and Cast," was recently developed to unambiguously cover all alloy types and product forms currently in widespread use.

ELECTRICAL COPPERS

Commercially pure copper is represented by UNS numbers C10100 to C13000. The various coppers within this group have different degrees of purity, and therefore different characteristics. Fire-refined tough pitch copper C12500 is made by deoxidizing anode copper until the oxygen content has been lowered to a value of 0.02 to 0.04%. Both the traditional method of "poling" (or "pitching") a bath of molten anode copper and the more modern method of deoxidizing with hydrocarbons produce metal with essentially the same high ductility and excellent electrical conductivity. Fire-refined tough pitch copper contains a small amount of residual sulfur, normally 10 to 30 ppm, and a somewhat larger amount of cuprous oxide, normally 500 to 3000 ppm.

Electrolytic tough pitch copper C11000 is made from cathode copper—that is, copper that has been refined electrolytically. C11000 is the most common of all the electrical coppers. It has high electrical conductivity, in excess of 100% IACS. It has the same oxygen content as C12500, but contains less than 50 ppm total metallic impurities (including sulfur).

Oxygen-free coppers C10100 and C10200 are made by induction melting prime-quality cathode copper under nonoxidizing conditions produced by a granulated graphite bath covering and a protective reducing atmosphere that is low in hydrogen. Oxygen-free coppers are particularly suitable for applications requiring high conductivity coupled with exceptional ductility, low gas permeability, freedom from hydrogen embrittlement or low out-gassing tendency.

If resistance to softening at slightly elevated temperature is required, C11100 is often specified. This copper contains a small amount of cadmium, which raises the temperature at which recovery and recrystallization occur. Oxygen-free copper, electrolytic tough pitch copper and fire-refined tough pitch copper are available as silver-bearing coppers having specific minimum silver contents. The silver, which may be present as an impurity in anode copper or may be intentionally alloyed to molten cathode copper, also imparts resistance to softening to cold worked metal. Silver-bearing coppers and cadmium-bearing coppers are used for applications such as automotive radiators and electrical conductors that must operate at temperatures above about 200 °C (400 °F).

If good machinability is required, C14500 (tellurium-bearing copper) or C14700 (sulfur-bearing copper) can be selected. As might be expected, machinability is gained at a modest sacrifice in electrical conductivity.

Addition of small amounts of elements such as silver, cadmium, iron, cobalt and zirconium to deoxidized copper imparts resistance to softening at times and temperatures encountered in soldering operations, such as those used to join components of automobile and truck radiators, and

in semiconductor packaging operations.

The thermal and electrical conductivities and the room-temperature mechanical properties are unaffected by small additions of these elements. However, the cadmium coppers and zirconium coppers work harden at higher rates than either silver-bearing copper or electrolytic tough pitch copper, as shown in Fig. 1.

Cold rolled silver-bearing copper is used extensively for automobile-radiator fins. Usually such strip is only moderately cold rolled, because heavy cold rolling makes silver-bearing copper more likely to soften during soldering or baking operations. Some manufacturers prefer cadmium copper C14300, because it can be severely cold rolled without making it susceptible to softening during soldering. Figure 2 illustrates the softening characteristics of C14300 and C11400 as measured for several temperatures and two tempers. As illustrated in Fig. 2(b), C14300 cold rolled to a tensile strength of 440 MPa (64 ksi) retains 91% of its strength after a typical core bake of 3 min at 345 °C (650 °F). Silver-bearing copper C11400 given the same cold reduction retains only 60% of its tensile strength after the same baking schedule.

Another application in which softening resistance is of paramount importance is leadframes for electronic devices, such as plastic dual in-line packages. During packaging and assembly, leadframes may be subjected to temperatures up to 350 °C for several minutes and up to 500 °C for several seconds. It is most desirable that the leads maintain good strength, because they are pressed into socket connectors, often by automated assembly machines, and softened leads collapse, causing spoilage.

Alloy C15100 (copper-zirconium), alloy C15500 (copper-silver-magnesium-phosphorus), alloy C19400 (copper-iron-phosphorus-zinc), and alloy C19500 (copper-iron-cobalt-tin-phosphorus) are popular alloys for these applications because they have good conductivity, good strength and good softening resistance. Figures 3 and 4 compare the softening resistance of these alloys with electrolytic copper C11000.

COPPER ALLOYS

The most common way to catalog copper and its alloys is to divide them into six families: coppers, dilute copper alloys, brasses, bronzes, copper nickels and nickel silvers. The first family, the coppers, is essentially commercially pure copper, which ordinarily is soft and ductile and contains less than about 0.7% total impurities. The dilute copper alloys contain small amounts of various alloying elements that modify one or more of the basic properties of copper. Each of the remaining families contains one of five major alloying elements as its primary alloying ingredient (see Table 2). All five of the major alloying elements have room-temperature solid solubility in copper of at least 8 at. %.

Solid-Solution Alloys. The most compatible alloying elements with copper are those that form solid-solution fields. These include all elements forming useful alloy families (see Table 2) plus manganese. Hardening in these systems is great enough to make useful objects without encountering brittleness associated with second phases or compounds.

Cartridge brass is typical of this group, consisting of 30% Zn in copper and exhibiting no beta phase except an occasional small amount due

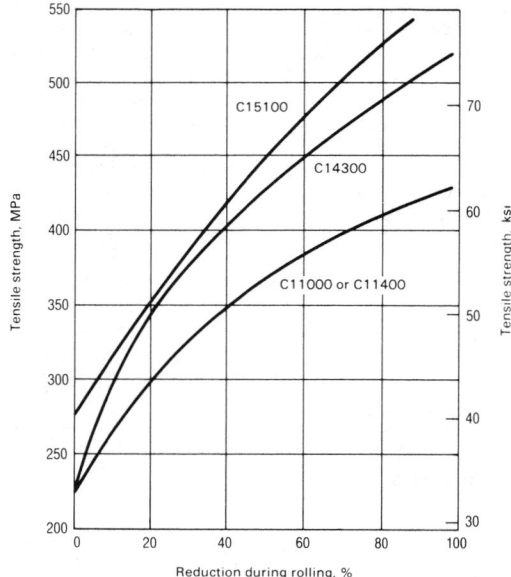

Fig. 1. Tensile strength versus reduction during rolling for cadmium copper (C14300), zirconium copper (C15100) and tough pitch copper (C11000)

to segregation, which normally disappears after the first anneal. Provided that there are no tramp elements, such as iron, cold working and grain growth relationships are easily reproduced in practice.

Modified Solid-Solution Alloys. The solid-solution-strengthened alloys of copper are noted for their strength and formability. Since they are single-phase and are not transformed by heating or cooling, their maximum strength is developed by cold working, such as cold rolling or cold drawing. Formability is reduced in proportion to the amount of cold work applied.

Modifications of some solid-solution alloys were developed, by adding elements which react to form dispersions of intermetallic particles. These dispersions have a grain refining and strengthening effect. As a result, higher strengths can be produced with less cold working, and consequently better formability at higher strength results. Since these modifications do not require large amounts of costly elements, the gains are reasonably economical.

Alloy C63800 (Cu 95, Al 2.8, Si 1.8, Co 0.4) is a high-strength alloy with an annealed tensile strength of 82 ksi (570 MPa) nominal, and nominal tensile strengths of 96 to 130 ksi (660 to 900 MPa) for the standard rolled tempers. Cobalt

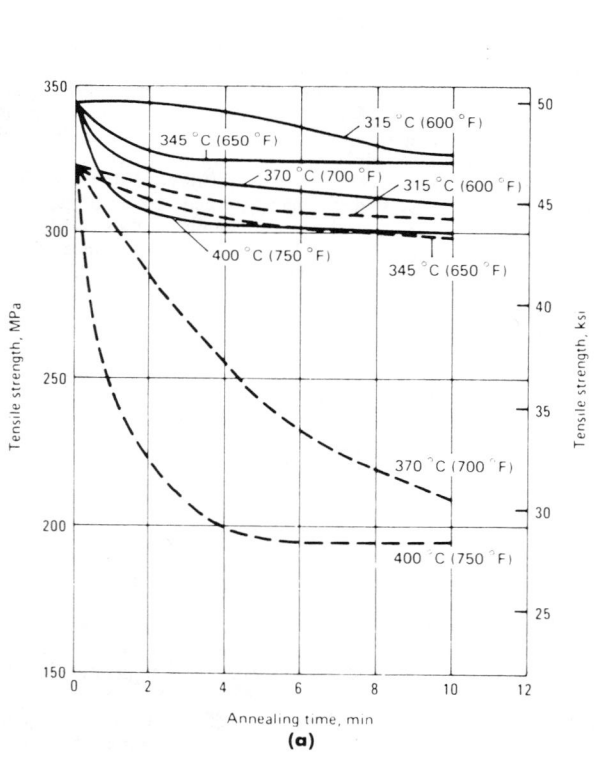

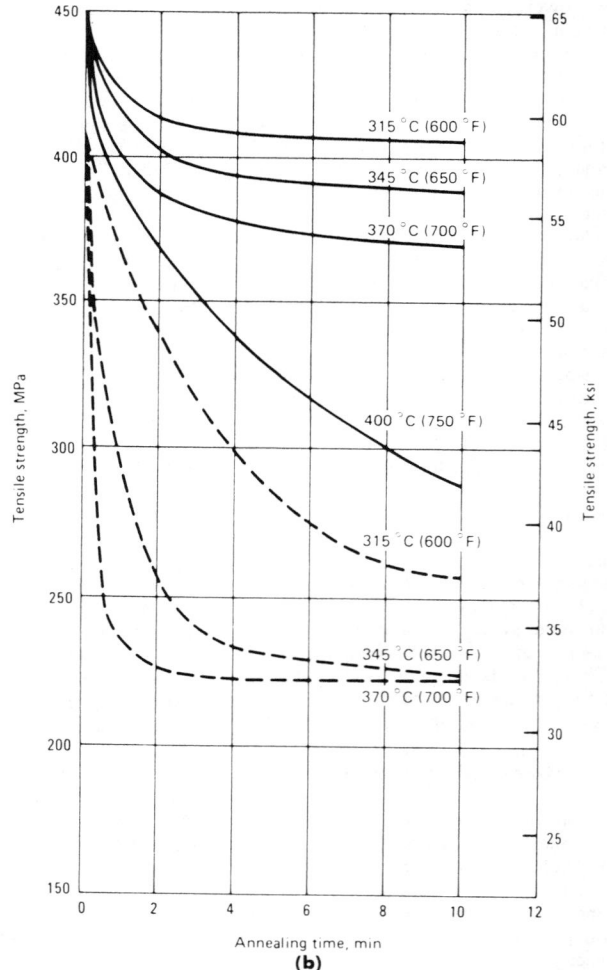

Solid curves are for C14300; dashed curves for C11400. (a) Softening curves for material cold reduced 21% in area from 0.0038 to 0.0030 in. in thickness. (b) Softening curves for material cold reduced 90% in area, from 0.0300 to 0.0030 in. in thickness.

Fig. 2. Softening characteristics of cadmium-bearing copper and silver-bearing tough pitch copper

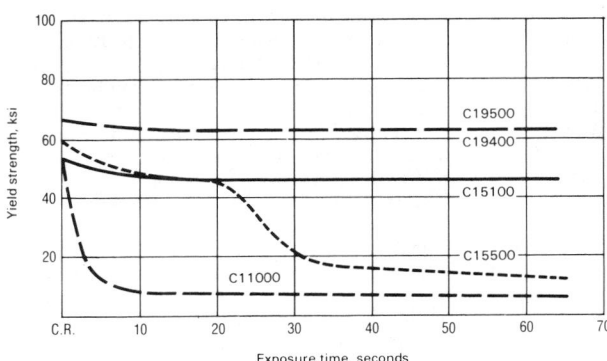

Fig. 3. Softening resistance of leadframe materials at the upper limit of temperature used (500 °C)

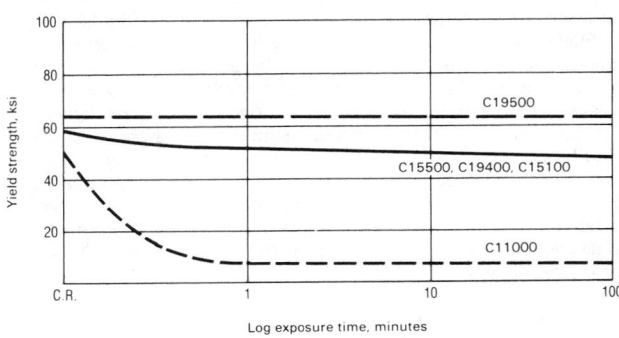

Fig. 4. Softening resistance of leadframe materials at an intermediate temperature level (350 °C)

Table 2. Classification of copper and copper alloys

Family	Principal alloying element	Solid solubility, at. %(a)	UNS numbers(b)
Coppers, high-copper alloys	(c)	...	C10000
Brasses	Zn	37	C20000, C30000, C40000, C66400 to C69800
Phosphor bronzes	Sn	9	C50000
Aluminum bronzes	Al	19	C60600 to C64200
Silicon bronzes	Si	8	C64700 to C66100
Copper nickels, nickel silvers	Ni	100	C70000

(a) At 20 °C (68 °F). (b) Wrought alloys. (c) Various elements having less than 8 at. % solid solubility at 20 °C (68 °F).

provides the dispersion of strengthening intermetallic particles.

Alloy C68800 (Cu 73.5, Zn 22.7, Al 3.4, Co 0.4) is a high-strength modified aluminum brass. Its bend formability parallel to the direction of rolling is outstanding relative to its strength. It also owes some of its unique properties to a dispersion of intermetallic particles occasioned by the presence of cobalt. Its strength range is essentially the same as that of alloy C63800.

Alloy C65400 (Cu 95.44, Si 3, Sn 1.5, Cr 0.06) is a very-high-strength alloy, which has excellent stress-relaxation resistance at temperatures up to 105 °C. Its nominal strength range in rolled tempers is 82 to 137 ksi (570 to 945 MPa). Electrical contact and connector springs are heat treated at 200 to 250 °C for one hour to stabilize internal stresses and maximize stress-relaxation resistance.

Alloy C66400 (Cu 86.5, Zn 11.5, Fe 1.5, Co 0.5) is a low-zinc brass modified by the addition of iron and cobalt. The dispersion of intermetallic particles resulting from these additions strengthens the alloy. At the same time, conductivity is only moderately reduced, and resistance to stress-corrosion cracking is very high. A high-zinc brass of the same strength and conductivity would be subject to stress-corrosion cracking unless plated for protection.

There are probably other modified solid-solution-strengthened alloys described in the literature of the brass-mill industries. The above should serve as examples of this additional class of copper alloys which is expanding through the development efforts of the producers of brass mill products throughout the world.

Age-Hardenable Alloys. Age hardening produces very high strengths, but is limited to those few copper alloys in which the solubility of the alloying element decreases sharply with decreasing temperature. The beryllium coppers can be con-

sidered typical of the age-hardenable copper alloys.

Wrought beryllium coppers can be precipitation hardened to the highest strength levels attainable in copper-base alloys. There are two commercially significant alloy families employing two ranges of beryllium with additions of cobalt or nickel. The so-called red alloys contain beryllium at levels ranging from approximately 0.2 to 0.7 wt % with additions of nickel or cobalt totaling 1.4 to 2.7 wt % depending on the alloy—e.g., C17500, C17510. These low-beryllium alloys achieve relatively high conductivity (e.g., 50% IACS) and retain the pink luster of other low-alloy coppers. The "red" alloys achieve yield strengths ranging from about 25 to 80 ksi (170 to 550 MPa), unheat treated, to greater than 130 ksi (895 MPa) after precipitation hardening, depending on degree of cold work.

The more highly beryllium-alloyed systems can contain from 1.6 to 2.0 wt % beryllium, and about 0.25 wt % cobalt—e.g., C17000 and C17200. These alloys are frequently termed the "gold" alloys due to the shiny luster imparted by the substantial amount of beryllium present, about 12 at. %. The "gold" alloys are the high-strength beryllium coppers and can attain yield strengths ranging from approximately 30 to 100 ksi (205 to 690 MPa) in the age-hardenable condition to above 200 ksi (1380 MPa) after aging. The conductivity of the "gold" alloys is lower than that of the "red" alloy family by virtue of the high beryllium content. However, conductivity ranging from about 20% to higher than 30% IACS is obtained in wrought products depending on cold work amount and heat treatment schedule. For enhanced machinability in rod and wire, lead is added (e.g., C17300).

Other age-hardenable alloys include C15000; C15100 (zirconium copper); C18200, C18400 and C18500 (chromium coppers); C19000 and C19100

(copper-nickel-phosphorus alloys); and C64700 (copper-nickel-silicon alloy).

Some age-hardening alloys have different desirable characteristics, such as high strength combined with better electrical conductivity than the beryllium coppers.

C71900 (copper-nickel-tin alloy) and other, similar alloys can be hardened by spinodal decomposition. By combining cold working with hot working these alloys can achieve high strengths, equivalent of the hardenable beryllium coppers. These alloys are unique in that their forming characteristics are isotropic, and thus do not reflect the directionality normally associated with wrought alloys.

Other Alloys. Certain aluminum bronzes, most notably those containing more than about 9% Al, can be hardened by quenching from above a critical temperature. The hardening process is a martensitic-type process, similar to the martensitic hardening that occurs when iron-carbon alloys are quenched. Mechanical properties of aluminum bronzes can be varied somewhat by temper annealing after quenching or by using an interrupted quench instead of a standard quench.

Aluminum bronzes alloyed with nickel or zinc utilize reversible martensitic transformations to provide "shape memory" effects.

Insoluble Alloying Elements. Lead, tellurium and selenium are added to copper and its alloys to improve machinability. They, along with bismuth, make hot rolling and hot forming nearly impossible and severely limit the useful range of cold working.

An exception here are the high-zinc brasses, which become fully beta phase at high temperature. The beta phase can dissolve lead—thus avoiding a liquid grain-boundary phase at hot forging or extrusion temperatures. Most free-cutting brass rod is made by beta extrusion. C37700, one of the leaded high-zinc brasses, is so readily hot forged that it is the standard alloy against which the forgeability of all copper alloys is judged.

DEOXIDIZERS

Li, Na, Be, Mg, B, Al, C, Si and P can be used to deoxidize copper. Ca, Mn and Zn can sometimes be considered deoxidizers, although they normally fulfill different roles.

The first requirement of a deoxidizer is that it have an affinity for oxygen in molten copper. Probably the second most important requirement is that it be relatively inexpensive compared to copper and any other additions. Thus, although zinc normally functions as a solid-solution strengthener, it is sometimes added in small amounts to function as a deoxidizer, because it has high affinity for oxygen and is relatively low in cost. In tin bronze, phosphorus has traditionally been the deoxidizer, hence the name "phosphor bronzes" for these alloys. Silicon instead of phosphorus is the deoxidizer for chromium coppers because phosphorus severely reduces electrical conductivity. Most deoxidizers contribute to hardness and other qualities, which often makes classification as a deoxidizer indistinct.

PRODUCTION OF COPPER METALS

The copper industry in the United States, broadly speaking, is composed of two segments: producers (mining, smelting and refining companies) and fabricators (wire mills, brass mills, foundries and powder plants). The end products of copper producers, the most important of which is refined cathode copper, are sold almost entirely to copper fabricators. The end products of copper fabricators may be generally described as mill products, and consist of wire and cable, sheet, strip, plate, rod, bar, mechanical wire, tubing, forgings, extrusions and castings. These products are sold to a wide variety of industrial users. Certain mill products—chiefly wire, cable and most tubular products—are used without further metalworking. On the other hand, most flat-rolled products, rod, bar, mechanical wire, forgings and castings go through multiple metalworking, machining, finishing and/or assembly operations before emerging as finished products.

Copper Producers. Figure 5 is a simplified flow diagram of the copper industry. The box at upper left represents mining companies which remove vast quantities of low-grade material, mostly from open-pit mines, in order to extract copper from the earth's crust. Approximately 2.5 t (tonnes) of overburden must be removed along with each tonne of copper ore. (The ratio of overburden to ore is sometimes as high as 3 to 1.) The ore itself averages only about 0.5% copper. The importance of efficient materials handling by copper producers is dramatized by the fact that, in the United States, more tonnage must be moved in mining copper than in mining all other metals combined.

Copper ore normally is crushed, ground and concentrated (usually by flotation) to produce a beneficiated ore containing 20 to 25% copper. Most copper ores are sulfide ores, composed chiefly of copper sulfide but also containing a significant amount of iron sulfide as an impurity plus recoverable trace amounts of silver and gold. Ore concentrates most often are reduced to the metallic state by a pyrometallurgical process. Traditionally, the concentrated ore is smelted in a reverberatory furnace to produce a copper sulfide–iron sulfide matte, and the matte is oxidized in a converter to convert the FeS to iron oxides, which separate out in a slag, and to burn off the sulfur from the CuS, leaving blister copper, which contains at least 98.5% copper. Fire refining of blister copper removes most of the oxygen and other impurities, leaving a product at least 99.5% pure, which is cast into anodes. Finally, most anode copper is electrolytically refined, usually to a purity of at least 99.95%. Gold and silver slimes are a byproduct of electrorefining, which helps defray the cost of the electrical energy used to refine the anode copper. The resulting cathodes are the normal end product of the producer companies, and are a common item of commerce. The consumption of refined copper (mostly cathodes) in the United States was about 2 million t (4.4 billion lb) in 1977, about 22% of the world total of 9 million t (19.8 billion lb).

Hydrometallurgical processing is an alternative to pyrometallurgical processing that has more recently become commercially important. Heap leaching followed by cementation is the most common hydrometallurgical process, although others such as vat leaching or agitation leaching, both of which require richer ores, have also become important. In heap leaching, sulfuric acid is percolated through waste dumps formed from the rejected material (tailings) of the flotation process to leach out (dissolve) most of the remaining copper. The copper in the pregnant leach liquor is recovered by a process called cementation; in this process, the liquor is passed over scrap steel, upon which the copper precipitates

from solution. Impure copper precipitates then are sent to a converter or smelting furnace for pyrometallurgical refining.

The liquor from heap leaching can be concentrated to produce an electrolyte suitable for use in electrowinning, whereas vat leaching and agitation leaching produce such electrolytes directly. In electrowinning, copper is extracted electrolytically from the electrolyte much as anode copper is electrorefined. The chief difference is that the copper is present in an enriched electrolyte instead of as impure copper anodes. Electrowon copper (about 99.9% pure) is not quite as pure as electrorefined copper (ordinarily about 99.97% pure).

The box at lower left in Fig. 5 represents the portion of copper supply provided by scrap. This portion is substantial: nearly half of the copper consumed in the United States each year is derived from recycled scrap. (Runaround scrap, which is scrap recycled within a particular plant, is not included in these statistics.) About 25 to 30% of the scrap is fed into the smelting or refining stream and thus quickly loses its identity. The remainder is consumed directly by brass mills, by ingot makers (whose main function is to melt scrap into alloy ingot for use by foundries) and by foundries themselves.

The box labeled "Copper consumed" in Fig. 5 represents the total tonnage of refined copper plus the copper content of scrap consumed directly by fabricator companies. To this sum are added the amounts of various alloying elements used in producing copper alloys, and the alloy content of the directly consumed scrap, to obtain "Metal consumed."

Copper Fabricators. The four classes of copper fabricators together consume about 98% of the total output of the copper producers. Other industries, such as the steel, aluminum and chemical industries, consume the remaining 2%. As shown in Fig. 5, wire mills and brass mills are roughly equal in output: on the average, each produces 40 to 50% of total mill products. Foundries account for roughly 10% of the fabricated mill products, and powder plants about 1%.

Wire mills make copper wire and cable. The starting material is refined copper cathodes. Tra-

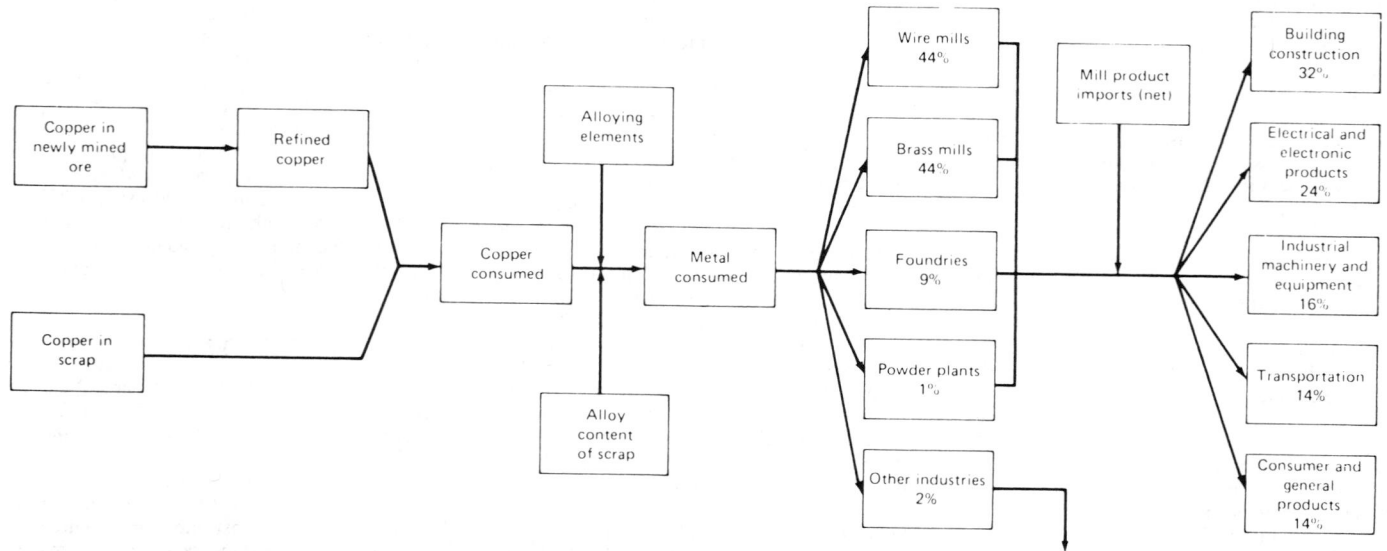

Fig. 5. Flow of copper from mine production and scrap collection through end use (percentages based on 1977 data)

Table 3. Major wrought copper and copper-base alloy systems

Copper or copper alloy group	UNS numbers	Approximate U.S. shipments in 1983, millions of pounds	Remarks
Wire Mill Shipments			
Coppers	C10000 to C15900	2600	C11000 is predominant material.
Brass Mill Shipments			
Coppers	C10000 to C15900, C16000 to C16900, C18000 to C18900	944	Includes modified coppers, cadmium copper, and chromium coppers.
Common brasses	C20000 to C29900	475	Of this, 89% is sheet, strip and plate.
Leaded brasses	C30000 to C39900	684	Of this, 95% is rod.
Tin bronzes	C50000 to C53900	36	Unleaded only; also known as phosphor bronzes.
Aluminum bronzes, silicon bronzes and manganese bronzes	C60000 to C68400	20	
Copper nickels	C70000 to C72900	81	
Nickel silvers	C73000 to C79900	11	
Others	C17000 to C17900, C19000 to C19900, C40000 to C49900, C54000 to C54900, C68500 to C69900	110	Includes beryllium coppers, copper-iron alloys, tin brasses, leaded tin bronzes, aluminum brasses, and silicon brasses.

ditionally, cathodes have been melted and then cast into wirebars, followed by hot rolling to wire rod. More recently, continuous casting has been chosen for new plants or for major renovation of older plants. Facilities for producing wire rod may be integrated into a refinery, be a separate operation or be integrated into a wiredrawing plant.

Wire rod is cold drawn to final dimensions through a series of dies. The cold drawn wire may or may not be annealed, depending on requirements. Wire may be used as a single conductor, but more often is fabricated into a stranded conductor; most copper wire is insulated. Various types of electrical cable are produced from individual conductors, each of which may be stranded and/or insulated separately before being incorporated into the finished cable.

Brass mills melt and alloy feedstock to make sheet, strip, plate, tubing, rod, bar, mechanical wire, forgings and extrusions. Of the feedstock used by brass mills, about half is scrap and about half is virgin metal. Fabricating processes such as hot rolling, cold rolling, extrusion and drawing are employed to convert the melted and cast feedstock into wrought mill products. Some brass mills are secondary mills that do not melt and cast the feedstock they fabricate into mill products, but merely reroll strip to thinner gages or redraw tubing or mechanical wire to smaller dimensions.

About one-third of the output of U.S. brass mills is unalloyed copper and high-copper alloys — chiefly in such forms as plumbing and air-conditioning tube, busbar and roofing sheet. Copper alloys comprise the remaining two-thirds, and are distributed approximately as indicated in Table 3. The several varieties of leaded brass rod (which exhibits outstanding machinability and good corrosion resistance) and unleaded brass strip (which has high strength, corrosion resistance, excellent formability and good electrical properties) constitute about three-fourths of the total tonnage of wrought copper alloys shipped from U.S. brass mills. Other alloy types of commercial significance include tin bronzes, which are noted for their excellent cold forming behavior; tin brasses, known for outstanding corrosion resistance; copper nickels, which are particularly resistant to seawater; nickel silvers, which combine a silvery appearance with good formability and corrosion

resistance; beryllium coppers, which can provide outstanding strength when hardened; and aluminum bronzes, which have high strength along with good resistance to both chemical attack and mechanical abrasion.

Foundries use prealloyed ingot, scrap and virgin metal as raw materials. Their chief products are shape castings for many different industrial and consumer goods, the most important of which are plumbing products, industrial valves and bearings.

Powder plants produce powder metallurgy parts, chiefly sintered bronze bushings.

APPLICATIONS OF COPPER AND ITS ALLOYS

The five major market categories at far right in Fig. 5 constitute the chief customer industries of the copper fabricators. Of the chief customer industries, the largest is building construction, which purchases large quantities of electrical wire, tubing and parts, for builders' hardware and for electrical, plumbing, heating and air-conditioning systems. Next are electrical and electronic products, including those for telecommunications, electronics, wiring devices, electric motors and power utilities. The industrial machinery and equipment category includes industrial valves and fittings; industrial, chemical and marine heat exchangers; and various other types of heavy equipment, off-road vehicles and machine tools. Transportation applications include road vehicles, railroad equipment and aircraft parts; automobile radiators and wiring harnesses are the most important products in this category. Finally, consumer and general products include electrical appliances, fasteners, ordnance, coinage and jewelry.

About 90% of the total tonnage of wrought copper alloys sold by U.S. fabricator plants is represented by the 15 application categories listed in Table 4. In the three categories that account for the greatest tonnage — telecommunications, automotive, and plumbing and heating — a continuing effort has been made to conserve materials and to manufacture products more efficiently. Most often this effort involves redesign of components, and is accomplished through reductions in material gage. In some instances, such as small motors for appliances and other devices, there is a trend toward using more copper for each unit to increase the energy efficiency of the end product.

Table 4 also shows the mill products used in each application category, plus the property or properties that dictate the use of copper or its alloys. High electrical conductivity is a major reason for choosing copper in ten of the fifteen categories, corrosion resistance a major reason in seven, ease of fabrication (including good machinability and good formability) in nine (although ease of fabrication is a significant factor in all fifteen categories), and good heat-transfer properties in five.

Table 4. Major end-use applications for copper and copper alloys in the United States — 1983

Application	% of total	Principal mill products	Principal reason(s) for using copper
Plumbing and heating	13.8	Copper tube, brass rod, castings	Corrosion resistance, machinability, heat transfer
Building wiring	12.8	Copper wire	Electrical properties
Telecommunications	10.5	Copper wire	Electrical properties
In-plant equipment	9.7	All	Corrosion resistance, wear resistance, electrical properties, heat transfer, machinability
Automotive: automobiles, trucks and buses	9.4	Brass and copper strip, copper wire	Heat transfer, electrical properties, corrosion resistance
Power utilities	7.2	Copper wire and bar	Electrical properties
Air conditioning and commercial refrigeration	6.5	Copper tube and wire	Heat transfer, formability, electrical properties
Electronics	5.9	Alloy strip, copper wire	Electrical properties, formability
Industrial valves and fittings	3.8	Brass rod, castings	Corrosion resistance, machinability
Lighting and wiring devices	2.8	Alloy strip, copper wire	Electrical properties
Appliances and extension cords	2.7	Copper wire and tube	Electrical properties, heat transfer
Military and commercial ordnance	2.0	Brass strip and tube	Ease of fabrication
Builders' hardware	1.6	Brass rod and strip	Corrosion resistance, formability, aesthetics
Fasteners and closures	1.2	Brass wire and strip	Corrosion resistance, formability, machinability
Coinage	1.1	Alloy and copper strip	Ease of fabrication, corrosion resistance, electrical properties, aesthetics
Other	9.0	All	Various

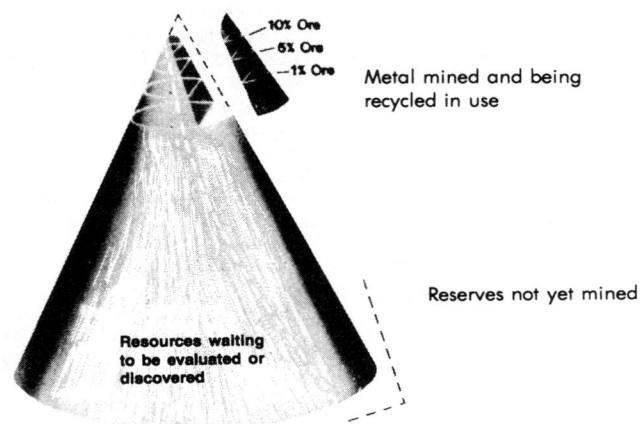

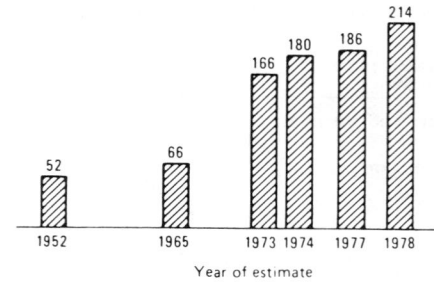

U.S. Bureau of Mines estimates in billions of pounds of copper content

Fig. 7. United States copper reserves

Fig. 6. United States copper reserves and resources

COPPER SUPPLY AND RESOURCES

All of the copper in the earth's crust is represented in Fig. 6 as a hypothetical "cone of resources." In this representation, the richest ores are at the tip of the cone. Up to now, man has discovered, extracted and used only a slice off the tip of the cone. Behind that slice are the known copper reserves that have been discovered but not yet mined. Below the known reserves are the rest of the earth's resources, only a fraction of which are known to be economically recoverable using current methods of extractive metallurgy. In 1983, the U.S. Bureau of Mines estimated that there were 199 billion lb of known copper reserves in the United States, plus another 634 billion lb of resources. By comparison, only about 2.3 billion lb of newly mined copper were consumed in 1978.

Not only are there substantial copper reserves in the United States (enough to ensure self-sufficiency for about the next half-century even if no new reserves are added), but there also are continual improvements in discovery and extractive metallurgy. As shown in Fig. 7, copper reserves in the United States increased from 52 billion lb to 214 billion lb from 1952 to 1978 — a rate of increase in reserves which greatly outstripped the total consumption of newly mined copper over the same period of time.

The U.S. copper industry relies heavily on scrap copper as a raw material. For instance, in 1983 the copper industry used 2.6 billion lb of scrap, or about 53% of the total amount of copper used in that year. Such reliance is typical of a mature economy, and recycling of products such as building wire and plumbing, even after a 40- to 50-yr service life, is not uncommon. In fact, by percentage, copper is recycled more than any other engineering metal.

Substantial reserves and extensive use of scrap both contribute to the self-sufficiency of the United States, and therefore to the reliability of supply of copper to manufacturing industries. Over the period 1970 to 1983, the United States was more than 90% self-sufficient in copper.

Copper Tubular Products

TUBE AND PIPE made of copper or copper alloys are used extensively for carrying potable water in buildings and homes. These products also are used throughout the oil, chemical and process industries to carry diverse fluids, ranging from various natural and process waters to seawater to an extremely broad range of strong and dilute organic and inorganic chemicals. In the automotive and aerospace industries, copper tube is used for hydraulic lines, heat exchangers (such as automotive radiators), air conditioning systems and various formed or machined fittings. In marine service, copper tube and pipe are used to carry potable water, seawater and other fluids, but their chief application is in tube bundles for condensers, economizers and auxiliary heat exchangers. Copper tube and pipe are used in food and beverage industries to carry process fluids for beet and cane sugar refining, for brewing of beer and for many other food processing operations. In the building trades, copper tube is used widely for heating and air conditioning systems in homes, commercial buildings, and industrial plants and offices. Table 1 summarizes the copper alloys that are standard tube alloys, and gives ASTM specifications and typical uses for each of the alloys.

Frequently, resistance to corrosion is a critical factor in selecting a tube alloy for a specific application. Information that can help determine the alloy(s) most suitable for a given type of service is given in the article "Corrosion Characteristics of Copper and Copper Alloys," (page 7·35).

Joints in copper tube and pipe are made in various ways. Permanent joints can be made by brazing or welding. Semipermanent joints are made most often by soldering, usually in conjunction with standard socket-type solder fittings, but threaded joints also can be considered semipermanent joints for pipe. Detachable joints are almost always some form of mechanical joint — flared joints, flange-and-gasket joints, and joints made using any of a wide variety of specially designed compression fittings are all common.

Properties of Tube. As with most wrought products, the mechanical properties of copper tube depend on prior processing. With copper, it is not so much the methods used to produce tube, but rather the resulting metallurgical condition that has the greatest bearing on properties. Table 2 summarizes tensile properties for the standard tube alloys in their most widely used conditions.

PRODUCTION OF TUBE SHELLS

Copper tubular products are produced by two different processes:

- By extruding or piercing billets
- By forming strip into tubular shape and welding.

Extrusion of copper and copper alloy tube shells is done by heating a billet of material above the recrystallization temperature, and then forcing material through an orifice in a die and over a mandrel held in position within the die orifice. The clearance between mandrel and die determines the wall thickness of the extruded tube shell.

In extrusion, the die is located at one end of the container section of an extrusion press; the metal to be extruded is driven through the die by a ram, which enters the container from the end opposite the die. Tube shells are produced either by starting with a hollow billet or by a two-step operation in which a solid billet is first pierced and then extruded.

Extrusion pressure varies with alloy composition. Alloy C36000 (61.5Cu-3Pb-35.5Zn) requires a relatively low pressure, whereas C26000 (70Cu-30Zn) and C44300 (71.5Cu-1Sn-27.5Zn-0.06As) require the highest pressure of all the brasses. Most of the coppers require an extrusion pressure intermediate between those for C26000 and C36000. C71500 (70Cu-30Ni) requires a very high extrusion pressure.

Extrusion pressure also depends on billet temperature, extrusion ratio (the ratio of the cross-sectional area of the billet to that of the extruded section), speed of extrusion and degree of lubrication. The flow of metal during extrusion depends on many factors, including copper content of the alloy, amount of lubricant, and die design.

Rotary piercing on a Mannesmann mill is another method commonly used to produce seamless pipe and tube from copper and certain copper alloys. Piercing is the most severe forming operation

Table 1. Copper tube alloys and typical applications

UNS number	Alloy type	ASTM specifications	Typical uses
C10200	Oxygen-free copper	B68, B75, B88, B111, B188, B280, B359, B372, B395, B447	Bus tube, conductors, wave guides
C12200	Phosphorus deoxidized copper	B68, B75, B88, B111, B280, B306, B359, B360, B395, B447, B543, B641	Water tubes; condenser, evaporator and heat exchanger tubes; air conditioning and refrigeration, gas, heater and oil burner lines; plumbing pipe and steam tubes; brewery and distillery tubes; gasoline, hydraulic and oil lines; rotating bands
C19200	Copper	B111, B359, B395, B469	Automotive hydraulic brake lines; flexible hose
C19400	...	B543, B586	Heat exchanger tube and water tube
C23000	Red brass, 85%	B111, B135, B359, B395, B543	Condenser and heat exchanger tubes, flexible hose; plumbing pipe; pump lines
C26000	Cartridge brass, 70%	B135	Plumbing brass goods
C33000	Low-leaded brass (tube)	B135	Pump and power cylinders and liners; plumbing brass goods
C36000	Free-cutting brass		Screw machine parts; plumbing goods
C43500	Tin brass		Bourdon tubes; musical instruments
C44300	Inhibited admiralty metal	B111, B359, B395	Condenser, evaporator and heat exchanger tubes; distiller tubes
C60800	Aluminum bronze, 5%	B111, B359, B395	Condenser, evaporator and heat exchanger tubes; distiller tubes
C65100	Silicon bronze B	B315	Heater exchange tubes; electrical conduits
C65500	Silicon bronze A	B315	Chemical equipment, heat exchanger tubes; piston rings
C68700	Arsenical aluminum brass	B111, B359, B395	Condenser, evaporator and heat exchanger tubes; distiller tubes
C70600	Copper nickel, 10%	B111, B359, B395, B466, B467, B543, B552	Condenser, evaporator and heat exchanger tubes; salt water piping; distiller tubes
C71500	Copper nickel, 30%	B111, B359, B395, B446, B467, B543, B552	Condenser, evaporator and heat exchanger tubes; distiller tubes; salt water piping

Table 2. Typical mechanical properties for copper alloy tube(a)

Temper	Tensile strength MPa	ksi	Yield strength(b) MPa	ksi	Elongation(c), %	Temper	Tensile strength MPa	ksi	Yield strength(b) MPa	ksi	Elongation(c), %
C10200						C43500					
OS050	220	32	69	10	45	OS035	315	46	110	16	46
OS025	235	34	76	11	45	H80	515	75	415	60	10
H55	275	40	220	32	25	C44300					
H80	380	55	345	50	8	OS025	365	53	150	22	65
C12200						C60800					
OS050	220	32	69	10	45	OS025	415	60	185	27	55
OS025	235	34	76	11	45	C65100					
H55	275	40	220	32	25	OS015	310	45	140	20	55
H80	380	55	345	50	8	H80	450	65	275	40	20
C19200						C65500					
H55(d)	290	42	205(e)	30(e)	35	OS050	395	57	...	...	70
C23000						H80	640	93	...	...	22
OS050	275	40	83	12	55	C68700					
OS015	305	44	125	18	45	OS025	415	60	185	27	55
H55	345	50	275	40	30	C70600					
H80	485	70	400	58	8	OS025	305	44	110	16	42
C26000						H55	415	60	395	57	10
OS050	325	47	105	15	65	C71500					
OS025	360	52	140	20	55	OS025	415	60	170	25	45
H80	540	78	440	64	8						
C33000											
OS050	325	47	105	15	60						
OS025	360	52	140	20	50						
H80	515	75	415	60	7						

(a) Tube size: 25 mm (1 in.) OD by 1.65 mm (0.065 in.) wall. (b) 0.5% extension under load. (c) In 50 mm or 2 in. (d) Tube size: 4.8 mm (0.1875 in.) OD by 0.76 mm (0.030 in.) wall. (e) 0.2% offset. (f) Tube size: 9.5 mm (0.375 in.) OD by 2.5 mm (0.097 in.) wall.

customarily applied to metals. The process takes advantage of tensile stresses that develop at the center of a billet when it is subjected to compressive forces around its periphery. In rotary piercing, one end of a heated cylindrical billet is fed between rotating work rolls that lie in a horizontal plane and are inclined at an angle to the axis of the billet, see Fig. 1. Guide rolls beneath the billet prevent it from dropping from between the work rolls. Because the work rolls are set at an angle to each other as well as to the billet, the billet is simultaneously rotated and driven forward toward the piercing plug, which is held in position between the work rolls.

The opening between work rolls is set smaller than the billet, and the resultant pressure acting around the periphery of the billet opens up tensile cracks, and then a rough hole, at the center of the billet just in front of the piercing plug. The piercing plug assists in further opening the axial hole in the center of the billet, smooths the wall of the hole and controls the wall thickness of the formed tube.

Coppers and plain alpha brasses can be pierced, provided the lead content is held to less than 0.01%. Alpha-beta brasses can tolerate higher levels of lead without adversely affecting their ability to be pierced.

When piercing brass, close temperature control must be maintained because the range in which brass can be pierced is narrow. Each alloy has a characteristic temperature range within which it is sufficiently plastic for piercing to take place. Below this range, the central hole does not open up properly under the applied peripheral forces. Overheating may lead to cracked surfaces. Suggested piercing temperatures for various alloys are given in the following table:

UNS number	Piercing temperature °C	°F
C11000	815-870	1500-1600
C12200	815-870	1500-1600
C22000	815-870	1500-1600
C23000	815-870	1500-1600
C26000	760-790	1400-1450
C28000	705-760	1300-1400
C46400	730-790	1350-1450

PRODUCTION OF FINISHED TUBES

Welded tube is made from clean strip in either cold rolled or annealed tempers. The strip is formed into a tubular shape on a precision forming mill.

For forge-welded tube, the edges of the strip are heated to the required welding temperature,

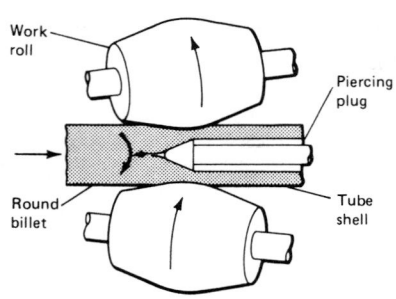

Arrows indicate direction of motion.
Fig. 1. Schematic diagram of metal piercing

usually by high-frequency electric current, and are pressed firmly together causing a forge-type joint to be formed with internal and external flash or bead. The external flash is always removed, providing a smooth external surface. The internal flash is also removed by scarfing, providing a smooth, clean internal surface.

For fusion-welded tube, the edges of the strip are brought together and welded, usually by a gas tungsten arc welding process, without the addition of filler metal, causing a fusion-type joint to be formed with no internal or external flash or bead.

Welded tube may be used in the as-welded condition or, subsequent to welding, may be further processed by the standard cold drawing, reducing and annealing processes.

Cold drawing of extruded or pierced tube shells to smaller sizes is done on draw blocks for coppers and on draw benches for brasses and other alloys. With either type of machine, the metal is cold worked by pulling the tube through a die that reduces the diameter. Concurrently, wall thickness is reduced by drawing over a plug or mandrel that may be either fixed or floating. Cold drawing increases the strength of the material and simultaneously reduces ductility. Tube size is reduced — outside diameter, inside diameter, wall thickness and cross-sectional area all are smaller after drawing. Because the metal work hardens, tubes may be annealed at intermediate stages when drawing to small sizes. However, coppers are so ductile that they frequently can be drawn to finished size without intermediate annealing.

Tube reducing is an alternative process for cold sizing of tube. In tube reducing, semicircular grooved dies are rolled or rocked back and forth along the tube while a tapered mandrel inside the tube controls the inside diameter and wall thickness. The process yields tube having very accurate dimensions and better concentricity than can be achieved by tube drawing.

The grooves in the tube reducing dies are tapered, one end of the grooved section being somewhat larger than the outside diameter of the tube to be sized. As the dies are rocked back and forth, the tube is pinched against the tapered mandrel, which reduces wall thickness and increases tube length. The tube is fed longitudinally, and rotated on its axis to distribute the cold work uniformly around the entire circumference. Feeding and rotating are synchronized with die motion and take place after the dies have completed their forward stroke.

Tube reducing may be used for all alloys that can be drawn on draw benches. Slight changes in die design and operating conditions may be required to accommodate different alloys. Small-diameter tubes may be produced by block or bench drawing following tube reducing.

PRODUCT SPECIFICATIONS

Copper and copper alloy tube and pipe are available in a wide variety of nominal diameters and wall thicknesses, from small-diameter capillary tube to 12-in. nominal diameter pipe. The standard dimensions and tolerances for several different kinds of copper tube and pipe are given in the ASTM specifications listed in Table 3, along with other requirements for the tubular products. Seamless copper tube for automotive applications ($\frac{1}{8}$ to $\frac{3}{4}$ in. nominal diameter) is covered by SAE J528. Requirements for copper tube and pipe to be used in condensers, heat exchangers, economizers and similar unfired pressure vessels are given in the ASME specifications listed in Table 3. (ASME materials specifications are almost always identical to ASTM specifications having the same numerical designation; for example, ASME SB111 is identical to ASTM B111.) Certain tube alloys are covered in AMS specifications, which apply to materials for aerospace applications. These are given below:

Table 3. ASTM and ASME specifications for copper tube and pipe

Tubular product	ASTM	ASME
Seamless pipe and tube, copper-nickel alloy . . .B466		SB466
Seamless pipe and tube, copper-silicon alloy B315		SB315
Seamless pipe and tube, for electrical conductors B188		. . .
Seamless pipe, standard sizes . .B42		. . .
Seamless pipe, threadless B302		. . .
Seamless tube B75		SB75
Seamless tube, brass B135		SB135
Seamless tube, bright annealed B68		. . .
Seamless tube, capillary, hard-drawn B360		. . .
Seamless tube, condenser and heat exchanger B111, B395		SB111, SB395
Seamless tube, condenser and heat exchanger, with integral fins B359		SB359
Seamless tube, for air conditioning and refrigeration service B280		. . .
Seamless tube, drainage B306		. . .
Seamless tube, general requirements B251		. . .
Seamless tube, rectangular waveguide B372		. . .
Seamless tube, water B88		. . .
Welded pipe and tube, copper-nickel alloy B467		SB467
Welded tube, C10800 and 12000 . . . . .B543		SB543
Welded tube, all other coppers. .B447		. . .

AMS specification	Product	Copper alloy
4555	Seamless brass tube, light annealed	C26000, C33000
4558	Seamless brass tube, drawn	C33200
4625	Phosphor bronze, hard temper	C51000
4640	Aluminum bronze	C63000
4665	Seamless silicon bronze tube, annealed	C65500

Copper Wire and Cable

WIRE made from copper and its alloys has been used since about 2000 to 3000 BC. According to archaeological evidence, the ancient Assyrians, Babylonians and Egyptians were skilled in producing copper wire for ornamental purposes. Drawing wire through a die is a much more modern development. Development of wiredrawing processes during the Middle Ages concentrated to a large extent on drawing iron and steel wires to make pins and instrument strings. But with the invention of the electric telegraph in 1847 came the requirement for long continuous lengths of electric conductor wire made of copper. In 1850, copper wire was used to make a submarine cable connecting England and France.

At the beginning of the twentieth century, wire was still being drawn through single dies — a process commonly known as "bull-block" drawing. Dies were made by punching a series of holes in a steel plate.

Multiple wiredrawing machines were introduced about 1900. As a result, chilled cast iron plates and dies that could be reamed to size replaced the punch-sized steel-plate dies. Lubricants were introduced because of the considerable heat generated by friction between the wire and draw-capstan and by successive reductions through progressively smaller dies. In turn, use of lubricants permitted wire to be drawn at faster speeds.

The prime development during the 1920's was the introduction of drawing dies made of tungsten carbide. High hardness and lack of porosity make tungsten carbide dies ideal for high-speed wiredrawing. Tungsten carbide dies are standard today. For very fine wire sizes, below about 1.3 mm (0.05 in.), diamond dies are used.

CLASSIFICATION OF COPPER FOR CONDUCTORS

Copper metals used for electrical conductors fall into three general categories: high-conductivity coppers, high-copper alloys, and electrical bronzes.

High-conductivity coppers are covered by ASTM specifications B4, B5, B170, B442 and B623. ASTM B4 covers both high-resistance Lake copper and low-resistance Lake copper. Lake copper is fire refined from local Lake Superior ore deposits. ASTM B5 covers copper electrolytically refined from blister copper, converter

Table 1. Characteristics of solid round copper wire: ASTM B1, B3, B258

| Conductor size, AWG | Conductor diameter, mils | Conductor area, circular mils | Net weight, lb/1000 ft | Soft (annealed) wire | | Hard drawn wire | | |
				Minimum elongation in 10 in., %	Nominal resistance, Ω/1000 ft	Nominal breaking strength, lb	Nominal tensile strength, ksi	Nominal resistance, Ω/1000 ft
4/0	460.0	211 600	640.5	35	0.0491	8143	49.0	0.05044
3/0	409.6	167 800	507.8	35	0.06180	6720	51.0	0.06361
2/0	364.8	133 100	402.8	35	0.07791	5519	52.8	0.08019
1/0	324.9	105 600	319.5	35	0.09821	4518	54.5	0.1011
1	289.3	83 690	253.3	30	0.1239	3888	56.1	0.1289
2	257.6	66 360	200.9	30	0.1563	3002	57.6	0.1625
3	229.4	52 620	159.3	30	0.1971	2439	59.0	0.2050
4	204.3	41 740	126.3	30	0.2485	1970	60.1	0.2584
5	181.9	33 090	100.2	30	0.3134	1590	61.2	0.3259
6	162.0	26 240	79.44	30	0.3952	1280	62.1	0.4110
7	144.3	20 820	63.03	30	0.4981	1030	63.1	0.5180
8	128.5	16 510	49.98	30	0.6281	826.1	63.7	0.6532
9	114.4	13 090	39.62	30	0.7923	660.9	64.3	0.8239
10	101.9	10 380	31.43	25	0.9991	529.3	64.9	1.039
11	90.7	8 230	24.9	25	1.26	423	65.4	1.31
12	80.8	6 530	19.8	25	1.59	337	65.7	1.65
13	72.0	5 180	15.7	25	2.00	268	65.9	2.08
14	64.1	4 110	12.4	25	2.52	214	66.2	2.62
15	57.1	3 260	9.87	25	3.18	170	66.4	3.31
16	50.8	2 580	7.81	25	4.02	135	66.6	4.18
17	45.3	2 050	6.21	25	5.06	108	66.8	5.26
18	40.3	1 620	4.92	25	6.40	85.5	67.0	6.66
19	35.9	1 290	3.90	25	8.04	68.0	67.2	8.36
20	32.0	1 020	3.10	25	10.2	54.2	67.4	10.6
21	28.5	812	2.46	25	12.8	43.2	67.7	13.3
22	25.3	640	1.94	25	16.2	34.1	67.9	16.9
23	22.6	511	1.55	25	20.3	27.3	68.1	21.1
24	20.1	404	1.22	20	25.7	21.7	68.3	26.7
25	17.9	320	0.970	20	32.4	17.3	68.6	33.7
26	15.9	253	0.765	20	41.0	13.7	68.8	42.6
27	14.2	202	0.610	20	51.4	10.9	69.0	53.4
28	12.6	159	0.481	20	65.2	8.64	69.3	67.8
29	11.3	128	0.387	20	81.0	6.96	69.4	84.3
30	10.0	100	0.303	15	104.0	5.47	69.7	108.0
31	8.9	79.2	0.240	15	131.0	4.35	69.9	136.0
32	8.0	64.0	0.194	15	162.0	3.53	70.2	169.0
33	7.1	50.4	0.153	15	206.0	2.79	70.4	214.0
34	6.3	39.7	0.120	15	261.0	2.20	70.6	272.0
35	5.6	31.4	0.0949	15	330.0	1.75	70.9	343.0
36	5.0	25.0	0.0757	15	415.0	1.40	71.1	431.0
37	4.5	20.2	0.0613	15	513.0	1.13	71.3	534.0
38	4.0	16.0	0.0484	15	648.0	0.898	71.5	674.0
39	3.5	12.2	0.0371	15	850.0	0.691	71.8	884.0
40	3.1	9.61	0.0291	15	1079.0	0.543	72.0	1122.0
41	2.8	7.84	0.0237	15	1323.0	0.443	72.0	1376.0
42	2.5	6.25	0.0189	15	1659.0	0.353	72.0	1726.0
43	2.2	4.48	0.0147	15	2143.0	0.274	72.0	2228.0
44	2.0	4.00	0.0121	15	2593.0	0.226	72.0	2696.0

copper, black copper or Lake copper. ASTM B170 covers oxygen-free electrical copper.

Some specialty coppers are produced by adding minimal amounts of hardening agents (such as chromium, tellurium, beryllium, cadmium or zirconium). These are used in applications where high anneal resistance is required.

A series of bronzes has been developed for use as conductors; these alloys are covered by ASTM B105. These bronzes are intended to provide better corrosion resistance and higher tensile strengths than standard conductor coppers.

CLASSIFICATION OF WIRE AND CABLE

Round Wire. Standard nominal diameters and cross-sectional areas of solid round copper wires used as electrical conductors are prescribed in ASTM B258. Wire sizes have almost always been designated in the American Wire Gauge (AWG) system. This system is based on fixed diameters for two wire sizes (4/0 and 36 AWG, respectively) with a geometric progression of wire diameters for the thirty-eight intermediate gages and for gages smaller than 36 AWG (see Table 1).

This is an inverse series in which a higher number denotes a smaller wire diameter. Each increase of one AWG number is approximately equivalent to a 20.7% reduction in cross-sectional area.

ASTM B1 specifies hard drawn round wire that has been reduced at least four AWG numbers (60% reduction in area) and ASTM B3 specifies soft (or annealed) copper wire. Table 1 gives the sizes and properties of wires specified in ASTM B1, B3 and B258.

Square and Rectangular Wire. ASTM B48 specifies soft (annealed) square and rectangular copper wire.

Stranded wire is normally used in electrical applications where some degree of flexing is encountered either in service or during installation. In order of increasing flexibility, the common forms of stranded wire are: concentric lay, unilay, rope lay and bunched.

Concentric-lay stranded wire and cable are composed of a central wire surrounded by one or more layers of helically laid wires, with the direction of lay reversed in successive layers, and with the length of lay increased for each succes-

sive layer. The outer layer usually has a left-hand lay.

ASTM B8 establishes five classes of concentric-lay stranded wire and cable, from AA (the coarsest) to D (the finest). Details of concentric-lay constructions are given in Table 2.

Unilay stranded wire is composed of a central core surrounded by more than one layer of helically laid wires, all layers having a common lay length and direction. This type of wire sometimes is referred to as "smooth bunch." The layers usually have a left-hand lay.

Rope-lay stranded wire and cable are composed of a stranded member (or members) as a central core, around which are laid one or more helical layers of similar stranded members. The members may be concentric or bunch-stranded. ASTM B173 and B172 establish five classes of rope-lay stranded conductors: classes G and H, which have concentric members; and classes I, K and M, which have bunched members. Construction details are shown in Tables 3 and 4. These cables are normally used to make large, flexible conductors for portable service, such as mining cable or apparatus cable.

Table 2. Characteristics of concentric-lay stranded copper conductors: ASTM B8

Conductor size, circular mils or AWG	Nominal weight lb/1000 ft	Nominal resistance (uncoated wire), Ω/1000 ft	Class AA		Class A		Class B		Class C		Class D	
			Number of wires	Diameter of individual wires, mils	Number of wires	Diameter of individual wires, mils	Number of wires	Diameter of individual wires, mils	Number of wires	Diameter of individual wires, mils	Number of wires	Diameter of individual wires, mils
5 000 000	15 890	0.002 178	…	…	169	172.0	217	151.8	271	135.8	271	135.8
4 500 000	14 300	0.002 420	…	…	169	163.2	217	144.0	271	128.9	271	128.9
4 000 000	12 590	0.002 696	…	…	169	153.8	217	135.8	271	121.5	271	121.5
3 500 000	11 020	0.003 082	…	…	127	166.0	169	143.9	217	127.0	271	113.6
3 000 000	9 353	0.003 561	…	…	127	153.7	169	133.2	217	117.6	271	105.2
2 500 000	7 794	0.004 278	…	…	91	165.7	127	140.3	169	121.6	217	107.3
2 000 000	6 175	0.005 289	…	…	91	148.2	127	125.5	169	108.8	217	96.0
1 900 000	5 886	0.005 568	…	…	91	144.5	127	122.3	169	106.0	217	93.6
1 800 000	5 558	0.005 877	…	…	91	140.6	127	119.1	169	103.2	217	91.1
1 750 000	5 403	0.006 045	…	…	91	138.7	127	117.4	169	101.8	217	89.8
1 700 000	5 249	0.006 223	…	…	91	136.7	127	115.7	169	100.3	217	88.5
1 600 000	4 940	0.006 612	…	…	91	132.6	127	112.2	169	97.3	217	85.9
1 500 000	4 631	0.007 052	…	…	61	156.6	91	128.4	127	108.7	169	94.2
1 400 000	4 323	0.007 556	…	…	61	151.5	91	124.0	127	105.0	169	91.0
1 300 000	4 014	0.008 137	…	…	61	146.0	91	119.5	127	101.2	169	87.7
1 250 000	3 859	0.008 463	…	…	61	143.1	91	117.2	127	99.2	169	86.0
1 200 000	3 705	0.008 815	…	…	61	140.3	91	114.8	127	97.2	169	84.3
1 100 000	3 396	0.009 617	…	…	61	134.3	91	109.9	127	93.1	169	80.7
1 000 000	3 088	0.010 88	37	164.4	61	128.0	61	128.0	91	104.8	127	88.7
900 000	2 779	0.011 75	37	156.0	61	121.5	61	121.5	91	99.4	127	84.2
800 000	2 470	0.013 22	37	147.0	61	114.5	61	114.5	91	93.8	127	79.4
750 000	2 316	0.014 10	37	142.4	61	110.9	61	110.9	91	90.8	127	76.8
700 000	2 161	0.015 11	37	137.5	61	107.1	61	107.1	91	87.7	127	74.2
650 000	2 007	0.016 27	37	132.5	61	103.2	61	103.2	91	84.5	127	71.5
600 000	1 853	0.017 63	37	127.3	37	127.3	61	99.2	91	81.2	127	68.7
550 000	1 698	0.019 23	37	121.9	37	121.9	61	95.0	91	77.7	127	65.8
500 000	1 544	0.021 16	19	162.2	37	116.2	37	116.2	61	90.5	91	74.1
450 000	1 389	0.023 51	19	153.9	37	110.3	37	110.3	61	85.9	91	70.3
400 000	1 235	0.026 45	19	145.1	19	145.1	37	104.0	61	81.0	91	66.3
350 000	1 081	0.030 22	12	170.8	19	135.7	37	97.3	61	75.7	91	62.0
300 000	926.3	0.035 26	12	158.1	19	125.7	37	90.0	61	70.1	91	57.4
250 000	771.9	0.042 31	12	144.3	19	114.6	37	82.2	61	64.0	91	52.4
4/0	653.3	0.049 99	7	173.9	7	173.9	19	105.5	37	75.6	61	58.9
3/0	518.1	0.063 04	7	154.8	7	154.8	19	94.0	37	67.3	61	52.4
2/0	410.9	0.079 48	7	137.9	7	137.9	19	83.7	37	60.0	61	46.7
1/0	326.0	0.100 2	7	122.8	7	122.8	19	74.5	37	53.4	…	…
1	258.4	0.126 4	3	167.0	7	109.3	19	66.4	37	47.6	…	…
2	204.9	0.159 4	3	148.7	7	97.4	7	97.4	19	59.1	…	…
3	162.5	0.201 0	3	132.5	7	86.7	7	86.7	19	52.6	…	…
4	128.9	0.253 4	3	118.0	7	77.2	7	77.2	19	46.9	…	…
5	102.2	0.319 7	…	…	…	…	7	68.8	19	41.7	…	…
6	81.05	0.403 1	…	…	…	…	7	61.2	19	37.2	…	…
7	64.28	0.508 1	…	…	…	…	7	54.5	19	33.1	…	…
8	50.98	0.640 7	…	…	…	…	7	48.6	19	29.5	…	…
9	40.42	0.808 1	…	…	…	…	7	43.2	19	28.2	…	…

Bunch stranded wire is composed of any number of wires twisted together in the same direction without regard to geometric arrangement of the individual strands. ASTM B174 provides for five classes (I, J, K, L and M); these conductors are commonly used in flexible cords, hookup wires and special flexible welding conductors. Typical construction details are given in Table 5.

Tin-coated Wire. Solid and stranded wires are available with tin coatings. These are manufactured to the latest revisions of ASTM B33, which covers soft or annealed tinned copper wires, and B246, which covers hard drawn or medium hard drawn tinned copper wires. Characteristics of tinned round solid wire are given in Table 6.

FABRICATION OF WIRE ROD

Wire rod is the intermediate product in the manufacture of wire. Although "wire rod" is the term used in the U.S. for the intermediate product, the term "drawing stock" is used in international standards and customs documents.

Rolling. The traditional process for converting prime copper into wire rod involves hot rolling of cast wirebar. Almost all drawing stock is rolled to 8 mm (0.32 in.) diameter. Larger sizes, up to 22 mm (0.87 in.) or more in diameter are available on special order.

Some special oxygen-free copper wirebar is produced by vertical casting, but most wirebar is produced by horizontal casting of tough-pitch copper into open molds. The oxygen content is controlled at 0.03 to 0.06% to give a level surface. Cast wirebars weigh about 110 to 135 kg (250 to 300 lb) each. Their ends are tapered to facilitate entry into the first pass of the hot rolling mill.

Prior to rolling, bars are heated to about 925 °C (1700 °F) in a neutral atmosphere and then rolled on a continuous mill through a series of reductions to yield round rod about 6 to 22 mm ($1/4$ to $7/8$ in.) in diameter. The hot rolled rod is coiled, water quenched, and then pickled to remove the black cupric oxide that forms during rolling. This method can produce rod at rates up to 7.5 kg/s (30 tons/h).

Disadvantages of this process include (a) high capital investment to achieve low operating cost; (b) relatively small coils that must be welded together for efficient production, where the welded junctions present potential sources of weakness in subsequent wiredrawing operations; and (c) unsuitability of rod rolled from cast wirebars for certain specialized wire applications.

Because of the disadvantages inherent in producing rolled rod from conventionally cast wirebars processes have been developed for continuously converting liquid metal directly into wire rod, thus avoiding the intermediate wirebar stage.

Continuous Casting. In 1963, the first plant for continuous casting and rolling of copper wirebar went into operation. This process, which was developed by the Southwire Co. in conjunction with Western Electric Co., was essentially an adaptation of a process developed by Ilario Properzi prior to 1950—a process that had been used for many years by aluminum and zinc producers for conversion of prime metal or scrap into wire rod.

Continuously cast wire rod has come to dominate the copper wire rod market and now accounts for more than 1.8 billion kg (2 million tons) annually, or about 50% of the total amount of wire rod produced.

Other casting systems that have been developed include the Properzi, Contirod, and General Electric (G.E.) dip-form systems. Smaller-capacity machines have been developed by Ou-

Table 3. Characteristics of rope-lay stranded copper conductors having uncoated or tinned concentric members: ASTM B173

Conductor sizes, circular mils or AWG	Class G				Class H				Conductor sizes, circular mils or AWG	Class G				Class H			
	Diameter of individual wires, mils	No. of ropes	No. of wires each rope	Net weight, lb/1000 ft	Diameter of individual wires, mils	No. of ropes	No. of wires each rope	Net weight, lb/1000 ft		Diameter of individual wires, mils	No. of ropes	No. of wires each rope	Net weight, lb/1000 ft	Diameter of individual wires, mils	No. of ropes	No. of wires each rope	Net weight, lb/1000 ft
5 000 000	65.7	61	19	16 052	53.8	91	19	15 057	600 000	37.5	61	7	1 908	29.2	37	19	1 923
4 500 000	62.3	61	19	14 434	51.0	91	19	14 429	550 000	35.9	61	7	1 749	28.0	37	19	1 768
4 000 000	58.7	61	19	12 814	48.1	91	19	12 835	500 000	43.9	37	7	1 579	34.2	61	7	1 587
3 500 000	55.0	61	19	11 249	45.0	91	19	11 234	450 000	41.7	37	7	1 425	32.5	61	7	1 433
3 000 000	50.9	61	19	9 635	41.7	91	19	9 647	400 000	39.3	37	7	1 265	30.6	61	7	1 271
2 500 000	59.6	37	19	8 012	46.4	61	19	8 006	350 000	36.8	37	7	1 109	28.6	61	7	1 110
2 000 000	53.3	37	19	6 408	41.5	61	19	6 405	300 000	34.0	37	7	947.1	26.5	61	7	953.0
1 900 000	52.0	37	19	6 099	40.5	61	19	6 100	250 000	31.1	37	7	792.4	24.2	61	7	794.8
1 800 000	50.6	37	19	5 775	39.4	61	19	5 773	4/0	39.9	19	7	666.6	28.6	37	7	670.1
1 750 000	49.9	37	19	5 617	38.9	61	19	5 627	3/0	35.5	19	7	527.7	25.5	37	7	532.7
1 700 000	49.2	37	19	5 460	38.3	61	19	5 455	2/0	31.6	19	7	418.1	22.7	37	7	422.2
1 600 000	47.7	37	19	5 132	37.2	61	19	5 146	1/0	28.2	19	7	333.0	20.7	37	7	334.3
1 500 000	59.3	61	7	4 772	46.2	37	19	4 815	1	25.1	19	7	263.8	18.0	37	7	265.4
1 400 000	57.3	61	7	4 456	44.6	37	19	4 487	2	36.8	7	7	206.9	22.3	19	7	208.2
1 300 000	55.2	61	7	4 135	43.0	37	19	4 171	3	37.8	7	7	164.4	19.9	19	7	165.8
1 250 000	54.1	61	7	3 972	42.2	37	19	4 017	4	29.2	7	7	130.3	17.7	19	7	131.2
1 200 000	53.0	61	7	3 814	41.3	37	19	3 847	5	26.0	7	7	103.3	15.8	19	7	104.5
1 100 000	50.8	61	7	3 502	39.6	37	19	3 537	6	23.1	7	7	81.52	14.0	19	7	82.06
1 000 000	48.4	61	7	3 179	37.7	37	19	3 206	7	20.6	7	7	64.83	12.5	19	7	65.42
900 000	45.9	61	7	2 859	35.8	37	19	2 891	8	18.4	7	7	51.72	11.1	19	7	51.59
800 000	43.3	61	7	2 544	33.7	37	19	2 562	9	15.3	7	7	40.59	9.9	19	7	41.04
750 000	41.9	61	7	2 383	32.7	37	19	2 412	10	14.6	7	7	32.57	...	...	...	...
700 000	40.5	61	7	2 226	31.6	37	19	2 252	12	11.5	7	7	20.20	...	...	...	...
650 000	39.0	61	7	2 064	30.4	37	19	2 085	14	9.2	7	7	12.93	...	...	...	...

Table 4. Characteristics of rope-lay stranded copper conductors having uncoated or tinned bunched members: ASTM B172

Conductor size, circular mils or AWG	Class of strand	Construction and wire size, AWG	Total No. of wires	Approx diameter, in.	Net weight, lb/1000 ft	Conductor size, circular mils or AWG	Class of strand	Construction and wire size, AWG	Total No. of wires	Approx diameter, in.	Net weight, lb/1000 ft
1 000 000	I	19×7×19/24	2 527	1.290	3306	250 000	K	7×7×61/30	2 499	0.638	802
	K	37×7×39/30	10 101	1.329	3272		M	19×7×48/34	6 384	0.658	821
	M	61×7×59/34	25 193	1.353	3239	4/0	I	19×28/24	532	0.569	683
900 000	I	19×7×17/24	2 261	1.217	2959		K	7×7×43/30	2 107	0.584	676
	K	37×7×35/30	9 065	1.255	2936		M	19×7×40/34	5 320	0.598	684
	M	61×7×53/34	22 631	1.279	2909	3/0	I	19×22/24	418	0.502	537
800 000	I	19×7×15/24	1 995	1.140	2611		K	7×7×34/30	1 666	0.516	535
	K	19×7×60/30	7 980	1.174	2585		M	19×7×32/34	4 256	0.532	547
	M	61×7×47/34	20 069	1.200	2580	2/0	I	19×18/24	342	0.452	439
750 000	I	19×7×14/24	1 862	1.099	2437		K	7×7×27/30	1 323	0.457	424
	K	19×7×57/30	7 581	1.143	2455		M	19×7×25/34	3 325	0.467	427
	M	61×7×44/34	18 788	1.160	2415	1/0	I	19×14/24	266	0.396	342
700 000	I	19×7×13/24	1 729	1.057	2262		K	19×56/30	1 064	0.408	338
	K	19×7×52/30	6 916	1.089	2240		M	7×7×54/34	2 646	0.414	337
	M	61×7×41/34	17 507	1.117	2251	1	I	7×30/24	210	0.350	267
650 000	I	19×7×12/24	1 596	1.014	2088		K	19×44/30	836	0.359	266
	K	19×7×49/30	6 517	1.056	2111		M	7×7×43/34	2 107	0.368	268
	M	61×7×38/34	16 226	1.074	2086	2	I	7×23/24	161	0.304	205
600 000	I	7×7×30/24	1 470	0.971	1906		K	19×35/30	665	0.319	211
	K	19×7×45/30	5 985	1.010	1938		M	7×7×34/34	1 666	0.325	212
	M	61×7×35/34	14 945	1.028	1921	3	I	7×19/24	133	0.275	169
550 000	I	7×7×28/24	1 372	0.936	1779		K	19×28/30	532	0.283	169
	K	19×7×41/30	5 453	0.961	1766		M	7×7×27/34	1 323	0.288	168
	M	61×7×32/34	13 664	0.981	1757	4	I	7×15/24	105	0.243	134
500 000	I	7×7×25/24	1 225	0.882	1588		K	7×60/30	420	0.250	132
	K	19×7×38/30	5 054	0.924	1637		M	19×56/34	1 064	0.257	134
	M	37×7×49/34	12 691	0.900	1631	5	I	7×12/24	84	0.216	107
450 000	I	7×7×23/24	1 127	0.845	1461		K	7×48/30	336	0.223	106
	K	19×7×34/30	4 522	0.871	1465		M	19×44/34	836	0.226	105
	M	37×7×44/34	11 396	0.892	1465	6	I	7×9/24	63	0.186	80
400 000	I	7×7×20/24	980	0.785	1270		K	7×38/30	266	0.197	84
	K	19×7×30/30	3 990	0.816	1292		M	19×35/34	665	0.201	84
	M	37×7×39/34	10 101	0.837	1298	7	K	7×30/30	210	0.174	66
350 000	I	7×7×18/24	882	0.743	1143		M	19×28/34	532	0.178	67
	K	19×7×26/30	3 458	0.757	1120	8	K	7×/30	168	0.155	53
	M	37×7×34/34	8 806	0.779	1132		M	7×60/34	420	0.158	53
300 000	I	7×7×15/24	735	0.675	953	9	K	7×19/30	133	0.137	42
	K	7×7×61/30	2 989	0.701	959		M	7×48/34	336	0.140	42
	M	19×7×57/34	7 581	0.720	975	10	M	7×37/34	259	0.122	33
250 000	I	7×7×13/24	637	0.626	826	12	M	7×24/34	168	0.097	21

Table 5. Characteristics of bunch stranded copper conductors having uncoated or tinned members: ASTM B174

Conductor size, AWG	Class of strand	Number and size of wire, AWG	Approx. diameter, in.	Approx. weight, lb/1000 ft	Conductor size, AWG	Class of strand	Number and size of wire, AWG	Approx. diameter, in.	Approx. weight, lb/1000 ft
7	I	52/24	0.168	64.9	16	J	16/28	0.057	7.84
8	I	41/24	0.148	51.1		K	26/30	0.058	8.03
9	I	33/24	0.132	41.2		L	41/32	0.059	8.10
10	I	26/24	0.117	32.4		M	65/34	0.059	7.97
	J	65/28	0.118	31.9	18	J	10/28	0.044	4.90
	K	104/30	0.120	32.1		K	16/30	0.045	4.94
12	J	41/28	0.093	20.1		L	26/32	0.046	5.14
	K	65/30	0.094	20.1		M	41/34	0.046	5.02
	L	104/32	0.096	20.6	20	J	7/28	0.038	3.43
14	J	26/28	0.073	12.7		K	10/30	0.035	3.09
	K	41/30	0.074	12.7		L	16/32	0.036	3.16
	L	65/32	0.075	12.8		M	26/34	0.037	3.19
	M	104/34	0.076	12.7					

Table 6. Characteristics of tinned solid round copper wire: ASTM B33, B246, B258

Conductor size, AWG	Net weight, lb/1000 ft	Soft (annealed) wire Nominal resistance, Ω/1000 ft	Soft (annealed) wire Minimum elongation in 10 in., %	Hard drawn wire Nominal resistance, Ω/1000 ft	Hard drawn wire Minimum breaking strength, lb	Conductor size, AWG	Net weight, lb/1000 ft	Soft (annealed) wire Nominal resistance, Ω/1000 ft	Soft (annealed) wire Minimum elongation in 10 in., %	Hard drawn wire Nominal resistance, Ω/1000 ft	Hard drawn wire Minimum breaking strength, lb
2	200.9	0.1609	25	...	...	18	4.92	6.66	20		
3	159.3	0.2028	25	...	...	19	3.90	8.36	20		
4	126.3	0.2557	25	0.2680	1773	20	3.10	10.6	20		
5	100.2	0.3226	25	0.3380	1432	21	2.46	13.3	20		
6	79.44	0.4067	25	0.4263	1152	22	1.94	16.9	20		
7	63.03	0.5127	25	0.5372	927.3	23	1.55	21.1	20		
8	49.98	0.6465	25	0.6776	743.1	24	1.22	26.7	15		
9	39.62	0.8154	25	0.8545	595.1	25	0.970	34.4	15		
10	31.43	1.039	20	1.087	476.1	26	0.765	43.5	15		
11	24.9	1.31	20	1.37	381.0	27	0.610	54.5	15		
12	19.8	1.65	20	1.73	303.0	28	0.481	69.3	15		
13	15.7	2.08	20	2.18	241.0	29	0.387	86.1	15		
14	12.4	2.62	20	2.74	192.0	30	0.303	110.0	10		
15	9.87	3.31	20	3.46	153.0	31	0.204	141.0	10		
16	7.81	4.18	20	4.37	121.0	32	0.194	174.0	10		
17	6.21	5.26	20			33	0.153	221.0	10		
						34	0.120	281.0	10		

tokumpu, Davy, Wertli, Lamitref and others.

Advantages of continuous casting and rolling include (a) large coil weights, up to 10 Mg (11 tons); (b) ability to reprocess scrap at considerable savings; (c) improved rod quality and surface condition; (d) homogeneous metallurgical conditions and close process control; and (e) low capital investment and low operating costs for moderate production rates.

The standard feed for continuous casting processes is cathode copper, which is charged directly into a melting furnace. An ASARCO shaft furnace is used for the Southwire, Properzi, and Contirod systems, but an electric furnace is preferred for the smaller systems, such as the G.E. dip-form process.

Southwire Continuous Rod System. In the Southwire system, the casting machine produces a cast copper bar 2500 to 3000 mm² (4 to 5 in.²) in cross-sectional area by pouring molten copper between a grooved wheel made of steel or copper and a steel band. The cast bar is cooled by water sprays as the wheel rotates, and is withdrawn by pinch rolls as it exits from the wheel. Next it goes to a bar-conditioning unit, and then to the rolling mill. After passing through a series of reductions in the mill, the rod enters an in-line pickling system, which quenches and cleans the rod prior to final coiling.

Properzi System. The Properzi system is basically the same as the Southwire process except for slight differences in design of the casting wheel and configuration of the rod mill.

Contirod Process. The Contirod process was developed jointly by Hazelett and Metallurgic Hoboken Overpelt (Belgium). Molten copper is passed between two water-cooled, counterrotating steel belts having specially cooled side dams. The resulting bar is sent through a conventional Krupp rolling mill similar to that used in the Southwire system. Production rates of about 6.3 to 7.5 kg/s (25 to 30 tons/h) have been achieved.

G.E. Dip-Form Process. The General Electric dip-form process was introduced in 1964. A seed rod approximately 9 mm (0.35 in.) in diameter is passed at a controlled rate through a bath of molten C10100. Copper freezes onto the seed rod, thickening its diameter to about 16.5 mm (0.65 in.). The rod emerging from the copper bath is hot rolled on a 2-stand mill, cooled and coiled. The entire operation is performed in a controlled atmosphere, from the time C10100 cathodes enter the melting furnace until the rod emerges. This rod is generally used for production of fine and ultrafine wires. Production rates up to 2.5 kg/s (10 tons/h) can be obtained.

Outokumpu Process. The Outokumpu process is a continuous casting process. Metal is pulled through a graphite die where it solidifies. One end of the die extends into the melt and the other is surrounded by a water-cooled jacket. The Outokumpu process differs from other continuous casting processes in that the direction of withdrawal is vertically upwards. A 12-strand plant can produce up to 12.7 Gg (14 000 tons) of oxygen-free rod annually.

WIREDRAWING AND WIRE STRANDING

Preparation of Rod. In order to provide a wire of good surface quality, it is necessary to have clean wire rod with a smooth, oxide-free surface. Conventional hot rolling cleaning must be a separate operation. With continuous casting, which provides better surface quality, the rod passes through a cleaning station as it exits from the rolling mill.

The standard method for cleaning copper wire rod is pickling in hot 20% sulfuric acid followed by rinsing in water. When fine wire is being produced, it is necessary to provide rod of even better surface quality. This can be achieved in a number of ways. One is by open-flame annealing of cold drawn rod—that is, heating to 700 °C (1300 °F) in an oxidizing atmosphere. This eliminates shallow discontinuities. A more common practice, especially for fine magnet-wire applications, is die shaving, in which rod is drawn through a circular cutting die made of steel or carbide to remove approximately 0.13 mm (0.005 in.) from the entire surface of the rod.

Wiredrawing. Single-die machines called "bull blocks" are used for drawing special heavy sections such as trolley wire. Drawing speeds range from about 1 to 2.5 m/s (200 to 500 ft/min). Tallow is generally used as the lubricant and the wire is drawn through hardened steel or tungsten carbide dies. In some instances, multiple-draft tandem bull blocks (in sets of 3 or 5 passes) are used instead of single-draft machines.

Tandem drawing machines having 10 to 12 dies

for each machine are used for breakdown of hot rolled or continuous cast copper rod. The rod is reduced in diameter from 8.3 mm (0.325 in.) to about 2 mm (0.08 in.) by drawing it through dies at speeds up to 25 m/s (5000 ft/min). The drawing machine operates continuously; the operator merely welds the end of each rod coil to the start of the next coil.

Intermediate and fine wires are drawn on smaller machines that have 12 to 20 or more dies each. The wire is reduced in steps of 20 to 25% in cross-sectional area. Intermediate machines can produce wire as small as 0.5 mm (0.020 in.) in diameter, and fine wire machines can produce wire in diameters from 0.5 mm to less than 0.25 mm (0.010 in.). Drawing speeds are typically 25 to 30 m/s (5000 to 6000 ft/min) and may be even higher.

All drawing is performed with a copious supply of lubricant to cool the wire and prevent rapid die wear. Traditional lubricants are soap and fat emulsions, which are fed to all machines from a central reservoir. Breakdown of rod usually requires a lubricant concentration of about 7%; drawing of intermediate and fine wires, concentrations of 2 to 3%. Today, synthetic lubricants are becoming more widely accepted.

Drawn wire is collected on reels or stem packs, depending on the next operation. Fine wire is collected on reels carrying as little as 4.5 kg (10 lb); large-diameter wire, on stem packs carrying up to 450 kg (1000 lb). To ensure continuous operation, many drawing machines are equipped with dual take-up systems. When one reel is filled, the machine automatically flips the wire onto an adjacent empty reel and simultaneously cuts the wire. This permits the operator to unload the full reel and replace it with an empty one without stopping the wiredrawing operation.

Production of Flat or Rectangular Wire. Depending on size and quantity, flat or rectangular wire is drawn on bull block machines or Turk's-head machines, or is rolled on tandem rolling mills with horizontal and vertical rolls. Larger quantities are produced by rolling, smaller ones by drawing.

Annealing. Wiredrawing, like any other cold working operation, increases tensile strength and reduces ductility of copper. Although it is possible to cold work copper up to 99% reduction in area, copper wire usually is annealed after 90% reduction.

In some plants, electrical-resistance heating methods are used to fully anneal copper wire as it exits from the drawing machines. Wire coming directly from drawing passes over suitably spaced contact pulleys that carry the electrical current necessary to heat the wire above its recrystallization temperature in less than a second.

In plants where batch annealing is practiced, drawn wire is treated either in a continuous tunnel furnace, where reels travel through a neutral or slightly reducing atmosphere and are annealed during transit, or in batch bell furnaces under a similar protective atmosphere. Annealing temperatures range from 400 to 600 °C (750 to 1100 °F) depending chiefly on wire diameter and reel weight.

Wire Coating. Four basic coatings are used on copper conductors for electrical applications:

1 Lead, or lead alloy
  (80Pb-20Sn) . . . .(ASTM B189)
2 Nickel . . . .(ASTM B355)
3 Silver . . . .(ASTM B298)
4 Tin . . . .(ASTM B33)

Coatings are applied to (a) retain solderability for hookup-wire applications, (b) provide a barrier between the copper and insulation materials such as rubber, that would react with the copper and adhere to it (thus making it difficult to strip insulation from the wire to make an electrical connection) or (c) prevent oxidation of the copper during high-temperature service.

Tin-lead alloy coatings and pure tin coatings are the most common; nickel and silver are used for specialty and high-temperature applications.

Copper wire can be coated by hot dipping in a molten metal bath, electroplating or cladding. With the advent of continuous process, electroplating has become the dominant process, especially because it can be done "on line" following the wiredrawing operation.

Stranded wire is produced by twisting or braiding several wires together to provide a flexible cable.

Different degrees of flexibility for a given current-carrying capacity can be achieved by varying the number, size and arrangement of individual wires. Solid wire, concentric strand, rope strand and bunched strand provide increasing degrees of flexibility; within the last three categories, a larger number of finer wires provides greater flexibility.

Stranded copper wire and cable are made on machines known as "bunchers" or "stranders." Conventional bunchers are used for stranding small-diameter wires (34 AWG up to 10 AWG). Individual wires are payed off reels located alongside the equipment and are fed over flyer arms that rotate about the take-up reel to twist the wires. The rotational speed of the arm relative to the take-up speed controls the length of lay in the bunch. For small, portable, flexible cables, individual wires are usually 30 to 34 AWG, and there may be as many as 150 wires in each cable.

A tubular buncher has up to 18 wire-payoff reels mounted inside the unit. Wire is taken off each reel while it remains in a horizontal plane, is threaded along a tubular barrel and is twisted together with other wires by a rotating action of the barrel. At the take-up end, the strand passes through a closing die to form the final bunch configuration. The finished strand is wound onto a reel that also remains within the machine.

Supply reels in conventional stranders for large-diameter wire are fixed onto a rotating frame within the equipment and revolve about the axis of the finished conductor. There are two basic types of machines. In one, known as a rigid-frame strander, individual supply reels are mounted in such a way that each wire receives a full twist for every revolution of the strander. In the other, known as a planetary strander, the wire receives no twist as the frame rotates.

These types of stranders are comprised of multiple bays, with the first bay carrying six reels and subsequent bays carrying increasing multiples of six. The core wire in the center of the strand is payed off externally. It passes through the machine center and individual wires are laid over it. In this manner, strands with up to 127 wires are produced in one or two passes through the machine depending on its capacity for stranding individual wires.

Normally, hard-drawn copper wire is stranded on a planetary machine so that the strand will not be as springy and will tend to stay bunched rather than spring open when it is cut off. The finished product is wound onto a power-driven external reel that maintains a prescribed amount of tension on the stranded wire.

INSULATION AND JACKETING

Of the three broad categories of insulation—polymeric, enamel and paper-and-oil—polymeric insulation is the most widely used.

Polymeric Insulation. The most common polymers are polyvinyl chloride (PVC), polyethylene, ethylene propylene rubber (EPR), silicone rubber, polytetrafluoroethylene (PTFE) and fluorinated ethylene propylene (FEP). Polymide coatings are used where fire resistance is of prime importance, such as in wiring harnesses for manned space vehicles. Until a few years ago, natural rubber was used, but this has now been supplanted by synthetics such as butyl rubber and EPR. Synthetic rubbers are used wherever good flexibility must be maintained, such as in welding or mining cable.

Many varieties of PVC are made, including several that are flame resistant. PVC has good dielectric strength and flexibility, and is one of the least expensive conventional insulating and jacketing materials. It is used mainly for communication wire, control cable, building wire and low-voltage power cables. PVC insulation is normally selected for applications requiring continuous operation at temperatures up to 75 °C (165 °F).

Polyethylene, because of its low and stable dielectric constant, is specified when better electrical properties are required. It resists abrasion and solvents. It is used chiefly for hookup wire, communication wire and high-voltage cable. Cross-linked polyethylene (XLPE), which is made by adding organic peroxides to polyethylene and then vulcanizing the mixture, yields better heat resistance, better mechanical properties, better aging characteristics, and freedom from environmental stress cracking. Special compounding can provide flame resistance in cross-linked polyethylene. Typical uses include building wire, control cables and power cables. The usual maximum sustained operating temperature is 90 °C (200 °F).

PTFE and FEP are used to insulate jet aircraft wire, electronic equipment wire and specialty control cables, where heat resistance, solvent resistance and high reliability are important. These electrical cables can operate at temperatures up to 250 °C (480 °F).

All of the polymeric compounds are applied over copper conductors by hot extrusion. The extruders are machines that convert pellets or powders of thermoplastic polymers into continuous covers. The insulating compound is loaded into a hopper that feeds it into a long, heated chamber. A continuously revolving screw moves the pellets into the hot zone, where the polymer softens and becomes fluid. At the end of the chamber, molten compound is forced out through a small die over the moving conductor, which also passes through the die opening. As the insulated conductor leaves the extruder it is water cooled and taken up on reels. Cables jacketed with EPR and XLPE go through a vulcanizing chamber prior to cooling to complete the cross-linking process.

Enamel Film Insulation. Film-coated wire, usually fine magnet wire, is composed of a metallic conductor coated with a thin, flexible enamel film. These insulated conductors are used for electromagnetic coils in electrical devices, and must be capable of withstanding high breakdown voltages. Temperature ratings range from 105 to 220 °C (220 to 425 °F), depending on enamel composition. The most commonly used enamels are based on polyvinyl acetals, polyesters and epoxy resins.

Equipment for enamel coating of wire often is custom built, but standard lines are available. Basically, systems are designed to insulate large numbers of wires simultaneously. Wires are passed through an enamel applicator that deposits a controlled thickness of liquid enamel onto the wire. Then the wire travels through a series of ovens to cure the coating, and finished wire is collected on spools. In order to build up a heavy coating of enamel, it may be necessary to pass wires through the system several times. In recent years, some manufacturers have experimented with powder-coating methods. These avoid evolution of solvents, which is characteristic of curing conventional enamels, and thus make it easier for the manufacturer to meet OSHA and EPA standards. Electrostatic sprayers, fluidized beds and other experimental devices are used to apply the coatings.

Paper-and-Oil Insulation. Cellulose is one of the oldest materials for electrical insulation and is still used for certain applications. Oil-impregnated cellulose paper is used to insulate high-voltage cables for critical power-distribution applications. The paper, which may be applied in tape form, is wound helically around the conductors using special machines in which six to twelve paper-filled pads are held in a cage that rotates around the cable. Paper layers are wrapped alternately in opposite directions, free of twist. Paper-wrapped cables then are placed inside special impregnating tanks to fill the pores in the paper with oil and to ensure that all air has been expelled from the wrapped cable.

The other major use of paper insulation is for flat magnet wire. In this application, magnet-wire strip (with a width-to-thickness ratio greater than 50 to 1) is helically wrapped with one or more layers of overlapping tapes. These may be bonded to the conductor with adhesives or varnishes. The insulation provides highly reliable mechanical separation under conditions of electrical overload.

Copper Alloy Castings

COPPER ALLOY CASTINGS are used in applications that require superior corrosion resistance, high thermal or electrical conductivity, good bearing-surface qualities or other special properties. Casting makes it possible to produce parts whose shape cannot be easily obtained by fabricating methods such as forming or machining. Often, it is more economical to produce a part as a casting than to fabricate it by other means.

Compositions of copper casting alloys may differ from those of their wrought counterparts for various reasons. Generally, casting permits greater latitude in the use of alloying elements, because the effects of composition on hot or cold working properties are not important. However, imbalances among certain elements, and trace amounts of certain impurities in some alloys, will diminish castability and may result in castings of questionable quality.

Many of the casting alloys have lead contents of 5% or more. Alloys containing such high percentages of lead are not suited to hot working, but are ideal for low- to medium-speed bearings, where the lead prevents galling and excessive wear under boundary-lubrication conditions.

The tolerance for impurities is normally greater in castings than in their wrought counterparts — again because of the adverse effects certain impurities have on hot or cold workability. On the other hand, impurities that inhibit response to heat treatment must be avoided in both castings and wrought products.

The choice of an alloy for any casting usually depends on four factors: metal cost, castability, properties and final cost.

Metal cost is a minor consideration if only a few castings are to be made. However, when the product is a mass-produced or highly competitive item, or when metal cost is a major portion of the final cost of the castings, this factor becomes of prime importance.

CASTABILITY

Castability should not be confused with "fluidity," which is the ability of a molten alloy to fill a mold cavity completely in every detail. Castability, on the other hand, is the ease with which an alloy responds to ordinary foundry practice, without requiring special techniques for gating, risering, melting, sand conditioning or any of the other factors involved in making good castings. High fluidity often ensures good castability, but is not solely responsible for that quality in a casting alloy.

Foundry alloys generally are classed as "high shrinkage" or "low shrinkage." To the former class belong the manganese bronzes, aluminum bronzes, silicon bronzes, silicon brasses, and some nickel silvers. They are more fluid than the low-shrinkage red brasses, more easily poured, and give high-grade castings in sand, permanent mold, plaster, die and centrifugal casting processes. With high-shrinkage alloys, careful design is necessary to (a) promote directional solidification, (b) avoid abrupt changes in cross section, (c) avoid notches (by using generous fillets), and (d) properly place gates and risers; all of which help avoid internal shrinks and draws. Turbulent pouring must be avoided to prevent formation of dross. Liberal use of risers or exothermic compounds ensures adequate molten metal to feed all sections of the casting. Table 1 presents foundry characteristics of seventeen standard alloys, including a comparative ranking for both fluidity and overall castability for sand casting; number 1 represents the highest castability or fluidity ranking.

All copper alloys can be successfully cast in sand. Sand casting is the most economical casting method and allows the greatest flexibility in casting size and shape.

Permanent mold casting is best suited for tin, silicon, aluminum and manganese bronzes, and for yellow brasses. Die casting is well suited for yellow brasses, but increasing amounts of permanent mold alloys are also being die cast. Size is a definite limitation for both methods, although large slabs weighing as much as 4500 kg (10 000 lb) have been cast in permanent molds. Brass die castings generally weigh less than 0.2 kg (0.5 lb) and seldom exceed 0.9 kg (2 lb). The limitation of size is due to reduced die life with larger castings.

Virtually all copper alloys can be cast successfully by the centrifugal casting process. Castings of virtually any size from less than 100 g to more than 22 000 kg (less than 0.25 to more than 50 000 lb) have been made.

Because of their low lead contents, aluminum bronzes, yellow brasses, manganese bronzes, low-nickel bronzes, and silicon brasses and bronzes are best adapted to plaster mold casting. For most of these alloys, lead should be held to a mini-

Table 1. Foundry properties for principal copper alloys for sand casting

UNS number	Common name	Shrinkage allowance, %	Approx liquidus temperature °C	Approx liquidus temperature °F	Castability rating(a)	Fluidity rating(a)
C83600	Leaded red brass	5.7	1010	1850	2	6
C84400	Leaded semi-red brass	2.0	980	1795	2	6
C84800	Leaded semi-red brass	1.4	955	1750	2	6
C85400	Leaded yellow brass	1.5 to 1.8	940	1725	4	4
C85800	Yellow brass	2.0	925	1700	4	4
C86300	Manganese bronze	2.3	920	1690	6	2
C86500	Manganese bronze	1.9	880	1615	6	2
C87200	Silicon bronze	1.8 to 2.0	...	...	8	3
C87500	Silicon brass	1.9	915	1680	7	1
C90300	Tin bronze	1.5 to 1.8	980	1795	3	6
C92200	Leaded tin bronze	1.5	990	1810	3	6
C93700	High-lead tin bronze	2.0	930	1705	1	6
C94300	High-lead tin bronze	1.5	925	1700	1	6
C95300	Aluminum bronze	1.6	1045	1910	8	5
C95800	Aluminum bronze	1.6	1060	1940	8	5
C97600	Nickel silver	2.0	1145	2090	5	7
C97800	Nickel silver	1.6	1180	2160	5	7

(a) Relative rating for casting in sand molds. The alloys are ranked from 1 to 8 in both over-all castability and fluidity; 1 is the highest or best possible rating.

Table 2. Typical properties of copper casting alloys

UNS number	Tensile strength		Yield strength(a)		Compressive yield strength(b)		Elongation, %	Hardness, HB(c)	Electrical conductivity, % IACS
	MPa	ksi	MPa	ksi	MPa	ksi			
ASTM B22									
C86300	820	119	570(d)	82(d)	490	71	18	177	9.05
C90500	275-345	40-50	140-160	20-23	...	...	24-43	75-85	10.5-11.5
C93700	270	39	125	18	125	18	30	67	10.0
ASTM B61									
C92200	280	41	110	16	105	15	45	64	14.5
ASTM B62									
C83600	240	35	105	15	100	14	32	62	15.0
ASTM B66									
C94300	160-205	23-30	75-105	11-15	80-95	12-14	7-16	42-55	...
ASTM B147									
C86200	625-670	91-97	315-345	46-50	345	50	19-25	170-195(e)	7-8
C86300	820	119	570(d)	82(d)	490	71	18	177	9.0
C86400	415-540	60-78	170-275	25-40	140-180	20-26	15-30	80-95	20-24
C86500	490	71	180	26	165	24	40	98	20.5
ASTM B148									
C95200	480-600	70-87	170-205	25-30	185-215	27-31	22-38	110-140(e)	12-14
C95300(f)	480-585	70-85	205-240	30-35	110-140	16-20	20-35	110-160(e)	12-15
C95300(g)	550-655	80-95	275-380	40-55	240-275	35-45	12-16	160-225(e)	13.8
C95400(f)	515-655	75-95	205-285	30-41	...	...	12-20	150-185(e)	13-15
C95400(g)	620-690	90-100	310-360	45-52	...	...	6-15	190-235(e)	...
C95500(f)	620-725	90-105	275-345	40-50	...	...	7-20	175-210(e)	8-9.5
C95500(g)	760-855	110-124	415-550	60-80	...	...	5-12	215-260(e)	...
ASTM B176									
C85800(h)	380	55	205(d)	30(d)	...	...	15	...	22
C87800(h)	620	90	205(d)	30(d)	...	...	25	...	...
C87900(h)	400	58	205(d)	30(d)	...	...	15	...	...
ASTM B584									
C83600	243	35	105	15	100	14	32	62	15
C83800	205-260	30-38	85-115	12-17	76-83	11-12	15-27	50-60	...
C84400	200-270	29-39	90-115	13-17	...	...	18-30	50-60	18
C84800	260	38	105	15	85	12	37	59	16.5
C85200	240-275	35-40	85-95	12-14	55-70	8-10	25-40	40-55	15-22
C85400	205-260	30-38	75-105	11-15	62	9	20-35	40-60	18-25
C85700	275-310	40-45	95-140	14-20	...	...	15-25	50-75	20-26
C86200	625-670	91-97	315-345	46-50	345	50	19-25	170-195(e)	7-8
C86300	820	119	573	82(d)	490	71	18	177	9.0
C86400	415-540	60-78	170-275	25-40	...	20-26	15-30	80-95	20-24
C86500	490	71	180	26	140-180	24	40	98	20.5
C86700	550	80	220	32	165	...	15	...	...
C87200	380-450	55-65	150-205	22-30	105-150	15-22	25-55	85-120	4.5-6.4
C87400	345-485	50-70	145-225	21-33	...	...	20-50	70-130	...
C87500	470	68	207	30	185	27	17	115	6.0
C87600	414 min	60 min	207 min	30 min	...	...	16 min	...	...
C90300	275-345	40-50	125-150	18-22	...	...	25-50	60-75	12-13
C90500	275-345	40-50	140-160	20-23	...	...	24-43	75-85	10.5-11.5
C92200	280	41	110	16	105	15	45	64	14.5
C92300	225-295	33-43	110-165	16-24	62-76	9-11	18-30	60-75	10-12
C93200	205-260	30-38	115-145	17-21	...	...	12-20	55-65	...
C93500	195-240	28-35	83-105	12-15	90	13	20-35	55-65	15
C93700	270	39	125	18	125	18	30	67	10.0
C93800	170-225	25-33	95-140	14-20	90-110	13-16	10-18	50-60	...
C94300	160-205	23-30	76-105	11-15	83-97	12-14	7-16	42-55	...
C94700(f)	310	45	140	20	...	...	25	...	...
C94700(j)	515	75	345	50	...	...	5	...	...
C94800(f)	275	40	140	20	...	...	20	...	...
C94900	262 min	38 min	97 min	15 min	...	...	15 min	...	...
C97300	205-275	30-40	105-140	15-20	...	...	10-25	50-60	5.7
C97600	325	47	180	26	168	24	22	85	4.8
C97800	345-450	50-65	180-275	26-40	...	...	15-25	120-150	4-5

(a) At 0.5% extension under load. (b) At permanent set of 0.1%. (c) 500 kg load; 10 mm diam ball. (d) At 0.2% offset. (e) 3000 kg load. (f) M01 temper. (g) TQ00 temper. (h) M04 temper. (j) TF00 temper.

mum because it reacts with the calcium sulfate in the plaster, resulting in discoloration of the surface of the casting and increased cleaning and machining costs. Size is a limitation on plaster mold casting, although aluminum bronze castings that weigh as little as 100 g (0.22 lb) have been made by the lost-wax process and castings that weigh more than 150 kg (330 lb) have been made by conventional plaster molding.

Control of Solidification. Production of consistently sound castings requires an understanding of the solidification characteristics of the alloys as well as knowledge of relative magnitudes of shrinkage. The actual amount of contraction during solidification does not differ greatly from alloy to alloy. Its distribution, however, is a function of the freezing range and the temperature gradient in critical sections. Manganese and aluminum bronzes are similar to steel in that their freezing ranges are quite narrow — about 40 and 14 °C (70 and 25 °F), respectively. Large castings can be made by the same conventional methods used for steel, as long as proper attention is given to placement of gates and risers — both those for controlling directional solidification and those for feeding the primary central shrinkage cavity.

Tin bronzes have wider freezing ranges (about 165 °C or 300 °F for C83600). Alloys with such wide freezing ranges form a mushy layer during solidification, resulting in interdendritic shrinkage or microshrinkage. Because feeding cannot take place properly under these conditions, open grain or porosity results in the affected sections. The only practical means of overcoming interdendritic shrinkage is to maintain close temperature control of the metal during pouring and to provide for rapid solidification. These requirements limit thickness of section and pouring temperatures, and require a gating system that will ensure directional solidification. Sections up to 25 mm (1 in.) in thickness are routinely cast. Sections up to 50 mm (2 in.) thick may be cast, but only with difficulty and under carefully controlled conditions. A bronze with a narrow solidification range and good directional solidification characteristics is recommended for castings having section thickness greater than about 25 mm.

It is difficult to achieve directional solidification in complex castings. The most effective and most easily used device is the chill, which can be employed to initiate or accelerate solidification, and thus promote soundness. For irregular sections, chills must be shaped to fit the contour of the section of the mold in which they are placed. Insulating pads and riser sleeves sometimes are effective in slowing down the solidification rate in certain areas to maintain directional solidification.

MECHANICAL PROPERTIES

Most copper-base casting alloys containing tin, lead or zinc have only moderate tensile and yield strengths, low to medium hardness, and high elongation. When higher tensile or yield strength is required, the aluminum bronzes, manganese bronzes, silicon brasses, silicon bronzes, and some nickel silvers are used instead. Most of the higher-strength alloys have better-than-average resistance to corrosion and wear. Mechanical and physical properties of copper-base casting alloys are presented in Table 2. (Throughout this discussion, as well as in Table 2, mechanical properties quoted are for test bars. Properties of the castings themselves are almost always lower and depend on section size and process variables.)

Tensile strengths for cast test bars of aluminum bronzes and manganese bronzes range from 450 to 900 MPa (65 to 130 ksi), depending on composition; aluminum bronzes attain maximum tensile strength only after heat treatment.

Although manganese and aluminum bronzes often are used for the same applications, the manganese bronzes are handled more easily in the foundry. As-cast tensile strengths as high as 800 MPa (115 ksi) and elongations of 15 to 20% can be obtained readily in sand castings, and slightly higher values in centrifugal castings. Stresses may be relieved at 175 to 200 °C (350 to 400 °F). Lead may be added to the lower-strength manganese bronzes to increase machinability, but at the expense of decreased tensile strength and elongation. Lead content should not exceed 0.1% in high-strength manganese bronzes. Although manganese bronzes range in hardness from 125 to 250 HB, they are readily machined with proper tools.

Tin is added to the low-strength manganese bronzes to enhance resistance to dezincification, but should be limited to 0.1% in high-strength manganese bronzes unless great sacrifices in strength and ductility can be accepted.

Manganese bronzes are specified for marine propellers and fittings, pinions, ball-bearing races, worm wheels, gearshift forks and architectural work. Manganese bronzes also are used for rolling-mill screwdown nuts and slippers, bridge trunnions, gears and bearings, all of which require high strength and hardness.

Various cast aluminum bronzes contain 9 to 14% Al and lesser amounts of iron, manganese or nickel. They have a very narrow solidification range and, because of the greater need for adequate gating and risering compared to most other copper casting alloys, are more difficult to cast. A wide range of properties is obtainable, especially after heat treatment, but close control of composition is necessary. Like the manganese bronzes, aluminum bronzes can develop tensile strengths well over 700 MPa (100 ksi).

Most aluminum bronzes contain from 0.75 to 4% Fe to refine grain structure and increase strength. Alloys containing from 8 to 9.5% Al cannot be heat treated unless other elements (such as nickel or manganese) in amounts over 2% are added as well. They have higher tensile strength and greater ductility and toughness than any of the ordinary tin bronzes. Applications include valve nuts, cam bearings, impellers, hangers in pickling baths, agitators, crane gears, and connecting rods.

The heat treatable aluminum bronzes contain from 9.5 to 11.5% Al, in addition to iron, with or without nickel or manganese. These alloys resemble heat treated steels in structure and in response to quenching and tempering; castings are quenched in water or oil from temperatures between 760 and 925 °C (1400 and 1700 °F) and tempered at 425 to 650 °C (800 to 1200 °F), depending on exact composition and required properties.

From the range of properties shown in Table 2, it can be seen that all the maximum properties cannot be obtained in any one aluminum bronze. In general, alloys with higher tensile strength, yield strength and hardness have lower values of elongation. Typical applications of the higher-hardness alloys are rolling-mill screwdown nuts and slippers, worm gears, bushings, slides, impellers, nonsparking tools, valves and dies.

Aluminum bronzes resist corrosion in many substances, including pickling solutions. When corrosion occurs, it often proceeds by dealuminification, a form of dealloying in which aluminum is lost preferentially. Duplex alpha-plus-beta aluminum bronzes are more susceptible to dealloying than the all-alpha aluminum bronzes.

Aluminum bronzes have a high fatigue limit, considerably greater than that of manganese bronze or any other cast copper alloy. Unlike those of Cu-Zn and Cu-Sn-Pb-An alloys, the mechanical properties of aluminum and manganese bronzes do not decrease much with increases in casting cross section. This is because of a narrow freezing range, which results in a denser structure when castings are properly designed and properly fed.

Whereas manganese bronzes become hot short above 230 °C (450 °F), aluminum bronzes can be used at temperatures as high as 400 °C (750 °F) for short periods of time without appreciable loss in strength. For example, room-temperature tensile strengths of 540 MPa (78 ksi) decline to 530 MPa (76.7 ksi) at 260 °C (500 °F), 460 MPa (67 ksi) at 400 °C (750 °F) and 400 MPa (58 ksi) at 540 °C (1000 °F). Corresponding elongation values change from 28% to 32, 35 and 25%, respectively.

Unlike manganese bronzes, many aluminum bronzes increase in yield strength and hardness, but decrease in tensile strength and elongation, on slow cooling in the mold. While some manganese bronzes precipitate a relatively soft phase during slow cooling, aluminum bronzes precipitate a hard constituent rather rapidly within the narrow temperature range 565 to 480 °C (1050 to 900 °F). Hence, large castings, or smaller castings that are cooled slowly, will have properties different from small castings cooled relatively rapidly. The same phenomenon occurs on heat treating the hardenable aluminum bronzes. Cooling slowly through the critical temperature range after quenching, or tempering at temperatures within this range, will decrease elongation. Addition of 2 to 5% Ni greatly diminishes this effect.

Nickel brasses, silicon brasses and silicon bronzes, although generally higher in strength than red metal alloys, are used more for their corrosion resistance.

Distributions of hardness and tensile-strength data for separately cast test bars of three different alloys are shown in Fig. 1.

Cast beryllium coppers achieve variations in properties principally by varying heat treatment conditions. The "red" beryllium copper alloys are exemplified by C82000 and C82200; the "gold" alloys include C82400, C82500, C82600 and C82800. The casting alloys typically contain larger amounts of beryllium than their wrought counterparts. The "gold" casting alloys, in particular, have excellent casting characteristics, and can be poured at relatively low temperatures into molds with intricate shapes and fine detail.

Properties of Test Bars. Mechanical properties of separately cast test bars often differ widely from those of production castings poured at the same time, particularly when the thickness of the casting differs markedly from that of the test bar.

Variation in casting section size particularly affects mechanical properties of tin bronzes. With increasing section size up to about 50 mm (2 in.), mechanical properties—both strength and elongation—of the castings themselves are progressively lower than the corresponding properties of separately cast test bars. Elongation is particularly affected; for some tin bronzes, elongation of a 50-mm section may be as little as $1/10$ that of a 10-mm (0.4-in.) section or of a separately cast test bar.

The metallurgical behavior of many copper alloy systems is complex. Cooling rate (a function of casting section size) directly influences grain size, segregation and intergranular shrinkage; these factors, in turn, affect the mechanical properties of the cast metal. Therefore, molding and casting techniques are based on metallurgical characteristics as well as casting shape.

DIMENSIONAL TOLERANCES

Typical dimensional tolerances are different for castings produced by different molding methods. A molding process involving two or more mold parts requires greater tolerances for dimensions that cross the parting line than for dimensions wholly within one mold part. For castings made in green sand molds, tolerance across the parting line depends on the accuracy of pins and bushings that align the cope with the drag.

Figure 2 shows variations in two important dimensions for 50 production castings of red brass. The 3.750-in. dimension presented the greatest difficulty: none of the 50 production castings had

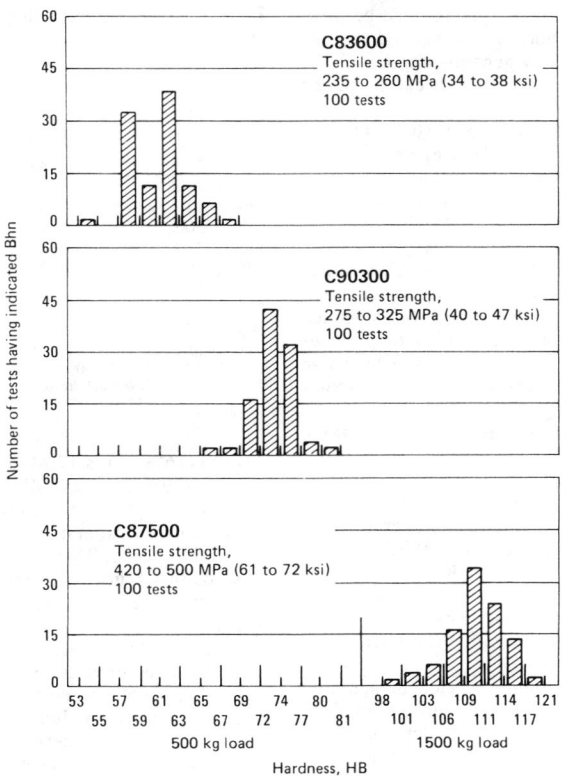

Fig. 1. Distribution of hardness for three copper casting alloys of different tensile strengths

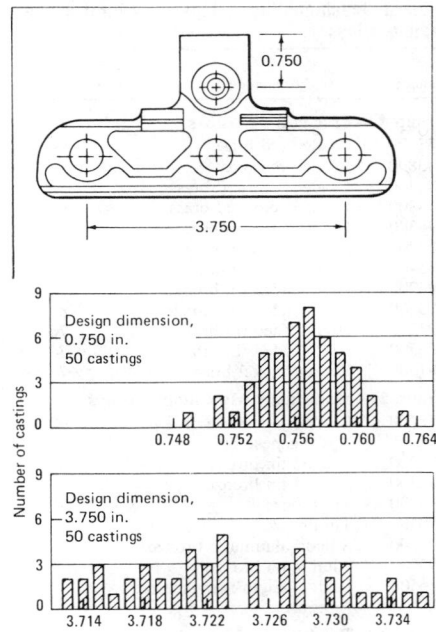

Parts were cast in green sand molds made using the same pattern.

Fig. 2. Variations from design dimensions for a typical red brass casting

an actual dimension as large as the nominal design value. Figure 3 shows dimensional variations in two similar cored valve castings. For each design, both the cores and the corresponding cavities in the castings were measured for about 100 castings. For both designs, the castings had actual dimensions less than those of the cores. This indicates that cores may need to have a slightly larger nominal size than is desired in the finished casting in order to ensure proper as-cast hole sizes.

MACHINABILITY

Machinability ratings of copper casting alloys are similar to those of their wrought counterparts. The cast alloys can be separated into three groups. The first group includes only those containing a single, copper-rich phase plus lead. Whether present for some other purpose or merely to improve machinability, lead facilitates chip breakage, thus allowing higher machining speeds with decreased tool wear and improved surface finishes.

Alloys of the second group contain two or more phases. Generally, the secondary phases are harder or more brittle than the matrix. Silicon bronzes, several aluminum bronzes and the high tin bronzes belong to this group. Hard and brittle secondary phases act as internal chip breakers, resulting in short chips and easier machining. Based on metallographic appearance, manganese bronzes would seem likely to produce a short chip, but they produce a long spiral chip, smooth on both sides, which does not break. Some aluminum bronzes, on the other hand, produce a long spiral chip that is rough on the underside and that breaks, thus acting like a short chip. Some of the alloys in the second group are classified "moderately machinable" because tools wear more rapidly when these alloys are machined, even though chip formation is entirely adequate.

The third group, the most difficult to machine, is composed mainly of the high-strength manganese bronzes and aluminum bronzes that are high in iron or nickel content. Relative machinability of alloys belonging to the three groups is shown in Table 3.

GENERAL-PURPOSE ALLOYS

General-purpose copper casting alloys are often classified as either red or yellow alloys. General properties of these alloys are shown in Table 2.

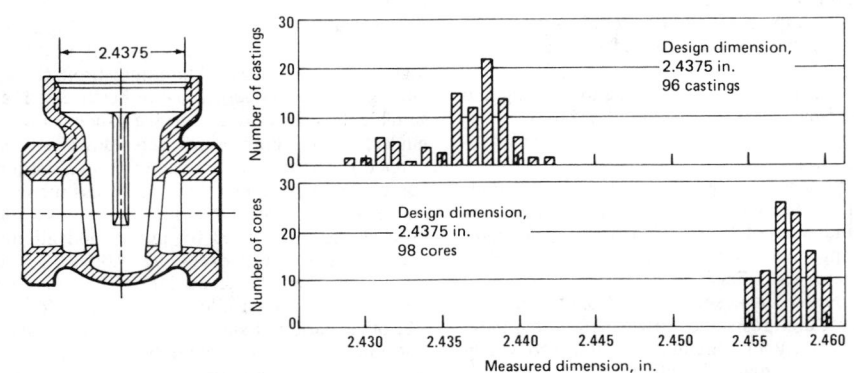
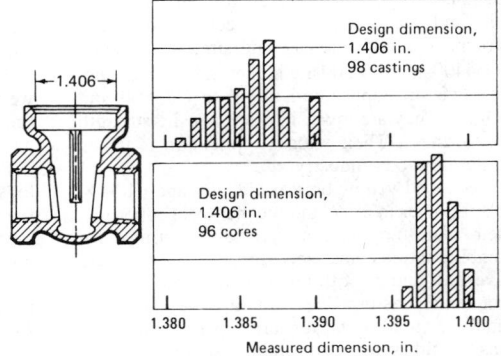

Valve bodies, similar in design but of different sizes, were made using dry sand cores to shape the internal cavities. Upper histograms indicate dimensional variations for the castings; lower histograms, variations for the corresponding cores.

Fig. 3. Variations from design dimensions for two typical cast red brass valve bodies

Table 3. Machinability ratings of several copper casting alloys

UNS number	Common name	Machinability rating(a), %
Group 1 — Free-cutting Alloys		
C83600	Leaded red brass	90
C83800	Leaded red brass	90
C84400	Leaded semi-red brass	90
C84800	Leaded semi-red brass	90
C94300	High-leaded tin bronze	90
C85200	Leaded yellow brass	80
C85400	Leaded yellow brass	80
C93700	High-leaded tin bronze	80
C93800	High-leaded tin bronze	80
C93200	High-leaded tin bronze	70
C93500	High-leaded tin bronze	70
C97300	Leaded nickel brass	70
Group 2 — Moderately Machinable Alloys		
C86400	Leaded high-strength manganese bronze	60
C92200	Leaded tin bronze	60
C92300	Leaded tin bronze	60
C90300	Tin bronze	50
C90500	Tin bronze	50
C95600	Silicon-aluminum bronze	50
C95300	Aluminum bronze	35
C86500	High-strength manganese bronze	30
Group 3 — Hard-to-Machine Alloys		
C86300	High-strength manganese bronze	20
C95200	9% aluminum bronze	20
C95400	11% aluminum bronze	20
C95500	Nickel-aluminum bronze	20

(a) Machinability rating expressed as a percentage of the machinability of C36000, free-cutting brass. The rating is based on relative speed for equivalent tool life. For instance, a material having a rating of 50 should be machined at about half the speed that would be used to make a similar cut in C36000.

The leaded red and leaded semi-red brasses respond readily to ordinary foundry practice and are rated very high in castability. C83600 is the best known of this group and usually is referred to by one of its common names — 85-5-5-5 or ounce metal. C83600 and its modification, C83800 (83-4-6-7), constitute the largest tonnage of copper-base foundry alloys. They are used where moderate corrosion resistance, good machinability, moderate strength and ductility, and good castability are required. C83800 has lower mechanical properties, but better machinability and lower initial metal cost than C83600.

Both C83600 and C83800 are used for plumbing goods, stopcocks, faucets, flanges, feed pumps, meter casings and parts, general household and machinery hardware and fixtures, paper machinery, hydraulic and steam valves, valve disks and seats, impellers, injectors, memorial markers, plaques, statuary and similar products.

C84400 and C84800 are higher in lead and zinc, and lower in copper and tin, than C83600 and C83800. They are lower in price, tensile strength and hardness. Their widest application is in the plumbing goods industry.

The leaded yellow brasses C85200 and C85400 are still lower in price and mechanical properties. Their main applications are die castings for plumbing goods and accessories, low-pressure valves, air and gas fittings, general hardware and ornamental castings. In general, they are best suited for small parts; for larger parts, thick sections should be avoided, and phosphorus may be added to increase fluidity if pressure tightness is required. Aluminum (0.15 to 0.25%) is sometimes added to yellow brasses to increase fluidity and to give a smoother surface, but aluminum

should not be added if the casting must be pressure-tight or if phosphorus is to be added as well.

All the red and yellow general-purpose alloys, when properly made and cleaned, can be plated with nickel or chromium.

Alloys that do not contain lead, such as the tin bronzes C90500 (Navy G bronze) and C90300 (modified Navy G bronze), are considerably more difficult to machine than leaded alloys. Alloys containing 10 to 12% Sn, 1 to 2% Ni and 0.1 to 0.3% P are known as gear bronzes. Up to 1.5% Pb is frequently added to increase machinability. Addition of lead to C90300 increases machinability, but a concurrent decrease in tin is needed to maintain elongation. The leaded tin bronzes include C92200 (known as steam bronze, valve bronze or Navy M bronze) and C92300 (commercial G bronze).

All of the tin bronzes are suitable wherever corrosion resistance, leak tightness or greater strength is required at higher operating temperatures than can be tolerated with leaded red or semi-red brasses. The limiting temperature for longtime operation of C92200 is 290 °C (550 °F); for C90300, C90500 and C92300, it is 260 °C (500 °F) because of embrittlement caused by precipitation of a high-tin phase. This reaction does not occur in tin bronzes with tin contents less than about 8%.

For elevated-temperature service in handling fluids and gases, the ASME Boiler and Pressure Vessel Code (in Table UNF-23) defines allowable working stresses for C92200 (leaded tin bronze ASTM B61) and C83600 (leaded red brass ASTM B2) at different temperatures, as given in Table 4.

Nickel frequently is added to tin bronzes to increase density and leak tightness. Alloys containing more than 3% Ni are heat treatable, but must contain less than 0.01% Pb for optimum properties; one example is C94700 (88 Cu – 5 Sn – 2 Zn – 5 Ni).

BEARING AND WEAR PROPERTIES

Copper alloys have long been used for bearings because of their combination of moderate to high strength, corrosion resistance, and either wear resistance or self-lubrication properties. The choice of an alloy depends on required corrosion resistance and fatigue strength, rigidity of backing material, lubrication, thickness of bearing material, load, speed of rotation, atmospheric conditions and other factors. Copper alloys may be cast into plain bearings, cast on steel backs, cast on rolled strip, made into sintered powder-metallurgy shapes, or pressed and sintered onto a backing material.

Three groups of alloys are used for bearing and wear-resistant applications: phosphor bronzes (Cu-Sn); copper-tin-lead (low-zinc) alloys; and manganese, aluminum and silicon bronzes.

Phosphor bronzes (Cu-Sn-P or Cu-Sn-Pb-P alloys) have residual phosphorus ranging from a few hundredths of 1% (for deoxidation and slight hardening) to a maximum of 1%, which imparts great hardness. Nickel often is added to refine grain size and disperse the lead. Copper-tin bearings have high resistance to wear, high hardness and moderately high strength. C90700 is so widely used for gears that it is called gear bronze.

Phosphor bronzes of higher tin content, such as C91100 and C91300, are used in bridge turntables, where loads are high and rotational movement is slow. The maximum load permitted for

Table 4. Allowable working stresses for C92200 and C83600 castings(a)

Temperature		ASTM B61		ASTM B62	
°C	°F	MPa	ksi	MPa	ksi
38	100	47	6.8	41	6.0
65	150	47	6.8	41	6.0
93	200	47	6.8	40	5.8
120	250	47	6.8	38	5.5
150	300	45	6.5	34	5.0
175	350	41	6.0	31	4.5
205	400	38	5.5	24	3.5
230	450	34	5.0	24	3.5
260	500	28	4.0	24	3.5
290	550	23	3.3	24	3.5

(a) Tensile strength of 235 MPa (34 ksi) min. is specified for C92200 in ASTM B61; 205 MPa (30 ksi) min. for C83600 in ASTM B62. (b) From ASME Boiler and Pressure Vessel Code, Table UNF-23.

C91100 (16% Sn) is 17 MPa (2500 psi), and for C91300 (19% Sn), 24 MPa (3500 psi). These bronzes are high in phosphorus (1% max) to impart high hardness, and low in zinc (0.25% max) to prevent seizing. They are very brittle, and because of this brittleness they are sometimes replaced by either manganese bronzes or aluminum bronzes.

High-leaded tin bronzes are used where a softer metal is required at low to moderate speeds and at loads not exceeding 5.5 MPa (800 psi). Alloys of this type include C93200, C93500 and C93700. The last of these alloys, also known as 80-10-10, is an excellent general bearing alloy, especially well suited for applications where lubrication may be deficient. C93700 is widely used in machine tools, electrical and railroad equipment, steel-mill machinery and automotive applications. C93200 and C93500 are less costly than C93700 and are used chiefly for replacement bearings in machinery. C93800 (15% Pb) and C94300 (24% Pb) are used where high loads are encountered under conditions of poor or nonexistent lubrication, or under corrosive conditions, such as in mining equipment (pumps and car bearings), or in dusty atmospheres, as in stone-crushing and cement plants. These alloys replace the tin bronzes or low-leaded tin bronzes where operating conditions are unsuitable for alloys containing little or no lead.

High-strength manganese bronzes have high tensile strength, hardness and resistance to shock. Large gears, bridge turntables (slow motion and high compression), roller tracks for antiaircraft guns, and recoil parts of cannons are typical applications.

Aluminum bronzes with 8 to 9% Al are used widely for bushings and bearings in light-duty or high-speed machinery. Aluminum bronzes containing 11% Al, either as cast or heat treated, are suitable for heavy-duty service (such as valve guides, rolling-mill bearings, screwdown nuts and slippers) and precision machinery. As aluminum content increases above 11%, hardness increases and elongation decreases to low values. Such bronzes are well suited for guides and aligning plates, where wear is excessive. Above 13% Al, aluminum bronzes exceed 300 HB in hardness but are brittle. Such alloys are suitable for dies and other parts not subjected to impact loads.

Aluminum bronze generally has a considerably higher fatigue limit and freedom from galling than manganese bronze. On the other hand, manganese bronze has great toughness for equivalent tensile strength and does not need to be heat treated.

Table 5. Composition and typical properties of heat treated copper casting alloys of high strength and conductivity

UNS number	Nominal composition	Tensile strength MPa	ksi	Yield strength MPa	ksi	Elonga- tion, %	Hardness	Electrical conductivity, % IACS
C81400	.....99Cu-0.8Cr-0.06Be	365	53	250	36	11	69 HRB	70
C81500	.....99Cu-1Cr	350	51	275	40	17	105 HB	85
C81800	.....97Cu-1.5Co-1Ag-0.4Be	705	102	515	75	8	96 HRB	48
C82000	.....97Cu-2.5Co-0.5Be	660	96	515	75	6	96 HRB	48
C82200	.....98Cu-1.5Ni-0.5Be	655	95	515	75	7	96 HRB	48
C82500	.....97Cu-2Be-0.5Co-0.3Si	1105	160	1035	150	1	43 HRC	20
C82800	.....96.6Cu-2.6Be-0.5Co-0.3Si	1140	165	1070	155	1	46 HRC	18

ELECTRICAL AND THERMAL CONDUCTIVITY

Electrical and thermal conductivity of any casting will invariably be lower than for wrought metal of the same composition. Copper castings are used in the electrical industry for their current-carrying capacity and as water-cooled parts of melting and refining furnaces for their high thermal conductivity. However, for a copper casting to be sound and have electrical or thermal conductivity of at least 85%, care must be taken in melting and casting. The ordinary deoxidizers (silicon, tin, zinc, aluminum and phosphorus) cannot be used, because small residual amounts lower electrical and thermal conductivity drastically. Calcium boride or metallic lithium will help to produce sound castings with high conductivity.

Cast copper is soft and low in strength. Increased strength and hardness and good conductivity can be obtained with heat treated alloys containing silicon, cobalt, chromium, nickel and beryllium in various combinations.

These alloys, however, are expensive and less readily available than the standardized alloys.

Table 5 presents some of the properties of these alloys after heat treatment.

COST CONSIDERATIONS

During design of a copper alloy casting, foundry personnel or the design engineer must choose a method of producing internal cavities. There is no general rule for choosing between cored and coreless designs. A cost analysis will determine which is the more economical method of producing the casting, although frequently the choice can be decided by experience.

In one instance, costs were compared for producing a small (1/2-in.) valve disk both as a cored casting and as a machined casting (internal cavities made without cores). The machined casting could be produced for about 78% of the cost of making the identical casting using dry sand cores—a savings of 22% in favor of the machined casting. In a similar instance, producing a larger (1 1/2-in.) valve disk as a cored casting requiring only a minimal amount of machining saved more than 8% in over-all cost compared to producing the same valve disk without cores. Thus, for two closely related parts, there may be a decisive difference in manufacturing economy when all cost factors are taken into account.

Stress Relaxation in Copper and Copper Alloys

COPPER and copper alloys are used extensively in structural applications in which they are subject to moderately elevated temperatures. Examples include automotive radiators, solar heating panels, communications cable and electrical connectors. At relatively low operating temperatures, these alloys can undergo stress relaxation (decrease in stress resulting from transformation of elastic strain into plastic strain in a constrained solid), which can lead to service failures. Because of the wide variations in composition and processing among commercial copper alloys, resistance to stress relaxation varies considerably. Of course, selection of an alloy for a given application is based not only on stress-time-temperature response but also on such factors as cost, basic mechanical and physical properties, operating temperature, service environment and formability. For many applications, electrical conductivity is a primary consideration.

STRESS-RELAXATION DATA

Unalloyed copper C11000 is probably the most inexpensive high-conductivity copper and is used extensively because of its ease of fabrication. The stress-relaxation behavior of this material is rather poor, as demonstrated in Fig. 1, in which relaxed stress is plotted as a function of time and temperature for 0.25-mm (0.010-in.) C11000 wire initially stressed in tension to 89 MPa (13 ksi). Comparison of stress values at a given time for different temperatures illustrates the very sharp dependence of stress relaxation on temperature for this copper. At 93 °C (200 °F), for example, no tension remains after 10^5 h (11.4 years), whereas 40% of the initial stress remains after 40 years at room temperature. For C11000 and for many other copper metals, stress relaxation in a given time period is inversely proportional to absolute temperature (Ref 1).

The stress-relaxation behavior of C10200 (oxygen-free copper) is somewhat better than that of C11000, as shown in Table 1, which also presents data for many other high-conductivity coppers. A more extensive comparison of the mechanical behavior of C10200 and C11000 has been presented by Opie, Taubenblat and Hsu (Ref 2).

Improvement in stress-relaxation behavior can be achieved by adding alloying elements that cause solid-solution strengthening, age hardening or dispersion hardening (Ref 3 and 4).

Table 2 provides bending-stress-relaxation data at 75 °C for solid-solution and dispersion-hardened copper alloys commonly used in electrical and electronic spring applications. Tensile strength and electrical conductivity values also are tabu-

Table 1. 10,000-hour tensile-stress-relaxation data for selected types of tinned copper wire

Alloy	Temper	Temper- ature °C	°F	Initial stress(a) MPa	ksi	Stress remaining, %
C10200	...O61	27	80	82	11.9	69
C11000	...O61	23	73	89	12.9	60
C15000	...H04(b)	23	73	203	29.5	93
C16200	...H04(c)	23	73	226	32.8	95

(a) Initial stress set at about 55% of 0.2% offset yield strength. (b) In-process strand anneal. (c) Batch annealed.

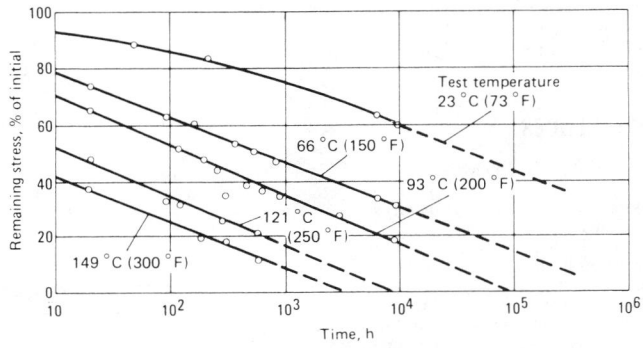

Data are for tinned 30 AWG (0.25-mm-diam) annealed ETP copper wire; initial elastic stress, 89 MPa (13 ksi).

Fig. 1. Tensile-stress-relaxation characteristics of C11000

Table 2. Bending-stress-relaxation data at 75 °C for commonly used electronic spring materials

Alloy	Conductivity, % IACS	Tensile strength MPa	Tensile strength ksi	Stress remaining, %(a) 1000 h	Stress remaining, %(a) 10^5 h
C15100	90	434	63	81	74
C19400	65	448	65	88	83
C19500	50	586	85	79	72
C26000	28	565	82	74	63
C42500	28	565	82	87	80
C51000	15	483	70	92	86
C65400	7	779	113	87	80
C68800	18	800	116	82	73
C72500	11	613	89	90	87
C75200	9	620	90	93	90

(a) The values for percentage of stress remaining at 10^5 h are extrapolated from 1000-h test data. Initial stress level was about 80% of the 0.2%-offset yield stress.

lated. In general, the resistance to stress relaxation, as measured by percentage of stress remaining, increases with solute hardening additions but with some sacrifice in electrical conductivity. It should be noted that stress-relaxation resistance can be expected to decrease with increasing tensile strength (temper) for a given alloy.

C17200 is a typical example of a copper alloy strengthened by precipitation hardening. Stress-relaxation data for this alloy are shown in Fig. 2 for two levels of cold work. C17200 has adequate resistance to stress relaxation at temperatures up to 120 °C (250 °F), provided the initial stress is below the elastic limit.

STRESS RELAXATION IN MECHANICAL COMPONENTS

A solderless wrapped connection such as the one shown in Fig. 3, in which electrical contact is made by wrapping a wire around a terminal, is a typical application for high-conductivity copper metals where stress relaxation is of concern. Typical operating temperatures can be as high as 85 °C (185 °F); generally, conductivities higher than 98% IACS at 20 °C (68 °F) are desirable. After the connection is made, it is maintained by elastic stresses in the two members. If the wire undergoes stress relaxation, electrical contact between the wire and the terminal may be lost.

A typical application of the lower-conductivity, high-strength copper alloys is the pressure-type, split-beam connector shown in Fig. 4. The knife edges of the connector first cut through the insulation on the conductor and then must maintain electrical contact with it. Materials used for connectors of this type, depending on operating stress and temperature, include some of the phosphor bronzes, nickel silvers, copper nickels, beryllium coppers and some of the newer copper-nickel-tin alloys (Ref 5) strengthened by spinodal decomposition.

REFERENCES

1. A. Fox, Stress-Relaxation Characteristics in Tension of High-Strength, High Conductivity Copper and High Copper Alloy Wires, *Journal of Testing and Evaluation,* Vol 2, No. 1 (Jan 1974), p 32-39
2. W. R. Opie, P. W. Taubenblat and Y. T. Hsu, A Fundamental Comparison of the Mechanical Behavior of Oxygen-Free and Tough Pitch Coppers, *Journal of the Institute of Metals,* Vol 98, 1970, p 245
3. A. Fox and J. J. Swisher, Superior Hook-Up Wires for Miniaturized Solderless Wrapped Connections, *Journal of the Institute of Metals,* Vol 100, 1972, p 30
4. J. Crane, "Performance and Fabricability Requirements for Copper Base Alloys in Electrical and Electronic Components," Technical Paper MS76-363, SME, Dearborn, MI 1976.
5. J. T. Plewes, Spinodal Cu-Ni-Sn Alloys Are Strong and Superductile, *Metal Progress,* July 1974, p 46

ADDITIONAL READING

E. Kula and V. Weiss, ed., *Residual Stress and Stress Relaxation,* Plenum Publishing Corp., 1982
A. Fox, ed., *Stress Relaxation Testing,* ASTM STP 676, American Society for Testing and Materials, Philadelphia, 1979.

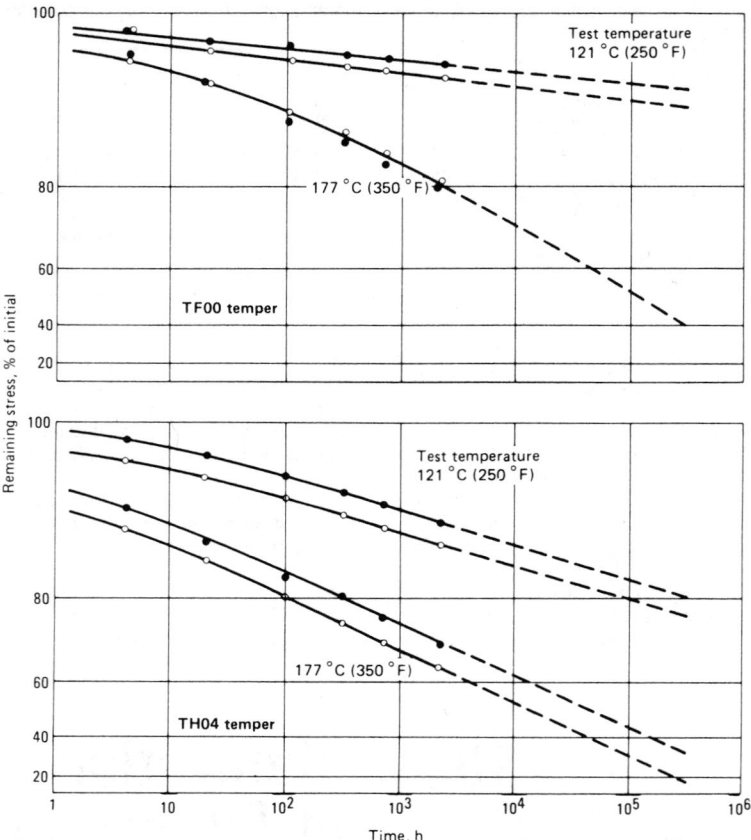

Data are for beryllium copper (1.9% Be) strip, 0.38 mm (0.015 in.) thick. Open symbols are for an initial test stress equal to 80% of the monotonic bending yield stress; solid symbols, for an initial stress 50% of the bending yield stress.

Fig. 2. Stress relaxation in C17200 at two levels of initial stress

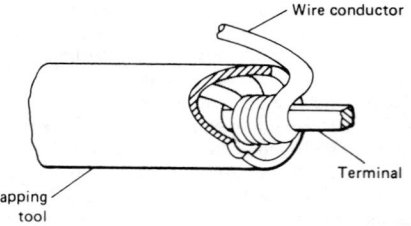

Wrapping tool is removed after connection is made.

Fig. 3. Typical solderless wrapped connection

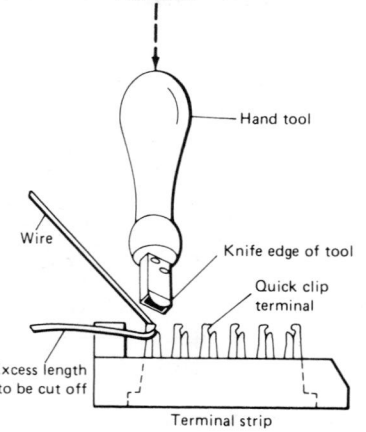
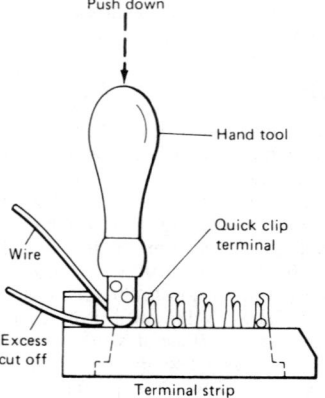

Fig. 4. Typical quick clip connection

Corrosion Characteristics of Copper and Copper Alloys

COPPER and its alloys are used in many applications that require service for extended periods in environments that can be aggressive to other metals. Copper metals resist the atmosphere, fresh and salt waters, alkaline solutions (except those containing ammonia) and many organic chemicals. Their resistance to corrosion by oxidizing acids depends mainly on the severity of oxidizing conditions in the acid solution. Copper metals are suitable for use with many salt solutions.

Copper reacts with sulfur and sulfides to form copper sulfide. As a result, coppers and copper alloys normally are not selected for service in environments known to contain sulfur or sulfides.

Coppers and copper alloys, like most other metals and alloys, are susceptible to several forms of corrosion, depending mainly on environmental conditions.

COMMON FORMS OF CORROSION

Table 1 presents identifying characteristics of the forms of corrosion that commonly attack copper metals, and the most effective means of combating each.

General Corrosion

General corrosion is well-distributed attack of an entire surface with little or no localized penetration, and it is the least damaging of all forms of attack. It is the only form of corrosion for which weight-loss data can be used to accurately estimate penetration rates.

General or uniform corrosion results from prolonged contact with environments in which the corrosion rate is very low, such as fresh, brackish and salt waters; many types of soil; neutral, alkaline and acid salt solutions; organic acids; and sugar juices. Many other substances that bring about uniform thinning at a faster rate include oxidizing acids, sulfur-bearing compounds, ammonia and cyanides.

Galvanic Corrosion

It two dissimilar metals are in electrical contact with each other and immersed in a conductive solution, a potential results which enhances corrosion of the more electronegative member of the couple (the anode) and partly or completely protects the more electropositive member (the cathode). Copper alloys are almost always cathodic to other common structural metals such as steel and aluminum. When steel or aluminum is put in contact with a copper metal, the corrosion rate of the steel or aluminum increases whereas that of the copper metal decreases. The common grades of stainless steel exhibit variable behavior; that is, copper metals may be either anodic or cathodic to the stainless steel, depending on conditions of exposure. Copper alloys usually corrode preferentially when coupled with high-nickel alloys, titanium or graphite.

Corrosion potentials of copper alloys generally range from -0.2 to -0.4 mV when measured against a saturated calomel reference electrode; the potential of pure copper is about -0.3 mV. Alloying additions of zinc or aluminum move the potential toward the anodic (more electronegative) end of the range; additions of tin or nickel move the potential toward the cathodic (less electronegative) end. Galvanic corrosion between two copper metals is seldom a significant problem, because the potential difference is so small.

Table 2 lists a galvanic series of metals and alloys valid for dilute aqueous solutions such as seawater and weak acids. The metals that are grouped together may be coupled to each other without significant galvanic damage. However, connecting metals from different groups leads to damage of the more anodic metal; the larger the difference in galvanic potential between groups, the greater the corrosion.

Accelerated damage due to galvanic effects is usually greatest near the junction, where electrochemical current density is the highest. Figure 1 shows a common example of galvanic corrosion of a steel coupling that joined two lengths of red brass pipe.

Another factor that affects galvanic corrosion is area ratio. An unfavorable area ratio exists when the cathodic area is large and the anodic area is small. The corrosion rate of the small anodic area may be several hundred times greater than if the anodic and cathodic areas were equal in size. Conversely, when a large anodic area is coupled to a small cathodic area, current density and damage due to galvanic corrosion are much less. For example, copper rivets (cathodic) used to fasten steel plates together lasted longer than 1.5 years in seawater, but steel rivets used to fasten copper plates were completely destroyed during the same period.

There are five major methods of eliminating or significantly reducing galvanic corrosion:

1. Selecting dissimilar metals that are as close as possible to each other in the galvanic series.
2. Avoiding coupling of small anodes to large cathodes.
3. Insulating dissimilar metals completely wherever practicable.
4. Applying coatings and keeping them in good repair, particularly on the cathodic member.
5. Using a sacrificial anode—that is, coupling the system to a third metal that is anodic to both structural metals.

Pitting

As with most commercial metals, corrosion of copper metals results in pitting under certain conditions. Sometimes pitting is general over the entire surface, giving the metal an irregular and roughened appearance. In other instances, pits are concentrated in specific areas and are of various sizes and shapes.

Localized pitting is the most damaging form of corrosive attack because it reduces load car-

Table 1. Guide to corrosion of copper metals

Form of corrosion	Identifying characteristics	Techniques for preventing corrosion
General thinning	Uniform metal removal	Select proper alloy for environmental conditions based on weight-loss data.
Galvanic corrosion	Preferential corrosion near a more cathodic metal	Avoid electrically coupling dissimilar metals. Maintain optimum ratio of anode area to cathode area. Maintain optimum concentration of oxidizing constituent in corroding medium.
Pitting:		
Concentration cell	Water-line pitting	
	Crevice corrosion	Design out crevices.
	Pitting under foreign objects or dirt	Keep metal clean
Impingement	Erosion attack from turbulent flow plus dissolved gases, generally as lines of pits in direction of fluid flow	Design for streamline flow. Keep velocity low. Remove gases from liquid phase.
Fretting	Chafing or galling, often occurring during shipment	Lubricate contacting surfaces. Interleave sheets of paper between sheets of metal. Decrease load on bearing surfaces.
Dealloying	Preferential dissolution of zinc or nickel, resulting in layer of sponge copper	Select proper alloy for environmental conditions based on metallographic examination of corrosion specimens
Stress-corrosion cracking	Cracking, usually intercrystalline but sometimes transcrystalline, that is often fairly rapid	Select proper alloy based on stress-corrosion tests. Reduce applied or residual stress. Remove mercury compounds or ammonia from environment.
Corrosion fatigue	Several transcrystalline cracks	Select proper alloy based on fatigue tests in service environment. Reduce mean or alternating stress.
Intergranular corrosion	Corrosion along grain boundaries without visible signs of cracking	Select proper alloy for environmental conditions based on metallographic examination of corrosion specimens.

Table 2. Galvanic series in seawater

Anodic End

Magnesium	Lead
Magnesium alloys	Tin
Zinc	Muntz metal (C28000)
Galvanized steel	Manganese bronze (C67500)
Aluminum 5052H	Naval brass (C46400)
Aluminum 3004	Nickel (active)
Aluminum 3003	Inconel (active)
Aluminum 1100	
Aluminum 6053	Cartridge brass (C26000)
Alclad aluminum alloys	Admiralty metal (C44300)
	Aluminum bronze (C61400)
Cadmium	Red brass (C23000)
Aluminum 2017	Copper (C11000)
Aluminum 2024	Silicon bronze (C65100)
	Copper nickel, 30% (C71500)
Low-carbon steel	Nickel (passive)
Wrought iron	Inconel (passive)
Cast iron	
Ni-Resist	Monel
Type 410 stainless steel (active)	Type 304 stainless steel (passive)
50Pb-50Sn solder	Type 316 stainless steel (passive)
Type 304 stainless steel (active)	Silver
Type 316 stainless steel (active)	Gold
	Platinum
	Cathodic End

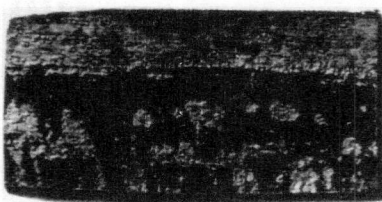

Flattened section of admiralty metal (C44300) tube taken from a ship condenser. High velocity and turbulence of the fluid stream caused impingement attack, which can be seen as a series of horseshoe-shape pits at top of flattened tube section. Shown at about 2×.

Fig. 2. Impingement attack of a condenser tube

are more erosion resistant than the brasses or tin brasses. When contaminated waters are involved, filtering or screening of the liquids and frequent cleaning of surfaces can be very effective in minimizing impingement attack.

Another form of attack, called *fretting* or *fretting corrosion,* appears as pits or grooves in the metal surface surrounded or filled with corrosion product. Fretting is sometimes referred to as *chafing, road burn, friction oxidation, wear oxidation* or *galling.*

The basic requirements for fretting are as follows:

1. Repeated relative (sliding) motion between two surfaces must occur. The relative amplitude of the motion is generally very small — motion of only a few tenths of a millimetre is typical.
2. The interface must be under load.
3. Both load and relative motion must be sufficient to produce deformation of the interface.
4. Oxygen and/or moisture must be present.

Fretting does not occur on lubricated surfaces in continuous motion such as axle bearings, but rather on dry interfaces subject to repeated small relative displacements. A classic type of fretting occurs during shipment of bundles of mill products having flat faces, such as hexagonal rod, octagonal rod and rectangular bar.

Fretting is not confined to coppers and copper alloys, but has been recognized on almost every kind of surface — steel, aluminum, noble metals, mica and glass.

Fretting can be controlled, and sometimes eliminated, by (*a*) lubricating with low-viscosity, high-tenacity oils to reduce friction at the interface between the two metals and to exclude oxygen from the interface; (*b*) separating the faying surfaces by interleaving an insulating material; (*c*) increasing the load to reduce motion between faying surfaces (may be difficult in practice, because only a minute amount of relative motion is necessary to produce fretting); or (*d*) decreasing the load at bearing surfaces to increase the relative motion between parts, which reduces fretting.

Dealloying

Dealloying is a corrosion process whereby the more active metal is selectively removed from an alloy, leaving behind a weak deposit of the more noble metal.

Copper-zinc alloys containing more than 15% Zn are susceptible to a dealloying process called dezincification. In dezincification of brass, selective removal of zinc leaves a relatively porous

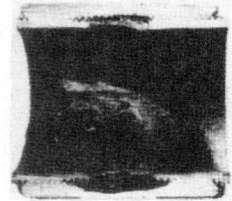

Galvanic corrosion of a 38-mm (1½-in.) steel coupling joining two sections of red brass (C23000) pipe. Note that the brass pipes (cathodic) have not corroded, whereas the steel coupling was severely attacked. About ⅔ size.

Fig. 1. Galvanic attack between steel and red brass

rying capacity and increases stress concentration by creating depressions or holes in the metal. Pitting is the usual form of corrosive attack at surfaces on which there are incomplete protective films, nonprotective deposits of scale or extraneous deposits of dirt or other foreign substances.

Crevice corrosion, water-line attack, deposit attack, impingement attack, concentration-cell action and *fretting* are some of the special types of localized attack. Pitting results from formation of local electrolytic cells due to differences in metal-ion or oxygen concentration at adjacent areas of the metal surface. Crevice corrosion results from a depletion of oxygen in crevices so that the metal in a crevice becomes anodic to metal outside the crevice, which is exposed to an oxygen-bearing solution. Water-line attack is a term used to describe pitting due to a differential oxygen cell functioning between the well-aerated surface layer of a liquid and the oxygen-starved layer immediately beneath it. The pitting occurs just below the water line.

Local cell action may also result from the presence of foreign objects or debris such as dirt, pieces of shell, or vegetation, or it may result from rust, permeable scales or uneven accumulation of corrosion product on the metallic surface. Such materials screen the affected area, causing oxygen deficiency by making it difficult for fresh oxygen-bearing solution to diffuse to the site — hence the name deposit attack. Sometimes this type of attack can be controlled by cleaning the surfaces. For instance, condensers and heat exchangers are cleaned periodically to prevent deposit attack.

Impingement attack (sometimes called erosion-corrosion) occurs where gases, vapors or liquids impinge on metal surfaces at high velocities, such as in condensers or heat exchangers. It is most often found with waters containing low compounds of sulfur and with polluted, contaminated or silty salt water or brackish water. The erosive action locally removes protective films, thereby contributing to the formation of concentration cells and to localized pitting of anodic sites.

Impingement attack is characterized by undercut grooves, waves, ruts, gullies and rounded holes; it usually exhibits a directional pattern. Pits are elongated in the direction of flow and are undercut on the upstream side. When the condition becomes severe, it may result in a pattern of horseshoe-shaped grooves or pits with their open ends pointing downstream. As attack progresses, the pits may join, forming fairly large patches of undercut pits (see Fig. 2). When this form of corrosion occurs in a condenser tube, it is usually confined to a region near the inlet end of the tube where fluid flow is rapid and turbulent. If some of the tubes in a bundle become plugged, the velocity is increased in the remaining tubes; therefore, the unit should be kept as clean as possible.

Impingement attack can be reduced, and the life of the unit extended, by decreasing fluid velocity, streamlining the flow and removing entrained air. This usually is accomplished by redesigning water boxes, injector nozzles and piping to reduce or eliminate low-pressure pockets, obstructions to smooth flow, abrupt changes in flow direction and other features that cause local regions of high-velocity or turbulent flow. Condensers and heat exchangers are less susceptible to impingement attack if they are made of one of the aluminum brasses or copper nickels, which

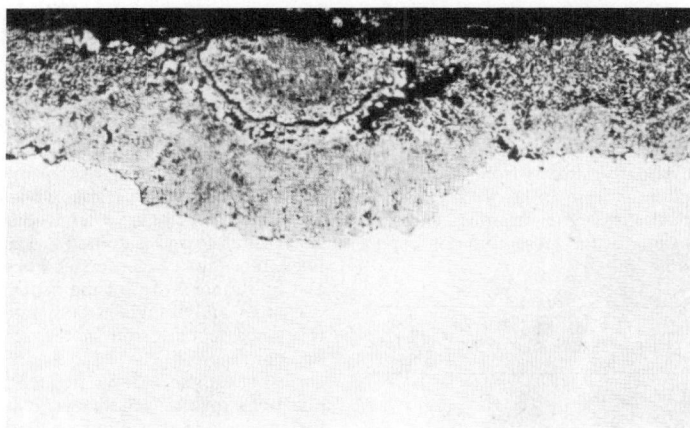

Combined uniform-layer and plug-type dezincification that occurred over 17 years of service; about 75×, unetched.

Fig. 3. Dezincification of C27000 innercooler tubes from an air compressor

and weak layer of copper and copper oxide. Corrosion of a similar nature continues beneath the primary corrosion layer, resulting in gradual replacement of sound brass by weak, porous copper. Unless arrested, dealloying eventually penetrates the metal, weakening it structurally and allowing liquids or gases to leak through the porous mass in the remaining structure. The term *plug-type dealloying* refers to the dealloying that occurs in local areas; surrounding areas usually are unaffected or only slightly corroded. An early stage of plug-type dezincification is illustrated in Fig. 3. In *uniform-layer dealloying,* the active component of the alloy is leached out over a broad area of the surface. Dezincification is the usual form of corrosion for uninhibited brasses in prolonged contact with waters high in oxygen and carbon dioxide. It is frequently encountered with quiescent or slowly moving solutions. Slightly acidic water, low in salt content and at room temperature, is likely to produce uniform attack, whereas neutral or alkaline water, high in salt content and above room temperature, often produces plug-type attack.

Brasses with copper contents of 85% or more resist dezincification. Brasses with two-phase structures usually dezincify in two stages: the high-zinc beta phase first and then the lower-zinc alpha phase.

Tin tends to inhibit dealloying, especially in cast alloys. C46400 (naval brass) and C67500 (manganese bronze), which are alpha-beta brasses containing about 1% Sn and are widely used for naval equipment, have reasonably good resistance to dezincification. Addition of a small amount of phosphorus, arsenic or antimony to admiralty metal (an all-alpha 71Cu-28Zn-1Sn brass) very effectively inhibits dezincification. Inhibitors are not entirely effective in preventing dezincification of the alpha-beta brasses, because they do not prevent dezincification of the beta phase.

Where dezincification is a problem, red brass, commercial bronze, inhibited admiralty metal and inhibited aluminum brass can be used successfully. In some instances, the economic penalty of avoiding dealloying by selecting a low-zinc alloy may be unacceptable. (Low-zinc alloy tube requires fittings that are available only as sand castings, whereas fittings for higher zinc tube can be die cast or forged much more economically.) Where selection of a low-zinc alloy is unaccept-

able, inhibited yellow brasses generally are preferred.

Attack similar to dezincification occurs in other alloys. Dealuminification occurs in some copper-aluminum alloys. Decobaltification of cobalt-base alloys has been reported, and denickelification of copper nickels may occur under special conditions. "Parting" of gold-silver alloys is a form of dealloying.

Intergranular Corrosion

Intergranular corrosion is an infrequently encountered form of attack that occurs most often in applications involving high-pressure steam. This type of corrosion penetrates the metal along grain boundaries—often to a depth of several grains (see Fig. 4)—which distinguishes it from surface roughening. Mechanical stress apparently is not a factor in intergranular corrosion. The alloys

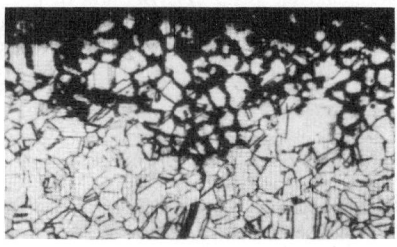

Longitudinal section of an arsenical admiralty metal tube; about 150×, etched.

Fig. 4. Intergranular corrosion of C44500

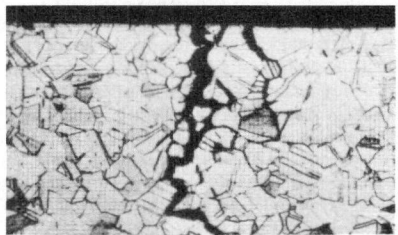

Intergranular cracking shown at about 60× in an etched specimen.

Fig. 5. Typical appearance of stress-corrosion cracking in copper alloys

that appear to be the most susceptible to this form of attack are Muntz metal, admiralty metal, aluminum brasses and silicon bronzes.

Stress-Corrosion Cracking

Stress-corrosion cracking and season cracking are two names for the same phenomenon—the apparently spontaneous cracking of stressed metal. Stress-corrosion cracking is largely intergranular (see Fig. 5), but sometimes transgranular cracking also may be detected. Stress-corrosion cracking occurs only if a susceptible alloy is subjected to the combined effects of sustained stress and certain chemical substances.

Ammonia and ammonium compounds are the corrosive substances most often associated with stress-corrosion cracking of copper alloys. Sometimes these compounds are present in the atmosphere; in other instances, they are in cleaning compounds or in chemicals used for the treatment of boiler water. Both oxygen and moisture must be present for ammonia to be corrosive to copper alloys; other compounds, such as carbon dioxide, are thought to accelerate stress-corrosion cracking in ammoniacal atmospheres. Moisture films on metal surfaces will dissolve significant quantities of ammonia, even from atmospheres that are low in ammonia concentration.

Susceptibility to stress-corrosion cracking can be mitigated by proper choice of microstructure and alloy composition. Microstructure and composition can be controlled most effectively by selecting the right combination of alloy, forming process, thermal treatment and metal-finishing process. Although test results may indicate that a finished part is not susceptible to stress-corrosion cracking, such an indication does not ensure complete freedom from cracking, particularly where service stresses are high.

Both applied and residual stresses can lead to failure by stress-corrosion cracking. Susceptibility is largely a function of stress magnitude. Stresses near the yield strength are usually required, but parts have failed under lower stresses. In general, the higher the stress, the weaker the corroding medium may be to cause cracking. The reverse also is true: the stronger the corroding medium, the lower the required stress.

Sources of Stress. Applied stresses result from ordinary service loading or from fabricating techniques such as riveting, bolting, shrink fitting, brazing and welding.

Residual stresses are of two types: differential-strain stresses, resulting from nonuniform plastic strain during cold forming; and differential-thermal-contraction stresses, resulting from nonuniform heating and/or cooling.

Residual stresses induced by nonuniform straining are influenced chiefly by the method of fabrication. In some fabricating processes, it is possible to cold work a metal extensively and yet produce only a low level of residual stress. For example, residual stress in a drawn tube is influenced by die angle and amount of reduction. Wide-angle dies (about 32°) produce higher residual stresses than narrow-angle dies (about 8°). Light reductions yield high residual stresses because only the surface of the alloy is stressed; heavy reductions yield low residual stresses because the region of cold working extends deeper into the metal. Most drawing operations can be planned so that residual stresses are low and susceptibility to stress-corrosion cracking is negligible.

Residual stresses resulting from upsetting, stretching or spinning are harder to evaluate and harder to control by varying tooling and process conditions. For these operations, stress-corrosion cracking can be prevented more effectively by selecting a resistant alloy or by treating the metal after fabrication.

Alloy Composition. Brasses containing less than 15% Zn are highly resistant to stress-corrosion cracking. Phosphorus-deoxidized copper and tough pitch copper rarely exhibit stress-corrosion cracking, even under severe conditions. On the other hand, brasses containing 20 to 40% Zn are highly susceptible to stress-corrosion cracking. Susceptibility increases only as zinc content is increased from 20 to 40%.

There is no indication that other elements commonly added to brasses increase the probability of stress-corrosion cracking. Phosphorus, arsenic, magnesium, tellurium, tin, beryllium and manganese are thought to decrease susceptibility under some conditions. Addition of 1.5% silicon is known to decrease the probability of cracking.

In Table 3, various copper alloys are ranked according to their relative stress-corrosion susceptibility in ammonia environments. Susceptibility in other environments may vary. In Table 3, the scale of 0 to 1000 covers the range from essential immunity to stress corrosion under normal service conditions to high susceptibility typified by C26000.

Table 3. Relative stress-corrosion susceptibility of representative copper alloys

Alloy	Susceptibility index	Alloy	Susceptibility index
C26000	1000	C51000	20
C35300	1000	C11000	0
C76200	300	C15100	0
C23000	200	C19400	0
C77000	175	C65400	0
C66400	100	C70600	0
C68800	75	C71500	0
C63800	50	C72200	0
C75200	40		

Altering the microstructure cannot make a susceptible alloy totally resistant to stress-corrosion cracking. However, the rapidity with which susceptible alloys crack appears to be affected by grain size and structure. All other factors being equal, rate of cracking increases as grain size increases. The effects of structure on stress-corrosion cracking are not sharply defined, chiefly because they are interrelated with effects of both composition and stress.

Control Measures. Stress-corrosion cracking can be controlled, and sometimes prevented, by (a) selecting copper alloys that have high resistance to cracking (notably those with less than 15% Zn); (b) reducing residual stress to a safe level by thermal stress relief, which usually can be applied without significantly decreasing strength; or (c) altering the environment, such as by changing the predominant chemical species present or introducing a corrosion inhibitor.

Residual and assembly stresses may be eliminated by recrystallization annealing after forming or assembly. Recrystallization annealing cannot be used when the integrity of the structure depends on the higher strength of strain-hardened metal, which always contains a certain amount of residual stress. Thermal stress relief (some-

times called relief annealing) can be specified when the higher strength of a cold worked temper must be retained. Thermal stress relief consists of heating the part for a relatively short time at low temperature. Specific times and temperatures depend on alloy composition, severity of deformation, prevailing stresses and size of load being heated. Usually, time is from 30 min to 1 h and temperature is from 150 to 425 °C (300 to 800 °F). Typical stress-relieving times and temperatures for some of the more common copper alloys follow.

UNS number	Common name	Temperature °C	°F	Time, h
C22000	Commercial bronze	205	400	1
C26000	Cartridge brass	260	500	1
C28000	Muntz metal	190	375	1/2
C44300, C44400, C44500	Admiralty metal	300	575	1
C51000, C52400	Phosphor bronze, 5 or 10%	190	375	1
C65500	Silicon bronze	370	700	1
C61300 C61400	Aluminum bronze	400	750	1
C71500	Copper nickel, 30%	425	800	1

The exact thermal treatment should be established by examining specific parts for residual stress, as described in the next section of this article. If such examination indicates that a thermal treatment is insufficient, temperature and/or time should be adjusted until satisfactory results are obtained. Parts in the center of a furnace load may not reach the desired temperature as soon as parts around the periphery. Because of this, it may be necessary to compensate for furnace loading when setting process controls or to limit the number of parts that can be stress relieved together.

Mechanical methods such as stretching, flexing, bending, straightening between rollers, peening and shot blasting also may be used to reduce residual stresses to a safe level. These methods depend on plastic deformation to either decrease dangerous tensile stresses or convert them to less objectionable compressive stresses.

Corrosion Fatigue

The combined action of corrosion (usually pitting corrosion) and cyclic stress may result in corrosion-fatigue cracking. Like ordinary fatigue cracks, corrosion-fatigue cracks generally propagate at right angles to the maximum tensile stress in the affected region. However, cracks resulting from simultaneous corrosion and alternating stress often propagate much more rapidly than cracks caused by alternating stress alone. Also, corrosion-fatigue failure usually involves several parallel cracks, whereas it is rare for more than one crack to be found in a part that has failed as a result of simple fatigue.

Ordinarily, corrosion fatigue can be readily identified by the presence of several cracks emanating from corrosion pits. Cracks not visible to the unaided eye or at low magnification can be made visible by deep etching or plastic defor-

mation or detected by eddy-current inspection. Corrosion-fatigue cracking is often transgranular, but there is evidence that certain environments induce intergranular fatigue cracking in copper metals.

In addition to effectively resisting corrosion, copper and copper alloys also resist corrosion fatigue in many applications that involve repeated stress and corrosion. These applications include such parts as springs, switches, diaphragms, bellows, aircraft and automotive gasoline and oil lines, tubes for condensers and heat exchangers, and fourdrinier wire for the paper industry.

Copper alloys high in both fatigue limit and resistance to corrosion in the service environment are more likely to have good resistance to corrosion fatigue. Alloys frequently used in applications involving both cyclic stress and corrosion include beryllium coppers, phosphor bronzes, aluminum bronzes and copper nickels.

EFFECT OF ALLOY COMPOSITION

Brasses are basically copper-zinc alloys and are the most widely used group of copper alloys. Resistance of brasses to corrosion by aqueous solutions does not change markedly as long as the zinc content does not exceed about 15%; above 15% Zn, dezincification may occur. Quiescent or slowly moving saline solutions, brackish waters and mildly acidic solutions are environments that often lead to dezincification.

As shown in Fig. 6, resistance to pitting is almost total when zinc content exceeds 15%. But the brasses that resist pitting are severely degraded by dezincification, which causes them to lose a substantial portion of their strength.

Where exposure to sulfur compounds is involved, brasses containing the highest amounts of zinc have the best resistance (Tables 4 and 5).

Susceptibility to stress-corrosion cracking is significantly affected by zinc content: alloys containing more zinc are more susceptible. Resistance increases substantially as zinc content decreases from 15% to zero; stress-corrosion cracking in commercial coppers is practically unknown.

Elements such as lead, tellurium, beryllium, chromium, phosphorus and manganese have little or no effect on corrosion resistance of coppers and binary copper-zinc alloys. These elements are added to enhance mechanical properties such as machinability, strength and hardness.

Tin-bearing Brasses. Tin additions significantly increase the corrosion resistance of some brasses, especially resistance to dezincification. Prime examples of this effect are two tin-bearing brasses: uninhibited admiralty metal (no active UNS number) and naval brass (C46400). Uninhibited admiralty metal was once widely used to make heat-exchanger tubes; it has largely been replaced by inhibited grades of admiralty metal (C44300, C44400 and C44500), which have even greater resistance to dealloying. Admiralty metal is a variation of cartridge brass (C26000) produced by adding about 1% Sn to the basic 70Cu–30Zn composition. Similarly, naval brass is the alloy resulting from addition of 0.75% Sn to the basic 60Cu–40Zn composition of Muntz metal (C28000).

Aluminum Brasses. An important constituent of the corrosion film on a brass that contains a few percent aluminum in addition to copper and zinc

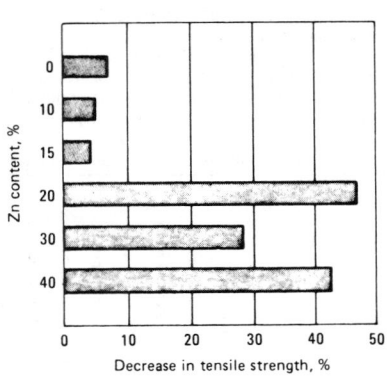
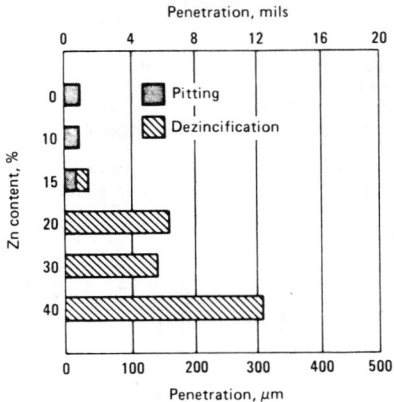

Brass strip, 0.8 mm (0.032 in.) thick, was immersed for 60 days in 0.01 *M* NH₄Cl solution at 45 °C (113 °F).

Fig. 6. Effect of zinc content on corrosion of brasses

is aluminum oxide, which markedly increases resistance to impingement attack in turbulent, high-velocity saline water. For example, the arsenical aluminum brass C68700 (76Cu-22Zn-2Al) is frequently used for marine condensers and heat exchangers where impingement attack is likely to pose a serious problem. Aluminum brasses are susceptible to dezincification and stress-corrosion cracking unless they are inhibited, which usually is done by adding 0.02 to 0.10% As.

Inhibited Alloys. Addition of phosphorus, arsenic or antimony (typically, 0.02 to 0.10%) to admiralty metal, naval brass or aluminum brass effectively produces high resistance to dezincification at alpha service temperatures. Inhibited alloys have been used extensively for components such as condenser tubes, which must accumulate several years of continuous service between shutdowns for repair or replacement.

Phosphor Bronzes. Addition of tin and phosphorus to copper produces good resistance to flowing seawater and to most nonoxidizing acids except hydrochloric. Alloys containing 8 to 10% Sn have high resistance to impingement attack. Phosphor bronzes are much less susceptible to stress-corrosion cracking than brasses and are similar to copper in resistance to sulfur attack.

Copper Nickels. C71500 (copper nickel, 30%) has the best general resistance to aqueous corrosion of all the commercially important copper alloys, but C70600 (copper nickel, 10%) is often selected because it offers good resistance at lower cost. Both of these alloys, although well suited to applications in the chemical industry, have been used most extensively for condenser tubes and heat-exchanger tubes in recirculating steam systems. They are superior to coppers and to other

Table 5. Corrosion of copper alloys in hot paper-mill vapor containing SO₂(a)

UNS number	Alloy type	Weight loss, g/m²·d
...	Bronze (90Cu-10Sn)	22.03
C61800	Aluminum bronze	26.42
C51100	Phosphor bronze	28.58
C73200	Nickel silver, 75-20	35.56
C52100	Phosphor bronze, 8% C	39.43
C65800	Silicon bronze	50.22
C77000	Nickel silver, 55-18	63.78
C75200	Nickel silver, 65-18	67.42
...	Nickel bronze (88.5Cu-5Sn-5Ni-1.5Si)	70.49

(a) Temperature, 200 to 220 °C (390 to 430 °F); atmosphere, 17 to 18% SO₂ plus 1 to 2% O₂; test duration, mainly 30 days, but some longer.

copper alloys in resisting acid solutions and are highly resistant to stress-corrosion cracking and impingement corrosion.

Nickel Silvers. The two most common nickel silvers are C75200 (nickel silver, 65-18) and C77000 (nickel silver, 55-18). They have good resistance to corrosion in both fresh and salt waters. Chiefly because their relatively high nickel contents inhibit dezincification, C75200 and C77000 usually are much more resistant to corrosion in saline solutions than are brasses of similar copper content.

Copper-silicon alloys generally have the same corrosion resistance as copper, but higher mechanical properties and superior weldability. These alloys appear to be much better than the common brasses in resistance to stress-corrosion cracking. Silicon bronzes are susceptible to embrittlement by high-pressure steam and should be tested for

suitability in the service environment before being specified for components to be used at elevated temperature.

Aluminum bronzes containing 5 to 12% Al have excellent resistance to impingement corrosion and high-temperature oxidation. Aluminum bronzes are used for beater bars and for blades in wood pulp machines because of their ability to withstand mechanical abrasion and chemical attack by sulfite solutions.

In most practical commercial applications, the corrosion characteristics of aluminum bronzes are primarily related to aluminum content. Alloys with up to 8% Al normally have completely face-centered-cubic alpha structures and good resistance to corrosion attack. Alloys C60600, C60800, C61000, C61300, C61400 and C61500 are of this type. As aluminum content increases above 8%, alpha-beta duplex structures appear. The beta phase is a high-temperature phase retained at room temperature on fast cooling from 565 °C (1050 °F) or above. Slow cooling, or long exposure at temperatures from 320 to 565 °C (610 to 1050 °F), tends to decompose the beta phase into a brittle alpha + gamma-2 eutectoid having either a lamellar or a nodular structure. The beta phase is less resistant to corrosion than the alpha phase, and eutectoid structures are even more susceptible to attack.

Depending on specific environmental conditions, beta phase or eutectoid structure in aluminum bronze can be selectively attacked by a mechanism similar to dezincification of brasses. The use of proper quench-and-temper treatments on duplex alloys such as C62400 and C95400 produces a tempered beta structure with reprecipitated acicular alpha crystals, a combination often superior in corrosion resistance to the normal annealed structures.

Iron-rich particles are distributed as small round or rosette particles throughout the structures of aluminum bronzes containing more than about 0.5% Fe. These particles sometimes impart a rusty tinge to the surface, but have no known effect on corrosion rates.

Nickel-aluminum bronzes are more complex in structure, with the introduction of the kappa phase. Nickel appears to alter corrosion characteristics of the beta phase to give greater resistance to dealloying and cavitation-erosion in most liquids. There is some evidence that for C63200, and perhaps for C95800, quench-and-temper treatments yield even greater resistance to dealloying. C95700, a high-manganese cast aluminum bronze, is somewhat inferior in corrosion resistance to C95500 and C95800, which are low in manganese and slightly higher in aluminum.

Aluminum bronzes are generally suitable for service in nonoxidizing mineral acids, such as phosphoric, sulfuric and hydrochloric: organic acids, such as lactic, acetic, or oxalic; neutral salt solutions, such as sodium or potassium chloride; alkalies, such as sodium hydroxide, potassium hydroxide and anhydrous ammonium hydroxide; and various natural waters, including sea, brackish and potable waters. Environments to be avoided include nitric acid, some metallic salts such as ferric chloride and chromic acid, moist chlorinated hydrocarbons, and moist ammonia. Aeration can result in accelerated corrosion in many media that appear compatible.

Exposure under high tensile stress to moist ammonia can result in stress-corrosion cracking. In certain environments, corrosion can lower the

Table 4. Corrosion of selected copper alloys in cracked oil containing 1.4% S

UNS number	Alloy type	Exposure time, days	Percent loss in tensile strength(a) at:			
			360 °C (680 °F)	315 °C (600 °F)	285 °C (545 °F)	255 °C (490 °F)
C23000	Red brass, 85%	27	100(b)	100(c)	100	100
C28000	Muntz metal	27	12(b)	7.5(d)	1	1.5
C46400	Naval brass	24	...	1.5	0	2
...	Uninhibited admiralty metal	27	13(b)	6(c)	3	2
C44400	Antimonial admiralty metal	27	16.5(b)	6(c)	4	2.5
...	Aluminum brass	24	...	7	16	10
C71500	Copper nickel, 30%	24	...	100	100	57
...	Silicon bronze, 3%	34	...	100	100	100

(a) Specimens 0.8 by 13 mm (0.032 by 0.50 in.) in cross section were exposed at different locations within a high-pressure fractionating column, each location having a characteristic average temperature. (b) 115-day exposure. (c) 26-day exposure. (d) Length of exposure unavailable.

Table 6. Corrosion ratings of copper and copper alloys in various corrosive media

This table is intended to serve only as a general guide to the behavior of copper and copper alloys in corrosive environments. It is impossible to cover in a simple tabulation the performance of a material for all possible variations of temperature, concentration, velocity, impurity content, degree of aeration and stress. The ratings are based on general performance; they should be used with caution, and then only for the purpose of screening candidate alloys.

The letters E, G, F and P have the following significance:

E—Excellent. Resists corrosion under almost all conditions of service.

G—Good. Some corrosion will take place, but satisfactory service can be expected under all but the most severe conditions.

F—Fair. Corrosion rates are higher than for the "G" classification, but the metal can be used if needed for a property other than corrosion resistance and if either the amount of corrosion does not cause excessive maintenance expense or the effects of corrosion can be lessened, such as by use of coatings or inhibitors.

P—Poor. Corrosion rates are high, and service is generally unsatisfactory.

Corrosive medium	Coppers	Low-zinc brasses	High-zinc brasses	Special brasses	Phosphor bronzes	Aluminum bronzes	Silicon bronzes	Copper nickels	Nickel silvers
Acetate solvents	E	E	G	G	E	E	E	E	E
Acetic acid(a)	E	E	P	P	E	E	E	E	G
Acetone	E	E	E	E	E	E	E	E	E
Acetylene(b)	P	P	(b)	P	P	P	P	P	P
Alcohols(a)	E	E	E	E	E	E	E	E	E
Aldehydes	E	E	F	F	E	E	E	E	E
Alkylamines	G	G	G	G	G	G	G	G	G
Alumina	E	E	E	E	E	E	E	E	E
Aluminum chloride	G	G	P	P	G	G	G	G	G
Aluminum hydroxide	E	E	E	E	E	E	E	E	E
Aluminum sulfate and alum	G	G	P	G	G	G	G	E	G
Ammonia, dry	E	E	E	E	E	E	E	E	E
Ammonia, moist(c)	P	P	P	P	P	P	P	F	P
Ammonium chloride(c)	P	P	P	P	P	P	P	F	P
Ammonium hydroxide(c)	P	P	P	P	P	P	P	F	P
Ammonium nitrate(c)	P	P	P	P	P	P	P	F	P
Ammonium sulfate(c)	F	F	P	P	F	F	F	G	F
Aniline and aniline dyes	F	F	F	F	F	F	F	F	F
Asphalt	E	E	E	E	E	E	E	E	E
Atmosphere:									
Industrial(c)	E	E	E	E	E	E	E	E	E
Marine	E	E	E	E	E	E	E	E	E
Rural	E	E	E	E	E	E	E	E	E
Barium carbonate	E	E	E	E	E	E	E	E	E
Barium chloride	G	G	F	F	G	G	G	G	G
Barium hydroxide	E	E	G	E	E	E	E	E	E
Barium sulfate	E	E	G	E	E	E	E	E	E
Beer(a)	E	E	G	E	E	E	E	E	E
Beet-sugar syrup(a)	E	E	G	E	E	E	E	E	E
Benzene, benzine, benzol	E	E	E	E	E	E	E	E	E

Corrosive medium	Coppers	Low-zinc brasses	High-zinc brasses	Special brasses	Phosphor bronzes	Aluminum bronzes	Silicon bronzes	Copper nickels	Nickel silvers
Benzoic acid	E	E	E	E	E	E	E	E	E
Black liquor, sulfate process	P	P	P	P	P	P	P	G	P
Bleaching powder (wet)	G	G	P	G	G	G	G	G	G
Borax	E	E	E	E	E	E	E	E	E
Bordeaux mixture	E	E	G	E	E	E	E	E	E
Boric acid	E	E	G	E	E	E	E	E	E
Brines	G	G	P	G	G	G	G	E	E
Bromine, dry	E	E	E	E	E	E	E	E	E
Bromine, moist	G	G	P	F	G	G	G	G	G
Butane(d)	E	E	E	E	E	E	E	E	E
Calcium bisulfate	G	G	P	G	G	G	G	G	G
Calcium chloride	G	G	F	G	G	G	G	G	G
Calcium hydroxide	E	E	G	E	E	E	E	E	E
Calcium hypochlorite	G	G	P	G	G	G	G	G	G
Cane-sugar syrup(a)	E	E	E	E	E	E	E	E	E
Carbolic acid (phenol)	F	G	P	G	G	G	G	G	G
Carbonated beverages(a)(e)	E	E	E	E	E	E	E	E	E
Carbon dioxide, dry	E	E	E	E	E	E	E	E	E
Carbon dioxide, moist(a)(e)	E	E	E	E	E	E	E	E	E
Carbon tetrachloride (dry)	E	E	E	E	E	E	E	E	E
Carbon tetrachloride (moist)	G	G	F	G	E	E	E	E	E
Castor oil	E	E	E	E	E	E	E	E	E
Chlorine, dry(f)	E	E	E	E	E	E	E	E	E
Chlorine, moist	F	F	P	F	F	F	F	G	F
Chloracetic acid	G	F	P	F	G	G	G	G	G
Chloroform, dry	E	E	E	E	E	E	E	E	E
Chromic acid	P	P	P	P	P	P	P	P	P
Citric acid(a)	E	E	P	E	E	E	E	E	E
Copper chloride	F	F	P	F	F	F	F	F	F
Copper nitrate	F	F	P	F	F	F	F	F	F
Copper sulfate	G	G	P	G	G	G	G	E	G

Corrosive medium	Coppers	Low-zinc brasses	High-zinc brasses	Special brasses	Phosphor bronzes	Aluminum bronzes	Silicon bronzes	Copper nickels	Nickel silvers
Corn oil(a)	E	E	G	E	E	E	E	E	E
Cottonseed oil(a)	E	E	G	E	E	E	E	E	E
Creosote	E	E	G	E	E	E	E	E	E
Dowtherm "A"	E	E	G	E	E	E	E	E	E
Ethanol amine	G	G	G	G	G	G	G	G	G
Ethers	E	E	E	E	E	E	E	E	E
Ethyl acetate (esters)	E	E	G	E	E	E	E	E	E
Ethylene glycol	E	E	G	E	E	E	E	E	E
Ferric chloride	P	P	P	P	P	P	P	P	P
Ferric sulfate	P	P	P	P	P	P	P	P	P
Ferrous chloride	G	G	P	G	G	G	G	G	G
Ferrous sulfate	G	G	P	G	G	G	G	G	G
Formaldehyde (aldehydes)	E	E	G	E	E	E	E	E	E
Formic acid	G	G	P	F	G	G	G	G	G
Freon, dry	E	E	E	E	E	E	E	E	E
Freon, moist									
Fuel oil, light	E	E	E	E	E	E	E	E	E
Fuel oil, heavy	E	E	G	E	E	E	E	E	E
Furfural	E	E	F	E	E	E	E	E	E
Gasoline	E	E	E	E	E	E	E	E	E
Gelatin(a)	E	E	E	E	E	E	E	E	E
Glucose(a)	E	E	E	E	E	E	E	E	E
Glue	E	E	G	E	E	E	E	E	E
Glycerin	E	E	G	E	E	E	E	E	E
Hydrobromic acid	F	F	P	F	F	F	F	F	F
Hydrocarbons	E	E	E	E	E	E	E	E	E
Hydrochloric acid (muriatic)	F	F	P	F	F	F	F	F	F
Hydrocyanic acid, dry	E	E	E	E	E	E	E	E	E
Hydrocyanic acid, moist	P	P	P	P	P	P	P	P	P
Hydrofluoric acid, anhydrous	G	G	P	G	G	G	G	G	G
Hydrofluoric acid, hydrated	F	F	P	F	F	F	F	F	F
Hydrofluosilicic acid	G	G	P	G	G	G	G	G	G
Hydrogen(d)	E	E	E	E	E	E	E	E	E
Hydrogen peroxide up to 10%	G	G	F	G	G	G	G	G	G

(continued)

fatigue limit to 25 to 50% of the normal atmospheric value.

PROTECTIVE COATINGS

Copper metals resist corrosion in many environments because on initial exposure they react with one or more constituents of the environment, thereby forming a surface layer of protective reaction products.

In certain applications, corrosion resistance of copper metals may be increased by applying metallic or organic protective coatings. Provided the coating material is able to resist corrosion adequately, service life may depend on impermeability, continuity, and adhesion to the basis metal. The electropotential relationship of the coating to the basis metal may be important, especially with metallic coatings and at uncoated edges.

Tin, lead and solder, used extensively as coatings, ordinarily are applied by hot dipping, although electroplating is used also.

Tin arrests corrosion caused by sulfur; it is most effective as a coating for copper wire and cable insulated by rubber that contains sulfur. Lead-coated copper is used chiefly for roofing applications where contact with flue gases or other products that contain dilute sulfuric acid is likely. Tin or lead coatings sometimes are applied to copper intended for ordinary atmospheric exposure, but this is done primarily for architectural effect; the atmospheric corrosion resistance of bare copper is excellent in rural, urban, marine and most industrial locations.

Electroplated chromium is used for decoration, for improvement of wear resistance or for reflectivity. Because it is somewhat porous, it is not effective for corrosion protection. Where corrosion protection is important, electroplated nickel is used most often as a protective coating under electroplated chromium.

Clear lacquer sometimes is applied to preserve a bright natural color for decorative reasons. For instance in one particular architectural application, a commercial lacquer specifically developed for use on copper effectively preserved the warm color of a copper roof.

SELECTION FOR SPECIFIC ENVIRONMENTS

Copper and copper alloys are not merely suitable but often superior for many applications included in the following broad classifications:

Table 6. (continued)

Corrosive medium	Coppers	Low-zinc brasses	High-zinc brasses	Special brasses	Phosphor bronzes	Aluminum bronzes	Silicon bronzes	Copper nickels	Nickel silvers
Hydrogen peroxide over 10%	P	P	P	P	P	P	P	P	P
Hydrogen sulfide, dry	E	E	E	E	E	E	E	E	E
Hydrogen sulfide, moist	P	P	F	F	P	P	P	F	F
Kerosine	E	E	E	E	E	E	E	E	E
Ketones	E	E	E	E	E	E	E	E	E
Lacquers	E	E	E	E	E	E	E	E	E
Lacquer thinners (solvents)	E	E	E	E	E	E	E	E	E
Lactic acid(a)	E	E	F	E	E	E	E	E	E
Lime	E	E	E	E	E	E	E	E	E
Lime sulfur	P	P	F	F	P	P	P	F	F
Linseed oil	G	G	G	G	G	G	G	G	G
Lithium compounds	G	G	P	F	G	G	G	E	E
Magnesium chloride	G	G	F	F	G	G	G	G	G
Magnesium hydroxide	E	E	G	E	E	E	E	E	E
Magnesium sulfate	E	E	G	E	E	E	E	E	E
Mercury or mercury salts	P	P	P	P	P	P	P	P	P
Milk(a)	E	E	G	E	E	E	E	E	E
Molasses	E	E	G	E	E	E	E	E	E
Natural gas(d)	E	E	E	E	E	E	E	E	E
Nickel chloride	F	F	P	F	F	F	F	F	F
Nickel sulfate	F	F	P	F	F	F	F	F	F
Nitric acid	P	P	P	P	P	P	P	P	P
Oleic acid	G	G	F	G	G	G	G	G	G
Oxalic acid(g)	E	E	P	P	E	E	E	E	E
Oxygen(h)	E	E	E	E	E	E	E	E	E
Palmitic acid	G	G	F	G	G	G	G	G	G
Paraffin	E	E	E	E	E	E	E	E	E
Phosphoric acid	G	G	P	F	G	G	G	G	G
Picric acid	P	P	P	P	P	P	P	P	P
Potassium carbonate	E	G	E	E	E	E	E	E	E
Potassium chloride	G	G	P	F	G	G	G	E	E
Potassium cyanide	P	P	P	P	P	P	P	P	P
Potassium dichromate (acid)	P	P	P	P	P	P	P	P	P
Potassium hydroxide	G	G	F	G	G	G	G	E	E
Potassium sulfate	E	E	G	E	E	E	E	E	E
Propane(d)	E	E	E	E	E	E	E	E	E
Rosin	E	E	E	E	E	E	E	E	E
Sea water	G	G	F	E	G	E	G	E	E
Sewage	E	E	F	E	E	E	E	E	E
Silver salts	P	P	P	P	P	P	P	P	P
Soap solution	E	E	E	E	E	E	E	E	E
Sodium bicarbonate	E	E	G	E	E	E	E	E	E
Sodium bisulfate	G	G	F	G	G	G	G	E	E
Sodium carbonate	E	E	G	E	E	E	E	E	E
Sodium chloride	G	G	P	F	G	G	G	E	E
Sodium chromate	E	E	E	E	E	E	E	E	E
Sodium cyanide	P	P	P	P	P	P	P	P	P
Sodium dichromate (acid)	P	P	P	P	P	P	P	P	P
Sodium hydroxide	G	G	F	G	G	G	G	E	E
Sodium hypochlorite	G	G	P	G	G	G	G	G	G
Sodium nitrate	G	G	P	F	G	G	G	E	E
Sodium peroxide	F	F	P	F	F	F	F	G	G
Sodium phosphate	E	E	G	E	E	E	E	E	E
Sodium silicate	E	E	G	E	E	E	E	E	E
Sodium sulfate	E	E	G	E	E	E	E	E	E
Sodium sulfide	P	P	F	F	P	P	P	F	F
Sodium thiosulfate	P	P	F	F	P	P	P	F	F
Steam	E	E	F	E	E	E	F	E	E
Stearic acid	E	E	F	E	E	E	E	E	E
Sugar solutions	E	E	G	E	E	E	E	E	E
Sulfur, solid	G	G	E	G	G	G	G	E	G
Sulfur, molten	P	P	P	P	P	P	P	P	P
Sulfur chloride (dry)	E	E	E	E	E	E	E	E	E
Sulfur chloride (moist)	P	P	P	P	P	P	P	P	P
Sulfur dioxide (dry)	E	E	E	E	E	E	E	E	E
Sulfur dioxide (moist)	G	G	P	G	G	G	G	F	F
Sulfur trioxide (dry)	E	E	E	E	E	E	E	E	E
Sulfuric acid 80-95%(j)	G	G	P	F	G	G	G	G	G
Sulfuric acid 40-80%(j)	F	F	F	P	F	F	F	F	F
Sulfuric acid 40%(j)	G	G	P	F	G	G	G	G	G
Sulfurous acid	G	G	P	G	G	G	G	F	F
Tannic acid	E	E	E	E	E	E	E	E	E
Tartaric acid(a)	E	E	E	E	E	E	E	E	E
Toluene	E	E	E	E	E	E	E	E	E
Trichloracetic acid	G	G	P	F	G	G	G	G	G
Trichlorethylene (dry)	E	E	E	E	E	E	E	E	E
Trichlorethylene (moist)	G	G	F	G	E	E	E	E	E
Turpentine	E	E	E	E	E	E	E	E	E
Varnish	E	E	E	E	E	E	E	E	E
Vinegar(a)	E	E	P	F	E	E	E	E	G
Water, acidic mine	F	F	P	F	G	F	F	P	F
Water, potable	E	E	G	E	E	E	E	E	E
Water, condensate(c)	E	E	E	E	E	E	E	E	E
Wetting agents(k)	E	E	E	E	E	E	E	E	E
Whiskey(a)	E	E	E	E	E	E	E	E	E
White water	G	G	E	E	E	E	E	E	E
Zinc chloride	G	G	P	G	G	G	G	G	G
Zinc sulfate	E	E	P	E	E	E	E	E	E

(a) Copper and copper alloys are resistant to corrosion by most food products. Traces of copper may be dissolved and affect taste or color of the products. In such cases, copper alloys often are tin coated. (b) Acetylene forms an explosive compound with copper when moisture or certain impurities are present and the gas is under pressure. Alloys containing less than 65% copper are satisfactory; when the gas is not under pressure, other copper alloys are satisfactory. (c) Precautions should be taken to avoid stress-corrosion cracking. (d) At elevated temperatures, hydrogen will react with tough pitch copper, causing failure by embrittlement. (e) Where air is present, corrosion rate may be increased. (f) Below 150 °C (300 °F), corrosion rate is very low; above this temperature, corrosion is appreciable and increases rapidly as the temperature rises. (g) Aeration and elevated temperature may increase corrosion rate substantially. (h) Excessive oxidation may begin above 120 °C (250 °F). If moisture is present, oxidation may begin at lower temperatures.

(j) Use of high-zinc brasses should be avoided in acids because of the likelihood of rapid corrosion by dezincification. Copper, low-zinc brasses, phosphor bronzes, silicon bronzes, aluminum bronzes and copper nickels offer good resistance to corrosion by hot and cold dilute sulfuric acid and to corrosion by cold concentrated sulfuric acid. Intermediate concentrations of sulfuric acid sometimes are more corrosive to copper alloys than either concentrated or dilute acid. Concentrated sulfuric acid may be corrosive at elevated temperatures due to breakdown of acid and formation of metallic sulfides and sulfur dioxide, which cause localized pitting. Tests indicate that copper alloys may undergo pitting in 90 to 95% sulfuric acid at about 50 °C (122 °F), in 80% acid at about 70 °C (160 °F), and in 60% acid at about 100 °C (212 °F). (k) Wetting agents may increase corrosion rates of copper and copper alloys slightly to substantially when carbon dioxide or oxygen is present, by preventing formation of a film on the metal surface and by combining (in some instances) with the dissolved copper to produce a green, insoluble compound.

1. Applications requiring resistance to atmospheric exposure, such as roofing and other architectural uses, hardware, building fronts, grille work, hand rails, butts, lock bodies, doorknobs and kick plates
2. Fresh-water supply lines and plumbing fittings, where superior resistance to corrosion by various types of water and soils are important
3. Marine applications—most often fresh-water and seawater supply lines, heat exchangers, condensers, shafting, valve stems and marine hardware—where resistance to seawater, hydrated salt deposits and biofouling from marine organisms are important.
4. Heat exchangers and condensers in marine service, in steam power plants and in chemical process applications, and liquid-to-gas or gas-to-gas heat exchangers where either process stream may contain a corrosive contaminant
5. Industrial and chemical-plant process equipment involving exposure to a wide variety of organic and inorganic chemicals

Selection of a suitably resistant material requires consideration of the many factors that influence corrosion.

Over the years, experience has been the best criterion for selecting the most suitable alloy for a given environment. The Copper Development Association has compiled much field experience in the form of the ratings shown in Table 6. The table should be used only as a guide: small changes in the environmental conditions sometimes change the performance of a given alloy from "suitable" to "not suitable."

Whenever there is a lack of operating experience, whenever reported test conditions do not closely match conditions for which alloy selection is being made, and whenever there is doubt as to the applicability of published data, it is always best to conduct an independent test program. Field tests are the most reliable. Laboratory tests can be equally valuable, but only if operating conditions are precisely defined and then accurately simulated in the laboratory. Long-term tests generally are preferred because the reaction that dominates the initial stages of corrosion may be quite different from the reaction that dominates later on. If short-term tests must be used as the basis for alloy selection, the test program should be supplemented with field tests so that the laboratory results can be re-evaluated in light of true operating experience.

Table 7. Atmospheric corrosion of selected copper alloys

| | Corrosion rates(a) at indicated locations | | | | | | | | | | | |
| | Altoona, PA | | New York, NY | | Key West, FL | | La Jolla, CA | | State College, PA | | Phoenix, AZ | |
Alloy	μm/yr	mils/yr	μm/yr	mils/yr	μm/yr	mils/yr	μm/yr	mils/yr	μm/yr	mils/yr	μm/yr	mils/yr
C11000	1.40	0.055	1.38	0.054	0.56	0.022	1.27	0.050	0.43	0.017	0.13	0.005
C12000	1.32	0.052	1.22	0.048	0.51	0.020	1.42	0.056	0.36	0.014	0.08	0.003
C23000	1.88	0.074	1.88	0.074	0.56	0.022	0.33	0.013	0.46	0.018	0.10	0.004
C26000	3.05	0.120	2.41	0.095	0.20	0.008	0.15	0.006	0.46	0.018	0.10	0.004
C52100	2.24	0.088	2.54	0.100	0.71	0.028	2.31	0.091	0.33	0.013	0.13	0.005
C61000	1.63	0.064	1.60	0.063	0.10	0.004	0.15	0.006	0.25	0.010	0.51	0.002
C65500	1.65	0.065	1.73	0.068	...	...	1.38	0.054	0.51	0.020	0.15	0.006
C44200	2.13	0.084	2.51	0.099	...	...	0.33	0.013	0.53	0.021	0.10	0.004
70Cu-29Ni-1Sn(b)	2.64	0.104	2.13	0.084	0.28	0.011	0.36	0.014	0.48	0.019	0.10	0.004

(a) Derived from 20-yr exposure tests. Types of atmospheres: Altoona, industrial; New York City, industrial marine; Key West, tropical rural marine; La Jolla, humid marine; State College, northern rural; Phoenix, dry rural. (b) This alloy is obsolete, but it indicates the corrosion resistance expected of C71500.

Erroneous conclusions based on laboratory results also may be reached by inaccurately measuring corrosion damage, especially when corrosion is slight. It has become common practice to express test results in terms of penetration or average reduction in metal thickness, even when corrosion was actually measured by weight loss. Weight-loss or average-penetration data are valid only when corrosion is uniform. When corrosion occurs predominantly by pitting or some other localized form, or when corrosion is intergranular or involves formation of a thick, adherent scale, direct measurement of the extent of corrosion provides the most reliable information. (A common technique is to measure the maximum depth of penetration observed on a metallographic cross section through the region of interest.) Statistical averaging of repeated measurements on one or more specimens may or may not be warranted. Despite the deficiencies in laboratory testing, information gained in this way serves as a useful starting point for alloy selection. Operating experience may later indicate the need for a more discriminating selection.

Atmospheric Exposure

Comprehensive tests conducted over a 20-year period under the supervision of The American Society for Testing and Materials, plus many service records, have confirmed the suitability of copper and copper alloys for atmospheric exposure (see Table 7). Copper and copper alloys resist corrosion by industrial, marine and rural atmospheres except atmospheres containing ammonia or certain other agents where stress-corrosion cracking has been observed in high-zinc alloys (greater than 20% Zn). The copper metals most widely used in atmospheric exposure are C11000, C22000, C23000, C38500 and C75200. C11000 is a most satisfactory material for roofing, flashings, gutters and downspouts.

Colors of different copper alloys are often important in architectural applications, and color may be the main criterion for selection of a specific alloy. After surface preparation such as sanding or polishing, different copper alloys vary in color from silver to yellow to gold to reddish shades. Different alloys having the same initial color may show differences in color after weathering under similar conditions. Therefore, alloys having the same or nearly the same composition are usually used together for consistent appearance in a specific structure.

Soils

Copper, zinc, lead and iron are the metals most commonly used in underground construction. Data

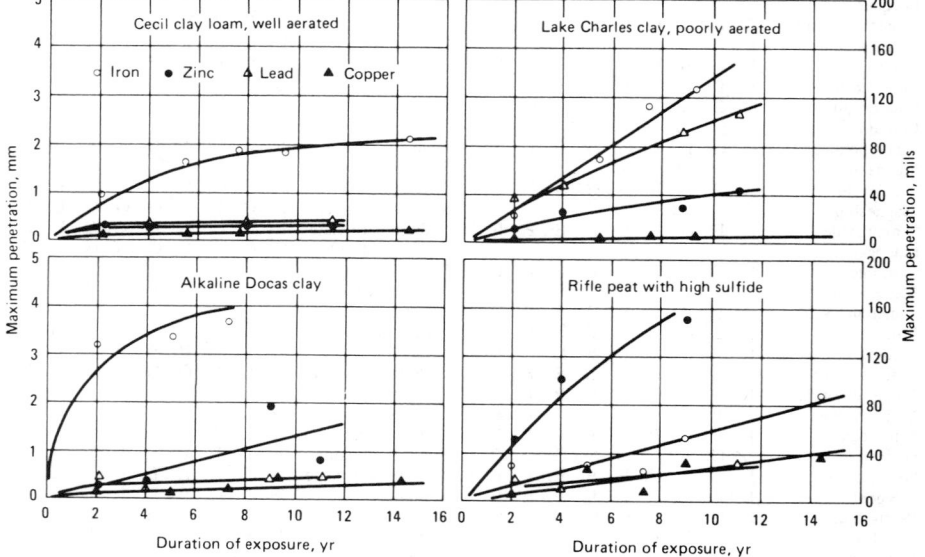

Fig. 7. Corrosion of copper, iron, lead and zinc in four different types of soil

compiled by the National Bureau of Standards (Circular 579) compare the behavior of these materials in soils of the following four types: (a) well-aerated acid soils low in soluble salts, (b) poorly aerated soils, (c) alkaline soils high in soluble salts (Docas clay), and (d) soils high in sulfides.

Corrosion data as a function of time for copper, iron, lead and zinc exposed to these four types of soil are given in Fig. 7. Copper shows high resistance to corrosion by these soils, which are representative of most soils found in the United States. Where local soil conditions are unusually corrosive, it may be necessary to employ some means of protection, such as cathodic protection, neutralizing backfill (limestone, for example), protective coating or wrapping.

For many years, the National Bureau of Standards has conducted studies on corrosion of underground structures to determine the specific behavior of metals and alloys when exposed for long periods in a wide range of soils. Results indicate that tough pitch coppers, deoxidized coppers, silicon bronzes, and low-zinc brasses behave essentially alike. Soils containing cinders with high concentrations of sulfides, chlorides, or hydrogen ions corrode these materials. In this type of contaminated soil, corrosion rates of copper-zinc alloys containing more than about 22% zinc increase as zinc content increases. Corro-

sion generally results from dezincification. In soils that contain only sulfides, corrosion rates of the copper-zinc alloys decline with increasing zinc content, and no dezincification occurs.

Exposure to Fresh Water

Copper is used extensively for handling fresh water. Copper tube in the K-gage range with flared fittings was designed for underground water service, and along with type L tube has now become standard for this application.

The greatest single application of copper tube is for hot- and cold-water distribution lines in homes and other buildings, though considerable quantities are also used in heating lines (including radiant heating lines for homes), drain tubes and fire safety systems.

Exposure to Steam. Copper and copper alloys are resistant to attack by pure steam, but if much carbon dioxide, oxygen or ammonia is present, the condensate is corrosive. Even though wet steam at high velocities can cause severe impingement attack, copper alloys are used extensively in condensers and heat exchangers. Copper alloys also are used for feedwater heaters, although their use in such applications is somewhat limited because of their rapid decline in strength and creep resistance at moderately elevated temperatures. Copper nickels are the preferred copper alloys for the higher temperatures and pressures.

Use of copper in systems handling hot water and steam is limited by working pressures of tubes and joints. For example, copper tube $1/4$ to 1 in. in nominal diameter joined with 50Sn-50Pb solder may be used at 120 °C (250 °F) and 585 kPa (85 psi). The working pressure at this temperature in tube of the same size can be increased to 1380 kPa (200 psi) when the system is joined using 95Sn-5Sb solder. When the joining material is a silver-base brazing alloy having a melting point above 540 °C (1000 °F), working pressure at 120 °C for tube in this size range can be increased to 2070 kPa (300 psi). A few copper alloys have shown a tendency to fail by stress-corrosion cracking when highly stressed and exposed to steam. Alpha aluminum bronzes that do not contain tin are among the susceptible alloys.

Steam condensate that has been properly treated so that it is relatively free of noncondensable gases, as in a power-generating station, is relatively noncorrosive to copper and copper alloys. Rates of attack in most such exposures are less than 2.5 μm/yr (0.1 mil/yr). Copper and its alloys are not attacked by condensate that contains a significant amount of oil (such as condensate from a reciprocating steam engine).

The rate of attack is significantly increased by dissolved carbon dioxide or oxygen, or both. For example, condensate containing 4.6 ppm of oxygen and 14 ppm carbon dioxide, having a pH of 5.5 at 68 °C (154 °F), caused an average penetration of 175 to 350 μm/yr (7 to 14 mils/yr) when in contact with C12200, C14200, C23000, C44300 to C44500 and C71000.

Steel tested under the same conditions was penetrated at about twice the rate given for the copper alloys listed above, whereas tin-coated copper proved much more resistant and was attacked at a rate of less than 25 μm/yr (1 mil/yr).

Steps that may be taken to attain the best life in condensate systems are (*a*) ensure that tubes are installed with enough slope to allow proper drainage, (*b*) reduce the quantity of corrosive agents (usually carbon dioxide and oxygen) at the source by either mechanical or chemical treatment of feedwater, or (*c*) chemically treat the steam.

The quantity of ammonia present, particularly in a condensing system, significantly influences service life of copper alloys. Boilerhouse operators should be cautioned about using ammonia to make boiler water more alkaline. This practice can lead to contamination and subsequent corrosion or stress-corrosion cracking in condensers and heat exchangers.

Salt Water

An important use of copper alloys is in handling seawater in ships and tidewater power stations. Copper itself, although fairly useful, is usually less resistant than C44300 to C44500 (inhibited admiralty metal), C61300 (aluminum bronze), C68700 (aluminum brass), C70600 (copper nickel, 10%) or C71500 (copper nickel, 30%). The superior performance of these alloys is partly the result of their inherent insolubility in seawater, but more because of their ability to form films of corrosion products that resist erosion by turbulently flowing seawater carrying entrained air.

Copper nickels are superior to copper and other copper alloys in resisting erosion in high-velocity salt water. Corrosion of copper alloys in a well-controlled desalting plant is described in a paper by A. Cohen and P. F. George: "Copper Alloys in the Desalting Environment" (NACE, 1974).

Corrosion rates of copper and its alloys in relatively quiescent seawater are typically less than 50 μm/yr (2 mils/yr). Copper nickels, aluminum bronzes and aluminum brass often show corrosion rates of less than 25 μm/yr (1 mil/yr). The instantaneous corrosion rate of all copper alloys tends to decrease as duration of exposure increases. Localized corrosion (pitting), with penetration rates on the order of 125 μm/yr (5 mils/yr), can occur under deposits or debris, or adjacent to crevices. These higher corrosion rates will also be seen if the velocity limits given above are exceeded. Plug-type dealloying of high-zinc brasses and certain aluminum bronzes can occur at rates often exceeding 250 μm/yr (10 mils/yr).

Biofouling. The copper alloys as a group tend to resist fouling by marine organisms, although the degree of resistance varies among the various alloys. The alloys most resistant to biofouling are those containing more than 85% copper (except the aluminum bronzes) and are the ones usually selected when this property is of prime importance. In order to maintain fouling resistance, copper alloys must be allowed to corrode freely. Traditionally, the minimum corrosion rate thought necessary to provide fouling resistance was about 25 μm/yr (1 mil/yr). However, copper nickels resist fouling when the instantaneous corrosion rate is on the order of 2.5 μm/yr (0.1 mil/yr), which implies that the corrosion-product film itself may resist fouling. The corrosion rates of copper alloys are such that they can protect only themselves from fouling. They do not release enough copper ions into seawater to prevent adjacent noncopper surfaces from becoming fouled.

Heat Exchangers and Condensers

The choice of material for condenser and heat-exchanger tubes requires a survey of service conditions, examination of tube previously used and evaluation of its service life, and a review of the type, form and location of corrosion experienced in the unit or in similar units. Types of water and operating conditions vary widely, and any estimate of probable tube performance must be based on specific operating factors. Tubes of various alloys, inhibited admiralty metal (C44300, C44400 and C44500), inhibited aluminum brass (C68700), aluminum bronzes (C61300 and C63200), copper nickels (C70600 and C71500) and phosphorus-deoxidized coppers (C12000 to C12300), have all been found to give satisfactory and economical performance in these applications. Further specific details are given in Volume 2 of Metals Handbook, Ninth Edition.

Drain Tubes. Copper has been used successfully for waste and vent lines in drains. The first such installations were made in the mid 1930's and since then many municipalities have approved the use of copper drain lines. Development of "Sovent" fittings now enables construction of a single-stack drain system in high-rise buildings instead of the two-stack system formerly used.

Acids

Copper is widely used for industrial equipment handling acid solutions. A fairly definite separation exists between those acids that can be handled by copper and those that cannot. In general, copper alloys are successfully used with nonoxidizing acids such as acetic, sulfuric, hydrochloric and phosphoric, as long as the concentration of oxidizing agents such as dissolved oxygen (air) and ferric or bichromate ions is low. Broadly speaking, a thoroughly agitated or stirred solution, or one into which a stream of air has been bubbled, approaches air saturation and thus is not a suitable acid medium for copper. Acids that are oxidizing agents in themselves (such as nitric, sulfurous and not concentrated sulfuric acids, and acids carrying oxidizing agents such as ferric salts, bichromate ions or permanganate ions) cannot be handled in equipment made of copper or its alloys.

In dilute solutions (up to 1% acid), the corrosive action of a nonoxidizing acid on copper is relatively low; corrosion rates are usually below 6 g/m²·d (60 mdd) or 250 μm/yr (10 mils/yr). This is only true of oxidizing acids when the concentration does not exceed 0.01%. At such low acid concentrations, aeration has little effect in either oxidizing or nonoxidizing acids.

Nonoxidizing acids with near-zero aeration have virtually no corrosive effect. Rates in 1.2N sulfuric, hydrochloric and acetic acids are less than 0.1 g/m²·d in the absence of air.

Except for hydrochloric acid, nonoxidizing acids that contain as much air as is absorbed in quiet contact with the atmosphere are weakly corrosive. Rates generally range from 0.5 to 6 g/m²·d, or an equivalent of about 20 to 250 μm/yr (0.8 to 10 mils/yr).

Air-saturated solutions of nonoxidizing acids are likely to be strongly corrosive, with rates of 5 to 30 g/m²·d, which is equivalent to 0.2 to 1.25 mm/yr (8 to 50 mils/yr). This rate is higher for hydrochloric acid. The actual corrosion in any aerated acid depends on acid concentration, temperature and other factors difficult to classify. Except in very dilute solutions, oxidizing acids corrode copper rapidly — at rates usually above 50 g/m²·d or 2.1 mm/yr (85 mils/yr). The reaction is independent of aeration.

Phosphoric, acetic, tartaric, formic, oxalic, malic and similar acids normally react in a manner comparable to sulfuric acid.

Alkalis

Copper and its alloys resist alkaline solutions except those containing ammonium hydroxide, or compounds that hydrolyze to ammonium hydroxide or cyanides. Ammonium hydroxide reacts with copper to form soluble complex copper cations whereas the cyanides react to form soluble complex copper anions.

Strong NH_4OH solutions attack copper and copper alloys rapidly, compared with the rates of attack by metallic hydroxides, because of the formation of a soluble complex copper-ammonium compound. However, in some applications the corrosion of copper exposed to dilute solutions of ammonium hydroxide is low. For example, copper specimens submerged in 0.01N NH_4OH solution at room temperature for one week lost weight at a rate of 1.5 g/m²·d, which is equivalent to about 60 μm/yr (2.5 mils/yr).

Ammonium hydroxide solutions also attack copper-zinc alloys. Alloys containing more than 15% zinc are susceptible to stress-corrosion cracking when stressed and exposed to ammonium hydroxide. The stress may be due to applied service loads or to unrelieved residual stresses.

In quiescent 2N solutions at room temperature, copper-zinc alloys corrode at 1.8 to 6.6 mm/yr

(70 to 260 mils/yr), copper-nickel alloys at 0.25 to 0.50 mm/yr (10 to 20 mils/yr), copper-tin alloys at 1.3 to 2.5 mm/yr (50 to 100 mils/yr) and copper-silicon alloys at 0.75 to 5 mm/yr (30 to 200 mils/yr).

Copper and its alloys are suitable for handling anhydrous ammonia provided that the ammonia remains anhydrous and is not contaminated with water and oxygen. In one test conducted for 1200 h, C11000 and C26000 each showed an average penetration of 5 μm/yr (0.2 mil/yr) in contact with anhydrous ammonia at atmospheric temperature and pressure.

Salts

Copper metals are widely used in equipment for handling salt solutions of various kinds, particularly those that are nearly neutral. Among these are the nitrates, sulfates, and chlorides of sodium and potassium. Chlorides usually are more corrosive than the other salts, especially in strongly agitated, aerated solutions.

The nonoxidizing acid salts, such as the alums and certain metal chlorides (magnesium and calcium chlorides) that hydrolyze in water to produce an acidic pH, behave essentially the same as dilute solutions of the corresponding acids. Corrosion rates generally range from 2.5 to 1500 μm/yr (0.1 to 60 mils/yr) at room temperature, depending on the degree of aeration and the acidity.

Neutral salt solutions can be handled successfully by copper alloys. Consequently, these alloys are used in heat-exchanger and condenser equipment exposed to seawater.

Such alkaline salts as sodium silicate, sodium phosphate and sodium carbonate attack copper alloys at low but different rates at room temperature. On the other hand, alkali cyanide is aggressive and attacks copper alloys fairly rapidly because it forms a soluble complex copper anion.

Oxidizing salts corrode copper and copper alloys rapidly; therefore, copper metals should not be used with oxidizing salt solutions except those that are very dilute. Aqueous sodium dichromate solutions can be safely handled by copper alloys, but the presence of a highly ionized acid such as chromic or sulfuric acid may increase the corrosion rate several hundred times, because the dichromate acts as an oxidizing agent in acidic solutions. In one test, a copper nickel corroded at 2.5 to 250 μm/yr (0.1 to 10 mils/yr), and a copper-tin alloy (phosphor bronze) at 5 μm/yr (0.2 mil/yr), when handling an aqueous sodium dichromate solution. The rate increased 200 to 300 times for both metals when chromic acid was added to the solution. In solutions containing ferric, mercuric or stannic ions, a copper nickel showed a corrosion rate of 75 μm per *day* (3 mils per *day*), while copper-zinc and copper-tin alloys showed a still greater rate of 625 μm per *day* (25 mils per *day*).

Salts of metals more noble than copper (such as nitrates of mercury and silver) corrode copper alloys rapidly, concurrently plating out the noble metal on the copper surface. Rate of attack is influenced by temperature and acidity. A film of

mercury on a high-zinc brass (more than 15% Zn) may cause intercrystalline cracking by liquid-metal embrittlement if the alloy is under tensile stress, either residual or applied.

Organic Compounds

Copper and many of its alloys resist corrosive attack by organic compounds such as amines, alkanolamines, esters, glycols, ethers, ketones, alcohols, aldehydes, naphtha and gasoline, and by most organic solvents.

Although corrosion rates of copper and copper alloys in pure alkanolamines and amines are low, they can be significantly increased if these compounds are contaminated with water, acids, alkalis or salts, or with combinations of these impurities, particularly at high temperatures.

Gasoline, naphtha and other related hydrocarbons in pure form will not attack copper or any of the copper alloys. However, in manufacture of hydrocarbon materials, process streams are likely to be contaminated with one or more substances such as water, sulfides, acids and various organic compounds. These contaminants attack copper and its alloys. Corrosion rates for C44300 and C71500 exposed to gasoline are low (see Table 7), and these two alloys are successfully used in equipment for refining gasoline.

Creosote. Copper and copper alloys are generally suitable for use with creosote, although creosote attacks some high-zinc brasses. C11000, C23000, C26000, C51000 and C65500 typically corrode at rates less than 500 μm/yr (20 mils/yr) when exposed to creosote at 24 °C (75 °F).

Linseed Oil. Copper and its alloys are fairly resistant to corrosion by linseed oil. All of the alloys show some attack, but none exhibits corrosion severe enough to make it unsuitable for this application. C11000, C51000 and C65500 showed corrosion rates less than 500 μm/yr (20 mils/yr) in linseed oil at 24 °C (75 °F). C26000 had a rate of 500 to 1250 μm/yr (20 to 50 mils/yr).

Benzol and Benzene. C11000, C23000, C26000, C51000 and C65500 tested in these two materials at 24 °C had corrosion rates less than 500 μm/yr (20 mils/yr).

Sugar. Copper is used successfully for vacuum-pan heating coils, evaporators, and juice extractors in manufacture of both cane and beet sugar. Inhibited admiralty metals, aluminum brass, aluminum bronzes and copper nickels are also used for tubes in juice heaters and evaporators. Bimetal tubes of copper and steel have been used by manufacturers of beet sugar to counteract stress-corrosion cracking of copper tubes caused by exposure to ammonia from beets grown in fertilized soil.

Beer. Copper is used extensively in brewing of beer. In one installation, the wall thickness of copper kettles thinned from an original thickness of 16 mm (^{5}/$_8$ in.) to 10 mm (^{3}/$_8$ in.) in a 30-year period. Brazing with BAg filler metals eliminates the possibility that the alkaline compounds used for cleaning copper equipment will destroy joints by attacking tin-lead solders. Steam coils require more frequent replacement than any other brewery equipment. They have a service life of 15 to 20 years. The service life of other copper items exposed to process streams in a brewery is 30 to 40 years.

Sulfur compounds free to react with copper (hydrogen sulfide, sodium sulfide or potassium sulfide) form copper sulfide. Reaction rates depend on alloy composition; the alloys of highest resistance are those of high zinc content.

Inhibited admiralty metals also are excellent alloys for use in heat exchangers and condensers handling sulfur-bearing petroleum products and using water as the coolant. C44300, C44400 and C44500, which are inhibited toward dezincification by addition of arsenic, antimony or phosphorus to the basic 70Cu-29Zn-1Sn composition, offer good resistance to corrosion from sulfur plus excellent resistance to the water side of the heat exchanger.

Table 8. Corrosion of C44300 and C71500 exposed to gasoline in a refinery

Service condition(a)	Temperature °C	Temperature °F	Average penetration rate μm/yr	Average penetration rate mils/yr
C44300				
Straight-run (untreated)				
Tower liquid(b)	121	250	1270 min	50 min
Storage(c)	4-27	40-80	63	2.5
Distilled tops from straight-run gasoline(d)	35	95	1270	50
Cracked gasoline (top tray in tower)(e)	204	400	15	0.6
Sweet gasoline vapor(f)	177	350	7.5	0.3
C71500				
Straight-run (untreated)				
Tower liquid(b)	121	250	180	7
Storage(c)	4-27	40-80	180	7
Distilled tops from straight-run gasoline(d)	35	95	1140	45
Cracked gasoline (top tray in tower)(e)	204	400	200	8
Sweet gasoline vapor(f)	177	350	10	0.4
Aviation gasoline (top of column)	121	250	2.5	0.1

(a) Gasoline or related hydrocarbons will not attack copper or its alloys. Attack depends on the type and amount of impurities in the gasoline, such as water, sulfides, mercaptans, aliphatic acids, naphthenic acids, phenols, nitrogen bases, and dissolved gases. (b) 100 lb of H_2S present per 1000 bbl of gasoline. (c) 0.02 to 0.03 g H_2S per litre of gasoline. (d) pH controlled by NH_3. (e) H_2S and HCl present. (f) Vacuum operation.

8 MAGNESIUM

Reviewed by Joseph H. Waibel, Dow Chemical Co.

This section was condensed from Metals Handbook, Ninth Edition, Volume 2, Properties and Selection: Non-ferrous Alloys and Pure Metals, pages 525 to 609. For more detailed information on the topics covered herein, the reader is referred to the larger work. Additional information on magnesium and magnesium alloys can be found in Parts III and IV of this Desk Edition.

Selection and Application

MAGNESIUM and magnesium alloys are used in a wide variety of structural and nonstructural applications. Structural applications include industrial, materials-handling, commercial and aerospace equipment. In industrial machinery, such as textile and printing machines, magnesium alloys are used for parts that operate at high speeds and thus must be lightweight to minimize inertial forces. Materials-handling equipment includes dockboards, grain shovels and gravity conveyors. Commercial applications include luggage and ladders. Good strength and stiffness at both room and elevated temperatures combined with light weight make magnesium alloys especially valuable for aerospace applications.

Magnesium is also employed in various nonstructural applications. It is used as an alloying element in alloys of aluminum, zinc, lead and other nonferrous metals. It is used as an oxygen scavenger and desulfurizer in manufacture of nickel and copper alloys, as a desulfurizer in the iron and steel industry, and as a reducing agent in production of beryllium, titanium, zirconium, hafnium and uranium. Another important nonstructural use of magnesium is in the Grignard reaction in organic chemistry. In finely divided form, magnesium finds some use in pyrotechnics, both as pure magnesium and alloyed with 30% or more aluminum. It is also used for cathodic protection of other metals from corrosion and in construction of dry-cell and reserve cell batteries. Gray iron foundries use magnesium and magnesium-containing alloys as ladle-addition agents introduced just before the casting is poured. The magnesium makes the graphite particles nodular and greatly improves the properties of the cast iron.

Because of its rapid but controllable response to etching as well as its light weight, magnesium is used increasingly in photoengraving.

Primary magnesium is furnished to ASTM B92, grade 9980A, with a specified minimum magnesium content of 99.8%. Also available are special grades of primary magnesium in which manganese, aluminum and iron impurities are held to especially low levels. These special grades are employed in chemical and metallurgical applications, such as preparation of uranium metal and other reactive metals.

Aluminum and zinc are relatively soluble in solid magnesium, but their solubilities decrease at low temperatures. The solubility of aluminum is 12.7% by weight at 437 °C (819 °F) and 3.0% at 93 °C (200 °F); solubility of zinc is 6.2% at 340 °C (644 °F) and 2.8% at 204 °C (400 °F). Solubilities of manganese, zirconium and cerium are less than 1.0% by weight at 482 °C (900 °F). At the eutectic temperature, 4.5% thorium is soluble in magnesium. Manganese is effective in improving corrosion stability of magnesium alloys that contain aluminum and zinc.

Designations. A standard system of alloy and temper designations, adopted in 1948, is explained in Table 1. As an example of how the system works, consider magnesium alloy AZ91C-T6, the nominal composition and typical properties of which are given in Table 2. The first part of the designation, AZ, signifies that aluminum and zinc are the two principal alloying elements. The second part of the designation, 91, means that aluminum and zinc are present in rounded-off percentages of 9 and 1, respectively. The third part, C, indicates that this is the third alloy standardized with 9% Al and 1% Zn as the principal alloying additions. The fourth part, T6, denotes that the alloy is solution treated and artificially aged.

CASTING ALLOYS

There are several systems of magnesium alloys for sand and permanent mold castings: magnesium-aluminum-manganese with and without silicon or zinc (AM, AS and AZ), magnesium-zirconium (K), magnesium-zinc-zirconium with and without rare earths (ZK, ZE and EZ), magnesium-thorium-zirconium with and without zinc (HK, HZ and ZH), and magnesium-silver-zirconium with rare earths or thorium (QE and QH). Nominal compositions and typical properties of these alloys are given in Table 2.

AZ91C and AZ81A have almost completely replaced AZ63A where good ductility and moderately high yield strength are required at tem-

Table 1. Standard four-part ASTM system of alloy and temper designations for magnesium alloys(a)(b)

First part	Second part	Third part	Fourth part
Indicates the two principal alloying elements	Indicates the amount of the two principal alloying elements	Distinguishes between different alloys with the same percentages of the two principal alloying elements	Indicates condition (temper)
Consists of two code letters representing the two main alloying elements arranged in order of decreasing percentage (or alphabetically if percentages are equal)	Consists of two numbers corresponding to rounded-off percentages of the two main alloying elements and arranged in same order as alloy designations in first part	Consists of a letter of the alphabet assigned in order as compositions become standard	Consists of a letter followed by a number (separated from the third part of the designation by a hyphen)
A-Aluminum E-Rare Earth H-Thorium K-Zirconium M-Manganese Q-Silver S-Silicon T-Tin Z-Zinc	Whole numbers	Letters of alphabet except I and O	F-As fabricated O-Annealed H10 and H11-Slightly strain hardened H23, H24 and H26-Strain hardened and partially annealed T4-Solution heat treated T-5-Artificially aged only T6-Solution heat treated and artificially aged T8-Solution heat treated, cold worked and artificially aged

(a) As an example of a typical four-part designation, AZ91C-T6 is explained in the text. (b) This system is now also used by SAE.

Table 2. Nominal compositions and typical room-temperature mechanical properties of magnesium alloys

Alloy	Al	Mn(a)	Th	Zn	Zr	Others	Tensile strength MPa	ksi	Yield strength Tensile MPa	ksi	Compressive MPa	ksi	Bearing MPa	ksi	Elongation in 50 mm or 2 in., %	Shear strength MPa	ksi	Hardness, HRB(b)
Sand and permanent mold castings																		
AM100A-T61	10.0	0.1	...	...	...	...	275	40	150	22	150	22	...	...	1	...	...	69
AZ63A-T6	6.0	0.15	...	3.0	...	...	275	40	130	19	130	19	360	52	5	145	21	73
AZ81A-T4	7.6	0.13	...	0.7	...	...	275	40	83	12	83	12	305	44	15	125	18	55
AZ91C-T6	8.7	0.13	...	0.7	...	...	275	40	195	21	145	21	360	52	6	145	21	66
AZ92A-T6	9.0	0.10	...	2.0	...	...	275	40	150	22	150	22	450	65	3	150	22	84
EZ33A-T5	...	...	...	2.7	0.6	3.3 RE	160	23	110	16	110	16	275	40	2	145	21	50
HK31A-T6	...	...	3.3	...	0.7	...	220	32	105	15	105	15	275	40	8	145	21	55
HZ32A-T5	...	...	3.3	2.1	0.7	...	185	27	90	13	90	13	255	37	4	140	20	57
K1A-F	...	...	...	...	0.7	...	180	26	55	8	...	...	125	18	1	55	8	...
QE22A-T6	...	...	...	...	0.7	2.5 Ag, 2.1 Di	260	38	195	28	195	28	...	...	3	...	...	80
QH21A-T6	...	...	60	...	0.7	2.5 Ag, 1.0 Di	275	40	205	30	...	...	...	...	4	...	...	...
ZE41A-T5	...	...	...	4.2	0.7	1.2 RE	205	30	140	20	140	20	350	51	3.5	160	23	62
ZE63A-T6	...	...	...	5.8	0.7	2.6 RE	300	44	190	28	195	28	...	...	10	...	...	60-85
ZH62A-T5	...	...	1.8	5.7	0.7	...	240	35	170	25	170	25	340	49	4	165	24	70
ZK51A-T5	...	...	...	4.6	0.7	...	205	30	165	24	165	24	325	47	3.5	160	23	65
ZK61A-T5	...	...	...	6.0	0.7	...	310	45	185	27	185	27	...	...	...	170	25	68
ZK61A-T6	...	...	...	6.0	0.7	...	310	45	195	28	195	28	...	...	10	180	26	70
Die castings																		
AM60A-F	6.0	0.13	...	...	...	...	205	30	115	17	115	17	...	...	6	...	...	...
AS41A-F(c)	4.3	0.35	...	...	...	1.0 Si	220	32	150	22	150	22	...	...	4	...	...	...
AZ91A and B-F(d)	9.0	0.13	...	0.7	...	...	230	33	150	22	165	24	...	...	3	140	20	63
Extruded bars and shapes																		
AZ10A-F	1.2	0.2	...	0.4	...	...	240	35	145	21	69	10	...	...	10	...	...	...
AZ21X1-F(c)	1.8	0.02	...	1.2	...	...	...	...	...	...	...	...	...	...	...	...	...	...
AZ31 B and C-F(e)	3.0	...	...	1.0	...	...	260	38	200	29	97	14	230	33	15	130	19	49
AZ61A-F	6.5	...	...	1.0	...	...	310	45	230	33	130	19	285	41	16	140	20	60
AZ80A-T5	8.5	...	...	0.5	...	...	380	55	275	40	240	35	...	...	7	165	24	82
HM31A-F	...	1.2	3.0	...	...	...	290	42	230	33	185	27	345	50	10	150	22	...
M1A-F	...	1.2	...	...	...	...	255	37	180	26	83	12	195	28	12	125	18	44
ZK21A-F	...	...	...	2.3	0.45(a)	...	260	38	195	28	135	20	...	...	4	...	...	...
ZK40A-T5	...	...	...	4.0	0.45(a)	...	276	40	255	37	140	20	...	...	4	...	...	...
ZK60A-T5	...	...	...	5.5	0.45(a)	...	365	53	305	44	250	36	405	59	11	180	26	88
Sheet and Plate																		
AZ31B-H24	3.0	...	...	1.0	...	...	290	42	220	32	180	26	325	47	15	160	23	73
HK31A-H24	...	...	3.0	...	0.6	...	255	33	200	29	160	23	285	41	9	140	20	68
HM21A-T8	...	0.6	2.0	...	...	...	235	34	170	25	130	19	270	39	11	125	18	...
PE(f)	3.3	...	...	0.7	...	...	...	...	...	...	...	...	...	...	...	...	...	...

(a) Minimum. (b) 500-kg load, 10-mm ball. (c) For battery applications. (d) A and B are identical except that 0.30% max residual Cu is allowable in AZ91B. (e) Properties of B and C are identical, but AZ31C contains 0.15 min Mn, 0.1 max Cu and 0.03 max Ni. (f) Photoengraving grade.

peratures up to 120 °C (250 °F). In similar fashion, AZ92A has virtually replaced AM100A. In any of these Mg-Al-Zn alloys, an increase in aluminum content raises yield strength but reduces ductility for comparable heat treatment. The castability of these alloys is good, and final selection of the specific composition may be based on tests of finished castings.

The difference between die casting alloys AZ91A and AZ91B is maximum copper content, which does not affect mechanical properties. Because of its higher maximum copper content (0.30%), AZ91B has lower resistance to corrosion. The reason for the higher maximum copper is to allow the alloy to be made from secondary metal, which reduces the cost of the alloy. Die castings are used in the as-cast condition.

Die cast alloy AM60A has better elongation and toughness, but lower tensile and yield strengths, than AZ91A or AZ91B. It is used in production of die-cast automotive wheels and in some archery equipment. Die cast alloy AS41A has creep strength much superior to that of AZ91A, AZ91B or AM60A up to 175 °C (350 °F), and good elongation, yield strength and tensile strength. One use of AS41A is in crankcases for air-cooled automotive engines.

KIA is primarily used where high damping ca-

pacity is required. It has low tensile and yield strength.

ZK and ZH alloys develop the highest yield strengths of all magnesium casting alloys and can be cast into complicated shapes. However, these grades are more costly than alloys of the AZ series.

The two ZK casting alloys in use are ZK51A and ZK61A. The latter, which has a slightly higher zinc content, has significantly greater strength than ZK51A (Table 2). Both alloys maintain high ductility after an artificial aging treatment (T5). The strength of ZK61A can be further increased (3 to 4%) by solution treatment plus artificial aging (T6), without impairing ductility. Both of these alloys have fatigue strengths equal to those of Mg-Al-Zn alloys, but they are more susceptible to microporosity and hot cracking and are less weldable. Addition of either thorium or rare-earth metals overcomes these deficiencies. The strength properties of ZE63A are equivalent to those of ZK61A, those of ZH62A are equivalent to or better than those of ZK51A, but those of ZE41A are somewhat lower than those of ZK51A (Table 2).

ZE41A alloy was developed to meet the growing need for an alloy with medium strength, good weldability, and better castability than that of

AZ91C or AZ92A. It has good fatigue and creep properties and maximum freedom from microshrinkage. Unlike in the AZ alloys, there is a very close relationship between separately cast test bar properties and those obtained from the casting itself, even where relatively thick cast sections are involved. ZE41A is used up to 160 °C (320 °F) in such applications as aircraft engines, helicopter and airframe components, and wheels and gear boxes.

ZE63A is a high-strength alloy with excellent tensile and yield strengths, which are achieved by heat treating in a hydrogen atmosphere. Because hydriding proceeds from the surface, heat treating time and penetrability are limiting factors. This alloy has excellent casting characteristics.

The Mg-RE-Zr alloys are used at temperatures from 175 to 260 °C (350 to 500 °F). Because their high-temperature strengths exceed those of Mg-Al-Zn alloys, a savings in weight is possible.

The Mg-RE-Zn-Zr alloy EZ33A has good strength stability when exposed to elevated temperatures. (Strength stability is the ability to resist deterioration of strength from extended exposure to elevated temperature.) EZ33A castings usually are quite free from porosity, but are more susceptible to inclusions of dross than are the Mg-

Al-Zn alloys. For these reasons, they are more difficult to cast in some designs than Mg-Al-Zn alloys. EZ33A castings have excellent pressure tightness. ZE41A, discussed earlier, is similar to EZ33A, but has higher tensile and yield strengths due to its higher zinc content. Some sacrifice is made in castability and weldability of ZE41A in return for higher mechanical properties.

When the operating temperature of an engine housing was increased from 120 to 205 °C (250 to 400 °F), alloy EZ33A-T5 was successfully substituted for AZ92A-T6. The change was based on creep tests of separately cast bars of the two alloys; stress values at three temperatures, for 0.1% creep in 1000 h, were as follows:

	Temperature		Stress	
Alloy	°C	°F	MPa	ksi
AZ92A-T6	205	400	.... 6.9	1.0
	260	500	.... 2.1	0.3
EZ33A-T5	205	400	 58	8.4
	260	500	 26	3.7
	315	600	.... 8.3	1.2

The Mg-Th-Zr alloys HK31A and HZ32A are intended primarily for use at temperatures of 200 °C (400 °F) and higher, for which properties superior to those of EZ33A are required. For full development of properties, HK31A requires the T6 treatment (solution heat treatment plus artificial aging), whereas HZ32A, which contains zinc, requires only the T5 treatment (artificial aging). HK31A and HZ32A castings have been used at temperatures as high as 345 to 370 °C (650 to 700 °F) in a few applications. The Mg-Zn-Th-Zr alloy ZH62A differs from other Mg-Th-Zr alloys in that it is intended primarily for use at room temperature.

Mg-Th-Zr alloys are more difficult to cast than EZ33A because they are more subject to formation of inclusions and defects as a result of gating turbulence. The tendency for inclusions to form in Mg-Th-Zr alloys is particularly marked in thin-wall parts that require high pouring rates. These alloys have adequate castability for production of complex parts of moderate to heavy wall thickness.

At 260 °C (500 °F) and slightly higher, HZ32A is equal to or better than HK31A in short-time and long-time creep strength at all extensions. HK31A has higher tensile, yield and short-time creep strengths up to 370 °C (700 °F). However, HZ32A has greater strength stability at elevated temperatures, and much better foundry characteristics, than does HK31A.

QE22A has high tensile and yield strengths as well as fairly good properties at temperatures up to 204 °C (400 °F). QH21A has properties similar to those of QE22A at room temperature but superior properties at temperatures from 204 °C (400 °F) to 260 °C (500 °F). Both QE22A and QH21A have good castability and weldability. They require solution and aging heat treatments to achieve their higher mechanical properties. Also, they are relatively expensive because of their silver contents.

WROUGHT ALLOYS

Wrought magnesium alloys are produced as bars, billets and shapes, wire, sheet, plate and forgings.

Extruded bars and shapes are made of several types of magnesium alloys (see Table 2). For normal strength requirements, one of the Mg-Al-Zn (AZ) alloys is usually selected. The strength of these alloys increases as aluminum content increases. AZ31B is a widely used, moderate-strength alloy with good formability. AZ31B is also used extensively for cathodic protection. AZ31C is a lower-purity, commercial variation of AZ31B for lightweight structural applications. M1A and ZM21A are alloys that can be extruded at higher speeds than AZ31B but have limited use due to their lower strength. Alloy AZ10A, because of its low aluminum content, has lower strength than AZ31B, but it can be welded without subsequent stress relief. AZ61A and AZ80A can be artificially aged for additional strength (with a sacrifice in ductility). AZ80A is not available in hollow shapes. AZ21X1 is an alloy designed specifically for use in battery applications.

Alloy ZK60A is used where high strength and good toughness are required. This alloy is heat treatable and normally is used in the artificially aged (T5) condition. ZK21A and ZK40A are lower in strength and more readily extrudable than ZK60A and have had limited use in hollow tubing with high strength requirements.

HM31A has moderate strength and is suitable for use in applications requiring good strength and creep resistance at temperatures from 150 to 425 °C (300 to 800 °F).

Forgings are made of alloys AZ31B, AZ61A, AZ80A, M1A and ZK60A, the compositions and properties of which are listed under extruded bars and shapes in Table 2. Alloy HM21A, which is listed under sheet and plate in Table 2, is also a good forging alloy. Alloys M1A and AZ31B may be used for hammer forgings, whereas the other alloys are almost always press forged. However, there has been a gradual decline in the use of the Mg-Mn alloy M1A. AZ80A has greater strength than AZ61A and requires the lowest rate of deformation of the Mg-Al-Zn alloys. ZK60A has essentially the same strength as AZ80A and greater ductility. To develop maximum properties, both AZ80A and ZK60A are heat treated to the T5 (artificially aged) condition. AZ80A may be given the T6 solution heat treatment, followed by artificial aging to provide maximum creep stability. HM21A is used in the T5 temper, and is useful at temperatures up to 370 to 425 °C (700 to 800 °F) where good creep resistance is needed.

Sheet and plate are rolled from Mg-Al-Zn (AZ and photoengraving grade alloy, or "PE") and Mg-Th (HK and HM) alloys (see Table 2).

AZ31B is the alloy most widely used for sheet and plate and is available in several grades and tempers. It can be used at temperatures up to 100 °C (200 °F). HK31A and HM21A are suitable up to 315 and 345 °C (600 and 650 °F), respectively. HM21A has superior strength and creep resistance. For example, an air impeller manufactured from thick plate of alloy HK31A failed as a result of excessive creep. A change to HM21A led to satisfactory performance and service life. Test coupons machined from the two materials gave the following values of stress for 0.1% creep in 100 h:

At 205 °C (400 °F):
    HM21A .... 86.2 MPa (12.5 ksi)
    HK31A .... 41 MPa (6.0 ksi)
At 260 °C (500 °F):
    HM21A .... 72.4 MPa (10.5 ksi)
    HK31A .... 28 MPa (4.0 ksi)
At 315 °C (600 °F):
    HM21A .... 52 MPa (7.5 ksi)
    HK31A .... 14 MPa (2.0 ksi)

Alloy PE is a special-quality sheet, with excellent flatness, corrosion resistance and etchability, that is used in photoengraving.

Good formability is an important requirement for most sheet materials. The approximate formability of magnesium alloy sheet is indicated by its ability to withstand 90° bending over a mandrel without cracking. The minimum-size mandrel (minimum radius) over which the sheet can be bent without cracking depends on alloy composition and temper, material thickness and temperature.

When correct temperatures and forming conditions are employed, all magnesium alloys can be deep drawn to about equal reduction.

MECHANICAL PROPERTIES

Mechanical properties of magnesium alloys are given in Table 2. These are typical values and, for castings, are obtained by testing separately cast specimens. Tensile strengths of investment mold and shell mold castings compare favorably with those of sand and permanent mold castings: yield strength, tensile strength and percentage elongation may vary with cooling rate and generally are lower than those of separately cast sand mold test bars.

Most magnesium alloys have ratios of tensile strength to density and tensile yield strength to density that are comparable to those of other common structural materials.

Compressive Strength. Compressive yield strength is defined as the stress required to produce a deviation or offset of 0.2% from the modulus line. For castings, compressive yield strength is approximately equal to tensile yield strength. For wrought alloys, however, yield strength in compression may be considerably less than yield strength in tension. The ratio of yield strength in compression to yield strength in tension varies from about 0.4 for alloy M1A to an average value of about 0.7 for the other wrought magnesium alloys. Typical compressive-yield-strength values for various magnesium alloys are given in Table 2.

Maximum design stresses for magnesium alloy columns that are loaded axially and that have sufficient stability to prevent local failure may be determined, for columns in the long-column range, by using the Euler column formula. (A long column is one whose length and cross section are such that the stress at which it will buckle does not exceed the elastic limit of the column material.) Maximum design stresses for magnesium alloy columns in the short-column range are dependent on strengths and forms of the alloys being tested. (A short column is any column of such length and cross section that it fails under compressive loading by plastic yielding and/or crushing, rather than by buckling.) In practical applications, the maximum design stress of a column is considered to be the minimum compressive yield stress of the material. Various formulas have been developed for deriving maximum design stresses for columns of intermediate length (those that fail by a combination of elastic buckling and plastic yielding and/or crushing).

Bearing strength is particularly important in design of bolted and riveted joints. Bearing yield strength is defined as the stress required to produce an offset from the initial straight portion of the curve equal to 2% of hole diameter. Bearing-strength values listed in Table 2 were determined using specimens with an edge distance (from the center of the hole) of $2^{1}/_{2}$ times the pin diameter and a width of 8 times the pin diameter. Increasing edge distance to more than about twice the pin diameter has little effect on bearing-strength values. Sheet thicknesses in a wide range have been tested, and no effect of the ratio of pin diameter to sheet thickness has been observed, except when buckling occurred. A pin diameter not greater than four times sheet thickness prevents buckling.

Shear strength is important in design of joints in magnesium parts, such as threaded joints and spot welds. Values for castings and extrusions given in Table 2 were obtained by the conventional double-shear method, using solid rods. Values for sheet and AZ80A-T5 structural shapes were obtained by the punch method, using flat specimens.

Hardness and Wear Resistance. Magnesium alloys have sufficient hardness for all structural applications except those involving severe abrasion. Hardness values are given in Table 2. Although rather wide variations in hardness are observed, resistance of the alloys to abrasion varies by only about 15 to 20%. When subjected to wear by rubbing, by frequent removal of studs or by heavy bearing loads, magnesium may be protected by inserts of steel, bronze or nonmetallic materials, attached as sleeves, liners, plates or bushings. Such inserts may be attached mechanically by pressing, shrinking, riveting, bolting or bonding; in castings, inserts may be cast in place. Magnesium alloys perform satisfactorily as bearing materials where loads do not exceed 14 MPa (2 ksi), shafts are hardened (350 to 600 HB), lubrication is ample, speeds are low (5 m/s, or 1000 ft/min, max) and operating temperature does not exceed 105 °C (220 °F).

Fatigue strength of magnesium alloys, as determined using laboratory test samples, covers a relatively wide scatter band such as is characteristic of other metals. The *S-N* curves have a gradual change in slope and become essentially parallel to the horizontal axis at 10 to 100 million cycles.

The fatigue strengths are higher for wrought products than for cast test bars. Increasing surface smoothness improves resistance to fatigue failure. For example, removing the relatively rough as-cast surfaces of castings by machining improves the fatigue properties of the castings. Sharp notches, small radii, fretting and corrosion are more likely to reduce fatigue life than are variations in chemical composition or heat treatment.

A specimen of alloy ZK60A-T5 (static yield strength, 290 MPa or 42 ksi) with a machined 60° notch with a radius of 0.025 mm (0.001 in.) has a fatigue limit of 28 MPa (4 ksi) at 500 million cycles, compared with 110 MPa (16 ksi) for the unnotched specimen. This is a notch factor of 0.25. For a shorter life of 100 000 cycles, the notch factor is about 0.48. As the severity of the notch decreases, its effect on fatigue limit decreases rapidly. For instance, a semicircular notch with radius of 1.2 mm (0.047 in.) reduces fatigue strength by only 20%, compared with 75% for the sharp V-notch cited above.

When fatigue is the controlling factor in design, every effort should be made to decrease the severity of stress raisers. Use of generous fillets in re-entrant corners and gradual changes of section greatly increase fatigue life. Situations in which the effects of one stress raiser overlap those of another should be eliminated. Further im-

provement in fatigue strength can be obtained by inducing stress patterns conducive to long life. Cold working of the surfaces of critical regions by rolling or peening to achieve appreciable plastic deformation produces residual compressive surface stress and increases fatigue life.

Surface rolling of radii is especially beneficial to fatigue resistance, because radii generally are the locations of higher-than-normal stresses. In surface rolling, size and shape of the roller, as well as feed and pressure, are controlled to obtain definite plastic deformation of the surface layers for an appreciable depth (0.25 to 0.38 mm, or 0.010 to 0.015 in.). In all surface working processes, caution must be exercised to avoid surface cracking, which decreases fatigue life. For example, if shot peening is used, the shot must be smooth and round. Use of broken shot or grit may result in surface cracks.

Low-Temperature Properties. With decreasing temperature, magnesium alloys increase in tensile strength, yield strength and hardness, but generally decrease in ductility (see Tables 3 and 4 for the results of tensile tests at both room and low temperatures).

Elevated-Temperature Properties. Elevated temperatures have adverse effects on tensile and yield strengths. The effects of elevated temperatures on the mechanical properties of magnesium alloys are evaluated by considering: (*a*) the strength as determined by bringing the test specimen up to temperature and testing immediately (short-time test); (*b*) the strength of temperature after prolonged heating at elevated temperature; (*c*) the effect on room-temperature properties of heating at elevated temperature for short and long times; and (*d*) the deformation produced by prolonged heating under load (creep test).

Data showing the effects of elevated temperatures on the mechanical properties of several magnesium alloys are presented in Tables 5 and 6.

Table 3. Low-temperature tensile properties of various wrought magnesium alloys(a)

Alloy	Thickness mm	in.	Tensile strength MPa	ksi	Yield strength MPa	ksi	Elongation, %
Transverse tests of plate alloys at 24 °C (75 °F)							
HK31A-H24	6.35	0.250	240	35.2	180	25.9	21.0
HK31A-O	6.35	0.250	200	29.0	125	18.0	30.5
HM21A-T8	6.35	0.250	240	35.0	170	24.8	13.7
Longitudinal tests of sheet and plate alloys at −54 °C (−65 °F)							
HK31A-H24	1.63	0.064	300	43.3	220	32.0	5.0
	6.35	0.250	280	40.8	230	33.4	9.0
HK31A-O	1.63	0.064	275	39.9	150	21.4	20.7
	6.35	0.250	265	38.3	150	21.5	18.0
HM21A-T5	(b)	(b)	270	39.5	110	15.8	9.3
HM21A-T8	1.63	0.064	275	39.6	175	25.6	6.2
	6.35	0.250	265	38.4	205	29.7	4.7
Longitudinal tests of sheet and plate alloys at −196 °C (−320 °F)							
HK31A-H24	1.63	0.064	370	54.0	225	33.0	6.2
HK31A-H24	6.35	0.250	365	52.9	240	34.7	8.0
Welded(c)	6.35	0.250	230	33.7	180	25.9	1.5
HK31A-O	1.63	0.064	330	47.9	170	24.3	12.7
HK31A-O	6.35	0.250	325	47.2	170	24.7	12.5
Welded(c)	6.35	0.250	205	29.7	150	21.6	2.2
HM21A-T5	(b)	(b)	320	46.6	125	18.1	8.0
HM21A-T8	1.63	0.064	330	47.6	170	24.9	4.0
HM21A-T8	6.35	0.250	325	47.3	210	30.6	4.2
Welded(c)	6.35	0.250	330	33.1	145	20.9	1.5

(a) Values are averages of two to four tests at room temperature (2-in. gage length). Values of duplicate tests at low temperatures are also averages (1-in. gage length). (b) Specimen machined from a forging. (c) Welding rod was EZ33A; weld bead intact.

Table 4. Low-temperature tensile properties of various magnesium casting alloys(a)

Alloy	Tensile strength MPa	ksi	Yield strength MPa	ksi	Elongation, %	Charpy impact Unnotched J	ft · lb	Notched J	ft · lb
At 24 °C (75 °F)									
AZ91C-T6	290	41.8	130	19.2	6.3	7.96	5.87	1.36	1.00
AZ92A-T6	290	41.8	160	23.4	4.0	7.62	5.62	0.68	0.50
EZ33A-T5	190	27.5	115	16.9	7.6	7.46	5.50	0.84	0.62
HK31A-T6	225	32.7	110	16.3	9.5	16.61	12.25	3.80	2.81
ZH62A-T5	275	39.9	190	27.9	5.7	15.02	11.08	1.02	0.75
At −78 °C (−109 °F)									
AZ91C-T6	305	44.3	150	21.6	5.1	6.26	4.62	1.36	1.00
AZ92A-T6	295	42.7	170	24.6	2.3	6.44	4.75	0.76	0.56
EZ33A-T5	190	27.6	125	18.0	3.1	4.83	3.56	0.68	0.50
HK31A-T6	300	43.3	120	17.5	8.6	16.43	12.12	3.21	2.37
ZH62A-T5	330	47.6	200	29.2	2.7	18.99	14.00	1.02	0.75
At −196 °C (−321 °F)									
AZ91C-T6	310	44.9	180	26.0	1.7	4.06	3.00	1.02	0.75
AZ92A-T6	320	46.5	195	28.5	0.8	4.57	3.37	0.68	0.50
EZ33A-T5	200	29.0	140	20.3	2.2	5.00	3.69	0.68	0.50
HK31A-T6	330	48.1	135	19.6	6.1	13.72	10.12	3.05	2.25
ZH62A-T5	320	46.6	235	34.1	1.0	8.56	6.31	1.02	0.75

(a) Values are averages of two to four tests on separately cast bars.

Table 5. Effects of elevated temperatures on tensile strengths of magnesium alloys

Alloy	Tested at exposure temperature — Exposed 10 min at 20 °C (70 °F) MPa	ksi	Exposed 10 min at 150 °C (300 °F) MPa	ksi	315 °C (600 °F) MPa	ksi	Exposed 1000 h at 205 °C (400 °F) MPa	ksi	315 °C (600 °F) MPa	ksi	Tested at room temperature — Exposed 1000 h at 205 °C (400 °F) MPa	ksi	315 °C (600 °F) MPa	ksi
Castings														
AZ63A-T6	275	40	165	24	55	8	110	16	...	...	255	37	...	...
AZ92A-T6	275	40	195	28	55	8	115	17	...	...	270	39	180	26
EZ33A-T5	160	23	145	21	83	12	130	19	76	11	170	25	180	26
HK31A-T6	215	31	195	28	125	18	180	26	62	9	240	35	180	26
HZ32A-T5	200	29	145	21	83	12	115	17	76	11	220	32	235	34
ZH62A-T5	290	42	195	28	69	10	...	...	...	...	...	...	...	...
QH21A-T6	275	40	235	34	97	14	...	...	...	...	...	...	...	...
Extrusions														
AZ80A-T5	380	55	235	34	69	10	...	...	...	...	...	...	...	...
ZK60A-T5	365	53	180	26	41	6	...	...	...	...	315	46	315	46
HM31A-F	275	40	195(a)	28(a)	115	17	...	...	...	...	...	...	...	...
Sheet														
AZ31B-H24 ...	285	41	145	21	48	7	90	13	62(a)	9(a)	255	37	260	38
HK31A-T6	255	37	180	26	115	17	...	...	55	8	255	37	215	31
HM21A-T8	235	34	140	20	97	14	...	...	...	...	...	...	...	...

(a) Tested at 260 °C (500 °F).

Table 6. Effects of elevated temperatures on values of creep stress and elastic modulus for magnesium alloys

Alloy	Creep stress(a) at 205 °C (400 °F) MPa	ksi	315 °C (600 °F) MPa	ksi	Elastic modulus at 205 °C (400 °F) GPa	10^6 psi	315 °C (600 °F) GPa	10^6 psi
Castings								
AZ92A-T6	3.4	0.5	...	...	31	4.5	21	3.0
EZ33A-T5	38	5.5	6.9	1.0	40	5.8	38	5.5
HK31A-T6	64	9.3	14	2.0	40	5.8	39	5.6
HZ32A-T5	52	7.5	22	3.2	40	5.8	39	5.6
ZH62A-T5	17	2.5	...	...	40	5.8	38	5.5
Extrusions								
ZK60A-T5	7	1.0(b)	...	...	...	...	...	...
HM31A-F	83	12.0	41	6.0	40	5.8	38	5.5
Sheet								
AZ31B-H24 ...	7	1.0(b)	...	...	30	4.3	17	2.5
HK31A-T6	69	10.0	17	2.5	40	5.8	25	3.6
HM21A-T8	76	11.0	34	5.0	40	5.8	34	5.0

(a) Stress to produce 0.2% total extension in 1000 h for cast alloys and 100 h for wrought alloys. (b) Tested at 150 °C (300 °F).

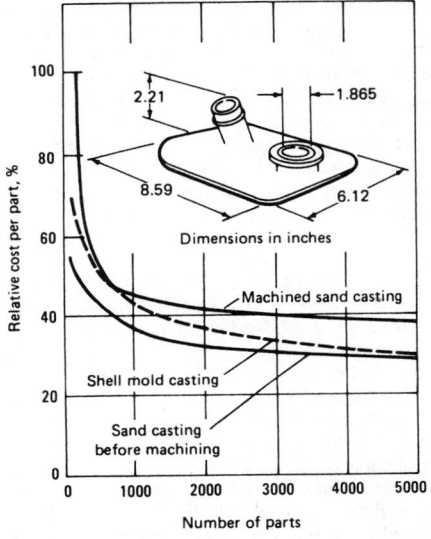

The sand casting required an extra machining operation to meet a dimensional limit that could be held in the shell mold casting without machining. Thus, the curve labeled "Machined sand casting" should be used in comparing total costs of the two casting methods.

Fig. 1. Comparison of cost-quantity relationships for two methods of casting a Mg-Al-Zn alloy part

Designs of many parts for use at elevated temperatures under continuous loads are based on maximum allowable deformation. The limiting creep-stress values given in Table 6 are based on 0.2% total extension. The alloys that contain thorium have the greatest resistance to creep at 205 and 315 °C (400 and 600 °F), and the Mg-Al-Zn alloys have the lowest resistance.

The decrease in modulus of elasticity with increase in temperature, a characteristic of Mg-Al-Zn alloys, is considerably less for thorium-containing alloys.

SELECTION OF PRODUCT FORM

Selection of a particular product form for a structural application is based on mechanical-property requirements and on cost, availability and fabricability. Requirements for production and design may change under operating conditions or as need arises. A part originally machined from bar stock may subsequently be made by extrusion or forging. Assemblies built up by joining sheets and extrusions may be redesigned as castings with equivalent performance at lower cost.

Castings. Parts too intricate to fabricate economically by other methods can be produced as castings. Sand, permanent mold and die castings are more widely used than investment and shell

Table 7. Comparison of costs for making a typical aircraft engine casting from three different magnesium alloys

Item	AZ91C	ZE41A	QE22A
Cores	$ 41.18	$ 41.18	$ 41.18
Molding	79.44	79.44	79.44
Metal	76.33	125.83	310.77
Cleaning	56.66	56.66	56.66
Heat treatment	18.28	1.39	8.03
Visual inspection	11.00	11.00	11.00
Nondestructive testing	94.68	94.68	94.68
Fixturing	8.80	8.80	8.80
Total cost per casting	$386.37	$418.98	$610.56

mold castings. The choice of casting method is determined primarily by size, shape, quantity, cost and desired mechanical properties of the casting. Cost of magnesium alloy castings is governed largely by ingot price, alloy castability and required heat treatment. Ingot price increases with additions of rare-earth metals, zirconium and thorium. Small changes in composition may affect cost of heat treatment. Comparative costs for casting an aircraft engine part from three different magnesium alloys are given in Table 7.

Magnesium alloys are cast by the permanent mold process when the number of parts required justifies the very high cost of equipment. The mechanical properties of sand and permanent mold castings are comparable, but the permanent mold process normally provides closer control of dimensions and produces better cast surfaces.

Cost of castings also is influenced by such factors as required tolerances, mold and die costs, and machining costs. The quantity of a part to be produced is an important factor affecting cost and must be considered in seeking the most economical method of production.

Figure 1 illustrates a casting that was produced at lower cost by shell mold casting than by sand casting and machining when more than 700 castings were made. The sand mold casting required one extra machining operation to obtain a dimension that could be held within the tolerance of the shell mold.

Die castings made of magnesium alloys may be selected in preference to aluminum die castings of the same design because of the savings in weight. Service requirements and size may govern whether a magnesium alloy is selected for use in a die casting, but quantity is the most important factor, because die castings are high-production items.

Magnesium alloy die castings, like castings in general, are always priced and purchased on a per piece basis. Cost per pound varies, depending primarily on complexity of design, wall thickness, number of cavities in the mold, and quality level.

Extrusions. Magnesium alloys are extruded as round rods and a variety of bars, tubes and shapes. A wide variety of special shapes also can be extruded. Extrusion is selected as a means of producing certain shapes when (a) several small extrusions or a combination of extrusions and sheet can be joined to form an assembly, (b) shapes are desired that are uneconomical to machine from castings, and (c) pieces cut from extrusions can replace individually cast or forged parts.

The extrusion process offers many design possibilities not economically attainable by other production methods. These include re-entrant angles and undercuts, thin-wall tubing of large diameter and variations in section thickness, almost without restriction. Probably the most important factor in determining whether a magnesium alloy shape will extrude well is good symmetry, preferably around both axes.

Very thin and wide sections with large circumscribing circles should be avoided. The optimum width-to-thickness (w/t) ratio for magnesium extrusions normally is less than 20. Parts with higher ratios can be extruded but require more generous tolerances. A thick section tapering to a thin wedge must always be modified by rounding the edge, or the die may not fill properly. A thin leg attached to a thick body of an extrusion should be limited to a length not exceeding ten times the leg thickness. Semiclosed shapes requiring long, thin die tongues should be avoided. For best extrudability, the length of the tongue should not exceed three times its width, although it is possible to extrude lengths five times the width. Similarly, shapes requiring unbalanced die tongues do not constitute good extrusion design. Hollow shapes that contain unsymmetrical voids, or voids separated by sections of inadequate thickness, are undesirable.

Sharp outside corners result in excessive stress concentration and die breakage and should be avoided. Inside corners should be filleted to reduce stress concentrations in the part and to ensure complete filling of the die during extrusion. Regardless of the shape being extruded, it is difficult to hold distances between thin sections to close tolerances.

Many shapes can be extruded economically. Extrusion dies are relatively inexpensive, and dimensions can be held closely enough so that machining often is unnecessary.

Impact extrusions are tubular parts of symmetrical shape. The impact type of hot extrusion is particularly applicable when:

- It is not practical to make the part by any other method, such as with parts requiring very thin walls, where thin walls having high strength are essential or where irregular profiles must be incorporated in the part.
- High production rates are required, where scrap loss from machining would be excessive if the

part were made by other means, where strength requirements cannot be met by die castings, where the number of manufacturing operations or the number of parts in an assembly can be reduced by the use of impact extrusions, where portions of the part require zero draft, and where closer tolerances are required.

In designing impact extrusions, the following factors should be taken into consideration:

- A wide variety of symmetrical shapes is possible.
- Variations in wall thickness are possible (thin sidewall, thick bottom; thick sidewall, thin bottom).
- Ribs, flanges, bosses and indentations can be incorporated.
- Length-to-diameter ratios may range from 1.5:1 to 15:1. Ratios from 6:1 to 8:1 are considered good working ratios.
- Reduction in area varies with the alloy being extruded and is limited by the size of available equipment. Parts with reduction in area up to 95% have been made. In general, extrusions of alloys M1A and AZ31B can have thinner walls than AZ61A, AZ80A and ZK60A extrusions.
- Sharp corner radii are possible in some areas of impact extrusions. This is not true of other product forms.
- Average properties of impact extrusions are slightly higher than typical properties of the hot extruded stock from which the parts are made.

Forgings. Magnesium forgings can be produced in the same variety of shapes and sizes as can forgings of other metals. Maximum size is limited primarily by size of available equipment. Tolerances can be held to the same values as in normal forging of other metals and vary somewhat with forging size and design.

Forgings have the best combination of strength characteristics of all forms of magnesium. They are used where light weight coupled with rigidity and high strength are required. Magnesium forgings are sometimes used because of their pressure tightness, machinability and lack of warpage rather than because of their high strength-to-weight ratio.

Forging is used for parts to be produced in quantities sufficient to amortize die costs and for parts requiring high strength and ductility and greater uniformity and soundness than can be obtained with castings. For small quantities, hand forgings may be used, but die forgings have better mechanical properties and are less expensive in larger quantities.

The ease with which magnesium can be worked greatly reduces the number of forging operations needed to produce finished parts. Many of the steps commonly required in forging brass, bronze and steel (such as punching, planishing, drawing and ironing, sizing and coining, and edging and rolling) are unnecessary in forging magnesium. Bending, blocking and finishing are the principal steps used in forging magnesium.

Die design and resulting metal flow cause variations in tensile properties at different sections of large forgings. An example involving aircraft wheels forged from magnesium alloys is illustrated in Fig. 2. Test specimens were taken at several locations in forgings made from alloy AZ80A-T5 and from alloy ZK60A in the T5 and T6 conditions.

Mechanical properties of magnesium alloy forgings or castings, as determined by testing of separately cast or forged bars, are useful for

evaluating certain characteristics on a comparative basis and serve as a means of control. However, test results for these bars may vary significantly from properties of specimens taken from various locations in production castings or forgings. The amount of this variation is affected by section thickness and direction of metal flow.

INSERTS

Magnesium surfaces that are subjected to heavy bearing loads or severe wear require protection. Inserts that provide such protection to the surfaces of holes can be made of various materials and can be attached in numerous ways.

Cast-in Inserts. Inserts in magnesium may be fixed in place by casting the magnesium around them. Cast-in inserts for use in magnesium may be made of steel, brass, bronze or other metals. Nonferrous inserts may be plated with chromium to prevent alloying with magnesium, although such plating is seldom used. Tinning of ferrous inserts prevents galvanic corrosion of the magnesium.

Cast-in inserts become securely fixed when the cast metal shrinks around them. The insert is even more securely fixed if the outside of the insert is knurled or grooved. Care should be taken to ensure that sharp corners or insufficient metal around the insert does not set up concentrations of stress.

Shrinkage of the magnesium alloy around cast-in inserts may cause high residual stress in the metal surrounding the insert. This possibility can be significantly reduced by preheating the insert, although inserts with wall thicknesses of 1.3 mm (0.050 in.) or less will be sufficiently preheated by contact with the hot die and the molten magnesium. If these designs and manufacturing conditions are not followed, the high residual stresses may lead to failure of the part in service due to stress-corrosion cracking. Safe design dimensions for cast-in steel inserts with wall thicknesses greater than 1.3 mm (0.050 in.) are shown in Fig. 3.

In highly stressed castings, it may be desirable to cast in a pilot to which the insert may be attached, thus avoiding high shrink stresses around the insert. Cast-in inserts may also be used for providing design details not otherwise feasible, such as lubrication lines of steel or copper, or appendages that could not be attached conveniently otherwise. Cast-in inserts complicate manufacture of castings and should be used only if other methods are not feasible.

Press-Fit and Shrink-Fit Inserts. Because of the low modulus of elasticity of magnesium, greater interference must be used for press-fit or shrink-fit inserts in magnesium than for inserts in other metals in order to obtain sufficient gripping force. An interference of 0.5 to 1.0 mm/m (0.0005 to 0.001 in./in.) is usually satisfactory, but this may be increased appreciably where high torque loads are likely to be encountered. Table 8 presents recommended interferences for steel and bronze inserts in normal service at various temperatures. The values given serve only as a guide, because service temperature, type of insert material, severity of service, thickness of insert, and sensitivity of the alloy to stress-corrosion cracking all influence the correct amount of interference. Differences in thermal expansion, yield strength and modulus of elasticity must also be considered for service at elevated temperatures.

Inserts are more easily assembled by shrinking than by pressing. It is relatively easy to heat a magnesium part and thus expand a hole in the part sufficiently to receive an insert. Where nec-

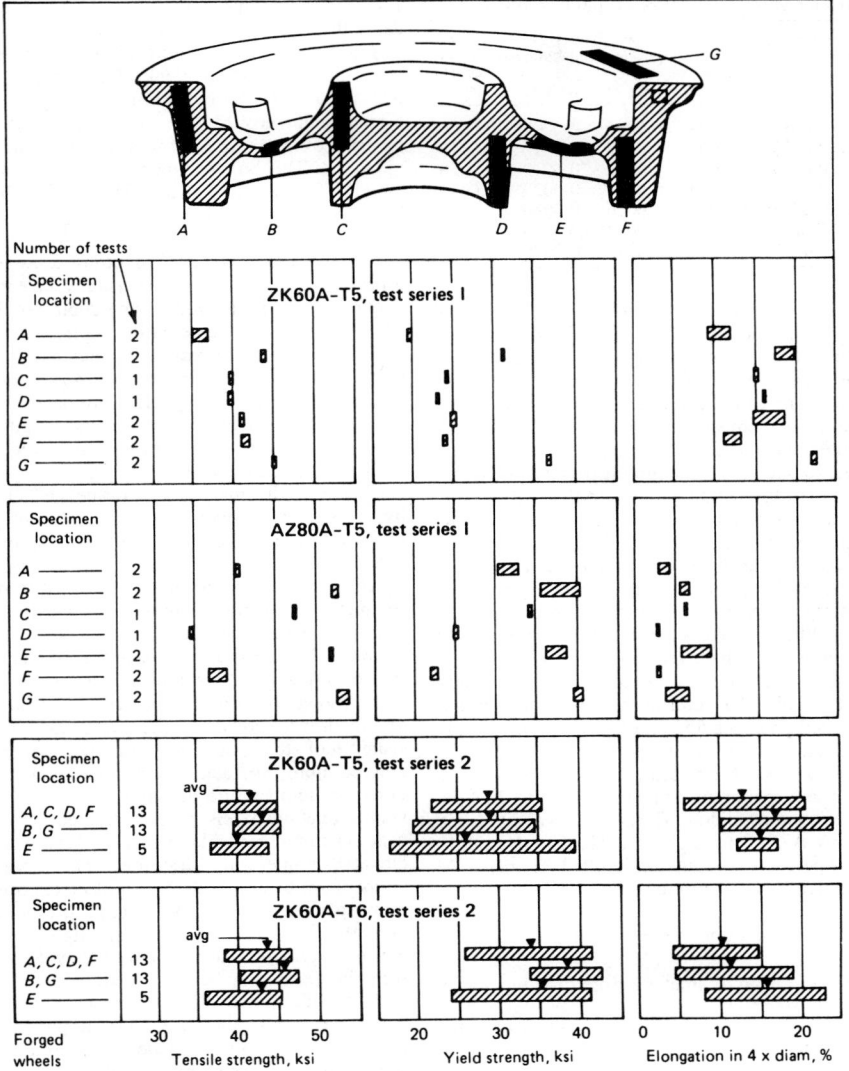

Fig. 2. Effects of alloy, heat treatment and specimen location on mechanical properties of forged magnesium aircraft wheels

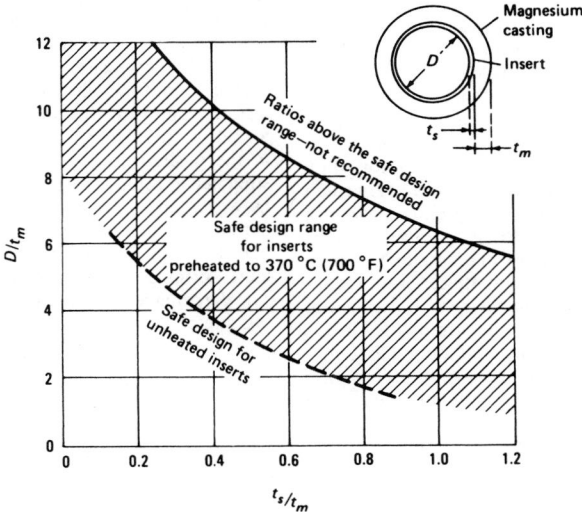

For inserts having wall thickness (t_s) greater than 1.3 mm (0.050 in.).

Fig. 3. Safe design dimensions for cast-in steel inserts in magnesium alloy castings

essary, the insert may be cooled to facilitate insertion. On the other hand, assembly by pressing requires careful machining of both insert and hole, and proper lubrication. When large interferences are required, it is best to specify shrink-fit inserts, because press-fits may score the magnesium.

Screwed-in Inserts. Various types of screwed-in (and other mechanically attached) inserts may also be used successfully in magnesium parts. When screwed-in inserts are used, locations of threaded holes must be chosen carefully in order to avoid stress concentrations and to provide sufficient engagement length for the thread.

FORMABILITY

Magnesium alloys, like other alloys with hexagonal crystal structures, are much more workable at elevated temperatures than at room temperature. Consequently, magnesium alloys usually are formed at elevated temperatures, and cold forming is used only for mild deformations around generous radii. The methods and equipment used in forming magnesium alloys are the same as those commonly employed in forming alloys of other metals, except for differences in tooling and technique that are required when forming is done at elevated temperatures.

Working of metals at elevated temperature has several advantages over cold working. Magnesium parts usually are drawn at elevated temperature in one operation without repeated annealing and redrawing, thus reducing the time involved for making the part and also eliminating the necessity of additional die equipment for extra stages. Hardened dies are unnecessary for most types of forming. Hot formed parts can be made to closer dimensional tolerances than cold formed parts because of less springback. Suggested maximum forming temperatures and times for various wrought magnesium alloys are given in Table 9. The column headed "Time" gives the maximum time the alloy can be held at temperature without adverse effects on properties.

Sheet and Plate. Rolled magnesium alloy products include flat sheet and plate, coiled sheet, circles, tooling plate and tread plate. These products are supplied in a variety of standard and nonstandard sizes.

The ability to use increased section thickness without weight penalty is of particular importance in designs that employ magnesium sheet. Thick-sheet construction provides the rigidity necessary in a structure, without the need for costly assembly of ribs and similar reinforcing members.

Rolled magnesium alloy products can be worked by most conventional methods. For severe forming, sheet in the annealed (O temper) condition is preferred. However, sheet in the partially annealed (H24 temper) condition can be formed to a considerable extent. Because heat has significant effects on properties of hard rolled magnesium, properties of the metal after exposure to elevated temperature must be considered in forming. The design curves shown in Fig. 4 give minimum values suitable for design use. Although the curves are based primarily on tests of sheet 1.63 mm (0.064 in.) thick or less, check tests indicate reasonable applicability for gages up to 6.35 mm (0.250 in.).

Figure 4 shows how the properties of AZ31B-H24 vary with exposure time at several temperatures. The curves have been extrapolated above the typical property levels of AZ31B-H24 sheet.

Table 8. Approximate interferences suggested for press-fit and shrink-fit inserts in magnesium

Outside diameter mm	in.	Wall thickness mm	in.	Steel inserts — Operating temperatures 21 °C (70 °F)	38 °C (100 °F)	93 °C (200 °F)	149 °C (300 °F)	Bronze inserts — Operating temperatures 21 °C (70 °F)	38 °C (100 °F)	93 °C (200 °F)	149 °C (300 °F)
12.70	0.50	1.59	0.06	....0.0004	0.0005	0.0009	0.0013	0.0004	0.0005	0.0007	0.0009
25.40	1.0	2.38	0.09	....0.0006	0.0010	0.0017	0.0025	0.0006	0.0008	0.0012	0.0016
50.80	2.0	3.17	0.13	....0.0010	0.0015	0.0031	0.0048	0.0010	0.0013	0.0019	0.0028
76.20	3.0	3.97	0.16	....0.0015	0.0024	0.0047	0.0072	0.0015	0.0019	0.0028	0.0044
101.6	4.0	4.76	0.19	....0.0021	0.0034	0.0062	0.0096	0.0021	0.0026	0.0039	0.0060
127.0	5.0	5.56	0.22	....0.0028	0.0044	0.0080	0.0121	0.0028	0.0034	0.0053	0.0079
152.4	6.0	6.35	0.25	....0.0036	0.0053	0.0098	0.0147	0.0036	0.0043	0.0070	0.0098

Table 9. Maximum forming temperatures and times for wrought magnesium alloys

Alloy	Temperature °C	°F	Time
Sheet			
AZ31B-O 288	550	1 h	
AZ31B-H24 163	325	1 h	
HK31A-H24 343	650	15 min	
	371	700	5 min
	399	750	3 min
Extrusions			
AZ61A-F 288	550	1 h	
AZ31B-F 288	550	1 h	
M1A-F 371	700	1 h	
AZ80A-F 288	550	½ h	
AZ80A-T5 193	380	1 h	
ZK60A-F 288	550	½ h	
ZK60A-T5 204	400	½ h	

Table 10. Maximum time at temperature to maintain properties of magnesium alloy sheet(a)

Maximum time, min	Temperature °C	°F
AZ31B-H24		
0.3 260	500	
1 224	435	
2 210	410	
3 202	395	
4 196	385	
5 188	370	
10 182	360	
30 174	345	
60 163	325	
HK31A-H24		
15 343	650	
5 371	700	
3 385	725	
3 399	750	
HM21A-T8(b)		
60 399	750	
10 427	800	

(a) Annealed sheet will endure much higher temperatures than heat treated sheet for short periods without significant reduction of properties at room temperature. (b) Based on limited data obtained in laboratory tests.

Thus, if the value selected from a curve exceeds the actual property level of the material before exposure, the actual figure must be used.

Tests indicate that the effects of multiple exposures at elevated temperature are cumulative. Using Fig. 4 as an example, suppose that a part is held at 195 °C (380 °F) for 10 min; the resultant design compressive yield stress is 159 MPa (23.1 ksi). Then suppose the part is subsequently exposed for 300 min at 170 °C (340 °F). Because 200 min at 170 °C and 10 min at 195 °C both result in a minimum compressive yield stress of 159 MPa, they are equivalent exposures. The total equivalent exposure time at 170 °C then is 300 plus 200, or 500, which according to the data given in Fig. 4 results in a compressive yield stress of 155 MPa (22.5 ksi).

AZ31B-H24 sheet is commonly hot formed at temperatures below 160 °C (325 °F) to avoid annealing it to room-temperature property levels lower than the specified minimums. Annealing is a function of both time and temperature of exposure; thus, temperatures higher than 160 °C can be tolerated if exposure is carefully controlled. Table 10 shows the maximum permissible combination of time and temperature that will ensure that the specified minimum room-temperature properties of AZ31B-H24, HK31A-H24 and HM21A-T8 can be retained. This table is used for establishing limits of time and temperature for single exposures in normal forming operations. Whenever the sheet must endure multiple exposures, or whenever time of exposure at a given temperature must exceed the value indicated in Table 10, refer to the data compilations presented on pages 553 and 595 in Metals Handbook, 9th Edition, Volume 2.

Deep Drawing. Magnesium alloys can be cold drawn to a maximum reduction of 15 to 25% in the annealed condition. The cold drawability limit of alloy AZ31B-O is about 20%. The drawability, or percentage reduction in blank diameter, is calculated by the formula:

$$\text{Percentage reduction} = \frac{D - d}{D} \times 100$$

where D is blank diameter before drawing and d is punch diameter.

The time and temperature for annealing AZ31B-O are 15 min at 260 °C (500 °F). Cold drawn parts are stress relieved at 150 °C (300 °F) for 1 h after the final draw, to eliminate the danger of cracking from residual stresses.

The technique for hot drawing of magnesium alloy sheet has been developed largely so that drawing can be completed in a single operation. Heating of magnesium alloys increases their drawability to such an extent that most parts can be made in a single draw. The amount of possible reduction increases as temperature increases up to about 230 °C (450 °F) for alloy AZ31B. It is usual practice to draw annealed AZ31B sheet to reductions as high as 68% in a single draw. When heated, magnesium can be drawn to higher reductions in a single draw than can other metals. Maximum single-draw reduction as a function of temperature is illustrated in Fig. 5. These values were determined for 1.63-mm (0.064-in.) sheet using a cupping die 38 mm (1½ in.) in diameter. The radii on the draw ring and punch were both 5t, and drawing speeds were 38 mm (1½ in.) and 508 mm (20 in.) per minute. Note in Fig. 5 that the amount of reduction possible at a given temperature also is influenced by drawing speed.

The possibilities inherent in two-step draws are illustrated by the following parts: In the first operation, 610-mm (24-in.) blanks of 0.64-mm (0.025-in.) annealed sheet were drawn to a cup 200 mm (8 in.) in diameter by 400 mm (16 in.) deep; they were redrawn to a cup 140 mm (5½ in.) in diameter by 585 mm (23 in.) deep. Starting with a rectangular blank of 1.3-mm (0.051-in.) AZ31B-O, 455 by 485 mm (18 by 19 in.), a rectangular box 111 by 273 by 165 mm (4³⁄₈ by 10³⁄₄ by 6½ in.) deep was drawn in the first operation. This box was then redrawn into a rectangular box 89 by 254 by 171 mm (3½ by 10 by 6³⁄₄ in.) deep, having 5.6-mm (⁷⁄₃₂-in.) corner radii.

For most parts, however, depth of draw is not a primary consideration, and usually no trouble is experienced in drawing to the depth required. More trouble is encountered in keeping the metal free from puckers in parts with rounded corners or contours. Temperatures above those required for maximum drawability often are necessary to eliminate these puckers. On unusual or difficult jobs, it may be necessary to vary the procedure to obtain minimum scrap.

Choice of die materials is influenced chiefly by severity of the operation and number of parts to be produced. For most applications, unhardened low-carbon steel boiler plate or cast iron is satisfactory. For runs of 10 000 parts or more, for maximum surface smoothness, or for close tolerances where no significant die wear can be tolerated, hardened tool steels are recommended. W1 or O1 tool steels are satisfactory for extremely long runs (one million parts). For the most severe draws, however, the more abrasion-resistant tool steels, such as A2 or D2, will probably be more satisfactory and more economical. For room-temperature drawing, it is usually desirable that die steels be heat treated to obtain near-maximum hardness in service. However, for elevated-temperature drawing, the maximum temperature to which the dies will be exposed in drawing must also be considered. In this situation, the dies must be tempered slightly above the maximum service temperature, even though some hardness may be sacrificed.

Parts with smooth bottoms are usually drawn with open dies or dies in which the punch does not bottom out on the female, or bottom, plate. These parts are formed by the pull exerted on the blank by the pressure ring and the draw ring. Mating dies are used with magnesium only in forming of parts having re-entrant portions that cannot be made by other means.

Stretch Forming. Both magnesium sheet and magnesium extrusions can be stretch formed. The temper of the alloy has no effect on the techniques employed. Sheet usually is heated to 165 to 290 °C (325 to 550 °F) and slowly stretched to the desired contour. Annealed sheet can be stretched at room temperature to a limited extent. However, the formabilities of alloys of any temper are so much greater at elevated temperatures than at room temperature that elevated-temperature forming is preferred for most operations. The percentage of differential stretching (P) during stretch forming is expressed as follows:

$$P = \frac{L - S}{S} \times 100$$

where L is the longest stretched length in the part and S is the shortest comparable length parallel to the longest stretched length. L and S are assumed to be of equal length before stretching.

Hot stretch forming results in minimum springback; the little springback that may occur is controlled by adding about 1% to the total

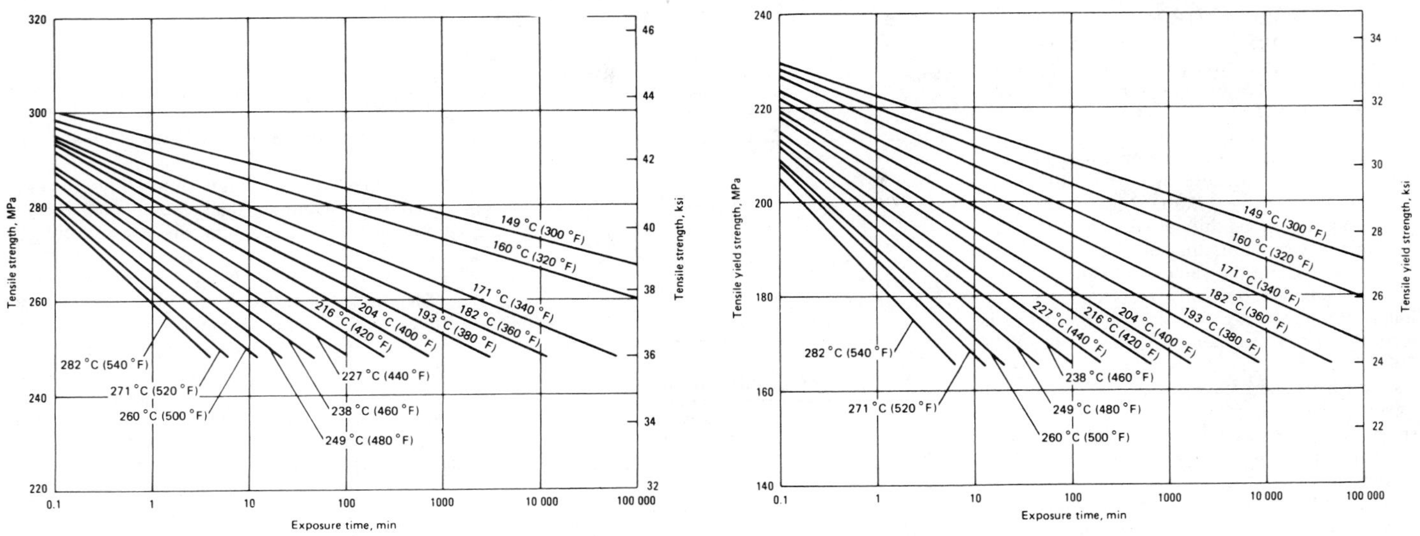

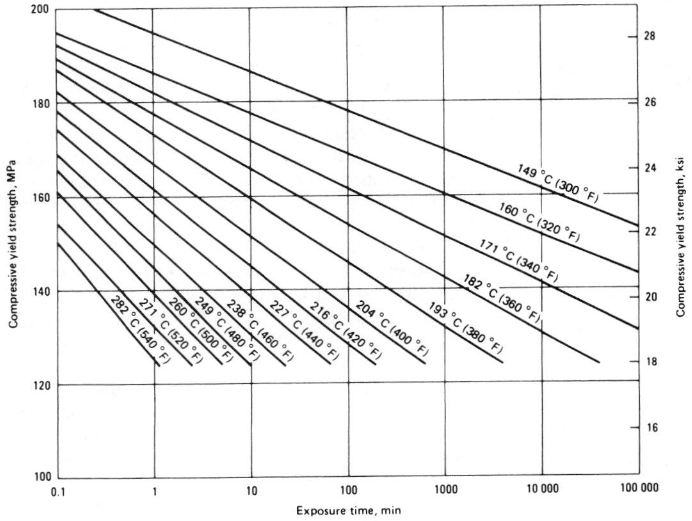

Data are based on sheet 1.63 mm (0.064 in.) thick. Check tests indicate reasonable applicability for thicknesses up to 6.35 mm (0.250 in.).

Fig. 4. Effects of exposure time at elevated temperature on mechanical properties of AZ31B-H24 at room temperature

stretch. Contour and springback control are good for positive contour radii curvature up to 6 m (20 ft).

For differential stretching of sheet over dies of low curvature, the maximum practical limit is about 15%. A 12% maximum is considered more desirable, however, because it permits enough overstretch for springback control.

Wrinkling often is a problem in stretch forming of magnesium sheet, particularly in making asymmetrical parts of low curvature. The best method of controlling wrinkling is by building the proper restraints in the dies.

Dies may be made of a variety of materials, including magnesium alloys, aluminum alloys, iron, steel or zinc alloys. One aircraft company has reported the use of dies made of concrete cast over a wire mesh and heated by electrical resistance. Zinc alloy blocks should not be used above 230 °C (450 °F). Grippers should not have sharp serrated edges, which tear magnesium alloys; the use of emery paper between grips and magnesium sheet helps to reduce the probability of tearing.

Heat control is important, and proper arrangement of heating units provides correct heat distribution. In addition to electrical resistance heating, infrared radiant heating units can be

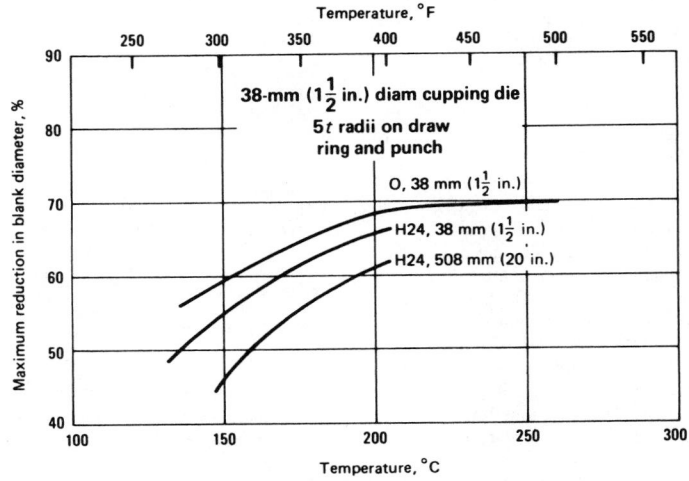

Fig. 5. Effect of temperature on the drawability of 1.63-mm (0.064-in.) AZ31B sheet

employed. Thermostatic control is essential and should be used on all heaters. The resistance heaters can be placed at various points where critical forming occurs. It is important that heating be sufficient to induce plastic flow, yet not

great enough to cause excessive elongation and rupture of the part.

Die temperature varies with the material being formed. AZ31B-O sheet is usually formed at 290 °C (550 °F) without loss of mechanical properties. Hard rolled AZ31B-H24 can withstand a temperature of 165 °C (325 °F) for 1 h and higher temperatures for shorter periods. The maximum time at temperature to maintain original properties is given in Table 10. Extruded magnesium alloys may be heated to 315 °C (600 °F) in most instances. For most contours, only heating of the die is necessary, because the sheet will pick up heat quickly from the hot die. More complicated parts may require additional heating of the sheet by radiant heat or thermal blankets. Thick material may require the same treatment.

Lubricants are normally recommended, colloidal graphite being best for high-temperature forming. Lower-temperature operation (up to 260 °C, or 500 °F) may permit use of heat-resisting waxes and greases, or a dry-film stearate-type lubricant instead of colloidal graphite. Some shops have used heat-resisting synthetic rubber or fiberglass cloth between the magnesium sheet and the forming block.

Annealed magnesium sheet may be shrunk satisfactorily at room temperature. The amount of shrinkage possible can be greatly increased by heating. Standard shrinking machines are employed.

Extrusion Formability. Production bending of extrusions may be done on standard angle rolls, in mating dies, in stretch-forming machines or in other specialized bending equipment. If the forming is severe, extrusions are heated to approximately 260 to 345 °C (500 to 650 °F) and formed hot.

Some bend-radii data for flat extruded magnesium strip, as well as maximum forming temperatures and times, are given in Table 11. Bending of more complex extrusions is more difficult, and the bend radii required must be established for a given shape.

The minimum bend radii for round magnesium tubing varies with the ratio of outer diameter to wall thickness (Table 12).

JOINING OF MAGNESIUM ALLOYS

Welding. Magnesium alloys are welded readily by gas metal-arc welding and by resistance spot welding. Rods of approximately the same composition as the base metal generally are satisfactory. With alloys HM21A and HM31A, EZ33A rods give higher joint efficiencies.

Butt and fillet joints are preferred in magnesium because they are the easiest to make by arc welding, and they provide more consistent results than other types of joints. Lap joints sometimes are used, but generally are less satisfactory than butt joints for load-carrying applications.

Arc-welded joints in annealed magnesium alloy sheet and plate have room-temperature tensile strengths less than 10% lower than those of the base metal (joint efficiencies greater than 90%). Tensile strengths of arc welds in hard rolled material, however, are significantly lower than those of the base metal (joint efficiencies of only 60 to 85%) as a result of the annealing effect of welding. Consequently, room-temperature strengths of arc-welded joints in magnesium alloy sheet and plate are about the same regardless of the temper of the base metal.

Joint efficiencies also are affected by service temperatures. For example, arc welds in HK31A-H24 sheet exhibit joint efficiencies of 75 to 80% at room temperature, but nearly 100% at 260 °C (500 °F). Joint efficiencies of arc welds in HM21A-T8 sheet range from about 80% at room

Table 11. Suggested limits for bending extruded flat magnesium alloy strip(a)

Alloy	Typical bend radius at 21 °C (70 °F)	Limits for hot bending — At temperature °C	At temperature °F	Time, h	Typical bend radius
AZ61A-F	1.9t	288	550	1	1.0t
AZ80A-F	2.4t	288	550	1/2	0.7t
AZ31B-F	2.4t	288	550	1	1.5t
M1A-F	4.8t	371	700	1	2.0t
AZ80A-T5 ...	8.3t	193	380	1	1.7t
ZK60A-F	12.0t	288	550	1/2	2.0t
ZK60A-T5 ...	12.0t	204	400	1/2	6.6t

(a) For extruded flat strip 2.29 by 22.2 mm (0.090 by 0.875 in.).

Table 12. Form bending of magnesium tubing

Alloy	Forming temperature °C	°F	Bend radius, D(a)
AZ31B-F	21	70	4D
....	93	200	3D
AZ61A-F	21	70	4D
....	−7	20	3D
M1A-F	21	70	6D
....	204	400	4D
ZK60A-F	21	70	5D

	Min bend radius at 21 °C (70 °F)(b)		
	D/t = 17	D/t = 6	D/t = 3
AZ61A-F(c)	5D	2½D	2D
AZ61A-F(d)	2½D	2½D	2½D
AZ31B-F(c)	6D	4D	3D
AZ31B-F(d)	3D	2D	2D
M1A-F(c)	6D	3D	2½D
M1A-F(d)	6D	6D	2½D

(a) D = tube outside diameter. Bend radius taken to axis of tube. (b) Minimum bend radius for various D/t ratios at 21 °C (70 °F). D = tube outside diameter; t = wall thickness. (c) Tubing unfilled before bending. (d) Tubing filled with low-melting alloy (50 Bi, 26.7 Pb, 13.3 Sn, 10 Cd) before bending.

Table 13. Times and temperatures for stress relief of arc welds in magnesium alloys

Alloy	Stress relief Temperature °C	°F	Time, min
Sheet			
AZ31B-H24 149		300	60
AZ31B-O 260		500	15
Extrusions(a)			
AZ31B-F 260		500	15
AZ61A-F 260		500	15
AZ80A-F 260		500	15
AZ80A-T5 204		400	60
Castings			
AZ63A 260		500	60
AZ81A 260		500	60
AZ91C 260		500	60
AZ92A 260		500	60

(a) When extrusions are welded to sheet, distortion may be minimized by using a lower stress-relieving temperature and longer time. For example, 60 min at 150 °C (300 °F) instead of 15 min at 260 °C (500 °F).

temperature to 100% at 200 °C (400 °F). HM31-T5 extrusions exhibit joint efficiencies of 75 to 85% at room temperature to about 370 °C (700 °F), and 100% at 425 °C (800 °F) and above. There are no appreciable differences in properties between welds made with alternating current and those made with direct current.

Stress Relieving. Arc welds in some magnesium alloys—specifically the Mg-Al-Zn series and alloys containing more than 1% Al—are subject to stress-corrosion cracking, and thermal treatment must be used to remove the residual stresses

that cause this condition. The parts are placed in a jig or clamping plate and heated at the temperatures indicated in Table 13 for the specified times. After heating, the parts are cooled in still air. The use of jigs is sometimes necessary so that relief of stresses does not result in warpage of the assembly.

The other types of magnesium alloys, including those containing manganese, rare earths, thorium, zinc or zirconium, are not sensitive to stress corrosion and normally do not require stress relief after welding.

Spot welds in magnesium have good static strength, but fatigue strength is lower than for either riveted or adhesive-bonded joints. Spot welded assemblies are used mainly for low-stress applications and are not recommended where joints are subject to vibration.

Recommended spot spacings and edge distances for spot welds are given in Table 14. Where magnesium sheets of unequal thickness are to be spot welded, the thickness ratio should not exceed 2½ to 1.

Seam welds of the continuous and intermittent types have strength properties comparable to those of spot welds. Shear strengths of 19.2 to 40.2 kg/linear mm (1075 to 2250 lb/linear in.) of welded seam can be obtained in AZ31B sheet from 1.0 to 3.2 mm (0.040 to 0.12 in.) thick.

Cost of weldments is less likely to vary significantly with quantity than cost of other methods of fabrication. Therefore, weldments are used most often where quantities are small or where fabrication of specific designs is impractical or impossible by other methods.

Adhesive bonding of magnesium has become an important fabrication technique. The fatigue characteristics of adhesive-bonded lap joints are better than those of other types of joints. The probability of stress-concentration failure in adhesive-bonded joints is minimal. Adhesive bonding permits the use of thinner materials than can be effectively riveted. The adhesive fills the spaces between the contacting surfaces and thus acts as an insulator between any dissimilar metals in the joint. It also permits manufacture of assemblies having surfaces smoother than those associated with riveting.

Adhesive bonding has been limited almost exclusively to lap joints. The following are a few of the general factors that should be considered when designing adhesively bonded joints:

• Joint strengths vary with lap width, metal thickness, direction in which loads are applied and type of adhesive used.
• The joint should be designed so that it provides a sufficiently large bonded area.
• The adhesive layer should be uniform in thickness.
• The adhesive layer should be as thin as possible, yet applied in sufficient quantity so that no joints are "starved."
• Joints should be designed so that pressure and heat can be readily applied.
• The curing temperatures of the common structural adhesives are below the temperatures at which the properties of hard rolled magnesium sheet are affected, and thus they do not significantly reduce the properties of magnesium alloys in the annealed (O) condition.

Riveting. Essentially the same procedures employed in riveting other materials are used in riveting magnesium alloys. Standard procedures are used for drilling and countersinking holes. Both

Table 14. Recommended spot spacings and edge distances for spot welds in magnesium alloy sheet

| Sheet thickness | | Spot spacing | | | | Edge distance | | | |
mm	in.	Minimum mm	in.	Nominal mm	in.	Minimum mm	in.	Nominal mm	in.
0.508	0.020	6.35	0.25	12.70	0.50	3.81	0.15	6.35	0.25
0.635	0.025	6.35	0.25	12.70	0.50	4.06	0.16	6.35	0.25
0.813	0.032	7.87	0.31	15.75	0.62	4.57	0.18	6.35	0.25
1.015	0.040	9.65	0.38	19.05	0.75	5.08	0.20	6.35	0.25
1.296	0.051	10.41	0.41	19.05	0.75	5.84	0.23	7.87	0.31
1.626	0.064	12.70	0.50	25.40	1.00	6.85	0.27	9.65	0.38
2.057	0.081	15.75	0.62	31.75	1.25	7.87	0.31	10.41	0.41
2.591	0.102	15.75	0.62	31.75	1.25	9.40	0.37	12.70	0.50
3.175	0.125	19.05	0.75	38.10	1.50	11.18	0.44	15.75	0.62

dimpling and machine countersinking are used in flush riveting. With machine countersinking, it is desirable to have a cylindrical land with a minimum depth of 0.38 mm (0.015 in.) at the bottom of the hole. Thus, machine countersinking is limited to sheet thick enough to permit lands of this depth with a given size of rivet. Dimpling of magnesium alloy sheet is a hot forming operation; to prevent reduction of properties during dimpling, the sheet must not be heated to excessively high temperatures or for long periods.

Only aluminum rivets should be used if galvanic incompatibility is to be minimized, and those up to 8 mm ($^5/_{16}$ in.) in diameter can be driven cold. The ease of driving rivets of alloy 5056 will vary with the temper. Quarter-hard temper (5056-H32) is satisfactory for all normal riveting.

MACHINABILITY

Magnesium and its alloys can be machined at extremely high speeds using greater depths of cut and higher rates of feed than can be used in machining other structural metals. There are no significant differences in machinability among magnesium alloys. Therefore, a specific magnesium alloy rarely, if ever, is selected in place of another magnesium alloy solely on the basis of machinability.

Because of the free-cutting characteristic of magnesium, chips produced in machining it are well broken. Dimensional tolerances of about ±0.1 mm (a few thousandths of an inch) can be obtained using standard operations.

The power required to remove a given amount of metal is lower for magnesium than for any other commonly machined metal. The tabulation below compares power requirements for various metals, based on volume of metal removed per minute:

Metal	Relative power
Magnesium alloys	1.0
Aluminum alloys	1.8
Brass	2.3
Cast iron	3.5
Low-carbon steel	6.3
Nickel alloys	10.0

An outstanding machining characteristic of magnesium alloys is their ability to acquire an extremely fine finish. Often, it is unnecessary to grind and polish magnesium to obtain a smooth finished surface. Surface smoothness readings of about 0.1 μm (3 to 5 μin.) have been reported for machined magnesium and are attainable at both high and low speeds, with or without cutting fluids.

Cutting Fluids ("Coolants"). In machining of magnesium alloys, cutting fluids provide far smaller reductions in friction than they provide in machining of other metals and thus are of little use in improving surface finish and tool life. Most machining of magnesium alloys is done dry, but cutting fluids sometimes are used for cooling the work.

Although less heat is generated during machining of magnesium alloys than during machining of other metals, higher cutting speeds and magnesium's low heat capacity and relatively high thermal-expansion characteristics may make it necessary to dissipate the small amount of heat that is generated. Generation of heat can be minimized by use of correct tooling and machining techniques, but sometimes cutting fluids are needed to reduce the possibilities of distortion of the work and ignition of fine chips. Because they are used primarily to dissipate heat, cutting fluids are referred to as "coolants" when used in machining of magnesium alloys.

Numerous mineral-oil cutting fluids of relatively low viscosity are satisfactory for use as coolants in machining magnesium. Suitable coolants represent a compromise between cooling power and flash point. Additives designed to increase wetting power are usually beneficial. Only mineral oils should be used as coolants; animal and vegetable oils are not recommended.

Water-soluble oils, oil/water emulsions or water solutions of any kind should not be used on magnesium. Water reduces the scrap value of magnesium turnings and introduces potential fire hazards during shipment and storage of machine-shop scrap.

Safe Practice. The possibility of chips or turnings catching fire must be considered when magnesium is to be machined. Chips must be heated close to their melting point before ignition can occur. Roughing cuts and medium finishing cuts produce chips too large to be readily ignited during machining. Fine finishing cuts, however, produce fine chips that can be ignited by a spark. Stopping the feed and letting the tool dwell before disengagement, and letting the tool or tool holder rub on the work, produce extremely fine chips and should be avoided.

Factors that increase the probability of chip ignition are (a) extremely fine feeds, (b) dull or chipped tools, (c) improperly designed tools, (d) improper machining techniques and (e) sparks caused by tools hitting iron or steel inserts. Feeds less than 0.02 mm (0.001 in.) per revolution and cutting speeds higher than 5 m/s (1000 ft/min) increase the risk of fire. Even under the most adverse conditions — with dull tools and fine feeds — chip fires are very unlikely at cutting speeds below 3.5 m/s (700 ft/min).

Distortion of magnesium parts during machining occurs rarely and usually can be attributed to excessive heating or improper chucking or clamping.

Heating of the work is increased by use of dull or improperly designed tools or very fine cuts. Because magnesium has a relatively high coefficient of thermal expansion, such excessive heating results in substantial increases in dimensions — particularly in thin sections, where heating causes relatively large increases in temperature. Use of sharp, properly designed tools, mineral-oil coolants and relatively coarse feeds and depths of cut reduces excessive heating. Wide variations in room temperature during machining also can cause sufficient dimensional change to affect machining tolerances.

Clamping should always be done on heavier sections of magnesium castings, and clamping

Table 15. Relative bending strength, stiffness and weight of some structural metals(a)

Material	Thickness	Bending strength	Stiffness	Weight
For equal thickness				
1025 steel	100	100.0	100.0	100.0
6061-T6 aluminum sheet and extrusions	100	97.2	34.5	34.5
AZ31B magnesium extrusions	100	47.2	22.4	22.5
ZK60A-T5 magnesium extrusions	100	88.9	22.4	22.5
AZ31B-H24 magnesium sheet	100	73.4	22.4	22.5
For equal bending strength				
1025 steel	100	100	100.0	100.0
6061-T6 aluminum sheet and extrusions	101	100	35.8	34.8
AZ31B magnesium extrusions	146	100	69.2	32.9
ZK60A-T5 magnesium extrusions	106	100	26.7	23.9
AZ31B-H24 magnesium sheet	117	100	35.6	26.3
For equal stiffness				
1025 steel	100	100	100	100.0
6061-T6 aluminum sheet and extrusions	143	199	100	49.3
AZ31B magnesium extrusions	165	129	100	37.2
ZK60A-T5 magnesium extrusions	165	242	100	37.2
AZ31B-H24 magnesium sheet	165	200	100	37.2
For equal weight				
1025 steel	100	100	100	100
6061-T6 aluminum sheet and extrusions	290	817	841	100
AZ31B magnesium extrusions	444	930	1962	100
ZK60A-T5 magnesium extrusions	444	1753	1962	100
AZ31B-H24 magnesium sheet	444	1451	1962	100

(a) Comparison made at room temperature for rectangular beams of constant width with the following minimum yield strengths: 1025 steel, 250 MPa (36 ksi); 6061-T6 aluminum, 240 MPa (35 ksi); magnesium alloys, average of minimum tensile yield and compressive yield strengths.

pressures should not be high enough to cause distortion. Special care should be taken with light parts that could be distorted easily by the chuck or by use of heavy cuts.

Distortion of magnesium parts seldom is caused by stresses during casting, forging or extruding, but may result from stresses caused by straightening or welding. Such stresses can be relieved prior to machining by heating at 260 °C (500 °F) for 2 h and slowly cooling. However, such treatment causes some loss of strength in AZ31B-H24 sheet products. If distortion of parts is observed after rough machining, the cutting tool should be inspected to ensure that it is sharp and properly ground. If so, size of cut should be decreased. With complex parts or parts machined to extremely close tolerances, it may be advisable to stress relieve or, if time permits, to store parts for two or three days, between rough machining and finishing.

DESIGN AND WEIGHT REDUCTION

By substitution of magnesium alloys for heavier metals such as steel and aluminum alloys, many structural parts can be substantially reduced in weight with little or no redesign. This is possible because manufacturing limitations make many parts heavier than necessary. For example, a casting, for successful filling of the mold, may require minimum wall thickness greater than that dictated by service requirements and the strength of the metal used. Similarly, forgings and extrusions sometimes must be made thicker than necessary, and the light weight of magnesium can be used to advantage with these product forms also. In many instances, a casting, forging or extrusion for which magnesium is substituted for a heavier metal can have adequate strength with no increase in wall thickness.

In other parts, substitution of magnesium may require greater wall thickness, and substantial redesign may be necessary in order to realize maximum saving of weight. Because strength and stiffness in bending of many structural sections increase approximately as the square and cube of the section depth, respectively, it is possible to obtain large increases in strength and stiffness with moderate increases in depth and cross-sectional area. When such increases in depth are permissible, it usually is economical to redesign the part for magnesium. The greater bulk of the redesigned part reduces local instability, and, although the saving in weight is less than maximum, the reduced instability allows design simplification and thus reduces manufacturing costs.

Magnesium alloys are compared with aluminum alloys and steel on the bases of thickness, bending strength (defined as the product of yield strength and section modulus), stiffness and weight, at room temperature, in Table 15.

Corrosion Resistance

CORROSION RESISTANCE of a magnesium or magnesium alloy part depends on environmental conditions and on the chemical composition, thermal and mechanical history and surface condition of the part. The factors affecting corrosion resistance that are discussed in this article are the presence of heavy-metal impurities (iron, copper, etc.) in the magnesium or magnesium alloy, the type of environment (rural atmosphere, marine atmosphere, elevated temperature, etc.), the surface condition of the part (bare, treated, painted, etc.), and whether the part is in contact with other parts.

In some environments, contact with parts made of dissimilar metals can lead to severe damage of the magnesium part due to galvanic corrosion, unless the joint is properly designed and protected. Therefore, a separate section on magnesium-to-dissimilar-metal assemblies has been included near the end of this article.

Unalloyed magnesium is not extensively used for structural purposes. Consequently, it is the corrosion resistance of the alloys of magnesium that usually are of concern. Magnesium alloys, when properly made and applied, are corrosion-resistant and are used successfully in a wide variety of commercial, industrial and aerospace applications. The corrosion that is encountered usually is a result of improper design or application, or inadequate protective finish.

METALLURGICAL FACTORS

Chemical Composition. Figure 1 shows the effects of 17 alloying elements on corrosion of magnesium-base binary alloys in 3% NaCl solution. Four elements—iron, nickel, cobalt and copper—have extremely deleterious effects on corrosion resistance; these elements are considered impurities in magnesium, with definite tolerance limits for good corrosion resistance.

Magnesium-manganese binary alloy M1A, which contains about 1.2% Mn, has comparatively good tolerance for both iron and nickel impurities; the iron tolerance limit is 0.017%, the same as for pure magnesium. Greater amounts of iron are less detrimental when manganese is

present. Manganese also increases the tolerance limit for nickel.

Magnesium-aluminum binary alloys have tolerance limits for iron that are significantly lower than that of unalloyed magnesium. With as little as a few hundredths percent aluminum, the tolerance limit for iron is decreased from 0.017% to a few thousandths percent. With 7% Al, it is about 0.0005%. With 10% Al, the limit is too low to be determined, possibly because of the formation of an iron-aluminum phase even more active than discrete iron particles alone. The addition of even a few hundredths percent manganese to the magnesium-aluminum binaries is sufficient to raise their tolerance limit for iron to 0.002%, probably because of the formation of the compound (Fe, Mn) Al_3.

Figure 2 shows how additions of 0.5 and 3% zinc affect the relationship of iron content to corrosion of Mg-6Al-0.2Mn alloys in 3% NaCl solutions. Addition of 0.5% Zn does not shift the

position of the tolerance limit for iron (0.002%); it does somewhat reduce corrosion rates at higher percentages of iron. Addition of 3% Zn raises the tolerance limit for iron to 0.003% and reduces the corrosion rate when iron content is higher than 0.003%. Zinc additions show somewhat similar effects in raising the tolerance limits for nickel and copper.

In outdoor urban exposures, corrosion of magnesium alloys containing aluminum decreases as aluminum content increases, up to about 9%. However, additions of cerium (mischmetal), thorium or zinc usually result in increased rates of atmospheric corrosion.

Heat Treatment and Cold Work. With controlled-purity Mg-Al-Mn alloys containing 0 to 1% Zn, there is little if any difference in corrosion rate between as-fabricated material and solution heat treated and aged material, provided that the heat treated material is air cooled from the solution heat treating temperature. With similar cooling,

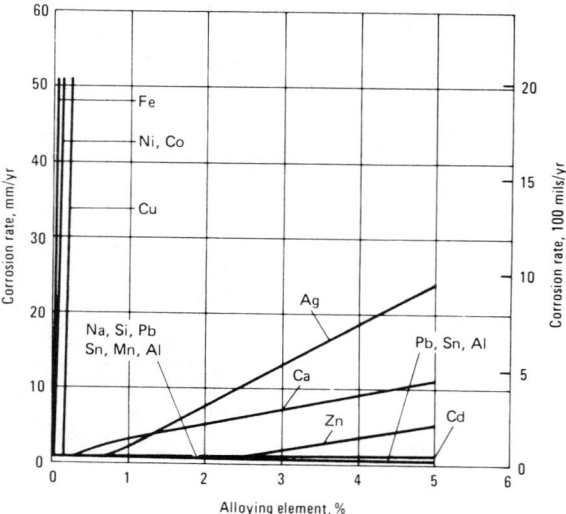

Fig. 1. Effects of alloy content on corrosion rates of magnesium-base binary alloys tested in 3% NaCl solution

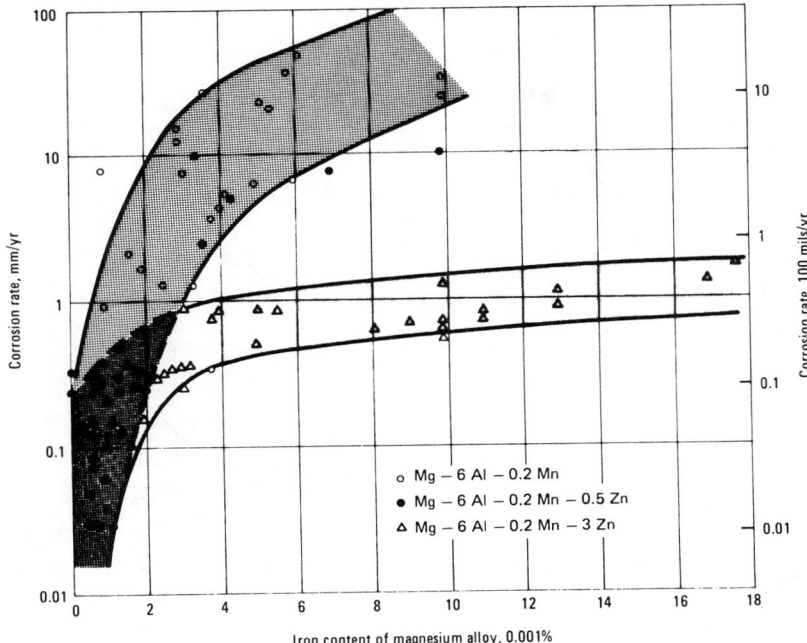

Fig. 2. Effect of zinc content on relationship between iron content and corrosion rate for Mg-6Al-0.2Mn alloys in 3% NaCl solution

controlled-purity alloys containing 2 to 3% Zn corrode at slightly higher rates in the solution heat treated or solution heat treated and aged condition than in the as-cast condition. In alloys containing iron above the tolerance limit, solution heat treatment will increase the corrosion rate by a factor of 2 to 5. This factor is lower if aging follows solution treatment.

The effect of heat treatment on corrosion rate in 3% NaCl is particularly noticeable adjacent to welds in alloys containing aluminum and zinc (severe attack may occur in the rapidly cooled heat-affected zone).

Cold working of magnesium alloys, such as stretching or bending, has no appreciable effect on corrosion rate. Corrosion rate is increased by shot or grit blasting because of surface contamination. Acid pickling to a depth of 0.04 to 0.05 mm (0.0015 to 0.0021 in.) often is used to remove the corrosion-active contamination, but reprecipitation of the contaminant is possible with acid pickling. Therefore, fluoride anodizing is used instead when complete removal of the contaminant is essential.

ENVIRONMENTAL FACTORS

Atmospheres. A clean unprotected magnesium-alloy surface exposed to indoor or outdoor atmospheres free from salt spray will develop a gray film (discussed later) that protects the metal from corrosion while causing only negligible losses in mechanical properties. Materials-handling equipment is an example of successful application of unfinished magnesium. Chlorides, sulfates and foreign materials that hold moisture on the surface will promote corrosion and pitting unless the metal is protected by properly applied coatings.

The surface film that ordinarily forms on magnesium alloys exposed to the atmosphere does not protect the metal from further attack. If the film is tight and adherent, however, it decreases the rate of further attack. Unprotected magnesium and magnesium alloy parts are resistant to rural atmospheres and moderately resistant to industrial and mild marine atmospheres, provided that they do not contain joints or recesses that entrap water and are not in contact with dissimilar metals. Magnesium alloys generally do lose strength in industrial and mild marine atmospheres, whether exposed in a stressed or an unstressed condition. However, except for exposure close to seawater, the conductivity of any entrapped water present during atmospheric exposure is low. Consequently, the rate of corrosion is low and is less influenced by alloy composition and impurities.

Corrosion of magnesium alloys increases with relative humidity. At 9.5% humidity, neither pure magnesium nor any of its alloys shows evidence of surface corrosion after 18 months. At 30% humidity, a small amount of an amorphous phase and slight corrosion are evident. At 80% humidity, an amorphous phase is present over about 30% of the surface, and the surface exhibits considerable corrosion. Crystalline magnesium hydroxide is formed only when relative humidity is at or above 93%.

In marine atmospheres heavily loaded with salt spray, magnesium alloys require protection for prolonged survival. Magnesium alloys are less resistant to atmospheric corrosion than aluminum alloys but considerably more resistant than low-carbon steel. Table 1 compares a magnesium alloy, an aluminum alloy and a low-carbon steel on the bases of loss in weight and loss in tensile strength in three different atmospheres. None of the metals was given surface protection.

Fresh Water. In stagnant distilled water at room temperature, magnesium alloys rapidly form a protective film that prevents further corrosive action. Small amounts of dissolved salts in water, particularly chlorides or heavy-metal salts, will break down the protective film locally, resulting in pitting.

Dissolved oxygen plays no major role in corrosion of magnesium, in either fresh water or salt solutions. However, agitation, or any other means of destroying or preventing formation of a protective film, leads to corrosion. When magnesium is immersed in a small volume of stagnant water, its corrosion rate is negligible. When the water is constantly replenished or circulated through an ion-exchange bed so that the solubility limit of magnesium hydroxide is never reached, the corrosion rate becomes significant. For example, AZ31B showed little attack after 35 days in stagnant distilled water at 52 °C (125 °F), but when the water was continuously replenished so that the pH remained at about 6.8, the rate of attack increased to 0.18 mm/yr (7 mils/yr).

Salt Solutions. Severe corrosion may occur in neutral solutions of salts of heavy metals such as copper, iron and nickel. Such corrosion occurs when the heavy metal or the heavy-metal basic salts, or both, plate out to form active cathodes on the anodic magnesium surface. Chloride solutions are corrosive because chlorides, even in small amounts, usually break down the protective film on magnesium, although sodium chloride is less corrosive than sodium iodide. Fluorides form insoluble magnesium fluoride and consequently are not appreciably corrosive. Oxidizing salts—especially those containing chlorine or sulfur atoms—are more corrosive than nonoxidizing salts; but chromates, vanadates, phosphates and many others are film-forming, and thereby retard corrosion, except at elevated temperatures.

The amounts and distribution of impurities in magnesium alloys strongly affect results of corrosion tests in salt solutions. For example, chloride impurities from included flux particles can cause catastrophic breakdown of a magnesium alloy part because magnesium chloride will react with water to form magnesium hydroxide and release chlorine for further chloride formation, which starts a chain reaction.

Acids and Alkalis. Magnesium is attacked rapidly by all mineral acids except hydrofluoric and chromic acids. Hydrofluoric acid does not attack magnesium to an appreciable extent because it forms an insoluble, protective magnesium fluoride film on the magnesium; however, pitting develops at low acid concentrations. With increasing temperature, the rate of attack increases at

Table 1. Results of a 2.5-year atmospheric exposure of an aluminum alloy, a magnesium alloy and a low-carbon steel

Alloy	Corrosion rate μm/yr	mils/yr	% change in tensile strength
Marine atmosphere(a)			
Aluminum, 2024-T3 . . .	2	0.06	2.5
Magnesium, AZ31B-H24	18	0.70	7.4
Low-carbon steel (0.27% Cu)	150	5.91	75.4
Industrial atmosphere(b)			
Aluminum, 2024-T3 . . .	2	0.08	1.5
Magnesium, AZ31B-H24	27.7	1.09	11.2
Low-carbon steel (0.27% Cu)	25.4	1.00	11.9
Rural atmosphere(c)			
Aluminum, 2024-T3 . . .	0.1	0.005	0.4
Magnesium, AZ31B-H24	13	0.53	5.9
Low-carbon steel (0.27% Cu)	15	0.59	7.5

(a) At Kure Beach, NC. (b) At Madison, IL. (c) Near Midland, MI.

the liquid line, but to a negligible extent elsewhere. Magnesium has been used commercially for float-control units, tanks and piping for closed units handling concentrated hydrofluoric acid.

Pure chromic acid attacks magnesium and its alloys at a very low rate. However, traces of chloride ion in the acid will markedly increase this rate. A boiling solution of 20% chromic acid in water is widely used to remove corrosion products from magnesium alloys without attacking the base metal.

Magnesium is resistant to dilute alkalis and even to 50% caustic liquors at temperatures as high as 60 °C (140 °F), but the corrosion rate increases rapidly with temperatures above 60 °C. Therefore, hot alkaline cleaners suitable for cleaning steel are recommended for cleaning magnesium. Such cleaners should have a pH of 11 to 13.

Organic Compounds. Aliphatic and aromatic hydrocarbons, ketones, ethers, glycols and higher alcohols are not corrosive to magnesium and its alloys. Ethyl alcohol causes slight attack. Anhydrous methyl alcohol causes severe attack. The rate of attack in the latter is reduced by the presence of water. Pure halogenated organic compounds do not attack magnesium at ambient temperatures. At elevated temperatures, or if water is present, such compounds may cause severe corrosion—particularly compounds whose final products are acidic.

Dry fluorinated hydrocarbons, such as the freon refrigerants, usually do not attack magnesium alloys at room temperature, but when water is present they may stimulate significant attack. At elevated temperatures, fluorinated hydrocarbons may react violently with magnesium alloys.

In aqueous organic acids, such as fruit juices, attack of magnesium is slow but measurable. Milk causes attack, particularly when souring.

At room temperature, ethylene glycol solutions produce negligible corrosion of magnesium used alone or galvanically connected to steel; at elevated temperatures (such as 115 °C, or 240 °F), the rate increases, and galvanic corrosion occurs unless inhibitors such as those used for aircraft engines are added. Alloy M1A has been used extensively for aircraft oil and gasoline tanks. However, tetraethyl lead and ethylene dibromide (added to raise the antiknock rating of gasoline) will cause severe pitting at the liquid line in gasoline-water mixtures, unless slushing compounds are used or special capsules containing slightly soluble chromates or other inhibitors are placed in the sump of the tank. Alloy AZ91C is used in applications involving contact with oil, such as accessory housings.

Gases. Dry chlorine, iodine, bromine and fluorine cause little or no corrosion of magnesium at room temperature. Even when it contains 0.02% water, dry bromine causes no more attack at its boiling temperature (58 °C, or 136 °F) than at room temperature. The presence of a small amount of water causes pronounced attack by chlorine, some attack by iodine, and negligible attack by bromine or fluorine. Dry, gaseous sulfur dioxide or ammonia causes no attack at ordinary temperatures; if water vapor is present, some corrosion may occur.

Soils. Except when used as galvanic anodes, magnesium alloys have good corrosion resistance in clay or nonsaline sandy soils but poor resistance in saline sandy soils. Bituminous paints over primed surfaces retard attack and provide protection against corrosion equal to or better than comparable treatments on low-carbon steel.

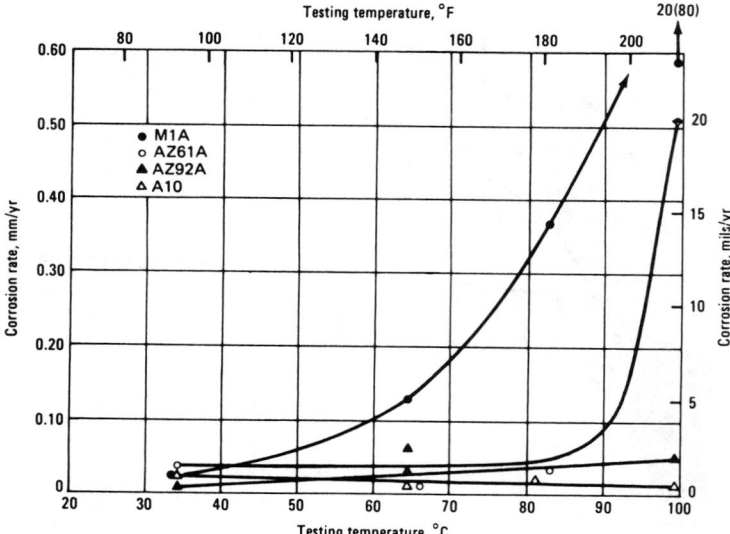

Fig. 3. Effects of temperature on corrosion rates of magnesium alloys M1A, AZ61A, AZ92A and A10 in tap water

Elevated Temperature. The effect of elevated temperature on corrosion of magnesium depends largely on the purity of the metal. The corrosion rates of controlled-purity magnesium and its alloys immersed in 3% sodium chloride solution at 100 °C (212 °F) are approximately twice the rates at room temperature. When cathodic impurities, such as iron and nickel, are present in amounts greater than their tolerance limits, corrosion and pitting increase as temperature increases.

Water vapor in air or in oxygen sharply increases the rates of oxidation of magnesium and its alloys at temperatures above 100 °C (212 °F), but boron trifluoride, sulfur dioxide and sulfur hexafluoride are effective in reducing them. The presence of boron trifluoride or sulfur hexafluoride in the ambient atmosphere is particularly effective in suppressing high-temperature oxidation up to and including the temperature at which the alloy normally ignites.

The corrosion rate of magnesium in ordinary tap water containing approximately 70 ppm chloride is low, but increases with temperature. Figure 3 shows comparative data for four alloys. Alloy M1A shows a greater tendency toward severe pitting with increasing temperature than magnesium alloys containing aluminum. Soluble fluorides are effective inhibitors of corrosion of magnesium in hot water. For example, addition of 0.10% sodium fluoride (by weight) reduced the

Table 2. Some protective chemical treatments for magnesium alloys

Common names	Military specifications	Type of treatment	Alloys on which treatment may be used	Type of solution	Time, min	Temperature °C	Temperature °F	Current density, A/m²	Remarks
Chrome pickle, Dow 1	(a)	Dip	All	Chromic acid, nitric acid, water	1-15	88-100	190-212	None	(b)
Dichromate, Dow 7	(c)	Dip	(d)	Sodium dichromate, calcium or magnesium fluoride, water	30	100	212	None	(e)
Dow 17	(f)	Anodic	All	Ammonium acid fluoride, sodium dichromate, phosphoric acid, water	1-30	71-82	160-180	55-540 at 60-90 v	(g)
HAE	(h)	Anodic	All	Potassium hydroxide, aluminum hydroxide, trisodium phosphate, potassium fluoride, potassium manganate, water	60-90	24-29	75-85	160 at 110 v	(g)

(a) MIL-M-3171C, type I. (b) Protection during shipment and storage; good paint base; slight dimensional etching loss. (c) MIL-M-3171C, type III (DOW 7). (d) All wrought alloys except those containing thorium. (e) Good combination of paint-base and protective qualities; requires acid fluoride pickle pretreatment. (f) MIL-M-45202B: type I, class C (thin); type II, class D (heavy). (g) Thin coatings (0.005 mm or 0.0002 in.) for flexibility and paint base; heavy coatings (0.03 mm or 0.001 in.) for maximum corrosion and abrasion resistance. (h) MIL-M-45202B: type I, class A (thin); type II, class A (heavy).

corrosion rate of alloy AZ31B in boiling distilled water from 0.41 to 0.02 mm/yr (16 to 0.9 mil/yr).

The rate of oxidation of magnesium in oxygen increases with temperature. At elevated temperature, the oxidation rate is a linear function of time. Most common alloying elements, such as aluminum and zinc, increase the rate of oxidation at a given temperature. Cerium, lanthanum, calcium, and beryllium reduce the rate of oxidation below that of pure magnesium. Beryllium additions have the most striking effects, protecting some alloys at temperatures up to the melting point over extended periods of time.

PROTECTIVE SURFACE TREATMENTS

Chemical treatments of magnesium alloys are used extensively to provide bases for paint and to protect the metal. The dip coatings (chrome pickle and dichromate treatments; see Table 2) are very thin coatings used primarily for protection during shipment and storage and as primers for subsequent painting. These coatings should not be heated above 260 °C (500 °F). Discretion should be exercised when using chromate coating without subsequent painting because some chromate coatings are pyrophoric — they spark when hit.

Anodic coatings are thicker and harder than dip coatings. Flash (thin) coats often are used as paint bases. Thicker (heavier) coats (without paint) give some protection against corrosion in moderately corrosive atmospheres, but sealing the coating by impregnation or painting is usually desirable. In marine atmospheres, or in any location conducive to galvanic corrosion, resin impregnation followed by painting with chromate-pigmented primer and then with a glossy-finish top coat is necessary for maximum serviceability. Several commercially available primers and finishing lacquers are satisfactory. Anodic coatings can be heated to 345 °C (650 °F) with no reduction in corrosion resistance. With suitable heat-resistant paints, exposure times of $^1/_2$ h at 425 °C (800 °F) have been survived by anodically treated ramjet parts made of alloys HK31A and HM21A.

GALVANIC CORROSION

Magnesium in electrical contact with other metals can suffer severe galvanic corrosion in high-conductivity environments, such as salt solutions, unless one or more of the following measures are taken:

- Selection of the dissimilar metal for galvanic compatibility with the magnesium, or electroplating the magnesium with a metal having galvanic compatibility with the dissimilar metal
- Protection of the magnesium and the dissimilar metal by suitable surface treatments
- Use of an insulating washer or gasket between the dissimilar metals to prevent the completion of an electrical circuit
- Inhibition of the galvanic cell by using chromates in the sealing compounds or primers.

High-purity (99.99%) aluminum is galvanically compatible with magnesium, but the small amounts of iron and copper normally present in commercial aluminum alloys effectively destroy such compatibility. In addition, the high pH (10.5) of water in contact with magnesium can lead to attack of the aluminum.

Figure 4 illustrates the effects of iron and magnesium in aluminum on the galvanic compatibil-

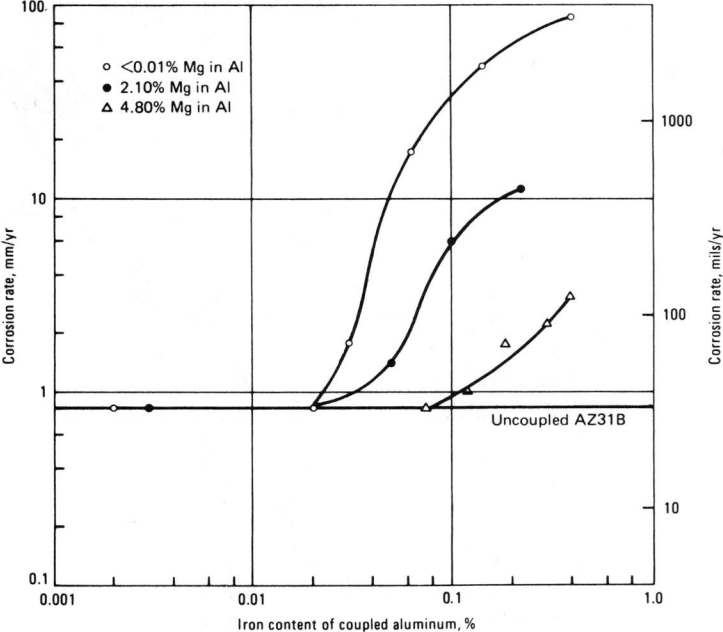

Corrosion rate of uncoupled AZ31B also shown for comparison.

Fig. 4. Corrosion rates of magnesium alloy AZ31B coupled with aluminum containing varying amounts of iron and magnesium in 3% NaCl solution

ity of aluminum with magnesium alloy AZ31B. The effects were determined using actual couples immersed in stagnant 3% NaCl solutions. High-purity aluminum causes almost no galvanic corrosion of magnesium, but when the iron content of the aluminum exceeds about 200 ppm (0.02%), cathodic activity becomes significant. Compatibility then diminishes rapidly with increasing iron content. Figure 4 also shows that addition of magnesium to aluminum markedly improves the tolerance limit for iron. This is why rivets made of aluminum alloy 5056, which contains about 5% Mg, have been used successfully in magnesium alloy structures without causing significant galvanic corrosion.

Magnesium ions in a sodium chloride solution may suppress the galvanic corrosion of magnesium–aluminum alloy couples in such a solution. However, addition of magnesium chloride to a sodium chloride solution actually increases the dissolution rate of uncoupled commercial-purity magnesium in the sodium chloride. Uncoupled magnesium is attacked less by natural seawater than by a sodium chloride solution of similar concentration, because of the sulfate content of the seawater, which probably stimulates flocculation of magnesium hydroxide sol on the metal surface to form a semiprotective film.

Figure 5 illustrates how suitable control of iron content can improve the compatibility of a commercial aluminum fastener alloy (6061) with magnesium alloy AZ31B. In seawater, 6061 is compatible with AZ31B, but in applications involving exposure to chlorides (such as road splash) AZ31B suffers severe galvanic corrosion. As shown in Fig. 5, aluminum alloy 6061 never becomes completely compatible with AZ31B even when the iron content of the 6061 is very low. An improvement is realized by holding the iron to a maximum of 0.1% (1000 ppm).

Ferrous alloys, nonferrous alloys containing nickel and/or copper, and titanium-base alloys have poor galvanic compatibility with magne-

sium alloys. The compatibility of steel fasteners with magnesium alloys is increased by electroplating the fasteners with tin, cadmium, zinc or chromium. Tin is the best coating for steel fasteners in contact with magnesium alloys, but better still are fasteners made of high-purity aluminum alloys containing magnesium. It should be noted that when magnesium-aluminum couples are immersed in seawater, the alkaline magnesium corrosion products induced by contact with aluminum in turn corrode the aluminum.

Metals compatible with one magnesium alloy may be incompatible with another. For example, reducing the iron content of aluminum alloy 6061 provides a significant reduction in galvanic corrosion when the 6061 is coupled with Mg-Al-Zn alloys, but a much smaller reduction when it is coupled with Mg-Th alloys. The improvement with Mg-Th alloys is small because these alloys, especially HK31A, develop appreciably higher potentials than do Mg-Al-Zn alloys and thus require a higher degree of galvanic compatibility for metals coupled with them.

In actual practice, magnesium alloy parts are most widely used under conditions of atmospheric exposure where the corrosive environment is a poor electrical conductor, resulting in very small galvanic current. For example, die cast magnesium alloy generator parts with steel inserts and bolted connections showed little galvanic attack after 15 years of exposure in an industrial atmosphere, whereas parts made entirely of steel rusted under the same conditions.

PROTECTION OF ASSEMBLIES

Contact of magnesium with magnesium, with dissimilar metals and with wood provides potential sources of corrosion of various types unless the magnesium is adequately protected. The protection required varies with the severity of the corrosive environment and the material with which the magnesium is coupled. Although corrosive

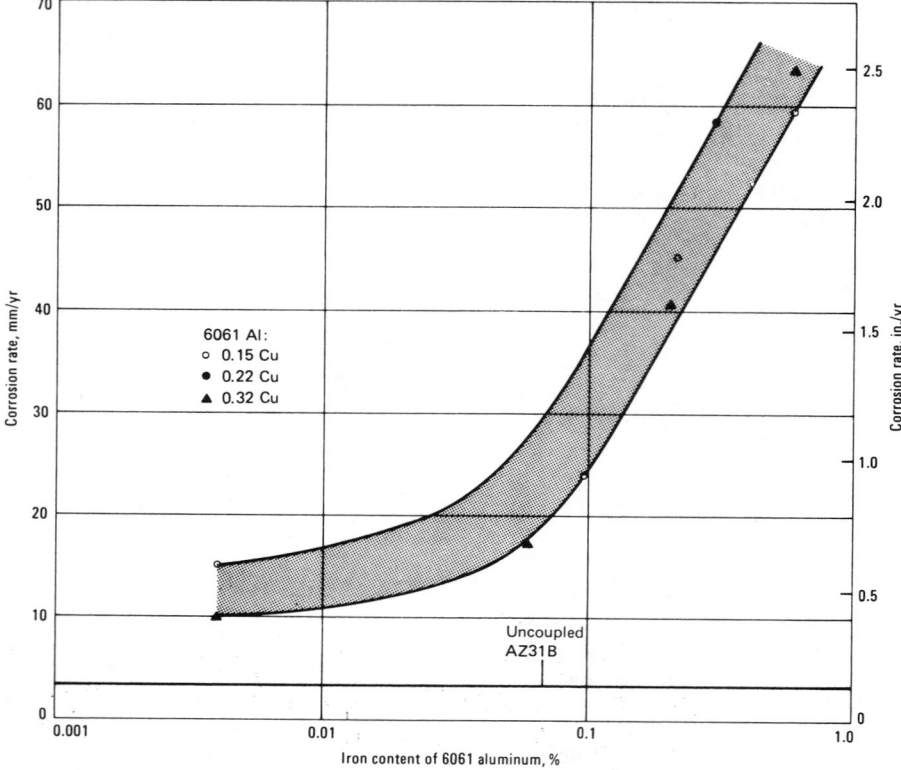

Corrosion rate of uncoupled AZ31B also shown for comparison.

Fig. 5. Effect of iron content on corrosion rate of magnesium alloy AZ31B coupled with aluminum alloy 6061-T6 in 3% NaCl solution

attack from any source can jeopardize the satisfactory performance of magnesium, attack resulting from metal-to-metal contact is probably the most detrimental. Applications conducive to severe attack include those involving continuous outdoor exposure or wetting of the magnesium with corrosive materials such as chloride-laden splash. For indoor use where condensation is not likely, no protection is necesary; there are many such applications. Ordinary paint coatings applied for decorative purposes provide adequate protection under normal indoor conditions.

Magnesium-to-Magnesium Assemblies. For all practical purposes, galvanic corrosion between magnesium alloys is negligible. However, good assembly practice dictates that the magnesium faying surfaces be given one or more coats of a chromate-pigmented primer.

Magnesium-to-wood assemblies present an unusual problem because of the absorbency of wood. If wood is wetted, it will hold the water in contact with the magnesium. As the water is absorbed, the natural acids of the wood tend to leach out, and continuous contact of the acids with the magnesium causes corrosion. To protect the magnesium from attack, the wood should be sealed with paint or varnish to prevent absorption of moisture and the faying surface of the magnesium should be treated as described above under "Magnesium-to-Magnesium Assemblies."

Magnesium-to-Dissimilar-Metal Assemblies. In order to minimize or prevent galvanic action between magnesium and dissimilar metals, it is necessary to employ one of the following procedures:

- Protect the dissimilar metal as well as the magnesium.

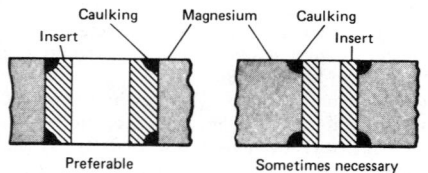

Fig. 6. Two methods of avoiding external contact between magnesium and dissimilar-metal inserts

- Use dissimilar metals that are compatible with magnesium.
- Separate one from the other so that the corroding medium cannot complete an electrical circuit.

Moisture-impervious films should be used to separate magnesium from dissimilar metals (often steel or aluminum). If either the dissimilar metal or the magnesium is coated with an unbroken film, no galvanic corrosion occurs.

Dissimilar metals that are compatible with magnesium, such as aluminum alloys 5052, 5056, 6053, 6061 and 6063, should be used for washers, shims, fasteners (rivets and bolts) and structural members.

Tin is the preferred coating material for steel and brass surfaces in contact with magnesium; it is more effective than cadmium and zinc coatings. However, tin, cadmium and zinc coatings only retard, and do not prevent, galvanic corrosion under the worst conditions. In some milder environments they are satisfactory.

Stainless steel, titanium, copper, Monel, and aluminum alloys such as 2024, Alclad 2024, 7075, Alclad 7075, and 3003 will corrode magnesium when coupled with it under corrosive conditions. Therefore, protection is required.

Proper design also calls for use of nonabsorbent tapes or sealing compounds in joints. Vinyl tapes as thin as 0.08 mm (0.003 in.) and other nonabsorbent tapes, such as rubber tape, can be used. (Cloth-supported tape is not recommended because the cloth acts as a wick.)

Fasteners and Inserts. Most assemblies require fasteners such as rivets, bolts or screws. Magnesium alloys are not used for making such fasteners, and thus choice of fastener material is important.

Very pure aluminum is almost completely compatible with magnesium. Alloy additions of magnesium, manganese or silicon to aluminum do not affect this compatibility. Alloy additions or impurities such as copper, nickel and iron, and to a lesser degree zinc, can adversely affect the compatibility. However, the adverse effect of these elements is suppressed to a large degree by the presence of magnesium in the aluminum alloy. As a consequence, aluminum alloy 5056 rivets are recommended for use with magnesium. Rivets of aluminum alloys 6053 and 6061 are almost as compatible with magnesium as 5056 rivets and may be used as substitutes (6053 rivets are preferable to 6061 rivets). Aluminum rivets should be anodized or chemically treated before use. Washers of alloy 5052 under cadmium-plated steel bolts are also recommended. Bolts and hardware of compatible aluminum alloys are preferred to coated steel bolts and hardware wherever it is possible to use them.

Steel and copper rivets, as well as steel, nickel, aluminum (other than alloys 5056, 6061 and 6053) and brass bolts and screws, should not be used bare in magnesium assemblies, because they are not compatible with the magnesium. In general, tin, zinc or cadmium plating of these parts is recommended, followed by chemical treatment to provide the plating with better paint-adhesion characteristics. However, such plating only retards, and does not prevent, galvanic corrosion.

Special organic coatings, such as baked vinyl plastisols, epoxies and high-temperature-resistant fluorinated hydrocarbon resin coatings, can be used for insulation of fasteners and inserts.

Most steel inserts used in magnesium parts should be plated with tin, cadmium or zinc. If severe service conditions are expected, annular grooves 3.2 mm ($^1/_8$ in.) or more in width should be provided and filled with caulking compound, as shown in Fig. 6. Preferably, these grooves should be formed by machining the inserts, but usually the small size of the inserts makes it necessary to counterbore the hole in the magnesium. Inserts that are well covered with grease or oil during service usually do not require additional protection.

The choice of materials for bimetallic joints is given in the following tabulation, listed in order of preference:

Magnesium-to-aluminum assemblies:
1 5056 (wire and rivets)
2 6061 (extrusions and sheet)
3 5052 (sheet)
4 6053 (extrusions and rivets)

Magnesium-to-steel assemblies:
1 Tin-coated steel
2 Cadmium-coated steel
3 Zinc-coated steel

9 TITANIUM

Edited by Matthew J. Donachie, Jr., The Hartford Graduate Center

This section was condensed from Metals Handbook, Ninth Edition, Volume 3, Properties and Selection: Stainless Steels, Tool Materials and Special-Purpose Metals, pages 351 to 417. For more detailed information on the topics covered herein, the reader is referred to the larger work.

Introduction and Overview

TITANIUM is a low-density element (approximately 60% of the density of steel) which can be highly strengthened by alloying and deformation processing. Titanium is nonmagnetic and has good heat-transfer properties. Its coefficient of thermal expansion is somewhat lower than that of steels and less than half that of aluminum. Titanium and its alloys have melting points higher than those of steels, but maximum useful temperatures for structural applications generally range from 427 to 538 °C (800 to 1000 °F). (Titanium aluminide alloys show promise for applications above these temperatures.) Titanium has the ability to passivate, and thereby exhibit a high degree of immunity to attack by most mineral acids and chlorides. Titanium is nontoxic and generally biologically compatible with human tissues and bones. The combination of high strength, stiffness, good toughness, low density and good corrosion resistance provided by various titanium alloys at very low to moderately elevated temperatures allows weight savings in aerospace structures and other high-performance applications. The excellent corrosion resistance and biocompatibility coupled with good strengths make titanium and its alloys useful in chemical and petrochemical applications, marine environments and biomaterials applications.

Titanium and its alloys are used primarily in two areas of application where the unique characteristics of these metals justify their selection: corrosion-resistant service and strength-efficient structures. For these two diverse areas, selection criteria differ markedly. Corrosion applications normally utilize low-strength "unalloyed" titanium mill products fabricated into tanks, heat exchangers or reactor vessels for chemical-processing, desalination or power-generation plants. In contrast, high-performance applications typically utilize high-strength titanium alloys in a very selective manner depending on factors such as thermal environment, loading parameters, available product forms, fabrication characteristics and inspection and/or reliability requirements. As a result of their specialized usage, alloys for high-performance applications normally are processed to more stringent and costly requirements than "unalloyed" titanium for corrosion service.

Historically, titanium alloys have been used instead of iron or nickel alloys in aerospace applications because titanium saves weight in highly loaded components that operate at low to moderately elevated temperatures. Many titanium alloys have been custom designed to have optimum tensile, compressive and/or creep strength at selected temperatures and, at the same time, to have sufficient workability to be fabricated into mill products suitable for specific applications. During the life of the titanium industry, various compositions have had transient usage, but one alloy, Ti-6Al-4V, has been consistently responsible for about 45% of industry application. Ti-6Al-4V is unique in that it combines attractive properties with inherent workability (which allows it to be produced in all types of mill products, in both large and small sizes), good shop fabricability (which allows the mill products to be made into complex hardware) and the production experience and commercial availability that lead to reliable and economic usage. Thus Ti-6Al-4V has become the standard alloy against which other alloys must be compared when selecting a titanium alloy (or custom designing one) for a specific application. Ti-6Al-4V also is the standard alloy selected for castings that must exhibit superior strength.

Applications

Strength efficiency, fatigue-crack-growth rate and fracture toughness, plus manufacturing considerations such as welding and forming requirements, normally provide the criteria that determine the alloy composition, structure (alpha, alpha-beta or beta), heat treatment (some variant of either annealing or solution treating and aging), and level of process control selected or prescribed for structural titanium alloy applications. However, for lightly loaded structures, where titanium is normally selected because it offers greater resistance to temperature effects than aluminum, commercial availability of required mill products and ease of fabrication may dictate selection. Here,

one of the grades of unalloyed titanium usually is chosen. In some cases, corrosion resistance, not strength or temperature resistance, may be the major factor in selection of a titanium alloy.

Rotating components in gas turbine engines require titanium alloys that maximize strength efficiency and metallurgical stability at elevated temperatures. These alloys also must exhibit low creep rates along with predictable behavior in stress rupture and low-cycle fatigue. To reproducibly provide these properties, stringent user requirements are specified to ensure controlled homogeneous microstructures and total freedom from melting imperfections such as alpha segregation, high-density or low-density tramp inclusions, and unhealed ingot porosity or pipe. Currently, Ti-6Al-4V, Ti-5Al-2.5Sn, Ti-6Al-2Sn-4Zr-2Mo and Ti-6Al-2Sn-4Zr-6Mo are some alloys used in gas turbine engine applications.

Aerospace pressure vessels similarly require optimized strength efficiency in addition to auxiliary properties such as weldability and predictable fracture toughness at cryogenic to moderately elevated temperatures. To provide this combination of properties, stringent user specifications require controlled microstructures and freedom from melting imperfections. For cryogenic applications, the interstitial elements oxygen, nitrogen and carbon are carefully controlled to improve ductility and fracture toughness. For these applications, the basic titanium alloy Ti-6Al-4V (or Ti-6Al-4V-ELI), processed to either the annealed or the solution treated and aged (STA) condition, is widely used. Ti-5Al-2.5Sn-ELI is an attractive alternative.

Aircraft structural applications as well as high-performance automotive and marine applications require corrosion resistance and high strength efficiency. The latter normally is achieved by judicious alloy selection combined with close control of mill processing; corrosion resistance is primarily a function of composition. Ti-6Al-4V and Ti-6Al-6V-2Sn are two of the alloys currently used for high-performance applications, but beta alloys have been proposed for springs. Titanium alloys are ideal spring materials because of the lower modulus and high strength possible.

Optic-system support structures are a little-known but very important structural application for titanium. Complex castings are used in surveillance and guidance systems for aircraft and missiles to support the optics where wide temperature variations are encountered in service. The chief reason for selecting titanium for this application is that the thermal-expansion coefficient of titanium most closely matches that of the optics.

Titanium alloys have been employed in submarine hulls and submersible research vehicles. Because of favorable heat-transfer behavior, thermal stability and corrosion resistance, titanium is being used in heat exchangers for power generation. Corrosion resistance also is a factor in the selection of titanium for use in outer containers for high-level solidified nuclear waste.

Economic considerations normally determine whether titanium alloys will be used for corrosion service. Capital expenditures for titanium equipment generally are higher than for equipment fabricated from competing materials such as stainless steel, brass, bronze, copper nickel or carbon steel. As a result, titanium equipment must yield lower operating costs, longer life or reduced maintenance to justify selection based on lower total-life-cycle cost. Commercially pure titanium satisfies the basic requirements for corrosion service.

Because of its unique corrosion behavior, titanium is used extensively in prosthetic devices such as artificial heart pumps, pacemaker cases, heart-valve parts and load-bearing bone or hip-joint replacements or bone splints. (In general, body fluids are chloride brines that have pH values from 7.4 into the acidic range and also contain a variety of organic acids and other components—media to which titanium is totally immune.)

PHYSICAL METALLURGY

Titanium undergoes an allotropic transformation at about 885 °C (1625 °F), changing from a close-packed hexagonal crystal structure (alpha phase) to a body-centered cubic crystal structure (beta phase). The transformation temperature (beta transus—completion of transformation to beta on heating) is strongly influenced by the interstitial elements oxygen, nitrogen and carbon (alpha stabilizers), which raise the transformation temperature; by hydrogen (beta stabilizer), which lowers the transformation temperature; and by metallic impurity or alloying elements, which may either raise or lower the transformation temperature.

Depending on their microstructure, titanium alloys fall into one of four classes: alpha, near-alpha, alpha-beta or beta. These classes denote the general type of microstructure after processing. An alpha alloy will form no beta phase. A near-alpha or super-alpha alloy will form limited beta phase on heating and may appear microstructurally similar to an alpha alloy. An alpha-beta alloy will consist of alpha and retained or transformed beta, and a beta alloy will tend to retain the beta phase on initial cooling to room temperature but will precipitate secondary phases during heat treatment.

Effects of Alloying Elements

The role of the interstitial elements oxygen, nitrogen and carbon was described above. The substitutional alloying elements also play an impor-

tant role in controlling the microstructure and properties of titanium alloys.

Tantalum, vanadium and niobium are beta isomorphous (i.e., have similar phase relations) with body-centered cubic titanium. Titanium does not form intermetallic compounds with the beta isomorphous elements. Eutectoid systems are formed with chromium, iron, copper, nickel, palladium, cobalt, manganese and certain other transition metals. These elements have low solubility in alpha titanium and decrease the transformation temperature. They usually are added to alloys in combination with one or more of the beta isomorphous elements to stabilize the beta phase and prevent or minimize formation of intermetallic compounds, which might occur during service at elevated temperature.

Zirconium and hafnium are unique in that they are isomorphous with both the alpha and beta phases of titanium. Tin and aluminum have significant solubility in both alpha and beta phases. Aluminum increases the transformation temperature significantly whereas tin lowers it slightly. Aluminum, tin and zirconium commonly are used together in alpha and near-alpha alloys. In alpha-beta alloys, these elements are distributed approximately equally between the alpha and beta phases. Almost all commercial titanium alloys contain one or more of these three elements because they are soluble in both alpha and beta phases, and particularly because they improve creep strength in the alpha phase.

Many more elements are soluble in beta titanium than in alpha. Beta isomorphous alloying elements are preferred as additions because they do not form intermetallic compounds. However, iron, chromium, manganese and other compound formers sometimes are used in beta-rich alpha-beta alloys or in beta alloys, because they are strong beta stabilizers and improve hardenability and response to heat treatment. Nickel, molybdenum and palladium improve corrosion resistance of unalloyed titanium in certain media.

Secondary Phases and Martensitic Transformations

Intermetallic compounds and other secondary phases are formed in titanium alloy systems. The more important phases, historically, have been omega (ω) and alpha-2 (α_2). Omega phase has not proven to be a constituent in commercial systems using present-day processing practice. Alpha-2 has been considered to be a factor in some cases of stress-corrosion cracking. Most present interest in α_2 centers on its use as a matrix for a high-temperature titanium alloy.

Beta phase decomposes, usually by martensitic transformations, in the alpha-beta alloys. The beta-to-martensite transition is responsible for the acicular structure in quenched and/or quenched and aged titanium alloys. Hardenability of a titanium alloy refers, generally, to the ability of an alloy to permit full transformation of the alloy to martensitic phases or to retain beta to room temperature. Alpha prime (α') and alpha double prime (α'') martensites are brought out by cooling and decompose on aging to alpha and beta phases.

ALLOY SYSTEMS

There are several grades of unalloyed titanium. The primary difference between grades is oxygen and iron content. Grades of higher purity (lower interstitial content) are lower in strength, hardness and transformation temperature than those higher in interstitial content. The high sol-

ubility of the interstitial elements oxygen and nitrogen makes titanium unique among metals, and also creates problems not of concern in most other metals. For example, heating titanium in air at high temperature results not only in oxidation but also in solid-solution hardening of the surface as a result of inward diffusion of oxygen (and nitrogen). A surface-hardened zone of "alpha-case" (or "air contamination layer") is formed. Normally, this layer is removed by machining, chemical milling or other mechanical means prior to placing a part in service, because the presence of alpha-case reduces fatigue strength and ductility.

Table 1 lists some commercial and semicommercial titanium grades and alloys currently available, which are subdivided into four groups: unalloyed (commercially pure) grades, alpha and near-alpha alloys, alpha-beta alloys, and beta alloys.

Ti-6Al-4V is the most widely used titanium alloy, accounting for about 45% of total titanium production. Unalloyed grades comprise about 30% of production, and all other alloys combined comprise the remaining 25%.

Selection of an unalloyed grade of titanium, an alpha or near-alpha alloy, an alpha-beta alloy or a beta alloy depends on desired mechanical properties, service requirements, cost considerations and the other factors that enter into any material-selection process.

Unalloyed Titanium

Unalloyed titanium usually is selected for its excellent corrosion resistance, especially in applications where high strength is not required. Yield strengths of unalloyed (commercially pure) grades (see Table 1) vary from 170 MPa (25 ksi) to 480 MPa (70 ksi) simply as a result of variations in the interstitial and impurity levels. Oxygen and iron are the primary variants in these grades; strength increases with increasing oxygen and iron contents.

Alpha and Near-Alpha Alloys

Alpha alloys that contain aluminum, tin and/or zirconium are preferred for high-temperature as well as cryogenic applications. Alpha-rich alloys generally are more resistant to creep at high temperature than alpha-beta or beta alloys. The extra-low-interstitial alpha alloys (ELI grades) retain ductility and toughness at cryogenic temperatures, and Ti-5Al-2.5Sn-ELI has been used extensively in such applications.

Unlike alpha-beta and beta alloys, alpha alloys cannot be strengthened by heat treatment. Generally, alpha alloys are annealed or recrystallized to remove residual stresses induced by cold working. Alpha alloys have good weldability because they are insensitive to heat treatment. They generally have poorer forgeability and narrower forging-temperature ranges than alpha-beta or beta alloys, particularly at temperatures below the beta transus. This poorer forgeability is manifested by a greater tendency for center bursts or surface cracks to occur, which means that small reduction steps and frequent reheats must be incorporated in forging schedules.

Alpha alloys that contain small additions of beta stabilizers (Ti-8Al-1Mo-1V or Ti-6Al-2Nb-1Ta-0.8Mo, for example) sometimes have been classed as "super-alpha" or "near-alpha" alloys. Although they contain some retained beta phase, these alloys consist primarily of alpha and behave more like conventional alpha alloys than alpha-beta alloys.

Table 1. Summary of commercial and semicommercial grades and alloys of titanium

Designation	Tensile strength (min) MPa	ksi	0.2% yield strength (min) MPa	ksi	N (max)	C (max)	H (max)	Fe (max)	O (max)	Al	Sn	Zr	Mo	Others
Unalloyed grades														
ASTM Grade 1	240	35	170	25	0.03	0.10	0.015	0.20	0.18	...	...	...	...	...
ASTM Grade 2	340	50	280	40	0.03	0.10	0.015	0.30	0.25	...	...	...	...	...
ASTM Grade 3	450	65	380	55	0.05	0.10	0.015	0.30	0.35	...	...	...	...	...
ASTM Grade 4	550	80	480	70	0.05	0.10	0.015	0.50	0.40	...	...	...	...	...
ASTM Grade 7	340	50	280	40	0.03	0.10	0.015	0.30	0.25	...	...	...	...	0.2 Pd
Alpha and near-alpha alloys														
Ti-0.3Mo-0.8Ni	480	70	380	55	0.03	0.10	0.015	0.30	0.25	...	...	...	0.3	0.8 Ni
Ti-5Al-2.5Sn	790	115	760	110	0.05	0.08	0.02	0.50	0.20	5	2.5	...	...	...
Ti-5Al-2.5Sn-ELI	690	100	620	90	0.07	0.08	0.0125	0.25	0.12	5	2.5	...	...	...
Ti-8Al-1Mo-1V	900	130	830	120	0.05	0.08	0.015	0.30	0.12	8	...	...	1	1 V
Ti-6Al-2Sn-4Zr-2Mo	900	130	830	120	0.05	0.05	0.0125	0.25	0.15	6	2	4	2	...
Ti-6Al-2Nb-1Ta-0.8Mo	790	115	690	100	0.02	0.03	0.0125	0.12	0.10	6	...	...	1	2 Nb, 1 Ta
Ti-2.25Al-11Sn-5Zr-1Mo	1000	145	900	130	0.04	0.04	0.008	0.12	0.17	2.25	11.0	5.0	1.0	0.2 Si
Ti-5Al-5Sn-2Zr-2Mo(a)	900	130	830	120	0.03	0.05	0.0125	0.15	0.13	5	5	2	2	0.25 Si
Alpha-beta alloys														
Ti-6Al-4V(b)	900	130	830	120	0.05	0.10	0.0125	0.30	0.20	6.0	...	...	...	4.0 V
Ti-6Al-4V-ELI(b)	830	120	760	110	0.05	0.08	0.0125	0.25	0.13	6.0	...	...	...	4.0 V
Ti-6Al-6V-2Sn(b)	1030	150	970	140	0.04	0.05	0.015	1.0	0.20	6.0	2.0	...	...	0.75 Cu, 6.0 V
Ti-8Mn(b)	860	125	760	110	0.05	0.08	0.015	0.50	0.20	...	...	...	...	8.0 Mn
Ti-7Al-4Mo(b)	1030	150	970	140	0.05	0.10	0.013	0.30	0.20	7.0	...	...	4.0	...
Ti-6Al-2Sn-4Zr-6Mo(c)	1170	170	1100	160	0.04	0.04	0.0125	0.15	0.15	6.0	2.0	4.0	6.0	...
Ti-5Al-2Sn-2Zr-4Mo-4Cr(a)(c)	1125	163	1055	153	0.04	0.05	0.0125	0.30	0.13	5.0	2.0	2.0	4.0	4.0 Cr
Ti-6Al-2Sn-2Zr-2Mo-2Cr(a)(b)	1030	150	970	140	0.03	0.05	0.0125	0.25	0.14	5.7	2.0	2.0	2.0	2.0 Cr, 0.25 Si
Ti-10V-2Fe-3Al(a)(c)	1170	170	1100	160	0.05	0.05	0.015	2.5	0.16	3.0	...	...	...	10.0 V
Ti-3Al-2.5V(d)	620	90	520	75	0.015	0.05	0.015	0.30	0.12	3.0	...	...	...	2.5 V
Beta alloys														
Ti-13V-11Cr-3Al(c)	1170	170	1100	160	0.05	0.05	0.025	0.35	0.17	3.0	...	...	...	11.0 Cr, 13.0 V
Ti-8Mo-8V-2Fe-3Al(a)(c)	1170	170	1100	160	0.05	0.05	0.015	2.5	0.17	3.0	...	...	8.0	8.0 V
Ti-3Al-8V-6Cr-4Mo-4Zr(a)(b)	900	130	830	120	0.03	0.05	0.020	0.25	0.12	3.0	...	4.0	4.0	6.0 Cr, 8.0 V
Ti-11.5Mo-6Zr-4.5Sn(b)	690	100	620	90	0.05	0.10	0.020	0.35	0.18	...	4.5	6.0	11.5	...

(a) Semicommercial alloy; mechanical properties and composition limits subject to negotiation with suppliers. (b) Mechanical properties given for annealed condition; may be solution treated and aged to increase strength. (c) Mechanical properties given for solution treated and aged condition; alloy not normally applied in annealed condition. Properties may be sensitive to section size and processing. (d) Primarily a tubing alloy; may be cold drawn to increase strength.

Alpha-Beta Alloys

Alpha-beta alloys contain one or more alpha stabilizers or alpha-soluble elements plus one or more beta stabilizers. These alloys retain more beta phase after solution treatment than do near-alpha alloys, the specific amount depending on the quantity of beta stabilizers present and on heat treatment.

Alpha-beta alloys can be strengthened by solution treating and aging. Solution treating usually is done at a temperature high in the two-phase alpha-beta field, and is followed by quenching in water, oil or other suitable quenchant. As a result of quenching, the beta phase present at the solution treating temperature may be retained or may be partly transformed during cooling by either martensitic transformation or nucleation and growth. The specific response depends on alloy composition, solution treating temperature (beta-phase composition at the solution temperature), cooling rate and section size. Solution treatment is followed by aging, normally at 480 to 650 °C (900 to 1200 °F), to precipitate alpha and produce a fine mixture of alpha and beta in the retained or transformed beta phase. Transformation kinetics, transformation products and specific response of a given alloy can be quite complex; a detailed review of the subject is beyond the scope of this article.

Solution treating and aging can increase the strength of alpha-beta alloys 30 to 50%, or more, over the annealed or over-aged condition. Response to solution treating and aging depends on section size; alloys relatively low in beta stabilizers (Ti-6Al-4V, for example) have poor hard-enability and must be quenched rapidly to achieve significant strengthening. For Ti-6Al-4V, the cooling rate of a water quench is not rapid enough to significantly harden sections thicker than about 25 mm (1 in.). As the content of beta stabilizers increases, hardenability increases; Ti-5Al-2Sn-2Zr-4Mo-4Cr, for example, can be through hardened with relatively uniform response throughout sections up to 150 mm (6 in.) thick. For some alloys of intermediate beta-stabilizer content, the surface of a relatively thick section can be strengthened, but the core may be 10 to 20% lower in hardness and strength. The strength that can be achieved by heat treatment is also a function of the volume fraction of beta phase present at the solution treating temperature. Alloy composition, solution temperature and aging conditions must be carefully selected and balanced to produce the desired mechanical properties in the final product.

Although the ability of alpha-beta alloys to be precipitation hardened has been studied in laboratory programs since the early days of the titanium industry, there have been relatively few production applications of solution treated and precipitation (age) hardened alloys. This situation appears to be changing, because alloys such as Ti-6Al-2Sn-4Zr-6Mo, Ti-5Al-2Sn-2Zr-4Mo-4Cr and certain high-hardenability beta alloys have been developed specifically to be age hardened for improved strength—about 30 to 40% above that of annealed alloys.

Beta Alloys

Beta alloys are richer in beta stabilizers and leaner in alpha stabilizers than alpha-beta alloys.

They are characterized by high hardenability, with beta phase completely retained on air cooling of thin sections or water quenching of thick sections. Beta alloys have excellent forgeability, and in sheet form can be cold formed more readily than high-strength alpha-beta or alpha alloys. After solution treating, beta alloys are aged at temperatures of 450 to 650 °C (850 to 1200 °F) to partially transform the beta phase to alpha. The alpha forms as finely dispersed particles in the retained beta, and strength levels comparable or superior to those of aged alpha-beta alloys can be attained. The chief disadvantages of beta alloys in comparison with alpha-beta alloys are higher density, lower creep strength and lower tensile ductility in the aged condition. Although tensile ductility is lower, the fracture toughness of an aged beta alloy generally is higher than that of an aged alpha-beta alloy of comparable yield strength.

In the solution treated condition (100% retained beta), beta alloys have good ductility and toughness, relatively low strength, and excellent formability. Solution treated beta alloys begin to precipitate alpha phase at slightly elevated temperatures, and thus are unsuitable for elevated-temperature service without prior stabilization or over-aging treatment.

Beta alloys, despite the name, actually are metastable, because cold work at ambient temperature or heating to a slightly elevated temperature can cause partial transformation to alpha. The principal advantages of beta alloys are that they have high hardenability, excellent forgeability, and good cold formability in the solution treated condition.

Processing

TITANIUM metal passes through three major steps during processing from ore to finished product: (a) reduction of titanium ore to a porous form of titanium metal called "sponge;" (b) melting of sponge to form ingot; and (c) remelting and casting into finished shape, or primary fabrication, in which ingots are converted into general mill products followed by secondary fabrication of finished shapes from mill products. Machining of cast structures or mill products is required, and welding invariably is needed in casting repair or build-up of structures from castings or mill products. Powder metallurgy processing is being considered for some designs. At each of these steps, the mechanical and physical properties of the titanium in the finished shape may be affected by any one of several factors, or by a combination of factors. Among the most important are (a) amounts of specific alloying elements and impurities, (b) melting process used to make ingot, (c) casting process and volume of cast article plus use of densification techniques such as hot isostatic pressing (HIP) to reduce casting porosity, (d) method for mechanically working ingots into mill products and (e) the final step employed in working, fabrication or heat treatment.

MELTING

Raw Materials

Control of raw materials is extremely important in producing titanium and its alloys, because there are many elements of which small amounts can have major effects on the properties of these metals in finished form. The raw materials used in producing titanium are: titanium in the form of sponge metal, alloying elements, and reclaimed titanium scrap (usually called "revert").

Titanium sponge must meet stringent specifications for control of ingot composition. Most important, sponge must not contain hard, brittle and refractory titanium oxide, titanium nitride or complex titanium oxynitride particles that, if retained through subsequent melting operations, could act as crack-initiation sites in the final product.

Carbon, nitrogen, oxygen, silicon and iron commonly are found as residual elements in sponge. These elements must be held to acceptably low levels because they raise the strength and lower the ductility of the final product.

Alloying-element purity is as important as purity of sponge and must be controlled with the same degree of care to avoid undesirable residual elements—especially those that can form refractory or high-density inclusions in the titanium matrix.

Basically, oxygen and iron contents determine strength levels of commercially pure titanium (ASTM and ASME grades 1, 2, 3 and 4) and the differences in mechanical properties between extra-low-interstitial (ELI) grades and standard grades of titanium alloys. (Table 1 illustrates this effect.) In higher-strength grades, oxygen and iron are intentionally added to the residual amounts already in the sponge to provide extra strength. On the other hand, carbon and nitrogen usually are held to minimum residual levels to avoid embrittlement.

Reclaimed Scrap. Use of reclaimed scrap makes production of ingot titanium more economical than production solely from sponge. If properly controlled, addition of scrap (commonly referred to as "revert") is fully acceptable—even in materials for critical structural applications, such as rotating components for jet engines.

All forms of scrap can be remelted—machining chips, cut sheet, trim stock and chunks. To be utilized properly, scrap must be thoroughly cleaned and carefully sorted by alloy and by purity before being remelted. During cleaning, surface scale must be removed, because adding titanium scale to the melt could produce refractory inclusions or excessive porosity in the ingot. Machining chips from fabricators who use carbide tools are acceptable for remelting only if all carbide particles adhering to the chips are removed; otherwise, hard high-density inclusions could result. Improper segregation of alloy revert would produce off-composition alloys and could degrade the properties of the resulting metal.

Ingot

Melting practice for most titanium and titanium alloy ingot is to melt twice in an electric-arc furnace under vacuum—a procedure known as the "double consumable-electrode vacuum-melting process." In this two-stage process, titanium sponge, revert and alloy additions are initially mechanically consolidated and then are melted together to form ingot. Ingots from the first melt are used as the consumable electrodes for second-stage melting. Processes other than consumable-electrode arc melting are used in some instances for first-stage melting of ingot for noncritical applications. Usually, all melting is done under vacuum, but in any event the final stage of melting must be done by the consumable-electrode vacuum-arc process.

Double melting is considered necessary for all applications to ensure an acceptable degree of homogeneity in the resulting product. Triple melting is used to achieve better uniformity. Triple melting also reduces oxygen-rich or nitrogen-rich inclusions in the microstructure to a very low level by providing an additional melting operation to dissolve them.

Segregation and other compositional variations directly affect the final properties of mill products.

Melting in a vacuum reduces the hydrogen content of titanium and essentially removes other volatiles. This tends to result in high purity in the cast ingot. However, anomalous operating factors such as air leaks, water leaks, arc-outs, or even large variations in power level affect both the soundness and the homogeneity of the final product.

Still another factor is ingot size. Normally, ingots are 650 to 900 mm (26 to 36 in.) in diameter and weigh 3600 to 6800 kg (8000 to 15 000 lb). Larger ingots are economically advantageous to use and are important in obtaining refined macrostructures and microstructures in very large sections, such as billets with diameters of 400 mm (16 in.) or greater. Ingots up to 1000 mm (40 in.) in diameter and weighing more than 9000 kg (20 000 lb) have been melted successfully, but there appear to be limitations on the improvements that can be achieved by producing large ingots due to increasing tendency for segregation with increasing ingot size.

Segregation in titanium ingot must be controlled because it leads to several different types of imperfections that cannot be readily eliminated either by homogenizing heat treatments or by combinations of heat treatment and primary mill processing.

Type I imperfections, usually called "high interstitial defects," are regions of interstitially stabilized alpha phase that have substantially higher hardness and lower ductility than the surrounding material, and that also exhibit a higher beta transus temperature. They arise from very high nitrogen or oxygen concentrations in sponge, master alloy or revert. Type I imperfections frequently, but not always, are associated with voids or cracks. Although type I imperfections sometimes are referred to as "low-density inclusions," they often are of higher density than is normal for the alloy.

Type II imperfections, sometimes called "high-aluminum defects," are abnormally stabilized alpha-phase areas that may extend across several beta grains. Type II imperfections are caused by segregation of metallic alpha stabilizers, such as aluminum, and contain an excessively high proportion of primary alpha having a microhardness only slightly higher than that of the adjacent matrix. Type II imperfections sometimes are accompanied by adjacent stringers of beta—areas low in both aluminum content and hardness. This condition is generally associated with closed solidification pipe into which alloy constituents of high vapor pressure migrate, only to be incorporated into the microstructure during primary mill fabrication. Stringers normally occur in the top portions of ingots and can be detected by macroetching or anodized blue etching. Material containing stringers usually must undergo me-

Table 1. Tensile properties of annealed titanium sheet as influenced by oxygen and iron contents

Material	Maximum impurity content, % Oxygen	Iron	Minimum tensile strength MPa	ksi	Minimum yield strength(a) MPa	ksi
Unalloyed Ti, grade 1	0.18	0.20	240	35	170	25
Unalloyed Ti, grade 2	0.25	0.30	345	50	275	40
Unalloyed Ti, grade 3	0.35	0.30	450	65	380	55
Unalloyed Ti, grade 4	0.40	0.50	655	95	485	70
Ti-6Al-4V	0.20	0.30	925	134	870	126
Ti-6Al-4V-ELI	0.13	0.25	900	130	830	120
Ti-5Al-2.5Sn	0.20	0.50	830	120	780	113
Ti-5Al-2.5Sn-ELI	0.12	0.25	690	100	655	95

(a) At 0.2% offset.

tallographic review to ensure that the indications revealed by etching are not artifacts.

Beta flecks, another type of imperfection, are small regions of stabilized beta in material that has been alpha-beta processed and heat treated. In size, they are equal to or larger than prior beta grains. Beta flecks are either devoid of primary alpha or contain less than some specified minimum level of primary alpha. They are caused by localized regions either abnormally high in beta-stabilizer content or abnormally low in alpha-stabilizer content. Beta flecks are attributed to microsegregation during solidification of ingots of alloys that contain strong beta stabilizers. They are most often found in products made from large-diameter ingots. Beta flecks also may be found in beta-lean alloys such as Ti-6Al-4V that have been heated to a temperature near the beta transus during processing.

Type I and type II imperfections are not acceptable in aircraft-grade titanium because they degrade critical design properties. Beta flecks are not considered harmful in alloys lean in beta stabilizers if they are to be used in the annealed condition. On the other hand, they constitute regions that incompletely respond to heat treatment, and for this reason microstructural standards have been established for allowable limits on beta flecks in various alpha-beta alloys. Beta flecks are more objectionable in beta-rich alpha-beta alloys than in leaner alloys.

PRIMARY FABRICATION

Primary fabrication includes all operations that convert ingot into general mill products—billet, bar, plate, sheet, strip, extrusions, tube and wire. These mill products can be readily utilized in secondary manufacture of parts and structures.

Primary fabrication is very important in establishing final properties, because many secondary fabrication operations may have little or no effect on metallurgical characteristics. Some secondary fabrication processes, such as forging and ring rolling, do impart sufficient reduction to play the major role in establishing material properties.

Reduction to Billet

Generally, the first breakdown of production ingot is a press cogging operation done in the beta temperature range. Modern processes utilize substantial amounts of working below the beta transus to produce billets with refined structures. These processes are carried out at temperatures high in the alpha region to allow greater reduction and improved grain refinement with a minimum of surface rupturing. Where maximum fracture toughness is required, beta processing (or alpha-beta processing followed by beta heat treatment) generally is preferred.

Final tensile properties of alpha-beta alloys are strongly influenced by the amount of processing in the alpha-beta field—both below the beta transus temperature and after recrystallization. Such processing increases the strength of high-alpha grades in large section sizes. With modern processing techniques, billet and forged sections readily meet specified tensile properties prior to final forging.

Properties of Rolled Bar, Plate and Sheet

Bars up to about 100 mm (4 in.) in diameter are unidirectionally rolled, and their properties commonly reflect total reduction in the alpha-beta

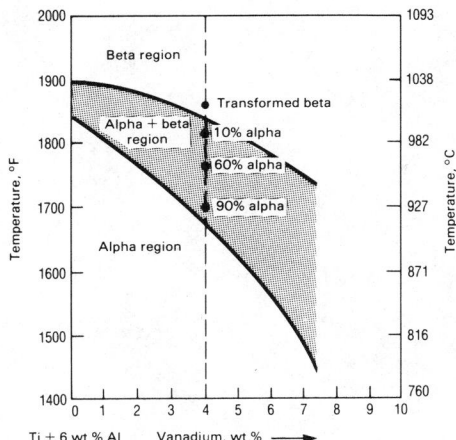

Fig. 1. Phase diagram that predicts results of forging or heat treatment practice

range. For example, a round bar 50 mm (2 in.) in diameter rolled from a Ti-6Al-4V billet 100 mm square typically is 140 to 170 MPa (20 to 25 ksi) lower in tensile strength than rod 7.8 mm ($^5/_{16}$ in.) in diameter rolled on a rod mill from a billet of the same size at the same rolling temperatures. For bars about 50 to 100 mm in diameter, strength may not decrease with section size, but transverse ductility and notched stress-rupture strength at room temperature do become lower. In diameters greater than about 75 to 100 mm (3 to 4 in.), annealed Ti-6Al-4V bars may not meet prescribed limits for stress-rupture at room temperature—1170 MPa (170 ksi) min to cause rupture of a notched specimen in 5 h—unless the material is given a special duplex anneal. Transverse ductility is lower in bars about 65 to 100 mm ($2^1/_2$ to 4 in.) in diameter because it is not possible to obtain the preferred texture throughout bars of this size.

Plate and sheet commonly exhibit the same tensile properties in both the transverse and longitudinal directions relative to the final rolling direction. With the precise control systems now available, proper texturing and directionality can be obtained in alpha-beta sheet by unidirectional rolling. These characteristics favorably affect tensile properties of Ti-6Al-4V sheet in various gages. Other properties, such as fatigue resistance, also are improved by this type of rolling. Directionality in properties is observed only as a slight drop in transverse ductility of plate greater

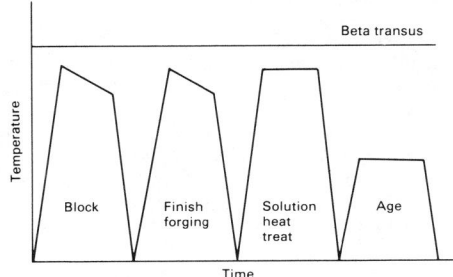

Fig. 2. Schematic diagram of a conventional forging and subsequent heat treatment sequence for producing alpha-beta structure

than 25 mm (1 in.) thick. For forming applications, some customers specify a maximum allowable difference between tensile strengths in the transverse and longitudinal directions.

SECONDARY FABRICATION

Secondary fabrication refers to manufacturing processes such as die forging, extrusion, hot and cold forming, machining, chemical milling and joining, all of which are used for producing finished parts from mill products and, in the case of machining, chemical milling and joining, for producing finished parts from castings. Each of these processes may strongly influence properties of titanium and its alloys, either alone or by interacting with effects of processes to which the metal has previously been subjected.

Die Forging

One of the main purposes of die forging, in addition to shape control, is to obtain a combination of mechanical properties that generally does not exist in bar or billet. Tensile strength, creep resistance, fatigue strength and toughness all may be better in forgings than in bar or other forms.

Forging is a common method of producing wrought titanium alloy articles. Forging sequences and subsequent heat treatment can be used to control the microstructure and resulting properties of the product. Forging is more than just a shapemaking process. The key to successful forging and heat treatment is the beta transus temperature. Figure 1 shows, schematically, the possible locations for temperature of forging and/or heat treatment of a typical alpha-beta alloy such as Ti-6Al-4V. The higher the processing temperature in the alpha-plus-beta region, the more beta is available to transform on cooling. On quenching from above the beta transus, a completely transformed, acicular structure arises. The exact form of the globular (equiaxed) alpha and the transformed beta structures produced by processing depends on the exact location of the beta transus, which varies from heat to heat, and the degree and nature of deformation produced. Section size is important, and the number of working operations can be significant. Conventional forging may require two or three operations, whereas isothermal forging may require only one. A schematic of a conventional forging and subsequent heat treatment sequence is shown in Fig. 2. The solution heat treatment offers a chance to modify or tune the as-forged microstructure, while the aging cycle modifies the transformed beta structures to an optimum dispersion.

Microstructural control is basic to successful processing of titanium alloys. Undesirable structures (grain-boundary alpha, beta fleck, "spaghetti" or elongated alpha) can interfere with optimum property development. Titanium ingot structures can carry over to affect the forged product. Beta processing, despite its adverse effects on some mechanical properties, can reduce forging costs, while isothermal forging offers a means of reducing forging pressures and/or improving die fill and part detail. Isothermal beta forging is finding use in the production of more creep-resistant components of titanium alloys.

Reheating of alpha-beta titanium alloys after hot working can substantially alter the microstructure. Careful attention must be paid to the development of microstructure right through the heat treatment steps. Superficially similar microstructures may not produce the same levels of

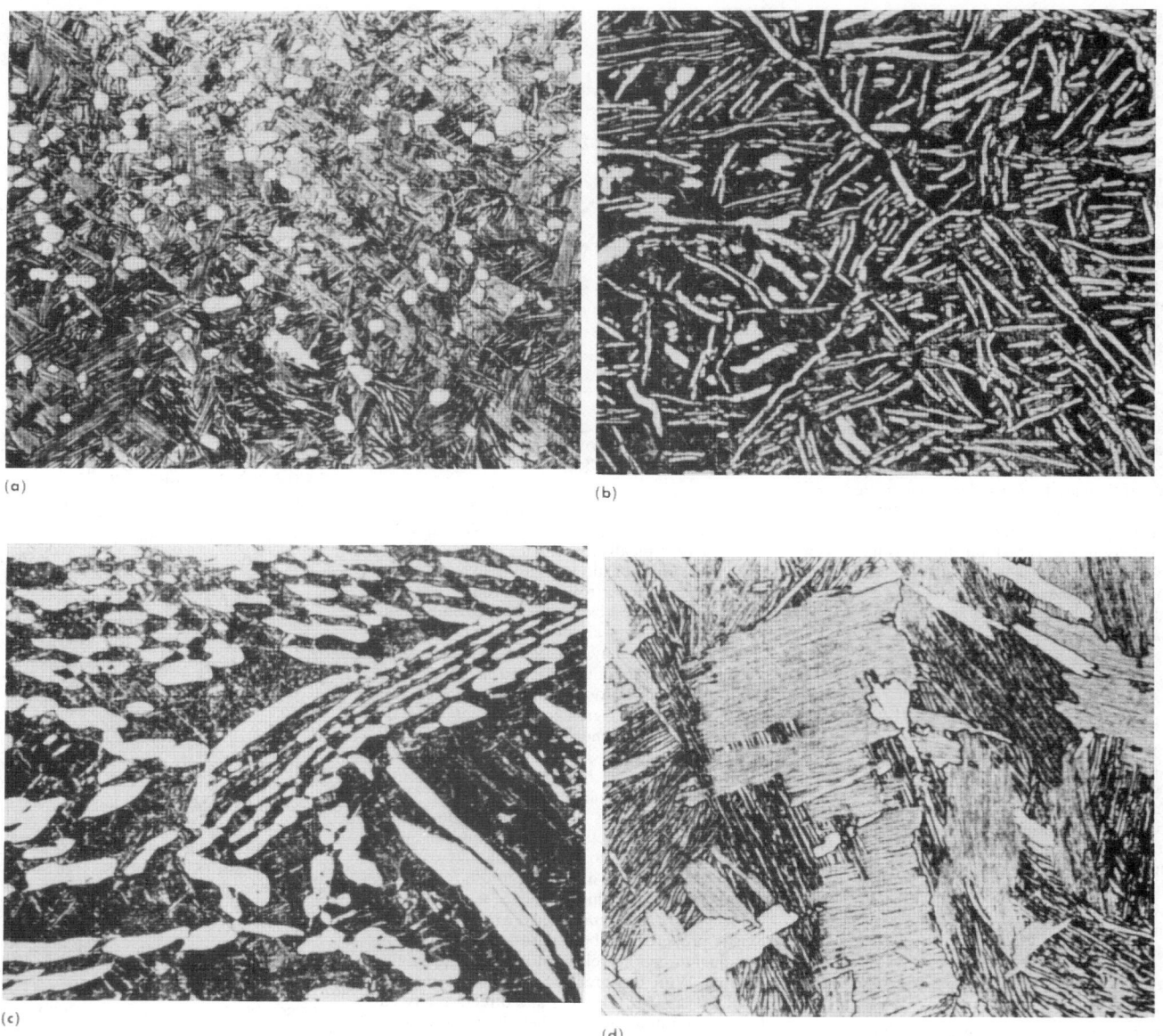

(a) 6% equiaxed primary alpha plus fine platelet alpha in Ti-6Al-4V alpha-beta forged, then annealed 2 h at 705 °C (1300 °F) and air cooled. (b) 23% elongated, partly broken up primary alpha plus grain-boundary alpha in Ti-6Al-4V, alpha-beta forged and water quenched, then annealed 2 h at 705 °C and air cooled. (c) 25% blocky (spaghetti) alpha plates plus very fine platelet alpha in Ti-6Al-4V alpha-beta forged from a spaghetti-alpha starting structure, then solution treated 1 h at 955 °C (1750 °F) and reannealed 2 h at 705 °C. (d) 92% alpha basket-weave structure in Ti-6Al-4V beta forged and slow cooled, then annealed 2 h at 705 °C. Structures in (a) and (b) produced excellent combinations of tensile properties, fatigue strength and fracture toughness. Structure in (c) produced very poor combination of mechanical properties. Structure in (d) produced good fracture toughness, but poorer tensile properties and fatigue resistance.

Fig. 3. Microstructures corresponding to different combinations of properties in Ti-6Al-4V forgings

mechanical properties. Solution heat treatment and aging of nonworked or insufficiently worked structures will not produce optimum strengths or toughness in titanium alloys.

The effects of different thermomechanical processing schedules on the mechanical properties and the corresponding structures of three titanium alloys—Ti-6Al-4V, Ti-6Al-6V-2Sn and Ti-6Al-2Sn-4Zr-6Mo—may be considered to illustrate the effects of processing schedules on properties. The microstructures of Ti-6Al-4V shown in Fig. 3 correspond to two of the schedules that produced good combinations of properties and two that produced inferior combinations. Note the substantial difference in microstructure in the same final product—which, in combination with the resulting properties (see description in the

caption for Fig. 3)—demonstrates that control of thermomechanical processing can control the microstructures and corresponding final properties of forgings.

Figure 4 summarizes the results of an extensive study of alpha-beta forging versus beta forging for several titanium alloys. Although yield strength after beta forging was not always as high as that after alpha-beta forging, values of notched tensile strength and fracture toughness were consistently higher for beta-forged material.

The beta-forged alloys tend to show a transformed beta or acicular microstructure (Fig. 3b), whereas alpha-beta-forged alloys show a more equiaxed structure (Fig. 3d). Tradeoffs are required for each structural type (acicular vs equiaxed) since each structure has unique capa-

bilities. Table 2 shows the relative advantages of equiaxed and acicular microstructures.

Extrusion

Extrusion is used to make rodlike products, as an alternative mill process to rolling. Properties are affected by processing conditions in much the same way as they are for rolled or forged products. The properties of extruded products, however, are not identical to those of die forged structures. Even where similar microstructures are produced, the thermomechanical working possible in open- and closed-die forging permits much more control over the resultant properties. One of the more unusual applications of extrusion has been in the production of tapered wing spars for a military aircraft.

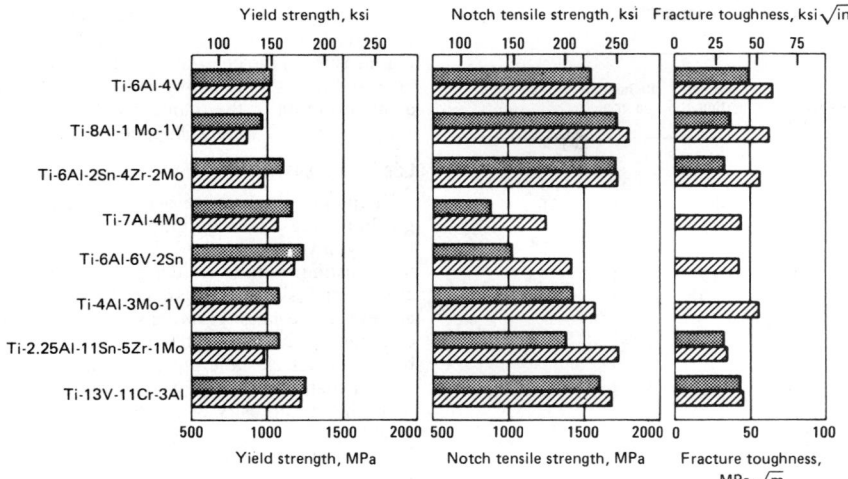

Shaded bars represent alpha-beta-forged material; striped bars represent beta-forged material.

Fig. 4. Comparison of mechanical properties of alpha-beta-forged and beta-forged titanium alloys

Forming

Titanium and titanium alloy sheet and plate are strain hardened by cold forming. This normally increases tensile and yield strengths, and causes a slight drop in ductility. Titanium metals exhibit a high degree of springback in cold forming. To overcome this characteristic, titanium must be extensively overformed or, as is done most frequently, hot sized after cold forming.

Hot forming does not greatly affect final properties. Forming at temperatures from 595 to 815 °C (1100 to 1500 °F) allows the material to slip more readily and simultaneously stress relieves the deformed material; it also minimizes springback. The true net effect in any forming operation depends on total deformation and actual temperature during forming. Titanium metals also tend to creep at elevated temperature; holding under load at the forming temperature (creep forming) is another alternative for achieving the desired shape without having to compensate for extensive springback.

In all forming operations, titanium and its alloys are susceptible to the Bauschinger effect — a drop in compressive yield strength in one loading direction accompanied by an increase in tensile strength in another direction due to strain hardening. The Bauschinger effect is most pronounced at room temperature: plastic deformation (1 to 5% tensile elongation) at room temperature always introduces a significant loss in compressive yield strength, regardless of the initial heat treatment or strength of the alloys. At 2% tensile strain, for instance, the compressive yield strengths of Ti-4Al-3Mo-1V and Ti-6Al-4V

Table 2. Relative advantages of equiaxed and acicular microstructures

Equiaxed:
- Higher ductility and formability
- Higher threshold stress for hot salt stress corrosion
- Higher strength (for equivalent heat treatment)
- Better hydrogen tolerance
- Better low-cycle fatigue (initiation) properties

Acicular:
- Superior creep properties
- Higher fracture-toughness values

drop to less than half the values for solution-treated material. Increasing the temperature reduces the Bauschinger effect; subsequent full thermal stress relieving completely removes it.

Temperatures as low as the aging temperature will remove most of the Bauschinger effect in solution-treated titanium alloys. Heating or plastic deformation at temperatures above the normal aging temperature for solution-treated Ti-6Al-4V will cause over-aging to occur and, as a result, all mechanical properties will decrease.

Machining

Machining of titanium alloys requires cutting forces only slightly higher than those needed to machine steels, but these alloys have metallurgical characteristics that make them somewhat more difficult to machine than steels of equivalent hardness. The beta alloys are the most difficult titanium alloys to machine. When machining conditions are selected properly for a specific alloy composition and processing sequence, reasonable production rates of machining can be achieved at acceptable cost levels. Table 3 shows machinability comparisons of several titanium alloys with other materials.

Joining

Adhesive bonding, brazing, mechanical fastening, metallurgical bonding and welding are all used routinely and successfully to join titanium and its alloys. The first three processes do not affect the properties of these metals as long as joints are properly designed. Metallurgical bonding includes all solid-state joining processes in which diffusion or deformation plays the major role in bonding the members together. Because these processes are performed below but close to the beta transus, metallurgical effects, either normally caused by heating at that temperature or resulting from contamination, should be anticipated. Properly processed joints have similar properties to the base metal and, because bonding is carried out at a temperature high in the alpha-beta field, material properties appear similar to those resulting from high-temperature annealing. With most alloys, a final low-temperature anneal will produce properties characteristic of typical annealed material.

Welding has the greatest potential for affecting material properties. In all types of welds, contamination by interstitial impurities such as oxygen and nitrogen must be closely controlled to maintain useful ductility in the weldment. Alloy composition, welding procedure and subsequent heat treatment are highly important in determining the final properties of welded joints.

Mechanical properties for representative alloys and types of welds can be summarized as follows:

- Welding generally increases strength and hardness.
- Welding generally decreases tensile and bend ductility.
- Welds in unalloyed titanium grades 1, 2 and 3 do not require postweld treatment unless the material will be highly stressed in a strongly reducing atmosphere. In such event, stress relieving or annealing may prove useful.
- Welds in beta-rich alloys such as Ti-6Al-6V-2Sn may have a high tendency to fracturing with little or no plastic straining. Weld ductility can be improved by postweld heat treatment consisting of slow cooling from a high annealing temperature.
- Rich beta-stabilized alloys can be welded, and such welds generally exhibit good ductility. Satisfactory properties are extremely difficult to produce in some beta alloy welds, however.

Titanium alloys can be welded by gas tungsten-arc welding in an inert atmosphere or can be electron beam or laser welded. Electron beam and laser welds are made without filler metal and weld beads have high ratios of depth to width. This combination allows excellent welds to be made in heavy sections, with properties very close to those of the base metal.

Welding must be done under strict environmental controls to avoid pickup of interstitials that can embrittle the weld metal. Small and moderate-size weldments ordinarily are enclosed within environmentally controlled chambers during welding. Larger weldments are made with the aid of portable chambers that only partly enclose the components, or with the aid of "trailers," either of which maintains a protective atmosphere on both front and back sides of the weld until it has cooled below about 480 °C (1000 °F).

Powder Metallurgy

Powder metallurgy (P/M) products having properties similar to those of other forms are now

Table 3. Machinability comparisons of several titanium alloys with other materials

Alloy	Condition(a)	Machinability rating(b)
Aluminum alloy 2017	STA	300
B1112 resulfurized steel	HR	100
1020 carbon steel	CD	70
4340 alloy steel	A	45
Commercially pure titanium	A	40
302 stainless steel	A	35
Ti-5Al-2.5Sn	A	30
Ti-6Al-4V	A	22
Ti-6Al-6V-2Sn	A	20
Ti-6Al-4V	STA	18
HS25 (Co-base)	A	10
René 41 (Ni-base)	STA	6

(a) STA = solution treated and aged; HR = hot rolled; CD = cold drawn; A = annealed. (b) Based on a rating of 100 for B1112 steel.

Table 4. Comparison of typical room-temperature properties of wrought, cast and P/M titanium products

Product and condition	Tensile strength MPa	ksi	Yield strength MPa	ksi	Elonga-tion, %	Reduction in area, %	Impact strength (a) J	ft·lb
Unalloyed Ti								
Wrought bar, annealed	550	80	480	70	18	33	35	26
Cast bar, as cast	635	92	510	74	20	31	26	19
P/M compact, annealed(b) ..	480	70	370	54	18	22	...	...
Ti-5Al-2.5Sn-ELI								
Wrought bar, annealed	815	118	710	103	19	34	...	...
Cast bar, as cast	795	115	725	105	10	17	...	...
P/M compact, annealed and forged(c)	795	115	715	104	16	27	...	...
Ti-6Al-4V								
Wrought bar, annealed	1000	145	925	134	16	34	22	16
Cast bar:								
As cast	1025	149	880	128	12	19	19	14
Annealed	1015	147	890	129	10	16	...	...
Solution treated and aged(d)	1180	171	1085	157	6	11	...	...
P/M compact:								
Annealed(b)	825-855	120-124	740-785	107-114	5-8	8-14	...	...
Annealed and forged(c)	925	134	840	122	12	27	...	...
Solution treated and aged(d)	965	140	895	130	4	6	...	...

(a) Charpy, at − 40 °C (−40 °F). (b) About 94% dense. (c) Almost 100% dense. (d) Aging treatment not specified.

being manufactured. Table 4 compares room-temperature properties of titanium and several titanium alloys in wrought, cast and powder forms. The processes for manufacture of titanium powders are slow and costly, however, and this has resulted in slow growth of powder metallurgy as a means of manufacturing titanium parts.

One of the most important considerations in manufacture of a titanium powder metallurgy product is control of oxygen content, because oxygen has the same undesirable effects on properties of P/M parts as it has on those of wrought products. Powders must be handled very carefully, because they have a very high affinity for oxygen.

CASTING

All titanium castings have compositions based on those of the common wrought alloys. There are no commercial titanium alloys developed strictly for casting applications. This is unusual, because in other metallic systems alloys have been developed specifically as casting alloys, often to overcome certain problems such as poor castability of a wrought-alloy composition. No unusual problems regarding castability or fluidity have been encountered in any of the titanium metals cast to date.

The major reason for selecting a titanium casting instead of a wrought titanium product is cost.

This cost advantage may be attained through increased design flexibility, better utilization of available metal or reduction in the cost of machining or forming parts.

Titanium castings are unlike castings of other metals in that they may be equal or nearly equal in tensile and creep-rupture strengths to their wrought counterparts. Strength guarantees in most specifications for titanium castings are the same as for wrought forms. Typical ductilities of cast products, as measured by elongation and reduction in area, are lower than typical values for wrought products of the same alloys. Fracture toughness and crack-propagation resistance may equal those of corresponding wrought material. However, the fatigue strength of cast titanium is inferior to that of wrought titanium. Fortunately, the fatigue strength of cast titanium can be enhanced by further processing and heat treatment.

Titanium castings, like wrought titanium products, are used primarily in three areas of application: aerospace products, marine service and industrial (corrosion) service. Commercially pure titanium is used for the vast majority of corrosion applications, whereas Ti-6Al-4V is the dominant alloy for aerospace and marine applications. Ti-6Al-2Sn-4Zr-2Mo-Si is being selected with increasing frequency for elevated-temperature service. Castings also have been supplied in alloys Ti-5Al-2.5Sn, Ti-8Al-1Mo-1V and Ti-6Al-6V-2Sn, as well as in several European alloys.

Titanium castings now are used extensively in the aerospace industry and to lesser but increasing measure in the chemical-process, marine and other industries. Titanium castings still represent a small portion of the titanium industry — about 1 to 2% of total weight shipped.

Casting Processes

Titanium castings have been cast in machined graphite molds, rammed graphite molds and proprietary investments used for precision investment casting. Significant design complexity, tolerance and surface-finish control have been achieved, and large parts may be cast. Porosity continues to be a potential problem. However, by use of hot isostatic pressing, internal soundness of titanium castings can be improved to the point that no porosity or small voids can be detected.

For large defects in noncritical locations, weld repair is common foundry practice. However, because titanium can become embrittled from pickup of oxygen, hydrogen and other contaminants during welding, weld repair of titanium castings must be carefully executed. Properly executed weld repair may not degrade the fatigue resistance of cast titanium, and a study has demonstrated that welding does not drastically affect the creep properties of cast Ti-6Al-4V.

Hot Isostatic Pressing

Hot isostatic pressing (HIP) of titanium castings became a production reality in the late 1970's. The HIP schedule that has become the industry standard is 2 h at 900 °C (1650 °F) under argon pressurized to 105 MPa (15 ksi).

Initially, hot isostatic pressing was used with excellent results to salvage parts that had been rejected after radiographic inspection. The effectiveness of the technique gave rise to plans to use HIP for routine parts, but high cost made such plans economically questionable. However, for certain casting configurations, adequate feeding by use of conventional risering is virtually impossible, and, in order to meet aerospace nondestructive inspection standards, shrinkage voids are closed by welding. In such instances, hot isostatic pressing becomes a means of avoiding weld

Table 5. Typical room-temperature tensile properties of several cast titanium alloys

Alloy	Condition	Tensile strength MPa	ksi	Yield strength(a) MPa	ksi	Elonga-tion(b), %	Reduction in area, %
Commercially pure titanium	As-cast or annealed	550	80	450	65	17	32
Ti-6Al-4V	As-cast or annealed	1035	150	890	129	10	19
Ti-6Al-2Sn-4Zr-2Mo	Duplex annealed	1035	150	895	130	8	16
Ti-5Al-2.5Sn-ELI	Annealed	805	117	745	108	11	...

(a) At 0.2% offset. (b) In 50 mm or 2 in.

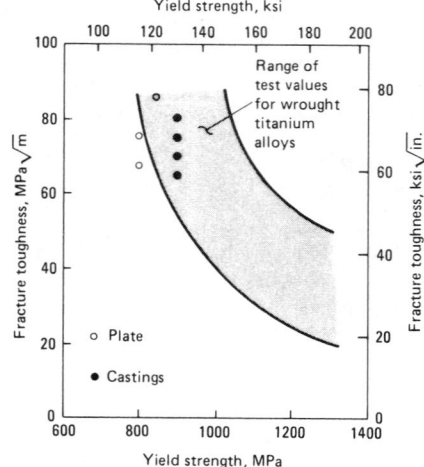

Fig. 5. Fracture toughness of Ti-6Al-4V castings compared with that of Ti-6Al-4V plate and of other Ti alloys

repair and its attendant extra handling and NDT costs.

From a technical viewpoint, hot isostatic pressing is a heat treatment, although some studies have claimed that HIP alone does little, if anything, to enhance mechanical properties of Ti-6Al-4V castings. Properties of hot isostatically pressed alloys are a function of the HIP temperature relative to the beta transus and the post-HIP heat treatment. With castings of marginal to substandard quality, hot isostatic pressing raises the lower limit of data scatter and raises the degree of confidence in the reliability of cast products.

HIP is considered by many to be a process that simplifies the problem of defining a standard for internal casting quality. Hot isostatic pressed parts

also are esthetically more acceptable. At the same time, use of HIP ensures that subsurface microporosity will be healed and therefore will not become exposed on a subsequently machined or polished surface to mar the finish or to act as a possible site for fatigue-crack propagation.

Properties

Cast titanium alloys are equal or nearly equal in strength to wrought alloys of the same compositions. However, typical ductilities are below the typical values for comparable wrought alloys, but still above the guaranteed minimum values for the wrought metals. Because castings of Ti-6Al-4V have been used in aerospace applications, the most extensive data have been de-

veloped for this alloy. Typical room-temperature tensile properties are given in Table 5 for cast commercially pure titanium and for three cast titanium alloys. Figure 5 compares plane-strain fracture-toughness values for Ti-6Al-4V castings with values for Ti-6Al-4V plate and for other wrought titanium alloys.

Generally, an improvement in fatigue resistance is gained by hot isostatic pressing of cast material. In addition, results of research suggest that substantial improvement in resistance to fatigue-crack propagation can be obtained by beta heat treating and over-aging of cast alloys.

Actual crack-growth rates will be influenced by casting quality and postcasting heat treatment, including hot isostatic pressing.

Corrosion Resistance

UNALLOYED TITANIUM is highly resistant to corrosion by many natural environments, including seawater, body fluids, and fruit and vegetable juices. Titanium is used extensively for handling salt solutions (including chlorides, hypochlorides, sulfates and sulfides), wet chlorine gas and nitric acid solutions. On the other hand, hot concentrated low-pH chloride salts (such as boiling 30% $AlCl_3$ and boiling 70% $CaCl_2$) corrode titanium. Warm or concentrated solutions of HCl, H_2SO_4, H_3PO_4 and oxalic acid also are damaging. In general, all acidic solutions that are reducing in nature corrode titanium unless they contain inhibitors. Strong oxidizers, including anhydrous red fuming nitric acid and 90% H_2O_2, also cause attack. Ionizable fluoride compounds, such as NaF and HF, activate the surface and can cause rapid corrosion; dry chlorine gas is especially harmful.

Titanium has limited oxidation resistance in air at temperatures above about 650 °C (1200 °F), and chlorides or hydroxides deposited on its surface can accelerate oxidation. Exposure to liquid or gaseous oxygen, nitrogen tetroxide or red fuming nitric acid can cause titanium to react violently under impact loading.

Titanium has been used to contain liquid or supercritical hydrogen at cryogenic temperatures, but above −100 °C (−150 °F) hydrogen may severely embrittle titanium. The potential for embrittlement is enhanced where hydrogen flow rates are high or where coatings on the titanium become damaged.

In unalloyed titanium and many titanium alloys, weld zones are just as resistant to corrosion as the base metal. Other fabrication processes (such as bending, forming and machining) also appear to have no influence on basic corrosion resistance.

GALVANIC CORROSION

Coupling of titanium to dissimilar metals usually does not accelerate corrosion of the titanium except in reducing environments, where titanium does not become passivated. Under reducing conditions, it has a galvanic potential similar to that of aluminum and will undergo accelerated corrosion when coupled to more-noble metals.

In most environments, titanium is the cathodic member of any galvanic couple. It may accel-

erate corrosion of the other member but in most instances will itself be unaffected. If the surface area of the titanium exposed to the environment is small in relation to the exposed surface area of the other metal, the effect of the titanium on the corrosion rate of the other metal will be negligible; but if the exposed area of titanium greatly exceeds that of the other metal, severe corrosion of the other metal may result.

Because titanium is nearly always the cathodic member of any galvanic couple, hydrogen may be evolved at its surface in an amount proportional to the galvanic current flow. This may result in formation of surface hydride films that generally are stable and cause no problems. At temperatures above 75 °C (170 °F), however, the hydrogen may diffuse into the titanium metal, causing embrittlement. In some environments, titanium hydride is unstable and decomposes or reacts, with a resultant loss of metal.

ALLOYING ADDITIONS

Anodic control of the corrosion reaction predominates when titanium is exposed to a reducing acid such as hydrochloric or sulfuric. Alloying with elements that reduce anodic activity therefore should improve corrosion resistance. This can be accomplished by using alloying elements that: (a) shift the corrosion potential of the alloy in the positive direction (cathodic alloying), (b) increase the thermodynamic stability of the alloy and thus reduce the ability of the titanium to dissolve anodically or (c) increase the tendency of titanium to passivate. The first group includes noble metals such as platinum, palladium and rhodium. The second includes nickel, molybdenum and tungsten. The third group includes zirconium, tantalum, chromium and possibly molybdenum. Considerable work has been done on the use of noble metals as alloying additions in titanium. An outgrowth of this work has been the development of Ti-0.2Pd, which has considerably greater resistance to corrosion in reducing environments than that of unalloyed titanium. In addition, work on alloying for thermodynamic stability has resulted in Ti-2Ni, which was developed for service in hot brine environments where crevice corrosion is sometimes a problem.

Various studies have shown that crevice cor-

rosion resistance of titanium is improved by addition of molybdenum. The commercial alloy Ti Code 12, which contains 0.3% Mo and 0.8% Ni, combines some of the favorable properties of nickel and molybdenum additions while avoiding the negative aspects. This alloy has excellent resistance to pitting and crevice corrosion in high-temperature brines, which sometimes attack commercially pure titanium, and also has better resistance to oxidizing environments such as nitric acid.

CREVICE CORROSION

Titanium is subject to crevice corrosion in brine solutions containing oxidizers. Although crevice corrosion of titanium is observed most often in hot chloride solutions, it also occurs in iodide, bromide and sulfate solutions. Susceptibility increases with increasing temperature, increasing concentration of chloride ions, decreasing concentration of dissolved oxygen and decreasing pH. In solutions with neutral pH, crevice corrosion of titanium has not been observed at temperatures below 120 °C (250 °F). At lower pH values, crevice corrosion sometimes is encountered at temperatures below 120 °C.

EROSION-CORROSION AND CAVITATION

For most materials there are critical velocities beyond which protective films are swept away and accelerated corrosion attack occurs. This accelerated attack is known as erosion-corrosion. The critical velocity differs greatly from one material to another and may be as low as 0.6 to 0.9 m/s (2 to 3 ft/s). For titanium, the critical velocity in seawater is more than 27 m/s (90 ft/s). Numerous erosion-corrosion tests have shown titanium to have outstanding resistance to this form of attack.

Erosion-corrosion can be greatly aggravated by the presence of abrasive particles (such as sand) in a flowing fluid. Titanium exhibited superior resistance to this type of attack in seawater containing fine sand that flowed through conventional titanium condenser tubes at the rate of 1.8 m/s (6 ft/s).

Cavitation-resistance tests have proved titanium to be one of the metals most resistant to cavitation damage.

STRESS-CORROSION CRACKING

Unalloyed titanium generally is immune to stress-corrosion cracking unless it has a high oxygen content (0.3% or more). For this reason, stress-corrosion cracking is of little concern in the chemical-process industries, where unalloyed titanium is most commonly used. On the other hand, certain alloys of titanium used principally in the aerospace industry are subject to stress-corrosion cracking.

One of the first reported instances of stress-corrosion cracking of unalloyed titanium occurred in red fuming nitric acid. It was also found that a pyrophoric surface deposit was formed on exposure to this acid. There was no evidence that these two phenomena are related, but addition of 1.5 to 2.0% water to the acid completely inhibited both reactions. Since then, stress-corrosion cracking has been demonstrated in hot dry sodium chloride, methanol, HCl solutions, seawater, chlorinated solvents, nitrogen tetroxide, mercury and cadmium.

One of the important variables affecting susceptibility to stress-corrosion cracking is alloy composition. Aluminum additions increase susceptibility to stress-corrosion cracking; alloys containing more than 6% Al generally are susceptible to stress corrosion. Additions of tin, manganese and cobalt are detrimental, whereas zirconium appears to be neutral. Beta stabilizers such as molybdenum, vanadium and niobium are beneficial. Susceptibility of titanium alloys to stress-corrosion cracking also can be affected by heat treatment.

ACCELERATED CRACK PROPAGATION IN SEAWATER

Titanium is known to be highly resistant to corrosion by seawater. However, for certain alloys, components containing very sharp notches or cracks exhibit accelerated crack propagation and thus lose resistance to fracture when exposed to seawater.

Failure of titanium due to loss of fracture resistance appears to be similar to delayed fracture of high-strength steels containing sharp notches or cracks on exposure to various liquid environments; it is not considered a form of stress-corrosion cracking.

Exposure to seawater does not appear to diminish service life of titanium alloys, such as Ti-8Mn and Ti-5Al-2.5Sn, that exhibit this phenomenon in laboratory testing. These two alloys have been employed successfully in aircraft for many years without reported failures. Apparently, the conditions leading to accelerated crack propagation (primarily, the existence of a crack) have not been encountered in service.

Accelerated crack propagation in seawater can be avoided by proper alloy selection. Alloys containing more than 6% aluminum are particularly susceptible. Additions of tin, manganese, cobalt and oxygen are detrimental, whereas beta stabilizers such as molybdenum, niobium and vanadium tend to reduce or eliminate susceptibility to this phenomenon. Unalloyed titanium is not susceptible unless it contains more than about 0.3% oxygen.

HOT SALT CORROSION

Titanium and titanium alloys can be damaged by halogenated compounds at temperatures above 260 °C (500 °F). Chloride salts—especially sodium chloride—are very detrimental. Residual salts cause surface pitting, or even cracking of certain alloys under high tensile loads. Although rarely encountered in service, cracking of titanium parts because of hot salt corrosion was encountered by fabricators during stress-relieving operations. Responsibility was traced to vapors of chlorinated cleaning fluids that were not completely removed prior to thermal processing, chloride traces from other process fluids (including tap water) and even salt residues from fingerprints.

The extent of damage by salts is directly related to temperature, exposure time and tensile-stress level. Processing history, alloy composition, salt composition and other environmental conditions also have important effects. Susceptibility to hot salt corrosion appears to be influenced considerably by processing and alloy additions. Therefore, control of these factors should make it possible to avoid this phenomenon in service.

LIQUID-METAL EMBRITTLEMENT

Some titanium alloys crack under tensile stress when in contact with liquid cadmium, mercury or silver-base brazing alloys. This type of embrittlement differs from stress-corrosion cracking although there are some similarities. Liquid-metal embrittlement appears to result from diffusion along grain boundaries and formation of brittle phases, which in turn produce the loss of ductility.

Titanium also can be embrittled by contact with certain solid metals (cadmium and silver, for example) when the titanium is under tensile stress. Service failures have occurred in cadmium-plated titanium alloys at temperatures as low as 65 °C (150 °F), and in silver-brazed titanium parts at temperatures above 315 °C (600 °F).

Silver-plated components should not be used in contact with titanium under stress at temperatures above 230 °C (450 °F). Cadmium-plated parts such as interference-fit fasteners or press-fit bushings should not be used in contact with titanium at any temperature. Other cadmium-plated parts and fasteners should not be used in contact with titanium at temperatures above 230 °C.

Properties, Compositions and Applications of Selected Alloys

PROPERTIES

Properties are functions of composition and processing. Typically, tensile elastic moduli lie in the range from 100 to 120 GPa (14.7 to 17 × 10^6 psi) and decrease with temperature. Texture control and heat treatment can effect changes in moduli. Physical properties vary with alloy composition, and some may vary with processing as it influences microstructure.

Tensile properties cover a wide range of values (see Table 1); moderate- to high-strength alloys typically show tensile-strength values in the range from 895 to 1065 MPa (130 to 155 ksi) (see Table 1). Generally, alloys such as Ti-6Al-6V-2Sn and Ti-6Al-2Sn-4Zr-6Mo tend to have the highest room-temperature strengths of all available alloys.

Creep properties of the highest-tensile-strength alloys usually drop off markedly with temperature. Figures 1 and 2 compare tensile strengths and 150-h creep strengths of a few titanium alloys. The fatigue strength of titanium alloys is of interest because it does not show a marked drop with temperature until temperatures in excess of 315 to 425 °C (600 to 800 °F) are reached. A related design property, fracture toughness (K_{Ic}), is of interest, particularly in applications of high-strength titanium alloys. Table 2 gives typical fracture-toughness values for three high-strength alloys as a function of microstructure. Toughness is influenced by texture and is a function of test direction (see Table 3). An additional property now frequently determined is the fatigue-crack-growth rate (da/dN), which is plotted as a function of stress-intensity-factor range (ΔK). Figure 3 indicates the great heat-to-heat scatter which can be found in this property.

COMPOSITIONS AND APPLICATIONS

Unalloyed Ti Grade 1

Common names. Unalloyed titanium; commercially pure titanium

UNS number. R50250

Composition limits. 0.03 max N; 0.10 max C; 0.015 max H (deviations from max H values: bar, 0.0125; billet and castings, 0.0100); 0.18 max O; 0.20 max Fe; 0.05 max others (each); 0.3 max others (total); rem Ti

Typical uses. Applications requiring high ductility for fabrication but relatively low strength in service; other uses where maximum ease of formability is required and where low iron and interstitial content might enhance corrosion resistance. Weldable.

Precautions in use. Hydrogen embrittlement of titanium can occur in pickling solutions (or other hydrogenating solutions) at room temperature and at elevated temperatures during air exposure or in exposures to reducing atmospheres. Elevated-temperature atmospheric exposure also results in oxygen and nitrogen contamination which increases in severity with increasing temperature and time of exposure. Violent oxidation reac-

Table 1. Typical mill-guaranteed room-temperature tensile properties of selected titanium alloys

Alloy	Tensile strength MPa	ksi	Yield strength MPa	ksi	Ductility Elongation, %	Reduction in area, %
Ti-6Al-4V	895	130	825	120	10	20
Ti-6Al-6V-2Sn	1065	155	995	145	10	20
Ti-6Al-2Sn-4Zr-6Mo	1030	150	965	140	10	20
Ti-6Al-2Sn-4Zr-2Mo	895	130	825	120	10	25
Ti-8Al-1Mo-1V	895	130	825	120	10	20

Table 2. Typical fracture-toughness values of high-strength titanium alloys

Alloy	Alpha morphology	Yield strength MPa	ksi	Fracture toughness (K_{Ic}) MPa·m$^{1/2}$	ksi·in.$^{1/2}$
Ti-6Al-4V	Equiaxed	910	130	44-66	40-60
	Transformed	875	125	88-110	80-100
Ti-6Al-6V-2Sn	Equiaxed	1085	155	33-55	30-50
	Transformed	980	140	55-77	50-70
Ti-6Al-2Sn-4Zr-6Mo	Equiaxed	1155	165	22-23	20-30
	Transformed	1120	160	33-55	30-50

Table 3. Effect of test direction on mechanical properties of textured Ti-6Al-2Sn-4Zr-6Mo plate

Test direction(a)	Tensile strength, MPa	Yield strength, MPa	Elongation, %	Reduction in area, %	Elastic modulus, GPa	K_{Ic} MPa·m$^{1/2}$	ksi·in.$^{1/2}$	K_{Ic} specimen orientation
L	1027	952	11.5	18.0	107	75	68	L-T
T	1358	1200	11.3	13.5	134	91	83	L-T
S	938	924	6.5	26.0	104	49	45	S-T

(a) High basal-pole intensities reported in the transverse direction, 90° from normal, and also intensity nodes in positions 45° from the longitudinal (rolling) direction and about 40° from the plate normal.

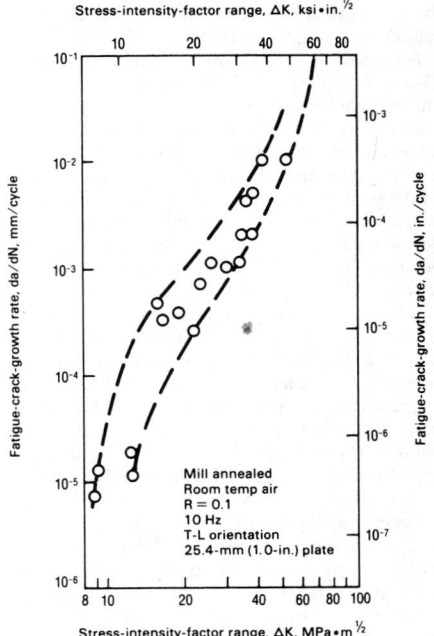

Fig. 3. Measurements of fatigue-crack growth for six heats of mill-annealed Ti-6Al-4V, showing the scatter that can occur in such measurements.

tions can occur between titanium and liquid oxygen or between titanium and red fuming nitric acid.

General corrosion behavior. Though highly reactive, titanium is extremely resistant to corrosion in many aggressive environments. For example, it is highly resistant to seawater and to nitric acid. Resistance to general corrosion has been ascribed to a thin, inert film that forms rapidly on the surface when titanium is exposed to air and to passive films that form on the surface in certain aggressive media.

Ti-5Al-2.5Sn
Ti-5Al-2.5Sn-ELI

Common names. A-110; Ti-5-2½ (standard grade); Ti-5-2½-ELI (extra-low-interstitial grade)

UNS numbers. R54520 (Ti-5Al-2.5Sn); R54521 (Ti-5Al-2.5Sn-ELI)

Composition limits. 0.035 to 0.05 max N; 0.05 to 0.10 max C; 0.01 to 0.02 max H; 0.12 to 0.20 max O; 0.25 to 0.50 max Fe; 0.05 to 0.10 max others (each); 0.30 to 0.40 max others (total); 4.0 to 6.0 Al; 2.0 to 3.0 Sn; rem Ti

Typical uses. Standard grade: gas turbine engine casings and rings, aerospace structural members in hot spots (near engines and wing leading edges), and chemical-processing equipment that requires both better elevated-temperature strength than that of unalloyed titanium and excellent weldability. Other applications requiring good weld fabricability and intermediate strength at service temperatures up to 480 °C (900 °F). High-purity grade: pressure vessels for liquified gases and other applications requiring better ductility and toughness (at somewhat lower strength) than those of

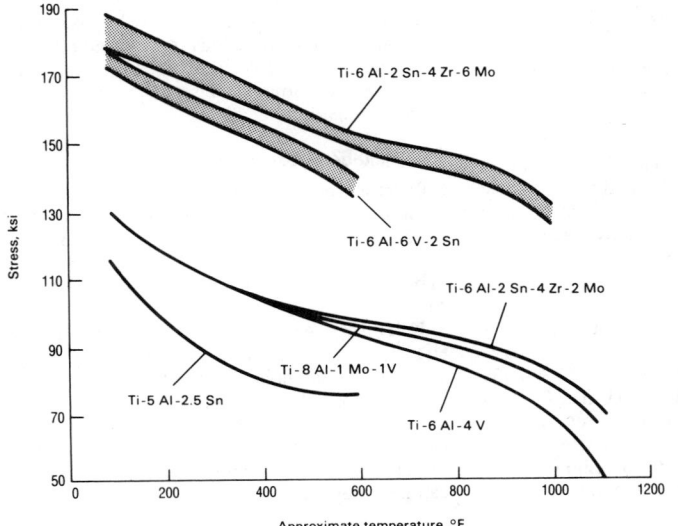

Fig. 1. Comparison of typical tensile strengths reported for various titanium alloys

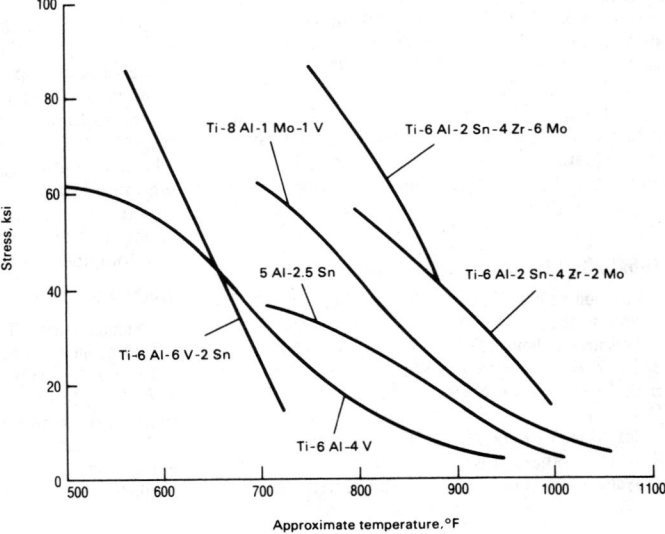

Fig. 2. Comparison of typical 150-h, 0.1% creep strengths reported for various titanium alloys

the standard grade, particularly in hardware for service at cryogenic temperatures

Precautions in use. Hydrogen embrittlement of Ti-5Al-2.5Sn (and its ELI modification) can occur in hydrogenating solutions at room temperature, in air or reducing atmospheres at elevated temperatures, and even in pressurized hydrogen at cryogenic temperatures. Oxygen and nitrogen contamination can occur in air at elevated temperatures; such contamination increases in severity with increasing exposure time and temperature. Ti-5Al-2.5Sn is susceptible to hot salt (particularly chloride) stress-corrosion cracking and to accelerated crack propagation in aqueous solutions at ambient temperatures. Environments in which Ti-5Al-2.5Sn is to be used should be carefully controlled to ensure against material degradation. Ti-5Al-2.5Sn-ELI is more tolerant of environments capable of degrading properties, but over-all immunity is not ensured by using the high-purity alloy. Ti-5Al-2.5Sn should never come in contact with liquid oxygen.

General corrosion behavior. Similar to grades 2 and 3 of unalloyed titanium. The standard alloy and its high-purity modification (Ti-5Al-2.5Sn-ELI) are highly resistant to many reactive substances (such as seawater and nitric acid solutions). Deep sea exposures (at depths of about 1500 m, or 1 mile) produced no corrosion in 751 days (rate estimated at <2.5 μm/yr or <0.1 mil/yr). In seawater flowing at about 36 m/s (120 ft/s), this alloy exhibited a corrosion rate of 5.5 μm/yr (0.22 mil/yr).

Resistance to specific corroding agents. Especially susceptible to corrosion by solid salts (such as chlorides) at elevated temperatures (hot salt stress corrosion). Exposure to salt causes pitting and cracking in 100 h at 315 °C (600 °F) at stress levels above about 207 MPa (30 ksi). Stress-corrosion cracking at lower temperatures is possible after longer exposure times at higher stress levels. Accelerated crack propagation at room temperature in the presence of a pre-existing crack is possible in seawater, chloride solutions and other active solutions.

Ti-8Al-1Mo-1V

Common names. Ti-8-1-1; Ti-811
UNS number. R54810
Composition limits. 0.05 max N; 0.08 max C; 0.015 max H; 0.12 max O; 0.30 max Fe; 0.10 max others (each); 0.40 max others (total); 7.35 to 8.35 Al; 0.75 to 1.25 Mo; 0.75 to 1.25 V; rem Ti
Typical uses. Sheet and forgings for high-speed aircraft structural components (primarily skin); turbine parts; forgings for compressor disks, plates and hubs; cargo flooring. Other applications where light, high-strength, highly weldable material with low density is required.

Ti-6Al-2Sn-4Zr-2Mo

Common names. Ti-6-2-4-2; Ti-6242
UNS number. R54620
Composition limits. 5.5 to 6.5 Al; 1.8 to 2.2 Sn; 3.6 to 4.4 Zr; 1.8 to 2.2 Mo; 0.25 max Fe; 0.05 max C; 0.05 max N; 0.12 max O; 0.0125 max H (bar and billet, 0.0100 max H; forgings and sheet, 0.0150 max H); 0.10 max others (each); 0.30 max others (total); rem Ti
Typical uses. Forgings and flat-rolled products

used in gas turbine engine and air-frame applications where high strength and toughness, excellent creep resistance, and stress stability at temperatures up to 595 °C (1100 °F) are required.

Ti-2.25Al-11Sn-5Zr-1Mo

Common name. Ti-679
UNS number. R54790
Composition limits. 2.00 to 2.50 Al; 10.50 to 11.50 Sn; 4.00 to 6.00 Zr; 0.80 to 1.20 Mo; 0.15 to 0.27 Si; 0.12 max Fe; 0.15 max O; 0.040 max C; 0.040 max N; 0.008 max H; rem Ti
Typical uses. Jet-engine blades and wheels, large bulkhead forgings, other applications requiring high-temperature creep strength plus stability and short-time strength.
Resistance to specific corroding agents. Superior to Ti-6Al-4V, Ti-8Al-1Mo-1V and Ti-5Al-5Sn-5Zr in resistance to hot-salt stress-corrosion cracking at 290 to 480 °C (550 to 900 °F)

Ti-6Al-4V
Ti-6Al-4V-ELI

Common names. Ti-6-4; Ti-64; Ti-6-4 ELI; Ti-64 ELI
UNS numbers. R56400 (Ti-6Al-4V); R56401 (Ti-6Al-4V-ELI)
Composition limits. Ti-6Al-4V: 5.5 to 6.75 Al; 3.5 to 4.5 V; 0.05 max N; 0.10 max C; 0.015 max H; 0.40 max Fe; 0.20 max O; 0.1 max others (each); 0.5 max others (total); rem Ti
Typical uses. Ti-6Al-4V is the most widely used titanium alloy. It is specially processed to both standard and ELI compositions to provide mill-annealed or beta-annealed structures, and is sometimes solution treated and aged. It is used for aircraft gas turbine disks and blades; is extensively used, in all mill product forms, for air-frame structural components and other applications requiring strength at temperatures up to 315 °C (600 °F); and is also used for high-strength prosthetic implants.

Ti-6Al-6V-2Sn

Common names. Ti-6-6-2; Ti-662
UNS number. R56620
Composition limits. 0.04 max N; 0.05 max C; 0.015 max H; 0.20 max O; 0.10 max others (each); 0.40 max others (total); 5.0 to 6.0 Al; 5.0 to 6.0 V; 1.5 to 2.5 Sn; 0.35 to 1.00 Fe; 0.35 to 1.00 Cu; rem Ti
Typical uses. Applications requiring high strength at temperatures up to 315 °C (600 °F). In the forms of sheet, light-gage plate, extrusions and small forgings, this alloy is used for airframe structures where strength higher than that of Ti-6Al-4V is required. Usage is generally limited to secondary structures, because attractiveness of higher strength efficiency is minimized by lower fracture toughness and fatigue properties.

Ti-6Al-2Sn-4Zr-6Mo

Common names. Ti-6-2-4-6; Ti-6246
UNS number. R56260
Composition limits. 5.5 to 6.5 Al; 3.6 to 4.4 Zr; 1.8 to 2.2 Sn; 5.5 to 6.5 Mo; 0.04 max N; 0.04 max C; 0.12 max O; 0.15 max Fe; rem Ti
Typical uses. Forgings in intermediate-temperature-range sections of gas turbine engines, particularly in disk and fan-blade components of

compressors. Available as billet and bar for forging stock, sheet and plate. Should be considered for long-time load-carrying applications at temperatures up to 400 °C (750 °F) and short-time load-carrying applications at temperatures up to 540 °C (1000 °F)

Ti-8Mn

Common name. None
UNS number. R56080
Composition limits. 6.5 to 9.0 Mn; 0.20 max C; 0.20 max O; 0.07 max N; 0.015 max H; rem Ti
Typical uses. Sheet, strip and plate in limited current usage for aircraft sheet components, structural parts

Ti-7Al-4Mo

Common names. Ti-7-4; Ti-74
UNS number. R56740
Composition limits. 6.5 to 7.3 Al; 3.5 to 4.5 Mo; 0.10 max C; 0.013 max H; 0.30 max Fe; 0.05 max N; 0.20 max O; rem Ti
Typical uses. Bar, forgings and forging stock in limited current usage for jet-engine disks, compressor blades and spacers, and sonic horns

Ti-10V-2Fe-3Al

Common name. Ti-10-2-3
Composition limits. 2.6 to 3.4 Al; 9.0 to 11.0 V; 1.8 to 2.2 Fe; 0.13 max O; 0.05 max C; 0.05 max N; 0.015 max H; 0.30 max others (total); rem Ti
Typical uses. Applications up to 315 °C (600 °F) where medium to high strength and high toughness are required in bar, plate or forged sections up to 125 mm (5 in.) thick. The combination of high strength and high toughness available with this alloy is superior to that in any other commercial titanium alloy. For applications requiring uniformity of tensile properties at surface and center locations

Ti-13V-11Cr-3Al

Common names. B-120VCA; Ti-13-11-3
UNS number. R58010
Composition limits. 12.5 to 14.5 V; 10 to 12 Cr; 2.4 to 4.0 Al; 0.35 max Fe; 0.08 max N; 0.10 max C; 0.015 max H; 0.20 max O; rem Ti
Typical uses. Missile applications such as solid rocket motor cases where extremely high strengths are required for short periods of time and for other structural applications in advanced manned and unmanned airborne systems. Springs for airframe applications

Ti-11.5Mo-6Zr-4.5Sn

Common name. Beta III
Composition limits. 10.0 to 13.0 Mo; 4.50 to 7.50 Zr; 3.75 to 5.25 Sn; 0.10 max C; 0.020 max H; 0.05 max N; 0.18 max O; 0.35 max Fe; 0.01 max others (each); 0.04 max others (total); rem Ti
Typical uses. Aircraft fasteners (especially rivets) and sheet metal parts where cold formability and strength potential can be used to greatest advantage. Possible use in plate and forging applications where high strength capability, deep hardenability and resistance to stress corrosion are required and somewhat lower aged ductility can be accepted

10 TIN

This section was condensed from articles entitled Tin and Its Alloys, and Properties of Tin and Tin Alloys, both by Joseph B. Long, Tin Research Institute, Inc., in Metals Handbook, Ninth Edition, Volume 2, Properties and Selection: Nonferrous Alloys and Pure Metals, pages 613 to 625. For more detailed information, the reader is referred to the larger work. Additional information on tin and tin alloys may be found in the selected references at the end of this section and elsewhere in the Desk Edition by consulting the index.

TIN was one of the first metals known to man. Throughout ancient history, various cultures recognized the virtues of tin in coatings, alloys and compounds, and use of the metal increased with advancing technology. Today, tin is an important metal in industry even though the annual tonnage used is much smaller than those of many other metals. One reason for the small tonnage is that, in most applications, only very small amounts of tin are used at a time.

Production of tin ore generally is centered in areas far distant from centers of consumption. The leading tin-producing countries (excluding the USSR and China) are, in descending order, Malaysia, Bolivia, Indonesia, Thailand, Australia, Zaire and Nigeria. These countries supply over 85% of total world production.

Cassiterite, a naturally occurring oxide of tin, is by far the most economically important tin mineral. The bulk of the world's tin ore is obtained from low-grade placer deposits of cassiterite derived from primary ore bodies or veins associated with granites or rocks of granitic composition.

Primary ore deposits can contain very low percentages of tin (0.01%, for example), and thus large amounts of soil or rock must be worked to provide recoverable amounts of tin minerals. Unlike ores of other metals, cassiterite is very resistant to chemical and mechanical weathering, but extended erosion of primary lodes by air and water has resulted in deposition of the ore as eluvial and alluvial deposits.

Underground lode deposits of tin ores are worked by sinking shafts and driving adits, the rock being broken from the working face by drilling and blasting. Cassiterite is recovered from eluvial and alluvial deposits by dredging, gravel pumping and hydraulicking. In open-pit mining, a much less widely employed mining method, mechanical and manual methods are used to move tin-bearing materials. After ball-mill concentration of the ore, a final culling is provided at dressing stations.

The final concentrates, which contain 70 to 77% tin, are then sent to the smelter where they are mixed with anthracite and limestone. This charge is heated in a reverberatory furnace to about 1400 °C (2550 °F) to reduce the tin oxide to impure tin metal, which is again heated in huge cast iron melting pots to refine the metal. Steam or compressed air is introduced into the molten metal, and this treatment, plus addition of controlled amounts of other elements which combine with the impurities, results in tin of high purity (99.75 to 99.85%). This high-purity tin often is treated again by liquating or electrolytic refining, which provides tin with a purity level approaching 99.99%.

After the tin is refined, it is cast into ingots weighing 12 to 25 kg (26 to 56 lb) or bars in weights of 1 kg (2 lb) and upwards. Tin normally is sold by brand name, and the choice of brand is determined largely by the amounts of impurities that can be tolerated in each end product. High-purity brands of tin may contain small amounts of lead, antimony, copper, arsenic, iron, bismuth, nickel, cobalt and silver. Total impurities in commercially pure tin rarely exceed 0.25%.

TIN IN COATINGS

Tinplate. The largest single application of tin is in manufacture of tinplate (steel sheet coated with tin), which accounts for about 40% of total world tin consumption. Since 1940, the traditional hot dip method of making tinplate has been largely replaced by electrodeposition of tin on continuous strips of rolled steel. Electrolytic tinplate can be produced with either equal or unequal amounts of tin on the two surfaces of the steel base metal. Nominal coating thicknesses for equally coated tinplate range from 0.38 to 1.54 μm (15 to 60 μin.) on each surface. The thicker coating on tinplate with unequal coatings (differential tinplate) rarely exceeds 2.0 μm (80 μin.). Tinplate is produced in thicknesses from 0.15 to 0.60 mm (0.006 to 0.024 in.).

Over 90% of world production of tinplate is used for containers (tin cans). Traditional tinplate cans are made of three pieces of tin-coated steel: two ends and a body with a soldered sideseam. Innovations in can manufacture have produced two-piece cans made by drawing and ironing. Tinplate cans find their most important use in packaging of food products, beer and soft drinks, but also are used for holding paint, motor oil, disinfectants, detergents and polishes. Other applications of tinplate include fabrication of signs, filters, batteries, toys, gaskets, and containers for pharmaceuticals, cosmetics, fuels, tobacco and numerous other commodities.

Electroplating accounts for one of the major uses of tin and tin chemicals. Tin is used in anodes, and tin chemicals are used in formulating various electrolytes, for coating a variety of substrates. Tin electroplating can be performed in either acid or alkaline solutions. Sodium or potassium stannates form the bases of alkaline tin plating electrolytes that are very efficient and capable of producing high-quality deposits. Advantages of these alkaline stannate baths are that they are not corrosive to steel and that they do not require additional agents. Acid electrotinning solutions operate at higher current densities and higher plating rates and require additions of organic compounds.

A number of alloy coatings can be electroplated from mixed stannate-cyanide baths, including coatings of tin-zinc and tin-cadmium alloys and a wide range of tin-copper alloys (bronzes). The bronzes range in tin content from 7 to 98%. Red bronze deposits contain up to 20% tin; high-tin bronzes, called speculum, usually contain about 40% tin.

Tin-nickel and tin-lead electrodeposits are plated from acid electrolytes and are important coatings for printed circuits and electronic components. Tin-cobalt plate is used in applications requiring an attractive finish and good corrosion resistance.

Two ternary alloy electrodeposits are used by industry. These are the copper-tin-lead alloy for bearing surfaces and the copper-tin-zinc alloy for coatings in certain electronic applications.

Hot Dip Coatings. Coating of steel with lead-tin alloys produces a material called terneplate. Terneplate is easily formed and easily soldered and is used as a roofing and weather sealing material and in construction of automotive gasoline tanks, signs, radiator header tanks, brackets, chassis and covers for electronic equipment and sheathing for cable and pipe.

Hot dip tin coatings are used on wire for component leads as well as food handling and processing equipment. In addition, hot dip tin coatings are used to provide the bonding layer for babbitting of bearing shells.

UNALLOYED TIN

There are only a few applications where tin is used unalloyed with other metals. Unalloyed tin is well recognized as the most practical lining material for handling high-purity water in distillation plants because it is chemically inert to pure water and will not contaminate the water in any way.

In the manufacture of plate glass, the molten glass is fed from the furnace onto the surface of a molten tin bath, which is protected from oxidation by an atmosphere of nitrogen containing some hydrogen. The natural forces of surface tension and gravity within the bath ordinarily

Table 1. Designations, chemical compositions and applications of commercially pure tins

ASTM B339	Designation	Class	Sn, max	Sb, max	As, max	Bi, max	Cd, max	Cu, max	Fe, max	Pb, max	Ni + Co, max	S, max	Zn, max	General applications
	Grade designation							Composition(a), %						
AAA	Electrolytic	Extra-high purity	99.98	0.008	0.0005	0.001	0.001	0.002	0.005	0.010	0.005	0.002	0.001	Analytical standards, research
AA	Electrolytic	High purity	99.95	0.02	0.01	0.01	0.001	0.02	0.01	0.02	0.01	0.01	0.005	Research, pharmaceuticals, fine chemicals
A(b)	A, Straits	High purity; commercial	99.80	0.04	0.05	0.015	0.001	0.04	0.015	0.05	0.01	0.01	0.005	Tinplate, foil, collapsible tubes, block tin products, pewter
B(c)	B	General purpose	99.80	...	0.05	...	...	...	...	...	...	...	...	Less exacting, general purpose
C	C	Intermediate grade	99.65	...	...	...	...	...	...	...	...	...	...	General-purpose alloys
D	D	Lower intermediate grade	99.50	...	...	...	...	...	...	...	...	...	...	General-purpose alloys
E	E	Common	99.00	(d)	...	...	...	(d)	...	(d)	...	...	...	Cast bronze, bearing metal, general-purpose alloys, lead-base alloys

(a) The maximum impurity limits listed here, which are from ASTM Standard Classification B339, are not specification limits but simply guides to the maximum impurity contents commonly found in the various brands of tin that fall into these grades. (b) ASTM Grade A includes about 80 to 90% of the refined tin produced. (c) Grade B is intended for those uses where the specific impurity limitations of Grade A are not critical. (d) Limits of these impurities may be specified for some uses.

produce plate glass about 6 mm (¼ in.) thick, but thickness of the glass can be varied by the speed at which the molten glass is drawn from the float bath and the temperature of the tin. Glass ribbons are formed with surfaces flat and parallel. Surfaces of the glass are so smooth that surface polishing is not required.

Chemical compositions of commercially pure tins are given in Table 1.

TIN IN ALLOYS

Chemical compositions of commercially available tin alloys are given in Table 2. Information on tin-base fusible alloys and tin alloys for sliding bearings can be found in Section 20 of this Handbook.

Solders account for the second largest use of tin (after tinplate). Tin is an important constituent in solders because it wets and adheres to many common base metals at temperatures considerably below their melting points. Tin is alloyed with lead to produce solders with melting points lower than those of either tin or lead. Small amounts of various metals, notably antimony and silver, are added to tin-lead solders to increase their strength. These solders can be used for joints that are subjected to high or even subzero service temperatures.

Both solder compositions and applications of joining by soldering are many and varied. Commercially pure tin is used for soldering sideseams of cans for special food products and aerosol sprays. The electronics and electrical industries employ solders containing 40 to 70% tin, which provide strong and reliable joints under a variety of environmental conditions. General-purpose solders (50Sn-50Pb and 40Sn-60Pb) are used for light engineering applications, plumbing and sheet metal work. Lower-tin solders (20 to 35% Sn, remainder Pb) are used in joining cable and in production of automobile radiators and heat exchangers. Low-tin solders are used in large amounts to fill crevices at seams and welds in automotive bodies, thereby providing smooth joints and contours. Solders containing about 2% tin (remainder lead) are used for can sideseams to provide hermetic seals. Tin-zinc solders are used to join aluminum, while tin-antimony and tin-silver solders are employed in applications requiring joints with high creep resistance.

Additional information on tin solders can be found in the article on soldering in Section 30 of this Handbook.

Alloys for Organ Pipes. Tin-lead alloys are used in the manufacture of organ pipes. These materials commonly are named "spotted metal" because they develop large nucleated crystals or "spots" when solidified as strip on casting tables. The pipes that produce the diapason tones of organs generally are made of alloys with tin contents varying from 20 to 90% according to the tone required. Broad tones generally are produced by alloys rich in lead; as tin content increases, the tone becomes brighter. Cold rolled tin-copper-antimony alloys (95% Sn) also have been used successfully in the manufacture of pipes, and adoption of these alloys has improved the efficiency and speed of fabrication of finished pipes. This composition provides for a bright appearance which is more tarnish resistant than the tin-lead alloys.

Pewter is a tin-base white metal containing antimony and copper. Originally, pewter was defined as an alloy of tin and lead, but to avoid toxicity and dullness of finish, lead is excluded from modern pewter. These modern compositions contain 1 to 8% antimony and 0.25 to 3.0% copper. Pewter casting alloys usually are lower in copper than pewters used for spinning hollowares and thus have greater fluidity at casting temperatures.

Pewter is malleable and ductile and is easily spun or formed into intricate designs and shapes. Pewter parts do not require annealing during fabrication. Much of the costume jewelry produced today is made of pewter alloys centrifugally cast in rubber or silicone molds.

Chemical-composition limits for modern pewter are given in Table 3.

Bearing Materials. Tin has a low coefficient of friction, which is the first consideration in its use as a bearing material. Tin is structurally a weak metal, and when used in bearing applications it is alloyed with copper and antimony for increased hardness, tensile strength and fatigue resistance. Normally, the quantity of lead in these alloys, called tin-base babbitts, is limited to 0.35 to 0.5% to avoid formation of the tin-lead eutectic, which would significantly reduce strength properties at operating temperatures.

Lead-base bearing alloys, called lead-base babbitts, contain up to 10% tin and 12 to 18% antimony. In general, these alloys are inferior in strength to tin-base babbitts, and this must be equated with their lower cost. Segregation of the constituents of these alloys may provide some difficulties during centrifugal casting of linings. During casting, careful selection of rotational speed in relation to bearing size is necessary. Additions of cerium, arsenic or nickel also assist in controlling segregation of these alloys.

In addition to the tin-base and lead-base babbitts, there is a series of intermediate lead-tin bearing alloys. These alloys have tin and lead contents between 20 and 65% plus various amounts of antimony and copper. Increasing the tin content of these alloys provides higher hardness and greater ease of casting. These alloys are less prone to segregation during melting than lead-base babbitts. Cast intermediate bearing alloys, however, exhibit lower strength values than either tin-base or lead-base babbitts.

Bearing alloys must maintain a balance between softness and strength. Aluminum-tin bearing alloys represent an excellent compromise between the requirements for high fatigue strength and the need for good surface properties such as softness, seizure resistance and embeddability. Aluminum-tin bearing alloys are usually employed in conjunction with hardened steel or ductile iron crankshafts and allow significantly higher

Table 2. Chemical-composition limits and typical uses of tin alloys

Name	Composition limits	Typical uses
Hard tin (99.6Sn-0.4Cu)	...	Collapsible tubes and foil
Antimonial tin solder (95Sn-5Sb)	95 Sn desired, 0.20 max Pb, 4.5 to 5.5 Sb, 0.15 max Bi, 0.08 max Cu, 0.04 max Fe, 0.005 max Al, 0.005 max Zn, 0.05 max As	Soldering of electrical equipment, joints in copper tubing and cooling coils for refrigerators. High-tin solders are used for joining parts of electrical apparatus because they have higher electrical conductivity than high-lead solders. High-tin solders are also used where lead may be a hazard — for instance, in contact with foodstuffs. Tin solders that contain 5% Sb (or 5% Ag) are suitable for use at higher temperatures than tin-lead solders.
Tin-silver solder (95Sn-5Ag)	...	Soldering of components for electrical and high-temperature service. See *Antimonial tin solder*.
Soft solder (70Sn-30Pb)	Grade 70A: 70 Sn desired, 30 Pb nominal, 0.12 max Sb, 0.25 max Bi, 0.08 max Cu, 0.02 max Fe, 0.005 max Al, 0.005 max Zn, 0.03 max As. Grade 70B: same as grade 70A except limits for Sb are 0.20 to 0.50. **Consequence of exceeding impurity limits.** See *Soft solder (60Sn-40Pb)*.	Joining and coating of metals (see *Antimonial tin solder*). Except in very special situations, grades 70A and 70B are used for the same applications. Grade 70B is specified when the presence of antimony is required to ensure that the change from beta tin to alpha tin (called the "tin pest"), with the accompanying change in volume and drastic loss in solder strength, does not occur.
Soft solder (63Sn-37Pb)	Grade 63A: 63 Sn desired, 37 Pb nominal, 0.12 max Sb, 0.25 max Bi, 0.08 max Cu, 0.02 max Fe, 0.005 max Al, 0.005 max Zn, 0.03 max As. Grade 63B: same as grade 63A except limits for Sb are 0.20 to 0.50. **Consequence of exceeding impurity limits.** See *Soft solder (60Sn-40Pb)*.	Soldering of electrical components (see *Antimonial tin solder*). Except in very special situations, grades 63A and 63B are used for the same applications. Grade 63B is specified when the presence of antimony is required to ensure that the change from alpha tin to beta tin (called the "tin pest"), with the accompanying change in volume and drastic loss in solder strength, does not occur.
Soft solder (60Sn-40Pb)	Grade 60A: 60 Sn desired, 40 Pb nominal, 0.12 max Sb, 0.25 max Bi, 0.08 max Cu, 0.02 max Fe, 0.005 max Al, 0.005 max Zn, 0.03 max As. Grade 60B: same as grade 60A except limits for Sb are 0.20 to 0.50. **Consequence of exceeding impurity limits.** Antimony is slightly detrimental to wetting properties; antimony is an intentional addition to grade 60B, as noted at right under typical uses. Bismuth causes some discoloration of solder surface. Copper levels above about 0.25% and iron levels above about 0.1% cause grittiness of solder. Excessive zinc causes oxidation of solder to be more noticeable. Excessive aluminum causes appreciable oxidation of solder. Even at the maximum allowable level of 0.03%, arsenic may cause dewetting problems in soldering of brass.	Solder for electronic and electrical work, especially mass soldering of printed circuits (see *Antimonial tin solder*). Except in very special situations, grades 60A and 60B are used for the same applications. Grade 60B is specified when the presence of antimony is required to ensure that the change from alpha tin to beta tin (called the "tin pest"), with the accompanying change in volume and drastic loss in solder strength, does not occur.
Tin babbitt alloy 1 (91Sn-4.5Sb-4.5Cu)	Bearings (ASTM B23): 90 to 92 Sn, 4 to 5 Sb, 4 to 5 Cu, 0.35 max Pb, 0.08 max Fe, 0.10 max As. Die castings (ASTM B102): 90 to 92 Sn, 4 to 5 Sb, 4 to 5 Cu, 0.35 max Pb, 0.08 max Fe, 0.08 max As, 0.01 max Zn, 0.01 max Al	Sleeve bearings, die castings
Tin babbitt alloy 2 (89Sn-7.5Sb-3.5Cu)	88 to 90 Sn, 7 to 8 Sb, 3 to 4 Cu, 0.35 max Pb, 0.08 max Fe, 0.10 max As, 0.08 max Bi, 0.005 max Zn, 0.005 max Al, 0.05 max Cd	Sleeve bearings
Tin babbitt alloy 3 (84Sn-8Sb-8Cu)	83 to 85 Pb, 7.5 to 8.5 Sb, 7.5 to 8.5 Cu, 0.35 max Pb, 0.08 max Fe, 0.10 max As, 0.08 max Bi, 0.005 max Zn, 0.005 max Al, 0.05 max Cd	Sleeve bearings
Bearing alloy (75Sn-12Sb-10Pb-3Cu)	74 to 76 Sn, 11 to 13 Sb, 9.3 to 10.7 Pb, 2.5 to 3.5 Cu, 0.08 max Fe, 0.15 max As	Sleeve bearings
Casting alloy (65Sn-18Pb-15Sb-2Cu)	64 to 66 Sn, 14 to 16 Sb, 17 to 19 Pb, 1.5 to 2.5 Cu, 0.08 max Fe, 0.15 max As, 0.01 max Zn, 0.01 max Al	Sleeve bearings, die castings
Tin die-casting alloy (82Sn-13Sb-5Cu)	80 to 84 Sn, 12 to 14 Sb, 4 to 6 Cu, 0.35 max Pb, 0.08 max Fe, 0.08 max As, 0.01 max Zn, 0.01 max Al	Die castings
Tin foil (92Sn-8Zn)	...	Foil for food packaging
White metal (92Sn-8Sb)	...	Costume jewelry

loading than tin- or lead-base bearing alloys.

Low-tin aluminum-base alloys (5 to 7% Sn) containing small amounts of strengthening elements, such as copper and nickel, are often used for connecting-rod and thrust bearings in high-duty engines. Strict dimensional tolerances must be adhered to, and oil contamination should be avoided. Alloys containing 20 to 40% tin, remainder aluminum, show excellent resistance to corrosion by products of oil breakdown and good embeddability, particularly in dusty environments. The higher-tin alloys have adequate strength and better surface properties, which make them useful for crosshead bearings in high-power marine diesel engines.

Type metals are cast alloys containing various proportions of lead, antimony and tin. They do not readily segregate on solidification from the melt, but they are subject to porosity in the central regions of type characters and slugs because

Table 3. Chemical-composition limits for modern pewter

Specification	Sn	Sb	Cu	Pb, max	As, max	Fe, max	Zn, max	Cd, max
ASTM B560:								
Type 1(a)	91-93	6-8	0.25-2.0	0.05	0.05	0.015	0.005	...
Type 2(b)	90-93	5-7.5	1.5-3.0	0.05	0.05	0.015	0.005	...
Type 3(c)	95-98	1.0-3.0	1.0-2.0	0.05	0.05	0.015	0.005	...
BS5140	Rem	5-7	1.0-2.5	0.5	...	...	...	0.05
	Rem	3-5	1.0-2.5	0.5	...	...	...	0.05
DIN17810	Rem	1-3	1-2	0.5	...	...	...	...
	Rem	3.1-7.0	1-2	0.5	...	...	...	...

(a) Casting alloy, nominal composition 92Sn-7.5Sb-0.5Cu. (b) Sheet alloy, nominal composition 91Sn-7Sb-2Cu. (c) Special-purpose alloy.

air in molds escapes with difficulty. Good fill of the mold should be ensured by rapid injection, and temperature of the metal should be high enough to avoid premature solidification and entrapment of gases.

Battery-Grid Alloys. Lead-calcium-tin alloys have been developed for storage-battery grids — largely as replacements for antimonial lead alloys. Use of ternary lead-base alloys containing up to 1.3% tin has substantially reduced gassing, and therefore batteries whose grids are made of these alloys do not require periodic water additions during their working life. Two chief methods of grid manufacture are casting and fabrication of wrought alloys including punching, roll forging and expanded metal processes.

Copper Alloys. Copper-tin bronzes were some of the first alloys used by man, and these alloys continue to be used for structural and decorative purposes. True bronzes contain tin in amounts up to 10% as well as very small amounts of phosphorus. Quaternary bronzes containing 5 Sn, 5 Zn, 5 Pb, remainder Cu are used for general-purpose castings for applications requiring reasonable strength and soundness, such as gears, pumps and automotive fittings. Special copper-base alloys with 20 to 24% tin have been used historically for cast bells of excellent tonal quality. Spinodal Cu-Ni-Sn alloys containing 2 to 8.5% tin have excellent elastic properties and have replaced tin-free Cu-Ni alloys in some spring and electrical-contact applications. In addition to these uses in copper-base alloys, small quantities of tin (0.75 to 1.0%) are added to copper-zinc alloys (brasses) for increased corrosion resistance. Cast leaded brasses may contain up to 4% tin.

Dental alloys for making amalgams contain silver, tin, mercury, and some copper and zinc. The copper increases hardness and strength and the zinc acts as a scavenger during alloy manufacture, protecting major constituents from oxidation. Most dental alloys presently available contain 25 to 27% tin and consist mainly of the intermetallic compound Ag_3Sn. When porcelain veneers are added to gold alloys for high-grade dental restoration, 1% tin is added to the gold alloy to ensure bonding with the porcelain.

Cast Irons. The presence of about 0.1% tin in flake or ductile iron castings ensures a completely pearlitic structure, and this pearlite is retained even at elevated temperatures. Commercially pure tin is added to the cast iron in the form of shot, bars or cast pieces; in cupola melting, the tin is commonly added to the ladle or to the cupola spout during tapping. Tin is also added to special mixing chambers along with suitable inoculant materials in the production of ductile iron castings. Because the mixing chambers are an integral part of the mold, this technique allows one-step treatment of the molten metal as it enters the mold and overcomes "fading" (loss of effectiveness of inoculating additions before the metal is cast). Also, the mixing chamber provides immediate dissolution of the tin in the iron

and ensures uniform distribution in the casting.

Titanium Alloys. Tin strengthens titanium alloys by forming solid solutions. Titanium can exist in the low-temperature alpha phase or the higher-temperature beta phase, which remains stable up to the melting point. In titanium alloys, relative amounts of alpha and beta phases present at the service temperature have profound effects on properties. Aluminum additions raise the transformation temperature and stabilize the alpha phase, but may cause embrittlement in amounts greater than 7%. However, with tin additions, increased strength without embrittlement can be obtained in aluminum-stabilized alpha titanium alloys. Optimum strength and workability can be obtained with 5% aluminum and 2.5% tin; in addition, this alloy has the advantage of being weldable. Alpha-beta titanium alloys contain aluminum as an alpha stabilizer and combinations of beta stabilizers (such as chromium, iron, molybdenum, manganese and vanadium), as well as tin and zirconium as substitutional solid-solution strengthening elements. Such alloys have good strength and creep resistance at elevated temperatures. Strength and forming properties of many of these alloys can be optimized by various heat treatments.

Zirconium alloys are similar to titanium alloys in that the elements they contain can be divided into two classes: alpha stabilizers, which raise the transformation temperature, and beta stabilizers, which lower it. Tin and aluminum are alpha stabilizers in zirconium alloys and enhance high-temperature strength. A commercial series of corrosion-resistant zirconium alloys containing 0.15 to 2.5% tin has been developed for nuclear service.

Powder Applications. Most of the supply of tin powders is used in making sintered bronze or sintered iron parts. However, tin powders are also employed in making paste solders and creams used in the plumbing and electronic manufacturing industries. A minor use of tin and tin alloy powders is in sprayed coatings for food-handling equipment, metallizing of nonconductors and bearing repairs. Tin particles are also used in food-can lacquers to decrease dissolution of iron and exposed lead-base solder by the food product.

Additions of 2% tin powder and 3% copper powder aid sintering of iron compacts. The tin provides a low-melting-point phase which in turn provides diffusion paths for the iron. Iron-tin-

copper compacts sintered at 950 °C (1740 °F) have mechanical properties comparable to those of iron-copper powder metallurgy parts containing 7 to 10% copper sintered at 1150 °C (2100 °F). In addition, closer control of finished dimensions is afforded by the iron-tin-copper mixture, which results in improved quality and cost effectiveness.

Sintered compacts made from mixtures of iron and tin-lead solder powders are suitable for certain low-stress engineering applications. Warm pressing of these compacts (at about 450 °C or 840 °F) provides cohesion of the iron-solder mixtures, but avoids recrystallization of the iron powder so that any work hardening obtained during compaction is retained. Different properties in pressed-and-sintered compacts can be obtained by varying the pressing conditions and the relative amounts of iron and solder powders.

TIN IN CHEMICALS

The manufacture of inorganic and organic chemicals containing tin constitutes one of the major uses of metallic tin. The use of tin compounds has grown so rapidly over the past quarter century that the tin chemicals industry has been transformed from one based mainly on recovered secondary tin to one that consumes significant amounts of primary ingot tin.

Tin chemicals are used for such widely diversified purposes as: electrolyte solutions for depositing tin and its alloys; pigments and opacifiers for ceramics and glazes; catalysts and stabilizers for plastics; pesticides, fungicides and antifouling agents in agricultural products, paints and adhesives; and corrosion-inhibiting additives for lubricating oils.

SELECTED REFERENCES

Extractive Metallurgy of Tin, 2nd Ed., by P. A. Wright: Elsevier Publishing Co., Amsterdam–London–New York, 1982, 228 pages

Tin Social and Economic History, 1st Ed., by E. S. Hedges: Edward Arnold Publishers, Ltd., London, 1964, 194 pages

Tin and Its Alloys, 1st Ed., edited by E. S. Hedges: Edward Arnold Publishers, Ltd., London, 1960, 424 pages

Tin, 2nd Ed., by C. L. Mantell: Hafner Publishing Co., Inc., New York, 1959, 573 pages

11 ZINC

By Hugh Morrow III, Zinc Institute Inc.

For additional information on zinc and zinc alloys, the reader is referred to Metals Handbook, Ninth Edition, Volume 2, pages 629 to 655, and to the index of the present work.

ZINC, its alloys and its chemical compounds represent the fourth most industrially utilized metal (behind iron, aluminum and copper). Slab zinc and zinc oxide are the primary materials for the vast majority of these applications, and the compositions of these materials, as specified by the American Society for Testing and Materials (ASTM), are given in Tables 1 and 2. Other grades of slab zinc, such as continuous galvanizing grade (CGG) and controlled lead grades (CLG), are not covered in ASTM specifications because they vary from one user to another. In addition, most zinc oxide producers market a number of different zinc oxide grades varying in particle size and shape, and in over-all composition, to conform to the specific needs of particular applications.

Zinc is used in five principal areas of application: in coatings and anodes for corrosion protection of irons and steels; in zinc casting alloys; as an alloying element in copper, aluminum, magnesium and other alloys; in wrought zinc alloys; and in zinc chemicals. In the corrosion-protection category, hot dip or continuous galvanizing accounts for the majority of zinc consumption. Almost all of the zinc used in zinc casting alloys is employed in die casting compositions. Among zinc-containing alloys, copper-base alloys such as brasses are the largest zinc consumers. Rolled zinc is the principal form in which wrought zinc alloys are supplied, although drawn zinc wire for metallizing is showing increasing usage. In the zinc-chemical category, zinc oxide is the major compound.

Recent statistics have shown that consumption of zinc in the United States conforms to the following pattern:

Application	Percent of total consumption
Galvanizing	44.5
Die casting alloys	25.6
Brasses	13.5
Rolled zinc products	5.7
Zinc oxide	9.1
Miscellaneous	1.6

For galvanizing applications, the less-pure or higher-lead grades are usually used. However, all of the impurities (Pb, Al, Fe) must be controlled for the various galvanizing processes even though they do not necessarily have to be kept as low as possible. Annual consumption of slab zinc used in various galvanized products in the

Table 1. Compositions of three grades of slab zinc (ASTM B6)

Grade	Composition(a), wt%			
	Pb, max	Fe, max	Cd, max	Zn, min (difference)
Special high grade (SHG)(b)	0.003	0.003	0.003	99.990
High grade (HG)	0.03	0.02	0.02	99.90
Prime Western (PW)(c)	1.4	0.05	0.20	98.0

(a) When specified for use in manufacture of rolled zinc or brass, aluminum shall not exceed 0.005%. (b) Tin shall not exceed 0.001%. (c) Aluminum shall not exceed 0.05%.

Table 2. Compositions of American Process and French Process zinc oxides (ASTM D79)

Process	Composition, %				
	Zinc oxide, min	Total S, max	Moisture and other volatile matter, max	Total impurities(a), max	Coarse particles(b), max
American	98	0.2	0.5	2.0	0.25
French	99	0.1	0.5	1.0	0.10

(a) Including moisture and other volatile matter. (b) Total residue retained on No. 325 (45-μm) sieve.

United States is summarized in Fig. 1.

The grade of zinc employed for die casting alloys and anodes is usually special high grade, whereas all grades are employed for brasses and rolled alloys, depending on the specific application. United States slab zinc consumption by grades for 1980, for example, was estimated to be as follows:

Grade	Short tons	Percent of total
Special high grade	390 568	43.7
High grade	83 950	9.4
Continuous galvanizing grade ...	20 415	2.3
Controlled lead grade	77 504	8.7
Prime western grade	319 947	35.8
Remelt grade	1 749	0.1
Total	894 133	100.0

Zinc oxide is employed in rubber, paints, ceramics, chemicals, agriculture, photocopying, floor coverings, and coated fabrics and textiles. Both French Process and American Process grades are used in most of these applications depending on the specific character of the end product. Zinc oxide's largest use is in rubber products, where it is employed as an activator for the accelerators used to speed up the vulcanization process. It is used as a pigment in paints and ceramics, as a soil nutrient in the agricultural field and as a stabilizer in plastics, and provides the photosensitive character for coated papers used in some photocopying techniques.

CORROSION PROTECTION

Almost half of the zinc consumed in the world is used for corrosion-protection coatings on irons, mild steels and low-alloy steels. Zinc corrodes at much lower rates than do steels in atmospheric exposure, and in addition will corrode sacrificially when the coated steel is exposed such as at scratches or cut ends. Zinc anodes are also used to provide galvanic sacrificial protection in underwater and underground applications.

Zinc Coatings

The specific coating techniques by which zinc is applied to provide corrosion protection include postfabrication hot dip galvanizing, continuous hot dip galvanizing, electrogalvanizing, electroplating, metallizing, zinc dust/zinc oxide painting and mechanical plating/sherardizing. The first three techniques are the most important commercially, and, according to a recent Zinc Development Association estimate, account for over 90% of the total zinc used for coatings. Nevertheless, the other methods also consume significant quantities of zinc:

Metallizing	25 000 metric tons
Zinc paint	50 000 metric tons
Electroplating	80 000 metric tons
Mechanical plating/ sherardizing	3 000 metric tons

Products and Applications. The major product forms of zinc-coated materials include hot dip post-

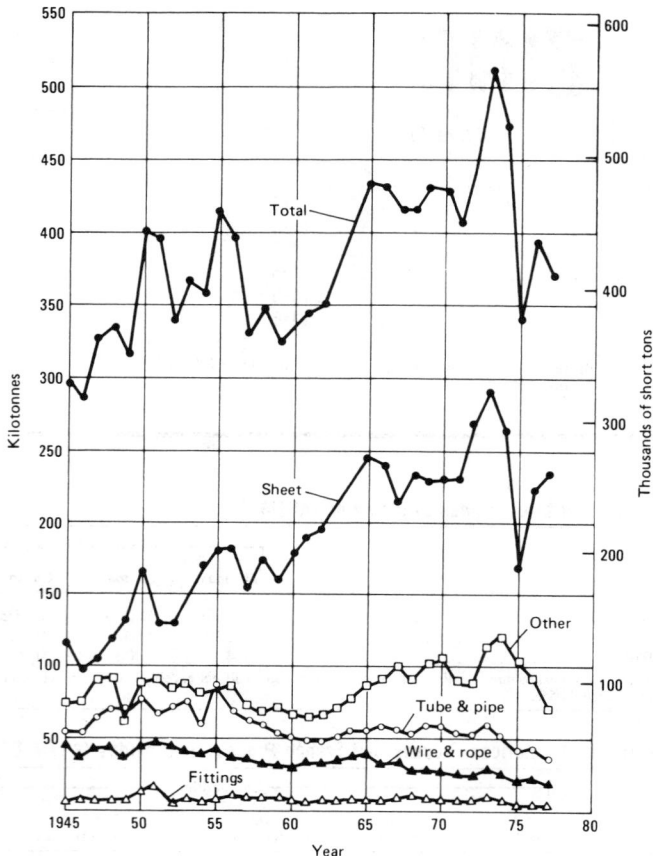

Fig. 1. Annual consumption of slab zinc used in galvanized products in the United States

fabrication products, sheet, strip, wire and tube. Galvanized sheet is used mainly for building and construction, automotive underbody panels and domestic and household appliances. Electroplating, mechanical plating and sherardizing are normally employed on fasteners and other relatively small objects for which thin, uniform coatings are required. Zinc spraying is used on large structures (such as bridges) which either are too large for convenient hot dip galvanizing or are already in place. Zinc-containing paints are effective for a wide range of products and are often used as primers for painting storage tanks, ships and other large structures. Galvanizing, painting with zinc-rich paints and zinc thermal-spray metallizing are often considered complementary zinc coating processes each of which has particular attributes that make it the most suitable zinc application method for a specific job.

Types of Zinc Coatings. The various zinc coating processes may also be differentiated by the type of zinc coating produced on the steel being protected. Conventional hot dip galvanized products exhibit a surface layer consisting of a thin coating of relatively pure zinc with three underlying iron-zinc alloy layers, progressively containing more iron and less zinc from the surface down to the ferrous base metal. Thus the zinc content decreases through the hot dip coating from the surface to the steel. Continuous hot dip sheet products, on the other hand, exhibit a thin pure zinc coating obtained by use of small aluminum additions to the liquid zinc bath and by running the sheet through the bath at a higher speed and in a much shorter time than for postfabrication hot dip galvanizing. The zinc coatings obtained by electrogalvanizing, electroplating, mechanical plating and sherardizing are also thin, relatively pure zinc layers. Metallizing and zinc-dust painting permit application of heavier coatings.

The nature of the zinc coating, whether it is relatively pure or is an iron-zinc alloy layer, is important in that it affects the formability and weldability of the precoated sheet products and, to a lesser degree, the corrosion resistance of the zinc-coated article. Where postfabrication hot dip galvanizing ensures that all parts of an assembly are fully covered with zinc and thus protected from corrosion, continuous hot dip galvanizing and electrogalvanizing allow production of differentially coated products which can readily be processed in mass-production operations such as those encountered in the automotive industry. In products coated with zinc-rich paints, the binder phase and other paint ingredients besides the zinc-dust pigment may provide extra corrosion protection by acting as a barrier between the steel and the corrosive environment. Metallizing, electroplating, mechanical plating and sherardizing (a higher-temperature form of mechanical plating) all produce relatively pure zinc coatings which are suitable for additional finishing operations. There are also several zinc alloy coatings which exhibit improved corrosion performance.

In recent years, a whole series of zinc coating alloys have been developed to meet a wide variety of needs. Most of these alloys are based on the Zn-Al system and vary in composition from 0.2 to 55% Al, sometimes containing small

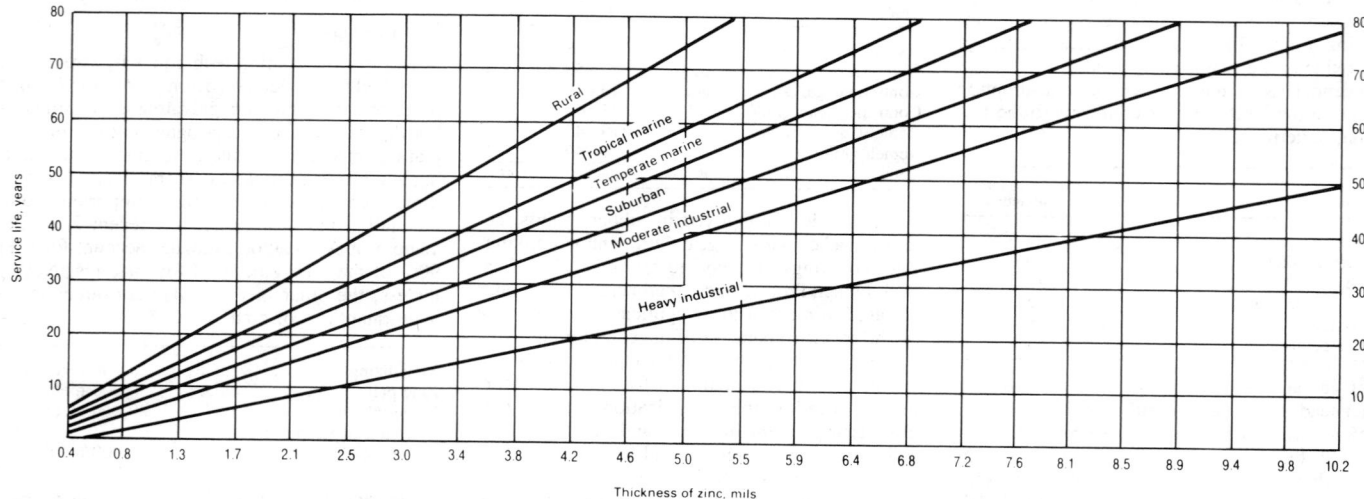

Fig. 2. Effect of zinc coating thickness on service life of hot dip galvanized coatings for various types of atmospheric exposure

amounts of a third alloying element. These alloys have generally been employed for continuous hot dip galvanizing, although Fe-Zn and Ni-Zn alloys are being studied for electrogalvanizing, a Zn-15Al alloy is used for metallizing and Zn-Cd alloys are employed in producing corrosion-resistant mechanically plated surfaces. In addition, the vast number of processes and products available for continuous galvanizing and electrogalvanizing, and the huge number of zinc-rich paint formulations available, suggest that zinc coatings for corrosion protection may be utilized in a virtually limitless number of ways.

Coating Thickness. The relationship between coating thickness and service life of zinc-coated steel is well known and is shown in Fig. 2 for hot dip galvanized coatings in six different types of atmospheric exposure. Service life increases directly with coating thickness for all types of environments. Thus, thicker zinc coatings such as those obtained by metallizing or zinc-rich painting give longer lives than those produced by continuous hot dip galvanizing or electrogalvanizing. On the other hand, steels with thick coatings are not readily plastically deformed or welded. At equivalent coating weights there is probably little difference between the corrosion protection afforded by the several zinc coating techniques. However, it must be remembered that the zinc in each of the various coatings may exist in a different form, which could result in differences in corrosion protection. Postfabrication hot dip galvanizing produces a thin zinc outer layer coupled with inner Fe-Zn alloy layers. Continuous hot dip galvanizing yields a thin, relatively pure zinc layer alloyed with a small amount of aluminum. Electrogalvanized sheet exhibits a dense pure zinc layer although there is an increasing tendency in the automotive industry towards use of Fe-Zn and Ni-Zn electrogalvanized coatings. Electroplated zinc and mechanically plated zinc are thin, dense coatings whereas metallized zinc is a thicker, more porous coating. Zinc-rich paints contain a number of other materials besides pure zinc dust.

The thickness of zinc coatings is normally specified in microns (μm) or mils except for hot dip galvanized products where coatings are specified in terms of the weight, in grams per square metre (g/m^2) or ounces per square foot (oz/ft^2), which can easily be converted to the correct thickness. However, for sheet products, the hot dip coating weight is reported per square metre or square foot of sheet, and thus the thickness of the zinc coating on each side of the sheet is only half that in the case of nonsheet products. Thus, a G90 coating on a beam represents a thickness of roughly 38 μm (1.5 mils) whereas on sheet it represents a thickness *on each side* of 19 μm (0.75 mil). The thicknesses of the various zinc coatings and the general requirements for their application are reported in the following specifications:

Technique	Specification(a)
Postfabrication galvanizing	ASTM A90, A123
Continuous sheet galvanizing	ASTM A525
Electrogalvanizing	No general specification
Electroplating	ASTM B633
Metallizing	CSA G189, BS 2569
Zinc-rich painting	SSPC PS Guide 12.00, SSPC Paint 20
Mechanical plating	ASTM B695
Sherardizing	BS 4921

(a) CSA stands for Canadian Standards Association, BS for British Standard and SSPC for Steel Structures Painting Council.

Coating Uniformity. For all zinc-coated products, the uniformity of the coating, as well as its thickness, is of great importance. The lead content of hot dip galvanizing baths is carefully controlled because it produces a strong effect on the surface tension of molten zinc and thus promotes formation of a uniform coating which rapidly covers all bare spots. The uniformity of electrocoated products, on the other hand, is governed by the geometry of the material being coated, the electrodes and the plating solutions. Metallizing and zinc-rich painting depend to a far greater extent on the applicator's skill. Mechanical plating techniques exhibit excellent uniformity because the zinc particles are essentially affixed one at a time to a constantly rotating workpiece.

Zinc Anodes

If a structure requiring protection is not, or cannot be, coated with zinc, corrosion still can be controlled by using zinc as a sacrificial galvanic anode in a cathodic-protection system.

Cathodic protection is effective in such diverse media as salt water, fresh waters and most soils. The structures most frequently protected by the cathodic method are underground pipelines and ship hulls. The source of the protective cathodic current can be either impressed (rectified ac or specially generated dc) or galvanic.

When a metal surface is in contact with an electrolyte, differences in electrical potential develop between local areas of the surface corresponding to their respective electrochemical reactivities; a more-reactive (less-noble) area is referred to as being "anodic" to a less-reactive (more-noble, or "cathodic") area. The difference in potential between two areas causes current to flow from the cathodic to the anodic area through the metal, and from the anodic to the cathodic area through the electrolyte to complete the circuit. Where the current enters the electrolyte, metal ions go into solution, causing anodic corrosion.

Such corrosion can be prevented, and the metal protected, if the metal (steel, for example) is connected electrically by a wire or rod to zinc that also is in contact with the electrolyte. Under these conditions, current flows from the external zinc anode through the electrolyte to the steel surface, and this current is of sufficient magnitude to afford cathodic protection to the steel. Zinc anodes are normally procured to ASTM B418 or MIL-A-18001.

CASTING ALLOYS

Zinc-base casting alloys fall into three categories: die casting alloys; foundry alloys; and other casting alloys. Of these three categories, die casting alloys are by far the most important, accounting for more than 25% of all zinc consumption in recent years. Zinc foundry alloys are a relatively new development, and have not yet achieved significant production although they are commercially available from numerous sources. Slush-casting and forming-die alloys are produced in only small quantities but have been utilized for some time.

Die Casting Alloys

The major use of zinc as a structural material is in alloys for pressure die casting. These alloys are commonly known as alloys 3, 5 and 7, and are based on the hypoeutectic Zn-4Al composition. The compositions of ingots for die casting are covered by ASTM B240 while those for the die castings themselves are found in ASTM B86; both are given in Table 3, along with the equivalent ASTM and UNS (Unified Numbering System) designations. Zinc alloy die castings have been in use for more than 50 years and are highly castable, easily finished, economical materials with good mechanical properties. They are used for a wide range of decorative and light structural parts and lend themselves readily to rapid mass-production techniques such as those required in the automotive industry. These alloys may be cast into very complex shapes, in intricate detail, to tight dimensional tolerances, with excellent surface finish and often in very thin section sizes. The low casting temperatures for these alloys result in longer die life and higher production rates than those of other die cast metals.

This excellent combination of processing characteristics and properties is due in large part to careful control of the impurities (Fe, Pb, Cd and Sn) in zinc alloy die castings. For this reason, special high grade is the usual form of slab zinc used for production of zinc die casting alloys. Dimensional stability, intergranular corrosion, machinability and mechanical properties are among a few of the characteristics which are sensitive to impurities in these alloys.

Alloy 3 is characterized by good impact strength and long-term dimensional stability. Alloy 5 exhibits somewhat higher tensile strength and creep resistance than alloy 3 but has lower impact strength at elevated temperatures. Alloy 7 is a high-purity version of alloy 3 and thus has sim-

Table 3. Compositions of zinc die casting alloy ingots and die castings

Grade	ASTM designation	UNS designation	Cu	Al	Mg	Fe, max	Pb, max	Cd, max	Sn, max	Ni	Zn
Die casting ingots (ASTM B240)											
Alloy 3	AG40A	Z33520	0.10 max	3.9-4.3	0.025-0.05	0.075	0.004	0.003	0.002	...	Rem
Alloy 5	AC41A	Z35531	0.75-1.25	3.9-4.3	0.03-0.06	0.075	0.004	0.003	0.002	...	Rem
Alloy 7	...	Z33522	0.10 max	3.9-4.3	0.010-0.020	0.075	0.0020	0.0020	0.0010	0.005-0.020	Rem
Die castings (ASTM B86)											
Alloy 3	AG40A	Z33520	0.25 max	3.5-4.3	0.020-0.05	0.100	0.005	0.004	0.003	...	Rem
Alloy 5	AC41A	Z35531	0.75-1.25	3.5-4.3	0.03-0.08	0.100	0.005	0.004	0.003	...	Rem

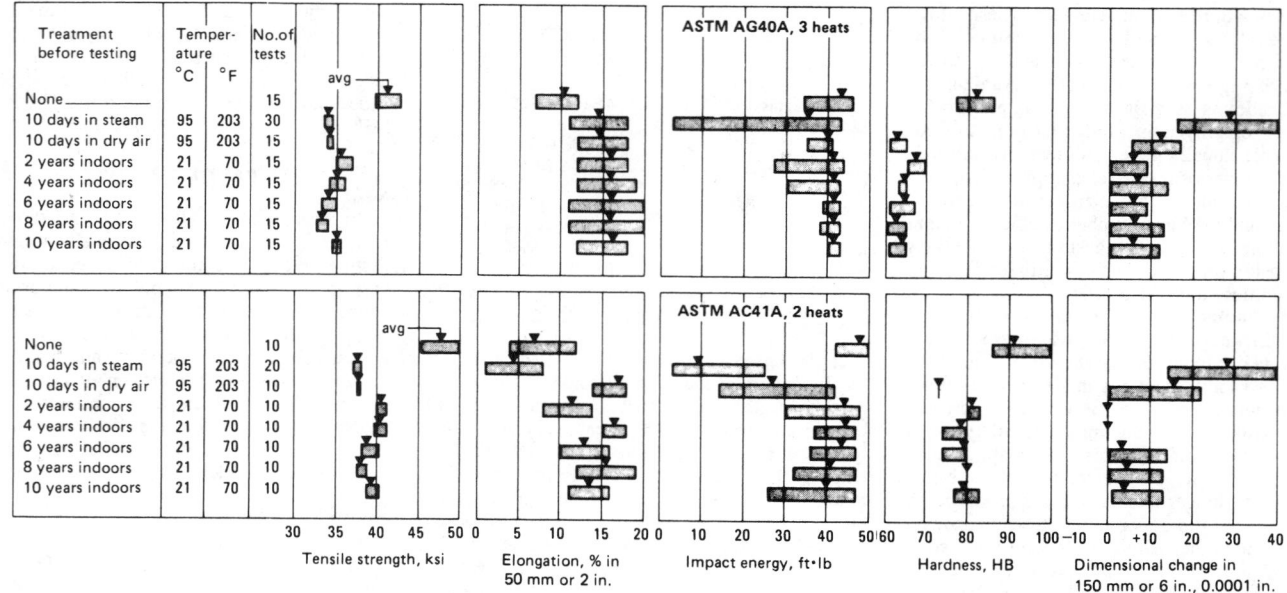

Treatment before testing	Temperature °C	°F	No.of tests
None			15
10 days in steam	95	203	30
10 days in dry air	95	203	15
2 years indoors	21	70	15
4 years indoors	21	70	15
6 years indoors	21	70	15
8 years indoors	21	70	15
10 years indoors	21	70	15

ASTM AG40A, 3 heats

None			10
10 days in steam	95	203	20
10 days in dry air	95	203	10
2 years indoors	21	70	10
4 years indoors	21	70	10
6 years indoors	21	70	10
8 years indoors	21	70	10
10 years indoors	21	70	10

ASTM AC41A, 2 heats

Tensile strength, ksi — Elongation, % in 50 mm or 2 in. — Impact energy, ft·lb — Hardness, HB — Dimensional change in 150 mm or 6 in., 0.0001 in.

Fig. 3. Mechanical properties and dimensional changes characteristic of zinc die casting alloys. (Source: New Jersey Zinc Co.)

ilar mechanical properties but exhibits superior casting and finishing characteristics.

Limiting service conditions for standard zinc-base die casting alloys are as follows. At temperatures slightly above 95 °C (200 °F), their tensile strength is reduced 30% and their hardness 40%. At subzero temperatures, some embrittlement occurs but impact strength is still in the same range as that of aluminum and magnesium die casting alloys at normal service temperatures. At room temperature, impact strength of zinc die castings is much higher than that of aluminum or magnesium die castings or iron sand castings. Tensile and other properties of alloys AG40A and AC41A after various aging treatments are given in Fig. 3.

The zinc die casting alloys, alloys 3, 5 and 7, are not intended for elevated-temperature service and will creep significantly under load at 75 °C (165 °F) or higher. Tensile strength is correspondingly decreased, and ductility increased, at these temperatures. Several compositional modifications of zinc die casting alloys have been developed to permit their use at higher temperatures. In alloy 5, additions of chromium (0.1 to 0.2%) and titanium (0.15 to 0.25%), coupled with an increase in the copper level to 1.0 to 1.5% and virtual elimination of aluminum (down to 0.01 to 0.04% from 3.9 to 4.3%), were employed to make a more creep-resistant material designated alloy 16 (ILZRO 16). A series of foundry alloys, with 8, 12 and 27% Al, have also been developed and will subsequently be discussed. At least one of these alloys, Zn-27Al, has been die cast and exhibits good creep properties.

Zinc alloy die castings will undergo aging and slight shrinkage if placed in service in the as-cast condition. The maximum shrinkage for alloy 3 is 0.0007 in./in., two-thirds of which occurs during the first four weeks of service. The maximum shrinkage for alloy 5 is 0.0009 in./in. If required, stabilization heat treatments may be applied to accelerate the aging process and thereby avoid it in service. The stabilization anneals may be carried out at temperatures from 25 to 150 °C (75 to 300 °F) for corresponding times of 5 weeks

to 30 min. It has been found that the optimum stabilization treatment is one of the following: 3 h at 100 °C (212 °F); 5 h at 85 °C (185 °F); 10 h at 70 °C (160 °F). Lower-temperature annealing is preferable because it results in less shrinkage after treatment, but the 100 °C anneal is more practical from a production standpoint.

Zinc alloy die castings are invariably cast within quite close dimensional limits, but some machining is commonly required to remove flash and to drill, ream or tap holes. Alloys 3, 5 and 7 may be machined by conventional tools and methods. Generally only light cuts at high speeds are necessary. High speed steel tools are usually satisfactory, but cemented tungsten carbide tools produce better finishes and permit closer dimensional control and shorter tool-grinding times.

Zinc die casting alloys are especially amenable to a wide range of subsequent finishing operations that are applied for decorative purposes, improved corrosion resistance or better abrasion performance. These finishing techniques basically comprise mechanical techniques, electrodeposition, chemical conversion, application of organic coatings, or use of other specialty coatings. Mechanical finishing techniques include buffing, polishing, brushing and tumbling. Metals electrodeposited on zinc die castings include copper, nickel, chromium, brass, silver and gold. The copper/nickel/chromium plating system is perhaps the one most widely used, but there is a huge variety of other electrodeposited coatings available to produce different colors and textures on zinc alloy die castings. The chemical conversion coatings include chromates, phosphates and other inorganics, which may be applied directly to the fresh zinc surface or to one of the other subsequent coatings to produce certain colors and surface effects and to improve corrosion resistance. The organic coatings include paints, enamels, lacquers, varnishes and plastics. There are many such organic coatings which can be applied successfully in a number of different ways. Some of the specialty techniques employed for finishing zinc die castings include vacuum metallizing, hot foil stamping, and anodizing, which has also

been called "Iridizing." It is the enormous range of finishing operations which can be carried out rapidly and economically on zinc alloy die castings that has made them so useful for decorative and some functional applications, although they may be used in light to moderate structural applications as well.

Zinc alloy die castings may be joined into assemblies or components by many of the techniques employed for other engineering materials. Most often, they are fastened by mechanical means such as rivets, tags, studs, rollovers, bosses or threaded fasteners. Adhesives have also been used to successfully join zinc die castings. Alloys 3, 5 and 7 are weldable and solderable, but these two joining techniques are not normally employed except in special circumstances. Soldering with the normal lead-tin solders may be done only if the die casting is first electroplated with nickel. Soldering may be performed on the bare metal by using an 82.5Cd-17.5Zn alloy, but in either case loss of strength and possible contamination may be encountered in die castings made from these zinc alloys in remelted form. Gas welding of zinc alloy die castings may be accomplished under reducing conditions using a filler rod similar in composition to the die casting alloy. Pulsed resistance arc welding has also been successfully used to join zinc alloy die castings.

Foundry Alloys

Zinc foundry alloys, commonly referred to as the "ZA" alloys, have been developed during the past 20 years and are now increasing in commercial usage. In contrast to hypoeutectic Zn-4Al die casting alloys, ZA alloys are hypereutectic compositions containing from 8 to 27% Al. The chemical requirements for these alloys are specified in ASTM B669 and are summarized in Table 4. The ZA alloys are suitable for casting by sand, permanent mold, graphite permanent mold, shell mold and high-pressure die casting methods. The graphite permanent mold process was specifically developed on the basis of the good castability and low casting temperature of the ZA alloys.

Table 4. Compositions of zinc foundry (ZA) alloy ingots (ASTM B669)

Grade	UNS designation	Cu	Al	Mg	Fe, max	Pb, max	Cd, max	Sn, max	Zn
ZA-8	Z25630	0.8-1.3	8.0-8.8	0.015-0.030	0.10	0.004	0.003	0.002	Rem
ZA-12	Z35630	0.5-1.25	10.5-11.5	0.015-0.030	0.075	0.004	0.003	0.002	Rem
ZA-27	Z35840	2.0-2.5	25.0-28.0	0.010-0.020	0.10	0.004	0.003	0.002	Rem

Table 5. Typical compositions of rolled zinc and zinc alloys (ASTM B69)

Alloy group	UNS designation	Pb	Fe, max	Cd	Cu	Mg	Zn
Unalloyed Zn	Z21210 ...0.05 max		0.010	0.005 max	0.001 max	...	Rem
Zn-Pb-Cd	Z21310 ...0.05-0.12		0.012	0.005 max	0.001 max	...	Rem
Zn-Pb-Cd	Z21540 ...0.30-0.65		0.020	0.20-0.35	0.005 max	...	Rem
Zn-Cu	Z44330 ...0.05-0.12		0.012	0.005 max	0.65-1.25	...	Rem
Zn-Cu	Z45330 ...0.05-0.12		0.015	0.005 max	0.75-1.25	0.007-0.02	Rem

Table 6. Typical compositions of zinc alloy sheet and strip (QQ-Z-100A)

Alloy group	Cu	Pb, max	Fe, max	Cr, max	Mn, max	Cd, max	Ti	Zn
Zn-Cu	0.2-1.2	0.25	0.015	0.02	0.01	0.07	...	Rem
Zn-Cu-Ti	0.5-0.8	0.20	0.015	0.02	0.01	0.01	0.08-0.16	Rem

ZA alloys exhibit mechanical properties equal to or exceeding those of conventional zinc die casting alloys and those of cast irons, aluminum alloys and copper alloys. In addition, they have excellent bearing properties, wear resistance and machinability. Advantageous foundry properties include low melting temperatures and hence low melting energy consumption, increased die life and mold stability. They may be readily cast in thin sections in sand molds and require no fluxing or degassing. The ZA-27 composition is normally indicated for higher-strength applications, while the ZA-12 alloy is the most widely used at present. All three materials have been successfully high-pressure die cast as well as being cast by the normal foundry methods.

Other Casting Alloys

Two other classes of zinc-base casting alloys find application, albeit in limited quantities. These are termed slush-casting alloys and forming-die or "Kirksite" alloys. Slush-casting alloys are Zn-5Al eutectic compositions containing 4.75 to 5.50% Al with or without small additions of copper or silicon. They are used for lighting fixtures, lamp bases, small statues, and casket hardware. Forming-die alloys usually contain about 4% Al and 3% Cu and may contain small magnesium additions. They are normally plaster mold or sand cast, and have been employed for dies and punches used in forming, blanking, trimming, drawing and stretching of large bodies and shapes from light-gage sheet aluminum, magnesium, stainless steel and low-carbon steel.

WROUGHT ALLOYS

Wrought zinc and zinc alloy products may be obtained in several different forms, such as rolled strip, sheet and foil; forged or extruded bar, rod and other shapes; and drawn wire and rod. These products are readily fabricated and welded, and exhibit good corrosion resistance in many types of service. Because pure zinc will creep under load at room temperature, alloying additions are necessary in alloys used for structural applications. Furthermore, zinc is anisotropic in its de-

Table 7. Typical compositions of rolled superplastic zinc alloys

Alloy group	Al	Cu	Mg	Zn
Zn-Al	20-24	...	...	Rem
Zn-Al	20-24	0.4-0.6	...	Rem
Zn-Al	20-24	0.4-0.6	0.005-0.03	Rem

formation behavior and will exhibit textures and preferred orientations.

Rolled Zinc Alloys

The products that account for the greatest usage of wrought zinc alloys are rolled sheet, strip and foil. They are employed for such applications as architectural and building materials, dry cell batteries, photoengraving plates, coinage, electrical contacts, superplastic alloy applications, and other miscellaneous uses similar to those for which wrought brasses are used. Rolled zinc products are popular because they can readily be formed by a number of secondary deformation techniques, exhibit good corrosion resistance, and are readily joined by conventional welding and soldering methods.

Commercially available rolled zinc alloys may be divided into five groups on the basis of their alloying elements:

1. Unalloyed zinc
2. Zinc-copper alloys
3. Zinc-copper-titanium alloys
4. Zinc-lead-cadmium alloys
5. Zinc-aluminum alloys.

Some of these compositions are covered in ASTM B69, "Rolled Zinc," while others are listed in Federal Specification QQ-Z-100A, "Zinc Alloy Sheet and Strip." A large number of compositions are available in the United States for rolled zinc alloys depending on the specific application requirements, and an even wider variety are available in Europe, where rolled zinc alloys have generally been more extensively utilized up to now. The compositions of some typical rolled zinc alloys are summarized in Tables 5, 6 and 7.

Included in the Zn-Al alloy group (Table 7) are the superplastic rolled zinc alloys. These materials contain roughly 22 wt % Al and may also have small additions of copper and/or magnesium. They are superplastic—i.e., are uniformly and plastically deformable in the same manner as plastics and molten glasses, at moderately elevated temperatures and low strain rates. They can be easily fabricated from sheet form into a desired shape, and then heat treated to restore their high strength properties. Additions of copper and magnesium to the Zn-22Al composition increase tensile, yield and creep strengths and decrease ductility, but have little effect on impact strength. Annealing at 350 °C (660 °F) followed by air cooling also produces increases in strength. The superplastic Zn-22Al alloys are normally processed into sheet by conventional methods and then fabricated into complex shapes by processes such as stretch forming, deep drawing, compression molding and thermoforming.

Forged or Extruded Zinc Alloys

The superplastic Zn-22Al alloys discussed above may also be processed by standard forging and extrusion methods. However, the two alloy groups most often used for forgings and extrusions are Zn-Al alloys with 11.0 to 14.5% Al and Zn-Cu-Ti compositions containing 1% Cu and 0.1% Ti. Both the Zn-Al and Zn-Cu-Ti alloys are recommended for forgings while the Zn-14.5Al alloy is utilized for applications requiring good impact strength at low temperatures. These alloys are readily machined; can be joined by soldering, welding or adhesives; and can be easily finished with paints, polymers or electroplated coatings. Secondary forming operations can easily be accomplished on Zn-Al and Zn-Cu-Ti alloys. The compositions of these forging and extrusion alloys are presented in Table 8.

Zinc Alloys for Drawn Wire and Rod

The drawn wire and rod products are used mainly for thermal-spray-metallizing wire, filler-metal wire for soldering, and wire which is subsequently fabricated into nails, screws, slate hooks and wire gauze. These alloys range from virtually pure zinc to multiple-alloying-element compositions, and some typical compositions are shown in Table 9. Other Zn-Al alloys with much higher aluminum contents (10 to 20%) have also been produced in wire and rod form for specialty applications.

ZINC-CONTAINING ALLOYS

Almost 15% of the zinc consumed is used in zinc-containing alloys. By far the largest group is the copper-zinc alloys known as brasses and bronzes. Zinc is also used, in varying degrees, as an alloying element in aluminum-base, magnesium-base and tin-base systems and is often one of the constituents in brazing and soldering alloys. The compositions of some of the typical systems in which zinc is employed as a major alloying element are summarized in Table 10.

ZINC DUST AND POWDER

Zinc dust refers to material condensed from zinc vapor, whereas zinc powder normally refers to atomized molten zinc. Zinc dust and powder are employed in production of zinc-rich paints and in powder-fed thermal spraying and coating op-

Table 8. Typical compositions of zinc alloys for forging and extrusion

Alloy group	Trade name(a)	Al	Mg	Cu	Ti	Mn	Zn
				Composition, wt %			
Zn-Al	Korloy 2570	11.0	0.02	0.75	...	...	Rem
Zn-Al	Korloy 2573	14.5	0.02	0.70	...	...	Rem
Zn-Cu-Ti	Korloy 3130	...	...	1.0	0.1	...	Rem
Zn-Cu-Ti	Korloy 3330	...	...	1.0	0.1	1.0	Rem

(a) Designations of Cominco Ltd.

Table 9. Typical compositions of zinc alloys for drawing of wire and rod

Alloy group	Commercial designation(a)	Fe, max	Cd, max	Pb, max	Cu, max	Ti, max	Al	Zn
				Composition, wt %				
Unalloyed Zn	Alloy 302	0.0025	0.0025	0.003	...	...	...	Rem
Unalloyed Zn	Alloy 308	0.013	0.007	0.004	...	...	...	Rem
Zn-Cu-Ti	Alloy 220	0.005	0.005	0.006	0.7	0.16	...	Rem
Zn-Al	Alloy 1311	0.003	0.06	0.006	0.005	...	0.3 max	Rem
Zn-Al	Alloy 1312	0.005	0.005	0.005	0.005	...	1.6-2.4	Rem

(a) Designations of Platt Brothers & Co.

Table 10. Nominal compositions of typical zinc-containing alloys

Alloy base	Generic name	Nominal composition, wt %
Copper	Gilding metal	Cu-5Zn
	Red brass	Cu-15Zn
	Cartridge brass	Cu-30Zn
	Muntz metal	Cu-40Zn
	Admiralty brass	Cu-28Zn-1Sn
	Aluminum brass	Cu-22Zn-2Al
	Architectural bronze	Cu-40Zn-3Pb
	Nickel silver B	Cu-27Zn-18Ni
Aluminum	AA 7075	Al-5.5Zn-2.5Mg-1.5Cu-0.3Cr
	AA 712	Al-5.8Zn-0.6Mg-0.5Cr-0.2Ti
	AA A380	Al-8.5Si-3.5Cu-2Zn-1Fe
Magnesium	ZK61A	Mg-6Zn-0.7Zr
	ZH62A	Mg-5.7Zn-1.8Th-0.7Zr
	ZE63A	Mg-5.7Zn-2.6RE-0.7Zr
Tin	Silver foil	Sn-10Zn
	Parson's white brass	Sn-35Zn-5Cu
	Queens metal	Sn-13Zn-17Pb-17Sb
Silver	Sterling silver solder	Ag-18Zn-2.5Cu
	French silver solder	Ag-17Zn-23Cu-18Cd
Gold	18-kt white gold	Au-18.5Ag-5.5Zn-1Cu
	10-kt white gold	Au-25Ni-25Cu-8.3Zn
Solders	Low temperature	70Sn-30Zn
	Intermediate temperature	40Cd-60Zn
	High temperature	95Zn-5Al

erations. Because of their reactivity they are employed in many chemical engineering applications, particularly in the pulp, paper, textile and linoleum industries. These materials are also used in several metal refining operations, in dry cell batteries, and in a large number of other minor applications.

ZINC CHEMICALS

The zinc chemicals of major importance are zinc oxide, zinc sulfide, zinc sulfate, zinc chloride, zinc chromate (zinc yellow), zinc stearate and zinc phosphate. Zinc oxide is by far the most important and accounts for about 10% of total zinc consumption. It is used in rubber, paints, ceramics, general chemical applications, agriculture and photocopying. Rubber applications represent about half of all zinc oxide usage. Here ZnO acts as an activating agent for the organic accelerators used to reduce the curing time for vulcanization. It is also present as a pigment in many paint and ceramic formulations. Zinc sulfide is also employed as a pigment, and, in combination with barium sulfate, is known as the pigment "lithopone." Zinc sulfide is used in phosphors for cathode-ray tubes and fluorescent lights, and is often modified with small amounts of cadmium and/or selenium to produce a wide range of colors. Zinc sulfate's principal applications are in the agricultural and textile industries. Zinc chloride is also used in processing and dyeing of textiles and fabrics.

SELECTED REFERENCES

Zinc—The Science and Technology of the Metal, Its Alloys and Compounds, by C. H. Mathewson: Reinhold Publishing Co., New York, 1964

Engineering Properties of Zinc Alloys, 2nd Ed.: International Lead Zinc Research Organization, Inc., New York, April 1981

Zinc Dust and Zinc Powder, by B. C. Hafford, W. E. Pepper and T. B. Lloyd: International Lead Zinc Research Organization, Inc., New York, May 1982

Zinc: Its Corrosion Resistance, by C. J. Slunder and W. K. Boyd (2nd Ed., revised and edited by T. K. Christman *et al*): Zinc Institute, Inc., New York, Aug 1983

Zinc Oxide—Properties and Applications, by H. E. Brown: International Lead Zinc Research Organization, Inc., New York, 1976

12 LEAD

By Jerome F. Smith, Lead Industries Association, Inc.

This section was adapted from Properties of Lead and Lead Alloys, Lead Industries Association, Inc., and Metals Handbook, Ninth Edition, Volume 2, Properties and Selection: Nonferrous Alloys and Pure Metals, pages 493 to 522. For more detailed information, the reader is referred to the larger works.

LEAD occurs in deposits widespread throughout the world. Although it is not considered a rare metal, it makes up only 0.0016% of the earth's crust. Lead ores do, however, tend to be concentrated in pockets, which makes mining of such ores economical. Lead's properties—high density, low melting point, corrosion resistance, malleability, unusual electrical properties, and ability to form useful alloys and chemical compounds—combined with its readily available forms and relative low cost, make it a unique material for solution of a variety of problems.

FORMS OF LEAD

The malleability of lead (and most of its alloys) allows it to be rolled to any desired thickness, from 2 to as little as 0.0005 in. Sheets are produced in standard widths up to 8 ft, with larger sizes available if necessary. Readily cut to desired size and shape, lead sheets can be fabricated into useful structures by lead welding (burning) or soldering. Sheet may be stamped to produce gaskets, washers and impact extrusion blanks. Tin-coated lead sheet can be produced by rolling lead and tin together.

Lead is extruded in the forms of pipe, rod, wire and practically any cross section, such as H-shape window cames, hollow stars, rectangular ducts, etc. Commercially available extrusions are available from 12-in. pipe down to 0.010-in.-diam solder wire. Lead is extruded around steel bars, as well as soft materials such as rubber- or plastic-covered power cables. Common flux-cored solders and collapsible tubes are typical of the variety of lead and lead alloy extrusions.

The low melting point of lead, 621 °F, makes it one of the simplest metals to cast at about 700 °F. It is used in massive counterweights, in sailboat keels, and as tiny die castings in instruments. Type metal is noted for its ability to produce fine detail, and storage-battery metal grids are examples of commercial lead castings. Small die castings, 0.05 in. in wall thickness and with as-cast dimensions reproducible to 0.001 in., are possible. Antimony, tin or arsenic, and other elements, may be alloyed with lead to produce certain properties such as castability, strength or greater hardness.

Lead shot is produced by taking advantage of the fact that the surface tension of lead is such that when molten lead is poured through a sieve, and allowed to free-fall, it forms perfect spheres before solidifying. The size of the shot is controlled by the sieve size. The shot is collected at the bottom of a shot tower and graded accurately for sizes, which are available from 0.04 up to

0.23 in. Larger-size shot is cast in permanent molds. Produced in abundance, the bulk of the shot is used as shotgun ammunition. Easily poured and free-flowing shot may be used when mass or shielding is required in an irregular enclosure.

Lead powder, particles and flakes are produced, usually by atomization, in diameters of 4 microns and up. The particles impart useful properties when added to grease and pipe-joint compounds. They are also used as constituents of powder metallurgy products, such as bearings, brake linings and clutch facings. Lead and lead alloy powders are also used to make solder pastes. As radiation-shielding and sound-attenuation products, lead powder is incorporated into rubber and plastics.

Lead wool, a loose rope of fibers weighing about 0.5 lb/ft, is produced by passing molten lead through a fine sieve and allowing it to solidify. The fibrous rope, when forced into a crevice under considerable force, cold welds into a homogeneous mass, forming a solid metal seal. This calking process is useful where temperature or explosion hazards prohibit the use of flame heating. The gaps between lead shielding sheets in nuclear submarines are often filled and calked with lead wool.

LEAD ALLOYS

Lead forms a wide range of low-melting alloys. It readily alloys with tin in all proportions, forming the tin-lead solders used widely in industry. Alloyed with antimony or calcium, lead is used both as castings and sheet in automotive and standby storage batteries. Bearing alloys consist of lead alloyed with combinations of antimony, arsenic, copper, etc., to yield suitable hardness and embeddability. Type metals are alloys containing antimony, tin and arsenic, with excellent casting and hardness as required by the graphic trades. When lead is added to other alloys, such as steel, brass and bronze, it promotes machinability, corrosion resistance or other special properties.

In addition to consideration of the physical and mechanical properties of lead, its chemical properties, mainly corrosion resistance, account for many of its uses. Lead is durable under varying weather conditions, exposure to most types of soil, marine and industrial atmospheres, and the action of many corrosive chemicals. Its resistance to sulfuric acid is used advantageously in the manufacture of the acid and in the most common method of storing electricity—the storage battery.

Alloyed with tin (from 7 to 15%), lead is used as a hot dip coating alloy to improve the corrosion resistance of steel sheet, known as terne-coated steel. Not only is the corrosion resistance enhanced, but the solderability and the drawing characteristics of the steel sheet are also improved. Electroplated lead and its alloys are used for the same purpose as are hot dip coatings.

Research and development programs taking advantage of the physical, mechanical and chemical properties of lead have yielded many alloys for specific industries. Typical compositions are shown in Tables 1 to 7.

The basic specification covering pig lead, as produced by most primary and secondary lead smelters in the United States, is the ANSI/ASTM B29-79 Standard Specification for Pig Lead (Table 1).

By agreement between purchaser and supplier, limits may be established for elements not shown in the specification, as well as variations in the specified limits—i.e., bismuth levels up to 0.15% may be allowed in common lead.

The ANSI/ASTM B29-79 specification is used widely in the plumbing trade and for chemical, radiation-protection and sound-attenuation applications.

The requirements for lead alloys for extrusion over paper-, plastic- or rubber-insulated copper telephone and power cables have led to the development of some typical compositions (Table 2). It is necessary that the sheathing have corrosion and fatigue resistance, along with suitable extrusion characteristics. In addition to the ANSI/

Table 1. Compositions of four grades of pig lead (ANSI/ASTM B29-79)

Grade of lead	UNS No.	Ag, max	Ag, min	Cu, max	Cu, min	Ag + Cu, max	As + Sb + Sn, max	Zn, max	Fe, max	Bi, max	Pb(a), min
Corroding	50042	0.0015	...	0.0015	...	0.0025	0.002	0.001	0.002	0.050	99.94
Common	50045	0.005	...	0.0015	...	...	0.002	0.001	0.002	0.050	99.94
Chemical	51120	0.020	0.002	0.080	0.040	...	0.002	0.001	0.002	0.005	99.90
Copper-bearing	51121	0.020	...	0.080	0.040	...	0.002	0.001	0.002	0.025	99.90

(a) By difference.

Table 2. Compositions of lead alloys for telephone and power cables

| UNS No. | Alloy designation | Nominal or preferred composition, wt % | | | | | | |
		As	Sn	Bi	Te	Sb	Cu	Ag
52605	1% antimony lead	...	...	0.04	...	1.0	0.0015 max	0.005 max
50310	F-3	0.15	0.10	0.10	...	Min	Min	...
	Gencalloy "A"	0.10-0.17	0.09-0.13	Min	...	Min	0.04-0.06	0.01 max
	Tellurium cable alloy	0.18-0.20	0.13-0.14	0.06-0.08	0.07-0.10	Min	Min	Min

Table 3. Compositions of antimonial lead storage-battery alloys

| UNS No. | Composition, wt % | | | | |
	Sb	Sn	As	Cu	Pb
52760	2.75	0.2	0.18	0.075	Rem
52765	2.75	0.3	0.3	0.075	Rem
52770	2.9	0.3	0.15	0.04	Rem
52840	2.9	0.3	0.15	0.05	Rem

Table 4. Compositions of calcium-lead storage-battery alloys

| UNS No. | Composition, wt % | | |
	Ca	Sn	Pb
50760	0.075	...	Rem
50770	0.10	...	Rem
50736	0.065	0.2	Rem
50737	0.065	0.5	Rem
50775	0.10	0.3	Rem
50780	0.10	0.5	Rem

Table 5. Compositions of type-metal alloys

| UNS No. | Designation | Composition, wt % | | |
		Sn	Sb	Pb
53425	Linotype	5	11	84
52830	Electrotype	3	3	94
53530	Stereotype	6	14	82
53570	Monotype	7	15	78
53750	Foundry type	12	24	64

Table 6. Compositions of lead casting alloys

| UNS No. | Composition, wt % | | | | |
	Sn	Sb	As, max	Cu, max	Pb
53560	4-6	14-16	0.15	0.5	79-81
53340	...	9.25-10.75	0.15	...	89-91
53470	0.75	12.75	0.4	0	86

Table 7. Compositions of lead-base bearing metals

| UNS No. | Composition, wt % | | | | |
	Sn	Sb	As	Cu	Pb
54727	25	13	...	3	59
53560	5	15	0.15	0.5	80
53620	1	15	1.4	0.6	82
53320	5	9	...	...	86

ASTM B29 unalloyed grades of lead, alloys in common use are shown also.

The highly developed lead acid storage battery is still the most economical means for storing electrical energy. The superior corrosion resistance of lead alloys in sulfuric acid, along with the required physical and electrical properties, have led to the development of a wide variety of alloys to meet certain battery requirements. Large communication standby batteries use virtually pure lead, while automotive batteries, where strength and vibration resistance are required, generally utilize sophisticated alloys. In general, the electrically conducting metal grid has been cast to

shape, but with the advent of the newer alloys for use in maintenance-free batteries (no need to add water), new production methods have been introduced—i.e., the slitting and stretching of rolled sheet.

The alloys now in use in automotive storage batteries are of two general types, antimonial leads (Table 3) and calcium-lead alloys (Table 4).

Alloys of lead, tin and antimony are used in the graphic arts as type metal, sleeve bearings, special castings and hollow slush castings. These alloys have fair mechanical and excellent antifrictional properties, both of which are useful in sleeve bearings. Its low melting point and ability to produce castings with very fine detail account for lead's use in the graphic arts. Table 5 gives compositions of typical type-metal alloys. The designations show the traditional uses for these alloys, but the same compositions are also used for many miscellaneous castings.

In order to obtain specific castings or mechanical properties, alloys such as those characterized in Table 6 have been developed for die casting or permanent mold gravity casting.

The excellent antifrictional properties, sufficient mechanical properties and low melting point account for the use of antimonial tin-lead alloys as sleeve-bearing materials. Arsenic and copper are sometimes added to improve the elevated-temperature and fatigue properties of the alloy, as shown in Table 7.

MECHANICAL PROPERTIES

Although the general properties of various grades of lead, UNS No. 50001 through 50090, are essentially the same, the properties of alloys with less than 99% lead may vary considerably. The reported mechanical properties of lead alloys are affected by a number of factors—e.g., grain size, phase distribution, thermal and fabricating history. The values reported in the literature rarely indicate all the test-specimen information and testing procedures. Alloys showing age hardening and some work hardening also may lose some of the property values on aging. Lead's low recrystallization and melting temperatures make it more difficult to determine property values than for aluminum, steel or copper, which melt at higher temperatures.

The mechanical properties of lead and its alloys are quite low compared to other metals when it is used as a structural member. The low properties have to be considered while taking advantage of the valuable properties, such as corrosion resistance. Quite often lead is used in conjunction with stronger structural members, such as lead sheet bonded to or supported by a steel structure in chemical process applications.

The limiting factors in the use of lead or its alloys as structural members are creep strength and fatigue resistance. Creep strength is generally accepted as the ability of a metal to withstand a sustained load without undue stretching or failure. For most design purposes, a limit of no more than $1/2$% elongation per year is ac-

cepted. For 99%+ lead, a value of about 200 psi per year has been accepted, although this represents only about 10% of the ultimate tensile strength.

Fatigue strength is the maximum cyclic stress that will not cause failure in a test specimen when stressed an indefinite number of cycles. Vibration or thermal cycling in the presence of oxygen are the chief causes of fatigue failure in lead or its alloys.

Typical mechanical properties of the alloys discussed in this section are presented in Table 8.

TIN-LEAD SOLDER ALLOYS

Tin-lead alloys are the most widely used of all solders. They have the advantage of a low melting range, which makes them ideal for joining most metals by convenient heating methods with little or no damage to heat-sensitive parts.

A pure metal always melts at a single temperature. Most solder alloys melt over a range of temperatures. The temperature at which a solder begins to melt is called the solidus. The temperature at which it is completely molten is the liquidus. Between these temperatures, part of the solder is molten and part is solid, thus the solder has a pasty consistency.

Care should be taken in specifying the correct solder for the job, since each alloy is unique with regard to its composition and, in general, its properties. Table 9 gives the melting characteristics of some tin-lead solders and lists their typical applications. When referring to tin-lead solders, the tin content is customarily given first—for example, 40/60 refers to 40% tin and 60% lead by weight.

The solders containing less than 5% tin are used for sealing precoated containers, for coating and joining metals, and for applications where service temperatures exceed 250 °F. At those temperatures, strength is taken care of by design, and the solder functions primarily as a seal. The 10/90, 15/85 and 20/80 solders are used for sealing cellular automobile radiators and filling seams and dents in automobile bodies.

General-purpose solders are 40/60 and 50/50. Soldering of automobile radiator cores, plumbing, electrical and electronic connections, roofing seams and heating units are but a few of the typical uses for these solders.

The 60/40 and 63/37 alloys are used where components are heat sensitive and where minimum heat should be used to make a solder joint. These alloys also provide the greatest ease and speed of joining. Electronic devices, computers and communications equipment are typical products using these solders.

For the electronics industry, silver is added to tin-lead solders to reduce the dissolution of silver from silver alloy coatings. Silver may also be added to improve creep resistance.

Tin/silver lead alloys exhibit good tensile, creep and shear strengths. Some are used for higher-temperature bonds in sequential soldering operations. Fatigue properties are increased by the addition of silver to the solder. The 1Sn-1.5Ag-97.5Pb solder is used in cryogenic equipment because it does not embrittle at low temperatures.

Table 10 gives compositions and melting temperatures of the most common silver-containing solders.

Lead alloyed with tin, bismuth, cadmium, indium or other elements, either alone or in com-

Table 8. Typical mechanical properties of selected lead alloys

UNS	Ultimate tensile strength pounds/square inch megapascal	Elongation %	Yield strength PSI megapascal	Shear strength PSI MPa	Hardness Brinell/ Rockwell (HR)	Creep MPa	Fatigue
50042	1740 PSI 12-13 MPa	30	7978 PSI 55 MPa	1810 PSI 12.5 MPa	3.2-4.5		3.2 MPa @ 10⁷ cycles
50132	5076 PSI 35 MPa 1.8 MPa @ 100 °C	28			13	19.5 MPa-1000 hrs. 7.5 MPa-1000 hrs. @ 100 °C	
50310	16-30 MPa	25-40			5-10	0.13%/yr @ 2.07 MPa	
50737	6500-7000 PSI 45 MPa	15					
50750	10,152 PSI 70 MPa	10	9570 PSI 66 MPa			28 MPa for 1000 hrs.	
50760	36-39 MPa	35-40			70-80 HR		
50770	5500-5900 PSI 38 MPa	30-45					
50775	5946 PSI 41-45 MPa	20-35			90-95 HR		
50780	6497 PSI 44.8-51.7 MPa	25-35			85-90 HR		
51110							
51120	2320 PSI 16-19 MPa	30-60	808 PSI 6-8 MPa		4-6	3% per year @ 2.07 MPa	4.3 MPa for 10⁷ cycles
51121							
52605	2900 PSI 20 MPa	50			7	2.4 MPa for 10⁻⁵ yr	
52830					12.4		
52901	4020 PSI 27.6 MPa	48			8.1		10.3 MPa for 10⁷ cycles
53425					22		
53530					23		
53565	10,000 PSI 69 MPa	5			20		27 MPa for 2 × 10⁷ cycles
53570					24		
53620	10,297 PSI 71 MPa	2			20		30 MPa for 2 × 10⁷ cycles
53750					33		
54210							4.7 MPa for 10⁷ cycles
54321	4060 PSI 218 MPa	55	1450 PSI 10 MPa		8		
54520	4350 PSI 30 MPa 8 MPa @ 100 °C	10 30% @ 100 °C			10	3.5 MPa for 1000 hrs. 1.1 MPa for 1000 hrs. (100 °C)	
54820	4930 PSI 34 MPa	18		4060 PSI 28 MPa	12	790 KPa for .01% per day	
54915	5367 PSI 37 MPa 6 MPa @ 100 °C	25 130		32 MPa	12	2.1 MPa for 1000 hrs.	
55030	4700 PSI 40.7 MPa	60	4790 PSI 33 MPa	5220 PSI 36 MPa	14		
55111	7610 PSI 52.5 MPa 19 MPa @ 100 °C	30-60 135-100 @ 100 °C		5380 PSI 37.1 MPa	16	2.9 MPa for 1000 hrs. .45 MPa for 1000 hrs. @ 100 °C	
55133	6235 PSI 43 MPa 19 MPa @ 100 °C	7				16 MPa for 1000 hrs. 2.7 MPa for 1000 hrs. @ 100 °C	

Table 9. Compositions, melting characteristics and typical uses of tin-lead solders

UNS No.	Composition, wt % Sn	Pb	Melting temperature, °F Solidus	Liquidus	Pasty range	Uses
54210	2	98	601	611	10	Side seams for can manufacturing.
54320	5	95	581	594	13	For automobile radiators.
54520	10	90	514	576	62	For coating and joining metals.
54560	15	85	440	551	110	For coating and joining metals. For filling dents or seams in automobile bodies.
54710	20	80	361	531	170	For machine and torch soldering.
54720	25	75	361	511	150	
54820	30	70	361	491	130	General purpose and wiping solder.
54850	35	65	361	477	116	Wiping solder for joining lead pipes and cable sheaths. For automobile radiator cores and heating units.
54915	40	60	361	460	99	
54950	45	55	361	441	80	For automobile radiator cores and roofing seams.
55030	50	50	361	421	60	For general purpose. Most popular of all.
13600	60	40	361	374	13	Primarily used in electronic soldering applications where low soldering temperatures are required.
13630	63	37	361	361	0	Lowest melting (eutectic) solder for electronic applications.

Table 10. Compositions and melting temperatures of common silver-containing solders

UNS No.	Composition, wt %			Melting temperature, °F	
	Sn	Ag	Pb	Solidus	Liquidus
50131 ...	1	1.5	97.5	588	588
55133 ...	62	2.0	36.0	354	372
50151		2.5	97.5	579	579

bination, forms alloys with particularly low melting points. Some of these alloys, which melt at temperatures even lower than the boiling point of water, are referred to as fusible alloys. They are used in automatic safety devices, such as fire-extinguishing sprinkler heads. Indium-containing alloys are used for low-temperature soldering or where gold scavenging must be avoided. Table 11 gives compositions of typical alloys, along with their melting points.

There are many other formulations for lead alloys that have been developed for very specific uses. The composition tables in this section show only typical formulas and are not meant to exclude other formulations, some of which are proprietary.

The following comprehensive tabulation of data has been compiled by Lead Industries Association, Inc., from selected sources believed to be the most reliable. The data shown are for unalloyed lead (i.e., 99.9% pure) with no appreciable alloying agents.

PROPERTIES OF UNALLOYED LEAD

Physical Properties

Color ....Bluish gray
Atomic number 82
Atomic arrangement
  (monomorphic) ......Face-centered cubic
Atomic weight 207.22
Isotopes of 203, 204, 205, 206, 207, 208, 209 and 210 have been determined, but common lead is considered to be a constant mixture of isotopes of 207.22 atomic weight.
Atomic volume at 20 °C
  (calculated) 18.27
Density:
  Ordinary lead at
    327.4 °C, just
    solid ....11.1005 g/cm³
  At 327.4 °C, just
    liquid ....10.686 g/cm³
  At 550 °C ....10.418 g/cm³
  At 800 °C ....10.132 g/cm³
  Rolled sheet at
    20 °C ....11.35 to 11.37 g/cm³
  Pure cast lead at
    20 °C ....11.34 g/cm³
Lead vapor is
  monatomic.
Hardness, Mohs
  scale 1.5
Density, vapor
  (hydrogen 1)
  (calculated) 103.6

Vapor at 1525 °C
  (calculated) ....1.67 mg/cm³
Weight:
  Pure cast lead at 20 °C
    (calculated) ....0.4092 lb/in.³
  Equivalent to ......707 lb/ft³
  Rolled (density, 11.37)
    (calculated) ......709 lb/ft³
  Liquid at 327.4 °C
    (calculated) ......666 lb/ft³
  Sheet lead, 1 ft² by
    1/64 in. thick ......1 lb (approx)
  Volume of 1 lb of cast
    lead at 20 °C
    (calculated) ....2.44 in.³

Thermal Properties

Melting point ....327.4 °C (621 °F)
Increase in melting point
  for each increase of
  150 atm pressure ....1.2 °C
Boiling point
  (temperature of visible
  ebullition) ....1525 °C (2777 °F)
Vapor pressure:
  At 808 °C ....0.08 mm Hg
  At 1000 °C ....1.77 mm Hg
  At 1200 °C ....23.29 mm Hg
  At 1365 °C ....166.0 mm Hg
  At 1525 °C ....760.0 mm Hg
  At 1870 °C ....6.3 atm
  At 2100 °C ....11.7 atm
Specific Heat:
  At −150 °C ....0.02805 cal/g
  At −100 °C ....0.02880 cal/g
  At −50 °C ....0.02949 cal/g
  At 0 °C ....0.0302 cal/g
  At 50 °C ....0.03054 cal/g
  At 100 °C ....0.03155 cal/g
  At 300 °C ....0.0338 cal/g
  At 327.3 °C (solid) ..0.0358 cal/g
  At 327.4 °C (liquid) ..0.0335 cal/g
  At 378 °C ....0.0338 cal/g
  At 418 °C ....0.0335 cal/g
  At 459 °C ....0.0335 cal/g
Atomic heat per °C
  (calculated) 6.26
Latent heat of fusion:
  Per gram ....5.47 g·cal
  Per mole ....1133 g·cal
To melt 1 lb of lead,
  heating from 20 °C,
  requires ....7100 g·cal,
                     or 28.4 Btu
Latent heat of
  vaporization, per
  gram ....223 g·cal
Thermal conductivity
  (silver = 100) 8.2
Thermal conductivity, per °C/s/cm³:
  At −247.1 °C ......0.117 g·cal
  At −160 °C ....0.092 g·cal
  At 0 °C ....0.083 g·cal
  At 100 °C ....0.081 g·cal
  At 200 °C ....0.077 g·cal

  At 300 °C ....0.074 g·cal
  At 400 °C ....0.038 g·cal
  At 500 °C ....0.037 g·cal
  At 600 °C ....0.036 g·cal
Coefficient of linear
  expansion at −190 to
  +19 °C ....0.0000265 per °C
Coefficient of linear
  expansion at 17 to
  100 °C ....0.0000293 per °C
  Equivalent to ......0.0000163 per °F
Coefficient of cubical expansion:
  Liquid at melting point
    to 357 °C ....0.0000129 per °C
  At 17 to 100 °C
    (calculated) ....0.0000879 per °C
Increase in volume from
  20 °C to liquid at
  melting point 6.1%
Decrease in volume on
  solidification 3.85%
Increase in volume on
  melting 4.01%
Shrinkage on casting
  taken in practice as ..7/64 to 5/16 in./ft
Shrinkage in volume
  calculated from liquid
  at melting point to
  20 °C 5.8%
Recrystallization,
  minimum possible
  temperature ....−33 °C (approx)

Electrical Characteristics

Specific electrical resistance:
  At − 183 °C ....6.02 μΩ/cm³
  At −78 °C ....14.1 μΩ/cm³
  At 0 °C ....20.4 μΩ/cm³
  At 90.4 °C ....28 μΩ/cm³
  At 196.1 °C ....36.9 μΩ/cm³
  At 318 °C ....94 μΩ/cm³
  At 600 °C ....107.2 μΩ/cm³
Temperature coefficient
  of electrical
  resistivity ....−0.00411 to 0.0043
                       per °C
Specific electrical conductance:
  At 0 °C ....5.05 × 10⁴ cm⁻¹·Ω⁻¹
  At 18 °C ....4.83 × 10⁴ cm⁻¹·Ω⁻¹
  At melting point ....1.06 × 10⁴ cm⁻¹·Ω⁻¹
Atomic electrical
  conductance
  (calculated) ....1.139 × 10⁶
Relative electrical
  conductance (copper =
  100) 7.82
Relative electrical
  resistance (copper =
  100) 1280
Electrolytic solution
  pressure:
  Ions of Pb⁺⁺ ....6.3 × 10⁻⁵ atm
  Ions of Pb⁺⁺⁺⁺ .....3.0 × 10⁻⁷⁴ atm
Electrochemical equivalence:

Valence	mg/C	C/mg	g/A·h
1	2.1473	0.4657	7.7302
2	1.0736	0.9314	3.8651
3	0.5368	1.8628	1.9326

Electrodeposition of 1
  lb of lead requires ..117 A·h
Potential of lead electrode:
  With normal calomel
  electrolyte ....0.415 V

Table 11. Compositions and melting temperatures of low-melting lead alloys

UNS No.	Composition, wt %					Melting temperature, °C	
	Sn	Bi	In	Cd	Pb	Solidus	Liquidus
50620	8.3	44.7	19.1	5.3	22.6	47	47
50640	12.0	49.9	21.0	...	18.0	58	58
50645	9.3	50.0	...	6.2	34.5	70	78
50665	15.5	52.5	...	...	32.0	95	95
50680	...	55.5	...	...	44.5	124	124

With normal
hydrogen
electrolyte ....0.132 V
On absolute scale ..0.145 V
Thermopotential vs
platinum ....+0.44 mV/°C
Magnetic susceptibility, per gram:
At 18 to 330 °C ...−0.12 × 10⁻⁶
At 300 to 600 °C ..−0.08 × 10⁻⁶

CORROSION RESISTANCE

Lead has such a successful record of service in exposure to the atmosphere and to water that its resistance to corrosion by these media is often taken for granted. Underground, thousands of kilometres of lead-sheathed cable and lead pipe give reliable, long-term performance all over the world. In the chemical industry, lead is a major constituent in corrosion-resistant equipment necessary for handling many chemicals (see Table 12).

Corrosion of lead in aqueous electrolytes is an electrochemical process. The metal either enters the solution at anodic sites as metallic cations or is converted anodically to solid compounds. Both corrosion reactions can be represented by the reaction:

$$Pb - 2e \rightarrow Pb^{++}$$

This oxidation reaction, which takes place at anodic sites, is accompanied by a reduction of some constituent in the electrolyte at cathodic sites. In neutral salt solutions, the cathodic reaction is the reduction of dissolved oxygen:

$$\tfrac{1}{2}O_2 + H_2O + 2e^- \rightarrow 2OH^-$$

In acid solutions free of oxygen, the corresponding cathodic reaction is:

$$2H^+ + 2e^- \rightarrow H_2$$

Rate of corrosion is a function of the current flowing between the anodes and cathodes of the corrosion cell. Many factors and conditions can initiate or influence this flow of current. In corrosion of a single metal such as lead, local anodes and cathodes may be set up as a result of inclusions, inhomogeneities, stress variations and differences in temperature. In bimetallic (galvanic) corrosion, the anodic and cathodic sites are on different metals, with the less noble metal (anode) corroding in preference to the more noble metal (cathode).

In most environments, lead is cathodic to steel, aluminum, zinc, cadmium and magnesium, and thus will accelerate corrosion of these metals. With titanium and passivated stainless steels, lead is the anode of the cell and suffers accelerated attack. In either instance, rate of corrosion is governed by the difference in potential between the two metals, the ratio of their areas, and their polarization characteristics.

The corrosion rate of lead usually is under anodic control, because the most important determinant usually is the solubility and other physical characteristics of the corrosion products formed at anodic sites. The majority of these products are relatively insoluble lead salts that are deposited on the lead surface as impervious films, which tend to stifle further attack. The formation of such insoluble protective films is responsible for the high resistance of lead to corrosion by sulfuric, chromic and phosphoric acids.

In general, anything causing injury to the protective film increases corrosion rate. Factors that

help create or strengthen the film reduce corrosion rate. Therefore, the life of lead-protected equipment can be extended, for example, by washing it with film-forming aqueous solutions containing sulfates, carbonates or silicates. This procedure is suggested for protecting lead when it will be in contact with corrosives that do not form protective films.

Corrosion in Water

Extensive service experience and laboratory testing have indicated that corrosion rate of lead generally is quite low in a wide variety of waters. The only major applications where lead cannot

Table 12. Corrosion of lead in hydrochloric acid at 24 °C (75 °F)

HCl concentration, %	Chemical lead μm/yr	Chemical lead mils/yr	6% antimonial lead μm/yr	6% antimonial lead mils/yr
1%	610	24	840	33
5%	410	16	510	20
10%	560	22	1 090	43
15%	790	31	3 810	150
20%	1880	74	4 060	160
25%	4830	190	5 080	200
35%(a)	8890	350	13 720	540

(a) Commercially concentrated HCl.

Table 13. Corrosion of chemical lead in industrial and domestic waters(a)

Type of water	Temperature °C	Temperature °F	Aeration	Agitation	Corrosion rate μm/yr	Corrosion rate mils/yr
Condensed steam, traces of acid	21-38	70-100	None	Slow	21.59	0.85
Mine water:						
pH 8.3, 110 ppm hardness	20	68	Yes	Slow	6.60	0.26
160 ppm hardness	19	67	Yes	Slow	7.11	0.28
110 ppm hardness	22	72	Yes	Slow	6.35	0.25
Cooling-tower water, oxygenated, from Lake Erie	16-29	60-85	Complete	None	134.6	5.3
Los Angeles aqueduct water, treated with chlorine and copper sulfate	Ambient		...	0.5 ft/s	9.65	0.38
Spray cooling water, chromate treated	16	60	Yes	...	9.4	0.37

(a) Total immersion.

Table 14. Corrosion of lead in natural waters

Location and type of water	Type of test	Agitation	Corrosion rate μm/yr	Corrosion rate mils/yr	Ref
Bristol Channel; seawater	Immersion about 93% of the time	...	12.7	0.50	1
Southhampton Docks; seawater	Half tide level	...	2.79	0.11	2
Gatun Lake, CZ; tropical fresh water	Immersion	None	2.03	0.08	3
Fort Amador, CZ; tropical Pacific Ocean	Immersion	Flowing(a)	9.14	0.36	
Fort Amador, CZ; tropical Pacific Ocean	Mean tide level	Flowing(a)	5.08	0.20	
San Francisco Harbor; seawater	Mean tide level	Flowing	10.67	0.42	
Port Hueneme Harbor, CA; seawater	Immersion	Flowing(b)	5.59	0.22	4
Kure Beach, NC; seawater	Immersion	...	15.24	0.60	

(a) At 150 mm/s (0.5 ft/s). (b) At 60 mm/s (0.2 ft/s).

Table 15. Corrosion of lead in various natural outdoor atmospheres

Location	Type of atmosphere	Duration of test, years	Type of lead	Corrosion rate μm/yr	Corrosion rate mils/yr	Ref
Altoona, PA	Industrial	10	Chemical	0.737	0.029	5, 6
			Pb-1Sb	0.584	0.023	5, 6
New York City	Industrial	20	Chemical	0.381	0.015	5, 6
			Pb-1Sb	0.330	0.013	5, 6
Sandy Hook, NJ	Seacoast	20	Chemical	0.533	0.021	5, 6
			Pb-1Sb	0.508	0.020	5, 6
Key West, FL	Seacoast	10	Chemical	0.584	0.023	5, 6
			Pb-1Sb	0.559	0.022	5, 6
LaJolla, CA	Seacoast	20	Chemical	0.533	0.021	5, 6
			Pb-1Sb	0.584	0.023	5, 6
State College, PA	Rural	20	Chemical	0.330	0.013	5, 6
			Pb-1Sb	0.356	0.014	5, 6
Phoenix, AZ	Semi-arid	20	Chemical	0.102	0.004	5, 6
			Pb-1Sb	0.308	0.012	5, 6
Kure Beach, NC (80 ft site)	East coast, marine	2	Chemical	1.321	0.052	7
			Pb-6Sb	1.041	0.041	7
Newark, NJ	Industrial	2	Chemical	1.473	0.058	7
			Pb-6Sb	1.067	0.042	7
Point Reyes, CA	West coast, marine	2	Chemical	0.914	0.036	7
			Pb-6Sb	0.660	0.026	7
State College, PA	Rural	2	Chemical	1.397	0.055	7
			Pb-6Sb	0.991	0.039	7
Birmingham, England	Urban	7	99.96%Pb	0.939	0.037	8
			Pb-1.6Sb	0.102	0.004	8
Wakefield, England	Industrial	1	99.995%Pb	1.879	0.074	8
Southport, England	Marine	1	99.995%Pb	1.778	0.070	8
Bourneville, England	Suburban	1	99.995%Pb	1.956	0.077	8
Cardington, England	Rural	1	99.995%Pb	1.422	0.056	8
Cristobal, CZ	Tropical, marine	8	Chemical	1.346	0.053	3
Miraflores, CZ	Tropical, marine	8	Chemical	0.762	0.030	3

Table 16. Corrosion of lead alloys in various soils(a)

Type of soil	Chemical lead(b) Corrosion rate μm/yr	Corrosion rate mils/yr	Max pit depth μm	Max pit depth mils	Tellurium lead(c) Corrosion rate μm/yr	Corrosion rate mils/yr	Max pit depth μm	Max pit depth mils	Antimonial lead(d) Corrosion rate μm/yr	Corrosion rate mils/yr	Max pit depth μm	Max pit depth mils
Cecil clay loam	<2.54	<0.1	457	18	<2.54	<0.1	406	16	<2.54	<0.1	229	9
Hagerstown loam	<2.54	<0.1	787	31	<2.54	<0.1	762	30	<2.54	<0.1	406	16
Lake Charles clay	7.62	0.3	2540	100	10.16	0.4	2718	107	10.16	0.4	2642	104
Muck	7.62	0.3	1321	52	7.62	0.3	1346	53	7.62	0.3	1295	51
Carlisle muck	5.08	0.2	508	20	5.08	0.2	533	21	2.54	0.1	305	12
Rifle peat	<2.54	<0.1	838	33	<2.54	<0.1	584	23	<2.54	<0.1	711	28
Sharkey clay	7.62	0.3	1778	70	7.62	0.3	1854	73	10.16	0.4	2261	89
Susquehanna clay	<2.54	<0.1	864	34	2.54	0.1	1016	40	2.54	0.1	356	14
Tidal marsh	<0.25	<0.01	305	12	<0.25	<0.01	203	8	<0.25	<0.01	152	6
Docas clay	<2.54	<0.01	635	25	<2.54	<0.1	432	17	<2.54	<0.1	483	19
Chino silt loam	<2.54	<0.1	381	15	<2.54	<0.1	508	20	<2.54	<0.1	178	7
Mohave fine gravelly clay	<2.54	<0.1	610	24	<2.54	<0.1	584	23	2.54	<0.1	406	16
Cinders	7.62	0.3	2159	85	7.62	0.3	1549	61	10.16	0.4	1168	46
Merced silt loam	<2.54	<0.1	610	24	<2.54	<0.1	406	16	<2.54	<0.1	229	9

(a) Maximum exposure time, 11 years. (b) 0.056 Cu, 0.002 Bi, 0.001 Sb. (c) 0.08 Cu, 0.01 Sb, 0.043 Te. (d) 0.036 Cu, 5.3 Sb, 0.016 Bi.

be used are those involving some pure waters containing oxygen and soft natural waters, especially if contamination is of concern. In contrast, as discussed above, addition of calcium and magnesium salts further enhances resistance of lead to corrosion by water.

Corrosion rates of chemical lead in several industrial and domestic waters are presented in Table 13. It should be noted that corrosion rate is relatively low, even where water hardness is below 125 ppm. (A corrosion rate for a fresh water is also included among the data for seawater in Table 14.)

Atmospheric Corrosion

Lead in most of its forms exhibits consistent durability in all types of atmospheric exposure, including industrial, rural and marine (see Table 15). These three atmospheric environments are distinct, because each involves different factors that promote corrosion. In rural areas, which are relatively free of pollutants, the only important environmental factors influencing corrosion rate are humidity, rainfall and air flow. However, near or on the sea, chlorides entrained in marine air often exert a strong effect on corrosivity. In industrial environments, sulfur oxide gases and the minerals in solid emissions considerably change patterns of corrosion behavior. However, the protective films that form on lead and its alloys are so effective that corrosion is insignificant in most natural atmospheres.

Corrosion in Underground Ducts

Lead is used extensively in the form of sheathing for power and communications cables because of its impermeability to water and its excellent resistance to corrosion in a wide variety of soil conditions. Cables are either buried directly in the ground or installed in ducts or conduits. In the United States, the preferred method is to lay cable in ducts or conduits made of materials such as cement, vitrified clay or wood.

Severe corrosion of lead in underground service (in ducts or directly in the soil) is the exception rather than the rule. However, because

Table 17. Solubility of lead compounds

Lead compound	Formula	Temperature °C	°F	Solubility(a), Kg/m³
Acetate	$Pb(C_2H_3O_2)_2$	20	68	433
Bromide	$PbBr_2$	20	68	8.441
Carbonate	$PbCO_3$	20	68	0.0011
Basic carbonate	$2PbCO_3, Pb(OH)_2$	...	...	Insoluble
Chlorate	$Pb(ClO_3)_2, H_2O$	18	64	0.513
Chloride	$PbCl_2$	20	68	9.9
Chromate	$PbCrO_4$	25	77	0.000058
Fluoride	PbF_2	18	64	0.64
Hydroxide	$Pb(OH)_2$	18	64	0.155
Iodide	PbI_2	18	64	0.63
Nitrate	$Pb(NO_3)_2$	18	64	565
Oxalate	PbC_2O_4	18	64	0.0016
Oxide	PbO	18	64	0.017
Orthophosphate	$Pb_3(PO_4)_2$	18	64	0.00014
Sulfate	$PbSO_4$	25	77	0.0425
Sulfide	PbS	18	64	0.1244
Sulfite	$PbSO_3$	...	...	Insoluble

(a) In water at room temperature.

repair or replacement of underground components is difficult and expensive, proper corrosion protection is recommended in any underground service.

Corrosion in Soil

Soils vary widely in physical and chemical characteristics and, consequently, in corrosive effect. A summary of the characteristics and properties of soils is given in the National Bureau of Standards circular on underground corrosion.

The data in Table 16 show that in most soils the average corrosion rate of lead is low—from less than 2.5 to 10 μm (0.1 to 0.4 mil) per year. It should be noted, however, that depth of pitting often is a more important measure of underground corrosion behavior than corrosion rate.

Resistance to Chemicals

The excellent resistance of lead and lead alloys to corrosion by a wide variety of chemicals is attributed to the polarization of local anodes caused by formation of a relatively insoluble surface film

of lead corrosion products. The extent of protection depends on the compactness, adherence and solubility of these films.

Solubilities of various lead compounds in water at room temperature are given in Table 17. These data are general indicators of the behavior of lead in solutions that promote formation of these compounds. The solubility of a lead corrosion product, however, depends on the solution in which the lead is immersed. Therefore, the solubility of that corrosion product in water is not always an adequate indicator of its behavior in another solution. This fact is illustrated by the variation in solubility of lead sulfate in sulfuric acid as acid concentration and temperature change. The lead sulfate film is less soluble in sulfuric acid solutions than it is in water. Solubility drops to a minimum value at acid concentrations of 30 to 60% and then increases at higher concentrations. At intermediate concentrations, the sulfate film is so insoluble that corrosion is negligible.

Increases in temperature generally increase corrosion rate. This effect is primarily due to increases in film solubility.

13 PRECIOUS METALS

By Alan A. Johnson and J. A. von Fraunhofer, University of Louisville

Additional and more detailed information on precious metals can be found in Metals Handbook, Ninth Edition, Volume 2, Properties and Selection: Nonferrous Alloys and Pure Metals, pages 657 to 705.

THE EIGHT PRECIOUS METALS are ruthenium, rhodium, palladium, silver, osmium, iridium, platinum and gold, and they occupy contiguous positions in the periodic table. Ruthenium, rhodium, palladium and silver have atomic numbers 44, 45, 46 and 47, respectively, and are in period 5. Osmium, iridium, platinum and gold have atomic numbers 76, 77, 78 and 79, respectively, and are in period 6. The six platinum-group metals—ruthenium, rhodium, palladium, osmium, iridium and platinum—are in group VIII, while silver and gold are in group Ib. Thus, together they occupy a 2 × 4 block of positions in the periodic table (see Table 1).

The platinum-group metals consist of two triads. The period 5 triad contains ruthenium, rhodium and palladium; the period 6 triad contains osmium, iridium and platinum. In some ways the members of a triad are similar, while in others corresponding members of the two triads seem to be related. Some of these relationships can be seen by examining Table 2, which shows the crystal structures and some of the properties of all eight precious metals.

Four of the platinum metals—rhodium, palladium, iridium and platinum—have face-centered cubic structures, while two—ruthenium and osmium—have hexagonal close-packed structures and are the first members of their respective triads. In each triad, the melting point decreases with increasing atomic number, and each metal in the second triad has a higher melting point than the corresponding metal in the first triad. In contrast to this behavior, the densities of all the metals in the first triad are close to 12 Mg/m³, while the densities of those in the second triad are all close to 22 Mg/m³. The first triad is sometimes referred to as the "light triad" and the second triad as the "heavy triad."

Although there is not much variation among the linear-expansion coefficients of the six metals, the metals of the first triad have somewhat higher values than the corresponding metals of the second triad. There is, however, a clear distinction between the specific-heat values for the two triads. The metals in the first triad all have specific heats close to 240 J/kg·K, while those in the second triad have values close to 130 J/kg·K.

In both triads, the second metal has the highest thermal conductivity. As would be expected from the Wiedemann-Franz law, electrical resistivity shows the inverse behavior. The second metal in each triad has the highest electrical resistivity. No over-all differences in thermal conductivity and electrical resistivity are apparent between the two triads.

Properties of silver and gold have been included in Tables 2 and 3. Like four of the platinum metals, silver and gold have face-centered cubic crystal structures. Silver comes after the first platinum-metal triad and gold after the second. Tables 2 and 3 show that at least some of their properties continue the trends established in the triads. For example, the melting points of silver and gold continue the trend of decreasing melting point with increasing atomic number exhibited by both triads. The density of silver (10.49 Mg/m³) is reasonably close to the density of the first-triad metals, which are all close to 12 Mg/m³, while the density of gold (19.32 Mg/m³) is reasonably close to the values for the second-triad metals, which are all close to 22 Mg/m³.

The linear-expansion coefficients of silver and gold are higher than those of the platinum-group metals. When the values for silver and gold are included next to their preceding triads, as in Table 2, it appears that there may be a systematic increase of the expansion coefficient with atomic number. The specific heats of silver and gold are closely similar to those of the metals in the triads that precede them. It is in the thermal conductivity and electrical resistivity of silver and gold that large differences between them and the platinum-group metals are apparent. The thermal conductivities of silver (428 W/m·K) and gold (318 W/m·K) are both at least twice the values for any platinum metal. Similarly, the electrical resistivities of silver (15.9 nΩ·m) and gold (23.5 nΩ·m) are substantially lower than those of the platinum-group metals. Silver has the lowest room-temperature electrical resistivity of all of the precious metals.

It should be realized that the property values given in Tables 2 and 3 are approximate. They may depend to some extent on impurity content and on lattice-defect content as determined by prior mechanical and thermal history.

Some of the mechanical properties of the precious metals, such as the elastic moduli, are sufficiently insensitive to impurity content that

Table 1. Positions of precious metals in the periodic table(a)

Period	Group VIII			Group Ib
5	Ruthenium (44)	Rhodium (45)	Palladium (46)	Silver (47)
6	Osmium (76)	Iridium (77)	Platinum (78)	Gold (79)

(a) Numbers in parentheses are atomic numbers.

Table 2. Crystal structures and physical properties of precious metals

Metal	Crystal structure	Melting point, °C	Density at 20 °C, Mg/m³	Coefficient of linear expansion at 20 °C, µm/m·K	Specific heat, J/kg·K	Thermal conductivity, W/m·K
Ruthenium	hcp	2310	12.45	5.05	240	110
Rhodium	fcc	1963	12.41	8.3	247	150
Palladium	fcc	1552	12.02	11.8	245	70
Silver	fcc	962	10.49	19.0	235	428
Osmium	hcp	~2700	22.57	2.6	130	60.5
Iridium	fcc	2447	22.65	6.8	130	147
Platinum	fcc	1769	21.45	9.1	134	71.1
Gold	fcc	1064	19.30	14.2	128	318

Table 3. Electrical properties of precious metals

Metal	Electrical resistivity at 20° C, nΩ·m	Superconducting transition temperature, K	Critical field, Oe
Ruthenium ...	76	2.04	66
Rhodium	45	...	...
Palladium ...	108	...	...
Silver	15.9	...	...
Osmium	95	0.655	65
Iridium	53	0.14	19
Platinum	106	...	...
Gold	23.5	...	...

meaningful values can be given for the nominally pure metals. Others, such as the yield stress, ultimate tensile strength and hardness, are so impurity sensitive that only typical values can be given. Values of the latter reported in the literature show a great deal of variation, as they do for all other nominally pure metals. The property values given in Table 4 should be interpreted accordingly. The values of tensile strength and hardness are for fully annealed specimens. The values for cold worked specimens are of course much greater and depend on the amount of cold work.

There is a qualitative difference between the two precious metals that have a close-packed hexagonal structure (ruthenium and osmium) and the others. These two metals are brittle, or nearly brittle, at room temperature, while the others exhibit limited ductility. Rhodium and iridium have limited ductility, usually about 5 to 10% elongation in a conventional tensile test, but the others have a high degree of ductility.

The literature on the phase diagrams of pairs of precious metals is confusing and the information presented is, in many cases, incomplete or contradictory. Seventeen of the twenty-eight diagrams are reasonably well established, and these are listed in Table 5. The information on which this list is based, and fragmentary information on most of the other eleven systems, can be found in the standard English language compilations of phase diagrams of Hansen, Elliot, Shunk and Moffatt, and in the book *Phase Diagrams of the Precious Metals,* by He Chunxiao *et al,* published in a bilingual format in Peking, China.

The established diagrams are remarkable in that they are all simple. The Ag-Au, Ag-Pd, Au-Pd, Os-Ru, Pd-Pt and Pt-Rh systems all exhibit a full range of solid state miscibility. The Au-Pt, Ir-Pt and Pd-Rh systems exhibit incomplete miscibility in the solid state in that they have phase diagrams each containing a miscibility gap. The Ag-Ru System exhibits substantial immiscibility in the solid state since its phase diagram has a simple eutectic reaction.

The remaining seven systems, Ag-Pt, Au-Ru, Ir-Ru, Os-Pd, Pd-Ru, and Rh-Ru, all exhibit simple peritectic reactions. Six of them are systems in which one component is close-packed hexagonal and the other face-centered cubic. It has been suggested that all such systems may have peritectic reactions.

The book by He Chunxiao contains at least partial information on more than two hundred ternary systems involving at least one precious metal. In nine systems, Ag-Au-Pd, Ag-Au-Pt, Ag-Ir-Pd, Ag-Pd-Pt, Ag-Pd-Rh, Au-Pd-Pt, Au-Pd-Ru, and Au-Pt-Rh, all three components are precious metals.

The platinum metals are obtained mainly from lode deposits in South Africa, the USSR and Canada, where they usually are associated with nickel and copper sulfides. The yield from these sources contains 80 to 85% platinum and palladium. Placer deposits, mainly in South Africa, produce mainly osmium and iridium. The total output from placer deposits accounts for only about 2% of world production of the platinum-group metals.

The United States is a net importer of silver. Domestic production is, however, substantial. It is, to some extent, derived from silver ores, but it is obtained principally as a by-product of the production of copper, lead, zinc and other met-

Table 4. Some room-temperature mechanical properties of the precious metals

Metal	Young's modulus, GPa	Tensile strength, MPa (typical)	Hardness, HV (typical)
Ruthenium	494	500	400
Rhodium	394	430	130
Palladium	126	200	50
Silver	71	130	27
Osmium	581	—	500
Iridium	549	500	200
Platinum	177	140	48
Gold	80	130	25

Table 5. Established binary phase diagrams for which both components are precious metals

System	Type of phase diagram
Ag-Au	Complete miscibility
Ag-Pd	Complete miscibility
Ag-Pt	Peritectic reaction
Ag-Ru	Eutectic reaction
Au-Pd	Complete miscibility
Au-Pt	Miscibility gap
Au-Ru	Peritectic reaction
Ir-Pt	Miscibility gap
Ir-Ru	Peritectic reaction
Os-Pd	Peritectic reaction
Os-Ru	Complete miscibility
Pd-Pt	Complete miscibility
Pd-Rh	Miscibility gap
Pd-Ru	Peritectic reaction
Pt-Rh	Complete miscibility
Pt-Ru	Peritectic reaction
Rh-Ru	Peritectic reaction

als. Enough gold is produced in the United States to satisfy about a quarter of the domestic demand. It comes from domestic and imported ores and is a by-product of the production of copper and other metals. Nearly one-half of all the gold that has been mined in the world is contained in the vaults of various governments and is thus unavailable for use.

The domestic production and demand for precious metals in the United States varies from year to year. Statistics are published periodically by the U.S. Bureau of Mines in *Mineral Commodity Profiles* and *Mineral Industry Surveys.* The reader is referred to the most recent issues of these publications for up-to-date information. What is quite apparent, however, is that the United States depends heavily on overseas sources for precious metals and is vulnerable to possible future shortages of them because of this dependence.

This dependency on overseas sources is all the more unfortunate because the use of these metals pervades our everyday life. Their uses are discussed in some detail below. Gold, silver and platinum alloys are used heavily in making jewelry; gold and other alloys are used in restorative dentistry; and sterling silver is used for making tableware. Less obvious, but equally pervasive, are the use of the platinum-group metals as catalysts in automotive pollution-control equipment and for making the spinnerets used in the manufacture of artificial fibers; the use of silver in photographic processes; the use of gold alloys as solders in solid-state circuits and electrical contacts; and many other applications.

Precious metals are used mainly for their exceptional oxidation and corrosion resistance. Some of these characteristics are shown in Tables 6 and 7. These two tables are intended to provide a general picture of how pure precious metals perform and should not be used as a basis for selecting a material for a specific application. For most applications, an alloy is more suitable than a pure metal because of the increased strength and other property changes brought about by alloying. Gold and silver are readily fabricated, but the platinum metals vary greatly in their fabricability (see Table 8).

The precious metals are bought and sold in kilograms and troy ounces. One kilogram is equal to 32.15 troy ounces. One troy ounce is equal to 20 pennyweights, 480 grains or 1.097 avoirdupois ounces. The term "fineness" is used to express the proportion by weight of gold or silver in a material in parts per thousand. Pure gold or silver is described as 1000 fine; sterling silver, which contains 92.5% Ag and 7.5% Cu, is 925 fine; the U.S. coin silver used to mint coins prior to 1965, which contains 90% Ag and 10% Cu, is 900 fine. The weight percentage of gold in an alloy can also be expressed in "karats." Pure gold is described as 24 karat, while the number of karats used to describe an alloy is the weight fraction of gold in that alloy multiplied by 24. Thus, an alloy containing 75% Au can be described as (0.75 × 24) karat—i.e., 18 karat.

Table 6. Oxidation characteristics of precious metals

Metal	Oxidation characteristics
Ruthenium ...	Forms a volatile oxide
Rhodium	Good oxidation resistance at all temperatures
Palladium ...	Forms an oxide film in the range 400 to 800 °C
Silver	Oxide is marginally stable near room temperature and unstable at elevated temperatures
Osmium ...	Forms a volatile oxide
Iridium	Forms an oxide film in the range 600 to 1000 °C
Platinum ...	Does not oxidize at any temperature
Gold	Does not oxidize at any temperature

Table 7. Corrosion characteristics of precious metals

Metal	Corrosion characteristics
Ruthenium ...	Resistant to aqua regia; resistant to most acids up to about 100 °C
Rhodium	Resistant to boiling aqua regia and hot HCl; resistant to hot Cl
Palladium ...	Corrosion resistance is less than that of platinum; poor resistance to highly oxidizing environments
Silver	Resistant to food products and hot concentrated solutions of many organic acids; resistant to hot caustic alkali solutions; resistant to wet Cl_2 gas at room temperature but not above
Osmium	Somewhat less corrosion resistant than other platinum-group metals
Iridium	Resistant to boiling aqua regia and hot HCl; resistant to hot Cl
Platinum	Resistant to HNO_3, HF and H_2SO_4 at high temperatures; resistant to perchlorates, persulfates and hypochlorites; poor resistance to Cl at high temperatures
Gold	Resistant to hot HNO_3 and H_2SO_4; resistant to dry HCl gas at high temperatures; attacked by HNO_3 + HCl and HNO_3 + H_2SO_4 mixtures; poor resistance to Cl above about 80 °C

In terms of the quantities used in the United States each year, the major uses of silver, in order of importance, are photography, electrical and electronic equipment and components, silverware, and appliance manufacture. For gold the important sources of demand are jewelry and the arts, various industrial applications, dental materials, and investment products. The platinum-group metals are used mainly for automotive applications, electrical equipment and components, various applications in the chemical industry, and dental materials.

INDUSTRIAL APPLICATIONS

Precious metals and their alloys are used extensively as catalysts. In terms of volume, the most important application is now the use of a Pt-Rh alloy in automobile pollution-control equipment. Examples of the many catalytic applications in the chemical industry are the use of silver for production of formaldehyde from methanol, and ethylene oxide from ethylene; and the use of a Pt-Rh alloy to produce nitric acid from ammonia and air, and hydrogen cyanide from ammonia, air and methane.

Pt-Rh alloys, or sometimes Au-Pt alloys, are also used for the spinnerets employed in manufacture of fibers of rayon, glass and other materials. Melting of glass is carried out in platinum crucibles if it is important to avoid contamination. Special-purpose crucibles also are made from iridium for melting lead, from ruthenium for melting bismuth, from silver for melting sodium hydroxide, and from gold for melting sulfur.

There are many uses for precious metals and their alloys in the electrical and electronic industries. Electrical contacts frequently are made from gold or silver alloys. Fuses may be made from gold or from a gold-silver alloy. Among high-temperature electrical applications are the use of a thoriated Pt-W alloy for spark-plug electrodes, a thoriated Rh-Pt alloy for jet-engine glow plugs, and Ag-Mg-Ni alloys for high-temperature wiring. In solid-state electronics, silver, gold, rhodium and palladium are all used as conductors. Eutectic, or near-eutectic, gold-silicon alloys are used as low-melting-point solders.

Electrochemical applications for precious metals include the use of platinum and Pt-Pd alloys as insoluble anodes for electrolytic protection; and the use of platinum and Pt-Ir alloys as insoluble anodes for production of persulfates and perchlorates, and for electroplating. Silver is used for the positive plates in primary and secondary batteries, and for tantalum capacitor containers. Various platinum-group metals are used in fuel-cell electrodes.

Platinum and Pt-Rh alloys are widely used in thermocouples, and some pairs of materials for which temperature/emf tables are available are shown in Table 9. At temperatures above the range for Pt-Rh alloys, iridium in combination with an Ir-Rh alloy can be used if the environment is oxygen-free.

For applications where high reflectivity is required, a metal that does not oxidize must be used. For over-all high reflectivity in the infrared, optical and ultraviolet parts of the spectrum, rhodium may be used. Gold has especially high reflectivity in the infrared region. It is used, for example, to coat the suits worn by astronauts when working outside a spacecraft.

Alloys of silver, gold, platinum and palladium are used for manufacture of jewelry. Silver jew-

Table 8. Fabricability of precious metals

Metal	Fabricability
Ruthenium	Difficult to work at any temperature
Rhodium	Can be cold worked only after hot working
Palladium	Readily fabricated at room temperature
Silver	Readily fabricated at room temperature
Osmium	Virtually impossible to work at any temperature
Iridium	Can be cold worked only after hot working
Platinum	Readily fabricated at room temperature
Gold	Readily fabricated at room temperature

Table 9. Frequently used Pt-Rh thermocouples

Type	Materials	Maximum temperature for extended service, °C
S	Pt-10Rh vs Pt	1400
R	Pt-13Rh vs Pt	1400
B	Pt-30Rh vs Pt-6Rh	1600

elry usually is made from sterling silver, which is a Ag-7.5Cu alloy, even though it has a tendency to tarnish. Most gold jewelry is made from alloys in the Au-Ag-Cu ternary system, although zinc or nickel sometimes is added, and occasionally cobalt or iron may be used to control grain growth. By varying the composition, a range of colors can be obtained. The platinum alloys used include Pt-10Ir and Pt-5Ru. The latter is known as "hard platinum." For uses where low density is required, such as earrings, an alloy of 4.5% Ru with palladium is used. Jewelry coated with rhodium is known for its exceptional whiteness and wear resistance.

METAL FINISHING

All of the precious metals, except osmium, have applications in the metal-finishing industry, either as decorative finishes or for industrial uses. Most commonly, these metals are applied by electrodeposition (electroplating), but other processes are also used in special cases.

Precious metals usually are electrodeposited from complexed solutions, commonly employing the cyanocomplex, with reasonably good current efficiencies, producing fine-grain deposits. Complexation of the precious-metal ions lowers their discharge potentials, which is necessary in order to prevent their discharge or precipitation by base metals. Additionally, in certain cases, such as with ferrous metals, a thin flash of metal is applied from a "strike" bath prior to electroplating to ensure that there is no precipitation or "plating out" of the precious metal from solution. The use of complexed ions also permits the codeposition of other metals to produce alloy deposits, with attendant improvements in mechanical properties such as surface hardness and wear resistance, as well as color changes, which are important with many decorative finishes.

Traditionally, electrodeposition of silver (silver plating) was performed primarily for decorative purposes, typically in making jewelry, ornaments, tableware and artifacts. In recent years, however, silver electrodeposition has become increasingly important in engineering, where it is

used for coating of bearing surfaces and a wide variety of electrical contacts and switches. In most cases of decorative plating, silver deposits are applied in thicknesses of 2.5 to 50 μm. Industrial applications, such as coating of bearing surfaces and electroforming, require thicker deposits, typically 0.1 to 1.5 mm thick. Further, the physical properties of electrodeposited silver are being continuously improved for many applications and to satisfy specific functional requirements.

Electrodeposition is performed from baths based on the argentocyanide ion. The basic or conventional cyanide bath, buffered with alkali-metal carbonates and/or hydroxides and free cyanide ion, will produce a matte white silver deposit that can be buffed or polished to a bright finish. The conventional bath is used mainly for decorative finishes and is operated at temperatures slightly above room temperature. Heavier deposits used for engineering applications utilize solutions containing higher silver contents, known as high-speed baths, which operate at higher solution pH levels and elevated temperatures. In general, a strike bath is used with both ferrous and nonferrous metals to ensure good adhesion of subsequent thicker silver electrodeposits. The strike bath has a lower silver content and a higher free cyanide level than the conventional silver electrodeposition system and, accordingly, deposits a thinner coating of silver.

Modern technology has developed a variety of additives that permit direct deposition of bright or shiny electrodeposits with little need for subsequent polishing. Bright electrodeposits, however, tend to have higher levels of internal stress, and some tend to tarnish more rapidly than the buffed matte finishes. The newer addition agents employed with silver electroplating baths are now mainly proprietary products which, in addition to improving deposit brightness, reduce internal stress and improve surface hardness. In many cases, a further increase in surface hardness can be achieved by means of heat treatment. Codeposition of hydrogen accompanies silver deposition, and this phenomenon necessitates a degassing procedure for sensitive components such as bearing surfaces in aircraft engines and other severe service applications.

Pure gold is electrodeposited for both industrial and decorative purposes. Pure gold electrodeposits are usually classified in terms of military specifications (MIL-G-455204B and its later amendments), which specify three types — type I (99.7% Au min), type II (99.0% Au min) and type III (99.9% Au min); and four hardness ranges — grade A (maximum hardness, 90 HK), grade B (91 to 129 HK), grade C (130 to 200 HK) and grade D (201 HK and greater). Generally, type I deposits conform to grades A to C, type II deposits to grades B to D, and type III deposits only to grade A. Other metals that are added in minor amounts for their hardening effects generally are not considered to be impurities under these specifications. Other specifications, such as ASTM and SAE aerospace material specifications, have comparable purity and hardness requirements for pure gold deposits.

All three types of pure gold deposits are obtained from the same basic hot cyanide bath, which contains gold in the form of potassium gold cyanide together with potassium cyanide, carbonate and phosphate in varying amounts. The gold content is varied depending on the required deposit thickness, with higher contents used for thicker electrodeposits. This type of electroplating bath is economical but is of high pH, tends to be unstable, and becomes contaminated through attack on copper and copper alloy substrates. It also is unsuitable for printed circuit boards due to attack on the adhesives used in their manufacture.

Gold electroplating solutions of lower pH, known as neutral and acid baths, have been developed to overcome the disadvantages of the hot cyanide bath. These solutions are based on phosphates or phosphoric acid and have pH values ranging from 2 to 7. A potassium citrate/citric acid bath having a pH of 3 to 6 also has been developed that is stable and provides thick deposits. All of these acid baths contain gold as potassium gold cyanide. They have lower cathode efficiencies than the cyanide bath, but have the advantages of bright electrodeposits, ease of electroplating, absence of staining and low electrodeposit porosity.

Because pure gold deposits are usually soft, ductile and somewhat porous, they often are unsuitable for service conditions where they may be subject to wear and/or abrasion. Harder electrodeposits and those with varied surface colorations are obtained by means of codeposition of other metals—commonly silver, copper, nickel, cadmium and cobalt, either singly or in combination—as well as polymers. In most cases, the alloying metal is added as the cyanide complex for the hot cyanide bath or as a metal salt or chelate for acid/neutral baths. Although alloyed electrodeposits are harder and generally brighter than pure gold, they tend to be less ductile.

Silver is added to gold baths as a cyanide complex to increase the hardness and brightness of the electrodeposit. Deposit hardness can be increased up to 150 HK, and deposits containing up to 8% Ag can be achieved. Gold-copper electrodeposits are also hard and tough; 18-karat Au-Cu electrodeposits have suitable hardness, wear resistance and ductility for many industrial applications. Nickel and cobalt additions often are made to increase hardness and wear resistance, particularly for electrical contacts. The alloying-element content of the deposit may range from 0.05 to 25% depending on the required hardness (120 to 450 HK), although purity has to be carefully controlled to ensure good solderability and low contact resistance. Other metals, such as indium, antimony and gallium, have been codeposited with gold for specific applications, such as in the semiconductor field; additions of cadmium or lead sometimes are made to achieve antiquing effects.

Electrodeposits of palladium are used primarily in the electronics field as an alternative to gold and rhodium. Palladium deposits have hardness and wear resistance comparable to those of hard golds but are inferior in wear characteristics to the much harder rhodium electrodeposits. Compared with gold deposited from a hot cyanide bath, palladium has two major advantages: the solution does not attack printed circuit boards, and, because palladium has a lower density than gold, comparable deposit thicknesses of palladium are inherently cheaper. Some of the advantages of palladium electrodeposits have been negated by the development of neutral and acid gold baths. Nevertheless, industrial applications of palladium electrodeposits are increasing, although palladium is little used as a decorative finish due to its dark color and tendency to tarnish.

Palladium, like other metals of the platinum

group, is electrodeposited from baths based on the so-called "P" salt, diamino-dinitrito-palladium, typically buffered with ammonium nitrate and sodium nitrite. Other palladium salts used for electrodeposition are the chelate formed with N,N'-cyclohexane diamine tetraacetic acid and tetramino-palladous bromide. Electrodeposited palladium is reasonably ductile, with low internal stress, but is darkish gray in color at deposit thicknesses greater than 1 μm. It can, however, be buffed to a bright finish.

The principal applications of electrodeposited platinum are in the production of platinized titanium anodes used for electroplating gold and rhodium and in cathodic protection systems. Platinum plating solutions are based either on the "P" salt or on sodium hexahydroxyplatinate. Electroplating solutions based on both salts tend to be somewhat unstable, but proprietary solutions based on sulfamates or on mixtures of sulfuric and phosphoric acids have greater stability. These deposits have an attractive appearance, are hard, and protect substrates such as zirconium and nickel from corrosion.

There have been few reports of electrodeposition of iridium from aqueous solutions; generally, the solutions used have been based on the bromide or chloride. The aqueous complexes appear to be unstable and tend to give irreproducible results. Most electrodeposition of iridium is performed using a molten cyanide bath and can produce deposits up to 10 μm thick. Generally, iridium is not electroplated, because it appears to offer little advantage over the more easily operated rhodium plating system.

Until comparatively recently, there has been little interest in electrodeposition of ruthenium, due to its complicated chemistry. However, it does have a number of attributes—notably, high hardness, good wear and corrosion resistance, and a lower cost than rhodium. Ruthenium electroplating solutions are based on sulfamate or nitroso salts. Thick deposits tend to crack but have excellent properties for electrical contacts.

Rhodium is the member of the platinum group most widely used for electrodeposition. It has good color, high hardness and excellent resistance to corrosion. Most electrodeposition solutions are based on phosphoric acid, phosphoric-sulfuric acids or sulfuric acid, although baths based on sulfamic and fluoboric acids have also been reported. Until comparatively recently, most applications of rhodium electrodeposition were in jewelry and artifacts. Typically, flash deposits 0.5 to 5 μm thick have been used to protect silverware from tarnishing. Thicker deposits are used in the electronics industry where advantage is taken of their high wear resistance and excellent corrosion resistance.

Thick electrodeposits of rhodium are subject to internal stresses, and the cathode efficiency of the plating process is low, with considerable gassing occurring at the cathode. This requires careful process control to avoid pitting of the electrodeposit. However, proprietary baths for rhodium electrodeposition are now available and produce excellent deposits.

SOLDERING ALLOYS

Most precious-metal alloys used for joining have relatively high melting points and thus are properly described as hard solders or brazing alloys. A notable exception is the use of gold-silicon al-

loys of eutectic or off-eutectic composition for die bonding and other applications in solid-state electronics. They are either silver- or gold-base alloys. Silver solders are used when the joint is expected to be subjected to elevated temperatures and/or significant stresses. Gold solders, which have better corrosion resistance than silver solders, are used mainly for dental applications and the manufacture of jewelry. Both types are used with a flux, often based on borax or boric acid together with a fluoride, which dissolves surface films and protects the surfaces being joined from oxidation.

The silver solders, which are often used to join stainless steels and other base-metal alloys, contain principally silver, copper and zinc. Small amounts of tin, cadmium or phosphorus are often added to modify the fusion temperature. The compositions of silver solders differ widely, but the ranges for silver, copper and zinc generally are within the limits 10 to 80%, 15 to 50% and 4 to 35%, respectively. The liquidus temperature is usually between 620 and 700 °C.

Gold solders contain principally gold, silver and copper. The solidus and liquidus temperatures tend to increase with gold content—i.e., with the solder's fineness. Some typical compositions are given in Table 10. In addition to the gold, silver and copper shown in Table 10, all gold solders contain 2 to 3% Sn and 2 to 4% Zn. These two elements lower the solidus and liquidus temperatures.

Table 10. Compositions and fusion ranges for several gold solders

Fineness	Approximate composition, wt %			Fusion range, °C
	Au	Ag	Cu	
450	45	30-35	15-20	680-760
600	60	12-32	12-22	720-835
650	65	16	13	765-800
729	72.9	12	10	755-835
800	80	3-8	8-12	745-870
809	80.9	8	7	800-870

The color of the solder usually is important for applications in dentistry and in the manufacture of jewelry. It can be controlled by varying the silver and copper contents. Whiteness can be imparted by using nickel instead of copper. In general, solders of high fineness are used to join alloys high in precious-metal content. As copper content is increased, the range of temperature over which melting occurs increases. Increasing the silver content has the opposite effect. Gold solders of high fineness tend to be softer and more ductile than those of low fineness. Joints that were quenched after soldering—i.e., from about 700 °C—are described as "softened." Age hardening occurs if the joint is allowed to cool slowly.

DENTAL APPLICATIONS

The science of dental materials is complex because it involves the use of a wide range of metals, alloys, plastics and ceramics, alone and in combinations, to achieve durable and esthetically pleasing restorations. The metals and alloys used change considerably with time in response to the appearance of new products on the market, fluctuations in the prices of the various metals, and changing esthetic values. The information offered below was drawn from manufacturers' catalogs early in 1984.

Dental amalgams, which are used for about four out of five dental fillings, are made by mixing particles of an alloy consisting primarily of silver and tin with liquid mercury, a process known as trituration, to form a plastic mass. This mixture is inserted into the prepared tooth cavity, where it sets in a few minutes. The restoration has good wear and abrasion characteristics, relatively high strength and acceptable corrosion resistance.

The alloy particles used in an amalgam may contain 40 to 75% Ag, 26 to 30% Sn, 2 to 30% Cu, 0 to 2% Zn and 0 to 3% Hg. A conventional amalgam contains 2 to 4% Cu; only the newer varieties have higher copper contents. The extra copper imparts improved long-term creep resistance, higher tensile and compressive strengths, and improved corrosion and tarnish resistance. When zinc is added, it acts as a scavenger to prevent oxidation of the silver and tin. Sometimes the particles are coated with mercury to facilitate trituration.

Conventional amalgams employ irregular lathe-cut particles. Newer alloys use spherical particles or mixtures of spherical and lathe-cut particles. Particle shape influences the setting and manipulative properties of the amalgam but has little effect on its clinical life. Particle types and chemical compositions of some typical amalgam particles are presented in Table 11.

Dental casting alloys are classified by the American Dental Association (ADA) as type I (soft), type II (medium), type III (hard) and type IV (extra hard). Some type III silver-palladium casting alloys are used for crowns, fixed bridgework and hard inlays, and some type IV alloys are used for high-stress bridgework, partial dentures, bars, clasps and extra-hard inlays. Some silver alloys are used for porcelain/metal restorations — i.e., restorations in which dental porcelain is fired onto a metal substrate. These silver alloys are less expensive than gold alloys, yet have reasonable strength and corrosion resistance and may be cast using the conventional lost wax process. The major chemical constituents of some silver alloys are given in Table 12. Table 13 pre-

Table 11. Typical particle types and compositions of dental amalgam alloys

| Type of alloy | Particle type | Composition(a), wt % | | |
		Ag	Sn	Cu
Conventional	Lathe cut	68-70	26-28	2-4
Conventional	Spherical	68-70	26-28	2-4
High copper	²/₃ lathe cut	68-70	26-28	2-4
	¹/₃ spherical	72	0	28
High copper	²/₃ lathe cut	40-60	27-30	13-30
	¹/₃ spherical	40-60	27-30	13-30

(a) Major components; some amalgams may also contain 0-2% Zn and 0-3% Hg.

Table 12. Compositions (major constituents) of several silver casting alloys

| Alloy type | Composition, wt % | | | | |
	Ag	Pd	Au	Cu	Co
Type III	70	25	...	...	...
Type IV	45	25	15	...	...
Type IV	50	30	3	...	...
Porcelain bonding	25-40	50-88	0-2	...	...
Porcelain bonding	35-40	5-8	...	20-25	12-14

Table 13. Typical physical properties of silver and gold dental casting alloys

Alloy type	Density, Mg/m³	Elongation, %	Hardness, HV	Melting range, °C
Type I (yellow gold)	16.6	36	80	940-960
Type II (yellow gold)	15.9	38	101	920-960
Type III (yellow gold)	15.5	39(a) 19(b)	121 182	930-960
Type III (low gold)	12.8	30(a) 13(b)	138 230	840-920
Type III (silver-palladium)	10.6	10(a) 8(b)	143 154	1020-1100
Type IV (yellow gold)	15.2	35(a) 7(b)	149 264	920-945
Type IV (low gold)	13.6	38(a) 2(b)	186 254	870-930
Type IV (silver-palladium)	11.3	10(a) 6(b)	180 270	930-1020
Porcelain bonding (yellow gold)	18.3	5	182	1150-1180
Porcelain bonding (white gold)	13.5-16.7	6-20	210-220	1270-1300
Porcelain bonding (silver-palladium)	10.7	20	189	1230-1300

(a) Quenched. (b) Hardened by slow cooling.

Table 14. ADA classification of gold casting alloys

Alloy type	Applications	Precious-metal content (min), %	Hardness(a), HV
I (soft)	Low-stress inlays	83	50-90
II (medium)	Moderate-stress inlays	78	90-120
III (hard)	Crowns, bridges and hard inlays	78	120-150
IV (extra hard)	High-stress bridgework, bars and clasps	75	>150

(a) In quenched state.

sents some typical physical properties, together with the corresponding properties for some gold casting alloys.

Pure silver is sometimes used for temporary crowns and for the splints used to maintain jaw alignment following oral/maxillofacial surgery. In the latter application, the germicidal action of silver reduces the risk of infection.

The ADA specifications for gold casting alloys are presented in Table 14. The concepts of fineness and karat rating are not used in these specifications. Instead, a minimum precious-metal content is specified for each type. In this context the term "precious metal" means gold or a platinum-group metal; silver is specifically excluded from the definition. Gold casting alloys may also be divided into yellow golds, white golds, low golds and porcelain-bonding golds. Economic pressures now limit the use of conventional yellow golds, which contain at least 60% Au. White golds, which derive their color from additions of palladium and/or silver, contain at least 50% Au. Low golds contain 42 to 55% Au and are light yellow in color. Their mechanical properties and casting characteristics are comparable to those of high golds, but they are somewhat subject to tarnishing, especially in the presence of sulfides.

Porcelain-bonding gold alloys have unusually high precious-metal contents. Their combined gold and platinum contents range from 85 to 95%. Small amounts of indium and tin are added to promote a strong porcelain-to-metal bond. The mechanism by which bonding occurs is not yet well understood, but it seems that indium and tin oxides form at the metal/porcelain interface and, perhaps by combined chemical and mechanical effects, enhance the bonding. These alloys do not contain either copper or silver, because these elements can diffuse into the porcelain and discolor it. In recent years, semiprecious silver-palladium alloys and various base-metal alloys have been used successfully in conjunction with porcelain.

Pure gold is used to a limited extent in dentistry. Thin sheets or hollow foil cylinders may be laid into a prepared cavity and worked into position using a manually or mechanically operated mallet. This type of restoration has been largely replaced by restorations using an amalgam, a plastic or a casting.

PHARMACOLOGY

Certain compounds of precious metals — notably those of silver, platinum and gold — have applications as pharmaceutical agents in medicine. Silver compounds have germicidal action and are used topically. Silver nitrate is an anti-infective that is used in dilute solutions as a germicide and astringent to the mucous membranes. Silver nitrate ophthalmic solution is used to prevent gonorrheal ophthalmia neonatorum. Silver sulfadiazine is a topical antimicrobial drug used as an adjunct for prevention and treatment of wound sepsis in patients with second- and third-degree burns.

The principal application of gold compounds in medicine is in the treatment of rheumatoid arthritis, where they produce symptomatic relief. Many gold compounds are used in adjunctive treatment, including gold sodium thiomalate, gold sodium thiosulfate, aurothioglucose and aurothioglycanide. These gold compounds are used as one part of a therapy program, rather than as complete treatments on their own. They are useful in the early active stage of rheumatoid arthritis if it cannot be adequately controlled by other anti-inflammatory agents. Gold Au(198) solution is used in cancer therapy where the primary interest is in the radiation effect of the radioisotope.

Cisplatin or cis-diaminedicloroplatinum is used as palliative therapy for metastatic ovarian and testicular tumors, usually in combination therapy with other chemotherapeutic agents following surgery and/or radiotherapy.

The pharmacological action of cisplatin is due to its action on DNA, but that of the gold compounds is poorly understood. The germicidal/antimicrobial action of silver salts appears to be an inherent property of silver. This characteristic is receiving attention in the orthopedic field, where percutaneous (through-the-skin) pins and other fixation devices have been electroplated or otherwise coated with metallic silver to prevent microbial invasion.

METALLOGRAPHY

The examination of the precious metals and their alloys by optical microscopy, scanning electron microscopy, transmission electron microscopy, and field ion microscopy has been discussed in the book *Physical Metallurgy of Platinum Metals,* by E. Savitsky *et al.* These authors present optical micrographs showing alloy microstructures, slip lines, low-angle boundaries, and annealing twins; scanning electron micrographs showing fracture surfaces; transmission electron micrographs showing dislocation structures; and several field ion micrographs. Some of the etching techniques which they recommend are shown in Table 15.

SELECTED REFERENCES ON PRECIOUS METALS

Metals Handbook, 9th Ed., Vol 2: American Society for Metals, Metals Park, OH, 1979, p 657-705 (various articles on precious metals) and p 714-833 (compilation of properties of pure metals)

Metals Handbook, 9th Ed., Vol 3: American Society for Metals, Metals Park, OH, 1980, p 662-695 (article on electric-contact materials) and p 696-720 (article on thermocouples)

Metals Reference Book, 5th Ed., by C. J. Smithells: Butterworths, 1976

Rare Metals Handbook, 2nd Ed., edited by C. A. Hampel: R. E. Krieger Publishing Co., 1971

Solders and Soldering, 2nd Ed., by H. H. Manko: McGraw-Hill, 1979

Basic Metal Finishing, by J. A. von Fraunhofer: Elek Science, 1976

Council on Dental Materials, Instruments and Equipment: Status Report on Low-Gold-Content Alloys for Fixed Prosthesis: *J. American Dental Assn.*, Vol 100, 1980, p 237-240

Council on Dental Materials, Instruments and Equipment: Classification and Definition of Alloys Used for Casting Substrates for Porcelain Veneering: *J. American Dental Assn.*, Vol 103, 1981, p 755-757

Phase Diagrams of the Precious Metals, by He Chunxiao, Ma Guangchen, Wang Wenna, Wang Yongli and Zhao Huaizhi: The Metallurgical Industry Press, Peking, China, 1978.

The Physical Metallurgy of the Platinum Metals, by E. Savitsky, V. Polyakova, N. Gorina and N. Roshan (Translated by I. V. Savin): M.I.R. Publishers, Moscow, 1975; Translation: Pergamon Press, Oxford, 1978.

Table 15. Some etching techniques used with precious-metal alloys

Alloy type	Reagents	Etching method
Ruthenium	10% soln. of oxalic acid	Electrochemical
	4 p. NaCl + 1 p. HCl	Electrochemical
Rhodium	1 p. $K_3Fe (CN)_6$ + 10 p. KOH	Electrochemical
	1 p. $FeCl_3$ + 4 p. HCl	Electrochemical
Palladium	Saturated soln. of Br in C_2H_5OH	Chemical
	48% HNO_3 + 2% HCl + 50% H_2O	Chemical
Silver	H_2O_2 + a few drops of NH_4OH	Chemical
Osmium	15% soln. of H_3PO_4	Electrochemical
	15% soln. of HNO_3	Electrochemical
Iridium	10% soln. of NaOH	Electrochemical
	5 to 30% soln. of HCl	Electrochemical
Platinum	Conc. aqua regia	Chemical, 80 °C
	Saturated soln. of KCN	Electrochemical
Gold	20% soln. of aqua regia	Chemical, 80 °C

14 PURE METALS

This section was condensed from Metals Handbook, Ninth Edition, Volume 2, Properties and Selection: Non-ferrous Alloys and Pure Metals, pages 707 to 833. Additional information on pure metals is located in the tables beginning on page 1·43 of this volume.

Preparation and Characterization of Pure Metals

By G. T. Murray, California Polytechnic State University

AS A RESULT of the constant quest for the true values of physical and chemical properties of metals, there has been continual improvement in the purity levels attainable and in the accuracy and completeness of techniques for measuring these purity levels. Therefore, the property values reported for "pure metals" in this section, which were determined at different times and in different laboratories, vary considerably in meaningfulness from one metal to another and from one property measurement to another.

The rapidly growing electronic microcircuit industry also has placed severe demands on metal suppliers for metals of the highest reproducible purity attainable. Trace impurity elements in concentrations below 1 ppm can prevent proper functioning of certain electronic devices.

The need for "ultrapure" metals for measurement of physical and chemical properties and for the electronic microcircuit industry poses two important problems: how to obtain such purity and how best to measure levels of trace impurity elements.

PREPARATION METHODS

What is commonly referred to as "commercial-purity" metal normally is used as the starting material in ultrapurification operations. Depending on the metal in question, "commercial purity" usually means a purity between 99.0 and 99.95%. Commercial-purity metal can be prepared by a variety of processes, of which electrolytic processes such as electrowinning and electrorefining are among the most common. In both electrowinning and electrorefining, metal is deposited by electroplating from a bath. In electrowinning, the starting material usually is in the form of a concentrated ore or compound; in electrorefining, it is in metallic form. Many different types of baths are employed. For titanium and vanadium, fused salt baths are used, whereas chromium sometimes is produced by electrolysis of an aqueous solution of chromium-alum or chromic acid. For applications such as semiconductors, material produced by electrolytic processes is of insufficient purity and must be subsequently ultrapurified by one of the methods described below.

Fractional crystallization is a liquid-phase method that relies on differences in solubility in a liquid solvent among the various solid phases present in the impure metal. Basically, the metal to be purified is dissolved in a hot, often organic, solvent. The solvent selected is such that the metal is much more soluble at elevated temperatures than at lower temperatures but that impurities are fairly soluble even at the lower temperatures. On subsequent cooling of the solvent, the "pure" metal precipitates out of solution whereas most of the impurities remain. This process can be repeated many times, using a new batch of solvent each time. Gallium has been purified to the 99.9999% level by this method. This purity is required for manufacture of semiconducting gallium arsenide, which is employed for light-emitting diodes in items such as small hand-held calculators, digital watches and a variety of electronic displays.

Ultrapure silver, gold, palladium and platinum also are produced by fractional crystallization. In some instances, the metal being refined is precipitated and impurities are left in the solvent (as described above); in others, the impurities are precipitated (as compounds). Maximum purity in these metals, however, is obtained by zone refining (see below) following fractional crystallization.

Zone refining, also a liquid-phase technique, probably is the most widely used of all preparation methods. The classic zone refining experiments by Pfann (Ref 1) led to production of germanium of sufficient purity for development of the first transistor.

In zone refining, a molten zone is made to move slowly from one end of a bar of impure metal to the other. During this "zone pass," redistribution of impurities takes place because of differences between the solubility limits of impurity elements (limiting impurity concentrations) in the liquid phase of the metal and the corresponding limits in the solid phase. Under equilibrium conditions, the resulting distribution is measured by the coefficient K_o, which is defined as follows:

$$K_o = C_s/C_l$$

where C_s is impurity concentration in the just-freezing solid phase and C_l is impurity concentration in the liquid phase. In practically all instances of freezing, equilibrium is not attained, and it is more appropriate to use an effective distribution coefficient, K_e, which is a function of freezing velocity, impurity diffusion and thickness of the diffusion layer, as well as the ratio C_s/C_l. When K_e is less than one, as in most instances, and with slow movement of the zone (for example, 10 mm/h or 0.39 in./h), the impurity concentration in the solid phase, C_s, at distance x from the starting end after a single pass of a liquid zone of length l, is as follows:

$$C_s/C_o = 1 - (1 - K_e) \exp [-K_e x/l]$$

where C_o is initial concentration in the liquid phase. Additional passes of the zone in the same direction cause further concentration of impurities at one end of the bar. After many zone passes, this end is removed and discarded.

Metals and semiconductors were first zone refined by placing a bar of the material in a long boat-type crucible. Later to be introduced was the floating-zone technique (Ref 2), in which the metal is suspended in a vertical position and the molten zone is held in place by its own surface tension; heat sources commonly used for this technique include an electron beam and an induction coil. In the floating-zone technique, the diameter of the bar is limited to approximately 15 mm (0.59 in.), but this method has the distinct advantage that the material being refined is not contacted, and thus not contaminated, by a crucible. This is particularly advantageous for the high-melting-temperature reactive metals such as titanium, zirconium, niobium, tantalum, vanadium, tungsten and molybdenum, whereas metals such as gold, silver, copper, aluminum, zinc, lead, tin and bismuth, which melt below about 1200 °C (2190 °F), are zone refined in a boat.

Vacuum Melting. Zone refining of materials often is conducted in a dynamic vacuum in order to enhance the degree of purification. However, many metals—particularly those with high melting points—can be purified to a significant degree by the vacuum melting process alone. Although vacuum melting may not produce the degree of purity attainable by zone refining, it is less expensive and yields material of sufficient purity for a wide variety of applications.

In vacuum melting, purification occurs by degassing—that is, removal of oxygen, nitrogen and hydrogen, as well as CO or CO_2 formed by side reactions of oxygen with carbon—and by vacuum distillation of high-vapor-pressure impurity elements.

Degassing takes place because the solubility of gaseous elements in the liquid decreases when the partial pressure of the same elements in the surrounding gaseous medium is decreased. This was experimentally verified for partial pressures of about 10 to 100 kPa (75 to 750 mm of mercury) in the early experiments of Seiverts, which led to the well-known relationship:

$$S \propto \sqrt{P}$$

where S is the solubility of a gas in the liquid phase and P is the partial pressure of the same gas in the surrounding medium.

This purification process is dependent on (a) the ability of the vacuum system to maintain a sufficiently low gas partial pressure near the molten surface, (b) diffusion of gas atoms through the liquid to the surface, (c) the presence or absence of any stirring action that might enhance transport of gas atoms in the liquid phase, and (d) the composition of the starting material.

Vacuum distillation during melting is a purification process based on preferential evaporation of solute. The degree of purification is dependent on the ratio of the vapor pressure of the solute to that of the solvent. For a high degree of purification, it is essential that solute vapor pressure be high relative to solute partial pressure in the gaseous medium in the immediate vicinity of the molten surface. As the solute content at the liquid/vapor interface becomes diminished, a concentration gradient is set up within the liquid. At this time, which may be very early in the melting operation, material transport in the liquid phase becomes the rate-controlling process. Thus, provided that vapor pressures are favorable and that the pumping speed of the vacuum system is sufficient to maintain a low partial pressure of the solute element, purification should proceed at a rate that depends on the diffusivity of the solute in the liquid.

Distillation. Straight distillation (in which heated material changes from solid to liquid to vapor), like vacuum-melting distillation, is an important vapor-phase purification process. If the distillation is conducted under conditions of near-equilibrium between the liquid and vapor phases, impurity elements will concentrate in the liquid phase. The vapor, or distillate, is then allowed to condense to form a solid of higher purity than that of the starting material. The most common distillation method is fractional distilllation, in which the metal is repeatedly vaporized and condensed to liquid on a series of plates placed in a vertical column. A high reflux ratio (ratio of amount of liquid returning to the column from the condenser to the amount of metal removed to the condenser) is desirable. Some metals, however, are purified in a single stage by simply condensing all the vapor produced by the still; this process has been used for alkali metals such as barium, calcium, lithium and sodium (Ref 3). Distilled magnesium is further purified by zone refinement.

A variation of straight distillation is sublimation, in which the metal passes directly from the solid phase to the vapor phase. Only metals that have high vapor pressure when in the solid state are suited to this process. Such a metal has a higher vapor pressure than most impurity metals, so that such impurities are left to concentrate in the remaining solid while the vapor is condensed, forming higher-purity metal.

Chemical Vapor Deposition. In purification by chemical vapor deposition, the starting material is reacted to form a compound, and the compound is subsequently decomposed in the vapor state. The metal vapor then is condensed to form a solid higher in purity than the starting material.

One of the more popular of the chemical vapor deposition processes is the iodide process, which has been used extensively to purify titanium, zirconium and chromium (Ref 4). For all three of these metals, the starting charge of the metal is reacted to form a volatile metal-iodide compound, which in turn is thermally decomposed to liberate iodine vapor. The pure metal is allowed to condense onto a suitably heated substrate (glass tubes and wires of the base metal have been used), while the iodine returns to the metal charge to form more iodide compound. Hence, the iodine acts as a carrier of the metal from the charge to the substrate.

In this process, some impurities are almost always carried over to the vapor phase along with the metal being purified. However, if a proper temperature is maintained, oxygen, nitrogen, hydrogen and carbon, as well as many metallic impurities, will not be carried over. Typical purities obtained are about 99.96% for titanium, 99.98% for zirconium (plus hafnium, which is present at about the 200-ppm level) and 99.995% for chromium. In all cases, the starting metal has a purity on the order of 99.9%. Chromium has been purified to its highest state to date by this method. Only iron is carried over with these metals to a significant extent. Thus, if a low-iron starting metal is used, the condensed vapor will approach a purity level of 99.999%.

Other metals that have been purified by chemical vapor deposition include hafnium, thorium, vanadium, niobium, tantalum, molybdenum and many less commercially important metals (Ref 4).

CHARACTERIZATION OF PURITY

The traditional system of describing metal purity is based on measurement of total impurity-element content and subtraction of this number from 100%. The result is reported in terms of "number of nines"—for example, "five nines," which indicates a purity of 99.999%. This system is quite adequate for many applications; however, unless the method of measurement and its sensitivity are reported, the system is meaningless and unacceptable for some fast-growing technical fields. In many analyses, certain elements (the gaseous elements, for example) are not measured, and often the method employed is not sensitive enough to detect impurity levels near the low end of the parts-per-million range. These impurity elements, referred to as "trace" elements, now can be detected by a variety of analytical methods.

Trace-Element Analysis. The sophisticated analytical methods now available include electron-beam microprobe analysis and the chromatographic methods. However, the most commonly used methods are:

• Emission spectroscopy—for simultaneous determination of metallic elements in the parts-per-million range and greater

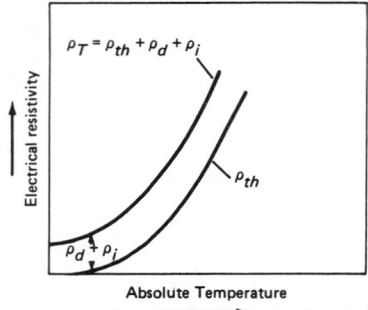

Fig. 1. Idealized graph of the components of total electrical resistivity of a metal near absolute zero

• Mass spectroscopy—for determination of metallic elements in the parts-per-billion range; the analytical results are as accurate as the reference standard used
• Neutron activation—for determination of metallic elements, and particularly oxygen, in the parts-per-million range
• Atomic absorption—for sequential determination of metallic elements in the parts-per-billion range
• Vacuum fusion—for determination of hydrogen and nitrogen in the parts-per-million range
• Conductometry—for determination of carbon in the parts-per-million range

Emission spectroscopy is the most common analytical method and normally is used for detecting trace elements in concentrations of 10 to 1000 ppm. It is relatively inexpensive and yields results for all metallic elements in one analysis. Mass spectroscopy is more sensitive (and more expensive) than emission spectroscopy, and can easily detect impurity levels as low as 0.01 ppm. However, accuracy is dependent on good standards, and the technique is not accurate above the 50- to 100-ppm level for some elements. Generally, it is the best method for verification of purity at the five nines level and for obtaining information on all residual elements in the sample. Neutron activation analysis is more sensitive than mass spectroscopy, but is unable to detect many elements because of their inherent radioactive characteristics. It is an expensive method, but for certain elements that are difficult to identify, it is extremely sensitive and accurate. Atomic absorption is excellent for concentrations of 0.1 to 10 ppm, when only a few elements are present. However, the specific elements being sought must be known, which generally requires that atomic absorption analysis be preceded by emission spectroscopic analysis. All factors consid-

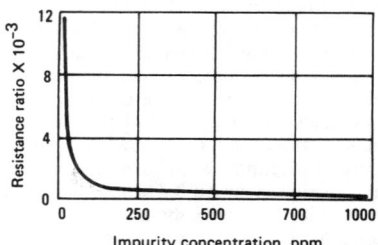

Fig. 2. Effect of interstitial impurity-atom concentration on resistance ratio of refractory metals

Table 1. Resistance ratios of samples of zone-refined metals(a)

Metal	Resistance ratio
Aluminum	40 000
Gold	2 000
Molybdenum	14 000
Nickel	3 000
Niobium	2 000
Rhenium	45 000
Tantalum	7 000
Tungsten	90 000
Vanadium	300
Zirconium	200

(a) See Table 2 for impurity contents of these samples.

Table 2. Impurity-atom concentrations of samples of zone-refined metals(a)

Impurity element	Concentration(b), ppm of impurity of zone-refined:									
	Al	Au	Mo	Ni	Nb	Re	Ta	W	V	Zr
C	5	<1	10	40	8	5	10	5	57	20
H	<1	<1	0.9	0.2	0.4	0.2	<0.1	0.1	3	3
O	5	2	4.3	25	23.4	0.5	3.5	0.8	250	200
N	<1	1	0.5	10	4	1	2.3	0.1	3	2
Ag	<0.08	4	<0.7	<0.002	<0.3	<0.001	<0.004	<0.12	<0.002	<0.4
Al	...	ND	<0.03	0.3	0.15	0.05	0.05	0.07	0.1	3
As	<0.05	ND	<0.01	<0.04	<0.01	<0.002	<0.002	<0.005	<0.05	<0.01
Au	<0.02	...	<0.02	<0.15	<0.03	<0.15	<0.2	<0.3	0.6	<0.2
Bi	<0.03	ND	<0.02	<0.01	<0.01	<0.12	<0.04	<0.12	<0.02	<0.007
Ca	0.15	ND	0.04	0.1	0.02	0.05	<0.008	0.02	0.1	0.04
Cd	<0.05	ND	<1.0	<0.08	<0.5	<0.02	<0.007	<0.025	<0.03	0.5
Cl	0.4	ND	0.4	0.1	0.03	0.1	0.01	0.2	0.1	2
Co	<0.1	ND	<0.06	<0.1	<0.01	0.06	0.3	0.1	<0.15	<0.007
Cr	0.2	ND	0.1	1.5	0.05	0.08	0.2	<0.001	<5	0.5
Cu	0.2	1	<0.02	<0.04	<0.01	0.005	0.02	0.005	<0.3	0.01
Fe	0.5	2	12	12	0.12	3	0.3	0.01	<20	30
Ga	<0.1	ND	<0.02	<0.4	<0.01	<0.004	<0.003	<0.01	20	<0.02
Ge	<0.1	ND	<0.02	<0.7	<0.01	<0.02	<0.005	<0.04	<0.6	<0.03
Hf	<0.06	ND	<0.03	<0.03	<0.02	<0.01	<0.4	<0.04	<0.03	40
In	<0.1	ND	<1	<1	<0.07	<0.2	<0.002	<0.03	<0.03	<0.08
Ir	<0.06	ND	<0.03	<0.02	<0.01	<0.2	<0.2	<0.15	<0.06	<0.03
K	0.4	ND	1	0.2	<0.04	0.01	0.02	<0.02	0.4	0.004
Li	<0.03	ND	<0.02	<0.02	<0.01	0.004	0.001	<0.001	<0.02	<0.001
Mg	<0.1	ND	<0.25	0.02	<0.05	0.02	0.006	0.15	<0.25	<0.05
Mn	<0.1	ND	0.06	0.03	0.03	0.02	0.01	0.03	<0.15	<0.03
Mo	<0.1	ND	...	0.5	<0.7	4	0.2	<0.1	0.08	<0.6
Na	<0.1	ND	<1	<0.04	<0.03	ND	0.015	<0.01	<0.05	<1
Nb	<0.04	ND	1	<0.02	...	1.2	25	<1	0.8	<0.5
Ni	<0.1	ND	0.1	...	0.15	0.2	1.5	<0.02	12	1.5
P	<0.06	ND	<0.03	<2	<30	0.02	<0.05	<0.05	0.2	0.1
Pb	<0.1	0.5	<0.03	<0.02	<0.02	<0.25	<0.08	<0.25	<0.003	<0.015
Pd	<0.1	ND	<1	<0.03	<0.5	<0.02	<0.08	<0.25	15	<0.8
Pt	<0.06	ND	<0.06	<0.03	0.02	<0.3	<0.4	<0.5	<0.04	<0.2
Re	<0.04	ND	<0.4	<0.02	<0.02	...	<0.2	<1	<0.5	<0.3
Rh	<0.04	ND	<0.1	<0.01	<0.06	<0.005	<0.002	<0.06	<0.5	<0.2
Ru	<0.1	ND	<0.3	<0.03	<0.4	<0.04	<0.006	<0.2	<0.5	<0.6
S	0.4	ND	1	<0.12	<0.07	1	0.02	0.07	0.1	<1
Sb	<0.03	ND	<0.2	<2	<0.04	<0.004	<0.004	<0.03	<0.02	<0.15
Si	<1	0.5	0.08	<0.2	0.6	0.5	<0.02	0.3	20	1.5
Sn	<0.04	ND	<0.4	<0.4	<0.3	<0.02	<0.006	<0.02	<0.03	<0.25
Ta	<0.02	ND	2	<0.5	50	3	...	5	<0.3	<0.2
Ti	<0.1	ND	<1	0.15	<0.02	<0.07	0.01	<0.01	6	1
V	<0.1	ND	<0.02	<0.01	<0.8	<0.001	0.01	<0.01	...	0.05
W	0.07	ND	20	1.5	6.4	15	1.2	...	7	<0.7
Zn	1	ND	<0.02	<0.4	<0.02	0.005	<0.004	<0.02	<0.4	<0.5
Zr	<0.08	ND	<0.04	<0.15	<0.3	<0.003	<0.1	<0.6	<0.12	...
RE(c)	<0.5	<0.1	<1	<0.5	<0.5	<0.1	<0.2	<0.2	<1	<0.5

(a) See Table 1 for resistance ratios of these samples. (b) ND, not detectable (<0.01 ppm). (c) Rare earths.

ered, mass spectroscopy is the preferred method for measuring trace elements in ultrahigh-purity metals.

In summary, the only way of describing purity that is both accurate and meaningful is to state the entire list of possible impurities, the findings, and the limits of detection applicable to the specific analytical procedure used for each element.

Resistance-Ratio Test. The amount of resistance to passage of electrons through a sample of high-purity metal, particularly at low temperatures, is extremely sensitive to the amount of trace elements present in the sample. This fact gives rise to the "resistance-ratio" test, which is an extremely sensitive qualitative method of measuring purities of 99.999% and higher. This test is valuable not only because it is extremely sensitive but also because measurement of electrical resistivity is relatively simple.

Making a resistance measurement at a single (low) temperature would require accurate dimensional measurements. To avoid this requirement, resistance measurements are made both at the low temperature and at room temperature, and the ratio of the room-temperature value to the low-temperature value is reported. Unless otherwise stated, it is assumed that the low-temperature measurement was made at liquid-helium temperature (4.2 K).

The electrical resistivity of a metal can be conveniently divided into three parts:

$$\rho_T = \rho_{th} + \rho_d + \rho_i$$

where ρ_T is total resistivity, ρ_{th} is resistivity due to thermal vibrations of the lattice, ρ_d is resistivity due to lattice imperfections (consisting primarily of vacancies, dislocations, and grain boundaries) and ρ_i is resistivity due to impurity atoms. Variation of ρ_T with temperature in terms of the components ρ_{th} and $(\rho_d + \rho_i)$ is depicted in Fig. 1. The sum $\rho_d + \rho_i$ is essentially temperature independent, whereas ρ_{th} is strongly temperature dependent ($\rho_{th} \propto T^5$), approaching zero at absolute zero temperature. Thus, resistivity near absolute zero affords a measure of $\rho_d + \rho_i$.

Point defects (vacancies) contribute to resistivity to about the same extent as impurity atoms. However, well-annealed metals contain far fewer point defects than impurity atoms. Dislocations of a typical density of 10^{11} per m^2 contribute an insignificant amount to the resistance ratio. In the highest-purity metals obtained to date (impurity concentrations of 10^{-5} to 10^{-6} atomic %), the contribution of ρ_d is still small compared with that of ρ_i. Total resistivity near 0 K, therefore, is a good measure of the ρ_i contribution.

Caution should be exercised in using the resistance ratio as a characterization of purity for several reasons. Most important is the fact that in resistance-ratio testing the impurity element in question is not determined (different impurity elements have vastly different effects on ρ_i). In addition, because only impurity atoms in solid solution are effective electron-scattering centers, nothing is learned about the impurity content in precipitate (compound) form. Finally, even for impurity atoms in solid solution, the resistance ratio is a sensitive measure of purity only when the impurity level is about 100 ppm or less. This is illustrated in Fig. 2, which is an estimate of variation in resistance ratio with concentration of interstitial atoms (O_2, N_2 and C) in refractory metals. This graph shows that the impurity level can be reduced from 500 to 250 ppm without appreciably affecting the ratio, whereas a reduction from 5 to 2.5 ppm has a marked effect.

Resistance ratios of zone-refined metals are listed in Table 1. Their corresponding chemical compositions, as measured by mass spectroscopy for metallic elements, vacuum fusion for gaseous elements and conductometric analysis for carbon, are listed in Table 2. Ratios higher than those shown in Table 1, and ratios for other metals, have been reported (Ref 5); however, they were not accompanied by chemical analyses. In fact, some of the ratios were so large that the impurity concentrations they indicated were too low to be detected by methods currently available.

REFERENCES

1. W. G. Pfann, *Transactions of the American Institute of Mining, Metallurgical and Petroleum Engineers*, Vol 194, 1952, p 861
2. H. C. Theuerer, *Transactions of the American Institute of Mining, Metallurgical and Petroleum Engineers*, Vol 206, 1956, p 1316
3. P. A. Schmidt, *Journal of the Electrochemical Society*, Vol 113, 1966, p 201
4. R. F. Rolsten, *Iodide Metals*, Wiley, New York, 1961
5. W. G. Pfann, *Zone Melting*, 2nd Ed., published jointly by Wiley, New York, and by the Cryogenics Laboratory, National Bureau of Standards, Boulder, CO

Selected Properties of Pure Metals*

ACTINIUM (Ac)

See "Properties of the Actinide Metals," at the end of this section.

ALUMINUM (Al)

See Section 6, "Aluminum," in this volume.

AMERICIUM (Am)

See "Properties of the Actinide Metals," at the end of this section.

ANTIMONY (Sb)

Antimony is used as an alloying element in lead alloys (for battery grids, printers' type, solder, bearings, cable sheathing and ammunition) and in tin alloys (such as pewter and costume jewelry alloys). Trace quantities are added to copper-base alloys, to prevent dezincification; to ductile iron, to assist in forming nodular graphite; and to gray iron, where the antimony acts as a powerful pearlite former. In the form of Sb_2O_3, antimony is used in enamels, glass, pigments, catalysts and flame retardants. Antimony is used as a component of III-V semiconductors such as InSb, AlSb and GaSb, and as an alloying ingredient in thermoelectric alloys.

Tensile strength. 11.40 MPa

Hardness. 30 to 58 HB

Elastic modulus. Tension, 77.759 GPa; shear, 19 GPa

ARSENIC (As)

In quantities of 0.5 to 2%, arsenic improves the sphericity of lead shot. Small percentages are added to lead alloys for battery grids and cable sheathing to improve their hardness. In amounts up to 3%, arsenic improves the properties of lead-base bearing alloys. Minor additions improve corrosion resistance and increase recrystallization temperature of copper and stabilize pearlite in ductile and gray cast irons. Arsenic is a component of III-V semiconductors such as GaAs, GaAsP and InAs.

Arsenic is normally available in the α-metallic and β-amorphous form. Other allotropes of arsenic have been reported but supportive evidence is meager.

The toxicity of arsenic is related to its chemical state and can be extremely high. The degree of toxicity of arsenic in the elemental state is relatively low. Metallic arsenic on exposure to the atmosphere will develop an oxide coating and for this reason care must be taken not to ingest or handle it. Inhalation of dust or fumes is to be avoided. Controlled exhaust ventilation of work area is required to comply with OSHA standard of 10 μg of arsenic per cubic metre of air.

Hardness. 3.5 Mohs

BERKELIUM (Bk)

See "Properties of the Actinide Metals," at the end of this section.

BERYLLIUM (Be)

Beryllium is used as an alloying addition to copper and nickel to produce an age-hardening alloy used for springs, electrical contacts, spot welding electrodes and nonsparking tools. Beryllium is also added to aluminum and magnesium to achieve grain refinement and oxidation resistance. Unalloyed beryllium is used in weapons, spacecraft, nuclear reactor reflector segments, neutron sources, windows for x-ray tubes and radiation detection devices, rocket nozzles, aircraft brake discs, precision instruments and mirrors. Inhalation or ingestion of beryllium and its compounds should be avoided. Users should comply with occupational safety and health standards applicable to beryllium in Title 29, Part 1910, Code of Federal Regulations.

Color. Steel gray

General corrosion behavior. Beryllium, a highly reactive metal, forms stable compounds with most other elements. It has excellent corrosion resistance at room temperaure (except for certain acids and alkalis) because of a thin protective oxide coating.

Resistance to specific corroding agents. Beryllium reacts appreciably with oxygen and nitrogen above 760 °C. Impure beryllium containing carbide or chloride reacts with moist air at room temperature. Beryllium does not react with hydrogen at any temperature; it reacts with fluorine at room temperature and with Cl, B, I, HCl, HF and CO_2 at elevated temperatures. Beryllium has excellent resistance to pure water up to 300 °C if carbide and chloride are absent and if grain size is fine. Beryllium is attacked by dilute HF, HCl, H_2SO_4, and HNO_3 at room temperature.

Machinability. Powder metallurgy material behaves well in most machining operations, provided that carbide tools are used. Chip formation is similar to that of cast iron. All machining, especially careless grinding, produces a damaged surface layer that must be removed by etching a minimum of 0.05 mm per surface for critical, highly stressed applications.

Recrystallization temperature. 725 to 900 °C, depending on amount of cold work and annealing time

Hot-working temperature. 800 to 1000 °C

Tensile properties, especially elongation, depend strongly on preferred orientation and grain size. Beryllium mill products are produced chiefly by powder metallurgy using different consolidation procedures. Wrought products are produced either from cast or powder metallurgy materials. Extreme anisotropy in elongation occurs in wrought material; hence, uniaxial tensile results are valueless as an indication of behavior under conditions of complex stress. Tensile properties of different forms may vary extremely, with the greatest uniformity occurring for vacuum hot-pressed powder. See Table 1 for some representative values.

Hardness. 75 to 85 HRB

Poisson's ratio. 0.02 to 0.075

Elastic modulus. 275 to 300 GPa

Impact strength. 1.4 to 5.4 J

Plane-strain fracture toughness. 9 to 13 MPa$\sqrt{m}$

BISMUTH (Bi)

Bismuth is used extensively in the production of fusible alloys (low-melting-point alloys), as a carbide stabilizer in the manufacture of malleable iron, and as an additive to low-carbon steel and aluminum to improve machinability. Compounds of bismuth are used for catalysts, in pharmaceutical and for semiconductor applications.

Hardness. 7.0 HB; 2.5 Mohs

Elastic modulus. Tension, 32 GPa

BORON (B)

Hot-wire boron is prepared by reduction of boron halides on a hot tungsten wire. The boron may form as wire or small chunks. It is black, the purest form of boron. Electrolytic boron, prepared by molten salt electrolysis, is a black powder of 40 to 325 mesh. Electrolytic boron has a

Table 1. Tensile properties of beryllium

Temperature, °C	Tensile strength, MPa	Yield strength(a), MPa	Elongation, %
Vacuum hot-pressed block			
Room temperature	228-352(b)	186-262	1-3.5
200	207-297	...	6-15
400	152-186	...	19-40
600	138-152	...	15-25
800	48	...	7-8
Hot extruded billet			
Room temperature	483-690	310	5-20
Cross-rolled sheet			
Room temperature	483-621	345-414	10-40
Hot-isopressed billet (high purity)			
Room temperature	345-414	242-276	3-6

(a) 0.2% offset. (b) Instrument grades have minimum tensile strength of 345 MPa with microyield strength of 35 to 83 MPa.

*The purity of the sample used in determining any of the properties listed in these compilations varies considerably from one metal to another and from one property to another. Furthermore, the technique used to measure a given property may vary. As a result, the accuracy and precision of the reported values vary. However, the values listed in these compilations are regarded to be the best values available. Additional data may be found in the compilations for commercially pure grades of metals, but these additional values may not apply to samples of laboratory purity. All temperatures related to thermodynamic data have been corrected to the International Practice Temperature Scale of 1968, unless otherwise noted. For a discussion of the 1948 and 1968 temperature scales, see T. B. Douglas, *Journal of Research of the National Bureau of Standards*, Vol 73A, 1969, p 451-470.

purity of 99% and above. Magnesium-reduced boron, prepared by reduction of boric acid (B_2O_3) with magnesium, results in a light brown powder in the purity range of 95 to 97%.

Boron formed on hot tungsten wire is used for reinforcement of metals and plastics and weight reduction. Electrolytic boron powder is used for preparation of borides for deoxidation of alloys. Isotropic B-10 and its compounds are used for neutron absorption. Boron in compound form is used for medicinal and cleaning purposes, as rocket fuels, diamond substitutes, and additives to aluminum alloys to improve electrical and thermal conductivity, and for grain refining of aluminum alloys.

Boron is nontoxic. No precautions are required for wire-form boron. Electrolytic product contains fine dust, which slowly oxidizes. Large particles are not affected, but the fine dust should be kept in closed containers and under a protective gas such as argon. Magnesium-reduced boron is a powder that requires only the normal precautions of air filtration and facial masks for the workers. Borides require no handling or storage precautions, but boron hydrides are very sensitive to shock and can detonate easily, and boron halides are corrosive and toxic.

Tensile properties. Tensile strength: 98.8% pure, amorphous, 1.6 to 2.4 MPa; fibers, 2.6 to 3.1 MPa

Compressive properties. Compressive strength: with B_2O_3 present, up to 0.5 MPa

Hardness. 99.9% crystalline: 3300 HK (with 100-g load), 9.3 Mohs

CADMIUM (Cd)

Cadmium is used for electroplating steel to improve its corrosion resistance, in powder form for mechanical plating of fasteners and other parts for corrosion protection, in low-melting-point alloys, brazing alloys, bearing alloys, nickel-cadmium batteries, and nuclear control rods, and as an alloying addition to copper to improve hardness. It also is used in the manufacture of pigments, plastic stabilizers, phosphors and semiconductor compounds. Cadmium with a minimum purity of 99.90% is available commercially; cadmium of this purity is covered by ASTM B440. Care must be taken to avoid creating toxic dust or fumes; melting or handling conditions that do create dust or fumes require capture at the source by an exhaust ventilation system. When capture of dust or fumes is not feasible, approved NIOSH respiratory protective equipment must be worn. Maximum threshold limit values are 0.2 mg/m³ for cadmium dust and 0.1 mg/m³ for cadmium oxide fumes.

General corrosive behavior. As a protective coating on steel and cast iron parts, cadmium offers corrosion protection in marine atmospheres, under alkaline conditions, and in damp indoor applications. Cadmium-plated steel fasteners resist galvanic attack when used with aluminum parts.

Tensile strength. 71 MPa

Elongation. 50% in 1 in.

Hardness. 16 to 23 HB

Poisson's ratio. 0.33 at room temperature

Elastic modulus. Tension, 55 GPa; shear, 19.2 GPa

Liquid surface tension. 0.564 N/m at 330 °C; 0.611 at 450 °C

CALCIUM (Ca)

Metallic calcium is used as a reducing agent in the preparation of thorium, zirconium, uranium, chromium, vanadium and the rare earths. It is also used as a deoxidizer, decarburizer or desulfurizer for various ferrous and nonferrous alloys. Calcium is used as an alloying or modifying agent for aluminum, beryllium, copper, lead, tin and magnesium alloys. Other uses for calcium include getters for residual gases in high vacuums and vacuum-tube applications and reagents for purification and scavenging of inert gases. Calcium reacts readily with atmospheric components, particularly water vapor, and is not inert to nitrogen. To avoid contamination it must be handled in a dry inert gas atmosphere or in a vacuum.

Tensile properties. Annealed: tensile strength, 48.0 MPa; yield strength, 13.7 MPa; elongation, 51 to 53%; reduction in area, 58 to 62%. As rolled: tensile strength, 115 MPa; yield strength, 84.8 MPa; elongation, 7%; reduction in area, 35%

Hardness. Annealed: 16 to 18 HB

CALIFORNIUM (Cf)

See "Properties of the Actinide Metals," at the end of this section.

CERIUM (Ce)

Cerium is used as an alloying additive to ferrous alloys to scavenge sulfur, oxygen, etc., and to nodulize cast iron; it improves high-temperature oxidation resistance of superalloys. It is also used in glass polishing compounds, petroleum cracking catalysts, lighter flints, glass decolorizing agents, carbon arc lights, ceramic capacitors, $CeCo_5$ permanent magnets and pyrophoric ordnance devices. Cerium readily oxidizes at room temperature in air. It should be stored in vacuum or inert atmosphere; storage in oil is not recommended. Turnings can be ignited easily and burn white hot. Finely divided cerium should not be handled in air.

Tensile properties. β-phase: yield strength, 87 MPa; reduction in area, 24% at 24 °C. γ-phase: tensile strength, 117 MPa; yield strength, 28 MPa; elongation, 22%, reduction in area, 34% at 24 °C

Hardness. 22 HV

CESIUM (Cs)

Cesium ignites immediately on contact with air if poured or sprayed and reacts explosively with water. Cesium may form peroxide compounds if allowed to oxidize in the absence of water normally present in atmosphere. The resulting per-

oxides may be shock sensitive with easily reduced compounds in the same manner that potassium superoxide is shock sensitive with mineral oil. Cesium metal must be contained under vacuum, inert gas, or anhydrous liquid hydrocarbons protected from oxygen or air exposure. Safety and handling information is available from suppliers.

Viscosity. 0.686 mPa·s at 28.64 °C

Surface tension. 0.0394 N/m at 28.64 °C

CHROMIUM (Cr)

Chromium, also known as chrome metal or electrolytic chromium, is an alloying agent used in steel and various nickel-base and cobalt-base superalloys, aluminum-base alloys, electrical resistance alloys, hard facing grains and powders, and for electroplating. No special precautions need be taken when using chromium. Chromium metal, produced by the electrolytic or pyrometallurgical processes, has chromium content in the range of 99.0 to 99.5%, with carbon at 0.050 max. In addition, an iodide process chromium is available with typical purity of 99.99% Cr.

Resistance to specific corroding agents. (A 10% solution at 12 °C was used unless otherwise noted.) Chromium is resistant to the following acids: acetic, aqua regia, benzoic (saturated), butyric, carbonic, citric, fatty, formic, hydrobromic, hydroiodic, lactic, nitric, oleic, oxalic, palmitic, phosphoric, picric, salicylic, stearic and tartaric. Chromium is not resistant to hydrochloric acid or other halogen acids. Chromium is resistant to the following agents: acetone, air, ethyl and methyl alcohol, higher alcohols, aluminum chloride, aluminum sulfate, ammonia, ammonium chloride, barium chloride, beer, benzyl chloride (saturated and 100%), calcium chloride, carbon dioxide, carbon disulfide, carbon tetrachloride (saturated and 100%), dry chlorine, chlorobenzene (saturated and 100%), chloroform, copper sulfate, ferric chloride, ferrous chloride, foodstuffs, formaldehyde, fruit products, glue, hydrogen sulfide (100%), magnesium chloride, milk, mineral oils, motor fuels, crude petroleum products, phenols, photographic solutions, printing ink, sodium carbonate, sodium chloride, sodium hydroxide, sodium sulfate, sugar, sulfur (100%), sulfur dioxide (100%), chlorinated, distilled or rain water, zinc chloride and zinc sulfate.

Tensile properties. Iodide chromium at room temperature, as-swaged: tensile strength, 413 MPa; 0.2% yield strength, 362 MPa; elongation, 44%; reduction in area, 78%. Iodide chromium at room temperature, swaged and recrystallized: tensile strength, 282 MPa; elongation, 0%; reduction in area, 0%. Electrolytic chromium, see Table 2.

Table 2. Tensile properties of recrystallized, swaged, arc-cast electrolytic chromium(a)

Temperature, °C	Tensile strength, MPa	Proportional limit, MPa	Elongation in 25 mm, %	Reduction in area, %	Modulus of elasticity, GPa
20	83	...	0	0	0.248
200	234	...	0	0	...
300	154(b)	11.7	3	4	0.290
350	197	105	6	8	0.168
400	225(c)	132	51	89	0.227
500	...	...	30	75	...
600	242	69	42	81	0.200
700	203	...	33	85	...
800	180	97	47	92	0.255

(a) Recrystallized at 1200 °C in hydrogen. Strain rate of testing, 0.017 m/m per min. (b) Yield strength, 131 MPa. (c) Yield strength, 140 MPa.

Hardness. As cast, forged: room temperature, 125 HB; 700 °C: 70 HB. Electrodeposited, annealed: 500 to 1250 HB, depending on the amount of hydrogen in the deposit. Electrodeposited and annealed: 70 to 90 HB. Extruded, annealed at 1100 °C: 110 HV. Extruded, annealed, rolled at 400 °C: 160 HV

Elastic modulus. Tension, see Table 2.

Impact strength. Unnotched Charpy, as arc-cast electrolytic chromium: room temperature, 2 J; 400 °C: 160 J

COBALT (Co)

Cobalt is used as an alloying element in (a) permanent and soft magnetic materials, (b) superalloys — high-temperature creep-resistant alloys, (c) hard facing and wear-resistant alloys, (d) sintered carbide cutting tools, (e) steels — high speed, tool, and others, (f) cobalt-base tool materials, (g) electrical-resistant alloys, (h) high-temperature spring and bearing alloys, (i) magnetostrictive alloys, and (j) special-expansion and constant-modulus alloys.

Tensile properties. See Table 3.

Poisson's ratio. 0.32

Elastic modulus. Tension, 211 GPa; shear, 826 GPa; compression, 183 GPa

Coefficient of thermal expansion. Linear, 13.8 μm/m·K near room temperature; 14.2 μm/m·K at 200 °C

COLUMBIUM (Cb)

See *Niobium.*

COPPER (Cu)

See Section 7, "Copper," in this volume.

CURIUM (Cm)

See "Properties of the Actinide Metals," at the end of this section.

DYSPROSIUM (Dy)

Dysprosium is used as a control rod in nuclear reactors, and in phosphors, catalysts and garnet microwave devices. It is also used to measure neutron fluxes. Dysprosium will remain shiny in air at room temperature. However, turnings can be ignited and will burn white hot. Finely divided metal should not be handled in air.

Tensile properties. Tensile strength, 132 MPa; yield strength, 39 MPa; elongation, 23%; reduction in area, 22%

Hardness. 44 HV

EINSTEINIUM (Es)

See "Properties of the Actinide Metals," at the end of this section.

ERBIUM (Er)

Erbium is used in lasers and in phosphors, garnet microwave devices, ferrite bubble devices and catalysts. Erbium will remain shiny in air at room temperature. However, turnings can be ignited and will burn white hot. Finely divided metal should not be handled in air.

Tensile properties. Tensile strength, 139 MPa; yield strength, 37 MPa; elongation, 14%; reduction in area, 14%

Hardness. 42 HV

Table 3. Mechanical properties of cobalt

Form and purity	Tensile strength, MPa	0.2% yield strength, MPa	Compressive yield strength, MPa
As cast (99.9)	234.4	...	291.0
Annealed (99.9)	255.1	...	386.8
Swaged (99.9)	689.5	...	...
Zone refined (99.8)	944.6	758.5	...

EUROPIUM (Eu)

Europium is used as control rods in nuclear reactors and as phosphors, especially as the red component in color television screens. Europium oxidizes rapidly in air at room temperature; therefore, it should be handled and stored under an inert atmosphere; storage in oil is not recommended. Finely divided europium can ignite spontaneously in air.

Hardness. 17 HV

GADOLINIUM (Gd)

Gadolinium is used as a burnable poison in shields and control rods in nuclear reactors, in host materials for rare earth phosphors, catalysts and garnet microwave devices. Gadolinium will tarnish slightly in air. Turnings can be ignited and burn white hot. Finely divided gadolinium should not be handled in air.

Tensile properties. Tensile strength, 122 MPa; yield strength, 17 MPa; elongation, 47%; reduction in area, 58%

Hardness. 37 HV for polycrystalline; 23 HV for {1010} prismatic face; 69 HV for {0001} basal plane

GALLIUM (Ga)

Gallium is used predominantly in the electronics industry where it is combined with elements of group III, IV or V of the periodic table to form semiconducting materials; most often, it is combined with arsenic and/or phosphorus for uses in light emitting diodes, laser diodes, solar cells, transistors, etc. Combined as the oxide with other oxides in garnets for magnetic bubble domain devices; as the metal for heat transfer medium, eutectic alloys, liquid seals, high temperature lubricant; in superconducting compounds such as GaV_3 and as compounds in organic reactions. Commercially available metal ranges from 99.5% pure to 99.9999 + %. The most common impurities are Hg, Pb, Sn, Zn and Cu. If certain impurity limits of high-purity gallium are exceeded, the optoelectric properties of electronic materials are degraded or destroyed. Gallium is tested for purity using emission spectrography, mass spectrography and by residual resistivity measurement. Gallium ordinarily is not considered to be hazardous but in compounds or alloys, it may be toxic, depending upon nature of the other components or ions. Gallium in aluminum causes severe intergranular corrosion of the aluminum.

Hardness. 1.5 to 2.5 Mohs

Elastic modulus. Compressibility, at 20 °C: 0.021 nm³/m³·Pa between 15 and 50 MPa

GERMANIUM (Ge)

See reference tables beginning on page 1•43.

GOLD (Au)

See Section 13, "Precious Metals," in this volume.

HOLMIUM (Ho)

Holmium is used in phosphors and ferrite bubble devices and will remain shiny in air at room temperature. Turnings can be ignited and will burn white hot. Finely divided holmium should not be handled in air.

Tensile properties. About the same as dysprosium and erbium

Hardness. 46 HV

INDIUM (In)

The major application of indium is in solders and fusible alloys (low-melting-point alloys). Other applications are in the manufacture of atomic reactor control rods, bearings, low-pressure sodium lamps and as a surface coating on aluminum wire conductors for making low-resistance contact and terminal joints. Some lesser applications are in dental alloys, semiconductors, radiation detector badges, surface lubricants and for protective finishes.

Tensile strength. 2.62 MPa

Compressive strength. 2.14 MPa

Hardness. 0.9 HB

IRIDIUM (Ir)

Small crucibles made of iridium have been used for studying high-temperature reactions. A major use for iridium is in crucibles for producing large, pure, defect-free man-made crystals for electronic and industrial applications. Single crystals so formed are used as substrates in magnetic bubble memory devices, solid-state lasers, insulating substrates for semiconductors, monoclinic filters and substitutes for natural gemstones in jewelry. Iridium also is used as an alloying element to harden platinum, electrodes in spark plugs for severe operating conditions such as jet engine igniters, thermocouple elements, and radioactive isotopes for industrial applications and cancer therapy.

Tensile properties. Properties of 0.5 mm wire. Tensile strength: annealed at 1000 °C, 1100–1240 MPa; hot drawn, 2070–2480 MPa. Elongation: annealed, 20–22%; hot drawn, 13–18%

Hardness. Annealed at 1000 °C, 200–240 HV; as cast, 210–240 HV; hot drawn, 600–700 HV

IRON (Fe)

Iron of sufficient purity so that its properties are essentially those of the element is commonly called high-purity iron. Iron is *not* an article of commerce; instead, it is employed almost exclusively in research. Iron of very high purity can be prepared by a variety of methods, but the last stage of the process is purification by floating zone refining, often combined with treatment in oxidizing and reducing atmospheres. In order to maintain this purity, it is essential to avoid contamination of the iron, which can occur by reactions with the atmosphere or with containers.

General corrosion behavior. Irons of high purity

show a remarkably high resistance to corrosion, sometimes remaining untarnished in laboratory atmospheres for months or years. It was shown that zone-refined iron corrodes at the same rate in hydrochloric acid whether cold worked or annealed and is not affected by any heat treatment schedule. Results reported indicate an orientation effect, as certain crystal faces are attacked more than others.

LANTHANUM (La)

Lanthanum is used as an alloying additive to ferrous alloys to scavenge sulfur, oxygen, etc.; it improves high temperature oxidation resistance of super-alloys. Lanthanum is also used in optical lenses, petroleum cracking catalysts, carbon-arc lights, lighter flints and ceramic capacitors. Lanthanum readily oxidizes at room temperature in air. It should be stored in vacuum or inert atmosphere; storage under oil is not recommended. Turnings can be ignited easily and burn white hot. Finely divided lanthanum should not be handled in air.

Tensile properties. Similar to neodymium
Hardness. 28 HV

LEAD (Pb)

See Section 12, "Lead," in this volume.

LITHIUM (Li)

Lithium is used as a scavenging agent for inert gases; as an alloying element with aluminum, magnesium, zinc and lead; in heat transfer applications; in tritium breeding; in the synthesis of organic compounds and in battery anode material. Compounds containing lithium are used as refrigerant dryers and catalysts, high temperature lubricants and as reagents in the ceramic and chemical industries. Lithium is very reactive and care must be taken to avoid reaction with air, water vapor or other reactive gases. Airtight containers should be used for containment. Niobium, tantalum and molybdenum containers are preferred for temperatures above 600 °C. Ferrous alloys are not recommended for long term use above 550 °C but are satisfactory below this temperature.

Hardness. 0.6 Mohs

LUTETIUM (Lu)

Lutetium is used in ferrite bubble devices. It will remain shiny in air at room temperature. Turnings can be ignited and will burn white hot. Finely divided lutetium should not be handled in air.

Tensile properties. About the same as erbium
Hardness. 44 HV

MAGNESIUM (Mg)

Primary magnesium has a minimum purity of 99.8% and must meet definite specifications limiting individual impurities. This purity is sufficient for most chemical and metallurgical uses. Most of the pure magnesium sold is produced electrolytically as primary magnesium. For applications requiring a minimum of specific impurities, special grades of electrolytic magnesium are available. Silicothermic magnesium is produced by thermal reduction of magnesium oxide. High-purity sublimed magnesium is produced by sublimation of primary electrolytic magnesium under vacuum.

Unless otherwise indicated, the properties listed here for pure magnesium were determined on metal of 99.98 + % purity.

Alloyed with small amounts of aluminum, manganese, rare earths, thorium, zinc or zirconium, magnesium yields alloys with high ratios of strength to weight at both room and elevated temperatures. The alloys have unexcelled machinability, workability by all common methods, stability in many atmospheres, and high damping capacity. Magnesium is an active chemical element and reacts with many common chemical oxidizing agents. A number of metals such as thorium, titanium, uranium and zirconium are prepared by thermal reduction with magnesium. As a catalyst, magnesium is useful for promoting organic condensation, reduction, addition and dehalogenation reactions. It is useful for the synthesis of complex and special organic compounds by the Gringnard process. Its use in pyrotechnics is well established. Magnesium powder can be dispersed in hydrocarbons and mixed in solid propellants for high-energy fuels. Magnesium alloyed with other metals such as aluminum, copper, cast iron, lead, nickel and zinc improves their properties. It also deoxidizes copper and brass, desulfurizes iron and nickel, and de-bismuthizes lead. As a galvanic anode, it provides effective corrosion protection for water heaters, underground pipe lines, ship hulls, ballast tanks and other underground and underwater structures. Small lightweight high-current-output primary batteries use magnesium alloy as the anode. Low capture cross section for thermal neutrons and low-level retention of induced radioactivity point to varied uses for magnesium in atomic energy.

General resistance to corrosion. Dependent on surface film formation; the rate of formation, solution, or chemical change of the film varies with the medium to which it is exposed and also with the alloying elements or impurities present in the metal. Magnesium has good resistance to both indoor and outdoor atmospheres and, in the absence of galvanic couples, even shows resistance to more aggressive environments such as seawater. Indoor tarnishing is controlled largely by the relative humidity. In mild marine and industrial inland atmospheres, the degree of corrosion resistance far exceeds that of mild steel. In stagnant distilled water at room temperature, magnesium forms a protective film that stops action.

Resistance to specific agents. The action of salt solutions on magnesium is dependent on both the anion and the cation of the dissolved salt. Neutral solutions of heavy-metal salts will generally cause severe attack. Magnesium suffers little, if any, attack in alkalis, chromates, fluorides, nitrates or phosphates; more vigorous corrosion occurs in solutions of chlorides, bromides, iodides and sulfates. Mineral acids, except hydrofluoric and chromic acids, dissolve magnesium rapidly. Aqueous solutions of organic acids attack magnesium, whereas fatty acids, hot or cold, dry or containing water, do not. Magnesium is not affected by aliphatic and aromatic hydrocarbons, ketones, ethers, glycols and alcohols, with the exception of anhydrous methyl alcohol. The latter reaction is inhibited, but not completely suppressed, by the presence of water in the methyl alcohol. Pure halogenated organic compounds do not attack magnesium at ordinary temperatures, but at elevated temperatures, or if water is present, corrosion can be severe. No marked reaction was found to occur between magnesium and methyl chloride, carbon tetrachloride, or chloroform, even after prolonged heating under increased pressures. Lower alkyl halides, up to amyl derivatives, have been shown to react with magnesium only under pressure and at temperatures in excess of 270 °C, but higher alkyl halides are reported to react with magnesium at their boiling points. In general, the presence of water greatly stimulates the reaction between magnesium and halogenated compounds at elevated temperatures. Fluorinated hydrocarbons are generally without action on magnesium when dry.

Casting temperature. 705 to 760 °C.

Type of flux. Open-pot melting, Dow No. 250; crucible melting, Dow No. 310

Precautions in melting. Molten metal must be protected from the atmosphere by the use of inert gas or protective fluxes. Molten magnesium does not react with carbon, silicon carbide, or combinations of these. There is little, if any, reaction with molybdenum, tungsten or tantalum. Low-carbon (welded or cast) steel crucibles are used as containers for molten magnesium of commercial purity (avoid nickel-bearing steels). Use a protective agent (Dow No. 181) to prevent magnesium from burning when it is being poured in an open atmosphere. The usual safety precautions observed with any molten metal should be observed. Preheat all tools or metal introduced in molten magnesium. Keep pot settings free from iron scale.

Precautions in fabrication. A supply of an approved extinguishing agent should be readily accessible to any machining, grinding or similar operations on magnesium. Good housekeeping and sharp machine tools are the best deterrents to magnesium fires. Heat treating furnaces should have a protective atmosphere, such as SO_2 or BF_3, when operating at high temperatures. Magnesium powder must be kept dry.

Machinability index. For pure magnesium and all magnesium alloys, 500 (free cutting brass = 100)

Hot working temperature range. 93 to above 482 °C for 99.98% Mg; 177 to above 482 °C for 99.80% Mg

Annealing temperature. 150 to 200 °C. Maximum reduction between anneals, 50 to 60% under suitable conditions

Forming temperature. 150 to 200 °C for best results

Joining. Rivet composition, aluminum alloy 5056. Oxyacetylene weld with pure magnesium welding rod, magnesium welding flux and neutral flame. Resistance welding is satisfactory. Helium-arc or argon-arc welding is preferred. Use pure magnesium welding rod and no flux

Recrystallization temperature. 93 °C for 1-h anneal after 30% cold reduction (99.98% Mg); 177 °C for 1-h anneal after 30% cold reduction; 93 °C for 1-h anneal after 60% cold reduction (99.80% Mg)

Tensile properties. See Table 4.

Compressive properties. See Table 4.

Hardness. See Table 4.

Elastic modulus. Tension at 20 °C. 99.98% Mg: dynamic, 44 GPa; static; 40 GPa. 99.80% Mg: dynamic, 45 GPa; static, 43 GPa

MANGANESE (Mn)

See reference tables beginning on page 1·43.

Table 4. Typical mechanical properties of magnesium at 20 °C

Form and section	Tensile strength, MPa	0.2% tensile yield strength, MPa	0.2% compressive yield strength, MPa	Elongation(a), %	Hardness	
					HRE	HB(b)
Sand cast, 13-mm (¹/₂-in.) diam	90	21	21	2-6	16	30
Extrusion, 13-mm (¹/₂-in.) diam	165-205	69-105	34-55	5-8	26	35
Hard rolled sheet	180-220	115-140	105-115	2-10	48-54	45-47
Annealed sheet	160-195	90-105	69-83	3-15	37-39	40-41
(a) In 50 mm or 2 in. (b) 500-kg load; 10-mm diam ball.						

MERCURY (Hg)

Mercury is the only common metal that is liquid at room temperature. It is rarely found in the free and uncombined state in nature, most often occurring as the ore *cinnabar* (HgS). Mercury is widely used for thermometers, barometers, diffusion pumps and other laboratory instruments. It is used commercially in mercury-vapor lamps, in lamps and lamp tubes for advertising signs, in switches for instruments and control devices, in dental preparations, and in batteries. Mercury chemicals are widely used for making pesticides, antifouling paints, high-grade paint pigments, explosives and medicines. Mercury cells are used for production of caustic chlorine.

Prime virgin mercury as commonly obtained by refining directly from mercury ores has a purity of at least 99.9%, and in many instances 99.99%. Metal of lesser purity is generally obtained by reclaiming discarded mercury.

Precautions in use. *Health hazard.* Mercury vapor is readily absorbed through the respiratory tract, the gastrointestinal tract or unbroken skin. Mercury acts as a heavy-metal poison, but its effect becomes known only after prolonged exposure. Acute poisoning from mercury vapor is extremely rare. Mercury absorbed from vapor is eliminated from the human body fairly quickly through the urinary and fecal tracts. Mercury levels resulting from exposure to mercury vapor or to inorganic mercury compounds (including aryl mercury compounds such as phenylmercuric acetate) are rapidly reduced and do not accumulate. On the other hand, mercury levels resulting from exposure to alkyl mercury compounds such as methyl mercury or ethyl mercury compounds cannot be eliminated quickly, and tend to accumulate.

Mercury is a very volatile element, and dangerous levels of mercury vapor are readily attained at room temperature in enclosed spaces that are not adequately ventilated. The present toxicity limit for mercury vapor in air is 0.1 mg/m³. At 20 °C, air saturated with mercury vapor contains a concentration more than 100 times this limit.

Because of the toxic nature of mercury and many mercury compounds, certain precautions are mandatory during handling and disposal. Containers should be securely covered. All operations involving mercury metal should be carried out in a well-ventilated area or in a closed system to prevent accumulation of mercury vapor in the workspace; this is of utmost importance if the operation involves heating mercury above room temperature. Workspaces should be continually monitored with special electronic instruments to detect any rise of mercury-vapor concentration above the established safe working limit. Workers should be provided with masks or special breathing devices, and the level of mercury in the body of every worker should be periodically monitored by specially trained medical personnel. Any spills of liquid, or escape of vapor from a closed heated system, must be countered by immediate decontamination of the affected workspace. Disposal is ordinarily accomplished by sending impure mercury or concentrated mercury compounds to reclamation centers where purified metal is produced from the discards. Mercury compounds such as methyl mercury are dangerous pollutants and are required to be removed from effluents before they are discharged into natural waters. Sludges and other solid wastes that are contaminated with small concentrations of mercury are sometimes buried at approved sites.

MISCHMETAL (MM)

Mischmetal is used as an alloying additive in ferrous alloys to scavenge sulfur, oxygen and other substances. Mischmetal is also added to magnesium-base alloys to improve high temperature strength and to ductile irons to nodularize graphite. Other uses of mischmetal include lighter flints, pyrophoric ordnance devices and mischmetal cobalt (MMCo₅) permanent magnets. Mischmetal oxidizes at room temperature in air. Turnings can be ignited easily and burn white hot. Finely divided mischmetal should not be handled in air. Because mischmetal is an indefinite mixture of rare earth metals, the properties of a particular mischmetal depend on its composition, which in turn depends on the mineral source for the mixture. Listed below are the properties of bastnäsite-derived mischmetal. All values are estimated.

UNS number. E21000

Composition limits. Total mixed rare earths: 99.0 min; mixture consists of 50.0 Ce; 38.0 La; 12.0 Nd; 4.0 Pr; 1.0 other rare earths

Tensile properties. At 24 °C: tensile strength, 138 MPa; yield strength, 48 MPa; elongation, 25%; reduction in area, 50%

Hardness. 35 HV

MOLYBDENUM (Mo)

Molybdenum is used as alloying additions and for electrical and electronic parts; missile and aircraft parts, high-temperature furnace parts; die-casting cores; hot working tools; boring bars; thermocouples; nuclear-energy applications; corrosion-resistant equipment; equipment for glass melting furnaces; metallizing. Molybdenum is not suitable for continued service at temperatures above 500 °C in an oxidizing atmosphere unless protected by adequate coating

Tensile properties. See Fig. 1 and 2.

Hardness. See Fig. 3.

Directional properties. If not cross rolled, the tensile strength of molybdenum sheet may be as much as 20% greater in the direction of rolling than when the inclination of the direction of tension to that of rolling is between 45 and 90°

Properties of single crystals. Tensile strength, 350 MPa

General corrosion behavior. Molybdenum has particulary good resistance to corrosion by mineral acids, provided oxidizing agents are not present. It is also resistant to many liquid metals and to most molten glasses. In inert atmospheres, it is unaffected up to 1760 °C by refractory oxides. Molybdenum is relatively inert in hydrogen, ammonia and nitrogen up to about 1100 °C, but a superficial nitride case may be formed in ammonia or nitrogen.

Consolidation. In most instances, molybdenum is consolidated from powder by compacting under pressure followed by sintering in the range from 1650 to 1900 °C. Some molybdenum is consolidated by a vacuum-arc-casting method in which a preformed electrode is melted by arc formation in a water-cooled mold.

Hot working temperature. Generally forged between 1180 and 1290 °C down to 930 °C

Annealing temperature. Normal stress-relieving temperature is 870 to 980 °C

Recrystallization temperature. Depends on prior working and condition; 1180 °C for full recrystallization in 1 h of a ⁵/₈-in. bar reduced 97% by rolling

Suitable forming methods. Conventional methods

Precautions in forming. Must be heated to the proper temperature relative to its thickness and forming speed

Heat treatment. Not hardenable by heat treatment but only by work hardening

Suitable joining methods. Can be brazed or joined mechanically, as well as welded by arc, resistance, percussion, flash and electron-beam methods. Arc-cast molybdenum is preferred to a powder metallurgy product for welding. Absolute cleanliness of surface is essential. Fusion welding must be carried out in closely controlled inert atmosphere

NEODYMIUM (Nd)

Neodymium is used as an alloying additive to ferrous alloys to scavenge sulfur, oxygen and other elements and to strengthen magnesium alloys. It is also used as a laser material, glass coloring agent and in petroleum cracking catalysts, carbon-arc lights, lighter flints, and ceramic capacitors. Neodymium oxidizes at room temperature in air. It should be stored in a vacuum or inert atmosphere; storage in oil is not recommended. Turnings can be ignited easily and will burn white hot. Finely divided neodymium should not be handled in air.

Tensile properties. Tensile strength, 169 MPa; yield strength, 71 MPa; elongation, 28%; reduction in area, 72%

Hardness. 18 HV

NEPTUNIUM (Np)

See "Properties of the Actinide Metals," at the end of this section.

NICKEL (Ni)

See the article "Nickel and Nickel Alloys," in Section 15.

NIOBIUM (COLUMBIUM) (Nb)

Niobium is used as an alloying element in nickel- and cobalt-based superalloys as well as some grades of stainless and low alloy steels. It is also used as an alloy base for various combi-

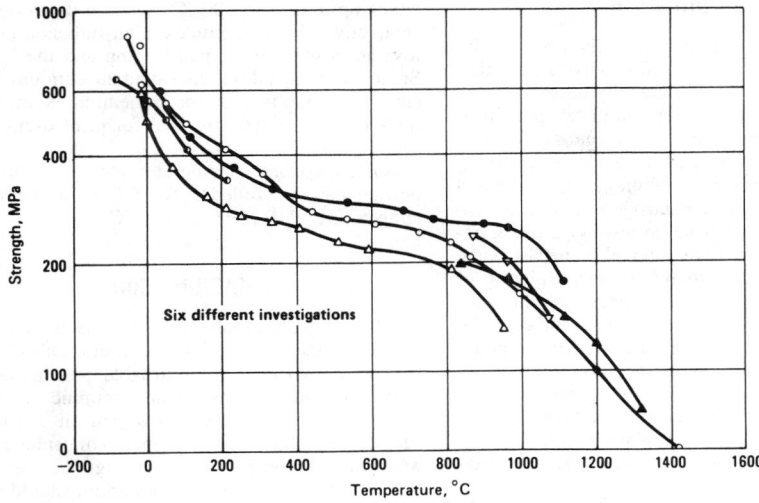

Fig. 1. Temperature dependence of the tensile strength of molybdenum

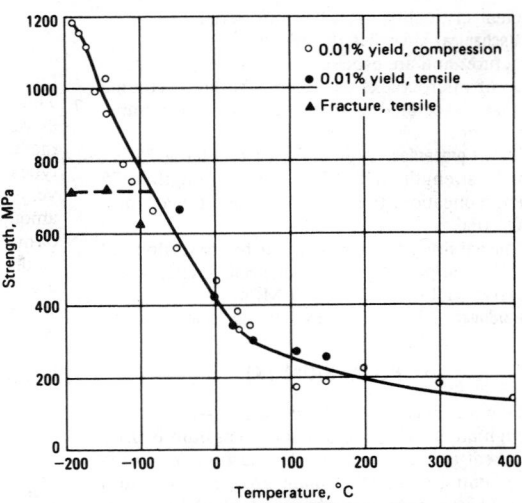

Fig. 2. Temperature dependence of the yield and fracture strengths of molybdenum

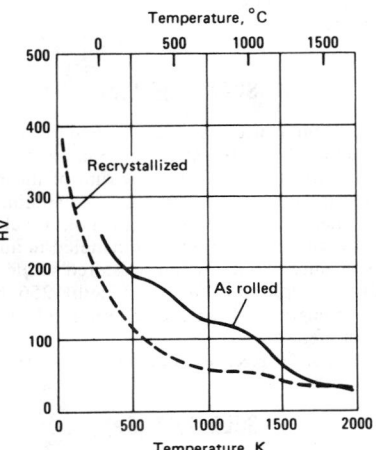

Fig. 3. Temperature dependence of the hardness of molybdenum

nations with zirconium, hafnium, tungsten, tantalum and molybdenum to increase high temperature mechanical properties. Niobium, niobium-titanium alloys and niobium-tin alloys are used as superconductors. Niobium oxidizes and becomes contaminated with absorbed oxygen rapidly above about 400 °C in oxygen-containing atmospheres, including atmospheres normally considered neutral or reducing; absorbs hydrogen at temperatures between about 250 and 950 °C from hydrogen-containing atmospheres. Contamination by interstitial elements results in loss of ductility at ambient temperature. Consequences of high impurity levels include impaired fabricability, increased ductile-to-brittle transition temperature, considerable low-temperature strengthening with attendant loss in ductility, intensified strain-aging effects at slightly elevated temperature, and slight strengthening at higher temperature.

General corrosion behavior. Niobium is moderately to highly resistant to corrosion in most aqueous media that are usually considered highly corrosive, such as dilute mineral acids, organic acids and organic liquids. Notable exceptions are dilute strong alkalis, hot concentrated mineral acids, and hydrofluoric acid, which attack the

metal rapidly. Gaseous atmospheres at high temperature attack niobium rapidly, primarily by oxidation, although oxygen contents may be very low. Niobium and its alloys are remarkably resistant to corrosion by certain liquid metals, notably lithium and NaK, and to high temperatures (900 to 1010 °C); this coupled with low-capture cross section for thermal neutrons renders niobium materials most attractive for reactor application.

Tensile properties. Highly dependent on purity, particularly the content of interstitial elements. Values listed are for material of good commercial purity (only 100 to 200 ppm interstitial contaminants). Wrought: tensile strength, 585 MPa; elongation, 5%. Annealed: tensile strength, 275 MPa; yield strength, 207 MPa; elongation, 30%; reduction in area, 80%

Hardness. Annealed: 80 HV. Wrought: 160 HV

Elastic modulus. At 25 °C: tension, 103 GPa; shear, 37.5 GPa. At 870 °C: tension, 90 GPa

Ductile-to-brittle transition temperature. <147 K; increases sharply with lower purity

Alloying practice. High-vacuum powder metallurgy techniques may be utilized effectively. Consumable-electrode vacuum arc melting and electron-beam furnace melting may be used for alloying purposes. High-vacuum techniques purify niobium at temperatures above 1980 °C, through volatilization of NbO_2.

Precautions in melting. Exclude atmospheric contaminants as completely as possible. Cold hearth techniques are required to prevent crucible reaction.

Recrystallization temperature. Material cold reduced 70 to 80% completely recrystallizes in 1 h at 1090 °C

Hot working temperature. 800 to 1100 °C may be necessary to break down the ingot structure of niobium. This requires conditioning of the breakdown product to remove the contaminated surface layer. Subsequent working is done cold.

Maximum reduction between anneals. Virtually unlimited.

Precautions in forming. Because of the high probability of seizure and galling, selection of lubricant and die material is important in extreme-pressure methods. Carbon tetrachloride (for machining) or sulfonated tallow or proprietary waxes (for spinning and drawing) are preferred lubricants. Polished aluminum bronze has been rec-

ommended as a die material for extreme-pressure processes.

Suitable joining methods. Welding processes capable of excluding interstitial contaminants from the hot zone are satisfactory.

OSMIUM (Os)

Osmium and its alloys are useful for their hardness and resistance to wear and corrosion. The resistance to rubbing wear is more than would be expected on the basis of hardness; alloys of equal hardness are inferior to osmium. Osmium is used in fountain-pen nibs, phonograph needles, electrical contacts, and instrument pivots. Osmium should not be heated in the presence of oxygen, since it has a toxic oxide OsO_4, that boils off at 130 °C.

Hardness. Approx 800 HV, arc-melted button

PALLADIUM (Pd)

See Section 13, "Precious Metals," in this volume.

PLATINUM (Pt)

See Section 13, "Precious Metals," in this volume.

PLUTONIUM (Pu)

The term plutonium usually implies ^{239}Pu of at least 95% purity (generally 99.7 to 99.99 wt %). Small amounts of δ-phase stabilizers, such as 0.1 wt % Al, may cause retention of the δ-phase at room temperature. The term plutonium, however, also implies ^{239}Pu sufficiently free of δ-phase stabilizers so that only the α-phase is present at room temperature.

Typical uses of plutonium include nuclear weapons, nuclear fuel, neutron sources, heat sources for thermoelectric generators, production of higher isotopes and transplutonic elements. Plutonium is a highly radioactive alpha emitter, extremely poisonous, and is properly handled in gloveboxes. It is about twice as poisonous as radium to the human system. The maximum permissible body burden is 0.6 μg. When handling quantities in excess of 300 g, the possibility of nuclear criticality must be considered.

Mechanical properties depend heavily on microstructure, and are especially sensitive to the presence of microcracks caused by the large volume change associated with the β-to-α phase transformation.

Tensile properties. Typical for cast alpha at 25 °C: tensile strength, 415 MPa; yield strength, 275 MPa; elongation, 0.2 to 0.5%; proportional limit, 160 MPa

Compressive properties. Typical for cast alpha at 25 °C: compressive strength, 830 MPa; compressive yield strength, 415 MPa

Hardness. 250 to 283 HV, 10-kg load

POTASSIUM (K)

Few uses have been found for potassium metal, though an alloy of sodium and potassium is used as a heat-transfer medium. Because potassium is an essential element in plant growth, potassium or K_2O is a main component of plant fertilizer. Potassium is also used as the super oxide, KO_2, to produce oxygen in gas masks. Potassium is highly reactive and must be handled with great care. Use of dry and oxygen-free inert gas atmosphere is essential if reactions are to be avoided.

PRASEODYMIUM (Pr)

Praseodymium is used as an alloying additive to ferrous alloys to scavenge sulfur, oxygen, etc. It is also used as a glass and ceramic coloring agent, and in petroleum cracking catalysts, carbon-arc lights and $PrCo_5$ permanent magnets. Praseodymium oxidizes at room temperature in air. It should be stored in vacuum or inert atmosphere; storage in oil is not recommended. Turnings can be ignited easily and will burn white hot. Finely divided praseodymium should not be handled in air.

Tensile properties. Approximately the same as neodymium

Hardness. 37 HV

PROMETHIUM (Pm)

Promethium is used as a lightly shielded radioisotope power source. It is also a highly radioactive β emitter (^{147}Pm).

Tensile properties. About the same as neodymium

Hardness. 63 HK

PROTACTINIUM (Pa)

See "Properties of the Actinide Metals," at the end of this section.

RHENIUM (Re)

Rhenium is used for electrical contacts, thermocouples, filaments for electronic devices, and to increase ductility in alloys of molybdenum and tungsten for use in electronics, thermocouples and welding rods. Rhenium begins to generate a white nonpoisonous vaporous oxide, Re_2O_7, at about 600 °C when heated in air.

Tensile properties. True stress, B, at unit strain is 2.53 GPa; strain hardening exponent, n, is 0.353

Elastic modulus. At 20 °C, 460 GPa

Hardness. Arc-melted button, 135 HK; annealed rod, 270 HK; rod swaged 40% in cross-sectional area, 825 HK

RHODIUM (Rh)

Pure rhodium is used principally where maintained high and uniform reflectivity is essential. This includes mirrors, principally those made by electrodeposition of rhodium on metal, plus some made by subliming rhodium on glass. Various types of light filters can be made by applying very thin coatings by subliming. A substantial amount of rhodium is electrodeposited as a nontarnishing finishing plate on jewelry articles, including white gold, silver, and other metals. Some use is also made of rhodium for the plating of sliding electrical contact surfaces; for this purpose, rhodium is sometimes applied over a heavier plate of palladium. Some pure wrought rhodium is used, but rhodium is most important as an alloying element with platinum. These materials find much use at elevated temperatures for crucibles, furnace windings, glass-working equipment, thermocouples, and particularly for catalysts.

Tensile properties. Tensile strength: annealed, 951 MPa; hard, 2068 MPa

Hardness. 101 HB, 122 HV

RUBIDIUM (Rb)

Current uses of rubidium are limited but include such applications as vacuum tubes and photoelectric cells. Potential uses include use as a heat transfer medium and as a fuel for ion propulsion engines. Chemical behavior of rubidium is similar to that of potassium; it reacts vigorously with air or water and should be handled and stored in a dry-inert environment.

Hardness. 0.3 Mohs

RUTHENIUM (Ru)

Ruthenium is used as a hardener for platinum and palladium for jewelry and other applications, including 10 to 11% Ru platinum for aircraft magneto contacts and similar contacts. Ruthenium can be used as an electric contact at temper-

Table 5. Resistance of silicon to specific corroding agents

Corrosive agent	Resistance
Air	Resistant
Ammonia	Resistant; reacts with vapors at bright red heat
Bromine	Resistant; burns at 930 °F
Carbon dioxide	Resistant
Chlorine	Resistant; burns at 640 °F
Copper sulfate	Resistant (10% solution)
Ferric chloride	Resistant (10% solution)
Hydrochloric acid	Resistant (dilute or conc, cold or boiling)
Hydrofluoric acid	Resistant (dilute or conc, cold or boiling)
Hydrogen sulfide	Resistant
Iodine	Resistant
Nitric acid	Resistant (dilute or conc, cold or boiling)
Oxygen	Resistant
Potassium hydroxide	Attacked
Sodium hydroxide	Attacked
Sulfur	Resistant; reacts at elevated temperatures
Sulfur dioxide	Resistant
Sulfuric acid	Resistant (dilute or conc, cold or boiling)
Water, distilled	Resistant
Water, rain	Resistant

atures up to around 500 °C, because its oxide is conductive. Hard, complex high-ruthenium alloys are also used for pen tipping and the like. Some of these alloys also contain osmium. A significant application for ruthenium is in the thick-film paste systems used for printed-circuit resistance elements.

Tensile properties. Tensile strength for compact powder bar, hot rolled 50%: 540 MPa at room temperature, 246 MPa at 3650 °C

SAMARIUM (Sm)

Alloyed with cobalt, samarium is used as a permanent magnet, $SmCo_5$. Samarium is also used in nuclear reactors as a burnable poison, as a phosphor and in catalysts and ceramic capacitors. This metal oxidizes slowly in air at room temperature. Storage in an inert atmosphere or vacuum is recommended. Turnings can be ignited easily. Finely divided samarium should not be handled in air.

Tensile properties. Tensile strength, 157 MPa; yield strength, 69 MPa; elongation, 22%; reduction in area, 34%

Hardness. 39 HV

SCANDIUM (Sc)

Scandium is used as a neutron window or filter in reactors. It is also used in high-intensity lamps because of the multilined spectrum of incandescent scandium vapor. Turnings of scandium can be ignited and will burn white hot. Finely divided scandium should not be handled in air. Ingots of pure scandium can be stored in air.

Tensile properties. Tensile strength, 256 MPa; yield strength, 174 MPa; elongation, 5%; reduction in area, 8%

Hardness. 50 HK, 36 HV, 40 HB, 85 HRH

SELENIUM (Se)

Selenium is used in rectifiers, in photovoltaic cells, in xerographic drums, as a colorizing and decolorizing agent in glass, as a color pigment used in paints, ceramics and plastics, as an additive to improve machinability of low-carbon steels, stainless steels, copper alloys and Invar, as an additive to lead-antimony battery grid metal to improve properties, and as a vulcanizing agent to improve temperature and abrasion resistance of rubber.

Hardness. 2.0 Mohs

SILICON (Si)

Silicon is used mainly in the primary and secondary aluminum industry, in silicones and in steels. The purity of the silicon for these purposes varies from 96.7 to 98.5 Si, 0.10 to 0.75 Al, and 0.03 to 0.40 Ca. The remainder is chiefly iron.

Resistance to specific corroding agents. See Table 5.

Compressive strength. Chill cast specimens, 25 by 25 by 90.2 mm: 92.87 MPa

Elastic modulus. Tension. Chill cast specimens, 25 by 25 by 90.2 mm: 112.7 GPa. Chill cast specimens, 6.5 by 6.5 by 76 mm: 106.8 GPa

Bending properties. Chill cast specimens, 6.5 by 6.5 by 76 mm on 75-mm span: deflection under 53-N load, 0.0457 mm; modulus of rupture, 62.37 MPa; breaking strength, 156 N

SILVER (Ag)

See Section 13, "Precious Metals," in this volume.

SODIUM (Na)

Sodium is used as a liquid metal heat-transfer medium, a working fluid for evaporative heat pipes, and an electrical conductor in homopolar generators. It is also used in vapor lamps for highway lighting, as an alloying addition for lead, zinc, and aluminum and as a reactant for deoxidation of metals and for reduction of metal fluorides. Sodium is highly reactive with water; hydrogen released by the reaction is potentially explosive. Molten sodium will burn in ambient air. Iron-based alloys are usually selected as containers in transport of liquid sodium. Argon, helium and nitrogen are used as cover gases to minimize sodium oxidation. Sodium fires are best extinguished by closing off air accesses or by blanketing with either nitrogen or inert solids, such as carbon granules. Commercial extinguishing media include sodium chloride, sodium carbonate and calcium phosphate. Carbon tetrachloride and solid carbon dioxide extinguishers should not be used on sodium fires.

STRONTIUM (Sr)

See reference tables beginning on page 1•43.

TANTALUM (Ta)

Tantalum provides a combination of properties not found in many refractory metals—excellent fabricability, low ductile-to-brittle transition temperature and high melting point. The largest use of tantalum at this time is in electrolytic capacitors. Sizeable quantities of tantalum also are used in chemical process equipment (such as heat exchangers, condensers, thermowells and lined vessels), notably for handling nitric, hydrochloric, bromic and sulfuric acids, and combinations of these acids with many other chemicals. Spinnerettes for extruding man-made fibers constitute another important application of tantalum. Because of its high melting point, tantalum is used for heating elements, heat shields and other components of vacuum furnaces. Tantalum has been used in specialized aerospace and nuclear applications. Tantalum also is used in prosthetic devices in contact with body fluids and as an alloy component in superalloys, and tantalum carbide is an important constituent of cemented carbide cutting tools made from mixtures of titanium, tungsten and tantalum carbides. Yield and ultimate strengths are increased, and ductility is reduced, by increases in the amount of interstitial elements (oxygen, nitrogen, carbon and hydrogen). Embrittlement of the tantalum can occur if contamination by these elements is sufficiently severe. High-purity tantalum (99.90% min) is available commercially with the following maximum impurity limits, in ppm: 500 Nb, 300 W, 100 to 200 O, 100 Fe, 100 Mo, 50 to 75 C, 50 to 75 N, 50 Ni, 50 Si, 50 Ti, 10 H.

General corrosion behavior. Tantalum oxidizes in air above 300 °C. It has excellent resistance to corrosion by a large number of acids, by most aqueous solutions of salts, by organic chemicals and by various combinations and mixtures of these agents. Also, tantalum exhibits good resistance to many corrosive as well as common gases and to many liquid metals.

Compacting pressure. 10 to 85 MPa depending on the physical properties of the powder

Sintering temperature. 2300 to 2600 °C in high vacuum will essentially remove all detrimental impurities contained in the powder.

Machinability. Fully recrystallized unalloyed tantalum has machinability similar to that of soft copper. Use chlorinated hydrocarbons, light oil or water-soluble oil as a cutting fluid, and high speed tool steel or cemented carbide tools. Tantalum can be successfully turned, bored, drilled, tapped, reamed, shaped, milled, sawed and ground to desired tolerances and surface finishes.

Joining. Gas tungsten-arc, gas metal-arc, resistance and electron beam welding can be used for joining tantalum. High-purity inert gas (argon or helium) or vacuum must be used in fusion welding. Resistance spot and seam welding can be done in air or under water with proper precautions. Silver brazing alloys, copper, and several specially developed refractory metal brazing alloys can be used to braze tantalum to itself or to dissimilar metals such as stainless steels. Brazing is done in vacuum or under an inert atmosphere (high-purity argon or helium). Tantalum also can be bonded to dissimilar metals by explosive cladding, and in some instances by roll bonding.

Cleaning. To avoid contamination of tantalum by interstitial elements and metallic impurities, it is mandatory that the material be chemically cleaned before any heating operation (such as annealing or welding). Such cleaning involves thorough degreasing (detergent or solvent); chemical etching in 20 vol % HF, 20 vol % H_2SO_4 and 60 vol % HNO_3; hot- and cold-water rinsing (deionized water recommended); and spot-free drying. The etching solution may be strengthened (by adding HF) or weakened (by adding water) to achieve the amount of stock removal necessary to ensure cleanness.

Precautions in melting. Exclude oxygen, hydrogen, nitrogen and carbon. Melt in vacuum or inert atmosphere.

Hot working temperature. None; it is worked cold.

Annealing temperature. Above 1050 °C in high vacuum for complete recrystallization, with resulting grain size as shown in Table 6

Maximum reduction between anneals. Greater than 95%

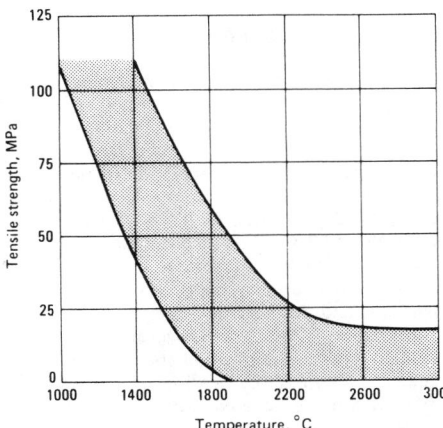

Note that upper portion of curve is characterized by high strain rates and high interstitial content whereas lower portion of curve is characterized by low strain rates and low interstitial content.

Fig. 4. High-temperature tensile strength of tantalum

Table 6. Tantalum grain size after annealing

Final annealing temperature(a), °C	Average ASTM grain size(b)
1200	5 to 6
1300	4
1400	3 to 4
1425	3 to 4
1600	2
1700	1
1800	0 to 1

(a) Material cold rolled 75% after intermediate annealing, then annealed 1 h at indicated temperature. (b) Determined by comparison with ASTM grain-size chart at 100 ×.

Suitable forming methods. Tantalum can be formed by spinning, deep drawing, bulging, bending, blanking, punching and stretch forming using conventional methods, equipment and tooling normally found in shops fabricating heat-resistant alloys.

Tensile properties. See Fig. 4.

Hardness. Electron-beam melted, 110 HV; P/M compact, 120 HV

Poisson's ratio. 0.35 at 20 °C

Elastic modulus. Tension: 186 GPa at 20 °C, 159 GPa at 750 °C. Shear: 69 GPa at 20 °C

TECHNETIUM (Tc)

Technetium is used as a radioactive tracer in medicine with potential uses arising from its favorable corrosion-inhibiting properties and its high superconducting transition temperature. There is contamination hazard due to its radioactivity. Classed as moderately toxic. All sample preparation, etc., which could disperse solid ^{99}Tc must be carried out in glove-box facilities. The data that follow are for ^{99}Tc only.

Tensile properties. Tensile strength. 1510 MPa as-rolled (46% reduction); 798.6 MPa after annealing 10 min at 950 °C; 0.2% offset yield strength, 1290 MPa as-rolled (46% reduction); 319 MPa after annealing 10 min at 950 °C (fully recrystallized); elongation in 1 in.: 4% as-rolled (46% reduction); 30% after annealing 10 min at 950 °C

Hardness. 46% cold worked, 394 HV, 442 HB; annealed at 950 °C: 151 HV, 112 HB

TELLURIUM (Te)

Tellurium is used as an additive to steel and copper to improve machinability, an additive to cast iron to control depth of chill, in the production of malleable cast iron as a carbide stabilizer, and in lead-base alloys to improve their properties. It is an important constituent of thermoelectric alloys. It is also used in fuses for explosives, as a vulcanizing agent in rubber, as a catalyst in chemical reactions, as a glass-forming agent in glasses, and as a colorizing agent in glass and ceramics.

Tensile strength. 10.8 to 11.25 MPa

Hardness. 25 HB, 2.3 Mohs

TERBIUM (Tb)

Terbium is used as a phosphor and in catalysts. It will remain shiny in air at room temperature. Turnings can be ignited and will burn white hot. Finely divided terbium should not be handled in air.

Tensile properties. About the same as gadolinium

Hardness. 38 HV for polycrystalline; 30 HV for

{10$\bar{1}$0} prismatic face; 80 HV for {0001} basal plane

THALLIUM (Tl)

See reference tables beginning on page 1·43.

THORIUM (Th)

Thorium, as a solid or fluid in elemental, intermetallic or oxide form, is used as a fuel for nuclear reactors since it is a fertile material for the generation of fissionable uranium-233. The oxide form of thorium is used for gas mantles. Thorium oxide additions control grain size in tungsten filaments and strengthen nickel alloys (TD nickel). Thorium metal is used as an alloying addition in magnesium technology and is used as a deoxidant for molybdenum, iron and other metals. Thorium has a variety of applications in electronic technology. Thorium is radioactive. Pure, fresh thorium is a weak α-emitter, but old thorium, with accumulated decay products, also emits β-particles and penetrating γ-rays. Thorium is also chemically quite reactive. In finely divided form, thorium can be pyrophoric, and in dust form, it may be explosive. Chemical toxicity of thorium and its compounds is generally low.

Tensile properties. As-cast thorium (0.02 to 0.08 wt% carbon): tensile strength, 219 MPa; yield strength (0.2% offset), 144 MPa; elongation, 34%; reduction in area, 35%

Compressive properties. Values closely comparable to tensile values

Hardness. 56 to 114 HV with 20-kg load

THULIUM (Tm)

Thulium is used in phosphors, ferrite bubble devices and catalysts. Irradiated thulium (^{169}Tm) is used as a portable radiographic source. Thulium will remain shiny in air at room temperature. Turnings can be ignited and will burn white hot. Finely divided thulium should not be handled in air; due to its high vapor pressure at its melting point, thulium should not be arc melted.

Tensile properties. About the same as erbium
Hardness. 48 HV

TIN (Sn)

See Section 10, "Tin," in this volume.

TITANIUM (Ti)

See Section 9, "Titanium," in this volume.

TUNGSTEN (W)

See the article "Electrical Contact Materials," in Section 20.

URANIUM (U)

Natural uranium nominally contains 0.006% ^{234}U, 0.72% ^{235}U, and the remainder ^{238}U. The term "enriched uranium" designates uranium containing higher-than-normal ^{235}U; "depleted uranium" designates lower-than-normal ^{235}U. Other designations include alpha uranium, beta uranium and gamma uranium for the three polymorphic forms.

The most common use is as enriched uranium in nuclear reactors. It is used either in unalloyed

Table 7. Typical mechanical properties of vanadium metal at room temperature

Condition	Tensile strength, MPa	Yield strength, MPa	Elongation in 50 mm, %	Reduction of area, %	Hardness HRA	Hardness HRB	Cold bend, deg
Bar, 25.4-mm diam(a)							
Hot rolled	472	439	27.0	54.4	...	85	...
Wire, 3.9-mm diam(b)							
Vacuum annealed	538	463	25.0	87.5	48	...	180
Cold drawn 80%	910.8	765	6.8	76.5	54	...	180
Sheet, 1.9-mm thick(c)							
Vacuum annealed	536	454	20.0	53.0	...	83	180
Cold rolled 84%	828	776.3	2.0	40.6	...	100	180

(a) Specimen size: 12.8-mm diam by 51 mm. (b) Specimen size: 3.9-mm diam by 51 mm. (c) Specimen size: 1.9 by 12.7 by 51 mm.

metallic form or as uranium oxide. In the latter case, reactor fuel elements usually are made of a mixture of uranium oxide and plutonium dioxide. Enriched uranium is also used as a nuclear explosive. ^{233}U, which is produced by neutron irradiation of ^{232}Th, is also fissionable and so can be used in nuclear reactors.

Massive uranium metal offers no substantial problem in handling and storage. It oxidizes slowly in air and forms an adherent oxide. Finely divided uranium metal, on the other hand, is pyrophoric and care must be exercised to prevent fires. Uranium is a controlled nuclear material and emits weak alpha radiation. It is also a heavy-metal poison and, like other heavy-metal poisons, must be processed under controlled conditions to avoid ingestion of fumes or dust by workers.

Wide variations exist in all mechanical properties of alpha uranium and depend markedly on a large number of parameters, most notably preferred orientation, grain size, fabrication history, heat treatment, and type and distribution of impurities. For instance, fracture stress decreases from approximately 600 MPa for a grain size of 1 μm to 130 MPa for a grain size of 10 μm.

Tensile properties. At 293 K (approximate). As cast: tensile strength, 400 MPa; 2% offset yield strength, 200 MPa; elongation, 4%; reduction in area, 10%. Beta annealed (grain size, 500 μm): tensile strength, 615 MPa. Wrought alpha uranium: tensile strength, 1150 MPa; yield strength, 740 MPa; elongation, 7%; reduction in area, 14%

Hardness. Coarse-grained α-U, 185 HV at 300 K; fine-grained α-U, 250 HV at 300 K. β-U, 30 HV at 950 K. γ-U, 1 HV at 1100 K

VANADIUM (V)

Commercial applications for pure vanadium are limited. The primary present use is as an alloying agent for steel. Other alloy applications of interest are in electronics, superconductivity, and nuclear power. Typical composition of pure vanadium is: 99.7 V (by difference); impurities: 0.03 Fe, 0.04 Al, 0.10 Si, 0.01 Ti, 0.03 Mo, 0.02 O, 0.02 Ni, 0.02 C, 0.001 H. Increased impurity levels, particularly of silicon, oxygen, nickel, carbon and hydrogen, have adverse effects on hardness and ductility.

Resistance to specific corroding agents. At room temperature, vanadium and its alloys have excellent resistance to corrosion in salt water and dilute hydrochloric acid; good corrosion resistance in sodium hydroxide solutions; and poor corrosion resistance in nitric acid solutions. Resistance to attack by liquid alkali metals is good.

Tensile properties. Typical at 1025 °C: tensile strength, 53 MPa; elongation, 37% in 1 in.

Hardness. 72 HB, electron beam ingot. See also Table 7.

Poisson's ratio. 0.36
Elastic modulus. Tension, 124 to 137 GPa; shear, 46.4 GPa
Creep-rupture characteristics. Limiting creep stress, 4.63 MPa for 1% deformation in 24 h at 1000 °C. Stress/density ratio at 1000 °C, 110
Recrystallization temperature. 800 to 1010 °C
Standard finishes. The machining of vanadium metal is similar to that of stainless steel and presents no special problem except where the metal surface has been severely contaminated with oxygen and nitrogen.
Suitable joining methods. Satisfactory electric welding of vanadium requires adequate protection of the weld pool and heat-affected zone with a neutral gas, such as argon or helium, to prevent or minimize contamination with oxygen, hydrogen and nitrogen. Flame welding is not practical because of the reactivity of any combustion-gas mixture with molten vanadium. Vanadium can be joined by welding to ferritic and austenitic stainless steels, titanium and titanium alloys, as well as to low-carbon steel.

YTTERBIUM (Yb)

Ytterbium is used for phosphors, ceramic capacitors, ferrite devices and catalysts. Ytterbium (^{170}Yb), which has been formed by neutron irradiation of thulium (^{169}Tm), is used as a portable radiograph source; ytterbium foils are used to measure pressure and as stress transducers. Ytterbium will tarnish slightly at room temperature in air. Massive ytterbium can be handled in air, but should be stored in an inert atmosphere or vacuum. Finely divided ytterbium should not be handled in air.

Tensile properties. Tensile strength, 59 MPa; yield strength, 6.9 MPa; elongation, 42%; reduction in area, 90%
Hardness. 17 HV

YTTRIUM (Y)

Yttrium is used in magnesium alloys and oxidation-resistant alloys, and it is also used in garnets and ferrites for electronic components. Yttrium is a host material for rare earth phosphors, including the red color (Eu) in color television screens. Some simulated diamonds (yttrium aluminum garnets) contain yttrium. Yttrium tarnishes slowly in air at room temperature. Turnings can be ignited quite easily and burn with great evolution of heat. Finely divided yttrium should be handled with great care and should be kept away from air and oxidizing agents.

Tensile properties. At 25 °C, annealed rod: tensile strength, 186 MPa; yield strength, 27 MPa; elongation, 17% in 1 in.; reduction in area, 24%
Hardness. 40 HV

ZINC (Zn)

See Section 11, "Zinc," in this volume.

ZIRCONIUM (Zr)

Zirconium is nontoxic and, consequently, does not require serious limitations on its use because of health hazards.

Zirconium is pyrophoric because of its heat-producing reaction with oxidizing elements such as oxygen. Large pieces of sheet, plate, bar, tube and ingot can be heated to high temperatures without excessive oxidation or burning, but small pieces with a high surface-area-to-mass ratio, such as machine chips and turnings, are easily ignited and burn at extremely high temperatures. It is recommended that large accumulations of chips and other finely divided material be avoided. Also, in storing the chips and turnings care should be taken to place the material in nonflammable containers and isolated areas. One storage method that works quite well is to keep the material covered with water in the containers and in turn, use oil on the water to keep it from evaporating. If a fire accidentally starts in zirconium, do not attempt to put it out with water or ordinary fire extinguishers, but use dry sand, powdered graphite, or commercially available Met-L-X powder. Large quantities of water can be used to control and extinguish fires in other flammables in the vicinity of a zirconium fire.

When zirconium is exposed to highly corrosive attack by concentrated acids, such as red fuming nitric acid, it is possible that in time exposed surfaces of the zirconium will be converted to a finely divided powder, which can ignite, possibly with explosive force. If the proper balance between water vapor and nitrogen dioxide above the liquid is maintained, this hazard can be eliminated.

Microstructure. Polishing and etching zirconium to observe the microstructure is not difficult when the proper techniques are used. Due to the tendency of zirconium to smear during polishing, an attack-polish technique is used. A solution of alumina is used in conjunction with a dilute acid solution to attack the sample surface chemically at about the same rate that it is being removed by abrasion. The right combination of wheel speed, hand pressure, acid solution, and abrasive will result in a true undisturbed microstructure. *Sanding* — the sample is sanded on abrasive cloth down to a $^{3}/_{0}$ grit. *Polishing* — the sample is then polished on a wheel using nylon cloth over met cloth (the nylon is fairly acid resistant), according to the following procedure:

Apply abrasive solution (5 g of 3 μm alumina/150 mL H_2O) to wheel
Apply acid solution (250 mL H_2O/22 mL

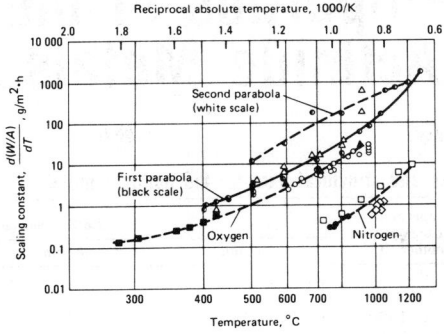

Fig. 5. Scaling rate of zirconium at elevated temperature

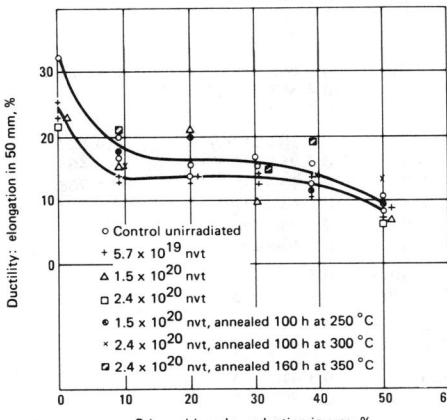

Fig. 6. Effect of irradiation and subsequent annealing on the ductility of sponge zirconium

HNO_3/3 mL HF) to wheel
Spin for several seconds (approx 1000 RPM) to allow an even film to form on the cloth
Reduce speed to approx 550 RPM and polish sample with light to moderate pressure between the wheel and the sample. When the sample is removed from the wheel, wash immediately with H_2O (squirt bottle works well) or overetching will occur. Repeat above polishing as necessary to obtain a surface with no disturbed metal

To maintain the proper acid concentration on the wheel, the polishing cloth should be thoroughly rinsed with water after 1 to 2 min use. New alumina and acid should be applied to continue polishing.
Etching — swab the sample surface with acid solution (22 mL H_2O/22 mL HNO_3/3 mL HF)

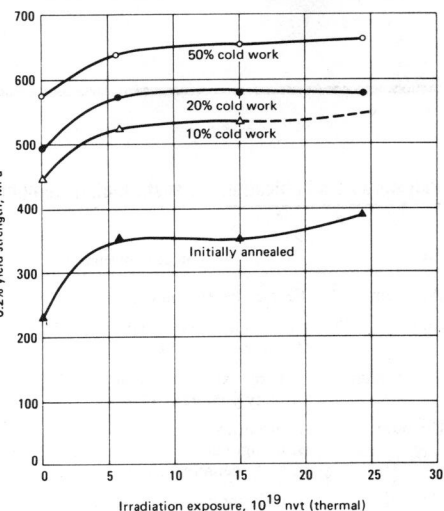

Fig. 7. Effect of irradiation on the yield strength of sponge zirconium. Exposure temperature, 50 to 60 °C.

for 5 to 10 s, rinse in water to prevent overetching.

Resistance to specific agents. Although zirconium is able to maintain a bright surface permanently at room temperature, to form a stable oxide of high melting point, and possibly to form a continuous oxide surface, oxidation resistance in air or oxygen at moderately high temperatures is poor, as indicated in Fig. 5.

Tensile properties. See Fig. 6 and 7.

Poisson's ratio. 0.35 at room temperature

Elastic properties. Tension, 99.284 GPa. Pressure dependency of specific volume (compressibility):

Pressure, MPa	Relative specific volume
0.10	1.000
2940	0.967
3920	0.956
5880	0.937
6860	0.929
7850	0.922
8830	0.916
9810	0.910

Specific damping capacity. 17×10^{-4} at 25 °C, decreasing to 6.2×10^{-4} at 260 °C with a sharp increase to 1×10^{-2} at 610 °C

Coefficient of friction. For zirconium sliding on zirconium, 0.42 at 20 °C

Velocity of sound. 4.62 km/s

Properties of the Actinide Metals

Polymorphic modifications, transformation temperatures and structural data for the actinide metals

Metal	Lattice symmetry	Temperature range of stability, °C	Lattice constants a, nm	b, nm	c, nm	β, deg	Density(a), Mg/m³	Atoms per unit cell
Actinium	Face-centered cubic		0.5 331	...	...	...	10.07	4
Thorium	α, face-centered cubic ....Below 1400		0.5 0843(25 °C)	...	...	...	11.724	4
	β, body-centered cubic 1400-1750		0.4 11(1450 °C)	...	...	...	11.10	2
Protactinium	α, body-centered tetragonal ......Below 1165		0.3 929	...	0.3 241	...	15.37	2
	β, body-centered cubic 1165-1575		0.3 81(1186 °C)	...	...	...	13.87	2
Uranium	α, orthorhombic ....Below 666		0.2 8537(298 °C)	0.5 8695	0.4 9548	...	19.07	4
	β, tetragonal 666-774		1.0 763 ± 5(720 °C)	...	0.5 652 ± 5	...	18.11	30
	γ, body-centered cubic 774-1132		0.3 524 ± 2(805 °C)	...	...	...	18.06	2
Neptunium	α, orthorhombic ....Below 279		0.6 663 ± 3	0.4 723 ± 1	0.4 887 ± 2	...	20.45	8
	β, tetragonal 279-574		0.4 897 ± 2(313 °C)	...	0.3 388 ± 2	...	19.36	4
	γ, body-centered cubic 574-637		0.3 518(600 °C)	...	...	...	18.04	2
Plutonium	α, monoclinic ....Below 118		0.6 183(21 °C)	0.4 822	1.0 963	101.79	19.86	16
	β, monoclinic 118-200		0.9 284(190 °C)	1.0 463	0.7 859	92.13	17.70	34
	γ, orthorhombic 200-312		0.3 159(235 °C)	0.5 768	1.0 162	...	17.14	8
	δ, face-centered cubic 312-458		0.4 6371(320 °C)	...	...	...	15.92	4
	δ', body-centered tetragonal 458-480		0.3 34(465 °C)	...	0.4 44	...	16.00	2
	ε, body-centered cubic 480-640		0.3 636(490 °C)	...	...	...	16.51	2
Americium	α, close-packed double hexagonal ...Below 1074		0.3 4(20 °C)	...	1.1 2	...	13.6	4
	β, face-centered cubic 1074-1176		0.4 894 ± 5	...	...	...	13.65	4
Curium	α, close-packed double hexagonal ...Below 1176		0.3 49	...	1.1 33	...	13.5	4
	β, face-centered cubic 1176-1330		...	...	...	...		4
Berkelium	α, close-packed double hexagonal ...Below 910		0.3 416 ± 5	...	1.1 068	...	14.79	4
	β, face-centered cubic 910-983		0.4 999 ± 5	...	...	...	13.24	4
Californium	α, close-packed double hexagonal		0.4 002	...	1.2 804	...	9.31	4
	β, face-centered cubic ....Below 940		0.5 74	...	...	...	8.72	4
Einsteinium	α, close-packed double hexagonal		7	...	7	...	7	4
	β, face-centered cubic ....Below 820		0.5 75	...	...	...	8.84	4

(a) Determined by x-ray methods.

Properties related to $N(E_F)$ for α-phase actinide metals

Metal	$\rho_{300} - \rho_4$(a), nΩ·m	χ_{300}(b), 10^4 emu/mol	γ(c), mJ/ (mol·K²)	n(d)	U_0/W(e)	Magnetic ordering or superconducting(f)	Metal	$\rho_{300} - \rho_4$(a), nΩ·m	χ_{300}(b), 10^4 emu/mol	γ(c), mJ/ (mol·K²)	n(d)	U_0/W(e)	Magnetic ordering or superconducting(f)
Thorium	140	0.8	4.3	3	0.3	S	Americium	670	7.8	6.0	5	50	S
Protactinium ...	180	2.7	...	2.8	(0.4)	S	Curium	860	119.0	...	2	>50	M
Uranium	310	3.9	9.1	1-4(g)	0.6	S	Berkelium	...	...	...	...	>50	M
Neptunium	1230	5.5	12.4	2	0.7	...	Californium	...	...	...	...	>50	M
Plutonium	1380	5.1	17.0	2	1.8	...							

(a) Electrical resistivity (difference in values at 300 and 4 K) $\propto N(E_F)$. (b) Magnetic susceptibility at 300 K (27 °C). (c) Coefficient of the electronic term in the specific heat, $\gamma \propto N(E_F)$. (d) Low temperature value in $\rho - \rho_0 = aT^n$. (e) Ratio of polar-state formation energy to band-width (for $U_0/W \leq 1$, bands are formed). (f) M, orders magnetically; S, superconductor. (g) A function of crystallographic direction.

Typical mechanical properties of pure actinide metals(a)

Metal	Tensile strength, MPa	Yield strength, MPa	Elongation, % in 50 mm	Hardness, HV	Metal	Tensile strength, MPa	Yield strength, MPa	Elongation, % in 50 mm	Hardness, HV
Thorium(b)	120	48	36	150	For comparison:				
Uranium(c)	585	240	10	220	Nickel(e)	315	60	30	70
Plutonium(d)	525	300	8	250					

(a) Measured at room temperature on high- or commercial-purity, polycrystalline material after prior plastic deformation and annealing. (b) Iodide (high-purity) grade; forged, hot and cold rolled, then annealed at 850 °C. (c) Magnesium-reduced (commercial purity) grade; hot rolled as α phase, annealed in β phase and water quenched. (d) Electrolytic (high-purity) grade; prior treatment consisted of extrusion at 108 °C without subsequent annealing. (e) Electrolytic grade; cold rolled, then annealed at 550 °C.

15 CORROSION-RESISTANT MATERIALS

This section was condensed from Metals Handbook, Ninth Edition, Volume 3, Properties and Selection: Stainless Steels, Tool Materials and Special-Purpose Metals, pages 1 to 185. For more detailed information the reader is referred to the larger work. Additional information on corrosion-resistant materials may be found in this Desk Edition by consulting the sections that deal with processing and with testing, inspection and quality control.

Stainless Steels

Edited by James D. Redmond, Climax Molybdenum Company

Review Committee: William J. Schumacher, ARMCO Research and Technology; Robert L. Caton, Carpenter Technology Corp.; John Ziemianski, Allegheny Ludlum Steel Corp.

Condensed from Metals Handbook, Ninth Edition, Volume 3, pages 3 to 124

Wrought Stainless Steels

STAINLESS STEELS are more resistant to rusting and staining than plain carbon and low-alloy steels. They have superior corrosion resistance because they contain relatively large amounts of chromium. Other elements, such as copper, aluminum, silicon, nickel and molybdenum, also increase corrosion resistance in specific environments and can be used to enhance oxidation resistance.

Stainless steels may be defined as alloy steels containing at least 10% chromium—with or without other elements. It has been customary in the United States to include with stainless steels those alloys that contain as little as 4% chromium. Together, these steels form a family known as stainless and heat-resisting steels. Few of these alloys contain more than 30% chromium or less than 50% iron. This section includes some information on alloys that contain from 4 to 10% chromium as well as on those with chromium contents of 30% or more. Steels in both categories are used mainly for elevated-temperature applications.

CLASSIFICATION OF STAINLESS STEELS

Stainless steels are commonly divided into five groups: (a) martensitic, (b) ferritic, (c) austenitic, (d) precipitation-hardening and (e) duplex stainless steels. Compositions of standard stainless steels of these five types are presented in Tables 1 to 5.

FACTORS IN SELECTION

The first and most important step toward successful use of a stainless or heat-resisting steel is selection of a type that is appropriate for the application. There are a large number of standard types that differ from one another in composition, corrosion resistance, physical properties and

Table 1. Compositions of standard martensitic stainless steels

| UNS number | Common designation | Composition(a), % | | | | | | | |
		C	Mn	Si	Cr	Ni	P	S	Others
S40300	403	0.15	1.00	0.50	11.5-13.0	0.60	0.04	0.03	...
S41000	410	0.15	1.00	1.00	11.5-13.5	0.75	0.04	0.03	...
S41008	410S	0.08	1.00	1.00	11.5-13.5	0.60	0.04	0.03	...
S41040	410Cb (XM-30)	0.18	1.00	1.00	11.5-13.5	...	0.04	0.03	0.05-0.30 Nb
S41400	414	0.15	1.00	1.00	11.5-13.5	1.25-2.50	0.04	0.03	...
S41600	416	0.15	1.25	1.00	12.0-14.0	...	0.04	0.15 min	0.6 Mo
S41610	416 Plus X (XM-6)	0.15	1.5-2.5	1.00	12.0-14.0	...	0.06	0.15 min	...
S41623	416Se	0.15	1.25	1.00	12.0-14.0	...	0.06	0.06	0.15 min Se
S42000	420	0.15 min	1.00	1.00	12.0-14.0	...	0.04	0.03	...
S42020	420F	0.30-0.40	1.25	1.00	12.0-14.0	...	0.06	0.15 min	0.6 Cu (optional)
S42200	422	0.20-0.25	1.00	0.75	11.5-13.5	0.5-1.0	0.04	0.03	0.75-1.25 Mo; 0.75-1.25 W; 0.15-0.3 V
S43100	431	0.20	1.00	1.00	15.0-17.0	1.25-2.50	0.04	0.03	...
S44002	440A	0.60-0.75	1.00	1.00	16.0-18.0	...	0.04	0.03	0.75 Mo
S44003	440B	0.75-0.95	1.00	1.00	16.0-18.0	...	0.04	0.03	0.75 Mo
S44004	440C	0.95-1.20	1.00	1.00	16.0-18.0	...	0.04	0.03	0.75 Mo
S44020	440F	0.95-1.20	1.25	1.00	16.0-18.0	0.75	0.04	0.10-0.35	0.40-0.60 Mo
S50100	501	0.10 min	1.00	1.00	4.0-6.0	...	0.04	0.03	0.40-0.65 Mo
S50200	502	0.10	1.00	1.00	4.0-6.0	...	0.04	0.03	0.40-0.65 Mo
S50300	503	0.15	1.00	1.00	6.0-8.0	...	0.04	0.04	0.45-0.65 Mo
S50400	504	0.15	1.00	1.00	8.0-10.0	...	0.04	0.04	0.9-1.1 Mo
...	414L	0.06	0.50	0.15	12.5-13.0	2.5-3.0	0.04	0.03	0.5 Mo; 0.03 Al
...	Trim Rite	0.15-0.30	1.00	1.00	13.5-15.0	0.25-1.00	0.04	0.03	0.40-1.00 Mo

(a) All single values are maximum values.

Table 2. Compositions of standard ferritic stainless steels

UNS number	Common designation	C	Mn	Si	Cr	Ni	P	S	Others
							Composition(a), %		
S18200	18-2 FM	0.08	2.50	1.00	17.5-19.5	...	0.04	0.15 min	1.50-2.50 Mo
S40500	405	0.08	1.00	1.00	11.5-14.5	0.60	0.04	0.03	0.10-0.30 Al
S40900	409	0.08	1.00	1.00	10.5-11.75	0.50	0.045	0.045	6 × %C min, 0.75 max Ti
S42900	429	0.12	1.00	1.00	14.0-16.0	0.75	0.04	0.03	...
S43000	430	0.12	1.00	1.00	16.0-18.0	0.75	0.04	0.03	...
S43020	430F	0.12	1.25	1.00	16.0-18.0	...	0.06	0.15 min	...
S43023	430FSe	0.12	1.25	1.00	16.0-18.0	...	0.06	0.06	0.15 min Se
S43035	439 (XM-8)	0.07	1.00	1.00	17.00-19.00	0.50	0.040	0.03	0.20 + 4(C+N) min, 1.10 max Ti; 0.015 max Al; 0.04 max N
S43036	430Ti	0.10	1.00	1.00	16.0-19.5	0.75	0.04	0.03	5 × %C min, 0.75 max Ti
S43400	434	0.12	1.00	1.00	16.0-18.0	...	0.04	0.03	0.75-1.25 Mo
S43600	436	0.12	1.00	1.00	16.0-18.0	...	0.04	0.03	0.75-1.25 Mo; 5 × %C min, 0.70 max (Nb+Ta)
S44200	442	0.20	1.00	1.00	18.0-23.0	0.60	0.04	0.04	...
S44400	444, 18-2	0.025	1.00	1.00	17.5-19.5	1.00	0.04	0.03	1.75-2.5 Mo; 0.035 N; 0.2 + 4(C+N) min, 0.80 max (Ti+Nb)
S44600	446	0.20	1.50	1.00	23.0-27.0	0.60	0.04	0.03	0.25N
S44627	E-Brite, 26-1 (XM-27)	0.01	0.40	0.40	25.0-27.5	0.50	0.02	0.02	0.75-1.5 Mo; 0.015 N; 0.2 Cu; 0.5(Ni+Cu); 0.05-0.20Nb
S44635	Monit	0.025	1.00	0.75	24.5-26.0	3.5-4.5	0.04	0.03	3.5-4.5 Mo; 0.20 + 4(C+N) min, 0.80 max (Ti+Nb); 0.035 N
S44660	Sea-cure/Sc-1	0.025	1.00	1.00	25.0-27.0	1.5-3.5	0.04	0.03	2.5-3.5 Mo; 0.2 + 4(C+N) min, 0.80 max (Ti+Nb); 0.035 N
S44700	AL 29-4	0.010	0.30	0.20	28.0-30.0	0.15	0.025	0.02	3.5-4.2 Mo; 0.15 Cu
S44735	AL 29-4C	0.025	1.00	0.75	28.0-30.0	0.50	0.040	0.03	3.50-4.50 Mo; 0.2 + 4(C+N) min, 0.080 max (Ti+Nb)
S44800	AL 29-4-2	0.010	0.30	0.20	28.0-30.0	2.0-2.5	0.025	0.02	3.5-4.2 Mo
...	18SR	0.08	0.60	0.75-1.25	17.0-19.0	...	0.04	0.03	1.75-2.50 Al; 6 × %C min, 0.75 max Ti
...	430FR	0.03	1.25	2.00	16.0-18.0	...	0.04	0.03	0.25-0.75 Mo

(a) All single values are maximum values.

Table 3. Compositions of standard austenitic stainless steels

UNS number	Common designation	C	Mn	Si	Cr	Ni	P	S	Others
						Composition(a), %			
S20100	201	0.15	5.5-7.5	1.00	16.0-18.0	3.5-5.5	0.06	0.03	0.25 N
S20200	202	0.15	7.5-10.0	1.00	17.0-19.0	4.0-6.0	0.06	0.03	0.25 N
S20500	205	0.12-0.25	14.0-15.5	1.00	16.0-18.0	1.0-1.75	0.06	0.03	0.32-0.40 N
S20910	Nitronic 50, 22-13-5	0.06	4.0-6.0	1.00	20.5-23.5	11.5-13.5	0.04	0.03	1.5-3.0 Mo; 0.2-0.4 N; 0.1-0.3 Nb; 0.1-0.3 V
S21400	Tenelon (XM-31)	0.12	14.0-16.0	0.3-1.0	17.0-18.5	1.00	0.045	0.03	0.35 N min
S21460	Cryogenic Tenelon (XM-14)	0.12	14.0-16.0	1.00	17.0-19.0	5.0-6.0	0.06	0.03	0.35-0.50 N
S21600	216 (XM-17)	0.08	7.5-9.0	1.00	17.5-22.0 *	5.0-7.0	0.045	0.03	2.0-3.0 Mo; 0.25-0.50 N
S21603	216L	0.03	7.5-9.0	1.00	17.5-22.0	5.0-7.0	0.045	0.03	2.0-3.0 Mo; 0.25-0.50 N
S21800	Nitronic 60	0.10	7.0-9.0	3.5-4.5	16.0-18.0	8.0-9.0	0.06	0.03	0.08-0.18N
S21900	Nitronic 40, 21-6-9	0.08	8.0-10.0	1.00	19.0-21.5	5.5-7.5	0.06	0.03	0.15-0.40 N
S24000	Nitronic 33, 18-3-12	0.08	11.50-14.50	1.00	17.0-19.0	2.25-3.75	0.060	0.030	0.20-0.40 N
S24100	Nitronic 32, 18-2-12	0.15	11.0-14.0	1.0	16.5-19.0	0.5-2.5	0.06	0.03	0.20-0.45 N
S28200	18-18 Plus	0.15	17.0-19.0	1.00	17.5-19.0	...	0.045	0.03	0.75-1.25 Mo; 0.75-1.25 Cu; 0.4-0.6 N
S30100	301	0.15	2.00	1.00	16.0-18.0	6.0-8.0	0.045	0.03	...
S30200	302	0.15	2.00	1.00	17.0-19.0	8.0-10.0	0.045	0.03	...
S30215	302B	0.15	2.00	2.0-3.0	17.0-19.0	8.0-10.0	0.045	0.03	...
S30300	303	0.15	2.00	1.00	17.0-19.0	8.0-10.0	0.20	0.15 min	0.6 Mo (optional)
S30310	303 Plus X (XM-5)	0.15	2.5-4.5	1.00	17.0-19.0	7.0-10.0	0.20	0.25 min	...
S30323	303Se	0.15	2.00	1.00	17.0-19.0	8.0-10.0	0.20	0.06	0.15 min Se
S30345	303 Al-Modified	0.15	2.00	1.00	17.0-19.0	8.0-10.0	0.05	0.11-0.16	0.40-0.60 Mo; 0.60-1.00 Al
S30400	304	0.08	2.00	1.00	18.0-20.0	8.0-10.5	0.045	0.03	0.10 N
S30403	304L	0.03	2.00	1.00	18.0-20.0	8.0-12.0	0.045	0.03	0.10 N
S30409	304H	0.04-0.10	2.00	1.00	18.0-20.0	8.0-10.5	0.045	0.03	...
S30430	18-9-LW	0.10	2.00	1.00	17.0-19.0	8.0-10.0	0.045	0.03	3.0-4.0 Cu
S30451	304N	0.08	2.00	1.00	18.0-20.0	8.0-10.5	0.045	0.03	0.10-0.16 N
S30452	304HN	0.08	2.00	1.00	18.0-20.0	8.0-10.5	0.045	0.03	0.16-0.30 N
S30453	304LN	0.03	2.00	1.00	18.0-20.0	8.0-12.0	0.045	0.03	0.10-0.16 N
S30500	305	0.12	2.00	1.00	17.0-19.0	10.5-13.0	0.045	0.03	...
S30800	308	0.08	2.00	1.00	19.0-21.0	10.0-12.0	0.045	0.03	...
S30815	253MA	0.10	0.80	1.40-2.00	20.0-22.0	10.0-12.0	0.040	0.030	0.03-0.08 Ce; 0.14-0.20 N
S30900	309	0.20	2.00	1.00	22.0-24.0	12.0-15.0	0.045	0.03	...
S30908	309S	0.08	2.00	1.00	22.0-24.0	12.0-15.0	0.045	0.03	...
S30940	309Cb	0.08	2.00	1.00	22.0-24.0	12.0-16.0	0.045	0.03	10 × %C min, 1.10 max (Nb+Ta); 0.10 N
S31000	310	0.25	2.00	1.50	24.0-26.0	19.0-22.0	0.045	0.03	...
S31008	310S	0.08	2.00	1.50	24.0-26.0	19.0-22.0	0.045	0.03	...
S31040	310Cb	0.08	2.00	1.50	24.0-26.0	19.0-22.0	0.045	0.030	10 × %C min, 1.10 max (Nb+Ta); 0.10 N
S31254	254 SMO	0.02	1.00	0.80	19.5-20.5	17.5-18.5	0.030	0.010	6.00-6.50 Mo; 0.50-1.00 Cu; 0.18-0.22 N

(continued)

Table 3. (continued)

UNS number	Common designation	C	Mn	Si	Cr	Ni	P	S	Others
						Composition(a), %			
S31400	314	0.25	2.00	1.5-3.0	23.0-26.0	19.0-22.0	0.045	0.03	...
S31600	316	0.08	2.00	1.00	16.0-18.0	10.0-14.0	0.045	0.03	2.0-3.0 Mo; 0.10 N
S31603	316L	0.03	2.00	1.00	16.0-18.0	10.0-14.0	0.045	0.03	2.0-3.0 Mo; 0.10 N
S31609	316H	0.04-0.10	2.00	1.00	16.0-18.0	10.0-14.0	0.045	0.03	2.0-3.0 Mo
S31620	316F	0.08	2.00	1.00	17.0-19.0	10.0-14.0	0.20	0.10 min	1.75-2.5 Mo
S31635	316Ti	0.08	2.00	1.00	16.0-18.0	10.0-14.0	0.045	0.030	5(C+N) min, 0.70 max Ti; 0.10 max N; 2.0-3.0 Mo
S31640	316Cb	0.08	2.00	1.00	16.0-18.0	10.0-14.0	0.045	0.030	10 × %C min, 1.10 max (Nb+Ta); 0.10 max N; 2.0-3.0 Mo
S31651	316N	0.08	2.00	1.00	16.0-18.0	10.0-14.0	0.045	0.03	2.0-3.0 Mo; 0.10-0.16 N
S31653	316LN	0.03	2.00	1.00	16.0-18.0	10.0-14.0	0.045	0.03	2.0-3.0 Mo; 0.10-0.16 N
S31700	317	0.08	2.00	1.00	18.0-20.0	11.0-15.0	0.045	0.03	3.0-4.0 Mo; 0.10 N
S31703	317L	0.03	2.00	1.00	18.0-20.0	11.0-15.0	0.045	0.03	3.0-4.0 Mo; 0.10 N
S32100	321	0.08	2.00	1.00	17.0-19.0	9.0-12.0	0.045	0.03	5(C+N) min, 0.70 max Ti; 0.10 N
S32109	321H	0.04-0.10	2.00	1.00	17.0-19.0	9.0-12.0	0.045	0.03	4(C+N) min, 0.70 max Ti; 0.10 N
S34700	347	0.08	2.00	1.00	17.0-19.0	9.0-13.0	0.045	0.03	10 × %C min, 1.10 max (Nb+Ta)
S34709	347H	0.04-0.10	2.00	1.00	17.0-19.0	9.0-13.0	0.045	0.03	8 × %C min, 1.10 max (Nb+Ta)
S34800	348	0.08	2.00	1.00	17.0-19.0	9.0-13.0	0.045	0.03	0.2 Co; 10 × %C min, 1.10 max (Nb+Ta)
S34809	348H	0.04-0.10	2.00	1.00	17.0-19.0	9.0-13.0	0.045	0.03	0.2 Co; 8 × %C min, 1.00 max (Nb+Ta)
S38100	18-18-2 (XM-15)	0.08	2.00	1.5-2.5	17.0-19.0	17.5-18.5	0.03	0.03	...
S38400	384	0.08	2.00	1.00	15.0-17.0	17.0-19.0	0.045	0.03	...
N06333	RA-333	0.08	2.00	0.75-1.50	24.0-27.0	44.0-47.0	0.030	0.030	2.5-4.0 Co; 2.5-4.0 Mo; 2.5-4.0 W
N08020	20Cb-3	0.07	2.00	1.00	19.0-21.0	32.0-38.0	0.045	0.035	2.0-3.0 Mo; 3.0-4.0 Cu; 8 × %C min Nb
N08026	20Mo-6	0.03	1.00	0.50	22.0-26.0	33.0-37.0	0.03	0.03	5.0-6.7 Mo; 2.00-4.00 Cu
N08028	Sanicro 28	0.020	2.50	1.0	26.0-28.0	29.5-32.5	0.030	0.030	3.0-4.0 Mo; 0.6-1.4 Cu
N08330	330	0.08	2.00	0.75-1.5	17.0-20.0	34.0-37.0	0.03	0.03	1.0 Cu
N08366	AL-6X	0.03	2.00	1.0	20.0-22.0	23.5-25.5	0.030	0.03	6.0-7.0 Mo
N08700	JS-700	0.04	2.00	1.00	19.0-23.0	24.0-26.0	0.04	0.03	4.3-5.0 Mo; 0.5 Cu; 8 × %C min, 0.50 max Nb; 0.005 Pb; 0.035 Sn
N08904	904L	0.02	2.00	1.00	19.0-23.0	23.0-28.0	0.045	0.035	4.0-5.0 Mo; 1.0-2.0 Cu
...	308L	0.03	2.00	1.00	19.0-21.0	10.0-12.0	0.045	0.03	...
...	317LM	0.03	2.00	1.00	18.0-20.0	13.0-17.0	0.045	0.03	4.0-5.0 Mo; 0.10 N
...	317LMN	0.03	2.00	1.00	17.0-20.0	13.5-17.5	0.045	0.03	4.0-5.0 Mo; 0.10-0.20 N
...	20Mo-4	0.03	1.00	0.50	22.5-25.0	35.0-40.0	0.035	0.035	3.5-5.0 Mo; 0.5-1.5 Cu; 0.15-0.35 Nb
...	302 HQ-FM	0.06	2.00	1.00	16.0-19.0	9.0-11.0	0.040	0.20	1.2-1.4 Cu
...	JS-777	0.04	2.00	1.00	19.0-23.0	24.0-26.0	0.045	0.035	4.0-5.0 Mo; 1.9-2.5 Cu
...	Cronifer 2328	0.04	0.75	0.75	22.0-24.0	26.0-28.0	0.030	0.015	2.5-3.5 Cu; 0.40-0.70 Ti; 2.5-3.0 N
...	Nitronic 20	0.28-0.38	1.5-3.5	0.60-0.90	22.0-24.0	7.0-9.0	0.04	0.015	0.28-0.40 N

(a) All single values are maximum values.

Table 4. Compositions of standard precipitation-hardening stainless steels

UNS number	Common designation	C	Mn	Si	Cr	Ni	P	S	Others
						Composition(a), %			
S13800	PH 13-8 Mo (XM-13)	0.05	0.20	0.10	12.25-13.25	7.5-8.5	0.01	0.008	2.0-2.5 Mo; 0.90-1.35 Al; 0.01 N
S15500	15-5 PH (XM-12)	0.07	1.00	1.00	14.0-15.5	3.5-5.5	0.04	0.03	2.5-4.5 Cu; 0.15-0.45 (Nb+Ta)
S15700	PH 15-7 Mo (632)	0.09	1.00	1.00	14.0-16.0	6.5-7.75	0.04	0.03	2.0-3.0 Mo; 0.75-1.5 Al
S17400	17-4 PH	0.07	1.00	1.00	15.0-17.5	3.0-5.0	0.04	0.03	3.0-5.0 Cu; 0.15-0.45 (Nb+Ta)
S17600	Stainless W (635)	0.08	1.00	1.00	15.0-17.5	6.0-7.5	0.04	0.03	0.4 Al; 0.4-1.2 Ti
S17700	17-7 PH (631)	0.09	1.00	1.00	16.0-18.0	6.5-7.75	0.04	0.03	0.75-1.5 Al
S35000	AM-350	0.07-0.11	0.5-1.25	0.50	16.0-17.0	4.0-5.0	0.04	0.03	2.5-3.25 Mo; 0.07-0.13 N
S35500	AM-355	0.10-0.15	0.5-1.25	0.50	15.0-16.0	4.0-5.0	0.04	0.03	2.5-3.25 Mo
S36200	Almar 362	0.05	0.50	0.30	14.00-14.50	6.25-7.00	0.04	0.03	0.55-0.90 Ti
S45000	Custom 450 (XM-25)	0.05	1.00	1.00	14.0-16.0	5.0-7.0	0.03	0.03	1.25-1.75 Cu; 0.5-1.0 Mo; 8 × %C min Nb
S45500	Custom 455 (XM-16)	0.05	0.50	0.50	11.0-12.5	7.5-9.5	0.04	0.03	0.5 Mo; 1.5-2.5 Cu; 0.8-1.4 Ti; 0.1-0.5 (Nb+Ta)
S66286	A286	0.08	2.00	1.00	13.50-16.00	14.0-27.0	0.04	0.03	0.35 Al; 0.0010-0.010 B; 1.00-1.50 Mo; 1.90-2.35 Ti; 0.10-0.50 V

(a) All single values are maximum values.

mechanical properties; selection of the optimum type for a specific application is the key to satisfactory performance at minimum total cost.

Below is a checklist of characteristics to be considered in selection of the proper type of stainless steel for a specific application.

- Corrosion resistance
- Resistance to oxidation and sulfidation
- Strength and ductility at ambient and service temperatures
- Suitability for intended fabrication techniques
- Suitability for intended cleaning procedures
- Stability of properties in service
- Toughness
- Resistance to abrasion and erosion
- Resistance to galling and seizing
- Surface finish and/or reflectivity
- Magnetic properties

Table 5. Compositions of standard duplex stainless steels

UNS number	Common designation	C	Mn	Si	Cr	Ni	P	S	Others
S31200	44LN	0.030	2.0	1.0	24.0-26.0	5.5-6.5	0.045	0.030	0.14-0.20 N; 1.2-2.0 Mo
S31500	3RE60	0.030	1.5-2.0	1.4-2.0	18.0-19.0	4.25-5.25	0.030	0.030	2.5-3.0 Mo
S31803	2205	0.030	2.0	1.0	21.0-23.0	4.5-6.5	0.030	0.020	2.50-3.5 Mo; 0.08-0.20 N
S32550	Ferralium 255	0.030	1.5	1.0	24.0-27.0	4.5-6.5	0.040	0.030	1.5-2.5 Cu; 0.10-0.25 N; 2.0-4.0 Mo
S32900	329, 7-Mo	0.08	1.0	0.75	23.0-28.0	2.5-5.0	0.040	0.03	1.0-2.0 Mo
...	DP-3	0.030	2.0	1.0	24.0-27.0	5.5-7.5	0.030	0.020	0.5-1.5 Cu; 0.08-0.20 N; 2.0-4.0 Mo

(a) All single values are maximum values.

Table 6. Wear compatibility of dissimilar corrosion-resistant alloy couples

Test conditions: $1/2$-in.-diam crossed (90°) cylinders; no lubricant; 16-lb load; 105 rpm; room temperature; 120-grit finish; 10 000 cycles; specimens degreased in acetone; duplicate tests; weight loss corrected for differences in density.

Stainless steel couples

Alloy	Hardness(a)	Type 304	Type 316	S17400	Nitronic 32	Nitronic 50	Nitronic 60	Type 440C
Type 304	99 HRB	12.8	...	...	...	...	...	...
Type 316	91 HRB	10.5	12.5	...	...	...	...	...
S17400	43 HRC	24.7	18.5	52.8	...	...	...	...
Nitronic 32	95 HRB	8.4	9.4	17.2	7.4	...	...	...
Nitronic 50	99 HRB	9.0	9.5	15.7	8.3	10.0	...	...
Nitronic 60	95 HRB	6.0	4.3	5.4	3.2	3.5	2.8	...
Type 440C	57 HRC	4.1	3.9	11.4	3.1	4.3	2.4	3.8

(a) Hardness values given apply to both axes.

Other dissimilar corrosion-resistant couples

Alloy	Hardness	Silicon bronze(a)	Chrome plate	Stellite 6B(b)	Monel K-500(c)	Waukesha 88(d)
Type 304	99 HRB	2.1	2.3	3.1	...	8.1
S17400	43 HRC	2.0	3.3	3.8	34.1	9.1
Nitronic 32	95 HRB	2.3	2.5	2.0	...	7.6
Nitronic 60	95 HRB	2.2	2.1	1.9	22.9	8.4
Type 316	91 HRB	...	...	...	33.8	9.6
Monel K-500	34 HRC	...	...	...	30.7	9.3

(a) Hardness, 93 HRB. (b) Hardness, 48 HRC. (c) Hardness, 34 HRC. (d) Hardness, 81 HRB.

- Thermal conductivity
- Electrical resistivity
- Sharpness (retention of cutting edge)
- Rigidity
- Dimensional stability.

Corrosion resistance frequently is the most important characteristic of a stainless steel, but often is also the most difficult to assess for a specific application. General corrosion resistance to natural conditions and to pure chemical solutions is comparatively easy to determine.

General corrosion is often much less serious than localized forms such as stress-corrosion cracking, crevice corrosion in tight spaces or under deposits, pitting attack, and intergranular attack in sensitized material such as weld heat-affected zones. Such localized corrosion can cause unexpected and sometimes catastrophic failure while most of the structure remains unaffected, and therefore must be considered carefully in design and in selection of the proper grade of stainless steel. Corrosive attack can also be increased dramatically by seemingly minor impurities in the medium that may be difficult to anticipate but that can have major effects, even when present in only parts-per-million concentrations; by heat transfer through the steel to or from the corrosive medium; by contact with dissimilar metallic materials; by stray electrical currents; and by many other subtle factors. At elevated temperatures, attack can be accelerated significantly by seemingly minor changes in atmosphere that affect scaling, sulfidation or carburization.

Laboratory corrosion data can be misleading in predicting service performance. Even actual service data have limitations, because "similar" corrosive media may differ substantially due to slight variations in some of the corrosion factors listed above.

Mechanical properties at service temperature obviously are important, but satisfactory performance at other temperatures must be considered also. Thus, a product for arctic service must have suitable properties at subzero temperatures even though steady-state operating temperature may be much higher; room-temperature properties after extended service at elevated temperature can be important for applications such as boilers and jet engines, which are intermittently shut down.

Fabrication and Cleaning. Frequently a particular stainless steel is chosen for a fabrication characteristic such as formability or weldability. Even a required or preferred cleaning procedure may dictate selection of a specific type. For instance, a weldment that is to be cleaned in a medium such as nitric-hydrofluoric acid, which attacks sensitized stainless steel, should be produced from stabilized or low-carbon stainless steel even though sensitization may not affect performance under service conditions.

Experiences in the use of stainless steels indicates that many factors can affect their corrosion resistance. Some of the more prominent factors are:

- Chemical composition of the corrosive medium, including impurities
- Physical state of the medium—liquid, gaseous, solid, or combinations thereof
- Temperature variations
- Aeration of the medium
- Oxygen content of the medium
- Bacteria content of the medium
- Ionization of the medium
- Repeated formation and collapse of bubbles in the medium
- Relative motion of the medium with respect to the steel
- Chemical composition of the metal
- Nature and distribution of microstructural constituents
- Continuity of exposure of the metal to the medium

Table 7. Galling resistance of stainless steels

Block material	Condition(a)	Nominal hardness(a), HB	410	416	430	440C	303	304	316	S17400	Nitronic 32	Nitronic 60
Type 410	Hardened and stress relieved	352	3	4	3	3	4	2	2	3	46	50+
Type 416	Hardened and stress relieved	342	4	13	3	21	9	24	42	2	45	50+
Type 430	Annealed	159	3	3	2	2	2	2	2	3	3	36
Type 440C	Hardened and stress relieved	560	3	21	2	11	5	3	37	3	50+	50+
Type 303	Annealed	153	4	9	2	5	2	2	3	3	50+	50+
Type 304	Annealed	140	2	24	2	3	2	2	2	2	30	50+
Type 316	Annealed	150	2	42	2	37	3	2	2	2	3	38
S17400	H950	415	3	2	3	3	2	2	2	2	50+	50+
Nitronic 32	Annealed	235	46	45	8	50+	50+	30	3	50+	30	50+
Nitronic 60	Annealed	205	50+	50+	36	50+	50+	50+	38	50+	50+	50

(Column header: Unlubricated threshold galling stress, 10^6 psi, in "button and blank" galling test(b) using button made of:)

(a) Condition and hardness apply to both blank and button materials. (b) Tests were terminated at 50×10^6 psi, and thus a value of 50+ indicates that the specimen did not gall.

- Surface condition of the metal
- Stresses in the metal during exposure to the medium
- Contact of the metal with one or more dissimilar metallic materials
- Stray electric currents
- Differences in electric potential
- Marine growths such as barnacles
- Sludge deposits on the metal
- Carbon deposits from heated organic compounds
- Dust on exposed surfaces
- Effects of welding, brazing and soldering.

Surface finish is important more often than any other characteristic except corrosion resistance, and stainless steels sometimes are selected because they are available in a variety of attractive finishes. Selection of surface finish may be made on the basis of appearance, frictional characteristics or sanitation. Selection of finish may in turn influence selection of alloy because of differences in availability or durability of the various finishes for different alloys.

Galling and Wear Resistance. A comparison of wear compatibility and galling resistance of several stainless steels is presented in Tables 6 and 7.

PRODUCT FORMS

Plate is a flat-rolled or forged product more than 254 mm (10 in.) in width and at least 4.76 mm (0.1875 in.) in thickness.

Stainless steel plate generally is produced in the annealed condition and is either blast cleaned or pickled. Blast cleaning generally is followed by further cleaning in appropriate acids to remove surface contaminants such as particles of steel picked up from the mill rolls.

Sheet is a flat-rolled product in coils or cut lengths at least 610 mm (24 in.) wide and less than 4.76 mm (0.1875 in.) thick.

Strip is a flat-rolled product, in coils or cut lengths, less than 610 mm (24 in.) wide and 0.13 to 4.76 mm (0.005 to 0.1875 in.) thick. Cold finished material 0.13 mm (0.005 in.) thick and less than 610 mm (24 in.) wide fits the definitions of both strip and foil, and may be referred to using either term.

Cold rolled stainless steel strip is manufactured from hot rolled, annealed and pickled strip (or from slit sheet) by rolling between polished rolls. Depending on desired thickness, various numbers of cold rolling passes through the mill are required for effecting the necessary reduction and for securing the desired surface characteristics and mechanical properties.

Hot rolled stainless steel strip is a semifinished product obtained by hot rolling slabs or billets, and is produced for conversion to finished strip by cold rolling.

Foil is a flat-rolled product, in coil form, up to 0.13 mm (0.005 in.) thick and less than 610 mm (24 in.) wide. Foil is produced in slit widths with edge conditions corresponding to No. 3 and No. 5 edge conditions for strip.

Bar is a product supplied in straight lengths; it is either hot or cold finished, and is available in various shapes, sizes and surface finishes. This category includes (*a*) small shapes whose dimensions do not exceed 76 mm (3 in.) and (*b*) hot rolled flat stock at least 3.2 mm (0.125 in.) thick and up to 254 mm (10 in.) wide.

Hot finished bar commonly is produced by hot rolling, forging or pressing ingots to blooms or billets of intermediate size, which are subsequently hot rolled, forged or extruded to final dimensions. Whether rolling, forging or extrusion is selected as the finishing method depends on several factors, including composition and final size.

Wire is a coiled product derived by cold finishing hot rolled and annealed rod. Cold finishing imparts excellent dimensional accuracy, good surface smoothness, fine finish and specific mechanical properties. Wire is produced in several tempers and finishes.

Wire is customarily referred to as round wire when the contour is completely cylindrical and as shape wire when the contour is other than cylindrical. Shape wire is cold finished either by drawing or by a combination of drawing and rolling.

Special Wire Commodities. There are many classes of stainless steel wire that have been developed for specific components or for particular applications. The unique properties of each of these individual wire commodities are developed by employing a particular combination of composition, steel quality, process heat treatment and cold drawing practice.

Cold heading wire is produced in any of the various types of stainless steel. In all instances, cold heading wire is subjected to special testing and inspection to ensure satisfactory performance in cold heading and cold forging operations.

Of the chromium-nickel group, S30430 is the all-purpose type used for cold heading wire and generally is necessary for severe upsetting. Other grades commonly cold formed include 303Se, 304, 316, 321, 347, 384, 410 and 430.

Spring wire is drawn from annealed rod. The types of stainless steel in which spring wire is commonly produced include 302, 304, 316 and nitrogen-strengthened grades.

Spring wire in large sizes can be furnished in a variety of finishes such as dry-drawn lead, copper, lime and soap, or oxide and soap. Fine sizes usually are wet drawn, although they can be dry drawn.

Rope wire is used to make rope, cable and cord for a variety of uses such as aircraft control cables, marine ropes, elevator cables, slings and anchor cables.

Rope wire is made of type 302 or type 304 unless a higher level of corrosion resistance is required, in which case type 316 is generally selected. Special nonmagnetic characteristics may be required, which necessitates selection of heats having little or no ferrite or martensite in the microstructure, and use of special drawing practices to limit or avoid deformation-induced transformation to martensite.

Weaving wire is used in weaving of screens for many different applications in coal mines, sand-and-gravel pits, paper mills, chemical plants, dairy plants, oil refineries and food-processing plants. Annealing and final drawing must be carefully controlled to maintain uniform temper and finish throughout each coil or spool. Because weaving wire must be ductile, it usually is furnished in the annealed temper with a bright annealed finish, or in the soft temper with either a lime-soap finish or an oil- or grease-drawn finish.

Semifinished Products. Blooms and slabs are hot rolled, hot forged or hot pressed to approximate cross-sectional dimensions, and generally have rounded corners. These semifinished products, as well as tube rounds, are produced in random lengths or are cut to specified lengths or to specified weights.

The nominal cross-sectional dimensions of blooms and slabs are designated in inches and fractions of an inch. The size ranges commonly listed as hot rolled stainless steel blooms and slabs include square sections 100 by 100 mm (4 by 4 in.) and larger, and rectangular sections at least 10 300 mm² (16 in.²) in cross-sectional area.

Pipe, tubes and tubing are made either by piercing rounds or by rolling and welding strip. They are used for conveying gases, liquids and solids, and for diverse mechanical and structural purposes. (Cylindrical forms intended for use as containers for storage and shipping purposes, and products cast to tubular shape, are not included in this category.)

Pipe is distinguished from tubes chiefly by the fact that is is commonly produced in relatively few standard sizes. Tubing is generally made to more exacting specifications regarding dimensions, finish, chemical composition and mechanical properties than either pipe or tubes.

Stainless steel tubular products are classified as follows according to intended service:

- Stainless steel tubing for general corrosion-resistant service
- Stainless steel pressure pipe
- Seamless steel pressure tubes
- Stainless steel sanitary tubing
- Stainless steel mechanical tubing
- Stainless steel aircraft tubing
- Aircraft structural tubing
- Aircraft hydraulic-line tubing.

MECHANICAL PROPERTIES

Mechanical properties of most stainless steels, especially ductility and toughness, are higher than the same properties of carbon steels. Strength and hardness can be raised by cold work for ferritic and austenitic types, and by heat treatment for precipitation-hardening and martensitic types. Certain ferritic stainless steels also can be hardened slightly by heat treatment.

Austenitic Types. Basic room-temperature properties of standard austenitic stainless steels and of several nonstandard austenitic stainless steels are given in Tables 8 and 9. Certain austenitic stainless steels—the so-called "metastable" types—can develop higher strengths and hardnesses than other "stable" types for a given amount of cold work. In metastable austenitic stainless steels, deformation triggers transformation of austenite to martensite.

Ferritic types of stainless steel are defined as those that contain at least 10% chromium and that have microstructures of ferrite plus carbides. Minimum mechanical properties of ferritic stainless steels are shown in Table 10. These steels are less ductile than the austenitic types. Strength is enhanced only moderately by cold working.

Martensitic types are iron-chromium steels with or without small additions of other alloying elements. They are ferritic in the annealed condition, but are martensitic after rapid cooling in air or a liquid medium from above the critical temperature. Steels in this group usually contain no more than 14% Cr—except types 440A, 440B and 440C, which contain 16 to 18% Cr—and an amount of carbon sufficient to promote hardening. They may or may not contain other elements; if they do, the total concentration usually is no more than 2 to 3%. Martensitic stainless

Table 8. Minimum room-temperature mechanical properties of annealed austenitic stainless steels

UNS number	Common designation	Tensile strength MPa	ksi	Yield strength(a) MPa	ksi	Elonga-tion, %	Reduction, in area, %	Hardness (max), HRB
S30100	301	515	75	205	30	40	...	88
S30200	302	515	75	205	30	40	...	88
S30215	302B	515	75	205	30	...	...	...
S30430	302Cu	450-585	65-85	...	...	...	...	...
S30300	303	585(b)	85(b)	240(b)	35(b)	50(b)	55(b)	...
S30323	303Se	585(b)	85(b)	240(b)	35(b)	50(b)	55(b)	...
S30400	304	515	75	205	30	40	...	88
S30403	304L	480	70	170	25	40	...	88
S30451	304N	550	80	240	35	30	...	...
S31651	316N	550	80	240	35	30	...	...
S30500	305	480	70	170	25	40	...	88
S30800	308	515	75	205	30	40	...	88
S32100	321	515	75	205	30	40	...	88
S34700	347	515	75	205	30	40	...	88
S34800	348	515	75	205	30	40	...	88
S30900	309	515	75	205	30	40	...	95
S30908	309S	515	75	205	30	40	...	95
S31000	310	515	75	205	30	40	...	95
S31008	310S	515	75	205	30	40	...	95
S31400	314	515	75	205	30	30	40	...
S31600	316	515	75	205	30	40	...	95
S31620	316F	585(b)	85(b)	240(b)	35(b)	40(b)	55(b)	...
S31700	317	515	75	205	30	35	...	95
S31703	317L	515	75	205	30	35	...	95
N08330	330	480	70	210	30	30	...	...
S38400	384	415-550	60-80	...	...	...	...	...
S38500	385	415-550	60-80	...	...	...	...	...
N08904	904L	490	71	220	31	35	...	95
N08366	AL-6X	515	75	205	30	30	...	...
S38100	18-18-2	515	75	205	30	40	...	96
N08700	JS-700	550	80	205	30	30	40	...
N08020	20Cb-3	585	85	275	40	30	...	95
...	304LN	515	75	205	30	...	...	...
...	308L	550(b)	80(b)	207(b)	30(b)	60(b)	70(b)	...
...	312	655	95	...	...	20	...	...
...	316LN	515(b)	75(b)	205(b)	30(b)	60(b)	70(b)	...
...	317LM	515	75	205	30	35	50	95
...	332	550(b)	80(b)	240(b)	35(b)	45(b)	70(b)	...
...	Crutemp 25	615(b)	89(b)	275(b)	40(b)	40(b)	...	...
...	JS-777	550	80	240	35	30	40	95

(a) At 0.2% offset. (b) Typical values.

Table 9. Minimum mechanical properties of annealed high-nitrogen austenitic stainless steels

UNS number	Common designation	Tensile strength MPa	ksi	Yield strength(a) MPa	ksi	Elonga-tion, %	Reduction, in area, %	Hardness (max), HRB
S20100	201	655	95	310	45	40	...	...
S20200	202	655	95	310	45	40	...	...
S20500	205	830(b)	120(b)	475(b)	69(b)	58(b)	62(b)	98(b)
S21600	216	690	100	415	60	40	...	100
S30451	304N	550	80	240	35	30	...	88
S30452	304HN	620	90	345	50	30	...	100
S31651	316N	550	80	240	35	30	...	95
S24100	Nitronic 32	690	100	380	55	30	50	...
S24000	Nitronic 33	690	100	415	60	40	...	...
S21900	Nitronic 40	690	100	415	60	40	...	...
S20910	Nitronic 50	825	120	515	75	30	...	...
S21800	Nitronic 60	655	95	345	50	35	55	...
S28200	18-18 Plus	760	110	415	60	35	55	...
S21400	Tenelon	860	125	485	70	40	...	...

(a) At 0.2% offset. (b) Typical values.

steels may be hardened and tempered in the same manner as alloy steels. They have excellent strength and are magnetic. Minimum mechanical properties of martensitic stainless steels are shown in Table 11.

Martensitic stainless steels harden when cooled off the mill after hot processing, so they often are given a process anneal at 650 to 760 °C (1200 to 1400 °F) for about 4 h. Process annealing differs from full annealing, which is done by heating at 815 to 870 °C (1500 to 1600 °F), cooling in the furnace at a rate of 40 to 55 °C/h (75 to 100 °F/h) to about 540 °C (1000 °F) and then cooling in air to room temperature. Occasion-

ally, martensitic types are purchased in the tempered condition; this condition is achieved by cooling directly off the mill to harden the steel and then reheating to a tempering temperature of 540 to 650 °C (1000 to 1200 °F), or by reheating the steel to a hardening temperature of 1010 to 1065 °C (1850 to 1950 °F), cooling it, and then tempering it.

In heat treating of martensitic stainless steels, temperatures up to about 480 °C (900 °F) are referred to as stress-relieving temperatures because little change in tensile properties occurs on heating hardened material to these temperatures. Temperatures of 540 to 650 °C are referred to as

tempering temperatures, and temperatures of 650 to 760 °C are called annealing temperatures.

Precipitation-hardening types generally are heat treated to final properties by the fabricator.

Duplex Types. The duplex stainless steels (see Table 5) are mixtures of austenite and ferrite. Their resistance to chloride stress-corrosion cracking is in between those of the ferritics and austenitics, and decreases with increasing cold working. Because duplexes have better toughness than do ferritics, they are available in plate thicknesses and may therefore be used for tubesheets, for example. The crevice-corrosion resistance of these steels is somewhat less than that of ferritic or austenitic grades of equivalent chromium and molybdenum contents.

A recent development in duplex stainless steels is the addition of nitrogen. These stainless steels can be cold worked to strengths unobtainable in ferritic and most austenitic grades. These alloys are characterized by reduced alloy-element segregation between the ferrite and austenite phases; consequently, they exhibit better as-welded corrosion resistance than do conventional duplex grades.

NOTCH TOUGHNESS AND TRANSITION TEMPERATURE

Notched-bar impact testing of stainless steels is likely to show a wide scatter in test results, regardless of type or test conditions. Because of this wide scatter, only general behavior of the different classes can be described.

Austenitic types have good notched-bar impact resistance. Charpy impact energies of 135 J (100 ft·lb) or greater are typical of all types at room temperature. Cryogenic temperatures have little or no effect on notch toughness; ordinarily, austenitic stainless steels maintain values exceeding 135 J even at very low temperatures. On the other hand, cold work lowers the resistance to impact at all temperatures.

Martensitic stainless steels exhibit decreasing resistance to impact with decreasing temperature, and the fracture appearance changes from a ductile mode at mildly elevated temperatures to a brittle mode at subzero temperatures. This fracture transition is characteristic of martensitic materials, although for martensitic stainless steels the upper-shelf energy is greater than for most other martensitic steels. Both the upper-shelf energy and the lower-shelf energy are not greatly influenced by heat treatment in martensitic stainless steels. However, the temperature range over which transition occurs is affected by heat treatment, minor variations in composition, and prior cold work. Heat treatments that result in high hardness move the transition range to higher temperatures, and those that result in low hardness move the transition range to lower temperatures.

Fracture-toughness data are not available for many of the standard types of stainless steels. Most testing has been concentrated on the high-strength precipitation-hardening stainless steels, because these materials have been used in critical applications where fracture-toughness testing has been found most useful for evaluating materials.

FATIGUE STRENGTH

Three types of fatigue tests are used to develop data on fatigue behavior of stainless steels. The most common of these tests is the rotating-beam test, which most closely approximates the kind

Table 10. Minimum mechanical properties of annealed ferritic stainless steels

UNS number	Common designation	Tensile strength MPa	ksi	Yield strength(a) MPa	ksi	Elongation, %	Reduction, in area, %	Hardness, HRB
S40500	405	415	60	170	25	20	...	88 max
S40900	409	415	60	205	30	22(b)	...	80 max
S42900	429	450	65	205	30	22(b)	...	88 max
S43000	430	450	65	205	30	22(b)	...	88 max
S43020	430F	585-860	85-125	...	...	...	...	...
S43400	434	530(c)	77(c)	365(c)	53(c)	23(c)	...	83 max
S43600	436	530(c)	77(c)	365(c)	53(c)	23(c)	...	83 max
S44200	442	515	75	275	40	20	...	95 max
S44400	444	415	60	275	40	20	...	95 max
S44600	446	515	75	275	40	20	...	95 max
S44625	E-Brite 26-1	450	65	275	40	22(b)	...	90 max
S44635	Monit	650	94	550	80	20	...	100 max
S44660	Sea-cure/SC-1	550	80	380	55	20	...	100 max
S44700	29-4	550	80	415	60	20	...	98 max
S44800	29-4-2	550	80	415	60	20	...	98 max
...	18SR	620(c)	90(c)	450(c)	65(c)	25(c)	...	90 min(c)

(a) At 0.2% offset. (b) 20% elongation for thicknesses of 1.3 mm (0.050 in.) or less. (c) Typical values.

Table 11. Minimum mechanical properties of martensitic stainless steels

UNS number	Common designation	Tensile strength MPa	ksi	Yield strength(a) MPa	ksi	Elongation, %	Reduction, in area, %	Hardness
S40300	403	485	70	205	30	25(b)	...	88 HRB max
S41000	410	450	65	205	30	22(b)	...	95 HRB max
S41008	410S	415	60	205	30	22	...	95 HRB max
S41040	410Cb	485	70	275	40	12	35	...
S41400	414	795	115	620	90	15	45	...
S41800	418(c)	1450(d)	210(d)	1210(d)	175(d)	18(d)	52(d)	...
S42000	420(e)	1720	250	1480(d)	215(d)	8(d)	25(d)	52 HRC(d)
S42200	422(f)	965	140	760	110	13	30	...
S43100	431(c)	1370(d)	198(d)	1030(d)	149(d)	16(d)	55(d)	...
S44002	440A	725(d)	105(d)	415(d)	60(d)	20(d)	...	95 HRB(d)
S44003	440B	740(d)	107(d)	425(d)	62(d)	18(d)	...	96 HRB(d)
S44004	440C	760(d)	110(d)	450(d)	65(d)	14(d)	...	97 HRB(d)
S50100	501	485(d)	70(d)	205(d)	30(d)	28(d)	65(d)	...
S50200	502	485(d)	70(d)	205(d)	30(d)	30(d)	70(d)	...
...	414L	795(d)	115(d)	550(d)	80(d)	20(d)	60(d)	...
...	416 Plus X	515	75	275	40	30	60	...

(a) At 0.2% offset. (b) 20% elongation for thicknesses of 1.3 mm (0.050 in.) or less. (c) Tempered at 260 °C (500 °F). (d) Typical values. (e) Tempered at 205 °C (400 °F). (f) Intermediate and hard tempers.

of loading to which shafts and axles are subjected. The flexural fatigue test is used to evaluate the behavior of sheet, and most closely simulates the action of leaf springs, which are expected to flex without deforming or breaking. The axial-load fatigue test subjects a fatigue specimen to unidirectional loading that can range from full reversal (tension-compression) to tension-tension loading, and can have virtually any conceivable ratio of maximum stress to minimum stress. In general, fatigue conditions involving tension-compression loading (stress ratio, R, between 0 and -1) lead to shorter fatigue lives than conditions involving tension-tension loading (stress ratio, R, between 0 and $+1$) at the same value of maximum stress.

ELEVATED-TEMPERATURE PROPERTIES

Many stainless steels—particularly the austenitic types 304, 309, 310, 316, 321 and 347 and certain precipitation-hardening types such as PH 15-7 Mo, 15-5 PH, 17-4 PH, 17-7 PH, AM-350 and AM-355—are used extensively for elevated-temperature applications such as high-temperature heat exchangers and superheater tubes for power boilers.

Extended service at elevated temperature can result in embrittlement of austenitic stainless steels or in "sensitization" which degrades the ability of the material to withstand corrosion, particularly in acid media. Most often, such degradation is caused by the precipitation of secondary phases

such as carbides and sigma phase. Precipitation depends on both time and temperature: longer times at temperature and higher temperatures both promote more extensive precipitation.

MILL FINISHES

Sheet, strip, plate, bar and wire made of stainless steel all have different designations of mill finish, each representing a standardized appearance that is characteristic of the process used to impart final mechanical properties. Although the various mill finishes are standardized, there is sufficient variability in mill processing so that exact matching of color and reflectivity cannot be expected from lot to lot. Even wider differences can be expected between mill products from different producers.

IN-SERVICE CARE

Despite the fact that stainless steel surfaces are generally considered nontarnishing, they still need a certain amount of care to maintain a given surface appearance under normal conditions of service.

Architectural Applications. For exposed exterior surfaces in inland, light industrial areas, minimum maintenance is needed. Ordinarily, normal rainfall is adequate to maintain the desired appearance, and only sheltered areas such as entryways need occasional washing with a scrub brush or a pressurized stream of water. In marine

atmospheres and heavy industrial areas, periodic cleaning with detergents and water to remove salt and dirt deposits is advisable. Heavy or stubborn deposits may have to be removed with strong industrial cleaners.

For interior surfaces, only occasional cleaning with detergent and water is required for maintenance of finish. Where fingerprints are a problem, a commercial glass cleaner or wax is suggested. Often, a No. 4 sheet finish is specified to minimize the effect of fingerprints on appearance.

Food-Handling Applications. Stainless steel is widely specified for food-handling equipment because of its excellent "bacterial cleanability." In many instances, strong sanitizing or sterilizing solutions are used for cleaning the equipment to prevent bacterial contamination of the food products being processed. Where this is done, it should be standard practice to monitor exposure time and thoroughly flush the cleaned surfaces with water. Burnt-on foods and grease spots can be removed by soaking in hot water and detergent. Stubborn spots can be removed by scrubbing with a nonabrasive cleanser and a fiber brush, a sponge, or a pad of stainless steel wool or nickel-silver wool.

Chemical, textile and drug applications often require high purity in the product being processed. Stainless steel is used in equipment for these industries not only because it is chemically inert to the products, which effectively eliminates corrosion as a possible source of low-level contamination, but also because the surfaces of stainless steel equipment are easy to clean and sterilize, which effectively eliminates bacteria and residues as sources of contamination. Equipment usually is cleaned with strong chemical cleaners, then repeatedly rinsed with water.

Fabrication of Wrought Stainless Steels

FABRICATION of wrought stainless steels differs from fabrication of carbon and low-alloy steels primarily because stainless steels (a) are stronger, harder and more ductile, (b) work harden more readily and (c) generally must present a corrosion-resistant surface in the finished product. These characteristics dictate use of greater power, more frequent repair or replacement of processing equipment, and application of procedures to minimize or correct surface contamination.

FORMING

The method chosen for forming stainless steel should be based on the characteristics of the type to be used and the thickness of the part to be formed. As indicated above, power requirements are higher for forming stainless steels than for forming carbon steels—particularly austenitic types, which work harden more rapidly than ferritic types. Warm or hot forming may be necessary for thicknesses that can be formed cold in carbon steel.

Because of their high ductility, austenitic types are the most readily formed stainless steels: austenitic stainless steel sheet can be severely drawn. Austenitic types 201 and 301 can be formed with biaxial stretching in excess of 35% because partial transformation to martensite during deformation helps the metal resist necking and deform more uniformly. (For unusually severe forming, composition may have to be adjusted to suit the particular job, and slow forming may be necessary to prevent buildup of heat and loss of the martensite effect.) Ferritic and lower-alloy martensitic types also can be cold formed extensively. However, they are less ductile than austenitic types, and thus forming of these alloys is more limited and intermediate annealing is more likely to be needed. The higher-carbon martensitic types such as 440A, 440B and 440C have only limited cold formability. Stainless steels generally gall more readily than other steels and thus require more attention to lubrication during forming. Straight mineral oils rarely provide adequate lubrication where sliding contact occurs in forming, and lubricants with extreme-pressure additives are often used.

Surface contamination during forming and handling should be kept to a minimum by thoroughly cleaning equipment and providing proper lubrication.

Cold formed stainless steel parts usually are used in the as-formed condition. However, for applications in which stress-corrosion cracking may occur, austenitic types susceptible to this failure process should be solution annealed after forming to remove residual stresses. Under all but the mildest service conditions, hot formed parts require postforming annealing to restore corrosion resistance and/or ductility.

FORGING

All standard types of stainless steel can be forged. However, as alloy content increases within a given group, forging becomes more difficult. Forging difficulties are most common in initial breakdown of high-alloy ingots, and precautions may be needed to avoid surface ruptures.

Working Temperatures. Typical forging-temperature ranges for most standard stainless steels are indicated in Fig. 1. A wide range of forging temperatures may be used for most of the common austenitic types because of the natural workability of austenite and the absence of allotropic transformation. The conventional 18-8 types often are forged at temperatures up to 1260 °C (2300 °F). However, the upper temperature limit is lower for the higher-alloy grades due to metallurgical changes at higher temperatures that can cause surface ruptures. Maximum temperature is lowest for types 309, 310 and 330. Adherence to maximum temperature limits is particularly important for ingot breakdown, where severe tearing along grain boundaries in the cast metal may occur if temperature is too high.

Small amounts of delta ferrite can impair the forgeability of some austenitic types—particularly in upset forging, where some of the tangential tensile forces are perpendicular to ferrite stringers. Types 304, 309, 316, 317 and 321 are especially likely to contain significant amounts of ferrite, and it may be advisable to limit ferrite content of these grades for severe forging applications. Ferrite can be particularly troublesome in initial ingot breakdown; a homogenization treatment at about 1150 °C (2100 °F), to transform some of the delta ferrite to austenite before heating to forging temperature, can be helpful.

Austenitic stainless steel forgings should be solution annealed to restore corrosion resistance

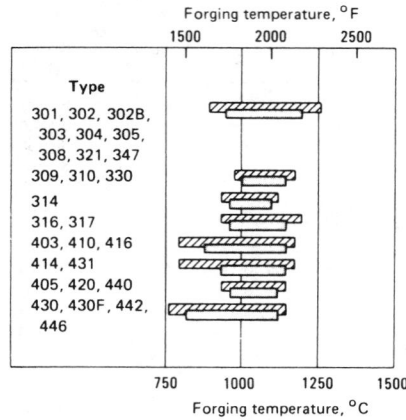

Crosshatched bars show temperature ranges that have been recommended by some but on which there is not general agreement. Solid bars are generally accepted.

Fig. 1. Typical temperature ranges for forging of stainless steels

and maximum ductility. For some applications, annealing may be omitted if working is finished above 870 °C (1600 °F) and the forging is rapidly cooled below 425 °C (800 °F) in order to prevent carbide precipitation. This approach should be used with caution, and should incorporate metallurgical checks to determine that detrimental carbide precipitation has actually been avoided. The stabilized types (321, 347 and 348) and extra-low-carbon types (304L and 316L) will not precipitate carbides as readily, and may be used in the as-forged condition with fewer precautions if maximum corrosion resistance and ductility are not essential.

The lower-carbon martensitic stainless steels can be hot worked over a wide temperature range. Finishing temperature is limited only by the allotropic transformation, which begins near 800 °C (1500 °F). The higher-carbon types (420 and 440) have a more limited forging-temperature range because of lower hot ductility. Because all martensitic types are highly hardenable, they require some type of heat treatment (generally annealing) after forging. Except for type 410, which may be used in the annealed condition, martensitic types should be further heat treated, if not for higher hardness then for best corrosion resistance, before being put into service.

A broad range of forging temperatures is shown in Fig. 1 for the ferritic steels. Forging is done at slightly lower temperatures than for austenitic types because the ferritic types tend to exhibit grain growth and structural weakness. Finishing temperatures are closely restricted only for types 405, 430 and 442. These alloys require special consideration because of grain-boundary weakness due to development of a small amount of austenite. The other ferritic types are commonly finished at temperatures as low as 700 °C (1300 °F). For fully ferritic types such as type 446, at least the final 10% reduction (and preferably more) should be performed below 870 °C (1600 °F) to achieve grain refinement and to develop optimum room-temperature ductility. Annealing after forging is recommended for ferritic types, because most of them contain substantial amounts of brittle martensite in the as-forged condition even though this may not be apparent from hardness measurements.

COLD WORKING

The cold working operations that can be successfully performed on most stainless steels include cold heading, cold drawing, cold extrusion and cold riveting. Cold working of stainless steel is more difficult than cold working of carbon steel because of differences in strength and work hardening, and power requirements are proportionately greater.

Cold Heading. As a result of the wide differences in work-hardening capabilities among stainless steels, some types are much more adaptable than others to cold heading. The surface finish of the part depends largely on the finish of the wire, which suggests the use of lightly drawn rather than annealed wire. Forming pressure increases with severity of shape, and lubrication of dies is difficult when upsetting is severe. Ferritic and lower-carbon martensitic types behave like carbon or low-alloy steels in cold heading.

Cold headed parts made of austenitic and ferritic stainless steels are most often used in the as-headed condition, although annealing may be needed if service conditions are likely to cause stress-corrosion cracking. Martensitic types are usually heat treated after heading.

MACHINING

Stainless steels as a class are more difficult to machine than carbon and low-alloy steels because of their higher strength and higher work-hardening rates ("gummy" nature). These characteristics require greater power and lower machining speed, shorten tool life, and sometimes lead to difficulty in obtaining a fine finish on the machined surface. Wide variations exist in these characteristics among the different stainless steels.

Procedures. In machining of stainless steels, special attention must be paid to equipment in order to control the effects of strength and work hardening. Rigid equipment and tooling are necessary to prevent chatter. Chip-curler tools are generally recommended because of the tough, stringy chips produced—particularly in machining austenitic and high-alloy ferritic types. Carbide cutting tools are preferred because they provide acceptable tool life at production machining speeds.

The following precautions should be instituted to avoid work hardening. Tools should never be permitted to ride or glaze without cutting, because the surface can work harden to the extent that cutting tools will become burned before they penetrate the surface. Care should be taken to ensure that hardening from one operation does not interfere with subsequent machining. For example, a tripod punch rather than a conventional center punch is preferred for hole location to prevent work hardening at the spot that will be touched first by the drill.

Heavier feeds and lower speeds than those used in machining low-alloy steels are used to minimize work hardening.

Type Variations. Low-alloy martensitic and ferritic stainless steels have machining characteristics much like those of low-alloy steels, whereas the higher-carbon martensitic stainless steels are among the most difficult metals to machine. Austenitic and precipitation-hardening stainless steels vary more widely in machining characteristics within each class than do the ferritic and martensitic grades. Most easily machined are the free-machining types. The unique characteristics of free-machining stainless steels should be considered, particularly as part of an effort to control strength and work hardening.

The structures of low-alloy martensitic and ferritic stainless steels make these types somewhat brittle, resulting in reasonably good chip breakage. However, hardness levels generally are higher than those of annealed low-alloy steels. The low-alloy martensitic stainless steels often must be machined in the hardened-and-tempered condition (up to 38 HRC), which produces excellent dimensional accuracy and surface finish.

Higher-carbon martensitic stainless steels such as types 420 and 440, and particularly type 440C, are progressively more difficult to machine because of their high annealed hardnesses (up to 240 HB) and the presence of hard, abrasive chromium carbides in their microstructures. The high-chromium ferritic stainless steels such as type 446 are difficult to machine because, like austenitic types, they are "gummy" and produce stringy chips.

Austenitic stainless steels such as types 304 and 316 have tensile strengths of 550 to 620 MPa (80 to 90 ksi) in the annealed condition—the same range of strength as that of annealed 1050 carbon steel. However, austenitic stainless steels exhibit much greater spreads between yield and ultimate strengths and much higher work-hardening rates—particularly the leaner alloys such as types 302 and 304.

Precipitation-hardening types vary considerably in machining characteristics because of differences in structure. They may be ferritic, martensitic, austenitic or two-phase, so machining characteristics will be characteristic of the structure that exists at the time of machining. Like martensitic types, precipitation-hardening stainless steels sometimes are machined after being heat treated to high strength in order to produce parts with closer tolerances than those obtainable by machining before heat treatment.

Free-machining stainless steels such as types 416, 430F and 303 are significantly more machinable than their non-free-machining counterparts because they contain small amounts of various free-machining additives. The most common additive is sulfur, which minimizes buildup of metal on cutting edges and promotes chip breakage, thereby permitting higher machining speeds and lower power consumption, and promoting longer tool life. The sulfur is present in nonmetallic inclusions, usually complex manganese sulfides. Selenium has beneficial effects similar to those of sulfur, but generally gives a better surface finish. Selenium also imparts greater ductility to free-machining stainless steels than does sulfur. Other free-machining additives are phosphorus, which reduces matrix "gumminess," and lead, which lubricates tools and facilitates chip breakage. Increases in machinability may be somewhat less for the higher alloys—particularly for the high-carbon martensitic stainless steels such as types 440A, 440B and 440C, for which abrasion from chromium carbides is a primary limiting factor.

Because of the high costs of labor and capital, the economic benefits of free-machining stainless steels can be substantial. The free-machining types are slightly more expensive, but this is more than offset by savings in machining costs when extensive machining is required. As a rule of thumb, a free-machining type may be cost-effective in any application where more than 10% of the material must be machined away.

The characteristics of free-machining types must be carefully considered to ensure that parts will perform satisfactorily in service. These steels have somewhat lower corrosion resistance than their unmodified counterparts, particularly under conditions where nonmetallic inclusions may initiate pitting. Free-machining stainless steels have a limited ability to be cold headed, but selenium-bearing types are more readily cold headed than other free-machining types.

HEAT TREATING

Stainless steels are subjected to various heat treatments depending on the type and on the requirements of the application. These treatments, which include annealing, hardening and stress relieving, restore desirable properties such as corrosion resistance and ductility to metal altered by prior fabrication operations. Heat treatment is often performed in controlled atmospheres to prevent detrimental surface effects.

Annealing. All types of stainless steels can be annealed. Annealing of austenitic types not only recrystallizes the grains and softens the metal, but also takes chromium carbides into solution in the austenite. Because of the latter effect, the process is sometimes referred to as solution annealing. Temperatures must exceed an intermediate range to avoid sensitization due to carbide precipitation along grain boundaries. Annealing temperatures usually are above 1040 °C (1900 °F), although some types may be annealed at closely controlled temperatures as low as 1010 °C (1850 °F) when fine grain size is important. Time at temperature is kept short to hold surface scaling to a minimum and to control grain growth, which can lead to "orange peel" in forming.

Annealing of austenitic stainless steels is occasionally called quench annealing because the metal must be cooled rapidly, usually by water quenching, to prevent sensitization (except for stabilized and extra-low-carbon types). Precipitation of chromium carbides can severely impair corrosion resistance because chromium is depleted in the matrix immediately adjacent to the carbides and/or because the carbides themselves may induce galvanic corrosion. Therefore, if water quenching is not used, thorough investigation is needed to ensure that sensitization does not occur. The investigation must take actual composition into account, because the rate of carbide precipitation varies markedly with composition: a heat of type 304 containing 0.05% carbon may be free of precipitation under cooling conditions that would produce heavy sensitization in the same alloy containing 0.08% carbon. Austenitic stainless steels are softened by recrystallization at the annealing temperature and, unlike most other steels, are not hardened by quenching.

A stabilizing anneal is sometimes performed after conventional annealing for types 321, 347 and 348. Most of the carbon content is combined with titanium in type 321 or with niobium in types 347 and 348 when these types are annealed in the usual manner. A further anneal at 870 to 900 °C (1600 to 1650 °F) for 2 to 4 h followed by rapid cooling precipitates all possible carbon as a titanium or niobium carbide and prevents subsequent precipitation of chromium carbide. It is believed that this special protective treatment is sometimes useful when service conditions are rigorously corrosive—especially when service also involves temperatures from about 400 to 870 °C (750 to 1600 °F).

Before annealing or other heat treating operations are performed on austenitic stainless steels, the steel should be cleaned to remove oil, grease and other carbonaceous residues. Such residues lead to carburization during heat treating, which degrades corrosion resistance.

All martensitic and most ferritic stainless steels can be subcritical annealed (process annealed) by heating into the upper part of the ferrite temperature range, or full annealed by heating above the critical temperature into the austenite range, followed by slow cooling. Usual temperatures are 760 to 830 °C (1400 to 1525 °F) for subcritical annealing, and 845 to 900 °C (1550 to 1650 °F) for full annealing. When material has been previously heated above the critical temperature, such as in hot working, at least some martensite is present even in ferritic stainless steels such as type 430. Relatively slow cooling at about 25 °C/h (50 °F/h) from full annealing temperature, or holding for one hour or more at subcritical annealing temperature, is required to produce the desired soft structure of ferrite and spheroidized carbides. However, parts that have undergone only cold working after full annealing can be subcritically annealed satisfactorily in less than 30 minutes.

The ferritic types that retain predominantly single-phase structures throughout the working temperature range (types 409, 442, 446 and 26Cr-1Mo) require only short recrystallization annealing in the range from 760 to 955 °C (1400 to 1750 °F). The higher-chromium types such as 446 and 26Cr-1Mo require rapid cooling through the range from 540 to 370 °C (1000 to 700 °F) to avoid "885 °F" embrittlement and consequent loss of ductility.

Hardening. Martensitic stainless steels are hardened by austenitizing, quenching and tempering much like lower-alloy steels. Austenitizing temperatures normally are 980 to 1010 °C (1800 to 1850 °F)—well above the critical temperature. As-quenched hardness increases with austenitizing temperature to about 980 °C (1800 °F) and then decreases due to retention of austenite. For some types, the optimum austenitizing temperature may depend on the subsequent tempering temperature.

Preheating before austenitizing is recommended to prevent cracking in high-carbon types and in intricate sections of low-carbon types. Preheating at 790 °C (1450 °F) and then heating to the austenitizing temperature is the most common practice, but very large or extremely intricate parts sometimes are successively preheated at 540 °C (1000 °F) and then at 790 °C (1450 °F) before austenitizing.

Martensitic stainless steels have high hardenability because of their high alloy content. Air cooling from the austenitizing temperature is usually adequate to produce full hardness. Oil quenching is sometimes used, particularly for larger sections. Tempering temperature must be chosen for the optimum combination of hardness, toughness and corrosion resistance. Parts should be tempered as soon as they have cooled to room temperature—particularly if oil quenching has been used—to avoid delayed cracking. Parts sometimes are refrigerated to −75 °C (−100 °F) before tempering to transform retained austenite—particularly where dimensional stability is important, such as in gage blocks made of type 440C. Tempering at temperatures above 510 °C (950 °F) should be followed by relatively rapid cooling to below 400 °C (750 °F) to avoid "885 °F" embrittlement.

Some precipitation-hardening stainless steels require more complicated heat treatments than standard martensitic types. For instance, a semi-austenitic precipitation-hardening type may require annealing, trigger annealing (to condition austenite for transformation on cooling to room temperature), subzero cooling (to complete the transformation of austenite) and aging (to fully harden the alloy). On the other hand, martensitic precipitation-hardening types often require nothing more than a simple aging treatment.

Stress Relieving. Stainless steel weldments generally are heated to temperatures below the usual annealing temperature to decrease high residual stresses when full annealing after welding is impossible. Most often, stress relieving is performed on weldments that are too large or intricate for full annealing or on dissimilar-metal weldments consisting of austenitic stainless steel welded to alloy steel. Stress relieving at temperatures below 400 °C (750 °F) is an acceptable practice but results in only modest stress relief.

Stress relieving at 425 to 925 °C (800 to 1700 °F) significantly reduces residual stresses that otherwise might lead to stress-corrosion cracking or dimensional instability in service. One hour at 870 °C (1600 °F) typically relieves about 85% of the residual stresses. However, stress relieving in this temperature range also precipitates grain-boundary carbides, resulting in sensitization that severely impairs corrosion resistance in many media. Sensitized austenitic stainless steels are susceptible to intergranular corrosion or stress-assisted intergranular corrosion even in some media that are considered mild. To avoid these effects, it is strongly recommended that a stabilized stainless steel (type 321, 347 or 348) or a low-carbon type (304L or 316L) be used, particularly when lengthy stress relieving is required.

When austenitic stainless steels have been cold worked to develop high strength, low-temperature stress relieving will increase the proportional limit and yield strength (particularly compressive yield strength). A two-hour treatment at 345 to 370 °C (650 to 700 °F) is normally used; temperatures up to 425 °C (800 °F) may be used if resistance to intercrystalline corrosion is not required for the application. Higher temperatures will reduce strength and sensitize the metal, and generally are not used for stress relieving cold worked products.

Stress relieving of martensitic or ferritic stainless steel weldments will simultaneously temper both weld and heat-affected zones, and for most types will restore corrosion resistance to some degree. However, annealing temperatures are relatively low for these grades, and normal subcritical annealing is the heat treatment usually selected if the weldment is to be heat treated at all.

JOINING PROCESSES

Stainless steels are commonly joined by welding, brazing and soldering. Arc welding is the overwhelming choice for joining stainless steel to stainless steel because it gives a relatively crevice-free joint of high joint efficiency. Resistance welding also is often used for austenitic types, because it produces a mechanically strong joint quickly and inexpensively. Procedures and precautions appropriate for the various types are important if optimum corrosion resistance and mechanical properties are to be attained in the completed assembly. Brazing usually is preferred for joining stainless steels to other metals.

Welding. All stainless steels can be readily welded by any common arc welding process. Major factors that must be considered are (a) corrosion resistance in the weld and heat-affected zones, (b) residual stress, which can lead to distortion, weld cracking or fissuring, and (c) for martensitic and ferritic types, mechanical properties in the weld and heat-affected zones.

Resistance welding of stainless steels also is quite common. In fact, austenitic stainless steels are resistance welded more often than any other metal except carbon steel. The welding currents used in welding austenitic stainless steels are significantly lower than those required for welding carbon steels because of the lower thermal conductivity, higher electrical resistivity and nonmagnetic character of the austenitic types. Distortion is somewhat more troublesome than in resistance welding of low-alloy steels because austenitic stainless steels have greater coefficients of thermal expansion. Corrosion resistance may not be seriously impaired by carbide precipitation, because welding times are too short to produce extensive sensitization. Crevice corrosion, a form of localized corrosion resulting from limited access of oxygen, can be a problem in spot welded sheets when they are exposed in certain media. Martensitic and ferritic types also can be joined by resistance welding, although these types are used for resistance welded components much less commonly than austenitic types. For martensitic types and those ferritic types that form some martensite on cooling after welding, a second heating pulse can be programmed into the welding cycle to temper the weld area.

Stainless steels are rarely joined by gas welding. In oxyacetylene welding of stainless steels, great skill in controlling the welding atmosphere is required to prevent either oxidation or carburization of the weld puddle. Gas welding with an atomic hydrogen torch is less demanding, but such equipment is no longer widely used.

Austenitic types are the most weldable stainless steels, but are also the ones most different in welding behavior from carbon and low-alloy steels. Probably the most important metallurgical factor to consider in planning for austenitic stainless steel weldments is susceptibility to grain-boundary carbide precipitation (sensitization) at moderately elevated temperatures. Material immediately adjacent to the weld will be heated to or above the annealing temperature and will be free of precipitation. At some distance away — perhaps 3 mm (1/8 in.) or more, depending on welding parameters — the base metal is heated to 650 to 870 °C (1200 to 1600 °F) and grain-boundary carbides can precipitate despite the short time at temperature. Carbide precipitation severely impairs corrosion resistance in many media, including the acids most often used for pickling to remove oxide. Nevertheless, because sensitization occurs in such a narrow region of the heat-affected zone, many austenitic stainless steel weldments are used in the as-welded condition without concern.

Free-machining stainless steels are quite difficult to weld, with porosity and segregation being common problems. However, these types can be welded successfully with special consumable electrodes (type 312 electrodes, for instance) if precautions are taken to exclude all traces of hydrogen.

Most precipitation-hardening stainless steels can be arc welded by techniques similar to those used for welding austenitic types. Normally, filler-metal composition is selected to match or nearly match base-metal composition so that response to heat treatment is similar. Welding usually is followed by full heat treatment, although aging alone may be adequate for single-pass welds in some types (such as 17-4 PH). In heat treated weldments, weld strength closely approaches base-metal strength, but toughness and ductility of the weld may be somewhat inferior, especially for material heat treated to high strength. Weldability of the fully austenitic precipitation-hardening stainless steels HNM and 17-10P is limited because these steels are hot short above 1180 °C (2150 °F) and therefore are susceptible to underbead cracking. HNM and 17-10P have been flash butt welded successfully.

Brazing. All stainless steels can be brazed, and this process is frequently used for joining stainless steels to other metals. All brazing techniques can be used, but furnace brazing is employed more often than any other technique because it permits brazing to be done in a protective environment (usually hydrogen or vacuum) that prevents oxidation of the stainless steel. Most brazing of austenitic types is performed at temperatures in the sensitization range, and solution annealing after brazing is impossible. Therefore, stabilized or extra-low-carbon types must be used if service conditions might lead to intergranular corrosion of sensitized material. In brazing of martensitic and ferritic types, a brazing filler metal that melts below the critical temperature of 830 °C (1525 °F) is normally used to avoid martensitic hardening. The heat of brazing will temper, and possibly soften, hardened martensitic stainless steels.

Virtually all types of brazing filler metals are used, including silver, nickel, gold and copper alloys. High-phosphorus copper-base filler metals should be avoided, however, because they have harmful effects on stainless steels. Certain austenitic types (21Cr-6Ni-9Mn, for instance) should not be brazed with any copper-base filler metal because molten copper attacks the stainless steel during brazing.

Destructive penetration of stainless steel is possible during brazing. A form of stress-corrosion cracking develops in some types if they are brazed while in a highly stressed condition. Penetration during brazing may be prevented by annealing before brazing, by heating parts slowly enough to relieve stresses before the brazing temperature is reached, or by selecting a brazing filler metal that inhibits such penetration. Grain-boundary penetration may also occur if high-temperature brazing alloys that melt above 980 °C (1800 °F) are used. This problem can be minimized by selecting a favorable brazing alloy, using as little brazing alloy as possible and keeping time at temperature short. Welding through or near a brazed joint can also cause penetration of stainless steel and should be avoided.

Use of a silver brazing alloy to which nickel has been added helps minimize crevice corrosion of brazed joints exposed to certain corrosive media such as seawater. Complete elimination of this problem in a type 430 brazement can be attained only by using a special filler metal that contains tin and nickel.

PRECAUTIONS IN WELDING

The most obvious problem associated with welding of stainless steel is maintenance of uniform resistance to corrosion across the weld zone and the adjacent base-metal zones. This problem is commonly overcome by close control of composition and welding conditions. Sometimes, postweld heat treatment is required to restore corrosion properties altered during welding—particularly in material subject to sensitization.

Sensitization (harmful carbide precipitation) can be avoided, where necessary, by any of three methods:

• *Solution annealing after welding* relieves any residual stresses and improves the structure and corrosion resistance of the weld metal itself. However, solution annealing and the postanneal pickling that it necessitates are costly and inconvenient. Distortion may be a serious problem, particularly if water quenching is used to ensure that general reprecipitation of carbides does not occur during cooling. Postweld annealing is impossible for large weldments such as tanks and pressure vessels.

• *Limitation of carbon content.* Carbides will not precipitate in a weld heat-affected zone if carbon content is low enough. Extra-low-carbon stainless steels such as types 304L and 316L (0.030% max C) are commonly used in the United States, and such low carbon contents are necessary for weldments that are to be stress relieved. Even the restriction of carbon content to 0.030% max may not completely protect against sensitization and loss of corrosion resistance in the higher-nickel alloys or in any austenitic stainless steel that has been mildly cold worked. Limitation of carbon to 0.05% max will prevent sensitization in the heat-affected zone under most welding conditions, particularly in light sections. When exposed to strongly oxidizing corrosive media such as hot nitric acid, molybdenum-bearing types 316 and 317 are susceptible to intergranular corrosion in the heat-affected zone due to formation of grain-boundary sigma phase, even when carbides are not precipitated. When weldments of molybdenum-bearing grades are to be used in oxidizing media, it is common practice to subject sensitized samples from each heat to corrosion testing for assessment of susceptibility to intergranular corrosion.

• *Stabilization of carbon.* Sensitization in the heat-affected zone can be prevented by tying up the carbon in titanium carbides (as in type 321) or in niobium carbides (as in types 347 and 348) so that detrimental chromium carbides cannot form. The stabilized grades are also recommended for applications involving long-time exposure to temperatures in the sensitization range, such as in prolonged stress relief or service at high temperatures. However, weldments of stabilized grades heated into the sensitization range and later exposed to certain corrosive media can suffer from severe localized corrosion called knifeline attack. The stable carbides are partly dissolved in a very narrow zone immediately adjacent to the weld, and during later exposure in the sensitization range chromium carbides can precipitate. This unusual type of sensitization can be prevented by annealing or stabilization annealing, even locally, after welding.

Weld Cracking. Austenitic stainless steel welds are extremely tough and ductile, and thus cold weld cracking is almost never a problem. However, austenitic stainless steels are susceptible to hot cracking or microfissuring as they cool from the solidus to about 980 °C (1800 °F). Microfissuring can be prevented or kept to a minimum by eliminating or reducing tensile stress imposed on the weld during cooling through this range. To some degree, microfissuring can be controlled by controlling concentrations of residual elements such as phosphorus. However, the most common control measure is to ensure the presence of at least 3 to 4% ferrite in the as-deposited weld. Small amounts of this phase seem to prevent the cracking that often occurs in fully austenitic weld metal. Ferrite content is usually estimated on the basis of composition by use of the DeLong diagram, which is a modification of the long-used Schaeffler diagram. DeLong's modification takes into account the potent austenitic stabilization effect of nitrogen. Because ferrite contents calculated in this manner are not completely precise, it is recommended that for critical applications actual ferrite content be determined by magnetic analysis of as-deposited weld metal. For production welds, measurement is especially preferred to calculation in the common instance where a high-ferrite welding electrode is used to weld lower-ferrite base metal. Weld composition then varies with the degree of dilution.

Control of ferrite content is not always an acceptable solution to microfissuring. Ferrite is a magnetic phase, reduces corrosion resistance in some media and may lead to embrittlement in long-time elevated-temperature service exposure due to precipitation of sigma phase. Ferrite content in the weld can be reduced significantly (typically by 2 to 4%) by annealing after welding; but where postweld annealing is not possible, fully austenitic welds may be required. Some steels such as type 310 are fully austenitic through the entire specified composition range. Weld cracking can be minimized in fully austenitic stainless steels by welding with low heat input, minimizing restraint, designing for low constraint and keeping residual elements at low concentrations.

Ferritic types are less ductile than austenitic types and therefore are more susceptible to weld cracking. Certain ferritic stainless steels (type 430, for instance) form significant amounts of martensite on cooling after welding, which increases susceptibility to cold cracking. Preheating at 150 to 230 °C (300 to 450 °F) is recommended to minimize weld cracking in all ferritic types.

In fully ferritic types such as 409, 446 and 26Cr-1Mo, welding causes grain coarsening in the base metal immediately adjacent to the weld. Toughness therefore is reduced, particularly in heavy sections, and cannot be restored by postweld heat treatment. Ferritic stainless steels that form austenite at elevated temperatures are not coarsened significantly, but postweld annealing is recommended to transform the resulting martensite and enhance ductility in the heat-affected zone.

Martensitic stainless steels are even more susceptible to weld cracking than ferritic types. Preheating at 200 to 300 °C (400 to 600 °F) generally is required. Postweld annealing is standard practice, particularly for steels with carbon contents greater than 0.20%.

CLEANING AND FINISHING

Proper cleaning and finishing of stainless steel parts are essential for maintenance of the corrosion resistance and appearance for which stainless steels are specified. The degree of care required depends on the nature of the application; the most stringent precautions (such as clean-room assembly and sophisticated postassembly cleaning) are used for critical applications such as nuclear-reactor cores, pharmaceutical and food-handling equipment, and some aerospace applications.

Stainless Steels in Corrosion Service

THE CORROSION RESISTANCE of stainless steels is believed, although not by all investigators, to result from the presence of a thin hydrous oxide film on the surface of the metal. The film varies in composition from alloy to alloy and with different treatments, such as rolling, pickling and heat treating. For any stainless steel, this film, stabilized by chromium, is considered to be continuous, nonporous, insoluble and self-healing. If broken, the film will repair itself when re-exposed to air or a suitable oxidizing agent.

Passivity is the corrosion-resistant behavior produced by the presence of a "passive" oxide film. It is not a constant state; it exists only in certain environments or under certain conditions. The range of conditions over which a stainless steel exhibits passivity may be broad or narrow, and passivity may be destroyed by slight changes in conditions. Under circumstances favorable to passivity, stainless steels have solution potentials approaching those of noble metals. When passivity is destroyed, the potential approximates that of ordinary iron.

Stainless steels are normally passive, but when exposed to mildly oxidizing corrosive solutions, they become active. Hence, oxygenating agents must be present and must be replenished constantly to maintain passivity. Otherwise, localized corrosion frequently results—for example, the crevice corrosion that occurs beneath barnacles in seawater.

An increase in the velocity of a corrosive solution increases the rate at which oxygen dissolved in the solution is brought in contact with the steel. Because of this, the rate at which a stainless steel undergoes electrochemical corrosion tends to decrease as velocity increases. Increased velocity, however, tends to increase mechanical actions such as erosion and cavitation, which can prevent formation of a passive film or remove a film originally present. Thus for most stainless steels, corrosion rates tend to decrease with increasing velocity up to some limiting velocity, then increase again as mechanical effects begin to compete with purely electrochemical corrosion. The limiting velocity varies not only with steel composition, but also with temperature, amount of suspended solids, type and concentration of corrodent, and other environmental factors.

Stainless steels would have a high solution tendency wherever the metal is in direct contact with a corrodent were it not for the presence and maintenance of an inert, passive film. The metal itself, with no covering, does not have good corrosion resistance. Thus, by nature, the resistance of stainless steel is usually either quite good or very bad. In corrosive service, it is seldom intermediate in performance.

The high solution tendency is also evident when the passive film is destroyed locally. If that happens, stainless steels can fail by localized mechanisms such as pitting, crevice corrosion, intergranular corrosion or stress-corrosion cracking. Localized corrosion can be catastrophic. Usually a very small portion of the stainless steel area is involved, and the damage may be difficult to detect before failure occurs.

EFFECT OF COMPOSITION

Stainless steels derive their resistance to corrosion from the presence of chromium. Increasing the chromium content in steel progressively enhances resistance to rusting in the atmosphere. The rate at which steel develops a passive film in the atmosphere depends on its chromium content. However, only the grades containing about 10% Cr or more develop passivity. Some grades, even in this passive range, may show evidence of superficial attack in severe marine and industrial atmospheres.

The presence of nickel in high-chromium stainless steels greatly improves their resistance to certain nonoxygenating media. Nickel may not be needed for corrosion resistance in some atmospheric environments, but will impart other desired properties.

Manganese is an effective austenite stabilizer that does not significantly alter the resistance to corrosion of high-chromium steels. In the 200-series stainless steels, manganese is used as a substitute for part of the nickel that would otherwise be required to make the steels austenitic at room temperature.

The presence of molybdenum in stainless steels greatly improves their resistance to solutions of halogen salts and to pitting in seawater. Additions of molybdenum strengthen the passive film in some media where it is otherwise likely to fail. Addition of molybdenum greatly reduces the incidence of pitting. However, when pits do occur—in type 316, for example—they become just as deep as those in molybdenum-free grades (such as type 304). As already mentioned, pitting corrosion will also develop under foreign deposits that prevent access of oxygen to the surface of the steel. Pitting, however, may occur in environments that are free from foreign deposits, if oxygenating conditions are borderline or halogen salts are present in sufficient concentration.

EFFECT OF HEAT TREATMENT

The changes in microstructure produced by different heat treatments have considerable influence on the corrosion resistance of stainless steels. These steels normally exhibit greater resistance when all the carbon is in solution, producing a homogeneous single-phase structure.

Unstabilized austenitic stainless steels become subject to severe attack along grain boundaries at room temperature in a number of corrosive media if the metal is heated to temperatures in the range from 550 to 850 °C (1000 to 1550 °F). This attack is known as intergranular corrosion and results from precipitation of chromium carbide and consequent depletion of chromium in the areas adjacent to the grain boundaries. Decreasing carbon content reduces susceptibility of the steel to carbide precipitation and subsequent intergranular attack.

Besides carbon content, time at the sensitizing temperature is important. Precipitation of carbides at the grain boundaries may take place very rapidly. For example, after welding, the metal adjacent to the weld may become subject to intergranular attack, whereas the weld metal itself and the base metal outside the heat-affected zone may remain virtually free from attack. Intergranular susceptibility may be removed by annealing, or may be avoided by using stabilized compositions (types 321 and 347) or extra-low-carbon grades. (Types 321 and 347 are subject to "knifeline attack" under certain conditions, as discussed in a later section of this article.)

Mechanical properties of stainless steels are not severely affected by sensitization. However, subsequent intergranular corrosion has a deleterious effect on the ability of a stainless steel structure to sustain a load.

Martensitic steels must be properly heat treated to give optimum resistance to corrosion, and normally they show maximum resistance to atmospheric corrosion in the fully hardened condition. Tempering at 375 °C (700 °F) or less relieves quenching stresses and greatly improves ductility and toughness without severely reducing resistance to atmospheric corrosion. However, tempering between 375 and 560 °C (700 and 1050 °F) should be avoided, because temperatures in this range impart low toughness as well as low resistance to corrosion.

Corrosion resistance of the ferritic grades may be adversely affected by some thermal treatments. Therefore, it is advisable to anneal the nonhardenable 10 to 29% Cr grades after welding, in order to obtain maximum resistance to corrosion and to obtain ductility in the heat-affected zones adjacent to welds. As with austenitic steels, these problems are minimized by using stabilized compositions such as types 321 and 347 or extra-low-carbon grades such as types 304L and 316L.

If properly solution heat treated, austenitic chromium-nickel steels exhibit passivity in a wide variety of corrosive environments. Austenitic steels resist corrosion most effectively when they are heated to temperatures from 1040 to 1150 °C (1900 to 2100 °F) and cooled rapidly; this heat treatment produces a homogeneous austenitic structure. The lower side of the above temperature range is usually preferred to minimize warpage and to avoid difficulty in scale removal. Rapid cooling from the heat treating temperature is essential in order to keep the carbides in solution. Very small parts may be air cooled, but larger parts must be water quenched to prevent sensitization. Specific guidelines for avoiding sensitization during heat treatment are difficult to establish because sensitization in a given part is affected by factors such as thickness and mass of the part, as well as by composition (especially carbon content and amounts of carbide formers other than chromium). With regard to cooling rate, it is not cooling rate *per se* that is important, but rather the amount of time that a part spends within the sensitization range. Here, a few minutes at a temperature near the middle of the range is equivalent to several hours at temperatures near the extremes of the range.

Cold work sometimes reduces the corrosion resistance of stainless steels. This effect is rather specific and depends on the composition of the steel, the extent and uniformity of cold work, and particularly the nature of the environment. For example, local cold work, such as that produced by imprinting identification numbers with metal stamps, seems to have a deleterious effect on the corrosion resistance of stainless steel.

EFFECT OF WELDING

The degree of sensitization induced by welding varies chiefly with heat input per unit length of weld. Low heat input, which is characteristic of arc welding processes and high travel speeds, produces lower amounts of sensitization than high heat input, which is characteristic of oxyfuel-gas welding and low travel speeds. Oxyfuel-gas welding is seldom used to weld stainless steels; it not only entails high heat input but also has a high potential for carburizing the stainless steel and thereby increasing the sensitization effect in the weld and heat-affected zones.

EFFECT OF SURFACE CONDITION

To ensure satisfactory service life, the surface condition of stainless steel must be given careful attention. Smooth surfaces, plus freedom from surface imperfections, blemishes and all traces of scale and other foreign material, reduce the probability of corrosion. Generally, a smooth, highly polished reflective surface has greater resistance to corrosion. Rough surfaces are more likely to catch dust, salts and moisture, which tend to localize corrosive attack.

Oil and grease can be removed by either hydrocarbon solvents or alkaline cleaners, but these cleaners must be removed before heat treatment. Hydrochloric acid formed from residual amounts of trichloroethylene, used for degreasing, has caused severe attack of stainless steels. Surface contamination may be caused by machining, shearing and drawing operations. Small particles of metal from tools become embedded in the steel surface and, unless removed, may cause localized galvanic corrosion. These particles are removed best by immersing the surface in a solution containing approximately 20% nitric acid, heated to a temperature between 50 and 60 °C (120 and 140 °F). Some believe that this treatment promotes development of passivity.

Shot blasting or sand blasting should be avoided unless iron-free silica sand is used; metal shot in particular will contaminate the stainless steel surface. If shot blasting or shot peening with metal grit is unavoidable, it is essential to clean the parts after blasting or peening by immersing them in a solution of about 20% nitric acid.

EFFECT OF DESIGN AND FABRICATION

Failure due to corrosion often can be eliminated by suitable changes in design without changing the type of steel. Factors to be considered include joint design, surface continuity and concentration of stress. Welds should be well spaced, and any applicable codes or standards should be followed for location and manner of making attachments. Weldments should be designed for economical cutting of plates and for good fit-up at joints. Butt joints are preferred over lap joints. If lap welds are necessary, they should be sealed completely against penetration by corrosive solutions; otherwise, crevice or concentration-cell corrosion may result. Use of certain attachments, such as doubling plates or reinforcing plates encircled by fillet welds, should be minimized; such attachments induce biaxial residual stresses that generally are difficult to relieve by heat treatment. In order to avoid diffusion of carbon into stainless steel tanks from low-carbon steel supporting legs, which can occur during service at elevated temperatures, the legs should be welded to a stainless steel saddle, and the saddle welded to the tank.

Wherever possible, welding should be done in the downhand position; this generally produces the most consistently sound welds. Stringer beads are preferred for manual welding because the heat input and residual stress are lower than for other types of beads. Welds less than 5 mm ($^3/_{16}$ in.) thick may be water cooled to lessen the effects of welding heat and to reduce distortion. Control of heat effects to avoid sensitization is generally unnecessary with stabilized stainless steels (types 321 and 347) or extra-low-carbon grades (types 304L and 316L), but freedom from residual fabricating stresses is important, particularly where there is a possibility that stress-corrosion cracking may occur in service. It is especially important to provide for good fit-up at weld joints; poor fit-up causes residual stresses, because either the parts have to be forced into position and then welded or welds are uneven and nonuniform due to gaps and unevenness at the joint. Residual stresses can also be kept to a minimum by using weld sequences that do not impose excessive constraint on the joint.

Stainless steel weldments should be thoroughly cleaned, by blast cleaning with iron-free grit or by pickling, to remove all contaminants such as oxides and iron dust, weld spatter, welding flux, dirt and organic matter. This should be followed by final cleaning with a solution of 10% nitric–1% hydrofluoric acid. The final acid cleaning step should be done with care because severe attack along the heat-affected zone can occur if the cleaning time is prolonged or the amount of hydrofluoric acid in the solution is above 1%.

Elimination of notches, threads, grinding or abrasive scratches, corners, grooves and similar stress raisers is essential. Generous fillets at corners, smooth contour welds without undercuts, ground and polished welds, rounded corners, ground and polished edges, and flat surfaces will eliminate causes of stress concentration.

Cold forming operations, such as rolling of tubes into tube sheets, should be held to a minimum. Severe forming operations, such as rolling of dished heads for tanks should be done by hot working above 870 °C (1600 °F). Castings, bolts, tank ends and other hot formed components should be annealed at 1100 °C (2000 °F) and quenched in water or in an air blast.

If heat treatment is required, consideration should be given to structural stability at the heat treating temperature. Surfaces must be clean and free from carbonaceous materials such as oil, grease and paint, which may be absorbed at high temperature and may thereby increase the carbon content. Availability of heat treating facilities in the locality where the stainless steel is to be fabricated will influence cost. Sometimes use of the higher-priced stabilized or extra-low-carbon grades may reduce total cost by eliminating heat treatment.

Hot gases that are innocuous to the stainless steel may form corrosive condensates on cold portions of a poorly insulated unit. Conversely, vapors from noncorrosive liquids may cause attack, and exhausts and overflows should be designed to prevent hot vapor pockets. Generally, open ends of inlets, outlets and tubes in heat exchangers should be flush with tank walls or tube sheets to avoid build-up of harmful corrodents, sludges and deposits.

INTERGRANULAR CORROSION

Improper heat treatment of stainless steels can lead to early failure under severely corrosive conditions and can greatly reduce service life in many relatively mild environments. Unstabilized austenitic stainless steels that contain more than 0.03% C become susceptible to intergranular corrosion because complex chromium carbides precipitate along grain boundaries when the steel is exposed to temperatures from about 550 to 850 °C (1000 to 1550 °F). Normal welding procedures also induce susceptibility. The fact that a stainless steel is susceptible to intergranular corrosion does not necessarily mean that it will be so attacked. However, many corrosive media, including some considered only mildly corrosive, will corrode susceptible steels intergranularly. The possibility of intergranular attack must be recognized unless ruled out by previous experience. Chromium carbide precipitation in austenitic stainless steels can be eliminated by:

1. Heating the steel, after final fabrication or processing, to a temperature high enough to dissolve the carbides (usually, 1040 to 1150 °C, or 1900 to 2100 °F), and cooling rapidly enough to avoid reprecipitation. Localized heat treatment of the area immediately adjacent to a weld is not satisfactory for prevention of chromium carbide precipitation. For effective heat treatment, the entire unit must be heated and quenched.
2. Using a stainless steel that has been stabilized with niobium or titanium, which combine with the carbon and thereby prevent harmful precipitation of chromium carbides.
3. Reducing carbon content to such a low level that difficulty is avoided. The availability of extra-low-carbon stainless steels (types 304L and 316L) has made this method of great practical importance.

Among the methods available for detecting susceptibility to intergranular corrosion are the boiling 65% nitric acid (Huey) test, the copper–copper sulfate–sulfuric acid test, the nitric acid–hydrofluoric acid test, the oxalic acid etch test and the ferric sulfate–sulfuric acid test. Details for performing all of these tests are given in ASTM A262. In tests on unstabilized stainless steels, specimens must represent the condition in which the steel is to be used. In tests performed on stabilized or extra-low-carbon grades to determine whether the steel as supplied is resistant to sensitization, specimens are sensitized by heat treatment in the temperature-sensitive range (for example, 1 h at 675 °C, or 1250 °F) to reveal any latent susceptibility.

The nitric acid test is not ordinarily used for routine evaluation of types 316L and 321 because phenomena other than chromium carbide precipitation can cause these steels to corrode intergranularly in boiling 65% nitric acid. Intermetallic compounds of iron and chromium known as sigma and chi phases (or perhaps transition phases intermediate in the formation of sigma and chi) may be responsible for this effect.

The danger of chromium carbide precipitation is now so widely recognized and guarded against that relatively few failures occur from this cause. Nevertheless, costly intergranular corrosion still is sometimes encountered.

In atmospheric exposures or mildly corrosive conditions where freedom from contamination is the primary objective, precautions against intergranular corrosion usually are not required.

"Knife-line attack" is a special form of intergranular corrosion sometimes encountered in type 321 and to a lesser degree in type 347. It usually occurs when weldments are subjected for a considerable period of time to temperatures within the sensitizing range after fabrication—as, for instance, in a stress-relieving treatment. After such a treatment, metal immediately adjacent to welds may be attacked by corrosive media. The explanation is that the affected area has been heated to a temperature high enough to decompose the titanium or niobium carbides. Consequently, part of the carbon combines with chromium during subsequent exposure within the temperature range for chromium carbide precipitation. The metal is therefore made susceptible to intergranular corrosion. A "stabilizing" treatment at 900 °C (1650 °F) after welding may be beneficial if such problems are encountered.

PITTING CORROSION

Because stainless steels are passive under almost all conditions in which they are normally used, any localized corrosion, under circumstances that prevent restoration of passivity, may cause rapid penetration at the point of initiation. This occurs because an active-passive electrolytic cell is formed between the large cathodic (passive) area and the small anodic area under attack; the surrounding oxygen serves as a depolarizer, and pitting proceeds.

Solutions containing chlorides are especially detrimental because they promote formation of active-passive electrolytic cells. Acid chlorides in their higher-valence state (such as cupric chloride and ferric chloride) are particularly severe, but any chloride in appreciable concentration is a possible source of trouble. Solutions of other halide salts and of some sulfates may cause pitting.

As shown in Fig. 2, molybdenum in stainless steels dramatically increases resistance to pitting.

CREVICE CORROSION

Crevices formed by joints and connections, or at points of contact between metals and nonmetals, are most frequently subject to attack. Similarly, deposits of foreign matter may promote local attack. When a stainless steel is immersed in seawater, fouling by growth of barnacles promotes local corrosion.

In a crevice, the oxygen supply is limited and cannot repair the passive oxide film, and a so-called "differential concentration cell" appears. Porous substances in the crevice may trap corrosive solutions such as seawater or moisture condensed from the atmosphere. A crevice stays damp longer than a fully exposed surface. Salts are likely to accumulate in crevices and under deposits, particularly if the area around the crevice is alternately wet and dry. Gasket materials containing sulfur or graphite contribute to this type of corrosion. On the other hand, corrosion will

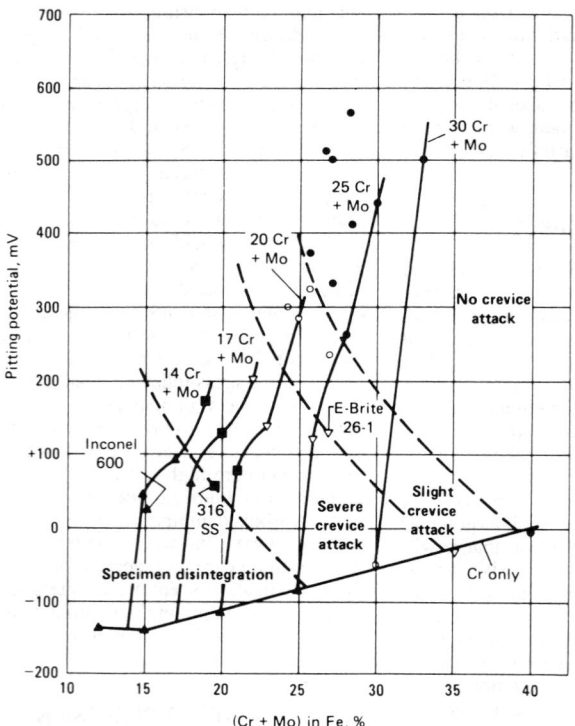 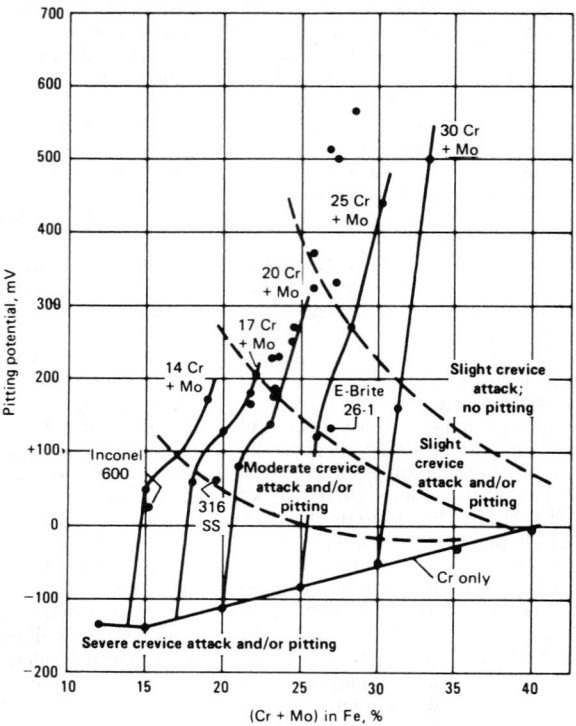

Pitting potentials were measured in deaerated synthetic seawater (pH of 7.2 ± 0.2) at 90 °C (194 °F) vs a saturated calomel electrode. Left: resistance to crevice attack after 6 days in oxygenated 10% ferric chloride at 21 ± 1 °C (70 ± 2 °F). Right: resistance to crevice attack after 14 days at 121 °C (250 °F) in synthetic seawater containing about 60 ppm oxygen.

Fig. 2. Relationship between critical pitting potential and resistance to crevice corrosion for Fe-Cr-Mo alloys

seldom occur if the crevice is sealed completely to prevent intrusion of moisture.

If the oxygen concentration decreases below a level necessary to maintain passivity in the anodic area, there is a double electrolytic effect. The difference in oxygen concentration alone will tend to accentuate attack on the anodic area; then the electrolytic potential between the active anode and the passive cathode is quite high, in itself, leading to a further degree of sustained continuous corrosion. There is a tendency for the cathode to become polarized with "plated hydrogen," but, because oxygen is immediately available, the hydrogen film is destroyed, permitting attack to proceed, often at an unacceptably high rate. Chlorides and other nonoxygenating salts, being electrolytes that will not contribute to passivity, assist in this type of corrosion, often leading to destructive pitting. Furthermore, the solution within the crevice becomes very acidic (pH of 1.2 has been reported), which adds to the accelerated attack.

GALVANIC AND CONCENTRATION-CELL CORROSION

The factors that influence galvanic corrosion include conductivity of the circuit, potential between anode and cathode, polarization, relative areas of cathode and anode, geometrical relationships between dissimilar-metal surfaces, and contact between metals. Of these, relative areas of anodic and cathodic surfaces have the most pronounced effect on the extent of damage produced by galvanic action, because a small anode and a large cathode result in an increase in current density at the anode with a great consequent increase in the rate of corrosion. Thus, small dif-

ferences in potential under these conditions may produce extensive corrosion because of increased current density in the anodic areas.

STRESS-CORROSION CRACKING

Stress-corrosion cracking in austenitic stainless steels has been reported in chloride solutions when high stresses (residual or applied) are present in the metal. Solutions of sodium, magnesium, calcium, zinc or lithium chlorides, and most ethyl chlorides, are among the most aggressive. Of these, a boiling concentrated solution of magnesium chloride is very aggressive and will cause austenitic steels to fail by stress-corrosion cracking in short periods of time ($^{1}/_{2}$ to 2 h) when the applied stress is near the yield strength of the steel. A boiling 42% magnesium chloride solution is especially severe in this respect and has been used for laboratory comparison testing.

Stress-corrosion cracking in annealed-and-quenched alloys is characterized by branching transgranular cracks, and this type of cracking is often characteristic of chloride-induced stress-corrosion cracking in sensitized material. However, sensitized material may display stress-assisted intergranular corrosion having a cracklike appearance under conditions where intergranular attack in unstressed material would be minor. This latter failure mode is often referred to as intergranular stress-corrosion cracking.

The corrosive environments that can cause stress-corrosion cracking are rather specific and are usually environments that cause little or no general attack. Conventional transgranular stress-corrosion cracking usually occurs only in chloride or caustic solutions. Intergranular stress-corrosion cracking occurs most often in polythionic

acid solutions or in high-temperature oxygenated water. Because failure is caused by the combined action of stress, temperature and the specific corrodent, it can be controlled or avoided by reducing or eliminating any one or more of the contributing factors.

Cracking has been observed even in hot water of relatively low chloride content, especially where cold worked parts were involved or where chloride was concentrated in crevices or pockets.

As-drawn or severely cold worked stainless steels crack readily in contact with aqueous hydrogen sulfide, but annealed stainless steels are not susceptible unless stressed well above their yield strength. Several media, including strong hot caustic solutions under pressure, have been reported to cause cracking, although most failures in hot caustic solutions may have been caused by unreported chloride impurities. Chloride compounds cause most stress-corrosion cracking in stainless steel.

Ferritic stainless steels are not considered susceptible to stress-corrosion cracking in chloride solutions but will pit badly. They also appear to resist cracking in caustic solutions. Martensitic grades are known to crack under stress when exposed to chlorides.

To avoid stress-corrosion cracking, some equipment fabricated from types 304L and 316L has been stress relieved at temperatures ordinarily used for carbon steel (540 to 650 °C, or 1000 to 1200 °F). Although this temperature range would be harmful to ordinary 18-8 because of carbide precipitation, it has no harmful effect on extra-low-carbon types. Table 12 shows the residual stresses in solid austenitic stainless steels after various stress-relieving treatments. Integrally bonded or attached stainless linings on car-

Table 12. Residual stresses in austenitic stainless steel after various treatments

Stress-relieving temperature °C	°F	Time, h	Residual stress MPa	ksi
After welding 9¹/₄-in.-OD, 6¹/₂-in.-ID pipe				
As welded	...	....	206-177	30.0-25.7
600	1100	16	138	20.0
600	1100	48	138	20.0
600	1100	72	159	23.0
650	1200	4	148-165	21.5-24.0
After welding 5-in.-OD, 4-in.-ID pipe				
As welded	...	....	127-101	18.5-14.7
650	1200	4	94.5-105	13.7-15.3
650	1200	12	110	16.0
650	1200	36	108	15.6
900	1650	2	nil	nil
1000	1850	1	nil	nil

bon steel cannot be stress relieved effectively by heating because of the large difference in coefficient of expansion.

Specimens of type 304L and 316L heated to 650 °C (1200 °F) for one to three days and then tested in the copper–copper sulfate–sulfuric acid test as prescribed in ASTM A262 have shown no intergranular corrosion after three 72-h immersions. Exposure of the low-carbon grades to temperatures in the carbide-precipitation range may create harmful effects in certain corrosive media. For example, type 316L may fail in the Huey test (boiling 65% nitric acid) after heating for only 1 h at 650 °C (1200 °F); this may be caused by formation of sigma phase at grain boundaries. The mechanism of sigma-phase formation is similar to carbide precipitation in that there is a depletion of chromium in the adjacent areas. Depletion caused by sigma-phase formation may be less severe than that caused by carbide precipitation. This would account for intergranular attack occurring in boiling nitric acid but not in acidified copper sulfate.

Although low-temperature stress relief may be adequate to avoid stress-corrosion cracking in certain corrosives, stresses are not completely removed. For complete relief of residual stress, a temperature of 900 °C (1650 °F) is required. If the threshold stress for stress-corrosion cracking is known to be higher than the residual stress remaining after stress relieving at a low temperature, the low-temperature stress relief should be practical. If the stress levels are not known, high-temperature stress relief should be considered.

Weld deposits exposed to a sensitizing temperature sometimes are subject to stress-corrosion cracking; the use of niobium-stabilized welding electrodes may prevent this. Use of extra-low-carbon weld metal (less than 0.03% C) is not as effective in preventing cracking. In many instances, residual stresses resulting from welding are sufficient to sustain stress-corrosion cracking, even in the absence of appreciable service stresses. Consequently, stress relieving of weldments is generally considered good practice for applications where the environment is known or suspected to contain a substance that induces stress-corrosion cracking.

CORROSION TESTING

Because actual service conditions are difficult or impossible to reproduce in standard laboratory tests, results of such tests usually can serve only as a guide. Chemical conditions, temperature,

velocity and aeration should parallel those in the proposed process; therefore, field tests in existing equipment in a comparable process should be used wherever possible, in order to duplicate anticipated conditions. If the installation will be expensive to fabricate and install, simulated service tests may be made in a pilot plant.

Sometimes, stainless steel may be so adversely affected during fabrication that it fails prematurely in service; tests to evaluate the effect of proposed fabrication methods should be included in the corrosion-test program. It is advisable to include welded specimens typical of the job, and also sensitized specimens, in order to evaluate weld deposits, heat effects in the weld zone, and the possibility of stress-corrosion cracking from residual stresses. Stress-corrosion specimens loaded to various levels of stress should be tested to evaluate stress-corrosion cracking susceptibility and to indicate whether fabricated equipment should be stress relieved. Specimens heat treated to represent job conditions should be tested. It is also important that surface finish be typical of the job and the same on all specimens. Frayed surfaces or cold worked edges produced in preparation of test specimens should be removed before testing.

Careful evaluation of test results is of utmost importance. Microscopic examination of the surface of the specimen for attack at grain boundaries is helpful when corrosion rates are low or available testing time is short. Grain-boundary attack generally indicates poor serviceability. Recently developed electrochemical tests also may be used for screening materials; as with most laboratory corrosion tests; these should be considered only for initial screening of candidate materials.

ATMOSPHERIC ENVIRONMENTS

Geographic location markedly affects the rate and type of corrosion of stainless steel and must be considered when selecting a specific grade.

Rural Atmospheres

All grades of stainless steel are suitable for exposure to rural or other atmospheres that are essentially free from contamination, even when relative humidity may approach 100%. Selection for such applications depends on cost, availability, mechanical properties, fabricability and appearance.

Selection for exposure in a dry rural area often involves merely choosing the grade that results in the lowest total cost. The lower-alloy grades may rust slightly and reveal a sensitivity to variations in original finish, but for most practical purposes all grades will serve satisfactorily in uncontaminated atmospheres, regardless of humidity.

Industrial Atmospheres

Most grades of stainless steel are suitable for use in industrial atmospheres, although lower-chromium grades may be unsuitable for more severely contaminated atmospheres. Application often depends on the appearance required. Lower-chromium grades may fulfill service requirements but will tarnish severely. If appearance is important, type 430 is the lowest-alloy grade that can be used, and a higher-alloy grade usually is required. Type 302 has given excellent service in most applications.

In atmospheres free from chloride contamination, stainless steels have excellent corrosion

resistance. Types 430, 302, 304 and 316 normally do not show even superficial rust. Some rusting may occur in marine atmospheres or in industrial exposures where surfaces become contaminated with chloride salts. Rusting is most likely to be severe on sheltered surfaces that are not well washed by rain.

In choosing a grade of stainless steel for a specific industrial atmospheric application, all possible abnormal local conditions should be considered. Experiences of other users of stainless steels in the same region should be evaluated.

Transportation Equipment

It is often difficult to choose grades of stainless steel for transportation equipment, such as trucks and trains, because the equipment may be exposed to the entire range of atmospheric conditions during service life, and different components on the same vehicle may have to meet different service or appearance requirements. The difficulty has become progressively more acute since the early 1950's, when use of de-icing salt and salt mixtures in the United States increased dramatically.

Several types of stainless steels—chiefly types 409, 430, 434, 201, 301 and 304—are used extensively in transportation service, in both exterior and interior applications.

Automobiles. Stainless steel types 409, 430, 434, 202, 301 and 304 are used for automotive parts. Type 434 usually is preferred for trim, partly because it more closely resembles chromium plate in color and partly because its corrosion resistance is better than that of type 430.

Due to the increased use of de-icing salts and general pollutants, use of type 430 for automotive trim has diminished considerably in recent years, having been replaced by type 434. Nonroping alloys such as type 436 also have replaced type 430 in some applications.

Type 301 is generally preferred for wheel covers because of its excellent corrosion resistance and because of its mechanical properties. During forming, this alloy develops a work-hardened condition that makes it an excellent spring retainer as well as a decorative, corrosion-resistant cover.

Since 1960, type 409 has found extensive use in automotive exhaust systems. Initially, mufflers made entirely of type 409 were used in some automobiles, with excellent service performance. In more recent years, noncritical structural parts or muffler parts that exhibit very little corrosion have been redesigned in less-corrosion-resistant materials. All catalytic converters, used as pollution-control equipment in automobiles, are constructed of type 409.

SEAWATER ENVIRONMENTS

Exposure to seawater introduces a number of factors not present to any great extent in atmospheric exposure. Selection of a stainless steel grade for seawater immersion is far more complex than selection for atmospheric service. Types 304 and 316 (especially 316) usually give the best service. However, in stagnant seawater (particularly badly contaminated harbor waters), all types are likely to pit severely from biofouling. Even type 316 may be completely unsatisfactory if water velocity is less than 1.5 m/s (5 ft/s). Recently developed highly alloyed ferritic and austenitic types such as AL 29-4-2, AL 29-4C, Sea-cure, Monit, 254 SMO, Ferralium 255 and AL-6X are

Table 13. Crevice corrosion of ferritic stainless steels in low-velocity seawater

	Nominal composition, %		Probability of initiation(a), %		Maximum depth			
					61 days		272 days	
Cr	Mo	Ni	61 days	272 days	mm	in.	mm	in.
18	2	...	13	...	0.64(b)	0.025(b)	...	...
26	1	...	0	...	0	0	...	...
26	1	...	3	...	0.43	0.017	...	...
28	2	...	0	0.8	0	0	0.14	0.006
28	2	4	0	7.5	0	0	0.2	0.008
25	3.5	...	0	0	0	0	0(c)	0(c)
25	3.5	2	...	0	...	...	0	0
25	3.5	4	0	0	...	...	0	0
29	4	...	0	0	0	0	0	0
29	4	2	1.4	0	<0.02	<0.0008	0	0

(a) At ambient temperature. (b) Perforated. (c) 186 days.

Table 14. Galvanic corrosion of medium-carbon steel in contact with type 304 stainless steel, titanium or copper in seawater

	Corrosion rate of steel		Increase caused by galvanic effect	
Couple(a)	μm/yr	mils/yr	μm/yr	mils/yr
Steel (uncoupled) ...	790	31	...	...
Steel-type 304	910	36	130	5
Steel-titanium	1170	42	280	11
Steel-copper	2540	100	1750	69

(a) Area of steel equaled area of other metal in couple. Seawater flow rate, 2.4 m/s (7.8 ft/s); temperature, 10 °C (50 °F).

extending the use of stainless steels in such highly corrosive applications.

To maintain good corrosion resistance, stainless steels must be kept in contact with constantly flowing seawater and must be free from deposits and biofouling. The latter factors interfere with the maintenance of a passive film. When stainless steels fail in condensers, it is from pitting attack rather than the corrosion-erosion common to many condenser alloys. Weldments in unstabilized grades are subject to localized attack in seawater because welding upsets the condition of heat treatment necessary for good corrosion resistance.

Test results on the crevice-corrosion behavior of some of the newly developed ferritic stainless steels in low-velocity seawater are given in Table 13. In general, both ferritic and austenitic alloys show improved resistance to pitting and crevice corrosion with increased amounts of chromium and/or molybdenum. AL-6X (20Cr-24Ni-6Mo) has been used for seawater condenser tubes with very few problems during ten years of service.

Galvanic couples with stainless steels may result in severe corrosion of the other metal when the couple is immersed in seawater. Although this is a good general rule, stainless steels generally cause less galvanic corrosion of less-noble metals than do other noble metals (Table 14).

Pollution. Impurities in water can result in conditions that definitely require stainless steel. Stainless steels may be especially suitable if con-

ditions are oxidizing. Other types of pollution may make use of stainless steels inadvisable.

Cavitation Erosion. Stainless steel has excellent resistance to cavitation erosion in seawater. Austenitic grades are recommended for uses such as pump impellers and ship propellers.

Because of good resistance to cavitation erosion, two proprietary stainless steels (type 384 and Carpenter Cb-3) have been found especially satisfactory for use in pumps that handle seawater.

CHEMICAL ENVIRONMENTS

The performance of stainless steels in the chemical process industries can be affected adversely by (a) general corrosion, (b) intergranular corrosion, (c) stress-corrosion cracking, (d) pitting, (e) crevice corrosion and (f) galvanic corrosion. Often, minor variations in environment markedly affect service life. Design and fabrication practices also exert an influence.

Acetic Acid. Austenitic stainless steels are fully resistant to attack by all concentrations of acetic acid at ambient temperatures. At higher temperatures, differences in resistance among austenitic types become apparent (see Table 15). Types 304 and 347 act similarly in acetic acid. Both show rates of 38 μm/yr (1.5 mils/yr) or less in all concentrations of refined acid up to 99% at temperatures below 50 °C (120 °F). Both can be used in refined acid at concentrations as high as about 50% at temperatures up to the boiling point of the solution. Tests should be made to determine the suitability of these two steels at concentrations above 50% and temperatures above 50 °C (Table 15).

Types 309 and 310 may be used safely in acid concentrations up to about 50% at temperatures up to the boiling point of the solution, or in all concentrations of refined acid up to 99% at temperatures below 75 °C (170 °F). Tests should be made to determine their suitability in acid concentrations above 50% and at temperatures greater than 75 °C.

Because they contain molybdenum, types 316 (Table 15) and 317, as well as the cast stainless steel CN-7M, have very good corrosion resistance. They may be used in all concentrations of acid up to 99% and at temperatures up to the boiling point. These data apply to pure, refined acid; contaminants—even in small amounts—limit the use of 18-8 stainless steels to temperatures below 50 °C (120 °F).

Type 317 is more resistant than type 316 to corrosive attack in hot acetic acid and is used considerably in distillation equipment, in spite of being somewhat difficult to fabricate. Cast stainless steels of the CN-7M type are used to make pumps and valves for handling hot acetic acid.

Molybdenum-containing ferritic stainless steels also show good corrosion resistance to acetic acid solutions. Alloys 18Cr-2Mo and 26Cr-1Mo have corrosion resistance at least as good as that of type 316.

As the concentration approaches that of glacial acetic acid at boiling temperatures and at superheated vapor temperatures, none of the stainless steels are sufficiently resistant. For these concentrations and conditions, nickel-base alloys must be used.

Types 304 and 347 are used for a wide variety of acetic acid equipment, including stills, base heaters, holding tanks, heat exchangers, pipelines, valves and pumps, under conditions ranging from dilute solutions (below 50%) at room temperature to concentrated solutions (up to 99%) at 135 °C (275 °F). Type 316, heat treated after fabrication, is suitable for fractionating equipment for acetic acid in concentrations of 30 to 99%; type 304 is not suitable. Halogen ions must be eliminated from the solutions because they have an adverse effect on corrosion rates. Type 316 is used for storage vessels, pumps and process piping to handle glacial acetic acid; type 304 is not satisfactory for these applications because it discolors the acid.

For temperatures above 50 °C (120 °F) and applications involving acetic acid that is mixed or contaminated with other substances, types 316 and 317 are generally satisfactory in equipment such as stills, heat exchangers and holding tanks. The superiority of the molybdenum-bearing types is illustrated by the results of plant tests in boiling 99% acetic acid.

Castings of CN-7M are used extensively as pumps, valves and related equipment for all concentrations of acetic acid at temperatures up to the boiling point. These alloys have good resistance to mixtures of acetic acid with small amounts of sulfuric or formic acid.

The presence of reducing agents, either as impurities or as necessary constituents in the process, destroys the passivity of stainless steels. Mixtures of acetic acid with other acids—especially sulfuric, hydrochloric and formic—may be more corrosive than acetic acid itself, particularly at high temperatures.

Slight changes in solution concentrations can have significant effects on corrosion rates and suitable corrosion tests must be made whenever any change in operating conditions is contemplated.

Stainless steels are not always satisfactory for contact with hot solutions of acetic acid in concentrations greater than about 25% and containing 2% or more reducing agent (such as formic acid). If oxidizing agents such as sodium dichromate may be added to the acetic acid, the useful lives of stainless steel components can be appreciably extended.

Table 15. Corrosion of three austenitic stainless steels in acetic acid solutions

			Corrosion rate					
Acetic	Temperature		Type 304		Type 316		Type 347	
acid, %	°C	°F	μm/yr	mils/yr	μm/yr	mils/yr	μm/yr	mils/yr
10	Boiling	Boiling	<38	<1.5	...	...	...	...
25	104	219	<130	<5.0	<130	<5.0	<130	<5.0
50	Boiling	Boiling	12.7	0.5	2.5	0.1	...	...
70	Boiling	Boiling	840	33.0	...	...	...	...
75	30	86	2.5	0.1	...	...	...	...
	145	293	400	15.8	104	4.1	...	...
99.5	140	284	...	...	5	0.2	287	11.3
	Boiling	Boiling	686	27.0	23	0.9	175	6.9

In some boiling liquors where severely corrosive conditions exist (for example, where acetic acid is contaminated with various chlorides), nickel-base alloys such as Hastelloy C have proved more resistant than stainless steels, and their use is justified economically by longer equipment life.

In equipment constructed of both wrought 18-8 stainless steels and stainless steel castings, CF-8M castings are generally used instead of CF-8 castings because CF-8M has greater resistance to pitting attack. To obtain service comparable to that of molybdenum-bearing wrought 18-8 stainless steels, the chromium content of cast alloys should be on the high side of the composition range. Castings should be fully annealed for best service.

Ammonia. Stainless steels show good resistance to all concentrations of ammonia up to the boiling point and find numerous applications in production of ammonia. For instance, in desulfurizers, types 304, 316 and 20Cb-3 are used for wire mesh screens and support grating. These are critical parts that must resist any corrosion.

Tubes for primary reformer furnaces usually are HK-40 castings (comparable to wrought type 310). For shrouds in secondary reformers, for crossover lines, for piping and for other equipment operating below 900 °C (1650 °F), types 304 and 321 have been used. In shift converters, types 304 and 410 are used for wire mesh screens and grating, while type 304 is used for piping, including tees and ells. Wire mesh screens and gratings in methanators are mostly of type 304.

Because of the possibility of stress-corrosion cracking, especially from chlorinated cooling water, type 430 tubes are used in water-cooled exchangers in the synthesis step. Pitting may be encountered if the chloride content of the cooling water is too high. Under such conditions, the more pit-resistant ferritics such as 18Cr-2Mo, 26Cr-1Mo and 29Cr-4Mo can be used. Catalyst cartridges and integral heat exchangers in synthesis converters are constructed of type 304.

Chlorinated Solvents. The halogen derivatives of methane, ethane, ethylene, propane and benzene are widely used in dry cleaning, metal cleaning, vapor degreasing and solvent extraction processes, and as chemical intermediates.

Stainless steels are not corroded by chlorinated solvents when water is absent; but when a water phase is present, the compounds hydrolyze to form hydrochloric acid and sometimes organic acids. Although the presence of metallic materials usually increases the rate of decomposition, contact with stainless steels does not, and their use in equipment for handling chlorinated solvents is generally satisfactory.

Consideration should be given to the use of type 316, type 317 or Carpenter 20Cb-3 for applications where pitting is encountered. Intergranular attack sometimes occurs at welded joints; tests should be made to evaluate this possibility. Stress-corrosion cracking also may be encountered in equipment handling chlorinated solvents.

Chromic Acid. Although chromic acid is a highly oxidizing acid, it will corrode stainless steels. Stainless steels can be used in low concentrations and/or at low temperatures.

Citric acid is slightly less corrosive than acetic acid; however, it also is a nonoxidizing acid. All stainless steels will resist attack by citric acid at low concentrations or temperatures. On the other hand, at elevated temperatures, at high concentrations or in the presence of contaminants such as chlorides, the more highly alloyed stainless steels are preferred. Chlorides can cause pitting,

Table 16. Corrosion of stainless steels in boiling 45% formic acid

Type	Corrosion rate	
	µm/yr	mils/yr
304	1220	48
316	280	11
26Cr-1Mo	76	3
29Cr-4Mo	50	2

crevice corrosion and possibly stress corrosion cracking of austenitic stainless steels.

Esters. Pure esters, such as methyl acetate, ethyl acetate, propyl acetate and vinyl acetate, are not corrosive toward stainless steels. However, in the esterification step during manufacture of esters, catalysts such as sulfuric acid, which can cause extensive corrosion, are used. Beyond this step, the process streams become less corrosive and stainless steels may be used.

Fatty Acids. Fatty acids of lower molecular weight, such as acetic and formic acids, require use of 18-8 stainless steels. The following discussion is concerned with the acids of higher molecular weight, such as lauric, myristic, palmitic and stearic acids, which are less corrosive. At temperatures up to 65 °C (150 °F), cheaper metals such as carbon steel and aluminum are moderately corroded, but if color and absence of contamination of the acid are important, 18-8 stainless steels should be used.

At temperatures below 175 °C (350 °F), all standard 18-8 types are satisfactory; above 175 °C, type 316 is needed to avoid pitting and general corrosion.

Corrosion in fatty-acid vapors is no greater than in liquid, except at high vapor velocities. Under these conditions, erosion-corrosion rates are lower in type 316 than in types 304, 321 and 347.

Pitting and loss of surface metal are caused by high-temperature plant processes. There are no reports of straight fatty acids having caused intergranular failures in 18-8 stainless steels.

Cast alloys, including type CN-7M, have been satisfactory. Wrought molybdenum-bearing stainless steels and the newer precipitation-hardening stainless steels have been used for pump and valve parts to prevent galling or to provide increased hardness. High-nickel cast iron has given satisfactory service in fatty acids at 260 °C (500 °F).

Fatty acids mixed with chlorides cause failure by stress-corrosion cracking. Acidulation of fatty acids by sulfuric acid produces wide variations in the corrosion rates of stainless steels. Factors that contribute to such variations include (a) unknown dilution of the concentrated sulfuric acid, (b) moisture inherent in the fatty acids, (c) temperature and (d) methods of agitation.

Fertilizers. Stainless steels are used extensively for handling fertilizers. For dry fertilizers, type 409 has found wide acceptance. Type 304 is preferred for liquid types. The major corrosion problems are encountered with potash. Depending on the source, potash sometimes contains substantial amounts of potassium chloride, which can cause extensive pitting of stainless steels.

Formic Acid. The behavior of stainless steels in contact with formic acid is similar to its reaction to acetic acid. In many instances, a slightly higher corrosion rate can be expected. Impurities such as formaldehyde in formic acid can cause pitting of stainless steels.

Any of the austenitic stainless steels can safely handle solutions of formic acid at room temperature. High-chromium ferritic stainless steels

containing molybdenum show some promise in hot formic acid solutions, as the laboratory test data in Table 16 indicate.

Hydrochloric Acid. Although types 316, 317 and 329; Carpenter 20Cb-3; and the cast alloys CN-7M and CF-8M find some use in very dilute aerated hydrochloric acid environments, stainless steels are not usually recommended for this service.

Solutions containing chloride salts at pH values below 7.0 are essentially hydrochloric acid environments. Pitting and stress-corrosion cracking are encountered at acid concentrations less than 1%, depending on temperature, aeration and agitation.

Bimetallic couples between stainless steels and other alloys should be avoided, because corrosion may be accelerated at the junction. In such couples, the stainless steel may become the anode in dilute hydrochloric acid, resulting in loss of passivity and rapid corrosion.

Hydrochloric acid at pH values of 2.0 to 4.0 and temperatures of 50 to 80 °C (120 to 180 °F) has caused pitting and subsequent failure of stainless steel heat-exchanger tubing and heating coils. Calcareous scale in hydrochloric acid vessels has induced pitting failures. Activated carbon that settled out of HCl solutions has caused pitting of heating coils and tank bottoms made of type 316 stainless steel.

Hydrofluoric Acid. At room temperature, stainless steels are extensively corroded by hydrofluoric acid except at very low and very high concentrations.

Lactic Acid. The use of types 304, 316 and 317; Carpenter 20Cb-3; and the cast alloys CN-7M and CF-8M is limited in lactic acid solutions. Molybdenum-containing varieties generally have greater corrosion resistance than type 304.

Purity, concentration, temperature, aeration and agitation are environmental factors that influence selection of a particular type of stainless steel for use in process equipment. The presence of chlorides or sulfates in lactic acid solutions increases the severity of corrosion. Impure solutions from which lactic acid is ultimately separated and concentrated are usually more corrosive than purified solutions. Stainless steels are not suitable for use with lactic acid above 95 °C (200 °F).

Heating coils or heat exchangers for lactic acid should be designed for use with hot water or low-pressure steam. Decomposition of lactic acid and subsequent formation of a carbonaceous deposit on heating coils can result in pitting and perforation under the deposit.

Monoethanolamine. Stainless steels have excellent resistance to corrosion by monoethanolamine, and by monoethanolamine saturated with carbon dioxide plus oxygen, at temperatures up to 95 °C (200 °F). Stainless steel is used in preference to carbon steel in process steps where carbon dioxide is stripped from monoethanolamine — for example, in reboilers, heat exchangers and parts of fractionating columns. For heat exchangers, a common practice is to specify stainless steel only for tube bundles rated for 1.0-MPa (150-psi) steam inside and monoethanolamine that is rich in carbon dioxide outside. Type 304 is adequate.

Nitric Acid. Stainless steels, first used commercially on a large scale in service involving nitric acid, continue to be used in such installations. These first applications involved an alloy containing 15 to 18% chromium (now type 430) and, soon thereafter, 18Cr-8Ni steel (now type 304). The necessity for full annealing to prevent accelerated corrosion and intergranular attack in nitric acid was soon demonstrated by service failures of improperly heat treated and as-welded equipment. These difficulties were eliminated by postfabrication heat treatments involving slow cooling from about 790 °C (1450 °F) for type 430 and rapid cooling from about 1095 °C (2000 °F) for type 304. Subsequently, for the austenitic grades, the use of stabilizing elements (particularly niobium in type 347) and, more recently, reduction of carbon content to 0.03% max (type 304L) have been effective in controlling this problem without the necessity for quenching fabricated equipment from a high-temperature heat treatment. In the as-welded condition, types 304L and 347 show satisfactory resistance to corrosion by nitric acid and are therefore suitable for field-erected equipment.

Corrosion by nitric acid in storage is slight for concentrations up to about 94%, but the acid condensate is of higher concentration, and attack becomes appreciable on parts of the tank exposed to the condensate. In hot concentrated solutions where attack is too severe to be tolerated, high-silicon iron can be used if its mechanical properties are suitable.

In reactions under pressure and at temperatures considerably above the atmospheric boiling point, corrosion rates of all stainless steels increase rapidly with both temperature and concentration, and only very dilute nitric acid solutions can be handled suitably in equipment made of stainless steel.

Type 304 in the annealed-and-water-quenched condition has essentially the same resistance to corrosion by nitric acid as types 304L and 347, but type 304 should be heat treated after fabrication to prevent intergranular corrosion. Types 316 and 316L in the annealed condition have similar resistance to nitric acid, but unless they are required for reasons other than resistance to corrosion, their use usually is not justified because of higher cost.

Type 309S-Cb is somewhat more resistant under the most severe conditions and is occasionally used where lower-alloy stainless steels are not satisfactory. If properly annealed and water quenched, types 309 and 310 have about the same resistance as type 309S-Cb. However, unless their carbon content is less than about 0.10%, these alloys cannot be cooled fast enough in commercial heat treatments to avoid susceptibility to intergranular corrosion.

Type 430 is still widely used for various kinds of equipment in the ammonia oxidation process for manufacture of nitric acid, and for tank cars, storage tanks, forged valves and other components. Type 430 costs less than the austenitic grades and, although temperature ranges are somewhat more limited at various nitric acid concentrations, this alloy is adequate for many applications. Its principal limitation is that it requires heat treatment after fabrication and is therefore not suitable for equipment erected in the field or for equipment repair.

The resistance of chromium steels to nitric acid is related directly to chromium content.

Type 446 is comparable to type 304L in resistance to nitric acid, but because it is more difficult to fabricate it is employed in only a few special applications. With the exception of type 430, none of the other chromium steels is used to any appreciable extent in contact with nitric acid.

Stainless steels are relatively insensitive to factors such as aeration, velocity and agitation, because nitric acid is oxidizing and tends to favor passivity. Neither pitting nor stress-corrosion cracking occurs under these circumstances. However, nitric acid causes intergranular attack in unstabilized stainless steels that contain more than 0.03% C, unless they have been properly heat treated. The presence of hydrofluoric acid in nitric acid, as in certain pickling solutions, increases such attack. Hydrofluoric acid also increases the rate of general corrosion, as do appreciable amounts of other halides.

In hot dilute mixtures of nitric acid and sulfuric acids, no appreciable attack of stainless steels occurs when the ratio of nitric acid to sulfuric acid is about 2 to 1, or higher. This is one of several examples where sufficient nitric acid will prevent attack that would otherwise occur. With very dilute hot mixtures of sulfuric acid and nitric acid (about 1 to 1.5% total acid), where the proportion of nitric acid will not maintain passivity for the austenitic grades, type 443 (20Cr-1Cu) has greater resistance. Cast grades such as CN-7M containing 3 to 4% copper, CF-8 and CF-8M are widely used for valves, pumps and other castings in nitric acid service. Addition of stabilizing elements (usually niobium) or restriction of carbon content to 0.03% is not ordinarily justified, because most stainless castings can be quenched readily in water, and they are seldom welded in place in the field. Types CF-8M and CN-7M, containing molybdenum or molybdenum and copper, are no more resistant to nitric acid than the CF-8 (18-8) grade but will handle a wider variety of process solutions. High molybdenum (and silicon) contents are somewhat detrimental to resistance to hot nitric acid in intermediate and high concentrations. Straight chromium cast stainless steels are seldom used in nitric acid service.

Phosphoric Acid. Resistance of stainless steels to corrosion by phosphoric acid depends on acid concentration, temperature, contamination and alloy composition. The higher-molybdenum special stainless steels, such as JS700 and alloy 904L, and the high-chromium-molybdenum ferritics appear to extend the usefulness of stainless steels into higher phosphoric acid concentrations.

Sodium Hydroxide. All stainless steels resist corrosion by all concentrations of sodium hydroxide at temperatures up to about 65 °C (150 °F). At higher temperatures, various stainless steels undergo varying degrees of corrosion.

In recent years, the high-chromium-molybdenum ferritics, particularly E-Brite 26-1, have found extensive use in production of caustics. E-Brite 26-1 tubing is used extensively for the first-, second- and third-effect evaporators handling 16% NaOH – 14% NaCl, 26% NaOH – 9% NaCl and 45% NaOH – 5% NaCl at temperatures as high as 150 °C (300 °F) with no reported problems.

Sulfation and Sulfonation Products. Austenitic stainless steels and carbon steels have low corrosion rates in oleum (fuming sulfuric acid) and sulfuric acid in concentrations higher than 80% at room temperature. At 100 to 103% there is a distinct rise in the corrosion rate for carbon steel. Above 103%, both stainless and carbon steels have satisfactory corrosion rates. Steels of the 300 series are satisfactory for sulfonation practice at room temperature with 78% sulfuric acid mixed with sulfonation products. At 60 °C (140 °F), corrosion of 300-series steels is excessive. Corrosion rates at these temperatures are reported in Table 17.

Sulfuric Acid. The 18-8 varieties of stainless steel

Table 17. Corrosion by 78% sulfuric acid mixed with sulfonation products

Alloy	Corrosion rate	
	μm/yr	mils/yr
At 27 °C (80 °F)		
Type 316	5.1	0.2
Carpenter 20Cb-3	None	None
Inconel	None	None
Mild steel	510	20
At 60 °C (140 °F)		
Type 316	510	20
Hastelloy B	15	0.6
Hastelloy C	38	1.5
Carpenter 20Cb-3	76	3.0
Inconel	205	8.0
Mild steel	3400	134

are resistant to corrosive attack by sulfuric acid within rather narrow ranges of concentration and temperature. Although stainless steels may be used safely in contact with 80 to 100% sulfuric acid at ambient temperature (carbon steel is ordinarily used in this range), they are attacked at slightly higher temperatures. Sulfuric acid in concentrations of 1 to 5% at ambient temperature should not be stored in vessels of molybdenum-free stainless steels. Type 316 may be used for this purpose; 317, with a higher molybdenum content, may be used safely in this range of acid concentration at temperatures as high as 65 °C (150 °F).

Alloys such as Carpenter 20Cb-3 and CN-7M resist all concentrations of sulfuric acid at temperatures up to 60 °C (140 °F), and concentrations as high as 10% up to the boiling point, but do not resist all concentrations over a wide range of temperatures.

The preceding data pertain to pure sulfuric acid. Addition of oxidizing agents (such as nitric acid, air and copper salts) will widen the range of applicability for all stainless steels; reducing agents (such as hydrogen) will narrow the range of usefulness. If other than pure sulfuric acid is used with stainless steel, corrosion tests must be made under service conditions in order to evaluate the usefulness of specific alloys. Applications should be restricted to only those concentrations of sulfuric acid and temperatures that have proved satisfactory in corrosion tests. Tests should include annealed, stressed and crevice-type specimens, as well as specimens sensitized by heating for 1 h or longer at 650 °C (1200 °F).

Agitation and aeration in stainless steel equipment and the velocities of sulfuric acid solutions in piping should be adequate to keep all solids suspended; 1.5 to 4.5 m/s (5 to 15 ft/s) is usually sufficient. Deposits such as charred organic matter or calcium sulfate scale may induce pitting and perforation. Surfaces should be kept clean during shutdown periods.

Oxidizing agents such as nitric acid, chromic acid and sodium dichromate have dramatic effects on corrosion of stainless steels in sulfuric acid. Substantial reductions in corrosion rates for types 304 and 316 occur when such agents are added even in relatively small amounts. Aeration of sulfuric acid solutions has a similar inhibiting effect. Advantage can be taken of these inhibiting effects when designing with stainless steels, as long as substantial safeguards are taken to maintain the proper inhibiting effect.

Sulfurous Acid and Sulfur Dioxide. Carpenter 20Cb-3 and types 316, 317, CF-8M and CN-7M have been used in equipment for sulfur dioxide (wet)

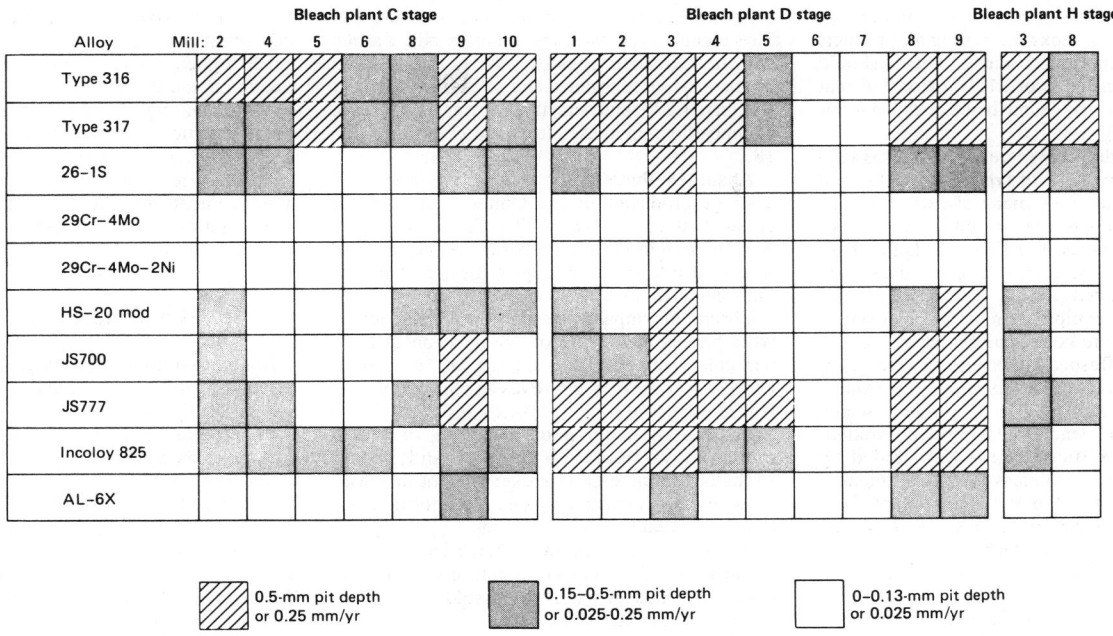

Charts represent selected results from corrosion tests in comparable stages of the bleach plants in ten different paper mills.

Fig. 3. Performance of ten stainless alloys in paper-mill bleach plant service

and sulfurous acid environments. The molybdenum in these alloys is responsible for their good resistance to the reducing environments of sulfurous acid. Wrought type 316 and cast CF-8M are the most widely used.

Complete suspension of any solids present is necessary to avoid crevice-type pitting. Crevice pockets, lapped joints, 90° corner intersections and similar obstructions should be avoided, and the surfaces should be clean and smooth.

Cold and hot working should be limited to minor forming operations that keep the hardness of the steel below 96 HRB. Stress-corrosion cracking can occur in steel exposed to sulfurous acid solutions containing 100 ppm or more of metal chlorides.

To ensure against failure of weld zones in severely corrosive environments (more than 1.0% sulfur dioxide), extra-low-carbon alloys should be used for equipment that requires significant amounts of welding.

Crevice-type pitting has caused failure of tank bottoms and perforation of heating coils. Solids such as filtering aids, activated carbon and bentonite, which settle on the bottoms of tanks and the tops of heating coils, cause pitting and perforation of these surfaces.

Severe erosion-corrosion of stainless steel pump impellers and valve bodies has been caused by sulfurous acid slurries that contain suspended solids. Pump impellers may fail from erosion-corrosion in a few weeks. Alloys such as 316, CF-8M and CN-7M generally give the longest service in erosion-corrosion environments. The useful service life of stainless steels can be predicted from data obtained in laboratory corrosion testing of circular specimens centered on a stainless steel shaft revolving at high speed (10 000 rpm) in the process slurry.

PHARMACEUTICALS AND FINE CHEMICALS

Stainless steels, principally the 18-8 types, are used by the fine chemical and pharmaceutical in-dustries for maintaining sanitary conditions in corrosive and noncorrosive environments, as well as for their basic resistance to corrosion. Sanitation requirements dictate selection of stainless steels for most installations.

Purity, color and stability of pharmaceutical products may be greatly affected by the presence of trace quantities of metallic ions. For example, traces of iron affect vitamin B_6 by forming a highly colored complex; therefore, stainless steel is unsuitable for handling vitamin B_6 hydrochloride, even though corrosion rates may be very low. Stainless steel has proved satisfactory for processing of vitamin C; however, all traces of copper must be eliminated, because copper in aqueous solutions accelerates the rate of decomposition of vitamin C by a factor of about 3000. Generally, Carpenter 20Cb-3 can be used with solutions acidified with hydrochloric acid even if the solution is more acidic than pH 5.0.

Historically, types 304 and 316 have been used for washers, evaporators, green- and white-liquor equipment, head boxes, save-alls, pipe and tubing. Types 316 and 317 have been used in pulp washers and rotary-drum filter equipment. Closed systems require materials more resistant to pitting and crevice corrosion. Recent corrosion tests conducted by the Metals Task Force of the TAPPI Corrosion and Materials Engineering Committee in bleach plant stages C, D and H of 10 different paper mills indicate the need for such alloys (see Fig. 3).

Of course, use of stainless steels in the pulp and paper industry is not confined to solving corrosion problems. Stainless steels often are selected because they resist scaling and sliming, prevent contamination and have good mechanical and physical properties.

FOODS

Stainless steels are used in the food industry for their corrosion resistance wherever metal comes in contact with food products during pro-cessing, and because they can be quickly and easily cleaned and made sanitary. Applications include pumps, tubing, tanks, kettles, filling machines, heat exchangers and vacuum tanks. Process temperatures may be as high as 150 °C (300 °F), and processing may be done in vessels operating under pressure or vacuum.

Stainless steels also are used where food-processing equipment is subjected to water spray, continual flow of water, or severe clean-up procedures; equipment includes bottle washers, continuous coolers, continuous cookers and conveyors. In such applications, the product does not contact the metal, but use of carbon steel may (a) cause rust to appear on the surface of the product container or (b) result in severe corrosion of equipment and lead to costly replacement or repair.

In the primary stages of processing, certain concentrated ingredient solutions that contain salt and vinegar, such as pickle liquors and sauces, are highly corrosive. Food products such as vegetables contain little acid, and their effects on stainless steels are negligible. Handling of sauces or pickle liquors has caused much difficulty due to pitting of tubing (type 304) at the level of the liquid. Type 316 is more resistant, but tubing of this alloy can become pitted in 4 to 12 months. Proprietary alloys such as JS700, AL-6X and 29-4, which are more resistant to pitting and crevice corrosion, should be considered for applications where type 316 becomes pitted.

Pumps of type CF-8 stainless steel are used for less-corrosive solutions, such as vegetables; for sauces, pickle liquors and items containing vinegar, type CF-8M is necessary. Because pumps and impellers may need to be interchanged between different processes, type CF-8M is preferred for such units. Type 304 is used for exhaust hoods over open kettles and tanks because of its ability to withstand corrosive fumes and vapors from sauces and liquors containing vinegar. Impellers for handling fumes from these tanks and kettles are made of type 316. Type 304 is also used for agitators in corrosive solutions and is recommended for tanks for storing all products except pickle liquors and sauces, which should be stored in tanks made of 316L, or of 316 that has been heat treated, sand blasted and passivated.

Drums for holding dry ingredients such as vegetables usually are made of type 304. Tanks used as filler bowls should be made of extra-low-carbon stainless steels to prevent pitting attack adjacent to welded areas. For filler bowls that handle food products less corrosive than sauces, type 304 is recommended.

Food-handling equipment must be designed so that the surfaces are kept clean by the product as it is continually flushed across the surface; there must be no crevices or corners where the product can become lodged. Welds in corners, which cannot be ground smooth, should be avoided. Many corrosion difficulties can be avoided by frequent cleaning of stainless steel equipment.

In equipment for the more-corrosive food products, extra-low-carbon stainless steels should be used wherever possible. Generally, it costs less to use these alloys than to use stainless steels that require annealing to combat sensitization.

For food-processing equipment, experience has shown a 2B finish to be superior to a No. 4 finish. The bright, cold rolled 2B finish has fewer grooves where tiny food particles can cling.

Corrosion-Resistant Steel Castings

CAST CORROSION-RESISTANT STEELS resist attack by aqueous solutions at or near room temperature, and by hot gases and high-boiling-point liquids at temperatures up to 650 °C (1200 °F).

All cast corrosion-resistant steels contain more than 11% chromium, and most contain from 1 to 30% nickel (a few have less than 1% Ni). Carbon content, especially important in its influence on both corrosion resistance and strength, is usually less than 0.20% and sometimes lower than 0.03%. Compositions of these steels are given in Table 18.

About two-thirds of the corrosion-resistant steel castings produced in the United States are of grades that contain 18 to 22% Cr and 8 to 12% Ni; the straight chromium compositions are also produced in considerable quantities, particularly the steel with 11.5 to 14.0% Cr.

Chromium imparts passivity to ferrous alloys when present in amounts of more than about 11%, particularly if conditions are strongly oxidizing. Corrosion resistance improves as chromium content is increased.

In general, addition of nickel to iron-chromium alloys improves ductility and impact strength. An increase in nickel content increases resistance to corrosion by neutral chloride solutions and weakly oxidizing acids.

Addition of molybdenum increases resistance to pitting attack by chloride solutions. It also extends the range of passivity in solutions of low oxidizing characteristics.

Addition of copper to duplex-structure Ni-Cr alloys produces alloys that can be precipitation hardened to higher strength and hardness. Addition of copper to single-phase austenitic alloys greatly improves their resistance to corrosion by sulfuric acid. In all Fe-Cr-Ni stainless alloys, resistance to corrosion by environments that cause intergranular attack can be improved by lowering the carbon content.

The type CA-15 alloy (11.5 to 14% Cr) can be hardened by quenching and tempering for wear applications. It is widely used for trim in carbon steel valves, for pumps handling acid mine water containing abrasives, and in paper-mill applications where wear resistance is as important as corrosion resistance. CA-6NM alloy has corrosion resistance similar to that of CA-15, plus a much higher impact strength and resistance to cavitation and the ability to be welded with little or no preheating. Type CB-30 alloy (18 to 22% Cr) is ferritic at all temperatures, and thus, as normally made, cannot be hardened by heat treatment. However, because of its higher chromium content, it resists acids in a wider range of concentrations over a wider range of temperatures than do the 12% Cr alloys.

IRON-CHROMIUM-NICKEL ALLOYS

Iron-chromium-nickel alloys have found wide acceptance and constitute about 60% of total production of high-alloy castings. They generally are both austenitic and slightly magnetic. However, they may vary from nonmagnetic to rather strongly magnetic depending on the composition of the heat or on effects of metal thickness (solidification rate) and heat treatment. The most popular alloys of this type are CF-8 and CF-8M. These alloys are nominally 18-8 stainless steels, and are the cast counterparts of wrought types 304 and 316, respectively. Their carbon contents are maintained at 0.08% max.

Alloys CF-3M and CF-8M are modifications of CF-3 and CF-8 containing 2 to 3% molybdenum to enhance general corrosion resistance. Their passivity under weakly oxidizing conditions is more stable than that of CF-3 and CF-8. They have good resistance to such corrosive media as sulfurous and acetic acids, and they are more resistant to pitting by mild chlorides. These alloys are suitable for use in flowing seawater, but will pit under stagnant conditions.

Alloy CG-8M, which is a modification of CF-8M that contains 3 to 4% Mo, has increased resistance to sulfurous and sulfuric acid solutions and to halogen compounds. This alloy is not suit-

Table 18. Standard designations and compositions of corrosion-resistant steel castings

ACI type(a)	Wrought alloy type(b)	C (max)	Mn (max)	Si (max)	Cr	Ni	Others(c)
CA-6NM	... .	0.06	1.00	1.00	11.5 to 14.0	3.5 to 4.5	0.40 to 1.0 Mo
CA-15	410	0.15	1.00	1.50	11.5 to 14.0	1.0 max	0.5 max Mo(d)
CA-40	420	0.40	1.00	1.50	11.5 to 14.0	1.0 max	0.5 max Mo(d)
CB-7Cu-1	... .	0.07	0.70	1.00	15.5 to 17.7	3.6 to 4.6	2.5 to 3.2 Cu; 0.20 to 0.35 Nb; 0.05 max N
CB-7Cu-2	... .	0.07	0.70	1.00	14.0 to 15.5	4.5 to 5.5	2.5 to 3.2 Cu; 0.20 to 0.35 Nb; 0.05 max N
CB-30	431	0.30	1.00	1.50	18.0 to 22.0	2.0 max	...
CC-50	446	0.50	1.00	1.50	26.0 to 30.0	4.0 max	...
CD-4MCu	... .	0.04	1.00	1.00	25.0 to 26.5	4.75 to 6.0	1.75 to 2.25 Mo; 2.75 to 3.25 Cu
CE-30	312	0.30	1.50	2.00	26.0 to 30.0	8.0 to 11.0	...
CF-3(e)	304L	0.03	1.50	2.00	17.0 to 21.0	8.0 to 12.0	...
CF-3M(e)	316L	0.03	1.50	2.00	17.0 to 21.0	8.0 to 12.0	2.0 to 3.0 Mo
CF-8(e)	304	0.08	1.50	2.00	18.0 to 21.0	8.0 to 11.0	...
CF-8C	347	0.08	1.50	2.00	18.0 to 21.0	9.0 to 12.0	Nb(f)
CF-8M	316	0.08	1.50	2.00	18.0 to 21.0	9.0 to 12.0	2.0 to 3.0 Mo
CF-12M	316	0.12	1.50	2.00	18.0 to 21.0	9.0 to 12.0	2.0 to 3.0 Mo
CF-16F	303	0.16	1.50	2.00	18.0 to 21.0	9.0 to 12.0	1.50 max Mo; 0.20 to 0.35 Se
CF-20	302	0.20	1.50	2.00	18.0 to 21.0	8.0 to 11.0	...
CF-105MnN	... .	0.10	9.00(g)	4.50(g)	16.0 to 18.0	8.0 to 9.0	0.08 to 0.18 N
CG-6MMN	... .	0.06	6.00(g)	1.00	20.5 to 23.5	11.5 to 13.5	1.5 to 3.0 Mo; 0.1 to 0.3 Nb; 0.1 to 0.3 V; 0.2 to 0.4 N
CG-8M	317	0.08	1.50	1.50	18.0 to 21.0	9.0 to 13.0	3.0 to 4.0 Mo
CH-20	309	0.20	1.50	2.00	22.0 to 26.0	12.0 to 15.0	...
CK-20	310	0.20	2.00	2.00	23.0 to 27.0	19.0 to 22.0	...
CN-7M	... .	0.07	1.50	1.50	19.0 to 22.0	27.5 to 30.5	2.0 to 3.0 Mo; 3.0 to 4.0 Cu
CN-7MS	... .	0.07	1.50	3.50(g)	18.0 to 20.0	22.0 to 25.0	2.5 to 3.0 Mo; 1.5 to 2.0 Cu

(a) Most of these standard grades are covered by ASTM A743 and A744. (b) Type numbers of wrought alloys are listed only for nominal identification of corresponding wrought and cast grades. Composition ranges of cast alloys are not the same as for corresponding wrought alloys; cast alloy designations should be used for castings only. (c) Phosphorus content is 0.04% max except in CF-16F, which has 0.17% max P; sulfur content is 0.04% max in all grades. (d) Molybdenum not intentionally added. (e) CF-3A, CF-3MA and CF-8A have the same composition ranges as CF-3, CF-3M and CF-8, respectively, but have balanced compositions so that ferrite contents are at levels that permit higher mechanical-property specifications than those for related grades. They are covered by ASTM A351. (f) Nb, 8 × %C min (1.0% max); or Nb + Ta, 9 × %C (1.1% max). (g) For CF-105MnN, manganese ranges from 7.00 to 9.00 and silicon ranges from 3.50 to 4.50. For CG-6MMN, manganese ranges from 4.00 to 6.00. For CN-7MS, silicon ranges from 2.50 to 3.50.

able for use in nitric acid or other strongly oxidizing media.

PRECIPITATION-HARDENING ALLOYS

Corrosion-resistant alloys capable of being hardened by low-temperature treatment to obtain improved mechanical properties usually are duplex-structure alloys with much more chromium than nickel. Addition of copper enables these alloys to be strengthened by precipitation hardening. These alloys are significantly higher in strength than the other corrosion-resistant alloys even without hardening.

CB-7Cu-1 and CB-7Cu-2 alloys have corrosion resistance between those of CA-15 and CF-8. They are widely used for structural components requiring moderate corrosion resistance, as well as for components requiring resistance to erosion and wear.

CD-4MCu alloy is widely used in many applications where its good corrosion resistance (which often equals or even exceeds that of CF-8M) and excellent resistance to erosion make it the most desirable alloy. CD-4MCu has outstanding resistance to nitric acid and mixtures of nitric acid and organic acids as well as excellent resistance to a wide range of corrosive chemical process conditions. This alloy normally is used in the solution-annealed condition, but it can be precipitation hardened for carefully selected applications where lower corrosion resistance can be tolerated and where there is no potential for stress-corrosion cracking.

IRON-NICKEL-CHROMIUM ALLOYS

For some types of service, extensive use is made of iron-nickel-chromium alloys that contain more nickel than chromium. Most important among this group is alloy CN-7M, which has a nominal composition of 28 Ni, 20 Cr, 3.5 Cu, 2.5 Mo and 0.07 max C. In effect, this alloy is made by adding 20% Ni and 3.5% Cu to alloy CF-8M, which greatly improves resistance to hot, concentrated, weakly oxidizing solutions such as sulfuric acid and also improves resistance to severely oxidizing media. Alloys of this type can withstand all concentrations of sulfuric acid at temperatures up to 65 °C (150 °F) and many concentrations up to 79 °C (175 °F). They are widely used in nitric-hydrofluoric pickling solutions, phosphoric acid, cold dilute hydrochloric acid, hot acetic acid, strong hot caustic solutions, brines, and many complex plating solutions and rayon spin baths.

CN-7MS alloy has corrosion resistance much the same as that of CN-7M. CN-7MS has outstanding resistance to corrosion by high-strength nitric acid (over 90%).

Influence of contaminants is one of the most important considerations in selecting an alloy for a particular process application. Ferric chloride in relatively small amounts, for example, will cause concentration-cell corrosion and pitting. Buildup of corrosion products in a chloride solution may increase the iron concentration to a level high enough to be destructive. Thus, chlorine salts, wet chlorine gas and unstable chlorinated organic compounds cannot be handled by any of the iron-base alloys, which creates a need for nickel-base alloys.

MECHANICAL PROPERTIES

Corrosion-resistant steel castings are somewhat lower in modulus of elasticity in tension than carbon or low-alloy steels, varying from about 165 GPa (24×10^6 psi) for high-nickel grades to 200 GPa (29×10^6 psi) for high-chromium, low-nickel alloys. This compares with 207 GPa (30 $\times 10^6$ psi) for carbon and low-alloy steels.

Heat Treatment. Stainless steel castings are almost always heat treated. For the hardenable ferritic-martensitic straight chromium compositions, heat treating is done primarily to obtain desired mechanical properties. Castings such as CA-15, a 12% Cr alloy, are air cooled or oil quenched from about 980 °C (1800 °F) and tempered between 540 and 760 °C (1000 and 1400 °F), depending on specified properties. Tempering between 370 and 540 °C (700 and 1000 °F) causes a marked loss in impact resistance and should be avoided.

FERRITE IN AUSTENITIC GRADES

Ferrite content is controlled by proper balance of composition. With other elements unchanged, ferrite content decreases as carbon content increases.

Chromium, molybdenum and silicon promote formation of ferrite (magnetic); carbon, nickel, nitrogen and manganese favor formation of austenite (nonmagnetic). For example, a cast extra-low-carbon grade (0.03 max C) cannot be completely nonmagnetic unless it contains 12 to 15% Ni. The wrought grades of these alloys normally contain about 13% Ni. They are made fully austenitic to improve rolling and forging characteristics.

Cast austenitic alloys usually have from 5 to 20% ferrite distributed in discontinuous pools throughout the matrix. In ordinary service, where these steels may be heated in the range from 425 to 650 °C (800 to 1200 °F), carbide precipitation occurs at the edges of the ferrite pools in preference to the austenite grain boundaries. When the steel is heated above 650 °C, the ferrite pools transform to chi or sigma phase. If these pools are distributed in such a way that a continuous network is formed, embrittlement or a network of corrosion penetration may result. Also, if the amount of ferrite is too great, it may form continuous stringers where corrosion can take place, producing a condition similar to grain-boundary attack.

Some solutions attack the austenite phase in heat treated alloys, whereas others attack the ferrite. For instance, calcium chloride solutions attack the austenite. On the other hand, a 10° Baumé cornstarch solution, acidified to a pH of 1.8 with sulfuric acid and heated to a temperature of 135 °C (275 °F), attacks the ferrite.

CORROSION

In alloys of the CF type, the effects of composition on rates of general corrosion attack have been studied, and certain definite relationships have been established. Through use of the Huey test it has been shown that, in this standardized environment, carbide-free quench-annealed alloys of various nickel, chromium, silicon, carbon and manganese contents have corrosion rates directly related to these contents.

Intergranular attack can be avoided in CF alloys (a) by addition of the stabilizing element niobium (columbium), (b) by use of an extra-low-carbon grade such as CF-3 or CF-3M, or (c) possibly by formation of small amounts of ferrite, which can be induced either by reducing some of the austenite-stabilizing elements such as nickel and carbon or by increasing such ferrite-stabilizing elements as molybdenum, silicon and chromium. Addition of niobium to molybdenum-containing type CF alloys has been found unsatisfactory for castings. When both niobium and molybdenum are present, the ferrite phase tends to form as an interconnected network and is especially likely to transform into the brittle sigma phase. As a result, castings in the as-cast condition become embrittled and have a tendency to crack.

In general, intergranular corrosion is of less concern in the straight chromium alloys, especially those containing 25% Cr or more.

GALLING

Corrosion-resistant steel castings are susceptible to galling and seizing when dry surfaces slide or chafe against each other. However, the surfaces of the castings can be nitrided so that they are hard and wear resistant. Tensile properties are not impaired. After being nitrided, the 19Cr-9Ni and straight chromium alloys are resistant to superheated steam (10 MPa at 500 °C, or 1.5 ksi at 930 °F), saturated steam, boiler feedwater and petroleum-base fuels; they are not resistant to halogen acids or salts, nor to any corrosive medium that will attack the untreated alloy. Nitriding reduces resistance to corrosion by concentrated nitric or mixed acids.

Parts such as gate disks for gate valves usually are furnished in the solution-treated condition but may be nitrided to reduce susceptibility to seizure in service. Similar results are obtained by hard facing these types of parts with Co-Cr-W alloys or by using CF-105MnN cast stainless steel (cast Nitronic 60).

MACHINING

The machinability of straight chromium alloys is as good as or better than that of annealed 19Cr-9Ni alloys. The machining characteristics of all iron-chromium-nickel alloys (chromium in excess of nickel) are about on a par with those of quench-annealed CF alloys. CE alloys, and alloys that contain niobium, are somewhat easier to machine; CH is slightly less machinable than CF.

WELDING

Corrosion-resistant steel castings can be welded by shielded metal-arc welding (SMAW), gas tungsten-arc welding (GTAW), gas metal-arc welding (GMAW) and electroslag (submerged-arc) welding. Austenitic castings normally are welded without preheat, and are solution annealed after welding. Martensitic castings require preheating to avoid cracking during welding, and are given an appropriate postweld heat treatment.

When welds are properly made, tensile and yield strengths of the welded joint are similar to those of the unwelded castings. Elongation generally is lower for specimens taken perpendicular to the weld bead.

Nickel and Nickel Alloys

Original contributors: Donald L. Pasquine, International Nickel Co.; John Gadbut and Donald E. Wenschhof, Huntington Alloys, Inc.; Robert B. Herchenroeder, Cabot Corp.; C. R. Bird, Stainless Foundry and Engineering, Inc.; D. L. Graver, Huntington Alloys, Inc.; Warren M. Spear, International Nickel Co., Inc.

Condensed from Metals Handbook, Ninth Edition, Volume 3, pages 125 to 182.

NICKEL is a hard, tough and malleable silver-white metal that has good resistance to oxidation and corrosion. It is used principally in association with other elements to impart strength, toughness and corrosion resistance to a wide range of ferrous and nonferrous alloys. Nickel, like iron and cobalt, is magnetic, and thus nickel alloys often are used in magnetic applications.

Tensile Properties. Typical values and ranges of tensile properties for selected wrought nickel alloys are presented in Tables 1 and 2 (Table 1 also gives hardness values). Typical tensile properties as a function of temperature are given in Table 3.

USES OF NICKEL

Of the total amount of nickel consumed, about 55% ordinarily is used as alloying additions to stainless and alloy steels, with the major portion going into austenitic chromium-nickel stainless steels. Another important market for nickel, comprising about 20% of total consumption, is in nonferrous alloys, where nickel is the major component in a large number of nickel-base, cobalt-base and copper-base alloys as well as other families of nonferrous alloys such as those containing substantial amounts of molybdenum. The major nonalloying use of nickel is in electroplating, which traditionally comprises about 10 to 15% of total demand. The balance of the nickel consumption is in the foundry industry and in a wide range of minor applications.

Stainless steels, principally the 18Cr-8Ni varieties that comprise the 300 series, are by far the largest consumers of nickel, accounting for 43% of total consumption. Nickel is used in about 70% of all stainless steel production. Nickel is added to stainless steels primarily to increase corrosion resistance, improve fabricability and improve strength at elevated temperatures. Austenitic stainless steels find widespread application in consumer products and in equipment for the process, chemical, energy and transportation industries.

Nonferrous alloys, comprising about 20% of total nickel consumption, are used primarily for resistance to corrosion and heat. Included in this group are nickel-base, copper-base, cobalt-base and molybdenum-base alloys. Nickel-base superalloys constitute one important class of nonferrous alloys and are used chiefly for jet-engine parts and other advanced propulsion systems.

High-nickel alloys also are important materials of construction for equipment used in chemical and petrochemical processing. Nonferrous alloys are used in a broad variety of applications in the transportation, power, process, chemical and electronics industries.

Electroplating, the major nonalloying use of nickel, accounts for about 12% of total nickel consumption. It is used primarily for decorative coatings in which nickel is overplated with chromium, although industrial coatings and electro-forming applications are significant. The principal objectives of nickel plating are to improve appearance, surface finish and corrosion resistance, but nickel also is used to build up worn or mismachined parts and to improve wear resistance. Nickel/chromium coatings are applied to such basis materials as steel, zinc alloy die castings, copper alloys, plastics, magnesium alloys and aluminum alloys. Nickel is electroplated on a wide variety of products such as automobile parts, consumer appliances and equipment for the food-processing and chemical industries.

Alloy steels other than austenitic stainless steels account for 10% of nickel consumption. Nickel in amounts up to 18% is an important alloying addition to steel. It normally is used in conjunction with one or more other alloying elements (such as chromium and molybdenum) to permit development of optimum hardness, strength, ductility and processing characteristics. Alloy steels are used in the automotive, construction and machinery industries for parts requiring high strength, toughness and wear resistance. Nickel improves the atmospheric corrosion resistance of steels, particularly when in combination with other elements such as copper and chromium. Nickel lowers the ductile-to-brittle transition temperature of steel, making nickel-containing steels useful for transporting and handling liquefied gases and for machinery and structures used at subzero temperatures.

Foundry products, both ferrous and nonferrous, account for 8% of nickel consumption. Nickel, in combination with chromium, is an important constituent in cast heat-resistant and corrosion-resistant stainless steels. In the cast heat-resistant alloys, its principal function is to strengthen and toughen the matrix. In cast corrosion-resistant alloys, nickel supplements the passivating effects of chromium under oxidizing conditions, and increases basic corrosion resistance under reducing conditions. Addition of molybdenum to the latter alloys reduces the tendency toward pitting in seawater and other chloride solutions. Nickel, in amounts up to 3.5%, is also used in engineering grades of steel castings, where it improves strength and ductility in much the same manner as it improves these properties in wrought products. Nickel in amounts up to about 35% is added to improve the properties of a wide range of cast irons, including gray iron, ductile iron, white iron and the various austenitic cast irons. Nickel is added to gray iron in amounts up to 3.5%, with or without other alloying elements, to improve strength, wear resistance and machinability. In several abrasion-resistant chilled and white cast irons, additions of up to 5% nickel, alone or in combination with chromium and molybdenum, improve wear resistance. Several important se-

Table 1. Typical tensile properties and hardness of five nickel alloys

Form and/or condition	Tensile strength MPa	ksi	Yield strength (0.2% offset) MPa	ksi	Elongation, %	Hardness
Nickel 270						
Rod and bar, hot finished	345	50	110	16	50	40 HRB
Strip:						
Cold rolled	655	95	621	90	4	95 HRB
Annealed	345	50	110	16	50	35 HRB
Sheet, annealed	345	50	110	16	45	30 HRB
Monel 502						
Hot rolled	586	85	255	37	47	74 HRB
Annealed	572	83	234	34	48	73 HRB
Annealed and aged	986	143	655	95	27	24 HRC
Inconel 600						
Rod:						
As rolled	672	97.5	307	44.5	46	86 HRB
Annealed	624	90.5	210	30.4	49	75 HRB
Plate:						
As rolled	682	99.0	346	50.2	42	87 HRB
Annealed	639	92.7	199	28.9	49	75 HRB
Tubing:						
As drawn	993	144.0	916	132.8	8	34 HRC
Annealed	693	100.5	279	40.5	43	83 HRB
Inconel 690						
Tube, annealed	731	106	365	53	41	97 HRB
Strip, annealed	758	110	372	54	40	88 HRB
Rod:						
As rolled	765	111	434	63	40	90 HRB
Annealed	710	103	317	46	49	90 HRB
Plate, as rolled	765	111	483	70	36	95 HRB
Hastelloy G-3						
Sheet(a)	690	100	325	47	50	79 HRB
Sheet(b)	685	99	305	44	53	83 HRB
Plate(c)	740	107	365	53	56	87 HRB
Bar(d)	695	101	295	43	59	80 HRB

(a) 0.63 to 0.97 mm (0.025 to 0.038 in.) thick. (b) 1.4 to 4.8 mm (0.056 to 0.187 in.) thick. (c) 6.4 to 19 mm (0.25 to 0.75 in.) thick. (d) 13 to 25 mm (0.5 to 1.0 in.) in diameter.

Table 2. Tensile properties(a) of four nickel alloys

Form and condition	Tensile strength MPa	ksi	Yield strength (0.2% offset) MPa	ksi	Elongation, %
Nickel 200					
Rod and bar					
Hot finished	414-586	60-85	103-310	15-45	55-35
Cold drawn	448-758	65-110	276-690	40-100	35-10
Annealed	379-517	55-75	103-207	15-30	55-40
Plate					
Hot rolled	379-690	55-100	138-552	20-80	55-35
Annealed	379-552	55-80	103-276	15-40	60-40
Sheet					
Hard	621-793	90-115	483-724	70-105	15-2
Annealed	379-517	55-75	103-207	15-30	55-40
Strip					
Spring temper	621-896	90-130	483-793	70-115	15-2
Annealed	379-517	55-75	103-207	15-30	55-40
Tubing					
Stress relieved	448-758	65-110	276-621	40-90	35-15
Annealed	379-517	55-75	83-207	12-30	60-40
Wire					
Annealed	379-586	55-85	103-345	15-50	50-30
Spring temper	862-1000	125-145	724-931	105-135	15-2
Monel 400					
Rod and bar					
Annealed	517-621	75-90	172-345	25-50	60-35
Hot finished	552-758	80-110	276-690	40-100	60-30
Cold drawn, stress relieved	579-827	84-120	379-690	55-100	40-22
Plate					
Hot rolled	517-655	75-95	276-517	40-75	45-30
Annealed	483-586	70-85	193-345	28-50	50-35
Sheet					
Annealed	483-586	70-85	172-310	25-45	50-35
Hard	690-827	100-120	621-758	90-110	15-2
Strip					
Annealed	483-586	70-85	172-310	25-45	55-35
Spring temper	690-965	100-140	621-896	90-130	15-2
Tubing, cold drawn					
Annealed	483-586	70-85	172-310	25-45	50-35
Stress relieved	586-827	85-120	379-690	55-100	35-15
Wire					
Annealed	483-655	70-95	207-379	30-55	45-25
Spring temper	1000-1241	145-180	862-1172	125-170	5-2
Inconel 600					
Rod and bar:					
Annealed	552-690	80-100	172-345	25-50	55-35
Cold drawn	724-1034	105-150	552-862	80-125	30-10
Hot finished	586-827	85-120	241-621	35-90	50-30
Plate:					
Hot rolled	586-758	85-110	241-448	35-65	50-30
Annealed	552-724	80-105	207-345	30-50	55-35
Sheet:					
Annealed	552-690	80-100	207-310	30-45	55-35
Hard	586-1034	120-150	621-862	90-125	15-2
Strip:					
Annealed	552-690	80-100	207-310	30-45	55-35
Spring temper	1000-1172	145-170	827-1103	120-160	10-2
Tubing:					
Hot finished	517-690	75-100	172-345	25-50	55-35
Cold drawn and annealed	552-690	80-100	172-345	25-50	55-35
Wire:					
Annealed	552-827	80-120	241-517	35-75	45-20
Spring temper	1172-1517	170-220	1034-1448	150-210	5-2
Incoloy 800					
Rod and bar:					
Annealed	517-690	75-100	207-414	30-60	60-30
Hot finished	552-827	80-120	241-621	35-90	50-25
Cold drawn	690-1034	100-150	517-862	75-125	30-10
Plate:					
Hot rolled	552-758	80-110	207-448	30-65	50-25
Annealed	517-724	75-105	207-414	30-60	50-30
Sheet:					
Annealed	517-724	75-105	207-379	30-55	50-30
Strip:					
Annealed	517-690	75-100	207-379	30-55	50-30
Tubing:					
Hot finished	517-724	75-105	172-414	25-60	50-30
Cold drawn, annealed	517-690	75-100	207-414	30-60	50-30
Wire:					
Annealed	552-758	80-110	241-448	35-65	45-25
Spring temper	965-1207	140-175	896-1172	130-170	5-2

ries of alloy cast irons can be either martensitic or austenitic depending on nickel content. Martensitic white irons, which contain up to 7% nickel in combination with chromium, are widely used in applications requiring resistance to wear and abrasion. The high-nickel alloy cast irons that contain sufficient nickel to produce austenitic structures are used for their unique ability to resist both wear and corrosion at elevated temperatures. The austenitic cast irons contain up to 36% nickel and varying amounts of chromium and copper; these alloys are used for gas turbines, exhaust manifolds, turbocharger components, and corrosion-resisting equipment such as pumps, valves and compressors. In production of ductile iron, magnesium, frequently in the form of a nickel-magnesium alloy, is the nodulizing agent added to molten cast iron that causes eutectic graphite to grow as spheres rather than flakes when the iron solidifies.

Miscellaneous uses, including chemicals, salts, catalysts, ceramics, coinage and permanent magnets, account for about 7% of total consumption of nickel.

SELECTED REFERENCES

Nickel and Its Alloys, Monograph 106, U.S. National Bureau of Standards, 1968
Minerals Yearbook, U.S. Bureau of Mines, Washington, 1977

Corrosion Resistance of Nickels and Nickel Alloys

NICKEL is highly resistant to a variety of corrosive media. It can be readily alloyed with elements such as copper, chromium, molybdenum and iron to produce alloys that are resistant to corrosion throughout a wide range of environments.

The common nickel alloy families are as follows: commercially pure nickel; binary systems, such as Ni-Cu, Ni-Si and Ni-Mo; ternary systems, such as Ni-Cr-Fe and Ni-Cr-Mo; and more complex systems, such as Ni-Cr-Fe-Mo-Cu (with other possible additions). Table 4 gives nominal compositions of alloys representative of each of these alloy families, with nickel contents varying from 32.5 to 99.5%.

Nickel alloys are used extensively in a variety of chemical-processing applications. These applications include production of ethylene, ammonia, nitric acid, caustic-chlorine and hydrofluoric acid, refining of salt; and processing of pulp and paper.

ALLOY CHARACTERISTICS

The corrosion characteristics of nickels and nickel alloys—notably, their corrosion resistance in a variety of media, which accounts for their wide industrial use—are described in the following paragraphs.

Commercially Pure Nickels. This family is represented in Table 4 by Nickel 200 and Nickel 201.

Footnotes for Table 2: (a) Values shown represent usual ranges for common section sizes. In general, values in the higher portions of the ranges are not obtainable with large section sizes, and exceptionally small or large sections may have properties outside the ranges.

Table 3. Typical tensile properties of five nickel alloys as a function of temperature

Temperature		Tensile strength		Yield strength (0.2% offset)		Elon-gation, %
°C	°F	MPa	ksi	MPa	ksi	
Nickel 200 (annealed)						
20	68	462	67.0	148	21.5	47.0
93	200	458	66.5	154	22.3	46.0
149	300	460	66.7	150	21.7	44.5
204	400	458	66.5	139	20.2	44.0
260	500	465	67.5	135	19.6	45.0
316	600	456	66.2	139	20.2	47.0
371	700	362	52.5	117	17.0	61.5
Nickel 201 (annealed)						
20	68	403	58.5	103	15.0	50
93	200	387	56.1	106	15.4	45
149	300	372	54.0	99	14.4	46
204	400	372	54.0	102	14.8	44
260	500	372	54.0	101	14.6	41
316	600	362	52.5	105	15.3	42
371	700	325	47.2	97	14.1	53
427	800	284	41.2	93	13.5	58
482	900	259	37.5	89	12.9	58
538	1000	228	33.1	83	12.1	60
593	1100	186	27.0	77	11.2	72
649	1200	153	22.2	70	10.2	74
Duranickel 301 (age hardened)						
21	70	1276	185	910	132	28
316	600	1158	168	827	120	29
371	700	1124	163	807	117	27
427	800	1069	155	786	114	24
482	900	972	141	752	109	11
538	1000	814	118	683	99	7
593	1100	648	94	517	75	5
649	1200	476	69	372	54	4
704	1300	290	42	234	34	8
760	1400	172	25	97	14	60
816	1500	117	17	62	9	98
Incoloy 825 (annealed)						
29	85	693	100.5	301	43.7	43
93	200	655	95.0	279	40.4	44
204	400	637	92.4	245	35.6	43
316	600	632	91.7	232	33.6	46
371	700	621	90.0	234	34.0	46
427	800	610	88.5	228	33.0	44
482	900	608	88.2	221	32.0	42
538	1000	592	85.9	229	33.2	43
593	1100	541	78.5	222	32.2	38
649	1200	465	67.5	213	30.9	62
760	1400	274	39.7	183	26.5	87
871	1600	135	19.6	117	17.0	102
982	1800	75	10.9	47	6.8	173
1093	2000	42	6.1	23	3.3	106
Hastelloy W (solution treated)						
425	800	725	105	260	38	56.0
650	1200	...	...	255	37	29.5
730	1350	465	67	...	...	16.0
815	1500	405	59	250	36	17.0
900	1650	353	52	220	32	14.5
980	1800	...	...	135	20	14.5
1065	1950	180	26	...	...	34.0

The latter is preferred for use at temperatures above 315 °C (600 °F) because its lower carbon content prevents graphitization and subsequent loss of ductility. The outstanding corrosion characteristics of Nickel 200 and Nickel 201 are their resistance to caustics; high-temperature halogens and hydrogen halides; salts other than oxidizing halides; and foods, for which these alloys are particularly well suited for maintaining product purity. Other important characteristics are high thermal and electrical conductivities and magnetic and magnetostrictive properties.

Ni-Cu Alloys. Monel 400 and Monel K-500 differ in that the strength and hardness of the latter may be increased by age hardening. Although these alloys share many of the corrosion characteristics of commercially pure nickel, they exceed nickel in resistance to sulfuric and hydrofluoric acids and to brine. Handling of waters, including seawater and brackish water, is a major area of application. Monel 400 and Monel K-500 are immune to chloride-ion stress-corrosion cracking, which often is a factor in their selection.

Ni-Mo and Ni-Si Alloys. Hastelloy B, noted for its superior resistance to hydrochloric acid, also has very good resistance to many nonoxidizing acids. As with commercially pure nickel and Ni-Cu alloys, corrosion rates of Hastelloy B in acid solutions can be greatly increased by the presence of strong oxidizers in the acid.

Hastelloy D is a cast alloy with outstanding resistance to sulfuric acid. At 66 °C (150 °F), it has exhibited a maximum corrosion rate of 0.15 mm/yr (6 mils/yr) at acid concentrations of from 2 to 96% (see Table 5).

Ni-Cr-Fe Alloys. Inconel 600 and Incoloy 800 are used primarily for their oxidation resistance and strength at elevated temperatures. In aqueous environments, these materials often are used to combat chloride-ion stress-corrosion cracking. Inconel 600 frequently is substituted for commercially pure nickel to resist caustic soda and halogens at high temperature. A common application of Incoloy 800 is sheathing for electric heating elements.

Ni-Cr-Mo Alloys. Significant additions of molybdenum make Hastelloy C-276 and Inconel 625 highly resistant to pitting. These alloys retain high strength and oxidation resistance at elevated temperatures, but they are used in the chemical industry primarily for their resistance to a wide variety of aqueous corrodents. In many applications, these alloys are selected because they are considered the only materials of this type capable of withstanding the severe corrosion conditions encountered.

Table 4. Nominal chemical compositions of nickels and nickel alloys

Alloy	C	Mn	S	Si	Cr	Ni	Mo	Cu	Ti	Fe	Others
							Nominal composition, wt %				
Nickel 200	0.08	0.18	0.005	0.18	...	99.5(a)	...	0.13	...	0.2	...
Nickel 201	0.01	0.18	0.005	0.18	...	99.5(a)	...	0.13	...	0.2	...
Monel 400	0.15	1.0	0.012	0.25	...	66.5(a)	...	31.5	...	1.25	...
Monel K-500	0.13	0.75	0.005	0.25	...	66.5(a)	...	29.5	0.60	1.00	2.73 Al
Hastelloy B	0.05 max	1.00	...	1.00	1.00	61.0	28.00	...	...	5.50	2.50 Co
Hastelloy D	0.12	0.90	...	9.25	1.00	82.0	...	3.00	...	2.00	1.50 Co
Inconel 600	0.08	0.5	0.008	0.25	15.5	76.0(a)	...	0.25	...	8.00	...
Incoloy 800	0.05	0.75	0.008	0.50	21.0	32.5	...	0.38	0.38	46.0	0.38 Al
Hastelloy C-276	0.01	0.5	0.02	0.03	15.50	57.0	16.00	...	...	5.5	0.02 P, 3.75 W, 0.2 V, 1.25 Co
Inconel 625	0.05	0.25	0.008	0.25	21.5	61.0(a)	9.0	...	0.2	2.5	0.2 Al, 3.65 Cb + Ta
Incoloy 825	0.03	0.50	0.015	0.25	21.5	42.0	3.0	2.25	0.90	30.0	0.10 Al
Hastelloy G	0.03	1.50	0.02	0.50	22.25	44.0	6.50	2.00	...	19.50	0.02 P, 0.50 W, 1.25 Co, 2.10 Cb + Ta
20Cb-3	0.04	1.00	0.02	0.50	20.0	34.0	2.50	3.5	...	...	0.02 P, 0.50 Cb + Ta

(a) Includes cobalt.

Table 5. Corrosion rates in sulfuric acid

Acid concentration, %	Corrosion rate			
	Hastelloy B		Hastelloy D	
	mm/yr	mils/yr	mm/yr	mils/yr
Tested at 66 °C (150 °F)				
2	0.13	5	0.15	6
5	0.10	4	0.13	5
10	0.08	3	0.13	5
25	0.03	1	0.05	2
50	0.03	1	0.03	1
60	0.03	1	0.15	6
77	...	...	0.05	2
80	...	...	0.05	2
85	...	...	0.05	2
90	...	...	0.05	2
96	...	...	0.03	1
Tested in boiling acid solution				
2	0.03	1	0.10	4
5	0.03	1	0.18	7
10	0.05	2	0.33	13
25	0.05	2	0.23	9
50	0.05	2	0.28	11
60	0.18	7	0.20	8
80	...	...	0.91	36
85	...	...	2.31	91
90	...	...	4.85	191
96	...	...	2.18	86

Ni-Cr-Fe-Mo-Cu Alloys. Carpenter 20Cb-3, Hastelloy G and Incoloy 825 offer good resistance to pitting, intergranular corrosion, chloride-ion stress-corrosion cracking and general corrosion in a wide variety of both oxidizing and reducing environments. These alloys frequently are used in applications involving sulfuric or phosphoric acid.

BEHAVIOR OF NICKEL ALLOYS IN CORROSIVE ENVIRONMENTS

All of the alloys referred to in this article have excellent resistance to corrosion in rural, industrial and marine atmospheres. Of these materials, only Monel 400 has been used to a significant degree solely for its resistance to atmospheric attack. Because of its low corrosion rate and the attractive patina that develops on its surface, Monel 400 is used in architectural applications such as roofs, gutters and flashings.

Corrosion by Waters. The major corrosion problems associated with industrial use of waters are crevice corrosion, pitting, stress-corrosion cracking and general corrosion. Many engineering materials have adequate resistance to general corrosion, but localized attack often is the life-limiting factor. In seawater, the high chloride content and fouling by marine organisms combine to produce severe crevice attack. Hastelloy C-276 and Inconel 625 are virtually immune to attack by seawater. Hastelloy G, Incoloy 825, Monel 400 and Monel K-500, although not immune to the detrimental effects of seawater, show good resistance to localized attack and excellent resistance to general corrosion. The other alloys listed in Table 4 are seldom used in seawater.

In distilled waters, fresh water and process waters, all of the alloys listed in Table 4 have good resistance to corrosion. Except for Carpenter 20Cb-3, Incoloy 800 and 825 and Hastelloy G, these materials are virtually immune to chloride-ion stress-corrosion cracking, and the excepted alloys are sufficiently resistant to it to be frequently employed specifically for this characteristic.

Corrosion by Sulfuric Acid. Commercially pure nickel adequately resists unaerated sulfuric acid, but would be selected for service in such an environment only if its use were dictated by other conditions, such as the presence of contaminants, exposure to alternating environments, or simultaneous exposure to caustics (encountered in some heat-exchanger applications).

Monel 400 is widely used for handling sulfuric acid under reducing conditions. At room temperature, its corrosion rate in air-free acid at concentrations up to 85% is less than 0.25 mm/yr (10 mils/yr). Tests in boiling sulfuric acid produced corrosion rates of 0.086 mm/yr (3.4 mils/yr) at 5% acid concentration, 0.061 mm/yr (2.4 mils/yr) at 10% and 0.19 mm/yr (7.5 mils/yr) at 19%. At 95 °C (203 °F), corrosion rates in unaerated acid at concentrations of up to 60% were less than 0.51 mm/yr (20 mils/yr).

Hastelloy B and Hastelloy D have exceptional resistance to sulfuric acid; results of tests at various acid concentrations are given in Table 5.

Inconel 600 and Incoloy 800 are used in low-concentration sulfuric acid at room temperature. Although they are rarely used in sulfuric acid service under any other conditions, Incoloy 800 has been employed in 99% acid at 120 °C (250 °F).

Hastelloy C-276 and Inconel 625 both exhibit good resistance to sulfuric acid; however, neither would be selected on this basis alone.

Carpenter 20Cb-3, Hastelloy G and Incoloy 825 have excellent resistance to sulfuric acid. Although the compositional differences among these alloys result in some variation in corrosion behavior, the alloys normally exhibit corrosion rates of less than 0.13 mm/yr (5 mils/yr) at all concentrations when solution temperature is below 50 °C (120 °F). Depending on composition, all three alloys exhibit maximum corrosion at acid concentrations between 60 and 80%.

Corrosion by Hydrochloric Acid. Nickel 200, Nickel 201, Monel 400 and Monel K-500 have room-temperature corrosion rates of below 0.25 mm/yr (10 mils/yr) in air-free hydrochloric acid at concentrations of up to 10%. Concentration of hydrochloric acid produced during hydrolysis of chlorides or chlorinated solvents usually is less than 0.5%; Nickel 200 and Monel 400 can withstand this environment at temperatures up to about 205 °C (400 °F). In air-saturated solutions, corrosion rate increases sharply. In boiling acid, Monel 400 has corroded at rates of 0.74 mm/yr (29 mils/yr) at 0.5% concentration, 1.07 mm/yr (42 mils/yr) at 1% and 1.12 mm/yr (44 mils/yr) at 5%. Rates for Nickel 200 are much higher.

Hastelloy B has outstanding resistance to hydrochloric acid, whereas Hastelloy D has moderate resistance. Hastelloy B corroded at a rate of 0.23 mm/yr (9 mils/yr) in 1, 2 and 5% HCl at 66 °C (150 °F). When acid concentration was increased to 37%, corrosion rate decreased to 0.05 mm/yr (2 mils/yr). In boiling HCl, corrosion rates were 0.05 mm/yr (2 mils/yr) at 1% concentration, 0.08 mm/yr (3 mils/yr) at 2%, 0.18 mm/yr (7 mils/yr) at 5%, 0.23 mm/yr (9 mils/yr) at 10%, 0.36 mm/yr (14 mils/yr) at 15% and 0.61 mm/yr (24 mils/yr) at 24%. Because chromium is rapidly attacked by hydrochloric acid, Inconel 600 and Incoloy 800 have little resistance to this acid. Because of their high molybdenum contents, both Hastelloy C-276 and Inconel 625 are resistant to all concentrations of hydrochloric acid at room temperature. At 66 °C (150 °F) in acid concentrations of from 5 to 37%, Hastelloy C-276 corrodes at rates of from 0.51 to 1.3 mm/yr (20 to 50 mils/yr). When tested at a 37% acid concentration at 66 °C (150 °F), Inconel 625 exhibited a corrosion rate of 0.38 mm/yr (15 mils/yr).

Carpenter 20Cb-3, Hastelloy G and Incoloy 825, although normally not considered candidate materials for hydrochloric acid service, exhibit useful room-temperature resistance at acid concentrations of up to 15%. When tested at room temperature, Hastelloy G corroded at a rate of 0.25 mm/yr (10 mils/yr) at 10% acid concentration, and Incoloy 825 exhibited corrosion rates of 0.12 mm/yr (4.9 mils/yr) at 5%, 0.18 mm/yr (7.2 mils/yr) at 10% and 0.19 mm/yr (7.3 mils/yr) at 15%.

Corrosion by Phosphoric Acid. Neither commercially pure nickel nor Ni-Cu alloys are used extensively in applications involving hot, concentrated phosphoric acid. At low temperatures or low acid concentrations, these materials show minor corrosion. Monel 400, for example, exhibits corrosion rates below 0.25 mm/yr (10 mils/yr) for all acid concentrations when tested at temperatures below 95 °C (200 °F).

Even in boiling 10 to 85% phosphoric acid, Hastelloy B and Hastelloy D show low corrosion rates. Inconel 600 and Incoloy 800 are resistant to all concentrations of pure phosphoric acid at room temperature. Corrosion rates increase rapidly with temperature.

Several contaminants, such as chloride, fluoride and silica, are present during manufacture of wet-process phosphoric acid. Because these contaminants increase susceptibility to pitting and crevice corrosion, resistance to general corrosion is not as important as resistance to local attack. In evaporators handling wet acid, Hastelloy C-276 and Inconel 625 have proved useful.

Carpenter 20Cb-3, Hastelloy G and Incoloy 825 have excellent resistance to all concentrations of phosphoric acid, even at boiling temperature. They also find application in the manufacture of superphosphoric acid.

Corrosion by Nitric Acid. Those alloys containing no chromium have poor resistance to nitric acid. Chromium-bearing nickel alloys show good resistance, with corrosion rates decreasing primarily as chromium content increases. Because boiling 65% nitric acid (Huey test) often is used to measure the resistance of an alloy to intergranular corrosion, it should be pointed out that alloying elements (such as niobium in 20Cb-3, Inconel 625 and Hastelloy G, and titanium in Incoloy 825) are added for preferential modification of carbide precipitation and for minimization of intergranular corrosion.

Corrosion by Organic Acids. Both commercially pure nickel and nickel-copper alloys find limited use in monocarboxylic acids. In glacial acetic acid at 110 °C (230 °F), Monel 400 has shown a corrosion rate of 0.33 mm/yr (13 mils/yr). Again, aeration normally increases corrosion rate.

Hastelloy B and Hastelloy D have very good resistance to most organic acids. In either acetic or formic acid at 70 °C (160 °F), the highest corrosion rate was 0.1 mm/yr (4 mils/yr).

Inconel 600 and Incoloy 800 have excellent resistance to hot, long-chain organic acids. Fat-splitting towers for stearic, oleic, linoleic and abietic acids are commonly fabricated from Inconel 600. Both Hastelloy C-276 and Inconel 625 display excellent resistance to organic acids. Formic acid normally is considered the most corrosive monocarboxylic acid. In all concentrations of boiling acid, both Hastelloy C-276 and Inconel 625 corrode at rates of 0.025 to 0.05 mm/

yr (1 to 2 mils/yr). These alloys are the preferred materials for construction of high-temperature distillation columns for glacial acetic acid.

Carpenter 20Cb-3, Hastelloy G and Incoloy 825 are highly resistant to organic acids and are adequate for most applications involving them.

Corrosion by Alkalis. Commercially pure nickel is unsurpassed by any common engineering material in resistance to corrosion by bases. An exception is ammonium hydroxide, which forms complexes with nickel and copper. Nickel 200 is not subject to corrosion by anhydrous ammonia or ammonium hydroxide in concentrations up to 1%. However, ammonium hydroxide in higher concentrations causes rapid attack.

Nickel 200 and Nickel 201 have excellent resistance to sodium hydroxide and potassium hydroxide at all concentrations and at all temperatures (even molten). In sodium hydroxide or potassium hydroxide at concentrations of less than 50%, Nickel 200 and Nickel 201 exhibit negligible corrosion rates (usually less than 0.005 mm/yr, or 0.2 mil/yr, even in boiling solutions). As concentration and temperature increase, corrosion rates increase slowly. Although stress cracking of nickel in molten anhydrous caustic soda has been reported, long-term laboratory and plant exposures of stressed specimens have not revealed any susceptibility to cracking. The presence of chlorates or oxidizable sulfur compounds increases the corrosive effect of caustics on nickel.

Nickel-copper alloys are not as resistant as pure nickel to the corrosive effects of alkalis. In boiling alkalis in concentrations up to 50%, however, corrosion rates for Ni-Cu alloys are still below 0.025 mm/yr (1 mil/yr), and thus these less-expensive materials are widely employed in such applications.

Hastelloy alloys B, C-276, D and G have excellent resistance to alkaline environments, but they are seldom employed unless other corrodents are involved. Also possessing excellent resistance to alkaline environments are 20Cb-3 and Incoloy 825; however, they, too, are seldom used in such environments.

In certain applications involving high-temperature caustics where sulfur is present or high strength is required, Inconel 600 is used instead of Nickel 201. The chromium content of Inconel 600 provides greater resistance to sulfur embrittlement. Inconel 600, in common with all nickel alloys except commercially pure nickels, is subject to stress-corrosion cracking when brought in contact with high-temperature, high-strength caustics. Thus, equipment should be fully stress relieved prior to use, and operating stresses should be minimized.

Corrosion by Salts. Except for halide salts, the corrosivity of a salt is based primarily on its oxidizing strength and on whether it hydrolyzes to an acid or a base. For example, materials that are resistant to nitric acid most likely are resistant to nitrates, including both sodium nitrate and ferric nitrate. These nitrate salts have high oxidizing strength and will readily hydrolyze to form nitric acid.

Halide salts, particularly chlorides, tend to promote localized attack such as pitting, crevice corrosion and stress-corrosion cracking. In general, high molybdenum contents help to control pitting and crevice corrosion, and high nickel contents resist chloride-ion stress-corrosion cracking.

Nickel 200 and Monel 400 are not subject to stress-corrosion cracking in any of the chloride salts. They have excellent resistance to all of the nonoxidizing halides. Oxidizing acid chlorides, such as ferric chloride and cupric chloride, are very corrosive to these alloys. Hypochlorites can cause pitting. A mixed group of very reactive and corrosive salts — phosphorus oxychloride, phosphorus trichloride, nitrosyl chloride, benzyl chloride and benzoyl chloride — is commonly contained in equipment made of Nickel 200.

Nickel 200 and Monel 400 have good resistance to solutions of neutral and alkaline salts such as carbonates, sulfates, nitrates and acetates. Even under severe conditions of concentration, temperature, agitation and aeration, corrosion rates normally are less than 0.1 mm/yr (5 mils/yr). Nickel 200 tubing is being used successfully in sodium chloride and sodium sulfate evaporators, and nickel-clad steel is used in construction of rotary salt driers. Monel 400 is widely used in salt plants for evaporators, crystallizers, filters, piping and similar equipment. In solutions of acid salts such as zinc chloride, ammonium sulfate and ammonium chloride, both Nickel 200 and Monel 400 have good resistance, but Monel 400 is more widely employed.

Hastelloy B has excellent resistance to nonoxidizing salts. Cupric chloride and ferric chloride are extremely corrosive to this alloy, whereas ammonium, aluminum and zinc chlorides are relatively harmless. Hastelloy B has little resistance to nitrates, chromates and other oxidizing salts. Typical use of this Ni-Mo alloy has been in connection with aluminum chloride-type catalysts, such as those used in alkylation of benzene during production of styrene. Corrosion rates in strong, boiling magnesium chloride are less than 0.05 mm/yr (2 mils/yr). This alloy also is resistant to pitting attack in chloride solutions.

The resistance of Inconel 600 to salts is very similar to that of Nickel 200 and Monel 400; however, in oxidizing acid salts, Inconel 600 is superior. This resistance does not apply to oxidizing acid chlorides. Inconel 600 has excellent resistance to silver nitrate, as used in photographic processing, and to strong, hot magnesium chloride. In nitrosyl chloride at temperatures above 43 °C (110 °F), this alloy is preferred over Nickel 200.

Incoloy 800 is subject to pitting in strong chloride solutions. It is highly resistant, although not immune, to stress-corrosion cracking. In salts other than halides, Incoloy 800 exhibits excellent resistance to a wide variety of both oxidizing and nonoxidizing media.

Inconel 625 and Hastelloy C-276 are resistant to all classes of salts, including oxidizing chlorides. Carpenter 20Cb-3, Hastelloy G and Incoloy 825 are less resistant to pitting than the higher-molybdenum alloys but much more resistant than Incoloy 800. These three alloys have excellent resistance to all classes of salts except the oxidizing halides.

Corrosion by Fluorine, Chlorine and Hydrogen Chloride. At room temperature, fluorine forms protective fluoride films on nickel, copper, magnesium and iron; thus, these metals are considered satisfactory for low-temperature service in fluorine. Nickel 201 and Monel 400 are preferred for high-temperature service in fluorine. All of the nickel-base alloys considered are resistant to dry chlorine and hydrogen chloride, most of them even at moderately high temperatures. Monel 400 is a standard material for trim on chloride cylinders and valves. Wet chlorine is successfully handled by Hastelloy C-276. Nickel 201 and Inconel 600 are the most widely used materials for service in chlorine and hydrogen chloride at elevated temperatures.

Corrosion-Resistant Nickel Alloy Castings

NICKEL-BASE ALLOY CASTINGS are widely used for handling corrosive media and are regularly produced by high-alloy steel foundries. The principal alloys are identified by Alloy Casting Institute (ACI) designations, and are included in ASTM A743, A744 and A494. In addition, several specialized proprietary grades are often specified for severe corrosion applications. Cast nickel-base corrosion-resistant alloys may be classified as follows:

- Nickel
- Nickel-copper alloys
- Nickel-chromium-iron alloys
- Nickel-chromium-molybdenum alloys
- Nickel-molybdenum alloys
- Nickel-base proprietary alloys.

Compositions of these alloys are given in Table 6.

The cast nickel-base alloys, with the exception of some high-silicon and proprietary grades, have wrought counterparts and frequently are specified as the cast components in systems built of both wrought and cast components made of a single alloy. Compositions of cast and equivalent wrought grades differ in minor elements because workability in wrought grades is a dominant factor, whereas castability and soundness are dominant factors in cast grades. The differences in minor elements do not result in significant differences in serviceability.

Nickel-base castings are employed most often in fluid-handling systems where they are matched with equivalent wrought alloys. They also are quite commonly used for pump and valve components or where crevices and high-velocity effects require a superior material in a wrought stainless steel system. Because of high cost, nickel-base alloys are usually selected only for severe service conditions where maintenance of product purity is of great importance and where less-costly stainless steels or other alternative materials are inadequate.

CAST NICKEL

CZ-100 is the ACI designation for the standard grade of cast nickel. A higher-carbon, higher-silicon grade is occasionally specified for greater resistance to wear and galling. Cast nickel is unsurpassed in handling concentrated and anhydrous caustic at elevated temperatures. It is also used in applications where product contamination by elements other than nickel cannot be tolerated.

The minor elements in CZ-100 provide for excellent castability and the production of sound, pressure-tight castings. Usual practice is to aim for 0.75% C and 1.0% Si. Carbon is present as finely distributed spheroidal graphite. A maximum carbon content of 0.10% or less is occasionally specified where castings are welded into wrought nickel systems. Low-carbon CZ-100, however, is a difficult material to cast and has no significant advantage over the higher-carbon CZ-100 under any known service conditions. CZ-100 is used in the as-cast condition.

Table 6. Compositions of cast nickel-base corrosion-resistant alloys

Designation	Composition limits(a), %								
	C	Si	Mn	Cu	Fe	Ni	Cr	Mo	Other
Nickel									
CZ-100 ....1.0		2.0	1.5	1.25	3.0	Rem	...	...	...
Ni-Cu alloys									
M-35-1 ....0.35		1.25	1.5	26-33	3.5	Rem	...	...	...
M-35-2 ....0.35		2.0	1.5	26-33	3.5	Rem	...	...	...
QQ-N-288:									
Grade A 0.35		2.0	1.5	26-33	2.5	62-68	...	...	...
Grade B 0.30		2.7-3.7	1.5	27-33	2.5	61-68	...	...	...
Grade C 0.20		3.3-4.3	1.5	27-31	2.5	60 min	...	...	...
Grade D 0.25		3.5-4.5	1.5	27-31	2.5	60 min	...	...	...
Grade E 0.30		1.0-2.0	1.5	26-33	3.5	60 min	...	...	1-3 (Nb+Ta)
Ni-Cr-Fe alloy									
CY-40 ....0.40		3.0	1.5	...	11	Rem	14-17	...	...
Ni-Cr-Mo alloy									
CW-12M-1 ....0.12		1.0	1.0	...	4.5-7.5	Rem	15.5-17.5	16-18	0.20-0.40 V, 3.75-5.25 W
CW-12M-2 ....0.07		1.0	1.0	...	3.0	Rem	17-20	17-20	...
Ni-Mo alloy									
N-12M-1 ....0.12		1.0	1.0	...	4-6	Rem	1.0	26-30	0.2-0.6 V
N-12M-2 ....0.07		1.0	1.0	...	3.0	Rem	1.0	30-33	...
Proprietary alloys									
Chlorimet 2 ....0.07		1.0	1.0	...	2.0	66 min	1.0	30-33	...
Chlorimet 3 ....0.07		1.0	1.0	...	3.0	60 min	17-20	17-20	...
Hastelloy B 0.12		1.0	1.0	...	4-6	Rem	1.0	26-30	0.2-0.6 V, 2.5 Co
Hastelloy C 0.12		1.0	1.0	...	4.5-7.0	Rem	15.5-17.5	16-18	0.2-0.6 V, 2.5 Co
Hastelloy D 0.12		8.5-10	0.5-1.25	2-4	2.0	Rem	1.0	...	1.5 Co
Illium 98(b) 0.05		1.0	1.0	5.5	...	Rem	28	8.5	...
Illium G(b) 0.20		1.0	1.0	6.5	5.0	Rem	22.5	6.5	...
Illium B(b) 0.05		3.5	1.0	5.5	...	Rem	28	8.5	...

(a) Where a single value is shown, that value is a maximum unless otherwise indicated. (b) Nominal composition.

Table 7. Mechanical-property requirements at room temperature for standard cast nickel-base alloys

Alloy designation	Min tensile strength		Min yield strength		Elongation, %	Hardness, HB
	MPa	ksi	MPa	ksi		
CZ-100	345	50	124	18	10	...
M-35-1	448	65	127	25	25	110-140
M-35-2	448	65	207	30	25	125-150
QQ-N-288, grade A	448	65	224	32.5	25	125-150
QQ-N-288, grade B	689	100	455	66	10	240-290
QQ-N-288, grade C(a)	825	120	550	80	10	250-300(b)
QQ-N-288, grade D(a)	...	...	...	...	...	300
QQ-N-288, grade E	448	65	221	32	25	125-150
CY-40	483	70	193	28	30	...
CW-12M-1	496	72	317	46	4	...
CW-12M-2	496	72	317	46	25	...
N-12M-1	524	76	317	46	6	...
N-12M-2	524	76	317	46	20	...

(a) Values are typical. (b) Minimum hardness requirement for solution-treated-and-age-hardened condition or cast-and-age-hardened condition.

Mechanical-property requirements for CZ-100 are listed in Table 7. Cast nickel has excellent toughness, thermal resistance and heat-transfer characteristics.

CZ-100 can be readily repair welded. It can be joined to other castings or to wrought forms using any arc or gas welding process; filler metal is nickel rod or wire. Joints must be prepared very carefully for welding, because small amounts of sulfur or lead will embrittle the welds.

The most common application for nickel castings is in the manufacture of caustic soda and in chemical processing with caustic. As temperature and caustic soda concentration increase, austenitic stainless steels are of limited usefulness. Ni-Cu (M-35) and Ni-Cr (CY-40) alloys often are applied at intermediate concentrations, and cast nickel is selected for higher caustic concentrations (including fused anhydrous NaOH). Minor amounts of elements such as oxygen and sulfur can have profound effects on the corrosion rate of nickel in caustic. Detailed corrosion data should be obtained before making a final selection.

NICKEL-COPPER ALLOYS

Cast 70Ni-30Cu alloys (Monels) are listed in ASTM A494, A743 and A744 as M-35-1 and M-35-2, and in Federal Specification, QQ-N-288 as compositions A, B, C, D and E.

The low-silicon grades M-35-1 (1.0% Si) and M-35-2 (1.5% Si) and compositions A and E (1.5% Si) are commonly used in conjunction with wrought nickel-copper and copper-nickel alloys in pumps, valves and fittings. A higher-silicon grade, composition B (3.5% Si), is used for rotating parts and wear rings because it combines corrosion resistance with high strength and wear resistance. Composition D (4.0% Si) is employed where exceptional resistance to galling is desired.

M-35-1, M-35-2, and QQ-N-288 compositions A and E are employed in the as-cast condition. Homogenization at 815 to 925 °C (1500 to 1700 °F) may slightly improve corrosion resistance, but under most corrosive conditions, alloy performance is not affected by the minor segregation present in as-cast metal.

At about 3.5% silicon, age hardening becomes possible. At approximately 3.8% Si, the solubility limit for silicon in the nickel-copper matrix is exceeded and hard, brittle silicides are formed. These effects are particularly evident in the high-silicon alloy composition D. The combination of aging during cooling to room temperature after casting, plus the hard silicides developed when silicon content exceeds about 3.8%, can cause considerable difficulty in machining. Softening of composition D is accomplished by solution heat treatment consisting of heating to 900 °C (1650 °F), holding at temperature 1 h for each 25 mm (1 in.) of section, and oil quenching. Maximum softening is obtained by oil quenching from 900 °C, but it is apt to result in quench cracks in complicated castings or castings with large differences in section size.

In solution heat treating of complicated castings, it is advisable to charge them into a furnace below 315 °C (600 °F) and heat them to 900 °C (1650 °F) at a rate that will limit the maximum temperature difference within any casting to about 55 °C (100 °F). After soaking, castings should be transferred to a furnace held at 730 °C (1350 °F), allowed to equalize in temperature, and then oil quenched. Alternatively, the furnace may be cooled rapidly to 730 °C (1350 °F), the casting temperature equalized and the castings subsequently quenched in oil.

Solution-treated castings are age hardened by placing them in a furnace at 315 °C (600 °F), heating uniformly to 600 °C (1100 °F), holding 4 to 6 h, and cooling in air.

Tensile properties of nickel-copper castings are controlled by the solution hardening effect of silicon or of silicon plus niobium. Increasing the copper content also has a minor strengthening effect. The tensile properties of M-35-1, M-35-2 and composition A are controlled by a carbon-plus-silicon relationship, and tensile properties of composition E are controlled by a silicon-plus-niobium relationship.

The combination of aging plus hard silicides in composition D results in an alloy with exceptional resistance to galling. As the silicon content is increased above 3.8%, the amounts of hard, brittle silicides in a tough nickel-copper matrix increase, ductility decreases sharply and tensile and yield strengths increase. Because of these effects, strength and ductility cannot be controlled readily, and thus minimum hardness is the only mechanical property specified for composition D.

The toughness of nickel-copper alloys decreases with increasing silicon content but all grades retain their room-temperature toughness down to at least −195 °C (−320 °F).

Weldability of nickel-copper alloys decreases with increasing silicon content but is adequate up to at least 1.5% Si. Niobium can enhance weldability of nickel-copper alloys, particularly when small amounts of low-melting-point residuals are present. Careful raw-material selection and good foundry practice, however, have largely eliminated any difference in weldability between niobium-bearing and niobium-free grades.

The high-silicon compositions B (3.5% Si) and D (4.0% Si) are considered not weldable. They can be brazed or soldered, as can the low-silicon grades.

Principal advantages of 70Ni-30Cu alloys are high strength and toughness coupled with excellent resistance to reducing mineral acids, organic acids, salt solutions, food acids, strong alkalis and marine environments.

The most common applications for 70Ni-30Cu alloy castings are in equipment for handling hydrofluoric acid, salt water, neutral and alkaline salt solutions, and reducing acids.

NICKEL-CHROMIUM-IRON ALLOY

The cast nickel-chromium-iron alloy CY-40 (Inconel) differs in composition from the corresponding wrought grade in having higher carbon, manganese and silicon contents, which impart the required qualities of castability and pressure tightness.

CY-40 is used in the as-cast condition because the alloy is insensitive to intergranular attack of the type encountered in as-cast or sensitized stainless steels. A modified cast nickel-chromium-iron alloy for nuclear applications (0.12% max carbon) is usually solution treated as an extra precaution.

The minimum mechanical properties in Table 7 are for a typical composition of 0.20 C, 1.50 Si, 1.0 Mn, 15.5 Cr, 8.0 Fe, rem Ni. Lower carbon and silicon contents for nuclear-grade castings result in a lower yield strength, but do not lower the minimum tensile strength.

CY-40 is frequently used for elevated-temperature fittings in conjunction with a wrought alloy of similar composition. Typical elevated-temperature properties are listed in Tables 8 and 9.

CY-40 castings can be repair welded or joined to other components using any of the standard

Table 8. Typical elevated-temperature tensile properties(a) of CY-40

Temperature		Tensile strength		Yield strength		Elonga-tion(b),
°C	°F	MPa	ksi	MPa	ksi	%
Room	....	486	70.5	293	42	16
480	900	...427	62	...	...	20
650	1200	...372	54.5	...	...	21
730	1350	...314	45.5	...	...	25
815	1500	...186	27	...	...	34

(a) Data are typical for investment cast test bars with a nominal composition of 0.20% C and 1.50% Si. (b) In 25 mm or 1 in.

Table 9. Typical elevated-temperature stress-rupture properties(a) of CY-40

Temperature		Stress to rupture in 100 h	
°C	°F	MPa	ksi
650	1200	 165	24
730	1350	 103	15
815	1500	 62	9
925	1700	 38	5.5

(a) Data are typical for investment cast test bars with nominal analysis of 0.20 C and 1.50 Si.

arc or gas welding processes. Rod and wire whose nickel and chromium contents match those of the castings should be used. Postweld heat treatment is not required after repair welding or fabrication unless residual stresses must be relieved.

CY-40 is commonly used to handle hot corrosives or corrosive vapors under moderately oxidizing conditions, where stainless steels might be subject to intergranular attack or stress-corrosion cracking. In recent years, CY-40 has been used extensively for components in systems handling hot boiler feedwater in nuclear power plants because it provides a greater margin of safety over stainless steels.

NICKEL-CHROMIUM-MOLYBDENUM ALLOY

Two grades of cast Ni-Cr-Mo alloy (ACI CW-12M-1 and CW-12M-2) are used in severe corrosion service—most often, service involving combinations of acids at elevated temperatures. These two versions of alloy CW-12M are also produced under the proprietary names Hastelloy C and Chlorimet 3.

The high chromium and molybdenum contents of CW-12M result in precipitation of carbides and intermetallic compounds during cooling in the mold. Because these precipitates adversely influence corrosion resistance, ductility and weldability, CW-12M should be solution treated at 1175 to 1230 °C (2150 to 2250 °F) and water or spray quenched.

The CW-12M grades have relatively high yield strengths due to the solution-hardening effects of chromium, molybdenum and silicon in CW-12M-2, and similar effects plus those of tungsten and vanadium in CW-12M-1. Ductility is excellent up to the limits of solid solubility. Inadequate heat treatment or improper composition balance, however, may result in formation of a hard brittle phase and in a rapid loss of ductility. Careful control within the specified composition range is necessary. Carbon and sulfur contents should be kept as low as practicable.

CW-12M can be arc or gas welded, using wire or rod of matching composition. For best weldability, carbon content of the base metal should be as low as practicable. The usual practice is to

solution treat and quench prior to repair welding. Heat treatment after welding generally is not necessary because the alloy is not subject to sensitization in the heat-affected zone.

CW-12M is probably the most common material for upgrading a system where service conditions are too demanding for standard or special stainless steels—most often, service involving combinations of acids and elevated temperatures. The cast alloy may be used in conjunction with similar wrought materials or it may be used to upgrade pump and valve components in a wrought stainless steel system.

NICKEL-MOLYBDENUM ALLOY

Two grades of cast Ni-Mo alloy (ACI N-12M-1 and N-12M-2) are frequently used for handling hydrochloric acid in all concentrations at temperatures up to the boiling point. These two versions of alloy N-12M are also produced commercially under the proprietary names Hastelloy B and Chlorimet 2.

Slow cooling in the mold is detrimental to corrosion resistance, ductility and weldability of N-12M. Because of this, the alloy should be solution treated, at a temperature of at least 1180 °C (2150 °F) for N-12M-1 and 1120 °C (2050 °F) for N-12M-2, and water quenched.

The N-12M grades have high yield strength due to the solution-hardening effect of molybdenum (Table 7). Ductility is controlled by the carbon and molybdenum contents. For best ductility, carbon content should be as low as practicable and molybdenum content should be adjusted to avoid formation of intermetallic phases.

N-12M can be arc or gas welded using wire or rod of matching composition. Castings should be solution treated and quenched prior to repair welding. Postweld heat treatment is not necessary because the alloy is not subject to sensitization in the heat-affected zone.

Applications of N-12M are specialized—mainly in the handling of hydrochloric acid. N-12M should not be used as a substitute for stainless steels in applications where the latter have proved inadequate; N-12M is not adequately resistant to most oxidizing solutions for which stainless steels are initially selected.

NICKEL-BASE PROPRIETARY ALLOYS

In addition to the standard ACI corrosion-resistant nickel-base alloys, there are a number of proprietary alloys that are widely used in corrosive service. Many of the proprietary alloys have excellent general corrosion resistance and are most commonly used where stainless steels are inadequate. Others, such as Chlorimet 2, Hastelloy B and Hastelloy D are used in specialized applications and should not be considered substitutes for stainless steels. Producers should be consulted before specifying these alloys, particularly for applications where there is no history of use.

In chemical processing where a mixture of corrosive solutions is involved, or where small amounts of a contaminant are present, corrosion rates of many common corrosion-resistant materials vary widely. It is under these more severe chemical-processing conditions that corrosion-resistant nickel-base alloys can be of greatest utility.

Materials for
Corrosion-Resistant Fasteners

By Joseph S. Orlando and William Ballantine, ITT Harper

Condensed from Metals Handbook, Ninth Edition, Volume 3, pages 183 to 185.

CORROSION-RESISTANT metallic fasteners are those made of stainless steels and nonferrous alloys. This broad definition could include hundreds of alloys, but in practice the materials actually used are limited to several stainless steels and several copper alloys, plus a few nickel, aluminum and titanium alloys. Fasteners can and have been made from unusual materials (tantalum, for example), but this discussion is primarily limited to those corrosion-resistant materials used in commercial fasteners that are readily available as standard (see Table 1).

STAINLESS STEELS

Over half of all industrial fasteners classified as corrosion resistant are made of stainless steels. This general designation covers austenitic, martensitic and ferritic stainless steels. Of all stainless steels, the 300 series austenitic types are the most popular for fastener use. Austenitic stainless steels are not hardenable by heat treatment and are nonmagnetic for all practical purposes. All alloys in this group have at least 8% nickel in addition to chromium. They offer a greater degree of corrosion resistance than martensitic and ferritic types, but offer a lesser degree of resistance to chloride stress-corrosion cracking.

Martensitic and ferritic stainless steels contain at least 12% chromium, but contain little or no nickel because it stabilizes austenite. Martensitic grades, such as types 410 and 416, are magnetic and can be hardened by heat treatment. Ferritic alloys, such as type 430, are also magnetic but cannot be hardened by heat treatment.

The fastener industry generally markets fasteners made of types 302, 303, 304 and 305 stainless steels as "18-8." These four alloys are similar in both corrosion resistance and mechanical properties. From the manufacturer's point of view, the choice of alloy depends on method of fastener production, which in turn depends on type and size of fastener and, to some extent, on production volume. Because no two manufacturers have identical equipment, the alloy selected for a given fastener will vary; as an indication, however, the alloys that a major fastener producer uses on orders for 18-8 are as follows:

- *Type 302* is used for machine and tapping screws.
- *Type 303* is used to make nuts machined from bar. It contains a small amount of sulfur, for improved machinability.
- *Type 304* is used for hot heading (examples: long bolts or large-diameter bolts beyond the range of cold heading equipment).
- *Type 305* is used for cold heading (examples: hex-head bolts and cold formed nuts).

Other 300-series stainless steels used in fasteners include the following:

Table 1. Standard corrosion-resistant fastener alloys

Commercial name	UNS No.	ASTM specifications
Stainless steels		
17-4 PH	S17400	
Type 302	S30200	
Type 303	S30300	
Type 304	S30400	
Type 305	S30500	F593: stainless steel bolts, hex cap screws, and studs
Type 309	S30900	
Type 310	S31000	F594: stainless steel nuts
Type 316	S31600	
Type 317	S31700	
Type 321	S32100	
Type 347	S34700	
Type 410	S41000	
Type 416	S41600	
Type 430	S43000	
Copper alloys		
ETP copper	C11000	
Yellow brass	C27000	
High leaded brass	C34200	
Free-cutting brass	C36000	
Naval brass, 63½%	C46200	
Naval brass, uninhibited	C46400	
Si-bearing aluminum bronze	C64200	
Low-silicon bronze B	C65100	
High-silicon bronze A	C65500	F468: nonferrous bolts, hex cap screws, and studs for general use
Nickel alloys		
Monel 400	N04400	
Monel 405	N04405	
Monel K-500	N05500	F467: nonferrous nuts for general use
Inconel 600	N06600	
Aluminum alloys		
Aluminum 1100	A91100	
Alloy 2024	A92024	
Alloy 6061	A96061	
Titanium alloys		
Commercial-purity titanium:		
ASTM grade 1	R50250	
ASTM grade 2	R50400	
ASTM grade 4	R50700	
Ti-6Al-4V (ASTM grade 5)	R56400	
Ti-0.2Pd (ASTM grade 7)	R52400	

- *Types 309 and 310* are higher in both nickel and chromium than the standard 18-8 alloys, and are used for high-temperature applications.
- *Types 316 and 317*, because they contain molybdenum, have better elevated-temperature strength and better resistance to pitting than 18-8 alloys.
- *Types 321 and 347* are similar to 18-8 alloys but are stabilized by addition of titanium (type 321) or niobium (type 347) to increase resistance to intergranular corrosion.

Ferritic and martensitic stainless steels for fasteners are largely limited to:

- *Types 410 and 416* are general-purpose corrosion and heat-resistant alloys; they are hardenable by heat treatment.
- *Type 430* has better corrosion- and heat-resistant qualities than type 410; it is not hardenable by heat treatment.

COPPER ALLOYS

Silicon bronzes have tensile strengths higher than those of low-carbon steels and are resistant to corrosion by the atmosphere, by fresh water and seawater, and by gases and sewage. Silicon bronzes are the copper alloys most commonly used for fasteners. They are nonmagnetic and have excellent machining and forming characteristics. C65100 is a low-silicon alloy suitable for cold heading; C65500 is suitable for hot forged fasteners.

Aluminum bronzes have better mechanical properties than silicon bronzes, but are much less frequently used in fasteners. Because of its good machinability, C64200 is the aluminum bronze most often used for fasteners.

Brasses, once the most commonly used materials for corrosion-resistant fasteners, now are specified less frequently than steels and silicon bronzes, which have higher mechanical properties. Brasses are still used in various applications, including electrical communications equipment, builders' hardware and many other consumer and industrial products. C27000 is used for cold headed fasteners, and C36000 is used for fasteners milled from bar.

Naval brasses are copper-zinc alloys containing small amounts of tin, which give them higher resistance to salt water and atmospheric corrosion. C46200 is used for cold headed fasteners, and C46400 is used for hot forged fasteners and for fasteners milled from bar.

NICKEL ALLOYS

Nickel-base alloys are characterized by good strength and good resistance to heat and corrosion. They are often specified for marine and chemical-plant uses.

- *Monel 400* is used for fasteners more often than any other nickel-base alloy.
- *Monel K-500* is heat treatable and, in effect, is a high-strength version (900-MPa, or 130-ksi, minimum tensile strength) of Monel 400.
- *Inconel 600* is used for fasteners that must retain both high strength and resistance to oxidation at temperatures as high as 870 °C (1600 °F).

ALUMINUM ALLOYS

Some aluminum alloys are used for industrial fasteners. They have good corrosion resistance and low weight. Typically, aluminum fasteners are used to join aluminum components.

- *2024-T4* is a heat treated alloy usually used for cold headed fasteners; its tensile strength is above 425 MPa (62 ksi).

- *6061-T6* is used for some nuts, both cold formed and machined.
- *1100* (commercial-purity aluminum) is used for some washers and rivets.

TITANIUM

Titanium and its alloys have excellent corrosion resistance and maintain their strength at moderately high temperatures. Most industrial titanium fasteners are made from commercial-purity titanium, and are used in chemical-equipment applications. Titanium aircraft fasteners, many of which are of proprietary design, are produced from titanium alloys of much higher strength.

INDUSTRY STANDARDS

The American Society for Testing and Materials (ASTM) has a working committee, "F16-Fasteners," with a series of subcommittees each of which deals with development of specific fastener standards. Subcommittee 4 works with nonferrous and stainless steel fastener standards. ASTM specifications in Table 1 (ASTM F467, F468, F593 and F594) are the four standards initially created by Subcommittee 4. They can be referred to for design criteria applicable to corrosion-resistant fasteners.

NONSTANDARD FASTENER ALLOYS

Previous sections of this article have dealt with those corrosion-resistant alloys that are used most often for fasteners whose designs are recognized as standard by the American National Standards Institute (ANSI). Not surprisingly, there are numerous other materials that are used, either for standard fasteners or for special parts, when dictated by strength considerations, corrosive conditions or temperature requirements. Some of these more specialized alloys are listed below.

- *Precipitation-hardening stainless steels,* such as 17-4 PH, 17-7 PH, PH 15-7 Mo, Custom 450 and Custom 455, are used to obtain higher strength than that available from 18-8 stainless steels.
- *Martensitic stainless steels* such as type 416 and type 420 are used to obtain better mechanical properties than can be achieved with types 410 and 430.
- *Carpenter 20Cb-3* is specified when greater corrosion resistance is required than can be offered by 18-8 stainless steels, such as for equipment handling hot sulfuric acid.
- *A-286,* a nonstandard stainless steel that has greater corrosion resistance than the 18-8 types, as well as good mechanical properties at elevated temperatures, has been used in applications requiring resistance to both heat and a corrosive substance, such as in specialized chemical-plant or petroleum-refinery applications.

16 HEAT-RESISTANT MATERIALS

Additional and more detailed information on heat-resistant materials can be found in Metals Handbook, Ninth Edition, Volume 3, Properties and Selection: Stainless Steels, Tool Materials and Special-Purpose Metals, pages 189 to 349. For additional information within the Desk Edition, the reader should consult the index.

Iron-Base Heat-Resistant Alloys

WROUGHT IRON-BASE HEAT-RESIS-TANT ALLOYS are used for applications involving metal temperatures above about 370 °C (700 °F) — the approximate upper limit for use of plain carbon steels under continuous load. The heat-resistant alloys include low-alloy steels that have been used at 370 °C or higher and for which high-temperature data are available. Also included are the austenitic stainless steels, high-temperature steels and precipitation-hardening alloys that have useful strengths at temperatures from 540 to 650 °C (1000 to 1200 °F) and higher while maintaining adequate oxidation resistance. Compositions of representative alloys dealt with in this article are presented in Table 1.

PRODUCT FORMS

Wrought heat-resistant alloys are manufactured in all the forms common to the metal industry: billet, bar, sheet, tubing and wire. Stainless steels are produced as tubing, wire, hot rolled sheet, plate, polished sheet, cold rolled strip, flat circles, hot rolled bar, cold finished bar, forging billet, tube rounds, structural shapes, pipe and rod. The elevated-temperature properties of any of these materials are influenced to some extent by the form of the product, depending largely on specific alloy characteristics such as oxidation resistance, type of oxide scale, thermal conductivity and thermal expansion. Duration and type of loading also may influence differences in properties among different product forms.

For alloys that form thin tenacious scales at elevated temperature, the stress-rupture properties of bar and sheet of the same alloy will be about the same. On the other hand, for alloys that are less resistant to oxidation, rupture values are likely to be significantly lower for sheet than for the same alloy in bar form because of the greater ratio of surface area to volume, which causes greater interaction between the environment and the substrate metal. In the case of oxidation, a fixed depth of oxidation (such as 3 to 5 mils) will more drastically affect sheet properties in 50-mil sheet than such a depth will affect properties of 250-mil bar stock.

Heat-resistant alloys are cold worked by hammering, swaging and rolling. Cold worked products such as bolts and studs also are produced.

ALLOWABLE DESIGN STRESSES

The maximum allowable stresses of the ASME Boiler Code are based on high-temperature laboratory data, and also reflect consideration of past operating experience. The two sections of the code that specify maximum allowable stresses deal with power boilers and unfired pressure vessels.

The procedure used by the Ferrous Section of the ASME Boiler Code Subcommittee on Allowable Stresses, as a basis for their allowable-stress tables, has been published in Section I of the Code for Power Boilers and in Section VIII of the Code for Unfired Pressure Vessels. The subcommittee established design criteria for unfired vessels that

Table 1. Nominal compositions of wrought iron-base heat-resistant alloys

Designation	UNS number	C	Cr	Ni	Mo	N	Nb	Ti	Others
Ferritic stainless steels									
405	S40500	0.15 max	13.0	...	...	...	...	...	0.2 Al
406	...	0.15 max	13.0	...	...	...	...	...	4.0 Al
409	S40900	0.08 max	11.0	0.5	...	...	...	6 × C min	...
430	S43000	0.12 max	16.0	...	...	...	...	...	...
434	S43400	0.12 max	17.0	...	1.0	...	...	...	...
439	S43027	0.07 max	18.25	...	...	...	...	0.2 + 4 (C + N)	...
18 SR	...	0.05	18.0	0.5	...	...	...	0.40	2.0 Al
18Cr-2Mo	...	...	18.0	...	2.0	...	...	...	...
446	S44600	0.20 max	25.0	...	...	0.25	...	...	...
E-Brite 26-1	S44627	0.01 max	26.0	...	1.0	0.015 max	0.1	...	...
26-1Ti	...	0.04	26.0	...	1.0	...	...	10 × C min	...
29Cr-4Mo	...	0.01 max	29.0	...	4.0	0.02 max	...	...	...
Quenched-and-tempered martensitic stainless steels									
403	S40300	0.15 max	12.0	...	...	...	...	...	...
410	S41000	0.15 max	12.5	...	...	...	...	...	...
416	S41600	0.15 max	13.0	...	0.6(a)	...	...	...	0.15 min S
422	S42200	0.20	12.5	0.75	1.0	...	...	...	1.0 W, 0.22 V
H-46	...	0.12	10.75	0.50	0.85	0.07	0.30	...	0.20 V
Moly Ascoloy	...	0.14	12.0	2.4	1.80	0.05	...	...	0.35 V
Greek Ascoloy	...	0.15	13.0	2.0	...	...	...	...	3.0 W
Jethete M-152	...	0.12	12.0	2.5	1.7	...	...	...	0.30 V
Almar 363	...	0.05	11.5	4.5	...	...	...	10 × C min	...
431	S43100	0.20 max	16.0	2.0	...	...	...	...	...

(continued)

Table 1. (continued)

Designation	UNS number	Composition, % C	Cr	Ni	Mo	N	Nb	Ti	Others
Precipitation-hardening martensitic stainless steels									
Custom 450		0.05 max	15.5	6.0	0.75	...	8 × C min	...	1.5 Cu
Custom 455		0.03	11.75	8.5	...	...	0.30	1.2	2.25 Cu
15-5 PH	S15500	0.07	15.0	4.5	...	...	0.30	...	3.5 Cu
17-4 PH	S17400	0.04	16.5	4.25	...	...	0.25	...	3.6 Cu
PH 13-8 MO	S13800	0.05	12.5	8.0	2.25	...	...	...	1.1 Al
Precipitation-hardening semiaustenitic stainless steels									
AM-350	S35000	0.10	16.5	4.25	2.75	0.10	...	...	...
AM-355	S35500	0.13	15.5	4.25	2.75	0.10	...	...	...
17-7 PH	S17700	0.07	17.0	7.0	...	...	...	...	1.15 Al
PH 15-7 Mo	S15700	0.07	15.0	7.0	2.25	...	...	...	1.15 Al
Austenitic stainless steels									
304	S30400	0.08 max	19.0	10.0	...	...	...	...	...
304L	S30403	0.03 max	19.0	10.0	...	...	...	...	...
304N	S30451	0.08 max	19.0	9.25	...	0.13	...	...	...
309	S30900	0.20 max	23.0	13.0	...	...	...	...	...
310	S31000	0.25 max	25.0	20.0	...	...	...	...	...
316	S31600	0.08 max	17.0	12.0	2.5	...	...	...	...
316L	S31603	0.03 max	17.0	12.0	2.5	...	...	...	...
316N	S31651	0.08 max	17.0	12.0	2.5	0.13	...	...	...
317	S31700	0.08 max	19.0	13.0	3.5	...	...	...	...
321	S32100	0.08 max	18.0	10.0	...	...	...	5 × C min	...
347	S34700	0.08 max	18.0	11.0	...	...	10 × C min	...	...
19-9 DL	K63198	0.30	19.0	9.0	1.25	...	0.4	0.3	1.25 W
19-9 DX	K63199	0.30	19.2	9.0	1.5	...	...	0.55	1.2 W
17-14-CuMo		0.12	16.0	14.0	2.5	...	0.4	0.3	3.0 Cu
202	S20200	0.09	18.0	5.0	...	0.10	...	...	8.0 Mn
216	S21600	0.05	20.0	6.0	2.5	0.35	...	...	8.5 Mn
21-6-9	S21900	0.04 max	20.25	6.5	...	0.30	...	...	9.0 Mn
Nitronic 32		0.10	18.0	1.6	...	0.34	...	...	12.0 Mn
Nitronic 33		0.08 max	18.0	3.0	...	0.30	...	...	13.0 Mn
Nitronic 50		0.06 max	21.0	12.0	2.0	0.30	0.20	...	5.0 Mn
Nitronic 60		0.10 max	17.0	8.5	2.0	...	...	...	8.0 Mn, 0.20 V, 4.0 Si
Carpenter 18-18 Plus		0.10	18.0	<0.50	1.0	0.50	...	...	16.0 Mn, 0.40 Si, 1.0 Cu

(a) Optional.

operate under the ASME Boiler Code, as follows:

- In the temperature range where elastic properties predominate, which varies from about 315 to 540 °C (600 to 1000 °F) depending on the alloy, maximum allowable stress shall not exceed $1/4$ of the tensile strength or $5/8$ of the yield strength at 0.2% offset. For bolting steels, maximum stress shall not exceed $1/5$ of the tensile strength or $1/4$ of the yield strength.
- In the temperature range where plastic properties predominate, maximum allowable stress shall not exceed the stress for a creep rate of 1% per 100 000 h or the rupture strength for 100 000-h life.
- In the intermediate range where both plastic and elastic properties predominate, maximum allowable stress shall be determined by a smooth curve joining the curves for the other two ranges.

CREEP AND STRESS-RUPTURE

Many of the published values for stress-rupture and creep are typical or average values. There is a spread above and below these values caused by differences among heats of metal, methods of processing, and variations in conducting the standard test procedures. A typical spread in 100 000-h rupture strength is shown at upper left in Fig. 1 for type 347 stainless steel. At 590 °C (1100 °F), the test values range from 105 to 205 MPa (15 to 30 ksi), and at 650 °C (1200 °F) from 41 to 145 MPa (6 to 21 ksi). The average curve is well above the stress levels allowed by the ASME Power Boiler Code. The graph at lower left in Fig. 1 shows the total range in stresses for a creep rate of 1% per 100 000 h. On a percentage basis, the spread is about the same as in

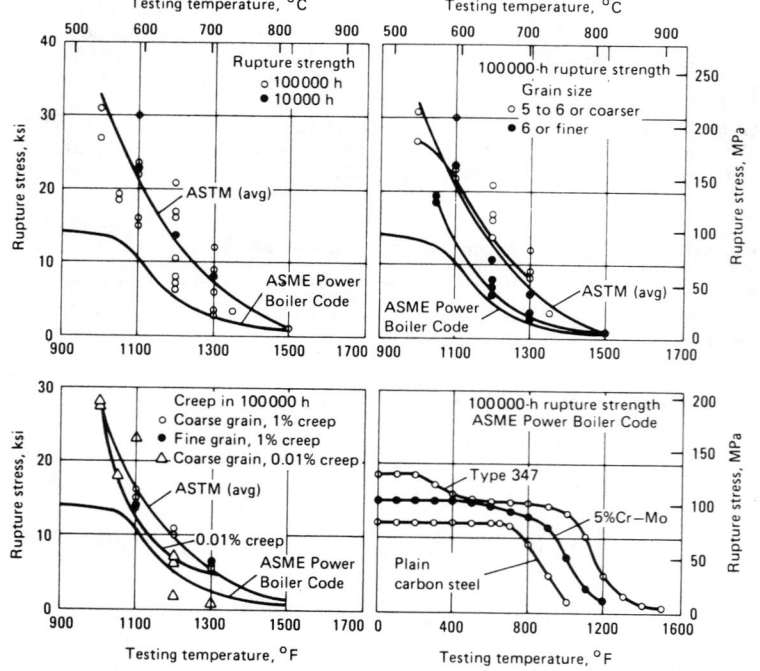

All charts except lower right are derived from ASTM STP 124, 1952.

Fig. 1. Rupture strength and creep properties of type 347 stainless steel compared with stresses permitted by the ASME Power Boiler Code

the upper left-hand chart. The stresses permitted by the ASME Power Boiler Code are below the average curve, except for low values for fine-grain material at 650 and 700 °C (1200 and 1300 °F). The results of an analysis of these tests are plotted in the upper right-hand chart in Fig. 1, classified according to grain size. Above 590 °C (1100 °F) the results fall into two distinct groups on the basis of grain size, with coarse-grain materials having higher rupture strengths.

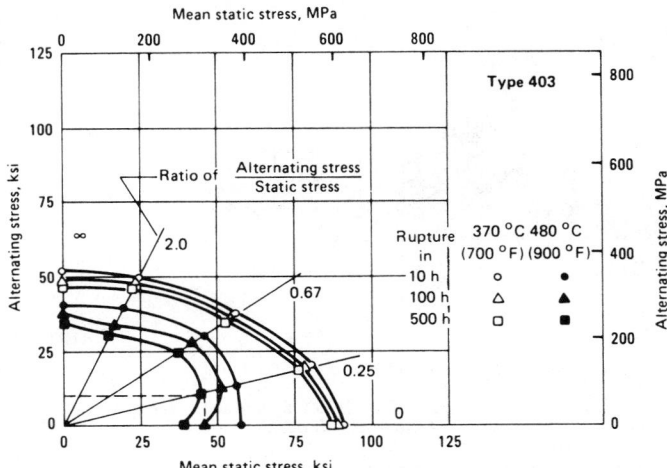

Relationship of static loads and superimposed alternating loads for type 403, oil quenched from 955 °C (1750 °F) and tempered at 540 °C (1000 °F). Axial alternating force up to ±2270 kg (±5000 lb) was produced by a 3600-rpm mechanical oscillator. A preload of 4540 kg (10 000 lb) was applied by calibrated springs and automatic follow-up.

Fig. 2. Effects of alternating loads on rupture life of type 403 stainless steel

DYNAMIC CREEP AND FATIGUE PROPERTIES

The relationship between static loads and superimposed fluctuating loads for type 403 stainless steel at two different temperatures is shown in Fig. 2. The points on the horizontal axes are the stresses that result in rupture under constantly applied static load at times indicated on the elliptical segments. The points of intersection of the elliptical segments with the vertical axes indicate the fatigue strengths for the indicated times when the mean stress is zero.

When a small fluctuating stress is superimposed on a static stress (for instance, $A = 0.25$), the static stress can be larger than the stress to rupture when no alternating stress is present, and the curve bulges to the right. This effect generally increases with time and temperature. A point in the bulge area also indicates longer life for a specimen with small alternating stress superimposed on the static stress.

Overheating. An alloy is selected for high-temperature service on the basis of its performance at some specific limiting temperature at which the design stresses can be maintained for the life of the equipment and within the limit of some predetermined amount of deformation. Static and dynamic tests on the alloy at this limiting temperature are satisfactory criteria for design if the component is not overheated or overstressed in service.

Relaxation of a bolted joint is a special case of tensile creep under gradually decreasing load with constant total elastic-plus-plastic deformation of the bolted assembly. The mechanical properties desired in a bolt are high yield strength, high relaxation resistance and satisfactory notch rupture strength for the expected life of the bolt.

FERRITIC STAINLESS STEELS

Many stainless steels of the 400 series have essentially ferritic structures at all temperatures. Types 405, 430, 434 and 446 form a certain amount of austenite when heated to high temperatures. Type 409 also may form some austenite, particularly if the titanium content is relatively low, while the other steels listed are completely ferritic at all temperatures.

The amount of chromium added for corrosion and oxidation resistance varies from 11% in type 409 to 29% in 29Cr-4Mo. Titanium is used to tie up carbon and nitrogen for structure control and resistance to intergranular corrosion.

An important structural characteristic of ferritic stainless steels is precipitation of alpha prime, a chromium-rich ferrite, when the steel is exposed to temperatures in the range from 370 to 540 °C (700 to 1000 °F). This precipitation results in an increase in hardness and a drastic reduction in room-temperature toughness, which is known as 475 °C (885 °F) embrittlement. This embrittlement occurs in all ferritic grades that have chromium contents above approximately 13%, and its severity increases at higher chromium levels. This characteristic has to be taken into consideration for ferritic grades intended for applications involving exposure to temperatures in the range from 370 to 540 °C (700 to 1000 °F), because subsequent room-temperature ductility will be severely impaired.

In the higher-chromium alloys such as type 446, 26-1 and 29-4, sigma phase is encountered at temperatures above 565 °C (1050 °F). Chi phase will also form in 26-1Ti and 29Cr-4Mo, if the titanium content is above 0.5%. Extensive formation of these phases can also result in severe embrittlement that persists up to at least 370 °C (700 °F).

The rupture strength and creep strength of types 430 and 446 are illustrated in Fig. 3. Long-time and short-time high-temperature strengths of ferritic steels are relatively low compared with those of austenitic steels.

The main advantage of ferritic stainless steels for high-temperature use is their good oxidation resistance, which is comparable to that of austenitic grades. In view of their lower alloy content and lower cost, ferritic steels should be used in preference to austenitic steels, stress conditions permitting.

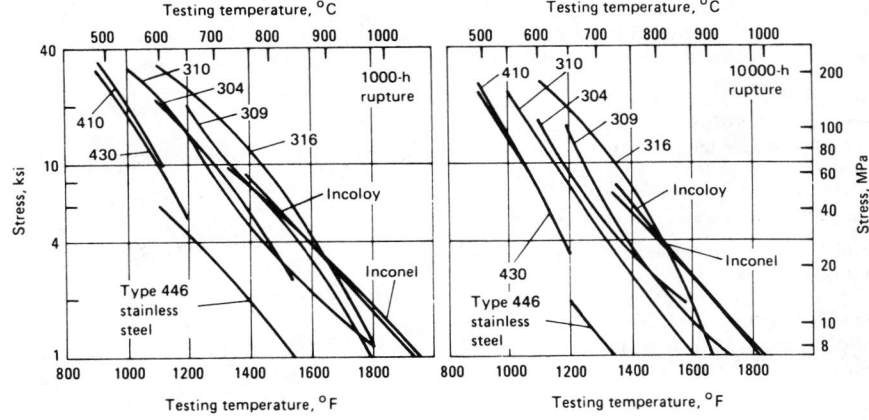

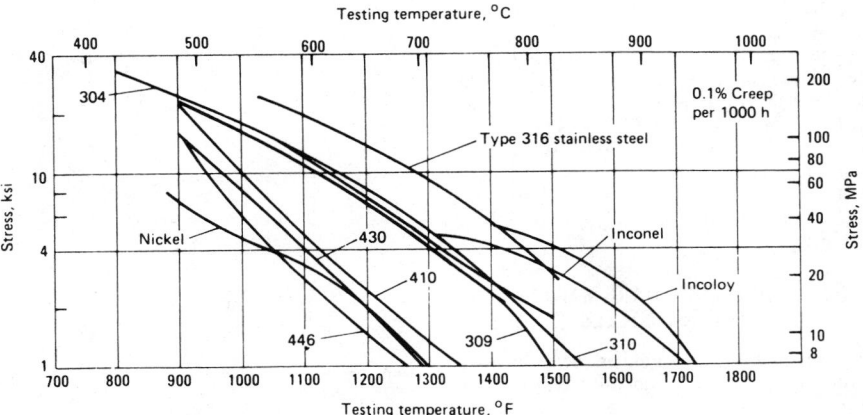

Fig. 3. Creep and stress-rupture comparisons for selected iron-base heat-resistant alloys

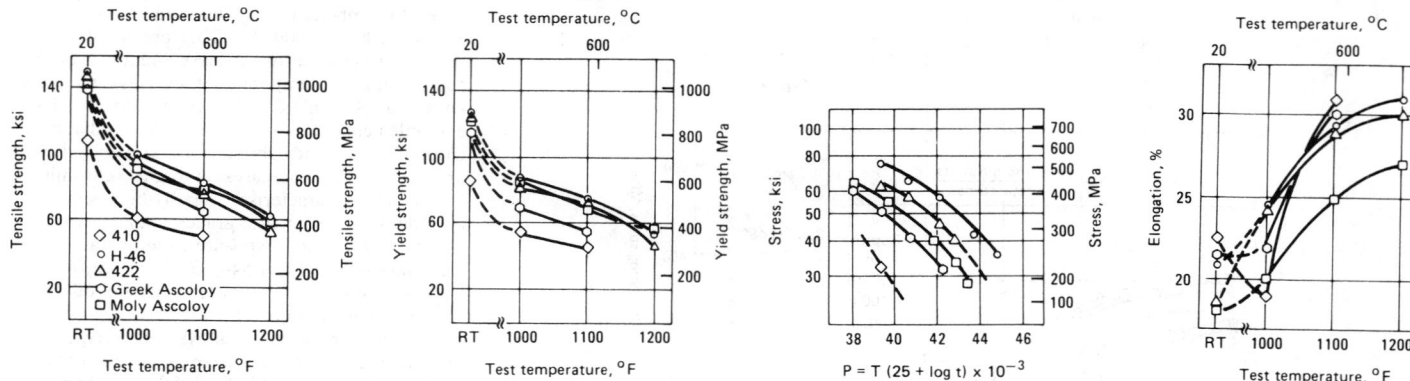

Heat treating schedules were as follows. Type 410: 1 h at 980 °C (1800 °F), oil quench; plus 2 h at 650 °C (1200 °F), air cool. Type 422: 1 h at 1040 °C (1900 °F), oil quench; plus 2 h at 650 °C, air cool. H-46: 1 h at 1150 °C (2100 °F), air cool; plus 2 h at 650 °C, air cool. Greek Ascoloy: 1 h at 955 °C (1750 °F), oil quench; plus 2 h at 650 °C, air cool. Moly Ascoloy: 30 min at 1050 °C (1925 °F), oil quench; plus 2 h at 650 °C, air cool.

Fig. 4. Comparison of mechanical properties of martensitic stainless steels

QUENCHED-AND-TEMPERED MARTENSITIC STAINLESS STEELS

Quenched-and-tempered martensitic stainless steels are essentially martensitic, and harden when cooled from the austenitizing temperature. These alloys offer good combinations of mechanical properties, with usable strength up to 590 °C (1100 °F), and relatively good corrosion resistance. The strength levels at temperatures up to 590 °C that can be attained in these alloys through heat treatment are considerably higher than those attainable in ferritic stainless steels, but the martensitic alloys have inferior corrosion resistance.

These alloys normally are purchased in the annealed or the fully treated (hardened-and-tempered) condition. They are used in the hardened-and-tempered condition. For best long-time thermal stability, these alloys should be tempered at a temperature 110 to 165 °C (200 to 300 °F) above the expected service temperature.

Properties. Quenched-and-tempered martensitic stainless steels can be grouped according to increasing strength and heat resistance as follows:

- Group 1 (lowest strength and heat resistance): types 403, 410 and 416
- Group 2: Greek Ascoloy and type 431
- Group 3: Moly Ascoloy (Jethete M152)
- Group 4 (highest strength and heat resistance): H-46 and type 422.

An actual comparison of mechanical-property data is presented in Fig. 4 for some of these alloys. The short-time tensile and rupture data shown in Fig. 4 were generated in tests of material that had been given austenitizing treatments typical for the specific alloys tested. Because these alloys normally are used at service temperatures near 540 °C (1000 °F), although they may be used up to 590 °C (1100 °F), data are shown for a relatively high tempering temperature of 650 °C (1200 °F), which results in good thermal stability in these alloys at 540 °C (1000 °F).

PRECIPITATION-HARDENING MARTENSITIC STAINLESS STEELS

The precipitation-hardening martensitic stainless (maraging) steels fill an important position between the chromium-free 18% Ni maraging steels and the 12% Cr, low-nickel quenched-and-tempered martensitic stainless alloys. These PH alloys contain 12 to 16% Cr for corrosion resis-

tance and scaling resistance at elevated temperatures and thus are intermediate in heat-resistance capability between the 18 Ni maraging steels and the 12% Cr, low-nickel quenched-and-tempered martensitic stainless alloys (such as type 422).

These alloys normally are purchased in the solution-annealed condition. Depending on the application, they may be used in the annealed condition or in the annealed-plus-age-hardened condition. In some cases, material will be supplied in an "over-aged" condition to facilitate forming of parts. The formed parts are then solution annealed following fabrication.

Properties. The following five maraging alloys are listed in Table 1 and can be divided into two groups on the basis of strength:

- Group 1 (lower strength): Custom 450, 17-4 PH and 15-5 PH
- Group 2 (higher strength): Custom 455 and PH 13-8 Mo.

Property data for these alloys are illustrated in Fig. 5. Data for 15-5 PH are not shown separately, because the properties of this alloy are very similar to those of 17-4 PH.

Short-time tensile data indicate that Custom 455 and PH 13-8 Mo have higher strengths than Custom 450, 17-4 PH and 15-5 PH. For all of these alloys, tensile and yield strengths drop rapidly at temperatures above 425 °C (800 °F), and tensile elongation is greater than 10% over the temperature range from ambient to 540 °C (1000 °F).

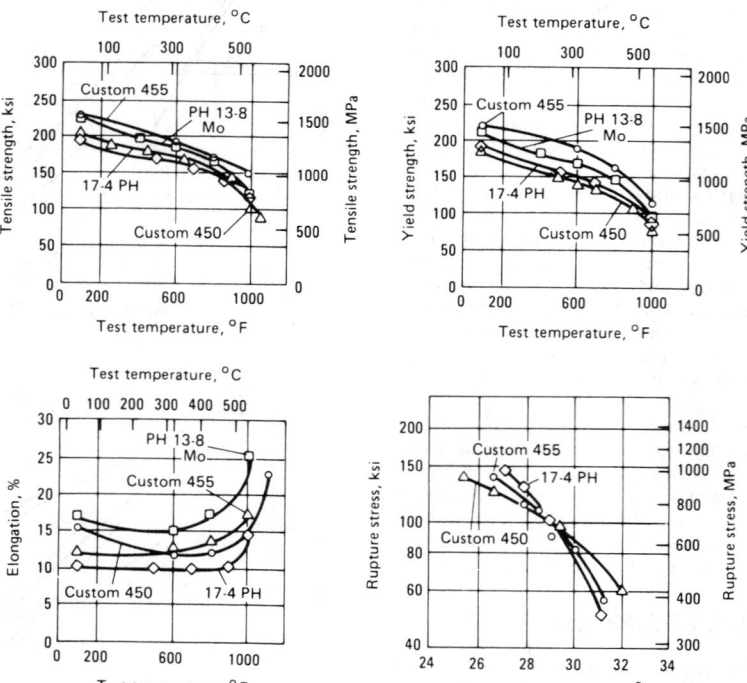

Heat treating schedules were as follows. Custom 450: 1 h at 1040 °C (1900 °F), water quench; plus 4 h at 480 °C (900 °F), air cool. 17-4 PH: 30 min at 1040 °C, oil quench; plus 4 h at 480 °C, air cool. Custom 455: 30 min at 815 °C (1500 °F), water quench; plus 4 h at 510 °C (950 °F), air cool. PH 13-8 Mo: oil quenched from 925 °C (1700 °F); plus 4 h at 540 °C (1000 °F), air cool.

Fig. 5. Comparison of mechanical properties of precipitation-hardening martensitic stainless steels

Stress-rupture data are compared in Fig. 5 by means of a master parameter plot. Data were developed at testing temperatures of 425 and 480 °C (800 and 900 °F) during the time periods of 100 and 1000 h.

PRECIPITATION-HARDENING SEMIAUSTENITIC STAINLESS STEELS

The precipitation-hardening semiaustenitic heat-resistant stainless steels are modifications of standard 18-8 austenitic stainless steels. Nickel contents are lower, and such elements as aluminum, copper, molybdenum and niobium are added. These alloys are solution annealed above 1040 °C (1900 °F), and in this condition can be formed, stamped, stretched and otherwise cold worked to about the same extent as can 18-8 alloys, although they are less ductile and may require intermediate annealing.

All the semiaustenitic stainless steels also can be used in the cold worked condition, in either sheet or wire form. Cold working causes partial transformation of the rather unstable austenite to martensite due to plastic deformation. Aging or tempering is performed after cold working.

AUSTENITIC STAINLESS STEELS

Austenitic stainless steels comprise a group of iron-base alloys that contain 16 to 25% Cr and residual to 20% Ni. Some alloys may contain as much as 18% Mn. These stainless steels are not hardenable by heat treatment but can be hardened by cold work. The effect of cold work on elevated-temperature mechanical properties depends on the recrystallization temperature of the alloy, the amount of residual stress resulting from

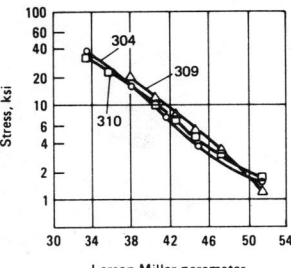 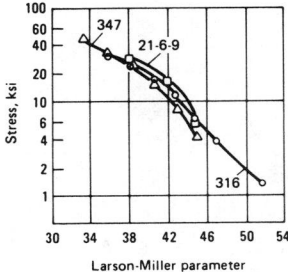 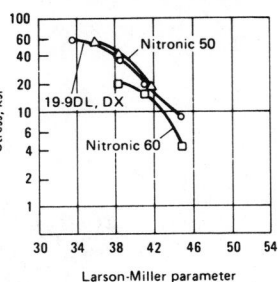

Heat treating schedules were as follows. Type 304: 1065 °C (1950 °F), water quench. Type 309: 1090 °C (2000 °F), water quench. Type 310: 1120 °C (2050 °F), water quench. Type 316: 1090 °C (2000 °F), water quench. Type 347: 1065 °C (1950 °F), water quench. 21-6-9: 1065 °C (1950 °F), water quench. 19-9 DX, DL: for tests above 705 °C (1300 °F), 1065 °C (1950 °F) and water quench, plus 705 °C and air cool; for tests below 705 °C, 705 °C and air cool. Nitronic 50: 1090 °C (2000 °F), water quench. Nitronic 60: 1065 °C (1950 °F), water quench. Larson-Miller parameter = $T/1000$ $(20 + \log t)$, where T is temperature in °R and t is time in h; all data taken from 1000-h tests.

Fig. 6. Stress-rupture plots for austenitic stainless steels

the cold work, and the duration of thermal exposure.

The austenitic stainless steels listed in Table 1 can be grouped into three categories, based primarily on composition, as follows:

- 300 *series alloys,* which are essentially Cr-Ni and Cr-Ni-Mo austenitic stainless steels to which small amounts of other elements have been added.
- *19-9 DL, 19-9 DX* and *17-14-CuMo,* all of which contain 1.25 to 2.5 Mo and 0.3 to 0.55 Ti. Other elements used include 1.25 W and 3 Cu in 17-14-CuMo.
- *Cr-Ni-Mn alloys,* which include types 202 and 216; 21-6-9; Nitronics 32, 33, 50 and 60; and Carpenter 18-18 Plus. These alloys contain 5 to 18 Mn and 0.10 to 0.50 N.

Stress-Rupture Properties. Type 316 stainless steel has the highest stress-rupture capability of all the 300 series alloys (see Fig. 6). Type 347 appears to be somewhat weaker than type 316, and weakens with increasing temperature at a higher rate. Types 304, 309 and 310 are inferior to types 316 and 347 in stress-rupture capability.

The 19-9 DL and 19-9 DX alloys have rupture strengths that are superior to those of all 300 series alloys over the temperature range for which rupture data are available (540 to 815 °C, or 1000 to 1500 °F).

The Cr-Ni-Mn alloys have higher stress-rupture capabilities than 300 series alloys, with the following exceptions: type 316 is superior to 21-6-9, and both 316 and 347 are stronger than Nitronic 60.

Superalloys

By Matthew J. Donachie, Jr., The Hartford Graduate Center

SUPERALLOYS are nickel-, iron-nickel- and cobalt-base alloys generally used at temperatures above about 540 °C (1000 °F). Iron, cobalt and nickel are transition metals with consecutive positions (Fe, Co, Ni) in the periodic table of the elements and are found in selected recoverable deposits in the earth's crust. The techniques of recovery of iron, nickel and cobalt have been well established for years and are not covered here.

The Fe-Ni-base superalloys are an extension of stainless steel technology and generally are wrought, whereas Co-base and Ni-base superalloys may be wrought or cast depending on the application/composition involved. Appropriate compositions of all superalloy base metals can be forged, rolled to sheet or otherwise formed into a variety of shapes. The more highly alloyed compositions normally are processed as castings. Fabricated structures can be built up by welding or brazing, but many highly alloyed compositions containing a high amount of hardening phase are difficult to weld. Properties can be controlled by adjustments in composition and by processing (including heat treatment), and excellent elevated-temperature strengths are available in finished products. Figure 1 compares stress-rupture behavior of the three alloy classes (Fe-Ni-, Ni- and Co-base).

IMPORTANT METAL CHARACTERISTICS

Pure iron has a density of 7.87 Mg/m³ (0.284 lb/in.³), and pure nickel and cobalt have densities of about 8.9 Mg/m³ (0.322 lb/in.³). Fe-Ni-base superalloys have densities of about 7.9 to 8.3 Mg/m³ (0.285 to 0.300 lb/in.³); Co-base superalloys, about 8.3 to 9.4 Mg/m³ (0.300 to 0.340

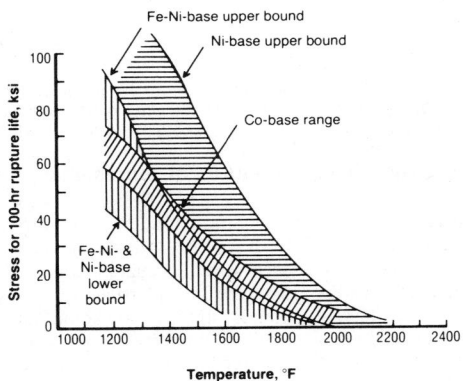

Fig. 1. Stress-rupture behavior of superalloys

lb/in.³); and Ni-base superalloys, about 7.8 to 8.9 Mg/m³ (0.282 to 0.322 lb/in.³). Superalloy density is considerably influenced by alloying additions: higher aluminum, titanium and chromium contents reduce density, whereas the solid-solution strengtheners tungsten and tantalum increase it. The degree of corrosion resistance of the pure metals depends strictly on the environment, but the corrosion resistance of superalloys depends primarily on the alloying elements added. The melting temperatures of the pure elements are as follows: nickel, 1453 °C (2647 °F); cobalt, 1495 °C (2723 °F); and iron, 1537 °C (2798 °F). Incipient melting temperatures and melting ranges of superalloys are functions of composition and prior processing. Generally, incipient melting temperatures are greater for Co-base than for Ni- or Fe-Ni-base superalloys. Ni-base superalloys may show incipient melting at temperatures as low as 1204 °C (2200 °F). Advanced Ni-base single-crystal superalloys having limited amounts of melting-point depressants tend to have incipient melting temperatures equal to or in excess of those of Co-base alloys. Iron and cobalt both undergo allotropic transformations: iron transforms from the body-centered-cubic (bcc) lower-temperature (alpha) form to the face-centered-cubic (fcc) high-temperature (gamma) form, and

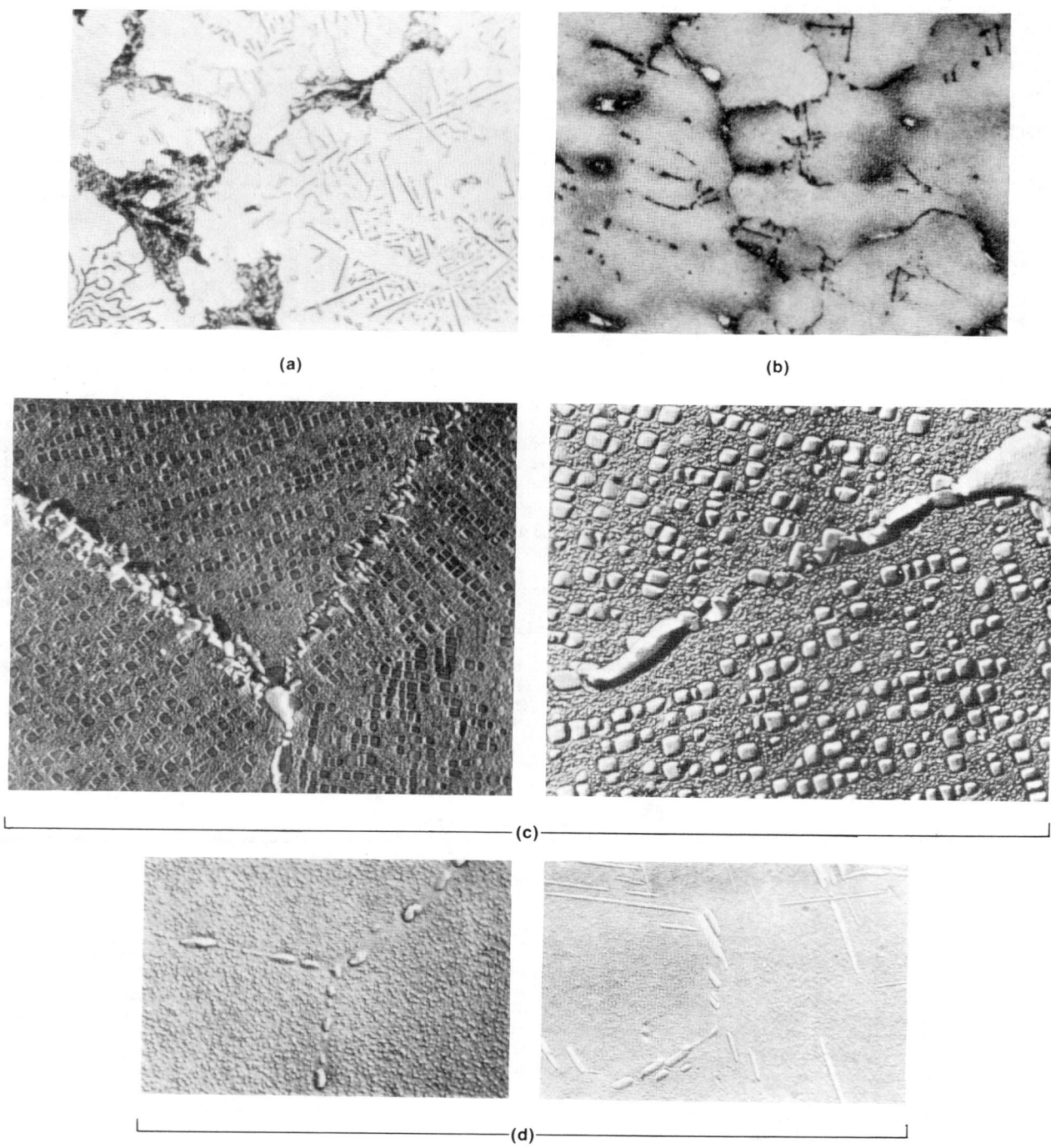

(a) Cast Co-base alloy (250×). (b) Cast Ni-base alloy (100×). (c) Wrought (left, 3300×) and cast (right, 5000×) Ni-base alloys. (d) Two wrought Fe-Ni-base alloys (left, 17 000×; right, 3300×). Note script carbides in (a) and (b) as well as eutectic carbide-cobalt grain-boundary structures in (a), spheroidal and cuboidal γ' as well as grain-boundary carbides in (c), and spheroidal γ' as well as grain-boundary carbides or grain-boundary and intragranular δ phase in (d). γ" not obvious but present in (d) (right). See also Fig. 6 and 9(a).

Fig. 2. Typical operating microstructures of representative superalloys

cobalt transforms from the hexagonal-close-packed (hcp) lower-temperature form to gamma fcc at higher temperatures. Nickel, on the other hand, is fcc at all temperatures. The fcc forms of iron and cobalt in superalloys based on either of these elements generally are stabilized by alloying elements. The upper limit of usage for superalloys is not restricted by the occurrence of allotropic transformation reactions but is a function of incipient melting temperature and dissolution of

strengthening phases. Some tendency toward transformation of the fcc phase to stable lower-temperature phases occasionally occurs in Co-base superalloys. The austenitic fcc matrices of superalloys have extended solubility for some alloying additions, excellent ductility, and favorable characteristics for precipitation of uniquely effective strengthening phases (Fe-Ni- and Ni-base superalloys).

Superalloys typically have moduli of elasticity

in the vicinity of 207 GPa (30×10^3 ksi), although moduli of specific polycrystalline alloys can vary from 172 to 241 GPa (25 to 35×10^3 ksi) at room temperature depending on the alloy system. Processing that leads to directional grain or crystal orientation can result in moduli of about 124 to 310 GPa (about 18 to 45×10^3 ksi) depending on the relation of grain or crystal orientation to testing direction. The physical properties, electrical conductivity, thermal conductivity and thermal expansion tend to be low (relative to other metal systems). These properties are influenced by the nature of the base metals (transition elements) and the presence of refractory-metal additions.

The superalloys are relatively ductile, although the ductilities of Co-base superalloys generally are less than those of Fe-Ni- and Ni-base superalloys. Fe-Ni- and Ni-base superalloys are readily available in extruded, forged or rolled form; the higher-strength alloys generally are found only in the cast condition. Hot deformation is the preferred process, cold forming usually being restricted to thin sections (sheet). Cold rolling may be used to increase short-time strength properties for applications at temperatures below the lower temperature level of 540 °C (1000 °F) established in this article for superalloy use.

PHASES AND STRUCTURES OF SUPERALLOYS

Superalloys consist of the austenitic fcc matrix phase gamma (γ) plus a variety of secondary phases. The principal secondary phases are the carbides MC, $M_{23}C_6$, M_6C and M_7C_3 (rare) in all superalloy types and gamma prime (γ') fcc ordered $Ni_3(Al,Ti)$ intermetallic compound in Ni- and Fe-Ni-base superalloys. The superalloys derive their strength from solid-solution hardeners and precipitating phases. Carbides may provide limited strengthening directly (e.g., through dispersion hardening) or, more commonly, indirectly (e.g., by stabilizing grain boundaries against excessive shear). In addition to those elements that produce solid-solution hardening and promote carbide and γ' formation, other elements (e.g., boron, zirconium, hafnium and cerium) are added to enhance mechanical or chemical properties. Some carbide- and γ'-forming elements may contribute significantly to chemical properties as well. Table 1 gives a generalized list of the ranges of alloying elements and their effects in superalloys. Typical operating microstructures of representative superalloys are shown in Fig. 2.

SUPERALLOY SYSTEMS

The three types of superalloys, Fe-Ni, Ni- and Co-base, are further subdivided into cast and wrought. A large number of alloys have been invented and studied, and many have been patented. A representative list of superalloys and compositions is given in Table 2.

Fe-Ni-Base. The most important class of Fe-Ni-base superalloys includes those alloys which are strengthened by intermetallic compound precipitation in a fcc matrix. The most common precipitate is γ', typified by A-286, V-57 or Incoloy 901, but some alloys precipitate gamma double prime (γ''), typified by Inconel 718. Another class of Fe-Ni-base superalloys is typified by the CRM series, which is hardened by carbides, nitrides and carbonitrides; some tungsten and molybdenum may be added to produce solid-solution hardening. Other Fe-Ni-base superalloys consist of modified stainless steels primarily strengthened by solid-solution hardening. Alloys in this last category vary from 19-9DL (18-8 stainless with slight Cr and Ni adjustments, additional solution hardeners and higher C) to Incoloy 800H (21 Cr, high Ni with small additions of Ti and Al, which yields some γ' phase). The Fe-Ni-base superalloys are used in the wrought condition although the CRM series was developed primarily for casting applications.

Ni-Base. The most important class of Ni-base superalloys is that strengthened by intermetallic-compound precipitation in a fcc matrix. The strengthening precipitate is γ', typified by Waspaloy or Udimet 700. Another class of Ni-base superalloys is represented by Hastelloy X, which is essentially solid-solution strengthened but probably also derives some strengthening from carbide precipitation produced through a working-plus-aging schedule. A third class includes oxide-dispersion-strengthened (ODS) alloys such as IN MA-754 and IN MA-6000E, which are strengthened by dispersions of inert particles such as yttria coupled in some cases with γ' precipitation (MA-6000E).

Ni-base superalloys are utilized in both cast and wrought forms, although special processing

Table 1. Common ranges of major alloying additions and their effects in superalloys

Element	Range, % Fe-Ni- and Ni-base	Range, % Co-base	Effect
Cr	5-25	19-30	Oxidation and hot corrosion resistance; carbides; solution hardening
Mo, W	0-12	0-11	Carbides; solution hardening
Al	0-6	0-4.5	Precipitation hardening; oxidation resistance
Ti	0-6	0-4	Precipitation hardening; carbides
Co	0-20	...	Affects amount of precipitate
Ni	...	0-22	Stabilizes austenite; forms hardening precipitates
Nb	0-5	0-4	Carbides; solution hardening; precipitation hardening (Ni-, Fe-Ni-base)
Ta	0-12	0-9	Carbides; solution hardening; oxidation resistance

Table 2. Compositions of selected superalloys

From *Metals Handbook*, 9th Edition, Vol 3, 1980, and *Metal Progress 1982 Materials and Processing Databook*, June 1982, except for IN MA-6000E and J-1570

Alloy	Cr	Ni	Co	Mo	W	Nb	Ti	Al	Fe	C	Other
Fe-Ni-base											
19-9DL 19.0		9.0	...	1.25	1.25	0.4	0.3	...	66.8	0.30	1.10 Mn; 0.6 Si
Incoloy 800 ....21.0		32.5	...	...	...	...	0.38	0.38	45.7	0.05	0.8 Mn; 0.5 Si
A-286 ....15.0		26.0	...	1.25	...	...	2.0	0.2	55.2	0.04	0.005 B; 0.3 V
V-57 ....14.8		27.0	...	1.25	...	...	3.0	0.25	48.6	0.08 max	0.01 B; 0.5 max V
Incoloy 901 ....12.5		42.5	...	6.0	...	...	2.7	...	36.2	0.10 max	...
Inconel 718 ....19.0		52.5	...	3.0	...	5.1	0.9	0.5	18.5	0.08 max	0.15 max Cu
Hastelloy X 22.0		49.0	1.5 max	9.0	0.6	...	...	2.0	15.8	0.15	...
Ni-base											
Waspaloy 19.5		57.0	13.5	4.3	...	...	3.0	1.4	2.0 max	0.07	0.006 B; 0.09 Zr
M252 ....19.0		56.5	10.0	10.0	...	...	2.6	1.0	<0.75	0.15	0.005 B
Udimet 500 ....19.0		48.0	19.0	4.0	...	...	3.0	3.0	4.0 max	0.08	0.005 B
Udimet 700 ....15.0		53.0	18.5	5.0	...	...	3.4	4.3	<1.0	0.07	0.03 B
Astroloy 15.0		56.5	15.0	5.25	...	...	3.5	4.4	<0.3	0.06	0.03 B; 0.06 Zr
René 80 ....14.0		60.0	9.5	4.0	4.0	...	5.0	3.0	...	0.17	0.015 B; 0.03 Zr
IN-100 ....10.0		60.0	15.0	3.0	...	...	4.7	5.5	<0.6	0.15	1.0 V; 0.06 Zr; 0.015 B
René 95 ....14.0		61.0	8.0	3.5	3.5	3.5	2.5	3.5	<0.3	0.16	0.01 B; 0.05 Zr
MAR-M 247 .... 8.25		59.0	10.0	0.7	10.0	...	1.0	5.5	<0.5	0.15	0.015 B; 0.05 Zr; 1.5 Hf; 3.0 Ta
IN MA-754 ....20.0		78.5	...	...	...	...	0.5	0.3	...	...	0.6 Y$_2$O$_3$
IN MA-6000E 15.0		68.5	...	2.0	4.0	...	2.5	4.5	...	0.05	1.1 Y$_2$O$_3$; 2.0 Ta; 0.01 B; 0.15 Zr
Co-base											
Haynes 25(L-605) 20.0		10.0	50.0	...	15.0	...	...	...	3.0	0.10	1.5 Mn
Haynes 188 ....22.0		22.0	37.0	...	14.5	...	...	...	3.0 max	0.10	0.90 La
S-816 ....20.0		20.0	42.0	4.0	4.0	4.0	...	...	4.0	0.38	...
X-40 ....22.0		10.0	57.5	...	7.5	...	...	...	1.5	0.50	0.5 Mn; 0.5 Si
WI-52 ....21.0		...	63.5	...	11.0	...	...	...	2.0	0.45	2.0 Nb + Ta
MAR-M 302 ....21.5		...	58.0	...	10.0	...	...	...	0.5	0.85	9.0 Ta; 0.005 B; 0.2 Zr
MAR-M 509 ....23.5		10.0	54.5	...	7.0	...	0.2	...	...	0.6	0.5 Zr; 3.5 Ta
J-1570 ....20.0		28.0	46.0	...	...	...	4.0	...	2.0	0.2	...

(powder metallurgy/isothermal forging) frequently is used to produce wrought versions of the more highly alloyed compositions (Udimet 700/Astroloy, IN-100).

An additional dimension of Ni-base superalloys has been the introduction of grain-aspect ratio and orientation as a means of controlling properties. In fact, in some instances, grain boundaries have been removed (see discussion of investment casting, below). Wrought powder metallurgy alloys of the ODS class and cast alloys such as MAR-M 247 have demonstrated property improvements owing to control of grain morphology by directional recrystallization or solidification.

Co-Base. The Co-base superalloys are invariably strengthened by a combination of carbides and solid-solution hardeners. The essential distinction in these alloys is between cast and wrought structures. Cast alloys are typified by X-40, and wrought alloys by Haynes 25. No intermetallic compound possessing the same degree of utility as the γ' precipitate in Ni- or Fe-Ni-base superalloys has been found to be operative in Co-base systems, although alloy J-1570 derives some of its strength from Ni_3Ti precipitation, and certain high-tungsten Co-base superalloys developed by NASA reportedly have been hardened by precipitation of Co_3W.

PROCESSING

Melting

A number of superalloys, particularly Co- and Fe-Ni-base alloys, are air melted by various methods applicable to stainless steels. However, for most Ni- or Fe-Ni-base superalloys, vacuum induction melting (VIM) is required as the primary melting process. The use of VIM reduces interstitial gases (O_2, N_2) to low levels, enables higher and more controllable levels of aluminum and titanium (along with other relatively reactive elements) to be achieved and results in less contamination from slag or dross formation than air melting. The benefits of reduced gas content and ability to control aluminum plus titanium are shown in Fig. 3 and 4. Vacuum arc remelting (VAR) and electroslag remelting (ESR) are the commonly employed secondary melting techniques.

The VIM process generally is used as the initial melting process for superalloys and may be the only melting process used when material for investment castings is being produced. However, for material which will be subjected to conventional wrought processing, particularly when the material is one of the higher-strength superalloys which must be hot worked to produce larger gas turbine parts, a secondary remelting operation is required. VIM ingots generally have coarse and nonuniform grain sizes, shrinkage and alloying-element segregation. Although these factors cause no problem in producing primary material which will be remelted for casting, such factors restrict the hot workability of forging alloys such as Incoloy 901, Waspaloy, Inconel 718 and Astroloy. The above problems are resolved by the use of VIM followed by VAR or ESR. In addition to alloy composition refinement, VAR and ESR act to refine the solidification structure of the resulting ingot. In some advanced Ni-base superalloys having high volume fractions (V_f) of γ', even VIM-VAR or VIM-ESR does not provide a satisfactory ingot structure for subsequent hot working. Such superalloys have been processed by powder metallurgy techniques (see "Powder Processing," below). Recent improvements in melting/ingot casting technology such as vacuum arc double electrode remelting (VADER) appear to offer the promise of sufficient structural refinement to permit such high-strength-alloy ingots to be processed in ingot form by appropriate extrusion and forging routes.

Electron-beam remelting/refining (EBR) has been evaluated as an alternative process for improving Ni- and Fe-Ni-base superalloy properties and processability through a further lowering of impurity levels and drastic reductions in dross/inclusion content. EBR can help to produce improved feedstock for casting operations or to provide more workable starting ingot for wrought processing. The expanded use of secondary melt processes such as EBR, argon-oxygen degassing (AOD) and VADER will be governed by the extent to which they each provide an economical means for Ni- and Fe-Ni-base superalloy processing.

Melting of Co-base superalloys generally does not require the sophistication of vacuum processing. An air induction melt (AIM) is commonly used, but VIM and ESR also have found application, the latter being used to produce stock for subsequent deformation processing. Alloys containing aluminum or titanium (J-1570) and tantalum or zirconium (MAR-M 302, MAR-M 509) must be melted by VIM. Vacuum melting

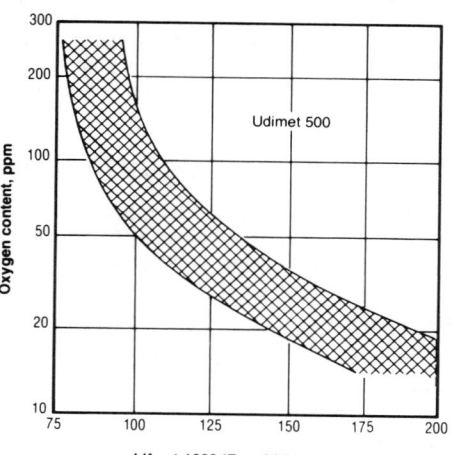

Fig. 3. Improvement of rupture life at 1600 °F and 25 ksi by reduced oxygen content produced by vacuum melting

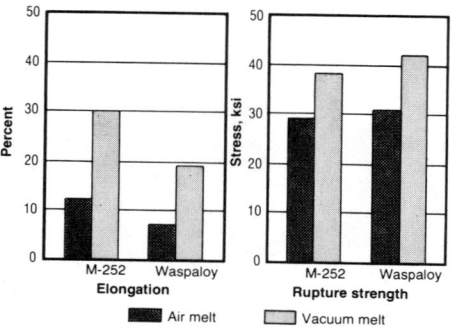

Fig. 4. Effects of vacuum melting, incorporating beneficial modifications in composition, on properties of two Ni-base superalloys

of other Co-base superalloys may enhance properties such as strength and ductility because of the improved cleanness and compositional control associated with this process.

Deformation Processing (Forging, Sheet Rolling)

The principal methods for deformation processing of superalloys are extrusion, extrusion followed by forging, and extrusion or cogging followed by sheet rolling. Processability is a direct function of alloy composition and cast ingot structure. Extrusion or cogging produces size reductions involving strains which lead to homogenization of the alloy during in-process anneals prior to final processing. Extrusion also may be used to set up grain size so that isothermal/superplastic forging can be performed. Not all alloy types are processed economically by all techniques. Forging is principally applied to Ni- and Fe-Ni-base superalloys and customarily is done by hammer (drop) forging or press forging. Traditionally, forging is carried out under falling temperature, because the workpiece is transferred from the preheat furnace into dies which are at lower temperatures. In recent years, isothermal forging has been carried out by using insulation and/or heating to maintain the part and forging dies at essentially the same constant temperature.

Forging processes have used open dies or closed dies, and generally, except in isothermal/superplastic forging, multiple forging steps are required. Rolling also is a multistep reduction process. Extrusion, if used to produce a commercial article (e.g., barstock), may be just a one-step operation.

Extrusion and rolling generally produce a final product of uniform grain size which is annealed, although cold finishing may be used for some superalloy sheet not meant for high-temperature operation. Forging operations may be conducted under one of two different philosophies. A prime objective, when cost is the determining factor, may be to produce a geometrical shape by the most economical means possible and rely on heat treatment to produce the required mechanical properties. A second approach is to aim processing at a preselected microstructure goal and achieve this goal through thermomechanical processing (TMP). Extensive studies of forging practice on Ni- and Fe-Ni-base superalloys have permitted development of full or partially recrystallized forged structures with varying amounts of stored energy. Use of the normal precipitation-hardening phases and other intermetallics — e.g., delta or eta (δ or η) — has been developed to control properties either through final structure in forging or through postforging heat treatment. Because phases such as γ' are effective inhibitors of grain growth, extrusion plus heat treatment at appropriate temperatures relative to the γ' solutioning temperature has been used to produce very fine grain structures which can be superplastically forged at reasonable temperatures. (However, such superplastically forged materials customarily are prepared by processing of powder, not ingot, superalloys.)

Composition affects superalloy forgeability. For example, cracking can occur as a result of detrimental trace elements such as sulfur; in that case, minor elements such as magnesium or cerium are added in some alloys to tie up such elements and thus improve forgeability. Nitrides or carbonitrides in stringer form or segregated compounds such as δ may lead to reduced properties and/or

Fig. 5. Continuous or nearly continuous MC film produced in grain boundaries of Waspaloy after high-temperature forging soak with no subsequent reduction, followed by normal solution treating and aging. 2700×.

cracking during forging. Detrimental carbide films have been produced by reheating without deformation (Fig. 5) and, occasionally, forgings fail because of overheating during the preheat or forging process. Table 3 gives qualitative forgeability ratings of cast-plus-forged superalloys.

For a given forged alloy composition, the size and morphology of the γ', the grain size, and the amount of residual cold/warm work affect the tensile strength attainable. Rupture life at a fixed composition is most likely to be a function of grain size. The microstructural control of the γ' (and the carbides) is associated with working below the solution temperature of γ'. Actual grain structure required and preferred is a function of the alloy system. A necklace structure (Fig. 6), in which small recrystallized grains surround larger warm worked grains, has been held desirable for some high-strength high-V_f γ' alloys. Other alloys rely on a fairly uniform fine grain size (ASTM 5 to 8) and appropriate bimodal γ' size distributions for strength.

Co-base superalloys have been extruded and forged (e.g., S-816), but most wrought Co-base superalloys (e.g., Haynes 25 and Haynes 188) are used in sheet form. ESR processing has been used to produce favorable ingot structures in more complex Co-base superalloys. TMP of Co-base superalloys is not practiced because of the more restricted ductility of these moderate-to-high-carbon alloys and the absence of useful intermetallic-compound strengtheners such as γ', which are needed to achieve control of grain size.

Table 3. Forgeability ratings of superalloys
From R. Galipeau and R. Sjoblad, *Materials Engineering*, Sept 1967.

Alloy	Forging temperature °C	Forging temperature °F	Forgeability
A-286	1065	1950	Excellent
Inconel 901 ...	1095	2000	Good to excellent
Hastelloy X ...	1095	2000	Excellent
Waspaloy	1080	1975	Good
Inconel 718 ...	1065	1950	Excellent
Astroloy	1095	2000	Fair to good

Powder Processing

Powder techniques are being used extensively in superalloy production. Principally, high-strength gas-turbine disk alloy compositions such as IN-100, which are difficult or impractical to forge by conventional methods, have been powder processed. Inert atmospheres are used in the production of powders, often by gas atomization, and the powders are consolidated by extrusion or hot isostatic pressing (HIP). HIP has been used either to produce shapes directly for final machining or to consolidate billets for subsequent forging. Extruded or HIP'ed billets often are isothermally forged to configurations for final machining. Minimal segregation, reduced inclusion sizes, ability to use very high-V_f γ' compositions and ease of grain-size control are significant advantages of the powder process. (Reduced costs, particularly through HIP formation of near-net-shape disks, may be possible with powder techniques, but the extent of actual cost savings is a function of alloy and part complexity.)

Powder techniques also have been used to produce turbine blade/vane alloys of the ODS type. Mechanical alloying is the principal technique for introducing the requisite oxide/strain energy combination to achieve maximum properties. Rapid-solidification-rate (RSR) technology has been applied to produce highly alloyed (very-high-V_f γ') superalloys and shows promise for advanced gas-turbine applications. RSR and ODS alloys can benefit from aligned crystal growth in the same manner as can directionally cast alloys. Directional recrystallization has been used in ODS alloys to produce favorable polycrystalline grain orientations with elongated (high-aspect-ratio) grains parallel to the major loading axis.

Investment Casting

Co-base and high-V_f γ' Ni-base superalloys are processed to complex final shapes by investment casting. Investment casting permits intricate internal cooling passages and re-entrant angles to be achieved and produces a near-net shape of very precise dimensions. Most investment castings are small polycrystalline articles ranging

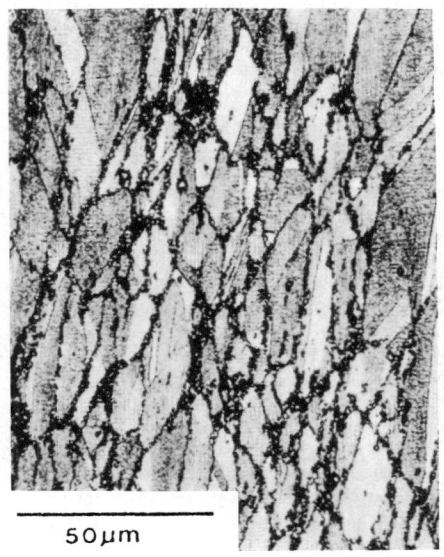

50µm

Fig. 6. Necklace structure in thermomechanically worked René 95. 300×.

in weight from less than a pound to a few pounds, but large investment castings are being made in configurations up to several feet in diameter and hundreds of pounds in weight. A cast alloy commonly used to obtain the economic benefits of large sections is Inconel 718; a wide range of alloys are cast as small parts in polycrystalline form. Grain size is controlled by an appropriate primary dip coat in the investment along with mold-metal pour temperatures and selective mold insulation. Inclusions are controlled by melting technology and the use of filters to eliminate dross. In the past two decades improved core materials and casting mold design plus control of grain size and inclusions have led to improvements in casting yield and, occasionally, improvements in cast part strength properties. Core shift in intricate gas-turbine airfoils and surface attack during chemical removal of cores can reduce casting yield and cause potential property-level reductions if not properly controlled.

Casting porosity has been a problem in parts having large cross sections and in small parts made of some high-V_f γ' alloys. Hot isostatic pressing (HIP) techniques used for powder processing have been applied successfully in many instances to eliminate nonsurface-connected porosity, particularly in large castings of Fe-Ni- and Ni-base superalloys. Improved fatigue and creep life generally result (Fig. 7), because casting quality is improved by HIP.

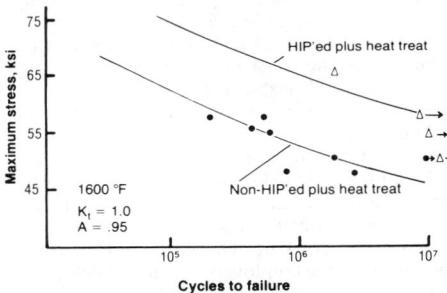

Fig. 7. Beneficial effect of HIP on high-cycle fatigue resistance of René 80

Directional-casting technology is a commonly accepted production process. Columnar grain structures have been produced by promoting unidirectional heat flow within the furnace during the solidification cycle. Substantial property improvements have resulted for many alloys: modulus parallel to the natural growth direction is lowered, leading to improvements in thermal-fatigue life; ductility generally is increased; and creep-rupture strength (life) is improved through removal of transverse grain-boundary segments. A logical extension of columnar-grain technology is the production of single crystals of aircraft gas-turbine airfoils. Removal of all grain boundaries and adjustments of alloy composition permitted by the absence of grain boundaries result in substantial improvements in strength capability. Directional casting of single-crystal articles provides composition flexibility and opens the possibility of additional alloy development for high-strength Ni-base superalloys.

Although Co-base superalloys can be directionally solidified in columnar grain structures, they are invariably cast as polycrystalline parts. Single-crystal manufacture of Co-base superalloys may be possible but has not been reported. It is doubtful that sufficiently significant property benefits would result from single-crystal or columnar-grain Co-base superalloys to warrant the expense of such processing.

Joining

Co-base superalloys are readily welded by gas metal-arc (GMA) or gas tungsten-arc (GTA) techniques. Cast alloys such as WI-52 and wrought alloys such as Haynes 188 have been extensively welded. Filler metals generally have been less highly alloyed Co-base alloy wire, although parent rod or wire (the same composition as the alloy being welded) have been used. Co-base superalloy sheet also is successfully welded by resistance techniques. Gas-turbine vanes which crack in service have been repair welded using the above techniques (e.g., WI-52 vanes using Haynes 25 filler rod and 540 °C, or 1000 °F, preheat). Appropriate preheat techniques are needed in GMA and GTA welding to eliminate tendencies for hot cracking. Electron-beam (EB) and plasma-arc (PA) welding can be used on Co-base superalloys but usually are not required in most applications because alloys of this class are so readily weldable.

Ni- and Fe-Ni-base superalloys are considerably less weldable than Co-base superalloys. Because of the presence of the strengthening phase, the alloys tend to be susceptible to hot cracking (weld cracking) and postweld heat treatment cracking, PWHT (strain age or delay) cracking. The susceptibility to hot cracking is directly related to the aluminum and titanium contents (γ' formers), as shown in Fig. 8. Hot cracking occurs in the weld heat-affected zone, and the extent of cracking varies with alloy composition and weldment restraint.

Ni- and Fe-Ni-base superalloys have been welded by GMA, GTA, EB, laser and PA techniques. Filler metals, when used, usually are weaker, more ductile austenitic alloys so as to minimize hot cracking. Occasionally base-metal compositions are employed as fillers. Welding is restricted to the lower-$V_f \gamma'$ (≤ 0.35) alloys, generally in the wrought condition. Cast alloys of high $V_f \gamma'$ have not been consistently welded successfully when filler metal is required, as in weld repair of service parts. However, EB welding can be used to make structural joints in such alloys. Friction or inertia welding also has been successfully applied to the lower-$V_f \gamma'$ alloys.

Because of their γ' strengthening mechanism and capability, many Ni- and Fe-Ni-base superalloys are welded in the solution-heat-treated condition. Special preweld heat treatments have been used for some alloys. Inconel 718 has unique welding characteristics in that its hardening phase, γ'', is precipitated more sluggishly at a lower temperature than is γ' so that the attendant welding-associated strains that must be redistributed are more readily accommodated in the weld metal and heat-affected zone. The alloy is welded in the solution-treated condition and then given a postweld stress-relief-and-aging treatment that causes γ'' precipitation. Some alloys—e.g., A-286—are inherently difficult to weld despite only moderate levels of γ' hardeners. There is some evidence that high-titanium alloys may be more difficult to weld than alloys of similar $V_f \gamma'$ relying on high Al/Ti ratios for their strength capabilities.

Weld techniques for superalloys must address not only hot cracking but PWHT cracking, par-

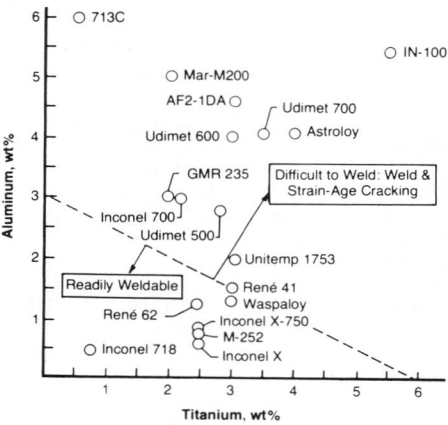

Fig. 8. Weldability diagram for some γ'-strengthened Fe-Ni- and Ni-base superalloys, showing influence of total Al + Ti hardeners

ticularly as it concerns microfissuring (microcracking), because it can be subsurface and therefore difficult to detect. Tensile and stress-rupture strengths may be hardly affected by microfissuring, but fatigue strengths can be drastically reduced.

In addition to being weldable by the usual fusion welding techniques discussed above, Ni- and Fe-Ni-base alloys can be resistance welded when in sheet form; brazing, diffusion bonding and transient liquid phase (TLP) bonding also have been employed to join these alloys. Brazed joints tend to be more ductility-limited than welds; diffusion bonding of superalloys has not found consistent application. TLP bonding has been found to be very useful, principally in turbine parts of aircraft gas-turbine engines. The distinguishing characteristic of TLP bonding which produces its excellent integrity is that, although a lower-temperature bond is made as in brazing, subsequent diffusion occurs at the bonding temperature, leading to a fully solidified joint which has a composition similar to that of the base metal and a microstructure indistinguishable from it. Consequently, the resultant joint can have a melting temperature and properties very similar to those of the base metal.

PROPERTIES AND MICROSTRUCTURE

The principal microstructural variables of superalloys are (*a*) the precipitate amount and its morphology, (*b*) grain size and shape, and (*c*) carbide distribution. Ni- and Fe-Ni-base superalloy properties are controlled by all three variables; the first variable is essentially absent in Co-base superalloys. Structure control is achieved through composition selection/modification and by processing. For a given nominal composition, there are property advantages and disadvantages of the structures produced by deformation processing and by casting. Cast superalloys generally have coarser grain sizes, more alloy segregation and improved creep and rupture characteristics. Wrought superalloys generally have more uniform, and usually finer, grain sizes and improved tensile and fatigue properties.

Ni- and Fe-Ni-base superalloys typically consist of γ' dispersed in a γ matrix, and the strength increases with increasing $V_f \gamma'$. The lowest V_f amounts of γ' are found in Fe-Ni-base and first-

generation Ni-base superalloys, where $V_f \gamma'$ is generally less than about 0.25 (25 vol %). The γ' is commonly spheroidal in lower-$V_f \gamma'$ alloys but often cuboidal in higher-$V_f \gamma'$ ($\gtrsim 0.35$) Ni-base superalloys. The inherent strength capability of such superalloys is controlled by the intragranular distribution; however, the usable strength in polycrystalline alloys is determined by the condition of the grain boundaries, particularly as affected by the carbide-phase morphology and distribution. Satisfactory properties are achieved by optimizing the γ' V_f and morphology (not necessarily independent characteristics) in conjunction with securing a dispersion of discrete globular carbides along the grain boundaries (Fig. 9). Discontinuous (cellular) carbide or γ' at grain boundaries (Fig. 9) increases surface area and drastically reduces rupture life even though tensile and creep strength may be relatively unaffected.

Wrought Ni- and Fe-Ni-base superalloys generally are processed to have optimum tensile and fatigue properties. At one time, when wrought alloys were used for creep-limited applications such as gas-turbine high-pressure turbine blades, heat treatments different from those used for tensile-limited uses were applied to the same nominal alloy composition in order to maximize creep-rupture life. Occasionally, the nominal compo-

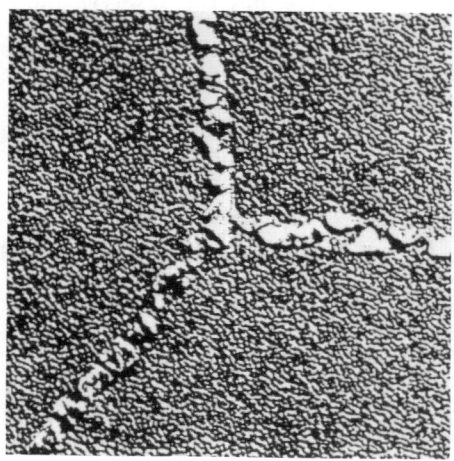

(a)

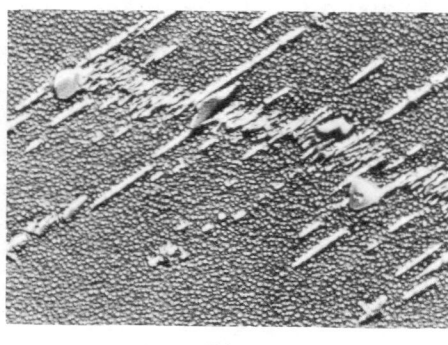

(b)

Fig. 9. Favorable discrete grain-boundary carbide precipitation (a), and less-favorable zipperlike, discontinuous carbide precipitation (b), in Waspaloy. Magnifications: (a), 10 000×; (b), 6800×.

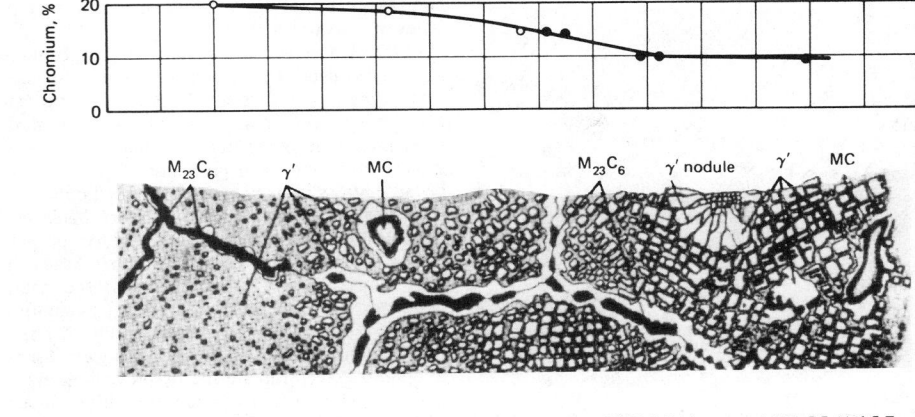

γ'-formers ⟶	2.5 Ti, 1.3Al	2.9 Ti, 2.9 Al	3.5 Ti, 4.3 Al	4.7 Ti, 5.5 Al	1.5 Ti, 5.5 Al, 1.5 Ta
Carbide formers ⟶	20Cr, 2.5 Ti	19 Cr, 4 Mo, 2.9 Ti	15 Cr, 5.2 Mo, 3.5 Ti	10 Cr, 3 Mo, 4.7 Ti, IV	9 Cr, 2.5 Mo, lOW, 1.5 Ta
Examples ⟶	Nimonic 80A	U-500	N-115/U-700/R-77	IN-100/R-100	Mar-M246

Adapted from *Heat Treatment, Structure and Properties of Nonferrous Alloys*, by C. R. Brooks: ASM, Metals Park, OH, 1982. Original source, "Nickel Alloys—The Heart of Gas Turbine Engines," by C. T. Sims: Paper 70-GT-24, ASME, New York, 1970.

Fig. 10. Qualitative description of the evolution of microstructure and chromium content of Ni-base superalloys

sition of an alloy such as IN-100 or Udimet 700/Astroloy varies according to whether it is to be used in the cast or the wrought condition.

Evolution of Microstructure

Superalloys contain a variety of elements in a large number of combinations to produce desired effects. Some elements go into solid solution to provide one or more of the following: strength (molybdenum, tantalum, tungsten, rhenium); oxidation resistance (chromium, aluminum); phase stability (nickel); and increased volume fractions (V_f) of favorable secondary precipitates (cobalt). Other elements are added to form hardening precipitates such as γ' (aluminum, titanium) and γ'' (niobium). Minor elements (carbon, boron) are added to form carbides and borides; these and other elements (cerium, magnesium) are added for purposes of tramp-element control. Some elements (boron, zirconium, hafnium) also are added to promote grain-boundary effects other than precipitation or carbide formation. Many elements (cobalt, molybdenum, tungsten, chromium, etc.), although added for their favorable alloying qualities, can participate, in some circumstances, in undesirable phase formation (sigma, mu, Laves, etc.).

Some of the elements mentioned above produce readily discernible changes in microstructure; other elements produce more subtle microstructural effects. The precise microstructural effects produced are functions of processing and heat treatment. The most obvious microstructural effects involve precipitation of geometrically close-packed (gcp) phases such as γ', formation of carbides and formation of topologically close-packed (tcp) phases such as sigma. Even when the type of phase is specified, microstructure morphology can vary widely; for example, script vs blocky carbides, cuboidal vs spheroidal γ', cellular vs uniform precipitation, acicular vs blocky sigma, and discrete γ' vs γ' envelopes. Typical Ni-base superalloy microstructures as they evolved from spheroidal to cuboidal γ' are sketched in Fig. 10.

The γ' phase is an ordered (Ll₂) intermetallic face-centered-cubic phase having the basic composition Ni₃(Al,Ti). Alloying elements affect γ'

mismatch with the matrix gamma (γ) phase, γ' antiphase-domain-boundary (APB) energy, γ' morphology and γ' stability. A related phase, eta (η), is an ordered (DO₂₄) hexagonal phase of composition Ni₃Ti which may exist in a metastable form as γ' before transforming to η. Other types of intermetallic phases such as delta (δ), orthorhombic Ni₃Cb or γ'', which is a body-centered-tetragonal ordered (DO₂₂)Ni₃Cb strengthening precipitate, have been observed.

Carbides also are an important constituent of superalloys. They are particularly essential in the grain boundaries of polycrystalline alloys for production of desired strength and ductility characteristics. Carbides also may provide some degree of matrix strengthening, particularly in Co-base alloys, and are necessary for grain-size control in some wrought alloys. Some carbides are virtually unaffected by heat treatment while others require such a step in order to be present. Various types of carbides are possible depending on alloy composition and processing. Some of the important types are MC, M₆C, M₂₃C₆ and M₇C₃, where M stands for one or more types of metal atom. In many cases the carbides exist jointly; however, they usually are formed by sequential reactions. The common carbide-reaction sequence for many superalloys is MC to M₂₃C₆, and the important carbide-forming elements are chromium (M₂₃C₆, M₇C₃), titanium, tantalum, niobium and hafnium (MC) and molybdenum and tungsten (M₆C). Boron may participate somewhat interchangeably with carbon and produces such phases as MB₁₂, M₃B₂ and others. One claim made for boron is that primary borides formed by adjustment of boron/carbon ratio are more amenable to morphological modification through heat treatment.

All superalloys contain some chromium plus other elements to promote resistance to environmental degradation. The role of chromium is to promote Cr₂O₃ formation on the external surface of an alloy. When sufficient aluminum is present, formation of the more protective oxide Al₂O₃ is promoted when oxidation occurs. A chromium content of 6 to 22 wt % generally is common in Ni-base superalloys, whereas a level of 20 to 30 wt % Cr is characteristic of Co-base superalloys

and a level of 15 to 25 wt % Cr is found in Fe-Ni-base superalloys. Amounts of aluminum up to about 6 wt % may be present in Ni-base superalloys.

A discussion of the function of alloying elements on microstructure would be incomplete without mention of the tramp elements. Elements such as silicon, phosphorus, sulfur, lead, bismuth, tellurium, selenium and silver, often in amounts as low as the parts-per-million (ppm) level, have been associated with property-level reductions, but they are not visible optically or with an electron microscope. Microprobe and Auger spectroscopic analyses have determined that grain boundaries can be decorated with tramp elements at high local concentrations. Elements such as magnesium and cerium tend to tie up some detrimental elements such as sulfur in the form of a compound, and titanium tends to tie up the element nitrogen as TiN. In such cases these and other similar compounds often are visible in the microstructure.

Function of Processing

Processing is considered to be the art/science of rendering the superalloy material into its final form. Processing and alloying elements are interdependent. The general microstructural changes brought about by processing result from the overall alloy composition plus the processing sequence. The role of heat treatment on phases will be referred to when prior microstructural effects are considered; the role of composition on phases has been discussed above.

The three most significant process-related microstructural variables other than those resulting from composition/heat treatment interactions are grain size, shape and orientation (in anisotropic structures). Grain size varies considerably from cast to wrought structure, generally being significantly smaller for the latter. Special processing—e.g., directional solidification or directional recrystallization—can effect changes not only in grain size but also in grain shape and orientation which significantly alter mechanical and physical properties. (Corrosion reactions, however, are primarily functions of composition.)

Effects of Prior Microstructure on Properties

γ' Precipitation. Ni- and Fe-Ni-base superalloys may be hardened by γ' precipitation (in an fcc γ matrix). The γ' in Fe-Ni-base and first-generation Ni-base alloys generally is spheroidal, whereas the γ' in later-generation Ni-base alloys generally is cuboidal (see Fig. 10 and 11). The V_f of γ' generally is about 0.2 or less in wrought Fe-Ni-base superalloys but may exceed 0.6 in Ni-base superalloys. Strengthening by γ' is related to many factors; the most direct correlations can be made with V_f of γ' and with γ' particle size. However, the correlation between strength and γ' size may be difficult to prove out in commercial alloys over the range of particle sizes available.

Before the age-hardening peak is reached during precipitation, the operative strengthening mechanism involves cutting of γ' particles by dislocations, and strength increases with γ' size (Fig. 12) at constant V_f of γ'. After the age-hardening peak is reached, strength decreases with continuing particle growth because dislocations no longer cut γ' particles but bypass them. This effect can be demonstrated for tensile or hardness behavior in low–V_f γ' alloys (A-286, Inconel 901, Waspaloy) but is not as readily apparent in high–

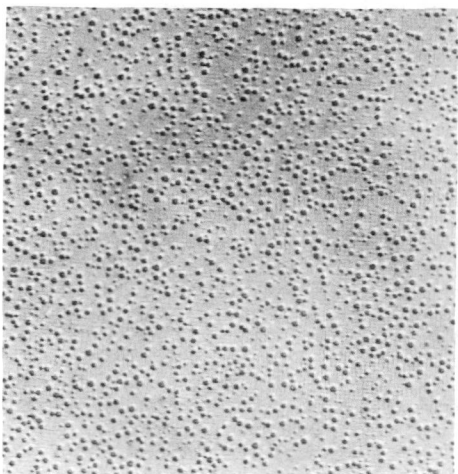

Note secondary (cooling) γ′ between primary cuboidal γ′ particles in Udimet 700. Original magnification, 6800×.

Fig. 11. Wrought Ni-base superalloys, showing spheroidal nature of early (low–V_f γ′) alloys (Waspaloy, left) and cuboidal nature of later (higher–V_f γ′) alloys (Udimet 700, right)

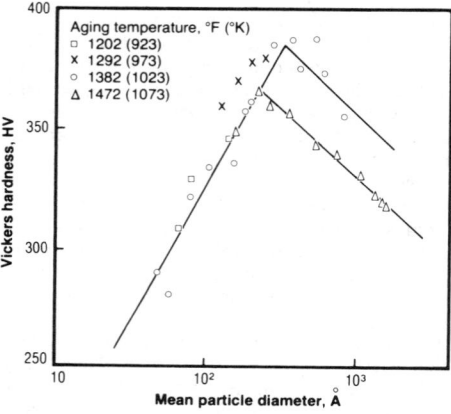

Cutting occurs at low particle sizes, bypassing at larger sizes. Note that aging temperature also affects strength in conjunction with particle size.

Fig. 12. Strength (hardness) vs particle diameter in a Ni-base superalloy

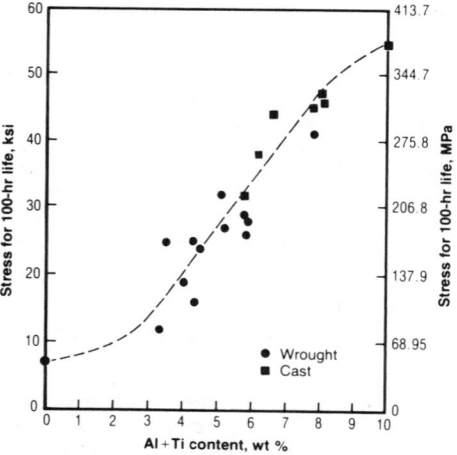

Fig. 13. Effect of Al + Ti content on strength of Ni-base superalloys at 870 °C (1600 °F)

V_f γ′ alloys such as MAR-M 246, IN-100, etc. For creep rupture the effects are less well defined; however, uniform fine-to-moderate γ′ sizes (0.25 to 0.5 μm) are preferred to coarse or hyperfine γ′ for optimum properties.

Alloy strength clearly depends on V_f γ′. The V_f γ′ and thus strength can be increased to a point by adding more hardener elements (aluminum, titanium). Alloy strengths increase as Al + Ti content increases (Fig. 13), and also as Al/Ti ratio increases. In wrought alloys the γ′ usually exists as a bimodal (duplex) distribution of fine γ′ and all of the Al + Ti contributes effectively to the hardening process. In cast alloys, the character of the γ′ precipitate developed can be extremely variable because of the effects of segregation and cooling rate. Large amounts of γ-γ′ eutectic and coarse γ′ may be developed during solidification. Subsequent heat treatments can modify these structures. Bimodal and trimodal γ′ distributions plus γ-γ′ eutectic can be found in cast alloys after heat treatment. Solution heat treatments at temperatures sufficiently high to homogenize the alloy and dissolve coarse γ′ and the eutectic γ-γ′ constituents for subsequent reprecipitation as a uniform fine γ′ have improved creep capability. For a columnar-grain Ni-base superalloy, a direct correlation has been found to exist between creep-rupture life at 980 °C (1800 °F) and the V_f of fine γ′ (Fig. 14).

In general, to achieve the greatest hardening in γ′-hardened alloys it is necessary to solution heat treat the alloys above the γ′ solvus. One or more aging treatments are employed in order to optimize the γ′ distribution and to promote transitions in other phases such as carbides. In some alloys, several intermediate and several lower-temperature aging treatments are used; in cast alloys or in very-high–V_f γ′ wrought alloys, a coating cycle or high-temperature aging treatment may precede the intermediate-temperature aging cycle. When multiple aging treatments are used, a superalloy may show the bimodal or trimodal γ′ distribution described above. An essential feature of γ′ hardening in Ni-base superalloys is that a temperature fluctuation which dissolves some γ′ does not necessarily produce permanent property damage, because subsequent

cooling to normal operating conditions reprecipitates γ′ in a useful form.

In the final analysis it is not possible to judge alloy performance by considering just the γ′ phase. The strength of γ′-hardened grains must be balanced by grain-boundary strength. If the γ′-hardened matrix becomes much stronger relative to grain boundaries, then premature failure occurs because stress relaxation becomes difficult.

Carbide Precipitation—Grain-Boundary Hardening. Carbides exert a profound influence on properties by their precipitation on grain boundaries. In most superalloys, $M_{23}C_6$ forms at the grain boundaries after a postcasting or postsolution treatment thermal cycle such as aging. A chain of discrete globular $M_{23}C_6$ carbides were found to optimize creep-rupture life by preventing grain-boundary sliding in creep rupture while concurrently providing sufficient ductility in the surrounding grain for stress relaxation to occur without premature failure.

In contrast, if carbides precipitate as a continuous grain-boundary film, properties can be severely degraded. $M_{23}C_6$ films were reported to reduce impact resistance of M252, and MC films were blamed for lowered rupture lives and ductility in forged Waspaloy. At the other extreme, when no grain-boundary carbide precipitate is present, premature failure also will occur because grain-boundary movement essentially is unrestricted, leading to subsequent cracking at grain-boundary triple points.

The role of carbides at grain boundaries in Fe-Ni-base superalloys is less well documented than for Ni-base alloys, although detrimental effects of carbide films have been reported. Studies of specific effects of grain-boundary carbides in Co-base alloys are even more sparse, because the carbide distribution in Co-base alloys arises from the original casting or on cooling after mill annealing for wrought Co-base alloys. The significantly greater carbon content of Co-base alloys leads to much more extensive grain-boundary carbide precipitation than in Ni- and Fe-Ni-base alloys. Carbides at grain boundaries in cast Co-base alloys appear as eutectic aggregates of M_6C, $M_{23}C_6$ and fcc γ-Co-base solid solution. No definitive study of the effects of varied carbide forms in grain boundaries on the mechanical behavior of Co-base superalloys has been reported. The lamellar eutectic (carbides–γ Co) nature

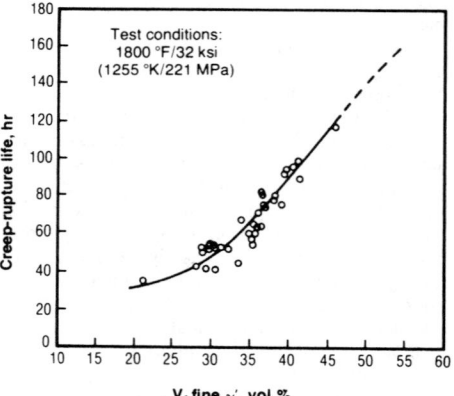

Fig. 14. Increase in creep-rupture life with increase in V_f of fine γ′, demonstrated in a columnar-grain, directionally solidified Ni-base superalloy

of carbides ($M_{23}C_6$-M_6C) in cast Co-base superalloys is interesting. A somewhat similar morphology of $M_{23}C_6$, occurring when it is precipitated in cellular form in Ni-base and Fe-Ni-base alloys, leads to mechanical-property loss in such alloys, but lamellar eutectic does not seem to degrade cast Co-base alloy properties. Cellular growth in Ni-base alloys was found to occur when a high supersaturation of carbon produced by solution heat treatment was not relieved by an intermediate precipitation treatment prior to aging at 705 to 760 °C (1300 to 1400 °F). The ductility of a Ni-base superalloy also was impaired by a different type of precipitation, namely Widmanstätten M_6C at grain and twin boundaries. However, formation of intragranular Widmanstätten M_6C after exposure of B-1900 Ni-base alloy did not appear to reduce properties.

Another effect produced by grain-boundary $M_{23}C_6$ carbide precipitation is the occasional formation, on either side of the boundary, of a zone depleted in γ' precipitate. These precipitate-free zones (PFZ) may have significant effects on rupture life of Ni- and Fe-Ni-base superalloys. If such zones should become wide or much weaker than the matrix, deformation would concentrate there, resulting in early failure. The more complex (higher–V_f γ') alloys do not show significant PFZ effects, probably because of their higher saturation with regard to γ'-forming elements. An effect seen concurrently with PFZ and not clearly separated from it is the γ' envelope produced by breakdown of TiC and consequent formation of $M_{23}C_6$ or M_6C + γ' (from the excess titanium). Not only is the role of the γ' envelope insufficiently established, but there is also the remote possibility that the excess Ti-rich area is really either η or a metastable γ' which could transform to η in use.

Carbide Precipitation—Matrix or General Hardening. Carbides affect the creep-rupture strengths of Co-base superalloys and some Ni- and Fe-Ni-base superalloys by formation within grains. In Co-base cast superalloys, script MC carbides are liberally interspersed within grains, causing a form of dispersion hardening which is not of a large magnitude owing to its relative coarseness. The distribution of carbides in cast alloys can be modified by heat treatment, but strength levels attained at all but the highest temperatures are substantially less than those of the γ'-hardened alloys. Consequently, cast Co-base alloys generally are not heat treated except in a secondary sense through the coating diffusion heat treatment of 4 h at 1065 to 1120 °C (1950 to 2050 °F) sometimes applied.

Wrought Co-base superalloys have carbide modifications produced during the fabrication sequence. Carbide distributions in wrought alloys result from the mill anneal after final working. Properties are largely a result of grain size, refractory-metal content and carbon level, which indicates the V_f of carbides available for hardening.

True solutioning, in which all minor constituents are dissolved, is not possible in most Co-base superalloys, because melting often occurs before all the carbides are solutioned. Some enhancement of creep-rupture behavior has been achieved by heat treatment. Rupture-time improvements can be gained by aging a modified Vitallium alloy (Fig. 15); larger increases have been produced by increasing the carbon content of the alloy. Solution treating and aging is not suitable for producing stable Co-base superalloys

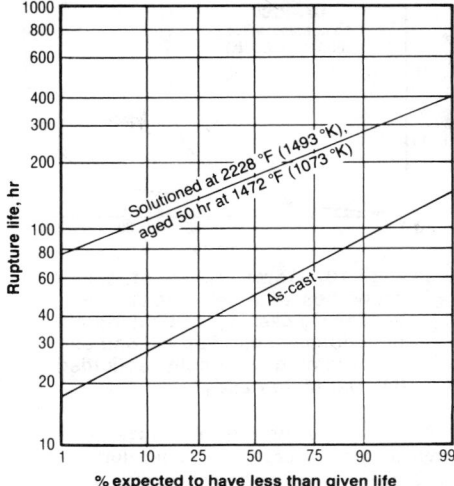

Fig. 15. Effect of heat treatment on a Co-base superalloy (HS-31), showing increase in strength resulting from carbide precipitation

for use above 815 °C (1500 °F) because of carbide dissolution or over-aging.

Matrix carbides in Ni-base and Fe-Ni-base superalloys also may be partially solutioned. MC carbides will not totally dissolve, however, without incipient melting of the alloy. Furthermore, MC carbides in such alloys tend to be unstable, decomposing to $M_{23}C_6$ at temperatures below about 815 to 870 °C (1500 to 1600 °F) or possibly converting to M_6C at temperatures of 980 to 1040 °C (1800 to 1900 °F) if the alloy has a sufficiently high molybdenum-plus-tungsten content. Matrix carbides generally contribute very small increments of strengthening to Ni- and Fe-Ni-base superalloys.

An interesting microstructural trend has taken place with the advent of single crystals of Ni-base superalloys. Because no grain boundaries exist there is little need for the normal grain-boundary strengtheners such as carbon. Consequently, very few matrix or subboundary carbides exist in such alloys.

Perhaps the most common other role of matrix carbides (also shared by grain-boundary carbides) is a negative one: they may participate in the fatigue-cracking process by premature cracking or by oxidizing at the surface of uncoated alloys to cause a notch effect. Oxidized carbides or precracked carbides from machining or thermal stresses can initiate fatigue cracks. Precracked carbides can be related to prior casting processes. Carbide size is important, and reduced carbide volumes and sizes in Ni-base alloys result in a reduction in precracked carbides. The longer solidification times and lower gradients of early directional solidification processes often resulted in moderately large carbides. However, improved gradients and the reduced carbon contents of single-crystal alloys (few or no carbides) have resulted in substantial improvements in fatigue resistance—particularly over similarly oriented columnar-grain alloys with normal carbon levels. This effect is most noticeable in LCF and TMF. No evidence is available to interpret the effect of the absence of carbides on HCF, but beneficial effects could be anticipated. Oxidized carbides can be minimized or prevented by several methods. Casting procedures and/or chemical com-

position may be modified to produce smaller primary carbides. Powder processing may be used to produce the same result. Carbon content may be reduced if it is not specifically required to enable the alloy to attain the desired strength levels. Reduced carbon is the rule in single-crystal and powdered superalloys. Of course, the alloy may be coated with an appropriate protective coating that leaves the carbides in a subsurface location.

Although there is limited documentation, it frequently is assumed that non-carbide-forming elements do influence the formation of carbides. Cobalt, for example, has been claimed to modify the carbides in Ni-base alloys, and phosphorus has produced a more general, more finely dispersed and smaller carbide precipitation than carbon alone in a heat-resisting Fe-base alloy. The modifying effect on carbides may be intragranular or intergranular depending on the modifier and the base-alloy system.

γ'' **Precipitation.** The practical use of γ'' precipitation is restricted to Fe-Ni-base alloys with niobium additions. Inconel 718 is the outstanding example of an alloy in which γ'' formation has been commercially exploited. The γ'' phase is disk shaped, and the V_f of γ'' in Inconel 718 is substantially in excess of that of γ'. Both γ'' and γ' phases will be found in alloys where γ'' is present, but γ'' can be the predominant strengthening agent. Although the strength of γ'' phase has not been studied, similar considerations to γ' behavior probably obtain. The most significant feature of γ'' is probably the ease with which it forms at moderate temperatures after prior solutioning by heat treatment or joining processes. Because of this behavior, an alloy can be aged, after welding, to produce a fully strengthened structure with exceptional ductility.

The γ'' phase, not normally a stable phase, can convert to γ' or to δ Ni_3Cb on long-time exposure. The strength of γ'' is additive to that of γ' phase. A lack of notch ductility in Inconel 718 has been associated with a γ'' PFZ; the γ'' PFZ can be eliminated and ductility restored by appropriate heat treatment. Alloys hardened with γ'' phase achieve high tensile strengths and very good creep-rupture properties at lower temperatures, but the conversion of γ'' to γ' or δ above about 675 °C (1250 °F) causes a sharp reduction in strength.

Effects of Boron, Zirconium and Hafnium. Within limits, significant improvements in mechanical properties can be achieved by additions of boron, zirconium and hafnium. However, only limited microstructural correlations can be made. The presence of these elements may modify the initial grain-boundary carbides or tie up deleterious elements such as sulfur and lead. Reduced grain-boundary-diffusion rates may be obtained, with consequent suppression of carbide agglomeration and creep cracking. Hafnium contributes to the formation of more γ-γ' eutectic in cast alloys; the eutectic at grain boundaries is thought (in modest quantities) to contribute to alloy ductility. The effects of these elements are limited to Ni-and Fe-Ni-base alloys; virtually no Co-base alloys contain them.

Effects of Processing

Three major processing techniques are used for controlling superalloy properties. Thermomechanical working is used for wrought Ni- and Fe-Ni-base alloys to store energy by producing a fine grain size and controlling dislocation density/

configuration. Improvements in tensile properties and LCF have resulted. A second processing route involves the use of powder metallurgy to produce reduced carbide size and greater homogeneity in materials with a resultant improvement in fatigue resistance. Furthermore, in conjunction with GATORIZING processing, alloy grain sizes of ASTM 8 to 12 are being routinely produced in very-high-strength alloys, resulting in additional fatigue-life benefits. (Major benefits result from the ability to form alloys such as IN-100, which are unforgeable by some standard procedures.)

The third area is casting control of grain size and morphology, especially by controlled solidification. Cast-alloy grain sizes have been made more uniform and, in some cases, have been reduced to enhance fatigue or tensile properties. Directional grain structures have improved strength (Fig. 16). Improved creep-rupture and fatigue resistance have resulted from the elimination of transverse grain boundaries and the favorable orientation of a low modulus direction to reduce strains. In the extreme, grains have been eliminated in single crystals with additional gains in creep-rupture behavior (Fig. 16).

Porosity in superalloys has led to fatigue and creep-rupture failure. Reduced porosity owing to hot isostatic pressing (HIP) has resulted in improved properties. Efforts to date have centered on Ni-base cast alloys, but the process should enhance properties of any cast alloy with nonsurface-connected casting porosity. (In the biomedical field, use of HIP has provided significant improvements in fatigue life of cast Vitallium alloy hip replacements.)

Effects of Thermal Exposure on Mechanical Properties

Superalloys generally behave much like other alloy systems on thermal exposure during testing or in service but with some differences due to the nature of the γ' precipitate. Most alloys with secondary phases undergo property degradation due to coalescence of secondary phases, which reduces their effectiveness. This behavior occurs in superalloys and is manifested by such phenomena as γ' agglomeration and coarsening; carbide precipitation and γ' envelope formation also occur. In addition, the superalloys may show a tendency to form less-desirable secondary phases such as σ. These detrimental phases generally reduce property levels of the superalloys in which they form because of their inherent properties and/or the consumption of elements intended for γ and γ'. Some of these phases can be prevented from forming by application of compositional control guided by the concept of the electron-vacancy number (N_v). Formation of topologically close-packed (tcp) phases such as sigma (σ), Laves and mu (μ) is found to be related to excess electron vacancies in the transition-element base metals (iron, nickel and cobalt). By ascribing N_v values to the alloying elements of the γ matrix, a weighted N_v can be calculated for γ and, by experiment, upper limits for N_v can be set for a given alloy composition to ensure the absence of tcp phases in reasonable exposure times. Unfortunately, simple calculations of this type have not been found applicable to δ or η formation, which can be detrimental in certain morphologies and/or in excess amounts. Trial-and-error adjustments of composition and processing generally are required to ensure that δ and η precipitation do not occur and cause a significant loss of properties.

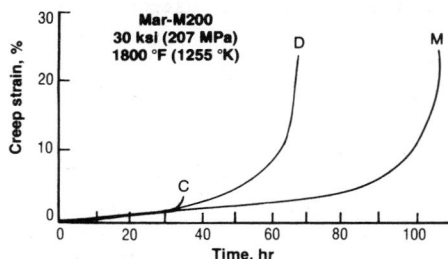

Fig. 16. Comparison of creep properties of polycrystalline conventionally cast (C), columnar-grain directionally solidified (D) and single-crystal directionally solidified (M) Mar-M 200 alloy

Table 4 presents typical values of tensile strength, yield strength and elongation, and Table 5 gives typical rupture strengths, of selected wrought superalloys at various temperatures.

Transformation of Hardening Phase. In Fe-Ni-base superalloys, the strengthening precipitate usually degrades in a moderate temperature regime, 650 to 760 °C (1200 to 1400 °F), forming structures and precipitate morphologies that are less effective in strengthening the alloy. In alloys hardened by large amounts of titanium, η phase replaces γ'. The precipitation of η may occur in two forms: at the grain boundaries as cellular product or intragranularly as Widmanstätten plates. The cellular precipitation is often associated with loss of mechanical properties, particularly notch stress-rupture strength (NSR). Intragranular plates also may cause some property loss, but data are not available. In some Ni-base superalloys with higher Ti/Al ratios (e.g., IN-738, IN-792, IN-939), η has been reported, but no information on property degradation is available.

In an alloy like Inconel 718, plates of orthorhombic δ phase will form on exposure after sufficiently long times at elevated temperatures. Because of its relatively coarse morphology, there is a deterioration of properties when excess amounts of δ form.

Formation of Transition-Element Phases. Of great concern in all superalloys has been the formation of detrimental secondary phases not directly associated with the primary hardening precipitate phases, γ' and γ''. Laves phases have been found in the Fe-Ni-base superalloys Inconel 718, Inconel 901 and A-286. Room-temperature yield strength and ductility of Inconel 718 are reduced by Laves formation, whereas Laves in A-286 does not affect properties. Laves in the Co-base alloy HA-25 severely degrades room-temperature ductility.

Phases such as Laves, σ and μ in acicular form generally are detrimental owing to their morphology, lack of ductility and the tieing up of some hardening elements. Sigma-phase formation in several Ni-base alloys has been shown to reduce stress-rupture life (see Fig. 17). Sigma generally forms with exposure and is controlled by establishing composition limits (phase control electron vacancy number, N_v) or by preventing an alloy from operating in the σ-forming temperature range. The extent of the effect that tcp and other secondary phases have on properties varies with the type of alloy (cast vs wrought), property being measured, initial microstructure, and environmental effects (coated vs uncoated, test atmosphere).

Morphological Changes in γ'. Changes in γ' during testing produce effects on properties which may

not be readily observed because of the simultaneous occurrence of other microstructural changes. Most commonly, γ' coarsens following well-established kinetics and agglomerates in a creep-rupture test (see Fig. 18), tending to form platelets of γ' on the [001] planes perpendicular to the applied stress. The coarsening of γ' can cause reduced rupture life (see Fig. 19). Overheating can cause accelerated coarsening as well as solutioning of some fine γ'. Properties may be reduced, but reprecipitation of fine γ' occurs in the case of mild overheating and property losses may be hard to detect.

Carbide Morphology/Type Changes. As exposure time increases, there is a tendency for further reduction in the primary MC carbide amounts and for further morphological changes together with a tendency to form increased amounts of the secondary carbides $M_{23}C_6$ and M_6C. If carbide films form and/or acicular carbides of the M_6C type form, alloy ductility and strength can be reduced. Agglomeration of carbides, however, can lead to increased ductility accompanied by a strength reduction. In cast Co-base superalloys, precipitation of additional intragranular carbides during service has led to sharply increased rupture strength and reduced rupture ductility. The precipitation of an intragranular acicular Widmanstätten carbide brought about by aging in another Co-base alloy degraded both rupture life and ductility.

Property Recovery After Thermal Exposure. As noted, changes in γ' occur with exposure to elevated temperatures. Most commonly, γ' coarsens and agglomerates, particularly under stress. Overheating can cause accelerated coarsening as well as solutioning of some fine γ'. Properties may be reduced in such circumstances, but when the overheating has been mild, reprecipitation of fine γ' occurs with a return to normal temperatures, and some property recovery occurs. Property recovery, however, does not occur in the case of additional thermal exposure of tcp, δ or η phases in superalloys. Neither is recovery achieved by thermal exposure after excessive carbide precipitation such as may occur in Co-base superalloys or after extensive γ' coarsening such as occurs in Ni- and Fe-Ni-base superalloys.

Recovery of properties is desired as a means of prolonging component life. If, as above, additional thermal exposure fails to promote better properties, then re-solution heat treatment and aging is required to restore property levels. While this process may be satisfactory in alloys degraded only by thermal exposure, service exposure under stress generally produces property losses (due in part to cavitation or wedge creep cracking), which cannot be recovered by simple heat treatment alone—at least for the more complex commercial superalloys. The use of hot isostatic pressing (HIP) plus re-solution and aging treatments of exposed superalloys has suggested that some improvements can be achieved under suitable circumstances, but the extent to which such salvage processing is economically viable is uncertain.

ENVIRONMENTAL EFFECTS

General Oxidation

Superalloys generally react with oxygen, and oxidation is the prime environmental effect on these alloys. At moderate temperatures—about 870 °C (1600 °F) and below—general uniform oxidation is not a major problem. At higher temperatures, commercial Ni- and Co-base superal-

Table 4. Typical mechanical properties of wrought cobalt-base and nickel-base superalloys

Cobalt-base alloys

Temperature °C	°F	Tensile strength MPa	ksi	Yield strength MPa	ksi	Elongation, %
Haynes 25 (L-605) sheet						
21	70	1010	146	460	67	64
540	1000	800	116	250	36	59
650	1200	710	103	240	35	35
760	1400	455	66	260	38	12
870	1600	325	47	240	35	30
Haynes 188, sheet						
21	70	960	139	485	70	56
540	1000	740	107	305	44	70
650	1200	710	103	305	44	61
760	1400	635	92	290	42	43
870	1600	420	61	260	38	73
S-816, bar						
21	70	965	140	385	56	30
540	1000	840	122	310	45	27
650	1200	765	111	305	44	25
760	1400	650	94	285	41	21
870	1600	360	52	240	35	16

Nickel-base alloys

Temperature °C	°F	Tensile strength MPa	ksi	Yield strength MPa	ksi	Elongation, %
Astroloy, bar						
21	70	1410	205	1050	152	16
540	1000	1240	180	965	140	16
650	1200	1310	190	965	140	18
760	1400	1160	168	910	132	21
870	1600	770	112	690	100	25
D-979, bar						
21	70	1410	204	1010	146	15
540	1000	1300	188	925	134	15
650	1200	1100	160	890	129	21
760	1400	720	104	655	95	17
870	1600	345	50	305	44	18
Hastelloy X, sheet						
21	70	785	114	360	52	43
540	1000	650	94	290	42	45
650	1200	570	83	275	40	37
760	1400	435	63	260	38	37
870	1600	255	37	180	26	50
IN-102, bar						
21	70	960	139	505	73	47
540	1000	825	120	400	58	48
650	1200	710	103	400	58	64
760	1400	440	64	385	56	110
870	1600	215	31	200	29	110
Inconel 600, bar						
21	70	620	90	250	36	47
540	1000	580	84	195	28	47
650	1200	450	65	180	26	39
760	1400	185	27	115	17	46
870	1600	105	15	62	9	80
Inconel 601, sheet						
21	70	740	107	340	49	45
540	1000	725	105	150	22	38
650	1200	525	76	180	26	45
760	1400	290	42	200	29	73
870	1600	160	23	140	20	92

Temperature °C	°F	Tensile strength MPa	ksi	Yield strength MPa	ksi	Elongation, %
Inconel 625, bar						
21	70	855	124	490	71	50
540	1000	745	108	405	59	50
650	1200	710	103	420	61	35
760	1400	505	73	420	61	42
870	1600	285	41	275	40	125
Inconel 706, bar						
21	70	1300	188	980	142	19
540	1000	1120	163	895	130	19
650	1200	1010	147	825	120	21
760	1400	690	100	675	98	32
Inconel 718, bar						
21	70	1430	208	1190	172	21
540	1000	1280	185	1060	154	18
650	1200	1230	178	1020	148	19
760	1400	950	138	740	107	25
870	1600	340	49	330	48	88
Inconel 718, sheet						
21	70	1280	185	1050	153	22
540	1000	1140	166	945	137	26
650	1200	1030	150	870	126	15
760	1400	675	98	625	91	8
Inconel X 750, bar						
21	70	1120	162	635	92	24
540	1000	965	140	580	84	22
650	1200	825	120	565	82	9
760	1400	485	70	455	66	9
870	1600	235	34	165	24	47
M-252, bar						
21	70	1240	180	840	122	16
540	1000	1230	178	765	111	15
650	1200	1160	168	745	108	11
760	1400	945	137	715	104	10
870	1600	510	74	485	70	18
Nimonic 75, bar						
21	70	750	109	...	...	41
540	1000	635	92	...	...	41
650	1200	538	78	...	...	42
760	1400	290	42	...	...	70
870	1600	145	21	...	...	68
Nimonic 80A, bar						
21	70	1240	179	620	90	24
540	1000	1100	160	530	77	24
650	1200	1000	145	550	80	18
760	1400	760	110	505	73	20
870	1600	400	58	260	38	34
Nimonic 90, bar						
21	70	1240	180	805	117	23
540	1000	1100	160	725	105	23
650	1200	1030	150	685	99	20
760	1400	825	120	540	78	10
870	1600	430	62	260	38	16
Nimonic 105, bar						
21	70	1140	166	815	118	12
540	1000	1100	160	775	112	18
650	1200	1080	156	800	116	24
760	1400	965	140	655	95	22
870	1600	605	88	365	53	25

Temperature °C	°F	Tensile strength MPa	ksi	Yield strength MPa	ksi	Elongation, %
Nimonic 115, bar						
21	70	1240	180	860	125	25
540	1000	1090	158	795	115	26
650	1200	1120	163	815	118	25
760	1400	1080	157	800	116	22
870	1600	825	120	550	80	18
Pyromet 860, bar						
21	70	1300	188	835	121	22
540	1000	1250	182	840	122	15
650	1200	1110	161	850	123	17
760	1400	910	132	835	121	18
René 41, bar						
21	70	1420	206	1060	154	14
540	1000	1400	203	1010	147	14
650	1200	1340	194	1000	145	14
760	1400	1100	160	940	136	11
870	1600	620	90	550	80	19
René 95, bar						
21	70	1620	235	1310	190	15
540	1000	1540	224	1250	182	12
650	1200	1460	212	1220	177	14
760	1400	1170	170	1100	160	15
Udimet 500, bar						
21	70	1310	190	840	122	32
540	1000	1240	180	795	115	28
650	1200	1210	176	760	110	28
760	1400	1040	151	730	106	39
870	1600	640	93	495	72	20
Udimet 520, bar						
21	70	1310	190	860	125	21
540	1000	1240	180	825	120	20
650	1200	1170	170	795	115	17
760	1400	725	105	725	105	15
870	1600	515	75	515	75	20
Udimet 700, bar						
21	70	1410	204	965	140	17
540	1000	1280	185	895	130	16
650	1200	1240	180	855	124	16
760	1400	1030	150	825	120	20
870	1600	690	100	635	92	27
Udimet 710, bar						
21	70	1190	172	910	132	7
540	1000	1150	167	850	123	10
650	1200	1290	187	860	125	15
760	1400	1020	148	815	118	25
870	1600	705	102	635	92	29
Unitemp AF2-1DA, bar						
21	70	1290	187	1050	152	10
540	1000	1340	194	1080	157	13
650	1200	1360	197	1080	157	13
760	1400	1150	167	1010	146	8
870	1600	830	120	715	104	8
Waspaloy, bar						
21	70	1280	185	795	115	25
540	1000	1170	170	725	105	23
650	1200	1120	162	690	100	34
760	1400	795	115	675	98	28
870	1600	525	76	515	75	35

loys are attacked by oxygen. The level of oxidation resistance at temperatures below about 980 °C (1800 °F) is a function of chromium content (Cr_2O_3 forms as a protective oxide); at temperatures above about 980 °C, aluminum content becomes more important in oxidation resistance (Al_2O_3 forms as a protective oxide). Chromium and aluminum can contribute in an interactive fashion to oxidation protection. The higher the chromium level, the less aluminum may be required to form a highly protective Al_2O_3 layer. However, the aluminum contents of many superalloys are insufficient to provide long-term Al_2O_3 protection, and so protective coatings are applied (see below). These coatings also prevent selective attack that occurs along grain boundaries and at surface carbides (see Fig. 20) and inhibit internal oxidation or subsurface interaction of O_2/N_2 with γ' envelopes, a process believed to occur in Ni-base superalloys.

Hot Corrosion

In lower temperature operating conditions, $\leq$ 870 °C (1600 °F), accelerated oxidation may occur in superalloys through the operation of selective fluxing agents. One of the better-documented accelerated oxidation processes is hot corrosion (sometimes known as sulfidation). The hot corrosion process is separated into two regimes: low temperature and high temperature. The principal method for combating hot corrosion is the use of a high chromium content ($\gtrsim$20 wt %) in the base alloy. Although Co-base superalloys and many Fe-Ni-base alloys have chromium levels in this range, most Ni-base alloys—especially those with high creep-rupture strengths—do not, because a high chromium content is not compatible with the high $V_f \gamma'$ required.

Higher titanium/aluminum ratios also seem to reduce attack on uncoated superalloys; and alloys with improved resistance to hot corrosion, based on slightly increased chromium contents and ap-

Table 5. Typical rupture strengths of selected wrought superalloys

Temperature °C	°F	100 h MPa	ksi	1000 h MPa	ksi
Incoloy 800					
650	1200	220	32	145	21
760	1400	115	17	69	10
870	1600	45	6.5	33	4.8
Incoloy 801					
650	1200	250	36	. . .	. . .
730	1350	145	21	. . .	. . .
815	1500	62	9	. . .	. . .
Incoloy 802					
650	1200	240	35	170	24
760	1400	145	21	105	15
870	1600	97	14	62	9
Inconel 600					
815	1500	55	8	39	5.6
870	1600	37	5.3	24	3.5
Inconel 601(a)					
540	1000	. . .	. . .	400	58
870	1600	48	7	30	4.3
980	1800	23	3.4	14	2.1
Inconel 617(b)					
815	1500	140	20	97	14
925	1700	62	9	. . .	5.5
980	1800	41	6	. . .	3.5
Inconel 625(a)					
650	1200	440	64	370	54
815	1500	130	19	93	13.5
870	1600	72	10.5	48	7
Inconel 718(c)					
540	1000	. . .	. . .	951	138
595	1100	860	125	760	110
650	1200	690	100	585	85
Inconel 751(d)					
815	1500	200	29	125	185
870	1600	120	175	69	10

Temperature °C	°F	100 h MPa	ksi	1000 h MPa	ksi
Inconel X750(e)					
540	1000	. . .	. . .	827	120
870	1600	83	12	45	6.5
925	1700	58	8.4	21	3.1
N-155, bar(f)					
650	1200	360	52	295	43
730	1350	195	28	150	22
870	1600	97	14	66	9.5
N-155(g)					
650	1200	380	55	290	42
N-155, sheet(f)					
980	1800	39	5.6	20	2.9
Nimonic 75(h)					
815	1500	38	5.5	24	3.5
870	1600	23	3.4	15	2.2
925	1700	14	2.1	10	1.5
980	1800	. . .	. . .	7.6	1.1
Nimonic 80A(j)					
540	1000	. . .	. . .	825	120
815	1500	185	27	115	17
870	1600	105	15	. . .	. . .
Nimonic 90(j)					
815	1500	240	35	155	22.5
870	1600	150	22	69	10
925	1700	69	10	. . .	. . .
Nimonic 105(k)					
815	1500	325	47	225	32
870	1600	210	30.2	135	19
Nimonic 115(m)					
815	1500	425	62	315	46
870	1600	315	46	205	30
925	1700	205	30	130	18.5
Nimonic 263(n)					
815	1500	170	24.5	105	15
870	1600	93	13.5	46	6.7
925	1700	45	6.5	. . .	. . .

(a) Solution treated 1150 °C (2100 °F). (b) Solution treated 1175 °C (2150 °F). (c) Heat treated to 980 °C (1800 °F) plus 720 °C (1325 °F) hold for 8 h, F.C. to 620 °C (1150 °F) hold for 8 h. (d) 730 °C (1350 °F) hold for 2 h. (e) Heat treat to 1150 °C (2100 °F) plus 840 °C (1550 °F) hold for 24 h, plus 705 °C (1300 °F) hold for 20 h. (f) Solution treated and aged. (g) Stress-relieved forging. (h) Heat treat to 1050 °C (1922 °F) hold for 1 h. (j) Heat treat to 1080 °C (1976 °F) hold for 8 h, plus 700 °C (1290 °F) hold for 16 h. (k) Heat treat to 1150 °C

(2100 °F) hold for 4 h, plus 1050 °C (1920 °F) hold for 16 h, plus 850 °C (1560 °F) hold for 16 h. (m) Heat treat to 1190 °C (2175 °F) hold for 1.5 h, plus 1100 °C (2010 °F) hold for 6 h. (n) Heat treat to 1150 °C (2100 °F) hold for 2 h, W.Q., plus 800 °C (1475 °F) hold for 8 h.

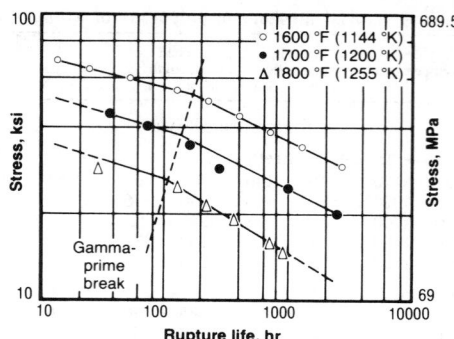

Fig. 19. Stress-rupture behavior of B-1900 Ni-base superalloy, showing break in slope believed to be caused by γ′ coarsening

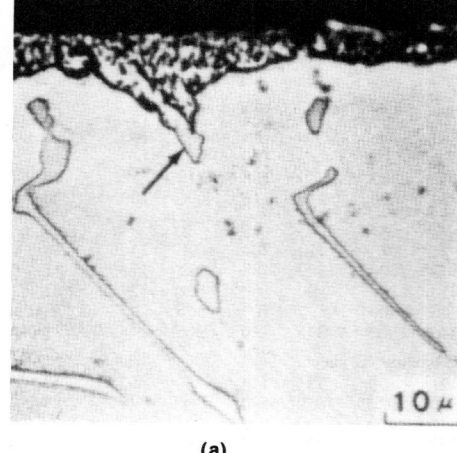

(a)

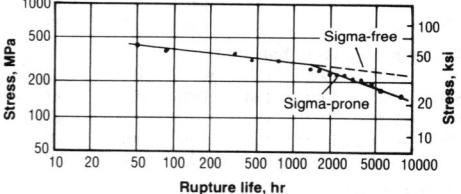

After M. J. Donachie, "Overheating, Creep and Alloy Stability": in *Metals Handbook,* 9th Ed., Vol 3, ASM, Metals Park, OH, 1980, p 220-229. Original source, "The Effect of Phase Instability on the High Temperature Stress Rupture Properties of Representative Nickel Base Superalloys," by D. M. Moon and F. J. Wall: in *International Symposium on Structural Stability in Superalloys,* Vol 1, High Temperature Alloys Committee, Met. Soc., Am. Inst. Min. Met. Pet. Engrs., New York, 1968, p 115-133.

Fig. 17. Stress-rupture behavior of Udimet 700, showing reduction in strength that occurs when sigma phase forms

propriate titanium/aluminum modifications, have been produced. For maximum uncoated hot-corrosion resistance, however, chromium contents in excess of 20 wt % appear to be required. Such alloys are not capable of achieving the strengths of the high–V_f γ′ alloys such as MAR-M 200. Consequently, coatings which protect the base

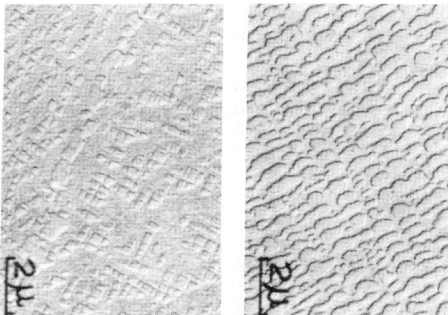

Left, as-heat-treated. Right, after 91.2 h at 252.3 MPa (36.6 ksi) and 893 °C (1640 °F). 4000×.

Fig. 18. Agglomeration of γ′ in Udimet 700 resulting from creep testing

metal (overlay coatings seem to provide the best surface protection), or sometimes environmental inhibitors, are used to suppress hot-corrosion attack in high-strength (high–V_f γ′) Ni-base superalloys.

Coatings

Development of increased strength (increased V_f γ′) in Ni-base superalloys led to reductions in

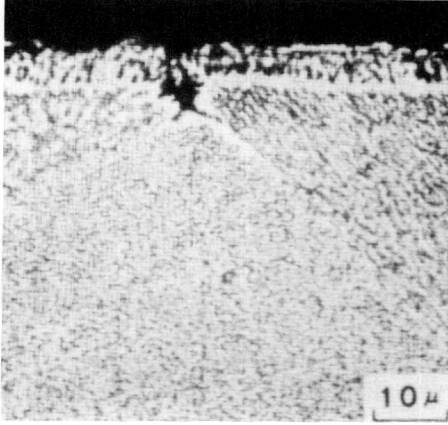

(b)

(a) Accelerated oxidation of MC carbide (arrow) at surface of MAR-M 200 at 925 °C (1700 °F). (b) Accelerated oxidation of grain boundary in Udimet 700 at 760 °C (1400 °F). Magnification, 1000×.

Fig. 20. Effects of oxidation on superalloys

chromium content and to greater oxidation attack and susceptibility to hot corrosion. Although the aluminum contents of some alloys were increased in order to enhance resistance to oxidation, intergranular and carbide attack worsened as operating temperatures escalated. Furthermore, some of the more oxidation-resistant alloys were found to be very poor in hot-corrosion resistance. In order to protect against local oxidation, and, later, against hot corrosion and similar fluxing reactions, coatings were applied to superalloys.

Use of coatings is a preferred method of protecting superalloy surfaces from environmental attack because:

- Coatings (at least overlay coatings) can be tailored to the hostile environment anticipated.
- Development of base alloys for strength is less inhibited, because significant protection (but not all of the protection) is provided by the coating.
- Coatings provide an opportunity to refurbish worn surfaces after exposure and environmental attack without causing significant degradation of the base metal.

Coatings are of two types: aluminide (diffusion) coatings and overlay coatings.

Aluminide (Diffusion) Coatings. The most common type of coating for environmental protection of superalloys is the aluminide diffusion coating, which develops an aluminide (CoAl or NiAl) outer layer with enhanced oxidation resistance. This outer layer is developed by the reaction of aluminum with the nickel or cobalt in the base metal.

Some use has been made of aluminides containing chromium or silicon, and, in recent years, noble metals such as platinum have been used to enhance the oxidation resistance of aluminides. The oxidation resistance of aluminide coatings is derived from formation of protective Al_2O_3 scales. Aluminide diffusion coatings generally are thin—about 50 to 75 μm (2 to 3 mils). They consume some base metal in their formation and, although deposited at lower temperatures, are invariably diffused at temperatures of about 1040 to 1120 °C (1900 to 2050 °F) prior to being placed in service.

Overlay coatings, generally referred to as MCrAl or MCrAlY coatings, are derived directly from the deposition process and do not require diffusion for their formation. The constituent denoted "M" in these designations has at various times been iron, cobalt, nickel, or combinations of nickel and cobalt. A high-temperature heat treatment (at 1040 to 1120 °C, or 1900 to 2050 °F) is performed to homogenize the coating and to ensure its adherence to the substrate.

MCrAl coatings are approximately twice as thick as aluminide coatings, and, through incorporation of yttrium, overlay coatings can be made to have improved corrosion resistance. An advantage of MCrAlY coatings is that their compositions can be tailored to produce greater or lesser amounts of chromium or aluminum within the coating, and thus the protectivity and mechanical properties of the coating can be balanced for optimum performance.

Effects of Coatings on Mechanical Behavior. Coatings seem to have little deleterious effect on tensile and creep-rupture properties. In fact, creep-rupture life may be enhanced by protection of the superalloy surface from oxidation or fluxing-agent attack. Thermomechanical fatigue (TMF) can be greatly affected by a coating because the ductilities of coatings tend to be low at low temperatures. Ductilities of aluminide coatings generally are lower than those of overlay coatings at lower temperatures (≤540 °C, or 1000 °F). However, ductilities of both overlay and diffusion coatings increase sharply at higher temperatures (≥650 °C, or 1200 °F). When TMF cycles have peak tensile strains at lower temperatures, coatings may crack within the first few cycles, and TMF response is related to the crack-propagation rate in the base metal. Some adjustment of ductility is possible, particularly in overlay coatings. However, because aluminum—the protective element—is a principal cause of reduced ductility, balances must be achieved between protectivity and resistance to TMF.

Coating Processes. Aluminide diffusion coatings generally are applied by pack processes, but slurry, electrophoretic and other techniques also have been used. Customarily, deposition is done at an intermediate temperature, followed by diffusion in a controlled-atmosphere furnace at about 1040 to 1120 °C (1900 to 2050 °F).

Overlay coatings are applied by physical vapor deposition (PVD) in vacuum chambers. They also may be applied by plasma spray (PS) techniques. Low-pressure PS techniques produce coatings with properties comparable to or better than those of PVD vacuum-produced coatings. (Argon is used in low-pressure plasma spraying, in opposition to the air PS process originally used.)

Overlay coatings tend to be line-of-sight coatings whereas pack-diffusion coatings do not. PS techniques afford more flexibility in application of overlay coatings than do PVD processes, because the angular relationship between the plasma and the part may be varied more or less in a large envelope so as to direct the coating to desired areas. PS also provides more opportunity for compositional flexibility than does PVD.

General Requirements of a Protective Coating. Coating selection will be based on knowledge of oxidation/corrosion behavior in laboratory, pilot-plant and field tests. Attributes that probaby will be required for successful coating selection include:

- High resistance to oxidation and/or hot corrosion
- Ductility sufficient to provide adequate resistance to TMF
- Compatibility with the base alloy
- Low rate of interdiffusion with the base alloy
- Ease of application and low cost relative to improvement in component life
- Ability to be stripped and reapplied without significant reduction of base-metal dimensions or degradation of base-metal properties.

Other Environmental Effects

Stress-corrosion cracking can occur in Ni- and Fe-Ni-base superalloys at lower temperatures. Hydrogen embrittlement at cryogenic temperatures has been reported for such alloys. Furthermore, so-called inert environments—for example, vacuum, or gases such as helium or argon—may produce mechanical behavior substantially different from baseline uncoated properties, which usually are determined in static-air tests.

APPLICATIONS

Superalloys have been used in cast, rolled, extruded, forged and powder-processed forms. Sheet, bar, plate, tubing, airfoils, disks and pressure vessels (cases) are but some of the shapes that have been produced. These metals have been used in aircraft, industrial and marine gas turbines, nuclear reactors, aircraft skins, spacecraft structures, petrochemical production and environmental-protection applications. Although developed for high-temperature applications, some are used at cryogenic temperatures. Orthopedic and dental prostheses have evolved from Co-base superalloy development. Applications continue to expand, although aerospace usage remains the predominant application.

SELECTED REFERENCES

General

Donachie, M. J., ed.: *Superalloys Source Book,* Am. Soc. for Metals, Metals Park, OH (1984).

Sullivan, C. P., Donachie, M. J., and Morral, F. R.: *Cobalt-Base Superalloys—1970,* Cobalt Information Center, Brussels (1970).

Sims, C. T., and Hagel, W. C., eds.: *The Superalloys,* Wiley, NY (1972).

Betteridge, W., and Heslop, J., eds.: *The Nimonic Alloys,* Second Edition, Crane, Russak and Co., NY (1974).

The proceedings of a series of COST-50 conferences published as follows:

Coutsouradis, D., et al., eds.: *High Temperature Alloys for Gas Turbines,* Applied Science Publishers, London (1978).

Brunetaud, R., et al., eds.: *High Temperature Alloys for Gas Turbines,* D. Reidel Pub. Co., Boston, MA (1982).

The proceedings of a continuing series of conferences held at Seven Springs Mountain Resort, Champion, PA, published as follows:

International Symposium on Structural Stability in Superalloys, Vol I, Vol II, High Temperature Alloys Committee, Met. Soc. Am. Inst. Min. Met. Pet. Eng., NY (1968).

Superalloys—Processing, Report MCIC-72-10, Metals and Ceramics Information Center, Columbus, OH (1972).

Kear, B. H., et al., eds.: *Superalloys: Metallurgy and Manufacture,* Claitor's Publishing Co., Baton Rouge, LA (1976).

Tien, J. K., et al., eds.: *Superalloys 1980,* Am. Soc. for Metals, Metals Park, OH (1980)

Microstructure

Johnson, J. L., and Donachie, M. J., Jr.: Microstructure of Precipitation Strengthened Nickel-Base Superalloys, Report System Paper C 6-18.1, Am. Soc. for Metals, Metals Park, OH (1966).

Kotval, P. S.: The Microstructure of Superalloys, Metallography, *1* (1969), 251-285.

Sabol, G. P., and Stickler, R.: Microstructure of Nickel-Based Superalloys, Phys. Stat. Solidi, *35* (1969), 11-52.

Kohlhaas, E., and Fischer, A.: The Metallography of Superalloys, Prakt. Metallog., *8* (1971), 3-25.

Radavich, J. F., and Couts, W. H.: Metallography of the Superalloys, Rev. High Temp. Mats., *1* (1971), 55-96.

Donachie, M. J., and Kriege, O. H.: Phase Extraction and Analysis in Superalloys—Summary of Investigations by ASTM Committee E-4 Task Group I, J. Mats., *7* (1972), 269-278.

Mihalisin, J. R.: Phase Chemistry and Its Relation to Alloy Behavior in Several Cast Nickel-Base Superalloys, Rev. High Temp. Mats., *2* (1974), 243-260.

Merrick, H. F.: Precipitation in Nickel-Base Alloys, in *Precipitation Processes in Solids,* Met. Soc. Am. Inst. Min. Met. Pet. Eng., NY (1978), 161-190.

Properties and Microstructure

Sullivan, C. P., and Donachie, M. J.: A series of articles in Mets. Eng. Qtly., *7*(2) (1967), 36-45; *9*(2) (1969), 16-29; *11*(4) (1971), 1-11.

Sims, C. T.: A series of articles appearing in J. Mets., *18*(1966), 1119-1134; *21*(12) (1969), 27-42.

Gell, M., et al.: The Fatigue Strength of Nickel-Base Superalloys, in *Achievement of High Fatigue Resistance,* STP 467, Am. Soc. Test. Mats., Philadelphia, PA (1970), 113-153.

Decker, R. F.: Strengthening Mechanisms in Nickel-Base Alloys, in *Steel Strengthening Mechanisms,* Climax Molybdenum Co., Greenwich, CT (1970), 147-170.

Woodford, D. A., and Mowbray, D.F.: Effect of Material Characteristics on Thermal Fatigue of Cast Superalloys: A Review, Mat. Sc. Eng., *16* (1974), 5-45.

Holt, R. T., and Wallace, W.: Impurities and Trace Elements in Nickel-Base Superalloys, Int. Met. Rev., *21* (1976), 1-24.

Processing

DeRidder, A. J., and Koch, R. W.: Controlling Variations in Mechanical Properties of Heat Resistant Alloys During Forging, Mets. Eng. Qtly., *5* (3) (1965), 61-64.

Rawson, J.: Effect of Temperature on Forgeability of Some Heat-Resistant Alloys, in *Reheating for Hot Working,* Iron and Steel Inst., London (1968), 65-69.

Heat Resistant Superalloys, Chapter 11 in *Forging Materials and Practices,* Reinhold, NY (1968), 254-293.

Muzyka, D. R.: Controlling Microstructures and Properties of Superalloys Via Use of Precipitated Phases, Mets. Eng. Qtly., *11* (4) (1971), 12-20.

Muzyka, D., and Maniar, G. N.: Microstructure Approach to Property Optimization in Wrought Superalloys, in *Metallography—A Practical Tool for Corre-*

lating the Structure and Properties of Materials, STP No. 557, ASTM, Philadelphia, PA (1974), 198-219.

Coyne, J. E.: Microstructural Control in Titanium- and Nickel-Base Forgings: An Overview, Met. Tech., *4* (1977), 337-345.

Ridder, A. J., and Koch, R.: Forging and Processing of High Temperature Alloys, in *MiCon 78,* STP No. 672, Am. Soc. Test. Mats., Philadelphia, PA (1979), 547-563.

Bartos, J. L.: Review of Superalloy Powder Metallurgy Processing for Aircraft Gas Turbine Applications, in *MiCon 78,* STP No. 672, Am. Soc. Test. Mats., Philadelphia, PA (1979), 564-577.

Gessinger, G. H.: Powder Metallurgy of Superalloys: Recent Developments, in *Powder Metallurgy Superalloys; Aerospace Materials for the 1980's,* Vol 2, MPR Publishing Services Ltd., Shrewsbury, England (1980).

VerSnyder, F. L., and Shank, M. E.: The Development of Columnar Grain and Single Crystal High Temperature Materials Through Directional Solidification, Mats. Sci. Eng., *6* (1970), 213-247.

Hallerberg, W. L.: Superalloy Investment Cast-

ings—Processing and Applications, Paper No. CM71-160, Soc. Mfg. Eng., Dearborn, MI (1971).

Stevens, R. A., and Flewitt, P. E. J.: Hot Isostatic Pressing to Remove Porosity and Creep Damage, Mats. in Eng., *3* (1982), 461-469.

Olofson, C., et al.: Machining and Grinding of Nickel- and Cobalt-Base Alloys, Report No. TM X-53446, NASA, Washington, DC (1966).

Vagi, J., et al.: Joining of Nickel and Nickel-Base Alloys, Report No. TM X-53447, NASA, Washington, DC (1966).

Heckman, G.: Stabilizing Heat Treatments for Gas Turbine Bucket Alloys, Paper No. 67-GT-55, Am. Soc. Mech. Engr., NY (Mar. 1967).

Burger, J. A., and Hanink, D. K.: Heat Treating Nickel-Base Superalloys, Met. Prog., *92*(1) (1967), 61-66.

Owczarski, W. A.: Process and Metallurgical Factors in Joining Superalloys and Other High Service Temperature Materials, in *Physical Metallurgy of Metal Joining,* Met. Soc. Am. Inst. Min. Met. Pet. Engrs., Warrendale, PA (1980), 166-189.

Heat-Resistant Castings

CASTINGS are classified as heat resistant if they are capable of sustained operation while exposed, either continuously or intermittently, to operating temperatures that result in metal temperatures in excess of 650 °C (1200 °F). Alloys used in castings for such service include iron-chromium ("straight chromium"), iron-chromium-nickel, iron-nickel-chromium, nickel-base and cobalt-base alloys. In application of heat-resistant alloys, considerations include: (*a*) resistance to corrosion at elevated temperatures; (*b*) stability (resistance to warping, cracking or thermal fatigue); and (*c*) creep strength (resistance to plastic flow).

Many alloys of the same general types are used also for their resistance to corrosive media at temperatures below 650 °C (1200 °F), and castings intended for such service are classified as corrosion resistant. Although there is usually a distinction between heat-resistant alloys and corrosion-resistant alloys, based on carbon content, the line of demarcation is vague—particularly for alloys used in the range from 480 to 650 °C (900 to 1200 °F).

GENERAL PROPERTIES

General characteristics of the five types of cast heat-resistant alloys are discussed below. For Fe-Cr, Fe-Cr-Ni and Fe-Ni-Cr alloys, Table 1 lists designations and compositions and Table 2 gives typical room-temperature properties. Compositions of low-iron nickel-base and low-iron cobalt-base alloys are given in Tables 3 and 4, respectively.

Fe-Cr alloys contain 10 to 30% Cr and little or no nickel. These alloys are useful chiefly for resistance to oxidation; they have low strength at elevated temperatures. Use of these alloys is restricted to conditions, either oxidizing or reducing, that involve low static loads and uniform heating. Chromium content depends on anticipated service temperature.

Fe-Cr-Ni alloys contain more than 13% Cr and more than 7% Ni (always more chromium than nickel). These austenitic alloys ordinarily are used under oxidizing or reducing conditions similar to those withstood by the ferritic iron-chromium al-

loys, but in service they have greater strength and ductility than the straight chromium alloys. They are used, therefore, to withstand greater loads and moderate changes of temperature. These alloys also are used in the presence of oxidizing and reducing gases that are high in sulfur content.

Fe-Ni-Cr alloys contain more than 25% Ni and more than 10% Cr (always more nickel than chromium). These austenitic alloys are used for withstanding reducing as well as oxidizing atmospheres, except where sulfur content is appreciable. (In atmospheres containing 0.05% or more hydrogen sulfide, for example, Fe-Cr-Ni alloys are recommended.) In contrast with Fe-Cr-Ni alloys, Fe-Ni-Cr alloys do not carburize rapidly or become brittle and do not take up nitrogen in nitriding atmospheres. These characteristics become enhanced as nickel content is increased, and in carburizing and nitriding atmospheres casting life increases with nickel content. Austenitic Fe-Ni-Cr alloys are used extensively under conditions of severe temperature fluctuations such as those encountered by fixtures used in quenching and by parts that are not heated uniformly or that are heated and cooled

intermittently. In addition, these alloys have characteristics that make them suitable for electrical-resistance heating elements.

Nickel-base alloys contain about 50% Ni and appreciable amounts of chromium, cobalt and refractory metals, but little or no iron; they may also contain aluminum and titanium (see Table 3). Originally, the high-chromium nickel-base alloys were developed for oxidation resistance, and those high in molybdenum for chemical corrosion resistance. Alloys with no more than 20% Cr became known as "superalloys" when strengths were remarkably increased by adding aluminum and titanium. Nickel-base heat-resistant alloys have high-temperature mechanical properties superior to those of other heat-resistant alloys. They are precision investment cast under vacuum in many configurations, but principally as turbine airfoils (blades and vanes). Nickel-base heat-resistant alloys are more costly than iron-base alloys but less expensive than cobalt-base alloys.

ASTM A560 describes two grades of chromium-nickel casting alloys: 50Cr-50Ni and 60Cr-40Ni. These two alloys (not strictly considered

Table 1. Compositions of ACI heat-resistant casting alloys

ACI designation	UNS number	ASTM specifications(a)	C	Composition(b), % Cr	Ni	Si (max)
HA	...	A217	0.20 max	8 to 10	...	1.00
HC	J92605	A297, A608	0.50 max	26 to 30	4 max	2.00
HD	J93005	A297, A608	0.50 max	26 to 30	4 to 7	2.00
HE	J93403	A297, A608	0.20 to 0.50	26 to 30	8 to 11	2.00
HF	J92603	A297, A608	0.20 to 0.40	19 to 23	9 to 12	2.00
HH	J93503	A297, A608	0.20 to 0.50	24 to 28	11 to 14	2.00
HI	J94003	A297, A567, A608	0.20 to 0.50	26 to 30	14 to 18	2.00
HK	J94224	A297, A351, A567, A608	0.20 to 0.60	24 to 28	18 to 22	2.00
HL	J94604	A297, A608	0.20 to 0.60	28 to 32	18 to 22	2.00
HN	J94213	A297, A608	0.20 to 0.50	19 to 32	23 to 27	2.00
HP	...	A297	0.35 to 0.75	24 to 28	33 to 37	2.00
HP-50WZ(c)	...	...	0.45 to 0.55	24 to 28	33 to 37	2.50
HT	J94605	A297, A351, A567, A608	0.35 to 0.75	13 to 17	33 to 37	2.50
HU	...	A297, A608	0.35 to 0.75	17 to 21	37 to 41	2.50
HW	...	A297, A608	0.35 to 0.75	10 to 14	58 to 62	2.50
HX	...	A297, A608	0.35 to 0.75	15 to 19	64 to 68	2.50

(a) ASTM designations are same as ACI designations. (b) Rem Fe in all compositions. Manganese content: 0.35 to 0.65% for HA, 1% for HC, 1.5% for HD and 2% for the other alloys. Phosphorus and sulfur contents: 0.04% max for all but HP-50WZ. Molybdenum is intentionally added only to HA, which has 0.90 to 1.20% Mo; maximum for other alloys is set at 0.5% Mo. HH also contains 0.2% max N. (c) Also contains 4 to 6% W, 0.1 to 1.0% Zr, and 0.035% max S and P.

nickel-base alloys) are cast into tube supports and other firebox fittings for certain stationary and marine boilers. The chief attribute of the Cr-Ni alloys is their resistance to hot-slag corrosion in boilers that fire oil high in vanadium content. Hot slag high in V_2O_5 content is extremely destructive to most other heat-resistant alloys. For example, iron-base alloys frequently last less than one-fourth as long as chromium-nickel alloys in regions of a firebox where the components are continually covered with molten slag containing V_2O_5.

Cobalt-base alloys contain about 50% or more cobalt plus appreciable amounts of chromium and refractory metals. The cobalt-base alloys were developed in the 1940's from Vitallium for turbocharger applications and became known as "superalloys" because of their superior high-temperature strengths compared with those of the Fe-Ni-Cr alloys then available. Cobalt-base alloys have good high-temperature mechanical properties and are precision investment cast in many configurations. They are not as strong as nickel-base alloys in short-time tests but are competitive in strength and corrosion resistance with nickel-base alloys at high temperatures and for long periods of operation. At present, the major uses of these alloys are furnace fixtures and turbine vanes.

MANUFACTURE

Iron-base alloys can be cast from heats melted in electric-arc furnaces that have either acid or basic linings. When melting is done in acid-lined furnaces, however, chromium losses are high and silicon content is difficult to control, and thus acid-lined furnaces are seldom used. All heat-resistant alloys can be melted in high-frequency induction furnaces. Initial melting also can be done in consumable-arc, electron beam or other furnaces with appropriate atmosphere control.

Iron-base alloy castings can be made by the static method, the centrifugal method or the investment process. The centrifugal method is used extensively in production of tubular parts such as radiant tubes for furnaces.

With the exception of a small tonnage of Fe-Cr heat-resistant alloy castings usually are not heat treated before being shipped. Because the 12Cr and 18Cr alloys are hardenable, castings of these alloys sometimes require full annealing for removal of casting stresses.

Nickel-base and cobalt-base alloy castings generally are made by investment casting. Cobalt-base alloys usually are melted and cast in air, although some of the more advanced alloys such as MAR-M 509 must be melted and cast under vacuum. Most nickel-base and cobalt-base castings are small, about 1 kg (2 lb) or less. Large nickel-base alloy castings, up to 1 m (3 ft) in diameter and about 0.3 m (1 ft) in length, have been produced.

For many applications, both nickel-base and cobalt-base alloy castings must be diffusion coated with a material high in silicon or aluminum to enable the alloy to resist the service atmosphere. Cobalt-base alloys are used as cast—with or without a diffusion coating, but without any other thermal treatment. Nickel-base alloys may be used as cast; in the cast, coated and aged condition; or in the cast, solution treated, coated and aged condition.

METALLURGICAL STRUCTURES OF IRON-BASE ALLOYS

The structures of Fe-Cr-Ni and Fe-Ni-Cr alloys must be wholly austenitic, or mostly austenitic with some ferrite, if these alloys are to be used for heat-resistant service. Depending on chromium and nickel contents, the structures of

Table 2. Typical room-temperature properties of ACI heat-resistant casting alloys

Alloy	Condition	Tensile strength MPa	ksi	Yield strength MPa	ksi	Elongation, %	Hardness, HB
HC	As cast.	760	110	515	75	19	223
	Aged(a)	790	115	550	80	18	...
HD	As cast.	585	85	330	48	16	90
HE	As cast.	655	95	310	45	20	200
	Aged(a)	620	90	380	55	10	270
HF	As cast.	635	92	310	45	38	165
	Aged(a)	690	100	345	50	25	190
HH, type 1	As cast.	585	85	345	50	25	185
	Aged(a)	595	86	380	55	11	200
HH, type 2	As cast.	550	80	275	40	15	180
	Aged(a)	635	92	310	45	8	200
HI	As cast.	550	80	310	45	12	180
	Aged(a)	620	90	450	65	6	200
HK	As cast.	515	75	345	50	17	170
	Aged(b)	585	85	345	50	10	190
HL	As cast.	565	82	360	52	19	192
HN	As cast.	470	68	260	38	13	160
HP	As cast.	490	71	275	40	11	170
HT	As cast.	485	70	275	40	10	180
	Aged(b)	515	75	310	45	5	200
HU	As cast.	485	70	275	40	9	170
	Aged(c)	505	73	295	43	5	190
HW	As cast.	470	68	250	36	4	185
	Aged(d)	580	84	360	52	4	205
HX	As cast.	450	65	250	36	9	176
	Aged(c)	505	73	305	44	9	185

(a) Aging treatment: 24 h at 760 °C (1400 °F), furnace cool. (b) Aging treatment: 24 h at 760 °C (1400 °F), air cool. (c) Aging treatment: 48 h at 980 °C (1800 °F), air cool. (d) Aging treatment: 48 h at 980 °C (1800 °F), furnace cool.

Table 3. Compositions of nickel-base heat-resistant casting alloys

Alloy designation	C	Ni	Cr	Co	Mo	Fe	Al	B	Ti	W	Zr	Others
B-1900	0.1	64	8	10	6	...	6	0.015	1	...	0.10	4 Ta(a)
Hastelloy X	0.1	50	21	1	9	18	...	...	...	1	...	...
IN-100	0.18	60.5	10	15	3	...	5.5	0.01	5	...	0.06	1 V
IN-738X	0.17	61.5	16	8.5	1.75	...	3.4	0.01	3.4	2.6	0.1	1.75 Ta, 0.9 Nb
IN-792	0.2	60	13	9	2.0	...	3.2	0.02	4.2	4	0.1	4 Ta
Inconel 713C	0.12	74	12.5	...	4.2	...	6	0.012	0.8	...	0.1	2 Nb
Inconel 713LC	0.05	75	12	...	4.5	...	6	0.01	0.6	...	0.1	2 Nb
Inconel 718	0.04	53	19	...	3	18	0.5	...	0.9	...	...	0.1 Cu, 5 Nb
Inconel X-750	0.04	73	15	...	...	7	0.7	...	2.5	...	...	0.25 Cu, 0.9 Nb
M-252	0.15	56	20	10	10	...	1	0.005	2.6	...	...	...
MAR-M 200	0.15	59	9	10	...	1	5	0.015	2	12.5	0.05	1 Nb(b)
MAR-M 246	0.15	60	9	10	2.5	...	5.5	0.015	1.5	10	0.05	1.5 Ta
MAR-M 247	0.15	59	8.25	10	0.7	0.5	5.5	0.015	1	10	0.05	1.5 Hf, 3 Ta
NX 188 (DS)	0.04	74	...	...	18	...	8	...	...	...	...	...
René 77	0.07	58	15	15	4.2	...	4.3	0.015	3.3	...	0.04	...
René 80	0.17	60	14	9.5	4	...	3	0.015	5	4	0.03	...
René 100	0.18	61	9.5	15	3	...	5.5	0.015	4.2	...	0.06	1 V
TRW-NASA VIA	0.13	61	6	7.5	2	...	5.5	0.02	1	6	0.13	0.4 Hf, 0.5 Nb, 0.5 Re, 9 Ta
Udimet 500	0.1	53	18	17	4	2	3	...	3	...	...	...
Udimet 700	0.1	53.5	15	18.5	5.25	...	4.25	0.03	3.5	...	...	...
Udimet 710	0.13	55	18	15	3	...	2.5	...	5	1.5	0.08	...
Waspaloy	0.07	57.5	19.5	13.5	4.2	1	1.2	0.005	3	...	0.09	...
WAZ-20 (DS)	0.20	72	...	...	...	...	6.5	...	...	20	1.5	...

(a) B-1900 + Hf also contains 1.5% Hf. (b) MAR-M 200 + Hf also contains 1.5% Hf.

Table 4. Compositions of cobalt-base heat-resistant casting alloys

Alloy designation	Nominal composition, %										
	C	Co	Cr	Ni	Al	B	Fe	Ta	W	Zr	Others
AiResist 13	0.45	62	21	...	3.4	...	...	2	11	...	0.1 Y
AiResist 213	0.20	64	20	0.5	3.5	...	0.5	6.5	4.5	0.1	0.1 Y
AiResist 215	0.35	63	19	0.5	4.3	...	0.5	7.5	4.5	0.1	0.1 Y
Haynes 21	0.25	64	27	3	...	...	1	...	...	...	5 Mo
Haynes 25; L-605	0.1	54	20	10	...	...	1	...	15	...	...
Haynes 151(a)	0.48	65	20	...	...	0.03	...	...	12.8	...	3 max Fe + Ni
J-1650	0.20	36	19	27	...	0.02	...	2	12	...	3.8 Ti
MAR-M 302	0.85	58	21.5	...	...	0.005	0.5	9	10	0.2	...
MAR-M 322	1.0	60.5	21.5	...	...	...	0.5	4.5	9	2	0.75 Ti
MAR-M 509	0.6	54.5	23.5	10	...	...	...	3.5	7	0.5	0.2 Ti
MAR-M 918	0.05	52	20	20	...	...	...	7.5	...	0.1	...
NASA Co-W-Re	0.40	67.5	3	...	...	...	...	...	25	1	2 Re, 1 Ti
S-816	0.4	42	20	20	...	...	4	...	4	...	4 Mo, 4 Nb, 1.2 Mn, 0.4 Si
V-36	0.27	42	25	20	...	...	3	...	2	...	4 Mo, 2 Nb, 1 Mn, 0.4 Si
WI-52	0.45	63.5	21	...	...	...	2	...	11	...	2 Nb + Ta
X-40	0.50	57.5	22	10	...	...	1.5	...	7.5	...	0.5 Mn, 0.5 Si

(a) Obsolete alloy, included for reference purposes.

these iron-base alloys can be austenitic (stable), ferritic (stable, but also soft, weak and ductile) or martensitic (unstable); therefore, chromium and nickel levels should be selected to achieve good strength at elevated temperatures combined with resistance to carburization and hot-gas corrosion.

A fine dispersion of carbides or intermetallic compounds in an austenitic matrix increases high-temperature strength considerably. For this reason, iron-base heat-resistant alloys are higher in carbon content than corrosion-resistant alloys of comparable chromium and nickel contents. By holding at temperatures where carbon diffusion is rapid (such as above 1200 °C, or 2200 °F) and then rapidly cooling, a high and uniform carbon content is established and up to about 0.20% C is retained in the austenite. Some chromium carbides are present in the structures of alloys with carbon contents greater than 0.20%, regardless of solution treatment.

Castings develop considerable segregation as they freeze. In standard grades either in the as-cast condition or after rapid cooling from a temperature near the melting point, much of the carbon is in supersaturated solid solution. Subsequent reheating precipitates excess carbides. The lower the reheating temperature, the slower the reaction and the finer the precipitated carbides. Fine carbides increase creep strength and decrease ductility. Intermetallic compounds such as Ni_3Al, if present, have a similar effect.

Reheating of material containing precipitated carbides in the range from 980 to 1200 °C (1800 to 2200 °F) will agglomerate and spheroidize the carbides, which reduces creep strength and increases ductility. Above 1100 °C (2000 °F), so many of the fine carbides are dissolved or spheroidized that this strengthening mechanism loses its importance. For service above 1100 °C, certain proprietary alloys of the Fe-Ni-Cr type have been developed. Alloys for this service contain tungsten to form tungsten carbides, which are more stable than chromium carbides at these temperatures.

Aging at a low temperature, such as 760 °C (1400 °F), where a fine, uniformly dispersed carbide precipitate will form, confers a high level of strength that is retained at temperatures up to those where agglomeration changes the character of the carbide dispersion (over-aging temperatures). Solution heat treatment or quench annealing, followed by aging, is the treatment generally employed to attain maximum creep strength.

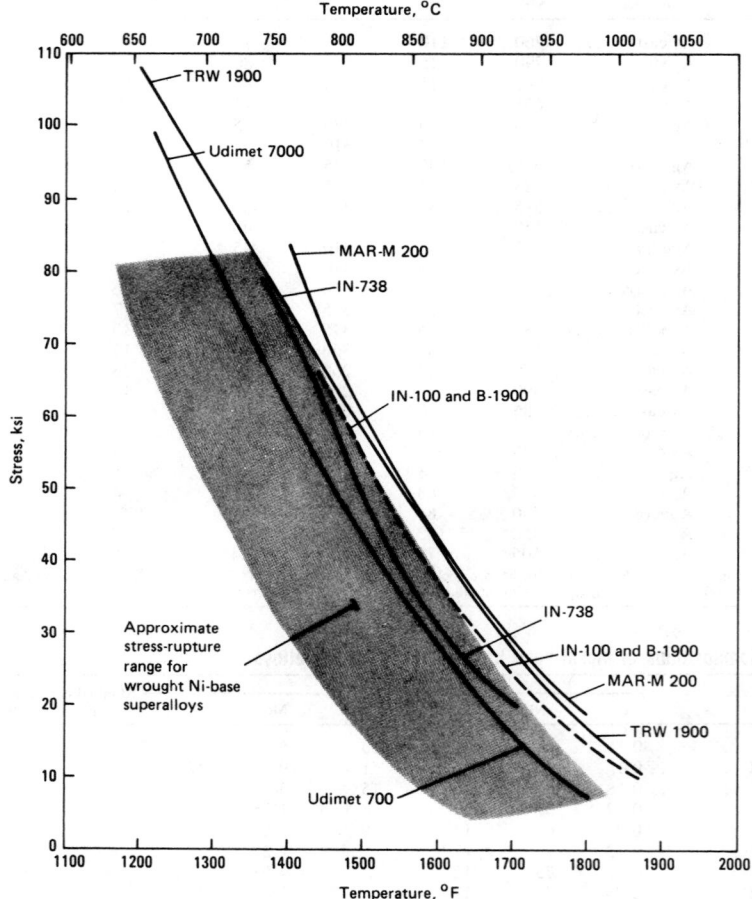

To convert stress values to MPa, multiply by 6.8948.

Fig. 1. Stress-rupture curves for 1000-h life of selected nickel-base alloys

Ductility usually is reduced when strengthening occurs; but in some alloys the strengthening treatment corrects an unfavorable grain-boundary network of brittle carbides, and both properties benefit. However, such treatment is costly and may warp castings excessively. Hence, this treatment is applied to heat-resistant castings only for the small percentage of applications where the need for premium performance justifies the high cost.

Carbide networks at grain boundaries are generally undesirable in iron-base heat-resistant alloys. Grain-boundary networks usually occur in very-high-carbon alloys or in alloys that have cooled slowly through the high temperature ranges where excess carbon in the austenite is rejected as grain-boundary networks rather than as dispersed particles. These grain-boundary carbide networks confer brittleness in proportion to their continuity.

Carbide networks also provide paths for selective attack in some atmospheres and in certain molten salts. Therefore, it is advisable in some salt-bath applications to sacrifice the high-temperature strength imparted by high carbon content and gain resistance to intergranular corrosion by specifying that carbon content be no greater than 0.08%.

METALLURGICAL STRUCTURES OF NICKEL-BASE ALLOYS

Nickel-base heat-resistant casting alloys generally contain substantial levels of aluminum and titanium (see Table 3). These elements strengthen the austenitic matrix through precipitation of $Ni_3(Al, Ti)$, an ordered fcc compound referred to as "gamma prime" (γ'). Various ratios of aluminum and titanium are used in the different nickel-base heat-resistant alloys; generally, titanium atoms can replace aluminum atoms up to a ratio of 3 Ti to 1 Al without altering the ordered fcc crystallographic structure of γ'. When excess titanium is present, Ni_3Ti, an ordered cph compound known as "eta phase" (η), precipitates. Because γ' is coherent with the matrix, precipitation of this phase has a greater strengthening effect than precipitation of η.

In addition to the strengthening imparted by γ' precipitation, solid-solution strengthening is conferred by addition of refractory elements, and grain-boundary strengthening by additions of boron, zirconium, carbon and hafnium. Hafnium also enhances grain-boundary ductility. Stress-rupture curves for various nickel-base alloys are shown in Fig. 1.

The strength of these alloys is complemented by superior corrosion resistance, which is conferred by chromium and aluminum (titanium may be more favorable than aluminum under hot-corrosion conditions). Coatings are used on most nickel alloys for temperatures exceeding about 815 °C (1500 °F) in order to provide adequate protection from oxidation and corrosion at these temperatures.

Nickel-base heat-resistant alloy castings are produced by investment casting under vacuum, and improvements in properties have been made not only through control of composition but also through more precise control of microstructure. A significant advance in microstructure control was the development of a columnar grain structure produced by directional solidification. Single crystals of nickel-base superalloys have been produced by directional solidification as well. The absence of grain boundaries permits elements such as carbon, zirconium and boron to be deleted from the composition. The resulting increase in melting point in turn provides improved flexibility in alloy composition and heat treatment.

Extensive use of nickel alloy castings essentially began with Inconel 713, and now alloys are available that can be used at temperatures up to about 1040 °C (1900 °F).

In addition to creep strength and corrosion resistance, two other properties — stability, and resistance to thermal fatigue — are important considerations in selection of nickel-base heat-resistant casting alloys. Thermal-fatigue resistance is partly controlled by composition, but it is also significantly affected by grain-boundary area and alignment relative to applied stresses. Crystallographic orientation of grains also influences thermal stresses, because the modulus of elasticity, which directly influences thermal stresses, varies with grain orientation. Stability of property values is directly influenced by metallurgical stability: any microstructural changes that take place during long-time exposure at high temperatures under stress cause attendant changes in properties. For instance, if the γ' phase coarsens, strength decreases. Also, potentially deleterious topologically close-packed (tcp) secondary phases, such as sigma, Laves and mu, may form. Coarsening of γ' can be controlled to some degree by adjusting alloy additions. Formation of tcp phases is controlled by adjusting the composition of the matrix to minimize the electron-vacancy number (N_v). A high N_v indicates a tendency toward formation of tcp phases. In general, an N_v value below 2.4 indicates minimal formation of deleterious phases; however, this relationship varies with base-alloy composition.

METALLURGICAL STRUCTURES OF COBALT-BASE ALLOYS

Cobalt-base heat-resistant casting alloys were first used in highly stressed gas-turbine blades during World War II. Although the initial alloy was used at temperatures no higher than those at which the Fe-Cr-Ni and Fe-Ni-Cr heat-resistant alloys were used, it gave satisfactory service under high stress and far surpassed the older alloys in creep and rupture properties. Compositions of selected cobalt-base casting alloys are given in Table 4. Stress-rupture properties of commonly used cobalt-base alloys are shown in Fig. 2.

Most high-temperature cobalt-base alloys contain appreciable amounts of carbon and derive their strength not only from solid-solution hardening by such elements as tungsten and chromium but also from carbide precipitation, which becomes less effective above about 815 °C (1500 °F). Nickel often is added to stabilize the high-temperature form of cobalt (face-centered cubic).

With the advent of vacuum melting, nickel-base alloys strengthened by precipitation of γ' phase promptly surpassed cobalt alloys in strength, and very few new cobalt-base casting alloys were developed for use in gas turbines after 1952. Generally, X-40 is used for latter-stage turbine airfoils of older gas turbines, and MAR-M 509, WI-52 and MAR-M 302 are employed for first-stage, and occasionally second-stage, turbine vanes in gas turbines. In terms of strength, cobalt alloys typically cannot compete with nickel alloys except at temperatures above 980 °C (1800 °F). However, because of the ease with which they can be repaired by welding, and because of their high chromium contents, which give them good corrosion resistance at these high temperatures, cobalt alloys find extensive use in high-pressure turbine vanes. Additional uses include cast components in burner cans and seals, and, for some cobalt alloys, heat-resistant castings for furnace hardware.

Five commonly used cobalt-base alloys are described below. All are investment cast, usually from air-melted stock. None of the commercial alloys is directionally solidified.

In application of cobalt-base alloys, stability and thermal-fatigue resistance must be considered, because cobalt-base alloys generally are less resistant to thermal fatigue than nickel-base alloys of comparable strength. Stability can be inferred from the electron-vacancy number (N_v). However, N_v numbers that delineate cutoff points for tcp-phase formation tend to be near 2.6, which is higher than corresponding numbers for nickel-base alloys. All cobalt-base alloys exhibit aging on exposure as a result of the strengthening effects of changes in the carbide structure.

Haynes 21 and *X-40* are used for turbine vanes in some gas turbines. They have good corrosion

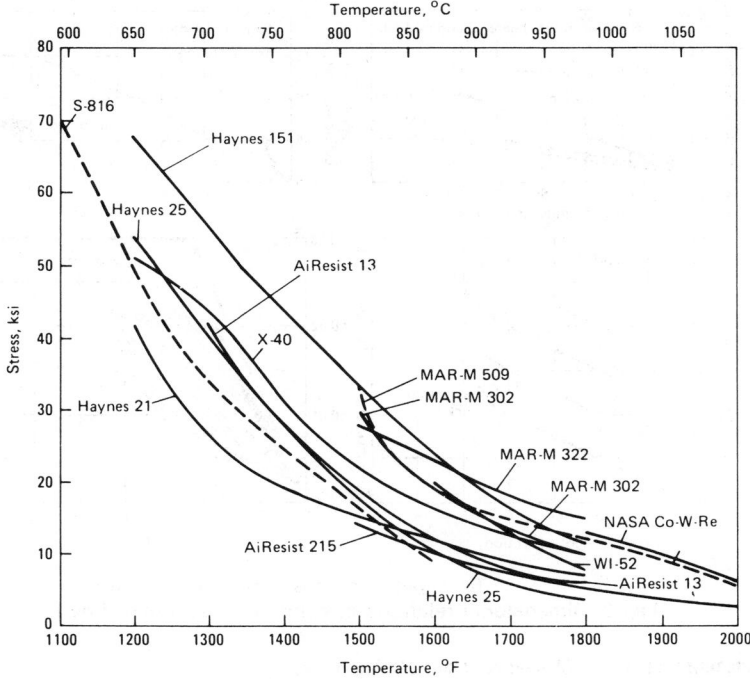

To convert stress values to MPa, multiply by 6.8948.

Fig. 2. Stress-rupture curves for 1000-h life of selected cobalt-base alloys

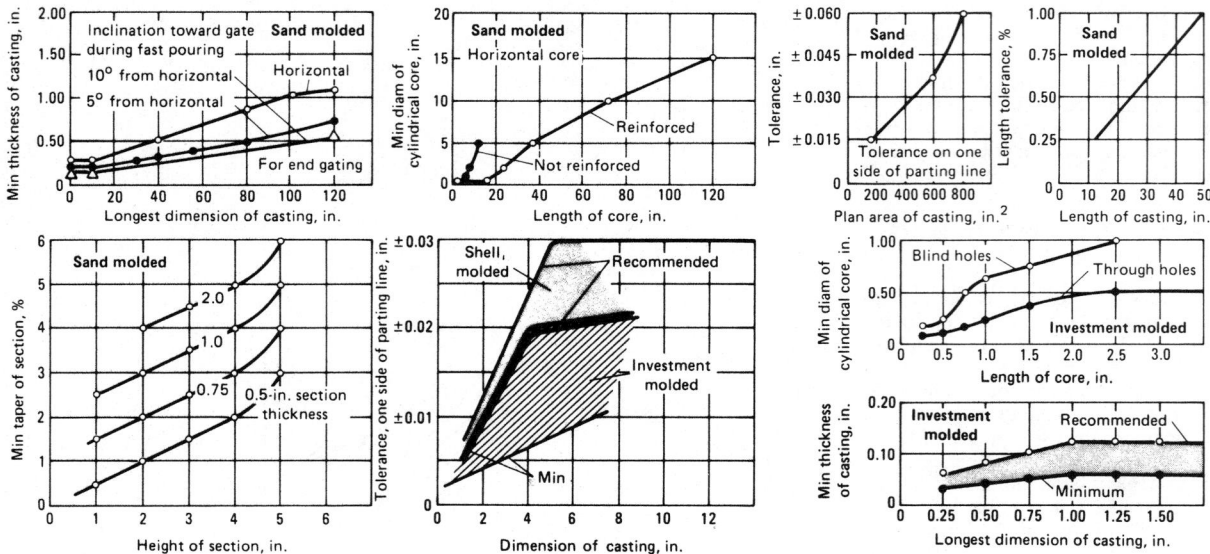

To convert dimensions in inches to equivalent values in millimetres, multiply by 25.

Fig. 3. Dimensional relations in sand, shell and investment molding for heat-resistant alloy castings

Table 5. Machining data for ACI heat-resistant casting alloys

ACI designation	Typical hardness, HB	Rough turning(a)		Finishing		Drilling speed(d), sfm(b)
		Speed, sfm(b)	Feed, ipr(c)	Speed, sfm(b)	Feed, ipr(c)	
HA	220	40 to 50	0.010 to 0.030	80 to 100	0.005 to 0.010	35 to 70
HC	220	40 to 50	0.025 to 0.035	80 to 100	0.010 to 0.015	40 to 60
HD	190	40 to 50	0.025 to 0.035	80 to 100	0.010 to 0.015	40 to 60
HE	270	30 to 40	0.020 to 0.025	60 to 80	0.005 to 0.010	30 to 60
HF	190	25 to 35	0.020 to 0.025	50 to 70	0.005 to 0.010	20 to 40
HH	200	25 to 35	0.015 to 0.020	50 to 70	0.005 to 0.010	20 to 40
HI	200	25 to 35	0.015 to 0.020	50 to 70	0.005 to 0.010	20 to 40
HK	190	25 to 35	0.020 to 0.025	50 to 70	0.005 to 0.010	20 to 40
HL	190	30 to 40	0.020 to 0.025	60 to 80	0.005 to 0.010	30 to 60
HN	160	35 to 45	0.020 to 0.025	70 to 90	0.005 to 0.010	40 to 60
HP	...	35 to 45	0.020 to 0.025	70 to 90	0.005 to 0.010	40 to 60
HT	200	40 to 45	0.025 to 0.035	80 to 90	0.005 to 0.010	40 to 60
HU	190	40 to 45	0.025 to 0.035	80 to 90	0.010 to 0.015	40 to 60
HW	200	40 to 45	0.025 to 0.035	80 to 90	0.010 to 0.015	40 to 60
HX	185	40 to 45	0.025 to 0.035	80 to 90	0.010 to 0.015	40 to 60

(a) Single-point high speed steel tools usually are ground to 4 to 10° side and back rake, 4 to 7° side relief, 7 to 10° end relief, 8 to 15° end cutting-edge angle, 10 to 15° side cutting-edge angle, and $1/32$- to $1/8$-in. nose radius. (b) To convert to m/s, multiply by 0.005. (c) To convert to mm/rev, multiply by 25. (d) Recommended drilling feeds are as follows: for drill diameters up to $1/8$ in., 0.001 to 0.002 ipr; $1/8$ to $1/4$ in., 0.002 to 0.004 ipr; $1/4$ to $1/2$ in., 0.004 to 0.007 ipr; $1/2$ to 1 in., 0.007 to 0.015 ipr; over 1 in., 0.015 to 0.025 ipr. Tapping speeds recommended for HA, HC, HD, HE and HL are 10 to 25 sfm; for HF, HH, HI, and HK, 10 to 20 sfm; and for HN, HT, HU, HW and HX, 5 to 15 sfm.

resistance and strength at temperatures from 700 to 815 °C (1300 to 1500 °F).

WI-52 is used for turbine vanes in some gas turbines. It can be used at temperatures up to about 950 °C (1750 °F) but must be coated for corrosion resistance.

MAR-M 302 is vacuum cast into turbine vanes for some gas turbines. This alloy is used at temperatures up to about 980 °C (1800 °F).

MAR-M 509 is vacuum cast into high-pressure turbine vanes and seals for gas turbines. The alloy has good corrosion resistance and strength

from 815 to 1010 °C (1500 to 1850 °F). MAR-M 509 generally is coated to provide the required corrosion resistance.

APPLICATIONS

In terms of tonnage, the most important use of heat-resistant castings is in metallurgical and other industrial furnaces. Iron-base alloys are most often used for this service, although significant amounts of nickel-base and cobalt-base alloys are used also. Other major applications for heat-resistant cast-

ings include turbochargers, gas turbines, power plant equipment, and equipment used in the manufacture of glass, cement, synthetic rubber, chemicals and petroleum products.

Alloy Selection. Heat-resistant alloys are selected on the basis of structural integrity in a specific application. Strength, creep resistance and corrosion resistance are prime factors influencing alloy selection. Next in importance is castability, though it is difficult to obtain a quantitative evaluation of this factor. The ability of the alloy to fill the mold must be assessed, as must the statistical distributions of product porosity and critical dimensions.

Because many castings must be machined to final dimensions, machinability of the as-cast metal often must be evaluated. Quantitative machining data are available for many cast heat-resistant alloys. Typical recommendations of speed, feed and depth of cut are summarized in Table 5 for iron-base casting alloys.

CASTING DESIGN FOR IRON-BASE ALLOYS

For most applications, ACI heat-resistant alloys are sand cast, although shell molding is also used. Sections in thicknesses of 4.8 mm ($3/16$ in.) and greater can be cast satisfactorily, and somewhat thinner sections also may be castable depending on casting design and pattern equipment. Dimensional tolerances for rough castings are influenced by quality of pattern equipment. In general, over-all dimensions and locations of cored holes can be held to 5.2 mm/m ($1/16$ in./ ft).

More specific information on casting design is presented in Fig. 3.

Refractory Metals and Alloys

A MAJORITY of the total tonnage of refractory metals produced is now used in aerospace applications, and much of the development of specific manufacturing techniques has been in that area.

Selection of a specific alloy from the refractory-metal group often is based on fabricability rather than on strength or corrosion resistance. Niobium, tantalum and their alloys are the most easily fabricated refractory metals. They can be formed, machined and joined by conventional methods. They are ductile in the pure state and have high interstitial solubilities for carbon, nitrogen, oxygen and hydrogen. Because of such high solubilities, these embrittling contaminants normally do not present problems in fabrication. However, tantalum and niobium dissolve sufficient amounts of oxygen at elevated temperatures to destroy ductility at normal operating temperatures. Therefore, elevated-temperature fabrication of these metals is used only when necessary. Protective coatings or atmospheres are mandatory unless some contamination can be tolerated. The allowable level of contamination, in turn, determines the maximum permissible exposure time in air at elevated temperature.

Table 1 lists nominal compositions of the refractory-metal alloys that are now commercially prominent.

Resistance of refractory metals to corrosion by liquid metals and aggressive acid solutions can cut maintenance and downtime if high initial cost can be accepted. Systems for containing liquid metals such as lithium and cesium at high temperature have been fabricated of Nb-1Zr alloy tubing; tantalum and tantalum-clad steel process equipment has performed well in high-temperature sulfuric acid service.

Most refractory metals and alloys are available as wire. Tungsten wire, for example, which comes in diameters as small as 0.0102 mm (0.0004 in.), is used as fiber reinforcement in composite materials in which the matrix is any one of various ductile alloys.

Table 1. Compositions of commercially important refractory-metal alloys

Designation	Nominal composition, %
Molybdenum alloys	
Mo-0.5Ti	Mo-0.5Ti-0.02W
TZM	Mo-0.5Ti-0.1Zr-0.02W
Mo-30W	Mo-30W
Niobium alloys	
Nb-1Zr	Nb-1Zr
FS-85	Nb-27.5Ta-11W-1Zr
SCb-291	Nb-10Ta-10W
Cb-752	Nb-10W-2.5Zr
B-66	Nb-5Mo-5V-1Zr
C-103	Nb-10Hf-1Ti
C-129Y	Nb-10W-10Hf-0.15Y
Tantalum alloys	
"63" Metal	Ta-2.5W-0.15Nb
Ta-10W	Ta-10W
T-111	Ta-8W-2Hf
T-222	Ta-10W-2.5Hf-0.01C
Tungsten alloys	
W-ThO$_2$	W-1ThO$_2$; W-2ThO$_2$
W-Mo alloys	Various Mo contents; W-2Mo and W-15 Mo are most common
W-Re alloys	Various Re contents up to 26%; W-1.5Re, W-3Re and W-25Re are most common
Doped W	50 ppm Si, 90 ppm K, 15 ppm Al, 35 ppm O

PRODUCTION

Primary production of refractory metals starts with consolidation by melting or powder metallurgy techniques. Hot forging or extrusion is used for breaking down ingots or powder compacts into sheet bar and solid rounds for processing into sheet, plate, foil, tubing and bar products. Table 2 gives typical mill-processing temperatures.

MACHINING

Compared with familiar structural alloys, refractory metals are considered difficult to machine. However, they can be machined using modifications of conventional machining practices.

Niobium alloys and tantalum alloys are readily machined using high speed steel or carbide tools. Machining and grinding characteristics vary from those of soft copper to those of annealed stainless steel.

Molybdenum is machined using carbide tools of the same configurations as those used for machining 1040 steel, the machining characteristics of these two metals being similar. Machining speeds for molybdenum alloys (TZM, for example) are about 40% higher than those for type 302 stainless steel. Finish grinding of molybdenum requires heavy coolant flow and use of alundum wheels to prevent heat checking. Tool configurations and grinding techniques are similar to those for grinding cast iron; conventional ma-

Table 2. Mill-processing temperatures for refractory metals

Metal or alloy	Forging Temperature(a) °C	Forging Temperature(a) °F	Typical total reduction, %	Extrusion Temperature(a) °C	Extrusion Temperature(a) °F	Typical reduction ratio	Rolling Temperature(a) °C	Rolling Temperature(a) °F	Typical total reduction (between anneals), %
Nb	980 to 650	1800 to 1200	50 to 80	1090 to 650	2000 to 1200	10:1	315 to 205 / 20	600 to 400 / 70	50 breakdown / 90 finish
Nb-1Zr	1200 to 980	2200 to 1800	50 to 80	1200 to 980	2200 to 1800	10:1	315 to 205 / 20	600 to 400 / 70	50 breakdown / 80 finish
FS-85	1320 to 980	2400 to 1800	50	1320 to 980	2400 to 1800	4:1	370 to 205 / 20	700 to 400 / 70	40 breakdown / 50 to 65 finish
SCb-291	1200 to 980	2200 to 1800	30	1320 to 980	2400 to 1800	4:1	370 to 260 / 20	700 to 500 / 70	50 breakdown / 60 to 75 finish
Cb-752	1200 to 980	2200 to 1800	30	1320 to 980	2400 to 1800	4:1	370 to 260 / 20	700 to 500 / 70	50 breakdown / 60 to 75 finish
B-66	1290 to 980	2350 to 1800	50	1320 to 980	2400 to 1800	4:1	1200 to 1090 / 20	2200 to 2000 / 500 to 70	50 breakdown / 25 to 50 finish
C-103	1320 to 980	2400 to 1800	50	1320 to 980	2400 to 1800	8:1	205 / 20	400 / 70	50 breakdown / 60 to 70 finish
C-129Y	1320 to 980	2400 to 1800	50	1320 to 980	2400 to 1800	4:1	425 / 20	800 / 70	50 breakdown / 60 to 70 finish
Ta	<500 / 20	<930 / 70	50 to 80 / Finish	1090	2000	10:1	370 to 260 / 20	700 to 500 / 70	80 breakdown / 90 finish
Ta-10W	1260 to 980 / 1090 to 815	2300 to 1800 / 2000 to 1500	50 / Finish	1650 to 1425	3000 to 2600	10:1	370 to 260 / 20	700 to 500 / 70	80 breakdown / 90 finish
T-222	1260 to 1200	2300 to 2200	50	2040 to 1650	3700 to 3000	10:1	370 to 260 / 20	700 to 500 / 70	75 breakdown / 50 to 75 finish
Mo	1320 to 1150 / 925 to 815	2400 to 2100 / 1700 to 1500	50 / Finish	1760 to 1370	3200 to 2500	8:1	1200 / 870	2200 / 1600	50 breakdown / 90 to 75 finish
Mo-0.5Ti	1425 to 1260 / 1320 to 1150	2600 to 2300 / 2400 to 2100	50 / Finish	1820 to 1480	3300 to 2700	8:1	1200 / 870	2200 / 1600	50 breakdown / 75 finish
TZM	1480 to 1320 / 1370 to 1200	2700 to 2400 / 2500 to 2200	50 / Finish	1820 to 1540	3300 to 2800	8:1	1350 to 1200 / 1000 to 980 / 315	2460 to 2200 / 1830 to 1800 / 600	50 breakdown / 60 / 10 finish
W	1820 to 1590 / 1320 to 1010	3300 to 2900 / 2400 to 1850	20 / Finish	1925 to 1650	3500 to 3000	9:1	1450 to 1400 / 1370 to 980	2640 to 2550 / 2500 to 1800	50 breakdown / 90 finish

(a) Where a range is given, the higher temperature is typical starting temperature and the lower temperature is minimum working temperature for that process.

Table 3. Recommended postweld annealing treatments for selected refractory alloys

Alloy	GTAW welds °C	GTAW welds °F	EB welds °C	EB welds °F
B-66	Not annealed		1040	1900
D-43	1315	2400	1315	2400
FS-85, C-129Y	1315	2400	1200	2200
Cb-752	1200	2200	1315	2400
SCb-291	1200	2200	Not annealed	
T-111, T-222	1315	2400	1315	2400
Ta-10W	Not annealed		Not annealed	

(a) One hour at temperature.

chines with standard feeds and speeds are satisfactory.

Turning is a problem only with tungsten. Carbide tools ground with negative back rake, 15° lead and 0° side rake are mandatory. All turning is done at room temperature.

For grinding tungsten, wheels of 60-grit silicon carbide or 46-grit alumina are recommended, and the use of normal grinding precautions, extra-light pressures and heavy coolant flow is required.

Tungsten and molybdenum must be punched and sheared above their ductile-to-brittle transition temperatures. Sheets over 1.3 mm (0.050 in.) thick must have excess thickness of 1.6 to 3.2 mm ($^1/_{16}$ to $^1/_8$ in.) to allow for belt sanding to final dimensions. Cutting can be done using abrasive (60-grit SiC) cutoff wheels. Tungsten and molybdenum are also suitable for electrical-discharge machining (EDM) and electrochemical machining (ECM).

FORMING

Sheet metal and tubing of tantalum or niobium alloys usually are formed in the annealed condition. Their forming behavior is similar to that of mild steel, except that they are more prone to galling, seizing and tearing. In thicknesses from 0.1 to 1.5 mm (0.004 to 0.060 in.), tantalum and niobium can be readily formed, blanked, punched, stamped or deep drawn at room temperature in steel dies (6% t clearance). Sheet must have a homogeneous, fine grain size (generally, ASTM No. 5 or finer) for satisfactory results. Coarse-grain sheet is likely to fail by localized necking during severe forming.

Both tungsten and molybdenum must have highly worked, fibrous microstructures to ensure adequate formability. Molybdenum disulfide and graphite lubricants facilitate forming. Sheet 0.5 mm (0.020 in.) or less in thickness is bent or

rolled at 21 to 93 °C (70 to 200 °F) for molybdenum and at 425 to 540 °C (800 to 1000 °F) for tungsten. Deep drawing of heavier sheets of molybdenum and tungsten is done at 425 °C (800 °F) and 925 °C (1700 °F), respectively.

JOINING

All refractory metals can be joined by electron beam (EB) welding, gas tungsten-arc welding (GTAW) or resistance welding. Two major problems are encountered: chemical changes due chiefly to atmospheric contamination, and microstructural changes resulting from thermal cycling. The latter changes include grain growth and different stages of precipitation hardening (solution, precipitation and over-aging). Preheating and postheating generally are required to minimize deleterious effects arising from precipitation hardening as well as from the residual stresses normally induced by welding. Although recrystallization and grain growth are unavoidable in weldments of wrought tungsten and molybdenum, proper choice of welding process and procedure can localize these effects. Electron beam welding has proved effective in achieving full weld penetration with an extremely narrow heat-affected zone.

All refractory metals suffer losses in ductility and increases in ductile-to-brittle transition temperature when welded, but niobium and tantalum alloys are less affected than molybdenum and tungsten alloys. Tantalum and niobium alloys generally retain greater than 75% joint efficiency after gas tungsten-arc welding. Preheating is not required, but postweld annealing can restore large amounts of ductility and toughness to commercial alloys. Table 3 summarizes recommended postweld annealing treatments for seven refractory-metal alloys, and Table 4 lists typical conditions for gas tungsten-arc and electron beam welding of refractory-metal sheet.

Tungsten is the most difficult refractory metal to join for satisfactory high-temperature service. Welding, and especially the EB process, offers the best compromise for joining tungsten for service at high temperatures. Mechanical joints are unsatisfactory unless molybdenum fasteners are used. Diffusion bonding is impractical because of severe tooling problems. Brazing for relatively low-temperature applications is done using precious metals (silver, palladium and platinum alloys) and transition metals (nickel and manganese alloys) as filler metals.

Table 5 lists typical brazing filler metals, and their maximum service temperatures, for all four refractory-metal systems.

Table 5. Typical brazing filler metals and service temperatures

Filler metal	Maximum service temperature °C	Maximum service temperature °F
For niobium alloys		
Si-Cr-Ni	980	1800
4Be-48Zr-48Ti	925	1700
Zr-6Be-19Nb	925	1700
Ti-0.5Si	1370	2500
Zr-0.1Be-16Ti-25V	1200	2200
V-35Nb	1200	2200
Ti-50Zr	1650	3000
Titanium	1760	3200
Ti-33Cr	1370	2500
Ti-3Al-11Cr-13V	1650	3000
For tantalum alloys		
Hf-7Mo	2090	3800
V-20Nb-20Ta	1870	3400
Ti-15Ta-25V	1650	3000
Ta-10Hf-(15-70)Nb	2200	4000
Nb-(30-50)Hf	2200	4000
Ta-10Hf	2200	4000
Nb-1.3B	1925	3500
Copper	980	1800
For molybdenum alloys		
Ti-3Be-25Cr	1590	2900
Pt-Mo	1650	3000
Zr-Ti	1230	2250
Cu-Au	815	1500
Ni-Cu	1200	2200
V-35Nb	1200	2200
Ti-30V	1370	2500
Ti-13Ni-25Cr	1760	3200
Co-10Ni-15W-20Cr	1320	2400
For tungsten		
Ag-Mn	870	1600
V-Nb-Ta	1925	3500
V-Ti-Ta	1925	3500
W-25Os	2200	4000
W-3Re-50Mo	2200	4000
Mo-5Os	1925	3500
Niobium	1650	3000
Tantalum	2200	4000

COATINGS

Surface protection is the most significant obstacle to widespread use of refractory metals in high-temperature oxidizing environments. The existing ceiling of about 1650 °C (3000 °F) is dictated by coating limitations: coatings have insufficient life at reduced pressures (below about 13 kPa, or 100 torr) and high temperatures (about 1370 °C, or 2500 °F) in oxidizing atmospheres and give unreliable protection, particularly at edges and corners.

Table 6 summarizes coating systems of current importance. Aluminide and silicide coatings with various modifications are available commercially. Much controversy exists on application methods; however, slurry techniques are usually preferred, because pack cementation processes can be used only for small parts.

Silicide coatings are used more often than aluminide coatings, and have been selected for radiative heat shields and leading edges on aerospace vehicles. Successful simulations and actual flight tests have lessened the concern over possible catastrophic failure of refractory-metal components due to localized defects in coatings. The vanadium-modified niobium disilicide coating system has proved outstanding.

Table 4. Typical conditions for welding 0.9-mm (0.035-in.) refractory-metal sheet

Alloy	Gas tungsten-arc welds Speed, in./min (a)	Gas tungsten-arc welds Clamp spacing, in.(a)	Gas tungsten-arc welds Current, A(b)	Gas tungsten-arc welds Arc gap, in.(a)	Electron beam welds Speed, in./min (a)	Electron beam welds Clamp spacing, in.(a)	Electron beam welds Deflection, in.(a)(c)	Electron beam welds Voltage, kV	Electron beam welds Current, mA
B-66	15	$^3/_8$	86	0.06	25	$^3/_{16}$	0.050	150	3.2
C-129Y	30	$^3/_8$	110	0.06	50	$^1/_2$	0.050	150	4.1
Cb-752	30	$^3/_8$	87	0.06	15	$^3/_{16}$	0.050	150	3.3
FS-85	15	$^3/_8$	90	0.06	50	$^3/_{16}$	0.050	150	4.4
T-111	15	$^3/_8$	115	0.06	15	$^1/_2$	0.050	150	3.8
T-222	30	$^1/_4$	190	0.06	15	$^1/_2$	0.050	150	4.5
Ta-10W	7.5	$^1/_4$	118	0.06	15	$^1/_2$	0.050	150	3.8

(a) To convert inch values to equivalent values in mm, multiply by 25. (b) Direct current, straight polarity. (c) Beam deflection at 60 cycles parallel to weld direction.

Table 6. Coatings for refractory metals

Coating designation	Method of application	Developer	Applicable substrate	Temperature limit(a) °C	°F
Aluminide coatings					
LB-2 (Al-Cr-Si)	Fused slurry	General Electric;	Nb	1425	2600
		McDonnell Douglas	Ta	1650	3000
Al-Si-Cr	Fused slurry	Sylvania	Nb	1425	2600
Sn-Al	Slurry dip or spray	GT&E	Mo	1480	2700
Ag-Si-Al	Hot dip	Sylvania	Nb	1540	2800
NAA-85 (Al$_2$O$_3$ + Al)	Slurry fusion	North American Rockwell	Nb	1425	2600
Silicide coatings					
Cr-Ti-Si (multilayered)	Vacuum pack and vacuum slip pack	TRW	Nb	1480	2700
			Ta	1480	2700
W Modified	Plasma spraying	TRW	W	1980	3600
Disil (Si + V, Cr, Ti)	Fluidized bed	Boeing	Nb	1540	2800
			Ta	1540	2800
L-7 (MoSi$_2$)	Slip pack	McDonnell Douglas	Mo	1650	3000
			W	1980	3600
PFR (Si + additives)	Pack cementation; fluidized bed	Pfaudler	Nb	1650	3000
PFR-5 (MoSi$_2$ + Cr)	Pack cementation	Pfaudler	Mo	1650	3000
R(Si-20Cr-5Ti)	Fusion	Sylvania	Nb	1650	3000
NS-4, TNV-7 (complex silicides)	Vacuum and high-pressure pack	Solar	Nb	1650	3000
Durak KA	Pack cementation	Chromizing	Nb	1425	2600
Durak B	Pack cementation	Chromizing	Mo	1650	3000
N-2 (Si + Cr, Al, B)	Pack cementation	Chromalloy	Nb	1425	2600
W-3	Pack cementation	Chromalloy	Mo	1650	3000

(a) Maximum temperature at which coating will give one hour of protection at atmospheric pressure.

Because operating temperatures must exceed 1650 °C (3000 °F) before use of tantalum and tungsten alloys can be justified for aerospace applications, coatings for niobium and molybdenum are more highly developed. Surface protection at temperatures above 1650 °C requires more complicated approaches, including new coating systems, new application techniques, and fresh concepts in materials design.

Some promising systems for protection of tantalum and tungsten from 1650 to 2200 °C (3000 to 4000 °F) include roll cladding with Ta-Hf alloys; slurry-type coatings of iridium-base alloys such as Ir-30Rh; and duplex and triplex silicide-base coating systems that combine slurry, slip, chemical vapor deposition and pack cementation processes.

MOLYBDENUM ALLOYS

There are two basic alloys of molybdenum in use today: TZM and Mo-30W. Nominal compositions of these alloys are given in Table 1.

TZM is used for high-temperature structural applications and in tooling for hot die forging. It has a higher recrystallization temperature, higher creep strength and higher tensile strength than pure molybdenum.

Mo-30W has outstanding resistance to corrosive attack by liquid metals, especially liquid zinc. Its melting point is 2830 °C (5125 °F), which is higher than the melting point of either TZM or pure molybdenum.

NIOBIUM (COLUMBIUM)

Niobium forms an oxide coating in most acid environments. This coating provides excellent corrosion resistance, especially to nitric and hydrochloric acids. Strong alkaline solutions and hydrofluoric acid attack niobium severely. At elevated temperatures the metal reacts with halogens, oxygen, nitrogen, carbon, hydrogen and sulfur. It forms high-melting-point refractory compounds with light elements such as carbon, boron, silicon and nitrogen.

Niobium can be used in contact with liquid lithium, sodium and sodium-potassium eutectic at temperatures well above 800 °C (1470 °F). Addition of 1% Zr increases the resistance of niobium to embrittlement caused by oxygen absorbed from the liquid metal. Typical compositions of niobium alloys are given in Table 1.

TANTALUM

Tantalum can be used for structural applications at service temperatures from 1370 to 1980 °C (2500 to 3600 °F), but for any exposure to an oxidizing environment at temperatures above 480 °C (900 °F) it requires a protective coating. It exhibits exceptional resistance to corrosion by acids (except HF and fuming H$_2$SO$_4$).

Tantalum's dense, dielectric oxide film makes it useful for miniature capacitors. Currently, the largest use of tantalum is in electrolytic capacitors; sintered P/M anodes are used in solid and wet electrolyte capacitors, and to a lesser extent precision tantalum foil is used in foil capacitors. Tantalum also is used extensively in chemical-process equipment such as heat exchangers, condensers, thermowells and lined vessels. Notably, it is used for condensing, reboiling, preheating and cooling of nitric, hydrochloric, bromic and sulfuric acids, and combinations of these acids with many other chemicals. Spinnerettes for extruding man-made textile fibers constitute another important use.

Because of its high melting point, tantalum is used for heating elements, heat shields and other components in vacuum furnaces. Tantalum and some tantalum alloys have been used in specialized aerospace and nuclear applications. Tantalum has been used in prosthetic devices in contact with body fluids and as an alloying element in superalloys. Tantalum carbide is an important constituent in complex cemented carbides used for cutting steel (see compositions of commercial tantalum alloys in Table 1).

TUNGSTEN

Tungsten's high melting point makes it the obvious choice for structural applications at very high temperatures. Design engineers must first overcome or accommodate its high density, brittleness at normal temperatures, poor oxidation resistance and poor fabricability.

Three types of tungsten alloys are produced commercially: W-ThO$_2$, W-Mo and W-Re alloys.

W-ThO$_2$ alloy is a dispersed second-phase alloy containing 1 to 2% thoria. The thoria dispersion enhances thermionic electron emission, which improves starting characteristics of GTAW welding electrodes. It also increases the efficiency of electron-discharge tubes and imparts creep strength to wire at temperatures above one-half the absolute melting point of tungsten.

W-Mo Alloys. Molybdenum forms a continuous solid solution with tungsten, the solidus of which is about 20 °C (36 °F) below the liquidus at 50% W. W-Mo alloys are used mostly for improved machinability where strength somewhat lower than those of tungsten and W-ThO$_2$ can be tolerated.

W-Re Alloys. Rhenium is soluble in tungsten up to 26%, above which the embrittling sigma phase begins to form. The W-1.5Re and W-3Re alloys are used to improve resistance to cold fracture in lamp filaments, especially for lamps exposed to vibrations. These alloys also contain the AKS dopants to improve creep strength in filament wires. Undoped W-1.5Re and W-3Re are used in thermocouple applications where strength is not a primary factor.

Machinable Tungsten Alloys

MACHINABLE TUNGSTEN ALLOYS are of interest for applications requiring material of high density or mass, such as counterweights for self-winding wristwatches and instruments, counterbalance weights for jet aircraft and helicopters, gyroscope rotors and similar inertial components, tool shanks for chatter-free machining tools, and shielding for use against x-rays and gamma rays. Because of the high melting point of tungsten, these components must be made by powder metallurgy techniques.

CLASSIFICATION OF ALLOYS

The specifications for machinable, high-density tungsten-base alloys usually divide them into four classes based on composition (Table 1).

Class 1 alloys may be basically tungsten-nickel-copper or tungsten-nickel-iron alloys. Tungsten-nickel-copper alloys of this class typically contain 90 W, 6 to 7 Ni and 3 to 4 Cu. Minor additions of other metals, such as molybdenum or cobalt, may be made to modify properties such as hardness. Class 1 tungsten-nickel-iron alloys usually contain 90 W, 5 to 7 Ni and 3 to 5 Fe.

Class 2, 3 and 4 alloys are usually tungsten-nickel-iron alloys with tungsten percentages in the ranges shown in Table 1, remainder Ni-Fe in a ratio of 7Ni:3Fe to 5Ni:5Fe. Sometimes a portion of the iron may be replaced with copper.

METHODS OF MANUFACTURE

The heavy-metal alloys usually are produced from a mixture of elemental, high-purity, fine-particle-size metal powders. The tungsten powder is about 2 to 3 μm (0.08 to 0.12 mil) in average particle size and is 99.99% pure. Fine high-purity nickel powder such as carbonyl nickel, fine electrolytic copper powder, and fine high-purity iron powder such as carbonyl iron are used. The powders are blended in a powder blender or ball mill for sufficient time to produce a homogeneous mixture and to achieve an apparent density compatible with the molding operation. Molding pressures of about 70 to 140 MPa (10 to 20 ksi) are used. The molded compact must be of such size as to allow for considerable shrinkage during the sintering operation.

Sintering. The molded parts are usually sintered in box-type electric sintering furnaces either by

plunging or stoking. The furnaces must have molybdenum or tungsten heating elements, because sintering temperatures range from about 1425 to 1650 °C (2600 to 3000 °F) depending on the exact composition of the alloy. In some instances, vacuum furnaces are used for sintering these materials, but the usual operation utilizes dry hydrogen or dissociated ammonia for the sintering atmosphere. Sintering times at temperature range from about 20 minutes for small parts to several hours for large blanks.

Hot Pressing. Some very large parts are produced by hot pressing rather than by cold pressing and sintering. This usually is done by leveling the powder mix in a graphite mold and heating in an induction coil while light pressure, sufficient to compact the mix to the required density at temperatures similar to the sintering temperatures mentioned above, is applied to the assembly. Hot pressed alloys of this type usually are more brittle and lower in strength than the cold pressed and sintered materials.

Machining and Finishing. Tungsten heavy alloys can be machined by the usual methods, such as turning, boring, shaping, drilling and tapping. For small runs, steel tools may be used, but generally carbide tools of the type used for machining cast iron are recommended.

PROPERTIES

Minimum mechanical properties of machinable heavy tungsten alloys are given in the specifications under which these materials are purchased. There are three specifications in general use: MIL-T-21014, ASTM B459 and AMS 7725. Typical mechanical properties of commercial machinable tungsten alloys are presented in Table 2.

Table 1. Classification of machinable tungsten alloys by composition, density and hardness

Class	Tungsten content, wt%	Density, Mg/m³	Hardness, HRC	Available in type
1	89-91	16.85-17.25	30-36	I
			32 max	II, III
2	91-94	17.15-17.85	33 max	II, III
3	94-96	17.75-18.35	34 max	II, III
4	96-98	18.25-18.85	35 max	II, III

Table 2. Typical mechanical properties of commercial machinable tungsten alloys

W-Ni-Cu alloy, class 1

Density	17.0 Mg/m³ (0.614 lb/in.³)
Tensile strength	785 MPa (114 ksi)
Yield strength(a)	605 MPa (88 ksi)
Elongation(b)	4%
Hardness	27 HRC
Proportional limit	205 MPa (30 ksi)
Modulus of elasticity	275 GPa (40 × 10⁶ psi)
Coefficient of thermal expansion	5.5 μm/m·°C (3.0 μin./in.·°F)
Magnetic properties	Virtually nonmagnetic

W-Ni-Fe alloy, class 1

Density	17.0 Mg/m³ (0.614 lb/in.³)
Tensile strength	895 MPa (130 ksi)
Yield strength(a)	615 MPa (89 ksi)
Elongation(b)	16%
Hardness	27 HRC
Proportional limit	260 MPa (38 ksi)
Modulus of elasticity	275 GPa (40 × 10⁶ psi)
Coefficient of thermal expansion	5.4 μm/m·°C (3.0 μin./in.·°F)
Magnetic properties	Slightly magnetic

W-Ni-Fe alloy, class 3

Density	18.0 Mg/m³ (0.650 lb/in.³)
Tensile strength	925 MPa (134 ksi)
Yield strength(a)	655 MPa (95 ksi)
Elongation(b)	6%
Hardness	29 HRC
Proportional limit	350 MPa (51 ksi)
Modulus of elasticity	310 GPa (45 × 10⁶ psi)
Coefficient of thermal expansion	5.3 μm/m·°C (2.9 μin./in.·°F)
Magnetic properties	Slightly magnetic

W-Ni-Fe alloy, class 4

Density	18.5 Mg/m³ (0.667 lb/in.³)
Tensile strength	795 MPa (115 ksi)
Yield strength(a)	690 MPa (100 ksi)
Elongation(b)	3%
Hardness	32 HRC
Proportional limit	450 MPa (65 ksi)
Modulus of elasticity	345 GPa (50 × 10⁶ psi)
Coefficient of thermal expansion	5.0 μm/m·°C (2.6 μin./in.·°F)
Magnetic properties	Slightly magnetic

(a) 0.2% offset. (b) In 25 mm or 1 in.

17 WEAR-RESISTANT MATERIALS

This section was condensed from Metals Handbook, Ninth Edition, Volume 3, Properties and Selection: Stainless Steels, Tool Materials and Special-Purpose Metals, pages 561 to 594. For more detailed information on the topics covered herein, the reader is referred to the larger work. Additional information within this volume can be located by consulting the index.

Hard Facing Materials

Condensed from Metals Handbook, Ninth Edition, Volume 3, pages 563 to 567.

HARD FACING is the process of applying, by welding, plasma spraying or flame plating, a layer, edge or point of wear-resistant metal onto a metal part to increase its resistance to abrasion, erosion, galling, hammering or other form of wear. Hard facing may be applied to new parts to improve their resistance to wear during service, or to worn parts for the purpose of restoring them to serviceable condition. It is frequently used in applications where systematic lubrication against abrasion is not feasible or is inadequate to give the desired service life. In general the wear-resistant coating is applied only to those critical surfaces of components where wear is maximum. Worn parts can be satisfactorily refaced or rebuilt many times before replacement becomes mandatory.

The economic success of the process often depends on selective application of relatively expensive hard facing alloys to comparatively inexpensive base metals. Heavy or bulky parts that are difficult and costly to move often can be repaired or rebuilt in the field or in the plant where they are installed, by welding with portable equipment.

MATERIAL CLASSIFICATIONS

Most hard facing alloys are marketed as proprietary materials. They are classified here in five major groups (1 to 5), primarily according to total alloy content (elements other than iron), with subdivisions based on major alloying elements (see Table 1). Usually, both wear resistance and cost increase as the group number increases. Choice of form and type of alloy depends on the application and the welding process to be used. Alloys for hard facing usually are available as bare cast or tubular rod, covered solid or tubular electrodes, solid wire, and powder.

Alloy Composition. The alloys in group 1A are low-alloy steels that, with few exceptions, contain chromium as the principal alloying element. The total alloy content (including carbon) is between 2 and 6%. These alloys often are used as buildup materials for support of harder, more highly alloyed hard facing alloys.

The iron-base alloys in group 1B are similar to those in group 1A except that they contain higher total alloy contents (6 to 12%), and in some instances carbon contents of 2% or more. Many tool steels and several alloy cast irons are included in this group.

Alloys in group 1 have the greatest shock resistance of all hard facing alloys except austenitic manganese steels, and have better wear resistance than low-carbon and medium-carbon steels, which are the base metals to which they are usually applied. These alloys are less expensive than the other hard facing alloys, and are extensively used where machinability is necessary and only moderate improvement over the wear properties of the base metal is required.

The alloys in group 2A are chromium-containing alloys with total alloy contents of 12 to 25%. Many of these alloys have appreciable molybdenum contents. This group also includes certain medium-alloy cast irons.

Molybdenum is the principal alloying element in nearly all group 2B alloys, most of which also contain appreciable amounts of chromium. These and the group 2C steels also have total alloy contents between 12 and 25%.

Table 1. Classification of hard facing materials by alloy groups

Group	Total alloy content, %	Principal alloying elements
Low-alloy ferrous materials		
1A .... 2 to 6		Cr, Mo, Mn
1B .... 6 to 12		Cr, Mo, Mn
High-alloy ferrous materials		
2A ....12 to 25		Cr, Mo
2B ....12 to 25		Mo, Cr
2C ....12 to 25		Mn, Ni
2D ....30 to 37		Mn, Cr, Ni
3A ....25 to 50		Cr, Ni, Mo
3B ....25 to 50		Cr, Mo
3C ....25 to 50		Co, Cr
Nickel-base and cobalt-base alloys		
4A ....50 to 100		Co, Cr, W
4B ....50 to 100		Ni, Cr, B
4C ....50 to 100		Cr, Ni, Co
Carbides		
5 ....75 to 96		WC, alone or in combination with other carbides such as TiC and TaC, all in a metal matrix.

Group 2C alloys are austenitic manganese steels. Although manganese content predominates, each of these alloys contains an appreciable amount of nickel or molybdenum as an austenite stabilizer.

The hard facing alloys in groups 2A and 2B are more wear resistant, less shock resistant and more expensive than those in group 1. Group 2C and 2D alloys are highly shock resistant, but have limited wear resistance unless subjected to work hardening. Group 2D alloys have total alloy contents of 30 to 37%, and carbon contents ranging from less than 0.10% to more than 1.0%.

Group 3 alloys have total alloy contents from 25 to 50%. They are high-chromium alloys, many of which contain nickel or molybdenum, or both. Carbon content ranges from about 1.75% to more than 5%. Group 3B alloys contain appreciable amounts of molybdenum and chromium, and group 3C alloys, appreciable amounts of cobalt and chromium. Group 3A, 3B and 3C alloys are characterized by massive hypereutectic alloy carbides that impart wear resistance and some degree of resistance to corrosion and heat. They are more expensive than the alloys in groups 1 and 2.

Cobalt-base and nickel-base alloys with total nonferrous metal contents from 50 to 99% are classified in group 4. The cobalt-base alloys (group 4A) generally are rated as the most versatile of the hard facing materials. They resist heat, abrasion, corrosion, impact, galling, oxidation, thermal shock, erosion and metal-to-metal wear. Some of these alloys retain useful hardness up to 825 °C (1500 °F) and resist oxidation temperatures up to 1100 °C (2000 °F). The nickel-base alloys (group 4B) are most effective for service involving both corrosion and wear. They are superior to other hard facing alloys where wear is caused by metal-to-metal contact, as in bearings. They retain useful hardness up to about 650 °C (1200 °F) and resist oxidation at temperatures up to 875 °C (1600 °F).

Group 5 materials consist of hard granules of carbide distributed in a metal matrix; they are extremely important for severe abrasion and cutting applications. Historically, tungsten carbides were used exclusively. Recently, however, carbides of certain other elements—notably titanium, tantalum and chromium—have been used with satisfactory results. Various matrix metals are employed, including iron, carbon steel, nickel-base alloys, cobalt-base alloys, and bronzes. Group 5 materials provide maximum abrasion resistance under service conditions involving low or moderate impact.

In addition to the materials in Table 1, copper-base alloys comparable in composition to bronzes are used as matrix metals for carbides or as overlays on less-expensive base metals, notably low-carbon steels. These overlays sometimes serve as bearing materials and provide resistance to corrosion and cavitation damage. They offer poor resistance to corrosion by sulfur compounds, to abrasive wear and to creep at elevated temperatures.

ALLOY SELECTION

In order to select the correct hard facing alloy for a particular wear application, a thorough analysis of anticipated service conditions must be carried out, and the factors that could potentially cause severe material degradation must be established. Many investigators acknowledge the following types of wear:

Adhesive wear
    Mild or oxidative wear
    Severe or metallic wear
Abrasive wear
    Low-stress scratching abrasion
    High-stress grinding abrasion
    Gouging abrasion
Erosive wear
    Impingement erosion
    Cavitation erosion

Table 2. General guide to selection of hard facing alloys

Service conditions	Hard facing materials
Metal-to-metal sliding, high contact stresses	Stellite 1, Tribaloy alloys
Metal-to-metal sliding, low contact stresses	Low-alloy hard facing steels
Metal-to-metal sliding combined with corrosion or oxidation	Cobalt-base or nickel-base alloys; selection of specific alloy depends on environmental conditions
Low-stress abrasion; particle impingement erosion at low angles	High-alloy cast irons
Severe low-stress abrasion; cutting-edge retention	Materials containing high proportions of carbides
Cavitation erosion; liquid-impingement erosion	Cobalt-base alloys
Heavy impact	High-alloy manganese steels
Heavy impact combined with corrosion of oxidation	Stellite 21, Stellite 6
Gouging abrasion	Austenitic manganese steels
Galling	Stellite 21, Stellite 6, Tribaloy T-400, Tribaloy T-800
Thermal stability and/or creep resistance at high temperatures	Cobalt-base alloys; carbide-type nickel-base alloys

Fretting

Once the service conditions for a particular hard facing application have been characterized, consideration should be given to interactions between candidate alloys and the base metal, and thus the process to be used for depositing the hard facing alloy becomes an important consideration. A general guide for selection of hard facing alloys, based on service conditions, is given in Table 2. In general, the following steps should be taken:

• Analyze service conditions to determine the types of wear resistance and environmental resistance required.

• Select a few candidate hard facing alloys.

• Analyze compatibility of these alloys with the base metal, including consideration of thermal stresses and possible cracking.

• Field test parts hard faced with candidate alloys.

• Select the optimum hard facing alloy, considering both cost and wear life.

• Select the hard facing process, considering deposition rate, degree of dilution, deposition efficiency, and over-all cost. The last should include cost of consumables and cost of processing.

Austenitic Manganese Steel

Condensed from Metals Handbook, Ninth Edition, Volume 3, pages 568 to 588.

THE ORIGINAL AUSTENITIC MANGANESE STEEL, containing about 1.2% C and 12% Mn, was invented by Sir Robert Hadfield in 1882. Hadfield's steel is unique in that it combines high toughness and ductility with high work-hardening capacity and, usually, good resistance to abrasion. Consequently, Hadfield's steel rapidly gained acceptance as a very useful engineering material.

However, austenitic manganese steel has certain properties that tend to restrict its use. It is difficult to machine and usually has a yield strength of only 345 to 415 MPa (50 to 60 ksi). Consequently, it is not well suited for parts that require close-tolerance machining or that must resist plastic deformation when highly stressed in service. However, hammering, pressing, cold rolling or explosion shocking of the surface raises the yield strength to provide a hard surface on a tough core structure.

COMPOSITION

Many variations of the original Hadfield's steel have been proposed, often in unexploited patents, but only a few have been adopted as significant improvements. These usually involve variations of carbon and manganese, with or without additional alloys such as chromium, nickel, molybdenum, vanadium, titanium and bismuth. The most common of these compositions, as listed in ASTM A128, are given in Table 1.

The available assortment of wrought grades is smaller and usually approximates ASTM composition B-3. Some wrought grades contain about 0.8% C and either 3% Ni or 1% Mo. A manganese steel foundry may have several dozen modified grades on its production list.

Carbon and Manganese. The ASTM A128 compositions in Table 1 do not permit any austenite transformation when the alloys are water quenched from above the A_{cm}, but this does not preclude lower ductility in heavy sections of the higher-carbon steels, because quenching is slowed by the heavy sections. The effect is due to formation of carbides along grain boundaries, and, in some degree, affects nearly all commercial castings except the very smallest.

The mechanical properties of austenitic manganese steel vary with both carbon and manganese contents. As the carbon content of cast 13% manganese steel is increased from 0.8 to 1.7%, there is a gradual increase in yield strength, whereas tensile strength and ductility reach maximum values at about 1.2% carbon and then decrease steadily. As carbon content is increased above about 1.1%, it becomes increasingly difficult to retain all of the carbon in solution in the austenite, which largely accounts for the reductions in tensile strength and ductility. Nevertheless, because abrasion resistance tends to increase as carbon content increases up to about 1.4% and possibly higher, carbon contents higher

Table 1. Standard composition ranges for austenitic manganese steel castings

ASTM A128 grade	C	Mn	Cr	Mo	Ni	Si (max)	P (max)
A	1.05-1.35	11.0 min	...	...	...	1.00	0.07
B-1	0.9-1.05	11.5-14.0	...	...	...	1.00	0.07
B-2	1.05-1.2	11.5-14.0	...	...	...	1.00	0.07
B-3	1.12-1.28	11.5-14.0	...	...	...	1.00	0.07
B-4	1.2-1.35	11.5-14.0	...	...	...	1.00	0.07
C	1.05-1.35	11.5-14.0	1.5-2.5	...	...	1.00	0.07
D	0.7-1.3	11.5-14.0	...	...	3.0-4.0	1.00	0.07
E-1	0.7-1.3	11.5-14.0	...	0.9-1.2	...	1.00	0.07
E-2	1.05-1.45	11.5-14.0	...	1.8-2.1	...	1.00	0.07
F	1.05-1.35	6.0-8.0	...	0.9-1.2	...	1.00	0.07

than 1.1% often are preferred even though the ductility of the steel may be lowered. Carbon levels over 1.4% are seldom used because of the difficulty of obtaining an austenite structure sufficiently free of grain-boundary carbides to avoid undesirably low values of both strength and ductility. The effect also may be observed in manganese steels containing less than 1.1% carbon because segregation may result in local variations of ±0.2% from the average carbon content determined by chemical analysis. The 0.7% C minimum of grades D and E-1 may be used to minimize carbide precipitation in heavy castings or in weldments, and similar low carbon contents are specified for welding filler metal. Carbides form in castings that are cooled slowly in the molds. In fact, carbides form in practically all as-cast grades containing more than 1.0% C, regardless of mold cooling rates. They often form in heavy-section castings during heat treatment if quenching is even somewhat ineffective in producing rapid cooling throughout the entire section thickness. Carbides can form during welding or during service at temeratures above about 275 °C (530 °F).

If carbon and manganese are lowered together, for instance to 0.53% C with 8.3% Mn or 0.62% C with 8.1% Mn, the speed of hardening by cold work is increased and strain-induced martensite may be produced. However, this does not provide enhanced abrasion resistance (at least to high-stress grinding abrasion) as is often hoped.

Manganese contributes the vital austenite-stabilizing effect of delaying transformation (but not eliminating it). Thus, in a simple steel that contains 1.1% Mn, isothermal transformation at 370 °C (700 °F) begins about 15 s after the steel is quenched to that temperature, whereas in a 13% Mn steel, transformation at 370 °C does not begin until after 48 h. Below 260 °C (500 °F), phase changes and carbide precipitation are so sluggish that for all practical purposes they may be neglected, in the absence of deformation, if manganese content exceeds 10%.

Manganese content has little effect on yield strength. In tensile testing, ultimate strength and ductility increase fairly rapidly with increasing manganese content up to about 12% and then tend to level off, although small improvements normally continue up to about 13% Mn.

Silicon and Phosphorus. As noted in Table 1, silicon and phosphorus are present in all ASTM A128 grades of austenitic manganese steel. Silicon is seldom added except for steelmaking purposes. Silicon contents exceeding 1% are not usual, because foundries do not like to have it pyramid in melts containing returned scrap. Silicon contents of 1 to 2% might be used to moderately increase yield strength, but other elements are preferred for this purpose. Loss of strength is abrupt above 2.2% Si, and Mn steel containing more than 2.3% Si may be worthless.

Since about 1960, low-phosphorus ferromanganese has been available, which has enabled steelmakers to greatly reduce phosphorus levels in manganese steel. The preferred practice is to hold phosphorus content below 0.04% even though a level of 0.07% is permitted by ASTM A128. Levels above 0.06%, which formerly were prevalent, contribute to hot shortness and low elongation at very high temperatures, and frequently are the cause of hot tears in castings and underbead cracking in weldments. In general, phosphorus in manganese steel tends to lower the ductility; the effect is more critical at or below

ambient temperatures. It is particularly advantageous to keep phosphorus at the lowest possible level in the grades that are welded and in manganese steel welding electrodes.

Special-Purpose Additives. Other elements are also added to certain grades of manganese steel to achieve desired properties. As can be seen in Table 1, chromium is added to grade C, molybdenum to grades E-1, E-2 and F, and nickel to grade D.

Chromium levels from 1.8 to 2.2% (and occasionally as high as 5%) are employed in manganese steel to moderately raise yield strength. Its effect on resistance to abrasion is unproven and appears to be inconsistent. Chromium reduces ductility by increasing the number of embrittling carbides in the austenite—particularly those containing more than 2.5% chromium.

Molybdenum additions, usually 0.5 to 2%, generally are made to improve the toughness and resistance to cracking of castings in the as-cast condition, and to raise the yield strength (and possibly toughness) of heavy-section castings in the solution treated and quenched condition. These effects occur because molybdenum in manganese steel is distributed partly in solution in the austenite and partly in primary carbides formed during solidification of the steel. The molybdenum in solution effectively suppresses formation of both embrittling carbide precipitates and pearlite, even when the austenite is exposed to temperatures above 275 °C (530 °F) during service or during welding. The molybdenum in primary carbides tends to change distribution of the carbides from continuous envelopes around the austenite dendrites to less-harmful nodular carbides, especially when molybdenum content exceeds about 1.5%.

Grade E-2, which contains about 2% Mo, may be given a special heat treatment to develop a structure of finely dispersed carbides in the austenite, which provides unusually high yield strength and, under some conditions, an improvement in abrasion resistance. Molybdenum is added to the lean-manganese steel, grade F, to partly suppress embrittlement in both as-cast and heat treated conditions.

Nickel, in amounts up to 4% or more, stabilizes the austenite because it remains in solid solution. It is particularly effective in suppressing precipitates of carbide platelets, which can form between about 300 and 550 °C (570 and 1020 °F). Therefore, the presence of nickel helps retain nonmagnetic qualities in the steel, especially in its decarburized skin, and lowers the rate of work hardening, thus providing high ductilities in lower-carbon grades. Abrasion resistance is lowered, but yield strength is not significantly affected. Nickel is used primarily in the lower-carbon or weldable grades of cast manganese steel and in wrought manganese steel products (including welding electrodes). In wrought products, nickel often is used in conjunction with molybdenum.

Because vanadium is a strong carbide former, it is added to manganese steels to substantially increase yield strength. However, a comparable decrease in ductility occurs. Vanadium is used on a limited basis in special-purpose compositions, such as a Mn-Ni-Mo-V austenitic alloy that can be age hardened to provide yield strengths of over 700 MPa (100 ksi). Tests in a jaw crusher have shown the abrasion resistance of this steel to be lower than those of regular Hadfield types.

Like nickel, copper in amounts of 1 to 5% is sometimes used in austenitic manganese steels to

stabilize the austenite. The effects of copper on mechanical properties have not been clearly established. Scattered reports indicate that it may have an embrittling effect, which may be due to the limited solubility of copper in austenite.

Other elements, such as bismuth and titanium, also are added to standard manganese steels. Bismuth can improve machinability.

Titanium can reduce carbon in the austenite by forming very stable carbides. The resulting properties may simulate those of a lower-carbon grade. Titanium may also somewhat neutralize the effect of too much phosphorus; some European practice is apparently based on this idea.

Sulfur. The sulfur content in manganese steel seldom influences its properties, because the scavenging effect of manganese operates to eliminate sulfur by fixing it in the form of innocuous rounded inclusions of manganese sulfide. Elongation of these inclusions in wrought steels may contribute to directional properties; in cast steel, such inclusions are harmless.

AS-CAST PROPERTIES

Although austenitic manganese steels in the as-cast condition generally are considered too brittle for normal use, Table 2 demonstrates that there are exceptions to this rule. Mechanical properties are listed for five grades of as-cast austenitic manganese steels in various thicknesses. These data indicate that lowering carbon content to less than 1.1% and/or adding about 1.0% Mo or about 3.5% Ni results in commercially acceptable as-cast ductilities in light and moderate section thicknesses. These data also apply to weld deposits, which normally are left in the as-deposited condition and therefore are essentially equivalent to material in the as-cast condition.

Commercial use of castings in the as-cast condition results in cost and energy savings and eliminates the problems of (*a*) decarburization of thin castings during solution treatment and (*b*) warpage during water quenching.

Heavy Sections. As section thickness increases, the rate at which castings cool in sand molds also slows, which increases the opportunity for embrittlement to occur by carbide precipitation. Shapes that tend to develop high residual stresses, such as cylinders and cones, can be particularly affected. These stresses most probably result from volume changes accompanying the carbide precipitation and austenite transformation that occur during normal cooling of castings.

Figure 1 shows the volume changes that occur during isothermal decomposition of a 1.25C-12.8Mn steel at temperatures between 850 and 500 °C (1560 and 930 °F), the principal range in which embrittlement occurs when a casting is cooled in its mold or reheated for reaustenitization. Between 850 and about 700 °C (1560 and 1300 °F), only carbides are precipitated, principally as envelopes around austenite grains and as lamellar-type patches within grains. The lamellar carbide patches have the appearance of coarse pearlite, but actually they are carbide plates in austenite. Below 700 °C, and particularly between 650 and 550 °C (1200 and 1020 °F), pearlite nodules, nucleated by previously precipitated carbides, grow relatively rapidly.

Transgranular acicular carbides also tend to precipitate below about 600 °C (1110 °F), especially in austenite containing more than about 1.1% C. This precipitation can continue down to about 300 °C (570 °F) in a 1.2C-12Mn steel. It

Table 2. Typical mechanical properties of as-cast austenitic manganese steels

C	Mn	Si	Other	Form	Section size mm	Section size in.	0.2% yield strength MPa	0.2% yield strength ksi	Tensile strength MPa	Tensile strength ksi	Elongation, %	Reduction in area, %	Impact strength(a) J	Impact strength(a) ft·lb	Hardness, HB
Plain manganese steels															
0.85	11.2	0.57	...	Round	25	1	...	...	440	64	14.5	...	...	...	...
1.11	12.7	0.54	...	Round	25	1	360	52	450	65	4	...	...	...	...
1.28	12.5	0.94	...	Keel block	100	4	...	...	330(b)	48(b)	1(b)	...	3.4	2.5	245
1.36	20.2	0.6	...	Y-block	50	2	...	...	425(b)	62(b)	1(b)	...	...	...	283
1Mo manganese steels															
0.61	11.8	0.17	1.10 Mo	Round	25	1	315	46	710	103	27.5	23	...	...	163
0.83	11.6	0.38	0.96 Mo	Round	25	1	345	50	695	101	30	29	...	...	163
1.16	13.6	0.60	1.10 Mo	Round	25	1	400	58	560	81	13	15	...	...	185
0.93	13.6	0.67	0.96 Mo	Plate	25	1	365	53	510	74	11	16	72	53	188
0.98	12.6	0.6	0.87 Mo	Plate	50	2	...	...	435(b)	63(b)	4(b)	...	...	...	...
1.30	13.1	0.78	0.99 Mo	Keel block	100	4	...	...	435(b)	63(b)	2(b)	...	8	6	230
1.33	19.8	0.6	0.99 Mo	Y-block	50	2	...	...	505(b)	73(b)	2.5(b)	...	...	...	231
2Mo manganese steels															
0.52	14.3	1.47	2.4 Mo	Round	25	1	370	54	600	87	15.5	13	...	...	220
0.75	14.1	0.99	2.0 Mo	Round	25	1	365	53	745	108	34.5	27	...	...	183
1.24	14.1	0.64	3.0 Mo	Round	25	1	440	64	600	87	7.5	10	...	...	235
1.34	12.0	0.43	2.2 Mo	Keel block	50	2	415	60	435	63	3.5	7	...	...	235
3.5Ni manganese steels															
0.75	13.0	0.95	3.65 Ni	Round	25	1	295	43	655	95	36	26	...	...	150
0.91	13.3	0.53	3.38 Ni	Round	25	1	...	...	510	74	24	...	...	...	...
6Mn-1Mo alloys															
0.90	5.8	0.37	1.46 Mo	Mill liner	100	4	325	47	340	49	2	...	9	7	181
1.00	6.0	0.43	1.03 Mo	Keel block	100	4	330	48	365	53	2	3	...	...	195
0.89	6.3	0.6	1.20 Mo	Plate	100	4	...	...	330(b)	48(b)	1(b)	...	...	...	...
1.27	6.1	0.42	1.07 Mo	Keel block	50	2	365	53	400	58	1	1	3	2	273

(a) Charpy V-notch. (b) Properties converted from transverse bend tests on 6-by-13-mm (1/4-by-1/2-in.) bars cut from castings and broken by center loading across 25-mm (1-in.) span.

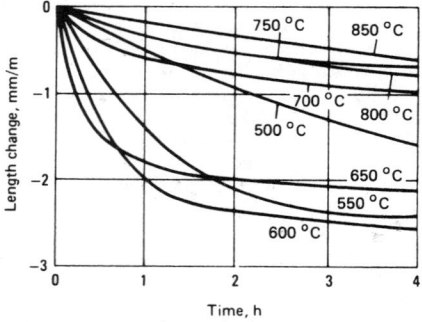

Adapted from Some Decomposition Structures of Austenitic Manganese Steels, by R. Castro and P. Garnier: *Revue de Metallurgie*, Vol 55, Jan 1958, p 17.

Fig. 1. Change in length of an austenitic 1.25C-12.8Mn steel during isothermal transformation

may be followed by transformation of some of the carbon-depleted austenite to martensite as the temperature approaches ambient.

HEAT TREATMENT

Heat treatment strengthens austenitic manganese steel so that it can be used safely and reliably in a wide variety of engineering applications. Solution annealing and quenching, the standard treatment that produces normal tensile properties and the desired toughness, involves austenitizing followed quickly by water quenching.

Variations of this treatment can be used to enhance specific desired properties such as yield strength and abrasion resistance.

Commercial heat treatment of manganese steel castings normally involves heating slowly to 1010

to 1090 °C (1850 to 2000 °F), soaking for 1 to 2 h at temperature and then quenching in agitated water. There is some tendency toward growth of austenite grains during soaking, although final austenite grain size in castings is determined largely by pouring temperature and solidification rate.

For grade E-2 manganese steel (see Table 1), a modified heat treatment often is specified or recommended. This treatment consists of heating castings to about 600 °C (1100 °F) and soaking them 8 to 12 h at temperature, which causes substantial amounts of pearlite to form in the structure. The castings then are further heated to about 980 °C (1800 °F) to reaustenitize the structure. This step converts the pearlitic areas to fine-grain austenite containing a dispersion of small carbide particles, which remain undissolved as long as the austenitizing temperature does not exceed about 1000 °C (1850 °F). Quenching then results in a dispersion-hardened austenite, which is characterized by higher yield strength, higher hardness and lower ductility than would be obtained if the same steel were given a full solution treatment at a higher austenitizing temperature. When this dispersion-hardening heat treatment is used, relatively high carbon contents are permissible, which in turn can improve abrasion resistance.

Precautions. Speed of quenching is important, but it is difficult to increase it beyond the rate of heat transfer from a hot surface to agitated water or the rate fixed by the thermal conductivity of the metal. As a result, heavy-section castings have lower mechanical properties at the center than do thinner castings. Figure 2 illustrates the cooling rates that can be expected when metal plates of four different thicknesses are quenched in water. Table 3 lists average properties observed in castings of 1.11C-12.7Mn-0.5Si-0.043P steel water quenched from about 1040 °C (1900 °F), which cooled the castings at the rates shown in Fig. 2.

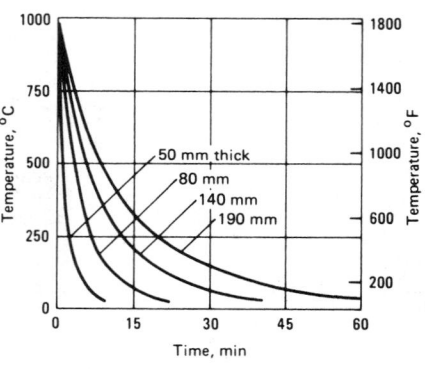

Cooling curves are approximately equivalent to those for plate of the thicknesses indicated.

Fig. 2. Cooling curves for austenitic manganese steel of various thicknesses

Residual stresses from quenching, coupled with the lower properties of heavy sections, establish the usual maximum thickness of commercial castings at about 125 to 150 mm (5 to 6 in.), although castings with sections up to 400 mm (16 in.) thick have been produced.

MECHANICAL PROPERTIES AFTER HEAT TREATMENT

As the section size of manganese steel increases, tensile strength and ductility decrease substantially in specimens cut from heat treated castings. This occurs because, except under specially controlled conditions, heavy sections do not solidify fast enough in the mold to prevent coarse grain size, a condition that is not altered by heat treatment. As shown in Table 3, fine-grain specimens may exhibit strength and elon-

Table 3. Average mechanical properties of 12.7Mn-1.1C-0.5Si-0.043P castings water quenched from 1040 °C (1900 °F)
Tension tests were performed on specimens 6.40 mm (0.252 in.) in diameter and 25 mm (1 in.) in gage length.

Plate thickness		Type of grain	Tensile strength		Elonga-tion(a), %	Reduction in area, %	Impact strength(b)	
mm	in.		MPa	ksi			J	ft · lb
50	2	Coarse	635	92	37.0	35.7	137	101
		Fine	820	119	45.5	37.4	134	99
140	5½	Coarse	545	79	22.5	25.6	115	85
		Fine	705	102	32.0	28.3	100	74

(a) In 25 mm or 1 in. (b) Izod V-notch.

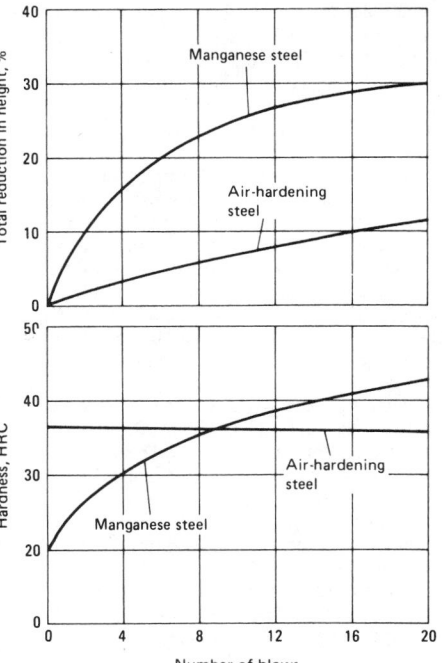

gation as much as 30% greater than those of coarse-grain specimens.

Grain size also accounts chiefly for the differences between cast and wrought manganese steels (the latter usually are of fine grain size). For cast grade B-2, the standard deviations for tensile strength and elongation are about 69 MPa (10 ksi) and 9%, respectively. The midrange values of 825 MPa (120 ksi) and 40% apply to sound, medium-grain cast specimens that have been properly heat treated. The scatter bands for this grade extend from 620 to 1035 MPa (90 to 150 ksi) for tensile strength and from 13 to 67% for elongation.

Mechanical properties vary with section size. Tensile strength, tensile elongation, reduction in area and impact strength are substantially lower in 100-mm-thick (4-in.-thick) sections than in 25-mm-thick (1-in.-thick) sections. Because section thicknesses of production castings often are from 100 to 150 mm (4 to 6 in.), this factor is an important consideration in specification of the proper grade.

Notched-bar impact test values can be exceptionally high. Charpy test specimens are sometimes bent and dragged through the machine rather than fractured. Sometimes, observed values are biased because of incorrect preparation of specimens. Notches should be cut by precision grinding to minimize work hardening at the apex of the notch.

Austenitic manganese steel remains tough at subzero temperatures above the M_s temperature. It is apparently immune to hydrogen embrittlement, although embrittlement has been produced in steels with low carbon contents (less than about 0.02%) and high manganese contents.

Resistance to crack propagation is high and is associated with very sluggish progressive failures. Because of this, any fatigue cracks that develop can be monitored, and the affected part(s) removed from service, before complete failure occurs — a capability that is a distinct advantage in railway trackwork. The fatigue limit of austenitic manganese steel has been reported as 270 MPa (39 ksi).

Yield strength and hardness vary only slightly with section size. Hardness of most grades is about 200 HB after solution annealing and quenching, but this value has little significance for estimating machinability or wear resistance. Hardness increases so rapidly due to work hardening during machining or in wear service that austenitic manganese steels must be evaluated on some basis other than hardness.

The true tensile characteristics of manganese steel are better revealed by the stress-strain curves in Fig. 3, which compare manganese steel with gray iron and with a tough ferritic steel of about the same nominal tensile strength. The low yield strength is significant and may prevent selection of this alloy where slight or moderate deforma-

tion is undesirable, unless the usefulness of the parts in question can be restored by grinding. However, if deformation is immaterial, the low yield values may be considered temporary — that is, deformation will produce a new, higher yield strength corresponding to the amount of strain that is absorbed locally.

WORK HARDENING

The approximate ranges of tensile properties produced in constructional alloy steels by heat treatment are developed in austenitic manganese steels by work hardening. In a tension test, yielding signifies the beginning of work hardening, and elongation is associated with its progress. Little or no reduction in area occurs in austenitic manganese steels by necking, because work hardening is greatest at the point of greatest deformation. The increase in strength due to cold work stops elongation, and deformation then occurs elsewhere until the hardening and reduction in area are substantially equalized throughout the specimen.

Manganese steels are unequalled in their ability to work harden, exceeding even the metastable austenitic stainless steels in this feature. For example, a standard grade of manganese steel containing 1.0 to 1.4% C and 10 to 14% Mn can work harden from an initial level of 220 HV to a maximum of more than 900 HV. After extended service, the hardness at the wearing surfaces of railway frogs typically ranges from 495 to 535 HB. Maximum attainable hardness depends on many factors, including specified composition, service limitations, method of work hardening and preservice hardening procedures. It appears that rubbing under heavy pressure can produce higher values of maximum attainable hardness than can be produced by simple impact.

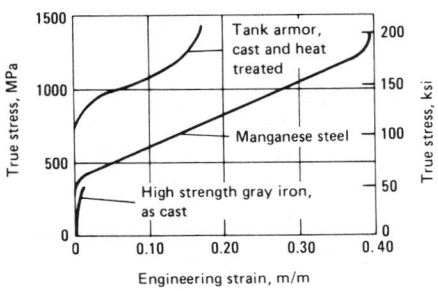

Alloy	C	Mn	Si	Cr	Other
Tank armor	0.29	1.30	0.52	0.37	0.36 Mo
Manganese steel	1.22	13.08	0.33	0.09	0.05 Al
Gray iron	2.79	0.75	1.32	0.10	...

Fig. 3. True stress vs engineering strain for manganese steel, cast ferritic tank armor of similar tensile strength and a high-strength gray iron

Specimens 25 mm (1 in.) in both diameter and length were struck repeatedly on one end by blows with an impact energy of 680 J (500 ft · lb). Composition and heat treatment of the manganese steel: 1.17C-12.8Mn-0.46Si; water quenched from 1010 °C (1850 °F). Composition and heat treatment of the air-hardening steel: 0.74C-0.88Mn-0.30Si-0.75Ni-1.40Cr-0.38Mo; air cooled from 900 °C (1650 °F), reheated to 700 °C (1300 °F) and air cooled.

Fig. 4. Plastic flow and work hardening of a manganese steel and an air-hardening steel under repeated impact.

Work-Hardening Methods. Work hardening usually is induced by impact, as from hammer blows. Light blows, even if they are of high velocity, cause shallow deformation with only superficial hardening even though the resulting surface hardness ordinarily is high. Heavy impact produces deeper hardening, usually with lower values of surface hardness. The course of flow under impact and the associated increase in hardness are illustrated in Fig. 4, which compares a standard 12% Mn steel with an air-hardening Cr-Ni-Mo alloy steel. Less well known is the fact that abrasion itself can produce work hardening.

Explosion hardening was developed as a substitute for hammer or press hardening that would achieve hardening with less deformation. Pentaerythritol tetranitrate in the form of plastic explosive sheet is cemented to the surface of the steel and then detonated. Several explosions may be required to attain the desired hardness.

Preservice Hardening. The low flow resistance and consequent low yield strength of manganese steel can be increased by several methods; preservice hardening by deformation is preferred. Special equipment for hammering or pressing is usually employed to induce deep-seated hardening. The depth of the affected zone usually ranges up to about 25 mm (1 in.). The superficial hardening that can be produced by high-velocity shot blasting is seldom satisfactory for service requirements.

Addition of alloying elements such as vanadium, chromium, silicon and molybdenum also is an effective means of raising yield strength, but vanadium and chromium reduce ductility.

Prehardening by deformation is particularly recommended for such operations as deep drawing of military helmets and cold forming of strip for use in body armor. The requirements of these operations necessitate careful control of the martensitic skin formed during solution annealing and quenching, because tensile deformation may produce cracks that render these thin products unsuitable for the intended use. The skin may not be thick enough to result in a noticeably defective product during manufacture but may contain minute cracks that lower resistance to further deformation. Solution and distribution of carbides are important in such applications.

Hardness Measurement. Methods of measuring and reporting the hardness of manganese steel differ from those used on other metals.

The common Brinell test reports hardness in terms of load divided by the curved area of the indentation, whereas the Meyer test (which uses the same indentor) reports hardness as load divided by the area of the indentation measured in the plane of the original surface. If a series of hardness numbers is obtained with various loads, it will be found that, for an annealed metal indented with a ball, Meyer hardness will increase with load, whereas for fully work-hardened metal, Meyer hardness is constant. The behavior of annealed metal—metal free of plastic strain or residual stress—can be used as a measure of the extent of work hardening.

This measurement technique generally employs a 10-mm-diam Brinell ball and a series of loads. Test load is plotted against diameter of the corresponding impression on logarithmic scales. The result is expected to be a straight line that fits the equation:

$$P = Ad^n \qquad \text{(Eq 1)}$$

where P is the applied load, d is the diameter of the impression, A is a constant for a given indenting ball size, and n is a measure of the tendency of the metal to strain harden. The exponent n, also called the Meyer index or Meyer exponent, is the slope of a plot of P versus d on log-log paper. It also can be expressed as:

$$n = \tan \phi \qquad \text{(Eq 2)}$$

where ϕ is the angle between the plot of impression diameters and its horizontal coordinate.

REHEATING

Before manganese steel parts are reheated in the field, the effects of such reheating must be seriously considered. Unlike ordinary structural steels, which become softer and more ductile when reheated, manganese steels suffer reduced ductility when reheated enough to induce carbide precipitation or some transformation of the austenite. As a general rule, manganese steels should never be heated above 260 °C (500 °F), either intentionally or accidentally, unless such heating

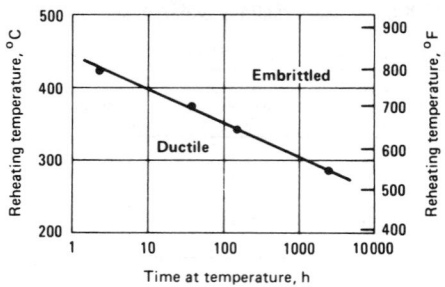

Prior to reheating, the alloy was annealed 2 h at 980 °C (2000 °F) and water quenched.

Fig. 5. Time-temperature relationship for embrittlement of 13Mn-1.2C-0.5Si steel

can be followed by standard solution annealing and quenching.

Time, temperature and composition are variables in the embrittlement process. At lower temperatures, embrittlement takes longer to develop. The time-temperature relationship in 13Mn-1.2C-0.5Si steel is illustrated in Fig. 5, which presents data based on metallographic examination for structural changes that indicate the beginning of embrittlement. At 260 °C (500 °F), transformation requires more than 1000 h; reheating to as high as 425 °C (800 °F), even with close control of temperature, may be done for no longer than 1 h if transformation is to be avoided. Figure 6 shows the effect of composition on the magnitude of embrittlement from reheating.

When 12 to 14% Mn steels are to be heated above about 290 °C (550 °F) during service or welding, it is recommended that carbon content be held below 1.0%, which will suppress embrittlement for at least 48 h at temperatures up to 370 °C (700 °F). Addition of 1.0% Mo will suppress embrittlement completely at temperatures up to 480 °C (900 °F) and will partly suppress it at temperatures of 480 to 600 °C (900 to 1100 °F). If carbon content is held below about 0.9%, addition of 3.5% Ni will completely suppress embrittlement up to 480 °C and will partly suppress it above this temperature. These rules can be expected to apply during heating periods of up to 100 h. For periods of 1000 h or more, limiting temperatures are substantially lower.

WEAR RESISTANCE

Compared with most other abrasion-resistant ferrous alloys, manganese steels are superior in toughness and moderate in cost, and it is primarily for these reasons that they are selected for a wide variety of abrasive applications. They usually are less resistant to abrasion than martensitic white irons and martensitic high-carbon steels, but often are more resistant than pearlitic white irons and pearlitic steels.

The type of abrasion to which a manganese steel is exposed has a major influence on how well it wears. Manganese steels have excellent resistance to metal-to-metal wear, as in sheave wheels, crane wheels and mine-car wheels; moderately good resistance to gouging abrasion, as in equipment for handling or crushing rock; intermediate resistance to high-stress (grinding) abrasion, as in ball-mill and rod-mill liners; and relatively low resistance to low-stress abrasion, as in equipment for handling loose sand or sand slurries.

Metal-to-Metal Contact. In applications involving metal-to-metal contact, work hardening of manganese steel is a distinct advantage because it decreases the coefficient of friction and confers resistance to galling if temperatures are not excessive. Compressive loads, rather than impact, provide the deformation required, producing a smooth, hard surface that has good resistance to wear but that does not abrade the contacting part. Sheaves, rails and castings for railway trackwork are common applications of this type. Manganese steel also has been used in some water-lubricated bearings.

Abrasion. In applications that involve heavy blows or high compressive and structural stresses, the very hard and abrasion-resistant martensitic cast irons may wear more slowly than manganese steel. However, these irons usually fail by early fracture with a considerable portion of the original cross section unworn, whereas manganese steel may wear almost to paper thinness before fracturing.

Pearlitic white cast iron, which has a hardness of about 400 to 450 HB, is equally brittle but less resistant to wear. Comparative tests on log washer lugs indicated that manganese steel was about 25% worn-out with no breakage, whereas in the same period white iron lugs wore to the point of uselessness with 14% breakage. In clay-crusher rolls, manganese steel lasted two to three times as long as white or chilled iron. In grinding-barrel liners, cast irons lasted two to three years compared with ten years for manganese steel. Part of this su-

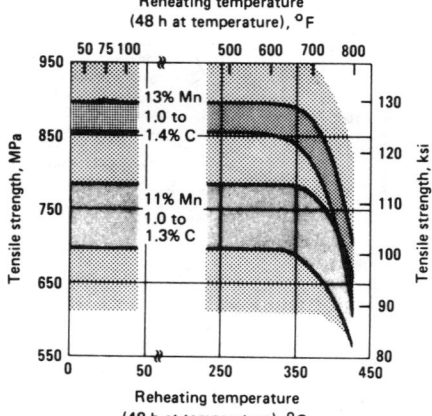

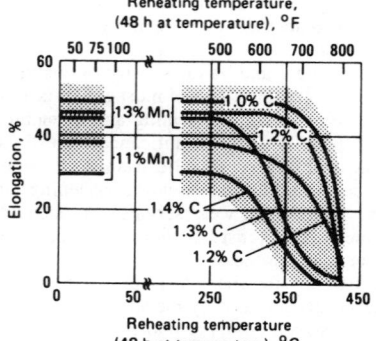

Cast bars 25 mm (1 in.) in diameter were reheated 48 h at the temperatures indicated, after solution annealing and quenching.

Fig. 6. Embrittlement from reheating manganese steel

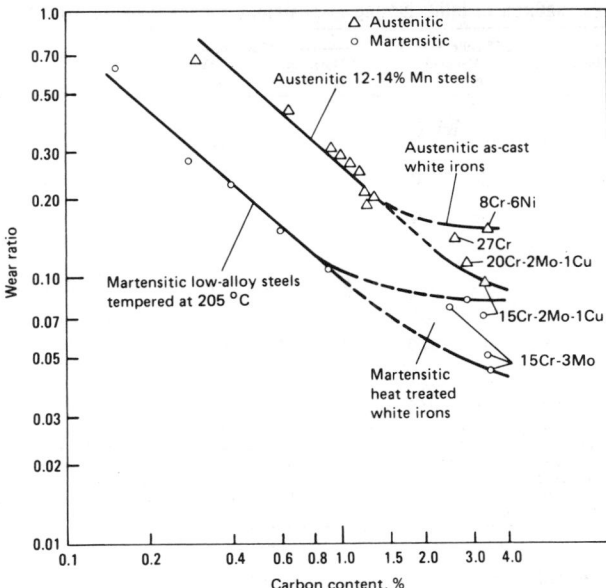

Adapted from Gouging Abrasion Test for Materials Used in Ore and Rock Crushing, Part II, by F. Borik and W. G. Scholz: *Journal of Materials*, Vol 6, No. 3, Sept 1971, p 590

Fig. 7. Relative wear ratios of ferrous alloys in jaw-crusher tests

periority of manganese steel over white cast iron is attributed to greater freedom from breakage and spalling, but some is probably due to better intrinsic wear resistance.

Manganese steel chain, with endless links cast in interlocking molds, also provides resistance to wear, lasting three to nine times as long as heat treated steel chain in certain applications. Manganese steel is valuable in conveyors as well as in dragline chain subjected to abrasion and used for carrying heavy loads.

Manganese steel is not satisfactorily resistant to wear by a stream of airborne abrasive particles (impingement erosion), such as in sandblasting or gritblasting equipment, and consequently should not be selected for such service.

Abrasion Testing. The abrasion resistance of austenitic 12 to 14% Mn steels with various carbon contents has been compared with the resistance of other steels and white irons in a jaw-crusher abrasion test (see Fig. 7). The wear rate for a quenched and tempered low-carbon, low-alloy steel (ASTM 517, type B at 269 HB), which was used as a comparative standard in each test, is shown also. When the relative wear rate (wear ratio) of each test material is plotted against increasing carbon content on a log-log scale, results for austenitic steels and irons tend to fall on a descending straight line, and results for martensitic steels and irons fall on a parallel line below the line for the austenitic alloys. A decrease in wear ratio represents a proportionate increase in abrasion resistance. Thus, Fig. 7 strongly supports the conclusions that abrasion resistance of both austenitic and martensitic steels improves with increasing carbon content and that, for a given carbon content, martensitic steels have better abrasion resistance than austenitic steels. However, martensitic steels and white irons have limited resistance to gouging abrasion due to their lack of toughness. The wear ratios of pearlitic steels, if plotted on Fig. 7, would lie above those of the austenitic steels. There is considerable scatter in the rates for pearlitic steels due to their wide variation in hardness for any given carbon content.

CORROSION

Manganese steel is not corrosion resistant. It rusts readily; and where corrosion and abrasion are combined, as they frequently are in mining and manufacturing environments, the metal may deteriorate or be dissolved at a rate only slightly lower than that of carbon steel. If the toughness or nonmagnetic nature of manganese steel is essential for a marine application, protection by galvanizing usually is satisfactory.

EFFECTS OF TEMPERATURE

The excellent properties of 13% Mn steel between −45 and +205 °C (−50 and +400 °F) make it useful for all ambient-temperature applications, even in arctic climates. It is not recommended for hot wear applications because of structural instability between 260 and 870 °C (500 and 1600 °F). At higher temperatures, it may lack the strength and ductility necessary to withstand severe welding stresses, and thus welding must be done under closely controlled conditions. It is not resistant to oxidation, and its creep-rupture properties are inferior to those of Fe-Cr-Ni austenites. Thus, it is unsuitable for structural applications in the red-heat range.

At −75 °C (−100 °F), cast manganese steels retain from 50 to 85% of their room-temperature impact resistance. They are considerably more brittle at liquid-air temperature (−185 °C, or −300 °F), but at all atmospheric temperatures encountered by railway trackwork and mining and construction equipment they have outstanding toughness that provides a valuable safety factor in comparison with ferritic steels at subzero temperatures.

Associated with the embrittlement produced by reheating above 260 °C (500 °F) are changes in physical properties stemming from the same transformations that cause the loss in toughness. Because both composition and time at temperature influence these changes, erratic behavior and a considerable range in such properties as thermal and electrical conductivity may be expected above 315 °C (600 °F).

Thermal-expansion characteristics of austenitic manganese steels are similar to those of other austenitic materials. The expected change in length on heating is about $1\frac{1}{2}$ times that of ferritic steels. A coefficient of linear thermal expansion of 18 μm/m · °C (10 μin./in. · °F) generally is precise enough near room temperature. Transformation to pearlite and precipitation of carbide influence values of the expansion coefficient in the range from 370 to 760 °C (700 to 1400 °F).

MAGNETIC PROPERTIES

The untransformed austenite of 13% Mn steel is virtually nonmagnetic, with a permeability of about 1.03 or less. This permits use of the material where a strong, tough, nonmagnetic metal is required, as in magnet cover plates, collector shoes for traveling cranes, stator core parts for generators and motors, liner plates for storage bins holding materials that are handled by lifting magnets, magnetic-separator parts, instrument-testing devices and furnace parts located in the magnetic fields of induction furnaces.

The changes that occur in composition of the surface during heat treatment may produce a magnetic skin, one that is either a martensitic surface layer or a magnetic oxide. Permeability values of 1.3± have been obtained on specimens that have this magnetic surface layer. Frequently, this layer does no harm, but if necessary it may be removed by grinding or pickling or may be prevented by suitable (although often expensive) corrective measures during heat treatment.

WELDING

Many of the common applications of austenitic manganese steel involve welding, either for fabrication or for repair. Consequently, it is important to understand that this material is unusually sensitive to the effects of reheating, often becoming embrittled to the point of losing its characteristic toughness. Oxyfuel-gas welding is so likely to produce embrittlement that it is not accepted as a practical method of welding this alloy. When properly done, electric-arc welding is the preferred method of joining manganese steels.

Arc Welding. Electrodes for arc welding austenitic manganese steel are commercially available in many compositions. They may be used for surfacing, for repair welds, and for joining manganese steel to itself or to carbon steels. They have a lowered carbon content to minimize carbon precipitation as they cool from welding heat. Though formulated to avoid embrittlement of the deposited filler metal, proper welding procedures still must be used to avoid damage in the heat-affected zone.

Electrodes of high manganese content, containing insignificant amounts of other alloying elements, are available also. Usually, these electrodes are recommended only for build-up of worn areas, because they are inherently lower in toughness than more highly alloyed grades. High-alloy, low-manganese electrodes generally have not been accepted as equivalent to high-manganese types.

Factors that are frequently overlooked are the losses in carbon, manganese and silicon that occur during welding. Although many electrode manufacturers compensate for these losses, improper welding techniques, such as use of excessive arc length and excessive puddling, may cause additional losses. The result is inferior properties in the weld deposit.

Carbon steels frequently are welded to high-manganese steel using austenitic stainless steel electrodes. Because the deposit tends to be a mixture or hybrid of the base and filler metals, it can have quite different properties. Often it is air-hardening, producing a martensitic zone as the weld cools. The ductility of martensite is low, but the strength is high and weldments are often satisfactory; the chief adverse factor may be cracks in the martensite. Cross-weld tensile properties of low-carbon 14Mn-1Mo steel plate welded to 1045 steel with EFeMn-A electrodes were 435 MPa (63 ksi) yield strength, 650 MPa (94 ksi) ultimate tensile strength, and 11% elongation with fracture in the 1045. These properties are superior to those of many weldments of carbon steel.

Filler metal from 0.75% carbon, 15% Mn, 3.5% Ni and 4.0% Cr seems to be superior to that of EFeMn-A and EFeMn-B.

The grades with chromium contents above 14% are also useful for joining manganese steel and for buildup of worn parts, but because of their low carbon content have poorer abrasion resistance. However, they are more machinable than the higher-carbon grades. If used to restore dimensions of crusher parts they should be overlaid with an effective hard facing alloy.

Precautions. The primary consideration in welding austenitic manganese steel is minimum heating of the parent metal to avoid embrittling transformations or carbide precipitation. This precludes preheating. Under the most favorable circumstances, some precipitation is expected, and the resulting heat-affected zones seldom attain the toughness of normal parent metal. Because manganese steel work hardens in service, it may be assumed that any worn area requiring repair or rebuilding will have a work-hardened surface. This surface must be removed before welding in order to prevent cracking in the heat-affected zone.

The low heat conductivity and high thermal expansion of manganese steel also cause difficulties, combining to produce steep thermal gradients and high residual stresses. Weld beads are subjected to tension as they cool; to minimize cracking, it is desirable to peen them while they are hot, producing plastic flow and changing the stress to compression. This hammering should be done promptly after 150 to 225 mm (6 to 9 in.) of weld bead has been deposited.

Table 4. Feed forces required in lathe turning of austenitic manganese steels(a)

Type of steel	Negative 7° rake Horizontal kg	lb	Vertical kg	lb	Flat tool Horizontal kg	lb	Vertical kg	lb	Positive 6° rake Horizontal kg	lb	Vertical kg	lb
1.12C-13Mn steel	68	150	161	355	79	175	161	355	116	255	170	375
3Ni-13Mn steel	63	140	156	345	...	...	...	...	127	280	172	380
1Mo-13Mn steel	82	180	168	370	...	...	...	...	111	245	175	385
1.12C-13Mn leaded steel(b)	59	130	152	335	50	110	141	310	127	280	172	380
Wrought type 304 stainless steel(c)	70	155	118	260	91	200	127	280	Welded to tool			
Cast CF-8 stainless steel(c)	68	150	116	255	...	...	...	...	229	505	261	575

(a) Specimens were 32-mm-diam (1.25-in.-diam) bars, toughened by water quenching. Roughing cuts 2.5 mm deep were taken using complex-carbide tools containing about 15% TiC + TaC (predominantly TiC) and about 7 to 10% Co. New cutting edges were used for each positive or negative rake. Cutting speed was 0.19 to 0.20 m/s (37 to 39 ft/min). (b) Recovery of 0.02% Pb from 0.35% added in ladle. The effect on machinability is inconclusive. (c) Stainless steels suffer in the comparison at this speed. They are more machinable at higher speeds and permit certain operations, such as drilling of 6.4-mm-diam ($^1/_4$-in.-diam) holes, that are very difficult with austenitic manganese steel. The type 304 stainless steel was cold finished.

MACHINING

Manganese steels are so tough, and work harden at the point of a cutting tool to such an extent, that frequently they are considered commercially unmachinable. However, these steels are regularly cut by adhering to generally accepted procedures. In addition, a new, more highly machinable grade of manganese steel has been developed and may be helpful in appropriate applications.

Procedures. Although details of practice and tool design differ, there is general agreement on the following procedures for machining manganese steels:

- Machine tools should be rigid and in good condition. Any factors that encourage chatter are undesirable.
- Tools should be sharp. Dull tools cause excessive work hardening of the cut surface and accentuate the difficulty of machining.
- Low speeds of about 9 to 12 m/min (30 to 40 ft/min) should be used. High speeds are likely to create red-hot chips and to cause rapid tool breakdown.
- Both cobalt high speed steel and cemented carbide tools can be used. The latter are preferred.
- Liberal use of a good grade of sulfur-bearing cutting oil is beneficial.
- In castings, holes should be formed by cores in the foundry, rather than by machining, whenever possible.

Various sources provide statements in favor of both positive-rake and negative-rake tools and both dry cutting and use of liquid coolants. Because high temperatures at the cutting edge are a large part of the problem, effective cooling seems desirable. Negative-rake tools are likely to require more force and thus to produce more heat. However, the thinner edge of a positive-rake tool is more vulnerable to heat. Comparative machining data are presented in Table 4.

SELECTED REFERENCES

The Equilibrium Diagram of Iron-Manganese-Carbon Alloys of Commercial Purity, by E. C. Bain, E. S. Davenport and W. S. N. Waring: *Transactions of AIME*, Vol 100, 1932, p 228

"Work Hardening and Martensite Formation in Austenitic Manganese Alloys," by C. H. Shih, B. L. Averbach and M. Cohen: Massachusetts Institute of Technology research report, 1953

Manganese Steel, by Hadfields Ltd.: Oliver and Boyd, London, 1956

Austenitic Manganese Steel Welding Electrodes, by H. S. Avery and H. J. Chapin: *The Welding Journal*, Vol 33, 1954, p 459

Gouging Abrasion Test for Materials Used in Ore and Rock Crushing, Part II, by F. Borik and W. G. Scholz: *Journal of Materials*, Vol 6, No. 3, Sept 1971, p 590

Some Decomposition Structures of Austenitic Manganese Steels, by R. Castro and P. Garnier: *Revue de Metallurgie*, Vol 55, Jan 1958, p 17

Austenitic Manganese Steel, by H. S. Avery: American Brake Shoe Company, 1949

Nonferrous Alloys for Wear Applications

Condensed from Metals Handbook, Ninth Edition, Volume 3, pages 589 to 594.

NONFERROUS ALLOYS most widely used for wear-resistant service are cobalt-base materials. Only wrought cobalt-base alloys are discussed in this article; hard facing alloys are the subject of a separate article in this volume. Cobalt-base materials are not the only class of nonferrous wear-resistant materials. For some types of applications, copper-beryllium alloys and certain aluminum bronzes are frequently specified—primarily in cast form and primarily for bearing applications.

COBALT-BASE ALLOYS

Wrought cobalt-base wear alloys are nearly identical in chemical composition with their hard facing alloy counterparts but with subtle differences in boron, silicon or manganese levels. Equally subtle differences in mill practice are used in producing the two classes of wear alloys. The significant difference between hard facing alloys and other cobalt-base wear alloys is microstructure, which varies with casting, sintering and rolling practices.

Wrought cobalt-base wear-resistant alloys are produced to standard production specifications designed to develop consistent microstructures and properties. The greatest quantity of these alloys is processed on typical hot and cold rolling equipment. However, there is a strong trend to-

Table 1. Compositions of principal cobalt-base wear alloys

Alloy	C	Cr	Ni	Mo	W	Fe	Mn	Si	Co
Stellite 6B	0.9-1.4	28-32	3.0	1.5	3.5-5.5	3.0	2.0	2.0	Rem
Stellite 6K	1.4-2.2	28-32	3.0	1.5	3.5-5.5	3.0	2.0	2.0	Rem
Haynes 25 (L-605)	0.05-0.10	19-21	9-11	...	14-16	3.0	1.0-2.0	1.0	Rem(b)
Tribaloy T-400(c)	0.06	8.5	(d)	28.5	...	(d)	...	2.6	Rem
Tribaloy T-800(c)	0.06	17.5	(d)	28.5	...	(d)	...	3.4	Rem

(a) Where a single value is shown instead of a range (except for the Tribaloy alloys), the value is a maximum limit. (b) Also contains 0.03 max P and 0.03 max S. (c) Nominal composition. (d) Ni + Fe content is 3.0 max.

Table 2. Typical wear data for selected alloys

Alloy	Condition	Volume loss, mm^3	Average wear coefficient(a)	Coefficient of friction(b)
Abrasive wear at room temperature(c)				
Stellite 6B	Mill annealed (38 HRC)	8.2	1.6×10^{-4}	...
Stellite 6K	Mill annealed (46 HRC)	13.3	3.2×10^{-4}	...
Haynes 25	Mill annealed (24 HRC)	53.0	6.7×10^{-4}	...
1090 steel	Hardened and tempered (55 HRC)(d)	37.2	2.7×10^{-3}	...
Type 316 stainless steel	As received (86 HRB)	81.4	6.7×10^{-4}	...
Type 304 stainless steel	As received (92 HRB)	102.1	1.0×10^{-3}	...
Adhesive wear at room temperature(e)				
Stellite 6B	Mill annealed (38 HRC)	0.293	1.2×10^{-5}	...
	Cold reduced 10% (44 HRC)	0.347	1.7×10^{-5}	...
Stellite 6K	Mill annealed (46 HRC)	0.561	2.4×10^{-5}	...
Haynes 25	Mill annealed (24 HRC)	0.285	0.8×10^{-5}	...
1090 steel	Hardened and tempered (55 HRC)(d)	0.293	2.0×10^{-5}	...
Adhesive wear at 540 °C (1000 °F)				
Tribaloy T-400		0.12(f)	4.0×10^{-5}(f)	0.25(f)
		0.07(g)	1.0×10^{-6}(g)	0.37(g)
		0.11(h)	1.0×10^{-6}(h)	0.44(h)
Tribaloy T-800		0.37(f)	3.0×10^{-5}(f)	0.19(f)
		0.03(g)	1.0×10^{-5}(g)	0.42(g)
		2.1(h)	1.0×10^{-5}(h)	0.62(h)
Stellite 6		0.07(f)	1.0×10^{-6}(f)	0.24(f)
		1.7(g)	1.7×10^{-5}(g)	0.62(g)
		2.3(h)	2.3×10^{-5}(h)	0.66(h)

(a) Wear coefficient, K, calculated from $K = Vh/PL$, where V is volume loss in mm^3, h is diamond pyramid (Vickers) hardness, P is test load in kg and L is sliding distance in mm. For elevated-temperature wear tests, hot hardness at the wear test temperature was used to calculate the wear coefficient. (b) Average final friction measurements on wear test specimens. (c) Rubber wheel test using dry sand abrasive. (d) Austenitized 1 h at 870 °C (1600 °F), water quenched, and tempered 4 min at 480 °C (900 °F). (e) Average of two or more tests of an alloy block rubbing against the edge of a rotating ring made of 4620 steel, case hardened to a surface hardness of 63 HRC. (f) 45-kg (100-lb) test load. (g) 90-kg (200-lb) test load. (h) 136-kg (300-lb) test load.

ward use of cobalt-base wear-resistant alloys in powder form. Powders are used either to produce powder metallurgy parts or to be applied as coatings by various commonly accepted methods.

Most wrought cobalt-base wear-resistant materials are proprietary and are not included under technical society, national standard or government specifications. Compositions of the five principal cobalt-base wear alloys are given in Table 1.

The primary alloys for severe wear applications are Stellite 6B, Stellite 6K and Haynes 25. These alloys have excellent resistance to most types of wear, as indicated in Table 2. The wear resistance is inherent and not the result of special heat treatment or of methods used to produce high surface hardness.

Cobalt-base wear alloys possess a number of other properties that significantly widen their range of application:

- Good resistance to impact and thermal shock.
- Good resistance to heat and oxidation: high temperatures have little effect on hardness, toughness and dimensional stability. These alloys are highly resistant to atmospheric oxidation at ordinary temperatures, and have good resistance to oxidation at elevated temperatures.

- Good corrosion resistance: these alloys resist attack by a variety of corrosive media. The combination of wear resistance and corrosion resistance makes Stellite 6B, for example, particularly useful in such applications as food-handling machinery and chemical-process equipment.
- High hot-hardness: wrought cobalt-base alloys retain high hardness even at red heat. Once cooled back to room temperature they recover full original hardness.

ALUMINUM BRONZES

The aluminum bronze family of alloys ranges from soft, ductile alpha alloys such as C60800, C61300 and C61400 to proprietary die alloys that are very hard and brittle. The hardness of aluminum bronzes in various forms ranges from 30 HRB to 40 HRC (67 to 375 HB), and there is a similarly large range of tensile and compressive strengths. Proper selection of an alloy from this family is the key to successful wear service, and is based on many complex factors.

As a group, the aluminum bronzes are not considered self-lubricating, and are therefore used where an adequate reliable source of lubrication can be maintained. These alloys are recommended for applications involving high loads and

moderate to low speeds. Natural surface oxidation during service forms a film that helps provide galling and seizing resistance in applications where the lubricating film might become temporarily marginal. General corrosion resistance is good, and the high strength and hardness result in relatively low conformability and embeddability. Abrasion resistance, particularly in the case of the hard die alloys, generally is superior to that of other copper alloys.

Some typical wear applications for aluminum bronzes are given in Table 3.

Wear Tests. The wear resistance and effective coefficient of friction for aluminum bronzes, as is true for most alloys, depend so much on specific conditions that generalizations can be misleading. One type of laboratory evaluation of wear involves the use of Amsler wear tests. In Amsler tests, thick disks of materials rotate against each other in a variety of modes, including counter-rotating slip, corotating slip, or one specimen rotating against a stationary specimen. A combination of rotational and axial slide may also be used. Tests are run dry or lubricated, and abrasive or corrosive substances may also be introduced. Wear rates are measured by weight loss and/or dimensional changes in the specimens. Table 4 summarizes results of some Amsler wear tests on six aluminum bronzes and two other copper alloys.

BERYLLIUM COPPER ALLOYS

Prior to the mid-1960's, few beryllium copper components were designed into equipment for service under conditions of sliding and rolling metal-to-metal contact. However, since that time, beryllium copper components have been produced for submarine telephone equipment, jet aircraft landing gear, wind tunnel apparatus and molds for producing plastic parts. There also has been tremendous growth in available forms of alloys for wear applications. They can now be cast and forged to shape and fabricated from tube, rod, bar or sheet.

During the late 1960's, beryllium copper alloys were used mostly to replace bronzes and low-alloy steels in applications where those alloys gave marginal service. Beryllium coppers were seldom specified in original bearing or bushing design until the 1970's. Beryllium coppers are among the hardest and strongest of all copper-base metals. Properly lubricated beryllium copper surfaces are more wear resistant than those of other copper-base alloys and many ferrous alloys. Wear rates for a beryllium copper sleeve bearing are compared with those of aluminum-nickel bronze in Fig. 1.

Table 3. Typical wear applications for aluminum bronzes

Alloy	Typical hardness	Applications
C61300, C61400	30-82 HRB	Wear plates, gibs, press guides
C62400, C95400	76-98 HRB	Gears, bushings, wear plates, mill slippers, rollers
C63000	90-100 HRB	Valve guides, bushings, rollers
C62500	27-32 HRC	Rollers, cams, bending dies
Die alloys	32-40 HRC	Automotive die inserts, dies for drawing stainless steel sinks, tube-bending dies

Table 4. Amsler wear test results for six aluminum bronzes, one nickel tin bronze and one aluminum brass

Alloy	Nominal composition	Typical hardness, HB	Wear rate(a) for load stress of:		Dynamic coefficient of friction for load stress of:	
			23 ksi	27.5 ksi	23 ksi	27.5 ksi
Lubricated with Mobil DTE-25						
AMS 4881	Cu-11Al-4.7Fe-5Ni	277	0.40	0.90	0.120	0.140
C61300	Cu-7Al-3.5Fe-0.3Sn	183	0.50	1.60	0.110	0.135
C62400	Cu-10.7Al-3.3Fe	200	1.10	1.30	0.120	0.140
C62500	Cu-13Al-4.2Fe	302	0.20	0.40	0.100	0.120
C86300	Cu-6.3Al-3Fe-25Zn-3.8Mn	223	0.85	0.90	0.125	0.140
C91700	Cu-12Sn-1.6Ni	106	2.50	4.30	0.130	0.140
C95400	Cu-10.7Al-4Fe	180	0.40	0.70	0.140	0.112
C95500	Cu-10.7Al-4Fe-3.7Ni-1Mn	228	0.25	0.60	0.105	0.140
Unlubricated						
AMS 4881		277	1.66	2.40	0.32	0.22
C61300		183	2.48	3.82	0.52	0.42
C62400		200	1.32	2.96	0.27	0.22
C62500		302	1.42	3.18	0.24	0.22
C86300		223	5.96	5.73	0.38	0.22
C91700		106	Erratic	5.18	0.88	0.69
C95400		180	1.76	4.85	0.28	0.22
C95500		228	1.20	2.20	0.42	0.26

(a) Wear rate, based on weight loss in mg for each 1000 revolutions of the test specimen, for a bronze disk 38 mm (1.5 in.) in diameter by 9.5 mm (0.375 in.) thick rotating against a similar specimen of 4340 steel hardened and tempered to 43 HRC under test conditions involving 110% slip and 4 mm (0.16 in.) axial slide. Wear surfaces were 0.6 μm (25 μin.) rms surface finish, or better.

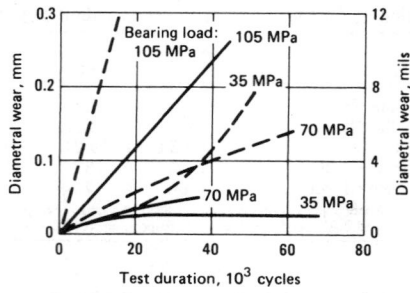

Solid lines are for C17200, beryllium copper; broken lines are for C63000, aluminum-nickel bronze. Shaft material: chromium-plated 4340 steel. Lubrication: AF17 (MIL-G-21164), applied at start of test only for 35-MPa test, and applied every 100 cycles in 70- and 105-MPa tests. Test conditions: bearing loads, 35, 70 and 105 MPa (5000, 10 000 and 15 000 psi); oscillation amplitude, ±45°; oscillation frequency, 2 Hz; velocity, 0.16 m/s (31.4 ft/min). Direction of loading reversed every 100 cycles of oscillation.

Fig. 1. Comparative wear data at three bearing loads for copper alloys C17200 and C63000

In addition, beryllium coppers exhibit excellent corrosion resistance in industrial and marine environments, have high electrical conductivity, are nonmagnetic and nonsparking, and resist anelastic behavior. Beryllium copper wear parts are almost always used with a lubricant, and preferably in a moist environment.

Alloy C17200 (98Cu-1.9Be-0.2Co), which has a tensile strength of 470 to 1460 MPa (68 to 212 ksi), is usually specified for wear applications. When well lubricated, no other copper-base alloy has equal load-carrying capacity as a bushing material. It resists vibration-induced fretting as well as wear in the presence of molten aluminum, zinc, cast iron, and non-chloride plastics. It is more resistant to sand blasting than low-alloy steels.

Sleeve bearing tests simulating airplane landing gear operation have shown that C17200 can sustain operating bearing loads up to about 700 MPa (100 ksi) when lubricated. Customarily, bearing stresses up to 415 MPa (60 ksi) are normal on beryllium copper aircraft bushings; overloads can be as high as 655 MPa (90 ksi). Even at such high stresses, no seizing or galling occurs as long as there is adequate lubrication, and wear rate is minimal for service lives exceeding 60 000 oscillatory cycles.

To get the most favorable wear rate from C17200 bushings, a complex BeO-containing oxide surface film must be developed and maintained. The thin, tenacious film probably is formed as a result of chemical reaction that takes place between the beryllium copper bearing and the lubricating grease under the conditions prevailing at the bearing interface. Breakthrough of the film in the absence of lubricating grease leads to galling and seizure.

Because well-designed lubrication systems are costly, design engineers usually seek applications where good lubrication is relatively easy to obtain. Typical situations that provide natural conditions for good lubrication are hammering action, low-speed reverse oscillation, and sliding motion between mating surfaces.

A linear actuator (jackscrew) for an aircraft hatch positioner, operating at bearing stresses over 205 MPa (30 ksi) at a low wear rate as well as at a high mechanical efficiency (high work output versus work input), is a good illustration of the optimum choice of design elements for a beryllium copper/steel wear couple. In this application, a C17200 beryllium copper nut (40 HRC) works against a nitrided AISI 4340 jackscrew (case hardness, 56 HRC) with a 29-mm (1⅛-in.) ACME 32-64 thread. A MIL-G-27617 high-tempera-

ture, Teflon-filled grease is used to obtain a coefficient of friction of 0.01 to 0.03. This combination provides high performance, low maintenance and long service life.

Techniques for Enhancing Wear Resistance. Self-lubrication of components has been accomplished by placing graphite in the beryllium copper surface. Wear properties can be augmented by casting beryllium coppers rather than machining components from wrought alloys. Wear properties also can be increased by orders of magnitude simply by oxidizing the surface of the alloy.

Structures with greater wear resistance than normal wrought C17200 have been developed by plasma depositing or detonation depositing atomized C17200 powder on suitable substrates. There is a thin BeO film on the surface of each spherical particle before plasma deposition. Thus, after deposition, a network of BeO films exists throughout the coating structure. Wear resistance superior to that of conventional cast and wrought C17200 also can be developed by hot compaction extrusion of atomized powders. Wear resistance can be further augmented by incorporating a dispersion of a hard phase of cobalt beryllide (CoBe) intermetallic particles in the 99.5Cu-1.9Be matrix.

18 TOOL MATERIALS

Reviewed and revised by Neil J. Culp, Carpenter Technology Corp.; Dennis D. Huffman, Timken Research; and R. J. Henry, University of Pittsburgh–Johnstown

This section was condensed from Metals Handbook, Ninth Edition, Volume 3, Properties and Selection: Stainless Steels, Tool Materials and Special-Purpose Metals, pages 419 to 559. For more detailed information on the topics covered in this section, the reader is referred to the larger work. Additional articles on tool materials can be found within this volume in the following sections: Machining (Section 27), Heat Treating (Section 28), Joining (Section 30) and Metallography (Section 35). The reader should also consult the index to locate information not otherwise categorized. Listings of selected references and additional reading for further research are presented at the end of this section.

Introduction and Overview

Tool Steels

A TOOL STEEL is any steel used to make tools for cutting, forming or otherwise shaping a material into a part or component adapted to a definite use. The earliest tool steels were simple, plain carbon steels, but beginning in 1868, and to a greater extent early in the 20th century, many complex, highly alloyed tool steels were developed. Although plain carbon tool steels were first used and still are employed occasionally, it is the alloy tool steels containing, among other elements, relatively large amounts of tungsten, molybdenum, manganese, vanadium and chromium which have made it possible to meet increasingly severe service demands and to provide greater dimensional control and freedom from cracking during heat treatment. Many alloy tool steels are also widely used for machinery components and structural applications where particularly severe requirements must be met.

In service, most tools are subjected to extremely high loads that are applied rapidly. They must withstand these loads a great number of times without breaking or undergoing excessive wear or deformation. In many applications, tool steels must provide this capability under conditions that develop high temperatures in the tool. No single tool material combines maximum levels of wear resistance, toughness, and resistance to softening at elevated temperatures. Consequently, selection of the proper tool material for a given application often requires a trade-off to achieve the optimum combination of properties.

Most tool steels are wrought products, but precision castings can be used to advantage in some applications. The powder metallurgy (P/M) process also is used in making tool steels, in both mill forms and near-net shapes. P/M tool steels may provide (a) more uniform carbide size and distribution in large sections and (b) special compositions that are difficult or impossible to produce by melting and casting and then mechanically working the cast product.

Tool steels are generally melted in small-tonnage electric-arc furnaces to economically achieve composition tolerances, good cleanness and precise control of melting conditions. Special refining and secondary remelting processes have been introduced to satisfy particularly difficult demands regarding tool steel quality and performance. Tool steels must have minimal decarburization held within carefully controlled limits. This requires that annealing be done by special procedures under closely controlled conditions.

The performance of a tool in service depends on proper design of the tool, accuracy with which the tool is made, selection of the proper tool steel and application of the proper heat treatment. A tool can perform successfully in service only when all four of these requirements have been fulfilled.

With few exceptions, all tool steels must be heat treated to develop specific combinations of wear resistance, resistance to deformation or breaking under high loads, and resistance to softening at elevated temperatures.

CLASSIFICATION AND CHARACTERISTICS

Table 1 gives composition limits for the tool steels most commonly used today. Each group of tool steels of similar composition, application or mode of quenching is identified by a capital letter; within each group, individual tool steel types are assigned code numbers.

High Speed Steels

High speed steels are tool materials developed largely for use in high speed cutting-tool applications. There are two classifications of high speed steels: molybdenum high speed steels (group M) and tungsten high speed steels (group T). Group M steels constitute about 95% of all high speed steel produced in the United States.

Group M and group T high speed steels are equivalent in performance; the main advantage of group M steels is lower initial cost (approximately 40% lower than that of similar group T steels).

Molybdenum high speed steels and tungsten high speed steels are similar in many other respects, including hardenability. Typical applications for group M and group T steels include cutting tools of all kinds. Some grades are satisfactory for cold work applications, such as cold-header die inserts, thread-rolling dies, punches and blanking dies. Steels of the M40 series are used to make cutting tools for machining modern, very tough, high-strength steels.

For die inserts and punches, high speed steels sometimes are underhardened—that is, quenched from austenitizing temperatures lower than those recommended for cutting-tool applications—as a means of increasing toughness.

Molybdenum high speed steels contain molybdenum, tungsten, chromium, vanadium, cobalt and carbon as principal alloying elements. Group M steels have slightly greater toughness than group T steels at the same hardness. Otherwise, mechanical properties of the two groups are similar.

Increasing the carbon and vanadium contents of group M steels increases wear resistance; increasing the cobalt content improves red hard-

Table 1. Composition limits of principal types of tool steels

AISI	SAE	UNS	C	Mn	Si	Cr	Ni	Mo	W	V	Co
Molybdenum high speed steels											
M1	M1	T11301...0.78-0.88	0.15-0.40	0.20-0.50	3.50-4.00	0.30 max	8.20-9.20	1.40-2.10	1.00-1.35	...
M2	M2	T11302...0.78-0.88; 0.95-1.05	0.15-0.40	0.20-0.45	3.75-4.50	0.30 max	4.50-5.50	5.50-6.75	1.75-2.20	...
M3, class 1	M3	T11313...1.00-1.10	0.15-0.40	0.20-0.45	3.75-4.50	0.30 max	4.75-6.50	5.00-6.75	2.25-2.75	...
M3, class 2	M3	T11323...1.15-1.25	0.15-0.40	0.20-0.45	3.75-4.50	0.30 max	4.75-6.50	5.00-6.75	2.75-3.75	...
M4	M4	T11304...1.25-1.40	0.15-0.40	0.20-0.45	3.75-4.75	0.30 max	4.25-5.50	5.25-6.50	3.75-4.50	...
M6	...	T11306...0.75-0.85	0.15-0.40	0.20-0.45	3.75-4.50	0.30 max	4.50-5.50	3.75-4.75	1.30-1.70	11.00-13.00
M7	...	T11307...0.97-1.05	0.15-0.40	0.20-0.55	3.50-4.00	0.30 max	8.20-9.20	1.40-2.10	1.75-2.25	...
M10	...	T11310...0.84-0.94; 0.95-1.05	0.10-0.40	0.20-0.45	3.75-4.50	0.30 max	7.75-8.50	...	1.80-2.20	...
M30	...	T11330...0.75-0.85	0.15-0.40	0.20-0.45	3.50-4.25	0.30 max	7.75-9.00	1.30-2.30	1.00-1.40	4.50-5.50
M33	...	T11333...0.85-0.92	0.15-0.40	0.15-0.50	3.50-4.00	0.30 max	9.00-10.00	1.30-2.10	1.00-1.35	7.75-8.75
M34	...	T11334...0.85-0.92	0.15-0.40	0.20-0.45	3.50-4.00	0.30 max	7.75-9.20	1.40-2.10	1.90-2.30	7.75-8.75
M36	...	T11336...0.80-0.90	0.15-0.40	0.20-0.45	3.75-4.50	0.30 max	4.50-5.50	5.50-6.50	1.75-2.25	7.75-8.75
M41	...	T11341...1.05-1.15	0.20-0.60	0.15-0.50	3.75-4.50	0.30 max	3.25-4.25	6.25-7.00	1.75-2.25	4.75-5.75
M42	...	T11342...1.05-1.15	0.15-0.40	0.15-0.65	3.50-4.25	0.30 max	9.00-10.00	1.15-1.85	0.95-1.35	7.75-8.75
M43	...	T11343...1.15-1.25	0.20-0.40	0.15-0.65	3.50-4.25	0.30 max	7.50-8.50	2.25-3.00	1.50-1.75	7.75-8.75
M44	...	T11344...1.10-1.20	0.20-0.40	0.30-0.45	4.00-4.75	0.30 max	6.00-7.00	5.00-5.75	1.85-2.20	11.00-12.25
M46	...	T11346...1.22-1.30	0.20-0.40	0.40-0.65	3.70-4.20	0.30 max	8.00-8.50	1.90-2.20	3.00-3.30	7.80-8.80
M47	...	T11347...1.05-1.15	0.15-0.40	0.20-0.45	3.50-4.00	0.30 max	9.25-10.00	1.30-1.80	1.15-1.35	4.75-5.25
Tungsten high speed steels											
T1	T1	T12001...0.65-0.80	0.10-0.40	0.20-0.40	3.75-4.00	0.30 max	...	17.25-18.75	0.90-1.30	...
T2	T2	T12002...0.80-0.90	0.20-0.40	0.20-0.40	3.75-4.50	0.30 max	1.00 max	17.50-19.00	1.80-2.40	...
T4	T4	T12004...0.70-0.80	0.10-0.40	0.20-0.40	3.75-4.50	0.30 max	0.40-1.00	17.50-19.00	0.80-1.20	4.25-5.75
T5	T5	T12005...0.75-0.85	0.20-0.40	0.20-0.40	3.75-5.00	0.30 max	0.50-1.25	17.50-19.00	1.80-2.40	7.00-9.50
T6	...	T12006...0.75-0.85	0.20-0.40	0.20-0.40	4.00-4.75	0.30 max	0.40-1.00	18.50-21.00	1.50-2.10	11.00-13.00
T8	T8	T12008...0.75-0.85	0.20-0.40	0.20-0.40	3.75-4.50	0.30 max	0.40-1.00	13.25-14.75	1.80-2.40	4.25-5.75
T15	...	T12015...1.50-1.60	0.15-0.40	0.15-0.40	3.75-5.00	0.30 max	1.00 max	11.75-13.00	4.50-5.25	4.75-5.25
Chromium hot work steels											
H10	...	T20810...0.35-0.45	0.25-0.70	0.80-1.20	3.00-3.75	0.30 max	2.00-3.00	...	0.25-0.75	...
H11	H11	T20811...0.33-0.43	0.20-0.50	0.80-1.20	4.75-5.50	0.30 max	1.10-1.60	...	0.30-0.60	...
H12	H12	T20812...0.30-0.40	0.20-0.50	0.80-1.20	4.75-5.50	0.30 max	1.25-1.75	1.00-1.70	0.50 max	...
H13	H13	T20813...0.32-0.45	0.20-0.50	0.80-1.20	4.75-5.50	0.30 max	1.10-1.75	...	0.80-1.20	...
H14	...	T20814...0.35-0.45	0.20-0.50	0.80-1.20	4.75-5.50	0.30 max	...	4.00-5.25	...	...
H19	...	T20819...0.32-0.45	0.20-0.50	0.20-0.50	4.00-4.75	0.30 max	0.30-0.55	3.75-4.50	1.75-2.20	4.00-4.50
Tungsten hot work steels											
H21	H21	T20821...0.26-0.36	0.15-0.40	0.15-0.50	3.00-3.75	0.30 max	...	8.50-10.00	0.30-0.60	...
H22	...	T20822...0.30-0.40	0.15-0.40	0.15-0.40	1.75-3.75	0.30 max	...	10.00-11.75	0.25-0.50	...
H23	...	T20823...0.25-0.35	0.15-0.40	0.15-0.60	11.00-12.75	0.30 max	...	11.00-12.75	0.75-1.25	...
H24	...	T20824...0.42-0.53	0.15-0.40	0.15-0.40	2.50-3.50	0.30 max	...	14.00-16.00	0.40-0.60	...
H25	...	T20825...0.22-0.32	0.15-0.40	0.15-0.40	3.75-4.50	0.30 max	...	14.00-16.00	0.40-0.60	...
H26	...	T20826...0.45-0.55(b)	0.15-0.40	0.15-0.40	3.75-4.50	0.30 max	...	17.25-19.00	0.75-1.25	...
Molybdenum hot work steel											
H42	...	T20842...0.55-0.70(b)	0.15-0.40	...	3.75-4.50	0.30 max	4.50-5.50	5.50-6.75	1.75-2.20	...

(continued)

ness but concurrently lowers toughness. High speed steels have unusually high resistance to softening at elevated temperatures (see Fig. 1) when compared with other tool steels; this is a result of their very high alloy contents.

Because group M steels readily decarburize and are easily damaged due to overheating under adverse austenitizing environments, they are more sensitive than group T steels to hardening conditions—particularly austenitizing temperature and atmosphere. This is especially true of the high-molybdenum, low-tungsten compositions.

Group M high speed steels are deep hardening. They must be austenitized at temperatures lower than those used for hardening group T steels, to avoid incipient melting. Group M high speed steels can develop full hardness when quenched from temperatures of 1175 to 1230 °C (2150 to 2250 °F). Type M10, which usually has slightly lower hardenability than other molybdenum high speed steels, must be oil quenched if section size is larger than about 40 to 50 mm (about 1½ to 2 in.).

The maximum hardness that can be obtained in group M tool steels varies with composition and section size. For those with lower carbon

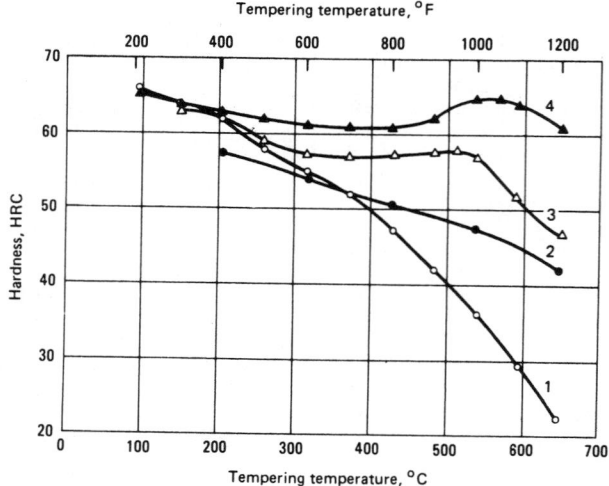

Curves are for 1 h at temperature. Curve 1 illustrates low resistance to softening at elevated temperature, such as is exhibited by group W and group O tool steels. Curve 2 illustrates medium resistance to softening, such as is exhibited by type S1 tool steel. Curves 3 and 4 illustrate high and very high resistance to softening, respectively, such as are exhibited by the secondary-hardening tool steels A2 and M2.

Fig. 1. Variation of hardness with tempering temperature for four typical tool steels

Table 1 (continued)

AISI	Designations SAE	UNS	C	Mn	Si	Cr	Composition(a), % Ni	Mo	W	V	Co
Air-hardening medium-alloy cold work steels											
A2	A2	T30102...0.95-1.05	1.00 max	0.50 max	4.75-5.50	0.30 max	0.90-1.40	...	0.15-0.50	...
A3	...	T30103...1.20-1.30	0.40-0.60	0.50 max	4.75-5.50	0.30 max	0.90-1.40	...	0.80-1.40	...
A4	...	T30104...0.95-1.05	1.80-2.20	0.50 max	0.90-2.20	0.30 max	0.90-1.40	...	...	...
A6	...	T30106...0.65-0.75	1.80-2.50	0.50 max	0.90-1.20	0.30 max	0.90-1.40	...	...	...
A7	...	T30107...2.00-2.85	0.80 max	0.50 max	5.00-5.75	0.30 max	0.90-1.40	0.50-1.50	3.90-5.15	...
A8	...	T30108...0.50-0.60	0.50 max	0.75-1.10	4.75-5.50	0.30 max	1.15-1.65	1.00-1.50	...	...
A9	...	T30109...0.45-0.55	0.50 max	0.95-1.15	4.75-5.50	1.25-1.75	1.30-1.80	...	0.80-1.40	...
A10	...	T30110...1.25-1.50(c)	1.60-2.10	1.00-1.50	...	1.55-2.05	1.25-1.75	...	...	...
High-carbon, high-chromium cold work steels											
D2	D2	T30402...1.40-1.60	0.60 max	0.60 max	11.00-13.00	0.30 max	0.70-1.20	...	1.10 max	1.00 max
D3	D3	T30403...2.00-2.35	0.60 max	0.60 max	11.00-13.50	0.30 max	...	1.00 max	1.00 max	...
D4	...	T30404...2.05-2.40	0.60 max	0.60 max	11.00-13.00	0.30 max	0.70-1.20	...	1.00 max	...
D5	D5	T30405...1.40-1.60	0.60 max	0.60 max	11.00-13.00	0.30 max	0.70-1.20	...	1.00 max	2.50-3.50
D7	D7	T30407...2.15-2.50	0.60 max	0.60 max	11.50-13.50	0.30 max	0.70-1.20	...	3.80-4.40	...
Oil-hardening cold work steels											
O1	O1	T31501...0.85-1.00	1.00-1.40	0.50 max	0.40-0.60	0.30 max	...	0.40-0.60	0.30 max	...
O2	O2	T31502...0.85-0.95	1.40-1.80	0.50 max	0.35 max	0.30 max	0.30 max	...	0.30 max	...
O6	O6	T31506...1.25-1.55(c)	0.30-1.10	0.55-1.50	0.30 max	0.30 max	0.20-0.30	...	...	...
O7	...	T31507...1.10-1.30	1.00 max	0.60 max	0.35-0.85	0.30 max	0.30 max	1.00-2.00	0.40 max	...
Shock-resisting steels											
S1	S1	T41901...0.40-0.55	0.10-0.40	0.15-1.20	1.00-1.80	0.30 max	0.50 max	1.50-3.00	0.15-0.30	...
S2	S2	T41902...0.40-0.55	0.30-0.50	0.90-1.20	...	0.30 max	0.30-0.60	...	0.50 max	...
S5	S5	T41905...0.50-0.65	0.60-1.00	1.75-2.25	0.35 max	...	0.20-1.35	...	0.35 max	...
S6	...	T41906...0.40-0.50	1.20-1.50	2.00-2.50	1.20-1.50	...	0.30-0.50	...	0.20-0.40	...
S7	...	T41907...0.45-0.55	0.20-0.80	0.20-1.00	3.00-3.50	...	1.30-1.80	...	0.20-0.30(d)	...
Low-alloy special-purpose tool steels											
L2	...	T61202...0.45-1.00(b)	0.10-0.90	0.50 max	0.70-1.20	...	0.25 max	...	0.10-0.30	...
L6	L6	T61206...0.65-0.75	0.25-0.80	0.50 max	0.60-1.20	1.25-2.00	0.50 max	...	0.20-0.30(d)	...
Low-carbon mold steels											
P2	...	T51602...0.10 max	0.10-0.40	0.10-0.40	0.75-1.25	0.10-0.50	0.15-0.40	...	...	...
P3	...	T51603...0.10 max	0.20-0.60	0.40 max	0.40-0.75	1.00-1.50	...	...	...	...
P4	...	T51604...0.12 max	0.20-0.60	0.10-0.40	4.00-5.25	...	0.40-1.00	...	...	...
P5	...	T51605...0.10 max	0.20-0.60	0.40 max	2.00-2.50	0.35 max	...	...	...	...
P6	...	T51606...0.05-0.15	0.35-0.70	0.10-0.40	1.25-1.75	3.25-3.75	...	...	...	...
P20	...	T51620...0.28-0.40	0.60-1.00	0.20-0.80	1.40-2.00	...	0.30-0.55	...	...	...
P21	...	T51621...0.18-0.22	0.20-0.40	0.20-0.40	0.20-0.30	3.90-4.25	...	...	0.15-0.25	1.05-1.25A1
Water-hardening tool steels											
W1	W108,W109, W110,W112	T72301...0.70-1.50(e)	0.10-0.40	0.10-0.40	0.15 max	0.20 max	0.10 max	0.15 max	0.10 max	...
W2	W209,210	T72302...0.85-1.50(e)	0.10-0.40	0.10-0.40	0.15 max	0.20 max	0.10 max	0.15 max	0.15-0.35	...
W5	...	T72305...1.05-1.15	0.10-0.40	0.10-0.40	0.40-0.60	0.20 max	0.10 max	0.15 max	0.10 max	...

(a) All steels except group W contain 0.25 max Cu, 0.03 max P and 0.03 max S; group W steels contain 0.20 max Cu, 0.025 max P and 0.025 max S. Where specified, sulfur may be increased to 0.06 to 0.15% to improve machinability of group H, M and T steels. (b) Available in several carbon ranges. (c) Contains free graphite in the microstructure. (d) Optional. (e) Specified carbon ranges are designated by suffix numbers.

contents—types M1, M2, M10 (low-carbon composition), M30, M33, M34 and M36—maximum hardness in typical tool cross sections is usually 65 HRC. For higher carbon contents—including types M3, M4 and M7—maximum hardness is about 66 HRC. A hardness of 66 HRC also can be developed in the lower-carbon, high-cobalt type M6. Maximum hardness of the higher-carbon cobalt-containing steels—types M41, M42, M43, M44 and M46—is 69 to 70 HRC.

Tungsten high speed steels contain tungsten, chromium, vanadium, cobalt and carbon as the principal alloying elements. Type T1 was developed partly as a result of the work of Taylor and White, who in the early 1900's found that certain steels with over 14% W, about 4% Cr and about 0.3% V resisted softening at temperatures high enough to cause the steel to emit radiation in the red part of the visible spectra—or in other words, they exhibited red hardness. In its earliest form, type T1 contained about 0.68 C, 18 W, 4 Cr and 0.3 V. By 1920, the vanadium content had been increased to about 1.0%. Carbon content was gradually increased over a 30-year period to its present level of 0.75%.

Group T tool steels are used primarily for cutting tools such as bits, drills, reamers, taps, broaches, milling cutters, and hobs. These steels are also used for making dies, punches, and high-load, high-temperature structural components such as aircraft bearings and pump parts. Type T15 is the most wear-resistant steel of this group.

Group T tool steels are characterized by high red hardness and wear resistance. They are so deep hardening that sections up to 75 mm (3 in.) in thickness or diameter can be hardened to 65 HRC or more by quenching in oil or molten salt.

Group T tool steels are all deep hardening when quenched from their recommended hardening temperatures of 1200 to 1300 °C (2200 to 2375 °F). They are seldom used to make hardened tools with section sizes greater than 75 mm (3 in.). Even very large cutting tools, such as drills 75 and 100 mm (3 and 4 in.) in diameter, have relatively small effective sections for hardening because metal has been removed to form the flutes. Some large-diameter solid tools are made from group T high speed steels; these include broaches and cold extrusion punches as large as 100 to 125 mm (4 to 5 in.) in diameter. For such tools, surface hardness is of primary importance.

The difference between surface hardness and center hardness varies with bar size. Section size and total mass of a given tool often have an effect on its response to a given hardening treatment that is equal to or greater than the effect of the grade of tool steel selected. For tools of extremely large diameter or heavy section, it is relatively common practice to use an accelerated oil quench to provide full hardness. This practice may yield values of Rockwell C hardness only one or two points higher than those obtainable through hot-salt quenching or air cooling, which ordinarily produce full hardness in tools smaller than about 75 mm (3 in.), but at such high hardnesses a one- or two-point increase in Rockwell hardness may prove quite significant.

Maximum hardness of tungsten high speed steels varies with carbon content, and to a lesser degree with alloy content. A hardness of at least 64.5 HRC can be developed in any high speed steel. Those types that have high carbon contents and hard carbides, such as T15, may be hardened to 67 HRC.

Hot Work Steels

Many manufacturing operations involve punching, shearing or forming of metals at high temperatures. Hot work steels (group H) have been developed to withstand the combinations of heat, pressure and abrasion associated with such operations.

Generally, group H tool steels have medium carbon contents (0.35 to 0.45%), and chromium, tungsten, molybdenum and vanadium contents totaling 6 to 25%. They are divided into three subgroups: chromium hot work steels (types H10 to H19), tungsten hot work steels (types H21 to H26) and molybdenum hot work steel (type H42).

Chromium hot work steels (types H10 to H19) have good resistance to heat softening because of their medium chromium contents and additions of carbide-forming elements such as molybdenum, tungsten and vanadium. The low carbon and low total alloy contents promote toughness at the normal working hardnesses of 40 to 55 HRC. Higher tungsten and molybdenum contents increase hot strength but slightly reduce toughness. Vanadium is added to increase resistance to washing (erosive wear) at high temperatures. An increase in silicon content improves oxidation resistance at temperatures up to 800 °C (1475 °F). The most widely used types in this group are H13, H12, H11 and, to a lesser extent, H19.

All of the chromium hot work steels are deep hardening. H11, H12 and H13 may be air hardened to full working hardness in section sizes up to 150 mm (6 in.); other group H steels may be air hardened in section sizes up to 300 mm (12 in.). The air-hardening qualities and balanced alloy contents of these steels result in low distortion during hardening. Chromium hot work steels are especially well adapted to hot die work of all kinds, particularly dies for extrusion of aluminum and magnesium, as well as die-casting dies, forging dies, mandrels and hot shears. Most of these steels have alloy and carbon contents low enough that tools made from them can be water cooled in service without cracking.

H11 tool steel is used to make certain highly stressed structural parts, particularly in aerospace technology. Material for such demanding applications is produced by vacuum-arc remelting of air-melted electrodes. Vacuum-arc remelting provides extremely low residual gas contents, excellent microcleanness and a high degree of structural homogeneity.

The chief advantage of H11 over conventional high-strength steels is its ability to resist softening during continued exposure to temperatures up to 540 °C (1000 °F) and at the same time provide moderate toughness and ductility at room-temperature tensile strengths of 1720 to 2070 MPa (250 to 300 ksi). In addition, because of its secondary hardening characteristic, H11 can be tempered at high temperatures, resulting in nearly complete relief of residual hardening stresses, which is necessary for maximum toughness at high strength levels. Other important advantages of H11, H12 and H13 steels for structural and hot work applications include ease of forming and working, good weldability, relatively low coefficient of thermal expansion, acceptable thermal conductivity and above-average resistance to oxidation and corrosion.

Tungsten Hot Work Steels. The principal alloying elements of tungsten hot work steels (types H21 to H26) are carbon, tungsten, chromium and vanadium. The higher alloy contents of these steels make them more resistant to high-temperature softening and washing than H11 and H13 hot work steels. However, high alloy content also makes them more prone to brittleness at normal working hardnesses (45 to 55 HRC) and makes it difficult for them to be safely water cooled in service.

Although tungsten hot work steels can be air hardened, they are usually quenched in oil or hot salt to minimize scaling. When air hardened, they exhibit low distortion. Tungsten hot work steels require higher hardening temperatures than chromium hot work steels, which makes them more likely to scale when heated in an oxidizing atmosphere.

Although these steels have much greater toughness, in many characteristics they are similar to high speed steels; in fact, type H26 is a low-carbon version of T1 high speed steel. If tungsten hot work steels are preheated to operating temperature before use, breakage can be minimized. These steels have been used to make mandrels and extrusion dies for high-temperature applications such as extrusion of brass, nickel alloys and steel, and are also suitable for use in hot forging dies of rugged design.

Molybdenum Hot Work Steel. There is only one active molybdenum hot work steel: type H42. This alloy contains molybdenum, chromium, vanadium and carbon, with varying amounts of tungsten. It is similar to tungsten hot work steels, having almost identical characteristics and uses. Although its composition resembles those of various molybdenum high speed steels, type H42 has a low carbon content and greater toughness. The principal advantage of type H42 over tungsten hot work steels is its lower initial cost. Type H42 is more resistant to heat checking than tungsten hot work steels but, in common with all high-molybdenum steels, requires greater care in heat treatment — particularly with regard to decarburization and control of austenitizing temperature.

Cold Work Steels

Because resistance to elevated temperatures (above 260 °C, or 500 °F) is not required for cold work tooling applications, the cold work die steels have alloy contents designed to provide good wear resistance and toughness in various combinations. There are three categories of cold work steels: air-hardening steels (group A); high-carbon, high-chromium die steels (group D); and oil-hardening steels (group O).

Air-hardening medium-alloy cold work steels (group A) contain enough alloying elements to enable them to achieve full hardness in sections up to about 100 mm (4 in.) in diameter on air cooling from the austenitizing temperature. (Type A6 through hardens in sections as large as a cube 175 mm, or 7 in., on a side.) Because they are air hardening, group A tool steels exhibit minimum distortion and the highest safety (least tendency to crack) in hardening. Manganese, chromium and molybdenum are the principal alloying elements used to provide this deep hardening. Types A2, A3, A7, A8 and A9 contain a high percentage of chromium (5%), which provides moderate resistance to softening at elevated temperatures (see curve 3 in Fig. 1 for a plot of hardness versus tempering temperature for type A2).

Types A4, A6 and A10 are lower in chromium content and higher in manganese content. They can be hardened from temperatures about 100 °C (about 200 °F) lower than those required for the high-chromium types, further reducing distortion and undesirable surface reactions during heat treatment at the expense of slightly lower abrasion resistance.

To improve toughness, silicon is added to type A8 and both silicon and nickel are added to types A9 and A10. Because of the high carbon and silicon contents of type A10, graphite is formed in the microstructure; as a result, A10 has much better machinability when in the annealed condition, and somewhat better resistance to galling and seizing when in the fully hardened condition, than other group A tool steels.

Typical applications for group A tool steels include shear knives, punches, blanking and trimming dies, forming dies and coining dies. The inherent dimensional stability of these steels makes them suitable for gages and precision measuring tools. In addition, the extreme abrasion resistance of type A7 makes it suitable for brick molds, ceramic molds and other highly abrasive applications.

The complex chromium or chromium-vanadium carbides in group A tool steels enhance the wear resistance provided by the martensitic matrix. Therefore, these steels perform well under abrasive conditions at less than full hardness. Although cooling in still air is adequate to produce full hardness in most tools, very massive sections should be hardened by cooling in an air blast or by interrupted quenching in hot oil.

High-carbon, high-chromium cold work steels (group D) contain 1.50 to 2.35% carbon and 12% chromium; with the exception of type D3, they also contain 1% molybdenum. All group D tool steels except type D3 are air hardening, and attain full hardness when cooled in still air. Type D3 is almost always quenched in oil (small parts can be austenitized in vacuum and then gas quenched); therefore, tools made of D3 are more susceptible to distortion and are more likely to crack during hardening.

Group D steels have high resistance to softening at elevated temperatures. These steels also exhibit excellent resistance to wear — especially type D7, which has the highest carbon and vanadium contents. All group D steels — particularly the higher-carbon types D3, D4 and D7 — contain massive carbides that make them susceptible to edge brittleness.

Typical applications for group D steels include long-run dies for blanking, forming, thread rolling and deep drawing; dies for cutting laminations; brick molds; gages; burnishing tools; rolls; and shear and slitter knives.

Oil-hardening cold work steels (group O) have high carbon contents, plus enough other alloying elements so that small to moderate sections can attain full hardness when quenched in oil from the austenitizing temperature. Group O tool steels vary in type of alloy, as well as in alloy content, even though they are similar in general characteristics and are used for similar applications. Type O1 contains manganese, chromium and tungsten. Type O2 is alloyed primarily with manganese. Type O6 contains silicon, manganese and molybdenum; it has a high total carbon content that includes free carbon as well as sufficient combined carbon to enable the steel to achieve maximum as-quenched hardness. Type O7 contains manganese and chromium, and has a tungsten content higher than that of type O1.

The most important service-related property of group O steels is high resistance to wear at normal temperatures, a result of high carbon content. On the other hand, group O steels have low resistance to softening at elevated temperatures.

The ability of group O steels to harden fully on relatively slow quenching yields lower distortion and greater safety (less tendency to crack) in hardening than is characteristic of the water-hardening tool steels. Tools made from these steels can be successfully repaired or renovated by welding if proper procedures are followed. In addition, graphite in the microstructure of type O6 greatly improves the machinability of annealed stock and helps reduce galling and seizing of fully hardened stock.

Group O steels are used extensively in dies and punches for blanking, trimming, drawing, flanging and forming. Surface hardnesses of 56 to 62 HRC, obtained through oil quenching followed by tempering at 175 to 315 °C (350 to 600 °F), provide a suitable combination of mechanical properties for most dies made from type O1, O2 or O6. Type O7, which has lower hardenability but better general wear resistance than any other group O tool steel, is more often used for tools requiring keen cutting edges. Oil-hardening tool steels are also used for machinery components (such as cams, bushings and guides) and for gages (where good dimensional stability and wear properties are needed).

The hardenability of group O steels can be measured effectively by the Jominy end-quench test. Hardenability bands for group O steels are shown in Fig. 2.

At normal hardening temperatures, group O steels retain greater amounts of undissolved carbides and thus do not harden as deeply as do steels that are lower in carbon but similar in alloy content. On the other hand, group O steels attain higher surface hardness. Raising the hardening temperature increases grain size, increases solution of alloying elements and dissolves more of the excess carbide, thereby increasing hardenability. However, raising the hardening temperature can have an adverse effect on certain mechanical properties—most notably ductility and toughness—and also can increase the likelihood of cracking during hardening.

Shock-Resisting Steels

The principal alloying elements in shock-resisting (group S) steels are manganese, silicon, chromium, tungsten and molybdenum, in various combinations. Carbon content is about 0.50% for all group S steels, which produces a combination of high strength, high toughness and low-to-medium wear resistance. Group S steels are used primarily for chisels, rivet sets, punches, driver bits and other applications requiring high toughness and resistance to shock loading. Types S1 and S7 are also used for hot punching and shearing, which require some heat resistance.

Group S steels vary in hardenability from shallow hardening (S2) to deep hardening (S7). In these steels of intermediate alloy content, hardenability is controlled to a greater extent by actual composition than by the incidental effects of grain size and melting practice, which are so important for group W steels. Group S steels require relatively high austenitizing temperatures to achieve optimum hardness; consequently, undissolved carbides are not a factor in control of hardenability. Type S2 is normally water quenched; types S1, S5 and S6 are oil quenched; type S7 is normally cooled in air, except for large sections, which are oil quenched.

Because group S steels exhibit excellent toughness at high strength levels, they often are considered for nontooling or structural applications.

Low-Alloy, Special-Purpose Steels

The low-alloy, special-purpose (group L) tool steels contain small amounts of chromium, vanadium, nickel and molybdenum. At one time, seven

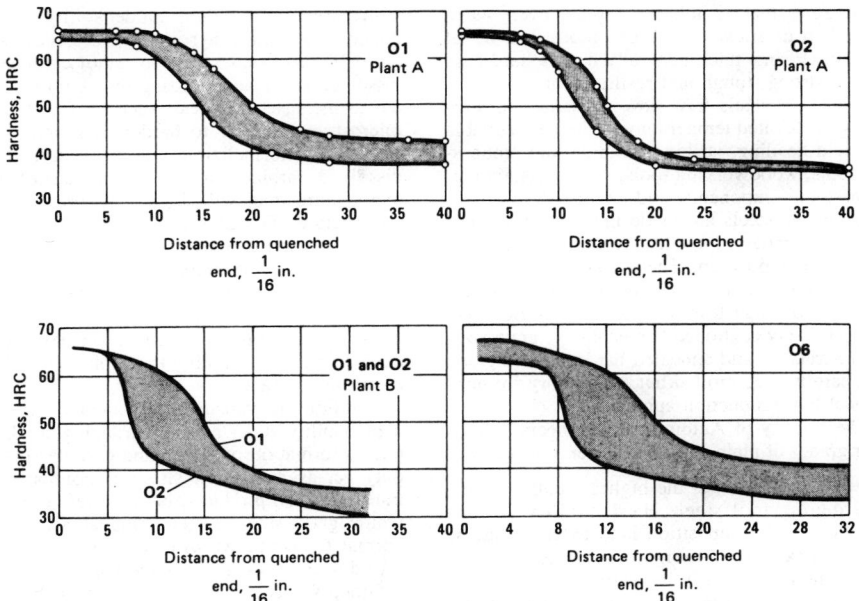

Hardenability bands from plant B represent the data from five heats each for O1 and O2 tool steels. Data from plant A were determined only on the basis of average hardness, not as hardenability bands. Data for O6 is for a spheroidized prior structure. O1 and O6 steels were quenched from 815 °C (1500 °F); O2 from 790 °C (1450 °F).

Fig. 2. End-quench hardenability bands for group O tool steels

steels were listed in this group, but because of falling demand, only types L2 and L6 remain. Type L2 is available in several carbon contents from 0.50 to 1.10%; its principal alloying elements are chromium and vanadium, which make it an oil-hardening steel of fine grain size. Type L6 contains small amounts of chromium and molybdenum, plus 1.50% nickel for increased toughness.

Although both L2 and L6 are considered oil-hardening steels, large sections of L2 are often quenched in water. A type L2 steel containing 0.50% carbon is capable of attaining about 57 HRC as oil quenched, but it will not through harden in sections more than about 13 mm (0.5 in.) thick. Type L6, which contains 0.70% carbon, has an as-quenched hardness of about 64 HRC; it can maintain a hardness above 60 HRC throughout sections 75 mm (3 in.) thick.

Group L steels generally are used for machine parts such as arbors, cams, chucks and collets, and for other special applications requiring good strength and toughness.

Mold Steels

Mold steels (group P) contain chromium and nickel as principal alloying elements. Types P2 to P6 are carburizing steels produced to tool steel quality standards. They have very low hardness and low resistance to work hardening in the annealed condition. These factors make it possible to produce a mold impression by cold hobbing. After the impression is formed, the mold is carburized, hardened and tempered to a surface hardness of about 58 HRC. Types P4 and P6 are deep hardening. In type P4, full hardness in the carburized case can be achieved by cooling in air.

Types P20 and P21 normally are supplied heat treated to 30 to 36 HRC—a condition in which they can be machined readily into large, intricate dies and molds. Because these steels are prehard-

ened, no subsequent high-temperature heat treatment is required, and distortion and size changes are avoided. However, when used for plastic molds, type P20 sometimes is carburized and hardened after the impression has been machined. Type P21 is an aluminum-containing precipitation-hardening steel and is supplied prehardened to 32 to 36 HRC. After machining and low-temperature aging, type P21 can reach 38 to 40 HRC in sections as large as it is practical to produce.

Nearly all group P steels have low resistance to softening at elevated temperatures, except for P4 and P21, which have medium resistance. Group P steels are used almost exclusively in low-temperature die-casting dies and in molds for injection or compression molding of plastics. Plastic molds often require very massive steel blocks up to 750 mm (30 in.) thick and weighing as much as 9 Mg (10 tons). Because these large die blocks must meet stringent requirements for soundness, cleanness and hardenability, electric-furnace melting, vacuum degassing and special deoxidation treatments have become standard practices in the production of group P tool steels. In addition, ingot casting and forging practices have been refined to achieve a high degree of homogeneity.

Water-Hardening Tool Steels

Water-hardening (group W) tool steels contain carbon as the principal alloying element. Small amounts of chromium and vanadium are added to most of the group W steels—chromium to increase hardenability and wear resistance, and vanadium to maintain fine grain size and thus enhance toughness. Group W tool steels are made with various nominal carbon contents (from about 0.60 to 1.40%); the most popular grades contain approximately 1.00% carbon.

Group W tool steels are very shallow hardening, and consequently develop a fully hardened zone that is relatively thin, even when quenched drastically. Sections more than about 13 mm (¹/₂ in.) thick generally have a hard case over a strong, tough and resilient core.

Group W steels have low resistance to softening at elevated temperatures. They are suitable for cold heading, striking, coining and embossing tools; woodworking tools; wear-resistant machine-tool components; and cutlery.

Group W steels are made in as many as four different grades or quality levels for the same nominal composition. These quality levels have been given various names by different manufacturers, and range from a clean carbon tool steel with precisely controlled hardenability, grain size, microstructure and annealed hardness to a grade less carefully controlled but satisfactory for noncritical low-production applications.

The Society of Automotive Engineers defines four grades of plain carbon tool steels as follows:

• *Special (grade 1)* is the highest-quality water-hardening tool steel. Hardenability is controlled, and composition held to close limits. Bars are subjected to rigorous testing to ensure maximum uniformity in performance.
• *Extra (grade 2)* is a high-quality water-hardening tool steel that is controlled for hardenability and is subjected to tests that ensure good performance in general applications.
• *Standard (grade 3)* is a good-quality water-hardening tool steel that is not controlled for hardenability and that is recommended for applications where some latitude in uniformity can be tolerated.
• *Commercial (grade 4)* is a commercial-quality water-hardening tool steel that is neither controlled for hardenability nor subjected to special tests.

Limits on manganese, silicon and chromium generally are not required for "special" and "extra" grades. The following Shepherd hardenability limits are prescribed instead:

Hardenability classification	Radial depth of hardening (P), ¹/₆₄ in.	Minimum fracture grain size (F)
Carbon content, 0.70 to 0.95%		
Shallow	10 max	8
Regular	9 to 13	8
Deep	12 min	8
Carbon content, 0.95 to 1.30%		
Shallow	8 max	9
Regular	7 to 11	9
Deep	10 to 16	8

The combined manganese, silicon and chromium contents of standard and commercial grades should not exceed 0.75%. Generally, both manganese and silicon are limited to 0.35% max in all standard and commercial grades; chromium is limited to 0.15% max in standard grades and 0.20% max in commercial grades.

The ability of a group W tool steel to perform satisfactorily in many applications depends on the depth of the hardened zone. Depth of hardening in these steels is controlled mainly by austenitic grain size, melting practice, alloy content, amount of excess carbide present at the quenching temperature and, to a lesser extent, initial structure of the steel prior to austenitizing for hardening.

Typical results in the Shepherd PF test indicate an increase in P value of 0.80 mm (²/₆₄ in.) for every increase in austenitic grain size of one ASTM number for the same grade. Increased amounts of undissolved carbides at the hardening

temperature will reduce hardenability. This is doubly important in hypereutectoid grades, which are deliberately quenched to retain carbides undissolved at the austenitizing temperature, in order to increase wear resistance. A fine lamellar microstructure prior to hardening, such as that obtained by normalizing, will result in fewer undissolved carbides at the normal austenitizing temperature than will a prior spheroidized microstructure. The presence of fewer carbides at the austenitizing temperature promotes deeper hardening because more carbon is dissolved in the austenite and because there are fewer carbides to act as nucleation sites for nonmartensitic transformation products. Thus, normalized bars have deeper hardenability than spheroidized bars of the same grade.

Addition of vanadium frequently decreases hardenability under normal hardening conditions due to formation of many fine carbides that not only act as nucleation sites for nonmartensitic transformation products, but also refine the austenitic grain size. Austenitizing at higher-than-normal temperatures will dissolve these excess carbides and thus increase the hardenability.

Group W steels with carbon contents lower than that of the eutectoid composition often have greater hardenability than hypereutectoid grades. Grain coarsening resulting from the higher austenitizing temperatures used for hypoeutectoid grades is one cause of this, but the main cause is the absence of excess carbides at the austenitizing temperature.

Figure 3 shows a typical relationship between bar diameter and case depth (60 HRC or above) for three W1 tool steels that are equal in carbon content (1% C) but differ in hardenability. Hardenability is varied by adjusting manganese and silicon contents and altering deoxidation procedure. This relationship illustrates the need for precise specification of hardenability in the selection of these grades: group W tool steels purchased without hardenability requirements could vary widely enough in this property to cause severe processing difficulties or actual tool failures.

With the very high cooling rates required for hardening of the W grades, there is a greater chance that the tool will crack during hardening. Consequently, most manufacturers prefer to use tool steels that can be hardened satisfactorily by

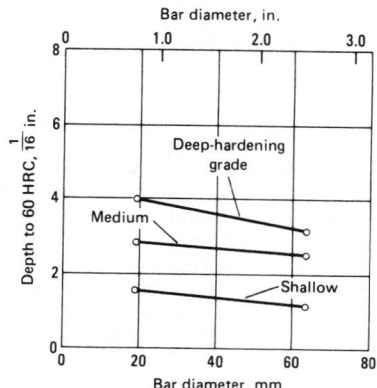

Fig. 3. Relation of bar diameter and depth of hardened zone for shallow-, medium- and deep-hardening grades of W1 tool steel containing 1% C

quenching in oil or cooling in air in order to attempt to avoid the expense involved if a tool cracks during heat treatment.

TYPICAL HEAT TREATMENTS AND PROPERTIES

Information on processing and service characteristics of tool steels is essential in understanding the problems involved in selection, processing and application of tool steels. A general guideline for these characteristics is presented in Table 2.

Technical representatives of tool steel producers can supply more specific information on the properties developed by specific heat treatments in the steels produced by their companies. They should be consulted as to the type of steel and heat treatment best suited to meet all service requirements at the least over-all cost.

The basic properties of tool steels that determine their performance in service are resistance to wear, deformation and breakage; toughness; and, in many instances, resistance to softening at elevated temperatures. Often, these characteristics can be measured by, or inferred from, direct measurement of hardness. Hardness of tool steels is most commonly measured and reported on the Rockwell C scale (HRC) in the United States and on the Vickers scale (diamond pyramid hardness, or HV) in the United Kingdom and Europe. It is significant that conversion from HRC to HV, or vice versa, is not linear (see Fig. 4).

For a given tool steel at a given hardness, wear resistance may vary widely depending on the wear mechanism involved and the heat treatment used. It is important to note also that among tool steels with widely differing compositions but identical hardnesses, wear resistance may vary widely under identical wear conditions.

The ability of a tool steel to withstand rapid application of high loads without breaking increases with decreasing hardness. With hardness held constant, wide differences can be observed among tool steels of different compositions, or among steels of the same nominal composition made by different melting practices or heat treated according to different schedules.

The ability of a tool steel to resist softening at elevated temperatures is related to (a) its ability to develop secondary hardening and (b) the amount of special phases, such as excess alloy carbides, in the microstructure. Useful information on the ability of tool steels to resist softening at elevated temperatures can be obtained from tempering curves such as those in Fig. 1.

Fabrication

The properties that influence the fabricability of tool steels include: machinability; grindability; weldability; hardenability; and extent of distortion, safety (freedom from cracking) and tendency to decarburize during heat treatment.

Machinability of tool steels can be measured by the usual methods applied to constructional steels. Results are reported as percentages of the machinability of water-hardening tool steels (see Table 3); 100% machinability in tool steels is equivalent to about 30% machinability in constructional steels, for which 100% machinability would be that of a free-machining constructional steel such as B1112.

Improving the machinability of a tool steel by altering either composition or preliminary heat treatment can be very important if a large amount

Table 2. Processing and service characteristics of tool steels
Adapted from "Tool Steels" (a Steel Products Manual): AISI, March 1978

AISI designation	Resistance to decarburization	Hardening response	Amount of distortion(a)	Resistance to cracking	Approximate hardness(b), HRC	Machinability	Toughness	Resistance to softening	Resistance to wear
Molybdenum high speed steels									
M1	Low	Deep	A or S, low; O, medium	Medium	60-65	Medium	Low	Very high	Very high
M2	Medium	Deep	A or S, low; O, medium	Medium	60-65	Medium	Low	Very high	Very high
M3 (class 1 and class 2)	Medium	Deep	A or S, low; O, medium	Medium	61-66	Medium	Low	Very high	Very high
M4	Medium	Deep	A or S, low; O, medium	Medium	61-66	Low to medium	Low	Very high	Highest
M6	Low	Deep	A or S, low; O, medium	Medium	61-66	Medium	Low	Highest	Very high
M7	Low	Deep	A or S, low; O, medium	Medium	61-66	Medium	Low	Very high	Very high
M10	Low	Deep	A or S, low; O, medium	Medium	60-65	Medium	Low	Very high	Very high
M30	Low	Deep	A or S, low; O, medium	Medium	60-65	Medium	Low	Highest	Very high
M33	Low	Deep	A or S, low; O, medium	Medium	60-65	Medium	Low	Highest	Very high
M34	Low	Deep	A or S, low; O, medium	Medium	60-65	Medium	Low	Highest	Very high
M36	Low	Deep	A or S, low; O, medium	Medium	60-65	Medium	Low	Highest	Very high
M41	Low	Deep	A or S, low; O, medium	Medium	65-70	Medium	Low	Highest	Very high
M42	Low	Deep	A or S, low; O, medium	Medium	65-70	Medium	Low	Highest	Very high
M43	Low	Deep	A or S, low; O, medium	Medium	65-70	Medium	Low	Highest	Very high
M44	Low	Deep	A or S, low; O, medium	Medium	62-70	Medium	Low	Highest	Very high
M46	Low	Deep	A or S, low; O, medium	Medium	67-69	Medium	Low	Highest	Very high
M47	Low	Deep	A or S, low; O, medium	Medium	65-70	Medium	Low	Highest	Very high
Tungsten high speed steels									
T1	High	Deep	A or S, low; O, medium	High	60-65	Medium	Low	Very high	Very high
T2	High	Deep	A or S, low; O, medium	High	61-66	Medium	Low	Very high	Very high
T4	Medium	Deep	A or S, low O, medium	Medium	62-66	Medium	Low	Highest	Very high
T5	Low	Deep	A or S, low; O, medium	Medium	60-65	Medium	Low	Highest	Very high
T6	Low	Deep	A or S, low; O, medium	Medium	60-65	Low to medium	Low	Highest	Very high
T8	Medium	Deep	A or S, low; O, medium	Medium	60-65	Medium	Low	Highest	Very high
T15	Medium	Deep	A or S, low; O, medium	Medium	63-68	Low to medium	Low	Highest	Highest
Shock-resisting steels									
S1	Medium	Medium	Medium	High	40-58	Medium	Very high	Medium	Low to medium
S2	Low	Medium	High	Low	50-60	Medium to high	Highest	Low	Low to medium
S5	Low	Medium	Medium	High	50-60	Medium to high	Highest	Low	Low to medium
S6	Low	Medium	Medium	High	54-56	Medium	Very high	Low	Low to medium
S7	Medium	Deep	A, lowest; O, low	A, highest; O, high	45-57	Medium	Very high	High	Low to medium
Low-alloy special-purpose steels									
L2	High	Medium	W, low; O, medium	W, high; O, medium	45-63	High	Very high(c)	Low	Low to medium
L6	High	Medium	Low	High	45-62	Medium	Very high	Low	Medium
Low-carbon mold steels									
P2	High	Medium	Low	High	58-64(c)	Medium to high	High	Low	Medium
P3	High	Medium	Low	High	58-64(c)	Medium	High	Low	Medium
P4	High	High	Very low	High	58-64(c)	Low to medium	High	Medium	High

(continued on the next page)

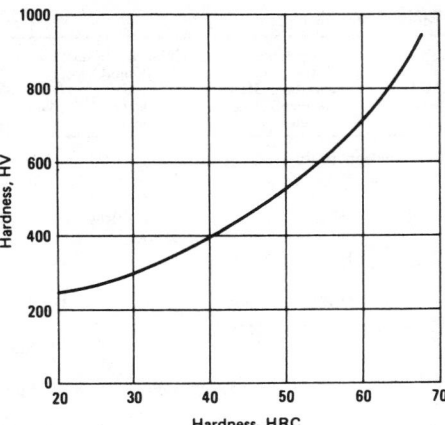

Fig. 4. Relation between Vickers and Rockwell C hardness scales

Table 3. Approximate machinability ratings for annealed tool steels

Type	Machinability rating
O6	125
W1, W2, W5	100(a)
A10	90
P2, P3, P4, P5, P6	75 to 90
P20, P21	65 to 80
L2, L6	65 to 75
S1, S2, S5, S6, S7	60 to 70
H10, H11, H13, H14, H19	60 to 70(b)
O1, O2, O7	45 to 60
A2, A3, A4, A6, A8, A9	45 to 60
H21, H22, H24, H25, H26, H42	45 to 55(b)
T1	40 to 50
M2	40 to 50
T4	35 to 40
M3 (class 1)	35 to 40
D2, D3, D4, D5, D7, A7	30 to 40
T15	25 to 30
M15	25 to 30

(a) Equivalent to approximately 30% of the machinability of B1112.
(b) For hardness range 150 to 200 HB.

of machining is required to form the tool and a large number of tools is to be made.

Grindability. One measure of grindability is the ease with which the necessary excess stock on heat treated tool steel can be removed using standard grinding wheels. The grinding ratio (grindability index) is the volume of metal removed per volume of wheel wear. The higher the grindability index the easier the metal is to grind. The index is valid only for specific sets of grinding conditions.

It should be noted that the grindability index does not indicate susceptibility to cracking during or after grinding, ability to produce the required surface (and subsurface) stress distribution, or ease of obtaining the required surface smoothness.

Weldability. The ability to construct, alter or repair tools by welding without causing the material to crack may be an important factor in selection of a tool material, especially if the tool is large. It is only rarely of importance in selecting materials for small tools. Weldability is largely a function of composition, but welding method and procedure also influence weld soundness. Generally, tool steels that are deep hardening and that are classified as having relatively high safety in hardening are among the more readily welded tool steel compositions. Never-

Table 2 (continued)

AISI designation	Hardening and tempering					Fabrication and service			
	Resistance to decarburization	Hardening response	Amount of distortion(a)	Resistance to cracking	Approximate hardness(b), HRC	Machinability	Toughness	Resistance to softening	Resistance to wear
Low-carbon mold steels (continued)									
P5	High	...	W, high; O, low	High	58-64(c)	Medium	High	Low	Medium
P6	High	...	A, very low; O, low	High	58-61(c)	Medium	High	Low	Medium
P20	High	Medium	Low	High	28-37	Medium to high	High	Low	Low to medium
P21	High	Deep	Lowest	Highest	30-40(d)	Medium	Medium	Medium	Medium
Water-hardening steels									
W1	Highest	Shallow	High	Medium	50-64	Highest	High(e)	Low	Low to medium
W2	Highest	Shallow	High	Medium	50-64	Highest	High(e)	Low	Low to medium
W5	Highest	Shallow	High	Medium	50-64	Highest	High(e)	Low	Low to medium
Chromium hot work steels									
H10	Medium	Deep	Very low	Highest	39-56	Medium to high	High	High	Medium
H11	Medium	Deep	Very low	Highest	38-54	Medium to high	Very high	High	Medium
H12	Medium	Deep	Very low	Highest	38-55	Medium to high	Very high	High	Medium
H13	Medium	Deep	Very low	Highest	38-53	Medium to high	Very high	High	Medium
H14	Medium	Deep	Low	Highest	40-47	Medium	High	High	Medium
H19	Medium	Deep	A, low; O, medium	High	40-57	Medium	High	High	Medium to high
Tungsten hot work steels									
H21	Medium	Deep	A, low; O, medium	High	36-54	Medium	High	High	Medium to high
H22	Medium	Deep	A, low; O, medium	High	39-52	Medium	High	High	Medium to high
H23	Medium	Deep	Medium	High	34-47	Medium	Medium	Very high	Medium to high
H24	Medium	Deep	A, low; O, medium	High	45-55	Medium	Medium	Very high	High
H25	Medium	Deep	A, low; O, medium	High	35-44	Medium	High	Very High	Medium
H26	Medium	Deep	A or S, low; O, medium	High	43-58	Medium	Medium	Very high	High
Molybdenum hot work steel									
H42	Medium	Deep	A or S, low; O, medium	Medium	50-60	Medium	Medium	Very high	High
Air-hardening medium-alloy cold work steels									
A2	Medium	Deep	Lowest	Highest	57-62	Medium	Medium	High	High
A3	Medium	Deep	Lowest	Highest	57-65	Medium	Medium	High	Very high
A4	Medium to high	Deep	Lowest	Highest	54-62	Low to medium	Medium	Medium	Medium to high
A6	Medium to high	Deep	Lowest	Highest	54-60	Low to medium	Medium	Medium	Medium to high
A7	Medium	Deep	Lowest	Highest	57-67	Low	Low	High	Highest
A8	Medium	Deep	Lowest	Highest	50-60	Medium	High	High	Medium to high
A9	Medium	Deep	Lowest	Highest	35-56	Medium	High	High	Medium to high
A10	Medium to high	Deep	Lowest	Highest	55-62	Medium to high	Medium	Medium	High
High-carbon, high-chromium cold work steels									
D2	Medium	Deep	Lowest	Highest	54-61	Low	Low	High	High to very high
D3	Medium	Deep	Very low	High	54-61	Low	Low	High	Very high
D4	Medium	Deep	Lowest	Highest	54-61	Low	Low	High	Very high
D5	Medium	Deep	Lowest	Highest	54-61	Low	Low	High	High to very high
D7	Medium	Deep	Lowest	Highest	58-65	Low	Low	High	Highest
Oil-hardening cold work steels									
O1	High	Medium	Very low	Very high	57-62	High	Medium	Low	Medium
O2	High	Medium	Very low	Very high	57-62	High	Medium	Low	Medium
O6	High	Medium	Very low	Very high	58-63	Highest	Medium	Low	Medium
O7	High	Medium	W, high; O, very low	W, low; O, very high	58-64	High	Medium	Low	Medium

(a) A, air cool; B, brine quench; O, oil quench; S, salt bath quench; W, water quench. (b) After tempering in temperature range normally recommended for this steel. (c) Carburized case hardness. (d) After aging at 510 to 550 °C (950 to 1025 °F). (e) Toughness decreases with increasing carbon content and depth of hardening.

theless, these "weldable" tool steels require careful preheating and postheating to ensure successful welding.

Hardenability includes both the maximum hardness obtainable when the quenched steel is fully martensitic and the depth of hardening obtained by quenching in a specific manner. In this context, depth of hardening must be defined — generally as a specific value of hardness or a specific microstructural appearance. As a very general rule, maximum hardness of a tool steel increases with increasing carbon content; increasing the austenitic grain size and the amount of alloying elements reduces the cooling rate required to produce maximum hardness (increases the depth of hardening). The Jominy end-quench test, which is used extensively for measurement of hardenability of constructional steels, has limited application for tool steels. This test gives useful information only for oil-hardening grades. Air-hardening grades are so deep hardening that the standard Jominy test is not sufficient to evaluate hardenability.

An air-hardenability test has been developed that is based on the principles involved in the Jominy test, but which uses only still air cooling and a 150-mm (6-in.) diameter end block to produce the very low cooling rates of large sections. Such tests provide useful information for research but are of limited use in devising production heat treatments. By contrast, water-hardening grades of tool steel are so shallow hardening that the Jominy test is not sensitive enough. Special tests, such as the Shepherd PF test, are useful for research and for special applications of water-hardening tool steels.

In the Shepherd PF test, a bar 19 mm ($^3/_4$ in.) in diameter, in the normalized condition, is brine quenched from 740 °C (1450 °F) and fractured; the case depth (penetration, P) is measured in 0.4-mm ($^1/_{64}$-in.) intervals, and the fracture grain size of the case (F) is determined by comparison with standard specimens. A PF value of 6–8 indicates a case depth of $^6/_{64}$ in. (2.4 mm) and a fracture grain size of 8. Fine-grain water-hardening tool steels are those with fracture grain sizes (F values) of 8 or more. Deep-hardening steels of this type have P values of 12 or more; medium-hardening steels, 9 to 11; and shallow-hardening steels, 6 to 8.

Distortion and Safety in Hardening. Minimal distortion in heat treating is important for tools that must remain within close size limits. For a more detailed discussion of this factor, see the article in this section on Distortion in Tool Steels. In general, the amount of distortion and the tendency toward cracking increase as the severity of quenching increases.

Resistance to decarburization is an important factor in determining whether or not a protective atmosphere is required during heat treating. In a decarburizing atmosphere, the rate of decarburization increases rapidly with increasing austenitizing temperature, and for a given austenitizing temperature the depth of decarburization increases directly with holding time. Some types of tool steels decarburize much more rapidly than others under the same conditions of atmosphere, austenitizing temperature and time.

MACHINING ALLOWANCES

The standard machining allowance is the recommended total amount of stock that the user should remove from the as-supplied mill form to provide a surface free from imperfections that might adversely affect response to heat treatment or the ability of tools to perform properly.

The decarburization resulting from oxidation at the exposed surfaces during forging and rolling of the tool steel is a major factor in determining the amount of stock that should be removed. Although extra care is used in producing tool steels, scale, seams and other surface imperfections may be present, and, if present, must be removed.

Besides the standard machining allowance, sufficient additional stock must be provided to allow for cleanup of any decarburization and distortion that may occur during final heat treatment. The amount of this allowance varies with the type of tool steel, the type of heat treating equipment and the size and shape of the tool.

Group W and group O tool steels are considered highly resistant to decarburization. Group M steels, cobalt-containing group T steels, group D steels and types H42, A2 and S5 are rated as poor in resisting decarburization.

Decarburization during final heat treatment is undesirable because it alters the composition of the surface layer, thereby changing the response to heat treatment of this layer and usually affecting adversely the properties resulting from heat treatment. Decarburization can be controlled or avoided by heat treating in a neutral salt bath or in a controlled atmosphere or vacuum furnace. When heat treating is done in vacuum, a vacuum of 100 to 200 μm Hg is satisfactory for most tools if the furnace is in good operating condition and has a very low leak rate. However, it is recommended that a vacuum of 50 to 100 μm Hg be used wherever possible.

If special heat treating equipment is not available, appreciable decarburization can be avoided by wrapping the tool in stainless steel foil. Type 321 stainless steel foil can be used at austenitizing temperatures up to about 1000 °C (1850 °F); either type 309 or type 310 foil is required at austenitizing temperatures from 1000 to 1200 °C (1850 to 2200 °F).

POWDER METALLURGY STEELS

In recent years, tool steels with improved properties have been produced by the powder metallurgy (P/M) process. In this process, a bath of prealloyed molten metal is gas atomized and quenched to produce a fine powder. The particles of this powder are screened and loaded into a steel container, which is then evacuated, and the particles are hot isostatically pressed to full density. The resulting compact is rolled or forged to size on conventional steel-mill equipment or, in some instances, is used in the as-compacted condition to make tools.

P/M tool steels have two major advantages: complete freedom from macrosegregation and uniform distribution of extremely fine carbides. These characteristics provide deeper hardening and faster response to hardening. The latter is important, particularly for molybdenum high speed steels, which tend to decarburize rapidly at austenitizing temperatures. P/M products also show less out-of-roundness distortion in large-diameter bars.

When sulfur is added to P/M tool steels, they exhibit a very fine homogeneous distribution of sulfides. This uniform sulfide distribution promotes better machinability. After heat treating, the refined microstructure of P/M tool steels promotes superior grindability and greater toughness compared with those of conventionally processed tool steels.

The following AISI types of high speed steels are available as P/M tool steels: M2, M2 with high sulfur and high carbon, M3 class 2 with sulfur, M4, M35 with sulfur, M42 and T15. P/M steels can be substituted for their conventional counterparts in all applications, and are particularly advantageous when heavy sections are required.

The freedom from gross segregation provided by the P/M process makes it possible for new higher-alloy tool steels to be readily fabricated. One type now available, which contains 1.50 C, 3.75 Cr, 3.00 V, 10 W, 5.25 Mo and 9.00 Co, is reported to have the highest hot hardness of any high speed steel. It has been used in cutting tools for critical applications such as machining of certain aerospace alloys and cutting applications where the highest speeds and feeds are required. Another type, CPM 10V, which contains 2.45 C, 5.25 Cr, 9.75 V and 1.30 Mo, is designed for extreme wear resistance in cold and warm work tooling. Its microstructure consists chiefly of a uniform distribution of hard, wear-resistant vanadium carbides in a tool steel matrix. The wear resistance of this steel at ordinary temperatures typically exceeds that of T15.

PRECISION-CAST TOOL STEELS

Precision casting of tools to nearly finished size offers important cost advantages through reductions in waste and machining. Casting is particularly advantageous when patternmaking costs can be distributed over a large number of tools.

Experience with cast forging and extrusion dies has shown that cast tools are more resistant to heat checking; minute cracks do occur, but they grow at much lower rates than in wrought material of the same grade and hardness. Slower propagation of thermal-fatigue cracks generally extends die life significantly. Mechanical testing of cast and wrought H13 indicates that yield and tensile strengths are virtually identical from room temperature to 600 °C (1100 °F), but that ductility is moderately lower in cast material. Hot hardness of cast H13 is higher than that of wrought H13 at temperatures above about 300 °C (about 600 °F); this hardness advantage increases with temperature and measures about eight points on the HRC scale at 650 °C (1200 °F).

Because cast dies exhibit uniform properties in all directions, no problem of directionality (anisotropy) exists. Dimensional control of castings is very consistent after an initial die is made and any necessary corrections are incorporated in the pattern. Reasonable finishing allowances are 0.25 to 0.38 mm (0.010 to 0.015 in.) on the impression faces, 0.8 to 1.6 mm ($^1/_{32}$ to $^1/_{16}$ in.) at the parting line of the mold, and 1.6 to 3.2 mm ($^1/_{16}$ to $^1/_8$ in.) on the back and outside surfaces. The hot work tool steels most commonly cast include hot work grades H12, H13, H21 and H25 and cold work grades D2, D5, A2 and D7.

SURFACE TREATMENTS

In many applications, service life of high speed steel tools can be increased by surface treatments.

Oxide coatings, provided by treatment of the finish-ground tool in an alkali-nitrate bath or by steam oxidation, prevent or reduce adhesion of the tool to the workpiece. Oxide coatings have doubled tool life—particularly in machining of gummy materials such as soft copper and non-free-cutting low-carbon steels.

Plating of finished high speed steel tools with 0.0025 to 0.0125 mm (0.1 to 0.5 mil) of chromium also prolongs tool life by reducing adhesion of the tool to the workpiece. Chromium plating is relatively expensive, and precautions must be taken to prevent tool failure in service due to hydrogen embrittlement.

Carburizing is not recommended for high speed steel cutting tools because the cases on such tools are extremely brittle. However, carburizing is useful for applications such as cold work dies that require extreme wear resistance and that are not subjected to impact or highly concentrated loading. Carburizing is done at 1035 to 1065 °C (1900 to 1950 °F) for short periods of time (10 to 60 min) to produce a case 0.05 to 0.25 mm (0.002 to 0.010 in.) deep. The carburizing treatment also serves as an austenitizing treatment for the whole tool. A carburized case on a high speed steel has a hardness of 65 to 70 HRC but does not have the high resistance to softening at elevated temperatures exhibited by normally hardened high speed steel.

Nitriding successfully increases life for all types of high speed steel cutting tools. However, gas nitriding in dissociated ammonia produces a case that is too brittle for most applications. Liquid nitriding for about 1 h at 565 °C (1050 °F) provides a light case, increasing both surface hardness and resistance to adhesion. For nitrided high speed steel taps, drills and reamers used in machining annealed steel, five-fold increases in life have been reported, with average increases of 100 to 200%. Obviously, if this nitrided case is removed when the tool is reground, the tool must then be retreated, which reduces the cost advantage of the process.

In addition, special surface-treatment processes, such as aerated nitriding baths, improve resistance to adhesive wear without producing excessive brittleness. Sulfur-containing nitriding baths provide a high-sulfur surface layer for additional resistance to seizing.

Sulfide Treatment. A low-temperature (190 °C; 375 °F) electrolytic process using sodium and potassium thiocyanate provides a seizing-resistant iron sulfide layer. This process can be used as a final treatment for all types of hardened tool steels without much danger of overtempering.

MARAGING STEELS

Certain high-nickel maraging steels are being used for special noncutting tool applications; 18Ni(250) is the type most frequently used. However, for the most demanding applications, the higher-strength 18Ni(300) is often preferred. For applications requiring maximum abrasion resistance, any of the maraging steels can be nitrided.

Maraging steels achieve full hardness—nominally 50 HRC for 18Ni(250), 54 HRC for 18Ni(300) and 58 HRC for 18Ni(350)—by a simple aging treatment, usually 3 h at about 480 °C (900 °F). Because hardening does not depend on cooling rate, full hardness can be developed uniformly in massive sections, with almost no distortion. Decarburization is of no concern in these alloys, because they do not contain carbon as an alloying element. If the long-time service temperature exceeds the aging temperature, maraging steels overage with a significant drop in hardness.

The 18Ni(250) grade is used for aluminum die-casting dies and cores, aluminum hot forging dies, dies for molding plastics, and various support tooling used in extrusion of aluminum. In die casting of aluminum, maraging steel dies can be used at higher hardness than is possible for dies made of H13 tool steel because maraging steel is not as prone to heat checking. Because the aging process results in very little size change, it is possible to machine the intricate impressions for plastic molding dies to final size prior to final hardening.

For molding extremely abrasive types of plastics, the higher surface hardness provided by 18Ni(300) maraging steel is desirable.

Superhard Tool Materials

SUPERHARD TOOL MATERIALS are exceptionally hard—far harder than any tool steel. Some superhard tool materials, such as hard alloys and cemented carbides, exhibit largely metallic characteristics, whereas others, such as ceramic tool materials, are considered nonmetallic.

Cemented carbides are the most widely used superhard tool materials. Their chief applications are in metalworking tools, mining tools and wear-resistant parts. Steel-bonded carbides have a heat treatable matrix, and are used extensively to make forming and stamping tools. Ceramics, boron nitride and diamond are used chiefly to make cutting tools. Diamond is also used for making small-diameter wiredrawing dies.

To take full advantage of the wear resistance of cemented carbides and other superhard tool materials in metalcutting applications, new types of machine tools having the required power and rigidity have been developed. The degree of improvement that can be obtained by using superhard tools in older existing equipment may be limited by a lack of sufficient power and rigidity.

CLASSIFICATION OF SUPERHARD TOOL MATERIALS

At present there is no universally accepted system for classifying carbides and other superhard tool materials. The systems most often employed by both producers and users to describe superhard tool materials are the SAE J1072 system, the ISO classification, the British Hard Metal Association system, and an informal application-oriented system known as the C-grade system. Close cooperation between user and producer is often the most fruitful means of selecting the proper grade.

CEMENTED CARBIDES

Cemented carbides are made by a powder metallurgy process in which finely divided compounds of refractory metals and carbon are bonded together to form a compacted solid of high strength and hardness. The first cemented carbide to be produced was tungsten carbide with a cobalt binder. Over the years, this original material has been modified in many ways to produce a variety of cemented carbides that can be used in a wide range of applications (see Table 4). These modifications consist mainly of varying the amount of metal used as the binder, varying the structure (grain size) of the carbide and substituting other metallic carbides for part of the tungsten carbide. Titanium carbide and tantalum carbide have been the metallic carbides used most widely in complex grades, but because of the limited world

supply of tantalum, niobium carbides are replacing tantalum carbides in increasing amounts. Cemented carbides containing only tungsten carbide are commonly referred to as "straight grades," whereas those containing other metallic carbides in addition to tungsten carbide are called "complex grades."

Straight grades of cemented carbides are used for cutting tools, drawing dies, forming-die inserts, punches and many other types of tools. Complex grades are used chiefly to make cutting tools and cutting-tool inserts for machining plain carbon and low-alloy steels. In these applications, tools made of complex grades exhibit less wear, because the crater formed on the top rake of the tool due to chip adherence is much smaller than that developed on tools made of straight grades. Also, the complex grades have better resistance to deformation at the temperatures developed along the cutting edge.

Cemented carbide is very hard, and holds its high hardness at temperatures well above those at which high speed steels begin to soften. Cemented carbide cutting tools originally were constructed by copper brazing small compacts into heavy steel shanks. These tools had to be sharpened by regrinding frequently for maximum tool efficiency. Although regrinding of brazed-tip tools is still done in many instances, it is much more common for cemented carbide to be used as relatively small indexable inserts that are mechanically held in a steel holder. These inserts are indexed until all available cutting edges have be-

come worn, and then are reground or recycled.

Manufacture by Powder Metallurgy. The conventional means of making cemented carbide tools is a powder metallurgy (P/M) process in which finely divided tungsten carbide powders are blended with cobalt powders, compacted by cold pressing to the desired shape (with allowance for shrinkage during sintering), and sintered in vacuum or under a controlled atmosphere at a temperature high enough to melt, or at least partly melt, the binder. One of the main limitations of this conventional process is the requirement that the parts be of uniform cross section along the direction of pressing, and be of limited length-to-diameter ratio, so that the compacting pressure can be applied relatively uniformly.

Large parts, and parts that cannot be conveniently compacted in dies, generally are made by cold isostatic pressing. The parts may or may not be presintered, and may or may not be machined to shape, before being liquid-phase sintered to final density. Large parts occasionally are compacted by hot pressing in graphite dies.

Hot isostatic pressing is commonly used to increase the soundness of sintered carbide parts. Hot isostatic pressing improves the surface finish and increases the average transverse rupture, compression and impact strengths of carbide tools. It is an essential step in the production of certain tools such as compressor plungers; grinder spindles; anvils and dies for very-high-pressure apparatus; cold extrusion punches; and swaging mandrels. Hot isostatic pressing also is preferred

Table 4. Typical applications of cobalt-bonded cemented carbides

Grade	Grain size	Application
Straight grades		
97WC-3Co	Medium	Machining of cast iron, nonferrous metals and nonmetallic materials; excellent abrasion resistance and low shock resistance; the most wear resistant of the straight WC-Co grades; maintains a sharp cutting edge and makes long finishing cuts to close tolerances possible; also used for fine wire dies and small nozzles
94WC-6Co	Fine	Machining nonferrous and high-temperature alloys
94WC-6Co	Medium	General-purpose machining of work materials other than steel; also used for small and medium-size compacting dies, coating dies, burnishing rings and nozzles
94WC-6Co	Coarse	Machining of cast iron, nonferrous metals and nonmetallic materials; also used for small wiredrawing dies, compacting dies, small drawing dies, and caps and rings. The hardest grade used in mining applications where impact is encountered, as in rotary percussive bits
90WC-10Co	Fine	Machining steel and milling high-temperature metals (including titanium and its alloys) at low feeds and speeds: face mills, end mills, form tools, cutoff tools and screw-machine tools
90WC-10Co	Coarse	Primarily used for mining roller bits and percussive drilling bits
84WC-16Co	Fine	Primarily used for mining and metalforming components
84WC-16Co	Coarse	Metalforming and mining components: medium and large dies where great toughness is required, blanking dies for punch presses, and large mandrels
75WC-25Co	Medium	Metalforming components for heavy impact applications, such as heading dies, cold extrusion dies, and punches and dies for blanking heavy stock
Complex grades		
71-74.5WC-10-12.5TiC-11-12.0TaC-4.5Co	Medium	Finishing, semifinishing and light roughing operations on plain carbon and alloy steels and alloy cast irons
72-73WC-7-8TiC-11.5-12TaC-8-8.5Co	Medium	Tough, wear-resistant grade for heavy-duty roughing cuts. Successfully withstands high temperatures encountered in heavy-duty machining, interrupted turning, scale cuts and milling of plain carbon and alloy steels and alloy cast irons
64TiC-28WC-2TaC-2Cr$_3$C$_2$-4Co	Medium	High-speed finishing of steels and cast irons
57WC-27TaC-16Co	Coarse	Cutting hot flash formed in the manufacture of welded tubing; also used to make dies for hot extrusion of aluminum wirebar and tubing

for parts that require a nearly perfect surface finish. Drawing dies and mandrels; extrusion punches; Sendzimer-mill rolls; strip-mill rolls; wire-flattening rolls; liners and plungers for pumps and compressors; burnishing rolls; and balls for burnishing, sizing and valve applications are among the types of tools and other parts that need exceptionally good surface finishes for optimum performance.

Parts that are long and slender, such as rod stock for circuit-board drills, are difficult to produce by either compaction in dies or isostatic pressing. Such parts are produced by extruding a mixture of carbide powders, metallic binders and a suitable organic vehicle, followed by sintering.

PROPERTIES OF CEMENTED CARBIDES

Specific properties of individual grades of cemented carbides depend not only on the composition of the carbide but also on its particle size and on the amount and type of binder.

The compositions and properties of nine straight grades and four complex grades of cobalt-bonded carbide are given in Table 5. Because properties are influenced by both composition and structure, both characteristics must be specified to define a specific grade.

Microstructure. Varying the structures of cemented carbides can improve tool performance in specific applications.

For straight tungsten carbides of comparable grain size (WC particle size), increasing cobalt content increases transverse strength and toughness but decreases hardness, compressive strength, elastic modulus and abrasion resistance. If we compare, for example, medium-grain carbides having 3, 6 and 25% cobalt, we find the 3% Co grade to have the greatest hardness and abrasion resistance—properties that make it well suited for wiredrawing dies and for cutting tools used in machining of cast iron and other abrasive or gummy materials. The 6% Co grade has moderate values for all properties, and is a good general-purpose carbide material. The 25% Co grade has the greatest toughness, and is used for applications involving heavy impact. Because of its relatively low hardness and abrasion resistance, it is not used for cutting tools. Similar parallels in properties and uses can be drawn both for the fine-grain grades and for the coarse-grain grades containing 6, 10 and 16% cobalt.

Another set of comparisons can be drawn for the grades containing 6% cobalt. All three grades—fine, medium and coarse—are used for cutting tools, but the applications to which they are applied involve different machining conditions and different work materials. The fine-grain material is used for finish to medium-rough machining of ductile, gray and chilled irons and of austenitic stainless steels, high-temperature alloys and nonmetallic materials; the medium-grain material for light to heavy machining of these same wrought work materials; and the coarse-grain grade for heavy to extremely heavy rough machining of such materials. The medium-grain material is widely employed for general-purpose machining because its properties have been found to offer a good practical balance between hardness and toughness. The coarse-grain grade, which has the lowest hardness and abrasion resistance and the best toughness of the three grades, is used where a combination of moderate hardness and high toughness is needed. Similar comparisons

can be made for the grades that contain 10 and 16% Co. In general, decreasing grain size improves abrasion resistance and makes it easier to retain the edge on a cutting tool; increasing grain size improves toughness and makes the cemented carbide more suitable for die applications.

For complex grades, comparisons similar to those drawn for the straight grades are not as readily made. Variations in carbide type, as well as in binder content, affect properties, which in turn influence suitability for specific types of service. In the microstructures the WC particles are angular, whereas those of TiC and TaC are more rounded.

The first two complex grades contain about the same amount of WC, but one contains twice as much binder. The lower-cobalt grade is used for lighter-duty cutting.

The complex grade high in TiC is relatively low in transverse strength and high in resistance to abrasion and cratering. It is used extensively for high-speed, light-duty finishing.

The complex grade highest in cobalt content and in TaC is preferred for hot work tools, in both cutting and shaping of metals.

Hardness. Following the practice of the producers and users of cemented carbides in the United States, Rockwell A hardness is reported in Table 5, and Fig. 5 graphically depicts changes in Rockwell A hardness as temperature is increased. The reduction in hardness between room temperature and about 800 °C (about 1475 °F) for 94WC-6Co with a very fine structure is 7 points on the Rockwell A scale—from 93 to 86 HRA. Lower hot hardness generally signifies lower resistance to deformation at high temperature. Nevertheless, cemented carbides with similar hot-hardness values sometimes show very significant differences in resistance to deformation at high temperature in service.

Abrasion Resistance. Most producers of cemented carbides and other superhard tool materials measure abrasion resistance by subjecting the materials to a test in which a specimen is held against a rotating wheel for a fixed number of revolutions while the specimen and wheel are immersed in a water slurry containing sharp aluminum oxide particles. Comparative rankings are reported, usually in terms of wear ratings based on the reciprocal of volume loss. Two standards apply to abrasive wear testing and the method of reporting results: ASTM B611 and CCPA P112. Not all producers have adopted the ASTM method of abrasive wear testing, so values of abrasion resistance cited by producers vary widely. Because of this variance, it is almost impossible to make valid comparisons among test results reported by different producers. It also is fallacious to use abrasion resistance as a measure of wear resistance of superhard tool materials when they are used for cutting steel or other materials: abrasion resistance in a standard test does *not* correspond directly to wear resistance in machining operations.

Values of comparative abrasion resistance are listed in Table 5. These are relative values only, and are based on a value of 100 for the most abrasion-resistant grade. Comparative abrasion resistance is lowered as cobalt content or grain size is increased. However, abrasion resistance is lower for complex carbides than for straight WC grades having the same cobalt content.

Corrosion resistance of cemented carbides is fairly good, and they may be employed advantageously in certain corrosive environments for applica-

tions where outstanding wear resistance is required. Resistance to water, to oils and other cutting fluids used in machining, and to alkaline attack is excellent; but the cobalt matrix in many cemented carbides is subject to attack by acids. When the cobalt is attacked, accelerated wear develops because of the rapid crumbling of unsupported carbide particles. Corrosion of the cobalt binder also may cause a drastic reduction in strength.

Cemented carbides used as tool materials begin to oxidize when heated in air above about 500 to 600 °C (900 to 1100 °F). However, as measured by weight changes or shape distortion, the rate of oxidation of grades containing large amounts of titanium carbide is much lower than that of straight grades at temperatures as high as 1000 °C (1800 °F).

Toughness. Cemented carbides are brittle materials; usually they will show less than 0.2% elongation in a tensile test.

Values of Charpy impact strength for cemented carbides have little significance and may be misleading. The energy absorbed during impact testing of very hard materials consists mainly of energy absorbed in elastic bending of the specimen and energy absorbed by the testing machine. The portion of total absorbed energy that is a measure of the toughness of a material—namely, the energy of plastic work and the energy necessary to create new surfaces—is only a few percent of the total energy measured.

Typical applications of cobalt-bonded cemented carbides are listed in Table 6.

COATED CEMENTED CARBIDES

Coated cemented carbides are materials consisting of substrates having compositions similar to those of conventional cemented carbides onto which thin coatings of very hard material are deposited, usually by chemical vapor deposition.

Coated cemented carbides have combinations of wear and breakage resistance superior to those of uncoated carbides. Coated cemented carbides became commercially available in 1970. In the United States, about 35% of the cemented carbide tools used for metalcutting in 1979-1980 were coated. Acceptance of coated carbides for metalcutting continues to increase steadily as improvements in the coating itself and in substrate materials are introduced. Coatings now are more uniform in thickness and have less porosity than earlier coatings; adhesion to the substrate has been improved, and undesirable interface reactions have been suppressed. New substrates designed to be coated are being used, which has resulted in improved resistance to breakage and thermal deformation, the two most important substrate properties.

Coated carbides are becoming more widely used because, in high-production machining operations, they permit cutting speeds to be significantly increased. For instance, compared with uncoated carbide tools, the same tool life can be obtained at as much as 50% greater cutting speed with a TiC-coated tool, and as much as 90% greater cutting speed with an Al_2O_3-coated tool. Selection of the proper coating should be based, however, not merely on the cutting speed, but also on the tool-wear mode (edge wear vs flank wear vs cratering). Tool configuration, relief angles and type of cut (roughing or finishing) also influence selection of the proper combination of substrate material and coating.

Table 5. Properties of representative cobalt-bonded cemented carbides

Nominal composition	Grain size	Hardness, HRA	Density Mg/m³	Density lb/in.³	Transverse strength MPa	ksi	Compressive strength MPa	ksi	Proportional limit, compression MPA	ksi	Modulus of elasticity GPA	10⁶ psi
97WC-3Co	Medium	92.5-93.2	15.3	0.55	1590	230	5860	850	2410	350	641	93
94WC-6Co	Fine	92.5-93.1	15.0	0.54	1790	260	5930	860	2550	370	614	89
	Medium	91.7-92.2	15.0	0.54	2000	290	5450	790	1930	280	648	94
	Coarse	90.5-91.5	15.0	0.54	2210	320	5170	750	1450	210	641	93
90WC-10Co	Fine	90.7-91.3	14.6	0.53	3100	450	5170	750	1590	230	620	90
	Coarse	87.4-88.2	14.5	0.52	2760	400	4000	580	1170	170	552	80
84WC-16Co	Fine	89	13.9	0.50	3380	490	4070	590	970	140	524	76
	Coarse	86.0-87.5	13.9	0.50	2900	420	3860	560	700	100	524	76
75WC-25Co	Medium	83-85	13.0	0.47	2550	370	3100	450	410	60	483	70
71WC-12.5TiC-12TaC-4.5Co	Medium	92.1-92.8	12.0	0.43	1380	200	5790	840	1170	170	565	82
72WC-8TiC-11.5TaC-8.5Co	Medium	90.7-91.5	12.6	0.45	1720	250	5170	750	1720	250	558	81
64TiC-28WC-2TaC-2Cr₃C₂-4.0Co	Medium	94.5-95.2	6.6	0.24	690	100	4340	630	...	...	...	...
57WC-27TaC-16Co	Coarse	84.0-86.0	13.7	0.49	2690	390	3720	540	1170	170	441	64

Nominal composition	Tensile strength MPa	ksi	Impact strength J	in.·lb	Relative abrasion resistance(a)	Thermal expansion μm/m·°C at 200 °C	at 1000 °C	μin./in.·°F at 400 °F	at 1800 °F	Thermal conductivity, W/m·K	Electrical conductivity, % IACS
97WC-3Co	...	...	1.13	10	100	4.0	...	2.2	...	121	5.3
94WC-6Co	...	...	1.02	9	100	4.3	5.9	2.4	3.3	...	...
	1450	210	1.36	12	58	4.3	5.4	2.4	3.0	100	7.8
	1520	220	1.36	12	25	4.3	5.6	2.4	3.1	121	10.0
90WC-10Co	...	...	1.69	15	22	...	...	...	...	...	...
	1340	195	2.03	18	7	5.2	...	2.9	...	112	11.4
84WC-16Co	...	...	3.05	27	5	...	...	...	...	...	...
	1860	270	2.83	25	5	5.8	7.0	3.2	3.9	88	9.2
75WC-25Co	1380	200	3.05	27	3	6.3	...	3.5	...	71	9.8
71WC-12.5TiC-12TaC-4.5Co	...	...	0.79	7	11	5.2	6.5	2.9	3.6	35	4.3
72WC-8TiC-11.5TaC-8.5Co	...	...	0.90	8	13	5.8	6.8	3.2	3.8	50	5.2
64TiC-28WC-2TaC-2Cr₃C₂-4.0Co	...	...	...	...	8	...	...	...	...	...	...
57WC-27TaC-16Co	...	...	2.03	18	3	5.9	7.7	3.3	4.3	...	...

(a) Based on a value of 100 for the most abrasion-resistant grade.

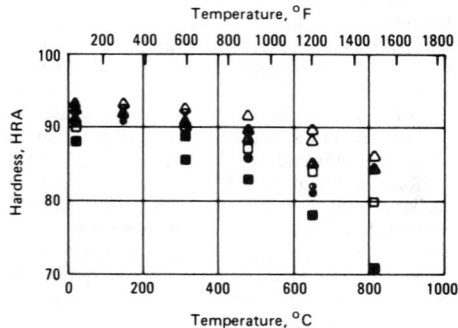

Symbol	Composition	Grain size
Key: o	94WC-6Co	Fine
•	94WC-6Co	Coarse
△	94WC-6Co	Very fine
▲	90WC-10Co	Very fine
□	85WC-15Co	Very fine
■	85WC-15Co	Coarse
◉	79WC-8TiC-4TaC-9Co	Medium
▲	72WC-8TiC-11.5TaC-8.5Co	Medium

Fig. 5. Hot hardness of cemented carbides

Table 6. C-grade classification system for cemented carbides

C grade	Application category
Machining of cast iron, nonferrous and nonmetallic materials	
C-1	Roughing
C-2	General-purpose machining
C-3	Finishing
C-4	Precision finishing
Machining of carbon and alloy steels	
C-5	Roughing
C-6	General-purpose machining
C-7	Finishing
C-8	Precision finishing
Wear-surface applications	
C-9	No shock
C-10	Light shock
C-11	Heavy shock
Impact applications	
C-12	Light impact
C-13	Medium impact
C-14	Heavy impact
Miscellaneous applications	
C-15	Hot weld-flash removal, light cuts
C-15A	Hot weld-flash removal, heavy cuts
C-16	Rock bits
C-17	Cold header dies
C-18	Wear at elevated temperatures and/or resistance to chemicals
C-19	Radioactive shielding, counterbalances and kinetic-energy devices

Titanium carbide coatings were the first coatings used on cemented carbide and are still the most widely used. TiC usually is deposited to a thickness of about 0.005 mm (0.0002 in.), but commercial TiC-coated tool materials vary greatly in thickness of coating and type of bond between coating and substrate. These variations can cause significant differences in metalcutting performance.

Some manufacturers use several different substrates and may offer as many as three or four grades of TiC-coated carbide that differ significantly in performance characteristics. A photomicrograph of the cross section of a typical cemented carbide coated with TiC is shown in Fig. 6. The TiC layer is at the top.

Commercial grades of TiC-coated carbide do not have the superior breakage resistance of uncoated heavy-duty roughing grades of cemented carbide, such as WC-TaC-8TiC-8.5Co, a roughing grade low in TiC and relatively high in cobalt. Coated grades are not as wear resistant as the most wear-resistant uncoated grades, such as 64TiC-32WC-4Co, a grade high in TiC and low in cobalt. Nevertheless, TiC-coated carbides perform well under a wide variety of machining conditions, often producing two to three times as many parts as can be produced using uncoated finishing or general-purpose grades of comparable breakage resistance. This comparison pre-

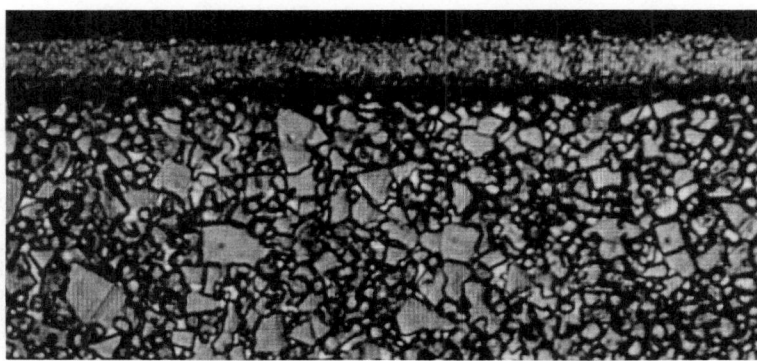

Etched with Murakami's reagent. Magnification, 1500×.
Fig. 6. Cross section of TiC-coated cemented carbide

sumes the use of similar machining conditions and similar criteria for determining the point at which the tool edge no longer can be used.

Aluminum oxide coatings are rapidly gaining acceptance for efficient metal removal at high cutting speeds. Carbide tools coated with aluminum oxide possess an edge strength much higher than that of solid ceramic cutting tools.

Commercial Al_2O_3-coated cemented carbides are available as single-coated or double-coated products. The former consist of an oxide coating 0.005-mm thick that is metallurgically bonded to a specially designed cemented carbide substrate. Double-coated products consist of a thin Al_2O_3 coating over a TiC-coated cemented carbide substrate.

Titanium nitride coatings are claimed to impart superior resistance to crater formation and flank wear in certain metal-cutting applications. TiN-coated products are easily recognized by their gold color.

Multiple coatings consisting of successive thin layers of titanium carbide, titanium carbonitride and titanium nitride are being used in increasing quantities. Multiple coatings are claimed to combine the desirable qualities of both nitride and carbide coatings, and impart to the tools better resistance to edge wear, flank wear and cratering, a combination that cannot be obtained with a single-layer coating.

NICKEL-BONDED TITANIUM CARBIDE

A satisfactory method of cementing titanium carbide using molybdenum carbide and nickel has been developed. The resultant material has good crater resistance, low coefficient of friction and low thermal conductivity. Although penetration hardness is high, abrasion resistance is lower than that of a cobalt-bonded tungsten carbide having equal or lower penetration hardness.

Typically, cemented titanium carbides contain 8 to 25% Ni, 8 to 15% Mo_2C and 60 to 80% TiC. Occasionally, tungsten carbide, cobalt, titanium nitride and other additives may be present in smaller amounts.

STEEL-BONDED CARBIDE

Steel-bonded carbide is a P/M tool material intermediate in wear resistance between tool steels and cemented carbides. Steel-bonded carbide consists of 40 to 55 vol % titanium carbide homogeneously dispersed in a steel matrix.

It is customary to follow tool steel practice and make the entire tool from steel-bonded carbide.

The tool can be joined to a supporting member either of the same material or of steel. This may be done by mechanical fastening, adhesive bonding or brazing. Brazing under vacuum using AWS BNi-8 (a nickel-manganese brazing filler metal) provides a ductile, high-strength joint with a remelt temperature in excess of the temperature recommended for heat treating grade C steel-bonded carbide.

Grade C is used for progressive stamping dies; lamination dies; dies for drawing, bending and curling sheet metal and wire; tube rolls; gages and fixtures. Grade CM is used to make tools for cold and warm forming of heavy-gage stock. Grade SK is used to make hot work rolls and forging dies, including dies for forging hot powder metals. Generally, steel-bonded carbide is not recommended for cutting tools to be used for machining ferrous metals because the hardness drops off too rapidly at the high temperatures developed at the cutting edge. Interface temperatures developed during cutting of nonferrous metals are not too high, on the other hand, and several grades of steel-bonded carbide have been used for machining these materials.

Steel-bonded carbide performs well in severe wear applications involving sliding friction. Its success has been attributed to exceptionally hard, extremely fine, rounded grains of titanium carbide exposed in slight relief at the surface. To provide this condition, the surface preparation of the tool after heat treatment requires lapping with a coarse compound (about 30-μm particle size) to remove any grinding marks and smeared metal, then lapping with a medium compound (15-μm particle size) and finally polishing with a fine abrasive (6-μm particle size) to a mirror finish. Aluminum oxide is satisfactory as the abrasive.

Other grades of steel-bonded carbide are available that have special-alloy matrices to provide corrosion resistance or nonmagnetic properties. Also, some grades can be surface treated to improve the wear resistance of the matrix in applications involving extremely fine abrasives. These surface treatments include nitriding and boriding.

CAST Co-Cr-W-Nb-C ALLOYS

Developed early in the 20th century to provide tools with better capabilities than high speed steel, cast cobalt-chromium-tungsten-niobium-carbon alloys are still produced. In general, as cutting tools, they bridge the gap between high speed steels and cemented carbides. Several cast Co-

Cr-W-Nb-C wear-resistant parts have been used in machinery applications as well.

Use of cast Co-Cr-W-Nb-C cutting tools should be considered:

- Where relatively low surface speeds cause build-up with cemented carbides
- Where machines lack the power or rigidity to use cemented carbides effectively
- Where higher production is desired than is possible with high speed steel tools
- For multiple-tool operations where surface speed of one or more operations falls between the recommended speeds for high speed steel and carbide tools
- For short runs on automatic equipment where form grinding of carbide tools is excessively costly
- For machining rough surfaces of castings where the surfaces contain abrasive materials such as residual sand, surface oxides, slag or refractory particles.

Tools made of cast Co-Cr-W-Nb-C alloys usually are not recommended for light, very fast finishing cuts.

CERAMICS

Ceramic cutting-tool materials use aluminum oxide as the base material. They are available as indexable inserts that can be used in the same holders as those used for cemented-carbide inserts. However, because ceramics are more brittle than carbides, extra care must be used to ensure that the inserts are firmly seated. Also, tool overhang should be kept at 50 to 100% of toolholder thickness.

The many available ceramic tools fall into three general groups:

- *Group A-1:*
  Al_2O_3 with up to about 10% of other oxides or carbides, primarily those of titanium, magnesium, molybdenum, chromium, nickel or cobalt. The mixtures are cold pressed and sintered.
- *Group A-2:*
  Essentially pure Al_2O_3, hot pressed
- *Group A-3:*
  Al_2O_3 plus 25 to 30% of a refractory carbide such as titanium carbide, hot pressed.

Typical properties of these groups are shown in Table 7. Also, the properties of a specific example of a Group A-1 material have been included, for which the microstructure is shown in Fig. 7.

Because ceramic tools are predominantly oxides, they are not subject to the oxidation that limits the usefulness of cemented carbides at high temperatures in air.

Ceramic tool materials retain their resistance to wear and deformation at much higher temperatures than the best cemented carbides. Consequently, ceramic tools generally can cut for acceptable periods of time at speeds much higher than those possible with cemented carbide tools. Coolants should not be used when cutting with ceramic tools.

POLYCRYSTALLINE CUBIC BORON NITRIDE

In 1973, polycrystalline composite cubic boron nitride (CBN) cutting tools were introduced. By use of high-pressure, high-temperature pro-

Table 7. Typical properties of ceramic tool materials

Property	Group A-1 General	Group A-1 Example(a)	Group A-2	Group A-3
Hardness: HRA	93-94	93-94	93-94	93-94
Density:				
Mg/m^3	3.96-3.98	4.1	4.0	4.24
$lb/in.^3$	0.142-0.143	0.148	0.144	0.153
Transverse strength:				
MPa	480-690	620	640	760
ksi	70-100	90	92.5	110
Compressive strength:				
MPa	3790-4480	2140(b)	4140	3930-4070
ksi	550-650	310(b)	600	570-590
Modulus of elasticity:				
GPa	390	400	390	...
10^6 psi	57	58	57	...
Impact strength:				
J	...	0.23	...	...
in. · lb	...	2	...	...
μm/m·°C	...	6.1(d)	7.2	7.7
Thermal expansion(c): μin./in.·°F	...	3.4(d)	4.0	4.3
Thermal conductivity:				
At room temperature:				
W/m·K	29	...	...	17-21
Btu/ft·°F	17	...	...	10-12
At 100 °C (212 °F):				
W/m·K	22	...	29	...
Btu/ft·h·°F	13	...	17	...
At 450 °C (850 °F):				
W/m·K	11	...	...	...
Btu/ft·h·°F	6.5	...	...	...
At 600 °C (1100 °F):				
W/m·K	...	...	...	14.7
Btu/ft·h·°F	...	...	...	8.4

(a) $89Al_2O_3$-11TiO, cold pressed and sintered. (b) Proportional limit. (c) At 21 to 200 °C (70 to 400 °F). (d) 8.3 μm/m·°C (4.6 μin./in.·°F) at 21 to 980 °C (70 to 1800 °F).

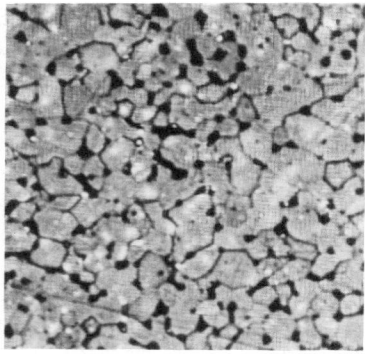

Phosphoric acid etch. Magnification, 750×. For material properties, see A-1 example in Table 7.

Fig. 7. Microstructure of $89Al_2O_3$-11TiO grade A-1 ceramic tool material

cesses, a layer of CBN is bonded to a cemented carbide substrate. The CBN is held together primarily by CBN-CBN intercrystalline bonds.

Most frequently, composite CBN tools are supplied in the same shapes as those used for cemented carbides. In many cases, CBN inserts can be brazed into a steel holder using the same procedures employed for cemented carbides, as long as special care is taken to avoid overheating the structure and to prevent molten flux from contacting the CBN layer.

Because the cost of a CBN tool is several times that of a cemented carbide tool, it is usually necessary to regrind CBN tools with diamond wheels and reuse them to minimize the cost per tool edge.

Because of their high hardness, their ability to hold this hardness and the fact that they resist oxidation at much higher temperatures than carbides, composite CBN tools are used effectively to cut difficult-to-machine superalloys at speeds several times higher than those possible with cemented carbides. The best conditions for their use are in the self-induced thermal machining mode.

In addition to nickel-base and cobalt-base high-temperature alloys, other ferrous materials—including chilled cast iron, Meehanite cast iron, tool steels (M2, M42, D2, A2, S5 and O1) and other steels (1055, 8620, 52100 and 4140) hardened to 50 to 70 HRC—have been machined successfully with CBN tools.

DIAMOND

Diamond, the tetrahedral form of carbon, is the hardest and most scratch-resistant material known. Its scratch hardness number is 10 on Mohs' scale. It will scratch all other materials and will be scratched by none. Similarly, its penetration hardness is 5000 to 12 000 HK, roughly twice that of the next hardest material.

These characteristics make diamond very attractive as a tool material. However, industrial-grade natural single-crystal diamonds are expensive even in small sizes. In addition, diamonds are very brittle and cleave easily along certain crystallographic planes. Also, diamond starts to oxidize rapidly at 650 °C (1200 °F), and at atmospheric pressure it reverts to graphite above 1500 °C (2700 °F).

These properties and the fact that carbon dissolves rapidly in iron at high temperatures make diamond unsatisfactory for machining ferrous alloys.

However, diamond tools are used effectively in machining high-silicon cast aluminum alloys, copper and its alloys, sintered cemented tungsten carbides, rubber impregnated with silica glass, glass-fiber/plastic and carbon/plastic composites, and high-alumina ceramics.

Diamonds are used extensively in resin-bonded and metal-bonded grinding wheels. In addition, diamond dies are used for drawing fine wire. Other common industrial uses of diamonds are as tools for dressing abrasive grinding wheels, as diamond cutoff saws and laps (when dispersed in a metal matrix) and as a polishing abrasive when in very finely divided form.

The relative abrasion resistance and Knoop hardness values of laminated diamond, natural diamond and cemented 94WC-6Co tool blanks are compared in the following table:

Tool material	Hardness, HK	Relative abrasion resistance
Laminated diamond/carbide composite	5500-8000	250
Natural diamond single crystal	8000-12 000	96 to 245
Cemented 94WC-6Co	1800-2200	2

The relative abrasion resistance is based on the time, in minutes, required to generate a specific size of wear land on the material when turning a siliceous hard rubber commonly used as a coating for steel rolls in the paper industry.

Distortion in Tool Steels

DISTORTION in tool steel parts includes all irreversible changes in size and shape that result from processing, from heat treatment, and from temperature variations and loading in service.

Changes in size or shape of tool steel parts may be either reversible or irreversible. Reversible changes are those caused by stressing in the elastic range or by temperature variations that neither cause changes in the metallurgical structure nor induce stresses that exceed the elastic range. Under such conditions, the initial dimensional values can be restored by a return to the original state of stress or temperature.

The upper limit of reversible dimensional change in a tool steel is determined by the stress required to initiate deformation (that is, the elastic limit corresponding to a preselected value of plastic strain), the elastic deformation per unit stress (modulus of elasticity), the effect of temperature on these properties, the coefficient of thermal expansion and the temperature-time combinations at which stress relief and phase changes occur.

For practical purposes the modulus of elasticity of all tool steels, regardless of composition or heat treatment, is 210 GPa (30×10^6 psi) at room temperature. Therefore, if a tool steel part deforms excessively under service loading but returns to its original dimensions when the load is removed, a change in grade or type of tool steel or in heat treatment will not be useful. To counteract excessive elastic distortion it is necessary to (a) reduce the applied stress by increasing the section size or (b) use a tool material with a higher modulus of elasticity (such as cemented tungsten carbide.)

Irreversible changes in size or shape of tool steel parts are those caused by stresses that exceed the elastic limit or by changes in metallurgical structure (most notably, phase changes). Such irreversible changes sometimes can be corrected by thermal processing (annealing, tempering or cold treating) or by mechanical processing to remove excess material or to redistribute residual stresses.

NATURE AND CAUSES OF DISTORTION

Distortion is a general term encompassing all irreversible dimensional changes. There are two main types: size distortion, which involves expansion or contraction in volume or linear dimensions without changes in geometrical form; and shape distortion, which entails changes in curvature or angular relations, as in twisting (warpage) or bending, and nonsymmetrical changes in dimensions. Usually, both size distortion and shape distortion occur during any heat treating operation.

SIZE DISTORTION IN TOOL STEELS

Typical volume percentages of martensite, retained austenite and undissolved carbides are given in Table 8 for four different tool steels quenched from their usual austenitizing temperatures.

Typical changes in linear dimensions for several tool steels are given in Table 9. As shown in this table, some tool steels, such as A10, show very little size change when hardened and tempered over the entire range from 150 to 600 °C (300 to 1100 °F).

Other types, such as the M2 and M41 high speed steels, expand about 0.2% (2 mm/m, or 0.002 in./in.) when hardened and tempered in the usual range of 540 to 595 °C (1000 to 1100 °F) to develop full secondary hardness. Although the information in Table 8 is useful in comparing size distortion in several tool steels, the factor of shape distortion makes it impossible to use these data alone to predict dimensional changes of a particular tool made from any of these steels.

SHAPE DISTORTION IN TOOL STEELS

In considering shape distortion, it is important to recognize that the strength of any tool steel decreases rapidly above about 600 °C (1100 °F). At the austenitizing temperature, the yield strength is so low that plastic deformation often occurs simply from the stresses exerted on the part by its own weight. Therefore, long parts, large parts and parts of complex shape must be properly supported at critical locations to prevent sagging at the hardening temperature.

Rapid heating increases shape distortion, especially in large tools and in complex tools containing both light and heavy sections. If the rate of heating is high, light sections will increase in temperature much faster than heavy sections. Likewise, the outer surfaces will increase in temperature much faster than the interior, especially in moderate to heavy sections. Differences in thermal expansion due to the differences in temperature will be sufficient to set up large stresses in the material. Under these stresses, the hotter regions will deform plastically, thereby relieving the thermally induced stress.

On continued heating, the hotter portions will begin to level off at the furnace temperature, while the cooler portions will continue to increase in temperature. This produces a decrease in thermal differential, which in turn causes at least a partial reversal in thermal stress because of the plastic deformation that took place when the temperature differential was high. This may or may not cause the part to undergo further plastic deformation, but if it does, the deformation will most likely be lesser in extent than the deformation that took place when the temperature differential was high, and will most likely be in a different direction.

Slow heating minimizes distortion by keeping temperature differentials low throughout the heating cycle. Ideally, all heat treatment of tool steel parts should start from a cold furnace to provide the greatest freedom from shape distortion during heating. Starting from a cold furnace is neither very practical nor energy efficient unless heat treating is being done in a vacuum furnace. For heat treating in fused salt or an atmosphere furnace, preheating of parts at an intermediate temperature prior to heating them to the austenitizing temperature provides a useful compromise.

On cooling to form martensite, large temperature differences between surface and interior, and between light and heavy sections, can cause severe shape distortion. This problem is most likely to arise if the hardenability of the steel is so low that a high cooling rate is required to obtain full hardness. In that event, it may be best to substitute a high-hardenability, air-hardening tool steel, especially when a large or complex part is being made.

However, if lower-hardenability steels that require liquid quenching are used, fixturing and pressure die quenching will help minimize distortion. Long, symmetrical parts should be fixtured, and should be quenched in the vertical position and agitated vertically while completely submerged in the quenching medium.

SPECIAL TECHNIQUES FOR CONTROLLING SHAPE DISTORTION

Besides being reduced through control of rates of heating and cooling, as discussed above, shape distortion can be reduced by quenching locally instead of quenching the entire part, or by using flame, induction, electron beam or laser methods to harden only that portion of the tool that must be hardened.

Special hardening procedures such as martempering and austempering may also be useful for controlling distortion.

Controlling out-of-roundness is important for certain precision applications, such as class C and D cutting hobs made of high speed steels. Class C and D hobs must be held to close size limits because they are not ground to size after heat treatment, but rather are used in the unground condition.

Normal size distortion in hardening and tempering can be accommodated by making the tool slightly oversize or slightly undersize, as required, and then heat treating. High speed steel bars, however, have been observed to go out-of-round as much as 0.05 mm (0.002 in.) when conventionally processed. This out-of-roundness problem can be combated by using specially processed wrought bars or bars made from hot isostatically pressed powders, which maintain the best possible symmetry during conventional heat treatment.

Stabilization involves reducing the amount of retained austenite that can slowly transform and thus produce distortion if the material is heated or subjected to stress. Stabilization also reduces internal (residual) stress, which in turn makes distortion in service due to stress relaxation less likely to occur. Stabilization is most important for tools

Table 8. Microconstituents in four tool steels after hardening

Steel	Hardening treatment	As-quenched hardness, HRC	Martensite, vol %	Retained austenite, vol %	Undissolved carbides, vol %
W1	790 °C (1450 °F), 30 min; WQ	67.0	88.5	9	2.5
L3	840 °C (1550 °F), 30 min; OQ	66.5	90	7	3.0
M2	1225 °C (2235 °F), 6 min; OQ	64	71.5	20	8.5
D2	1040 °C (1900 °F), 30 min; AC	62	45	40	15

Table 9. Typical dimensional changes in hardening and tempering

Tool steel	Hardening treatment °C	°F	Quenching medium	Total change in linear dimensions, %, after quenching	°C / °F	150 / 300	205 / 400	260 / 500	315 / 600	370 / 700	425 / 800	480 / 900	510 / 950	540 / 1000	565 / 1050	595 / 1100
O1	816	1500	Oil	0.22		0.17	0.16	0.18	...	...	...	...	...	...	...	...
O1	788	1450	Oil	0.18		0.09	0.12	0.13	...	...	...	...	...	...	...	...
O6	788	1450	Oil	0.12		0.07	0.10	0.14	0.10	0.00	−0.05	−0.06	...	−0.07	...	...
A2	954	1750	Air	0.09		0.06	0.06	0.08	0.07	...	0.05	0.04	...	0.06	...	...
A10	788	1450	Air	0.04		0.00	0.00	0.08	0.08	0.01	0.01	0.02	...	0.01	...	0.02
D2	1010	1850	Air	0.06		0.03	0.03	0.02	0.00	...	−0.01	−0.02	...	0.06	...	...
D3	954	1750	Oil	0.07		0.04	0.02	0.01	−0.02	...	...	...	...	...	...	...
D4	1038	1900	Air	0.07		0.03	0.01	−0.01	−0.03	...	−0.4	−0.03	...	0.05	...	...
D5	1010	1850	Air	0.07		0.03	0.02	0.01	0.00	...	0.3	0.03	...	0.05	...	...
H11	1010	1850	Air	0.11		0.06	0.07	0.08	0.08	...	0.3	0.01	...	0.12	...	...
H13	1010	1850	Air	−0.01		...	...	...	...	...	...	0.00	...	0.06	...	...
M2	1210	2210	Oil	−0.02		...	...	...	...	...	...	...	−0.06	0.10	0.14	0.16
M41	1210	2210	Oil	−0.16		...	...	...	...	...	...	...	−0.17	0.08	0.21	0.23

that must retain their size and shape over long periods of time.

If the tool steel chosen provides the required hardness after tempering at a relatively high temperature, it is possible to reduce the amount of retained austenite and the internal stress by multiple tempering. Initial tempering reduces internal stress and conditions the retained austenite so that it can transform to martensite on cooling from the tempering temperature. Usually, a second or third retempering treatment is necessary to reduce the internal stress set up by the transformation of retained austenite.

Single or repeated cold treatment to a temperature below M_f will cause most of the retained austenite to transform to martensite in plain carbon or low-alloy tool steels that must be tempered at low temperatures to achieve the hardness required for the application. Cold treatment may be applied either before or after the first temper, but if the tools tend to crack because of the additional stress induced by dimensional expansion during cold treatment, it is generally prudent to apply cold treatment after the tools have been tempered the first time. When cold treatment is applied after the first temper, the amount of retained austenite that transforms on cold treatment may be considerably less than would be expected, because some of the austenite may be stabilized by tempering prior to cold treating.

Cold treatment is usually done in a commercial refrigeration unit capable of attaining temperatures of −70 to −95 °C (−100 to −140 °F). Tools must be retempered promptly after returning to room temperature following cold treatment, to reduce internal stress and increase the toughness of the newly formed martensite.

For some tools, a small percentage of retained austenite is desirable to improve toughness and provide a favorable internal stress pattern that will help the tool withstand service stresses. For these tools, little or no stabilization may be preferred, and a full stabilizing treatment may actually result in tools that are unfit for service or only marginally able to perform their required functions.

Tool Materials for Special Applications

Cutting Tools

CUTTING TOOLS include all of the various styles of cutters used in machining of metals, plastics, woods and other machinable structural materials. The most common styles include single-point tools, drills, reamers, taps, threading dies, milling cutters, end mills, broaches, saws and hobs. For many of these styles, the actual cutting edge is on a detachable portion of the tool called an "insert." Inserts generally are made of materials different from those of the tool holders in which they are affixed.

SINGLE-POINT CUTTING TOOLS

Single-point cutting tools are those types of cutters having essentially only one cutting edge and/or one corner in contact with the workpiece throughout a given cutting cycle. Single-point cutting tools are used for turning, boring, shaping, planing and threading.

A substantial portion of all single-point tools used in the United States consist of superhard tool bits (inserts) mounted in tool holders usually made of carbon steel or alloy steel.

High speed steels are used for single-point cutting tools in the form of both solid tool bits and inserts. Molybdenum types M2 and M4 generally are recommended for solid tool bits used for general-purpose machining of metals whose hardness is below about 250 HB. For machining harder metals, M42 and T15 high speed steels, which have better hot hardness, are generally preferred. Types T4 and T5 are relatively common alternatives for types M2 and M4 in single-point tools for machining cast irons and copper alloys.

Cemented carbides are used largely for inserts. Single-point carbide tools are generally preferred for high-volume production machining, where productivity is significantly enhanced by high machining speeds, relatively deep cuts and relatively high feed rates, and where it is highly desirable to change tool bits only at infrequent intervals. With single-point carbide tools it is most important to establish optimum machining speed, because this factor has the greatest effect on tool life.

Specific types of carbide are preferred for single-point turning of certain metals. Straight grades of tungsten carbide are intended primarily for use in machining cast iron and nonferrous metals. When used for machining steel, the straight grades tend to crater rapidly, and thus it is often better to select a coated carbide or a complex grade containing titanium carbide instead.

Ceramic inserts, such as those of solid aluminum oxide, permit very high machining speeds to be used with no sacrifice in tool life. They also produce finer finishes than those obtained with other insert materials. Ceramics are weaker than carbides or coated carbides, and can be used only for applications where impact loading is low.

Diamond inserts may be either single-crystal natural diamonds or carbide substrates under a layer of randomly oriented fine-grain polycrystalline synthetic diamond. Single-crystal natural diamond inserts have outstanding wear resistance, but they cannot withstand high shock loading. Diamond-carbide composite inserts combine excellent wear resistance with good resistance to shock loading.

Diamond tools are preferred for machining soft, abrasive nonferrous metals, and abrasive nonmetallic materials such as cemented tungsten carbide, unfired ceramics, filled plastics, rubber, carbon and graphite.

Boron nitride bonded to a cemented carbide substrate offers exceptionally high wear resistance and edge life in the machining of high-temperature alloys and hardened ferrous metals. Next to diamond, cubic boron nitride is the hardest known material. This accounts for its cutting properties, and use of a cemented carbide substrate gives the insert satisfactory resistance to shock.

DRILLS

Drilling of holes for assembly of metal parts is performed principally with twist drills made of high speed steel. Carbide drills of several designs are sometimes used for drilling cast iron, high-silicon aluminum alloys and abrasive materials, as well as certain ceramic and plastic materials of high hardness. For drilling holes of large diameter—about 25 mm (1 in.) or greater—in-dexable carbide insert-style drills are sometimes used. The number of holes drilled with all types of carbide drills is still a small percentage of the total number of holes drilled.

High Speed Steel Drills. High speed steels M1, M2, M7 and M10, which have the highest strength and toughness, are used for most of the drills manufactured in the United States. Each of these steels has a specific range of carbon content normally associated with it. Steels with carbon contents at the upper end of the range normally are used for drills requiring less toughness and more resistance to abrasion. Drills subject to shock during use often are produced from steels with carbon contents at the lower end of the normal range, to provide better toughness.

Cemented Carbides for Drills. Drilling of holes in concrete, glass and various ceramic materials with carbide-tipped "masonry" drills is a well-known application for cemented carbide tool materials. Solid cemented carbide twist drills of special design are being used in large quantities for drilling printed circuit boards made of glass-filled epoxy that is abrasive and comparatively low in strength.

Carbide-tipped die drills have been used successfully on some difficult-to-machine heat-resistant materials, especially where the machine setup is very rigid.

Straight tungsten carbides containing up to 6% cobalt are used successfully in tipped twist drills for drilling cast iron and nonferrous metals.

REAMERS

Reamers are cutting tools used for finishing holes that must meet stringent finish and size-tolerance requirements that cannot be satisfied by simple drilled holes.

High speed steel, with its high resistance to softening at elevated temperatures, has replaced carbon steel for all except hand reamers.

In choosing materials for machine reamers, high hardness and high abrasion resistance are the most important properties. The general-purpose high speed steels (M1, M2, M7, M10 and T1) at high hardness levels, and the high-vanadium types (M3, M4 and T15), have been used successfully. The latter steels have greater resistance to abrasion than the lower-vanadium types. The very high resistance to softening at elevated temperatures provided by cobalt-containing high speed steels such as M33, M42, M6 and T5 is less often required in reaming than in drilling, because heat is more easily controlled in reaming.

Solid cemented carbide reamers and carbide-tipped reamers have been used widely and successfully in reaming almost all metals. General-purpose and harder grades of carbide are used most commonly.

TAPS

Taps are tools used for cutting or forming internal screw threads. Selection of tool materials for taps usually is done by the tap manufacturer. Three families of materials generally are considered: (a) carbon and alloy tool steels, (b) high speed steels and (c) cemented carbides.

Carbon and alloy tool steels are suitable for taps used for hand tapping or other low-speed, light-duty applications.

High speed steels are used for taps that must cut efficiently at high speeds and thus must resist softening under the extreme heat generated at the cutting edge of the tool. Most tap manufacturers use M1, M2, M7 and M10 for general-purpose taps.

For special tap designs that require greater resistance to abrasion, type M3 (class 2) or type M4 is generally used. Where requirements include high abrasion resistance and/or very high resistance to softening at elevated temperatures, types T15 and M42 are the most popular choices. Three common surface treatments — nitriding, oxide coating and hard chromium plating — are often used to improve tap life.

Cemented carbides can be considered for extremely abrasive applications, such as tapping of filled plastics and certain grades of cast iron.

END MILLS

An end mill is a shank-type milling cutter with cutting edges on its periphery and end surface. The shank may be either straight or tapered. These tools are available in diameters ranging from 0.8 to 75 mm ($^1/_{32}$ to 3 in.), and may have one or more cutting teeth (most have 2, 4 or 6 teeth). Larger sizes are made as special items on customer request. End teeth may be of the noncenter type or may be designed to cut to the axis of the tool to permit plunge cutting.

End-Mill Materials. The majority of end mills larger than 16 mm ($^5/_8$ in.) in diameter are made from high speed steel bars, which in the annealed state may be easily machined to shape, heat treated and finished by grinding. End mills smaller than 16 mm in diameter are commonly made from hardened cylindrical blanks of high speed steel into which the flutes are ground.

The majority of workpieces machined with high speed steel end mills have hardnesses below 300 HB. For these applications, end mills made from the more popular and less costly general-purpose tool steels such as M1, M2, M7 and M10 generally have proved satisfactory on the bases of both economy and performance.

For work materials that are difficult to machine and that have hardnesses from 350 to 450 HB, end mills made of cobalt high speed steels such as T15, M33 and M42 are most effective.

Cemented carbide end mills have the ability to withstand higher cutting temperatures and greater abrasion than high speed steel end mills but are somewhat less resistant to shocks caused by interrupted cuts or voids. Carbide end mills are used most often for machining nonferrous alloys; nonmetallic materials; work materials with hardnesses exceeding 450 HB; and materials with highly abrasive scaled surfaces, such as sand castings.

The straight grade of tungsten carbide containing 6% cobalt is used most commonly for solid carbide end mills and for brazed-tip tools, both of which are used for cutting cast iron, titanium alloys, nonferrous metals, stainless steels and plastics. A complex grade containing 72 to 73% WC, 7 to 8% TiC, 11.5 to 12% TaC and 8 to 8.5% cobalt binder is used for end mills in steel-cutting applications.

MILLING CUTTERS

Milling cutters that are relatively small and of complex configuration generally are made of high speed steels. The cutting teeth on tools of this type often have helical cutting edges, deep radial and/or axial gashes, irregular profiles or thin web sections. These complex configurations can be most readily obtained by machining the tool as an integral unit from annealed high speed steel, heat treating it to a suitable hardness, and then grinding it to size.

There are many special applications and operating conditions for which milling cutters are constructed of materials other than solid high speed steel. For economy, large-diameter cutters, and cutters of simple configuration used in high-production applications, often comprise high speed steel blade inserts attached to low-cost alloy steel bodies.

Because of their high cost, cemented carbides are used predominantly in those milling cutters into which carbide tips can be brazed or carbide blades inserted.

For general-purpose cutting of steel, the grades containing complex carbides are widely used in all types of milling cutters. Other types of workpiece materials, such as cast irons, various brasses, aluminum alloys and fiber composites, are most productively cut using the straight tungsten carbide grade containing 6% Co. This straight tungsten carbide also performs well in cutting stainless steels of the 300 series.

Tool steel milling cutters used for cutting plain carbon and low-alloy steels, cast irons and nonferrous alloys, where workpiece hardness does not exceed 30 HRC, normally are made of high speed steels such as M1, M2, M7 and M10. T1 cutters are available from most cutter manufacturers, but they are high in initial cost.

At intermediate hardness levels from 30 to 35 HRC, the high-vanadium grades M3 and M4 provide increased tool performance. As the hardness of the workpiece increases beyond 35 HRC, tool life of the general-purpose grades drops very rapidly. For workpiece hardness levels from 35 to about 45 HRC, cobalt high speed steels such as M42 and T15 are recommended. No high speed steel, however, is recommended for cutting metals with hardnesses much above 45 HRC; for these high-hardness materials, carbide tools must be used. In general, carbide milling cutters can be used at cutting speeds three to six times those permitted with high speed steel cutters.

As a group, nickel and cobalt high-temperature alloys are very difficult to machine, and best results are obtained by using cobalt-bearing high speed steel cutters, low cutting speeds and heavi-er-than-normal feed rates. Carbide cutters normally are not recommended because they are prone to chipping of the cutting edge.

HOBS

A hob is a type of milling cutter used to generate a repeating form about a center, such as in cutting of gear teeth, spline teeth and serrations. The hob and workpiece must rotate and mesh in a specific timed relationship, and thus the cutting teeth on the hob follow a helical or thread pattern around the periphery of the tool. This is the feature that distinguishes a hob from a milling cutter.

The majority of hobs are made from high speed steels, although hobs for special applications have been manufactured from cemented tungsten carbide, cast cobalt-chromium-tungsten alloys, and even some low-alloy tool steels. Virtually all types of high speed steel have been used, but the most widely used is type M2 in either the standard or the high-carbon version.

General-purpose types like M2 offer good combinations of edge strength and wear resistance, but high speed steels such as M3 and M4, which contain more carbon and vanadium, are used for hobbing harder and more abrasive materials. For applications that require increased resistance to softening at elevated temperatures, cobalt-bearing grades such as M42 and T15 are more suitable.

A recent innovation has been the production of high speed steel by powder metallurgy. A broad range of grades is available, and the same application guidelines apply to them as apply to wrought tool steels.

Shearing and Slitting Tools

MOST SHEAR BLADES are solid, one-piece blades made of tool steel. However, some are composite tools that consist of tool-material inserts in heat treated medium-carbon or low-alloy steel backings.

COLD SHEARING AND SLITTING OF METALS

Blade materials recommended for cold shearing of various metals are presented in Table 1, and blade materials for rotary slitting are given in Table 2.

Tool materials vary in toughness and wear resistance, and the metals being sheared vary in hardness and resistance to shearing. If the material to be sheared is very thin and of relatively low hardness, the shear-blade material can be low in toughness but must have optimum wear resistance. For shearing material of greater thickness and higher hardness, it may be necessary to decrease blade hardness, or change to a less wear-resistant blade material or to a shock-resistant tool steel having a hard case over a tough core, to obtain the toughness needed to resist edge chipping.

HOT SHEARING OF METALS

Shearing is done at elevated temperatures when the work material is thick and resistant to shearing or when hot shearing is otherwise desirable as part of the manufacturing process. The strong

Table 1. Recommended blade materials for cold shearing of flat metals

Material to be sheared	Blade material(s) for work-metal thickness of:		
	6 mm (¹/₄ in.) or less	6 to 13 mm (¹/₄ to ¹/₂ in.)	13 mm (¹/₂ in.) and over
Carbon and low-alloy steels up to 0.35% C	D2, A2	A2, A9	S2, S5, S6, S7
Carbon and low-alloy steels, 0.35% C and over	D2, A2	A9, S5, S7	S2, S5, S6, S7
Stainless steels and heat-resisting alloys	D2, A2	A2, A9, S2	S2, S5, S6, S7
High-silicon electrical steels	D2, T15, cemented carbide inserts(a)	S2, S5, S7	(b)
Copper and aluminum alloys	D2, A2	A2	S2, S5, S6, S7
Titanium alloys	D2	...	...

(a) Carbide inserts usually are brazed to heat treated medium-carbon or low-alloy steel backings. (b) Seldom sheared in these thicknesses.

Table 2. Recommended blade materials for rotary slitting of flat metals

Material to be sheared	Blade material(s) for work-metal thickness of:		
	4.5 mm (³/₁₆ in.) or less	4.5 to 6.5 mm (³/₁₆ to ¹/₄ in.)	6.5 mm (¹/₄ in.) or more
Carbon, alloy and stainless steels	D2	D2, A2, A9	A9, S5, S6, S7
High-silicon electrical steels	D2, M2	D2	...
Copper and aluminum alloys	A2, D2	A2, D2	A2, S5, S6, S7
Titanium alloys	D2, A2	...	...

secondary hardening of group H tool steels provides sufficient resistance to softening to make them useful for shear blades operating at temperatures up to 425 °C (800 °F).

Hardness of blades for hot shearing varies considerably with conditions such as thickness and temperature of the metal being sheared, and type and condition of available equipment. However, hardness is usually kept within the range from 38 to 48 HRC.

MATERIALS FOR MACHINE KNIVES

Unlike materials for metal-slitting applications, materials for knives used in cutting papers, films and foils usually are selected on the basis of cost and wear resistance, without consideration of toughness.

Score cutters are most commonly made of 52100 steel hardened to 60 to 62 HRC—a hardness level that is adequate for most operations and that will not result in scoring of an opposing platen sleeve or hardened roll. Such sleeves and rolls are made of 52100 or carburized steel, hardened to not less than 60 HRC. Alternative materials for score cutters are O7 and D2 tool steels.

Shear cutters can be produced from a much wider array of alloys, the choice being influenced by such factors as tool design, material to be cut, machine design, and maintenance limitations. The entire range of tool steels (including the popular 52100 and other low-alloy types plus the higher-alloy group D, group M and group T steels), as well as specialty tool steels and cemented carbides, can be considered. For standard applications, 52100 and O1 are selected most often, chiefly because they are readily available as sheet, bar stock or tubing.

Burst cutters are, by design and function, fairly thin knives. Material selection thus is restricted to alloys that are readily available in thin sheet, a group that includes 52100 steel, 1075 steel and razor-blade stock. Except in those applications where resistance to elevated temperature is required (as in core cutting with a stationary blade), use of high speed steels is rarely economical.

Single-knife cutters are used to reduce wide rolls of paper, foam or textile to narrower rolls without unwinding and rewinding. These knives have long, thin, one-sided or two-sided bevels and are kept sharp by means of one or more grinding wheels situated on the machine, which are activated either automatically (for grinding at specific intervals) or manually (for grinding as required).

Under normal circumstances, such knives are made from L2, L6 or D2 tool steel. Certain applications of single knives, including slicing of foam or impregnated fabrics, cause the thin bevel to heat up considerably as it penetrates the material being slit. In these cases, an alloy that has higher resistance to elevated temperatures, such as M2 or T1 tool steel, may be required.

A standard paper-trimmer knife consists of a carbon steel backing and an insert that provides the actual cutting edge. The insert material can range from O1 to M2 to one of the group T high speed steels. For most applications, O1 is ideal with respect to initial cost, ease of maintenance, and adequate performance between resharpenings.

Sheeter and cutoff knives are similar in function to trimmer knives and frequently are made with cutting-edge inserts of M2 high speed steel hardened to 62 to 64 HRC. Under optimum conditions, life of these blades is very long—50 to 100 million cuts between resharpenings.

Some sheeter or cutoff blade designs are too narrow in cross section to allow the use of inserts; such knives are most often made of solid 52100, O1 or D2 tool steel.

Blanking and Piercing Dies

BLANKING AND PIERCING DIES include the punches, dies and related components used to blank, pierce and shape metallic and nonmetallic sheet and plate in a stamping press. The primary measure of the performance of a die material in blanking or piercing service is the number of acceptable parts that can be produced.

Sectional views of the blanking dies and the blanking and piercing punches used for making simple parts are shown in Fig. 1. More complex parts require notching and compound dies.

MATERIALS FOR SPECIFIC TOOLS

Punches and Dies. Typical materials for punches and dies used for blanking parts of different sizes and degrees of severity from several different work materials about 1.3 mm (0.050 in.) thick, in various quantities, are given in Table 3. (Sketches of typical parts are presented in Fig. 2.) Typical materials for the punches and dies used to shave several work materials of this same thickness in various quantities are given in Table 4.

Tables 3 and 4 may be used to select punch and die materials for parts made of sheet thicker or thinner than the 1.3 mm used in the examples. For sheet of greater thickness, use the punch and die material recommended for the next greater production quantity than the quantity actually to be made (the column to the right of the actual production quantity in the table). Similarly, for sheet of lesser thickness, use the punch and die material recommended for the next lower production quantity (shift one column to the left of the actual production quantity).

Typical materials for perforator punches used on several different work materials are given in Table 5. The usual limiting slenderness ratio (punch diameter to sheet thickness) for piercing aluminum, brass and steel is 2.5:1 for unguided punches and 1:1 for guided punches. The limiting slenderness ratio for piercing spring steel and stainless steel ranges from 3:1 to 1.5:1 for unguided punches and from 1:1 to 0.5:1 for accurately guided punches.

Table 6 gives typical materials for perforator bushings of all three types (punch holder, guide or stripper, and perforator or die). These recommendations are particularly applicable to precision bushings—for instance, where the outside diameter is ground to a tolerance of −0, +0.008 mm (−0, +0.0003 in.) and is concentric with the inside diameter within 0.005 mm (0.0002 in.) TIR. The hardness of W1 bushings should be 62 to 64 HRC, and that of D2 bushings, 61 to 63 HRC.

Die plates and die parts that hold inserts normally are made of gray iron, alloy or tool steel. For stamping thick sheet or hard materials, either class 50 gray iron, or 4140 steel heat treated to a hardness of 30 to 40 HRC, should be used. For long-run die plates for stamping thick or hard materials, steels such as 4340 and H11 are preferred when inserts are pressed into the die plates, and 4340 is nearly always used when inserts are screwed in. Die plates for stamping thin or soft sheet may be made of class 25 or class 30 gray iron or of mild steel.

Secondary Tooling. Punch holders and die shoes for carbide dies are made of high-strength gray iron or low-carbon steel plate. Yokes for retaining carbide sections usually are made of O1 tool steel hardened to 55 to 60 HRC. Backup plates for carbide tools are preferably made of O1 hardened to 48 to 52 HRC. Strippers ordinarily can be made of low-carbon or medium-carbon steel (1020 or 1035) plate. Where a hardened plate is used for medium-production work, 4140 flame hardened, W1 conventionally hardened or W1 cyanided and oil quenched is often preferred. Hardened strippers for carbide dies and high-production D2, D4 or CPM 10V dies are made of O1 or A2, hardened to 50 to 54 HRC.

Custom-made hardened guides and locator pins usually are made of W1 or W2 for most medium- or long-run dies, or of alloy steels such as 4140 for low-cost short-run dies.

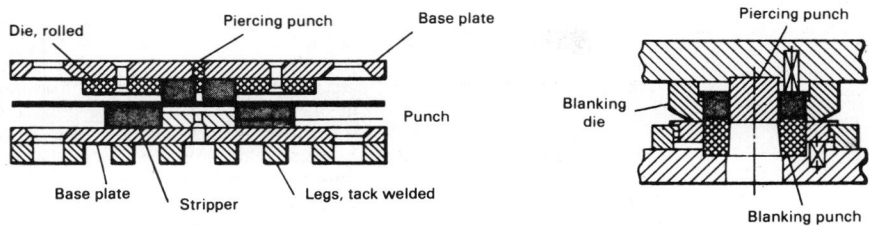

Tools at left are for short-run production of parts similar to parts 1 and 2 in Fig. 2 from relatively thin-gage sheet metal; tools at right are for longer runs. Refer to Table 3 for tooling recommendations.
Fig. 1. Sectional views of typical tools used for blanking and piercing simple shapes

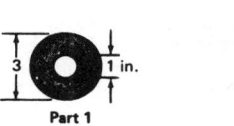 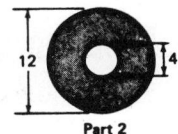  

Dimensions are in inches; to find equivalent metric values (mm), multiply listed values by 25. Parts 1 and 2 are relatively simple parts, and require dies similar to those illustrated in Fig. 8. Parts 3 and 4 are more complex, and require notching and compound or progressive dies.
Fig. 2. Typical parts of varying severity commonly produced by blanking and piercing

Table 3. Typical punch and die materials for blanking 1.3-mm (0.050-in.) sheet
For sketches of typical parts, see Fig. 2.

Work material	Tool material for production quantity of:				
	1000	10 000	100 000	1 000 000	10 000 000
Part 1 and similar 75-mm (3-in.) parts					
Aluminum, copper and magnesium alloys	Zn (a), O1, A2	O1, A2	O1, A2	D2	Carbide
Carbon and alloy steel, up to 0.70% C, and ferritic stainless steel	O1, A2	O1, A2	O1, A2	D2	Carbide
Stainless steel, austenitic, all tempers	O1, A2	O1, A2	A2, D2	D4	Carbide
Spring steel, hardened, 52 HRC max	A2	A2, D2	D2	D4	Carbide
Electrical sheet, transformer grade, 0.6 mm (0.025 in.)	A2	A2, D2	A2, D2	D4	Carbide
Paper, gaskets, and similar soft materials	W1 (b)	W1 (b)	W1 (c), A2 (d)	W1 (d), A2 (d)	D2
Plastic sheet, not reinforced	O1	O1	O1, A2	D2	Carbide
Plastic sheet, reinforced	O1 (e), A2	A2 (f)	A2 (f)	D2 (f)	Carbide
Part 2 and similar 300-mm (12-in.) parts					
Aluminum, copper and magnesium alloys	Zn (a), 4140 (g)	4140 (h), A2	A2	A2, D2	Carbide
Carbon and alloy steel, up to 0.70% C, and stainless steels up to quarter hard ...	4140 (h), A2	4140 (h), A2	A2	A2, D2	Carbide
Stainless steel, austenitic, over quarter hard	A2	A2, D2	D2	D2, D4	Carbide
Spring steel, hardened, 52 HRC max	A2	A2, D2	D2	D2, D4	Carbide
Electrical sheet, transformer grade, 0.6 mm (0.025 in.)	A2	A2, D2	A2, D2	D2, D4	Carbide
Paper, gaskets, and similar soft materials	4140 (j)	4140 (j)	A2	A2	D2
Plastic sheet, not reinforced	4140 (j)	4140 (h), A2	A2	D2	Carbide
Plastic sheet, reinforced	A2 (e)	A2 (e)	D2 (e)	D2 (e)	Carbide

(continued on the next page)

Press Forming Dies

PRESS FORMING is a process in which sheet metal is made to conform to the contours of a die and punch — largely by being bent or moderately stretched, or both. The suitability of a tool material for a press forming die is determined by the number of parts that can be produced using that die. This number is influenced by the size and shape of the part, the type and thickness of the metal being formed, lubrication practice, and the allowable variation in dimensions.

Typical Tool Materials. Tooling for the part shown in Fig. 3 consists of a punch and a lower die. In operation, the punch pushes the blank through the lower die, which causes wear of the lower die. The metal closely envelops the punch, with little sliding, and in that event a punch generally produces about ten times as many parts as a lower die made of the same material. However, at areas where the part shrinks against the punch during forming, wear (and possibly galling) of the punch surface will occur, particularly when the forming is done in single-action dies. For a small die and punch, the cost of steel is of minor importance, and type D2 tool steel may be used for production quantities as low as 10 000. If galling occurs during preproduction trials, the tool can be nitrided.

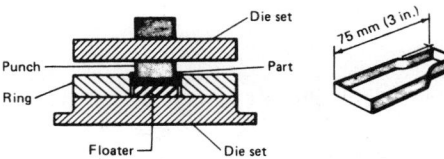

For typical die materials, see Table 7.
Fig. 3. Cross section of die used for small part of mild severity

Tooling for the part in Fig. 4 consists of a punch, an upper die and a lower die. Without the upper die, excessive wrinkling would occur at the shrink flanges. As for the part shown in Fig. 3, a less wear-resistant material is required for the punch and upper die than for the lower die. Under conditions for which the tooling is typically made of tool steel (see Table 8), the tooling is in the form of inserts in a lower die made of cast iron, as indicated in Fig. 4, and the punch is made of a tool steel such as D2.

Typical lower-die materials for press forming large parts similar to that shown in Fig. 4 are given in Table 8. For quantities less than 100 000 pieces, the entire lower die is typically made of the material indicated in the selection table, without inserts. The punch is made of a less wear-resistant material, which usually is the same as the lower-die material in the first column to the left of the quantity being considered.

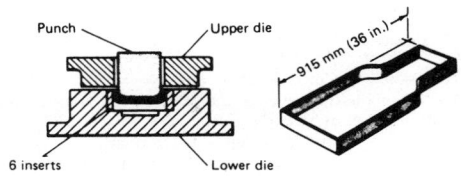

For typical die materials, see Table 8.
Fig. 4. Cross section of die used for large part of mild severity

Table 3. (continued)

Work material	\multicolumn{5}{c}{Tool material for production quantity of:}
	1000	10 000	100 000	1 000 000	10 000 000
Part 3 and similar 75-mm (3-in.) parts					
Aluminum, copper and magnesium alloys	O1, A2	O1, A2	O1, A2	A2, D2	Carbide
Carbon and alloy steel, up to 0.70% C, and ferritic stainless steel	O1, A2	O1, A2	O1, A2	A2, D2	Carbide
Stainless steel, austenitic, all tempers	A2	A2, D2	A2, D2	D2, D4	Carbide
Spring steel, hardened, 52 HRC max	A2	A2, D2	D2, D4	D2, D4	Carbide
Electrical sheet, transformer grade, 0.6 mm (0.025 in.)	A2	A2, D2	D2, D4	D2, D4	Carbide
Paper, gaskets and other soft materials	W1 (b)	W1 (b)	W1 (k), A2	W1 (k), A2	D2
Plastic sheet, not reinforced	O1	O1	A2	A2, D2	Carbide
Plastic sheet, reinforced	O1 (m)	A2 (f)	A2 (f)	D2 (f)	Carbide
Part 4 and similar 300-mm (12-in.) parts					
Aluminum, copper and magnesium alloys	A2	A2	A2, D2	A2, D2	Carbide
Carbon and alloy steel, up to 0.70% C, and ferritic stainless steel	A2	A2	A2, D2	A2, D2	Carbide
Stainless steel, austenitic, up to quarter hard	A2	A2	A2, D2	D2, D4	Carbide
Stainless steel, austenitic, over quarter hard	A2	D2	D2	D2, D4	Carbide
Spring steel, hardened, 52 HRC max	A2	A2, D2	D2	D2, D4	Carbide
Electrical sheet, transformer grade, 0.6 mm (0.025 in.)	A2	A2, D2	D2	D2, D4	Carbide
Paper, gaskets, and other soft materials	W1 (b)	W1 (b)	W1 (n)	W1, A2	D2
Plastic sheet, not reinforced	A2	A2	A2	A2, D2	Carbide
Plastic sheet, reinforced	A2 (f)	A2 (f)	D2 (f)	D2 (f)	Carbide

Note: Although carbide is recommended in this table only for 10 million pieces, it should usually be considered also for runs of 1 to 10 million pieces.

(a) Zn refers to a die made of zinc alloy plate and a punch of hardened tool steel. (b) For punching up to 10 000 parts, the W1 punch and die would be left soft and the punch peened to compensate for wear if necessary. (c) For punching 10 000 to 1 000 000 pieces, the W1 punch can be soft so that it can be peened to compensate for wear, or it can be hardened and ground to size. (d) Of the two alternatives listed, A2 tool steel is preferred if compound tooling is to be used for quantities of 10 000 to 1 000 000. (e) This O1 punch may have to be cyanided 0.1 to 0.2 mm (0.004 to 0.008 in.) deep to make even 1000 pieces. (f) For the application indicated, the punch and die should be gas nitrided 12 h at 540 to 565 °C (1000 to 1050 °F). (g) Soft. (h) Working edges are flame hardened in this application. (j) May be soft or flame hardened. (k) For punching 10 000 to 1 000 000 pieces, the punch would be W1, left soft so that it can be peened to compensate for wear, and the die would be O1, hardened. (m) Cyaniding of the punch is advisable, even for 1000 pieces. (n) For punching 10 000 to 1 000 000 pieces, the W1 die would be hardened and the W1 punch would be soft, so that it could be peened to compensate for wear.

Table 4. Typical punch and die materials for shaving 1.3-mm (0.050-in.) sheet

Work material	\multicolumn{4}{c}{Tool material for production quantity of:}
	1000	10 000	100 000	1 000 000
Aluminum, copper and magnesium alloys	O1 (a)	A2	A2	D4 (b)
Carbon and alloy steel, up to 0.30% C, and ferritic stainless steel	A2	A2	D2	D4 (b)
Carbon and alloy steel, 0.30 to 0.70% C	A2	D2	D2	D4 (b)
Stainless steel, austenitic, up to quarter hard	A2	D2	D4 (b)	D4 (b)
Stainless steel, austenitic, over quarter hard, and spring steel hardened to 52 HRC max	A2	D2	D4 (b)	M2 (b)

(a) Type O2 is preferred for dies that must be made by broaching. (b) On frail or intricate sections, D2 should be used in preference to D4 or M2. Carbide shaving punches may also be practical for this quantity.

Tables 7 and 8 may be used to select lower-die materials for parts made of sheet thicker or thinner than the 1.3 mm used for the examples, or for parts of greater or lesser severity than those shown in Fig. 3 and 4. For parts of greater severity or sheet of greater thickness, use the die material recommended for the next greater production quantity than the quantity actually to be made (the column to the right of the actual production quantity in the table). Similarly, for parts of lesser severity or sheet of lesser thickness, use the die material recommended for the next lower production quantity (shift to the next column to the left of the actual production quantity).

Deep Drawing Dies

DEEP DRAWING is a process in which sheet metal is formed into round or square cup-shape parts by making it conform to a punch as it is drawn through a die. In conventional deep drawing, successive draws are made in the same direction. The types of dies and other tooling used for conventional deep drawing are illustrated in Fig. 5.

For economy in manufacture, a drawn part should always be produced in the fewest steps possible. Ironing—that is, thinning the walls of the part being drawn by using a reduced clearance between punch and die—is used almost universally in multioperation deep drawing. Ironing helps produce deep draws and uniform wall thickness in the fewest operations. Each operation is designed for maximum practical reduction of the metal being drawn. Accordingly, the information given here is predicated on use of reductions near the maximum of about 35%.

MATERIALS FOR SPECIFIC TOOLS

Draw Rings. Table 9 gives typical materials for draw rings (both dies and backup rings) used in drawing and ironing cups of various diameters and lengths from the three basic types of sheet

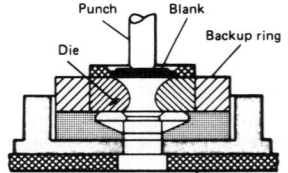

First operation in drawing

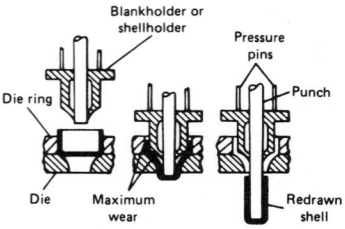

Conventional redrawing of thin-wall shells

Fig. 5. Tooling components used in conventional drawing operations

Table 5. Typical materials for perforator punches

Work material	Punch material for production quantity of:		
	10 000	100 000	1 000 000
Punch diameters up to 6.4 mm ($^1/_4$ in.)			
Aluminum, brass, carbon steel, paper and plastics	M2	M2	M2
Spring steel, stainless steel, electrical sheet and reinforced plastics ...	M2	M2	M2
Punch diameters over 6.4 mm ($^1/_4$ in.)			
Aluminum, brass, carbon steel, paper and plastics	W1	W1	D2
Spring steel, stainless steel, electrical sheet and reinforced plastics ...	M2	M2	M2

Table 6. Typical materials for perforator bushings

Work material	Bushing material for production quantity of:		
	10 000	100 000	1 000 000
Aluminum, brass, carbon steel, paper and plastics	W1 (a)	W1 (a)	D2
Spring steel, stainless steel, electrical sheet and reinforced plastics	D2	D2	D2 or carbide

(a) When bushings are of a shape that cannot be ground after hardening, an oil-hardening or air-hardening steel is recommended to minimize distortion.

metal listed in Table 10. The data in Table 9 are given for both round and square cups drawn from stock 1.6 mm (0.062 in.) thick in three typical production quantities. Similar data for a large square cup and a large pan are given also. Design dimensions for all seven parts referred to in Table 9 are given in Fig. 6. The square parts have liberal corner radii consistent with favorable die life.

Punches and Blankholders. Typical materials for punches and for blankholders (pressure pads) or shellholders (pressure sleeves) are given in Table 11. The materials listed in Table 11 are for punches and blankholders used in drawing and

ironing round and square steel cups similar to parts 2 through 7 in Fig. 6.

More wear-resistant materials are required not only for the tools used in drawing and ironing harder or thicker stock or for those used for longer runs, but also for tools used to achieve greater percent reductions during ironing.

Metalworking Rolls

ROLLS are used to reduce the cross section of metal or to change its shape, or both. Cylindrically shaped, rolls are placed in a mill housing

supported on journal bearings against which screwdown pressure is exerted. Rolls rotate in opposite directions, the metal bar, plate or sheet passing between them longitudinally (Fig. 7a). Cross rolling is used for making seamless tube or for straightening round bar (Fig. 7b). Cross rolling is also used for widening billet (to make wide sheet or plate products), as well as for producing a more homogeneous microstructure.

CAST IRON ROLLS

Cast iron rolls are used in the as-cast condition or after stress relief. Some high-alloy iron rolls are heat treated by holding at high temperature followed by several lower-temperature treatments. Cast irons used for rolls are metastable and may be white or gray depending on composition, inoculation (if any), cooling rate and other factors.

In American practice, cast iron rolls are classified as (a) chilled iron rolls, (b) grain rolls, (c) sand iron rolls, (d) ductile iron rolls and (e) composite rolls.

Chilled iron rolls (hardness, 50 to 90 HSc) have a definitely formed, clear, homogeneous chilled white iron body surface and a fairly sharp line of demarcation between the chilled surface and the gray iron interior portion of the body. Clear chilled iron rolls can be made in unalloyed or alloyed grades.

Alloy chilled iron rolls have hardnesses that range from 60 to 90 HSc and that are controlled by carbon and alloy contents. Customary maximum percentages of alloying elements are 1.25

Table 7. Typical lower-die materials for forming a small part of mild severity from 1.3-mm (0.050-in.) sheet
For die cross section and part shape, see Fig. 3.

Metal being formed	Quality requirements			Lubri-cation(b)	Lower-die materials(a) for total production quantity of:				
	Finish	Tolerance mm	in.		100	1 000	10 000	100 000	1 000 000
1100 aluminum, brass, copper(c)	None	None	None	Yes	Epoxy-metal, mild steel	Polyester-metal, mild and 4140 steel	Polyester-glass(d), mild and 4140 steel	O1, 4140	A2, D2
1100 aluminum, brass, copper(c)	None	±0.1	±0.005	Yes	Epoxy-metal, mild and 4140 steel	Polyester-metal, mild and 4140 steel	Polyester-glass(d), mild and 4140 steel	4140, O1, A2, D2	A2, D2
1100 aluminum, brass, copper(c)	Best	±0.1	±0.005	Yes	Epoxy-metal, mild steel	Polyester-metal, mild and 4140 steel	Polyester-glass(d), mild and 4140 steel	4140, O1, A2	A2, D2
Magnesium or titanium(e)	Best	±0.1	±0.005	Yes	Mild steel	Mild and 4140 steel	A2	A2	A2, D2
Low-carbon steel, to $^1/_4$ hard	None	None	None	Yes	Mild and 4140 steel	Mild and 4140 steel	4140, mild steel chromium plated, D2	A2	D2
Type 300 stainless, to $^1/_4$ hard	None	None	None	Yes	Mild and 4140 steel	Mild and 4140 steel	Mild and 4140 steel	A2, D2	D2
Low-carbon steel	Best	±0.1	±0.005	Yes	Mild and 4140 steel	Mild and 4140 steel	Mild and 4140 steel	A2, D2, nitrided D2	D2, nitrided D2
High-strength aluminum or copper alloys	Best	±0.1	±0.005	No(f)	Mild and 4140 steel	Mild and 4140 steel	Mild steel chromium plated and 4140	Cr plated O1; A2	D2, nitrided D2
Type 300 stainless, to $^1/_4$ hard	None	±0.1	±0.005	Yes	Mild and 4140 steel	Mild and 4140 steel	Mild and 4140 steel	Cr plated O1; A2	D2
Type 300 stainless, to $^1/_4$ hard	Best	±0.1	±0.005	Yes	Mild and 4140 steel	Mild and 4140 steel	Mild steel chromium plated, D2	D2, nitrided D2	D2, nitrided D2
Heat-resisting alloys	Best	±0.1	±0.005	Yes	Mild and 4140 steel	Mild and 4140 steel	Mild steel chromium plated, D2	D2, nitrided D2	D2, nitrided D2
Low-carbon steel	Good	±0.1	±0.005	No(f)	Mild and 4140 steel	Mild and 4140 steel	Mild steel chromium plated	D2, nitrided D2	D2, nitrided D2

(a) When more than one material for the same conditions of tooling is given, the materials are listed in order of increasing cost; however, final choice often depends on availability rather than on small differences in cost or performance. Where mild steel is recommended for forming fewer than 10 000 pieces, the dies are not heat treated. For forming 10 000 pieces and more, such dies should be carburized and hardened. Where 4140 is recommended for fewer than 10 000 pieces, it should be pretreated to a hardness of 28 to 32 HRC. Flame hardening of high-wear areas is recommended for quantities greater than 10 000 pieces. (b) Specially applied lubrication, rather than mill oil. (c) Soft. (d) With inserts. (e) Heated sheet. (f) Use lubrication to make 1 to 100 parts.

Table 8. Typical lower-die materials for forming a large part of mild severity from 1.3-mm (0.050-in.) sheet
For die cross section and part shape, see Fig. 4.

Metal being formed	Finish	Tolerance mm	Tolerance in.	Lubrication(b)	100	1 000	10 000	100 000	1 000 000
1100 aluminum, brass, copper(c)	None	None	None	Yes	Epoxy-metal, polyester-metal, zinc alloy	Polyester-metal, zinc alloy	Epoxy or polyester-glass(d), zinc alloy	Alloy cast iron	Cast iron or A2(e)
1100 aluminum, brass, copper(c)	None	±0.1	±0.005	Yes	Epoxy-metal, polyester-metal, zinc alloy	Polyester-metal, zinc alloy	Alloy cast iron	Alloy cast iron	Alloy cast iron
1100 aluminum, brass, copper(c)	Best	±0.1	±0.005	Yes	Epoxy-metal, polyester-metal, zinc alloy	Polyester-metal, zinc alloy	Alloy cast iron	Alloy cast iron	Alloy cast iron, A2(e)
Magnesium or titanium(f)	Best	±0.1	±0.005	Yes	Cast iron, zinc alloy	Cast iron, zinc alloy	Cast iron	Alloy cast iron	Alloy cast iron, A2(e)
Low-carbon steel, to ¼ hard	None	None	None	Yes	Epoxy-metal, polyester-metal, zinc alloy	Epoxy-glass, polyester-glass, zinc alloy	Epoxy or polyester-glass(d), cast iron	Alloy cast iron	
Type 300 stainless, to ¼ hard	None	None	None	Yes	Epoxy-metal, polyester-metal, zinc alloy	Epoxy-glass, polyester-glass, zinc alloy	Epoxy or polyester-glass(d), alloy cast iron	A2(e)	D2(e)
Low-carbon steel	Best	±0.1	±0.005	Yes	Zinc alloy	Epoxy-glass polyester-glass, zinc alloy	Alloy cast iron	D2, nitrided A2(e)	D2, nitrided D2(e)
High-strength aluminum or copper alloys	Best	±0.1	±0.005	No(g)	Zinc alloy	Polyester-glass, zinc alloy	Alloy cast iron	Alloy cast iron	Nitrided A2(e), nitrided D2(e)
Type 300 stainless, to ¼ hard	None	±0.1	±0.005	Yes	Zinc alloy	Zinc alloy	Alloy cast iron	D2, nitrided A2(e)	D2(e), nitrided D2(e)
Type 300 stainless, to ¼ hard	Best	±0.1	±0.005	Yes	Zinc alloy	Zinc alloy	Alloy cast iron	Nitrided D2	Nitrided D2(e)
Heat-resisting alloys	Best	±0.1	±0.005	Yes	Zinc alloy	Zinc alloy	Alloy cast iron	Nitrided D2	Nitrided D2(e)
Low-carbon steel	Good	±0.1	±0.005	No(g)	Zinc alloy	Zinc alloy	Alloy cast iron	Nitrided D2	Nitrided D2(e)

(a) When more than one material for the same conditions of tooling is given, the materials are listed in order of increasing cost; however, final choice often depends on availability rather than on small differences in cost or performance. Where mild steel is recommended for forming fewer than 10 000 pieces, the dies are not heat treated. For forming 10 000 pieces and more, such dies should be carburized and hardened. Where 4140 is recommended for fewer than 10 000 pieces, it should be pretreated to a hardness of 28 to 32 HRC. Flame hardening of high-wear areas is recommended for quantities greater than 10 000 pieces. (b) Specially applied lubrication, rather than mill oil. (c) Soft. (d) With inserts. (e) Use as inserts in cast iron body. (f) Heated sheet. (g) Use lubrication to make 1 to 100 parts.

Mo, 1.00 Cr and 5.5 Ni. Many different combinations are used to produce desired properties. Rolls of this type, particularly in the harder grades, are used chiefly for rolling flat work, both hot and cold. The softer, machinable grades are used for rolling rod and small shapes.

Grain rolls are "indefinite chill" iron rolls (hardness, 40 to 90 HSc) that have an outer chilled face on the body. These rolls have high resistance to wear and good finishing qualities, to considerable depths. The harder grades are used for hot and cold finishing of flat rolled products, and the softer grades are for deep sections (even with small rolls).

Sand iron rolls (no chill; hardness, 35 to 45 HSc) are cast in sand molds, in contrast to chilled iron rolls and grain rolls, the bodies of which are cast directly against chills. Sand iron rolls are used chiefly for intermediate and finishing stands on mills that roll large shapes. They are also used for roughing operations in primary mills.

Ductile iron rolls (hardness, 50 to 65 HSc) are made of iron of restricted composition to which magnesium or rare earth metals are added under controlled conditions to cause the graphite to form, during solidification, as nodules instead of the flakes common to gray iron. The resulting iron has strength and ductility properties between those of gray iron and those of steel.

Composite rolls, sometimes called double-pour rolls (hardness: bodies, 70 to 90 HSc; necks, 40 to 50 HSc), are rolls in which the body surface is made of a richly alloyed, hard cast iron resistant to wear, and the necks, wabblers, and central areas of the body are of a tougher and softer material. The metals are bonded firmly together during casting to form an integral structure. The chief applications of composite rolls in rolling of steel have been work rolls for four-high hot and cold strip mills and for plate mills; in rolling of nonferrous metals, the chief application has been rolls for hot breakdown and cold reduction of sheet and strip.

CAST STEEL ROLLS

Differentiation between cast iron rolls and cast steel rolls cannot be made strictly on the basis of carbon content. Iron rolls usually are of compositions that produce free graphite in unchilled portions; in contrast, steel rolls do not exhibit free graphite.

Cast steel rolls have higher hardness than cast iron rolls, and the superior strength of cast steel rolls often makes them preferable.

Composition. Alloy steel rolls have almost entirely superseded carbon steel rolls. Compositions of most alloy steel rolls are within the following limits: 0.40 to 2.60 C; less than 0.12 S, usually 0.06 max; less than 0.12 P, usually 0.06 max; up to 1.25 Mn; up to 1.50 Cr; up to 1.50 Ni; and up to 0.60 Mo. Higher carbon contents increase hardness and wear resistance. Some rolls have higher alloy contents, but these usually are for special purposes.

Applications. Cast steel rolls are graded according to carbon content. The general applications of these rolls are listed in Table 12.

FORGED STEEL ROLLS

Hardened forged steel rolls are used principally for cold rolling various metals in the form of coiled sheet and strip. Extremely high pressures are used in cold rolling, and forged rolls have sufficient strength, surface quality and wear resistance for cold rolling operations. Forged rolls sometimes are employed in nonferrous hot mills in preference to iron rolls because of their higher bending strength and resistance to metal pickup.

Composition. The most commonly used composition for forged steel rolls, sometimes known as regular roll steel, averages 0.85 C, 0.30 Mn, 0.30 Si, 1.75 Cr and 0.10 V. About 0.25% Mo sometimes is added to this basic composition, and the chromium content may be varied to obtain specific characteristics. For rolling nonferrous metals, a forged steel containing 0.40 C and 3.00 Cr is preferred. In Sendzimir mills, the work rolls and first and second intermediate supporting and drive rolls usually are made from high-carbon high-chromium steel with 1.50 or 2.25% C and 12.00% Cr (D1 or D4). For more severe service, work rolls of M1 molybdenum high speed steel are used. Special P/M alloys produced through

Table 9. Typical materials for draw rings used in drawing and ironing both round and square parts

For part designs and over-all dimensions, see Fig. 6.

Metal to be drawn	Total number of parts to be drawn		
	10 000	100 000	1 000 000
Cups up to 76 mm (3 in.) across, drawn from 1.6-mm (0.062-in.) sheet (parts 1, 2 and 3)			
Drawing quality aluminum and copper alloys	W1; O1	O1; A2	A2; D2
Drawing quality steel	W1; O1	O1; A2	A2; D2
300-series Stainless steel	W1 chromium plated; aluminum bronze	Nitrided A2; aluminum bronze	Nitrided D2 or D3; cemented carbide
Cups 305 mm (12 in.) or more across, drawn from 1.6-mm (0.062-in.) sheet (parts 4 and 5)			
Drawing quality aluminum and copper alloys	Alloy cast iron(a)	Alloy cast iron(a); A2 inserts(b)	A2 or D2 inserts(b)
Drawing quality steel	Alloy cast iron(a)	Alloy cast iron(c); A2 inserts(b)	A2 or D2 inserts(b)
300-series stainless steel	Alloy cast iron(d); aluminum bronze inserts(b)	A2 or aluminum bronze inserts(b)	Nitrided A2 or D2 inserts(b)
Square cups similar to part 6, drawn from 1.6-mm (0.062-in.) sheet			
Drawing quality aluminum and copper alloys(e)	W1	O1; A2	A2; D2
Drawing quality steel(e)	W1	O1; A2	A2; D2; nitrided A2 or D2
300-series stainless steel(f) . . .	W1; aluminum bronze	Nitrided A2; aluminum bronze	Nitrided A2 or D2
Large pans similar to part 7; drawn from 0.8-mm (0.031-in.) sheet			
Drawing quality aluminum and copper alloys	Alloy cast iron(a)	Alloy cast iron(a); A2 corner inserts(b)	Nitrided A2 or D2 inserts(b)
Drawing quality steel	Alloy cast iron(a)	Alloy cast iron(a); A2 corner inserts(b)	Nitrided A2 or D2 inserts(b)
300-series stainless steel	Alloy cast iron(d); aluminum bronze	Nitrided A2 or aluminum bronze inserts(b)	Nitrided A2 or D2 inserts(b)

(a) Wearing surfaces flame hardened. (b) In flame hardened alloy cast iron. (c) Quenched and tempered for part 4; flame hardened for part 5. (d) Flame hardened on wearing surfaces to not over 420 HB. (e) For drawing aluminum, copper and steel, the tool material would be used as corner inserts. (f) For drawing stainless steel, inserts would be used for all wear surfaces.

Table 10. Sheet metals that require similar drawing-die materials

Type of sheet metal	Maximum hardness	Metals that require similar drawing-die materials
Drawing quality aluminum and copper alloys	64 HRF(a)	All aluminum and clad aluminum alloy sheet, copper and alloys, zinc and alloys, silver, pewter and Monel
Drawing quality steel	70 HRB	Carbon steel, grades 1008 to 1020
	75 HRB	Carbon steel, grades 1021 to 1030
Austenitic stainless steel	95 HRB	301, 302, 304, 305, 308, 310, 316, 317 steel; 410 and 430 carbon steel clad with stainless steel; copper clad with stainless steel; magnesium drawn at 200 to 300 °C (400 to 600 °F) with no ironing of sides; 17-4 PH, 17-7 PH and PH 15-7 Mo stainless steels

(a) Roughly equivalent to 58 HB (500-kg load) or 24 HR30T.

Table 11. Typical materials for punches and blankholders

For part designs and over-all dimensions, see Fig. 6.

Die component	Total number of parts to be drawn		
	10 000	100 000	1 000 000
For round steel cups like part 2			
Punch(a)	Carburized 4140; W1	W1; carburized S1	A2; D2
Blankholder(b)	W1; O1	W1; O1	W1; O1
For square steel cups like part 3			
Punch(a)	Carburized 4140; W1	W1; carburized S1	A2; D2
Blankholder(b)	W1; O1	W1; O1	W1; O1
For round steel cups like parts 4 and 5			
Punch(a)	Alloy cast iron(c)	O1 (d)	A2 (c); D2 (c)
Blankholder(b)	Alloy cast iron(c)	Alloy cast iron(e)	O1; A2
For square steel cups like parts 6 and 7			
Punch(a)	Carburized 4140 (f)	W1; O1 (d)	Nitrided A2; D2 (d)
Blankholder(b)	Alloy cast iron(c)	W1; O1	O1; A2

(a) Chromium plating is optional on punches, to reduce friction between part and punch and thus facilitate removal of the part. Cast iron, however, should not be plated. (b) Also applies to shellholder, pressure pad or pressure sleeve. (c) Flame hardening not necessary. (d) The punch holder is flame hardened alloy cast iron with a nose insert of the indicated tool steel. (e) For part 4, this blankholder is quenched and tempered; for part 5, it is flame hardened. (f) The punch holder is alloy cast iron with a nose insert of the indicated steel.

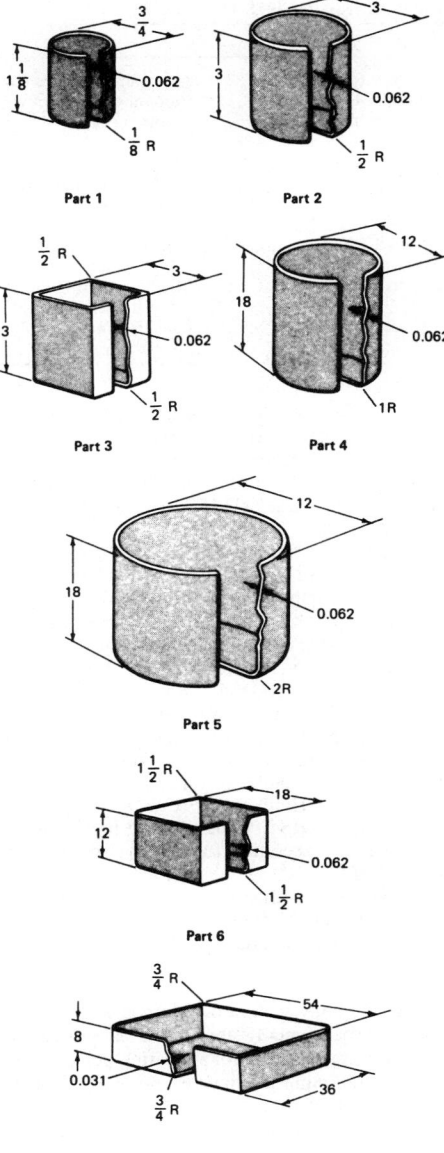

Part 1   Part 2   Part 3   Part 4   Part 5   Part 6   Part 7

Dimensions are given in inches; to find equivalent metric dimensions (mm), multiply listed dimensions by 25. Corner radii comply with standard commercial practice. For typical deep drawing materials, see Table 9.

Fig. 6. Seven typical deep drawn parts

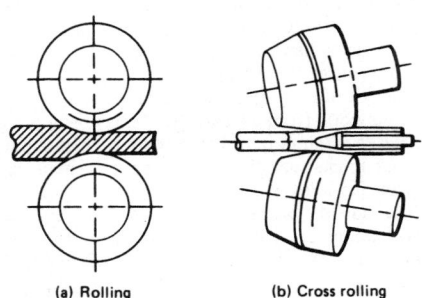

(a) Rolling   (b) Cross rolling

Fig. 7. Typical arrangements of metalworking rolls

Table 12. Applications of cast steel rolls

Carbon, %	Applications
0.50 to 0.65	Applications where strength is the prime and only requirement
0.70 to 0.85	Blooming mills; roughing stands in jobbing, plate and sheet mills; muck mills
0.90 to 1.05	Blooming mills; slab mills; roughing stands in continuous bar mills; backing rolls
1.10 to 1.25	Blooming and slab mills where breakage is not great; piercing mills; roughing stands in billet, bar, rail and structural mills
1.35 to 1.55	Intermediate stands for rail mills; structural, continuous billet and continuous bar mills
1.60 to 1.80	Intermediate stands for continuous bar and billet mills; middle rolls for three-high mills
1.85 to 2.05	Middle rolls for rail and structural mills; finishing mills where housing design is too limited for iron rolls
2.10 to 2.60	Finishing rolls for unusual conditions
2.65 and up	Special applications

application of the P/M process to high-alloy tool steel have wear resistance approaching that of carbide, which makes them attractive for some special forged steel rolls. Other high-wear alloys in the MHO series, or T15, also are applied on these occasions.

Hardness. The hardness range for forged steel rolls varies with the specific application and is developed with the cooperation of mill operators.

Hardness of work rolls for rolling thin strip averages about 95 HSc; lower hardnesses are employed for rolling thicker strip. In temper and finishing mills, work-roll hardness sometimes is higher than 95 HSc, and for special applications such as foil rolls, can be as high as 100 HSc. In nonferrous rolling, especially in aluminum plate mills, work-roll hardness generally ranges from 60 to 80 HSc. Hardness of backing rolls varies from 55 to 95 HSc; values on the high side of this range are specified for rolls in small mills and foil mills.

For Sendzimir mills, customary hardness is 61 to 64 HRC for D1 and D4 steel work rolls and 64 to 66 HRC for high speed steel work rolls. Customary hardness of intermediate rolls is 58 to 62 HRC.

Only the body section of a forged roll is hardened. Journals usually are not hardened, except those for direct-contact roller-bearing designs, for which a minimum hardness of 80 HSc is specified. In normal practice, the journals of forged rolls range in hardness from 30 to 50 HSc.

SLEEVE ROLLS

Use of forged and hardened sleeve-type backing rolls in certain hot strip and cold reduction mills has become a common practice because such rolls are more economical.

Sleeves are forged from high-quality alloy steel. Chromium-molybdenum-vanadium and nickel-chromium-molybdenum-vanadium compositions are generally used. Sleeves are heat treated by liquid quenching in either oil or water and are tempered to hardnesses of 50 to 85 HSc, depending on application.

The mandrel over which the sleeve is slipped may be made from a cast roll that has been worn below its minimum usable diameter, from a new casting made specifically for use as a mandrel, or from an alloy steel forging.

The outside diameter of the mandrel and the inside diameter of the sleeve are accurately machined or ground for a shrink fit. Mounting is accomplished by heating the sleeve to obtain the required expansion and then either slipping the sleeve over the mandrel or inserting the mandrel into the sleeve. This operation is performed with

the mandrel in a vertical position. A locking device prevents lateral movement of the sleeve. Final machining is done after the sleeve is mounted.

MISCELLANEOUS FORGED ROLLS

Auxiliary rolls such as leveler rolls and pinch rolls are employed in processing and handling equipment associated with rolling mills. These rolls are characterized by their long, slender shape. They are made from forged or rolled bars of 52100 steel or carburizing grades, and are processed to a hardness of approximately 95 HSc. Rolls used for various types of straightening machines generally are of sleeve design. One roll may be concave in body shape while its mating roll is straight. Standard compositions for forged steel rolls may be employed, and bodies may be hardened to 85 to 90 HSc.

CEMENTED TUNGSTEN CARBIDE ROLLS

Cemented tungsten carbide rolls have been used for rolling metals under a wide variety of conditions and in many types of rolling mills. They are used for both cold and hot rolling, and are made in all sizes from 6 to 400 mm (¹/₄ to 16 in.) in diameter. Rolls for slitters and trimmers also have been made of cemented tungsten carbide.

Use of cemented tungsten carbide rolls for applications varying from conventional cold rolling of flat sheet and wire to continuous hot rolling of rod in a wide range of sizes is continuing to expand, and technology of design and composition for these rolls is changing rapidly. Therefore, consultation with experienced carbide manufacturers is advised in selecting materials for specific applications.

Coining Dies

IN COLD COINING operations, the surface metal being worked is free to flow only to the small extent required by the coining operation. Flow to a larger extent constitutes cold extrusion.

DECORATIVE COINING

In dies used for decorative coining, materials that can be through hardened to produce a combination of good wear resistance, high hardness and high toughness are preferred.

Typical Die Materials. For dies up to 50 mm (2 in.) in diameter, consumable-electrode vacuum-melted or electroslag remelted 52100 steel pro-

vides the clean microstructure necessary for development of critical polished die surfaces. When heat treated to a hardness of 59 to 61 HRC, 52100 steel provides excellent die life. This steel also is suitable for photochemical etching, a process used in place of mechanical die sinking for engraving many low-relief dies. L6 tool steel at a hardness of 58 to 68 HRC is suitable for dies up to 100 mm (4 in.) in diameter. It can be through hardened, has enough toughness for long-life applications, and is suitable for photochemical etching of low-relief patterns. Air-hardening tool steels are preferred for coining and embossing dies greater than 100 mm (4 in.) in diameter. One of the chief reasons for choosing air-hardening tool steels is their low degree of distortion during heat treatment. Type A6 is a nondeforming, deep-hardening tool steel that often is used for large dies that must be hardened to 59 to 61 HRC. Air-hardening hot work steels such as type H13 are used at a hardness of 52 to 54 HRC for applications requiring especially high toughness.

For dies containing high-relief impressions, lowest die cost is obtained by machining the impressions directly into the dies when the number of pieces to be coined is less than the anticipated die life. For longer runs that require two or more identical dies, it is less expensive to produce the impressions by hubbing.

O1 and A2 tool steels are alternative choices for machined dies in production quantities up to about 100 000 pieces. The small additional cost of A2 is often justified because A2 gives longer life, especially when aluminum alloys, alloy steels, stainless steels or heat-resisting alloys are being coined.

Dies for Coining Silverware. Probably the greatest amount of industrial coining is done with drop hammers in the silverware industry, in producing highly embossed designs on surfaces such as teaspoon handles. Water-hardening steels such as W1 are almost always used for making such coining dies, whether the product is made of silver, a copper alloy or stainless steel. Water-hardening grades are selected because die blocks made of these steels can be reused repeatedly. After a die block fails, the block is annealed, the impression is machined off, and a new impression is hubbed before the die is rehardened. Dies made of deep-hardening tool steels such as O1, A2 and D2 are not reused (as are W1 dies), because they fail by deep cracking.

For ordinary designs requiring close reproduction of dimensions, dies may be made of A2, or of the high-carbon high-chromium steels D2, D3 and D4, to obtain greater compression resistance. For coining of designs with deep configurations and either coarse or sharp details, where dies usually fail by cracking, a deep-hardening carbon tool steel may be used at lower hardness, or O1, or S5 or S6, may be selected. In some instances, it may be desirable to select an air-hardening type such as A2, which would provide improved dimensional stability and wear resistance. A hot work steel such as H11, H12 or H13 may prove to be best where extreme toughness is the predominant requirement. When die failure occurs by rapid wear, a higher-hardness steel, or a more highly alloyed wear-resistant steel such as A2, may solve the problem.

For articles coined on drop hammers from series 300 austenitic stainless steels, it has sometimes been found advantageous to use steels of the S1, S5, S6 and L6 types, oil quenched and tempered to 57 to 59 HRC. Because the carbon contents of these grades are between 0.50 and 0.70%, they are less resistant to wear than W1, A2 and D2, but are tougher and more resistant to chipping and splitting. If necessary, the wear resistance of S5 tool steel dies can be improved slightly by carburizing to a depth of 0.13 to 0.25 mm (0.005 to 0.010 in.).

COINING IN PROGRESSIVE DIES

Tool steels recommended for coining a cup-shape part to final dimensions in the last stages of progressive stamping are shown in Table 13. This press coining operation involves partial confinement of the entire cup within the die. This produces high radial die pressures and thus requires pressed-in inserts on long runs, to prevent die cracking.

The punch material can be the same as the die material, except that O1 should be substituted for W1 in applications where W1 might crack during quenching.

P/M STEELS FOR COINING DIES

The application of hot isostatic processing to P/M production of high speed steels and special high-alloy steels has expanded the range of tool steel grades available for long-run coining dies. Dramatic increases in toughness and grindability have been achieved. Type M4 is an excellent example. When made by P/M processing, M4 has approximately twice the toughness and two to three times the grindability of conventionally processed M4.

Cold Heading Tools

THE MANY TYPES of fasteners that are used in large quantities are manufactured in cold heading machines. Most of the tools used in these machines require maximum surface hardness for wear resistance, along with maximum strength and toughness to enable them to withstand high service pressures without breaking.

Materials used for tools in single-die, two-die-three-blow and multistation cold heading machines are listed in Table 14.

TOOL STEEL DIES

Solid cold heading dies made of W1 and W2 tool steels are used for short production runs. These tools are hardened by flush quenching, which provides the desired combination of high hardness, high strength and high toughness. Most long production runs require dies made of the more highly alloyed tool steels M1, M2 and D2, or of cemented carbides. Some of the new P/M alloys have also given excellent performance and are economical alternatives to cemented carbides on long runs. These materials are most commonly used as inserts in H13 tool steel cases; hardness of such cases usually is held to about 48 HRC to provide the best combination of backup strength and freedom from hazardous breakage. However, for applications involving high-interference fits between insert and case, the greater high-strength fracture toughness of maraging steels such as 18Ni(300) makes them very desirable as materials for cases, although at some sacrifice in wear resistance.

Table 13. Typical tool steels for coining a preformed cup to final size on a press

Metal to be coined	Die material(a) for total quantity of:		
	1000	10 000	100 000
Aluminum and copper alloys	W1	W1	D2
Low-carbon steel	W1	O1	D2
Stainless steel, heat-resistant alloys and alloy steels	O1	A2	D2

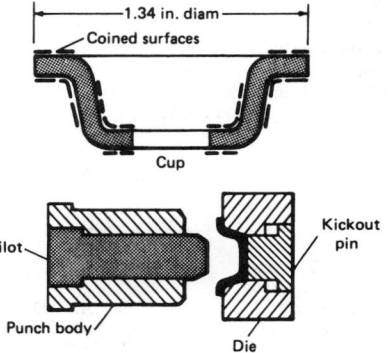

(a) For quantities over 10 000, the materials given are for die inserts. All selections shown are for machined dies. The same material would be used for the punch, except that O1 should be substituted for W1 in applications where W1 might crack during heat treating.

Table 14. Typical materials for tools in cold heading machines

Tool	Material
Cutoff quill (die)	M4, or cemented carbide insert
Cutoff knife	M4, or cemented carbide insert
Upset, cone or spring punch	W1, S1, M1, or cemented carbide insert
Cone-punch knockout pin	M2, or CPM 10V
Backing plug	O1
Finish-punch case	H13
Finish-punch insert	M1, M2, CPM 10V, or cemented carbide
Die case	H13
Die insert	M1, D2, M2, CPM 10V, or cemented carbide
Die knockout pin	M2, or CPM 10V

CEMENTED CARBIDE TOOLS

As a rule of thumb, cemented carbide tools properly designed and utilized should provide roughly ten times the life of steel tools. Although tungsten carbide tooling is higher in initial cost than tool steel tooling and lower in impact resistance, the longer life and superior dimensional integrity of carbide result in far lower cost per thousand pieces produced. For best results, cemented carbide tools should be used in cold heading machines that are in good condition, with tight rams, minimum vibration and accurate alignment.

Cold Extrusion Tools

IN COLD EXTRUSION, neither the tooling nor the work is preheated. However, the heat generated by plastic deformation of the workpiece under steady and nearly uniform pressure may be sufficient to require tool steels with relatively high resistance to softening at elevated temperatures.

In the cold extrusion process, backward displacement from a closed die progresses in the direction opposite that of punch travel, as shown at left in Fig. 8. Parts often are cuplike in shape and have wall thicknesses equal to the clearance between punch and die.

In forward extrusion, the metal is forced in the direction of punch travel (Fig. 8, center). One end of the die recess is just large enough to receive the starting slug, and the other end has a small orifice of the shape required for the final part.

Sometimes the two methods of extrusion are combined so that some of the metal flows backward and some flows forward, as shown at right in Fig. 8.

Compressive strength of the punch and tensile strength of the die are among the most important factors influencing the selection of materials for cold extrusion tools. Thus, almost without exception and particularly for extrusion of steel, the primary tools in contact with the work must be made from steels that will through harden in the section sizes involved.

The primary mechanism of tool deterioration in cold extrusion is abrasive and adhesive (galling) wear of both punch and die.

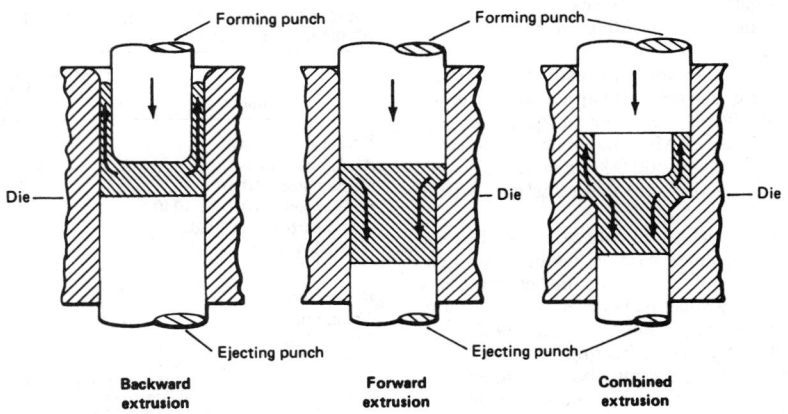

Fig. 8. Backward, forward and combined forward and backward displacement in cold extrusion

Lubrication, at least theoretically, is the prime factor controlling tool wear, because there is no metal-to-metal contact between the tools and the work when lubrication is ideal. In practice, the stresses imposed on tooling can vary by more than 100% with changes in lubrication.

PUNCH AND DIE MATERIALS

Cold extrusion of a part from 1018 or 1021 steel requires about 10% more extrusion pressure than extrusion of the same part from 1010 steel. For low-alloy steels, forming loads are about 20 to 30% greater than those for 1010. Medium-carbon steels such as 1030 and 1040 also require forming loads about 20 to 30% higher than those for 1010 steel.

Dies made of W1 tool steel generally are satisfactory for extruding the softer alloys of aluminum. Steels such as A2 and D2 are preferred for tools used in extrusion of the stronger aluminum alloys, because enough heat sometimes is generated to soften tools made of W1. In extrusion of aluminum, tool wear is roughly proportional to the yield strength of the work metal. Thus, the common impact extrusion alloys 1100, 6061, 2014 and 7075 cause progressively more wear on tools in the order listed.

Tables 15 to 20 list typical tool steels used in tools for cold extrusion of steels and aluminum alloys, in two quantities, for the series of hypothetical parts shown in Fig. 9. These simple parts are seldom encountered in practice; however, the principles described can be related to actual production components of comparable severity.

SECONDARY TOOLING COMPONENTS

Table 21 lists the constructional and tool steels used for the secondary tooling components for cold extrusion of parts 1 to 7 (see Fig. 9) from steel. (Tooling for forward or backward extrusion of aluminum consists of only a die, a die holder, a punch and an ejector.) The steels listed in Table 21 reflect the moderate to high extrusion severity involved in producing parts 1 to 7.

Cemented Carbides. The grades of cemented carbides used most frequently in cold extrusion punches and dies are shown in Table 22. In general, the quantity of parts to be produced and the required dimensional tolerances are more important than tool size in deciding whether or not to use carbide.

As a specific example, it is estimated that carbide tooling could not be justified for extruding part 5 (see Fig. 9) from steel in quantities less than 100 000 pieces. Much larger quantities would be required (perhaps 500 000 or more) to justify use of carbide tooling for extruding part 5 from aluminum.

In addition to its use in cold extrusion dies and die inserts, cemented carbide has been used extensively for cold extrusion punches for many years, producing small parts for automotive, farm-equipment and other high-production industries. Typical of such parts are bearing cups, valve lifters, wrist pins and spark-plug bodies. Larger parts now are being produced by cold extrusion. Primarily, these are cuplike shapes that either are used as is or are subsequently punched out to produce bushings.

Table 15. Typical tool steels for backward extrusion of parts 1 and 2
For designs of parts, see Fig. 9.

Metal to be extruded(a)	Total quantity of parts to be extruded	
	5 000	50 000
Punch material(b)		
Aluminum alloys A2	A2, D2, M4 (c)	
Carbon steel, up to 0.40% C A2	D2, M2 (b), M4 (c)	
Carburizing grades of alloy steel A2	M2 (d), M4 (c)	
Die material(b)		
Aluminum alloys ......W1 (e)	W1 (e)	
Carbon steel, up to 0.40% C O1, A2	A2 (f)	
Carburizing grades of alloy steel O1, A2	A2 (f)	
Knockout material(b)		
Aluminum alloys A2	D2	
Carbon steel, up to 0.40% C, and carburizing grades of alloy steel A2	A2, D2	

(a) For part 1, starting with a solid slug; for part 2, starting with part 1. In aluminum, part 2 can be made directly from a cylindrical blank. (b) Where two or more tool materials are recommended for the same conditions, they are given in order of cost, with the less or least expensive shown first. (c) First choice in automotive parts processing. (d) Liquid nitrided. (e) The 1.00% C grade is recommended. (f) Gas nitrided on the inside surface only.

Table 16. Typical tool steels for drawing part 3
For design of part, see Fig. 9.

Metal to be drawn(a)	Total quantity of parts to be extruded	
	5 000	50 000(b)
Punch material(c)		
Aluminum alloys A2	D2, M4 (d)	
Carbon and alloy steel, up to 0.40% C O1	A2, M4 (d)	
Die material(c)		
Aluminum alloys ....W1 (e)	W1 (e), A2	
Carbon and alloy steel, up to 0.40% C O1, A2	A2 (f), D2	

(a) Starting with part 2 (Table 15) for steel. In aluminum, the part would be made in one backward extrusion from a cylindrical slug. (b) For quantities greater than about 100 000 parts in steel, carbide punches and dies should be considered, especially if close tolerances must be maintained. (d) First choice in automotive parts processing. (e) The 1.00% C grade is recommended. (f) Gas nitriding is recommended on the inside surface. F2 tool steel may be used in place of A2.

Table 17. Typical tool steels for drawing part 4
For design of part, see Fig. 9.

Metal to be drawn(a)	Total quantity of parts to be drawn	
	5 000	50 000
Punch material(b)		
Aluminum alloys A2	D2, M4 (c)	
Carbon steel, 1010 ....M2	M2, M4 (c)	
Carbon steel, 1020 to 1040, and carburizing grades of alloy steel ....M2 (d)	M2 (d), T15, M4 (c)	
Die material(b)		
Aluminum alloys A2	A2, D2	
Carbon and alloy steel A2	A2 (d), D2	

(a) In steel, a part would be made in two operations with an intermediate process anneal (see text). In aluminum, it would be made in one backward extrusion. (b) Where two or more tool materials are recommended for the same conditions, they are given in order of cost, with the less or least expensive shown first. (c) First choice in automotive parts processing. (d) Nitriding treatment is recommended.

Table 18. Typical tool steels for forward extrusion of part 5
For design of part, see Fig. 9.

Metal to be extruded(a)	Total quantity of parts to be extruded	
	5 000	50 000(b)
Punch material(c)		
Aluminum alloys A2	D2, M4 (d)	
Carbon steel, 1010 ....A2	D2, M4 (d)	
Carbon steel, 1020 and 1040, and carburizing grades of alloy steel A2	M2 (e)	
Die material(c)		
Aluminum alloys ....W1 (f)	A2, D2	
Carbon and alloy steel, up to 0.40% C A2	A2 (g)	

(a) Starting with part 2 (Table 15) for steel. Aluminum would be extruded from a cylindrical slug. (b) For quantities greater than about 100 000 parts in steel, carbide punches and dies should be considered, especially if close tolerances must be maintained. (c) Where two tool materials are recommended for the same conditions, they are given in order of cost, with the less expensive shown first. (d) First choice in automotive parts processing. (e) Nitrided. (f) The 1.00% C grade is recommended. (g) Liquid nitrided.

Table 19. Typical tool steels for forward extrusion of part 6
For design of part, see Fig. 9.

Metal to be extruded	Total quantity of parts to be extruded	
	5 000	50 000
Punch material(a)		
Aluminum alloys A2	D2, M4 (b)	
Carbon and alloy steels, up to 0.40% C A2, D2	M2 (c), M4 (b)	
Die material(a)		
Aluminum alloys A2	A2	
Carbon and alloy steels, up to 0.40% C ......(d)	(d)	

(a) Where two tool materials are recommended for the same conditions, they are given in order of cost, with the less expensive shown first. (b) First choice in automotive parts processing. (c) Liquid nitrided. (d) No tool steel can be recommended without qualification. Medium-carbon alloy tool steels such as H12, H21 and 6F5 have given the best results.

Table 20. Typical tool steels for forward extrusion of part 7
For design of part, see Fig. 9.

Metal to be extruded(a)	Total quantity of parts to be extruded	
	5 000	50 000
Punch material(b)		
Aluminum alloys A2	D2, M4 (c)	
Carbon and alloy steel, up to 0.40% C, and series 300 stainless steels A2	D2, M4 (c)	
Die material(b)		
Aluminum alloys A2	D2	
Carbon and alloy steel, up to 0.40% C, and series 300 stainless steels ....(d)	(d)	
Knockout material(b)		
Aluminum alloys, steels and series 300 stainless steels ....1020 (e)	1020 (e)	
	O1 (f)	O1 (f)

(a) Starting with a ring-shaped blank. (b) Where two tool materials are recommended for the same conditions, they are given in order of cost, with the less expensive shown first. (c) First choice in automotive parts processing. (d) No tool steel can be recommended without qualification. Medium-carbon alloy tool steels such as H12, H21, 6F5 and 6H2 have given the best results. (e) Or other low-carbon or low-alloy steels for knockout pins. (f) Knockout heads.

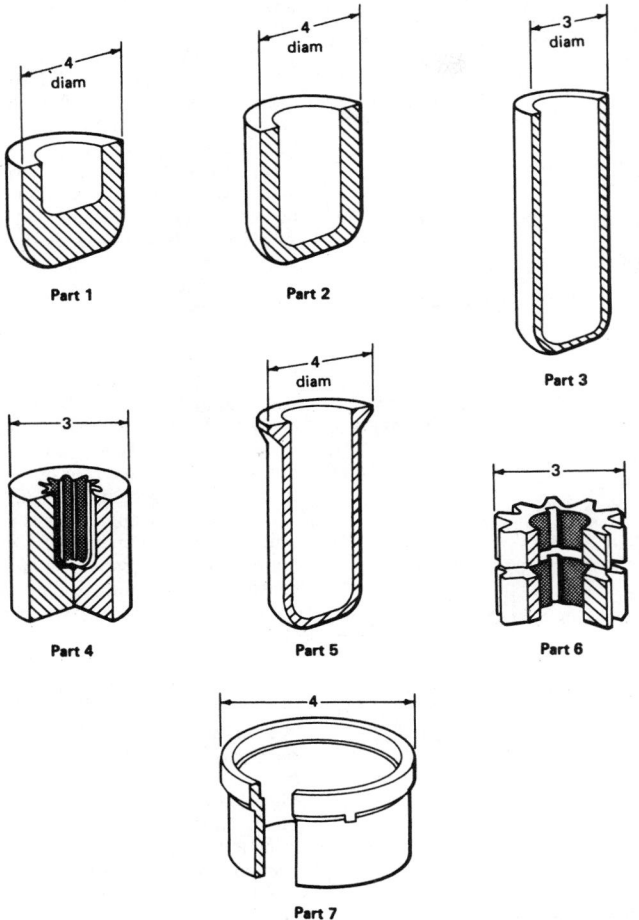

Part 1   Part 2   Part 3

Part 4   Part 5   Part 6

Part 7

Dimensions are in inches; for equivalent metric sizes (mm), multiply by 25.4. For typical die materials used in extruding these parts from low-carbon steel and aluminum, see Tables 15 to 20.

Fig. 9. Seven hypothetical cold extruded parts

Table 21. Typical steels for secondary tooling components used in extruding steel parts

Part number	Related table	Type of operation	Upper die plate	Punch backup plate	Inner shrink ring	Outer shrink ring	Primary support block	Secondary support block	Lower die plate
1 and 2	Table 15	Backward . . .	1040	S1	S4, S5	1040	W2	W2	1040
3	Table 16	Drawing	1040	S1	S4, S5	1040	W2	W2	1040
4	Table 17	Drawing	W2	8620 (a)	4340 (a)	W2	S4, S5	D4	W2
5	Table 18	Forward	1040	S1	S4, S5	1040	W2	W2	1040
6	Table 19	Forward	W2	8620 (a)	4340 (a)	W2	S4, S5	D4	W2
7	Table 20	Forward	W2	8620 (a)	4340 (a)	W2	S4, S5	D4	W2

Note: Part designs are shown in Fig. 9.

(a) Or other alloy steel having hardenability appropriate for the component.

Table 22. Cemented carbides used most frequently for cold extrusion punches and die inserts

Type of service	Composition, %		Grain size
	Tungsten carbide	Cobalt binder	
Punches			
High impact	84	16	Fine
Medium impact	88	12	Fine
Dies and die inserts			
High impact	75	25	Medium coarse
Medium impact	84	16	Medium fine
Light impact, maximum wear	88	12	Fine

Tools for Drawing Wire, Bar and Tubing

SELECTION of tool materials for cold drawing of metal into continuous forms such as wire, bar and tubing depends primarily on the size, composition, shape, stock tolerance and quantity of the metal being drawn. Cost of the tool material is also important and may be decisive.

Dies and mandrels used for cold drawing are subjected to severe abrasion. For this reason, most wire, bar and tubing is drawn through dies having diamond or cemented tungsten carbide in-serts, and tube mandrels usually are fitted with carbide nibs. Small quantities, odd shapes and large sizes are more economically drawn through hardened tool steel dies.

WIREDRAWING DIES

Table 23 gives recommended materials for wiredrawing dies. For round wire, dies made of diamond or cemented tungsten carbide are always recommended, without regard to the composition or quantity of the metal being drawn. For short runs or special shapes, hardened tool steel is less costly, although carbide gives superior performance in virtually any application.

Diamond Dies. The use of diamond dies is restricted only by limitations on the sizes of available industrial diamonds and by cost, which is extremely high for diamonds in larger sizes. These tools can outperform cemented tungsten carbide dies by 10 to 200 times, depending on the alloy being drawn, and thus are cost effective despite their high unit cost.

Diamond dies are available in two types: natural and synthetic. Natural diamond dies, which have been used longer, are made from single crystals and produce exceptional surface finishes in the drawn product. Generally, natural diamond is less expensive than synthetic diamond for smaller hole sizes. For a hole size of 0.66 mm (0.0259 in.), the two materials are about the same in cost. For larger sizes, synthetic diamond is more economical.

Cemented tungsten carbide is economical for wiredrawing dies in most applications above the range of size where diamond can be used. The softer cemented carbides, which contain about 8% cobalt, are less brittle and can withstand greater stock reductions without breaking, but wear more rapidly than lower-cobalt grades.

Tool steel used for wiredrawing dies should have near-maximum hardness (62 to 64 HRC) for reductions below about 20%. For greater reductions, because of the possibility of breakage, hardness should be decreased to 58 to 60 HRC, even though the rate of wear will increase.

DRAWING BARS AND TUBING

Table 24 shows die and mandrel materials recommended for drawing bars, tubing and complex shapes. Diamond is virtually never used in larger sizes; cemented tungsten carbide is recommended for three-fourths of all applications. Tool steels are rarely used to make tools for drawing commercial-quality round bars less than 90 mm (3.5 in.) in diameter. Cemented tungsten carbide is used to draw stainless steel tubes as large as 280 mm (11 in.) in outside diameter.

Mandrels. Either carbide or hardened tool steel is satisfactory for mandrels used in drawing tubes, but carbide is more economical for tubes less than 125 mm (5 in.) in diameter. Carbide nibs are available in lengths sufficient for this purpose. Mandrel nibs are available in either a braze-type design or a shell design that permits mechanical attachment to the shank for easy replacement. Carbide tips are also recommended for mandrels used in drawing of shapes, but with reservations on the use of carbide similar to those that apply to the use of carbide in dies for drawing of shapes. Tool steel mandrels are recommended for drawing of tubes over 125 mm (5 in.) in inside diameter.

Table 23. Recommended materials for wiredrawing dies

Metal to be drawn	Wire size mm	Wire size in.	Recommended die materials for: Round wire	Recommended die materials for: Special shapes
Carbon and alloy steels	<1.57	<0.062	Diamond, natural or synthetic	M2 or cemented tungsten carbide
	>1.57	>0.062	Cemented tungsten carbide	
Stainless steels; titanium, tungsten, molybdenum and nickel alloys	<1.57	<0.062	Diamond, natural or synthetic	M2 or cemented tungsten carbide
	>1.57	>0.062	Cemented tungsten carbide	
Copper	<2.06	<0.081	Diamond, natural or synthetic	D2 or cemented tungsten carbide
	>2.06	>0.081	Cemented tungsten carbide	
Copper alloys and aluminum alloys	<2.5	<0.100	Diamond, natural or synthetic	D2 or cemented tungsten carbide
	>2.5	>0.100	Cemented tungsten carbide	
Magnesium alloys	<2.06	<0.081	Diamond, natural or synthetic	
	>2.06	>0.081	Cemented tungsten carbide	

Table 24. Recommended tool materials for drawing bars, tubing and complex shapes

Metal to be drawn	Round bars and tubing(a) Common commercial sizes Bar and tube dies	Round bars and tubing(a) Common commercial sizes Tube mandrels(b)	Maximum commercial size(c): dies and mandrels	Complex shapes: dies and mandrels(a)(b)
Carbon and alloy steels	Tungsten carbide	W1 or carbide	D2	CPM 10V or carbide
Stainless steels; titanium, tungsten, molybdenum and nickel alloys	Diamond or carbide(d)	D2 or carbide	D2 or M2(a)	F2 or carbide(e)
Copper, aluminum and magnesium alloys	W1 or carbide	W1 or carbide	D2 or CPM 10V	O1, CPM 10V or carbide

(a) Tool steels for both dies and mandrels are usually chromium plated. (b) "Carbide" indicates use of cemented carbide nibs fastened to steel rods. (c) 10-in. OD by 3/4-in. wall. (d) Under 1.6 mm (0.062 in.), diamond; over 1.6 mm (0.062 in.), tungsten carbide. (e) Recommendations for large tubes or complex shapes apply to stainless steel only.

Closed-Die Hot Forging Tools

THE CLOSED-DIE FORGING TOOLS discussed here are restricted to die blocks, die inserts and trimming tools used for hot forging in vertical presses and hammers. In hammer forging, the hammer—whether a gravity, steam, airdrop or counterblow hammer—strikes a sudden blow, imposing a shock load on the forging dies. In press forging, the working pressure is applied as a fast push rather than a blow, so that the dies are subjected to less shock. However, because a press is much slower acting than a hammer, the dies in a press absorb more heat from the hot blank during the forging cycle.

The size and shape of the part being forged influence the force and energy required to reshape the hot plastic metal—from the initial shape of the forging stock (usually round, square or flat) to that of the finished forging. The force and energy required are further influenced by the composition of the metal being forged. For example, as the alloy content of a steel increases, hot strength and subsequent resistance to flow also increase, and more energy is required to forge the same shape. Similarly, some copper alloys and aluminum alloys are considerably easier to forge than others.

The basic causes of premature die failure are excessive force, abrasion, and excessive temperature. In addition, dies may break in a brittle manner if used cold, and thus preheating to 260 to 300 °C (500 to 600 °F) is recommended. This may be accomplished by placing "warmers" (pieces of hot steel) between the dies or by installing gas-fired or electrical heating devices to maintain temperature during idle periods.

DIE MATERIALS

Prehardened die blocks suitable for making forging dies are available in a range of compositions and hardnesses. Prehardened tool steels also are available in other forms for making small die blocks, die inserts and trimming tools. All steels available in a prehardened condition also are available in the annealed condition for ease of machining; once machined to the desired contours, they can be hardened and finished by methods ordinarily used for standard tool steels.

Prehardened die-block steels usually are purchased on the basis of hardness and proprietary name. Typical tool steels, and their Brinell hardnesses, for die blocks and die inserts used in hammer and press forging are shown in Tables 25A and 25B. In these tables, recommended die materials are listed for the six hypothetical shapes of increasing severity illustrated in Fig. 10. In-

formation is included for two production quantities and four types of work metal.

The recommendations for steel and hardness level in Tables 25A and 25B are based on lowest over-all cost (which includes both material and fabrication costs), and on avoiding breakage. Lower-alloy tool steels at lower hardnesses are acceptable when relatively few parts are to be made from carbon or low-alloy steel. Somewhat higher hardness is required in dies for forging stainless steels and heat-resisting alloys.

The relatively expensive, high-alloy H11 and H12 tool steels are desirable for forging copper alloys, because copper alloys are quite resistant to flow at their maximum forging temperatures. Copper oxide is abrasive, and copper alloy forgings usually are made to closer tolerances than steel forgings; both of these characteristics demand high wear resistance in the die material.

As shown in Tables 25A and 25B, higher hardnesses can be used for press-forging dies than for hammer-forging dies because the former are not subjected to impact. Because of the longer times in contact with hot work metal, press-forging dies must be made of a steel having greater resistance to softening at elevated temperatures.

The compositions of certain nonstandard tool steels used for die blocks are shown in Table 26.

TOOLS FOR TRIMMING

Usually, trimming of metal flash at temperatures below 150 °C (300 °F) is referred to as cold trimming, and trimming at 500 °C (1000 °F) or higher is termed hot trimming. For the purpose of selecting materials for trimming tools, trimming at temperatures between 150 and 500 °C is also considered hot trimming.

Whether trimming is done hot or cold depends chiefly on whether the trim is to be normal or close and on the composition of the metal being trimmed. Table 27 gives typical materials for both hot and cold trimming tools. Carbon and alloy steels may be trimmed either hot or cold, but close trimming is generally done hot. Stainless steels and heat-resisting alloys are usually trimmed hot, whereas nonferrous alloys may be trimmed either hot or cold.

Cold Trimming. Punches for cold trimming carbon and alloy steel forgings are commonly made from discarded die blocks. Because of this practice, prehardened low-alloy die steels are the materials most widely used for punches and are considered satisfactory for many different types of punches. Although high-carbon alloy tool steels such as A2 have been used successfully as blade materials for normal cold trimming of carbon and alloy steel forgings, D2 is usually a better choice because of its longer life.

Hardened and tempered 6150 steel has been used successfully in punches for normal cold trimming of nonferrous forgings. Other alloy steels similar to 6150 in hardenability are also satisfactory for this application. For close trimming, the edge-holding properties of D2 make it more desirable as a punch material. It is also recommended for blades used in close cold trimming of nonferrous forgings.

Hot Trimming. Prehardened die steels are satisfactory punch materials for hot trimming of carbon, alloy and stainless steel forgings. Punch materials for hot trimming of nonferrous forgings are less critical, and 1020 steel, either as rolled or as annealed, is widely used for reasons of economy.

Table 25A. Typical tool steels for die blocks and die inserts
For illustration of forged parts, see Fig. 10.

| Work metals | Typical die materials and hardness ranges | | | |
| | Hammer forging | | Press forging | |
	100 to 10 000 parts	10 000 parts or more	100 to 10 000 parts	10 000 parts or more
For making parts of severity no greater than part 1				
Carbon and alloy steels	6F2 or 6G at 341 to 375 HB	6F2 or 6G at 388 to 429 HB	6F2 or 6G at 388 to 429 HB	6F3 at 369 to 388 HB or H11 or H12 at 388 to 405 HB(a)
Stainless steels and heat-resisting alloys	6F2 or 6G at 388 to 429 HB	6F2 or 6G at 388 to 429 HB	6F2 or 6G at 388 to 429 HB	H11 or H12 insert at 477 to 543 HB or H26 at 514 to 577 HB(b)
Aluminum and magnesium alloys	6F2 or 6G at 302 to 331 HB	6F2 or 6G at 341 to 375 HB	6F2 or 6G at 341 to 375 HB	6F3 at 375 to 405 HB or H11 or H12 at 448 to 477 HB(a)
Copper and copper alloys	6F2 or 6G at 341 to 375 HB or H11 or H12 at 405 to 433 HB	H11 or H12 at 405 to 448 HB	6F2 or 6G at 341 to 375 HB or H11 or H12 at 477 to 514 HB	H11 or H12 at 477 to 514 HB
For making parts of severity no greater than part 2				
Carbon and alloy steels	6F2 or 6G at 341 to 375 HB	6F2 or 6G at 341 to 375 HB	6F2 or 6G at 388 to 429 HB	6F3 at 369 to 388 or H11 or H12 at 388 to 405 HB(a)
Stainless steels and heat-resisting alloys	6F2 or 6G at 341 to 375 HB	6F2 or 6G at 341 to 375 HB with 6F3 insert at 405 to 448 HB	...	...
Aluminum and magnesium alloys	6F2 or 6G at 302 to 331 HB	6F2 or 6G at 341 to 375 HB or H11 or H12 at 405 to 448 HB(a)	6F2 or 6G at 341 to 375 HB	6F2 or 6G at 341 to 375 HB with 6F3 insert at 405 to 448 HB or H11 or H12 at 448 to 477 HB(a)
Copper and copper alloys	6F2 or 6G at 341 to 375 HB	6F2 or 6G at 341 to 375 HB or H11 or H12 at 405 to 448 HB(a)	6F2 or 6G at 341 to 375 HB	H11 or H12 at 477 to 514 HB
For making parts of severity no greater than part 3				
Low-alloy steels, stainless steels and heat-resisting alloys	6F2 or 6G at 302 to 331 HB	6F2 or 6G at 302 to 331 HB with insert of same steel at 341 to 375 HB	...	...
Aluminum, magnesium and copper alloys	6F2 or 6G at 269 to 293 HB	6F2 or 6G at 302 to 331 HB	...	...

(a) Recommended for long runs—for example, 50 000 forgings. (b) Recommended for forging higher-alloy heat-resisting materials, such as nickel-base and cobalt-base alloys.

Hard faced carbon steels are often preferred for use as trimming edges (see Table 27). One advantage of hard faced blades is that chipped or broken edges are easily repaired. The cobalt-base alloy indicated in Table 27 (type 4A) is extremely high in resistance to shock, heat and abrasion. Most forging shops prefer to use this alloy (or a similar one) for facing all trimming blades, either hot or cold. Information on alloy 4A and other hard facing alloys may be found elsewhere in this volume.

TOOLS FOR ISOTHERMAL FORGING

The temperatures typically used for isothermal forging of titanium alloys (870 to 980 °C, or 1600 to 1800 °F) and of nickel-base superalloys (925 to 1100 °C, or 1700 to 2000 °F) impose severe hot-strength, creep and oxidation-resistance requirements on the die materials. Dies for isothermal forging of titanium alloys typically are made of cast nickel-base superalloys that were initially developed for blades in gas turbine engines. Of these superalloys, Inconel 713C and IN-100 have been used with great success for isothermal forging dies. However, in the most severe environments of high temperature and stress, excessive creep can be a major cause of failure for IN-100 dies. In severe environments, replacement of IN-100 by alloys of even higher creep strength, such as TRW-NASA VIA, has proved effective.

Isothermal forging of nickel-base superalloys often is performed with dies made of molybdenum alloys. The susceptibility of molybdenum alloy dies to oxidation at working temperatures requires that forging be performed in vacuum.

Hot Upset Forging Tools

THE TOOLS discussed in this article are restricted to header dies, gripper dies and auxiliary tools used in forging machines or upsetters operating in a horizontal plane.

Forgings made by the hot upset method vary in size from 13-mm (1/2-in.) bolts to 305-mm (12-in.) flanged pipe sections for the oil industry. Production rate varies with size of forging and amount of automation. Use of automatic feeds allows rates as high as 7200 pieces per hour in such applications as boltmaking, and such rates require tool materials that will serve continuously at high temperatures for long periods of time. For medium-size parts such as automotive forgings, which are produced at rates of 120 to 150 pieces per hour, the dies cool off enough between blows so that die materials with lower hot strength can be used. The high-alloy, high-hardness tools used in the bolt industry are not suitable for medium-size automotive upset forgings because such high-hardness tools are too susceptible to breakage. Dies for still larger upset forgings made at lower rates may require higher strength at forging temperature and higher alloy content because of the longer sustained contact between the hot workpiece and the tools.

Complexity of die shape also influences selection of die steels for hot upset forging. Sharp corners and edges greatly increase stress concentration, and thin sections may be subject to extreme loads and high thermal stresses. Internal punches and mandrels are subjected to high impact loads and sliding abrasive wear, and often are designed to be replaceable because of their short life. Replaceable inserts may be used for areas of gripper dies subject to short life and for parts requiring close tolerances.

The material to be forged is important in selecting a die material. At forging temperature, carbon and low-alloy steels have lower strength than stainless steels and heat-resisting alloys and can be forged with less costly tool steels. In upset forging of heat-resisting steels and titanium alloys, even the best die steels may have a short life. The compositions and properties of the AISI grades and types of tool steel discussed in this article are presented elsewhere in this volume. The nominal compositions of six nonstandard tool steels used to make tools for hot upset forging are shown in Table 28.

HEADING TOOLS AND GRIPPER DIES

Tools for hot upset forging generally are constructed in the form of insert dies. Table 29 summarizes the typical materials used for hot upset forging tools for making parts of the various degrees of severity shown in Fig. 11. Part 1 in Fig. 11 represents straight upset forging; parts 2 and 3 represent two degrees of severity for pierced and upset forgings. Each of the three shapes is made in one blow. For all three, the original unsupported length of stock required to make the part is 2.5 times the diameter of the starting stock.

In hot upset forging of simple flanged shapes from low-carbon and alloy carburizing steels, the heading tool usually wears faster than the gripper die. For example, in hot upsetting of hexagonal bolt heads on 1020, 1045 or 4140 steel shanks 13 to 25 mm (0.5 to 1 in.) in diameter, H13 tool steel is used for both the heading tool and the gripper die, but the heading tool has greater hardness. Under the same wear conditions, the higher hardness could have been expected to provide a lower rate of wear, but forging of about 28 000 pieces produced the same amount of wear (0.15 mm, or 0.006 in.) in the heading tool as forging of about 40 000 pieces produced in the softer gripper tool.

Table 25B. Typical tool steels for die blocks and die inserts
For illustration of forged parts, see Fig. 10.

Work metals	Hammer forging 100 to 10 000 parts	Hammer forging 10 000 parts or more	Press forging 100 to 10 000 parts	Press forging 10 000 parts or more
For making parts of severity no greater than part 4				
Carbon and alloy steels	6F2 or 6G at 341 to 375 HB, solid or with H11 or H12 plug(a) at 369 to 388 HB	6F2 or 6G at 341 to 375 HB with H11 or H12 plug at 369 to 388 HB or H11 or H12 at 405 to 433 HB	6F2 or 6G at 388 to 429 HB, solid or with H11 or H12 plug(a) at 405 to 433 HB	6F2 or 6G at 388 to 429 HB with H11 or H12 plug at 405 to 433 HB
Stainless steels and heat-resisting alloys	6F2 or 6G at 341 to 375 HB, solid or with H11 or H12 plug(a) at 429 to 448 HB	6F2 or 6G at 341 to 375 HB with H11 or H12 insert at 429 to 448 HB	6F2 or 6G at 388 to 429 HB, solid or with H11 or H12 plug(a) at 429 to 448 HB	6F2 or 6G at 341 to 375 HB with H11 or H12 plug at 429 to 448 HB
Aluminum and magnesium alloys	6F2 or 6G at 341 to 375 HB or H11 or H12 at 405 to 433 HB	6F2 or 6G at 341 to 375 HB with H11 or H12 plug at 405 to 433 HB or H11 or H12 at 405 to 433 HB	6F2 or 6G at 341 to 375 HB or H11 or H12 at 405 to 433 HB	6F2 or 6G at 341 to 375 HB with H11 or H12 plug at 429 to 448 HB
Copper and copper alloys	H11 or H12 at 405 to 433 HB	H11 or H12 at 405 to 433 HB	H11 or H12 at 405 to 433 HB	6F2 or 6G at 341 to 375 HB with H11 or H12 plug at 429 to 448 HB
For making parts of severity no greater than part 5				
Carbon and alloy steels	6F2 or 6G at 302 to 331 HB	6F2 or 6G at 302 to 331 HB, solid or with 6F3 plug(b) at 369 to 388 HB	6F2 or 6G at 341 to 375 HB	6F3 at 369 to 388 HB with H11 or H12 plug at 369 to 388 HB
Stainless steels and heat-resisting alloys	6F2 or 6G at 302 to 331 HB	6F2 or 6G at 302 to 331 HB with H11 or H12 plug at 369 to 388 HB	. . .	. . .
Aluminum and magnesium alloys	6F2 or 6G at 269 to 293 HB	6F2 or 6G at 269 to 293 HB with plug of same steel at 302 to 331 HB	6F2 or 6G at 341 to 375 HB	6F2 or 6G at 341 to 375 HB with H11 or H12 plug(c) at 429 to 448 HB
Copper and copper alloys	6F2 or 6G at 302 to 331 HB	6F2 or 6G at 302 to 331 HB with H11 or H12 plug at 405 to 448 HB	6F2 or 6G at 341 to 375 HB	H11 or H12 at 477 to 514 HB
For making parts of severity no greater than part 6				
Low-alloy steels, stainless steels and heat-resisting alloys	6F2 or 6G at 269 to 293 HB(d)	6F2 or 6G at 269 to 293 HB with plug of same steel at 341 to 375 HB	. . .	. . .
Aluminum, magnesium and copper alloys	6F2 or 6G at 269 to 293 HB	6F2 or 6G at 269 to 293 HB	. . .	. . .

(a) Recommended for 1000 to 10 000 forgings. (b) Recommended for long runs—for example, 50 000 forgings. (c) For long runs—for example, 50 000 forgings—a solid block made from H11 or H12 tool steel at 477 to 514 HB is recommended. (d) For quantities over 1000, a plug of the same material at 341 to 375 HB is recommended.

Table 26. Nominal compositions of nonstandard tool steels for die blocks

Steel(a)	C	Mn	Si	Cr	Ni	Mo	V
6G	0.55	0.80	0.25	1.00	. . .	0.45	0.10(b)
6F2	0.55	0.85	0.25	1.00	1.00	0.40	0.10(b)
6F3	0.55	0.60	0.85	1.00	1.80	0.75	0.10(b)

(a) Neither AISI nor SAE has assigned type numbers to these tool steels. (b) Optional.

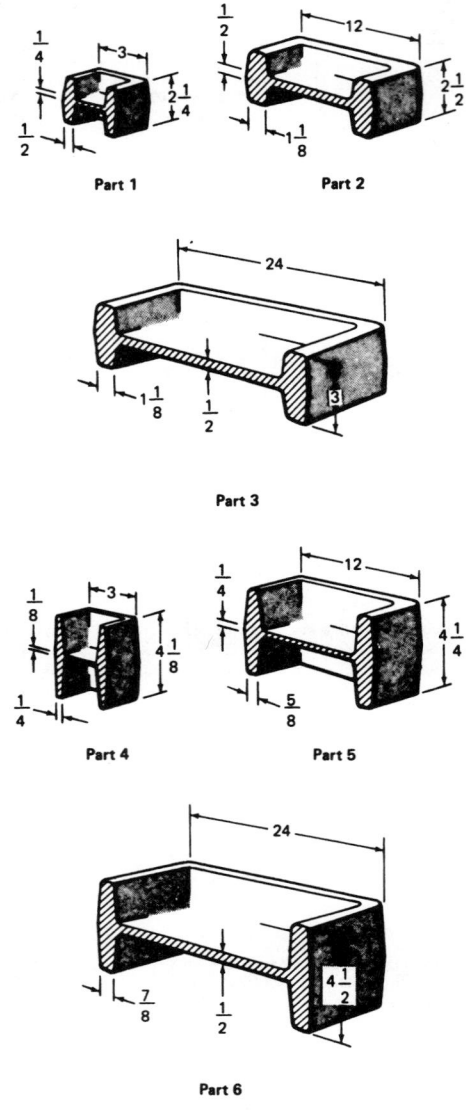

Dimensions are in inches; to find equivalent metric units (mm), multiply listed values by 25. For typical die materials and hardness ranges, see Tables 25A and 25B.

Fig. 10. Six hypothetical parts of progressively increasing severity

The reverse may be true under certain severe conditions of forming. For instance, an unsymmetrical steering-gear forging was made with the segment stock all gathered on one side in the gripper die, and this resulted in greater metal movement, with correspondingly heavier die loads and increased abrasion, in that area.

The quantity of water used as a coolant for hot upset forging dies ordinarily does not affect selection of a die steel for forging a given part. In the bolt industry, however, water-cooled tools made of tungsten-bearing tool steels are susceptible to cracking and heat checking at high production rates (on the order of 100 pieces per min-

ute). At high production rates such as these, the hot bolt stock is in nearly constant contact with the dies, allowing them little chance to cool, and under these conditions water cooling imposes severe thermal shock. Low-carbon (0.58 to 0.65% C) M10 high speed steel heat treated to 58 HRC performs well in such applications, as does T1 heat treated to the same hardness.

When a lower-alloy tool steel such as 6G or 6F2 is used for inserts, wear resistance can be improved by adjusting water flow to keep die temperatures below 200 °C (400 °F). At such temperatures, little improvement can be gained by changing to a different low-alloy die steel.

Lubrication of dies impairs speed and is not used extensively for upset forging of steel. Some lubrication may be used in deep punching and piercing, but only enough to prevent the workpiece from sticking to the punch. Resistance to wear and abrasion can be improved by use of die

Table 27. Typical materials for trimming dies

Material to be trimmed	Cold trimming				Hot trimming(a)	
	Normal trim		Close trim			
	Punch	Blade	Punch	Blade	Punch	Blade
Carbon and alloy steels	6F2 or 6G, at 341 to 375 HB	D2 at 54 to 56 HRC	Generally hot trim		6F2 or 6G at 341 to 375 HB	Hard facing alloy 4A on 1035 steel(b); or D2 at 58 to 60 HRC
Stainless steels and heat-resisting alloys	Generally hot trim		Generally hot trim		6F2 or 6G at 388 to 429 HB	D2 at 58 to 60 HRC
Aluminum, magnesium and copper alloys	6150 at 461 to 477 HB	Hard facing alloy 4A on 1020 steel(b); or O1 at 58 to 60 HRC	D2 at 58 to 60 HRC	D2 at 58 to 60 HRC	1020 soft	Hard facing alloy 4A on 1020 steel(b)

(a) Both normal and close trimming. (b) Hard facing alloy 4A has nominal composition as follows: 1 C, 30 Cr, 3 Ni, 4.5 W, 60 Co, rem Fe. For greater detail, refer to the article on Hard Facing Materials in this volume.

Table 28. Nominal compositions of nonstandard tool steels used in tools for hot upset forging

Type(a)	Composition, %						
	C	Mn	Si	Cr	Ni	Mo	V
6F	0.55	0.80	0.25	1.00	...	0.45	0.10(b)
6F2	0.55	0.75	0.25	1.00	1.00	0.30	0.10(b)
6F3	0.55	0.60	0.85	1.00	1.80	0.75	0.10(b)
6F4	0.20	0.70	0.25	...	3.00	3.35	...
6H1	0.55	...	...	4.00	...	0.45	0.85
6H2	0.55	0.40	1.10	5.00	1.50	1.50	1.00

(a) UNS, AISI and SAE have not assigned type numbers to these tool steels. (b) Optional.

Table 29. Typical tool materials for hot upset forging
For part designs and over-all dimensions, see Fig. 11.

Material forged	Tool material types and hardness ranges for total production quantity of:					
	100		1000 to 10 000		50 000 and up	
	Gripper die	Heading tool	Gripper die	Heading tool	Gripper die	Heading tool
For parts of maximum outside upsetting severity (part 1)(a)						
Carbon and low-alloy steels	4150 at 38 to 42 HRC or 4340 insert at 38 to 42 HRC	W1 with 0.70 C at 42 to 46 HRC or 4340 insert at 38 to 42 HRC	6H1 or H11 at 46 to 50 HRC or 4340 insert at 38 to 42 HRC	6H1 or H11 at 44 to 48 HRC or 6G (b) insert at 41 to 45 HRC	6H1 or H11 at 46 to 50 HRC or 4340 insert at 38 to 42 HRC	H11 at 46 to 50 HRC or 6H2 at 52 to 56 HRC or 6G (b) insert at 41 to 45 HRC
Stainless steels and heat-resistant alloys (up to type 310)	6G (b) insert at 38 to 42 HRC	6F3 insert at 42 to 46 HRC	6F3 insert at 42 to 46 HRC	H11 at 46 to 50 HRC or same for insert	6H2 at 52 to 56 HRC or H11 insert at 44 to 48 HRC	6H2 at 52 to 56 HRC or H11 insert at 48 to 52 HRC
For parts requiring both upsetting and piercing (parts 2 and 3)						
Carbon and alloy steels	4340 insert at 38 to 42 HRC or 6G (b) insert at 36 to 40 HRC	4340 insert at 42 to 46 HRC or 6G (b) insert at 41 to 45 HRC	4340 insert at 38 to 42 HRC or 6G (b) insert at 36 to 40 HRC	H11 at 42 to 46 HRC or H11 insert at 46 to 48 HRC	H11 or H11 insert at 42 to 46 HRC	H12 or M10 at 50 to 52 HRC or H11 insert at 46 to 50 HRC
Stainless steels and heat-resistant alloys (up to type 310)	(c)	(c)	(c)	(c)	(c)	(c)

(a) All heads are round and made in one blow with relative dimensions shown. (b) 6F2 die steel may be used interchangeably with 6G. (c) The same tool materials are recommended for upsetting part 2 as are shown for part 1. Part 3 is too severe to be made from a stainless steel or a heat-resistant alloy.

lubricants, but there is no known correlation between lubrication and die performance that would make lubrication a factor in die-steel selection.

Hot Extrusion Tooling

TOOLING for hot extrusion must operate under severe conditions of temperature, pressure and abrasive wear. Fundamentally, the extrusion process consists of forcing material in a plastic condition through a suitable die under high pressure to form a long, continuous shape. Some of the softer metals such as lead and aluminum are sufficiently plastic to be extruded at or near room temperature, but most other metals and alloys are extruded only at elevated temperatures.

The prevailing commercial hot extrusion process, based on tonnage of product, is single-charge direct extrusion with butt discard. The essential parts of the tooling for this process are illustrated in Fig. 12.

With the ram retracted, a hot billet or slug is placed in the container. A dummy block is inserted between the ram and billet; then the hot billet is pushed into the container liner and advanced under high pressure against the die. The metal is squeezed through the die opening, assuming the desired shape, and is severed from the remaining stub by sawing or shearing.

Tool Materials. Table 30 lists typical materials and hardnesses for tools used in hot extrusion. Hot extrusion of aluminum and magnesium is in many respects similar, the major difference being the pressure required. Often, the same tooling materials can be used for extrusion of either aluminum or magnesium.

In addition to the typical materials listed in Table 31, special insert materials and surface treatments have been specified (particularly for tools used in extruding complex shapes) where better resistance to wear at higher temperatures is required. Special insert materials include special grades of cemented tungsten carbide, nickel-bonded titanium carbides, and aluminum oxide ceramics. Special surface treatments include nitriding, aluminide coating, and application of proprietary materials by vapor deposition or sputtering.

Die-Casting Dies

MATERIALS for die-casting dies must have good resistance to thermal shock and good resistance to softening at elevated temperatures. Resistance to softening is necessary because dies must be able to withstand the erosive action of molten metal under high injection velocity. Other properties that influence selection of materials for die-casting dies are hardening characteristics, machinability, resistance to heat checking, and weldability.

Performance of die-casting dies is directly related to casting temperature, injection pressure, thermal gradients within the dies, and frequency of exposure to high temperature. These variables are the principal ones used in the selection tables in this article. Tool steels of relatively high alloy content are required for components in direct contact with molten metal, as indicated in Table 32.

Die hardness is less critical for die casting of zinc than for die casting of alloys of higher cast-

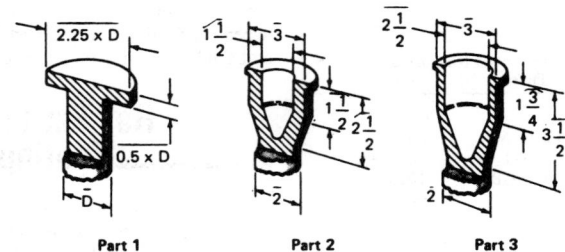

Part 1 **Part 2** **Part 3**

Dimensions for parts 2 and 3 are given in inches; to find equivalent metric dimensions (mm), multiply listed dimensions by 25. Corner radii comply with standard commercial practice. For typical header-tool and gripper-die materials, see Table 29.

Fig. 11. Three typical parts made by hot upset forging

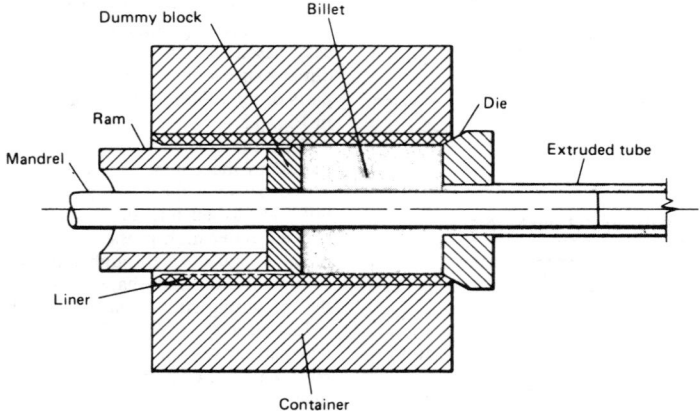

In this arrangement, the mandrel is attached to the press piercer and operates through a hollow ram.

Fig. 12. Typical tooling for extrusion of seamless tube

Table 30. Typical materials and hardnesses for tools used in hot extrusion

Tooling application	Aluminum and magnesium Tool material	Hardness, HRC	Copper and brass Tool material	Hardness, HRC	Steel Tool material	Hardness, HRC
Dies, for both shapes and tubing	H11, H12, H13	47-51	H11, H12, H13 H14, H19, H21	42-44 34-36	H13 Cast H21 inserts	44-48 51-54
Dummy blocks, backers, bolsters, and die rings	H11, H12, H13	46-50	H11, H12, H13 H14, H19 Inconel 718	40-44 40-42 ...	H11, H12, H13 H19, H21 Inconel 718	40-44 40-42 ...
Mandrels	H11, H13	46-50	H11, H13	46-50	H11, H13	46-50
Mandrel tips and inserts	T1, M2	55-60	Inconel 718	...	H11, H12, H13 H19, H21	40-44 45-50
Liners	H11, H12, H13	42-47	A-286, V-57	...	H11, H12, H13	42-47
Rams	H11, H12, H13	40-44	H11, H12, H13	40-44	H11, H12, H13	40-44
Containers	4140, 4150, 4340	35-40	4140, 4150, 4340	35-40	H13	35-40

Table 31. Typical compositions of superalloys used for extrusion tools

Alloy	C	Mn	Si	Cr	Ni	Mo	Nb	Ti	Al	Fe	V	B
A-286	0.05	1.35	0.50	15.00	26.00	1.25	...	2.00	0.20	53.6	0.30	0.015
V-57	0.08	0.35	0.75	14.80	27.00	1.25	...	3.00	0.25	52.0	0.30	0.010
Inconel 718	0.04	0.20	0.30	18.60	53.00	3.10	5.0	0.90	0.40	18.5	...	...

ing temperature. Consequently, prehardened insert steels (29 to 34 HRC) are often used to make tooling for die casting of zinc.

Hot work tool steels are almost always used to make tooling for casting higher-melting-point alloys, such as aluminum, magnesium and copper. For casting aluminum and magnesium alloys, H13 steels are hardened to about 44 to 48 HRC; for casting copper alloys, H20, H21 and H22 steels generally are used at hardnesses of 38 to 45 HRC.

Injection components for both zinc and aluminum die-casting dies are subjected to considerable contact with molten metal and to severe erosion. Recommended materials and heat treatments for such dies are given in Table 33.

Slides, Guides, Cores and Pins. The moving parts of a die-casting die must have the same general characteristics as those of the stationary parts, and also must provide resistance to wear. Table 34 lists materials recommended for slides, guides, cores and ejector pins.

Die Lubricants. Most commercial die lubricants or mold-release agents reduce soldering or sticking of the casting to the die. The use of an effective lubricant results in easier cleaning of dies, less wear and longer runs between die polishes.

However, some of these die lubricants attack the die steel as well as the particles of the casting alloy adhering to the die surface. If used frequently, such lubricants will shorten die life.

Preheating and Cooling. Water cooling of cores and slides must be considered where significant wear is likely. Die-casting dies used for casting aluminum or copper are water cooled in service and should be preheated to about 175 °C (350 °F) before a run is started. Preheating should be done with water flowing slowly through the die to prevent the serious damage that would result from introduction of cold water into a hot die.

Ejection components are not subjected to high temperatures. Consequently, hot rolled 1020 steel in the as-received condition is used for making ejector boxes, rails, ejector plates, support posts and support blocks for tooling employed in die casting of aluminum and zinc parts.

Trim dies are used in secondary operations and are made of wear-resistant materials, as indicated in Table 35.

Powder-Compacting Tools

THE COMPACTING PROCESS most widely used to convert metallic powders into components is the closed-die process. The basic tooling for this process includes a precision-machined die, an upper and lower punch and, if the part is to be hollow, a core rod. These components are attached to the platens of a mechanical or hydraulic press by means of adapters, which often enclose the basic tooling and give it backup support.

The materials generally used for these tools are (a) cemented carbides and (b) tool steels with or without special surface treatments.

Dies. Die inserts for compaction of carbide, ceramic or ferrite powder most frequently are the medium- or coarse-grain 94WC-6Co grades of cemented carbide. Cemented tungsten carbide containing 12 to 16% cobalt may be used to make inserts for compacting metal powders in medium-to-long production runs.

Cemented carbides are relatively expensive, and shaping of parts to the required form must be done either by electrical discharge machining or by specialized methods of grinding.

Wear-resistant tool steel inserts are sometimes used instead of carbide inserts. Crucible CPM 10V is frequently chosen for medium-to-long production runs. Other wear-resistant tool steels, usually D2 or a high speed steel such as M2 or M4, have been used for short-run applications, as have high-wear P/M alloys. Tool steel inserts generally are heat treated to a working hardness of 62 to 64 HRC. For increased wear resistance, a nitrided case may be specified for dies made of CPM 10V or D2. For certain part designs, a solid die rather than an insert die is a more practical choice; an air-hardening 5%-Cr tool steel such as A2 is generally used for such applications.

Punches. The stresses imposed on punches during service are such that toughness is a much more important material requirement than wear resistance, although wear resistance cannot be ignored. Type A2, and sometimes the shock-resisting type S7, is preferred for punches.

For applications in which A2 or S7 punch faces become severely abraded, a more wear-resistant grade such as D2, D3, high-wear P/M alloys or M2 should be considered. Cemented carbides are too brittle to perform successfully as punches or punch-face inserts.

Core Rods. Both toughness and wear resistance are important criteria in selection of core-rod materials, but generally the primary consideration is wear resistance. For particularly abrasive conditions, CPM 10V has been used successfully, as have D2, M2 and A2 tool steels that have been nitrided or coated with tungsten carbide.

Molds for Plastics and Rubbers

IN MOLDS for parts made of plastic materials (including rubbers), the type of material to be molded is a major factor governing the choice of mold material. For example, some plastics require mold materials that resist abrasion, corrosion, heat, or high compression loads.

This article covers typical mold materials, and methods of producing molds, for injection, blow, transfer and compression molding of the chief groups of plastic materials. The following definitions describe the basic steps in each molding process, and the principal tooling involved.

Injection molding is a method of forming plastic objects from granular or powdered thermoplastics by (*a*) using heat and pressure to plasticize the material in a chamber and then (*b*) forcing part of the fluid mass into a cooler chamber, where it solidifies. Injection molding requires pressures of 70 to 140 MPa (10 000 to 20 000 psi).

Blow molding is a method of forming hollow objects in which a thermoplastic in the form of a molten or softened tube (parison) is inserted into a cool mold. The parison is then pressurized at low internal pressure so that it expands against the sides of the mold, where it solidifies. Blow molding requires pressures of only 0.2 to 0.7 MPa (25 to 100 psi). The primary advantage of blow molding is the ability to form re-entrant curves.

Transfer molding is a method of forming plastic objects from granular, powdered or pre-formed thermosets by softening the material in a heated chamber and then forcing essentially the entire mass into another hot chamber, where it cures.

Compression molding is a method of forming objects from either thermosets or thermoplastics by placing the material in a heated mold cavity open at the top, and then simultaneously applying heat and compressing the material with a "force."

Cavity refers to the female portion of the mold, which forms the outer surface of the molded article. (The cavity frequently is also called the *die*.)

Plunger refers to the ram or piston used to displace the fluid or semifluid material in transfer molding and injection molding. (In compression molding, the ram or piston usually is called the *force*.)

SELECTION OF MOLD MATERIALS

Table 36 lists typical materials used for making machined molds. The choice of mold material depends on the type of plastic to be molded, and on the quantity and shape of the parts to be made. The mold materials given in Table 36 are for several hypothetical parts representing different degrees of molding severity; these hypothetical parts are illustrated in Fig. 13.

Machined Molds. Large cavities usually are produced by machining. For these large cavities, steels that require high-temperature heat treatment and rapid cooling may exhibit an unacceptable amount of distortion; it may be necessary to use prehardened die steels, or steels (such as nitriding steels or maraging steels) that require only a relatively low-temperature final heat treatment to develop acceptable wear resistance.

Hubbed Molds. In many instances, small multiple-cavity molds can be produced at lowest cost by hubbing. This involves pressing a male master plug, known as the hub, into the metal block. (This process is also called "hobbing.") If forming a mold by hubbing is to be economical, the mold steel must be very soft and have a low rate of work hardening. Such steels must be carburized and heat treated to provide the required wear resistance and surface hardness.

Satisfactory hubbing of mold cavities depends not only on the steel to be hubbed but also on selection of the proper steel for the hub. Steels for master hubs should have good machinability and workability in the annealed condition, high compressive strength, and resistance to abrasion in the heat treated condition. Hub steels also should exhibit minimum distortion and size change in heat treatment, minimum scaling, and the ability to be polished to a high finish.

Table 32. Recommended materials for die-casting dies and die inserts

Components	Typical material(s) and hardness
Cavity inserts for Al and Mg castings	H13 hardened to 45 to 48 HRC
Cavity inserts for Zn castings	P20 prehardened to 300 HB
Cavity inserts for long-run Zn castings	H13 hardened to 44 to 46 HRC
Cavity inserts for Cu castings	H20, H21, H22 hardened to 44 to 48 HRC
Holder blocks	4140 prehardened to 300 HB

Table 33. Typical materials for injection components

Component	Metal being cast	Material and condition
Sprue spreader	Zinc	H13 hardened to 250 to 290 HB and nitrided
Sprue bushing	Zinc	Nitralloy hardened to 250 to 300 HB and nitrided
Nozzle and adapter	Zinc	H13 hardened to 46 to 48 HRC and nitrided
Shot sleeve	Aluminum	H13 hardened to 46 to 48 HRC and nitrided
Shot pad	Aluminum	H13 hardened to 46 to 48 HRC and nitrided
Plunger tip	Aluminum	Beryllium copper hardened to 38 to 42 HRC

Table 34. Materials for slides, guides, cores and ejector pins

Component	Material(s)	Condition
Slide carrier	4130, 4140, 6150	Hardened to 46 to 50 HRC
Slide lock	4140, 6150	Hardened to 46 to 50 HRC
Leader pin	1117	Carburized and hardened to 58 to 62 HRC
Guide bushing	1018	Carburized and hardened to 58 to 62 HRC
Guide block	4140	Carburized and hardened to 56 to 60 HRC
Guide plate	4140	Carburized and hardened to 56 to 60 HRC
Ejector pin	H13	Hardened to 34 to 40 HRC and nitrided
Return (surface) pin	H13	Hardened to 34 to 40 HRC and nitrided
Core	H13	Hardened to 48 to 55 HRC

Table 35. Typical materials for trim dies

Component	Material	Condition
Base, holder	1020 (a)	As received
Trim ring, plate	A2	Hardened to 56 to 58 HRC
Punch, pad	1020 (a)	As received
Guide pin, guide bushing	C1117	Carburized and hardened to 58 to 62 HRC
Slide block	A2	Hardened to 56 to 58 HRC
Cam	1020 (a)	Carburized and hardened to 46 to 48 HRC
(a) Hot rolled.		

Table 36. Typical materials for machined molds

Material to be molded	Parts 1, 4 and 7 — 10 000 to 100 000	Parts 1, 4 and 7 — 1 000 000 to 10 000 000	Parts 2, 5 and 8 — 10 000 to 100 000	Parts 2, 5 and 8 — 1 000 000 to 10 000 000	Parts 3, 6 and 9 — 10 000 or more
Mold materials for thermoplastics					
Group 1: general-purpose plastics and rubbers	P20 or P21, prehardened(b); 414L, prehardened	O1, 53 to 57 HRC; P6 or P20, carburized; S7, 51 to 57 HRC; 420 stainless, 45 to 50 HRC	P20 or P21, prehardened(b); 414L, prehardened	P6, carburized, 54 to 58 HRC; P20 or P21, prehardened(b); 414L, prehardened	P20 or P21, prehardened(b); 414L, prehardened
Group 2: fluid plastics	P6 or P20, carburized, 54 to 58 HRC	O1, 53 to 57 HRC; S7, 51 to 57 HRC	P6, carburized, 54 to 58 HRC	P6, carburized, 54 to 58 HRC; H13 (c), 48 to 52 HRC; S7, 51 to 57 HRC; 420 stainless, 45 to 50 HRC	P20 or P21, prehardened(b); 414L, prehardened
Group 3: corrosive and high-temperature plastics	P20 or P21, prehardened(b) and nickel plated(d); 414L, prehardened; 420 stainless, 45 to 50 HRC	O1, 53 to 57 HRC, nickel plated(d); 420 stainless, 45 to 50 HRC; S7, 51 to 57 HRC	P20 or P21, prehardened(b) and nickel plated(d); 414L prehardened; 420 stainless, 45 to 50 HRC	P6, carburized, 54 to 58 HRC, nickel plated(d); 420 stainless, 45 to 50 HRC	P20 or P21, prehardened(b) and nickel plated(d); 414L, prehardened
Mold materials for thermosets					
Group 4: general-purpose plastics	L2, 53 to 57 HRC; P20, carburized, 54 to 58 HRC; S7, 51 to 57 HRC	L2, carburized, 53 to 57 HRC; A2, 53 to 57 HRC; P20, carburized, 54 to 58 HRC; S7, 51 to 57 HRC	P20 or P6, carburized, 54 to 58 HRC	P20 or P6, carburized, 54 to 58 HRC	P20 or P6, carburized, 50 to 55 HRC
Group 5: plastics requiring high-temperature curing	H13, 48 to 52 HRC; S7, 51 to 57 HRC	H13, 48 to 52 HRC; S7, 51 to 57 HRC	P4, carburized, 52 to 56 HRC; H13, 48 to 52 HRC	P4, carburized, 52 to 56 HRC; H13, 48 to 52 HRC; S7, 51 to 57 HRC	P4, carburized, 52 to 56 HRC
Group 6: rubbers	Class 30 gray iron(e)	Class 30 gray iron; 1020, soft, chromium plated(f); A2, 53 to 57 HRC(g)	1020, soft, chromium plated(f)	1020, soft, chromium plated(f)	1020, soft, chromium plated(f)
Group 7: low-pressure and abrasive plastics	P20, carburized, 54 to 58 HRC; L2, 53 to 57 HRC; S7, 51 to 57 HRC	P20, carburized, 54 to 58 HRC; L2, 53 to 57 HRC; S7, 51 to 57 HRC	P20, carburized, 54 to 58 HRC; L2, flame hardened	P20, carburized, 54 to 58 HRC; L2, flame hardened; S7, 51 to 57 HRC	P20 or P6, carburized, 50 to 55 HRC

(a) Where more than one mold material is given for a specific set of conditions, they are arranged in order of preference unless otherwise noted. (b) Hardness of prehardened steels should be 300 HB minimum. (c) Preferred for molding parts 5 and 8. (d) Recommended thickness of plating, 0.005 to 0.025 mm (0.0002 to 0.0010 in.). (e) Cast 356-T6 aluminum recommended for quantities up to 10 000 parts. (f) Recommended thickness of plating, 0.005 to 0.015 mm (0.0002 to 0.0005 in.). (g) Provides increased resistance to handling.

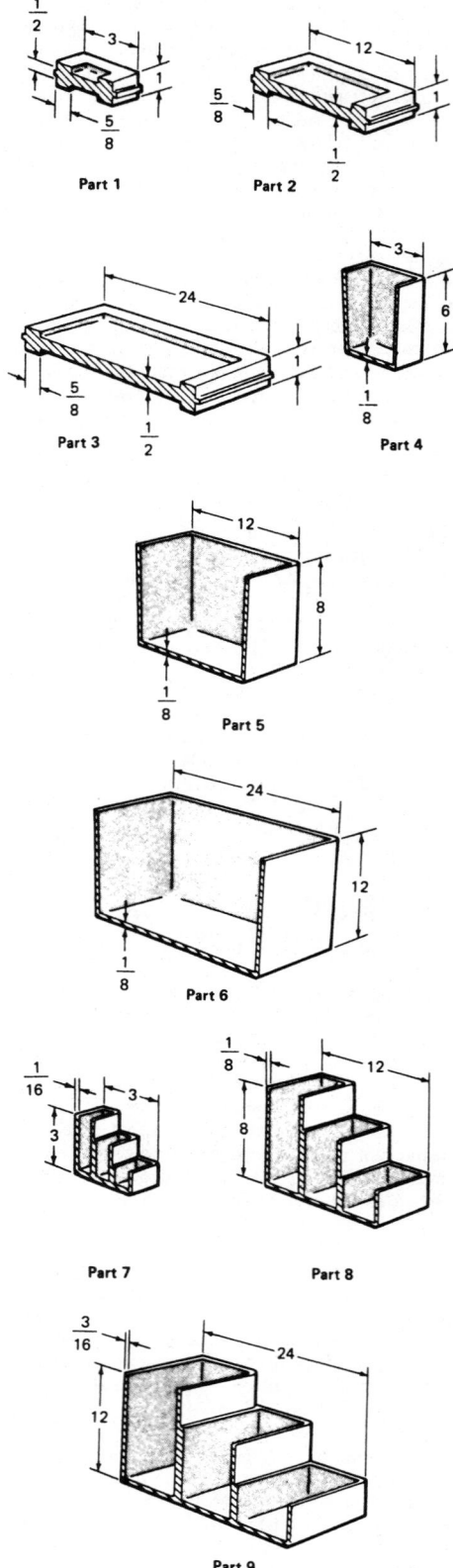

Dimensions are in inches; for equivalent metric dimensions (mm), multiply listed values by 25. See Table 36 for typical mold materials used in producing parts similar in general configuration and degree of severity to those shown above.

Fig. 13. Hypothetical shapes typical of molded plastic products

For relatively simple hubs containing sharp detail but no feather edges (which are susceptible to edge wear, edge breakdown, or loss of detail), S1 is usually recommended. Hardnesses of 59 to 61 HRC are recommended, and slight carburization of the surface during heat treatment is often beneficial. Most hubs fall into this general category. Therefore, S1 is the tool steel most commonly used in hubbing.

Steels O1, A2 and D2, which have high hardness and good abrasion resistance, are used for high-production hubs of simple design for which long life and resistance to bulging are important considerations but for which high toughness is relatively unimportant. A hardness in the range from 60 to 63 HRC is recommended for this type

of hubbing. Hubs of intricate or complex design must be notch tough and often are made of 6F5 tool steel. Such hubs are heat treated to hardnesses of 58 to 59 HRC. However, 6F6 tool steel is more suitable for hubs of complex design that incorporate feather edges, which require exceptional resistance to brittleness and edge failure. Intentional carburization during heat treatment of this type of steel is often helpful.

L6 tool steel is used principally for inexpensive hubs that are pressed into the mold material with relatively low hubbing pressures.

Cast Molds. Many small-cavity and multiple-cavity molds are pressure cast using a hub as the pattern for the die cavity.

Hubs used for pressure casting of beryllium

Table 37. Typical materials and hardnesses for common types of production gages

Types of gage	Gaging tolerance 0.01 mm (0.0005 in.) Part hardness up to 350 HB — Gage material(a)	Hardness, HRC	Gaging tolerance 0.01 mm (0.0005 in.) Part hardness over 350 HB — Gage material(a)	Hardness, HRC	Gaging tolerance 0.05 mm (0.002 in.) Part hardness up to 350 HB — Gage material(a)	Hardness, HRC	Gaging tolerance 0.05 mm (0.002 in.) Part hardness over 350 HB — Gage material(a)	Hardness, HRC
Occasional gaging of dimensions up to 100 mm (4 in.)								
Gage blocks, gaging pins, anvils, buttons	W1, O1, O2	61-64						
Cylindrical ring and plug gages	1212 (b), W1, O1, O2	61-64						
Threaded ring and plug gages	W1, O1, O2	61-64						
Height and length gages	W1, O1, O2	61-64						
Spline gages	W1, O1, O2	61-64						
Snap gages(c), thread rolls	O1, O2, L7	61-64						
Feeler gages	W1, O1, O2	45-52						
Alignment bars	1212 (b)	...						
Frequent, long-term gaging of dimensions up to 12 mm (1/2 in.)								
Gage blocks, gaging pins, anvils, buttons	L7 (d)	61-64	D2 (e)	57-64				
	D2 (e)	57-64	M2 (f)	62-65				
			Carbide(g)	...				
Cylindrical ring and plug gages	M2 (f)	62-65	M2 (f)	62-65				
			Carbide(g)	...				
Threaded ring and plug gages	A2 (e)	56-64	M2	62-65				
Height and length gages	M2	62-65	M2	62-65				
Spline gages	O6	61-64	A2 (e)	56-64				
Snap gages(c), thread rolls	L7 (d)	61-64	L7 (d)	61-64				
Feeler gages	L7 (d)	45-50	D2	45-50				
Frequent, long-term gaging of dimensions from 12 to 100 mm (1/2 to 4 in.)								
Gage blocks, gaging pins, anvils, buttons	A2 (e)	56-64	Carbide(g)	...	A2	62-64	M2 (f)	62-65
	D2 (e)	57-64			D2	62-64		
	M2	62-65			M2	62-65		
Cylindrical ring and plug gages	M2 (f)	62-65	Carbide(g)	...	D2	62-64	M2 (f)	62-65
					M2	62-65		
Threaded ring and plug gages	A2 (e)	56-64	M2 (f)	62-65	A2	62-64	M2	62-65
	D2 (e)	57-64						
Height and length gages	A2 (e)	56-64	M2	62-65	A2	62-64	M2	62-65
	D2 (e)	57-64						
Spline gages	O6	61-64	A2 (e)	56-64	O6	61-64	A2, D2	62-64
			D2 (e)	57-64				
Snap gages(c), thread rolls	L7 (d)	61-64	D2 (e)	57-64	L7	61-64	D2	62-64
			M2	62-65			M2	62-65
Alignment bars	1212 (b)	...	8620 (b)	...	1212 (b)	...	8620 (b)	...
	8620 (b)	...	4140 (h)	...	8620 (b)	...	4140 (h)	...

(a) Where more than one tool material is listed for a specific set of conditions, the last material listed is usually preferred for large sections. (b) Carburized, with a case not more than 1/5 the section thickness and having a minimum surface hardness equivalent to 61 HRC. (c) Snap-gage bodies generally are made of stress-relieved cast iron, ASTM A48, class 20, 30 or 35. (d) 52100 steel has proved a satisfactory substitute for L7. (e) For close tolerances, this steel must be tempered in the secondary hardening range for maximum stability. (f) Liquid nitriding after full hardening is recommended to produce a surface hardness of about 1100 HV. (g) Cemented tungsten carbide is usually selected, but chromium carbide or boron carbide also can be used. (h) Heat treated to 26 to 30 HRC, then gas nitrided for 24 h.

copper molds must resist softening at the elevated temperatures involved. Therefore, a hot work steel, such as H13, is usually chosen for this application. For some applications, such as gate inserts, machinable steel-bonded carbides can be shaped into the insert, and then heat treated to hardnesses on the order of 68 HRC.

Thread-Rolling Dies

THE PROPERTIES that are most significant in selecting materials for thread-rolling dies are hardness, toughness and wear resistance. Hardness and toughness must be high enough to enable the dies to withstand the forces exerted on them in service; good wear resistance is necessary because the prime cause for removing thread-rolling dies from service is spalling of the crests of the die threads, which is allowed to continue until the contours of the threads being rolled no longer meet dimensional or functional requirements.

The materials most commonly used to make thread-rolling dies are M1 and M2 high speed tool steels; D2 high-carbon, high-chromium tool steel; and A2 medium-alloy cold work tool steel. The steel chosen should be adequately annealed

before hardening and should have a sufficiently uniform carbide-particle distribution.

In most applications, D2, M1 and M2 are about equal in performance, whereas the service life of A2 dies is somewhat lower. In general, D2, M1 and M2 should be selected for long production runs and for rolling larger parts, coarser threads and alloys of higher hardness. The hardness ranges given in the following table have been found satisfactory for most applications:

Die material	Recommended hardness, HRC Flat dies	Circular dies
For threading aluminum, copper or soft steel blanks		
A2	57 to 60	56 to 58
D2	60 to 62	58 to 60
M2	58 to 60	58 to 60
For threading ferritic steel (hardness, >95 HRB) or austenitic stainless steel blanks		
A2	57 to 59	56 to 58
D2	59 to 61	58 to 60
M2	59 to 61	58 to 60

Hardness values indicated above should be achieved by double tempering after quenching. Early failure is more likely if double tempering is not used.

If diameter and lead tolerances of the rolled

part permit, the dies can be ground or machined before hardening. The recommended steels for that practice, given in descending order of preference, and their average distortion during heat treatment necessary for obtaining the required hardness, are as follows:

Type of steel	Approx average distortion, %
A2	0.04
D2	0.05
M1 or M2	0.11

If tolerances require lower average distortions, dies must be ground after heat treatment. Die life usually is not decreased if grinding is done after heat treatment, as long as proper (nonabusive) grinding techniques are used.

Flat dies are used for producing most standard threaded fasteners and most wood screws. Flat dies made of D2 are usually ground before hardening, because D2 is susceptible to grinding cracks if improperly ground after hardening.

Gages

SELECTION OF MATERIALS for gages depends to a large extent on tolerance to be checked, number of items to be gaged, composition and hardness of the material being gaged, size and complexity of the item and cost of the gage material. Abrasive wear is the predominant factor in determining the useful life of production gages. Therefore, gage surfaces that contact the workpiece must be hard enough to provide adequate resistance to abrasion. The actual hardness required depends on the hardness and abrasive characteristics of the workpiece surfaces, number of pieces to be gaged and the tolerance to be checked. Production gages that must be used in hostile environments may have to be corrosion resistant. Also, if the gage must operate over a range of temperatures, thermal expansion must be considered when making final material selection.

GAGE MATERIALS

Typical materials and hardnesses for the more common types of production gages are given in Table 37. In this table, it should be noted that feeler gages, which are thin flexible strips of precisely controlled thickness, must be tempered back to lower hardness so that they will not break during use.

Precision Gages. Making gages to extra-close tolerances often requires special gage materials. In applications that require extremely close gage tolerances, fine surface finishes, exceptional wear characteristics and outstanding dimensional stability, boron carbide and even jewels make excellent gage elements.

Combination Gages. Frames, bodies and bases of combination gages commonly are made of cast iron. The cast iron used for gages is generally a class 20, 30 or 35 gray iron that has been stress relieved at 450 to 480 °C (850 to 900 °F).

Aluminum is also used in combination gages for handles, bodies and bases. Aluminum handles and bodies may be made either of soft grades such as 1100 or, if more strength is required, of harder grades such as 2017 or 2024.

Materials used for wear surfaces in combination gages must have the same properties as those required for the solid gages discussed above. Wear inserts generally are made from hardened tool steel or cemented carbide. However, when the insert is a relatively simple shape that is not very susceptible to distortion and cracking in heat treatment, it may be possible to use an inexpensive tool steel instead.

Inspection fixtures are gages, but belong to a different category from those discussed so far. They are used mainly for checking dimensions and contours of large stampings such as automobile body components.

Plastics or plastic-faced cast iron composites are frequently used for inspection fixtures (checking fixtures) because these materials are lower in cost and lighter in weight than solid cast iron.

Master gages require high accuracy, good wear resistance and maximum stability. They are used in checking other gages and are expected to hold their accuracy over long periods of time.

No one material has the ideal combination of properties desired for master gages, although high-carbon high-chromium steels such as D2 and D4 have most of the desired characteristics.

—— Selected References on Tool Materials ——

The Metallurgy of Tool Steels, by P. Payson: John Wiley & Sons, Inc., 1962

Metallurgy and Heat Treatment of Tool Steels, by R. Wilson: McGraw-Hill, London, 1975

"Tool Steels" (a Steel Products Manual): American Iron and Steel Institute, March 1978

Tool Steels, by G. A. Roberts and R. A. Cary: American Society for Metals, 1980

Tool Steel Simplified, Revised Ed., by F. R. Palmer *et al.*: Chilton Book Co., Radnor, PA, 1978

ADDITIONAL READING

World Directory and Handbook of Hard Metals, 2nd Ed., by Kenneth J. A. Brooks: Engineers' Digest Limited, London, 1979

"Engineering Properties of Ceramics," by J. F. Lynch, C. G. Rederer and W. H. Duckworth: Technical Report AFML-TR 66-52, Air Force Materials Laboratory, 1966

Some Plain Talk About Carbides, by H. S. Kalish: *Manufacturing Engineering and Management,* July 1973

"A System of Classification of Hard Metal Grades for Machining": Technical Publication No. 1, The British Hard Metal Association, Sheffield, England, 1967

An Analysis of Charpy Impact Testing as Applied to Cemented Carbide, by R. C. Lueth: in *Instrumented Impact Testing,* STP 563, American Society for Testing and Materials, 1974

Where Solid Titanium Carbide Stands, by H. S. Kalish: *American Machinist,* 7 Jan 1974, p 50-52

"Machinable Carbides for High Performance Tooling and Wear Parts," by S. E. Tarkan and M. K. Mal: Technical Paper MR 73-927, Society of Manufacturing Engineers

Tool and Die Failures Source Book, edited by Serope Kalpakjian: American Society for Metals, 1982

"Tantung: The Premiere Cast Alloy": Bulletin 72-1, VR/Wesson Div. of Fansteel, Inc., 2 June 1972

"Ceramic Tools," by E. D. Whitney: Technical Paper TE 73-205, Society of Manufacturing Engineers

Cutting Performance and Practical Merits of Carbide Ceramics, by K. Ogawa, M. Furukawa and Y. Hara: *Nippon Tungsten Review,* Vol C, 1973

"Borazon and Diamond Cutting Tools," by R. E. Hanneman and L. E. Gibbs: Technical Information Series, Report No. 73 CRD 182, General Electric Co., June 1973

Some Experiments to Compare Diamond and Diamond Compact Cutting Tools, by M. Casey and J. Wilks: *Sixteenth International Tool Design and Research Conference,* McMillan Press Ltd., Sept 1975

Distortion in Tool Steels, by B. S. Lement: American Society for Metals, 1959

19 POWDER METALLURGY MATERIALS

By Leander F. Pease III, Powder-Tech Associates, Inc.

Additional and more detailed information on powder metallurgy materials can be found in Metals Handbook, Ninth Edition, Volume 7, Powder Metallurgy. Additional information within the Desk Edition can be located in Section 25 and by consulting the index.

THIS SECTION contains typical mechanical-property data, at various densities, for the following P/M materials:

1. Ferrous materials (Fe, Fe-C, Fe-Cu-C, Fe-Ni-C), medium to high density
2. Stainless steels (AISI types 303, 316, 410), medium density
3. Nonferrous materials (bronze, brass, nickel silver), low to medium density
4. P/M forged low-alloy steels
5. Stainless steels, fully dense
6. Injection molded materials, high density
7. Tool steels, fully dense
8. Aluminum alloys, medium density
9. Nickel- and cobalt-base alloys
10. Superalloys
11. Titanium and titanium-base alloys.

FERROUS MATERIALS, MEDIUM TO HIGH DENSITY

Ferrous materials of medium to high density are made by blending iron powder with copper, carbon and nickel, pressing, and sintering for 20 to 30 min at about 1120 °C (2050 °F). The graphite (carbon) dissolves rapidly to form steel. Copper melts and partially diffuses into the iron particles; nickel diffuses somewhat less, and is left as nickel-rich islands. For densities above about 7.1 g/cm³, the materials are often pressed, presintered at 845 °C (1550 °F), re-pressed to high density (7.2 to 7.6 g/cm³) and final sintered.

Most of these materials have been standardized by the Metal Powder Industries Federation (MPIF), ASTM and SAE. The various grades, densities, compositions and cross references are given in Tables 1 and 2. MPIF Standard 35 (Ref 1) is the source of the mechanical-property data given in Table 3 for iron-base alloys containing carbon, copper and nickel. These are typical properties, not minimums. MPIF is, however, revising Standard 35 to include minimum yield strengths for as-sintered materials and minimum ultimate tensile strengths for heat treated materials.

Table 3 shows the general trend of increasing mechanical properties with increases in density. It must be emphasized that the data in Table 3 apply for the base iron employed (electrolytic for Ni steels) and for the processing used (presintering and re-pressing for all materials above 6.6 g/cm³). Other iron powders and other processing procedures give different results. Under the new MPIF system of specified minimum tensile prop-

Table 1. Designations and compositions of ferrous P/M structural materials

| Material | Designation(a) | | | MPIF composition limits and ranges(b), % | | | |
	MPIF	ASTM	SAE	C	Ni	Cu	Fe
P/M iron	F-0000	B 310, class A	853, class 1	0.3 max	...	...	97.7-100
P/M steel	F-0005	B 310, class B	853, class 2	0.3-0.6	...	...	97.4-99.7
	F-0008	B 310, class C	853, class 3	0.6-1.0	...	...	97.0-99.1
P/M copper iron	FC-0200	...	...	0.3 max	...	1.5-3.9	93.8-98.5
P/M copper steel	FC-0205	...	...	0.3-0.6	...	1.5-3.9	93.5-98.2
	FC-0208	B 426, grade 1	864, grade 1, class 3	0.6-1.0	...	1.5-3.9	93.1-97.9
	FC-0505	...	...	0.3-0.6	...	4.0-6.0	91.4-95.7
	FC-0508	B 426, grade 2	864, grade 2, class 3	0.6-1.0	...	4.0-6.0	91.0-95.4
	FC-0808	B 426, grade 3	864, grade 3, class 3	0.6-1.0	...	6.0-11.0	86.0-93.4
	...	B 426, grade 4	864, grade 4, class 3	0.6-0.9	...	18.0-22.0	75.1 min
P/M iron-copper	FC-1000	B 222; B 439, grade 3	862	0.3 max	...	9.5-10.5	87.2-90.5
P/M iron-nickel	FN-0200	B 484, grade 1, class A	...	0.3 max	1.0-3.0	2.5 max	92.2-99.0
P/M nickel steel	FN-0205	B 484, grade 1, class B	...	0.3-0.6	1.0-3.0	2.5 max	91.9-98.7
	FN-0208	B 484, grade 1, class C	...	0.6-0.9	1.0-3.0	2.5 max	91.6-98.4
P/M iron-nickel	FN-0400	B 484, grade 2, class A	...	0.3 max	3.0-5.5	2.0 max	90.2-97.0
P/M nickel steel	FN-0405	B 484, grade 2, class B	...	0.3-0.6	3.0-5.5	2.0 max	89.9-96.7
	FN-0408	B 484, grade 2, class C	...	0.6-0.9	3.0-5.5	2.0 max	89.6-96.4
P/M iron-nickel	FN-0700	B 484, grade 3, class A	...	0.3 max	6.0-8.0	2.0 max	87.7-94.0
P/M nickel steel	FN-0705	B 484, grade 3, class B	...	0.3-0.6	6.0-8.0	2.0 max	87.4-93.7
	FN-0708	B 484, grade 3, class C	...	0.6-0.9	6.0-8.0	2.0 max	87.1-93.4
P/M infiltrated steel	FX-1005	...	...	0.3-0.6	...	8.0-14.9	80.5-91.7
	FX-1008	...	...	0.6-1.0	...	8.0-14.9	80.1-91.4
	FX-2000	B 303, class A	870	0.3 max	...	15.0-25.0	70.7-85.0
	FX-2005	B 303, class B	...	0.3-0.6	...	15.0-25.0	70.4-84.7
	FX-2008	B 303, class C	872	0.6-1.0	...	15.0-25.0	70.0-84.4

(a) Designations listed are nearest comparable designations; ranges and limits may vary slightly between comparable designations. (b) MPIF standards require that the total amount of all other elements be less than 2.0%, except that the total amount of other elements must be less than 4.0% in infiltrated steels.

Table 2. Typical density designations and ranges of ferrous P/M materials
Density of pure iron is 7.87 g/cm³.

MPIF density suffix	Designation ASTM type(a)	Designation SAE type	Density, g/cm³
N	I	1(b)	Less than 6.0
P	II	2	6.0 to 6.4
R	III	3	6.4 to 6.8
S	IV	4	6.8 to 7.2
T	V(c)	5(c)	7.2 to 7.6
U	...	...	7.6 to 8.0

(a) ASTM B 426 only; different density ranges used in ASTM B 310 and B 484. (b) Density range of 5.6 to 6.0 g/cm³ is specified. (c) Minimum density of 7.2 g/cm³ is specified.

erties, suppliers will select the correct powders, density and processing to meet the minimum properties.

The values of Young's modulus, Poisson's ratio and the linear coefficient of thermal expansion depend mainly on the density of the low-alloy ferrous parts, when compared under conditions of average sintering. These data are given in Table 4, for a range of densities.

The compressive strengths and hardnesses of sintered ferrous materials can be increased by steam treating for 1 to 4 h at 480 to 595 °C (900 to 1100 °F). This process fills the surface pores with black iron oxide (Fe_3O_4). The increase in hardness is shown in Table 5. The process can also make parts pressure-tight to 690 kPa (100

psi) and provide increased corrosion resistance. The oxide may lower the impact resistance although it increases the abrasive wear resistance.

STAINLESS STEELS, MEDIUM DENSITY

Medium-density stainless steels are typically produced by pressing at 414 to 827 MPa (30 to 60 tsi) and sintering at 1120 to 1150 °C (2050 to 2100 °F) in dissociated ammonia, hydrogen or vacuum. The data presented in Table 6 are representative of such processing and have been selected from MPIF Standard 35 (Ref 1). Better elongation, impact resistance and corrosion resistance are obtained by sintering at 1260 to 1315 °C (2300 to 2400 °F). These advanced ma-

Table 3. Typical mechanical properties of ferrous P/M materials

Designation	MPIF density suffix(a)	Condition(b)	Tensile strength MPa	ksi	Yield strength MPa	ksi	Elongation in 25 mm (1 in.), %	Fatigue strength MPa	ksi	Impact energy(c) J	ft·lb	Apparent hardness	Elastic modulus GPa	10⁶ psi
F-0000	N	AS	110	16	75	11	2.0	40	6(d)	4.1	3.0	10 HRH	70	10.5
	P	AS	130	19	95	14	2.5	50	7(d)	6.1	4.5	70 HRH	90	13
	R	AS	165	24	110	16	5	60	9(d)	13	9.5	80 HRH	110	16
	S	AS	205	30	150	22	9	80	11(d)	20	15	15 HRB	130	19
	T	AS	275	40	180	26	15	105	15(d)	34	25	30 HRB	160	23
F-0005	N	AS	125	18	105	15	1.0	45	7(d)	3.4	2.5	5 HRB	70	10.5
	P	AS	170	25	140	20	1.5	65	10(d)	4.7	3.5	20 HRB	90	13
	R	AS	220	32	160	23	2.5	85	12(d)	6.8	5.0	45 HRB	110	16
		HT	415	60	395	57	0.5	155	23(d)	...	...	100 HRB	110	16
	S	AS	295	43	195	28	3.5	110	16(d)	12	9.0	60 HRB	130	19
		HT	550	80	515	75	0.5	210	30(d)	...	...	25 HRC	130	19
F-0008	N	AS	200	29	170	25	0.5	75	11(d)	2.7	2.0	35 HRB	70	10.5
		HT	290	42	...	...	<0.5	110	16(d)	...	...	90 HRB	70	10.5
	P	AS	240	35	205	30	1.0	90	13(d)	4.1	3.0	50 HRB	90	13
		HT	400	58	...	...	<0.5	150	22(d)	...	...	100 HRB	90	13
	R	AS	290	42	250	36	1.5	110	14(d)	4.7	3.5	65 HRB	110	16
		HT	510	74	...	...	<0.5	195	28(d)	...	...	25 HRC	110	16
	S	AS	395	57	275	40	2.5	150	22(d)	9.5	7.0	75 HRB	130	19
		HT	650	94	625	91	<0.5	245	36(d)	...	...	30 HRC	130	19
FC-0200	P	AS	160	23	115	17	2.5	60	9(d)	7.5	5.5	80 HRH	90	13
	R	AS	205	30	145	21	4	80	11(d)	9.5	7.0	15 HRB	110	16
	S	AS	255	37	160	23	7	95	14(d)	23	17	30 HRB	130	19
FC-0205	P	AS	275	40	235	34	1.0	105	15(d)	4.7	3.5	35 HRB	90	13
	R	AS	345	50	260	38	1.5	130	19(d)	7.5	5.5	60 HRB	110	16
		HT	585	85	560	81	<0.5	220	31(d)	...	...	30 HRC	110	16
	S	AS	425	62	310	45	3.0	160	24(d)	13	9.5	75 HRB	130	19
		HT	690	100	655	95	<0.5	260	38(d)	...	...	35 HRC	130	19
FC-0208	N	AS	225	33	205	30	<0.5	85	13(d)	3.4	2.5	45 HRB	70	10.5
		HT	295	43	...	...	<0.5	110	16(d)	...	...	95 HRB	70	10.5
	P	AS	310	45	280	41	<0.5	115	17(d)	4.1	3.0	50 HRB	90	13
		HT	380	55	...	...	<0.5	145	21(d)	...	...	25 HRC	90	13
	R	AS	415	60	330	48	1.0	155	23(d)	6.8	5.0	70 HRB	110	16
		HT	550	80	...	...	<0.5	210	30(d)	...	...	35 HRC	110	16
	S	AS	550	80	395	57	1.5	210	30(d)	11	8.0	80 HRB	130	19
		HT	690	100	655	95	<0.5	260	38(d)	...	...	40 HRC	130	19
FC-0505	N	AS	240	35	205	30	0.5	90	13(d)	4.1	3.0	50 HRB	70	10.5
	P	AS	345	50	290	42	1.0	130	19(d)	6.1	4.5	60 HRB	90	13
	R	AS	455	66	380	55	1.5	170	25(d)	6.8	5.0	75 HRB	116	16
FC-0508	N	AS	330	48	295	43	<0.5	125	18(d)	4.1	3.0	60 HRB	70	10.5
	P	AS	425	62	395	57	1.0	160	24(d)	4.7	3.5	65 HRB	90	13
		HT	480	70	480	70	<0.5	185	27(d)	...	...	30 HRC	90	13
	R	AS	515	75	480	70	1.0	195	29(d)	6.1	4.5	85 HRB	116	16
FC-0808	N	AS	250	36	...	...	<0.5	...	...	...	...	55 HRB	...	...
FC-1000	N	AS	205	30	...	...	0.5	...	...	...	...	70 HRF	...	...
FN-0200	R	AS	195	28	125	18	4	75	11	19	14	38 HRB	115	17
	S	AS	260	38	170	25	7	105	15	43	32	42 HRB	145	21
	T	AS	310	45	205	30	11	125	18	68	50	51 HRB	160	23
FN-0205	R	AS	255	37	160	23	3.0	105	15	14	10	50 HRB	115	17
		HT	565	82	450	65	0.5	225	33	8.1	6	32 HRC	115	17
	S	SS	345	50	215	31	3.5	140	20	24	18	70 HRB	145	21
		HT	760	110	605	88	1.0	305	44	22	16	42 HRC	145	21
	T	SS	420	61	255	37	4.5	165	24	43	32	85 HRB	160	23
		HT	925	134	725	105	2.0	370	54	38	28	46 HRC	160	23

(continued)

(a) For density range, see Table 2. (b) AS, as sintered; SS, sintered and sized; HT, heat treated, typically austenitized at 870 °C (1600 °F), oil quenched and tempered 1 h at 200 °C (400 °F). (c) Unnotched Charpy test. (d) Estimated as 38% of tensile strength. (e) X indicates infiltrated steel; see Table 1.
Source: Ref 1.

Table 3. (continued)

Designation	MPIF density suffix(a)	Condition(b)	Tensile strength MPa	ksi	Yield strength MPa	ksi	Elongation in 25 mm (1 in.), %	Fatigue strength MPa	ksi	Impact energy(c) J	ft·lb	Apparent hardness	Elastic modulus GPa	10⁶ psi
FN-0208 .... R		AS	330	48	205	30	2.0	130	19	11	8	62 HRB	115	17
		HT	690	100	650	94	0.5	275	40	8.1	6	34 HRC	115	17
	S	AS	450	65	280	41	3.0	180	26	19	14	79 HRB	145	21
		HT	930	135	880	128	0.5	370	54	16	12	45 HRC	145	21
	T	AS	545	79	345	50	3.5	220	32	30	22	87 HRB	160	23
		HT	1105	160	1070	155	0.5	415	60	24	18	47 HRC	160	23
FN-0400 .... R		AS	250	36	150	22	5	95	14	22	16	40 HRB	115	17
	S	AS	340	49	205	30	6	140	20	47	35	60 HRB	145	21
	T	AS	400	58	250	36	6.5	160	23	68	50	67 HRB	160	23
FN-0405 .... R		AS	310	45	180	26	3.0	125	18	14	10	63 HRB	115	17
		HT	770	112	650	94	0.5	310	45	8.1	6	27 HRC	115	17
	S	AS	425	62	240	35	4.5	165	24	20	15	72 HRB	145	21
		HT	1060	154	880	128	1.0	415	60	14	10	39 HRC	145	21
	T	AS	510	74	295	43	6.0	205	30	41	30	80 HRB	160	23
		HT	1240	180	1060	154	1.5	450	65	19	14	44 HRC	160	23
FN-0408 .... R		AS	395	57	290	42	1.5	160	23	8.1	6	72 HRB	115	17
	S	AS	530	77	390	57	3.0	215	31	14	10	88 HRB	145	21
	T	AS	640	93	470	68	4.5	255	37	22	16	95 HRB	160	23
FN-0700 .... R		AS	560	52	205	30	2.5	145	21	16	12	60 HRB	115	17
	S	AS	490	71	275	40	4	195	28	28	21	72 HRB	145	21
	T	AS	585	85	330	48	6	240	34	35	26	83 HRB	160	23
FN-0705 .... R		AS	370	54	240	35	2.0	150	22	12	9	69 HRB	115	17
		HT	705	102	550	80	0.5	280	41	11	8	24 HRC	115	17
	S	AS	525	76	330	48	3.5	205	30	23	17	83 HRB	145	21
		HT	965	140	760	110	1.0	385	56	20	15	38 HRC	145	21
	T	AS	620	90	390	57	5.0	250	36	33	24	90 HRB	160	23
		HT	1160	168	895	130	1.5	500	65	27	20	40 HRC	160	23
FN-0708(e) .... R		AS	395	57	280	41	1.5	160	23	8	6	75 HRB	115	17
	S	AS	550	80	380	55	2.5	220	32	16	12	88 HRB	145	21
	T	AS	655	95	455	66	3.0	260	38	22	16	96 HRB	160	23
FX-1005(e) .... T		AS	570	83	440	64	4.0	...	...	19	14	75 HRB	135	20
		HT	830	120	740	107	1.0	...	...	9.5	7.0	35 HRC	135	20
FX-1008(e) .... T		AS	620	90	515	75	2.5	...	...	16	12	80 HRB	135	20
		HT	895	130	725	105	60.5	...	...	9.5	7.0	40 HRC	135	20
FX-2000(e) .... T		AS	450	65	...	...	1.0	...	...	20	15	60 HRB	...	...
FX-2005(e) .... T		AS	515	75	345	50	1.5	...	...	12.9	9.5	75 HRB	125	18
		HT	790	115	655	95	<0.5	...	...	8.1	6.0	30 HRC	125	18
FX-2008(e) .... T		AS	585	85	515	75	1.0	...	...	14	10	80 HRB	125	18
		HT	860	125	740	107	<0.5	...	...	6.8	5.0	42 HRC	125	18

(a) For density range, see Table 2. (b) AS, as sintered; SS, sintered and sized; HT, heat treated, typically austenitized at 870 °C (1600 °F), oil quenched and tempered 1 h at 200 °C (400 °F). (c) Unnotched Charpy test. (d) Estimated as 38% of tensile strength. (e) X indicates infiltrated steel; see Table 1.
Source: Ref 1.

Table 4. Effects of density on elastic modulus, Poisson's ratio, and coefficient of thermal expansion of P/M steels

MPIF density suffix(a)	Density, g/cm³	Elastic modulus GPa	10⁶ psi	Poisson's ratio	Coefficient of thermal expansion, 10⁻⁶/K
N 5.6-6.0		72	10.5	0.18	8.1
P 6.0-6.4		90	13	0.20	8.7
R 6.4-6.8		110	16	0.21	9.2
S 6.8-7.2		130	19	0.23	9.8
T 7.2-7.6		160	23	0.26	10.4
Theoretical 7.86		205	30	0.28	11-12

(a) See Table 2 for density designations.

Table 5. Effects of steam treating on density and apparent hardness of ferrous P/M materials

MPIF designation	MPIF density suffix(a)	Density, g/cm³ Sintered	Steam treated	Apparent hardness Sintered	Steam treated
F-0000 .... N		5.8	6.2	7 HRF	75 HRB
	P	6.2	6.4	32 HRF	61 HRB
	R	6.5	6.6	45 HRF	51 HRB
F-0008 .... M		5.8	6.1	44 HRB	100 HRB
	P	6.2	6.4	58 HRB	98 HRB
	R	6.5	6.6	60 HRB	97 HRB
FC-0700 .... N		5.7	6.0	14 HRB	73 HRB
	P	6.35	6.5	49 HRB	78 HRB
	R	6.6	6.6	58 HRB	77 HRB
FC-0708 .... N		5.7	6.0	52 HRB	97 HRB
	P	6.3	6.4	72 HRB	94 HRB
	R	6.6	6.6	79 HRB	93 HRB

(a) For density range, see Table 2.

terials are in the process of becoming standardized by MPIF.

NONFERROUS MATERIALS, LOW TO MEDIUM DENSITY

Bronzes generally are made by mixing copper and tin powders, pressing at 138 to 207 MPa (10 to 15 tsi) and sintering for 10 to 20 min at 815 to 845 °C (1500 to 1550 °F). The resultant microstructure is predominantly alpha bronze, with little or no Cu-Sn intermetallic compounds. The properties of interest are radial crushing strength and pore volume (%), for holding oil in bearing applications. Brasses and nickel silvers are pressed-and-sintered prealloyed powders. The mechanical properties for all three kinds of material are given in Table 7.

P/M FORGED LOW-ALLOY STEELS

P/M forging and hot forming are processes that can be applied to many different powders. The powders are preformed into medium-density shape, sintered, reheated, and restruck while hot, to bring the material to near-theoretical density. If the preform is re-pressed without substantial lateral flow, the process is called hot forming. If there is a 10 to 50% increase in area through lateral flow, the process is called hot forging. The final properties of the forged parts depend on density and heat treatment, and the dynamic properties (impact and fatigue) depend on powder cleanness and low final oxygen content. The latter is determined by starting powder, alloying elements and method of sintering. High-temperature sintering (at 1205 to 1315 °C, or 2200 to 2400 °F) in a reducing atmosphere can lead to low oxygen contents (100 to 300 ppm) and superior dynamic properties. Lateral flow can break up powder oxide films and partially substitute for highly reductive sintering. The data given in Table 8 for selected low-alloy steels illustrate the

Table 6. Typical mechanical properties of medium-density P/M stainless steels

| MPIF designation | MPIF density suffix(a) | Composition, % | | | | | Tensile strength | | 0.2% yield strength | | Elongation in 25 mm (1 in.), % | Density, g/cm³ |
		Cr	Ni	Mo	Si	Fe	MPa	ksi	MPa	ksi		
SS-303	P	17	12	···	0.7	rem	241	35	220	32	1	6.2
	R	17	12	···	0.7	rem	358	52	324	47	2	6.6
SS-316	P	16	13	2	0.7	rem	262	38	220	32	2	6.2
	R	16	13	2	0.7	rem	372	54	275	40	4	6.6
SS-410	N	12	···	···	0.8	rem	289	42	283	41	<1	5.8
	P	12	···	···	0.8	rem	379	55	372	54	<1	6.2

Note: All materials sintered in dissociated ammonia.
(a) For density range, see Table 2.
Source: Ref 1

Table 7. Typical mechanical properties of copper-base P/M materials

| MPIF designation | MPIF density suffix(a) | Composition, % | | | | | Tensile strength | | 0.2% yield strength | | Elongation in 25 mm (1 in.), % | 0.1% compressive yield strength | | Apparent hardness, HRH | Density, g/cm³ |
		Sn	Zn	Ni	Pb	Cu	MPa	ksi	MPa	ksi		MPa	ksi		
Bronzes															
CT-0010	N	10	···	···	···	rem	55	8	···	···	1	48	7	···	5.6-6.0
	R	10	···	···	···	rem	96	14	···	···	1	76	11	···	6.4-6.8
	S	10	···	···	···	rem	124	18	···	···	2.5	121	17.5	···	6.8-7.2
Brasses															
CZ-0010	T	···	10	···	···	rem	138	20	62	9	13	···	···	57	7.2-7.6
	U	···	10	···	···	rem	186	27	69	10	18	···	···	70	7.6-8.0
CZ-0030	T	···	30	···	···	rem	214	31	89	13	20	···	···	76	7.2-7.6
	U	···	30	···	···	rem	255	37	103	15	26	···	···	85	7.6-8.0
CZP-0210	T	···	10	···	2	rem	124	18	48	7	14	···	···	46	7.2-7.6
	U	···	10	···	2	rem	176	25.5	55	8	20	···	···	60	7.6-8.0
CZP-0220	T	···	20	···	2	rem	165	24	76	11	13	···	···	55	7.2-7.6
	U	···	20	···	2	rem	193	28	89	13	19	96	14	68	7.6-8.0
	W	···	20	···	2	rem	221	32	103	15	23	110	16	75	8.0-8.4
CZP-0230	T	···	30	···	2	rem	193	28	76	11	22	···	···	65	7.2-7.6
	U	···	30	···	2	rem	234	34	89	13	27	···	···	76	7.6-8.0
Nickel silvers															
CZN-1818	U	···	18	18	···	rem	206	30	···	···	10	110	16	75	7.6-8.0
	W	···	18	18	···	rem	255	37	···	···	12	124	18	85	8.0-8.4
CZNP-1818	U	···	18	18	1.5	rem	206	30	···	···	10	110	16	75	7.6-8.0
	W	···	18	18	1.5	rem	241	35	···	···	12	117	17	85	8.0-8.4

(a) For density range, see Table 2.
Source: Ref 1

Table 8. Typical mechanical properties of P/M forged low-alloy steels
All materials are in the hardened-and-tempered condition unless otherwise indicated.

| Material | Processing | Ultimate tensile strength | | 0.2% yield strength | | Elongation in 25 mm (1 in.), % | Reduction in area, % | Charpy V-notch impact energy | | Hardness | Fracture toughness (K_{Ic}) | | Density, % of theoretical | Ref |
		MPa	ksi	MPa	ksi			J	ft·lb		MPa √m	ksi √in.		
Fe-2MCM-0.67C(a)(b)	···	960	139.3	590	86	···	12	···	···	98 HRB	···	···	···	2
Fe-2MCM-0.67C(a)	···	1900	275.6	1500	218	···	4.5	···	···	49 HRC	···	···	···	···
4120	Sintered at 1315 °C (2400 °F), re-pressed	701	101.7	616	89.4	14	46	38	28	20-25 HRC	···	···	100	3
1520	Sintered at 1315 °C (2400 °F), re-pressed	936	135.7	···	···	9	13	39	29	20-25 HRC	···	···	100	···
4130	Gas atomized, −65 mesh	1586	230	1303	189	5	3	10	7.5	46 HRC	49	45	100	4
4640	Gas atomized, −65 mesh	···	···	···	···	···	···	7	5	55 HRC	36	33	100	4
	Water atomized	···	···	···	···	···	···	7	5	42 HRC	37	34	100	4
	Sintered at 1200 °C (2190 °F)	1040	150.8	1000	145	20	40	36	26	310-350 HV	···	···	99	6
Fe-2Ni-0.35C	Mixed elemental powders	938	136	600	87	13	44	···	13	31 HRC	···	···	99	5
Fe-0.55Ni-0.32Mo-0.47Mn-0.23Cr-0.30C	Sintered at 1200 °C (2190 °F)	1020	147.9	970	141	17	37	46	34	···	···	···	···	6
Fe-3Cu-0.5C-0.3S	···	873	127	···	···	6.5	···	···	···	274 HV	···	···	99	7
Fe-9Cu-0.34Mn-0.43Ni-0.65Mo-0.31C	···	1675	245	1410	205	13	31	19	14	49 HRC	···	···	99	8
Fe-0.35Mn-0.57Mo-1.95Ni-0.5C	···	1200	174	1120	162	10	19	30	22	475 HV	···	···	99	9
4630 modified	Sintered at 1205 °C (2200 °F)	148	215	1331	193	6	10	8	6	42 HRC	···	···	98	10

(a) MCM is a master alloy containing 20% Mn, 20% Cr, 20% Mo, and 7% C. (b) As-sintered condition

Table 9. Typical mechanical properties of nearly dense P/M stainless steel
Based on high-temperature sintering.

Alloy	Condition	Ultimate tensile strength MPa	ksi	0.2% yield strength MPa	ksi	Elongation in 25 mm (1 in.), %	Hardness	Impact strength J	ft·lb	Density, g/cm³	Theoretical density, g/cm³
Ultimet 04, 304	Sintered	593	86	248	36	36	80 HRB	10.8(a)	8(a)	7.8	7.9
Ultimet 16, 316	Sintered	687	99.6	308	44.7	26	94 HRB	8.1(a)	6(a)	7.7	7.8
	Solution treated and quenched	684	99.3	329	47.7	45	90 HRB	5.4(a)	40(a)	7.7	7.8
Ultimet 40C, 440C	Sintered	...	...	...	...	...	20-30 HRC	2.7(b)	2(b)	7.6	7.7
	Hardened and tempered	...	...	...	...	...	50-60 HRC	2.7(b)	2(b)	7.6	7.7

(a) Charpy V-notch. (b) Unnotched. Source: Ref 11

Table 10. Typical mechanical properties of fully dense stainless steel

Property	P/M material	Wrought material
Extruded 0.3- by 15.5-mm (0.1- by 0.61-in.) 317LM tube(a)		
Ultimate tensile strength, MPa (ksi)	693 (100)	693 (100)
0.2% yield strength, MPa (ksi)	324 (47)	353 (51)
Reduction in area, %	71	73
Elongation in 25 mm (1 in.), %	47	50
Hot isostatically pressed type 316		
Ultimate tensile strength, MPa (ksi)	579 (84)	...
0.2% yield strength, MPa (ksi)	288 (42)	...
Elongation in 25 mm (1 in.), %	58	...

(a) Gas-atomized powder, canned, cold isostatically pressed, and extruded. Source: Ref 12

Table 11. Typical as-sintered mechanical properties of injection molded P/M materials

Material	Density, g/cm³	Tensile strength MPa	ksi	0.2% yield strength MPa	ksi	Elongation in 25 mm (1 in.), %
Fe-2Ni	7.7	380	55	241	35	25
316L	...	517	75	345	50	18
17-4 PH	7.63	1028	149	966	140	12
IN-100	...	1083	157	904	131	12

Source: Ref 13

effects of processing variables on final properties. The fatigue ratio (endurance limit/ultimate tensile strength) is typically 0.3 to 0.4 for as-forged material and 0.4 to 0.5 for heat treated material.

STAINLESS STEELS, FULLY DENSE

High-density stainless steels have been produced by P/M forging, but that is not the usual method. At high temperatures (1315 to 1425 °C, or 2400 to 2600 °F), austenitic and ferritic stainless steels densify during sintering, and it is possible, with minor chemical variations, to achieve near-theoretical density in pressed-and-sintered parts. The resulting data for this process are shown in Table 9.

An alternative process for making the stainless steel in mill shapes (e.g., tubing) is to start with clean gas-atomized powder. The powders are packed in cans, sealed, cold isostatically pressed to reduce bulk, and extruded into tubing. The CIP'ed cans can also be hot isostatically pressed (HIP'ed) to full density, thus avoiding the extrusion step. Table 10 summarizes data for extruded and HIP'ed materials, and compares them with data for materials made by wrought (ingot) metallurgy.

INJECTION MOLDED MATERIALS, HIGH DENSITY

Finely divided (1- to 10-μm) powders can be mixed with organic binders and injection molded like plastics. After very careful binder removal, the metal preform can be sintered to near-full density. The reaction is driven by the high surface area and elevated temperatures (1150 to 1315 °C, or 2100 to 2400 °F). The densification is accompanied by over 10% linear shrinkage. The high final densities (often 95 to 99%) result in better dynamic properties than those of conventionally pressed-and-sintered materials. The injection process allows for considerable geometric freedom in the design of the part. This process is presently being used to make smaller parts (less than 75 mm, or 3 in., in diameter).

Table 11 summarizes properties for four different alloys.

TOOL STEELS, FULLY DENSE

Fully dense P/M tool steels are made either from gas-atomized tool steel powders, by HIP'-ing, or from water-atomized tool steel powders, by cold compaction and sintering. In both cases the resulting parts have a fine, uniform carbide distribution with no segregation. This results in a material with excellent grindability—that is, a high ratio of volume removed to volume of wheel worn. These materials also have improved hot workability and exhibit less distortion during heat treating than their ingot metallurgy counterparts. It is also convenient to make special grades with higher volume fractions of carbide than can be made conventionally.

The compositions of several grades are given in Table 12. Mechanical properties and comparisons with ingot metallurgy counterparts are presented in Tables 13A and 13B.

ALUMINUM ALLOYS, MEDIUM DENSITY

Pressed-and-sintered aluminum alloys consist of aluminum powder admixed with minor amounts of magnesium, silicon and copper. The compacts are pressed at 138 to 276 MPa (10 to 20 tsi) and

Table 12. Nominal compositions of P/M tool steels

AISI designation	Composition, wt% C	Mn	Si	Cr	V	W	Mo	Co
M2	0.85	0.30	0.30	4.0	2.0	6.0	5.0	...
M4	1.30	0.30	0.30	4.0	4.0	5.5	4.5	...
M42	1.10	0.30	0.30	3.75	1.15	1.5	9.5	8.0
CPM Rex 20	1.30	0.30	0.30	3.75	2.0	6.25	10.5	...
T15	1.55	0.30	0.30	4.0	5.0	12.25	...	5.0
CPM Rex 25	1.80	0.30	0.35	4.0	5.0	12.5	6.5	...
CPM Rex 76	1.50	0.30	0.30	3.75	3.1	10.0	5.25	9.0
CPM 10V	2.45	0.50	0.90	5.25	9.75	...	1.30	...
D2	1.55	0.35	0.45	11.5	0.9	...	0.8	...
ASP 23	1.3	...	...	4.2	3.1	6.4	5.0	...
ASP 30	1.3	...	...	4.2	3.1	6.4	5.0	8.5
ASP 60	2.3	...	...	4.0	6.5	6.5	7.0	10.5

(a) 12.7-mm (1/2-in.) radius notch
Source: Ref 16, 17

Table 13A. Mechanical properties of tool steels made from cold compacted and vacuum sintered water-atomized powders

Grade	Hardness, HRC	Ultimate tensile strength(a) MPa	ksi	Elongation(b), %	Impact strength(c) J	ft·lb	Ref
M2	62-65	750-2000	109-290	12-14	9-12	7-9	14, 15
M35	63-66	770-2000	112-290	6-9	8-11	6-8	14, 15
T15	64-67	770-2000	112-290	3-6	8-11	6-8	14, 15

(a) Values depend on heat treatment. Lowest values are for fully annealed condition. (b) Fully annealed. (c) Triple temper. Izod unnotched impact bar (ASTM E 23)

Table 13B. Mechanical properties of gas-atomized and hot isostatically pressed P/M tool steels and ingot metallurgy tool steels

Alloy	Hardness, HRC	Bend fracture stress MPa	ksi	Bend deflection at fracture mm	in.	Charpy V-notch impact strength(a) J	ft·lb	Fracture toughness MPa√m	ksi√in.	0.2% compressive yield strength MPa	ksi	Ref
Ingot metallurgy alloys												
M2	64-65	3819	554	···	···	23	17	···	···	···	···	16
M4	64-65	3585	520	···	···	16	12	···	···	···	···	16
M42	67-68	2565	372	···	···	7	5	···	···	···	···	16
T15	66-67	2151	312	···	···	5	4	···	···	···	···	16
D2	62	2068	300	···	···	23	17	···	···	···	···	16
P/M alloys												
M2 CPM	64-65	4991	724	···	···	41	30	···	···	···	···	16
M4 CPM	64-65	5377	780	···	···	43	32	···	···	···	···	16
M42 CPM	67-68	4005	581	···	···	16	12	···	···	···	···	16
T15 CPM	66-67	4674	678	···	···	···	14	···	···	···	···	16
CPM Rex 20	67-68	4005	581	···	···	19	12.5	···	···	···	···	16
CPM Rex 25 (M61)	66-67	4323	627	···	···	15	11	···	···	···	···	16
CPM Rex 76 (M48)	69	4088	593	···	···	14	10	···	···	···	···	16
CPM 10V	63	4240	615	···	···	23	17	···	···	···	···	16
ASP 23	66	4800	696	···	···	···	···	13	12	3500	508	17
ASP 23	62	···	···	···	···	···	···	19	17	2800	406	17
ASP 30	66	5100	740	2.1	0.083	···	···	···	···	3600	522	18
ASP 60	67	4600	667	···	···	···	···	···	···	···	···	19

sintered at 595 to 620 °C (1100 to 1150 °F) to a density of 90 to 95% of theoretical. Mechanical properties of alloys 601AB and 201AB are given in Table 14.

NICKEL- AND COBALT-BASE ALLOYS

Most small- to medium-size parts made from these wear- and corrosion-resistant alloys traditionally have been castings. Prealloyed powders are now used as the starting materials. Powders are cold pressed and high-temperature sintered to near-full density. Parts that are free from alloy segregation can be produced. Table 15 summarizes compositions and mechanical properties of typical nickel- and cobalt-base P/M alloys.

SUPERALLOYS

These highly alloyed materials are produced from powders to prevent alloy segregation and to provide better forgeability (freedom from cracking). Several consolidation methods may be used, but all require the use of very clean powder that is free of oxides and inclusions. One consolidation method consists of canning; extrusion to a fully dense, fine grain structure; and superplastic forming of the part (gatorizing). In another consolidation method, powder is canned in a vessel of predetermined larger dimensions, evacuated, sealed, and hot isostatically pressed to near-net shape. An alternative consolidation method adds a final forging step to the second consolidation process. Powder metallurgy processing provides

segregation-free high alloying and a near-net shape that increases material utilization.

Nominal compositions of selected P/M superalloys are given in Table 16; typical mechanical properties are summarized in Table 17. In addition to room-temperature tensile properties, superalloys are characterized by superior elevated-temperature properties (stress-rupture and creep) and low-cycle fatigue.

TITANIUM AND TITANIUM-BASE ALLOYS

Titanium P/M parts are favored for superior corrosion resistance and high strength-to-weight ratio (for aircraft applications). Selection of the P/M process depends on the method of powder production (blended elemental or prealloyed) and desired final density. Medium- to high-density parts with close tolerances are made by pressing and sintering. Starting powder is usually sponge fines with master alloy additions of vanadium and aluminum to make Ti-6Al-4V. These are referred to as blended elemental powders. Sintering causes the vanadium and aluminum to diffuse into the titanium. Table 18 compares properties of wrought and P/M titanium materials. For aircraft applications, which require fully dense parts, the starting powder should be a high-purity, prealloyed material. Such powder is often made by the plasma rotating electrode process. These powders are canned in oversize containers and hot isostatically pressed to near-net shape, resulting in full density with reduced material utilization and machining.

REFERENCES ON POWDER METALLURGY MATERIALS

1. Metal Powder Industries Federation Standard 35, "P/M Materials Standards and Specifications," Metal Powder Industries Federation, Princeton, NJ

2. Huppmann, W. J. and Albano-Muller, L., Production of Powder Forged Parts of Complex Geometry, in *Modern Developments in Powder Metallurgy*, Vol 12, Metal Powder Industries Federation, Princeton, NJ, 1981, p 631

3. Hanejko, F., Mechanical Properties of Powder Forged 4100 and 1500 Type Alloy Steels, in *Modern Developments in Powder Metallurgy*, Vol 12, Metal Powder Industries Federation, Princeton, NJ, 1981, p 689

4. Pilliar, R. M. *et al.*, Fracture Toughness Evaluation of Powder Forged Parts, in *Modern Developments in Powder Metallurgy*, Vol 7, Metal Powder Industries Federation, Princeton, NJ, 1974, p 51

5. Badia, F., Heck, F., and Tundermann, J., Effect of Composition and Processing Variations on Properties of Hot Formed Mixed Elemental P/M Nickel Steels, in *Modern Developments in Powder Metallurgy*, Vol 7, Metal Powder Industries Federation, Princeton, NJ, 1974, p 255

6. Lindskog, P., Reduction of Oxide Inclusions in Powder Preforms Prior to Hot Forming, in *Modern Developments in Powder Metallurgy*, Vol 7, Metal Powder Industries Federation, Princeton, NJ, 1974, p 285

7. Tsumuki, C. *et al.*, Connecting Rods by P/M Hot Forging, in *Modern Developments in Powder Metallurgy*, Vol 7, Metal Powder Industries Federation, Princeton, NJ, 1974, p 385

8. Mocarski, S. and Hall, D. W., Properties of Hot Formed Mo-Ni-Mn P/M Steels with Admixed Copper, in *Modern Developments in Powder Metallurgy*, Vol 9, Metal Powder Industries Federation, Princeton, NJ, 1977, p 467

(continued on page 19·10)

Table 14. Room-temperature mechanical properties of aluminum alloys

Alloy designation	Nominal composition	Oxygen content, %	Temper designation	Test direction	Density, % of theoretical	Tensile strength MPa	ksi	Yield strength MPa	ksi	Elongation, %	Fracture toughness (K_{Ic}) MPa √m	ksi √in.	Ref
Extrusions													
X7090	0.12 Si, 0.15 Fe, 0.6-1.3 Cu, 2-3 Mg, 7.3-8.7 Zn, 1.0-1.9 Co	0.2-0.5	T6E192	Longitudinal	100	676	98	641	93	10(a)	...	...	25
				Long transverse	100	648	94	600	87	10(a)	...	...	25
	0.12 Si, 0.15 Fe, 0.6-1.3 Cu, 2-3 Mg, 7.3-8.7 Zn, 1.0-1.9 Co	0.2-0.5	T7E71	Longitudinal	100	621	90	579	84	9(a)	30.8	28	25
				Short transverse	100	558	81	496	72	8(a)	19.8	18	25
X7091	0.12 Si, 0.15 Fe, 1.1-1.8 Cu, 2-3 Mg, 5.8-7.1 Zn, 0.2-0.6 Co	0.2-0.5	T6E192	Longitudinal	100	614	89	558	81	11(a)	...	...	25
				Long transverse	100	586	85	538	78	13(a)	...	...	25
	0.12 Si, 0.15 Fe, 1.1-1.8 Cu, 2-3 Mg, 5.8-7.1 Zn, 0.2-0.6 Co	0.2-0.5	T7E69	Longitudinal	100	593	86	545	79	11(a)	46.2	42	25
				Long transverse	100	545	79	496	72	9(a)	33.0	30	25
				Short transverse	100	524	76	455	66	9(a)	26.4	24	25
	0.12 Si, 0.15 Fe, 1.1-1.8 Cu, 2-3 Mg, 5.8-7.1 Zn, 0.2-0.6 Co	0.2-0.5	T7E70	Longitudinal	100	538	78	483	70	11	...	...	25
				Long transverse	100	510	74	462	67	10	47.3	43	25
Pressed bars													
601AB	1.0 Mg, 0.6 Si, 0.25 Cu, rem Al	...	T1	Longitudinal	91.1	110	16	48	7	6(b)	...	...	26
			T4	Longitudinal	91.1	141	20.5	97	14	5(b)	...	...	26
			T6	Longitudinal	91.1	183	26.5	176	25.5	1(b)	...	...	26
			T61	Longitudinal	...	241	35.0	237	34.4	2(b)	...	...	26
			T1	Longitudinal	93.7	121	17.5	55	8	7(b)	...	...	26
			T4	Longitudinal	93.7	148	21.5	100	14.5	5(b)	...	...	26
			T6	Longitudinal	93.7	224	32.5	214	31	2(b)	...	...	26
			T61	Longitudinal	...	252	36.5	247	35.8	2(b)	...	...	26
			T1	Longitudinal	96.0	124	18.0	59	8.5	8(b)	...	...	26
			T4	Longitudinal	96.0	152	22.0	103	15.0	5(b)	...	...	26
			T6	Longitudinal	96.0	252	36.5	241	35.0	2(b)	...	...	26
			T61	Longitudinal	...	255	37.0	248	36.2	2(b)	...	...	26
201AB	0.5 Mg, 0.8 Si, 4.4 Cu, rem Al	...	T1	Longitudinal	91.0	169	24.5	145	21.0	2(b)	...	...	26
			T4	Longitudinal	91.0	210	30.5	179	26.00	3(b)	...	...	26
			T6	Longitudinal	91.0	248	36.0	...	...	...	...	...	26
			T61	Longitudinal	...	343	49.7	339	49.2	0.5(b)	...	...	26
			T1	Longitudinal	92.9	201	29.2	170	24.6	3(b)	...	...	26
			T4	Longitudinal	92.9	245	35.6	205	29.8	3.5(b)	...	...	26
			T6	Longitudinal	92.9	322	46.8	...	...	...	...	...	26
			T61	Longitudinal	...	349	50.6	342	49.6	0.5(b)	...	...	26
			T1	Longitudinal	97.0	209	30.3	181	26.2	3.0(b)	...	...	26
			T4	Longitudinal	97.0	262	38.0	214	31.0	5.0(b)	...	...	26
			T6	Longitudinal	97.0	332	48.1	327	47.5	2.0(b)	...	...	26
			T61	Longitudinal	...	356	51.7	354	51.3	2.0(b)	...	...	26

Note: T1, as sintered; T4, solution heat treated for 30 min at 520 °C (970 °F) for 601AB or 505 °C (940 °F) for 201AB, quenched, aged 4 weeks at room temperature; T6, solutionized and quenched as T4, aged 18 h at 160 °C (320 °F); T61, re-press bars at 345 MPa (25 tsi), temper as T6
(a) Elongation in 50 mm (2 in.). (b) Elongation in 25 mm (1 in.)

Table 15. Typical mechanical properties and compositions of nickel- and cobalt-base P/M alloys

Alloy designation	Room temperature Ultimate tensile strength, MPa (ksi)	Elon- gation, %	Hardness, HRC	540 °C (1000 °F) Ultimate tensile strength, MPa (ksi)	Elon- gation, %	Hardness, HRC	650 °C (1200 °F) Ultimate tensile strength, MPa (ksi)	Elon- gation, %	Hardness, HRC	760 °C (1400 °F) Ultimate tensile strength, MPa (ksi)	Elon- gation, %	Hardness
Stellite 3	863 (125)	<1	54	725 (105)	<1	40	690 (100)	<1	39	621 (90)	1	28 HRC
Stellite 6	897 (130)	<1	40	828 (120)	1	37	766 (111)	1	30	518 (75)	10	15 HRC
Stellite 19	1035 (150)	<1	49	... (...)	...	...	... (...)	...	...	... (...)	...	...
Stellite 31	828 (120)	4	...	676 (98)	14	...	614 (89)	13	...	... (...)	...	...
Stellite 190	621 (90)	<1	58	518 (75)	<1	54	518 (75)	<1	46	518 (75)	<1	34 HRC
Star J Metal	523 (76)	0.1	56	539 (78)	0.1	52	569 (82)	0.1	43	573 (83)	0.1	31 HRC
Stellite 98 M2	794 (115)	0.3	58	725 (105)	0.3	...	690 (100)	0.5	...	656 (95)	0.5	...

(continued)

Table 15. (continued)

Alloy designation	Room temperature			540 °C (1000 °F)			650 °C (1200 °F)			760 °C (1400 °F)		
	Ultimate tensile strength, MPa (ksi)	Elongation, %	Hardness, HRC	Ultimate tensile strength, MPa (ksi)	Elongation, %	Hardness, HRC	Ultimate tensile strength, MPa (ksi)	Elongation, %	Hardness, HRC	Ultimate tensile strength, MPa (ksi)	Elongation, %	Hardness
Haynes 208	690 (100)	<1	44	552 (80)	<1	41	552 (80)	<1	37	483 (70)	1	25 HRC
Haynes N-6	656 (95)	2	30	545 (79)	3	25	545 (79)	4	20	428 (62)	7	82 HRB
Haynes 711	559 (81)	<1	50	490 (71)	<1	43	490 (71)	<1	43	504 (73)	<1	27 HRC

Alloy	Composition, %											
	Ni	Si	Fe	Mn	Cr	Mo	W	C	V	B	Co	Other (total)
Stellite 3	3(a)	1(a)	3(a)	1(a)	31	...	12.5	2.4	...	1(a)	rem	1(a)
Stellite 6	3(a)	1.5(a)	3(a)	1(a)	29	1.5(a)	4.5	1.2	...	1(a)	rem	2(a)
Stellite 19	3(a)	1(a)	3(a)	1(a)	31	...	10.5	1.9	...	1(a)	rem	2(a)
Stellite 31	10.5	1(a)	2(a)	1(a)	25.5	...	7.5	0.5	...	...	rem	2(a)
Stellite 190	3(a)	1(a)	5(a)	1(a)	26	1(a)	1.4	3.1	...	1(a)	rem	2(a)
Star J Metal	3(a)	1(a)	3(a)	1(a)	32.5	...	17.5	2.5	...	1(a)	rem	2(a)
Stellite 98 M2	3.5	1(a)	5(a)	1(a)	30	0.8(a)	18.5	2	4.2	1	rem	2(a)
Haynes 208	rem	1(a)	12.5	0.75(a)	26	10	10	2.6	...	1(a)	10	2(a)
Haynes N-6	rem	1.5(a)	3(a)	1(a)	29	5.5	2	1.1	...	0.6	3	...
Haynes 711	rem	...	2.3	...	27	7	3	2.7	...	1(a)	12	2

(a) Maximum
Source: Ref 20

Table 16. Nominal compositions of P/M superalloys

Alloy	Composition, wt%													
	C	Cr	Mo	Fe	Co	Al	Ti	B	Nb	V	Hf	W	Zr	Ni
IN-100	0.1	10.0	3.5	1.0	14.0	4.5	5.5	0.01	...	1.0	...	...	0.05	rem
René 95	0.1	14.0	3.5	...	8.0	3.5	2.5	0.01	3.5	...	...	3.6	0.05	rem
Astroloy	0.05	15.0	5.0	...	18.0	4.0	3.5	0.03	...	...	...	...	0.05	rem
MERL 76	0.025	12.5	3.0	...	18.5	5.0	4.3	0.02	1.4	...	0.4	...	0.06	rem
AF-115	0.05	10.5	2.8	...	15.0	3.8	3.9	0.02	1.8	...	0.8	5.9	0.06	rem
Udimet 100	0.03	14.72	4.90	0.62	17.72	3.86	3.53	0.026	...	...	...	0.04	0.03	rem

Table 17. Typical mechanical properties of P/M superalloys

Property	René 95(a) (Ref 21)	Low-carbon Astroloy(b) (Ref 22)	Low-carbon Astroloy(c) (Ref 22)	Low-carbon Astroloy(d) (Ref 22)	IN-1000(e) (Ref 23)	MERL 76(f) (Ref 23)	Udimet 700(g) (Ref 24)	Udimet 700(a) (Ref 24)
0.2% yield strength at 210 °C (410 °F), MPa (ksi)	1257 (182)	973 (141)	928 (135)	994 (143)	1095 (159)	1188 (172)	860 (125)	1115 (162)
Ultimate tensile strength at 210 °C (410 °F), MPa (ksi)	1671 (242)	1376 (200)	1338 (194)	1359 (197)	1594 (231)	1674 (243)	1355 (197)	1515 (219)
Elongation, %	20	22	26	28	26	21	25	18.5
Reduction in area, %	20.3	23	28	32	27	22	27	18.5
Creep at 595 °C (1110 °F) at 1034 MPa (150 ksi), 100 h/% strain	0.15	...	...	...	...	...	...	...
Stress rupture at 650 °C (1200 °F) at 1034 MPa (150 ksi), service life (hours)/% elongation	29.5/5.4 28.4/4.7	...	...	...	...	...	...	...
Strain-controlled low-cycle fatigue at 535 °C (1000 °F), strain/cycles to failure	0.78/26 948 0.66/94 447	...	...	...	...	...	...	...
Stress rupture at 621 MPa (90 ksi) at 730 °C (1350 °F) at 151 h, % elongation/% reduction in area	...	14/17	17/22	16/21	...	...	...	...
0.2% yield strength at 705 °C (1300 °F), MPa (ksi)	...	...	...	...	1044 (151)	...	...	...
Ultimate tensile strength at 705 °C (1300 °F), MPa (ksi)	...	...	...	...	1265 (184)	...	...	...
% elongation/% reduction in area at 705 °C (1300 °F)	...	...	...	...	19/23	...	...	...
Stress rupture at 730 °C (1350 °F) at 655 MPa (95 ksi), hours to failure/% elongation	...	...	...	...	35.6/16.6 25.5/11.0 37.0/14.5	...	...	...

(continued)

Table 17. (continued)

Property	René 95(a) (Ref 21)	Low-carbon Astroloy(b) (Ref 22)	Low-carbon Astroloy(c) (Ref 22)	Low-carbon Astroloy(d) (Ref 22)	IN-1000(e) (Ref 23)	MERL 76(f) (Ref 23)	Udimet 700(g) (Ref 24)	Udimet 700(a) (Ref 24)
Creep at 705 °C (1300 °F) at 551 MPa (80 ksi), time for 0.1%/time for 0.2%	...	...	...	...	140.5/193.5	...	...	...
	...	...	...	...	100.0/142.0	...	...	...
	...	...	...	...	91.0/125.0	...	...	...
0.2% yield strength at 620 °C (1150 °F), MPa (ksi)	...	...	...	...	...	1136 (165)	...	...
Ultimate tensile strength at 620 °C (1150 °F), MPa (ksi)	...	...	...	...	...	1492 (216)	...	...
% elongation/% reduction in area at 620 °C (1150 °F)	...	...	...	...	...	18.5/17	...	...

(a) Hot isostatically pressed and hardened and tempered. (b) Produced by rapid omnidirectional compaction. Consolidated at 811 MPa (58.8 tsi); 0.5-s dwell in composite of copper-nickel fluid dies. Preheated to 1075 °C (1970 °F); held at temperature 1 h. Powder was electrodynamically degassed prior to vacuum filling. Post-consolidation solution treated at 1165 °C (2125 °F) for 4 h, fan air cooled. (c) Hot isostatically pressed at 1150 °C (2100 °F) at 104 MPa (15 ksi) for 4 h. Hot loaded at 925 °C (1700 °F). Hot unloaded at 980 °C (1800 °F). Post-consolidation solution treated at 1120 °C (2050 °F) for 2 h, fan air cooled. (d) Hot isostatically pressed and forged. Forging conditions: open die side upset at 1095 °C (2000 °F). Average reduction is 52%. Aging heat treatment for all processes: 650 °C (1200 °F) for 24 h, air cooled plus holding at 760 °C (1400 °F) for 8 h, air cooled. All tensile specimens tested normal to the forging direction. (e) Hot isostatically pressed and gatorized billet. (f) Hot isostatically pressed and gatorized. (g) As hot isostatically pressed

Table 18. Mechanical properties of P/M and wrought titanium and titanium-base alloys

Alloy	Processing	Density, %	Ultimate tensile strength, MPa (ksi)	Yield strength MPa (ksi)	Elongation, %	Reduction in area, %	Elastic modulus, GPa (10^6 psi)	Fatigue limit, notched MPa (ksi)	Fracture toughness, MPa $\sqrt{m}$ (ksi $\sqrt{in.}$)	Ref
Wrought commercial purity titanium grade II	...	100	345 (50)	344 (50)	5	35	103 (14.9)	...	...	28
	...	95.5	414 (60)	324 (47)	15	14	103 (15)	...	...	28
Sponge commercial-purity P/M titanium(a)	...	94	427 (62)	338 (49)	15	23	...	...	...	29
	Forged	100	455 (66)	365 (53)	23	30	...	...	...	29
Wrought Ti-6Al-4V (AMS 4298)	...	100	896 (130)	827 (120)	10	20	114 (16.5)	427 (62)	55(e) (50)(e)	28
P/M Ti-6Al-4V	Blended elemental alloy, cold pressed	95.5	876 (127)	786 (114)	8	14	117 (17)	193 (28)	45(e) (40)(e)	28, 30
	...	98+	919 (133)	839 (121.6)	10.9	19.0	...	262 (38)	56(e) (51)(e)	30, 31
	Blended elemental alloy, forged preforms or vacuum hot pressed	99 min	937 (136)	862 (125)	12-18	15-40	116 (16.8)	414 (60)	61(e) (56)(e)	28, 30
	Hot isostatically pressed prealloy	100	947 (137.4)	868 (125.9)	18.8	43.2	117 (17)	414 (60)	...	28
	Solution treated and aged	99	1103 (160)	1013 (147)	4.9	7.6	...	...	...	32
	Rapid omnidirectional compacted(c)	100	1014 (147)	944 (137)	18.4	40.9	...	...	...	22
Plasma rotating electrode processed Ti-6Al-6V-2Sn	Hot isostatically pressed	100	1053 (152.7)	1008 (146.3)	18	36.5	110 (16)	448 (65)	...	33
Plasma rotating electrode processed Ti-6Al-4V	Hot isostatically pressed	100	951 (138)	910 (132)	15	39	...	414(d) (60)(d)	83(f) (76)(f)	34
P/M Ti-6Al-4V(a)	...	94	827 (120)	738 (107)	5	8	...	...	...	29
	Forged	100	920 (133.5)	841 (122)	11.5	25	...	...	...	29
P/M Ti-6Al-4V(b)	Hot isostatically pressed	100	917 (133)	827 (120)	13	26	...	...	...	29
P/M Ti-6Al-6V-2Sn	...	99	1067 (155)	977 (142)	10	14	...	...	...	32

(a) 0.12% oxygen. (b) 0.2% oxygen. (c) Consolidated at 811 MPa (58.8 tsi), 0.5-s dwell in low-carbon steel fluid dies. Preheat temperature was 940 °C (1725 °F), held at temperature $^3/_4$ h. Powder was vacuum filled into fluid dies following cold static outgassing for 24 h. (d) K_t = 3. (e) K_c. (f) K_{Ic}

9. Saritas, S., James W. B., and Davies, T. J., Influence of Preforging Treatments on the Mechanical Properties of Two Low Alloy Powder Forged Steels, *Powder Metall.,* Vol 3, 1981, p 131

10. Pietrocini, T. W. and Gustafson, D. A., Fatigue and Toughness of Hot Formed Cr-Ni-Mo and Ni-Mo Prealloyed Steel Powders, in *Modern Developments in Powder Metallurgy,* Vol 4, Plenum Press, New York, 1971, p 431

11. High Technology Materials, Amstead Research Laboratories, Bensenville, IL, private communication, 1983

12. Aslund, C., "Fully Dense Stainless Steel Products Compete Successfully With Forged Products," Metal Powder Industries Federation National Powder Metallurgy Conference, New Orleans, 1983

13. Billiet, R., "Plastic Metals from Fiction to Reality With Injection Molded P/M Materials," Metal Powder Industries Federation National Powder Metallurgy Conference, Montreal, June 1982

14. Consolidated Metallurgical Industries Inc., Farmington Hills, MI, private communication, 1983

15. Beiss, P. and Huppman, W. S., "Sintering of P/M Tool Steels to Full Density," presented at Powder Metallurgy Short Course "Fully Dense P/M Materials for High Performance Applications," sponsored by Metal Powder Industries Federation, New Orleans, Feb 1983

16. Stasko, W., Crucible Research Center, Pittsburgh, private communication, 1983

17. Hellman, P., "Potential of High Strength P/M High Speed Steel," in *Processing and Properties of High Speed Tool Steels,* Ferrous Metallurgy Committee of the Metallurgical Society of American Institute of Mining, Metallurgical and Petroleum Engineers 109th Annual Meeting, Las Vegas, 26 Feb 1980

18. Uddeholm High Speed Steel, UHB ASP 30, Uddeholm Steel Corp., Totowa, NJ, private communication, 1983

19. Powder Metallurgy High Speed Steel, ASP 60, Uddeholm Corp., Totowa, NJ, private communication, 1983

20. Stellite Powder Metallurgy Products, Cabot Wear Technology Division, Cabot Corp., Kokomo, IN, private communication, 1983

21. Coyne, J. E. *et al.,* "Superalloy Powder Engine Components: Controls Employed to Assure High Quality Hardware," in *Powder Metallurgy Superalloys,* Metal Powder Report Conference, Zurich, Nov 1980

22. Kelto, C. A., Kelsey-Hayes Powder Technology Center, Traverse City, MI, private communication

23. Coyne, J. E. *et al.,* "Superalloy Turbine Components, Which Is the Superior Manufacturing Process, As HIP'd, HIP'd plus Isoforged, or Gatorizing℗ of Extrusion Consolidated Billets," in *Powder Metallurgy Superalloys,* Metal Powder Report Conference, Zurich, Nov 1980

24. Moser, G., *et al.,* Hot Isostatic Pressing of Superalloys, presented at Powder Metallurgy Superalloys, Aerospace Materials for the 1980's, Vol 2, Metal Powder Report Conference, Zurich, Nov 1980

25. Hart, R. M., "Wrought P/M Aluminum Alloys X7090 and X7091," Aluminum Company of America, Alcoa Technical Center, Alcoa Center, PA, Aug 1981, unpublished

26. "Aluminum Powder for Powder Metallurgy Parts," Aluminum Company of America, General Information, Pittsburgh, 1971

27. Properties of Powders and Powder Metallurgy Products, in *Aluminum,* Vol 1, *Properties and Physical Metallurgy,* Kent R. Van Horn, Ed., American Society for Metals, 1967

28. Garriott, R. E. and Thellmann, E. L., Titanium Powder Metallurgy—The Commercial Reality, 1976 International Powder Metallurgy Conference, Metal Powder Industries Federation, Princeton, NJ, 1976

29. Abkowitz, S., Isostatic Pressing of Complex Shapes from Titanium and Titanium Alloys, in *Titanium Alloys Science and Technology,* Proceedings of the 4th International Conference on Titanium, Kyoto, Japan, 19 May 1980

30. Anderson, T. J. *et al.,* "Fracture Behavior of Blended Elemental P/M Titanium Alloys," United States Air Force contracts F 33615-76-5227, F 33615-79-C5151, F33615-77-C-5008, 1979

31. Toaz, M. W., Blended Titanium P/M Alloys—Current Developments, *Metal Powder Rep.,* Vol 37 (No. 2), Feb 1982, p 85

32. Eloff, P. C., "Development of Low Cost Titanium Structures Using Blended Elemental Powder Metallurgy," Winter Annual Meeting of American Society of Mechanical Engineers, Washington, DC, 15 Nov 1981

33. Witt, R. H. and Bruce, J. S., Progress on Hot Isostatic Pressing of Titanium, 28th National Society for Advancement of Material and Process Engineering Symposium, 12 April 1983

34. Bruce, J. S., *et al.,* Evaluation of Titanium P/M Technology for Naval Aircraft Components, 28th National Society for Advancement of Material and Process Engineering Symposium, 12 April 1983

20 MATERIALS FOR SPECIAL APPLICATIONS

This section was condensed from Metals Handbook, Ninth Edition, Volume 3, Properties and Selection: Stainless Steels, Tool Materials and Special-Purpose Metals, pages 595 to 822. For more detailed information on the topics covered herein, the reader is referred to the larger work.

Magnetically Soft Materials

Condensed from Metals Handbook, Ninth Edition, Volume 3, pages 597 to 614.

MAGNETICALLY SOFT MATERIALS are ferromagnetic materials that have little or no retentivity—that is, if they are magnetized in a magnetic field and then are removed from that field, they lose most, if not all, of the magnetism they exhibited while in the field.

The most important characteristics of magnetically soft materials are: (a) low hysteresis loss (easy domain movement during magnetization); (b) low eddy-current loss from electric currents induced by flux changes; (c) high magnetic permeability, and sometimes constant permeability at low field strengths; (d) high magnetic saturation induction; and (e) minimum or definite change in permeability with temperature in special applications. Cost, availability and ease of processing also influence final choice of material.

Magnetically soft materials made in large quantities include high-purity iron, low-carbon steels, silicon steels, iron-nickel alloys, iron-cobalt alloys, and ferrites.

Impurities. Ferromagnetic properties depend on crystal structure and composition. Carbon, sulfur, nitrogen and oxygen are especially deleterious to ferromagnetic properties because they distort the lattice of the crystal structure and even in small amounts may greatly interfere with easy movement of magnetic domains, which is the basis of such properties.

A similar disturbance caused by carbon and nitrogen, known as "aging," occurs when low solubility at room temperature causes the excess solute to precipitate slowly as small particles within grains. In irons and iron-silicon alloys,

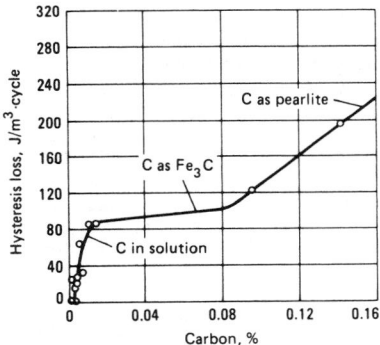

Induction $B = 1$ tesla (10 000 gausses).

Fig. 1. Relationship between carbon content and hysteresis loss for unalloyed iron

carbon content preferably should be less than 0.003% for best permeability, low hysteresis loss and minimal aging. Figure 1 shows the relationship between carbon content and hysteresis loss for iron. Hysteresis loss is similarly related to sulfur and oxygen contents.

Grain Size. For most applications, grain size should be as large as possible for nonoriented materials (Fig. 2). In oriented grades of silicon steel, optimum magnetic properties are usually obtained with grain sizes of 2 to 10 mm depending on the degree of crystal orientation: increases in grain size above 10 mm are accompanied by significant increases in both domain-wall spacing and eddy-current losses.

Grain Orientation. All ferromagnetic crystals are magnetically anisotropic — that is, they have different magnetic properties in different crystallographic directions. In nickel, the direction of easiest magnetization is the cube diagonal $\langle 111 \rangle$; in iron it is the cube edge $\langle 100 \rangle$. Crystal orien-

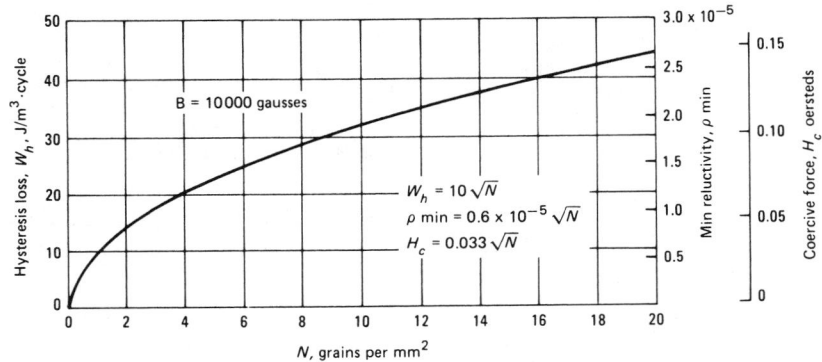

$$W_h = 10\sqrt{N}$$
$$\rho\text{ min} = 0.6 \times 10^{-5}\sqrt{N}$$
$$H_c = 0.033\sqrt{N}$$

Fig. 2. Effect of grain size on magnetic properties of pure iron and silicon iron

tation, therefore, is a basic factor in the determination and control of magnetic properties.

Alloying Elements. Pure iron is very soft magnetically and has a high saturation induction, but requires special handling during manufacture due to its low mechanical strength. Pure iron is used extensively in dc applications and in some ac relays. However, its low electrical resistivity makes it unsuitable for use in ac circuits, which constitute a great majority of all industrial applications of magnetic materials. Addition of alloying elements to iron increases its electrical resistivity and results in alloys that can be used in ac circuits. Figure 3 shows the changes in resistivity that result from additions of silicon, aluminum and other elements to iron.

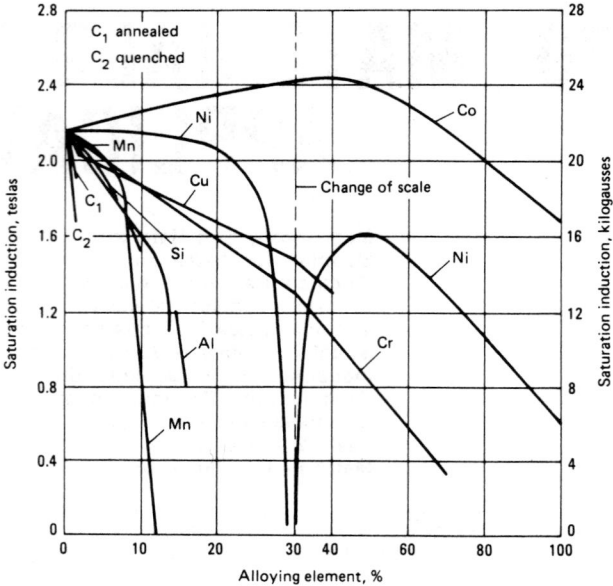

Fig. 4. Effect of alloying elements on room-temperature saturation induction of iron

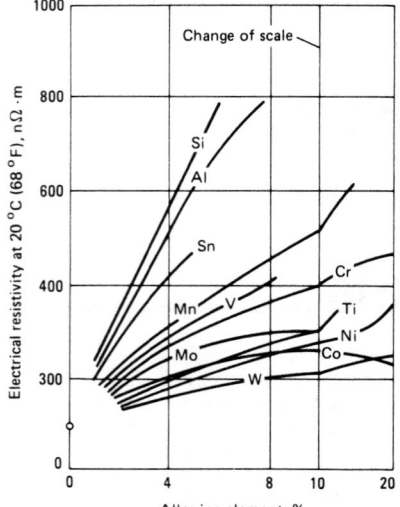

Fig. 3. Effect of alloying elements on electrical resistivity of iron

Addition of silicon in sufficient amounts eliminates the allotropic transformation in iron. Consequently, Si-Fe alloys can be annealed at high temperature to promote grain growth, thus facilitating development of preferred grain orientation. However, room-temperature saturation induction is reduced by alloy additions other than cobalt (see Fig. 4).

HIGH-PURITY IRONS

For experimental uses, 99.99% pure iron can be produced; by high-temperature annealing for prolonged times, a maximum permeability of about 100 000 with a hysteresis loss of about 10^{-5} J/cm³ per cycle at a flux density of 1.0 tesla (10 000 gausses) has been obtained (Fig. 5).

The saturation induction of iron, based on a density of 7.878 g/cm³, is given as $4\pi I_s = 2.1580 \pm 0.001$ tesla (21 580 ± 10 gausses), where I_s is the intensity of magnetization or magnetic moment per unit volume. The electrical resistivity is 98 nΩ·m (59 Ω·circ mil/ft) at 20 °C (68 °F), and the temperature coefficient is 0.0065 per °C (0.0036 per °F).

Commercial irons with purities of 99.6 to 99.8% are available in a variety of shapes. These irons have a saturation induction of about 2.15 teslas (21 500 gausses), a specific gravity of 7.85 and a resistivity of 107 nΩ·m (64 Ω·circ mil/ft) at 20 °C (68 °F).

LOW-CARBON STEELS

For many applications that require less than superior magnetic properties, low-carbon steels (type 1010, for example) are used. Frequently, higher-than-normal phosphorus and manganese contents are used to increase electrical resistivity. Such steels are not purchased to magnetic specifications. Although low-carbon steels exhibit power losses higher than those of silicon steels, they have better permeability at high flux density. This combination of magnetic properties, coupled with low price, makes low-carbon steels especially suitable for applications such as fractional-horsepower motors, which are used intermittently.

NONORIENTED SILICON STEELS

Except for saturation induction, the magnetic properties of iron containing a small amount of silicon are better than those of pure iron. Few commercial steels contain more than 3.5% silicon, because at levels above 4% the steel becomes brittle and difficult to process with cold rolling methods.

The commercial grades of silicon steel in common use (0.5 to 3.5% Si) are made mostly in electric or basic-oxygen furnaces. Nonoriented grades are melted with careful control of impurities; better grades have sulfur contents of about 0.01% or less. Continuous casting and vacuum degassing may be employed. After hot rolling,

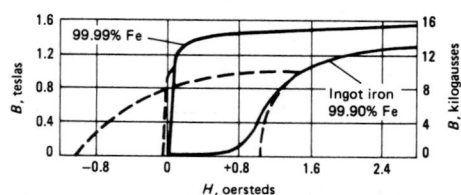

Fig. 5. Partial hysteresis loops and *B-H* curves for two types of iron

the hot bands are annealed, pickled and cold rolled to final thickness as continuous coils.

Semiprocessed grades of strip are not sufficiently decarburized for general use, and so decarburization and annealing to develop potential magnetic quality must be done by the user. This procedure is practical for small laminations accessible to the annealing atmosphere. Fully processed grades are strand annealed in moist hydrogen at about 825 °C (1520 °F) to remove carbon. The final annealing operation is very important and is carried out at a higher temperature — up to 1100 °C (2000 °F) for continuous strip — to cause grain growth and development of magnetic properties. Use of a protective atmosphere is vital. The steel frequently is coated with organic or inorganic materials after annealing, to reduce eddy currents in lamination stacks.

The vast majority of finished nonoriented silicon steel is sold in either full-width coils (860 to 1220 mm; 34 to 48 in.) or slit-width coils, but some is sold as sheared sheets. All coils are sampled and tested according to ASTM A343, and graded as to quality.

ASTM and AISI specify standard grades of electrical steel. The AISI designations were adopted in 1946 to eliminate the wide variety in nomenclature formerly used. When originally adopted, the AISI designation number approximated ten times the maximum core loss, in watts per pound, exhibited by 29-gage samples when tested at a flux density of 1.5 teslas (15 000 gausses) and a magnetic circuit frequency of 60 Hz.

More specific information is given by the present ASTM designation, as follows: the first two digits indicate thickness in mm × 100; following these digits is a letter (F, S, G or H) that indicates material and test conditions; the last three digits indicate maximum core loss (in watts per pound) × 100. In the present designation, code letter F denotes fully processed nonoriented electrical steel whose core loss is determined at 1.5 teslas and 60 Hz on a 50/50 Epstein specimen consisting of as-sheared test strips, 50% sheared parallel to the grain and 50% sheared across the grain. Code letter S indicates semiprocessed non-oriented electrical steel whose core loss is determined at 1.5 teslas and 60 Hz on a 50/50 Epstein specimen that has been annealed 1 h at 845 °C (1550 °F). Code G denotes fully processed grain-oriented electrical steel whose core loss is determined at 1.5 teslas and 60 Hz on a parallel-grain Epstein specimen stress relieved 1 h at 790 °C (1450 °F). Code H indicates fully processed grain-oriented electrical steel whose core loss is determined at 1.7 teslas and 60 Hz on a parallel-grain Epstein specimen stress relieved 1 h at 790 °C.

Table 1 lists some typical applications. Fully processed grades require only stress-relief annealing after fabrication or stamping. Semiprocessed grades must be decarburized by the customer to develop full magnetic properties. Suppliers' recommendations on annealing should be followed carefully.

Best permeability at high induction is obtained in steels with lower silicon contents. Low core loss is obtained with higher silicon contents, larger grains, lower impurity levels and thinner gages. The absence of preferred crystal orientation is actually helpful, because isotropic properties are desired in punchings for rotating machinery.

ORIENTED SILICON STEELS

Grain size is as important in silicon steel as it is in iron with regard to core losses and low-flux-density permeability. For high-flux-density permeability, however, crystallographic orientation is the deciding factor. Like iron, silicon steels are more easily magnetized in the direction of the cube edge, $\langle 100 \rangle$. For special compositions, rolling and heat treating techniques are used to promote secondary recrystallization in the final anneal at about 1175 °C (2150 °F) or higher, which results in a well-developed texture with the cube edge parallel to the rolling direction $\{110\}\langle 001 \rangle$. Conventional oriented grades contain about 3.15% Si.

About 1970, improved $\{110\}\langle 001 \rangle$ texture was developed by modification of composition and processing. The improved high-permeability material usually contains about 2.9 to 3.2% Si. Conventional grain-oriented 3.15% silicon steel has grains about 3 mm (0.12 in.) in diameter. The high-permeability silicon steel tends to have grains about 8 mm (0.31 in.) in diameter. Ideally, grain diameter should be less than 3 mm to minimize excess eddy-current effects from domain-wall motion. Special coatings provide electrical insulation and induced tensile stresses in the steel substrate. These induced stresses lower core loss and minimize noise in transformers.

High-grade silicon electrical steel does not age significantly as received from the mill, because its carbon content has been reduced to about 0.003% or less. With higher carbon contents, core loss can increase with time because of carbide precipitation. Also, silicon steel may age appreciably if not correctly heat treated in a manner that completely stabilizes its physical structure.

IRON-ALUMINUM ALLOYS

Although aluminum and silicon have similar effects on electrical resistivity and some magnetic properties of iron, aluminum is seldom substituted for silicon because of the resulting difficulties in fabrication. Aluminum is used most commonly as small (<0.5%) additions to the better grades of nonoriented silicon steel to increase

Table 1. Silicon contents, densities and applications of electrical steel sheet and strip

AISI type	Nominal Si + Al content, %	Assumed density, Mg/m³	Characteristics and applications
Lamination steel			
...	0	7.85	High magnetic saturation; magnetic properties may not be guaranteed; intermittent-duty small motors
Nonoriented electrical steels			
M47	1.05	7.80	Ductile, good stamping properties, good permeability at high inductions; small motors, ballasts, relays
M45	1.85	7.75	Good stamping properties, good permeability at moderate and high inductions, good core loss; small generators, high-efficiency continuous-duty rotating machines, ac and dc
M43	2.35	7.70	
M36	2.65	7.70	Good permeability at low and moderate inductions, low core loss; high reactance cores, generators, stators of high-efficiency rotating machines
M27	2.80	7.70	
M22	3.20	7.65	Excellent permeability at low inductions, lowest core loss; small power transformers, high-efficiency rotating machines
M19	3.30	7.65	
M15	3.50	7.65	
Oriented electrical steels			
M6	3.15	7.65	Grain-oriented steel has highly directional magnetic properties with lowest core loss and highest permeability when flux path is parallel to rolling direction; heavier thicknesses used in power transformers, thinner gages generally used in distribution transformers. Energy savings improve with lower core loss
M5	3.15	7.65	
M4	3.15	7.65	
M3	3.15	7.65	
High-permeability oriented steel			
...	2.9-3.15	7.65	Low core loss at high operating inductions

electrical resistivity and thereby reduce eddy currents. Ternary alloys of iron, silicon and aluminum have high resistivity and good permeability at low flux density. Increases in silicon and aluminum reduce saturation induction. At low flux densities, the magnetic properties of these alloys can be made to approach those of some iron-nickel alloys.

IRON-NICKEL ALLOYS

The effects of nickel content in iron-nickel alloys on saturation induction after annealing and on initial permeability are illustrated in Fig. 6 and 7, respectively. Various amounts of other elements, particularly molybdenum, often are added to these alloys in order to develop or accentuate specific characteristics.

In iron-nickel alloys, as in all magnetic materials, magnetic properties are controlled by saturation magnetization and magnetic anisotropy

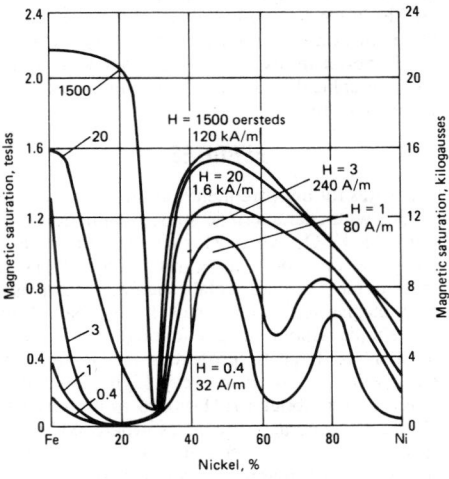

All specimens were annealed at 1000 °C (1830 °F) and cooled in the furnace.

Fig. 6. Magnetic saturation of iron-nickel alloys at various field strengths

energies. Of the various types of magnetic anisotropy energy, the magnetocrystalline and magnetostrictive anisotropies are the most important in this system. Two broad classes of alloys have been developed in the Fe-Ni system. The high-nickel alloys (about 79% Ni) have high initial permeability (Fig. 7) but low saturation induction (approximately 0.9 tesla, or 9000 gausses), whereas the low-nickel alloys (about 50% Ni) are lower in initial permeability but higher in saturation induction (about 1.6 teslas, or 16 000 gausses).

The data plotted in Fig. 8 are from early laboratory studies and illustrate the effects of both composition and heat treatment on initial permeability (μ_0). Values of μ_0 above 12 000 are now obtained commercially in 50% Ni alloys, and values above 60 000 are obtained in 79% Ni alloys containing 4% Mo.

To obtain high initial permeability, both magnetocrystalline anisotropy (K_1) and magnetostrictive anisotropy (λ_s) must be minimized. Cooling rate—or, in other words, the degree of atomic ordering achieved in the critical temperature range from 760 to 315 °C (1400 to 600 °F)—has a profound influence on the ability to minimize K_1. Consequently, cooling of this alloy from heat treating temperatures must be precisely controlled to obtain $K_1 = 0$. For commercially practical cooling rates of about 100 °C/min (180 °F/min), optimum composition for achieving $K_1 = 0$ and $\lambda_s = 0$ is about 4% Mo and 80% Ni.

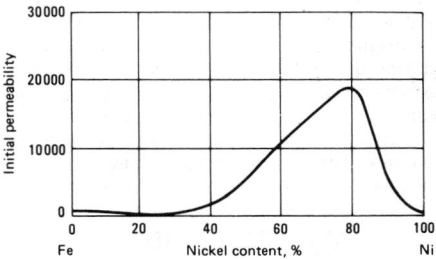

Fig. 7. Initial permeability at 2 mT (20 gausses) for annealed Fe-Ni alloys

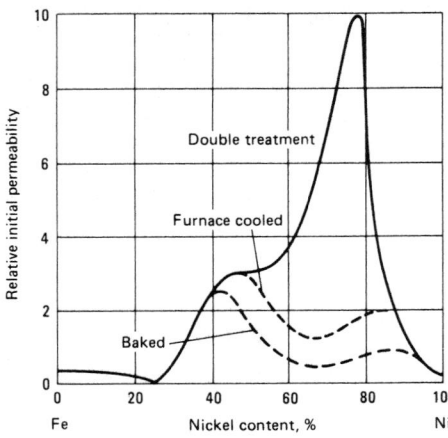

Treatments were as follows: furnace cooled—1 h at 900 to 950 °C (1650 to 1740 °F), cooled at 100 °C/h (180 °F/h); baked—furnace cooled plus 20 h at 450 °C (840 °F); double treatment—furnace cooled plus 1 h at 600 °C (1110 °F) and cooled at 1500 °C/min (2700 °F/min).

Fig. 8. Relative initial permeability at 2 mT (20 gausses) for Fe-Ni alloys given various heat treatments

Because $K_1 = 0$ and $\lambda_s = 0$ for commercial alloys such as Supermalloy, Moly Permalloy and Hymu 80, all of which contain about 4% Mo and 80% Ni, grain orientation is not critical for these alloys. Purity, however, influences permeability and core loss. Interstitial impurities such as carbon and nitrogen must be minimized by special melting procedures and by careful final annealing of laminations and other core configurations. Oxygen and sulfur are also objectionable. Best magnetic properties in Fe-Ni alloys are obtained by annealing in pure dry hydrogen (dew point less than −50 °C, or −58 °F) at 1000 to 1200 °C (1830 to 2200 °F) for several hours to reduce carbon, nitrogen, sulfur and oxygen contents. Sulfur contents higher than several ppm and carbon in excess of 20 ppm are detrimental to final annealed magnetic properties.

Most applications of 50% Ni alloys are based on requirements for high saturation induction. Nickel content is not critical near the middle of the iron-nickel series (50% Ni) and may be varied from 45 to 60%, but for highest saturation induction it should be held close to 50% (Fig. 6). Although $K_1 = 0$, the value of λ_{100}, which is the magnetostrictive constant in the $\langle 100 \rangle$ easy direction of magnetization, is close to zero for these alloys. Therefore, the initial permeability is still reasonable and in fact reaches a small maximum (Fig. 7). In some applications, such as converters, a high squareness ratio is desired—that is, a high ratio of remanence to saturation induction, B_r/B_s. In 50% Ni alloys, where the magnetocrystalline anisotropy is not zero, excellent squareness can be achieved by careful development of a cube texture during the final high-temperature anneal. In applications of Fe-Ni alloys it is thus important to control magnetic anisotropy energies, purity and texture, depending on what combination of properties is desired.

AMORPHOUS MATERIALS

Suitable alloys of iron prepared in amorphous, noncrystalline form have the attractive combination of high permeability and high volume re-

sistivity. In the preferred method of production, the metal is rapidly quenched from the melt onto cooled rotating drums to form long ribbons about 40 μm (1.6 mils) thick. Attractive compositions contain 40 to 80% Fe with various additions of carbon, cobalt, boron, silicon, nickel and phosphorus. These materials are characterized by low hysteresis loss and low coercive force. However, Curie temperature is limited to about 400 °C (750 °F), and magnetic saturation to about 1.6 teslas (16 000 gausses).

Potential applications include substitution for conventional nickel-iron alloys in electronic devices. A major goal is to utilize the low core loss of amorphous material, which is less than one-half that of conventional grain-oriented silicon steel 0.26 mm (0.010 in.) thick, in distribution transformers (provided that production costs are competitive).

IRON-COBALT ALLOYS WITH HIGH MAGNETIC SATURATION

Pure iron has a saturation induction of 2.16 teslas (21 600 gausses). Higher values require cobalt additions of up to 65%; the highest values (about 2.42 teslas, or 24 200 gausses) are obtained with cobalt contents of about 35% (Fig. 4). Use of alloys containing 25 to 50% Co is limited by low resistivity and high hysteresis loss, the high cost of cobalt and the brittleness of alloys containing more than 30% Co. However, with small additions of vanadium and special treatment in processing, 50% Co alloys can be cold rolled commercially to any gage, and strip is ductile enough to be punched and sheared. The 27% Co alloy is more easily fabricated, and less subject to degradation by stresses, than the 50% Co alloy. Furthermore, proper annealing of the 27% Co alloy produces magnetic properties suitable for both dc and ac applications.

Alternating-current applications require low eddy-current and hysteresis losses. Eddy-current losses can be minimized by a proper combination of composition and thickness.

AUSTENITIC STAINLESS STEELS

Austenitic stainless steels usually are not considered to be magnetic materials. However, increases in tensile strength due to cold working of these alloys are accompanied by increases in intrinsic permeability. This phenomenon is illustrated graphically in Fig. 9 for nine austenitic stainless steels.

FERRITES

Ferrites for high-frequency applications are ceramics with characteristic spinel-magnetic structures ($M \cdot Fe_2O_4$, where M is a metal) and usually comprise solid solutions of iron oxide and one or more oxides of other metals such as manganese, zinc, magnesium, copper, nickel and cobalt. They are unique among magnetic materials in their outstanding magnetic properties at high frequencies, which result from very high resistivities ranging from about 10^8 $\Omega \cdot$ cm to as high as 10^{14} $\Omega \cdot$ cm. Hence, at frequencies where eddy-current losses for metals become excessive, ferrites make ideal soft magnetic materials. Because ferrites have inherently high corrosion resistance, parts made of these materials normally do not require protective finishing.

Disadvantages of ferrites include low magnetic saturation, low Curie temperature, and relatively

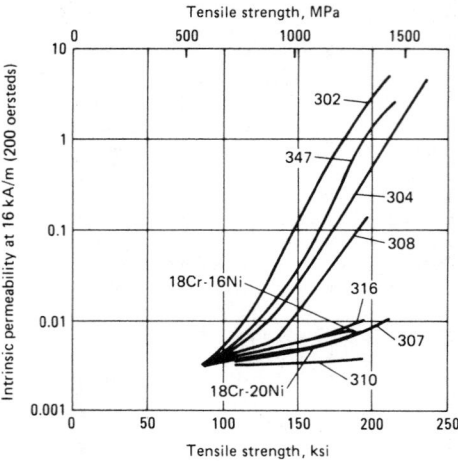

Annealed hot rolled strips 2.4 to 3.2 mm (0.095 to 0.125 in.) thick before cold reduction. For normal permeability values, add unity to the numbers given on vertical scale.

Fig. 9. Correlation of increased tensile strength from cold working and the permeability of cold worked austenitic stainless steels

poor mechanical properties compared with those of metals. Ferrites are produced from powdered raw materials by mixing, calcining, ball milling, pressing to shape, and firing to the desired magnetic properties. The final product is hard, brittle and unmachinable, and thus close dimensional tolerances must be obtained by grinding.

Many different types of ferrites are available for magnetic use. They can be classified into three general types: square-loop ferrites for computer memories, linear ferrites for transformers and for inductors in filters, and microwave ferrites for microwave devices. In recent years, due to increasing usage of semiconductors for computer memories, square-loop ferrites have decreased in importance.

Microstructure and composition have much stronger influences on the magnetic properties of ferrites than on those of metals. Hence, properties of finished ferrite parts can vary drastically with purity and structure of raw materials, with the nature of binders used and with the ceramic-processing technique employed. In general, lithium ferrites, Mn-Mg-Zn ferrites and Mn-Mg-Di ferrites are used for computer memories. Lithium ferrite is higher in Curie temperature and saturation magnetization, but lower in switching speed, than Mn-Mg-Zn and Mn-Mg-Di ferrites. Linear ferrites comprise Mn-Zn and Ni-Zn ferrites. Mn-Zn ferrite is higher in saturation magnetization, but lower in resistivity, than Ni-Zn ferrite. Mn-Zn ferrite is preferred for frequencies up to about 1 MHz. For microwave applications, Ni-Zn, Mg-Mn-Al and Mg-Mn-Cu ferrites are used, as well as garnets of the type $M_{3+x}Fe_{5-x}O_{12}$ (where M = Y + Al or Y + Gd + Al).

CONSTANT PERMEABILITY WITH CHANGING TEMPERATURE

In all magnetic materials, magnetic properties change with temperature. Proper selection and preparation of materials, and proper circuit design, can minimize these changes.

Change in flux density with temperature for iron tested at four different values of magnetizing force is plotted in Fig. 10. Operation of a device at a

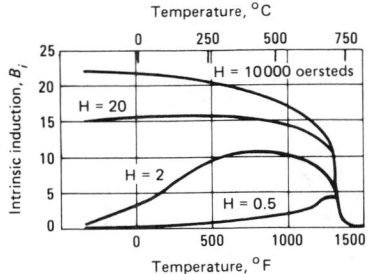

Fig. 10. Variation of induction with temperature for iron, at four different values of magnetizing force

flux density of 1.5 teslas (15 000 gausses) would be only slightly affected by variations in operating temperature near ambient. There is a similar minimized temperature effect for all materials, except that the flux density for optimum operation depends on the materials, given the proper flux density and temperature range. Great changes occur at temperatures approaching the Curie temperature (Fig. 10).

For many reasons, it is not always possible to operate a material at the best flux density for temperature stability. One way of obtaining better temperature stability is to use magnetic materials in insulated powder form, such as pressed Permalloy powder cores, which have good temperature stability due to the presence of many built-in air gaps. Another method involves use of Fe-Ni alloys, such as Isoperm or Conpernik, that have been drastically cold rolled and then underannealed to produce a partly strained alloy less sensitive to temperature changes.

In a third method, two alloy powders with opposite temperature coefficients are combined for use in the desired temperature range. Special Permalloy powder cores are combined with small amounts of Fe-Ni-Mo powder having a low Curie temperature and a negative temperature coefficient near room temperature. The 30% Ni irons are of the low-Curie-temperature type.

ALLOYS FOR MAGNETIC TEMPERATURE COMPENSATION

The variation in magnetic permeability of iron and iron alloys that occurs with temperature can seriously affect the accuracy of various electrical measuring instruments. To compensate for such changes, a certain amount of the magnetic flux can be shunted around the moving part of the instrument by using an alloy with high negative

magnetic temperature coefficient between −18 and +100 °C (0 and 212 °F). The amount of shunted flux therefore decreases with increases in ambient temperature, forcing more flux through the moving part than would otherwise be possible. Nearly complete compensation for temperature changes can be made by correct design of parts. A watt-hour meter is one example. Another example is an automobile speedometer, where a temperature compensator compensates for the change in pole strength of the permanent magnet and the change in electrical resistivity of the aluminum drag cup, over a temperature range.

Iron-nickel alloys are frequently used commercially for this purpose.

Magnetic induction and losses decrease, and permeability at low to intermediate induction increases, as operating temperature increases. However, temperature instability becomes greater as the operating temperature approaches the Curie temperature, or nonmagnetic point.

COMPRESSED POWDERED IRON

For applications in which complicated magnetic circuits would otherwise require considerable machining, it may be helpful to press iron powder in a mold and sinter the part in vacuum or in a reducing atmosphere. Use of P/M techniques to make magnetically soft components can eliminate all machining operations and may save as much as 50% of total cost.

The densities of most iron-base P/M parts range from 6.2 to 7.0 Mg/m³. High densities and superior magnetic properties can be obtained by using higher molding pressures or higher sintering temperatures, or by making the parts from high-purity iron-base powders (with or without small additions of alloying elements such as silicon and phosphorus). The cost of such parts is higher and, unless magnetic flux density exceeds 1 tesla (10 000 gausses), performance is no better than that of less-expensive iron-base P/M parts. This is especially true in magnetic circuits with short flux paths.

Selection of a rolled or sintered iron for a specific application depends on die cost, required finish, tolerances and number of parts. For example, design of soft iron pole pieces in the magnetic circuits of small panel-type motors permits the use of sintered P/M parts.

HEAT TREATMENT

Magnetic materials such as iron-nickel alloys in the cold rolled condition must be annealed to

develop the desired grain structure and magnetic properties.

Annealing conditions required for silicon steels depend on processing carried out by the supplier and on cost-versus-performance factors. Semi-processed grades must be annealed near 840 °C (1550 °F), after stamping of laminations, for removal of carbon and development of magnetic properties. Fully processed grades of nonoriented or grain-oriented steels require annealing in the range 750 to 875 °C (1375 to 1600 °F) only for removal of fabrication stresses. For wide laminations used in a flat condition, as in large power transformers, low-stress strip requiring no annealing at all is available.

STRESS EFFECTS

Magnetic properties such as power (core) loss and permeability are very sensitive to stresses. Plastically strained materials usually must be annealed unless the volume of strained material is only a small fraction of the total. In oriented silicon steels, a compressive stress of as little as 3450 to 6900 kPa (500 to 1000 psi) can increase core loss by 50 to 100%. Thus, great care must be exercised in punching and assembling electrical steels to ensure that stresses are relieved before the unit is assembled and that new stresses are not introduced during assembly. In some instances, however, small tensile stresses induced by coatings can improve the properties of grain-oriented silicon steel.

SELECTED REFERENCES

Soft Magnetic Materials, by R. Ball: Heyden & Son Ltd., London, 1979. [Handbook of soft magnetic materials with comprehensive commercial data.]

Magnetism and Metallurgy of Soft Magnetic Materials, by C. W. Chen: North Holland, New York, 1977. [General text on soft magnetic materials.]

Magnetic Materials and Their Applications, by C. Heck: Crane, Russak & Co., Inc., New York, 1974. [Comprehensive treatment of all magnetic materials in terms of component applications.]

Ferromagnetic Materials—A Handbook on the Properties of Magnetically Ordered Substances, edited by E. P. Wohlfarth: Elsevier, New York, 1980. [Volume 1 contains an article on "Amorphous Ferromagnets" by F. E. Luborsky. Volume 2 contains articles on "Soft Magnetic Metallic Materials" by G. Y. Chin and J. H. Wernick, "Ferrites for Non-Microwave Applications" by P. L. Slick and "Microwave Ferrites" by J. Nicolas.]

"Steel Products Manual—Flat Rolled Electrical Steel": American Iron and Steel Institute, March 1978. [Properties of low-carbon steel and silicon steel as established by ASTM.]

Materials for Permanent Magnets

Condensed from Metals Handbook, Ninth Edition, Volume 3, pages 615 to 639.

THE TERM "permanent magnet" is used to describe materials that are normally used in a single magnetic state and that have sufficiently high resistance to demagnetizing fields and sufficiently high magnetic flux output to provide useful and stable magnetic fields. This implies

insensitivity to temperature effects, mechanical shock and demagnetizing fields. This article does not consider magnetic memory or recording materials in which the magnetic state is altered during use. It does include, however, hysteresis alloys used in motors.

Permanent magnet materials include a variety of metals, intermetallics and ceramics. Commonly included are certain steels, Alnico, Cunife, Fe-Co alloys containing V or Mo, Pt-Co, hard ferrites, and cobalt – rare earth alloys. Each type of magnet material possesses unique magnetic and

mechanical properties, corrosion resistance, temperature sensitivity, fabrication limitations and cost. These factors provide designers with a wide range of options in designing magnetic parts.

Permanent magnet materials are based on the cooperation of atomic and molecular moments within a magnet body to produce a high magnetic induction. This induced magnetization is retained because of a strong resistance to demagnetization. These materials are classified as ferromagnetic, and do not include diamagnetic or paramagnetic materials. The natural ferromagnetic elements are iron, nickel and cobalt. Other elements, such as manganese and chromium, can be made ferromagnetic by alloying to induce proper atomic spacing. Ferromagnetic metals combine with other metals or with oxides to form ferrimagnetic substances; ceramic magnets are of this type. Although scientific literature lists many magnetic substances, relatively few have gained commercial acceptance because of the commercial requirement for low cost and high efficiency.

Permanent magnet materials are marketed under a variety of trade names and designations throughout the world. The United States designations will be used here.

Permanent magnet materials are developed for their chief magnetic characteristics: high induction, high resistance to demagnetization, and maximum energy content. Magnetic induction is limited by composition; the highest saturation induction is found in binary Fe-Co alloys. Resistance to demagnetization is conditioned less by composition than by shape or crystal anisotropies and the mechanisms that subdivide materials into microscopic regions. Precipitations, strains and other material imperfections, and fine particle technology are all used to obtain a characteristic resistance to demagnetization.

Maximum energy content is most important because permanent magnets are used primarily to produce a magnetic flux field (which is a form of potential energy). Maximum energy content and certain other characteristics of materials used for magnets are best described by a hysteresis loop. Hysteresis is measured by successively applying magnetizing and demagnetizing fields to a sample and observing the related magnetic induction.

FUNDAMENTALS OF MAGNETISM

For understanding a permanent magnet, Faraday's concept of representing a magnetic flux field by "lines of force" is very useful. The lines of force radiate outward from a "north pole" and return at a "south pole." The lines of force can be revealed by a powder pattern made by sprinkling iron powder on a paper placed above a bar magnet. The number of lines per unit area is the magnetic induction and is designated B. Induction consists of lines of force due to the magnetic field and lines of magnetization due to the ferromagnetism of the magnet:

$$B = H + B_i \qquad \text{(Eq 1)}$$

where H is the magnetic field strength and B_i is the intrinsic induction.

Magnetic Hysteresis. A hysteresis loop is a common method of characterizing a permanent magnet. The intrinsic induction is measured as the magnetizing field is changed (see Fig. 1). Starting with a virgin state of the material at the origin O, induction increases along curve I to the point marked $+S$ as the field is increased from zero to maximum. The point $+S$ is the point at which induction no longer increases with higher magnetizing field, and is known as the saturation induction. When the magnetizing field is reduced to zero in permanent magnets, most of the induction is retained. In Fig. 1, when the field is reduced through zero to $-S$, the induction decreases from $+S$ to B_r, to $-S$. At zero field, there is a remanent magnetization in the sample, defined as B_r; the value of B_r approaches the saturation induction in well-prepared permanent magnet materials.

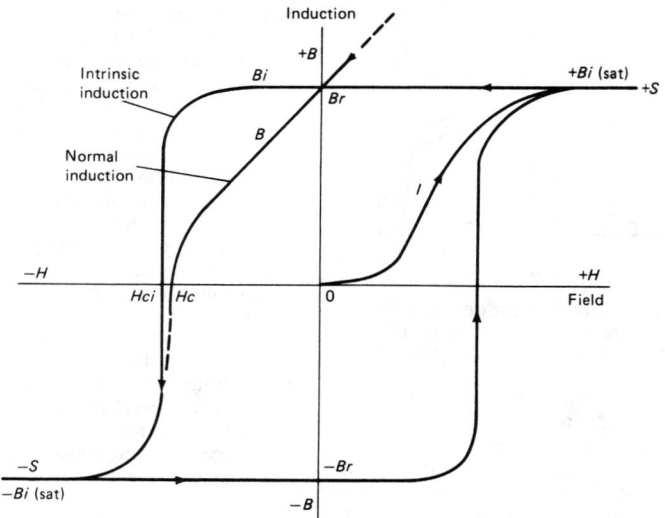

Fig. 1. Major hysteresis loop for a permanent magnet material

If the field is increased again in the positive direction, the induction passes through $-B_r$ to $+S$ as shown, and not through the origin. Thus, there is a hysteresis between the magnetization and demagnetization curves, and this plot is called the hysteresis loop. The two halves of the loop are generally symmetrical and form a major loop, which represents the maximum energy content, or the amount of magnetic energy that can be stored in the material. Innumerable minor loops can be measured within the major loop, measurements being made to show the effects of lesser fields on magnets under operating conditions.

Demagnetization. The particular value of the demagnetizing field is called the intrinsic coercive force H_{ci}. Figure 1 includes the normal demagnetization curve derived from the intrinsic curve. The field required to reduce induction B to zero is the normal coercive force H_c. The important practical features of the curve for application to permanent magnet materials are the numerical values of B_r and H_c, and the area within the hysteresis loop.

Because a permanent magnet most often is used to provide a flux field in a space outside itself, the material rests within its own field, which is a self-demagnetizing field. Therefore, for practical applications, a magnet designer is interested primarily in the second quadrant of the hysteresis loop, called the demagnetization curve (see Fig. 2). This curve represents the resistance to demagnetization and, in an affirmative sense, the ability of a material to establish a magnetic field in an air gap or adjoining magnetic material.

Magnetic Energy. The maximum magnetic energy available for use outside the magnet body is proportional to the largest rectangle that fits inside the normal demagnetization curve. It is indicated by the product $(B_d H_d)_{max}$ and is usually cited as the figure of merit for determining the quality of permanent magnet materials.

A characteristic useful in selecting permanent magnet materials subjected to varying demagnetizing conditions is the permeability at the operating point: $\mu = B_d/H_d$. For example, a straight-line demagnetization curve where $B_r = H_c$ would have the ideal permeability of 1.0; a magnet of such a material would recover spontaneously all flux when a partial demagnetizing field is re-

moved. The corresponding intrinsic curve would be flat out to the knee, and the material would retain maximum energy. Cobalt – rare earth magnets come closest to ideal permanent magnet behavior.

Figure 3 is the product of B and H at each point along the demagnetization curve, plotted against B. On the demagnetization curve, each value of B or H involves the other as a coordinate variable. The maximum value of their product, designated as $(B_d H_d)_{max}$, represents the maximum magnetic energy that a unit volume of the material can produce in an air gap. The most efficient design for a magnet is that which employs the magnet at the flux density corresponding to the $(B_d H_d)_{max}$ value.

The amount of total external magnetic flux available from a magnet operating in an open-circuit condition depends on its shape. This relation is shown in Fig. 4 for one specific shape. The permeability μ is the ratio of the total external permeance B_d to the permeance of the space occupied by the magnet, H_d, and is equal to the slope of the demagnetization curve.

An enlarged plot of the first and second quadrants of the intrinsic induction curve is given in

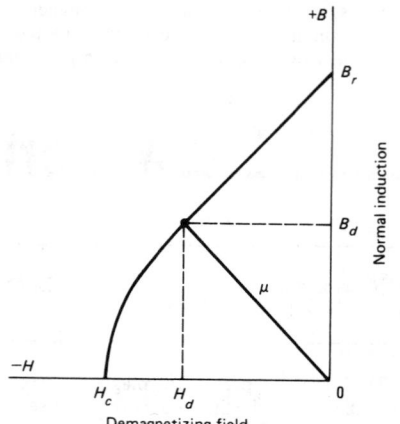

Fig. 2. Normal demagnetization curve for a permanent magnet material

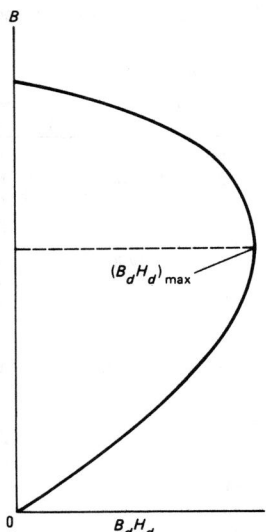

Fig. 3. Typical energy-product curve for a permanent magnet material

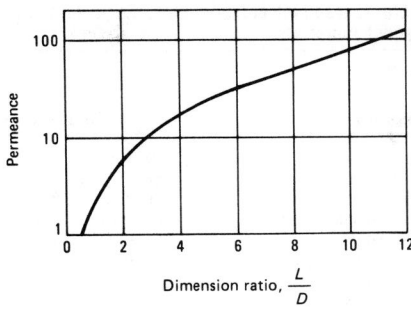

Fig. 4. Relation between magnetic properties and dimensions of straight bar magnets of circular cross section

Fig. 5. The intrinsic induction curve is primarily of interest to the materials scientist, who is concerned about the effect of composition and processing on the various intrinsic parameters of the material: B_{is}, B_r, and H_{ci}. The applications engineer, on the other hand, is concerned about the flux density in an air gap due to both B_i and H. Accordingly, he is likely to be more interested in the induction curve and in the values of B_r, H_c and $(BH)_{max}$.

Lines of constant energy product $(B_d H_d)$ usually are plotted in the second quadrant area and, as illustrated in Fig. 6, are a series of hyperbolic curves superimposed on the rectangular B-H grid of the demagnetization curves. The maximum values of external energy are therefore readily available in relation to the demagnetization curve. In this form, the grid constitutes an efficient guide for the design engineer. In practice, a magnet with a fixed air gap would have one fixed B_d/H_d operating point on the demagnetization curve corresponding to the material being used. For variable air gaps, such as are produced by relative movement between the armature and field poles of electrical machinery, the external energy available at the air gap changes continuously, resulting in a so-called minor loop with minimum and maximum values. In practice, the minor loop is plotted on the demagnetization curve to determine the location of the loop on the curve, and to evaluate the extent of flux variation within the minor loop cycle. Efficient design of equipment using permanent magnets, such as magnetos, small generators and motors, requires that the minor loop operate near the $(B_d H_d)_{max}$ point.

Magnetically soft materials differ from permanent magnet materials not only in their higher permeabilities, but also, and more significantly, in their much lower resistance to demagnetization. The best magnetically soft materials have H_c values of virtually zero. The hysteresis loop of such a material retraces itself through or near the origin point with each cycle.

Conversely, permanent magnet materials have wide hysteresis loops, characterized by high values of H_{ci}, which range from about 100 to over 20 000 Oe.

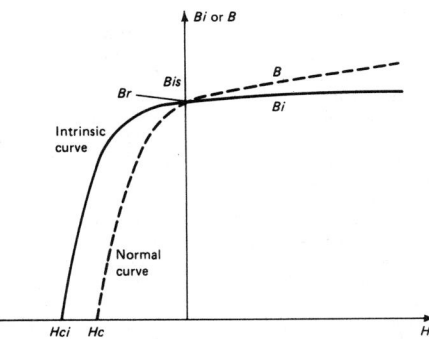

Fig. 5. Intrinsic magnetization curve B_i in the first and second quadrants compared with the curve for B

COMMERCIAL PERMANENT MAGNET MATERIALS

Table 1 lists most of the permanent magnet materials commercially available in the U.S. and their nominal compositions. Magnetic properties are presented in Table 2. Physical and mechanical properties are summarized in Table 3. Gen-

erally, the production of permanent magnet materials is controlled to achieve magnetic characteristics, and other properties are allowed to vary according to the manufacturing process used. The selection of materials and the design of permanent magnets for particular applications comprise a well-defined engineering art; design assistance is available from most major producers.

Permanent magnets are superior to electromagnets for many uses because they maintain their fields without an expenditure of electrical power and without the generation of heat.

Tables 2 and 3 give nominal properties only. Even under the most carefully controlled manufacturing conditions, some variation from these nominal values must be expected and considered in practical application.

STABILIZATION AND STABILITY

There is an important group of permanent magnet applications where the accuracy or performance of the device is drastically affected by very small changes (1% or less) in the strength of the magnet. These applications include braking magnets for watt-hour meters, magnetron magnets, special torque motor magnets, and most dc panel and switchboard instrument magnets. Operation of these devices requires extreme accuracy over a moderate range of conditions, or moderate accuracy over an extreme range of conditions.

If the nature and magnitude of the conditions are known, it often is possible to predict the flux change. It also may be possible, by exposing the magnet to certain influences in advance, to render the magnet insensitive to subsequent changes in service. For many years, permanent magnets in instruments have exhibited long-term stability on the order of one part per thousand (0.1%). More recently, investigations in conjunction with inertial guidance systems for space vehicles have shown that long-term stability on the order of 1 to 10 ppm (0.0001 to 0.001%) can be achieved. This incredible stability of a magnetic field achieved with modern permanent magnets contrasts sharply with the stability of very early per-

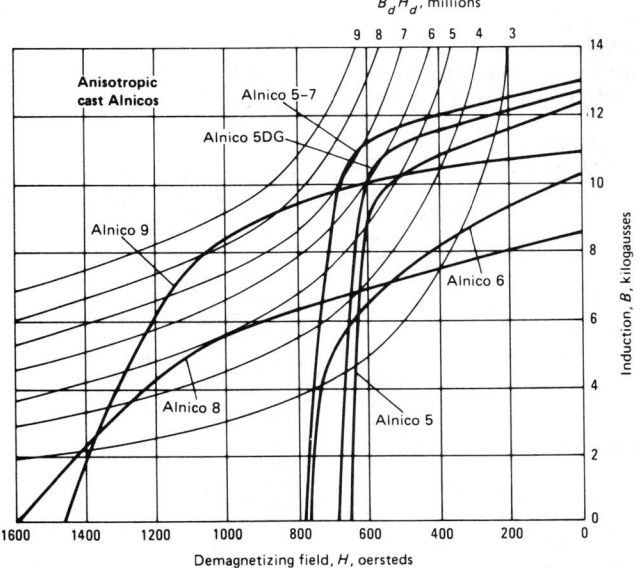

Fig. 6. Demagnetization curves for anisotropic cast Alnicos

manent magnets, in which both structural and magnetic changes caused a significant loss of magnetization with time.

IRREVERSIBLE CHANGES

Losses in magnetization with time can be classified as either reversible or irreversible. Irreversible changes are defined as changes where the affected properties remain altered after the influence responsible for the change has been removed. For example, if a magnet loses field strength under the influence of elevated temperature, and if the field strength does not return to its original value when the magnet is cooled to room temperature, the change is considered irreversible.

Changes in Metallurgical State. Irreversible changes begin to occur at different temperatures for different alloys. These changes usually depend on both time and temperature, and thus short exposures above the recommended temperatures may be tolerated. These changes may take the form of growth of the precipitate phase, such as in Alnico, Cunife and Cunico; precipitation of another phase, such as γ precipitation in Alnico; an increase in the amount of an ordered phase, such as in PtCo; stress-relief effects, such as in quenched steels and Vicalloy; an increase in grain size, as in $BaO \cdot 6Fe_2O_3$; oxidation, as occurs with metals, or reduction, as occurs with oxides; radiation damage; cracking; or changes in dimensions.

Irreversible metallurgical changes often can be counteracted, and original properties restored, by a suitably chosen thermal treatment. For example, if Alnico 5 has become degraded by exposure to 700 °C (1290 °F), it may be solution treated at 1300 °C (2370 °F), cooled in a magnetic field and aged at 600 °C (1110 °F) to re-attain the optimum metallurgical structure.

A nuclear environment is known to cause changes in metallurgical structure, and thus may cause changes in magnetic properties. Permanent magnet materials tested were not affected by neutron irradiation at levels below about 3×10^{17} neutrons/cm^2. Results of later work at levels up to 10^{20} n/cm^2 showed some degradation. The Alnicos are not affected by radiation up to 5×10^{20} n/cm^2 at neutron energies greater than 0.4 eV, and up to 2×10^{19} n/cm^2 for neutron energies greater than 2.9 MeV. Radiation effects were found to be independent of temperature, but high temperatures tended to counteract radiation effects.

Changes in magnetic state may be caused by temperature effects, such as ambient temperature changes or statistical local temperature fluctuations within the material; mechanical effects, such as mechanical shock or acoustical noise; or magnetic field effects, such as external fields, circuit reluctance changes or magnetic surface contacts. In all of these situations, the loss in magnetization may be restored by remagnetizing.

Mechanical shock and vibration add energy to a permanent magnet, and decrease the magnetization in the same manner as discussed for the case of thermal energy. The only difference is that energy imparted thermally to the magnet is precisely kT, whereas the energy imparted mechanically usually is not known. Thus, repetitive shocks or continual vibration should decrease the magnetization by the same logarithmic relations as for thermal effects, but where time is re-

Table 1. Nominal compositions, Curie temperatures and magnetic orientations of selected permanent magnet materials

Designation	Nominal composition	Approximate Curie temperature °C	°F	Magnetic orientation(a)
3¹⁄₂% Cr steel	Fe-3.5Cr-1C	745	1370	No
6% W steel	Fe-6W-0.5Cr-0.7C	760	1400	No
17% Co steel	Fe-17Co-8.25W-2.5Cr-0.7C	...	...	No
36% Co steel	Fe-36Co-3.75W-5.75Cr-0.8C	890	1630	No
Cast Alnico 1	Fe-12Al-21Ni-5Co-3Cu	780	1440	No
Cast Alnico 2	Fe-10Al-19Ni-13Co-3Cu	810	1490	No
Cast Alnico 3	Fe-12Al-25Ni-3Cu	760	1400	No
Cast Alnico 4	Fe-12Al-27Ni-5Co	800	1475	No
Cast Alnico 5	Fe-8.5Al-14.5Ni-24Co-3Cu	900	1650	Y,H
Cast Alnico 5DG	Fe-8.5Al-14.5Ni-24Co-3Cu	900	1650	Y,H,C
Cast Alnico 5-7	Fe-8.5Al-14.5Ni-24Co-3Cu	900	1650	Y,H,C
Cast Alnico 6	Fe-8Al-16Ni-24Co-3Cu-2Ti	860	1580	Y,H
Cast Alnico 7	Fe-8Al-18Ni-24Co-4Cu-5Ti	840	1540	Y,H
Cast Alnico 8	Fe-7Al-15Ni-35Co-4Cu-5Ti	860	1580	Y,H
Cast Alnico 9	Fe-7Al-15Ni-35Co-4Cu-5Ti	...	...	Y,H,C
Cast Alnico 12	Fe-6Al-18Ni-35Co-8Ti	...	...	No
Sintered Alnico 2	Fe-10Al-17Ni-12.5Co-6Cu	810	1490	No
Sintered Alnico 4	Fe-12Al-28Ni-5Co	800	1475	No
Sintered Alnico 5	Fe-8.5Al-14.5Ni-24Co-3Cu	900	1650	Y,H
Sintered Alnico 6	Fe-8Al-16Ni-24Co-3Cu-2Ti	860	1580	Y,H
Sintered Alnico 8	Fe-7Al-15Ni-35Co-4Cu-5Ti	860	1580	Y,H
Bonded ferrite A	BaO-6Fe₂O₃ + organics	450	840	No,P
Bonded ferrite B	BaO-6Fe₂O₃ + organics	450	840	No
Sintered ferrite 1	BaO-6Fe₂O₃	450	840	No,P
Sintered ferrite 2	BaO-6Fe₂O₃	450	840	Y,A
Sintered ferrite 3	BaO-6Fe₂O₃	450	840	Y,A
Sintered ferrite 4	SrO-6Fe₂O₃	460	860	Yes
Sintered ferrite 5	SrO-6Fe₂O₃	460	860	Yes
Lodex 30	9.9Fe-5.5Co-77.0Pb-8.6Sb	980	1800	Y,A
Lodex 31	16.0Fe-9.0Co-67.5Pb-7.5Sb	980	1800	Y,A
Lodex 32	19.2Fe-10.8Co-63.0Pb-7.0Sb	980	1800	Y,A
Lodex 33	21.9Fe-12.3Co-59.2Pb-6.6Sb	980	1800	Y,A
Lodex 36	9.9Fe-5.5Co-77Pb-8.6Sb	980	1800	No,E
Lodex 37	16Fe-9Co-67.5Pb-7.5Sb	980	1800	No,E
Lodex 38	19.2Fe-10.8Co-63Pb-7.0Sb	980	1800	No,E
Lodex 40	9.9Fe-5.5Co-77Pb-8.6Sb	980	1800	No,P
Lodex 41	16Fe-9Co-67.5Pb-7.5Sb	980	1800	No,P
Lodex 42	19.2Fe-10.8Co-63.0Pb-7.0Sb	980	1800	No,P
Lodex 43	21.9Fe-12.3Co-59.2Pb-6.6Sb	980	1800	No,P
P-6 alloy	45Fe-45Co-6Ni-4V	...	...	No
Cunife	20Fe-20Ni-60Cu	410	770	Y,R
Cunico	29Co-21Ni-50Cu	860	1580	No
Vicalloy I	39Fe-51Co-10V	855	1570	No
Vicalloy II	35Fe-52Co-13V	855	1570	Y,R
Remalloy 1	17Mo-12Co-71Fe	900	1650	No
Remalloy 2	20Mo-12Co-68Fe	900	1650	No
Platinum cobalt	76.7Pt-23.3Co	480	900	No
Cobalt–rare earth 1	Co₅Sm	725	1340	Y,A
Cobalt–rare earth 2	Co₅Sm	725	1340	Y,A
Cobalt–rare earth 3	Co₅Sm	725	1340	Y,A
Cobalt–rare earth 4	(Co,Cu,Fe)₇Sm	...	...	Y,A

(a) Y, yes; H, orientation developed during heat treatment; C, columnar crystal structure developed; P or E, some orientation developed during pressing or extrusion; R, orientation developed by rolling or other mechanical working; A, orientation developed predominantly by magnetic alignment of powder prior to compacting but alignment influenced by pressing forces also.

Table 2. Nominal magnetic properties of selected permanent magnet materials
For nominal compositions, see Table 1; for mechanical and physical properties, see Table 3.

Designation	H_c, Oe	H_{cf}, Oe	B_r, G	B_{is}, G	$(BH)_{max}$, MG·Oe	B_d, G	H_d, Oe	Required magnetizing field, Oe	Permeance coefficient at $(BH)_{max}$	Average recoil permeability
3¹⁄₂% Cr steel	66	...	9 500	...	0.29	...	...	...	...	...
6% W steel	74	...	9 500	...	0.33	...	...	...	...	...
17% Co steel	170	...	9 500	...	0.65	...	...	...	...	...
36% Co steel	240	...	9 750	...	0.93	...	...	...	...	...
Cast Alnico 1	440	455	7 100	10 500	1.4	4 500	305	2 000	14	6.8
Cast Alnico 2	550	580	7 250	10 900	1.6	4 500	350	2 500	12	6.4

Table 2 (continued)

Designation	H_c, Oe	H_{cf}, Oe	B_r, G	B_{is}, G	$(BH)_{max}$, MG·Oe	B_d, G	H_d, Oe	Required magnetizing field, Oe	Permeance coefficient at $(BH)_{max}$	Average recoil permeability
Cast Alnico 3	470	485	7 000	10 000	1.4	4 300	320	2 500	13	6.5
Cast Alnico 4	730	770	5 350	8 600	1.3	3 000	420	3 500	8.0	4.1
Cast Alnico 5	620	625	12 500	13 500	5.25	10 200	525	3 000	18	4.3
Cast Alnico 5DG	650	655	12 900	14 000	6.1	10 500	580	3 500	17	4.0
Cast Alnico 5-7	730	735	13 200	14 000	7.4	11 500	640	3 500	17	3.8
Cast Alnico 6	750	...	10 500	13 000	3.7	7 100	525	4 000	13	5.3
Cast Alnico 7	1050	...	8 570	9 450	3.7	...	...	5 000	8.2	...
Cast Alnico 8	1600	1 720	8 300	10 500	5.0	5 060	950	8 000	5.0	3.0
Cast Alnico 9	1450	...	10 500	...	8.5	...	...	7 000	7.0	...
Cast Alnico 12	950	...	6 000	...	1.7	3 150	540	5 000	5.6	...
Sintered Alnico 2 ..	525	545	6 700	11 000	1.5	4 300	345	2 500	12	6.4
Sintered Alnico 4 ..	700	760	5 200	...	1.2	3 000	400	3 500	...	7.5
Sintered Alnico 5 ..	600	605	10 400	12 050	3.60	7 850	465	3 000	18	4.0
Sintered Alnico 6 ..	760	790	8 800	11 500	2.75	5 500	500	4 000	12	4.5
Sintered Alnico 8 ..	1550	1 675	7 600	9 400	4.5	4 600	1000	8 000	5.0	2.1
Bonded ferrite A ..	1940	...	2 140	...	1.0	1 160	...	12 000	1.3	1.1
Bonded ferrite B ...	1150	...	1 400	...	0.4	...	...	8 000	1.2	1.1
Sintered ferrite 1 ...	1800	3 450	2 200	...	1.0	1 100	900	10 000	1.2	1.2
Sintered ferrite 2 ...	2200	2 300	3 800	...	3.4	1 850	1650	10 000	1.1	1.1
Sintered ferrite 3 ...	3000	3 650	3 200	...	2.5	1 600	1600	10 000	1.1	1.1
Sintered ferrite 4 ...	2200	2 300	4 000	...	3.7	2 150	1700	12 000	1.2	1.05
Sintered ferrite 5 ...	3150	3 590	3 550	...	3.0	1 730	1730	15 000	1.0	1.05
Lodex 30	1250	1 470	4 000	4 400	1.6	2 200	750	6 000	3.4	1.5
Lodex 31	1140	1 180	6 300	7 000	3.4	4 400	770	6 000	5.3	1.9
Lodex 32	940	960	7 350	8 300	3.5	5 400	650	5 000	8.2	2.6
Lodex 33	865	875	8 000	9 200	3.2	5 850	545	5 000	10.5	3.0
Lodex 36	1210	1 380	3 500	4 400	1.5	1 850	800	5 000	2.0	2.0
Lodex 37	1000	1 080	5 450	7 000	2.1	3 150	670	5 000	5.8	3.0
Lodex 38	850	890	6 200	8 300	2.2	3 700	600	5 000	7.0	3.5
Lodex 40	1100	1 400	2 700	4 400	0.8	1 400	600	5 000	2.0	2.5
Lodex 41	990	1 100	4 350	7 000	1.4	2 400	600	5 000	3.8	3.2
Lodex 42	845	920	5 300	8 300	1.4	2 750	510	5 000	7.6	3.5
Lodex 43	710	750	6 000	9 200	1.3	3 300	400	5 000	10	3.8
P-6 alloy	58	...	14 000	19 000	0.5	10 500	48	300	220	23
Cunife	550	555	5 400	5 900	1.5	4 000	325	2 500	12	3.7
Cunico	680	750	3 400	4 500	0.8	1 950	390	3 000	5.0	3.2
Vicalloy I ...	240	242	8 400	12 900	0.9	5 600	160	1 000	...	...
Vicalloy II ...	415	420	9 050	...	2.3	7 000	325	2 000	...	...
Remalloy 1	250	...	9 700	14 200	1.0	6 100	155	1 000	40	13
Remalloy 2	340	345	8 550	...	1.2	5 400	220	2 000	...	...
Platinum cobalt	4450	5 400	6 450	...	9.2	3 500	2700	20 000	1.2	1.2
Cobalt – rare earth 1	9000	20 000	9 200	9 800	21	...	...	30 000	...	...
Cobalt – rare earth 2	8000	>25 000	8 600	...	18	4 400	4100	30 000	...	1.05
Cobalt – rare earth 3	6700	>15 000	8 000	...	15	4 000	3700	30 000	...	1.1
Cobalt – rare earth 4	5700	6 500	9 400	...	21	4 600	4600	>15 000	...	...

placed, for example, by number of impacts.

Little work has been done regarding stabilization to minimize mechanical effects because it is seldom found necessary after thermal and field stabilization. There is limited information that suggests that both thermal and alternating field exposure will minimize, but not entirely eliminate, the change in magnetization due to shock.

REVERSIBLE CHANGES

A loss in magnetization caused by a disturbing influence, such as temperature or an external magnetic field, is considered reversible if the original properties of the magnet return when the disturbing influence is removed.

Temperature Effects. The properties of a magnet vary with temperature in a manner that often can be predicted. The variation of B_{is} with temperature can be calculated from theory, provided detailed knowledge of the crystallographic and magnetic structure of the magnetic phase is available. In many other instances, such information is not yet available, but direct measurements of B_{is} vs T have been made.

Changes in H_{ci} with temperature can be predicted from the changes with temperature of anisotropy and magnetization. This assumes knowledge of the physical origin of all anisotropies contributing to H_{ci}. For a case where uniaxial anisotropy predominates, as in $BaO \cdot 6Fe_2O_3$, quite good agreement between calculated and experimental results is obtained. In a case where shape anisotropy is dominant, as for Lodex elongated particles with various Co contents, calculated and experimental results also are in good agreement, especially when the small crystal anisotropy contributions are considered. The case of Alnico is similar to that of Lodex, but crystal anisotropy is more in evidence. In addition, there is greater uncertainty as to the effect of the so-called nonmagnetic phase, especially at lower temperatures where the nonmagnetic phase may contribute appreciable magnetization. In the case of steels, the temperature dependence based on the inclusion mechanism is difficult to predict.

Demagnetization curves may change in both shape and peak values with changes in temperature.

Time Effects at Constant Temperature. In ferromagnetic materials, the intensity of magnetization does not instantly attain its equilibrium value when the applied field is suddenly changed. This time dependence may be due to eddy current effects or to reversible or irreversible magnetic viscosity. In general, eddy current effects are important only for a very short time — normally, less than a second after a change in the applied field. Such effects are not considered here. "Reversible" magnetic viscosity has been shown to be due to ionic diffusion in the crystal lattice and thus has a time-temperature dependence characteristic of diffusion processes. The time constant is

$$\tau = \tau_\infty \exp (E/RT)$$

where τ_∞ is the time constant at infinite temperature and E is the activation energy, normally 0.1 to 1 eV. The time constant appears to be important only in magnetically soft materials, and only at high frequencies.

"Irreversible magnetic viscosity is important to the stability of permanent magnets. Irreversible magnetic viscosity is due to the influence of thermal fluctuations on magnetization or the domain process responsible for magnetization. The effect of thermal agitation has been considered in terms of the energy required to activate irreversible domain processes. The time-temperature dependence of magnetization was shown to be given by

$$M(t) = S \ln t$$

where $S = \lambda N M_s kT$. Here, N is the number of blocks, or regions of magnetization M_s per unit volume; λ is the constant probability density of energy E of all these blocks; and k is Boltzmann's constant. Because these factors are all relatively independent of temperature (except near the Curie temperature), S is nearly directly proportional to T. The results of experiments are in agreement with this equation. Aging at room temperature results in losses in magnetization for many materials.

Effects of Temperature Variations. Various permanent magnet materials undergo changes in magnetization as the temperature is cycled above and below room temperature. For a long bar operating above $(BH)_{max}$, the change in M is reversible. For a shorter bar operating below $(BH)_{max}$, the first cooling cycle results in a substantial loss in magnetization. After the initial low-temperature exposure, the changes in M are reversible, but at a level below the initial magnetization. These data suggest that by proper choice of dimensions, a reversible coefficient of approximately zero could be achieved over a limited range of temperature.

DESIGN CONSIDERATIONS

Stability can have a significant influence on choice of magnet material, as well as on component shape and magnetic circuit arrangement. For example, the rather drastic change in coercive force of oriented barium ferrite with temperature requires special considerations in design. Here, the lowest permeance coefficient (B/H) that can be used is established by stability considerations rather than by magnetic circuit analysis.

For the more widely used permanent magnet materials, reversible changes in magnetization are encountered by cooling below room temperature. Because the reversible remanence changes are closely approximated by a straight line, a reversible temperature coefficient is listed. The values of the coefficient are very small and may be of different sign for different magnet shapes. Consequently, it is often possible to carefully design magnet shape to yield very small variations in remanence with temperature. Similar changes may result upon heating above room temperature. It is important to distinguish between irreversible losses and reversible changes. It is common practice prior to use to cycle a magnet between the temperature extremes to be encountered in service. Nearly all of the irreversible loss is encountered in one temperature cycle, but in some instances four or five cycles may be necessary.

In applications that are extremely sensitive to magnetization changes, it is very common to use a temperature compensating circuit to counteract reversible changes over the operating temperature range. Temperature-sensitive iron-nickel alloys are used as magnetic shunts for this purpose. A shunt is mounted beside the permanent magnet and simply diverts flux from the air gap

Table 3. Nominal mechanical and physical properties of selected permanent magnet materials
See Table 1 for compositions, Curie temperatures and magnetic orientations; see Table 2 for nominal magnetic properties.

Designation	Density, Mg/m³	Tensile strength MPa	ksi	Transverse modulus of rupture MPa	ksi	Hardness, HRC	Coefficient of linear expansion, μm/m·K	Electrical resistivity, nΩ·m	Maximum service temperature °C	°F
3½% Cr steel	7.77	…	…	…	…	60-65	12.6	290	…	…
6% W steel	8.12	…	…	…	…	60-65	14.5	300	…	…
17% Co steel	8.35	…	…	…	…	60-65	15.9	280	…	…
36% Co steel	8.18	…	…	…	…	60-65	17.2	270	…	…
Cast Alnico 1	6.9	28	4.1	96	14	45	12.6	750	540	1004
Cast Alnico 2(a)	7.1	21	3.1	52	7.5	45	12.4	650	540	1004
Cast Alnico 3	6.9	83	12	157	23	45	13.0	600	480	896
Cast Alnico 4	7.0	63	9.1	167	24	45	13.1	750	590	1094
Cast Alnico 5(a)(b)	7.3	37	5.4	73	11	50	11.4	470	540	1004
Cast Alnico 5DG	7.3	36	5.2	62	9.0	50	11.4	470	…	…
Cast Alnico 5-7	7.3	34	4.9	55	8.0	50	11.4	470	540	1004
Cast Alnico 6(a)	7.4	157	23	314	46	50	11.4	500	540	1004
Cast Alnico 7	7.3	108	16	…	…	60	11.4	580	…	…
Cast Alnico 8	7.3	64	9.3	…	…	56	11.0	500	540	1004
Cast Alnico 9	7.3	48	6.9	55	8.0	56	11.0	…	…	…
Cast Alnico 12	7.4	275	40	343	50	58	11.0	620	480	896
Sintered Alnico 2	6.8	451	65	480	70	43	12.4	680	480	896
Sintered Alnico 4	6.9	412	60	588	85	…	13.1	680	590	1094
Sintered Alnico 5	7.0	343	50	392	57	44	11.3	500	540	1004
Sintered Alnico 6	6.9	382	55	755	110	44	11.3	530	540	1004
Sintered Alnico 8	7.0	…	…	382	55	43	…	…	…	…
Bonded ferrite A(c)	3.7	4.4	0.63	…	…	…	94	~10¹³	95	203
Sintered ferrite 1(d)	4.8	49	7.1	…	…	…	10	~10¹³	400	752
Sintered ferrite 2	5.0	…	…	…	…	…	10	~10¹³	400	752
Sintered ferrite 3	4.5	…	…	…	…	…	18	~10¹³	400	752
Sintered ferrite 4	4.8	…	…	…	…	…	…	10¹³	400	752
Sintered ferrite 5	4.5	…	…	…	…	…	…	10¹³	…	…
Lodex 30	10.1	…	…	31	4.5	…	18	1200	200	392
Lodex 31	9.6	6.9	1.0	31	4.5	…	18	1200	200	392
Lodex 32	9.3	6.9	1.0	31	4.5	…	18	1200	200	392
Lodex 33	9.2	…	…	31	4.5	…	18	1200	200	392
Lodex 36	10.2	…	…	108	16	…	18	1200	200	392
Lodex 37	9.7	…	…	108	16	…	18	1200	200	392
Lodex 38	9.6	…	…	108	16	…	18	1200	200	392
Lodex 40	10.2	…	…	27	3.9	…	18	1200	200	392
Lodex 41	10.1	6.9	1.0	27	3.9	…	18	1200	200	392
Lodex 42	9.8	6.9	1.0	27	3.9	…	18	1200	200	392
Lodex 43	9.4	…	…	27	3.9	…	18	1200	200	392
P-6 alloy	7.9	2160	313	1 180	170	65	11	300	…	…

Table 3 (continued)

Designation	Density, Mg/m³	Tensile strength		Transverse modulus of rupture		Hardness, HRC	Coefficient of linear expansion, μm/m·K	Electrical resistivity, nΩ·m	Maximum service temperature	
		MPa	ksi	MPa	ksi				°C	°F
Cunife	8.6	686	99	. . .	. . .	95 HRB	12	180	350	662
Cunico	8.3	588	85	. . .	. . .	95 HRB	14	240	500	932
Vicalloy I	8.2	2060	299	. . .	. . .	62	7	630	450	842
Remalloy 1	8.2	882	128	. . .	. . .	60	9.3	450	500	932
Platinum cobalt	15.5	1370	199	1 570	230	26	11	280	350	662
Cobalt – rare earth(e) . . .	8.2	3430	498	13 730	1 990	50	511; 131	500	250	482

(a) Specific heat; 460 J/kg·K (0.11 Btu/lb·°F). (b) Thermal conductivity: 25 W/m·K (170 Btu·in./ft²·h·°F) at room temperature. (c) Thermal conductivity: 0.62 W/m·K (4.3 Btu·in./ft²·h·°F). (d) Thermal conductivity: 5.5 W/m·K (38 Btu·in./ft²·h·°F). (e) Specific heat: 375 J/kg·K (0.09 Btu/lb·°F). Thermal conductivity: 15 W/m·K (104 Btu·in./ft²·h·°F).

as the temperature decreases. Temperature compensation by shunting requires overdesign of the magnet to allow for the loss in flux through the shunt at low operating temperatures.

Stress Effects. Some magnets subjected to tension or compression show large changes in properties. This is especially true of Vicalloy and Cunife. The changes are reversible, often even after considerable deformation has occurred. The changes are due to the contribution that stress makes to the total anisotropy of the system.

Magnetization Prior to Use. Magnets are magnetized in applied fields supplied by dc or pulsed-current electromagnets. Where practical, saturating magnetizing fields are recommended to gain full use of magnetic potential energy. Magnets are demagnetized by heating to Curie temperature or by applying an ac or dc field to reduce the measured induction to zero.

Electrical Resistance Alloys

Condensed from Metals Handbook, Ninth Edition, Volume 3, pages 640 to 661.

ELECTRICAL RESISTANCE ALLOYS include both the types used in instruments and control equipment to measure and regulate electrical characteristics and those used in furnaces and appliances to generate heat. In the former applications, properties near ambient temperature are of primary interest; in the latter, elevated-temperature characteristics are of prime importance. In common commercial terminology, electrical resistance alloys used for control or regulation of electrical properties are called *resistance alloys*, and those used for generation of heat are referred to as *heating alloys*. Electrical resistance materials of a third class, used in applications where heat generated in a metal resistor is converted to mechanical energy, are termed *thermostat metals*. All three classes of electrical resistance alloys are discussed in this article.

RESISTANCE ALLOYS

The primary requirements for resistance alloys are uniform resistivity, stable resistance (no time-dependent aging effects), reproducible temperature coefficient of resistance, and low thermoelectric potential, versus copper. Properties of secondary importance are coefficient of expansion, mechanical strength, ductility, corrosion resistance, and ability to be joined to other metals by soldering, brazing or welding.

Nominal compositions and physical properties of metals and alloys used to make resistors for instruments and controls are listed in Table 1.

Resistance alloys must be ductile enough so that they can be drawn into wire as fine as 0.01 mm (0.0004 in.) in diameter or rolled into narrow ribbon from 0.4 to 50 mm ($^1/_{64}$ to 2 in.) wide and from 0.025 to 6.4 mm (0.001 to 0.25 in.) thick.

Alloys must be strong enough to withstand fabrication operations, and it must be easy to procure an alloy that has consistently reproducible properties. For instance, successive batches of wire must have closely similar electrical characteristics: if properties vary from lot to lot, resistors made of wire from different batches may cause a given model of instrument to exhibit widely varying performance under identically reproduced conditions or may cause large errors in a given instrument when a resistor from one batch is used as a replacement part for a resistor from another batch.

Coefficients of expansion of both the resistor and the insulator on which it is wound must be considered, because stresses can be established that will cause changes in both resistance and temperature coefficient of resistance. It is equally important that consideration be given to the choice between single-layer and multiple-layer wound resistors, because of the difference in rate of heat dissipation between the two styles.

In design of primary electrical standards of very high accuracy, cost of resistance material is not a consideration. For ordinary production components, however, cost may be the deciding factor in material selection.

RESISTORS

Resistors for electrical and electronic devices may be divided into two arbitrary classifications on the basis of permissible error: those employed in precision instruments in which over-all error is considerably less than 1%, and those employed where less precision is needed. The choice of alloy for a specific resistor application depends on the variation in properties that can be tolerated.

In many electronic devices, resistors whose error in resistance value is 5 to 10% are entirely satisfactory. Most resistors for this classification are made of carbon. Carbon resistors are not discussed in this article. Here, we are concerned chiefly with metallic resistors such as wirewound precision resistors and potentiometers, resistance thermometers, and ballast resistors.

Some applications of resistance materials require devices with large thermal coefficients of resistance, either positive or negative. A device of this type is called a thermistor. Thermistors are made almost exclusively of ceramic semiconductor materials.

Precision resistors (those with less than 1% error) require careful material selection. The ideal material for a precision resistor should have a thermal coefficient of resistance equal to zero for the temperature range over which the resistor will operate. In addition, to ensure freedom from thermoelectric effects, it should have a small or negligible thermoelectric potential versus copper, which is the material normally used for the connecting conductor. Temperature differentials may exist among various junctions between a resistance wire and a connecting wire, resulting in a network of thermocouples that can cause parasitic electromotive forces in the circuit; this effect is especially critical in precise dc circuits. In an apparatus where extreme precision is required, it is advisable to make the connecting wires of the same material as the resistors or to design the apparatus so that all dissimilar-metal junctions are at the same operating temperature.

Selection of a material for, and specific dimensions of, a precision resistor must include consideration of equipment size and heat-dissipation characteristics. Temperature excursions from the ambient or from a specified operating temperature may be undesirable, because they may cause net changes in resistance that will affect the stability or accuracy of the instrument. The magnitude of the change in resistance can be calculated using the temperature coefficient of resistance. For example, a resistor made of a low-resistivity material could be several times larger than one made of a higher-resistivity material and yet achieve the same total resistance. The large resistor would have a much greater surface area and therefore could dissipate much more heat, and thus, despite its low resistivity, would attain a lower steady-state temperature than would be possible for a small, high-resistivity resistor operating under the same conditions. Alloys used for precision resistors generally have resistivities ranging from 500 to 1350 nΩ·m (300 to 800 Ω·circ mil/ft).

Table 1. Typical properties of electrical resistance alloys

Basic composition	Resistivity(a), nΩ·m(b)	TCR, ppm/°C(c)	Thermoelectric potential vs Cu, μV/°C	Coefficient of thermal expansion(d), μm/m·°C	Tensile strength(a) MPa	Tensile strength(a) ksi	Density(a) Mg/m³	Density(a) lb/in.³
Radio alloys								
98Cu-2Ni	50	+1350 (25-105 °C)	−13 (25-105 °C)	16.5	205-410	30-60	8.9	0.32
94Cu-6Ni	100	+550 (25-105 °C)	−13 (25-105 °C)	16.3	240-585	35-85	8.9	0.32
89Cu-11Ni	150	+430 (25-105 °C)	−25 (25-105 °C)	16.1	240-515	35-75	8.9	0.32
78Cu-22Ni	300	+160 (25-105 °C)	−36 (0-75 °C)	15.9	345-690	50-100	8.9	0.32
Manganins								
87Cu-13Mn	480	±15 (15-35 °C)	+1 (0-50 °C)	18.7	275-620	40-90	8.2	0.30
83Cu-13Mn-4Ni	480	±15 (15-35 °C)	−1 (0-50 °C)	18.7	275-620	49-90	8.4	0.31
85Cu-10Mn-4Ni(e)	380	±10 (20-45 °C)	−1.5 (0-50 °C)	18.7	345-690	50-100	8.4	0.31
Constantans								
57Cu-43Ni	490	±20 (25-105 °C)	−43 (25-105 °C)	14.9	410-930	60-135	8.9	0.32
55Cu-45Ni	500	±40 (20-1000 °C)	−42 (0-75 °C)	14.9	455-860	66-125	8.9	0.32
Nickel-chromium-aluminum alloys								
75Ni-20Cr-3Al-2(Cu, Fe or Mn)	1330	±20 (−55 - +105 °C)	−0.1 (25-105 °C)	12.6	690-1380	100-200	8.1	0.29
72Ni-20Cr-3Al-5Mn	1355	±20 (−55 - +105 °C)	−0.1 (25-105 °C)	13	690-1380	100-200	7.1	0.26
Nickel-base alloys								
94Ni-3Mn-2Al-1Si	315	+2400 (20-100 °C)		12.3	550-1035	80-150	8.5	0.31
80Ni-20Cr	1125	+85 (−55 - +100 °C)	+5 (0-100 °C)	13	690-1380	100-200	8.4	0.31
78.5Ni-20Cr-1.5Si	1080	+85 (25-105 °C)	+3.9 (25-105 °C)	13.5	690-1380	100-200	8.3	0.30
76Ni-17Cr-4Si-3Mn	1330	±20 (−55 - +150 °C)	−1 (20-100 °C)	15	900-1380	130-200	7.8	0.28
71Ni-29Fe	200	+4500 (25-105 °C)	−40 (25-105 °C)	15	480-1035	70-150	8.4	0.31
68.5Ni-30Cr-1.5Si	1180	+90 (25-105 °C)	−1.2 (25-105 °C)	12.2	825-1380	120-200	8.1	0.29
60Ni-16Cr-24Fe	1120	+150 (25-105 °C)	+0.9 (25-105 °C)	13.5	655-1200	95-175	8.4	0.30
35Ni-20Cr-45Fe	1015	+400 (25-105 °C)	−1.1 (25-105 °C)	15.6	550-1200	80-175	8.1	0.29
Iron-chromium-aluminum alloys								
73.5Fe-22Cr-4.5Al	1350	±50	−3.0 (0-100 °C)	11	690-965	100-140	7.25	0.262
73Fe-22Cr-5Al	1390	±50	−2.8 (0-100 °C)	11	690-965	100-140	7.15	0.258
72.5Fe-22Cr-5.5Al	1450	±50	−2.6 (0-100 °C)	11	690-965	100-140	7.1	0.256
81Fe-15Cr-4Al	1250	±50	−1.2 (0-100 °C)	11	620-900	90-130	7.43	0.268
Pure metals								
Aluminum (99.99+)	26.55	+4290(a)	−3.4 (0-50 °C)	23.9(a)	50-110	7-16	2.70	0.098
Copper (99.99)	16.73	+4270 (0-50 °C)	0	16.5(a)	115-130	17-19	8.96	0.324
Gold (99.999+)	23.50	+4000 (0-100 °C)	+0.2 (0-100 °C)	14.2(a)	130	19	19.32	0.698
Iron (99.94)	970	+5000(a)	+12.2 (0-100 °C)	11.7(a)	180-220	26-32	7.87	0.284
Molybdenum (99.9)	52	+3300(a)	+6.9 (0-100 °C)	4.9	690-2140	100-310	10.22	0.369
Nickel (99.8)	80	+6000 (20-35 °C)	−22 (0-75 °C)	15	345-760	50-110	8.90	0.322
Platinum (99.99+)	106	+3920 (0-100 °C)	+7.6 (0-100 °C)	8.9(a)	125	18	21.45	0.775
Silver (99.99)	16	+4100(a)	−0.2 (0-100 °C)	19.7	125	18	10.49	0.379
Tantalum (99.96)	135	+3820 (0-100 °C)	−4.3 (0-100 °C)	6.5(a)	690-1240	100-180	16.6	0.600
Tungsten (99.9)	55	+4500(a)	+3.6 (0-100 °C)	4.3(a)	1825-4050	265-590	19.25	0.695

(a) At 20 °C (68 °F). (b) To convert to Ω·circ mil/ft, multiply by 0.6015. (c) Temperature coefficient of resistance is $(R - R_0)/R_0(t - t_0)$, where R is resistance at t °C and R_0 is resistance at the reference temperature t_0 °C. (d) At 25 to 105 °C. (e) Shunt manganin.

Resistance thermometers are commonly made of copper, nickel or platinum; these devices are precision resistors whose resistance change with temperature is stable and reproducible over specified ranges of temperature. For resistance thermometers, the larger the temperature coefficient of the material, the greater the accuracy and ease of measurement. Temperature coefficients of relatively pure metals are greatly affected by small amounts of impurities. In fact, one of the most sensitive tests of the purity of a metal is measurement of its temperature coefficient of resistivity, which decreases sharply with increasing impurity or alloy content.

Ballast resistors are used extensively in industrial circuits to maintain constant currents over long periods of time. In such an application, a ballast resistor must be able to dissipate energy in such a way as to control current over a wide range of voltages. Wires with the proper temperature coefficient of resistance can be made to change resistance rapidly with changes in current, due to self heating, in such a manner that the current in the circuit will remain nearly constant even when there are fluctuations in voltage across the circuit. Because ballast resistors operate at elevated temperatures, mechanical properties are important also. Typical materials used in ballast resistors are pure iron, pure nickel, and nickel-iron alloys such as 71Ni-29Fe (see Table 1).

Reference resistors and virtually all other applications of resistance alloys demand temperature coefficients of resistance lower than ±20 ppm/°C (±20 μΩ/Ω·°C). This requirement stems from the fact that, for these applications, resistance errors resulting from the small changes in ambient temperature that are continually taking place cannot be tolerated. In the most demanding of these applications, resistors often are mounted in thermally insulated containers and are carefully maintained at a temperature slightly above the maximum anticipated ambient temperature.

The most important requirement of a resistor used as a reference standard is that its value be predictable within narrow limits over long periods of time. Many reference resistors exhibit a nearly linear change in resistance with time. Hence, resistance between dates of calibration can be determined by interpolation; resistance at future points in time can be determined by extrapolation, but undue reliance should not be placed on extrapolated values. Figure 1 shows the change in resistance with time for a 10-kΩ resistor made of a Ni-Cr-Al-Cu alloy.

Stability, or the ability to maintain a specific value of resistance within narrow limits over a long period of time, is an important requirement of materials for precision resistors and reference resistors. Principal sources of instability are: (a) relief of residual stresses during service; (b) time-de-

Table 2. Properties of thermostat metals frequently selected for some common service temperatures

Temperature range of maximum sensitivity °C	Temperature range of maximum sensitivity °F	Composition High-expanding side	Composition Low-expanding side	Resistivity at 24 °C (75 °F) nΩ·m	Resistivity at 24 °C (75 °F) Ω·circ mil/ft	Flexivity(a) μm/m·°C	Flexivity(a) μin./in.·°F
−20 to +150	0 to 300	75Fe-22Ni-3Cr	64Fe-36Ni	780	470	26.3	14.6
−20 to +200	0 to 400	75Fe-22Ni-3Cr	Pure Ni	160	95	8.3	4.6
		72Mn-18Cu-10Ni	64Fe-36Ni	1120	675	38.5	21.4
120 to 290	250 to 550	67Ni-30Cu-1.4Fe-1Mn	60Fe-40Ni	565	340	16.6	9.2
150 to 450	300 to 850	66.5Fe-22Ni-8.5Cr	50Fe-50Ni	580	350	11.2	6.2

(a) At 40 to 150 °C (100 to 300 °F). See ASTM B106 for standard test method for determining flexivity of thermostat metals.

Table 3. Typical properties of resistance heating materials

Basic composition	Resistivity(a), nΩ·m(b)	Average change in resistance(c), %, from 20 °C to:				Thermal expansion, μm/m·°C, from 20 °C to:			Tensile strength		Density	
		260 °C	540 °C	815 °C	1095 °C	100 °C	540 °C	815 °C	MPa	ksi	Mg/m³	lb/in.³
Nickel-chromium and nickel-chromium-iron alloys												
78.5Ni-20Cr-1.5Si (80-20) ...	1080	+4.5	+7.0	+6.3	+7.6	13.5	15.1	17.6	690 to 1380	100 to 200	8.41	0.30
77.5Ni-20Cr-1.5Si-1Nb	1080	+4.6	+7.0	+6.4	+7.8	13.5	15.1	17.6	690 to 1380	100 to 200	8.41	0.30
68.5Ni-30Cr-1.5Si (70-30) ...	1180	+2.1	+4.8	+7.6	+9.8	12.2	...	...	825 to 1380	120 to 200	8.12	0.29
68Ni-20Cr-8.5Fe-2Si	1165	+3.9	+6.7	+6.0	+7.1	...	12.6	...	895 to 1240	130 to 180	8.33	0.30
60Ni-16Cr-22.5Fe-1.5Si	1120	+3.6	+6.5	+7.6	+10.2	13.5	15.1	17.6	655 to 1205	95 to 175	8.25	0.30
35Ni-30Cr-33.5Fe-1.5Si	1055	+7.95	+14.9	+18.0	+22.0	14.6	17.5	16.0	895 to 1380	130 to 200	7.90	0.29
35Ni-20Cr-43.5Fe-1.5Si	1015	+8.0	+15.4	+20.6	+23.5	15.7	15.7	...	550 to 1205	80 to 175	7.95	0.29
35Ni-20Cr-42.5Fe-1.5Si-1Nb	1015	+8.0	+15.4	+20.6	+23.5	15.7	15.7	...	550 to 1205	80 to 175	7.95	0.29
Iron-chromium-aluminum alloys												
83.5Fe-13Cr-3.25Al	1120	+7.0	+15.5	...	...	10.6	...	...	515 to 1035	75 to 150	7.30	0.26
81Fe-14.5Cr-4.25Al	1245	+3.0	+9.7	+16.5	...	10.8	11.5	12.2	550 to 1170	80 to 170	7.28	0.26
79.5Fe-15Cr-5.2Al	1370	+1.9	+5.5	+8.9	+9.6	11.3	12.6	...	550 to 895	80 to 130	7.12	0.26
73.5Fe-22Cr-4.5Al	1355	+0.3	+2.9	+4.3	+4.9	10.8	12.6	13.1	725 to 1205	105 to 175	7.15	0.26
72.5Fe-22Cr-5.5Al	1455	+0.2	+1.0	+2.8	+4.0	11.3	12.8	14.0	760 to 1205	110 to 175	7.10	0.26
Pure metals												
Molybdenum	52	+110	+238	+366	+508	4.8	5.8	...	690 to 2160	100 to 313	10.2	0.369
Platinum	105	+85	+175	+257	+305	9.0	9.7	10.1	345	50	21.5	0.775
Tantalum	125	+82	+169	+243	+317	6.5	6.6	...	345 to 1240	50 to 180	16.6	0.600
Tungsten	55	+91	+244	+396	+550	4.3	4.6	4.6	3380 to 6480	490 to 940	19.3	0.697
Nonmetallic heating-element materials												
Silicon carbide	995 to 1995	−33	−33	−28	−13	4.7	...	...	28	4	3.2	0.114
Molybdenum disilicide	370	+105	+222	+375	+523	9.2	...	...	185	27	6.2	0.212
MoSi₂ + 10% ceramic additives	270	+167	+370	+597	+853	13.1	14.2	14.8	...	...	5.6	0.202
Graphite	9100	−16	−18	−13	−8	1.3	...	...	1.8	0.26	2.3	0.057

(a) At 20 °C (68 °F). (b) To convert to Ω·circ mil/ft, multiply by 0.6015. (c) Changes in resistance may vary somewhat, depending on cooling rate.

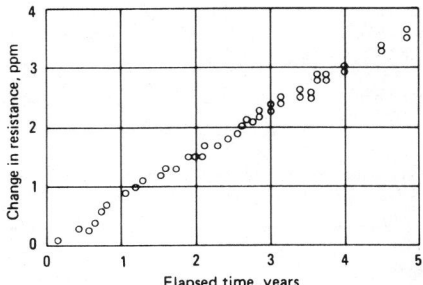

Fig. 1. Change in resistance of a 10-kΩ resistor with time

pendent or time-temperature-dependent metallurgical changes, such as precipitation of a second phase; and (c) corrosion or oxidation.

Residual stresses often are relieved at room temperature over long periods of time through a process known as stress relaxation. Stress relaxation alters the resistance of a coil at a rate of change that increases with the original level of residual stress. For this reason, only carefully preannealed wires are used for precision resistors. Stresses induced during winding, weaving or other operations in fabrication of resistors from preannealed wire must be kept to a minimum. Thorough annealing of finished resistors is not always possible, because the wires may be enameled or may be coated with a textile insulation of only moderate resistance to heat. Either type of coating limits to about 140 °C (285 °F) the temperature that can be used for stress relieving finished resistors.

Figure 2 shows the effect of residual stress on the stability of a manganin alloy subjected to different amounts of cold work. The top curve illustrates that a low-temperature stress-relieving treatment substantially eliminates the stresses that would, in time, have been eliminated due to natural relaxation at room temperature.

Table 4. Recommended maximum furnace operating temperatures for resistance heating materials

Basic composition	Approximate melting point		Maximum furnace operating temperature in air	
	°C	°F	°C	°F
Nickel-chromium and nickel-chromium-iron alloys				
78.5Ni-20Cr-1.5Si (80-20)	1400	2550	1150	2100
77.5Ni-20Cr-1.5Si-1Nb	1390	2540	...	...
68.5Ni-30Cr-1.5Si (70-30)	1380	2520	1200	2200
68Ni-20Cr-8.5Fe-2Si	1390	2540	1150	2100
60Ni-16Cr-22.5Fe-1.5Si	1350	2460	1000	1850
35Ni-30Cr-33.5Fe-1.5Si	1400	2550	...	...
35Ni-20Cr-43.5Fe-1.5Si	1380	2515	925	1700
35Ni-20Cr-42.5Fe-1.5Si-1Nb	1380	2515	...	...
Iron-chromium-aluminum alloys				
83.5Fe-13Cr-3.25Al	1510	2750	1050	1920
81Fe-14.5Cr-4.25Al	1510	2750	...	...
79.5Fe-15Cr-5.2Al	1510	2750	1260	2300
73.5Fe-22Cr-4.5Al	1510	2750	1280	2335
72.5Fe-22Cr-5.5Al	1510	2750	1375	2505
Pure metals				
Molybdenum	2605	4730	400(a)	750(a)
Platinum	1770	3216	1500	2750
Tantalum	2975	5390	500(a)	930(a)
Tungsten	3375	6116	300(a)	570(a)
Nonmetallic heating-element materials				
Silicon carbide	2410	4370	1600	2900
Molybdenum disilicide	2080	3775	1700 to 1900	3100 to 3270
MoSi₂ + 10% ceramic additives	1800	3270	...	...
Graphite	3650 to 3695(b)	6610 to 6690(b)	400(c)	400(c)

(a) Recommended atmospheres for these metals are a vacuum of 10⁻⁴ to 10⁻⁵ mm Hg, pure hydrogen, and partly combusted city gas dried to a dew point of +4 °C (+40 °F). In these atmospheres the recommended temperatures would be:

	Vacuum	Pure H₂	City gas
Mo	1650 °C (3000 °F)	1760 °C (3200 °F)	1700 °C (3100 °F)
Ta	2480 °C (4500 °F)	Not recommended	Not recommended
W	1650 °C (3000 °F)	2480 °C (4500 °F)	1700 °C (3100 °F)

(b) Graphite volatilizes without melting at 3650 to 3695 °C (6610 to 6690 °F). (c) At approximately 400 °C (750 °F) (threshold oxidation temperature), graphite undergoes a weight loss of 1% in 24 h in air. Graphite elements can be operated at surface temperatures up to 2205 °C (4000 °F) in inert atmospheres.

Resistors represented by the top curve were stress relieved at 140 °C for 48 h to stabilize their resistance within about 20 ppm of the nominal value. Resistors not stress relieved, as represented by the other curves, continue to change in resistance almost indefinitely. For most mod-

ern, hermetically sealed precision resistors annealed at 140 °C, the change in resistance does not exceed 10 ppm/yr, and for many it does not exceed 5 ppm/yr. However, resistors made of manganin that are used as reference standards require greater stability, and stress relief at 140 °C

Table 5. Comparative life of heating-element materials in various furnace atmospheres
See Table 6 for atmosphere compositions.

Element material	Oxidizing: (air)	Reducing: dry H₂ or type 501	Reducing: type 102 or 202	Reducing: type 301 or 402	Carburizing: type 307 or 309	Reducing or oxidizing, with sulfur	Reducing, with lead or zinc	Vacuum
Nickel-chromium and nickel-chromium-iron alloys								
80Ni-20Cr	Good to 1150 °C	Good to 1175 °C	Fair to 1150 °C	Fair to 1000 °C	Not recommended(a)	Not recommended	Not recommended	Good to 1150 °C
60Ni-16Cr-24Fe	Good to 1000 °C	Good to 1000 °C	Good to fair to 1000 °C	Fair to poor to 925 °C	Not recommended	Not recommended	Not recommended	...
35Ni-20Cr-45Fe	Good to 925 °C	Good to 925 °C	Good to fair to 925 °C	Fair to poor to 870 °C	Not recommended	Fair to 925 °C	Fair to 925 °C	...
Iron-chromium-aluminum alloys								
Fe-23Cr-4.5Al-1Co	Good to 1150 °C	Fair to poor to 1150 °C(b)	Not recommended	Not recommended	Not recommended	Fair in oxidizing atmosphere	Not recommended	...
Fe-37Cr-7.5Al	Good to 1320 °C	Fair to poor to 1300 °C(b)	Not recommended	Not recommended	Not recommended	Fair in oxidizing atmosphere	Not recommended	...
Pure metals								
Molybdenum	Not recommended(c)	Good to 1650 °C	Not recommended	Not recommended	Not recommended	Not recommended	Not recommended	Good to 1650 °C
Platinum	Good to 1400 °C	Not recommended	Not recommended	Not recommended	Not recommended	Not recommended	Not recommended	...
Tantalum	Not recommended	Not recommended	Not recommended	Not recommended	Not recommended	Not recommended	Not recommended	Good to 2500 °C
Tungsten	Not recommended	Good to 2500 °C(d)	Not recommended	Not recommended	Not recommended	Not recommended	Not recommended	Good to 1650 °C
Nonmetallic heating element materials								
Silicon carbide . . .	Good to 1600 °C	Fair to poor to 1200 °C	Fair to 1375 °C	Fair to 1375 °C	Not recommended	Good to 1375 °C	Good to 1375 °C	Not recommended
Graphite	Not recommended	Fair to 2500 °C	Not recommended	Fair to 2500 °C	Fair to poor to 2500 °C	Fair to 2500 °C in reducing atmosphere	Fair to 2500 °C	...
Molybdenum disilicide	Good to 1800 °C	...	...	...	...	...	...	...

Inert atmosphere of argon or helium can be used with all materials. Nitrogen recommended only for the nickel-chromium group. Temperatures listed are element temperatures, not furnace temperatures.

(a) Special 80Ni-20Cr elements with ceramic protective coatings designated for low voltage (8 to 16 V) can be used. (b) Must be oxidized first. (c) Special molybdenum heating elements with MoSi₂ coating can be used in oxidizing atmospheres. (d) Good with pure H₂ only.

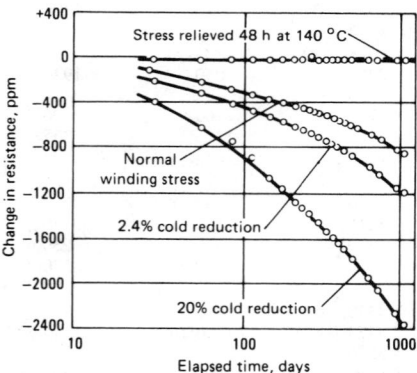

Fig. 2. Change in resistance of manganin resistors upon aging at room temperature

Table 6. Types and compositions of standard furnace atmospheres
See Table 5 for comparative life of heating elements in these atmospheres.

Type	Description	Composition, vol %					Typical dew point	
		N₂	CO	CO₂	H₂	CH₄	°C	°F
Reducing atmospheres								
102(a)	Exothermic unpurified	71.5	10.5	5.0	12.5	0.5	+27	+80
202	Exothermic purified	75.3	11.0	...	13.0	0.5	−40	−40
301	Endothermic	45.1	19.6	0.4	34.6	0.3	+10	+50
402	Charcoal	64.1	34.7	...	1.2	...	−29	−20
501	Dissociated ammonia	25	...	...	75	...	−51	−60
Carburizing atmospheres								
307	Endothermic + hydrocarbon	No standard composition					...	...
309	Endothermic + hydrocarbon + ammonia	No standard composition					...	...

(a) This atmosphere, refrigerated to obtain a dew point of +4 °C (+40 °F), is widely used.

is not adequate. One-ohm resistors of the best grade (the double-wall type) are treated as follows. A coil of wire is wound on a steel mandrel and annealed at about 500 °C (930 °F) in a protective atmosphere for 6 h or more. The coil is removed and slipped over an insulated tube of the same diameter as the mandrel, and then is hermetically sealed using a second tube slightly greater in diameter. In most resistors of this type, the change in resistance does not exceed 1 ppm/yr.

The second factor affecting stability of precision resistors is the metallurgical stability of the alloy being used as the resistance element; any metallurgical change will be detrimental. All resistance alloys are single-phase solid-solution alloys; thus, the changes in resistance that occur are relatively small but not insignificant. Changes in resistance are caused by internal changes such as long-range order-disorder reactions in 71Ni-29Fe alloys, short-range order or clustering in quaternary nickel-chromium alloys, and even minor ordering in manganin alloys. Accordingly, resistance of these alloys is affected by heat treatment and by rates of cooling from heat treating temperatures. Power resistors that can oper-

ate at temperatures as high as 300 °C (570 °F) can in effect be heat treated during service. The net effect during service can be an increase in resistance for nickel-chromium alloys, a decrease for manganin and either an increase or a decrease for nickel-iron alloys.

The third factor affecting stability of resistors is corrosion and/or oxidation. Corrosion of the resistance element will decrease its effective cross section, resulting in a corresponding increase in resistance. If the corrosion attack is selective, changes will occur in temperature coefficient of resistance and thermal emf as well as in resistivity. These corrosive effects may be minimized by protecting the wire with an enamel or plastic coating. One relatively common source of corrosive attack, but one that is often overlooked, is flux residue at soldered or brazed joints. Another less obvious cause of instability is the presence of tin-containing solder. Intergranular stress corrosion, believed to originate during thermal stress-relieving treatments, may cause open circuits.

Combinations of these three factors — residual stresses, metallurgical instability, and corrosion or oxidation — account for the complex changes in resistance that often occur in resistors.

The ease with which alloys can be soldered, brazed or welded is an important consideration in selection of materials for precision resistors. Improperly brazed or soldered joints frequently cause resistance instability in the circuit. Metals to be soldered must be cleaned prior to tinning so that solder can completely wet the surfaces and maintain electrical continuity. For copper-nickel alloys this is relatively simple, because protective oxide coatings are not formed on these alloys. Nickel-chromium alloys must be tinned immediately after cleaning and before an inherent protective oxide forms.

The resistance value of a resistor may change if the hydrostatic pressure on the resistance element is changed; for manganin this change is about 22.5 p$\Omega/\Omega \cdot$Pa (0.155 $\mu\Omega/\Omega \cdot$psi). Sealed resistors may also be affected by changes in external pressure. In a double-wall one-ohm resistor, for example, a change in pressure on the inner tube will cause a change in tube diameter, thus altering the length of wire wound on the tube. The magnitude of the resistance change depends in part on the thickness of the wall, and for commercial resistors is typically less than the hydrostatic pressure coefficient (PCR) of manganin. Unsealed resistors wound on mica cards containing air bubbles may have pressure coefficients several times greater than that predicted from the hydrostatic pressure coefficient of the alloy. This effect is important only if there is a large change in pressure, which would be most likely if there were a large change in elevation above sea level.

THERMOSTAT METALS

A thermostat metal is a composite material (usually in the form of sheet or strip) that consists of two or more materials bonded together, of which one may be a nonmetal. Because the materials bonded together to form the composite differ in thermal expansion, the curvature of the composite is altered by changes in temperature: this is the fundamental characteristic of any thermostat metal. A thermostat metal is, therefore, a complete, self-contained transducing system capable of transforming heat directly into mechanical energy for control, indicating or monitoring purposes.

In applications such as circuit breakers, thermal relays, motor overload protectors, and flashers, the change in temperature necessary for operation of the element is produced by the passage of current through the element itself — in other words, the change is produced by I²R heating. In certain other applications, any increase in the temperature of the thermostat element caused by I²R heating is objectionable, and a thermostat metal with low electrical resistivity is required.

For circuit breakers and similar devices, there are thermostat metals that differ in electrical resistivity but that are similar in other properties.

This allows a manufacturer to design a complete series of circuit breakers of different ratings in which the thermostat elements are all of the same size but have different electrical resistances. Resistivity is varied by incorporating a layer of a low-resistivity metal between outer layers of two other metals that have high resistivities and that differ widely in expansion coefficient.

In one series of commercial thermostat metals with resistivities ranging from 165 to 780 n$\Omega \cdot$m (100 to 470 $\Omega \cdot$circ mil/ft) at 24 °C (75 °F), high-purity nickel is used for the intermediate layer. In a series with resistivities from 33 to 165 n$\Omega \cdot$m (20 to 100 $\Omega \cdot$circ mil/ft), high-conductivity copper alloys are employed for the intermediate layer.

The use of a manganese-copper-nickel alloy having a resistivity of 1745 n$\Omega \cdot$m (1050 $\Omega \cdot$circ mil/ft) for one of the outer layers has extended the practical upper resistivity limit of thermostat metals to 1620 n$\Omega \cdot$m (975 $\Omega \cdot$circ mil/ft) at 24 °C.

Tolerances on resistivity at a standard temperature vary from ±3 to ±10%, depending on the type of thermostat metal and its resistivity.

About 30 different alloys are used to make over 50 different thermostat metals. Most of these 30 alloys are nickel-iron, nickel-chromium-iron, chromium-iron, high-copper and high-manganese alloys.

Thermostat metals are available as strip or sheet in thicknesses ranging from 0.13 to 3.2 mm (0.005 to 0.125 in.) and widths from 0.5 to 300 mm (0.020 to 12 in.). They are easily formed into the required shapes. Thermostat metals usually are selected on the basis of the temperature range in which they are required to operate. They are available for various operating ranges between −185 and +540 °C (−300 and +1000 °F). Properties and typical bimetal combinations for several temperature ranges are given in Table 2.

HEATING ALLOYS

Resistance heating alloys are used in many varied applications — from small household appliances to large industrial furnaces. In appliances, heating elements are designed for intermittent short-term service at about 100 to 1090 °C (200 to 2000 °F). In industrial furnaces, elements often must operate continuously at temperatures as high as 1300 °C (2350 °F) for furnaces used in metal treating industries, as high as 1700 °C (3100 °F) for kilns used for firing ceramics, and occasionally as high as 2000 °C (3600 °F) for special applications.

The primary requirements of materials used for heating elements are high melting point, high electrical resistivity, reproducible temperature coefficient of resistance, good oxidation resistance in furnace environments, absence of volatile components, and resistance to contamination. Other desirable properties are good elevated-temperature creep strength, high emissivity, low thermal expansion and low modulus (both of which help minimize thermal fatigue), good resistance to thermal shock, and good strength and ductility at fabrication temperatures.

Table 3 gives physical and mechanical properties, and Table 4 presents recommended maximum operating temperatures, for resistance heating materials. Of the four groups of materials listed in these tables, the first group (Ni-Cr and Ni-Cr-Fe alloys) serves by far the greatest number of applications.

The ductile wrought alloys in the first group

have properties that enable them to be used at both low and high temperatures in a wide variety of environments. The Fe-Cr-Al compositions (second group) are also ductile alloys. They play an important role in heaters for the higher temperature ranges, which are constructed to provide more effective mechanical support for the element. The pure metals that comprise the third group have much higher melting points. All of them except platinum are readily oxidized and are restricted to use in nonoxidizing environments. They are valuable for a limited range of application, primarily for service above 1370 °C (2500 °F). The cost of platinum prohibits its use except in small, special furnaces.

The fourth group, nonmetallic heating-element materials, are used at still higher temperatures. Silicon carbide can be used in oxidizing atmospheres at temperatures up to 1650 °C (3000 °F); two varieties of molybdenum disilicide are effective up to maximum temperatures of 1700 and 1800 °C (3100 and 3270 °F) in air. Molybdenum disilicide heating elements are gaining increased acceptance for use in industrial and laboratory furnaces. Among the desirable properties of molybdenum disilicide elements are excellent oxidation resistance, long life, constant electrical resistance, self-healing ability, and resistance to thermal shock.

DESIGN OF RESISTANCE HEATERS

Regardless of which heating alloy is selected, design of the heating element is important. One of the most important rules is to allow for unhindered expansion and contraction so as to avoid concentration of stresses as the temperature changes.

For service at lower temperatures, particularly from 400 to 600 °C (750 to 1100 °F), formed heating elements are used in ovens. In this construction, a heater support is made of two high-alloy rods spaced approximately 300 mm (12 in.) apart in a frame made of angle sections. The rods, whose length is determined by rated electrical input, contain spool insulators around which is wound a ribbon element made of a heating alloy. In a similar alternative construction, the ribbon element is replaced by a continuous helical coil of 5 gage or smaller wire.

Ribbon sizes for oven heaters range from 0.09 to 0.20 mm (0.0035 to 0.008 in.) thick, and from 9.5 to 16 mm (³/₈ to ⁵/₈ in.) wide. Oven heaters are rated to give maximum output at a watt density of approximately 8 kW/m² (5 W/in.²). (Watt density is obtained by dividing total power input to the elements by total surface area of the heater.) For 110- or 220-V oven heaters operating under normal conditions, expected life of Ni-Cr elements in air is three to five years.

For furnace temperatures up to 1175 °C (2150 °F), sinuous loop elements generally are formed from ribbon having a width-to-thickness ratio of about 12 to 1 and dimensions varying from 0.76 to 3.2 mm (0.030 to 0.125 in.) in thickness and from 13 to 38 mm (¹/₂ to 1¹/₂ in.) in width. Round rods of resistance material also may be formed into elements. Rod elements have been used by several furnace manufacturers.

ATMOSPHERES

Based on element temperature, Table 5 rates serviceabilities of various heating-element materials as good, fair or not recommended for the temperatures and atmospheres indicated. (See Table 6 for atmosphere compositions.) Element temperatures are always higher than furnace control temperatures; the difference depends on watt-density loading on the element surface. Thus, when furnaces are operated near maximum element temperature in the more active atmospheres, watt-density loading should be lower and element cross-sectional area should be higher.

With the exception of molybdenum, tantalum, tungsten and graphite, commonly used resistor materials have satisfactory life in air and in most other oxidizing atmospheres.

Atmosphere Contamination. Sulfur, if present, will appear as hydrogen sulfide in reducing atmospheres and as sulfur dioxide in oxidizing atmospheres. Sulfur contamination usually comes from one or more of the following sources: high-sulfur fuel gas used to generate the protective atmosphere; residues of sulfur-base cutting oil on the metal being processed; high-sulfur refractories, clays or cements used for sealing carburizing boxes; and the metal being processed in the furnace. Sulfur is destructive to Ni-Cr and Ni-Cr-Fe heating elements. The higher the nickel content, the greater the attack.

Lead and zinc contamination of a furnace atmosphere may come from the work being processed. This is a common occurrence in sintering furnaces for processing powder metallurgy parts. In the presence of a reducing atmosphere, lead will vaporize from leaded bronze powders (such as those used to make sintered bronze bushings) and attack the heating elements, forming lead chromate. Metallic lead vapors are even more harmful than sulfur to Ni-Cr alloys, and will cause severe damage to a heating element in a matter of hours if unfavorable conditions of concentration and temperature exist.

Electrical Contact Materials

Condensed from Metals Handbook, Ninth Edition, Volume 3, pages 662 to 695.

IF AN IDEAL ELECTRICAL CONTACT MATERIAL could be found, it would have high electrical conductivity to minimize the heat generated during passage of current; high thermal conductivity to dissipate both the resistive and arc heat developed; high reaction resistance to all environments in which it was to be used to avoid formation of insulating oxides, sulfides and other compounds; and immunity to arcing damage on the making and breaking of electrical contact. The force required to close a contact made of this material would be low, as would the electrical resistance between mating members. The melting point of the material would be high enough to limit arc erosion, metal transfer, and welding or sticking, but it would also be low enough to increase resistance to re-ignition in switching. (When the melting point is high, contacts continue to heat gas in the contact gap after the current drops to zero, thus facilitating re-ignition.) The vapor pressure would be low to minimize arc erosion and metal transfer. Hardness would be high to provide good wear resistance, and yet ductility would be high enough to ensure ease of fabrication. Purity of the material would be maintainable at a level that ensures consistent performance. Neither the material nor any process step necessary to fabricate it would present an environmental hazard. Finally, the material would be available at low cost in any desired form.

Because no metal has all the desired properties, a wide variety of contact materials is required to accomplish the objectives of different contact applications. The economic choice of materials is usually a compromise between the various processing variables and the application requirements. Load conditions, service requirements, and ambient conditions present during the life of the unit must be considered in the selection of contact materials.

FAILURE MODES OF MAKE-AND-BREAK CONTACTS

In an electric make-and-break switching device, the contact points, which usually consist of a pair of thin shaped slabs, perform the actual duty of making, carrying and breaking the current. Make-and-break contacts differ from slid-

ing contacts in that the moving member of the switching device travels perpendicular to the contact surfaces. As a result, arcs generated during opening and closing actions always strike and consequently damage the conducting surfaces.

Arcing, except in a circuit with an extremely low potential or low current, is a major factor — if not the main factor — causing failure of contact points.

When a pair of contacts opens in a live circuit, an arc is often generated between the contact pair, which remains until they are separated by a certain gap. Relatively less severe arcing occurs when the contacts close. Arcing also occurs when the moving contact bounces away from the stationary contact during closing. The arc causes contact erosion by blowing away the molten metal droplets, vaporizing the material, and transforming the metal to ion jets. Sometimes the material vaporizes from one contact and then condenses onto the other contact, thereby altering the surface configuration of both contacts. This is known as material transfer.

Oxidation of the contact surfaces, which may be accelerated by the heat from arcing, is a serious problem, because most metallic oxide films are nonconductive or semiconductive. The oxide film may easily increase contact resistance. In high-current circuits, this may cause excessive contact heating. In low-voltage and low-current circuits, the oxide films can grow so thick that they completely insulate the contact surfaces before the contact bodies erode. This happens more frequently when a pair of contacts operates in a hostile environment such as a polluted industrial atmosphere. Condensed organic polymers also play a role in precious metal contacts at light loads. These polymers come from monomers which evaporate from resins and are polymerized on the active catalytic metal surfaces of contacts.

Welding. To make a pair of contacts more conductive, a mechanical load is always applied on the contact pairs. Theoretically, the load could make two rigid contact surfaces touch at no more than three points. However, the touching points at both surfaces yield either elastically or plastically, resulting in larger areas of contact. These constricted regions carry the current through the contacts and form regions of high current density. Heat is generated in these areas and, if the temperature becomes high enough, the two contact points eventually are welded together.

Another kind of welding occurs as contacts close. The arcs generated during closing and bouncing of a moving contact melt a small por-

tion of both contacts. On reclosure, solidification of the molten material welds the contact pair. Occasionally, the strength of the weld exceeds the opening force of the switching device, resulting in catastrophic failure of the entire electrical system because the contacts fail to open on command.

Bridge Formation. When a pair of contacts opens, the contact area gradually decreases because of the gradual lessening of contact pressure. The continuous opening action causes the contact areas to reach a stage at which the current density of the constricted areas is so great that it melts the material in these regions. Continuous separation of the contact points now pulls the molten metal, forming a current-carrying bridge. The temperature of the molten bridge continues to rise as the contact points pull apart. It may become high enough to evaporate the material and finally break the circuit. This "bridge" phenomenon during the opening of a pair of contacts slightly damages the surfaces of the contacts and evaporates some of the bridge material. This generally results in pitting of one contact surface and buildup of material on the other; an uneven continuous transfer may eventually erode one of the contacts. Furthermore, the surface asperities from the continuous bridge formation may interlock the contact pair and interfere with their mechanical separation.

PROPERTY REQUIREMENTS FOR MAKE-AND-BREAK CONTACTS

The four failure modes discussed above determine the requirements of materials for make-and-break contacts. The most important requirements are listed below. In selecting a material, it is often necessary to reach a compromise that provides adequate properties without jeopardizing essential qualities of the component as a whole, such as reliability, life and cost.

- **Electrical conductivity:** Because the conduction of electricity between the pair of contacts depends on only a few constricted spots, the higher the electrical conductivity, the less the amount of heat that will be generated by high current density in these spots.

- **Thermal properties:** High melting and boiling points decrease evaporation loss caused by high arcing heat. High thermal conductivity disperses the heat rapidly and quenches the arc.

- **Chemical properties:** Contact materials should be corrosion resistant so that insulating films (either oxides or other compounds) do not form easily when the contacts operate in a hostile environment.
- **Mechanical properties:** The major loads applied to a contact pair are the closing force and the impact between movable and stationary contact points during closing. An induced relative movement between two contact surfaces always exists during closing. In some devices, such as certain types of relays, a wiping motion is purposely designed into the device to destroy any oxide films that form. However, friction between wiping surfaces produces wear of the contacts upon repeated opening and closing. Generally, hard materials are more resistant to wear. However, hard materials often have high contact resistances and low thermal conductivities, both of which contribute to a greater tendency toward contact welding. Hard materials also have high tensile strengths, which may or may not be advantageous in electrical contact applications.
- **Fabrication properties:** Contact materials should have the capability of being welded, brazed or otherwise joined to backing materials. In addition, they should have sufficient malleability to enable them to be shaped, or they should be capable of being formed by P/M techniques.

SLIDING CONTACTS

The applications of sliding contacts are usually quite different from those of make-and-break contacts. Friction, contact temperature, mechanical considerations and wear also are different.

The fundamental difference between make-and-break contacts and sliding contacts is that sliding contacts require films on the contact faces to facilitate sliding without seizure or galling; shear must take place within this film with only minor disturbance of both materials. A lubricant of some kind is always necessary. This can be provided by graphite if there is moisture present—such as in an environment having a dew point of about −20 °C (−4 °F) or higher. Alternatively, lubrication can be provided by very thin oil films, although excessive oil vapor causes overfilming. It can also be provided by molybdenum disulfide, and by other chalcogenides of molybdenum, tungsten and niobium. Oxygen, sulfur and other contaminants cause increased filming.

In applications in air, a drop in voltage can result from an equilibrium between oxidation and filming (which tend to increase the drop) and fretting or film breakdown and cleaning action (which tend to decrease the drop). In the absence of lubricants, fretting and oxidation are most important. In inert or reducing gases, oxidation is largely eliminated, and the voltage drop decreases until counteracted by mechanical factors. Noble metals that are properly lubricated also minimize voltage drops in air.

Brush contacts generally contain an appreciable amount of metal if they are intended for use in low-voltage (<24 V) applications. Large quantities of brush contacts are used in automotive and related industries as starter brushes and auxiliary motor brushes; copper-graphite is the principal material. Silver-graphite brushes are used primarily in instruments and in outer-space applications. Some silver-graphite brushes are used in seam welders and similar equipment.

Oxidation of sliding contacts is similar to that of make-and-break contacts, except that the surface disturbed by friction oxidizes more rapidly. In most applications in air, the metal surface generates a film that is a complex mixture of graphite, oxide, sulfide and water, which tends to decrease the conducting area.

The surfaces generated on metal-graphite brushes as they wear are effective cleaning agents in that they abrade films and keep larger areas available for conduction. Even so, it is sometimes advisable to have additional abrasive material in the brushes to prevent overfilming in critical atmospheres.

Because the major factor in friction is the shear strength of any film that is present, the composition of this film, as affected by atmospheric contaminants, is important.

Brush Materials. Considering the range of commercially available metal powders, graphites, other lubricants, and processing variables, there are unlimited possibilities for development of suitable brush materials. However, only a limited number of commercial grades have been developed.

More brush contacts are made from copper and its alloys than from any other class of material. In applications where copper metals undergo substantial oxidation, silver metals may be used. Tungsten or, more rarely, molybdenum is used where a high melting point is required. Platinum, palladium and gold are used where reliable closure with low force is required. Brushes clad or electroplated with precious metals, and brushes made of sintered alloys, are important for general applications in power switching relays. Silver/cadmium oxide is the most widely used contact material for medium- to high-energy applications. Recently, aluminum has become more popular for insert-type contacts because of its good machinability, formability and light weight. However, it is not as good as some other metals in contact properties—for instance, corrosion can be a limiting factor. When contact properties are important, aluminum should be plated or clad with copper or silver. More recently, tin platings have been accepted for certain applications. Protective lubricants have been added to the joints in some instances.

Interdependence Factors. When contacts are attached to a carrier, which is usually a copper alloy, the properties of the carrier material and the properties of the interface between contact and carrier (that is, the area of bond and the conductivity across the interface) are critical to ultimate performance. The contact carrier serves as a heat sink as well as a structural member and electrical conductor. The over-all efficiency of the system depends on the contact, the contact carrier and the method of attachment, all of which affect the size of the contact required for a specific application. To conserve precious metal, the contact materials, carrier material and method of attachment must be optimized. Some high-strength, high-conductivity copper alloys are used for carriers because of their structural properties and resistance to softening at brazing temperatures.

The attachment method that provides minimum interface alloying, minimum softening of the carrier and maximum bond area generally produces the best combination of properties for the contact system as a whole. The methods that best satisfy these criteria are percussion welding and diffusion bonding. Percussion welding can be utilized for round, square or rectangular con-

tacts and does not require a special backing for attachment purposes. This is especially helpful in minimizing the cost of Ag-CdO contacts, because the fine silver backing adds about 20% to the cost. Percussion welding does not soften the carrier as brazing does, and can be controlled to provide a high level of reliability.

COMMERCIAL CONTACT MATERIALS

Commercial materials for electrical contacts are divided into two categories based on their manufacturing methods: (a) wrought materials, which include both pure metals and alloys; and (b) composite materials, which include powder metallurgy products and internally oxidized silver alloys.

Copper Metals

High electrical and thermal conductivities, low cost, and ease of fabrication account for the wide use of copper in electrical contacts. The main disadvantage of copper contacts is low resistance to oxidation and corrosion. In many applications, the voltage drop resulting from the film developed by normal oxidation and corrosion is acceptable. In some circuit breaker applications, the contacts are immersed in oil to prevent oxidation. In other applications, such as in drum controllers, sufficient wiping occurs to maintain fairly clean surfaces, thus providing a circuit of low resistance. In some applications, such as knife switches, plugs and bolted connectors, contact surfaces are protected with grease or coatings of silver, nickel or tin. In power circuits, where oxidation of copper is troublesome, contacts frequently are coated with silver. Vacuum-sealed circuit breakers use oxygen-free copper contacts (wrought or powder metal) for optimum electrical properties.

In air, copper does not provide high resistance to arcing, welding or sticking. Where these characteristics are important, copper-tungsten or copper-graphite mixtures are used. However, when used in a helium atmosphere, a Cu-CdO contact performs similarly to an Ag-CdO contact. Copper alloys are used for high currents in vacuum interrupters.

Pure copper is relatively soft, anneals at low temperatures, and lacks the spring properties sometimes desired. Some copper alloys, harder than pure copper and having much better spring properties, are listed in Table 1. The annealing temperature of copper can be increased by additions of 0.25% Zn, 0.5% Cr, 0.03 to 0.06% Ag (10 to 20 oz per ton) or small amounts of finely dispersed metal oxides, such as Al_2O_3, with little loss of conductivity. On the other hand, improved spring properties are obtained only at the expense of electrical conductivity. Precipitation-hardened alloys, dispersion-hardened alloys, and powder metal mixtures can provide a wide range of mechanical and electrical properties.

Applications. Copper-base metals are commonly used in plugs, jacks, sockets, connectors and sliding contacts. Because of tarnish films, the contact force and amount of slide must be kept high to avoid excessive contact resistance and high levels of electrical noise. Yellow brass (C27000) is preferred for plugs and terminals because of its machinability. Phosphor bronze (C50500 or C51000) is preferred for thin socket and connector springs and for wiper-switch blades because of its strength and wear resistance. Nickel silver is sometimes preferred over yellow brass for re-

Table 1. Properties of copper metals used for electrical contacts

UNS number	Solidus temperature °C	°F	Electrical conductivity, % IACS	Hardness OS035 temper	H02 temper	Tensile strength OS035 temper MPa	ksi	H02 temper MPa	ksi
C11000	1065	1950	100	40 HRF	40 HRB	220	32	290	42
C16200	1030	1886	90	54 HRF	64 HRB(a)	240	35	415(a)	60(a)
C17200	865	1590	15 to 33(b)	60 HRB(c)	93 HRB(d)	495(c)	72(c)	655(d)	95(d)
C23000	990	1810	37	63 HRF	65 HRB	285	41	395	57
C24000	965	1770	32	66 HRF	70 HRB	315	46	420	61
C27000	905	1660	27	68 HRF	70 HRB	340	49	420	61
C50500	1035	1900	48	60 HRF	59 HRB	276	40	365	53
C51000	975	1785	20	28 HRB	78 HRB	340	49	470	68
C52100	880	1620	13	80 HRF	84 HRB	400	58	525	76

(a) H04 temper. (b) Depends on heat treatment. (c) TB00 temper. (d) TD02 temper.

lay and jack springs because of its high modulus of elasticity and strength, and also for its resistance to tarnishing and better appearance. Sometimes, copper alloy parts are nickel plated to improve surface hardness, reduce corrosion and improve appearance. However, nickel carries a thin but hard oxide film that has high contact resistance; very high contact force and long slide are necessary to rupture the film. To maintain low levels of resistance and noise, copper metals should be plated or overlaid with a precious metal.

Silver Metals

Silver, in pure or alloyed form, is the most widely used material for a considerable range of make-and-break contacts (1 to 600 A). Mechanical properties and hardness of pure silver are improved by alloying, but its thermal and electrical conductivities are adversely affected. Figure 1 shows the effect of different alloying elements on the hardness and electrical resistivity of silver. Properties of the principal silver metals used for electrical contacts are given in Table 2. Silver is widely used in contacts that remain closed for long periods of time and, in the form of electroplate, is widely used as a coating for connection plugs and sockets. It is also used on contacts subject to occasional sliding, such as in rotary switches, and to a limited extent for low-resistance sliding contacts, such as slip rings.

Electrical and Thermal Conductivity. Silver has the highest electrical and thermal conductivities of all metals at room temperature and, as a result, will carry high currents without excessive heating, even when dimensions of the contacts are only moderate. Although good thermal conductivity is desired once the contact is in service, such conductivity increases the difficulty of assembly welding.

In component assemblies, migration of silver through electrical insulation may cause failure of the insulation. When in contact with certain materials such as phenol fiber and under electric potential, silver migrates ionically through or across the insulating material, producing thread-like connections that lower the resistance across the insulation. This reduces insulating qualities, and the reduction is even greater if moisture is present in the atmosphere. Insulators must be designed with care to avoid this hazard.

Oxidation Resistance. Silver is used instead of copper chiefly because of its resistance to oxidation in air. In general, silver oxide is not a problem on silver contacts, whether or not the contacts make and break the circuit. However, silver oxide can be produced by exposure to ozone, as well as by other methods. This oxide has high resistivity, is decomposed slowly on heating at about 175 °C (350 °F), is decomposed rapidly at about 350 °C (650 °F) and is removed by arcing.

Silver is vulnerable to attack by sulfur or sulfide gases in the presence of moisture. The resulting sulfide film may produce significant contact resistance, particularly where contact force, voltage, or current is low. Direct current brings silver ions from the matrix into the sulfide where they form connecting bridges. Therefore, particularly at high direct current, the film becomes somewhat conducting. The resistance of a silver sulfide film decreases as temperature increases — Ag_2S decomposes slowly at 360 °C (680 °F) and more rapidly at higher temperatures. In addition, the film may increase erosion and entrap dust.

Limitations of Silver Contacts. Silver will provide a fairly long contact life for make-and-break con-

tacts and will handle up to 600 A. In pure silver contacts, difficulties sometimes arise from transfer of metal from one electrode to the other, which leads to the formation of buildups on one contact surface and holes in the other. When used in dc circuits, silver contacts are subject to ultimate failure by mechanical sticking as a direct result of metal transfer. The direction of transfer is generally from the positive contact to the negative, but under the influence of arcing, the direction may be reversed. With high currents or inductive loads, it may be desirable to shunt the load with a resistance-capacitance (RC) protection network to reduce erosion.

When arcing produces a glow discharge in air, the rate of erosion of silver is unusually high because of a chemical interaction with air to form $AgNO_2$.

For low resistance and low noise levels, the design of the contact device must provide sufficient force and slide to break through any silver sulfide film and maintain film-free metal-to-metal contact at the interface. Connectors should have high slide force and several newtons normal force. Rotary switches that have up to 490 mN normal force and considerable slide should have a protective coating of grease to reduce sulfiding and to remove abrasive particles. In low-noise transmission circuits, silver should not be used on relay and other butting contacts that have less than 196 mN force; other precious-metal coatings, such as gold or palladium, should be used instead of silver.

A silver sulfide film has a characteristic voltage drop of several tenths of a volt. Where this drop is tolerable, silver contacts will provide re-

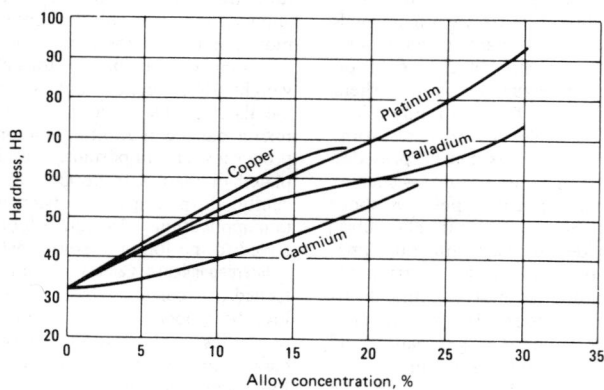

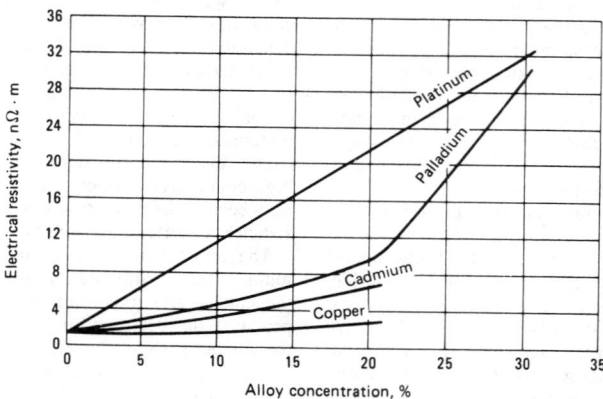

Fig. 1. Hardness and electrical resistivity versus alloy content for silver alloy contacts

Table 2. Properties of silver metals used for electrical contacts

Alloy	Solidus temperature °C	°F	Electrical conductivity, % IACS	Hardness, HR15T Annealed	Cold worked	Tensile strength Annealed MPa	ksi	Cold worked MPa	ksi	Density, Mg/m³	Elongation in 50 mm or 2 in., % Annealed	Cold worked
99.9Ag	960	1760	104	30	75	172	25	310	45	10.51	55	5
99.55Ag-0.25Mg-0.2Ni	...	...	70	61	77	207	30	345	50	10.34	35	6
99.47Ag-0.18Mg-0.2Ni-0.15Cu	...	...	75	64	84	...	...	...	...	10.38	...	...
99Ag-1Pd	...	...	79	44	76	179	26	324	47	10.14	42	3
97Ag-3Pd	977	1790	58	45	77	186	27	331	48	10.53	37	3
97Ag-3Pt	982	1800	45	45	77	172	25	324	47	10.17	37	3
92.5Ag-7.5Cu	821	1510	88	65	81	269	39	455	66	10.34	35	5
90Ag-10Au	971	1780	40	57	76	200	29	317	46	11.03	28	3
90Ag-10Cu	777	1430	85	70	83	276	40	517	75	10.31	32	4
90Ag-10Pd	999	1830	27	63	80	234	34	365	53	10.57	31	3
86.8Ag-5.5Cd-0.2Ni-7.5Cu	...	...	43	72	85	276	40	517	75	10.10	43	3
85Ag-15Cd	877	1610	35	51	83	193	28	400	58	10.17	55	5
77Ag-22.6Cd-0.4Ni	...	...	31	50	85	241	35	469	68	10.31	55	4
75Ag-24.5Cu-0.5Ni	...	...	75	78	85	310	45	552	80	10.00	32	4
72Ag-28Cu	777	1430	84	79	85	365	53	552	80	9.95	20	5
60Ag-23Pd-12Cu-5Ni	...	...	11	86	93	538	78	758	110	10.51	22	3

Table 3. Properties of gold metals used for electrical contacts

Alloy	Solidus temperature °C	°F	Electrical conductivity, % IACS	Hardness, HR15T Annealed	Cold worked	Tensile strength Annealed MPa	ksi	Cold worked MPa	ksi	Density, Mg/m³
99Au	1085	1985	74	40	65	...	...	...	...	19.36
90Au-10Cu	932	1710	16	76	91	400	58	705	102	17.18
75Au-25Ag	1029	1885	17	50	77	...	...	...	...	15.96
72.5Au-14Cu-8.5Pt-4Ag-1Zn	954	1750	10	88	96	...	...	...	...	16.11
72Au-26.2Ag-1.8Ni	...	...	14	61	81	230	33	345	50	15.56
71Au-5Ag-9Pt-15Cu	...	...	8	88.5	75(a)	700	101	1170	170	16.02
69Au-25Ag-6Pt	1029	1885	10	70	84	275	40	415	60	15.92
50Au-50Ag	...	...	...	...	...	...	...	...	...	13.59

(a) Rockwell 15N.

Table 4. Properties of platinum and palladium metals used for electrical contacts

Alloy	Solidus temperature °C	°F	Electrical conductivity(a), % IACS	Hardness, HR15T Annealed	Cold worked	Tensile strength Annealed MPa	ksi	Cold worked MPa	ksi	Density, Mg/m³	Elongation(a) in 50 mm or 2 in., %
99.9Pt	1770	3220	15	60	73	138	20	241	35	21.45	35
95Pt-5Ru	1775	3230	5	84	89	414	60	793	115	20.57	18
92Pt-8Ru	...	...	4	86	91	483	70	896	130	20.27	15
90Pt-10Ir	1780	3240	7	87	92	379	55	620	90	21.52	12
89Pt-11Ru	1815	3300	4	91	96	586	85	1034	150	19.96	12
86Pt-14Ru	1843	3350	3	93	99	655	95	1172	170	19.06	10
85Pt-15Ir	1787	3250	6	90	95	517	75	827	120	21.52	12
80Pt-20Ir	1808	3290	5	93	97	689	100	1000	145	21.63	12
75Pt-25Ir	1819	3310	5	95	98	862	125	1172	170	21.68	10
73.4Pt-18.4Pd-8.2Ru	...	...	4	90	92	517	75	862	125	17.77	12
65Pt-35Ir	1899	3450	4	97	99	965	140	1344	195	21.80	8
99.9Pd	1554	2830	16	62	78	193	28	324	47	12.17	28
95Pd-5Ru	1593	2900	8	79	89	372	54	517	75	12.00	15
89Pd-11Ru	1649	3000	6	85	92	483	70	689	100	12.03	13
72Pd-26Ag-2Ni	1382	2520	4	82	90	469	68	689	100	11.52	13
60Pd-40Ag	1338	2440	4	65	91	372	54	689	100	11.30	28
60Pd-40Cu	1199	2190	8	82	92	565	82	1331	193	10.67	20
35Pd-9.5Pt-9Au-14Cu-32.5Ag	1085	1985	5	90	94	689	100	1034	150	11.63	18

(a) For material in annealed condition.

liable contact closure. Failure to close, however, may be greater than with other precious metal contacts because of impacted dirt, with a sulfide film acting as a dirt catcher.

For many applications, silver is too soft to give acceptable mechanical wear. Alloying additions of copper, cadmium, platinum, palladium, gold and other elements are effective in increasing the hardness and modifying the contact behavior of silver.

Gold Metals

Pure gold has unsurpassed resistance to oxidation and sulfidation, but a low melting point and susceptibility to erosion limit its use in electrical contacts to situations where the current is not more than 0.5 A. Although oxide and sulfide films do not form on gold, a carbonaceous deposit is sometimes formed when a gold contact is operated in the presence of organic vapors. The resistance of this film may be several ohms.

When gold is used in contact with palladium or rhodium, very low contact resistances have been reported.

The low hardness of gold can be increased by alloying with copper, silver, palladium and platinum, but usage is necessarily restricted to low-current applications because of the low melting point.

Properties of gold and its alloys are listed in Table 3. If low tarnish rates and low contact resistance are to be preserved, the gold content should not be less than about 70%.

Precious Metals of the Platinum Group

Platinum and palladium are the two most important metals of the platinum group. These metals have a high resistance to tarnishing, and therefore provide reliable contact closure for relays and other devices having contact forces of less than 490 mN. Their high melting points, low vapor pressure, and resistance to arcing make them suitable for contacts that close and open the load, particularly in the range up to 1 A. The low electrical and thermal conductivities of these metals, as well as their cost, generally exclude them from use at currents above about 5 A.

Palladium has an arcing limit only slightly less than that of platinum and gives comparable performance in relays for telephones and similar services handling 1 A or less. Palladium is a satisfactory substitute for platinum in these applications.

Properties of platinum and palladium metals are presented in Table 4.

Chemical Properties. Platinum has a high resistance to corrosion, including resistance to oxidation, sulfidation and salt water. It will not form a stable oxide at any temperature.

Palladium is resistant to oxidation at ordinary temperatures. If heated above 350 °C (660 °F), it will oxidize slowly to form an oxide that is stable at room temperature. However, the oxide is decomposed promptly on heating to 800 °C (1470 °F) or by arcing. The oxide is not considered to be a significant factor in the reliability of closure of telephone-type relays.

The presence of organic vapors in the contact area can seriously influence the life and reliability of electrical contacts, particularly the low-force precious metal contacts universally employed in high-reliability low-noise circuits. The damaging organic vapors may arise from coil forms, wire coatings, insulation, soldering flux, potting and sealing compounds, and other organics in associated electrical equipment, as well as from external sources.

Organic contamination may produce two distinctly different forms of contact damage: activation and polymer formation.

Activation is the development of a carbon deposit on the contacting surfaces, formed by the decomposition of the organic contaminant in the arc. This deposit markedly increases arc erosion. The carbon deposits decrease the current needed to sustain an arc and prolong the arcing time. A 95% reduction in contact life may result from activation brought about by the presence of organic vapors. Activation can be reduced or eliminated by using insulating materials that are not sources of organic vapors, by adequate ventilation and, perhaps, by absorbing the organic vapors in a getter such as carbon.

Polymer formation is the development of a polymer-like insulating brown powder on contacts in dry circuits (those not carrying current on make or break), and may lead to transient open circuits. The insulating brown powder is believed to result from the adsorption of the organic vapor on the contact surface, followed by its polymerization by the friction associated with contact operation. The sliding motion both forms the polymer and pushes it outside the slide area, where it builds up as a brown powder. A transient open circuit occurs when some of the built-up powder falls into the contact area.

Controlled experiments have shown that the type of contact metal influences the amount of polymer formed. The greatest amount of polymer is formed on the platinum metals; lesser amounts are formed on gold and some base metals; polymer does not form on silver.

Elimination of materials that give rise to organic vapors is a possible solution to polymer formation, but one that is difficult to carry out. From a practical standpoint, the problem has been solved in telephone circuits by cladding one of a mating pair of palladium contacts with a very thin layer of gold. In dry circuits, the one gold surface significantly reduces polymer formation, although in working circuits, the gold soon wears off and exposes the palladium base.

Erosion and Sticking. The arcing current limit for platinum group metals is about 1 A, and contact life is long if the current is kept below this value. With currents higher than 1 A or with inductive loads, it may be desirable to shunt the load or contact with a resistance-capacitance network to reduce erosion, and to reduce failures caused by snagging of pits and buildups. In general, for equal volumes of contact metal, the life of platinum or palladium contacts is about ten times the life of silver contacts.

Resistance and Noise. Palladium is used almost universally on relays and relay-type switch contacts in telephone systems within the United States for talking circuit transmission. In this service, palladium is essentially noise-free, and is used in preference to platinum or gold alloys because it is more economical.

In a few isolated instances, where the palladium talking circuit contacts have been subject to vibration in service, noise troubles have developed because of polymer formation. In these few instances, the difficulty has been met by the use of gold alloys, which greatly reduces the production of polymer.

Tungsten and Molybdenum

Most tungsten and molybdenum contacts are made in the form of composites with silver or copper as the other principal component. Tungsten, which was one of the earliest metals other than copper and silver adopted for electrical contact applications, has the highest boiling point (5930 °C, or 10 700 °F) and melting point (3110 °C, or 5625 °F) of all metals; it also has very high hardness at both room and elevated temperatures. Therefore, as a contact material, it offers excellent resistance to mechanical wear and electrical erosion. Its main disadvantages are low corrosion resistance and high electrical resistance. After a short period of operation, an oxidized film will build up on tungsten contacts, resulting in very high contact resistance. Considerable force is required to break through the film, but high pressure and considerable impact cause little damage to the underlying metal because of its high hardness. Tungsten contacts are used in switching devices with closing forces of more than 19.6 N and in circuits with high voltages and currents not more than 5 A, such as automotive ignitions, vibrators, horns, voltage regulators, magnetos, and electric razors. In low-voltage dc devices, tungsten is always used as the negative contact, and is paired with a positive contact made of precious metal.

Tungsten rods or strips that are consolidated by swaging or rolling from sintered powder compacts have very poor ductility. They cannot be cold worked, in contrast to other contact materials. Tungsten disks are usually cut from rods or punched from strips and then brazed directly to functional parts such as breaker arms, brackets or springs.

Properties such as grain size, grain configuration, and the degree of fibrous structure, which affect contact behavior, are controlled by using special swaging methods and annealing cycles. Tungsten disks usually are supplied with a ground finish but they can also be electrochemically polished to obtain high-luster surfaces.

The high boiling and melting points of molybdenum—5560 °C (10 040 °F) and 2610 °C (4730 °F), respectively—are second only to those of tungsten and rhenium. Molybdenum is not used as widely as tungsten because it oxidizes more readily and erodes faster on arcing than tungsten. Nevertheless, because the density of molybdenum (10.2 Mg/m^3 or 0.369 lb/in.^{3}) is about half that of tungsten (19.3 Mg/m^3 or 0.697 lb/in.^{3}), use of molybdenum is advantageous where mass is important. Its cost by volume is also lower.

In addition to its use in make-and-break contacts, molybdenum is widely used for mercury switches because it is not attacked, but only wetted, by mercury. Like tungsten, molybdenum strips and sheets are made by swaging or rolling sintered powder compacts. Disks made from rods or sheets are brazed to blanks or other structural components. Table 5 lists the properties of tungsten and molybdenum.

Aluminum

In recent years, because of its good electrical and mechanical properties, ready availability and favorable cost, aluminum has gained importance as a conductor material. It has replaced copper in many applications, such as rectangular, tubular and channel bus conductors. It has advantages over copper in density, mass, electrical conductivity, availability and cost; it is lighter, and the same mass will conduct more current for the same voltage drop.

Aluminum 1350 is preferred for contact materials because of its high conductivity (61.8% IACS), but it is low in strength and, for some designs, requires additional support. Where strength and resistance to joint relaxation are important, heat treated 6101 is better suited and is used to a considerable extent, although there is some sacrifice in electrical conductivity (57 to 60% IACS for 6101).

As a contact metal, aluminum is generally poor because it oxidizes readily. Where aluminum is used in contacting joints, it should be plated or clad with copper, silver or tin. Aluminum should never be used for power applications where arcing is present. For instance, if aluminum contacts were substituted for silver in a motor starter, an explosion due to noninterruption of current on motor-starter de-energization would probably occur on load interruption.

Composite Materials

There are three major groups of composite contact materials made by powder metallurgy

Table 5. Typical properties of tungsten and molybdenum(a)

Tungsten

Hardness	70 HRA, 385 HV
Modulus of elasticity:	
At 20 °C (68 °F)	405 GPa (59 × 10^6 psi)
At 1000 °C (1830 °F)	325 GPa (47 × 10^6 psi)
Density	19.3 Mg/m^3 (0.697 lb/in.^{3})
Melting point	3410 °C (6170 °F)
Boiling point	5900 °C (10 650 °F)
Specific heat at 20 °C (68 °F)	140 J/kg (0.033 Btu/lb·°F)
Thermal conductivity at 20 °C (68 °F)	130 W/m·K (75 Btu/ft·h·°F)
Coefficient of linear thermal expansion at 20 °C (68 °F)	4.43 μm/m·°C (2.46 μin./in.·°F)
Specific resistance at 20 °C (68 °F)	5.5 nΩ·m
Electrical conductivity at 20 °C (68 °F)	31% IACS

Molybdenum

Hardness	58 HRA, 210 HV
Modulus of elasticity:	
At 20 °C (68 °F)	325 GPa (47 × 10^6 psi)
At 1000 °C (1830 °F)	270 GPa (39 × 10^6 psi)
Density	10.22 Mg/m^3 (0.369 lb/in.^{3})
Melting point	2622 °C (4750 °F)
Boiling point	4800 °C (8672 °F)
Specific heat at 20 °C (68 °F)	270 J/kg (0.065 Btu/lb·°F)
Thermal conductivity at 20 °C (68 °F)	155 W/m·K (89 Btu/ft·h·°F)
Coefficient of linear thermal expansion at 20 °C (68 °F)	5.53 μm/m·°C (3.07 μin./in.·°F)
Specific resistance at 20 °C (68 °F)	5.2 nΩ·m
Electrical conductivity at 20 °C (68 °F)	33% IACS

(a) Some of the physical properties of tungsten and molybdenum vary considerably with cross-sectional area and grain structure.

methods: refractory and carbide-base, silver-base, and copper-base.

Because manufacturing methods affect properties of materials with the same composition, the most common methods of producing composite electrical contact materials are discussed below.

Infiltration is used exclusively for making refractory metal and carbide-base composite contact materials. Metal powder or carbide powder is first blended to the desired composition with or without a small amount of binder to impart green strength, then is pressed and sintered into a skeleton of the required shape. Silver or copper is then infiltrated into the pores of the skeleton. This method produces the most densified composites, generally 97% or more of theoretical density. Complete densification is not possible because of the presence of some closed pores in the sintered skeleton. After infiltration, the contact is sometimes chemically or electrochemically etched so that only pure silver appears on the surface. The contact thus treated has better corrosion resistance and performs better in the early stages of use.

Press-Sinter. For small refractory-metal contacts (not exceeding about 25 mm, or 1 in., in diameter), a high-density material can be obtained by pressing a blended powder of exact final composition into shape and then sintering it at the melting temperature of the low-melting-point component (liquid-phase sintering). In some cases, an activating agent such as nickel, cobalt or iron is added to improve the sintering effect on the refractory metal particles. For this process, powders of much finer particle size are required so that more bonding surface exists. However, the

skeleton formed by this process is weaker than that formed by the infiltration process. Formation of the skeleton usually shrinks the apparent volume of the refractory portion of the composition, thus bleeding out the molten component onto the surface of the finished contact.

Press-Sinter-Re-press. The press-sinter-re-press process is used for all categories of contact materials, especially those in the silver-base category. Blended powders of the correct composition are compacted to the required shape and then sintered. Afterward, the material is further densified by a second pressing (re-pressing). Sometimes the properties can be modified by a second sintering or annealing. The versatility of this process makes it applicable for contacts of any configuration and of any material. However, it is difficult to obtain material with as high a density as is obtained with other processes. Material thus produced also may have weak bonding between particles.

Press-Sinter-Extrude. Blended powder of final composition is pressed into an ingot and sintered. The ingot is then extruded into wires, slabs or other desired shapes. The extruded material may be subsequently worked by rolling, swaging or drawing. Material made by this method is usually fully dense.

The press-sinter-extrude process is used mostly for silver-base composites. Other processes used for manufacturing silver-base composite contacts are direct extrusion or direct rolling of loose powder. Although they appear to be uncommon, they are economically feasible if the equipment is properly designed and built.

Preoxidize-Press-Sinter-Extrude. This method is used

exclusively for making silver/cadmium oxide (Ag-CdO) material. Alloys are reduced to small particles in the shape of flakes, slugs or shredded foil. These particles are oxidized and then consolidated with the press-sinter-extrude process. Material made by this method is more uniform than the same material made by conventional internal oxidation. Mechanical properties are superior to those of the same material made by the press-sinter-re-press method.

Coprecipitation. Conventional blending or mechanical mixing of silver and cadmium oxide powders begins by dissolving the proper amounts of silver and cadmium metals in nitric acid. Compounds of silver and cadmium coprecipitate from the solution when the pH value of the solution is changed by adding either hydroxide or carbonate solutions. During subsequent calcination at about 500 °C (930 °F), the compound mixture decomposes to form a mixture of silver and cadmium oxide. Alkali-metal content can be controlled in the ppm range by adequate washing. Controlled amounts of sodium, potassium and lithium may enhance electrical life. Excessive amounts of these elements can lead to rapid erosion, restrike, and generally poor electrical life. Depending on device design, the range may be from 10 to 300 ppm. Contacts are consolidated from this mixture by conventional P/M methods. The microstructure of contacts made by this method displays a finer particle size and a more uniformly dispersed CdO phase than material made by conventional blending. The fine particle dispersion results in good contact welding resistance, presumably because of the formation of slaglike inclusions.

Thermocouples for Industrial Applications

Condensed from Metals Handbook, Ninth Edition, Volume 3, pages 696 to 720.

ACCURATE MEASUREMENT of temperature is one of the most common and vital requirements in science, engineering and industry. Measurement of temperature is generally thought to be one of the simplest and most accurate measurements that can be made. This is a misconception. Unless proper techniques are employed, highly inaccurate readings can occur and either useless data can be generated or materials can be misprocessed. Also, under certain conditions, it may be difficult or impossible to obtain accurate temperature measurements regardless of whether or not proper techniques are employed.

Seven types of instruments, under appropriate conditions and within specific operating ranges, may be used for measurement of temperature: thermocouple thermometers, radiation pyrometers, resistance thermometers, liquid-in-glass thermometers, filled-system thermometers, optical pyrometers and bimetal thermometers. The success of any temperature-measuring system depends not only on the capacity of the system but also on how well the user understands the principles, advantages and limitations of its application.

The thermocouple thermometer is by far the most widely used device for measurement of

temperature. Its favorable characteristics include good accuracy, suitability over a wide temperature range, fast thermal response, ruggedness, high reliability, low cost and great versatility of application.

Essentially, a thermocouple thermometer is a system consisting of (a) a temperature-sensing element called a thermocouple, which produces an electromotive force (emf) that varies with temperature, (b) a device for sensing emf, which may include a printed scale for converting emf to equivalent temperature units, and (c) an electrical conductor (extension wires) for connecting the thermocouple to the sensing device. Although any combination of two dissimilar metals and/or alloys will generate a thermal emf, only seven thermocouples are in common industrial use today. These seven have been chosen on the basis of such factors as mechanical and chemical properties, stability of emf, reproducibility and cost.

Measurement of Temperature by a Thermocouple. A setup for measurement of temperature by use of a thermocouple is illustrated schematically in Fig. 1. The welded junction of thermocouple PN is inserted into an electric furnace the temperature of which is to be measured. The ice-point cold junction is provided by two mercury U-tubes embedded in a Dewar flask packed with shaved ice. The legs of the thermocouple are inserted into the mercury U-tubes and connected to the positive and negative terminals of a potentiometer by insulated copper wires. The temperature

of the furnace then can be obtained by measuring the emf generated by the thermocouple and referenced to the established emf table for that particular thermocouple. Commercially available automatic compensating cold junctions can be used in place of the above-mentioned mercury U-tubes to achieve a 0 °C reference junction. These may be built into an indicating or recording instrument used to measure the emf developed by the thermocouple or external of the measuring instrument.

In the absence of an ice junction, the thermocouple wires may be connected directly to the terminals of the potentiometer. The ambient temperature of the terminals is measured by a thermometer and converted to emf in millivolts from the emf table. The total emf generated by the thermocouple between the hot junction and the ice point is the sum of the emf thus measured by the potentiometer and this ambient-temperature correction factor. The temperature of the hot junction can be obtained by referring this total emf to the established table.

For additional information, see ASTM E563, "Standard Recommended Practice for Preparation and Use of Freezing Point Reference Baths."

Preparation of the Measuring Junction. The two dissimilar thermoelements must be joined at the temperature-measuring junction to form the thermocouple. The joint must have good thermal and electrical conductivity without adversely affecting the mechanical and electrical properties of the thermocouple wires at this joint.

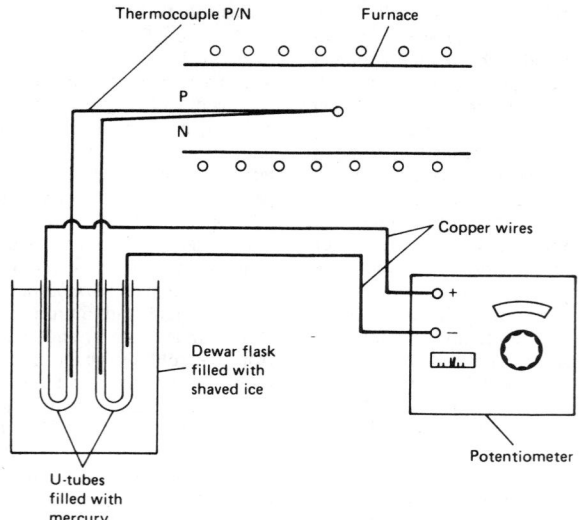

Adapted from Ref 1.

Fig. 1. Schematic diagram of the experimental setup for measuring temperature using a thermocouple and an ice-point reference junction

Prior to being joined, the thermoelements are straightened to facilitate insertion into hard-fired ceramic insulators. In this operation, care should be taken to avoid excessive cold working of the wires, which has a deleterious effect on the emf of the couple. After being cut to the desired length, the thermocouple wires are cleaned carefully (to remove lubricant residue, fingerprints and other contaminants) with a suitable solvent such as methyl ethyl ketone, Freon TF or isopropyl alcohol prior to joining.

For applications below about 500 °C (about 1000 °F), base-metal thermocouple wires may be silver brazed using borax as a flux. Above this temperature, thermocouple junctions usually are prepared by welding. Noble-metal thermocouples should always be joined by welding. Thermocouples are usually welded using gas, electric-arc, resistance, tungsten-inert gas and plasma-arc processes. In gas welding, a neutral flame is required (preferably oxidizing for noble metals). Prior to gas or arc welding, the ends of types E, J, K and T thermocouple wires are first twisted one and a half turns.

Effecting a hot junction in a sheathed thermocouple requires a higher degree of skill, special equipment and considerable care. After the sheath has been stripped away, joining usually is done by gas tungsten-arc or plasma-arc welding. A clean, dry and well-lighted work area is required to produce a finished element of good integrity. An oven capable of continuous operation at 90 °C (200 °F) should be available for storage of unsealed sheathed thermocouples during unavoidable delays in forming of junctions. Use of such an oven will minimize pickup of airborne moisture and other contaminants.

THERMOCOUPLE MATERIALS

Commercially available thermocouples are grouped according to material characteristics (base metal or noble metal) and standardization. At present, four base-metal thermocouples and three noble-metal thermocouples have been standardized and given letter designations by ANSI (American National Standards Institute), ASTM (American Society for Testing and Materials) and ISA (Instrument Society of America). Among the remaining thermocouples in use, some have not been assigned letter designations because of limited usage, and some are being considered for standardization.

Standard Thermocouples. The base compositions, melting points and electrical resistivities of the individual thermoelements of the seven standard thermocouples are presented in Table 1. Maximum operating temperatures, and limiting factors in environmental conditions, are listed also.

The relations between emf and temperature for the individual thermoelements with reference to Platinum 67 and for the seven standard thermocouples are shown in Fig. 2 and 3, respectively. Tolerances for initial calibration of standard thermocouples (those meeting established tables within a specified tolerance) are listed in Table 2.

THERMOCOUPLE EXTENSION WIRES (REF 3, 4)

Thermocouple extension wires, also known as extension wires or lead wires, are electrical conductors used for connecting the thermocouple wires to the temperature measuring and control instrument. Extension wires usually are supplied in cable form, with positive and negative wires electrically insulated from each other. The chief reasons for using extension wires are economy and mechanical flexibility.

- *Economy.* Base-metal thermoelements, which cost less than $10 per pound in 1980, are always used as extension wires for the noble-metal thermocouple wires, which in 1980 cost about $700 per troy ounce. For base-metal thermocouples, use of extension wires permits periodic replacement of the thermocouple, which is exposed to elevated temperatures, without replacing the insulated extension-wire cables.
- *Mechanical Flexibility.* Insulated solid or stranded wires in sizes from 14 to 20 gage are used as extension wires. This lends mechanical sturdiness and flexibility to the thermocouple circuitry while permitting the use of larger-di-

ameter (usually 3.2 mm, or ⅛ in.) base-metal thermocouples for improved oxidation resistance and service life, or smaller-diameter (usually 0.51 mm, or 0.020 in.) noble-metal thermocouple wire to save cost.

COLOR CODING OF THERMOCOUPLE WIRES AND EXTENSION WIRES (REF 5)

For many years the Instrument Society of America has coordinated an effort to standardize color coding of thermocouple and extension wires in the United States. The main objective has been to establish uniformity in designation of various types of thermocouples and extension wires to provide, by means of insulation color, identification of wires by type or composition as well as by polarity when used as part of a thermocouple system. The present color designations are indicated in ANSI MC96.1(1975). Color coding is not uniform throughout the world and presently is being evaluated by the International Electrotechnical Commission in an attempt at world standardization.

THERMOCOUPLE CALIBRATION

The temperature/emf relationship for a specific thermocouple combination is a definite physical property and thus does not depend on details of the apparatus or method used for determining this relationship. Consequently, thermocouples can be calibrated by any of several methods, the choice of which depends on type of thermocouple, temperature range, accuracy required, size of wires, apparatus available and personal preference.

Calibration of a thermocouple is achieved through determination of its electromotive force (emf) at a series of known temperatures, which when coupled with a standardized means of interpolation will give values of emf over the entire temperature range in which it will be used. A standard thermometer that indicates temperatures on a universally acceptable scale is required, as well as a means of measuring the emf of the thermocouple and a controlled heat source wherein the thermocouple and the standard can be brought to the same temperature.

Temperature Scales (Ref 5 and 6). Meaningful measurement of temperature requires a scale with appropriate units, just as measurement of length requires a yardstick or metre stick with all of its subdivisions. The ideal temperature scale is known as the thermodynamic scale. However, measurement of temperature on this scale (using a gas thermometer) is extremely difficult even under laboratory conditions. For many years prior to 1927, the need for a more practical temperature scale had been apparent.

In 1927, such a scale, named the International Temperature Scale (ITS 27), was adopted by the Seventh General Conference on Weights and Measures. Among other advantages, this scale served to unify the existing national temperature scales (Germany, UK, USA, etc.). The scale was revised in 1948, and in a 1960 modification the word "Practical" was inserted in the name of the Scale, which now became the International Practical Temperature Scale. The Scale was revised again in 1968, and was amended in 1975.

The present scale, the "International Practical Temperature Scale of 1968 (amended in 1975)," or "IPTS 68 (amended 1975)," was designed in such a way that the temperature measured on it

Table 1. Properties of standard thermocouples

Type	Thermo-elements	Base composition	Melting point, °C	Resistivity, nΩ·m	Recommended service	Max temperature °C	Max temperature °F
J	JP	Fe	1450	100	Oxidizing or reducing	760	1400
	JN	44Ni-55Cu	1210	500			
K	KP	90Ni-9Cr	1350	700	Oxidizing	1260	2300
	KN	94Ni-Al, Mn, Fe, Si, Co	1400	320			
T	TP	OFHC Cu	1083	17	Oxidizing or reducing	370	700
	TN	44Ni-55Cu	1210	500			
E	EP	90Ni-9Cr	1350	700	Oxidizing	870	1600
	EN	44Ni-55Cu	1210	500			
R	RP	87Pt-13Rh	1860	196	Oxidizing or inert	1480	2700
	RN	Pt	1769	104			
S	SP	90Pt-10Rh	1850	189	Oxidizing or inert	1480	2700
	SN	Pt	1769	104			
B	BP	70Pt-30Rh	1927	190	Oxidizing, vacuum or inert	1700	3100
	BN	94Pt-6Rh	1826	175			

REFERENCE TABLES FOR THERMOCOUPLES

Practical use of thermocouples requires that the selected thermocouple meet an established or standardized temperature/emf relationship within acceptable tolerance limits. Because the thermocouple in a thermoelectric thermometer system is replaced periodically due to drift, failure or other reasons, conformance to an established temperature/emf relationship is necessary in order to permit interchangeability when commercially available readout equipment is used. Such widely acceptable reference tables contain information on S and T thermocouples and are available in NBS Monographs 124 (Cryogenic) and 125 (Standard Couples), ANSI MC96.1 and ASTM E230. Less detailed versions of these tables (at intervals of 10 °C, or 18 °F) usually may be obtained from producers or distributors of these thermocouples.

CHANGE OF CALIBRATION DURING SERVICE

Any thermocouple can be subject to failure (of a type that creates an open circuit) during service. Failure can be caused by localized melting of the thermoelements as a result of overheating, by vibration resulting in fatigue failure, or by gradual reduction of wire diameter through high-temperature oxidation. Prior to failure, the emf calibration of a thermocouple will change, primarily as a result of the individual or combined changes in chemical composition, homogeneity and structure that take place in the thermoelements. The magnitudes and directions of these changes are dependent on temperature, time, wire diameter and environmental conditions.

THERMOCOUPLE ASSEMBLIES

Conventional Thermocouples. Some typical thermocouple assemblies employed in industrial applications are shown in Fig. 4. In the assembly shown at the top of this figure, a closed-end pipe protection tube may be substituted for the nipple

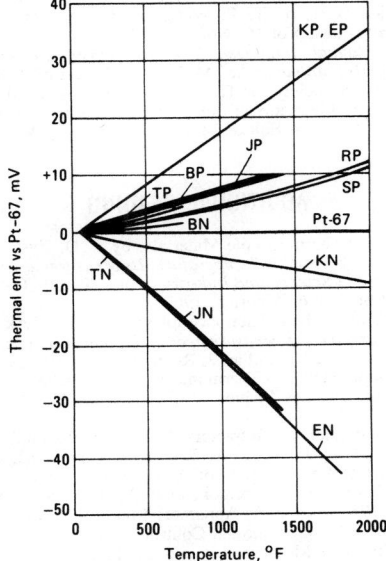

Adapted from Ref 2.

Fig. 2. Thermal emf of standard thermoelements

- Pt-10Rh/Pt thermocouple 630.74 °C to 1064.43 °C (gold point)
- Above 1064.43 °C, defined in terms of the Planck radiation law using 1064.43 as a reference temperature (optical pyrometer).

Methods of Thermocouple Calibration. Initial calibration of a thermocouple can be done by any of the following methods:

- Freezing-point calibration
- Direct thermoelement emf measurement vs platinum
- Thermoelement comparison method
- Calibration of thermocouples by comparison methods.

In the freezing-point method of calibration, the emf output of the thermocouple as a whole is measured during the cooling cycle of molten pure metals. In the second and third methods, the emf of both the positive and negative thermoelements are individually measured versus platinum or another calibrated standard.

closely approximates the thermodynamic temperature; the difference is within the limits of the present accuracy of measurement.

The IPTS 68 (amended 1975) is based on the assigned values of the temperatures of 13 reproducible equilibrium states (defining fixed points) and on standard instruments calibrated at these temperatures. Interpolation is provided by formulas used to establish the relations between indications on standard instruments and values of International Practical Temperature.

The IPTS 68 uses both International Practical Kelvin Temperature, symbol T_{68}, and International Practical Celsius Temperature, symbol t_{68}. The relation between T_{68} and t_{68} is the same as that between T and t on the Thermodynamic Scale—that is, $t_{68} = T_{68} - 273.15$ K. The units of T_{68} and t_{68} are the kelvin symbol, K, and the degree Celsius symbol, °C, as in the case of thermodynamic temperature T and Celsius temperature t. The standard instruments used are:

- Platinum resistance thermometer 13.81 K to 630.74 °C

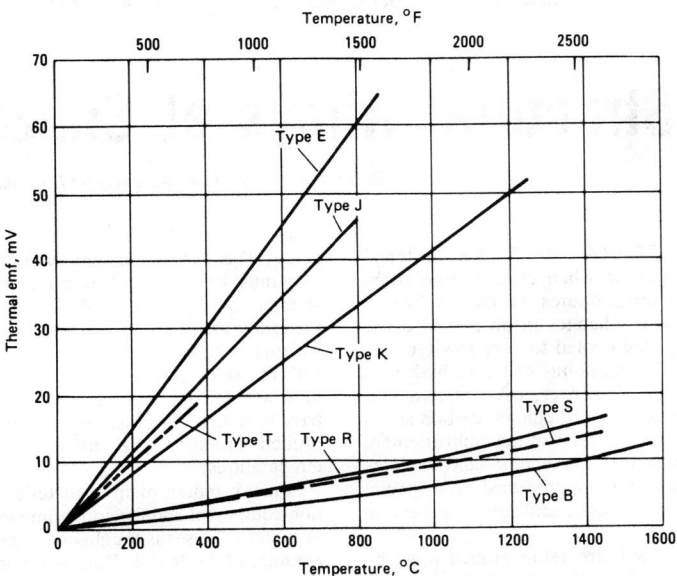

Thermal emf plots are based on IPTS-68 (1974).

Fig. 3. Thermal emf curves for ISA standard thermocouples

Table 2. Initial calibration tolerances for thermocouples when the reference junction is at 0 °C
Adapted from ANSI MC96.1.

Thermocouple type	Temperature range, °C	Initial calibration tolerance Standard (whichever is greater)	Special (whichever is greater)
T	0 to 350	±1 °C or ±0.75%	±0.5 °C or ±0.4%
J	0 to 750	±2.2 °C or ±0.75%	±1.1 °C or ±0.4%
E	0 to 900	±1.7 °C or ±0.5%	±1 °C or ±0.4%
K	0 to 1250	±2.2 °C or ±0.75%	±1.1 °C or ±0.4%
R or S	0 to 1450	±1.5 °C or ±0.25%	±0.6 °C or ±0.1%
B	800 to 1700	±0.5%	. . .
T(a)	−200 to 0 °C	±1 °C or ±1.5%	(b)
E(a)	−200 to 0 °C	±1.7 °C or ±1%	(b)
K(a)	−200 to 0 °C	±2.2 °C or ±2%	(b)

(a) Thermocouples and thermocouple materials are normally supplied to meet the limits of error specified in the table for temperatures above 0 °C. The same materials, however, may not fall within the subzero limits of error given in the second section of the table. If materials are required to meet the subzero limits, the purchase order must so state. Selection of materials usually will be required. (b) Little information is available to justify establishment of special limits of error for subzero temperatures. Limited experience suggests the following limits for types E and T thermocouples: Type E, −200 to 0 °C ± 1 °C or ± 0.5%; Type T, −200 to 0 °C ± 0.5 °C or ± 0.8%. These limits are given only as a guide for discussion between purchaser and supplier. Due to the characteristics of the materials, subzero limits of error for type J thermocouples and special subzero limits for type K thermocouples are not listed.

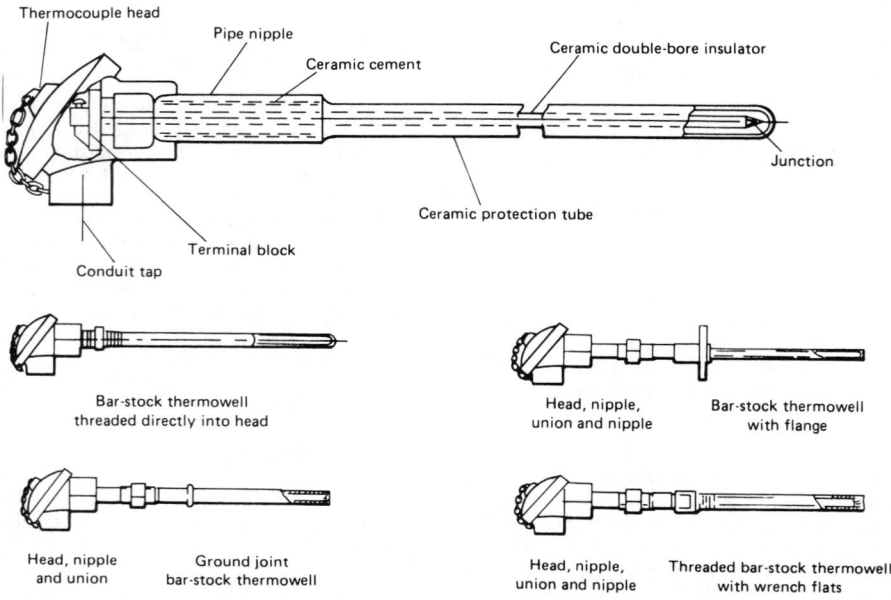

Adapted from Ref 5.
Fig. 4. Typical industrial thermocouples insulated with hard-fired ceramics

and ceramic protection tube in base-metal thermocouple applications. For additional details, see also ANSI MC96.1, "Temperature Measurement Thermocouples," and suppliers' literature.

Metal-Sheathed Thermocouples. In metal-sheathed couples, the wires are insulated from each other and from the sheath by means of compressed pure refractory oxide powder. The resulting assembly (thermocouple wires, oxide powder and integral sheath) is flexible enough to be formed around a diameter equal to four times that of the assembly, without damage.

REFERENCES

1. "EMF Measurements," by T. P. Wang: Technical Paper MF77-958, Society of Manufacturing Engineers, 1977

2. Temperature Sensors, by T. P. Wang: *Instrument and Control Systems*, Vol 40, 1967, p 100

3. A New Stable Nickel-Base Thermocouple, by C. D. Starr and T. P. Wang: *ASTM Journal of Testing and Evaluation*, Vol 21, 1976, p 42

4. The HI BX, A New Type B Thermocouple Extension Wire, by T. P. Wang and C. D. Starr: *ISA Transactions*, Vol 16, No. 3, 1977, p 85

5. *Manual on the Use of Thermocouples in Temperature Measurement*: ASTM STP 470B (revised 1980), American Society for Testing and Materials, 1980

6. The International Practical Temperature Scale of 1968, Amended Edition of 1975: *Metrologia*, Vol 12, 1976

ADDITIONAL READING

Newer Thermocouple Materials, by E. D. Zysk and A. R. Robertson: in *Temperature, Its Measurement and Control in Science and Industry*, Vol 4, Part 3, Instrument Society of America, 1967

Platinum Metal Thermocouples, by E. D. Zysk: in *Temperature, Its Measurement and Control in Science and Industry*, Vol 3, Part 2, Reinhold, New York, 1962

"Noble Metals in Thermometry," by E. D. Zysk: Engelhard Industries Technical Bulletin, Vol 5, No. 3, Dec 1964

"Calibration of Refractory Metal Thermocouples," by E. D. Zysk and D. A. Toenshoff: Paper #12, 11-4-66, Instrument Society of America, Oct 1966.

"Thermocouple Thermometers": PMC Standard No. 8-10-1963, Scientific Apparatus Makers Association, Process Measurement and Control Section, 1963

"Precision Measurement and Calibration Temperature": NBS Special Publication, Vol II, National Bureau of Standards, 1968

Structural Alloys at Subzero Temperatures

By James E. Campbell, Metallurgical Consultant, Columbus, Ohio

ALL STRUCTURAL METALS undergo changes in properties when cooled from room temperature to temperatures in the "subzero" range. The greatest changes in properties occur when the metals are cooled to very low temperatures near the boiling points of liquid hydrogen and liquid helium. However, even at temperatures encountered in arctic regions, carbon steels become embrittled. Because of the embrittlement problem, components of structures and vehicles for arctic regions are often produced from special steels that do not become embrittled at service temperatures.

The effects of subzero temperatures must be considered in selection of materials for aircraft, missiles and space vehicles that are exposed to the temperatures of upper altitudes and outer space. These structures are "weight limited," so they must be fabricated from materials with high strength-to-weight ratios. At the same time, these materials are required to retain high levels of fracture toughness at all exposure temperatures for "fail safe" service. Certain aluminum alloys, cold worked stainless steels, and titanium alloys have been used successfully in fabricating weight-limited structures that are exposed to subzero temperatures.

A major requirement of materials for liquefaction equipment and for containment and transport of liquefied gases is toughness at the boiling temperature of the liquid. Boiling points for some of the liquefied gases are: methane, −161 °C (−258 °F); oxygen, −183 °C (−297 °F); nitrogen, −196 °C (−320 °F); hydrogen, −253 °C (−423 °F); and helium, −269 °C (−452 °F). Temperatures below −150 °C (−238 °F) often are identified as cryogenic temperatures. Aluminum alloys, copper and copper alloys, nickel-alloy steels, stainless steels and high-nickel alloys have been used successfully for liquefaction, containment and transport of these liquids.

Certain metallic materials become superconductors at temperatures below about −260 °C (−436 °F). To achieve these temperatures, superconducting devices must be cooled with liquid helium. Therefore, structural materials selected for cryogenic components of superconducting machinery, magnets and transmission systems must be suitable for use at liquid-helium temperature. Furthermore, these components are subjected to high stresses in service, and thus safeguards must be employed to minimize service failures. In order to obtain the required strength and toughness along with a reasonable degree of fabricability, certain austenitic stainless steels and high-nickel alloys (superalloys) usually are designated for highly stressed components of structures that will be cooled with liquid helium.

In this article, tables of typical room-temperature and subzero-temperature mechanical properties are presented for:

- Aluminum alloys
- Copper and copper alloys
- High-nickel alloys, including superalloys
- Ferritic alloy steels
- Stainless steels
- Titanium and titanium alloys.

In general, yield strengths and tensile strengths of structural alloys increase as the exposure temperature is decreased. The effect of low-temperature exposure on the ductility and toughness of these alloys depends on alloy composition.

Values of Young's modulus generally increase, and values of Poisson's ratio generally decrease, as the testing temperature for the above alloys is decreased.

The numerical values presented in the tables have been averaged from available property data from the literature and represent state-of-the-art information. Because only limited amounts of test data are available for most metals and alloys, the numerical values represent only "composite" averages and are not statistical mean values or design values. Because of space limitations, tabulated data are limited to values obtained in tests of longitudinal specimens if longitudinal data are available. Corresponding data obtained on transverse specimens usually do not vary significantly from the longitudinal data for these alloys. Additional data on all of the metals discussed in this article and on other metals are presented in Ref 1, which gives 99 references to sources of original data.

ALUMINUM ALLOYS

Aluminum alloys represent a very important class of structural metals for subzero-temperature applications. Aluminum and aluminum alloys have face-centered-cubic (fcc) crystal structures. Most fcc metals retain good ductility at subzero temperatures. Aluminum can be strengthened by alloying and heat treatment while still retaining good ductility along with adequate toughness at subzero temperatures. Nominal compositions of aluminum alloys that have been considered for subzero service are presented in Table 1.

Tensile Properties. Typical tensile properties of aluminum alloys at room temperature and at subzero temperatures are presented in Tables 2 to 7. Data are presented for various product forms and tempers because these variables influence properties.

As shown in Table 2, aluminum alloy 2014-T6 has relatively high strength at room temperature and at subzero temperatures. It retains about the same ductility at liquid-hydrogen temperature as at room temperature. Aluminum alloy 2024 (Table 2) also has relatively high strength at both room and subzero temperatures.

Tensile properties of aluminum alloy 2219 are somewhat lower than for 2014-T6, but 2219 has better toughness at room and subzero temperatures. Tensile properties of alloy 2219 weld-

ments are presented in Table 3. Because of its favorable properties at cryogenic temperatures, alloy 2219-T87 plate has been used for the liquid oxygen and liquid hydrogen tanks for the space shuttle. Sufficient testing has been conducted on the 2219-T87 alloy to permit an evaluation of the design-allowable properties for sheet, plate and weldments (Ref 2).

Aluminum alloy 3003 is used in fabrication of brazed heat exchangers and other equipment employed in gas-liquefaction plants. Tensile properties of alloy 3003 as annealed, as processed to two H tempers, and as welded are given in Table

4. For this alloy, ease of fabrication, along with good ductility and notch toughness at room and subzero temperatures, are the most significant properties.

Tensile properties of aluminum alloy 5083 plate and weldments are presented in Table 5. This alloy is not heat treatable; for maximum toughness, it is used in the annealed (O) condition. It is readily weldable, and the yield strength of the weld metal is nearly equal to that of the base metal. Design properties of 5083-O sheet, plate and weldments to −162 °C (−260 °F) are given in Ref 3.

Table 1. Nominal compositions of aluminum alloys

Alloy designation	Nominal composition, %								
	Si	Cu	Mn	Mg	Cr	Zn	Ti	Zr	Others
Wrought alloys									
2014	0.8	4.4	0.8	0.5	...	...	...	...	...
2024	...	4.4	0.6	1.5	...	...	...	...	...
2219	...	6.3	0.3	...	...	...	0.06	0.18	0.1V
3003	...	0.12	1.2	...	...	...	...	...	...
5083	...	...	0.7	4.4	0.15	...	...	...	...
5456	...	...	0.8	5.1	0.12	...	...	...	...
6061	0.6	0.28	...	1.0	0.20	...	...	...	...
7005	...	...	0.45	1.4	0.13	4.5	0.04	0.14	...
7039	0.1	0.05	0.25	2.8	0.20	3.0	0.05	...	0.2Fe
7075	...	1.6	...	2.5	0.23	5.6	...	...	...
Cast alloys									
355	5.0	1.2	...	0.5	...	...	...	...	...
C355	5.0	1.3	...	0.5	...	...	...	...	...
356	7.0	...	...	0.3	...	...	...	...	...
A356	7.0	...	...	0.3	...	...	...	...	...

Table 2. Typical tensile properties of 2000 series aluminum alloys

Temperature		Tensile strength		Yield strength		Elongation, %	Reduction in area, %
°C	°F	MPa	ksi	MPa	ksi		
2014 sheet, T6 temper, longitudinal orientation							
24	75	490	71.1	440	63.6	9.6	...
−78	−108	515	74.7	475	68.8	9.6	...
−196	−320	580	84.0	510	74.2	11.4	...
−253	−423	685	99.7	565	82.0	12.2	...
−269	−452	670	97.0	570	82.5	10.4	...
2014 plate, T651 temper, longitudinal orientation							
24	75	465	67.6	430	62.2	11.2	33
−196	−320	580	84.0	525	76.0	12.0	22
−253	−423	660	95.6	555	80.2	15.0	23
−269	−452	660	95.4	565	81.8	12.8	20
2024 sheet in T4 temper, longitudinal orientation							
24	75	465	67.7	295	42.8	19	...
−78	−108	480	69.8	300	43.7	22	...
−196	−320	585	84.9	375	54.1	27	...
−253	−423	740	107	505	73.3	16	...
2024 plate in T351 temper, longitudinal orientation							
24	75	465	67.1	345	50.3	22	28
−253	−423	740	107	530	77.2	22	20
2024 plate in T4 temper, longitudinal orientation							
24	75	465	67.7	365	53.3	17	17
−78	−108	480	69.6	375	54.7	17	4
−196	−320	555	80.8	460	66.5	11	11
−253	−423	650	94.4	555	80.6	8	9
2024 plate in T851 temper, longitudinal orientation							
24	75	495	72.0	455	65.8	8	17
−78	−108	535	77.8	490	71.3	6	14
−196	−320	690	99.8	575	83.3	8	13
−269	−452	715	104	625	90.7	9	14
2219 sheet in T62 temper, longitudinal orientation							
24	75	415	60.2	290	41.9	10	...
−78	−108	460	66.5	345	50.2	11	...
−196	−320	545	78.8	370	53.6	14	...
−253	−423	650	94.5	390	56.3	23	...
2219 plate in T87 temper, longitudinal orientation							
24	75	465	67.8	380	55.2	11	26
−78	−108	515	74.8	420	61.0	14	27
−196	−320	585	84.5	460	66.6	13	25
−253	−423	675	97.6	490	71.0	14	21
−269	−452	675	97.8	510	74.2	15	23

Table 3. Typical tensile properties of alloy 2219 weldments using 2319 filler metal

Temperature °C	°F	Tensile strength MPa	ksi	Yield strength MPa	ksi	Elongation, %	Reduction in area, %
Sheet in T62 temper, longitudinal orientation(a)(b)							
24	75	315	45.4	210	30.5	3	...
−196	−320	370	54.0	275	39.7	2	...
−253	−423	435	63.2	345	50.0	6	...
Sheet in T62 temper, longitudinal orientation(b)(c)							
24	75	415	60.5	300	43.5	8	...
−196	−320	520	75.2	355	51.8	8	...
−253	−423	565	81.6	405	58.5	4	...
Plate in T87 temper, longitudinal orientation(a)(d)							
24	75	280	40.9	170	24.6	5	27
−78	−108	285	41.4	170	25.0	4	...
−196	−320	395	57.0	205	29.6	8	18
−253	−423	425	61.9	270	38.8	4	...
Plate in T87 temper, longitudinal orientation(a)(b)							
24	75	275	40.0	165	23.8	6	27
−78	−108	285	41.0	170	24.8	4	...
−196	−320	410	59.3	190	27.8	9	22

(a) As welded. (b) Gas tungsten-arc welding process. (c) Reheat treated to T62. (d) Gas metal-arc welding process.

Table 4. Typical tensile properties of alloy 3003 base metal and weldment

Temperature °C	°F	Tensile strength MPa	ksi	Yield strength MPa	ksi	Elongation, %	Reduction in area, %
Plate, O temper, longitudinal orientation							
24	75	110	16.1	41	6.0	40	...
−78	−108	140	20.1	48	7.0	43	...
−196	−320	230	33.7	59	8.5	46	...
−253	−423	365	53.0	69	10.0	48	...
Plate, H12 temper, longitudinal orientation							
24	75	130	19.0	110	16.0	...	...
−78	−108	150	22.0	125	18.0	...	...
−196	−320	235	34.0	160	23.0	...	...
Bar, H14 temper, longitudinal orientation							
24	75	160	22.9	145	21.1	17	68
−78	−108	175	25.3	155	22.3	18	59
−196	−320	250	36.6	180	25.9	32	56
−269	−452	400	58.1	210	30.1	32	49
Plate, H112 temper, as welded(a)							
24	75	110	16.1	52	7.6	24	67
−78	−108	135	19.3	57	8.3	26	66
−196	−320	230	33.7	74	10.8	31	52
−269	−452	350	51.1	130	18.5	28	25

(a) Gas metal-arc welded, 1100 filler.

Table 5. Typical tensile properties of 5000 series aluminum alloy plate and weldments

Temperature °C	°F	Tensile strength MPa	ksi	Yield strength MPa	ksi	Elongation, %	Reduction in area, %
5083-H113 plate, longitudinal orientation							
24	75	335	48.6	235	34.2	15	23
−196	−320	465	67.1	275	39.6	31	31
−253	−423	620	90.0	305	43.9	30	24
−269	−452	590	85.8	280	40.5	29	28
GMA weld in 5083-H113 plate, 5183 filler metal, longitudinal specimen(a)							
24	75	295	42.5	145	20.9	16	30
−196	−320	420	60.8	175	25.1	21	25
−253	−423	445	64.2	185	27.1	11	14
5456-H321 plate, longitudinal orientation							
24	75	355	51.2	235	34.1	14	11
−78	−108	380	55.0	235	34.3	20	31
−196	−320	490	71.3	275	39.9	26	28
−253	−423	635	92.2	305	43.9	18	17
−269	−452	640	92.6	320	46.5	24	25
GMA weld in 5456-H321 plate, 5556 filler metal, longitudinal specimen(a)							
24	75	310	44.6	155	22.5	13	23
−78	−108	310	45.2	155	22.5	16	28
−196	−320	405	59.0	180	26.2	14	18

(a) As welded.

Ships of current design for transport of liquefied natural gas (LNG) contain five spherical tanks made of aluminum alloy 5083-O built into the structure of each ship.

Aluminum alloy 5456 (Table 5) is another non-heat-treatable alloy that has good welding characteristics and good ductility and toughness at cryogenic temperatures. It is an alternative to alloy 5083.

Aluminum alloy 6061 is usually used in the T6 temper. Specimens from sheet, plate, bar and forgings, all in T6 temper, have about the same strength and ductility at the same subzero temperature (data for sheet and plate are given in Table 6).

Aluminum alloy 7005-T5351 extrusions have higher strength than 6061-T6 plate, but about the same ductility at the same temperatures (Table 7).

Aluminum alloy 7039 plate in the T6 temper retains good ductility and notch toughness at cryogenic temperatures. Specimens from weldments also retain good ductility and notch toughness (Table 7). Alloy 7039-T6 has been recommended for cryogenic pressure vessels.

Aluminum alloy 7075 is representative of the high-strength nonweldable 7000 series alloys. It is primarily an aircraft and aerospace alloy and has relatively low fracture toughness in the T6 temper. Normally it is not used in applications involving cryogenic temperatures, but the decision to use 7075-T6 plate for the intertank skirt between the liquid oxygen and liquid hydrogen sections of the external tank of the space shuttle was probably based on the high strength of this alloy. Mechanical fasteners are used in joints between the skirt and the tanks. Tensile properties of 7075 sheet and plate at subzero temperatures are presented in Table 7. Processing of alloy 7075 to the T7351 temper improves its ductility and notch toughness at cryogenic temperatures.

Aluminum alloy castings often are specified for pump parts and for other components that are more readily produced by casting than by other fabricating processes. A variety of alloys, casting methods and heat treatments can be applied in producing aluminum alloy castings for subzero applications. With suitable designs and quality-control standards, cast aluminum components can be produced that are satisfactory for many subzero applications, but the ductility of these alloys (355 and 356 series) is relatively low at room temperature and at subzero temperatures.

Fracture Toughness. Data on fracture toughness of several aluminum alloys at room and subzero temperatures are summarized in Table 8. The room-temperature yield strengths for the alloys in this table range from 142 to 536 MPa (20.6 to 77.7 ksi), and room-temperature plane-strain fracture-toughness values for both bend and compact (CT) specimens range from 22.3 to 39.9 MPa·m$^{1/2}$ (20.3 to 36.3 ksi·in.$^{1/2}$).

Of the alloys listed in Table 8, 5083-O has substantially greater toughness than the others and its toughness increases as exposure temperature decreases. Of the other alloys, 2219-T87 has the best combination of strength and fracture toughness, both at room temperature and at −196 °C (−320 °F), of the alloys that can be readily welded.

Alloy 6061-T651 has good fracture toughness at room temperature and at −196 °C (−320 °F), but its yield strength is lower than that of alloy 2219-T87. Alloy 7039 also is weldable and has a good combination of strength and fracture toughness at room temperature and at −196 °C. Alloy 2124 is similar to 2024, but with a higher-purity base and special processing for improved fracture toughness. Tensile properties of 2124-T851 at subzero temperatures can be expected to be similar to those of 2024-T851.

Fatigue Properties. Data on rates of fatigue-crack growth in aluminum alloys at subzero temperatures are limited, but the available data show that fatigue-crack-growth rates at subzero temperatures are lower than those at room temperature at the same ΔK levels.

Results of axial and flexural fatigue tests at 10^6 cycles on aluminum alloy specimens at room temperature and at subzero temperatures indicate that, for a fatigue life of 10^6 cycles, fatigue strength is higher at subzero temperatures than at room temperature for each alloy. This trend is not necessarily valid for tests at higher stress levels and shorter fatigue lives, but, at 10^6 cycles, results are consistent with the effect of subzero temperatures on tensile strength.

COPPER AND COPPER ALLOYS

Copper and copper alloys have fcc crystal structures and retain high degrees of ductility and toughness at subzero temperatures. Nominal compositions of several copper alloys that have been considered for use at subzero temperatures are presented in Table 9.

The major use of copper at cryogenic temperatures is for the stabilizer components of the windings in superconducting magnets, solenoids and power cables. High-purity high-conductivity copper is usually specified for this application. The purpose of such a stabilizer is to stabilize the superconducting condition in the windings.

Tensile Properties. Tensile properties of copper and copper alloys at subzero temperatures are summarized in Table 10. These alloys have good strength, ductility and notch toughness even at liquid-helium temperature.

Fatigue Properties. Results of flexural fatigue testing of several higher-strength copper alloys at subzero temperatures show that fatigue strengths of both unnotched and notched specimens increase as testing temperature is decreased.

HIGH-NICKEL ALLOYS

Nickel is another fcc metal that retains good ductility and toughness at subzero temperatures. Compositions of selected high-nickel alloys are given in Table 11. Some of these alloys exhibit excellent combinations of strength, ductility and toughness over the entire range of subzero temperatures. These alloys have been selected for some of the most critical structural components in recent designs for large superconducting motors, generators and other energy-related equipment. They also are suitable for service at elevated temperatures and may be used in applications involving exposure to both subzero and elevated temperatures.

Tensile Properties. Typical tensile properties of high-nickel alloys and of high-nickel alloy weldments, at room temperature and at subzero temperatures, are presented in Table 12. All of the alloys in Table 12 retain good ductility at the lowest testing temperature. Notch tensile data indicate that they also retain good notch toughness at the lowest testing temperature. Ductility of heat treated weldments is lower, but is not affected by extended exposure to subzero temperatures.

Table 6. Typical tensile properties of aluminum alloy 6061 sheet, plate and weldments

Temperature °C	°F	Tensile strength MPa	ksi	Yield strength MPa	ksi	Elongation, %	Reduction in area, %
Sheet, T6 temper, longitudinal orientation							
24	75	320	46.3	290	42.2	12	...
−78	−108	350	50.8	310	44.8	12	...
−196	−320	425	61.8	340	49.6	18	...
−253	−423	495	72.0	365	52.6	26	...
Plate, T6 temper, longitudinal orientation							
24	75	320	46.5	295	43.0	16	36
−78	−108	360	52.5	325	46.8	21	44
−196	−320	445	64.2	360	52.3	25	40
−253	−423	525	76.2	375	54.3	26	36
Plate, T651 temper, longitudinal orientation							
24	75	310	44.9	290	42.2	16	50
−196	−320	400	58.3	335	48.9	23	48
−269	−452	485	70.1	380	55.0	26	42
GMA weld in T6 temper sheet, 4043 filler metal, longitudinal specimen(a)							
24	75	220	32.2	160	23.2	5	...
−196	−320	325	47.4	200	28.8	10	...
−253	−423	450	65.5	220	32.0	10	...
GMA weld in T6 temper plate, 5356 filler metal, longitudinal specimen(a)							
24	75	225	32.7	155	22.6	8	31
−78	−108	255	37.1	170	24.7	9	36
−196	−320	325	47.0	190	27.3	14	39
−269	−452	400	57.7	245	35.3	14	24
GMA weld in T6 temper plate, 5356 filler metal, longitudinal specimen(b)							
24	75	280	40.5	200	29.3	10	33
−78	−108	320	46.4	240	35.1	12	44
−196	−320	395	57.1	235	33.9	20	29
−269	−452	475	69.1	305	44.5	19	24

(a) As welded. (b) Heat treated to T6.

Table 7. Typical tensile properties of 7000 series aluminum alloys and weldments

Temperature °C	°F	Tensile strength MPa	ksi	Yield strength MPa	ksi	Elongation, %	Reduction in area, %
7005 extrusion, T5351 temper, longitudinal orientation							
24	75	425	62.0	380	55.0	15	43
−78	−108	465	67.8	405	58.6	14	30
−196	−320	580	83.9	465	67.5	17	27
−253	−423	705	102	505	73.4	18	28
−269	−452	675	97.6	520	75.6	17	22
7039 sheet, T6 temper, longitudinal orientation							
24	75	455	66.0	410	59.8	11	...
−78	−108	510	73.9	460	66.7	12	...
−196	−320	575	83.2	495	72.0	15	...
−253	−423	665	96.2	535	77.4	14	...
7039 plate, T6 and T61 tempers, longitudinal orientation							
24	75	450	65.2	400	57.7	13	...
−78	−108	490	71.0	440	63.6	18	...
−196	−320	575	83.5	480	69.6	14	...
−253	−423	660	95.9	525	76.0	13	...
GMA weld in 7039 sheet, T6 temper, 5183 filler metal(a)							
24	75	365	53.2	240	34.8	7	...
−196	−320	390	56.7	300	43.4	4	...
GMA weld in 7039 plate, T61 temper, 5039 filler metal(a)							
24	75	355	51.8	215	31.1	9	23
−196	−320	400	58.3	250	36.1	9	12
−253	−423	375	54.5	...	...	2	4
7075 sheet, T6 temper, longitudinal orientation							
24	75	555	80.8	500	72.4	10	...
−73	−100	585	85.1	525	76.2	10	...
−196	−320	680	98.6	610	88.7	12	...
−253	−423	785	114	705	102	7	...
7075 plate, T651 temper, longitudinal orientation							
24	75	580	84.2	530	77.1	10	14
−78	−108	595	86.2	595	86.0	10	11
−196	−320	705	102	650	94.4	7	10
−253	−423	770	112	710	103	5	...
−269	−452	825	120	770	112	6	9
7075 plate, T7351 temper, longitudinal orientation							
24	75	525	76.2	455	66.2	10	22
−78	−108	575	83.4	500	72.3	10	17
−196	−320	675	98.2	570	82.5	11	14
−269	−452	760	110	605	88.1	11	12

(a) Naturally aged for 15 to 30 days.

Table 8. Fracture toughness of aluminum alloy plate

Alloy and condition	Room-temperature yield strength MPa	ksi	Specimen design	Orientation	24 °C (75 °F) MPa·m$^{1/2}$	ksi·in.$^{1/2}$	−196 °C (−320 °F) MPa·m$^{1/2}$	ksi·in.$^{1/2}$	−253 °C (−423 °F) MPa·m$^{1/2}$	ksi·in.$^{1/2}$	−269 °C (−452 °F) MPa·m$^{1/2}$	ksi·in.$^{1/2}$
2014-T651	432	62.7	Bend	T-L	...23.2	21.2	28.7	26.1	...	...	...	...
2024-T851	444	64.4	Bend	T-L	...22.3	20.3	24.4	22.2	...	...	...	...
2124-T851(a)	455	66.0	CT	T-L	...26.9	24.5	32.0	29.1	...	...	...	...
	435	63.1	CT	L-T	...29.2	26.6	35.0	31.9	...	...	...	...
	420	60.9	CT	S-L	...22.7	20.7	24.3	22.1	...	...	...	...
2219-T87	382	55.4	Bend	T-S	...39.9	36.3	46.5	42.4	52.5	48.0	...	...
			CT	T-S	...28.8	26.2	34.5	31.4	37.2	34.0	...	...
	412	59.6	CT		...30.8	28.1	35.9	32.7	...	...	...	...
5083-O	142	20.6	CT	T-L	...27.0(b)	24.6(b)	43.4(b)	39.5(b)	...	...	48.0(b)	43.7(b)
6061-T651	289	41.9	Bend	T-L	...29.1	26.5	41.6	37.9	...	...	...	...
7039-T6	381	55.3	Bend	T-L	...32.3	29.4	33.5	30.5	...	...	...	...
7075-T651	536	77.7	Bend	T-L	...22.5	20.5	27.6	25.1	...	...	...	...
7075-T7351	403	58.5	Bend	T-L	...35.9	32.7	32.1	29.2	...	...	...	...
7075-T7351	392	56.8	Bend	T-L	...31.0	28.2	30.9	28.1	...	...	...	...

(a) 2124 is similar to 2024, but with higher-purity base and special processing to improve fracture toughness. (b) $K_{Ic}(J)$.

Table 9. Nominal compositions of copper and copper alloys

UNS number	Common name	Composition, % Cu	Zn	Others
C10200	Oxygen-free copper	99.95(Cu + Ag)	...	...
C12200	Phosphorus deoxidized, high residual phosphorus copper	99.90	...	0.02 P
C17200	Beryllium copper	98.1	...	1.9 Be
C26000	Cartridge brass, 70%	70.00	30.00	...

Table 10. Typical tensile properties of copper and copper alloys(a)

Temperature °C	°F	Tensile strength MPa	ksi	Yield strength MPa	ksi	Elongation, %	Reduction in area, %
C10200 bar, O61 temper(b)							
24	75	 220	32.2	75	10.9	54	86
−78	−108	 270	39.0	80	11.6	53	84
−196	−320	 360	52.2	88	12.8	60	84
−253	−423	 420	60.7	90	13.1	69	83
C12200 bar, O61 temper(b)							
24	75	 215	31.3	46	6.7	45	76
−78	−108	 265	38.3	46	6.6	56	87
−196	−320	 350	50.6	51	7.4	62	84
−253	−423	 440	63.8	58	8.4	68	83
−269	−452	 415	60.4	54	7.9	65	81
C26000 bar, H03 temper(c)							
24	75	 655	95.2	420	60.9	14	58
−78	−108	 695	101	445	64.3	17	62
−196	−320	 805	117	475	68.6	28	63
−253	−423	 910	132	505	73.4	32	58
C17200 sheet, TD02 temper(d)							
24	75	 620	90	550	80	15	...
−78	−108	 655	95	600	87	20	...
−196	−320	 805	117	690	100	37	...
−253	−423	 945	137	750	109	45	...
C17200 sheet, TH02 temper(e)							
24	75	 1320	191	1140	166	2.8	...
−253	−423	 1640	238	1230	178	3.5	...

(a) Longitudinal orientation. (b) Annealed. (c) $^3/_4$ hard. (d) Solution treated, cold worked to $^1/_2$ hard. (e) Aged 2 h at 315 °C (600 °F).

Table 11. Nominal compositions of high-nickel alloys

UNS number	Alloy designation	Ni	Cr	Fe	Mn	Si	C	Others
N05500	Hastelloy C	......Rem	15.5	5.5	0.5	0.5	0.07	1.5Co, 16Mo, 4.0W
N06600	Inconel 600	......Rem	15.8	7.2	0.2	0.2	0.04	0.10Cu
N09706	Inconel 706	......39 to 44	16	Rem	0.10	0.10	0.04	0.35 max Al, 3.0 (Nb + Ta), 1.7Ti
N07718	Inconel 718	......Rem	18.6	18.5	...	...	0.04	0.4Al, 0.9Ti, 5.0Nb, 3.1Mo
N07750	Inconel X-750	....Rem	15.0	6.8	0.7	...	0.04	0.8Al, 2.5Ti, 0.85Nb

Fracture Toughness. Fracture-toughness data for several high-nickel alloys and weldments at room temperature and at subzero temperatures are presented in Table 13. Because of the high toughness of most of these alloys, all of the fracture-toughness data except that for Inconel 718 solution treated and double aged (STDA) were obtained by the J-integral method and converted to $K_{Ic}(J)$.

These alloys normally retain a high degree of toughness at temperatures as low as −269 °C (−452 °F). Fracture toughness of fusion and heat-affected zones of welds in heat treated weldments tends to be lower than that of the base metal.

Fatigue Properties. For the high-nickel alloys, fatigue-crack-growth rates at subzero temperatures are either equal to or lower than the rates at room temperature for the same ΔK values. For design purposes, the use of room temperature fatigue-crack-growth-rate data for subzero applications is feasible.

Results of fatigue-life tests at 10^6 cycles on axial and flexural specimens of several high-nickel alloys at room temperature and at subzero temperatures indicate that, at 10^6 cycles, the fatigue strengths of these alloys are higher at subzero temperatures that at room temperature.

FERRITIC ALLOY STEELS

Ferritic steels exhibit a transition in toughness from ductile, high-energy fracture at temperatures above the transition temperature to brittle, low-energy fracture at temperatures below the transition temperature. This is characteristic of metals with body-centered-cubic (bcc) crystal structures. The usual method of measuring transition temperature in ferritic steels is the Charpy V-notch impact test, in which specimens are tested over a range of subzero temperatures (ASTM method A370). The transition temperatures determined by this method do not indicate at what temperatures brittle fracture will occur in structural components or threaded fasteners, but they do provide a means of rating the notch toughness of steels at subzero temperatures. This method provides a means of setting minimum standards for steels intended for use at subzero temperatures. These minimum standards are presented in ASTM A20 (standard specification for pressure vessels). ASTM specifications A203, A353, A442, A516, A517, A537, A553, A612, A645, A662 and A724 describe steel plates with minimum Charpy V-notch energy or lateral expansion requirements at testing temperatures from −26 to −196 °C (−15 to −320 °F). ASTM specifications A353, A553, and A645 describe alloy steel plates with minimum Charpy energy requirements at testing temperatures from −170 to −196 °C (−275 to −320 °F).

The following ASTM specifications describe other steel products for subzero service:

A333 Seamless and welded steel pipe
A334 Seamless and welded carbon and alloy steel tubes
A350 Forged or rolled carbon and alloy steel flanges, forged fittings, and valves
A352 Ferritic steel castings
A420 Pipe and fittings of wrought carbon steel and alloy steel
A522 Forged or rolled 8 and 9% nickel steel flanges, fittings and valves
A671 Electric-fusion-welded steel pipe

In addition to the steels listed above, a number of proprietary ferritic steels have been developed by several steel producers to meet certain requirements for service at subzero temperatures as low as about −112 °C (−170 °F).

ASTM specification A203 covers ferritic alloy steel plates with nominal nickel contents of 2.25 and 3.50%. For 1-in. plate, minimum Charpy impact requirements are applicable for the 2.25% Ni grades at −68 °C (−90 °F) and for the 3.50% Ni grades at −101 °C (−150 °F).

ASTM specification A645 covers ferritic alloy steel plates containing 4.75 to 5.25% Ni and 0.20 to 0.35% Mo. For 1-in. plate, minimum Charpy impact requirements are designated at −170 °C (−275 °F) for hardened, temperized and reversion-annealed plate.

Double normalized and tempered 9% nickel steel is covered by ASTM A353, and quenched and tempered 8% and 9% nickel steels are covered by ASTM A553 (types I and II). For quenched and tempered material, the minimum lateral expansion in Charpy V-notch impact tests is 0.38 mm (0.015 in.). Charpy tests on 9% Ni steel (type I) are conducted at −195 °C (−320 °F); for 8% Ni steel (type II), they are conducted at −170 °C (−275 °F).

Carbon and alloy steel castings for subzero-temperature service are covered by ASTM Standard Specification A757. These specifications recognize the effect of nickel content in reducing the transition temperatures of nickel steels.

In order to ensure that a given steel structure will have a reasonable degree of toughness at the minimum service temperature, it is recommended that the appropriate ASTM specification (or ASME SA20) be applied in designating the steels to be used. Minimum strength levels also are given in these specifications.

For applications involving exposure to temperatures from 0 to −196 °C (+32 to −320 °F), the ferritic nickel steels usually are considered first, if they have sufficient corrosion resistance. Such applications include storage tanks for liquefied hydrocarbon gases and structures and machinery designed for use in cold regions.

Tensile Properties. Typical tensile properties of 5% and 9% Ni steels at room temperature and at subzero temperatures are presented in Table 14. Yield and tensile strengths increase as testing temperature is decreased. These steels remain ductile at the lowest testing temperatures.

Fracture Toughness. Ferritic nickel steels are too tough at room temperature for valid fracture-toughness (K_{Ic}) data to be obtained on specimens of reasonable size, but limited fracture-toughness data have been obtained on these steels at subzero temperatures by the J-integral method. Results of these tests are presented in Table 15. The 5% Ni steel retains relatively high fracture toughness at −162 °C (−260 °F), and the 9% Ni steel retains relatively high fracture toughness at −196 °C (−320 °F). These temperatures approximate the minimum temperatures at which these steels may be used.

Fatigue-Crack-Growth Rates. For the 3.5% Ni steel, fatigue-crack-growth rates at −78 °C (−108 °F) and at −101 °C (−150 °F) are about the same as the rate at room temperature. However, in the stress-intensity-factor range (ΔK) above about 20 MPa·m$^{1/2}$ (18 ksi·in.$^{1/2}$), fatigue-crack-growth rates are substantially greater at −196 °C (−320 °F) than at room temperature. For the 5% Ni steel, fatigue-crack-growth rates also were substantially higher at −196 °C (−320 °F) than at room temperature or at −162 °C (−260 °F) in the stress-intensity-factor range above 30 MPa·m$^{1/2}$ (27 ksi·in.$^{1/2}$). For the 9% Ni steel, fatigue-crack-

growth rates were substantially higher at −269 °C (−452 °F) than at higher testing temperatures in the stress-intensity-factor range above 24 MPa·m$^{1/2}$ (22 ksi·in.$^{1/2}$). The fatigue-crack-growth-rate data for the ferritic nickel steels are consistent with results of other tests.

STAINLESS STEELS

Austenitic stainless steels have been used extensively for subzero applications to −269 °C (−452 °F). These steels contain sufficient amounts of nickel and manganese to depress the M_s temperature into the subzero range. Thus they retain a face-centered-cubic crystal structure on cooling

from hot working or annealing temperatures.

Compositions of austenitic stainless steels of interest are presented in Table 16. Small amounts of nitrogen are added to some of these steels to increase their strength. Manganese additions are used in some steels to replace some of the nickel.

Yield and tensile strengths of austenitic stainless steels increase substantially as testing temperature is decreased, and the stable grades retain good ductility and toughness at −269 °C (−452 °F).

Strengths of these steels can be increased by cold rolling or cold drawing. Cold working at −196 °C (−320 °F) is more effective in increasing strength than cold working at room temper-

Table 12. Typical tensile properties of high-nickel alloys and weldments

Temperature °C	°F	Tensile strength MPa	ksi	Yield strength MPa	ksi	Elongation, %	Reduction in area, %
Hastelloy C sheet, cold rolled 20%, longitudinal orientation							
24	75	1140	165	1000	145	13	...
−196	−320	1520	220	1280	186	32	...
−253	−423	1740	252	1380	200	33	...
Inconel 600 bar, cold drawn, longitudinal orientation							
24	75	940	136	890	129	15	56
−78	−108	985	143	910	132	20	58
−196	−320	1160	168	1030	150	26	62
−253	−423	1250	181	1100	160	30	56
−257	−430	1280	186	1210	176	20	56
Inconel 706 forged billets(a)							
24	75	1260	183	1050	152	24	33
−196	−320	1570	228	1200	174	29	33
−269	−452	1680	243	1250	181	30	33
Inconel 718 sheet, longitudinal orientation(b)							
24	75	1330	193	1090	158	18	...
−78	−108	1490	216	1190	172	17	...
−196	−320	1730	251	1310	190	21	...
−253	−423	1740	252	1340	194	16	...
Inconel 718 bar, longitudinal orientation(c)							
24	75	1410	204	1170	170	15	18
−196	−320	1650	239	1340	197	21	20
−269	−452	1810	263	1410	204	21	20
Inconel 718 forgings, longitudinal orientation(c)							
24	75	1340	194	1150	167	24	35
−78	−108	1350	196	1190	172	29	45
−196	−320	1630	237	1300	188	26	34
−253	−423	1680	244	1320	192	28	42
−269	−452	1810	263	1410	204	21	20
GTA weld in Inconel 718 sheet, no filler metal(d)							
24	75	1320	191	1150	167	5	...
−196	−320	1560	226	1290	187	4	...
−253	−423	1730	251	1410	205	5	...
GTA weld in Inconel 718 forging, 718 filler metal(e)							
24	75	1260	183	2000	159	2	6
−196	−320	1430	208	1280	186	2	4
−269	−452	1650	239	1280	185	28	33
Inconel X-750 sheet, longitudinal orientation(f)							
24	75	1220	177	815	118	24	...
−78	−108	1320	192	875	127	28	...
−196	−320	1500	217	905	131	32	...
−253	−423	1590	230	940	136	32	...
Inconel X-750 bar, longitudinal orientation(f)							
24	75	1340	194	985	143	25	49
−196	−320	1570	228	1050	152	32	45
−253	−423	1700	246	1090	158	33	42
−257	−430	1720	249	1080	157	33	46
GTA weld in Inconel X-750 sheet, X-750 filler metal(g)							
24	75	1290	187	860	125	22	...
−78	−108	1340	195	915	133	24	...
−196	−320	1540	224	945	137	30	...
−253	−423	1660	241	1020	148	28	...
GTA weld in Inconel X-750 forged billet, F69 filler metal(e)							
24	75	1100	159	855	124	9	12
−196	−320	1110	161	930	135	6	9
−269	−452	1120	163	960	139	6	9

(a) Aged 1 h at 980 °C (1800 °F), AC, 8 h at 730 °C (1350 °F), FC to 620 °C (1150 °F), held 8 h, AC. (b) Aged 1 h at 955 °C (1750 °F), AC, 8 h at 720 °C (1325 °F), FC to 620 °C (1150 °F), held 10 h, AC. (c) Aged $^{3}/_{4}$ h at 980 °C (1800 °F), AC, 8 h at 720 °C (1325 °F), FC to 620 °C (1150 °F), held 10 h, AC. (d) Weldment aged 8 h at 720 °C (1325 °F), FC to 620 °C (1150 °F), held 10 h, AC. (e) Weldment aged 1 h at 980 °C (1800 °F), AC, 8 h at 730 °C (1350 °F), FC to 620 °C (1150 °F), held 8 h, AC. (f) Annealed and aged 20 h at 700 °C (1300 °F). (g) Weldment aged 20 h at 700 °C (1300 °F), AC.

Table 13. Fracture toughness of high-nickel alloys and weldments

Alloy and condition(a)	Form	Room-temperature yield strength MPa	ksi	Specimen design	Orientation	Fracture toughness, K_{Ic} or $K_{Ic}(J)$, at: 24 °C (75 °F) MPa·m$^{1/2}$	ksi·in.$^{1/2}$	−196 °C (−320 °F) MPa·m$^{1/2}$	ksi·in.$^{1/2}$	−269 °C (−452 °F) MPa·m$^{1/2}$	ksi·in.$^{1/2}$
Inconel 706 (VIM-VAR) STDA	Forging	1065	154	CT	C-R	133(b)	121(b)	...	...	157(b)	143(b)
Inconel 706 (VIM-VAR) GTA weld, ST/W/STDA	Forging-weldment	1065	154	CT	...	...	...	...	...	58.2(b)(c)	53.0(b)(c)
Inconel 718 STDA	Bar	1170	170	CT	T-S	96.4	87.8	103	94.0	112	102
Inconel 718 (VIM-VAR) STDA	Forging	1165	169	CT	C-R	61.1(b)	55.6(b)	...	...	75.0(b)	68.2(b)
Inconel 718 (VIM-VAR) GTA weld, ST/W/STDA	Forging-weldment	1165	169	CT	...	...	...	51.1(b)(c) 66.2(b)(d)	46.5(b)(c) 60.2(b)(d)	51.7(b)(c) 60.9(b)(d)	47.1(b)(c) 55.4(b)(d)
Inconel X-750 (VIM-VAR) STDA	Forging	825	120	CT	C-R	...	...	...	...	76.1(b)(e)	69.2(b)(e)
Inconel X-750 (VIM) STDA	Forging	920	133	CT	C-R	...	...	...	...	145(b)	132(b)
Inconel X-750 (AAM-VAR) STDA	Forging	850	123	CT	C-R	...	...	...	...	237(b)	216(b)
Inconel X-750 (VIM-VAR) ST/W/STDA	Forging-weldment	825	120	CT	...	...	...	...	...	134(b)(c)(f) 176(b)(c)(g)	122(b)(c)(f) 160(b)(c)(g)

(a) VIM, vacuum induction melted; VAR, vacuum arc remelted; AAM, air arc melted; ST, solution treated; W, welded. STDA for Inconel 706 and Inconel X-750: 980 °C (1800 °F) 1 h, AC, 730 °C (1350 °F) 8 h, FC to 620 °C (1150 °F) 8 h, AC. STDA for Inconel 718: 980 °C (1800 °F) 1 h, AC, 720 °C (1325 °F) 8 h, FC to 620 °C (1150 °F), hold 8 h, AC. Filler metals: F-718 for Inconel 706 and Inconel 718; Inco F69 for Inconel X-750. (b) $K_{Ic}(J)$. (c) Fusion zone. (d) Heat-affected zone. (e) This heat of Inconel X-750 had carbide precipitates at the grain boundaries, which caused abnormally low fracture toughness. (f) Gas tungsten-arc weld. (g) Vacuum electron beam weld.

Table 14. Typical tensile properties of ferritic nickel steels

Temperature °C	°F	Tensile strength MPa	ksi	Yield strength MPa	ksi	Elongation, %	Reduction in area, %
A645 plate, longitudinal orientation(a)							
24	75	715	104	530	76.8	32	72
−168	−270	930	135	570	82.9	28	68
−196	−320	1130	164	765	111	30	62
A353 plate, longitudinal orientation(b)							
24	75	780	113	680	98.6	28	70
−151	−240	1030	149	850	123	17	61
−196	−320	1190	172	950	138	25	58
−253	−423	1430	208	1320	192	18	43
−269	−452	1590	231	1430	208	21	59
A553-I plate, longitudinal orientation(c)							
24	75	770	112	695	101	27	69
−151	−240	995	144	885	128	18	42
−196	−320	1150	167	960	139	27	38

(a) Quenched, tempered, reversion annealed. (b) Double normalized and tempered: 900 °C (1650 °F) 1 h/in. of thickness, AC; 790 °C (1450 °F) 1 h/in. of thickness, AC; 570 °C (1050 °F) 1 h/in. of thickness, AC or WQ. (c) Quenched and tempered: 800 °C (1475 °F), WQ; 570 °C (1050 °F) 30 min/in. of thickness, AC or WQ.

Table 15. Fracture toughness of 5% and 9% Ni steel plate for compact tension specimens in T-L orientation

Alloy and condition	Yield strength(a) MPa	ksi	Fracture toughness, $K_{Ic}(J)$, at: −162 °C (−260 °F) MPa·m$^{1/2}$	ksi·in.$^{1/2}$	−196 °C (−320 °F) MPa·m$^{1/2}$	ksi·in.$^{1/2}$	−269 °C (−452 °F) MPa·m$^{1/2}$	ksi·in.$^{1/2}$
5Ni steel (A645) quenched, temperized, reversion annealed	535	77.5	196	178	87.1	79.3	58.4	53.2
9Ni steel (A553 type I) quenched and tempered	690	99.9	...	...	184	167	...	...

(a) At room temperature.

ature. For metallurgically unstable stainless steels such as 301, 304 and 304L, plastic deformation at subzero temperatures causes partial transformation to martensite, which increases strength.

Type 301 has been used in the form of extra-hard cold rolled sheet to provide high strength in such applications as the liquid oxygen and liquid hydrogen tanks for the Atlas and Centaur rockets.

Maximum strength is obtained in A-286 alloy by solution treating and aging. Type 416 is a ferritic chromium stainless steel that is usually used in the quenched-and-tempered condition. It is included in this series because there are applications in rotating pumps and other machinery in which a magnetic material is needed to activate counters.

Type 304 stainless steel usually is used in the annealed condition for tubing, pipes and valves employed in transfer of cryogens; for Dewar flasks and storage tanks; and for structural components that do not require high strength.

Types 310 and 310S are considered to be metallurgically stable for all conditions of cryogenic exposure. Therefore, these steels are used for structural components in which maximum stability and a high degree of toughness are required at cryogenic temperatures. Type 316 stainless steel is less stable than type 310.

For higher-strength components of cryogenic structures, there are several stainless steels (such as 21-6-9, Pyromet 538, Nitronic 40, Nitronic 60 and Kromarc 58) that contain significant amounts of manganese in place of some of the nickel, along with small additions of nitrogen and other elements that increase strength.

Cast stainless steels of the compositions given in Table 16 have been used for bubble chambers,

Table 16. Nominal compositions or composition ranges for selected austenitic stainless steels

UNS number	Alloy designation	C	Mn	P max	Si	S max	Cr	Ni	Mo	Others
							Nominal composition, %			
Wrought alloys										
S30100	AISI 301	0.15 max	2.0 max	0.045	0.03	1.0 max	16 to 18	6 to 8	...	...
S30400	AISI 304	0.08 max	2.0 max	0.045	0.03	1.0 max	18 to 20	8 to 12	...	...
S30403	AISI 304L	0.03 max	2.0 max	0.045	0.03	1.0 max	18 to 20	8 to 12	...	...
S31000	AISI 310	0.25 max	2.0 max	0.045	0.03	1.5 max	24 to 26	19 to 22	...	...
S31008	AISI 310S	0.08 max	2.0 max	0.045	0.03	1.5 max	24 to 26	19 to 22	...	...
S31600	AISI 316	0.08 max	2.0 max	0.045	0.03	1.0 max	16 to 18	10 to 14	2.0 to 3.0	...
S32100	AISI 321	0.08 max	2.0 max	0.045	0.03	1.0 max	17 to 19	9 to 12	...	(5 × C) Ti min
S34700	AISI 347	0.08 max	2.0 max	0.045	0.03	1.0 max	17 to 19	9 to 13	...	(10 × C)(Nb + Ta)
S41600	AISI 416	0.15 max	1.25 max	0.06	0.15 min	1.0 max	12 to 14	...	...	...
	21-6-9	0.04 max	8 to 10	...	...	1.0 max	19 to 21.5	5.5 to 7.5	...	0.2 to 0.35N
	Pyromet 538	0.02	9.5			0.15	20	7.0	...	0.2N
	Nitronic 40	0.03	9.2			0.6	19.5	7.0	...	0.3N
	Nitronic 60	0.07	8.0			3.5	17	8.5	...	0.1N
K66286	A-286	0.05	1.4			0.4	15	26	1.25	0.2Al, 2.0Ti, 0.3V, 0.005B
	Kromarc 58	0.03	9.3	0.005	0.005	0.05	15.5	23	2.2	0.02Al, 0.16V, 0.17N, 0.008Zr, 0.016B
Cast alloys										
	CF-8	0.08 max	1.5 max	...	...	2.0 max	18 to 21	8 to 11	...	...
	CF-8M	0.08 max	1.5 max	...	...	2.0 max	18 to 21	9 to 12	2 to 3	...
	Kromarc 55	0.04	9.5	...	...	0.3	16	20	2.25	...

(a) Other specifications for 310 give lower limits on carbon content.

for cylindrical magnet tubes for superconducting magnets, for valve bodies, and for other components that are cooled to −269 °C (−452 °F) in service.

Tensile Properties. Typical tensile properties of annealed 300 series austenitic stainless steels at room temperature and at subzero temperatures are presented in Table 17, and tensile properties of cold worked 300 series stainless steels are given in Table 18. Cold working substantially increases yield and tensile strengths and reduces ductility, but ductility and notch toughness of the cold worked alloy often are sufficient for cryogenic applications. Tensile properties of other stainless steels are presented in Table 19. For the annealed alloys, the greatest effect of the nitrogen addition is to produce an increase in yield strength at cryogenic temperatures. Kromarc 58 has been used for several structural applications in prototype superconducting generators. Other nitrogen-strengthened stainless steels have comparable properties. The data for cold worked Kromarc sheet indicate how these alloys can be strengthened by cold working. Solution treating and aging of A-286 alloy develops good strength with good ductility and notch toughness in the cryogenic range. Because of its low ductility, alloy 416 is not recommended for use below −196 °C (−320 °F) except in nonstressed applications.

Results of tensile tests on stainless steel weldments at subzero temperatures, given in Table 20, may be significant in selecting stainless steels for cryogenic applications. For annealed plate tested in the as-welded condition, Kromarc 58 has the most favorable properties of the alloys in Table 20. Tensile properties of A-286 weldments can be improved by age hardening; however, there is a significant advantage in being able to use weldments in the as-welded condition, as can be done with Kromarc weldments.

Tensile properties of cast stainless steels indicate that alloys CF8 and CF8M have yield strengths at subzero temperatures comparable to those of equivalent wrought alloys. Yield and tensile strengths of as-cast Kromarc 55, however, are considerably lower than those of the wrought alloy Kromarc 58.

Table 17. Typical tensile properties of annealed type 300 austenitic stainless steels

Temperature		Tensile strength		Yield strength		Elongation, %	Reduction in area, %
°C	°F	MPa	ksi	MPa	ksi		
304 sheet, longitudinal orientation							
24	75	660	95.5	295	42.5	75	...
−196	−320	1625	236	380	55.0	42	...
−253	−423	1800	261	425	62.0	31	...
−269	−452	1700	247	570	82.5	30	...
304 plate, longitudinal orientation							
24	75	590	85.9	330	47.6	64	...
−253	−423	1720	250	410	59.4	...	...
304L sheet, longitudinal orientation							
24	75	660	95.9	295	42.8	56	...
−78	−108	980	142	250	36.0	43	...
−196	−320	1460	212	275	39.6	37	...
−253	−423	1750	254	305	44.5	33	...
−269	−452	1590	230	405	58.5	29	...
310 sheet, longitudinal orientation							
24	75	570	83.0	240	35.0	50	...
−196	−320	1080	156	545	79.1	68	...
−253	−423	1300	188	715	104	56	...
−269	−452	1230	178	770	112	58	...
310S forging, transverse orientation							
24	75	585	84.8	260	37.9	54	71
−196	−320	1100	159	605	87.6	72	52
−269	−452	1300	189	815	118	64	45
316 sheet, longitudinal orientation							
24	75	595	86.4	275	39.8	60	...
−253	−423	1580	229	665	96.6	55	...
321 sheet, longitudinal orientation							
24	75	620	89.6	225	32.4	55	...
−196	−320	1380	200	315	45.6	46	...
−253	−423	1650	239	375	54.5	36	...
347 sheet, longitudinal orientation							
24	75	650	94	255	37	52	...
−196	−320	1365	198	420	61	47	...
−253	−423	1610	234	435	63	35	...
347 bar							
24	75	670	97.4	340	49.3	57	76

Fracture Toughness. Fracture-toughness data for stainless steels are limited because steels of this type that are suitable for use at cryogenic temperatures have very high toughness. The fracture-toughness data that are available were obtained by the J-integral method and converted to $K_{Ic}(J)$ values. Such data for base metal and weldments are shown in Table 21. Fracture-toughness values for base metals are relatively high even at −269 °C (−452 °F); fracture-toughness values for the fusion zones (FZ) of welds may be lower or higher than that of the base metal.

Fatigue Properties. Fatigue-crack-growth rates for austenitic stainless steels and weldments are generally lower at subzero temperatures, or about equal at room temperature and at subzero temperatures, except for 21-6-9 stainless steel. For 21-6-9, fatigue-crack-growth rates are higher at

Table 18. Typical tensile properties of cold worked type 300 austenitic stainless steel sheet

Temperature		Tensile strength		Yield strength		Elongation, %
°C	°F	MPa	ksi	MPa	ksi	
301, hard cold rolled (42 to 60% reduction), longitudinal orientation						
24	75	1310	190	1200	174	18
−78	−108	1560	226	1130	164	23
−196	−320	2020	293	1380	200	19
−253	−423	2110	306	1610	233	14
304, hard cold rolled, longitudinal orientation						
24	75	1320	191	1190	173	3
−78	−108	1470	213	1300	188	10
−196	−320	1900	276	1430	208	29
−253	−423	2010	292	1560	226	2
304L, 70% cold reduced, longitudinal orientation						
24	75	1320	192	1080	156	3
−196	−320	1770	256	1530	222	14
−253	−423	1990	288	1770	256	2
310, 75% cold reduced, longitudinal orientation						
24	75	1180	171	1100	160	3
−78	−108	1410	204	1290	187	4
−196	−320	1720	249	1540	223	10
−253	−423	2000	290	1790	259	10

Table 19. Typical tensile properties of stainless steels other than 300 series

Temperature		Tensile strength		Yield strength		Elongation, %	Reduction in area, %
°C	°F	MPa	ksi	MPa	ksi		
21-6-9 plate, longitudinal orientation(a)							
24	75	705	102	385	55.9	54	80
−78	−108	895	130	590	85.4	60	75
−196	−320	1510	219	970	141	41	33
−253	−423	1660	241	1220	177	16	26
−269	−452	1700	247	1350	196	22	30
Pyromet 538 plate, longitudinal orientation(b)							
24	75	675	97.9	340	49.0	75	81
−196	−320	1370	199	800	116	76	73
−269	−452	1490	216	1010	147	52	59
Nitronic 40 plate, electroslag remelted; as rolled							
24	75	1010	146	840	122	35	72
−73	−100	1170	169	945	137	36	71
−196	−320	1830	266	1540	223	31	64
Nitronic 60 bar, annealed, longitudinal orientation							
24	75	750	109	400	58.1	66	79
−73	−100	1020	148	535	77.9	70	81
−196	−320	1500	218	695	101	60	66
−253	−423	1410	204	860	125	24	27
Kromarc 58 sheet, annealed, longitudinal orientation(c)							
24	75	695	101	285	41.5	62	. . .
−78	−108	825	120	395	57.6	59	. . .
−196	−320	1280	185	695	101	82	. . .
−253	−423	1450	210	880	128	56	. . .
Kromarc 58 sheet, cold rolled 80%, longitudinal orientation(d)							
24	75	1280	186	1210	175	6	. . .
−78	−108	1510	219	1430	207	7	. . .
−196	−320	1880	272	1740	252	9	. . .
−253	−423	2100	304	1990	288	1	. . .
A-286 sheet, longitudinal orientation(e)							
24	75	860	125	410	59.9	36	. . .
−196	−320	1230	179	615	89.4	44	. . .
−253	−423	1460	212	745	108	34	. . .
416 bar, longitudinal orientation(f)							
24	75	1400	203	1200	174	15	53
−78	−108	1500	218	1260	183	15	52
−196	−320	1800	261	1600	232	9	24
−253	−423	2020	293	2020	293	0.4	2

(a) Annealed 1 h at 1065 °C (1950 °F), WQ. (b) Annealed 1 h at 1095 °C (2000 °F), WQ. (c) Annealed 1 h at 1065 °C (1950 °F). (d) Annealed at 1065 °C (1950 °F), cold rolled 80%. (e) Heat treatment: 1/2 h at 980 °C (1800 °F), WQ, aged 16 h at 595 °C (1100 °F), AC. (f) Heat treatment: 1 h at 980 °C (1800 °F), OQ, tempered 4 h at 370 °C (700 °F), AC.

−269 °C (−452 °F) than at room temperature.

Results of flexural and axial fatigue tests at 10^6 cycles on austenitic stainless steels at room temperature and at subzero temperatures show that fatigue strength increases as exposure temperature is decreased.

TITANIUM AND TITANIUM ALLOYS

Unalloyed titanium and alpha titanium alloys have close-packed hexagonal (cph) crystal structures, which accounts for the fact that the properties of these metals do not follow the same trends at subzero temperatures as do the properties of metals with fcc or bcc structures.

Many of the available titanium alloys have been evaluated at subzero temperatures, but most service experience at such temperatures has been limited to Ti-5Al-2.5Sn and Ti-6Al-4V alloys. These alloys have very high strength-to-weight ratios at cryogenic temperatures and have been the preferred alloys for special applications at temperatures of −196 to −269 °C (−320 to −452 °F). Among these applications are spherical pressure vessels that are parts of the propulsion and reaction-control systems for the Atlas and Centaur rockets, the Apollo and Saturn launch boosters, and the lunar module.

Commercially pure titanium may be used for tubing and other small-scale cryogenic applications that involve only low stresses in service.

The Ti-5Al-2.5Sn alloy usually is used in the mill-annealed condition and has a 100% alpha microstructure. The Ti-6Al-4V alloy may be used in the annealed condition or in the solution-treated-and-aged condition, but for maximum toughness in cryogenic applications the annealed condition is preferred. The Ti-6Al-4V alloy is an alpha-beta alloy that has significantly higher yield and ultimate tensile strengths than those of the all-alpha alloy.

Interstitial impurities such as iron, oxygen, carbon, nitrogen and hydrogen tend to reduce the toughness of these alloys at both room and subzero temperatures. Therefore, for maximum toughness, extra-low-interstitial (ELI) grades are specified for critical applications. The composition limits for these alloys are given in Table 22. Note that the iron and oxygen contents of the ELI grades are substantially lower than those of the normal interstitial (NI) grades. Iron is a strong stabilizer of the beta phase, which has a bcc crystal structure. The NI grades are suitable for service to −196 °C (−320 °F); for temperatures below −196 °C, ELI grades generally are specified. For ELI grades, reduced creep strength at room temperature must be considered in design for pressure-vessel service.

There are two precautions that should be emphasized in considering titanium and titanium alloys for service at cryogenic temperatures: titanium and titanium alloys must not be used for transfer or storage of liquid oxygen, and titanium must not be used where it will be exposed to air while below the temperature at which oxygen will condense on its surfaces. Any abrasion or impact of titanium that is in contact with liquid oxygen will cause ignition.

Tensile Properties. Typical tensile properties of titanium and of titanium alloys Ti-5Al-2.5Sn and Ti-6Al-4V at room temperature and at subzero temperatures are presented in Table 23. Marked increases in yield and tensile strengths are evident for commercial titanium and for titanium alloys as test temperature is reduced from room temperature to −253 °C (−423 °F). In the cryogenic temperature range, these alloys have the highest strength-to-weight ratios of all fusion-weldable alloys that retain nearly the same strength in the weld metal as in the base metal. Yield and tensile strengths of an electron-beam weldment of Ti-5Al-2.5Sn(ELI) sheet are also given in Table 23.

The recrystallization annealing treatment used for the Ti-6Al-4V(ELI) forging was developed as a means of improving fracture toughness in large forgings and thick plate.

Values of Young's modulus for titanium alloys increase substantially as test temperature is decreased.

Fracture Toughness. Data on plane-strain fracture toughness (K_{Ic}) at subzero temperatures for alloys Ti-5Al-2.5Sn and Ti-6Al-4V are presented in Table 24 along with corresponding data for weldments in Ti-6Al-4V(ELI). These data indicate that there is a modest reduction in fracture toughness as test temperature is reduced from room temperature to subzero temperatures. However, the ELI grades have better toughness than the corresponding normal interstitial grades at subzero temperatures. The limited data for electron-beam weldments indicate that at −196 °C (−320 °F) there is a slight reduction in toughness in both fusion and heat-affected zones in Ti-6Al-4V(ELI) weldments when compared with the base metal.

Fatigue Properties. Data on fatigue-crack-growth rates for Ti-5Al-2.5Sn and Ti-6Al-4V alloys indicate that the exposure temperature has no effect on the fatigue-crack-growth rates for Ti-5Al-2.5Sn and Ti-6Al-4V(NI). However, over part of the ΔK range, the rates for Ti-6Al-4V(ELI) are higher at cryogenic temperatures than at room temperature at the same ΔK values.

Values of fatigue strength at 10^6 cycles for titanium alloys show that for unnotched specimens, fatigue strength increased substantially when the test temperature was reduced from room temperature to −196 °C (−320 °F). When the test temperature was reduced to −253 °C (−423 °F), the fatigue strengths for some series of alloys were lower than at −196 °C. Fatigue strengths were much lower in the notched specimens than in the corresponding unnotched specimens. At −196 and −253 °C, the welded specimens had lower fatigue strengths than the base-metal specimens. Therefore, in designing welded structures of titanium alloys that will be subjected to fatigue loading at subzero temperatures, the weld areas usually should be thicker than the remaining areas. Hemispheres for spherical pressure vessels are machined so that the butting sections for the equatorial welds are thicker than the remaining sections, excluding inlet and discharge ports.

REFERENCES

1. Alloys for Structural Applications at Subzero Temperatures, by James E. Campbell: in *Metals Handbook*, 9th Ed., Vol 3, American Society for Metals, 1980, p 721-772
2. "Determination of Design Allowable Properties, Fracture of 2219-T87 Aluminum Alloy," by W. L. Engstrom: NASA CR-115388, The Boeing Company, Seattle, Mar 1972
3. Design Stresses for Aluminum Alloy 5083-O and 5183 Welds at Cryogenic Temperatures, by K. O. Bogardus and R. C. Malcolm: in *Properties of Materials for Liquefied Natural Gas Tankage*, STP 579, American Society for Testing and Materials, Philadelphia, Sept 1975, p 190-204

ADDITIONAL READING

Materials at Low Temperatures, edited by R. P. Reed and A. F. Clark: American Society for Metals, 1983
Cryogenic Materials Data Handbook, Vol I and II, by F. R. Schwartzberg, S. H. Osgood, and R. G. Herzog: AFML-TDR-64-280, Martin Marietta Corporation, Denver, Aug 1978
Handbook on Materials for Superconducting Machinery, by K. R. Hanby, *et al.*: MCIC-HB-04, Metals and Ceramics Information Center, Battelle Columbus Laboratories, Columbus, OH, Jan 1977
New Data on Aluminum Alloys for Cryogenic Applications, by J. G. Kaufman and E. W. Johnson: in

Table 20. Typical tensile properties of stainless steel weldments

Alloy and condition	Welding process	Filler	Form	Base-metal orientation	Test temperature °C	°F	Yield strength MPa	ksi	Tensile strength MPa	ksi	Elongation, %	Reduction in area, %
Type 310, ³/₄ hard; tested as welded	GTA	310	Sheet	L	24	75	380	55	530	77	4	...
					−78	−108	525	76	725	105	4	...
					−196	−320	750	109	1025	149	4	...
Type 310S, annealed	SMA	310S	Plate	...	24	75	330	48	580	84	40	76
					−196	−320	670	97	1070	155	46	67
					−269	−452	825	120	1105	160	26	24
21-6-9, annealed	SMA	Inconel 625	Plate	Weld(a)	−269	−452	875	127	1260	183	31	27
				HAZ(a)	−269	−452	1730	251	1875	272	21	33
	GTA	Inconel 625	Plate	Weld(a)	−269	−452	950	138	1220	177	18	20
				HAZ(a)	−269	−452	1740	252	1925	279	17	37
	GMA	Inconel 625	Plate	Weld(a)	−269	−452	835	121	1090	158	19	27
				HAZ(a)	−269	−452	1690	245	1870	271	15	27
Pyromet 538, annealed	GTA	Pyromet 538	Plate	...	24	75	415	60	725	105	51	74
					−196	−320	1005	146	1455	211	48	61
					−269	−452	1240	180	1650	239	31	24
	GMA	In-182	Plate		24	75	415	60	730	106	53	75
					−196	−320	800	116	1050	152	6	37
					−269	−452	805	117	1090	158	6	40
Kromarc 58 annealed plate; tested as welded	GTA	Kromarc 58	Plate	L	24	75	495	72	915	133	36	61
					−196	−320	855	124	1325	192	46	41
					−269	−452	1060	154	1440	209	33	40
A-286, annealed sheet; welded and age hardened	GTA	A-286	Sheet	L	24	75	600	87	860	125	11	...
					−78	−108	615	89	930	135	13	...
					−196	−320	745	108	1145	166	16	...
					−253	−423	870	126	1280	186	15	...

(a) Weld parallel with specimen axis; weld specimens were all weld metal; HAZ specimens contained HAZ plus some weld metal and some base metal.

Table 21. Fracture toughness of austenitic stainless steels and weldments

Alloy and condition(a)	Form	Room-temperature yield strength MPa	ksi	Orientation	24 °C MPa·m¹ᐟ²	(75 °F) ksi·in.¹ᐟ²	−196 °C MPa·m¹ᐟ²	(−320 °F) ksi·in.¹ᐟ²	−269 °C MPa·m¹ᐟ²	(−452 °F) ksi·in.¹ᐟ²
Type 310S, annealed	Plate	260	37.9	T-L	...	...	...	...	259	236
	Weldment(b)	...	...	...	...	...	...	...	116(c)	106(c)
Pyromet 538, STQ	Plate	340	49	T-L	...	...	275(b)	250(b)	181	165
	Weldment(d)	...	...	...	...	...	...	...	82(c)	74(c)
	Weldment(b)	...	...	...	...	...	...	...	175(c)	159(c)
Kromarc 58, STQ	Plate	370	53.8	T-L	...	...	...	...	214	195
	Weldment(d)	...	...	...	...	...	...	...	155(c)	141(c)
A-286, STA	Bar	610	88.2	T-S	125	114	123	112	118	107
	Plate	820	119	L-T	161	146	...	...	179	163
	Weldment(d)	...	...	...	...	...	...	...	247(c)	225(c)

(a) STQ = solution treated and quenched. STA for A-286: 5 h at 900 °C (1650 °F), OQ; age 20 h at 720 °C (1325 °F), AC. Filler wires: for 310S, E310-16; for Pyromet 538, 21-6-9; for Kromarc 58, K-58; for A-286, Inconel 92. (b) Shielded metal-arc weld. (c) In weld fusion zone, as welded. (d) Gas tungsten-arc weld.

Table 22. Compositions of titanium alloys

Alloy	Al	Sn	V	Fe max	O max	C max	N max	H max	M max
Ti-75A	...	...	...	0.40	0.20		0.07	0.0125	...
Ti-5Al-2.5Sn	4.0-6.0	2.0-3.0	...	0.50	0.20	0.15	0.07	0.020	0.30
Ti-5Al-2.5Sn(ELI) (a)	4.7-5.6	2.0-3.0	...	0.20	0.12	0.08	0.05	0.0175	...
Ti-6Al-4V	5.5 to 6.75	...	3.5 to 4.5	...	...	...	...	...	...
Ti-6Al-4V(ELI) (a)	5.5 to 6.5	...	3.5 to 4.5	0.15	0.13	0.08	0.05	0.015	...

(a) Extra low interstitial.

Advances in Cryogenic Engineering, edited by K. D. Timmerhaus, Vol 6, Plenum Press, 1961, p 637–649

Aluminum Alloys for Cryogenic Service, by J. E. Campbell: *Materials Research and Standards*, Vol 4, No. 10, Oct 1964, p 540–548

Effects of Low Temperatures on the Mechanical Properties of Structural Metals, by H. L. Martin, *et al.*: NASA SP-5012 (01), Office of Technology Utilization, National Aeronautics and Space Administration, Washington, DC, 1968

LNG Materials and Fluids: A Users Manual of Property Data in Graphic Format, First Edition, edited by D. Mann: National Bureau of Standards, Boulder, CO, 1977

Low-Temperature Mechanical Properties of Welded and Brazed Copper, by R. P. Reed: in *Advances in Cryogenic Engineering*, edited by K. D. Timmerhaus, Vol 14, Plenum Press, 1969, p 83–87

Tensile and Impact Properties of Selected Materials from 20° to 300 °K, by K. A. Warren and R. P. Reed: National Bureau of Standards Monograph 63, 28 June 1963

Tensile Properties of Copper, Nickel, and Some Copper-Nickel Alloys at Low Temperatures, by G. W. Geil and N. L. Carwile: NBS Circular 520, National Bureau of Standards, Washington, DC, 7 May 1952, p 67–96

The Fatigue Behavior of Certain Alloys in the Temperature Range from Room Temperature to −423 F, by D. N. Gideon, *et al.*: in *Advances in Cryogenic Engineering*, edited by K. D. Timmerhaus, Vol 7, Plenum Press, 1962, p 503–508

Mechanical Properties of Several Nickel-Base Alloys at Room and Cryogenic Temperatures, by J. L. Christian: in *Advances in Cryogenic Engineering*, edited by K. D. Timmerhaus, Vol 12, Plenum Press, 1967, p 520–531

Fracture of Structural Alloys at Temperatures Approaching Absolute Zero, by R. L. Tobler: *Proceedings of the Fourth International Conference on Fracture*, University of Waterloo, Ontario, Canada, June 1977

Titanium Alloys for Cryogenic Service, by R. G. Broadwell and R. A. Wood: *Materials Research and Standards*, Vol 4, No. 10, Oct 1964, p 549–554

Mechanical Properties of Titanium Alloys at Cryogenic Temperatures, by J. L. Christian and A. Hurlich: in *Advances in Cryogenic Engineering*, edited by K. D. Timmerhaus, Vol 13, Plenum Press, 1968, p 318–333

Table 23. Typical tensile properties of titanium and two titanium alloys

Temperature °C	°F	Tensile strength MPa	ksi	Yield strength MPa	ksi	Elongation, %	Reduction in area, %
Ti-75A sheet, annealed, longitudinal orientation							
24	75	580	84	465	68	25	...
−78	−108	750	109	615	89	25	...
−196	−320	1050	152	940	136	18	...
−253	−423	1280	186	1190	173	8	...
Ti-5Al-2.5Sn sheet, nominal interstitial annealed, longitudinal orientation							
24	75	850	123	795	115	16	...
−78	−108	1080	156	1020	148	13	...
−196	−320	1370	199	1300	188	14	...
−253	−423	1700	246	1590	231	7	...
Ti-5Al-2.5Sn (ELI) sheet, annealed, longitudinal orientation							
24	75	800	116	740	107	16	...
−78	−108	960	139	880	128	14	...
−196	−320	1300	188	1210	175	16	...
−253	−423	1570	228	1450	210	10	...
Ti-5Al-2.5Sn (ELI) sheet/weldment, annealed, EB weld							
24	75	815	118	785	114	...	...
−196	−320	1300	189	1210	176	...	...
−253	−423	1510	219	1380	200	...	...
Ti-6Al-4V (ELI) sheet, annealed, longitudinal orientation							
24	75	960	139	890	129	12	...
−78	−108	1160	168	1100	160	9	...
−196	−320	1500	217	1420	206	10	...
−253	−423	1770	256	1700	246	4	...
Ti-6Al-4V (ELI) plate, annealed, longitudinal orientation							
24	75	890	129	840	122	15	37
−253	−423	1640	238	1600	232	...	8
Ti-6Al-4V (ELI) forgings, as forged, longitudinal orientation							
24	75	970	141	915	133	14	40
−78	−108	1160	168	1120	163	13	31
−196	−320	1570	227	1480	214	11	31
−253	−423	1650	239	1570	227	11	24
Ti-6Al-4V (ELI) forging, recrystallization annealed(a)							
24	75	890	129	825	120	14	41
−196	−320	1430	207	1370	198	10	16

(a) Recrystallization annealing treatment: 4 h at 930 °C (1700 °F), FC to 760 °C (1400 °F) in 3 h, cooled to 480 °C (900 °F) in 3/4 h, AC.

Table 24. Fracture toughness of two titanium alloys and weldments

Alloy and condition(a)	Form	Room-temperature yield strength MPa	ksi	Specimen design	Orientation	24 °C (75 °F) MPa·m^{1/2}	ksi·in.^{1/2}	196 °C (−320 °F) MPa·m^{1/2}	ksi·in.^{1/2}	−253 °C (−423 °F) MPa·m^{1/2}	ksi·in.^{1/2}	−269 °C (−452 °F) MPa·m^{1/2}	ksi·in.^{1/2}
Ti-5Al-2.5Sn(NI), annealed	Plate	875	127	CT	L-T	71.8	65.4	53.4	48.6	...	...	...	...
		875	127	Bend	L-T	...	...	...	...	51.4	46.8	...	...
		875	127	Bend	L-S	...	...	...	...	50.2	45.7	...	...
Ti-5Al-2.5Sn(ELI), annealed	Plate	705	102	CT	L-T	...	...	111	101	...	...	...	...
		705	102	Bend	L-T	...	...	...	...	89.6	81.5	...	...
Ti-5Al-2.5Sn(ELI), as forged	Forging	760	110	CT	R-L	...	...	...	...	79.4	72.3	...	...
					R-C	...	...	...	...	58.5	53.2	...	...
Ti-5Al-2.5Sn(ELI)	Forging(b)	780	113	CT	...	...	...	...	...	54.4 to 75.3	49.5 to 68.5	...	...
Ti-6Al-4V (NI), annealed	Bar	940	136	CT	T-L	47.4	43.2	38.8	35.3	...	...	38.5	35.1
Ti-6Al-4V (ELI), as forged	Forging	830	120	CT	T-L	...	...	61.0	55.5	...	...	54.1	49.2
Ti-6Al-4V (ELI), RA	Forging	830	120	CT	M-L(c)	...	...	62.8	57.2	...	...	...	...
					M-R(c)	...	...	62.0	56.4	...	...	...	...
Ti-6Al-4V (ELI), RA	Forging	830	120	CT	M-R(c)	...	...	61.1(d)	55.6(d)	...	...	...	...
Electron beam welded, SR	Weldment	...	...	...	M-L(c)	...	...	56.9(d)	51.8(d)	...	...	...	...
					M-R(c)	...	...	57.1(e)	52.0(e)	...	...	...	...
					M-R(c)	...	...	51.0(f)	46.4(f)	...	...	...	...

(a) NI = normal interstitial content. ELI = extra low interstitial content. RA = recrystallization annealed: 4 h at 930 °C (1700 °F), FC to 810 °C (1400 °F) in 3 h, cooled to 480 °C (900 °F) in 3/4 h, AC. SR = stress relieved: 50 h at 540 °C (1000 °F), AC. (b) Range for 18 tests. (c) M-L and M-R are specific orientations in a specific forging. (d) Fusion zone. (e) Heat-affected zone. (f) Heat-affected-zone boundary.

Industrial Uses of Depleted Uranium

Condensed from Metals Handbook, Ninth Edition, Volume 3, pages 773 to 780.

DEPLETED URANIUM (sometimes referred to as DU) is a by-product of the process by which the fissionable isotope U-235 is extracted from natural uranium, and thus can be considered a by-product of the nuclear industry. From an engineering standpoint, the most singular property of uranium is its great density—almost twice that of lead, and nearly as great as those of gold and tungsten. Because of this high density, thin layers of uranium are capable of absorbing as much penetrating radiation, such as gamma rays, as could be absorbed by much thicker layers of less dense metals such as lead and iron.

Uranium is much easier to fabricate than dense metals such as tungsten and rhenium, and is much less costly than heavy metals such as gold and platinum. These qualities make uranium a good candidate for objects that must be small, yet very heavy for their size. Depleted uranium is relatively abundant and available. Consequently, industrial non-nuclear usage has increased steadily during recent years.

APPLICATIONS

There are three principal non-nuclear uses of depleted uranium: radiation shielding; counterweights in airplanes, helicopters and missiles; and armor-piercing projectiles for military ordnance. Unalloyed uranium is used mainly in shielding and counterweight applications. Heat treated uranium alloys are used in ballistic or armor-piercing ordnance applications. Besides these three chief uses, depleted uranium has been used in several specific applications where its combination of great density, good fabricability and relatively good mechanical properties give it an advantage over alternative materials.

PROPERTIES OF DEPLETED URANIUM

Uranium is an allotropic metal having three phases in the solid state. Below 688 °C, the metal is in the alpha phase (orthorhombic), which exhibits increasing ductility from room temperature to the phase-transformation temperature. From 688 to 775 °C, the metal is in the beta phase (tetragonal), which is brittle and unworkable. From 775 °C to the melting point, it is in the gamma phase (body-centered cubic), which exhibits great ductility and very low strength.

Uranium is a highly anisotropic material. Properties can vary extensively, depending on fabrication history and orientation with respect to the direction of working. Impurities such as carbon, iron, silicon and aluminum have strong effects on mechanical properties. The properties given in this section are typical of production material for unalloyed depleted uranium of standard purity and for U-0.75Ti and U-2Mo, the two depleted uranium alloys most extensively produced for non-nuclear applications.

- **Unalloyed uranium (as cast)**
  Melting point: 1130 °C
  Density: 19 Mg/m^3
  Tensile strength: 450 MPa (65 ksi)
  Yield strength (0.2% offset): 207 MPa (30 ksi)
  Modulus of elasticity (tension): 172 GPa (25 × 10^6 psi)
  Elongation: 1 to 5%
  Reduction in area: 1 to 10%
  Hardness: 50 to 100 HRB

Hardness and strength of unalloyed uranium can be increased by warm or cold working.

- **Uranium alloys U-0.75Ti and U-2Mo**

	Melting point, °C	Density, Mg/m^3
U-0.75Ti	1200	18.6
U-2Mo	1150	18.5

Mechanical properties of these two alloys vary widely with heat treatment. The standard heat treatment for ordnance applications consists of heating into the gamma phase (about 850 °C), quenching in water or oil, and then aging at any of various temperatures in the range 350 to 450 °C. Strength increases and ductility decreases with increasing aging temperature until the material reaches a peak-aged condition at about 450 °C. Above this temperature, the material becomes overaged and loses strength but gains in ductility.

Health Hazards. Depleted uranium is only mildly radioactive (specific activity of 3.6 × 10^{-7} Ci/g vs 6.77 × 10^{-7} Ci/g for natural uranium) and is listed with natural uranium and thorium as a "low specific activity" (LSA) material in shipping regulations. Like lead and like metals with atomic numbers higher than that of lead, depleted uranium is a heavy-metal poison that can be lethal if a sufficient amount of dust or fumes is ingested.

Licensing. Possession of more than 6.8 kg (15 lb) of depleted uranium in any form requires a license from the U.S. Nuclear Regulatory Commission. Title 10, Part 40, of Federal Regulations describes the requirements and the necessary steps for obtaining such a license. In addition, other local, state and federal regulations may apply to possession and use of uranium objects.

SELECTED REFERENCES

"Trends in the Use of Depleted Uranium," Report NMAB-275, National Materials Advisory Board, National Academy of Sciences, Washington, June 1971

K. H. Eckelmeyer, Aging Phenomena in Dilute Uranium Alloys, *Physical Metallurgy of Uranium Alloys*, Proceedings of the Third Army Materials Technology Conference, Vail, CO, 1974, 2nd Ed., edited by J. J. Burke, D. A. Colling, A. E. Gorum and J. Greenspan: Metals and Ceramics Information Center, Columbus, OH, and Brooke Hill Publishing Co., Chestnut Hill, MA, 1976, p 463-509

Zirconium and Zirconium Alloys

Condensed from Metals Handbook, Ninth Edition, Volume 3, pages 781 to 791.

ZIRCONIUM and most of its alloys exhibit strong anisotropy because of two characteristics of the metal—zirconium has a close-packed hexagonal crystal structure at room temperature and it undergoes allotropic transformation to a body-centered-cubic structure at about 870 °C (1600 °F). The strong anisotropy profoundly influences the engineering properties of zirconium and its alloys and must be taken into account when selecting and processing a zirconium metal. The most common alloys are rather dilute alpha alloys whose characteristics are generally similar to those of unalloyed zirconium.

THE ALLOTROPIC TRANSFORMATION

In zirconium, the low-temperature alpha phase has a close-packed hexagonal crystal structure. This phase transforms to a body-centered-cubic structure at about 870 °C (1600 °F). Small amounts of impurities, particularly oxygen, strongly affect the transformation temperature, as shown in Table 1.

The transformation on cooling generally results in a Widmanstätten structure of alpha zirconium; beta phase cannot be retained even by rapid quenching. The platelets of the Widmanstätten structure are finer, the more rapid the cooling rate. Other elements affect phase stability as follows:

Alpha-stabilizing elements raise the temperature of the allotropic alpha-to-beta transformation. These elements include Al, Sb, Sn, Be, Pb, Hf, N, O and Cd. Phase diagrams for many of the binary alloy systems formed between these various elements and zirconium exhibit a peritectic or a peritectoid reaction at the Zr-rich end.

Beta-stabilizing elements lower the alpha-to-beta transformation temperature. Typical beta stabilizers include Fe, Cr, Ni, Mo, Cu, Nb, Ta, V, Th, U, W, Ti, Mn, Co and Ag. For binary alloy systems between zirconium and these elements, there usually is a eutectoid, and often a eutectic reaction as well, at the Zr-rich end of the phase diagram.

Low-solubility intermetallic compound formers carbon, silicon and phosphorus have very low solubility in zirconium, even at temperatures in excess of 1000 °C (1830 °F). They readily form stable intermetallic compounds and are relatively insensitive to heat treatment.

Most alloying elements and impurities are soluble in beta zirconium but relatively insoluble in

Table 1. Variation in allotropic transformation temperature with oxygen content of unalloyed zirconium

Temperature °C	°F	Phases present at oxygen content of: 1640 ppm	1370 ppm	970 ppm
954	1750	.... β	β	β
932	1710	....α + β	β	β
927	1700	....α + β	β	β
921	1690	....α + β	β	β
915	1680	.... α	α + β	β
910	1670	.... α	α + β	β
904	1660	.... α	α + β	β
893	1640	.... α	α + β	α + β
888	1630	.... α	α	α + β
885	1625	.... α	α	α + β
865	1590	.... α	α	α
857	1575	.... α	α	α

alpha zirconium, where they exist primarily as intermetallic compounds. Size and distribution of these secondary phases are largely governed by reactions that take place during the last transformation from beta to alpha and by subsequent mechanical working at lower temperatures.

Heating at temperatures near the alpha-beta transus, or in the alpha-plus-beta region, causes migration of many impurities to grain boundaries, which impairs ductility and corrosion resistance, particularly in zirconium alloys.

COLD WORK AND RECRYSTALLIZATION

The degree to which unalloyed zirconium may be cold worked depends on both metal purity and method of reduction. Zirconium work hardens rapidly, reaching maximum hardness and strength after cold reduction of only about 20%. But during cold rolling, reductions of about 50% are common and reductions of 80% can be accomplished in some instances. Maximum reductions of 80% are obtained by starting with very soft metal and using machines that feature multiaxial loading (such as cold Pilger machines or Sendzimir rolling mills). Reductions during cold drawing are generally about 15 to 30%. Initially, deformation results in twinning, which reorients the lattice for slip — the primary mechanism for cold working.

Recrystallization is a function of amount of cold work, temperature and time, with time playing a relatively small role. In heavily cold worked material, recrystallization commences at about 510 °C (950 °F), and process annealing of such material usually is conducted at 620 to 790 °C (1150 to 1450 °F). Recrystallization will occur in times as short as 15 min, but much longer times normally are used to ensure that the entire furnace load reaches temperature. Grain growth is nearly nonexistent at the usual annealing temperatures; times of 100 h or more are required to effect grain growth of 2 to 3 ASTM sizes.

Large grains can be grown by annealing after cold reduction of about 5 to 8%. The most common source of large grains is reannealing after a straightening or forming operation that imparts only a small amount of cold work.

ANISOTROPY AND PREFERRED ORIENTATION

The relationships among preferred orientation of zirconium crystals, the working practice that caused it and the properties that result from it are complex and have been studied in great detail. In most engineering applications, it is important to understand that wrought forms of zirconium

and its alloys have different tensile properties in the rolling (or longitudinal) direction than they have in the transverse direction. Yield strength is higher in the transverse direction; tensile strength is slightly higher in the rolling direction.

THE ROLE OF OXYGEN

Originally, oxygen was considered a troublesome impurity in zirconium, and considerable effort was devoted to its elimination. But when oxygen levels were finally reduced below 1000 ppm, it was found that required strength levels in Zircaloy could no longer be met. The status of oxygen then changed to one of a controlled solid-solution alloying agent. Early methods for determining oxygen content were crude and relatively imprecise, so hardness (which is roughly related to oxygen content but much easier to measure) became the controlled attribute and is still widely used to express purity or grade for unalloyed zirconium and zirconium alloys.

The oxygen content of Kroll process sponge varies from about 500 to 2000 ppm, depending on the number of purification steps and the effectiveness of each step. A hardness of 125 HB indicates "soft" sponge with an oxygen content of about 800 ppm, whereas 165 HB is considered "hard" and indicates an oxygen content of about 1600 ppm. Crystal bar zirconium generally has hardness below 100 HB and contains less than 100 ppm oxygen.

Oxygen is a potent strengthener at room temperature, but much of its effectiveness is lost at elevated temperature.

CORROSION RESISTANCE OF UNALLOYED ZIRCONIUM

Unalloyed zirconium is generally more resistant to corrosion than stainless steels, but less resistant than tantalum, in many chemical media. Zirconium resists most inorganic and organic acids and is totally resistant to alkalis. It is attacked, however, by fluoride ions, wet chlorine, aqua regia, concentrated sulfuric acid, and ferric and cupric chlorides. Zirconium and its alloys derive their corrosion resistance from a regenerating, adherent oxide film that forms in most media.

The alloys of zirconium were developed specifically for use in nuclear reactors because they have good mechanical strength and ductility at reactor service temperatures; low neutron absorption; sufficient corrosion resistance to withstand pressurized water and steam; and adequate heat conductivity. As an added benefit, these properties are relatively stable even after extensive irradiation in a reactor core.

Stress-Corrosion Cracking. Zirconium and its alloys are generally resistant to stress-corrosion cracking in seawater and in most aqueous chemical media. Stress-corrosion cracking has been reported, however, in (*a*) concentrated methanol, (*b*) solutions containing heavy-metal chlorides, (*c*) organic solutions with small quantities of chlorides and (*d*) gaseous iodine or fused salts containing iodine. The related phenomenon of liquid-metal embrittlement has been reported for zirconium in contact with molten cesium and with liquid sodium or cadmium.

Pitting. The regenerating oxide film on zirconium usually keeps pitting to a minimum. Still, severe pitting has been observed in hydrochloric acid solutions containing ferric or cupric ions.

Crevice Corrosion. Zirconium is highly resistant to accelerated corrosion under gaskets and fas-

Table 2. Corrosion rates of zirconium in air, oxygen and nitrogen

Temperature °C	°F	Weight gain, g/m²·h, in: Air	Dry air	Oxygen	Nitrogen
425	800	... 0.76	...	...	...
500	930	... 3.6	3.5	1.9	...
600	1110	... 9.5	6.3	3.8	...
700	1290	...16.9	16.0	7.5	0.47
800	1470	... ...	...	...	0.72
900	1650	... ...	...	...	1.5
1000	1830	... ...	...	...	2.7
1100	2010	... ...	...	...	6.5
1200	2190	... ...	...	...	10.2

teners or in corners and overlaps.

Oxidation. Zirconium forms a visible oxide film at about 200 °C (about 400 °F) and begins to exhibit a loose white scale with prolonged exposure at over 425 °C (800 °F). Oxidation rates at several temperatures from 425 to 1200 °C (800 to 2190 °F) are given in Table 2. When heated in air, zirconium reacts primarily with oxygen rather than with nitrogen.

CORROSION RESISTANCE OF ZIRCONIUM ALLOYS

The most common alloys, Zircaloy-2 and Zircaloy-4, contain the strong alpha stabilizers tin and oxygen, plus the beta stabilizers iron, chromium and nickel. There is an extensive alpha-plus-beta field from about 790 to 1010 °C (1450 to 1850 °F). Most alloying elements form intermetallic compounds, and distribution of these compound phases is critical to corrosion resistance in steam and hot water. Generally these alloys are forged in the beta region, then solution treated at about 1065 °C (1950 °F) and water quenched. Subsequent hot working and heat treating is done in the alpha region (below 790 °C) to preserve the fine, uniform distribution of intermetallic compounds that results from solution treating and quenching.

Except for being somewhat stronger and less ductile than unalloyed grades, the Zircaloys are quite similar to unalloyed zirconium in all aspects of metallurgical behavior.

Zr-2.5Nb is the only other alloy of zirconium that has significant commercial importance. In zirconium, niobium is a mild beta stabilizer, and induces a eutectoid reaction when the niobium content exceeds about 1%. (The eutectoid point occurs at 20% Nb.) The mechanical and physical properties of Zr-2.5Nb are very similar to those of the Zircaloys; its corrosion resistance is slightly inferior to that of the Zircaloys.

Zirconium ores generally contain substantial amounts of zirconium's sister element, hafnium. Consequently, zirconium metal produced directly from sponge may contain several percent hafnium. Hafnium has chemical and metallurgical properties similar to those of zirconium, but its nuclear properties are markedly different: hafnium is a neutron absorber whereas zirconium is not. As a result, there are nuclear and non-nuclear grades of zirconium and zirconium alloys, the nuclear grades being essentially hafnium free and the non-nuclear grades containing as much as 4.5% hafnium. Properly speaking, Zircaloy-2, Zircaloy-4 and Zr-2.5Nb are alloy names that apply only to nuclear-grade materials. ASTM specifications for non-nuclear grades list R60704 as the alloy corresponding approximately to Zircaloy-2 and R60705 as the alloy corresponding approximately to Zr-2.5Nb.

Low-Expansion Alloys

Condensed from Metals Handbook, Ninth Edition, Volume 3, pages 792 to 798.

LOW-EXPANSION ALLOYS of iron and nickel, and their various modifications, are used for absolute standards of length, such as rods and tapes for geodetic work. Other applications include:

- Compensating pendulums and balance wheels for clocks and watches
- Moving parts that require control of expansion, such as pistons for some internal-combustion engines
- Bimetal strip
- Glass-to-metal seals
- Thermostatic strip
- Vessels and piping for storage and transportation of liquefied natural gas
- Superconducting systems in power transmission
- Integrated-circuit lead frames
- Components for radios and other electronic devices.

Alloys of iron and nickel have many inconsistent properties, depending on the relative proportions of the two elements. The coefficients of linear expansion range from a small negative value (-0.5 $\mu m/m \cdot K$) to a large positive value (20 $\mu m/m \cdot K$).

Alloys that contain less than 36% nickel have much higher coefficients of expansion than alloys containing 36% nickel or more. Alloys that contain less than 36% nickel are of the so-called "irreversible" type and are excluded from the present discussion.

One alloy, containing 36% nickel with small quantities of manganese, silicon and carbon amounting to a total of less than 1%, has a coefficient of expansion so low that its length is almost invariable for ordinary changes in temperature. For this reason, the alloy was named "Invar."

After the discovery of Invar, an intensive study was made of the thermal and elastic properties of several similar alloys. Those with higher nickel contents were found to have higher coefficients of expansion. The alloy containing 39% nickel has a coefficient of expansion corresponding to that of low-expansion glasses. The 46Ni alloy has a coefficient equivalent to that of platinum (9.0 $\mu m/m \cdot K$) and has been named "Platinite." "Dumet wire" is an alloy containing 42% nickel. Covered with copper to prevent gassing at the seal, it is used to replace platinum as the "seal-in" wire in incandescent lamps and vacuum tubes. The 56Ni alloy has a coefficient approaching that of ordinary steel (11 $\mu m/m \cdot K$). Nilvar is identical with Invar (36% Ni).

Elinvar, containing 36% nickel and 12% chromium, has an invariable modulus of elasticity over a considerable range of temperature. It also has low thermal expansivity.

There is an advantage in replacing some of the nickel with cobalt in the 36Ni alloy. Substitution of 5% cobalt for 5% nickel provides an alloy with an expansion coefficient even lower than that of Invar. The 31Ni-5Co alloy is also less susceptible to variations in heat treatment.

EFFECT OF COMPOSITION ON EXPANSIVITY

The effect of variation in nickel content on linear expansivity is shown in Fig. 1. Minimum expansivity occurs at about 36% Ni. Small additions of other metals have considerable influences on the position of this minimum. The effects of additions of manganese, chromium, copper and carbon are shown in Fig. 2.

Minimum expansivity shifts toward higher nickel contents when manganese or chromium is added and toward lower nickel contents when copper or carbon is added. The value of the minimum expansivity for any of these ternary alloys is, in general, greater than that of a typical Invar.

Additions of silicon, tungsten and molybdenum produce effects similar to those caused by additions of manganese and chromium; the composition of minimum expansivity shifts toward higher contents of nickel.

Addition of carbon is said to produce instability in Invar. This instability is attributed to the changing solubility of carbon in the austenitic matrix during heat treatment.

INVAR

Invar and related alloys have low coefficients of expansion only over a rather narrow range of temperature (see Fig. 3). At low temperatures, in the region from A to B in Fig. 3, the coefficient of expansion is high. In the interval between B and C, the coefficient decreases, reaching a minimum in the region from C to D. With increasing temperature, the coefficient begins again to increase in the range from D to E, and

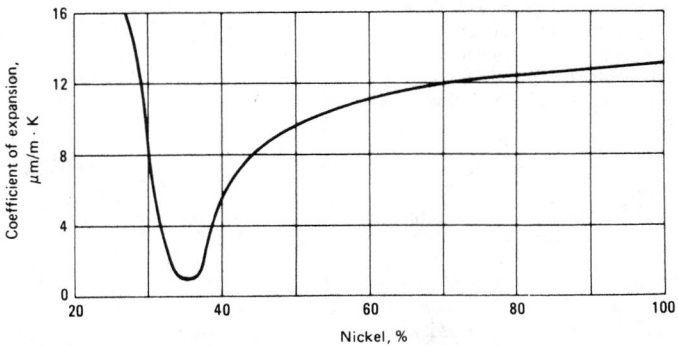

Fig. 1. Coefficient of linear expansion at 20 °C vs Ni content for Fe-Ni alloys containing 0.4% Mn and 0.1% C

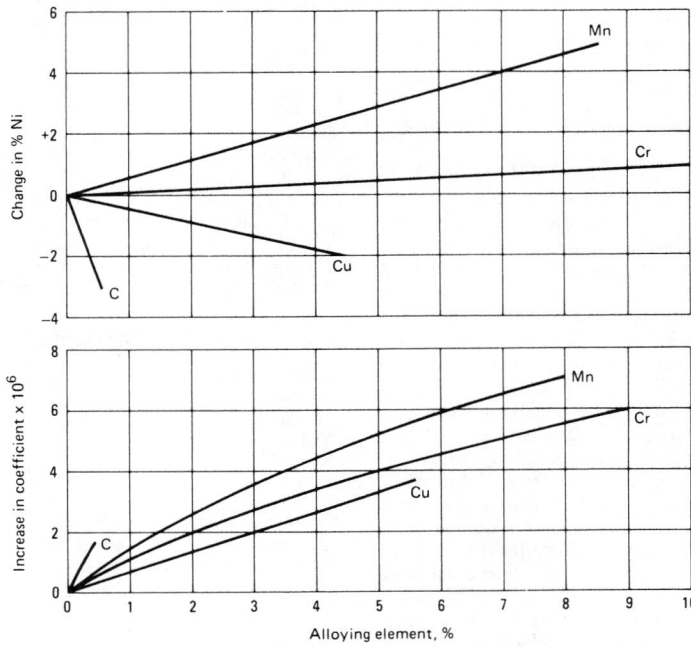

Top: Displacement of nickel content caused by additions of manganese, chromium, copper and carbon to alloy of minimum expansivity. Bottom: Change in value of minimum coefficient of expansion caused by additions of manganese, chromium, copper and carbon.

Fig. 2. Effect of alloying elements on expansion characteristics of Fe-Ni alloys

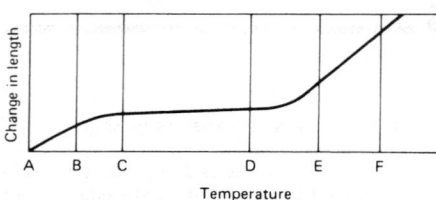

Fig. 3. Change in length of a typical Invar over different ranges of temperature

thereafter (from *E* to *F*) the expansion curve follows a trend similar to that of the nickel or iron of which the alloy is composed. The minimum expansivity prevails only in the range from *C* to *D*.

Between *D* and *E*, the coefficient is changing rapidly to a higher value. The temperature limits for a well-annealed 64 alloy are 162 and 271 °C (324 and 520 °F). These temperatures correspond to the initial and final losses of magnetism in the material. The slope of the curve between *C* and *D* is then a measure of the coefficient of expansion over a limited range of temperature.

Table 1 gives coefficients of linear expansion of iron-nickel alloys between 0 and 38 °C (32 and 100 °F). The expansion behavior of several Fe-Ni alloys over wider ranges of temperature is represented by curves 1 to 5 in Fig. 4; for comparison, Fig. 4 also includes data (curve 6) for an ordinary carbon steel.

Heat treatment and cold work change the expansivity of Invar (or Nilvar) considerably. The effect of heat treatment is shown in Table 2. The expansivity is greatest in well-annealed material and least in quenched material.

Cold drawing also decreases expansivity. The values for the coefficients in the following table are from experiments conducted on two heats of Invar:

Material condition	Expansivity, $\mu m/m \cdot K$(a)
Direct from hot mill	1.4
	1.4
Annealed and quenched	0.5
	0.8
Quenched and cold drawn(b)	0.14
	0.3

(a) Individual measurements for two heats of Invar. (b) 3.2 to 6.4 mm (0.125 to 0.250 in.) diam.

By cold working after quenching, it is possible to produce material with a zero, or even a negative, coefficient of expansion. A negative coefficient may be increased to zero by careful annealing at a low temperature. However, these artificial methods of securing an exceptionally low coefficient produce instability in the material. With lapse of time and variation in temperature, exceptionally low coefficients usually revert to normal values. For special applications (geodetic tapes, for example), it is essential to stabilize the material by cooling it slowly from 100 to 20 °C (212 to 68 °F) over a period of many months, followed by prolonged aging at room temperature. However, unless the material is to be used within the limits of normal atmospheric variation in temperature, such stabilization is of no value. Although these variations in heat treating practice are important in special applications, they are of little significance for ordinary uses.

Magnetic Properties. Invar and all similar iron-nickel alloys are ferromagnetic at room temper-

Table 1. Thermal expansion of Fe-Ni alloys between 0 and 38 °C

Ni, %	Mean coefficient, $\mu m/m \cdot K$
31.4	3.395 + 0.00885 *t*
34.6	1.373 + 0.00237 *t*
35.6	0.877 + 0.00127 *t*
37.3	3.457 − 0.00647 *t*
39.4	5.357 − 0.00448 *t*
43.6	7.992 − 0.00273 *t*
44.4	8.508 − 0.00251 *t*
48.7	9.901 − 0.00067 *t*
50.7	9.984 + 0.00243 *t*
53.2	10.045 + 0.00031 *t*

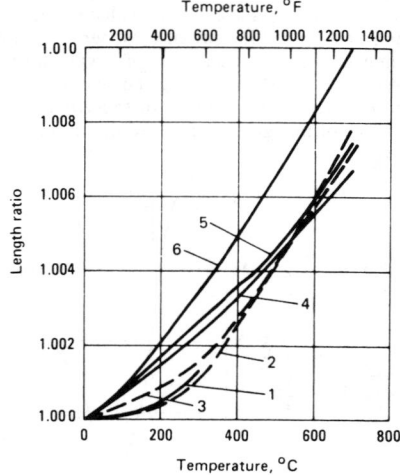

Curve 1, 64Fe-31Ni-5Co; curve 2, 64Fe-36Ni (Invar); curve 3, 58Fe-42Ni; curve 4, 53Fe-47Ni; curve 5, 48Fe-52Ni; curve 6, carbon steel (0.25% C).

Fig. 4. Thermal expansion of Fe-Ni alloys

ature. They become paramagnetic at higher temperatures. The points of inflection in the curves in Fig. 4 indicate the loss of magnetism. The loss of magnetism in a well-annealed sample of a true Invar begins at 162 °C (324 °F) and ends at 271 °C (520 °F). In a quenched sample, the loss begins at 205 °C (400 °F) and ends at 271 °C.

Slow cooling through this range of temperature eliminates to a large extent the troublesome variability in properties of materials of this class.

Electrical Properties. The electrical resistivity of Invar is between 750 and 850 m$\Omega \cdot$m at ordinary temperatures. The temperature coefficient of electrical resistivity is about 1.2 n$\Omega/\Omega \cdot$K over the range of low expansivity. The thermoelectric potential versus pure copper is about 10 μV/K.

Other Physical and Mechanical Properties. Table 3 presents data on miscellaneous properties of Invar in the hot rolled and forged conditions. The effects of temperature on mechanical properties of forged 66Fe-34Ni are illustrated in Fig. 5.

Processing. Considerable care must be used in hot working of iron-nickel alloys, because at hot working temperature they have a tendency to check and break up when carelessly handled. Invar and related alloys should be annealed in a reducing atmosphere. Because they are susceptible to intercrystalline attack during annealing, they should be processed in an atmosphere that contains a large percentage of a neutral gas (such as nitrogen) and a small percentage of a reducing gas. Cold rolling and drawing of iron-nickel alloys are quite similar to corresponding processing procedures for nickel.

Heat Treatment. Annealing of Invar is done at 750 to 850 °C (1380 to 1560 °F). When the alloy is quenched in water from these temperatures, expansivity is decreased, but instability is induced both in actual length and in coefficient of expansion. To overcome these deficiencies and to stabilize the material, it is necessary to anneal Invar at a low temperature such as 95 to 150 °C

Table 2. Effect of heat treatment on coefficient of thermal expansion of Invar

Condition	Temperature °C	°F	Mean coefficient, $\mu m/m \cdot K$
As forged	17 to 100	63 to 212	1.66
	17 to 250	63 to 480	3.11
Quenched from 850 °C (1560 °F)	18 to 100	65 to 212	0.64
	18 to 250	65 to 480	2.53
Quenched from 850 °C and tempered	16 to 100	60 to 212	1.02
	16 to 250	60 to 480	2.43
Cooled from 850 °C to room temperature in 19 h	16 to 100	60 to 212	2.01
	16 to 250	60 to 480	2.89

Table 3. Physical and mechanical properties of Invar

Solidus temperature	1425 °C (2600 °F)
Density	8.0 Mg/m³ (0.29 lb/in.³)
Tensile strength	450 to 585 MPa (65 to 85 ksi)
Yield strength	275 to 415 MPa (40 to 60 ksi)
Elastic limit	140 to 205 MPa (20 to 30 ksi)
Elongation	30 to 45%
Reduction in area	55 to 70%
Scleroscope hardness	19
Brinell hardness	160
Modulus of elasticity	150 GPa (21.4 × 10⁶ psi)
Thermoelastic coefficient	500 $\mu m/m \cdot K$
Specific heat	515 J/kg·K (0.123 Btu/lb·°F) at 25 to 100 °C (78 to 212 °F)
Thermal conductivity	11 W/m·K (6.4 Btu/ft·h·°F) at 20 to 100 °C (68 to 212 °F)
Thermoelectric potential (against copper)	9.8 μV/K at −96 °C (−140 °F)

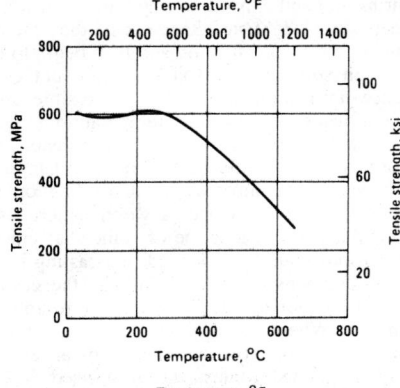

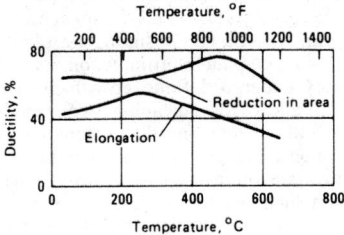

Alloy composition: 0.25 C, 0.55 Mn, 0.27 Si, 33.9 Ni, rem Fe. Heat treatment: annealed at 800 °C (1475 °F) and furnace cooled.

Fig. 5. Mechanical properties of a forged 34% Ni alloy

(200 to 300 °F) and then cool it to room temperature over a period of several months. Slow cooling through the magnetic transformation range also may be satisfactory.

Where exacting specifications call for definite and invariable properties, alloys with low coefficients of expansion are being made by powder metallurgy. Where free machining is desirable, addition of 0.25% selenium produces satisfactory results.

IRON-NICKEL ALLOYS OTHER THAN INVAR

Iron-nickel alloys that have nickel contents higher than that of Invar retain to some extent the expansion characteristics of Invar. Because

Table 4. Expansion characteristics of Fe-Ni alloys

Composition, %			Inflection temperature		Mean coefficient of expansion, from 20 °C to inflection temperature, $\mu m/m \cdot K$
Mn	Si	Ni	°C	°F	
0.11	0.02	30.14 155		310	9.2
0.15	0.33	35.65 215		420	1.54
0.12	0.07	38.70 340		645	2.50
0.24	0.03	41.88 375		705	4.85
. . .	. . .	42.31 380		715	5.07
. . .	. . .	43.01 410		770	5.71
. . .	. . .	45.16 425		795	7.25
0.35	. . .	45.22 425		795	6.75
0.24	0.11	46.00 465		870	7.61
. . .	. . .	47.37 465		870	8.04
0.09	0.03	48.10 497		925	8.79
0.75	0.00	49.90 500		930	8.84
. . .	. . .	50.00 515		960	9.18
0.25	0.20	50.05 527		980	9.46
0.01	0.18	51.70 545		1015	9.61
0.03	0.16	52.10 550		1020	10.28
0.35	0.04	52.25 550		1020	10.09
0.05	0.03	53.40 580		1075	10.63
0.12	0.07	55.20 590		1095	11.36
0.25	0.05	57.81	None		12.24
0.22	0.07	60.60	None		12.78
0.18	0.04	64.87	None		13.62
0.00	0.05	67.98	None		14.37

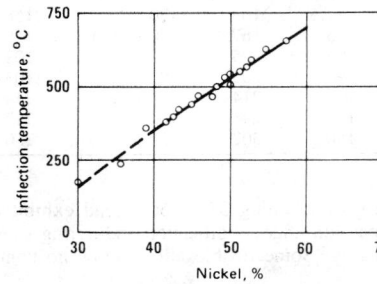
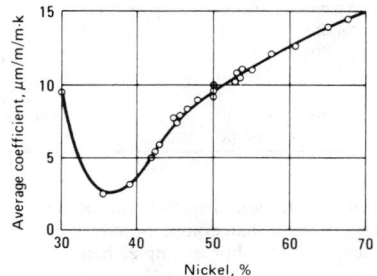

Left: Variation of inflection temperature. Right: Variation of average coefficient of expansion between room temperature and inflection temperature.

Fig. 6. Effect of nickel content on expansion of Fe-Ni alloys

further additions of nickel raise the temperature at which the inherent magnetism of the alloy disappears, the inflection temperature in the expansion curve rises with increasing nickel content. Although this increase in range is an advantage in some circumstances, it is accompanied by an increase in coefficient of expansion.

Table 4 and Fig. 6 present information on the coefficients of expansion of iron-nickel alloys at temperatures up to the inflection temperature. They also give data on alloys containing up to 68% Ni.

Fusible Alloys

Condensed from Metals Handbook, Ninth Edition, Volume 3, pages 799 to 801.

THE TERM "fusible alloys" refers to any of the more than 100 white metal alloys that melt at relatively low temperatures. Most commercial fusible alloys contain bismuth, lead, tin, cadmium, indium and antimony, and special alloys of this class may also contain significant amounts of zinc, silver, thallium or gallium.

Many of the fusible alloys used in industrial applications are based on eutectic compositions (see Table 1). Use of such alloys is important in automatic safety devices such as fire sprinklers,

boiler plugs and controls for furnaces. Under ambient temperature, such an alloy has sufficient strength to hold parts together, but at a specific elevated temperature the fusible-alloy link will melt, thus disconnecting the parts. In fire sprinklers, the links melt when dangerous temperatures are reached, releasing water from piping systems and extinguishing the fire.

In addition to eutectic alloys, each of which melts at a specific temperature, there are numerous noneutectic alloys, which melt over a range of temperatures. Some of the more important noneutectic alloy compositions are listed in Table 2.

A fusible alloy with a wide melting range is useful in staking rods and tubing in assemblies

because the alloy is distributed around part surfaces while still molten and provides a firm anchorage after it solidifies.

Properties. Most fusible alloys are heavy, bright, silvery, nontarnishing metals that can be melted repeatedly with negligible loss of elemental constituents. They are ageable alloys and thus their mechanical properties often are dependent on the period of time that has elapsed since casting, as well as on conditions of casting and rate of solidification. The test conditions employed also affect mechanical-property values. For example, many fusible alloys may appear brittle when subjected to sudden shock but exhibit high ductility under low rates of strain.

In certain alloys, normal thermal contraction

Table 1. Compositions and melting temperatures of several eutectic fusible alloys

Bi	Pb	Composition, % Sn	Cd	Other	Melting temperature °C	°F
44.70	22.60	8.30	5.30	19.10 In 46.8		116
49.00	18.00	12.00	...	21.00 In 58		136
50.00	26.70	13.30	10.00	 70		158
51.60	40.20	...	8.20	... . 91.5		197
52.50	32.00	15.50	...	 95		203
54.00	...	26.00	20.00	... . 102.5		217
55.50	44.50	...	...	 124		255
58.00	...	42.00	...	... . 138.5		281
...	30.60	51.20	18.20	 142		288
60.00	...	...	40.00	 144		291
...	...	67.75	32.25	 177		351
...	38.14	61.86	...	 183		361
...	...	91.00	...	9.00 Zn 199		390
...	...	96.00	...	3.50 Ag 221.3		430
...	79.7	...	17.7	2.60 Sb 236		457
...	87.0	...	...	13.00 Sb 247		477

Table 2. Compositions, yield temperatures and melting-temperature ranges for several noneutectic fusible alloys

Bi	Pb	Composition, % Sn	Cd	Other	Yield temperature °C	°F	Melting-temperature range °C	°F
50.50	27.8	12.40	9.30	... . 70.5		159	70 to 73	158 to 163
50.00	34.5	9.30	6.20	... . 72.0		162	70 to 79	158 to 174
50.72	30.91	14.97	3.40	... . 72.5		163	70 to 84	158 to 183
42.50	37.70	11.30	8.50	... . 72.5		163	70 to 90	158 to 194
35.10	36.40	19.06	9.44	 75		167	70 to 101	158 to 214
56.00	22.00	22.00	...	 96		205	95 to 104	203 to 219
67.00	16.00	17.00	...	 96		205	95 to 149	203 to 300
33.33	33.34	33.33	...	 101		214	101 to 143	214 to 289
48.00	28.50	14.50	...	9.00 Sb 116		241	103 to 227	217 to 441
40.00	...	60.00	...	 150		302	138 to 170	280 to 338

due to cooling after solidification can be partly, completely, or more than compensated for by expansion due to aging. For example, bismuth al-

loys containing 33 to 66% lead exhibit net expansion after solidification and during subsequent aging. Some fusible alloys show no contraction

(shrinkage) and expand rapidly while still warm; others show slight shrinkage during the first few minutes after solidification and then begin to expand; in still others, expansion does not commence until some time after the fusible alloy casting has cooled to room temperature.

All three characteristics — net expansion, net contraction, and little or no volume change — provide specific advantages, depending on the application. For instance, a wood pattern used for making molds must be of somewhat greater dimensions than those desired in a casting in order to compensate for shrinkage of the casting on solidification and during cooling to room temperature. Where metal patterns are cast from a master wood pattern, two such allowances will have to be made unless the alloy used for the metal patterns possesses zero shrinkage. Fusible alloys with eutectic compositions are often used for casting metal patterns from wood masters because they undergo definite growth that is sufficient to allow for "cleaning" of production castings without reducing dimensions below required values. The growth characteristics of fusible alloys are often used to advantage when a metal part such as a turbine blade is to be firmly anchored in a lathe chuck. After the part is machined, the fusible alloy is melted away.

Growth of fusible alloys that exhibit this characteristic varies from 0.01 to 0.7% in linear dimensions, depending primarily on alloy composition.

Generally, load-bearing capacity of fusible alloys is good, although some deformation will occur under prolonged stress. In addition, hardness and other mechanical properties of many fusible alloys change gradually with time, probably due to the same microstructural changes that are responsible for growth or shrinkage on aging.

Materials for Sliding Bearings

Condensed from Metals Handbook, Ninth Edition, Volume 3, pages 802 to 822.

A SLIDING BEARING (plain bearing) is a machine element designed to transmit loads or reaction forces to a shaft that rotates relative to the bearing. Journal bearings are cylindrical (full cylinders or segments of cylinders) and are used when the load or reaction force is essentially radial—that is, perpendicular to the axis of the shaft. Thrust bearings are ring-shape bearings (full rings or segments of rings) and are used when the load or reaction force is parallel to the direction of the shaft axis. Both radial and axial loads can be accommodated by flange bearings, which are journal bearings constructed with one or two integral thrust-bearing surfaces. The sliding movement of the shaft surface or thrust-collar surface relative to the bearing surface is characteristic of all plain bearings. In many applications, plain bearings offer advantages over rolling-contact bearings—advantages such as lower cost, smaller space requirements, ability to operate with marginal lubricants, resistance to corrosion and ability to sustain high specific loads.

CLASSIFICATIONS

Functions and Applications. When considered in terms of function or location in a machine, bearings are commonly called by the following names, which describe their use:

- Connecting-rod bearing
- Main bearing
- Piston-pin bushing
- Camshaft bearing or bushing
- Crankshaft thrust washer
- Leaf-spring bushing
- Front-axle bushing
- Idler-gear bushing
- Brake-pedal shaft bushing
- Rocker-arm bushing
- Transmission-gear bushing
- Reverse-pinion bushing
- Differential-trunnion bushing
- Intermediate gear thrust washer
- Countershaft thrust washer
- Electric-motor shaft bushing.

The terms "bearing" and "bushing" are used interchangeably, and do not have meanings that are significantly different in terms of function or location in a machine.

A major distinction can be drawn, however, between (a) connecting-rod and main bearings and (b) the remaining items in the list. In a reciprocating engine the connecting-rod and main bearings must support the entire firing load on the pistons, which is transmitted through the connecting rods to the crankshaft. These bearings therefore are subject to exceptionally severe cyclic loading, shaft-deflection effects, and variations in lubricant-film thickness. The conditions imposed on the other bearings listed differ mainly in type of load (steady or intermittent), magnitude of load, speed of operation, direction of operation (unidirectional or oscillating) and operating temperature.

Structural Characteristics. Bearings are also frequently classified according to material construction as solid (single-metal), bimetal (two-layer) or trimetal (three-layer) bearings.

In current practice, connecting-rod and main bearings of reciprocating engines are usually of either bimetal or trimetal construction, but single-metal bearings are employed occasionally. Bearings for other machinery applications are most often of single-metal or bimetal construction, although trimetal construction is sometimes required.

Size, Configuration and Manufacturing Method. With respect to size, bearings commonly are classified as either thin-wall or heavy-wall bearings; in general, bearings greater than about 5 mm (0.2 in.) in wall thickness and not less than 150 mm (6 in.) in diameter are considered heavy-wall bearings. Configuration may be further described as half round, full round, flanged or washer. SAE standards classify bearings used in mass-produced machinery (virtually all thin-wall bearings) into three groups: sleeve-type half bearings, split-type bushings and thrust washers.

OPERATING CONDITIONS

It is necessary to analyze both the mechanical design and the material requirements of a bearing in terms of the system in which it will be used. The important mechanical components of a system include the bearing housing, the lubricant film, the surface and subsurface of the bearing itself, and the surface of the journal or thrust collar. The interactions among these components in an operating machine can be exceedingly complex, and they are not subject to rigorous analysis on the basis of any established general theory.

However, a substantial body of empirical knowledge and a growing fund of theoretical understanding exist, permitting engineering decisions to be made on questions of bearing design and material selection with a minimum of trial-and-error testing.

The technical factors that are most important from the standpoint of material selection are discussed briefly below.

Loads and Speeds. The approximate magnitude of the specific load to which a bearing will be subjected should be known, so that materials of insufficient strength can be eliminated from consideration. However, precise values of actual bearing unit loads in operating machinery are not easily obtained. The most common approximation is the mean unit load, P, which is calculated from the equation

$$P = \frac{F}{L \times D} \qquad \text{(Eq 1)}$$

where F is maximum total force acting on the bearing, L is axial bearing length and D is bearing diameter.

Total force usually can be estimated with reasonable accuracy from known machine design parameters, and sometimes can be determined experimentally. Values of P for bearings in commercial machinery vary from nearly zero to 105 MPa (15 ksi). Where the loading on a bearing is steady or varies in a simple pattern, the P value is a satisfactory reference for selecting bearing materials of adequate strength.

Lubrication and Friction. In order to minimize the friction associated with the sliding movement of plain bearings, a fluid lubricant is almost invariably used. Oils and greases are the most commonly used lubricants; but other fluids, including hydraulic fluids, water and even air, may be used in special applications.

Every effort should be made to design any bearing so that a continuous film of lubricant will be maintained between the sliding surfaces during operation.

Heat and Temperature. Even in the absence of metallic contact, generation of some frictional heat is unavoidable in machinery bearings. In any heat engine, additional heating occurs by conduction from the working fluid. Artificial cooling of bearings generally is accomplished by use of a heat exchanger installed in the lubricant system, by means of which excessively high oil and bearing temperatures can be avoided. In machinery other than heat engines, it is not usually necessary to employ artificial lubricant cooling.

In general, successful lubrication and bearing operation require that the inlet lubricant temperature be maintained below about 135 °C (275 °F) if the lubricant is a hydrocarbon oil. Under these conditions, bearing-back temperatures of 175 °C (350 °F) or less usually can be maintained. Temperatures as high as 260 °C (500 °F) can be tolerated by some synthetic lubricating fluids, with correspondingly higher bearing-back temperatures. Actual surface temperatures of bearing liners in operation are rarely known, but must be assumed to be higher than corresponding bearing-back temperatures.

Corrosion. Except in pumps that handle corrosive fluids, plain bearings usually are not required to operate in extremely hostile chemical environments. It is possible, however, for corrosive problems to develop in lubricating oils as a result of oxidation and/or by reaction with engine coolants and combustion products. Fatty acids, alcohols, aldehydes and ketones formed in this way can cause corrosion of bearing metals. Acidic sulfur compounds in lubricating oils, which may be initially present or which may result from oxidation or from contamination by combustion products, also can act as corrodents.

PROPERTIES OF BEARING MATERIALS

Surface and Bulk Properties. The nature of the conditions under which plain bearings must operate and the wide ranges over which these conditions can vary lead to concern for bearing-material properties of two kinds: (*a*) surface properties (those associated with the bearing surface and immediate subsurface layers) and (*b*) bulk properties.

Conventional engineering definitions of material properties are not sufficient to characterize all the essential attributes of bearing materials. Although there is no universally accepted system of nomenclature, measurement or testing for these properties, they can be defined and studied in terms of the following characteristics:

- *Compatibility:* the antiwelding and antiscoring characteristics of a bearing material when operated with a given mating material
- *Conformability:* the ability of a material to yield to and compensate for slight misalignment and to conform to variations in the shape of the shaft or of the bearing-housing bore
- *Embeddability:* the ability of a material to embed dirt or foreign particles and thus prevent them from scoring and wearing shaft and bearing surfaces
- *Load capacity:* the maximum unit pressure under which a material can operate without excessive friction, wear and fatigue damage
- *Fatigue strength:* the ability of a material to function under cyclic loading below its elastic limit without developing cracks or surface pits
- *Corrosion resistance:* the ability of a material to withstand chemical attack by uninhibited or contaminated lubricating oils
- *Hardness:* the ability to resist plastic flow under high unit compressive loads, convention-

ally measured by indentation hardness testing
- *Strength:* the ability to resist elastic and plastic deformation under load, conventionally measured by compression, shear and tensile testing.

Compatibility can be regarded as a purely surface characteristic. Conformability and embeddability involve the surface and immediate subsurface, but are strongly related to the bulk properties of strength and hardness. The other characteristics relate principally to bulk properties.

Measurement and Testing. Of all the characteristics in the list above, only hardness and strength can be measured satisfactorily by standard laboratory test methods. Many special dynamic test rigs and test methods have been developed in the plain-bearing industry to evaluate and measure the other characteristics and their interactions. Although much useful information has been developed through laboratory rig testing, it is still often necessary to test bearing materials and designs in full-size operating machines in order to clearly establish their over-all suitability. Such testing is necessarily expensive and time-consuming. It should be undertaken only after careful study of the conditions under which the bearing will operate and of prior experience in similar applications.

Bearing-Material Structures. All commercially significant bearing metals except silver are used as polyphase alloys. Typical microstructures are shown beginning on page 232 in Volume 7 of the 8th Edition of this Handbook. Four general microstructural types can be recognized:

- *Soft Matrix With Discrete Hard Particles.* Lead babbitts and tin babbitts are of this type.
- *Interlocked Soft and Hard Continuous Phases.* Many copper-lead alloys are of this type.
- *Strong Matrix With Discrete Soft-Phase Pockets.* Leaded bronzes, aluminum-tin alloys and aluminum-lead alloys are of this type.

Laminated Construction. One of the most useful concepts in bearing-material design came with the recognition in 1941 that the effective load capacities and fatigue strengths of lead and tin alloys were sharply increased when these alloys were used as thin layers intimately bonded to strong bearing backs of bronze or steel. Use is made of this principle (Fig. 1) in both two-layer and three-layer constructions, in which the surface layer is composed of a lead or tin alloy, usually no more than 0.13 mm (0.005 in.) thick.

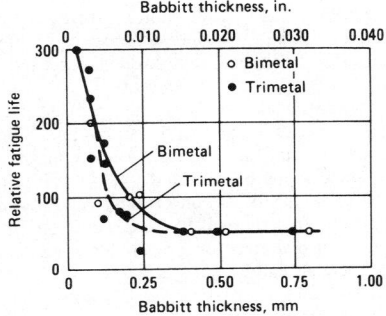

Bearing load: 14 MPa (2000 psi) for all tests.
Fig. 1. Variation of bearing life with babbitt thickness for lead or tin babbitt bearings

Unimpaired compatibility is provided by such a layer, together with reasonably high levels of conformability and embeddability. Other useful compromises can be effected between surface and bulk properties by employing an intermediate copper alloy or aluminum alloy layer between the surface alloy layer and a steel back. In these three-layer constructions, use of surface-layer thicknesses as low as 0.013 mm (0.0005 in.) offers even more favorable compromises between surface and bulk properties than are possible with two-layer constructions.

BEARING-MATERIAL SYSTEMS

Because of the widely varying conditions under which bearings must operate, commercial bearing materials have evolved as specialized engineering materials systems rather than as commodity products. They are used in relatively small tonnages and are produced by a relatively small number of manufacturers. Much proprietary technology is involved in alloy formulation and processing methods. Successful selection of a bearing material for a specific application often requires close technical cooperation between the user and the bearing producer.

Single-Metal Systems

Virtually all single-metal sliding bearings are made of either copper alloys or aluminum alloys. Considerable ranges of compositions and properties are available in the older copper group.

Single-metal systems do not exhibit outstandingly good surface properties, and their tolerance of boundary and thin-film lubrication conditions is limited. As a result, the load-capacity rating for a single-metal bearing usually is low relative to the fatigue strength of the material from which it was made. Because of their metallurgical simplicity, these materials are well suited to small-lot manufacturing from cast tubes or bars, using conventional machine-shop processes.

Copper Alloys. Except for commercial bronze and low-lead tin bronze, copper alloys in single-metal systems are almost always used in cast form. This provides thick bearing walls (3.20 mm, or 0.125 in.) strong enough so that the bearing is retained in place when press fitted into the housing.

Commercial bronze and low-lead tin bronze (alloys C22000 and C47600) are used extensively in the form of wrought strip for thin-wall bushings, which are made in large volumes by high-speed press forming. The poor compatibility of these alloys can be improved by embedding a graphite-resin paste in rolled or pressed-in indentations, so that the running surface of the bushing consists of interspersed areas of graphite and bronze. Such bushings are widely used in automotive engine starting motors.

The lead in leaded tin bronzes is present in the form of free lead, dispersed throughout a copper-tin matrix so that the bearing surface consists of interspersed areas of lead and bronze. This improves compatibility, conformability and embeddability. In general, the best selection from this group of materials for a given application will be the highest-lead composition that can be used without risking excessive wear, plastic deformation or fatigue damage.

Low-lead and unleaded bronzes are also used in porous bushings produced by powder metallurgy methods. The sintered bushings are impregnated with oil, which provides a built-in supply of lubricant. Such bushings are widely used in applications involving light loads and requiring self-lubricating properties.

Aluminum Alloys. Virtually all solid aluminum bearings used in the United States are made from alloys containing from 5.5 to 7% tin, plus smaller amounts of copper, nickel, silicon and magnesium. Starting forms for bearing fabrication include cast and wrought tubes as well as rolled plate and strip, which can be press formed into half-round shapes. As in the case of solid bronze bearings, relatively thick bearing walls are employed.

The tin in these alloys is present in the form of free tin, dispersed throughout an aluminum matrix so that the bearing surface consists of interspersed areas of aluminum and tin. Surface properties are enhanced by the free tin in much the same way that those of bronze are improved by the presence of free lead.

The high thermal expansion of aluminum poses special problems in maintaining press-fit and running clearances. Various methods are employed for increasing yield strength, through heat treatment and cold work, to overcome plastic flow and permanent deformation under service temperatures and loads.

Bimetal Systems

All bimetal systems employ a strong bearing back to which a softer, weaker, relatively thin layer of a bearing alloy is metallurgically bonded. Low-carbon steel is by far the most widely used bearing-back material, although alloy steels, bronzes, brasses and (to a limited extent) aluminum alloys are also used. When steel bearing backs are employed, load-capacity ratings for both copper and aluminum alloys are sharply increased above those of the corresponding single metals, without degrading any other properties.

Bronze-back bearings do not exhibit combinations of performance characteristics substantially different from those of steel-back bearings. The practical advantages of bronze as a bearing-back material lie partly in the economics of small-lot manufacturing and partly in the relative ease with which worn bronze-back bearings can be salvaged by rebabbitting and remachining. From the standpoint of performance, the advantage of bronze over steel as a bearing-back material is the protection bronze affords against catastrophic bearing seizure in case of severe liner wear or fatigue.

Although the surface properties of bronze bearing-back materials are not impressive, they are superior to those of steel, and these "reserve" bearing properties can be of considerable practical importance in large, expensive machinery used in certain critical applications.

Trimetal Systems

All trimetal systems employ a steel bearing back, an intermediate layer of relatively high fatigue strength, and a tin alloy or lead alloy surface layer. Most trimetal systems are derived from steel-back bimetal systems by addition of a surface layer.

The strengthening effects of thin-layer construction are notable in those systems that incorporate electroplated lead alloy surface layers approximately 0.025 mm (0.001 in.) thick. Comparison of fatigue-strength and load-capacity ratings of these systems with those of corresponding bimetal systems shows that the thin lead alloy surface layer upgrades not only surface properties but also fatigue strength. The increase in fatigue strength can be attributed at least in part to the elimination of surface stress raisers, from which fatigue cracks can propagate.

Trimetal systems with electroplated lead-base surface layers and copper or aluminum alloy intermediate layers provide the best available combinations of cost, fatigue strength and surface properties. Such bearings have high tolerances for boundary and thin-film lubrication conditions, and thus can be used under higher loads than can any of the bimetal systems.

TIN ALLOYS

Tin-base bearing materials (babbitts) are alloys of tin, antimony and copper that contain limited amounts of zinc, aluminum, arsenic, bismuth and iron. The compositions of tin-base bearing alloys, according to ASTM B23 and SAE specifications, are given in Table 1.

The presence of zinc in these bearing metals generally is not favored. Arsenic increases resistance to deformation at all temperatures; zinc has a similar effect at 38 °C (100 °F), but causes little or no change at room temperature. Zinc has a marked effect on the microstructures of some of these alloys. Small quantities of aluminum (even less than 1%) will modify their microstructures. Bismuth is objectionable because, in combination with tin, it forms a eutectic that melts at 137 °C (279 °F). At temperatures above this eutectic, alloy strength is decreased appreciably.

In high-tin alloys, such as ASTM grades 1, 2 and 3, and SAE 11 and 12, lead content is limited to 0.50% or less because of the deleterious effect of higher percentages on the strength of these alloys at temperatures of 149 °C (300 °F) or above. Lead and tin form a eutectic that melts at 183 °C (361 °F). At higher temperatures, bearings become fragile as a result of formation of a liquid phase within them. Mechanical properties of ASTM grades 1 to 3 are given in Table 2.

Table 1. Compositions of tin-base bearing alloys

Designation	Sn(a)	Sb	Pb (max)	Cu	Fe (max)	As (max)	Bi (max)	Zn (max)	Al (max)	Others (max total)
ASTM B23 alloys										
Alloy 1	91.0	4.5	0.35	4.5	0.08	0.10	0.08	0.005	0.005	...
Alloy 2	89.0	7.5	0.35	3.5	0.08	0.10	0.08	0.005	0.005	...
Alloy 3	84.0	8.0	0.35	8.0	0.08	0.10	0.08	0.005	0.005	...
Alloy 11	87.5	6.8	0.50	5.8	0.08	0.10	0.08	0.005	0.005	...
SAE alloys										
SAE 11	86.0	6.0–7.5	0.50	5.0–6.5	0.08	0.10	0.08	0.005	0.005	0.20
SAE 12	88.0	7.0–8.0	0.50	3.0–4.0	0.08	0.10	0.08	0.005	0.005	0.20

(a) Desired in ASTM alloys; specified minimum in SAE alloys.

Table 2. Properties of selected ASTM B23 tin-base bearing alloys

Desig-nation	Specific gravity	Compressive yield strength(a)(b) At 20 °C (68 °F) MPa	ksi	At 100 °C (212 °F) MPa	ksi	Compressive ultimate strength(a)(c) At 20 °C (68 °F) MPa	ksi	At 100 °C (212 °F) MPa	ksi	Hardness(d), HB At 20 °C	At 100 °C	Solidus temperature °C	°F	Liquidus temperature °C	°F	Pouring temperature °C	°F
Alloy 1	7.34	30.3	4.40	18.3	2.65	88.6	12.85	47.9	6.95	17.0	8.0	223	433	371	700	440	825
Alloy 2	7.39	42.1	6.10	20.7	3.00	102.7	14.90	60.0	8.70	24.5	12.0	241	466	354	669	425	795
Alloy 3	7.46	45.5	6.60	21.7	3.15	121.3	17.60	68.3	9.90	27.0	14.5	240	464	422	792	490	915

(a) The compression-test specimens were cylinders 1½ in. long and ½ in. in diameter, machined from chill castings 2 in. long and ¾ in. in diameter. (b) Values for yield point were taken from stress-strain curves at a deformation of 0.125% reduction of gage length. (c) Values for ultimate strength were taken as the unit load necessary to produce a deformation of 25% of the length of the specimen. (d) Tests were made on the bottom face of parallel machined specimens cast at room temperature in a steel mold 2 in. in diameter by ⅝ in. deep. The Brinell hardness values listed are the averages of three impressions on each alloy, using a 10-mm ball and applying a 500-kg load for 30 s.

The mechanical-property values obtained from massive cast specimens are dependent on temperature. Hardness and compression tests are sensitive also to duration of the load because of the plastic nature of these materials. Bulk properties may be of some value in initial screening of materials, but they do not accurately predict behavior of the material in the form of a thin layer bonded to a strong backing, which is the manner in which the babbitts are normally used. The relationship that exists between bearing life and thickness of babbitt is shown in Fig. 1, and the marked influence of operating temperature is shown in Fig. 2.

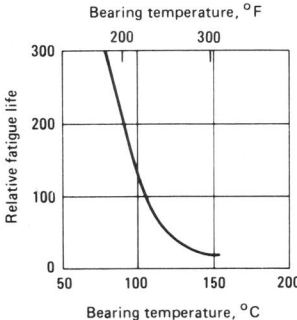

SAE 12 alloy lining, 0.05 to 0.13 mm (0.002 to 0.005 in.) thick, on steel backing. Bearing load: 14 MPa (2000 psi).

Fig. 2. Variation of bearing life with temperature for SAE 12 bimetal bearings

Compared with other bearing materials, tin alloys have low resistance to fatigue, but their strength is sufficient to warrant their use under low-load conditions. These alloys are commercially easy to bond and handle, and they have excellent antiseizure qualities. Furthermore, they are much more resistant to corrosion than lead-base bearing alloys.

LEAD ALLOYS

There are two types of lead babbitts: (a) alloys of lead, tin, antimony and in many instances arsenic; and (b) alloys of lead, calcium, tin and one or more of the alkaline earth metals. Many alloys

Table 3. Nominal compositions of lead-base bearing alloys

Designation	Pb	Sb	Sn	Cu (max)	Fe (max)	As	Bi (max)	Zn (max)	Al (max)	Cd (max)	Others
ASTM B23 alloys											
Alloy 7 (a)	Rem	15.0	10.0	0.50	0.1	0.45	0.10	0.005	0.005	0.05	...
Alloy 8	Rem	15.0	5.0	0.50	0.1	0.45	0.10	0.005	0.005	0.05	...
Alloy 13 (b)	Rem	10.0	6.0	0.50	0.1	0.25(a)	0.10	0.005	0.005	0.05	...
Alloy 15 (c)	Rem	16.0	1.0	0.50	0.1	1.10	0.10	0.005	0.005	0.05	...
Other alloys											
SAE 16	Rem	3.5	4.5	0.10	...	0.05(a)	0.10	0.005	0.005	0.005	...
AAR M501 (d) ..	Rem	8.75	3.5	0.50	...	0.20(a)	...	...	...	...	...
SAE 19	Rem	...	10.0	...	...	...	...	...	...	...	...
SAE 190	Rem	...	7.0	3.0	...	...	...	...	...	...	...
Proprietary alloys											
A	95.65	...	3.35	0.08	...	...	...	...	...	...	0.67 Ca
B	83.30	12.54	0.84	0.10	...	3.05	...	...	...	...	...
C	Rem	10.0	3.0	0.20	...	...	...	...	...	...	2.0 Ag

(a) Also SAE 14. (b) Also SAE 13. (c) Also SAE 5. (d) Association of American Railroads, Specification M501; also ASTM B67.

of the first group have been used for centuries as type metals, and were probably employed as bearing materials because of the properties they were known to possess. The advantages of arsenic additions in this type of bearing alloy have been generally recognized since 1938. Alloys of the second type were developed early in the 20th century.

Nominal compositions of lead-base bearing alloys covered by ASTM specifications are listed in Table 3, along with compositions of other proprietary alloys. Additional information on the mechanical properties of some of these alloys is given in Table 4.

In the absence of arsenic, the microstructures of these alloys comprise cuboid primary crystals of SbSn, or of antimony embedded in a ternary mixture of Pb-Sb-SbSn in which lead forms the matrix. The number of these cuboids per unit volume of alloy increases as antimony content increases. If antimony content is more than about 15%, the total amount of the hard constituents increases to such an extent that the alloys become too brittle to be useful as bearing materials.

Arsenic is added to lead babbitts to improve their mechanical properties, particularly at elevated temperatures. All lead babbitts are subject to softening or loss of strength during prolonged exposure to the temperatures (95 to 150 °C; 200 to 300 °F) at which they serve as bearings in internal-combustion engines. Addition of arsenic minimizes such softening. Under suitable casting conditions, the arsenical lead babbitts—for example, SAE 15 (ASTM grade 15)—develop remarkably fine and uniform structures. They also have better fatigue strength than arsenic-free alloys.

Arsenical babbitts give satisfactory service in many applications. Use of these alloys increased greatly during the second world war, particularly in the automobile industry and in diesel engines. The alloy most widely used is SAE 15 (ASTM grade 15), which contains 1% arsenic. Automobile bearings of this alloy usually are made from continuously cast bimetal (steel/babbitt) strip. When properly handled, this alloy can withstand the considerable strain that results from forming the bimetal strip into bearings.

Diesel-engine bearings often are cast as individual bearing shells by either centrifugal or gravity methods. In applications where higher hardness is required and where formability requirements are less severe (rolling-mill bearings, for instance), an alloy that contains 3% arsenic

Table 4. Properties of selected ASTM B23 lead-base bearing alloys

Designation	Specific gravity	Compressive yield strength(a)(b) At 20 °C MPa	ksi	At 100 °C MPa	ksi	Compressive ultimate strength(a)(c) At 20 °C MPa	ksi	At 100 °C MPa	ksi	Hardness(d), HB At 20 °C	At 100 °C	Solidus temperature °C	°F	Liquidus temperature °C	°F	Pouring temperature °C	°F
Alloy 7	9.73	24.5	3.55	11.0	1.60	107.9	15.65	42.4	6.15	22.5	10.5	240	464	268	514	338	640
Alloy 8	10.04	23.4	3.40	12.1	1.75	107.6	15.60	42.4	6.15	20.0	9.5	237	459	272	522	340	645
Alloy 15	10.05	...	...	...	...	...	...	...	...	21.0	13.0	248	479	281	538	350	662

(a) The compression-test specimens were cylinders 1.5 in. long, 0.5 in. in diameter, machined from chill castings 2 in. long, 0.75 in. in diameter. (b) Values were taken from stress-strain curves at a deformation of 0.125% reduction of gage length. (c) Values were taken as the unit load necessary to produce a deformation of 25% of the length of the specimen. (d) Tests were made on the bottom face of parallel-machined specimens that had been cast at room temperature in a steel mold, 2 in. in diameter by 0.625 in. deep. Values listed are the averages of three impressions on each alloy, using a 10-mm ball and applying a 500-kg load for 30 s.

has been used successfully (alloy B in Table 3).

For many years, lead-base bearing alloys were considered to be only low-cost substitutes for tin alloys. However, the two groups of alloys do not differ greatly in antiseizure characteristics, and when lead-base alloys are used with steel backs and in thicknesses below 0.75 mm (0.03 in.), they have fatigue resistance that is equal to, if not better than, that of tin alloys. Bearings of any of these alloys remain serviceable longest when they are no more than 0.13 mm (0.005 in.) thick (see Fig. 1). The superiority of lead alloys over tin alloys becomes more marked as operating temperature increases. For this reason, automotive engineers generally favor lead-base alloys of compositions that approximate ASTM alloys 7 and 15 and SAE alloy 16. SAE alloy 16 is cast into and on a porous sintered matrix, usually copper-nickel, bonded to steel. The surface layer of babbitt is 0.025 to 0.13 mm (0.001 to 0.005 in.) thick.

The fatigue resistance of bearing materials depends to a great extent on the design of the bearing. The strength and rigidity of the supporting structure, the thickness of the backing metal (steel or bronze), the thickness of the bearing material and the character of the bond between the bearing material and the backing are all factors of consequence in bearings for use in high-speed reciprocating engines, such as the main and connecting-rod bearings of automobile and aircraft engines.

Resistance to fatigue is somewhat less important in bearings that operate under static load—for example, journal bearings in traction-motor supports for diesel locomotives and in railway freight cars. In such bearings, antiseizure characteristics, conformability, compressive strength, and resistance to abrasion and corrosion are of greater significance. The lining metal generally employed in such journal bearings is the low-arsenic AAR alloy (ASTM B67) cast onto a leaded bronze back.

Pouring temperature and rate of cooling markedly influence the microstructures and properties of lead alloys, particularly when they are used in the form of heavy liners for railway journals. High pouring temperatures and low cooling rates, such as result from use of overly hot mandrels, promote segregation and formation of a coarse structure. A coarse structure may cause brittleness, low compressive strength and low hardness. Therefore, low pouring temperatures (325 to 345 °C; 620 to 650 °F) usually are recommended. Because these alloys remain relatively fluid almost to the point of complete solidification (about 240 °C, or 465 °F, for most compositions), they are easy to manipulate and can be handled with no great loss of metal from drossing.

Use of lead babbitts containing calcium and alkaline earth metals is confined almost entirely to railway applications, although these babbitts also are employed to some extent in certain diesel-engine bearings. One of the more widely used alloys contains 1.0 to 1.5% tin, 0.50 to 0.75% calcium, and small amounts of various other elements. The strength of this alloy approximates that of a tin alloy containing 90% Sn, 8% Sb and 2% Cu. Hardness of this lead alloy is about 20 HB, and the solidus is 321 °C (610 °F). The liquidus is probably near 338 °C (640 °F). The pouring temperature, which varies from 500 to

Table 5. Designations and nominal compositions of copper-base bearing alloys

UNS number	SAE	Other	Former SAE	Cu	Sn	Pb	Zn	Other	Form	Use
Commercial bronze										
C22000	795	...	...	90	0.5	...	9.5	...	Wrought strip	Solid bronze bushings and washers
Unleaded tin bronzes										
C90300	C90300	...	620	88	8	0	4	...	Cast tubes	Solid bronze bearings
C90500	C90500	...	62	88	10	0	2	...	Cast tubes	Solid bronze bearings
C91100	...	...	...	84	16	0	0	...	Cast tubes	Solid bronze bearings
C91300	...	...	...	81	19	0	0	...	Cast tubes	Solid bronze bearings
Low-lead tin bronzes										
C92200	C92200	...	622	88.5	6	1.5	4	...	Cast tubes	Solid bronze bearings
C92300	C92300	...	621	87.0	8.5	0.5	4	...	Cast tubes	Solid bronze bearings
C92700	C92700	...	63	87.5	10	2	0.5	...	Cast tubes	Solid bronze bearings
Medium-lead tin bronzes										
C54400	791	...	...	88	4	4	4	...	Wrought strip	Solid bronze bushings and washers
...	...	F32/62	...	87	4	4	3	2 Fe	Cast on steel back	Bimetal bushings and washers; trimetal intermediate layer
C83600	C83600	...	40	85	5	5	5	...	Cast tubes	Solid bronze bearings and bronze bearing backs
C93200	C93200	...	660	83	7	7	3	...	Cast tubes	Solid bronze bearings
C93600	793	...	...	85	4	8	3	...	Cast on steel back	Bimetal surface layer
...	798	...	...	88	4	8	...	...	Sintered on steel back	Bimetal surface layer
C93700	C93700	...	64	80	10	10	...	...	Cast tubes	Solid bronze bearings and bronze bearing backs
...	792	...	...	80	10	10	...	...	Cast on steel back	Bimetal surface layer and trimetal intermediate layer
...	797	...	...	80	10	10	...	...	Sintered on steel back	Bimetal surface layer
High-lead tin bronze										
C93800	C93800	...	67	78	6	16	...	...	Cast tubes	Solid bronze bearings and bronze bearing backs
...	...	AMS 4825	...	74	10	16	...	...	Cast on steel back	Bimetal surface layer
...	794	...	...	71.5	3.5	23	2	...	Cast on steel back	Bimetal surface layer
...	799	...	...	74	3	23	...	...	Sintered on steel back	Bimetal surface layer
...	...	AMS 4824	...	75	1	24	...	...	Cast on steel back	Trimetal intermediate layer
...	...	F780	...	74	2.5	23.5	...	...	Sintered on steel back	Trimetal intermediate layer

Table 5 (continued)

UNS number	SAE	Other	Former SAE	Cu	Sn	Pb	Zn	Other	Form	Use
High-lead tin bronze (continued)										
...	...	F15/112	...	72	3	25	...	...	Cast on steel back	Bimetal surface layer and trimetal intermediate layer
...	...	AMS 4840	...	70	5	25	...	...	Cast tubes	Solid bronze bearings
Copper-lead alloys										
...	49	...	...	75.5	0.5	24	...	...	Cast on steel back	Trimetal intermediate layer
...	49	...	...	75.5	0.5	24	...	...	Sintered on steel back	Trimetal intermediate layer
...	48	...	...	70	...	28	...	1.5 Ag	Cast on steel back	Bimetal surface layer and trimetal intermediate layer
...	482	...	...	67	5	28	...	...	Sintered on steel impregnated with Pb-Sn	Bimetal surface layer
...	480	...	...	65	...	35	...	...	Cast on steel back	Bimetal surface layer
...	480	...	...	65	...	35	...	...	Sintered on steel back	Bimetal surface layer
...	481	...	...	55	0.25	40	...	5 Ag	Cast on steel back	Bimetal surface layer
...	484	...	...	55	3	42	...	(a)	Sintered on steel back, infiltrated with Pb	Bimetal surface layer
...	485	...	...	48	1	51	...	(a)	Sintered on steel back, infiltrated with Pb	Bimetal surface layer
				98	2	...	...	(b)	Sintered on steel back, infiltrated with Pb	Bimetal surface layer
...	...	F510	...	41	2	48	...	7 Ni, 2 Sb(a)	Sintered on steel back, infiltrated with Pb-Sn-Sb alloy (SAE 16)	Bimetal surface layer and trimetal intermediate layer
				86	2	...	...	12 Ni(b)	Sintered on steel back, infiltrated with Pb-Sn-Sb alloy (SAE 16)	Bimetal surface layer and trimetal intermediate layer
...	...	M100A	...	40	1	48	...	9 Ni, 2 Sb(a)	Sintered on steel back, infiltrated with Pb-Sn-Sb alloy (SAE 16)	Bimetal surface layer and trimetal intermediate layer
				85	...	...	...	15 Ni(b)	Sintered on steel back, infiltrated with Pb-Sn-Sb alloy (SAE 16)	Bimetal surface layer and trimetal intermediate layer

(a) Composition of dense alloy after infiltration. (b) Composition of open grid before infiltration.

520 °C (930 to 970 °F), is relatively high and accounts for the formation of a larger volume of dross than that encountered in melting of Pb-Sb-Sn alloys. Care must be taken to avoid contamination of the alloy with antimonial lead babbitts, and vice versa. Deformability and resistance to wear are of the same order as those of the other lead babbitts. Most alloys of this type are subject to corrosion by acidic oils.

OVERLAYS

The improvement in fatigue life that can be achieved by decreasing babbitt-layer thickness has already been noted. Economically as well as mechanically, it is difficult to consistently achieve very thin uniform babbitt layers bonded to bimetal shells by casting techniques. Therefore, the process of electroplating a thin precision babbitt layer on a very accurately machined bimetal shell was perfected. A specially designed plating rack allows the thickness of the coplated babbitt layer to be regulated so accurately that machining usually is not required.

Coplated tin babbitts were found to be inferior in performance to lead babbitts. Plated babbitts are somewhat different in structure and composition from their cast counterparts. SAE alloy 190 is the most widely used overlay plate. The tin content of this alloy gives it better wear resistance than that of pure lead, and is necessary to protect the lead from corrosion; the copper content increases fatigue life.

When an SAE 190 overlay is plated directly onto a copper-lead bimetal surface, the tin has a tendency to migrate to the copper-lead interface, forming a brittle copper-tin intermetallic compound and/or diffusing into the lead phase. This decreases the corrosion resistance of the overlay and causes embrittlement along the bond line. To avoid this deterioration, a thin, continuous barrier layer, preferably nickel about 1.3 μm (0.05 mil) thick, is plated onto the copper-lead surface just prior to plating of the overlay. In addition to providing better surface behavior, overlays improve fatigue performance of the intermediate layer by preventing cracking in this layer. Plated overlays generally range in thickness from 0.013 to 0.05 mm (0.0005 to 0.002 in.), with fatigue life increasing markedly as overlay thickness decreases. In order to take full advantage of the improved fatigue life achieved with thin overlays, it is necessary to minimize assembly imperfections (such as misalignment) and to maintain close tolerances on machined shafts and bearing bores. Engine components must be thoroughly cleaned before assembly, and adequate air and lubricant filtration must be maintained if the overlay is to survive during the useful life of the bearing. Under adverse wear conditions, however, premature removal of the overlay will not necessarily impair operation of the bearing, because the exposed intermediate bearing alloy layer should continue to function satisfactorily.

COPPER ALLOYS

Copper-base bearing alloys comprise a large family of materials with a considerable range of properties. They include commercial bronze, copper-lead alloys, and leaded and unleaded tin bronzes. They are used alone in single-metal bearings, as bearing backs with babbitt surface layers, as bimetal layers bonded to steel backs, and as intermediate layers in steel-backed tri-metal bearings.

Pure copper is a relatively soft, weak metal. The principal alloying element used to harden and strengthen it is tin, with which it forms a solid solution. However, when tin content is higher than about 8%, a hard constituent (the alpha-delta eutectoid) develops in cast copper-tin alloys because of deviation from true equilibrium. This constituent is quite hard, and its presence causes a considerable improvement in wear resistance. Lead is present in all cast copper-base bearing alloys as a nearly pure, discrete phase, because its solid solubility in the matrix is practically nil. The lead phase, which is exposed on the running surface of a bearing, constitutes a site vulnerable to corrosive attack under certain operating conditions.

The antifriction behavior of copper-base bearing alloys improves as lead content increases, although at the same time strength is degraded because of increased interruption of the continuity of the copper alloy matrix by the soft, weak lead. Thus, through judicious control of tin content, lead content and microstructure, an entire family of bearing alloys has evolved to suit a wide variety of bearing applications.

Table 5 gives specification numbers and nominal compositions of copper-base bearing alloys, as well as the forms in which the alloys are used and general notations on typical applications.

Commercial Bronze. Lead-free copper alloys are characterized by poor antifriction properties but fairly good load-carrying ability. Wrought commercial bronze strip (SAE 795) with 10% zinc can be readily press formed into cylindrical bushings and thrust washers. Strength can be increased by cold working this inexpensive material.

Unleaded Tin Bronzes. The unleaded copper-tin alloys are known as phosphor bronzes because they are deoxidized with phosphorus. They are used principally in cast form as shapes for specific applications, or as rods or tubes from which solid bearings are machined. They have excellent strength and wear resistance, both of which improve with increasing tin content, but poor surface properties. They are used for bridge turntables and trunnions in contact with high-strength steel, and in other slow-moving applications.

Low-Lead Tin Bronzes. The inherently poor machinability of tin bronzes can be improved by adding small amounts of lead. Such additions do not significantly improve surface properties such as lubricity, however, and applications for these alloys are essentially the same as those for unleaded tin bronzes.

Medium-Lead Tin Bronzes. The only wrought strip material in this group of alloys is SAE 791, which is press formed into solid bushings and thrust washers. C83600 is used in cast form as bearing backs in bimetal bearings. SAE 793 and 798 are chemically similar low-tin, medium-lead materials that are cast or sintered on steel backs and used as surface layers for medium-load bimetal bushings. SAE 792 and 797 are higher in tin and slightly higher in lead, are cast or sintered on steel backs, and are used for heavy-duty applications such as wrist pin bushings and heavy-duty thrust surfaces.

High-lead tin bronzes contain medium to high amounts of tin, and relatively high lead contents to markedly improve antifriction characteristics. SAE 794 and 799 are widely used as bushings for rotating loads, and have the same chemical

matrix as 793 and 798 but with three times as much free lead. Both are generally cast or sintered on steel backs and are used for somewhat higher speeds and lower loads than alloys 793 and 798. The 3Sn-25Pb alloy cast on a steel back provides a much stronger bronze matrix than plain 75-25 copper-lead alloy, and is used with a plated overlay as the intermediate layer in heavy-duty trimetal bearing applications such as main and connecting-rod bearings in diesel truck engines. This construction provides the highest load-carrying ability available at the present time in copper alloy trimetal bearings.

Copper-lead alloys are used extensively in automotive, aircraft and general engineering applications. These alloys are usually cast or sintered to a steel backing strip from which parts are blanked and formed into full-round or half-round shapes depending on final application. Copper-lead alloys continuously cast on steel strip typically consist of copper dendrites perpendicular and securely anchored to the steel back, with an interdendritic lead phase. In contrast, sintered copper-lead alloys of similar composition are composed of more equiaxed copper grains with an intergranular lead phase.

The high-lead alloys (28 to 40% Pb) may be used bare on steel or cast iron journals as medium-duty automotive bearings. Tin content in these alloys is restricted to a low value to maintain a soft copper matrix, which along with the higher lead improves the antifriction/antiseizure properties. Bare bimetal copper-lead bearings are used less frequently today than they were some years ago because the lead phase, present as nearly pure unalloyed lead in all cast copper alloys, is susceptible to attack by corrosive products that can form in the crankcase lubricant during the longer oil-change periods now in use. Therefore, many of the copper-lead alloys with lead contents near 25% are used as bases for plated overlays in trimetal bearings for automotive and diesel engines.

Other alloys included in this group are the special sintered and infiltrated or impregnated materials SAE 482, 484, and 485. The last two items in Table 5 consist of an open copper-nickel or copper-nickel-tin grid, which is sintered onto a steel back, then infiltrated with a Pb-Sn-Sb alloy (SAE 16) to make bimetal grid bearings. Alternatively, the lead-base alloy may be overcast so that it completely covers the grid. Excess babbitt can then be machined off, leaving a very thin layer covering the grid, to make a medium-duty trimetal bearing.

ALUMINUM ALLOYS

Successful commercial use of aluminum alloys in plain bearings dates back to about 1940, when low-tin aluminum alloy castings were introduced to replace solid bronze bearings for heavy machinery. Production of steel-backed strip materials by roll bonding became commercially successful about 1950, permitting the development of practical bimetal and trimetal bearing-material systems using aluminum alloys in place of babbitts and copper alloys.

The ready availability of aluminum and its relatively stable cost have provided an incentive for continuing development of its use in plain bearings. Aluminum single-metal, bimetal and trimetal systems now can be used in the same load ranges as babbitts, copper-lead alloys and high-lead tin bronzes. Moreover, the outstanding cor-

rosion resistance of aluminum has become an increasingly important consideration in recent years, and has led to widespread use of aluminum alloy materials in automotive engine bearings in preference to copper-lead alloys and leaded bronzes.

Designations and Compositions. Alloy designations and nominal compositions of the commercial aluminum-base bearing alloys used most extensively in the United States are listed in Table 6. In these alloys, additions of silicon, copper, nickel, magnesium and manganese function to strengthen the aluminum through solid-solution and precipitation mechanisms. Fatigue strength and the opposing properties of conformability and embeddability are largely controlled by these elements. Tin, cadmium and lead are instrumental in upgrading the inherently poor compatibility of aluminum. Silicon has a beneficial effect on compatibility in addition to a moderate strengthening effect. Although not well understood theoretically, this compatibility-improving mechanism is of considerable practical value and is utilized effectively in the high-lead and tin-free alloys.

Mechanical Properties and Alloy Tempers. Conventional mechanical properties, somewhat like microstructural features, are of more value in understanding fabrication behavior of aluminum-base bearing alloys than in predicting their bearing performance. With the exception of solid aluminum alloy bearings, in which there is no steel back and where press-fit retention depends entirely on the strength of the aluminum alloy, mechanical properties of finished bearings are rarely specified — and then usually for control purposes only. Consideration of some of these properties (Table 7) does, however, contribute to an understanding of these alloys as a family of related engineering materials, and of their relationship to the better-known structural aluminum alloys. The wrought alloys as a group are low in hardness and strength compared with conventional aluminum structural alloys. With one exception (No. 14, Table 6), no use is made of heat treatment or cold working for increasing mechanical strength.

Cast aluminum-base bearing alloys are low in hardness and strength compared with conventional cast aluminum alloys, but are heat treated and cold worked to increase their yield-strength levels above as-cast values.

The majority of current commercial applications of aluminum-base bearing alloys involve steel-backed bimetal or trimetal bearings. To determine the most cost-effective aluminum material for any specific application, consideration should be given to the economic advantages of bimetal versus trimetal systems. The higher cost of the high-tin and high-lead alloys usually is offset by eliminating the cost of the lead alloy overlay plate. If the higher load capacity of a trimetal material is required, it then becomes important to select an aluminum liner alloy that provides adequate but not excessive strength, so that conformability and embeddability are not sacrificed unnecessarily. The tin-free alloy group (alloys 10 to 14 in Table 6) offers a wide range of strength properties, and the most economical choice usually is found in this group.

SILVER ALLOYS

Use of silver in bearings is largely confined to unalloyed silver (AMS 4815) electroplated on steel shells, which then are machined to very close di-

Table 6. Designations and nominal compositions of aluminum-base bearing alloys

No.	SAE	AA	Other	Al	Si	Cu	Ni	Mg	Sn	Cd	Other	Form	Typical applications
High-tin aluminum alloy													
1	783	8081	...	79	...	1	...	...	20	...	...	Wrought strip, O temper, bonded to steel back	Bimetal surface layer
High-lead aluminum alloys													
2	...	...	F-66	85	4	1	...	...	1.5	...	8.5 Pb	Powder rolled and sintered strip, O temper, bonded to steel back	Bimetal surface layer
3	...	...	AL-6	88	4	0.5	...	0.5	1	...	6 Pb	Wrought strip, O temper, bonded to steel back	Bimetal surface layer
Low-tin aluminum alloys													
4	770	850.0	...	91.5	0.7	1	1	...	6.5	...	...	Cast tubes, T101 temper(a)	Solid aluminum bearings; aluminum bearing backs
5	...	A850.0	...	89.5	2.5	1	0.5	...	6.5	...	...	Cast tubes, T101 temper(a)	Solid aluminum bearings; aluminum bearing backs
6	...	B850.0	...	89.5	...	2	1.2	0.8	6.5	...	...	Cast tubes, T5 temper(b)	Solid aluminum bearings; aluminum bearing backs
7	...	...	MB-7	89	0.7	1	1.7	1	7	...	...	Cast tubes, T5 temper(b)	Solid aluminum bearings; aluminum bearing backs
8	780	828.0	...	90.5	1.5	1	0.5	...	6.5	...	...	Wrought strip and plate, H12(c) temper	Solid aluminum bearings; aluminum bearing backs
												Wrought strip, O temper, bonded to steel back	Bimetal surface layer; trimetal intermediate layer
9	...	...	A300	91	1	2	...	...	6	...	...	Wrought strip, O temper, bonded to steel back	Trimetal intermediate layer
Tin-free aluminum alloys													
10	781	4002	...	95	4	0.1	...	0.1	...	1	...	Wrought strip, O temper, bonded to steel back	Bimetal surface layer; trimetal intermediate layer
11	782	...	...	95	...	1	1	...	...	3	...	Wrought strip, O temper, bonded to steel back	Bimetal surface layer; trimetal intermediate layer
12	...	...	A250	...	...	1	1	...	...	3	1.5 Mn	Wrought strip, O temper, bonded to steel back	Trimetal intermediate layer
13	...	...	AS78	88	11	1	...	...	...	...	...	Wrought strip, O temper, bonded to steel back	Trimetal intermediate layer
14	...	4002	F-154	95	4	0.1	...	0.1	...	1	...	Wrought strip, T6 temper(d), bonded to steel back	Trimetal intermediate layer

(a) Artificially aged and cold pressed. (b) Artificially aged. (c) Strain hardened, approximately 25% cold reduction. (d) Solution treated and artificially aged.

Table 7. Approximate mechanical properties of aluminum-base bearing alloys

Classification	Tensile strength		Yield strength		Hardness, HB(a)
	MPa	ksi	MPa	ksi	
High-tin aluminum strip	114	16.5	41	6	30
High-lead aluminum strip	117	17	62	9	32
Low-tin aluminum strip	117 to 138	17 to 20	48 to 124	7 to 18	32 to 40
Tin-free aluminum strip	124 to 207	18 to 30	62 to 138	9 to 20	38 to 48
Low-tin aluminum castings	159 to 234	23 to 34	117 to 172	17 to 25	54 to 74

(a) 500-kg load, 10-mm ball.

mensional tolerances and finally precision plated to size with a thin overlay of soft metal. The overlay may be of a lead-tin-copper or lead-tin alloy. In some aircraft applications, the overlay consists of a plated layer of lead with a final layer of indium. Such bearings are then heat treated to diffuse the indium into the lead.

OTHER METALLIC BEARING MATERIALS

Gray Cast Irons. Cast irons are standard materials for certain applications involving friction and wear, such as brake drums, piston rings, cylinder liners and gears. Cast irons perform well in such applications, and thus should be given consideration as bearing materials.

Cemented Carbides. Hard materials such as cemented tungsten carbide have been used successfully for various specialized bearing and seal applications. With proper design and materials selection, performance of sleeve-type antifriction bearings, mechanical rubbing-face seals, and seals employing packings can be improved by making them from carbide.

NONMETALLIC BEARING MATERIALS

Today, nonmetallic bearing materials are widely used. They have many inherent advantages over metals, including better corrosion resistance, lighter weight, better resistance to mechanical shock, and the ability to function with very marginal lubrication or with no lubricant at all. The major disadvantages of most nonmetallics are their high coefficients of thermal expansion and low thermal conductivity characteristics.

A wide variety of plastic composites is now being used very successfully in bearing applications. Addition of fiber reinforcements and fillers such as solid lubricants and metal powders to the resin matrix can significantly improve the physical, thermal and tribological properties of these plastics. This is illustrated in Table 8, where a few of the more promising materials are listed. The PV values (stress × velocity) are for dry operation. Even with marginal lubricants, such as water, some of these compounds have truly outstanding load-carrying capacities. It should be noted, however, that certain plastics wear at a

higher rate when a lubricant is present. When the effectiveness of a lubricant depends on formation of a transferred film, its use may inhibit material transfer, resulting in higher wear.

The following paragraphs present more detailed discussions of some nonmetallic materials typically used for bearings.

Nylon. The low melting point of nylon limits its use to temperatures below about 150 °C (300 °F). To obtain dimensional stability, nylon should be stress relieved at a temperature at least 28 °C (50 °F) above the maximum temperature expected in service. This is usually accomplished by heat treating the nylon in oil or some other suitable liquid. Graphite, molybdenum disulfide and other fillers are added to nylon to improve its bearing properties. The static coefficient of friction for nylon against nylon is more than twice the kinetic friction. The friction values for steel against nylon are lower than for nylon against nylon.

Teflon, a polytetrafluoroethylene resin, is a thermoplastic material. It is used as a bearing material mainly for two reasons: (a) chemical inertness; and (b) at low speeds, an extremely low coefficient of friction with sliding metals (about 0.05). Use of Teflon as a bearing material is limited, however, because of its low thermal conductivity, high thermal expansion, thermal instability and poor resistance to cold flow. In designing Teflon bearings, consideration must be given to the fact that Teflon has a transition point at 21 °C (70 °F), which results in a linear increase of 4 mm/m (0.004 in./in.). When Teflon is heated to about 340 °C (650 °F) or higher, it gives off toxic fumes, and therefore it must be

Table 8. Typical PV values for plastics sliding unlubricated on steel at a velocity of 100 ft/min

Type of plastic	Operating temperature limit(a)		Typical PV(b) values
	°C	°F	
Teflon (unfilled)	230 to 260	450 to 500	1800
Nylon (unfilled)	120 to 150	250 to 300	2000 to 4000
Acetal (unfilled)	82	180	3000 to 4500
Polysulfone (unfilled)	150 to 175	300 to 350	4000 to 5000
Teflon + 30% bronze powder	230 to 260	450 to 500	28 000
Acetal + 30% glass fiber + 15% Teflon fiber	82	180	12 000
Polyester + 30% glass fiber + 15% Teflon fiber	135	275	30 000
Formulated polyphenylene sulfide (proprietary)	175	350	35 000 to 45 000
Formulated polyamide-imide (proprietary)	205	400	12 000 to 12 900
Polyimide + 15% graphite powder	260	500	>300 000

(a) Continuous use. (b) PV, stress in psi × sliding velocity in ft/min.

cooled during fabrication operations such as machining. It can be used at service temperatures (260 °C; 500 °F) higher than those for nylon, and it is not hygroscopic.

Teflon is used in bearings in several ways: (*a*) as a film applied by water dispersion and then cured; (*b*) as an impregnant in a metal matrix; and (*c*) as a woven layer, supported by a woven layer of glass bonded to a metal surface. Teflon applied by water dispersion generally is used as a means of preventing fretting corrosion where there is intermittent oscillation. Woven Teflon is recommended for oscillating or low-speed use, although it has been used successfully at loads as high as 400 MPa (60 ksi). However, a rule of thumb for application of this material is to work

to a maximum PV value of 30 000 (load in pounds per square inch multiplied by velocity in feet per minute). Laboratory tests on metal-filled Teflon (60% Teflon; 40% metal) proved that bearings with bronze as the metal filler were greatly superior in wear resistance to those with either lead or aluminum. At low speeds, the dynamic coefficient of friction of unmodified Teflon is lower than that of the filled types; this is not true at high speeds. However, because of the increased strength provided by the fillers, it may be desirable to accept a slightly higher coefficient of friction to gain strength.

Carbon-graphite is used extensively in bearing and brush applications. Its excellent performance as a brush material confirms its desirable bearing

qualities. Its service temperature, usually limited to about 370 °C (700 °F), has recently been increased by processing techniques. Carbon-graphite has good wear resistance at temperatures too high for conventional lubricants. Carbon-graphite can also operate as a bearing in water, gasoline and other nondestructive solvents. It has a low coefficient of expansion, and its thermal conductivity is between those of copper and cast iron. Although it possesses reasonable strength, its edges are likely to chip or crack during machining or installation. It is not usually considered for applications involving high impact loads. It is used in packing rings, seals, instrument bearings, and sleeve and thrust bearings.

Wood. Lignum vitae, one of the hardest woods known, has been used for centuries as a lining for various underwater bearings in ships, where metal corrodes severely. It is inexpensive and readily obtainable. Oil-impregnated wood is also used in some bearing applications.

A composition material that has a base of either paper, fabric or asbestos may be substituted for wood in applications such as ship-rudder bearings, liners for rolling mills, inking-roll bearings, bushings and pump sleeves.

Rubber often is used in bearings for devices that operate in water. Rubber can absorb shock and has fairly good resistance to abrasion and other types of wear. Rubber bearings usually are backed by a metal shell that provides additional strength. If the rubber bearing is properly designed, much of the solid contaminating material in the water can be washed out through longitudinal passages fabricated in the bearing surface.

Part III

PROCESSING

21 EXTRACTIVE METALLURGY

By A. W. Schlechten (deceased), W. J. Kroll Institute for Extractive Metallurgy,
and C. A. Natalie, Colorado School of Mines

Introduction and Overview

EXTRACTION and refining of metals was once the major division of the subject of metallurgy. Procedures and equipment had evolved mainly by trial and error over long periods of time, and the study of metallurgy was largely a detailed listing of the current practices and equipment, with due regard for previous methods and devices.

Later on, extraction and refining were displaced from their pre-eminent position, particularly in academic and research circles, by an increasing interest in the properties and the processing of metals—subjects that seemed to lend themselves more to scientific investigation and theoretical considerations.

This shift led to great advances in the knowledge of metal properties along with the development of new and improved methods of processing. Many of these advances were based on greater knowledge of basic properties and better understanding of the mechanisms and chemistry of processing.

Unfortunately, the growing interest in so-called "physical" metallurgy left extractive metallurgy with very few adherents. Government support, industrial efforts and academic interest were directed toward studies of properties and processing, while extraction and refining were hardly considered.

This neglect was particularly true in the United States, and it was soon apparent that most of the progress in extraction and refining was being made in other countries. Examples of this trend can be pointed out:

- Extraction and refining of lead was improved greatly in Australia at the Port Pirie plant of the Broken Hill Associated Smelters (Ref 1) with the invention and development of (a) up-draft sintering; (b) larger blast furnaces with multi-level tuyeres; (c) continuous lead refining; and (d) vacuum dezincing.
- Simultaneous smelting of zinc and lead in a blast furnace—the Imperial Smelting Process (Ref 2). This development, based on thermodynamic studies, showed that zinc oxide could be smelted and reduced in the same chamber in which fuel was being burned; this procedure was previously thought to be theoretically and practically impossible.
- Continuous smelting of copper was developed at Noranda in Canada (Ref 3) and at Mitsubishi in Japan (Ref 4).
- The electrolytic zinc process was greatly im-

proved by developments in Norway and Tasmania, particularly by improved precipitation of iron from zinc sulfate solutions (Ref 5).

The above examples are only a partial list and do not imply that there were no developments in the United States during this period. Uranium production and the development of commercial production of titanium and zirconium were largely United States achievements.

Fortunately, the pendulum is now swinging the other way. Government agencies have been aware that there may be strategic metal deficiencies and that the production of common metals must be made more efficient to be competitive with other countries and to avoid pollution and health hazards. Industrial concerns are establishing larger and better-equipped laboratories for the same reasons: to gain economic benefits and to avoid pollution and health hazards. Academic research is being revived by an increase in available funds, an increase in the amount of fundamental data on record, a better understanding of the process mechanisms and chemistry, and more sophisticated equipment for the study of processes.

SOURCES OF METAL

The Earth's Crust. Except for very small contributions from outer space (see below), the ultimate source of almost all metals is the earth's crust. The formal description of the lithosphere characterizes it as consisting of an outer granite layer from 10 to 32 km (6 to 20 miles) thick resting on a somewhat thicker stratum probably composed of basaltic rock, both igneous and metamorphic. These two layers form the earth's crust, according to the theories of Washington (Ref 6).

The average chemical composition of the uppermost portion of the lithosphere is about the same as that of igneous rocks. According to the calculations of Clarke and Washington, eight elements—oxygen, silicon, aluminum, iron, calcium, sodium, potassium and magnesium—comprise 98.3% (by weight) of the igneous rocks. This would seem discouraging to the extractive metallurgist, but fortunately there are occurrences known as ore deposits.

Uneven distribution of the elements in the earth's crust has resulted in concentrations of valuable elements and minerals to the extent that these deposits can be mined and the ore pro-

cessed to recover metals at a profit. The origins and occurrences of ore deposits are very complex and have been the subjects of much theorizing. Examples of the processes which formed ore deposits are: (a) magmatic segregation; (b) solution and precipitation; and (c) physical phenomena, such as erosion and concentration.

Outer Space. Very small amounts of metals have come from outer space in the forms of meteorites and moon rocks. Meteorites can be roughly divided into irons and stones; there are many more specimens of the former, although it is thought there are actually more falls of stone meteorites.

Iron meteorites are easily distinguished by their chemical composition, which averages 90.8% Fe, 8.6% Ni, 0.60% Co, and lesser amounts of other elements. Their microstructure is distinctive also, exhibiting the banded Widmanstätten pattern.

Meteorite size varies from the size of a pea or even less to the huge 35 900-kg (79 200-lb) specimen found in Greenland or the 59 900-kg (132 000-lb) mass found in South West Africa. There is evidence that meteorites were used as sources of iron in many civilizations, and it was generally known that they had fallen from the sky.

Moon rocks, brought back to earth by the astronauts, have a composition mainly of calcium aluminum silicate, which is not very interesting as a commercial source of metals, although some thought is being given to the nature of the equipment and the source of energy that would be necessary for smelting of material from the moon, and for recovery of metals, in outer space.

Obviously, present-day recovery of metals from space is minute in comparison with total production. However, it is interesting that neither meteorites nor moon rocks have contained any elements that are not common in the earth's crust (although some mineral forms are unique).

ORES AND ORE MINERALS

Ores. One definition of an ore is *a naturally occurring aggregate of minerals from which a metal or metals may be extracted at a profit.* The economic implication of the phrase "at a profit" encompasses many factors, such as metal prices, labor costs, cost of supplies, extent of the deposit, environmental considerations and other factors, so that evaluation is most difficult. There are many examples of properties that have been developed and put into production only to find that the operation was not economical. Other examples describe deposits that could not be considered as ore at one time but became economical to mine and process when some of the factors, either economic or technical, changed.

Ore Minerals. The ore minerals that are most simple to treat are those where the valuable metal occurs in its elemental or native form. Not many metals occur naturally as elements; the chief examples are gold, silver, copper, and the platinum metals. Other metals have been found in elemental form, but not to the extent needed for economic exploitation. Most metals occur in the earth's crust combined with other elements in the form of minerals. The possible combinations are very great, so there are many minerals, but those of economic importance are relatively few.

The components of the earth's crust can roughly be divided into oxidized minerals and sulfide minerals, with infinite gradations between the two groups. In regard to naturally occurring ores, the metallurgist asks the following questions: What is the valuable mineral? Is the mineral oxidized, or is it a sulfide? What is the composition of the worthless material accompanying the valuable

mineral? Is it high in SiO_2 (acidic), or do bases, such as $CaCO_3 \rightarrow CaO$, predominate? With this information the metallurgist can plan the sequence of processes for extraction of the metal from the ore mineral and can calculate the flux additions needed to remove the gangue material in a high-temperature process in the form of a proper slag.

Thus, an ore can be relatively simple or exceedingly complex, but will consist in essence of: (*a*) the primary valuable mineral, usually a compound; (*b*) the greatly predominant worthless gangue; (*c*) valuable by-products; and (*d*) detrimental impurities. As mentioned before, the composition of the valuable mineral will dictate to some extent the type of reduction process used. The composition of the gangue will also influence the reduction processes that can be used. Valuable by-products may increase the economic return from a given ore; good examples are the occurrence of gold in copper ores, the presence of silver in lead ores, and the recovery of molybdenum from copper ores. Finally, the presence of certain impurities may increase the cost

of treatment or even make processing unfeasible; examples are arsenic in copper ores and bismuth in lead ores.

REFERENCES

1. "The Port Pirie Smelters," by F. A. Green: B. H. A. S. Pty., Ltd., Melbourne, 1977, p 170
2. Combined Zinc-Lead Smelting, by B. G. Perry and D. A. Temple: *Erzmetall*, Vol 12, 1959, p. 479-486
3. Design and Operation of the Noranda Process Continuous Smelter, by L. A. Mills, G. D. Hallett and C. J. Newman: in *Extractive Metallurgy of Copper*, edited by J. C. Yannopoulos and J. C. Agarwal, Metallurgical Society of AIME, 1976, p 458-487
4. Commercial Operation of Mitsubishi Continuous Copper Smelting and Converting Process, by T. Nagano and T. Suzuki: in *Extractive Metallurgy of Copper*, edited by J. C. Yannopoulos and J. C. Agarwal, Metallurgical Society of AIME, 1976, p 439-457
5. The Jarosite Process, by V. Arregui, A. R. Gordon and G. Steintveit: in *Lead-Tin-Zinc*, edited by J. M. Cigan, T. S. Mackey and T. J. O'Keefe, Metallurgical Society of AIME, 1979, p 97-123
6. "The Data of Geochemistry," by F. W. Clarke and H. S. Washington: Bulletin 770, U.S. Geological Survey, 1924

Mineral Processing

THE TERM "mineral processing" covers a somewhat greater area than the older names "ore dressing," "mineral dressing" and "mineral beneficiation." Whatever the name, this branch of metallurgy includes those operations needed to separate valuable minerals from worthless rock by what are essentially physical means involving no more than superficial chemical changes. This rather elaborate definition is needed to cover both the obviously physical separations based on density, magnetic properties and electrostatic properties, as well as the most widely used flotation process, which utilizes the differences in surface properties that can be altered by surface chemistry.

Mineral processing has as its goal the elimination of most of the worthless gangue in natural ores by operations which do not change the mineralogical status of the valuable metal but rather increase the percentage of the valuable mineral by eliminating valueless rock. The resulting product with higher metal content is called a "concentrate." The rejected material is referred to as "tailings" or "tails."

Mineral processing is economically justified by the early elimination of worthless rock, so that a smaller weight and volume of concentrate can be shipped rather than the bulky low-grade ore. In the majority of installations, the mineral-processing mill (or concentrator) is near the mine, while the reduction plant may be at some distance in a more central location. Hence, it is much preferable to ship concentrates to the reduction plant. Direct-smelting ores existed in the past when the upper portions of veins were being mined and the ore was rich and relatively low in gangue.

COMMINUTION OPERATIONS AND DEVICES

The crude ore must first be broken down in particle size to release the valuable mineral from its physical bonds with the worthless gangue. The

extent of the subdivision of the ore to obtain a reasonable degree of liberation will depend on the characteristics of the individual ore. Ores are referred to as fine-grain or coarse-grain ores to describe the average natural particle size of the valuable mineral constituents as well as that of the worthless gangue.

Laboratory tests can be made in which ore samples are crushed and ground to varying extents and the resultant product separated into size fractions by screening on a set of laboratory screens with elutriation of the subsieve fraction. The size fractions can be examined under the microscope to determine the extent of liberation, and concentration tests can be run on size fractions to determine the degree of concentration that can be obtained. The decision as to the extent of comminution must balance the higher degree of concentration theoretically possible with more liberation against the actual cost of increased grinding as well as the fact that, when the particle size is very small, concentration processes do not work as efficiently and tailing losses are more likely.

Particle-size reduction from large to very fine in one operation, such as achieved by a stamp mill, has been replaced by two- or three-stage procedures referred to as crushing and grinding; the combination of the two is called "comminution."

Crushing of ore reduces mine-size pieces one or two feet in diameter, or even larger, to pieces averaging 100 to 150 mm (4 to 6 in.) in diameter. Of course, a certain proportion of all other sizes are made characteristic of the crusher as has been shown by several investigators, particularly by Schuhmann (Ref 1), who showed a linear relationship between successive screen sizes and the logarithm of cumulative percent retained on each size of screen.

Many devices have been proposed for coarse crushing, but the choice of crushers for the nor-

mal, natural ore has been narrowed down to jaw crushers or gyratory crushers, as illustrated in Fig. 1.

Intermediate crushing to reduce the size of the gyratory- or jaw-crusher product is sometimes introduced into more elaborate flowsheets. The favorite device is a short-head cone crusher, also shown in Fig. 1.

Grinding. Following coarse crushing and possibly intermediate crushing, the next step is grinding to take the ore from a maximum size of a few inches to the range of small sizes needed to achieve reasonable liberation.

Again, many devices and procedures have been invented for this step, but the average plant uses a tumbling, grinding mill in which the grinding media are steel or ceramic balls, steel rods, or pieces of the ore itself. This latter process is called "autogenous" grinding.

Figure 2 shows three basic crushing-and-grinding circuits using the devices mentioned above.

The grinding operation can be conducted either wet or dry, but normally in ore mills wet grinding is used because the subsequent concentration process is also wet. Other advantages of wet grinding are the avoidance of severe dust problems and the merits of wet classification for particle-size control.

A mill can be operated in the open-circuit mode, where the ore is added at the feed end and the entire product accepted at the discharge end. This means that the reduction in particle size for all of the ore must be essentially completed in one pass — a procedure which produces considerable overgrinding of most of the feed.

An alternative scheme is to put the grinding mill in the closed-circuit mode with a classifying device. The mill is operated so that not all of the ore is sufficiently reduced in size, but the mill discharge is fed to a classifier which returns the oversize particles to the mill feed and permits the

			Size (mm)	Power (kW)	Speed (r.p.m.)	Reduction Ratio	Characteristics and Applications
JAW CRUSHERS	Blake (Double-Toggle)	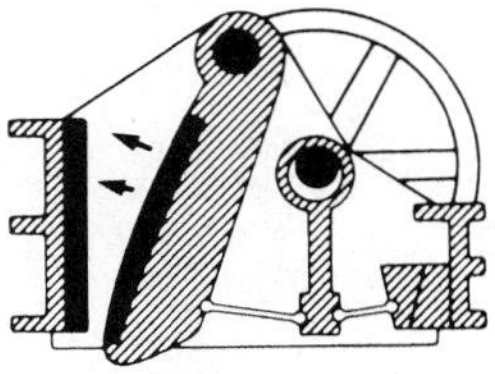	125 (gape) x 150 (width) to 1600 x 2100	2.25 to 225*	300 to 100	Average 7:1 Range 4:1 to 9:1	Originally the standard jaw crusher used for primary and secondary crushing of hard, tough abrasive rocks. Also for sticky feeds. Relatively coarse slabby product, with minimum fines. Flywheel evens power draft.
	Overhead Pivot (Double Toggle)	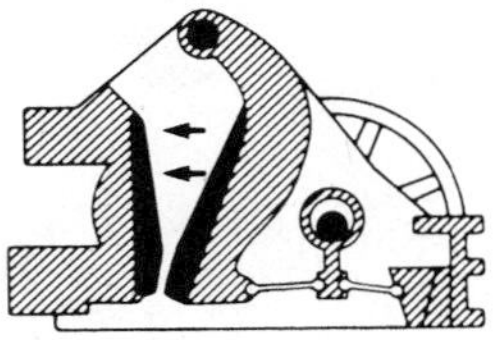	180 x 305 to 1220 x 1525	11 to 150*	390 to 257	Average 7:1 Range 4:1 to 9:1	Similar applications to Blake. Overhead pivot; reduces rubbing on crusher faces, reduces choking, allows higher speeds and therefore higher capacities. Energy efficiency higher because jaw and charge not lifted during cycle.
	Overhead Eccentric (Single-Toggle)		125 x 150 to 1600 x 2100	2.25 to 400*	300 to 120	Average 7:1 Range 4:1 to 9:1	Originally restricted to smaller sizes by structural limitations. Now in same sizes as Blake, which it has tended to supersede, because overhead eccentric encourages feed and discharge, allowing higher speeds and capacity, but with higher wear and more attrition breakage and slightly lower energy efficiency. Unsuitable for very hard, tough abrasive rock. Sometimes made with twin swing jaws.
	Dodge	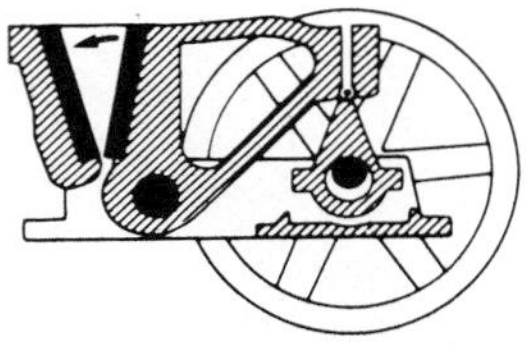	100 x 150 to 280 x 380	2.25 to 11*	300 to 250	Average 7:1 Range 4:1 to 9:1	Bottom pivot gives closer sized product than Blake, but Dodge is difficult to build in large sizes, and is prone to choking. Generally restricted to laboratory use.
GYRATORY CRUSHERS	True Gyratory	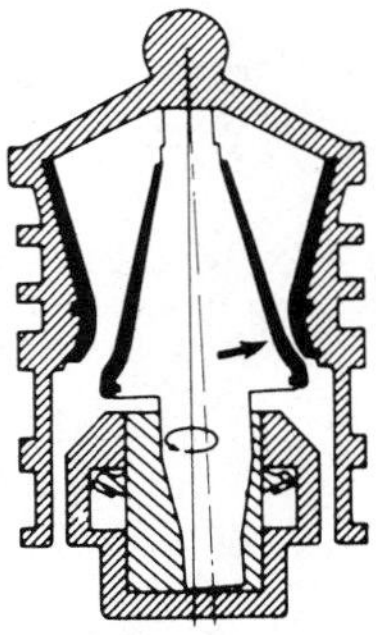	760 (opening width) x 1400 (mantle maximum diameter) to 2135 x 3300	5 to 750	450 to 110	Average 8:1 Range 3:1 to 10:1	True gyratory crushers characterised by diverging crushing surfaces (outer surface or bowl has inward slope towards bottom). Used for primary and secondary rock, with minimum fines. Taller, higher capacity, and more suitable for slabby feeds than jaw crusher.
	Cone	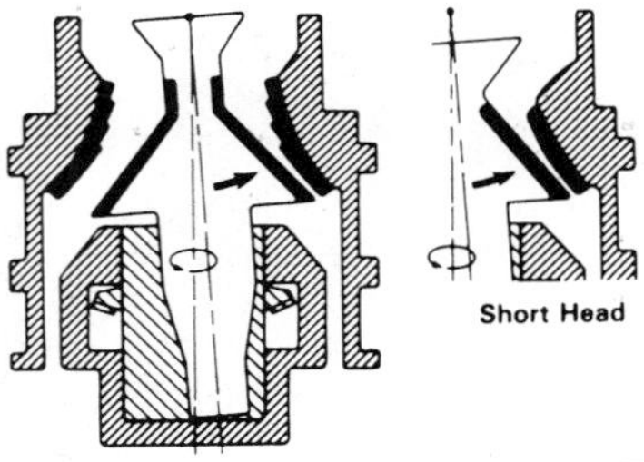 Short Head	600 (cone diameter) to 3050	22 to 600	290 to 220	Secondary crushing 6:1 to 8:1 Tertiary crushing 4:1 to 6:1	Cone gyratories are characterised by converging crushing surfaces (outer surface tends to parallel mantle surface). Used for secondary and tertiary crushing. Generally as the particle size decreases (e.g. tertiary crushing) the outer crushing surface is made straighter and more parallel to a steeper mantle (often called a "Short Head" crusher). Tertiary crushers are often choke fed.

* For very hard rock, power may be up to 50% higher provided machine is strengthened.

Source: Introduction to Mineral Processing, by E. G. Kelly and D. J. Spottiswood: John Wiley, New York, 1982.

Fig. 1. The major types of crushers

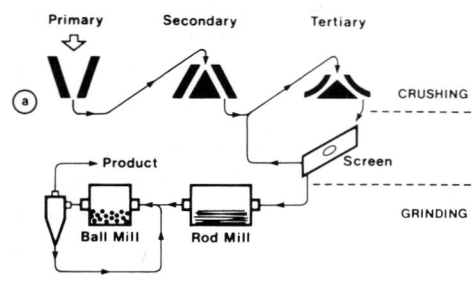

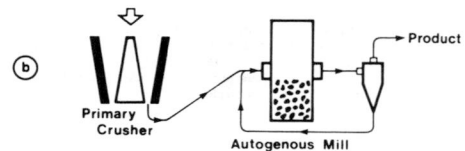

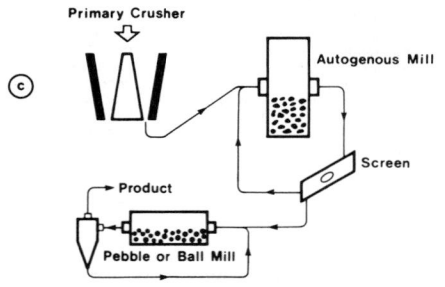

(a) Conventional. (b) Autogenous. (c) Autogenous with separate fine grinding. Source: Same as for Fig. 1.
Fig. 2. Three basic types of crushing-and-grinding circuits

Courtesy of Mine and Smelter Corp., division of Kennedy Van Saun Corp.
Fig. 3. Spiral (mechanical) classifier

undersize particles to continue on to the next step in the process.

Classification has been defined as separation of particles according to their settling rates in a fluid; normally, in mineral processing, this refers to settling in water, although dry classifiers are common for collecting dust from crushing operations. Wet classifiers used in conjunction with grinding can be subdivided into mechanical and hydraulic classifiers.

Figure 3 shows a spiral (mechanical) classifier to which a pulp from a grinding mill is being fed; the larger and heavier particles settle to the bottom of the pool at the lower right end and are conveyed by the spiral to the upper left end to be returned for more grinding. The smaller and lighter particles overflow the lower right end and are ready for the next step in the process.

A hydraulic classifier (hydrocyclone) is shown in Fig. 4. The fine particles in the overflow are discharged from the top, and the coarse particles in an underflow are discharged at the bottom.

CONCENTRATION OPERATIONS AND DEVICES

The concentration of valuable minerals and the discard of worthless rock (gangue) must be based on a difference in some property of the two chief constituents. Perhaps the simplest example would be hand sorting based on the appearance of the mixed pieces of ore and gangue. This method requires cheap labor and hand-size pieces of material; attempts to mechanize the method with a

power-driven ejector activated by observation of optical properties have not been very successful.

Gravity Devices. Most valuable minerals are higher in density than waste rock, and thus various devices can be used to achieve separation. The miner's pan and the rocking cradle are examples of simple and antiquated gravity devices. A shaking table or a jig can be mechanized to handle larger quantities of material and to achieve better separation, but they have both been supplanted in most installations because of their inability to treat the fine particles generated by the extent of the grinding necessitated by fine ore–waste rock textures.

Magnetic Devices. Separation based on magnetic properties is widely used for concentration of iron ores. If the iron mineral is magnetite, the magnetite particles can be attracted by permanent magnets or electromagnets so as to separate them from the gangue particles, as shown in Fig. 5. If the iron ore is predominantly hematite (Fe_2O_3), a preliminary treatment using partial reduction will convert the hematite to the magnetic Fe_3O_4, making magnetic separation again feasible.

Electrostatic Devices. A somewhat similar method of separation is based on differences in electrostatic properties. The ore, preferably in a limited size range that is best described as a sand, is fed onto a conducting and grounded rotor and is exposed to an ionizing electrode. The particles take a charge and then those that are conducting lose the charge to the rotor and are thrown off. The nonconducting particles retain their charge, cling to the rotor, and are eventually brushed off into a separate container. This procedure has been successful—especially in the treatment of beach sands—and, in conjunction with magnetic and gravity separation, can make separate rutile, ilmenite, chromite, garnet, magnetite and silica fractions.

Flotation is by far the most widely used method of concentrating nonferrous ores; it is most suc-

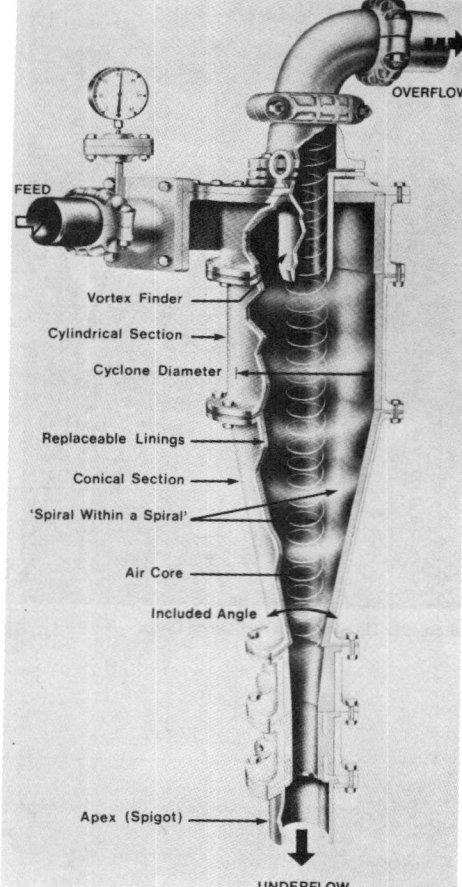

Courtesy of Krebs Engineers.
Fig. 4. Hydraulic classifier (hydrocyclone)

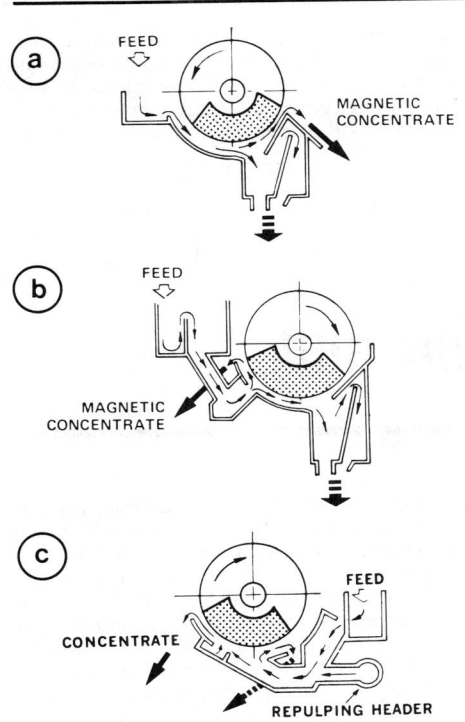

(a) Concurrent. (b) Counterrotation. (c) Countercurrent. Courtesy of Eriez Magnetics.

Fig. 5. Wet-drum low-intensity magnetic separator tanks

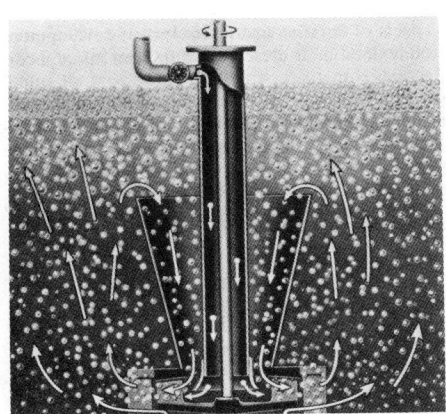

Courtesy of Joy Manufacturing Co.

Fig. 6. Denver flotation cell mechanism

Courtesy of Environmental Equipment Div., FMC Corp.

Fig. 7. Cutaway view of thickener

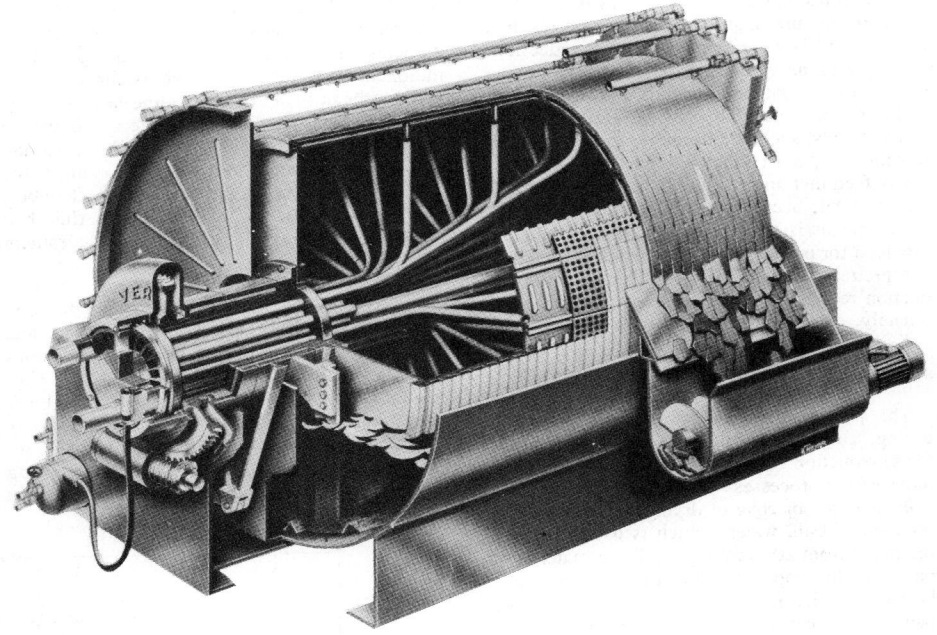

Courtesy of Filters Vernay.

Fig. 8. Cutaway view of drum filter with scraper discharge

cessful when used on sulfide ores, but has application to oxidized forms of minerals as well. The separation is based on whether or not mineral surfaces are wettable. The finely ground ore in an aqueous pulp is fed to a flotation machine which has an impeller that provides vigorous agitation. Air is introduced into the cell either from a compressed-air line or by the pumping action of the impeller. The air bubbles, distributed by the agitation, rise through the pulp and form an attachment to the nonwetted mineral particles and carry them to the surface, where a froth is formed that can be scraped from the cell.

The wetted particles, usually the gangue, do not attach to the bubbles, and hence remain in the pulp to be discharged as tailings. A single one-pass separation is usually not sufficient, so that the first, rougher concentrate may be cleaned or even recleaned. Likewise, the tailings from the first separation may be scavenged to avoid discarding valuable minerals. These retreatments plus the possibility of regrinding or even more elaborate steps can lead to very complex flowsheets; modern techniques of mathematical modeling have helped to achieve optimum flowsheets, and sensitive instrumentation has resulted in a high degree of control and thus greater efficiency.

Figure 6 shows a cross section of a flotation machine of one type, although there is almost no end to the types of flotation machines that have been invented, manufactured and used.

The nonwettable surface necessary for particles to adhere to bubbles and be carried to the surface is native only to graphite and molybdenite. Usually, reagents must be added to the pulp

to change the surface properties as desired. At one time, oils were used which would give sulfides an oily surface that would make them float, but would not affect the gangue remaining in the pulp. Later, more selective reagents were used which could distinguish among various sulfides. These reagents, called "collectors," are added along with other agents that will stabilize the froth, depress or activate selected minerals, control pH, and generally improve the collection and selection of desired minerals.

DEWATERING AND DISPOSAL OF TAILINGS

The concentrate products from any of the concentrating processes mentioned above, if operated wet, which most of them are, require dewatering. The usual first step is thickening in a large cylindrical tank; the stream of concentrate and water enters the tank and in the quiescent surroundings the mineral particles settle to the bottom where slowly revolving rakes move the mudlike deposit to a center discharge where it is pumped to the next step. The overflow, practically devoid of solids, can be returned to the process if dissolved reagents are not detrimental.

The thickened underflow from the concentrate thickeners is normally sent to a vacuum filter of a drum or disk type which reduces the moisture content of the cake to about 10% or less. The cake is then sent to the smelter, usually by truck

or rail, where it is received, sampled, and analyzed for moisture content and key elemental composition. A thickener and a drum filter are shown in Fig. 7 and 8.

The tailings from the mill can be thickened if there is a need to recover most of the mill water; otherwise the tailings discharge can be run through pipes or flumes to the tailing pond. In the usual procedure, a circumferential piping system with numerous outlets is used to discharge the tailings

toward the center of the pond. The larger particles quickly settle out and build up the outer walls of the pond; the finer particles settle as the flow approaches a central discharge, and hopefully the overflow will be clear enough to be discarded.

REFERENCE

1. Energy Input and Size Distribution in Comminution, by R. S. Schuhmann, Jr.: *Trans AIME/SME,* Vol 217, 1960, p 22-25

Pyrometallurgical Reduction and Refining of Metals

THE MINERAL CONCENTRATE from milling and concentration of crude ores is usually a complex mixture of minerals and some residual gangue, and is not suitable for direct reduction. Some preliminary treatment is necessary, and the extent of this treatment depends on the reactivity of the major metal. The raw feed of highly reactive metals, such as aluminum, magnesium, titanium, etc., must be pretreated to give a high-purity feed that upon reduction will yield a high-purity metal, because the refining of highly reactive metals is a difficult task. In contrast, the raw feed for production of low-reactive metals is not pretreated extensively and hence upon reduction results in an impure metal, which, fortunately, can be refined effectively.

PREPARATION PROCESSES AND DEVICES

The pyrometallurgical pretreatments include drying, calcination, roasting to oxides or sulfates, reduction roasting, chlorination, and even more exotic processes.

Drying. The objective of drying operations is the removal of bulk water, which is usually within the range from 20 to 30 vol %. For a material to be dried, the vapor pressure of the water must be raised so that it is higher than the surrounding partial pressure of water. This evaporation of water can be accomplished either by heating the solid material or by using a vacuum system. Because of their cost, vacuum systems are usually used only in special cases such as the removal of organics which would decompose at high temperatures.

Because the evaporation of water is endothermic (the heat of vaporization of water is 44 kJ/mole or 10.5 kcal/mole), energy must be supplied for the evaporation of water as well as to bring the material up to drying temperature. In most cases, the energy required for evaporation is greater than any other energy requirement in the process. Drying processes are usually operated at temperatures close to the boiling point of water. This low temperature allows the use of low-quality fuels, such as hot combustion gases from other processes or cheap producer-gas fuels. Also, the low temperature used for the drying processes poses no serious materials problem for consideration of equipment.

The most common furnaces used for drying are fixed-bed furnaces, rotary kilns and fluidized-bed furnaces. Fixed-bed furnaces or shaft furnaces are most suitable for drying coarse materials. Rotary

kilns are usually used with feed material of a mixed size. Fluidized-bed furnaces are used when the material contains a high percentage of fine particles. If the concentrate is to be roasted on multihearth roasters, drying will occur on the upper hearths utilizing the ascending heat from the roasting hearths. Some flash roasters also dry the concentrate on the upper hearths, take it out and grind it to eliminate lumps, and then inject the fine, dry material into the combustion chamber. Wet concentrates can be fed into a fluid-bed roaster and will rapidly dry in the hot, turbulent bed.

Wet concentrates can be fed into reverberatory furnaces, but care must be taken that the feed does not build up on the sides of the furnace and then topple into the hot bath, causing explosions violent enough to damage the furnace roof. Calculations show that drying concentrates in a high-temperature furnace is wasteful of fuel, but may still be cheaper than drying in a separate step requiring additional equipment and labor. Feeding wet concentrates into a converter is definitely a

strenuous procedure and is to be avoided if possible. The question that arises, particularly with the treatment of cement copper from precipitation plants, is whether it is better to dilute the high-grade material by adding it to roaster feed or reverberatory feed or to risk the problems of adding it to the converter. The over-all conclusion is that most drying is achieved in connection with other processes, although there are examples of independent drying steps.

Calcination differs from drying in that it is a higher-temperature operation for removal of chemically bound water or for decomposition of carbonates to remove carbon dioxide gas. The temperature of the operation depends on the material that is to be treated.

By far the most common objective of calcination processes in the metallurgical industry is the decomposition of metal carbonates to form metal oxides. Figure 1 shows the decomposition pressures and temperatures of some metal carbonates. At one atmosphere, the temperatures range from 1000 to 1500 °C (1830 to 2730 °F). De-

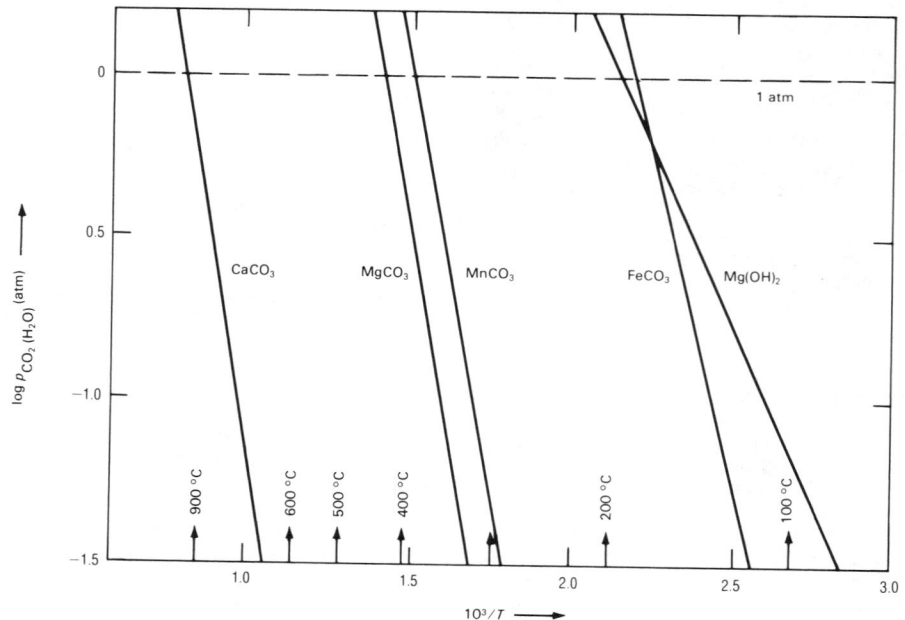

Fig. 1. Decomposition pressures (logarithmic) of various carbonates and hydrates as functions of inverse temperature

composition of carbonates is done by heating the material until a reaction such as

$$CaCO_{3(s)} = CaO_{(s)} + CO_{2(g)}$$

can occur. For calcium carbonate at one atmosphere, the reaction occurs at approximately 1000 °C (1830 °F).

Researchers have found that the rate at which decomposition of carbonates will occur is controlled by the heat input (Ref 1). Because calcination is a gas-solid reaction with an oxide-product layer, the rate of the process is controlled by conduction of heat through the oxide layer that surrounds the unreacted carbonate. If the particles of feed material are round, the conduction of heat, and thus the rate of calcination, is proportional to the square of the diameter of the particles.

To increase calcination rates in industrial furnaces, large amounts of excess heat are commonly used. The excess heat increases the calcination rate, but it also causes rather high temperatures in the oxide-product layer. For some metal carbonates, this may cause significant changes in the physical properties of the oxide product. For example, calcium oxide sinters readily at temperatures just above the temperature required for calcination. Research has shown that the more excess heat that is used in the calcination process, the lower the pore volume, the specific surface area, and the reactivity of the calcium oxide product (Ref 2).

Fixed-bed or shaft furnaces are commonly used for calcination of lump carbonate material. These furnaces are popular because of their low maintenance, operation and capital costs as well as their efficient fuel utilization. However, because of the long residence times and high surface temperatures that are achieved, the oxide products tend to be less reactive, or "hard-burned." Rotary kilns, on the other hand, produce oxides that are more reactive, or "soft-burned," but rotary kiln furnaces have lower fuel efficiency and higher maintenance, operation and capital costs. In strictly pyrometallurgical plants, calcination usually is achieved during heating of the mineral mixture for other purposes.

Roasting is a process in which the mineral mixtures are heated to temperatures just below the melting or sintering point of the mineral in the presence of a gas, usually air, to achieve a chemical change. Roasting reactions are carried out at temperatures below the melting points of the sulfides, oxides and sulfates but high enough to achieve significant rates of reaction. Therefore, temperatures for roasting are commonly between 500 and 900 °C (930 and 1650 °F).

The roasting process is carried out by contacting the sulfide with air at elevated temperatures. Once the roasting has started it is autogenous, meaning that no external fuel is required. This is possible because of the exothermic nature of the roasting reactions. The burning of the sulfides provides heat that is sufficient, and often more than sufficient, to run the process.

The equilibrium relationships of the metal-oxygen-sulfur system are used to determine which solid phase, oxide, sulfide or sulfate will be present as a product. The equilibrium calculations for such a system are complex. However, it is possible to show the equilibrium relationships of the system graphically, as shown in Fig. 2 for the Cu-O-S and Fe-O-S systems at 700 °C (1300 °F).

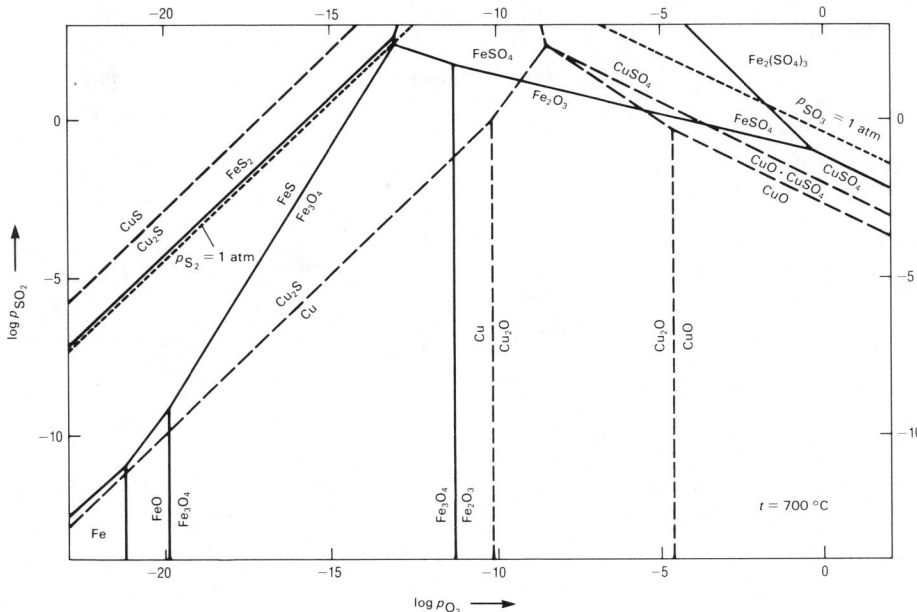

Whole lines: Fe compounds. Dashed lines: Cu compounds. Dotted lines: gases. Ternary compounds are disregarded.

Fig. 2. Predominance area diagram for roasting of iron and copper sulfides at 700 °C (1300 °F)

This figure is commonly called a predominance area diagram. Specifying two gas compositions, such as oxygen and sulfur dioxide, will fix the solid phase that is stable.

Pyrometallurgical plants usually have as their goal the production of oxides from sulfides for subsequent reduction by carbon or other reducing agents. Production of sulfates by roasting would be desirable if subsequent leaching were to take place. The production of metal is a phenomenon observed in sinter-roasting of lead concentrates where appreciable amounts of "windbox" lead are produced.

Roasting in a reducing atmosphere is normally used for partial reduction, as in the case of hematite (Fe_2O_3) ores being converted to magnetite (Fe_3O_4) which is then concentrated with magnetic separators.

Roasting in a chlorine atmosphere or in the presence of chlorides is a method for producing metal chlorides which can be volatilized or leached.

Roasting Devices. Originally, sulfide ores were roasted in heaps exposed to the atmosphere. Subsequently, a great variety of roasters was invented to achieve temperature control, atmosphere control, heat conservation, and greater exposure of mineral surfaces to the gas by frequent rabbling or by dispersion of the solids into a particle/gas suspension.

Many varieties of roasters have become obsolete, so that present-day installations are mostly limited to multihearth, flash and fluidized-bed roasters. Multihearth roasters, as shown in Fig. 3, are cylindrical in plan with 4 to 12 superimposed hearths. The feed enters at the top and is rabbled first to the center, where it drops to the second hearth to be rabbled to the outer edge, where it drops to the third hearth, and so on until the roasted material exits from the bottom hearth of the roaster. The gases rise countercurrent to the material movement and are usually removed

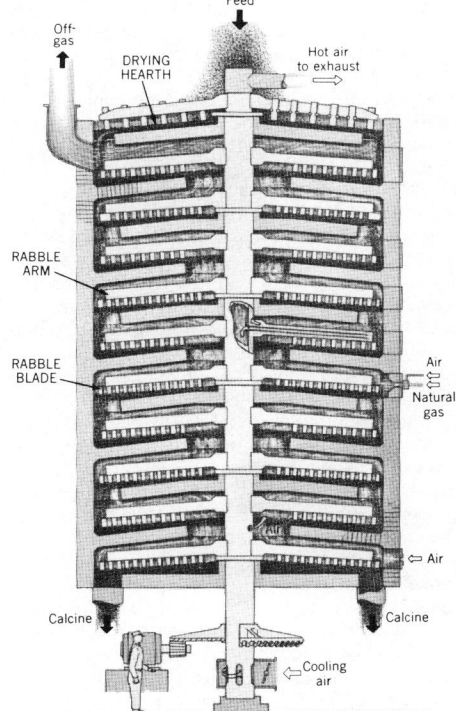

Fig. 3. Cutaway view of typical multihearth roasting furnace

at the second hearth and sent to flues, dust collectors or possibly an acid plant.

A flash roaster, as shown in Fig. 4, has a larger capacity than a multihearth roaster of similar dimensions, because of greater exposure of mineral surfaces to the reacting gas. The example shown reveals that the earlier models of the flash

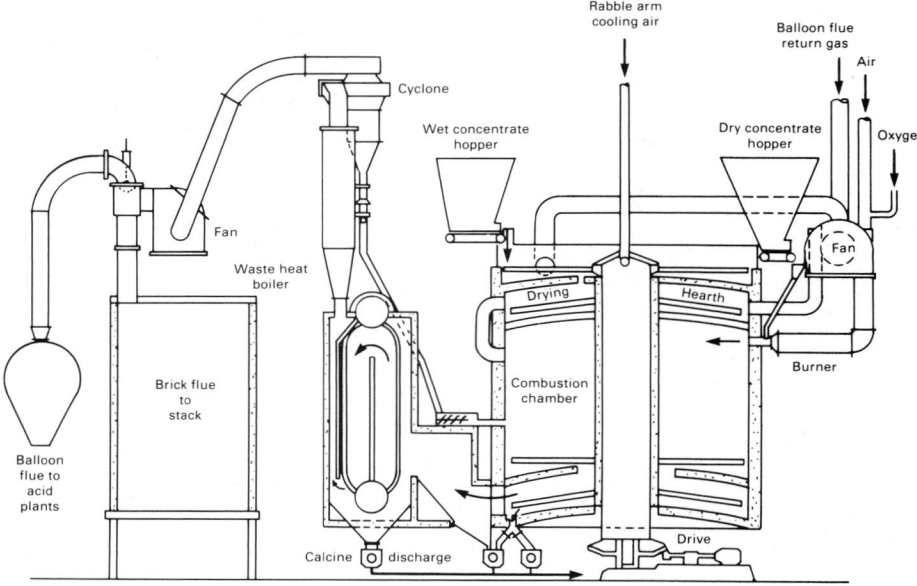

Source: R. E. Eyre, *Can Min Metall Bull*, Vol 54, No. 592, 1961, p 589.
Fig. 4. Flash roaster with drying hearth above combustion chamber

roaster were made by removing most of the hearths of a multihearth roaster. The two upper hearths dried the wet feed, and the resulting dry material was injected into the combustion chamber where rapid oxidation occurred. In later models the drying hearths are at the bottom of the cylindrical structure.

A fluid-bed roaster reverses the concept of a flash roaster by blowing the gases through a turbulent bed of mineral particles, as shown in Fig. 5.

Sintering, sinter-roasting or blast sintering is a variation of roasting carried out at a temperature high enough to obtain partial fusion of the particles so that they adhere to each other to form "sinter" or a "sinter-cake." Sintering of iron ore or iron concentrates has as its primary purpose the coalescence of small particles into larger agglomerates that will not blow out of the furnace. At the same time, some drying, calcination and possibly roasting will be achieved.

Sintering of nonferrous sulfide concentrates of lead, copper, zinc and other metals has the dual function of particle-size control and simultaneous roasting, but the roasting is of even more importance because the final sulfur content may be critical.

A sintering machine can be no more than a grate on which the charge is spread and then ignited; air is either blown up through the grate and the bed, or sucked down through the bed and grate. High-sulfide ores generate enough heat to reach the temperature needed for sintering. Low-sulfur ores, or iron ores with no fuel value, require fuel additions—usually in the form of carbon.

Continuous sintering machines, such as the Dwight-Lloyd type shown in Fig. 6, can be controlled to a greater extent and also simplify feeding of the charge and discharging of the finished sinter. Most continuous sintering machines run on a downdraft principle, but it has been found in lead smelting that updraft sintering will make a sinter with higher lead content and eliminate most of the windbox lead.

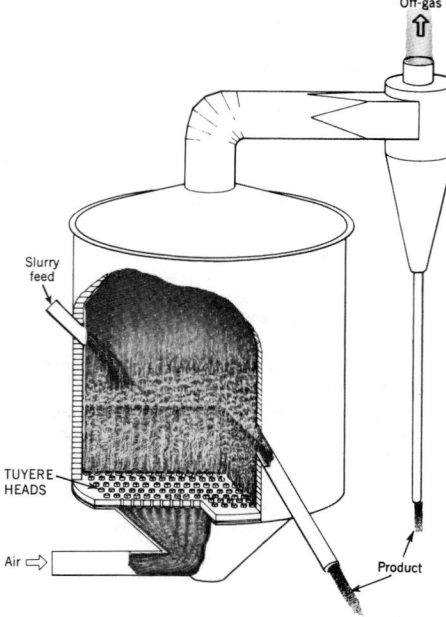

Fig. 5. Cutaway view of fluid-bed roaster (Source: Ref 3)

Sinter-roasting is the customary method of preparing feed for lead blast furnaces; at one time it was also used for preparation of feed for copper blast furnaces, but such furnaces are no longer used in North America. It is also the method for treating zinc sulfide concentrates preparatory to smelting either in electrothermic furnaces or in the Imperial Smelting furnace.

PYROMETALLURGICAL REDUCTION

As has been pointed out before, the metal contents of ores and concentrates are rarely in the form of elemental metals, but are mostly metal compounds, such as the original minerals found in the ore, or compounds derived from the mineral after preliminary treatment. Therefore, the chief objective of smelting is to reduce the desired metal from the compound in the feed.

This procedure can be represented by the general equation

$$MX + R = M + Rx$$

where M is any metal, X is the element or elements originally combined with M, and Rx is a reaction product. The use of a reducing agent, R, is not required in every instance, but there are a few examples of metal compounds decomposing to yield the elemental metal just by thermal decomposition. A well-known laboratory experiment is the heating of mercuric sulfide (cinnabar) in air to yield mercury and sulfur dioxide. Another example is the refining of zirconium or hafnium by thermal decomposition of the tetraiodide.

Reducing Agents. As can be seen from the above equation, a reducing agent with a strong affinity for the element or elements (X) originally combined with the metal will react with the X with an evolution of free energy, which helps to force the reaction to the right.

Theoretically, a reducing agent can be any element that can take the X from the MX under the circumstances of the reduction reaction. However, considerations of cost, the nature of the reaction by-product, the solubility of the reducing agent in the metal produced and the possibility of reaction of the metal with the reducing agent all narrow the field of choice, so that commercial reducing agents are few in number.

Metal oxides are usually reduced with carbon for several reasons: carbon is relatively cheap (despite the increasing cost of coke); the resulting CO/CO_2 by-product escapes from the reacting mass; and, most important of all, the change in the standard free energy of reaction of

$$MO + C = M + CO$$

becomes more negative with increasing temperature. This means that even the most refractory oxides will be decomposed by carbon if the mixture is heated to a high enough temperature. In a practical sense there are other possible chemical effects that keep this simple reaction from being a universal procedure for metal production. First is the possibility that the metal is a carbide former and will react with the carbon so as to yield an impure metal. The best example is in the production of pig iron at high temperatures in a blast furnace, where the metal will contain several percent carbon in contrast to almost no carbon in iron produced by a low-temperature process. Other examples are molybdenum, tungsten and vanadium, which have appreciable carbon contents if reduced from metal oxide–carbon mixtures.

Another interfering phenomenon arises if oxygen (the metal oxide) is soluble in the metal being produced. Such solubility reduces the activity of the oxygen so that even the most highly reducing conditions cannot reduce the oxygen content below an equilibrium condition. This has been shown particularly for titanium and zirconium, which cannot be reduced from their respective oxides using carbon or carbide agents to obtain a low enough oxide content to meet commercial requirements.

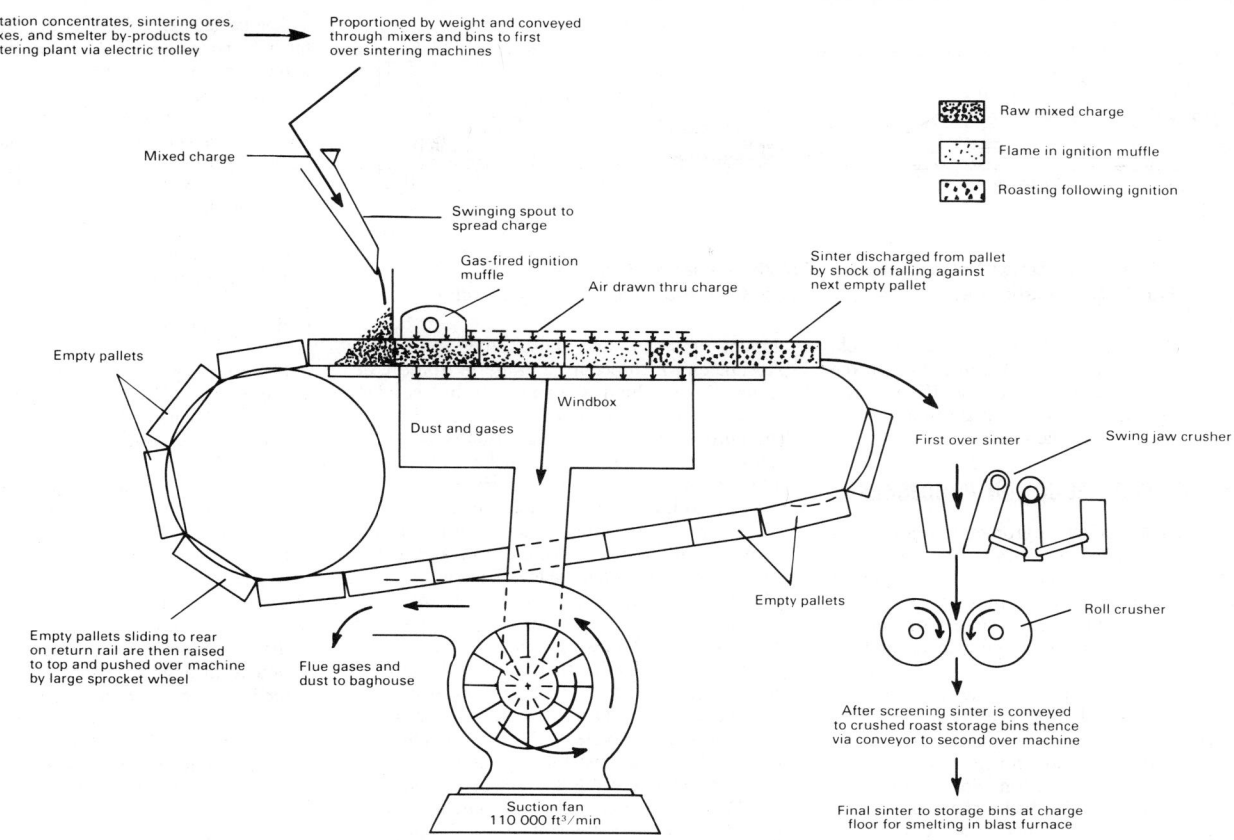

Source: United States Smelting, Mining and Refining Co.

Fig. 6. Dwight-Lloyd downdraft-type sintering plant

Hydrogen can be used to reduce metal oxides if carbon reduction is not suitable; for example, hydrogen-reduced tungsten can meet specifications for low-carbon metal. The cost of hydrogen will be higher than carbon if the metal oxide has a large negative value of free energy of formation, which in turn would lead to an equilibrium composition of the reaction gas high in hydrogen and low in water vapor. Either the moisture content of the reaction gas must be reduced prior to recirculation or some other use must be found for the hydrogen–water vapor mixture in order to profit from the residual hydrogen content.

Other metals. There are a few well-known examples of metal oxides being reacted with another metal having a greater affinity for oxygen. The famous Goldschmidt reaction of iron oxide mixed with aluminum powder produces molten iron and a cloud of finely divided alumina; it was used to weld railroad rails and large drive shafts before the development of effective portable welding outfits.

Vacuum reduction of calcined dolomite with ferrosilicon depends on the magnesium oxide being reduced by the silicon, followed by the reaction of the resulting silica with the calcium oxide content of the calcined dolomite to form a calcium silicate and thus avoid reaction of some of the magnesium oxide with the silica.

In general, the reduction of metal oxides with other metals is adopted only when there are objections to the use of carbon.

Reduction of Metal Compounds Other Than Oxides. Nonferrous metals occur as sulfides in many nat-ural deposits, and so there is reason to consider direct reduction of sulfides. In the past, lead sulfide was reduced with scrap iron to form iron sulfide and metallic lead; this process suffered from the relatively high cost of the iron used as the reducing agent, and also from the tendency of the heavy and viscous iron sulfide to entrap excessive amounts of lead.

As mentioned before, some metal oxides, such as titanium and zirconium oxides, will not yield up all of their oxygen to even the most powerful reducing agents because the oxygen is dissolved in the metal and has very low activity. As a result, these metals are best reduced from their chlorides with sodium or magnesium, both of which are stronger chloride formers, as the reducing agents.

Reduction of Metals from Metal Compounds. It was found early that galena (PbS) could be mixed into a strongly burning fire which, if stirred frequently to expose the galena to both oxidizing and reducing conditions, would result in some metallic lead being produced. First, there was a roasting reaction in which lead oxide and lead sulfate were produced, followed by a reaction between these oxidation products and some of the remaining lead sulfide to produce metallic lead and SO_2 gas. An examination of the reaction equation

$$2\,PbO + PbS = 3\,Pb + SO_2$$

shows that the free energy of formation of the SO_2 must exceed the free energy required to decompose the lead oxide and the lead sulfide.

This type of reaction has been used to produce lead in so-called ore hearths or mechanical ore hearths, but has been abandoned because of high labor costs and loss of lead to the slag formed.

Copper converting utilizes this same sequence of partial oxidation followed by reaction to produce blister copper from copper matte. Converting of copper sulfide is conducted at temperatures high enough so that all constituents are liquid and the reactions are rapid. Attempts to convert lead sulfide at high temperatures result in high volatilization losses. However, lead bullion containing 3% S is converted so as to oxidize the sulfur at Boliden's Ronnskar Works.

Nickel sulfide can be converted to metallic nickel by using high temperatures which can be achieved in a top-blown converter using oxygen.

Slags. The reduction reactions previously referred to, if performed on impure feed material, usually require formation of a slag as a means of disposing of gangue material left in the feed — that is, pure PbO reduced with carbon would yield metallic lead and a CO/CO_2 mixture that would escape from the reacting mixture. If the PbO is accompanied by gangue, as is usually the case, the gangue probably would not melt and would remain as a solid which would be difficult to remove cleanly from the metal. It would be preferable to add fluxes (reagents) which will react with the gangue to produce a fluid slag. Other desirable features for the slag are: low value, low melting point, low viscosity, low tendency to react with refractories or to dissolve the metal, and low density so as to obtain good separation.

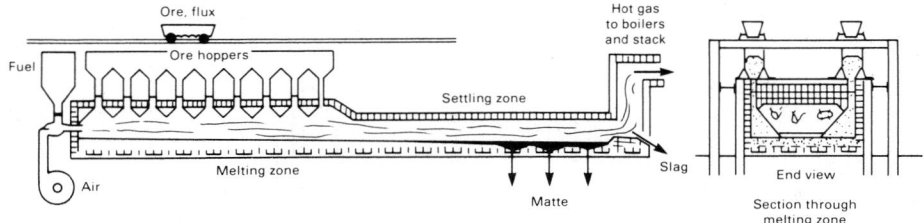

Source: *Extraction Metallurgy*, by J. D. Gilchrist, Pergamon Press, Oxford, 1967.

Fig. 7. Schematic views of copper matte smelting reverberatory furnace

In that gangue is most likely to be siliceous, the usual fluxes are lime and iron oxide. In the iron blast furnace, conditions are reducing to iron oxides, and so the main flux is lime resulting from limestone additions to the charge.

DEVICES FOR REDUCTION PROCESSES

Reverberatories. Many pyrometallurgical processes are conducted in so-called reverberatory furnaces which are rectangular in plan with an arched roof. The fuel, which may be gas, oil, or powdered coal, combusts over the hearth and the heat "reverberates" down onto the hearth and the charge on the hearth; schematic views of such a furnace are shown in Fig. 7. The reverberatory furnace is very suitable for melting a finely divided charge, such as is encountered in matte smelting of copper sulfide concentrates, but has some problems with reduction. For example, attempts to reduce liquid baths by charcoal or coke additions encounter difficulty in achieving contact with the bath, because of the slag. However, some new smelting processes propose close control of furnace conditions by blowing mixtures of gases of various reducing or oxidizing powers through protected tuyeres submerged in the bath or by top blowing. These furnaces can be of the reverberatory type, or can be more like kilns with a mechanism for partial rotation.

Blast Furnaces. A blast furnace is the most appropriate device for reducing the oxides of metals that are not highly reactive. The best-known and largest example is the iron blast furnace, but smaller versions are used in lead, copper and tin metallurgy.

Iron Blast Furnaces. The iron blast furnace as it is used today is the result of hundreds of years of evolution and development. The blast furnace has developed from a simple hole in the ground during the iron age to a modern furnace with continuous tapping, oxygen injection, high top pressures and other modern improvements.

Figure 8 shows a schematic diagram of an iron blast furnace and the temperatures and chemical reactions that exist in a working furnace. The furnace consists of a cylindrical, refractory-lined shaft that can range from 8 to 15 m (25 to 50 ft) in diameter at its widest point. The furnace is usually narrower at the top and expands from top to bottom to allow for expansion of material as it mixes through the furnace.

At the bottom of the furnace is the hearth, which is lined with carbon and collects the molten product of the blast furnace. Just above the hearth area, air is injected through water-cooled copper nozzles referred to as tuyeres.

Operation of the furnace consists of continuously charging the solid material into the top of the stack. This solid material includes ore which consists mainly of the oxides of iron, a small

amount of silica, and those other oxides and phosphates which occur naturally in iron ore deposits. Coke, which is mainly carbon, is added as a fuel and reductant. Also, fluxes are added. The purpose of fluxes, which usually consist of limestone and dolomite, is to make a slag of suitable composition with the impurities.

As the solid material or burden moves down the furnace it comes in contact with the air that is injected at the bottom, and chemical reactions (see Fig. 8) start to take place. The carbon burns to provide heat and to produce carbon monoxide to reduce the oxides. As the burden approaches the hearth it becomes molten and the slag and metal collect in the hearth. The slag is removed and discarded, and the metal or pig iron is removed and sent to stations where it is further purified and made into steel. Pig iron, which is the final product of a blast furnace, consists of iron with approximately 4% C, 0.8 to 2% Si, 0.01 to 0.03% S, 0.3 to 0.5% P, and some manganese depending on the content of the feed material.

Lead blast furnaces (see Fig. 9) are much smaller than iron blast furnaces, because the volume of material treated at a lead smelter would not be sufficient to justify use of a furnace as large as a 30-m (100-ft) iron furnace. The plan of the lead furnace is rectangular rather than circular so that the relatively low-pressure blast can penetrate to the center of the charge. Figure 9 shows a lead blast furnace and shows how the shaft is constructed of water-cooled jackets resting on the crucible made of refractory brick. The tops of the jackets may fit into a section of brick shaft that completes the height of the furnace.

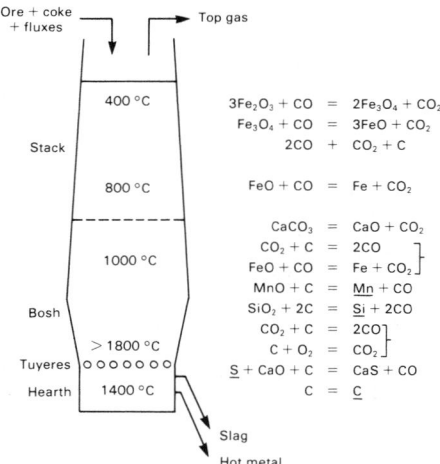

Fig. 8. Schematic illustration of an iron blast furnace, showing temperatures and chemical reactions

The furnace is charged at the top with lead sinter, which also contains most of the fluxes, coke, and additional fluxes if needed to adjust the final slag composition. As the charge descends in the shaft it moves countercurrent to the ascending hot reducing gas generated by the combustion of coke in the tuyere zone. The charge is dried, the carbonates are calcined and the metal oxides are reduced by the CO in the gases as the descending charge gets hotter. At the same time the siliceous gangue reacts with the lime and iron oxide flux to form droplets of slag. Finally, the whole charge is melted with the molten lead lying on the bottom of the crucible and the slag floating on the lead.

If there is appreciable sulfur in the charge, it will combine with any copper, with any elemental iron and with lead to form an artificial mixture of sulfides known as "matte." Disposal of matte can be a problem, because it probably will contain appreciable amounts of precious metals. The logical method of disposal is to sell the matte to a copper smelter, but the copper smelter will not wish to pay anything for the lead content and only fractional amounts for the precious metals.

The older practice was to let the slag accumulate in the furnace and to tap it at intervals; at the same time the lead discharged almost continuously through a siphon tap extending down to the bottom of the crucible. More recently, both slag and metal are tapped continuously into a forehearth where the slag overflows and the lead accumulates to be tapped periodically.

Recently, experiments have been conducted to determine if a heated blast or an oxygenated blast would provide benefits in fuel economy. The relatively low temperatures of the lead blast furnace reactions are not improved by blast pretreatment to the extent found in iron furnaces, and so there is some doubt as to whether the additional cost is warranted.

Exhaust gases from lead blast furnaces are conducted through flues to obtain some cooling and to permit larger particles to precipitate out. The gases then go to bag filters to remove finely divided fume, which is eventually returned to the furnace charge. Gases are discharged from a stack.

Copper Blast Furnaces. Furnaces similar to lead blast furnaces have been used in copper metallurgy, but there are no examples in North America at this time. Reduction of copper oxide ores in a blast furnace suffered from simultaneous reduction of some of the iron oxides in the charge to yield a mixed iron-copper alloy referred to as "black copper." If reducing conditions were made less intense in the furnace to avoid the iron contamination, then the copper content of the slag became excessive.

More extensive use was made of blast furnaces for simple melting of copper sulfides to form a matte. Originally, the feed was lump ore of sufficient copper content to be suitable for direct smelting. Later, when only concentrates were available for feed, it was necessary to sinter the feed before smelting. This practice has been abandoned in North America but is still practiced in Japan.

Imperial Furnaces (Lead-Zinc Blast Furnaces). It was generally thought that ZnO could not be smelted in a blast furnace, because the reaction ZnO + C = Zn + CO is readily reversible and because, if combustion occurs in the same vessel as the reduction, the amount of CO will be increased, favoring the back reaction, so that the product in the condenser is high in ZnO. Thermodynamic

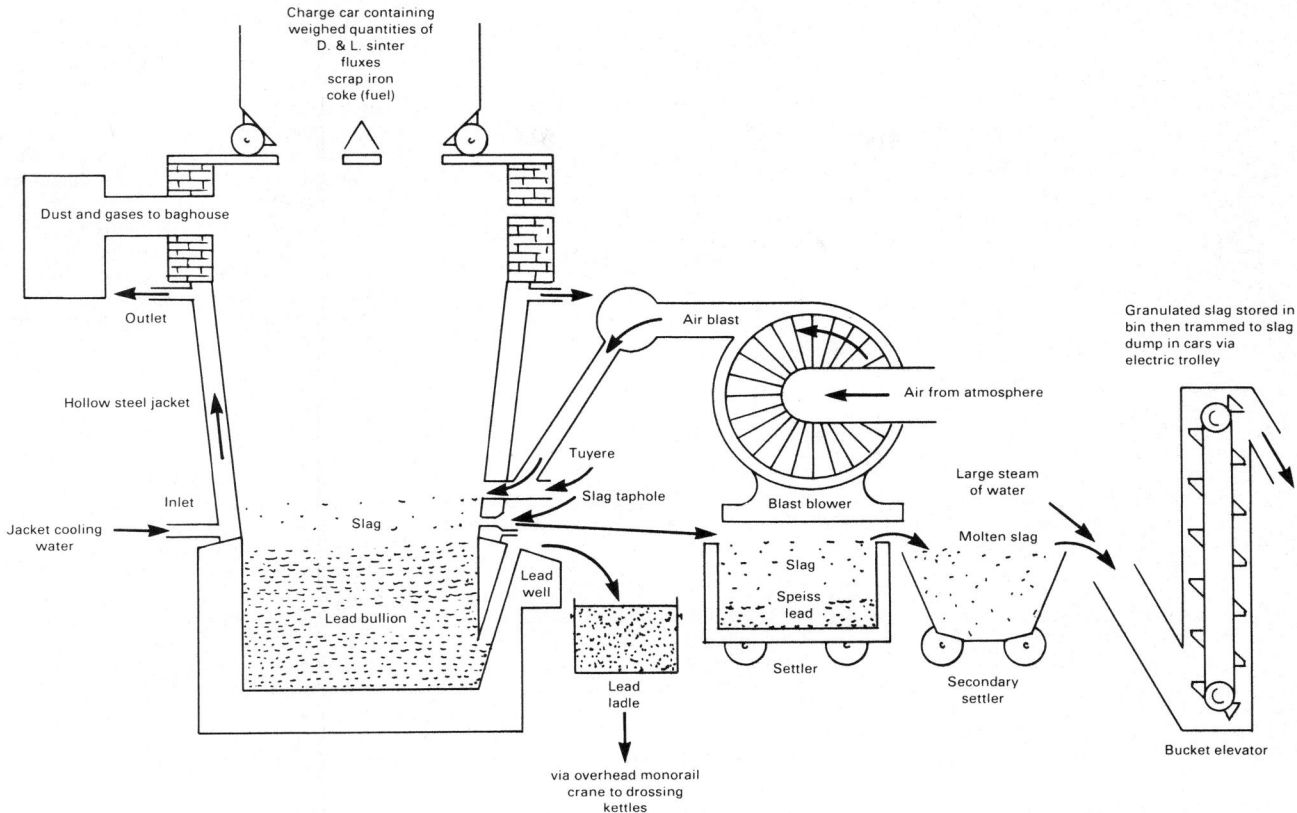

Source: United States Smelting, Mining and Refining Co.
Fig. 9. Lead blast furnace, open-top type

studies by the metallurgists of Imperial Smelting showed that the reduction would not reverse if the gas mixture was kept at a high enough temperature. The Imperial Furnace then was built to have no cold zones, and the charge was preheated before entering the furnace and the gases were kept hot in the condensers. Probably most important of all, the zinc was captured from the gas mixture, not by cooling and condensation, but by absorption on small drops of liquid lead created by violent mixing of a lead bath in the condenser. Subsequent cooling of the zinc-lead alloy yielded two immiscible liquid layers — the upper one high in zinc, which can be sent to a refining column, and the lower one high in lead, which is recirculated to the condenser.

It was found, if the furnace was fed a mixture of lead and zinc oxides, that the lead could be reduced and tapped from the furnace while the zinc vapors issued from the upper part of the furnace to the condenser and that the lead reduction would not require additional reducing agent.

Imperial Furnaces have been installed in many countries, but there are no examples in the United States.

Converters. A converter is essentially a refractory-lined vessel holding a molten bath of metal or metal sulfide into which air or oxygen can be blown to achieve rapid oxidation. The details of shape and construction depend on the material to be treated.

Historically, the Bessemer converter changed the steel industry because it provided a cheap and rapid method of oxidizing carbon, silicon and manganese from pig iron, so that the content of

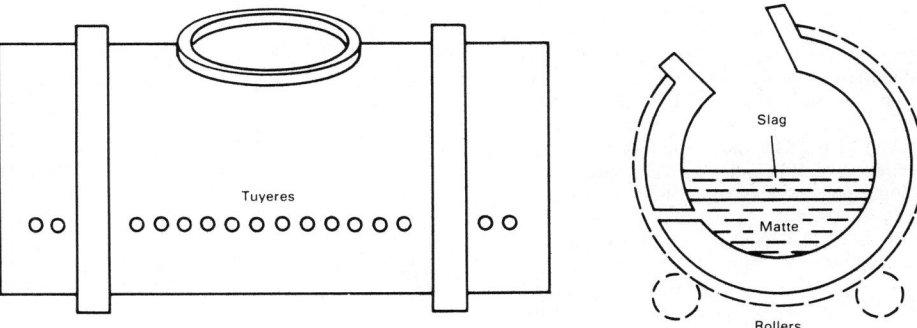

Fig. 10. Pierce-Smith converter (schematic)

the impurities could be readjusted to obtain a desirable steel.

The success of the Bessemer converter attracted the attention of copper metallurgists, because the partial oxidation of copper matte to eliminate iron as an oxide and to produce blister copper from the molten copper sulfide was a long and laborious task. The first attempts, with bottom-blown converters, failed because metallic copper froze and blocked the tuyeres. Subsequently, side-blown converters with provision for tuyere-punching made the process a success. Some copper converters have kept the pear shape of the original Bessemer converter, but today most installations for copper converting are horizontal cylinders, as shown in Fig. 10.

Retorts. Some metals can be reduced only at temperatures above their vaporization points, so

that they issue from the reacting mass in the vapor state and must be condensed to liquid or solid. The best example of this procedure was smelting of zinc oxides plus carbon in horizontal retorts which were refractory cylinders closed at one end and heated from the outside (see Fig. 11). The charge of oxide sinter and coke or coal was charged in the open end, which was then closed with a refractory condenser. Upon heating, the zinc vapors and CO entered the condenser where the temperature was low enough to condense the zinc vapors to liquid zinc, which was then retained in the condenser while the CO issued to the atmosphere. The zinc was removed several times during the reduction run of 24 to 48 h.

The excessive labor and many small units of the horizontal retort furnaces led to the development of the New Jersey Vertical Retort (see

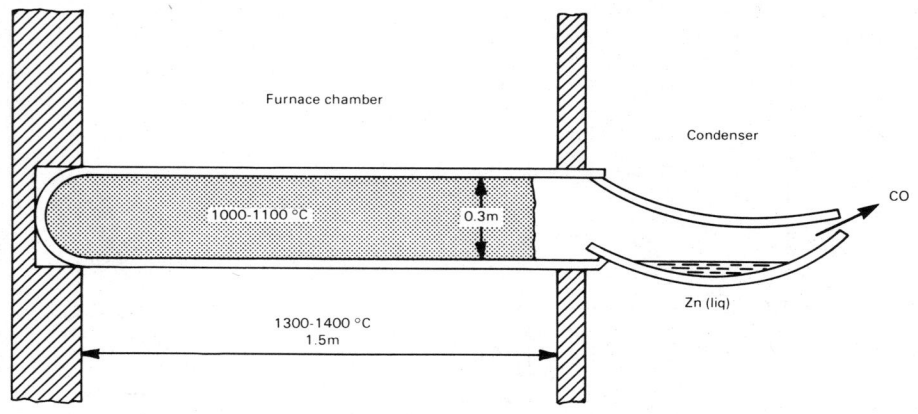

Fig. 11. Horizontal zinc retort (Source: Ref 4)

Fig. 12), a much larger unit with semicontinuous operation utilizing periodic feeding of hot batches of sinter and coke. The vapors are condensed in a large spray condenser and the zinc is tapped periodically.

Reduction of calcined dolomite with ferrosilicon is usually carried out in alloy steel retorts heated at one end and cooled at the other end with a water jacket. A vacuum is maintained in the retort to assist migration of the magnesium vapor to the cold zone, where the metal condenses to a solid "muff" of magnesium.

PYROMETALLURGICAL REFINING

Those metals reduced from impure mixtures are bound to contain impurities, because of the impossibility of controlling the reducing conditions to be sharply selective. In addition, there may be contamination from the reducing agent or from the vessel or furnace in which the reduction was made. The purpose of refining is to remove various impurities until the host metal can meet the specifications for the particular grade of metal desired. In some instances, particularly the removal of gold and silver as impurities, the refining process may be profitable and more than pay for the cost.

Liquation. The freezing out of impurities by reducing the temperature of the bath suggests itself as a simple method of refining but actually has limited application unless accompanied by a reagent treatment.

For example, in a binary alloy system, such as gold and silver showing a continuous solid solution, cooling of the melt will freeze out an alloy somewhat higher in the higher-melting-point element (gold), but there is no clear-cut separation of the two metals.

In a eutectic system of two metals, such as silver and lead, the original alloy will be on the lead side of the eutectic and cooling of the mixture will freeze out almost pure lead, except for mechanically entrapped alloy, but the liquid phase cannot go above the eutectic composition of 2.5% Ag.

Alloy systems (such as lead-zinc) which form two immiscible layers upon cooling can be used on occasion to yield two usable products, as has been explained in the discussion of the Imperial Furnace.

Refining with Gaseous Reagents. Removal of impurities from metals often can be made more effective and more selective by the use of reagents; these can be divided into gaseous and solid, with oxygen being the most prominent of the gaseous group.

Refining with oxygen presumes that the impurity is more readily oxidized than the host metal and that the resulting impurity oxide is not soluble in the host metal. Examples of this procedure include converting of pig iron with air or oxygen to remove carbon, silicon and manganese. In the open hearth these reactions are altered to some extent by using iron oxide ore as a source for some of the oxygen and carrying active slags which react with some of the impurity oxides and promote elimination.

Another example is high-temperature oxidation of impure lead by injection of air into a molten bath, a process which eliminates arsenic, antimony and (to a lesser extent) tin as oxides which combine with PbO to form a slag that can be skimmed. This process shows dramatically the selectivity of the oxidation in that gold, silver and bismuth contents are not oxidized.

In copper converting, despite the strongly oxidizing conditions, the resulting blister copper usually still carries some iron and sulfur, which are eliminated in the subsequent fire refining process where air is blown into the molten blister and the remaining iron and sulfur are oxidized. Additions of silica sand are added to the bath to react with the iron oxide; this slag contains a high percentage of copper and must be returned to the flowsheet at an earlier step.

Oxidation can be achieved with air, oxygen or steam injections. These represent oxygen at different activities and may be the basis for obtaining more selectivity in oxidation.

Chlorine as a Gaseous Reagent for Refining. Although zinc can be selectively oxidized from molten lead by injection of air, a fairly high temperature is necessary, which requires a furnace rather than a kettle, and the mixed zinc-lead oxide skimmings are not easily treated. Removal of zinc as a chloride can be achieved at a lower temperature by injecting chlorine gas into the lead-zinc alloy in a closed kettle. Not only does the

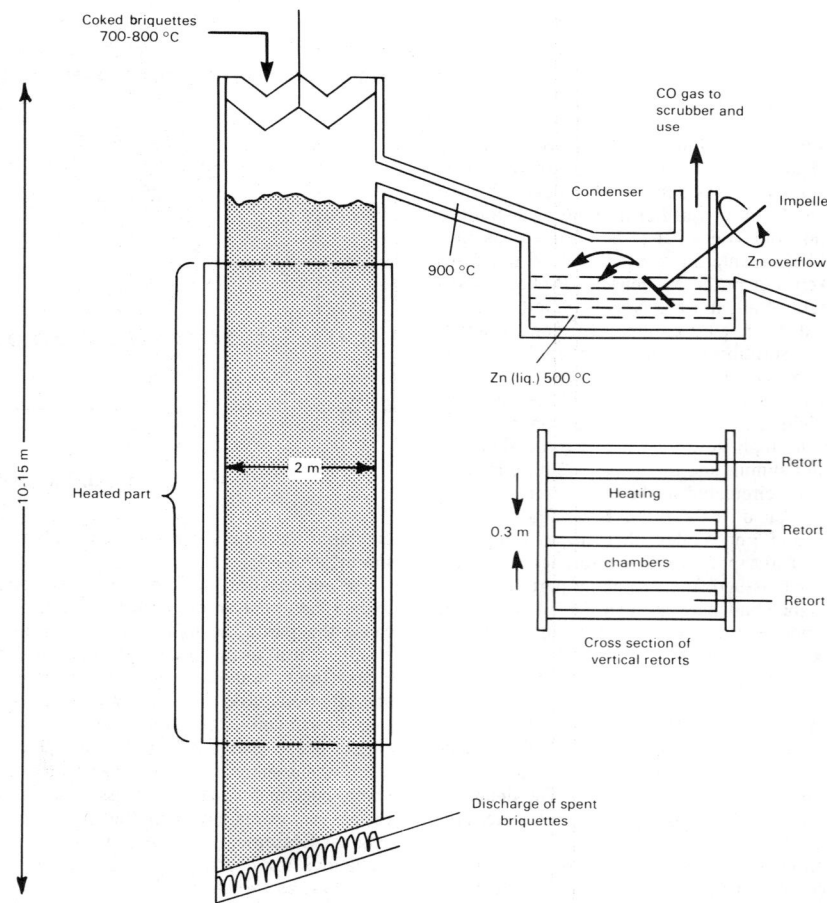

Fig. 12. Vertical (New Jersey) continuous zinc retort (Source: Ref 4)

reaction occur at a lower temperature, but the mixed chloride slag may have commercial value as well.

Chlorine has been proposed as a means of refining aluminum wherein the aluminum is removed from its impurities in the form of volatile aluminum monochloride (AlCl), which is then disproportionated in another vessel to yield high-purity aluminum and aluminum trichloride; the latter is returned to the reaction chamber to form more monochloride.

In a similar way, iodine (I_2) can be reacted with impure titanium, zirconium or hafnium to form a volatile iodide that leaves the impurities behind. The volatile iodide is brought into contact with a hot filament which decomposes the iodide, leaves the pure metal to deposit on the filament and releases the iodine to return and react again with the feed metal.

Reagents for promotion of impurity removal depend on their ability to react selectively with the impurities and to form an insoluble phase that can simply be removed from the parent metal. Many examples can be given, such as: addition to sulfur to remove copper from lead; addition of zinc to remove gold and silver from lead; and addition of calcium and magnesium to remove bismuth from lead.

REFERENCES

1. C. N. Satterfield and F. Feakes: *Amer Inst Chem Eng J*, Vol 5, 1959, p 115-122
2. Influence of Lime Properties on Rate of Dissolution, by C. A. Natalie and J. W. Evans: *Ironmaking and Steelmaking*, No. 3, 1979, p 101-109.
3. The Winning of Nickel, by J. R. Boldt, Jr.: International Nickel, Inc., New York, 1967
4. Principles of Extractive Metallurgy, 2nd Ed., by T. Rosenqvist: McGraw-Hill, New York, 1983

Hydrometallurgical Reduction and Refining Processes

IN MANY CASES, ores or concentrates are not immediately ready for processing by hydrometallurgical techniques, and pretreatment processes are used to prepare such ores or concentrates. Three typical preparation processes are drying, calcination and roasting. These procedures have been discussed in a previous article and will not be repeated here.

LEACHING PROCESSES

Objectives and Principles. Leaching is essentially a separation process which is accomplished with the use of aqueous solutions. The success of a leaching process is determined by the ability to select a suitable aqueous environment that will decompose the mineral that contains the valuable metal or compound.

In general, the objectives of leaching can be subdivided into the following broad categories:

- Production of a pure compound for further processing by pyrometallurgical techniques
- Production of a metal from impure metal or metal compounds which have been prepared by a pyrometallurgical process
- Direct production of a metal from an ore or concentrate.

In development of a total process for recovery of a particular metal or compound, economic factors are often important in determining which of the above objectives is selected. Once the objective is selected, thermodynamic and kinetic conditions of the system determine the limit of success that can be expected. The theoretical success is determined by thermodynamic constraints, which describe the maximum limits, whereas kinetic constraints relate to the over-all time required and affect the reactor size and design.

The thermodynamics of leaching are concerned with the ability to decompose a particular compound so that it will dissolve and become stable in the aqueous solution used for leaching. Thermodynamics is a rather complex branch of science that allows one to predict the maximum extent to which a particular reaction will proceed. A complete discussion of thermodynamics is beyond the scope of this article. Therefore, it will have to be sufficient to say that proper calculations can predict the maximum amount of mineral that can be leached if no kinetic prob-

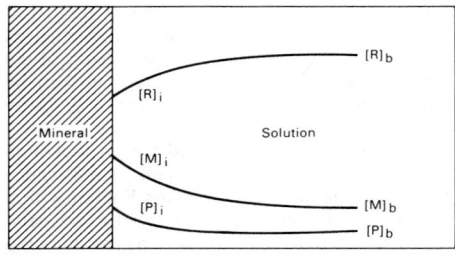

Fig. 1. Dimensional sketch of a mineral surface that is dissolving completely

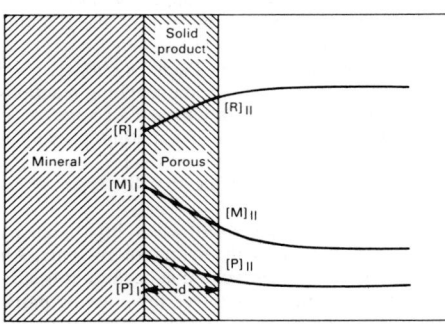

Fig. 2. Dimensional sketch of a mineral surface that is decomposing and leaving a porous residue on the surface

lems are involved. This is often referred to as a system that reaches equilibrium.

The study of how fast a mineral will decompose in a given environment is referred to as kinetics. Although a mineral may often be easy to decompose based on thermodynamic considerations, kinetically the process may be so slow that it does not occur within a reasonable time period.

Figures 1 and 2 are simple schematic drawings of what a mineral/water interface is believed to look like on a microscopic scale. In these diagrams, the concentration of the active chemical in the bulk of the solution is greater than the concentration near the surface of the mineral. This occurs because the active chemical is removed from solution by reaction at the mineral surface. In Fig. 1, the rate at which the reaction or decomposition of the mineral will occur is determined by the following steps: (*a*) diffusion of reactant R from the bulk of the solution to the surface of the mineral, (*b*) reaction of the reagent with the surface to form a soluble species and (*c*) diffusion of the product metal species M away from the surface. In Fig. 2, two additional possibilities exist: (*d*) diffusion of reagent R through the porous product layer and (*e*) diffusion of the product metal species M through the product layer.

Extensive research has been done to develop mathematical models that will describe the process given in steps (*a*) to (*e*) above (Ref 1). In many cases, these mathematical treatments of kinetics allow calculation of the time required for a given mineral to decompose. However, in some cases the mineral is so complex that the kinetic information can only be obtained through experimentation.

Leaching chemistry can be divided into some rather broad classifications, as follows:

- Dissolution of a salt in water—for example:
$CuSO_{4(s)} + nH_2O = CuSO_4 \cdot nH_2O_{(aq)}$
- Acid dissolution—for example:
$ZnO_{(s)} + 2 H^+_{(aq)} = Zn^{++}_{(aq)} + H_2O$
- Alkali dissolution—for example:
$Al_2O_{3(s)} + 2 OH^-_{(aq)} = 2 AlO^-_{2(aq)} + H_2O$
- Base exchange—for example:
$CaWO_{4(s)} + CO^=_{3(aq)} = CaCO_{3(s)} + WO^=_{4(aq)}$
- Leaching with complex ion formation:
$CuO_{(s)} + 2 NH^+_4 + 2 NH_3 = Cu(NH_3)^{++}_{4(aq)} + H_2O$
- Leaching with oxidation:
$CuS_{(s)} + 2 Fe^{+++}_{(aq)} = Cu^{++}_{(aq)} + 2 Fe^{++}_{(aq)} + S^0_{(s,l)}$
- Leaching with reduction:
$MnO_{2(s)} + SO_{2(g)} = Mn^{++}_{(aq)} + SO^=_{4(aq)}$

The above reactions are only a few examples of the many leaching reactions that describe recovery of mineral resources.

Simple Leaching Processes. The method used to carry out a particular leaching chemistry may vary depending on physical as well as economic considerations. Some simple methods used for leaching include *in situ* leaching, heap leaching and agitation leaching.

In situ leaching is mineral dissolution that is done with an ore "in place" underground at the time of treatment. *In situ* leaching is usually done in worked-out stops of high-grade mines, support pillars left behind after mining, or low-grade deposits. This type of treatment requires the surrounding rock to be tight and impermeable to flow of solution in order to contain the leaching solution.

The procedure for *in situ* leaching begins with breaking of the rock by block-cave mining or blasting. Breaking or fracturing of the rock must be done in order to provide for exposure of the mineral values to the leaching solution. After the fracturing procedure, the area is flooded with leaching solution. This type of leaching is often slow and requires that the leaching solution remain in contact with the ore for a long period of time (for years in some cases) as the mineral dissolves. Once the mineral of interest has dissolved to a reasonable level, the solution is pumped out and contained in storage ponds. Fine solid particles from the ore that have collected in the leaching solution settle out in the storage ponds. After settling has occurred, clear solution from the storage ponds is pumped to precipitation tanks where the metal values are recovered and the leaching solution is returned to the mine area.

The main advantages of *in situ* leaching are low treatment costs, low equipment requirements, low capital costs, and the ability to treat low-grade ores at a profit. *In situ* leaching has been practiced at Miami, Arizona (Ref 2), at Mammoth, Arizona (Ref 3), at Winnemucca, Nevada (Ref 4), at Mountain City, Nevada (Ref 5) and at other mines.

The heap-leaching process is similar to *in situ* leaching in that it does not require extensive leaching equipment, such as tanks, slurry pumps, and thickeners. This, in turn, keeps capital and maintenance costs down. Therefore, the principal feed materials for heap leaching are low-grade ores, ores not amenable to flotation, and discarded waste rock from previous processing with metal values below milling grade.

Heap leaching begins with construction of waterproof pads. The pads are usually made of concrete, packed clay, or plastic sheets, and are properly sloped for drainage and collection of leaching solution. Once the pads are ready, the feed material is heaped onto the pad with the coarse material (100-to-200-mm, or 4-to-8-in., chunks) on the bottom if possible. In any case, if excessive amounts of fine material are included in the heap, it may cause reduced flow of solution through the material. To overcome this problem, the U.S. Bureau of Mines has developed a low-cost agglomeration process that gives the heap uniform permeability by ensuring uniform distribution of solid material (Ref 6). The agglomeration pretreatment has increased precious-metal recovery from 37 to 90% in some cases.

The leaching solution is brought in contact with the solids by pumping and evenly spraying the solution over the top of the heap. As the solution percolates through the heap, it dissolves the metal values. This pregnant solution then drains from the leach pad, into a clay or plastic lined storage pond. From the storage pond the solution is pumped to a precipitation processing station where the metal values are recovered and the barren solution is returned to the heap for leaching. Figure 3 shows a typical heap-leaching process.

Agitation leaching is usually used with high-grade ores where the metal values are of fine grain size and are well disseminated in the host rock. This situation requires extensive crushing and grinding before leaching to expose the solution to the minerals that are to be dissolved. Also, the

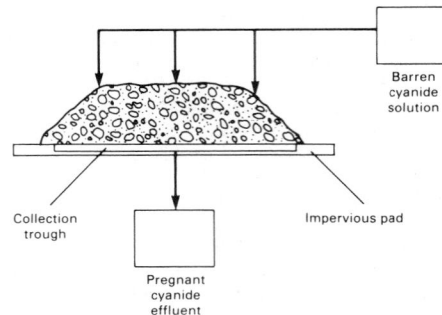

Fig. 3. Heap-leaching cyanidation

solid/liquid mixture of ore and solution is agitated to increase the rate at which the dissolution will take place. Because of the extensive amount of equipment that is required for agitation leaching, high recoveries (approximately 90% or greater), short residence times (hours instead of days as with heap leaching) and smaller tonnages of ores to be handled are all requirements of the process. Agitation is usually accomplished by either bubble action using compressed air or by mechanical agitation using impellers.

The standard equipment used for agitation with compressed air is the Pachuca tank. The Pachuca tank is a tank with a cone-shape bottom with a 60° incline. A central pipe through the bottom of the tank acts as an air lift which pushes the pulp in an upward motion. Figures 4 and 5 show typical Pachuca tank design. The average size of these tanks is 12 m (40 ft) in height and 3 m (10 ft) in diameter, with a 250-mm (10-in.) central lift pipe. Pachuca tanks are made of wood or steel, and are lined with inert material if the leaching solution is extremely corrosive.

A distinct advantage of Pachuca tanks is good aeration of the solution. Some leaching reactions require oxygen for oxidation, and air agitation is well suited as a method of agitating as well as aerating the solution. Lamont (Ref 7) has found that aeration is best in the free-lift type of Pachuca tank.

Pachuca tanks are used when high-intensity agitation is not one of the main requirements. Air agitation is a popular technique because it is simple and efficient and involves few moving parts. In general, Pachuca tanks have lower operating costs and fewer maintenance problems than mechanically agitated tanks.

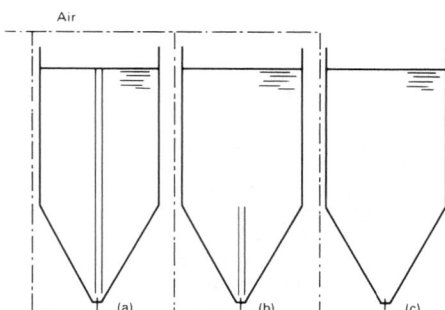

(a) Full-center-column tank. (b) Stub-column tank. (c) Free-airlift tank.

Fig. 4. Three types of Pachuca tanks

When high-intensity agitation is required, mechanical agitation with impellers is usually used. Figure 6 shows a typical agitation tank with mechanical agitation. The main difference among mechanically agitated tanks is the type of impeller used for agitation.

Three major types of impellers are popular: marine propellers, paddles, and turbine impellers. Marine propellers produce an axial flow pattern, paddles produce a tangential flow pattern and turbine impellers produce a combined radial and axial flow pattern. These flow patterns are depicted in Fig. 7.

The flow patterns in mechanically agitated leaching are also affected by the type of tank used. In an unbaffled tank, the flow pattern is circular with a deep vortex caused by the rotation of the pulp. A typical flow pattern in an unbaffled agitated tank is shown in Fig. 8. This circular flow pattern can often cause separation of solid and liquid instead of mixing.

To overcome the problems with unbaffled tanks, baffles are installed. Baffles are vertical strips that are mounted radially on the tank walls. The width of the baffles is usually one-tenth to one-twelfth the diameter of the tank. Figures 9 and 10 show the flow patterns in baffled tanks. As can be seen, the circular motion and the deep vortex are eliminated. For mixing slurries, the baffles are usually located at a distance equal to one-half their width from the bottom of the tank to avoid accumulation of solid material behind them.

An excellent treatment of the theory, design

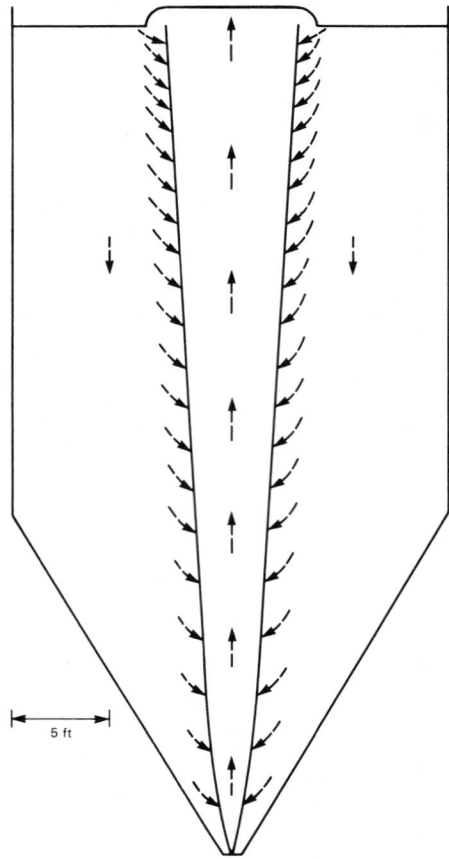

Fig. 5. Representation of flow in a free-airlift Pachuca tank

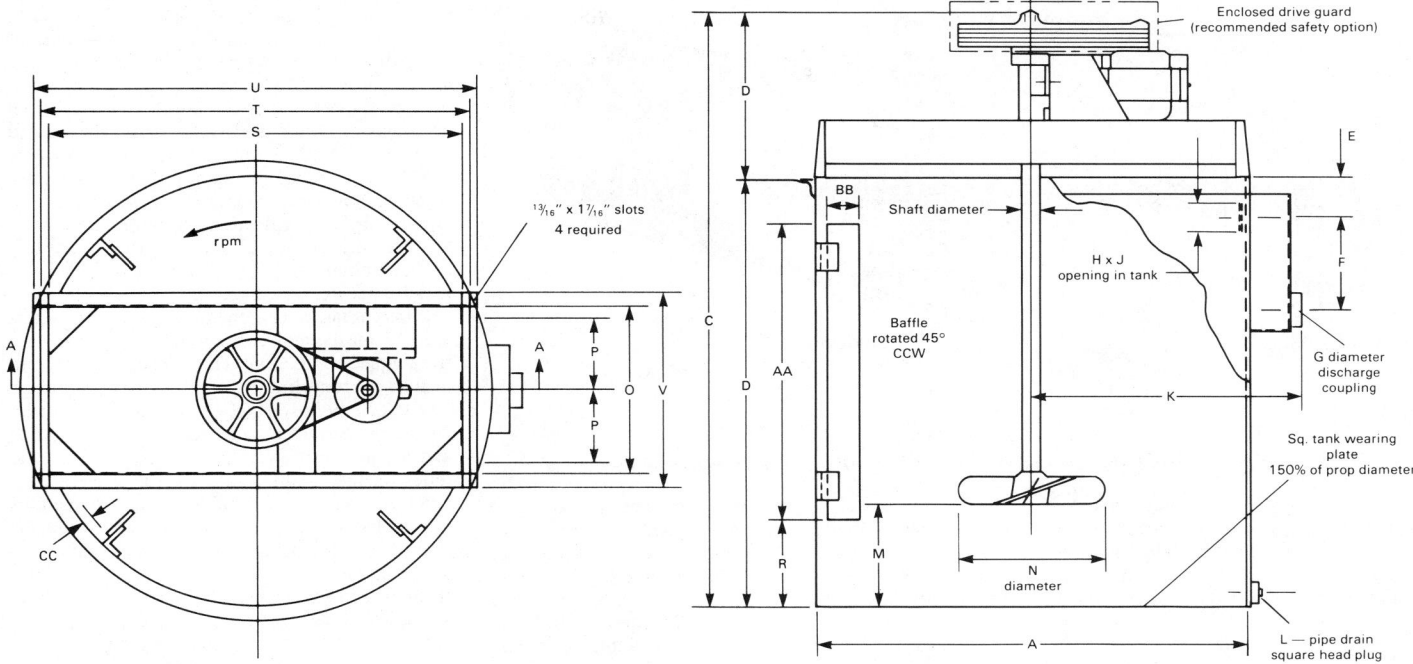

Courtesy of Denver Equipment Div., Joy Industrial Equipment Co.
Fig. 6. Agitation-type leaching tank

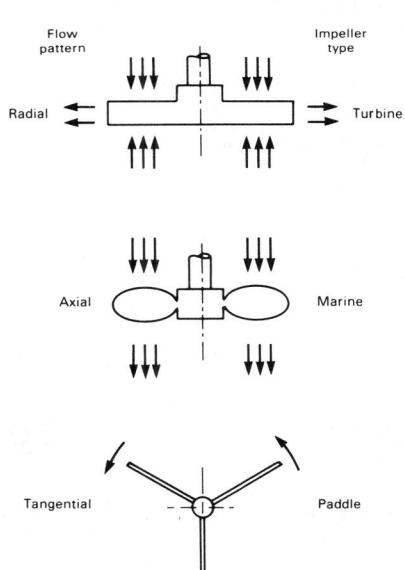

Fig. 7. Classification of impellers by their flow patterns

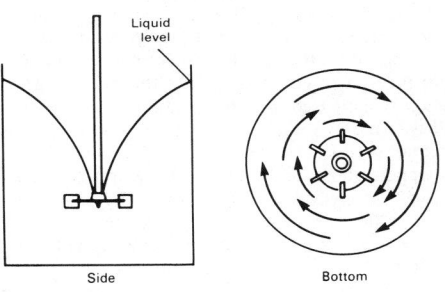

Fig. 8. Typical flow pattern for either axial- or radial-flow impellers in unbaffled tank

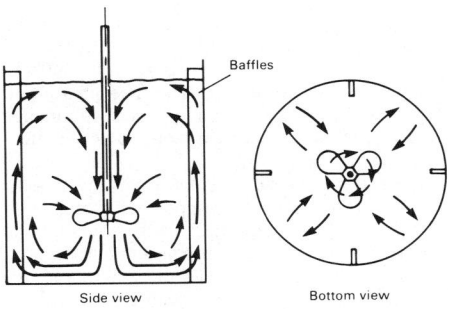

Fig. 9. Typical flow pattern in baffled tank with propeller or axial-flow turbine positioned on center

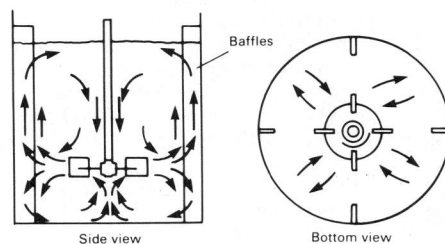

Fig. 10. Typical flow pattern in baffled tank with turbine positioned on center

and scale-up of mechanically agitated tanks can be found in a book by Oldshue (Ref 8).

Pressure leaching is similar to agitation leaching except that the process is done at elevated pressures and temperatures. Agitation leaching, in open tanks at sea level, is limited to 100 °C (212 °F), which is the boiling point of water at one atmosphere. As pressure on the solution is increased, the boiling point of water is also increased. Thus, higher leaching temperatures can be employed by increasing the pressure at which the operation takes place. Therefore, the main

objective of pressure leaching is to speed dissolution of metal values by permitting higher operating temperatures and increasing the solubility of gaseous species that may take part in the leaching reaction.

An example of the need for increased gas solubility is the oxidation of sulfides in aqueous solutions. At a pressure of one atmosphere, metal sulfides are insoluble even in strong acidic solutions. Increasing the temperature and pressure of the system increases the solubility of oxygen in solution. These increases in oxygen solubility and temperature cause rather rapid oxidation of some metal sulfides to sulfates which are quite soluble in acid solutions. This ability to achieve oxidation in the leaching step by increasing the pressure can eliminate the need for pretreatment steps, such as oxidation roasting of sulfide ores.

The standard piece of equipment used for pressure leaching is the autoclave. Autoclaves usually are made of metal for strength and often are of stainless steel or titanium because of the severe corrosion that can occur at high temperatures and pressures. If the attack on the autoclave

is extreme, linings of glass, lead or refractory brick can be used. Most autoclaves have agitators for mixing combined with heating and cooling coils for close temperature control of the process.

Both vertical and horizontal autoclaves are used in hydrometallurgy. In either case, they are four to five times greater in length than in diameter.

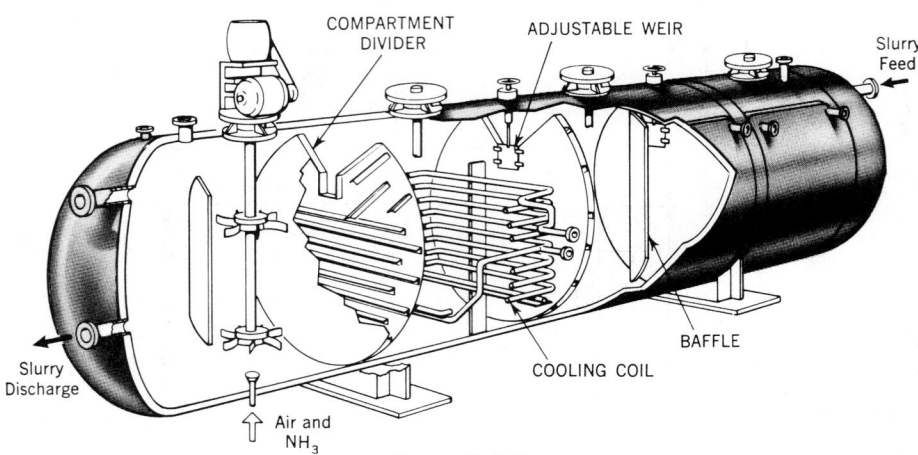

COMPARTMENT DIVIDER

ADJUSTABLE WEIR

Slurry Feed

Slurry Discharge

Air and NH₃

COOLING COIL

BAFFLE

Source: Ref 21.

Fig. 11. Cutaway view of a Sherritt-Gordon pressure autoclave

Baffle plates are used to divide the autoclave into approximately four compartments, each of which has mechanical agitation. Heating and/or cooling coils run through the system to allow for addition or extraction of heat when required. A typical autoclave is shown in Fig. 11.

SOLUTION PURIFICATION

Because ores are quite complex and contain many minerals, leaching with strong solutions usually produces a solution that contains the desired metal values as well as impurities. It is, therefore, often necessary to purify the leaching solution before metal recovery. The most common methods for solution purification include chemical and physical treatment, solvent extraction and ion exchange.

Chemical and Physical Treatment. Removal of ionic species from solution as compounds, which is referred to as precipitation, is a very common and efficient technique used for solution purification. Precipitation is accomplished by making adjustments to the solution which cause formation of compounds that are no longer soluble in the solution.

The concept of the solubility product is used to predict, and to perform calculations concerning, precipitation from solution. The solubility product for a given compound is defined as the product of the concentrations of cation and anion of the compound of interest, each raised to the power of its proportion in the compound. For example, the solubility product for a hypothetical compound $MX_{2(s)}$ would be

$$[M^{++}][X^-]^2 = K_{s0}$$

based on the reaction

$$M^{++} + 2X^- = MX_{2(s)}$$

If the product of the concentrations (left-hand portion of first equation above) exceeds the value of the solubility constant (K_{s0}) in a solution, precipitation of the compound occurs. Table 1 gives a list of pK_{s0} values of some common compounds ($pK_{s0} = -\log_{10} K_{s0}$).

When used as a purification technique in hydrometallurgy, precipitation can be initiated by one of a few different methods. For example, precipitation can be done by addition of chemicals. If the appropriate cation or anion is added to a solution, it will force the precipitation of a specific compound or compounds by exceeding the solubility products. A good example of this type of precipitation would be precipitation of metal cations by addition of H_2S to a solution. Because metallic sulfides in general have very small solubility products, addition of sulfide anions causes precipitation of insoluble metal sulfides.

Precipitation can also be accomplished by evaporation of water from the solution. As the water evaporates, the concentrations of all ionic species increase until one or more solubility products are exceeded and precipitation occurs.

Changes in the pH of solutions can also be used to precipitate compounds. As the pH of a solution is increased and the solution becomes more basic, the hydroxide ion (OH^-) concentration increases and solid hydroxides precipitate. Iron, copper, cobalt and nickel are precipitated selectively as hydroxides in solutions by raising the pH with milk of lime. Iron precipitates at approximately pH 2.5, copper at 5.8, cobalt at 8.3 and nickel at 9.4.

Precipitation can be used in a process for both removal of impurities as well as concentration of metal values in the form of a compound. An example of elimination of an impurity is the removal of iron from copper sulfate solutions. The iron content of copper sulfate solutions which will undergo electrolysis for copper should not be above 1.2 g/L (0.16 oz/gal). To remove iron without removing valuable copper, the ferrous iron (Fe^{++}) in solution is first oxidized to ferric iron (Fe^{+++}) by adding manganese dioxide and

then is precipitated as ferric hydroxide. This is carried out by raising the pH to 2.2 by partly neutralizing the acid solution with additions of copper hydroxide filter cake. The iron hydroxide solid is removed by thickening.

Desired metal values can also be removed from an impure solution and concentrated in a solid compound. A good example of this is recovery of sulfides of nickel, copper, lead and zinc from leaching solutions. In general, these precipitates require further purification, but there are advantages to this technique. Treatment costs are low and solutions with very low concentrations of metal values can be treated.

Solvent extraction is a chemical process which is used to purify and concentrate a given species from aqueous solution. This objective is accomplished by recycling an organic solution, which will selectively exchange the metal species of interest between an impure aqueous feed solution and a pure fresh aqueous solution. The process relies on the immiscibilities of the aqueous solution and the organic solution as well as the ability to establish conditions which favor the chemical stability of the metal species in the organic solution or the aqueous solution as desired. Purification is then achieved by extracting a metal species from the impure aqueous solution to the organic solution and then stripping the metal species from the organic solution back to a fresh aqueous solution.

A typical solvent-extraction flowsheet is shown in Fig. 12. The key to successful solvent extraction is proper selection of the organic. The or-

Table 1. Solubility products at 25 °C (77 °F) for salts forming only two principal ionic species in dilute aqueous solution

Ions of equal charge		Ions of unequal charge	
Salt	pK_{s0}(a)	Salt	pK_{s0}(a)
TlCl	3.72	Ag_2SO_4	4.80
$AgBrO_3$	4.28	$Ag_2C_2O_4$	11.30
Hg_2SO_4(b)	6.17	BaF_2	5.76
$SrSO_4$	6.55	$Cu(IO_3)_2$	7.13
$AgIO_3$	7.52	MgF_2	8.18
$PbSO_4$	7.80	SrF_2	8.54
AgCl	9.75	CaF_2	10.40
$BaSO_4$	9.96	$Mg(OH)_2$	10.74
AgSCN	12.00	$Pb(IO_3)_2$	12.59
AgBr	12.28	Hg_2Cl_2(b)	17.88
AgI	16.08	$Ce(IO_3)_3$	9.50
		$La(IO_3)_3$	11.21

(a) K_{s0} has been extrapolated to zero ionic strength. (b) Hg_2^{2+} is a single ion.

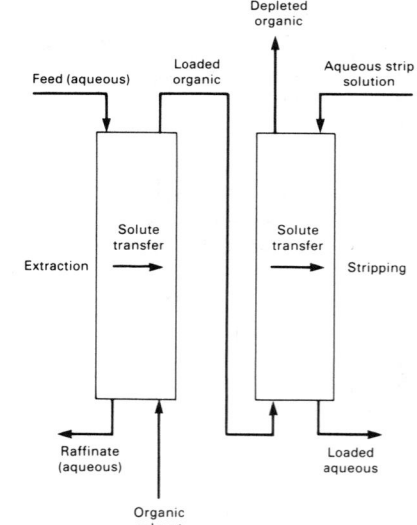

Depleted organic

Feed (aqueous)

Loaded organic

Aqueous strip solution

Solute transfer

Solute transfer

Extraction

Stripping

Raffinate (aqueous)

Loaded aqueous

Organic solvent

Fig. 12. Typical flowsheet for solvent extraction

ganic must be selective to the species being purified, and the reaction must be reversible so that the metal species can be transferred from impure aqueous to fresh aqueous via the organic phase. Solvent-extraction chemistry can roughly be separated into two types of reactions: solvation and exchange reactions.

Solvation involves transfer of neutral molecular species between aqueous and organic. In the transfer from the aqueous to the organic phase, the neutral species simply dissolves in the organic solution. The capacity of the organic phase for solvation reactions is governed by solubility considerations. The organic phases used for solvation can be alcohols, ethers, esters, ketones, or phosphorus-containing compounds such as trialkyl phosphates and trialkyl phosphine oxides.

Exchange reactions involve the formation of specific bonds between metal species and active compounds in the organic phase. To be extracted into the organic phase, the metal species forms an organic salt with the active organic compounds, which then dissolves in the organic phase. Exchange reactions can be cationic or anionic depending on the system involved. A more complete description of solvation and exchange reactions is given in an article by Bailes, Hanson and Hughes (Ref 9).

The organic solutions typically used for exchange reactions are made up of a combination of a carrier, an extractant and a modifier. The carrier or diluent is the inert organic that makes up the bulk of the solution. The carrier comprises approximately 90% of the solution and acts as a vehicle for carrying the active compound referred to as the extractant. Common diluents are kerosine, rykue and naphthene. The extractant or active agent is the compound that contains the functional group that is capable of chemically reacting with the particular metal species of interest in the aqueous phase. The modifier, which is usually an alcohol, is added because it increases extracting power, increases selectivity, improves phase separation and prevents formation of solid organic compounds.

The choice of equipment for possible use in solvent extraction is outlined in a paper by Reissinger and Schröter (Ref 10). However, the most common equipment used in solvent extraction is the mixer-settler (see Fig. 13). A combination of mixer-settler units is set up in series to allow for the countercurrent flow of organic and aqueous solutions from stage to stage. Countercurrent staged systems are further explained in an article by Bridges and Rosenbaum (Ref 11), including the methods used to do calculations in staged countercurrent flow.

Ion Exchange. Purification of solutions by ion exchange is accomplished by interchange of metal ions between aqueous solutions and a solid, insoluble resin. The chemistry of the ion-exchange process is very similar to the reactions in solvent-extraction systems except that the organic is a solid resin in the case of ion exchange. In fact, ion exchange is sometimes used as a substitute for solvent extraction when it becomes necessary to avoid problems of emulsion formation and solvent loss due to entrainment.

The ion-exchange process involves adsorption followed by elution. Adsorption is the removal of metal ionic species from an aqueous solution when that solution is passed through a bed of ion-exchange resins. Elution is the recovery of the metal ionic species in fresh solution by passing a suitable fresh solution through the previously loaded resins.

Ion-exchange resins are classified as cationic resins and anionic resins. Cationic resins, which exchange cationic species, are made of strong acid or weak acid groups and exchange H^+ ions. Anionic resins are strong bases, such as quaternary ammonium group bases, and weak bases, such as secondary or tertiary amine group bases. In most cases, chloride ions exchange with anions in solution.

Ion-exchange resins are made by first synthesizing a cross-linked polymer structure and then adding functional groups to this hard, organic

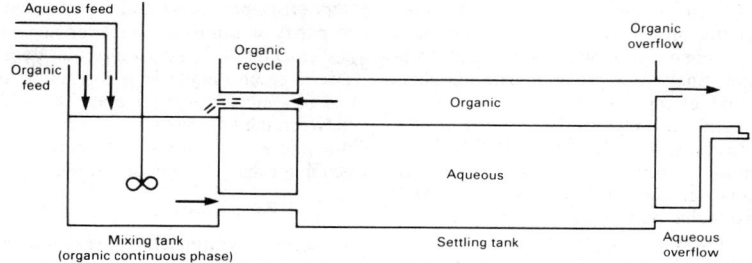

Fig. 13. Schematic diagram of a mixer-settler unit for solvent extraction

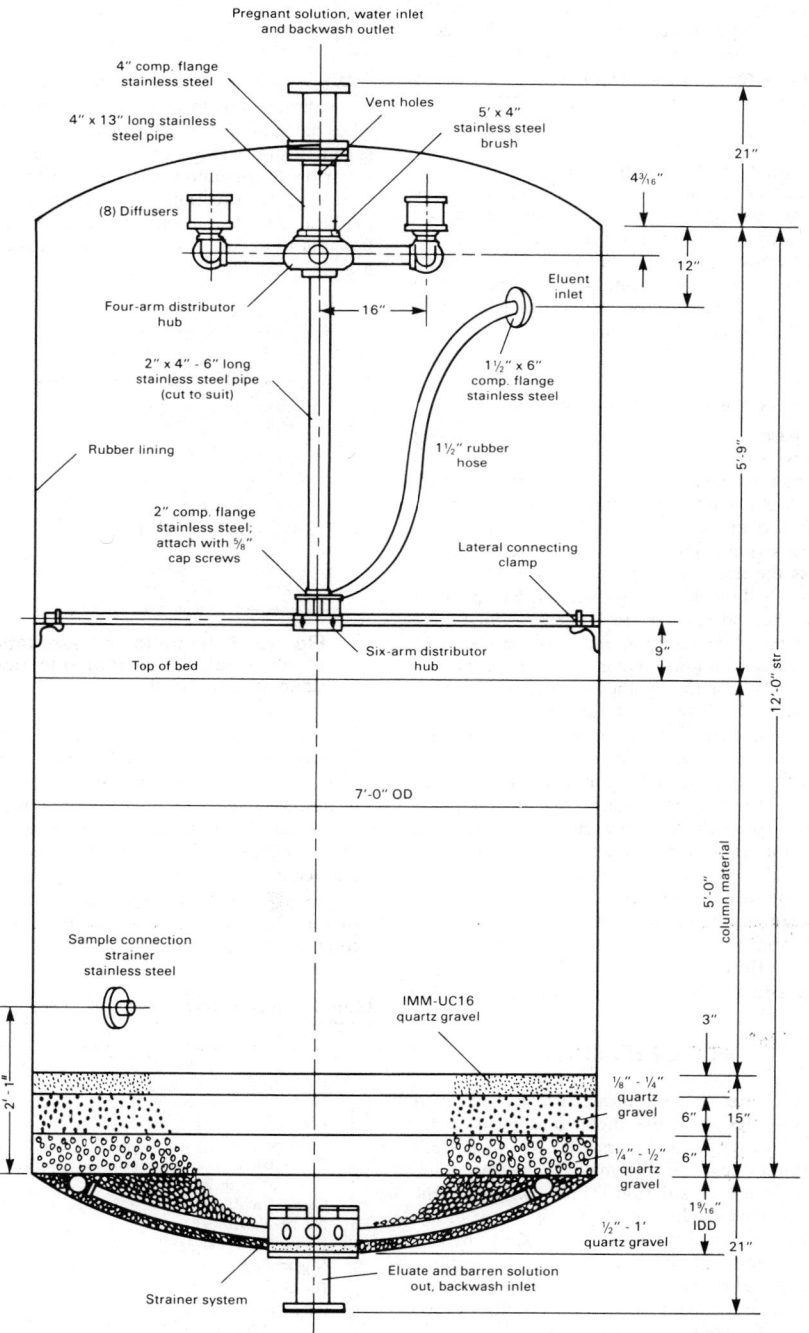

Source: W. Q. Hull and E. T. Pinkney, *Ind Eng Chem*, Vol 49, No. 1, 1957, p 7.

Fig. 14. Ion-exchange column for uranium extraction

structure. The amount of cross-linking that is used in the manufacture of the resin has a dramatic effect on the chemical as well as physical properties of the final resin product. The chemical reactions and effects of cross-linking are discussed in detail in books by Dorffner (Ref 12), Lydersen (Ref 13) and Helfferich (Ref 14).

The important properties of ion-exchange resins from a design standpoint can be summarized as capacity, selectivity and mechanical properties. The capacity of a particular resin is the amount of a specific inorganic group that the resin will hold per unit of resin. For engineering purposes, this is described as either weight capacity or volume capacity. The weight capacity is the weight of a specific inorganic group that is contained in a specific weight of resin. The specific amount of resin is the weight when the resin is completely converted to a standard form, such as H^+ or Cl^-, and the weight capacity is expressed as milliequivalents per gram of resin. The volume capacity is the amount of an inorganic group per unit volume of a packed bed and is usually expressed as capacity per litre of swollen bed.

The affinity of a resin for different ions in solution varies. Therefore, the result is a certain degree of selectivity of one ion over the next in solution. This selectivity can be described with a distribution coefficient such as

$$K = \frac{(\% \text{ equivalent of ion in resin})}{(\% \text{ equivalent of ion in solution})}$$

For a particular resin, the selectivity coefficient varies with the particular species of interest. This phenomenon makes it possible to purify a particular metal ion from a complex solution by selecting it over the other ions in solution.

Because ion-exchange resins are used over and over as the transfer media for purification, they require good mechanical properties. Resins must be durable and resistant to breakage, must have low chemical degradation and must be insoluble in aqueous solutions. Their physical properties are the result of the synthesis procedure, which is explained well by Helfferich (Ref 14).

The ion-exchange process usually involves fairly high capital costs and a large plant area as a result of the sophisticated equipment and the large amounts of in-process material that are required. However, properly run ion-exchange facilities can result in up to 99% efficiency during normal operation. The ion-exchange process is usually done using columns, as shown in Fig. 14. The columns are arranged in a staged countercurrent system similar to solvent-extraction circuits. Examples of flowsheets and calculations for ion-exchange circuits are given by Lydersen (Ref 13).

METAL DEPOSITION

Once an ore or concentrate has been leached to remove the valuable metal and purified by chemical or physical treatment, solvent extraction or ion exchange, the metal of interest must be recovered from solution. Three common techniques of metal reduction from aqueous solution are metallic replacement (cementation), hydrogen reduction and charcoal (carbon) adsorption.

Metallic Replacement (Cementation). Metallic replacement or cementation is one of the oldest methods of recovering metals from aqueous solution. The first known reference to cementation is from the 16th century (Ref 15). Despite the long time it has been in use, cementation is a very efficient process and is still commonly used to purify solutions and recover metals.

Cementation is essentially the precipitation or discharge of a noble or less-reactive metal in favor of a more-reactive metal. The basic reaction between the two metals is electrochemical in nature and can be represented by the following reaction for the reduction of cadmium by zinc metal:

$$Cd_{(aq)}^{++} + Zn_{(s)} = Cd_{(s)} + Zn_{(aq)}^{++}$$

The above electrochemical reaction can, in turn, be separated into two half-reactions:

$$Zn_{(s)} = Zn_{(aq)}^{++} + 2e^-$$

and

$$Cd_{(aq)}^{++} + 2e^- = Cd_{(s)}$$

When an active metal, such as zinc, is added to a solution containing cadmium ions, reduction takes place in microcells such as the one depicted schematically in Fig. 15. The half-reactions that occur do so separately at an arbitrary, finite distance and require transfer of electrons between the locations where each half-reaction takes place. For this to occur, it is essential that the solid is a good conductor of electricity.

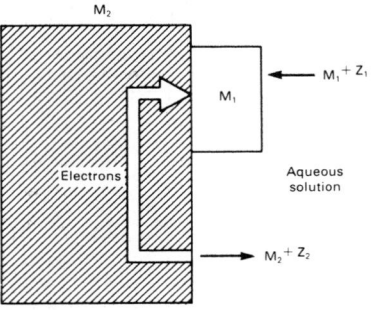

Fig. 15. Schematic representation of an electrochemical microcell used in cementation

The tendency of one metal to displace or reduce another metal from solution is based on the electromotive series of metals. Table 2 shows an electromotive series of some common metals. The table is based on thermodynamic considerations which rank the metals according to their potential to do reduction. An electropositive or active metal, such as zinc, will tend to reduce a less-active or noble metal, such as copper or silver, from a solution. In general, when two metals are considered, the one that is more electropositive will tend to reduce a less-electropositive metal from solution. Also, the greater the difference in potential between the two metals the more complete the reaction will be at equilibrium. However, even a small difference in potential will result in a very extensive degree of reduction.

A good example of cementation used on an industrial scale is reduction of copper ions from solution by metallic iron. The reaction is:

$$Cu_{(aq)}^{++} + Fe_{(s)} = Cu_{(s)} + Fe_{(aq)}^{++}$$

The simplest and most common method of copper cementation is the use of open launders. Copper sulfate leaching solution is fed through open launders containing steel scrap, where the above displacement reaction occurs and produces a very pure copper which is recovered in the bottom of the launders; however, the copper may be contaminated with residual iron.

Recently, research has led to the development of more efficient equipment for carrying out cementation reactions. The most significant development from an industrial point of view has been the development of cementation cones (Ref 16). Cementation cones have been very successful and have been accepted and used by industry.

A great deal of effort has also been directed in the last decade toward an understanding of the basic principles of this old process. For example, it has been determined that the rate at which cementation reactions occur depends on initial concentrations, temperature, agitation, polarization characteristics of different metals, and addition agents. Nadkarni and Wadsworth (Ref 17 and 18), Rickard and Fuerstenau (Ref 19), and Miller (Ref 20) give excellent descriptions of the theory and principles involved in cementation reactions.

Gaseous Reduction. Reduction of metals from aqueous solution can be done with reducing gases instead of by metallic reduction. Reducing gases, such as hydrogen, carbon monoxide and sulfur dioxide, can be used. However, hydrogen is the most widely used because it is simple and relatively inexpensive, because it is produced on a large scale and because the reaction products (H^+ or OH^-) are simple. The reaction products from carbon monoxide and sulfur dioxide must be further treated after the reduction step.

The use of reducing gases is similar to cementation in that it is a replacement reaction, such as

$$Cu_{(aq)}^{++} + H_{2(g)} = Cu_{(s)} + 2\,H_{(aq)}^+$$

The tendency of the above reaction to produce

Table 2. Electromotive series of some common metallic elements

Strongest reducing action	$E°$		$E°$	Strongest oxidizing action
$Li \rightleftarrows Li^+ + e^-$	3.05	$Cd \rightleftarrows Cd^{++} + 2e^-$	0.40	
$Cs \rightleftarrows Cs^+ + e^-$	2.92	$Co \rightleftarrows Co^{++} + 2e^-$	0.28	
$Rb \rightleftarrows Rb^+ + e^-$	2.92	$Ni \rightleftarrows Ni^{++} + 2e^-$	0.25	
$K \rightleftarrows K^+ + e^-$	2.92	$Sn \rightleftarrows Sn^{++} + 2e^-$	0.14	
$Ba \rightleftarrows Ba^{++} + 2e^-$	2.90	$Pb \rightleftarrows Pb^{++} + 2e^-$	0.13	
$Sr \rightleftarrows Sr^{++} + 2e^-$	2.89	$H_2 \rightleftarrows 2H^+ + 2e^-$	0.00	
$Ca \rightleftarrows Ca^{++} + 2e^-$	2.87	$Cu \rightleftarrows Cu^{++} + 2e^-$	−0.34	
$Na \rightleftarrows Na^+ + e^-$	2.71	$2I^- \rightleftarrows I_2 + 2e^-$	−0.53	
$La \rightleftarrows La^{+++} + 3e^-$	2.52	$Ag \rightleftarrows Ag^+ + e^-$	−0.80	
$Mg \rightleftarrows Mg^{++} + 2e^-$	2.34	$Hg \rightleftarrows Hg^{++} + 2e^-$	−0.85	
$Be \rightleftarrows Be^{++} + 2e^-$	1.85	$2Br^- \rightleftarrows Br_{2(l)} + 2e^-$	−1.06	
$Al \rightleftarrows Al^{+++} + 3e^-$	1.67	$Pt \rightleftarrows Pt^{++} + 2e^-$	−1.2	
$Mn \rightleftarrows Mn^{++} + 2e^-$	1.18	$2H_2O \rightleftarrows O_2 + 4H^+ + 4e^-$	−1.23	
$Zn \rightleftarrows Zn^{++} + 2e^-$	0.76	$2Cl^- \rightleftarrows Cl_2 + 2e^-$	−1.36	
$Cr \rightleftarrows Cr^{+++} + 3e^-$	0.74	$Au \rightleftarrows Au^+ + e^-$	−1.68	
$Fe \rightleftarrows Fe^{++} + 2e^-$	0.44	$2F^- \rightleftarrows F_2 + 2e^-$	−2.65	

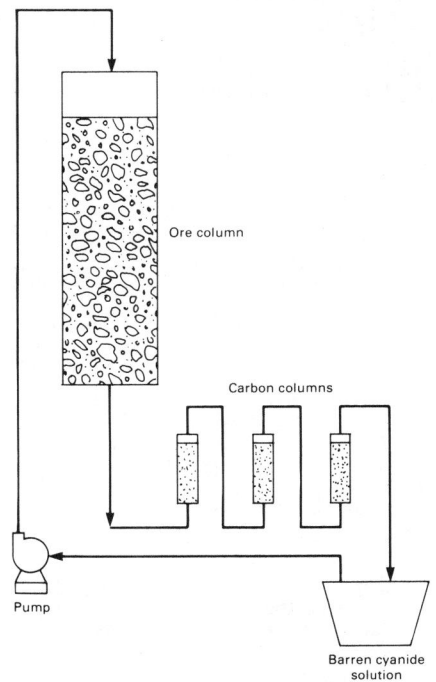

Fig. 16. Laboratory column leaching unit

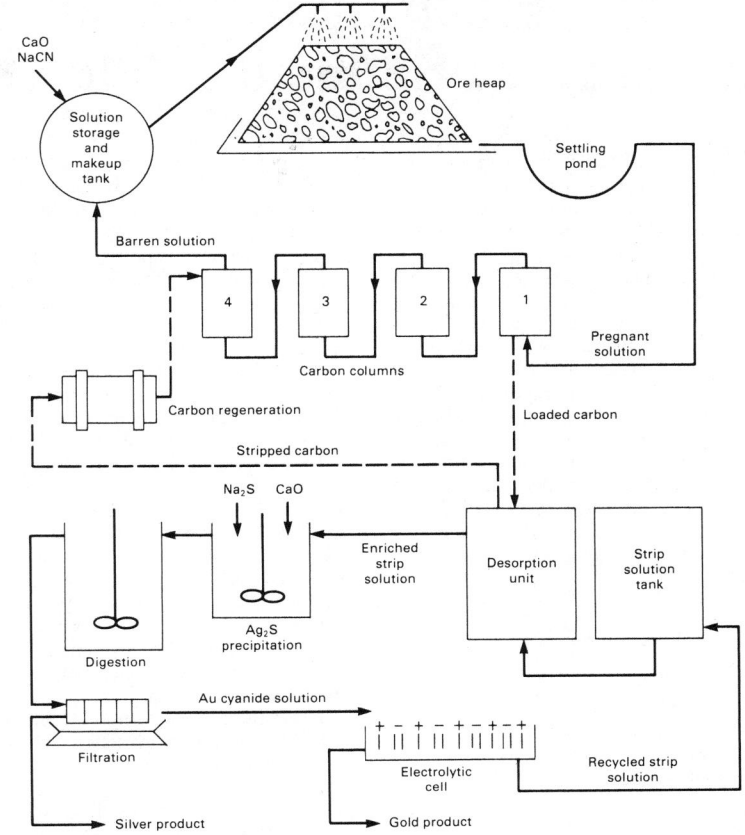

Fig. 17. Heap leaching–charcoal adsorption process for gold ores low in silver

copper can be increased by increasing the pressure of hydrogen gas. This fact is used to reduce nickel from solution according to the reaction

$$Ni_{(aq)}^{++} + H_{2(g)} = Ni_{(s)} + 2 H_{(aq)}^{+}$$

At room temperature and one atmosphere, the above reaction will not produce nickel as indicated by the potentials listed in Table 2. If the reaction is carried out in an autoclave that permits high hydrogen pressures and high temperatures, nickel can be reduced from solution. Although the mechanism by which nickel is reduced from ammoniacal solutions is more complex than the reaction listed above, nickel is reduced commercially at high temperature and pressure in the presence of ammonia (Ref 21). The principle of increasing pressure to force the reduction to occur is the same.

Charcoal Adsorption. Reduction and adsorption on carbon of metals from aqueous solution has been used almost exclusively to recover the noble precious metals, such as gold and silver. This process is based on the principle that these easily reduced metals can be reduced by solid carbon at low temperatures and deposited in metallic form on the carbon.

In a typical charcoal-adsorption process, the metal leaching solution is fed to carbon columns as shown in Fig. 16, and the metal is almost completely removed from the solution by adsorption on the solid carbon. After the carbon is loaded, it is removed from the circuit and the metal is stripped away. The carbon can also be added to the leaching liquor and agitated without the requirement of carbon columns. This process is referred to as the "carbon in pulp" process and is in commercial operation in the United States (Ref 22 and 23).

De-adsorption or stripping can be done by passing a hot, caustic stripping solution over the column. Stripping is followed by electrowinning from solution of a very pure gold or silver product. An example of a typical flowsheet, including charcoal adsorption, is shown in Fig. 17. The loaded carbon can also be burned, which leaves a gold- or silver-rich ash. However, this procedure destroys the carbon, and in many cases is too expensive.

REFERENCES

1. Hydrometallurgical Processes, by M. E. Wadsworth and J. D. Miller: in *Rate Processes of Extractive Metallurgy*, edited by H. Y. Sohn and M. E. Wadsworth, Plenum Press, New York, 1979, p 133-241

2. In-Place Leaching at Miami Mine, Miami, Arizona, by J. B. Fletcher: *Trans SME/AIME*, Vol 250, No. 4, Dec 1971, p 310-316

3. Rancher's Big Blast Shatters Copper Ore Body for In-Situ Leaching: *Engineering and Mining J*, Vol 173, No. 4, 1972, p 98-100

4. Rancher's Development Sets Off Blast: Will Leach at Big Mike: *Mining Engineering*, Vol 25, No. 8, 1973, p 10

5. Developments and In-Place Leaching of Mountain City Chalcocite Ore Body, by C. B. Catanach: in *Extractive Metallurgy of Copper*, Vol II, edited by J. C. Yannapoulos and J. C. Agarwal, TMS/AIME, 1976, p 849-872

6. "Agglomeration—Heap Leaching Operations in the Precious Metals Industry": U.S. Bureau of Mines Information Circular 8945, 1983

7. Air Agitation and Pachuca Tanks, by A. G. W. Lamont: *Can J Chem Eng*, Aug 1958, p 153

8. *Fluid Mixing Technology*, by J. Y. Oldshue: Chemical Engineering Division of McGraw-Hill, New York, 1983

9. Liquid-Liquid Extraction-Metals, by P. J. Bailes, C. Hanson and M. Hughes: *Chem Eng*, Aug 30, 1976

10. K. Reissinger and J. Schröter: *Chem Eng*, Nov 6, 1978

11. D. W. Bridges and J. B. Rosenbaum: U.S. Bureau of Mines Information Circular 8193, 1962

12. *Ion Exchangers, Properties, and Applications*, by K. Dorffner: Ann Arbor Science Publishing Co., Ann Arbor, MI, 1973

13. Chapter 7 in *Mass Transfer in Engineering Practice*, by A. L. Lydersen: John Wiley, New York, 1983

14. *Ion Exchange*, by F. Helfferich: McGraw-Hill, New York, 1962

15. *De re Metallica*, by G. Agricola: Dover, New York, 1950

16. Cone-Type Precipitates for Improved Copper Recovery, by R. H. Spedden, E. E. Malouf and J. D. Prater: *J Metals*, Vol 18, 1966, p 1137-1141

17. A Kinetic Study of Copper Precipitation on Iron—Part I, by R. M. Nadkarni *et al: Trans Met Soc AIME*, Vol 239, 1967, p 581-585

18. A Kinetic Study of Copper Precipitation on Iron—Part II, by R. M. Nadkarni and M. E. Wadsworth: *Trans Met Soc AIME*, Vol 239, 1967, p 1066-1074

19. An Electrochemical Investigation of Copper Concentration by Iron, by R. S. Rickard and M. C. Fuerstenau: *Trans Met Soc AIME*, Vol 242, 1968, p 1487-1493

20. Cementation, by J. D. Miller: Section 3.3 in *Rate Processes of Extractive Metallurgy*, edited by H. Y. Sohn and M. E. Wadsworth, Plenum Press, New York, 1979

21. The Winning of Nickel, by J. R. Boldt, Jr., and P. Queneau: International Nickel, Inc., New York, 1967

22. D. Jackson: *Engineering and Mining J*, 1974, p 65-70

23. S. J. Hussey, H. B. Salisburgy and G. M. Potter: U.S. Bureau of Mines Report of Investigation 8268, 1978

Electrolytic and Electrothermal Reduction and Refining

ELECTROMETALLURGY is the area of extractive metallurgy that deals with production of metals from ions by application of electrical energy. This article will discuss the basic concepts of electrochemistry, electrowinning, electrorefining and fused salt electrolysis as they relate to metal production and purification. The term "electrowinning" will be used to describe the recovery of metals from aqueous solutions obtained from prior leaching and purification procedures. Electrorefining will be discussed as a purification method which is accomplished by dissolution of an impure metal (anode) and deposition of a pure metal (cathode) at different sites. Fused salt electrolysis is production of metals which cannot be electrolyzed from aqueous solution due to their relative positions in the electromotive series and which are reduced from molten salts.

BASIC CONCEPTS

If a metal is placed in a solution of metal ions, the first metal may dissolve while the second metal deposits from its ions. A common example of this is the cementation reaction between zinc metal and copper ions:

$$Zn_{(s)} + Cu^{++}_{(aq)} = Zn^{++}_{(aq)} + Cu_{(s)}$$

If the reverse procedure is tried — that is, if copper metal is placed in a solution containing zinc ions — no reaction will take place to any measurable degree. For example, if the solution containing zinc ions has initially no copper ions, the reaction will occur to a very small extent and will stop when a certain very small concentration of copper ions has been produced. In the opposite case, zinc metal will react with copper ions almost to completion, stopping only when the concentration of copper ions is very small.

The above experiment can be repeated with many combinations of metals, and the ability of one metal to replace ions of another metal from solution can be used to arrange the metals in a series. The series so formed would show the abilities of the metals to reduce each other. Such an electromotive series is shown in Table 2 in the previous article. The potentials listed in that table are measured values which will be described later in this article.

The reactions described in establishing an electromotive series are referred to as electrochemical reactions. Electrochemical reactions are those reactions which involve oxidation (increase in valence) and reduction (decrease in valence) and can be arranged in an electrochemical (galvanic) cell as shown in Fig. 1.

For the example of copper metal deposition using zinc metal, the oxidation reaction is

$$Zn_{(s)} = Zn^{++}_{(aq)} + 2e^-$$

which results in two electrons. These electrons are consumed by copper ions according to the following reduction reaction:

$$Cu^{++}_{(aq)} + 2e^- = Cu_{(s)}$$

An electrochemical cell can be set up as in Fig. 1 by using a copper electrode as M_1 in a solution of copper sulfate and a zinc electrode as M_2 in a solution of zinc sulfate. If the external circuit is short circuited, electrons will flow, as zinc dissolves, from the zinc electrode (anode) to the copper electrode (cathode), which causes deposition of copper metal. By placing a voltmeter between the cathode and the anode, the potential difference between the oxidation half-reaction and reduction half-reaction can be measured.

Because measurement of potential requires that it be measured between the electrodes, an absolute potential of any half-reaction cannot be measured. Therefore, reactions are measured to a defined standard electrode. The potentials given in Table 2 (p 21•18) are specifically those measured relative to a standard hydrogen electrode at 25 °C (77 °F) when all ions are in $1M$ concentrations, all gases are at a pressure of one atmosphere, and solid phases are pure. Changes in concentration, temperature and pressure may change the position of a particular metal in the series. Calculation of potentials for conditions other than the specific ones mentioned is covered by Maron and Prutton (Ref 1).

Returning to the example of an electrochemical cell with copper and zinc electrodes, it is apparent that the chemical energy that exists between the copper and zinc electrodes can be converted to electrical energy. This is, of course, the concept of a battery. However, the external circuit can be replaced with a dc power supply which can be used to force electrons to go in a direction opposite to that in which they tend to go naturally.

As the potential of the power supply is gradually increased, the electron flow from the zinc electrode to the copper electrode becomes less and less. Eventually, if enough power is supplied, the electrons reverse direction and travel from the copper electrode to the zinc electrode. This now causes oxidation (dissolution of copper) to occur at the copper electrode and reduction (zinc plating) to occur at the zinc electrode. The cell is now operating as an electrolytic cell — that is, by application of electrical energy, the reduction of zinc is being forced to occur. The reversible or ideal potential, which must be overcome to accomplish this (decomposition potential), is the difference between the reduction potential of zinc and the oxidation potential of copper

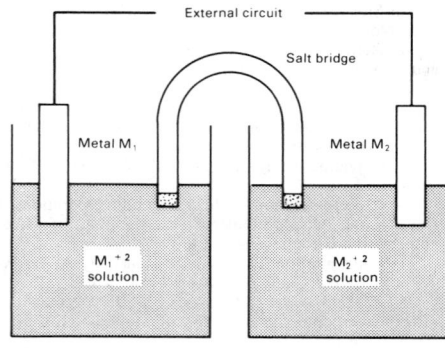

Fig. 1. Electrochemical (galvanic) cell

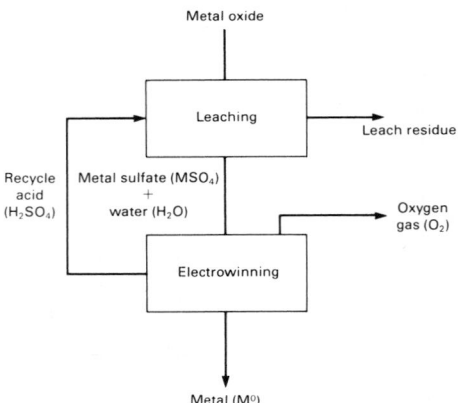

Fig. 2. Simplified block diagram showing the cyclical nature of the leaching/electrowinning process

(for a given metal, the reduction potential is the negative of the oxidation potential).

ELECTROWINNING

The purpose of electrowinning is to recover metal values from leaching solutions. The process usually is done with a two-step leaching/electrowinning arrangement that recycles the acid between the leaching operation and the electrowinning operation. A simplified diagram of the leaching/electrowinning process is presented in Fig. 2.

An example of electrowinning that can be used to describe the reactions in each step is sulfuric acid leaching of a metal oxide MO. In the leaching step the sulfuric acid is consumed as the metal dissolves, according to the reaction

$$MO_{(s)} + H_2SO_{4(aq)} = H_2O + MSO_{4(aq)}$$

The leaching solution is transferred to the electrowinning step for deposition of the metal. The leaching solution is fed into cells which consist of inert anodes (usually made of lead) and cathodes (which can be made of stainless steel, aluminum, or starter sheets of metal M), arranged as parallel plates with cathodes and anodes alternating. Examples of plant cell layouts are shown in Fig. 3.

In the electrowinning step, the power source forces electrons to the cathode, and metal is reduced to produce the very pure metal:

$$MSO_{4(aq)} + 2e^- = M_{(s)} + SO^=_{4(aq)}$$

At the anode, oxygen is produced by decomposition of water:

$$H_2O = 2 H^+_{(aq)} + \tfrac{1}{2} O_{2(g)} + 2e^-$$

Also, the water decomposition and metal plating cause regeneration of acid:

$$SO^=_4 + 2 H^+ = H_2SO_4$$

which is recycled back to the leaching step.

The potential required between cathode and anode in electrowinning is on the order of 1.25

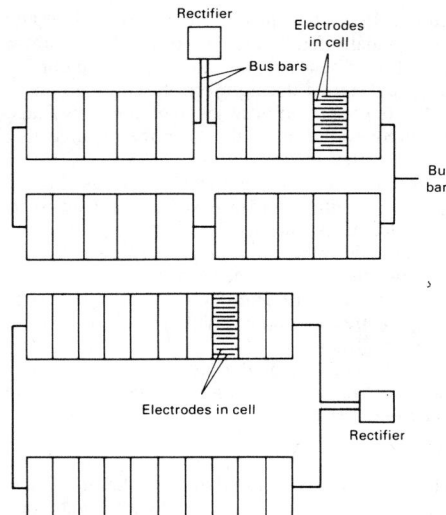

Fig. 3. Examples of plant cell layouts

to 1.75 V. This potential is the combination of the decomposition voltage (the reversible potential between the cathode and the anode), the ohmic drop due to the resistance of the electrolyte, the ohmic drop due to connections and conductors, and polarization of the cell.

Polarization is a phenomenon that results from the nature of the current flow in the system. The decomposition potential is a potential that is measured only when the cell reactions take place under conditions of infinitely low current or net zero current. In electrolysis, where current flows at a finite rate, additional voltage drops occur as a result of phenomena that take place near the electrode solution boundaries. These extra voltages are called overvoltages and are caused by reactants not being supplied to electrodes as fast as products are removed (concentration overvoltage) or by molecular phenomena (activation overvoltage). The complex subject of polarization is covered in more detail by other authors (Ref 2 and 3).

One of the problems associated with electrowinning (and electrorefining) is development of acid mists above the electrolytic cells. These mists require proper ventilation in all cell houses to avoid mist buildup. Also, methods such as the use of a blanket of foam on the surface of the cell have been used to depress these mists.

The resistance of the electrical connections and the resistance of the electrolyte cause generation of heat as current passes through the cell. In order to avoid excessive heating of the cell, this heat must be removed. Some heat is lost in evaporation, but is often not enough to avoid accumulation of heat and excessive electrolyte temperatures. The optimum solution temperature in most electrowinning plants is 30 to 45 °C (85 to 115 °F). Temperatures much higher than this can cause deterioration of the cathode deposit as well as increased corrosion of the leaching anodes. Temperature control usually can be accomplished by incorporating a heat exchanger in the circuit, as shown in Fig. 4.

Once the metal has been deposited on the cathode it must be recovered. This usually involves peeling the plated metal away from the cathode so that the cathode can be recycled back to the cells. For this reason, it is important that the metal

not be tightly bonded to the cathode. Depending on the metal being plated, different cathode materials can be used. For copper electrowinning, titanium, stainless steel and copper itself have been used successfully. Aluminum is used to recover zinc and titanium, and stainless steel has been used for electrowinning manganese and cobalt.

ELECTROREFINING

The purpose of electrorefining is further refinement of metal to a very pure metal product by electrolysis. The advantages of electrorefining over other refinement techniques are the high purity which can be obtained in the product and the recovery of secondary metals, which often are a series of precious metals, in a concentrated state. The secondary metals are then recovered from the concentrated form in a separate processing step.

Metals made by smelting of sulfides, such as copper, lead and nickel, usually are prime candidates for refinement by electrolysis. The impurities that contaminate these metals are those metals that are most often associated with sulfides. The precious metals gold, silver, platinum and selenium, as well as iron, bismuth, arsenic and antimony, are usually present. The metal from the smelters, which is from 90 to 98.5% pure, is cast into anodes prior to shipment to the electrorefining cell house.

When the impure anodes reach the cell house, they are put into cells that contain an acid solution of the anode metal ion. This solution is referred to as the electrolyte. Cathode sheets are placed between the anodes and are the locations where the pure metal will be deposited. A dc power source is wired to the anodes and cathodes in a manner such that electrons are forced into the cathode sheets, causing reduction, and electrons are stripped from the anodes, causing oxidation.

At the anode, the oxidation reaction results in the dissolution of the metal by the reaction

$$M = M^{++} + 2e^-$$

and the electrons flow back through the power source to the cathode. As the anode dissolves, the impurities either dissolve into the electrolyte with the metal or remain in the metallic state.

The metals that remain in the metallic state are those metals which are more noble in the electromotive series than the metal that is being refined. As the anode dissolves away, these noble metals fall to the bottom of the cell and collect in a concentrated product that is commonly called "anode slime" or "anode mud." Anode mud is very valuable because it is usually made up of precious metals, such as gold, silver, platinum, selenium, etc. In copper electrorefining, for example, the slimes can be over 90% silver. The metals that are less noble on the electromotive series of metals will dissolve along with the metal that makes up the anode and, if not removed, will build up and contaminate the electrolyte.

The sequence of events in electrorefining begins with placing of the anodes in the cells, followed by dissolution during electrolysis. Approximately 85% of the anode is dissolved before it is removed from the tank. If more than 85% of the anode is dissolved, it loses its structure and falls into the cell, causing problems of removal and short circuiting. The unused portion is sent back to the smelter, where it is used as scrap and recast into more anodes. The cycle time

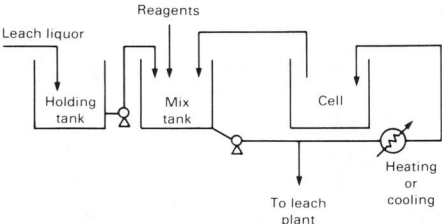

Fig. 4. Electrowinning plant circulation system

for anodes in a typical electrorefining operation is 14 to 28 days.

The cathodes in electrorefining cells are the final locations of the metal to be recovered. As the electrons are stripped from the anode and forced to the cathode, metal-ion reduction and metal deposition occur on the cathode according to the reaction

$$M^{++} + 2e^- = M$$

Because the less-noble metals remain in the metallic state as the anode dissolves, the only metal that plates on the cathode is the metal that made up the bulk of the anode. Those impurities that are less noble and that dissolve in the electrolyte are not plated on the cathode and remain in solution as metallic ions. The cathode sheets that are originally put into electrorefining cells are called either "starter sheets" or "mother sheets" and can be either reusable or used only once. Reusable mother sheets are made of a metal other than the metal to be reduced and form a very thin surface oxide layer. This oxide layer prevents tight bonding of the product metal, and, once the cathodes are loaded, the product metal is stripped off, allowing the reusable sheet to be returned to the electrorefining cells. Typical metals employed as reusable mother sheets are titanium, stainless steel and aluminum. Starter sheets can also be thin sheets of the product metal; when the cathodes are removed, the whole sheet is washed and shipped. The cycle time for cathodes is somewhat longer than the anode cycle time but usually is not more than twice as long. Current densities on cathodes are kept low to ensure quantity deposits and often are approximately 215 A/m^2 (20 A/ft^2).

The voltage required between anode and cathode during electrorefining is significantly lower than that required for electrowinning. For example, electrorefining of copper requires only about 0.25 V. This low voltage is the result of the fact that the reversible decomposition voltage between two copper electrodes is essentially zero. The other voltage drops, such as ohmic drop of the electrolyte and polarization (overvoltage), exist in electrorefining as they do in electrowinning.

The cells used for electrorefining are most commonly rectangular boxes that allow for easy placement and removal of anodes and cathodes. They are typically 100 to 200 mm (4 to 8 in.) wider, and 150 to 300 mm (6 to 12 in.) deeper, than the electrodes. In copper electrorefining, the anodes are approximately 1.0 by 1.0 by 0.05 m (39 by 39 by 2 in.). The bottoms of the tanks are usually sloped to allow for collection and flushing of the anode slimes. Since most electrolytes are sulfuric acid based (with the exception of lead electrorefining), the lining material is PVC, rubber or lead. Figure 5 is a drawing of typical electrorefining cells including electrodes.

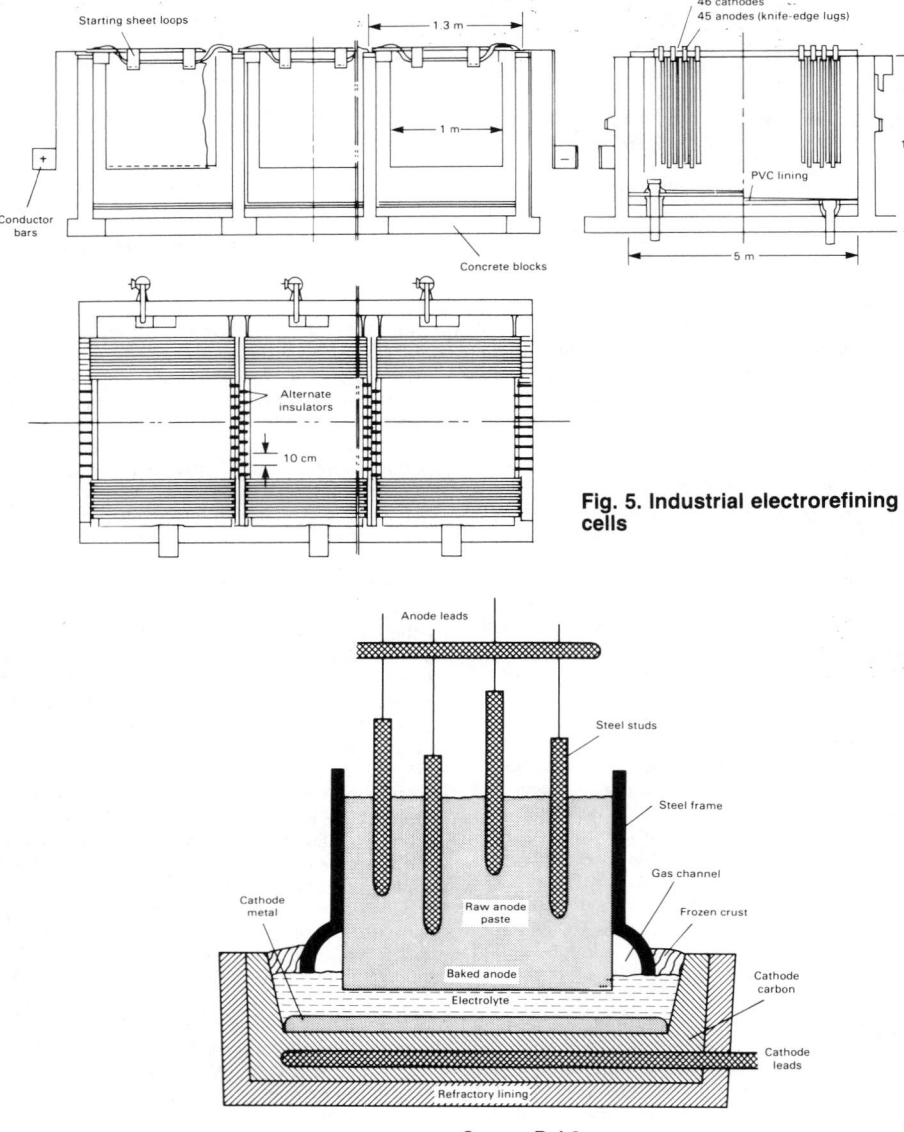

Starting sheet loops

1.3 m

1 m

Conductor bars

Concrete blocks

46 cathodes
45 anodes (knife-edge lugs)

PVC lining

1.5 m

5 m

Alternate insulators

10 cm

Fig. 5. Industrial electrorefining cells

Anode leads

Steel studs

Steel frame

Gas channel

Cathode metal

Raw anode paste

Frozen crust

Baked anode

Cathode carbon

Electrolyte

Cathode leads

Refractory lining

Source: Ref 2

Fig. 6. Hall-Héroult aluminum production cell with self-baking anode

current flow. The anodes are also of carbon and are gradually fed into the top of the cell because, as will be discussed later, the anodes are continually consumed during electrolysis. A group of cells are connected in series to obtain the voltage required by the particular dc power source that is being used.

For aluminum, the electrolyte used is cryolite (Na_3AlF_6) with 8 to 10% Al_2O_3 dissolved in it. Other additives, such as CaF_2 and AlF_3, are added to obtain desirable physical properties. The melting point of the electrolyte is approximately 940 °C (1725 °F), and the Hall-Héroult cell operates at temperatures of approximately 960 to 1000 °C (1760 to 1830 °F).

At the cathode of the aluminum cell, aluminum is reduced from an ionic state to a metallic state — for example:

$$Al^{+++} + 3e^- = Al_{(l)}$$

This is a very simplified representation of the complex reactions that take place at the cathode. However, it does represent the over-all production of molten aluminum, which forms a molten pool in the bottom of the cell. Periodically, the molten pool of aluminum metal is drained or siphoned from the bottom of the cell and cast.

At the anode, oxygen is oxidized from its ionic state to oxygen gas. The oxygen gas in turn reacts with the carbon anode to form carbon dioxide gas, which gradually consumes the anode material. Two types of anodes are in use: prebaked and self-baking. Prebaked anodes are individual carbon blocks that are replaced one after another as they are consumed. Self-baking anodes, as shown in Fig. 6, are made up of a carbon paste which is fed into a steel frame above the cell. As the anode descends in the cell it hardens and new carbon paste is fed continually into the top of the steel frame.

Impurities in the aluminum oxide raw material which are more noble than aluminum are reduced at the cathode along with the aluminum. Examples of two common metals associated with aluminum ores that fit this description are iron and silicon. It is, therefore, very important that raw materials be as free of these metal oxides as possible. By careful control of raw materials, aluminum with a purity of 99% or higher may be produced.

Magnesium is also produced by molten salt electrolysis. Magnesium does not alloy with iron, and so the cells used consist of iron pots which act as cathodes and carbon anodes. The electrolyte used is a magnesium chloride – sodium chloride – calcium chloride mixture which allows the process to be run at approximately 750 °C (1380 °F). The raw material for magnesium may be either magnesium chloride or magnesium oxide. Magnesium electrolysis differs from aluminum electrolysis in that the magnesium metal formed at the cathode is less dense than the electrolyte and rises to float on top of the electrolyte, and this requires special precautions when the molten metal is recovered from the cell.

Soluble impurities which dissolve with the anode and are not reduced at the cathode must be removed to avoid concentration in the electrolyte. If not removed, these impurities lower the conductivity of the electrolyte, which in turn increases the voltage required and thus increases the energy required. An example of this type of impurity is nickel in copper electrorefining. Nickel is an impurity in the anode which dissolves during electrolysis and collects in the electrolyte. Most copper refiners keep the concentration of nickel below 10 to 20 g/L (1.3 to 2.7 oz/gal) by continuously bleeding off a small portion of the electrolyte, removing nickel by chemical means and returning the nickel-free electrolyte to the cells.

Insoluble or more-noble impurities that remain in the metallic state collect in the bottom of the electrorefining cells. This anode mud (slimes) collects until the anode-cathode cycles are complete. At that time, the electrolyte is pumped from the cell, and the slimes are flushed out and collected by filtration. Because the slimes are the final destination of the precious metals, they are further refined to recover gold, silver, platinum and other associated precious metals.

MOLTEN SALT ELECTROLYSIS

Reactive metals more electropositive in the electromotive series than manganese cannot be electrowon or electrorefined from aqueous solutions because water will decompose at the cathode to form hydrogen gas before reduction of the metal occurs. Therefore, electrolysis of very electropositive metals is done from molten salt electrolytes. The most common metal recovered electrolytically from a molten salt electrolyte is aluminum.

Present practice for aluminum electrolysis involves the use of the Hall-Héroult cell as pictured in Fig. 6. The cell is lined with carbon, which acts as the cathode; steel bars are embedded in the cathode lining to provide a path for

REFERENCES

1. *Principles of Physical Chemistry*, by S. H. Maron and C. F. Prutton: MacMillan, London, 1965

2. *Principles of Extractive Metallurgy*, 2nd Ed., by T. Rosenqvist: McGraw-Hill, New York, 1983, p 438-441

3. *Electrochemical Methods*, by A. J. Bard and L. R. Faulkner: Wiley, New York, 1980

22 IRON AND STEELMAKING

Edited by Robert D. Pehlke, University of Michigan

Basic Technology

THIS ARTICLE presents a brief description of the production of steel, including the processes utilized and the most common operating practices. Initially, the processes utilized in production of direct-reduced iron ore are outlined. This technology is slowly being adopted as a source of iron units in addition to the more common practice of using pig iron produced in blast furnaces and/or iron-base scrap.

The iron blast-furnace process is described in detail, as are the most common steelmaking processes. In the articles that follow the present article, refining of steel in the ladle, and secondary refining and remelting processes, are outlined. Finally, casting of steel utilizing standard ingot practices or continuous casting machines is described, followed by a brief summary of ingot breakdown and hot rolling practices.

Direct-Reduction Processes

DIRECT-REDUCTION PROCESSES (hereafter termed "DR" processes) produce metallic iron from ores by removing most of the oxygen at temperatures below the melting temperatures of the materials in the process. Historically, DR processes were used before the advent of the blast furnace. However, these early processes could not be carried out on a large scale and were eventually supplanted by higher-capacity indirect processes based on blast-furnace smelting.

Products from DR processes (generally referred to as "direct-reduced iron" or "DRI") may be refined to steel in any commercial steelmaking process, or used to increase the productivity of blast-furnace or other melting processes.

Attempts to develop large-scale DR processes have embraced practically every known type of apparatus for contacting the reactants. However, out of the many projects and intensive activity, fluidized beds, moving-bed shafts, fixed-bed retorts and rotary-kiln processes have emerged as the successful commercial operations. Each of these processes has its own set of special characteristics, requirements and advantages for a particular situation.

Chemical, Physical and Process Concepts

Reduction of iron ore in any DR process is accomplished by the same reactions that occur in a blast furnace, including reaction by CO and H_2 through successive oxide-reduction reactions to metallic iron as:

$$Fe_2O_3 \rightarrow Fe_3O_4 \rightarrow FeO \rightarrow Fe^o$$

When the reduction reactions take place below about 1000 °C (1830 °F) and the reducing agents are restricted to CO and H_2, the DR product will be porous and of essentially the same size as the original iron ore particle. When the reducing temperature is above 1000 °C, the reduced material begins to sinter. At about 1200 °C (2190 °F), a pasty mass forms. This temperature represents the upper limit for the DR process, because metallic iron forms above 1200 °C and will absorb carbon with the resultant fusing or melting of the solid. The lower temperature limit for final reduction in any DR process is established by the fact that when iron oxide is reduced below about 650 °C (1200 °F), it is usually pyrophoric when cooled to ambient temperature, which presents handling and storage problems.

DESCRIPTIONS OF COMMERCIAL PROCESSES

The four principal DR processes are the fluidized bed, moving-bed shaft, fixed-bed retort and rotary-kiln processes. The effective performance of these processes depends on process design to ensure that good contact is made between the reacting gases and the solids. Ultimately, their performance depends on how closely they can approach the theoretical energy requirements for the system at outputs high enough to provide economical plant productivity.

Fluidized-Bed Processes. In the fluidized-bed process, solids are preheated in a separate fluidized bed by *in-situ* combustion of a suitable fuel with air, and reducing gases are preheated in an indirect-fired heat exchanger. In this way, the endothermic heat of reduction is introduced as sensible heat in the entering solids and gases. The rate of the reduction reactions will then be a function of the temperature and pressure in the reduction beds, the porosity and size distribution of the ore, the composition of the reducing gas, the ore/gas ratio, the depth of the reduction beds and the effectiveness of gas/solid contact. The reduction rate generally increases with increasing temperature and pressure up to about 500 kPa (73 psi). When reduction is done below about 650 °C (1200 °F), the DRI will be pyrophoric. Above about 800 °C (1470 °F), the beds tend to defluidize because of sticking of the reduced material. The fluidized-bed processes are therefore operated between these temperature limits.

Moving-Bed Shaft Processes. The moving-bed shaft DR processes comprise two distinct processing sections—reducing-gas generation and ore reduction. The iron-bearing raw materials can be fired pellets and/or lump ore, screened just prior to introduction to the top of the shaft reducer to minimize the fines and attendant problems with gas-flow distribution and solids movement. Excessive fines in the shaft, either in the feed or from decrepitation during reduction, usually will result in poor gas-flow distribution and erratic descent of solids in the shaft reducer. The shaft furnace process comprises an upper reduction zone and a lower cooling zone.

The spent reducing gas leaving the shaft is cleaned and cooled in high-efficiency scrubbers and packed cooling towers, and then used as a fuel and reforming gas in the natural-gas reformer. The reformer is a catalytic system wherein natural gas is reformed with a stoichiometric quantity of recycled off-gas (the mole ratio of the CO_2 and H_2O in the off-gas to the carbon in the hydrocarbon portion of the natural gas is close to unity). The reformer is equipped with a convection section for preheating the feed-gas mixture. A portion of the reformed gas may be cooled to adjust the temperature of the reducing gas entering the shaft.

Fixed-Bed Retort DR Processes. In these processes, the efficient countercurrent contact afforded by the blast-furnace stack is approached by arranging fixed-bed reactors (e.g., four) in series so that preheating, reduction and cooling take place concurrently with continuous flow of the process-gas stream through the reactors. In this way the sensible energy in the hot reduced product is transferred to the reducing gas before it reaches the reduction stage, and sensible energy in the hot spent reducing gas is transferred to the ore charge before the spent gas is exhausted.

As in moving-bed processes, the iron-bearing raw materials can be fired pellets and/or lump ore screened just prior to introduction to the first reactor to minimize the fines and attendant problems with gas-flow distribution and discharge of the bed after reduction and cooling. Fresh feed is charged to a reactor that has just been emptied.

22•1

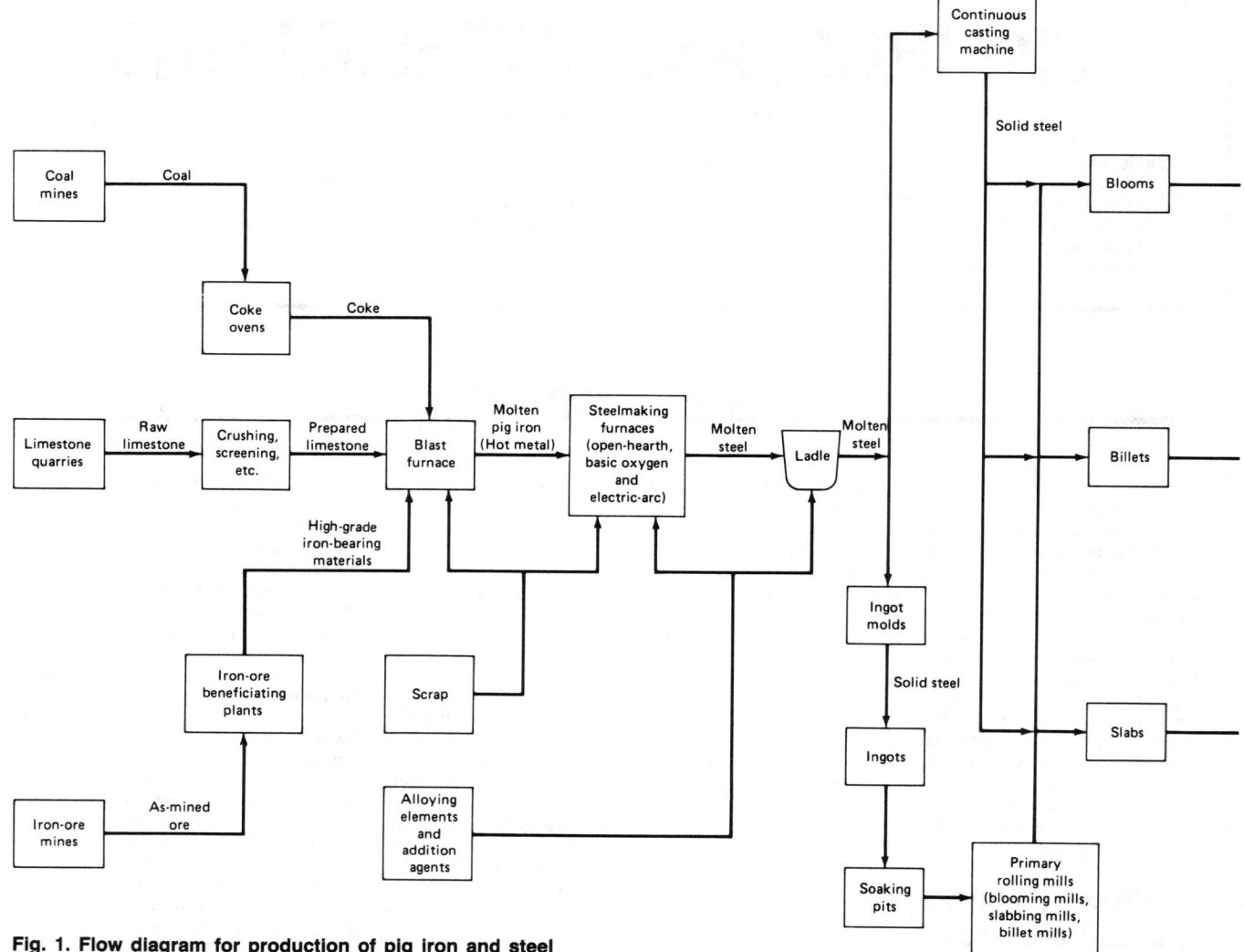

Fig. 1. Flow diagram for production of pig iron and steel

This discharge/charging stage, during which the reactor is not connected to the gas-distribution manifold, comprises several hours of the total treatment cycle. During this time the other reactors are undergoing preheating and prereduction, primary reduction, and final cooling, respectively. Each reactor passes through these stages successively.

Rotary-Kiln DR Processes. In rotary kilns, the ore, coal, recycle char, limestone and/or dolomite for sulfur control are charged together at the feed end and heated to the reaction temperature by burning volatiles and CO generated during reduction with controlled amounts of air introduced at appropriate locations through the kiln length.

Kilns are attractive because they represent a simple but effective means of contacting ores and agglomerates with solid reductants at temperatures high enough to provide relatively rapid reduction of FeO with concomitant gasification of fixed carbon. Because the heat-transfer rates and reactant-contact effectiveness are not as good as for the other DR processes discussed previously, energy consumption per unit of DRI is generally higher for kiln processes. This accounts for the fact that the rotary-kiln processes based on high-grade feeds and highly volatile noncoking coals

have still not been widely used in commercial operation.

COMMERCIAL USES FOR DRI

Major uses for DRI are as metallic iron units for steelmaking in electric-arc furnaces, open hearths, basic oxygen furnaces and induction furnaces. Minor uses are in foundries and as a superagglomerate for increasing productivity and decreasing energy consumption in blast furnaces, or in submerged-arc smelting furnaces, as a chemical reagent for copper cementation, and for use in making iron powder.

Production of Pig Iron and Steel

A SIMPLE flow diagram based on blast-furnace operations is shown at the left-hand side of Fig. 1. The right-hand side of Fig. 1 presents a flow diagram for conventional steelmaking processes. Thus, the total process of conversion of iron ore into steel products is covered in Fig. 1. Principal processes involved in converting raw

steel into the various mill-product forms are presented in Fig. 2.

Iron Blast Furnaces

THE MODERN blast furnace, as shown in Fig. 3, is an extremely complex piece of equipment in construction as well as in operation.

Iron is supplied to the furnace as ore in the form of iron oxide, either hematite (Fe_2O_3) or magnetite (Fe_3O_4). Pure hematite contains about 70% Fe, but ores presently being mined throughout the world may contain as little as 50% Fe, the difference being represented by gangue, which consists mostly of silica and alumina, and up to 10% moisture. Taconite ores containing as little as 25% Fe are also being mined. These ores are crushed and ground to a very fine particle size. The iron oxide is then separated and rolled into pellets on a balling disk. These pellets are then heated to provide a high-grade iron oxide pellet that can be handled without breakage. Prepared burdens, containing fluxes as in self-fluxing sinter or pellets, are now used in most blast furnaces. Iron is also added by charging mill scale,

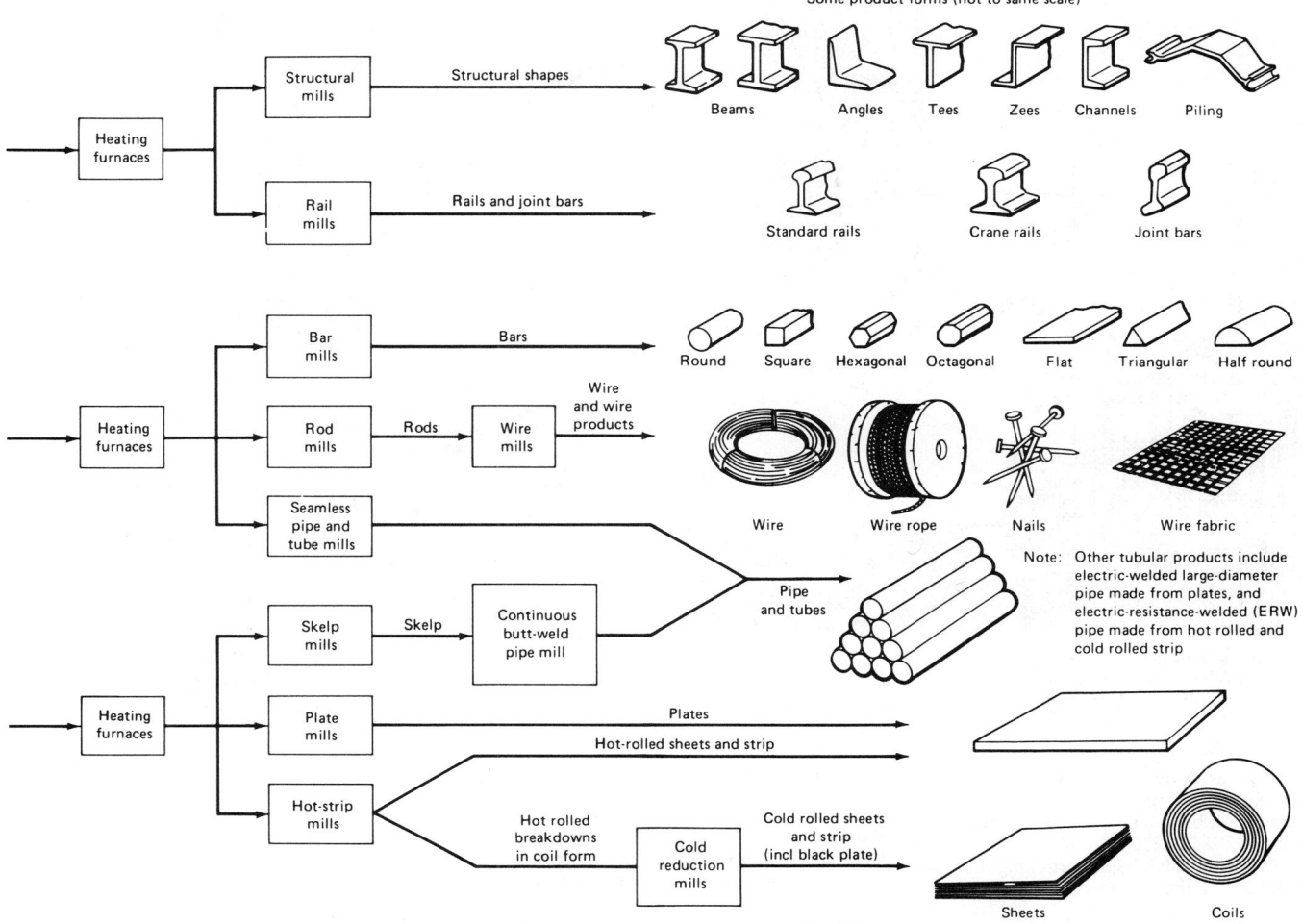

Fig. 2. Flow diagram showing the principal processes involved in converting raw steel into mill-product forms (excluding coated products)

sinter, slag from steelmaking furnaces, and scrap. The fuel used to provide temperature and also the source of reducing gases is coke, the solid product of the destructive distillation of coal. Limestone is added to form a fluid slag and to restrict the amounts of silicon and sulfur entering the pig iron. The blast, which is heated air, sometimes enriched with natural gas or oxygen, supplies oxygen to the process, reacting with the carbon in the coke to form carbon monoxide, the principal gaseous reducing agent.

As indicated in Fig. 1, the blast furnace charge is comprised of iron ore, coke and limestone. The product of the blast furnace is pig iron, which may be cast into "pigs" and allowed to solidify. More often, however, it is retained in the liquid condition and transferred to a steelmaking furnace. In the operation of the blast furnace, a blast of preheated air (hence the name of the furnace) passes upward through a permeable bed of solid raw materials; the products of the series of reactions that occur produce stack gases at the top of the furnace and molten slag and pig iron at the bottom. Pig iron contains greater amounts of carbon (typically 4 to 5%) and phosphorus (typically 0.1 to 0.5%), which is higher than permit-

ted by most steel specifications. Thus, carbon and phosphorus must be removed in the steelmaking process.

A coke-oven battery is usually an integral part of such a plant, but not necessarily so, because it may be located elsewhere and coke may be transported to the blast-furnace plant. Thus, as far as the furnace proper is concerned, the facilities start with the ore yard, where iron-bearing raw materials are stored, often blended, and ultimately transferred either to the sinter plant for production of sinter or to the stockhouse for charging to the furnace.

Burden material is raised to the top of the furnace via either a skip hoist or a belt conveyor. Large, modern furnaces, requiring a flow on the order of 5 tons per minute to the furnace top, most often use belt conveyors.

Two types of tops are commonly used on blast furnaces. One is the bell-less top shown in Fig. 4. At the top of the furnace, the charge materials that have been placed on the belt sequentially are discharged into the receiving hopper, which feeds into one of two lock hoppers. The solids are distributed within the furnace by a rotating chute. By controlling the rotational speed and angle, the

materials can be layered in any desired pattern within the furnace. With this system, charging is essentially continuous, permitting furnace pressure at the top to be maintained at levels up to 205 kPa (30 psi).

An alternative top, the two-bell top, also provides a means for distributing the burden in the furnace shaft and maintaining a closed system so that the furnace can be operated at pressures above atmospheric, as shown in Fig. 5.

The modern furnace shown in Fig. 3 and 4 is more than 45 m (150 ft) high. The refractory lining is alumina-silica brick, and the stock line is further protected by several rows of fixed cast armor plates. Cooling water is used to protect various structural components of the furnace. Most modern blast furnaces use a carbon block hearth with an air-duct system for cooling.

The blast for a typical large blast furnace is furnished by axial-flow compressors with flow rates of 150 000 scfm or greater at pressures up to 450 to 485 kPa (65 to 70 psi). The heat evolved in the compression stage is retained in the "cold" blast by insulation of the cold blast main.

Stoves are external combustion units that are heated by burning a mixture of blast-furnace and

Source: Iron Blast Furnace, by G. H. Geiger, in *Metallurgical Treatises*, TMS-AIME, Warrendale, PA, 1981.

Fig. 3. Side view of a modern blast furnace

Source: Iron Blast Furnace, by G. H. Geiger, in *Metallurgical Treatises*, TMS-AIME, Warrendale, PA, 1981.

Fig. 4. Detailed view of the top section of a typical blast furnace

coke-oven gases. Stoves typically are on gas for about an hour and on blast for about 30 min. Blast temperatures can be as high as 1315 °C (2400 °F), and average 1205 °C (2200 °F), with mixer valves at the stoves automatically proportioning the correct flow through the stove and the bypass.

CHEMISTRY OF THE PROCESS

The principal reaction taking place in the blast-furnace process is reduction of iron oxide by carbon. The actual mechanism for this process involves gaseous reduction by carbon monoxide according to the following reaction:

$$3\,CO + Fe_2O_3 = 2\,Fe + 3\,CO_2 \qquad \text{(Reaction 1)}$$

In the presence of an excess of carbon at a high temperature, CO_2 is at once reduced to CO:

$$CO_2 + C = 2\,CO \qquad \text{(Reaction 2)}$$

The actual reduction of Fe_2O_3 by CO may take place in three steps, the Fe_2O_3 being successively reduced to Fe_3O_4, FeO and finally Fe. In addition to reducing the oxides of iron, carbon monoxide also reduces the oxides of manganese, silicon and phosphorus according to the following reactions:

$$MnO + CO = Mn + CO_2 \qquad \text{(Reaction 3)}$$

$$SiO_2 + 2\,CO = Si + 2\,CO_2 \qquad \text{(Reaction 4)}$$

$$P_2O_5 + 5\,CO = 2\,P + 5\,CO_2 \qquad \text{(Reaction 5)}$$

The water vapor in the blast also plays a role in the process:

$$H_2O + C = CO + H_2 \qquad \text{(Reaction 6)}$$

The hydrogen liberated by the above reaction may react with iron oxide, reducing it:

$$FeO + H_2 = H_2O + Fe \qquad \text{(Reaction 7)}$$

The water so formed is again decomposed. It should be noted that relatively few of the reactions involved furnish the heat required for the process, but that it is the oxidation of carbon and some of the reduction reactions involving carbon monoxide that furnish the heat to dry the raw materials, decompose the limestone, melt the iron and slag, and replace the heat losses.

The reactions occurring in iron and steelmaking primarily involve adjustment of the oxygen potential, first to reduce the iron ore to an impure pig iron, and then to increase the oxygen potential to oxidize impurities from that iron in the steelmaking operation. In the iron blast furnace the oxygen potential is extremely low in the hearth zone and continually increases as the reducing gases move up the furnace, reacting with iron oxide and coke according to Reactions 1 and 2. The temperature profile of the furnace shows a continually decreasing temperature from about 1650 °C (3000 °F) in the bosh zone just above the tuyeres down to about 175 °C (350 °F) at the top of the furnace.

IMPROVEMENTS IN BLAST-FURNACE OPERATION

In addition to design improvements, the principal advancements in blast-furnace operation during the past two decades have been control of the blast humidity, oxygen enrichment of the blast, natural gas or other fuel injection, pressurized operation, and increased use of beneficiated materials.

Large modern furnaces operating at high top pressures, above 205 kPa (30 psi), often utilize

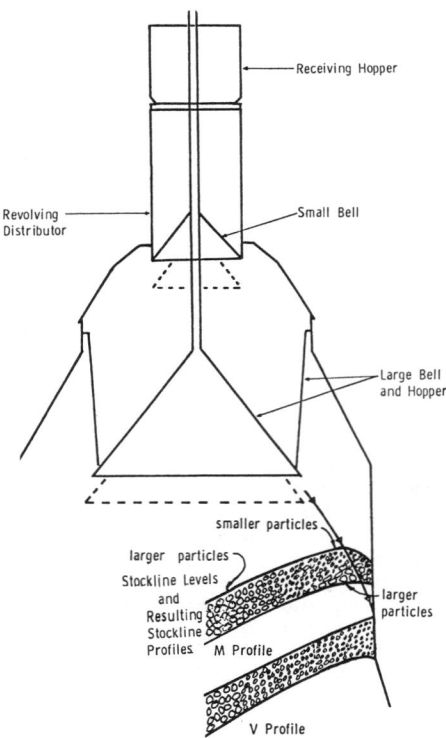

Source: Iron Blast Furnace, by G. H. Geiger, in *Metallurgical Treatises*, TMS-AIME, Warrendale, PA, 1981.

Fig. 5. Two-bell top, with burden trajectory and two different burden distributions

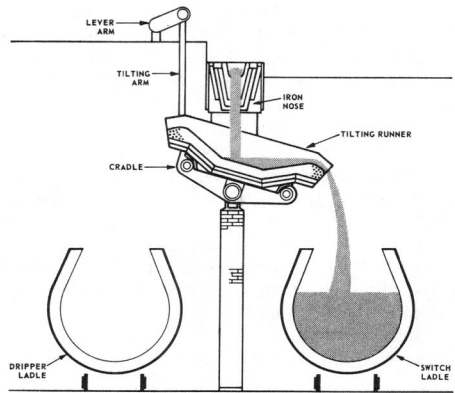

Source: Iron Blast Furnace, by G. H. Geiger, in *Metallurgical Treatises*, TMS-AIME, Warrendale, PA, 1981.

Fig. 6. Distribution mechanism for diverting a continuous tapping stream from one ladle to another

the high-pressure gas to drive an expansion turbine which generates electric power. The gases are then passed through a dust catcher and scrubber, which cool and clean the gases. The blast-furnace gas is then blended with coke-oven gas for use throughout the integrated steelmaking plant.

CASTING OF METAL AND SLAG FROM TAP HOLES

Very large furnaces may have two cast houses, on opposite sides of the furnace, and two tap holes in each cast house. The metal and slag are directed from the tap holes to the transfer ladles in clay-lined runners. Since the production rate from such a large furnace is on the order of 5 tons per minute, if the flow of hot metal in the runner is to be kept under control, it must be almost continuous. A given runner, or pair of opposing runners, is used until the clay lining is eroded to the point where tapping must be started from the adjacent hole. This changeover typically takes place every three or four days. Slag is skimmed from hot metal in the runners, and diverted to slag pits, where it is water cooled. Recently, techniques have been developed either to granulate the slag for road-fill or to recover the latent heat in the slag by running water through pipes buried in the slag, creating steam.

The hot metal is collected in ladles for transportation to steelmaking shops. In some recent installations, hot-metal flow is diverted from one ladle to the next using a tilting device as illustrated in Fig. 6.

Steelmaking Furnaces and Processes

THE MOST commonly used steelmaking furnaces are the basic open-hearth furnace, acid or basic electric furnaces, and the basic oxygen furnace. The distinction between acid and basic is made according to the character of the furnace lining and the flux used in the furnace. A basic furnace is required to reduce the phosphorus con-

tent from that of typical pig iron to the maximum amount allowable in most grades of steel.

The Open-Hearth Process. Historically, the open-hearth process was the principal steelmaking process for high-tonnage operation. However, the open-hearth process now has been superseded largely by the faster and more efficient basic oxygen process.

The open-hearth furnace, shown schematically in Fig. 7, is a regenerative furnace that includes a large, shallow basin lined with either basic or acid refractory material, an arched roof lined with high-temperature refractory, gas or fuel oil burners, and regenerators (called checkers) for preheating the combustion air. The furnace is charged with a mixture of liquid pig iron, solid pig iron and steel scrap (in almost any desired proportion, although 50% scrap is common), and fluxes; some iron ore is often used in the flux. Fuel burned in the furnace is necessary to melt solid scrap and pig iron and to maintain the proper operating temperature during the period of 4 to 10 h required to produce a heat of steel. Shorter heat times are achieved with oxygen injection. Almost any grade of carbon or low-alloy steel can be made in an open-hearth furnace. The size and expense of the open-hearth furnace usually limits its use to large steel-producing facilities.

Electric Furnaces. Any one of several types of electric furnaces may be used in the manufacture of steel products. However, the most commonly used types are the electric-arc furnace and the induction furnace. Of these two types, the electric-arc furnace is used far more extensively. Induction melting generally is used for producing relatively small quantities of specialty steels or for remelting and refining special alloys (see subsequent article).

The electric-arc furnace usually is built to operate in air; the other furnaces can be built to operate in air or as vacuum furnaces. Because electric furnaces can be conveniently built to a small scale, they can be used in small-scale steelmaking operations. Because they do not require molten pig iron in the charge, they also are used in facilities that are not located near blast furnaces. In certain installations, electric furnaces

are less costly to install and operate than basic oxygen or open-hearth equipment. Electric furnaces can be used for producing specialty steels, such as tool steels, stainless steels and aircraft-quality steels, or for manufacturing the common grades of carbon and alloy steels.

The electric-arc furnace, shown schematically in Fig. 8, is the most widely used type of electric furnace; for steelmaking, a basic refractory lining is usually used. Three carbon electrodes extending through the roof contact the metal charge; three-phase alternating current flows through the metal between the electrodes. The charge usually includes cold pig iron, scrap and virgin materials such as ingot iron and ferroalloys. The acid-lined electric-arc furnace is most often used in the production of steel castings.

Induction furnaces, which often have basic linings, are used in both steelmaking and steel foundry operations. The furnace consists of a crucible or refractory lining to contain the metal, a coil of an electrical conductor around the crucible, and a source of alternating current of the desired frequency, which is usually between line frequency and about 4 kHz. Alternating current passing through the coil induces currents in the charge inside the crucible; these induced currents heat the charge, which may include pig iron, steel scrap and virgin materials such as ingot iron and ferroalloys. Vigorous stirring of the molten steel is a characteristic of induction melting.

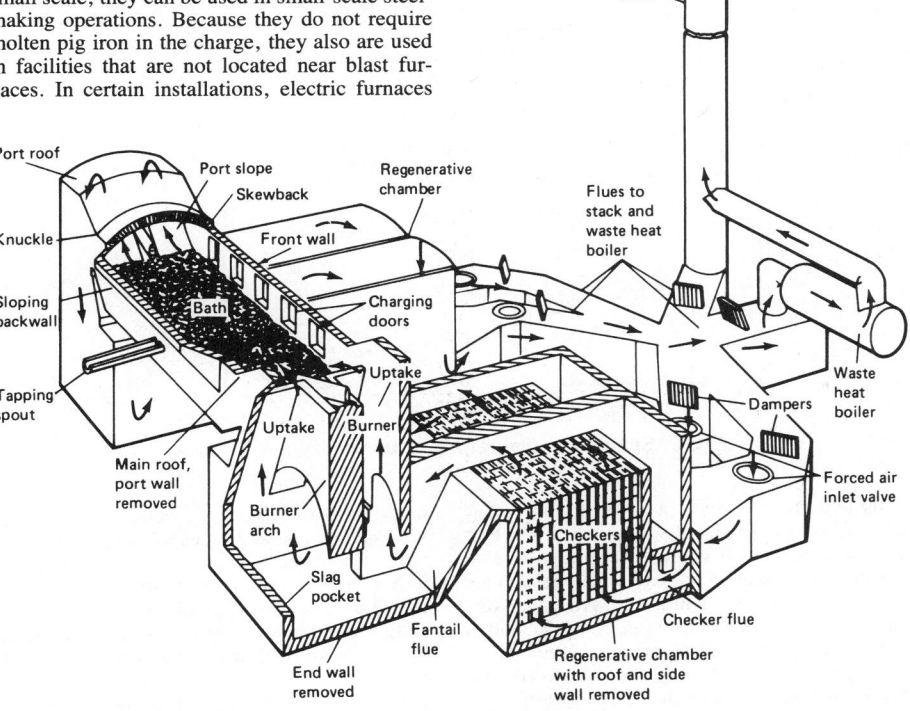

Fig. 7. Schematic diagram of a basic open-hearth furnace

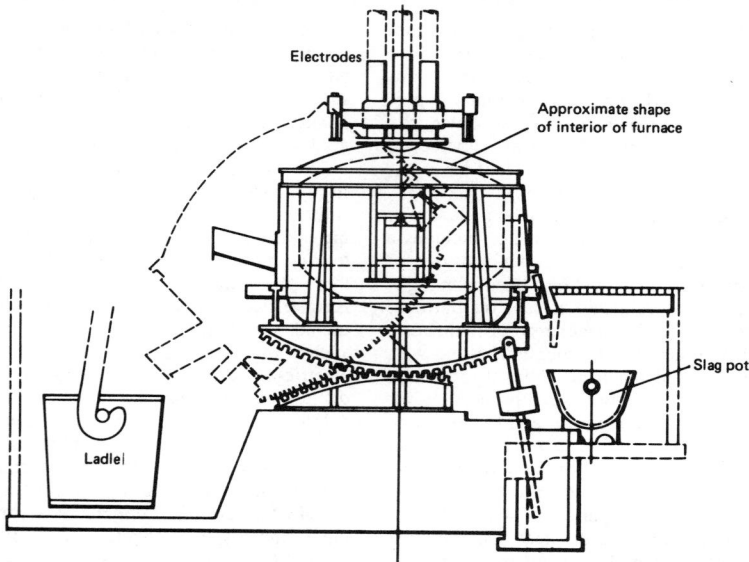

Fig. 8. Schematic diagram of an electric-arc furnace

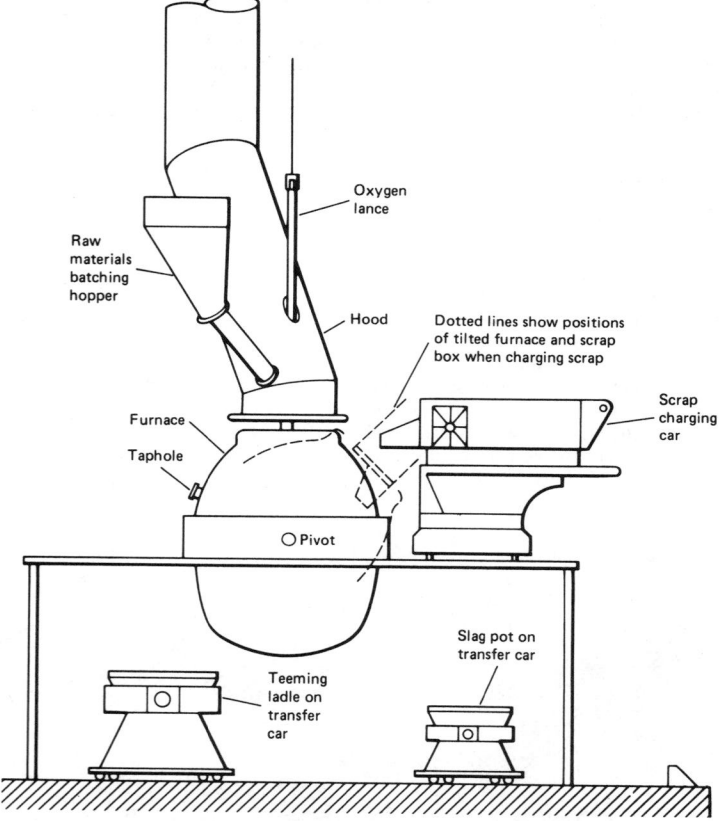

Fig. 9. Schematic diagram of a basic oxygen furnace

The Basic Oxygen Process. A schematic representation of a basic oxygen furnace is shown in Fig. 9. This arrangement consists of a large tiltable vessel lined with basic refractory material and the necessary accessory equipment. After the furnace is charged with molten pig iron (which usually comprises 65 to 75% of the charge), scrap steel and fluxes, a lance is brought down near the surface of the molten metal and a jet of high-velocity oxygen impinges upon the metal. The oxygen reacts with carbon and other impurities in the steel to form liquid compounds that dissolve in the slag and gases that escape from the top of the vessel. A heat of steel can be produced in less than 1 h. No fuel or electric power is required by the basic oxygen furnace (except to operate accessory equipment), but large quantities of oxygen are required. Almost any grade of carbon or low-alloy steel can be made in this type of furnace. Because it uses small amounts of scrap, the amount of tramp alloying elements introduced by the scrap is minimized. Furthermore, some alloying elements, such as chromium, are oxidized preferentially to iron during the oxygen "blow," so that there may be a significant loss of alloying elements during production of alloy steels.

In a bottom-blown basic oxygen furnace, the oxygen is introduced through an opening in the bottom of the furnace, rather than through a lance. In all other respects, it is similar to the top-blown basic oxygen furnace.

Both types of basic oxygen furnaces require large amounts of molten pig iron; thus, they are normally used in conjunction with blast furnaces in large steel-manufacturing facilities.

The basic oxygen process is now the principal method of high-tonnage steel production.

The Submerged Injection (Q-BOP) Process. The bottom-blown oxygen steelmaking process uses a converter with a replaceable bottom containing tuyeres through which oxygen is introduced. The number of tuyeres depends on the size of the furnace, the number increasing with the size. Tuyere placement is critical if advantage is to be taken of the stirring capability of bottom injection, particularly under conditions where oxygen flow rates are decreased. Replacement of the bottom refractory plug for the vessel which contains the tuyeres was required four to five times as often as replacement of the remaining vessel lining in the early stages of the process. However, with improved operating and refractory practices, this ratio has been decreased to unity with recent operating practice indicating up to 1000 heats for the bottom plug and working lining of the vessel. The absence of the lance in the bottom-blown process has offered opportunities to decrease total plant height and has also allowed relatively easy conversion of older open-hearth plants.

With the flexibility in plant design and lower conversion costs on a brown-field site, and particularly considering the potential metallurgical advantages, the bottom-blown process has shown rapid adoption.

The rapid growth of the submerged tuyere processes is also reflected in the parallel growth of postrefining processes. One specific example is the argon oxygen decarburization process (AOD), which uses an inert gas in the outer tuyere and an oxidizing gas in the inner tuyere. This process has rapidly achieved adoption for duplex refining of stainless steel and more recently for alloy grades. In addition, the process has gained rapid adoption throughout the foundry industry as a means for completing refining of small heats of a wide range of steels and alloys.

Comparing the top- and bottom-blown processes from an operational viewpoint brings out certain advantages for each. Because of lower plant height required for the bottom-blown process, the minimum cost is less either for a green-field site or for conversion of an open-hearth or Bessemer shop. However, the scrap-melting capability of the bottom-blown process is less, compared to the top-blown process, because of the lower iron oxide content of the slag, a factor which can be looked upon as representing an increase in process yield.

Combined top and bottom blowing is currently being explored to combine the advantages of each type of oxygen steelmaking. These early tests have shown higher yield, improved manganese recovery, better control of oxygen content and lower oxygen concentrations, as well as improved dephosphorization.

Secondary Refining Processes

Ladle Metallurgy

REGARDLESS of which steelmaking process is used, removal of reaction products from the molten steel during the deoxidation process is most critical. Nonmetallic inclusions in the form of oxides of silicon, manganese and aluminum may influence final properties adversely.

Sulfur in steel has presented a problem in steelmaking from the very beginning. Improved surface quality can be achieved by lowering sulfur levels to 0.015% or less. The achievement of low sulfur contents in steelmaking — for example, by use of two slag practices in electric furnaces — has presented operating difficulties and production delays.

In killed steels with low oxygen contents, as when aluminum is used for deoxidation and grain-size control, sulfur combines with manganese in the form of highly deformable manganese sulfides. These manganese sulfides have a low melting point and, as the last liquid to solidify in the steel, collect as films at grain boundaries. During hot rolling, the manganese sulfides are plastically deformed into elongated stringers extending parallel to the rolling direction. This shape and distribution of sulfides can have a marked negative effect on the directional properties of steel products. Two methods can be utilized to minimize the influence of the total volume of sulfide inclusions in the steel product. This can be accomplished by decreasing the total sulfur input, treating the hot metal to remove sulfur and controlling the scrap charge; by adjusting melting practices, including the use of additional quantities of flux; and by direct treatment of the liquid steel with desulfurizing agents, such as calcium or magnesium. A second approach involves modifying the shape or morphology of sulfide inclusions by producing a relatively nondeformable complex sulfide and/or oxysulfide by addition of calcium or rare earth metals.

Ladle metallurgy was first used to produce high-quality steels, but is now being utilized for many grades of steel because of the economic advantages of higher productivity. Ladle treatments of steel can be described generally in five categories:

1. Synthetic slag systems
2. Gas stirring or purging
3. Direct immersion of reactants, such as rare earths
4. Lance injection of reactants
5. Wire feeding of reactants.

These are often used in combination to produce synergistic effects — e.g., synthetic slag and gas stirring for desulfurization followed by direct immersion, injection or wire feeding for control of inclusion shape.

Synthetic Slag Systems. Addition of a prepared flux layer to the steel ladle during or immediately following tapping offers an opportunity to accomplish specific refining reactions. To prevent reversion of undesirable elements, such as phosphorus, sulfur and oxygen, from the steelmaking slag, as little slag as possible should be allowed to escape from the furnace into the ladle. However, prevention of heat loss requires that some thermal insulation cover the steel in the ladle. This can be achieved with considerable advantage using a fluid, sulfur-free, highly reactive synthetic slag. The basicity can be controlled and desulfurization achieved. Materials currently in use generally are based on the calcia-alumina system with additions of fluorspar and perhaps a small amount of a reducing agent such as carbon, aluminum or calcium silicide. Promotion of desulfurization can be achieved by good mixing and stirring of the system by a gas stream or by induction stirring. These materials must be dry to minimize hydrogen pickup.

Gas Stirring and Purging. Bubbling of a gas such as nitrogen, or in appropriate circumstances the more expensive gas argon, through the steel ladle will promote flotation of inclusions and provide homogeneity of temperature and composition following the addition of ladle additives and deoxidizers. Gas or induction stirring also creates local metal currents flowing vertically downward, aiding in recovery of buoyant additions made by plunging or wire injection.

Immersion. Reactants can be added to molten steel — for example, by plunging a cannister containing the reagent.

Lance Injection. Several processes are currently in use for injection of solid particulate metallic reactants such as magnesium, calcium, calcium-silicon, ferrosilicon, aluminum, or elements for microalloying; nonmetallic fluxes, such as CaO or CaO-Al$_2$O$_3$; or mixtures of these additions. One such process is illustrated in Fig. 1. These systems generally use a refractory-coated lance immersed deep in the ladle, and are based on dense-phase transport. They utilize less than 30% gas by volume, which causes relatively little disturbance of the bath for injection of a substantial amount of reactant. The capital and operating costs for lance injection are generally higher than those for alternative systems, and this technique may show lower efficiencies because the lighter reactants tend to float out in the gas-bubble stream.

Wire Feeding. Solid reactants in wire form or as a wire core can be fed directly into the steel ladle, into the casting tundish or, under some circumstances, directly into the mold — all yielding a high recovery of the reactant. The operating costs for these systems are relatively low, depending primarily on the economics of reactant preparation.

A basic or neutral (high-alumina) ladle lining is required for desulfurization to low levels, because the reactants used will usually attack acidic linings, such as fireclay. High-alumina refractories (70 to 90% alumina) are typically used. A suggested installation is a 70% alumina lining with fine-grain magnesite refractory at the slag line.

Adoption of steel refining by ladle metallurgy permits steelmaking processes to operate at lower cost and with higher productivity while simultaneously ensuring production of high-quality steels.

Protection of Pouring Streams. Recognizing that oxygen from the environment will react readily with liquid steel, the steelmaker can avoid oxygen pickup by shrouding the liquid steel pouring stream, and can minimize reoxidation by maintaining a smooth, compact pouring stream, in contrast to one which is rippled, ragged or spraying.

There are several approaches to minimization of oxidation during pouring. One method is a shrouding system such as that shown in Fig. 2.

Refining and Remelting Processes

THE PRINCIPAL refining and remelting processes used in production of special alloys are argon-oxygen decarburization (AOD), electroslag refining (ESR), vacuum arc remelting (VAR) and vacuum induction melting (VIM).

Argon-Oxygen Decarburization (AOD). The AOD process was originally developed for production of stainless steels but rapidly found use in production of many nonstainless grades. Almost immediately it was applied to production of high-chromium, nickel-base superalloys. Since then it has been used for production of almost all classes of steels, including low-chromium steels, cobalt-base alloys, chromium-free nickel grades, and alloys used for their expansion, resistive and

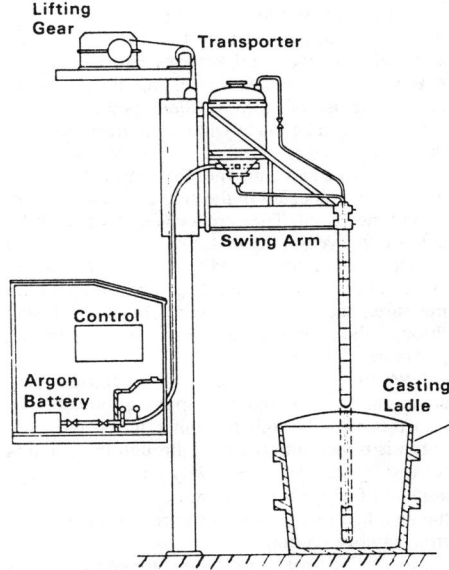

Fig. 1. Schematic illustration of a lance-injection process

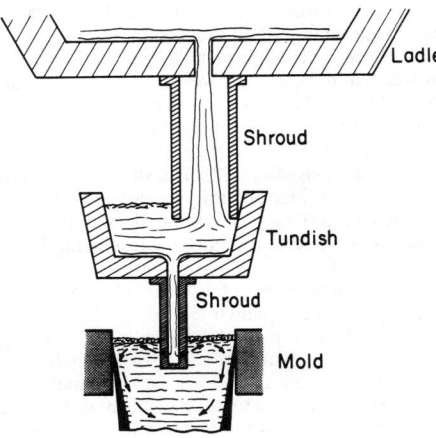

Fig. 2. Schematic illustration of a shrouding system for minimizing oxidation during pouring

magnetic properties. The incentive for use of this process for production of low-chromium steels is its good degassing characteristics, permitting in some cases the pouring of grades sensitive to hydrogen cracking without the need for more expensive vacuum degassing, and its excellent sulfur-removal and good inclusion-removal characteristics, which provide mechanical properties superior to those obtained when only arc-furnace melting is used.

The excellent carbon-removal characteristics of this process, even in the presence of high chromium, have permitted economical production of a whole class of low-carbon alloys. Originally this included extra-low-carbon versions of conventional stainless steels and nickel-base, corrosion-resistant superalloys.

AOD Equipment and Processing. Liquid metal produced in an electric-arc furnace is refined in a pear-shape, refractory-lined vessel. The metal and slag fill the bottom one-third to one-half of the vessel, the large space being needed in order to contain the metal and slag during rapid injection of argon-oxygen mixtures into the bath. The vessel is mounted on a trunnion ring, and the entire assembly can be tilted for tapping or sampling. Attachment of the vessel to the ring is such that a vessel with worn refractories can be quickly removed and a relined vessel rapidly put in its place. The use of multiple shells permits refractory work to take place without interrupting production.

Gas is injected into the bath through two or more tuyeres located in the side and near the bottom of the vessel. They are located along the back side of the vessel so that, when the top of the vessel is tilted forward, the tuyeres are uncovered. This permits sampling and temperature measurement without blowing metal and slag through the mouth of the vessel onto operating personnel.

Each tuyere consists of two concentric tubes. An argon-oxygen mixture or pure argon is blown into the bath through the inner tube, and pure argon is blown into the bath through the annulus formed by the two tubes. Only a relatively small amount of argon is blown into the vessel through the annulus; its function is to prevent the tuyeres from rapidly eroding.

Most AOD installations are equipped with gas-handling systems capable of delivering various gas mixtures as desired. Nitrogen may be used in place of argon with substantial savings, provided that nitrogen pickup from the blown gas is not deleterious to the product.

A pit large enough to hold the teeming ladle is situated in front of the vessel. A movable platform can be placed across the pit to permit sampling and temperature measurement through the mouth of the vessel.

Electroslag Remelting (ESR). The success achieved in producing high-integrity weldments by the electroslag process led to the development of electroslag remelting (ESR) for producing high-purity metals.

A schematic view of a basic electroslag remelting furnace is shown in Fig. 3. The material (electrode) to be refined is remelted by passing a current through it into a molten slag which is resistively heated and which, in turn, melts the electrode. Molten metal droplets form on the end of the electrode and fall through the slag, forming an ingot in a water-cooled copper crucible. The process continues until the electrode is consumed and the ingot is formed. ESR is employed

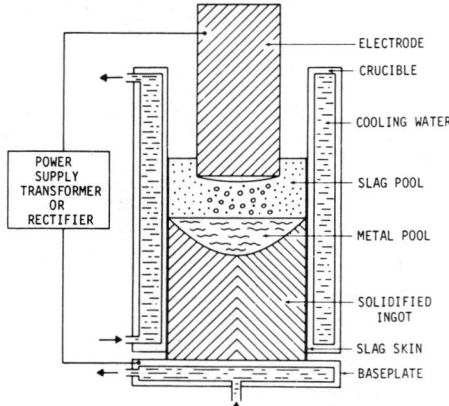

Source: The Argon Oxygen Decarburization and Electroslag Remelting Processes, by H. J. Klein *et al.*, in *Metallurgical Treatises*, TMS-AIME, Warrendale, PA, 1981.

Fig. 3. Schematic illustration of a basic electroslag remelting (ESR) furnace

in order to improve material quality through (*a*) removal of oxide and sulfide inclusions; (*b*) control of the solidification rate, and thus elimination of shrinkage porosity and minimization or elimination of segregation; and (*c*) improvement of surface quality. In general, the ESR process has been shown to be capable of accomplishing these objectives, and materials produced by the ESR technique normally exhibit better property levels and, more significantly, less scatter in these properties. In addition, the hot workability of the materials is significantly improved and, in some cases, superalloys which were not previously forgeable can be processed in that manner after ESR.

Vacuum arc remelting (VAR) is a casting process carried out in a vacuum with the objective of remelting the consumable electrode to produce an ingot which has improved chemical and physical homogeneity.

The energy that promotes remelting of the electrode is supplied by an arc which occurs between the bottom surface of the electrode (usually the cathode) and the top surface of the newly formed ingot (usually the anode). A typical VAR setup is depicted schematically in Fig. 4.

Originally, VAR was used to melt refractory metals such as tungsten, tantalum and molybdenum, all of which required water-cooled metal molds because these metals react with ceramic molds. Electrodes of reactive metals such as titanium and zirconium are prepared by consolidating powders or metallic chunks into forms of required shape. Sometimes the initial VAR melting is followed by a second (and even a third) VAR step to ensure the required properties.

Electrodes of iron- or nickel-base superalloys are cast from air-electric, air-induction or AOD furnaces. If the superalloys contain appreciable amounts of reactive metals, the electrodes are cast from vacuum induction melts.

Vacuum induction melting (VIM) is the most flexible of the vacuum melting processes, because it allows for independent control of temperature, pressure, and mass transport by means of stirring. As a consequence, VIM provides the greatest degree of control over alloy composition of all the melting processes, which is probably its most important metallurgical justification.

Production VIM furnaces range in size from less than one ton to 60 tons in melt capacity. The basic components for these units are essentially the same, as shown schematically in Fig. 5.

The basic charge is melted in an induction furnace with a rammed or brick lining. Most furnaces with melting capacities greater than 4 tons are brick-lined. The frequency and power requirements are similar to those of air-melting counterparts, but every attempt is made to keep coil voltage as low as practical so as to avoid corona and arcing in certain ranges of reduced pressure. Special insulating techniques also are used for this purpose.

A water-cooled metal tank is required to enclose the melting furnace. The configuration varies widely from furnace to furnace, depending on production requirements. Most large-scale furnaces employ a two-chamber arrangement — one to enclose the melting furnace and the other to house the molds and their handling mechanism (as shown in Fig. 5). These chambers usually are connected by means of large-scale vacuum valves, and each chamber has its own access doors and vacuum pumps. In this way, melting can proceed

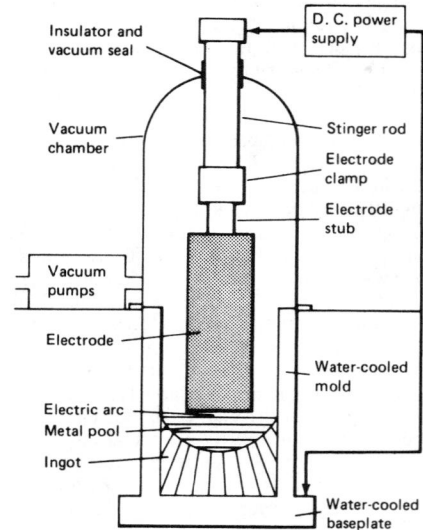

Fig. 4. Schematic diagram of a typical vacuum arc remelting (VAR) furnace

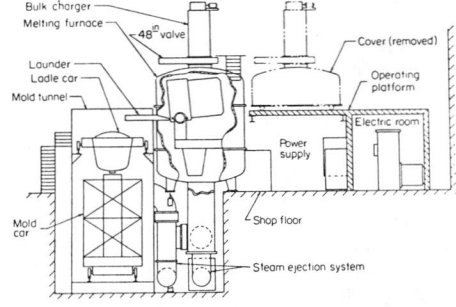

Source: Principles and Practices of Vacuum Induction Melting and Vacuum Arc Remelting, by J. W. Pridgeon *et al.*, in *Metallurgical Treatises*, TMS-AIME, Warrendale, PA, 1981.

Fig. 5. Schematic view of a 15-ton vacuum induction melting (VIM) furnace and associated equipment

while the molds are being set up or removed, so that "turnaround time" can be minimized.

In the early days of production vacuum melting, the pumping systems consisted of mechanical pumps to rough down the chambers from 760 to 1 torr. Ejector-type diffusion pumps then reduced the pressure to the operating range of approximately 10^{-2} torr. The ultimate pressure was a dynamic level resulting from the quantity of the evolving gases and the throughput characteristics of the pumps.

As furnaces increased in size, steam ejectors were used to handle the larger gas loads, and installations of twin six-stage systems with inlet sizes of 1.2 to 1.5 m (about 4 to 5 ft) were not uncommon. Because of the very low thermal efficiency of the steam ejector, however, operating costs have risen sharply as energy costs have increased, and as a result there is a trend back to the mechanical booster combination using high-speed blowers.

The mold configuration, arrangement, and means for introduction and removal depend on the types of ingots required. It is not uncommon to see two furnaces side-by-side with the product from one being scores of small ingots 75 mm (3 in.) round by 750 mm (30 in.) long and the product from the other being two ingots 600 mm (24 in.) round by 3 m (118 in.) long.

Cars frequently are used in large furnaces to transport the molds into the furnace, to shift them under the tundish during pouring, and to remove them after solidification (see Fig. 5).

Casting and Rolling

Casting of Steel

WHEN METAL is teemed into molds, it becomes either an ingot or a casting. In reality, an ingot *is* a casting, because it is prone to all the structural characteristics of castings (shrinkage, porosity, presence of inclusions, segregation, and nonhomogeneous grain size). However, if the "casting" is to be mechanically worked (or wrought) by rolling, forging or hammering, it is known as an ingot. If the "casting" is in the desired shape and requires only machining, it is a true casting and a typical foundry product.

As an alternative, steel may be continuously cast into any one of several product forms.

Ingot Practice. There are four basic types of ingot structures, the differences among them being based on the degree of deoxidation performed on the liquid steel. These four basic ingot types, in order of decreasing deoxidation, are: (*a*) killed, (*b*) semikilled, (*c*) capped and (*d*) rimmed.

The ingots sketched in Fig. 1 were poured in bottle-top molds to the same height as indicated by the dotted lines. A fully killed (deoxidized) steel ingot is shown in Fig. 1(a). No gas evolution is evident. The cavity in the top section is caused by shrinkage as the metal solidifies, and is known as "pipe." Killed steel usually is poured in big-end-up molds equipped with refractory hot tops which retard the solidification of the top of the ingot and tend to confine the pipe to the hot top section of the ingot.

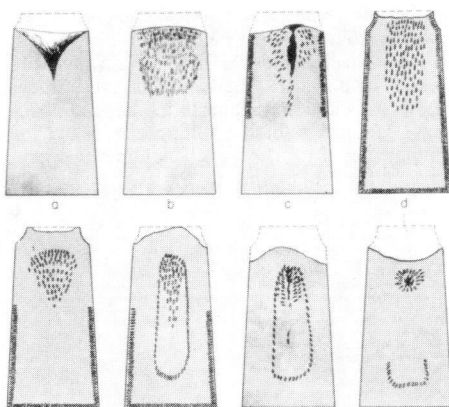

Fig. 1. Cross sections of ingots, showing the varying degrees of rimming action (see text)

The structure shown in Fig. 1(b) identifies a semikilled ingot. Here some gas evolution is shown, and in this case it was sufficient to overcome the shrinkage which normally occurs during solidification of the steel. Scattered blowholes in the upper half of the ingot replace the pipe usually found in fully killed steel ingots. This is caused by the pressure of gas evolving on solidification, which causes the top surface of the ingot to bulge upward during solidification. There are no blowholes in the lower part of the ingot. This is due to the ferrostatic pressure, which prevented evolution of gas in the lower half of the ingot.

The structure of the ingot shown in Fig. 1(c) shows evidence of much more gas evolution than is shown in Fig. 1(b). This excess of gas caused formation of honeycombed blowholes in the upper half of the ingot and ruptured the frozen top surface of the ingot. The gas pressure forced molten steel up through the ruptured surface. This is known as a "bleeding" ingot.

The evolution of gas in the ingot shown in Fig. 1(d) was such that blowholes were formed close to the surface and from the top to the bottom of the ingot. The top surface was kept from freezing by the gas evolution until a heavy metal cap was placed over the top of the ingot. This results in the term "capped steel." The structure of a typical capped ingot is shown in Fig. 1(e). In this instance there was a much greater degree of gas evolution. Here the gas bubbles, which would otherwise form blowholes, were swept away by the strong current of gas traveling upward along the sides of the ingot. Blowholes could not form even in the lower half of this ingot until the gas evolution had decreased considerably. By the time the gas evolution had decreased to this extent, the metal along the sides of the ingot had solidified to such a distance in from the surface that a thick skin was formed. This skin serves to protect the blowholes from contact with air and gases during the subsequent heating and rolling operations. The blowholes, being "deep-seated" and thus protected from oxidation, will weld shut during the rolling operation.

Typical structures of rimmed steel are shown in Fig. 1(f), (g) and (h), differing only in degree of gas evolution. Figure 1(f) shows more gas evolution than does Fig. 1(e), but not enough to prevent the formation of blowholes from exceeding the shrinkage. As a result, the top surface of the ingot rose slightly during solidification. Figure 1(g) shows a strong evolution of gas, causing a honeycomb formation of blowholes which very

closely balanced the shrinkage tendency so that the top center of the ingot did not rise or fall appreciably as it solidified. Figure 1(h) shows a violent rimming (or "wild") action, and is typical of low-metalloid steel. There was no chance for a honeycomb of blowholes to form, and the top surface fell during solidification, causing pipe in the top section of the ingot.

Continuous Casting of Steel. In the continuous casting process, molten steel is poured at a steady rate into a water-cooled mold.

Depending on the size of the product cast, a shell 3.2 to 12.7 mm ($\frac{1}{8}$ to $\frac{1}{2}$ in.) thick forms quickly by solidification at the mold wall. As molten metal is poured into the top of an oscillating mold to a predetermined level, the casting is withdrawn at a controlled rate from the bottom of the mold (see Fig. 2). The solidified shell acts as a container for the still-molten metal inside the casting. As the casting is withdrawn from the mold, the cast billet, bloom or slab is sprayed with water to cool and solidify it further. A flying shear or torch cuts the casting off at a level below which the metal has solidified.

Nearly all steels continuously cast today are killed, with typical compositions ranging from low-carbon grades for sheet and strip to medium-carbon grades in the range of 0.2 to 0.6% C, with manganese contents over 0.6% for plate applications. Extensive experiments for casting rimming steels were conducted at several steel-producing plants, but the cost and the precise control necessary to produce a sound rim zone made this process noncompetitive.

Continuous casting has a greater metal yield than conventional ingot-mold practice, which results in reduced fuel requirements and greater productivity. The hot metal yield in conventional ingot practice is about 87%; in continuous casting of billets, the yield is 94 to 98%, depending on the tonnage of the heat and the cross-sectional shape of the product cast.

The process allows casting of special cross-sectional shapes, such as dumbbell or dogbone shapes for rolling into structural sections.

The quality of the cast product is uniform and good to excellent. Scarfing is reduced, and often eliminated. The product is comparatively free from macrosegregation, grain size is easy to control, steel cleanness is equal to or better than that obtained in ingot-mold practice and the product withstands severe hot reduction. The cost of equipment is relatively low; small-tonnage mills are feasible, and there is a major savings in large-tonnage plants.

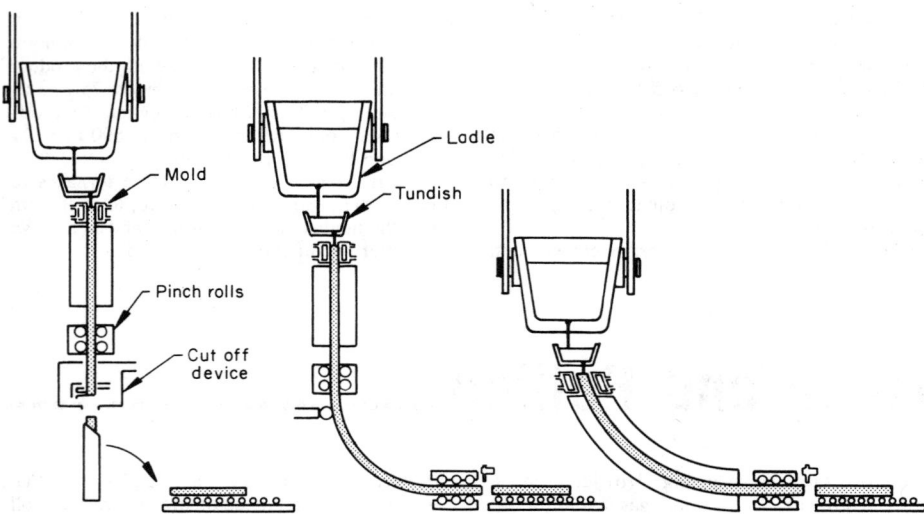

Source: A Decade of Development in the Continuous Casting of Special Steels, by R. S. Wagstaff and G. E. Stock, in *Continuous Casting of Steel*, Special Report No. 89, Iron and Steel Institute, London, 1965, p 122.

Fig. 2. Vertical, bent-strand and curved-mold continuous casting machines

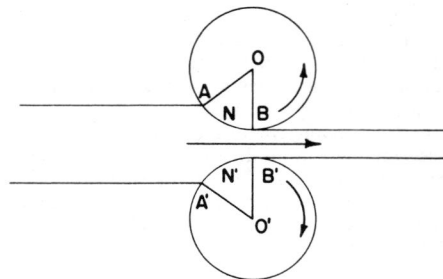

Fig. 3. Schematic illustration of the mechanical deforming action that occurs during hot rolling

Ingot Breakdown

FROM THE FORM of an ingot, steel must be reduced in cross-sectional area and formed into a useful shape. This is accomplished in several stages by hot deformation of the metal. If the steel is to be used for carving knives it will be rolled into strip; if it is to be used for automobile bodies it will be rolled into sheet. Other products, such as large axles and turbine rotors, are produced by drop hammers or forging presses. The most economical method for making the desired product is determined, and the selected process is then carried out in such a manner that the finished steel will have the required size, shape and mechanical properties.

Ingot structures are generally coarse and fragile, and are subject to tearing when rolled or forged. The ingot should be deformed slightly and slowly in initial hot working to produce a newly recrystallized structure which is more resistant to cracking and tearing as hot working proceeds. It is for this reason that most high-carbon and/or high-alloy steels (tool steels, for example) are cogged in forging machines before they are subjected to rolling. As a rule, the ingots of specialty steels are small and can easily be handled in forging hammers or presses.

The higher the temperature of the steel prior to hot working, the more easily it can be deformed, regardless of composition.

Production Practice. For the most part, production practice employs a sequence like that shown in Fig. 1 and 2 (pages 22•2 and 22•3) in the article on basic technology (ingot to finished product form, from left to right in the flow diagram).

The ingots are first heated uniformly to 1120 to 1345 °C (2050 to 2450 °F) in soaking pits prior to rolling. The first rolling operations are performed in blooming or slabbing mills, usually located in close proximity to the soaking pits.

Blooms, billets and slabs are the semifinished products of the blooming and slabbing mills. These products are then converted into the various product forms—structurals, bar, strip, sheet and wire—as required.

Hot Rolling

BY FAR the greater part of modern steel production from ingots is accomplished by rolling—principally by hot rolling, although some product forms are further finished by cold rolling or other cold working processes.

Rolling, as it applies to shaping of steel, consists of passing the steel ingot (or bloom, slab or billet) between two rolls revolving in opposite directions at the same speed and so spaced that the distance between them is somewhat less than the thickness of the piece of steel entering them. The rolls grip the steel, reduce it in cross-sectional area, and deliver it increased in length. The spread (the extent of lateral spreading) depends chiefly on the amount of reduction and the shape of the section entering the rolls. In this instance, re-

duction means the percentage decrease in cross-sectional area of the steel—that is, the portion of cross-sectional area eliminated. For example, reduction of an 8-by-8-in. cross section to a 4-by-4-in. cross section is a 75% reduction. The calculations are:

$$\frac{(8 \times 8) - (4 \times 4)}{8 \times 8} \times 100 = 75\%$$

The mechanical deforming action that occurs during hot rolling is presented in Fig. 3. The rotating rolls are in contact with the metal through the arcs AB and A′B′. Frictional force acts along these arcs and is at a maximum at points N and N′. The frictional force pulls the metal into the space between the rolls. The metal enters the rolls at a speed somewhat lower than the speed of the roll surface and is delivered at a higher speed. The degree of forward slip is the ratio of the delivery speed to the surface speed of the rolls. The points on the arcs AB and A′B′ where the speed of the roll surface is equal to the speed of the metal is known as the neutral point and coincides with the points of maximum pressure, N and N′. The arcs AB and A′B′ are called the contact arcs, and the angles AOB and A′O′B′ are the contact angles or rolling angles. The maximum contact angle at which the metal will enter the rolls without being pushed is the angle of bite. The maximum angle of bite is about 30° for smooth rolls and 35 to 42° for ragged rolls. Ragged rolls are those whose surfaces have been purposely roughened by cutting shallow grooves in the roll surface parallel to the roll axis. The contact area is the area of the steel under the contact arc.

The temperature of the steel has a decided effect on the changes in physical dimensions effected by rolling. Slight increases in temperature give the steel additional plasticity, which decreases the power required and increases the degree of plastic deformation in the desired directions. The nature of the rolling action and the composition of the particular steel limit the maximum and minimum rolling temperatures.

23 CASTING

Edited by P. J. Mikelonis, General Casting Corp.

This section was condensed from Metals Handbook, 8th Edition, Volume 5, Forging and Casting, pages 149 to 448. For more detailed information on the topics covered herein, the reader is referred to the larger work. To locate related information on casting in this volume, the reader should consult the index.

Molding and Casting Processes

Patterns for Sand Molding

A PATTERN is a form made of wood, metal or other suitable material, such as wax, polystyrene, polyurethane or epoxy resin, around which molding material is packed to shape the casting cavity of a mold. The pattern is constructed to casting design with the addition of: (*a*) shrinkage allowance, to compensate for contraction of liquid metal in cooling; and (*b*) an allowance, called taper or draft, on the vertical sides of the pattern, to facilitate removal from the sand or other molding medium. Patterns for some castings must include, in addition, projections called core prints, to support cores used to produce shapes, such as internal cavities, that cannot be molded directly from the pattern. Some patterns incorporate provisions for gating or other projections, to facilitate the flow of metal.

Sand cores are made in core boxes (negative patterns) to the same standards of accuracy, and with the same problems of service life under abrasion, as patterns. Core boxes are illustrated and discussed in the article on Sand Cores and Coremaking.

MATERIALS FOR PATTERNS

The materials of which patterns are usually made differ greatly in their characteristics and therefore in the applications to which they are best suited. The decision as to what material to use for a specific pattern depends on:

- Expected production quantity (initial and possible repeat orders)
- Dimensional accuracy required
- Molding process to be used in the foundry, including the type and size of molding machine

if a molding machine is to be used
- Size and shape of casting.

Table 1 presents a comparison of the important characteristics of five commonly used pattern materials—wood, aluminum, steel, plastic and cast iron.

Over-all utility often requires that patterns be made of a combination of materials, such as wood with plastic inserts on wear surfaces, wood with metal inserts, or aluminum alloy with steel inserts.

Table 1. Comparison of characteristics of five pattern materials

SOURCE: *Introduction to Foundry Technology*, by D. C. Ekey and W. P. Winter, McGraw-Hill, New York, 1958, p 46

Charac-teristic	Ratings(a) for:				
	Wood	Alumi-num	Steel	Plas-tic	Cast iron
Machin-ability	E	G	F	G	G
Wear resistance	P	G	E	F	E
Strength	F	G	E	G	G
Weight(b)	E	G	P	G	P
Repair-ability	E	P	G	F	G
Resistance to:					
Corrosion(c) ..	E	E	P	E	P
Swelling(c) ..	P	E	E	E	E

(a) E = excellent; G = good; F = fair; P = poor. (b) As a factor in operator fatigue. (c) By water.

TYPES OF PATTERNS

The type of pattern used in a specific production application is determined by the number of castings required, the stage of development of the design of the casting (if the casting is likely to be revised, an inexpensive pattern should be used), the complexity of the shape of the casting, and the molding process used in the foundry.

The types of patterns most commonly used are one-piece patterns and two-piece patterns (match-plate or cope-and-drag); these are described below. Figure 1 shows a one-piece wood pattern, an aluminum match-plate pattern, and a metal or plastic cope-and-drag pattern for a water-pump casting. The one-piece wood pattern on a follow board (Fig. 1a), which is used for jobbing work only, can be used to make molds for 20 to 30 castings. Accuracy of the casting dimensions depends greatly on the molder's skill. The cast aluminum match-plate pattern shown in Fig. 1(b) can be used to produce up to 50 000 castings, to greater dimensional accuracy than is possible with the wood pattern. With the cope-and-drag setup (Fig. 1c), if a plastic pattern is used, pattern life expectancy is 20 000 castings, although patterns produced from polyurethanes can have life expectancies equaling that of aluminum. If the pattern is made of metal, it can be used to make 100 000 to 110 000 castings.

Two-piece patterns (also called split patterns) are used for castings of more intricate design or for castings required in greater quantities. They are split along a parting surface determined by the shape of the casting: each pattern section must, of course, be able to be withdrawn from its half of the mold. Sometimes, the two halves of the pattern are mounted on opposite sides of a common board, called a match board; sometimes, one half of the pattern is mounted on the cope board and the other half on the drag board. Patterns of the first type are match-plate patterns; those of the second type are cope-and-drag patterns.

Match-plate patterns are split patterns the halves of which are separated by a plate of the required thickness and shape. "Match plate" commonly refers to metal patterns, but wood patterns mounted on a board are used in the same man-

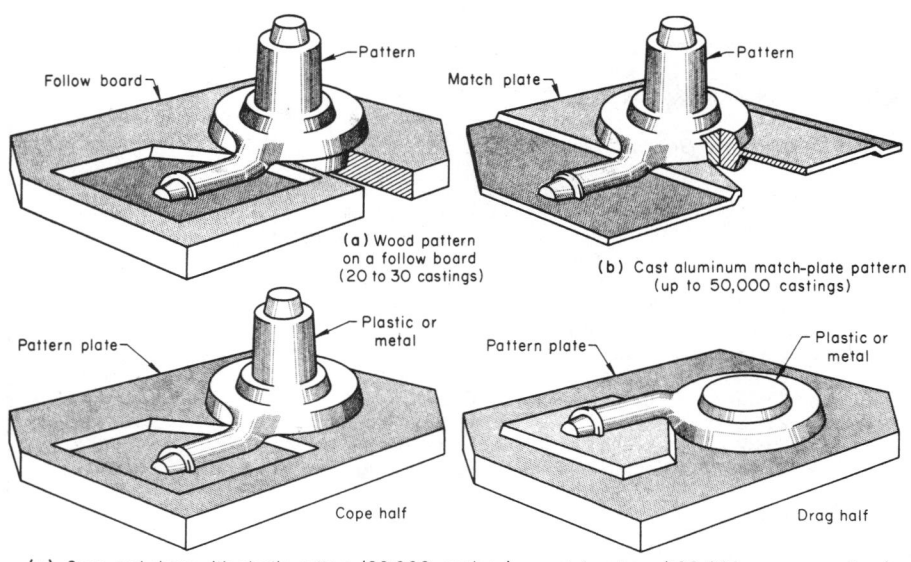

(a) Wood pattern on a follow board (20 to 30 castings)

(b) Cast aluminum match-plate pattern (up to 50,000 castings)

(c) Cope and drag with plastic pattern (20,000 castings) or metal pattern (100,000 or more castings)

Fig. 1. Three types of patterns for producing a water-pump casting in various quantities

ner, and plastic patterns can be mounted on a plastic match plate.

If the pattern has a flat parting surface, halves can be made separately and fastened to the plate. However, most *metal* match-plate patterns, whether they have a flat parting surface or not, are cast integrally with the plate—usually in a plaster mold, but sometimes in a sand mold. In integral casting, the cope and drag mold halves are shaped from a master pattern and then are aligned so that the mold cavities face each other but remain separated by a distance equal to the thickness specified for the metal plate. The open edges are sealed with metal strips, and when molten metal is poured it fills the pattern halves of the mold and the space that separates them. The resulting pattern is a metal plate having, on one side, the cope half of the pattern and, on the other, the drag half.

Cope-and-drag patterns are split patterns the halves of which are mounted on separate plates or boards. These patterns may be made of either wood or metal. For machine molding, metal is used, to withstand the higher pressures exerted, and the heavier jolt imparted, by the molding machine. In high-wear areas, either plastic or metal inserts can be used.

Cope-and-drag patterns are adapted to high-volume production and to production of large castings. If made of metal, they are more expensive than metal match-plate patterns; but total molding cost per casting is less when production volume is high, because the cope and drag pattern sections can be molded at the same time. Two or more patterns can be mounted on one plate or board for simultaneous production of more than one casting per mold.

Multipiece patterns are required when a one-piece or two-piece pattern that can be withdrawn from the mold cannot be designed. The additional pieces can be extra body pieces (molded in the cheek), loose pieces or loose core prints.

PATTERNS FOR GATES AND RISERS

A pattern often includes gates and risers, although they can be cut into the molding sand by hand or can be made of loose pieces. Patterns, including gates and risers, for producing a casting from one-piece wood, metal match-plate, and cope-and-drag patterns, are shown in Fig. 2. Figure 2(a) shows a wood pattern for which gating is cut by hand when one to ten castings are to be produced. Figure 2(b) shows a gating system cut into the follow board, for use with the wood pattern when a greater number of castings is to be produced. The patterns shown in Fig. 2(c) and (d) include built-in gating systems: Fig. 2(c) shows gating and risering for producing castings with metal match-plate patterns; Fig. 2(d) shows gating and risering for producing castings with cope-and-drag patterns. When a metal match-plate pattern has a flat parting surface, the gating system, like the pattern sections, can be cast integrally or can be made separately ("loose") and fastened to the plate.

SELECTION OF PATTERN TYPE

The number of castings to be produced and the accuracy required are primary considerations in the choice of an appropriate type of pattern. As mentioned previously, metal patterns, especially those with harder-metal or plastic inserts at points of high wear, will retain accuracy longer. The number of castings to be produced also determines the molding equipment that will be used, and the equipment available also affects pattern choice.

Figure 3 shows different types of pattern equipment that can be used for making a simple gray iron casting (Fig. 3a) in small to large quantities.

In the development stage, when the cored connection boss on the side may require resizing, only a few castings will be produced. Therefore,

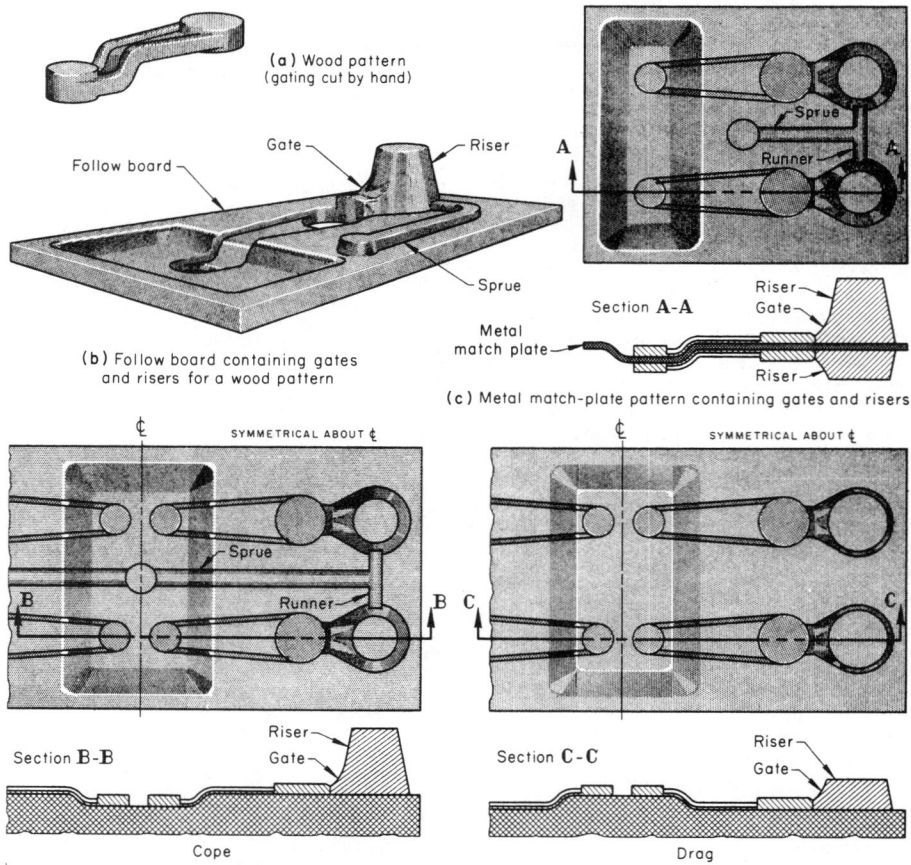

(a) Wood pattern (gating cut by hand)

(b) Follow board containing gates and risers for a wood pattern

(c) Metal match-plate pattern containing gates and risers

(d) Cope-and-drag pattern containing gates and risers

Fig. 2. Patterns for a sand casting and its gating and risering systems, for four different methods of mold production

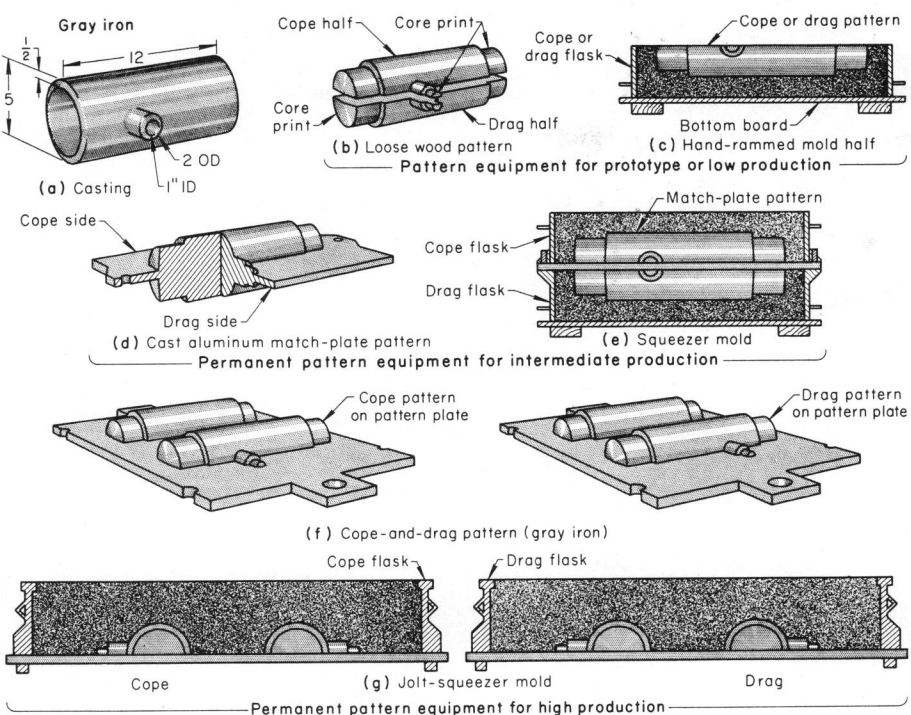

Fig. 3. **Different types of pattern equipment and molding methods for making a simple casting in small or large quantities. (Dimensions are in inches.)**

a well-constructed wood pattern (Fig. 3b) and hand ramming of molds (Fig. 3c) will be satisfactory, and will enable engineering changes to be made quickly and at minimum cost.

After the design and quality of the casting have been approved, a permanent pattern is selected, based on annual production requirements. If annual production requirements are moderate, so that the casting can be produced economically on a squeezer line or on the molding floor, a cast aluminum match-plate pattern (Fig. 3d) and a squeezer mold (Fig. 3e) are used.

If production requirements for the casting increase, so that jolt-squeezer or sand-slinger molding is economical (Fig. 3g), a more rugged pattern will be required, because of the higher pressures exerted against the pattern and the heavier jolt to which it is subjected. The pattern, of the cope-and-drag type, will consist of two or more gray iron patterns on gray iron pattern plates (Fig. 3f). The drag pattern is mounted in a molding machine, and the drag half of the mold is made and placed on a moving conveyor. Cores, for the hollow center of the casting, are set in the drag half of the mold. The cope half is made in another molding machine and, as the conveyor moves toward the pouring station, this half is placed on top of the drag half, making a complete mold, ready for pouring.

Sand Molding

SAND, combined with a suitable binder, can be packed rigidly about a pattern so that, when the pattern is removed, a cavity corresponding to the shape of the pattern remains. Molten metal poured into this cavity and solidified develops a cast replica of the pattern. The sand that forms the mold cavity is friable after the metal is cast, and can readily be broken away for removal of the casting.

TYPES OF MOLDS

Molds for sand casting are broadly classified as: (*a*) green sand molds, (*b*) skin-dried molds, (*c*) dry sand molds, (*d*) dry sand core molds and (*e*) other types of sand molds.

Green sand molds are the most widely used of all sand molds. They are made of sand, clay, water and other materials and are used without further conditioning. ("Green" means that the molded sand mixture is allowed to remain moist.) Both ferrous and nonferrous castings are produced in these molds. The molds are prepared, metal is poured, and castings are shaken out in rapid production cycles. Green sand molding is adaptable to machine molding, as in squeeze molding machines, automatic molding machines, and sand slingers.

Advantages of green sand molds are:

- Green sand molding is the least expensive method of producing a mold.
- There is less distortion than in dry sand molds, because no baking is required.
- Flasks are ready for reuse in minimum time.
- Dimensional accuracy is good across the parting line.
- There is less danger of hot tearing of castings than in other types of molds.

Disadvantages of green sand molds are:

- Sand control is more critical than in dry sand molds.
- Erosion of the mold is more common in the production of large castings.
- Surface finish deteriorates as the weight of the casting increases.
- Dimensional accuracy decreases as the weight of the casting increases.

Skin-Dried Molds. Large molds and molds for pit work are usually skin dried. In dry climates, an appreciable amount of moisture is removed from

the surface of molds merely by storing them for a time. Torches are also used to remove moisture. Molds that are skin dried to a depth of less than 13 mm (¹/₂ in.) are still classified as green sand molds.

Dry sand molds are oven dried to a depth of 13 mm (¹/₂ in.) or more. (The dried layer is usually deeper than 13 mm, depending on section thickness, and may extend clear through the section.) The molds are baked at 150 to 375 °C (300 to 700 °F) for 8 to 48 h, depending on the binders used in the sand mixture and the amount of sand surface to be dried, and on the requirements of production cycles. Dry sand molds generally are used in preference to green sand molds for making medium-size to large castings, such as large rolls, housings, gears and machinery components.

Advantages of dry sand molds are:

- They are stronger than green sand molds, and thus are less susceptible to damage in handling.
- Over-all dimensional accuracy of the mold is better than for green sand molds.
- Surface finish of castings is better, mainly because dry sand molds are coated with a wash.

Disadvantages of dry sand molds are:

- Castings are more susceptible to hot tears.
- Distortion is greater than for green sand molds, because of the baking.
- More flask equipment is needed to produce the same number of finished pieces, because processing cycles are longer than for green sand molds.
- Production is slower than for green sand molds.

Dry Sand Core Molds. When flasks are too large to fit in an oven, or when it costs too much to dry a large mass of sand, molds are made from assemblies of sand cores. A core mold is usually prepared from core-sand mixtures.

The principal advantages of dry sand core molds are:

- Exceptionally good dimensional accuracy can be maintained.
- Dry sand core molding is adaptable to green sand foundries that do not have large drying facilities.

The main disadvantage of dry sand core molding is the extreme care that must be used in setting the cores.

Other sand molds include furan no-bake, oil no-bake, oil-oxygen, carbon dioxide – sodium silicate, shell, and cement molds. All are made by modifications of dry sand molding methods, and have a very low water content—usually less than 1%.

Advantages of the processes for making these types of molds are:

- Sands are free-flowing, and therefore ramming is eliminated or reduced.
- Tensile strengths of molds are higher than those of conventional sand molds. This permits reduction of mold weight and easier handling of large molds.
- Molds can be made without flasks.
- Production rates are high.
- Most of these molds can produce castings to closer tolerances than are obtainable in green sand molds.

The major disadvantage of most of these processes is that the sands must be used immediately. Shell mold sands are exceptions, because they can be stored indefinitely. Some of these processes—for example, the carbon dioxide–sodium silicate process—are better adapted to jobbing-type production than to long production runs.

PROPERTIES AND SELECTION OF SANDS

Silica sands are used for most sand molding operations. Silica sands are of two general types: naturally bonded and synthetic. Both types exist in nature, but have been formed under different conditions.

Naturally bonded sands (commonly referred to as bank sands) contain up to 20% of clay-base contaminants. The suitability of these sands for molds and cores depends largely on the total amount of contaminant, composition of the contaminant, and metal being cast. Naturally bonded sands containing more than 5% clay-base contaminant are unsuitable for making oil-sand cores, because the clay impairs the effectiveness of the oil binder. Some bank sands are formed by disintegration of igneous rocks and contain substantial amounts of feldspar. Such sands are usually satisfactory for molds used for casting of aluminum alloys, which have low melting points, but are not usually suitable for casting of ferrous alloys, especially steel. Sands that contain low-melting-point granules such as mica, sodium salts or potassium salts are usually objectionable for sand molding.

Synthetic sands (also called lake sands or sharp sands) are made up of a base sand grain. This is usually silica, although zircon, olivine or chromite may be used for certain applications. For instance, olivine is used extensively in the production of austenitic manganese steel castings, because it is not attacked by manganese oxide slags. Binder materials (mostly clay) are added to the base sand, which is nearly pure, to make the desired molding mixture.

Most foundries that produce castings from metals that melt at high temperatures use synthetic sands, because their compositions can be more closely controlled. Various mixtures of naturally bonded and synthetic sands are commonly used for malleable iron and gray iron. Naturally bonded sands are generally satisfactory for the lower-melting-point metals.

SAND RECLAMATION

When a casting is poured in a green sand mold, the layer of sand adjacent to the poured metal is heated to nearly the temperature of the molten metal. If this temperature is higher than about 600 °C (1100 °F), the sand will transform and change volume, and the clay binder will lose its combined water. The burnt clay-sand mixture no longer has desirable molding properties and is usually separated as coarse lumps from the remaining sand at shakeout. If the burnt sand is to be reused, it must be processed mainly for removal of the burnt clay, which will no longer form a satisfactory bond with the addition of water. Also, additives such as sugars, wood fiber and sea coal are removed to yield a clean sand.

The burnt sand that has been treated for reuse is said to be reclaimed, while the partially burnt sand that is reused after being reconditioned is called system sand.

The partially burnt (system) sand is either unaffected during the casting cycle, or at most has lost the temper water required to develop the clay bond.

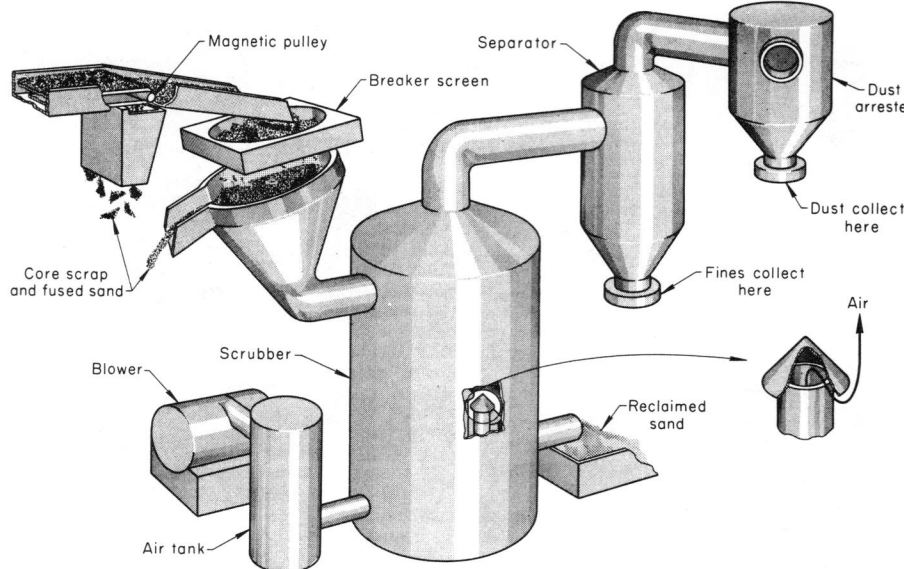

Fig. 4. Pneumatic scrubbing system used for dry reclamation of used molding sand

The amount of clay that is burnt depends largely on the size, weight and composition of the castings being poured. For instance, in pouring small, relatively thin-section gray iron castings in molds such as are made on a squeezer line, very little clay is burnt. The sand may be used after retempering (adding water). Addition of little or no new sand is required except to make up for the normal losses. Conversely, when heavy castings are poured from the same metal, a greater amount of clay is burnt. Under these conditions, the system sand must either be diluted with new sand and clay or the system sand must be reclaimed.

If cores are used extensively in making a casting, two shakeout operations may be necessary in order to keep the burnt core sand from contaminating the recirculating system sand.

Reclamation Methods. The three methods of sand reclamation commonly used are dry reclamation, wet reclamation and thermal reclamation.

Dry reclamation employs air separation to remove fines, such as silica flour and clay. Pneumatic scrubbing systems of the type shown in Fig. 4 have two to eight compartments in which abrasion of grain against grain takes place to remove the spent binder.

The size and shape of sand grains have a bearing on scrubbing effectiveness and output rates in pneumatic scrubbing systems.

Wet Reclamation. In wet reclamation, used sand is slurried with water on a mining slab (an inclined concrete apron surrounded by a concrete wall), or in a rotary-barrel tumbling unit. The slurry then goes to a scrubber (Fig. 5); after scrubbing, it is pumped into a classifier (Fig. 6) for removal of clay and other fine material. Finally, the sand is dried.

Depending on the shape of the grains and, to some extent, on the amount of scrubbing, wet-reclaimed sand usually appears gray. This is a result of carbon deposits retained in surface imperfections in the sand grains.

Benefits of a wet-sand reclamation system include:

• Casting finish equal to that obtained with new sand

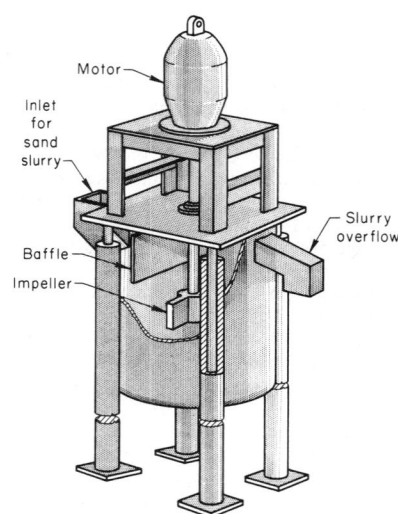

Fig. 5. Scrubbing system used for wet reclamation of used molding sand

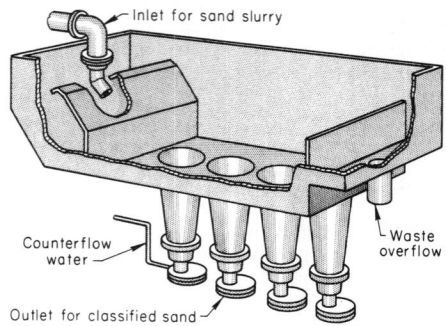

Fig. 6. Hydraulic counterflow classifier used in wet reclamation of used molding sand

- Acceptable removal of spent clays and fine materials
- Sand cost per ton substantially lower than that of new sand, which may have to be shipped a great distance
- Closer control of grain distribution than is obtained with new sand
- No lowering of the sintering point.

Thermal reclamation involves heating the sand to 650 to 825 °C (1200 to 1500 °F) in air to remove carbonaceous material; some clay may be removed also, by abrasion of the sand as it travels through the furnace. The efficiency of clay removal depends on the type of clay used; for example, western bentonite is more readily removed than fireclay. Thermal reclamation alone cannot completely remove a clay binder. However, the small amount of carbon left by the wet reclamation process can be effectively removed by thermal reclamation. The resulting sand is essentially equal to new sand, except for a reddish-yellow color caused by pickup of a small amount of iron, which is not harmful.

When the only binder used is organic, as in shell molding, thermal reclamation is completely effective in removing the used binder.

CLAY BONDING AGENTS

Clay, the most commonly used bonding agent for molding sand, is primarily aluminum silicate, but because it is a decomposition product of many types of igneous rock, it varies widely in composition.

Clays have three common characteristics: (a) they can be made plastic and adhesive by being mixed with the correct amount of water; (b) they can be dried, and then replasticized by the addition of water, provided the drying temperature is not too high; and (c) they become calcined, or "dead," if heated to too high a temperature, and then cannot be replasticized with water. (The calcining temperature varies with the type of clay.)

Clays are of three general types:

- Montmorillonite, or bentonite, of which there are two types: sodium, or western bentonite; and calcium, or southern bentonite. Southern bentonite produces molds with higher green strength and lower hot and dry strengths than are obtained with western bentonite.
- Kaolinite, commonly called fireclay
- Illite, which is derived from the decomposition of certain shales.

When mixed with water, all clays develop both adhesive and cohesive properties; the amount of each depends on water content. When the water content of the clay-water mixture is low, clays cohere (stick to themselves) rather than adhere (stick to a foreign substance), whereas the reverse is true at high water contents.

SAND MIXTURES

Sand for molds is a mechanical mixture containing as much as 98% silica sand, plus clay, carbonaceous material, and water. Fireclay, western bentonite and southern bentonite are used either singly or in combination to bond the sand grains. The carbonaceous material may be ground bituminous coal (sea coal), cellulose (cereal or wood flour), or any of various proprietary materials, including natural and synthetic complex resin products.

Mold coatings (or washes) are used to obtain a better finish on castings, including a smoother surface, less scabbing and buckling, and less metal penetration in the mold. The coating is applied to increase the refractory characteristics of the surface by sealing the mold at the sand/metal interface.

Mold coatings resemble paints and are generally compounded from three basic ingredients: a refractory filler, a vehicle, and a suspension agent. The filler material for mold coatings for steel castings is usually silica flour, zircon flour or chromite flour. Any one or a mixture of these flours, which are prepared by grinding the sands to 200-mesh or finer, can be used, depending on the severity of the application. The vehicle is either water or a commercial grade of alcohol. If the mold is to be dried, water is used. When the mold is produced in green sand, or by one of the processes where water is not included in the molding mixture, an alcohol-vehicle coating is used. Bentonite or sodium alginate is generally used as the suspension agent. Only enough is used to keep the mixture in reasonable suspension under constant agitation.

MOLDING METHODS

Hand ramming is the oldest and slowest method of making a mold. It is usually necessary for loose-pattern molding, and it is used also on mounted patterns when sample castings are being produced.

Jolt-type molding machines (Fig. 7) operate with the pattern mounted on a pattern plate, which in turn is fastened to the machine table. The table is fastened to the top of the operating air piston. A flask is placed on the table or on the pattern plate, and is located by pins so that the pattern is centered in the flask cavity. The flask is filled with sand, and the jolt valve to the air supply is opened. This allows compressed air at about 620 MPa (90 psi) to enter under the piston and lift it a few inches, at which point a port is uncovered and all of the air is released. The piston, together with the table, pattern, sand and flask, falls by gravity until the bottom of the piston strikes on the bottom of its cylinder with a sharp jolt. Air re-enters and the action repeats, at several times a second, until the air supply is turned off. The number of jolts can be preset to suit the pattern, sand, and mold hardness required. Because the sand is compacted only by its own weight, it is harder near the pattern plate, and a tall pattern thus can have appreciable decrease in mold hardness vertically.

When ramming is complete, push-off pins, bearing against the bottom edges of the flask and operated by a separate air cylinder, lift or strip the flask and contained sand off the pattern. Various mechanisms are used to accomplish stripping and then also to rotate the flask 180° so that the mold cavity will be upward.

Ramming of the sand on the pattern by jolting eliminates much of the manual ramming labor. It also increases production rate and provides more uniformly rammed molds.

Jolt-squeeze molding machines use the same pattern equipment as jolt machines, but after the jolting operation, the sand is squeezed by air pressure, producing a more uniform and harder rammed mold. Such equipment takes flasks up to approximately 750 mm by 1 m by 375 mm (30 by 40 by 15 in.), and is widely used because of its simplicity and low capital cost.

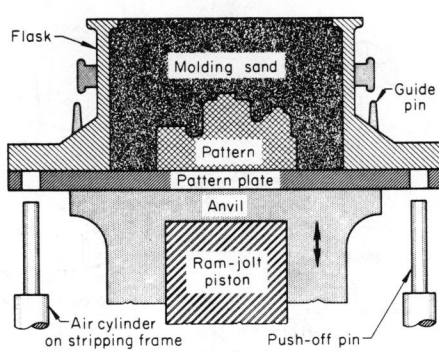

Fig. 7. Essential components of jolt-type molding machine. (See text for description of operation.)

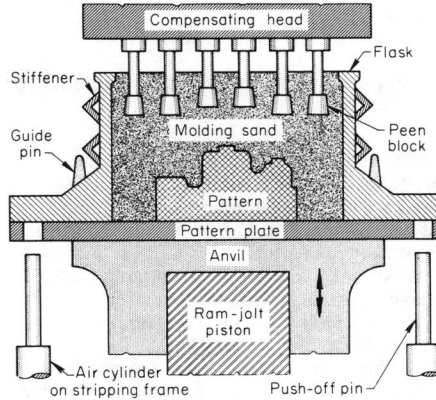

Fig. 8. Essential components of a high-pressure jolt-squeeze molding machine. (See text for description of operation.)

A high-pressure jolt-squeeze machine (Fig. 8) uses a hydraulic system for the squeeze operation. The compensating head equalizes the pressure applied to each floating peen block as the sand-filled flask is hydraulically raised against the peen blocks. This develops a uniformly dense packing of molding sand against the entire surface of the pattern. The mold may or may not be jolted, depending on requirements.

GATING SYSTEMS

The function of a gating system is to fill a mold cavity completely, avoid or prevent pickup or entrapment of loose sand or slag, and feed liquid shrinkage. The gating system should be designed to promote progressive solidification from the point most distant from the gate toward the gate. Where risers and a properly designed gating system do not compensate for all conditions of shrinkage and sectional variation in castings, changes in the design of the casting may be required. Although it is sometimes possible to effect the required directional metal solidification through the use of chills, this is not usually the preferred solution to a metal-feeding difficulty, mainly because the use of chills increases molding cost.

Shrinkage is not the only problem arising from improper gating. Failure to trap slag, or pickup of other types of nonmetallics from the gate, will cause inclusions in the casting. Inadequate rate

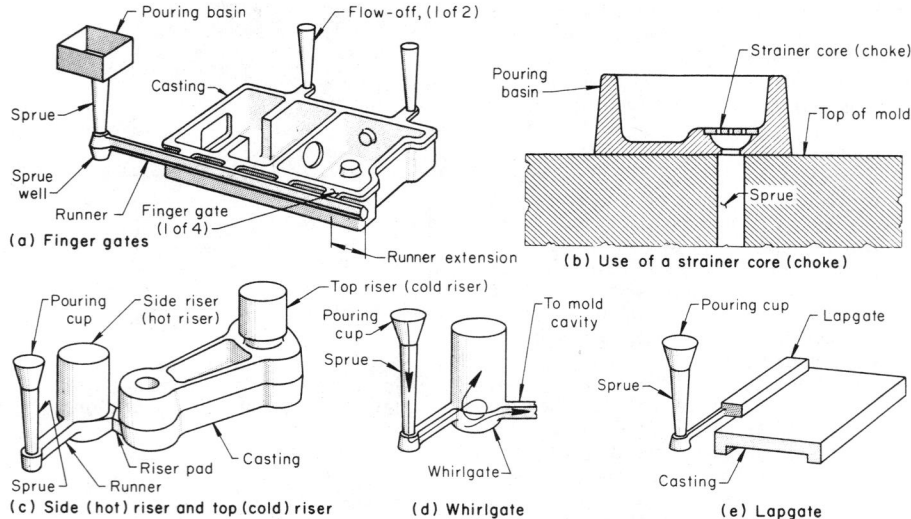

Fig. 9. Five types of gating systems for regulating the flow of metal into a mold. (See text.)

of metal entry will result in cold shuts and misruns, or in incomplete filling of thin sections.

Types of Gating Systems. Five types of gating systems are illustrated in Fig. 9. In a system using gates off the runner (called "finger" gates), molten metal is delivered from the pouring ladle to the pouring basin, which is suitably elevated to develop the required pressure for complete filling of the mold cavity (see Fig. 9a).

A strainer core that serves as the choke in a gating system can be located at any point in the system ahead of the runner, although it is conveniently located either in the pouring basin, as shown in Fig. 9(b), or at the bottom of the sprue. A tapered sprue may provide enough choke so that no strainer core is needed.

Figure 9(c) shows a gating system with a side (hot) riser and a top (cold) riser. The side riser receives the molten metal directly from the runner before it enters the casting cavity, and is more effective than the top riser, which fills up with the coldest metal and is likely to solidify before the casting.

Figure 9(d) shows a whirlgate, which carries slag or dirt to the center of the riser by centripetal force, thus preventing dirt from entering the gate and the casting.

Figure 9(e) shows the use of a lapgate to allow molten metal to enter the mold cavity. A lapgate is easily removed from a casting (often, it will simply fall off), because of the narrow connection (2.4 to 4.8 mm, or $3/32$ to $3/16$ in.) between the gate and the casting. Consequently, after the casting is cleaned for removal of burned-on sand, the gating area needs little, if any, grinding. (For this reason, lapgates are known also as "no-grind" gates or "kiss" gates.)

Shell Molding

SHELL MOLDING is a process in which a mold is formed from a mixture of sand and a thermosetting resin binder that is placed against a heated metal pattern. When the mixture is heated in this manner, the resin cures, causing the sand grains to adhere to each other, forming a sturdy shell that constitutes half of a mold. After the shell has been cured and stripped from the pattern, any cores required are set, the cope and drag

halves of the mold are secured together and placed in a flask, and backup material is added; then the mold is ready to be poured.

APPLICABILITY

Shell molding is used for making production quantities of castings that range in weight from a few ounces to approximately 180 kg (400 lb) apiece, in both ferrous and nonferrous metals. Castings weighing at least 450 kg (1000 lb) apiece have been made in shell molds, but not in production quantities.

The shell mold process is particularly suited to castings for which:

- The greater dimensional accuracy attainable with shell molding (as compared with conventional green sand molding) can reduce the amount of machining required for completion of the part.
- As-cast dimensions may not be critical, but smooth surfaces (smoother than can be obtained in green sand) are the primary objective.

In addition to producing castings that have greater accuracy and smoother surfaces, shell molding has two other advantages over green sand molding: (a) less sand required, and (b) fewer restrictions on casting design.

The limitations or disadvantages of shell mold casting are:

- Maximum casting size and weight are limited (see above).
- High cost of patterns, which must be machined from metal
- High cost of resin binder
- Relative inflexibility in gating and risering. Gates and risers must be incorporated, at least in part, into the shell mold pattern.
- Shrinkage factors vary with casting practice. (Two foundries using the same pattern may pour castings with different dimensional variations.)
- More equipment and control facilities are needed, such as for heated metal patterns.

PROPERTIES AND SELECTION OF MOLDING SANDS

The silica and zircon contents of these sands are high, which suggests that the sands are gen-

erally low in organic material, and clay content is low.

Low percentages of organic material and clay, which are obtained by washing (if necessary), minimize resin consumption. When excessive amounts of organic material and clay are present, more resin is required to strengthen the mold. Resin consumption increases also as the sand grain size becomes finer; therefore, the coarsest grain size that will produce the casting finish required should be used.

Grain Size and Shape. In shell mold casting, the greater strength available in the resin bond allows the use of a finer grain size for a given casting weight than is possible in sand casting. Except for aluminum castings, best results are obtained with sand having a narrow screen distribution.

Moisture Content. Dry sand is necessary for the cold resin-coating process. Moisture can cause dry resin to ball up, thereby decreasing the amount of resin available to coat the sand grains. In the hot coating process, moisture is generally not a problem, because it is driven off when the sand is heated to the minimum temperature of 120 °C (250 °F) before coating.

Breakage of molds or cores during molding reflects low hot strength. This can be caused by the use of a finer sand screen distribution without a proportionate increase in resin content, by excessive amounts of clay or organic material, and by round sand grains.

RESINS

The synthetic resins used in shell molding are thermoplastic resins to which setting or polymerizing agents have been added to produce thermosetting characteristics.

Phenol-formaldehyde resins, produced with an insufficiency of formaldehyde so they develop thermoplastic characteristics, can be made to harden permanently, or to set, on heating by the addition of hexamethylenetetramine (commonly called hexamine or hexa). The storage life of these resins is limited, and care must be taken to avoid any increase in temperature during storage. Frequently the resin is refrigerated to extend storage life.

Phenol-formaldehyde resins are the most commonly used in shell molding, because, when combined with sand, they have high strength, resistance to heat and moisture, and good flowing and curing properties. Generally, the phenolic resins used for shell molding are of the thermosetting, two-stage type, often referred to as novolak resins. The two-stage designation indicates that the resin was formed with a molar excess of phenol in the presence of an acid catalyst and is like a thermoplastic. To obtain thermosetting properties, the catalyst hexa is added during coating.

When the resin is formed with a molar excess of formaldehyde with a basic catalyst, it is referred to as resol, a one-stage resin. This resin is reactive as made, and does not require the addition of a catalyst during coating. Because of this, the resin will gradually age, and if it is to remain suitable for use in shell molding it must be refrigerated. Also, the one-stage and urea-modified phenolics absorb moisture, and therefore are not widely used to coat sands that will be used for shell molds and cores. Sand that is properly coated with phenolic novolak resins has excellent resistance to moisture absorption and remains free-flowing.

Phenolic resins used in shell molding are: (*a*) novolak varnishes with 60 to 70% solids, (*b*) water-borne novolaks with 75 to 80% solids and (*c*) flake and lump resins. Novolak varnishes are only slightly soluble in water and are furnished in organic solvents, mainly alcohol. Water-borne novolaks must be used either with hot sand or with air heated to 190 to 260 °C (375 to 500 °F); flake and lump resins require hot sand.

Resin powders are purchased according to the properties that are important in the formulation of resin-sand mixtures: melting point, cure rate, flow rate, particle size and hexa content. Table 2 lists properties of ten synthetic resins used in shell molding.

Solids content of a resin is the amount or percentage of binder or resinous material that is available for bonding the sand. The remainder of the resin is alcohol and water, which are removed in the coating cycle. The only limit to the maximum solids content of a resin is that imposed by an excessive increase in viscosity. The highest solids content is 80%.

Coating formulations are usually based on resin solids, and hexa and wax are added as weighed in their solid form. Any change in the solids content of the resin requires an appropriate adjustment in the hexa-to-wax ratio. If the proper adjustment is not made, either brittle shells or soft (spongy) shells will be produced.

The foundryman cannot control the solids content of a resin except by diluting it with alcohol. However, resins with a range of controlled solids levels are commercially available. These resins can be used to maintain the correct solids content.

Viscosity of a resin varies with resin temperature; therefore, the temperature at which viscosity is measured must always be recorded along with the viscosity. If the viscosity of a resin is too high, sand coating may be incomplete, which will result in low-strength shell molds that will break and crack during pouring. The viscosity of resins depends on the solids content, as shown in the following tabulation:

| | Viscosity at 24 °C (75 °F) | |
Solids, %	Pa·s	P
58 to 62	0.25 to 0.60	2.5 to 6.0
68 to 72	1.80 to 4.50	18.0 to 45.0
75 to 77	8.00 to 14.00	80.0 to 140.0

The viscosity of a resin can be reduced by diluting the resin with an organic solvent, such as alcohol.

PREPARATION OF RESIN-SAND MIXTURES

A resin-sand mixture for shell molding can be prepared by either of two methods: (*a*) mixing resin and sand according to conventional dry mixing techniques or (*b*) coating the sand with resin. The dry resin-sand mixes are subject to segregation and dusting, which makes them unsuitable for mold blowing; they can be used only for dump-box production of molds. The resin-coating method has largely superseded dry resin-sand mixing.

Cold coating is a process in which sand and liquid resin, both at or near room temperature, are blended together with other materials to produce a resin-coated sand. The liquid resin can be either a one-stage or a two-stage resin, having either water or alcohol, or both, as a carrier. The carrier must be subsequently removed from the resin-sand mixture, leaving a dry, free-flowing sand,

Table 2. Properties of ten synthetic resins used in shell molding

Resin No.	Melting point °C	°F	Curing time at 150 °C (300 °F), s	Flow in 17 min at 125 °C (257 °F), mm	Particle size, % on 200-mesh screen	Hexa concentration, %
1	100-105	210-225	25-35	10-17	0.5-1.0	13-15
2	105-115	220-235	35-45	16-22	0.5-1.0	11-13
3	95-100	205-215	35-45	10-17	0.5-1.0	13-15
4	105-115	225-240	40-47	10-15	8.0-10.0	13-15
5	105-110	220-230	45-50	16-22	1.0-1.5	13-15
6	105-115	225-240	50-55	22-28	5.0-10.0	11-13
7	95-100	205-215	55-60	16-22	0.1-0.2	11-13
8	100-105	210-225	55-60	22-34	0.5-1.0	13-15
9	100-105	210-225	65-75	22-28	0.1-0.5	8-10
10	110-120	230-250	80-100	24-32	0.1-0.2	8-10

each grain of which is covered with a thin coating of resin.

The main advantage of the cold coating method is that the sand is used at room temperature, thus eliminating the necessity of equipment to heat the sand. Also, the melting point and tensile strength of the coated sand are easier to control by this method.

The main disadvantages of the cold method of coating sand are: (*a*) the necessity of handling large amounts of liquids that are not incorporated into the final product; (*b*) the need for removing these liquids from the sand mix, which greatly increases the over-all mixing cycle time; and (*c*) the explosion hazard associated with the evaporation of low-molecular-weight hydrocarbons.

Hot coating is a process in which resin and sand, in combination, are heated to 120 to 135 °C (250 to 275 °F). At this temperature, the thermoplastic novolak resin will coat the sand grains during mixing, but will not set. The catalyst (hexa) is then added, usually in water solution, and distributed throughout the mixture, which is then cooled.

There are at least three advantages of hot coating: (*a*) the tensile strength of hot-coated sand is about 25% higher than that of cold-coated sand; (*b*) hot-coated sand does not dust, as does cold-coated sand; and (*c*) there is no explosion hazard in the hot coating process, as there may be in cold coating, because alcohol is not used in the hot coating process.

The principal disadvantage is the necessity for heating the sand prior to mixing and mulling.

Equipment for Coating and Handling. Resin coating of sand can be done at high speed on a nearly continuous basis by use of the equipment shown

in Fig. 10. In this equipment, the air heater heats the low-speed muller to 150 to 175 °C (300 to 350 °F). The sand charge and granulated novolak resin are mixed in the low-speed muller until the sand is coated. The resin-coated hot sand is then discharged into the running high-speed muller, which is water cooled. When all of the hot sand has been discharged, a water solution of the required amount of hexa is sprayed onto it. The sand is then discharged immediately through the vibratory screen into the oscillating conveyor for transfer to storage bins.

Various methods, depending on the size of the foundry operation, can be utilized to transfer coated sand from the muller or storage hopper to the shell molding machine. In small foundries, the sand is generally loaded into fiber drums or tote boxes and transferred by hand carts or fork trucks to the feed hopper on the machine. Larger plants use automated transfer systems; for example, conveyor-belt, bucket-monorail or pneumatic systems, such as shown in Fig. 11.

Because prolonged and excessive handling of coated sand causes resin to separate from the sand grains, a sand-transfer system should keep travel distance to a minimum. A conveyor-belt system should be constructed to limit the number of directional transfers, not only to minimize separation of sand and resin, but also to eliminate segregation of sand grains. Automated bucket-monorail conveyors are normally used when several different sand mixtures are required or when the number of directional transfers cannot be reduced. If a pneumatic transfer system is selected, a low-pressure system is preferable.

Coated sand does not pick up moisture readily unless the surrounding air is extremely humid;

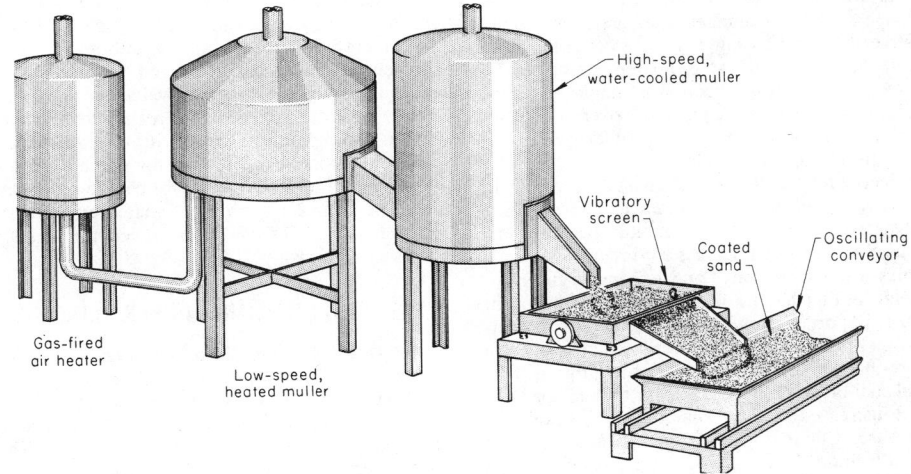

Fig. 10. Two-muller system for resin coating of shell sand by the hot process

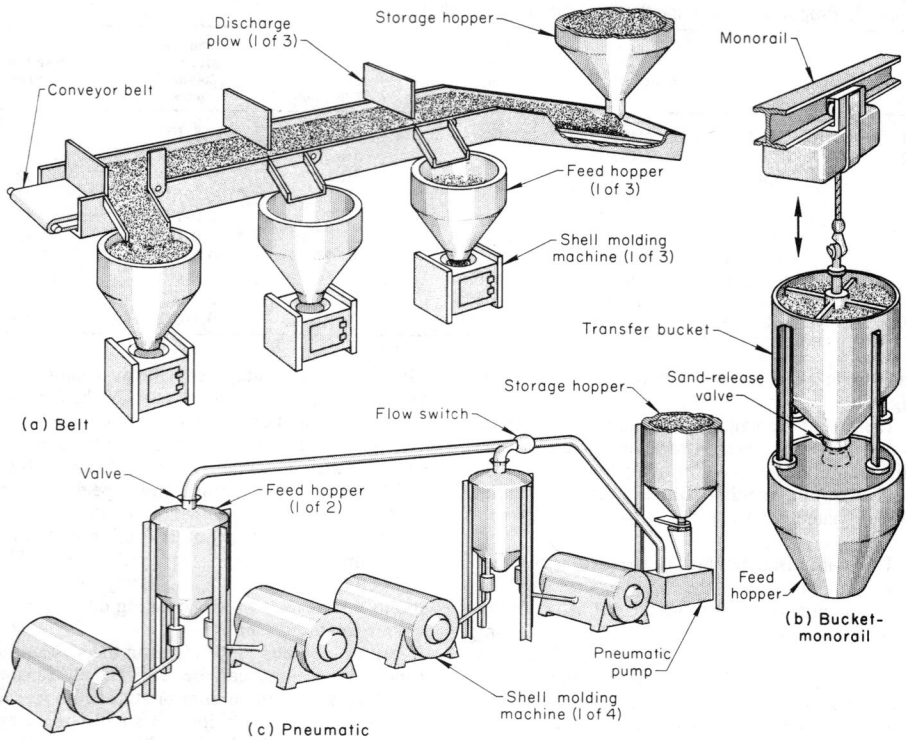

Fig. 11. Three types of systems for automatic transfer of coated sand from muller to shell molding machine

therefore, transfer time from muller or coating unit is not critical. (In fact, coated sand can be stored for several months if kept in a fairly dry area.) Temperature variations that result from changes in weather have no effect on coated sand, provided that it was thoroughly dry when removed from the muller.

Phenolic Urethane Resins for Core and Mold Making

By R. L. Naro, Ashland Chemical Co.

SAND/BINDER SYSTEMS based on phenolic urethane chemistry provide a multiplicity of operational advantages compared with alternative chemical-curing resin systems. Phenolic urethane resins cure at room temperature. The nature of this curing reaction is unique and provides many significant advantages over older heat-curing resin systems as well as other room-temperature curing systems.

Heat-curing systems, which are declining in popularity, suffer from a number of serious disadvantages and drawbacks, mainly because heat is used to initiate curing. As a consequence, cycle times usually are long, and expensive driers or metal tooling having limited life must be used. Other important drawbacks associated with heat-curing binder systems are decreased efficiency in core-machine uptime because of lengthy heat-up and cool-down times required for tooling changes, and increased tooling maintenance. Likewise, because of their slower cycles, these systems are usually more labor-intensive than resin systems that cure without heat. Conversely, alternative

room-temperature curing resin systems do not provide the desirable and flexible curing properties and characteristics inherent in phenolic urethane systems.

Phenolic urethane resins can be used in one of two forms: (1) as a liquid catalyzed no-bake system or (2) as a vapor-cured cold-box system. The no-bake system employs phenolic urethane binders and cures without baking; curing takes place at room temperature and is achieved only after the catalyst is added to the sand/resin mixture. The vapor-catalyzed or cold-box system also uses phenolic urethane resin binders, and curing is achieved only after a gaseous catalyst is passed through the sand/binder aggregate in a vented core box.

Depending on the method of catalysis, cure times ranging from a few seconds to as long as 1 h are readily achievable. Tensile-strength development is extremely rapid with both systems; 1-h tensile values easily exceed 1.4 MPa (200 psi), and fully cured tensile values will exceed 2.1 MPa (300 psi) with 1.5% resin levels (based on sand weight). An almost infinite variety of molds and cores is made with the process. Cores weighing up to 135 kg (300 lb) can be produced in a one-minute cycle on automated equipment. Even faster cycles can be achieved for smaller cores and molds.

PHENOLIC URETHANE NO-BAKE BINDERS

Resins and Catalysts. Two liquid-resin components and a liquid catalyst comprise the system. The part I phenolformaldehyde resin is a clear, amber, organic polyol and the source of active hydroxyl groups (OH) necessary for bonding. The part II resin component is a dark liquid that provides active isocyanate groups (NCO) in the form

of polymers of the MDI type (methylene bis phenylisocyanate). Part I storage life is one year or more at temperatures between 16 and 27 °C (60 and 80 °F). Part II has essentially unlimited storage stability in tightly closed containers.

Both part I and part II resins are diluted with aliphatic and aromatic solvents. The primary purpose of these solvents is to reduce binder viscosity. Typically, the viscosities of the part I and part II resins are adjusted to 200 cps or lower to provide good pumpability, rapid and efficient sand-coating qualities and good flowability of mixed sand. A second purpose of the solvents is to enhance resin reactivity.

For most ferrous-metal casting applications, total binder levels of 0.8 to 2% (based on sand weight) provide good sand-bonding properties. The two resins are normally used in offset proportions such as 55/45 or 60/40 of part I to part II. Unbalanced ratios are often used to minimize the nitrogen contribution from the polymeric isocyanate resin component (part II). In aluminum applications, binder levels ranging from 0.8 to 1% are used to facilitate core shakeout.

Several catalyst series based on amine derivatives have been developed for use with PUN (phenolic urethane no-bake) binders. During use, the catalyst is precisely metered into the part I resin stream just prior to entry into the mixer. Precatalyzing the resin to avoid using a third pump limits the storage life of the resin, and variations in cure time can occur. If the resins are precatalyzed, they must be used within a one-week period.

Standard catalysts provide strip times ranging from 2 to 15 min at a sand temperature of 24 °C (75 °F). Newer catalysts are capable of providing either slower or faster core/mold strip times. For example, catalysts that provide strip times as short as 10 s have been developed for utilization in no-bake core-blowing machines. In addition, slower catalyst systems have also been introduced to accommodate those foundries requiring extended working times. Bench lives ranging from 15 to 60 min are possible with these catalysts.

A typical curing curve for a phenolic urethane no-bake binder and two alternate no-bakes is shown in Fig. 12. The curing reaction is unique compared with other systems, because a latent catalysis occurs. The compressive strength, a measure of the sand-mix flowability, remains very low during the working life or bench life of the mix. However, once curing has been initiated, curing rates and compressive-strength buildup are extremely rapid. Since the resultant crosslinking reaction evolves no side products, e.g., condensation reactions, curing proceeds at the same rate throughout the sand body. There is little to no difference in curing rate between compressive test cores cured in air and those sealed in their ramming sleeves. Such uniform strength development takes the guesswork out of determining actual strip times in foundry applications. The ratio of work time to strip time for PUN binders varies between 0.75 and 0.85, depending on catalyst level. As this ratio approaches unity, the more desirable are the curing characteristics for any given system. Higher productivity and core-box turnover also result.

The tensile properties attainable with phenolic urethane no-bake binders are directly related to the binder percentage used. Using a reasonably good sand, 1-h and 24-h (fully cured) tensile strengths are typically 1.6 and 2.2 MPa (225 and 325 psi), respectively, with 1.5% binder (based on sand).

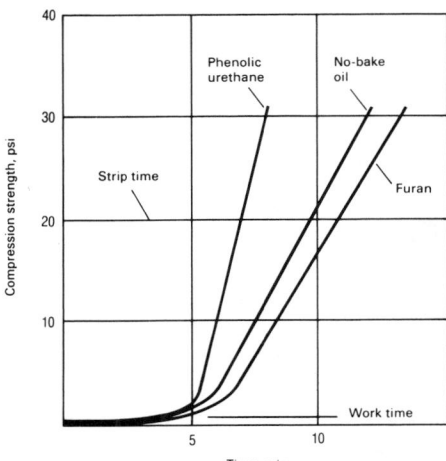

All systems were catalyzed to provide 5 min of working time (bench life of the mixed sand).

Fig. 12. Curing curve comparison for phenolic urethane no-bake binders compared with a typical acid-catalyzed no-bake (furan) binder and an alkyd oil no-bake binder

Sands. Sand properties, both physical and chemical, exert a profound effect on the performance characteristics of all no-bake binders. PUN binders are somewhat more tolerant to varying sand properties, and most grades of silica, lake and bank sands can be used successfully. Most foundries using PUN binders employ either subangular lake or round-grain silica sands with grain distributions ranging from 50 to 70 GFN. Specialty sands, such as zircon, chromite and olivine, are also used. Important considerations in selecting a sand are grain shape and distribution, incoming temperature, moisture content and surface chemistry. It is important that these parameters be controlled to provide curing consistency.

Chemically, the acid demand value and pH of the base sand provide a good indication of binder/sand reactivity. The ADV and pH values indicate the extent or presence of alkaline impurities such as calcium carbonate (lime) or other various metallic compounds of iron, aluminum, sodium, etc., on the sand-grain surfaces which may affect curing speed. Generally, basic sands tend to accelerate the curing rate while acidic sands retard it. The use of various additives, such as iron oxide or clays, to the sand mix will also affect the reaction rate and mechanical properties of the mix.

Other important factors that affect binder performance are sand temperature and moisture content. Higher temperatures accelerate the curing rate for a given catalyst level whereas lower temperatures retard curing. Control of sand temperature is critical in attempts to maintain a consistent work time/strip time relationship. Sand moisture also has a significant effect on curing rate. Typically, sand moisture levels are maintained below 0.25%. Sand should be stored in dry, covered silos. To minimize sand segregation, silo design should incorporate multiple discharge openings feeding into a single outlet.

PUN bonded sands are readily reclaimable. Foundries using PUN binders for molding will typically reclaim sand to reduce sand costs. Foundries that have had the most success in reclaiming sand utilize reclamation equipment that not only breaks the sand down into individual grains but also scrubs the sand in a separate operation.

Mixing and Process Applications. To utilize the rapid curing speed of phenolic urethane no-bake binders for producing cores and molds, it is important to choose an efficient, high-speed, zero-retention continuous mixer. With such mixers, strip times as short as one minute from core-box filling can easily be achieved. Although it is possible to utilize batch mixing, catalyst additions must be reduced to decrease curing rate and avoid premature setting. This can also be accomplished through the use of a less-reactive catalyst. Because most batch mixers require 2- to 4-min mix cycles, strip times approaching 10 min or longer must be employed. With continuous mixers, the design should incorporate accurate resin and catalyst metering and consistent sand delivery as well as providing adequate mixing to ensure uniform resin coverage on the sand grains. Most important are accurate blending and dispersion of the resin and catalyst on all sand-grain surfaces in a uniform manner.

Because of the short strip times achievable with PUN binders, most foundries automate core and molding loops to speed production. Loops may vary in design, based on individual requirements, but usually consist of a high-speed mixer, a vibration or compaction table, one or more rollover draw units, a core- or mold-removal conveyor and a return conveyor for empty core boxes. Examples of a basic no-bake layout loop and a highly automated loop are shown in Fig. 13 and 14, respectively. Further improvements in productivity can be achieved by using PUN binders on newly developed no-bake core blowers. With designs such as the one shown in Fig. 15, blow-to-blow cycles of 30 s can be achieved for cores weighing up to 9 kg (20 lb).

PHENOLIC URETHANE COLD-BOX BINDERS

Resins and Catalysts. The basic resin chemistry employed for the phenolic urethane cold-box version is essentially identical to that used for the no-bake system; the part I resin differs only in solvent composition. Depending on solvent selection, resin systems can be tailored to provide high out-of-the-box tensile properties, improved humidity resistance and improved bench life. Part II resins contain polymeric MDI (methylene diphenyl isocyanate), also dissolved in solvents. Because of the large variety of sands and sand chemistries, several part II resin components are available. Small additions of organic-acid extenders are added to these part II resins to neutralize alkaline impurities (present in some sands) and thus extend bench life.

Two catalysts are used to cure phenolic urethane resins: triethylamine (TEA) and dimethylethylamine (DMEA). These catalysts have flammability limits of 1.3 to 8.0 vol % in air and must be handled with extreme care. The actual amount of catalyst used has been found to be very dependent on generator design and core complexity. Experience has generally shown that target amounts of catalyst range anywhere from 0.05 to 0.08% (0.14 to 0.23 cm^3 per kilogram, or 0.010 to 0.017 oz per pound, of sand).

The curing mechanism involves rapid reaction of hydroxol groups from the part I resin component with isocyanate groups from the part II component as an amine catalyst vapor is passed through the core. The formation of a rigid urethane film that bonds the sand occurs within sec-

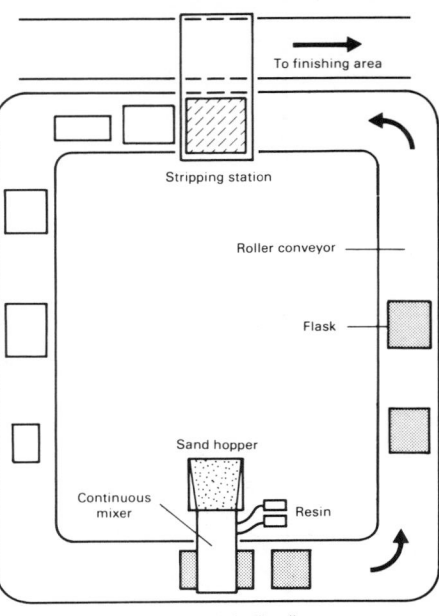

Fig. 13. Typical "no-bake loop," illustrating essential components

onds after contact with the catalyst vapor. Because curing occurs at room temperature, its speed is only limited by the amount of time it takes to sweep the catalyst vapor through the core. As an aid to a better understanding of how the process works, Fig. 16 schematically illustrates the various steps involved in coremaking.

Bench life of mixed sand may vary anywhere from 1 to 5 h. The quality of sand, sand temperature and other foundry variables exert major influences on bench-life performance. High levels of alkaline impurities present in certain lake and bank sands may have a catalytic effect on the resin system and shorten bench life. Several part II resins are available that improve bench life of the cold-box mixes made with such sands.

Total binder level is dependent on sand characteristics and core-strength requirements. Using a reasonably good sand, 30-s, 1-h and fully cured (24-h) tensile values of 1.28, 2.07 and 2.24 MPa (185, 300 and 325 psi), respectively, are attainable with resin levels of 1.5% (based on sand).

Sands. Selection of sands for phenolic urethane cold-box binders is very similar to selection for the no-bake version. Lake, bank and silica sands work well with the process as long as acid-demand value, moisture and temperature are within specified limits (see no-bake specifications).

Mixers for Sand Preparation. There are three basic types of mixers used for mixing cold-box resins with sand: (1) zero-retention, high-speed continuous mixers, (2) low-speed, auger-type through continuous mixers and (3) batch mixers.

Zero-retention, high-speed mixers offer the advantage of obtaining maximum bench life by limiting the amount of sand that is prepared at any given time. Ideally, these mixers are mounted or located so as to discharge directly into a coreblower sand hopper. Swivel mounting can be used to accommodate two or three core blowers.

Through-type continuous mixers have the same general characteristics as the high-speed models. However, such low-speed continuous mixers tend to demonstrate better mixing efficiency in pro-

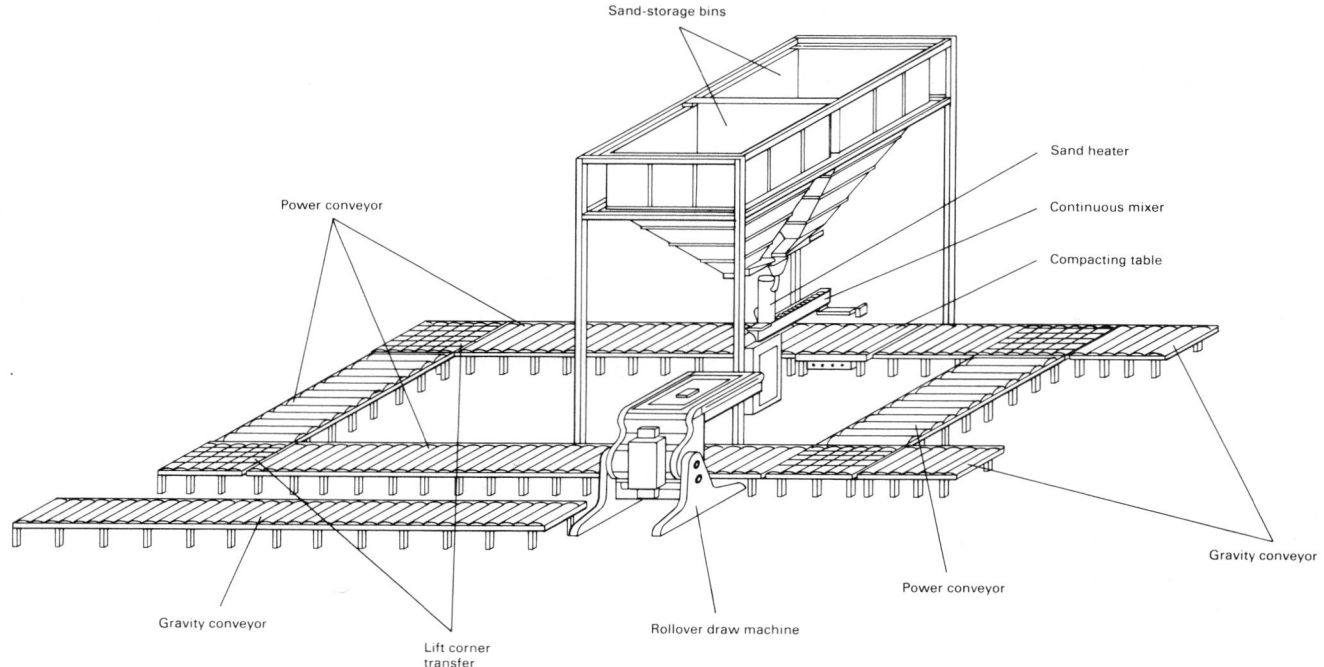

Strip times as short as 1 min for 45-kg (100-lb) cores and/or molds can be achieved using this concept.

Fig. 14. Highly automated no-bake loop layout

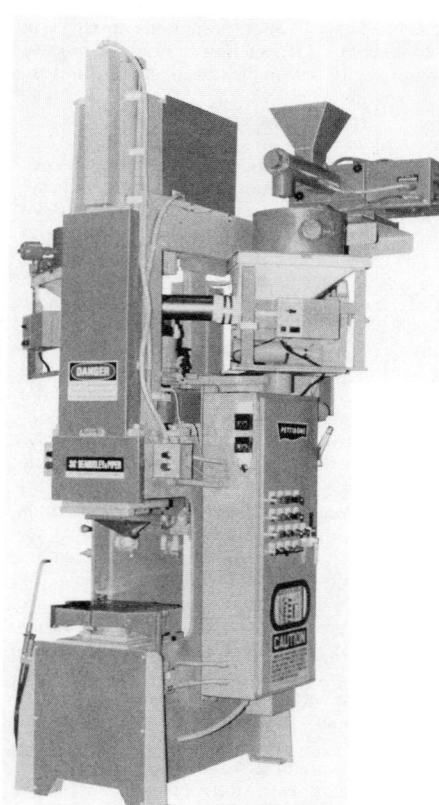

The machine illustrated can cycle twice per minute, producing 9-kg (20-lb) core assemblies.

Fig. 15. No-bake core blower designed to handle phenolic urethane no-bake binder systems

duction, primarily due to fewer maintenance problems.

Batch-type mixers with S-blade designs are also commonly used. Mixing time may vary from 2 to 4 min. Although batch mixers tend to produce mixes having somewhat higher tensile strengths, their satisfactory use depends on tailoring the batch size to core-blower consumption. With batch-type mixers it is important to prevent heat buildup that may shorten bench life.

Catalyst Generators. The preferred method of generating the gaseous catalytic vapor is through use of a vaporizer. The simplest method for accomplishing catalyst vaporization is bubbling of a suitable carrier gas (CO_2 or N_2) through a bath of liquid amine. Figure 17 illustrates the basic concept, and such a unit can be readily constructed on site with shop maintenance.

Equipment and Core Blowers. The density and general quality of blown cores can be affected greatly by the blowing and curing unit on which the particular job is run. For instance, a lacy, elongated cure is usually more easily made on a machine which blows sand perpendicular to the parting line of the box. Curing of such a core also is more easily accomplished with this orientation.

As a guide to matching the core machine to the core, three principles of cold-box operation have been formulated from the experience of foundrymen over the years. First, the tooling should move as little as possible and as smoothly as possible from the start of the blow to the finish of the curing cycle. Sudden stops, jolting, or excessive jiggling of the tooling after filling may cause "sticking" to the cope side of the box. Second, tight parting-line contact must be maintained during and between blowing and curing operations. Lastly, the travel path of sand during blowing and the travel path of catalyst and air during curing should be the same in most cases. For instance, when an edge-blow machine is used to produce a thin, elongated core, the gassing direction should be edgewise, not side-to-side. Venting of such a box for side-to-side gassing

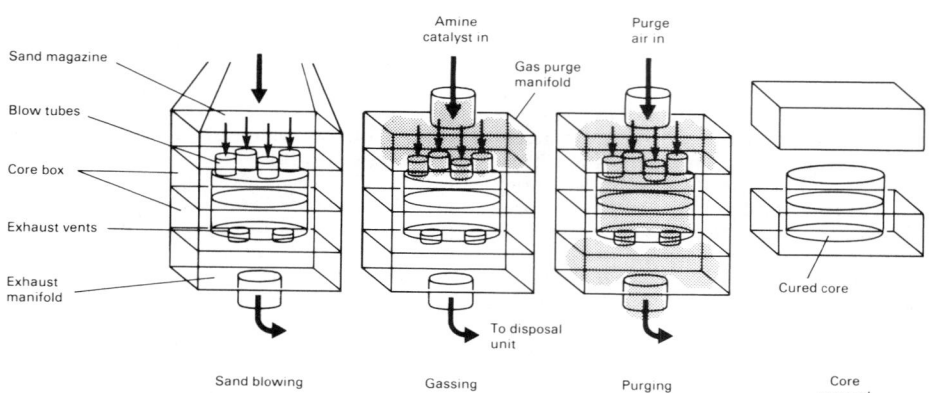

Fig. 16. Schematic illustration of essential steps in cold-box process coremaking

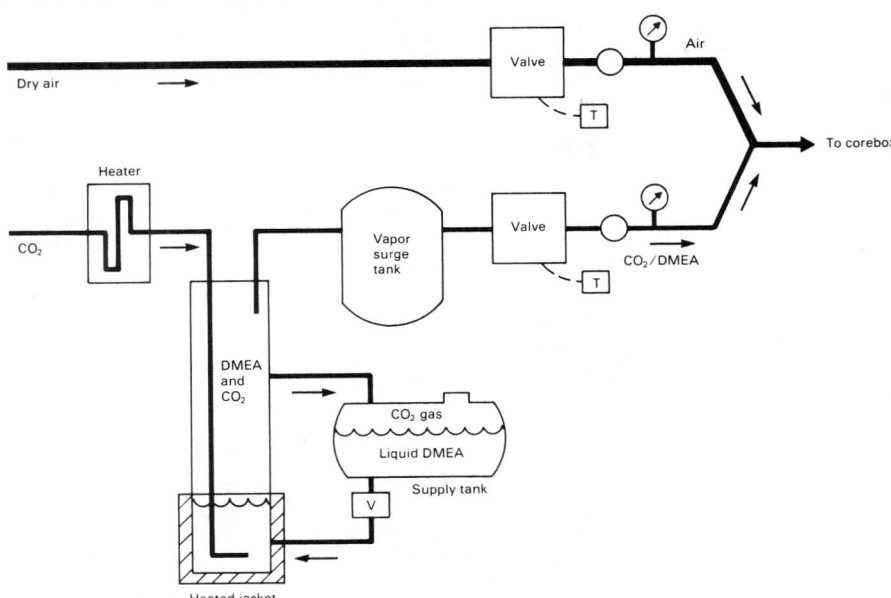

Fig. 17. Vaporizer or bubbler-type generator used to generate catalytic amine vapor for curing

will usually interfere with blowing of a tightly compacted core and will lead to high catalyst usage.

Many styles of core blowers, equipped with integral curing mechanisms, are available for cold-box core production. Regardless of specific core-blower design, all core blowers can be divided into two broad design categories: top blow with horizontal parting line (Fig. 18) and edge blow with vertical parting line (Fig. 19). Both have advantages and disadvantages, depending on core configuration.

Horizontally parted boxes can be rigged to distribute sand into cavities quite evenly. Conse-

quently, a more intricate core can be blown and cured. Blow-tube-location arrangements are unlimited. Inserts or segmented cavities can be utilized in large horizontally parted boxes and for making a combination of different cores in one cycle. For example, if a casting requires five cores per casting, two of one kind and one each of the other three, the tooling can be built to blow and cure this requirement with each cycle.

There are several disadvantages associated with horizontally parted boxes. It is harder to keep core vents clean because sand is blown at 90° to them. Core finishing, such as parting-line fins and blow-tube scars, is more severe than on vertically parted boxes. To keep core finishing to a minimum, blow-tube locations should be kept in core-print areas with a minimum located in casting areas. Setting up a horizontally parted box in a core machine usually is more time-consuming than setting up a vertically parted box, because of size, ejector-plant adjustment, pick-off fingers, and positioning and alignment of blow and gassing heads.

Vertically parted core boxes are ideal for cylindrically shaped cores because of the limited zone for blowing. Vents stay clean because sand is being blown parallel to them. Venting is usually more critical for core designs other than cylindrical shapes. One approach for venting a vertically parted box is to (a) vent to blow a sound or dense core, and (b) add additional vents if needed for gassing. Care must be taken so as not to overvent this type of core box. Higher blow pressures are often required for the intricate cores on an edge-blow machine, and wear is more predominant because of the limited blow area. Most edge-blow machines can be modified to include,

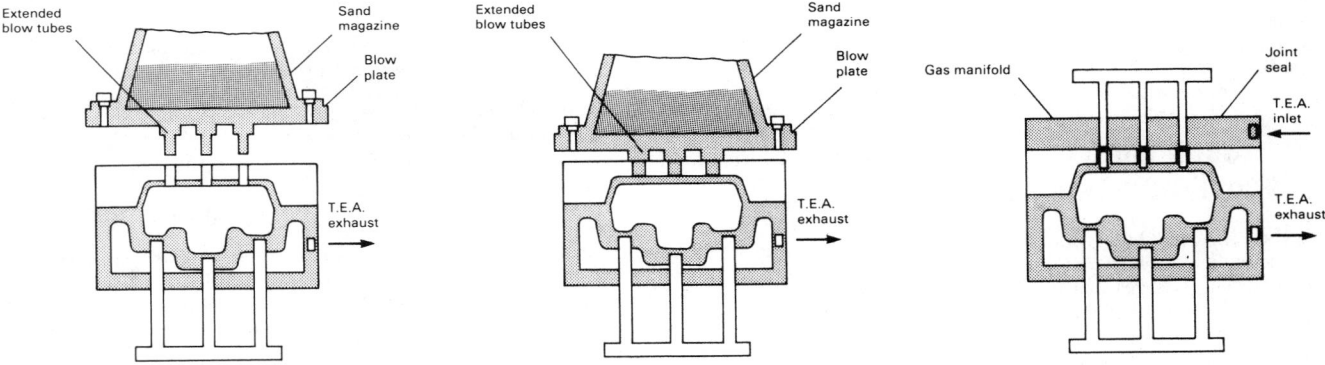

Fig. 18. Schematic illustration of a horizontally parted, top-blow core box with blow tubes

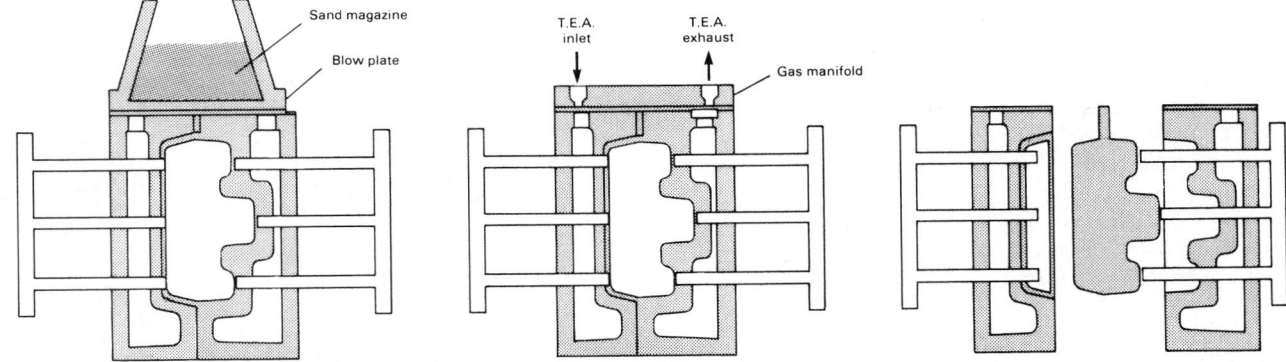

Note: The preferred method of gassing is edge-wise, in the core-blowing direction.
Fig. 19. Schematic illustration of a vertically parted, edge-blow core box with side-to-side gassing manifolds

or purchased with, a bottom-draw mandrel plate. Consequently, three-part core boxes can be designed, and this is especially desirable for lightening out larger cores. The resulting sand savings, shorter curing times and lighter core weight will be reflected in lower casting cost. Three-part designs may be accomplished with horizontally parted boxes, but with much added tooling expense. Vertically parted core boxes are usually cheaper than horizontally parted boxes because of their more simplified rigging.

A great variety of materials is available for corebox construction. Whereas cast iron construction is most commonly used in production foundries in the U.S., some production and experimental boxes are fabricated from aluminum, wood, epoxy, urethanes, etc. Solid aluminum and wood are seldom used in high production due to low wear resistance. Urethane-lined tooling offers the advantages of low cost and long wear; however, its sensitivity to solvent cleaners and certain release agents has proved to be a drawback.

It is possible to convert existing tooling from oil sand, hot box, shell and CO_2 to cold box with minimum reworking. Dimensional variations among the various processes must be checked to allow for any variations in cavity dimensions and operating temperatures as opposed to cold processes. Existing CO_2 boxes have proved to be extremely easy to convert.

Amine-Catalyst Disposal. Several methods are currently being used for disposal of amine catalysts. In the most popular method, exhaust gases are piped to a chemical absorption scrubber. Units of the type shown in Fig. 20 have attained wide popularity. Other methods of amine disposal include incineration by piping the curing mixture into a cupola, oven stack, or gas incineration disposal unit. Smaller jobbing foundries that employ the process often use only local exhaust ventilation in the area of the core box and pipe amine exhaust to the outside atmosphere since smaller operations often do not use much amine.

Sand Cores and Coremaking

CORES are separate shapes of sand that are placed in the mold to provide castings with contours, cavities and passages that are not otherwise practical or physically obtainable by means of the mold. Many objects are made at reasonable cost only because cores are used.

Cores are composed mainly of sand, but also contain one or more binder materials. Most of the principles that apply to making a sand mold apply also to making a sand core.

The basic principles of coremaking are illustrated in Fig. 21. In the four views in the top row of Fig. 21, core boxes are shown in section after ram-up; in each of the middle and lower views, the core box has been inverted on a plate and lifted, leaving the core on the plate, ready for baking.

Figure 21(a) shows a core that is semicylindrical—one of the most common core shapes. No problem is presented in removing this core from the core box; natural draft is provided by core shape, and no obstructions exist.

In Fig. 21(b), two bosses are formed by the core box. Boss A is simple to produce and adds nothing to the cost of producing the core, although it does add a small amount to the cost of the core box. Boss B, however, creates an un-

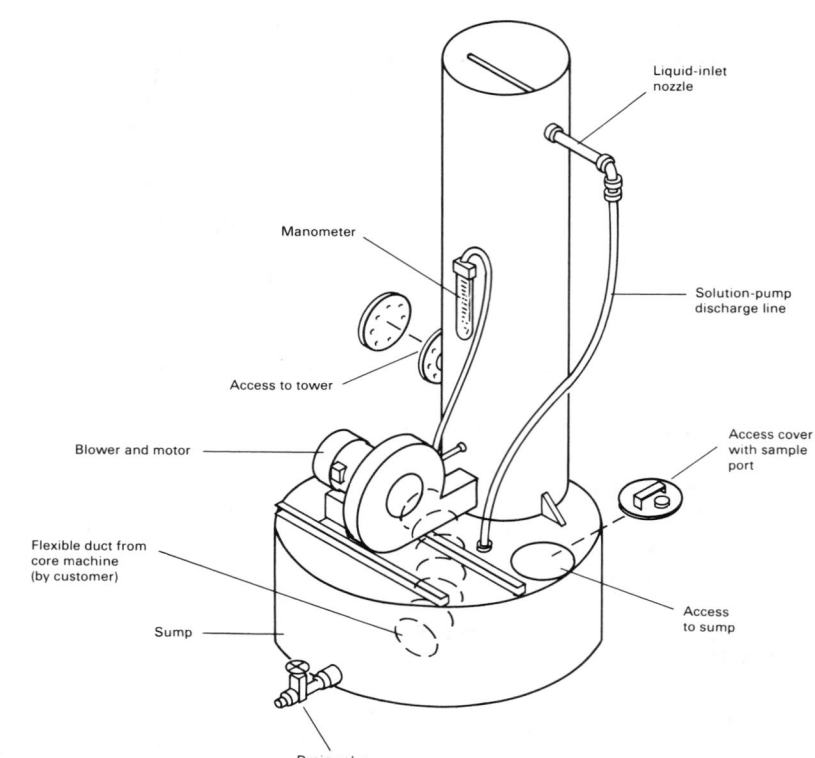

Fig. 20. Typical amine-catalyst scrubbers used for catalyst disposal

dercut that requires a loose piece, as shown, to permit the core to be freed without destroying the desired shape. This adds to the cost of the core box. It also adds to the cost of producing the core, and may increase the dimensional variations in the location of the boss.

Figure 21(c) shows a core box in which two more conditions are incorporated that require loose pieces. Loose piece A permits the side of the core to be produced without draft. Bosses or ribs could be appended to this face without adding to molding problems, provided that they could be drafted to permit easy withdrawal of the loose piece. Loose piece B forms a cavity in the core that assists in producing a rib on the casting. The overhang of the core box would lock the core in the box unless a loose piece were provided as shown.

Figure 21(d) shows the additional problem that arises when the loose piece at the face of the core box projects excessively into the core. When the core box is inverted and the loose piece removed, excessive overhand of the core could cause the sand to sag and lose its dimensional accuracy. To prevent this, support must be provided for the core. In the core box shown, the loose piece is removed and replaced with bedding sand before the core box is inverted.

CORE COMPOSITION

The compositions of cores vary considerably, although sand is the basic ingredient. Binders and fillers are the other ingredients.

Core sands are usually silica sands, but zircon, olivine, chromite, carbon and chamotte sands are

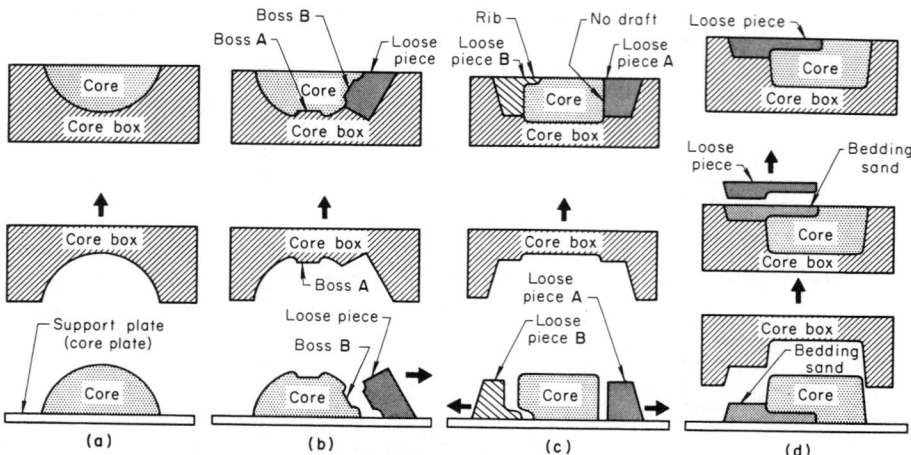

Fig. 21. Basic principles of coremaking. (See text for discussion.)

used also. Coarser sands permit greater permeability and are preferred for cores. Sand that contains more than 5% clay cannot be used for cores, because of the larger amount of additives required. In selecting a core sand, it must be considered that most of the sand from burned-out cores enters the molding-sand system and may well be a significant portion of the system sand. Under these conditions, a sand must be selected that will be suitable for both molds and cores.

Core binders most commonly used are of three general types: (a) binders that harden at room temperature, (b) binders that require heat to harden, and (c) clays. Cores often contain more than one binder.

REINFORCING OF CORES

Cores often must be reinforced to give them the necessary strength for handling and for withstanding the lifting force exerted on the core by metal under it in the mold. Cast iron weighs approximately $4\frac{1}{3}$ times as much as core sand. Hence, for iron castings, it is more often necessary to reinforce a core for the resistance to the lifting force of the liquid metal than for resistance to handling.

Small cores can be reinforced by rods or wires placed in the cores, but in large cores, where rods may not give sufficient strength, iron or steel frames of bolted, cast or welded construction are used. These frames are called core arbors. Figure 22 shows a cross-sectional view of a mold for casting a cored shell. Note the core arbor, wrapped with burlap, on which the sand has been rammed to form the core. Here the arbor is used to support the core in an upright position and to center it precisely and rigidly in the mold.

Whenever practical, an arbor should be used in large cores, because it eliminates the need for rods and pipe.

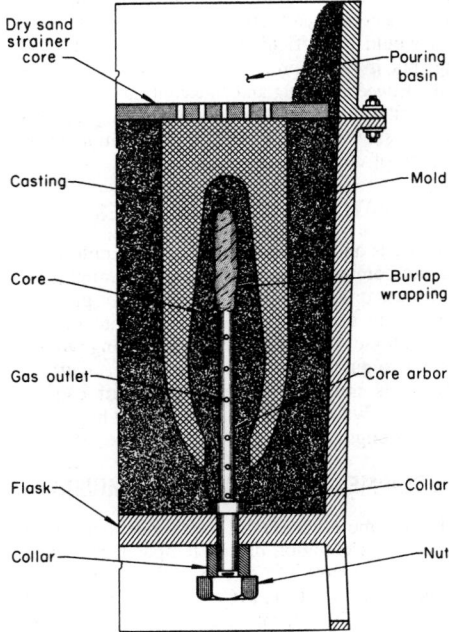

Fig. 22. Details of a mold-and-core assembly for making a cast iron shell, showing the use of a core arbor

CORE RAMMING

The degree of compactness necessary to produce a satisfactory core depends on the type of binder used and on the size and shape of the core. Core sands bonded with air-drying binders, such as sodium silicate or portland cement, must be rammed well to bring the sand grains into proper contact. Oil/sand core mixtures do not require hard ramming; many cores are made from oil/sand mixtures by jolting, blowing or squeezing.

Core blowing is done by filling the core-box cavity with sand that is suspended in a stream of air and introduced into the cavity at high velocity through blowing holes. The sand is retained by the cavity walls and the air is exhausted through vent holes in the core box and core plate. Core blowing is a high-speed operation; a core-blowing machine, under proper conditions, will deliver a blown core in a core box within $3\frac{1}{2}$ to $6\frac{1}{2}$ s, depending on the size of the machine. More time may be necessary before the core is removed from the box, depending on whether or not the binder system being used requires hardening or curing of the core while it is in the box. A typical setup for core blowing is shown in Fig. 23.

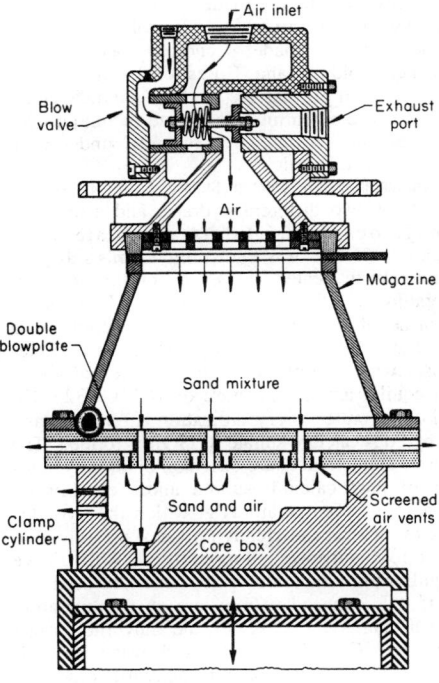

Fig. 23. Typical core-blowing setup

A pneumatic core-blowing machine is a self-contained unit, consisting principally of a sand reservoir and blowplate arranged to travel back and forth from the filling position to the blowing position. That mechanism is mounted on a carriage in the upper structure, which is supported by a frame and a base. The base houses the machine table and vertical clamping mechanism. In the filling position, the upper throat of the sand reservoir is open to receive sand from the hopper. After filling, the reservoir and blowplate are moved into position directly over the core box.

The machine table and core box are raised by a vertical clamp diaphragm, thus forming a seal between the sealing ring in the head of the ma-

chine and the upper throat of the reservoir, and between the blowplate and the core box itself. Usually, a working clearance of about 0.4 mm ($\frac{1}{64}$ in.) is allowed between the blowplate and the core box when adjusting the machine. Compressed air is applied when the elements of the machine are in this position.

Horizontal clamps, when used, function in unison with the vertical clamp. After the blow has taken place, the clamps are released, the table is dropped, and the reservoir is returned to the filling position. That sequence of operations usually is controlled by a combination of hand and foot operating valves. It also is quite common to use a fully automatic operating valve, which requires only a single motion from the operator to start the machine through its sequence of operations.

STOCK CORES

Cores made in quantities and stored for future use are called stock cores. Most stock cores are of uniform cross section, and are cut to required lengths. Round cores ranging in diameter from 9.5 to 75 mm ($\frac{3}{8}$ to 3 in.) are the most common, although almost any shape, of a size within a 75-mm-diam (3-in.-diam) circle, can be produced (see Fig. 24).

Many of the shapes shown in Fig. 24 are made to fixed lengths (usually 20 or 24 in.) in multiple core boxes on molding machines, but the most uniform cores are obtained by the continuous molding process, in which the core is formed by forcing sand through a die of the shape and size desired.

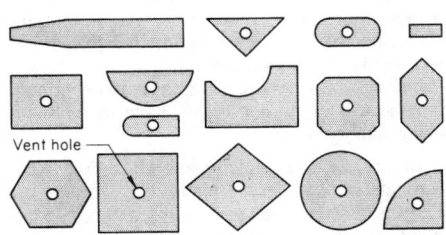

Fig. 24. Typical sectional shapes of stock cores produced as straight lengths

CORE BAKING

Cores are heated to remove moisture and to develop the strength of the binder.

Ovens. Various sizes and types of ovens are available for baking cores, the temperature being generally under 260 °C (500 °F). Core type and size, and the binders used, usually govern the type of oven selected.

Temperature and Time. The baking temperature and the length of time that a core must be heated to develop optimum properties depend on the following factors:

- Type of oven used
- Size of the core
- Moisture content of the core
- Type of binder
- Shape, fineness distribution, and permeability of sand grains
- Thermal conductivity, heat capacity, and density of the sand
- Atmospheric humidity.

Core size. Large cores require longer baking times than small cores.

Plaster Mold Casting

PLASTER MOLD CASTING is a specialized casting process used to produce nonferrous castings that have greater dimensional accuracy, smoother surfaces and more finely reproduced detail than can be obtained in sand molds or coated permanent molds. The four generally recognized plaster mold processes are:

1. Conventional plaster mold casting
2. Match-plate-pattern plaster mold casting
3. The Antioch process
4. The foamed-plaster process.

Applicability. Castings made in plaster molds are more expensive than those made in sand molds. Therefore, plaster mold processes are used only when acceptable results cannot be obtained by sand casting or another less-expensive casting process. Intricately designed impellers, components of electronics gear, and molds for some types of rubber tires are examples of parts that are produced by plaster mold casting. If these parts were cast in most other types of molds, machining costs would be exceedingly high.

For many parts, only a small portion of the total area requires the capabilities of plaster molding. For such parts, composite molds are more suitable (see the article on Composite Mold Casting).

Although about 90% of the plaster mold castings weigh less than 9 kg (20 lb) each, castings weighing as much as 34 kg (75 lb) are produced in substantial quantities, and some weigh considerably more then 34 kg (75 lb). An aluminum alloy casting weighing 1815 kg (4000 lb) has been successfully cast in a plaster mold.

Advantages. Most of the advantages to be gained from the use of plaster molds have to do with casting quality:

- Castings having especially smooth surfaces and intricate designs can be made.
- Dimensional accuracy of castings is good (about equal to the accuracy of castings made in investment molds).
- Because the castings normally cool quite slowly, thinner walls can be cast then by most other processes.
- Slow cooling minimizes warpage and promotes uniformity of structure and mechanical properties in the castings.
- Plaster molds are especially well-suited to the use of chills, which permits close control of thermal gradients in the molds.

Disadvantages in using plaster molds are largely related to molding and casting procedures:

- Cost is high, partly because of the lengthy processing procedures, and partly because the mold materials are not reusable.
- Mold processing requires more equipment than is required for most other molding processes.
- Because of the lengthy processing procedures needed to make plaster molds, duplicate (or multiple) pattern equipment, as well as duplicates of other processing equipment, may be required in order to meet production schedules.
- Permeability of plaster molds is inherently low. Thus, it is usually necessary to apply pressure or vacuum during pouring, or to provide greater permeability by a special procedure.
- Close control over every phase of the process is mandatory, in order to obtain good reproducibility.

Each of the disadvantages listed above is associated, to some extent, with all of the plaster mold processes, but the degree to which it is a limitation varies among the processes. The second item in the list, for example, is particularly true of the Antioch process. The fourth is less true of Antioch-process and foamed-plaster molds than of conventional molds; in fact, the Antioch and foamed-plaster processes are the "special procedures" developed to provide greater mold permeability.

PLASTER MOLD COMPOSITIONS

In all of the processes for making plaster molds and cores, the principal mold ingredient is calcium sulfate. Moldene talc, another important ingredient of plaster mold compositions, is a fibrous talc that serves to increase the strength of the mold in both the green and dry states, increase permeability, and help control expansion characteristics. Small amounts of two or more other ingredients are generally used to attain specific properties in the mold. Silica or zircon sands comprise up to 50% of the material for some plaster molds. Water is used for converting the dry mixture into a slurry.

Cores typically are made of the same material and by the same process as are molds, but sometimes cores are made of another material, such as shell molding sand. Cores for the match-plate plaster mold process are made of sand. Specific mold and core formulations, and techniques for making the slurries, are discussed under each process in this article.

Characteristics of Calcium Sulfate. Calcium sulfate exists as two different hydrates and also in an anhydrous form. In its dihydrate form ($CaSO_4 \cdot 2H_2O$), it is known by chemists as gypsum. When heated above an equilibrium temperature of 128 °C (262 °F), gypsum loses three-fourths of its water to form the hemihydrate ($CaSO_4 \cdot {}^1\!/_2H_2O$), which is plaster of paris. Likewise, when the plaster of paris is heated above an equilibrium temperature of 163 °C (325 °F), all of the water of crystallization is removed and anhydrous calcium sulfate ($CaSO_4$) is formed. The above temperatures are based on laboratory studies of pure calcium sulfate under equilibrium conditions. Temperatures at which the hydrates are formed depend greatly on vapor pressure. For practical purposes, temperatures well above equilibrium temperatures are generally used.

If it is exposed to moisture at a temperature lower than 100 °C (212 °F), the anhydrite formed near 163 °C (325 °F) will absorb water and will revert to the hemihydrate; the anhydrite produced by heating to about 540 °C (1000 °F) will absorb water less rapidly. Similarly, when plaster of paris (the hemihydrate) is mixed with water, gypsum (the dihydrate) is re-established, yielding a coherent solid in a few minutes. This reaction is called "setting." Prolonged exposure of the hemihydrate to moist air at temperatures below 100 °C (212 °F) results in a slow conversion to the dihydrate.

Commercially, all three forms of calcium sulfate are sometimes referred to as gypsum.

The transitions of calcium sulfate in response to water and to heat are the basis of all of the plaster mold processes. In the production of a plaster mold, plaster of paris is mixed with water to form a slurry. This slurry is immediately poured over a pattern or into a core box, where it sets, forming a solid mold or core composed of gypsum with free water distributed throughout it.

The next stage is to dry the plaster to remove the excess or free water. This is done in ovens. Various drying temperatures are used, depending on the plaster molding process. The most practical means of removing free water is to heat the gypsum for a sufficient length of time at some temperature between 100 and 128 °C (212 and 262 °F) — usually, at about 120 °C (250 °F).

After the free water has been removed, the heating may be continued to remove the chemically combined water at the surface or through the entire mold.

At the time the metal is poured, the plaster mold may be composed entirely of the dihydrate (although this is uncommon), or, at the opposite extreme, it may be completely anhydrous. Most often the composition ranges from the anhydrous form at the surface to the dihydrate form in the center of the mold section.

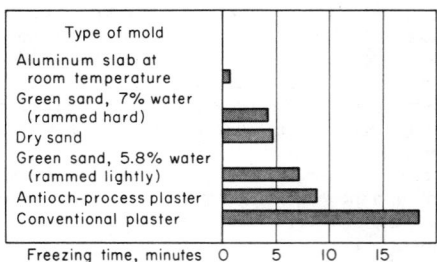

Fig. 25. Comparison of freezing times for identical test castings poured in various types of molds

Plaster molds have low heat capacity. Therefore, cooling rates for castings made in such molds are low. Figure 25 shows that the freezing time for a test casting made in a conventional plaster mold is more than four times as long as for one made in a hard-rammed green sand mold.

Slow cooling has both advantages and disadvantages. It permits better feeding, particularly in thin sections, and replication of intricate detail that would be difficult to obtain with fast cooling; but it slows production by lengthening the time between pouring and shakeout, and, for most alloys, it results in castings of lower strength. (This latter disadvantage can usually be eliminated by the use of chills in specific areas.)

PATTERNS AND CORE BOXES

Patterns and core boxes for plaster mold casting are commonly made of metal (aluminum alloy, brass or zinc alloy). Flexible rubber patterns are widely used for producing intricate molds. For example, in making molds for casting wheels having angular vanes, flexible rubber patterns having as much as 30° negative draft can be withdrawn without damaging molds that have high green strength.

EQUIPMENT FOR DRYING THE MOLDS

Plaster molds made by any process must be dried. Most common drying temperatures range from 120 to 260 °C (250 to 500 °F), but temperatures up to 870 °C (1600 °F) have been used under carefully controlled conditions. Once the optimum temperature for a specific application has been determined, it must be closely maintained — usually within ±6 °C (±10 °F).

After this initial requirement has been met, the choice of equipment depends largely on the temperature to be used and the number of molds to be processed. For drying at 120 to 260 °C (250 to 500 °F), either batch-type or conveyor core ovens are satisfactory.

Higher-temperature drying requires furnaces similar to those used for heat treating of steel. Either batch-type or conveyor furnaces can be used, but conveyor furnaces are generally preferred, especially for drying large quantities of molds. The molds must be cooled to at least 205 °C (400 °F) before they are removed from the furnace.

Cast epoxy resin patterns are useful for some purposes. For instance, when several patterns are required, they can be made quickly by molding them from epoxy resin in metal master dies.

Because patterns for production casting by plaster mold processes must withstand repeated exposure to liquid slurries, wood patterns usually are not adequate. However, they are used for producing a few castings; they are suitable, in particular, as patterns for match-plate pattern molds. Wood patterns must be sealed or they will swell and distort.

FLASKS

Flasks are usually made of low-carbon steel. They vary in size in accordance with the size of the pattern, the number of identical molds to be produced, and the number of patterns in a flask.

When the flask is to hold only one pattern half, especially when only a few molds are required, a simple, bottomless flask is placed on a mold board and the pattern half is positioned within it on the board, ready to receive the slurry. When many identical molds are required, especially when two or more patterns are placed in a single flask, a flask with a flat bottom that serves as a mold board is used; the pattern halves are arranged on the flask bottom. In the Antioch process, for small single-pattern molds, pattern and flask often are an integral unit.

The flask for a given application should be large enough to allow space for the same thickness of plaster surrounding the pattern as would be needed if the mold material were dry sand.

Standardization of flask sizes is important in production operations, because this simplifies tooling, especially when vacuum pouring is used. Standardization of flask size is also desirable for pouring the slurry and drying the molds. In some foundries, it is common practice to select as standard a flask size that is suitable for the largest casting that normally will be poured. When smaller castings are to be made, "family mold" practice is often used. In this practice, two or more patterns (sometimes as many as 16) are positioned in the flask.

METALS CAST

Only nonferrous metals are cast in plaster molds. Ferrous alloys are not suitable because most of them are poured at temperatures that would melt the calcium sulfate of which the molds are made. (Calcium sulfate undergoes a phase transformation at 1195 °C, or 2180 °F, and melts at 1450 °C, or 2640 °F.)

Aluminum. All of the aluminum alloys that can be cast successfully in sand molds are suitable for casting in plaster molds. The more readily castable alloys—43, A344, 355 and 356—are preferred.

Copper. Most of the coppers and copper alloys that can be cast successfully in sand molds can be cast in plaster molds; again, the more castable alloys are preferred (see the discussion on pour-

ing, near the beginning of the article on Production of Copper Alloy Castings in this volume). Copper alloys containing more than about 5% lead are not generally recommended for casting in plaster molds, because the higher-lead alloys react with some mold compositions, resulting in poor surfaces on the castings, defeating one of the objectives of plaster mold casting.

Magnesium alloys are not recommended for plaster mold casting. Reaction between magnesium alloys and the mold material is likely. In particular, magnesium alloys will react with any free water that remains in the mold, and will cause an explosion.

Zinc alloys are frequently cast in plaster molds, most often for prototype castings. The die-casting alloys AG40A and AC41A are often used, but a proprietary alloy whose coefficient of thermal expansion is very close to that of aluminum alloys is frequently cast. Master patterns appropriate for this zinc alloy or for aluminum can be made according to a single shrinkage rule.

Nominal compositions of the three zinc alloys mentioned are:

Alloy	Composition, %			
	Al	Cu	Mg	Zn
AG40A	4.10	0.10 max	0.035	Rem
AC41A	4.10	1.00	0.055	Rem
Proprietary	12.00	0.87	0.02	Rem

Except for allowances for shrinkage, the practice for making the molds is the same as for aluminum and zinc alloys.

Melting practice for metal to be poured into plastic molds is the same as that used for preparing the metal for pouring into other types of molds.

CONVENTIONAL PLASTER MOLD CASTING

A brief outline of conventional plaster molding is given in Table 3.

Table 3. Sequence of operations for producing conventional plaster molds

1 Mix dry ingredients
2 Add dry ingredients to water
3 Soak (2 to 4 min)
4 Mix (2 to 5 min)
5 Coat patterns (or core boxes)
6 Pour slurry
7 Set at room temperature (15 min)
8 Remove pattern
9 Dry molds (or cores)
10 Assemble cores and mold halves

Composition and Preparation of Dry Ingredients. Dry ingredients ready to mix with water are available as proprietary compositions. For small operations, purchasing these ready-to-mix dry ingredients is usually economical. Compositions vary considerably, but the gypsum content (typically, a blend of gray and white gypsum) is commonly 70 to 80%, by weight, of the dry mixture. Moldene talc (usually 20 to 30%) is added to control several other characteristics of the plaster. Small amounts of two or more other ingredients are usually added to control set time, mold strength, and expansion. (In the conventional process, no special procedures or ingredients are used to increase the permeability of the mold.) Lime, portland cement, high-strength gypsum cement, asbestos, and terra alba are some of the additives that are most frequently used. Table 4 lists the dry ingredients in a formulation that has proved successful in a foundry that pours a variety of

Table 4. Typical composition of dry material for conventional plaster molds

Ingredient	Weight	
	kg	lb
Metal casting base (gray gypsum)	45.4	100
White molding plaster	45.4	100
Moldene talc	34.0	75
High-strength gypsum cement	1.4	3
Asbestos	0.5	1
Terra alba	0.5	1

NOTE: Slurry is made by mixing 1 part of the above dry mixture with 1½ parts water (parts by weight).

aluminum alloy, copper, and copper alloy castings. Although several different formulations may produce acceptable results for a given application, once a foundry gets good results with a particular formulation, best reproducibility (of dimensions, in particular) is obtained by adhering to that formulation for all applications. Standard mixes are available.

Mixing the Slurry. Equipment requirements for mixing the slurry are more precise. The principal components of a batch-type mixer are a bucket and a propeller (see Fig. 26). The bucket height should be equal to or slightly greater than the top diameter, and the bottom diameter should be approximately two-thirds of the top diameter (see Fig. 26).

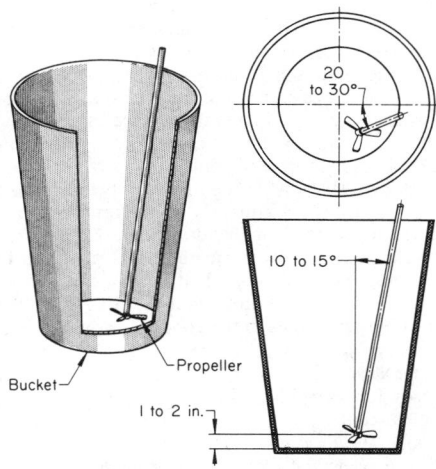

Fig. 26. Preferred shape of bucket and position of propeller for batch mixing of plaster slurries for conventional plaster molds

Preparing the Pattern. Because the slurry begins to set as soon as it is mixed, it must be poured over the pattern (or into the core box) almost immediately, and the pattern must be ready to receive the slurry.

Pouring the Slurry. When the slurry is poured into the flask, the lip of the bucket should be kept close to the pattern. The slurry should be poured at a constant rate and made to flow over the pattern rather than to splash on it.

Set time varies somewhat, depending on the slurry composition, but 15 min is usually the maximum set time. After the slurry has set, the pattern and flask are removed and the drying cycle is started as soon as is practical.

Drying the Molds. All conventional plaster molds must be dried enough to expel free water. In addition, it is usually desirable to remove the water of crystallization (that is, to calcine) to a depth of at least 13 mm (½ in.) below the surface. Some foundries prefer to have the molds completely calcined.

Oven drying should begin as soon as is practical after the mold has been removed from the flask. Molds that have become partly dried by standing at room temperature are more susceptible to cracking than those that are oven dried immediately after setting. If the molds must stand at room temperature for some time (such as over a weekend, or even overnight), they should be covered with a damp cloth or stored in a humid atmosphere.

During drying, the mold should be uniformly supported on its edge or face. Common practice is to place the mold on a perforated flat metal plate, a rigid metal grid, or some other type of level support. If smooth plates are used for support, they should be covered with a thin layer of talc or similar material, to allow movement of the mold during drying, without danger of its cracking or warping.

Time and temperature cycles used for drying conventional plaster molds vary widely among foundries. Temperature may vary from about 175 to 870 °C (350 to 1600 °F), and time from 45 min to 72 h. The fact that furnaces are operated at high temperatures (760 to 870 °C, or 1400 to 1600 °F) does not mean that all areas of the mold must reach this temperature range, but the interior of the mold must be at a temperature of at least 105 °C (220 °F), to ensure removal of the free water. (Mold temperatures are measured by thermocouples in the center of mold sections.) Specific time and furnace temperature required for drying a specific mold must usually be determined by experimentation. Once established, the time-temperature cycle must be rigidly controlled for best reproducibility.

Mold Assembly. After the mold has been allowed to cool to approximately room temperature, cores are placed in the drag half. Ceramic or plaster pins are placed in the holes provided in the drag half. The cope half is then lowered so that the pins protruding from the drag enter the matching pin holes in the cope, as shown in Fig. 27.

Preheating. Following assembly, the mold is ready for preheating. Some foundries preheat all conventional plaster molds to a pre-established temperature (commonly, 120 °C, or 250 °F) before pouring the metal. Other foundries preheat

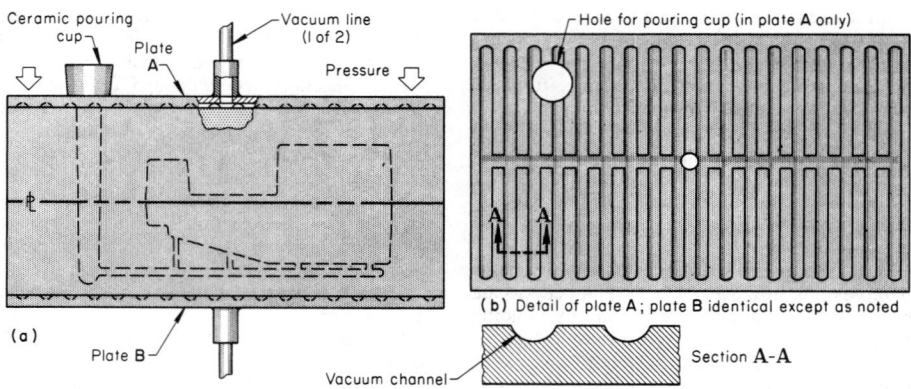

(a) Side view of conventional plaster mold positioned between upper and lower plates for pouring with vacuum assist. (b) Details of a top plate, showing vacuum channels.

Fig. 28. Typical setup for vacuum-assist pouring of a conventional plaster mold

the molds only in specific applications for which preheating has proved advantageous. Preheating of the mold can help to minimize defects or to obtain better replication of fine detail in the casting.

Pouring Practice. A dried plaster mold made by the conventional process has extremely low permeability—about 1 to 2 AFS, compared with 80 and upward for sand molds. Because of this low permeability, either vacuum assist or pressure is usually required for pouring of molds. Simple gravity pouring is unusual.

Most foundries use vacuum assist for pouring conventional plaster molds. A typical setup for vacuum-assist pouring is shown schematically in Fig. 28(a). In this setup, the mold is supported by a fixed bottom plate (plate B in Fig. 28a).

PRACTICE FOR CASTING OF MATCH-PLATE PATTERNS

Match-plate patterns are cast by a particular adaptation of the conventional method of plaster molding. Changes in details of the conventional method ensure high accuracy and smooth surface finish, which are required of metal match-plate patterns.

Common dimensional requirements for match-plate patterns are: match between the cope and drag, within 0.25 mm (0.010 in.); and parallelism between the two halves, within 0.5 mm (0.020 in.). Surface finish of 2.5 μm (100 μin.) or better is obtained.

Sizes and Weights of Match-Plate Patterns. A match-plate pattern consists of a cope section and a drag section separated by a plate. These three components are cast as an integral unit. A typical match-plate pattern is shown in Fig. 29.

Metal match-plate patterns vary considerably in size and weight. Patterns weighing as much as 180 kg (400 lb) each have been cast, although this is unusually large. Match-plate patterns sel-

dom weigh more than 45 kg (100 lb) each, and the majority weigh from 8 to 16 kg (18 to 35 lb).

A single plate may hold one pattern or many patterns. Thickness of the plate for small patterns having one pattern per plate is usually 9.5 mm (3/8 in.). The plate thickness increases as the size and weight of the pattern increase: a thickness of 19 mm (3/4 in.) would be used for a plate 1.1 by 1.1 m (44 by 44 in.). Plate thickness is also increased to provide adequate stiffness and bend resistance in designs involving a stepped parting line separating cope and drag mold sections.

Mold Composition. Materials for molds for match-plate patterns are available as proprietary dry mixtures ready for mixing with water. However, some foundries prefer to make their own mixtures. A typical mixture for match-plate pattern molds is given in Table 5.

A comparison of Tables 4 and 5 shows a significant difference between the compositions of conventional plaster molds and those of molds for match-plate patterns. In conventional molds, both white and gray gypsum are used. In match-plate-pattern molds, only a white molding plaster (a pure grade of gypsum) is used. This ensures a smooth mold finish, which, in turn, produces the smooth surface required on match-plate patterns.

Table 5. Typical composition of dry material for molds for metal match-plate patterns

Ingredient	Weight	
	kg	lb
White molding plaster	45.4	100
Moldene talc	18.6	41
Hydrated lime	0.7	1.5
Portland cement	0.2	0.5

NOTE: Slurry is made by mixing 1 part of the above dry mixture with 1.65 parts water (parts by weight).

Master patterns for match-plate patterns are usually made of wood (for economy), but they can be made of metal. Wood patterns must be lacquered or otherwise coated, to prevent absorption of water from the slurry.

Separate master patterns are made for the cope and the drag. The plate portion is developed by the technique described under "Mold Assembly," below, rather than by means of a pattern.

Before the slurry is poured, the patterns are coated with a release agent.

Flasks for casting of match-plate-pattern molds are bottomless boxes. They differ from flasks for

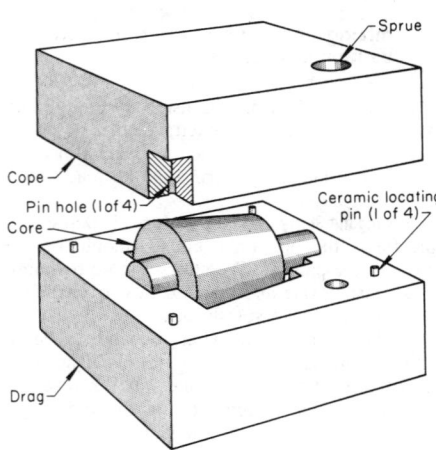

Fig. 27. Plaster mold and core, showing locating pins for matching the cope and drag sections

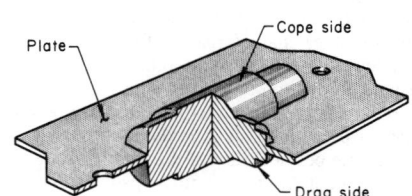

Fig. 29. A typical metal match-plate pattern

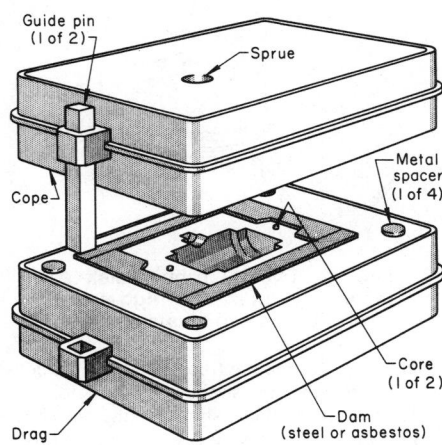

Fig. 30. Assembly details of a match-plate mold. (See text for discussion.)

conventional plaster mold casting in two respects. First, standardization of flask size is seldom feasible in match-plate-pattern mold work, because pattern sizes vary widely. But standardization is less important, because molds are not required in production quantities. Second, because the mold remains in the flask through the drying and metal-pouring operations, provision for matching the cope and drag sections must be incorporated in the flask, rather than in the pattern as in conventional operations. Use of guide pins for the matching of mold sections is indicated in Fig. 30.

Mixing and Pouring the Slurry. Equipment and procedures for mixing the slurry and pouring it over the patterns are essentially the same as for the conventional process (see Fig. 26 and accompanying text). Common practice is to make several vents in the mold with a nail or wire immediately after the slurry has been poured. The vents serve two purposes. Vent holes traversing the entire mold sections of dry molds provide openings for the escape of steam and other gases when metal is poured. Through-vents in wet molds facilitate separation of mold halves and removal of patterns, by acting as channels for injection of compressed air.

Set Time. A slurry of the type defined in Table 5 will set in 14 to 16 min. After the slurry sets, the patterns are removed and the molds are dried.

Drying. Molds for match-plate patterns, like molds for conventional plaster castings, should be dried as soon as possible after the plaster has set. High-temperature drying cannot be used for match-plate molds, because it results in unacceptable distortion and size change. Match-plate molds are usually dried at 120 to 205 °C (250 to 400 °F) for 12 to 72 h. Size and section thickness of the molds determine the length of time in the drying ovens. The center of the thickest section of the mold should reach at least 105 °C (220 °F) before drying is stopped. Permeability of a match-plate mold at this stage is approximately the same as that of a conventional mold (1 to 2 AFS).

Mold Assembly. After the mold halves have reached room temperature, the cope and drag sections are matched as shown in Fig. 30. Assembly of a match-plate mold is significantly different from that of a conventional plaster mold. Cores, if used, are positioned in the drag mold section. (Cores are used if holes are

required in the plate. Internal cores are used only in large match-plate molds, so that the patterns will be lighter and can be handled more easily. Cores are made of sand.) The cope section, aligned by guide pins, is lowered onto the drag (Fig. 30). It is then raised for a distance equal to the desired thickness of the plate. Metal spacers of this thickness are inserted at each corner, and dams of steel or asbestos are placed so as to form the desired outer contour of the plate (Fig. 30). When the metal is poured, it flows outward in the space between the cope and drag sections and forms the plate portion of the match-plate pattern.

Metals Used. Match-plate patterns are cast from aluminum alloys, most frequently alloys 355 and 356. One foundry using a blend of equal quantities of 356 and 319 reported greater ductility and better machinability than could be obtained if either alloy were used alone. Ductility is important in making match-plate patterns, because patterns often require straightening.

Pouring Practice. Because of the low permeability of a match-plate mold, some assistance is required to fill the mold quickly and completely. Pressure assist is used rather than vacuum, in part because pressure equipment is more adaptable to a variety of flask sizes than is vacuum equipment.

The equipment for pressure casting is illustrated in Fig. 31. The first step in its use is to place a diaphragm of sheet asbestos (about 3.2 mm, or 1/8 in., thick) over the sprue in the mold. A ceramic-lined cylinder is then placed on the diaphragm, as shown in Fig. 31. A predetermined amount of metal is poured into the cylinder, and a cap is placed on the cylinder and clamped tight. The cap is attached to a source of compressed air. After the cap has been secured in position, the air valve is opened. Air pressure against the molten metal ruptures the asbestos diaphragm and forces the metal into the mold cavity. A pressure of 10 to 17 kPa (1 1/2 to 2 1/2 psi) is kept on the sprue for about 1/2 h.

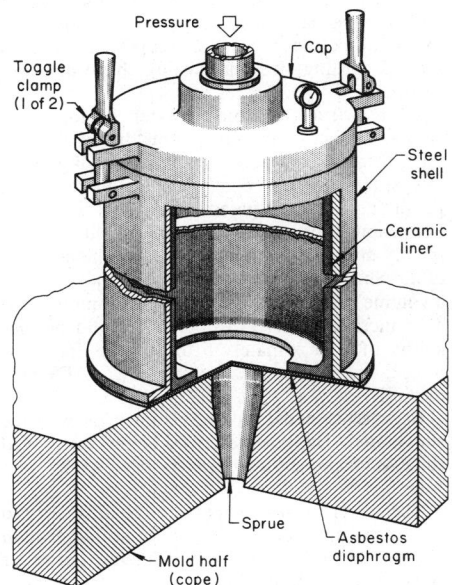

Fig. 31. Cylinder used for pouring a casting with pressure assist. (See text for discussion.)

THE ANTIOCH PROCESS

The Antioch process (U.S. Patent 2 220 703) was developed to overcome the principal limitations of conventional plaster molds and cores, without sacrificing the advantages of the plaster mold process.

If undried molds are partially dehydrated and then allowed to rehydrate without being disturbed, gypsum crystals slowly recrystallize into granules about the size of sand grains, and the mold acquires a porous structure of relatively high permeability. Permeability is held within a range of 15 to 30 AFS, compared with 1 to 2 AFS for conventional plaster molds. Recrystallization does not take place at the surface of the mold, because not enough water is present. Therefore, the surface remains smooth.

In addition to the greater permeability developed by the dehydration-rehydration process, the molds produced have greater heat capacity than conventional plaster molds, because they are composed of approximately 50% sand. Figure 25 shows that the freezing time for a casting in an Antioch-process mold is only about 20% longer than the freezing time for an identical casting in a lightly rammed green sand mold, and less than half the time required for a casting in a conventional plaster mold.

Unlike conventional molds, Antioch-process molds do not shrink. In fact, they expand slightly—0.001 to 0.0025 in./in.—during processing.

Because of their porous structure, the molds have low dry strength. This characteristic, in promoting early collapse of cores as the casting cools, minimizes hot tears in the castings. For very large molds, the low dry strength sometimes necessitates the use of internal reinforcement, which is achieved with hardware cloth or core rods such as those used in making sand cores. When possible, reinforcement is avoided, because of the difference in expansion between the reinforcing metal and the molding material.

After setting, but before the dehydration-rehydration treatment, Antioch-process molds have relatively high green strength. When flexible rubber patterns are used, this high green strength permits withdrawal of patterns having a severe back draft, without damage to the mold. This makes the Antioch process particularly well suited to the production of molds for parts having angular, bladelike sections—rotors and nozzles, for example.

In addition to the cost, which is high for all plaster molds, the major disadvantage of Antioch-process molds is the long time required for processing them. This ties up expensive equipment for long periods.

Mold and Core Materials. The dry mixture for Antioch-process molds and cores consists of silica sand, white molding plaster, moldene talc, and a small amount of material, such as portland cement, for expansion control. The typical mixture given in Table 6 varies somewhat among different foundries. However, once a formulation has been established in a specific foundry, it is retained for all castings, regardless of their size and shape. Best results are obtained by weighing all ingredients accurately. Only by consistent use of a specific formulation is it possible to obtain maximum reproducibility.

Processing. The sequence of operations for producing Antioch-process molds is given in Table 7. A comparison of Tables 7 and 3 shows that

Table 6. Typical composition of dry material for Antioch-process molds

Ingredient	Weight kg	lb
Washed silica sand (AFS 50 is typical)	22.7	50
White molding plaster	19.1	42
Moldene talc	3.4	7.5
Portland cement	0.2	0.5

NOTE: Slurry is made by mixing 45 kg (100 lb) of the above dry mixture with 24 kg (54 lb) of water.

Table 7. Sequence of operations for producing Antioch-process molds

1 Mix dry ingredients
2 Add dry ingredients to water
3 Soak (1 to 3 min)
4 Mix (2 to 4 min)
5 Coat patterns (or core boxes)
6 Pour slurry
7 Set at room temperature (15 to 20 min)
8 Remove pattern
9 Dehydrate in autoclave (6 to 12 h)
10 Rehydrate in air (14 h)
11 Dry molds (or cores)
12 Assemble cores and mold halves

the chief difference in making Antioch-process molds and conventional plaster molds is in the dehydration and rehydration treatments (steps 9 and 10 in Table 7).

Set time for a slurry formulated from a composition such as that shown in Table 6 will be approximately 15 to 20 min. Set time can be decreased by adding up to 3% terra alba and heating the water. For instance, the minimum set time of 6 to 7 min is achieved by adding 3% terra alba and using water at 32 °C (90 °F). Temperature and humidity of the surrounding atmosphere have very little influence on set time, although an atmospheric temperature of 21 to 27 °C (70 to 80 °F) is preferred.

Dehydration. The time between setting of the slurry and the beginning of the dehydration cycle is not extremely critical if steps are taken to prevent the mold from drying out. If the set molds are covered with damp cloths, they can be held overnight, or sometimes even over a weekend, without significant effect on subsequent dehydration. If the molds are placed in humidity cabinets, they can be stored for longer periods before dehydrating. However, the dehydration cycle should begin soon after the pattern is removed if the mold cannot be kept moist.

For dehydration, the molds are placed on suitable racks in a standard autoclave. The autoclave is sealed, and steam is admitted. The autoclave is operated with a steam pressure of 105 kPa (15 psi) for 6 to 8 h. For extremely large molds, it is operated for 12 h. The autoclave is then opened, and the molds are removed.

Rehydration. The mold is permitted to remain at room temperature for 14 h. After rehydration, the mold is ready for drying.

Drying temperature ranges from 175 to 230 °C (350 to 450 °F), and drying time from 1 to 70 h. Drying time depends mainly on the size of the mold and the temperature used. The center of the mold must reach a temperature of at least 120 °C (250 °F). This can be accomplished considerably more quickly at an oven temperature of 230 °C (450 °F) than at 175 °C (350 °F).

Regardless of the cycle used, it is important that the same cycle be used for all molds of the same size. Only by close control of the cycle can maximum reproducibility (of dimensions, in particular) be achieved.

Mold assembly is essentially the same as described for conventional plaster molds. After the molds have cooled to room temperature, cores (if used) are placed in the drag and the cope is placed over the drag-and-core assembly. Matching is done by means of locating pins. Pins used for matching Antioch-process molds are usually 13 to 19 mm (¹/₂ to ³/₄ in.) in diameter. Even when molds are permitted to remain in their flasks, guide pins on the sides of the flasks are seldom used for matching.

Metals Cast. All of the aluminum alloys that can be cast in other types of plaster molds can be cast in Antioch-process molds. Most copper-base alloys can be cast in Antioch-process molds. Yellow brass is the copper alloy most often cast. The Antioch process is seldom used for alloys that must be poured at temperatures above about 1040 °C (1900 °F).

Pouring Practice. It is generally possible to pour castings in Antioch-process molds by gravity, using gating systems that are similar to those used for sand molding. Molds are usually at room temperature when pouring begins.

Where difficulty is encountered in replicating fine detail, or where thin sections have not filled properly, vacuum assist can be applied. The technique is generally the same as for pouring conventional molds with vacuum assist.

FOAMED-PLASTER MOLDS

The foamed-plaster process offers a means for obtaining greater mold permeability than can be obtained in conventional plaster molds. This gain is achieved by adding a foaming agent, such as alkyl aryl sulfonate, either to the dry ingredients before mixing or to the liquid slurry, as a separately generated foam mix. A special method of mixing foams the slurry with many fine air bubbles, thereby decreasing the density and increasing the volume of the slurry.

In general, the applicability of foamed-plaster molds is the same as that of plaster molds made by other procedures, as far as composition of metal poured, casting size and casting shape are concerned.

Characteristics. Foamed-plaster molds have smooth surfaces with air cells just below the surface. During setting and subsequent drying of the molds, these air cells become interconnected, thus permitting escape of gases as the metal is poured. Permeability of a foamed-plaster mold depends mainly on the volume increase from the addition of air when the slurry is mixed. For most molds, a volume increase of 50 to 75% is recommended. This increase usually results in a mold permeability of approximately 5 to 15 AFS for dried molds; many foundries get 20 to 25 AFS permeability.

Equipment for mixing will vary somewhat with the slurry used. The type described below has proved suitable for mixing a proprietary dry mixture with water.

The mixer must be capable of beating air into the slurry and producing air cells no larger than about 0.25 mm (¹/₁₀₀ in.) in diameter. Large air cells are not permitted, because they break under pressure from molten metal, resulting in casting defects. Proper mixing can be accomplished with several types of mixers. Regardless of the type of mixer used, the greater the power input, the

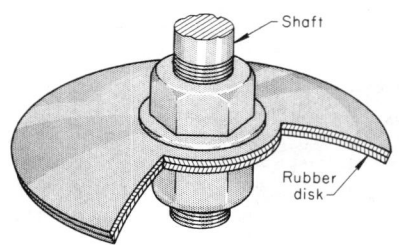

Fig. 32. Two-ply rubber disk attached to shaft for high-speed mixing of foamed-plaster slurries

finer the mold structure (the smaller the air cells) and the smoother the surface of the mold.

The bucket should be similar to the one shown in Fig. 26, but the mixing device is a round, two-ply, 3-mm-thick (¹/₈-in.-thick) rubber disk (which can be made from 3-mm two-ply rubber belting) attached to a shaft, as shown in Fig. 32. Diameter of the shaft is not critical.

Composite Mold Casting

COMPOSITE MOLDS are complete molds assembled from several components among which at least one component varies from the others in the process by which it was made and in the molding material.

The goal in the use of composite mold assemblies is to derive the specific advantages of each molding material without being penalized by its disadvantages. The use of three or four different types of mold or core materials in one mold assembly is common, and the use of as many as seven different types of mold materials in a single mold assembly has been reported.

Mold and Core Combinations. Any specific component of a composite mold assembly is usually made of one of the following materials or products:

* Silica sand with air-set binder
* Zircon sand with air-set binder
* CO_2 sand (sand bonded with sodium silicate and treated with carbon dioxide)
* Shell (sand bonded with resin)
* Plaster or a mixture of plaster and sand
* Permanent mold (metal).

The most common combinations are:

* Permanent mold – plaster
* Shell – plaster
* Permanent mold – shell – plaster
* Sand with air-set binder – plaster
* CO_2 sand – plaster
* Shell – permanent mold.

Metals Cast. Because of demand, composite molds are mainly used for casting of aluminum alloys. Aluminum alloy castings made in composite mold assemblies are variously known by terms such as "premium-quality castings" and "engineered castings."

Many aluminum alloy castings now made in composite mold assemblies are of the type formerly made in all-plaster molds because of the capabilities of the plaster mold process to fulfill precise requirements for dimensions, surface finish and thin walls.

However, because of the disadvantages of

plaster molds (poor heat transfer, low permeability and high cost), a primary objective is to restrict the use of plaster to portions of the mold assembly where it is essential. Other materials are then used for portions of the mold where dimensional accuracy and surface-finish requirements are less stringent or where faster cooling rates (to develop specific mechanical properties) are required.

REASONS FOR USE OF COMPOSITE MOLDS

Composite molds are used for one or more of the following reasons:

• Decreased cost of mold material
• Decreased processing time
• Increased accuracy of castings
• Improved casting finish
• Increased strength (often in specific areas)
• Decreased amount of evolved gas.

Decreased Cost of Mold Material. A wide variation in cost exists among the various mold materials; generally, the greatest dimensional accuracy and smoothest surface finish are obtained only by the use of the most expensive materials, such as plaster. For small castings the cost of the mold material is less important, but for larger castings the cost of the mold material contributes significantly to over-all cost. Partly because of the high cost of material, all-plaster molds are seldom used when the total volume of the mold exceeds one cubic foot, and often all-plaster molds having a volume of much less than one cubic foot have been replaced by composite molds.

For most castings, stringent requirements for accuracy and finish do not prevail over the entire surface of the casting—sometimes because some as-cast surfaces are noncritical, and sometimes because one or more surfaces will be subsequently machined. Under these conditions, a less costly molding material is used for the noncritical areas.

Decreased Processing Time. For some castings, the time required for processing of mold components has a larger influence on the cost of the casting than does the material cost. Large mold components can be made from shell, air-set sand or CO_2 sand in two hours or less, whereas many hours are required for the processing of plaster mold components.

Increased Accuracy of Castings. Mold components made from any of the materials listed under "Mold and Core Combinations," above, can produce castings to a higher degree of dimensional accuracy than can be achieved with green sand molds. Plaster components, however, are capable of producing somewhat greater accuracy than the others. Therefore, plaster components are used for those portions of a mold where the ultimate in accuracy or replication of fine detail is required.

Improved Casting Finish. Mold components made from the six materials listed under "Mold and Core Combinations" are capable of providing castings that have smoother surfaces than can be obtained in green sand molds; plaster components generally provide the smoothest surfaces. Therefore, the use of composite molds permits attainment of the smoothest surfaces where needed without incurring a penalty in cost for the same degree of smoothness where it is not needed.

Increased Strength of Castings. The strength of aluminum alloy castings is affected significantly by mold materials because of the differences among

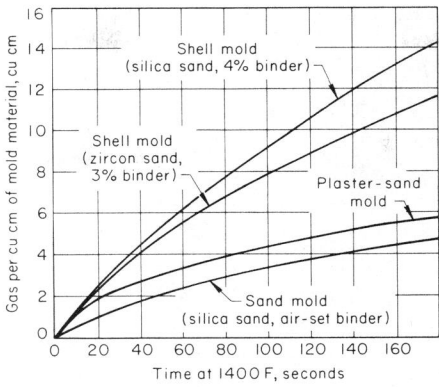

Fig. 33. Amount of gas evolved by four mold materials, as a function of time at 760 °C (1400 °F)

these materials in rate of heat absorption. Whereas plaster components can provide the greatest accuracy and smoothest surface finish, they are, when used without chills, the poorest mold material for heat absorption and may result in cooling rates too slow to develop required strength in the castings.

Decreased Amount of Evolved Gas. Evolved gases sometimes cause defective castings. Because there are considerable differences among the amounts of gas evolved by different mold materials within a given time, this is sometimes a factor in the selection of material for a specific mold component (most often, the core material).

Data on the rate and amount of gas evolution of four mold materials at 760 °C (1400 °F) are given in Fig. 33. Here it can be seen that within 160 s only about one-third the amount of gas is evolved from silica sand with an air-set binder as from silica-sand shell material with 4% binder. Thus, if internal porosity in castings is a problem with shell cores, changing to plaster or air-set sand cores would be one approach to correcting the difficulty.

SELECTION OF MOLD MATERIALS

Selection of materials for the components of any composite mold is based initially on meeting requirements of dimensional accuracy, surface finish, strength and soundness.

The quantity of castings required is the second basis for selection of mold materials. For prototype work or very limited production, every attempt is made to use mold and core materials that require minimum tooling—for instance, CO_2 or air-set sand. If quantity requirements increase, materials of mold components may be changed to permanent (metal) or shell; either of these involves increased tooling cost.

Sometimes, however, side considerations restrict flexibility in material selection. For example, there are many applications in which it is preferred to glue plaster cores in their green state to another mold component. This means that the material in the other component must withstand the processing cycle used for the plaster cores. Shell material can withstand such a cycle, whereas most other mold materials cannot. Under these conditions, therefore, shell will be used instead of CO_2 or air-set sand, regardless of other factors.

The practical use of composite mold casting is illustrated by the two examples which follow.

Impeller castings with a trimmed weight of 19.5 kg (43 lb) each (38.5 kg, or 85 lb, of metal poured for each casting) were originally poured from aluminum alloy C355 in a composite mold comprised of a permanent mold drag, a plaster cope (with metal chills), and plaster cores. The plaster cope and the cores were made by the Antioch process. Acceptable castings were made in this mold, but because the cope was massive, the cost of the plaster material was high and the time required for processing the mold was excessive (far longer than was required for the plaster cores).

With an improved composite mold (Fig. 34), the permanent mold drag and plaster cores were retained, but the cope was changed to air-set silica sand (with metal chills), which could be prepared in 2 h or less. The improved mold provided the required surface smoothness, dimensional accuracy and mechanical properties. In both the original and the improved cope, fiber glass was used to insulate the six risers.

Torque Converter. Originally, a torque converter was cast from aluminum alloy C355 in a mold assembly comprised entirely of plaster components (Antioch process), with metal chills being used in the drag. Then, with no change in the position of the cavity in the mold, three changes were made in mold material, resulting in the composite mold shown in Fig. 35. In this composite mold, the drag was made as two components; the outer component of the two was metal (permanent mold), and the inner component was shell molded. The one-piece cope was also shell molded.

Production quantities warranted the tooling cost for shell components. However, regardless of quantity requirements, neither CO_2 nor air-set sand would have been considered for the inner component of the drag, because it was necessary to glue the individual plaster cores to the inner drag component while the cores were still green, processed cores being too fragile. (CO_2 and air-set sand components cannot withstand the processing cycle required for Antioch cores, whereas shell molded materials suffer only a slight loss of strength in the processing cycle used for the Antioch-process cores.)

Acceptable castings were produced in both the all-plaster mold and the composite mold. In the composite mold, however, the permanent mold outer drag component not only provided the required cooling rate and reduced the amount of expendable mold material, but also saved the processing time formerly required for the plaster and saved the labor cost for placing the metal chills in the drag. Changing from a plaster to a shell cope saved the long processing time required for the former plaster cope. Likewise, processing time was greatly reduced by changing the inner drag component to shell material. The shell and permanent mold components provided acceptable accuracy and surface finish on the portions of the casting that were shaped by these mold materials.

Cleaned castings weighed 10 kg (22 lb) each; 12 kg (27 lb) of metal was required to pour each casting.

ASSEMBLY OF MOLD COMPONENTS

Because composite molds are used mainly for producing castings to precise specifications, assembling the mold components requires more care and more numerous inspection operations than does assembling the components of a more conventional mold, such as a green sand mold.

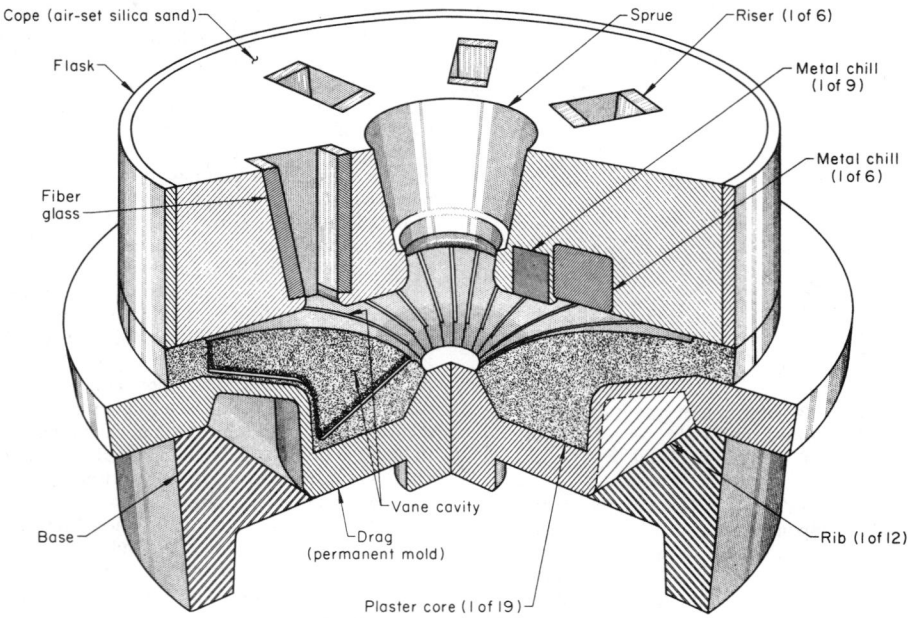

Changing to the air-set sand cope from a plaster cope saved material cost and processing time.
Fig. 34. Composite mold (permanent mold drag, air-set sand cope, plaster cores) for an impeller cast from aluminum alloy C355

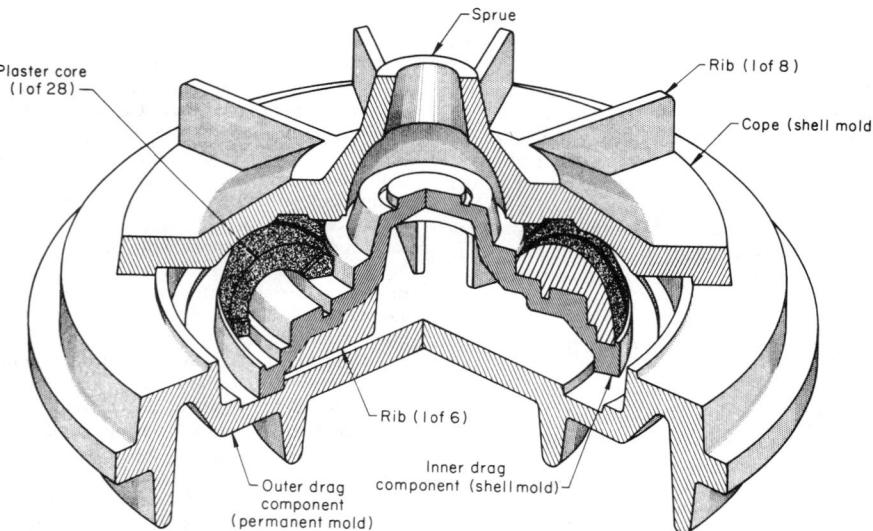

Fig. 35. Composite mold (permanent and shell drag, shell cope, plaster cores) for an aluminum alloy C355 torque converter previously cast in an all-plaster mold

For instance, permanent mold (metal) components of composite assemblies, although similar to those used in conventional permanent mold casting, require closer attention. They are inspected each time they are assembled with other components, so as to keep close control of the mold coating.

Drying. Some mold components, such as those made of CO_2 sand or sand with an air-set resin binder, may contain moisture at the time of assembly. Only experience under specific foundry conditions and with similar assemblies can indicate whether or not drying just before assembly is required. For some castings, a slight mold blow caused by moisture is of no consequence, but for others minute blows are detrimental. To ensure freedom from moisture, the components are dried immediately before assembly.

Mismatch between cope and drag sections of composite molds is sometimes controlled by guide pins in a flask (the same as in green sand molding). However, guide pins are not accurate enough for most applications. A more frequently used method for preventing mismatch is to mold accurately located metal bushings into both the cope and drag sections. When the mold components are assembled, metal pins are inserted in the bushings of the drag, and then the cope is placed on the drag so that the pins also enter the bushings in the cope.

Damage and Repair. Before assembly, mold components are inspected for broken corners or other damage they may have suffered in handling. When a component has been damaged, it is a matter of judgment whether it will cost less to make repairs or to replace the damaged component.

Investment Casting

INVESTMENT CASTING uses a mold that has been produced by surrounding an expendable pattern with a refractory slurry that sets at room temperature. The pattern (usually of wax or plastic) is then melted or burned out, leaving the mold cavity. Investment casting is also known as the "lost-wax process" and as "precision casting."

In sand casting, wood or metal patterns are used to make the impression in the molding material. The pattern can be reused, but the mold is expendable. In investment casting, a metal pattern die is used to produce the patterns, which, in turn, are used to produce ceramic molds. Both the patterns and the molds are expendable. Ceramic cores are used, as required, and these also are expendable.

Advantages of the process are:

- The process permits mass production of complex shapes that are difficult or impossible to produce by the more conventional casting processes or by machining.
- The mold materials and the technique used permit reproduction of fine detail, greater dimensional accuracy, and smoother surfaces than can be obtained by other processes.
- The process is adaptable to almost any metal that can be melted and poured. Castings containing more than one metal, such as an integrally cast steel hub in a copper rotor, are also feasible.
- With modification of the basic process, production of castings weighing up to 23 kg (50 lb) is not uncommon. The production of very large castings weighing as much as 450 kg (1000 lb) is sometimes feasible, although such castings are rarely produced by the investment method. (For typical weights, see *Investment Casting Handbook,* Investment Casting Institute, Chicago, 1968.)
- Castings can be produced that require little or no finishing for completion, thus minimizing the importance of selecting easy-to-machine metals.
- The process permits close control of metallurgical properties such as grain size, grain orientation, and directional solidification, which results in close control of mechanical properties.
- The process is adaptable to melting and casting of alloys that must be poured in a vacuum or under protection of an inert atmosphere.
- The castings produced do not vary in dimensions across parting lines, as do castings made by other processes. In fact, parting lines are normally eliminated because of the monolithic molds used in investment casting. (When a "parting line" shows, it has been caused by flash on the pattern that was not removed before molding.)

Limitations of the process include:

- Size and weight of castings that can be produced by the process are usually limited by physical and economic considerations, as well as by the capacity of available equipment. In general, the process can be applied most advantageously to castings weighing not more than 4.5 kg (10 lb).

- The initial tooling costs for larger castings (4.5 to 23 kg, or 10 to 50 lb) are generally high.

PROCESS CLASSIFICATION

Two distinct processes, differing in method of mold preparation, are used in the production of investment castings—namely, the shell investment process and the solid investment process. In general, the two processes do not differ appreciably, if at all, in pattern preparation or pattern assembly. However, patterns for the shell process are always precoated, whereas precoating of patterns for the solid process is generally not required unless the properties of the backup refractory are inadequate for the specific application. Precoating methods for both processes are similar: the pattern is dipped in a fine slurry, and a granulated refractory is applied by sprinkling or some other suitable method.

In the shell investment process, after precoating, the pattern assembly is alternately dipped in a coating slurry and "stuccoed" with granulated refractory, either by sprinkling or by suspending in a fluidized bed, until the shell is built up to desired thickness. Usually, the refractory grain ranges in size from 20 to 100-mesh, the fine material being used for the initial coat and progressively coarser grains for subsequent coats. Each coat of slurry and grain is air dried before subsequent coats are applied.

In contrast, in the solid investment process, the pattern assembly (precoated, if it is to be used for casting alloys with a melting point above 1100 °C, or 2000 °F) is encircled by a flask, which in turn is filled with a refractory mold slurry. The mold slurry, called a "backup slurry," hardens in air, forming a solid mass in which the pattern assembly is encased.

The steps in making a casting in shell and solid investment molds are shown schematically in Fig. 36 and 37 and are considered in detail in subsequent sections of this article.

Molds made by the two processes are not necessarily equivalent. Therefore, both economic and technical factors must be considered in deciding between the two processes. Melting, pouring and finishing operations are generally the same for both molding processes.

PATTERN DIES

For extremely limited production, and in the development of procedures for high production, patterns for investment molding can be machined from an expendable material, usually a plastic such as polystyrene. When many castings are required, however, the patterns are produced by injecting the expendable wax or plastic material into cast or (more commonly) machined pattern molding dies. The dimensional tolerances of these dies (including allowances for shrinkage in the pattern-die material, pattern material, and the metal to be investment cast) are closely controlled.

Machined Pattern Dies. Cavities can be machined directly, without the need for a master pattern, although a master pattern sometimes is used to aid in reproduction when machining is done with tracing equipment. Methods used for machining

the cavities are essentially the same as those used for diesinking.

Selection of material for pattern dies depends mainly on the pattern material that will be used, injection pressures (largely dictated by the pattern material), and quantity of patterns to be produced.

For producing patterns from liquid or semisolid waxes, using pressures no greater than about 2.4 MPa (350 psi), pattern dies can be cast in bismuth-tin alloy, zinc alloy, or some other metal that has a relatively low melting temperature and is easily worked. These pattern dies are sometimes feasible for production of small quantities of patterns. If pattern dies are produced by machining, a low-carbon steel (unhardened), an aluminum alloy, a magnesium alloy or a brass is used. Machining of pattern dies is most feasible when larger quantities of patterns are to be produced, especially if they are of complex shape.

For pattern materials that require high injection pressures (solid waxes and plastics), the pattern die is in effect a plastic mold. Thus, the die materials used are generally the same as those used for conventional molding of plastics, depending primarily on production quantity requirements. Mold steels such as P20 and P21 in the prehardened condition (300 HB or slightly higher) are commonly used for up to 100 000 patterns. When they are in the prehardened condition, these steels can be machined and put in service without heat treatment, thus eliminating distortion. When a larger number of patterns is to be produced and longer pattern-die life is required, the dies can be made from P6 or P20 mold steel, carburized and hardened to 54 to 58 HRC, or from a tool steel such as O1, hardened to 53 to 57 HRC.

WAX PATTERNS

Wax patterns are prepared by injecting liquid or semisolid wax into a pattern die. The wax is then allowed to solidify, the pattern die is parted, and the patterns are lifted out.

Injection waxes are usually blended from selected waxes, resins and modifiers of the types listed in Table 8. Typical proportions of components for a workable formulation are shown in Table 9. This formulation does not contain a modifier, but does contain an antioxidant. Fillers, such as water mixed with granulated polystyrene, are sometimes added to formulations of this type, to reduce cavitation and contraction of the pattern.

Only a few formulations of waxes for patterns have been disclosed in patent specifications. Published material on waxes for patterns deals mainly with the properties of waxes and how they can be measured and controlled.

Depending on the characteristics of the wax, it may be introduced into the pattern die in the liquid state, the mushy or slush state (between liquidus and solidus), or the solid state (at a temperature just below the mushy state).

A typical liquid injection wax is comprised of a hard vegetable wax and paraffin into which is incorporated a high percentage of resin (rosin or rosin ester) to reduce shrinkage. It congeals at 60 to 65 °C (140 to 150 °F), and is injected at a temperature of 70 to 75 °C (160 to 170 °F) and a pressure of 345 kPa to 1.03 (50 to 150 psi).

PLASTIC PATTERNS

Plastic patterns are made in one of two ways, depending primarily on the quantity of patterns

Table 8. Materials used in formulating injection waxes for patterns

Type of material	Melting point (M) or softening point (S): °C	°F
Hard waxes		
Vegetable wax (candelilla) ....M:	67	152
Vegetable wax (carnauba) ....M:	89	192
Mineral wax (montan) ....M:	86	187
Synthetic wax, nonchlorinated .....M:	86	186
Synthetic wax, chlorinated ....M:	93	199
Microcrystalline waxes		
Petroleum origin .....M:	79	175
Petroleum origin .....M:	63	145
Insect origin (beeswax, USP) .....M:	64	147
Soft resinous plasticizers		
Rosin derivatives ....S:	43-71	110-160
Terpene resins ......S:	40	104
Coal-tar resins ......S:	25-75	77-167
Petroleum hydro-carbon resins ......S:	70	158
Chlorinated resins ....S:	60-73	140-163
Elastomer polymers ..S:	Wide range	
Hard resins		
Rosin derivatives ....S:	138-193	280-380
Terpene resins (from plants).......S:	85	185
Coal-tar resins ......S:	70-107	158-224
Petroleum hydro-carbon resins ......S:	95-100	203-212
Chlorinated resins ....S:	98-110	208-230
Modifiers		
Synthetic wax, nonchlorinated .....S:	70-105	158-221
Elastomer polymers ...S:	149	300
Polyethylene resins....S:	99-108	210-226

Table 9. Typical formulation of an injection wax for patterns

Hard wax(a) 40%	
Microcrystalline wax 25%	
Soft resinous plasticizers(b) 15%	
Hard resins(b) 20%	
Antioxidant 0.05%	

(a) Microcrystalline, amorphous. (b) Amorphous.

required. Patterns for prototype or experimental castings, or for the production of a few castings, are machined and assembled from molded or extruded plastic products, such as rod and sheet. For production quantities large enough to justify molding costs, patterns are made by injecting a semisolid plastic material into a pattern mold (or die) that will serve to produce the desired shape, using conventional plastic molding techniques.

Pattern Materials. A widely used material for plastic patterns is the "all-purpose" grade of polystyrene. Other types of thermoplastic plastics, including polyethylene, may be used to obtain specific properties, but under normal conditions polystyrene provides the required properties at lowest cost. The basic requirements of plastic patterns and pattern materials are: (a) adequate strength; (b) ability to be molded or machined to a smooth surface finish; and (c) ability to be burned out of the investment mold without leaving a residue. Defects in the molded material, such as air bubbles, are not objectionable provided that they do not occur at the surface.

Advantages and Limitations. As a pattern material, polystyrene has a number of advantages, the most

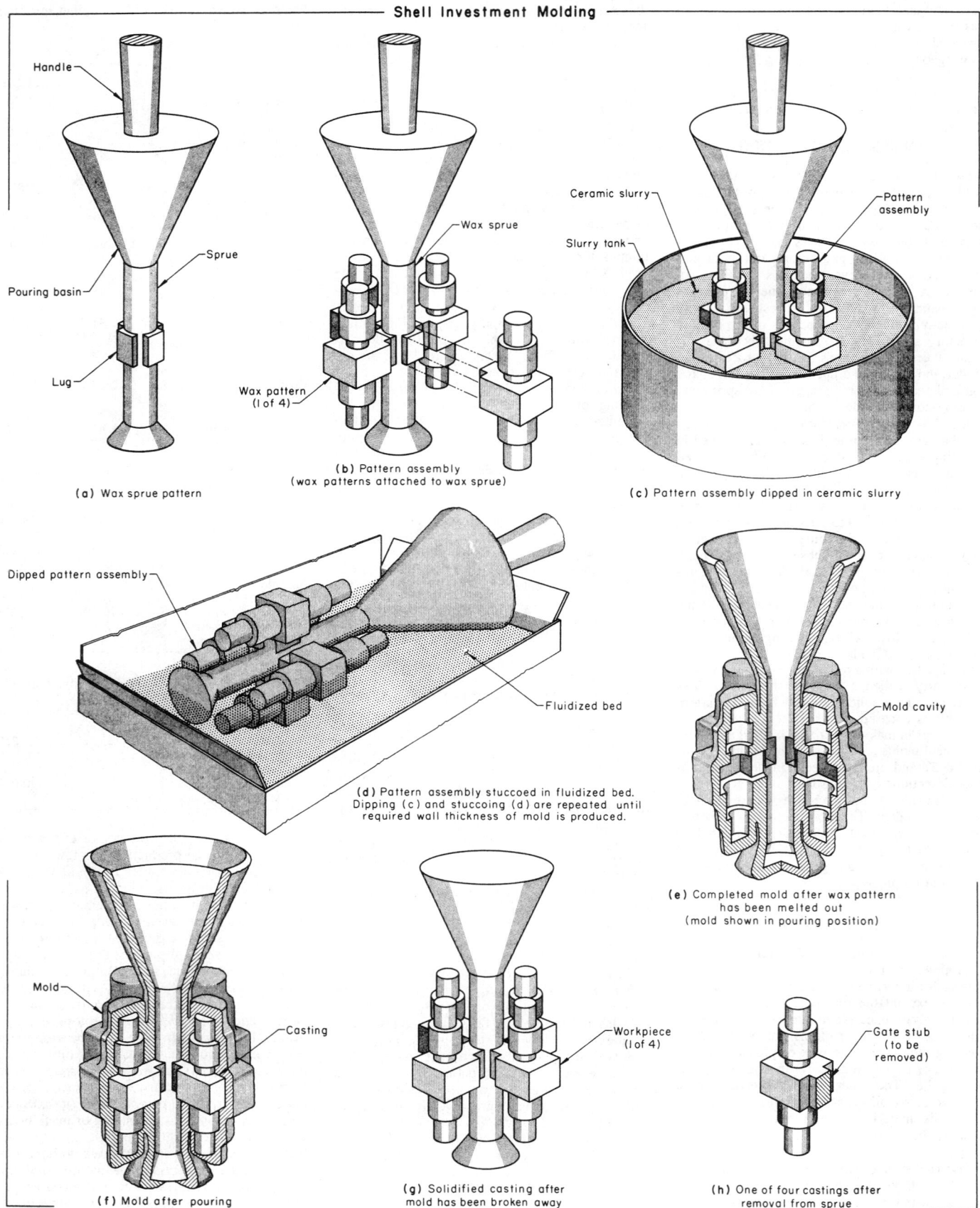

Shell Investment Molding

Handle

Pouring basin

Sprue

Lug

(a) Wax sprue pattern

Wax sprue

Wax pattern (1 of 4)

(b) Pattern assembly (wax patterns attached to wax sprue)

Ceramic slurry

Pattern assembly

Slurry tank

(c) Pattern assembly dipped in ceramic slurry

Dipped pattern assembly

Fluidized bed

(d) Pattern assembly stuccoed in fluidized bed. Dipping (c) and stuccoing (d) are repeated until required wall thickness of mold is produced.

Mold cavity

(e) Completed mold after wax pattern has been melted out (mold shown in pouring position)

Mold

Casting

(f) Mold after pouring

Workpiece (1 of 4)

(g) Solidified casting after mold has been broken away

Gate stub (to be removed)

(h) One of four castings after removal from sprue

Fig. 36. Steps in the production of a casting by the shell investment molding process, using a wax pattern

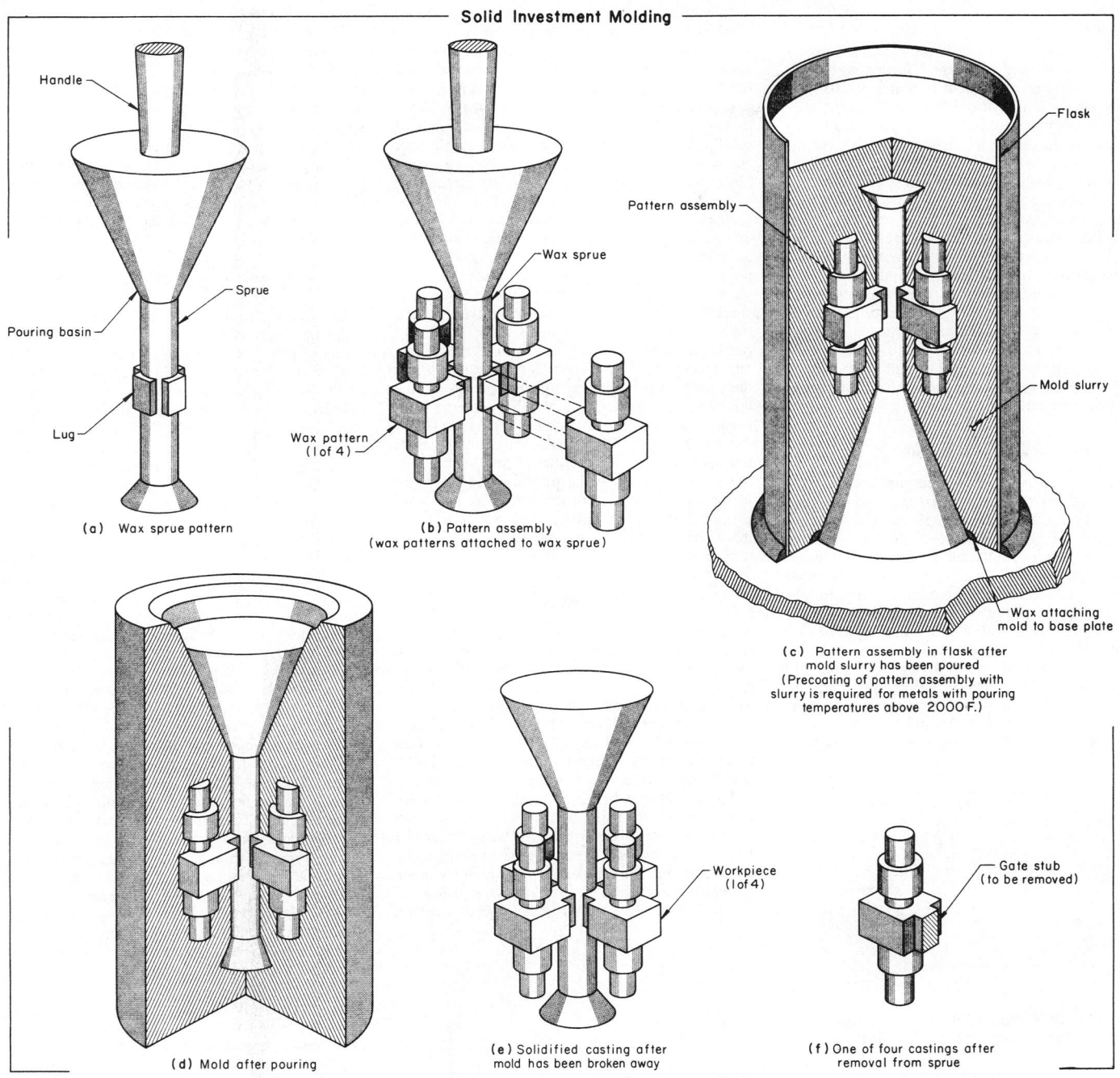

— Solid Investment Molding —

(a) Wax sprue pattern

Handle
Pouring basin
Sprue
Lug

(b) Pattern assembly
(wax patterns attached to wax sprue)

Wax sprue
Wax pattern
(1 of 4)

(c) Pattern assembly in flask after
mold slurry has been poured
(Precoating of pattern assembly with
slurry is required for metals with pouring
temperatures above 2000 F.)

Flask
Pattern assembly
Mold slurry
Wax attaching
mold to base plate

(d) Mold after pouring

(e) Solidified casting after
mold has been broken away

Workpiece
(1 of 4)

(f) One of four castings after
removal from sprue

Gate stub
(to be removed)

Fig. 37. Steps in the production of a casting by the solid investment molding process, using a wax pattern

outstanding being its high mechanical strength compared with that of wax. For this reason, polystyrene patterns do not require the special handling that must be accorded wax patterns, to avoid damage prior to and during investment. Of greater importance, the higher mechanical strength of polystyrene permits automatic ejection of patterns from the pattern mold, thereby shortening pattern production time and increasing efficiency. Because of other inherent properties of polystyrene, and the higher injection pressures it requires, patterns can be made with thinner sections, finer definition, sharper corners and better surface finish than can be achieved with wax.

Several patterns can be injection molded simultaneously and ejected as a complete unit. In a typical patternmaking application for small (1-lb) steel castings, an assembly of 10 patterns and their common runner system was injection molded and automatically ejected every 18 s from a multiple-cavity, water-cooled, low-carbon steel mold.

A major disadvantage in the use of certain plastics, including polystyrene, is related to their thermal expansion characteristics. During burnout, polystyrene patterns may expand enough to crack ceramic shell molds. Consequently, they are unsuitable for use with shell investments. Plastics that liquefy and flow at sufficiently low

temperatures can be used with both shell and solid investments.

MERCURY PATTERNS

Mercury patterns are prepared by pouring mercury into pattern dies that have been cooled to −57 °C (−70 °F) or lower. The filled dies are held at −57 °C or lower (usually in a medium of acetone) until the mercury is frozen; then the patterns are removed from the pattern dies.

All subsequent processing, including pattern inspection and mold development up to pattern meltout, is completed at temperatures below the freezing point of mercury (about −38 °C, or −37 °F).

After investment, the melted mercury is drained from the mold and is cleaned successively in a 20% solution of nitric acid in water, a 20% solution of sodium hydroxide in water, and acetone. These cleaning treatments remove all contamination, rendering the mercury suitable for reuse.

Advantages and Limitations. Possibly the most important advantage of mercury as a pattern material is that it does not expand in changing from the solid (frozen) to the liquid state. Thus, meltout does not adversely affect the condition or dimensional accuracy of the surrounding mold.

A major disadvantage of mercury patterns is the requirement for making and keeping them at extremely low temperature. Other disadvantages, which account for the diminishing use of mercury patterns, include the high cost of mercury, its toxicity, and the special handling problems that are entailed in its use.

CORE PRACTICE

Some investment castings require complex internal cavities (holes, air passages, and vents). Openings that are large enough and that have no undercuts or backdrafts are formed in the patterns by means of movable metal inserts (cores) in the pattern dies. These cores are withdrawn after the pattern is made, leaving the appropriate internal cavity in the pattern. The cavity in the pattern can then be filled with either a preformed ceramic body or with mold material.

For more complicated shapes, the pattern-die cores that form portions of the pattern cannot be withdrawn. Typical examples of such pattern shapes are shown in Fig. 38 and 39. To produce such complex patterns, an alternative method must be used.

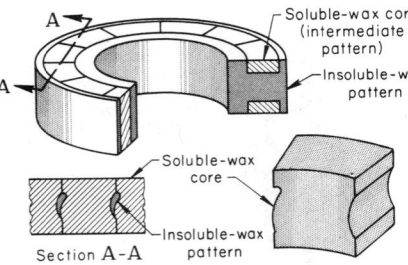

Fig. 38. Cross section of a pattern in which soluble and insoluble waxes were used

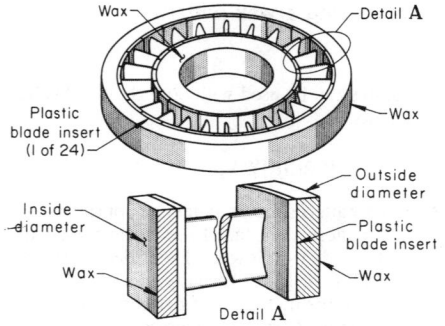

Fig. 39. Cored pattern developed with plastic blade inserts and injected wax

One method, illustrated in Fig. 38, is to use pattern-die cores that are made from a wax that will dissolve in water or a weak acid. The soluble-wax cores (or "intermediate patterns") are produced in a die, assembled (in a ring in Fig. 38), and placed in the pattern die. An ordinary pattern wax (which is not soluble in water or acid solutions) is injected around the soluble-wax cores, which are then leached out, leaving a one-piece wax pattern with the necessary cavities.

Another method, illustrated in Fig. 39, is to prepare the complex pattern by assembling several pattern pieces, which have been produced in a separate die or dies. The assembly can be composed entirely of wax pieces, or of both wax and plastic pieces, as in Fig. 39. For large complex shapes with heavy sections, several wax pattern pieces are assembled. For smaller, more delicate parts, the usual practice is to combine plastic pattern pieces into a subassembly and then to inject the pattern wax around the plastic pieces to produce the final pattern shape. Plastic pieces are used because they are stronger and more stable dimensionally, and do not distort in handling. Wax is usually injected around the plastic to compensate for expansion of the plastic during burnout of the assembled pattern from the ceramic mold (to prevent mold cracking). The wax also seals the joints created by assembly of the plastic pattern pieces.

PATTERN ASSEMBLY

As much of the gating as possible is included in wax patterns—particularly when more than one pattern is produced in the pattern die at one time. Examples of pregating of patterns are shown in Fig. 36(b) and 37(b). In these applications, the gates were injected as integral parts of the individual patterns. This technique greatly simplifies assembly and can be used to make a mold cluster containing a large number of individual patterns.

The use of standardized, or universal, feeders also simplifies the assembly of wax patterns. Two such feeders are shown in Fig. 40. The feeder in Fig. 40(a) consists of a pouring basin and a main cross runner. The runner can be made with or without the slots, and the size and number of slots can be varied. The feeder in Fig. 40(b) consists

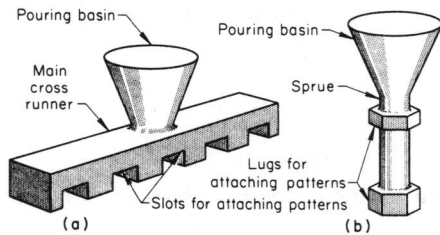

Fig. 40. Two types of universal feeders to which wax patterns are attached

of a pouring basin, a sprue, and multisided lugs. The number of sides on the gate lugs, as well as the size and location of the lugs, can be varied depending on the patterns to be assembled. This type of feeder pattern may also be produced with round or rectangular sprues.

Patterns are frequently joined to the feeding system by dip sealing. The gate of the pattern can be dipped in molten wax and held against a wax runner or sprue until the wax sets, which

takes only a few seconds. Figure 41(a) shows the assembly of a pattern and sprue where the gate is dipped in molten wax and is inserted in the slot of a universal feeder. In Fig. 41(b), the gates of the patterns are arranged on a horizontal runner and the assembly is placed, upside down, in molten wax to cover the runner and seal the patterns to it. The dipping operation must be done carefully, to avoid splashing of the individual patterns. Also, the pattern assembly must not be allowed to remain in the molten wax for a prolonged period of time, lest the patterns and runner melt or distort.

CERAMIC SHELL INVESTMENT PROCESS

Ceramic shell molds are used primarily for investment casting of carbon and alloy steels, stainless steels, heat-resistant alloys, and other alloys with melting points above 1100 °C (2000 °F). The molds are built up by dipping pattern assemblies into slurries of ceramic powders suspended in a liquid, draining the excess, stuccoing the wetted surface with a dry refractory grain, either by sprinkling or by immersion in a fluidized bed, and then drying the resultant coating. This process is repeated until the required mold thickness is achieved. The initial coating is termed a precoat and usually employs a slurry that is made of finely ground particles, to provide a smooth surface. The smoothness of the precoat largely determines the smoothness of the cast surface. Subsequent coatings usually contain increasingly coarser refractory grains. A method for applying the coatings to a pattern assembly is shown in Fig. 36(c) and (d).

The number of coatings applied depends on the required thickness of the shell. The thickness of ceramic shells usually ranges from 6.5 to 13 mm ($^1/_4$ to $^1/_2$ in.), depending mainly on casting shape and weight, cluster size, and type of ceramic and binder. To some extent, shell thickness depends also on the temperature and humidity of the environment in which the shell is made. To maintain optimum mold permeability, in addition to saving dipping time and cost, shell thickness should be the minimum that will do the job.

Ceramic shells are dried or cured after each dipping operation, to allow bonding of the individual layers. In drying, moisture removal is regulated by control of wet and dry bulb temperatures, air flow, and time. Final drying and bonding of the shell usually occur during the dewaxing operation and during firing of the mold before casting. Drying time depends on the shape of the patterns, and on air temperature, humidity and circulation. Shells have been completely fabricated and dewaxed in 8 h; generally, however, the completed shell is allowed to dry for a minimum of 24 h prior to dewaxing.

The ceramic shell process can be mechanized, and machines are available to apply as many as eight dip coats per hour. Rapid drying between coats is made possible through the use of an ethyl silicate–bonded dip coating, which can be gelled (chemically set) in an ammonia atmosphere.

SOLID INVESTMENT PROCESS FOR FERROUS ALLOYS

Mold materials for investment casting of ferrous alloys (and nonferrous alloys with high pouring temperatures) must be capable of withstanding temperatures above 1100 °C (2000 °F) without deteriorating. Gypsum, therefore, is not

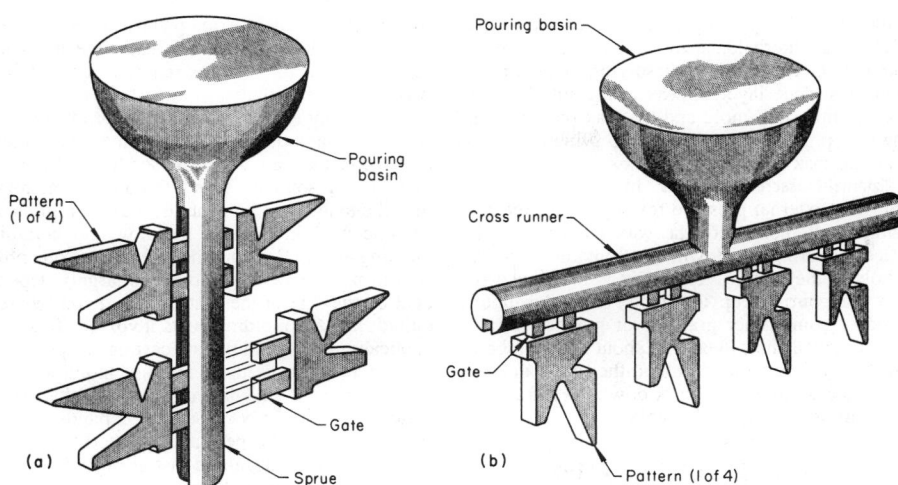

Fig. 41. Two assemblies for which different methods of dip sealing are used to join patterns to a universal feeding system. (See text for discussion.)

a suitable bonding or refractory material. Precoating is standard practice in the preparation of solid investments for ferrous castings. Precoating of patterns with a fine-grain slurry makes it possible to use coarser-grain (and therefore less expensive) slurries in the mold backup without affecting surface finish.

Precoating. The slurries used for precoating patterns in the solid investment process are similar to those used in the shell investment process. Sodium silicate, ethyl silicate and colloidal silica are the most widely used binders.

Precoats must be properly dried or cured. High humidity during fabrication (for example, in excess of 80%) will lower the dry strength of the precoat. Precoats that have not been dried sufficiently are weak and will slough off into the mold cavity, resulting in dirty castings. The quality of the precoat is important, because it determines the quality of the casting surface.

Control of stuccoing following precoating is essential. A coarse stucco will give less trouble than one of 150-mesh or finer. The coarser grain permits more open space for better adhesion of subsequent layers. A fine grain produces a more uniform surface but prevents optimum bonding of subsequent coats. Usually, two layers of slurry coating and stucco are applied.

When their thermal expansions are different, the precoat usually expands more than the backup investment, and buckles away from it. The thermal expansion of the precoat can be reduced by substituting milled zircon for some of the silica flour in the precoat slurry. Mixtures of one to two parts (by weight) silica flour to one part zircon have been used to correct buckling.

Investment Materials. Backup slurries for solid investments, in common with precoat slurries, are composed of two basic ingredients: a refractory and a binder. The refractory is usually composed of a graded aggregate of powders. Ordinarily, the basic ingredient of the refractory powders is silica sand; alumina and silicates comprise the remainder. This graded aggregate is designed to give the required mold strength and permeability, combined with needed expansion characteristics and resistance to incipient melting, at minimum cost. The silica sand is usually a mixture of selected different particle sizes.

Ethyl silicate is the most widely used binder. It relies on the hydrolysis of the silicate in a non-

aqueous solution, such as alcohol, to form a relatively stable solution or colloid containing a form of silica. By changing the pH of the hydrolyzed ethyl silicate solution, the silica precipitates as a gel. Gelling is obtained by hydrolyzing the ethyl silicate–alcohol mixture by adding water and a catalyst such as hydrochloric acid. Then, if an alkaline material such as magnesium oxide is added, the pH is raised until a gel is produced. The setting time is controlled by varying the rate of pH increase through additions of a weakly alkaline material such as magnesium oxide.

Close control is mandatory when ethyl silicate is used as a binder. Impurities in the mixture and improper hydrolization can cause premature gelling or, at the opposite extreme, complete lack of gelling. Mixing temperature should be maintained at $24 \pm 1\,°C$ ($75 \pm 2\,°F$). Mixing temperature is so important that many plants measure the temperature of each mix after one minute of stirring. This temperature is recorded as part of the process-control information.

Investment materials can also be bonded with a mixture of monomagnesium phosphate, monoammonium phosphate and magnesium oxide. A typical mixture, consisting of 9 kg (20 lb) of monoammonium phosphate, 9 kg of magnesium oxide and 3.6 kg (8 lb) of monomagnesium phosphate, when blended with 160 kg (350 lb) of refractory, will set to a solid with the addition of water in 20 to 30 min.

Phosphate-bonded investments cost significantly less than those bonded with ethyl silicate. With phosphate as a bonding material, setting is gradual and continuous, starting from the time water is mixed with the dry ingredients and continuing until a solid material is formed. With ethyl silicate, the material is completely liquid until the pH reaches the point of instability, and then it solidifies rapidly.

Phosphate-bonded materials contract considerably when they are heated to 650 °C (1200 °F) and cooled to room temperature. Figure 42 shows the effect of heating and cooling on linear expansion of a graded-silica, phosphate-bonded mixture having a water-to-powder ratio of 18 to 100.

Mixing and Vacuuming. After suitable additions of water have been made, both gypsum-bonded and phosphate-bonded investment mixtures are mechanically stirred for 3 to 4 min and vacuumed.

The vacuumed investment is next poured into the flasks and again vacuumed. Vacuuming removes air bubbles and voids, thus providing a good solid mold around the patterns.

The investment rises during vacuuming; therefore, a rubber collar or other flask extension is provided to allow for this expansion. During vacuuming, vibration for 30 to 45 s at the height of the "boiling action" also helps to remove air bubbles.

It is absolutely necessary to have rapid vacuum pumpdown capacity to obtain a vigorous boil in the investment within 30 to 45 s. This produces a stirring action, prevents settling, breaks up any froth that might form at the investment/wax or investment/precoat interface, and helps to produce a bubble-free mold. The vacuum system, when dry, should be capable of pumping down to a vacuum of 10 mm mercury (1.3 kPa) or less within 20 to 30 s. The vacuum line from the pump to the investing station should be as short as possible for efficient pumpdown, and the pipe diameter should be in accordance with the pump manufacturer's specifications. Pipe diameters smaller than specified produce a throttling effect and reduce pumping efficiency. An approved filter should be provided in the line to protect the pump from abrasive investment material. For best results, valves specially designed for vacuum service are recommended and should be used.

SOLID INVESTMENT PROCESS FOR NONFERROUS ALLOYS

Solid investment practice for casting nonferrous alloys with pouring temperatures below 1100 °C, or 2000 °F (aluminum-base alloys and many copper-base alloys) differs from that used for ferrous alloys, primarily because of the effects of pouring temperature on bonding materials and refractories.

The mold material for nonferrous castings generally consists of alpha gypsum—which serves as both a binder and refractory—and a blend of other refractories, such as various grades of silica, to control mold permeability. The refractories used are fine-grained, and precoating is not required.

As shown in Fig. 37(c), the pattern assembly is placed on a plate and encircled by a metal flask that is open at both ends. The investment slurry is poured directly into the flask, which is vibrated or subjected to vacuum, or both, to remove air bubbles. The slurry is then permitted to solidify by chemical action.

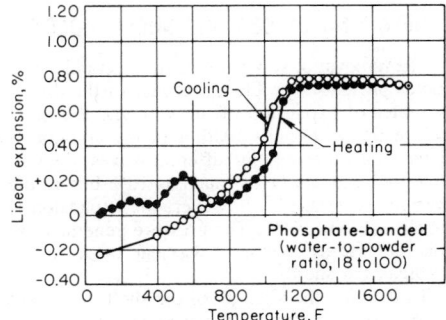

Fig. 42. Effect of temperature on linear expansion of graded-silica, phosphate-bonded investment mold material used in producing steel castings

Alpha gypsum (calcium sulfate hemihydrate) is a high-strength, relatively low-cost, commercially available product that makes use of the least-expensive liquid vehicle (water). A typical gypsum-bonded material contains about 35.5% cristobalite, 29% silica flour, 33.5% calcium sulfate, and 2% of a highly absorbent diatomaceous earth (to provide permeability). Small amounts of commercial accelerators and retarders also may be added, to control setting and thermal expansion. Commonly used accelerators are sodium chloride, potassium sulfate, and powdered calcium sulfate dihydrate. Citric acid and sodium citrate are used as retarders.

The recommended consistency for the investment slurry is determined by the ratio of water to powder. The correct weight of water required for any desired weight of powder is calculated from the water-to-powder ratio that is marked on each container by the manufacturer. Small variations in water content will produce large variations in properties of the investment. Therefore, the proportions of water and powder must be accurately measured.

Gypsum-bonded mixtures contract in linear dimensions after heating and cooling. Typical expansion-contraction curves for a cristobalite, gypsum-bonded mixture having a water-to-powder ratio of 38 to 100 are shown in Fig. 43. (Mixing and vacuuming of gypsum-bonded mixtures are described in the preceding section.)

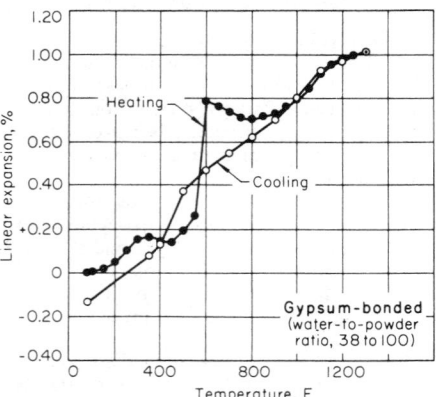

Fig. 43. Effect of temperature on linear expansion of a typical cristobalite, gypsum-bonded investment mold material used in producing nonferrous castings

DEWAXING OF CERAMIC SHELL MOLDS

The removal of wax patterns from ceramic shell molds presents a problem, because of the different rates of expansion of the wax and the mold before the wax melts. Although the amount of expansion varies among different waxes, the volume increase can be as much as 10% before the wax melts. (Usually, however, the expansion is much less than 10%.) The pressure generated by this increase in volume of wax can be sufficient to break the shell.

To increase the strength of the shell to prevent it from breaking under the wax pressure would be self-defeating, because the collapsibility of the mold would be poorer and heat dissipation would be less, which might lead to hot tears in the casting. The only practical methods for preventing mold breakage when removing wax from a shell

mold are: (*a*) supplying external pressure to the shell to counterbalance the internal pressure of the wax; and (*b*) rapidly dissolving or melting a skin or surface layer of wax at the interface of the ceramic shell mold and the wax pattern, and thereby producing a cavity into which the remaining wax can expand.

External-Pressure Techniques. In one method of applying external pressure for dewaxing, the ceramic shell mold with the wax pattern in it is placed, pouring cup down, in a container that has a hole in the bottom. The mold is positioned so that the pouring cup is directly over this hole. Sand or some other granular or powder refractory (or metal shot), heated to about 300 °C (about 600 °F), is then poured around the shell, and the container is vibrated to pack down the refractory medium as quickly as possible.

REMOVAL OF PLASTIC PATTERNS

Most plastic patterns are burned out during the forepart of the preheating or firing cycle. During burnout, polystyrene patterns may expand enough to crack ceramic shell molds, and so they are commonly used only with solid investments. However, certain synthetic thermoplastics that liquefy and flow at temperatures well below their flash points can be melted out by the same methods used to melt wax.

Except for the types that flow out of the mold instead of burning, plastics are not recoverable.

DEWAXING AND PREHEATING OF SOLID INVESTMENT MOLDS

Mold cracking caused by expansion of the wax is not a problem in dewaxing of solid molds, because of the backing provided in the solid mold. In practice, solid molds are dewaxed, burned out (fired), and preheated for pouring in the same cycle.

Best results are obtained when molds are dewaxed and burned out on the same day that they are invested. Molds that are allowed to air dry for several days are likely to become brittle and less able to withstand the burnout cycle. This results in a high rejection rate because of flash, core failure, and dirt. Slow drying at 80 to 90 °C (180 to 200 °F) in a meltout oven also results in a high rejection rate from the same causes. Therefore, it is recommended that molds be loaded into a preheated burnout furnace.

MOLD PREHEATING TEMPERATURE

Mold preheating temperatures for investment casting depend mainly on the composition of the metal being cast and on the size and complexity of the casting. Although many investment molds are poured at room temperature, more often the molds are preheated for pouring. Preheating temperatures range from 90 to 1050 °C (200 to 1900 °F).

Ceramic Mold Casting

CERAMIC MOLD CASTING is most widely used for production of precision castings that require patterns too large and unwieldy for molding with expendable wax or plastic patterns, or for production of castings in limited quantities

for which permanent wood patterns may be more economical and require less lead time to make than the metal pattern dies required for molding wax or plastic patterns.

Ceramic molding is intended for production of castings of high quality, not only in terms of their dimensional accuracy and surface finish, but also in terms of soundness and freedom from nonmetallic inclusions. In general, the capabilities of ceramic-shell investment molding and ceramic molding are similar, and the selection of one process in preference to the other is largely dependent on the size of the casting, the quantities required, and the molding costs involved. In some applications, depending on casting shape, permanent patterns used in ceramic molding may provide greater dimensional accuracy than wax patterns, primarily because wax expands during melt-out. Also, the permanent patterns are less susceptible to damage and distortion in handling than wax or plastic patterns.

SUITABLE WORK METALS

Both ferrous and nonferrous alloys are cast in ceramic molds, but ferrous applications are more numerous and account for the major tonnage of castings produced. Alloys of aluminum, copper (especially beryllium copper), nickel and titanium are nonferrous alloys suitable for ceramic mold casting; the ferrous alloys that are suitable include ductile iron, carbon and low-alloy steels, stainless steels and tool steels.

SHAW PROCESS

The Shaw process relates to two distinctly different types of ceramic molds: a one-piece, all-ceramic mold; and a composite ceramic mold consisting of an inexpensive fireclay backup material with a relatively thin facing of ceramic slurry. (A composite mold is shown in Fig. 44.) Selection of mold type depends almost exclusively on the size of the casting and the cost of mold material. Many small castings can be produced economically in one-piece, all-ceramic molds, because the amount of expensive ceramic slurry needed for the mold is moderate, and the additional pattern and labor costs for composite molding cannot be justified.

Patterns. Two sets of patterns are commonly required for production of composite molds—a set of oversize preform patterns for molding the coarse backup material, and a second set of patterns, representative of the dimensional accuracy desired in the casting, for molding the ceramic facing. Thus, for cope-and-drag molding, a total of four patterns may be required.

The preform patterns are made 2.4 to 9.5 mm ($^3/_{32}$ to $^3/_8$ in.) oversize to allow for the thickness of the ceramic facing. In some applications, the preform pattern can be eliminated by using the final pattern, backed with a sheet of plastic, cardboard or felt of suitable thickness, to mold the preform impression in the coarse fireclay backup. Because solid ceramic molds are made without backup material, only the final patterns are required for molding.

Preparing the Backup. The backup refractory commonly consists of a coarse-grain chamotte (an aluminous fireclay that has been calcined at a high temperature) and a sodium silicate binder. A typical screen analysis for the chamotte is as follows:

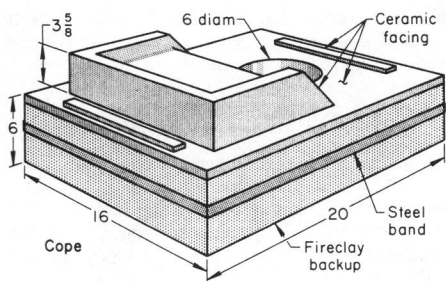

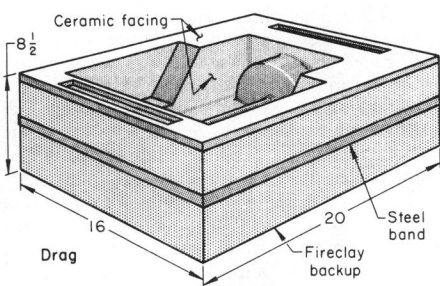

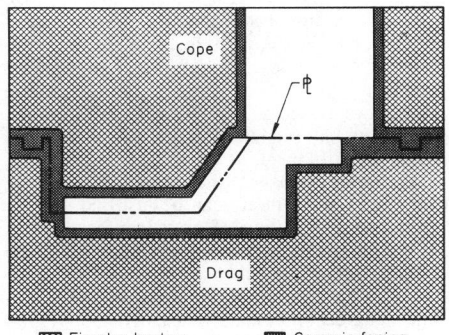

☒☒ Fireclay backup   ■ Ceramic facing

Section through cope and drag

This Shaw-process mold was used for casting H13 tool steel dies for use in hot upset forging of steel axles. Dimensions are in inches.

Fig. 44. Typical composite ceramic mold

Cumulative percentage on 6 mesh . . . . . . Trace
On 8 mesh . . . . 5 to 9%
On 12 . . . . 27 to 41
On 16 . . . . 60 to 68
On 30 94
On 70 99

Carbon dioxide gas is diffused into the mold material for hardening. One method of doing this consists of piercing the mold material in several locations with small-diameter rods, thus creating channels into which the gas can be introduced. More uniform gassing can be accomplished in a vacuum chamber. After the gassing and hardening operation, the preform pattern is removed from the mold.

Preparing the Ceramic Facing. The principal ingredient of the ceramic facing usually is finely comminuted zircon or calcined mullite, or a mixture of both, although fused silica, magnesium oxide, and other refractory flours have also been used. Typical screen analyses for two calcined mullite flours are:

Mullite flour	Cumulative % on mesh size of:		
	100	200	325
A	3	35 to 65(a)	. . .
B	2	17 to 32	30 to 45(a)
(a) Remainder passes through this mesh size.			

Pouring an All-Ceramic Mold. An unbacked, all-ceramic mold is made entirely from slurry of the type used for facing a composite ceramic mold. The procedure for pouring an all-ceramic mold is shown in Fig. 45. The pattern, affixed to a flat plate, is surrounded by a flask into which the ceramic slurry is poured. After the chemical gelling action is completed the green mold is stripped from the pattern, and the flask, which is built with a small amount of draft, is removed from the mold. The mold is then ready for burn-off.

Burn-off (Mold Stabilization). After gelling, the ceramic mold or mold facing is ignited with a torch (Fig. 45) and burns until most volatiles are consumed. During burn-off, the ceramic develops a microcrazed pattern — a three-dimensional network of microscopic cracks induced by the rapid evaporation of the alcohol in the slurry and by solid-phase reactions. Craze cracking can be

avoided by allowing the mold material to air dry for 1 h or more prior to ignition.

Baking is the final stage of processing, in which all remaining volatiles are removed and the colloidal silica left by the hydrolyzed ethyl silicate binder forms a high-temperature bond of SiO_2, stable up to temperatures approaching its melting point (1710 °C, or 3110 °F), thus providing enough mold strength to resist washout, or erosion, by molten metal. To perform these functions satisfactorily, baking must be done at not less than 480 °C (900 °F). Baking composite molds at more than 650 °C (1200 °F) may cause differential expansion between the facing and backup layers, due to contraction of the soda-rich silica bond in the backing material. (Use of backup and facing materials that have similar thermal-expansion characteristics will prevent separation of the two materials on firing.) This differential expansion can produce distortion in the mold cavity. Nevertheless, some foundries bake composite molds at as high as 815 to 980 °C (1500 to 1800 °F) without harmful effects.

Pouring. Prior to pouring, molds are usually checked for cleanness and temperature. Mold temperature at the start of pouring ranges from 40 to 540 °C (100 to 1000 °F), depending on casting shape and the alloy being poured. Cope and drag halves are assembled, and usually a layer of cement is troweled on to seal the mold around the parting line. For safety, heavier molds are reinforced by fastening a steel band around the outside of the mold (see Fig. 44). This assists in handling the mold, and prevents seepage of hot metal in the event that the mold wall cracks during pouring. Finally, ceramic pouring tubes and exothermic or insulating riser sleeves are set up in the cope.

The mold is clamped firmly, and metal is poured either directly from the melting furnace or from a ladle. Pouring must proceed as rapidly as possible, to avoid cold shuts and gas voids. After the metal has been poured, additional exothermic material is sprinkled on the risers to keep them fluid and prevent shrinkage tears.

UNICAST PROCESS

The Unicast process differs from the Shaw process principally in the method of mold sta-

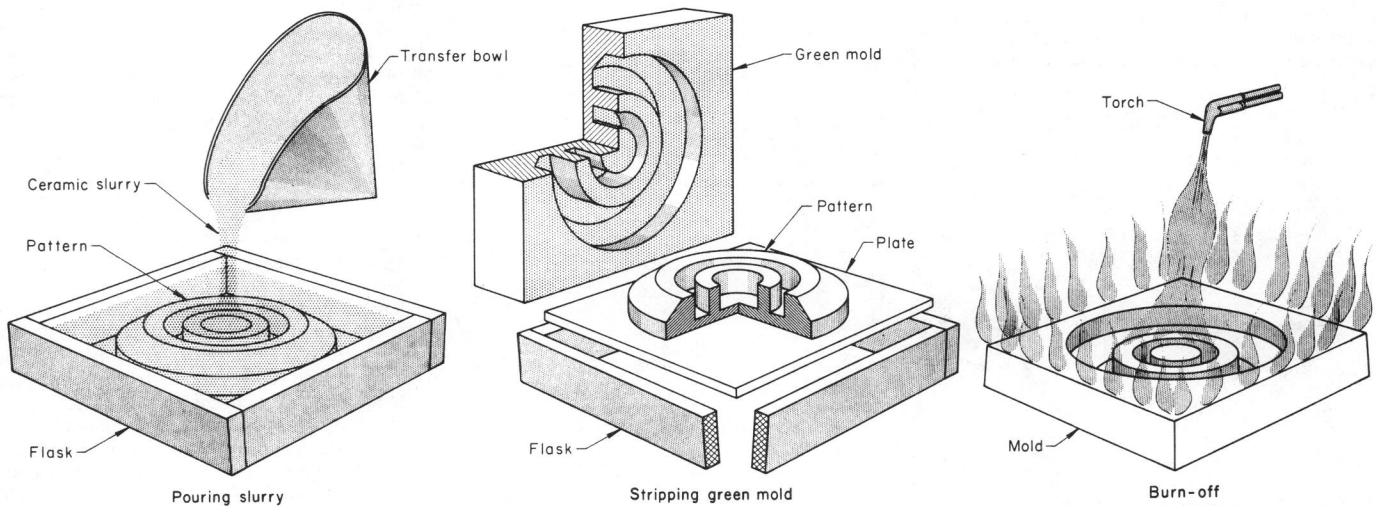

Fig. 45. Pouring, stripping and burn-off of a one-piece, all-ceramic mold. (See text for discussion.)

bilization employed. Mold stabilization refers to the treatment given the fine ceramic facing of a composite mold, or the total mass of an all-ceramic mold, shortly after it has set and while it is still green.

Another difference between the Shaw and Unicast processes relates to the sequence followed in preparing composite molds. In the Shaw process, preparation of the coarse mold backup precedes pouring of the fine ceramic facing; in the Unicast process, this sequence is reversed.

In a typical molding and casting sequence, a pattern, mounted on a baseplate, is enclosed within a flask. A thin coating (4.8 to 9.5 mm, or $^3/_{16}$ to $^3/_8$ in.) of fine-grain ceramic slurry is applied to exposed pattern surfaces by spraying or similar means. The coating becomes tacky almost on contact and is ready to receive backing material. An inexpensive, coarse backup slurry is poured rapidly over the facing coat until the flask is filled. The slurry is chemically controlled so as to set to a semirigid state within 2 to 3 min, whereupon the upper surface of the mold is leveled with the flask edges with a striking tool. The flask is removed, and a vacuum clamping plate is placed in position on top of the mold. The entire assembly is inverted on a stripping machine, the pattern is stripped from the mold, and the clamping vacuum is released.

The stripped mold is transferred to the chemical hardening tank or chamber, where it is immersed in, or sprayed with, hardening fluid. It is then cured by heating with direct flame impingement or in a furnace.

Permanent Mold Casting

IN PERMANENT MOLD CASTING, a metal mold consisting of two or more parts is used repeatedly for producing many castings of the same form. The liquid metal enters the mold by gravity. (The process does not, however, include pouring of ingots in metal molds.) Simple cores are made of metal, but more complex cores are made of sand or plaster. When sand or plaster cores are used, the process is called semipermanent mold casting.

Permanent mold casting is particularly suitable for high-volume production of small, simple castings that have fairly uniform wall thickness and no undercuts or intricate internal coring. The process can also be used to produce moderately complex castings; however, production quantities should be high enough to justify the cost of the molds.

Compared with sand casting, permanent mold casting permits production of more uniform castings, with closer dimensional tolerances, superior surface finish and improved mechanical properties.

Permanent mold casting has the following limitations: (*a*) although no maximum size has been established, the process is most practical for production of small castings; (*b*) not all alloys are suited to permanent mold casting; (*c*) the process can be prohibitively expensive for low production; and (*d*) some shapes cannot be made by the process because of the location of the parting line or because of difficulty in removing the casting from the mold.

SUITABLE CASTING METALS

Metals that can be cast in permanent molds include aluminum, magnesium, zinc and copper alloys, and hypereutectic gray iron.

Aluminum alloys have low density, which, combined with their oxide-film-forming characteristics, makes them flow somewhat sluggishly.

The shrinkage of aluminum alloys during solidification is relatively large, and provision must be made for ample metal feed during solidification. After solidification, aluminum alloys are soft at elevated temperature, and castings may distort during removal from the mold.

Magnesium alloys are less castable than aluminum alloys, and have relatively poor feeding characteristics in thin-wall castings. Also, the castings are more sensitive to hot shortness (brittleness at elevated temperature) than are aluminum alloy castings. Generous fillets are required when the casting contains large bosses or when one section of the casting is much larger than another. Sharp casting detail cannot be obtained with magnesium alloys, and shapes that shrink onto mold sections are susceptible to cracking and should be avoided.

Copper alloys solidify at high temperatures, and some have narrow solidification ranges. They shrink onto cores and other mold elements, and must be ejected from molds as soon as possible.

Zinc alloys can be cast in permanent molds, but because the castings are usually made in large quantities, they are more often die cast.

Gray iron is used successfully in high-volume production of small (28 g to 13.5 kg, or 1 oz to 30 lb), simple castings. However, more complex gray iron castings, with internal coring and marked changes in section, have also been successfully made by the permanent mold process.

MAXIMUM SIZE OF CASTINGS

Practical sizes of permanent mold castings are limited by cost. The maximum sizes that have been cast differ among the casting alloys.

Aluminum Alloys. In high production, permanent mold castings weighing up to 13.5 kg (30 lb) are made from aluminum alloys in casting machines. However, much larger castings can be produced. For instance, aluminum alloy castings of relatively simple design with a trimmed weight of 354 kg (780 lb) were produced in a three-section permanent mold that had a vertical parting line.

Magnesium alloys, despite their comparatively low castability, have been cast in permanent or semipermanent molds to produce relatively large and complex castings. For instance, an 8-kg (17.7-lb) housing for an emergency power unit was poured from alloy AZ91C in a semipermanent mold. The mold utilized vertical parting and an oil-sand core to develop the vanes and internal surfaces of the casting. Surface finish of the casting varied from 6.4 to 12.7 μm (250 to 500 μin.).

In another application, 24-kg (53-lb) spoolhead castings, 760 mm (30 in.) in diameter, were produced from alloy AZ92A in a two-segment permanent mold with vertical parting. These castings were used as ends for fiber rolls, which have heavy hub sections and thin peripheral rims and function like spools for thread.

Copper alloy permanent mold castings weighing over 9 kg (20 lb) rarely can be justified.

Gray Iron. Production of gray iron castings in permanent molds is seldom practical when the castings weigh more than 13.5 kg (30 lb). One reason is that the largest mold that available machines can accommodate is usually 455 by 510 by 150 over 150 mm (18 by 20 by 6 over 6 in.). Also, the longest cycle time for these machines is 7 to 8 min. This limits cooling time, and hence limits the size and weight of castings that can be produced.

CASTING MACHINES

Manually operated permanent mold casting machines may consists of a simple "book" mold arrangement, such as that shown in Fig. 46; or, for castings with high ribs or walls that require mold retraction without rotation, the machine shown in Fig. 47 can be used. With either type of machine, after the casting has solidified, the mold halves are separated by manually releasing the eccentric mold clamps.

Automatic Machines. For high-volume production, the manual drives are replaced by two-way hydraulic mechanisms. These can be programed to open and close in a preset cycle. Thus, except for pouring of the metal and removal of castings, the operation is automatic.

A method of permanent mold casting has been developed in which the metal is not ladled by hand. This is called the Wessel process, the equipment for which is shown in Fig. 48. In this method, the permanent mold is mounted on rails against the end face of a tilting reverberatory furnace. As the furnace is tilted about an axis near its center of gravity, metal flows through a pouring hole in the wall of the furnace into the mold. The assembly remains in its tilted position for a predetermined interval, then returns to the start-

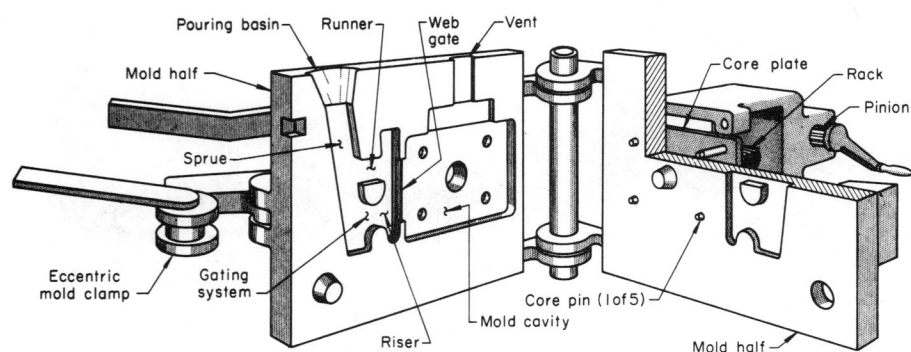

Fig. 46. Book-type manually operated permanent mold casting machine, used principally with molds having shallow cavities

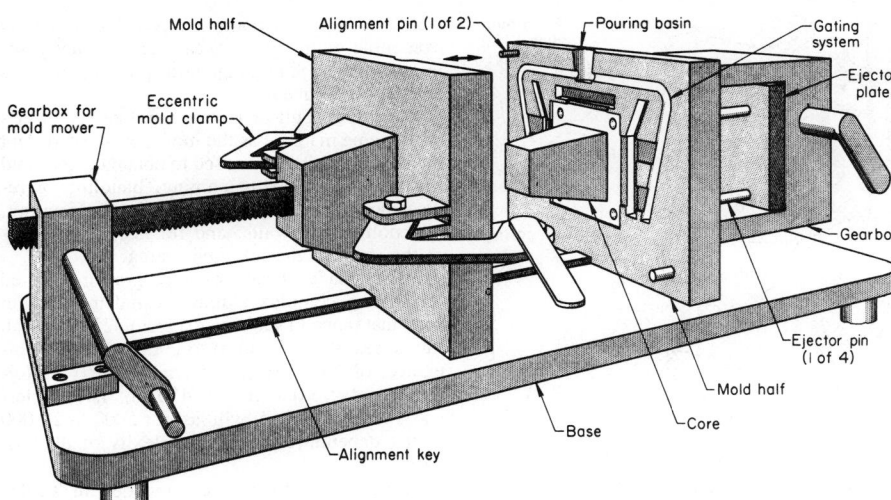

Fig. 47. Manually operated permanent mold casting machine with straight-line retraction (required for deep-cavity molds)

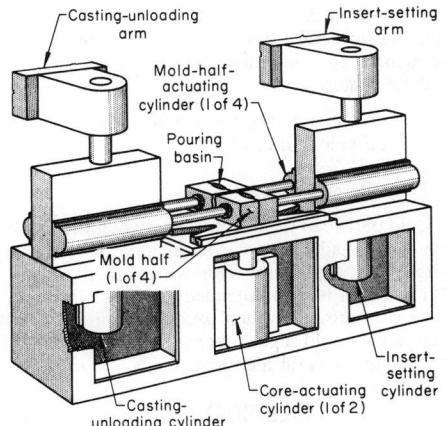

Fig. 48. Automatic permanent mold casting machine, with a setter for cast-in inserts

ing position. Tilting is done by means of a hydraulic cylinder.

Molds are parted vertically, parallel to the direction of metal flow. One mold half slides on the mounting rails; the other, which is hinged, swings away from the mounting rails, pulling the casting and sprue with it to leave the pouring hole clear for the next cycle. Core manipulation and casting ejection are the same as in conventional practice.

MOLD DESIGN

A simple permanent mold is shown in the book-type casting machine illustrated in Fig. 46. Here the two mold halves are hinged on a pin and aligned. The mold cavity with the mold halves closed determines the shape of the casting. The casting is poured by means of the sprue and runners to the riser, which is provided with a web gate to the mold cavity. The cavity is vented to allow escape of air.

The plate-shape cavity shown in Fig. 46 required five core pins, which were moved by means of the manually driven gearbox mounted on the

back of the right-hand mold half.

In operation, the mold halves are closed and locked. Metal is then poured to fill the gating system and the mold cavity. After the metal has solidified, the mold is opened, leaving the casting on the core pins. The core pins are withdrawn and the casting is removed manually.

The mold shown in Fig. 46 is designed with the parting vertical and in a single plane. This mold also could be designed for horizontal parting, or with parting in two or more planes. Also, instead of side gating, it could have bottom gating (see Fig. 49).

The mold shown in Fig. 47 is also designed for vertical parting and side gating. However, because of the deep cavity required (and correspondingly long core), a hinged type of mold cannot be used. The mold shown is opened and closed by straight-line movement of one mold half to and away from the other mold half, which remains fixed.

Undercuts on the outside of a casting complicate mold design and increase casting cost, because additional mold parts or expendable cores are needed. Complicated and undercut internal sections are usually made more easily with expendable cores than with metal cores, although collapsible steel cores or loose metal pieces can sometimes be used instead of expendable cores.

Number of castings per mold is a major consideration in designing the mold; the objective is to have the optimum number of cavities per mold that will yield acceptable castings at the lowest cost. Except for very small and thin castings, as the weight of the metal being cast per mold increases, the cycle time of the machine also increases. These increases, however, are not directly proportional. A mold with the maximum number of cavities often will produce more castings per unit of time than a mold with a smaller number of cavities that was designed to operate on a shorter cycle. This is because there is a minimum solidification time for every casting, regardless of the number of cavities in the mold. Sometimes, as the number of cavities is increased, the number of rejects increases, but this is usually offset by the greater productivity.

Progressive Solidification. Alloys should be cast so that solidification takes place progressively toward the risers, which are generally to one side of the casting. To achieve this solidification pattern, thinner sections of the casting should be away from the gating system, and heavy sections should be adjacent to it. Rib sections and thin walls vent and fill more easily when they are vertical, but filling a vertical mold cavity promotes turbulence in the molten metal, resulting in excessive dross; consequently, the mold should be tilted when being poured.

For relatively simple castings, the cavities may be placed one above the other (see Fig. 49); the metal then flows through the lower cavities to fill those above. This permits maximum utilization of the face area of the mold. However, for more complex castings, especially those for which there are significant projections in the cavities, it is usually necessary to gate each cavity individually.

In design of a permanent mold, the part is laid out in the orientation decided on and the mold is designed around it, allowing sufficient space for gating, the necessary seal to prevent metal leakage, and coring and mold inserts. It is common practice to contour the back of the mold so that its exterior conforms roughly with the cavity. This permits more even temperature distribution and heat dissipation. For castings with heavy sections, the adjacent mold sections are generally heavier. For aluminum castings, a ratio of three or four mold-wall thicknesses to one casting-wall thickness is often used, but a mold wall this thin cannot always be used in making thin-wall castings without jeopardizing mold stability. Ribbing is often used to stiffen the mold structure, but

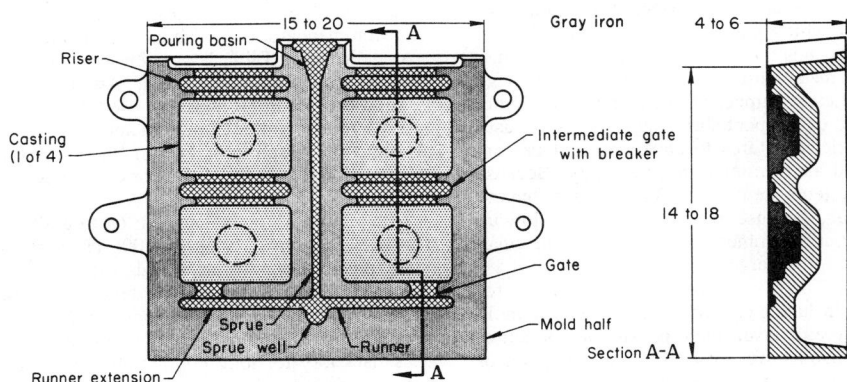

Fig. 49. Bottom-gated permanent mold with stacked cavities for four castings. (Dimensions are in inches.)

excessive ribbing can cause distortion by increasing the temperature differential between inner and outer mold surfaces.

GATING SYSTEMS

Permanent mold castings can be gated from the top, side or bottom, and single or multiple gates can be used.

Top Gating. In this gating arrangement, the sprue and the riser are usually the same. Thin sections are placed farthest from the gate, so that directional solidification is toward the gate. After pouring, the gate functions as a riser. The metal in the riser solidifies last, thus ensuring sound metal throughout the casting.

Side gating is frequently used, particularly for aluminum castings. In this gating system, the riser is at the top of the casting. The gate extends up the side of the casting to nearly 75% of its height, which ensures that the metal at the top of the casting and in the riser is hotter than the first metal to enter the mold. Thin sections should be placed remote from the gate and riser. The direction of solidification is from the mold cavity toward the gate and riser, so that porosity caused by shrinkage is minimized.

Bottom gating is always used for pouring gray iron castings and may be used for other metals. A typical gating system, as shown in Fig. 49, consists of a pouring basin, a choked sprue, a sprue well, a horizontal bottom runner, a runner extension, and vertical gates. For the mold shown in Fig. 49, the two bottom die cavities lead into breakers and intermediate gates that feed the two top die cavities. These lead into risers. Risering is a minor consideration in gating systems for pouring gray iron, because a hypereutectic cast iron does not shrink during solidification.

Gating systems for permanent molds are less flexible than those for sand molds, and are nearly always located in parting planes. Gating must supply metal fast enough to fill all sections of the casting, with a minimum of turbulence. Gating systems that permit minimum turbulence are especially important in casting of aluminum and magnesium alloys, because turbulence creates excessive amounts of oxide, which may cause defective castings. Molds for aluminum and magnesium alloys are poured in the vertical position or tilted from vertical. By this method, air is readily displaced and vented off at the mold parting.

Besides supplying liquid metal to compensate for casting shrinkage, risers reduce the velocity of the metal before it enters the cavity, and help sustain the mold temperature. The number of points at which metal is admitted to the mold cavity depends on section thickness and the distance the metal must flow. Excessive flow through an insufficient number of inlets may result in hot spots and consequent shrinkage. Sprues are usually restricted in area to choke off and prevent dross and air from entering the cavity. Because gating systems are often easier to enlarge than to reduce, and because oversize gating can slow up the cycle, it is common practice to start with small gates, and to enlarge them if necessary.

Misruns. There are several possible causes for misruns, including entrapped air, insufficient mold temperature or insufficient pouring temperature. Often, misruns are caused by a combination of two or more such conditions. Adequate venting is the simplest way to eliminate entrapped air. Increasing the mold temperature or the pouring

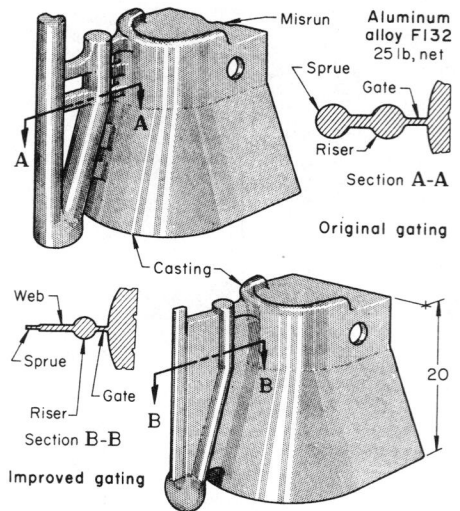

In redesigned mold for the new gating, provision was made for application of external heat to prevent misruns. In original mold, misrun area was impractical to heat, and an insulating mold coating was used to prevent misruns.

Fig. 50. Original gating system for a transmission-case casting, and improved gating than saved 3.2 kg (7 lb) of metal

temperature, or both, may eliminate misruns, but is likely to cause other defects such as porosity. One common approach is to increase the mold temperature at the critical locations by applying heat from an external source (antichill), or by using an insulating type of mold coating in the specific location of the mold to prevent the liquid metal from chilling in a specific area. Causes and the corrective steps taken to eliminate misruns that apply in some cases are illustrated in Fig. 50.

SELECTION OF MOLD AND CORE MATERIALS

Four principal factors affect selection of materials for permanent molds and cores: (*a*) the pouring temperature of the metal to be cast, which determines the amount of heat that must be removed by the mold; (*b*) the size of the casting and (*c*) the number of castings per mold, both of which limit the selection of materials to those that will withstand the effects of the high temperature and the casting cycles; and (*d*) the cost of the mold material. These variables form the basis for the recommendations in Tables 10, 11 and 12.

As indicated in Table 10, gray iron is the mold material most commonly used. Aluminum or graphite molds are sometimes used for small-quantity production of aluminum and magnesium castings, and graphite or carbon liners on steel are sometimes used for molds for casting copper alloys.

With aluminum or magnesium casting alloys, it is not unusual to obtain 100 000 castings, or more, per mold; however, molds for copper or gray iron casting alloys have a shorter life, because of the higher pouring temperatures required. Gray iron molds without tool steel inserts are satisfactory for long production runs of aluminum and magnesium castings that will be machined extensively and for which surface finish is not a major consideration. In casting of zinc,

well over 100 000 pours are possible in a gray iron mold (although die casting is usually selected when such large quantities of zinc castings are to be produced).

Mold Inserts. Full or partial mold-cavity inserts of the same material as the mold, or of a different material, are sometimes used to obtain longer mold life, or to simplify machining, handling or replacement. Inserts can be used also for venting, for cooling thin walls, and for heating portions of the mold or the full cavity area.

Inserts made of cast-to-shape gray iron, used for casting complex aluminum and magnesium parts that range in surface area from 325 cm^2 (with tolerances of ±0.75 mm) to 2900 cm^2 (with tolerances of ±1.5 mm), or from 50 in.^{2} (with tolerances of ±0.030 in.) to 450 in.^{2} (with tolerances of ±0.060 in.), will last for 5000 to 20 000 pours, depending on the complexity of the cast parts.

Materials for cores are recommended in Tables 11 and 12 on the basis of performance over a wide range of coring requirements for small and large cores.

An expendable core is used when the location or shape of the core would not permit it to be removed from the casting, or where intricate design can be obtained at less cost with materials for such cores. The order of preference in the use of these materials is:

1. Sand (oil-bonded or resin-bonded, shell, carbon dioxide–silicate)
2. Plaster
3. Graphite and carbon.

Mold parts can be cast to shape when the castings to be made are not complicated. These are suitable for applications for which gray iron or H11 die steel is recommended in Table 10. Such cast molds are most often used when production is large and mold replacement or shipping schedules require several identical molds.

MOLD COATINGS

A mold coating is applied to mold and core surfaces to serve as a barrier between the molten metal and the surfaces of the mold while a skin of solidified metal is formed. Mold coatings are used for four purposes:

1. To prevent premature freezing of the molten metal
2. To control the rate and direction of solidification of the casting and thus its soundness and structure
3. To minimize thermal shock to the mold material
4. To prevent soldering of molten metal to the mold.

Types. Mold coatings are of two general types—insulating and lubricating. Some coatings perform both functions. A good insulating coating can be made from (by weight) one part sodium silicate to two parts colloidal kaolin in sufficient water to permit spraying. The lubricating coatings, usually include graphite in a suitable carrier. Fifteen typical compositions of mold coatings are listed in Table 13. Coatings are available as proprietary materials.

The various requirements of a mold coating are not always obtained with one coating formulation. Often, these requirements are met by applying different coatings to various locations in the mold cavity.

Table 10. Recommended materials for permanent molds

Alloy to be cast	Number of pours		
	1000	10 000	100 000
For casting small parts (maximum dimension: 25 mm, or 1 in.)			
Zinc	Gray iron; 1020 steel	Gray iron; 1020 steel	Gray iron; 1020 steel
Aluminum; magnesium	Gray iron; 1020 steel	Gray iron; 1020 steel; H11 die steel	Gray iron with H14 inserts; 1020 steel
Copper	Gray iron	Gray iron	2½% beryllium copper
Gray iron	Gray iron(a)	Gray iron(a)	(Quantity not poured)
For casting medium-size and large parts (maximum dimension: up to 915 mm, or 36 in.)			
Zinc	Gray iron; H11 (b)	Gray iron; H11 (b)	Gray iron; H11 (b)
Aluminum; magnesium	Gray iron	Gray iron	Gray iron; H11; H14 (c)
Copper	Alloy cast iron	Alloy cast iron	Alloy cast iron(d)
Gray iron	Gray iron(a)	Gray iron(a)	(e)

(a) Same composition as being poured. (b) H11 is used where polish is required on parts of medium size. (c) Recommendations are for medium-size parts; for large parts, recommended materials are gray iron with H11 inserts or solid H11 die steel. (d) Recommendations are for medium-size parts; large parts are not poured in this quantity. (e) Parts are not poured in this quantity.

Table 11. Recommended materials for small cores (less than 75 mm, or 3 in., in diameter by 250 mm, or 10 in., long) for permanent molds

Alloy to be cast	Recommended core material(a)
Zinc	Sand; plaster; gray iron; 1010 steel
Aluminum or magnesium	1010 or 1020 steel; sand; plaster; H11 die steel or equivalent(b); carbon(c)
Copper	Sand; 1020 steel; gray iron; plaster(d); graphite(c)
Gray iron	Sand; graphite; carbon; gray iron

(a) Core materials are listed in descending order of preference. (b) Hardened to 42 to 45 HRC. (c) For use with relatively few pours. (d) For casting of aluminum bronzes.

Table 12. Recommended materials for large cores (more than 75 mm, or 3 in., in diameter by 250 mm, or 10 in., long) for permanent molds

Alloy to be cast	Number of pours		
	1000	10 000	100 000
Zinc	Gray iron; 1020 steel	Gray iron; 1020 steel	Gray iron; 1020 steel
Aluminum; magnesium	Gray iron(a); gray iron with 1020 steel inserts(a); sand, plaster(a)	Gray iron(a); gray iron with 1020 or H11 inserts(a); sand; plaster(a)	Gray iron(a); gray iron with H11 inserts(a); H11 die steel
Copper	Sand	Sand	(b)
Gray iron	Sand; graphite; carbon; gray iron	Sand; graphite; carbon; gray iron	(b)

NOTE: Core materials are listed in descending order of preference.

(a) Except for openings with complex shapes, which require the use of expendable sand cores. (b) Parts are not poured in this quantity.

Table 13. Typical compositions of coatings for permanent molds

Coating No.	Composition, wt % (remainder, water)									
		Insulators					Lubricants			
	Sodium silicate	Whiting	Fire-clay	Metal oxide	Diatomaceous earth	Soapstone(a)	Talc(a)	Mica(a)	Graphite	Boric acid
1	2	...	4	...	...	...	...	...	1	...
2	8	...	4	...	...	...	...	...	...	...
3	...	7	...	...	...	...	...	...	...	7
4(b)	12	9	...	...	...	...	...	...	...	...
5	5	11	...	2	...	...	5	...	...	...
6	9	...	4	...	...	14	...	...	...	...
7	11	...	...	17	...	...	...	...	...	...
8	...	...	4	...	...	23	...	5	...	...
9	7	...	1	...	...	23	...	2	...	...
10	23	...	...	...	...	...	20	...	...	...
11	30	...	...	...	5	...	10	...	...	...
12	18	...	...	...	41	...	...	...	...	...
13	8	...	...	60	...	...	...	...	...	...
14	7	...	...	...	...	...	62	...	...	...
15	20	53	...	...	...	...	...	...	...	...

(a) Serves also as an insulator. (b) Plus 2 wt % silicon carbide, for wear resistance.

CONTROL OF MOLD TEMPERATURE

Optimum mold temperature is that temperature which will produce a sound casting in the shortest time. For an established time cycle, control of temperature is achieved largely by the use of auxiliary cooling or heating and by control of coating thickness.

Auxiliary cooling is often achieved by forcing air or water through passages in mold sections adjacent to the heavy sections of the casting. Water is more effective, but scale will soon coat the passages in the areas where heat extraction is needed most, making frequent adjustments in water flow rates necessary. Eventually, the flow of water becomes so restricted that the scale deposit must be removed. The problem of scale formation has been solved in some plants by use of recirculating systems containing either demineralized water or another fluid such as ethylene glycol. However, such practice is infrequently used.

Mold Coating. A mold coating of controlled thickness can equalize solidification rates between thin and heavy sections. Chills and antichills can be used to further adjust solidification rates, so that freezing proceeds rapidly from thin through intermediate and into heavy sections, and finally into the feeding system.

Chills are used to accelerate solidification in a segment of a mold. This can be done by directing cooling air jets against a chill inserted in the mold or, more simply, by using a metal insert without auxiliary cooling. Alternatively, chilling can be achieved by removal of some or all of the mold coating in a specific area to increase thermal conductivity. Chills may be used to increase production rate, improve metal soundness, and increase mechanical properties.

Antichills. The function of an antichill is to slow the rate of cooling in a specific area. Heat loss in a segment of a permanent mold may be reduced by directing an external heating device, such as a gas burner, against an antichill inserted in the mold.

POURING TEMPERATURE

Generally, permanent mold castings are poured with metal that is maintained within a relatively narrow temperature range. This range is basically established by the composition of the metal being poured, although size and weight of the casting, mold-cooling practice, mold coating, and gating systems also are considered in establishing the optimum pouring temperature for a specific casting.

Low Pouring Temperature. If pouring temperature is lower than optimum, the mold cavity will not fill, inserts (if used) will not be bonded, the gate or riser will solidify before the last part of the casting, and thin sections will solidify too rapidly and interrupt directional solidification. Low pouring temperature consequently results in misruns porosity, poor casting detail and cold shuts.

High pouring temperature causes shrinkage of the casting and mold warpage. Warpage leads to loss of dimensional accuracy. Also, variations in metal composition may develop if the casting metal has components that become volatile at a high pouring temperature. High pouring temperature also increases solidification time (thus decreasing production rate) and almost always shortens mold life.

REMOVAL OF CASTINGS FROM MOLDS

After a casting has solidified, the mold is opened and the casting is removed. To facilitate release of the casting from the mold, a lubricant is often added to the mold coating. The use of as much draft as permissible on all portions of the casting makes ejection easier. For many castings, ejector pins or pry bars must be used.

Core pins and cores should be designed so as not to interfere with removal of castings from the mold.

Aluminum alloy castings require at least a 1° draft for easy mechanical ejection from the mold prior to manual removal (the more draft, the better for ejection). The mold coating should contain a lubricating agent (usually graphite) to prevent sticking.

Magnesium alloy castings are subject to cracking when removed from the mold, because the metal is hot short. Thus, the use of adequate draft is mandatory. On ribs, a draft of 5° is an absolute minimum. However, 10° is recommended and will result in fewer ejection difficulties.

Copper alloy castings will stick in the molds for any of several reasons, but insufficient draft is usually the main one.

Gray iron castings, with very few exceptions, are manually removed from permanent molds. Automatic or semiautomatic devices are ineffective and impractical because of the high metal temperatures and because of the differences in expansion and contraction rates of the material of the ejection devices and the gray iron mold.

Die Casting

Reviewed by W. T. Andresen, American Die Casting Institute, Inc.

DIE CASTINGS are produced by forcing molten metal under pressure into permanent steel dies. Die casting involves metal flow at high velocities induced by the application of pressure. Because of this high-velocity filling, die casting can produce shapes that are more complex than shapes that can be produced by permanent mold casting. (In Europe, die castings are generally called "pressure die castings.")

In die casting, after the die has been closed and locked, molten metal is delivered to a piston pump, which may be cold or may be heated to the temperature of the molten metal. The pump plunger is advanced to drive the metal quickly through the feeding system while the air in the die escapes through vents. Sufficient metal is introduced to overflow the die cavities, fill overflow wells and develop some flash. As the extraneous metal solidifies, pressure is applied to the remaining metal and is maintained through a specified dwell time to allow the casting to solidify. The die opens and the casting is then ejected. While the casting die is open, it is cleaned, cooled and lubricated as required. Then the die is closed and locked, and the cycle is repeated.

The principal advantages of the die casting process are:

- More complex shapes can be made by die casting than by other metalforming processes.
- Because the dies are filled by pressure, castings with thinner walls, greater length-to-thickness ratio, and greater dimensional accuracy can be produced by die casting than by most other casting processes.
- Production rates are higher in die casting, especially when multiple-cavity dies are used, than in other casting processes.
- Because die castings are produced as almost completely finished parts, the investment in inventory and factory floor space is reduced to a minimum.
- Dies for die casting (like molds for permanent mold casting) can produce many thousands of castings without significant change in casting dimensions.
- Metal cost is often lower than in other casting processes, because die casting permits casting of thinner sections, and because metal savers are often designed into the part.

- Many die castings can be plated (finished) with minimum surface preparation.
- Some aluminum alloy die castings can develop higher strength than comparable sand castings.

The principal limitations of the die casting process are:

- Casting size is limited; casting weight seldom exceeds 23 kg (50 lb) and normally is less than 4.5 kg (10 lb).
- Depending on casting contours and gating, difficulty may be encountered with air entrapped in the die. Entrapped air is one cause of porosity.
- The facilities, consisting of the machine, the auxiliary equipment and the dies, are relatively expensive. Quantities of castings in excess of 1000 pieces per year are required for the process to be economical.
- With few exceptions, commercial use of the process is limited to metals having melting temperatures no higher than those of the copper-base alloys.

MACHINES

All die casting machines have one of two different metal-pumping systems: either a hot-chamber system or a cold-chamber system.

If the metal being cast melts at a low temperature or does not attack the injection-pump material, the pump can be placed directly in the molten metal bath (hot-chamber machines). If the molten metal attacks the pump material at casting temperature, the pump must not be placed in the metal bath, and a cold-chamber machine must be used. Lead, zinc and magnesium usually are cast by this process.

(In this section of this article, only the metal-pumping systems of die casting machines are discussed. For information on the metal-melting and

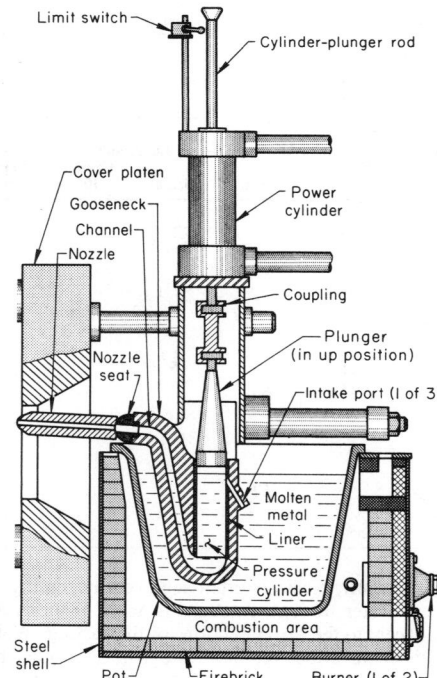

Fig. 51. Principal components of the shot end of a hot-chamber die casting machine

metal-holding furnaces and the metal-transfer systems used in connection with die casting machines, see subsequent sections of this article.)

Hot-Chamber Machines. The metal-pumping system shown in Fig. 51, which consists essentially of pressure and power cylinders, plunger, gooseneck and nozzle, is typical of hot-chamber injection (shot) systems. The gooseneck containing the pressure cylinder and plunger is submerged in the molten casting metal and is thus at the temperature of the metal bath. This arrangement allows the metal to be injected into the die cavities in minimum time and with minimum decrease in temperature.

With the plunger in the up position as shown in Fig. 51, molten metal flows from the pot into the pressure cylinder through the intake ports. With the die closed and locked, the power cylinder is energized to move the plunger downward. This seals off the intake ports. With further downward movement of the plunger, the molten metal is forced through the gooseneck channel and the nozzle into the die cavity. After a preset time to allow the metal to solidify in the die cavity, the power cylinder is activated in the reverse direction, thus pulling the plunger up. This uncovers the intake ports, and metal flows from the pot into the pressure cylinder. The machine is now ready for the next cycle. The power cylinder can be actuated either by oil or air (machines of recent or current vintage are activated by oil).

Castings weighing from a few grams to about 23 kg (about 50 lb) can be produced in hot-chamber machines. The amount of metal cast with one stroke can be varied by using different sizes of gooseneck assemblies. The weight of castings that can be made depends on the alloy being cast, the projected area of the shot, and the locking pressure.

Depending on degree of mechanization, process variables, and the part being cast, hot-chamber machines generally operate at rates of 50 to 500 shots per hour. Very small special machines greatly exceed these rates, ranging from 2000 to 5000 shots per hour up to 18 000 shots per hour for a zipper-casting machine.

The gooseneck is made of gray, alloy or ductile iron, or of cast steel. The choice of material is dictated by operating pressure, casting metal, and cost. Common practice is to insert a replaceable liner in the bore of the pressure cylinder (Fig. 51). The material for the liner should have good wear resistance and good resistance to softening at operating temperature.

Liners and nozzle seats usually are made from H13 or a high speed tool steel, nitrided alloy steel, or stainless steel. Nozzles must be able to resist the washing action of the molten metal and the scaling action of externally applied heat, and must be strong enough at operating temperatures to resist the pressure of molten metal. Alloy cast iron, H13 tool steel, and stainless steel are among the materials used for nozzles. A recent development of the Commonwealth Scientific Industrial Research Organization of Australia, called the PQZ theory, relates the important time element to the pressure on, and quantity of, the metal being pumped. (This research was sponsored by the International Lead Zinc Research Organization.)

Horizontal Cold-Chamber Machines. The shot end of a typical cold-chamber machine is shown in Fig. 52. In a cold-chamber machine, the shot chamber is unheated except for the heat from the molten metal ladled into it for casting, and the plunger

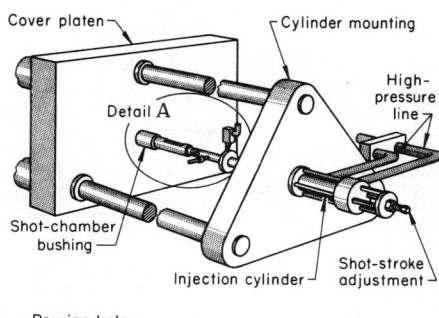

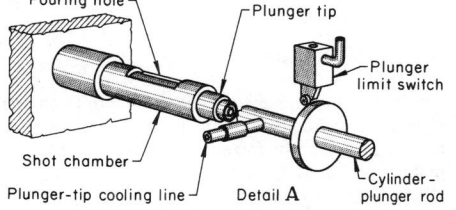

Fig. 52. Principal components of the shot end of a horizontal cold-chamber die casting machine

tip is water cooled to prevent it from overheating. To facilitate pouring of the metal into it, the shot chamber is mounted horizontally with a pouring hole in the top of the chamber wall (Fig. 52).

Figure 53 shows the operating cycle of a horizontal cold-chamber shot system. With the die closed and locked and the cylinder-plunger rod retracted, the metal is ladled through the pouring hole of the shot chamber (position 1 in Fig. 53). Then the injection cylinder is energized and moves the plunger rod forward. The plunger tip first closes off the pouring hole and then forces the metal into the die (position 2 in Fig. 53). After a preset dwell time, which permits the metal to solidify, the die is opened and the plunger is moved in the direction of initial travel to complete its full stroke (position 3 in Fig. 53). This final forward travel of the plunger tip forces the biscuit, or excess metal, free of the shot chamber and, at the same time, aids in stripping the cast-ing from the cover die. The plunger rod retracts, the casting is ejected from the ejector die, and the shot end is ready for the next cycle.

Cold-chamber injection systems can be used for all metals that can be die cast, but they are normally used for aluminum alloys, copper alloys, and the high-aluminum zinc (ZA) alloys.

The main advantages of cold-chamber machines are the relative freedom from attack of molten metal on equipment (because the shot chamber and plunger are not submerged in the molten metal), and high injection pressures. Injection pressures range from 55 to 207 MPa (8 to 30 ksi). Pressures up to 345 MPa (50 ksi) have been used in special or experimental applications.

The main disadvantages of cold-chamber machines are: (a) the need for an auxiliary method of feeding metal, (b) longer cycle time than is needed for hot-chamber operation and (c) the possibility of metal defects due to loss of superheat.

Vertical Cold-Chamber Machines. There are two basic types of vertical cold-chamber machines. In one type, the die parting is horizontal; in the other, it is vertical. The commonest type of vertical cold-chamber die casting machine with the die parting line in the horizontal plane is shown in Fig. 54. In this machine, metal is injected from below the die. Air is evacuated from the die and molten metal is drawn into the shot chamber by vacuum. Pressures for clamping the dies and injecting the metal are controlled from a single accumulator source, which provides and maintains a balanced pressure even in the event of a malfunction. The horizontal position of the die, and the location of the shot sleeve below the die, eliminate the need for gating upward to prevent premature metal flow from the shot sleeve. For good balance in multiple-cavity dies, the sprue may be located in the center of the die, as in Fig. 54. Metal runners may be gated from any point on the sprue, and they may follow a direct path to the cavities. Some of the larger vertical machines are equipped with center and off-center shot positions.

A vertical cold-chamber die-casting machine with vertically parting dies is shown in Fig. 55. The injection system includes a vertical shot chamber connected directly to the cover die half by a sprue bushing. A hydraulically actuated lower plunger covers the bushing hole in the vertical cold chamber while metal is being ladled into the chamber from above (position 1 in Fig. 55). After the metal has been ladled, the upper, or shot, plunger is actuated. As the pressure builds up, the lower plunger retracts so that the metal is forced through the sprue bushing into the die (position 2 in Fig. 55). After a dwell cycle for metal solidification, the upper plunger is withdrawn while the lower plunger rises and shears off the remaining slug of metal (commonly called a metal biscuit) and ejects it (position 3 in Fig. 55).

Machine Selection. Hot-chamber die casting machines are used mainly for casting of low-melting metals (such as zinc, tin and lead alloys). Although cold-chamber machines can be used for all die casting metals, they are normally used for aluminum, aluminum-zinc and copper alloys. Beyond this division, machine selection for making any casting should be based principally on clamping force and opening stroke, metal-pumping capacity, length of shot stroke, maximum shot pressure, maximum die opening and die size, maximum and minimum die height, clearance between beams, over-all size, and cost.

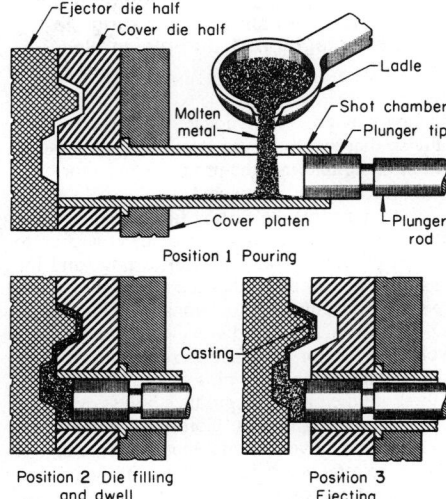

Fig. 53. Operating cycle of a horizontal cold-chamber die casting machine

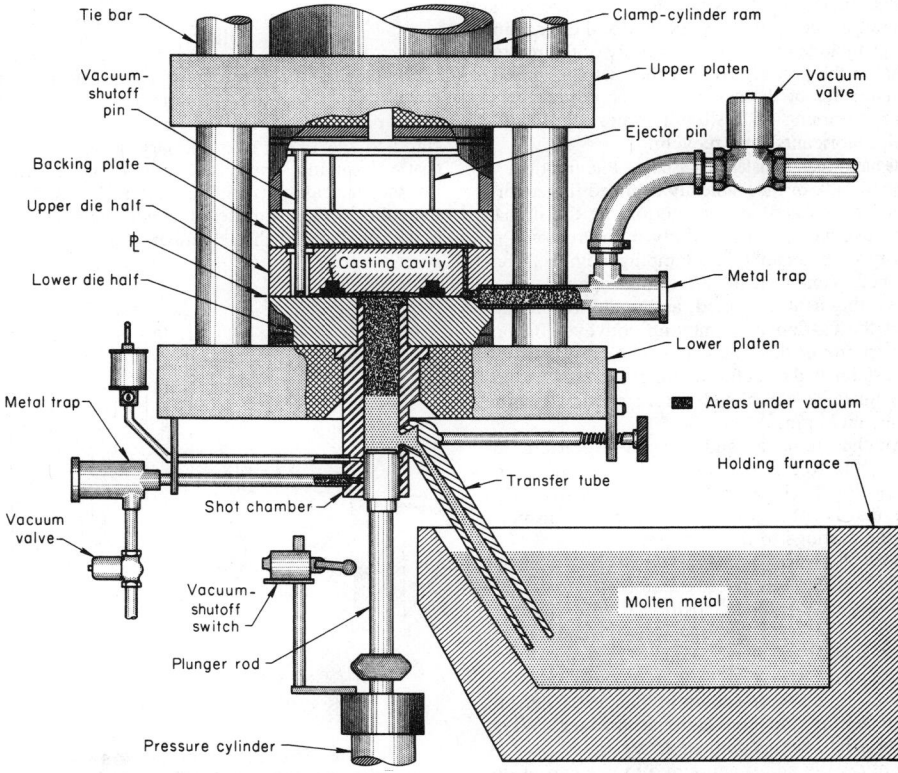

Fig. 54. Principal components of a vertical cold-chamber die casting machine with the die parting line in the horizontal plane

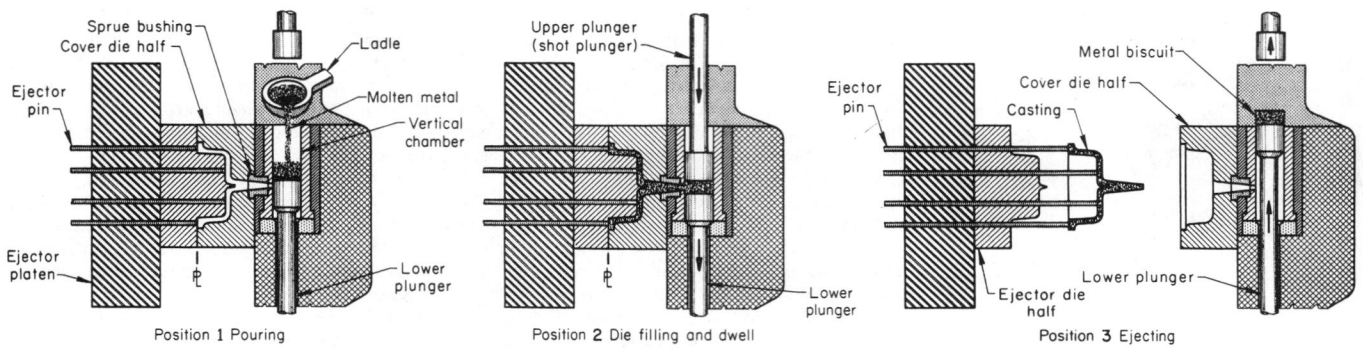

Fig. 55. Operating cycle of a vertical cold-chamber die casting machine with vertical die parting

The rule of thumb for selection of die casting machine size is to use the smallest machine that will do the job. This will ensure fundamental economy of operation, since the larger the machine, the slower its cycle. For example, a 365-t (400-ton) machine can cycle about twice as fast as a 725-t (800-ton) machine. Commercially available machines range in size (clamping force) from 23 to 2270 t (25 to 2500 tons).

Clamping force is not necessarily the deciding factor in selection of a die casting machine. Die dimensions must be considered. A machine of adequate tonnage for casting a part may have insufficient platen area or tie-rod spacing for the die, or the opening stroke may not be sufficient for removal of the casting.

AUXILIARY EQUIPMENT FOR MECHANIZATION

Die casting machines are frequently mechanized to provide greater consistency, to decrease the number of castings rejected to reduce costs. Operations performed by mechanized die casting setups include control of the supply of the molten metal in the pot, removal of the casting from the die, transfer of the casting to succeeding operations, trimming of flash and sprues, application of die lubricants, and recycling.

Removal and Transfer of Castings. Pneumatically, hydraulically or mechanically operated fingers that duplicate the actions and motions of the human arm have worked successfully in the removal of castings. A finger-type clamping device is positioned to enter the die after it has opened, grasp the casting as it is ejected, and withdraw and deposit the casting on a moving conveyor belt or in a trimming die. Recycling is controlled by a pre-established circuit. Action of the removal device must be completed before another casting cycle can begin.

Another removal and transfer procedure involves the free fall of castings into water. The die casting machine is located over a water-filled tank that is fitted with a conveyor for transporting the castings to the next operation.

In-the-die trimming uses an indexing hub adjacent to the die; the hub has two unloading arms containing socket recesses positioned 180° apart, as shown in Fig. 56. Each time a shot is made, metal is cast into one of these rotatable recesses, forming a carrier for the shot (which consists of two castings in the setup shown in Fig. 56). The shot is ejected in the conventional manner. When ejection has been completed, the entire shot is free of the die and is supported by the end of the unloading arm on the indexing hub. The hub then rotates 180° and swings the arm and the shot out

of the die into a cutoff area, to the position shown in Fig. 56. At the same time, the other unloading arm moves into position for the next shot. As the die closes, the shot (still attached to the unloading arm on which it was cast) moves between cutoff blades on the side of the die, and castings are separated from the runner and sprue and fall into an appropriate chute. The runner, still attached to the unloading arm, is then stripped from the unloading arm by a hydraulically operated stripper (not shown in Fig. 56). The runner and sprue fall into another chute and are returned to the furnace for remelting.

If the index is reduced to 120°, three operations (casting, piercing and trimming) can be combined in one die. This device can be used only with small castings, because of maintenance problems and the hazard of nicks and dents in the castings. Programable controllers, or microprocessors, for proper and consistent sequencing of the casting machine cycle are now in fairly common use.

DIES

Die-casting dies consist of two sections — a cover die half and an ejector die half — that meet at the die parting line. The cover die half is secured to the front, or stationary, platen of the casting machine. The sprue, or shot hole, for filling the die cavity is in this half and is aligned with the injector nozzle of a hot-chamber machine, or with the shot chamber of a cold-cham-

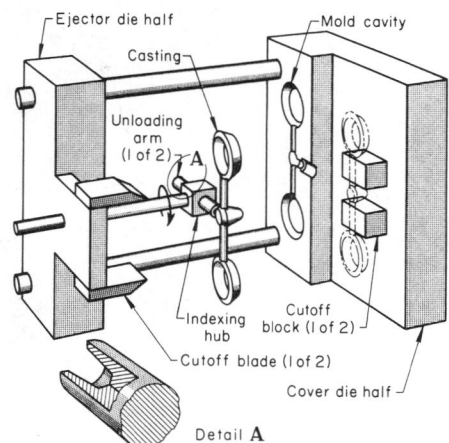

Fig. 56. Die that incorporates an indexing hub and rotary unloading arms for handling die castings during in-the-die trimming

ber machine. The ejector die half contains the ejector mechanism and, in most cases, the runner. It is normally fastened to the movable die platen, which is coupled with the actuating or clamping mechanism of the die-casting machine. A cross section of a simple hot-chamber die-casting die is illustrated in Fig. 57.

The die cavity is machined into hot work die steel inserts that are fitted into the two halves of the die block. The location of the parting line must ensure that, on opening of the die, the casting will pull away from the cover half and remain in the ejector half. Thus, when shapes having interior surfaces are to be cast, as much of

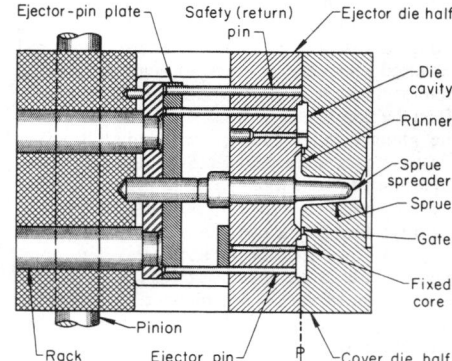

Fig. 57. Principal components of a simple hot-chamber die-casting die with integral rack-and-pinion ejection

the cavity as possible is located in the cover die half, so that when the die is opened, the casting will have shrunk onto the protruding ejector half, from which it can be ejected.

Ejector pins free the casting from the die. There must be a sufficient number of ejector pins to prevent the hot casting from distorting, and the pins must be placed so that the marks they leave are not in objectionable locations. When the casting is removed and the die is closed for the next cycle, the ejector-pin plate is forced back, thus withdrawing the ejector pins from the cavity.

Cores are die parts that produce holes, openings and other casting details. Cores that extend parallel with die movement and that therefore do not have to be removed before ejection of the casting are called fixed cores (see Fig. 57). The ejector pins force the casting from fixed cores as the casting is pushed out of the die cavity. Cores with axes not parallel with die movement are called movable cores; these cores must be withdrawn

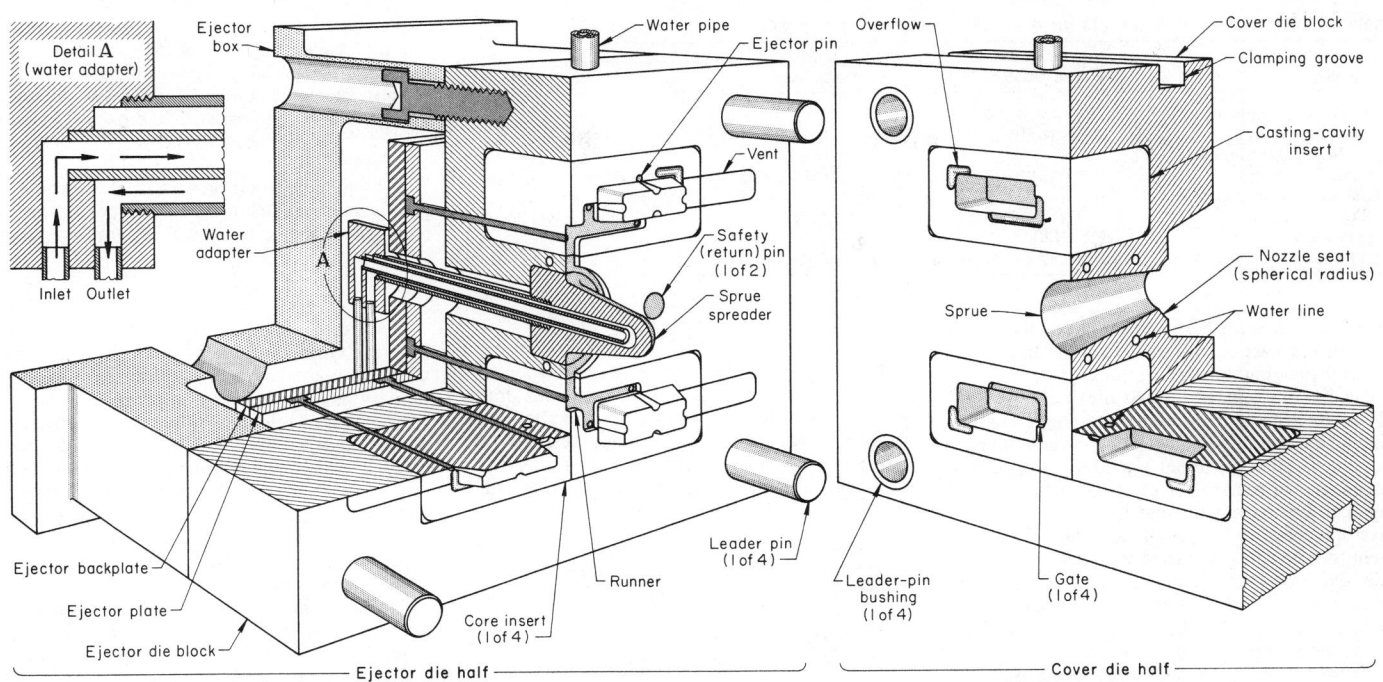

Fig. 58. Multiple-cavity die for casting four automotive lamp bezels per shot

by a separate mechanism before the casting is ejected from the die.

Slides are the movable die parts needed to build up die surfaces. Slides are used when it is impossible to avoid undercuts in a casting. The part of the die-cavity wall that forms the undercut portion is made on the face of a slide that is accurately fitted in a guide cut in the die block. The slide must be retracted before the casting can be ejected. A separate locking mechanism must be provided for each slide. The use of slides adds considerably to die cost.

Runners and Gates. The metal entering the sprue is directed into one or more passages, or runners. Near the die cavity, the cross-sectional area of the runner decreases to form a gate designed to direct the metal into the die cavity (Fig. 57). By decreasing the area of the metal-feeding system up to the gate orifice, increased velocity of the liquid metal (to minimize turbulence) is ensured.

Overflows. For some castings, the die cavity may feed into an overflow (see Fig. 58). Overflows perform several important functions:

• They serve as receptacles for the first metal entering the die cavity during each shot. Normally, this metal will be cooler than the metal behind it. In addition, it will oxidize to some extent as it forces the air that is in the cavity out through the vents. Feeding this cooler, oxidized metal to overflows both avoids cold shuts and traps any included oxides in appendages to the casting, which are later trimmed off, remelted and recovered.

• They provide additional mass to small castings, thereby helping to maintain a satisfactory and stable die temperature.

• They serve as locations for contact of ejector pins when the castings themselves are not permitted to have ejector-pin marks on any surface. (This theory, however, is challenged by many knowledgeable die casters.)

Cooling. In die casting, there is hot metal in the die cavity during a large portion of each cycle, and cycling is always rapid; measured in milliseconds, heat is dissipated to and retained by the dies, especially around the sprue and around heavy cross sections in the casting. To prevent overheating of a die, sections of the die are cooled to a controlled temperature by water circulating through passages drilled in these sections. Temperature control is achieved by control of the amount of water in circulation. The location of water-cooling lines in production die sets is shown in Fig. 58. Systems utilizing calorific oil for both heating and cooling, especially in die casting of magnesium, are now coming into use.

Single-cavity dies are used when: (a) castings to be produced have so large a surface area that the die for one casting fills the die space in the machine; (b) the volume of metal required for the casting is near the shot capacity of the machine; (c) the production quantity does not warrant a larger machine for multiple-casting production; (d) a suitable machine for multiple-cavity production is not available; or (e) the complexity of the casting requires movable slides and cores that preclude the use of more than one cavity.

Multiple-cavity dies consist of two or more duplicate cavities for making identical castings in one die (see Fig. 58). A multiple-cavity die, compared with a single-cavity die, reduces the unit cost per casting by producing a greater number of castings per machine cycle. In addition, multiple cavities may sometimes produce better castings, because the symmetrical spacing of cavities often provides a more uniform thermal balance in the die and a better distribution of mechanical forces. On the side of disadvantages, the increased number of cavities increases the potential operating problems and sometimes results in a slightly higher scrap rate. The number of shots per hour is slightly reduced for multiple-cavity dies, and these dies may require a larger die casting machine.

The number of castings required generally determines the type of tooling, since a certain quantity must be produced to justify the use of more elaborate tooling. The design of the casting is also a factor; the possibility of lowering costs by using multiple-cavity tooling decreases as size and complexity of the casting increase, and as tolerances become more stringent.

Combination dies, or "family" dies, consist of a series of cavities in one die for casting two or more different parts. The parts cast in a combination die are generally components of a single end-product assembly.

The development of a suitable die-cavity arrangement and an optimum gating system is usually more difficult in production with combination dies than in production with multiple-cavity dies. This is because differences in size and shape of castings cause an imbalance in the mechanical and thermal performance of the die. Two or more castings differing in volume and size by a ratio of 10 to 1 can be produced in a combination die, but a smaller ratio is preferred.

Combination dies require development of the tooling to achieve the required dimensional, visual and physical properties. The dimensional development work is usually simpler than for multiple-cavity tooling, because there is only one cavity for each component in the combination die.

One disadvantage of a combination die is that it must be operated under the conditions that will produce the most difficult casting of the group the most efficiently. The remaining castings in the group ordinarily could be produced at higher speeds and at less cost. Another disadvantage of combination tooling involves the fact that, typically, the castings produced in a combination die are intended for assembly into a single unit. This means that all of the castings will, in effect, have scrap rates equal to that of the casting having the highest scrap rate, since no unit can contain a defective casting. Consequently, if the volume required is high, a separate die for each different

Table 14. Recommended materials for die-casting dies and die inserts

Type of casting alloy	Die or insert material for:		
	Up to 50 000 shots	250 000 shots	1 million shots
Zinc alloys (25-mm, or 1-in., die cavity)	P20 (a)(b)	P20 (a)(b)	P20 (b), H13 (b)
Zinc alloys (100-mm, or 4-in., die cavity)	P20 (a)(b)	P20 (a)(c), 4150 mod (d)	4150 mod, H13 (b)
Aluminum and magnesium alloys	H11, H13	H13, H11	H13, H11
Copper alloys	H21, H20, H22	...	...

(a) Prehardened to 280 to 320 HB. (b) Recommended for die inserts. (c) Hardenability barely adequate. (d) Hardened to 44 to 46 HRC.

casting, or at least a separate die for the most troublesome casting, will generally be more economical than combination tooling.

Unit dies are separate, small dies, usually single-cavity, that are inserted in a single master holding die and operated several-at-a-time in large machines. Since different unit dies may be interchanged in the master die, a variety of castings can be made with one master die. Master dies are generally regarded as a capital investment, because they are used with many different unit dies for many different customers — very often over a period of many years. They consist of cover and ejector halves, which are mounted in the die casting machine.

The replaceable unit dies consist of cover and ejector unit blocks that contain the die cavity. Cavities can be precut inserts, or they can be made by cutting directly into the unit blocks. Casting economy dictates the type of cavity used.

The runner openings in the unit align with those in the center section of the master die. Suitable clamping devices are provided, to permit easy removal and installation of the unit dies while the master holder is mounted in the machine. The nozzle or shot chamber is just tangent to the cover half of the master die; it is untouched during die changes. The ejector mechanism, too, generally requires no adjustment when the unit dies are changed.

A unit die can be used advantageously to minimize the cost of tooling for producing a series of different small castings in small quantities. With unit dies, as with standard combination dies, it is difficult to achieve and maintain proper operating conditions for all the castings being made. This is true particularly in production of castings of different wall thicknesses and shapes, which normally require different machine-cycle times.

Die Materials. Selection of materials for die-casting dies depends mainly on the type of metal being cast and on production quantities. Table 14 gives recommendations for materials for dies and die inserts, based on the above variables. Recommended materials for cores, slides, and ejector pins are listed in Table 15.

Table 15. Recommended materials for cores, slides and ejector pins for die-casting dies

Casting alloy	Recommended material
Cores and slides(a)	
Zinc	440B (b), H13, H11, H12
Al or Mg	H13
Copper	H21, H20, H22
Ejector pins(a)	
Zinc	Nitriding steel, H13
Al or Mg	Nitriding steel, H13
Copper	H21, H20, H22

(a) Sliding components, except those used with dies for casting copper alloys, are nitrided for improved wear resistance. (b) Cores only.

DIE DESIGN

The shape of a die cavity must correspond to the dimensions of the part to be cast, plus draft, or taper, to facilitate ejection, plus an allowance for shrinkage. For close-tolerance castings, allowance must also be made for thermal expansion of the die cavity. The placement of the cavity in the die block is governed by:

- Parting-line requirements
- Requirements for moving cores and slides
- Location of a noncritical area in the casting for gating
- The need for positioning the gate so that the initial flow of metal is not obstructed by cores or other members

Experience with similar castings is most frequently the basis of die design. Regardless of past experience, however, changes in design ranging from minor to major are often required before optimum results are obtained. An excellent example of good design is shown in Fig. 59.

DIE WEAR

The rate of die wear is influenced chiefly by the temperature of the casting metal and by the design of the die. When the metal has a casting temperature no higher than that of zinc alloys, and the die is of simple design, it is not unusual to obtain more than 500 000 shots before there is a significant amount of die wear. As metals with higher and higher casting temperatures are used, progressing from zinc to aluminum and thence to copper alloys, die wear increases rap-

idly, regardless of die design. As the configuration of the casting and gating system becomes more complex, wear in localized portions of the die also increases, especially as the temperature of the casting metal rises. Die erosion (wash) is likely to be severe when the hot metal goes around a corner.

Although die wear generally does not vary greatly when metals of the same type and casting temperature are being cast, there are exceptions.

FURNACES

Melting of metal for die casting entails three different types of operations: (a) melting down of virgin ingot; (b) remelting of scrap from the foundry and from trimming operations, and reconstitution of this melt by means of suitable additions; and (c) holding of quantities of molten metal at a closely controlled temperature, adjacent to the die casting machine. Of these operations, the first and second involve continuous bulk *melting*, whereas the third operation is usually restricted to maintaining molten metal at a desired temperature.

Zinc Alloys. The furnaces used for melting and alloying of zinc alloys are generally of the pot type, although immersion-tube and induction furnaces also are used. The construction of some melting furnaces permits tilting to aid in pouring of the metal. In practice, a furnace designed specifically for melting should, in general, have a bath capacity of five to seven times the amount of metal required per hour.

Aluminum Alloys. For melting of aluminum alloys, gas furnaces have the advantages of low initial cost and easy repairability.

The pot, or crucible, furnace, which is similar to that used for zinc alloys, is constructed with a sheet-metal shell surrounding layers of insulation, insulating brick, and firebrick. A semispherical ceramic pot holds the metal.

The reverberatory furnace is a rectangular or cylindrical iron shell, refractory-lined throughout, in which the heat source is above the metal bath, as shown in Fig. 60. Flame or radiant tubes provide the heat. Flames from the burners may contact the aluminum, but heat transfer is mainly by radiation from the walls. Reverberatory fur-

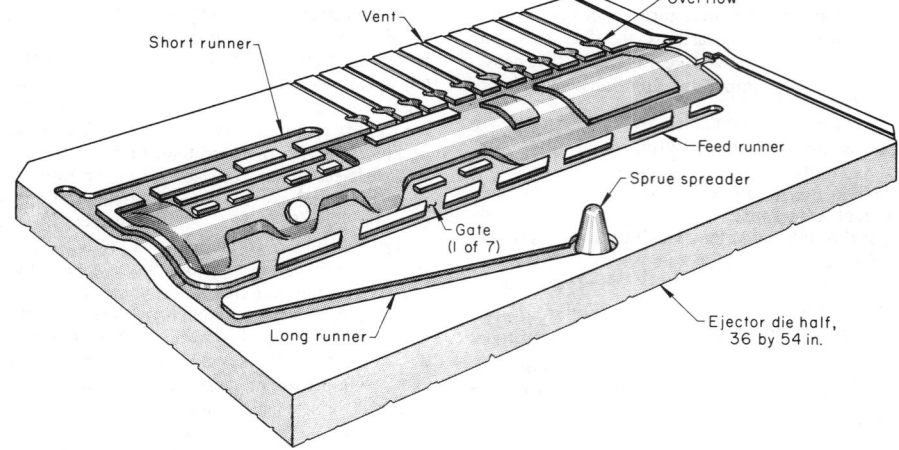

Dimensions of grill component: 1.07 m (42 in.) long by 305 mm (12 in.) wide, with an average wall thickness of 1.3 mm (0.050 in.). This system represents good design for metal feeding.

Fig. 59. Arrangement of gates, runners, vents and overflows in an ejector die half for making an automotive grill component

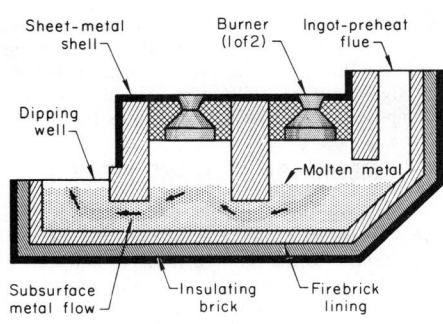

Fig. 60. Cross section of a gas-fired reverberatory furnace used for melting aluminum die casting alloys

Table 16. Compositions of zinc die casting alloys

	Alloy designation			Composition, %							
ASTM	Commercial	SAE	Covering ASTM specification	Cu	Al	Mg	Pb, max	Cd, max	Sn, max	Fe, max	Zn
AG40A	...	...	B240, ingot(a)	0.10 max	3.9-4.3	0.025-0.05	0.004	0.003	0.002	0.075	rem
AG40A	3	903	B86, die castings	0.25 max	3.5-4.3	0.020-0.05	0.005	0.004	0.003	0.100	rem
AC41A	...	...	B240, ingot(a)	0.75-1.25	3.9-4.3	0.03-0.06	0.004	0.003	0.002	0.075	rem
AC41A	5	925	B86, die castings	0.75-1.25	3.5-4.3	0.03-0.08	0.005	0.004	0.003	0.100	rem
...	7	903	....	0.25 max	3.5-4.3	0.005-0.02(b)	0.003	0.002	0.001	0.075	rem

(a) Zinc alloy ingot is sold to more rigid limits to allow for pickup of impurities in the casting process. (b) Alloy also contains 0.005 to 0.02% Ni.

naces are charged through doors in the wall, or through an external well (not shown in Fig. 60). They are divided into sections for heating or melting the metal before it enters the dipping well for ladling. A reverberatory furnace is generally used for melting rather than for holding.

Electric furnaces may be heated by induction, radiation or conduction. Those in general use are channel, low-frequency induction furnaces. Electric melting and holding prevents segregation of the various chemical elements in the alloy.

Magnesium Alloys. Gas-fired pot furnaces, stationary or tilting, and coreless induction furnaces are commonly used for melting magnesium alloys; reverberatory furnaces are used only to a limited extent. For a small operation, a stationary open-pot furnace generally is satisfactory. For a medium-size operation, it is better to have a tilting furnace to facilitate metal transfer and desludging of the pot. Coreless induction furnaces are economically feasible. Modern energy-efficient furnaces, both electric and gas-fueled, employing superb insulating materials, are used. These furnaces operate on less than 25% of the energy required for furnaces of older design and construction.

Melting and holding furnaces for magnesium alloys are similar in general construction to those for zinc alloys, but the combustibility of magnesium requires that the following precautions respecting furnace design and materials be observed:

• Molten magnesium reacts violently with some refractories. High-alumina refractories and high-density, superduty firebrick with approximately 57% silica and 43% alumina have given satisfactory service.
• Because molten magnesium reacts violently with iron oxide, an access door should be provided to permit removal of scale from the pot and furnace.
• A runout pan should be provided to hold the magnesium in the event of pot failure.
• A protective cover must be kept over the molten magnesium to prevent burning. The cover can be a proprietary flux or sulfur dioxide gas. Adequate ventilation must be provided to remove the strong, corrosive fumes of the flux or the gas. When sulfur dioxide gas is used as a cover, a hood with a small ladling door is placed over the holding pot so that the gas provides a protective curtain at the door of the furnace.

Copper Alloys. Gas-fired pot-type furnaces are the type most widely used for melting and hold-

ing copper alloys for die casting. The furnaces are similar to gas-fired pot furnaces used for zinc alloys. The pots are usually made of silicon carbide, and range in capacity from 30 to 100 kg (70 to 225 lb), depending on the installation. Gas consumption ranges from 0.19 to 0.31 m³/kg (3 to 5 ft³/lb) for melting, and from 0.09 to 0.14 m³/kg (1.5 to 2.3 ft³/lb) for holding. Gas consumption varies with the alloy being melted, because of differences in melting temperature. Maintenance requirements for these furnaces are not great. Normally, the lining must be patched after about 80 h of operation; other components must be repaired or replaced periodically, and the furnace must be completely rebuilt after about 24 months of operation.

Induction furnaces are sometimes used for melting copper alloys. Compared with pot furnaces, induction furnaces offer the advantages of less drossing, faster melting, better control of composition and temperature, less absorption of gases in the molten metal, and greater operator comfort. For some installations, high-frequency coreless induction furnaces are preferred because of speed of melting and flexibility in alloy changeover. A coreless furnace can be emptied and shut down during nonworking hours, and it can be started cold. Low-frequency core or channel-type induction furnaces are not practical for heats of less than 225 kg (500 lb).

CONTROL OF COMPOSITION

The compositions of casting alloys strongly influence the die casting process and the final product, and should be checked periodically as the castings come from the machine.

The greatest source of impurities is return scrap from rejected castings, especially those that have been plated, and trimmings (sprues, runners, and flash). Scrap materials should be remelted in separate facilities, and they should be analyzed before being recast.

Zinc Alloys. Compositions of zinc die casting alloys are given in Table 16. Good melting practice with minimum agitation and close temperature control will cause little change in these compositions. Overheating will result in a loss of aluminum through oxidation, and in an increase in iron due to a decrease in the scavenging action provided by aluminum. Return scrap for remelting must be clean and free of moisture. Clean scrap may be charged into the furnace in unlimited proportions, provided that it is of acceptable chemical composition. If there is doubt about the purity of the scrap, it should not be used until analyzed.

Aluminum in amounts of 3.5 to 4.3% is added to zinc alloys to increase strength and to improve

fluidity and castability. Casting of alloys containing less than 3.5% Al should be avoided. If aluminum content is too low (especially when it gets below 2%), the attack on dies and machinery becomes more aggressive and the castability of the alloy decreases. When aluminum content exceeds 4.3%, each increase in aluminum content up to the eutectic composition (5% aluminum) results in a corresponding decrease in ductility and shock resistance.

As shown in Table 16, small amounts of magnesium are present in zinc die castings to counteract the tendency of the impurities in the zinc alloys to promote subsurface corrosion. Magnesium also provides the hardness necessary for the handling of zinc die castings. Preferably, the magnesium content should be close to the minimum values given in Table 16. When the magnesium content of a zinc alloy casting exceeds 0.05%, the susceptibility of the casting to hot cracking increases.

Copper in small amounts helps minimize the adverse effects of impurities and moderately increases hardness and strength. However, zinc alloys containing more than 1.25% Cu are not dimensionally stable, and age with a severe decrease in ductility.

Iron in the amounts usually present has no detrimental effect on the properties of zinc alloys. However, excessive iron pickup can cause trouble. The iron picked up in melting is removed with the surface dross; thus, good skimming practice is mandatory.

Lead, cadmium and tin are controlled by the magnesium addition. If they are present in amounts above permissible limits, they promote hot shortness and susceptibility to intergranular corrosion.

Nickel and chromium contamination results from remelting of plated castings. Small amounts of silicon and manganese are sometimes introduced with aluminum alloy additions. However, the concentration of any of the above elements is seldom high enough to cause any difficulty.

Aluminum Alloys. Nominal compositions of aluminum die casting alloys are given in Table 17.

Table 17. Nominal compositions of aluminum die casting alloys

Alloy	Composition, %		
	Cu	Mg	Si
13	...	...	12.0
43	...	...	5.0
218	...	8.0	...
360	...	0.5	9.5
380	3.5	...	8.5
384	3.8	...	12.0

Aluminum alloys are extremely susceptible to iron pickup, especially under conditions of overheating (heating above 665 °C, or 1225 °F). Excessive iron pickup not only impairs the quality of the casting, but also shortens the life of metal components in the melting and casting equipment.

Iron is present in aluminum die casting alloys up to a maximum concentration of 2%, but normally in the range of 0.8 to 1.2%. At these concentrations, iron reduces the tendency of the casting metal to become soldered to dies, increases hot strength and minimizes hot cracking. At higher concentrations, iron decreases the fluidity of the alloy and may, in the presence of chromium and manganese, cause sludging. A complex AlFeMn(Cr)Si constituent that reduces ductility and machinability of aluminum alloys is formed at relatively low iron levels (above 0.8% Fe) when the other contributing elements are present.

Manganese is normally specified in aluminum die casting alloys to a maximum of 0.5%. Because manganese can form sludge when mixed with iron, manganese content must always be considered in relation to iron content, particularly if metal temperature is lower than about 650 °C (1200 °F).

Chromium content is normally less than the specified maximum of 0.25%.

Nickel content is usually limited to 0.5%, although some special alloys for which improved elevated-temperature properties are desired may contain as much as 3% Ni.

Zinc, when present in amounts up to 3%, mainly affects the specific gravity of the aluminum alloy. As zinc content increases above 4%, Brinell hardness increases approximately five points for each 1% increase in zinc. In the range of 4 to 8% Zn, aluminum alloys are hot short.

Magnesium content is limited to 0.10% (except in alloys 218 and 360). Excessive amounts of magnesium decrease the fluidity of molten aluminum alloys and also increase the hardness and decrease elongation and impact strength of the castings.

Magnesium Alloys. Alloys AZ91A and AZ91B are the only magnesium alloys normally used for die casting. Their compositions are identical (9 Al, 0.13 min Mn, 0.7 Zn), except that 0.3% copper is permitted in AZ91B.

Preferred practice is to premelt and refine the magnesium alloys, even when the melting stock consists entirely of clean ingots. This preliminary treatment is mandatory when any portion of the casting stock consists of scrap metal. Refining is accomplished by heating the metal to approximately 700 °C (1300 °F), adding a suitable flux, stirring the molten metal thoroughly, and then letting it stand for 10 to 15 min. The flux is generally a mixture of magnesium chloride, potassium chloride and other chlorides, plus a smaller percentage of calcium fluoride. The amount of flux needed varies from 1 to 3% (by weight) of the molten metal, depending on the cleanness of the melting stock.

Contamination of the molten metal can consist, on the one hand, of flux, charred lubricants, and oxides, and, on the other, of metallic elements in quantities above or below specified ranges. Flux, charred lubricants, and oxides result from improper melting, refining or metal-handling techniques. Flux inclusions cause corrosion. Charred lubricants and oxides affect the casting characteristics of the alloy and the mechanical properties of the casting. Excessive

burning or higher-than-normal viscosity can be an indication of poorly refined metal. Corrosion tests and metallographic examination of a section of the casting can be used to evaluate metal refining procedures.

Off-analysis alloys result from admixture of other alloys, contamination from the vessels used to contain or transfer the molten metal, and losses through remelting. Nickel-bearing alloys should not be used for equipment components that are submerged in the molten casting alloy—not only because components containing nickel will be short-lived, but also because they will contaminate the magnesium casting alloy. Beryllium and aluminum are lost on remelting, and compensatory additions of these elements may be necessary. Composition can be checked by wet-chemical or spectrographic methods. The frequency and extent of chemical analyses depend on operating conditions and quality requirements.

Aluminum is added to magnesium alloys for strength, hardness and castability. Excessive amounts of aluminum promote segregation.

Additions of zinc improve castability and corrosion resistance and, in combination with aluminum, improve mechanical properties. Hot shortness and the likelihood of cracking increase with increasing zinc content.

Manganese additions, in quantities up to the limits of solubility (0.25%), improve corrosion resistance.

Beryllium is sometimes added in quantities up to 0.001%, to reduce burning of the molten alloy.

Iron, nickel and copper are harmful impurities; they should be kept within the specified limits for residuals, because they adversely affect corrosion resistance.

Copper Alloys. The compositions of brass die casting alloys and of other copper-base die casting alloys are listed in Table 18. Controlling the compositions of these alloys depends on the makeup of the charge. Prepared alloys in pig or ingot form, or raw materials, may be used, but the charge normally consists of 50% new metal and 50% mill scrap. Scrap segregation is instrumental in preventing contamination. Proper alloy identification by tag, color code or part number is essential.

Casting scrap and scrap from secondary operations may be melted together and cast into pigs or small ingots that are individually analyzed. If the material can be adjusted to meet specifica-

tions, this is done at the time of melting for casting. When 100% scrap is used for the charge, care should be taken to avoid contamination, excessive dross, and oily chips. Usually, a charge composed entirely of scrap requires fluxing.

METAL TRANSFER AND FEEDING

Metal transfer generally includes two operations: transfer of the molten metal from the melting furnace to the holding furnace, and transfer from the holding furnace to the shot chamber of the die casting machine. Sometimes it is feasible to receive aluminum in liquid form rather than in ingot form, which requires a laundering system between the transport unit and the holding furnace.

Melting Furnace to Holding Furnace. In almost all die casting plants, melting and holding are separate operations. A single large-volume melting furnace may be used to supply molten metal to smaller holding furnaces at the die casting machines. The transfer mechanism must be able to provide a continuous supply of molten metal, heated to the approximate casting temperature, to all the holding furnaces being fed at a given time.

Metal should be transferred at, or slightly above, the holding temperature. When this is done, wide fluctuations in metal temperature are avoided. Furthermore, metal should be transferred to the holding furnace at regular intervals and in small amounts, so that there is a minimum of change in the level of metal in the holding furnace. This reduces buildup of dross on the walls and permits easier ladling.

A problem often unrecognized in the transfer of molten metal is the damaging effect of metal agitation. Aluminum is particularly susceptible to damage from agitation, because of its strong affinity for oxygen. Aluminum oxides having a specific gravity nearly equal to that of aluminum are formed during metal transfer, and are entrained in the form of film and of relatively massive particles. Nonmetallic inclusions can lead to a decrease in fluidity, which causes poor surface finish, excessive flow lines, incomplete filling of the die cavity, cold shuts, or porosity. When the oxide content exceeds 0.005% by weight, excessive tool wear and breakage may be expected in machining operations. The formation of dross by agitation reduces metal yield, thus increasing costs. As dross is generated, it should be removed, not transferred.

Table 18. Compositions of copper-base die casting alloys

Casting alloy	Composition, %										
	Cu	Si	Pb	Sn	Mn	Al	Fe	Mg	Ni	Other	Zn
Brass die casting alloys covered by ASTM B176 (a)											
Z30A . . . . . . .57 min	0.25	1.50	1.50	0.25	0.25	0.50	. . .	. . .	0.50(b)	30.0 min
ZS331A 63-67	0.75-1.25	0.25	0.25	0.15	0.15	0.15	. . .	. . .	0.50(b)	Rem
ZS144A 80-83	3.75-4.25	0.15	0.25	0.15	0.15	0.15	0.01	. . .	0.25(b)	Rem
Other copper-base die casting alloys											
Aluminum bronze (alloy 9B) . . . . . .86 min	. . .	. . .	. . .	. . .	9-11	0.75-1.5	. . .	. . .	. . .	. . .
Aluminum bronze (alloy 9D) . . . . . .78 min	. . .	. . .	. . .	3.5	10-11.5	3-5	. . .	3-5.5	. . .	. . .
Silicon bronze (alloy 13B) . . . . .Rem	3-5	. . .	. . .	. . .	. . .	. . .	. . .	. . .	. . .	12-16

(a) Percentages given are maximum values unless otherwise designated. (b) Arsenic, antimony and sulfur: 0.05% (max) each. Phosphorus, 0.01% max.

Pumping is a convenient and rapid method of transferring metal, although the ceramic pump used is rather fragile and somewhat costly to maintain. Agitation from this high-speed transfer process results in more dross formation than is desirable; however, dross can be prevented from entering the pressure cylinder or the shot chamber in excessive quantities by use of large (450- to 900-kg, or 1000- to 2000-lb) transfer ladles and by intermittent settling, fluxing and skimming of the molten metal to remove the dross.

Minimum turbulence occurs and, therefore, a minimum amount of dross forms when a siphon or a bottom-pour transfer ladle is used. Neither of these methods is as fast as pumping, and the siphon or ladle must be of a size that is compatible with the melting equipment that is to be used.

If hand ladling is used, the dross-covered surface should be pushed back gently to allow the ladle to fill with minimum dross inclusion.

Holding Furnace to Shot Chamber. Methods used to move the molten metal from holding furnace to shot chamber are:

- Ladling—manual or mechanized
- Pressure systems—including pressure on the surface of the molten metal, reduced pressure ahead of the metal (vacuum), and gravity
- Pumps—positive displacement and positive pressure.

Hand ladling is sometimes used for feeding the molten metal to the machine in cold-chamber operations, and is the least expensive from the standpoint of initial outlay and maintenance costs. However, hand ladling has some disadvantages. Because the metal is ladled from the top surface, metal contamination with oxides and other impurities occurs readily. In addition, it is difficult for an operator to maintain the same pace hour after hour.

Operator fatigue is minimized by the use of a mechanized ladle (Fig. 61). The mechanized ladle consists of a trough-and-ladle assembly, which pivots to permit filling of the ladle on the downstroke and pouring of the metal into the shot chamber on the upstroke. The source of movement is an air cylinder located above the apparatus shown in Fig. 61 and connected to the cylinder rod. Actuation of mechanized ladles can be correlated with the cycle of the casting machine. Cleaner metal can be obtained by taking the metal from below the surface of the metal bath.

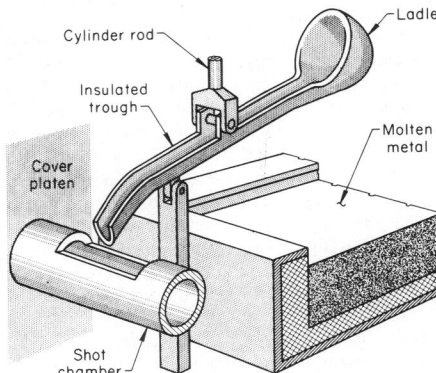

Fig. 61. Mechanized, pivoted ladle for transfer of molten metal from a holding furnace to the shot chamber of a cold-chamber die casting machine

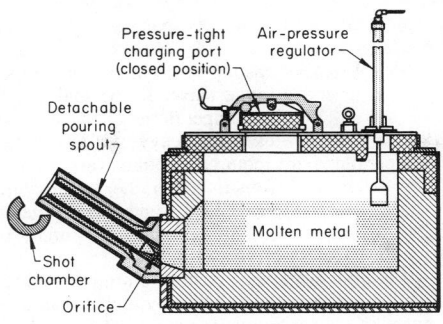

Fig. 62. Air-pressure system for supplying clean molten metal to the shot chamber of a cold-chamber die casting machine

In the system shown in Fig. 62, the molten metal is in a teakettle type of container. Air pressure is applied to the surface of the metal, forcing the metal through an orifice, up the pouring spout, and into the shot chamber. The weight of the shot is controlled by varying the volume and pressure of the air that is applied. Either by an accuracy ensured by the design of the pressure system, or by an external mechanical compensator, the weight of the shot can be controlled within ±5%. Shots weighing up to 45 kg (100 lb) can be made. This system is frequently used in casting of aluminum alloys.

A vacuum-feed system depends on pressure difference to force the metal from the holding furnace to the shot chamber. Details of vacuum systems vary considerably, but the fundamentals are illustrated in Fig. 63. In the system shown in Fig. 63, a vacuum pump evacuates the air from

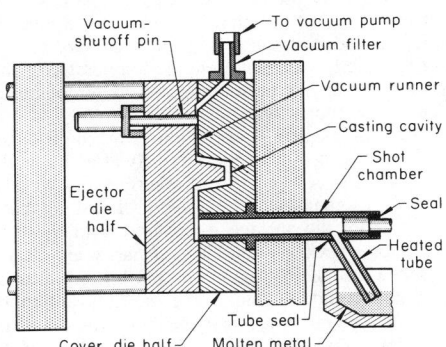

Fig. 63. Vacuum-feed system for supplying molten metal to the shot chamber of a cold-chamber die casting machine

the shot chamber and the heated tube (usually by way of the casting cavity and through a filter or trap). The metal from the holding pot then flows through the heated tube and into the shot chamber, to fill the vacuum. The amount of metal delivered is controlled by the evacuating system. After a predetermined interval, the plunger is advanced to cut off the metal flow and inject the metal that is in the shot chamber into the die cavity. Because actuation of the plunger is generally determined by time elapsed rather than by the volume of metal in the shot chamber, a change in the rate of metal delivery will result in a variation in the weight of the shot. Partial plugging of the tubes or leakage of air are the two most likely causes of variation in shot weight. A major

advantage of a vacuum-feed system is the speed of delivery of the metal, which results in a shorter cycle.

Vacuum-feed systems can be used with all casting alloys; however, for satisfactory tube life, special consideration must be given to material for the heated tube in relation to the casting metal. Black iron pipe or a series 400 stainless steel provides reasonable tube life for use with magnesium or zinc alloys; generally, a coated or ceramic tube is needed for aluminum alloys. In the design of the tube seal, consideration should be given to making the tube readily replaceable, and to keeping flash and plunger-lubricant from being deposited in the tube.

Gravity-feed systems are not used extensively, although they are sometimes employed in die casting of magnesium alloys. In the gravity-feed system illustrated in Fig. 64, the molten metal flows from the pot through the heated transfer pipe and seeks its own level in the chamber surrounding the valve stem. At the proper time in the casting sequence, a signal actuates the air cylinder, which moves the valve stem off the valve seat, thus allowing the metal to flow into the shot chamber. The amount of metal delivered is controlled by the length of time the valve is open. For magnesium alloys, a sulfur dioxide purge is provided to reduce burning and to help keep the valve seat clean. (The valve seat must be clean to get a good seal.) The control of heat on the transfer pipe is also important for good operation.

A third method of feeding molten metal to cold-chamber machines is the use of pumps. Pumps are used mainly for magnesium alloys, but some types have been developed for use with aluminum alloys. Most piston-type pumps have been replaced by constant-pressure centrifugal pumps of the type shown in Fig. 65. For this pump, the following operating cycle is used: the casting machine is closed and locked; the pump impeller goes into high speed; the valve of the delivery tube is opened by the air cylinder, allowing metal to flow up the tube and into the shot chamber; and, after a preset time (controlled by a timer), the valve closes and the impeller returns to low speed.

Most problems with a constant-pressure centrifugal pump have to do with the delivery tube. Improper heating and air infiltration cause the tube to become plugged. To reduce the downtime re-

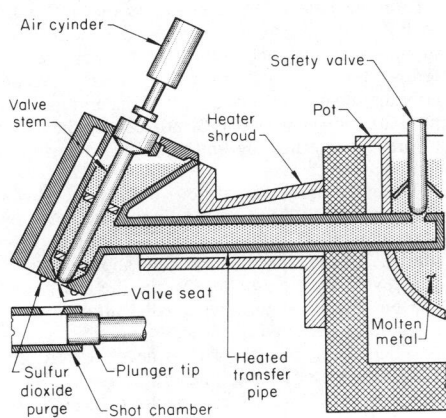

Fig. 64. Gravity-feed system for supplying molten magnesium alloy to the shot chamber of a cold-chamber die casting machine

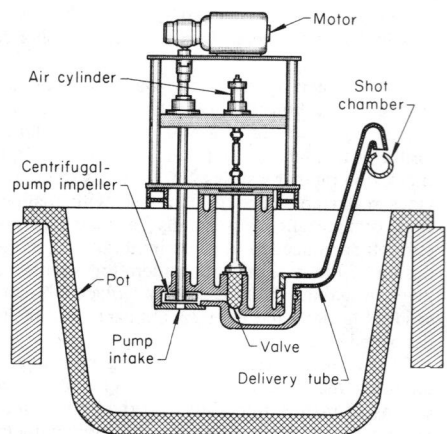

Fig. 65. Constant-pressure centrifugal pump for feeding molten magnesium alloy to the shot chamber of a cold-chamber die casting machine

quired for tube maintenance, the tubes are generally designed with slip fits. Tubes must be designed so that they can be removed when desired, but will not loosen during operation.

DIE TEMPERATURE

Die-casting dies are basically similar to heat exchangers, and thermal conditions must be balanced according to section thicknesses and the configuration of the die.

The temperature that a die will reach and hold during continuous operation depends on the temperature of the casting metal, the weight of the shot, cycle speed, surface area and shape of the die cavity, and provisions for die cooling. Optimum die temperature for a specific casting is determined by section thicknesses and by the type of finish required.

When optimum die temperature has been established, it should be maintained within ±6 °C (±10 °F). Specific areas of a die can be heated by using runners, copper inserts, and strip heaters, or calorific oil. These heating devices are either embedded in the die or mounted externally on the die. It is often necessary to add overflows on perimeters of castings when thin sections are far removed from the main runner (as in the die in Fig. 59). Overflows increase the metal flow in thin sections, and thus help to increase die temperature in these areas. Conversely, water-cooling channels are frequently concentrated behind the runner immediately adjacent to the sprue to prevent overheating and possible soldering at this location.

When the die temperature is too low, the overflows fail to fill, and the castings may have excessive internal porosity, cold shuts and flow marks. A study of the first few shots of a run will reveal the flow pattern in the die cavity and can be used to determine needed alterations in the gate and the overflow wells.

When a die is hotter than is necessary to provide complete filling and good finish, heat checking may occur. Heat checking shortens die life and promotes "hangup," or poor ejection. Failure to eject disrupts the casting cycle, and cycle disruptions contribute to fluctuations in die temperature.

Shrinkage caused by hot spots may occur when sufficient cooling cannot be provided or cores

cannot be built into the die, particularly in heavy sections.

Zinc Alloys. For casting zinc alloys, die temperatures generally range between 165 and 245 °C (325 and 475 °F). Temperatures at the low end of the range are used for heavy castings; higher temperatures are required for thin-wall castings. For a hardware finish, higher die temperatures (at least 220 °C, or 425 °F, and usually nearer 245 °C, or 475 °F) are generally required, regardless of section thickness.

Aluminum Alloys. For casting aluminum alloys, die temperatures are appreciably higher than those for zinc alloys; the usual range is 220 to 315 °C (425 to 600 °F), with the average near 290 °C (550 °F). Because of high heat requirements and the oxidation characteristics of aluminum alloys, good casting finish is considerably more difficult to obtain than with zinc alloys. A surface finish comparable to a hardware finish is rarely obtained on aluminum die castings.

Attack of aluminum on the steel dies becomes more pronounced as die temperatures increase. Cores that are located where heat extraction is difficult are more susceptible to attack and, subsequently, to soldering than are other die components. Several surface treatments of steel cores and cavity inserts help prevent attack by the molten metal.

Magnesium Alloys. Optimum die temperatures for casting magnesium alloys range from 245 to 275 °C (475 to 525 °F). Depending on the shape of the casting and the quality requirements, die temperatures may range from 230 to 290 °C (450 to 550 °F).

Copper Alloys. Die temperatures for copper alloys vary from 315 to 700 °C (600 to 1300 °F), depending on die size and location of the casting in the die. To prolong die life, a temperature near the low end of this range is preferred, but to obtain sound castings it may be necessary to sacrifice die life and increase die temperature. This accounts for the broad range of die temperatures used for copper alloys.

Dies for casting copper alloys are highly susceptible to heat checking. Shortly after a new die is put into service, the die surface shows evidence of checking; after several thousand shots, the entire die surface is covered with thermal-fatigue check marks. These marks are reproduced on the castings and must be removed by grinding and buffing if the castings are to be plated. To minimize heat checking, the die is run as cool as possible. Sometimes, preheating of the die increases its resistance to heat checking. Infrared die heaters can be used for preheating.

DIE LUBRICANTS

Lubricants prevent a casting from adhering to the die and provide the casting with a better finish. A correctly chosen lubricant will allow metal to flow into cavities that otherwise cannot be filled.

Selection of lubricant is based on the temperature of the metal being poured, the operating temperature of the die, and the alloy being cast. No lubricant will perform equally well for all casting alloys.

When the molten metal contacts an oil-containing lubricant, some of the lubricant decomposes and forms a carbonaceous powder, which remains on the surface of the casting after ejection from the die. Any carbonaceous residue on the die can be removed with an air jet. If the lubricant has the correct fire point, enough of it will remain on the die to allow at least five or

six shots before another application is necessary. When the carbonizing temperature of the lubricant is too high, the lubricant will be absorbed into the surface of the casting and will appear as an oil stain. When the carbonizing temperature is too low, all of the lubricant will be used on the first shot. The lubricant must carbonize slowly. In doing so, it must form a minimum amount of gas, but it must not actually burn.

Moving die parts, such as ejectors and cores, must be treated regularly with a high-temperature lubricant to prevent seizure. Oil suspensions of colloidal graphite are generally used for this purpose. Heavy graphite in grease (flake graphite is often used) is applied to plungers and shot chambers. In most plants, die lubricants are purchased as proprietary concentrates and mixed to the desired consistency.

Oil-base lubricants are being replaced by water-base lubricants. A mixture of approximately 50 parts water to one part lubricant is used both to cool the die and to act as a release agent.

Selection for Zinc Alloys. Lubricants for use with zinc alloys are of two general types: solvent-soluble compounds and water-mixed compounds.

Most solvent-soluble lubricants are mixtures of mineral spirits and oil, or kerosine and oil, with graphite sometimes added. These give excellent die release and, at the same time, lubricate the ejector pins. They do not, however, provide cooling for the die.

Water-mixed die lubricants (colloidal graphite or silicone emulsions) are excellent for cooling the die and provide good die release, but do not adequately lubricate the ejector pins. When water-mixed lubricants are used, it is desirable to use a separate pin lubricant and to apply it regularly.

Selection for Aluminum Alloys. Four types of lubricants are used with aluminum alloys: pigmented compounds, graphite greases, colloidal graphite in oil, and water-mixed compounds.

Pigmented compounds are mixtures of pigment and oil. The pigment has a higher melting point than the metal being cast, and will not embed itself in the surface of the casting. The oil provides the carbonaceous deposit that allows easy ejection. Frequently, when a die is worn or checked or has insufficient draft, a good, unstained casting can be obtained with this type of compound. A disadvantage of pigmented compounds is that they adhere to the dies and thus impair dimensional accuracy. This can be prevented by frequent cleaning of the dies with a caustic solution.

Graphite greases are best for dies that operate at temperatures ranging from cool to normal. The waxes in these compounds act as the wetting agents, and the oil and graphite are the lubricants. These compounds are usually mixed with kerosine in a ratio of fifteen parts kerosine to one part lubricant, and then are sprayed on the die. The mixture should be agitated frequently to prevent the graphite from settling out.

Colloidal graphite in oil is used when die temperatures range from normal to hot. Carbonizing takes place slowly with this type of compound, and if the dies are not hot enough, oil stains will show on the surface of the casting.

Water-mixed die sprays are colloidal-graphite or silicone emulsions that cool the die and act as a lubricant. Proprietary concentrates are usually mixed with water in ratios from 1:30 to 1:100.

To lubricate the plunger or ram, the above lubricants, with the exception of the water-mixed types, are used. The lubricant most commonly used for this purpose is colloidal graphite in oil, which can also be used on ejector pins, slides and cores.

Selection for Magnesium Alloys. The die cavity requires very little lubrication for magnesium alloy castings (very often none is required). The most common fault is overlubrication. If the die cavity is lubricated, the lubricant must be in the form of a fine mist generated by a spray gun. Dies can be sprayed with commercially available die lubricants either automatically (using fixed spray guns) or by hand.

Sliding components are lubricated with the same compounds as those discussed above for aluminum.

Selection for Copper Alloys. Sliding components are lubricated with the same materials as those described above for aluminum alloys. For copper alloys, lubricants must be used sparingly or not at all, because they will burn into the casting. An oil-and-graphite lubricant is often used around the sprue or on specific areas of the cavity where sticking is a problem. When a lubricant is used, it is applied by swabbing or spraying.

VARIABLES THAT DETERMINE CYCLE TIME

The optimum casting cycle is established on the basis of past experience and a trial production run. Changes may be required in dwell time, shot-injection time, pouring temperature, or methods of die cleaning and lubrication. Adjustments in die cooling are usually made by varying the flow rate of the cooling water or by using external cooling. Sometimes, changes must be made in the sizes and locations of cooling passages, gates and overflows.

The cycle time for producing a die casting (or for making a shot that yields more than one casting) depends largely on the size and weight of the shot. These basic factors are interrelated with machine capability, die-opening and closing time, pouring time, injection time, dwell time, metal temperature, extraction time, time for die cleaning, and time for application of lubricant.

Die-opening and closing time depends on the travel distance and the average speed of travel of the ejector half of the die, on the number and complexity of movable components used in the die, and on the speed and travel distance of the individual movable components. Die-opening time is usually slightly less than die-closing time.

Pouring time depends on the type of pouring system used and on the amount of metal per shot. Time for hand ladling is influenced by operator efficiency. Automatic-pouring time varies for the different systems.

Injection time varies with the volume of the metal cast (longer time for larger-volume castings) and the configuration of the gate used in the die. Generally, a longer time is required for a more constricted gate.

Dwell time depends on the cooling rate of the die (lower cooling rates require longer dwell times, and vice versa), casting weight (especially if walls are thick), and metal temperature (the hotter the metal, the longer the dwell time).

Extraction time depends on the type of actuation, speed and travel distance of the die, and whether extraction can partly overlap the die-opening operation. Extraction time also varies with operator efficiency in withdrawing the ejected casting from the machine (if withdrawal is manual). Location, size and number of ejector contacting surfaces, together with size, weight and configuration of the casting, influence the ejection portion of the extraction time.

Die-cleaning time is affected by the method and equipment used, the amount of residue left in the die, operator efficiency (manual operation), and cavity size and configuration.

Lubricant-application time is influenced by the method and equipment used, the amount of lubricant required, the area and configuration of the die cavity to be lubricated, operator efficiency (in manual application), and the condition of die-cavity surfaces.

TRIMMING

Runners, gates, overflows, and parting-line flash must be removed from a casting before it is ready for further processing. For small pro-duction quantities, removal can be done manually with a mallet and other hand tools; but when the production volume is large, trimming dies are used.

Trimming dies for die castings are often similar in design to the blanking and piercing dies used for sheet metal, and sometimes they resemble those used for trimming of closed-die forgings (see the article on Closed-Die Forging in Hammers and Presses, in this volume).

Trimming dies are mounted in a mechanical or hydraulic press. Presses used for trimming of castings are usually larger in bed area in relation to their rated tonnage than are presses used for trimming of forgings, because the forces required for trimming even relatively large die castings are low. The metal to be trimmed is ordinarily 0.75 to 1.3 mm (0.030 to 0.050 in.) thick.

The degree of complexity that is built into a trimming die depends mainly on production quantities. For instance, when quantities are small, a cored hole may be trimmed by a simple punch; the outside flash is removed in a second operation. For larger quantities, or for continuous high production, the two operations (or more than one of each kind) can be done in a single press stroke with a more complex die. For trimming complex castings, side slides may be incorporated to permit trimming in more than one direction in a single stroke of the press.

Die materials for trimming dies depend on the number of castings to be trimmed.

For example, for trimming a small quantity of castings, a simple punch may be made of low-carbon unhardened steel. Dies for production trimming are usually made from hardened tool steels such as W1, O1, A2 or D2. High-carbon high chromium tool steels, like D2 hardened to 58 to 60 HRC, give the longest life and are the most economical, especially for complex, expensive dies. In some plants, the established practice is to make all cutting components of the dies from a carbon steel such as 1020 or 1030 and then to hard face the edges with a cobalt-base alloy.

Melting and Casting of Ferrous Metals

Melting of Gray Iron

GRAY IRONS are characterized by the presence of most of the contained carbon as flakes of free graphite in the as-cast iron. Because of the presence of carbon dissolved in the molten iron in amounts of about 2.8 to 4.0%, gray iron has the lowest casting temperature, the least shrinkage, and the best castability of all ferrous metals.

Tensile strengths of 140 to 415 MPa (20 to 60 ksi) are readily attained (the higher strengths in this range usually require alloying), and most gray irons are used as cast. Heat treatment is effective in softening gray irons for better machinability or in hardening them for increased wear resistance and strength.

CLASSIFICATION BY TENSILE STRENGTH

The tensile strength of gray iron is affected both by the amount of free graphite present and by the size, shape and distribution of the graphite flakes. Flake size, shape and distribution are influenced strongly by metallurgical factors in the melting of the iron and in its subsequent treatment while molten, and by the rates of solidification and cooling in the mold. For this reason, a required tensile strength can be achieved with more than one chemical composition, and composition alone does not determine properties.

Gray iron is usually specified entirely by the mechanical properties of standard separately cast test bars, and composition is mentioned only when some special property is required. Relations between the properties of test bars and of castings poured from the same iron have been established on the basis of relative rates of solidification and cooling in the mold, which are the determining factors for a given iron as poured. The foundry selects the raw material, composition and treatment of the iron necessary for the melting equipment used, to meet the properties specified in a standard test bar.

Tensile strength obtained with a given gray iron as poured decreases as solidification rate and cooling rate in the mold decrease. These rates depend on the ratio of surface area to volume of the casting. Relationships are given for ten irons in Table 1 (see also Fig. 1).

CARBON EQUIVALENT

Cast iron containing only iron and carbon would contain little free graphite as cast. The presence of silicon in amounts of 0.5 to 3.0% is necessary to cause free graphite to separate during cooling of the iron in the liquid state.

MELTING METHODS

Gray iron is usually melted in a cupola, an induction furnace or an electric-arc furnace; the major but decreasing percentage is melted in cupolas. The composition of the melt is determined by the charge materials used; it is adjusted by operation of the melting equipment and by the addition of inoculants and alloying elements, to

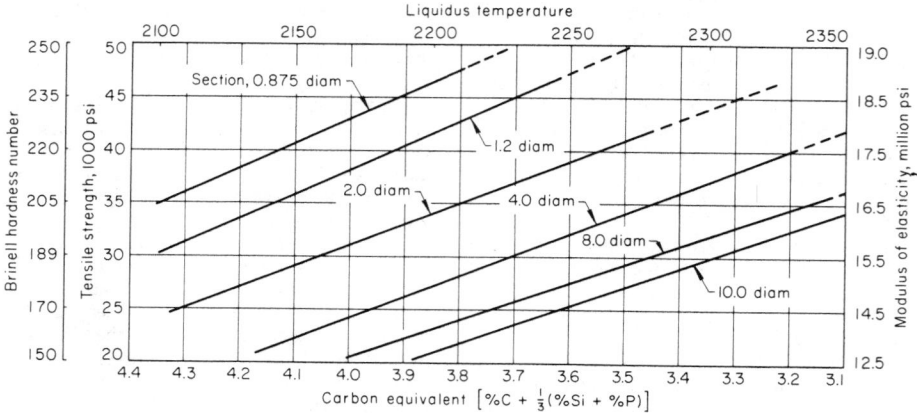

Fig. 1. Interrelation of mechanical properties, section diameter, carbon equivalent and liquidus temperature of gray iron

Table 1. Typical relations of minimum tensile strength to section thickness for gray iron castings

	Minimum tensile strength for casting thickness of:							
	Up to 13 mm (1/2 in.)		Over 13 mm (1/2 in.) to 25 mm (1 in.)		Over 25 mm (1 in.) to 50 mm (2 in.)		Over 50 mm (2 in.) to 75 mm (3 in.)	
Iron	MPa	ksi	MPa	ksi	MPa	ksi	MPa	ksi
1 ...	186	27	138	20	(a)	(a)	(a)	(a)
2 ...	228	33	179	26	131	19	(a)	(a)
3 ...	248	36	193	28	152	22	(a)	(a)
4 ...	262	38	207	30	165	24	(a)	(a)
5 ...	276	40	207	30	193	28	(a)	(a)
6 ...	276	40	228	33	186	27	124	18
7 ...	290	42	234	34	193	28	138	20
8 ...	310	45	255	37	214	31	159	23
9 ...	345	50	290	42	241	35	172	25
10 ...	359	52	296	43	255	37	186	27

(a) Casting to this thickness is not recommended.

produce castings that are within specified property limits.

CUPOLAS

The primary function of a cupola is to melt iron to a specified tapping temperature and chemical composition. A cupola is a vertical steel shaft into which coke, flux and metal are charged in alternating layers. Conventional, or refractory-lined, cupolas are the most widely used, although water-cooled (also called water-wall) cupolas have also come into extensive use.

The essential elements of conventional and water-cooled cupolas are shown in Fig. 2. Of particular concern are the bottom doors, sand bed, iron trough, wind box, tuyeres, charging doors, and linings.

Cupolas such as those shown in Fig. 2 are pre-

pared for operation by closing the bottom doors, supporting them by a prop, then placing on them a bed of rammed sand 150 to 250 mm (6 to 10 in.) thick. The taphole for iron is at the edge of the sand surface; the row of tuyeres for entry of air is 300 to 900 mm (12 to 36 in.) above the sand, depending on the cupola size and whether tapping is intermittent or continuous. A bed of coke is placed on the sand bottom and ignited, and then made up to a height of 1.0 to 1.5 m (40 to 60 in.) above the tuyeres. The shaft is filled through the charging door with alternate layers of coke, flux and metallic charge.

Air supply or blast is then introduced through the tuyeres, and intense heat of combustion is thereby developed in the coke bed. The metal at the surface of the bed melts and trickles down through the hot coke to collect on the sand bottom in the well below the tuyeres. The column of charge material descends to replace the metal melted, and a fresh layer of coke replenishes the coke burned in the bed to melt one charge. This process continues as long as air supply is continued and coke and metallic charge are added. Molten slag, which is coke ash and nonmetallics in the charge (fluxed usually by limestone), is also formed and floats on the surface of the molten iron in the well.

If the cupola is continuously tapped, as most medium-size and large cupolas are, the iron and slag flow continuously through the same taphole and are separated in a small basin in the spout, the slag floating and being discarded. For intermittent tapping, there is both an iron taphole at the front of the cupola and a slag taphole at the rear some 300 to 600 mm (12 to 24 in.) higher. The iron taphole is closed with a fireclay plug, so that iron and slag are accumulated in the well; as the level rises, the molten slag floating on the iron reaches the slag hole and flows out. When the iron level is near or at the slag hole, the iron

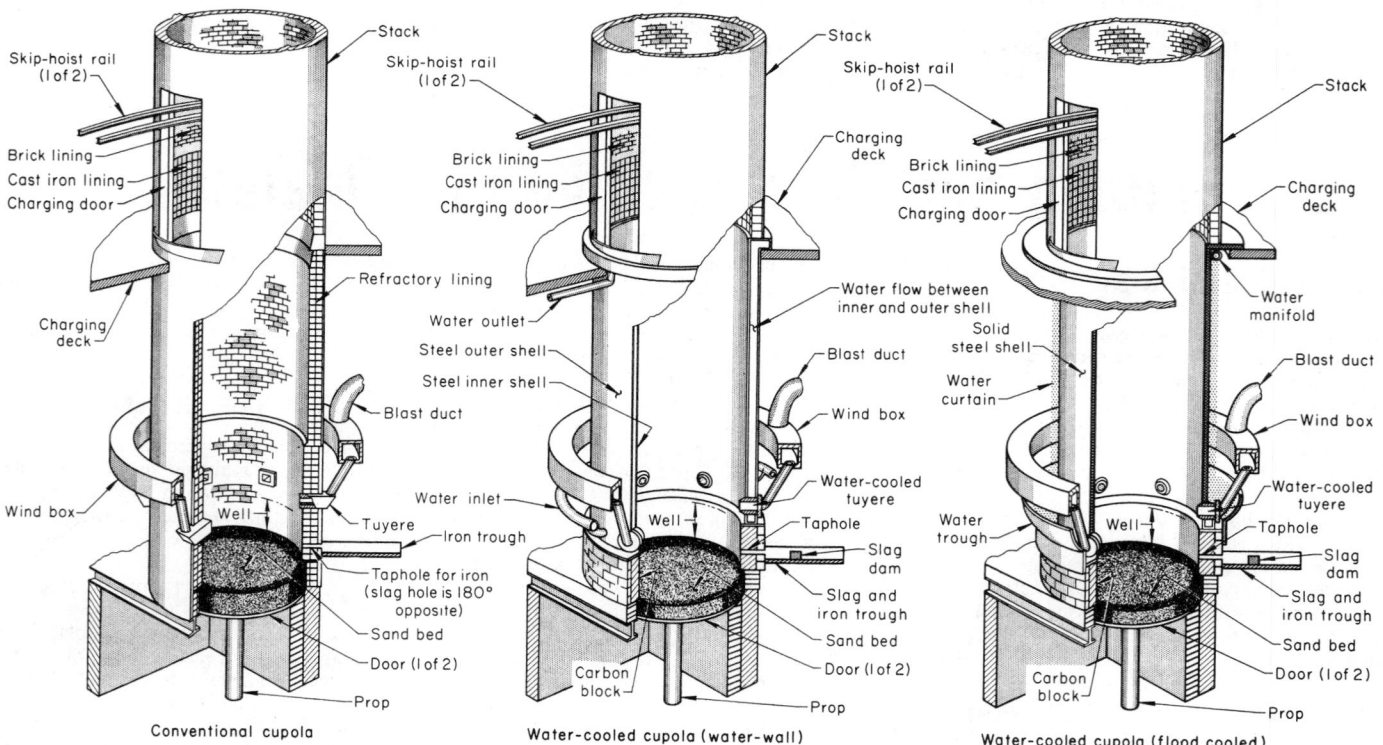

Fig. 2. Sectional views of conventional and water-cooled cupolas

taphole is opened by removing the fireclay plug, and most of the iron is drained out at a rate much higher than the melting rate. Then the iron taphole is reclosed and the cycle is repeated.

Charging. The charge is introduced into a cupola through the charging door, located 3.7 to 5.5 m (12 to 18 ft) above the tuyeres. Although many small cupolas are charged by hand, mechanical charging is always preferable because labor requirements are decreased.

Metal Tapping. Metal is usually tapped from the cupola to a transfer ladle or to a tilting forehearth, called a mixing ladle, or to an electric induction or resistance holding furnace, located in front of the cupola. The ladles must be heated before receiving the molten metal, in order to maintain iron temperature. The forehearth may have an auxilliary system for continuous heating.

Forehearth. The forehearth serves as a place to mix molten metal, thus making it more uniform in composition; to store molten metal temporarily; and to condition the metal, as by alloying or desulfurizing.

Slag Disposal. The slag formed in the cupola is removed continuously or intermittently. If the volume of slag is small, it may be run out on a dry sand bed and carried out with the bottom drop. Large volumes of slag are run into suitable slag pots and cooled, and then taken to a slag dump after the metal buttons that form at the bottom of the pots are removed. Slag also can be run into water tanks where it fritters and then is sluiced off to a container for disposal.

Advantages of cupola melting include flexibility in using a variety of low-cost materials to produce gray iron, high melting rates, low fixed costs per unit of output, continuous operation (for a week or more), and minimum downtime.

The cupola provides the foundry with a single melting unit that can produce large amounts of molten iron on a continuous basis. With a cupola 2.5 m (100 in.) in diameter, for example, a melting rate of 36 t/h (40 tons/h) can be obtained. Melting rates can be further increased by heating the blast.

Disadvantages of Cupola Melting. With improvements such as water cooling, hot blast, skip-hoist charging, humidity control and emission control, a cupola represents a large investment.

Cupola melting is best suited to making one type of iron. With good metallurgical control, the composition of the iron can be changed in the cupola during the course of operation, but even with the best control, this technique produces a certain amount of transition iron, which must either be pigged or be poured into noncritical castings.

Because the metals are subject to oxidation from the air blast, losses from oxidation are appreciable.

CUPOLA CHARGES

A cupola charge is composed of metal, fuel (coke), and a flux. Because of the changes in composition that take place during melting, the make-up of the charge is based largely on experience. As shown in Table 2, most elements undergo a loss during melting, although some, such as sulfur, increase when melting is done in an acid cupola. Carbon also increases during melting, and when a charge of high steel content is melted under a basic slag a net melting loss near zero is possible.

The make-up of the metal portion of a charge is based on the estimated losses and on the re-

Table 2. Approximate loss or gain of elements during acid melting in a cupola

Element or constituent of charge	Percentage of weight charged	
	Loss	Gain
Silicon(a)	7 to 12	...
Manganese(a)	10 to 20	...
Ferrosilicon (lump)	10 to 15	...
Ferromanganese (lump)	15 to 25	...
Spiegeleisen	15 to 25	...
Phosphorus	...	Trace
Ferrochromium (lump)	10 to 20	...
Nickel (shot or ingot)	2 to 5	...
Copper(b)	2 to 5	...
Alloys in briquets	5 to 10	...
Sulfur	...	40 to 60

(a) In pig iron and scrap. (b) As shot or as scrap having 4.8-mm ($^3/_{16}$-in.) minimum thickness.

quirements for the castings. Table 3 shows metal-charge compositions and final analyses for a class 30 iron and a class 40 iron. These compositions are typical and provide for minimum tensile strength only. Compositions for gray iron of a given tensile strength are likely to vary considerably among foundries.

COMPOSITION CONTROL OF CUPOLA IRON

In addition to being affected by variations within the charge, the composition of the iron can vary as a result of the distribution of the charge in the cupola, variations in blast, variations in start-up, and water leaks.

Adequate analytical facilities must be provided for accurately checking the composition of the iron by wet-chemical or spectrochemical methods. Direct-reading spectrometers permit quick, accurate chemical analyses. Vacuum-type spectrometers can be used to determine carbon, sulfur and phosphorus contents. Spectrochemical samples used for analysis of carbon should be poured in water-cooled copper molds and must be thin enough to solidify as white iron. A small amount of bismuth or tellurium in pellet form may be added to prevent formation of any free carbon in the sample.

Regardless of whether the wet-chemical or the spectrochemical method is used for control, the carbon-equivalent method is desirable also. In the carbon-equivalent method, a thermocouple is used to measure the liquidus temperature of the iron. This is recorded as a thermal-arrest line on a suitable chart. This method offers a quick and accurate method of detecting changes from the desired carbon equivalent, because carbon equivalent is related to liquidus temperature.

Carbon-equivalent determinations should be made at frequent intervals, so that indicated variations in iron composition can be immediately corrected. For cross checking, carbon equivalent should be determined on the same sample of iron that is used for chemical analysis.

Chill Control. White iron in gray iron castings is called chill. It occurs on initial solidification of the iron (most often in thin sections), and it is usually caused or aggravated by one of the following conditions:

• Low carbon or silicon content
• Excessive amount of carbide-stabilizing ele-

Table 3. Typical cupola charge compositions and final analyses for class 30 and class 40 gray iron castings

Material charged	Quantity %	Quantity lb	Total carbon %	Total carbon lb	Silicon %	Silicon lb	Phosphorus %	Phosphorus lb	Sulfur %	Sulfur lb	Manganese %	Manganese lb
Class 30 gray iron												
Foundry scrap (returns)	30.0	600	3.35	20.10	1.88	11.30	0.16	0.96	0.11	0.66	0.70	4.20
Purchased cast iron scrap	22.5	450	3.25	14.62	2.20	9.90	0.22	0.99	0.10	0.45	0.60	2.70
Rail steel	27.5	550	0.60	3.30	0.30	1.65	0.03	0.16	0.03	0.16	0.80	4.40
Pig iron	15.0	300	4.00	12.00	2.40	7.20	0.12	0.36	0.03	0.09	0.91	2.73
Silvery pig iron	5.0	100	2.50	2.50	10.48	10.48	0.10	0.10	0.05	0.05	0.88	0.88
Manganese briquets		3	...	...	...	...	...	...	...	...	...	2.00
Inoculant(b)	...	...	...	...	...	1.20	...	...	...	...	...	...
Total		2003		52.52		41.73		2.57		1.41		16.91
Charge analysis			2.62		2.09		0.13		0.07		0.85	
Change in melting	−5.0		+0.73		−0.21		+0.03		+0.04		−0.17	
Final analysis			3.35		1.88		0.16		0.11		0.68	
Class 40 gray iron												
Foundry scrap (returns)	27.5	550	3.15	17.34	1.45	7.97	0.09	0.50	0.08	0.44	0.80	4.40
Rail steel	60.0	1200	0.60	7.20	0.30	3.60	0.03	0.36	0.03	0.36	0.80	9.60
Pig iron	5.0	100	4.00	4.00	2.40	2.40	0.12	0.12	0.03	0.03	0.91	0.91
Silvery pig iron	7.5	150	2.50	3.75	10.48	15.72	0.10	0.15	0.05	0.07	0.88	1.32
Manganese briquets		6	...	...	...	...	...	...	...	...	...	4.00
Silicon briquets		4	...	...	...	2.00	...	...	...	...	...	...
Inoculant(b)	...	...	...	...	...	0.90	...	...	...	...	...	...
Total		2010		32.29		32.59		1.13		0.90		20.23
Charge analysis			1.62		1.63		0.06		0.05		1.01	
Change in melting	−5.0		+1.53		−0.16		+0.03		+0.03		−0.20	
Final analysis			3.15		1.47		0.09		0.08		0.81	

(a) These compositions provide only for minimum tensile strength; the proportion of charged components and the desired final analyses may vary considerably among different foundries. (b) For both class 30 and class 40 irons, the inoculant was added at the spout during tapping.

ments such as chromium, molybdenum and vanadium
- High sulfur content when not balanced by adequate amounts of manganese
- High gas content
- Low pouring temperature in conjunction with one of the above conditions.

Castings that have chilled edges are not machinable, and are usually scrapped unless the chill is shallow enough to be removed by grinding. Most types of chill can be broken down by long-cycle annealing, but this treatment is expensive and may reduce mechanical properties to an unacceptable level. Chill control can be accomplished by testing samples, then conditioning the molten iron as required.

Figure 3 shows a standard wedge block used to determine depth of chill. Such a test specimen may be poured in a sand mold and quenched in water as soon as it has solidified, or it may be poured in a special water-cooled mold. It is then broken, and the depth of the chill is measured in thirty-seconds of an inch by means of a special scale that fits over the chilled edge of the wedge block. The size of the chill sample varies with the type of iron produced. The sample should permit accurate measurement of the chill depth, but it should not be so sensitive that it produces

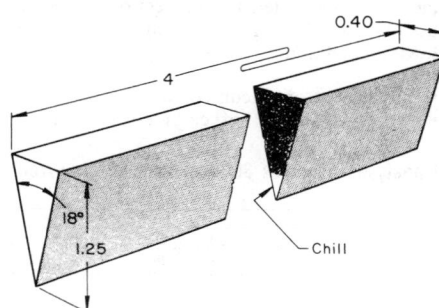

Fig. 3. Standard W2 wedge block used for measuring depth of chill (ASTM A367-60)

an entirely white section unless extreme conditions are experienced. A maximum chill depth on the sample should be established that can be tolerated without producing chill on the castings. If this maximum depth is exceeded, chill-reducing inoculants must be added to the molten iron before the castings are poured.

INDUCTION FURNACES

Types of Furnaces. Induction furnaces are classed as (a) coreless furnaces, which appear to consist of a simple crucible surrounded by a water-cooled copper coil, and (b) core or channel furnaces, in which molten metal forms a loop or channel around one leg of a transformer core. These two types of furnaces are based on different ways of converting electrical energy into heat energy by induction, and their operating characteristics are quite different. A coreless furnace is better adapted to and more widely used for melting and superheating, whereas a channel furnace is better suited to superheating, holding and duplexing.

Channel furnaces are simpler electrically than coreless furnaces and have greater electrical efficiency. They work readily on 60-cycle current and have been used for many years in smaller

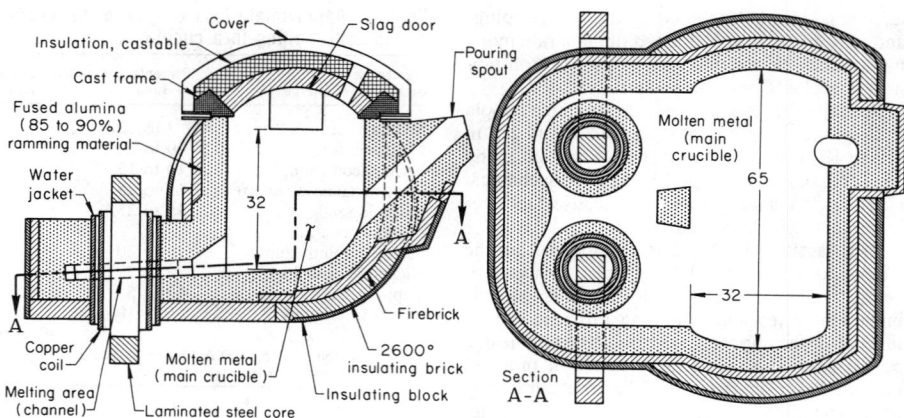

Fig. 4. Sectional views of one type of channel (core) induction furnace used for superheating, holding and duplexing of gray iron. (Dimensions are in inches.)

sizes in nonferrous foundries. The main development necessary for use with iron was better refractories for the channel. One design of channel furnace used for iron is shown in Fig. 4; there are several variations by different furnace manufacturers.

Inherent in the channel concept is the fact that only the relatively tiny amount of metal in the loop receives energy, and so that metal must have both a considerable temperature rise and rapid motion through the channel. Temperature rise is limited by refractory life, and the motion is provided inherently by a motor effect, but the rate of energy input is limited. This is one reason that channel furnaces are not used much for melting. Another reason for limited use in melting is that stirring contributed to the bath by the channels is relatively mild, so that additions of cold charge or of ferrosilicon or carbon to the surface of the bath do not mix or dissolve rapidly.

A channel furnace must generally be started with a supply of molten metal, and enough power should be left on over nights or weekends to keep the iron molten. For furnaces larger than about 18 t (20 tons), power rating in kilowatts is typically 30 times bath capacity in net tons, plus 400.

A coreless furnace is shown in Fig. 5; these furnaces vary little in general construction among manufacturers. The metal to be heated forms both core and short-circuited secondary, the copper coil being the primary of the transformer. For electromagnetic reasons, small coreless furnaces are not satisfactory at line frequency. Furnaces with capacities of 450 kg (1000 lb) or less are best at 180 or 540 cycles, which is obtained by frequency doubling in static equipment.

The primary coil is of copper tubing of special cross section to give good coupling, rigidly held to withstand electromagnetic forces, and with provision for thermal expansion. For good electrical coupling, the bath is usually deeper than it is wide; hence, the pressure of molten metal near the base of the coil can be considerable. Vertical bars of laminated transformer iron, which form the magnetic yokes, are spaced evenly around the outside of the coil and add to its strength in a major way. The supporting framework must be strong and stiff, because the entire furnace must rotate through about 100° to empty the crucible.

There is a strong stirring action in line-frequency coreless furnaces, up the center and down the walls in the upper half and down the center

and up the walls in the lower half (note arrows in Fig. 5). Mixing is therefore excellent, and both alloying elements and fresh charge are absorbed very rapidly. Stirring action is directly related to power input per unit volume, and is a major limitation on the power that can be applied to a bath of given size. Typically, the power rating in kilowatts is 200 times the bath capacity in net tons.

CHARGE MATERIALS AND PRACTICE FOR INDUCTION FURNACES

Since melting losses are low and recovery of alloying additions is high in induction furnace melting of gray iron, control of final composition is very good and a wide range of charge materials can be used. Charges are usually made up of steel scrap, cast iron scrap, foundry returns, and ferrosilicon and carbon in suitable amounts to adjust composition. Pig iron is seldom used. In moderate quantity, steel and iron chips or borings are readily absorbed in a coreless furnace bath because of the stirring action. These materials are economical, especially if available from other operations of the same company, but in most localities they are not practical as a major pur-

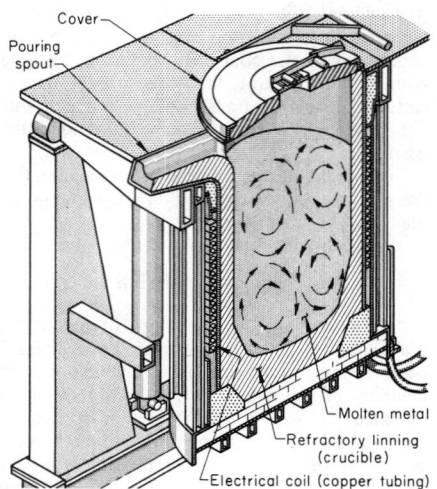

Fig. 5. Sectional view of a coreless induction furnace. (Arrows in crucible show direction of stirring action.)

chased component of the charge because there is a limited supply of clean borings of known composition at a satisfactory price. When available, chips or borings can comprise a maximum of about 50% of the charge in a coreless furnace. In channel furnaces, however, borings are not very useful, because the stirring action is mild and absorption into the bath is slow.

A typical charge for class 30 iron of 4.20 carbon equivalent is:

Steel scrap 41.3%
Gray iron borings 10.0
Foundry returns 45.0
Carbon 1.7
50% ferrosilicon. 2.0

Two charge compositions and charging sequences used in a high-production foundry for melting class 30 iron in induction furnaces are given in detail in Table 4.

Carbon equivalent can readily be adjusted to any desired level by balancing steel scrap content and additions of carbon and ferrosilicon. Alloying additions are absorbed rapidly at 90 to 95% recovery, and the ease of meeting chemical and mechanical specifications, including those difficult or impractical in cupola melting, is valuable. Carbon additions are made with pelletized petroleum coke of controlled grain size, to give high and reproducible recoveries. Because little sulfur is introduced in induction melting, the sulfur content of the melted iron is considerably lower

Table 4. Typical charge compositions and sequences for producing class 30 gray iron in an induction furnace

Charge material	Weight of material and sequence of charge(a)					
	First bucket		Second bucket		Third bucket	
	kg	lb	kg	lb	kg	lb
3.30C-1.80Si-0.60Mn(b) (44% returns, 32% turnings, 12% plate, 10% borings)						
Steel plate					860	1900
Steel turnings . .	2315	5100	. . .	. . .	. . .	. . .
Cast iron returns			1635	3600	1540	3400
Coke	115	250	. . .	. . .	. . .	. . .
Ferro- silicon			100	220	. . .	. . .
Cast iron borings			680	1500	. . .	. . .
Total weight of charge . . . . . 7245 kg (15 970 lb)						
3.30C-1.85Si-0.60Mn(c) (55% returns, 32% turnings, 10% plate)						
Steel plate			365	800	365	800
Steel turnings . .	2315	5100	. . .	. . .	. . .	. . .
Cast iron returns			1995	4400	1995	4400
Coke	115	250	. . .	. . .	. . .	. . .
Ferro- silicon			45	100	45	100
Total weight of charge . . . . . 7235 kg (15 950 lb)						

(a) Charge is made after three transfer ladles (7210 kg, or 15 900 lb) of hot metal have been tapped from the furnace. No hot metal is tapped during the charging sequence. All three buckets are charged in sequence, and the charge is completely melted before another charge is added. Maximum power is used after charging of the first bucket, to help dissolve the coke. (b) Elements may vary by ±0.10% each. Buckets are loaded in the following order. First bucket: steel turnings, coke, steel turnings. Second bucket: gray iron borings, ferrosilicon, gray iron returns. Third bucket: gray iron returns, steel plate. (c) Elements may vary by ±0.10% each. Buckets are loaded in the following order. First bucket: steel turnings, coke, steel turnings. Second bucket: steel plate, ferrosilicon, gray iron returns. Third bucket: gray iron returns, ferrosilicon, steel plate.

than that from a cupola, and thus less manganese is required for balance with sulfur.

Charging Practice. Charge materials are made up in weighed proportions by the same equipment and methods as those described for cupola melting. Accurate charge make-up is the essential basis of good composition control. The principal additives are carbon and ferrosilicon, both usually added directly to the furnace. Occasionally, a small furnace is charged by hand, but general practice is to use a bucket (see Fig. 6).

Charge Preparation. The desirable features of charge material for a coreless induction furnace are that the pieces not be too long, the bulk density be reasonably high, and the material be essentially dry.

The space receiving the charge in an induction furnace is a cylinder; to avoid bridging and hangup, the maximum dimension for pieces of charge material should be substantially smaller than the inside diameter of the lining. In a 14.5 t (16-ton) furnace with a melting rate of about 5.5 t/h (6 tons/h), for example, the inside diameter of the lining is about 1.2 m (48 in.), and standard prepared scrap plate (600 mm, or 24 in., or smaller in maximum dimension) is an ideal charge. Large foundry returns must be broken, and large plate and structural steel scrap must be cut.

High bulk density is important because the fresh charge must be contained in the space left by the molten iron tapped out, and if density is too low, extra charging may be necessary. Again, foundry returns are helpful, because they are relatively dense compared with much steel scrap. Steel turnings should be crushed to give acceptable density if used in quantity. A compensating factor with low-density material is that it is thin and dissolves rapidly in the molten bath.

Charge materials should be nearly free of oil or moisture; otherwise, a serious blowout of molten iron or an explosion can occur. The amount tolerable depends on the method of, and care in, charging, but any amount is a potential hazard. Borings and turnings should be degreased, not only to decrease this risk, but also to decrease the amount of smoke formed and uncertainty in carbon control.

Charge Preheating. Charge materials are preheated to dry and de-oil them, and to add heat to reduce the electrical energy input to the induction furnace.

Simple drying and de-oiling can be done by heating the material to a few hundred degrees Fahrenheit, and is readily done in simple equipment ranging from large hotplates to rotary drums. The economic justification is in negative terms, in freedom from unexpected trouble in charging.

If enough heat is added to the charge to replace an appreciable amount of electric power, an average temperature of at least 540 °C (1000 °F) should be used, and drying and de-oiling are then done incidentally. The net economic gain, including write-off of more expensive equipment, increases with preheat temperature.

INDUCTION MELTING PRACTICE

Induction melting is characterized by the close control that can be exercised over iron composition, temperature and melting rate. Operation can be on a batch or a continuous basis, the latter accounting for most of the tonnage melted.

Batch Melting. In jobbing shops particularly, molten iron is needed discontinuously and the composition may need to be changed frequently.

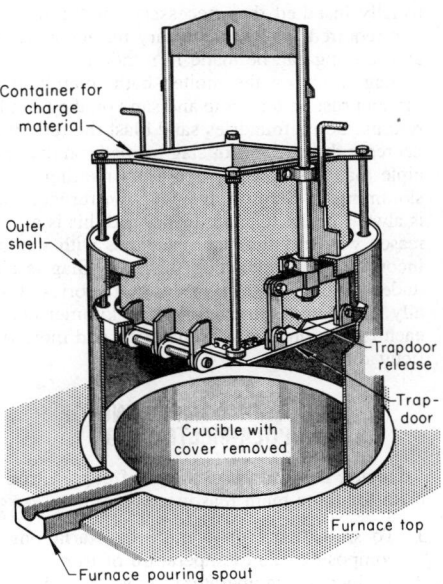

Fig. 6. Bottom-release bucket for charging an induction furnace

A coreless induction furnace is better suited to such requirements than is a cupola, because a coreless furnace is more readily started and stopped; also, control over composition is much better, because in a coreless furnace a bath changed in composition can be held until it is right, and tapping temperature is under close control. Operating costs are in practice similar to those for a cupola under the same conditions; the main disadvantage is capital cost and amortization.

Usually, a heel of one-third furnace capacity is retained to maintain good power input, and its composition is taken into account in the make-up of fresh charges if the next iron to be tapped is to be different. When the whole furnace capacity is necessary to pour a large casting, a starter block is necessary.

An example of batch operation is a roll foundry that replaced cupolas with a 36-t (40-ton) coreless induction furnace. The furnace is designed with a relatively low ratio of power input to crucible capacity, so that two heats can be made per shift, retaining adequate stirring action. Excellent control over iron composition and chill depth is obtained, and a particular advantage is that whole scrap rolls can be used for charge make-up without expensive breaking.

Batch melting is usually a single-shift operation; to prevent cracking of the lining, either the furnace should be kept about one-third full overnight or a torch should be used to keep the inside face of the lining above about 750 °C (about 1400 °F).

Continuous melting is the major use of induction melting; a very considerable tonnage of iron is so melted to feed mold production lines. General practice is to use several furnaces, tapping 10 to 30% of the capacity of each in rotation and immediately adding the same amount of fresh charge. In this way, transfer or pouring ladles are kept full in pace with the production line. A holding furnace is not necessary, and many foundries operate without one; but it does add flexibility, and several recent installations have included a channel holding furnace. One more furnace body is usually installed than necessary for the melting rate required, so that necessary refractory repair and relining can be made in rotation.

Slag forms on the molten bath, mainly from dirt and rust on the scrap and sand on the foundry returns. Some foundries sand blast the returns to decrease the quantity of slag formed, on the principle that this is less costly than melting it and skimming it. Slag on an induction-furnace bath is always cooler than the metal, and this is a main reason why metallurgical operations with slag are inconvenient in induction furnaces. Slag is also undesirable because it erodes refractories. Usually, the slag is removed with a skimmer before each tap, to prevent accumulation and more difficult removal.

INDUCTION COMPLEXING OF CUPOLA-MELTED IRON

Duplex melting (duplexing) of cupola iron is done for three main reasons:

1. To smooth out or average the variations in composition and temperature of iron
2. To increase temperature above that obtained from the cupola
3. To provide an active or heated holder where adjustments in temperature and composition can be made for a batch operation, as in a jobbing foundry.

The smoothing or averaging function is commonly done in any case in a sufficiently large forehearth or holder, and this in itself is not duplexing. However, heat losses from such holders are appreciable, and if temperature is not to be lost, some energy or heat must be added, which may be defined as duplexing. It is then simple to increase the energy input to give whatever temperature control or increase is desired. Induction furnaces are peculiarly well-suited to this function, because of their high efficiency and good mixing action. Data in Fig. 7 compare the carbon and silicon contents of a gray iron from a water-cooled cupola before and after duplexing in an 18-t (20-ton) induction furnace.

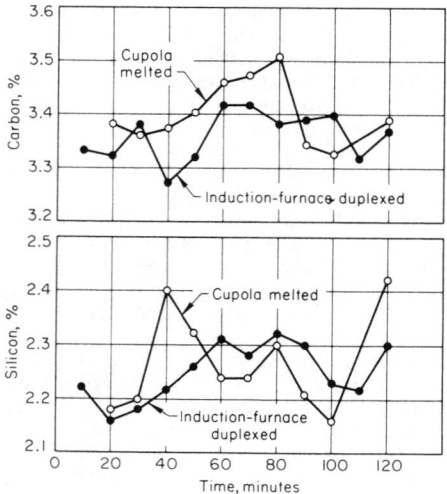

Fig. 7. Carbon and silicon contents in gray iron from a water-cooled cupola and after duplexing in an 18-t (20-ton) induction furnace

ARC-FURNACE MELTING

In the past, arc furnaces have been used in gray iron foundries for special irons and for batch or jobbing operations; very few have been used as holders or superheaters for cupolas. However, as it has become apparent that electric melting can compete with cupola melting on a straight economic basis, interest in the arc furnace as a primary melter has grown. In 1969 the first installation of arc furnaces in a major high-production foundry was put into operation, and others are in planning or erection stages.

An arc furnace is essentially a refractory hearth on which material can be melted by heat from electric arcs. The molten bath is relatively wide compared to its depth, so that bulky charge material can be handled. Reactions between slag and molten metal are efficient, because the interface area is large and the slag is at least as hot as the metal. Thermal efficiency on melt down is high, about 80%, because the arcs are surrounded by charge material. Efficiency of superheating molten metal is relatively low, about 20 to 30%, because the bath receives heat only from the surface.

Arc furnaces are sized in terms of shell diameter; this determines the bath capacity, depending on the thickness of refractory used. Typically, a 2.7-m (9-ft) furnace would hold a 9-to-11-t (10-to-12-ton) bath, and a 3.4-m (11-ft) furnace would hold a 20-to-23.5-t (22-to-26-ton) bath. Melting or production rate in tons per hour is a function of power input, and examination of over 100 furnaces making tonnage plain carbon steel shows that, on the average, production of one ton per hour requires 1000 kVA in transformer capacity. Power factor depends on the design of the electrical circuit and on the mode of operation of the power input; typically, it is 0.80.

Production of Gray Iron Castings

GRAY IRON CASTINGS are produced by several molding processes, but green sand molding is the most widely used. Significant tonnages of castings are also produced in dry sand molds, core molds, shell molds, molds made by the cold-set process, and permanent molds. Die casting of gray iron is restricted to a few highly specialized applications, because of the lack of a suitable die material. Gray iron can be cast into investment molds, but such castings are required only rarely.

Melting. Gray iron is melted in a cupola or an electric furnace, as described in the preceding article, "Melting of Gray Iron" (p 23•41).

SAND MOLDING

Molding-sand mixtures for gray iron castings must have: high green compressive strength, with correspondingly high green shear strength; permeability sufficient to vent mold gases; sufficient clay to absorb expansion; sufficient moisture to activate the clay; and sand of grain size, shape and distribution that will give the desired casting finish.

Sand. The sand component of the mixture is made up of various amounts of system sand and new sand. It is common practice to make continuous additions of clean new sand to a system.

Table 5. Typical screen analyses for bank sand and sharp sand

Bank sand (AFS 80 to 100)
Retained on 50-mesh screen . . . . 5% max
Retained on 70-mesh screen . . . . 20% max
Through 100-mesh screen . . . . 35% min

Sharp sand (AFS 43 to 51)
Retained on 30-mesh screen . . . . 1% max
Retained on 50-mesh screen . . . . 30% min
Through 70-mesh screen . . . . 20% max

The amount of new sand that is periodically added depends largely on the amount that is lost, which, in turn, depends on the nature of the foundry operation.

Specifications for bank sand usually require that it be low in all impurities, with no evidence of lime. Typical grain-size requirements for bank sand and sharp sand are listed in Table 5.

Clay Binders. The clay in a molding-sand mixture serves primarily as the binder for the sand particles. The clay also provides cushion space for the expanding silica grains during casting, by giving up moisture and contracting while continuing to maintain grain-to-grain adhesion.

The two principal clays used in gray iron foundries are western bentonite and southern bentonite. Southern bentonite produces sand molds with higher green strength and lower hot and dry strengths than are obtained with western bentonite.

Some foundries use 100% western bentonite and others use 100% southern bentonite, but most use a combination of the two. The proportions vary among foundries (and sometimes within a foundry) from three parts southern to one part western to the reverse ratio of three parts western to one part southern. The exact ratio is usually arrived at through experience. Typical composition and property specifications for bentonite are given in Table 6.

Dry Sand Molds. Sometimes, skin-dried or dry sand molds are used for casting of gray iron. When the surface of a mold is dried by torches or other means to a depth of less than 13 mm ($\frac{1}{2}$ in.), the mold is said to be skin-dried. A mold that has been oven dried to a depth of 13 mm or more is called a dry sand mold.

SHELL MOLDING

Shell molding, in which the molds are formed from thermosetting resin–sand mixtures, is used extensively for production of gray iron castings. The process has been employed for castings weighing up to 450 kg (1000 lb), although most gray iron castings produced in shell molds weigh less than 45 kg (100 lb).

The shell process provides better surface finish and greater dimensional accuracy than it is possible to obtain in green sand molding, and it often provides these improvements at lower cost—particularly when one or more machining operations can be minimized or eliminated because of the greater accuracy of the casting. For example, the casting shown in Fig. 8, which is a muff for cooling boiler components, depended on dimensional accuracy of the fins for cooling efficiency. When the muff was cast in a green sand mold, the fins required machining, but when it was cast in a shell mold, dimensional accuracy was obtained without machining.

For some castings, the shell process permits a saving in weight, in addition to reducing the amount of machining required. For instance, the

Table 6. Typical composition and property specifications for western and southern bentonites used in molding-sand mixtures

Type of bentonite	Silica	Alumina	Iron oxide	Calcium oxide	Magnesia	Potassium oxide	Sodium oxide	Titania	Phosphorus pentoxide	Loss on ignition
Western ...	62.4	24.6	3.2	0.4	2.9	0.2	0.6	...	...	6.1
Southern ...	55.4	15.6	3.7	0.9	1.5	1.4	...	0.5	0.3	...

Item	Western bentonite		Southern bentonite	
	Powdered(a)	Granular(b)	Powdered(a)	Semipowdered(b)
Sieve analyses and properties				
Sieve analysis	Thru 200-mesh, 80% min	On 20-mesh, 2% max; on 40-mesh, 25-40%; on 200-mesh, 50-70%; thru 200-mesh 2% max	On 200-mesh, 10% max	On 140-mesh, 20% max; on 200-mesh, 40% max; thru 200-mesh, 70% max
Fusion temp (min), °C (°F) ..	1260 (2300)	1260 (2300)	1315 (2400)	1315 (2400)
Green compressive strength(c) (min), kPa (psi)	34 (5)	28 (4)	52 (7.5)	45 (6.5)

(a) Used for dry additions. (b) Used for slurry additions. (c) Samples for testing green compressive strength are made by mixing 1920 g of AFS testing sand (50 to 70-mesh) with 80 g of bentonite (dried for 2 h at 105 °C, or 220 °F) for 2 min, adding 45 cm³ of water, and mixing for 3 min.

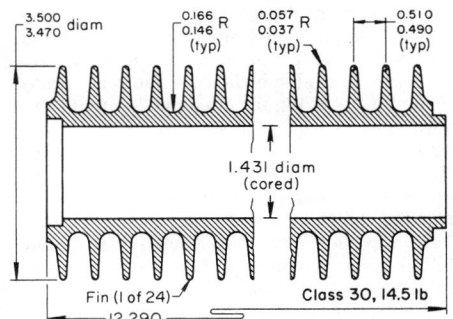

Fig. 8. Cooling muff that did not require machining for dimensional accuracy of fins when cast in a shell mold instead of a sand mold. (Dimensions are in inches.)

bushing shown in Fig. 9 weighed almost 3 lb less when cast in a shell mold than when cast in a green sand mold. In addition, fewer areas of this shell mold casting required machining (note shaded and unshaded areas in Fig. 9).

COLD-SET MOLDING

Cold-set molding using phenolic urethane resin binders with an acid catalyst allows setting up of sand at room temperature. Molds produced from phenolic urethane binders have high strength, good dimensional accuracy, good casting surface finish and good shakeout properties. Some disadvantages are higher sand-mix cost than that of green sand, plasticity problems when large, rangy molds are produced, and some odor problems. This process, which is not too abusive of tooling, is used primarily for low-production and prototype work allowing the use of relatively inexpensive wood or plastic patterns.

PERMANENT MOLD CASTING

Permanent mold casting of gray iron is used for relatively high production of small castings (usually, no heavier than 13.5 kg, or 30 lb). The rapid cooling rate obtained in permanent mold casting makes it particularly suitable for production of pressure castings required for compressor and hydraulic cylinders. Gray iron used for this process should be hypereutectic, so as to obtain maximum fluidity and minimum chill.

Most permanent molds are made of gray iron. Frequently, the molds are mounted on multiple-station turntables, but single-head stationary machines are also used. Most molds are designed with vertical parting and are gated into the bottom of the mold cavity. Permanent molds must be vented. Prior to first use, the molds are coated with a mixture of sodium silicate and china clay. After each casting cycle, the molds are blackened with lampblack that is generated by burning acetylene gas at low pressure without the addition of oxygen.

The complete casting cycle consists of setting cores, depositing lampblack, pouring the iron, ejecting the casting, and cleaning the mold with compressed air or by brushing. Molds are usually cooled by air during the entire cycle. Cycle time varies considerably, depending on the weight and thickness of the casting, but usually ranges from 3 to 7 min. Cycle time must be fast enough to maintain the desired mold temperature, but the casting must be solid at the time of ejection. When

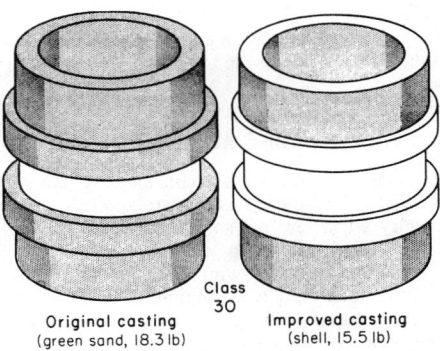

Fig. 9. Bushing that weighed less and required less machining as a shell mold casting than as a sand mold casting

the cycle is set up for a multiple-head machine, all of the castings being produced must be of similar weight and section thickness, because the casting cycle will be the same for each head.

CORES

All of the common types of cores—oil-sand, shell, hot-box, carbon dioxide and cold-set—are used for producing gray iron castings. The type of core need not be closely related to the mold material. For instance, the use of shell cores in green sand molds is common practice. Selection of core material depends mainly on core size, complexity, time permitted for making the core, required dimensional accuracy, and cost.

Oil-sand cores are usually the least costly type of core, in small sizes. For larger sizes, however, shell cores often cost less than oil-sand cores, because shell cores are hollow and do not require as much material. Oil-sand cores are generally less accurate in dimensions than are cores made by the shell or carbon dioxide process. Also, oil-sand cores must be baked, which consumes time, whereas hot-box, shell, carbon dioxide, and cold-set cores can be prepared in a few minutes.

POURING TEMPERATURE

Gray iron melted in a cupola is poured into the forehearth or the ladle at approximately 1565 °C (2850 °F). As delivered from an induction furnace, the metal is at approximately 1480 °C (2700 °F). Each time the metal is transferred, it loses 42 °C (75 °F) or more; heat losses occur in unheated mixing, and in transfer and pouring ladles, even when they are covered.

The fluidity of gray iron is proportional to the metal temperature in excess of the liquidus temperature for the given iron. However, addition of inoculants and alloying elements may change fluidity.

The temperature of a given iron for pouring a small, thin-section casting must be higher than the temperature for pouring a larger, thick-section casting. The composition of the iron also affects the optimum pouring temperature. Table 7 gives typical pouring temperatures for four classes of iron in making large and small castings with thin or thick sections.

Pouring temperatures that are too low are likely to cause misruns, because of insufficient fluidity. Pouring temperatures that are too high may result in very hard castings or in scabbing of castings. Scabbing occurs when pouring temperature is too high or pouring rate is too low. For example, when flywheel castings 19 to 50 mm (³/₄ to 2 in.) thick and up to 450 mm (18 in.) in diameter, were poured at 1425 °C (2600 °F) from class 30 gray iron, the surface of the cope half of the mold deteriorated, causing scabs on the casting. When the pouring temperature was reduced to 1370 °C (2500 °F), and the gates were enlarged to increase the rate of metal flow to the cavity, no scabbing occurred. Gates were enlarged on a trial-and-error basis until the desired results were obtained.

GATING AND FEEDING

Gating and feeding systems for casting gray iron serve the same purposes as for casting other metals. The important functions are:

• To fill the mold cavity rapidly but without turbulence

Table 7. Typical pouring temperatures for gray iron

ASTM class	Approximate liquidus temperatures		Pouring temperature							
			Small castings				Large castings			
			Thin sections		Thick sections		Thin sections		Thick sections	
	°C	°F	°C	°F	°C	°F	°C	°F	°C	°F
30	1150	2100	1400	2550	1370	2500	1345	2450	1315	2400
35	1175	2150	1425	2600	1400	2550	1370	2500	1345	2450
40	1200	2190	1450	2640	1420	2590	1395	2540	1365	2490
45	1220	2230	1470	2680	1445	2630	1415	2580	1390	2530

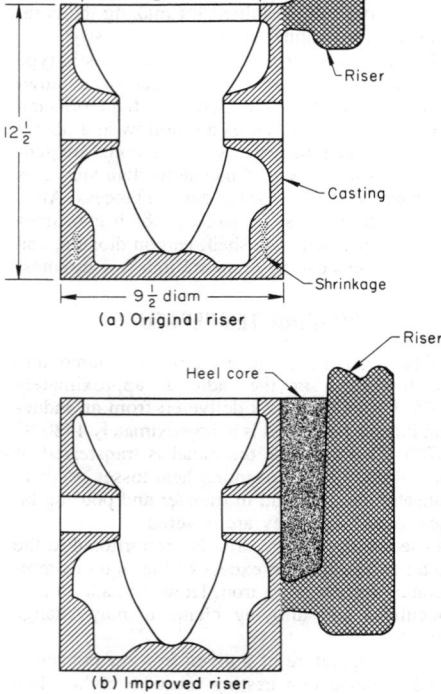

Composition of iron .... 3.00 to 3.20 total C, 1.40 to 1.80 Si, 0.12 max P, 0.09 max S, 0.80 to 1.00 Mn, 0.20 to 0.40 Cr, 0.20 to 0.40 Mo, and 0.40 to 0.60 Ni

Weight of trimmed casting .... 44.5 kg (98 lb)

Number of castings per mold .... One

Weight of metal poured per mold .... 68 kg (150 lb)(a)

Pouring temperature .... 1370 to 1400 °C (2500 to 2550 °F)

Shakeout method .... Mechanical

(a) Original riser system required a weight of 59 kg (130 lb).

Fig. 10. Relocation of riser that prevented solidification shrinkage in a compressor-piston casting. (Dimensions are in inches.)

- To prevent slag, dross or mold material from entering the mold
- To prevent the introduction of air or mold gases into the stream of metal
- To produce heat-transfer characteristics that will aid in the progressive solidification of the casting
- To enable production of the casting with the use of a minimum amount of metal.

Gating and feeding practice for gray iron is less critical than that for other metals. This is because graphite precipitates during solidification, and the expansion of the graphite compensates for solidification shrinkage. A class 20 gray iron precipitates enough graphite so that it requires no feed metal and may expand slightly rather than shrink; a class 50 iron will exhibit 4% volumetric shrinkage. Consequently, the grade of the iron poured will govern to some extent the design of the gating and feeding system.

Casting defects such as shrinks, scabs and misruns may be caused by some deficiency in the gating system. Some common defects and the changes that were made in gating and feeding to correct them are illustrated in Fig. 10.

DIMENSIONAL CONTROL

The main factors in dimensional control of gray iron castings are: allowance for shrinkage, compensation for distortion in certain large castings, and control of variables in green sand molding to minimize mold-wall movement. The machining layout often determines which dimensions of a casting must be closely controlled.

Shrinkage. Solidification shrinkage occurs when the metal changes from liquid to solid, and is a function mainly of the carbon content and the form in which the carbon is present (graphite or carbide). Because the form of the carbon also depends on the silicon content, the solidification shrinkage is influenced by the carbon equivalent. Solid contraction is offset by graphitization, which is also a function of carbon equivalent. The following represents a typical relationship between pattern shrinkage allowance and class of iron:

| ASTM class | Carbon equivalent, % | Pattern shrinkage allowance | |
		mm/m	in./ft
55	...	13.0	5/32
50	3.3 to 3.6	13.0	5/32
45	3.45 to 3.8	10.4	1/8
40	3.5 to 3.9	10.4	1/8
35	3.7 to 4.1	10.4	1/8
30	...	8.3	1/10

By determining carbon equivalent, corrections in the composition of the iron can be made in the ladle.

MACHINING ALLOWANCES

As-cast tolerances vary considerably among foundries. The tolerances that are applied in one foundry are given in Table 8. On castings that require some machining, enough stock must be allowed for cleanup, with the amount of machining stock held to a minimum so as not to waste metal and labor. Castings of different design must be considered separately, because of variations in size, shape, and casting technique.

The machining allowance for cylinder bores is a function of the inside diameter of the cylinder, although this is subject to some variation. Often, a plain cylinder will contract the normal amount

Table 8. As-cast tolerances, machining allowances and setup tolerances used by one manufacturer of gray iron sand castings

As-cast tolerances

8-in. max dimension .... Tolerance, ±1/32 in.	
14	±3/64
18	±1/16
24	±5/64
30	±3/32
36	±1/8

Machining allowances for single-bore cylinders

4-in. bore diam ...... Allowance, 0.12 to 0.20 in.	
4 to 8	0.12 to 0.24
8 to 12	0.20 to 0.32
12 to 20	0.25 to 0.40

Setup tolerances

12-in. OD .... Tolerance, 1/64 to 1/32 in.	
24	1/32 to 1/16
48	1/16 to 1/8
96	1/8 to 3/16

in length but, because of core restraint, will not contract across the bore. Suggested machining allowances for single-bore cylinders are given in Table 8.

The machining allowance for surfaces in the cope side of a casting should be twice as great as for those in the drag side. A small casting may have allowances of 3.2 mm (1/8 in.) in the drag and 6.4 mm (1/4 in.) in the cope. On large castings, such as engine bedplates, allowances may be 13 mm (1/2 in.) in the drag and 25 mm (1 in.) in the cope.

For a cylinder that is cast vertically, machining allowance normally will be twice as great in the cope as in the drag portion of the casting on surfaces parallel to the mold parting.

For a cylinder that is cast horizontally, core shift or crush is often a problem, and the rough casting should be checked for concentricity of cavities formed by cores. Minimum stock removal should be allowed for the drag surfaces, and maximum stock removal for the cope surfaces.

REPAIR WELDING

Welding as a means of repairing defective castings is generally accepted in most foundries. Very often, material specifications state the extent to which weld repairs may be made and the conditions under which welding must be done. In some applications, the amount and type of welding for specific castings are negotiated between the foundry and the customer. As an example of repair welding, thousands of tons of blocks and other castings for engines are repaired annually by welding.

Types of defects repaired by welding include porosity, sand holes, cold shuts, misruns, hot tears, and some types of cracks.

Welding methods most often used for repairing defective castings are:

- Shielded metal-arc (stick) welding
- Oxyacetylene gas welding
- Braze welding (arc or gas).

Other arc welding methods, such as gas metal-arc (MIG) welding, have been used successfully for repairing castings, but the shielded metal-arc method is usually preferred because of its versatility. For instance, areas of a casting can be reached with a stick electrode that would not be accessible with a MIG torch.

Arc welding is usually preferred over oxyacetylene gas welding if equipment and skilled operators are available. Arc welding is faster and costs less than oxyacetylene welding, partly because less preheating and postheating are required for arc welding. In addition, according to reports from a test program, higher strength is obtained in an arc-welded area than in one welded by oxyacetylene. However, the oxyacetylene method is successfully used in many gray iron foundries and is preferred for welding of surfaces that are to be machined. In comparison with electric welding, well-controlled oxyacetylene welding results in minimal carbide formation in the heat-affected zone.

Braze welding is used extensively for repair of gray iron castings.

Production of Compacted Graphite Iron Castings

COMPACTED GRAPHITE IRON is a new cast iron having a chemical composition close to that of ductile iron. It has been produced commercially for engineered castings since 1976. Its carbon and silicon contents are relatively high, with all other elements preferably kept low as in ductile iron.

The graphite form is the main difference between the two irons: ductile iron has a spheroidal form of free graphite, whereas the graphite in compacted graphite iron is in the form of short, blunt flakes that are interconnected.

The graphite in the compacted form gives this iron mechanical properties that are higher than those of gray iron although lower than those of ductile iron. The same reason—graphite form—allows compacted graphite iron to have higher thermal conductivity and better machineability than ductile iron.

Production of compacted graphite iron is done by processes similar to those used in producing ductile iron. Much of the current production involves treatment of a molten iron with a titanium- and cerium-bearing calcium-magnesium-ferrosilicon alloy. The titanium acts as a restricting element to the formation or retention of graphite in the spheroidal form, which would occur if only a magnesium ferrosilicon were used. Other alloys, such as cerium alloys, have been used to obtain the compacted graphite form. Whatever treatment alloy the producer uses must be used carefully and accurately as a function of the base-iron sulfur content, metal-treatment temperature and treatment method. For example, in the sandwich or ladle-pocket method, a base iron having 0.025% S would require a 1.2% addition of a titanium-bearing 5% Mg ferrosilicon for a metal temperature of 1525 °C (2775 °F). An increase of 0.005% S would require an added 0.1% of alloy; an increase of 28 °C (50 °F) in metal temperature would also require an added 0.1% of alloy.

Because the graphite can fade into a flake form or be exceptionally high in spheroidal graphite, the metallurgical disciplines for producing good compacted graphite iron are severe.

The procedures and production techniques used for ductile iron (regarding raw materials, melting furnaces, desulfurization, and treatment methods for introduction of graphite-control alloying additions) are also used for production of compacted graphite iron.

TREATMENT ALLOYS

Several cerium and magnesium alloys have been used to produce a compacted form of graphite. The range of sulfur content in the iron to be treated, the critical section size of the part to be cast, and the total chemical composition of the iron to be treated may dictate which alloy will give the most latitude in successfully producing the required graphite form. When sulfur content is below 0.01%, the iron may be treated with cerium-silicon, rare-earth silicides or mischmetal. A cerium content of 0.025 to 0.045% in the treated metal will result in compacted graphite formation.

Cerium-bearing magnesium ferrosilicons have been successfully used in treating irons with sulfur contents as high as 0.05% to produce compacted graphite. The amount of addition has to be closely controlled in relation to the sulfur content of the base metal for optimum results.

The cerium-titanium-calcium-magnesium-ferrosilicon alloy has been the most widely used proprietary alloy for the production of compacted graphite cast iron in the United States. When residual contents of 0.015 to 0.035% Mg and 0.08 to 0.14% Ti are retained in the iron, the compacted form of graphite will predominate in the structure. Extremes in casting section size and solidification rate may require some changes in composition ranges. Thinner sections will require less magnesium, and thicker sections will require more magnesium.

As with ductile iron, compacted graphite iron can be produced with a predominantly ferritic matrix or a predominantly pearlitic matrix produced by heat treatment, controlled mold shakeout, mold inoculants and pearlite-stabilizing alloys.

A ferritic matrix can be produced without heat treatment if the iron is relatively free of pearlite-stabilizing and carbide-forming elements. The use of a mold inoculant and slow cooling of the casting in the mold to below the lower critical temperature will enhance the formation of a ferritic matrix.

Pearlitic compacted graphite irons are produced by adding pearlite stabilizers such as copper, manganese or tin. Shaking out the castings from the mold at a temperature above the lower critical temperature may be helpful where the section size is large. Normal ladle inoculation or mold inoculation can be used in production of pearlitic irons.

Compacted graphite iron may also be alloyed to obtain special properties for any given service application. Use in elevated-temperature service may require addition of chromium, nickel, molybdenum, copper or vanadium to provide elevated-temperature strength or wear resistance.

MATERIAL PROPERTIES

At 80% minimum compacted form, mechanical properties of graphite will fall in the following ranges:

Tensile strength	260 to 450 MPa (38 to 65 ksi)
Yield strength	205 to 345 MPa (30 to 50 ksi)
Elongation	1.0 to 6.0%
Hardness	137 to 255 HB
Modulus of elasticity	145 to 159 GPa (21 to 23 × 10⁶ psi)
Impact strength	7 to 20 J (5 to 15 ft•lb)

Production of Ductile Iron Castings

DUCTILE IRON (also known as nodular iron, spherulitic iron, spherulitic graphite iron, and SG iron) is of gray iron composition with respect to carbon and silicon contents. Also, the melting equipment, handling temperatures and general metallurgy for ductile iron and gray iron are very similar.

The important difference between ductile iron and gray iron is that graphite separates during solidification of ductile iron as spheroids (instead of as flakes, as in gray iron), under the influence of the presence of a few hundredths of a per cent of magnesium. Because the presence of minute quantities of elements such as sulfur, lead, titanium and aluminum can interfere with or prevent this nodulizing effect, the molten iron for conversion to ductile iron must be purer than if gray iron is to be made. A small quantity of cerium added with the magnesium minimizes the effects that inhibit nodule formation, and thus makes it possible to produce the iron from raw materials of moderate cost.

Sulfur in the iron consumes magnesium uselessly, forming magnesium sulfide, which in turn forms a troublesome dross. Therefore, it is important that the iron be low in sulfur, preferably less than 0.02%, when magnesium is added. Also, because good ductility is an important property of ductile iron, pearlite and carbide stabilizers such as chromium, vanadium, manganese, tin and phosphorus must be low. Phosphorus content, in particular, should not exceed 0.06% if good low-temperature impact properties are required.

RAW MATERIALS

To gain control over tramp-element content, ductile iron is generally made from melts comprising selected steel scrap, special grades of pig iron, and foundry ductile iron return scrap. With the use of correct amounts of magnesium and cerium, good inoculation practice, rapid spectrographic analysis, and control of the quality of the metal structure by microscopic examination, the amounts of the raw materials entering the charge can be varied as economics, the metallurgy of the product mix, and the quality of the steel scrap allow.

Purchased cast iron scrap cannot be used, because the sulfur content is too high and other elements detrimental to ductile iron are usually present in objectionable quantities. Most steel scrap is higher in manganese than is desirable for best mechanical properties of the as-cast metal, especially in light sections. Cost and availability of melt materials, compared with the cost of heat treatment for correction of improper as-cast structure of the iron for a given product mix, are the deciding factors in shop practice.

FURNACES

Various melting units can be used for producing ductile iron if good control of the temperature and composition of the melt is maintained. Facilities employed are: (a) cupola melting with an acid slag, followed by desulfurization; (b) cupola melting with a basic slag; (c) duplex melting in an acid or basic cupola, followed by composition and temperature adjustments in an electric furnace; and (d) melting in an electric furnace, such as a coreless induction furnace or an arc furnace with a channel induction furnace used for holding.

DESULFURIZATION

Desulfurization practice outside the melting furnace consists essentially of introducing a highly basic material such as sodium carbonate, calcium carbide or quicklime to the molten iron in a suitable vessel and then agitating the bath to develop a chemical reaction between the basic material and the sulfur in the iron. Agitation can be done by using a shaking ladle, by injecting calcium carbide with nitrogen into the melt through a lance, by mechanically stirring calcium carbide into the metal, or by injecting nitrogen into the melt with a lance or through a porous plug after adding calcium carbide. Sulfur contents of less than 0.02% can be obtained consistently with these methods.

MAGNESIUM ADDITIONS

The boiling point of magnesium is below the melting temperature of cast iron, so for reasonable recovery, either magnesium must be added with a carrier alloy or the pressure of the system must be maintained above the vapor pressure of magnesium at the melting temperature (about 1.4 MPa, or 200 psi). Various commercial alloys are available for addition at atmospheric pressure. Typical are nickel-magnesium with 20 to 60% Mg; iron-silicon-magnesium with 5 to 50% Mg but most commonly 5 to 10%; magnesium-impregnated coke; and combination nickel-iron-silicon-magnesium alloys. Cerium, in amounts from 0.5 to 1.5% in the iron-silicon-magnesium alloys, is added to the melt in this form or separately as mischmetal containing about 50% cerium.

Methods of addition of magnesium to the melt differ in complexity and in recovery. Three that are widely used are the immersion-basket or plunging method, open-ladle methods, and the ladle-pocket method.

In the immersion-basket or plunging method, the magnesium alloy is anchored in a perforated refractory basket that is fastened to the end of a plunger or ram. As the magnesium alloy is forced rapidly under the surface of the melt by the ram, a cover or shield moves into place over the ladle, to contain splash without making a seal and building up pressure.

Open-ladle methods are similar to the immersion-basket method, except that the magnesium is added as magnesium coke. Typical are the perforated-basket plunging method, and one that uses a specially constructed ladle with a removable bottom and a perforated refractory separation between the coke and the metal to hold the coke under the surface of the metal as it is tapped into the ladle.

In the ladle-pocket method, base iron is poured rapidly over magnesium alloy that has been charged into a refractory well or pocket built into the bottom of an open receiving ladle. All of the magnesium alloy is charged into the well and covered with clean, small pieces of steel scrap or part of the ferrosilicon used as the late inoculation. The treating ladle is specially designed, having a height at least two times its diameter.

In-mold treatment of ductile iron has become very popular. In this procedure, a weighed quantity of carefully sized cerium-magnesium ferrosilicon alloy is placed in the runner system of the mold. As the metal is poured into the mold and flows through the runner system, the nodularizing treatment takes place. This procedure allows for relatively high magnesium recovery compared with ladle treatments.

INOCULATION

When magnesium is added to the base iron, the free graphite that forms on solidification is spheroidal instead of in flake form. Magnesium also strongly promotes formation of carbides. A silicon addition to the melt after the magnesium has been added promotes formation of well-shaped nodules and increases their number, so that less carbide and pearlite are formed. This results in a considerable increase in as-cast ductility.

The silicon content of the base iron is kept low enough so that the silicon from the magnesium alloy and from the inoculation adjusts final silicon content to the desired range. Depending on the source of the magnesium, the silicon content of the base iron ranges from 1.00 to 1.80%. This allows for the addition of silicon that may be contained in the magnesium alloy and for the subsequent addition of up to 0.60% Si as an inoculant.

Alloys used for inoculation include ferrosilicons containing either 75 or 85% Si; a calcium-bearing ferrosilicon with 85% Si; a calcium-silicon metal; or various combinations of these grades. A ferrosilicon containing 75% Si and controlled amounts of both calcium and aluminum is widely used to inoculate the melt.

Effective inoculation can be made by (a) reladling the metal after the magnesium addition, and adding the inoculant as the metal is poured; (b) adding magnesium to the first half of the tap or pour, and then adding the inoculant to the latter half; (c) stirring the inoculant into the melt; or (d) a combination of the last two methods. Sometimes the metal passing into the mold is inoculated by placing a pellet of finely pulverized inoculant in the basin directly under the sprue.

MOLDING AND CASTING PRACTICE

The technology of molding and pouring in a ductile iron foundry is similar in general to that employed in foundries that produce gray iron castings in similar sizes and quantities. The molding processes employed are those discussed in the articles on Sand Molding, Shell Molding, and Sand Cores and Coremaking.

Production of Malleable ———— Iron Castings ————

MALLEABLE IRONS are produced from base metal in the following ranges of composition:

Carbon	2.00 to 3.00%
Silicon	1.00 to 1.80
Manganese	0.20 to 0.50
Sulfur	0.02 to 0.17
Phosphorus	0.01 to 0.10
Boron	0.0005 to 0.0050
Aluminum	0.0005 to 0.0150

Each foundry uses an established melting practice, which permits it to produce iron within the above composition ranges with characteristics suitable to the requirements of the type of castings produced and the heat treating equipment utilized in that foundry.

Foundries having low tonnage requirements generally charge, melt down, refine and superheat the iron in batches. This practice is commonly referred to as "cold" melting. Such installations utilize electric induction, electric arc, or reverberatory ("air") furnaces for melting a properly proportioned charge of white iron returns, scrap iron and steel, pig iron, and ferroalloys. Air furnaces are fired with powdered coal, natural gas, or oil. The charge is laid in the hearth of the furnace and is melted by passing the flame over it. After the melt has been superheated and a sample has been chemically analyzed and the bath corrected by alloy additions, if necessary, the melt is tapped and transferred by ladle to molds. The tapping temperature, usually in the range of 1475 to 1600 °C (2700 to 2900 °F), depends on the composition of the iron, the fluidity of the iron required to fill the thinnest sections to be cast, and the facilities for metal transfer. Ordinarily, only one heat per day is melted in batch air-furnace operations.

Where larger tonnages are needed, or where a continuous supply of molten iron is required to pour conveyorized molding lines, electric melting furnaces or duplexing systems are employed. With electric installations, either multiple batch heats can be arc melted or line-frequency coreless induction furnaces can be used. In coreless induction melting, a portion of the superheated melt is periodically tapped and replenished at once with new charge.

Choice of an electric melting system for a malleable iron foundry involves the same considerations as for a gray iron foundry.

Cupola furnaces are also used to melt iron without duplexing for some malleable castings, especially pipe fittings. Furnace linings are generally either silica or superduty refractories. Metal superheating temperatures range from 1475 to 1600 °C (2700 to 2900 °F), depending on the necessity for reladling, the composition of the iron, and the section thickness of the castings to be produced.

Duplexed malleable iron is first melted in cupolas or in arc or induction furnaces, and then is refined in arc, induction or air furnaces. Usually, primary melting units are operated to provide molten metal for duplexing at relatively low temperatures (1425 to 1500 °C, or 2600 to 2750 °F), because primary melters generally are not efficient superheaters. Final superheating temperatures vary from 1475 to 1600 °C (2700 to 2900 °F), depending on the needs of individual foundries. Minor corrections in composition can be made in the duplexing furnace, but basic process control is generally accomplished in the primary melting unit.

CONTROL OF MELTING

Metallurgical control of the melting operation is based on the requirement for a molten iron of a certain composition that will:

- Solidify white in the castings to be produced
- Anneal on an established time-temperature cycle set to minimum values in the interest of economy
- Produce the desired hardenability after annealing.

Changes in melting practice or composition that would satisfy the first and third of these requirements generally oppose the satisfaction of the second, while attempts to improve annealability beyond an optimum level may result in difficulty with mottle (primary graphite) and low hardenability. One high-production malleable iron foundry operates to a composition standard of:

Carbon	2.50 ± 0.10%
Silicon	1.52 ± 0.08%
Manganese	[(% S × 1.7) + 0.15%] + 0.10, −0.05%
Sulfur	0.17% max
Chromium	0.09% max
Boron	0.0030 ± 0.0005%
Aluminum	0.0030 ± 0.0020%

A minimum carbon content is required in the interest of mechanical quality and annealability, because decreasing carbon content reduces fluidity of the molten iron, increases shrinkage during solidification, and reduces annealability. A maximum carbon content is imposed by the requirement that the casting be white as-cast. The silicon content is limited to ensure proper annealing during a short-cycle high-production annealing process and to avoid formation of primary graphite during solidification. Manganese and sulfur contents are balanced to ensure that all sulfur is combined with manganese and that only a safe, minimum quantity of excess manganese is present in the iron. An excess of either sulfur or manganese will retard annealing in the second stage and therefore increase annealing costs or decrease casting quality. Chromium content is kept to a low limit because chromium is a carbide stabilizer and because it retards second-stage annealing. Boron and aluminum are required for two reasons: to provide rapid annealing and to control nitrogen and oxygen activity.

CONTROL OF NODULE COUNT

Proper annealing in short time cycles and attainment of high levels of casting quality require that controlled graphite-particle distribution be obtained during the first-stage heat treatment. With low nodule count (few graphite particles per unit area or volume), mechanical properties are reduced from optimum values, and second-stage annealing time is unnecessarily long as a consequence of long diffusion distances during the low-temperature second-stage heat treatment. Too high a nodule count may be undesirable because of a reduction in the hardenability required for proper heat treatment of high-strength grades. Generally, a nodule count of 100 discrete graph-

ite particles per square millimetre appears to be optimum. This produces random particle distribution, with short distances between particles. Lower nodule counts are desirable for production of heavy-section castings that are to be hardened and tempered; higher nodule counts are desirable for fully annealed ferritic castings.

CONTROL OF ANNEALING

The rate of annealing for a hard iron casting depends on chemical composition, nucleation tendency as discussed above, and annealing temperature. With proper balance of boron content, base composition and melting practice, optimum numbers and distribution of graphite nuclei are developed in the early part of the first-stage anneal, and growth of the temper-carbon particles proceeds rapidly at any annealing temperature. An optimum iron will anneal completely through the first-stage reaction in approximately 1 h at 950 °C (1750 °F). Irons with lower silicon content or less than optimum nodule count may require as much as 20 h for completion of first-stage annealing. Longer times are required for castings more than 25 mm (1 in.) thick.

The temperature of first-stage annealing exercises considerable influence on the rate of annealing and the number and shape of graphite particles produced. Increasing annealing temperature accelerates the rate of decomposition of primary carbide and produces more graphite particles per unit of area or volume. However, high first-stage annealing temperatures result in excessive distortion of castings during annealing and the need for straightening operations after heat treatment. Annealing temperatures are adjusted to provide maximum practical annealing rates and minimum distortion, and therefore are controlled within the range of 900 to 975 °C (1650 to 1775 °F). Lower temperatures result in excessively long annealing times; higher temperatures provide excessive distortion and deterioration of nodule shape.

After first-stage annealing, the castings are cooled as rapidly as practical to about 750 °C (1375 °F) in preparation for the second stage of the annealing heat treatment. The fast cooling cycle requires 2 to 6 h, depending on the equipment used. For second-stage annealing of castings to become ferritic malleable iron, they are then cooled slowly at a rate of 3 to 17 °C/h (5 to 30 °F/h); during cooling, the carbon dissolved

in the austenite is converted to graphite on the existing temper-carbon particles, and a ferritic matrix results.

The furnace atmosphere for production of ferritic malleable iron in continuous controlled-atmosphere furnaces is controlled so that the ratio of carbon monoxide to carbon dioxide is in the range between 2-to-1 and 3-to-1. In addition, any sources of water vapor or hydrogen are eliminated, since the presence of hydrogen produces excessive decarburization of the surface of the castings. Proper control of the gas atmosphere is important to avoid undesirable surface structure. A high ratio of carbon monoxide to carbon dioxide causes retention of an excessively high content of combined carbon on the surface of the casting and produces a pearlitic rim, or "picture frame," on the final heat treated part. A low ratio of carbon monoxide to carbon dioxide permits excessive decarburization, which causes formation of a ferritic skin on the casting with an underlying rim of pearlite. This latter condition is produced when a significant portion of the subsurface metal is decarburized so much that no temper-carbon nodules can be developed during first-stage annealing (Fig. 11). When this occurs, the dissolved carbon cannot precipitate from the austenite, except in pearlite form.

METALLURGICAL CONTROL OF PEARLITIC MALLEABLE IRON

Melting and foundry controls for production of pearlitic malleable iron castings are the same as those required for ferritic malleable iron, since the base metal is the same for both materials (except for alloyed types) and the metallurgical requirements of the hard iron castings are identical. However, after first-stage annealing has been completed, the heat treating processes diverge, because of the requirements that ferritic malleable iron be produced with a carbon-free matrix and that pearlitic malleable iron be produced with a matrix containing a controlled amount of carbon in the combined form. The annealing equipment required for production of pearlitic malleable iron is similar in design and operation to that used for ferritic malleable iron, except that furnaces for heat treating pearlitic malleable do not require the fast-cool and slow-cool zones required for second-stage annealing of ferritic malleable but do require a temperature-equalization zone at the discharge end.

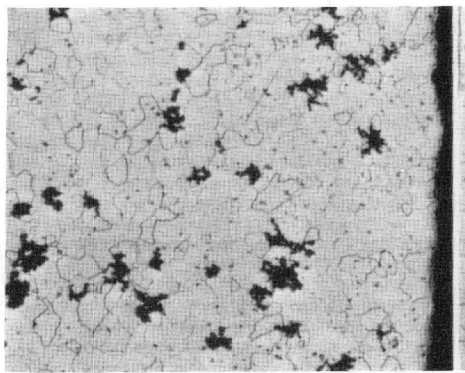 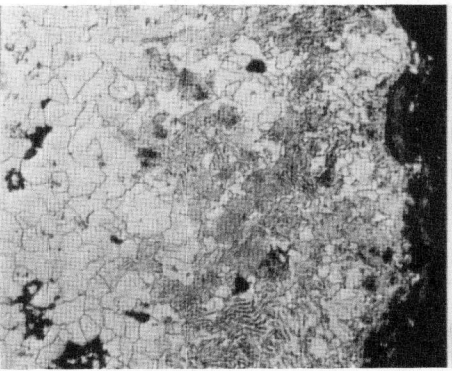 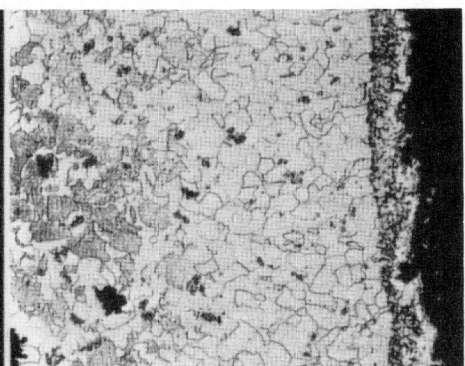

Proper atmosphere (15.6% CO, 5.4% CO₂)   Furnace atmosphere high in CO   Furnace atmosphere high in CO₂

Fig. 11. Influence of annealing furnace atmosphere on surface structure of malleable iron. (All specimens etched. Magnification, 100×.)

Production of Steel Castings

MELTING AND POURING of steel for production of sand castings is similar to melting and pouring of steel for casting into ingots that are subsequently forged or rolled. Also, the technology of molding in steel foundries is similar in a general way to that employed in foundries that produce castings from metals other than steel. This article is concerned primarily with production practices that are different for steel castings than for other castings, particularly iron castings.

MELTING

One major difference between melting practice in a steel foundry and in a steel mill is the higher tapping temperature used for foundry melting, to attain better fluidity of the molten steel. The producer of ingots for rolling is less concerned with fluidity, because mold filling is simpler in ingot molds than in sand molds for producing shapes having relatively thin sections.

The melting furnaces used in steel foundries are essentially the same as those used for production of steel ingots except that most foundry melting units are smaller. Although the equipment is the same, the processes are often different. Steel ingots may be made as rimming, semikilled or killed steel. Only thoroughly killed steel is used for steel foundry products. The method of production of the killed steel used for castings may differ from that used for wrought products because of the fluidity requirement. However, the salient features of making steel in a foundry are the same as those used for producing fully killed steel ingots.

Electric-Arc Melting

The direct-arc furnace consists essentially of a metal shell lined with refractories. This lining forms a melting chamber, of which the hearth is bowl-shaped. Three carbon or graphite electrodes carry the current into the furnace. Steel, either solid or molten, is the common conductor for the current flowing between the electrodes. Melting of the metal is achieved by arcs from the electrodes to the metal charge; the metal is melted both by direct impingement of the arcs and by radiation from the roof and walls. The electrodes are controlled automatically so that an arc of proper height may be maintained.

Acid Practice. In acid electric practice, the furnace hearth is composed of silica sand or ganister rammed into place. The furnace is charged with selected scrap low in phosphorus and sulfur content, because the acid process is not able to eliminate these elements. About 40% of the charge is usually made up of foundry scrap (gates and risers). The general practice is to charge small pieces first, in order to form a compact mass in the furnace, thus aiding electrical conductivity. The heavy and lumpy portion of the charge is placed over the smaller pieces, followed by the lightest portion.

Basic Practice. For basic electric-furnace melting, the furnace lining is a basic refractory such as magnesite or dolomite. The charge is usually composed of purchased scrap steel and foundry returns. During the melting period, small quantities of lime are added from time to time to form a protective slag over the molten metal. Iron ore is added to the bath just as melting is complete.

The slag is then highly oxidizing and in the correct condition to take up phosphorus from the metal. Shortly after all of the steel has melted, this first slag is taken off (if a two-slag process is to be used) and a new slag composed of lime, fluorspar, and sometimes a little sand, is added.

As soon as the second slag is melted, the current is reduced, and at intervals pulverized coke, carbon or ferrosilicon, or a combination of these, is spread over the surface of the bath. This period of furnace operation is known as the refining period, and its purpose is to reduce the oxides of iron and manganese in the slag and to form a calcium carbide slag, which is essential to the removal of sulfur from the metal. The refining slag has approximately the following composition: 45 to 55% CaO; 15 to 20% SiO_2; 0.50 to 1.5% FeO; and 5 to 15% CaF.

Adjustments are made in the carbon content of the bath by the addition of a low-phosphorus pig iron. After the proper bath temperature is obtained, ferromanganese and ferrosilicon are added and the furnace is tapped. Aluminum generally is added in the ladle as a final deoxidizer.

The basic electric-arc furnace is indispensable in the manufacture of high-alloy steels, including stainless steels. Alloy steels containing easily oxidized elements, such as chromium and manganese, can be remelted in the basic electric furnace without loss of chromium and manganese. For these applications, a single reducing slag is used to minimize oxidation.

Induction Melting

The high-frequency induction furnace is essentially an air transformer in which the primary is a coil of water-cooled copper tubing and the secondary is the metal charge. Inside the shell is placed the circular winding of copper tubing. Firebrick is placed on the bottom of the shell, and the space between that and the coil is rammed with grain refractory. The furnace chamber may be a refractory crucible or it may consist of a rammed and sintered lining. General practice is to use ganister rammed around a steel shell that melts down with the first heat, leaving a sintered lining. Basic linings are often preferred; a rammed lining of magnesia grain, or a clay-bonded magnesia crucible, may be used.

The process consists of charging the furnace with steel scrap and then passing a high-frequency current through the primary coil, thus inducing a much heavier secondary current in the charge, which heats it to the desired temperature. As soon as a pool of liquid metal has been formed, a pronounced stirring action takes place in the molten metal, which helps to accelerate melting. In this process, melting is rapid and there is only a slight loss of the easily oxidized elements. If a capacity melt is required, steel scrap is added continually during the melting-down period. As soon as melting is complete, the desired superheat temperature is obtained and the metal is deoxidized and tapped.

Gas Content in Conventional Melting

Proper melting practice and, to a lesser degree, proper heat treatment can limit the gas content of conventionally melted steel to acceptable levels, regardless of the type of furnace equipment employed. Failure to control oxygen, hydrogen and nitrogen contents may result in porosity or a severe decrease in ductility, or both.

Gas content is largely adjusted during the oxygen boil. After the cold charge is melted and the bath is in the temperature range of 1510 to 1540 °C (2750 to 2800 °F), oxygen is introduced into the molten metal, usually by means of a piping arrangement. The oxygen combines with the dissolved carbon in the steel to form bubbles of carbon monoxide. As the bubbles form, dissolved hydrogen and nitrogen are caught up in the bubbles in much the same way that dissolved oxygen finds its way into bubbles of boiling water. Thus, the bubbles of gas contaminants are boiled out.

Vacuum Degassing

Vacuum degassing reduces gas content and nonmetallic inclusions in steel to a minimum.

In the vacuum degassing process, steel from the furnace ladle is poured into a "pony" ladle. The steel from the pony ladle enters the vacuum chamber and flows into another ladle. Hydrogen and other gases are removed from the pouring stream as the metal enters the vacuum. The vacuum chamber is brought to atmospheric pressure after the heat is degassed, the ladle of degassed steel is removed from the vacuum chamber, and the castings are poured in the usual manner.

MOLDING

Green sand, dry sand, shell, and core molding are the processes most widely used for preparing molds for production of steel castings. Investment and ceramic molds are less frequently used. Steel cannot be cast into plaster molds because the high pouring temperature will destroy the plaster.

POURING, GATING AND RISERING

Pouring, gating and risering are three significant process variables that influence the surface condition, internal soundness and mechanical properties of steel castings.

Pouring

Three types of ladles are used for pouring steel castings: bottom-pour, teapot and lip-pour. Ladle capacity normally ranges from 45 kg to 36 t (100 lb to 40 tons), although ladles having much larger capacities are available.

The bottom-pour ladle has an opening in the bottom that is fitted with a refractory nozzle (Fig. 12). A stopper rod, suspended inside the ladle, pulls the stopper head up from its seat in the nozzle, allowing the molten steel to flow from the ladle. When the stopper head is returned to the position shown in Fig. 12, the flow is cut off. Position of the stopper head is controlled manually by the slide-and-rack mechanism shown at the left in Fig. 12.

The teapot ladle incorporates a ceramic wall, or baffle, that separates the bowl of the ladle from the spout. The baffle extends almost four-fifths of the distance to the bottom of the ladle (Fig. 13). As the ladle is tipped, hot metal flows from the bottom of the ladle up the spout and over the lip. Since the metal is taken from near the bottom of the ladle, it is free of slag and pieces of eroded refractory, although it may pick up foreign materials in the spout section or at the lip. The teapot design is feasible in various sizes, generally covering the entire range of casting sizes that are below the minimum size for which the bottom-pour ladle is used.

Lip-pour ladles are essentially similar in their external form to the teapot type. Because lip-pour

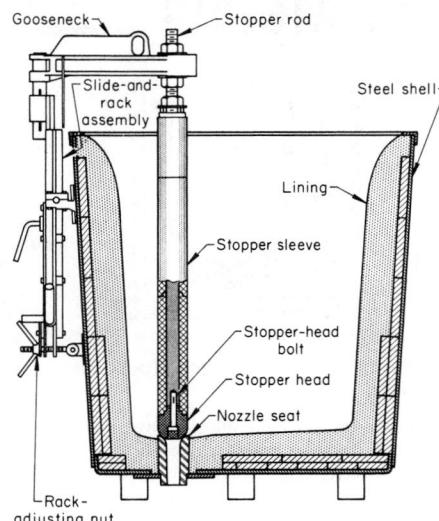

Fig. 12. Design of a bottom-pour ladle used for pouring large steel castings

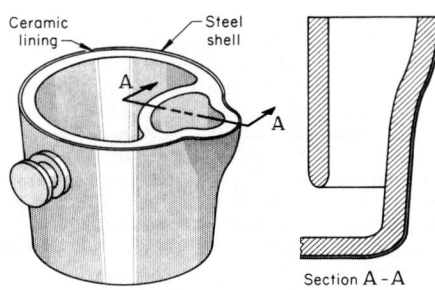

Fig. 13. Typical teapot-type ladle used for pouring small to medium-size steel castings

ladles have no baffles to hold back the slag, and because the hot metal is not taken from the bottom of the ladle, this type of ladle pours a dirtier steel and is seldom used to pour steel castings. Nevertheless, it is widely used as a tapping ladle (at the melting furnace) and as a transfer ladle to feed smaller ladles of the teapot type.

Gating

An effective gating system for pouring steel, as well as other metals, into sand molds is one that fills the mold as rapidly as possible without developing pronounced turbulence. It is essential that the mold be filled rapidly, mainly because heat is radiated to the walls and top surface of the mold as the molten steel rises in the mold cavity, and this heat can destroy the binder in the molding sand, causing the mold to collapse, unless the metal rises rapidly to provide the necessary support.

Preferred Metal Flow. According to preferred practice, the pourer directs the metal stream toward the pouring cup at the top of the mold, controlling the pouring rate to keep the cup full of molten steel throughout the pouring cycle. The opening in the bottom of the cup is directly over the sprue, or downgate, which is tapered with the large end up, thereby reducing the diameter of the stream of descending metal. The taper prevents the stream from pulling away from the walls

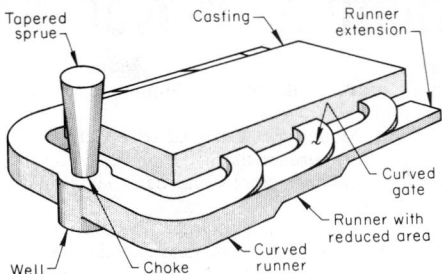

Fig. 14. Gating system for good metal flow

In this system, desired metal flow is obtained by proportioning the cross-sectional area of the choke of the sprue to all of the runners emanating from the sprue, and to all of the gates in accordance with a 1 : 4 : 4 ratio.

and drawing air into the gating system. The descending metal impinges on the sprue well at the bottom of the sprue, and the direction of flow changes from vertical to horizontal, with the metal flowing along runners to gates (ingates), and thence to the main body of the casting. A gating system that incorporates these features is shown in Fig. 14.

Risering

Molten steel contracts 0.9% per 55 °C (per 100 °F) as it cools from the pouring temperature to the solidification temperature. It then undergoes solidification contraction of 3% during freezing, and finally the solidified metal contracts 7.2% during cooling to room temperature. Therefore, in casting of steel, an ample supply of molten metal must be available from risers (reservoirs) to compensate for the volume decrease, or shrinkage cavities will develop in the locations that solidify last.

Because feeding from the riser depends on gravity, risers are usually located at the top of the casting. Riser forms are placed on the pattern and molded into the cope half of the mold. The riser cavity is usually open to the top of the mold, although blind risers are sometimes used.

Size and Shape. Formulas based on surface area, volume and freezing time of the casting are used to determine riser size. Most risers are cylindrical in shape, with their height approximately equal to their diameter. This configuration provides a low ratio of surface area to volume, which prolongs the time the steel remains liquid. A spherical riser would constitute an optimum design, but spherical shapes are difficult to mold.

Placement of a riser, in conjunction with its size, determines its effectiveness. The thicker sections of a casting act as reservoirs for feeding the thinner sections, which solidify first. Thus, risers are placed over thick sections that cannot be fed by other areas of the casting. Demonstrating this principle, the steel gear blank casting shown in Fig. 15 is provided with a large riser over the central hub and six smaller risers, equally spaced around the rim of the gear, to ensure adequate feeding. Metal enters the mold at the two gates, 180° apart.

Feeding Distance. Castings of uniform thickness present a different problem. Studies have established the feeding distances of a riser for various rectangular shapes in both the horizontal and vertical planes, with and without an end effect. (An end effect is the extra cooling provided by the sand cover of an end surface.)

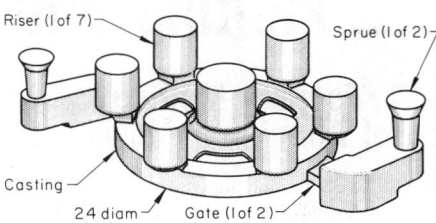

Fig. 15. Gating and feeding system used for casting a gear blank

The maximum feeding distance can be extended for a uniform section by adding a taper. The progressively thicker section solidifies in a progressively longer time, so that a favorable temperature gradient is established from the end of the section to the riser. A tapered pad of exothermic material placed in the mold along the length of the casting will also produce a favorable temperature gradient.

Cleaning

The choice of equipment and processing method for operations in the cleaning room of a steel foundry depends largely on the desired quality level of the castings, with respect to surface finish and dimensional accuracy, and on the cost that can be justified for the particular casting.

Small castings are usually shaken out, or removed from their molds, by a semiautomatic operation in which the molds are placed on a conveyor and are taken to a vibrating grate where the sand from the broken molds and some of that adhering to the castings falls through the grate while the casting vibrates to the far end of a table and is removed. Certain very fragile castings may require special processing to prevent mechanical damage at this and subsequent stages in the cleaning operation.

Castings are then allowed to cool until they are below about 205 °C (400 °F). At this point, the castings and the attached gates and risers still contain some adhering sand and a thin layer of scale.

Large Castings. For cleaning of large castings, portable equipment must be used while the casting rests on a worktable. These castings may be moved to the shakeout deck on a conveyor, or the individual flasks may be brought to the shakeout site with an overhead crane. When large castings must be shaken out while still at a red heat, the flask may be emptied by setting it on the edge of the shakeout deck so that the casting falls free of the deck and is allowed to cool before being placed on the shakeout deck for more complete removal of adhering sand. This practice helps to prevent mechanical damage to the castings while they are red-hot and soft.

Abrasive blasting follows rough cleaning on the shakeout deck. The blasting equipment may be of the same type as is used for small castings, but of larger capacity.

Influence of Casting Design on Cleaning. Cleaning of steel castings often represents a large proportion of the total cost of producing the casting. Frequently, designers of castings can contribute to a reduction in cleaning costs. For example, some castings require the addition of pads of metal at specific locations to enable proper feeding of liquid metal from risers. Such pads are costly to remove, but often they can be incorporated in the design of the casting and retained in the finished part.

Melting and Casting of Nonferrous Metals

Foundry Melting of Aluminum Alloys

FURNACES for melting of aluminum alloys in the foundry may be broadly classified into three types:

1. Direct fuel-fired furnaces
2. Indirect fuel-fired furnaces
3. Electrically heated furnaces.

DIRECT FUEL-FIRED FURNACES

Reverberatory furnaces (Fig. 1) generally are used to melt large amounts of aluminum to supply holding furnaces, or used to remelt scrap metal, and may be large enough to hold up to 80 t (90 tons) of molten aluminum alloy. The larger furnaces are disproportionately wider and longer than smaller furnaces, because bath depth is held to a maximum of about 750 mm (30 in.) regardless of the size of the furnace. Roof height above the molten metal depends on the height of the charging door, and that height depends on the kind of charge used. Roof height also depends on the heat-release factor relating the furnace volume to the heat input. In general, furnace builders prefer not to exceed 1.12 kJ per cubic metre (30 000 Btu per cubic foot) of space above the bath of molten metal.

Most reverberatory furnaces utilize a nozzle-mix burner that will throw a long flame, making use of "double-pass firing." This begins with a luminous or semiluminous flame, relatively high in the combustion chamber, that radiates heat to the refractory walls and roof. As the walls and roof become incandescent, they reradiate heat to the bath. On the return path to the flue, which is in the same wall as the burners, convective heat is transferred from the gases. This provides a double transfer of heat: radiation on the outgoing path and convection on the return path.

The exhaust port of a reverberatory furnace should have a cross-sectional area that will provide a slight positive pressure in the furnace during melting.

Charging. A wet-hearth furnace is charged by placing the charge material (solid or liquid) directly into the chamber or into a well. This procedure minimizes oxidation of the charge material. Molten metal may be charged through a launder (a refractory trough).

Dry-hearth furnaces are charged by placing solid material on the hearth, which slopes at an angle of 10 to 15°, on which the metal melts. Oxides, tramp iron and other nonmelting materials stay on the hearth as the molten aluminum runs into the bath.

Tapping. Molten metal may be removed from a reverberatory furnace through tapholes or by tilt pouring. Siphons, pumps and hand ladles may also be used.

Tapholes are plugged with fiber refractory cones, clay, or mixtures of sand and clay. Some furnaces have a taphole in a sump for removal of sludge. In addition, an upper taphole is used for pouring. Some furnaces use air-displacement pumps to deliver molten metal in measured quantities.

INDIRECT FUEL-FIRED FURNACES

In an indirect fuel-fired furnace, a barrier of some sort prevents contact of the hot combustion gases with the metal to be melted. Thus, there can be no pickup of the products of combustion by the metal charge, such as occurs in a direct fuel-fired furnace.

Crucible (pot) furnaces are the most typical of indirect fuel-fired furnaces. In a crucible furnace, there is a wall of silicon carbide or metal between the combustion gases and the metal charge.

Crucible furnaces are much used in aluminum foundries for melting or holding, or both, because of their versatility for alloy change. The major components of crucible furnaces for melting aluminum alloys are essentially the same as those of crucible furnaces used for melting other metals.

In the simplest form of crucible furnace, the pot is stationary, and the molten metal is ladled from it for casting. In a lift-out crucible furnace (Fig. 2), the pot has a pouring spout, and it is removed from the furnace by means of tongs and used as a pouring ladle.

ELECTRIC FURNACES

Most electric-furnace melting of aluminum alloys for casting is done in low-frequency induction furnaces, of the channel (core) and the coreless types, although high-frequency induction furnaces and electric-resistance crucible furnaces are used also and sometimes prove to be more suitable for specific applications.

Channel (core-type) low-frequency induction furnaces are used to supply molten aluminum for all methods of casting. Most channel induction furnaces for melting aluminum are 60-cycle furnaces, and range from 20 to 200 kW for capacities of 320 to 1360 kg (700 to 3000 lb) of aluminum, with melting rates of 45 to 450 kg/h (100 to 1000 lb/

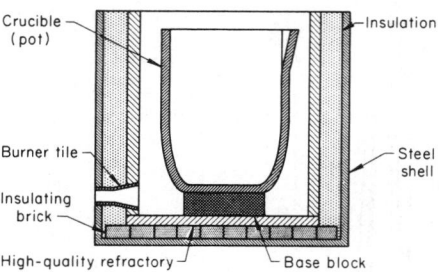

Fig. 2. Lift-out crucible (pot) furnace

h). These furnaces are controlled by voltage-regulated transformers and contactors.

The powerful electromagnetic field present in the channels causes the molten metal to flow from the sides to the center of each channel, from which it is forced in opposite directions out the ends of the channels. This flowing action stirs and mixes the molten metal enough so that mechanical stirring rarely is needed; it also helps to maintain uniform temperature throughout the molten metal.

Core-type low-frequency induction furnaces will melt about 2.3 kg (5 lb) of aluminum per kilowatt-hour; thus, they are more efficient than any burner-type furnace. Thermal efficiency per unit weight of aluminum melted and heated to pouring temperature can be 50% for 8-h operation, and as high as 70% for 24-h operation. The loss by oxidation is low, because the heat is generated within the metal. Also, the metal is cleaner than when melted in burner-type furnaces.

For greater capacity or for continuous operation, a holding hearth may be connected to the melting hearth, with separate controls for each hearth. In these combination units, metal melted in one side goes by gravity to the holding side. When such units are used only as holding furnaces for purposes such as large die-casting operations, they are charged with hot metal.

Coreless low-frequency induction furnaces are used mostly to melt turnings, foil and other fines. Coreless furnaces have the same melting and stirring action as channel (core-type) furnaces, and so there is little loss of metal by oxidation.

Electromagnetic forces in a coreless furnace produce intense stirring, which lifts the center of the surface of the molten metal to a higher level than the level at the crucible wall. Nonmetallics gather as a ring around the furnace walls in the plane of the center of the power coil. The circulation washes a fresh charge quickly into the molten bath, resulting in high recovery of metal. The bath remains at a lower temperature than would be practical for melting in the well of a reverberatory furnace.

A coreless furnace has no channels to clean, and it is not necessary to keep a molten heel in the furnace. It is thus adaptable to batch-type operation. The initial charge should consist of large shapes, to at least 15% of furnace capacity. Following meltdown of the large shapes, subsequent charges may consist of any convenient forms, and can be comprised entirely of fines.

MATERIALS OF CONSTRUCTION FOR FURNACES AND ACCESSORIES

Silicon carbide bonded with carbon is the most commonly used crucible material for melting and

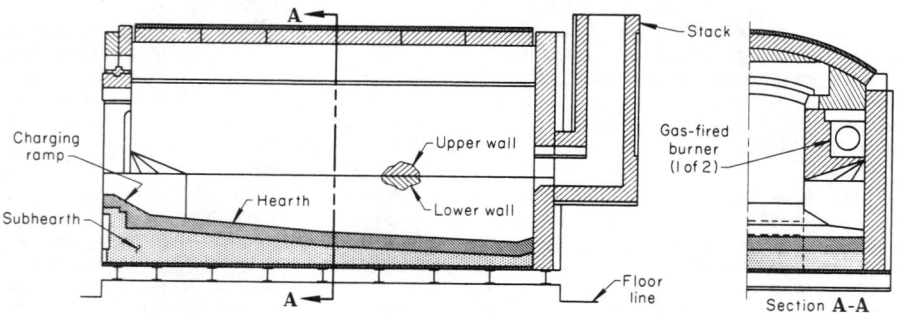

Fig. 1. Large reverberatory furnace charged through a ramp

holding aluminum alloys. Bonding the silicon carbide with silicon nitride provides a better crucible than is obtained with carbon bonding, but nitride-bonded crucibles cost more and are less commonly used. Silicon carbide crucibles bonded with carbon can, with good cleaning and maintenance practice, last for long periods of operation (up to 18 months in some foundries).

Crucibles of cast iron are not used for melting or holding aluminum that must not pick up iron, because there are no protective washes or coatings to make cast iron pots acceptable without excessive maintenance.

Ladles with capacities of up to 9 kg (20 lb) of molten aluminum generally are made of clay and graphite and are held with hand tongs. With care to avoid abrasion, such ladles have good life, but they will not last as long as ladles made of silicon carbide.

CHARGING PRACTICE

The furnace charge may be made up of any one or a combination of: returned gates and risers, returns from machining operations, prealloyed ingot, primary metal together with alloying elements or hardeners, and molten metal. The selection of metal for the charge depends on melting equipment and capacity, composition and quality of the alloy needed, cost, and available analytical equipment.

Direct charging into a holding furnace is never as desirable as separate melting, because there is less opportunity to clean the metal in direct charging. Direct charging of scrap is a source of oxide. The presence of water, oil or dirt on direct-charged metal will result in defective castings besides being a hazard to personnel. If solid metal is charged directly in a holding furnace, the chilling effect will promote the formation of sludge.

Purchased scrap may be in the form of briquetted chips, castings, clippings or other solid scrap that has been analyzed and classified.

Clean scrap is scrap that has been subjected to abrasive blasting to remove dirt, heavy oxide, and embedded sand, then heated above 260 °C (500 °F) for at least 4 h to remove contaminants such as water, oil and wax.

Machining scrap, such as borings and turnings, should be subjected to treatment in centrifuges, drum dryers, magnetic separators, and screens. Generated scrap of this type is sometimes briquetted, although it is usually fed into reverberatory melting furnaces without briquetting. All scrap must be clean and dry to prevent excessive metal loss and to prevent explosions.

Foundry Scrap. Floor sweepings and dross can constitute a costly loss of metal if not recovered. However, to prevent contamination of other charging materials, or an explosion, such material should be melted in a separate furnace and treated to remove dirt and oxide. The recovered metal is cast into ingots for remelting in small quantities with the normal charge for regular production.

Scrap-to-Ingot Ratio. The proportionate amounts of scrap (such as gates and risers) to new ingot in the charge depend to some extent on the requirements of the castings to be poured, but usually to a much greater extent on the control facilities in a particular foundry. For instance, in some foundries where requirements are extremely high (such as for the highest-quality aircraft castings) and facilities for reprocessing are minimal, only virgin ingot is used; gates, risers

and sprues are sold. In most aluminum foundries, however, the yield is less than 50% (percent by weight of salable castings compared to the amount of metal melted); thus, the practical approach is to reprocess the accumulated scrap.

DEGASSING

When molten metal is cooled from a higher temperature to a lower temperature for pouring (for instance, from 825 to 700 °C, or 1500 to 1300 °F), dissolved hydrogen is given up in conformance with the equilibrium conditions. However, the dissolved hydrogen is given up slowly. Thus, in foundry practice, additional degassing may be required. For instance, in one foundry casting various aluminum alloys, the number of castings rejected for over-all porosity increased to alarming proportions during normal operation. Tests showed that the porosity resulted from hydrogen gas that was introduced into the molten metal by melting and holding the metal at 825 °C (1500 °F) for $\frac{1}{2}$ to 3 h. Three corrective steps were taken: (a) metal temperature was reduced to 700 °C (1300 °F) before castings were poured; (b) the ratio of air to oil in the melting furnace was changed to provide oxidizing flames, thus reducing the possibility of hydrogen pickup from this source; and (c) the metal was degassed with a mixture of nitrogen and chlorine. These steps were effective in eliminating porosity.

Degassing fluxes to remove hydrogen are recommended for use after the surface of the bath has been fluxed for removal of oxides. The degassing fluxes also help to lift fine oxides and particles to the top of the bath. Removal of hydrogen by degassing is a mechanical action; hydrogen does not combine with the degassing agents.

Degassing agents include chlorine gas, nitrogen-chlorine mixtures, and hexachloroethane.

Regardless of which gas is used, it is admitted to the bath through fluxing tubes. The tubes, usually made of graphite or of porcelain enameled steel, are inserted in the bath so that they nearly reach the bottom. Apparatus for degassing is available in both stationary and portable units. A typical setup for degassing in a ladle is illustrated in Fig. 3.

CONTROL OF SLUDGE

Sludge, also called "sand," "sugar" and "silicon dropout," forms and settles out of aluminum alloy baths that contain about 5% or more silicon. However, the sludge itself does not contain silicon, but is a compound of iron, manganese and chromium.

Sludge formation is more common in die casting than in sand or permanent mold casting, because: (a) lower temperatures are used in die casting, and (b) the higher silicon and iron contents of aluminum die-casting alloys make them more susceptible to sludge formation.

Sludge in castings generally is first revealed by a reduction in the life of tools used in machining the castings. The reduced tool life is caused by hard spots in the castings, which appear as small shiny spots on the machined surface. Metallographic examination reveals the sludge particles.

Chemical analysis can also be used to establish the presence of excessive amounts of sludging elements in the metal. In one foundry, for example, the analysis of a casting that contained sludge was: 3.36 Cu, 1.24 Fe, 9.23 Si, 0.41 Mn,

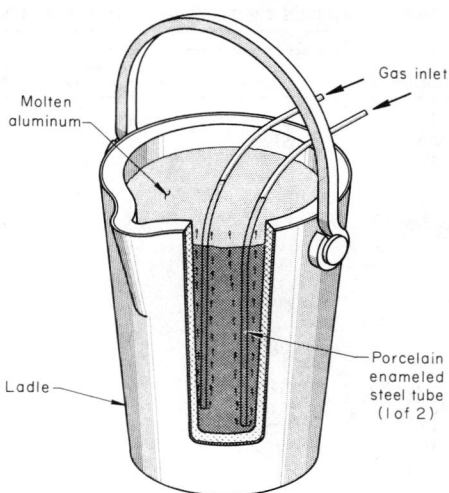

Fig. 3. Typical setup for degassing of molten aluminum alloy in a ladle

0.02 Mg, 2.59 Zn, 0.09 Ni, 0.14 Cr, rem Al. In this composition, the total of % Fe + 2(% Mn) + 3(% Cr) was 2.48%, which is well above the normally allowable 1.90%.

In addition to causing difficulty in the castings, sludge formation can deplete the molten bath of specified alloying elements.

Sometimes sludge originates in the central melting furnace and is then transferred to the holding furnace, although the temperature of a melting furnace is usually high enough that little or no sludge is formed. The source of sludge is usually the holding furnace. This furnace can contain as much as 150 mm (6 in.) of sludge segregated at the bottom. If this sludge is stirred to ladling depth, it will be poured into the castings.

Production of Aluminum Alloy Castings

ALUMINUM ALLOYS can be cast by any of the commercial casting processes. Sand casting, permanent mold casting and die casting are the processes most often used.

Nominal compositions of the casting alloys mentioned in this article are given in Table 1, along with their commercial designations and corresponding ASTM designations (where they exist).

Designations for commercial aluminum alloys often carry a letter prefix to denote an impurity level or the presence of a secondary alloying element. For example, alloy 356 is a 7% silicon, 0.3% magnesium alloy; alloy A356 has the same basic composition, but a limit of 0.2% iron is imposed. Also, alloy 214 can be purchased in three modifications—A214, B214 and F214. Although each of these alloys has about 4% magnesium as the major alloying element, A214 is alloyed with 1.8% zinc, B214 with 1.8% silicon, and F214 with 0.5% silicon.

Castability of aluminum alloys varies with composition. The aluminum-silicon alloys, such as alloys 13, 43, 355 and 356, are the easiest to cast—providing the greatest fluidity for casting of deep, thin sections. Generally, the higher the silicon content up to the eutectic composition

Table 1. Nominal compositions of aluminum casting alloys mentioned in this article

Commercial	ASTM B26, B85 and B108	Casting process(a)	Cu	Si	Mg	Zn	Other(b)
13	S12B	D	...	12.0	...	...	...
A13	S12A	D	...	12.0	...	...	1.3 Fe max
43	S5A,B,C	S,P,D	...	5.3	...	...	...
D132	...	P	3.5	9.0	0.8	...	0.8 Ni
F132	SC103A	P	3.0	9.5	1.0	...	...
195	C4A	S	4.5	0.8	...	...	...
B195	...	P	4.5	2.5	...	...	...
214	G4A	S	...	...	4.0	...	...
A214	GZ42A	P,D	...	...	4.0	1.8	...
B214	GS42A	S,P	...	1.8	4.0	...	...
F214	...	S	...	0.5	4.0	...	...
218	G8A	D	...	...	8.0	...	...
B218	...	S	...	...	7.0	...	...
Almag 35	GM70B	S,P,D	...	...	7.0	...	0.2 Mn
220	G10A	S	...	...	10.0	...	...
319	SC64D	S,P	3.5	6.0	...	...	...
333	SC94A	P	3.5	9.0	...	...	...
A344	...	P	...	7.0	...	...	0.2 Fe max
355	SC51A	S,P	1.3	5.0	0.5	...	...
C355	SC51B	P	1.3	5.0	0.5	...	0.2 Fe max
356	SG70A	S,P	...	7.0	0.3	...	...
A356	SG70B	P	...	7.0	0.3	...	0.2 Fe max
360	SG100B	D	...	9.5	0.5	...	...
A360	SG100A	D	...	9.5	0.5	...	1.3 Fe max
364	...	D	...	8.5	0.3	...	0.4 Cr, 0.03 Be
380	SC84B	D	3.5	8.5	...	...	...
A380	SC84A	D	3.5	8.5	...	...	1.3 Fe max
384	SC114A	D	3.8	12.0	...	...	...
A612	ZG61B	S	0.5	...	0.7	6.5	...
C612	ZC60A	S,P	0.5	...	0.4	6.5	1.0 Fe
D612, 40E	ZG61A	S,P	...	...	0.6	5.8	0.5 Cr
RR350	...	S	5.0 Cu, 1.5 Ni, 0.25 Mn, 0.25 Co, 0.25 Zr, 0.25 Sb, 0.2 Ti				

(a) S = sand casting, P = permanent mold casting, D = die casting. (b) About 0.15% Ti is specified in some alloys, but 0.25% Ti max usually is noted.

(about 12.5% Si), the easier an alloy is to cast. Hypereutectic alloys are more difficult to cast. Alloys that contain substantial amounts of magnesium, such as alloys 214, 218 and 220, are the most difficult to cast.

SAND CASTING

Sand casting is the most versatile method for casting aluminum alloys, providing the greatest latitude for size, shape, and alloys cast. It is preferred for making large castings, because the limits on maximum size and maximum section thickness are higher; section thicknesses up to 150 mm (6 in.) are feasible. Also, sand casting is usually selected for the production of small quantities of castings of almost any size, because of the relatively low tooling cost.

Disadvantages of sand casting, compared with other casting methods, are: greater cost per casting (omitting cost of tooling), rougher surfaces, and greater dimensional variation.

Aluminum alloys are generally easier to cast in sand molds than are iron and steel, for the following reasons: (a) because casting temperatures are lower, there is usually no mold burn-in; (b) no mold or core coatings or washes are required; (c) because of lower casting temperatures, less gas forms from mold and core components; and (d) because aluminum alloys weigh less, fewer castings are defective because of mold failure.

Sand. The principal differences between sand mixtures for casting aluminum alloys and those for casting ferrous or copper-base alloys are: (a) sands of finer grain size are used for aluminum alloys; (b) sand mixtures can have lower green compressive strength, thereby permitting easier shakeout; and (c) only a minimum of additives is needed.

The selection of molding sand depends on: (a) the surface finish specified for the casting, (b) the size of the casting, (c) the alloy to be cast, and (d) the method used and the equipment available for the preparation of the sand mixture.

Green vs Dry Sand Molds. Green sand molds are ordinarily used for casting aluminum alloys. Dry sand molds may be used because they provide greater accuracy and smoother casting surfaces, but dry sand molding is most often selected when complexity of the casting is a factor. Lack of strength in a green sand mold that allows core shift is sometimes a reason for using dry sand molds, as in Fig. 4.

Hot Tears. Aluminum alloy castings have very low hot strength and are thus more susceptible to hot tearing than are ferrous metal castings. Sometimes hot tearing can be prevented by careful selection of core materials, but more often minor redesign of the casting is the more practical approach. A redesign of casting struts that eliminated a problem with hot tearing is described in Fig. 5.

Shakeout and Cleaning. Mechanized shakeout can be used for most aluminum alloy castings, but some require special handling. At shakeout, aluminum alloy castings will bend under relatively low stress. Dropping of molds may be enough to cause bending. Castings with thin sections are more likely to bend than thick, chunky ones. Proper handling at shakeout will reduce or eliminate subsequent straightening operations. Castings that are subject to warpage should be allowed to remain in the mold for a longer time before shakeout.

The removal of internal cores is often a problem, especially when only a thin shell of metal surrounds a heavy core. On large castings and cores, pneumatic chipping hammers are used to break up the cores. For castings produced in squeezer-type molds, a core knockout machine is recommended. These machines have a pneumatic chipping hammer mounted on a frame directly above an anvil. The gate or riser on the casting is placed between the hammer and the anvil, and the hammer is started. Vibration is transmitted through the gates and casting, to break up the cores and thus facilitate their removal. For castings that are to be heat treated, if cores cannot be removed easily, they may be left in the casting. The heat treating temperature will completely burn out the binder, and the core sand can be poured out. However, this technique may contaminate both the heat treating furnace and the quench tank.

Gates and risers are removed from aluminum castings by band sawing, friction sawing, shearing or breaking. Band sawing is most widely used. The saws operate at 4.6 to 20.3 m/s (900 to 4000 ft/min). Small gates can often be removed by shearing in a mechanical press.

Snag grinding for casting cleanup is done on both coated abrasive belts and grinding wheels. Under many conditions, coated abrasive belts offer improved production, better surface finish, and better working conditions. Grit sizes used range from 24 to 150 mesh. Longer belt life and faster cutting will result if a wax-type lubricant is used.

Fins and lumps that, because of their location, cannot be ground off may be removed by car-

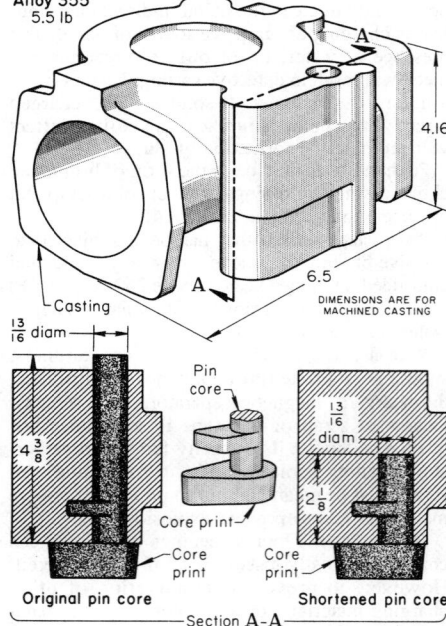

Alloy 355
5.5 lb

4.16
6.5
DIMENSIONS ARE FOR MACHINED CASTING

Casting
$\frac{13}{16}$ diam
$4\frac{3}{8}$
Pin core
Core print
Core print
Original pin core

$\frac{13}{16}$ diam
$2\frac{1}{8}$
Core print
Shortened pin core

— Section A-A —

Item	Green sand mold	Dry sand mold
Castings per mold	One	Two
Metal poured per mold, lb	15	25
Shakeout method	Manual	Jolting table
Production, castings per hr	2	2.7
Total castings produced	500	5000

Fig. 4. Housing cast in a dry sand mold to prevent core shift that occurred when a green sand mold was used. (Dimensions are in inches.)

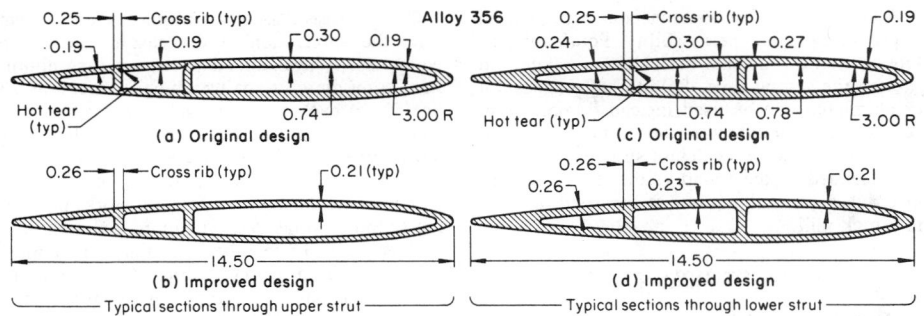

Fig. 5. Original design of cross sections of struts of a turbine air intake casting, and redesign that eliminated hot tears. (Dimensions are in inches.)

bide-tipped rotary files in high-speed air-powered rotary hand tools. Speeds for the rotary files can vary from 2500 to 18 000 rpm. Small chipping hammers and hand files are also used for removing fins and lumps.

SHELL MOLDING

The shell molding process is used in aluminum foundries to obtain greater dimensional accuracy and better surface finish on castings than can be obtained by sand casting, and, occasionally, to decrease cost. Equipment, materials and procedures for making shell molds and cores for aluminum alloy castings are generally the same as those for gray iron and other metals. However, because of the characteristics of aluminum alloys, special considerations (as discussed subsequently) are required for molds and cores.

Precision metal patterns and a considerable amount of special equipment are required for shell molding. Because of the cost of patterns and equipment, shell molding is most applicable to high-volume production, but shell molds and cores are also used for low-volume production when green sand molding fails to produce good castings.

The aluminum alloys that can be cast in sand molds can be cast in shell molds.

Molds. Because aluminum alloys solidify over a range of temperature and go through a mushy stage instead of forming a skin, heat-transfer requirements of the shell are quite different for aluminum alloys than for iron. Aluminum gives up its heat three times as fast as iron, which causes greater thermal shock to the mold. Before the metal solidifies, stresses are set up in the mold that can result in cracking or buckling of the mold, and damage to the casting.

Consequently, careful selection of sand, resin and processing procedure is important. Only subangular sands should be used, particularly for molds that have large flat surfaces (up to 500 by 750 mm, or 20 by 30 in.). A four-screen sand with about 5% pan fines of less than 325 mesh is recommended. The sand must be dry and free from clay.

Cores. Aluminum alloys have very low strength at temperatures just below the solidification range; therefore, it is mandatory that cores have low hot strength so that they will not cause hot tears in the castings by restricting metal contraction. Cores must be strong enough to resist breaking in handling and assembling, and yet be weak enough to collapse readily as the casting cools. This combination of properties can be obtained by: (*a*) use of additives, such as iron oxide or

Table 2. Suggested dimensional tolerances for aluminum alloy plaster mold castings

Type A dimension: between two points in same part of mold, not affected by parting plane or core

Specified dimension, in.	Tolerance, in.	
	Critical	Noncritical
Up through 1	±0.005	±0.010
Over 1	±0.005, +0.001 in. per in. over 1 in.	±0.010, +0.002 in. per in. over 1 in.

Type B dimension: across parting plane. Type A dimension plus the following:

Projected area of casting, $A_1 \times A_3$, sq in.	Added tolerance for parting plane, in.
Up through 10	0.005
Over 10 to 50	0.010
Over 50 to 100	0.020
Over 100	0.030

Type C dimension: affected by core. Type A dimension plus the following:

Projected area of casting affected by core, $A_3 \times G$, sq in.	Added tolerance for core, in.
Up through 10	0.005
Over 10 to 50	0.020
Over 50 to 100	0.030
Over 100	0.045

D dimension: draft

Critical locations........................	0°
Noncritical locations	2°

E dimension: minimum wall thickness: 0.060 in.
F dimension: allowance for finish

Maximum dimension, in.	Nominal allowance, in.
Up through 5	0.020
Over 5 to 12	0.030
Over 12 to 18	0.040

Minimum diameter of cored holes: 0.250 in.

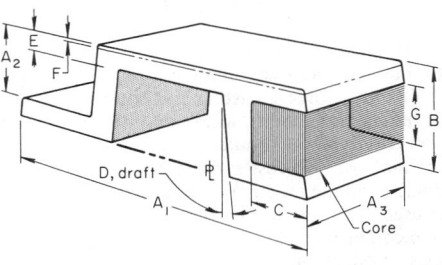

wood flour; (*b*) reduction of binder (resin) to the lowest possible level; and (*c*) use of round-grain sands. The most common practice is to make cores from round-grain sands (mixtures made with round grains have lower strength than those made with subangular grains). The sands are cold

or hot coated with 2 to 4% resin. These cores are satisfactory for most applications.

Gating and Pouring. As with sand castings, an important objective in the design of the gating system for shell mold castings of aluminum alloys is to minimize turbulence and prevent dross from entering the mold cavity.

The cope and drag halves of shell molds are held together by bonding with resin, by stapling the edges, or by bolting; these methods also can be used in combination.

Pouring practice is essentially the same as for pouring sand castings. Because aluminum alloys weigh only about one-third as much as the same volume of gray iron or steel, mold backup is seldom required. Most molds are poured in the horizontal position.

Dimensional Accuracy. The close tolerances that can be held when casting in shell molds is a major reason for their use. Dimensional tolerances for shell mold casting are closer than for casting in permanent molds and are approximately equal to those for plaster mold casting (see Table 2).

PLASTER MOLD CASTING

Plaster mold casting will produce aluminum alloy castings that have greater dimensional accuracy and smoother surfaces than can be obtained by sand casting or permanent mold casting.

Applicability. Plaster mold casting is used extensively for casting of aluminum alloy impellers such as those for air-conditioning equipment; it is the most economical method for obtaining the required accuracy and surface finish of the blades. Other applications of plaster mold casting include match plate patterns, molds for rubber and plastics, and precision electronics components such as waveguides.

Alloys Cast. The aluminum alloys that can be cast successfully in sand molds are suitable also for casting in plaster molds. However, alloys 43, A344, 355 and 356 are preferred, because they have high fluidity and resist hot cracking. Alloy 43 is satisfactory when high mechanical properties are not required, whereas the heat treatable alloys such as A344, 355 and 356 are used when high mechanical properties are required.

COMPOSITE MOLDS

Composite mold assemblies are molds constructed of several components, at least one of which differs from the others in regard to the process by which it was made and in molding material. Although the use of composite mold assemblies for casting aluminum alloys is not new, stringent requirements by the aerospace, electronics and air-conditioning industries have greatly expanded applications.

INVESTMENT CASTING

Aluminum alloys may be cast in either ceramic-shell or solid investment molds. Molds can be poured by gravity, by gas or metal pressure, by centrifugal force, or by the vacuum-assist method. Procedures for investment casting of aluminum alloys are generally the same as those employed for other metals (see the article on Investment Casting, in this section).

Applicability. Castings produced by investment molding are costly compared with those produced by the sand molding, permanent molding, or die-casting processes. The investment process should not be considered for castings that can be produced to acceptable standards by one of these other processes. The investment process is used when required shape, reproduction of surface detail, or dimensional accuracy is beyond the capabilities of the more conventional casting processes, and when machining to meet such requirements is impossible or prohibitively expensive. Components of electronics equipment, aerospace parts, and intricate parts for various instruments are the major applications of aluminum alloy investment castings.

Size Limitations. Although there are no theoretical limitations on the size of an investment casting, there are practical limitations; a dimension of 500 mm (20 in.) is considered maximum for most castings. Section thicknesses are usually no greater than 13 mm ($^1/_2$ in.), although sections 100 mm (4 in.) thick have been successfully cast in aluminum alloys.

Alloys C355 and A356 are most commonly used for investment castings, because they have good fluidity and also because their properties are generally acceptable for a wide range of applications. Alloys 40E, A612 and D612 are frequently specified, even though their foundry characteristics are near the opposite extreme from C355 and A356. For alloys 40E, A612 and D612, it is necessary to use more elaborate gating systems for late-stage feeding, which results in a lower yield per unit of metal poured; also, higher preheat temperatures for the molds are usually required for producing sound castings of these alloys.

Mechanical properties of investment castings, when properly engineered, are similar to those of sand and permanent mold castings of the same alloy. The investment process permits considerable versatility in control of solidification pattern and rate.

Mold Materials. In all commercially important investment materials for investment casting of aluminum alloys, various grades of silica comprise the refractory portion, and calcium sulfate hemihydrate is the bonding agent. The materials are available as dry blends formulated to individual specifications. A typical composition of a dry blend is as follows:

Silica:
Cristobalite	35.5%
Silica flour	29.0%
Calcium sulfate	33.5%
Diatomaceous earth	2.0%
Potassium sulfate	0.15 to 0.30%

Proportions of the refractory and the bond are designed to control the amount of expansion during green setting (initial drying) and firing. Typical thermal expansion and contraction characteristics are shown in Fig. 6.

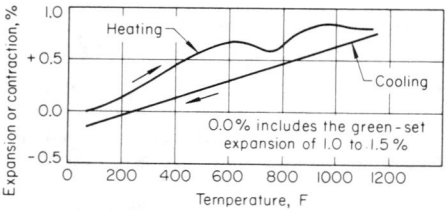

Fig. 6. Typical thermal expansion and contraction of an investment mold material bonded with calcium sulfate

A highly absorbent diatomaceous earth is used to provide adequate permeability. Potassium sulfate is used to nucleate the calcium sulfate and shorten the setting time. Sulfates of magnesium or aluminum can be used instead of potassium sulfate. The amounts used are influenced by the properties of the raw calcium sulfate.

If blended in the foundry, the dry refractory materials are usually ground to specified fineness in a rock mill.

PERMANENT MOLD CASTING

More aluminum alloy castings are made by the permanent mold process than by any other process except die casting. Semipermanent mold casting, wherein molds are made of metal but cores are expendable, is also used for aluminum alloys.

Aluminum alloy castings weighing as much as 350 kg (780 lb) each have been produced in permanent molds, but the majority weigh less than 14 kg (30 lb) each.

Alloys identified with the letter P in Table 1 are those most commonly used for casting in permanent molds. The castability of these alloys is by no means equal. The alloys with high silicon contents have best castability, and those with high magnesium contents have poorest castability.

Surface finish of permanent mold castings is smoother than that of sand castings. Surfaces of 7.0 to 12.7 μm (275 to 500 micro-in.) are normal. Type and thickness of mold coating, condition of the mold surface, and metal pouring temperature are the major variables that influence surface finish. With close control of these variables, finishes smoother than 7.0 μm (275 micro-in.) can be produced.

Mechanical properties are often better for permanent mold castings than for similar sand castings produced from the same alloy, because of the more rapid cooling in metal molds.

Applicability. Because tooling cost for permanent mold casting is higher than for sand casting, the permanent mold process is most often restricted to use for intermediate to high production. Where the permanent mold process is used to make small quantities, it is most often to obtain greater dimensional accuracy, which will minimize or eliminate machining operations.

A common practice is to cast in green sand in the prototype or limited-production stage and, as production requirements increase, to determine whether permanent mold or semipermanent mold casting will be the more practical.

DIE CASTING

Die casting is used for mass production of small to intermediate-size aluminum alloy castings. Ordinarily, when a casting can be made by any of the major methods, sand mold casting is used for prototype or limited production (because tooling cost is lowest and design changes can be made at minimum cost), permanent mold casting is used when the design is fixed and production is medium to high, and die casting is used for mass production.

Alloys Cast. The alloys in Table 1 that are designated by the letter D are those most commonly used for die casting. Castability of alloys follows the general rule that high-silicon alloys have the best castability and that high-magnesium alloys have the poorest. Alloys 13 and A13 are extensively used for die casting.

Size and Weight Limitations. Casting size and weight limit the application of die casting. Although equipment is available that can produce aluminum alloy die castings weighing as much as 23 kg (50 lb) each, most weigh considerably less. Components that have relatively large plan areas are routinely produced by die casting (see Fig. 59 in the article on Die Casting), but, as plan area increases, especially for thin work, producing sound castings becomes more difficult.

Machines. Cold-chamber machines are almost universally used for die casting of aluminum alloys. Hot-chamber machines are generally unsatisfactory, mainly because the molten metal is constantly in contact with some metal component in a hot-chamber machine, and severe contamination of the aluminum by iron is likely to take place.

Horizontal machines are more widely used than vertical machines, although vertical machines are sometimes preferred, as when center gating is used.

Dies for die casting aluminum alloys may be single-cavity, multiple-cavity, combination, or unit-type, depending mainly on the size and shape of the casting and on production requirements.

Dies used for casting aluminum alloys are subjected to more rigorous service than are those used for casting lower-melting metals, such as zinc alloys. They must be made of a material that will resist softening at the casting temperature (620 to 700 °C, or 1150 to 1300 °F). Hot work tool steels (such as H13) hardened to 44 to 48 HRC are most commonly used. Large dies are commonly constructed as unit dies—that is, as a master holding block with various dies inserted. The master holding block is made of an alloy steel such as 4140, frequently in the form of a casting, and the inserted dies are made of a hot work tool steel. Moving parts, such as slides and cores, are made of nitrided 7140 steel or of a hot work tool steel, which can also be nitrided to prolong service life.

Heat checking, from repeated stressing at elevated temperature, is the most common cause of failure of die-casting dies used for casting aluminum alloys. Heat checks in the dies result in poor surfaces on the castings.

SELECTION OF CASTING PROCESS

The appropriate casting process is selected on the basis of the size, weight and shape of the casting, the quantity to be produced, the mechanical properties and dimensional accuracy required, and the cost of making the casting to these requirements by the various processes. Cost is usually related closely to quantity and the weight of the casting.

Foundry Melting of Copper Alloys

MELTING TEMPERATURES of copper alloys are considerably higher than those of aluminum, magnesium, zinc or lead alloys; tapping temperatures for copper alloys are often as high as 1300 °C (2400 °F).

In the liquid state, copper alloys behave much like ferrous alloys of similar density. Molten copper alloys are susceptible to contamination from refractories as well as from the atmosphere. Copper casting alloys are subject to fuming from

Table 3. Melting temperatures of principal copper casting alloys

Nominal composition	Common name	ASTM B30 alloy No.	Start of melting (solidus) Copper phase °C	°F	Lead phase °C	°F	End of melting (liquidus) °C	°F
88Cu-6Sn-1.5Pb-4.5Zn	Navy M bronze	2A	826	1518	...	...	988	1810
80Cu-10Sn-10Pb	High-lead tin bronze	3A	762	1403	314	598	929	1705
85Cu-5Sn-5Pb-5Zn	Leaded red brass	4A	853	1568	316	601	1009	1849
76Cu-2.5Sn-6.5Pb-15Zn	Leaded semi-red brass	5B	831	1527	319	607	953	1748
57.5Cu-39.25Zn-1.25Fe-1.25Al-0.25Mn	Manganese bronze (65 000 psi)	8A	862	1583	...	...	880	1616
64Cu-26Zn-3Fe-5Al-4Mn	Manganese bronze (110 000 psi)	8C	885	1625	...	...	923	1693
64Cu-4Sn-4Pb-8Zn-20Ni	Nickel silver (20% Ni)	11A	1108	2027	309	588	1143	2089
81Cu-4Si-15Zn	Silicon brass	13B	821	1510	...	...	917	1683

vaporization of zinc, which is a major alloying element in about three-fourths of the copper casting alloys. With a few exceptions, such as beryllium copper and 1% Cr copper, the copper casting alloys contain at least 10% alloying additions and frequently these additions exceed 40%. Alloying additions have a marked effect on the temperature at which the metal begins to melt and the temperature at which melting is completed (solidus and liquidus). Temperatures at the beginning and at the end of melting are given in Table 3 for eight of the principal copper casting alloys. For any copper alloy, the temperature at which the metal is poured into the mold is higher than the liquidus temperature (for pouring temperatures, see Table 4, in the article on Production of Copper Alloy Castings).

MELTING FURNACES

Furnaces for melting the copper casting alloys are either fuel fired or electrically heated. They are broadly classified into four categories:

1. Crucible furnaces (lift-out or tilting)
2. Reverberatory (open-flame) furnaces
3. Induction furnaces (core or coreless)
4. Indirect-arc furnaces.

Selection of a furnace depends principally on the quantity of metal to be melted, the degree of purity required, and the variety of alloys to be melted.

CRUCIBLE FURNACES

Crucible melting by combustion heating is capable of providing high-quality metal at a relatively low initial investment. Crucible furnaces are available in a wide range of sizes. Crucible furnaces may be of either the lift-out or the tilting type (see Fig. 7 and 8).

REVERBERATORY FURNACES

Reverberatory furnaces are open-flame fuel-fired furnaces in which the charge is melted by radiation from the hot walls and roof and by convection from the movement of hot gases.

Reverberatory furnaces are available in a variety of sizes and designs. Capacities range from about 25 kg (about 50 lb) to many tons. Tilting reverberatory furnaces (Fig. 9) are the type most often used for melting copper alloys in the foundry. This type of furnace is charged by placing the charge material in the well. The burners for a tilting reverberatory furnace are on the same

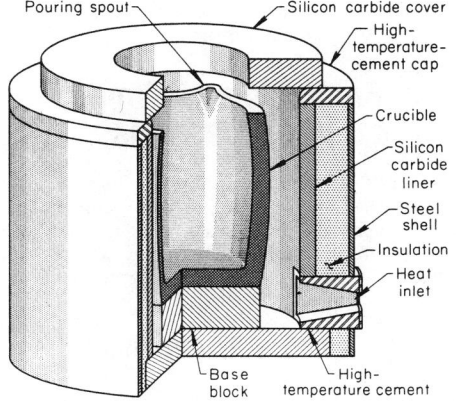

Fig. 7. Typical lift-out crucible furnace especially well-adapted to foundry melting of small quantities of copper alloys (usually less than 135 kg, or 300 lb)

side as the exhaust. This design feature serves to increase thermal efficiency, because the hot exhaust gases must return across the surface of the melt, which increases the amount of heating from convection.

Because in a reverberatory furnace heat is transferred to the melt directly, rather than through the containing walls as in crucible melting, reverberatory furnaces are capable of rapid melting with higher thermal efficiency than is obtained in crucible melting.

INDUCTION FURNACES

Induction furnaces possess several advantages for foundry melting of copper alloys. Such furnaces are clean and easy to control, cannot contaminate the melt with products of combustion, and are extremely flexible in their operation. Electrical characteristics (primarily, power input) can be related directly to the amount of heat generated. The melt can be held indefinitely at any desired temperature for charges down to about 10% of the rated furnace capacity. Electromagnetic stirring in these furnaces ensures homogeneous composition and uniform temperature. The main disadvantage of induction melting is the cost of the furnace.

Induction furnaces are of two general types: core (channel), using low-frequency power; and coreless, using either low-frequency or high-fre-

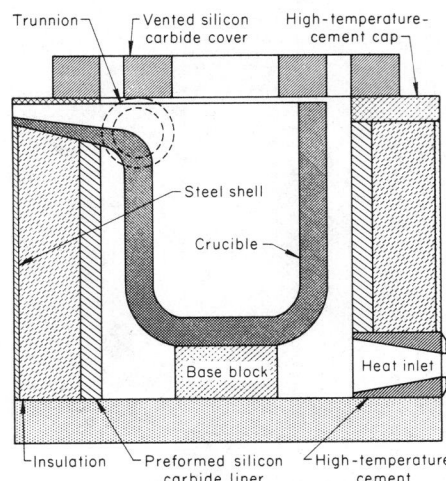

Fig. 8. Typical lip-axis tilting crucible furnace used for foundry melting of copper alloys. (Similar furnaces are available that tilt on a central axis.)

quency power. Low-frequency current is usually line frequency, generally 60 cycles per second. High-frequency current is supplied by motor generators, spark gaps, or electronic oscillators and ranges in frequency from one thousand to many thousand cycles per second.

Core (Channel) Furnaces. In a core induction furnace, a channel of molten metal acts as a shorted secondary loop coupled as a step-down transformer. Heat from the metal in the loop is transferred to the charge above. Core furnaces may have either a single or double secondary channel.

Coreless Furnaces. In the coreless type of induction furnace, the crucible that contains the charge is surrounded by the induction coil, and either the charge or the crucible acts as a shorted secondary.

Coreless induction furnaces are available as drop-coil units (Fig. 10), in which the coil is positioned around the crucible during melting. After melting, the coil and shell are lifted off as a unit and the crucible is used as a pouring ladle. Coreless furnaces are available also as tilting types, in which the crucible is permanently installed.

INDIRECT ARC FURNACES

Indirect arc furnaces may be used for high-production melting. An indirect arc furnace is

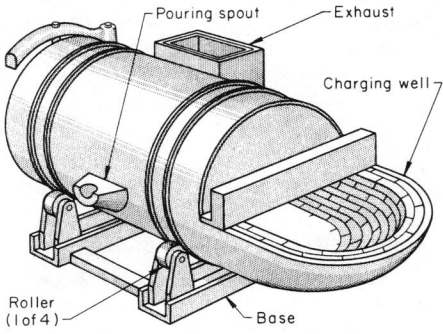

Fig. 9. Simplified view of a typical tilting reverberatory furnace

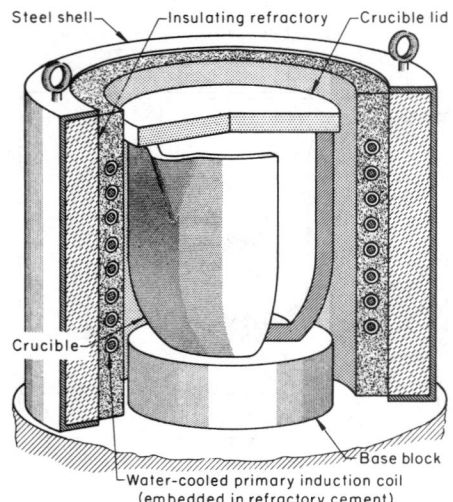

Fig. 10. Typical drop-coil coreless induction furnace used for melting copper alloys

barrel-shaped with a horizontal axis. As shown in Fig. 11, a graphite electrode enters each end of the furnace; one electrode is stationary and the other is movable. An arc is struck between the two electrodes at the center. Heat is transferred to the charge by direct radiation from the electric arc, from reflected radiation off the lining, and by conduction to the molten charge as it washes over heated refractory. Because the energy of the arc must be transferred to the charge and furnace walls by radiation from an arc at an extremely high temperature, there is a problem of vaporization of alloying constituents and difficulty in temperature control of the melt.

Indirect arc furnaces are rocked during the melting cycle. Rocking promotes faster melting, produces a more homogeneous melt, and minimizes wear of the refractory. The furnace is mounted on rollers and is rocked mechanically about a horizontal axis. The angle of rocking can be varied as needed, according to the location of the pouring spout and the height of molten metal in the furnace. The furnace is tilted (on the same axis on which it is rocked) for pouring.

Melting of aluminum bronze in indirect arc furnaces is not recommended, because of the excessive dross formation. High-zinc alloys are subject to loss of zinc by volatilization.

CHARGING

Generally, no attempt is made to refine the metal during melting in the foundry. Prealloyed

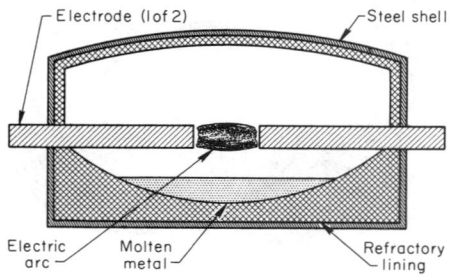

Fig. 11. Schematic view of an indirect arc furnace

ingot is available in nearly any desired alloy composition. Thus, alloying in the melting furnace from base materials should not be attempted. For best control, however, routine chemical analyses should be performed—principally because of melting losses, which vary considerably, depending on the type of melting furnace used, melting atmosphere, and preheating practice (if used).

The entire charge for a heat should be carefully weighed, recorded so that melting losses may be calculated and compensated for, and placed in a suitable container before being transferred to the melting furnace. Ideally, the entire charge is put into the furnace at one time. Breaking up a weighed charge is conducive to errors and results in lack of composition control. Crucible and reverberatory furnaces should be at a red heat before charging, so that the melting time is kept to a minimum. This practice minimizes contamination from the atmosphere.

Production of Copper Alloy Castings

COPPER ALLOY castings are produced by sand, shell, investment, plaster, ceramic, and permanent mold casting; and by die casting. Most of the articles in this volume that deal individually with these processes contain information on the casting of copper alloys.

POURING TEMPERATURE

Temperature ranges for pouring the principal copper casting alloys are given in Table 4. It should not be inferred from the breadth of most of these ranges that pouring temperature is not critical; as noted in Table 4, the ranges are intended for pouring various section thicknesses. For castings with minimum section thickness, the metal should be poured at a temperature near the high side of the range. Conversely, for castings that have all heavy sections, pouring temperatures should be near the low side of the range. Under any conditions, identical castings should be poured at the same temperature, insofar as possible. It is generally advisable to allow a variation of no more than 55 °C (100 °F) during the pouring of a specific mold, or when pouring several molds from the same ladle. The casting process used also influences the pouring temperature for a specific alloy. In die casting, for instance, a temperature near the low side of a given range is used, in the interest of longer die life.

Temperature Control. The only practical method of measuring the pouring temperature of copper

alloys is the use of a pyrometer with a calibrated thermocouple.

The gas content of liquid metal increases as the temperature increases; consequently, metal poured at too high a temperature may produce castings with gas porosity.

POURING PRACTICE

On the basis of the degree of care required when they are poured into molds, copper alloys can be classified into two groups:

1. Alloys that form tight, adherent, nonfluid slags or oxides. Typical are aluminum bronze and manganese bronze.
2. Alloys that form fluid slags or oxides. These include most of the alloys in general use—those containing various combinations of copper, tin, lead and zinc. Typical are high-leaded tin bronze (alloy 3A) and leaded red brass (alloy 4A).

Copper alloys in the first group (those that form tight, adherent, nonfluid slags) require great care in pouring. Their general behavior can be compared to that of aluminum casting alloys, and similar pouring techniques are recommended.

Good pouring practice for this first group of alloys includes attention to the following:

- Molten alloys with tight oxide films should never be stirred. After the ladle is filled or the crucible is removed from the furnace, and before pouring, the metal should be carefully skimmed but not stirred or mixed—thus minimizing oxide entrapment.
- In addition to avoiding stirring of molten aluminum bronze and manganese bronze, other forms of agitation should also be avoided. If the metal is melted in a tilting furnace and must be transferred to a ladle for pouring, the distance the metal must drop should be minimized by holding the ladle close to the furnace lip.
- Pouring should be smooth and even, to avoid splashing and separated metal streams. With careful pouring of aluminum bronze and manganese bronze, it is possible to form an aluminum oxide "glove" around the metal stream, which will protect the molten metal from further oxidation.

Alloys in this first group have a very narrow freezing range, so that they solidify in much the same way as does a pure metal. The total shrinkage is concentrated in the region of the casting that solidifies last. These alloys are thus prone to piping and gross shrinkage cavities. Risering is commonly used to prevent shrinkage from occurring in the casting. The metal is poured well

Table 4. Pouring temperatures of principal copper casting alloys

Composition	Alloy Name	No.	Pouring temperature(a) °C	°F
88Cu-6Sn-1.5Pb-4.5Zn	Leaded tin bronze (Navy M bronze)	2A	1075 to 1250	1950 to 2300
80Cu-10Sn-10Pb	High-leaded tin bronze	3A	1000 to 1225	1850 to 2250
85Cu-5Sn-5Pb-5Zn	Leaded red brass	4A	1075 to 1300	1950 to 2350
76Cu-2.5Sn-6.5Pb-15Zn	Leaded semi-red brass	5B	1075 to 1250	1950 to 2300
57.5Cu-39.25Zn-1.25Fe-1.25Al-0.25Mn	Manganese bronze (65 000 psi)	8A	950 to 1100	1750 to 2000
64Cu-26Zn-3Fe-5Al-4Mn	Manganese bronze (110 000 psi)	8C	975 to 1150	1800 to 2100
88Cu-3Fe-9Al	Aluminum bronze	9A	1100 to 1200	2000 to 2200
64Cu-4Sn-4Pb-8Zn-20Ni	Nickel silver (20% Ni)	11A	1225 to 1425	2250 to 2600
81Cu-4Si-15Zn	Silicon brass	13B	975 to 1150	1800 to 2100
96.5Cu-2.5Be-1.1Ni	Beryllium bronze	...	1000 to 1225	1850 to 2250

(a) Use the high side of the temperature range for castings with thin sections, and the low side for castings with thick sections.

above the liquidus so that the entire mold cavity is filled and so that solidification occurs from the bottom to the top, with feeding from a riser.

Copper alloys in the second group (those that form fluid slags) are generally less affected by turbulence in pouring than are those in the first group. Although turbulence in pouring can cause casting defects in any alloy, the fact that the oxides that are formed with this group of alloys separate readily from the molten metal means less likelihood of oxide entrapment in the casting, and greater likelihood of the escape of entrained air bubbles.

SAND MOLDING

Molding Sands. Selection of molding sands for copper alloys depends on the same factors as selection for other metals.

Natural sand can be used for most molds for the casting of copper alloys. Selection of grade depends largely on the type of casting being poured and the end requirements. For instance, in foundries that produce such items as bronze tablets and statues, fine sand with very little clay is used, and the sand mixture is conditioned to a velvetlike texture. The molds are sprayed with a water-soluble wash, and are torch dried to remove excess moisture. This practice produces the best surface conditions, but with some sacrifice of internal soundness. However, internal soundness is not usually important in producing this type of casting. In contrast, the plumbing goods industry uses coarse sand, which provides better internal soundness (pressure tightness) but rougher surfaces than does the fine sand.

Synthetic (compounded) sand is used to overcome the deficiencies of natural sand. The preferred arrangement of grain distribution for sand used in casting of copper alloys is essentially the same as in casting of other metals. Synthetic sand composed of fine-screen base sand requires the least clay and water to produce the strongest bond. Further, it requires less ramming and squeezing to produce the densest mold.

GATING AND FEEDING SYSTEMS FOR SAND CASTING

It has been estimated that up to 50% of foundry rejects are attributable to poor gating and feeding practice. Rejects can result from dross or slag, or misruns and porosity.

The gating system for a mold should ensure that the metal enters the mold as quietly as possible, to minimize the formation of dross from metal agitation. Strainer cores are not used extensively, but a choke in the system may be used to avoid washing, which results in dirty castings. Too much choke may result in such slow pouring that a wormy surface is produced; this is particularly likely to occur during pouring of heavy castings from alloys that contain large amounts of zinc. The size of the sprue should be related to the size of the runner, so that the sprue can be kept filled by pouring at a fairly uniform rate. Further, the total cross-sectional area of the gates should be related to that of the runner. Runners should be free of sharp edges, and gates should be filleted.

A gating system designed according to best foundry principles will: (*a*) maintain an adequate reservoir of molten metal; (*b*) fill the sprue quickly and keep it full; (*c*) keep the runner full of metal; (*d*) keep velocities of metal down so that metal

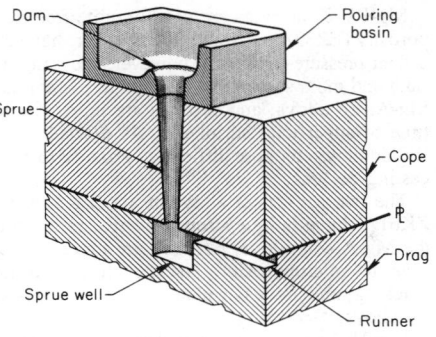

Fig. 12. Section of a typical sand mold, showing pouring basin with a dam, tapered sprue, sprue well, and a portion of the runner

does not squirt into the mold, causing eroded surfaces; (*e*) make provision for keeping the first metal poured (which is likely to be contaminated) from reaching the casting; and (*f*) maintain a temperature gradient to ensure proper feeding.

Pouring basins should be used, rather than permitting the metal to be poured straight down the sprue. A typical pouring basin is illustrated in Fig. 12. Ordinarily, this type of basin is prepared separately (usually from dry sand) and placed over the sprue prior to pouring. For some molds, however, the shape of such a pouring basin is cut in the cope. A pouring basin is usually made with a dam, as shown in Fig. 12. The dam permits some molten metal to accumulate in the basin before it starts down the sprue, thus promoting a more even flow. In addition to a dam, a skim core is sometimes incorporated in the pouring basin. A skim core consists of a section molded into the top of the pouring basin that acts as a skimmer to hold back any floating dross or dirt and to prevent it from being washed down the sprue.

Runners. Runners should be rectangular, and their maximum cross-sectional area should be four to six times that of the sprue exit. Inadequate cross section of a runner can cause premature chilling of the metal and result in misruns. When the runner has a larger cross section than the sprue exit, metal velocity is reduced in the runner. Also, turbulence is reduced and dross and dirt are permitted to rise or sink in the runner, depending on their density.

Runners are most commonly placed in the drag. With the runner in the drag and the gates in the cope, the metal flows for the full length of the runner before it begins to flow through the gates. This promotes uniform flow into the casting cavity. In gating systems like the one illustrated in Fig. 13, where two or more gates are in line, preferred practice is to decrease the cross section of the runner after each gate, and to extend the runner beyond the last gate. This runner extension serves as a dross trap, so that the first metal poured, which is likely to be dirty, does not enter the casting cavity. A riser may be placed near the sprue end of the runner if needed for additional feeding, although for pouring a simple flat casting, such as that shown in Fig. 13, risers are seldom needed.

Gates are usually in the cope, regardless of whether the casting is in the cope or the drag. Multiple gating as shown in Fig. 13 and 14 is preferred, particularly for those alloys that solidify over a broad temperature range. The total

cross-sectional area of the gates should be at least equal to the maximum cross-sectional area of the runner. If the total cross-sectional area of the gates is less than that of the runner, a pressurized system is developed whereby the metal squirts into the casting cavity.

Sprues. The smallest cross-sectional area of a sprue (the sprue exit) should be no larger than is required to achieve the desired flow rate during pouring.

Sprues less than about 100 mm (4 in.) high (including the height of the pouring basin) usually are not tapered. When a sprue is more than about 100 mm high, it should be tapered so that its shape conforms to the natural taper of a falling stream of metal. This avoids aspiration of air, which can cause oxide inclusions and entrapped air bubbles. The amount of taper depends on the height of the sprue.

SHELL MOLD CASTING

Copper alloys can be cast successfully in shell molds. Advantages of the shell mold process for copper alloys are the same as for steel and gray iron—namely, close dimensional control, smooth surface finish, and low cost in high-volume production.

PLASTER MOLD CASTING

Most copper alloys that can be cast successfully in sand molds can be cast with equal success in plaster molds. However, copper alloys that contain more than about 5% lead are not recommended for casting in plaster molds because the high-lead alloys react unfavorably with some mold compositions, resulting in rough surfaces

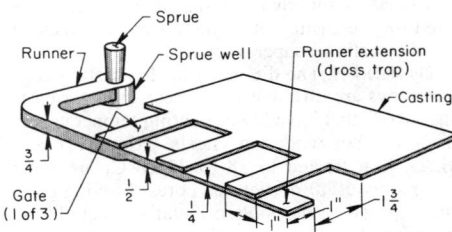

Source: Improved Copper-Base Alloy Gating Using Aluminum Techniques, by D. L. LaVelle, *Trans Am Foundrymen's Soc*, Vol 72, 1964, p 575-581. (Runner thicknesses have been added.)

Fig. 13. Typical gating and feeding system for casting a simple flat plate

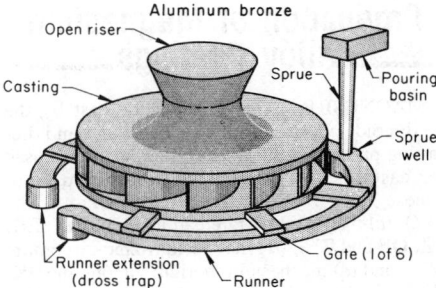

Fig. 14. Gating and feeding system with a riser attached to a thicker section of a casting having nonuniform sections

and thus defeating one of the primary purposes of casting in plaster molds.

Sometimes pouring temperature imposes a limitation on plaster mold casting of copper alloys. A plaster mold will be destroyed if heated above about 1425 °C (2600 °F). Therefore, maximum pouring temperature must be substantially below 1425 °C.

PERMANENT MOLD CASTING

Although the permanent mold casting process is applicable for a number of copper alloys, it is best suited to casting of tin, silicon, aluminum and manganese bronzes, and yellow brasses.

The advantages of permanent mold casting over sand casting are essentially the same for copper alloys as for aluminum alloys or other metals that can be cast by both processes. Likewise, for copper alloys as for other metals, the only major disadvantage of the permanent mold process is the higher tooling cost.

DIE CASTING

Copper alloys that are commonly cast in permanent molds can also be die cast; yellow brass is best suited to the die casting process. Copper alloy die castings seldom weigh more than 0.9 kg (2 lb), and most weigh less than 0.2 kg ($^1/_2$ lb).

The advantages derived from die casting of copper alloys are the same as those derived from die casting of other metals: close dimensional control, good surface finish, and high rate of production.

The main limitation in die casting copper alloys is short die life, which is caused by the high temperatures at which copper alloys must be cast.

Machines. Cold-chamber machines of the type used for die casting other metals are satisfactory for die casting copper alloys.

Die Materials. The die materials for casting copper alloys are different from those used for casting metals that have lower melting temperatures. Tungsten hot work tool steels (H20, H21 and H22), heat treated to 38 to 45 HRC, are most commonly used for dies and cores. Moving parts not exposed to the molten metal—coreholders, for example—are made of an alloy steel such as 6145, heat treated to 40 to 50 HRC.

Die Life. The high temperatures required in casting copper alloys shorten die life even though high-alloy die steels are used. In fact, the die casting process is not used extensively for copper alloys primarily because of the rapid wear of dies.

Production of Magnesium Alloy Castings

MAGNESIUM ALLOYS can be cast by the sand mold, permanent mold, shell mold and die-casting processes. Alloys in five systems are used for castings: (*a*) magnesium-aluminum-manganese (AM); (*b*) magnesium-aluminum-zinc (AZ); (*c*) magnesium–rare earth–zirconium (EK, EZ, QE and ZE); (*d*) magnesium-zinc-zirconium (ZK); and (*e*) magnesium-thorium-zirconium (HK, HZ and ZH). Compositions of the most frequently cast alloys in these systems are given in Table 5.

The Mg-Al-Mn and Mg-Al-Zn alloys are easily cast and are preferred when their properties meet application requirements.

Mg-RE-Zr castings are usually quite free from porosity (EZ33A castings, in particular, have excellent pressure-tightness); however, they are less fluid and more susceptible to dross formation than Mg-Al-Zn alloys, and castings are more susceptible to superficial shrinkage. As a result, Mg-RE-Zr alloys are more difficult to produce to some casting designs.

The two Mg-Zn-Zr alloys in use (ZK51A and ZK61A) have fatigue strengths equal to those of the Mg-Al-Zn alloys, but they are more susceptible to microporosity, surface sinks and hot cracking, and are less weldable, than the Mg-Al-Zn alloys.

Mg-Th-Zr alloys are more difficult to cast than Mg-RE-Zr alloys, because oxide inclusions and defects caused by gating turbulence are harder to control. Inclusions are most likely to occur in thin-wall castings that require a rapid pouring rate. Mg-Th-Zr alloys have adequate castability for production of complex parts having moderate to heavy wall thickness.

FURNACES AND OTHER EQUIPMENT FOR MELTING

Furnaces for melting and holding magnesium alloys are generally the same in design and construction as those used for melting and holding zinc or aluminum, but the nature of molten magnesium requires differences in equipment, refractories, and protection for molten surfaces:

- Crucibles can be made of wrought or cast steel, because molten magnesium does not attack ferrous metals.
- Refractories should be selected with care, because molten magnesium reacts violently (explosively) with some refractories. High-alumina refractories and high-density superduty firebrick with 57% silica and 43% alumina give satisfactory service.
- Because molten magnesium reacts explosively with iron oxide, a door should be provided through which scale dropping from the crucible or furnace parts can be removed, and a runout pan should be positioned under the crucible to contain molten metal in the event of a crucible failure.
- A protective flux or a protective gas must cover

Table 5. Nominal compositions of magnesium casting alloys

Alloy	Composition, %					
	Al	Mn. min	Zn	Th	Zr	Rare earths
Sand and permanent mold casting alloys						
Am100A	10.0	0.1	...	...	...	...
AZ63A	6.0	0.15	3.0	...	...	...
AZ81A	7.6	0.13	0.7	...	...	...
AZ91C	8.7	0.13	0.7	...	...	...
AZ92A	9.0	0.10	2.0	...	...	...
EK30A	...	...	...	...	0.3	3.3
EK41A	...	...	...	...	0.6	4.0
EZ33A	...	...	2.7	...	0.6	3.3
QE22 (a)	...	...	...	...	0.7	2.0
ZE41A	...	...	4.2	...	0.7	1.2
ZK51A	...	...	4.6	...	0.7	...
ZK61A	...	...	6.0	...	0.7	...
HK31A	...	...	...	3.3	0.7	...
HZ32A	...	...	2.1	3.3	0.7	...
ZH42	...	...	4.0	2.0	0.7	...
ZH62A	...	...	5.7	1.8	0.7	...
Die casting alloys						
AZ91A, B (b)	9.0	0.13	0.7	...	...	...

(a) Contains 2.5% silver. (b) Alloys AZ91A and AZ91B are identical except that residual copper to 0.35% max is allowable in AZ91B.

the molten magnesium to prevent burning. When sulfur dioxide is used, a hood with a small ladling door is placed over the crucible. The sulfur dioxide forms a protective curtain at the door. With either the flux or the sulfur dioxide gas, adequate ventilation must be provided to remove fumes.

MELTING PRACTICE

Only clean metal should be charged in melting furnaces. Dirty metal introduces gas and contamination into the melt. Metal should be kept free from moisture and oil. All scrap should be thoroughly cleaned by abrasive blasting, because sand adhering to foundry scrap causes silicon contamination. Gates and risers should be remelted as soon as possible after they are removed from castings.

Melting. Flux equal to about $1^1/_2$% of the capacity of the crucible should be placed in the bottom of the empty crucible. Dry ingot or scrap is then added. The metal should be placed so that it will not "bridge over" during melting. The charge is dusted with flux as often as is necessary to prevent oxidation of the molten magnesium. Additional solid metal is charged to the semiliquid metal until the crucible or pot is full.

Sludge Removal. If there is an appreciable amount of sludge in the crucible after the charge has melted, it should be removed. Metal temperature for desludging should be about 650 °C (1200 °F). A sludge-removal ladle is moved along the bottom of the crucible in such a manner as to collect the sludge. Metal and flux usually float on the top of the sludge as the ladle is brought to the surface, and they should be decanted. The sludge is dumped into a hot, dry, clean sludge pan. Desludging is repeated as long as significant amounts of sludge are picked up in the ladle. The crucible should be desludged as often as necessary, or at least once during each shift. Excess flux should also be removed.

Because sludge is mainly magnesium oxide, which is a good insulator, an excessive accumulation causes the bottom of the crucible to become overheated, resulting in abnormal scaling of the crucible and short crucible life.

Cleaning Molten Metal. The molten metal at 705 to 720 °C (1300 to 1325 °F) is stirred thoroughly but carefully, and flux equal to $1^1/_2$% of the charge is sprinkled on its surface. The melt is then allowed to settle for 10 to 15 min to precipitate nonmetallic impurities and flux. The condition of the ladle in open-crucible operations can be used as a good indication of the cleanness of the metal. If the ladle is clean and free from lumpy deposits of sludge, the metal is in good condition for casting.

Gases and nonmetallic impurities can be removed by chlorination (bubbling chlorine gas through the molten metal). The chlorine gas is introduced into the metal through a graphite tube. The magnesium chloride formed during chlorination is molten above 708 °C (1307 °F) and removes oxide particles and other suspended impurities. Too much magnesium chloride will form if the temperature of the melt is too high. The optimum temperature range for chlorination is 720 to 760 °C (1325 to 1400 °F), and the length of treatment is usually between 5 and 15 min.

SAND CASTING

Most magnesium alloy castings are made in sand molds. Sand castings vary in size from a

Table 6. Compositions and properties of molding sands used in sand casting of magnesium alloys

Mixture No.	AFS fineness of base sand	Binder Western bentonite	Southern bentonite	Sulfur	Boric acid	Diethylene glycol	Potassium fluoborate	Water	Green compression strength kPa	psi	AFS green permeability
1	65	2.0	...	1.5-2.0	1.5-2.0	1.5-2.0	0.5	2.5-3.2	62-83	9-12	80-140
2	65	4.0	...	1.5-2.0	1.5-2.0	0.75-1.25	0.5	2.0	55-83	8-12	80-120
3	65	2.0	2.0	1.5-2.0	1.5-2.0	0.75-1.25	0.5	2.0	55-83	8-12	80-120
4	65	4.0	...	...	0.75	...	(b)	4.0	48-62	7-9	125
5	65	2.5	2.5	3.0	...	2.5	1.5	2.5	62	9	100
6	80	3.0	...	1.0	1.00	1.5	...	2.5	48	7	100
7	90	2.5-4.0(c)(d)	...	1.0-3.0(d)	...	...	0.5-1.5(d)	(e)	62-76	9-11	40-65

(a) Remainder is sand. (b) Ammonium fluosilicate, 3.25%. (c) Modified bentonite containing oil as a mixing agent. (d) Low end of range is used for light to medium-weight castings; high end for heavy castings. (e) Oil, 2.5 to 4.0%.

few ounces to 1360 kg (3000 lb); most weigh less than 23 kg (50 lb).

Inhibitors. Practice for sand casting of magnesium alloys differs from practice that is applicable to other metals, because of the precautionary measures that must be taken to prevent metal-mold reactions. Inhibitors, such as sulfur, boric acid, potassium fluoborate and ammonium fluosilicate, are mixed with the sand to prevent these reactions. In the presence of boric acid, diethylene glycol acts as an inhibitor, although its primary purpose is to keep the sand from drying out and reduce the amount of water added to the sand. The amount of inhibitor added depends on:

- *Moisture Content of the Molding Sand.* Other factors being equal, the less moisture in the sand, the less inhibitor is required.
- *Pouring Temperature of the Metal.* The higher the pouring temperature of the metal, the greater the reactivity, and correspondingly greater amounts of inhibitor will be required.
- *Cooling Rate of the Casting in the Mold.* The larger the casting (or the thicker the sections), the slower the casting will cool. Volatile inhibitors are soon lost from the surface of the mold, requiring replenishment by inhibitors at a distance from the surface of the mold until the casting surface has solidified.

Molding Sands. In most magnesium alloy foundries, castings of various sizes are produced. Because it is desirable to use the same sand mixture for all the castings, the molding sand is adjusted to the needs of the largest casting produced.

Molding sands for magnesium alloys must have high permeability (large pores) to allow free flow of mold gases to the atmosphere. Fine-grain sands, clay and the inhibiting agents decrease permeability. Because coarse-grain sands produce rough surfaces on castings, a compromise between good surface finish and proper venting is necessary to produce quality castings. The use of natural molding sands, with their high clay content and lack of uniformity, is restricted to small castings. Grain size of molds for magnesium alloy castings is AFS 65 to 90 (see Table 6).

PERMANENT MOLD CASTING

Casting of magnesium alloys in permanent molds should be considered when permitted by part design and when the cost of permanent mold equipment is justified by the number of castings required. In general, alloys that are suitable for casting in sand molds can also be cast in permanent molds (see Table 5). For detailed information on the permanent mold process, see the article on Permanent Mold Casting. With good

foundry technique, permanent mold castings are equal in strength and performance to their sand cast counterparts. Compared with sand casting, permanent mold casting provides better surface finish and permits closer control of dimensions from casting to casting. If the part to be made is subject to frequent redesign, however, it would be uneconomical to use permanent mold casting.

Shape Limitations. Because magnesium alloys have relatively poor feeding characteristics (compared with those of aluminum alloys, for example) and are hot short at temperatures near their solidus temperatures, there are restrictions on the shapes that can be successfully cast. Adequate draft is necessary to prevent cracking when a casting is removed from the mold. Also, extreme care must be used in removing retractable cores, or the hot casting will be damaged. Cores must be retracted without side thrust, and when two or more cores are used they should be pulled simultaneously. Intricate shapes, such as deep ribs or sections requiring complex cores, are not practical to cast in magnesium alloys.

Mold Coatings. An effective mold coating must be reasonably insulating, not wettable by molten magnesium, capable of adhering tightly to the mold surface, and porous enough to allow venting. The most acceptable cast surfaces are obtained when the coating has a matte or textured structure. Extremely smooth coatings should be avoided, because they increase the depth of oxide skin.

Mold coatings consist of refractory pigments suspended in water. The suspension also contains a suitable binder, usually sodium silicate (water glass). Small quantities of graphite are sometimes added to assist in removing castings from the molds.

Several proprietary coatings are in use. A typical mold coating consists of 1 part (by volume) sodium silicate, 2 parts finely powdered kaolin, and 5 to 20 parts water. The amount of water used depends on the thickness of coating desired and whether the coating is to be applied by spray or brush.

Mold Temperature. The quality of permanent mold castings is strongly influenced by the pattern of temperature distribution throughout the mold. Generally, because of the low heat content of magnesium, temperatures throughout a permanent mold must range from 260 to 425 °C (500 to 800 °F). For some castings, the temperature pattern that is developed by pouring in regular production will provide an acceptable product. More often, however, the establishment of satisfactory thermal conditions in the mold will require auxiliary heating and cooling. Molds can be locally heated with suitably shaped gas burners. Packing the outside of the die with asbestos

insulation will help retain heat in the mold. Local cooling of the molds can be done with high-conductivity inserts or with air or water. The use of air-cooled fins in certain areas of the mold is effective in chilling, and so is the removal of the mold coating in small areas.

DIE CASTING

Magnesium alloys AZ91A and AZ91B (see Table 5) are most frequently used for die casting. These two alloys are identical except that up to 0.35% copper is permitted in AZ91B.

Die casting of magnesium alloys offers the same advantages as for other metals (see the article on Die Casting). The process has sometimes been used for making a few castings because of some special requirement, but die casting of magnesium alloys is usually applied to high-volume production.

The size of magnesium casting that can be produced in a die is subject to the same limits as those for aluminum alloy die castings.

Machines. The high-pressure cold-chamber die-casting machine is the most practical for magnesium alloys. The only special requirements for machines used in die casting of magnesium alloys are: (a) the materials of construction must be selected carefully, to prevent reaction with the molten metal; and (b) extra precautions must be taken to ensure that no water comes in contact with the molten metal.

Dies. Hot work tool steels such as H11 and H13 heat treated to 44 to 48 HRC are ordinarily used for dies and cores. For high-production operation, cores are sometimes nitrided to prolong their life. Heat checking of the die can be postponed by providing the die with a surface finish of 0.38 μm (15 micro-in.) and by stress relieving and reworking the surface after every 25 000 shots.

Very little die wash occurs in the production of magnesium alloy die castings. The only significant changes occur across parting lines, where a high injection speed can increase dimensions. The magnitude of wash is a function of ram speed, projected area of the casting, and locking tonnage of the machine.

Lubrication of the die face is required, but use of excessive amounts of lubricant should be avoided. A fine mist should be sprayed manually or automatically.

Die Temperature. Optimum die temperature for casting magnesium is 260 ± 14 °C (500 ± 25 °F). Die temperature is controlled by production rate, amount of cooling water used, and sometimes by auxiliary heaters. If die temperature is too low, misruns and surface defects result; if it is too high, die soldering, hot tears, and shrinkage defects occur.

Control of Composition. Production of sound die castings depends greatly on close control of metal cleanness and of melting and handling of molten metal. Premelting and refining of the metal is customary when clean ingots are charged and is mandatory when scrap metal is used.

Refining of a magnesium die casting alloy (AZ91B is most frequently used) consists of heating the metal to approximately 700 °C (1300 °F), adding flux, stirring the metal thoroughly, and letting the melt stand for 10 to 15 min. The flux generally used is a mixture of magnesium chloride, potassium chloride, and other chlorides plus a smaller amount of calcium fluoride.

Design Limits. Good casting design features ample fillets, rounded corners and blended sections.

The use of remote heavy sections should be avoided in die castings, because such sections are likely to be porous. Thin sections may be strengthened by the use of ribbed construction. Section thicknesses from 1.6 to 4.8 mm ($^1/_{16}$ to $^3/_{16}$ in.) are preferred, because maximum mechanical properties are obtained in that range.

Casting Finish. The best surface finish on magnesium alloy die castings is obtained by close control of die temperature, cavity filling, die lubrication, metal temperature, holding time in the die, and smoothness of die surfaces. Magnesium alloy die castings are seldom plated, so the term "hardware finish" does not apply.

Casting Defects. When magnesium alloys are melted, some of the impurities settle to the bottom of the melting or holding unit, forming a sludge. In die casting, buildup of sludge in the holding crucible or pot must be avoided, because the metal pump intake is down in the melt and,

at a certain level, sludge may be mixed with the metal. Improper stirring or settling of the metal in the holding pot also results in poor metal quality. Hand ladling through flux is not recommended. Instead of having a flux cover, a flux baffle or a sulfur dioxide dome, depending on the size of the holding pot, should be used.

Inclusions that indicate their presence by a white bloom after the casting is exposed to air suggest the presence of a chloride flux. Inert inclusions such as nitrides and oxides indicate a faulty handling technique and suggest that insufficient flux has been used.

Misruns are caused by too-slow filling of the die, overlubrication, wrong die or metal temperature, dirty metal, and improper gating. Too little metal in the shot well also causes misruns.

A cold shut occurs when two streams of molten metal meet but do not fuse completely. Cold shuts can be seen by visual inspection and may

be cause for rejection. This type of defect is most likely to occur behind cores and in the extremities of the casting. Sometimes it is caused by too-slow filling of the cavity. Cold shuts can be prevented by operating the die-casting machine at a greater speed, thus allowing less time for the metal to cool. In extreme cases, it may be necessary to rework the gating system to eliminate cold shuts.

Internal porosity indicates that gas has been trapped in the casting. When this condition is encountered, the venting, gating, and lubrication practices should be examined. As a rule, the total vent area should be at least half the gate area. Position of the vents is also important. Often, a change of vent position will solve the problem without increasing total area of the venting system. An increase in injection pressure may decrease internal porosity.

Magnesium is sensitive to overlubrication, which often causes black swirls.

Selected References on Casting

In addition to the references listed below, the reader may wish to contact the following organizations for specific information on foundry technology and management: American Foundrymen's Society, Des Plaines, IL; Investment Casting Institute, Dallas, TX; Iron Castings Society, Rocky River, OH; Society of Die Casting Engineers, River Grove, IL; Steel Founders' Society, Rocky River, OH.

Metallurgical Treatises, edited by J. K. Tien and J. F. Elliott, TMS-AIME, Warrendale, PA, 1981.

The Physical Metallurgy of Cast Iron, I. Minkoff, John Wiley, New York, 1983.

Principles of Solidification, B. Chalmers, John Wiley, New York, 1964.

Solidification and Casting, M. C. Flemings, McGraw-Hill, New York, 1974.

Principles of Metal Casting, 2nd ed., R. W. Heine, C. R. Loper, and P. C. Rosenthal, McGraw-Hill, New York, 1967.

The Structure of Cast Iron, A. Boyles, American Society for Metals, 1949.

Cupola Handbook, American Foundrymen's Society, 4th ed., 1975.

Casting Design Handbook, American Society for Metals, 1962.

Foundry Technology Source Book, edited by Paul J. Mikelonis, American Society for Metals, 1982.

Casting Defects Bibliography, Metals Information, American Society for Metals, 1984.

Precision Castings Bibliography, Metals Informa-

tion, American Society for Metals, 1984.

Continuous Casting, Ferrous Bibliography, Metals Information, American Society for Metals, 1984.

Continuous Casting, Non-Ferrous Bibliography, Metals Information, American Society for Metals, 1984.

Die Casting of Aluminum Bibliography, Metals Information, American Society for Metals, 1984.

Casting of Magnesium Alloys Bibliography, Metals Information, American Society for Metals, 1984.

Cast Bronze, H. J. Roast, American Society for Metals, 1953.

24 FORGING

Reviewed and revised by Taylan Altan and Raghu Raghupathi, Battelle-Columbus Laboratories

This section was condensed from Metals Handbook, Eighth Edition, Volume 5, Forging and Casting, pages 1 to 148. For more detailed information on the topics covered in this section, the reader is referred to the larger work.

Hammers and Presses for Forging

HAMMERS AND PRESSES for forging may be considered in two groups: those for closed-die forging and those for open-die work. Some simple open-die forging can be done in a closed-die forging hammer, and occasionally an open-die forging hammer is used with closed dies; however, this is uncommon. Two or more pieces of equipment may be used to produce a specific forging — for instance, a power hammer using flat dies for pancaking and then another hammer or a press for forging in closed dies.

CLOSED-DIE FORGING HAMMERS

With the exception of the counterblow hammer, forging hammers have a weighted ram, which, when it moves vertically in a downward stroke, exerts a striking force against the anvil near the base of the hammer. The upper half of a pair of closed dies is fastened to the weighted ram, and the lower half to the anvil cap. The work metal (in the form of a heated bar, billet, bloom or ingot) is placed on the lower die, and the striking force is imposed on the work metal by the upper die and ram, causing it to deform plastically with each successive blow. Although all hammers operate on the principle of high impact, designs vary, depending on the method of actuation. Common types of hammers include gravity drop hammers (board, hydraulic or air-lift) and power drop hammers (steam, hydraulic or air-driven).

BOARD DROP HAMMERS

Board drop hammers are widely used, especially for producing forgings weighing no more than a few pounds. In the board drop hammer, the ram is lifted by one or more boards keyed to it and passing between two friction rolls at the top of the hammer. The boards are rolled upward and are then mechanically released, permitting the ram to drop from the desired height. Power for lifting the ram is supplied by one or more motors. The hammers have falling weights, or rated sizes, of 400 to 10 000 lb (about 180 to

4550 kg); standard sizes range from 1000 to 5000 lb (about 455 to 2270 kg), in increments of 500 and 1000 lb (about 225 and 455 kg). The height of fall of the ram varies with hammer size, ranging from about 35 in. (890 mm) for a 400-lb (180-kg) hammer to about 75 in. (1.9 m) for a 7500-lb (3400-kg) hammer. The height of fall, and thus

the striking force, of the hammer is approximately constant for a given setting and cannot be altered without stopping the machine and adjusting the length of stroke. Anvils on board drop hammers are 20 to 25 times as heavy as the rams. Components of a typical board drop hammer are shown in Fig. 1.

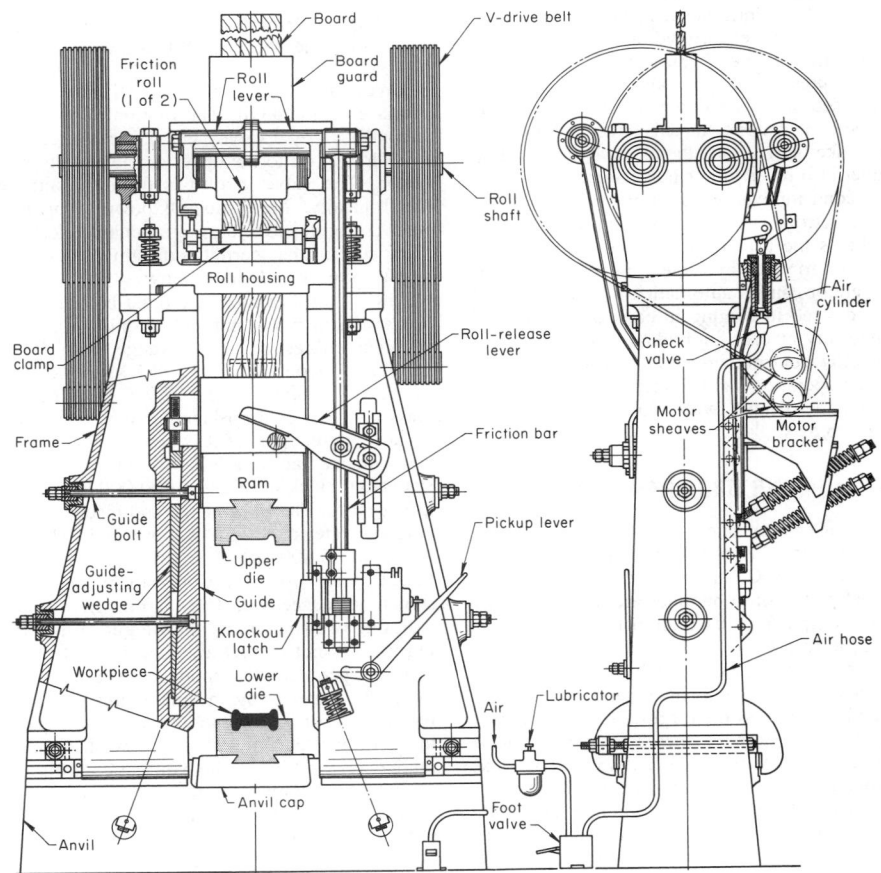

Fig. 1. Principal components of a board drop hammer

In using the board drop hammer for closed-die forging, the heated metal for forging is placed over the cavity in the lower die, and the ram and top die drop when the operator releases the board clamp (usually by means of a foot treadle). When the treadle is depressed and released by the operator, the ram drops once and returns to its raised position, where it remains until it is again released. The operator can cause successive blows to be struck by keeping the treadle depressed. In practice, especially in the production of small, simple closed-die forgings, usually the hammer is allowed to strike successive blows while the operator moves the workpiece from one impression to another without stopping the hammer. Since the stroke length is fixed, to provide the proper height of fall for the specific dies in the hammer, light and heavy blows to the workpiece cannot be interspersed.

AIR- AND HYDRAULIC-LIFT GRAVITY DROP HAMMERS

Air- and hydraulic-lift gravity drop hammers are similar to board drop hammers in that the forging force is derived from the weight of the falling ram assembly and upper die. They differ from board drop hammers in that the ram is raised by air, steam or hydraulic power.

The range of sizes generally available in air- or hydraulic-lift hammers is 500 to 10 000 lb (about 225 to 4550 kg). The weight of forging that can be produced in a drop hammer of any size is about the same as can be produced in a corresponding board drop hammer.

During recent years two significant innovations have been introduced in hammer designs. The first is the electrohydraulic gravity drop hammer. In this type of hammer, the ram is lifted with oil pressure against an air cushion. The compressed air slows down the upstroke of the ram and contributes to its acceleration during the downstroke blow. Thus, the electrohydraulic drop hammer also has a minor power-hammer action. The second innovation in hammer design is the use of electronic blow-energy control. Such control allows the user to program the drop height of the ram for each individual blow. As a result, the operator can set automatically the number of blows desired in forging in each die cavity and the intensity of each individual blow. The electronic blow control increases the efficiency of the hammer operations and decreases the noise and vibration associated with unnecessarily strong hammer blows.

POWER DROP HAMMERS

In a power drop hammer, the ram is accelerated during the downstroke by air, steam or hydraulic pressure. Components of a steam- or air-actuated power drop hammer are shown in Fig. 2. This equipment is used almost exclusively for closed-die forging.

The steam- or air-powered drop hammer is the most powerful machine in general use for the production of closed-die forgings by impact pressure. In a power drop hammer, a heavy anvil block supports two frame members that accurately guide a vertically moving ram; the frame also supports a cylinder that, through a piston and piston rod, motivates the ram. In its lower face, the ram carries an upper die, which contains one part of the impression that shapes the forging. The lower die, which contains the remainder of the impression,

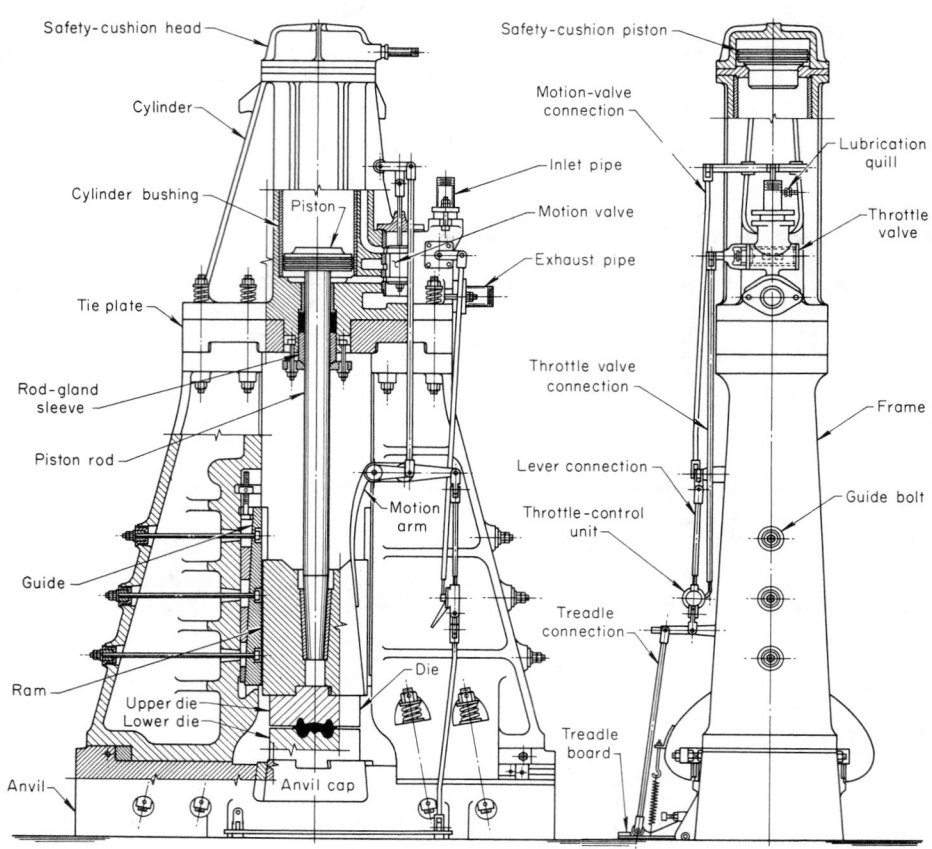

Fig. 2. Principal components of a power drop hammer with foot control to regulate the force of the blow

is keyed into an anvil cap that is firmly wedged in place on the anvil. The motion of the piston is controlled by a valve, which admits steam, air or hydraulic oil to the upper or lower side of the piston. The valve in turn is usually controlled electronically. Most modern power drop hammers are equipped with programmable electronic blow control that permits adjustment of the intensity of each individual blow.

Power drop hammers are rated by the weight of the striking mass, not including the upper die. Hammer ratings commonly range from 500 to 35 000 lb (about 225 to 15 875 kg), and occasionally to 50 000 lb (about 22 700 kg). The large mass of a power drop hammer is not apparent, because a great part of it is beneath the floor. A hammer rated at 50 000 lb will have a sectional steel anvil block weighing 1 000 000 lb (453 600 kg) or more. The ram, piston and piston rod will have an aggregate weight of approximately 45 000 lb (about 20 400 kg). The striking velocity obtained by the downward pressure on the piston sometimes exceeds 25 ft/s (7.6 m/s).

Rating of hammers by the weight of the striking mass is not correct although it has been the common practice. The more realistic method of rating hammers is by the maximum energy, in foot-pounds or joules, that the ram can impart to the hot metal during a single blow at the maximum energy setting of the hammer controls. The useful energy supplied to the forged metal by the hammer ram depends on (a) the hammer design (weight of the ram and the pressure on the top of the piston), (b) the ratio of the anvil weight versus the ram weight and (c) the hammer foundation design.

Apart from the size of power drop hammers and the force they make available for the production of large forgings (forgings commonly produced in power drop hammers range in weight from 50 lb to several tons), another important advantage is that the striking intensity is entirely under the control of the operator or is preset by the electronic blow-control system. Consequently, effective use can be made of auxiliary impressions in the dies to preform the billet to a shape that will best fill the finishing impressions in the dies, and result in proper grain flow, soundness and metal economy, with minimum die wear. When adequate preliminary impressions cannot be incorporated in the same set of die blocks, two or more hammers are used to produce adequate shaping or blocking before the final die is used.

Although they are generally advantageous, the greater striking forces that are developed with power drop hammers give rise to several disadvantages. As much as 15 to 25% (and, in hard finishing blows, up to 80%) of the kinetic energy of the ram is dissipated in the anvil block and foundation, and therefore does not contribute to deformation of the workpiece. This loss of energy is most critical when finishing blows are struck and the actual deformation per stroke is relatively slight. The transmitted energy imposes a high stress on the anvil block and may even break it. The transmitted energy also develops violent, and potentially damaging, shocks in the surrounding floor area, necessitating the use of shock-absorbing materials, such as timber or iron felt, in anvil-block foundations, and adding appreciably to foundation cost.

COUNTERBLOW HAMMERS

The counterblow hammer, a variation of the power drop hammer, is widely used in Europe. These hammers develop striking force by the movement of two rams, simultaneously approaching from opposite directions and meeting at a midway point. Some hammers are pneumatically or hydraulically actuated; others incorporate a mechanical-hydraulic or a mechanical-pneumatic system.

A vertical counterblow hammer with a steam-hydraulic actuating system is shown schematically in Fig. 3 (air-hydraulic systems are avail-

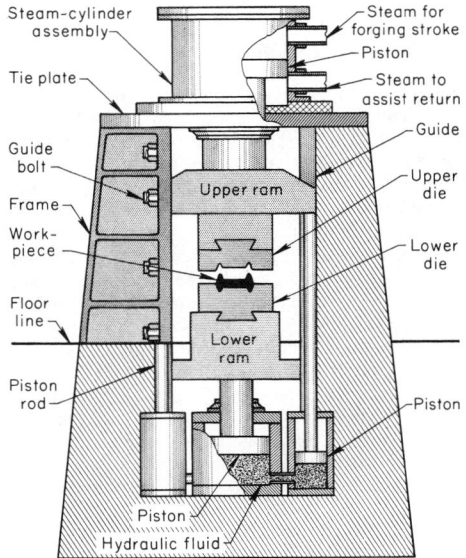

Fig. 3. Essential components of a vertical counterblow hammer with a steam-hydraulic actuating system

able also). In this hammer, steam is admitted to the upper cylinder and drives the upper ram downward. At the same time, pistons connected to the upper ram act through a hydraulic linkage in forcing the lower ram upward. Since the weight of the lower ram and piston assembly is greater than that of the upper assembly, the rams retract automatically after the blow. Retraction speed is increased by steam (or air) pressure acting upward on the piston. Through proper design relative to weights (including tooling and workpiece) and hydraulics (slower lower-assembly velocities), the kinetic energy of the upper and lower assemblies may be balanced at impact.

The rams of a counterblow hammer are capable of striking repeated blows; they develop combined velocities of 16 to 20 ft/s (4.9 to 6.1 m/s). Compared to single-action hammers, vibration of impact is reduced, and approximately the full energy of each blow is delivered to the workpiece, without loss to an anvil. As a result, the wear of moving hammer parts is minimized, contributing to longer operating life. At the time of impact, forces are canceled out, and no energy is lost to foundations. In fact the large inertia blocks and foundations of conventional power drop hammers are not required.

A horizontal counterblow hammer has two opposing, die-carrying rams that are moved horizontally by compressed air. Heated stock is positioned automatically at each die impression by a preset pattern of accurately timed movements of a stock-handling device. A 90° rotation of stock may be programmed between blows.

HIGH-ENERGY-RATE FORGING MACHINES

High-energy-rate forging (HERF) machines are of three basic designs—ram-and-inner-frame, two-ram, and controlled-energy-flow. Each differs from the others in engineering and operating features, but all are essentially very-high-velocity, single-blow hammers that require less mov-

ing weight than do conventional hammers to achieve the same impact energy per blow. All the designs employ counterblow principles, to minimize foundation requirements and energy losses. All the designs use inert high-pressure gas controlled by a quick-release mechanism for rapid acceleration of the ram. In none of the designs is the machine frame required to resist the forging forces.

Ram-and-Inner-Frame Machines. The machine shown schematically in Fig. 4(a) has a frame consisting of two units: an inner or working frame connected to a firing chamber, and an outer or guiding frame within which the inner frame is free to move vertically. As the trigger-gas seal is opened, high-pressure gas from the firing chamber acts on the top face of the piston and forces the ram and upper die downward. Reaction to the downward acceleration of the ram raises the inner frame and lower die.

The machine is made ready for the next blow by means of hydraulic jacks that elevate the ram until the trigger-gas seal between the upper surface of the firing chamber and the ram piston is re-established. Venting of the seal gas, and gas pressure on the lower lip of the piston, then hold the ram in the elevated position.

This type of machine is produced in several sizes, ranging in capacity from 12 500 to 550 000 ft·lb (16.95 to 745.7 kJ) of impact energy.

Two-Ram Machines. In a two-ram machine (Fig. 4b) the counterblow is achieved by means of an upper ram and a lower ram. An outer frame (not shown in Fig. 4) provides vertical guidance for the two rams. Vertical movement of the trigger permits high-pressure gas to enter the lower chamber and the space beneath the drive piston. This forces the drive piston, rod, lower ram and lower die upward. The reaction to this force drives the floating piston, cylinder, upper ram and upper die downward. The rods provide relative guidance between the moving upper and lower assemblies.

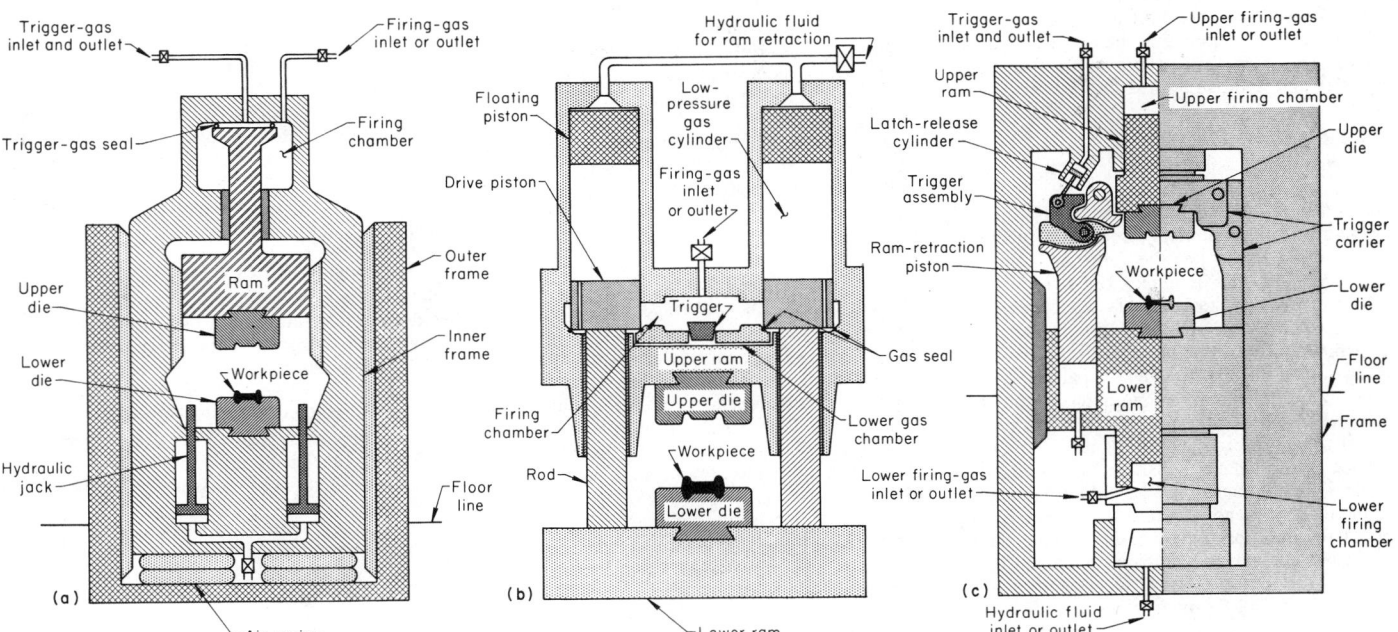

Triggering and expansion of gas in the firing chamber cause the upper and lower rams to move toward each other at high velocity. An outer frame provides close guiding surfaces for the rams. See text for descriptions of the mechanics of operation.

Fig. 4. Three types of machines for high-energy-rate forging: (a) ram-and-inner-frame machine, (b) two-ram machine and (c) controlled-energy-flow machine

After the blow, hydraulic fluid enters the cylinder, returning the upper and lower rams to their starting positions. The gas is recompressed by the floating pistons, and the gas seals at the lower edges of the drive pistons are re-established. When the trigger is closed, the hydraulic pressure is released, the high-pressure gas in the lower chamber expands through the drive-piston ports and forces the floating pistons up, and the machine is ready for the next blow.

Several different sizes of these machines are made; the largest has a rating of 300 000 ft·lb (406.7 kJ) of impact energy.

Controlled-energy-flow machines (Fig. 4c) are counterblow machines from the standpoint of having separately adjustable gas cylinders and separate rams for the upper and lower dies; however, self-reacting principles are not employed. The lower ram has a hydraulically actuated vertical-adjustment cylinder so that different stroke lengths may be preset.

The trigger, although pneumatically operated, is a massive mechanical latch that returns and holds the rams through mechanical support of the upper ram and hydraulic connection with the lower ram. With this arrangement, simultaneous release of the two rams is assured.

Controlled-energy-flow forging machines have been made in two sizes, with ratings of 73 000 and 400 000 ft·lb (99.0 to 542.3 kJ) of maximum impact energy.

Applicability. High-energy-rate forging machines basically are limited to fully symmetrical or concentric forgings such as wheels and gears or for coining applications where little metal movement but high die forces are required. The practical use of these machines has been limited and is expected to remain so. Parts that can be forged in high-energy-rate forging machines can also be forged in counterblow hammers or mechanical presses. High-velocity impact during the forging blow reduces die life and increases machine maintenance requirements. Therefore, lower-intensity-stroke machines — e.g., hammers and presses — are more economical to use in most applications. Exceptions may be in cases where the extremely high forging rate imparts unusual properties and microstructures to the forged materials.

OPEN-DIE FORGING HAMMERS

Open-die forging hammers, commonly known as general forging hammers, are used to make a large percentage of open-die forgings. In size, they range from small hammers rated at 25 to 50 lb (11.3 to 22.7 kg) to hammers with ratings as high as 24 000 lb (about 10 900 kg). The smaller hammers — those rated at less than 1000 lb (about 455 kg) — are usually found in repair or maintenance departments of manufacturing plants, where they are used to forge various repair parts, to dress tools, and to shape the many items that are made in maintenance departments. These small hammers are not considered part of a forge shop, and therefore will not be discussed here. Commercial forge shops use open-die forging hammers that have rated sizes ranging from about 1000 to 24 000 lb (about 455 to 10 900 kg).

A general forging hammer is operated by steam or compressed air, usually at pressures of 100 to 120 psi (690 to 825 kPa) for steam and 90 to 100 psi (620 to 690 kPa) for compressed air. These conditions are similar to those of the power drop hammer used for closed-die forging.

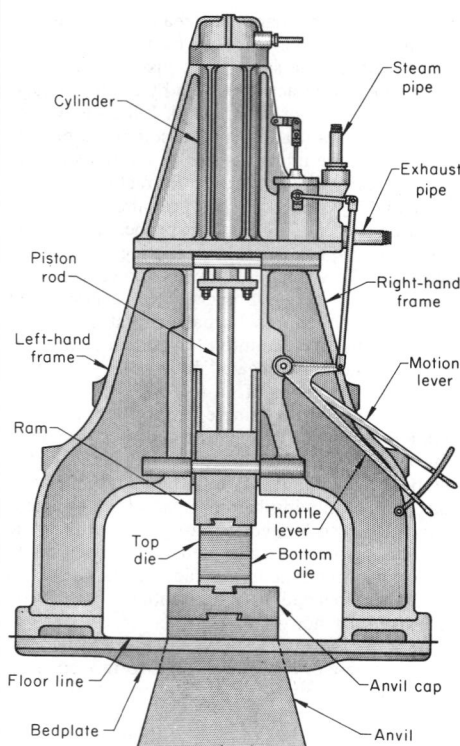

Fig. 5. Double-frame power hammer used for open-die forging

There are two basic differences between power drop hammers used for closed-die forging and the hammers used for open-die forging. First, a modern power drop hammer has blow-energy control to assist the operator in setting up the intensity of each blow. In this case the hammer stroke is limited by the upper die surface touching the surface of the lower die face. In open-die forging, the upper and lower dies do not "kiss" or touch each other; therefore, an additional stroke-position control is provided through control of the air or steam valve that actuates the hammer piston.

The second difference in hammer design is that the anvil of the general forging hammer is separate and independent of the hammer frame that contains the striking ram and top die. Separation of the anvil from the frame allows the anvil to give way under a heavy blow or a series of blows without disturbing the frame. The anvil may rest on a cushion of oak timbers, which absorbs the hammering shock.

Open-die forging hammers are made either with a single frame (these are often known as C-frame or single-arch hammers) or with a double frame (often called double-arch hammers).

Double-frame general forging hammers (see Fig. 5) usually come in rated sizes from 6000 to 24 000 lb (about 2720 to 10 900 kg). An advantage of a double-frame hammer is that the ram is rigidly guided. This makes it desirable for production work and general forging where rigidity is needed, and for forging of high-strength steels, heat-resisting alloys and other high-strength metals. Although hammers are inexpensive and versatile, they do not allow open-die forging with close tolerances. As a result, the increasing trend is to employ hydraulic presses for open-die forging work.

CLOSED-DIE FORGING PRESSES

Forging presses generally incorporate a ram that moves in a vertical direction to exert a squeezing action on the workpiece, in contrast with the repeated blows characteristic of hammer forging. Depending on their source of power, forging presses are classified as mechanical or hydraulic. Maximum capacities exceeding those of the largest power drop hammers are developed by hydraulic presses. In general, presses can produce all of the types of forgings produced by hammers and, in addition, can forge some alloys of moderate ductility that would shatter under the blows of a hammer.

MECHANICAL PRESSES

Driven by a motor and controlled by means of an air clutch, mechanical presses have a full-eccentric type of drive shaft that imparts a constant-length stroke to a vertically operating ram. The ram carries the top (or moving) die, whereas the bottom (or stationary) die is clamped to the die seat of the main frame. The ram stroke is shorter than that of a forging hammer or a hydraulic press. Ram speed is greatest at the center of the stroke, but force is greatest at the bottom of the stroke. Because of the short stroke, mechanical presses are best suited for low-profile forgings. Capacities of these forging presses are rated on the maximum force they can apply, and range from about 300 to 16 000 tons (about 270 to 14 500 tonnes).

Mechanical forging presses have principal components that are similar to those of eccentric-shaft, straight-side, single-action presses used for forming sheet metal. In detail, however, mechanical forging presses are considerably different from mechanical presses that are used for forming sheet. The principal differences are:

1. Forging presses, particularly their side frames, are built stronger than are presses for forming sheet metal.
2. Forging presses deliver their maximum force within 1/8 in. (3.2 mm) of the end of the stroke, because maximum pressure is required to form the flash.
3. The slide velocity in a forging press is faster than in a sheet-metal deep drawing press, because in forging it is desirable to strike the metal and retrieve the ram quickly to minimize the time the dies are in contact with the hot metal.

Unlike the blow of a forging hammer, a press blow is more of a squeeze than an impact, and is delivered by uniform stroke length. The character of the blow in a forging press resembles that of an upsetting machine, thus combining some features of hammers and upsetters. Mechanical forging presses use drive mechanisms similar to those of upsetters, although an upsetter generally is a horizontal machine.

Some advantages of mechanical forging presses are:

1. Higher production rates are possible with presses than with hammers. Forging presses have stroking rates from about 30 strokes per minute, for a 16 000-ton (14 000-tonne) press, to about 100 strokes per minute, for a 500-ton (455-tonne) press.
2. Because the impact is less in presses than in hammers, dies can be less massive, thus requiring less tool steel to make the dies. If de-

sired, cast dies can be used. Also because presses deliver a less severe impact blow, dies can be operated at higher hardness, which prolongs die life.

3. Presses require less operator skill than is required for hammers.
4. Presses are easier to automate than hammers and can produce more precise parts.
5. Presses generate less noise and vibration than hammers generate.

Some disadvantages of mechanical forging presses are:

1. Initial cost is high — as much as three times the cost of a hammer that will produce the same forging.
2. A press delivers consecutive strokes of equal force, and therefore is less suitable for preliminary shaping operations such as fullering or rollering.

Knuckle-joint presses are one type of mechanical press operated by an eccentric shaft or by a crankshaft. The use of the knuckle mechanism results in an increase in available forging force, but in a loss of ram-stroke length. Thus, knuckle-joint presses are used in the forging industry primarily for trimming of flash and for coining or sizing.

HYDRAULIC PRESSES

The ram of a hydraulic press is driven by hydraulic cylinders and pistons, which are part of a high-pressure hydraulic or hydropneumatic system. The principal components of a hydraulic press are illustrated in Fig. 6. Hydraulic presses are used for both open-die and closed-die forging.

There are two types of drives for hydraulic presses: direct drives and accumulator drives. In direct-driven presses, a hydraulic pump feeds oil under pressure into the cylinder of the press continuously during the stroke. Thus, the pump capacity, in terms of volume of oil per minute at a given pressure, determines the maximum available press speed and press load. In accumulator-

driven presses, a water emulsion that is used as the working medium is stored in an accumulator under nitrogen pressure. When the press stroke is initiated, the pressurized water emulsion is fed into the press cylinder. During the press stroke, the water emulsion expands and loses some of its initial pressure. In automatic programmed forging operations, direct drive offers the advantage of very close control of ram speed and ram position.

Capacities of hydraulic presses range from 300 to 85 000 tons (about 270 to 77 100 tonnes).

The principal advantages of hydraulic presses are:

1. Pressure can be changed, as desired, at any point in the stroke by adjusting the pressure control valve.
2. Rates of deformation can be controlled, and even varied during the stroke if required. This is especially important in forging of metals that will rupture if subjected to high deformation rates.
3. By use of split dies, many parts are made with offset flanges, projections, back draft, and other design features that are extremely difficult, if not impossible, to incorporate in hammer forgings.
4. When excessive heat transfer (from the hot workpiece to the dies) is not a problem, or can be eliminated, the gentle squeezing action of a hydraulic press results in lower maintenance cost and increased die life because of less shock compared with other types of forging equipment.
5. Maximum press load can be limited so as to protect the tooling.

Some of the disadvantages of hydraulic presses are:

1. The initial cost of a hydraulic press is higher than that of a mechanical press of equivalent capacity.
2. The action of a hydraulic press is slower than that of a mechanical press.
3. The slower action of a hydraulic press permits longer contact between the dies and the workpiece, and thus die life is shortened in forging of high-temperature materials (steels, titanium, nickel alloys) because of heat transfer from the hot workpiece to the dies.

SCREW PRESSES

Screw presses are widely used in Europe for (a) job-shop-type hardware forging, (b) forging of brass and aluminum parts, (c) precision forging of turbine and compressor blades and (d) gearlike parts. Recently, screw presses have also been introduced in North America for a wide range of applications, notably for forging steam-turbine and jet-engine compressor blades and diesel-engine crankshafts.

The screw press uses a friction, gear, electric or hydraulic drive to accelerate the flywheel and the screw assembly, and it converts the angular kinetic energy into the linear energy of the slide or ram. Figure 7 shows two basic designs of screw presses.

In the friction-drive press (Fig. 7a), the driving disks are mounted on a horizontal shaft and are rotated continuously. For a downstroke, one of the driving disks is pressed against the flywheel by a servomotor. The flywheel, which is connected to the screw either positively or by a fric-

tion slip clutch, is accelerated by this driving disk through friction. The flywheel energy and the ram speed continue to increase until the ram hits the workpiece. Thus, the load necessary for forming is built up and transmitted through the slide, the screw and the bed to the press frame. When the entire energy in the flywheel is used in deforming the workpiece and elastically deflecting the press, the flywheel, the screw and the slide stop. At this moment, the servomotor activates the horizontal shaft and presses the upstroke driving disk wheel against the flywheel. Thus, the flywheel and the screw are accelerated in the reverse direction and the slide is lifted to its top position.

In the direct-electric-drive press (Fig. 7b), a reversible electric motor is built directly on the screw and on the frame, above the flywheel. The screw is threaded into the ram or slide and does not move vertically. To reverse the direction of flywheel rotation, the electric motor is reversed after each downstroke and upstroke.

In a screw press the load is transmitted through the slide, screw and bed to the press frame. The available load at a given stroke position is supplied by the energy stored in the flywheel. At the end of a stroke, the flywheel and the screw come to a standstill before reversing the direction of rotation.

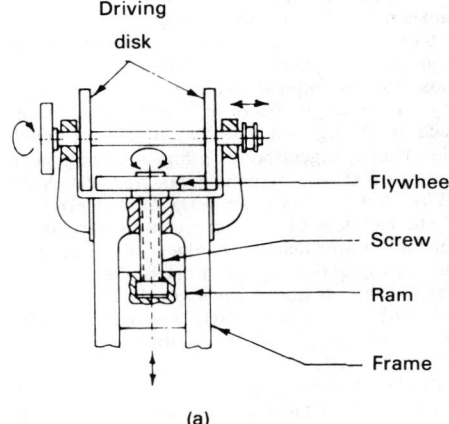

(a)

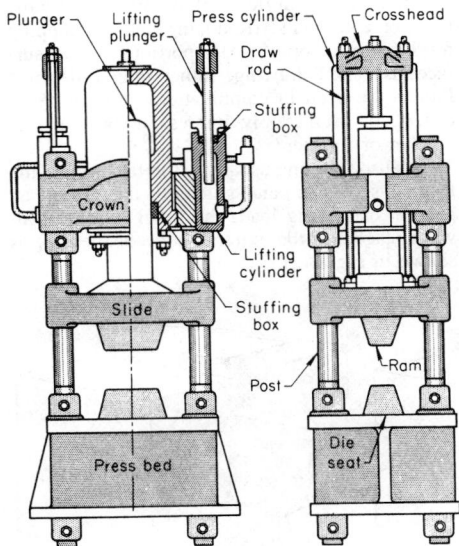

Fig. 6. Components of a four-post hydraulic press for closed-die forging

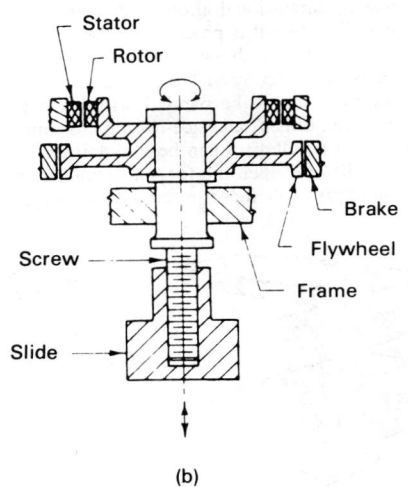

(b)

(a) Friction-drive press. (b) Direct-electric-drive press.
Fig. 7. Two basic designs of screw presses

The number of strokes per minute largely depends on the energy required by the specific forming process and on the capacity of the drive mechanism to accelerate the screw and the flywheel. In general, however, the production rate of a screw press is lower than that of a mechanical press, especially in automated high-volume operations.

During a downstroke, the ram velocity increases until the slide hits the workpiece. After the actual deformation starts, the ram velocity decreases depending on the energy requirements of the process. Thus, the ram velocity is greatly influenced by the geometry of the stock and of the part.

A screw press is operated like a hammer—i.e., the top and bottom dies "kiss" at each blow. Therefore the stiffness of the press, which affects the load and energy characteristics, does not influence the thickness tolerances of the forged part. As a result, die setup in a screw press is relatively simple because, as in hammer and hydraulic-press operations, no adjustments for flash are necessary.

Dies and Die Materials for Hammer and Press Forging

THE TOOLS discussed in this article are for use in vertical presses or hammers.

DIES FOR OPEN-DIE FORGING

Most open-die forgings are produced in a pair of flat dies—one attached to the hammer or to the press ram, and the other to the anvil. Swage dies (semicircular), V-dies, and V-die and flat-die combinations are also commonly used. These four types of die sets are shown in Fig. 1. In some applications, forging is done with a combination of a flat die and a swage die.

Steels used for dies for open-die forging differ from plant to plant, but are often the same as those used for impression dies in closed-die forging—for example, 6G or 6F2 steels. Alloy steels such as 4150 give satisfactory results for small dies. Some forgers prefer a higher-carbon steel and use a 0.70% carbon, 4300 grade (such as 4370, although this is not a standard steel).

The hardness of dies for open-die forging is generally lower than the hardness of impression dies for closed-die forging. If one of the die-block steels such as 6G or 6F2 is used, the usual hardness range is 302 to 331 HB. Dies made of 4150 or a similar alloy steel are usually heat treated to 277 to 321 HB.

Parallelism. If the faces of a set of dies mounted in a hammer or press are not parallel with each other, the deviation from parallelism will cause taper on the forgings, which may then be outside established dimensional allowances. The degree of parallelism that it is practical to maintain varies to some extent with the size of the dies. Fairly large dies (dies 965 by 510 mm, or 38 by 20 in., for example) should be parallel within 1.6 mm ($^1/_{16}$ in.) front-to-back and side-to-side. For smaller dies, closer parallelism can be maintained.

Life of dies for open-die forging is longer than that of impression dies for closed-die forging.

Because of the wide variety of forgings produced on the same open dies, life is usually expressed in production hours rather than in number of forgings.

It is not unusual for dies to operate in a hammer for 600 h before they require re-dressing. Re-dressing consists of planing or milling the die faces to remove deteriorated or damaged metal and to restore parallelism that may have been lost through wear. As little metal as possible is removed from each die face. Often, removal of less than 6.4 mm ($^1/_4$ in.) is sufficient for cleanup. Dies are usually designed to permit eight to ten re-dressings, which is equivalent to 4800 to 6000 h total life.

"LOOSE" TOOLING FOR BLOCKER-TYPE FORGINGS

Blocker-type forgings can be produced in metal die blocks containing cavities of simple shape, or with stops and gages, that the forging producer has in stock. These are called "loose" tools since they are not attached to the ram or anvil, but are held in position on a bottom flat die by a handle. Loose tooling can be used to produce a variety of simple shapes, on short notice, without the cost of dies.

Because they lack the detail and the dimensional accuracy of forgings completed in finisher dies by conventional closed-die forging methods, forgings produced in loose tooling usually require substantially more machining, to provide a finished part. Specific examples are shown in Fig. 2.

TYPES OF IMPRESSIONS IN DIES FOR CLOSED-DIE FORGING

Several different types of impressions can be used in a forging die, each type being designed

to serve a specific function. In a forging sequence that incorporates several types of impressions, each impression should be considered for the specific function that it is to perform, both by itself and in relation to the preceding and succeeding impressions. In particular, the design of one impression should provide for location of the workpiece in the succeeding die impression.

Finishers. The finisher impression gives the final over-all shape to the workpiece. It is in this impression that any excess work metal is forced out into the flash. Despite its name, the finisher impression is not necessarily the last step in the production of a forging. A bending or hot coining operation is sometimes used to give the final shape or dimensions to a forged part after it has passed through the finisher impression and the trimming die.

Preforming or roughing impressions are usually used in hammer forging. These include fullers, edgers, rollers, flatteners, benders, splitters and blockers. Each kind of preform impression is used as needed to shape the forging stock so that it can be formed efficiently and completely in the finisher impression.

Fullers. A fuller is a die impression used to reduce the cross section and to lengthen a portion of the forging stock. In longitudinal cross section, the fuller is usually elliptical or oval, to obtain optimum metal flow without producing laps, folds or cold shuts. (Reducing and lengthening forging stock between flat portions of die surfaces is called drawing, rather than fullering.) Fullers are used in combination with edgers or rollers, or as the only impression prior to the blocker or finisher.

Because fullering usually is the first step in the forging sequence, and generally uses the least amount of forging load, the fuller is almost always placed on the extreme edge of the die, as shown in Fig. 3(a).

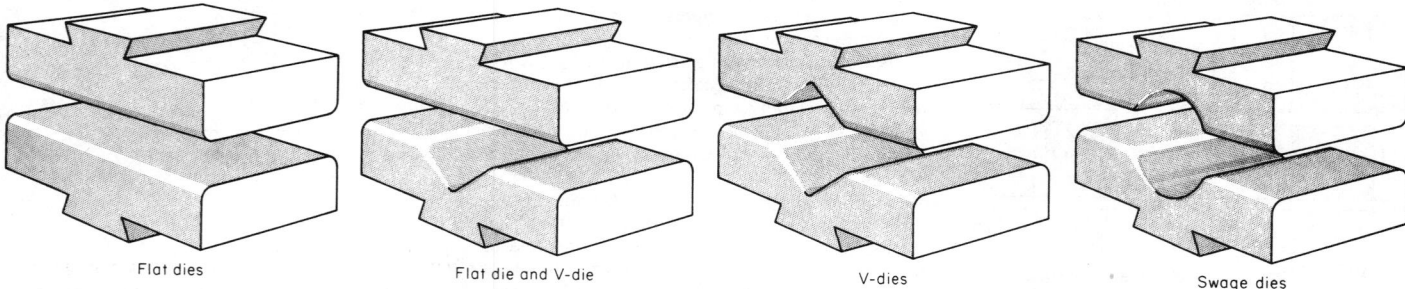

Flat dies   Flat die and V-die   V-dies   Swage dies

Fig. 1. Four types of die sets commonly used in open-die forging

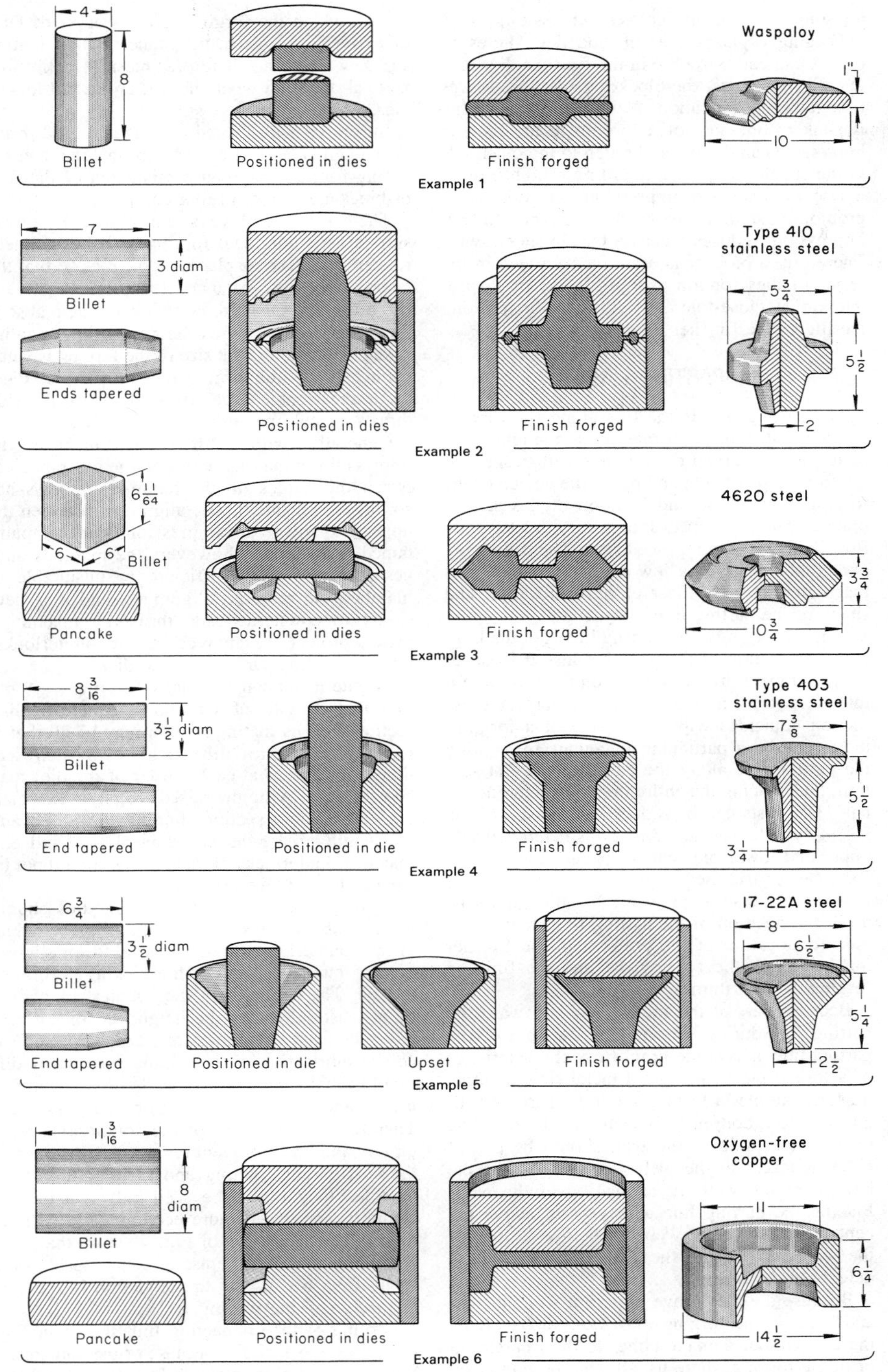

Example 1

Waspaloy

Example 2

Type 410 stainless steel

Example 3

4620 steel

Example 4

Type 403 stainless steel

Example 5

17-22A steel

Example 6

Oxygen-free copper

Fig. 2. Production of six blocker-type forgings with loose tooling in hammers. Dimensions are in inches.

Item	Example 1	Example 2	Example 3	Example 4	Example 5	Example 6
Stock size, in.	4 diam by 8 long	3 diam by 7 long	6 by 6(a) by $6^{11}/_{64}$ long	$3^1/_2$ diam by $8^3/_{16}$ long	$3^1/_2$ diam by $6^3/_4$ long	8 diam by $11^3/_{16}$ long
Stock weight, lb	$29^3/_4$	14	63	22	$18^1/_2$	180
Shipping weight, lb	$26^3/_4$	12	60	20	$10^1/_2$	135(b)
Forging temperature (max), °F	2100	2100	2200	2200	2150	1550
Size of hammer	6000 lb	3000 lb	6000 lb	3000 lb	3000 lb	1500 ton(c)
Men in forging crew	5	4	5	4	4	5

(a) Round-corner square. (b) After rough machining. (c) Hydraulic press.

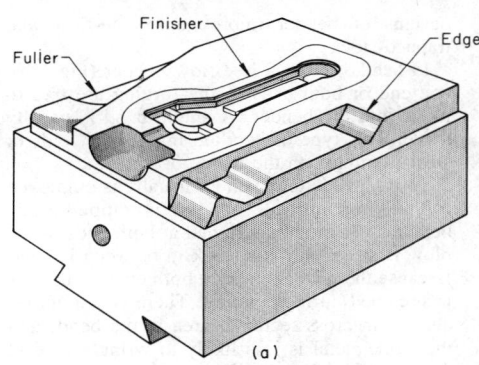

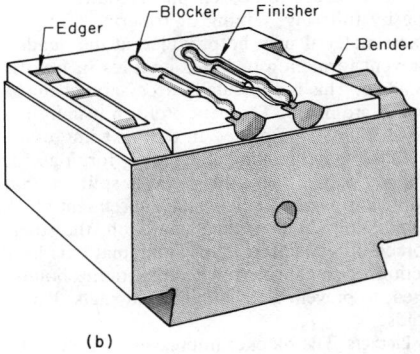

Fig. 3. Typical multiple-impression hammer dies for closed-die forging

Edgers are used to redistribute and proportion stock for heavy sections that will be further shaped in blocker or finisher impressions. Thus, the action of the edger is opposite to that of the fuller. A connecting rod is an example of a forging where stock is first reduced in a fuller to prepare the slender central part of the rod and then worked in an edger to proportion the ends for the boss and crank shapes.

The edger impression may be open at the side of the die block, as in Fig. 3(a), or confined, as in Fig. 3(b). An edger is sometimes used in combination with a bender in a single die impression, to reduce the number of forging blows necessary to produce a forging.

Rollers are used to round the stock (for example, from a square billet to a round barlike shape) and often to cause some redistribution of mass in preparation for the next impression. The stock usually is rotated, and two or more blows are needed to roller the stock.

The operation of a roller impression is similar to that of an edger, but the metal is partially confined on all sides, with shapes in the top and bottom dies resembling a pair of shallow bowls. Because of the cost of sinking the die impressions, rollering is more expensive than edging, provided that both operations can be done in the same number of blows.

Flatteners are used to widen the work metal, so that it more nearly covers the next impression or, with a 90° rotation, to reduce the width to within the dimensions of the next impression. The flattener station can be either a flat area on the face of the die or an impression in the die to give the exact size required.

Benders. A portion of the die can be used to bend the stock, generally along its longitudinal axis, in two or more planes. There are two basic designs of bender impressions: free-flow and trapped-stock.

In bending with a free-flow bender (Fig. 3b), one end or both ends of the forging are free to move into the bender. A single bend is usually made. This type of bending may cause folds or small wrinkles on the inside of the bend.

The trapped-stock bender usually is employed for making multiple bends. In trapped-stock bending, the stock is gripped at both ends as the blow is struck and the stock in between is bent. Because the metal is held at both ends, it usually is stretched during bending. There is a slight reduction in cross-sectional area in the bend, and the work metal is less likely to wrinkle or fold than in a free-flow bender.

Stock that is to be bent may require preforming by fullering, edging or rollering. Bulges of extra material may be provided at the bends, to prevent formation of kinks or folds in free-flow bending. This is particularly necessary when sharp bends are made. The bent preform usually is rotated 90° as it is placed in the next impression.

Splitters. In making fork-type forgings, frequently part of the work metal is split, so that it conforms more closely to the subsequent blocker impression. In a splitting operation, the stock is forced outward from its longitudinal axis by the action of the splitter. Generous radii should be used, to prevent formation of cold shuts, laps and folds.

Blockers. The blocker impression is used in both hammer and press forging. The blocker immediately precedes the finisher impression and serves to refine the shape of the metal preparatory to forging to final shape in the finisher. Usually, the blocker imparts the general final shape to the forging, omitting those details that restrict metal flow in finishing, and including those details that will permit smooth metal flow and complete filling in the finisher impression.

A blocker may be a streamlined model of the finisher, to provide a smooth transition from partially finished to finished forging. Streamlining helps the metal to flow around radii, reducing the possibility of cold shuts or other defects.

Sometimes, the blocker impression is made by duplicating the finisher impression in the die block and then rounding it off as required for smooth flow of metal. When this practice is used, the volume of metal in the blockered preform is greater than will be needed in the finisher impression. Also, the blocker impression is larger at the parting line than is the finisher impression. The excess metal causes the finisher impression to wear at the flash land—where the excess metal must be extruded as flash—and around the top of the impression. With wear, the finisher will produce forgings that cannot be properly trimmed or that are out of tolerance. The impression must be reworked more frequently or the die scrapped prematurely.

It is better practice to make the blocker impression slightly narrower and deeper than the finisher impression, with a volume that is equal to, or only slightly greater than, that of the finisher. The use of a blocker impression having this narrower design minimizes die wear at the parting line in the finisher impression. Moreover, it eliminates the occurrence of the type of lap that is likely to be produced in a finished forging made from a blockered preform of the rounded, finisher-duplicate sort described above—namely, the lap made when the finisher shaves excess metal from the sides of the blockered preform. An added benefit of the narrower design is that it allows for some wear of the blocker impression.

Forging of parts that include deep holes or bosses can cause trouble in the finisher. For producing such parts, the blocker sometimes serves as a gathering operation: A volume of metal that is sunk to one side of a forging in the blocker impression can be forced through to the other side in the finisher impression, filling a high boss.

Use of a blocker impression, in addition to promoting smooth metal flow in the finisher impression, reduces wear on the finisher cavity. Therefore, a blocker is often incorporated in the series of dies for long forging runs, or in production of close-tolerance forgings, to prolong the life of the finisher impression.

PARTING LINE

The parting line is the line along the forging where the dies meet. It may be in a single plane or it may be curved or irregular with respect to the forging plane, depending on the design of the forging. The shape and location of the parting line determine die cost, draft requirements, grain flow and trimming procedures. The following paragraphs describe a few of the considerations that determine the most effective location and shape of the parting line.

In most forgings, the parting line is at the largest cross section of the part, because it is easier to spread metal by forging action than to force it into deep die impressions. If the largest cross section coincides with a flat side of a forging, there may be a particular advantage in locating the parting line along the edges of the flat section, thus placing the entire impression in one die half. Die costs can be reduced, because one die is simply a flat surface. Also, mismatch between upper and lower dies cannot occur, and forging flash can be trimmed readily.

When a die set having one flat die cannot be used, the position of the parting line should provide for location of the preform in the finisher impression of the forging die and the finished forging in the trimming die.

Because part of the metal flow is toward the parting line during forging, the location of the parting line affects the grain-flow characteristics of a forged piece. For good metal-flow patterns in, for example, a forging having a vertical wall adjacent to a bottom web section, a parting line on the outer side of the wall should be placed either adjacent to the web section and near the bottom of the wall, or at the top of the wall. Placing the parting line at any point above the center of the bottom web but below the top of the wall may disrupt the grain flow and cause defects in the forging.

Because the dies move only in a straight line, and because the forging must be removed from the die without damage either to the impression or to the forging, there usually can be no undercuts in the die impressions. Frequently, the forging can be inclined, with respect to the forging plane, to overcome the effect of an undercut. In press forging, it is possible to use "split dies" that are confined in a holder during forging and opened up for removal of the forged part. With such dies it is possible to forge parts with undercuts and complex shapes.

LOCKS AND COUNTERLOCKS

Many forgings require a parting line that is not flat and, correspondingly, die parting surfaces that are neither planar nor perpendicular to the direction in which the forging force is applied. Dies that have a change in the plane of their mating surfaces, and that, therefore, mesh ("lock") in a vertical direction when closed, are called locked dies.

In forging with locked dies, side or end thrust is frequently a problem. A strong lateral thrust during forging may cause mismatch of the dies or breakage of the forging equipment.

There are several ways to eliminate or control side thrust. Individual forgings can be inclined, rotated or otherwise placed in the dies so that the lateral forces are balanced (see Fig. 4c). Flash can be used to cushion the shock and help absorb the lateral forces. When the production quantity is large enough and the size of the forging is small enough to permit forging in multiple-part dies, the impressions can be arranged so that the side thrusts cancel one another.

Generally, with optimum placement of the impression in the die, and with the clearance between the guides on the hammer or press absorbing some side thrust, alignment between the upper and lower die impressions can be maintained. Sometimes, however, the methods suggested above are insufficient or unsuitable to maintain the required alignment, and it is necessary to counteract side thrust by machining mating projections and recesses, or counterlocks, into the parting surfaces of the dies.

Counterlocks can be relatively simple. A pin lock that consists of a round or square peglike section with its mating section may be all that is required to control mismatch. Two such sections, or sections at each corner of the die, may be necessary. A simple raised section with a mating countersunk section running the width and the length of the die can control side and end match. Counterlocks of these types should not be used in long production runs.

Counterlocks in high-production dies should be carefully designed and constructed. The height of the counterlock usually is equal to, or slightly greater than, the depth of the locking portion of the die. The thickness of the counterlock should be at least 1.5 times the height, so that it will have adequate strength to resist side thrust. Adequate lubrication of the sliding surfaces is difficult to maintain, because of the temperature of the die and the heat radiated from the workpiece. Therefore, the surfaces of the counterlock wear rapidly and need frequent reworking. Because of the cost of constructing and maintaining counterlocks, they should be used only if a forging cannot be produced more economically without them. (The discussion of Fig. 7 shows that, in a specific application, the use of counterlocks could be justified because of the reduction in the cost of machining the forging.)

Forging of the connecting link shown in Fig. 4 requires a locked die, because of the part shape. With the die design shown in Fig. 4(a), side thrust is particularly large, because of the angle at which the die faces meet the inclined portion of the work metal. Because no means is provided to counteract side thrust, it is impossible to avoid mismatch of the upper and lower dies. The position of the forging in the die in Fig. 4(b) is the same as in Fig. 4(a), but a counterlock is machined into the die to counteract side thrust. With this arrangement, the possibility of mismatch is eliminated, but the cost of making and maintaining the dies is high. Figure 4(c) shows a position of the forging in the die that is preferable for production. The workpiece has been rotated so that the side thrusts produced when forging the ends

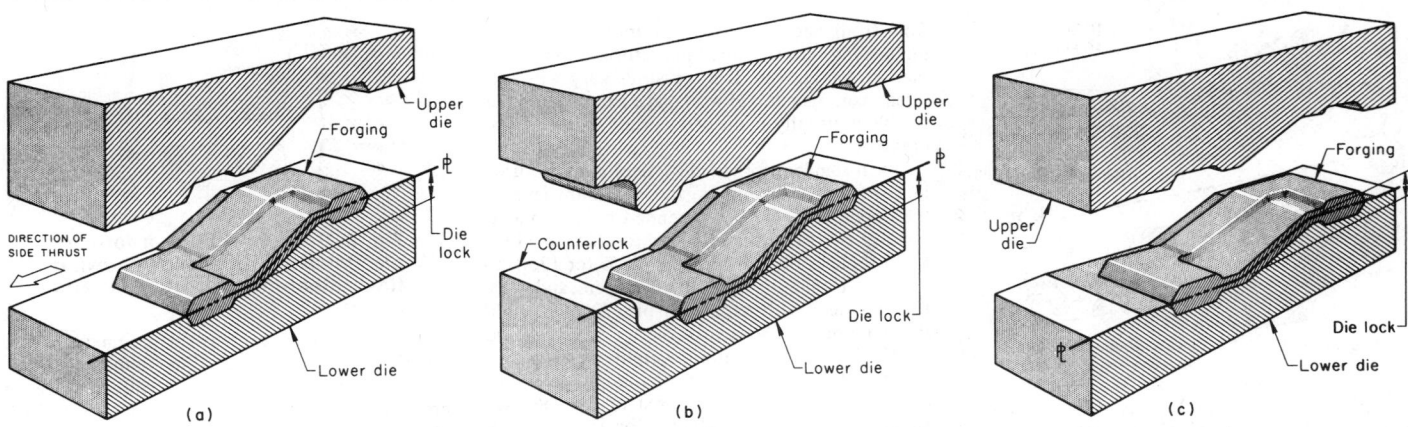

Locked dies (a) with no means of counteracting side thrust, (b) with counterlock and (c) requiring no counterlock because the forging has been rotated to minimize side thrust.

Fig. 4. Locked and counterlocked dies

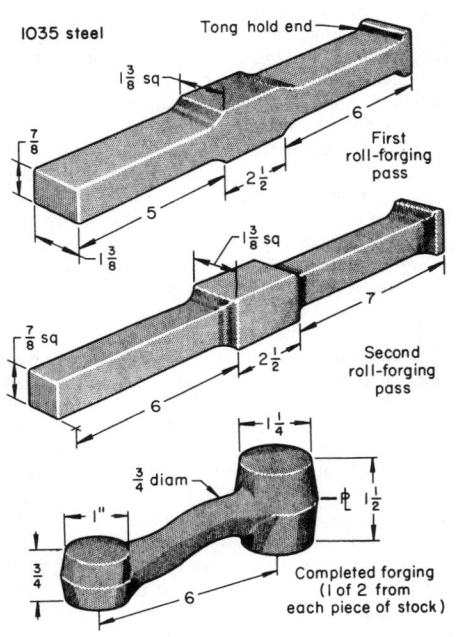

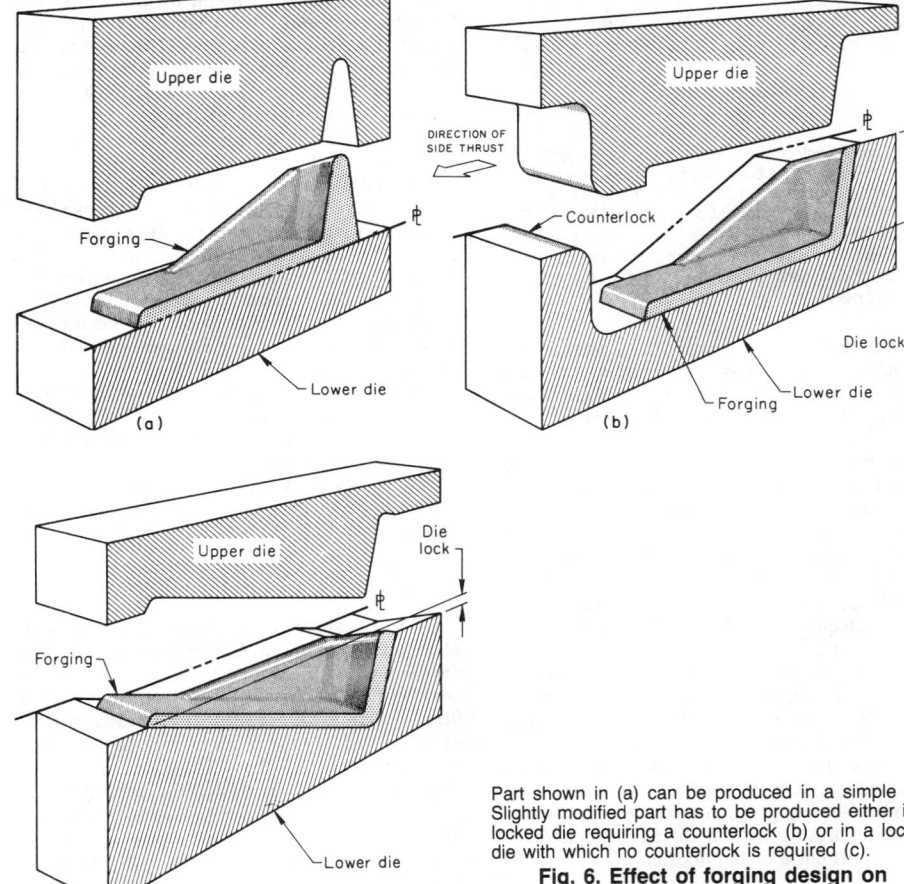

1035 steel

Tong hold end

$1\frac{3}{8}$ sq

$\frac{7}{8}$

6

First roll-forging pass

5

$2\frac{1}{2}$

$1\frac{3}{8}$

$1\frac{3}{8}$ sq

$\frac{7}{8}$ sq

7

Second roll-forging pass

6

$2\frac{1}{2}$

$1\frac{1}{4}$

$\frac{3}{4}$ diam

$1''$

$1\frac{1}{2}$

$\frac{3}{4}$

6

Completed forging (1 of 2 from each piece of stock)

Stock preparation .... Cold shearing
Stock size .... $1^3/_8$-in. round-corner square,
                              $10^1/_4$ in. long
Stock weight .... 5.5 lb(a)
Heating method ...... Oil-fired pusher-type furnace
Heating temperature .... 2200 °F
Forging-press capacity .... 2000 tons
Trimming-press capacity .... 125 tons
Forging-die material ...... 6F4 at 41 to 45 HRC
Forging-die life:
   Before recutting .... 19 500 forgings
   Total (one recut) .... 39 000 forgings
Setup time .... 2 h
Lubricant .... Graphite-oil
Production rate .... 250 forgings per hour(b)
  (a) Yielded two forgings. (b) 125 peices.

Fig. 5. Steering arm forged two at a time in closed dies from a pre-form produced by roll forging in two passes. Dimensions are in inches.

and the web cancel each other. No counterlock is required, and accurate forgings can be produced.

Another arrangement of impressions in locked dies that requires no counterlock is shown in Fig. 5. The design of the workpiece did not permit

inclining the forging with respect to the forging plane in order to minimize side thrust. Instead, since the part was small and production justified forging two parts at a time, side-thrust forces were balanced in a multiple-part die.

A slight change in the design of a forging can change the parting line from one that is contained in a single plane to one requiring a locked die. As shown in Fig. 6(a), when the outside corner at the intersection of the end and the base is square (not radiused), a straight parting line can be used and the entire impression can be contained in one die. Draft requirements cause the end of the forg-

Part shown in (a) can be produced in a simple die. Slightly modified part has to be produced either in a locked die requiring a counterlock (b) or in a locked die with which no counterlock is required (c).

Fig. 6. Effect of forging design on die requirements

ing to be relatively heavy and necessitate machining to maintain a right angle between the end and base surfaces. (If forgings of the design shown in Fig. 6a were made two-at-a-time by forging end-to-end and parted after forging, the production rate would be increased, and sinking of the impression in one die only would still be economical.) When a modification in forging design calls for radiusing of the outside corner, the parting line has to be changed and a locked die must be used. One possible arrangement of such a die is shown in Fig. 6(b). With this arrangement, a counterlock is needed to counteract side thrust,

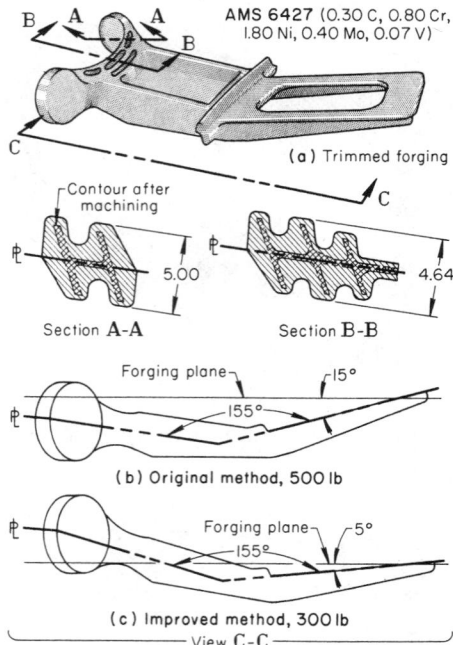

AMS 6427 (0.30 C, 0.80 Cr, 1.80 Ni, 0.40 Mo, 0.07 V)

(a) Trimmed forging

Contour after machining

5.00

Section A-A

4.64

Section B-B

Forging plane — 15°
155°

(b) Original method, 500 lb

Forging plane — 5°
155°

(c) Improved method, 300 lb

View C-C

Engine-mount fitting (a), and the two positions, (b) and (c), in which it could be produced. Position shown in (b) permitted production in a simple locked die. Position shown in (c) allowed a reduction in weight of the forging but required a counterlock in the die.

Fig. 7. Relation of forging design and die design. Dimensions are in inches.

and it is difficult to forge a 90° angle between the end and bottom surfaces. Inclining the forging in the die as shown in Fig. 6(c) eliminates the need for a counterlock, and the natural draft permits forging of a 90° corner.

The engine-mount fitting shown in Fig. 7(a) had to be forged in locked dies. It was too large to forge two-at-a-time, but lateral forces could be balanced when the part was forged in the inclined position shown in Fig. 7(b). When produced in this position, the forging weighed 225 kg (500 lb). Much of the metal was required to provide adequate draft for forging and had to be machined off.

Changing the angle of the parting line with respect to the forging plane from 15° to 5°, as shown in Fig. 7(c), placed the machined ribs in a more nearly vertical position. This permitted less draft at the ribs, and the weight of the forging could be reduced by 90 kg (200 lb). However, changing the die-cavity position upset the balance between the striking and lateral forces, making it necessary to put a counterlock in the die.

SIZE OF THE DIE BLOCK

The size of the die block is determined by the width of the finished "platter," including allowance for flash and gutters, and allowance for the preforming impressions and for sprues and gates for the blocker and finisher impressions. ("Platter" is defined as the entire workpiece on which the forging equipment performs work, including the flash, sprue, tonghold, and as many forgings as are made at one time.)

The impressions in a block should be spaced in such a way that the size of the die face is suitable for the size of the hammer, that flash cannot

flow from one impression to another, and that the forging is not pinched in preform impressions that are too narrow. Die pressures vary with different work metals and workpiece shapes. A larger die block usually is needed as die pressure is increased.

When standard forged alloy steel rams are used in gravity drop hammers, the minimum area of the upper die face recommended by one manufacturer is 30% of the ram area for 225- to 1135-kg (500- to 2500-lb) hammers, 35% for 1360- to 2270-kg (3000- to 5000-lb) hammers and 40% for hammers with capacities of 2720 kg (6000 lb) or more. The weight of the upper die should be 25 to 30% of the falling weight. Heavier dies are not recommended for regular practice. For power drop hammers, the area of the die face should be 50% of the ram area for 455- to 1360-kg (1000- to 3000-lb) hammers, 60% for 1815- to 3630-kg (4000- to 8000-lb) hammers and 70% for hammers with capacities of 4535 kg (10 000 lb) or more.

The minimum shut height of the dies should be at least 50 mm (2 in.) greater than the shut height of the hammer or press. The height of the die block determines the maximum impression depth, since adequate die material must remain between the bottom of the impression and the bottom face of the die block to provide strength in the die.

Relatively small "die inserts" usually are used in mechanical presses. This saves expensive die material and the machining required on large die blocks. The dies are set in recesses in holders fastened to the ram and bed of the press. The dies are held in the recesses by clamps, and screws extending through the holders into the recesses provide for adjustment of die position and hold the die in position.

In modern press-forging operations, quick die-change mechanisms are available. Thus, die inserts can be held by hydraulic clamps that hold or release very quickly. Another method is to set up the inserts in an extra die holder outside the press and to change the entire die holder before starting a new production run.

DRAFT

Draft, or taper, is added to straight sidewalls of a forging to permit easier removal from the die impression. Forgings having round or oval cross sections or slanted sidewalls form their own draft. Forgings having straight sidewalls, such as square or rectangular sections, can be forged by parting them across the diagonal and tilting the impression in the die so that the parting line is parallel to the forging plane. Another method is to place the parting line at an angle to the forging plane, and to machine a straight-wall cavity and a counterlock in each die.

The draft used in die impressions normally varies from 3° to 7° for external walls of the forging. Surfaces that surround holes or recesses have draft angles ranging from 5° to 10°. More draft is used on walls surrounding recesses, to prevent the forging sticking in the die as a result of natural shrinkage of the metal as it cools.

FLASH ALLOWANCE

Flash, or excess metal extruded from the finisher impression during forging, acts as a cushion for impact blows and as a pressure-relief valve for the almost incompressible work metal. Also,

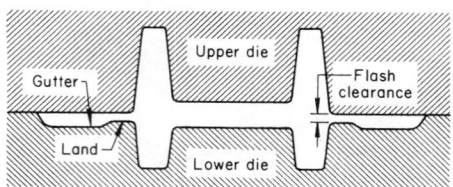

Fig. 8. Section through a forging-die finisher impression showing flash clearance, flash land, and gutter

it restricts the outward flow of the metal so that thin ribs and bosses can be filled in the upper die.

The finisher impression generally includes a provision for flash. A typical arrangement (flash clearance, flash land, and gutter) is shown in Fig. 8.

If the amount of flash clearance in the dies is too small and there is an excess volume of work metal in the impression, much greater forging load or extra hammer blows are required to bring the forging to size. This creates excessive wear on the flash land and produces extreme pressures within the impression and may cause dies to break. Conversely, if the flash clearance is too great, work metal needed to fill the impression flows out through the flash land and the forging is not properly filled. A balanced condition is needed, with just enough volume of metal to ensure that the flash clearance provided will force the work metal to fill the impression properly without causing excess wear and pressure.

The volume, the complexity and the height-to-weight ratio of the forging, as well as the type of work metal, have an effect on flash thickness. General practice has been to use smaller flash clearance for small forgings than for large forgings. The amount of clearance varies from a minimum of about 0.51 mm (0.020 in.) up to a maximum of about 9.52 mm (0.375 in.) for forgings weighing up to 90 kg (200 lb). About 3% of the maximum forging thickness is a reasonable guide for flash clearance.

The amount of excess work metal extruded from the impression can be too great to permit complete closing of the dies. To ensure complete closing, a gutter is provided for the excess metal after it has passed through the flash land.

LOCATION OF IMPRESSIONS

The preform and finisher impressions should be positioned across the die block so that the forging force is as near the center of the striking force (ram) as possible. This will minimize tipping of the ram, reduce wear on the ram guides and help to maintain the thickness dimensions of the forging. When the forging is transferred manually to each impression, the impression for the operation requiring the greatest forging force (usually, the finisher) is placed at the center of the die block, and the remaining impressions are distributed as nearly equally as possible on each side of the die block.

Symmetrical forgings usually have their centerline along the front-to-back centerline of the die block. For asymmetrical forgings, the center of gravity can be used as a reference for positioning the preform and finisher impressions in the die block.

The center of gravity of a forging does not necessarily correspond to the center of the forging force, because of the influence of thin sections on the forging force. Because the increase in force is not always directly proportional to the decrease in thickness, both the flash and the location of the thin sections must be considered when locating the impressions in a die block. Evenly distributed flash has little effect on an out-of-balance condition; very thin sections have a marked effect. Often it is necessary to make some calculations in order to determine the center of loading of the finisher impression.

DIE INSERTS

Die inserts are used for economy in production of some forgings. In general, they prolong the life of the die block into which they fit. Use of inserts can decrease production costs when several inserts can be made for the cost of making one solid die. The time required for changeover or replacement of inserts is brief, because a second set of inserts can be made while the first set is being used. Finally, more forgings can be made accurately in a die with inserts than in a solid die, because steel of higher alloy content and greater hardness can be used in inserts than would be safe or economical to use in solid dies.

Some commercial forge shops in which most of the forging units are drop hammers make only limited use of die inserts. However, nearly all press shops use die inserts.

Inserts can contain the impression of only the portion of a forging that is subject to greatest wear, or they can contain the impression of a whole forging. An example of the first type of insert is the plug-type insert used for forging deep cavities. Examples of the second type include master-block inserts that permit forging of a variety of shallow parts in a single die block, and inserts for replacement of impressions that wear the most rapidly in multiple-impression dies.

A plug-type insert (Fig. 9) is usually a projection in the center of the die, such as would be required for making a hub or cup forging. In some impressions, the plug may not be in the center, and more than one plug can be used in a single impression.

Although plugs are used in either shallow or deep impressions, the need is usually greater in

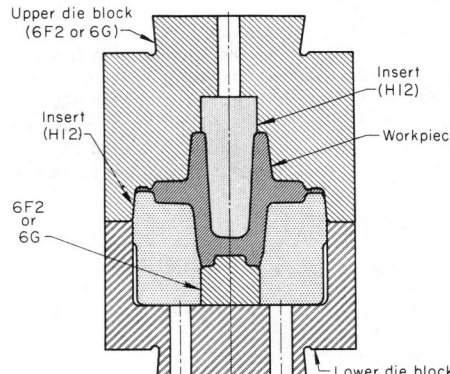

For forging this workpiece (an automotive axle housing), shown in cross section, an H12 tool steel plug in the upper die block is used in combination with a nearly complete H12 insert in the lower die block. **Fig. 9. Use of a plug-type insert in combination with a nearly complete insert in the lower die block for making a forging of extreme severity**

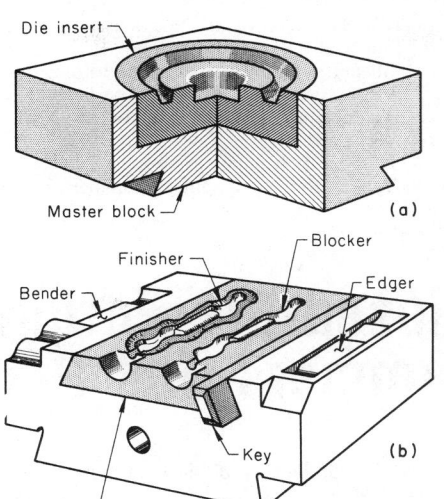

(a) Full insert and master block for use in forging of gear blanks in hammers. (b) Multiple-impression insert for use when wear is excessive on one or more impressions. Such an insert is usually secured by a key.

Fig. 10. Two types of die inserts used in hammer forging

deep impressions. For impressions of moderate depth, an insert is advantageous if medium or large quantities of forgings are required. For deep, narrow impressions like that shown in Fig. 9, a plug-type insert is always recommended. Sometimes it is advantageous to use a plug in combination with a complete or nearly complete insert, as in Fig. 9, where a long H12 steel plug is used in the upper die and an almost complete female insert is used in the lower die.

Plug inserts can be made either from prehardened die steel at a higher hardness than the main part of the die or, for still longer life, from one of the hot work tool steels. Where wear is extremely high, the plug can be hard faced.

Plugs are held in place by press fitting, by shrink fitting (by packing in dry ice before insertion) or by the use of plug keys (Fig. 10b).

Full inserts are generally used for making relatively shallow forgings. They offer one or more of the following advantages: (*a*) the insert can be of high hardness with less danger of breakage, because it has the softer block as a backing; (*b*) a higher-alloy steel can be used for the insert portion without a large increase in cost; (*c*) changes in forging design are less costly when inserts are used; (*d*) the same die block can be used for slightly different forgings by changing inserts; and (*e*) inserts can be readily replaced if breakage occurs. Full inserts are used in many commercial forge shops, where a set of standard master blocks is kept available for use. Such a master block with an insert is shown in Fig. 10(a).

On either side of the insert cavities, preforming can take place.

Another type of insert is for use in multiple-impression dies in which the impressions wear at different rates. Fuller, edger or bender impressions seldom are used for close-tolerance work and may wear slowly compared with other impressions. Inserts are used for only the impressions that wear most rapidly. In the multiple-impression die shown in Fig. 10(b), the blocker and finisher impressions are in the same insert.

This type of insert is not necessarily limited to shallow impressions. If the insert contains a single impression, the impression can be of any practical depth. However, if it contains several impressions, the impression depth is limited to about 64 mm (2$^1/_2$ in.) or less. Width of the insert must be considered: sufficient wall thickness must be allowed between the edge of the impression and the edge of the insert, so that the die-block walls are not weakened too greatly.

DIE STEELS

Prehardened die blocks are available in a range of compositions and hardnesses. Other tool steels are also available for use in small die blocks, die inserts and trimming tools.

Table 1 lists five hardness ranges, together with the approximate maximum sizes and weights in which prehardened die blocks (hardened and tempered by the manufacturer) are commercially available. Although hardness ranges are tabulated in both Brinell and Rockwell units, hardness measurements are usually made with a Brinell tester, because of the size of the block. Almost all of the die blocks listed are too large for testing on a Rockwell machine without special facilities. The Brinell test is standard, because it is less sensitive to the minute structural dissimilarities that are present in massive blocks and because design features of the Rockwell tester limit its application in testing large sections. A carbide ball, rather than a steel ball, is used in the Brinell tester, because carbide has better resistance to deformation at high hardness levels, and better dimensional stability, than steel.

Many forge shops take hardness measurements with the scleroscope, because of its portability. Such readings can be converted and reported in some other hardness scale.

Prehardened die-block steels are usually purchased on a basis of hardness and proprietary name. Additional information on properties and selection of tool materials for dies can be found in Section 18. Information on heat treating of dies can be found in Section 28.

FACTORS IN SELECTION OF DIE MATERIALS

Die steels are selected on the basis of the following characteristics:

Table 1. Hardness ranges and normal size and weight limits for hardened-and-tempered die blocks

| Hardness(a) | | Cross section | | Length | | Approximate weight | |
HB	HRC	mm	in.	mm	in.	kg	lb
444 to 477	47 to 50	. . .255 by 380	10 by 15	510	20	410	900
388 to 429	42 to 46	. . .255 by 430	10 by 17	510	20	455	1000
341 to 375	37 to 40	. . .380 by 510	15 by 20	915	36	1475	3250
302 to 331	32 to 36	. . .380 by 710	15 by 28	1220	48	2720	6000
269 to 293	28 to 31	. . . (b)	(b)	(b)	(b)	(b)	(b)

(a) Hardened-and-tempered blocks are available at lower hardnesses than those listed. (b) Limited only by facilities of vendor.

1. Ability to harden uniformly
2. Ability to resist the abrasive action of the hot metal while it is being forged
3. Ability to withstand high pressure and heavy shock loads
4. Ability to resist cracking and checking caused by heat.

   Selection of the most suitable combination of steel and hardness for die blocks and die inserts is influenced by:

1. Shape, size and weight of the forging
2. Composition of the metal to be forged
3. Temperature at which the work metal is to be forged
4. Number of forgings to be made
5. Type of forging equipment (hammer or press)
6. Cost of the die steel
7. Sequence of machining the die impressions (before or after hardening)
8. Forging tolerances (including those specified for draft angles)
9. Established plant practice and previous experience with similar applications
10. Availability of auxiliary equipment.

Closed-Die Forging in Hammers and Presses

CLOSED-DIE FORGING is the shaping of hot metal completely within the walls or cavities of two dies that come together to enclose the workpiece on all sides. The impression for the forging can be entirely in either die or divided between the top and bottom dies.

The forging stock, which is generally round or square bar, is cut to length to provide the volume of metal needed to fill the die cavities, plus an allowance for flash and sometimes for a projection for holding the forging. The flash allowance is, in effect, a relief valve for the extreme pressure produced in closed dies. Flash also acts as a brake to slow the outward flow of metal, to permit complete filling of thin sections.

This article discusses closed-die forging of carbon and alloy steels in hammers and presses. Related forging processes are described in other articles in this volume.

CAPABILITIES OF THE PROCESS

With the use of closed dies, complex shapes and heavy reductions can be made in hot metal within closer dimensional tolerances than are usually feasible with open dies. Open dies are used primarily for forging of simple shapes, or for making forgings that are too large to be contained in closed dies. Closed-die forgings are usually designed to require minimum subsequent machining.

Closed-die forging is adaptable to low-volume or high-volume production. In addition to producing final, or nearly final, metal shapes, closed-die forging allows control of grain-flow direction, and it often improves mechanical properties in the longitudinal direction of the workpiece.

Size of forgings produced in closed dies can range from a few ounces to several tons. The maximum size that can be produced is limited only by the equipment that is available for handling and for forging. Steel forgings weighing as much as 15 000 kg (33 000 lb), and with maximum dimensions of 890 mm (35 in.) wide by 2.9 m (115 in.) long, have been successfully forged in closed dies, although more than 70% of the closed-die forgings produced weigh 0.9 kg (2 lb) or less.

Shapes. Complex nonsymmetrical shapes that require a minimum number of operations for completion can be produced by closed-die forging. In addition, the process can be used in combination with other processes to produce parts having greater complexity or closer tolerances than are possible by forging alone. Cold coining and the assembly of two or more closed-die forgings by welding are examples of other processes that

Table 1. Classification of alloys in order of increasing forging difficulty

Alloy group	Approximate forging-temperature range	
	°C	°F
Aluminum alloys (least difficult)	400 to 550	750 to 1020
Magnesium alloys ...	250 to 350	480 to 660
Copper alloys	600 to 900	1110 to 1650
Carbon and low-alloy steels	850 to 1150	1560 to 2100
Martensitic stainless steels	1100 to 1250	2010 to 2280
Maraging steels	1100 to 1250	2010 to 2280
Austenitic stainless steels	1100 to 1250	2010 to 2280
Nickel alloys	1000 to 1150	1830 to 2100
Semiaustenitic PH stainless steels	1100 to 1250	2010 to 2280
Titanium alloys	700 to 950	1290 to 1740
Iron-base superalloys	1050 to 1180	1920 to 2160
Cobalt-base superalloys	1180 to 1250	2160 to 2280
Niobium alloys	950 to 1150	1740 to 2100
Tantalum alloys	1050 to 1350	1920 to 2460
Molybdenum alloys ..	1150 to 1350	2100 to 2460
Nickel-base superalloys	1050 to 1200	1920 to 2190
Tungsten alloys (most difficult)	1200 to 1300	2190 to 2370

can extend the useful range of closed-die forging.

FORGING MATERIALS

In closed-die forging, a material must satisfy two basic requirements: (*a*) the material strength (or flow stress) must be low so that die pressures are kept within the capabilities of practical die materials and constructions, and (*b*) the capability of the material to deform without failure (its forgeability) must allow the desired amount of deformation. By convention, impression- and closed-die forging refer to hot working. In Table 1, various alloy groups and their respective forging-temperature ranges are given in order of increasing forging difficulty. The forging material influences the design of the forging itself as well as the details of the entire forging process. For example, Fig. 1 shows that, owing to difficulties in forging, nickel alloys allow for less shape definition than aluminum alloys.

For a given metal, both the flow stress and the forgeability are influenced by (*a*) the metallurgical characteristics of the billet material and (*b*) the temperatures, strains, strain rates and stresses which occur in the deforming material.

FORGEABILITY OF STEEL

Forgeability varies considerably among the different carbon and alloy steels, depending on carbon and alloy content, forging temperature, and strength at forging temperature. Low-carbon steels are the most forgeable, partly because they can be safely forged at higher temperatures than high-carbon steels, and partly because of their low strength at elevated temperatures. As carbon and alloy contents increase, the strength of the metal at any temperature increases, as do the forging load requirements. The effects of steel composition on the forging loads and forging pressures required to effect upset reductions of increasing severity at different forging temperatures are shown in Fig. 2.

In most practical hot forging operations, the temperature of the workpiece material is higher than that of the dies. Metal flow and die filling are largely determined by (*a*) the resistance and the ability of the forging material to flow—i.e., flow stress and forgeability, (*b*) the friction and cooling effects at the die/material interface, and (*c*) the complexity of the forging shape. Of the two basic material characteristics, flow stress represents the resistance of a metal to plastic deformation, and forgeability represents the ability of a metal to deform without failure, regardless of the magnitude of load and stresses required for deformation.

The concept of forgeability has been used vaguely to denote a combination of both resistance to deformation and the ability to deform without fracture. A diagram illustrating this type of information is presented in Fig. 3. Since the resistance of a metal to plastic deformation is essentially determined by the flow stress of the ma-

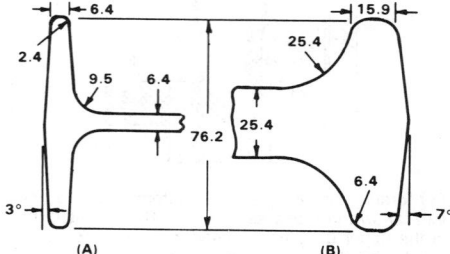

(A) Aluminum alloys. (B) Nickel-base superalloys. Dimensions are in millimetres.

Fig. 1. Comparison of typical design limits for rib-web structural forgings

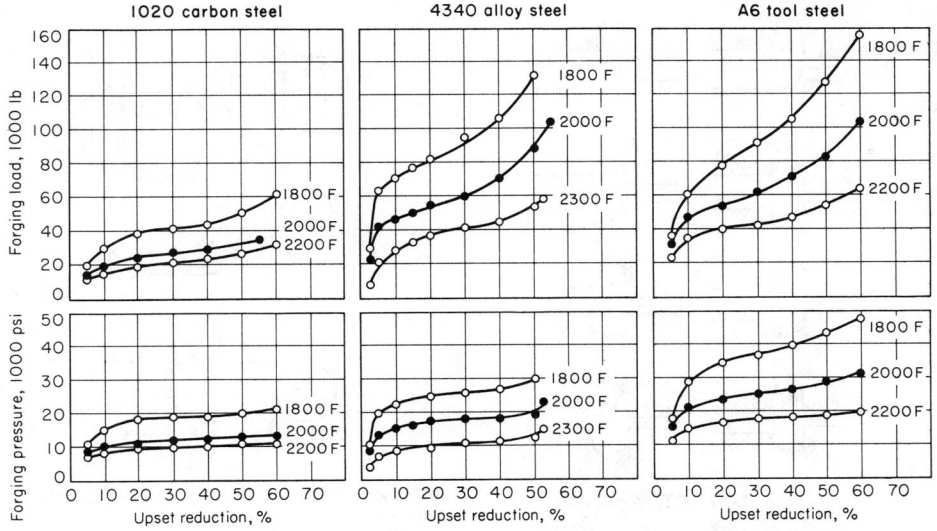

Fig. 2. Effects of compositions of three different steels on loads and pressures required for upset reductions of increasing severity at various temperatures

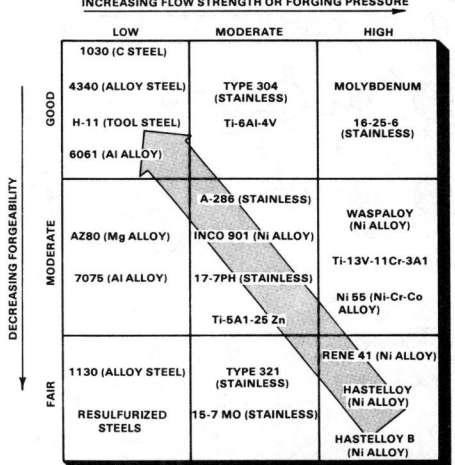

Fig. 3. Generalized diagram of influence of forgeability and flow strength on die filling. Shaded arrow denotes increasing ease of die filling.

terial at given temperature and strain-rate conditions, it is more appropriate to define forgeability as the capability of the material to deform without failure, regardless of pressure and load requirements.

In general, the forgeability of metals increases with increasing temperature. However, as temperature increases, grain growth occurs and, in some alloy systems, forgeability decreases with increasing grain size. In other alloys, forgeability is greatly influenced by the characteristics of second-phase compounds. The state of stress in a given deformation process significantly influences forgeability. In upset forging at large reductions, for instance, cracking may occur at the outside fibers of the billet, where excessive barreling occurs and tensile stresses develop. In certain extrusion-type forging operations, axial tensile stresses may be present in the deformation zone and may cause centerburst cracking. As a general and practical rule, it is important to provide compressive support to those portions of a less forgeable material that are normally exposed to the tensile and shear stresses.

The forgeability of metals at various deformation rates and temperatures can be evaluated by using various tests such as torsion, tension and compression tests. In all these tests, the amount of deformation prior to failure of the specimen is an indication of forgeability at the temperature and deformation rates used during that particular test.

FRICTION AND LUBRICATION IN FORGING

In forging, friction greatly influences metal flow, pressure distribution, and load and energy requirements. In addition to lubrication effects, the effects of die chilling or heat transfer from the hot material to colder dies must be considered. For example, for a given lubricant, friction data obtained in hydraulic-press forging cannot be useful in mechanical-press or hammer forging even if die and billet temperatures are comparable.

In forging, the lubricant is expected to:

- Reduce sliding friction between the dies and the forging in order to reduce pressure requirements, to fill the die cavity and to control metal flow.
- Act as a parting agent and prevent local welding and subsequent damage to the die and workpiece surfaces.
- Possess insulating properties so as to reduce heat losses from the workpiece and minimize temperature fluctuations on the die surface.
- Wet the surface uniformly so that local lubricant breakdown and uneven metal flow are prevented.
- Be nonabrasive and noncorrosive so as to prevent erosion of the die surface.
- Be free of residues that would accumulate in deep impressions.
- Develop a balanced gas pressure to assist quick release of the forging from the die cavity; this characteristic is particularly important in hammer forging, where ejectors are not used.
- Be free of polluting or poisonous components and not produce smoke on application to the dies.

No single lubricant can fulfill all the requirements listed above, and therefore a compromise must be made for each specific application.

CLASSIFICATION OF CLOSED-DIE FORGINGS

Closed-die forgings are generally classified as: (*a*) blocker-type, (*b*) conventional and (*c*) close-tolerance.

Blocker-type forgings are produced in relatively inexpensive dies, but their weight and dimensions are somewhat greater than those of counterpart conventional closed-die forgings. A blocker-type forging approximates the general shape of the final part, with relatively generous finish allowance and radii. Such forgings are sometimes specified when only a small number of forgings are required and the cost of machining parts to final shape is not excessive.

Conventional closed-die forgings are the most common type, and are produced to comply with commercial tolerances. These forgings are characterized by design complexity and tolerances that fall within the broad range of general forging practice. They are made closer to the shape and dimensions of the final part than blocker-type forgings are, and thus are lighter and have more detail.

Close-tolerance forgings usually are held to smaller dimensional tolerances than conventional forgings. Little or no machining is required after forging, because close-tolerance forgings are made with less draft and with thinner walls, webs and ribs. These forgings cost more, and require higher forging pressures per unit of plan area, than conventional forgings. However, the higher forging cost is sometimes justified by a reduction in machining cost.

SHAPE COMPLEXITY IN FORGING

Metal flow in forging is greatly influenced by part or die geometry. Often, several operations (preforming or blocking) are needed to achieve gradual flow of the metal from an initially simple shape (cylinder or round-cornered square billet) into the more complex shape of the final forging. In a general sense, spherical and blocklike shapes are the easiest to forge in impression or closed dies. Parts with long, thin sections or projections (webs and ribs) are more difficult to forge because they have more surface area per unit volume. Such variations in shape maximize the effects of friction and temperature changes and, hence, influence the final pressure required to fill the die cavities. There is a direct relationship between the surface-to-volume ratio of a forging and the difficulty of producing that forging.

The ease of forging more complex shapes depends on the relative proportions of vertical and horizontal projections on the part. Figure 4 is a schematic representation of the effects of shape on forging difficulties. Parts (c) and (d) would require not only higher forging loads but also at least one more forging operation than parts (a) and (b) to ensure die filling.

As shown in Fig. 5, the majority of forgings can be classified into three main groups. The first group consists of the so-called "compact shapes," whose three major dimensions (length, l, width, b, and height, h) are approximately equal. The number of parts that fall into this group is rather small. The second group consists of "disk shapes," for which two of the three dimensions (length, l, and width, b) are approximately equal and are

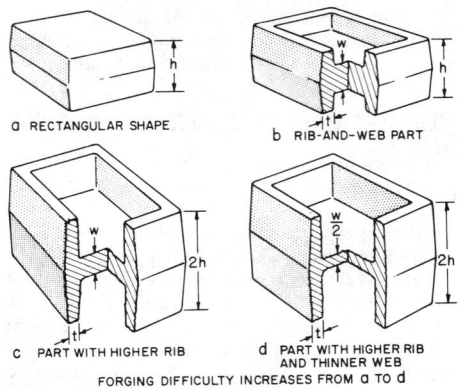

a RECTANGULAR SHAPE

b RIB-AND-WEB PART

c PART WITH HIGHER RIB

d PART WITH HIGHER RIB AND THINNER WEB

FORGING DIFFICULTY INCREASES FROM a TO d

Fig. 4. Rectangular shape and three modifications, showing increasing forging difficulty with increasing rib height and decreasing web thickness

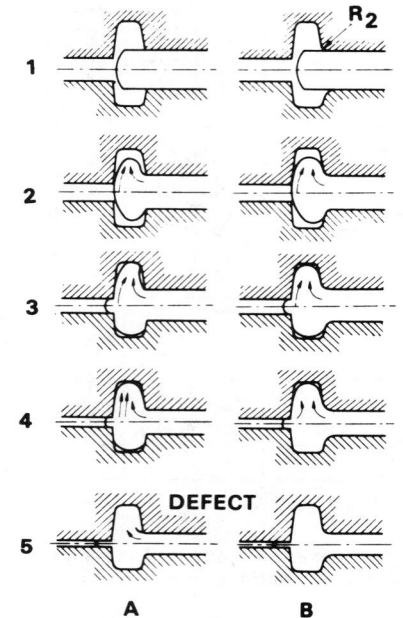

DEFECT

A   B

(A) Insufficient fillet radii. (B) Adequate fillet radii.

Fig. 6. Defect formation in forging due to insufficient fillet radii

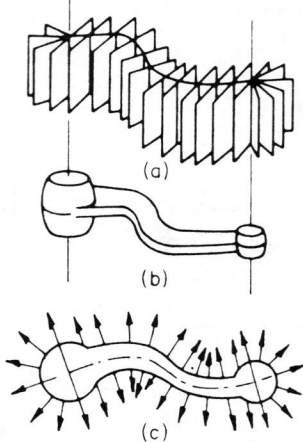

(a) Planes of flow. (b) Finish forging shape. (c) Directions of flow.

Fig. 7. Planes and directions of metal flow during forging of a relatively complex shape

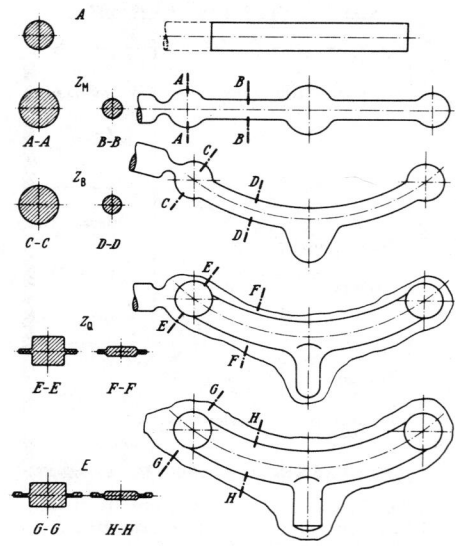

Fig. 8. Preforming, blocking and finish forging operations for an example forging

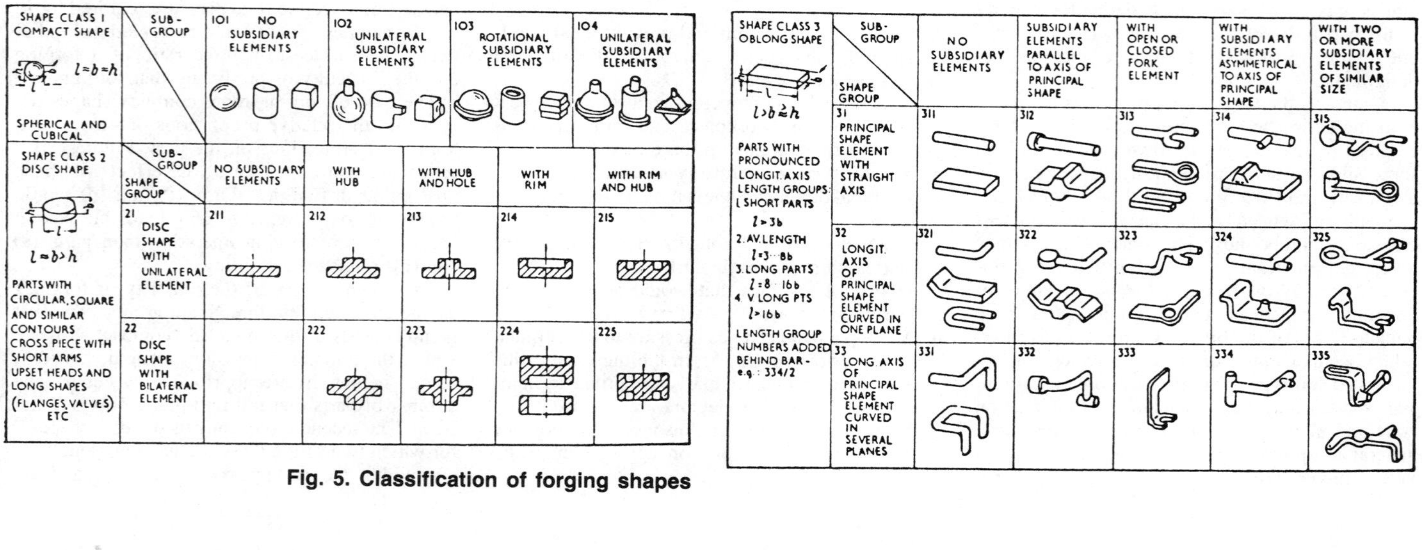

greater than the height, h. All round forgings belong to this group, which includes approximately 30% of all commonly used forgings. The third group consists of long shapes which have one major dimension significantly greater than the other two ($l > b \geq h$). These three basic groups are further divided into subgroups depending on the presence and type of elements subsidiary to the basic shape.

This "shape classification" can be useful for practical purposes, such as for estimating costs and for predicting preforming steps. However, this method is not entirely quantitative and requires some subjective evaluation based on past experience.

DESIGN OF BLOCKER (PREFORM) DIES

One of the most important aspects of impression- and closed-die forging is proper design of preforming operations and of blocker dies to achieve adequate metal distribution. Thus, in the finish forging operation, defect-free metal flow and complete die filling can be achieved, and metal losses into the flash can be minimized. In preforming, round or round-cornered square stock with constant cross section is deformed in such a manner that a desirable volume distribution is achieved prior to impression-die forging. In

blocking, the preform is die forged in a blocker cavity prior to finish forging.

The main objective of preforming is to distribute the metal in the preform so as to:

- Ensure defect-free metal flow and adequate die filling: for example, Fig. 6 illustrates how a defect can form as a result of an insufficient fillet radius in an H-shape cross section.
- Minimize the amount of material lost into flash.
- Minimize die wear in the finish forging cavity by reducing metal movement in this direction.
- Achieve desired grain flow and control mechanical properties.

Common practice in preform design is to consider planes of metal flow — i.e., selected cross sections of the forging — as shown in Fig. 7. The example presented in Fig. 8 illustrates the various preforming operations necessary to forge the part shown in that figure. The round bar from rolled stock is (a) rolled in a reducer roller for volume distribution, (b) bent in a die to provide

Fig. 5. Classification of forging shapes

the appropriate shape, (c) blocked in a blocker die cavity and (d) finish forged. In determining the various forging steps, it is first necessary to obtain the volume of the forging, based on the areas of successive cross sections throughout the forging.

Figure 9 shows two examples of obtaining a volume distribution through the following procedure:

1. Lay out a dimensioned drawing of the finish configuration, complete with flash.
2. Construct a baseline for area determination parallel to the centerline of the part.
3. Determine maximum and minimum cross-sectional areas perpendicular to the centerline of the part.
4. Plot these area values at proportional distances from the baseline.
5. Connect these points with a smooth curve (in instances where it is not clear how the curve

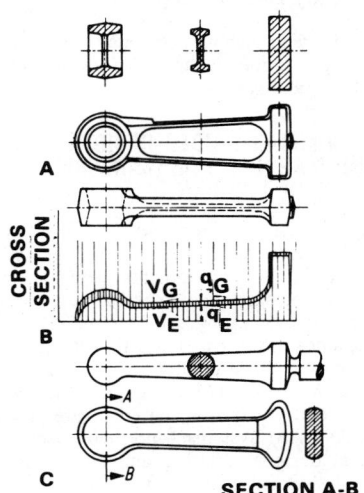

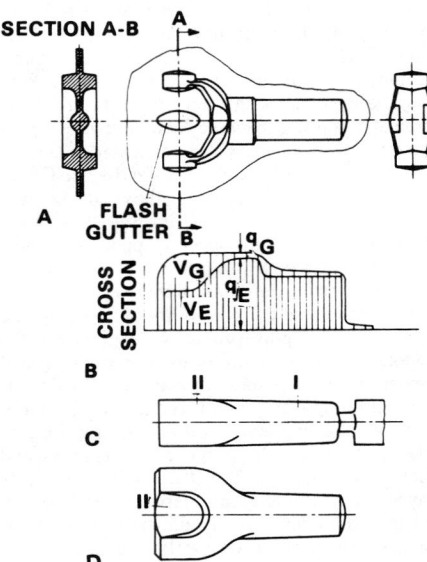

In both examples, A = forging, B = cross-sectional area vs length, C and D = ideal preform, V_E and q_E = volume and cross section of the finished forging, and V_G and q_G = volume and cross section of the flash.

Fig. 9. Preform designs for two example parts

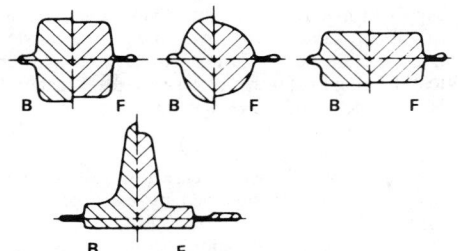

B = blocker; F = finished forging.

Fig. 10. Suggested blocker cross sections for various steel forgings

would best show the changing cross-sectional areas, additional points should be plotted to assist in determining a smooth representative curve).

6. Above this curve, add the approximate area of the flash at each cross section, giving consideration to those sections where the flash should be widest. The flash will generally be of a constant thickness but will be widest at the narrower sections and smallest at the wider sections (the proportional allowance for flash is illustrated by the examples in Fig. 9).
7. Convert the maximum and minimum area values to rounds or rectangular shapes having the same cross-sectional areas.

In designing the cross sections of a blocker (preform) die impression, several basic rules must be followed:

- The area of each cross section along the length of the preform must be equal to the area of the finish cross section augmented by the area necessary for flash. Thus, the initial stock distribution is obtained by determining the areas of cross sections along the main axis of the forging.
- All the concave radii (including fillet radii) of the preform should be larger than the radii of the forged part.
- Whenever practical, the dimensions of the preform should be greater than those of the finished part in the forging direction so that metal flow is mostly of the upsetting rather than the extrusion type. During the finishing operation, the material then will be squeezed laterally toward the die cavity without additional shear at the die/material interface. Such conditions minimize friction and forging load and reduce wear along the die surfaces.

Application of these three principles to steel forgings is illustrated in Fig. 10 for some solid cross sections. The qualitative principles of preform design are well known, but quantitative information is rarely available. For forging of rib-web-type aluminum- and titanium-alloy parts, blocker dimensions recommended by various companies fall into the ranges given in Table 2.

For forging of complex parts, empirical guidelines may not suffice, and trial-and-error procedures may be time-consuming and expensive; a more systematic and well-proven method for developing preform shapes is physical modeling, using a soft material such as lead, plasticine or wax as a model forging material and hard plastic or mild steel dies as tooling. Thus, with relatively low-cost tooling and with some experimentation, preform shapes can be determined.

FLASH DESIGN

The influences of flash thickness and flash-land width on forging pressure are reasonably well understood from a qualitative viewpoint (see Fig. 11). Essentially, forging pressure increases with (a) decreasing flash thickness and (b) increasing flash-land width because of combinations of increasing restriction, increasing frictional forces and decreasing metal temperatures at the flash gap.

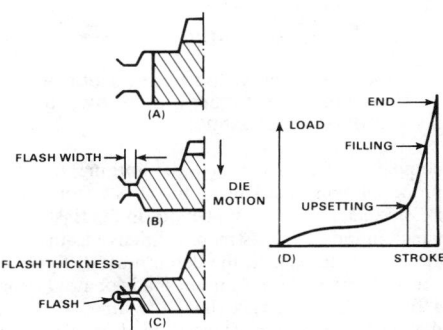

(A) Upsetting. (B) Filling. (C) End. (D) Load-stroke curve.

Fig. 11. Metal flow and load-stroke curve in impression-die forging

A typical load-versus-stroke curve for a closed-die forging is shown in Fig. 12. Loads are relatively low until the more difficult details are partly filled and the metal reaches the flash opening (Fig. 11). This stage corresponds to point P_1 in Fig. 12. For successful forging, two conditions must be fulfilled when this point is reached: (a) a sufficient volume of metal must be trapped within the confines of the die to fill the remaining cavities and (b) extrusion of metal through the narrowing gap of the flash opening must be more difficult than filling of the more intricate detail in the die.

As the dies continue to close, the load increases sharply to point P_2, the stage at which the die cavity is filled completely. Ideally, at this point, the cavity pressure provided by the flash geometry should be just sufficient to fill the entire cavity, and the forging should be completed. However, P_3 represents the final load reached in normal practice for ensuring that the cavity is

Table 2. Recommended preform (or blocker) dimensions for forging of rib-web-type aluminum and titanium alloy parts

Dimension in finish forging	Preform dimensions(a)	
	Aluminum alloys	Titanium alloys
Web thickness, t_F	$t_P = 1\text{-}1.5\ t_F$	$t_P = 1.5\text{-}2.2\ t_F$
Fillet radii, R_{FF}	$R_{PF} = 1.2\text{-}2\ R_{FF}$	$R_{PF} = 2\text{-}3\ R_{FF}$
Corner radii, R_{FC}	$R_{PC} = 1.2\text{-}2\ R_{FC}$	$R_{PC} = 2\ R_{FC}$
Draft angle, α_F	$\alpha_P = \alpha_F\ (2\text{-}5°)$	$\alpha_P = \alpha_F\ (3\text{-}5°)$
Rib width, W_F	$W_P = W_F - 0.8$ mm	$W_P = W_F - 1.6\text{-}3.2$ mm

(a) The first subscript of each dimension indicates either finish forging (F) or preform (P).

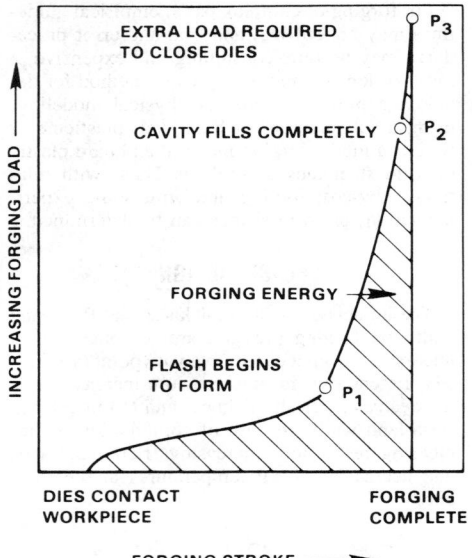

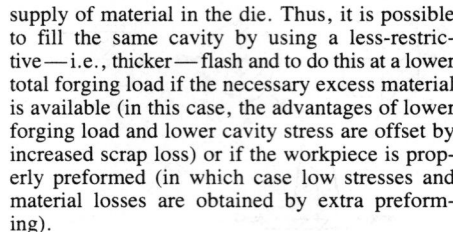

Fig. 12. Typical load-stroke curve for closed-die forging, showing three distinct stages

completely filled and that the forging has the proper dimensions. During the stroke from P_2 to P_3, all metal flow occurs near or in the flash gap, which in turn becomes more restrictive as the dies close. In that respect, the detail most difficult to fill determines the minimum load for producing a fully filled forging. Thus, the dimensions of the flash determine the final load required for closing the dies. Formation of the flash, however, is greatly influenced by the amount of excess material available in the cavity, because that amount determines the instantaneous height of the extruded flash and, therefore, the die stresses.

A cavity can be filled with various flash geometries provided that there is always a sufficient supply of material in the die. Thus, it is possible to fill the same cavity by using a less-restrictive—i.e., thicker—flash and to do this at a lower total forging load if the necessary excess material is available (in this case, the advantages of lower forging load and lower cavity stress are offset by increased scrap loss) or if the workpiece is properly preformed (in which case low stresses and material losses are obtained by extra preforming).

The "shape classification" (Fig. 5) has been utilized in systematic evaluation of flash dimensions in steel forgings. The results for shape group 224 are presented in Fig. 13 as an example.

In general, the flash thickness is shown to increase with increasing forging weight, while the ratio of flash width to flash thickness (w/t) decreases to a limiting value.

PREDICTION OF FORGING PRESSURE

It is often necessary to predict forging pressure so that a suitable press can be selected and so that die stresses can be prevented from exceeding allowable limits. In estimating the forging load empirically, the surface area of the forging, including the flash zone, is multiplied by an average forging pressure known from experience. The forging pressures encountered in practice vary from 550 to 965 MPa (40 to 70 tsi), depending on the material and the geometrical configuration of the part. Figure 14 gives forging pressures for parts made of various carbon steels (up to 0.6% C) and low-alloy steels. In these trials, flash ratios w/t (where w is flash-land width and t is flash thickness) from 2 to 4 were used. The variable which most influences forging pressure is the average height of the forging. The lower curve in Fig. 14 relates to relatively simple parts, and the upper curve to more difficult-to-forge parts.

Most empirical methods, summarized in terms of simple formulas or nomograms, are not sufficiently general for prediction of forging loads for a variety of parts and materials. Lacking a

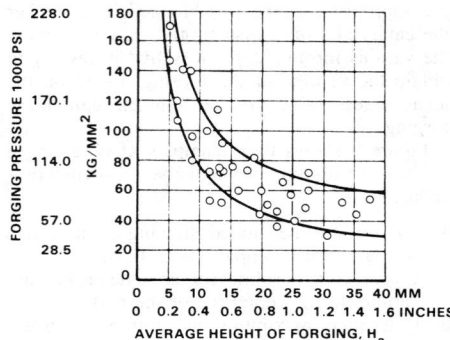

Lower curve relates to relatively simple parts; upper curve, to more difficult-to-forge parts. Data are for flash ratios from 2 to 4.

Fig. 14. Forging pressure vs average forging height for carbon and low-alloy steel forgings

suitable empirical formula, one may use analytical or computer-aided techniques for calculating forging loads and stresses.

COMPUTER-AIDED DESIGN AND MANUFACTURE (CAD/CAM) OF FORGING DIES

During the last decade, computers have been used to an increasing extent for forging applications. The initial developments were concentrated on NC machining of forging dies and of models for copy milling. In the mid-1970's, computer-aided drafting and NC machining were also introduced for structural forgings and for forging steam-turbine blades. During the early 1980's, several companies started to use stand-alone CAD/CAM systems—normally used for mechanical designs, drafting and NC machining—for design and manufacture of forging dies.

Stand-alone CAD/CAM systems are commercially available and have the necessary software for computer-aided drafting and NC machining. A typical CAD/CAM system consists of a minicomputer, a graphics display terminal, a keyboard, a digitizer with menu for data entry, an automatic drafting machine, and hardware for information storage and NC tape punching or floppy-disk preparation. Although such stand-alone CAD/CAM systems are expensive, they can increase productivity in mechanical drafting by 2 to 5 times, depending on the specific application. Such CAD/CAM systems also allow, at various levels of automation, three-dimensional representation of the forging and the possibility of zooming and rotating the forging-geometry display on the graphics terminal screen for the purpose of visual inspection. Ideally, these systems should also allow sectioning of a given forging—i.e., the description, drawing and display of desired forging cross sections for the purpose of die-stress and metal-flow analyses. Thus, the results can be displayed for easy interaction between the designer and the computer system, and modifications of die design can be made easily and alternatives can be explored.

The ultimate advantage of computer-aided design in forging is achieved when reasonably accurate and inexpensive computer software is available for simulating metal flow throughout a forging operation (see Fig. 15). Thus, forging "experiments" can be run on a computer by simulating the finish forging that would result from

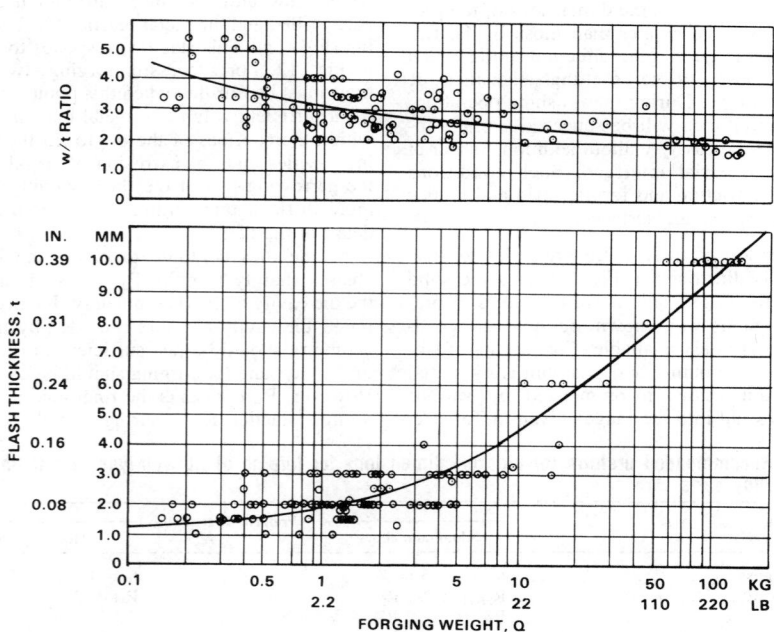

Fig. 13. Variations in flash-land-to-thickness ratio and in flash thickness, t, with weight, Q, for forgings of group 224 in Fig. 5 (materials: carbon and alloy steels)

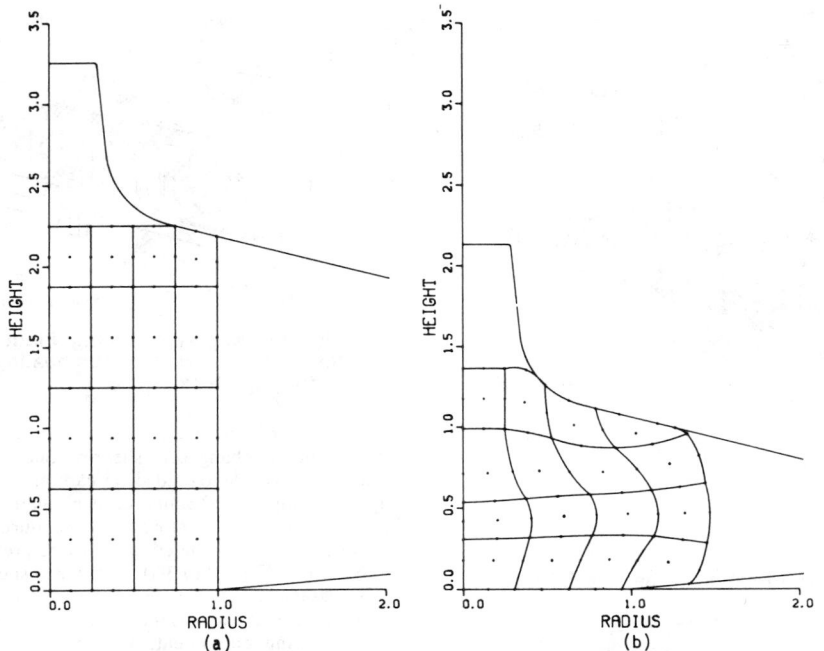

(a) Undeformed grid. (b) Deformation at a die stroke of one-half initial billet height. Temperature, 955 °C (1750 °F); m = 0.3.

Fig. 15. Simulation of axisymmetric spike forging

an "assumed" or "selected" blocker design, and the results can be displayed on a graphics terminal. If the simulation indicates that the selected blocker design would not fill the finisher die or that too much material would be wasted, then another blocker design can be selected and the computer simulation, or "trial," can be repeated. Such computer-aided simulations reduce the required number of expensive die tryouts.

HEATING EQUIPMENT

Furnaces for heating steel to forging temperatures may be batch-type or continuous. Batch furnaces are usually loaded and unloaded by hand, or by a robot, and heating time is determined on the basis of measurements or empirical guidelines. Continuous furnaces provide mechanical means of feeding the stock through the heating zone of the furnace, thus providing greater assurance of uniform heating.

Batch Furnaces. Most batch furnaces used for heating forging stock are of the slot type. Instead of a door, these furnaces have a horizontal opening across the front through which the forging stock is inserted and withdrawn. These furnaces are usually built on a steel frame and have a refractory hearth and firebrick sidewalls and roof. Slot-type batch furnaces may be oil-fired or gas-fired; burner units are normally placed in the sidewalls of the furnace. A gas curtain may be provided to minimize entry of oxygen into the furnace. Temperature can be controlled with a platinum/platinum-rhodium thermocouple, although radiation-type pyrometers are used more often. Such furnaces will heat 245 to 490 kg of steel per hour per square metre (50 to 100 lb per hour per square foot) of hearth.

Box furnaces of similar construction are also used for heating and may incorporate a door to exclude air, thereby minimizing heat losses.

Continuous furnaces are in two categories: (*a*) those that heat the entire piece of forging stock and (*b*) those that heat only part of the stock (localized heating). The common types of furnaces for heating the entire piece of stock are the continuous rotary-hearth and pusher furnaces. Those used for localized heating include screw-feed and conveyor types that hold the stock horizontally or vertically, or sometimes in an angular position.

Rotary-hearth furnaces can be almost any size, varying from a few feet to nearly 30 m (100 ft) in diameter. The hearth is usually doughnut-shaped, and can be rotated in either direction by means of a drive mechanism beneath the furnace. Hearth speed can be varied, as required, for heating stock of different sizes during one rotation. Stock may be manually or automatically loaded and unloaded in the same area. Rotary-hearth furnaces are widely used, especially for heating the entire piece of stock.

Pusher furnaces are customarily designed to heat steel for a particular type of part. The hearth can be either horizontal or sloped to assist movement of the stock through the furnace. The hearth brick will often be grooved to hold the circular or rectangular stock. Water-cooled skid rails are sometimes added to this type of furnace.

Pusher furnaces are rugged in construction and can usually accommodate large pieces of stock of more than one size. The stock is placed in compact rows on a loading table that is an extension of the hearth. The pusher mechanism, which can be hydraulic, mechanical or air-operated, is usually arranged to push one row at a time. At the discharge end, the stock can be removed by several methods: continuous conveyors outside or inside the furnace may be used to carry the stock away, or the stock can be pushed into discharge chutes, or removed manually.

Conveyor furnaces are similar to the pusher type to the extent that the stock is fed in at one end and removed at the other. In conveyor furnaces, conveyor chains, screw mechanisms, or hooks made of a heat-resisting alloy carry the stock through the heating zone. Conveyor-type furnaces are used extensively when stock requires only localized heating—as in hot upset forging. Two types of conveyor furnaces for localized heating are illustrated in Fig. 16 and 17. Bars are heated in the vertical position (Fig. 17) when floor space is limited. Automatic loading mechanisms can be used with either horizontal or vertical furnaces.

Atmosphere Control. An endothermic-gas generator is normally used to produce a gas that will be neutral to the steel being heated. Fuel-fired furnaces used with controlled or neutral atmospheres may be either of the full muffle type or heated by radiant tubes.

Most heating for forging is done with direct gas-fired or oil-fired furnaces. Gases present are principally carbon monoxide, carbon dioxide and water vapor (resulting from combustion of the fuel) and nitrogen. Carbon dioxide and water vapor are oxidizing agents, which cause decarburization and scaling of the steel. Carbon monoxide is a carburizing agent.

An optimum ratio of fuel to air is desired. Such a mixture will burn to produce a balance of gases that will result in an atmosphere essentially neutral (neither carburizing nor decarburizing) to the steel. Too little air results in inefficient heating and a thin tenacious scale on the steel; too much air also reduces heating efficiency and allows excessive scale formation.

Temperature Control. Best results in forging depend greatly on temperature control in heating the stock. The furnace should be equipped with pyrometric controls that can maintain temperature within ±5.6 °C (±10 °F). As the stock is heated to the pre-established temperature and is discharged, it should be checked occasionally (or checked automatically for each piece) with an optical pyrometer to determine whether the temperatures of the furnace and the stock are the same.

Induction heating is an efficient and economical method of heating stock for some high-volume applications. The lower the frequency, the greater the depth of flux penetration, and hence the greater the depth of heating. Higher frequencies cause the current to travel closer to the work surface, resulting in shallower heating. The heat always travels from about the surface toward the center, and for any specified cycle and power input the heating rate will depend on the electrical characteristics and thermal conductivity of the work material.

Induction heaters are built to a variety of sizes and shapes, determined by the shape of the workpiece, or of the portion of the workpiece, to be heated. An inductor can be a simple coil, or, for mass production, several heaters or coils can be connected and automated to provide a continuous supply of heated stock (Fig. 18). Electric current in the inductor is carried by water-cooled copper tubing. The copper inductor is insulated from the workpiece by a refractory material or by fitted brickwork. When feasible, the inductor should be placed near the source of electric power to minimize current losses.

The nature of the work determines the specific tools or equipment that will be needed, taking into consideration that the larger the workpiece to be heated, the lower the frequency.

Some advantages of the use of induction heating are:

- Accurate temperature control
- Negligible scale formation

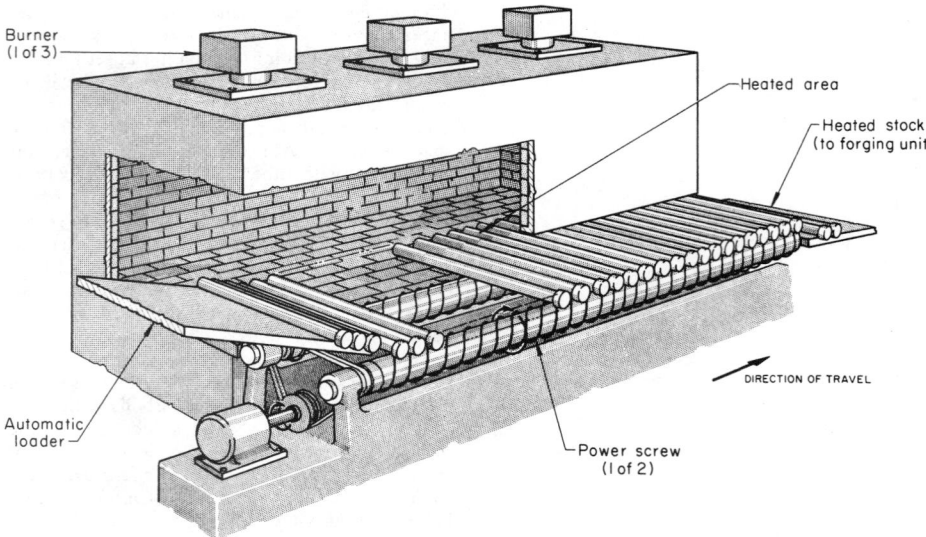

Fig. 16. Screw-feed conveyor furnace for localized heating of bars to be forged

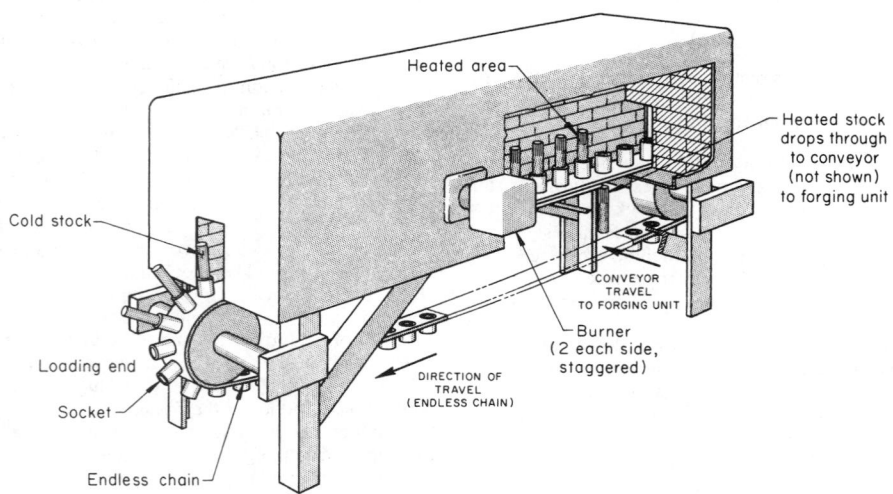

Fig. 17. Vertical chain-type conveyor furnace for localized heating of bars to be forged

- Ease of adapting to automated operation
- Better working conditions (cleaner, with less noise and heat)
- High efficiency per unit weight of steel heated
- Compactness of equipment.

The main disadvantages of induction heating are:

- High capital investment compared with fuel-fired furnaces
- Induction-heating installations are best adapted to mass production and consequently are less flexible than furnaces.
- Coil configurations are limited by practical considerations and by relative efficiency of power input.

An indication of the relative speed and efficiency of heating steel slugs by induction heating, compared with conventional furnace heating, was provided by a forge plant specializing in automotive components. At this plant, it was found that 34.1-mm, 2.9-kg (1^{11}/$_{32}$-in., 6.4-lb) and 35.7-mm, 3.6-kg (1^{13}/$_{32}$-in., 8-lb) round-corner square billets could be heated to forging temperature in 48 s by induction heating, but re-quired a 20-min heating cycle in conventional furnaces.

Complete automation can be achieved. Bar stock is loaded into a stock feeder that automatically sends the bars through a series of induction coils to the forging machine.

PREPARATION OF STOCK

Selection of the method for cutting the stock is based on several factors, which include: (*a*) number of blanks required, (*b*) thickness and hardness of the stock, (*c*) accuracy of cutting required and (*d*) whether the part will be forged from a flat position or pancaked (upset). The usual methods of preparing stock are described below.

Cold shearing of mill-length hot rolled bars is the most common method of stock preparation. It is usually economical to shear more than one bar in each cut, using multigrooved shear blades (provided that section thicknesses permit). It is common practice to use multiple shearing on carbon steel up to 75 mm (3 in.) round-corner square (rcs). Powered feed rolls with bar hold-down devices are usually needed for a good operation.

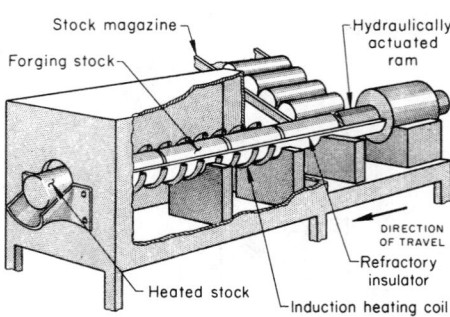

Fig. 18. Magazine-loading induction heater for continuous heating of forging stock

As the hardness of the work metal increases, the practice is changed to shearing one bar at a time. It is possible to cold shear carbon steel bars up to 125 mm (5 in.) round-corner square. However, when bars are stored outside during the winter, it is recommended that they be preheated to 65 to 150 °C (150 to 300 °F) prior to shearing.

Hot shearing is used to prepare stock of a size or hardness that exceeds the capacity of available cold shearing equipment. In most plants, hot shearing is required when bar or billet cross section (rcs) exceeds 89 mm (3^{1}/$_{2}$ in.).

Despite the advantages of shearing (cold or hot), the cut ends lack squareness. This may cause problems if the pieces of stock are forged by pancaking (forging by striking initial blows on the ends), depending on the ratio of length to cross section, and the degree of out-of-squareness of the sheared stock.

Cold sawing can be used for cutting of sizes not readily shearable, or when exact weight control is required. Sawing is relatively slow, and has high maintenance cost, compared with shearing.

Abrasive cutting is often used to prepare high-alloy and austenitic steels for forging. It is fast, it leaves a good finish, and kerf loss is normally less than in circular sawing. However, abrasive cutting is more limited in depth of cut than are band and circular sawing, because of the limits of wheel size—about 760-mm (30-in.) diameter. With a straight cutoff wheel, if the work metal does not revolve, the thickness of stock to be cut is limited to about one-third of the diameter of the abrasive wheel. If the work metal revolves, its thickness can be approximately two-thirds of the diameter of the cutoff wheel.

FORGING TEMPERATURES FOR STEEL

Maximum safe forging temperatures for carbon and alloy steels are given in Table 3, which

Table 3. Maximum safe forging temperatures for carbon and alloy steels of various carbon contents

| Carbon content, % | Maximum safe forging temperature | | | |
| | Carbon steels | | Alloy steels | |
	°C	°F	°C	°F
0.10	1290	2350	1260	2300
0.20	1275	2325	1245	2275
0.30	1260	2300	1230	2250
0.40	1245	2275	1230	2250
0.50	1230	2250	1230	2250
0.60	1205	2200	1205	2200
0.70	1190	2175	1175	2150
0.90	1150	2100	...	...
1.10	1110	2025	...	...

Table 4. Forging temperatures for tool steels

	Preheat slowly to:		Forging temperatures — Start forging at(a):		Do not forge below:	
Steels	°C	°F	°C	°F	°C	°F
Water-hardening tool steels						
W1 to W5 790	1450		980 to 1095(b)	1800 to 2000(b)	815	1500
Shock-resisting tool steels						
S1, S2, S4, S5 815	1500		1040 to 1150	1900 to 2100	870	1600
Oil-hardening cold work tool steels						
O1 815	1500		980 to 1065	1800 to 1950	845	1550
O2 815	1500		980 to 1040	1800 to 1900	845	1550
O7 815	1500		980 to 1095	1800 to 2000	870	1600
Medium-alloy air-hardening cold work tool steels						
A2, A4, A5, A6 870	1600		1010 to 1095	1850 to 2000	900	1650
High-carbon, high-chromium cold work tool steels						
D1 to D6 900	1650		980 to 1095	1800 to 2000	900	1650
Chromium hot work tool steels						
H11, H12, H13 900	1650		1065 to 1175	1950 to 2150	900	1650
H14, H16 900	1650		1065 to 1175	1950 to 2150	925	1700
H15 845	1550		1040 to 1150	1900 to 2100	900	1650
Tungsten hot work tool steels						
H20, H21, H22 870	1600		1095 to 1205	2000 to 2200	900	1650
H24, H25 900	1650		1095 to 1205	2000 to 2200	925	1700
H26 900	1650		1095 to 1205	2000 to 2200	955	1750
Molybdenum high speed tool steels						
M1, M10 815	1500		1040 to 1150	1900 to 2100	925	1700
M2 815	1500		1065 to 1175	1950 to 2150	925	1700
M4 815	1500		1095 to 1175	2000 to 2150	925	1700
M30, M34, M35, M36 815	1500		1065 to 1175	1950 to 2150	955	1750
Tungsten high speed tool steels						
T1 870	1600		1065 to 1205	1950 to 2200	955	1750
T2, T4, T8 870	1600		1095 to 1205	2000 to 2200	955	1750
T3 870	1600		1095 to 1230	2000 to 2250	955	1750
T5, T6 870	1600		1095 to 1205	2000 to 2200	980	1800
Low-alloy special-purpose tool steels						
L1, L2, L6 815	1500		1040 to 1150	1900 to 2100	845	1550
L3 815	1500		980 to 1095	1800 to 2000	845	1550
Carbon-tungsten special-purpose tool steels						
F2, F3 815	1500		980 to 1095	1800 to 2000	900	1650
Low-carbon mold steels						
P1	...		1205 to 1290	2200 to 2350	1040	1900
P3	...		1040 to 1205	1900 to 2200	845	1550
P4 870	1600		1095 to 1230	2000 to 2250	900	1650
P20 815	1500		1065 to 1230	1950 to 2250	815	1500
Other alloy tool steel						
6G 815	1500		1040 to 1150	1900 to 2100	845	1550

(a) The temperature at which to start forging is given as a range, the higher side of which should be used for large sections and heavy or rapid reductions, and the lower side for smaller sections and lighter reductions. As the alloy content of the steel increases, the time of soaking at forging temperature increases proportionally. Likewise, as the alloy content increases, it becomes more necessary to cool slowly from the forging temperature. With very-high-alloy steels, such as high speed steels and air-hardening steels, this slow cooling is imperative in order to prevent cracking and to leave the steel in a semisoft condition. Either furnace cooling of the steel or burying it in an insulating medium (such as lime, mica, or diatomaceous earth) is satisfactory. (b) Forging temperatures for water-hardening tool steels vary with carbon content. The following temperatures are recommended: for 0.60 to 1.25% C, the range given; for 1.25 to 1.40% C, the low side of the range given.

shows that forging temperature decreases as carbon content increases. The higher the forging temperature, the greater the plasticity of the steel, which results in easier forging (see Fig. 2) and less die wear; however, the danger of overheating and excessive grain coarsening is increased. If a steel that has been heated to its maximum safe temperature is forged rapidly and with large reduction, the energy transferred to the steel during forging can substantially increase its temperature, thus causing overheating.

The effect of carbon content on forging temperature is the same for most tool steels as for carbon and alloy steels. However, the complex alloy compositions of some tool steels have different effects on forging temperature. Forging temperatures for tool steels are listed in Table 4.

Heating Time. For any steel, the heating time must be sufficient to bring the center of the forging stock to forging temperature. A longer heating time than necessary results in excessive decarburization, scale and grain growth. For stock up to 75 mm (3 in.) in diameter, the heating time per inch of section thickness should be no more than 5 min for low-carbon and medium-carbon steels, or than 6 min for low-alloy steel. For stock 75 to 230 mm (3 to 9 in.) in diameter, the heating time should be no more than 15 min per inch of thickness. For high-carbon steels (0.50% C and higher), and for highly alloyed steels, slower rates are required, and sometimes preheating at temperatures from 650 to 760 °C (1200 to 1400 °F) is necessary to prevent cracking.

Finishing temperature should always be well above the transformation temperature of the steel being forged, to prevent cracking of the steel and excessive wear of the dies, but should be low enough to prevent excessive grain growth. For most carbon and alloy steels, 980 to 1095 °C (1800 to 2000 °F) is a suitable range for finish forging.

DESCALING BEFORE FORGING

Prevention of scale formation during heating, or removal of scale between heating and forging, will result in longer die life, smoother surfaces on the forging, and improved dimensional control. In addition, a scaly forging makes hot inspection unreliable and increases cleaning cost.

When controlled heating methods that minimize scale formation are not available, other methods can be used to remove scale from the heated metal prior to forging. These are: (a) busting and blowoff, (b) wire brushing or other mechanical methods, and (c) spraying a jet of water on the hot piece.

Busting and Blowoff. Busting is a preliminary operation in which the scale is broken. For instance, in flat forging, the edging operation breaks off the scale, which is then blown away by steam or air jet, manually or automatically. In pancake forging of round-corner squares, scale is first broken off each end by lightly flat forging two diagonally opposite edges; this is also known as a busting operation. In pancake forging of round stock, the scale is broken from the sides by an initial upsetting blow.

Mechanical Descaling. In one effective method, the heated stock is brushed with rotating wire brushes. Another form of mechanical descaling, applicable to round stock, is to shape a knifelike section to fit the heated stock; the stock is scraped across the knife edge to dislodge and remove the scale. For example, if a round bar is to be descaled, a knife section with the concave shape of a half circle is used. The heated bar is placed in the half-circle knife section and drawn through the knife to remove the scale from half of the bar. The bar is then rotated 180°, and the operation is repeated. This method is economical but less effective than wire brushing.

Water-jet descaling is another effective method. Four (or more) high-pressure nozzles are used, so that the water strikes simultaneously on all sides of the heated stock. The nozzles are usually in a cabinet, which is shielded at the opening into which the heated stock is inserted. Water is supplied to the nozzles at a pressure of 8.3 to 12.4 MPa (1200 to 1800 psi). The nozzles vary in size: 0.76 by 1.27 mm (0.030 by 0.050 in.) is usual for stock diameters from 38 to 75 mm (1½ to 3 in.). A 35° angle of the water stream (from the surface on which the stream is directed) usually provides the most efficient descaling. Water is used for only a fraction of a second, to prevent excessive cooling of the workpiece.

CONTROL OF DIE TEMPERATURE

Dies should be heated to at least 120 °C (250 °F), and preferably to 205 to 315 °C (400 to 600 °F), before forging begins. Heating is done by placing "warmers" (pieces of hot metal) between the die faces, or by using gas torches. Dies are sometimes heated in ovens before being placed in the hammer or press. Temperature-indicating crayons can be used to measure surface temperature. Failure to warm the dies is likely to result in die breakage when the hot billet is placed on them.

Operating Temperature. Normal hammer-forging and press-forging practices do not include special methods for cooling the dies—their mass and the lubricant usually provide natural cooling, and keep them within a safe operating range (usually 315 °C, or 600 °F, maximum). However, maximum operating temperature depends greatly on the die-steel composition, on the amount of lubricant used and on the stroking rate. Higher temperatures may be permitted for the higher-alloy steels such as H11. In no event should any portion of the die be operated at a temperature higher than that at which it was tempered. Most dies are tempered at 540 to 595 °C (1000 to 1100 °F), and sometimes higher; thus, the danger of exceeding the temperature is not great. However, the hardness *at working temperature* varies a great deal for different steels. For instance, high-alloy steels such as H11 and H21 retain a greater percentage of their room-temperature hardness at working temperature than do low-alloy die steels such as 6G.

TRIMMING

Method of trimming depends mainly on the quantity of forgings to be trimmed, the size of the forgings, and available equipment. A specific trimming procedure can sometimes eliminate a machining operation.

For small quantities or for large forgings, sawing or other machining operations are frequently used to remove the flash. For large quantities, the cost of trimming dies can usually be justified. Most forgings produced in closed dies are die trimmed.

With respect to die trimming, forging materials can be divided into two groups: those that can be trimmed cold and those that should be trimmed hot. Almost all materials can be cold trimmed, but some must have special treatment after forging and prior to cold trimming. Generally, a forging can be cold trimmed satisfactorily if the work metal in the "to be trimmed" condition has a tensile strength of not more than 690 MPa (100 ksi) or a hardness of not more than 207 HB.

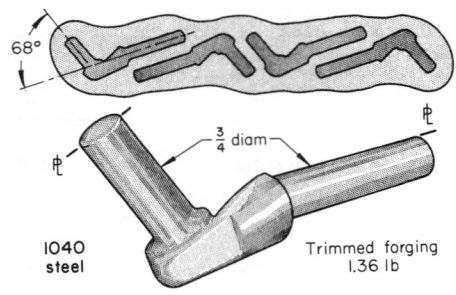

Stock size	$1^5/_{32}$-in. round-corner square, $21^3/_4$ in. long
Stock weight	8.16 lb(a)
Heating method	9-ft rotary furnace
Heating temperature	2250 °F
Descaling	Water spray, 1800 psi
Forging-press capacity	2000 tons
Forging-die material	6F3 at 40 to 42 HRC
Forging sequence	Edge, block and finish
Trimming-press capacity	150 tons(b)
Die material, hot trimming	1045 steel, hard faced
Die life, hot trimming	40 000 pieces
Trimmed weight	1.36 lb per forging
Lubricant	Graphite-water
Production rate	725 pieces per hour

(a) Yielded four forgings. (b) A 6-in. stroke was used for cold trimming one part at a time, and an 8-in. stroke for hot trimming four parts at a time.

Fig. 19. Arrangement of impressions for four-at-a-time forging and hot trimming of control-arm shafts which formerly had been cold trimmed one at a time

Cold trimming usually refers to trimming of metal flash at a temperature below 150 °C (300 °F). This method is extensively used, especially for small forgings. An advantage of cold trimming is that it can be done at any time — it need not be a part of the forging sequence, and no reheating of the forgings is needed. Another advantage is that the trimming blades can be adjusted so that the flash is sheared to a smooth surface that does not require machining.

A disadvantage of cold trimming is the power required; for thick flash, hot trimming often must be used because of a lack of press capacity for cold trimming.

Hot trimming is done at temperatures as low as 150 °C, or 300 °F (for nonferrous alloys) and as high as 980 °C (1800 °F) or above for steels and other ferrous alloys.

For some applications, there are reasons other than power consumption for using hot trimming, as is described in Fig. 19.

COOLING PRACTICE

Cooling in still air or in factory tote boxes is common practice and is usually satisfactory for carbon steel or low-alloy steel forgings when cross sections are no greater than approximately 64 mm (2 1/2 in.). Flaking may occur on larger forgings when they are air cooled. Flakes (also called "shatter cracks" or "snowflakes") are short, discontinuous internal fissures attributed to stresses produced by localized transformation and decreased solubility of hydrogen during cooling. In a fractured surface, flakes appear as bright silvery areas; on an etched surface, they appear as short cracks. Flaking indicates the need for cooling to at least 175 °C (350 °F) in a furnace, or by burying the piece in sand or slag. An alternative method for treating large forgings made of alloy steels such as 4340 consists of cooling in air to about 540 °C (1000 °F), followed by isothermal annealing at 650 °C (1200 °F).

Forgings of alloy tool steel should always be cooled slowly, as is recommended above for larger forgings of carbon and alloy steels.

Open-Die Forging

OPEN-DIE FORGING is variously known as hand, smith, hammer and flat-die forging. Forgings are made by this process when: (*a*) the forging is too large to be produced in closed dies; (*b*) mechanical properties of the work metal can be developed by open-die forging that cannot be obtained by machining from a bar or billet; (*c*) the quantity required is too small to justify the cost of closed dies; or (*d*) the delivery date is too close to permit the making of dies for closed-die forging. All forgeable metals can be forged in open dies.

Size and Weight. The size of a forging that can be produced in open dies is limited only by the capacity of available equipment for heating, handling and forging. Items such as marine propeller shafts, which may be several feet in diameter and as much as 75 ft long, are forged by open-die practice. Conversely, forgings no more than a few inches in maximum dimension are also produced in open dies.

Although an open-die forging may weigh as little as a few pounds or as much as 300 tons, probably 80% of all open-die forgings (number of forgings, not total weight) weigh between 30 and 1000 lb each.

Shapes. Highly skilled hammer and press operators, with the aid of various auxiliary tools, can produce relatively complex shapes in open dies. However, because forging of complex shapes is time-consuming and expensive, such forgings are produced only under unusual circumstances. Most open-die forgings have the following shapes:

- Rounds, squares, rectangles, hexagons and octagons forged from billet stock, either to develop mechanical properties that are superior to those of rolled bars or to provide these shapes in compositions for which the shapes are not readily available as rolled products
- Simple pancake forgings, made by upsetting a length of stock. Finished parts made from these forgings include gears, wheels, milling cutters, and tube sheet blanks.
- Hub forgings that have a small diameter adjacent to a large diameter and that are produced in small quantities
- Spindle, pinion gear, and rotor forgings. These forgings are for shaftlike parts and have their major or functional diameters either in the center or at one end with one or more smaller diameters extending from one or both sides of the major diameter in shaftlike extensions.
- Forged and pierced blanks, for subsequent conversion to rolled rings
- Various basic shapes that are developed between open dies with the aid of "loose" tooling. Depending on the design of the tooling, these forgings may be of the open-die type, or they may be closed-die blocker-type forgings.

HAMMERS AND PRESSES

The principles of operation of hammers and presses are discussed in the article "Hammers and Presses for Forging." In open-die forging, it is necessary to control the stroke, the position and the speed of the ram in order to obtain acceptable precision and quality in the forging parts. Therefore, in general, only air- or oil-driven power hammers and hydraulic presses are suitable for open-die forging. Modern open-die forging installations use direct-driven hydraulic presses with pulldown frame design and with quick die-changing mechanisms. In this type of press, the cylinder crosshead, which is located below floor level, is rigidly connected to the press columns. This assembly is movable and is well guided. The

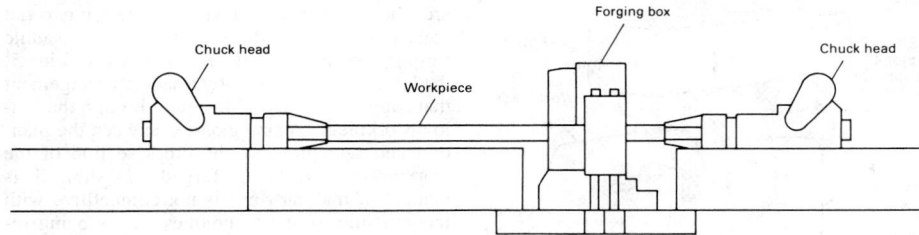

Fig. 1. Radial forging machine with two chuck heads (manipulators) for automatic operation under numerical control

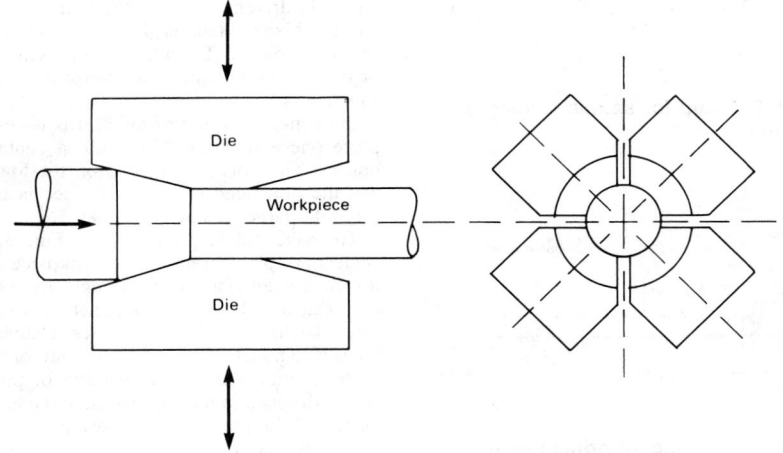

Fig. 2. Principles of radial forging process for producing tubes and shafts with variations in diameter along the axis

for producing large forgings in small quantities. When large-quantity production is required—such as for ingot reduction for manufacture of titanium billets or tool steel shapes—then it is justified to use fully automated radial forging machines (see Fig. 1). Such machines use two, four or six dies to produce solid or hollow round, square, rectangular or profiled sections (see Fig. 2). Such a typical machine has a forging box (Fig. 1) that contains the drive mechanism, which consists of eccentric shafts that drive the dies at rates of 250 to 1800 strokes per minute, depending on machine size. The stroke length is relatively short (a few inches). The stroke position of the opposite hammers and the axial and rotational motions of the manipulators are computer controlled during forging so that shafts with varying diameters along the axis can be easily produced.

DIES

Most open-die forgings are produced in a pair of flat dies—one attached to the hammer or to the press ram, and the other to the anvil. Swage dies (curved), V-dies, and V-die and flat-die combinations are also used. These four types of die sets are shown in Fig. 3. In some applications, forging is done with a combination of a flat die and a swage die. Dies are attached to platens and rams by either of the methods shown by the two views at the left in Fig. 3. Figure 3 also shows several types of dies that are held on the anvil manually, by means of handles.

Information on die materials, die parallelism, and die life for open-die forging is presented in the article "Dies and Die Materials for Hammer and Press Forging."

AUXILIARY TOOLS

To assist in forging production, mandrels, saddle supports, sizing blocks (spacers), ring tools, bolsters, fullers, punches, drifts (expansion tools),

center of gravity of such a press is low, and its stiffness is relatively great. Direct-driven hydraulic presses offer better control of ram speed and ram position. In modern installations this feature is very significant, because the press is usually integrated with a manipulator. The stroke

position of the press and the motions of the manipulator are computer controlled. As a result, the entire open-die forging operation can be programmed for given initial and final stock shape and workpiece material.

Integrated open-die forging installations are used

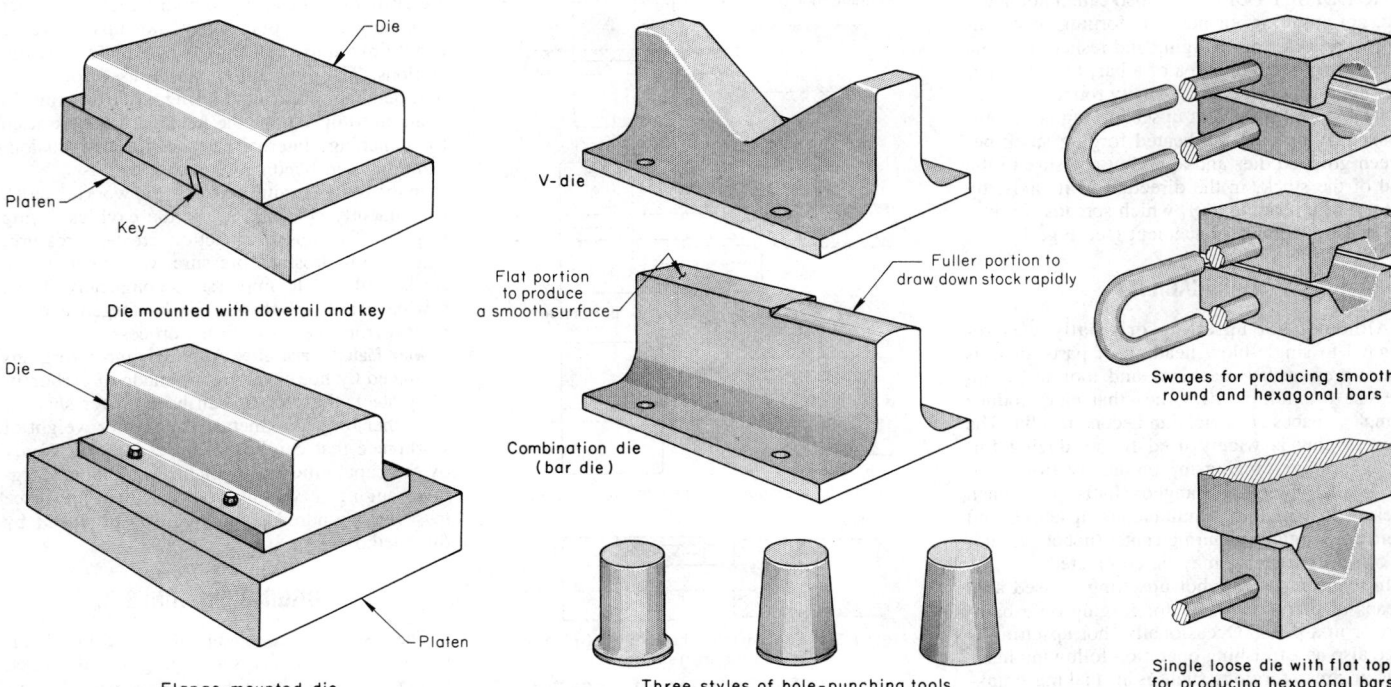

Fig. 3. Typical dies and punches used in open-die forging

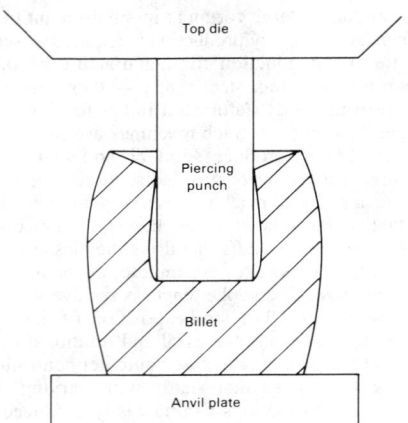

Fig. 4. Illustration of technique for piercing a hole in an upset forging using a straight punch (in preparation of a preform for ring rolling)

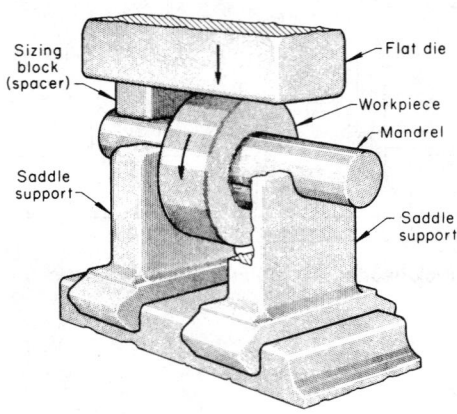

Fig. 5. Setup for saddle forging of a ring

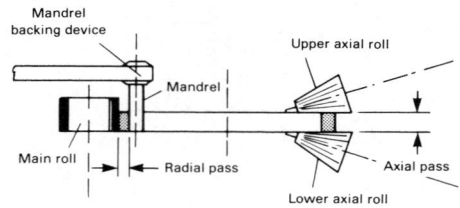

Fig. 6. Operational principles of a horizontal ring-rolling mill

and a wide variety of special tools (for producing special shapes) are used. Because most auxiliary tools are exposed to heat, they are usually made from the same steels as the dies.

An open-die forging can be made with an upper die that is flat, while the lower die is replaced with another type of tool. Often two or more hammers or presses and die setups are needed to complete a shape (or operations are done at different times in the same hammer or press by changing the tooling). For example, large rings are made by upsetting the stock between two flat dies, punching out the center, and then saddle forging (see Fig. 4 and 5). As shown in Fig. 5, the lower die is replaced by a saddle arrangement that supports a mandrel inserted through the hollow workpiece. A sizing block between the mandrel and ram prevents the cross section of the workpiece from being forged too thin. This method of making rings is not competitive with modern ring-rolling techniques and is being replaced by the use of horizontal and vertical ring-rolling mills. In a horizontal mill (see Fig. 6), the pierced blank is placed over a mandrel. The mandrel moves laterally toward the main roll, which is driven. As a result, both the mandrel and the blank rotate as the cross section of the blank is reduced. The axial rolls provide support of the deforming ring and control its width and squareness.

In open-die forging of hubs, it is necessary to place one end of the billet into a container or bolster. The bolster is placed on the lower die, and the other end of the workpiece is upset to forge the larger-diameter flange.

To make holes, punches (see Fig. 3, lower center) are placed on the hot workpiece and are driven through, or partly through, by a ram. A hole can also be made by punching from both sides. The method used to produce relatively deep holes is to punch from both sides until only a thin center section remains. A handful of powdered coal is dropped in the depression in the hot workpiece and the punch is again driven in. The gas created by the powdered coal increases the pressure and helps to force out the center slug.

Hot Upset Forging

HOT UPSET FORGING (also called hot heading, hot upsetting, or machine forging) is essentially a process for enlarging and reshaping some of the cross-sectional area of a bar, tube or other product form of uniform, usually round, section. In its simplest form, hot upset forging is accomplished by holding the heated forging stock between grooved dies and applying pressure to the end of the stock, in the direction of its axis, by the use of a heading tool, which spreads (upsets) the end by metal displacement (see Fig. 1).

APPLICABILITY

Although hot upsetting originally was restricted to single-blow heading of parts such as bolts, present-day machines and tooling permit the use of multiple-pass dies that can produce complex shapes accurately and economically. The process now is widely used for producing finished forgings ranging in complexity from simple headed bolts or flanged shafts to wrench sockets that require simultaneous upsetting and piercing. Forgings requiring center (not at bar end) or offset upsets also may be completed.

In many instances, hot upsetting is used as a means of preparing stock for forging on a hammer or in a press. Occasionally, hot upsetting is used also as a finishing operation following hammer or press forging, such as in making crankshafts.

Because the transverse action of the moving die and longitudinal action of the heading tool

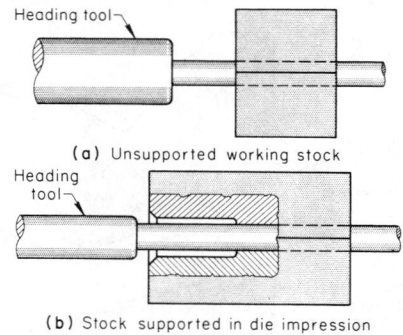

(a) Unsupported working stock

(b) Stock supported in die impression

(c) Stock supported in heading tool recess

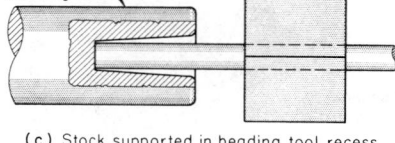

(d) Stock supported in heading tool recess and die impression

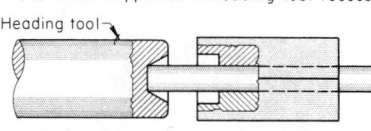

Fig. 1. Basic types of upsetter heading tools and dies, showing the extent to which stock is supported

are available for forging in both directions, either separately or simultaneously, hot upset forging is not limited to simple gripping and heading operations. The die motion can be used for swaging, bending, shearing, slitting and trimming. In addition to upsetting, the heading tools are used for punching, internal displacement, extrusion, trimming and bending.

In the upset forging process, the working stock is frequently confined in the die cavities during forging. The upsetting action creates pressure, similar to hydrostatic pressure, which causes the stock to fill the die impressions completely. Thus, a wide variety of shapes can be forged and removed from the dies by this process.

Work Material and Size. Although most forgings produced by hot upsetting are made of carbon or alloy steel, the process can be used for shaping any other forgeable metal. The size or weight of workpiece that can be hot upset is limited only by the capabilities of available equipment; forgings ranging in weight from less than an ounce to several hundred pounds can be produced by this method.

FORGING MACHINES

The essential components of a typical horizontal machine for hot upset forging are illustrated in Fig. 2. These machines are mechanically operated from a main shaft with an eccentric drive that operates a main slide, or header slide, horizontally. Cams drive a die slide, or grip slide, which moves horizontally at right angles to the header slide, usually through a toggle mechanism. The action of the header slide is similar to that of the ram in a mechanical press. Power is supplied to a machine flywheel by an electric motor. A flywheel clutch provides for "stop motion" operation, placing movement of the slides under operator control.

Forging takes place in three die elements: two gripper dies (one stationary and one moved by the die slide), which have matching faces with horizontal grooves to grip the forging stock and hold it by friction; and a heading tool, or "header," which is carried by the header slide in the plane of the work faces of the gripper dies and aligns with the grooves in these dies (see Fig. 3). The travel of the moving die is designated as the "die opening," and its timed relation to the movement of the header slide is such that the dies close during the early part of the header-slide stroke. The part of the forward header-slide stroke that takes place after the dies are closed is known as the "stock gather," and the amount that the returning header slide travels before the moving die starts to open is called the "hold-on," or the "hold."

The die opening determines the maximum diameter of upset that can be formed on a given machine, transferred between the dies and withdrawn through the throat, without pushing the workpiece forward and lifting it out over the top. The diameter of the stock, rather than the stock gather, determines the amount of stock that can be upset; the stock gather, however, has an important bearing on the depth to which internal displacement can be carried. The height of the die determines the number of progressive operations that can be accommodated in one set of dies.

Operation. The basic actions of the gripper dies and the header tools of an upsetter can be demonstrated by the three-station setup shown in Fig. 3. The stock is positioned in the first (topmost) station of the stationary die of the machine.

During the upset forging cycle, the movable die slides against the stationary die to grip the stock. The header tool, fastened in the header slide, advances toward and against the forging stock to spread it into the die cavity. When the header punch retracts to its back position, the movable die slides to the open position to release the forging. This permits the operator to place the partly forged piece into the next station, where the cycle of the movable die and header tool is repeated. Many forgings can be produced to final shape in a single pass of the machine. Others may require as many as five passes for completion.

TOOLS

The four basic types of upsetter heading tools and dies, shown schematically in Fig. 1, differ in operating principle as follows:

1. Tooling does not support exposed working stock (Fig. 1a). Stock is held by the gripper dies, and the heading tool advances to upset the exposed stock.
2. Stock is supported in the gripper-die impression (Fig. 1b). Using repeated blows, great lengths of stock can be upset by this method. The diameter of the preceding upset becomes the diameter of the working stock for the next pass.
3. Stock is supported in a recess in the heading tool, which is shaped like the frustum of a cone (Fig. 1c). Stock is gathered in the re-

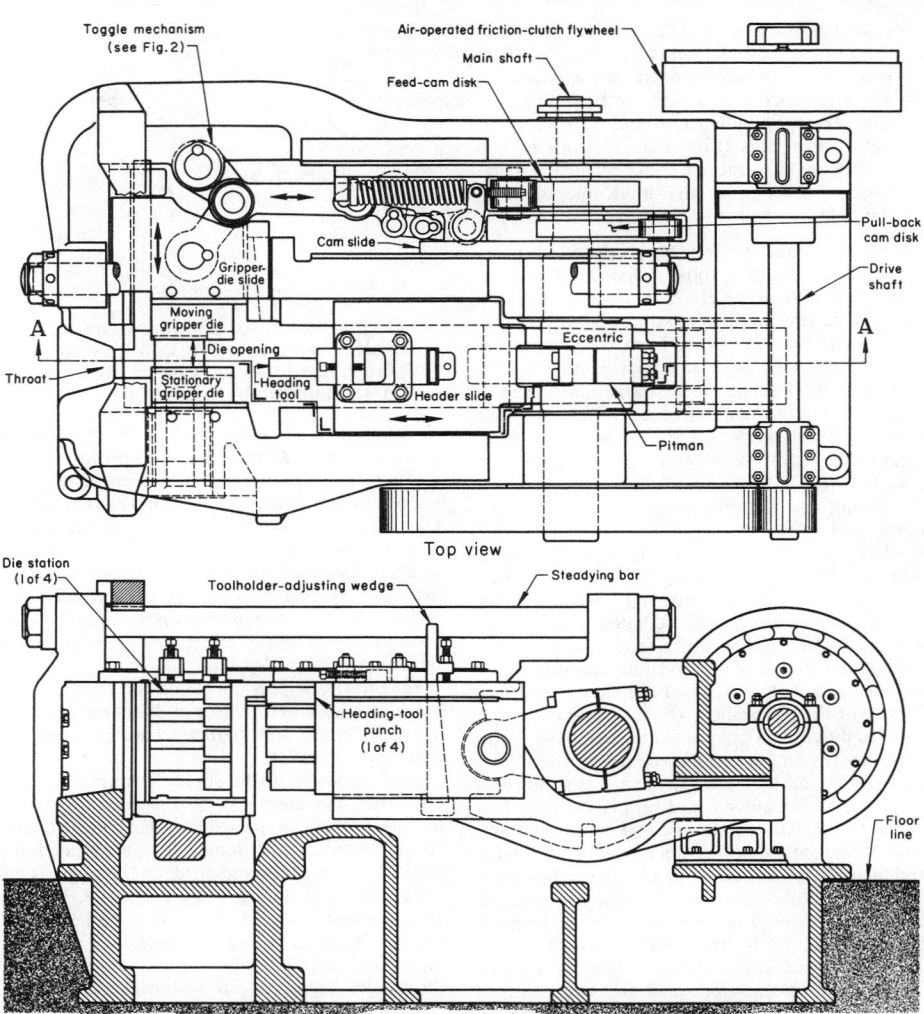

Fig. 2. Principal components of a typical horizontal machine for hot upset forging with a vertical four-station die. See text for description of operation

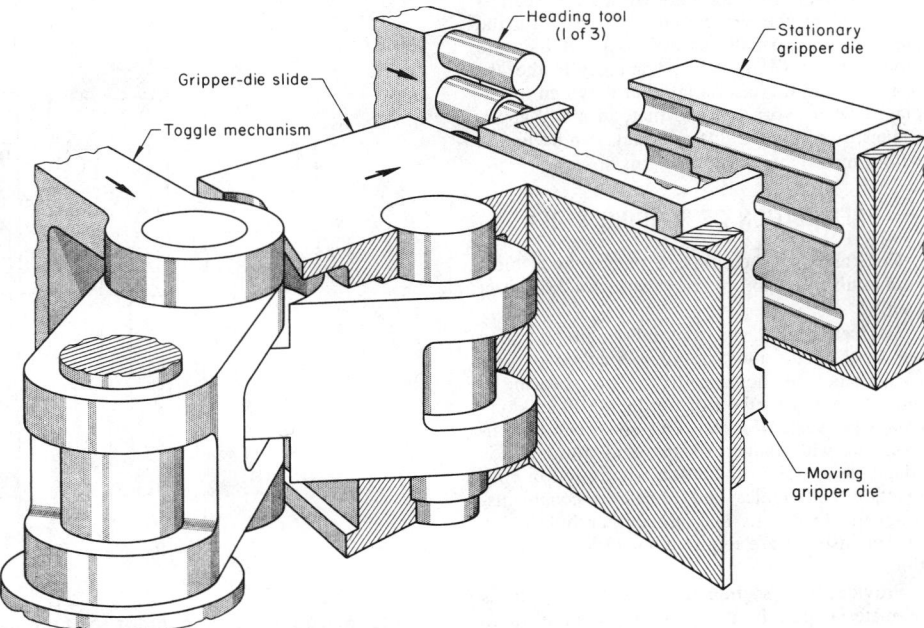

Fig. 3. Basic actions of the gripper dies and heading tools of an upsetter

cessed heading tool. This method is widely used when large amounts of stock must be gathered, as in forging of transmission shafts.

4. Stock is supported in both the frustum-shape recess of the heading tool and in recesses of the gripper dies (Fig. 1d). This method is widely used to achieve a better balance of metal displacement, especially in development of intricate, difficult-to-forge shapes.

Although some forgings are produced by a single stroke of the ram, most shapes require more than one pass to complete. The upsetter dies may incorporate several different impressions or "stations." The stock is moved from one impression (or station) to the next in sequence, to give the forging a final shape. Each move constitutes a "pass." Three or more passes are commonly used to complete the upset, and if flash removal (trimming) is a part of the forging operation, another pass is added.

Piercing and shearing passes can also be incorporated in the dies. In single-blow solid-die machines, the gripper dies are replaced by a shear arm and shear blade. A long heated bar of forging stock is placed in a slot and pushed against a stop. As the machine is actuated—either automatically or by means of a foot pedal—a motion similar to that of a conventional upsetter occurs, except that, instead of the dies closing, a section of the bar is sheared off. While the shear slide is moving, a cam actuates a transfer arm, which moves until it contacts the stock. The stock, now positioned between the shear blade and the transfer arm, is moved into proper position between the punch and the die. As the punch advances and contacts the stock, the shear blade and the transfer arm move apart. The punch continues its advance, and the forging is produced in a single blow. Ejector pins push the forging from the die and it drops onto an underground conveyor. Another heated bar of forging stock is pushed against the stop, and the cycle is repeated.

Tool Materials. For short runs, it is common practice to use solid dies made of lower-alloy steels such as 4340, 6G or 6F3. For runs of about 1000 pieces, higher-alloy hot work steels such as H11, 6H1 or 6H2 are commonly used—for the die if dies are small, or for inserts if dies are large. The two important advantages in the use of punch and die inserts are that they can be replaced when worn out and that, in many applications, two or more different parts can be forged with a master block by changing inserts.

PREPARATION OF FORGING STOCK

Cold and hot shearing are the methods most commonly used to prepare blanks for hot upset forging.

Cold shearing of blanks from mill-length hot rolled bar stock is the most common method of preparing stock for hot upsetting. Cold shearing is the most rapid method of producing blanks and entails no waste of metal. One shear can accommodate a wide range of sizes, and equipment is adaptable to mass production when used in conjunction with tables and transfer mechanisms. Magnetic feed rolls and proper bar hold-down devices usually are required for efficient operation.

Provided that section thickness and hardness of material permit, it is usually economical to shear as many bars in one cut as possible, using

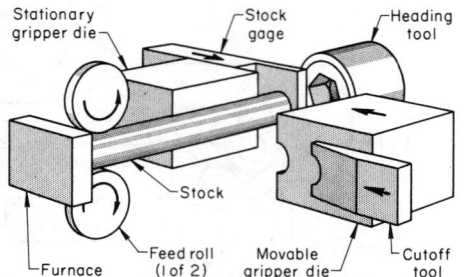

Fig. 4. Setup for simultaneous upsetting and cutoff of continuously fed heated mill lengths of stock in a semiautomatic header

multiple-grooved shear blades. It is common practice to use multiple shearing on low-carbon steel up to 50 mm (2 in.) in diameter.

For small-diameter blanks, it is often advantageous to use coiled cold drawn wire. This is straightened and cut off, and the blanks are stacked, by means of high-speed machines. The use of blanks made from wire is especially beneficial when shank diameter on the upset forging must be held to closer tolerances than can be obtained with hot rolled bars.

Hot shearing is recommended for cutting bars more than 89 mm ($3\frac{1}{2}$ in.) in diameter, and may be used for smaller-diameter bars in semiautomatic operations.

For diameters up to about 29 mm ($1\frac{1}{8}$ in.), and when the upset can be made in one blow, the preliminary preparation of individual blanks can be avoided. Mill-length bars are heated and fed into a semiautomatic header. The blank is cut off at the same time the upset is made. A stock gage between the gripper dies and the header die locates the stock before it is held by the gripper dies. The gage, mounted on a slide actuated by the header slide, retracts as the header tool ad-

vances. A typical tooling arrangement is shown in Fig. 4.

HEATING METHODS

The methods and temperatures used for heating the stock prior to hot upsetting are the same as those used in other forging processes. Temperatures required for forging various metals are given in articles in this volume dealing with the forging of specific metals.

Induction heating has been successfully adapted for use in upset forging, and offers the following advantages over other heating methods:

1. The fast heating rate obtained with induction (heating times of 10 to 20 s are common for stock under 50 mm, or 2 in., diameter) results in a minimum amount of scaling.
2. Induction heating permits accurate temperature control.
3. Induction heating is readily adaptable to localized heating.
4. Induction equipment requires less floor space than furnaces.

Although all types of furnaces (box, pusher, rotary-hearth, slot and conveyor furnaces) are used for heating for upset forging, accuracy of temperature control and adaptability to localized heating make induction heating particularly useful in forging of similar parts on a mass-production basis.

DIE COOLING AND LUBRICATION

Normal practice is to keep dies below 205 °C (400 °F) during operation. In some low-production operations, no coolant is required for keeping dies below this temperature. In most applications, however, a water spray (sometimes containing a small amount of salt or graphite) is used as a coolant.

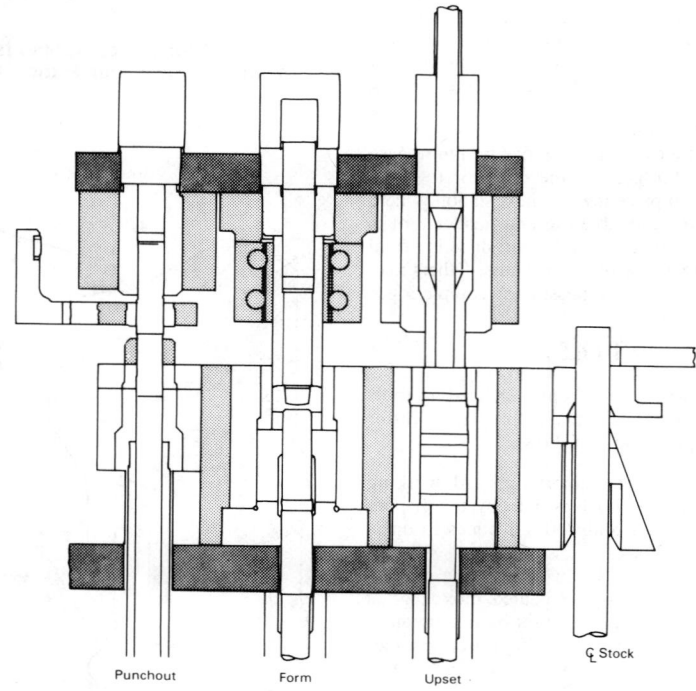

Fig. 5. Tooling arrangement for nutmaking in a three-die hot former (courtesy of National Machinery Co., Tiffin, OH)

Die lubrication slows production, and is not widely used in upsetting of steel. Because of the die action in upsetting, parts are less likely to stick than in hammer or press forging. In deep punching and piercing, however, sticking may be encountered, thus requiring a lubricant. An oil-graphite spray is an effective lubricant and may simultaneously provide adequate cooling. A recirculated suspension of alumina in water is used in some high-producton operations. In one high-production automotive shop, a recirculated suspension of water and graphite is sprayed onto the dies after each forging cycle and serves as both coolant and lubricant.

AUTOMATIC UPSET FORGING

For high production rates, fully automated hot forging machines ("hot formers") are used. These machines, originally developed for making nuts, can be used for producing a large variety of round, or nearly round, parts at very high production rates (60 to 150 parts per minute, depending on size).

A typical sequence of hot forming operations for nutmaking is shown in Fig. 5. The long bar is induction heated and sheared to length. The first blow upset forges and sizes the blank. The second blow forms a web and further sizes the material. During the last blow, the web is punched out and the part is ejected. The stock is transferred from one forming station to the next automatically by means of a transfer mechanism and "fingers" that are cam driven.

Forging of Stainless Steels, Heat-Resisting Alloys and Nonferrous Alloys

Forging of Stainless Steels

TECHNOLOGY for forging of stainless steels (i.e., billet heating, die temperatures, die-lubrication techniques and forging machines) is essentially the same as that for forging of carbon and alloy steels. The main difference is that stainless steels have higher flow stresses at forging temperatures; as a result, the amounts of pressure, load and energy required for forging a part of given configuration are greater for stainless steels than for carbon and alloy steels.

FORGING CHARACTERISTICS

Stainless steels can be divided into three broad categories: (a) martensitic stainless steels, including types 410, 414, 416, 420, 440A, 440B, 440C, Greek Ascoloy and Lapelloy C; (b) austenitic stainless steels, or 300 series alloys; and (c) precipitation-hardening stainless steels, such as AM-350, AM-355, 17-7 PH and 17-4 PH.

High-chromium martensitic stainless steels have forging characteristics similar to those of low-alloy steels. Because they are higher in chromium

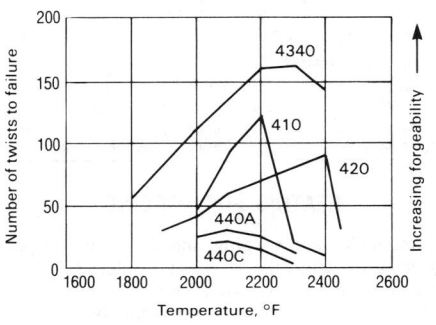

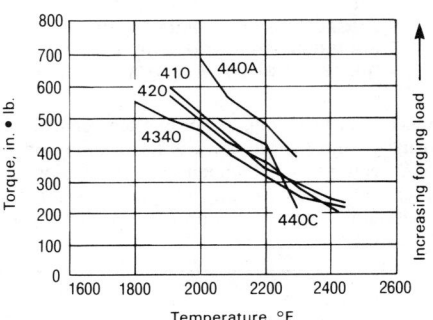

Source: A Laboratory Evaluation of the Hot Working Characteristics of Metals, by C. L. Clark and J. J. Russ, *Trans. AIME*, Vol 167, 1946, p 736–748.

Fig. 2. Forging characteristics of several martensitic stainless steels and of 4340 steel, as measured in hot-twist tests

content, however, their forging-load requirements are 30 to 50% higher. Maximum forging temperatures are generally 55 to 165 °C (100 to 300 °F) lower than those for low-alloy steels. Forging-temperature ranges recommended for various stainless steels are given in Fig. 1. Figure 2 compares hot-twist properties (twist ductility and torque) of several martensitic stainless steels with those of 4340 steel. This comparison is an indication of relative forgeability. Forging temperatures and relative forging loads (relative to 4340) are given for various martensitic alloys in Table 1. Martensitic stainless steels are characterized by high hardenability, and therefore they are subject to cracking and must be cooled slowly from the forging temperature.

Table 1. Forging temperatures and relative forging loads (relative to type 4340 steel) for various martensitic stainless steels

Material	Forging temperature °C	Forging temperature °F	Relative forging load
4340	1260	2300	1.0
410	1175	2150	1.2 to 1.5
414	1175	2150	1.3 to 1.5
416	1175	2150	1.3 to 1.5
Greek Ascoloy ...	1205	2200	1.5 to 1.7
420	1205	2200	1.1 to 1.2
440A	1150	2100	2.0
440B	1150	2100	2.0
440C	1120	2050	2.0
Lapelloy C	1230	2250	2.0

Austenitic stainless steels are more difficult to forge than carbon and low-alloy steels: they require greater loads, and their forging-temperature ranges are narrower. For example, forging pressures for type 304 stainless steel are given in Fig. 3 as a function of upset reduction and in Fig. 4 as a

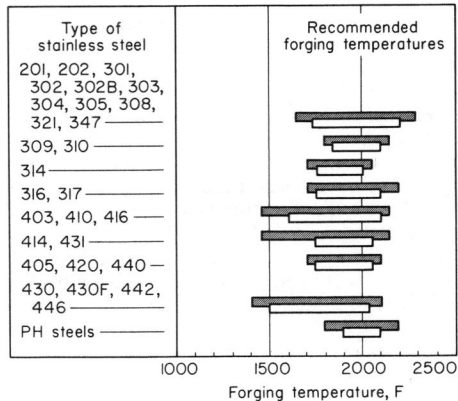

Crosshatched bars show temperature ranges that have been recommended by some plants but on which there is not general agreement. Open bars indicate generally accepted ranges.

Fig. 1. Recommended forging and finishing temperatures for stainless steels

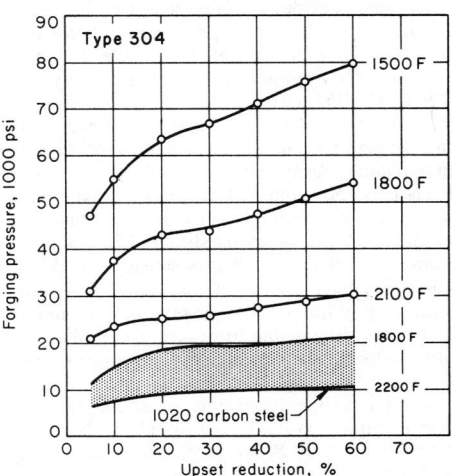

Source: *Forging Materials and Practices*, by A. M. Sabroff, F. W. Boulger and H. J. Henning, Reinhold, New York, 1968.

Fig. 3. Effect of upset reduction on forging pressure for various temperatures

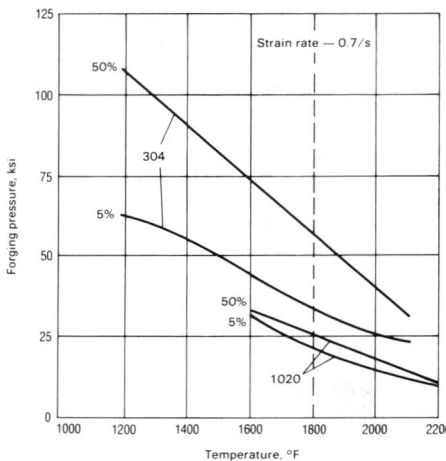

Source: Same as for Fig. 3.

Fig. 4. Comparison of forging-pressure requirements for type 304 stainless steel and type 1020 carbon steel at several temperatures

function of temperature. Forging temperatures are usually adjusted downward for forging operations requiring small amounts of deformation. For example, typical forging temperatures for type 304 stainless steel for various operations would be as follows:

Operations	Forging temperature °C	°F
Severe reductions (ingot breakdown, roll forging, drawing, blocking, back extrusion, etc.)	1260	2300
Moderate reductions (finish forging, upsetting, etc.)	1205	2200
Slight reductions (coining, restriking, end upsetting, etc.)	1120	2050

Similar adjustments are made for other austenitic grades. The temperatures used for slight reductions seldom exceed 1150 °C (2100 °F), regardless of the alloy being forged.

Precipitation-Hardening Stainless Steels. The relative forging behavior of precipitation-hardening stainless steels is illustrated in Table 2, which compares the forging characteristics of alloys 17-7 PH, AM-355 and 17-4 PH with those of 4340 steel. Because of the combination of lower forging temperature and greater stiffness, forging of precipitation-hardening stainless steels requires loads 30 to 50% higher than those for type 4340 steel, and, accordingly, heavier equipment is needed. On the other hand, the PH grades of stainless steel are much less sensitive to decarburization than are the higher-carbon low-alloy steels. In addition, they do not scale as much, and thus it is possible to design some precipitation-hardening stainless steel forgings for use with as-forged surfaces.

Forging temperatures recommended for several precipitation-hardening stainless steels are given in Table 3. When forging is done rapidly with large reductions in a single operation, it is good practice to lower the forging temperature by 55 °C (100 °F) or more in order to prevent formation of additional delta ferrite resulting from increases in temperature during forging. This is particularly true when forging is done in high-energy-rate machines.

Table 2. Comparison of forging characteristics of three precipitation-hardening stainless steels with those of type 4340 steel

Characteristic	4340	17-7 PH	AM-355	17-4 PH
Forging temperature, °C (°F)	1260 (2300)	1175 (2150)	1175 (2150)	1175 (2150)
Decarburization	High	Low	Low	Low
Scaling	High	Low	Low	Low
Grain-size control	Excellent	Fair	Fair	Good
Forgeability	Excellent	Fair	Good	Good
Forging pressure (relative)	1.0	1.4	1.4	1.4
Thermal cracking	Low	None	Low	Medium
Die wear	Low	Medium	Medium	Medium

Table 3. Recommended forging temperatures for precipitation-hardening stainless steels
Source: Same as for Fig. 3.

Alloy	Maximum forging temperature °C	°F	Light(a) °C	°F	Moderate(b) °C	°F	Severe(c) °C	°F	Variable(d) °C	°F
Semiaustenitic grades											
AM-350	1175	2150	1150	2100	1175	2150	1175	2150	1150	2100
AM-355	1205	2200	1095	2000	1175	2150	1175	2150	1095	2000
17-7 PH	1205	2200	1120	2050	1175	2150	1205	2200	1065	1950
PH 15-7 Mo	1230	2250	1095	2000	1150	2100	1175	2150	1095	2000
Martensitic grades											
Stainless W	1230	2250	1120	2050	1205	2200	1205	2200	1120	2050
17-4 PH	1205	2200	1150	2100	1175	2150	1175	2150	1150	2100

(a) Up to 15%. (b) 15 to 50%. (c) Over 50%. (d) Refers to forgings receiving widely differing reductions. End upsets, for example, receive large reductions on the upset end while the shaft may remain essentially undeformed.

HEATING FOR FORGING

Recommended forging and finishing temperatures for most of the standard stainless steels are indicated in Fig. 1, which represents temperatures used in a number of different forge plants. The crosshatched bars show temperature ranges that have been used in some plants, but on which there is not general agreement. Open bars represent the generally accepted ranges.

The thermal conductivity of stainless steels is lower than that of carbon or low-alloy steels. Therefore, stainless steels take longer to reach the forging temperature. However, they should not be soaked at forging temperature, but should be forged as soon as possible after reaching it. The exact time required for heating stock of a given thickness to the established forging temperature depends on the type of furnace used.

Equipment. Gas-fired and electrically heated furnaces are used with equal success for heating the stock. The gas employed should be essentially free from hydrogen sulfide and other sulfur-bearing contaminants. Oil-fired furnaces are widely used for heating the series 400 stainless steels and the 18-8 varieties, but because of the danger of contamination from sulfur in the oil they are considered unsafe for heating the high-nickel grades.

Although not absolutely necessary, heating of stainless steel is preferably done in a protective atmosphere. When gas heating is used, an acceptable protective atmosphere can usually be obtained by adjusting the fuel-to-air ratio. When the furnace is heated by electricity, the protective atmosphere (if used) must be separately generated.

As is the case with all steels, induction heating is used whenever it can be justified economically.

DIE LUBRICATION

Die-lubrication practice in forging of stainless steels is very similar to that used in forging of alloy steels.

Glass is sometimes used as a lubricant in press forging, being applied either by dipping the heated forging in molten glass or by sprinkling it with glass frit. Glass is an excellent lubricant, but its viscosity must be compatible with the forging temperature used. For best results, the viscosity of the glass should be maintained at 0.1 Pa·s (100 cP). Thus, when different forging temperatures are used, a variety of glass compositions must be stocked. Another disadvantage of glass is that it will accumulate in deep cavities, solidify, and impair metal flow. Therefore, the use of glass is generally confined to making shallow forgings that require maximum lateral flow.

TRIMMING

When production quantities justify the cost of tools, forgings are trimmed in dies. Hot trimming is preferred for all types of stainless steel, because less power is required and because there is less danger of cracking than in cold trimming. The precipitation-hardening stainless steels *must* be hot trimmed to prevent flash-line cracks, which can penetrate the forging.

Frequently, it is practical to hot trim immediately after the forging operation, before the workpiece temperature falls below a red heat. Less often, forgings are reheated to 900 to 955 °C (1650 to 1750 °F) and then trimmed.

Tool Materials. Punches for hot trimming of closed-die forgings are often made of 6G or 6F2 die block steel at 388 to 429 HB, and the blades are made of a high-alloy tool steel, such as D2, at 58 to 60 HRC. In some forge shops, both punches and blades for hot trimming are made of a carbon or low-alloy steel (usually with less than 0.30% C) and then hard faced, generally with a cobalt-base alloy (a typical composition is: 1.10 C, 30 Cr, 3 Ni, 4.50 W, rem Co).

Upset forgings can be hot trimmed in a final pass in the upsetter, or in a separate press. For trimming in the upsetter, H11 tool steel at 46 to 50 HRC has performed successfully on a variety of forgings with a normal flash thickness. For

Table 4. Cycle for sodium hydride (reducing) descaling of annealed stainless steel forgings

Operation sequence	Bath composition	Bath temperature °C	°F	Treatment time, min
Descale	1.5 to 2.0% NaH	400 to 425	750 to 800	20
Quench	Water (circulated in tank)	Cold	Cold	1 to 3
Acid clean	10% H_2SO_4	63	145	20
Acid brighten	10% HNO_3 – 2% HF	63	145	30
Rinse	Water (high-pressure spray)	Ambient	Ambient	2
Rinse	Water	79	175	1 to 2

trimming of heavy flash in the upsetter, H21 at 50 to 52 HRC is recommended. Tools for hot trimming in a separate press usually are made of a 0.30%-C carbon or low-alloy steel, and are hard faced with a cobalt-base alloy (typical composition: 1.10 C, 30 Cr, 3 Ni, 4.50 W, rem Co).

CLEANING

Stainless steels do not form as much scale as carbon or alloy steels, especially when a protective atmosphere is provided during heating. However, the scale that does form is tightly adherent, hard and abrasive. It must be removed prior to machining, or tool life will be severely impaired.

Mechanical or chemical methods, or a combination of both, can be used to remove scale. Abrasive blast cleaning is an efficient method, and is applicable to forgings of various sizes and shapes, in large or small quantities. When surfaces will not be machined or passivated, blasting must be done only with silica sand; the use of steel grit or shot will contaminate the surfaces and impair corrosion resistance.

Abrasive blast cleaning is usually followed by acid pickling. The forgings are then thoroughly washed in water.

Barrel finishing (tumbling) is sometimes used for descaling. Acid pickling is recommended after tumbling.

Wire brushing is sometimes used for removing scale from a few forgings. Brushes with stainless steel wire must be used unless the forgings will be machined or passivated.

Salt bath descaling followed by acid cleaning and brightening is an efficient method of removing scale. A typical procedure is detailed in Table 4.

Forging of Heat-Resisting Alloys

HEAT-RESISTING ALLOYS, because of their greater strengths at elevated temperatures, are more difficult to forge than most other metals. Some iron-base heat-resisting alloys, such as 19-9 DL and A-286, are similar to austenitic stainless steels in forgeability. Most heat-resisting alloys, however, are more difficult to forge than stainless steels, although they can be forged by open-die or closed-die forging, upsetting, extrusion forging, roll forging or ring rolling. Often, two or more of these methods are used in sequence.

FORGING ALLOYS

The heat-resisting alloys most commonly forged are listed in Table 5, together with their forging temperatures and forgeability ratings.

Forging temperatures given in Table 5 are the temperatures of the billets as they are removed

Table 5. Forging temperatures and forgeability ratings for heat-resisting alloys

Alloy	Upset and breakdown °C	°F	Finish forging °C	°F	Forge-ability rating(b)
Iron-base alloys					
A-286	1095	2000	1040	1900	1
V-57	1095	2000	1040	1900	1
16-25-6 ...	1095	2000	1095	2000	1
19-9 DL ..	1150	2100	1095	2000	1
Nickel-base alloys					
Alloy R-235 ..	1205	2200	1205	2200	3
Astroloy ..	1120	2050	1120	2050	5
Hastelloy W	1205	2200	1040	1900	4
Hastelloy X	1175	2150	1175	2150	3
Inconel 600	1150	2100	1040	1900	1
Inconel 700	1120	2050	1105	2025	4
Inconel 718	1095	2000	1040	1900	2
Inconel X-750 ..	1175	2150	1120	2050	2
Inconel 751	1150	2100	1150	2100	3
Incoloy 901	1150	2100	1095	2000	2
M-252	1150	2100	1095	2000	3
René 41 ...	1150	2100	1120	2050	4
U-500	1175	2150	1175	2150	3
U-700	1120	2050	1120	2050	5
Waspaloy	1165	2125	1040	1900	3
Cobalt-base alloys					
J-1570	1175	2150	1175	2150	2
J-1650	1150	2100	1150	2100	2
HS-25 (L-605)	1230	2250	1230	2250	3
S-816	1150	2100	1150	2100	4

(a) Lower temperatures are often used for specific forgings when structural uniformity is a requirement. (b) Based on considerations discussed in text. As the rating increases, forgeability decreases.

from the furnace. Forging should begin immediately, with a loss in temperature of no more than 42 °C (75 °F). Forging can be continued until the stock has cooled 110 °C, or 200 °F (or more for some alloys) below the temperatures given in Table 5, without damage to the work metal. However, because greater pressures are required, forging is seldom done at temperatures substantially lower than those in Table 5.

Forgeability Rating. In the forgeability ratings listed in Table 5, alloy A-286 is assigned an arbitrary value of 1, because it is one of the most forgeable of the heat-resisting alloys. The other alloys listed are assigned values that are multiples of 1, depending on their forgeability (as the arbitrary number increases, forgeability decreases). In establishing the forgeability ratings, the power required is a minor consideration. Forgeability is determined by the ease with which a given shape can be formed. The alloys that are difficult to forge generally require more blows and, conse-

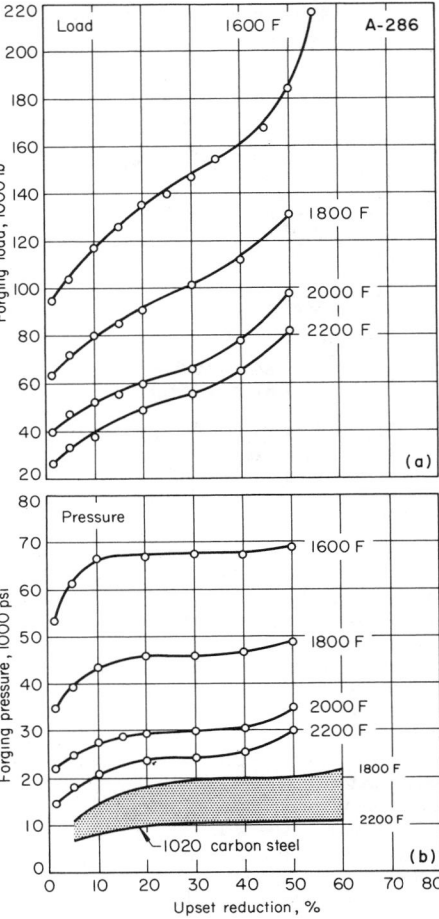

Source: "A Study of Forging Variables," by H. J. Henning, A. M. Sabroff and F. W. Boulger, U.S. Air Force Report ML-TDR-64-95, 1964, p 32.

Fig. 5. Effect of upset reduction at four temperatures on (a) forging load in forging of A-286 and (b) forging pressure for A-286 compared with that for 1020 carbon steel

quently, more operations than the more easily forgeable alloys. The values of 1 to 5 in Table 5 are based on the difficulties that are encountered in finishing the forging operation.

Rate of die deterioration, number of rejected forgings, and number of blows required for producing a given shape increase as forgeability decreases. These factors have been considered in establishing the ratings in Table 5.

Iron-Base Alloys. Stock for forgings of the iron-base alloys is generally furnished as press-forged squares or hot rolled rounds, depending on size. As-cast ingots are sometimes used.

Temperature has an important effect on forgeability. The optimum temperature range for forging A-286 and similar iron-base alloys is narrow. Forgeability of A-286, based on forging load required for various upset reductions at four forging temperatures, is shown in Fig. 5(a). Figure 5(b) shows that, on the basis of forging pressure, A-286 is considerably more difficult to forge than 1020 steel, even though A-286 is among the most forgeable of the heat-resisting alloys (Table 5). For instance, as shown in Fig. 5(b), 1020 steel at 1205 °C (2200 °F) requires only about 70 MPa

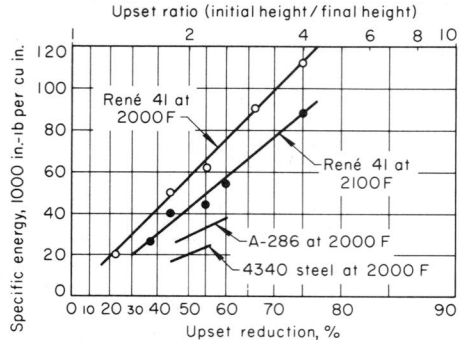

Source: *Forging Materials and Practices*, by A. M. Sabroff, F. W. Boulger and H. J. Henning, Reinhold, New York, 1968.

Fig. 6. Effects of upset reduction and forging temperature on specific energy required for forging two heat-resisting alloys and 4340 steel

(10 ksi) for an upset reduction of 30%, whereas for the same reduction, A-286 at 1205 °C requires approximately 170 MPa (25 ksi).

Figure 6 shows that, on the basis of the specific energy required for various percentages of upset reduction, A-286 needs approximately 50% more energy than 4340 steel when both are forged at 1095 °C (2000 °F). However, A-286 requires only about half the specific energy that René 41 requires, for the same upset reduction and the same forging temperature (Fig. 6).

Yield strengths—and, as a result, forging pressures—are highly dependent on forging temperature, as shown in Fig. 7 for several wrought nickel-base superalloys. Figure 7 also illustrates that, in the temperature range from 815 to 980 °C (1500 to 1800 °F), yield stress increases very rapidly as a result of slight decreases in temperature.

Strain rates also influence forging pressures. Figure 8 shows that as strain rate increases, more energy is required in presses and hammers.

Nickel-base alloys initially consisted of a few simple nickel-chromium alloys hardened by small additions of titanium and aluminum, for service up to 760 °C (1400 °F). With the development of production vacuum-melting techniques, workable alloys can be produced containing relatively large amounts of titanium, aluminum, zirconium, niobium and other reactive elements. The levels of nitrogen and oxygen are reduced by vacuum melting, which eliminates most of the nitrides and oxides that contribute to poor forgeability. Hence, the "second generation" of nickel-base alloys consists of numerous compositions containing larger amounts of hardening elements.

As shown in Table 5, all but one of the nickel alloys are less forgeable than the iron-base alloys—almost all require more force for producing a given shape. Astroloy and U-700 are the two most difficult-to-forge nickel-base alloys. For a given percentage of upset reduction at a forging temperature of 1095 °C (2000 °F), these alloys require about twice the specific energy needed for the iron-base alloy A-286 (see Fig. 6).

In the forgeability ratings listed in Table 5, Astroloy and U-700 alloys have about one-fifth the forgeability of Inconel 600. However, these ratings reflect only a relative ability to withstand deformation without failure; they do not indicate the energy or pressure needed for forging, nor

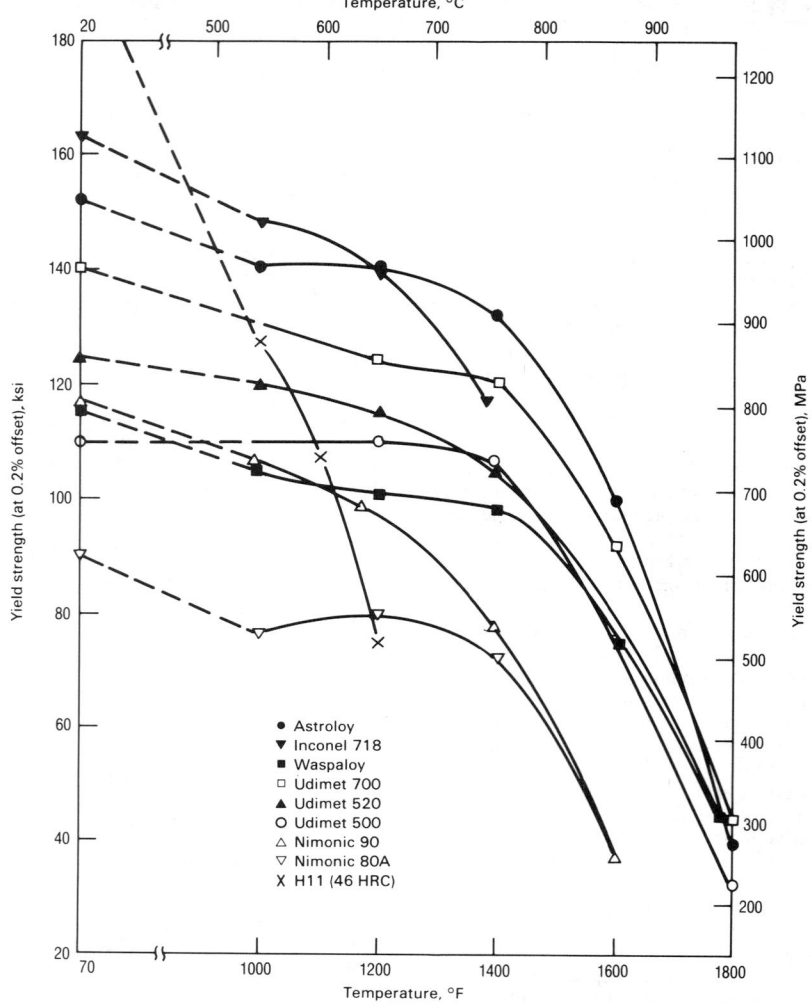

Source: "High Temperature, High Strength Nickel-Base Alloys," The International Nickel Co., Inc., 1977.

Fig. 7. Yield strengths of several wrought nickel-base superalloys (H11 tool steel included for comparison)

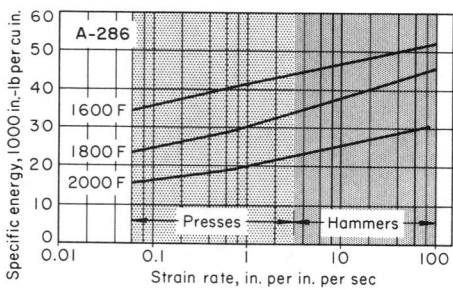

Source: Same as for Fig. 6.

Fig. 8. Effect of strain rate on specific-energy requirements in press and hammer forging of A-286 at three different temperatures

can the ratings be related to low-alloy steels and other alloys that are considerably more forgeable.

Forging of nickel-base alloys requires close control over both metallurgical and operational conditions. Particular attention must be given to control of work-metal temperature. Observers are usually required, to record data on transfer time, soaking time, finishing temperature and percentage reduction. Critical parts frequently are num-

bered, and precise records are kept. These records are useful in determining the cause of defective forgings, and they permit metallurgical analysis so that defects can be avoided in future products.

The nickel-base alloys are sensitive to minor variations in composition, which can cause large variations in forgeability, grain size and final properties. In one instance, wide heat-to-heat variations in grain size occurred in parts forged from Incoloy 901 in the same sets of dies. For some parts, optimum forging temperatures had to be determined for each incoming heat of material, by making sample forgings and examining them after heat treatment for variations in grain size and other properties.

Fine-grain-size nickel-base alloys produced by powder metallurgy techniques exhibit uniform structures. These alloys are particularly sensitive to variations in strain rate and temperature, as illustrated in Fig. 9 for alloy IN-100. Comparison of these data with data for cast IN-100 illustrates that the PM product has flow stresses lower than those of the cast product. In addition, the PM alloy exhibits very large values of elongation at low strain rates, which make this material suitable for slow deformation under isothermal conditions—i.e., dies at the same

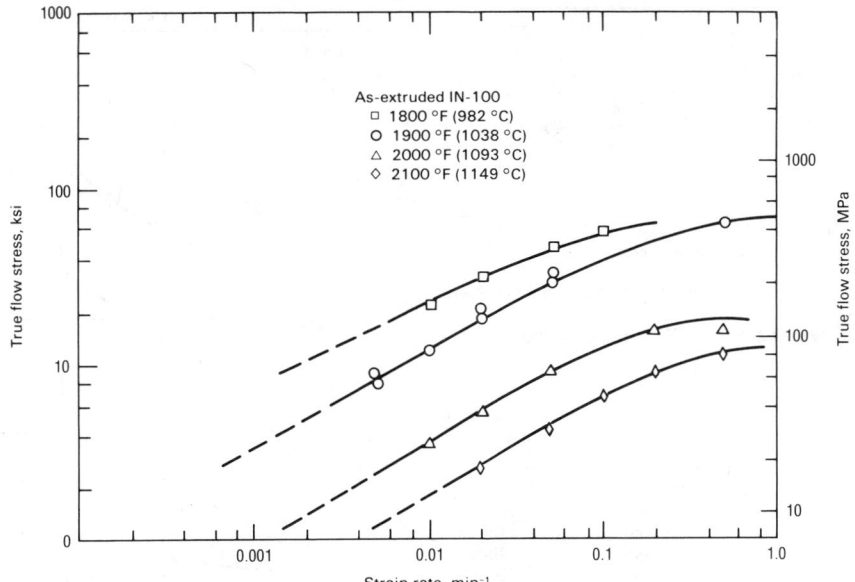

Source: Properties of IN-100 Processed by Powder Metallurgy, by L. N. Moskowitz, R. M. Pelloux and N. Grant, in *Superalloys – Processing*, Proc. 2nd Int. Conf., MCDC Report 72-10, Battelle, Columbus, OH.

Fig. 9. Flow-stress values for as-extruded IN-100 at various temperatures, plotted as log true flow stress vs log strain rate

temperature as that of the workpiece, to avoid die chilling.

Cobalt-Base Alloys. Many of the cobalt-base alloys cannot be forged successfully, because they ordinarily contain more carbon than the iron-base alloys and, therefore, greater quantities of hard carbides, which impair forgeability.

The four cobalt-base alloys listed in Table 5 are forgeable. The strengths of these alloys at elevated temperatures, including the temperatures at which they are forged, are considerably higher than those of iron-base alloys; hence, the pressures required in forging them are several times greater than those required for iron-base alloys.

Even when forged at their maximum forging temperature, alloys S-816 and HS-25 work harden; thus, forging pressure must be increased with greater reductions. Accordingly, these alloys generally require frequent reheating during forging, to promote recrystallization and lower the forging pressure for succeeding steps.

Forging conditions (temperature and reduction) have a significant effect on the grain size of cobalt-base alloys. Because low ductility, notch brittleness and low fatigue strength are associated with coarse grains, close control of forging and of final heat treatment is important.

Cobalt-base alloys are susceptible to grain growth when heated above about 1175 °C (2150 °F). They heat slowly and require a long soaking time for temperature uniformity. Forging temperatures and reductions, therefore, depend on the forging operation and on part design.

CONVENTIONAL FORGING

In conventional forging of heat-resisting alloys, well-known forging presses and hammers are used. The die materials and die-heating practices are the same as those used in forging high-alloy steels or stainless steels. However, more power and higher forging pressures are needed to forge the same shape from a heat-resisting alloy than from an alloy steel. Thus, required die pressures often may be so high that the desired

part definition may not be obtainable without risking die breakage. Die wear can be a major problem, and die lubrication is very critical— especially in conventional press forging.

ISOTHERMAL FORGING

Heat-resisting alloys, especially the fine-grain-size PM alloys based on nickel, are suitable for isothermal forging because they exhibit (a) a strong dependence of flow stress on temperature and (b) high strain-rate sensitivity at low strain rates. Isothermal forging eliminates die chilling; and thus forging can be carried out at low ram speeds (or strain rates). Under such forging conditions the material has high forgeability and low flow stress, which result in enhanced die filling and lower die pressures. Thus, fine details of the forged part are formed under creeplike conditions. For isothermal forging, hydraulic presses with accurate speed controls are used; typical ram speeds are on the order of 0.04 to 0.4 mm/s (0.1 to 1.0 in./min). At the beginning of the forging operation, the part usually is thick and has a small plan area, and relatively little pressure is required to deform it. Consequently, forging can be started at a high ram speed, which can be lowered at later stages of the ram stroke in order to reduce forging pressure and prevent plastic deformation of dies.

In isothermal forging, dies must maintain their strength at temperatures from 980 to 1095 °C (1800 to 2000 °F). The most widely used material in this application is the molybdenum alloy TZM, which can be used at temperatures up to 1205 °C (2200 °F). This alloy, however, is very reactive with oxygen at these high temperatures and thus requires the use of a protective atmosphere (such as argon or nitrogen) or vacuum around the die system during forging. As expected, the costs of the TZM die material and of equipment for oxidation protection are quite high. However, these costs are often justifiable because isothermal forging allows closer part definition—i.e., near-net-shape forging—than is possible with conventional forging.

HEATING FOR FORGING

Atmosphere protection for heating the forging stock is desirable but not essential, because heat-resisting alloys have high resistance to oxidation at elevated temperature. Protective atmospheres provide cleaner surfaces on finished forgings, and therefore minimize subsequent cleaning problems.

Furnaces. Electrically heated furnaces are often preferred for heating forging stock, because their temperatures can be closely controlled and the possibility of contaminating the work metal is minimized. Fuel-fired furnaces are used, but to a lesser extent than those heated by electrical resistance. If fuel-fired furnaces are used, it is mandatory that the fuel have an extremely low sulfur content, especially when heating nickel-base alloys, or contamination may occur.

TRIMMING

Small closed-die forgings of heat-resisting alloys are usually die trimmed, particularly if quantities are large. Die trimming is invariably done hot, because hot trimming requires less power than cold trimming. Also, the danger of cracking the forgings is minimized when trimming is done hot. As long as the forging temperature is not exceeded, there will be no adverse effects from hot trimming. Forgings are frequently hot trimmed immediately after the finish forging operation.

Most punches for hot trimming are made of 6G or 6F2 tool steel at a hardness of about 388 to 429 HB. Most trimming blades are made of a high-alloy tool steel, such as D2 at 58 to 60 HRC.

CLEANING

Scale and lubricants can be removed by chemical methods. Salt bath descaling, followed by acid pickling and finally sand blasting, is the procedure most widely used.

Mechanical methods, such as sand blasting, are used for cleaning heat-resisting alloy forgings, but usually as a supplement to the chemical methods. A final sand blasting operation helps to produce an attractive finish. It is mandatory that sand be used rather than steel grit.

Forging of Aluminum Alloys

ALUMINUM ALLOYS can be forged into essentially the same shapes as are forgeable from low-carbon steel. However, for a given forging shape, the pressure requirements vary over a wide range, depending primarily on chemical composition of the alloy and forging temperature. This is illustrated in Fig. 10, where forging-pressure requirements, as they are related to forging temperature and upset reduction, are compared for two aluminum alloys and for 1020 steel.

FORGING ALLOYS AND TEMPERATURES

Fifteen aluminum alloys most commonly used for forgings are listed in Table 6. Generally, all of these alloys can be forged to the same severity, although more power or more blows are required for some than for others.

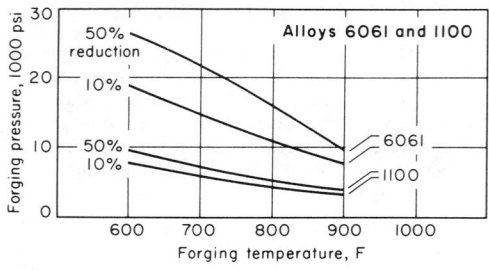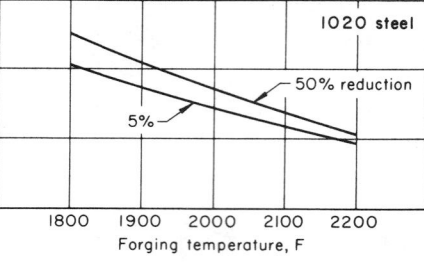

Source: *Forging Materials and Practices,* by A. M. Sabroff, F. W. Boulger and H. J. Henning, Reinhold, New York, 1968

Fig. 10. Forging pressures for two aluminum alloys and for 1020 steel at various upset reductions and forging temperatures

Table 6. Forging-temperature ranges for aluminum alloys

| Alloy | Forging temperature | | Alloy | Forging temperature | |
	°C	°F		°C	°F
1100	315-405	600-760	4032	415-460	780-860
2014	420-460	785-860	5083	405-460	760-860
2025	420-450	785-840	6061, 6151	430-480	810-900
2218	405-450	760-840	7039	380-440	720-820
2219	425-470	800-880	7075	380-440	720-820
2618	410-455	770-850	7079	405-455	760-850
3003	315-405	600-760	X7080	370-440	700-825

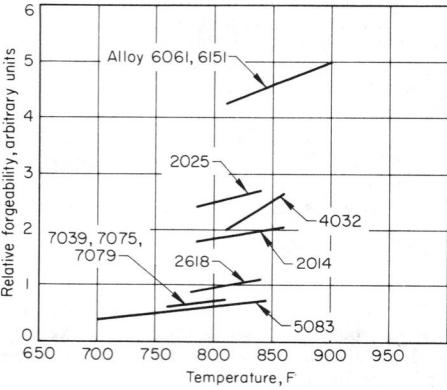

Forgeability increases as the arbitrary unit increases.

Fig. 11. Forgeabilities and forging temperatures of ten aluminum alloys

Temperature ranges for forging these alloys are included in Table 6. The recommended forging-temperature range for most of the alloys is relatively narrow (less than 55 °C, or 100 °F, for several alloys; for no alloy is the range greater than 90 °C, or 160 °F). Maintaining temperature is not usually a problem, because dies are heated to about the same temperature as that of the stock, and this minimizes temperature loss.

FORGEABILITY

Figure 11 illustrates the relative forgeability—i.e., the ability to deform without fracture—of the ten alloys that comprise the major tonnage of aluminum alloy forgings. The arbitrary units shown on the vertical axis of Fig. 11 are based principally on deformation per unit of energy absorbed at the various temperatures ordinarily used for forging these alloys. The difficulty of forging the alloys to specific degrees of severity was also considered in establishing these arbitrary units.

Alloys 1100 and 3003 are the most easily forged and would be rated higher than 5 units in Fig.

11 (forgeability increases as the arbitrary unit increases), but both of these alloys have limited use as forgings, because they cannot be strengthened by heat treatment.

There is considerable variation in the effect of temperature on forgeability; alloy 4032 shows the most marked response. The effect of temperature on forging load and pressure for alloy 6061 is shown in Fig. 12. The nearly twofold increase in load between 480 °C or 900 °F (the top of the forging range for 6061) and 400 °C or 750 °F (33 °C, or 60 °F, below the recommended range) indicates the principal reason why the recommended forging-temperature ranges for aluminum alloys, especially the high-strength alloys, are narrow.

In practice, forging is usually begun with the forging stock at the high side of the temperature range for the alloy being forged. Forging continues until the part is finished or until the work metal is too cold.

Aluminum alloys are rarely reheated for further forging unless intermediate operations, especially those requiring separate die sets, are required. Sometimes, minor increases in corner and fillet radii or draft permit production in one heating of a forging that otherwise would require reheating. On the other hand, intermediate punchout or trimming, although reducing the amount of power required for forging, may necessitate a reheat.

MACHINES AND DIES

The types of hammers and presses that are used for forging of aluminum alloys are essentially the same as those used for forging of other alloys (see the article entitled "Hammers and Presses for Forging," in this section). Hydraulic presses are very widely used in forging of aluminum alloys, because heating of the dies to a temperature near the workpiece temperature—i.e., to about 425 °C (800 °F)—eliminates die chilling. As a result, the time of contact between the forging and the dies, which usually is extensive in hydraulic presses, does not affect the forging pro-

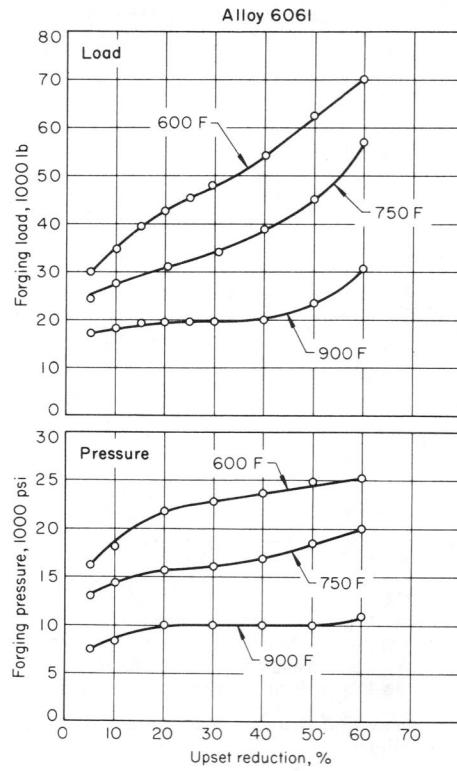

Source: "A Study of Forging Variables," by H. J. Henning, A. M. Sabroff and F. W. Boulger, U.S. Air Force Report ML-TDR-64-95, 1964.

Fig. 12. Forging-load and forging-pressure curves for upset reduction of alloy 6061 at a strain rate of 0.6 per second

cess. In addition, the easy control of the hydraulic press allows forging to proceed at a moderate speed and/or permits the forging operation to be stopped at any time during the stroke. Thus, a part can be forged in the same die, using several forging strokes. Isothermal forging—i.e., forging with dies at the same temperature as that of the workpiece being forged—also allows the use of split or wrapped dies, for production of parts with undercuts and precision-forged surfaces, as illustrated in Fig. 13.

Forging of aluminum alloys requires the use of dies specially designed for these alloys, for at least three reasons:

1. Aluminum alloys are seldom fullered or bent in the forging sequence; two forging stages, preforming and finishing, are most commonly used.
2. Allowances for shrinkage are greater than for steel.
3. Temperature control of dies for forging aluminum is critical; thus, facilities for heating dies and controlling die temperature during forging must be considered in die design.

Finish on dies used for forging of aluminum alloys is more critical than that on dies used for steel. Cavities must be highly polished to obtain acceptable surface finish on the forgings.

Table 7 lists materials for dies and die inserts for forging of aluminum alloys. Recommendations are based on the quantity to be forged, type of equipment used (press or hammer), and severity and size of the forging to be produced.

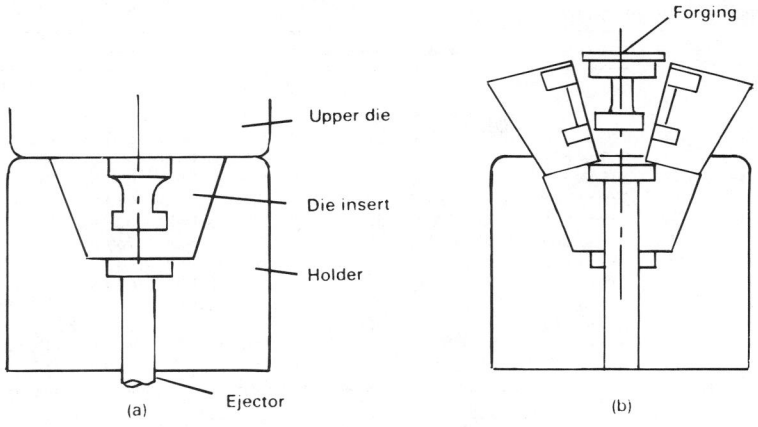

(a) Forging. (b) Ejection.

Fig. 13. Use of split dies for forging in a single die set

STOCK PREPARATION AND HEATING

The two methods most used for cutting stock into lengths for forging are sawing and shearing. Abrasive cutoff may be used, but it is slower than sawing for cutting aluminum, and, like sawing, produces burrs.

Gas-fired semimuffle furnaces are the most widely used for heating aluminum alloys for forging, mainly because gas is widely available and is usually the least expensive source of heat. Furnace design and construction necessarily vary with the requirements of the operation.

Oil-fired furnaces may be used if gas is not available. The oil must be low in sulfur content to avoid high-temperature oxidation, especially if used in a semimuffle furnace rather than a full-muffle furnace.

Electric furnaces are entirely satisfactory for heating aluminum alloys, but in most areas they cost more to operate than fuel-fired types, and hence they are seldom used.

Time at temperature is not critical for aluminum alloys. Long soaking times provide no advantage but, except for the high-magnesium alloys such as 5083, are seldom harmful. Long soaking times are sometimes difficult to avoid, as when a press breaks down. When forging a high-magnesium alloy, if the operation is delayed so that soaking time will be more than 4 h, the billets should be removed from the furnace or the furnace should be shut down. Before forging is resumed, the billets should be reheated.

LUBRICANTS

Dies are always lubricated for forging of aluminum alloys, because aluminum alloys have a tendency to adhere to steel at elevated temperatures. Lubrication practice is generally the same as that for forging of other metals: spraying with colloidal graphite mixed with water. If metal flow is a problem, as in forging metal into narrow rib sections, soap is added to the graphite mixture.

Excess lubricant may become a problem in forging of aluminum alloys, especially in dies that have intricate cavities. The preferred practice is to blow off excess lubricant with an air hose.

TRIMMING

Aluminum alloy forgings are usually cold trimmed. The method used depends largely on the size, quantity and shape of the forging. Large forgings, especially in small quantities, are usually trimmed by sawing off the flash. Web sections are removed by punchout or machining.

Trimming tools are ordinarily used for trimming large quantities, especially of small forgings that are relatively intricate and require several punchouts.

For normal trimming, both the punch and the die are often made of 6G or 6F2 die block steel (Table 7) at a hardness of about 444 to 477 HB. Tools of these steels are cheaper because they are often made from pieces of worn-out or broken forging dies. Blades for normal trimming are sometimes made by hard facing of low- or medium-carbon steels.

Forging of Copper and Copper Alloys

THE MOST FORGEABLE copper alloy (forging brass, alloy 377) can be forged into a given shape with substantially less force than would be required to forge the same shape from low-carbon steel. A less forgeable copper alloy, such as an aluminum bronze, can be forged with approximately the same force as that required for forging low-carbon steel.

Closed-Die Forging. Most copper alloy forgings are produced in closed dies.

Upset forging is used less for copper alloys than for steel, mainly because copper alloys are so easily extruded. A part having a long shaftlike section and a larger-diameter head often can be made at less cost by extruding the smaller cross section from a larger one than by starting with a small cross section and upsetting to obtain the head.

FORGING ALLOYS

Copper 122 and the copper alloys most commonly forged are listed in Table 8 in order of decreasing forgeability. They comprise at least 90% of all commercially produced copper alloy forgings. Forging brass, the least difficult alloy to forge, has been assigned an arbitrary forgeability rating of 100.

Some copper alloys cannot be forged to any significant degree because they will crack. Leaded copper-zinc alloys, such as architectural bronze,

Table 7. Recommended steels and hardnesses for dies and die inserts for hammer and press forging of aluminum alloys

| Maximum severity(a) | Total quantity to be forged | | | |
| | 100 to 10 000 | | 10 000 and over | |
	Tool steel	Hardness, HB	Tool steel	Hardness, HB
Hammer forging				
Part 1	6G, 6F2	302-331	6G, 6F2	341-375
Part 2	6G, 6F2	302-331	6G, 6F2	341-375
			H12(b)	405-448
Part 3	6G, 6F2	269-293	6G, 6F2	302-331
Part 4	6G, 6F2	341-375	6G, 6F2(c)	341-375
	H11	405-433	H11	405-433
Part 5	6G, 6F2	269-293	6G, 6F2(d)	269-293
Press forging				
Part 1	6G, 6F2	341-375	6F3	375-405
			H12(b)	448-477
Part 2	6G, 6F2	341-375	6G, 6F2(e)	341-375
			H12(b)	448-477
Part 3	6G, 6F2	302-331	6G, 6F2(f)	302-331
Part 4	6G, 6F2	341-375	6G, 6F2(g)	341-375
	H11	405-433		
Part 5	6G, 6F2	341-375	6G, 6F2(h)	341-375

Steel	Nominal composition
6F2	0.55 C, 0.75 Mn, 0.25 Si, 1.00 Ni, 1.00 Cr, 0.10 V (optional), 0.30 Mo
6G	0.55 C, 0.80 Mn, 0.25 Si, 1.00 Cr, 0.10 V, 0.45 Mo
H11	0.35 C, 5.00 Cr, 0.40 V, 1.50 Mo
H12	0.35 C, 5.00 Cr, 0.40 V, 1.50 W, 1.50 Mo

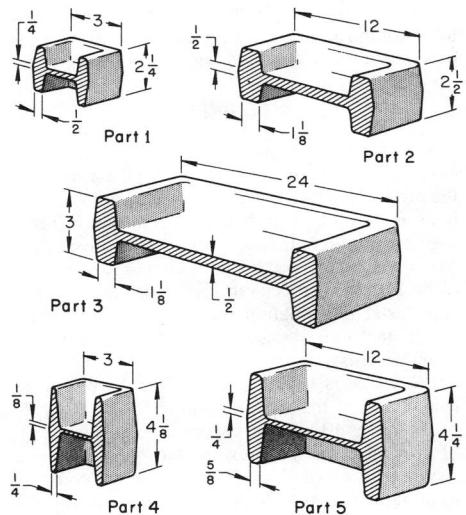

(a) Dimensions in illustration are in inches. (b) Recommended for long runs—for example, 50 000 forgings. (c) With either steel, use inserts of H11 at 405 to 433 HB. (d) With either steel, use inserts of 6G or 6F2 at 302 to 331 HB. (e) With either steel, use inserts of 6F3 at 405 to 448 HB. (f) With either steel, use inserts of 6G or 6F2 at 341 to 375 HB. (g) With either steel, use inserts of H12 at 429 to 448 HB. (h) With either steel, use inserts of H12 at 429 to 448 HB. For long runs (50 000 forgings), a solid block made of H12 at 477 to 514 HB is recommended.

which may contain more than 2.5% lead, are seldom recommended for hot forging. Although lead content improves metal flow, it promotes cracking in those areas of a forging, particularly deep-extruded areas, that are not completely supported by, or enclosed in, the dies. This does not mean that the lead-containing alloys cannot be forged, but rather that the design of the forging may have to be modified to avoid cracking.

The solubility of lead in beta brass at forging temperatures is about 2% maximum, whereas lead

Table 8. Forgeabilities and forging-temperature ranges for copper and copper alloys

Alloy No.	Alloy name	Relative forge-ability(a)	Single-blow press or hammer forging °C	°F	Flattening °C	°F	Blocking °C	°F	Finishing °C	°F
377	Forging brass	100	650 to 760	1200 to 1400	620 to 730	1150 to 1350	595 to 705	1100 to 1300	650 to 730	1200 to 1350
464	Naval brass	90	595 to 705	1100 to 1300	565 to 675	1050 to 1250	595 to 705	1100 to 1300	595 to 705	1100 to 1300
674	Mn-Si bearing brass	85	595 to 730	1100 to 1350	595 to 705	1100 to 1300	595 to 705	1100 to 1300	620 to 730	1150 to 1350
673	Mn-Si bearing brass	80	595 to 730	1100 to 1350	595 to 705	1100 to 1300	595 to 705	1100 to 1300	620 to 730	1150 to 1350
670	Manganese-aluminum bronze	75	595 to 705	1100 to 1300	595 to 675	1100 to 1250	595 to 705	1100 to 1300	620 to 730	1150 to 1350
624	Aluminum bronze, 11%	70	705 to 815	1300 to 1500	705 to 760	1300 to 1400	705 to 790	1300 to 1450	730 to 815	1350 to 1500
642	Aluminum-silicon bronze	70	730 to 900	1350 to 1650	730 to 845	1350 to 1550	730 to 870	1350 to 1600	760 to 900	1400 to 1650
675	Manganese bronze (A)	70	595 to 705	1100 to 1300	595 to 675	1100 to 1250	595 to 705	1100 to 1300	620 to 705	1150 to 1300
122	Phosphorus-deoxidized copper	65	730 to 845	1350 to 1550	705 to 790	1300 to 1450	730 to 845	1350 to 1550	760 to 845	1400 to 1550
616	Aluminum bronze, 9%	65	760 to 870	1400 to 1600	730 to 815	1350 to 1500	760 to 815	1400 to 1500	790 to 870	1450 to 1600
182	Chromium copper	60	650 to 760	1200 to 1400	620 to 730	1150 to 1350	650 to 760	1200 to 1400	675 to 760	1250 to 1400
628	Aluminum bronze, 10%	60	815 to 900	1500 to 1650	760 to 845	1400 to 1550	815 to 900	1500 to 1650	815 to 900	1500 to 1650

(a) This factor is based on the ability of the alloy to fill a given die impression under a force of 550 MPa (40 tsi), on extrudability, on optimum forging temperature, on permissible forging speed and on the abrasiveness of the alloy in causing die wear. (b) Actual slug temperature.

is insoluble in alpha brass at all temperatures. Consequently, although a lead content of up to 2.5% is permissible in 60Cu-40Zn alpha-beta brasses, lead in excess of 0.10% in a 70Cu-30Zn alpha brass will contribute to catastrophic cracking.

Other copper alloys can be forged only with greater difficulty and at higher cost. Cupro-nickel is an outstanding example. This alloy, mainly because of its higher forging temperature, is sometimes heated in a controlled atmosphere, thus complicating the process. The silicon bronzes, because of their high forging temperatures and their compositions, cause more rapid die deterioration than the common forging alloys.

MACHINES

Most copper alloy forgings are produced in crank-type mechanical presses. With these presses, the production rate is high, and less operator skill is needed and less draft is required than in forging copper alloys in hammers. Screw presses are also very widely used because they (a) allow parts to be forged to precision tolerances and (b) ensure faster deformation rates and shorter billet/die contact times than those characteristic of mechanical presses.

Press size is normally based on the projected (plan) area of the part, including flash. The rule of thumb is 40 tons of capacity per square inch of projected area. Therefore, a forging having a projected area of 5 in.² (32 cm²) will require a minimum of 200 tons (180 tonnes) capacity for forgings of up to medium severity. If the part is complicated (for example, with deep, thin ribs), the capacity must be increased.

DIES

Dies designed for forging copper or copper alloys are usually different from those designed for forging the same shapes from steel in three ways: (a) the draft angle can be decreased for forging copper (3° max and often less than 3°); (b) the die cavity is usually machined to dimensions that are 0.005 in. per in. less than for forging steel; and (c) the die cavity usually is polished to a better surface finish for forging copper.

Brass—the copper alloy that is the easiest to forge—can be forged using split and cored tooling to produce parts with undercuts, gate-valve bodies, water-meter bodies, elbows and cored T-pieces.

Die materials and hardnesses commonly used for forging copper alloys are shown in Table 9.

Table 9. Recommended steels and hardnesses for dies and die inserts for hammer and press forging of copper and copper alloys

Maximum severity(a)	100 to 10 000 Tool steel	Hardness, HB	10 000 and over Tool steel	Hardness, HB
Hammer Forging				
Part 1	H11	405-433	H12	405-448
	6G, 6F2	341-375		
Part 2	6G, 6F2	341-375	6G, 6F2	341-375
			H12(b)	405-448
Part 3	6G, 6F2	269-293	6G, 6F2	302-331
Part 4	H11	405-433	H11	405-433
Part 5	6G, 6F2	302-331	6G, 6F2(c)	302-331
Press Forging				
Part 1	H12	477-514	H12	477-514
	6G, 6F2	341-375		
Part 2	6G, 6F2	341-375	H12	477-514
Part 3		Part usually not press forged from copper alloys		
Part 4	H11	405-433	6G, 6F2(d)	341-375
Part 5	6G, 6F2	341-375	H12	477-514

(a) Dimensions in illustration are in inches. (b) Recommended for long runs—for example, 50 000 forgings. (c) With either steel, use insert of H12 at 405 to 448 HB. (d) With either steel, use insert of H12 at 429 to 448 HB.

Whether the dies are made entirely from a hot work steel such as H11 or H12 or whether inserts are used depends largely on the size of the die. Common practice is to make the inserts from a hot work steel and to press them into rings or holders made from a low-alloy die block steel (see Table 9) or L6 tool steel. Hardness of the ring or holder is seldom critical; a range of 341 to 375 HB is typical.

HEATING OF BILLETS OR SLUGS

Optimum forging temperatures for 12 alloys are given in Table 8.

Atmosphere protection during billet heating is not required for most alloys, especially when forging temperatures are below 705 °C (1300 °F). For temperatures toward the top of the range in Table 8, a protective atmosphere is desirable, and is sometimes required. An exothermic atmosphere is usually the least costly, and is satisfactory for heating copper alloys at temperatures above 705 °C.

Gas-fired furnaces are almost always used. Design of the furnace is seldom critical. Open-fired conveyor chain or belt types are the most commonly used.

HEATING OF DIES

Dies are always heated for forging copper and copper alloys, although, because of the good forgeability of copper alloys, die temperature is generally less critical than for forging aluminum or aluminum alloys.

Dies are seldom preheated in ovens. Heating is usually accomplished by ring burners. Optimum die temperatures vary from 150 to 315 °C (300 to 600 °F), depending on the forging temperature of the specific alloy. For alloys having low forging temperatures, a die temperature of 150 °C is sufficient. Die temperature is increased to as much as 315 °C for those copper alloys that have the highest forging temperatures shown in Table 8.

Die temperature is controlled by the use of temperature-sensitive crayons or surface pyrometers.

LUBRICANTS

Dies should be lubricated before each forging operation. A spray of colloidal graphite and water is usually adequate. Many installations include a spray that operates automatically, timed with the press stroke. For deep cavities, however, the spray is often inadequate, and so it is supplemented by swabbing with a conventional forging oil.

TRIMMING

Brass forgings are nearly always trimmed at room temperature. Because the forces imposed on the trimming tools are less than for trimming steel forgings, trimming of brass forgings seldom poses problems.

Large forgings, especially in small quantities, are commonly trimmed by sawing off the flash and punching or machining the web sections.

CLEANING

Scale and excess lubricants are easily removed from copper and copper alloy forgings by chemical cleaning. Pickling in dilute sulfuric acid is the most common method for cleaning brass and most other copper alloy forgings, although hydrochloric acid may also be used. Makeup of sulfuric and hydrochloric acid solutions, pickling procedures, and typical uses are given in Table 10.

Table 10. Typical solutions and conditions for cleaning copper and copper alloy forgings

Sulfuric acid
H_2SO_4 (1.83 sp gr) . . . . 4 to 15% by volume
Water . . . . Remainder
Temperature of solution . . . Room to 60 °C (140 °F)
Immersion time . . . . $^1/_2$ to 15 min
USES: (a) Removal of black copper oxide scale from brass forgings. (b) Removal of oxide from copper forgings.

Hydrochloric acid
HCl (20° Bé) . . . . 40 to 90% by volume
Water . . . . Remainder
Temperature of solution . . . . Room
Immersion time . . . . 1 to 3 min
USES: (a) Removal of scale and tarnish from brass forgings. (b) Removal of oxide from copper forgings.

Aluminum bronzes form a tough, adherent aluminum oxide film during forging. An effective method of cleaning aluminum bronze forgings is first to immerse them in a 10% solution (by weight) of sodium hydroxide in water, at 75 °C (170 °F) for 2 to 6 min. After rinsing in water, the forgings are pickled in acid solutions in the same way as brasses.

Table 11. Forging temperatures for the most commonly forged magnesium alloys

Alloy	Principal alloying elements(a), %	Initial forging temperature(b) Stock °C	Stock °F	Dies °C	Dies °F
Commercial alloys					
ZK21A	2.3 Zn, 0.45 Zr	330 to 290	625 to 550	315 to 260	600 to 500
AZ31B	3.0 Al, 1.0 Zn	330 to 290	625 to 550	315 to 260	600 to 500
AZ61A	6.5 Al, 0.9 Zn	355 to 315	675 to 600	345 to 290	650 to 550
High-strength alloys					
ZK60A	5.5 Zn, 0.45 Zr	355 to 260	675 to 500	290 to 205	550 to 400
AZ80A	8.5 Al, 0.5 Zn	400 to 315	750 to 600	290 to 205	550 to 400
Elevated-temperature alloys					
HM21A	2.0 Th, 0.8 Mn	510 to 400	950 to 750	425 to 370	800 to 700
EK31A	3.0 RE, 0.6 Zr	480 to 370	900 to 700	400 to 345	750 to 650
Special alloys(c)					
ZE42A	4.0 Zn, 2.0 RE	370 to 290	700 to 550	345 to 300	650 to 575
ZE62	6.0 Zn, 1.6 RE	355 to 300	675 to 575	345 to 300	650 to 575
QE22A	2.5 Ag, 2.0 RE	385 to 345	725 to 650	370 to 315	700 to 600

(a) RE = rare earths. (b) The strain-hardening alloys must be processed on a declining temperature scale within the range given, to preclude recrystallization. (c) These three alloys contain 0.6 or 0.7% zirconium (nominal) in addition to the principal elements listed.

Forging of Magnesium Alloys

MAGNESIUM ALLOYS most commonly forged are listed in Table 11, along with their forging temperatures.

The commercial alloys ZK21A, AZ31B and AZ61A listed in Table 11 are low-cost magnesium alloys suitable for applications in which slightly lower mechanical properties are acceptable. AZ31B is slightly more difficult to forge than the other two alloys. ZK21A has good weldability.

Of the two high-strength alloys in Table 11, ZK60A is more widely used, because it is inherently fine-grained and more workable.

Of the elevated-temperature alloys in Table 11, both EK31A and HM21A are readily forgeable.

To ensure good workability, only forging-grade billet or bar stock should be used. This type of product has been conditioned and inspected to eliminate surface defects that could open up during forging, and has been homogenized by the supplier to make it forgeable.

MACHINES AND DIES

Machines. Hydraulic presses or slow-action mechanical presses are the most commonly used machines for open-die and closed-die forging of magnesium alloys. In these machines, magnesium alloys can be forged with small corners and fillets, and thin web or panel sections. Corner radii of 1.6 mm ($^1/_{16}$ in.), fillet radii of 4.8 mm ($^3/_{16}$ in.), and panels or webs 3.2 mm ($^1/_8$ in.) thick are not uncommon. Draft angles required for extraction of the forgings from the dies may be held to 3° or less.

Magnesium alloys are not usually hammer forged, because they exhibit better hot ductility at lower strain rates.

Dies. Because forging temperatures for magnesium alloys are relatively low (see Table 11), conventional low-alloy hot work tool steels are satisfactory materials for forging dies.

Magnesium alloys are good conductors of heat, and therefore are readily chilled by cold dies, causing the alloys to crack. Because die contact during forging is extensive and is maintained for a prolonged period, dies must be heated to temperatures not much lower than those used to heat the stock (see Table 11). Thus, magnesium alloys are forged isothermally.

HEATING FOR FORGING

In most instances, the mechanical properties developed in magnesium forgings depend on the strain hardening induced during forging. Strain hardening is accomplished by keeping the forging temperature as low as is practical; however, if temperatures are too low, cracking will occur.

The temperature ranges generally used in the initial heating of magnesium alloys for forging are listed in Table 11. The temperature selected depends on the nature and size of the part, the amount of working that is required, and the number of operations involved. In a multiple-operation process, the forging temperature should be adjusted downward for each subsequent operation, to avoid recrystallization and grain growth. In addition to controlling grain growth, the reduction in temperature allows for residual strain hardening following the final operation.

Heating can be done with fuel-fired or electrically heated furnaces. Inert or reducing atmospheres are not needed at temperatures below 480 °C (900 °F).

Because forging temperatures are well below the melting points of the various alloys, no fire hazard exists when temperatures are controlled with reasonable accuracy. It is important, however, that uniformity of temperature be maintained — at least throughout the final heating zone — and that large gradients and hot spots be avoided in the preliminary heating zones. Furnaces that are equipped with fans for recirculating the air within the furnace provide the greatest uniformity of heating.

FORGING PRACTICE

Forging pressures for upsetting magnesium alloy billets between flat dies are shown in Fig. 14. At normal press forging speeds, the forging pressure first increases and then decreases slightly with increased upset reduction, probably because work-metal temperature increases during forging.

Forging temperature has a marked effect on forging-pressure requirements. Figure 15 shows the magnitude of this effect for magnesium alloy AZ31B in comparison with aluminum alloy 6061. As Table 12 shows, at normal forging temperatures, AZ31B requires greater forging pressure than carbon steel, alloy steel or aluminum, and requires less than stainless steel. Magnesium alloys flow less readily than aluminum into deep

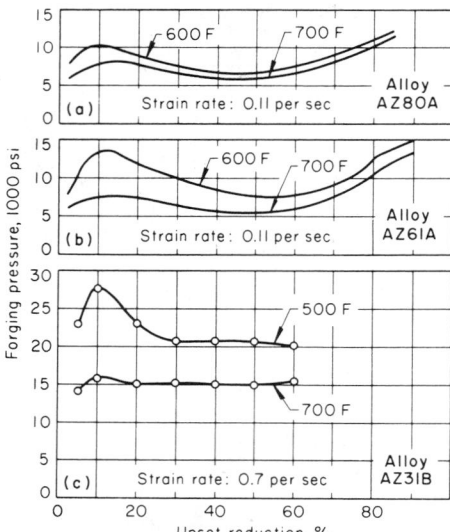

Source of data: For AZ80A, and AZ61A, R. L. Dietrich and G. Ansel, *Trans. ASM*, Vol 38, 1947, p 709-727; for AZ31B, "A Study of Forging Variables," by H. J. Henning, A. M. Sabroff and F. W. Boulger, U.S. Air Force Report ML-TDR-64-95, 1964.

Fig. 14. Forging pressures for upsetting of magnesium alloy billets between flat dies

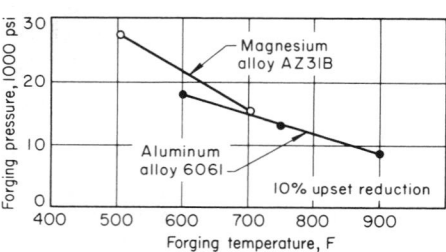

Source: Same as for AZ31B in Fig. 14.

Fig. 15. Effect of forging temperature on forging pressure required for upsetting to a 10% reduction at hydraulic-press speeds

vertical die cavities. If two dies are needed for a typical aluminum structural forging, the same part in a magnesium alloy may require three dies for successful forging. The lubricants and lubrication conditions used are similar to those used in forging of aluminum alloys.

TRIMMING

Magnesium alloy forgings are usually trimmed by band sawing, because: (*a*) sawing is usually cheaper than die trimming, especially for short-run operation; and (*b*) die trimming must be done hot, which often is inconvenient in a sequence of operations. However, die trimming is sometimes preferred because of high-production operation or intricate design of the forging. Under these conditions, the forging is either heated to 205 to 260 °C (400 to 500 °F) or is trimmed while still hot from the forging operation.

Whether the trimming is normal or close can affect the optimum trimming temperature. For most applications the trimming tools are adjusted for a closer trim than for similarly shaped forgings of other metals. If the trimming temperature is too high, the flash will tear and have a smeared

Table 12. Approximate forging pressure required for a 10% upset reduction at normal forging temperature in flat dies
Source: Same as for AZ31B in Fig. 14.

Work metal	Forging temperature °C	°F	Forging pressure MPa	ksi
1020 steel	1260	2300	55	8
4340 steel	1260	2300	55	8
6061 aluminum alloy	455	850	69	10
AZ31B magnesium alloy	370	700	110	16
304 stainless steel	1205	2200	152	22

appearance; if the temperature is too low, flash-line cracks will occur that go deep into the forging. Trimming in dies at or near room temperature should never be attempted. Hot trimming can be done before or after the forgings are quenched, whichever produces best results.

CLEANING

After the forgings have been sandblasted to remove the die lubricant, they are dipped in a solution of 8% nitric acid and 2% sulfuric acid, and then rinsed. This treatment is adequate for most surface inspection requirements. After inspection, the forgings can be protected for shipping purposes by dipping in dichromate.

Forging of Titanium Alloys

TITANIUM ALLOYS can be forged into essentially the same shapes as are forgeable from steel and other metals. For the same amount of metal flow, however, more power is required than for alloy steels such as 4340. This difference between titanium alloys and alloy steel is illustrated in Fig. 16, which summarizes the effect of strain rate and temperature on forging pressure for several titanium alloys and 4340 steel at their usual forging temperatures. Forging pressure increases in an approximately linear relation with the log-

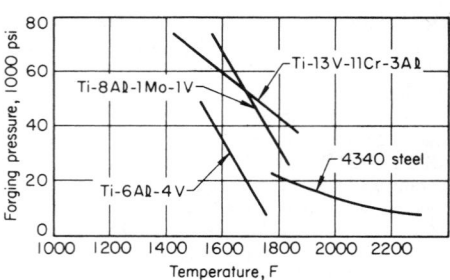

Source: Same as for Fig. 16.

Fig. 17. Effect of forging temperature on forging pressure for three titanium alloys and 4340 steel

arithm of the strain rate, as shown by the behavior of the Ti-13V-11Cr-3Al (beta) alloy. Thus, forging pressure for a given strain rate can be estimated by interpolating or extrapolating data available from other strain rates. The comparative data are in agreement with forging experience, which indicates that an alloy such as Ti-6Al-4V requires $1\frac{1}{2}$ to 2 times the equipment capacity needed for forging low-alloy steels to comparable shapes.

As Fig. 17 shows, the pressure required for forging titanium alloys rises at a faster rate as the temperature of the work metal decreases than does the pressure required for forging alloy steel.

FORGING ALLOYS

The titanium alloys that are forged most often are given in Table 13. This table also gives the range of forging temperatures, as well as the beta transus temperature, for each alloy. The beta transus temperature establishes the maximum temperature to which an alloy can be heated without developing a new microstructure that may result in reduced final properties. Alpha-beta alloys can be strengthened by heat treatments that cause some of the beta phase to transform into alpha, finely dispersed in the beta phase. Recently, several beta and near-beta alloys have been developed which have excellent properties. These alloys can be forged at temperatures lower than those required for alpha-beta alloys.

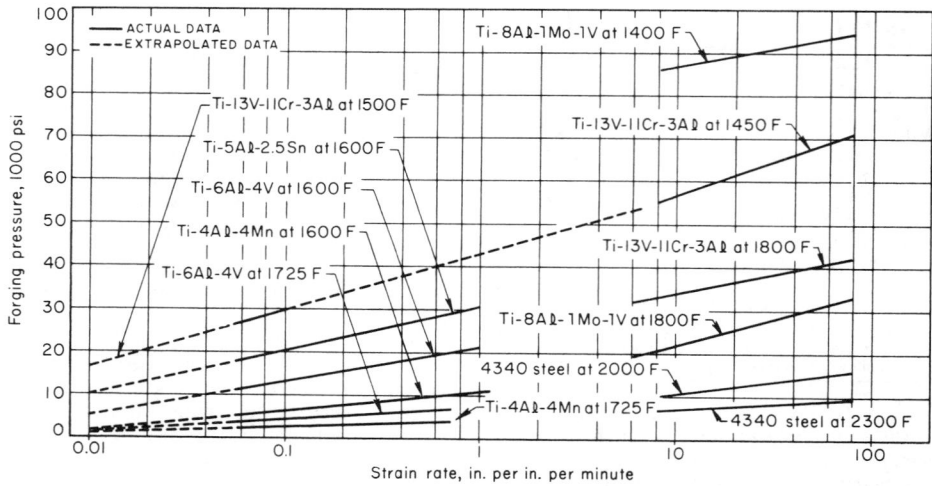

Source: *Forging Materials and Practices,* by A. M. Sabroff, F. W. Boulger and H. J. Henning, Reinhold, New York, 1968.

Fig. 16. Effect of strain rate on forging pressure for several titanium alloys and 4340 steel at various forging temperatures

Table 13. Approximate forging temperatures for titanium alloys
Source: "Forging Equipment, Materials, and Practices," by T. Altan *et al*, MCIC Report HB-03, Battelle-Columbus Laboratories, Columbus, OH, 1973.

Alloy	Beta transus temperature °C	°F	Forging temperature °C	°F
Commercially pure titanium	960	1760	870 to 925	1600 to 1700
Alpha alloys				
Ti-5Al-2.5Sn	1040	1900	940 to 1010	1725 to 1850
Ti-8Al-1Mo-1V	1015	1860	955 to 1010	1750 to 1850
Beta alloys				
Beta III	760	1400	845 to 900	1550 to 1650
Ti-13V-11Cr-3Al	720	1325	870 to 980	1600 to 1800
Ti-3Al-8V-6Cr-4Mo-4Zr	800	1475	815 to 870	1500 to 1600
Ti-10V-2Fe-3Al	800	1475	760 to 870	1400 to 1600
Alpha-beta alloys				
Ti-6Al-4V	995	1820	845 to 980	1550 to 1800
Ti-6Al-6V-2Sn	945	1735	845 to 915	1550 to 1675
Ti-6Al-2Sn-4Zr-2Mo	995	1825	925 to 980	1700 to 1800
Ti-6Al-2Sn-4Zr-6Mo	955	1750	885 to 925	1625 to 1700

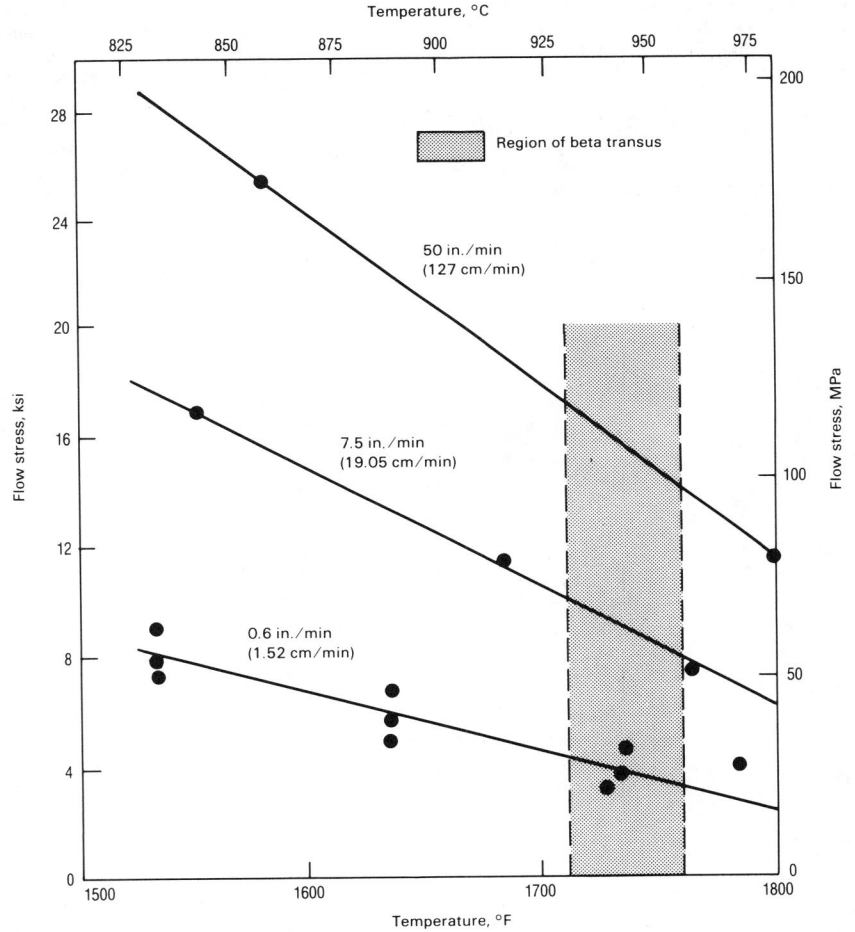

Source: Titanium Precision Forgings, by D. K. Fix, in *Titanium Science and Technology*, Vol. 1, Plenum Press, New York, 1972, p 441.

Fig. 18. Effects of deformation rate and temperature on flow stress of alloy Ti-6Al-6V-2Sn under isothermal forging conditions

Alpha, Near-Alpha and Alpha-Beta Alloys. This group includes the majority of commercial titanium alloys as well as the "workhorse" of titanium alloys, Ti-6Al-4V. Values of flow stress for these alloys are highly dependent on temperature and strain rate, as illustrated in Fig. 18. For a given composition, the flow stress also depends on microstructure. Especially at low strain rates—from 0.001 to 0.1 per second—a difference in alpha grain size of a factor of 2.5 can lead to differences in flow stress of a factor of about 2.

Beta Alloys. Development and commercial application of beta titanium alloys have occurred relatively recently (i.e., over the last 15 years). Some beta alloys have flow-stress values at temperatures of about 815 °C (1500 °F) that are comparable to those of alloy Ti-6Al-4V at temperatures of about 925 °C (1700 °F). This is due to the fact that the beta transus temperatures of these alloys are lower than those of alpha-beta alloys. Forging of beta alloys requires very close control of temperature, which is difficult to maintain using conventional forging practices. Beta forging is still not widely practiced and requires some changes in generally accepted processing specifications.

CONVENTIONAL FORGING

The forging machines, die design and die-heating practices used in conventional forging of titanium alloys are similar to those used in forging of alloy steels except that the forging pressures are higher. High degrees of dependency of flow stress on temperature and strain rate also require that these alloys not be allowed to lose temperature during forging in relatively low-speed machines (i.e., in hydraulic presses). In high-rate forging (i.e., in hammers), there is a danger of overheating as a result of heat generation due to large amounts of plastic deformation. Thus, depending on the forging machine being used, the initial stock temperature must be properly selected so as not to increase the temperature of the forging during deformation above the beta transus temperature. Thus, conventional forging of titanium alloys requires much closer temperature and process control than are necessary in forging of alloy steels.

DIES

Forging of titanium alloys requires the use of specially designed dies, for the following reasons:

1. Shrinkage allowance in die sinking is 0.100 in. per foot for forging titanium alloys, compared with 0.187 in. per foot for dies used for forging steel.
2. Titanium alloys fill die contours less readily than alloy steel, stainless steel or aluminum alloys; therefore, the die impressions for forging titanium alloys must have larger radii and fillets. For intricate or close-tolerance forgings, more forging steps, and therefore more dies, are usually needed for titanium alloys than for steel.
3. Dies for forging titanium alloys must be more rugged than for steel, because greater force is required. Dies for forging titanium alloys are commonly about 50% thicker than dies used for the same depth and severity of impression for forging steel. If the die is not made considerably thicker for titanium alloys, the number of resinkings that can be made without risk of die breakage will be fewer.
4. Maintaining a good surface finish on dies is more important for forging titanium than for steel, because of the relatively poor flow characteristics of titanium.

Practices for die-material selection, die hardening and die heating are similar to those used in forging of alloy steels.

PREPARATION AND HEATING OF STOCK

Production of sound forgings from titanium alloys requires that all surface defects on stock be removed. Therefore, all billets should be rough machined or rough ground on all surfaces, and bars should be turned or centerless ground to ensure that all surfaces are free from defects. Sawing is the preferred method for cutting stock to length.

Heating of titanium alloys is critical, both in terms of providing protection against harmful contaminants, such as oxygen, nitrogen, carbon and hydrogen, and of controlling temperature within narrow limits. Although heating in vacuum or in a suitably inert atmosphere is feasible, billets are often precoated with a glass-type coating material and heated in the usual manner.

Both a soda barium glass and a borosilicate glass are reported to provide complete fusion and excellent surface protection. This coating is retained during forging, where it acts as a lubricant. The working temperature of the glass coating is taken as the temperature at which the viscosity is 0.1 Pa·s (100 cP). Glass coatings soften between 675 and 760 °C (1250 and 1400 °F).

Prolonged heating should be avoided for all titanium alloys. If an unscheduled delay such as a press breakdown occurs, stock should be removed from the furnace, and then it should be reheated when forging is resumed.

CLEANING

Cracks or surface imperfections should be removed by grinding after forging operations.

Grinding normally is done at room temperature after the metal is air cooled from the forging temperature. Liquid-penetrant inspection or etching can be used as an aid to visual inspection of the surface for defects.

The oxide layer that forms on titanium at forging temperatures is brittle and can promote cracking of the underlying metal in subsequent bending or forging. Consequently, it is often desirable to remove the oxide between successive forging operations.

Grit blasting is a reasonably effective method of removing this oxide layer, which can vary in thickness from 0.13 to 0.76 mm (0.005 to 0.030 in.). The blasting is done with zircon sand (100 to 150 mesh) at an air pressure of 275 kpa (40 psi), although other types of grit are also used.

Salt bath descaling followed by acid pickling is an effective method for removing scale from titanium alloys. The process must be closely controlled, however, or the work metal may become embrittled. Also, titanium racks must be used; with steel racks, an electric potential would be generated, resulting in erosion of the titanium work metal.

ISOTHERMAL AND HOT-DIE FORGING

Because the forging behavior of titanium alloys is highly temperature-dependent, as illustrated in Fig. 16 for various alloys, it is desirable to forge these alloys under isothermal conditions (dies at the same temperature as the forging) or under hot-die conditions (dies slightly below forging temperature). Thus, the part can be forged at low speeds to reduce flow stress and forging pressure, without much danger of die chilling. The effect of strain rate on flow stress is shown in Fig. 18. Isothermal or hot-die forging, however, requires expensive die-heating systems and die materials that can maintain their strength at temperatures from 815 to 925 °C (1500 to 1700 °F). Die materials that are suitable for this purpose are heat-resisting alloys such as Waspaloy, Udimet 700, Astroloy, IN-100 and Inconel 713C. The effect of temperature on the strength of these alloys is discussed in the article "Forging of Heat-Resisting Alloys."

In isothermal and hot-die forging, stringent demands are placed on forging lubricants and stock coatings as well as on die lubrication. Various glass mixtures are used for coating the billets. In addition, it is often necessary to apply an additional "die-separation agent" to the stock prior to forging so that the forged part can be removed from the die without distortion. Several isothermal forging lubricants/coatings are commercially available, and many others have been developed by various companies active in this technology and are considered proprietary.

25 POWDER METALLURGY

Edited by Peter W. Lee, The Timken Co.

Additional and more detailed information on powder metallurgy can be found in Metals Handbook, Ninth Edition, Volume 7, Powder Metallurgy. For additional information within the Desk Edition, the reader should consult the index.

Ferrous Powder Metallurgy

By Peter W. Lee, The Timken Co., and B. Lynn Ferguson, Deformation Control Technology

Condensed from Powder Metallurgy: Applications, Advantages and Limitations, edited by Erhard Klar, American Society for Metals, 1983, pages 55 to 122.

Pressed-and-Sintered Parts

THE POWDER METALLURGY (P/M) process is defined as a process for producing a discrete shape from either metallic or nonmetallic particulate matter. The first commercial P/M parts were porous bronze bushings produced in the 1920's. However, the first commercial application of ferrous P/M parts was not developed until sintered oil-pump gears were produced in the late 1930's.

Approximately 80% of all ferrous powder produced is used in P/M part applications, and the automotive industry is the leading powder user. Other important areas are business machines, hand tools, lawn and garden equipment, and appliances.

The P/M process provides economical as well as technical benefits over conventional processing of wrought material. Some of the advantages of P/M processing are:

* *Reduction of Manufacturing Cost*. Generally, P/M processing provides higher material utilization, minimizes machining and reduces energy consumption. Consequently, over-all manufacturing cost is reduced.
* *Improved Performance*. Generally, the P/M process has the potential to improve performance through uniform properties, fine grain structures, and chemical homogeneity.
* *Design Flexibility*. P/M offers the designer material/property combinations that permit production of engineered structures for optimized mechanical and physical properties.
* *Production of Unique Materials*. P/M is the only viable process for manufacturing parts from materials with very poor workability, such as tungsten, beryllium, ceramics and carbides.

P/M PART FABRICATION PROCESSES

Manufacturing processes for pressed-and-sintered (P/S) P/M parts consist of powder production, cold compaction and sintering. The parts are used in applications requiring low mechanical-property levels. Secondary processes, such as re-pressing and infiltration, can be used to improve the mechanical properties of the pressed-and-sintered P/M parts. For high-stress, critical applications, pressed-and-sintered P/M parts are hot forged into full-density parts.

Production of Ferrous Metal Powders

Several commercial methods are available for production of ferrous powders. Among these, the most important processes are direct reduction of oxides and atomization of liquid metal.

Direct Reduction of Oxides. There are two commercially available processes for direct reduction of oxides: reduction with carbon (Hoeganaes Corp.) and reduction with hydrogen gas (Pyron Corp.).

In the Hoeganaes process, sponge iron is produced by direct reduction of magnetite (Fe_3O_4)

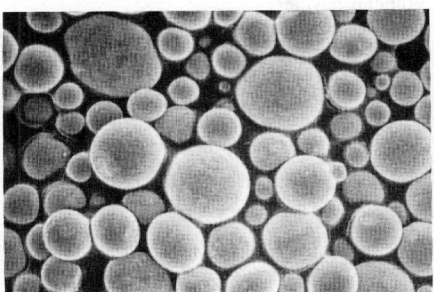

Fig. 1. Scanning electron micrograph of gas-atomized powder

ore. Cleaned ore is then charged into silicon carbide tubes with a crushed lime and coke mixture and heated at 1205 °C (2200 °F) for 24 to 36 h in a tunnel kiln. The caked sponge iron is then removed from the container, ground and screened. The cleaned powder is charged into a strip belt furnace at 925 °C (1700 °F) under a dissociated ammonia atmosphere in order to reduce the oxygen level of the powder.

In the Pyron process, iron powder is produced by reducing selected iron oxides under a hydrogen atmosphere. In this process, mill scale is ground, magnetically separated, and screened. The oxides are then reduced in a belt furnace under a hydrogen atmosphere at approximately 980 °C (1800 °F). The reduced iron is then ground, screened, and blended to specifications.

Atomization of Liquid Metal. Another major commercial process for production of powder is atomization of liquid metal by either gas or water.

In the *gas atomization* process, a liquid-metal stream is impinged upon by a subsonic or supersonic jet of inert gas. Atomization is produced by the kinetic energy of the atomizing fluid, typically nitrogen or argon for ferrous powders.

Generally, gas atomization produces smooth and spherical powders, as shown in Fig. 1. It is employed for highly alloyed materials which contain significant amounts of readily oxidized alloying elements. Because of the spherical shape of gas-atomized powder particles, they can be consolidated only by applying pressure and heat simultaneously. Hot isostatic pressing, hot extrusion or containerized hot pressing is used for consolidation of gas-atomized powder.

In *water atomization*, a high-pressure stream of water is forced through a nozzle to form a dispersion of droplets which impact the metal stream. Generally, water-atomized powders have highly irregular particle shapes and high oxygen levels. The formation of metal droplets during water atomization is the result of impact, not shear, of the water droplets on the surface of the metal stream.

A scanning electron micrograph of water-atomized powder is shown in Fig. 2. Because of

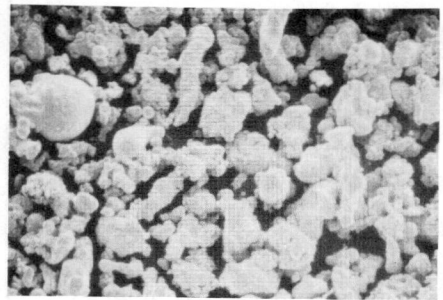

Fig. 2. Scanning electron micrograph of water-atomized powder

its highly irregular particle shape, water-atomized powder can be compacted at room temperature by either die pressing or cold isostatic pressing. Prealloyed powders are produced by atomization processes and are used for high-stress and high-performance applications.

Compaction and Sintering Technology

Blending Process. Solid-lubricant and graphite powders are blended with ferrous powder. The lubricant reduces the required ejection force, and the graphite provides carbon for reduction of oxides and the final carbon level of the sintered part. Because the solid lubricant tends to reduce the green strength of the compacts and to weaken the final sintered parts, a minimum amount of lubricant is recommended. Die-wall lubrication can be used instead of a lubricant blended into the powder mass to minimize degradation of properties by the presence of residual lubricant.

Compaction Process. After blending, the compaction of powder into green parts is the next major step of the P/M process. The Metal Powder Industries Federation (MPIF) has classified powder metallurgy parts into four groups, depending on their design complexity. Class I parts are one-level parts less than 6.35 mm (0.25 in.) thick which can be pressed with a force from one direction. Class II parts are one-level parts of any thickness that must be pressed with forces from two opposing directions. Class III parts are two-level parts of any thickness that must be pressed with forces from two opposing directions. Class IV parts are multilevel parts of any thickness that must be pressed with forces from two opposing directions.

The reduced iron powders and water-atomized powders are compacted into preforms by either die compaction or cold isostatic pressing. The irregularly shaped powder particles are mechanically interlocked during compaction and provide substantial green strength in the compact. Die compaction is probably the most widely used technique for P/M part production because of its speed and ability to produce fairly complex parts. However, highly complex P/M parts are produced by cold isostatic pressing.

During die compaction, friction develops as a result of the relative motion between die sidewalls and powder particles, between powder particles within the powder mass, and between powder particles and punch faces. Because of this friction, the local pressure within the powder mass caused by the upper punch load is not uniform, which results in density variations in the compact. For thick P/M parts, forces should be applied from both top and bottom directions to control the neutral axis. The density variations for both single- and double-action compaction are

shown schematically in Fig. 3. In addition, because metal powder does not flow under pressure, separate pressing forces must be provided for each level in multilevel parts. Even for single-level parts, uniform filling of the die cavity is important for obtaining compacts with uniform density. Uniform density in the compact is necessary to obtain uniform density in the final part after subsequent sintering operations.

The life of the compaction tooling is influenced by the required tolerances of the part. Close tolerances tend to decrease tool life by decreasing the amount of tool wear that can be accommodated. In addition, closer tolerances may require secondary operations such as coining or sizing, which increase costs. Therefore, it is recommended that tolerances of P/M parts be as liberal as possible.

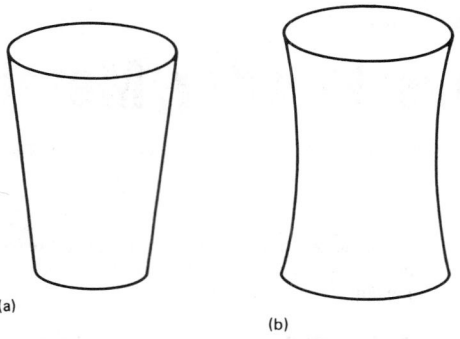

(a)

(b)

Fig. 3. Sintered green compacts made by (a) single-action and (b) double-action pressing, showing shape distortions that occurred during sintering as a result of density variations in the green compacts

Sintering Process. Sintering is probably one of the most complex phenomena in P/M processing. For powder metallurgy, sintering may be generally defined as the process wherein powder particles develop metallurgical bonding and densify under the influence of heat. During sintering of ferrous P/M parts, in particular, the following reactions take place:

- Metallurgical bonds are developed from the mechanical interlocks between powder metal particles in the compact.
- Metal oxides in the powder compact are reduced by reaction with the carbon from the blended graphite powders.
- The desired final carbon level of the P/M compact is obtained by diffusion of carbon from graphite powder.
- Densification of P/M compacts can be achieved during sintering. The degree of densification depends on the sintering parameters and the alloy(s) involved. Generally, higher sintering temperatures and longer sintering times promote densification of sintered parts.

A typical sintering cycle for ferrous powder consists of preheating or delubrication at about 705 °C (1300 °F) and sintering at 1120 °C (2050 °F) for 30 min in a reducing atmosphere, followed by cooling under an inert atmosphere.

The commercially used sintering atmospheres are endothermic gas, exothermic gas, dissociated ammonia, pure hydrogen, and nitrogen-base atmospheres. Among these atmospheres, endo-

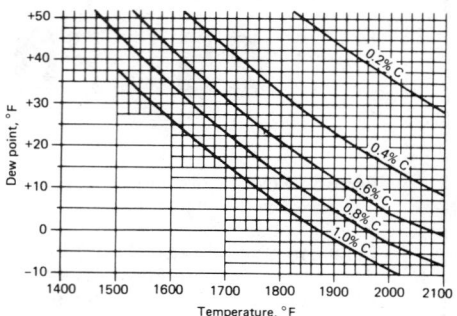

Fig. 4. Carbon potential of endothermic gas as a function of dew point and temperature

thermic gas is probably the most widely used, followed by dissociated ammonia. Pure hydrogen atmospheres have rather limited use (mainly for aerospace applications) due to their high cost. Nitrogen-base atmospheres are fairly new for powder metallurgy applications; however, their application increases steadily because of low cost and extremely low dew point.

Endothermic gas is readily produced from natural gas at reasonable cost. However, control of carbon level is difficult because the carbon potential of endothermic gas varies with temperature and dew point, as shown in Fig. 4. The carbon potential of endothermic gas with a dew point of −1 °C (+30 °F) varies from approximately 0.2% at 1120 °C (2050 °F) to 1.0% at about 850 °C (1560 °F).

A typical sintering furnace consists of three zones: a preheating or delubricating zone, a high-heat or sintering zone, and a cooling zone. For conventional sintering at 1120 °C (2050 °F), mesh-belt conveyor furnaces are the most widely used. Green compacts are carried on a conveyor mesh belt made from nickel-chromium alloy wire. Mesh belt furnaces can only be used at temperatures up to 1150 °C (2100 °F), due to the temperature limitation of the belt material.

TYPICAL PROPERTIES OF PRESSED-AND-SINTERED PARTS

In porous P/M materials, both static and dynamic mechanical properties are strongly dependent on the degree of porosity. The pores act as stress raisers under the influence of an external stress, which enhances crack propagation between pores and reduces the effective cross-sectional area of solid material. Generally, static properties such as tensile strength, yield strength, ductility and elastic modulus increase with decreasing porosity. Variations in the mechanical properties of sintered ferrous compacts made from six different powders are given as a function of density in Fig. 5 and 6. Because of the presence of porosity, the apparent hardness of porous P/M parts is lower than the actual particle hardness. Relationships between apparent hardness and particle hardness for the various density levels are given in Fig. 7.

Another common technique for providing P/M parts for high-stress applications is infiltration. Infiltration is a process in which interconnected pores in a P/M compact are filled with a liquid metal or alloy whose melting point is much lower than that of the compact. The driving force of infiltration is the minimization of the total surface-free energy of the system.

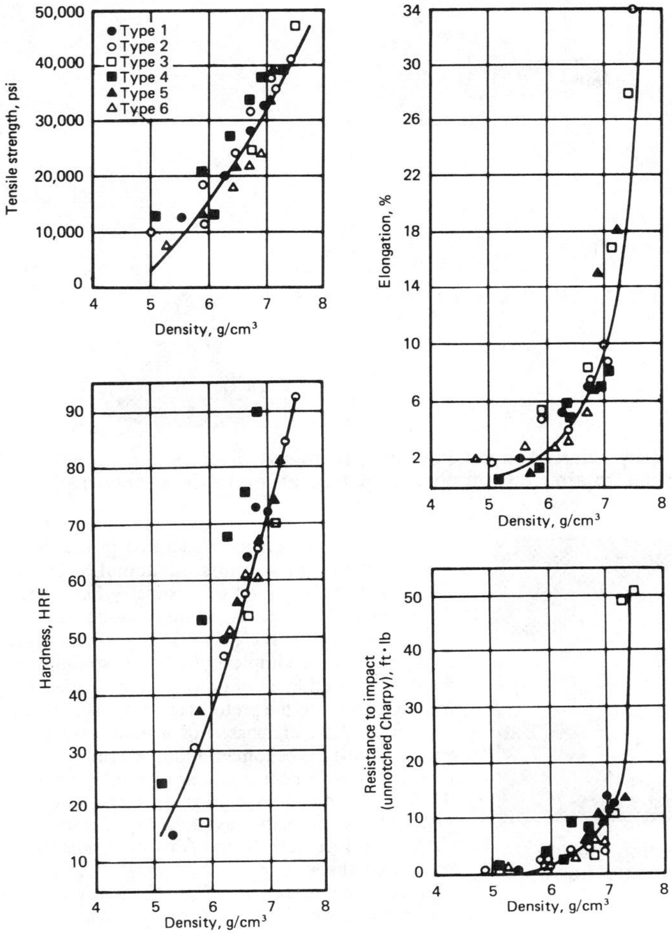

Fig. 5. Mechanical properties of iron compacts as a function of sintered density

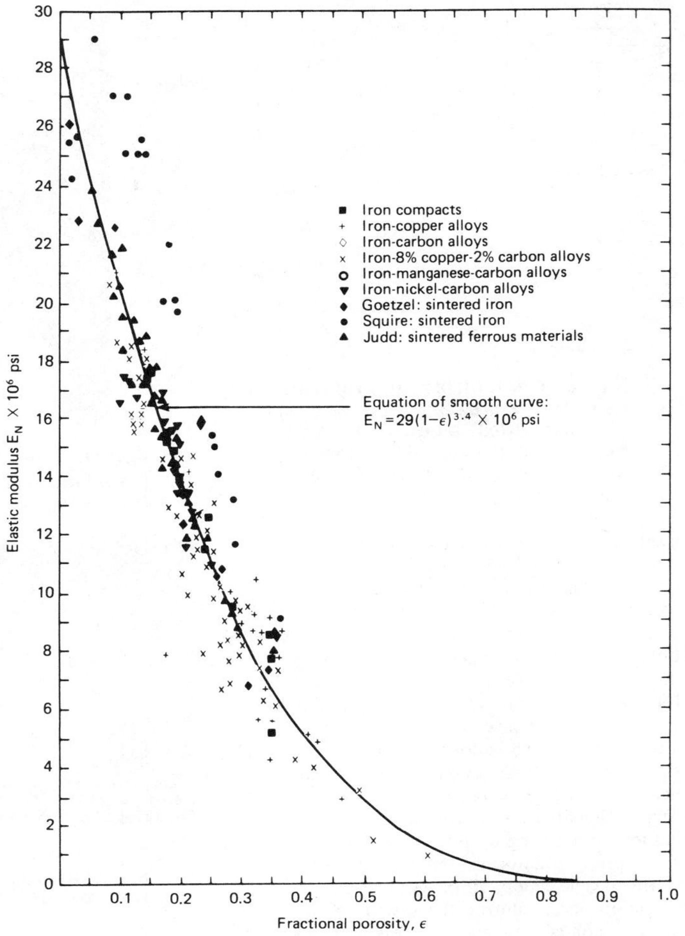

Fig. 6. Elastic modulus of ferrous sintered materials as a function of density

Equation of smooth curve:
$E_N = 29(1-\epsilon)^{3.4} \times 10^6$ psi

- Iron compacts
- Iron-copper alloys
- Iron-carbon alloys
- Iron-8% copper-2% carbon alloys
- Iron-manganese-carbon alloys
- Iron-nickel-carbon alloys
- Goetzel: sintered iron
- Squire: sintered iron
- Judd: sintered ferrous materials

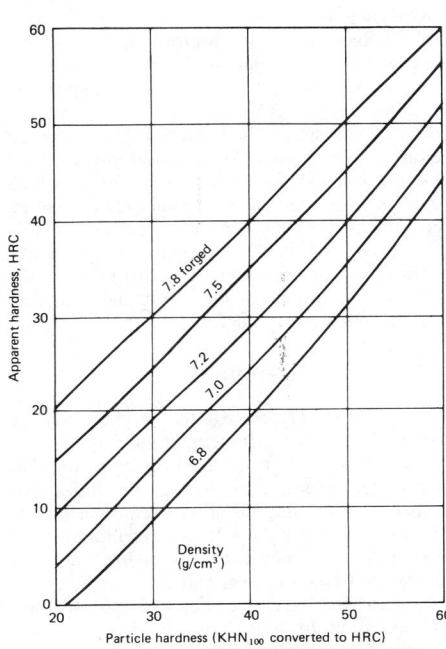

Fig. 7. Relationships between apparent hardness and particle hardness at various density levels

Some of the advantages of infiltration of P/M parts are:

- Increased mechanical properties such as tensile strength, hardness, impact resistance and fatigue strength
- Uniform density in parts
- Production of high-density parts (up to 7.2 g/cm³) by a single pressing and sintering operation
- Elimination or sealing of surface pores for secondary operations such as pickling and plating
- Production of P/M parts with selective density variations
- Assembly of multiple parts by sintering the individual pieces together and bonding the pieces into one part through common infiltration.

Fully Dense Parts

FERROUS PARTS of near and full theoretical density can be produced by a variety of thermomechanical methods. These methods involve various amounts of heat and mechanical work to aid densification of the powder mass. For production of low-alloy steel parts, powder forging has received the most commercial attention. This section presents the basic powder forging process steps and the fundamental relationships that have

been discovered between process variables and mechanical properties of products forged from sintered preforms.

Powder forging may be defined as the process of converting a porous preform into a fully dense part by forging in a trapped die cavity. Material utilization is high because flash is avoided, and complex shapes with net or near-net surfaces may be forged.

The basic steps of the powder forging process are:

1. Production of a preform
2. Preheating of the preform and tooling (for warm and hot forging)
3. Forging
4. Part finishing, including heat treatment, machining, coating, etc.

The first step is identical to conventional press/sinter practices. The shape and density of the preform are important to final properties, and are discussed in the section on preform design. Because these parts are fully dense after forging, finishing is similar to finishing of conventionally produced parts.

The two forms of powder forging are re-pressing, which is also known as minimum deformation processing or hot coining, and forging. The difference between these methods lies in the

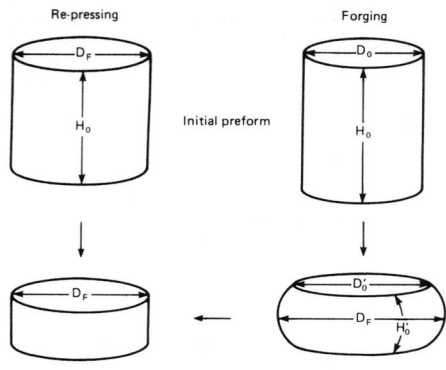

Fig. 8. Geometries of preforms used for re-pressing and forging of a cylindrical slug. In re-pressing, only the height, H_0, changes; in forging, both diameter and height change.

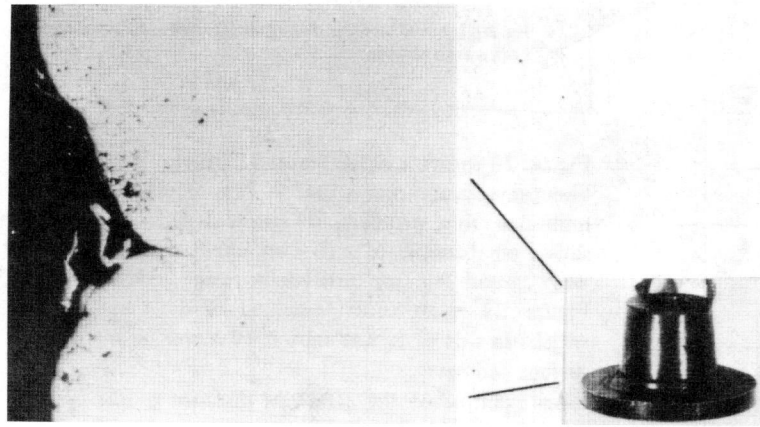

Fig. 10. Defect caused by internal metal flow in a two-level preform during re-pressing. Preform corner rotated and moved onto hub as metal flow occurred from flange into hub.

mode of pore closure. Preforms for re-pressing are similar to the final forged shape. Re-pressing involves densification by axial collapse of porosity (see Fig. 8). Over-all deformation of the preform is restricted to strain in the axial direction, with die-wall contact occurring very early in the densification process to limit lateral deformation. Forging involves deformation of a simple preform shape into a final forged shape. The preform shape is not as complex as the final shape, and a large amount of deformation is needed to change the preform shape into the final shape. Densification occurs as a result of shear deformation during upsetting, which takes place as the preform deforms both axially and laterally to fill the die cavity (Fig. 8). The bulk of densification occurs during the upsetting stage. As die-wall contact is made, re-pressing takes over, but in this case die fill rather than densification is the major consequence of the final re-pressing stage.

PREFORM DESIGN

Selection of a preform to be forged to a final part shape is the critical step in powder forging. Preform shape and density directly affect the magnitude of mechanical properties that are achieved in the final part. Porosity in the preform drastically reduces the workability of the preform material, which means that cracking during forg-

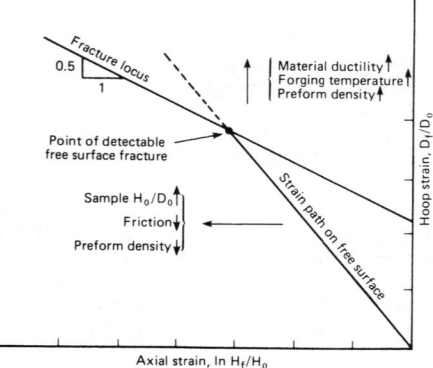

Fig. 9. Qualitative effects of forging variables on formability of porous preforms during hot forging

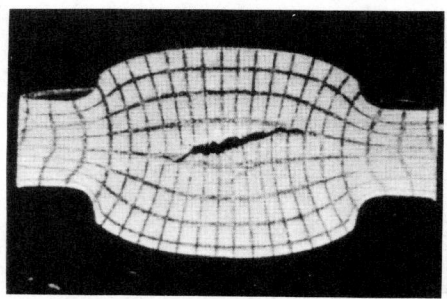

Fig. 11. Internal bursting during simultaneous forward and backward extrusion

ing can be a serious problem. Historically, successful preform designs have been developed through iterative trial-and-error procedures, using prior experience as a guide. Computer-aided design techniques are emerging that minimize trial and error and shorten forging development time and costs significantly.

A method of characterizing workability where free surface fracture is dominant, as is the case during upsetting, has been developed for wrought materials and has been successfully applied for porous preforms. Figure 9 shows a typical workability line (labeled "fracture locus"), a typical strain path, which is the local surface strain state generated during forging, and the effects that process variables have on these lines. Parts that can be forged with strain levels that remain below the line (slope = -0.5) can be forged without cracking. If strain levels above the line (unsafe region) are required to form the part, surface cracking is probable. As shown in Fig. 9, the position of the workability line is dependent on preform density and forging temperature. The strain to which a preform is subjected is dependent on aspect ratio (geometry), friction and preform density. Surface fracture may be avoided by adjusting the deformation temperature, preform density, preform shape, or friction conditions during forging.

In forging of complex shapes from simple preforms, defects can be formed by internal cavitation and by surface fracture at die-contact areas. An example of internal bursting, or cavitation, is shown in Fig. 10. In Fig. 11, high friction gradients are shown to cause surface fracture under

die faces. These defects, generated in the laboratory, accurately simulated actual defects. Workability rules based on geometry have been developed in order to avoid these defects.

Preform design for producing a single-level shape such as a simple cylinder is straightforward. A desired level of deformation is selected, and dimensions of the preform are calculated based on conservation of mass. For a multilevel part, preform design becomes more complex. Re-pressing is no longer a simple task of matching the final shape and preform shape through an adjustment for porosity, as it is for single-level shapes. Surface defects and porosity gradients in re-pressed shapes can be problems. An example of a lap defect is shown in Fig. 12. Several techniques are available to overcome these problems, including the use of preforms with different densities in different levels, and the use of multiple punches. These methods control local metal flow at junctions between part levels. Special preform shapes may also be used for this purpose.

Guidelines concerning preform design of complex shapes can be gathered from forging of generic shapes. By reducing complex shapes into combinations of generic shapes, preform sections can be designed, with the final preform shape being a combination of several preform sections. Figure 13 shows four possible preforms which could be forged into an axisymmetric part having cup and hub sections. Preform geometries in parts (b), (c) and (d) result in defective forgings due to cracking along the center bore off the hub, the internal cup surface, or the top of the cup. Preform (a) resulted in a defect-free forging.

FORGING

The critical variables of the forging step are forging temperature, die temperature, transfer of the heated preform from the furnace to the die, ejection temperature, and lubrication. Press type is also an important variable, because it affects time in the die and the rate of consolidation and deformation. Dimensional precision and part quality are based on press and die considerations and on forging variables. For steel powder preforms, the preform is typically coated with a graphite lubricant, preheated under a protective atmosphere to a preselected forging temperature (normally 980 to 1120 °C, or 1800 to 2050 °F), transferred in air to the forging die, and forged

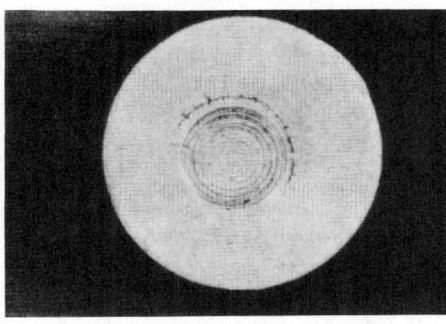

(a)

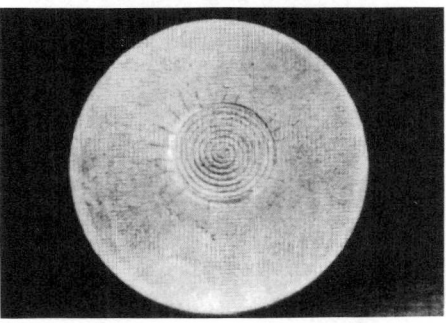

(b)

Both radial and circumferential cracks are visible in the vicinity of the transition (a and b). In addition to the cracks, interconnected pores were opened in the smooth outer region.

Fig. 12. Top views of disks compressed between dies having a grooved central region and a smooth outer region with grease lubrication

in one blow to final shape and full density. Tooling is heated to 150 to 370 °C, or 300 to 700 °F), to minimize die chill, and it is lubricated with a graphite lubricant.

Die design centers on the use of trapped dies to produce a flashless part. Die dimensions must take into account the forging temperature, the die preheat temperature, the forging load, the elastic strain of the die under load and on release of the forging load, the elastic strain of the forging on ejection from the die, and the temperature of the part at ejection. Part dimensional tolerances can be met only when these features are satisfactory. Once the die dimensions have been selected, the dimensions of the forged part can be controlled somewhat by adjusting the die and forging temperatures. For example, a part that is consistently undersize for a given set of forging conditions may be brought into size specification by increasing the die temperature and/or by decreasing the forging temperature. Similarly, if a part being forged is consistently oversize, the die temperature may be decreased and/or the forging temperature may be increased.

Transfer of the heated preform to the die must be rapid, because cooling and internal oxidation are to be avoided. Depending on forging temperature and density, there is a finite time period, ranging from 4 to 20 s, during which the preform is protected by residual furnace atmosphere. Transfer must be accomplished within this safe time period. Transfer time must also be repeatable for achievement of dimensional uniformity.

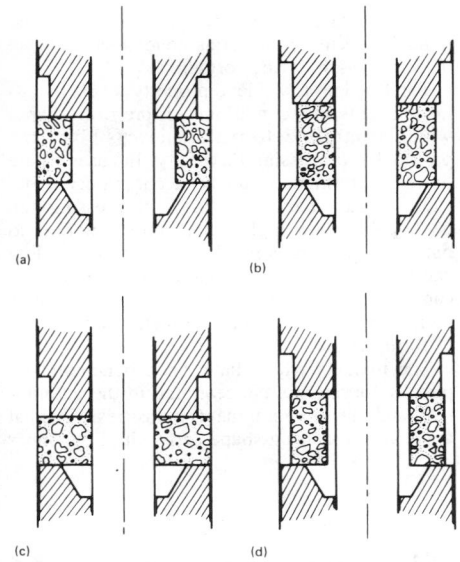

Fig. 13. Schematic illustration of the four ring-preform options for production of a complex part having cup and hub sections

Press speed affects directly the duration of contact between the part being forged and the die, the rate of pore closure, the rate of deformation, and the dwell time under load. Contact time dictates the amount of chilling of the part by the die and the amount of heating of the die by the part. Chilling of the preform can lead to residual porosity in surface sections. This surface-related porosity can act as a stress raiser and reduce toughness and fatigue resistance. Excessive die heating due to long contact times can cause overtempering of the tooling and can shorten die life. Because of these two negative features of contact time, fast-acting presses are desirable. Most commercial powder forging is performed on crank presses. Screw presses are seeing increased use due to even faster cycle times and lower contact

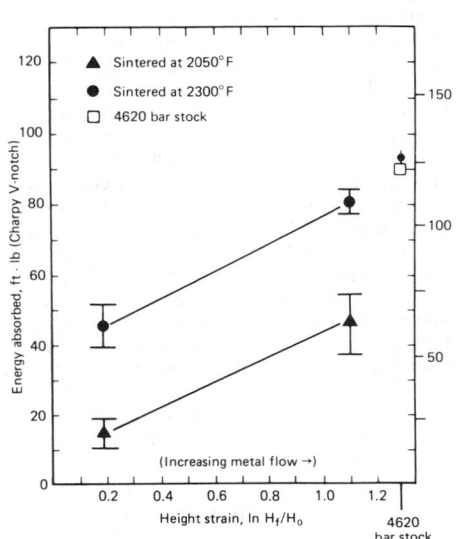

Fig. 14. Toughness of 4620 P/M forgings as a function of flow during forging and of sintering temperature

times, and improved control of part thickness.

Warm and cold forging of ferrous powder preforms are also suitable manufacturing processes. Preform workability is reduced as the working temperature is reduced, but control of part size increases as thermal expansions and contractions are eliminated. Shapes that can be formed under compressive strain states, such as bearing cups, are suitable for cold forging. Other low-carbon ferrous parts for automotive applications have been cold forged on a commercial basis.

MECHANICAL-PROPERTY RELATIONSHIPS OF FORGED POWDER PARTS

For powder forged parts, final mechanical properties are dependent on sintering conditions and deformation during part shaping. By control of these two features, the mechanical properties of a part may be optimized.

Strength, ductility, toughness and fatigue resistance increase as porosity is eliminated. The last three properties are especially sensitive to porosity, and they increase dramatically as the last traces of porosity are removed. This effect is shown for toughness in Fig. 14. Elimination of porosity in highly stressed regions of a part is essential for achievement of property levels comparable to wrought property levels.

As the level of upsetting is increased during the forging process, the magnitude of structure-sensitive properties increases. Figure 15 shows improvement in values of absorbed Charpy V-notch impact energy as deformation level increases. Figure 16 shows a similar effect for fatigue resistance. An interesting and useful feature of powder forgings is that deformation does not significantly degrade through-thickness properties as it does for wrought forgings. For repressed parts, through-thickness toughness is slightly lower than longitudinal toughness. Upsetting increases toughness in the longitudinal direction while toughness in the through-thickness direction remains relatively constant.

Figure 14 shows that sintering variables affect toughness. In this case, the higher sintering temperature of 1260 °C (2300 °F) results in a lower oxygen content and higher toughness than for material sintered at 1120 °C (2050 °F). Particle bonding during forging is enhanced due to cleaner

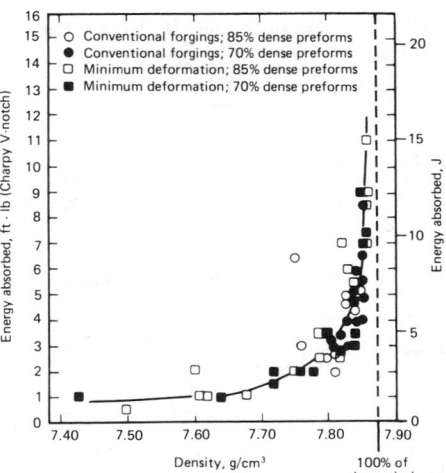

Fig. 15. Charpy V-notch strength of P/M forgings as a function of density

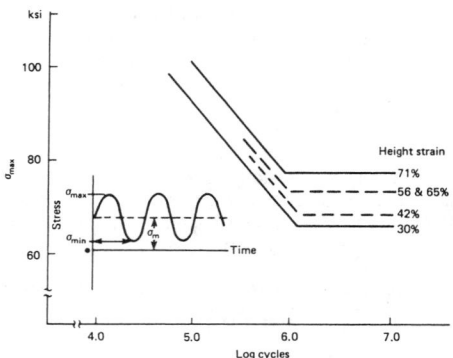

Fig. 16. Axial fatigue of P/M forged 4620 steel as a function of height strain during forging

particle surfaces. Toughness can be controlled through selection of sintering conditions and level of deformation during forging.

The freedom to achieve property levels by adjusting sintering conditions or preform shape, which controls deformation level, gives the partsmaker processing flexibility. In cases where limited deformation can be accommodated because of part-shape constraints, re-pressing can be coupled with high-temperature sintering to maximize properties for that part. If lateral flow can be accommodated, high dynamic properties can be achieved at normal sintering temperatures by maximizing the amount of metal flow needed to form the part.

The markets which have been penetrated by powder forging are concentrated in the automotive and garden-equipment industries. Typical applications are ring-shape parts with large inner

diameters, for which powder forging offers high material utilization, net or near-net surfaces, and high quality. Bearings, transmission components and other ring-shape components have proved successful economically and from a performance standpoint. The automotive market is viewed as the most viable market because of the high volume requirements considered necessary for economical implementation of powder forging. Typically, production runs of 50 000 or more parts are thought to be needed to make powder forging cost-competitive. Advances in methods for fast tooling changes will reduce this volume level. With innovative tooling concepts, fast tooling changes and increased confidence in performance of powder forgings, the volume criterion should drop to levels around 5000 to 10 000 parts per run. This will open powder forging to a much wider range of markets.

Stainless Steel Powder Metallurgy

By Donald L. Dyke, SSI Technologies, Inc.

STAINLESS STEEL powder metallurgy parts represent an important and interesting segment of the P/M industry. The volume of stainless steel parts produced doubled from 1970 to 1980. Most of these parts are used in environments that are only mildly corrosive, and they have performed well. However, stainless steel powder metallurgy is still a young and emerging segment of the P/M industry.

POWDERS

All commercial compacting-grade stainless steel powders are produced by water atomization. This manufacturing technique was developed in the early 1950's and opened the door to the growth of stainless steel powder metallurgy. This process produces a powder that is fully alloyed and that has an irregular particle shape and sufficient compressibility and green strength to be suitable

Table 1. Typical sieve analysis of P/M stainless steel powder

U.S. standard sieve size		Particle size, μm	Percentage (wt %)
	+100	>149	0 to 3
−100	+140	105 to 149	10 to 15
−140	+200	74 to 105	15 to 25
−200	+325	44 to 74	20 to 30
−325		<44	35 to 50

Table 2. Nominal compositions of stainless steel powders

Grade	Composition(a), %						
	Ni	Cr	Mo	Si	Mn	S	C
Austenitic:							
303L	. . . 12.5	17.5	. . .	0.7	0.35	0.2	0.02
304L	. . . 10.5	18.5	. . .	0.7	0.25	. . .	0.02
316L	. . . 13.0	17.0	2.1	0.7	0.20	. . .	0.02
Ferritic:							
434L	. . . . . .	17.0	1.0	0.7	0.25	. . .	0.02
Martensitic:							
410L	. . . . . .	13.0	. . .	0.7	0.40	. . .	0.02
(a) Remainder Fe.							

for P/M fabrication. A typical sieve analysis of such a powder is presented in Table 1.

The chemical compositions of stainless steel powders (see Table 2) generally meet the AISI limits for wrought alloys, but in practice are held to closer tolerances. Low carbon contents, and high nickel in the austenitic alloys, contribute to good compressibility. The austenitic grades are the most widely used, with type 316L alone accounting for more than one-third of all stainless steel powder.

COMPACTION

Design and processing of stainless steel P/M parts are subject to the same basic considerations as for other P/M materials. Compared with low-alloy ferrous compositions, however, stainless steel P/M parts require higher compacting pressures and exhibit lower green strength. Typical compaction behavior is illustrated in Fig. 1. Commercial compacting pressures range from 552 to 827 MPa (40 to 60 tsi). For stainless steel compacts, as for compacts of other ferrous powders, green strength is influenced by compacting pressure and type of lubricant. At best, however, green strength of P/M stainless steel is only about half that for P/M irons, and complex parts require special care in ejection and handling.

As with other powders, lubricant choice involves a compromise. Those lubricants which give high green strength generally cause lower compactibility, and vice versa. Lubricant selection is more important to successful fabrication of stainless steel than to fabrication of any other metal powder. Lithium stearate and synthetic waxes are popular lubricants.

SINTERING

Stainless steels for some applications are sintered at 1120 to 1150 °C (2050 to 2100 °F). When improved mechanical properties and corrosion resistance are required, however, temperatures as high as 1315 °C (2400 °F) are used. Contin-

uous mesh-belt furnaces are suitable up to about 1150 °C. Manual or automatic pushers, walking beam and vacuum furnaces are used at higher temperatures.

The most common commercial atmosphere for sintering stainless steel is dissociated ammonia. A dew point of −45 to −50 °C (−50 to −60 °F) at the dissociator is usually adequate for oxide-free work, but to ensure that no discoloration occurs during cooling and to allow some latitude in the sintering process, the dissociated ammonia is sometimes dried to a dew point of −60 °C (−80 °F) or better before being introduced into

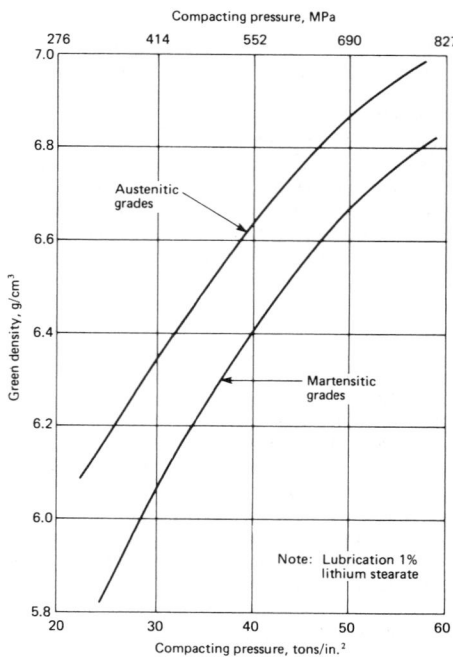

Fig. 1. Typical compaction behavior of stainless steel powders

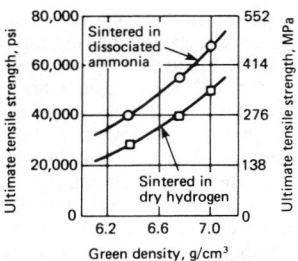
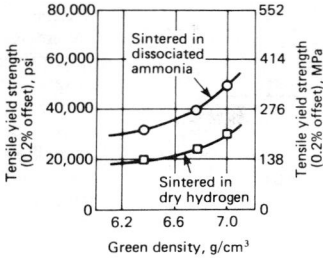
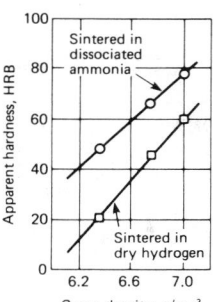

Fig. 2. Effects of green density and sintering atmosphere on tensile strength, yield strength and apparent hardness of sintered type 316L stainless steel powder

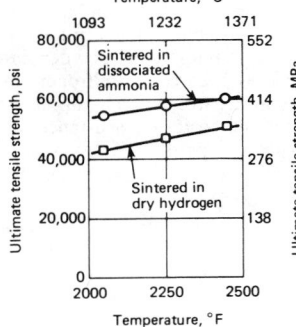
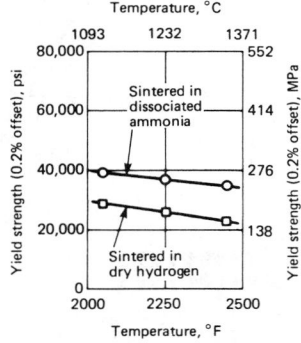
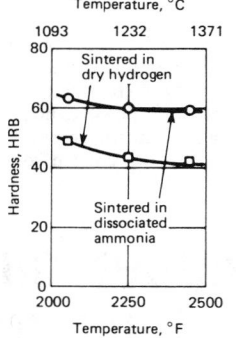

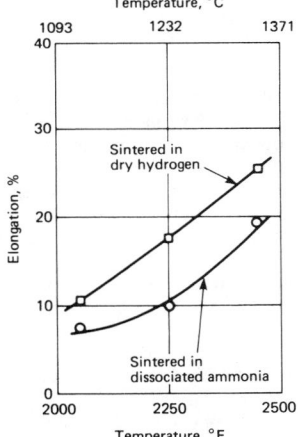

Fig. 3. Effects of sintering temperature and sintering atmosphere on tensile strength, yield strength, apparent hardness and elongation of type 316L stainless steel powder pressed to a density of 6.85 g/cm³ and sintered for 30 min

the furnace. The nitrogen in the dissociated ammonia is definitely not neutral to stainless steel. The effects of nitrogen on the mechanical properties of sintered type 316L can be seen in Fig. 2 and 3 by comparing the properties after sintering in hydrogen and dissociated ammonia. The higher nitrogen contents do not appear to be detrimental in most applications.

Although considerable data on hydrogen-sintered stainless steel has been developed, there is little commercial use of hydrogen because of its high cost. The principal alternative to dissociated ammonia is vacuum. Conventional cold-wall vacuum furnaces with mechanical pumping systems are used. A pressure of 200 to 1000 μm of nitrogen or argon frequently is maintained at the sintering temperature to prevent loss of elements such as chromium through vaporization.

Various mixtures of nitrogen and hydrogen gases offer another alternative to dissociated ammonia as a protective atmosphere for sintering stainless steel. It has been demonstrated that many parts can be successfully sintered in production with as little as 5% hydrogen in the atmosphere, in contrast with the 75% in dissociated ammonia.

PROPERTIES

The properties of sintered stainless steel increase with increasing density. Figure 2 shows the mechanical properties and hardness of type 316L as a function of green density and sintering atmosphere. Densities of 6.2 to 7.0 g/cm³ cover a large percentage of parts produced. Higher densities can be achieved with high-temperature sintering or multiple pressing.

The effects of sintering temperature on the mechanical properties of type 316L are shown in Fig. 3. Properties of "vacuum" sintered 316L are listed in Table 3, and properties of ferritic type 434L are given in Table 4.

Martensitic type 410 powder contains less than 0.03% carbon, but sufficient hardening is obtained by sintering in dissociated ammonia to satisfy many applications. When sintering is done in an atmosphere without nitrogen, it is necessary to add approximately 0.15% graphite to the powder. Table 5 shows the effects of sintering atmosphere, graphite additions and heat treating conditions on the strength and hardness of type 410 stainless steel at a density of 6.7 g/cm³ sintered at 1120 °C (2050 °F) for 30 min.

CORROSION RESISTANCE

The primary reason for selecting a stainless steel composition is usually its corrosion resistance. Because the corrosion behavior of P/M materials is not as well documented as that of comparable wrought materials, it is advisable to perform functional environmental tests on P/M parts. The corrosion resistance of sintered stainless steel depends on the processing of the material, on its porosity and on the environment to which it is exposed. Some generalities pertaining to corrosion resistance can be stated:

• High-temperature sintering usually improves corrosion resistance, especially in nitrogen-containing atmospheres.

• The carbon content must be maintained as low as possible. With proper housekeeping and control of furnace atmospheres, it can be held below 0.03%.

• Contamination with iron particles accounts for

Table 3. Properties of type 316L stainless steel sintered for 140 min at 1300 °C (2375 °F) in a vacuum furnace under a pressure of 300 μm of nitrogen and quenched in nitrogen

Compacting pressure		Density, g/cm³	Tensile strength		Yield strength(a)		Elongation(b), %
MPa	tsi		MPa	ksi	MPa	ksi	
552	40	6.9	548	79.5	359	52.0	16.0
690	50	7.1	590	85.6	383	55.6	16.6
827	60	7.2	592	85.9	381	55.3	18.2

(a) At 0.2% offset. (b) In 25 mm or 1 in.

Table 4. Properties of type 434L stainless steel sintered for 30 min at 1315 °C (2400 °F) in dissociated ammonia

Compacting pressure		Density, g/cm³	Tensile strength		Yield strength(a)		Elongation(b), %
MPA	tsi		MPa	ksi	MPa	ksi	
552	40	7.0	400	58.0	253	36.7	17
690	50	7.1	425	61.6	270	39.1	19
827	60	7.2	454	65.8	284	41.2	21

(a) At 0.2% offset. (b) In 25 mm or 1 in.

Table 5. Properties of sintered type 410 stainless steel

Processing treatment	Graphite added, %	Sintering atmosphere	Tempering temperature °C	°F	Tensile strength MPa	ksi	Apparent hardness, HRB
As sintered and cooled	0	NH₃	...		724	105	102
in water-jacketed	0.10	NH₃ 205	400		683	99	103
zone of furnace	0	H₂	...		393	57	68
	0.10	H₂ 175	350		710	103	95
Reheated in NH₃ and	0	NH₃ 205	400		627	91	106
oil quenched from	0.10	NH₃ 220	430		703	102	102
950 °C (1750 °F)	0	H₂	...		752	109	106
	0.10	H₂ 220	430		717	104	105
Reheated in H₂ and oil	0	NH₃ 205	400		731	106	104
quenched from 950 °C	0.10	NH₃ 205	400		745	108	104
(1750 °F)	0	H₂ 205	400		641	93	95
	0.10	H₂ 220	430		800	116	101

many instances of development of rust spots in atmospheric exposure or in other relatively mild environments. Isolation of the equipment used to process stainless steel is a reliable method of solving this problem.

Recent studies have shown that corrosion resistance increases with increasing part density in acid environments and decreases with increasing density in chloride environments. The reason for the latter phenomenon is that higher density results in smaller pores, which, in turn, are more sensitive to crevice corrosion.

Other studies have shown that some of the discrepancies among the effects of sintering atmospheres reported in the literature may be due to differences in oxygen and nitrogen contents. It has been shown, for instance, that the major surface oxide present on atomized powder particles as well as on sintered parts is not chromium oxide but silicon oxide. In nitrogen-containing sintering atmospheres, it is now well established that fast cooling after sintering is essential for preventing additional nitrogen absorption and precipitation of chromium nitrides. This process is analogous to the well-known carbide sensitization which also leads to chromium depletion and loss of corrosion resistance.

Prealloyed austenitic stainless steel powders containing small amounts of tin recently have become commercially available. Nitrogen has a low solubility in tin which is concentrated in the outermost layers of the stainless steel powder particles. The tin acts as a barrier to nitrogen absorption and chromium carbide precipitation and thus provides improved corrosion resistance, especially in lower-density parts.

Copper and Copper Alloy Powder Metallurgy

By Erhard Klar, SCM Metal Products

POROUS BRONZE BEARINGS, which were developed in the 1920's as the first major copper-base P/M application, still account for about 50% of all copper-base P/M materials. Other important copper-base P/M applications include friction materials, structural parts, copper-carbon brushes, and filters. Of more recent origin is the development of oxide-dispersion-strengthened copper for high-performance applications. Non-P/M applications of copper and copper alloy powders, and of copper flake, include uses in the chemical industry and as catalysts, coatings, metallic greases and decorative paints.

The following is a description of the production of copper and copper alloy powders and of the manufacture of the major types of P/M parts made from these powders. For more details on powder production, on manufacture of parts and on applications, see the corresponding articles in Metals Handbook, 9th Edition, Volume 7, Powder Metallurgy.

POWDER PRODUCTION

There are four major methods for production of plain copper powders: oxide reduction, atomization, electrolysis and hydrometallurgy. Atomization is also used to produce copper alloy powders. Over the years, these methods have varied in relative importance due to differences in cost and powder properties. Although copper powders made by different processes may be used interchangeably in some applications, powders made by oxide reduction and by atomization presently account for the bulk of the powders used.

Production of Copper Powder by Oxide Reduction

Sources of raw materials for production of copper powder by oxide reduction include copper scale, cement copper, particulate copper scrap

and atomized copper. The last two are initially roasted to form copper oxide. The particulate copper oxide is then reduced on a stainless steel belt in a continuous belt furnace. Suitable reducing atmospheres are hydrogen, dissociated ammonia, water-reformed natural gas, and similar gas mixtures. Reduction temperatures range from about 425 to 650 °C (800 to 1200 °F). The reduced copper oxide forms a more or less sintered porous cake of copper, which is ground into powder.

Oxide-reduced copper powders can be tailored so as to develop the properties needed in all major applications of copper powders. Oxide-reduced copper powders are of irregular particle shape, and the individual powder particles are porous (see Fig. 1). It is largely through control of particle size and particle pore structure that compacting, sintering, and engineering properties of the sintered parts can be controlled over

a wide range. Some powder grades are treated with antioxidants to stabilize them against oxidation. Table 1A lists properties of various grades of copper powders made by oxide reduction. Product brochures and data sheets from powder producers generally provide details on characteristics and performance properties of powders as well as recommendations for specific uses.

Production of Copper Powder by Atomization

Disintegration of a liquid stream by means of an impinging jet of liquid or gas is known as atomization. Nearly spherical powders, of high apparent density and unsuitable for die compaction, result from atomization of liquid copper or liquid copper alloys with inert gas (see Fig. 2). Spherical copper powders are used in copper flake production by ball milling.

Most atomized copper and copper alloy powders are made by water atomization. The particle shape of such powders is irregular (see Fig. 3 and 4). In the case of plain copper, the copper is melted and superheated to about 1150 to 1200 °C (2100 to 2200 °F) and atomized with pressurized water at a pressure of 10 to 14 MPa (1500 to 2000 psi). This results in a predominantly −100 mesh powder. Variation of water pressure permits control of average particle size over a wide range. Powders of which 60 to 80% comprise −325 mesh particles are feasible. Particle shape can be controlled to a lesser degree through addition of small amounts of magnesium to the molten copper prior to atomization. Properties of atomized copper powders are given in Table 1B.

Production of Copper Powder by Electrolysis

In production of copper powders by electrolysis the processing variables are adjusted so that a spongy deposit is formed rather than a smooth deposit as in electroplating. This is achieved pri-

Fig. 1. Scanning electron micrograph of copper-oxide-reduced copper powder

Table 1A. Properties of commercial grades of copper powder produced by the copper oxide process

Copper	Tin	Graphite	Lubricant	Hydrogen loss	Acid insolubles	Apparent density, g/cm³	Hall flow rate, s/50 g	+100	+150	+200	+325	−325	Green density, g/cm³	Green strength, MPa (psi) at: 165 MPa (12 tsi)	6.30 g/cm³
99.53	...	...	...	0.23	0.04	2.99	23	0.3	11.1	26.7	24.1	37.8	6.04	6 (890)	...
99.64	...	...	...	0.24	0.03	2.78	24	...	0.6	8.7	34.1	56.6	5.95	7.8 (1140)(a)	...
99.62	...	...	...	0.26	0.03	2.71	27	...	0.3	5.7	32.2	61.8	5.95	9.3 (1350)(a)	...
99.36	...	...	...	0.39	0.12	1.56	...	0.1	1.0	4.9	12.8	81.2	5.79	21.4 (3100)(a)	...
99.25	...	...	...	0.30	0.02	2.63	30	0.08	7.0	13.3	16.0	63.7	...	...	8.3 (1200)(a)
90	10	...	0.75	...	...	3.23	30.6	0.0	1.4	9.0	32.6	57.0	6.32	...	3.80 (550)
88.5	10	0.5	0.80	...	...	3.25	12(b)	...	...	...	...	...	...	...	3.6 (525)

(a) Measured with die-wall lubricant only. (b) Carney flow.

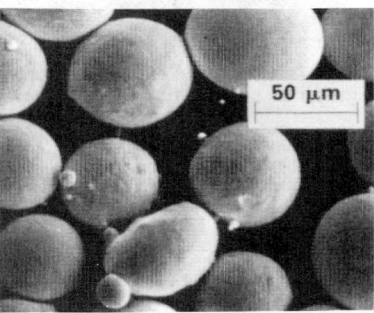

Fig. 2. Scanning electron micrograph of nitrogen-atomized copper powder

Fig. 3. Scanning electron micrograph of water-atomized copper powder (apparent density, 3.04 g/cm³)

Table 1B. Properties of commercial grades of water- and gas-atomized copper powders

Copper, %	Hydrogen loss	Acid insolubles	Hall flow rate, s/50 g	Apparent density, g/cm³	+100	−100+150	−150+200	−200+325	−325
99.65(a)	0.28	...	...	2.65	Trace	0.31	8.1	28.2	63.4
99.61(a)	0.24	...	...	2.45	0.2	27.3	48.5	21.6	2.4
99.43(a)	0.31	...	...	2.70	Trace	0.9	3.2	14.2	81.7
>99.1(b)	<0.35	<0.2	~50	2.4	<8	17-22	18-30	22-26	18-38
99.1	0.77	...	No flow	4.8	Trace	3	...	...	...
99.2	<0.7	...	9-13	4.9-5.5	7-14	←20-30→	←20-30→	15-30	30-50

(a) Water atomized plus reduced. (b) Contains magnesium.

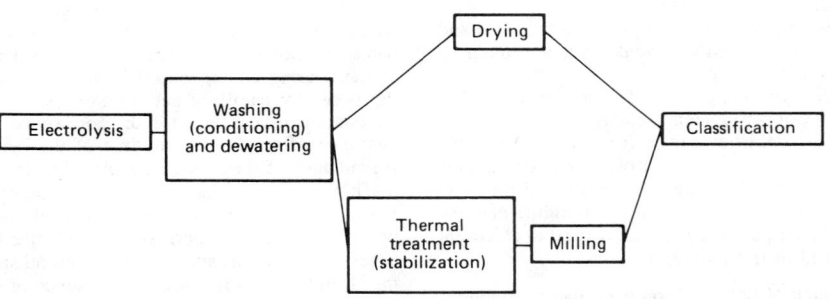

Fig. 5. Processing stages in production of copper powder by electrolysis

marily through use of a very high current density.

The processing stages of electrolytic copper powder production are shown schematically in Fig. 5. Typical electrolysis conditions include an electrolyte concentration of 5 to 8 g/L (0.7 to 1.1 oz/gal) of copper and 100 to 160 g/L (13 to 21 oz/gal) of sulfuric acid, a bath temperature of about 50 °C (120 °F), a current density of 540 to 1080 A/m² (50 to 100 A/ft²) and a cell voltage of about 1 V. Generally, the anodes are of electrolytically refined copper, and the cathodes are of lead alloy sheet. A broad range of particle sizes can be obtained through control of processing conditions, including the use of addition agents to the electrolyte. The particle shape of

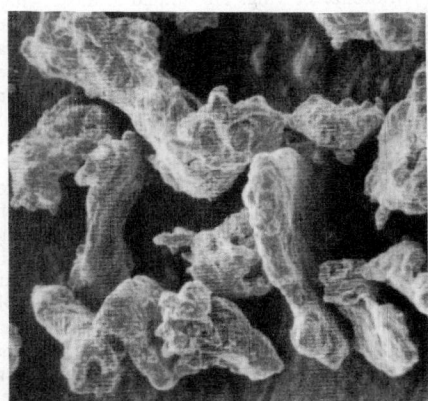

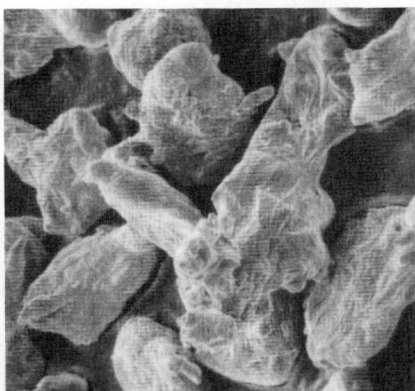

Fig. 4. Photomicrographs, at 165×, of prealloyed, air-atomized copper alloy powders. Left: brass (80Cu-18Zn-2Pb). Center: nickel silver (63Cu-18Ni-17Zn-2Pb). Right: bronze (89Cu-9Sn-2Zn).

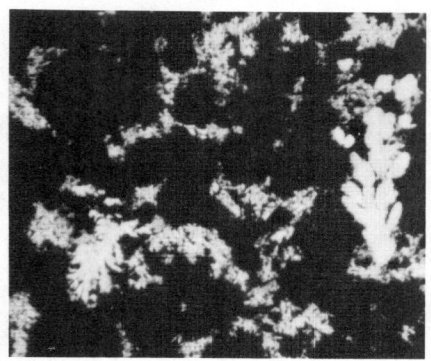

Source: "Technical Data—AMAX Metal Powders," AMAX Copper, Inc., 1968.

Fig. 6. Photomicrograph, at 85×, of electrolytic copper powder, showing dendritic structure

as-deposited powder is dendritic or fernlike (Fig. 6). Thorough washing of the deposited powder and treatment with surfactants are essential in order to prevent the powder from becoming oxidized. After being washed, the powder may be dried and used, but for most applications it is subjected to low-temperature reduction (at 480 to 760 °C, or 900 to 1400 °F). This furnace operation permits further control over particle size and shape. The reduced sinter cake is then broken up and ground into powder.

Electrolytic copper powders are of very high purity. Copper contents easily exceed 99.5%. Apparent densities range from 1 to 4 g/cm^3. Properties of typical electrolytic powder blends are summarized in Table 2. Because of their excellent purity and high green strength, electrolytic copper powders are widely used in electrical parts and in friction applications.

Production of Copper Powder by Hydrometallurgy

In this method, the basic processing steps consist of preparing a pregnant liquor by leaching copper ore or another suitable raw material, followed by precipitation of copper from this solution. The most important precipitation methods are cementation with iron, reduction with hydrogen or sulfur dioxide, and electrolysis. Use of several leach-precipitation steps, or the inclusion of flotation, solvent extraction or ion exchange, improves the purity of the final material. Due to economic factors, the only copper powder produced hydrometallurgically in the United States

today is cement copper, an impure copper powder precipitate obtained from copper sulfate solution by the addition of iron. Cement copper, like other hydrometallurgically produced copper powders, has low apparent density and high specific surface area (about 1 m^2/g). The powder particles are spongy and of irregular shape (Fig. 7). Cement copper often has higher green strength than most other copper powders, but its sintering activity is inferior because of the presence of finely divided unreducible oxides. Its primary use is in composite friction materials. Table 3 presents chemical analyses of cement copper from various sources.

MANUFACTURE OF PARTS AND APPLICATIONS

Self-Lubricating Porous Bronze Bearings

The major use of P/M bronze bearings is in fractional-horsepower electric motors. Industries that use these bearings include the automotive, home-appliance, consumer-electronics, business-machine, and industrial-equipment industries.

P/M oil-impregnated bronze bearings are popular because of their self-lubricating characteristics, which are made possible by an interconnecting pore system that acts as a lubricating oil reservoir. During use, as the bearing gets hot from friction between journal and bearing, the oil expands and flows to the bearing surface. After use, as the bearing cools, the oil is drawn back into the pores by capillary action. For many applications, the oil contained in the pore space of the bearing suffices for the lifetime of the bearing. Advantages of these bearings include simplified machine design, quiet operation, use in vertical positions for which solid bearings would be impractical because of lubricant run-out, use in inaccessible positions and freedom from oil splashing. Limitations arise from the presence of pores which lower strength and thermal conductivity, which in turn makes the bearings less useful under conditions of impact or fatigue loading and reduces the allowable frictional losses.

The most common compositions are 90Cu-10Sn with or without small amounts of graphite and/or lead. Graphite improves lubricity and is beneficial for interrupted use and for conditions of starved lubrication. Lead improves the tolerance toward misalignment of the bearing. Additions of iron reduce cost and increase strength but degrade corrosion resistance and general bearing properties.

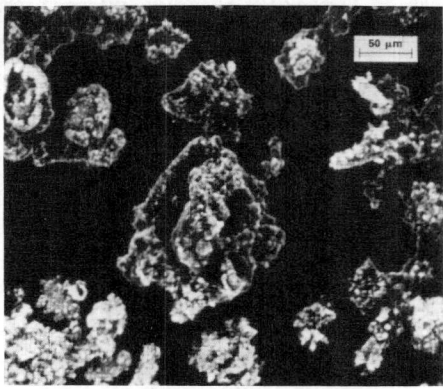

Fig. 7. Scanning electron micrograph of cement copper

Table 3. Chemical analyses of cement copper from various locations (dry basis)

Component	Composition, wt %, for location:			
	A	B	C	D
Total copper	75	83.0	87.4	85.0
Iron	6	2.4	0.7	10.0
Sulfur	1	0.5	...	1.1
Nitric acid insolubles	2	...	0.7	1.9
Hydrogen loss	16	...	...	...
Calcium oxide		0.08	...	...
Aluminum oxide (alumina)		1.2	0.5	...
Silicon dioxide		0.4	...	...
Lead		...	0.2	...
Oxygen		...	9.5	...

Most P/M bronze bearings are made from elemental blends of their constituents. Such blends provide compacting and sintering properties superior to those of prealloyed powders. Preblended and lubricated powder mixtures are available from powder producers. Typical lubricants, in amounts of 0.5 to 0.75%, include stearic acid, zinc stearate, and mixtures of these compounds.

Basic manufacturing procedures consist of compaction of the premixed powder to the appropriate shape and density (typically 70 to 82% of theoretical, corresponding to compacting pressures from 138 to 414 MPa, or 10 to 30 tsi), sintering to achieve a homogeneous metallurgical alpha bronze structure, oil impregnation, and sizing for dimensional precision. The shapes of P/M bronze bearings are limited to those which can be formed by axial compression. There are limits regarding maximum length-to-diameter ratios as well as length-to-wall-thickness ratios because of the requirement of reasonably uniform porosity along the axis of compression. Sizes of bushings range from 0.8 to 76 mm (0.03 to 3 in.).

Sintering of regular bronze bearings in a reducing atmosphere is accomplished at relatively low temperatures (815 to 870 °C, or 1500 to 1600 °F) and for short periods of time (3 to 8 min at temperature), made possible by the presence of a transient liquid phase during sintering. Diluted (i.e., iron-containing) and iron-base bearings require higher sintering temperatures and longer sintering times. After sintering, the bearings are sized for dimensional accuracy and then vacuum impregnated with high-grade turbine oil. Force fitting of the bearing into a housing is easily accomplished.

Table 2. Physical properties of typical copper powder blends
Source: "Technical Data—AMAX Metal Powders," AMAX Copper, Inc., 1968.

Apparent density, g/cm^3	Maximum flow, s/50 g	Screen analysis (mesh size), %				
		+100	−100+150	−150+200	−200+325	−325
2.4-2.6	32	0.5 max	5-15	25-35	25-42	22-32
2.5-2.6	32	0.2 max	1-11	13-23	20-37	43-53
2.45-2.55	33	0.2 max	3-13	17-27	23-40	33-43
2.5-2.6	35	0.2 max	1-10	9-19	24-31	55-65
2.7-2.8	32	0.2 max	1-10	7-17	15-32	54-64
2.5-2.6	40	0.2 max	1-6	5-15	11-26	65-75
2.1-2.5	...	0.1 max	0.5 max	4 max	8 max	90 min
High-conductivity powder						
2.5-2.6	32	0.2 max	1-11	13-23	20-37	43-53
Friction-grade powder						
1.7-2.0	...	0.5 max	1-6	5-15	10-26	60-80
High-density powder						
3.25-4.00	24	0.8 max	7-17	17-27	19-35	35-45

Table 4A. Permissible loads for sintered bronze self-lubricating bearings(a)
Source: ASTM B438.

Shaft velocity(b)			Permissible loads, grades 1 and 2							
			Type I		Type II		Type III		Type IV	
m/s	ft/min		MPa	psi	MPa	psi	MPa	psi	MPa	psi
Slow, intermittent	Slow, intermittent	...	22	3200	28	4000	28	4000	28	4000
0.13	25	...	14	2000	14	2000	14	2000	14	2000
0.26-0.51	50-100	...	3.4	500	3.4	500	3.9	550	3.9	550
Over 0.51-0.77	Over 100-150	...	2.2	365	2.2	325	2.5	365	2.5	365
Over 0.77-1.02	Over 150-200	...	1.7	280	1.7	250	1.9	280	1.9	280

(a) With a shaft velocity of less than 0.255 m/s (50 ft/min) and a permissible load greater than 6.89 MPa (1000 psi), an extreme-pressure lubricant should be used. With good heat-dissipation and heat-removal techniques, higher PV ratings can be obtained. (b) For shaft velocities greater than 1.02 m/s (200 ft/min), permissible loads may be calculated as follows: P = 50 000/V (P = 1.75/V).

Table 4B. Permissible loads for sintered iron-base self-lubricating bearings
Source: ASTM B439.

Shaft velocity(a)		Permissible loads			
		Grades 1 and 2		Grades 3 and 4	
m/s	ft/min	MPa	psi	MPa	psi
Slow, intermittent	Slow, intermittent	25	3600	55	8000
0.13	25	12	1800	20	3000
0.26-0.51	50-100	3.1	450	4.8	700
Over 0.51-0.77	Over 100-150	2.1	300	2.8	400
Over 0.77-1.02	Over 150-200	1.6	225	2.1	300

(a) For shaft velocities greater than 1.02 m/s (200 ft/min), permissible loads may be calculated using P = 50 000/V.

Table 4C. Permissible loads for iron-bronze sintered self-lubricating bearings
Source: ASTM B612.

Shaft velocity(a)		Permissible load	
m/s	ft/min	MPa	psi
Slow, intermittent	Slow, intermittent	28	4000
0.13	25	14	2000
0.26-0.51	50-100	2.8	400
Over 0.51-0.77	Over 100-150	2.1	300
Over 0.77-1.02	Over 150-200	1.4	200

(a) For shaft velocities greater than 1.02 m/s (200 ft/min), permissible loads may be calculated using P = 40 000/V.

Values of radial crushing strength for 89Cu-10Sn-1C bearings (ASTM B438) range from 103 to 183 MPa (15 to 26.5 ksi) for densities from 6.0 to 6.7 g/cm³. Industry standards (ASTM, SAE, MPIF) for P/M bearings provide information on compositional density, amount of interconnected porosity, tensile strength, compressive strength, elongation, and the K-factor, which is a measure of radial crushing strength. In addition to these properties, standards exist on dimensional tolerances, running clearances, and guideline values for the PV factor—i.e., for permissible loads as a function of shaft velocity (see Tables 4A, 4B and 4C).

Bearing life may be extended through the use of an oil-saturated felt or through wicking.

Friction Materials

Sintered metal friction materials were developed in the 1920's and commercialized in the 1930's. They are used in applications involving transmission of motion through friction (clutches) and for deceleration and braking. Braking involves conversion of mechanical energy into heat. Metal-base friction materials are stronger and more heat- and wear-resistant than organic-base friction materials. They also have higher energy capacity—i.e., they can absorb more energy for the same size clutch or brake.

Applications of sintered metal friction materials may be classified into dry and wet. Under wet conditions, the friction components, such as clutches in power-shift and automatic transmissions, are immersed in oil. Under dry conditions,

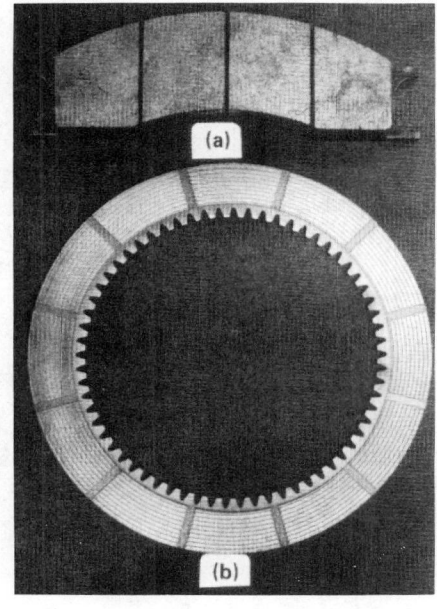

Fig. 8. Disc-brake pad (a) and clutch plate (b)

the friction components make direct contact without oil, such as in aircraft brakes and standard clutches. Figure 8 shows typical disc-brake-pad and clutch-plate designs.

Metallic friction materials are manufactured by compacting and sintering mixes of metal powders and friction-producing ceramic materials such as silicon dioxide or aluminum oxide. The nonmetallic components form a dispersion within a metallic matrix. The metallic matrix may be cop-

per-base (bronze, brass) or iron-base. In wet applications the matrix is always copper-base; in dry applications, the percentage of copper is reduced and iron is increased. Typical composition ranges are given in Table 5.

The powder mixtures are compacted at pressures of about 165 to 276 MPa (12 to 20 tsi). An important consideration is surface parallelism of these thin parts. For clutch-plate facings, the green facings are placed on copper-plated supporting steel backing plates, stacked in a belt-type sintering furnace and sintered in a protective atmosphere under pressure and at temperatures of 550 to 1000 °C (1020 to 1830 °F). Disc-brake pads may be sintered in conventional furnaces. Figure 9 shows the structure of a copper-base friction material, including the copper-plated layer and steel backing, after sintering at 650 °C (1200 °F) for 2 h.

After sintering, parts are machined to meet dimensional specifications. For friction facings operating in oil, the surfaces contain grooves which help remove heat and oil from the surface, thus increasing the coefficient of friction. In dry systems, grooves can prevent the crushing effect of thermal expansion, remove debris, and transfer water and other liquids if the surface becomes wet.

The coefficient of friction, μ (both static and dynamic values), is the single most important property in the selection of a facing material. It depends on composition, rubbing speed, pressure and temperature. With increasing rubbing speeds of the two surfaces, μ tends to drop (see Fig. 10). Applied pressure has a similar effect on the stability of μ. In wet applications, an increase in the temperature of the two rubbing surfaces to above 150 °C (300 °F) will cause a substantial drop in μ.

Wear rates are lower in wet applications, which permits the use of thinner facings. Wear rate depends on temperature, number of engagements, surface finish and composition of coupling plate and facing material.

P/M Copper-Base Structural Parts

Applications of copper-base P/M materials which rely mainly on the load-bearing capacities of the sintered parts are commonly classified as structural applications. The most common compositions include brass, nickel silver, and bronze.

Brass and Nickel Silver. Prealloyed brass and nickel silver powders are available in both leaded and nonleaded compositions. The range of brass alloys includes 90Cu-10Zn through 65Cu-35Zn. However, leaded versions of 80Cu-20Zn and 70Cu-30Zn are most commonly used for manufacture of sintered structural parts that may require secondary machining operations. The only nickel silver powder common to the industry has a nominal composition of 65Cu-18Ni-17Zn, which is modified by addition of lead for the purpose of improving machinability. The powders are usually blended with mixtures of lithium stearate and zinc stearate in amounts of 0.5 to 1.0 wt %.

Sintering is performed in a reducing atmosphere at temperatures from 760 to 925 °C (1400

Table 5. Nominal compositions of copper-base and iron-base friction materials

Premix	Composition, %						
	Copper	Iron	Lead	Tin	Zinc	Silicon dioxide	Graphite
Copper base	65-75	...	2-5	2-5	5-8	2-5	10-20
Iron base	10-15	50-60	2-4	2-4	...	8-10	10-15

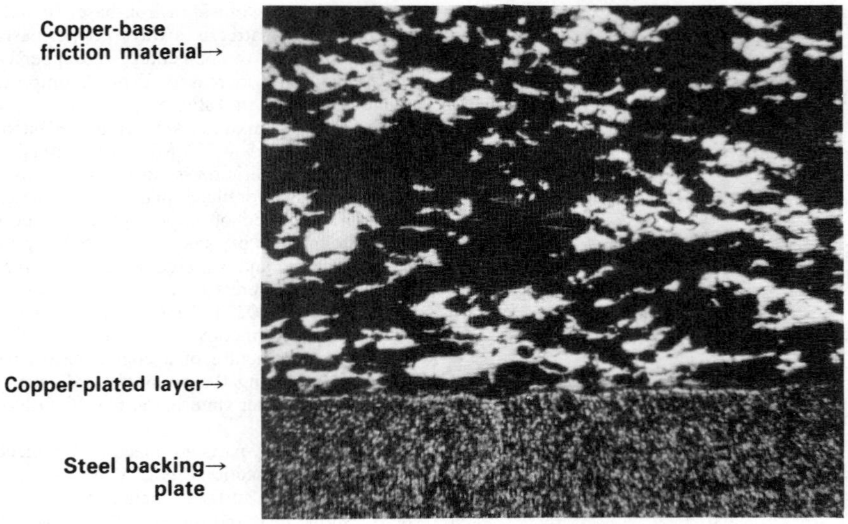

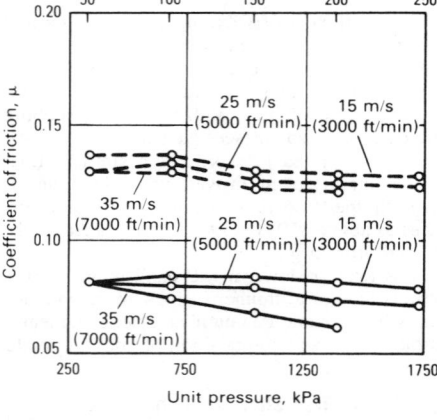

Copper-base friction material→

Copper-plated layer→

Steel backing→ plate

Fig. 9. Structure of a sintered copper-base friction material

sities in excess of 7.0 g/cm³ are preferably made from prealloyed powders because of better control of dimensions during sintering within the higher density range.

The most common composition is 90Cu-10Sn. Sintering requires certain precautions in order to develop optimum microstructures. Figure 11 shows typical strength-versus-density curves for 90Cu-10Sn bronze.

Copper-Infiltrated Parts

In copper infiltration, the pores of a sintered iron or steel part are filled with liquid copper. A so-called infiltrant slug (usually pressed from a copper alloy powder) is placed on top of the green steel part (top infiltration) or beneath it (bottom infiltration). Sintering of the steel part and its infiltration with copper are combined in a single process ("sintration"). Typical "sintrating" conditions are 30 min at 1120 °C (2050 °F) in endothermic gas. Dissociated ammonia or nitrogen-base atmospheres are also used. After infiltration, a small nonadhering residue can be disposed of easily. The infiltrated steel part is al-

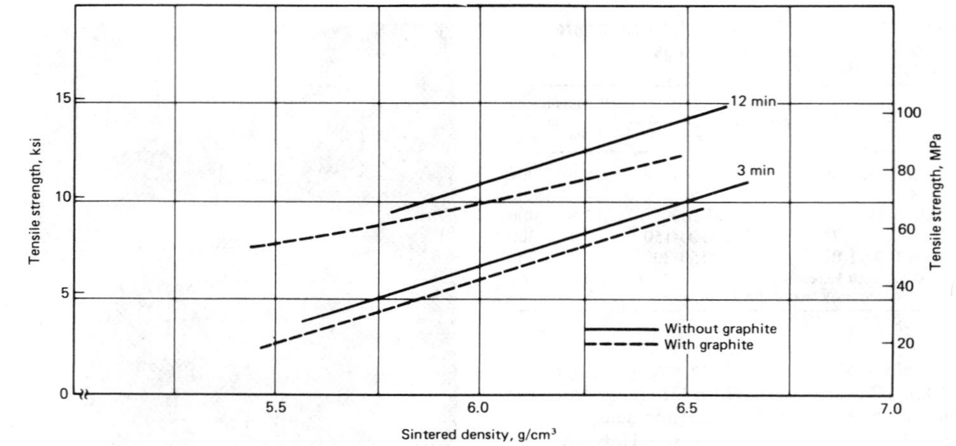

Solid lines indicate dynamic coefficient of friction. Dotted lines indicate static coefficient of friction.

Fig. 10. Effects of unit pressure and rubbing speed on the coefficient of friction in a wet system

Fig. 11. Effect of sintered density on tensile strength of 90/10 bronze compacts with and without graphite

to 1700 °F) for brasses, and from 870 to 980 °C (1600 to 1800 °F) for nickel silver. Variation of the length of sintering permits good control over dimensional change and ductility. Excellent ductility permits re-pressing, which provides further improvements in mechanical properties and dimensional precision.

Both brass and nickel silver compacts sinter to bright matte surface finishes that may be burnished to high luster for decorative and aesthetic purposes.

Applications of brass and nickel silver parts include latch bolts and cylinders for locks; shutter-mechanism components for cameras; gears; cams; actuator bars for timing and small generator drive assemblies; and decorative trim and medallions. Often, free-machining properties and corrosion resistance are critical to the selection of these compositions.

Bronze. Most P/M bronze parts are made from elemental premixes of tin and copper because of the superior compacting and sintering behavior of such premixes in comparison with prealloyed bronze powder. However, parts requiring den-

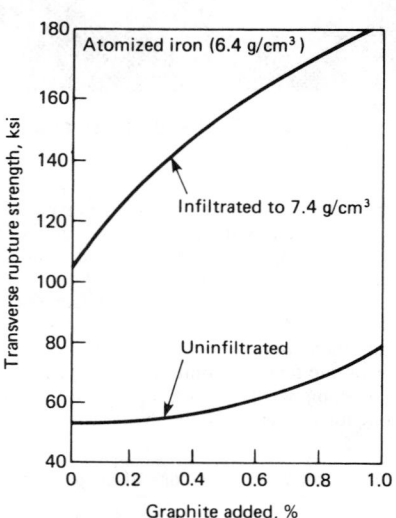

Fig. 12. Transverse rupture strength of copper-infiltrated steel part as a function of carbon content

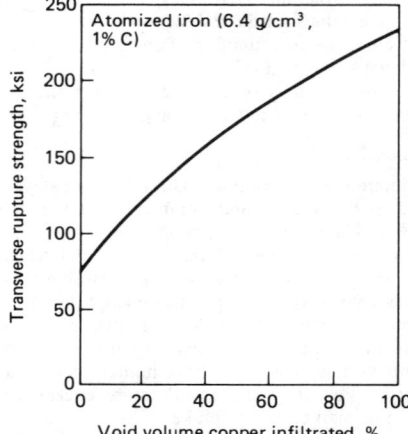

Fig. 13. Transverse rupture strength of copper-infiltrated steel part as a function of void volume infiltrated

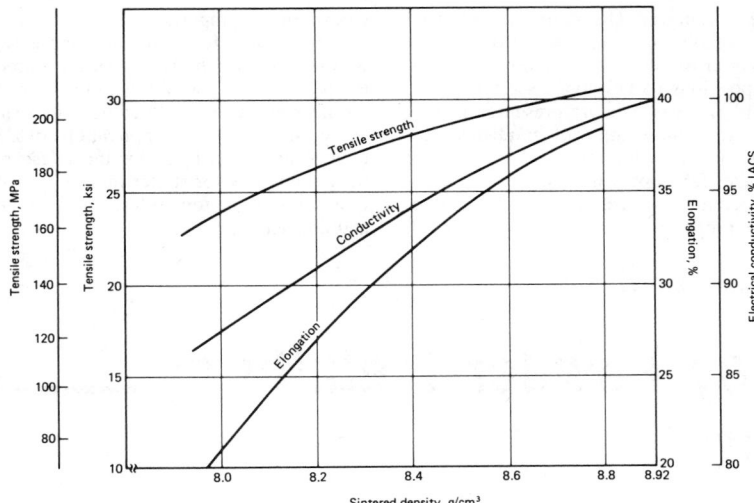

Source: Effect of Foreign Electrolytes on the Production of Copper Powder by Contact Deposition, by V. P. Artamonov and A. V. Pomosov: *Isv. V.U.Z. Tsvetn. Metall.*, No. 2, 1976, p 30-34 (in Russian); *Metall. Abstracts*, No. 54-0503, Nov 1976.

Fig. 14. Effect of sintered density on electrical conductivity of P/M copper

Electrical Parts and Oxide-Dispersion-Strengthened Copper

For the use of P/M parts in electrical and electronic applications, high electrical conductivity is of prime importance, and it is, therefore, essential to use very pure copper powder. It is then possible to obtain the values of strength and ductility shown in Fig. 14. Compaction and sintering require certain precautions to achieve complete sintering throughout the part. Full-density properties are approached or reached by double pressing or forging. Typical applications include commutator rings, contacts for circuit breakers, armature bearing blocks, shading coils for contactors, nose cones, switchgear components for use in switchboxes, and electrical twist-type plugs. Heat sinks for diodes used in the bases of silicon rectifiers for the alternating-current systems in automobiles, and tools for electrical-discharge

Table 6. Compositions of oxide-dispersion-strengthened (DS) coppers

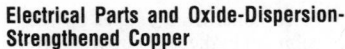

Grade	Copper		Al_2O_3	
	Wt %	Vol %	Wt %	Vol %
C15715	99.7	99.3	0.3	0.7
C15760	98.9	97.3	1.1	2.7

most fully dense. Figures 12 and 13 illustrate the effects of graphite in the steel skeleton, and the amount of void volume filled with copper, on transverse rupture strength. The total amount of copper may range from 12 to 30%. Major reasons for the use of copper infiltration include:

• Enhancement of mechanical properties
• Higher and more uniform density
• Sealing of porosity for secondary operations such as pickling, plating, brazing and painting
• Development of selective properties through localized infiltration
• Production of multiple-component assemblies
• Improvement of machinability
• Improvement of conductivity
• Noise reduction in moving components.

Copper-infiltrated parts find wide use in load-bearing applications in many types of machines and engines. P/M infiltrated gears are particularly well suited to copper infiltration, because infiltration adds strength to the teeth of the gear, which often suffer from lower density.

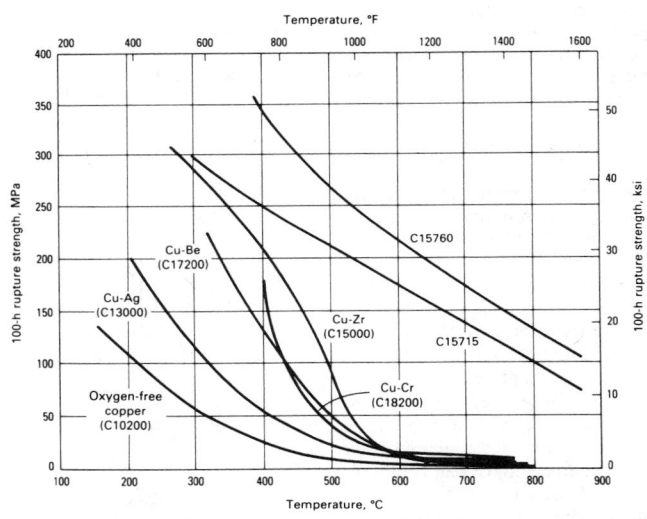

Fig. 16. Stress-rupture properties of oxide-dispersion-strengthened (DS) copper

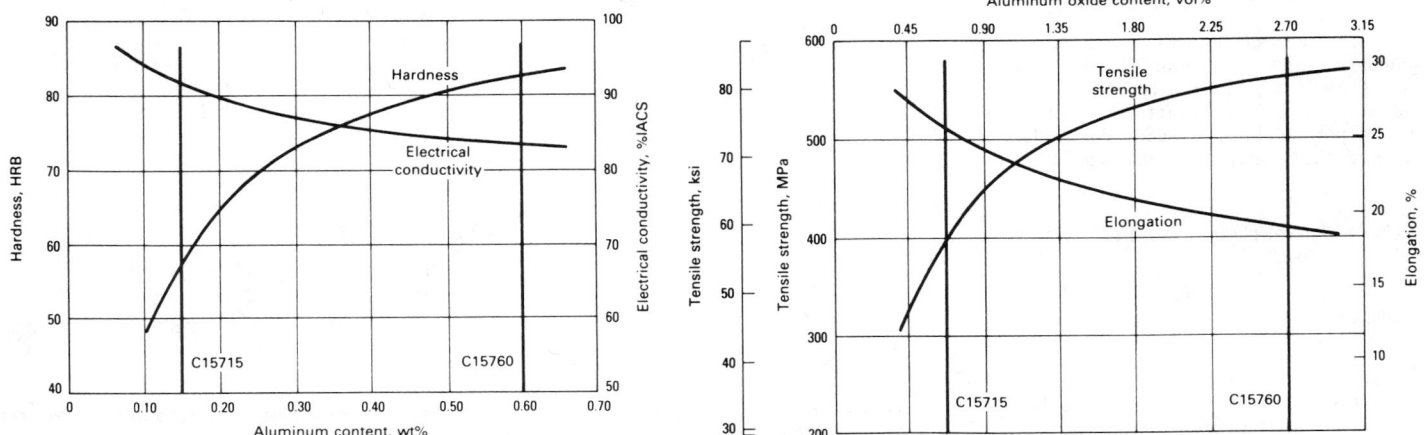

Fig. 15. Properties of oxide-dispersion-strengthened (DS) copper

machining, are also made from copper powder.

Oxide-dispersion-strengthened copper differs from the abovementioned pure copper in that it contains a fine dispersion of a refractory oxide, typically Al_2O_3. Commercial DS copper is made by internal oxidation of a copper-aluminum alloy powder. Methods for consolidation of the powder to full density include hot extrusion, hot forging and hot isostatic pressing. DS copper withstands high-temperature exposure without appreciable softening. The fully consolidated material is available in a wide range of sizes of rod and other cross sections, including strip and wire. Compositions of DS coppers are given in Table 6, and property data are presented in Fig. 15 and 16. The combination of good electrical conductivity and excellent elevated-temperature stability makes DS copper uniquely useful in elevated-temperature applications and where retention of these properties after exposure to elevated temperature is important. A major application is electrode caps for nonconsumable resistance welding, for which DS copper has outperformed RWMA class 2 type welding electrodes by a factor of seven, with the added benefit of substantial energy savings. Other applications of DS copper include its use in leads for incandescent lamps, in commutators for starter motors, in submerged fuel pumps in automobiles, and in relay blades and contact supports.

Aluminum Alloy Powder Metallurgy

By Byron S. Henderson, Alcoa

COMMERCIAL production of precision parts by powder metallurgy (P/M) techniques represents an important dimension in the fabrication of aluminum alloys. The exceptional properties of high strength, light weight, corrosion resistance, high thermal and electrical conductivities and response to a variety of finishing processes make aluminum a strong candidate in the P/M industry for application in automobiles, appliances, business machines, farm and garden equipment and many other areas. This article is devoted to the technology of conventional P/M press-and-sinter processing of aluminum alloys.

GENERAL ATOMIZING PROCESS

Atomized aluminum powders are produced by atomization of molten aluminum. The molten aluminum is drawn through an atomizing nozzle; the lower end of the nozzle dips into the molten aluminum, and the upper end terminates in a small orifice. This orifice is surrounded by a small chamber into which a jet of pressurized air is introduced. The air jet impinges on the stream of molten aluminum and disintegrates it into small particles (see Fig. 1). These irregularly shaped (nodular) particles, together with a substantial volume of cooling air, are drawn through a chiller chamber and into a cyclone and/or bag filter collecting system. The atomized powders are then graded by screening. After product sampling and inspection, the powders are ready for packing.

PHYSICAL PROPERTIES

Typical physical properties of atomized aluminum powders are given in Table 1. Aluminum reacts readily with moisture or free oxygen in the air during manufacture to form an oxide coating on the powder. Finer powders, due to their increased surface area, show the highest percent-

Table 1. Typical physical properties of atomized aluminum powder

Density, g/cm^3	2.7 (metal)
Melting point, °C	660
Boiling point, °C	2427
Surface tension at 800 °C, dyne/cm	865
Apparent density, g/cm^3	0.8 to 1.3
Tapped density, g/cm^3	1.2 to 1.5
Melting point of oxide, °C	2045
Al_2O_3 content, wt %	0.1 to 1.0

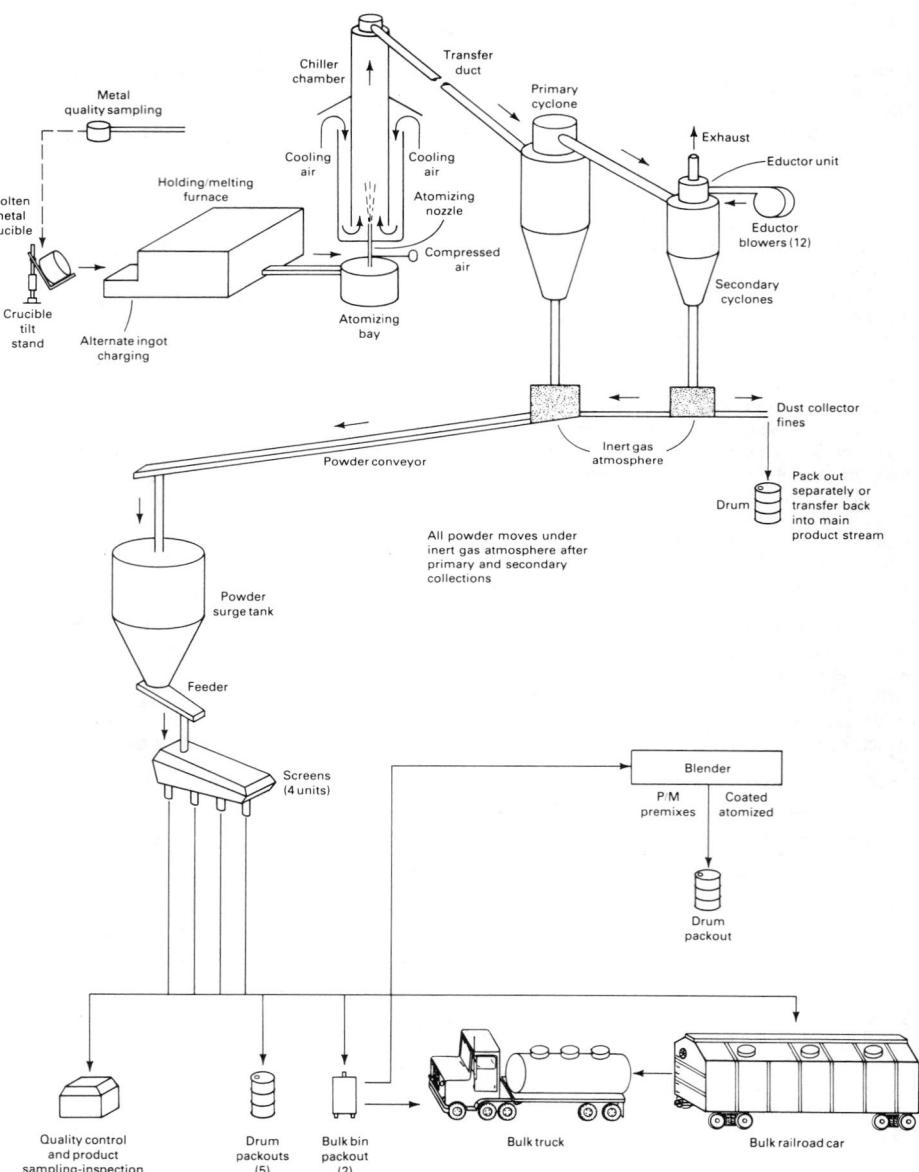

Entire operation is under computer control. Powder is packed in drums or bins or is loaded for bulk shipment in trucks or railroad cars.

Fig. 1. Alcoa process for production of aluminum powder by atomization

Table 2. Typical compositions of unalloyed atomized aluminum powders

| Type of powder | Composition of molten metal or starting ingot, wt % | | | Other metallics | |
	Al	Fe	Si	Each	Total
Regular atomized powders:					
Typical	99.7	...	...	...	...
Maximum	...	0.25(a)	0.15(a)	0.05	0.15
High-purity atomized powders:					
Minimum	99.97	...	...	...	...
Typical	99.976	0.007	0.008	...	0.009

(a) Fe + Si, 0.30% max.

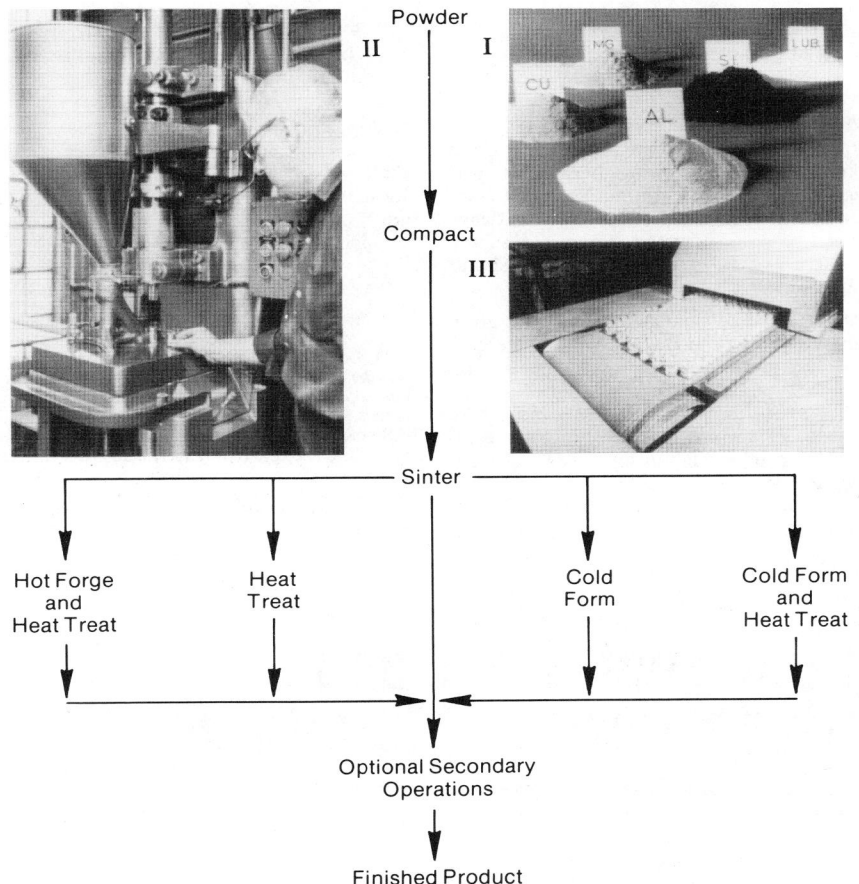

Fig. 2. Flow chart of the aluminum P/M process

age of oxide by weight. Because the aluminum powder surface is hydroscopic, it also contains moisture in the form of physically absorbed H_2O and hydrated aluminum oxide.

The thickness of the oxide film on the aluminum powder is relatively constant even when atomizing conditions vary. Experiments employing air, argon and nitrogen as atomizing gases, and gas temperatures ranging from ambient to 595 °C (1100 °F), have shown that the oxide thickness is 5.05 nm (50.5 Å) independent of the composition and temperature of the atomizing gas.

CHEMICAL PROPERTIES

Chemical compositions of unalloyed atomized aluminum powders are presented in Table 2. Iron and silicon are the major contaminants for both regular and high-purity powders.

Aluminum is stable in air because of its thin, natural oxide film. In finely divided powder form, however, aluminum is more chemically reactive than it is in massive form. Aluminum powders can oxidize further after atomization and can react with water to liberate hydrogen and form aluminum hydroxide. Atomized aluminum exposed to moisture-saturated air at room temperature slowly forms beta alumina trihydrate and hydrogen.

THE ALUMINUM P/M PROCESS

The aluminum P/M process consists of three basic steps (see Fig. 2). First, aluminum powders of controlled purity and particle size are mixed with alloying metal powders in precisely controlled quantities. Generally, a powdered lubricant is added to this premix to facilitate pressing

and ejection of parts and to minimize tool wear. Second, the premix is compacted at 97 to 345 MPa (7 to 25 tsi) in precision metal dies on specially designed mechanical or hydraulic P/M presses. In the third step, sintering, the green compacts are heated to 560 to 650 °C (1040 to 1200 °F) in a nitrogen or inert-atmosphere furnace at closely regulated temperatures. The sintering process metallurgically bonds the powder particles and develops the desired physical and mechanical properties in the part. For some applications, no further processing is required. Other parts may require one or more secondary operations such as sizing, coining, cold forming, hot forging, heat treating and finishing. These secondary operations improve the dimensions, properties or appearance of the finished part.

SECONDARY OPERATIONS

Secondary operations performed on aluminum P/M parts are similar to those for most other P/M materials. These include sizing or coining, heat treatment, machining, joining and finishing. Cold and hot forming are options which can be added when optimum properties are desired.

Sizing and coining of aluminum P/M parts are repressing operations that are performed either to control dimensional tolerances or to increase density, or both. Due to the distortion that may occur during sintering, either sizing or coining is usually necessary except for parts having relatively low profiles and generous tolerances or parts that will undergo secondary machining operations. Both mechanical properties and surface finish are improved by the coining operation, and detail quite often can be imparted to the surface of a part by coining rather than by molding.

Cold forming is an upsetting operation in which a significant amount of lateral flow of material occurs. Increases in mechanical properties can be realized beyond those obtainable by conventional coining. Upsetting of up to 20% may be involved, and density may be increased to 96 to 99% of theoretical.

Heat Treatment. Solution heat treating and aging after sintering will improve mechanical properties and increase ductility of aluminum P/M parts. The same thermal treatment applied to wrought products may be used for P/M parts. Parts should be heated in air for 30 min at 505 to 540 °C (940 to 1000 °F), then cold-water quenched. Depending on final property requirements, parts can be naturally aged (T4) or artificially aged (T6). The T4 temper provides good strength with maximum ductility. The T6 temper provides the maximum strength properties available through heat treating. The additional cost and decrease in ductility of the T6 temper have made the T4 temper the most popular. Heat treatment also improves corrosion resistance and machinability.

Machining. The machinability of P/M aluminum is one of its most attractive features. P/M aluminum provides excellent chip characteristics; chips are small and easily broken, with none of the long stringer buildup that occurs with some wrought aluminum alloys (see Fig. 3). This means good tool life, increased speeds and feeds, and hence higher production rates.

Drilling, tapping and milling can be performed by following the procedures used for wrought aluminum. Good lubricant flow, sharp tools, high speeds and rapid heat removal are the important parameters.

Fig. 3. Illustration of the excellent chip characteristics of P/M aluminum (left), compared with the long stringer buildup that occurs with some wrought aluminum alloys (right)

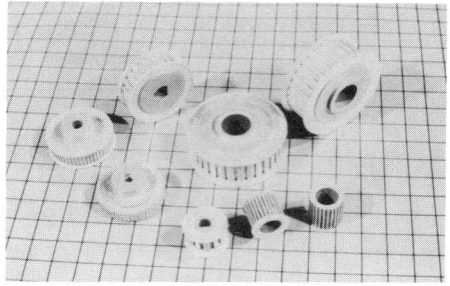

Fig. 4. Dual-flange pulleys made by staking or pressing stampings onto aluminum P/M pulley hubs

Fig. 5. Illustration of the variety of finishes that can be applied to aluminum P/M parts

Joining. Common methods employed for joining of aluminum P/M parts include press fitting, staking, adhesive bonding, and joining with threaded fasteners. Joining during sintering also works except where the shape of the assembled parts restricts the ability to properly size the structure.

Joining of aluminum P/M parts by copper infiltration is not possible, because the melting point of copper is higher than the sintering temperature of aluminum. Normal welding and brazing techniques are unsatisfactory for joining of aluminum P/M parts. Soldering can be used if the strength of such a bond is adequate.

Where part design allows, press fitting and staking are practical and economical methods of joining two or more parts together. These operations also may be used to join non-P/M components to sintered aluminum parts. Dual-flange pulleys (Fig. 4) are common examples of stampings staked or pressed onto aluminum P/M pulley hubs.

Finishing. Sintered aluminum parts can be given a variety of surface finishes. Many of the decorative and protective treatments currently employed for wrought and cast aluminum alloys can also be applied to aluminum P/M parts. These treatments include chemical cleaning, mechanical finishing and etching to achieve textures; anodizing for increased resistance to corrosion, wear and abrasion; coloring for decorative and functional purposes; electroplating; and painting. A hard-type anodic coating also can be applied in many applications. Thus, the designer has a wide choice of decorative and functional finishes that are not possible with other P/M materials. Figure 5 illustrates the variety of finishes that can be applied to aluminum P/M parts.

SELECTED REFERENCES

Aluminum P/M — Properties and Applications, by J. D. Generous and W. C. Montgomery: Chapter 8 in *Powder Metallurgy — Applications, Advantages and Limitations,* edited by E. Klar, American Society for Metals, Metals Park, OH, 1983

Production of Aluminum Powder by Atomization, by J. E. Williams, Jr.: in *Metals Handbook,* 9th Ed., Vol 7, *Powder Metallurgy,* American Society for Metals, Metals Park, OH, 1984

"Alcoa Aluminum Powders in Powder Metallurgy," F36-13713, Aluminum Company of America

High Speed Tool Steel and Cemented Carbide Powder Metallurgy

By Mark L. Shoenberger, Universal-Cyclops Specialty Steel Div., and Roy V. Leverenz, Kennametal, Inc.

High Speed Tool Steels

HIGH SPEED TOOL STEELS are a family of iron-base materials that contain 20 to 30% chromium and refractory-metal alloying constituents which are added to create carbides. Their most important properties are hardness, wear resistance and the ability to resist softening at elevated temperatures. Conventional ingot casting techniques for manufacturing these alloys produce heavy eutectic carbide segregation on cooling. The resulting microstructure is nonuniform and coarse grained, exhibiting variations in composition throughout the cross section. Over the years, to minimize these detrimental effects, extra processing steps have been added, including heavy hot work reductions (90 to 98%) and casting of smaller ingots for faster solidification.

The powder metallurgy (P/M) approach to production of high speed steels has been researched since the early 1960's and was first commercialized in the early 1970's. Use of P/M processing permits rapid solidification of fully alloyed molten metal into droplets that are free of segregation and that have a uniform distribution of fine carbides (see Fig. 1). When proper densification parameters are utilized, P/M high speed steel processing yields a product that is greatly superior to corresponding conventional cast and wrought alloys (see Fig. 2). The uniform structural homogeneity yields a finer carbide size and distribution, which pay benefits by promoting finer grain size, improved response to hardening, more uniform hardening, less out-of-round distortion in bar products, improved impact properties and superior grindability.

Modern P/M high speed steel processes all consist of essentially the same manufacturing steps. All commercial processes start with atomized powder created from a fully alloyed molten bath. Following melt/atomization, a powder-processing sequence is followed which may include screening, blending and/or chemical treatment of the powder. Subsequently, the processed powder is either containerized in molds of whatever configuration the manufacturer desires or cold pressed in dies. Containers have typically been made of metal or borosilicate glass, both of which lend themselves to vacuum integrity and the ability to be hermetically sealed. Consolidation of the containerized powder may then be accomplished by hot isostatic pressing (HIP), cold isostatic pressing (CIP), consolidation by atmospheric pressure (CAP), high-temperature vacuum sintering, or hot mechanical working such as rolling or extrusion. Postcompaction processing is utilized to impart the desired mechanical properties and/or increase densification of sintered products. This final step may include any combination of mechanical working, heat treatment, machining or grinding that is required to manufacture a finished product. Because all commercial P/M processes use these common manufacturing steps, each step is discussed briefly in the pages that follow.

Magnification (both), 1000×.

Fig. 1. Microstructures of P/M (top) and cast (bottom) high speed steels

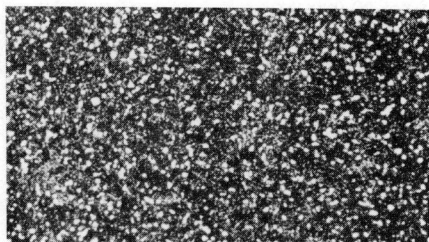

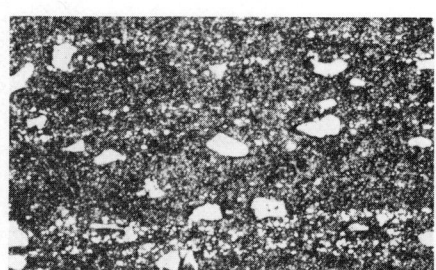

Note finer carbide size, and more uniform carbide distribution, in P/M structure. Magnification (both), 500×.

Fig. 2. Microstructures of CAP P/M (top) and conventional (bottom) 38-mm (1¹/₂-in.) diam M4 high speed steel bar

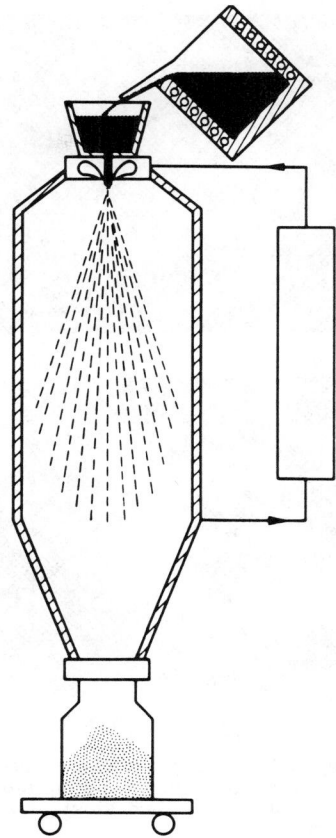

Fig. 3. Schematic illustration of a vertical gas-atomization unit

ATOMIZATION OF POWDER

Gas atomization of high speed steels is normally accomplished with nitrogen gas. Molten metal is poured from the furnace into a tundish that has a refractory nozzle affixed to its base. As the molten stream exits from the nozzle, it is impacted by a high-pressure atomization gas that disperses the stream into fine droplets which are rapidly solidified by gas or liquid quenchants. The nitrogen gas provides an atmosphere cover that prevents heavy oxidation of the powder.

Figures 3 and 4 are schematic illustrations of vertical and horizontal gas-atomization units. Although they produce similar products, their primary difference lies in the expenses associated with facility construction. Vertical atomization requires tall atomization chambers to cool the powder during its descent. This height can be reduced by using liquid nitrogen or water as a quenchant in the base of the chamber. Horizontal atomization provides a more compact unit which uses gas jets to cool the powder after it has been collected on a vibratory table and fluidizing bed.

Figure 5 illustrates a water-atomization setup wherein the molten stream exiting from the nozzle is impacted by high-pressure water jets. A short distance later, the metal particulate is quenched by an agitated water bath where the powder may be pumped off to a draining and drying station.

Figure 6 shows typical gas- and water-atomized high speed steel powders. Gas-atomized powders have attached satellites of smaller particles welded to larger particles due to collisions of atomized particles prior to solidification. Properties of vertically and horizontally gas-atomized powders are similar. Such powders typically have packing densities of 65 to 70% and

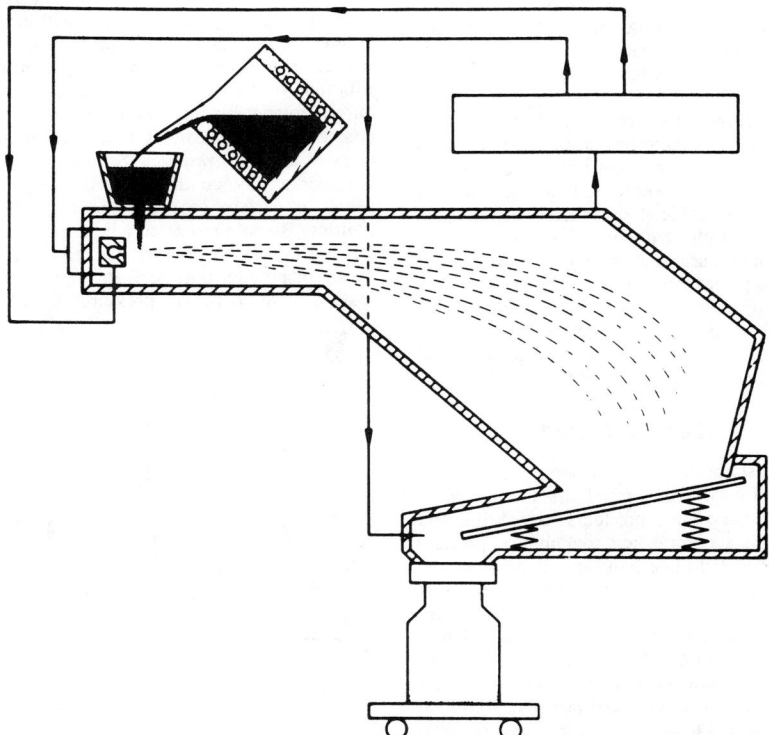

Fig. 4. Schematic illustration of a horizontal gas-atomization unit

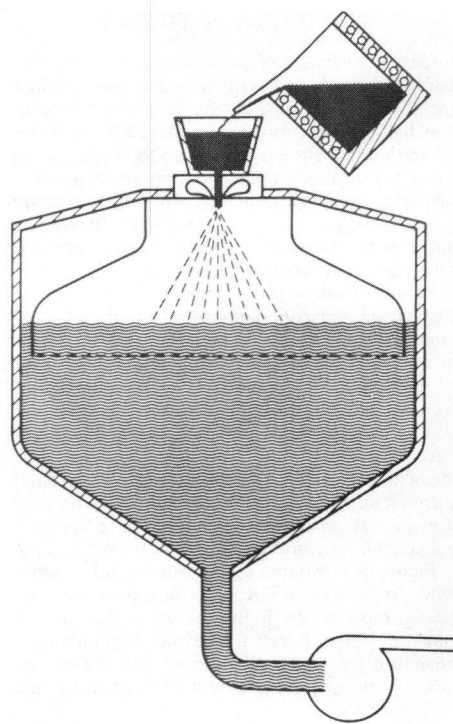

Fig. 5. Schematic illustration of a water-atomization unit

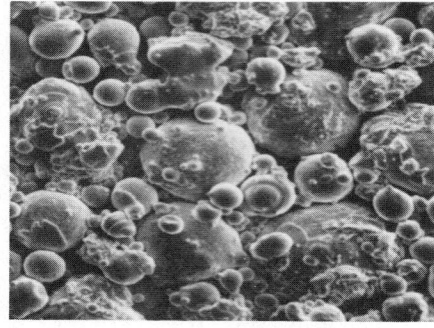

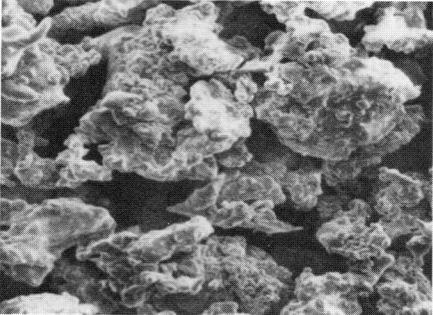

Magnifications: gas-atomized powder, 200×; water-atomized powder, 100×.

Fig. 6. Typical appearances of gas-atomized (top) and water-atomized (bottom) high speed steel powders

oxygen/nitrogen levels below 500 ppm. Care must be exercised to ensure cleanness of the powder by verifying the pressure integrity of the atomization chamber and controlling the level of metal in the tundish to prevent slag from entering the atomization stream. Heavily oxidized powder surfaces degrade material impact strength, and slag particulate will impart surface imperfections from the tool cutting edge to the part being machined.

Water-atomized powder is much more irregular in shape, thereby providing higher green strength for cold compaction processes. Oxygen content varies from 1000 to 3000 ppm for typical material, and an oxide shell typically encases the particle to a depth of about 0.2 μm (7.9 μin.). Due to the high particulate hardness, powder must be annealed prior to compaction. Most producers use vacuum annealing for reducing the surface oxide shell to less than 1000 ppm. Water-atomized powder parts may then be cold pressed and vacuum sintered to near net shape.

POWDER PROCESSING

Some postatomization powder processing is performed by all commercial manufacturers of P/M high speed steel products. Screening and blending are accomplished to enhance product quality and to facilitate handling of powder.

Screening is used to eliminate undesirable oversize atomization product which may prevent proper densification or contain detrimental metallurgical structures. Coarse-mesh, gas-atomized powder contains a higher frequency of hollow particles than finer-mesh powder. Removal of this particle fraction eliminates large residual porosity in the final product. Also, because large particulates have lower solidification rates, their

coarser microstructures would not be carried into the final product.

Powder blending generally is used to homogenize the particle distribution within a given atomization lot or to combine lots into a larger master blend to minimize handling. Uniform distribution of particulate is especially important during powder loading to preclude segregation of coarse and fine particles that may create striations in the final product. Striations create poor hot workability conditions that manifest themselves as surface cracking and/or duplex grain regions. Blending also may be used to introduce additives into the powder that enhance densification or powder part performance. The consolidation at atmospheric pressure (CAP) process

utilizes such a treatment by introducing a chemical additive to promote bonding of gas-atomized powders during subsequent high-temperature sintering.

CONSOLIDATION OF POWDER

The majority of P/M high speed tool steels on the commercial market are hot isostatically pressed (HIP), consolidated by atmospheric pressure (CAP) or sintered. Common manufacturing steps include vertical nitrogen gas atomization and post-compaction processing that includes thermomechanical working. Schematics typical of these processes are illustrated in Fig. 7 and 8.

High speed steel powder for the HIP process is screened and blended, then containerized in mild steel cans. The containers are fabricated by welding and checked for pressure integrity. Some manufacturers employ cold isostatic pressing (CIP) at this point in the process to provide a larger area of particle contact and thus reduce preheating time for HIP. Other producers place the containers in preheating furnaces and commence degassing as the material is heated. Once the container has reached the HIP temperature and an acceptable vacuum level exists, the tubulation is sealed off and the can is charged into the autoclave. Applied pressures vary from 145 to 207 MPa (21 to 30 ksi) depending on the material condition. Hot isostatic pressing time ranges from 15 min to 4 h depending on the equipment or processing sequence. The consolidated product is fully dense and subsequently undergoes thermomechanical working via rolling or forging. The container may remain during hot working but eventually is removed by scaling. Any residuals are removed by pickling or grinding. The resulting microstructure is fully homogeneous, with fine grain and carbide sizes. Material quality is high, but manufacturing costs related to the expensive processing equipment may result in higher prices in the marketplace. Billets manufactured in this manner may weigh from 400 to 9100 kg (880 to 20 000 lb).

The consolidation by atmospheric pressure (CAP) process is a type of vacuum sintering operation. Nitrogen-atomized powder is screened and blended. However, during the blending step a chemical additive is combined with the powder. The purpose of the additive is to act as a fluxing agent at powder-particle surfaces during sintering to enhance densification. Treated powder is loaded into borosilicate glass containers,

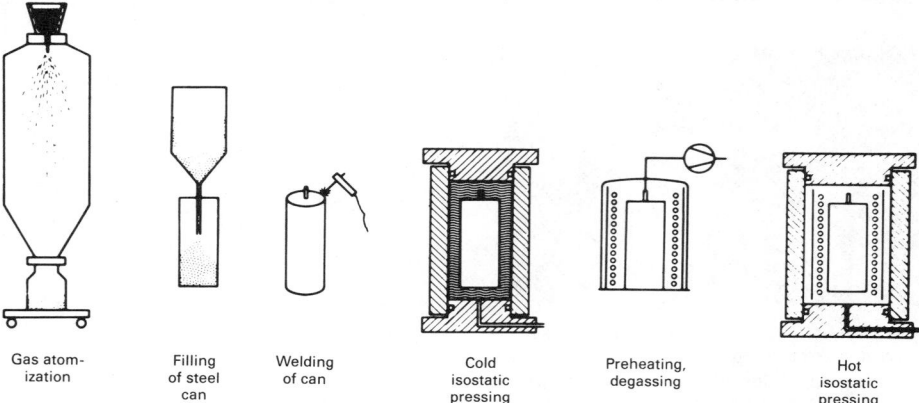

Gas atom-ization   Filling of steel can   Welding of can   Cold isostatic pressing   Preheating, degassing   Hot isostatic pressing

Fig. 7. Schematic illustration of the hot isostatic pressing (HIP) process. See text for discussion.

then subsequently degassed and sealed. The loaded glass containers are fixtured in an outer refractory support container and surrounded with sand as shown at top right in Fig. 8. They are then charged into the sintering furnace. As the temperature increases, the glass softens and plastically deforms under atmospheric pressure following the contour of the sintered preform. Due to the negative pressure inside the container, the external atmosphere presses the powder particles closer together to promote interparticle diffusion bonding during the thermal cycle. On cooling, the glass container fractures and spalls off, leaving only residual glass that can be removed by a light sandblasting operation. Preforms have densities of 95 to 99% of theoretical and require some hot working to achieve full density. The primary advantages of the CAP process are low capital investment and low manufacturing cost with no sacrifice in structural quality of the final product.

POSTCOMPACTION PROCESSING

Subsequent to consolidation, the majority of high speed tool steels require further processing to yield a saleable product. This processing may include annealing and hot working such as forging, extrusion or rolling. If none of these steps is required, direct machining or grinding may be necessary to finish the product. Normally these processing steps are important for developing final material properties. In general, lower processing temperatures (i.e., less than 1150 °C, or

2100 °F) provide a fine grain structure with small carbides. However, this can be permanently destroyed through improper heat treatment. Higher processing temperatures result in coarser structures. Because carbide and grain sizes are functions of time and temperature, the optimum processing route will be dictated by the material application and required performance. Because of their homogeneous nature, P/M tool steels provide better response to heat treatment and better grindability than corresponding conventional high speed steels. The lack of carbide banding and stringers provides predictable distortion during heat treatment. However, each application must be weighed individually in order to optimize cost and product fitness for use. This outcome will dictate which postcompaction processing route best fits the need.

Cemented Carbides

CEMENTED CARBIDES constitute a family of hard carbides of tungsten, tantalum, titanium and niobium in a binder of cobalt or nickel. Because they are some of the most versatile powder metallurgy (P/M) materials made, they play a prominent role in modern high speed production. Almost every product made requires a cemented carbide tool. Because cemented carbides are very hard and resist deformation, they are used to cut and shape most metals. Applications include turning, milling, drilling, back extrusion, draw-

ing dies, heading dies, coal mining, road planing, oil-well drilling bits and rod-mill rolls. Many materials are made by powder metallurgy to reduce costs, but cemented carbides can be made only in this way. Carbides are an extension of the tungsten industry. The tungsten is made by a P/M process, and carbon is added and reacted to form tungsten carbide. Titanium, tantalum and niobium carbides are added to obtain unique properties. Cobalt is the cementing agent that makes the material practical. Cobalt contents of available commercial grades range from 3 to 30%.

Tungsten carbide (WC) is made from scheelite and wolframite ores through the following processing steps:

1. *Manufacture of Ammonium Paratungstate (APT)*. The ore is converted to ammonium paratungstate by dissolving and reprecipitating the ore several times so that the resulting APT has almost no impurities. An alternative method is to convert the ore to WC in a thermit, where the WC crystals are precipitated out of a molten steel menstruum. This material will be used in step 5.

2. *Manufacture of Blue Oxide (WO_3)*. The ammonia is driven off from the APT by heating at about 900 °C (1650 °F) to produce blue oxide (WO_3). These individual steps are necessary to control the ultimate cemented carbide grain-size distribution.

3. *Reduction of Blue Oxide to Tungsten*. The blue oxide is reduced by hydrogen at a temperature of about 900 °C (1650 °F) to produce a fine tungsten powder. A rotary or stoking tube-type furnace is usually used. Time, temperature and moisture must be controlled closely.

4. *Production of Tungsten Carbide*. The tungsten is blended with carbon and heated in a controlled-atmosphere furnace at approximately 1550 °C (2820 °F) to form the WC. This blend is then crushed to −200 mesh.

5. *Manufacture of Grade Powder*. The tungsten carbide is mixed with varying amounts of cobalt, tantalum carbide, titanium carbide and niobium carbide to make the various grades of cemented carbide. The mixing is usually done in some type of ball mill with carbide balls in a solvent such as hexane, naphtha, alcohol or acetone.

6. *Powder Granulation*. A pressing lubricant (usually 2% paraffin) is added, and the powder is made free-flowing by pelletizing. Free-flowing powder is required in order to maintain size during pressing. Pellets are made by many different methods, including spray drying, compacting and granulating, and tumbling.

7. *Pressing Parts to Required Shapes*. There are many methods for making final cemented carbide parts. All of these methods include pressing the powder at approximately 205 MPa (15 tsi). Most parts, including cutting inserts, are made in automatic presses that may press parts to shape at rates up to 1000 pieces per hour. Very large parts are isostatically pressed in rubber molds. Pressure is applied from all sides through a water-and-oil mixture. These parts usually require additional machining such as turning, drilling and grinding. Other methods of forming are extrusion and injection molding.

8. *Sintering* is a process of heating the carbide compact to approximately 1480 °C (2700 °F) in a protective atmosphere or vacuum, which

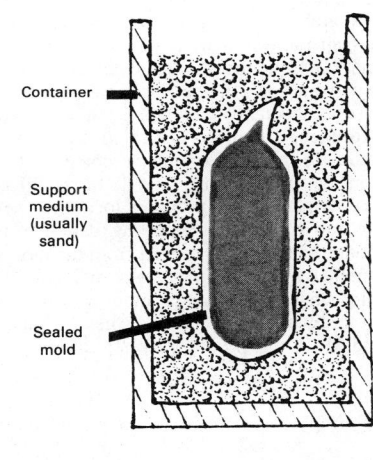

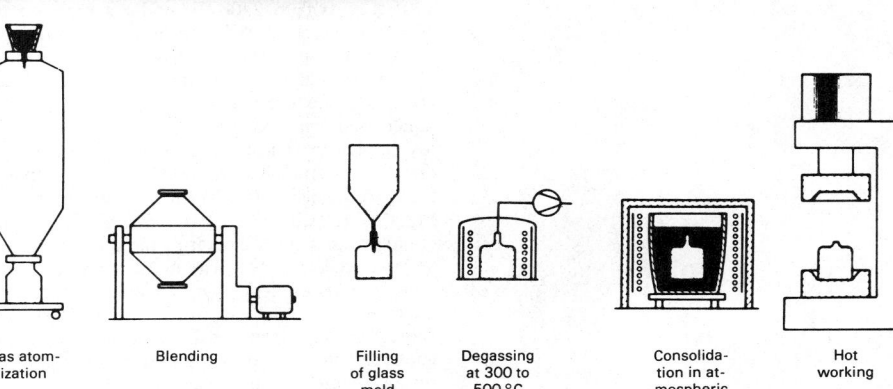

| Gas atom-ization | Blending | Filling of glass mold | Degassing at 300 to 500 °C | Consolidation in atmospheric furnace | Hot working |

Container

Support medium (usually sand)

Sealed mold

Fig. 8. Schematic illustration of the consolidation by atmospheric pressure (CAP) process. See text for discussion.

causes the carbide to fuse together without losing its shape. Sintering is accomplished by the formation of a liquid phase (eutectic) that produces an almost porosity-free body. The cobalt and WC eutectic dissolves and reprecipitates the WC particles, causing grain growth, which is necessary to achieve the unique combination of toughness and wear resistance that is characteristic of cemented carbides. During this process, the carbide shrinks about 20%. Carbon control is essential, and carbon content must be held within 0.02%. Hot pressing, which combines pressing and sintering in a graphite mold, is some-times used for large pieces.

9. *Finishing* is normally done by diamond turning or grinding. Due to the hard nature of cemented carbides, special shapes are made by electrodischarge machining with either electrode or wire. The trend for metal turning is toward unground and coated inserts.

Superalloy Powder Metallurgy

By John H. Moll, Colt Industries

SUPERALLOY PARTS made by advanced powder metallurgy (P/M) techniques are being used in advanced turbine engines. The advantages of the P/M process, as applied to turbine-engine hardware, are as follows:

- Ability to produce near-net shapes, which results in reduced material input and less machining to produce a finished part
- Improved property uniformity and alloy-development flexibility due to the elimination of macrosegregation
- Reduced energy requirements and shorter delivery time because the P/M process requires fewer processing steps than conventional ingot technology.

The superalloy engine shaft shown schematically in Fig. 1 is a good example of the material and fabrication savings attainable with P/M technology (Ref 1). It is noteworthy that the P/M process utilized markedly fewer processing steps and reduced the material input weight from 205 kg (452 lb) to 49 kg (108 lb).

The following is a brief description of the P/M superalloy process. A more complete summary of this subject is available in the literature (Ref 2 and 3).

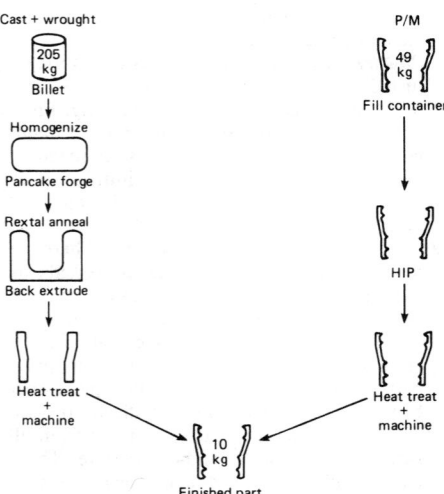

Fig. 1. Schematic comparison of processes required for manufacture of turbine engine shaft by cast-plus-wrought and P/M procedures (Ref 1)

P/M SUPERALLOY PROCESS

The process for producing P/M superalloy hardware generally involves production of spherical prealloyed powder, screening to remove oversize particles, blending the powders to homogenize the powder size distribution, loading the powder into containers, vacuum outgassing and sealing the containers, and then consolidating the powder to full density.

Powder Production

The principal commercial powdermaking processes used for superalloys are inert-gas atomization (Ref 4), vacuum atomization (Ref 5) and the plasma rotating-electrode process, or PREP (Ref 4, 6 and 7). In the gas-atomization process used for superalloys, metal is vacuum induction melted and subsequently atomized using an inert gas, usually argon. In the vacuum-atomization process for superalloys, molten metal is saturated with hydrogen. The metal is then streamed into a vacuum. As the gas expands, it comes out of solution and atomizes the metal. In the PREP process, a plasma arc provides localized melting at the end of a rapidly rotating bar. The rotating motion spins off the powder from the molten pool. Another atomization process that is still in the pilot stage is centrifugal atomization employing a rapidly spinning crucible, disk or plate (Ref 7 and 8). In this process, molten alloy streams are

Fig. 2. Argon-atomized superalloy powder

poured onto a spinning disk, crucible or plate. Droplets are ejected from the spinning base.

Most superalloy powder produced currently is spherical and has a tap density of about 65%. Tap density is important for reliable production of P/M shapes. Figure 2 shows powder produced by argon-gas atomization. In addition to being spherical, large particles often have smaller satellite particles attached to them. Also, particles may have a partial coating of splat, which occurs when a solid particle collides with a liquid particle. Following powder production, powders are screened to remove oversize particles. For superalloys, powder sizes range from −60 mesh (−250 μm) to −325 mesh (−44 μm) depending on the application. With superalloys, screening is utilized to limit the maximum nonmetallic inclusion size in the final product. This is a unique feature of the P/M process. After being screened, powders are blended to obtain a uniform size distribution. The powders are then ready for loading into containers for subsequent consolidation.

Powder Consolidation

Superalloy powders are consolidated to full density using a combination of temperature and pressure. The two principal techniques used to consolidate superalloy powders for aerospace applications are (*a*) hot isostatic pressing (HIP) and (*b*) extrusion followed by isothermal forging. For HIP, powder-filled containers are placed in an autoclave which is subsequently heated and pressurized. Superalloys are normally HIP'ed to full density at temperatures ranging from 1095 to 1205 °C (2000 to 2200 °F) under a pressure of 103 MPa (15 ksi). HIP can be used to produce parts of simple or very complex shapes depending on the container used for the powder. Metal containers are currently the main type of container used for production of HIP P/M superalloy hardware. These range from simple cylindrical containers made from steel pipe to complex-shape containers made from formed sheet (see Fig. 3). A unique process for producing very complex parts using a ceramic mold has been developed (Ref 9 to 12). The mold is produced by the lost wax process using specially developed materials. The mold is filled with powder and subsequently placed inside a cylindrical metal container surrounded by a granular ceramic. The metal container is then sealed. During consolidation, the granular ceramic transmits pressure from the outer container to the ceramic container. The process is capable of producing a variety of very complex shapes (see Fig. 4).

Fig. 3. Sectioned metal container for a HIP'ed powder multicomponent compressor spool

Fig. 4. Assorted parts made by the ceramic-mold process (courtesy of Crucible Research Center)

Extrusion plus isothermal forging is also used commercially to consolidate and form superalloy powders into useful shapes. Powders are loaded into cylindrical steel containers, which are evacuated, sealed and then extruded at high temperature to consolidate the powder and develop a fine uniform grain structure. This structure is capable of superplastic behavior at elevated temperatures and low strain rates. As such, the material can be formed to relatively complicated shapes using isothermal forging.

Several other techniques are being developed for consolidation of superalloy powders. These include the consolidation by atmospheric pressure (CAP) process and the fluid die process. In the CAP process (Ref 13 to 15), powder is loaded into a glass container which is subsequently evacuated and sealed. The assembly is then heated to a temperature slightly below the liquidus of the alloy being consolidated for 4 to 24 h depending on the compact size. Consolidation results from the pressure difference between the inside and outside of the container—i.e., ~100 kPa (~15 psi)—and from sintering mechanisms. The resulting product is not fully dense and requires subsequent hot isostatic pressing and/or hot working to attain full density and optimum properties.

The fluid die process (Ref 16 and 17) uses a shaped cavity that is machined, cast or forged into mating metal blocks. (The fluid die process is also referred to as the rapid omnidirectional compaction, or ROC, process; see Ref 2.) The cavity is filled with powder, evacuated and sealed. Consolidation is accomplished by applying pressure to the mold by either HIP or forging at high temperatures. After consolidation, the mold is removed by machining, leaching or melting. When melting is used, the metal mold is of a composition which melts at a temperature below that of the P/M superalloy part.

Once consolidated, P/M materials undergo conventional finishing processes such as heat treatment, nondestructive inspection and machining. In many instances, however, modified heat treatments have been developed for P/M products. Also, machining requirements are often markedly reduced and inspectability of the P/M product is often enhanced. For example, the uniform structure obtained with P/M products is ideally suited to ultrasonic inspection.

REFERENCES

1. Review of Superalloy Powder Metallurgy Processing for Aircraft Gas Turbine Engines, by J. L. Bartos: ASTM STP 672, 1979
2. *Metals Handbook*, 9th Ed., Vol 7, *Powder Metallurgy:* American Society for Metals, Metals Park, OH, 1984
3. Powder Metallurgy Parts for Aerospace Applications, by J. H. Moll, V. C. Petersen and E. J. Dulis: in *Powder Metallurgy: Applications, Advantages and Limitations,* edited by E. Klar, American Society for Metals, Metals Park, OH, 1983
4. A Review of the Techniques Available for the Production of Superalloy Powders, by J. W. Eggar and R. J. Siddall: in *P/M Superalloy Technology and Applications,* 1980 International Powder Metallurgy Conference
5. U.S. Patent 3 510 546, Homogeneous Metals, Inc.
6. Specialty Powders by the Rotating Electrode Process, by P. Lowenstein: *Progress in Powder Metallurgy,* Vol 37, MPIF, Princeton, NJ, 1982
7. Rapid Solidification Processing: An Overview, by D. J. Looft and E. C. Van Reuth: in *Rapid Solidification Processing,* Claitor's, Baton Rouge, 1980
8. Rapid Solidification Effects on Alloy Structures, by P. R. Holiday, A. R. Cox and R. J. Patterson II: in *Rapid Solidification Processing,* Claitor's, Baton Rouge, 1980
9. Hot Isostatic Pressing of Large Titanium Shapes, by V. C. Petersen, V. K. Chandhok and C. A. Kelto: in *Powder Metallurgy of Titanium Alloys,* AIME, 1980
10. Progress in P/M Superalloy and Titanium for Aircraft Applications, by E. J. Dulis, J. H. Moll, V. K. Chandhok and J. C. Hebeisen: in *The 1980's—Payoff Decade for Advanced Materials,* Vol 25, SAMPE, 1980
11. Outlook for As-HIPed Near-Net Shapes, by J. H. Moll: paper presented at Workshop on Conservation and Substitution Technology for Critical Materials, Nashville, TN, June 15–17, 1981 (to be published by National Bureau of Standards)
12. "Method for Making Powder Metallurgy Shapes": U.S. Patent 3 700 862, Crucible, Inc., Oct 24, 1972
13. "Powder Metallurgy": U.S. Patent 4 227 927, Oct 14, 1980
14. "Process for Compacting Metal Powders": U.S. Patent 3 704 508, Dec 5, 1972
15. Processing Effects on the Properties of P/M Rene 95 Near-Net Shapes, by J. D. Buzzanell and L. W. Lherbier: in *Superalloys 1980,* American Society for Metals, Metals Park, OH, 1980
16. "Container for Hot Consolidating Powder": U.S. Patent 4 142 888, March 6, 1979
17. Fluid Die Press Consolidation, by J. R. Lizenby, W. J. Rozmas, L. J. Barnard and C. A. Kelto: in *Powder Metallurgy Superalloys: Aerospace Materials for the 1980's,* MPR Publishing Services, Ltd., Shrewsbury, England, 1980

ADDITIONAL READING

Superclean Superalloy Powders, by P. Lowenstein: in *Powder Metallurgy Superalloys: Aerospace Materials for the 1980's,* MPR Publishing Services, Ltd., Shrewsbury, England, 1980

Nickel-Base Superalloy Powder Metallurgy—State-of-the-Art, by M. M. Allen, R. L. Athey and J. B. Moore: *Progess in Powder Metallurgy,* Vol 31, MPIF, Princeton, NJ, 1975

Superalloy Turbine Components—Which is the Superior Manufacturing Process: As-HIP, HIP + Isoforge or GATORIZING of Extrusion Consolidated Billet, by J. E. Coyne, W. H. Couts, C. C. Chen and R. P. Roehm: in *Powder Metallurgy Superalloys: Aerospace Materials for the 1980's,* Vol 1, MPR Publishing Services, Ltd., Shrewsbury, England, 1980

Development of Hot Isostatically Pressed (As-HIP) Powder Metallurgy Rene 95 Turbine Hardware, by J. L. Bartos and P. S. Mathur: in *Superalloys—Metallurgy and Manufacture,* AIME, Claitor's, Baton Rouge, 1976

Effect of Processing Variables on Powder Metallurgy Rene 95, by J. F. Barker and E. H. VanDerMolen: in *Superalloys—Processing,* AIME, Battelle, Columbus, OH, 1972

Influence of Atomization Stock on Properties of Hot Isostatically Pressed, Low-Carbon Astroloy, by D. J. Evans and G. M. Judd: *Modern Developments in Powder Metallurgy,* Vol 11, MPIF, Princeton, NJ, 1977

Effect of Heat Treatment and Slight Chemistry Variations on the Physical Metallurgy of Hot Isostatically Pressed, Low-Carbon Astroloy Powder, by M. T. Podob: *Modern Developments in Powder Metallurgy,* Vol 11, MPIF, Princeton, NJ, 1977

Development of a High Strength Hot-Isostatically-Pressed Disk Alloy, MERL 76, by D. J. Evans and R. D. Eng: *Modern Developments in Powder Metallurgy,* Vol 14, MPIF, Princeton, NJ, 1980

Application of Superalloy Powder Metallurgy for Aircraft Engines, by R. L. Dreshfield and R. V. Miner, Jr.: in *P/M Superalloy Technology and Application,* 1980 International Powder Metallurgy Conference

Production of P/M Near-Net Shape Superalloy Hardware, by F. J. Rizzo, J. Lane and J. H. Moll: SAE Paper No. 821515, Aerospace Congress and Exposition, Anaheim, CA, 1982

P/M Dual-Property Wheels for Small Engines, by J. H. Moll, J. H. Schwertz and V. K. Chandhok: *Progress in Powder Metallurgy,* Vol 37, MPIF, Princeton, NJ, 1982

"Development of Materials and Process Technology for Dual Alloy Disks," by C. S. Kortovich and J. M. Marder: NASA CR-165224, Oct 1981

"Nickel-Base Superalloy Articles and Method for Producing the Same": U.S. Patent 3 902 862, Crucible, Inc., Sept 2, 1975

Rapidly Solidified Powders

By Robert S. Carbonara, Battelle Columbus Laboratories

CLASSIFYING a powder as rapidly solidified (RS) can be rather arbitrary. Normally, rapidly solidified materials are quenched at rates greater than 10^4 °C/s. However, quench rates are not usually measured directly, but are inferred from the microstructure of the material. In fact, the most important feature of RS powders is their microstructure. The properties of the consolidated powder metal parts, to a large extent, depend on the RS microstructure. Of course, the consolidation of the powders also plays a significant role in their final properties.

POWDER PRODUCTION

There are several types of atomization processes for producing rapidly solidified powders. The cooling rate for the powder will depend on both the process and the powder particle size — that is, powder particles of a given size will have been quenched at different rates depending on the type of process and the processing conditions used. Two of the most widely used processes are *water and gas atomization* (Fig. 1). Both processes use a high-pressure fluid to disintegrate a continuous stream of molten metal into droplets. Gas-atomization pressures usually range from 2 to 8 MPa (300 to 1200 psi), and gas velocities from 50 to 200 m/s (165 to 655 ft/s). Use of inert gas as the atomizing medium produces smooth, spherical powder particles from 50 to 100 μm (2 to 4 mils) in diameter. Water-atomization pressures are higher, ranging from 4 to 20 MPa (600 to 3000 psi), and water velocities range from 40 to 150 m/s (130 to 490 ft/s). Water-atomized powder particles are larger in diameter (75 to 200 μm, or 3 to 8 mils), and are more irregular in shape, than gas-atomized powder particles. Even though water-atomized particles are larger, they can and usually do have a finer microstructure than gas-atomized particles of the same composition.

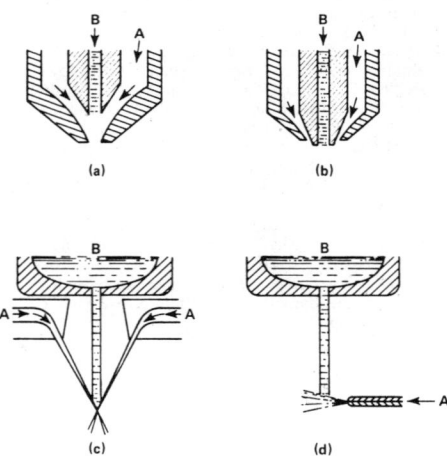

(a) Fully closed system with internal mixing. (b) Semiclosed system with external mixing. (c) Free-fall open system with V- or cone-jet impingement. (d) Free-fall open system with cross-jet impingement. A = atomizing fluid; B = liquid to be atomized.

Fig. 1. Basic nozzle designs for jet atomization

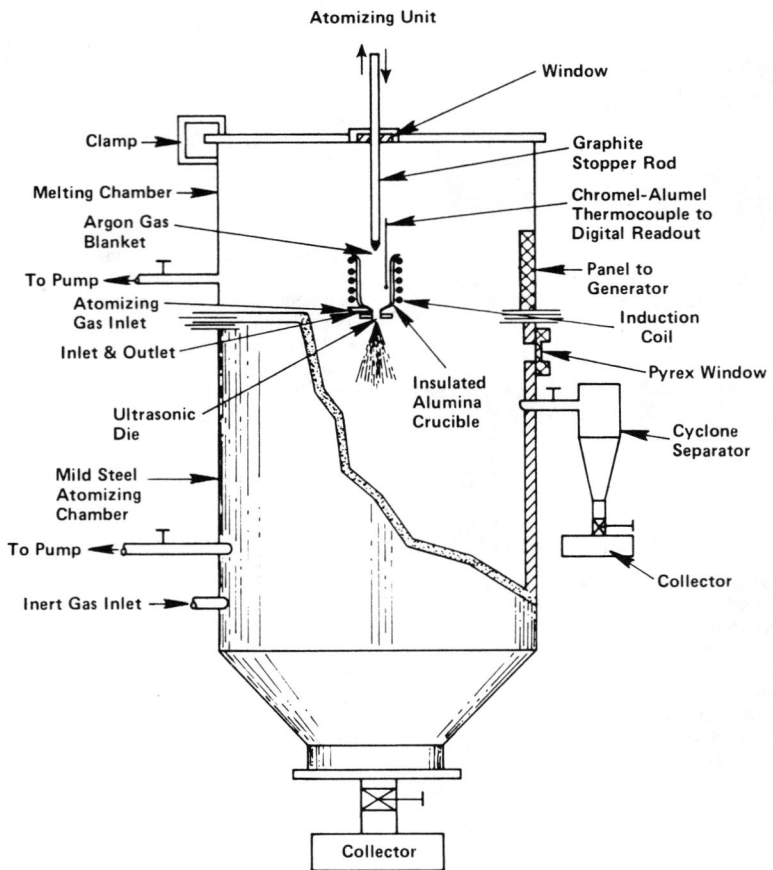

Fig. 2. Schematic illustration of an ultrasonic gas atomization (USGA) chamber

A variation of the gas-atomization process is the *ultrasonic gas atomization* (USGA) process (see Fig. 2). This process uses a high-velocity pulsed gas stream to break up the molten metal stream. Modified Hartman shock tubes are used to produce the high-frequency gas pulses with velocities up to 4600 m/s (15 100 ft/s). This process can produce high yields of powder with particle diameters less than 20 μm (0.8 mil). In spite of the very small particle sizes, some USGA powder has a larger grain microstructure than powders of equivalent size and composition but made by conventional gas atomization.

Gas atomization can also be used in conjunction with substrate cooling by directing the atomized droplets onto the outside of a rotating drum where they are "splat cooled" by conduction (Fig. 3). This process, called *drum splat quenching*, has been used to produce rapidly solidified aluminum alloy flakes, usually a few millimetres in diameter and 50 to 100 μm (2 to 4 mils) thick. Although the process is designed to have each "splat" leave the drum before another one impacts over it, such is not always the case, and multiple splats are often produced.

A recently developed liquid-atomization technique is the *rapid spinning cup* (RSC) process (see Fig. 4). In this process, a stream of molten metal is extruded and impacts a thick layer of

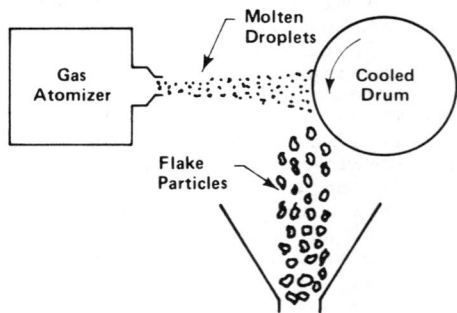

Fig. 3. Schematic representation of the drum splat quenching process

rotating liquid which is on the vertical interior wall of the spinning cup. The cup is designed with an annular lip on the top to contain the quenchant. The liquid in the cup atomizes the molten metal stream into droplets and, as the droplets form, they enter this liquid and are quenched. The centrifugal force on the particles accelerates them through the thick liquid layer. This movement of the particles through the liquid strips the vapor jacket that forms around the liquid-quenched particles and enhances the heat-

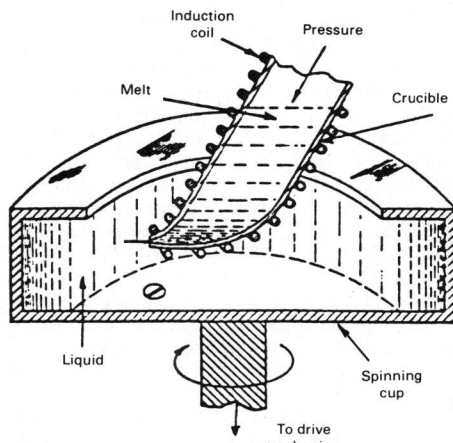

Fig. 4. Schematic illustration of the rapid spinning cup (RSC) apparatus

transfer conditions and produces a very high quench rate. Rotational cup speed can vary up to 16 000 rpm. At these high rotational speeds, the forces produce a large hydrostatic pressure in the liquid, which raises the boiling point of the liquid quenchant and reduces the vapor formation around the particles, which also enhances heat transfer. The RSC process can produce a wide variety of rather narrow size distributions. Particle-size distributions with mean values of about 400 to 500 μm (16 to 20 mils) can be easily produced, as can distributions in which 90% of the particles are less than 10 μm (0.4 mil) in diameter. Powders of superalloys, stainless steels, and alloys of aluminum, copper, zinc and tin, as well as powders of many other alloys, have been made by the RSC process.

There are several other atomization methods that do not use either liquids or gases for atomization. These are the *centrifugal atomization* processes in which liquid-metal droplets are thrown or dispersed from the lip or edge of a flat or concave rotating disk, dish, cup or rod.

In the *rapid solidification rate* (RSR) process (Fig. 5), liquid metal is fed from a bottom-pour crucible onto a concave dish that is rotating at speeds up to 35 000 rpm. As the droplets fly off the rotating dish, they pass through flowing helium gas, which is a high-heat-transfer gas, and this enhances the cooling rate of the particles. Because of the large amount of costly helium used, the economics of the process must be carefully considered. The process appears to be best suited

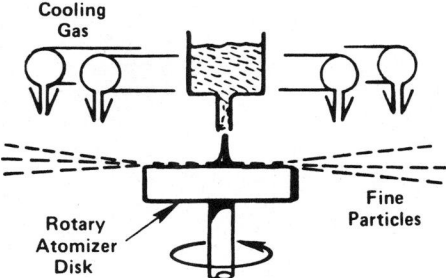

Fig. 5. Schematic representation of the rapid solidification rate (RSR) process

to superalloy and titanium alloy powders with particle diameters ranging from 20 to 200 μm (0.8 to 8 mils), both of which require the protection afforded by helium.

The molten-metal feed system used in this type of process is not limited to a bottom-pour crucible—in fact, both electron beam melting and electric-arc consumable-electrode melting have been used to feed molten metal into the rotating dish or cup.

A slight variation of this type of atomization process involves the use of a perforated cup into which the molten metal is fed. The molten metal droplets emerge from the cup through the perforation and form a rather coarse powder, with particles usually about 1 mm (0.04 in.) in diameter and 1 to 5 mm (0.04 to 0.2 in.) long. This process is best suited to lower-melting point alloys.

An alloy can also be centrifugally atomized by precision machining a rod or bar of the alloy (about 50 mm, or 2 in., in diameter) and rotating it about its long axis while melting one end. The end of this bar can be melted in a variety of ways. If it is melted by an electric arc, the process is called the *rotating electrode process,* or REP (see Fig. 6); if it is melted by a plasma arc, the pro-

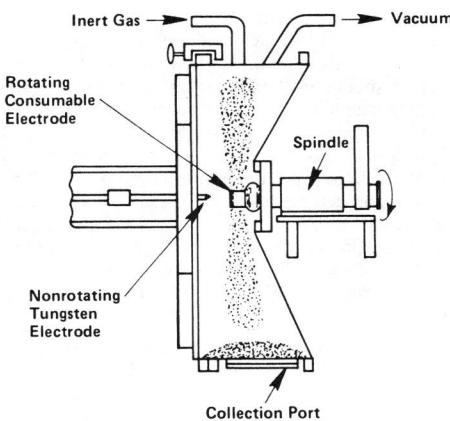

Fig. 6. Schematic representation of the rotating electrode process (REP)

cess is called the *plasma rotating electrode process* (PREP); and if the bar is melted by a laser, the process is called the *laser rotating electrode process* (LREP). It is also possible to use electron beam melting. The melting technique used depends primarily on the need for cleanness of the powder and on the type of atmosphere required for the alloy. As the electrode rotates at high speed, droplets fly off the molten end and form powder particles in sizes ranging from 50 to 400 μm (2 to 16 mils), which are *not* considered to be very rapidly solidified.

The *pulverisation sous vede* (PSV) process uses the same type of rod as in REP, but in PSV the rod is rotated about its vertical axis and the consumable electrode is electron beam melted. The rotation speeds are much lower than in REP, and the particles are much larger (usually greater than 125 μm or 5 mils) and are not rapidly solidified. This process uses a screen to prevent molten particles from hitting the chamber walls. The screen can lead to powder contamination, especially in titanium alloys.

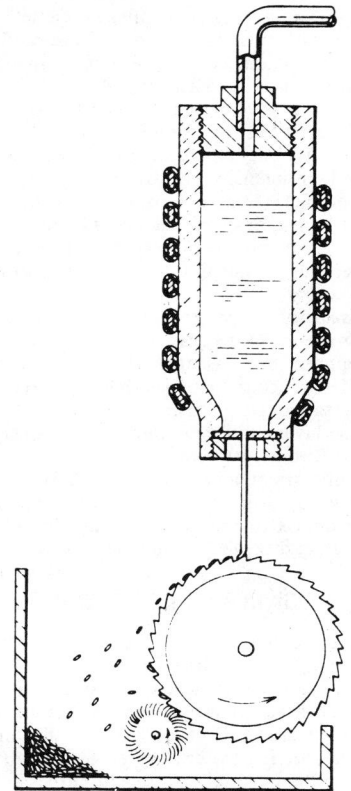

Fig. 7. Schematic representation of the melt spinning (MS) process

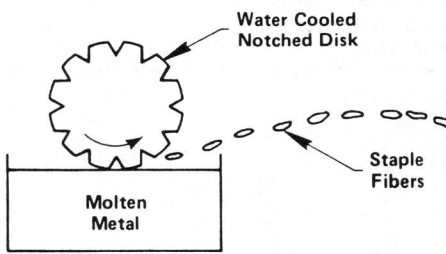

Fig. 8. Schematic representation of the melt extraction (ME) process

The *melt spinning* (MS) and *melt extraction* (ME) processes (see Fig. 7 and 8) are methods that use substrate cooling to produce RS powder. The conventional melt spinning process has been modified to produce RS flakes about 1 mm (0.04 in.) square and 50 to 100 μm (2 to 4 mils) thick. In the modified MS process, molten metal is extruded in a thin stream (about 20 μm, or 0.8 mil, in diameter) onto a rotating drum with a notched surface. It is the notches that produce the flakes, as opposed to the narrow ribbons produced in conventional MS. Solidification rates that are high enough to produce amorphous flakes are possible with this process.

In the ME process, the molten metal solidifies on the edge of a water-cooled disk, and flies off. By notching this wheel, RS powder can be produced with a very specific size. The distribution width can be kept to nearly zero for ME powder. However, it is possible to produce several sizes of powder particles if so desired. Particles as small as 30 to 50 μm (1.2 to 2 mils) in diameter and 500 μm (20 mils) long, and as large as 5 mm (0.2 in.) in diameter and 200 mm (8 in.) long, have been made by ME.

There are several other powdermaking processes which can produce RS powders. The *electrohydrodynamic atomization* (EHDA) process uses a high-intensity electric field to form very small droplets (from less than 1 μm, or 0.04 mil, up to 100 μm, or 4 mils) at the end of a capillary orifice. This produces very rapidly solidified powder, but is limited to the production of a few grams a day.

Plasma spray deposition is also considered by some to be an RS process. If the deposited layers are kept very thin (less than 250 μm, or 10 mils), RS microstructures are possible. However, unless precautionary steps are taken, successive deposition layers will transform the previously deposited RS microstructure.

Finally, irregularly shaped RS flakes can be produced by drip melting from a stationary rod or bar onto a rotating disk. As the drops strike the rotating disk, they flatten out into RS flakes.

POWDER CONSOLIDATION

The other aspect of RS powders is their consolidation. To consolidate a powder, it is necessary to use either heat or pressure, or both. However, heat can have a very detrimental affect on RS materials. Because many of the microstructures of RS materials are metastable, heat added to the material can transform the material out of its RS microstructure and, in many cases, undo the property enhancements achieved through RS. This is especially true of RS amorphous alloys, most of which have glass transition temperatures (above which they become crystalline) of about 400 to 600 °C (750 to 1100 °F). However, there are many alloys which moderate heating does not affect.

To eliminate or reduce the amount of heat needed for consolidation, several techniques are available. By using *dynamic compaction* methods, such as explosive, gas-gun or rail-gun compaction, very high pressures (several GPa) can be achieved for short periods of time (a few milliseconds). These high pressures are sufficient to cause metallurgical bonding between the powder particles and to develop fully dense parts. The shortcomings of this technique are the high costs of the special facilities and materials needed for explosive compaction and of the special dies and equipment required for gas- and rail-gun compaction, very limited shape capability, and of course the low production rate of all of these dynamic compaction methods. Also, it is necessary to consider the effects of the adiabatic heating caused by the sudden rise and fall in pressure in the material. This effect can be rather significant for many materials. In spite of their shortcomings, dynamic compaction methods may be very useful in special cases.

Hot isostatic processing (HIP) is another technique of powder consolidation that can be used for RS powders. In HIP, the pressures are much lower (about 200 MPa, or 30 ksi) but can be sustained for much longer periods of time (several hours). In addition, the temperature of the parts can be closely controlled to ensure maintenance of the desired microstructure and properties. However, HIP also has its limitations. The pressure/temperature combinations available in HIP and needed to retain RS properties are sometimes not adequate to cause sufficient metallurgical bonding between powder particles and/or full densification. But there are enough situations where the temperature and pressure are sufficient, which is why HIP is one of the major consolidation processes for RS powders. As a better understanding of RS materials develops, HIP will become more widely used for production of RS powder parts.

Recent advances in *powder forging* (PF) technology have made PF an alternative technique for consolidation of RS powders. In PF, a precise preform is produced from RS powders. Normally, the preform is of a simpler shape so that the desired final shape is produced to closely controlled dimensions in the hot forging step. PF differs from HIP in that densification occurs rapidly and involves lateral flow of the material mass, which causes high shear stresses at powder and pore surfaces and ruptures surface oxide films present on the powder particles. This promotes excellent metallurgical bonding between powder particles and collapsed pore surfaces and "fiberizes" inclusions in the lateral direction. Hence, preform design is critical to the success of the PF process.

Another consolidation process that relies more on pressure than on temperature is *extrusion*. One of the major features of extrusion that makes it attractive for consolidation of RS powders is that it involves large deformation and movement of the powder. Large reductions in the cross section of the extrusion billet are possible (as large as 30:1 for RS aluminum alloys). This causes new or fresh metallic surfaces to be exposed to one another and leads to metallurgical bonding between powder particles.

As with the other consolidation processes, extrusion has limitations. These are primarily related to the mechanical properties of the RS powder and how the powder yields and flows under deformation.

At the present time, HIP and extrusion appear to be the most attractive processes for consolidation of RS powders. There are, however, many RS powders that can be consolidated by pressing, sintering and other conventional P/M consolidation processes. The best method for a given RS powder depends on many factors — some of which are not yet known — that are unique to the specific RS powder.

26 FORMING

Edited by Roger N. Wright, Rensselaer Polytechnic Institute

Review Committee: Richard D. Campbell, Rensselaer Polytechnic Institute; Thomas A. Kircher, Rensselaer Polytechnic Institute; and Joseph T. Strauss, Rensselaer Polytechnic Institute

This section was condensed from Metals Handbook, Eighth Edition, Volume 4, and Ninth Edition, Volume 6, pages 896 to 920. For more detailed information on the topics covered in this section, the reader is referred to the larger works.

Forming of Steel Sheet, Strip and Plate

Presses and Auxiliary Equipment for Forming of Sheet Metal

PRESSES described in this article are mechanically or hydraulically powered machines used for producing parts from sheet metal.

Power presses can be classified according to the following characteristics: source of power, type of frame, method of actuation of slides, and number of slides in action. Presses in any of these classes are available in a range of capacities (tonnage or bed area), although the range is not necessarily the same for all types of presses. Characteristics of 18 types of presses are summarized in Table 1.

JIC Identification System. The Joint Industry Conference (JIC) system of identifying press characteristics is in general use. In a typical sample:

S4-750-96-72

the press is identified by the S as a single-action model (D is used for double-action, T for triple-action, and OBI for open-back inclinable); by the 4 as having four-point suspension; by the 750 as being rated at 750-ton capacity; and by the 96 and 72 as having a bed measuring 96 in. left-to-right and 72 in. front-to-back. Any other press can be so identified, by substitution of appropriate numerals for number of suspension points, tonnage rating, and bed dimensions. These characteristics are discussed later in this article.

The JIC also recommends that a metal tag be attached permanently to the press, stating the stroke length, shut height, kind and length of adjustment, strokes per minute, size and weight. If a die cushion is provided, an additional tag should describe it.

SOURCE OF POWER

Power presses for sheet-metal work can be driven hydraulically or mechanically. The performance characteristics and other operational features of hydraulic and mechanical presses are compared in the following list.

1. Force is exerted constantly throughout the stroke of a hydraulic press. Force developed by a mechanical press varies with the position of the slide.
2. The length of stroke is easily adjusted and controlled in a hydraulic press. In a mechanical press, the stroke is fixed by the throw of the crank or eccentric.

Table 1. Characteristics of 18 types of presses

Type of press	Type of frame							Position of frame				Action			Method of actuation								Type of drive				Suspension			Ram		Bed		
	Open-back	Gap	Straight-side	Arch	Pillar	Solid	Tie rod	Vertical	Horizontal	Inclinable	Inclined	Single	Double	Triple	Crank	Front-to-back crank	Eccentric	Toggle	Screw	Cam	Rack and pinion	Piston	Over direct	Geared, overdrive	Under direct	Geared, underdrive	One-point	Two-point	Four-point	Single	Multiple	Solid	Open	Adjustable
Bench	X	X				X		X		X	X	X			X		X		X		X	X	X				X			X		X	X	X
Open-back inclinable	X	X				X		X		X	X	X	X		X		X			X		X	X	X			X	X		X	X		X	
Gap-frame	X	X				X	X	X	X	X	X	X	X		X	X	X	X	X			X	X	X			X	X		X	X	X	X	X
Adjustable-bed horn		X				X		X			X	X										X	X	X			X	X		X		X	X	X
End-wheel		X				X		X				X					X	X					X	X			X	X		X		X		X
Arch-frame			X	X		X	X	X				X			X		X		X				X	X			X			X		X		X
Straight-side			X	X		X	X	X	X			X	X	X	X	X	X	X				X	X	X	X	X	X	X	X	X	X	X	X	X
Reducing	X	X	X			X		X	X			X	X					X					X	X			X	X		X	X	X	X	X
Knuckle-lever			X			X	X	X				X	X				X						X	X			X			X		X	X	X
Toggle-draw			X			X	X	X					X	X			X						X	X			X	X		X	X		X	
Cam-drawing	X	X	X			X	X	X	X				X				X			X			X	X			X				X		X	
Two-point single-action			X	X		X	X	X				X			X	X	X						X	X			X		X	X	X	X	X	X
High-production			X			X	X	X				X			X		X						X	X	X		X			X	X	X	X	X
Dieing machine			X			X		X				X			X	X							X	X	X		X			X		X		X
Transfer			X	X		X	X	X				X			X	X	X	X					X	X			X			X		X	X	X
Flat-edge trimming			X	X		X		X				X								X			X				X			X		X		X
Hydraulic			X	X		X	X	X	X	X		X	X	X								X					X	X		X	X	X	X	X
Press brake	X	X				X		X				X			X							X		X			X			X	X	X		X

3. The speed of a hydraulic press is adjustable over a wide range, whereas the speed of a mechanical press is limited by the type of drive.

4. A hydraulic press cannot be overloaded. It can deliver only a preset force, and slide motion stops when that force is reached. A mechanical press can be overloaded, resulting in damage to the press, if it is not equipped with overload protection.

5. Mechanical presses cycle faster and are better suited to high production than are hydraulic presses.

6. Because energy is stored in the flywheel, a mechanical press can use a smaller motor. For some applications, the size of the motor in a hydraulic press may be as much as $2^1/_2$ times as large as in an equivalent mechanical press.

7. Ram velocity in a mechanical press is higher, making this equipment more useful in operations such as blanking and piercing, in which a high-impact blow is needed. Blanking and piercing can be done in a hydraulic press, but the shock of the punch breaking through the metal can cause damage to the hydraulic system.

Mechanical Presses. In most mechanical presses, a flywheel is the major source of energy that is applied to the slides by cranks, gears, eccentrics or linkages during the working part of the stroke. During operation, the flywheel runs continuously and is engaged by the clutch only when a press stroke is needed. In some very large mechanical presses the drive motor is connected directly to the press shaft, thus eliminating the need for a flywheel and a clutch.

Two basic types of drive are used to transfer the rotational force of the flywheel to the main shaft of the press — the nongeared drive and the gear drive (Fig. 1).

Hydraulic Presses. Hydrostatic pressure against one or more pistons provides the power for a hydraulic press. Most hydraulic presses have a variable-volume, variable-pressure, concentric-piston pump to provide them with a fast slide opening

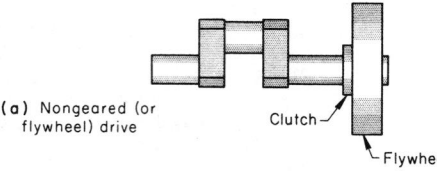

(a) Nongeared (or flywheel) drive

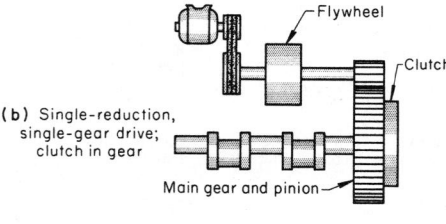

(b) Single-reduction, single-gear drive; clutch in gear

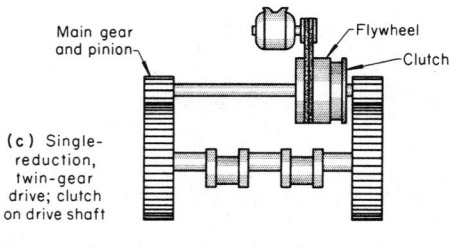

(c) Single-reduction, twin-gear drive; clutch on drive shaft

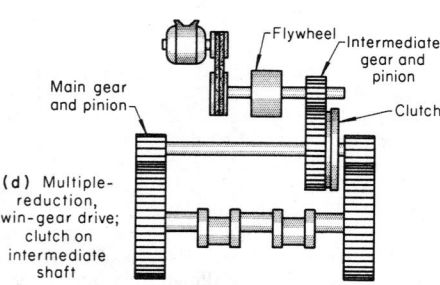

(d) Multiple-reduction, twin-gear drive; clutch on intermediate shaft

Fig. 1. Four types of drive-and-clutch arrangements for mechanical presses

and closing speed, as well as with a slow working speed at high forming pressure.

Pumps, reservoirs and other components of the hydraulic system usually are housed in the frame and in the crown of the press, although in some presses these components are housed below the bed. If all components are readily accessible for maintenance, their location is unimportant.

SLIDE ACTUATION IN MECHANICAL PRESSES

Rotary motion of the motor shaft on a mechanical press is converted into reciprocating motion of the slides by one of the following:

Crankshaft
Eccentric shaft
Eccentric-gear drive
Knuckle-lever drive
Rocker-arm drive
Toggle mechanism

Crankshafts. The most common mechanical drive for presses with capacities up to 300 tons is the crankshaft drive (Fig. 2).

Eccentric Shafts are similar to crankshafts. The eccentric completely fills the space between the supporting bearings of the press crown, thereby eliminating the deflection commonly caused by the unsupported portion where the crank cheeks normally would be. Eccentric drives (Fig. 2) are often used in high-speed short-stroke straight-side presses with progressive dies.

Eccentric-gear drives (Fig. 2) are used almost universally for large straight-side presses that operate at speeds of less than 50 strokes per minute. In place of a crankshaft, an eccentric is built as an integral part of the press drive gear. The eccentric gear permits strokes as long as 50 in.

Knuckle-lever drives combine the motions of a crank and a knuckle lever to drive the press slide (Fig. 2). Their use is limited to operations such as coining or embossing.

Rocker-arm drives apply crank or eccentric motion to a rocker arm that is connected to the press slide (Fig. 2). In this mechanism, the linkage is

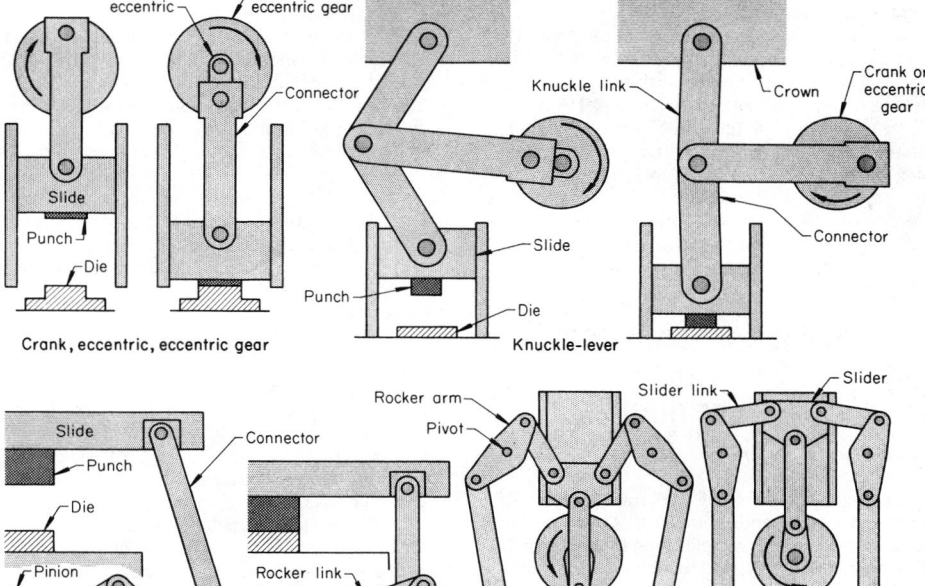

Fig. 2. Principles of operation of drives for mechanical presses

driven by an eccentric gear and a connecting rod. The rocker-arm drive is a variation of the knuckle-lever drive. However, a press with rocker-arm drive is not limited to coining operations, but can be used also for drawing or forming operations.

Toggle mechanisms are the most widely used means of providing the second action in double-action mechanical presses. The toggles operate an outer slide, which clamps the blank against the die, while the punch, operated by the inner slide directly from the crankshaft, performs the draw operation. Principal components of a toggle mechanism are shown in Fig. 2.

AUXILIARY EQUIPMENT

Most primary press operations are automated, so that equipment for feeding and unloading is used even for fairly short runs. Hand feeding, with its attendant hazards, is often confined to second operations on partly completed workpieces.

Planning for automated operations should include the following goals:

1. Maximum safety to the operator and to the equipment
2. High or nearly continuous production
3. Improved quality of the product and minimum scrap
4. Reduction in cost of the finished parts.

The shape and position of the part before and after each operation must be carefully studied to determine whether design changes, such as providing tabs or extra stock on the blank, will facilitate handling.

Automatic handling equipment can be divided into the following categories: feeding equipment, unloading equipment, and equipment for transferring the work from one press operation to the next.

Straighteners are used to remove coil kinks and to flatten stock before feeding into a die. Roller levelers improve formability of the stock by plastic working. Coil-handling equipment moves coiled stock to the press area and uncoils it with a minimum of damage to the stock and danger to the tools and operator.

An example of a modern sheet-forming system is shown in Fig. 3, including computer numerical controls (CNC) for feeding, punching, and unloading of stock.

Selection of Presses for Forming of Sheet Metal

SELECTION of press equipment for making a sheet metal part involves three steps:

1. Analyzing the part
2. Determining the type of tooling
3. Relating part and tooling requirements to press features.

Although these steps can be considered individually, the final decision is usually a set of compromises based on costs, lead time, and availability or utilization of equipment.

ANALYSIS OF PART

A drawing of the part and a sample part, if available, are carefully studied to determine what basic sheet metal operations, such as blanking, piercing, bending, forming or drawing, are necessary to make the part. The effect of one operation on another, relations of part features to one another, and the dimensional tolerances, influence the sequence of operations and die design. How the basic operations can be combined into a die or die station, size and shape of the part after each die operation, production rate and total quantity, and the composition and thickness of the work metal, all have an effect on the type of press selected.

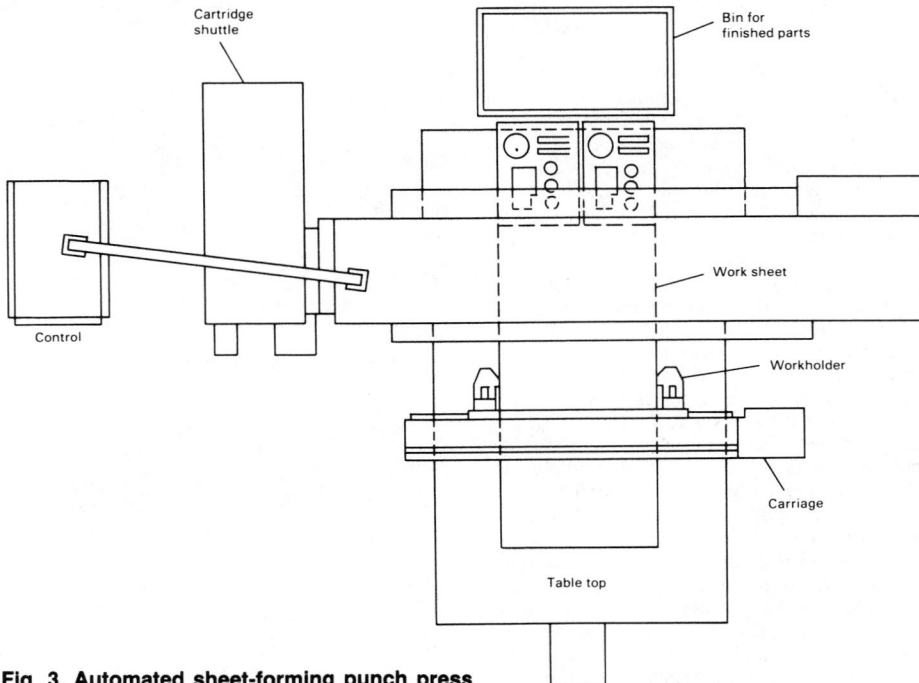

Fig. 3. Automated sheet-forming punch press

TYPE OF TOOLING

The operations necessary to make a part can be incorporated into a series of individual dies, a progressive die or a transfer die. Individual dies may be either single-operation or compound.

A turret-punch press carries the punches and dies to produce many jobs in a rotating turret. Each tool is an individual station. The turrets automatically rotate and firmly lock in position with two seconds' average time per station. Complicated hole patterns are made using single stations and combinations of different sizes and shapes.

A modern turret-punch press may carry as many as 36 different punches and dies and make from 75 to 175 hits per minute.

Press size and tonnage rating depend on the size of die and how much work it has to perform. The length of a progressive or transfer die is determined by the size of the part and the number of operations to be performed. The number of die stations is not necessarily the same as the number of basic operations, but depends on how these operations are combined and divided when the strip layout is designed. The layout should include the necessary idle stations to make a strong die, provide space for transfer from station to station and provide extra stations, if needed, to make the die function more smoothly. For a difficult part it is good practice, if time permits, to make temporary dies to determine by trial how the part will form and how many stations are necessary.

SELECTION OF A MECHANICAL PRESS

Mechanical presses are suitable for all forming operations, and also for blanking and piercing.

Tonnage Capacity. The force needed to do a given job depends on the strength and thickness of the work metal and on the perimeter of the surfaces to be worked. Forces required to do a given operation can be estimated by the formulas given in the articles in this volume that deal with the specific types of operations.

Mechanical presses develop their rated tonnage at a specific distance above the bottom of the stroke. Therefore, the point on the stroke at which work begins is important. Forming and drawing begin at a distance above the bottom of the stroke equal to or greater than the depth of draw. At this point the force exerted by the press is less than the rated tonnage.

Blanking and piercing are done near the bottom of the stroke, where press tonnage is maximum. Blanking in compound blank-and-form or blank-and-draw dies is done above the bottom of the stroke, where the tonnage is less than the full rating.

Restriking, coining and embossing operations are done at the bottom of the stroke and usually require more press tonnage than forming.

Selection of a Hydraulic Press. Table 1 in the preceding article gives specifications that should be considered in the selection of hydraulic presses.

Hydraulic presses are not used extensively for forming sheet metal, because of high initial and maintenance costs and difficulties in relocating in a press line. Also, cycle time for a hydraulic press is greater than for a comparable mechanical press.

Hydraulic presses, however, are particularly suited for making deep and intricate draws in all types of sheet metal. Although blanking can be done in the press, precut blanks are preferred for better control of the drawing operation and to

avoid damaging the hydraulic system as the punch breaks through the metal.

Metal flow is well controlled, because both the force on the slide and the velocity of the slide are constant throughout the working stroke. These features are an advantage when cupping and reverse redrawing in the same press stroke. The amount of impact between the work metal and the blankholder or punch is minimized, because the press ram is slowed automatically before die contact and then does its work at whatever speed and force are needed to produce the best results. This permits the use of a small punch nose.

Selection and Use of Lubricants in Forming of Sheet Metal

LUBRICATION is of two main types: fluid (hydrodynamic) and boundary. (Extreme-pressure lubrication is a special type of boundary lubrication.) These types are illustrated schematically in Fig. 4.

Fluid lubrication is typified by metal surfaces separated by a continuous film of lubricant having a thickness considerably greater than the height of the surface asperities of the metal (Fig. 4a).

Boundary lubrication is typified by metal surfaces separated by a lubricant film only a few mole-

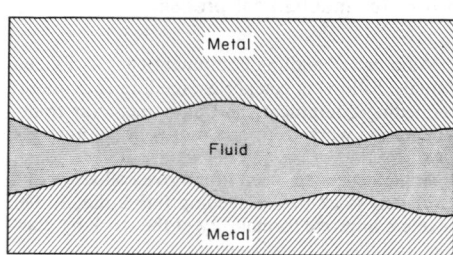

(a) Fluid (hydrodynamic) lubrication

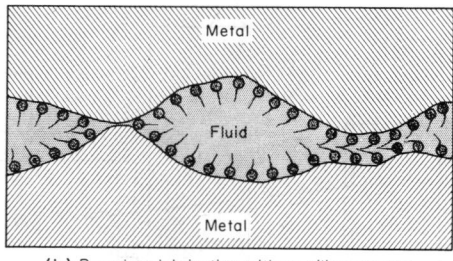

(b) Boundary lubrication with an oiliness agent

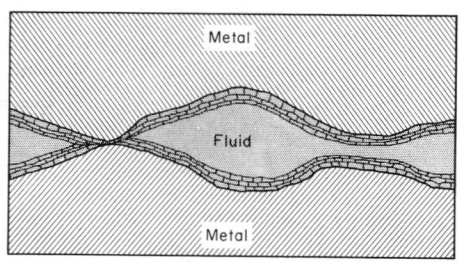

(c) Extreme-pressure boundary lubrication with a pigment (mechanical EP agent)

Fig. 4. Schematic representation of the types of lubrication

cules thick, with considerable metal-to-metal contact between asperities of the two surfaces (Fig. 4b). In most metal-forming operations, because of the high pressures and low speeds involved, lubrication is of the boundary type.

Extreme-Pressure (EP) Lubrication. Substances that have boundary-lubrication properties in an optimum combination are called extreme-pressure (EP) lubricants. One type contains finely divided inorganic solids (pigments, or mechanical EP agents) that physically separate workpieces from the tools. The use of extreme-pressure boundary lubrication with a pigment is illustrated in Fig. 4(c).

SELECTION FACTORS

The proper lubricant for sheet metal forming will depend on the severity of the forming operation. This in turn is governed by the type of sheet, its condition, the change of shape required, and the tool design.

The type of sheet will influence the expected difficulties. A ductile metal such as copper will tend to gall against the tool, a problem not so common with steel sheet. The surface reactivity of the sheet is important. For example, conventional and EP additives are not as efficient with stainless steels.

The condition of the sheet is primarily its strength (annealed or cold worked), and the presence of rust or scale. Change of shape is important, as friction is most critical in deep drawing and stretch forming, as opposed to blanking. Additionally, small radii in tool design may increase the severity of the operation, particularly in deep drawing.

Secondary considerations may be equally critical in lubricant selection. The lubricant must be nontoxic, it must not react unfavorably with the metal surface, it must be stable at working conditions, and it must be easily removable after forming.

Tables 2 and 3 give commonly recommended lubricants for forming of various sheet metals and their alloys.

Blanking of Low-Carbon Steel

A BLANK is a shape cut from flat or preformed stock. Ordinarily, a blank serves as a starting workpiece for a formed part; less often, it is a desired end product. This article deals with the production of blanks from low-carbon steel (such as 1008 and 1010) sheet and strip, in dies in a mechanical or hydraulic press.

METHODS OF BLANKING IN PRESSES

Cutting operations that are done by dies in presses to produce blanks include cutoff, parting, blanking, notching and lancing. The first three of these operations can produce a complete blank in a single press stroke. In progressive dies, two or more of these five operations are done in sequence to develop the complete outline of the blank and to separate it from the sheet, strip or coil stock.

Trimming, or cutting off excess material from the periphery of a workpiece, usually is done in dies, and is similar to blanking. Often it is the final operation on a formed or drawn part.

Table 2. Lubricants commonly used in forming of various sheet metals

Operation	Carbon and low-alloy Steels	Stainless steels	Copper and copper alloys	Aluminum and aluminum alloys
Blanking	Residual fatty oil from light mineral oil Mineral or fatty oil emulsions	Fatty oil emulsions (very light work) Chlorinated mineral oils Soap-fat emulsions Dry soap films (special applications)	Soap solutions Straight emulsions Mineral oils (blended with fatty oil or noncorrosive EP agent for heavier work) Soap-fat emulsions	Light mineral oils Soap-fat or mineral oil emulsions Mineral–fatty-oil blend
Light presswork	Residual fatty oil from rolling Soap-fat emulsions Mineral oils with EP agents Chlorinated oils		Same as above	Same as above
Deep drawing and heavy pressing	Undiluted chlorinated oils Pigmented soap-fat compounds Sulfochlorinated oils Dry soap films	Chlorinated oil (possibly undiluted) Pigmented soap-fat compounds Dry soap films Dry polymer coatings (Usually chlorinated and used with chlorinated oil)	Soap-fat emulsions (possibly pigmented) Undiluted fatty oils Noncorrosive EP oils	Mineral–fatty-oil blends Undiluted fatty oils Undiluted chlorinated oils Pigmented soap-fat compounds Dry soap or wax films
Stretch forming				Emulsions (often pigmented) Heavy mineral–fatty-oil blends

Table 3. Lubricants used in forming of titanium and magnesium and their alloys

Operation	Ti and Ti alloys	Mg and Mg alloys
Cold working	Oxide film with soap Fluoride-phosphate with soap or MoS_2 Dry polymer film	EP oils, waxes, lanolin
Warm working	Graphite or MoS_2-containing lubricants	Graphite in tallow Hard wax Colloidal graphite in naphtha
Hot working	Glasses	

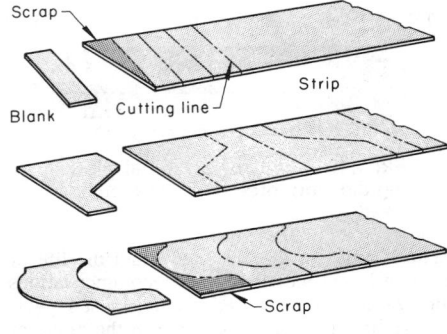

Fig. 5. Nested layouts for making blanks by cutoff

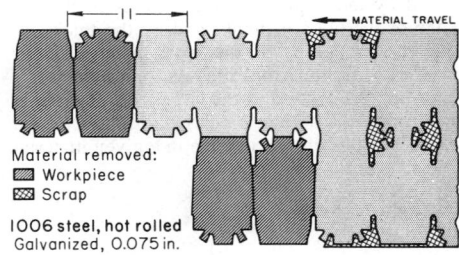

Material removed:
▨ Workpiece
▧ Scrap

1006 steel, hot rolled
Galvanized, 0.075 in.

Fig. 6. Layout for cutoff of four blanks at each press stroke from notched and seminotched strip

Cutoff is cutting along a line to produce blanks, without generating any scrap in the cutting operation, most of the part outline having been developed by notching or lancing in preceding stations. The cutoff line may have almost any shape—straight, broken or curved. After being cut off, the blanks fall onto a conveyor or into a chute or container.

A cutoff die may be used to cut the entire outline of blanks whose shape permits nesting in a layout that uses all of the material (except possibly at the ends of the strip), as shown in Fig. 5. Alternating positions can sometimes be used in nesting, as shown in the middle strip in Fig. 5, to avoid the production of scrap except at strip ends.

Cutoff also is used to cut blanks from strip that has already been notched to separate the blanks along part of their periphery, as shown in Fig. 6.

Parting (Fig. 7) is the separation of blanks by cutting away a strip of material between them.

Like cutoff, it can be done after most of the part outline has been developed by notching or lancing. It is used to make blanks that do not have mating adjacent surfaces for cutoff (Fig. 7); or blanks that must be spaced for ease of handling, to avoid distortion, or to allow room for sturdy tools. Some scrap is produced in making blanks by parting, which is therefore less efficient than cutoff in the use of material.

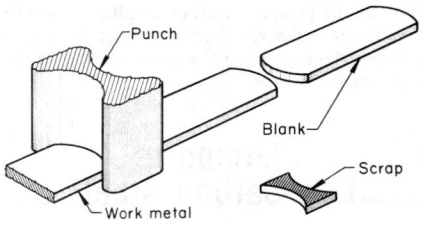

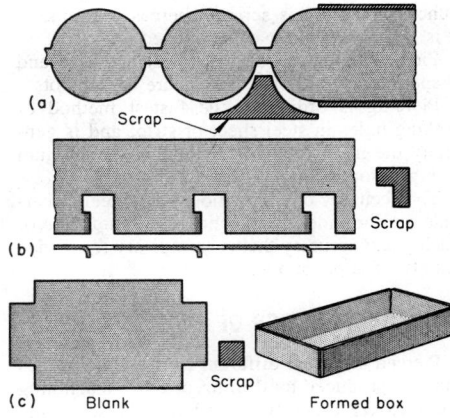

Fig. 8. Notched work illustrating use of notching or freeing of metal before drawing (a) and before forming (b), and for removing excess metal before forming (c)

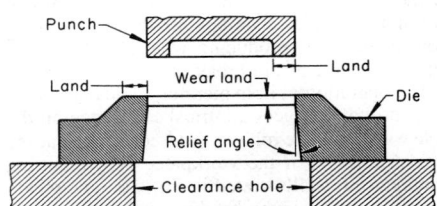

Fig. 9. Relief in a blanking die

Blanking (also called punching) is the cutting of the complete outline of a workpiece in a single press stroke. Usually a scrap skeleton is produced, so that blanking causes some waste of material. However, blanking is usually the fastest and most economical way to make flat parts, particularly in large quantities.

Notching is an operation in which the individual punch removes a piece of metal from the edge of the blank or strip (Fig. 8).

DIE CLEARANCE

The terms clearance, die clearance, and punch-to-die clearance are used synonymously to refer to the space between punch and die. Clearance is important for reliable operation of the blanking equipment, quality and type of cut edges, and life of punch and die.

Relief in a blanking die (Fig. 9) is the taper that is provided so that the severed blank can fall free. The relief angle may range from $1/2°$ to $2°$ from the vertical wall of the die opening. Sometimes, relief in a die is called draft or angular clearance. In some dies, the relief may start at the top of the die surface and have a taper of only 0.002 in. per inch per side.

Piercing of Low-Carbon Steel

PIERCING is the cutting of holes in sheet metal, generally by removing a slug of metal, with a punch and die. Piercing is like blanking, except that in piercing, the work metal that surrounds the piercing punch is the workpiece and the punched-out slug is scrap, whereas in blanking it is the workpiece that is punched out.

Pierced holes can be of almost any size and shape; elongated holes usually are called slots.

Piercing is ordinarily the fastest method of making holes in steel sheet or strip, and is generally the most economical method for medium to high production.

The accuracy of conventional tool steel or carbide dies provides pierced holes with a degree of quality and accuracy that is satisfactory for a wide variety of applications.

CHARACTERISTICS OF PIERCED HOLES

Pierced holes are different from through-holes that are produced by drilling or other machining methods. A properly drilled or otherwise machined through-hole has a sidewall that is straight for the full thickness of the work metal, with a high degree of accuracy in size, roundness and straightness. The sidewall of a pierced hole generally is straight and smooth for only a portion of the thickness, beginning near the punch end of the hole; the remainder of the wall is broken out in an irregular cone beyond the straight portion of the hole, producing what is called fracture, breakout, or die break (see Fig. 10).

The operation of hole piercing typically begins as a cut that produces a burnished surface on the hole wall and some rollover (curved surface caused by deformation of the workpiece before cutting commenced) as illustrated in Fig. 10. The punch completes its stroke by breaking and tearing away the metal that was not cut during the initial part of the piercing operation.

SELECTION OF DIE CLEARANCE

Clearance, or space between the punch and the sidewall of the die, affects the reliability of operation of piercing (and blanking) equipment, the

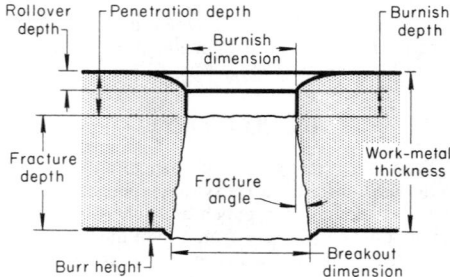

Curvature and angles are exaggerated for emphasis.
Fig. 10. Characteristics of a pierced hole

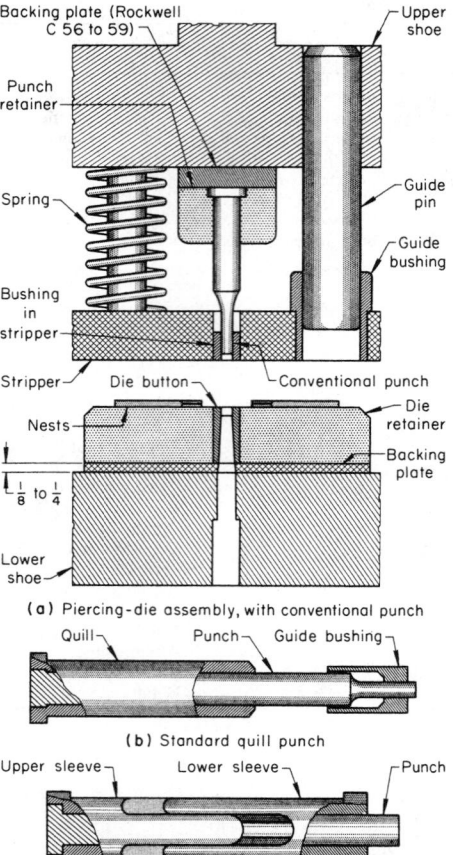

(a) Piercing-die assembly, with conventional punch

(b) Standard quill punch

(c) Telescoping-sleeve quill punch

Fig. 11. Layout of a typical piercing die, and three types of punches used

characteristics of the cut edges, and the life of the punch and die. Published recommendations for clearances have varied widely; most suggest a clearance per side somewhere in the range of 3 to 12.5% of the stock thickness for steel.

Establishment of the clearance to be used for a given piercing or blanking operation is influenced by the required characteristics of the cut edge of the hole or blank, and by the thickness and the properties of the work metal. Larger clearances prolong tool life. An optimum clearance may be defined as the largest clearance that will produce a hole or blank having the required characteristics of the cut edge in a given material and thickness.

PRESSES

Presses used in piercing are the same as those used in other pressworking operations. Open-back gap-frame presses of the fixed upright, fixed inclined, or inclinable types are common. The stock can be fed from the side with minimum interference from the press frame, and the parts can be removed from the front by the operator or ejected out the back by gravity or air jets.

TOOLS

A typical piercing die consists of: upper and lower die shoes, to which punch and die retainers

are attached; punches and die buttons; and a spring-actuated guided stripper (see Fig. 11).

Punches. Figure 11 shows three types of punches used for piercing: (a) conventional, (b) standard quill, and (c) telescoping-sleeve quill.

Fine-Edge Blanking and Piercing

FINE-EDGE BLANKING (also known as fine blanking, smooth-edge blanking, or fine-flow blanking) produces precise blanks in a single operation without the fractured edges characteristically produced in conventional blanking and piercing. In fine-edge blanking, a V-shape impingement ring (Fig. 12) is forced into the stock to lock it tightly against the die and to force the work metal to flow toward the punch, so that the part can be extruded out of the strip without fracture or die break. Die clearance is extremely small, and punch speed much slower than in conventional blanking.

Fine-edge piercing can be done either separately or at the same time as fine-edge blanking. In piercing small holes, an impingement ring may not be needed.

No further finishing or machining operations are necessary to obtain blank or hole edges comparable to machined edges, or to those that are conventionally blanked or pierced and then shaved. A quick touchup on an abrasive belt or a short treatment in a vibratory finisher may be used to remove the small burr on the blank.

PROCESS CAPABILITIES

Holes can be pierced in low-carbon steel with a diameter as small as 50% of stock thickness. In high-carbon steel, the smallest hole diameter is about 75% of stock thickness. Holes can be spaced as close to each other, or to the edge of the blank, as 50 to 70% of stock thickness. Total tolerances obtainable are: 0.0005 in. on hole diameter and for accuracy of blank outline; 0.001 in. on hole location with respect to a datum surface, and 0.001 in. on flatness.

No die break shows on the sheared surface of the hole. Blank edges may be rough for a few thousandths of an inch. Since cracking separation at the blank edge is prevented and shearing occurs throughout, the punch unloads gradually and there is no "snapthrough" of the punch. This reduces wear and damage at the tool faces.

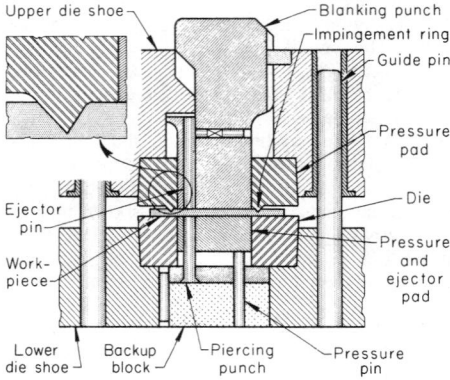

Fig. 12. Typical tooling setup for fine-edge blanking a simple shape

TOOLS

The design of tools for fine-edge blanking is based on the shape of the part, the method of making the die, the required load, and the extremely small punch-to-die clearance. The considerable loading and intended accuracy require that the press tools be sturdy and well supported to prevent deflection. The small clearance presupposes precise alignment of the punch and die.

Design. A basic tool comprises three functional components: the die, the punch, and back-pressure components. To produce good-quality blanks, the punch-to-die clearance must be uniform along the entire profile and must be suitable for the thickness and strength of the work metal. The clearance varies between 0.0002 and 0.0004 in. (see Fig. 12). A dulled die edge will produce the best surface finish on the sheared edge.

Blanking and Piercing of Magnetically Soft Materials (Electrical Sheet)

MAGNETICALLY SOFT materials are used for various static and rotating electrical devices. The majority of finished parts must be laminated—that is, composed of flat parts of a particular shape that are stacked to a given height and fastened together by riveting, bolting or welding. The stock is flat rolled metal and is usually called "electrical sheet." This sheet is available in coils or cut-to-length. For most applications, stock thickness ranges from 29 gage (0.0135 in.) to 24 gage (0.0239 in.). For some special purposes, sheet less than 0.001 in. thick is used.

Silicon steels are manufactured to specific magnetic, rather than mechanical, properties and are used in a wide range of applications. Table 4 lists the types most used, and their approximate silicon contents. Note that as the M-number decreases, the silicon content increases, except for the grain-oriented types, which all contain from 2.8 to 3.5% Si. Carbon content is not specified in silicon steels, but is preferably held to less than 0.005%. In most silicon steels, elements other than silicon and iron are present only in residual amounts, although small amounts of aluminum or other elements sometimes are added to obtain specific magnetic characteristics.

Size and Shape. Flat laminations of a wide variety of shapes and sizes are blanked and pierced from electrical sheet. Most, however, have shapes like those in Fig. 13 and 14. Laminations similar to those shown in Fig. 13 can range in diameter

Table 4. Designations and nominal silicon contents of flat rolled electrical steels

AISI type	Nominal silicon, %	AISI type	Nominal silicon, %
M-50	0 to 0.6	M-4(a) ...	2.8 to 3.5
M-45	0 to 0.6	M-5(a) ...	2.8 to 3.5
M-43	0.6 to 1.3	M-6(a) ...	2.8 to 3.5
M-36	1.4 to 2.2	M-7(a) ...	2.8 to 3.5
M-27	1.7 to 3.0	M-8(a) ...	2.8 to 3.5
M-22	2.5 to 3.5	M-15	2.8 to 5.0
M-19	2.5 to 3.8	M-14	4.0 to 5.0
(a) Grain-oriented			

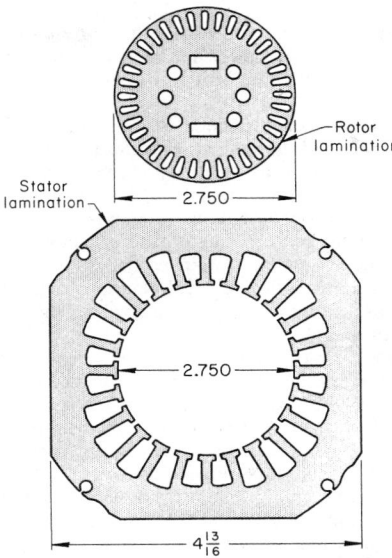

Laminations for rotating electrical machinery are blanked and pierced by single-station dies (sequence shown in Fig. 15).

Fig. 13. Typical rotor and stator laminations blanked and pierced from electrical sheet

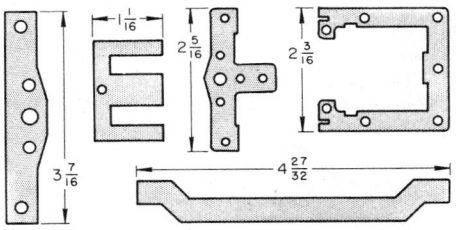

Fig. 14. Typical laminations blanked and pierced from electrical sheet for application in other than rotating machines

from less than 1 in. to 50 in. or more, and laminations like those shown in Fig. 14 can range in length from less than 1 in. to 12 in. or more.

PRESSES

A general-purpose punch press in good mechanical condition is acceptable for stamping laminations, but large-volume production of laminations by progressive-die methods requires the use of "high-productivity" or turret-punch presses.

DIES

Single-station dies and progressive dies are used for making laminations.

Single-Station Dies. Each single-station die performs one operation, and a set of dies for a lamination can be mounted in one press or different presses. Simple laminations like those shown in Fig. 14 usually are produced in one operation. More complex parts may require several operations. Figure 15 shows a typical sequence for the production of stator and rotor laminations in four operations.

Single-station dies can be used for punching any lamination, regardless of size, composition, shape, or quality requirements. However, production with single-station dies is relatively slow, so that the cost per piece is high for mass production. Laminations such as shown in Fig. 13 and 14 can be produced in large quantities at a lower cost in progressive dies.

Coining

COINING is a closed-die squeezing operation, usually performed cold, in which all surfaces of the workpiece are confined or restrained, resulting in a well-defined imprint of the die on the workpiece. It is also a restriking operation (called, depending on the purpose, sizing, or bottom or corner setting) used to sharpen or change a radius or profile.

APPLICABILITY

In coining, the surface of the workpiece copies the surface detail in the dies with dimensional accuracy that is seldom obtained by any other process. (It is because of this that the process is used for the minting of metallic coins.)

Decorative items, such as patterned tableware, medallions, and metal buttons, as well as coins, are produced by coining. When articles with a design and a polished surface are required, coining is the only practical production method to use. Also, coining is well suited to the manufacture of extremely small items, such as interlocking-fastener elements.

Dimensional accuracy equal to that available only with the very best machining practice can often be obtained in coining. Many automotive components are sized by coining. Sizing is usually done on semifinished products, frequently effecting significant savings in material and machining costs.

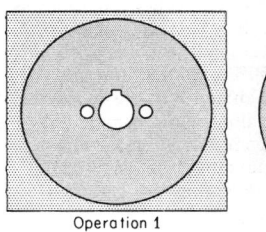
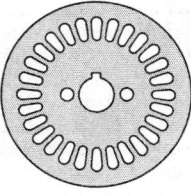
Operation 1
Operation 2

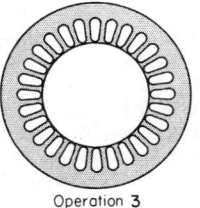
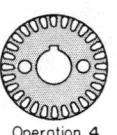
Operation 3
Operation 4

Operation 1: Stock blanked and pierced. **Operation 2:** Stator lamination notched. **Operation 3:** Rotor lamination separated from stator lamination. **Operation 4:** Rotor lamination notched.

Fig. 15. Sequence of operations for producing stator and rotor laminations from one blank using single-station dies

HAMMERS AND PRESSES

Drop Hammers. Gravity drop hammers in the size range of 900 to 2000 lb (weight of the ram) are used extensively in the tableware industry. Board hammers can be used, although pneumatic-lift hammers predominate for this type of coining. In producing tableware, reproduction of detail and finish are more important than dimensional control.

Mechanical presses with capacities of a few tons to several hundred tons are widely used in coining. The larger presses are usually of the knuckle type with production rates up to about 7500 pieces per hour. Small, specially built eccentric-driven presses are used for coining tiny parts at a rate of 2000 per minute.

Hydraulic presses are used extensively for sizing operations, especially for workpieces with large surfaces to be coined. Spacers ("kissing blocks") are required for maintaining close tolerances on the final dimensions of the part being sized. Hydraulic presses are sometimes favored because they are readily equipped with limiting devices that prevent overloading and possible breaking of the dies.

LUBRICANTS

Whenever possible, coining without a lubricant is preferred. If entrapped in the coining dies, lubricants can cause flaws in the workpieces. For example, under conditions of constrained plastic flow, an entrapped lubricant will be loaded in hydrostatic compression and will interfere with transfer of die detail to the workpiece. In many coining operations, however, because of work-metal composition or severity of coining, or both, the use of some lubricant is mandatory to prevent galling or seizing of the dies and the work metal.

For coining teaspoons, medallions or similar items from sterling silver, no lubricant is used. Some type of lubricant is ordinarily used for coining copper and aluminum and their alloys, and for coining stainless, alloy and carbon steels.

When coining items that do not require transfer of intricate detail, the type and amount of lubricant are less critical than for items of fine detail. A mixture of 50% oleum spirits and 50% medium-viscosity machine oil has been successful for prevention of galling and seizing for a large variety of coining operations. When coining involves maximum metal movement and high pressure, a commercial deep-drawing compound is sometimes used.

DIES

Dies for coining may be either single-station or progressive and usually are made from W1, O1, A2 or D2 tool steels. The use of these steels is indicated in Tables 5 and 6, which give tool steel recommendations for coining specific shapes. Recommended working hardnesses of coining dies made of the tool steels listed in Tables 5 and 6 are:

W1 .... Rockwell C 59 to 62
O1 .... Rockwell C 58 to 60
A2 and D2 .... Rockwell C 56 to 58

COINABILITY OF METALS

Limits to coining are established mainly by the unit loads in compression that the coining dies will withstand before deforming. Deformation of the dies results in dimensions that are out of tolerance in the workpiece and premature failure of the dies through the action of low-frequency fatigue.

In coining, deformation of the work metal is accomplished largely in a compression strain cycle, which leads to a progressive increase in compression flow strength as deformation progresses. This deformation cycle results in a product that has good bearing properties and wear resistance in service. Continued coining can raise the yield strength to a level that approaches the permissible maximum die load, where the coining must stop.

Sizing to close dimensional tolerances on several nonparallel surfaces can be achieved readily in the manufacture of small parts, such as interlocking-fastener elements. For large workpieces, ingenuity may be required to develop a coining process for sizing — ingenuity in the design of tooling to minimize the effect of distortion in the press, and ingenuity in the preparation of the workpiece to make necessary a minimum of metal flow during coining.

Table 5. Recommended tool steels for dies to coin a ¹/₂-in.-diam emblem or similar part

Use of die	Punch and die material for coining a total quantity of(a):		
	1000	10,000	100,000
Machined dies			
Drop hammers	W1	W1	O1(b), A2
Presses	O1	O1, A2	O1, A2
Hubbed dies			
Drop hammers	W1	W1	W1(c)
Presses	O1	O1, A2	A2, D2(d)

(a) For coining the emblem from aluminum, copper, gold or silver alloys, or from low-carbon, alloy or stainless steel. (b) O1 is recommended only for coining low-carbon steel, and copper, gold or silver alloys. (c) The average life of W1 dies in coining copper, gold and silver alloys softer than Rockwell B 60 would be about 40,000 ± 10,000 pieces. The life of W1 dies coining harder metals would be about half as great; thus, more than one set of dies would be needed for 10,000 parts. (d) Hot hubbed.

Table 6. Recommended tool steels for coining a preformed cup to final size on a press

Metal to be coined	Die material for coining a total quantity of(a):		
	1000	10,000	100,000
Low-carbon steel	W1	O1	D2
Alloy steel, stainless steel, heat-resisting alloys	O1	A2	D2
Al and Cu alloys	W1	W1	D2

(a) For quantities over 10,000, the die material refers to insert material. The materials shown are for dies made by machining. The punch material would be the same except that where heat treating of W1 would be hazardous, O1 would be safer in hardening.

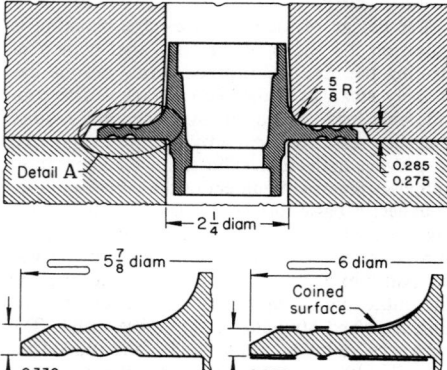

Fig. 16. Coining the flange on a forged wheel hub to final size, which cost less than sizing the flange by machining

For the flange of an automobile front-wheel hub (Fig. 16), the sizing operation required redistribution of the metal in the workpiece; this is possible in coining only over relatively short distances (a few multiples of the section thickness). Restricted metal flow, therefore, is the major limiting factor in a coining operation.

Press Bending of Low-Carbon Steel

PUNCH PRESSES are used for bending, flanging and hemming of low-carbon steel when production quantities are large, when close tolerances must be met, or when the parts are relatively small. Press brakes ordinarily are used for small lots, uncritical work, and long parts.

To estimate the press capacity needed for bending in V-dies, the bending load in tons can be computed from:

$$L = \frac{lt^2 kS}{s}$$

where L is press load, in tons; l is length of bend (parallel to bend axis), in inches; t is work-metal thickness, in inches; k is a die-opening factor (varying from 1.2 for a die opening of $16t$ to 1.33 for a die opening of $8t$); S is tensile strength of the work metal, in tons per square inch; and s is width of die opening, in inches.

For U-dies, the constant k should be twice the values shown above.

BENDABILITY AND SELECTION OF STEELS

Temper of the metal affects the bendability. Figure 17 shows the bend limitations for the standard AISI tempers of cold rolled carbon steel strip. Stock of No. 1 temper is not recommended for bending, except to large radii. Stock of No. 2 temper can be bent 90° over a radius equal to strip thickness, perpendicular to the rolling direction. Stock of No. 3 temper can be bent 90° over a radius equal to strip thickness, parallel to the rolling direction; it can also be bent 180° around a strip of the same thickness when the bend is perpendicular to the rolling direction.

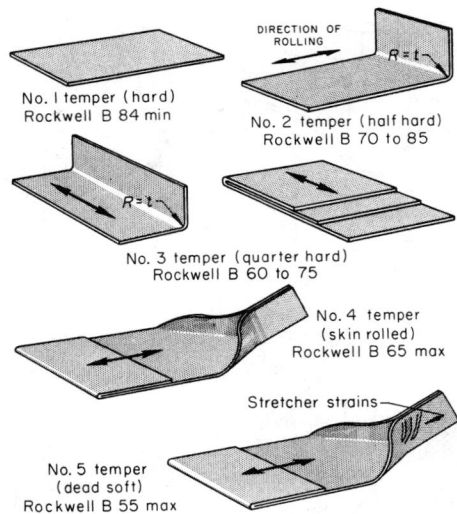

No. 1 temper (hard)
Rockwell B 84 min

No. 2 temper (half hard)
Rockwell B 70 to 85

No. 3 temper (quarter hard)
Rockwell B 60 to 75

No. 4 temper (skin rolled)
Rockwell B 65 max

Stretcher strains

No. 5 temper (dead soft)
Rockwell B 55 max

Stock of No. 1 (hard) temper sometimes is used for bending to large radii; each lot should be checked for suitability, unless furnished for specified end use by prior agreement.

Hardnesses shown are for steel containing 0.25% max C in the three hardest tempers and 0.15% max C in the No. 4 and 5 tempers. Hardness for No. 1 temper applies to thicknesses of 0.070 in. and greater; for thinner sheet, hardness would be Rockwell B 90 min.

Fig. 17. The most severe bend that can be tolerated by each of the standard tempers of cold rolled carbon steel strip (AISI Steel Products Manual on Carbon Steel Strip)

Stock of No. 4 or No. 5 temper can be bent 180° flat on itself in any direction. The No. 5 temper stock may develop stretcher strains and should not be used if these markings are objectionable.

Carbon content of the steel also affects the bendability. Increasing the carbon content of steel increases its strength but reduces its formability. The amount of carbon in steel sheet is generally limited to 0.10% or less. Table 7 shows minimum bend radii for several sheet materials.

DIE CONSTRUCTION

Although the same types of bending dies as those used in press brakes generally can be used in presses, there are major differences, as indicated below:

1. Because presses ordinarily are not long and narrow like press brakes, more consideration has to be given to clearance for removing the finished workpiece when the press is open, as well as clearance for the legs of the bend when the piece is being formed. The bed dimensions of a press also limit the size of workpiece that can be bent.

2. Presses cycle rapidly, and shut height is not as easy to change; therefore, fewer pieces are bent in air, as described in the article on Press-Brake Forming in this volume. More frequently, pieces are formed by bottoming the dies. This has the advantage of decreasing springback.

3. Usually (but not always), presses are used for workpieces less than 2 ft long, and press brakes for pieces longer than 2 ft. However, the au-

Table 7. Minimum bend radii for selected plain carbon and low-alloy steel sheet materials

Product	Quality temper or strength level	Thickness	Parallel to rolling direction	Across rolling direction
Cold rolled				
1008/1010	CQ	. . .	0.25 mm (0.01 in.)	. . .
1008/1010	DQ	. . .	0.25 mm (0.01 in.)	. . .
1008/1010	No. 3(b)	. . .	1t	0.5t
1008/1010	No. 2(c)	. . .	NR	1t
1008/1010	No. 1(d)	. . .	NR	NR
Hot rolled				
1008/1010	CQ	<2.25 mm (<0.09 in.)	0.75t	0.5t
		>2.25 mm (>0.09 in.)	1.5t	1t
1008/1010	DQ	<2.25 mm (<0.09 in.)	0.5t	0.25t
		>2.25 mm (>0.09 in.)	0.75t	0.5t
Annealed				
1020/1025	. . .	. . .	1t to 2t	. . .
4130, 8630	. . .	. . .	1.5t to 2t	. . .
1070, 1095	. . .	. . .	2t to 3t	. . .
ASTM A607 (HSLA)	345 MPa (50 ksi)	. . .	1.5t	1t
	415 MPa (60 ksi)	. . .	3t	2t
	480 MPa (70 ksi)	. . .	4t	3t

(a) CR, cold rolled; HR, hot rolled; CQ, commercial quality; DQ, drawing quality; t, sheet thickness; NR, not recommended; HSLA, high-strength low-alloy. (b) Quarter hard. (c) Half hard. (d) Full hard.

tomotive industry bends very large sheet-metal structural members on large presses by mass-production methods.

V-dies are composed of a V-block for a die and a wedge-shaped punch (see Fig. 18a). The width of the opening in the V is ordinarily at least eight times stock thickness. In bending, the workpiece is laid over the V in the die, and the punch descends to press the workpiece into the V to form the bend.

Wiping Dies. Another type of bending die is the wiping die (Fig. 18b). A pressure pad that is either spring loaded or attached to a fluid cylinder clamps the workpiece to the die before the punch makes contact. The punch descends and wipes one side of the workpiece over the edge of the die. The bend radius is on the edge of the die.

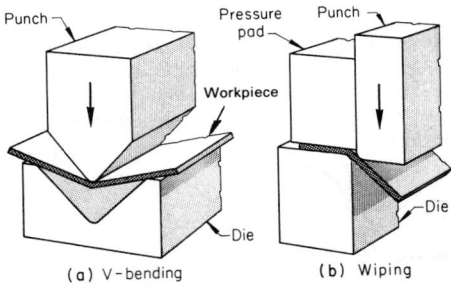

Fig. 18. Bending in a V-die and a wiping die

Rotary bending dies (Fig. 19) are used to make bends or twists in bars or strip. These dies use cam action to rotate the workpiece.

Figure 19 shows a rotary bending die in which a metal strip is being twisted 90° to make a connecting link. An inner cylinder has a 90° helical cam groove in its outer surface which engages a pin in an outer cylinder. The end of the workpiece fits in a slot in the bottom end of the inner cylinder. The ram forces the inner cylinder to bottom, and then causes the spring to compress, forcing the outer cylinder to move down over the inner cylinder. The pin in the groove makes the inner cylinder rotate, giving the workpiece a 90° twist.

LUBRICATION

Lubrication is less important for most bending operations than it is for other types of forming. In many bending operations, no lubricant is used; in others, mill oil remaining on the stock or a light mineral oil applied before forming is sufficient to prevent galling.

Exceptions to this practice are hole flanging, compression and stretch flanging, and severe bending in which wiping, ironing or drawing of the work metal may call for more effective lubrication.

──Press-Brake Forming──

PRESS-BRAKE FORMING is a process in which the workpiece is placed over an open die and is pressed down into the die by a punch. The punch is actuated by the ram portion of a ma-

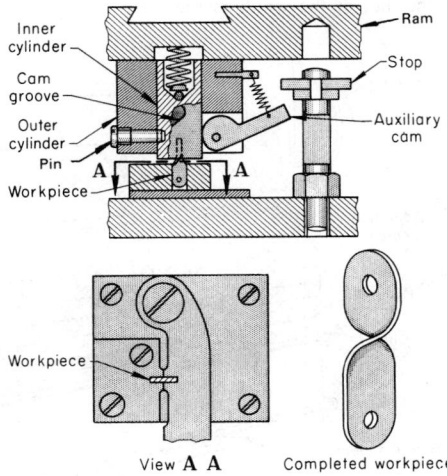

Die is shown in closed position; inner cylinder has rotated to give workpiece a 90° twist. The auxiliary cam prevents rotation of the inner cylinder until it is free of workpiece.

Fig. 19. Rotary bending die used for 90° twisting of strip metal

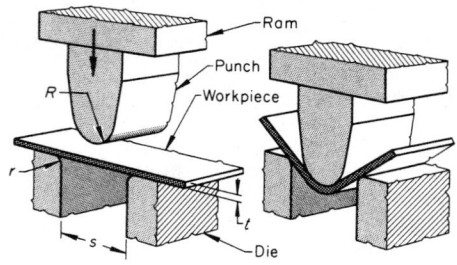

R = punch radius s = span width
r = die radius t = metal thickness

Fig. 20. Typical setup for press-brake forming in a die with a vertical opening

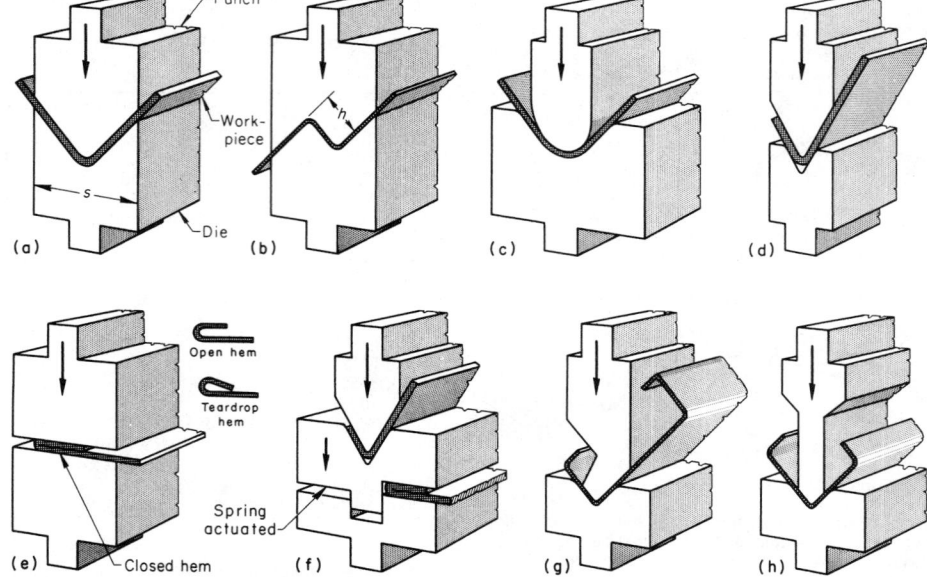

(a) 90° V-bending. (b) Offset bending. (c) Radiused 90° bending. (d) Acute-angle bending. (e) Flattening, for three types of hems. (f) Combination bending and flattening. (g) Gooseneck punch for multiple bends. (h) Special clearance punch for multiple bends.

Fig. 22. Dies and punches most commonly used in press-brake forming

chine called a press brake. The process is most widely used for forming of relatively long, narrow parts that are not adaptable to press forming, and for applications in which production quantities are too small to warrant the tooling cost for contour roll forming.

Simple V-bends or more intricate shapes can be formed in a press brake. Operations such as blanking, piercing, lancing, shearing, straightening, embossing, beading, wiring, flattening, corrugating and flanging can also be performed in a press brake.

PRINCIPLES

In press-brake forming, as in other forming processes, when a bend is made, the metal on the inside of the bend is compressed or shrunk, and that on the outside of the bend is stretched. The applied forces create a strain gradient across the thickness of the work metal in the area of die contact. Tensile strain occurs in the outer fiber, and compressive strain in the inner fiber.

The setup and tooling for press-brake forming (Fig. 20) are relatively simple. The distance the punch enters the die determines the bend angle and is controlled by the shut height of the machine (the distance between ram and bed at the bottom of the stroke; see Fig. 21). The span width of the die, or the width of the die opening, af-

fects the force needed to bend the workpiece. The minimum width is determined by the thickness of the work and sometimes by the punch-nose radius.

APPLICABILITY

Press-brake forming is most widely used for producing shapes from ferrous and nonferrous metal sheet and plate.

The length of plate or sheet that can be bent is limited only by the size of the press brake. For instance, a 600-ton press brake can bend a 10-ft length of $^3/_4$-in.-thick low-carbon steel plate to a 90° angle, with an inside radius of the bend equal to stock thickness. A lower-capacity press can be used if the included angle of the bend is greater than 90°, or if the bend radius is larger than stock

thickness, or if the length of bend is less than the bed length.

Work Metals. Press-brake forming is applicable to any metal that can be formed by other methods, such as press forming and roll forming. Low-carbon steels, high-strength low-alloy steels, stainless steels, aluminum alloys, and copper alloys are commonly formed in press brakes. High-carbon steels and titanium alloys are less frequently formed in a press brake, because they are more difficult to form.

PRESS BRAKES

The main advantages of press brakes are versatility, the ease and speed with which they can be changed over to a new setup, and low tooling costs. A press brake is basically a slow-speed punch press that has a long, relatively narrow bed and a ram mounted between end housings (Fig. 21). Rams are actuated mechanically or hydraulically.

Mechanical Press Brakes. The ram of a mechanical press brake is actuated by a crank or an eccentric through a gear train in which there is a clutch and a flywheel. The gear train is usually designed to provide fast movement of the ram. Shut height is adjustable by means of a powered screw in the pitman, or link, at each end of the ram. Length of ram stroke, however, is constant.

Hydraulic Press Brakes. The ram of a hydraulic press brake is actuated by two double-acting cylinders, one at each end of the ram. Force supplied by the hydraulic mechanism will not exceed the press rating; therefore, it is almost impossible to overload a hydraulic press brake. (When thicker metal is inadvertently used, the ram stalls.) For this reason, frames can be lighter and less costly than those for mechanical press brakes, which are subject to overloading.

Machines with numerical controls (NC) automatically adjust settings, thus increasing productivity and quality.

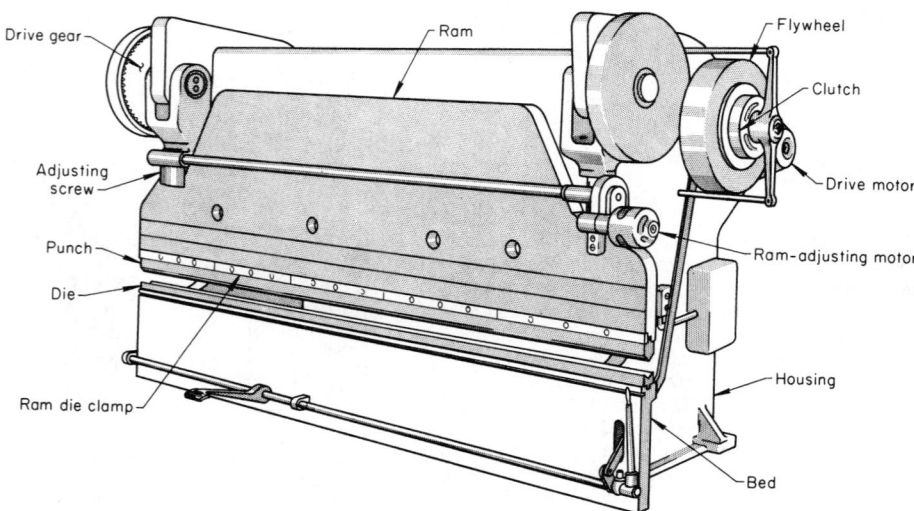

Fig. 21. Principal components of a mechanical press brake

DIES AND PUNCHES

V-bending dies and their corresponding punches (Fig. 22a and d) are the tools most commonly used in press-brake forming. The width of the die opening (shown as *s* in Fig. 22a) is usually a minimum of 8*t* (eight times the thickness of the work metal).

Offset Dies. Punch-and-die combinations like the one shown in Fig. 22(b) are often used to produce offset bends. Because an offset bend requires about four times as much force as a 90° V-bend, offset bending is usually restricted to relatively light-gage metal (0.125 in. or less). The depth of offset (*h* in Fig. 22b) should be a minimum of six times work-metal thickness, to provide stability at the bends.

Radius forming is done with a 90° die and a punch, each having a large radius (Fig. 22c). When the punch is bottomed, the inside radius of bend in the workpiece conforms to the radius of the punch over a part of the curve. The harder the punch bottoms, the more closely the work metal wraps around the punch nose, resulting in a smaller radius of bend and less springback.

Acute angles are formed by the die and punch shown in Fig. 22(d). The air-bending technique, which uses a die deep enough so the workpiece does not bottom out on the die, is often used for producing acute angles.

Flattening dies, shown in Fig. 22(e), are used to produce three types of hems (also shown in Fig. 22(e) after the metal has been formed into an acute angle. The combination die shown in Fig. 22(f) produces an acute angle on one workpiece and a hem on another, so that a piece is begun and a piece completed with each stroke of the press brake.

Gooseneck punches (Fig. 22g) and narrow-body or special clearance punches (Fig. 22h) are used to form workpieces to shapes that cannot be formed using punches having conventional width (two such workpiece shapes also are shown in Fig. 22g and h).

TOOL MATERIAL

Tool material used for press-brake bending and forming depends on work-metal composition. Most tool materials are low-carbon steel, low-alloy steel or gray iron. Sometimes, inserts of tool steel or carbides are placed in cast iron dies for making severe bends in less-formable work metal (such as high-strength low-alloy steel).

Press Forming of Low-Carbon Steel

PRESS FORMING is a metalworking process in which the workpiece takes the shape imposed by the punch and die. The applied forces may be tensile, compressive, bending or shearing, or various combinations of these. In some applications, the metal requires appreciable stretching in order to retain the shape of the formed part.

PRESSES

Restriking, coining and embossing are usually done in presses with more available tonnage than is needed for simple forming of similarly sized areas because in these operations the metal is

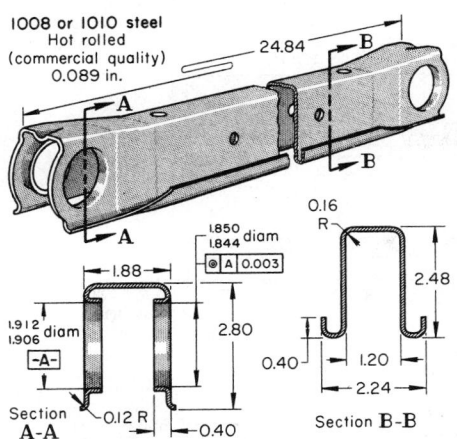

1008 or 1010 steel
Hot rolled (commercial quality) 0.089 in.

The two holes in the top of the control arm are 0.500-in.-diam tooling holes that were pierced during the blanking operation.

Fig. 23. Automotive control arm that was formed and pierced in a transfer press in five operations

confined and is plastically flowing. Progressive dies are used in presses that have sufficient tonnage to accommodate the total demands of the various stations.

Transfer presses are generally long-bed straight-side presses. The mechanism for moving the workpiece from station to station is a part of the machine to which transfer fingers are attached. The control arm shown in Fig. 23 was formed in five operations in a transfer press.

A **dial feed** is a type of transfer mechanism that moves the workpiece from die to die in a circular path rather than in a straight line.

Multiple-slide machines are designed for automatic, complete production of a variety of small formed parts. Flat stock is fed into a straightener, then to a feed mechanism and through one or more presses incorporated in the multiple-slide machine, for operations such as piercing, notching and bending—often in a progressive die. The feed mechanism then moves the metal into the multiple-slide forming area, where it is severed by a cutoff mechanism to predetermined lengths. The piece is usually formed around a center post by four sets of tools mounted 90° apart around the forming post. Finally, the part is stripped off the center post and dropped through a hole in the bed.

SPEED OF FORMING

Speed of forming has little effect on formability of steels used for simple bending or flanging, or moderate stretching. The maximum velocity of the punch when it contacts the blank in such conventional press forming is usually not greater than about 200 ft per min.

FORMABILITY

Formability of low-carbon steel is determined partly by the steelmaking practice. Hot rolled steel often forms stretcher strains on the surface after forming. This steel should be temper rolled with about 1% cold reduction before forming to avoid the stretcher strains. Cold rolled steel has better formability. Rimmed steel has better surface quality than killed steel. However, killed steel is

preferred where the part must have good mechanical properties which are uniform throughout the part.

In-process annealing is sometimes done after some press forming operations. This is to remove the effects of cold working and increase the formability for subsequent operations.

LUBRICATION

The type of lubricant usually has little effect on the grade of steel selected to form a given part. The main effects of a lubricant are to prevent die galling and die wear and to reduce the friction over critical areas, thus allowing proper flow of metal and possibly a reduction in severity class. The thicker gages and higher forming speeds require increasingly effective lubricants.

In progressive dies, a light oil sprayed on the strip as it enters the die often is enough to keep the stock lubricated through all stages. Generally, the oil is applied to the stock between the feeding device and the die.

DIES

Dies for press forming of low-carbon steel are made from a wide range of materials, including plastics, cast iron, tool steel and sintered carbide. Severity of forming, number of parts to be produced, workpiece shape, work-metal hardness, specified surface condition, and tolerances affect selection of the die material. These factors are discussed under "Press Forming Dies" in the article "Tool Materials for Special Applications" in Section 18.

Single-operation dies perform one operation at a time and are individually loaded and unloaded. They are usually set up in a press, and the operation is performed on a specific lot size. The die is then removed from the press, and the next die in the sequence is set up. For continuous production, a line of presses, each operating a single die, can produce finished pieces from raw stock without interruption for change in setup. Occasionally, more than one die is set up in a press at a time, and the parts are moved manually from one die to the next (a part formed using this practice is shown in Fig. 24).

Compound dies are one-station dies in which more than one operation is done on a workpiece in one press stroke without relocating the workpiece in the die. The operations done must be such that their inclusion does not weaken the die elements or restrict other operations. Generally the operations are done successively in the course of the press stroke, rather than simultaneously.

Typical combinations of operations are: cutting a blank from a strip, then forming; lancing and forming a tab or louver; or forming a flange and embossing a stiffening bead. When a die is used for blanking and forming a part, holes often can be pierced in the bottom with the same die. The combination of lancing and forming is common. Continued travel of the lancing punch does the forming. Flanging can be combined with forming or embossing, provided that no metal flow is necessary after the flange has been formed.

Several operations can be performed successively on a workpiece in a press, using two or more compound or single-operation dies. The parts can be manually transferred from die to die, eliminating storage and transfer between presses.

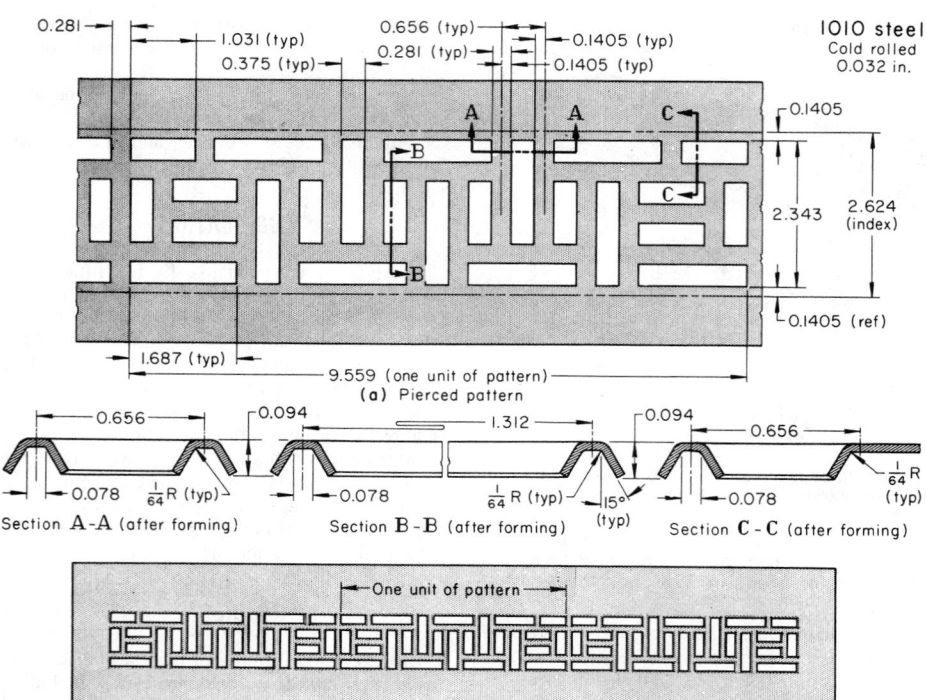

Section A-A (after forming) Section B-B (after forming) Section C-C (after forming)

(b) Completed workpiece, formed with one press stroke after piercing

Fig. 24. Decorative grill with a repetitive pattern, pierced and formed in single-operation dies

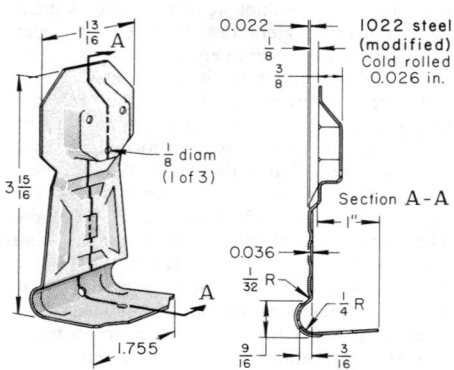

Several hooded rectangular openings like the one shown at the right end of the wrapper were made on the blind side of the piece by lancing and forming tabs.

Fig. 25. Housing wrapper that was formed in one press in six stages in compound and single-operation dies

Fig. 26. Spring end of a paper-towel holder that was bent, pierced, embossed, drawn and cut off in a progressive die

The capacity of a large-bed press may be more fully utilized by performing several operations during each press stroke (see Fig. 25).

Progressive dies perform a series of operations at two or more die stations during each press stroke as the stock is moved through the die. One or more operations are done on the workpiece at each die station. As the outline of the workpiece is developed in the trimming or forming stations, connecting tabs link the workpiece to the strip until the workpiece reaches the last station, where it is cut off and ejected from the die (an example is shown in Fig. 26). Pilot holes that are engaged by pilot pins in the die keep the workpieces aligned and properly spaced as they progress through the die.

Although a progressive die runs more slowly than a single-operation or a compound die for similar work, over-all production is usually higher, because the die is operated more continuously.

Press Forming of High-Carbon Steel

HIGH-CARBON STEEL strip (including spring steel and tool steel) is blanked, pierced and formed to make a variety of parts. The practices, precautions, presses and tools used in making high-carbon steel parts are like those used to produce similar parts of low-carbon steel (see the preceding articles in this volume). Differences that must be considered in blanking, piercing and forming high-carbon rather than low-carbon steel are: (a) more force is required for high-carbon steel because of its higher strength; (b) greater clearance between the punch and die is necessary in blanking and piercing; and (c) a more wear-resistant

tool material may be required before acceptable tool life can be obtained.

BLANKING AND PIERCING

The most important difference between blanking and piercing of high-carbon and of low-carbon steels is that greater clearance between punch and die is required for high-carbon steels.

Tool materials for blanking and piercing high-carbon steel are often carbide dies for high production (see Fig. 27) or tool steels.

Die Life in the blanking and piercing of high-carbon steel varies with different applications, depending greatly on the dimensional accuracy that must be maintained and the burr height that can be tolerated on the blanked parts.

FORMING PRETEMPERED STEEL

Mild forming of high-carbon steel in the quenched and tempered ("pretempered") condition (usually Rockwell C 47 to 55) is common practice (note Fig. 28). The severity of forming that can be done without cracking of the work metal depends mainly on thickness. When metal thickness is no more than about 0.015 in. it is possible to make relatively severe bends without fracturing the work metal. However, as metal thickness increases, the amount of forming that can be done on pretempered steel decreases rapidly.

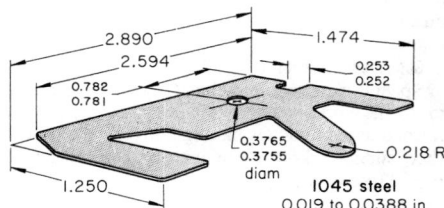

Fig. 27. Blanked and pierced textile-machine part for which carbide compound dies had 28 times the total life of those made of D2 tool steel

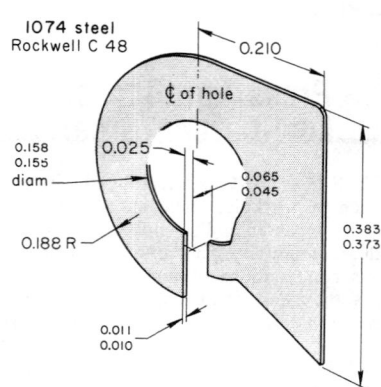

Pretempered stock eliminated the need for heat treating after forming.

Fig. 28. Clutch spring that was blanked, pierced and formed in an eight-station progressive die

Press Forming of Coated Steel

COATED STEEL sheet or strip is formed in the same presses as are used for forming uncoated steel. Forming procedures, however, must sometimes be modified, depending on the type of coating. During processing, scratching or breaking the coating (sometimes, only marring the surface) must be avoided, because these defects could cause rejection of the finished part.

Resistance to forming forces varies directly with the thickness and hardness of the steel base, so that the coating on thicker or harder steel is subjected to greater abrasion, surface shear, and die pressure. The coating can be sheared off or the dies can cause scratches, gouge marks or increased gloss in pressure areas on the coated product.

In many applications, special handling and processing techniques in forming are needed. Formability is usually less than for the same metal uncoated. Further restrictions are imposed on severity of forming by the need to avoid fracture or excessive porosity in corrosion-resistant surfaces, to avoid roughening or "orange peel" on decorative surfaces, or to avoid flaking of hot dip zinc or aluminum coatings.

ZINC-COATED STEEL

Most of the zinc-coated steel used in forming applications is hot dip galvanized low-carbon steel sheet and strip. A layer of metallic zinc on the surface of the work metal prevents galling during forming by eliminating direct contact of the steel against the die. This generally increases die life, because of the softness and lubricity of the zinc. The need for lubrication during forming is reduced by the presence of the zinc coating.

Table 8 gives formability limits of galvanized steel for exposed and unexposed parts, and Table 9 shows the effect of coating thickness on the diameter at which no flaking of the coating occurs at the outside of the bend in a bend test.

Tool Design. Tools for forming zinc-coated parts are of conventional design, and are made of cast iron and standard tool steels. However, parts formed of commercial quality continuous-annealed steel, and of steel over 0.060 in. thick, require more compensation for springback than conventional box-annealed steel and uncoated steel.

HOT DIP ALUMINUM-COATED STEEL

Moderately severe forming is done on hot dip aluminum-coated steel ("aluminized" steel). The same dies and pressworking practices are applicable to aluminum-coated stock as for similar uncoated stock. Drawing compounds are recommended for severe forming.

The aluminum coating influences the formability of the steel base in the same way as does the zinc coating on galvanized steel.

Requirements for resistance to corrosion in service often limit the permissible severity of forming. Hairline cracks that develop in the aluminum coating lead to lower service life at high temperature or in atmospheric exposure. Table 10 gives minimum diameters for 180° bends for different thicknesses of steel sheet with two types of aluminum coating.

Table 8. Formability limits for parts made from galvanized steel up to 0.060 in. thick

Type of operation	Maximum severity of forming	Grade of steel(a) Unexposed parts	Grade of steel(a) Exposed parts
90° bend	1*t* radius, minimum	CQ, CA or BA	CQ(b), CA or BA
Up to 180° bend	0.01-in. radius(c)	CQ, CA or BA	CQ(b), CA or BA
Drawing	10 to 20% elongation	DQ, CA(d) or BA	DQ(b), CA(d) or BA
Drawing	20 to 30% elongation	DQ, CA(d) or BA(e)	DQSK(b), CA(d) or BA
Drawing(f)	30 to 35% elongation	DQSK, CA(d) or BA(e)	DQSK(b), BA(e)

SOURCE: P. G. Nelson, *Metal Progress*, 82, 104 to 108 (Oct 1962)

(a) CQ = commercial quality; DQ = drawing quality; DQSK = drawing quality, special killed; CA = continuous annealed; BA = box annealed. (b) Must be temper rolled after galvanizing. (c) For 10% maximum elongation. (d) Heat treated after galvanizing. (e) Dead soft or annealed before galvanizing. (f) With possible buckling during drawing.

Table 9. Bend test requirements for hot dip galvanized steel sheet(a)

Coating designation	Bend diameter for sheet thickness range(b) 0.33-0.97 mm (0.0131-0.0381 in.)	0.97-1.90 mm (0.0382-0.0747 in.)	1.90-4.46 mm (0.0748-0.1756 in.)
G 235	2*t*	3*t*	3*t*
G 210	2*t*	2*t*	2*t*
G 185	2*t*	2*t*	2*t*
G 165	2*t*	2*t*	2*t*
G 140	*t*	*t*	2*t*
G 115	0	0	*t*
G 90	0	0	*t*
G 60	0	0	0
G 01	0	0	0

(a) From ASTM A525. Table does not apply to structural (physical) quality sheet. (b) Value listed is the minimum diameter of rod (or mandrel), in multiples of the galvanized sheet thickness (*t*) around which the galvanized sheet can be bent 180° in any direction at room temperature without flaking of the coating on the outside of the bend.

Table 10. Minimum diameters for corrosion-resistant 180° bends in various thicknesses of steel sheet with AISI types 1 and 2 hot dip aluminum coating

Sheet thickness (*t*), in.	Minimum bend diameter Type 1 coating(a)	Minimum bend diameter Type 2 coating(b)
0.0635	3*t*	5*t*
0.0516	3*t*	4*t*
0.0396	3*t*	3*t*
0.0336	...	2*t*
0.0276, 0.0217 and 0.0187	1*t*	1*t*

(a) Coating containing about 9% silicon and weighing 0.5 oz per sq ft. Minimum diameters are for no rusting at outside of bend after exposure in air to 25 cycles consisting of 30 min at 1100 F and 30 min of cooling. (b) Coating of commercially pure aluminum and weighing 1.15 oz per sq ft. Minimum bend diameters for a type 2 coating are for no rusting at the outside of the bend after one year of exposure to a mild industrial atmosphere.

TIN-COATED AND TERNE-COATED STEEL

Nearly all of the common forming methods are used on tin-coated and terne-coated low-carbon steel. Spinning is not ordinarily done on these materials, because of the likelihood of excessively thinning the coatings or of fusing the coatings.

The amount of tin on electroplated mill products ordinarily ranges from 0.25 to 1.5 lb per base box (31,360 sq in. of sheet), corresponding to a coating thickness per surface of 15 to 90 micro-in. Hot dip products have coating weights of 1.10 to about 2.5 lb per base box (66 to 150 micro-in. of thickness per surface).

Terne-coated steel products are hot dip coated with a lead-tin alloy that contains 10 to 25% by weight of tin. Terne coatings range from the thinnest coating that gives complete coverage of the steel to about 8 lb per double base box (about 40 to 170 micro-in. of thickness per surface).

The steel to which tin and terne coatings are applied varies in composition for different products and among different manufacturers, but is generally similar to 1008 or 1010.

NICKEL-PLATED AND CHROMIUM-PLATED STEEL

Press forming and roll forming are done on steel that has been electroplated in the coil with decorative copper-nickel or copper-nickel-chromium.

Conventional lubricants may be used in press forming of this material, particularly in high-volume production. Sometimes, however, no lubricant is used in making decorative parts. Instead, surface contact between the work metal and tools is prevented by the use of strippable plastic coatings or adhesive-backed paper on the work metal, or of loose paper between the work metal and the punch or the die. These materials protect the decorative finish on the preplated steel, prevent galling, and provide a controlled amount of friction for forming.

Forming of Steel Strip in Multiple-Slide Machines

MULTIPLE-SLIDE FORMING is a process in which flat stock or wire is progressively formed in one or more presses incorporated in a multiple-slide machine. A large variety of simple and intricately shaped parts can be formed this way.

Operations such as straightening, feeding, trimming, blanking, embossing, coining, lettering, forming to shape, and ejecting can all be done in one cycle of a multiple-slide machine. Forming generally is limited to bending operations, but the four slides and center post permit making some very complex parts. Deep drawing is not generally done in the forming or press stations of a multiple-slide machine.

APPLICABILITY

Multiple-slide forming is used to produce shapes from coiled strip or wire. The maximum size of workpiece that can be formed from strip metal in a multiple-slide machine is 3 in. wide by 14 in. long. Parts made from wire to 24-in. lengths (or longer if a special machine is used) and up to $^3/_8$ in. in diameter can be formed automatically from coil stock.

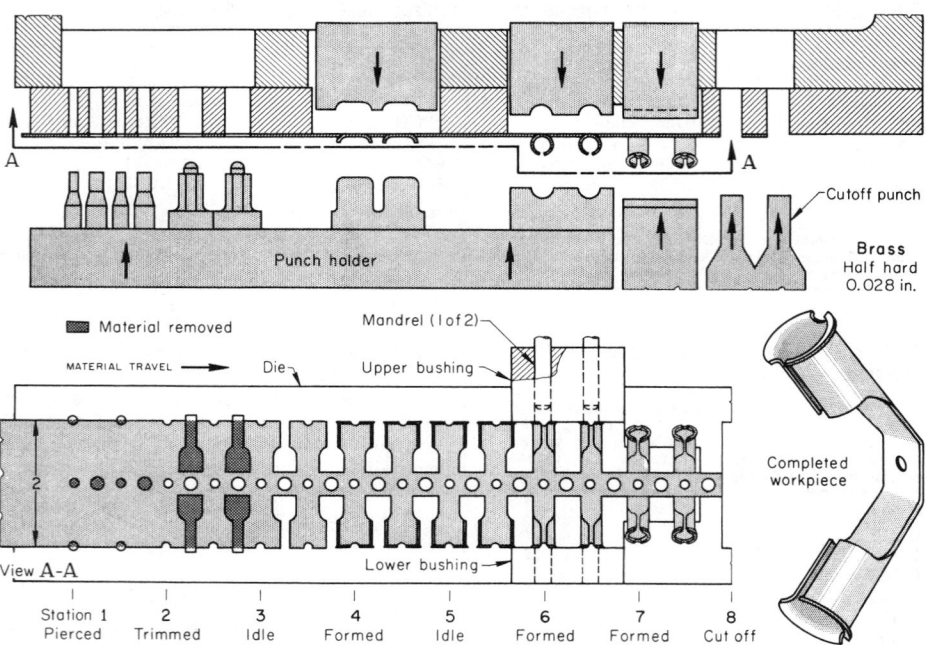

Fig. 29. Extended die that was used to form two double-socket contacts per cycle in eight die stations

If the work metal is comparatively thin and the bending is not severe, tempered strip material can be formed. Plated or otherwise coated materials can be formed, but it is usually better to coat after forming, because it is difficult to avoid marring coated surfaces during forming. However, nonmetallic inserts at appropriate points in the straightener, feeder and forming tools can be used to reduce tool marks.

In bending materials such as stainless steel, phosphor bronze, certain grades of brass and beryllium copper, or high-carbon steel, springback must be considered. Adjustments can be made in the forming tools to provide the amount of overbending required for the accuracy of the finished work.

More than one piece can be made in each cycle of a multiple-slide machine. For instance, Fig.

29 shows two double-socket contacts that were made in each cycle of a multiple-slide machine.

MULTIPLE-SLIDE MACHINES

Multiple-slide machines are made in a range of sizes, all similar in construction and principle. The larger machines have a longer die space, which enables more die stations to be used for the manufacture of complicated components. Generally, the number of strokes per minute decreases and the horsepower increases as the machine size increases.

The multiple-slide machine is an efficient machine for automatically producing workpieces that would require a number of dies or die stations in conventional single-action or double-action mechanical presses.

The four forming slides of a multiple-slide machine are generally sufficient for ordinary part-forming needs. However, complex parts can be formed at two or three levels around the center post, thereby doubling or tripling the number of forming positions available.

Figure 30 shows a plan view of the main units of a medium-size multiple-slide machine that uses a floor space of 12 by 5 ft, including the stock reel. Four shafts (A, B, C and D), mounted to a flat-top bedplate, are driven at equal speed through spur gearing E by an electric motor. Each of the four shafts is fitted with a positive-action cam F that drives a slide G (only 2 of 4 identified) on which the forming tools may be secured. In the center of the machine is a vertical post H into which the center post or former is fixed, and around which the work material is bent. The formed workpiece is removed from the center post by a stripper mechanism, and dropped through a hole in the bed. The stripper (a hardened steel plate) is secured to a vertical rod connected to a bell crank J. The bell crank is operated by a cam

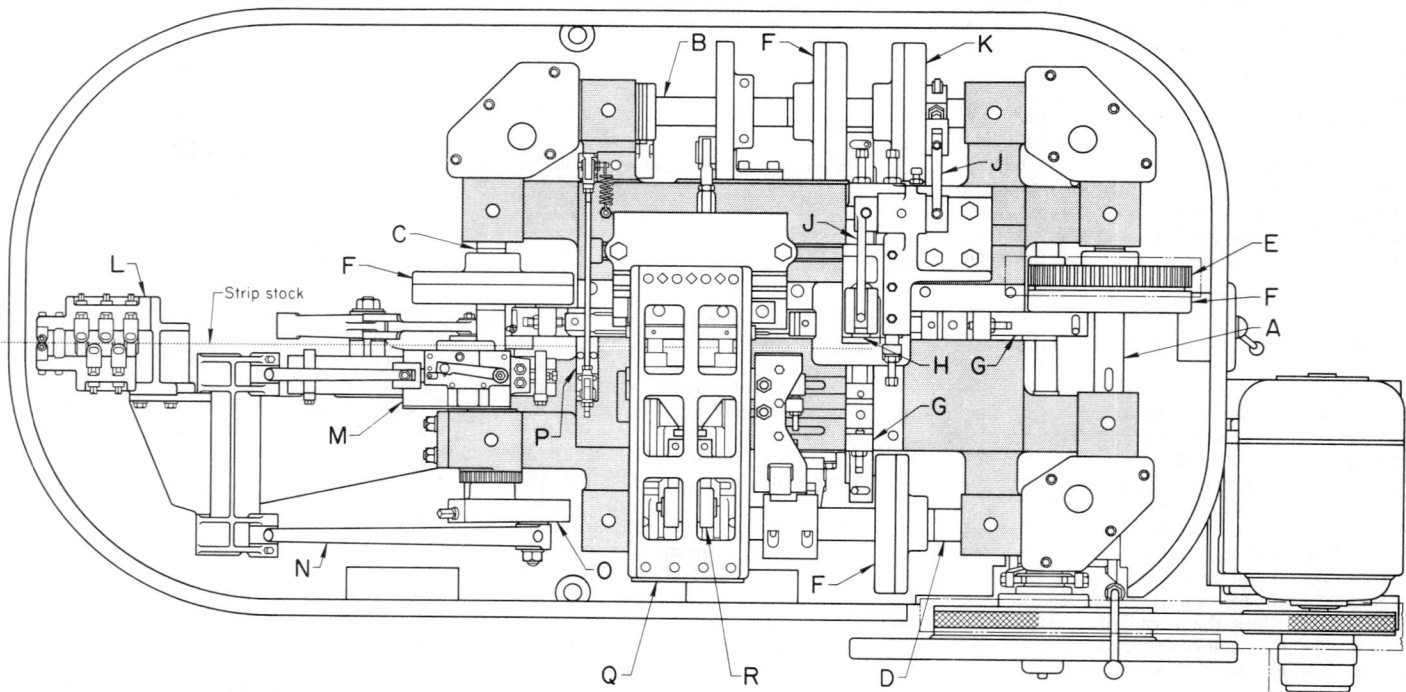

A, B, C and D—Integrated shafts. E—Spur gearing. F—Positive-action cam. G—Slide. H—Vertical post. J—Bell crank. K and R—Cams. L—Stock straightener. M—Automatic gripper in a feed slide. N—Links. O—Adjustable crank. P—Stationary gripper with cam-operated jaws. Q—Horizontal press containing dies. R—Cam.

Fig. 30. Plan view of a multiple-slide machine, showing major components

K on the rear shaft to give up-and-down motion to the stripper. All these parts comprise the "forming station" of the machine.

To the left of the machine proper is a stock straightener L, shown in working position with strip stock passing through it. Intermittent feeding of the work metal is accomplished by an automatic gripper in feed slide M. The gripper is actuated by link N and adjustable crank O, which is attached to shaft C. A separate gripper P is provided with cam-operated jaws, which grip the strip on the return stroke of the feed slide to prevent backward motion of the strip.

Deep Drawing

DRAWING is a process in which sheet metal is formed into round or square cup-shape parts. The work metal is placed over a shaped die and is pressed into the die with a punch (Fig. 31).

The distinction between shallow drawing and deep drawing is arbitrary, although shallow drawing generally refers to the forming of a cup no deeper than half its diameter, with little thinning of the metal. In deep drawing, the cup is deeper than half its diameter, and wall thinning, although not necessarily intentional, may be more than in shallow drawing.

The fundamentals of drawing discussed in this article are generally applicable to all ductile metals, although drawing practice varies for different work metals. Most of the examples of practice presented here are concerned with the processing of flat rolled low-carbon steels such as 1006, 1008 and 1010. For drawing other metals, the reader may consult the articles in this volume that deal with forming of specific metals.

Product Shape. Drawing is used to produce both shallow and deep straight-wall shells (or cups) from flat stock. These shells can be used as formed, like some pots and pans, or they may need attachments such as handles and pouring spouts. Drawn shells can also be thread rolled and edge curled to form a variety of products (such as screw ends for light bulbs).

FUNDAMENTALS OF DRAWING

Progressive stages of metal flow in deep drawing of a cylindrical cup are shown in Fig. 32. During the first stage, the punch contacts the blank (Fig. 32a), which rests on a die. The metal section denoted as 1 in Fig. 32 is bent and wrapped around the punch nose (Fig. 32b). The outer sections of the blank (denoted as 2 and 3 in Fig. 32) move radially toward the center of the blank until they flow over the die radius. The radial movement of the metal increases the thickness of the blank as the metal flows toward the die radius. As the metal flows over the die radius, this thickness decreases (because of tension in the cup wall and the clearance between the punch and die) and a straight-wall cup is formed (Fig. 32c and d). During drawing, the center of the blank (punch area) is essentially unchanged as it forms the bottom of the drawn cup. The areas that become the sidewall of the cup change from annular segments to long parallel-side cylindrical elements as they are drawn over the die radius.

A blankholder often is used in a draw die (see Fig. 31a), especially for thinner sheets to prevent the formation of wrinkles.

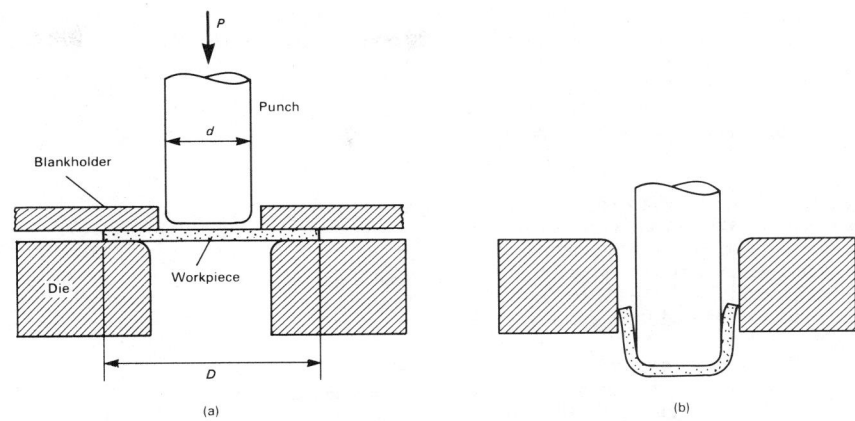

From George Dieter, "Mechanical Metallurgy," 2nd Ed., McGraw-Hill, p 688
Fig. 31. Deep drawing of a cylindrical cup: (a) before drawing; (b) after drawing

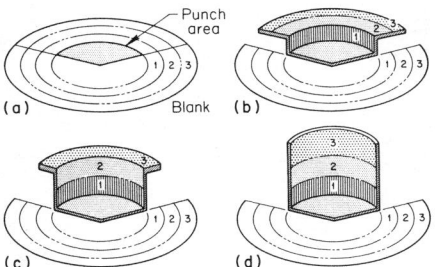

Fig. 32. Progression of metal flow in drawing a cup from a flat blank

DIRECT REDRAWING

Shells that are too deep to be drawn in a single operation are completed by one or more redraws.

In direct redrawing the drawn cup is slipped over the punch and is loaded in the die, as shown in Fig. 33(a). At first, the bottom of the cup is wrapped around the punch nose without reducing the diameter of the cylindrical section. Then the sidewall section enters the die and is gradually reduced to its final diameter. Metal flow takes place as the cup is drawn into the die, so that the wall of the redrawn shell is parallel to and deeper than the wall of the cup at the start of the redraw.

In a single-action redraw, the metal must be thick enough to withstand the compressive forces set up in reducing the cup diameter without wrinkling. Wrinkling can be prevented by the use of an internal blankholder (see Fig. 33a) and a double-action press, which usually permits a shell to be formed in fewer operations than by single-action drawing without the use of a blankholder.

REVERSE REDRAWING

Reverse redrawing is used for making deeper draws than is possible with direct redrawing. In reverse redrawing, the cupped workpiece is placed over a reversing ring and redrawn in the direction opposite to that used for drawing the initial cup (see Fig. 33b). Reverse redrawing can be done with or without a blankholder. The blankholder serves the same purposes as in direct redrawing.

Usually, metals that can be direct redrawn can be reverse redrawn. All of the carbon and low-alloy steels, austenitic and ferritic stainless steels, aluminum alloys, and copper alloys can be reverse redrawn.

Reverse redrawing requires more closely controlled processing than does direct redrawing. This control must begin with the blanks, which should be free from nicks and scratches, especially at the edges.

Ironing is the intentional thinning of the wall of

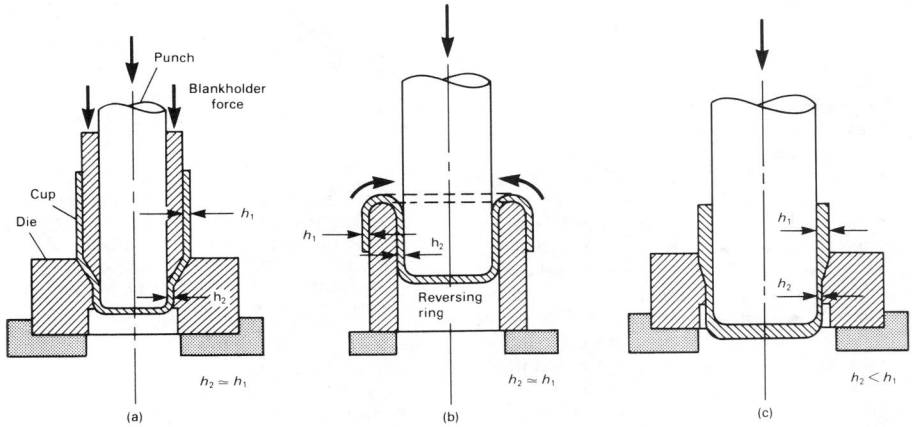

From John A. Schey, "Introduction to Manufacturing Processes," McGraw-Hill, 1977, p 174, Fig. 5.19
Fig. 33. Deformation of a cup-shaped part by (a) direct redrawing, (b) reverse redrawing, (c) ironing

a drawn shell (see Fig. 33c). The operation provides uniform wall thickness for either the full height or only the bottom part of the cup. It is often used to produce thin-wall, thick-bottom parts. The clearance between the punch and the die is less than metal thickness; this reduces wall thickness and increases the height as the shell is drawn through the die by the punch.

Reducing differs from redrawing in that the diameter is reduced at the mouth of the shell rather than at the bottom or for the full length. The top of the shell is pushed or compressed into the die rather than pulled through the die by the punch. Reducing is often referred to as necking, tapering, nosing or closing.

Expanding or bulging is the forming of irregular contours or surfaces of revolution on shells or rings by expanding or reducing the diameter for a portion of its cylindrical length. Such shapes cannot be removed from a regular punch or die. The change in shape is produced by a wedge-action punch or die, or by the use of a fluid or a rubber punch.

Sizing is used for forming workpieces to specific dimensions. The final forming of sharp radii should be combined with a small reduction in the cross-sectional area of the part. The metal should be crowded and not stretched into sharp corners. When the sizing operation is designed to eliminate oil-canning in the sidewall of a box or to establish a depth, the metal is stretched.

DRAW RATIOS

The reduction in drawing cylindrical shells is generally expressed in terms of the diameters of the blank and the cup. Strain depends only slightly on the blank thickness. The drawability (limiting drawing ratio) of a metal is expressed as the largest ratio of blank to punch diameter, D/d (see Fig. 31a), that can be drawn successfully.

Blank Size vs Cup Diameter. Reduction of drawn cups is usually expressed as the percentage reduction from the diameter of the blank to the inside diameter of the cup, or from the inside diameter of one shell to that of the next in redrawing. Percentage reduction is calculated using the formula $100(1 - d/D)$. Thus, the drawing of a 6-in.-diam cup from a 10-in.-diam blank results in a reduction of 40%. Nominal cup heights resulting from diameter reductions of 20, 30, 40 and 50% for blanks 2, 5, 10 and 20 in. in diameter are given in Table 11.

A chart that can be used to estimate cup size from blank size or cup size from an earlier draw is shown in Fig. 34.

Sample Calculation Using Fig. 34. To illustrate the use of Fig. 34, assume the problem of determining whether a cup of 7¹⁄₂-in. ID can be made in three draws of 40, 20, and 15% reduction, respectively, from an 18-in.-diam blank. To find the diameter of the cup after the first draw, trace the line for 18-in. blank diameter horizontally to its intersection with the diagonal for 40% reduction. From this intersection, draw a vertical line to the top of the chart and read 10.8 in. for the inside diameter of the cup after the first draw. Next, trace a horizontal line from 10.8 on the vertical axis until it intersects the diagonal for 20% reduction. From this point draw a vertical line to the bottom of the chart and read 8.6 in., which is the inside diameter of the cup after the second draw. The inside diam-

Table 11. Nominal height of cup calculated from blank diameter, cup diameter, and percentage reduction

Reduction, %(a)	Cup diam (d), in.	Cup height (h), in.(b)	Ratio h/d
Blank diameter, 2 in.			
20	1.6	0.23	0.14
30	1.4	0.37	0.26
40	1.2	0.54	0.45
50	1.0	0.75	0.75
Blank diameter, 5 in.			
20	4.0	0.55	0.14
30	3.5	0.92	0.26
40	3.0	1.35	0.45
50	2.5	1.87	0.75
Blank diameter, 10 in.			
20	8	1.15	0.14
30	7	1.87	0.26
40	6	2.7	0.45
50	5	3.75	0.75
Blank diameter, 20 in.			
20	16	2.2	0.14
30	14	3.6	0.26
40	12	5.3	0.44
50	10	7.4	0.74

(a) Reduction = $100(1 - d/D)$. (b) $h = (D^2 - d^2)/4d$.

Developed from Charts IV and V, "Computations for Metal Working Presses," published by the E. W. Bliss Co.

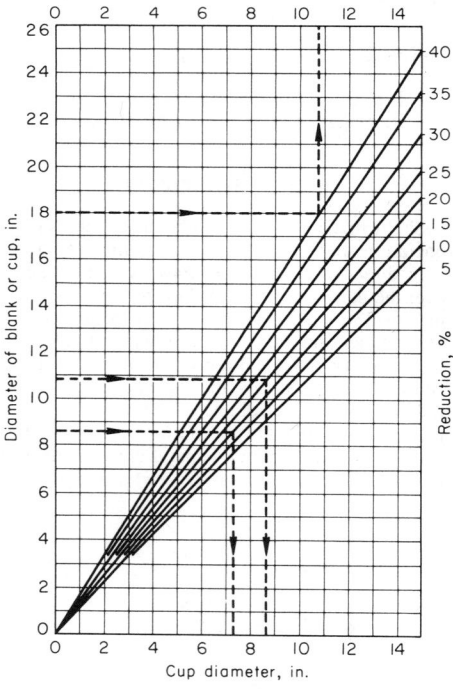

Fig. 34. Chart for checking percentage reduction in the drawing of cups. See text for sample calculation. The inside diameter is ordinarily used for the cup diameter.

eter of the cup after the third draw, assumed to be 15% reduction, is found using the same procedure, by drawing a line horizontally from 8.6 to its intersection with the diagonal for 15% reduction and from there to the bottom of the chart, which gives a reading of 7¹⁄₄ in. Accordingly, it is concluded that a 7¹⁄₂-in.-

ID cup can be drawn from an 18-in.-diam blank in the three assumed reductions.

Draw-reduction values for cupping and redrawing cylindrical shells in a double-action press or a die provided with a blankholder are given in Table 12.

It is emphasized that the values in Table 12 are approximate; they vary with work-metal composition, thickness and hardness, and with shell contour. For instance, a metal with maximum drawability, such as copper, can be reduced initially as much as 55%, but a refractory metal may allow an initial reduction of only 25%.

The first, or cupping, reduction percentage is given in Table 12 as 40%. In general, flaws introduced into the cup during the first draw cannot be removed in subsequent processing. Rather, they are aggravated, especially if they have resulted from strain instability, which can lead to wrinkles, puckers or metal thinning.

Some metals cannot be redrawn without first being annealed. This is especially true of metals (such as stainless steels) that initially possess high strength or that cold work or strain harden rapidly to high strength levels. High work-metal strength causes excessive force on the drawing tools, which can cause tool breakage. In addition, excessive force can cause lubrication breakdown.

Annealing for redrawing, if not closely controlled, can impair drawability. This is especially true if grain coarsening results from the annealing. This characteristic varies widely among different work metals.

Table 12. Typical relations of shell depth to draw reduction for ductile metal(a)

Depth of shell, in diameters	Number of draws	Cupping	Reduction, %(b) First redraw	Second redraw	Third redraw
¹⁄₂	1	40	...	...	...
1	2	40	25	...	...
1¹⁄₂	3	40	25	15	...
2	4	40	25	15	11

(a) Values are based on deep drawing of steel in a double-action press or a die provided with a blankholder. These values can serve as a guide for drawing of most ductile metals. Values are based on annealing between operations. (b) Percentage reduction = $100(1 - d/D)$.

Rectangular Shells. The drawing of oval, square or rectangular shells involves true drawing at the corners only; at the sides and ends, metal movement is more accurately described as bending. Stresses in the metal in the corner areas are severe, compared with the sidewall areas where only bending occurs. A portion of the surplus metal in the corner areas tends to move into the sidewall areas. If the punch-to-die clearance at the corners or in the sidewalls adjacent to the corners is too great, the metal may wrinkle rather than shrink or be compressed.

For square and rectangular boxes, except those with a width slightly more than twice the corner radius, the depth that the box can be drawn in one operation depends more on the size of the corner radius and on the stock thickness than on the width of the box.

When the ratio of the depth to the corner radius is 6 or less, the shell can usually be drawn in one operation. When this ratio is up to 12, two draws are required. Three draws are required for a ratio of 17 and four draws for a ratio of 22.

Square (or nearly square) cups and nearly cylindrical oval cups can be drawn in one operation if the area of the blank does not exceed the cross-sectional area of the punch by four times. This ratio is 4$\frac{1}{2}$ for boxes with a length-to-width ratio of 3, and decreases as the length-to-width ratio increases. Square boxes with a small corner radius (about 3% of box width) can be drawn to a maximum depth of about 80% of the width from low-carbon steel or some stainless steels, and to 70 to 75% of the width from aluminum alloys 3003-O and 5052-O.

Other factors that influence the producibility of a rectangular shell are work-metal composition, hardness and thickness, and radius at the bottom of the shell. Soft brass can easily be drawn to a depth of six times the corner radius in one operation, but some stainless steels can be drawn to a depth of only three times the corner radius. Likewise, thin material and small corner radii at the bottom of the shell reduce the maximum depth of draw. Shells that have a depth of more than six times the corner radius must be drawn in two or more operations.

PRESSES

Double-action presses are required for most deep drawing, because a more uniform blankholding force can be maintained for the entire stroke than is possible with a spring-loaded blankholder. Double-action hydraulic presses with a die cushion are often preferred for deep drawing, because of their constant drawing speed, stroke adjustment, and uniformity of clamping pressure. Regardless of the source of power for the slides, double-action straight-side presses with die cushions are best for deep drawing. Straight-side presses provide a wide choice of tonnage capacity, bed size, stroke and shut height.

Factors in Selection. Tonnage requirements, die space, and the length of stroke are the most important considerations in selecting a press for deep drawing.

DIES

Single-action dies (Fig. 35a) are the simplest of all drawing dies and have only a punch and a die. A nest or locator is provided to position the blank. The drawn part is pushed through the die and is stripped from the punch by the counterbore in the bottom of the die. The rim of the cup expands slightly to make this possible. Single-action dies can be used only when the forming limit permits cupping without the use of a blankholder.

Double-action dies have a blankholder. This permits greater reductions and the drawing of flanged parts. Figure 35(b) shows a double-action die of the type used in a double-action press. In this design, the die is mounted on the bottom, the punch is attached to an inner slide and the blankholder is attached to an outer slide. A pressure pad is used to hold the blank firmly against the punch nose during the drawing operation, and to lift the drawn cup from the die.

Figure 35(c) shows an inverted type of double-action die, which is used in single-action presses. In this design, the punch is mounted on the bottom and the die on the top. A die cushion can supply the blankholding force, or springs or air or hydraulic cylinders are incorporated in the die to supply the necessary blankholding force. The drawn cup is removed from the die on the up-

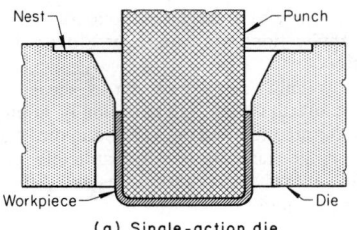

(a) Single-action die

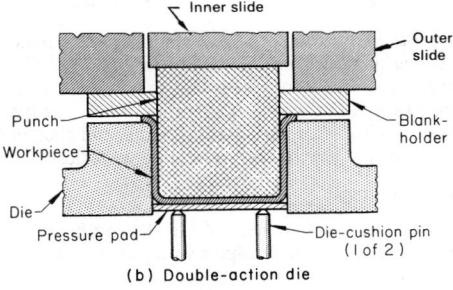

(b) Double-action die

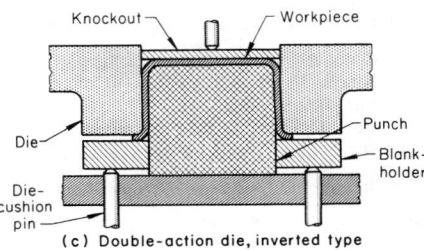

(c) Double-action die, inverted type

Fig. 35. Components of three types of simple dies, shown in setup for drawing a round cup. See text for discussion.

stroke of the ram, when the pin-like extension of the knockout strikes a stationary knockout bar attached to the press frame.

Compound Dies. When the initial cost is warranted by production demands, it is practical to combine several operations in a single die. Blanking and drawing are two operations commonly done in compound dies. Figure 36 shows how a cup is blanked, drawn, pierced and pinch trimmed in a single operation using a compound die. By this means, workpieces can be produced up to four times as fast as by the simple dies shown in Fig. 35.

MATERIAL FOR DIES AND PUNCHES

The selection of material for dies, punches and other tools for drawing sheet metal depends on workpiece composition, size and surface finish. Severity of the draw and production size are also important factors.

Most punches and dies are made of carburized, nitrided or chromium-plated low-alloy and tool

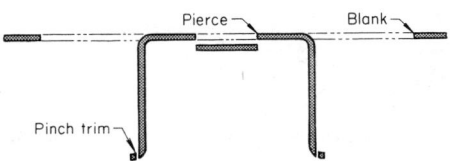

Fig. 36. Progressive blanking, drawing, piercing and pinch trimming of a cup in a compound die

steels. Alloy cast irons, aluminum bronze and even plastics are used for drawing certain materials. Cemented carbide inserts are placed in steel dies for drawing hard materials.

For further information on dies and punches, see "Deep Drawing Dies" in the article "Tool Materials for Special Applications" in Section 18.

LUBRICATION

Lubricants are used in all deep drawing operations. They range from ordinary machine oil to pigmented compounds.

Selection of lubricant is based primarily on ability to prevent galling, wrinkling or tearing during deep drawing. It is also influenced by ease of application and removal, corrosivity and other factors.

The type of lubricant also depends on severity of drawing and material being drawn. Aluminum and aluminum alloys are commonly drawn with mineral oil. For carbon steel, mineral oil, lard oil, or a soap solution are used. Castor oil is often used when drawing stainless steels.

Spinning

SPINNING is a method of forming sheet metal or tubing into seamless hollow cylinders, cones, hemispheres or other circular shapes by a combination of rotation and force. (Spinning of tube is covered under "Tube Spinning" in the article "Forming of Bars, Tube and Wire"). The method may be divided into two categories: manual spinning (with or without mechanical assistance to increase the force) and power spinning.

MANUAL SPINNING

Any metal ductile enough to be cold formed by other methods can be spun. Most spinning is done without applying heat to the work metal, although sometimes the metal is preheated to achieve one of two objectives: (*a*) to increase the ductility of hard-to-form metals such as beryllium, refractory metals, or magnesium; or (*b*) to reduce the strength of work metals, thus permitting greater thicknesses to be spun.

Applicability. Manual spinning is used for forming flanges, rolled rims, cups, cones and double-curved surfaces of revolution such as bells. Several typical shapes formed by manual spinning are shown in Fig. 37.

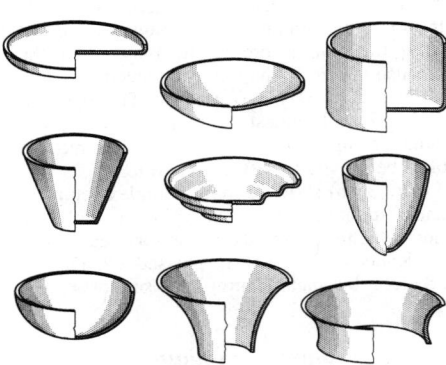

Fig. 37. Typical conical, cylindrical and dome shapes formed by manual spinning

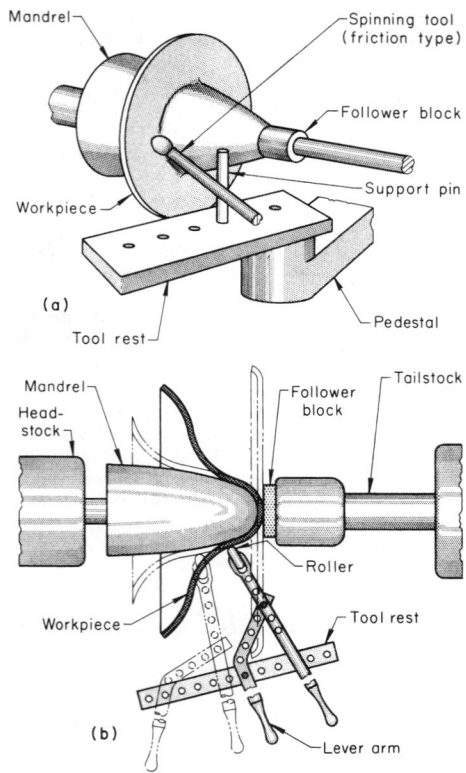

(a) Setup using a simple hand tool, applied like a pry bar. (b) Setup using scissorlike levers and a roller spinning tool.

Fig. 38. Manual spinning in a lathe

Manual spinning of low-carbon steel $\frac{1}{8}$ in. thick or aluminum $\frac{1}{4}$ in. thick is possible. Thicker materials can be spun by using various mechanical devices for applying the tool force. Manual spinning is good for low-volume production, but quality of work varies from one operator to the next.

EQUIPMENT FOR MANUAL SPINNING

A typical tool and workpiece setup for manual spinning is shown in Fig. 38(a). A mandrel is mounted on the headstock of a lathe. A circular blank (workpiece) is clamped to the mandrel by a follower block attached to the tailstock of the lathe. The blank is rotated by the headstock, while a friction-type spinning tool is manually pressed against the blank, forcing the blank over the preshaped mandrel. The tool is mounted on a tool rest by a support pin (fulcrum). The operator presses the tool against the blank by pivoting it around the support pin. The tool rest permits the tool to be moved to different positions.

Figure 38(b) shows a more complex setup for manual spinning. Here, the spinning tools are rollers which are mounted in the fork sections of long levers. The roller is pressed against the workpiece by manipulating the scissorlike levers.

POWER SPINNING

Virtually all ductile metals are processed by power spinning. Products range from small hardware items made in large quantities (metal tum-

blers, for instance) to large components for aerospace applications in unit or low-volume production.

Power spinning is also known as shear spinning, because the metal is intentionally thinned. Roller tools are mounted on compound tool rests, which are mechanically or hydraulically powered. This allows larger workpieces and different metals to be spun.

MECHANICS OF CONE SPINNING

The application of shear spinning to conical shapes is shown schematically in Fig. 39. The metal deformation is such that forming is in accordance with the sine law, which states that the wall thickness of the starting blank and that of the finished workpiece are related as follows:

$$t_2 = t_1 (\sin \alpha)$$

Where t_1 is the thickness of the starting blank, t_2 is the thickness of the spun workpiece, and α is one-half the apex angle of the cone.

Reducing wall thickness by 50% in accordance with the sine law is illustrated in Fig. 39, where:

D = diameter (the same in starting blank and cone)
t_1 = flat plate thickness
t_2 = wall thickness of side of spun cone
α = 30° (which is half the included angle).

Using the sine law for Fig. 39,

$$t_2 = t_1 (\sin \alpha) = 0.500 \times 0.5 = 0.250 \text{ in.}$$

When spinning in accordance with the sine law,

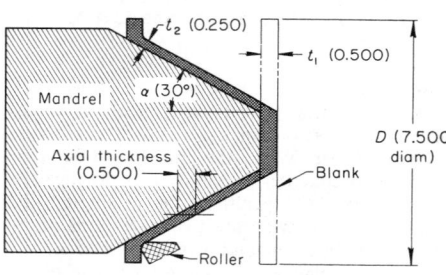

Fig. 39. Setup and dimensional relations for one-operation power spinning of a cone. See text for application of sine law in relation to this illustration.

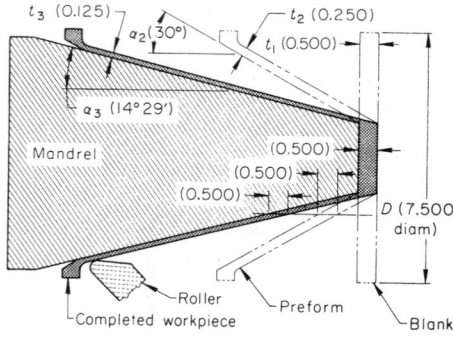

Fig. 40. Setup and dimensional relations for two-operation spinning of a cone to a small angle (less than 35° included angle)

the axial thickness is the same as the thickness of the starting blank (Fig. 39).

When spinning cones to small angles (less than 35° included angle), the best practice is to use more than one spinning pass with a different cone angle for each pass to form preforms, as illustrated in Fig. 40. When using this technique, the workpiece is annealed or stress relieved between passes.

MACHINES FOR POWER SPINNING

Power spinning machines can be horizontal or vertical. Components of a vertical machine are shown in Fig. 41. Automatic spinning machines are now available which have templates to guide the tool, or computer numerical controls (CNC) to automatically perform the spinning operation. These provide fast, accurate and reproducible results.

TOOLS FOR POWER SPINNING OF CONES

Mandrels are made of steel, cast iron, or hardwood. The tools are usually made of hardened carbon or low-alloy tool steels.

Rubber-Pad Forming

RUBBER-PAD FORMING employs a rubber pad on the ram of the press and a form block on the platen. The form block usually is similar to the punch in a conventional die, but it can be the die cavity. The blank is placed over the form block and during forming the rubber presses the blank around the form block, forming the workpiece. The rubber exerts nearly equal pressure on all workpiece surfaces.

GUERIN PROCESS

The Guerin process is the oldest and most basic of the production rubber-pad forming processes. Its advantages are simplicity of equipment, adaptation to small-lot production, and ease of changeover.

Some metals commonly formed by the Guerin process are listed in Table 13.

Presses and Tools. Almost any hydraulic press can be used in the Guerin process.

The main tools are the rubber pad and the form block (Fig. 42). The rubber pad is fairly soft (about Durometer A 60 to 75) and is usually three times as deep as the part to be formed. The pad can consist of a solid block of rubber, or of laminated slabs cemented together and held in a retainer, as shown in Fig. 42.

The form block is loosely mounted on a platen (see Fig. 42), which fits closely into the rubber-pad retainer to avoid extrusion of the rubber during the forming process. The form block usually contains locating pins to hold the blank in place.

Blanking. With the Guerin process, rubber pads can also be used for blanking and piercing. Rubber pads produce better edges on the workpiece than band sawing, and almost as good edges as those made by routing. An edge radius up to the thickness of the metal can be produced on some heavy-gage metals. The rubber-pad method can blank aluminum alloy 2024-O up to 0.032 in. thick. Minimum hole diameter or width of cutout

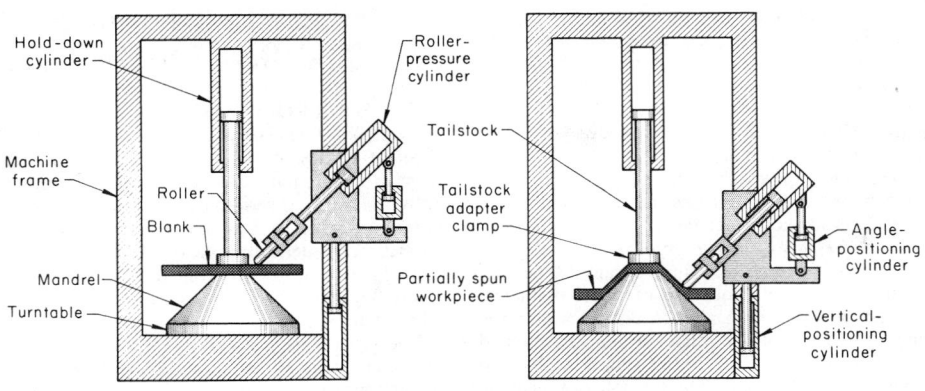

Fig. 41. Schematic illustration of power spinning in a vertical machine

Table 13. Metals commonly formed by the Guerin rubber-pad process

Metal	Maximum thickness, in.(a)
Mild forming	
Aluminum alloys:	
2024-O, 7075-W	0.187
2024-T4	0.064
Austenitic stainless steels:	
Annealed	0.050(b)
Quarter hard	0.032(c)
Titanium alloys	0.040(d)
Stretch flanging	
Aluminum alloy 2024-T4	0.064
Austenitic stainless steels:	
Annealed	0.050
Quarter hard	0.030

(a) Typical; varies with type of equipment and part design. (b) Up to 0.078 in. when compression dams are used. (c) Only very mild forming. (d) When heated to 600 °F.

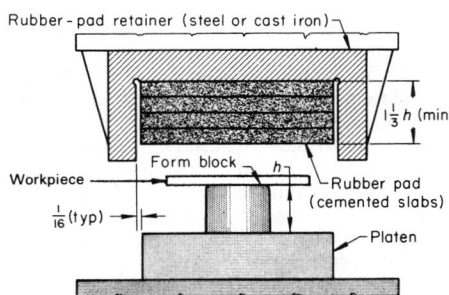

Fig. 42. Tooling and setup for rubber-pad forming by the Guerin process

is 2 in. A minimum of 1½-in. trim is needed for external cuts.

The form block is provided with a sharp cutting edge where the blank is to be sheared. In hard metal blocks, this edge can be cut into the form block, as shown in Fig. 43(a) and (b). Form blocks of soft metal, plastic, or wood need a steel shear plate for the cutting edge (Fig. 43c). The shearing edge should be undercut 3° to 6°.

The trim metal beyond the line of shear must be clamped firmly, so that the work metal will break over the sharp edge instead of forming around it. This clamping is done by a lock ring (Fig. 43a), by a grip plate (Fig. 43b), or by a raised extension of the form block (Fig. 43c).

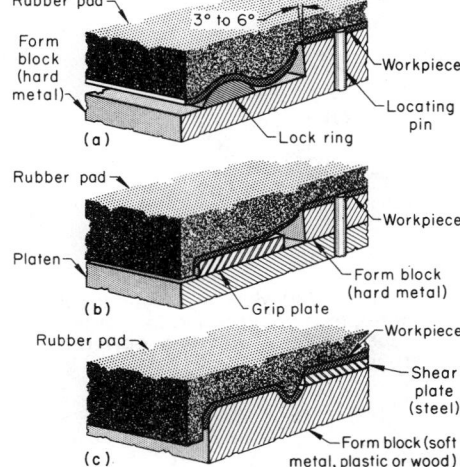

Fig. 43. Three techniques for blanking by the Guerin process. See text for discussion.

VERSON-WHEELON PROCESS AND HYDROFORMING

Several other techniques have been developed from the Guerin process for specific uses. The Marform process uses a pressure regulator to regulate the rubber-pad pressure. It is used for deep drawing and forming of wrinkle-free shrink flanges. Two other techniques are the Verson-Wheelon process and the Hydroforming process. Both use a rubber pad backed by a flexible hydraulic fluid cell. The fluid cell provides a higher and more uniform hydrostatic pressure on the workpiece than does rubber-pad forming alone.

The Verson-Wheelon press has a horizontal cylindrical steel housing that contains the hydraulic fluid cell above a sliding table on which the form block is mounted (Fig. 44). Hydraulic fluid is pumped into the cell, causing it to expand. The expansion creates the force needed to flow the rubber-pad downward, over and around the form block and the metal to be formed. This process is used for forming shallow parts.

In the Hydroform press, the hydraulic fluid is contained in the upper die cavity and is retained by a 2½-in.-thick rubber diaphragm. This cavity is called a pressure dome (see Fig. 45). A wear sheet is placed between the diaphragm and the workpiece. The punch pushes the workpiece

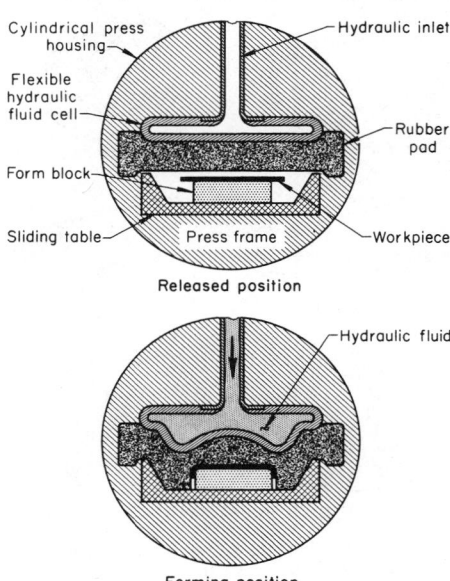

Released position

Forming position

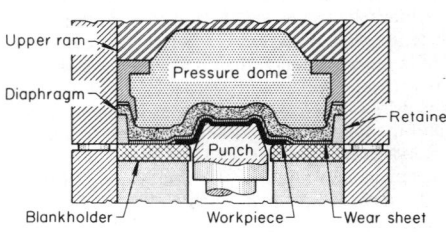

Fig. 45. Rubber-diaphragm forming in a Hydroform press

up into the diaphragm, forming the part. Deep drawing of complicated parts is possible because the rubber diaphragm applies a uniform pressure, especially to the sides of the part.

Both the Verson-Wheelon process and Hydroforming are used for forming aluminum, low-carbon steel, stainless steel, and heat-resisting alloys. Titanium can be formed with the Verson-Wheelon process. Copper alloys are commonly Hydroformed.

Three-Roll Forming

THREE-ROLL FORMING (also known as roll bending) is a process for forming plate, sheet, bars, beams, angles, or pipe into various shapes by passing the work metal between three properly spaced rolls. Sheet and plate are the mill products most often formed by the three-roll process.

Applicability. Any metal that can be formed by other cold forming processes can be three-roll formed. Some of the shapes commonly produced by three-roll forming from flat stock are illustrated in Fig. 46.

MACHINES

There are two basic types of three-roll forming machines: the pinch-roll type and the pyramid-

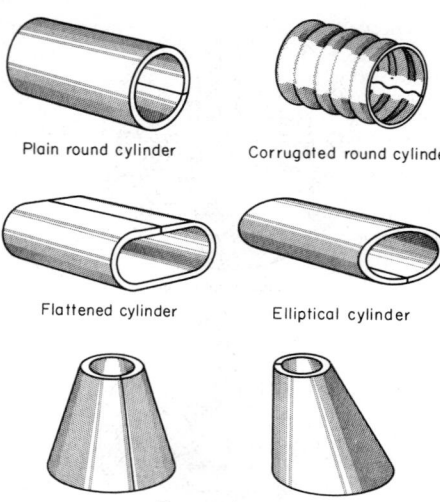

Fig. 46. Typical shapes produced from flat stock by three-roll forming

roll type. The rolls on most three-roll machines are positioned horizontally; a few vertical machines are used, mainly in shipyards. Vertical machines have one advantage in forming scaly plate: loose scale is less likely to become embedded in the work metal than when bending is done in horizontal rolls. With vertical rolls, however, it is difficult to handle wide sections that require careful support to avoid skewness in rolling.

Conventional pinch-type machines have the roll arrangement shown in Fig. 47. For rolling flat stock up to about 1 in. thick, each roll is of the same diameter. However, on larger machines, the top rolls sometimes are smaller in diameter to maintain approximately the same surface speed on both the inside and outside surfaces of the plate being formed. These heavier machines are supplied also with a slip-friction drive on the front roll to permit slip, because of the differential in surface speed of the rolls. Thus, as work-metal thickness increases, the diameter of the top roll is decreased in relation to the diameter of the lower rolls.

The position of the top roll is fixed, whereas the lower front roll is adjustable vertically to suit the thickness of the blank. The bending roll is adjusted angularly; this determines the diameter of the cylinder to be formed.

Shoe-Type Pinch-Roll Machines. One important modification of the conventional three-roll pinch-type machine is the shoe-type machine. This ma-

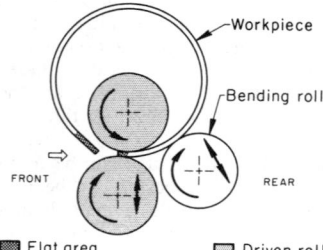

Fig. 47. End view of a cylindrical workpiece being rolled in a conventional pinch-type machine. Note large flat area on leading end, and smaller flat area on trailing end.

chine uses the pinch principle but in addition incorporates a forming shoe, as shown in Fig. 48. Because of the relation of the two front rolls and the forming shoe to the workpiece, the flat areas on the ends become barely discernible compared with the length of flat area obtained when rolling in a conventional machine.

The shoe-type machine is often used for manufacture of transformer cases and small tanks, such as jackets for hot-water tanks.

Pyramid-Type Machines. Figure 49 illustrates the arrangement of the rolls in a pyramid-type machine. The bottom rolls are of equal diameter, but are about 50% smaller in diameter than the top roll. The bottom rolls are gear-driven and (normally) fixed. Each roll is supported by two smaller rolls (Fig. 49). The top roll is adjustable vertically to control the diameter of the cylinder formed. The top roll, which rotates freely, depends on friction with the work metal for rotation. Backup rolls are not used on the top roll.

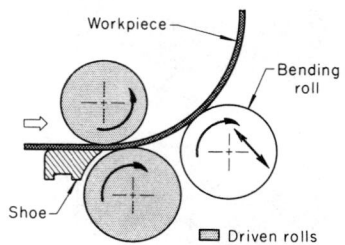

Fig. 48. End view of a cylindrical workpiece being rolled in a shoe-type machine with two powered rolls

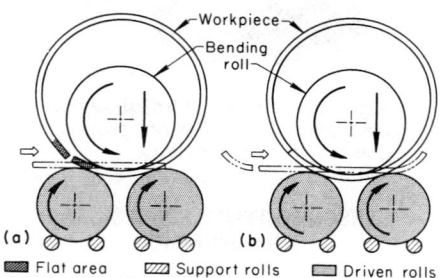

(a) Entrance of flat workpiece and shape of a nearly finished workpiece, including the flat areas on the leading and trailing ends. (b) Similar, except that the workpiece was prebent to minimize the flat areas on the ends.

Fig. 49. Arrangement of rolls in a pyramid-type machine

As shown in Fig. 49, the work metal is placed on the bottom rolls while the top roll is in a raised position. The top roll is then lowered to contact and bend the work metal a predetermined amount, depending on the diameter of the workpiece to be formed. To minimize flat areas on the ends, prebending of the ends often is done (see Fig. 49b).

ROLLS

Rolls used in three-roll forming machines are machined from steel forgings having a carbon content of 0.40 to 0.50% and a Brinell hardness of 160 to 210. Plain carbon steel such as 1045 has often been used; when greater strength is needed, rolls are forged from an alloy steel such as 4340.

Contour Roll Forming (Cold Roll Forming)

CONTOUR ROLL FORMING (also known as cold roll forming) is a process for forming metal sheet or strip stock into desired shapes of uniform cross section. The stock is fed longitudinally through a series of roll stations, each equipped with two or more contoured rolls (sometimes called roller dies). Most contour roll forming is done by working the stock progressively through two or more stations until the finished shape is produced.

The process is particularly suited to the production of large quantities and long lengths, with a minimum of handling. Auxiliary operations, such as notching, slotting, punching, embossing, curving and coiling, can be easily combined with contour roll forming.

APPLICABILITY

All metals that can be shaped by any of the common forming processes can be contour roll formed. The formability of the work metal controls the permissible speed of roll forming and the degree (severity) to which the metal can be formed. For instance, the speed at which the softest grade of aluminum strip can be contour roll formed may be as much as 400 times the speed permitted for rolling titanium strip into a similar shape.

Thickness Range of Work Metal. The range of work-metal thicknesses that can be shaped by contour roll forming can only be approximately given. Items such as steel measuring tape and brass radiator tubing are usually produced from strip 0.004 to 0.005 in. thick. At the opposite extreme, channels and Z-sections are produced from steel up to 0.312 in. thick. The maximum thickness of section that can be contour roll formed is usually limited by the size of machinery available and the amount of force the bearings and spindles can withstand. The length of section that can be roll formed is limited by the facilities for handling formed sections.

Shape. Under the most favorable conditions, it is possible to contour roll form almost any shape. However, formability of the work metal and the number of forming stations available may impose limitations on complexity of shape or severity of forming.

MACHINES

Machines used for contour roll forming are available with roll-shaft diameters of 1 to 15 in. and width capacities of 4 to 60 in. The number of roll stations varies from 1 to 40 for most machines. Most of these machines are built with individual units (roll stations), so that the initial installation can be limited to immediate needs, and additional units can be added when required. The different types of machines have many features of construction in common.

Overhung-spindle machines are the simplest type used for contour roll forming. An overhung-spindle machine with one roll station is shown in Fig. 50. The rolls and roll shafts (spindles) have no outboard support. The greatest advantage of an overhung-spindle machine is the convenience it affords in changing rolls, because there are no outboard housings to remove. Machines of this

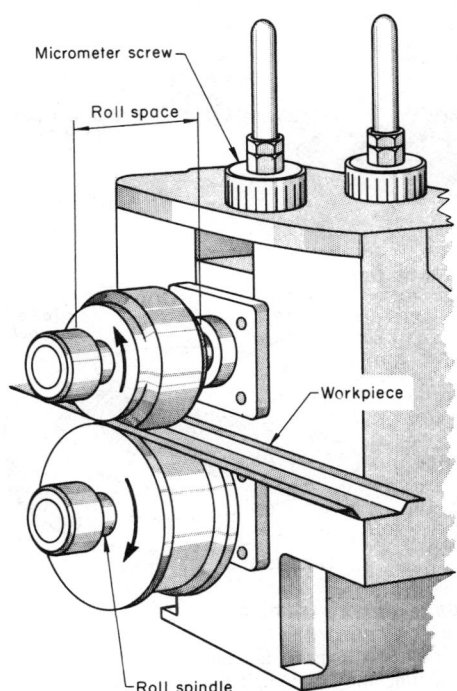

Fig. 50. Overhung-spindle machine (one roll station) for contour roll forming

type are available with roll-shaft diameters of 1 to 2 in. Their principal disadvantage is a lack of rigidity, which limits their capacity to a strip width of 6 in. and thickness of about 0.040 in. (for work metal not harder than low-carbon steel).

Outboard supports on the roll shafts are required for rolling strip widths greater than about 6 in. and thicknesses greater than about 0.040 in. A universal type of rolling machine having roll shafts supported at both ends is shown in Fig. 51. These machines are available with roll-shaft diameters ranging from 1 1/2 to 3 in. and various shaft lengths as required. These machines are best adapted to rolling strip no thicker than about 3/16 in.

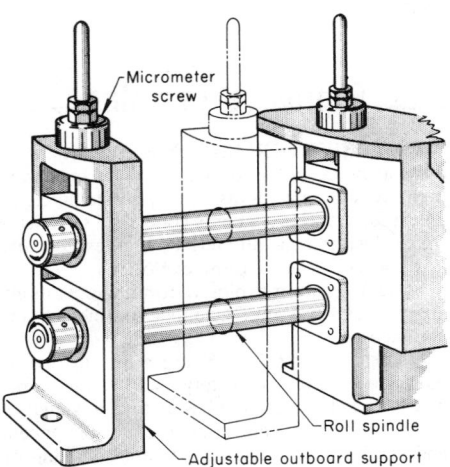

Fig. 51. Universal contour roll forming machine, with outboard support for roll shafts (rolls not shown)

ROLL DESIGN

As shown in Fig. 50 and 51, the principal forming rolls are mounted on two horizontal shafts. Usually both rolls are positively driven, although for some applications one roll is driven and the other is an idler. The efficiency of contour roll forming can often be increased by adding side rolls, which may either be idlers located between or beside forming stations, or be positively driven as separate forming stations.

Driven side rolls sometimes are the best means of forming complex shapes, particularly those shapes that have a number of sharp bends and will subsequently be completely closed (complex tubular shapes). The most common method of forming with driven side rolls is to remove the upper roll shaft from a station and replace it with an assembly similar to the one shown in Fig. 52. The rolls are mounted to pinions that are bushed and anchored to the large bracket on the housing, and the entire side-roll assembly is bolted to the machine. Driving is accomplished through bevel

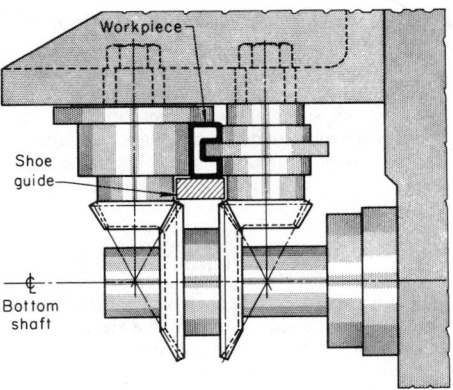

Fig. 52. Driven side rolls used in forming a complex shape

gears that mesh with other bevel gears mounted on the lower roll shaft. Side-driven rolls are usually smaller than vertical rolls; consequently their speed of rotation must be greater to match the surface speed of the other rolls. This is done by proportioning the sizes of the bevel gears to give the correct speed ratio.

SELECTION OF ROLL MATERIAL

The materials that are most commonly used for contour rolls are:

1. Low-carbon steel, turned and polished—not hardened
2. Gray iron (such as class 30), turned and polished—not hardened
3. Low-alloy tool steel (such as O1 or L6), hardened to Rockwell C 60 to 63 and sometimes chromium plated
4. High-carbon, high-chromium tool steel (such as D2), hardened to Rockwell C 60 to 63 and sometimes chromium plated
5. Bronze (usually aluminum bronze).

Quantity to be rolled is usually the major factor in choosing the most appropriate roll material. Split rolls, made of several sections, are often used to roll complex shapes.

Stretch Forming

STRETCH FORMING is the forming of sheet, bars, and rolled or extruded sections over a form block of the required shape while the workpiece is held in tension. The work metal is stretched just beyond its yield point (generally 2 to 4% total elongation) to retain the contour of the form block.

The four methods of stretch forming are: stretch draw forming (Fig. 53a and b); stretch wrapping, also called rotary stretch forming (Fig. 53c); compression forming (Fig. 53d); and radial-draw forming (Fig. 53e).

APPLICABILITY

Almost any shape that can be produced by other sheet forming methods can be produced by stretch forming. Drawn shapes that involve metal flow, particularly straight cylindrical shells, and details that result from compression operations such as coining and embossing, cannot be made. However, some embossing is done by the mating-die method of stretch draw forming (Fig. 53b).

Stretch forming is used to form aerospace parts from steel, nickel and aluminum, and from titanium alloys and other heat-resisting and refractory metals.

Stretch forming is used also to shape automotive body panels, both inner and outer, and frame members that could be formed by other processes but at higher cost. An example is the automobile roof shown in Fig. 54.

MACHINES AND ACCESSORIES

Stretch wrapping, compression forming, and radial-draw forming use rotary tables on which are mounted the form block, a ram gripping and tensioning or wiping device, and a mechanically or hydraulically actuated table gripper (Fig. 53c, d and e). In stretch wrapping, the workpiece is pulled from one end, wrapping the metal around the form block. In compression forming, a wiper shoe presses the workpiece against the form block. Radial-draw forming combines both pulling and pressing of the metal.

Stretch draw forming is done in three types of machines. In one type, the form block is mounted on a hydraulic cylinder and is pushed into the blank, which is held in tension by a pair of pivoting grippers. In another type, the form block is fixed to the table and the blank is drawn around it by a pair of grippers actuated by slides or a hydraulic cylinder. The third type of machine is a single-action hydraulic press equipped with a two-piece mating die. Grippers pull the blank over the lower die, then the upper die descends to produce the workpiece (Fig. 53b).

Accessory Equipment. Grippers and wiping shoes or rollers are made to conform to the rolled or extruded shape that is to be stretch formed. Jaws used for gripping sheet in stretch draw forming can be segmented or contoured to apply equal stretch to all parts of the sheet as it is formed.

Drop Hammer Forming

DROP HAMMER FORMING is a process for producing shapes by the progressive deformation of sheet metal in matched dies under the repetitive blows of a gravity drop hammer or a power

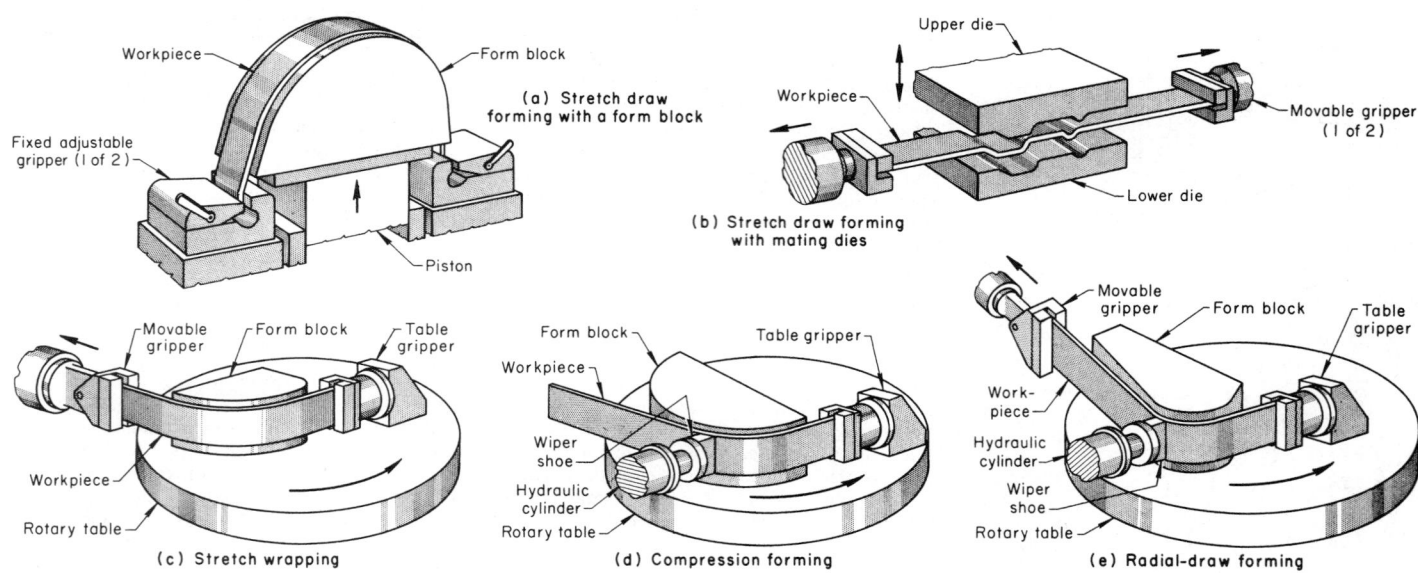

(a) Stretch draw forming with a form block

(b) Stretch draw forming with mating dies

(c) Stretch wrapping

(d) Compression forming

(e) Radial-draw forming

Fig. 53. Fundamentals of the techniques employed in the four methods of stretch forming

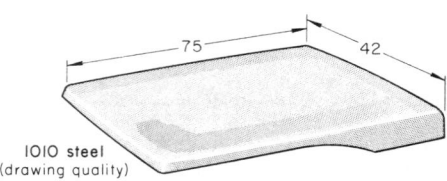

Fig. 54. Stretch draw formed automobile roof. (Dimensions are in inches.)

drop hammer. Configurations most commonly formed by the process include shallow, smoothly contoured, double-curvature parts; shallow-beaded parts; and parts with irregular and comparatively deep recesses. Small quantities of cup-shape and box-shape parts, curved sections, and contoured flanged parts also are formed.

Advantages and Limitations. The main advantages of drop hammer forming are: (*a*) low cost for limited production; (*b*) relatively low tooling costs; (*c*) dies that can be cast from low-melting alloys and that are relatively simple to make; (*d*) short delivery time of product, because of simplicity of toolmaking; and (*e*) the possibility of combining coining with forming.

Against these advantages, the following limitations must be weighed: (*a*) probability of forming wrinkles; (*b*) need for skilled operators, specially trained for this process; (*c*) restriction to relatively shallow parts with generous radii; (*d*) restriction to relatively thin sheet, from about 0.024 to 0.064 in. (thicker sheet can be formed only if the parts are shallow and have generous radii).

Drop hammer forming is not a precision forming method; tolerances of less than $1/32$ to $1/16$ in. are not practical.

HAMMERS FOR FORMING

Gravity drop hammers and power drop hammers are comparable to a single-action press. However, they can be used to do the work of a press equipped with double-action dies through the use of rubber pads, beads in the die surfaces, draw rings, and other auxiliary equipment.

Because they can be controlled more accurately and because their blows can be varied in intensity and speed, power drop hammers, particularly the air-actuated types, have virtually replaced gravity drop hammers. A typical air drop hammer, equipped for drop hammer forming, is shown in Fig. 55.

TOOLING

In general, a tool set consists of a die that conforms to the outside shape of the desired part, and a punch that conforms to the inside contour (see Fig. 55).

Tool Materials. Dies are cast from zinc alloy (3.5 Cu, 4 Al and 0.04 Mg), aluminum alloy, beryllium copper, ductile iron or steel. The wide use of zinc alloy as a die material stems from the ease of casting it close to the final shape desired. Its low melting point (717 F) also is advantageous. All dies, regardless of die material, are polished.

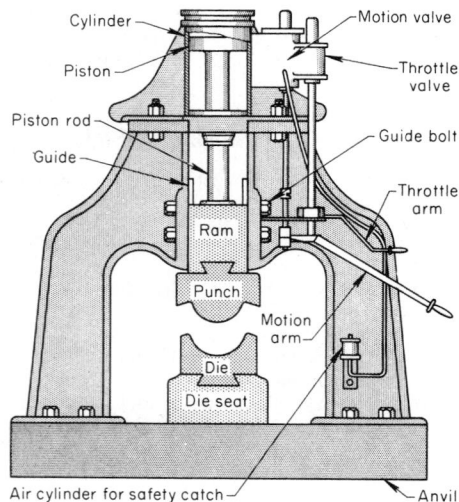

Fig. 55. Air-actuated power drop hammer equipped for drop hammer forming

Punches usually are made of lead or a low-melting alloy, although zinc or a reinforced plastic also may be used. The sharpness of the contours to be formed, the production quantity, and the accuracy desired primarily govern the choice of punch material. As a punch material, lead has the advantage of not having to be cast accurately to shape, because it deforms to assume the shape of the die during the first forming trial with a blank.

LUBRICANTS

Lubricants are used in drop hammer forming to facilitate deformation by reducing friction and minimizing galling and sticking, and to preserve or improve surface finish. Selection of a lubricant depends primarily on type of work metal, forming temperature, severity of forming, and subsequent processing. Recommendations for lubricants used with steels, and aluminum, magnesium and titanium alloys are given in the sections of this article that deal with processing of those metals.

Explosive Forming

EXPLOSIVE FORMING changes the shape of a metal blank or preform by the instantaneous high pressure that results from the detonation of an explosive. This article is concerned only with the explosives generally termed high explosives, and not with so-called low explosives.

Systems used for explosive forming operations are generally classified as either confined or unconfined.

Confined systems (Fig. 56) use a die, in two or more pieces, that completely encloses the workpiece. The closed system has distinct advantages for the forming of thin stock to close tolerances, and has been used for close-tolerance sizing of thin-wall tubing. However, confined systems are generally used only for forming of comparatively small workpieces, because economic feasibility decreases as the size of the workpiece and corresponding die increases.

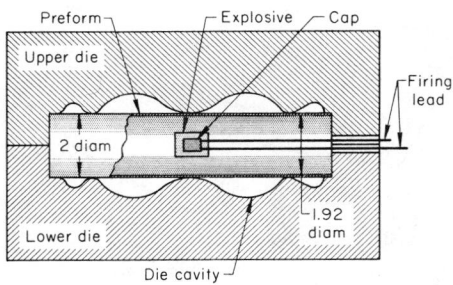

Fig. 56. Confined system for explosive forming

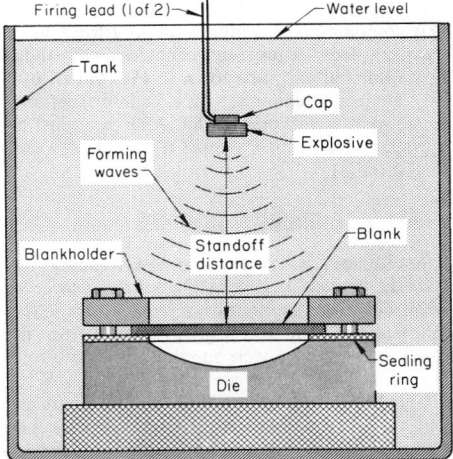

Fig. 57. Unconfined system for explosive forming

Unconfined Systems. In an unconfined system (Fig. 57), the shock wave from the explosive charge takes the place of the punch in conventional forming. A single-element die is used with a blank held over it, and the explosive charge is suspended over the blank at a predetermined distance (the "standoff" distance). The complete assembly can be immersed in a tank of water, as shown in Fig. 57, or a plastic bag filled with water can be placed over the blank.

The unconfined system is inherently inefficient, because only a small part of the total energy released by the explosion is effective as forming energy. The medium within which the explosion occurs plays an important part in determining the efficiency of the system. As the density of the medium increases, efficiency increases. For this reason, most explosive forming of large pieces is done in a medium more dense than air. Water is the most commonly used medium for ambient-temperature explosive forming. Molten aluminum has been used as the medium in explosive forming at elevated temperature.

Under normal operating conditions, it is best to detonate the explosive charge as far below the surface of the water as possible. This reduces the amount of water that is thrown by the explosion, and reduces the amount of energy lost by venting to the atmosphere the gas bubble that results from the detonated charge.

The amount of energy or peak pressure delivered can be calculated from standard formulas. Generally, a cylindrical charge (or point charge) is located near the centerline of the part and at a standoff distance that is related to the span of the workpiece over the die cavity. For large parts, it is generally impractical to use a point charge; for example, in forming large hemispheres or end closures for rocket motors, Primacord, shaped in a large loop and located close to the outer periphery of the part, is ordinarily used instead of a point charge. (Primacord is a cordlike detonating fuse that consists of a filament of explosive material covered by a protective, water-repellent coating.)

EQUIPMENT

The primary equipment for explosive forming in an unconfined system consists of a water tank, a crane, a vacuum pump and a detonator.

Water Tank. The water tank must be able to withstand the repeated impacts of the explosive shock without rupturing. Many tanks are designed to be large enough so that the shocks reaching the walls from centrally placed charges are considerably reduced.

Crane. A crane is usually necessary to move material around the facility and also in and out of the water tank. Ideally, the crane should be air operated, to avoid having electric power lines within the firing area. The required capacity of the crane will depend on the size and weight of the dies to be handled.

Vacuum Pump. A vacuum pump will probably be needed for most explosive forming operations when parts are formed under water. If the firing area is to be maintained with a minimum of electric lines, a venturi pump operating on water pressure will work satisfactorily. A mechanical pump driven by an electric motor can be used, with the vacuum lines being brought into the firing area from a pumping site remote from the firing area. An electrically driven mechanical pump is preferred, because it has a considerably greater capacity than a venturi pump and is probably more economical to operate.

Detonator. Under ideal conditions, the detonator for the electric blasting caps is the only electric circuit that should be permitted in the area where explosives are handled.

A detonator should be constructed on the fail-safe principle, so that any malfunction will immediately cause the circuit to be disarmed.

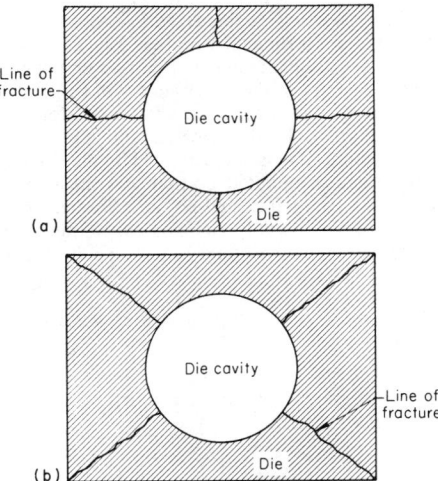

Fig. 58. Modes of fracture of a rectangular die under (a) static load and (b) shock load

DIE SYSTEMS AND MATERIALS

Basic differences between tooling for explosive forming and for conventional forming arise from the type of loading that the die material must be able to withstand. In explosive forming, high impact loads transmit shock waves through the metal that cause unusual stress patterns within the die material, and therefore, corners and other sites of stress concentrations should be eliminated where possible. Shock loading causes the die to fracture along lines from the corners, rather than through the thinnest section as in static fracture. Figure 58 shows the modes of fracture for conditions of static loading and dynamic (shock) loading.

Materials for Solid Dies. Solid dies made from heat treated alloy steel maintain contour, surface finish and dimensional accuracy for a relatively long time. To avoid brittle fracture under overloads, a maximum hardness of Rockwell C 50 is desirable.

Electromagnetic Forming (EMF)

ELECTROMAGNETIC forming, also known as magnetic pulse forming, is a process for forming metal by the direct application of an intense, transient magnetic field. The workpiece is formed without mechanical contact by the passage of a pulse of electric current through a forming coil.

The major application of EMF is for the single-step assembly of tubular parts to each other or to other components; it is used to a lesser extent for the shaping of tubular parts and the shallow forming of flat stock. Metals with high electrical conductivity are formed directly; poorly conductive metals are formed with the aid of highly conductive "drivers."

PROCESS DESCRIPTION

A very intense magnetic field that lasts only a few microseconds is produced by the discharge of a bank of capacitors into a coil called the forming coil. The resulting eddy currents induced in a conductive workpiece interact with the magnetic field to cause mutual repulsion between the workpiece and the forming coil. The force of this repulsion accelerates the workpiece against a die or a mandrel with enough stored energy to plastically deform the workpiece. The reaction to this shaping force is sustained by the forming coil, which must therefore be stiff and strong enough to withstand such forces.

Basic Circuit. Figure 59 shows the basic circuit for EMF, as used for compression forming of a tubular workpiece. It consists of a forming coil, an energy-storage capacitor, switches, and a power supply of nearly constant current to charge the capacitor.

Figure 59(a) shows the flux-density pattern of the magnetic field produced by discharging the capacitor through the forming coil in the absence of an electrically conductive workpiece. The evenly spaced flux lines indicate a uniform flux density within the coil.

Figure 59(b) shows the change in field pattern that results when the capacitor is discharged through a forming coil in which a tubular workpiece of highly conductive metal has been inserted. The magnetic field is distorted and the flux density intensified (flux lines are more closely

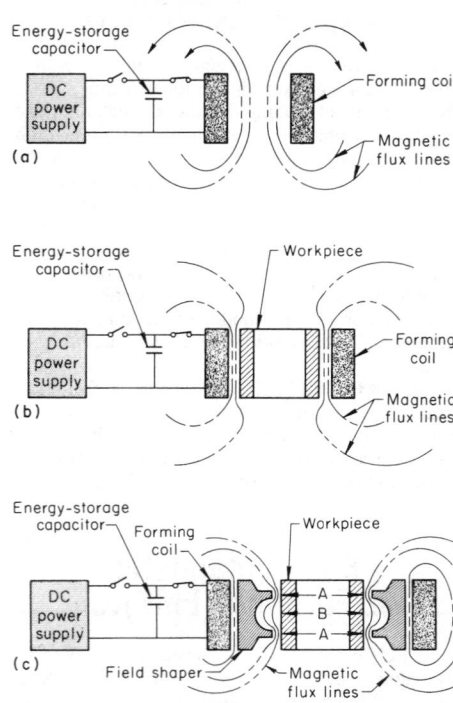

(a) Field pattern in absence of workpiece. (b) Field pattern with workpiece in forming coil. (c) Field pattern when field shaper is used.

Fig. 59. Basic circuit and magnetic field patterns for electromagnetic compression forming of a tubular workpiece

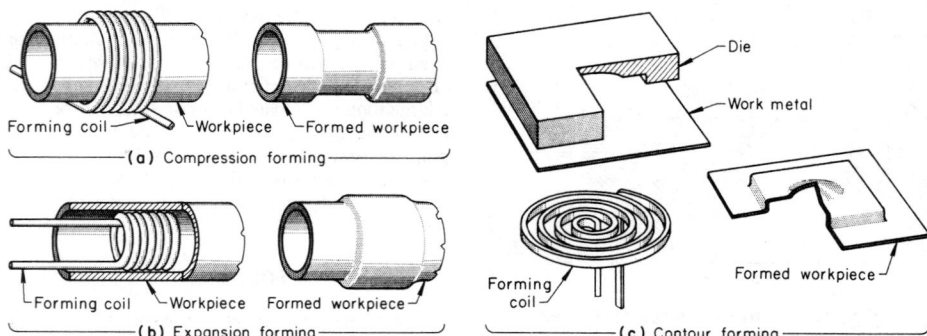

Fig. 60. Three basic methods of electromagnetic forming

spaced) by confinement to the small annular space between the coil and the workpiece.

Field Shapers. For efficient use of stored energy, coils are designed to minimize stray inductance and to avoid flux concentrations. Field patterns are ordinarily controlled with field shapers, which are massive current-carrying conductors. Field shapers need not be directly connected to the basic coil, but they may be inductively coupled to it.

Figure 59(c) illustrates the use of a field shaper to concentrate the force in certain regions of the workpiece. This technique not only produces high local forming pressures in desired areas, but also lengthens the life of the forming coil by preventing high pressures on weaker parts of the coil.

Forming Methods and Coil Types. Typical forming with the three basic types of coil used for compression, expansion and contour forming is shown schematically in Fig. 60. A tubular workpiece is compressed by an external coil, as shown in Fig. 60(a), usually against a grooved or suit-

ably contoured insert, plug, tube, or fitting inside the workpiece. A tubular workpiece is expanded by an internal coil, as shown in Fig. 60(b), usually against a collar or other component surrounding the workpiece. Flat stock is almost always contour formed against a die, as indicated in Fig. 60(c).

ENERGY SOURCES

In ordinary applications, the power supply may need to deliver almost instantaneous power of 1000 megawatts to the forming coil. Because this much power cannot be drawn directly from the usual industrial power supplies, specially designed energy sources are needed.

Flywheel-generator combinations, inductors and batteries have been used for energy storage and pulse discharge. For EMF applications, however, capacitor banks have been the most satisfactory means of energy storage.

Shearing, Slitting, and Gas and Arc Cutting

Shearing of Plate and Flat Sheet

SHEARING of sheet and plate is broadly classified by the type of blade (cutter) used— namely, straight or rotary. Straight-blade shearing is used for squaring and cutting flat stock to required shape and size. It is most often used for square and rectangular shapes, although triangles and other straight-sided shapes are also sheared with straight blades. Rotary shearing (not to be confused with slitting) is used for producing circular or other contoured shapes from sheet or plate.

STRAIGHT-BLADE SHEARING

In straight-blade shearing, the work metal is placed between a stationary lower blade and a movable upper blade. As the upper blade is forced down, the work metal is penetrated to a specific portion of its thickness, after which the unpenetrated portion fractures and the work metal separates (Fig. 1). The amount of penetration de-

pends largely on the ductility and thickness of the work metal. The blade will penetrate 30 to 60% of the work-metal thickness for low-carbon steel, depending on thickness. For a more ductile metal like copper, the penetration will be greater. Conversely, the penetration will be less for metals that are harder than low-carbon steel.

A sheared edge is characterized by the smoothness of the penetrated portion and the relative roughness of the fractured portion. Sheared edges cannot compete with machined edges, but when blades are kept sharp and in proper adjustment,

it is possible to obtain sheared edges that are acceptable for a wide range of applications.

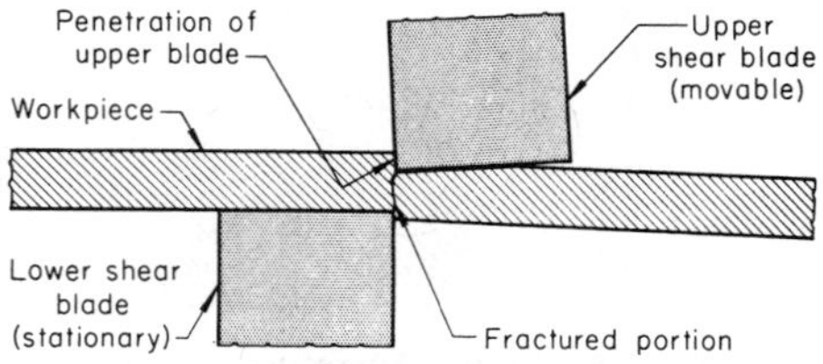

Applicability. Straight-blade shearing is the most economical method of cutting straight-sided blanks from stock no more than 2 in. thick. The process is also widely used for cutting sheet into blanks that subsequently will be formed or drawn on a punch press. Blanks can be cut to size within ±0.005 in., and strips can be cut to width within ±0.005 in. in lengths up to 12 ft.

Straight-blade shearing is seldom used for shearing metal harder than about Rockwell C 30.

MACHINES FOR STRAIGHT-BLADE SHEARING

Squaring Shears. Trimming and cutting of sheet or plate to specific size are usually done in a squaring shear (see Fig. 2). Squaring shears (also called resquaring or guillotine shears) are available in a wide range of sizes and designs. Some types permit slitting by moving the work metal a predetermined amount in a direction parallel with the cutting edge of the blade after each stroke of the shear.

Fig. 1. Mechanics of straight-blade shearing

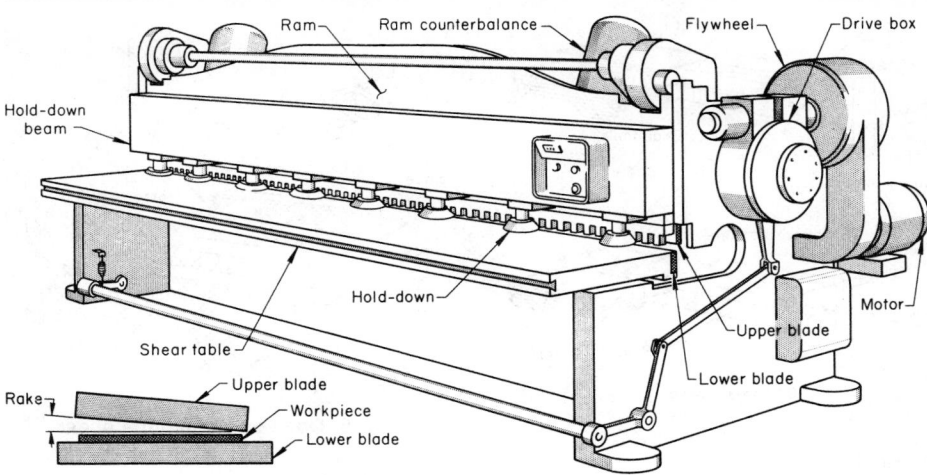

Fig. 2. Principal components of a squaring shear, and detail showing rake angle of blades

Table 1. Recommended blade materials for rotary shearing of flat metal

Metal to be sheared	Thickness to be sheared		
	3/16 in. or less	3/16 in. to 1/4 in.	1/4 in. or more
Carbon, alloy and stainless steels	D2	A2	S4; S5
High-silicon electrical steels	M2; D2	D2	...
Copper and aluminum alloys	A2; D2	A2; D2	A2
Titanium and titanium alloys	D2; A2	...	...

The sheet or plate is held rigid by hold-down devices while the upper blade moves down past the lower blade. Most sheet or plate is sheared by setting the upper blade at an angle, as shown in the lower left corner of Fig. 2. The scimitar guillotine shear has a curved cutting edge and cuts with a rolling, back-and-forth motion.

STRAIGHT SHEAR BLADES

Most shear blades are made in one piece from tool steel; some are made of carbon or alloy steel with hard-faced cutting edges or with inserts of tool steel or carbide. The composition, thickness and quantity of metal being sheared are the most important factors in the selection of blade material.

ROTARY SHEARING

Rotary shearing, or circle shearing (not to be confused with slitting), is a process for cutting sheet and plate in a straight line or in contours

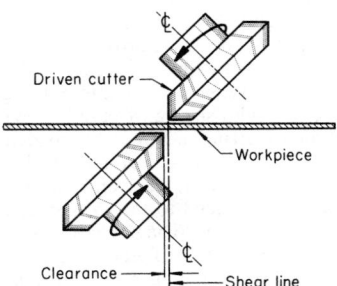

Fig. 3. Conventional arrangement of cutters in a rotary shearing machine, for production of a perpendicular edge

by means of two revolving tapered circular cutters. Recommended cutter materials are given in Table 1.

For conventional cutting, to produce a perpendicular edge, the cutters approach each other and line up vertically at one point (Fig. 3). The point of cutting is also a pivot point for the workpiece; because of the round shape of the blades, they offer no obstruction to movement of the workpiece to the right or left. This feature permits cutting of circles and irregular shapes that have small radii, and cutting in straight lines.

Slitting and Shearing of Coiled Sheet and Strip

COILED sheet or strip is cut to size for further processing, by slitting, for dividing it into narrower coils, and by shearing, for cutting it into flat pieces of specified length.

SLITTING

A slitting line for cutting wide coiled stock into narrower widths consists essentially of an un-

coiler for holding the coil, one or more slitters, and a re-coiler for simultaneous coiling of all slit strips (Fig. 4). Other equipment can be added to the line for scrap disposal, coil handling and packaging, leveling and edge conditioning.

Slitting is done by circular blades mounted on two arbors of the slitter.

Slitting lines are broadly classified as driven or pull-through types. In the driven type, uncoiler, slitter and re-coiler are each driven by a separate motor. The motors are synchronized to maintain constant speed of the metal as it travels through the slitting line.

In a pull-through slitting line, the drive motors on the slitter and uncoiler are used only to feed the coil stock far enough through the slitter to permit attaching the slit strips to the re-coiler. After the strips are attached, the motors for the uncoiler and slitter are disconnected and the driven re-coiler pulls the strip from the uncoiler through the slitter.

The choice between driven and pull-through lines depends largely on work-metal strength and thickness, number of slits, and slitting speed. In general, for slitting metal less than 0.010 in. thick, a driven slitter is preferred, because thin-gage metal is likely to tear in a pull-through slitter. The Clustercoil slitting process (Fig. 5) leaves interconnecting tabs between the multiple strips of metal.

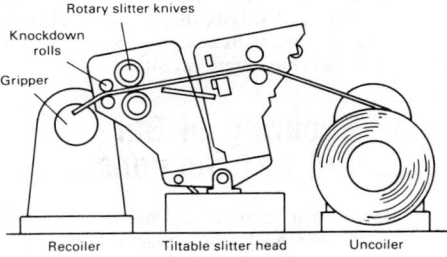

Fig. 5. The Clustercoil slitting process

SHEARING

Shearing lines (also called cut-up lines, or cut-to-length lines) are high-production setups for producing accurately cut-to-length sheets from coil stock. There are two basic types of shearing lines: the stationary-shear type and the flying-shear type. These machines uncoil the strip and cut it to the required lengths.

Stationary-shear lines (Fig. 6) consist of an uncoiler, rolls that remove the set from the uncoiled strip and feed it over a hump table, a stationary shear of the square-shear type, a gage table with a retractable stop, and a stacker that stacks the

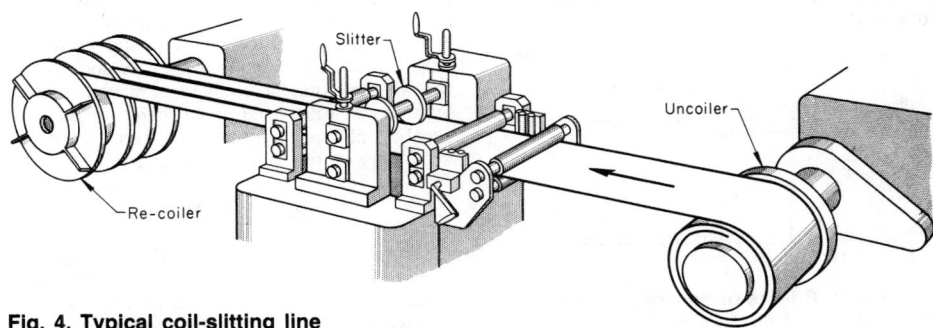

Fig. 4. Typical coil-slitting line

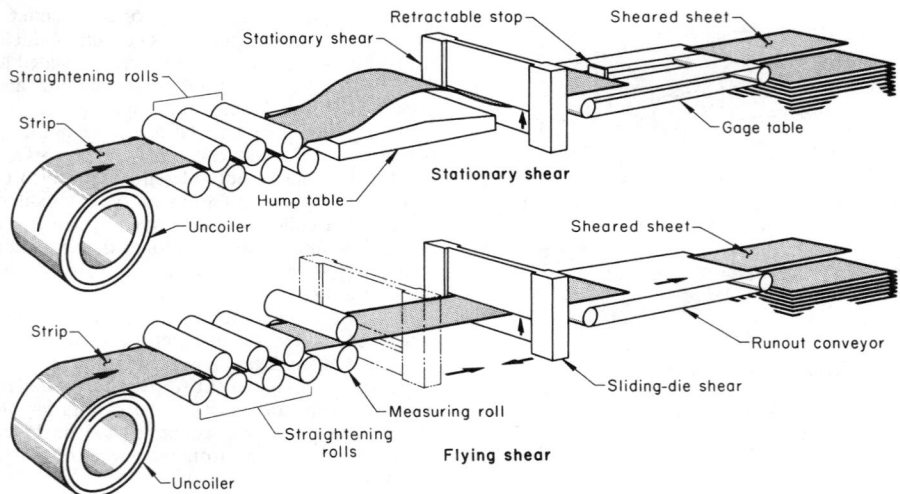

Fig. 6. Two types of shearing lines for cutting coiled strip into flat sheets

cut sheets as they are delivered from the gage table.

Flying-Shear Lines. One type of flying-shear cut-up line (Fig. 6) incorporates an uncoiler, straightening rolls, measuring rolls, a sliding-die shear, a runout conveyor, and a device for piling the sheared lengths. This type of flying-shear line can be operated at speeds that compare closely with those obtained with a hump-table stationary shear, for the same length of cut.

Shearing of Bars and Bar Sections

BARS and bar sections are sheared between lower and upper blades of a machine in which only the upper blade is movable. As the upper blade is forced down, the work metal is distorted and caused to fracture. Figure 7 shows the appearance of a sheared round bar. The burnished area, or depth of shear action by the blade, is usually one-fifth to one-fourth the diameter of the bar. In visual examination of a sheared edge, the burnished portion appears smooth, whereas the fractured portion is comparatively rough.

Applicability. In general, any metal that can be machined can be sheared, but power requirements increase as the strength of the work metal increases. Further, blade design is more critical and blade life decreases as the strength of the work metal increases.

Equipment is available for shearing round, hexagonal or octagonal bars up to 6 in. in diameter or thickness, rectangular bars and billets up to 3 by 12 in. in cross section, and angles up to 8 by 8 by $1\frac{1}{2}$ in.

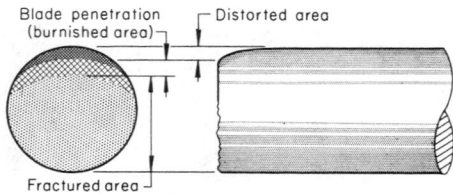

Fig. 7. Effects of shearing a round bar with a straight blade

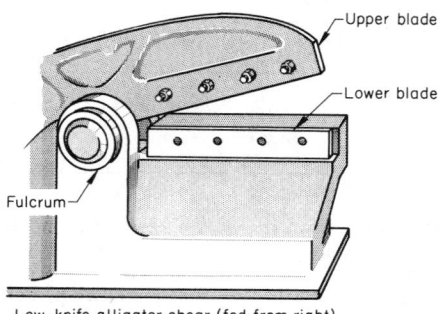

Low-knife alligator shear (fed from right)

Fig. 8. Alligator shear

MACHINES

Alligator shears are so named because the action resembles that of an alligator's jaw. In an alligator shear, the lower blade is stationary and the upper blade, held securely in an arm, moves in an arc around a fulcrum pin (Fig. 8). The shearing action is similar to that of a pair of scissors. A crankshaft transmits power to the shearing arm, and the leverage applied produces the force for shearing. Maximum shearing force is obtained closest to the fulcrum, and the mechanical advantage decreases as the distance between the point of shearing and the fulcrum increases.

Guillotine shears are designed for cutting bars and bar sections to desired lengths from mill stock, and are used extensively throughout the fabricating industry. Two general types are available — open-end (Fig. 9) and closed-end. The shear illustrated in Fig. 9 is called "open-end" because it has a C-frame construction with one end open and unsupported. Open-end shears are either single-end, for one operator, or double-end, for two operators. On double-end machines, both ends can be right-hand or left-hand, or one end can be right-hand and the other end left-hand, depending on the type of shearing to be done.

BLADE DESIGN

Conforming Blades. One method of minimizing distortion in sheared bars employs two hardened blades mounted face-to-face, with identical holes through each blade. The holes should conform to

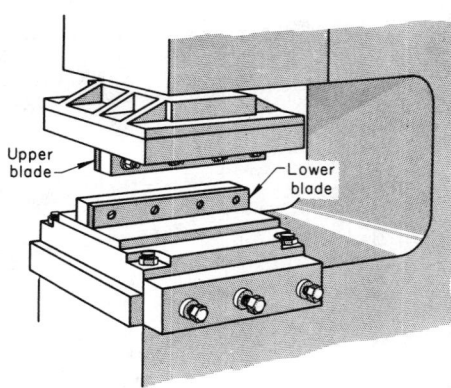

Fig. 9. Open-end guillotine shear

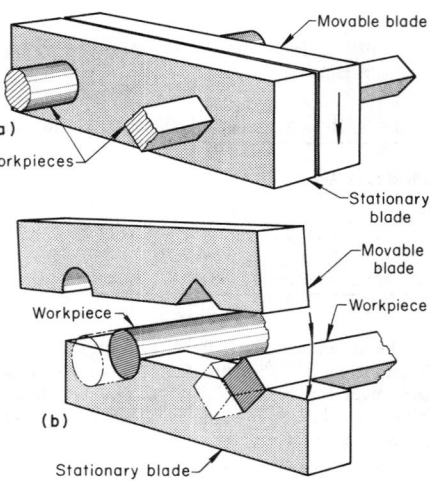

Fig. 10. Two types of blades for shearing of bars

the shape of the work metal and should be large enough to allow easy passage through the blades (Fig. 10a). One blade is movable vertically and one is stationary. Relatively little movement of the machine is required when blades of this type are used.

Shearing of round and square bars is more frequently done with the open-type blades illustrated in Fig. 10(b).

Oxyfuel Gas Cutting

OXYFUEL GAS CUTTING (OFC) is accomplished through a chemical reaction in which preheated metal is cut, or removed, by rapid oxidation in a stream of pure oxygen. A fuel-gas/oxygen flame heats the workpiece to ignition temperature, and a stream of pure oxygen feeds the cutting (oxidizing) action. With oxidation-resistant materials, such as stainless steels, either a metal powder (iron or iron-aluminum) or chemical flux is added to the oxygen stream to help the cutting action. The process is sometimes referred to as burning or flame cutting.

The simplest oxyfuel gas cutting equipment consists of two cylinders (one for oxygen and one for the fuel gas), gas flow regulators and gages, gas supply hoses, and a cutting torch with a set of exchangeable cutting tips. Such manually operated equipment is portable and inexpensive.

PRINCIPLES OF OPERATION

Oxyfuel gas cutting begins by heating a small area on the surface of the metal to the ignition temperature of 1400 to 1600 °F with an oxyfuel gas flame. Upon reaching this temperature, the surface of the metal will appear bright red. A cutting oxygen stream is then directed at the preheated spot, causing rapid oxidation of the heated metal and generating large amounts of heat. This heat supports continued oxidation of the metal as the cut progresses. Combusted gas and the pressurized oxygen jet flush the molten oxide away, exposing fresh surfaces for cutting. The metal in the path of the oxygen jet burns. The cut progresses, making a narrow slot, or kerf, through the metal.

During cutting, oxygen and fuel gas flow through separate lines to the cutting torch at pressures controlled by pressure regulators, adjusted by the operator. The cutting torch contains ducts, a mixing chamber, and valves to supply an oxyfuel gas mixture of the proper ratio for preheat and a pure oxygen stream for cutting to the torch tip. By adjusting the control valves on the torch handle or at the cutting machine controller, the operator sets the precise oxyfuel gas mixture desired. Depressing the cutting oxygen lever on the torch during manual operation initiates the cutting oxygen flow. For machine cutting, oxygen is normally controlled by the operator at a remote station or by numerical control. Cutting tips have a single cutting oxygen orifice centered within a ring of smaller oxyfuel gas exit ports. The operator changes the cutting capacity of the torch by changing the cutting tip size and by resetting pressure regulators and control valves. Because different fuel gases have different combustion and flow characteristics, the construction of cutting tips, and sometimes of mixing chambers, varies according to the type of gas.

Oxyfuel gas flames initiate the oxidation action and sustain the reaction by continuously heating the metal at the line of the cut. The flame also removes scale and dirt that may impede or distort the cut.

The rate of heat transfer in the workpiece influences the heat balance for cutting. As the thickness of the metal to be cut increases, more heat is needed to keep the metal at its ignition temperature. Increasing the preheat gas flow and reducing the cutting speed maintains the necessary heat balance.

Oxygen flow also must increase as the thickness of the metal to be cut increases. The jet of cutting oxygen must have sufficient volume and velocity to penetrate the depth of the cut and still maintain its shape and effective oxygen content.

MATERIALS

Oxyfuel gas cutting processes are used primarily for severing carbon and low-alloy steels. Other iron-based alloys and some nonferrous metals can be oxyfuel gas cut, although process modification may be required, and cut quality may not be as high as is obtained in cutting the more widely used grades of steel. High-alloy steels, stainless steels, cast iron, and nickel alloys do not readily oxidize and therefore do not provide enough heat for a continuous reaction. As the carbon and alloy contents of the steel to be cut increase, preheating or postheating, or both, often are necessary to overcome the effect of the heat cycle, particularly the quench effect of cooling.

Some of the high-alloy steels, such as stainless steel, and cast iron can be cut successfully by injecting metal powder or a chemical flux into the oxygen jet. Metal powders, such as iron, or iron and aluminum mixtures, burn with a high evolution of heat. Addition of these powders to the combustion stream aids in the cutting of stainless steel and cast iron, and even allows cutting of copper and brass. Fluxes combine with refractory oxides to form a slag of lower-melting-temperature compounds. This slag is driven out, enabling oxidation of the metal to proceed. This is mainly used in cutting of stainless steels.

Thickness Limits. Gas can cut steel less than $1/8$ in. thick to over 60 in. thick, though some sacrifice in quality occurs near both ends of this range. With very thin material, operators may have some difficulty in keeping heat input low to avoid melting the kerf edges and to minimize distortion. Steel under $1/4$ in. thick often is stacked for cutting of several parts in a single torch pass.

Applications. Large-scale applications of oxyfuel cutting are found in shipbuilding, structural fabrication, manufacture of earthmoving equipment, machinery construction, and in the fabrication of pressure vessels and storage tanks. Many machine structures, originally made from forgings and castings, can be made at less cost by redesigning them for OFC and welding with the advantages of quick delivery of plate material from steel suppliers, low cost of oxyfuel gas cutting equipment, and flexibility of design.

Oxyfuel gas cutting is used for beveling and weld preparation of plates. It is often used in cutting of scrap metal, removal of gates and risers from castings, and severing of bolts and rivets.

CHEMISTRY OF CUTTING

Carbon or low-alloy steel heated to ignition temperature (1600 °F) and supplied with pure (99.5%) oxygen oxidizes rapidly. The principal oxidation product is iron oxide (Fe_3O_4). The following can be used as a guide for oxygen requirements and heat emission:

$$3Fe + 2O_2 \rightarrow Fe_3O_4 + 267\,000 \text{ cal}$$

By weight of solid material, the reaction can be expressed as:

$$1 \text{ lb Fe} + 4.6 \text{ ft}^3 O_2 \rightarrow$$
$$1.38 \text{ lb } Fe_3O_4 + 2870 \text{ Btu}$$

For the complete oxidation of iron (with oxygen volumes at 68 °F and at 1 atm):

- 1 lb of iron requires 0.38 lb or 4.6 ft³ of oxygen.
- 1 in.³ of iron requires 0.109 lb or 1.31 ft³ of oxygen.
- 1 lb of oxygen requires 2.62 lb or 9.19 in.³ of iron.
- 1 ft³ of oxygen requires 0.22 lb or 0.76 in.³ of iron.

In OFC, all of the iron removed from the cut is not oxidized. Some of it (up to 30 or 40%) blows out of the cut as molten iron, along with the oxide, and becomes part of the slag.

Oxygen consumption and flow rates vary depending on whether economy, speed, or accuracy of cut is desired. For average straight-line cutting of low-carbon steel, consumption of cutting oxygen per pound of metal removed varies with thickness of the metal and is lowest at a thickness of 4 to 5 in.

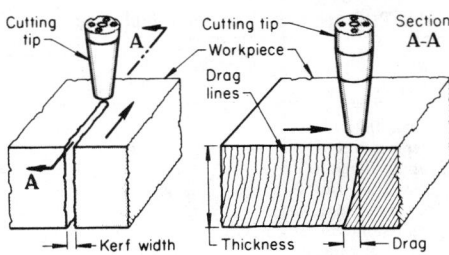

Fig. 11. Cross section of work metal during OFC showing drag on cutting face

By assuming that for every unit mass of iron oxidized an equal mass of iron melts, one can calculate the amount of heat generated by the cutting reaction—heat emitted is 2870 Btu/lb of iron oxidized. Melting of 1 lb of iron takes 680 Btu, based on a melting point of 2800 °F, 0.2 Btu/lb · °F as the specific heat, and 117 Btu/lb as the heat of fusion. Only a small amount of the heat melts the iron; most of it, about 2100 Btu/lb, goes into the reaction. Some of this superheats the molten metal, some soaks into the workpiece, and some leaves by radiation and convection. Most of it leaves with the slag and hot exhaust gases.

As cutting oxygen flows down through the cut, the quantity available for reaction decreases. If the flow of oxygen is large and well collimated, the rate of cutting through the depth of the cut is approximately constant. The cutting face remains vertical if the oxygen is in excess and if cutting speed is not excessive.

If oxygen flow is insufficient, or cutting speed too high, the lower portions of the cut react more slowly, and the cutting face curves behind the torch. The horizontal distance between point of entry and exit is called drag (see Fig. 11). Drag is often expressed as a percentage of the metal thickness.

FUEL GASES

Combustion of fuel gas produces the preheat flame, which initiates cutting action and helps sustain the operation. These flames heat the surface of the work to ignition temperature to (1) initiate cutting, (2) descale and clean the work surface, (3) supply heat to the work and the cutting oxygen stream to maintain the heat needed for continuous cutting, and (4) shield the cutting oxygen stream from surrounding air.

The common fuel gases used for preheating are listed in Table 2, along with their properties. The most important properties of the gases are the temperature and heat content (value) of the flame. The flame usually consists of an inner (primary) cone and an outer (secondary) cone. A gas whose inner flame has a high temperature and high heat release provides the most concentrated heat. These gases are superior for fast starts in flame cutting of high-alloy steels that are difficult to cut.

When low heat release is accompanied by high flame temperature, even though it is above the melting point of steel, heat is diffused and gives slow starts in flame cutting. Gases, such as natural gas, that release most of their heat in the outer flame are well suited to heating and heavy cutting.

Acetylene is the most widely used cutting gas. It burns hotter than any of the other common fuel

Table 2. Properties of common fuel gases

Properties	Fuel gases				
	Acetylene (C$_2$H$_2$)	MPS(a) (C$_3$H$_4$)	Propane (C$_3$H$_8$)	Propylene (C$_3$H$_6$)	Natural gas(b)
Neutral flame temperature, °F	5600	5200	4580	5200	4600
Primary flame heat emission, Btu/ft^3	507	517	255	433	11
Secondary flame heat emission, Btu/ft^3	963	1889	2243	1938	989
Total heat value(c), Btu/ft^3	1470	2406	2498	2371	1000
Combustion ratio (neutral flame), vol oxygen/vol fuel	2.5 to 1	4 to 1	5 to 1	4.5 to 1	2 to 1
Oxygen supplied through torch (neutral flame), ft^3 oxygen/lb fuel (60 °F)	18.9	22.1	37.2	31.0	44.9
Maximum allowable regulator pressure, psi	15	Cylinder	Cylinder	Cylinder	Line
Volume-to-weight ratio, ft^3/lb (60 °F)	14.6	8.85	8.66	8.25	23.6
Specific gravity(d)	0.906	1.48	1.52	1.48	0.62

(a) Stabilized methylacetylene propadiene. (b) Principally methane (CH$_4$). (c) Total heat value after vaporization. (d) At 60 °F; air = 1.

Table 3. Recommended conditions for manual OFC-A of carbon steel plate

Plate thickness, in.	Diameter of cutting orifice, in.	Oxygen pressure, psi	Cutting speed(a), in./min	Gas consumption, ft^3(b)			
				Per hour		Per linear foot	
				Oxygen	Acetylene	Oxygen	Acetylene
1/8	0.0380-0.0400	15-23	20-30	45-55	7-9	0.37-0.45	0.06-0.07
1/4	0.0380-0.0595	11-20	16-26	50-93	9-11	0.63-0.72	0.08-0.11
3/8	0.0380-0.0595	17-25	15-24	60-115	10-12	0.80-0.96	0.10-0.13
1/2	0.0465-0.0595	20-30	12-22	66-125	10-13	1.10-1.14	0.12-0.17
3/4	0.0465-0.0595	24-35	12-20	117-143	12-15	1.43-1.95	0.15-0.20
1	0.0465-0.0595	28-40	9-18	130-160	13-16	1.78-2.89	0.18-0.29
1 1/2	0.0595-0.0810	35-48	6-14	143-178	15-18	1.96-3.18	0.21-0.33
2	0.0670-0.0810	22-50	6-13	185-231	16-20	3.55-6.16	0.31-0.53
3	0.0670-0.0810	33-55	4-10	240-290	19-23	5.80-12.00	0.46-0.95
4	0.0810-0.0860	42-60	4-8	293-388	21-26	9.70-14.64	0.65-1.05
5	0.0810-0.0860	53-70	3.5-6.4	347-437	24-29	13.66-19.83	0.91-1.37
6	0.0980-0.0995	45-80	3.0-5.4	400-567	27-32	21.00-26.70	1.19-1.80
8	0.0995	60-77	2.6-4.2	505-615	31.5-38.5	29.30-38.84	1.83-2.42
10	0.0995	75-96	1.9-3.2	610-750	36.9-45.1	46.90-64.20	2.57-3.84
12	0.1200	69-86	1.4-2.6	720-880	42.3-51.7	67.70-103.00	3.98-6.05

Note: Values do not necessarily vary in exact proportion to plate thickness, because straight-line relations do not exist among pressure, speed, and orifice sizes.
(a) Lowest speeds and highest gas consumptions are for inexperienced operators, short cuts, dirty or nonuniform material. Highest speeds and lowest gas consumptions are for experienced operators, long cuts, clean and uniform material. (b) Pressure of acetylene for the preheating flames is more a function of torch design than of the thickness of the part being cut. For acetylene pressure data, see charts of manufacturers of apparatus.

gases, making it indispensable for certain jobs. Combustion occurs in stages. In the small inner cone at each preheat port at the tip of the torch, acetylene burns with feed-line oxygen:

$$2C_2H_2 + 2O_2 \rightarrow 4CO + 2H_2$$

This reaction gives off a blistering amount of heat; the tip of the inner cone is the hottest part. Burning continues in the outer envelope of the flame, in a cooler, blue-colored region:

$$4CO + 2H_2 + 3O_2 \rightarrow 4CO_2 + 2H_2O$$

Most oxygen for this reaction comes from the air surrounding the flame. Of the 2 1/2 parts of oxygen needed for acetylene to burn completely, about 1 1/2 comes from the air and 1 part from line oxygen. A neutral flame, recommended for manual cutting, consumes equal volumes of line oxygen and acetylene. The sharp distinction between the two flames helps in adjusting the oxygen-to-acetylene ratio for a reducing (carburizing), neutral, or oxidizing flame.

Recommended conditions for manual oxyacetylene cutting (OFC-A) of carbon steel plate of various thicknesses are shown in Table 3.

Stabilized methylacetylene propadiene (MPS), commonly called MAPP gas, combines the high flame temperature of acetylene with the high heat content of propane and propylene. It provides the most even heat distribution of any gas, and it competes with acetylene for almost every job.

Propane has a low flame temperature but higher heat content than natural gas. Propylene has a higher flame temperature than propane and is less expensive than acetylene. Natural gas is cheap, but it has a low flame temperature and low heat content and thus takes longer to preheat a plate. It is good for light- to heavy-gage materials. Selection of fuel gases depends on availability and cost.

EQUIPMENT

Manual gas cutting equipment consists of gas regulators, gas hoses, cutting torches, cutting tips, and multipurpose wrenches. Auxiliary equipment may include a hand truck, tip cleaners, torch ignitors, and protective goggles. Machine cutting equipment varies from simple railmounted "bug" carriages to large bridge-mounted torches that are

driven by computer-directed drives. Cutting machines, employing one or several cutting torches guided by solid template pantographs, optical line tracers, numerical controls, or computers, improve production rates and provide superior cut quality. Machine cutting is important for profile cutting—the cutting of regular and irregular shapes from flat stock.

Cutting torches, such as the one shown in Fig. 12, control the mixture and flow of preheat oxygen and fuel gas and the flow of cutting oxygen. The cutting torch discharges these gases through a cutting tip at the proper velocity and flow rate. Pressure of the gases at the torch inlets, as well as size and design of the cutting tip, limits these functions, which are operator controlled.

Oxygen inlet control valves and fuel gas inlet control valves permit operator adjustment of gas flow. Fuel gas flows through a duct and mixes with the preheat oxygen; the mixed gases then flow to the preheating flame orifices in the cutting tip. The oxygen flow is divided—a portion of the flow mixes with the fuel gas, and the remainder flows through the cutting oxygen orifice in the cutting tip. A lever-actuated valve on the manual torch starts the flow of cutting oxygen; machine cutting starts the oxygen from a panel control.

Fuel gases supplied at low pressure, such as natural gas tapped from a city line, require an injector-mixer (Fig. 12b) to increase fuel gas flow above normal operating pressures. Optimum torch performance relies on proper matching of the mixer to the available fuel gas pressure.

Cutting tips are precision-machined nozzles, produced in a range of sizes and types. Figure 13(a) shows a single-piece acetylene cutting tip. A two-piece tip used for natural gas (methane) or LPG is shown in Fig. 13(b). A tip nut holds the tip in the torch. For a given type of cutting tip, the diameters of the central hole, the cutting oxygen orifice, and the preheat ports increase with the thickness of the metal to be cut. Cutting tip selection should match the fuel gas; hole diameters must be balanced to ensure an adequate preheat-to-cutting oxygen ratio. Preheat gas flows through ports that surround the cutting oxygen orifice. Smoothness of bore and accuracy of size and shape of the oxygen orifice are important to efficiency. Worn, dirty bores reduce cut quality by causing turbulence in the cutting oxygen stream.

The size of the cutting tip orifice determines the rate of flow and velocity of the preheat gases and cutting oxygen.

Electric-Arc Cutting

ARC CUTTING melts metal by heat generated from an electric arc between an electrode and the base metal. Because extremely high temperatures are developed, arc cutting can be used to cut almost any metal and is generally faster than oxyfuel gas cutting. Modifications of the process include the use of compressed gases to cause rapid oxidation (or to prevent oxidation) of the workpiece, thus incorporating aspects of the gas cutting process. Arc cutting methods include air carbon arc, gas metal arc, gas tungsten arc, shielded metal arc, plasma arc, and oxygen arc cutting. This article covers the methods of industrial im-

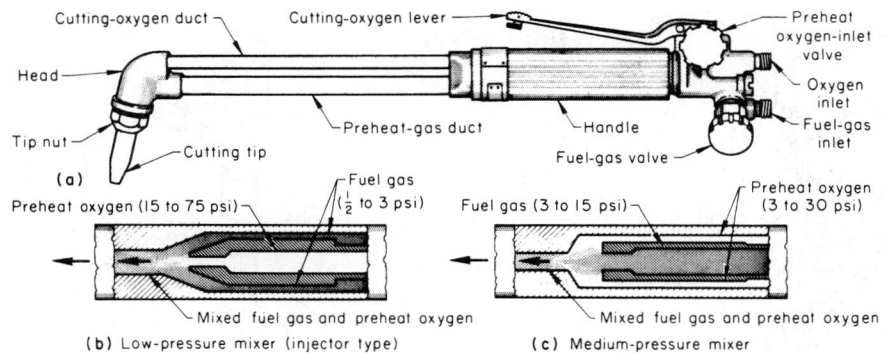

After the workpiece is sufficiently preheated, the operator depresses the lever to start the flow of cutting oxygen. Valves control the flow of oxygen and fuel gas to achieve required flow and mixture at the cutting tip. (b) and (c) Sections through preheat gas duct showing two types of mixers commonly used with the torch shown.

Fig. 12. (a) Typical manual cutting torch in which preheat gases are mixed before entering torch head.

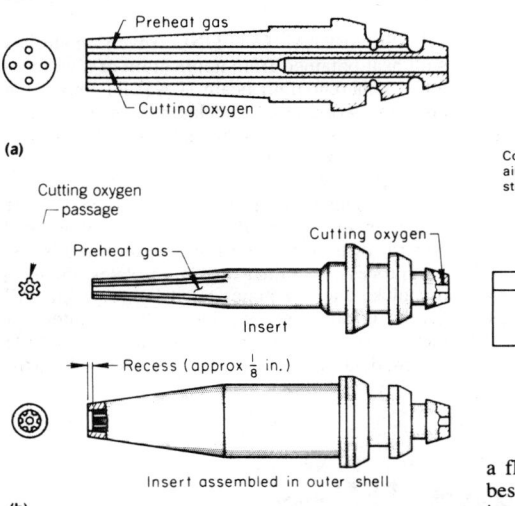

(a) Single-piece acetylene cutting tip. (b) Two-piece tip for natural gas or LPG. Fuel gas and preheat oxygen mix in tip. Recessed bore helps promote laminar flow of gas.

Fig. 13. Types of cutting tips

portance, which are air carbon-arc, plasma-arc, and oxygen-arc cutting.

AIR CARBON-ARC CUTTING AND GOUGING

Air carbon-arc cutting (AAC) and gouging severs or removes metal by melting it with the heat of an arc struck between an electrode and the base metal. The electrode is held in a holder similar to a shielded metal-arc welding (SMAW) electrode holder. The electrode is held at an angle sloping back from the direction of travel and an arc is struck by touching the electrode to the workpiece. A stream of compressed air flows from orifices in the holder and strikes the metal behind the arc, blowing away the molten metal (see Figure 14). The holder also contains an air flow control valve, an air hose connected to a source of compressed air, and a cable connected to a welding power supply.

Electrodes are made from mixtures of carbon and graphite, and most are copper-coated to improve conductivity.

Power Supply. Constant-voltage direct current with

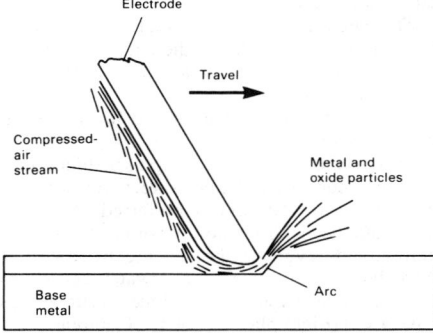

From "Welding Handbook," Vol. 2, AWS

Fig. 14. Cutting action in the air carbon-arc process

a flat to slightly rising voltage characteristic is best for most AAC applications. Direct current is preferred; copper alloys, however, cut better with alternating current. Recommended current levels for various sizes of electrodes are given in Table 4.

Air Supply. Compressed air from a shop line or a compressor at 80 to 100 psi should be used; pressure as low as 40 psi is suitable for light work. Deep grooves in thick metal require pressures up to 125 psi.

Applications. Air carbon-arc cutting is commonly used for: (1) weld-joint preparation (for example, gouging a U-groove along a joint between two plates as shown in Figure 15); (2) removal of defective welds; and (3) removal of gates, risers and defects from castings.

Since metal removal is by melting and not oxidation, AAC can cut most metals such as all steels; cast irons; and aluminum, magnesium, copper, and nickel alloys. It cannot cut metals that react rapidly with air.

Table 4. Recommended current levels for graphite electrodes of various sizes used in carbon-arc cutting

Electrode diameter, in.	Current, amp	Electrode diameter, in.	Current, amp
5/32	80 to 150	3/8	300 to 550
3/16	110 to 200	1/2	400 to 800
1/4	150 to 350	5/8	600 to 1000
5/16	200 to 450	3/4	800 to 1600

The ACC process can cut plate up to 1/2 in. thick or cut a groove up to 1 in. deep in one pass. Greater thicknesses require more passes. Rough cutting is done manually, but accurate work requires electrode holders mounted on motor-driven carriages. These torches have feedbacks to maintain constant arc length.

PLASMA-ARC CUTTING

Plasma-arc cutting (PAC) uses a high-velocity jet of high-temperature ionized gas to sever the metal. The plasma arc is between an electrode contained within the torch, and the base metal. The arc is concentrated by a nozzle onto a small area of the workpiece, where it melts the metal. The plasma jet then forces away the melted material, forming a narrow slot, or kerf.

Principles of Operation. At the temperature in the arc, the gas partially ionizes and exists as a plasma—a mixture of free electrons, positively charged ions, and neutral atoms. As this plasma passes through the arc, it is heated, expands, and accelerates toward the workpiece.

The PAC torch, as shown in Fig. 16, is similar to a gas tungsten-arc welding (GTAW) torch, but the tungsten electrode is recessed into a nozzle with a small opening. A high-frequency pulsing potential initiates a pilot arc between the tungsten electrode (cathode, negative) and the copper nozzle (anode, positive), both of which are water cooled. This pilot or nontransferred arc initiates an external transferred arc between the torch electrode and the workpiece, which is connected as the anode (positive). The pilot arc shuts off, and the external arc supplies the energy to sustain cutting.

The gases used are usually argon, nitrogen, hydrogen, oxygen and mixtures of these. Most steels are cut with nitrogen, a nitrogen-hydrogen mixture, or a nitrogen-oxygen mixture. Stainless steels and nonferrous metals are usually cut with inert-gas mixtures. Some torches provide a secondary gas shield or water spray surrounding the arc to shield it from the atmosphere.

Application. Plasma-arc cutting is frequently used for beveling, weld preparation, and shape cutting.

Because PAC generates very high temperatures (50 000 °F compared to 5500 °F for oxyfuel), it can be used on almost any material that conducts electricity, including those that are resistant to OFC. It can cut stainless steel up to 6 in. thick, aluminum alloys up to about 8 in. thick, and carbon steel up to 3 in. thick. It also works well on copper, magnesium and nickel, and their alloys.

Almost all PAC is done automatically. Computer numerically controlled equipment is being

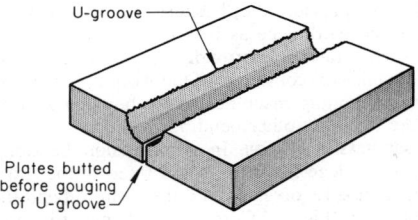

Fig. 15. Butted plates in which U-groove was produced by air carbon-arc gouging, in preparation for welding

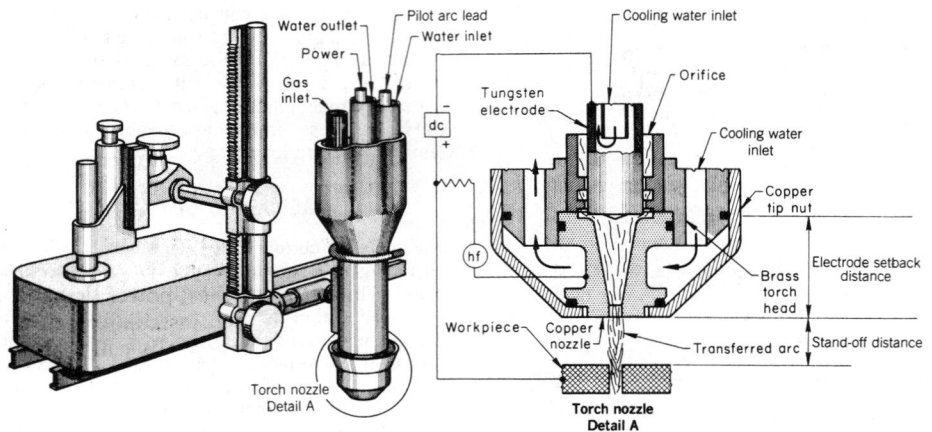

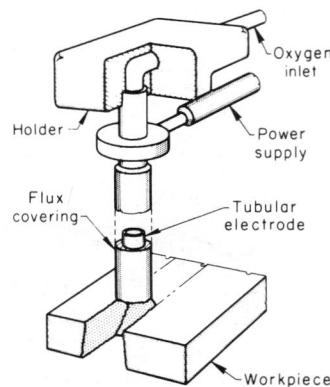

Fig. 17. Components of an oxygen-arc cutting electrode

Fig. 16. Essential components of a plasma-arc cutting torch

used more frequently. This equipment automatically controls the arc, gases and travel, and greatly increases the cutting speed and capabilities. The past decade has seen a great increase in the use of PAC, because of its high cutting speed (100 in./min for 1-in.-thick low-carbon steel). Most automatic cutting is done over a water-filled table. The water, which just touches the bottom of the plate, traps the fumes and slag.

OXYGEN-ARC CUTTING

Oxygen-arc cutting is very similar to air carbon-arc cutting. It uses a consumable flux-covered tubular electrode. The arc heats the material to kindling temperature and a stream of oxygen flows down the bore of the electrode and burns the material away.

The electrode covering insulates the electrode from arcing to the sides of the cut. The covering also acts as a flux to break up oxides on the workpiece surface.

The process finds greatest use in underwater cutting, mainly of steel.

Equipment. Oxygen-arc cutting uses direct or alternating current, although direct current electrode negative (DCEN) is preferred. The electrode and the electrode holder convey the electric current and oxygen to the arc. Electrode holders must be fully insulated; underwater cutting requires a flashback arrester, and the electrode must have a watertight plastic coating. Components of an oxygen-arc cutting electrode are shown in Fig. 17.

Laser-Beam Cutting

LASER-BEAM CUTTING severs materials with heat obtained from application of a concentrated coherent light beam (LASER) impinging upon the workpiece. The laser beam melts and vaporizes a small area of material. Most times, a high-velocity gas jet is used to blow the molten and vaporized material from the kerf.

Laser-beam cutting can be used to cut intricate shapes in metal, wood, and plastics. It is used also for drilling cooling holes in turbine blades. The process is limited to cutting plates about $1/2$ in. thick, and the plates must be very flat. All cutting is done by machine with computer numerical controls. Laser-beam cutting has little commericial use now, but will probably be used more in the future.

Forming of Bars, Tube and Wire

Bending of Bars and Bar Sections

BARS are bent by four basic methods: draw bending, compression bending, roll bending, and stretch bending.

Draw Bending. In draw bending, the workpiece is clamped to a rotating form and is drawn by the form against a pressure die, as shown in Fig. 1. The pressure die can be either fixed or movable along its longitudinal axis. A fixed pressure die must be able to withstand abrasion caused by the sliding of the work metal over its surface. A movable pressure die, because it moves forward with the workpiece as it is bent, is less subject to such abrasion. It provides better guidance and more uniform constraint of the work material. On power bending machines, draw bending is used more than any other bending method.

Compression Bending. In compression bending, the workpiece is clamped to a fixed form, and a wiper shoe revolves around the form to bend the workpiece (Fig. 1). Compression bending is most useful in bending rolled and extruded shapes. A bend can be made close to another bend in the workpiece without the need for the compound dies required in draw bending. Although compression

bending does not control the flow of metal as well as draw bending, compression bending is widely used in bending presses and in rotary bending machines.

Roll bending uses three or more parallel rolls. In one arrangement using three rolls, the axes of the two bottom rolls are fixed in a horizontal plane.

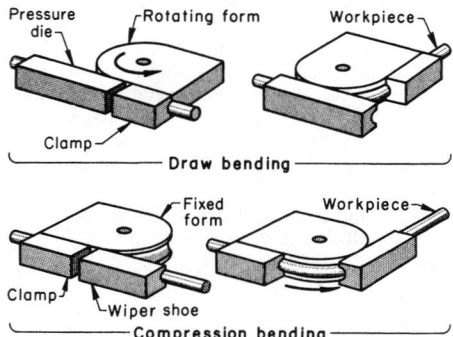

Fig. 1. Essential components and mechanics of draw bending and compression bending of bars and bar sections

The top roll (bending roll) is lowered toward the plane of the bottom rolls to make the bend, as shown in Fig. 2. The bottom rolls are power driven; the top roll is an idler and is moved up or down by a hydraulic cylinder.

Stretch bending is used for bending large irregular curves. The workpiece is gripped at the ends, stretched, and bent as it is stretched around a form. Usually, less springback occurs when the work is bent while it is stretched. The gripped ends are customarily trimmed off. This method can do in one operation what would otherwise take several operations, and sometimes does it better. The result is a possible saving in time and labor, even though stretch bending is a slow process. The tools, form blocks, or dies for stretch bending are simpler in design and less costly than conventional press tooling.

BENDING MACHINES

The machines used for bending of bars include the following: devices and fixtures for manual bending; press brakes; conventional mechanical and hydraulic presses; horizontal bending machines; rotary benders; and bending presses. Shapers have also been used to perform specific bending operations.

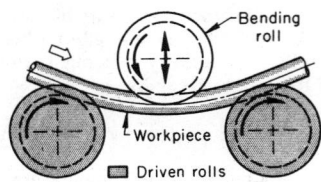

Fig. 2. Operating essentials in one method of three-roll bending

TOOLS

Tools for draw and compression bending are shown in Fig. 1. The form used in both processes is shaped to the contour of the bend. Usually, it is grooved to fit the work. Often, the form is part of a right cylinder whose straight portion (frequently, an insert) provides the surface against which the work is clamped. Hydraulic or mechanical pressure holds the clamp against the workpiece. Annular grooves or roughened surfaces grip the bar or bar section.

LUBRICATION

Successful bending depends to a large extent on the type of lubricant used. No one lubricant works equally well on all materials. Selection of a lubricant varies among different shops. Typical lubricants for bending specific metals are listed in Table 1.

Bending and Forming of Tubing

THE PRINCIPLES for bending tubing are much the same as for bending bars (see the preceding article, "Bending of Bars and Bar Sections"). Two important additional features in bending of tubes are that internal support is often needed, and sometimes support is needed on the inner side of a tube bend.

Wall thickness of the tubing affects the distribution of tensile and compressive stresses in bending; a thick-wall tube will usually bend more readily to a small radius than a thin-wall tube.

Table 2 gives the minimum practical inside radii for cold draw bending of round steel or copper tubing, with and without various supports against flattening and wrinkling.

SELECTION OF BENDING METHOD

The four most common methods of bending tubing are basically the same as those used in the bending of bars: compression bending, stretch bending, draw bending, and roll bending. The method selected for a particular application depends on the equipment available, number of parts required, the size and wall thickness of the tubing, the work metal, the bend radius, the number of bends in the workpiece, the accuracy required,

Table 2. Minimum practical inside radii for cold draw bending of annealed steel or copper round tubing to 180°(a)

	Minimum practical inside radius, in.		
	Grooved bending tools		Cylindrical bending block without mandrel; ratio, <30(b) (poor conditions)
Tubing OD, in.	With mandrel; ratio, <15(b) (best conditions)	With mandrel or filler; ratio, <50(b) (normal conditions)	
1/8	1/16	1/4	1/2
1/4	1/8	5/16	1
3/8	3/16	3/8	2
1/2	1/4	7/16	3
5/8	5/16	9/16	4
3/4	7/16	11/16	6
7/8	1/2	3/4	8
1	9/16	7/8	10
1 1/4	11/16	1	15
1 1/2	13/16	1 1/8	20
1 3/4	15/16	1 1/4	27
2	1 1/16	1 3/8	35
2 1/2	1 3/8	1 5/8	. . .
3	1 5/8	1 7/8	. . .
3 1/2	1 7/8	2 1/8	. . .
4	2 1/8	2 3/8	. . .

(a) Radii can be slightly less for a 90° bend, but must be slightly larger for 360°. (b) Ratio of outside diameter to wall thickness of tubing.

and the amount of flattening that can be tolerated.

TOOLS

Tools used for the bending of tubes are similar to those used for the bending of bars. One important difference is that tools for tubes need carefully shaped guide grooves to support the sidewalls and preserve the cross section during the bend.

For round tubes, the depth of the groove in the form block should be half the outside diameter of the tube, to provide sufficient sidewall support. The block becomes the template for holding the shape of the bend. Form blocks can be made of wood, plastic or hardboard; if they are to be used for an extensive production run, they can be made of tool steel and hardened.

BENDING TUBING WITH A MANDREL

Mandrels are sometimes used in bending to prevent collapse of the tubing or uncontrolled flattening in the bend. A mandrel cannot correct failure in bending after the failure has begun, nor can it remove wrinkles.

Five types of mandrels used in the bending of tubing are shown in Fig. 3. The plug mandrel and the formed mandrel are rigid, but the three other types shown are flexible or jointed to reach farther into the bend.

The largest diameter of the rigid portion of the mandrel should reach a short distance into the bend, the distance that it extends (past the tangent straight portion) depending on the kind of mandrel and the size of tube, and usually being

Fig. 3. Five types of mandrels used in bending of tubing. Broken vertical lines are points at which bends should be tangent to mandrel centerlines

established by trial. If the mandrel extends too far, it can cause a bulge in the bend. Conversely, if the mandrel does not extend far enough, wrinkles may form or the outer tube surface may flatten in the bend area.

The need for a mandrel depends on the tube and bend ratios. The tube ratio is D/t, where D is outside diameter and t is wall thickness. The bend ratio is R/D, where R is the radius of bend measured to the centerline.

Table 3 can be used to determine whether or not a mandrel is needed for bending steel tubing.

Table 3. Minimum centerline radii for bending steel tubing without a mandrel

Tubing OD. in.	Minimum centerline radius, in., for tubing with wall thickness, in., of:					
	0.035	0.049	0.065	0.083	0.093	0.120
3/16	5/16	1/4	3/16	. . .	. . .	. . .
1/4	1/2	3/8	5/16	. . .	. . .	. . .
5/16	7/8	3/4	5/8	. . .	. . .	. . .
3/8	1 1/2	1 1/4	1 1/8	1	. . .	. . .
1/2	2 1/4	2	1 3/4	1 1/2	. . .	. . .
3/4	4	3	2 1/2	2	. . .	. . .
1	8	6	4	3	2	2
1 1/2	. . .	. . .	12	10	8	6
2	. . .	. . .	. . .	24	20	16
2 1/2	. . .	. . .	. . .	. . .	24	20
3	. . .	. . .	. . .	. . .	. . .	25

BENDING TUBING WITHOUT A MANDREL

It is cheaper to bend tubing without a mandrel. Trial bending is generally necessary to find what bends can be made. Tubing with thick walls is more likely to be bendable without a mandrel than thin-wall tubing. Bends with large radii are more likely to be formable without a mandrel than those with small radii. Slight bends are more feasible than acute bends. Wide tolerances on permissible flattening make a bend easier to form without a mandrel.

Springback is greater without a mandrel, but it can be compensated for by overbending, or lessened by increasing force on the pressure die.

Table 1. Typical lubricants for bending various metals

Work metal	Lubricant
Low-carbon steel	Water-soluble, vegetable-oil-base drawing oil
Stainless steel and other high-alloy iron-base alloys	Mineral-oil-base drawing oil
Aluminum alloys and copper alloys	Mineral oil
Brass (severe bends)	Soap solution
Hot bending of carbon, alloy and stainless steels	Molybdenum disulfide

MACHINES

Powered rotary benders are commonly used to bend tubing as large as 8 in. OD. Capabilities exist for bending tubing as large as 12 in. OD with a ¼-in. wall, and special power benders can bend 18-in. pipe.

Bending presses are hydraulic machines made especially for bending both bars and tubes, but most often for tubes. The ram of a bending press can be stopped at any point in the stroke. Wing dies and a cushioning device help to wrap the work around the ram die, as shown in Fig. 4. When the ram moves down, it causes the wing dies to pivot by a sort of camming action and wrap the workpiece around the ram die. The wing dies wipe the work to control the flow of metal; a compression bend is made on each side of the ram die, without wrinkles or distortion.

Roll benders for bending tubes are similar to those used for bending bars, as described in the preceding article, but tolerances are more critical on the rolls and spacing. The contour of the rolls must match that of the tube to minimize wrinkling or flattening. Tubes of sizes up to 8-in. OD by 0.240-in. wall can be bent into arcs, circles or helixes. An example of coiling a helix in a three-roll bender is shown in Fig. 5.

LUBRICATION FOR TUBE BENDING

Where a mandrel is used, both the mandrel and the interior of the tube are heavily coated with a thick lubricant. Pigmented lubricants are useful for adding body between the mandrel and the tube. Sometimes, thick lubricants are heated to 250 F and sprayed onto the inner surface of the tube. An oil hole in a mandrel can be used to lubricate the inside of a tube during bending.

——Tube Spinning——

TUBE SPINNING is a rotary-point method of extruding metal, much like cone spinning, except that the sine law does not apply. Because the half angle of a cylinder is zero, tube spinning follows a purely volumetric rule, depending on the practical limits of deformation that the metal can stand without intermediate annealing.

Applicability. Spinning is one method of reducing the wall thickness of tubular shapes and increasing their strength, particularly for aircraft and aerospace applications. Producing specific shapes from tubing is a major function of tube spinning. For instance, one or more flanges can be spun at selected areas on a tube, often at savings in the cost of labor and material when compared with other processes, such as machining. Tube spinning also has been used because ring forgings having the desired relation between wall thickness and length were not available.

All ductile work metals are suitable for tube spinning; the practical ranges of compositions and strengths are approximately the same as for power spinning of cones. Metals as hard as Rockwell C 35 have been successfully spun. Most tube spinning is accomplished without heating the workpieces.

Preform Requirements. Preform is the name commonly applied to a tube or a tubular shape before it is spun. A preform may be a straight, symmetrical tube, or it may have been changed in

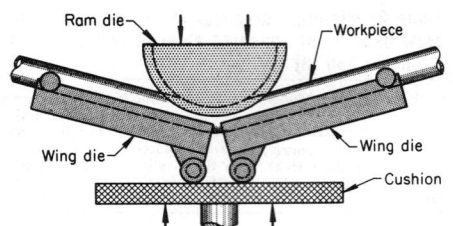

Fig. 4. Essential components and mechanics of a bending press

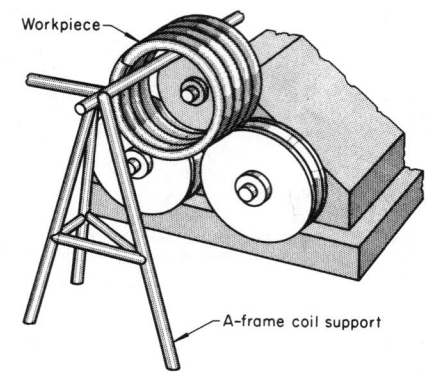

Fig. 5. Coiling a helix from round steel tubing by three-roll bending

shape by the addition of an internal flange for clamping.

Tubular shapes used for spinning include forged or centrifugally cast tubes (both of which are completely machined before spinning), welded tubing, seamless tubing, and extruded tubing.

METHODS OF TUBE SPINNING

Two distinctly different methods or techniques are used for tube spinning—namely, backward and forward. They are so termed because of the directional relations of metal flow and tool travel. In both methods, the workpiece is fixed in one position at one end, and the remaining length is free to slide along the mandrel.

Backward Spinning. In backward spinning, the workpiece is held against a fixture on the headstock, the roller advances toward the fixed end of the workpiece, and the work metal flows in the opposite direction, as illustrated in Fig. 6(a). Two advantages of backward spinning over forward spinning are: (*a*) the preform is simpler for backward spinning, because it slides over the mandrel and does not require an internal flange for clamping; and (*b*) the roller traverses a distance equal to only 50% of the length of the finished tube in making a 50% reduction of the tube wall, and a distance equal to only 25% of the final length for a 75% reduction.

Forward Spinning. In forward spinning, the roller moves away from the fixed end of the workpiece, and the work metal flows in the same direction as the roller, usually toward the headstock, as shown in Fig. 6(b). The main advantage in forward spinning compared with backward spinning is that it will overcome the problem of distortion. In forward spinning, closer control of length is possible because as metal is formed under the rollers it is not required to move again,

and any variation caused by variable wall thickness of the preform is continually pushed ahead of the rollers, eventually becoming trim metal beyond the finished length.

MACHINES FOR TUBE SPINNING

Machines used for tube spinning are usually the same as those used for power spinning of cones or other shapes. The few special features that are required for tube spinning are normally specified and can be supplied on all power spinning machines. A power spinning machine will have the same size capacity for tube spinning as for cone spinning.

Most tube spinning is done on machines with two opposed rollers as in Fig. 6(b). This practice minimizes the deflection caused by spinning with one roller, when the length-to-diameter ratio of the mandrel and workpiece is large. Even on machines employing opposed rollers, when the length-to-diameter ratio is excessively large, deflection of the mandrel is often a problem because the mandrel and workpiece are pushed off center. To counteract this problem, machines have been built with more than two rollers.

TOOLS FOR TUBE SPINNING

Tools required for tube spinning are: a mandrel, rollers (two are usually required), a puller ring (for removing the workpiece from the mandrel), a drive ring (which can also be used as a puller ring), tracer styluses (two are required when a tracer is used), and tracer bars (two are required).

Mandrels. Figure 7 shows a typical mandrel for tube spinning, and gives usual ranges of dimen-

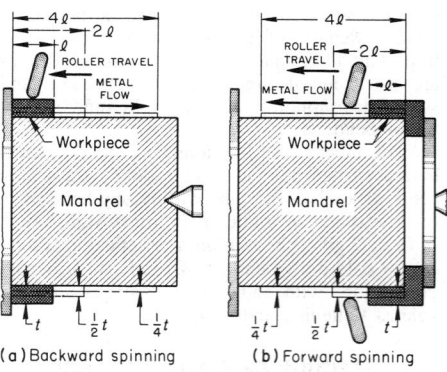

Fig. 6. Metal flow and roller travel in backward and forward spinning of tube

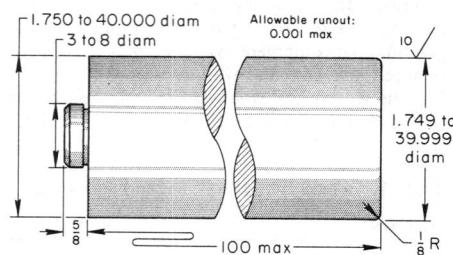

Fig. 7. Typical mandrel for tube spinning, showing usual ranges of dimensions

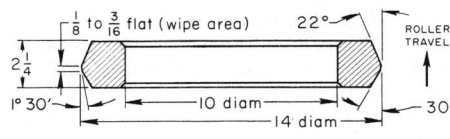

sions. Many mandrels are made solid, although as size increases and weight becomes excessive, the usual practice is to hollow them out, either by coring, if made from castings, or by boring if made from forgings or bars.

Rollers used for tube spinning are subjected to rigorous service. A typical roller used for tube spinning is illustrated in Fig. 8. The lead angle of the roller shown in Fig. 8 is given as 22°, but this angle is often larger. A surface finish of 10 micro-in. or better is preferred. Most rollers used for tube spinning are made from D2 or D4 tool steel, hardened to Rockwell C 60 or slightly higher. Rollers made in accordance with this practice have been known to last for 4000 to 5000 hr when spinning hot rolled tubes of 1020 to 1025 steel. The higher-vanadium types of high speed steels, such as M4 hardened to approximately Rockwell C 62, have also proved to be acceptable roller materials for production spinning.

Staggered rollers that have relatively large radii have been successfully used for forward spinning of workpieces such as missile cases. The two rollers shown in Fig. 9 are staggered radially so that each takes a portion of the total "bite." When this practice is used, the lead roller takes approximately 30% of the total "bite."

Straightening of Bars, Shapes and Long Parts

BARS, bar sections, structural shapes and long parts are straightened by bending, twisting or stretching.

Deviation from straightness in round bars may be expressed either as camber (deviation from a straight line) or as total indicator reading (TIR), per foot or for some other convenient length. Total indicator reading, which is twice the camber, is measured by rotating a round bar on its axis on rollers or centers and recording the needle travel on a dial gage placed in contact with the bar surface, generally midway between the supports. The indicator reading divided by the distance in feet between the supports gives the straightness in TIR per foot. Alternatively, the deviation is expressed in terms of the distance between the supports. The effect that changing the distance between supports has on the reading is illustrated in Fig. 10.

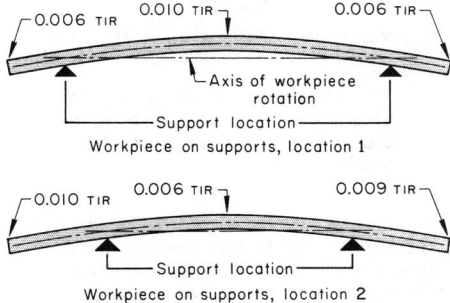

Fig. 10. Effect of distance between supports on straightness readings for round bars

MANUAL STRAIGHTENING

The original method of hand straightening is still used where capital investment is to be minimized, when accuracy and precision are required or when the shape of the bar or part makes machine straightening impractical.

The tools used in manual straightening include hammers and mallets, anvils, surface tables, vises, levers, grooved blocks, grooved rolls, twisting devices, various fixtures, and heating torches. The use of a grooved block (Fig. 11a) illustrates the basic principle of manual straightening by bending.

STRAIGHTENING IN PRESSES

Round bars up to 2 in. in diameter and from 2 to 10 ft long are often straightened in an arbor press. Larger workpieces are similarly straightened in power presses, which may have power rolls and hoists to move the work.

The principle of press straightening is illustrated in Fig. 11(b). The bar to be straightened is supported at points A and B with the convex side of the bow or kink toward point C. Sufficient force is applied at C to cause the bar to become bowed in the opposite direction.

ROTARY AND ROLL STRAIGHTENING

Rotary and roll straightening are cold finishing mill processes by which bars and structural shapes are provided with straightness adequate for most applications. Several rotary and roll straightening methods are illustrated in Fig. 11(c) to (h). For bars and shapes on which close tolerances must be maintained, rotary and roll straightening may be followed by press straightening.

In one type of roll straightening, square, flat, hexagonal and other flat-sided bars are passed continuously between sets of parallel-axis rolls (see Fig. 12). Uniform bends are introduced in such a way that the bar is straight when it leaves the rolls. By varying the distance between roll centers and the amount of offset, the degree of

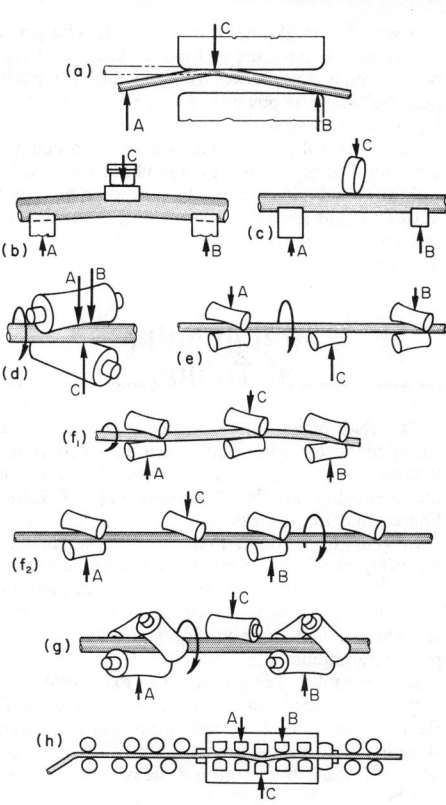

(a) Manual straightening with a grooved block. (b) Straightening in a press. (c) Simplest form of rotary straightening. (d) Two-roll straightening. (e) Five-roll straightening. (f_1 and f_2) Two arrangements of rolls for six-roll straightening. (g) Seven-roll straightening. (h) Wire straightening.

In all methods shown, the bar is supported at points A and B, and force at C on the convex side causes straightening.

Fig. 11. Principles of straightening by bending

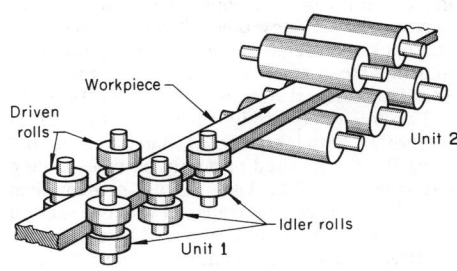

Fig. 12. Arrangement of vertical-shaft and horizontal-shaft rolls in a roll straightener for straightening a rectangular-section bar

bend can be adjusted according to the section size and yield strength of the metal being straightened.

STRETCH STRAIGHTENING

Many bars and shapes that are difficult to straighten by other methods can be straightened easily by stretching. However, this technique is usually confined to straightening of shapes that are uniform in cross section and length.

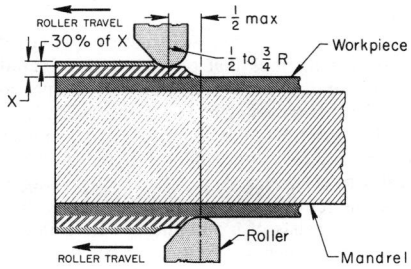

Fig. 9. Large-radius rollers staggered radially for forward spinning of tube, showing how each roller takes a portion of the total bite ("X" in illustration)

A stretch straightener has two heads with grips that clamp on the ends of the bar. One head can be adjusted to suit the workpiece length. The other head (tailstock) is powered for stretching and also for rotation to correct twist in the workpiece.

Most hot rolled bars can be straightened by stretching. Straightening by stretching also works well on rolled and extruded bars of aluminum and austenitic and ferritic stainless steels, but not on martensitic stainless steels unless they are first annealed.

Straightening of Tubing

TUBING of any cross-sectional shape can be straightened by using equipment and techniques that are basically the same as those discussed in the preceding article, "Straightening of Bars, Shapes and Long Parts."

In general, a round tube that has warped in annealing or other heat treatment is given a rough-straightening pass in a press or a roll straightener, followed by one or more passes in a rotary straightener and, if required, a finish pass in a press straightener.

Rough straightening eliminates excessive bow that would cause whipping in the inlet trough. Because the tube does not rotate during rough straightening, almost any amount of bow can be accepted; thus, both roll and press straightening can be used. If the tube is only slightly bowed, initial rough straightening can be omitted.

Straightening of tubing having shapes other than round is done in roll or press straighteners with much the same considerations governing the choice as those for similar solid shapes.

PRESS STRAIGHTENING

The straightening press can rough straighten tubing prior to roll straightening or rotary straightening, or can completely straighten tubing by removing end hooks that cannot be removed easily by any other method. Sometimes end hooks and large cambers must be removed to permit the tubing to enter the rotary straightener and to prevent dangerous whipping of the work.

Two-roll rotary straighteners (see Fig. 11 in the article "Straightening of Bars, Shapes and Long Parts") are used primarily on tubes having a diameter-to-wall thickness ratio of no more than 15 to 1. The machine is equipped with two skewed rolls, between which two guide shoes are mounted. One roll has a concave contour; the other is straight or convex. The rolls may be arranged in a horizontal or a vertical plane.

The tube is held between the guide shoes while the straight or convex roll bends the tube between the ends of the concave roll. The maximum deflection depends on the depth and the skew angle of the concave roll.

TWO-ROLL ROTARY STRAIGHTENING

The principle employed in rotary straightening of round tubes is basically the same as for solid round bars. The driven rolls, set at a predetermined angle, rotate the tube while conveying it in a lineal direction. The crest of the bow is stressed to, or beyond, the elastic limit once during each revolution, and the maximum stress point is repeated spirally along the length of the tube.

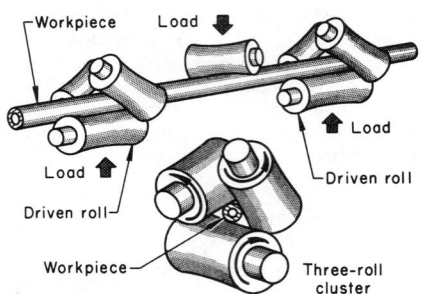

Fig. 13. Arrangement and principles of operation of three-roll clusters in a seven-roll rotary straightener

The distance between each stress point depends on the lineal travel for each revolution of the tube. Approximate values for lineal travel can be determined by multiplying the tube circumference by the tangent of the angle of the rolls.

MULTIPLE-ROLL ROTARY STRAIGHTENING

Rotary straighteners with five, six or seven rolls are also used in straightening of tubing (see Fig. 11 in the article "Straightening of Bars, Shapes and Long Parts").

A seven-roll rotary straightener has two three-roll clusters — one at the entry end and one at the exit end of the straightener — and a middle deflecting roll (Fig. 13). Normally, the two bottom rolls are driven and the five others are idlers. The middle roll (deflecting roll) moves vertically, and the four end idler rolls move in a circular path about pivot points in the base and apply pressure to the tube for feeding and straightening.

Rotary Swaging of Bars and Tubes

ROTARY SWAGING is a process for reducing the cross-sectional area or otherwise changing the shape of bars, tubes or wires by repeated radial blows with one or more pairs of opposed dies. The work is elongated as the cross-sectional area is reduced. Usually the starting workpiece is round, square or otherwise symmetrical in cross section, although other forms, such as rectangles, can be swaged.

Most as-swaged workpieces are round, the simplest being formed by reduction in diameter. However, swaging can also produce straight and compound tapers, produce contours on the inside diameter of tubing, and change round to square or other shapes.

APPLICABILITY

Swaging has been used to reduce tubes up to 14 in. in initial diameter and bars up to 4 in. in initial diameter.

Low strength (or hardness) and high ductility are required for good swageability.

Work Metals. Of the plain carbon steels, those with a carbon content of 0.20% or less are the most swageable. These grades can be reduced up to 70% in cross-sectional area by swaging. As carbon content or alloy content is increased, swageability is decreased.

Figure 14 shows the relation of hardness to carbon content for pearlitic and spheroidized microstructures and also shows three zones of swageability, indicating that a hardness of Rockwell B 85 is the maximum preferred for carbon steels, and that when hardness exceeds Rockwell B 102 swaging is impractical.

METAL FLOW DURING SWAGING

Metal flow during rotary swaging is not confined to one direction. As shown in Fig. 15, more metal moves out of the taper in a direction opposite to that of the feed than through the straight portion (blade). Some metal flow also occurs in the transverse direction, but it is restricted by the oval or side clearance in the dies.

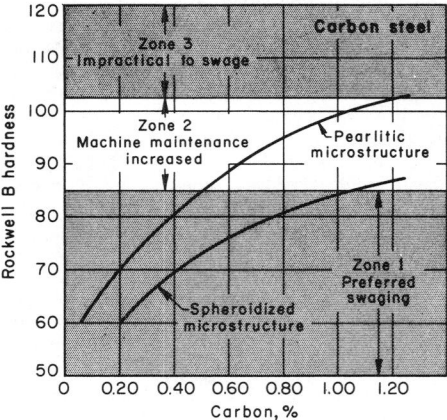

Fig. 14. Swageability of carbon steel, as a function of microstructure, hardness and carbon content

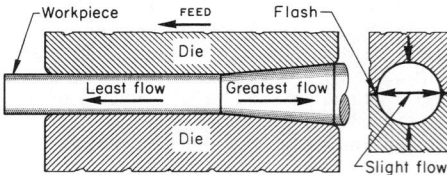

Fig. 15. Direction of metal flow in workpiece during rotary swaging

MACHINES

Rotary swaging machines are classified as standard rotary (see Fig. 16), die-closing, and stationary-spindle types. All these machines are equipped with dies that open and close rapidly to provide the impact action that shapes the workpiece. Die-closing and stationary-spindle type swaging machines can be altered to produce a slower, "squeeze-type" action of the dies.

Selection of Machine. A standard rotary swager is recommended for bar stock or tubing when swaging consists of a straight reduction in stock diameter or when it is used for tapering round workpieces.

When swaging bars or tubes that cannot be fed into the machine in the usual manner (see workpiece in Fig. 17), a die-closing swager is recommended.

A stationary-spindle machine is used when swaging involves changing of the cross-sectional

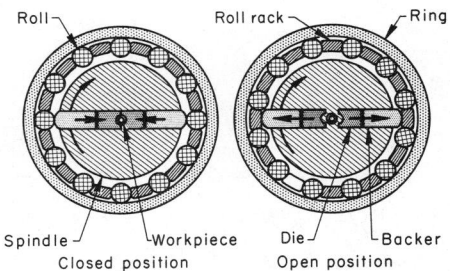

Fig. 16. Principal components and directions of movement in a standard two-die rotary swager

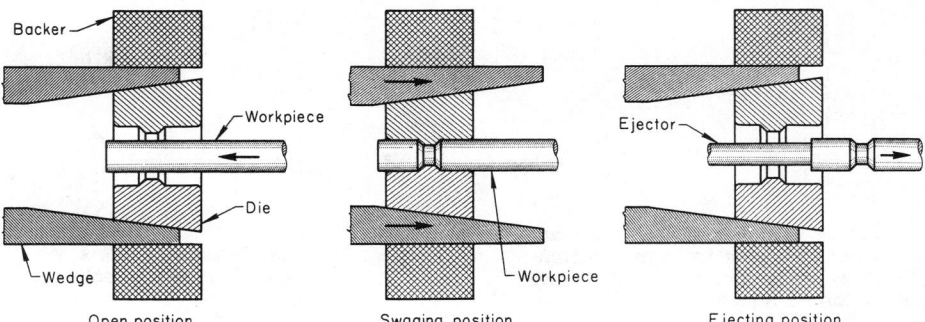

Fig. 17. Principal components and directions of movement in a die-closing swager

shape of a bar or tube to a different shape (such as a round to a square, or a square to a triangle), or when swaging the end of coiled wire. Stationary-spindle machines are sometimes called inverted swagers, because the spindle, dies and work remain stationary while the head and roll rack rotate. These machines are used for swaging shapes other than round.

The reciprocating action of the dies is the same as in swagers in which the spindle is rotated and the roll rack remains stationary. Principal components of a stationary-spindle machine are shown in Fig. 18.

Rolls and backers used for cold swaging are made from tool steel. The grade of tool steel used varies considerably, although many rolls and backers are made from one of the shock-resisting grades, such as S2, and then hardened and tempered to Rockwell C 55 to 60.

Almost all rolls and backers become work hardened, to a degree that depends on the severity of reduction of the swaged workpiece, the swageability of the work metal, the material used for the rolls and backers, total operating time, and adjustment of the machine. To reduce the effects of work hardening and to prolong service life, rolls and backers used in cold swaging are stress relieved periodically at 350 to 450 F for 2 to 3 hr. The stress-relieving temperature used must not be higher than the original tempering temperature, or softening will result. The frequency of stress relieving depends on the severity of swaging. Under normal conditions, rolls and backers should be stress relieved after every thirty hours of operation.

SWAGING DIES

Resistance to shock and wear are the primary requirements for cold swaging dies. Sometimes it is necessary to sacrifice some wear resistance to prevent die breakage from lack of shock resistance. In many applications, shallow-hardening tool steels such as W1 are used, because it is possible to develop extremely high surface hardness and a softer, tougher core with these steels. Another advantage in using the W grades of tool steel for swaging dies is that they are available in different carbon ranges (0.60 to 1.40%). It is usual to vary the carbon content in accordance with die size. For instance, steels for small dies (½ by ¾ in. in cross section) have a carbon content of about 1.4%, those for medium-size dies (2 by 2 in. in cross section) have about 0.95% carbon, and a 0.60% carbon content is used for large dies (5 by 8 in. in cross section).

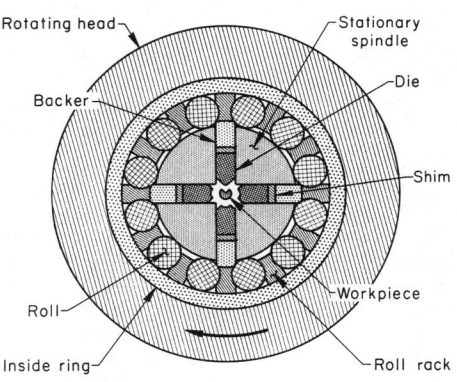

Fig. 18. Principal components of a stationary-spindle rotary swager

When using the practice outlined above, a zone about 9/64 in. in depth hardened to Rockwell C 62 to 65 is recommended. This can be obtained by water or brine quenching the W grade tool steels.

Types of Dies. Depending on the shape, size and material of the workpiece, dies range from the simple, single-taper, straight-reduction type to those of special design. Figure 19 illustrates nine typical die shapes.

LUBRICATION

The adverse effect that lubricants have on feeding conditions eliminates lubricants from use in many swaging operations (except between mandrels and workpieces). If a lubricant can be used, a better finish and longer tool life generally result.

Lubricants used include oils specifically prepared for swaging operations, phosphate coatings, molybdenum disulfide, kerosine, and Stoddard solvent, which is especially useful for swaging aluminum.

The main disadvantage in using lubricants is that excessive feedback can occur, especially when dies have a large entrance angle. Feedback is less likely to occur when large machines are used (about 200 tons or more), because they are equipped with automatic feed units capable of exerting several tons of force on the end of the blank.

Forming of Wire

WIRE FORMS are used to give high strength-to-weight ratio, an open construction (as in fan guards or baskets), resilience to absorb shock, and economy of automatic production of formed parts. Where production quantities are small or the size of the finished article is large, the wire may be straightened and cut to length as a preliminary operation before the individual pieces are fed into hand benders, kick presses, power presses equipped with appropriate dies, or coiling devices. For large quantities, the wire is straightened as it comes from the coil and is fed directly and continuously into power presses, automatic forming or spring-coiling machines, multiple-slide machines, or special machines that are actuated by cams or by air or hydraulic cylinders.

Tools used for forming wire should be made of tool steel hardened to Rockwell C 56 to 61. Water-hardening tool steels, such as W1, are usually adequate. For more severe forming and for longer tool life, D2 tool steel is recommended. Surfaces contacting the wire should be polished to prevent marking. They can usually be hardened after tryout in the soft state.

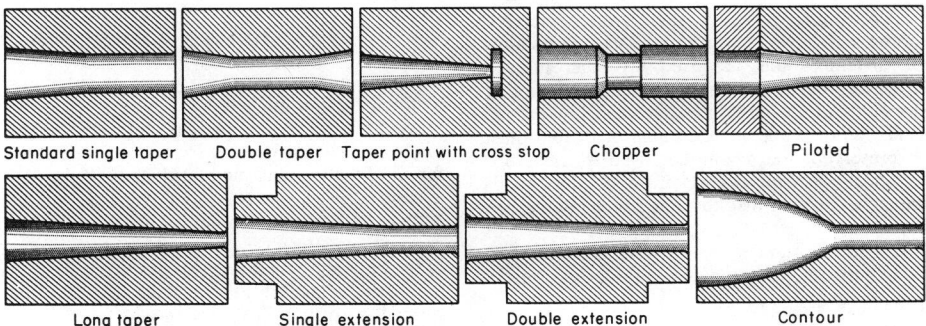

Standard single taper   Double taper   Taper point with cross stop   Chopper   Piloted

Long taper   Single extension   Double extension   Contour

Fig. 19. Typical die shapes used in rotary swaging

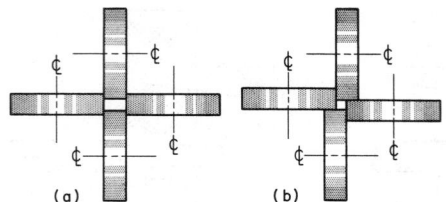

Fig. 20. Turk's-head rolls: (a) positioned in line to form a rectangular cross section, and (b) offset to form a square section

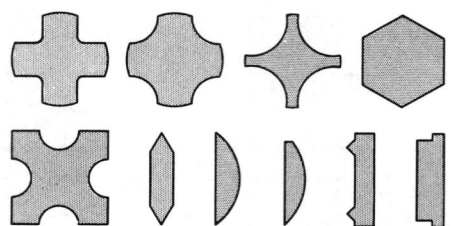

Fig. 21. Typical cross sections of wire formed in Turk's-head rolls

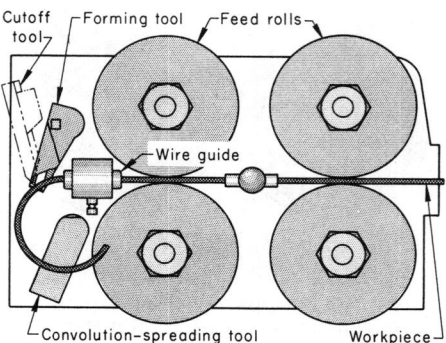

Fig. 22. Mechanism for winding springs that have coils of varying diameters

ROLLING OF WIRE IN A TURK'S-HEAD MACHINE

A Turk's-head machine generally has four rolls that will accommodate wire of one cross-section (generally, round) and cold roll it to another shape. Uses of the machine are:

1. To make accurate square and narrow rectangular wire directly from round wire
2. To finish special shapes from round or pre-formed rough shapes
3. To put edge contours on flat metal ribbon.

Operation of Machine. The machine has a cluster of four rolls whose axes are in the same plane and at right angles to each other as shown in Fig. 20. In operation, a coil of wire is supported in a payoff reel; the wire is pulled through the rolls by a capstan and then is re-coiled. A drawbench can be used for pulling short lengths (up to 100 ft) through the rolls.

The narrow rolls can be centered (opposed) to form some shapes, as in Fig. 20(a), or they can be offset for some other shapes, as in Fig. 20(b). Although the rolls shown are plain cylinders, these can be replaced with rolls ground to any shape that will form a section, simple or complex. Some sections are formed in several passes. It may take two passes to roll an accurate sharp-cornered square wire, and three or more passes through

the Turk's-head machine may be needed to make a complex section fill properly.

Simple or complex shapes may be drawn through a Turk's-head machine as fast as 600 ft per min, depending on the force and speed available in the drawbench, and on the amount of heating in the operation.

Some of the cross sections formed in Turk's-head rolls are shown in Fig. 21.

SPRING COILING

Small quantities of springs may be coiled in a lathe. The arbor around which the spring is wound is held in the chuck, and two wooden friction blocks are mounted on the cross slide. Numerous hand-operated devices are also used. Production coiling is done in single-purpose automatic spring coilers.

In a standard spring-coiling machine a pair of feed rolls pushes a calculated length of straightened wire through restricting guides against a coiling point and around a fixed arbor into a coil. At the end of the coiling cycle, the feed rolls stop and a cutoff mechanism actuates a knife, which severs the completed spring against the arbor as an anvil. A mechanism for winding springs whose coils vary in diameter is shown in Fig. 22. A flying knife separates the completed spring from the wire strand.

LUBRICANTS

Requirements of lubricants for wire-forming operations are more severe than for most other metalworking operations. The exceptionally high working pressures that may be reached require special lubricants to prevent galling, seizure or fracture of the wire, and to prevent excessive tool wear. Improper lubricating oils or compounds interfere with close-tolerance work and cause variations in the finished parts. The lubricant varies with the type of wire. Aluminum, copper alloys, basic steel wire, and steel spring wire each require a different lubricant.

Lubricants for wire forming can generally be classed in three groups: inorganic fillers, soluble oils, and boundary lubricants.

Inorganic fillers include solids such as white lead, talc, graphite, and molybdenum disulfide in a vehicle such as a neutral oil or paraffin oil.

Soluble oils include mineral oils to which agents such as sodium sulfonates have been added to make the oil emulsifiable in water. Soluble oils are good for cooling and corrosion prevention.

Boundary lubricants are thin adsorbed films and are usually subjected to high unit pressures.

Forming of Stainless Steel and Heat-Resisting Alloys

Forming of Stainless Steel

STAINLESS STEELS are blanked, pierced, formed and drawn in basically the same press tools and machines that are used for other metals. However, because stainless steels have higher strength and are more abrasive than low-carbon steels, and have a surface finish that often must be preserved, the techniques used in the fabrication of sheet-metal parts from stainless steels are more exacting than those used for low-carbon steels.

In general, stainless steels have the following characteristics, as compared with carbon steels:

1. Greater strength
2. Greater susceptibility to work hardening (austenitic grades)
3. More abrasiveness
4. Higher probability of welding and galling
5. Lower heat conductivity.

General ratings of the relative suitability of the commonly used austenitic, martensitic and ferritic types of stainless steels to various methods of forming are given in Table 1. These ratings are based both on formability and on the power required for forming.

As the table shows, the austenitic and ferritic steels are, almost without exception, well suited for all of the forming methods listed. Of the martensitic steels, however, only types 403, 410 and 414 are generally recommended for cold forming applications. Because the higher carbon content of the remaining martensitic types severely limits their cold formability, these steels are sometimes formed warm. (Warm forming also may be employed to good advantage with other stainless steels in difficult applications.)

FORMABILITY

Characteristics of stainless steel that affect its formability include yield strength, tensile strength, and ductility, and the effect of work hardening on these properties. These properties are, in turn, sensitive to chemical composition.

Figure 1 compares the effect of cold work on the tensile strength and yield strength of type 301 (an austenitic alloy), type 430 (a ferritic alloy) and 1008 low-carbon steel sheet.

Stress-Strain Relations. Figure 2 shows load-elongation curves for six types of stainless steel: four austenitic (202, 301, 302 and 304), one martensitic (410), and one ferritic (430). The information was obtained by drawing cups of the shape shown. The type of breakage in failure of the austenitic types was different from that of types 410 and 430, as shown in Fig. 2. The austenitic types broke in a fairly clean line near the punch

Table 1. Relative suitability of stainless steels for various methods of forming(a)
(A = Excellent; B = Good; C = Fair; D = Not generally recommended)

Steel	0.2% yield strength, 1000 psi	Blanking	Piercing	Press-brake forming	Deep drawing	Spinning	Roll forming	Coining	Embossing
Austenitic steels									
201	55	B	C	B	A-B	C-D	B	B-C	B-C
202	55	B	B	A	A	B-C	A	B	B
301	40	B	C	B	A-B	C-D	B	B-C	B-C
302	37	B	B	A	A	B-C	A	B	B
302B	40	B	B	B	B-C	C	...	C	B-C
303, 303(Se)	35	B	B	D(b)	D	D	D	C-D	C
304	35	B	B	A	A	B	A	B	B
304L	30	B	B	A	A	B	A	B	B
305	37	B	B	A	B	A	A	A-B	A-B
308	35	B	...	B(b)	D	D	...	D	D
309, 309S	40	B	B	A(b)	B	C	B	B	B
310, 310S	40	B	B	A(b)	B	B	A	B	B
314	50	B	B	A(b)	B-C	C	B	B	B-C
316	35	B	B	A(b)	B	B	A	B	B
316L	30	B	B	A(b)	B	B	A	B	B
317	40	B	B	A(b)	B	B-C	B	B	B
321, 347, 348	35	B	B	A	B	B-C	B	B	B
Martensitic steels									
403, 410	40	A	A-B	A	A	A	A	A	A
414	95	A	B	A(b)	B	C	B	C	C
416, 416(Se)	40	B	A-B	C(b)	D	D	D	D	C
420	50	B	B-C	C(b)	C-D	D	C-D	C-D	C
431	95	C-D	C-D	C(b)	C-D	D	C-D	C-D	C-D
440A	60	B-C	...	C(b)	C-D	D	C-D	D	C
440B	62	...	...	...	...	D	...	D	D
440C	65	...	...	...	...	D	...	D	D
Ferritic steels									
405	40	A	A-B	A(b)	A	A	A	A	A
430	45	A	A-B	A(b)	A-B	A	A	A	A
430F, 430F(Se) . . .	55	B	A-B	B-C(b)	D	D	D	C-D	C
442	...	A	A-B	A(b)	B	B-C	A	B	B
446	50	A	B	A(b)	B	B-C	C	B	B

(a) Suitability ratings are based on comparison of the steels within any one class; thus it should not be inferred that a ferritic steel with an A rating is more formable than an austenitic steel with a C rating for a particular method. (b) Severe sharp bends should be avoided.

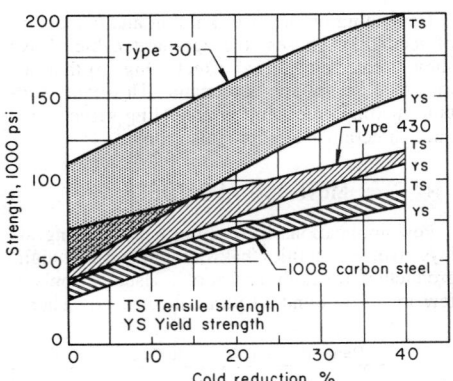

Fig. 1. Comparison of work-hardening qualities of type 301 austenitic stainless steel, type 430 ferritic stainless steel, and 1008 low-carbon steel

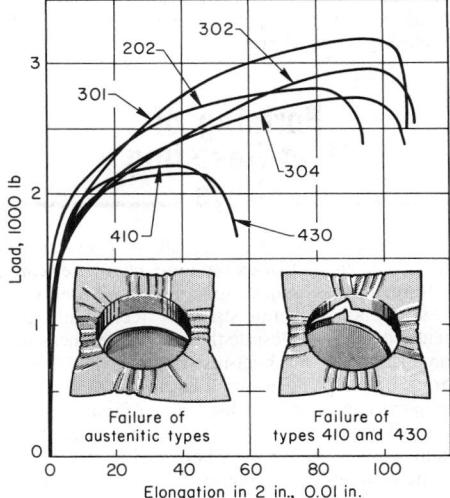

Fig. 2. Comparison of the ductility of six stainless steels, and of the types of failure resulting from deep drawing

nose radius, almost as if the bottom of the drawn cup were blanked out; types 410 and 430 broke in the sidewall in sharp jagged lines, showing extreme brittleness as a result of the severe cold work.

As shown in Fig. 2, the power required to form type 301 exceeds that required by the other austenitic alloys; also, it will withstand maximum elongation before failing. Types 410 and 430 require considerably less power to form but fail at comparatively low elongation levels.

Power requirements for forming of stainless steel, because of the high yield strength, are greater than for low-carbon steel; generally, twice as much power is used in forming stainless. Because the austenitic steels work harden rapidly in cold forming operations, the need for added power after the start of initial deformation is greater than that

for the ferritic steels. The ferritic steels behave much like plain carbon steels once deformation begins, although higher power also is needed to start plastic deformation.

LUBRICATION

Lubrication requirements are more critical in forming stainless steels than in forming carbon and alloy steels, because it usually is necessary to preserve the high-quality surface on stainless steels, and because stainless steels have higher strength, greater hardness, lower thermal conductivity, and higher coefficient of friction. In forming stainless steels, galling and spalling occur more readily, and higher temperatures are reached in a larger volume of the workpiece. Local or general overheating can change the properties of the work metal and the lubricant.

Table 2 lists the lubricants ordinarily used in forming stainless steel by various processes. Except for the special-purpose lubricants graphite and molybdenum disulfide, they are listed in the approximate order of increasing ability to reduce galling and friction. The ratings in Table 2 also consider other suitability factors such as cleanliness and ease of removal.

PRESS FORMING

Stainless steels are press formed with the same kind of equipment as is used in the forming of low-carbon steel. However, although all stainless steels are not the same in strength or ductility, they all need more power to form than do carbon steels. In general, presses should have the capacity for 60% more ram force than is needed for equivalent work in low-carbon steel, and frames should have the rigidity and bulk to withstand this greater force.

Dies. Besides wearing out faster, dies may fracture more readily when used with stainless steel than when used with low-carbon or medium-carbon steel, because of the greater forces needed to work stainless steel.

For the longest service in mass production, the wearing parts of the dies should be made of carbide, D2 tool steel, or high-strength aluminum bronze. Carbide can last ten times as long as most tool steels.

DEEP DRAWING

Percentages of reduction (blank diameter to inside diameter of cup) obtainable in deep drawing range from 40 to 60% for the chromium-nickel (austenitic) stainless steels of best drawability and from 20 to 30% for the straight-chromium (ferritic) grades. The amount of reduction obtainable varies greatly with the radius of the die and to a lesser extent with the radius of the punch nose. As the die radius decreases, the drawability decreases, as shown in Table 3 for austenitic stainless steel. Typically used punch and die radii are five to ten times metal thickness. With the ferritic grades, the drawability and ductility may decrease with increasing chromium content (unless carbon and nitrogen content is kept quite low). To offset this, such steels are often warmed moderately before drawing.

Presses used for deep drawing stainless steel differ only in power and rigidity from those used for low-carbon steel; because of the higher work-hardening rate of stainless steel and its inherent higher strength, presses used for deep drawing

Table 2. Suitability of various lubricants for use in forming of stainless steel
(A = Excellent; B = Good; C = Acceptable; NR = Not recommended)(a)

Lubricant	Blank-ing and piercing	Press-brake forming	Press forming	Mul-tiple-slide forming	Deep draw-ing	Spin-ning	Drop ham-mer form-ing	Con-tour roll form-ing	Emboss-ing
Fatty oils and blends(b)	C	B	C	A	C	A	C	B	B
Soap-fat pastes(c)	NR	NR	C	A	B	B	C	B	C
Wax-base pastes(c)	B	B	B	A	B	B	C	B	A
Heavy-duty emulsions(d)	B	NR	B	A	B	B	NR	A	B
Dry film (wax, or soap plus borax)	B	B	B	NR	B	A	B	NR	A
Pigmented pastes(c)(e)	B	NR	A	B	A	C	NR	NR	NR
Sulfurized or sulfochlorinated oils(f)	A	A	B+	A	C	NR	A	B	A
Chlorinated oils or waxes(g):									
High-viscosity types (h)	A(j)	NR	A	NR	A	NR	A(k)	A	NR
Low-viscosity types(m)	B+	A	A	A	B	NR	A(k)	A	A
Graphite or molybdenum disulfide(n)	NR	(p)	(p)	NR	(p)	NR	(p)	NR	NR

(a) Ratings consider effectiveness, cleanliness, ease of removal, and other suitability factors. (b) Vegetable or animal types; mineral oil is used for blending. (c) May be diluted with water. (d) Water emulsions of soluble oils; contain a high concentration of EP sulfur or chlorine compounds. (e) Chalk (whiting) is commonest pigment; others sometimes used. (f) EP types; may contain some mineral or fatty oil. (g) EP chlorinated mineral oils or waxes; may contain emulsifiers for ease of removal in water-base cleaners. (h) Viscosity of 4000 to 20,000 SUS. (j) For heavy plate. (k) For cold forming only. (m) Viscosity (200 to 1000 SUS) is influenced by base oil or wax, degree of chlorination, and additions or mineral oil. (n) Solid lubricant applied from dispersions in oil, solvent or water. (p) For hot forming applications only.

Table 3. Effect of die radius on reduction obtainable in deep drawing of austenitic stainless steel(a)

Die radius(b)	Reduction in drawing, %
15t	50 to 60
10t	40 to 50
5t	30 to 40
2t	0 to 10

(a) Per cent reduction = $100(1 - d/D)$, where D is the diameter of blank, and d is the inside diameter of the drawn piece. (b) t = stock thickness.

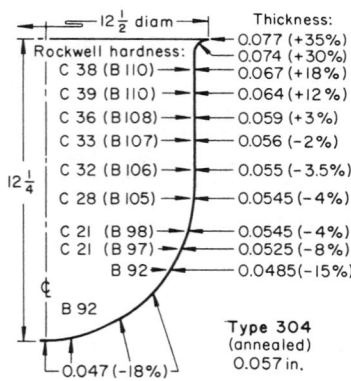

Fig. 3. Profile of a shell drawn from an austenitic stainless steel, showing variations in hardness and thickness produced by drawing

stainless steel often need 100% more ram force and the necessary frame stiffness to support this greater force.

Dies for drawing stainless steel must be able to withstand the high force and to resist galling. For ordinary service, D2 tool steel dies give a good combination of hardness and toughness. On long runs, carbide draw rings have exceptionally long life. Where friction and galling are the chief problems, draw rings are sometimes made of high-strength aluminum bronze.

Die clearance for heavy draws is 35 to 40% greater than the original metal thickness for aus-

tenitic alloys; for the ferritic alloys, which thicken less, 10 to 15% is generally adequate.

Figure 3 shows a profile of an austenitic stainless steel drawn part that illustrates the thickening pattern observed in drawing a cup from this material. If the process is one of stretching more than of drawing, the clearances do not have to compensate for natural thickening.

Clearances of less than the metal thickness generally are not used with stainless steel, because they result in "ironing" or squeezing of the metal between the male and female dies. The austenitic stainless steels are not suited to ironing, because their high rate of work hardening promotes scoring and rapid wear of the dies. Also, any substantial ironing in drawing austenitic stainless steels greatly increases the likelihood of fracturing the workpiece.

Forming of Heat-Resisting Alloys

WROUGHT heat-resisting alloys (iron-base, nickel-base and cobalt-chromium-nickel-base) can be formed by techniques similar to those used for the series 300 austenitic stainless steels, but with greater difficulty. Despite the wide differences in composition among heat-resisting alloys, all are strongly susceptible to work hardening. Figure 4 compares degree of work hardening of four nickel-base alloys with A-286 (iron-base), type 304 stainless steel, and low-carbon ferritic steel.

The differences in composition of the various heat-resisting alloys cause differences in their formability. Alloys that contain the greatest amounts of cobalt, such as HS-25 and Elgiloy, are more difficult to form than iron-base or low-cobalt nickel alloys. Most alloys that contain substantial amounts of molybdenum or tungsten for strengthening, such as Hastelloy B or René 41, are more difficult to form than alloys having low amounts of these elements.

Most of the iron-base and nickel-base alloys contain less than 0.15% carbon; more carbon than this causes excessive carbide precipitation (19-9 DL with 0.30% carbon is an exception). Small

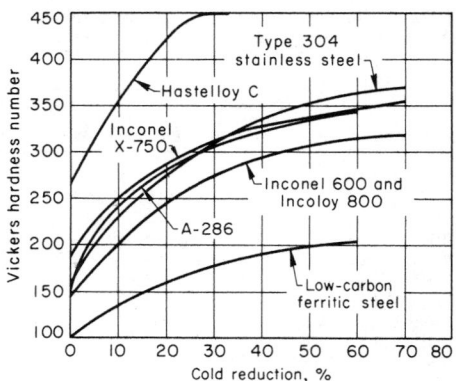

Fig. 4. Effect of cold reduction on hardness of heat-resisting alloys, type 304 stainless steel, and low-carbon ferritic steel

amounts of boron are used in some of the heat treatable nickel-base alloys, such as René 41 and U-700, to prevent precipitation of carbides at grain boundaries; too much boron causes cracking during forming.

Sulfur causes hot shortness of nickel-base alloys. Silicon should be less than 0.60%, and preferably less than 0.30%. More than 0.60% silicon causes cracking of cold drawn alloys, and may cause cracking of welds in some others. Usually, alloys with less than 0.30% silicon present no difficulties in forming.

Cold forming is preferred for heat-resisting alloys, especially in thin sheets; most of these alloys can be hot formed effectively only in a narrow temperature range.

ROLLING DIRECTION

Depending on the size, amount and dispersion of secondary phases, the age-hardenable alloys show greater directional effects (Fig. 5) than alloys that are not age-hardenable. However, vacuum melting and solution annealing serve to reduce directional effects (anisotropy)

METHODS AND TOOLS

Few applications in forming heat-resisting alloys involve quantities that warrant the use of high-production methods and tools. Usually, only a few to a few hundred parts are needed. There-

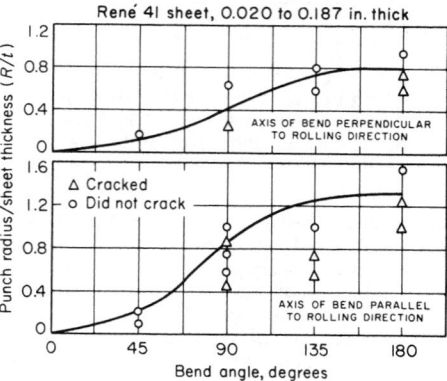

Fig. 5. Effect of direction of rolling on the formability of René 41 in press-brake bending

Table 4. Nominal compositions of refractory alloys available as sheet(a)

Alloy	Zr	Ti	V or Hf	W	Others
Columbium-base alloys					
FS-80	1.0	...	...	...	...
FS-82	1.0	...	...	...	33.0 Ta
FS-85	1.0	...	...	11.0	28.0 Ta
C-103	...	1.0	10.0 Hf	...	...
C-129Y	...	...	10.0 Hf	10.0	0.10 Y
B-66	1.0	...	5.0 V	...	5.0 Mo
D-14	5.0	...	...	...	...
D-36	5.0	10.0	...	...	...
D-43	1.0	...	...	10.0	0.10 C
Cb-752 ...	2.5	...	...	10.0	...
Cb-753 ...	1.2	...	5.0 V	...	...
Tantalum-base alloys					
Ta-10W ...	...	...	...	10.0	...
Ta-8W-2Hf	...	...	2.0 Hf	8.0	...
Ta-30Cb-					
7.5 V ...	...	...	7.5 V	...	30.0 Cb
Molybdenum-base alloys					
Mo-0.5Ti	...	0.5	...	...	0.03 C
TZM	0.1	0.5	...	...	0.03 C

(a) Commercially pure Cb, Ta, Mo and W sheets are also available.

Table 5. Conditions for press-brake forming of refractory metal sheet 0.020 to 0.050 in. thick(a)

Metal or alloy(b)	Forming temperature	Minimum bend radius(c) — Test data	Minimum bend radius(c) — Preferred	Spring-back
Columbium-base alloys (annealed)				
FS-82, C-103,				
C-129Y	Room	<1t	1t	2°-6°
D-36	Room	1t-2t	2t	3°-10°
Tantalum-base alloys(annealed)				
Tantalum	Room	<1t	1t	...
Ta-10W	Room	<1t	2t	1°-5°
Molybdenum-base alloys (stress relieved)				
Mo-0.5Ti,				
TZM	300 F	2t-5t	5t	3°-8°
Tungsten (stress relieved)				
Tungsten	600 F	2t-5t	5t	2°-8°

(a) Formed to a 120° bend angle in a 60° V-die at a ram speed of 10 to 120 ipm. (b) See Table 1 for compositions. (c) t = sheet thickness.

fore, methods that require a minimum of tooling, such as press-brake forming, drop-hammer forming, spinning and explosive forming, have been used more than other methods. Presses or other machines are the same as those used for forming steel, but more power is needed to form heat-resisting alloys because of their higher strength.

LUBRICATION

Some lubrication is usually required for best results in drawing, stretch forming, or spinning. In press-brake forming, lubrication is seldom needed. Sulfurized or sulfochlorinated oils may be used if the work is carefully cleaned afterward in a degreaser or an alkaline cleaner. Work that has been formed in zinc alloy dies should be flash-pickled in nitric acid before heat treatment, to prevent the possibility of zinc embrittlement.

Forming of Refractory Metals

REFRACTORY METALS are generally worked in small quantities, production rates are low, each piece is handled separately, and the forming process is closely controlled.

Table 4 shows the compositions of refractory alloys available as sheet. Typical conditions for bending sheet 0.020 to 0.050 in. thick are given in Table 5. Tensile forming parameters for sheet materials are summarized in Table 6.

FORMABILITY

Columbium and tantalum alloys generally are formed at room temperature in the annealed (recrystallized) condition, although the stress-relieved alloys are sufficiently ductile for most forming operations. Work hardening, especially of the stronger alloys, often necessitates annealing after severe forming.

Strong alloys of columbium, which are made in limited quantity, are not listed in Table 4. These

Fig. 6. Effect of temperature on the formability of Mo-0.5Ti sheet, as indicated by ratio of bend radius to sheet thickness

alloys have varying degrees of brittleness at low temperature, but can be formed by the same procedures used for molybdenum.

Molybdenum and tungsten are more difficult to form than columbium and tantalum, but if they are heated and certain precautions are taken, even complex parts can be formed. The greatest difficulty in forming these metals is their tendency toward brittle fracture (cracks and ruptures that occur with little or no plastic deformation), and delamination (a type of brittle behavior that produces cracks or ruptures parallel to the plane of the sheet). Tungsten can be hot formed only; it is brittle at room temperature.

At slow strain rates in tension and in bending, molybdenum, Mo-0.5Ti and TZM are ductile at room temperature, becoming brittle at lower temperatures. However, because of the high variable strain rates and tri-axial stresses in the usual forming processes, these metals are usually hot formed, to decrease the probability of brittle fracture. Molybdenum and tungsten blanks must have prepared edges to prevent cracking and splitting during forming.

Molybdenum and tungsten are generally stress relieved before forming. Recrystallization raises the transition (ductile-to-brittle failure) temperature.

EFFECT OF FORMING TEMPERATURE

Annealed columbium and tantalum alloys are formed at room temperature. Heating these alloys would reduce their formability, because of strain aging, and would cause oxidation and possible surface contamination.

Tungsten is brittle at room temperature, so thin tungsten sheet is formed at 600 to 1000 F, and thicker sheet or complex shapes are formed at 1000 to 1500 F after stress relief.

Molybdenum and molybdenum alloys, in thin sheets, can be formed cold to some extent, but heating helps to prevent fracture and delamination.

The TZM alloy is most ductile at 200 F (just above the ductile-to-brittle transition temperature). Further rise in temperature decreases the ductility, because of strain aging. Most molybdenum is formed at 200 to 600 F, but thicker metal or complex shapes are formed at 600 to 1200 F.

Figure 6 shows the effect of heating on the bending of Mo-0.5Ti sheet. It is most formable at 200 to 400 F. Forming in other zones may crack it.

Table 6. Elongation and true strain in forming refractory metal sheet of various thicknesses and orientations to rolling direction(a)

Alloy	Condition(b)	Thickness, in.	Forming temperature, F	Orientation(c)	Elongation, % in: 1 in.	Elongation, % in: 2 in.	True strain(d) ϵ_m	True strain(d) ϵ_c
Columbium-base alloys								
FS-82	Ann	0.010	70	L	32.0	26.0	0.180	0.230
				T	30.0	25.0	0.170	0.230
C-103	SR	0.030	70	L	18.0	14.0	0.122	0.145
				T	6.0	4.0	0.041	0.046
	Ann	0.030	70	L	30.0	24.0	0.152	0.232
				T	26.0	21.0	0.150	0.197
Tantalum-base alloy								
Ta-10W	Ann	0.040	70	L	39.0	30.5	0.180	0.283
				T	38.0	30.0	0.197	0.282
Molybdenum-base alloys								
Mo-0.5Ti	SR	0.020	70	L	19.0	15.0	0.102	0.164
				T	11.0	9.0	0.052	0.089
TZM	SR	0.035	70	L	19.0	15.0	0.074	0.130
		0.040	70	L	16.0	11.5	0.060	0.075
Tungsten								
Tungsten	SR	0.035	1100	T	4.0	2.5	0.019	0.022
			1800	T	3.5	...	0.021	0.023

(a) Based on testing one heat of material for each alloy. (b) Ann = annealed; SR = stress relieved. (c) L = longitudinal; T = transverse. (d) ϵ_m = true strain at maximum load; ϵ_c = maximum true strain at maximum true stress.

LUBRICANTS

The types of lubricants used in the forming of refractory metals include oils, extreme-pressure lubricants, soaps, waxes, silicones, graphite, molybdenum disulfide, copper plating, and an acrylic enamel coating that is produced by suspending powdered copper in acrylic resin.

Ordinary oils and greases are commonly used in the forming of columbium and tantalum, because these metals are generally formed at room temperature. For severe forming operations petrolatum is frequently used.

Solid lubricants and suspensions of suitable pigments, such as molybdenum disulfide with or without colloidal graphite, are used in the hot forming of molybdenum and tungsten.

Chlorinated lubricants and other lubricants that decompose upon heating to form toxic or noxious fumes must not be used without proper safety precautions.

——————— Forming of Nonferrous Metals ———————

Forming of —— Aluminum Alloys ——

ALUMINUM and its alloys are among the most readily formable of the commonly fabricated metals. There are, of course, differences between aluminum alloys and other metals in the amount of permissible deformation, in some aspects of tool design, and in details of procedure. These differences stem primarily from the lower tensile and yield strengths of aluminum alloys, and from their comparatively slow rate of work hardening. The compositions and tempers of aluminum alloys also affect their formability.

FORMABILITY

Non-Heat-Treatable Alloys. Alloys 1100 and 3003 are frequently used in forming applications, because of their excellent workability and low cost. If somewhat higher strength is required, alloys containing magnesium are commonly used (for example, in order of increasing strength, alloys 3004, 5052, 5154 and 5086).

If superior finishing characteristics are needed in addition to higher strength, an alloy containing a small amount of manganese in addition to magnesium (alloy 5053, 5252 or 5457) can be used. Holding impurities at a low level in alloys used for decorative and finishing purposes also helps in developing bright, uniform finish.

Heat treatable alloys are used in applications for which a high strength-to-weight ratio is required. These include alloys 6061, 2014, 2024, 7075 and 7178, in approximate order of increasing strength.

The annealed temper (O) is the most workable condition for forming, but it entails the greatest expense in subsequent heat treating and straightening. Alloys that have been freshly solution heat treated and quenched (W temper) are nearly as formable as when annealed, and can be given increased strength after forming by natural or artificial aging, without reheating and consequent exposure of the finished part to warping. Alloys can be stored in the W temper for a reasonable period at a low temperature. (Almost no aging occurs in most alloys at −20 F.)

Material that has been solution heat treated at the mill, but not artificially aged (T3, T4 or W temper), is generally suitable only for mild forming operations such as bending, mild drawing, or moderate stretch forming.

Solution heat treated and artificially aged (T6 temper) alloys are seldom used for forming, other than bending to standard radii and forming of very shallow shapes. Although alloys in the T6 temper are much stronger, they have lost so much ductility in hardening that they are apt to fracture in even moderately severe forming.

EQUIPMENT AND TOOLS

Most of the equipment used in the forming of steel and other metals is suitable for use with aluminum alloys. Because of the generally lower yield strength of aluminum alloys, press tonnage requirements are usually lower than for comparable operations on steel and higher press speeds can be used. Similarly, equipment for roll forming, spinning, stretch forming, and other forming operations on aluminum need not be so massive or rated for such heavy loading as for comparable operations on steel.

Tools. Total wear on tools used in forming aluminum is somewhat less than with steel. This results in part from the lower force levels involved, and in part from the smoother surface condition that is characteristic of aluminum alloys.

However, a higher-quality surface finish is generally required on tools used with aluminum alloys, to avoid marking. The oxide film on the surface of aluminum alloys is highly abrasive, and for this reason many forming tools are made of hardened tool steels. As a rule, these tools, even if otherwise suitable, should not be used interchangeably to form steel parts, because this could destroy the high finish on the tools.

Most aluminum alloys require smaller clearances between punches and dies in blanking and piercing than do steels. For drawing operations, similar tool radii but larger clearances are required, with respect to steel, to allow for the free flow of the metal and avoid excessive stretching during drawing.

The amount of springback is roughly proportional to the yield strength of the metal; therefore, the amount of springback in forming aluminum alloys is generally less than in forming low-carbon steels.

Table 1. Typical lubricants used in the forming of aluminum alloys
(Listed in approximate order of increasing effectiveness)(a)

1 Kerosine
2 Mineral oil (viscosity, 40 to 300 SUS at 100 F)
3 Petroleum jelly
4 Mineral oil plus 10 to 20% fatty oil
5 Tallow plus 50% paraffin
6 Tallow plus 70% paraffin
7 Mineral oil plus 10 to 15% sulfurized fatty oil plus 10% fatty oil
8 Dried soap films or wax films(a)
9 Fat emulsions in aqueous soap solutions plus finely divided fillers(b)
10 Mineral oil plus sulfurized fatty oil plus fatty oil plus finely divided fillers(b)

(a) For some applications, dried soap or wax films (lubricant No. 8) are less effective than lubricants No. 5, 6 and 7. (b) Typical fillers are chalk, lithopone, white lead, talc, mica, zinc oxide, clay sulfur and graphite.

The slower rate of work hardening of aluminum alloys permits a greater number of successive draws than is possible with steel.

Aluminum alloys have a fairly high coefficient of thermal expansion (14.4 micro-in./in./°F), and this must be accounted for in hot work tooling design.

LUBRICANTS

Lubricants must be selected specifically for their compatability with aluminum alloys and their suitability for the particular forming operation. A lubricant suitable for use on a steel part will not necessarily be suitable for use in the forming of a similar aluminum alloy part.

Properly formulated lubricants take into account the special requirements of regulation of moisture content in nonaqueous systems, corrosion inhibitors, and pH control, in order to prevent staining or corrosion of aluminum alloys and to make duration of contact with the workpiece less critical.

The lubricants most widely used in the forming of aluminum alloys are listed in Table 1 in approximate order of increasing effectiveness. The use of various special-purpose lubricants is discussed in sections of this article that deal with individual forming processes.

BLANKING AND PIERCING

Blanking and piercing of aluminum alloy flat stock are ordinarily done in punch presses, because of their high production rates and ability to maintain close tolerances. Press brakes are sometimes used, particularly for experimental or short-run production.

Because of the generally lower shear strength of aluminum alloys, lower-tonnage presses or press brakes are usually required than for comparable operations with steel.

Tool Materials. Aluminum alloys are classed with other soft materials, such as copper and magnesium alloys. In general, for a given tool material, tool life will be longer for blanking and piercing aluminum alloys than for steel. In some applications, a less expensive die can be used than with steel parts, particularly for relatively short runs.

Low-carbon steel or cast iron dies sometimes replace hardened tool steel dies, even for long runs. Punches are usually made from annealed or hardened tool steel, depending on the size and complexity of the part and on the length of the run. Carbide tools are seldom required, even for extremely long runs.

Clearance between punch and die must be controlled in blanking and piercing, in order to ob-

Table 2. Punch-to-die clearances for blanking and piercing aluminum alloys

Alloy	Temper	Clearance per side, %t(a)
1100	O	5.0
	H12, H14	6.0
	H16, H18	7.0
2014	O	6.5
	T4, T6	8.0
2024	O	6.5
	T3, T36, T4	8.0
3003	O	5.0
	H12, H14	6.0
	H16, H18	7.0
3004	O	6.5
	H32, H34	7.0
	H36, H38	7.5
5005	O	5.0
	H12, H14, H32, H34	6.0
	H36, H38	7.0
5050	O	5.0
	H32, H34	6.0
	H36, H38	7.0
5052	O	6.5
	H32, H34	7.0
	H36, H38	7.5
5083	O	7.0
	H323, H343	7.5
5086	O, H112	7.0
	H32, H34, H36	7.5
5154	O, H112	7.0
	H32, H34, H36, H38	7.5
5257(b)	O	5.0
	H25	6.0
	H28	7.0
5454	O, H112	7.0
	H32, H34	7.5
6061	O	5.5
	T4	6.0
	T6	7.0
7075	O	6.5
	W, T6	8.0
7178	O	6.5
	W, T6	8.0

(a) *t* = thickness of sheet. (b) Also alloys 5357, 5457, 5557 and 5657.

tain a uniform shearing action. Clearance is usually expressed as the distance between mating surfaces of punch and die (per side) in percentage of work thickness.

Correct clearance between punch and die depends on the alloy as well as the sheet thickness. Suggested punch-to-die clearances for blanking and piercing the common alloys are listed in Table 2.

PRESS-BRAKE FORMING

The press-brake forming techniques used with aluminum alloys are similar to those used with steel and other metals, differing only in some details of tool design.

Tolerances in press-brake forming are larger than those in punch-press operations. For simple shapes that are relatively long and narrow, a tolerance of $\pm\frac{1}{32}$ in. can usually be maintained. On larger parts of more complex cross section, the tolerance may be as much as $\pm\frac{1}{16}$ in.

Radii to which bends can be made depend on the properties of the metal, and on the design, dimensions and condition of tools. For most metals, the ratio of minimum bend radius to thickness is approximately constant, because ductility is the primary limiting factor on minimum bend radius. This is not true of aluminum alloys, for which the ratio increases with the thickness.

Table 3 shows the experimentally determined variation of minimum bend radius with alloy, temper and thickness for most of the commonly used aluminum alloys, in conventional bending operations with rigid dies.

Lubricants are needed for nearly all press-brake forming of aluminum alloys. The light protective film of oil sometimes present on mill stock is often adequate for mild bending.

CONTOUR ROLL FORMING

Aluminum alloys are readily shaped by contour roll forming, using equipment and techniques similar to those used for steel (see the article "Contour Roll Forming," in this volume). Operating speeds can be higher for the more ductile aluminum alloys than for most other metals. Speeds as high as 800 ft per min have been used in mild roll forming sections 50 to 100 ft long made of $\frac{1}{32}$-in.-thick alloy 1100-O coil stock.

Power requirements for roll forming of aluminum alloys are generally lower than for comparable operations on steel, because of the lower yield strength of most aluminum alloys.

Tooling. The design of rolls and related equipment, as well as the selection of tool materials, is discussed in the article "Contour Roll Forming." The most commonly used material is L6 tool steel, a low-alloy nickel-chromium grade with excellent toughness, wear resistance, and hard-enability. For extremely severe forming operations or exceptionally long runs, a high-carbon high-chromium grade such as D2 is preferred because it has superior resistance to galling and wear. These tool steels are hardened to Rockwell C 60 to 63.

DEEP DRAWING

Equipment, tools and techniques used for deep drawing aluminum are similar to those used for other metals. This section deals with those aspects of deep drawing that are peculiar to aluminum alloys. It is restricted to procedures using a rigid punch and die.

Equipment. Punch presses are used for nearly all deep drawing; press brakes are sometimes used for experimental or very short runs. Presses used for steel are also suitable for aluminum.

Tonnage requirements, determined by the same method as used for steel, are generally lower for comparable operations because of the lower tensile strength of aluminum alloys.

Press speeds are ordinarily higher than for steel. For mild draws, single-action presses are usually operated at 90 to 140 ft per min.

Tool Design. Tools for deep drawing are the same in general construction as those used with steel, but there are some significant differences. Aluminum alloy stock must be allowed to flow without undue restraint or excessive stretching. The original thickness of the metal is changed very little. This differs from deep drawing of stainless

Table 3. Minimum recommended radii for 90° cold bends in aluminum alloy sheet(a)

Alloy	Temper	Minimum bend radius in $\frac{1}{32}$, for sheet thickness, in., of:									
		0.016	0.025	0.032	0.040	0.050	0.063	0.090	0.125	0.190	0.250
1100	O	0	0	0	0	0	0	0	0	0	0
	H12	0	0	0	0	0	0	0	0	3	6
	H14	0	0	0	0	0	0	0	0	3	6
	H16	0	0	0	0	1	2	3	4	8	16
	H18	1	1	2	2	3	4	6	8	16	24
2014	O	0	0	0	0	0	0	0	0	3	6
	T6	2	4	4	5	7	8	15	20	36	64
2024 & alclad 2024	O	0	0	0	0	0	0	0	0	3	6
	T3	2	3	4	5	7	8	15	20	30	48
3003, 5005, 5357 and 5457, at tempers listed at right	O	0	0	0	0	0	0	0	0	0	0
	H12 or H32	0	0	0	0	0	0	0	0	3	6
	H14 or H34	0	0	0	0	0	0	1	2	4	8
	H16 or H36	0	0	1	2	2	3	5	6	12	24
	H18 or H38	1	1	2	2	3	5	9	12	24	40
3004, alclad 3004, 5154, 5254 and 5454, at tempers listed at right	O	0	0	0	0	0	0	0	2	3	8
	H32	0	0	0	1	1	2	3	4	9	18
	H34	1	1	1	2	2	3	5	6	12	24
	H36	1	1	1	2	3	4	6	9	18	24
	H38	1	1	2	3	4	6	9	16	30	40
5050	O	0	0	0	0	0	0	0	0	0	0
	H32	0	0	0	0	0	0	0	2	3	8
	H34	0	0	0	0	0	1	2	4	6	12
	H36	1	1	1	2	2	3	6	8	16	24
	H38	1	1	2	3	4	6	9	12	24	40
5052 and 5652	O	0	0	0	0	0	0	0	2	3	4
	H32	0	0	0	0	1	2	3	4	6	12
	H34	0	0	0	1	1	2	4	5	9	16
	H36	1	1	1	2	3	4	5	8	18	24
	H38	1	1	2	3	4	6	9	12	24	40
5086 and 5155	O	0	0	0	0	0	1	2	3	6	8
	H32	1	1	1	2	2	3	5	6	12	16
	H34	1	1	2	2	3	3	6	8	18	24
6061 & alclad 6061	O	0	0	0	0	0	0	0	2	3	4
	T6	1	1	2	2	3	4	6	9	18	28
7075 & alclad 7075	O	0	0	0	1	1	2	2	3	5	18
	T6	2	2	4	8	10	12	18	24	36	64
7178 & alclad 7178	O	0	0	0	1	1	2	3	5	9	18
	T6	2	3	4	8	10	12	21	28	42	80

(a) These radii represent average values for forming in conventional equipment with tools of good design and condition. The minimum permissible radii in a forming operation on a specific part are subject to several variables and can be determined only by forming under shop conditions.

Table 4. Typical clearances between punch and die for successive drawing operations

Draw	Clearance per side, % of stock thickness
Cylindrical shells	
First	110
Second	115
Third and subsequent	120
Final (tapered shells only)	100
Rectangular shells	
First and subsequent	110
Final	100

steel or brass sheet, which may be reduced as much as 25% in thickness in a single draw.

Clearances between punch and die are usually equal to the metal thickness plus about 10% per side for drawing alloys of low or intermediate strength. An additional 5 to 10% clearance may be needed for the higher-strength alloys and harder tempers. Typical clearances for multiple operations in drawing cylindrical and rectangular shells are given in Table 4.

Radii on Tools. Tools used for drawing aluminum alloys are ordinarily provided with draw radii equal to four to eight times the stock thickness. Punch nose radius is sometimes as large as ten times the stock thickness.

Tool Materials. The selection of materials for deep drawing tools is discussed under "Deep Drawing Dies" in the article "Tool Materials for Special Applications" in Section 18. Materials for small dies are chosen almost entirely on the basis of performance, but cost becomes a significant factor for large dies. Local variation in wear on tools is an important factor in tool life. A twentyfold variation in rate of wear can be observed on the die radius.

Lubricants for deep drawing aluminum alloys are usually commercial products based on the compositions listed in Table 1.

Practical limits for single-operation deep drawing of cylindrical cups and rectangular boxes have been expressed in terms of dimensional ratios as shown in Fig. 1. (Reverse redrawing can be used

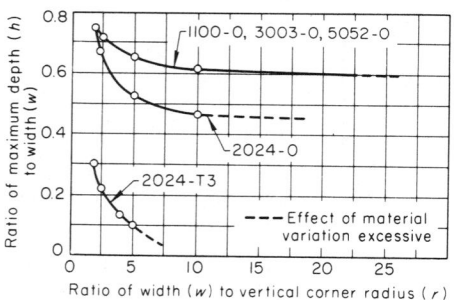

For cylindrical cups, width *w* equals the diameter, and vertical corner radius *r* equals half the diameter. Thus, the *w*/*r* ratio is 2, and values for *h*/*w* can be obtained from the graph.

For rectangular boxes, width *w* equals the square root of the projected bottom area (width times length). If length is more than three times width, drawing limits will be more severe than limits shown in the above graph. For flanged boxes, the width of the flange must be included in depth *h*.

Fig. 1. Drawing limits for one-operation forming of cylindrical cups or rectangular boxes from aluminum alloy sheet 0.026 to 0.064 in. thick

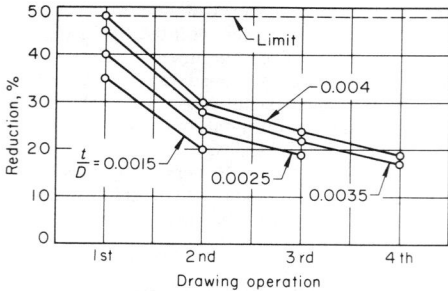

t = metal thickness. *D* = blank diameter.

Fig. 2. Effect of the relation of metal thickness to blank diameter on percentage reduction for successive drawing operations without intermediate annealing, for low-strength alloys such as 3003-O

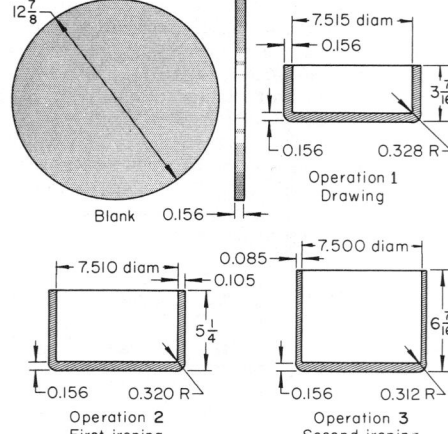

Fig. 3. Progression of shapes in producing a shell with a thick bottom and thin sides in one draw and two ironing operations

to obtain a deeper shell than indicated by the limits in Fig. 1 for conventional drawing methods.)

The relation of the metal thickness (*t*) to the blank diameter (*D*) is an important factor in determining the percentage reduction for each drawing operation. As this ratio decreases, the probability of wrinkling increases, requiring more blankholding pressure to control metal flow and prevent wrinkles from starting. Figure 2 shows the effect of this ratio on percentage reduction of successive draws, without intermediate annealing, for low-strength alloys such as 3003-O.

Ironing is avoided in most deep drawing applications with aluminum alloys, but can be used to produce a shell with a heavy bottom and thin sidewalls.

The shell is first drawn to approximately the final diameter. The drawing lubricant is then removed, and the shell is annealed, bringing it to temperature rapidly to minimize the formation of coarse grains in areas that have been only slightly cold worked.

The sidewalls can then be reduced in thickness by 30 to 40% in an ironing operation. By repeating the cleaning, annealing and ironing steps, an additional reduction of 20 to 25% can be obtained, with good control over wall thickness.

A typical use of ironing is shown in Fig. 3. Here a cylindrical shell is produced with a thick bottom and thin sidewalls by a single deep draw and two successive ironing operations. The approximate final diameter and about half the final depth are obtained in the drawing operation. Wall thickness is reduced 33% in the first ironing step and 19% in the second.

Forming of Beryllium

BERYLLIUM and a beryllium alloy containing 38% aluminum have been successfully formed by bending, three-roll forming, joggling, deep drawing, creep forming, and spinning. The following are required:

1. Equipment that can be controlled at slow speeds and that can withstand the use of heated dies
2. Dies that can withstand the temperature at which beryllium is commonly formed
3. Facilities for preheating and controlling the temperature of dies and workpieces
4. In some applications, facilities for stress relieving the work at 1300 to 1450 F
5. Special lubrication
6. Safety precautions when grit blasting is required for cleaning after forming.

FORMABILITY

The formability of beryllium is low compared with that of most other metals. The crystal structure is hexagonal close-packed; thus there are relatively few slip planes. Plastic deformation is limited and plastic flow is highly anisotropic. The recrystallization temperature for beryllium is around 1400 °F. Cold working is performed above 1000 °F and hot working above 1400 °F. Beryllium is oxidation resistant; thus no special atmosphere is required at these working temperatures. Beryllium is strain-rate sensitive, and thus forming must be performed at low strain rates. Sheet beryllium is formed by working from as-cast ingot or by the P/M technique of canned powder processing.

The Be-38Al alloy has somewhat better formability than ingot or powder sheet beryllium, because this alloy has lower strength, higher ductility, and a cubic crystal structure. The Be-38Al alloy also is less notch sensitive at room temperature than pure beryllium, and some mild forming can be done without heating the work. The maximum working temperature for this alloy is about 1000 °F; production forming is usually done between 400 and 800 °F.

Effect of temperature on formability (in terms of bend angle at fracture) of ingot sheet, two grades of powder sheet, and the Be-38Al alloy is shown in Fig. 4.

EQUIPMENT AND TOOLING

Presses operated by air or hydraulic systems are usually used for forming beryllium, because of the slow speeds required. Standard mechanical presses or other fast forming presses are not suitable.

Because the tools used for forming beryllium will be heated, allowances must be made for thermal expansion, high-temperature strength, and

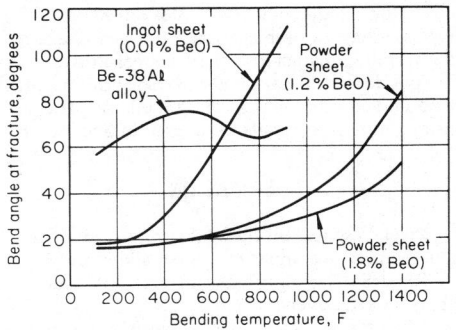

Fig. 4. Effect of temperature on bend angle of beryllium sheet, using a 2t bend radius. Angle plotted is the angle through which the sheet was bent before fracture occurred.

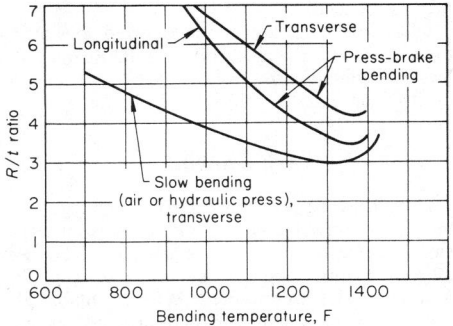

Fig. 5. Minimum bend limits for press-brake versus slower bending of cross-rolled beryllium powder sheet 0.060 to 0.068 in. thick. (R: bend radius; t: sheet thickness.)

oxidation when selecting tool material and designing tools.

Heating Dies and Workpieces. In most forming applications, both the die and the workpiece must be preheated. Dies are specially constructed to permit heating; heat may be supplied by either electrical elements or gas burners.

LUBRICATION

Lubrication or coating of some type is needed in a majority of forming operations. For less severe operations, such as bending, powdered mica has been used. For operations like joggling, forming in matched dies, or deep drawing, colloidal graphite in oil is commonly used. Colloidal graphite in oil is used also as a lubricant for spinning.

BENDING

To achieve consistent results in bending beryllium, slow speeds are required; thus, rapid methods such as press-brake bending cannot be used. As indicated in Fig. 5, the minimum bend radius (ratio of radius, R, to sheet thickness, t) is far lower for low-velocity methods of bending. Although the R/t ratio changes to some extent for different thicknesses, the data in Fig. 5 can be used as a guide.

Table 5. Compositions, bendability and formability ratings of the most commonly formed copper alloys

Alloy No.	Alloy name	ASTM specification(a)	Nominal composition, %					Bendability(b)	Formability(c)
			Cu	Zn	Pb	Sn	Ni		
110	Electrolytic tough pitch copper	B152	99.9	…	…	…	…	70	E
170	Beryllium copper, 1.7%	B194	98		(1.7 Be) (d)			90(e)	G
172	Beryllium copper, 1.9%	B194	97.8		(1.9 Be) (d)			90(e)	G
210	Gilding, 95%	B36(1)	95	5	…	…	…	100	E
220	Commercial bronze, 90%	B36(2)	90	10	…	…	…	90	E
226	Jewelry bronze, 87.5%	…	87.5	12.5	…	…	…	90	E
230	Red brass, 85%	B36(3)	85	15	…	…	…	70	E
240	Low brass, 80%	B36(4)	80	20	…	…	…	70	E
260	Cartridge brass, 70%	B36(6)	70	30	…	…	…	60	E
268	Yellow brass, 65%	B36(8)	65	35	…	…	…	60	E
280	Muntz metal, 60%	…	60	40	…	…	…	40	F
335	Low-leaded brass, 64.5%	B121(2)	64.5	35	0.5	…	…	50	G
340	Medium-leaded brass, 64.5%	B121(3)	64.5	34.5	1	…	…	40	G
342	High-leaded brass, 64.5%	B121(5)	64.5	33.5	2	…	…	20	F
353	High-leaded brass, 62%	B121(4)	62	36	2	…	…	20	F
356	Extra-high-leaded brass, 62%	B121(6)	62	35.5	2.5	…	…	20	P
408	Gilding bronze, 95%	…	95	3	…	2	…	80	E
413	Tin brass, 1%	…	90	9	…	1	…	70	E
430	Spring bronze, 2%	…	86	12	…	2	…	60	E
434	Spring bronze, 0.8%	…	85	14.2	…	0.8	…	60	E
464	Naval brass	B171	60	39.2	…	0.8	…	40	F
510	Phosphor bronze, 5% - A	B103(A)	95	…	…	5	…	60	E
521	Phosphor bronze, 8% - C	B103(C)	92	…	…	8	…	50	G
544	Free-cutting phosphor bronze	B103(B2)	88	4	4	4	…	20	G
614	Aluminum bronze, D	B169(D)	91	(7 Al, 2 Fe)				40	G
651	Low-silicon bronze, B	B97(B)	98.5	(1.5 Si)				70	E
655	High-silicon bronze, A	B97(A)	97	(3 Si)				70	E
706	Copper nickel, 10%	B171	88.7	(1.3)Fe			10	90	G
715	Copper nickel, 30%	B122(5)	70	…	…	…	30	80	G
735	Nickel silver, 72-18	B122(1)	72	10	…	…	18	50	E
745	Nickel silver, 65-10	B122(3)	65	25	…	…	10	50	E
752	Nickel silver, 65-18	B122(2)	65	17	…	…	18	50	E
754	Nickel silver, 65-15	…	65	20	…	…	15	50	E
757	Nickel silver, 65-12	…	65	23	…	…	12	50	E
770	Nickel silver, 55-18	B122(4)	55	27	…	…	18	40	G

(a) Nearest applicable specification. Number or letter in parentheses designates a specific alloy composition of several covered by the indicated ASTM specification. (b) Based on alloy 210 as 100. (c) E = excellent; G = good; F = fair; P = poor. (d) Nickel and/or cobalt, 0.2% min; nickel plus cobalt plus iron, 0.60% max. (e) Solution heat treated condition.

Forming of Copper and Copper Alloys

COPPER AND COPPER ALLOYS are readily formed into complicated shapes, even in foil thickness. The copper alloys commonly formed are characterized by strength and work-hardening rates intermediate between those of steel and of aluminum alloys.

ALLOYS

Compositions of the copper alloys most commonly formed are listed in Table 5 together with their bendability and formability ratings. Grouping by major alloying elements reflects the effect of composition on physical and mechanical properties, and thus on forming behavior. Alloys with the same general rating in Table 5 may nevertheless have differences in bendability or formability that are significant in specific applications.

Temper Designations. Copper alloys are supplied in annealed (soft) tempers and in cold worked (hard) tempers. The standard temper designations are defined in Table 6.

The forming behavior of material in the annealed tempers depends on grain size, which is controlled in production by the temperature of the final anneal. Higher annealing temperatures produce larger grain sizes, which usually correspond to lower hardness, lower strength, and increased ductility.

Table 6. Standard temper designations for copper alloy flat stock

	Average grain size, mm(a)	
Nominal	Minimum	Maximum
Annealed (soft) tempers(b)		
0.015	(c)	0.025
0.025	0.015	0.035
0.035	0.025	0.050
0.050	0.035	0.090
0.070	0.050	0.100
0.120	0.070	(d)

Standard designation	Nominal reduction in thickness, %
Rolled (hard) tempers	
Quarter hard	11
Half hard	21
Three-quarters hard	29
Hard (full hard)	37
Extra hard	50
Spring	60
Extra spring	68

(a) Usual range; other limits can be specified. (b) The larger the grain size, the lower the hardness. (c) Although no minimum grain size is required, the material must be fully recrystallized. (d) Not usually specified.

Material in the annealed temper is classified by average grain size into the six grades shown in Table 6. Special annealed tempers can sometimes be produced to meet particularly critical forming requirements by modifying standard mill procedures. For example, the standard brasses are available in strip of extremely fine grain size, intended for bending and shallow drawing applications, in addition to the standard annealed tempers. This material has an unusual combination of ductility, smoothness and strength, but the small grain size (below 0.010 mm) reduces ductility enough to limit the use of the material to shallow forming applications.

Cold worked material is classified on the basis of the amount of cold reduction in rolling after annealing. The seven standard temper designations, and the corresponding reduction in thickness for each, are listed in Table 6.

WORK HARDENING

Work-hardening effects for copper and yellow brass are shown in Fig. 6. The curves of tensile strength in Fig. 6(a) indicate the rate at which the two metals harden during cold working. On this basis, copper work hardens at a rate equivalent to an increase of about 400 psi in the tensile strength for each 1% reduction of area by drawing. In comparison, yellow brass work hardens at a rate equivalent to an increase of about 1000 psi for a 1% reduction in area.

FORMABILITY OF COPPER ALLOYS VS OTHER METALS

Minimum draw-die radius, as determined in an empirical test procedure, is used to compare the forming characteristics of metals. Table 7 shows the results obtained in forming a cup 1 in. deep and 1 in. in diameter, from 2-in.-diam annealed blanks of different metals in thicknesses of 0.010 to 0.060 in. Alloy 260 (cartridge brass, 70%), 1010 steel, and aluminum alloy 1100 all show the same minimum draw-die radii, which are about 40% smaller than the radii that can be used for annealed type 18-8 stainless steel.

EQUIPMENT AND TOOLING

Equipment that is used to form other commonly formed metals is used also for copper alloys. High-speed production, long runs, and small part size are characteristic of the production of commercial items from copper alloy flat stock. Size and tonnage requirements are generally lower than those for the forming of steel or aluminum alloys. Small, high-speed eyelet presses are widely used for forming of copper alloys.

Copper alloys can be formed in dies designed for other metals of approximately the same strength and formability. Dies designed for deep drawing or severe bending of low-strength aluminum alloys or low-carbon steels can be used with little or no modification for copper and the readily formable brasses, although for shallow drawing or mild bending it may be necessary to make die adjustments to compensate for the smaller amount of springback with these copper alloys. To use the dies for high-strength copper alloys, additional forming stages and die modifications may be required in order to maintain dimensions; completely new dies may be needed in some instances. Dies designed for high-strength aluminum alloys or high-strength steels are generally suitable for use on high-strength copper alloys.

Most drawing dies designed for aluminum alloys do not reduce stock thickness by more than 10%, and redesign of dies and forming procedures for a reduction in thickness of 40 to 50% is sometimes desirable to take full advantage of the forming characteristics of copper alloys.

Tool design is described in articles in this volume on the individual forming methods. The selection of tool materials is covered in Section 18.

LUBRICATION

Water-base and oil-base lubricants used in various forming operations on copper alloys are given in Table 8. They are listed in order of their increasing effectiveness.

Water-Base Lubricants. The soap-plus-fat paste compounds have the widest range of usefulness of the water-base lubricants, because they are readily modified by adjustment of the extent of dilution and by the addition of pigments. These pigments are of particular help in severe draws with copper-nickel alloys or heavy-gage brass, and in other heavy-duty applications where some mechanical separation between tools and workpiece is needed.

Oil-base lubricants are often preferred to water-base lubricants, because oil-base lubricants are less likely to stain the work if allowed to remain on it for an extended time.

DEEP DRAWING

Deep drawing, or drawing to a depth greater than the diameter or the smallest lateral dimension, can be done readily on nearly all copper alloys. In most applications of deep drawing, multiple draws are required. With a suitable alloy and temper, and with favorable processing conditions, several draws can be made in succession without intermediate annealing.

Drawing procedures, equipment and tools are generally the same as those used in the deep drawing of steel and aluminum. The emphasis is on mass production of relatively small articles, chiefly using eyelet presses. Deep drawing procedures for copper alloys emphasize ironing and thinning of sidewalls to a much greater degree

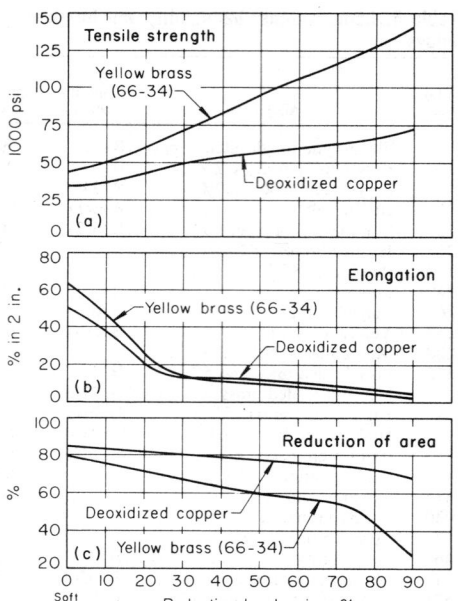

Fig. 6. Effect of cold drawing on tensile strength, elongation, and reduction of area of copper and yellow brass

Table 7. Minimum draw-die radii for copper alloy 260 and various other work metals(a)

Work-metal thickness, in.	Minimum draw-die radius, in.	
	Alloy 260(b), 1010 steel(c) & Al alloy 1100(d)	18-8 stainless steel(e)
0.010	0.05	0.08
0.020	7/64	3/16
0.040	7/32	3/8
0.060	5/16	1/2

(a) Minimum die radii for drawing a cup 1 in. deep by 1 in. in diameter from a 2-in.-diam annealed blank. (b) Cartridge brass, 70% (tensile strength, 49,000 psi). (c) Tensile strength, 47,000 psi. (d) Tensile strength, 13,000 psi. (e) Tensile strength, 90,000 psi.

Table 8. Lubricants commonly used in forming of copper alloys
(Listed in order of increasing effectiveness)

Water-base lubricants	Oil-base lubricants
Blanking and Piercing, Mild Bending, and Shallow Drawing (Less than 10% reduction in diameter) 1 Dilute soap solutions (0.3 to 2%) 2 Emulsions of 5 to 10% soluble oil(a) or fat(b) in water **Medium Bending and Drawing** (10 to 30% reduction in diameter) 3 Dilute soap solutions (1 to 2%) plus 1% fat plus 0.25% free fatty acid(c) 4 Emulsions of soluble oils (may contain fats and other additives) 5 Soap solution (5%) **Severe Bending and Drawing, and Spinning** (30 to 50% reduction in diameter) 6 Diluted soap-plus-fat pastes (2 to 20% final fat content; may contain mineral oil or free fatty acids) 7 Pigmented soap-plus-fat pastes (chalk, talc, mica) **Maximum-Severity Deep Drawing, and Spinning** (More than 50% reduction in diameter) 8 Dry soap (applied from hot 10 to 15% solution)	**Blanking and Piercing, Mild Bending, and Shallow Drawing** (Less than 10% reduction in diameter) 1 Volatile solvents (mineral spirits, isopropanol, chlorinated solvents)(d) 2 Volatile solvents plus 1 to 10% fatty oil(d,e) or wax 3 Mineral oil (40 to 100 sus at 100 F) **Medium Bending and Drawing** (10 to 30% reduction in diameter) 4 Mineral oil (100 to 300 sus at 100 F) plus 10 to 20% fatty oil 5 Mineral oil (250 to 300 sus at 100 F) plus 2 to 5% free fatty acid **Severe Bending and Drawing, and Spinning** (30 to 50% reduction in diameter) 6 Fatty oils containing 2 to 5% free fatty acid **Maximum-Severity Deep Drawing, and Spinning** (More than 50% reduction in diameter) 7 Pigmented mineral oils, fatty oils, or blends of these 8 Stabilized chlorinated oils (30 to 50% combined chlorine; may be blended with up to 20% mineral oil)

(a) Soluble oil is a mineral oil to which has been added an agent that makes it emulsifiable in water. (b) Fats are fatty oils, such as tallow, that are made water soluble by the addition of an emulsifier. (c) Fatty acid is usually stearic, oleic or palmitic acid, or a mixture of these. (d) Precautions must be taken in handling, storage and use, to avoid hazards of fire and toxicity. (e) Fatty oil is usually prime lard oil with less than 2% free fatty acid; also lanolin, sperm oil, castor oil or rapeseed oil.

than for steels, in order to minimize the number of draws and annealing operations and to improve the surface finish of the sidewall.

Forming of Magnesium Alloys

THE PRINCIPAL DIFFERENCE between forming magnesium alloys and forming steel, aluminum and copper is forming temperature. Magnesium has a hexagonal close-packed structure; thus the allowable deformation at low temperatures is very limited. The majority of production forming is thus carried out at elevated temperature.

COLD FORMING

Cold forming of magnesium alloys is restricted to mild deformation with a generous bend radius. Alloys AZ31B-O (special bending sheet) and LA141A-O are exceptions; they have much better room-temperature formability than most other magnesium alloys.

Bend Radii. Cylinders and cones can be formed from magnesium alloys at room temperature by using standard power rolls. Simple flanges can be press formed at room temperature. Table 9 gives minimum radii for fast bending at room temperature, as in a press brake. Slightly smaller bend radii than given in Table 9 may be used when forming speeds are slower, as in a hydraulic press, or when proved by trial.

HOT FORMING

Magnesium alloys are hot formed at temperatures of 250 to 800 F. Tables 10 and 11 give minimum bend radii for forming magnesium alloys at various temperatures, and list minimum and maximum temperatures and total time of exposure to maximum temperatures. Total time, as shown in Tables 10 and 11, is the cumulative total of the time intervals during which the work is at the specified temperature for the particular alloy and temper. Exceeding the temperature-time relation usually reduces mechanical properties, thus increasing the probability that premature failure will occur in service.

Magnesium alloys have a high coefficient of

Table 9. Recommended minimum bend radii for fast forming of magnesium alloys at room temperature

Alloy and temper	Min bend radius	Alloy and temper	Min bend radius
Sheet(a) (0.020 to 0.249 in. thick)		**Extruded flat strip** (0.875 in. by 0.090 in. thick)	
AZ31B-O, special bending sheet	3.0*t*	AZ31C-F	2.4*t*
AZ31B-O	5.5*t*	AZ31B-F	2.4*t*
AZ31B-H24 ...	8.0*t*	AZ61A-F	1.9*t*
HK31A-O	6.0*t*	AZ80A-F	2.4*t*
HK31A-H24 ...	13.0*t*	AZ80A-T5	8.3*t*
HM21A-T8	9.0*t*	HM31A-T5	11.0*t*
HM21A-T81 ...	10.0*t*	ZK21A-F	15.0*t*
LA141A-O	3.0*t*	ZK60A-F	12.0*t*
ZE10A-O	5.5*t*	ZK60A-T5	12.0*t*
ZE10A-H24	8.0*t*	*t* = work-metal thickness	

(a) Minimum bend radii are based on bending a 6-in.-wide specimen through 90°.

Table 10. Minimum bend radii and hot forming conditions for magnesium alloy sheet 0.020 to 0.249 in. thick

Alloy and temper	Recommended minimum bend radius (*t* = sheet thickness) for forming at:							Minimum temperature, F(a)	Maximum temperature, F	Total time at maximum temperature, minutes(b)
	70 F	200 F	300 F	400 F	500 F	600 F	700 F			
AZ31B-O(c) ...	5.5*t*	5.5*t*	4*t*	3*t*	2*t*	...	...	250	550	60
AZ31B-H24(c) ..	8*t*	8*t*	6*t*	...	...	...	...	250	345	30
									300	60
HK31A-O	6*t*	6*t*	6*t*	5*t*	4*t*	3*t*	2*t*	400	750	60
HK31A-H24 ...	13*t*	13*t*	13*t*	9*t*	8*t*	5*t*	3*t*	400	700	10
									650	60
HM21A-T8	9*t*	9*t*	9*t*	9*t*	9*t*	8*t*	6*t*	400	800	10
									750	60
HM21A-T81 ...	10*t*	9*t*	9*t*	9*t*	9*t*	8*t*	6*t*	400	700	10
									650	60
LA141A-O	3*t*	2*t*	1*t*	...	...	...	...	...	300	60
ZE10A-O	5.5*t*	5.5*t*	4*t*	3*t*	2*t*	...	...	...	525	60
ZE10A-H24	8*t*	8*t*	6*t*	...	...	...	...	...	350	3
									300	60

(a) Alloy strength increases and ductility decreases sharply at forming temperatures lower than minimum. (b) Mechanical properties of completed parts are adversely affected if total time (including preheating, handling and press time) at maximum indicated temperature is exceeded. (c) AZ alloys formed at temperatures below the minimum hot forming temperature must be stress relieved.

Table 11. Minimum bend radii and hot forming conditions for extruded flat magnesium alloy strip(a)

Alloy and temper	Minimum bend radius (*t* = strip thickness)	Minimum temperature, F(b)	Maximum temperature, F	Total time at maximum temperature, minutes(c)
AZ31C-F, AZ31B-F(d)	1.5*t*	250	550	60
AZ61A-F(d) ...	1.0*t*	400	550	60
AZ80A-F(d) ...	0.7*t*	285	550	30
AZ80A-T5(d) ..	1.7*t*	285	450	6
			380	60
ZK60A-F(e) ...	2.0*t*	300	550	30
ZK60A-T5(e) ..	6.6*t*	300	500	6
			400	30
HM31A-T5(e)	6.0*t*	550	800	60
			650	24 hr
ZK21A-F(e) ...	5.0*t*	300	600	30

(a) Applicable for 90° bends and for stretch forming. (b) Strength of the alloy increases and ductility decreases sharply at forming temperatures lower than minimum. (c) Mechanical properties of completed parts are adversely affected if total time (including preheating, handling and press time) at maximum indicated temperature is exceeded. (d) AZ alloys formed at temperatures below the minimum hot forming temperature must be stress relieved. (e) HM and ZK alloys should be stress relieved if straightening is required.

thermal expansion (16.1 micro-in./in./°F), which must be accounted for in the design of hot work tooling.

LUBRICANTS

Generally, lubrication is more important in hot than in cold forming of magnesium alloys, because these alloys are more likely to gall as the forming temperature increases.

Lubricants for forming magnesium alloys include mineral oil, grease, tallow, soap, wax, molybdenum disulfide, colloidal graphite in a volatile vehicle, colloidal graphite in tallow, and thin sheets of paper or fiber glass.

Selection of lubricant depends primarily on forming temperature. For temperatures up to 250 F, oil, grease, tallow, soap and wax are generally used.

When forming is done at temperatures higher than 250 F, the selection of lubricant is narrowed; ordinary oil, grease and wax are eliminated. Colloidal graphite can be applied at any

temperature that is used for forming magnesium alloys, but because graphite is difficult to remove and interferes with subsequent surface treatments, it is used as little as possible.

For some work, lubricants cannot be tolerated at any forming temperature, and thin sheets of paper or fiber glass (depending on temperature) are placed between the work metal and the tools instead of a conventional lubricant.

Forming of Nickel Alloys

NICKEL ALLOYS are strengthened primarily by (a) substitutional solid-solution effects, (b) precipitation hardening, and (c) strain hardening. The ductility of some nickel alloys in the annealed condition makes them adaptable to virtually all methods of cold forming.

Strain Hardening. Because strain hardening is related to the solid-solution strengthening afforded by alloying elements, strain-hardening rate generally increases with the complexity of the alloy. Accordingly, strain-hardening rates range from moderately low for nickel and nickel-copper alloys to moderately high for the nickel-chromium and nickel-iron-chromium alloys. Similarly, the age-hardenable alloys have higher strain-hardening rates than their solid-solution equivalents. Figure 7 compares the strain-hardening rates of six nickel alloys, in terms of the increase in hardness with increasing cold reduction, with those of four other materials. Note that the strain-hardening rates of the nickel alloys are greater than that of 1020 steel and most are less than that of 304 stainless steel.

Nickel alloys are strain-rate sensitive. Low forming rates are required for these alloys in order to avoid excessive forming pressures and reduced ductility.

Some of the higher-strength nickel alloys require forming at elevated temperatures. The hot forming temperature of these alloys is dependent on the melting temperature and/or the precipitation-reaction temperature of the particular alloy. Although high temperature does not significantly diminish the strength of some of these alloys, increased resistance to deformation due to strain hardening is reduced or eliminated.

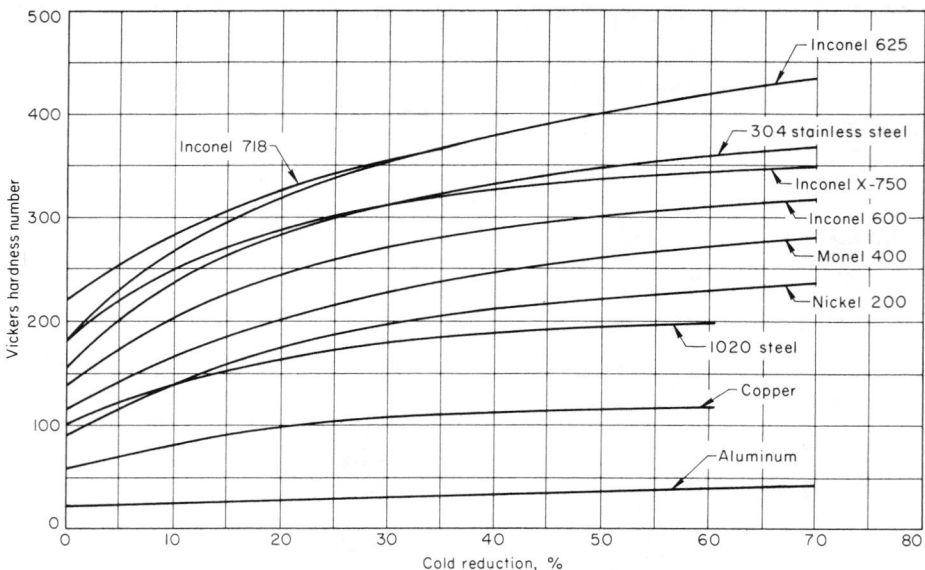

Fig. 7. Effect of cold work on the hardness of sheet metals

TOOLS AND EQUIPMENT

Nickel alloys do not require special equipment for cold forming. However, the physical and mechanical properties of nickel alloys frequently necessitate modification of tools and dies used for cold forming other metals. These modifications are discussed in this section. Information applying to specific cold forming operations is presented in the sections covering those operations.

Die materials used in cold forming austenitic stainless steel are suitable for similar operations on nickel alloys.

Soft die materials such as aluminum bronze, nickel-aluminum bronze, and zinc alloy are used when superior surface finishes are desired. However these materials have a relatively short service life. Parts formed with zinc alloy dies should be flash-pickled in dilute nitric acid to remove any traces of zinc picked up from the dies during forming. Zinc can cause embrittlement of nickel alloys during heat treatment or high-temperature service. For similar reasons, parts formed with brass or bronze dies should be pickled if the dies impart a bronze color to the workpiece.

Dies used for hot working the higher-strength nickel alloys must be made of material that retains its hardness and strength at elevated temperatures. An alternative to the use of such dies is to use cooled dies with heat insulation around the workpiece.

Tool Design. Because nickel alloys are likely to gall, and because of the high pressures developed in forming, tooling should be designed with liberal radii, fillets and clearances. The radii and clearances used in cold forming nickel alloys are usually larger than those used for brass and low-carbon steel, and about equal to those used for the austenitic stainless steels.

Nickel alloys, particularly the nickel-chromium alloys, have high yield strengths and strain-hardening rates. This necessitates stronger and harder dies and more powerful equipment than are required for low-carbon steel. Generally, 30 to 50% more power is required for nickel alloys than for low-carbon steel.

LUBRICANTS

Heavy-duty lubricants are required in most cold forming of nickel alloys.

Although sulfur and chlorine can improve lubricants, they can also have harmful effects if not completely removed after forming. Sulfur will embrittle nickel alloys at elevated temperatures such as might be encountered in annealing or age hardening, and chlorine can cause pitting of the alloys after long exposure.

Pigmented oils and greases should be selected with care as the pigment might be white lead (lead carbonate), zinc oxide, or similar metallic compounds that have low melting points. These elements can embrittle nickel alloys if the compounds are left on the metal during heat treatment. Inert fillers such as talc or flour can be used safely.

Maximum film strength can be obtained by using a coating of copper. However, because application and removal are expensive, metallic coatings are used as lubricants only in severe cold forming operations and then only when they can be properly removed.

Ordinary petroleum greases are seldom used in forming nickel alloys. These greases do not necessarily have the film strength indicated by their viscosity, and they do not have a strong polar attraction for metals. Molybdenum disulfide is seldom recommended for use with nickel alloys because of the difficulty in removing it.

Phosphates do not form usable surface compounds on nickel alloys and cannot be used as lubricant carriers.

Light-bodied mineral oils and water-base lubricants have limited film strength and lubricity and can be used only in light forming operations.

Forming of Titanium Alloys

TITANIUM ALLOYS can be formed in standard machines to tolerances similar to those obtained in the forming of stainless steel. However, in order to lessen the effect of springback vari-

ation on accuracy and to gain the advantage of increased ductility, the great majority of formed titanium parts are made by hot forming or by cold preforming and then hot sizing.

Characteristics of titanium and titanium alloys that must be considered in forming are:

1. Variation in mechanical properties from heat to heat
2. Notch sensitivity, which may cause cracking and tearing, especially in cold forming
3. Galling (worse than with stainless steel)
4. Poor ability to shrink (a disadvantage in some flanging operations)
5. Embrittlement from overheating and from absorption of gases, principally hydrogen (Scale and the surface layer adversely affected by the slower penetration of oxygen can be removed readily.)
6. Narrow spread (workability range) between yield strength and tensile strength.

The formability of annealed titanium alloys in six forming operations, at room temperature or at elevated temperatures, is given in Table 12.

Springback. In general, springback in forming titanium and titanium alloys varies directly with the ratio of bend radius to work-metal thickness, and inversely with forming temperature. These effects are illustrated for press-brake forming of alloy Ti-6Al-4V, in Fig. 8. Springback generally is reduced by increasing the forming pressure.

Springback in titanium alloys is more difficult to predict than springback in steel, although it depends on the same principles. Differences in the yield strength of various heats of titanium can cause differences in springback; higher ratios of yield strength to tensile strength generally result in greater springback.

TOOL MATERIALS AND LUBRICANTS

Tool materials for forming titanium are chosen to suit the forming operation, forming temperature, and expected quantity of production.

Cold forming can be done with epoxy-faced aluminum or zinc tools; hot forming, with ceramic, cast iron, tool steel, stainless steel, and nickel alloy tools.

Galling is the severest problem to be overcome in hot forming. Lubricants may react unfavorably with titanium when it is heated. Suspensions of graphite or molybdenum disulfide have been used successfully. If the lubricant reacts with oxidation products to produce a tenacious surface soil, it must be removed by sand blasting with garnet grit or 120-mesh aluminum oxide.

Temperature-resistant lubricants for hot forming have a graphite or molybdenum disulfide base. Zinc phosphate conversion coatings are sometimes first produced on the work-metal surface to aid in the retention of lubricants during severe forming.

Lubricants for cold forming of titanium are generally similar to those used for severe forming of aluminum alloys.

Tool materials and lubricants for cold and hot forming of titanium alloys are given in Table 13.

COLD FORMING

Commercially pure titanium and the most ductile titanium alloys can be formed cold to a limited extent. Alloy Ti-8Al-1Mo-1V sheet can be cold formed to shallow shapes by standard meth-

Table 12. Formability of annealed titanium alloys in six forming operations at room temperature or elevated temperature(a)

Press brake (minimum bend radius), 70 F	Guerin rubber-pad process(b)		Stretch wrap (maximum stretch), 70 F	Skin stretch (maximum stretch), 850 to 950 F	Drop hammer (max stretch), 900 to 1450 F	Joggle (Length/depth ratio),	
	Stretch (maximum stretch), 600 to 700 F	Shrink (maximum shrink), 600 to 700 F				70 F	600 to 700 F
Ti-13V-11Cr-3Al (1.5t)	Ti-13V-11Cr-3Al (10%)	Ti-13V-11Cr-3Al (6%)	Ti-8Mn (8%)	Ti-8Mn (18%)	Ti-13V-11Cr-3Al (16%)	Ti-13V-11Cr-3Al (1.25)	Ti-13V-11Cr-3Al (1)
Ti-8Mn (3t)	Ti-8Mn (7.5%)	Ti-8Mn (5%)	Ti-5Al-2.5Sn (8%)	Ti-6Al-4V (17%)	Ti-8Mn (16%)	Ti-8Mn (4)	Ti-8Mn (3)
Ti-5Al-2.5Sn (3.5t)	Ti-6Al-4V (5%)	Ti-6Al-4V (4%)	Ti-13V-11Cr-3Al (5.5%)	Ti-13V-11Cr-3Al (13.5%)	Ti-5Al-2.5Sn (13%)	Ti-5Al-2.5Sn (4)	Ti-6Al-4V (3)
Ti-8Al-2Cb-1Ta (4t)	Ti-5Al-2.5Sn (<5%)	Ti-5Al-2.5Sn (3%)	Ti-6Al-4V (3.5%)	Ti-5Al-2.5Sn (12.5%)	Ti-6Al-4V (13%)	Ti-6Al-4V (4.5)	Ti-5Al-2.5Sn (4.5)
Ti-4Al-3Mo-1V(c)(4.5t)							
Ti-2.5Al-16V(c) (4.5t)							
Ti-6Al-4V (4.5t)							
Ti-5Al-2.75Cr-1.25Fe (6.2t)							

(a) Alloys are listed in order of forming ease, the most formable alloy being at the top of the list. Numbers in parentheses following alloy designations are laboratory test values for the indexes of formability shown in parentheses at the top of each list. Laboratory index values shown should be relaxed at least 25% when designing for production. (b) The rubber-pad process is the least suitable of all processes listed for forming at elevated temperature. (c) Solution treated condition.

Fig. 8. Effect of ratio of punch radius to work-metal thickness on springback in press-brake forming alloy Ti-6Al-4V at 70 F and 1000 F

ods, but the bends must be of larger radii than in hot forming and must have shallower stretch flanges. Cold forming of other alloys generally results in excessive springback, requires stress relieving between operations, and requires more power. Titanium and titanium alloys are commonly stretch formed without being heated, although sometimes the die is warmed to 300 F.

For cold forming of all titanium alloys, formability is best at low forming speeds.

HOT FORMING

Heating titanium increases the formability, reduces springback, takes advantage of a lesser variation in yield strength and allows for maximum deformation with minimum annealing between forming operations. Hot sizing is used in almost 90% of the products made of titanium and its alloys. Severe forming must be done in hot dies, generally with preheated stock.

The greatest improvement in the ductility and uniformity of properties for most titanium alloys is at temperatures above 1000 F. However, contamination is also more severe at the higher temperatures.

Certain titanium alloys lend themselves to unique hot forming techniques such as diffusion bonding and superplastic forming. Diffusion bonding entails the contact of two parts under elevated temperature and pressure to form a metallurgical bond between them. Superplastic forming is performed under controlled conditions of temperature, strain rate and microstructure to enable greatly extended deformation limits.

Forming of Platinum Metals

FOUR OF THE PLATINUM METALS— platinum, palladium, rhodium and iridium—have the face-centered cubic crystal structure, which is usually associated with ductility. Yet only platinum and palladium can be cold worked from the cast condition. Rhodium must be broken down

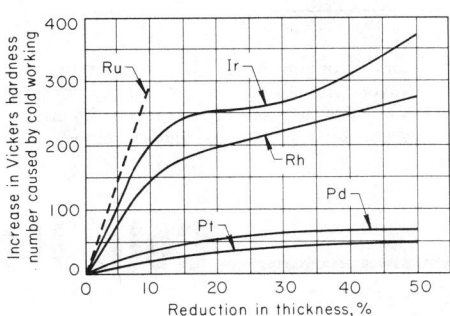

Fig. 9. Effect of cold work in increasing the hardness of the platinum metals

at a high temperature before it can be cold worked, and iridium can be cold worked, with difficulty, only after a fibrous structure has been imparted by careful hot working.

Ruthenium and osmium have a close-packed hexagonal structure. Osmium is completely unworkable and ruthenium very nearly so.

In general, the only problems special to working of the platinum metals are those resulting from surface contamination derived from rolls, swaging dies, and other tools. Base-metal impurities such as iron, which may be smeared on the surface or picked up as slivers or fine dust during

Table 13. Tool materials and lubricants for cold and hot forming of titanium alloys

Operation	Tool material	Lubricant
Cold forming		
Press forming, drawing, drop hammer forming	Cast zinc die or lead punch, with stainless steel caps	Graphite suspension(a)
Press-brake forming	4340 steel (Rc 36 to 40)	Graphite suspension(a)
Contour roll forming, three-roll forming	O2 tool steel	SAE 60 oil
Stretch forming:		
Sheet	Cast aluminum with epoxy face	Grease-oil mixture or wax
Sections	Cast zinc, cast bronze	Wax and graphite(b)
Extrusions	Low-carbon steel, 4130 steel	Molybdenum disulfide or graphite suspension(a)
Hot forming		
Press forming, drawing, drop hammer forming	High-silicon cast iron, RA-330 stainless steel, Inconel X-750, Incoloy 802	Graphite suspension(a)
Sizing	Low-carbon steel, high-silicon gray or ductile iron, H13 tool steel, 310 or RA-330 stainless steel, Inconel X-750, Hastelloy X, Incoloy 802	Graphite suspension(a)
Press-brake forming	H11, H13 tool steel; Incoloy 802	Graphite suspension(a)
Contour roll forming, three-roll forming	H11 or H13 tool steel	Graphite suspension(a)
Stretch forming:		
Sheet	Cast ceramic	Graphite suspension(a)
Sections	H11 or H13 tool steel, high-silicon gray iron	Wax and graphite(b)
Extrusions	4130 steel, 310 stainless steel	Molybdenum disulfide or graphite suspension(a)

(a) In a suitable volatile solvent. (b) Ten parts wax to one part graphite, by volume.

Table 14. Influence of cold work on the hardness of platinum, palladium, and the more important platinum alloys, with recommended annealing temperatures(a)

Reduction of area, %	Pt	Pd	10 Rh 90 Pt	20 Rh 80 Pt	40 Rh 60 Pt	10 Ir 90 Pt	20 Ir 80 Pt	25 Ir 75 Pt	10 Ru 90 Pt
Brinell hardness									
0	53	48	110	128	130	116	192	220	190
10	70	80	145	176	236	136	226	270	242
20	80	88	165	190	264	154	242	286	265
30	86	96	178	200	284	168	252	298	280
40	93	100	185	212	292	176	259	308	286
50	99	106	190	222	308	180	264	316	295
60	103	110	195	234	320	182	272	324	310
70	112	120	200	244	334	185	284	332	325
80	122	135	220	260	356	195	300	339	335
Recommended annealing temperature									
°F	1830	1560	2010	2010	2280	2010	2010	2190	2010
°C	1000	850	1100	1100	1250	1100	1100	1200	1100

(a) Values for hardness and annealing temperature will vary, because of differences in working procedures and in degree of purity of the alloy.

hot working or annealing, will alloy with the surface layers and diffuse inward. Thus, physical characteristics such as electrical resistivity are affected and surface cracking may develop.

PLATINUM

Hot Working. Platinum ingots are normally broken down by hot forging or rolling. Ingots are heated to 2200 to 2750 F (1200 to 1500 C), usually in a gas-fired furnace, supported on high-grade alumina.

Cold Working. Platinum responds readily to cold working and can be reduced 98% or more by rolling or wiredrawing. The rate of work hardening is slow, as shown in Fig. 9 and Table 14.

PLATINUM ALLOYS

The alloys of platinum with up to about 40% rhodium, 30% iridium, or 10% ruthenium comprise those of chief industrial use. All are worked by the same general methods as are used for platinum, allowance being made for the greater stiffness and hardness of the alloys. They can be forged, hot rolled, and hot swaged, usually at temperatures higher than for platinum. All respond to cold working by rolling, swaging, and wiredrawing.

In wiredrawing, platinum and platinum alloys are handled almost exactly like copper. Solid lubricants are used for drawing to about 3/32 in.; for smaller diameters, water-base lubricants of the soluble-oil type are suitable.

Cold Heading and Cold Extrusion

Cold Heading

COLD HEADING is a cold forging process in which the force developed by one or more strokes (blows) of a heading tool is employed to upset, or displace, the metal in a portion of a wire or rod blank to form a section of different contour or, more commonly, of larger cross section than the original. The process is widely used to produce a variety of small and medium-size hardware items—for example, bolts and rivets. However, the process is not limited to the cold deformation of the ends of a workpiece nor to conventional upsetting; metal displacement may be imposed at any point, or at several points, along the length of the workpiece and may incorporate extrusion in addition to upsetting. Advantages of the process over machining of the same parts from suitable bar stock include:

1. Almost no waste material
2. Increased tensile strength from cold work
3. Controlled metal flow.

SUITABLE WORK METALS

Most cold heading is done on low-carbon steel wire with hardness ranging from Rockwell B 75 to 87. This is the type of material for which most machines are rated. Copper, aluminum, stainless steel, and some nickel alloys are also cold headed. Titanium, beryllium, magnesium, and refractory metals are less formable at room temperature and are likely to crack when cold headed; these metals are sometimes warm headed.

Headability decreases as carbon and manganese content increase. Many head splits can be attributed directly to high carbon or high manganese, or both. Carbon steel wire that contains 0.25 to 0.44% carbon should be process annealed or spheroidize annealed, or both.

Quality Levels of Steel. Steel wire for cold heading is generally available in five quality levels (pertaining to surface quality), which are listed below, in order of increasing quality and cost:

1. Industrial quality
2. Cold heading quality
3. Recessed head or scrapless nut quality
4. Special head quality
5. Coil-turned, ground, or shaved wire (seam-free).

The difference in cost between items 1 and 5 above is usually about 30%.

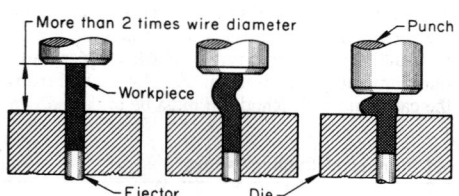

Fig. 1. Typical folding effect with a flat-end punch when heading low-carbon steel wire with unsupported length of more than two diameters

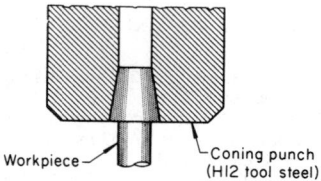

Fig. 2. Use of a coning punch in the first blow of a two-blow heading operation, which enables a low-carbon steel workpiece to be upset to a length of up to six diameters in two strokes

Selection of quality level depends largely on severity of the upset and the magnitude and number of defects that can be tolerated in the headed product.

Length of Upset. Metals are rated for cold heading on the basis of the length of stock, in terms of diameter, that can be successfully upset (compressed into a head). Using flat-end punches, most cold headers can upset up to approximately two diameters of low-carbon steel wire per stroke. If this unsupported length is increased, the stock is likely to buckle or fold on itself, as shown in Fig. 1. With more formable metals such as copper alloys, the length of upset per stroke may be three or more diameters—except for the leaded copper alloys, which are usually limited to 1 1/2 diameters because of the danger of splitting. Punches and dies can be designed to increase the headable length of any stock. For example, with a coning punch (Fig. 2), or a bulbing punch, it is possible to head as much as six diameters of low-carbon steel in two strokes.

MACHINES

Standard cold headers are classified according to whether the dies open and close to admit the work metal or are solid, and according to the number of strokes (blows) the machine imparts to the workpiece during each cycle. The die in a single-stroke machine has one mating punch; in a double-stroke machine, the die has two punches. The two punches usually reciprocate, so that each contacts the workpiece during a machine cycle. Figure 3 shows a doublestroke header with a reciprocating punch holder.

TOOLS

Tools used in cold heading consist principally of punches or hammers and dies. The dies can be made as one piece (solid dies) or as two pieces

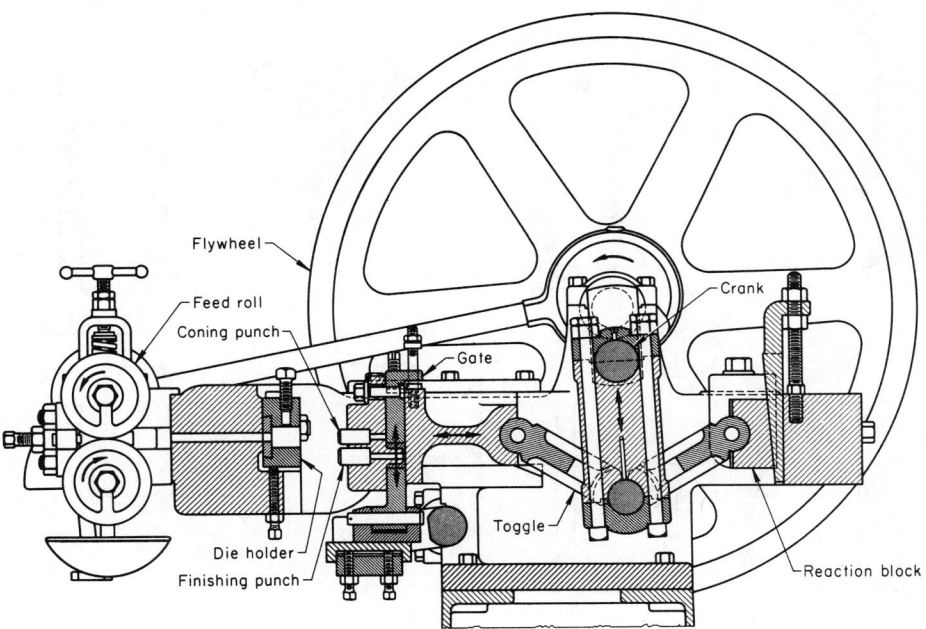

Fig. 3. Essential components and mechanism of a double-stroke solid-die toggle header

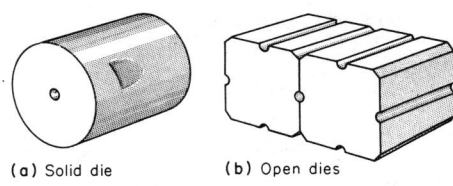

(a) Solid die   (b) Open dies

Fig. 4. Solid (one-piece) and open (two-piece) cold heading dies

(open dies), as shown in Fig. 4.

Solid dies (known also as closed dies) consist of a cylinder of metal with a hole through the center (Fig. 4a). Solid dies may be made entirely from one material, or may be made with the center portion surrounding the hole as an insert of a different material.

Open dies (also called two-piece dies) consist of two blocks with matching grooves in their faces (Fig. 4b). When the grooves in the blocks are put together, they match to form a die hole as in a solid die. The die blocks have as many as eight grooves on various faces, so that as one wears, the block can be turned to make use of a new groove.

TOOL MATERIALS

The shock loads imposed upon cold heading tools must be considered in selecting tool materials. For optimum tool life it is essential that both punches and dies have hard surfaces (preferably Rockwell C 60 or higher). However, with the exception of tools for cold heading of hard materials, the interior portions of the tools must be softer (Rockwell C 40 to 50, and sometimes as low as Rockwell C 35 for larger tools) or breakage is likely.

To meet these conditions, shallow-hardening tool steel such as W1 or W2 is used extensively for punches and open dies, and for solid dies made without inserts.

LUBRICATION

Although some of the more ductile metals can be successfully cold headed to moderate severity without a lubricant, most metals to be cold headed are lubricated to prevent galling of the work metal or the dies, sticking in the dies, and excessive die wear.

Lubricants used include lime coating, phosphate coating, stearates and oils, and plating with softer metals such as copper, tin or cadmium.

The ultimate in lubrication for steel to be cold headed is a coating of zinc phosphate with stearate soap—the same as is used for cold extrusion of steel.

Cold Extrusion

COLD EXTRUSION is so called because the slug or preform enters the die at room temperature or at a temperature appreciably below the recrystallization temperature. Any subsequent rise in temperature, which may amount to several hundred degrees Fahrenheit, is caused by the thermomechanical effects of plastic deformation and friction. Cold extrusion involves backward or forward, or combined backward-and-forward, displacement of metal by plastic flow under steady, though not uniform, pressure. Backward displacement from a closed die is in the direction opposite to punch travel, as shown in Fig. 5(a). Workpieces are often cup-shaped and have wall thickness equal to the clearance between the punch and die. In forward extrusion, the work metal is forced in the direction of the punch travel, as shown in Fig. 5(b). Sometimes these two basic methods of extrusion are combined so that some of the work metal flows backward and some forward, as shown in Fig. 5(c).

Metals Cold Extruded. Aluminum and aluminum alloys, copper and copper alloys, low-carbon and medium-carbon steels, modified carbon steels, low-alloy steels, and stainless steels are the met-

als most commonly cold extruded. The above listing is in the order of decreasing extrudability.

EXTRUSION RATIO

Extrusion ratio is determined by dividing the original area undergoing deformation by the final deformed area. The following calculations show how the ratio is determined for both backward and forward extrusion.

Calculation 1. If a 1-in.-diam slug is forward extruded into a $\frac{1}{2}$-in.-diam product, the extrusion ratio is determined by dividing the area of a 1-in. diameter (0.7854 sq in.), which is the area undergoing deformation, by the area of a $\frac{1}{2}$-in. diameter (0.1963 sq in.), which is the deformed area, thus giving an extrusion ratio of 4 to 1 (usually expressed as 4).

Calculation 2. In backward extruding a 1-in.-diam solid cylindrical slug to produce a tubular section having 1-in. OD and $\frac{1}{2}$-in. ID ($\frac{1}{4}$-in. wall), the extrusion ratio is calculated by dividing the area of the 1-in. OD (0.7854 sq in.) by the annular area of the 1-in. OD (the area undergoing deformation), which is determined by subtracting the area of a $\frac{1}{2}$-in. ID (0.1963 sq in.) from the area of the 1-in. OD, or 0.5891 sq in. Thus, the extrusion ratio is 1.33 to 1 (0.7854 sq in. divided by 0.5891 sq in.).

Calculation 3. Assuming that a tubular section of 1-in. OD and $\frac{1}{2}$-in. ID ($\frac{1}{4}$-in. wall) is to be extruded to a section having $\frac{1}{2}$-in. OD and $\frac{1}{4}$-in. ID ($\frac{1}{8}$-in. wall), the extrusion ratio is calculated by dividing the annular area of the 1-in.-OD tube (0.7854 sq in. minus 0.1963 sq in., or 0.5891 sq in.) by the annular area of the $\frac{1}{2}$-in. tubular section (0.1963 sq in. minus 0.0492 sq in., or 0.1471 sq in.). The extrusion ratio is thus 4 to 1.

PRESSES AND HEADERS

Hydraulic presses, mechanical presses, special knuckle-joint presses for cold extrusion, special cold forging machines, and cold heading machines are employed for cold extrusion. Most presses used for cold extrusion are essentially the same as those used for sheet metal forming.

Most cold extrusion operations are performed on mechanical presses or cold heading machines. Of the two, mechanical presses are used more often, because of their adaptability to other types of operations. Mechanical presses are generally less costly and are capable of higher speeds than are hydraulic presses of similar capacity.

The ram pressure that must be borne during the stroke is a function of the workpiece strength (as affected by composition, state of cold work, anneal, etc.) and the extrusion ratio, as shown in Fig. 6 and 7.

TOOLING

The components of a typical tool assembly used for backward extrusion of steel parts are identified in Fig. 8. There is considerable variation in tooling practice and in the design details of tool-assembly components.

TOOL MATERIALS

Compressive strength of the punch and tensile strength of the die are important considerations when selecting material for cold extrusion tools.

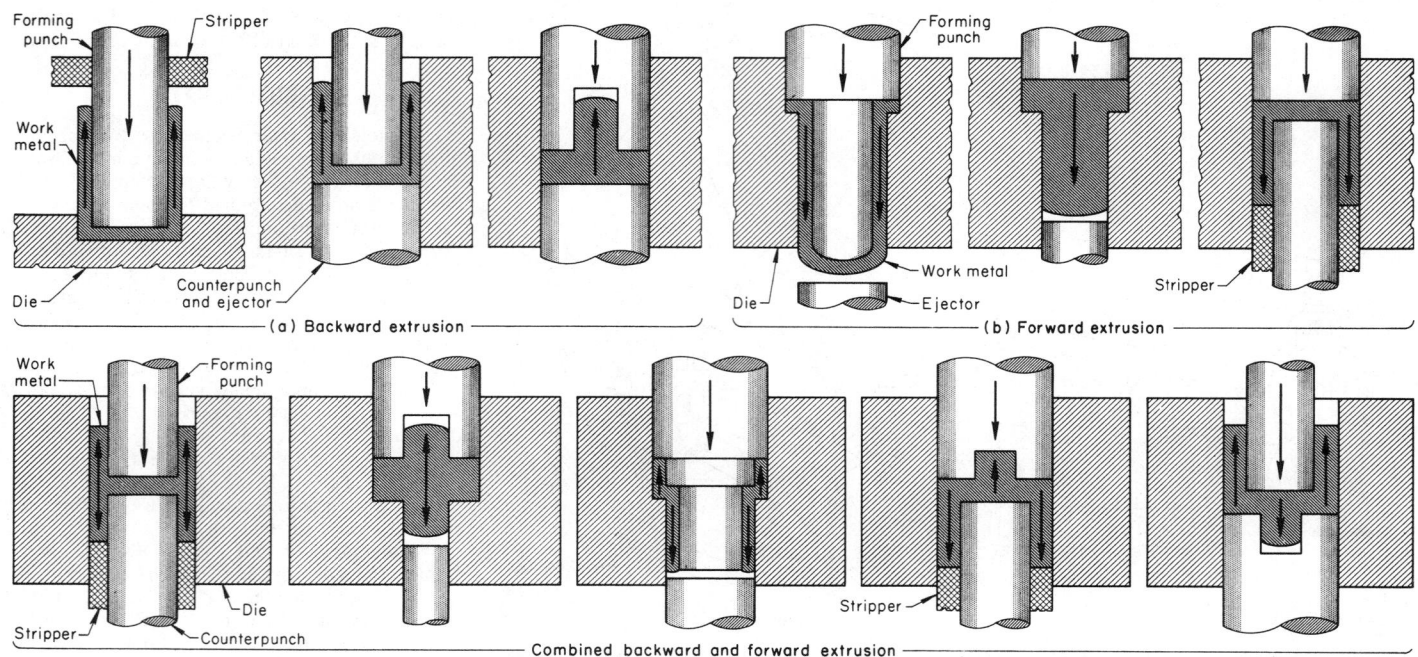

Fig. 5. Displacement of metal in cold extrusion: (a) backward, (b) forward, and (c) combined backward and forward

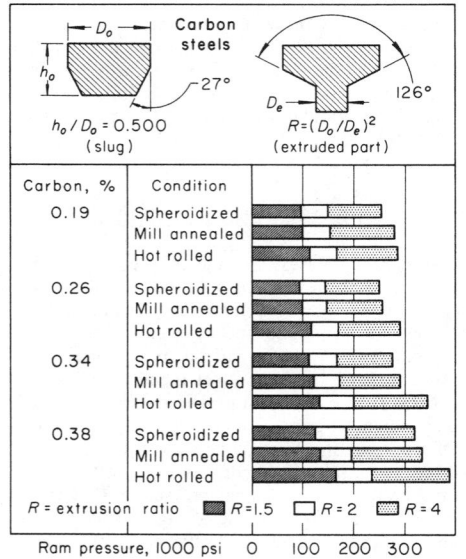

Fig. 6. Effect of carbon content, type of annealing treatment, and extrusion ratio on maximum ram pressure in forward extrusion of the carbon steel part shown from the preformed slug shown

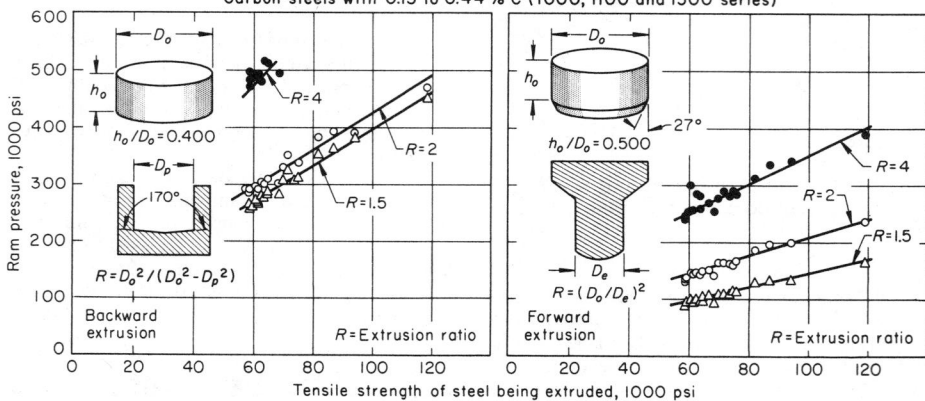

Fig. 7. Effect of tensile strength of steel being extruded on ram pressure required for backward and forward extrusion at different ratios

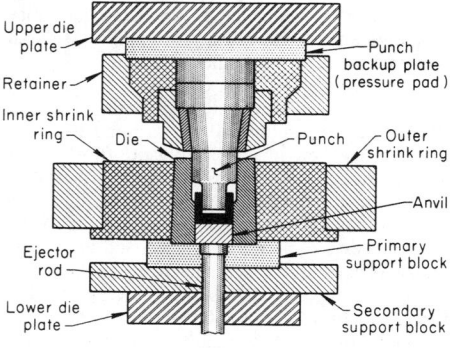

Fig. 8. Nomenclature of tools comprising a typical setup for backward extrusion of steel parts

Because the die is invariably prestressed in compression by the pressure of the inner and outer shrink rings, the principal requirement for a satisfactory die is a combination of tensile yield strength and prestressing that will prevent failure. Punches must have enough compressive strength to resist upsetting without being hazardously brittle. Thus, almost without exception, and particularly for extruding steel, the primary tools in contact with the workpiece must be made of steels that will harden through the section in the sizes involved. (This is notably different from cold heading tools, for which a hard case and soft core are usually desired.) Among the relatively few exceptions are small dies made of a water-hardening tool steel and bore quenched. As the bore hardens, the remainder of the die cools and shrinks, placing the bore in compression.

The degree of strength required for the tools is influenced by workpiece shape, composition and hardness of the metal being extruded, and production requirements.

PREPARATION OF SLUGS

Preparation of slugs often represents a substantial fraction of the cost in producing cold extruded parts. Table 1 lists the methods commonly used for cutting or otherwise producing the slug.

LUBRICANTS FOR STEEL

In most instances, the starting metal surface is given a conversion coating such as zinc phosphate to facilitate lubrication. A soap lubricant gives best results. Slugs are immersed in a dilute

Table 1. Advantages and disadvantages of commonly used methods for producing slugs for cold extrusion

Method of producing	Advantages	Disadvantages
Sawing from bar stock	Ends are square Outside diameter is symmetrical Work hardening is nominal	Loss of metal Burrs must be removed Operation is slow Much floor space required
Cutting off in an automatic bar machine	Ends are square Outside diameter is symmetrical Work hardening is nominal	Loss of metal Burrs must be removed Operation is slow Much floor space required
Blanking from plate	Useful when ratio of length to diameter is small	Metal loss is high Some work hardening occurs
Using a cast preform	Desired shape can be cast	Generally less extrudable than wrought metals.
Shearing in a press	No loss of metal Operation is fast	Size is difficult to control Slug is distorted Ends are work hardened Length must exceed diameter (for minimum distortion)
Shearing and upsetting in a press	No loss of metal Operation is fast	Slug is work hardened Slug is imperfectly filled out
Shearing and heading in a header	No loss of metal Operation is fast Wire or cold drawn bars usable	Slug is work hardened Diameter is limited to $1\frac{1}{4}$ in.

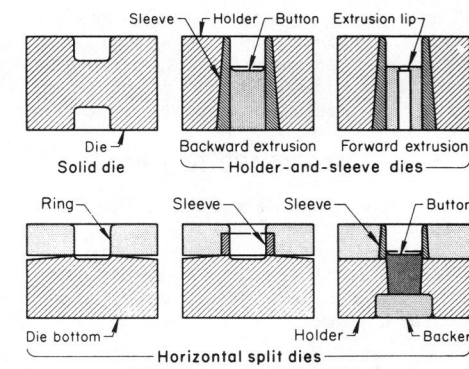

Fig. 9. Three types of dies used in the cold extrusion of aluminum alloy parts

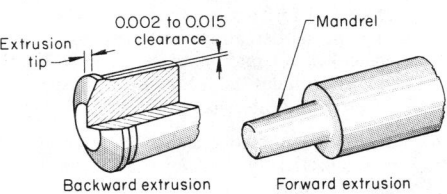

Fig. 10. Typical punches for backward and forward extrusion of aluminum alloy parts

(6 to 16 oz per gallon) soap solution at 145 to 190 F for 3 to 5 min. Some soaps are formulated to react chemically with a zinc phosphate coating, resulting in a layer of water-insoluble metal soap (zinc stearate) on the surfaces of the slugs. This coating has a high degree of lubricity and maintains a film between the work metal and tools at the high pressures and temperatures developed during extrusion.

Other soap lubricants, such as high-titer sodium tallow soaps, with or without filler additives, can be used effectively for mild extrusion of steel. This type of lubricant is absorbed by a phosphate coating, rather than reacting with it.

Although the lubricant obtained by reaction of soap and zinc phosphate is best for extruding steel, its use demands precautions. If soap builds up in the dies, workpieces will not completely fill out. Best practice is to vent all dies so that the soap can escape, and also to keep a coating of mineral seal oil (applied as an air-oil mist) on the dies to prevent adherence of the soap.

When steel extrusions are produced directly from coiled wire (similar to cold heading), the usual practice is to coat the coils with zinc phosphate. This practice, however, has one deficiency; because only the outside diameter of the work metal is coated, the sheared ends are uncoated at the time of extrusion. This deficiency is partly compensated for by constantly flooding the work with sulfochlorinated oil.

Cold (Impact) Extrusion of Aluminum Alloy Parts

ALUMINUM ALLOYS are well adapted to cold extrusion (often called impact extrusion). The lower-strength, more ductile alloys such as 1100 and 3003 are easiest to extrude. When higher mechanical properties are required in the final product, heat treatable grades are used, but extrusions from these metals are more susceptible to defects, such as laps or cracks, than are those from the lower-strength alloys.

Table 2. Relative pressure requirements for cold extruding annealed slugs of five aluminum alloys (alloy 1100 = 1.0)

Alloy	Relative extrusion pressure
1100	1.0
3003	1.2
6061	1.6
2014	1.8
7075	2.3

Although nearly all aluminum alloys can be cold extruded, the five alloys listed in Table 2 are most commonly used.

PRESSES

Presses for extruding aluminum alloys are not necessarily different from those used for extruding steel. There are, however, two considerations that enter into the selection of a press for aluminum: (a) because aluminum extrudes easily, the process is often applied to the forming of deep cuplike or tubular parts, and for this the press should have a long stroke; and (b) also because aluminum extrudes easily, the process is often used for mass production, which requires that the press be capable of high speeds.

The press must have a stroke long enough to permit removal of the longest part to be produced. Except for single-purpose equipment for relatively short extrusions, a stroke of at least 24 in. is required in large mechanical presses. A stroke of 36 in. or more is desirable in intermediate and large hydraulic presses.

TOOLING

Tools designed especially for extruding aluminum may be different from those used for steel, because aluminum extrudes more easily. For instance, a punch used for backward extrusion of steel should not have a length-to-diameter ratio greater than about 3 to 1, whereas this ratio, under favorable conditions, can be as high as 17 to

1 for aluminum (although a 10-to-1 ratio is usually the practical maximum).

Dies. Three basic types of dies for extruding aluminum are shown in Fig. 9.

Punches. Typical punches for forward and backward extrusion are shown in Fig. 10. In backward extruding of deep cuplike parts, specially designed punches may be used to facilitate stripping.

LUBRICANTS

Aluminum and aluminum alloys can be successfully extruded with lubricants such as high-viscosity oil, grease, wax, tallow, and sodium-tallow soap. Zinc stearate, applied by dry tumbling, is an excellent lubricant for extruding aluminum.

The lubricant should be applied to clean metal surfaces, free from foreign oil, grease and dirt. Preliminary etching of the surfaces increases the effectiveness of the lubricant.

For the most difficult aluminum extrusions (less extrudable alloys or greater severity, or both), the slugs should be given a phosphate treatment followed by application of a soap that reacts with the surface to form a lubricating layer similar to that formed when extruding steel.

Cold (Impact) Extrusion of Copper and Copper Alloy Parts

OXYGEN-FREE COPPER (alloy 102) is the most extrudable of the coppers and copper-base alloys. Other grades of copper and most of the copper-base alloys can be cold (impact) extruded, although there are wide differences in extrudability among the different compositions. For example, the harder copper alloys, such as aluminum-silicon bronze and nickel silver, are far more difficult to extrude than are the softer, more ductile alloys, such as cartridge brass (alloy 260).

EQUIPMENT AND TOOLING

Coppers and copper alloys can be extruded in hydraulic or mechanical presses, or in cold heading machines.

Tooling procedures and tool materials are essentially the same for extruding copper alloys as for extruding steel.

PREPARATION OF SLUGS

Surface Preparation. In applications involving minimum to moderate severity, copper slugs are often extruded with no special surface preparation before the lubricant is applied. However, for extruding the harder alloys (aluminum bronze, for instance) or for maximum severity, or both, best practice includes the following surface preparation before the lubricant is applied:

1. Cleaning in an alkaline cleaner to remove oil, grease and soil
2. Rinsing in water
3. Pickling in 10% (by vol) sulfuric acid at 70 to 150 F to remove metal oxides
4. Rinsing in cold water
5. Rinsing in a well-buffered solution, such as carbonate or borate, to neutralize residual acid or acid salts.

Lubrication. Zinc stearate is an excellent lubricant for extruding copper alloys. Common practice is to etch the slugs as described above, and then to coat them by dry tumbling in zinc stearate. An alternative procedure is to dip the slugs (preferably, etched slugs) in a solution of lanolin, zinc stearate, and trichlorethylene.

Impact Extrusion
of Lead Alloys

THE METHOD of producing collapsible tubes and other containers by impact extrusion involves plastic deformation of a flat blank by forcing it through an orifice under hydrostatic pressure. Extrusion time in a mechanical press is about $1/20$ sec; average production rate is 60 tubes per minute.

Tube Extrusion. Blanks for collapsible tubes are

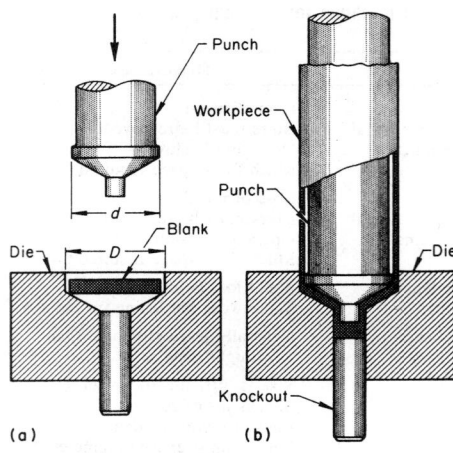

Fig. 11. Impact extruding a collapsible tube

stamped from rolled sheet. After the blank has been lubricated by tumbling in a lubricating medium such as zinc stearate, it is placed in an extrusion die having an inside diameter (D) equal to the outside diameter of the tube to be extruded (see Fig. 11a). A punch is centered above the die. This punch has an outside diameter (d) equal to the inside diameter of the tube. The press moves the punch downward to deform the slug, until all the void below the punch and die has been filled and the metal is completely trapped. As the punch descends farther, an annular orifice between the punch and die will be formed. The width of the orifice is equivalent to the desired thickness of the wall of the tube, conventionally 0.004 to 0.005 in. The punch descends still farther, displacing the metal, and since the annular opening between the punch and the die is the only place through which the metal can escape, the metal is forced to climb up the punch, as shown in Fig. 11(b). The tube is completed when the punch has descended to the desired depth, leaving a predetermined bottom thickness. After the first part of the cycle has been completed, the punch withdraws, the knockout pushes the tube out of the die, and the tube stays on the punch until removed by hand or by an automatic stripper. The tube is thus extruded in one stroke of the press. Trimming and threading, if required, can be done in a second operation. Pure lead, lead alloys (with alloying elements such as antimony, copper, silver, tellurium or tin) and lead sheathed with tin (duplex metal) can be processed in this manner.

Table 3. Pressures required at various temperatures to impact extrude four magnesium alloys to a reduction in area of 85%

| Alloy | Extrusion pressure (tons/sq. in.) at: | | | | | | |
	450 F	500 F	550 F	600 F	650 F	700 F	750 F
AZ31B	33	33	30	27	26	25	23
AZ61A	35	34	33	32	31	30	29
AZ80A	36	35	34	33	32	31	30
ZK60A	34	33	32	31	29	27	26

Impact Extrusion
of Magnesium Alloys

IMPACT EXTRUSION is used for producing symmetrical tubular workpieces, especially those with thin walls or irregular profiles for which other methods are not practical. As applied to magnesium alloys, the extrusion process cannot be referred to as "cold," because both blanks and tooling must be preheated to not less than 350 F.

Equipment and Tooling. Because mechanical presses are faster, they are more widely used for impact extrusion than are hydraulic presses, except when long strokes are needed.

Presses with a capacity of 100 tons and a stroke of 6 in. are adequate for most extrusion applications. Up to 100 extrusions per minute have been produced. Extrusion rate is limited only by press speed.

Dies for impact extruding magnesium alloys differ from those used for other metals, because magnesium alloys are extruded at elevated temperature (usually 500 F). Common practice is to heat the die with tubular electric heaters. The die is insulated from the press by asbestos cloth, and an insulating shroud is built around the die. The top of the die is also covered, except for punch entry and the feeding and ejection devices. The punch is not heated, but during continuous operation becomes hot, so that the punch should be insulated from the ram by asbestos.

Punches and dies usually are made of a hot work tool steel, such as H12 or H13, heat treated to Rockwell C 48 to 52.

Pressures for impact extruding magnesium alloys are about half those required for aluminum and depend mainly on alloy composition, amount of reduction, and operating temperature. Table 3 shows the pressures required to extrude several magnesium alloys to a reduction in area of 85% at seven temperatures from 450 to 750 F.

27 MACHINING

Edited by John F. Kahles, Metcut Research Associates Inc., and based upon machining data and information compiled by the staff of the Machinability Data Center of Metcut Research Associates Inc.

This section in large part was condensed from the Machining Data Handbook, Third Edition, published by Metcut Research Associates Inc., Cincinnati, Ohio. For more detailed information, the reader is referred to the larger work.

Types of Machining Processes

MACHINING is a technology concerned with all of the many different types of processes which remove material in order to manufacture useful goods. Application of the word "machining" has often been limited to include only those mechanical methods used to precisely shape a workpiece by removing some of the workpiece material with tools having sharp and geometrically precise cutting edges. Such processes are differentiated from grinding. Grinding is more widely used to describe processes which accomplish material removal mechanically with hard and brittle grains of abrasive materials, such as aluminum oxide, silicon carbide, and diamond. Abrasive grains of these materials are held together by various bonding materials, including resins, rubber, and vitreous bonds—in the form of abrasive wheels, belts, stones and papers.

The technology of machining and grinding, such as turning, milling, drilling, tapping, surface grinding, cylindrical grinding, and belt grinding, has been universally well established in industry throughout the world for many years. Its development is still being continued at a good rate with notable improvements being made in basic machine tools, in application of numerical and adaptive control, and in the utilization of new tool materials and cutting fluids.

In recent years, many new manufacturing methods—also with the objective of shaping objects by removing material—have been invented and have been developed well enough to assume an important role in industry. These methods—electrical discharge machining (EDM), electrochemical machining (ECM), electrochemical grinding (ECG), chemical milling (CHM), ultrasonic machining (USM), and others—have been variously termed "nonconventional," "nontraditional," "alternate," and "alternative." Such generic terms are no longer sufficiently descriptive and are even misleading in relation to modern

developments and the commercial importance of processes such as EDM, CHM, and others. Thus, it is logical under the heading "Machining" to include not only machining and grinding operations but all of the manufacturing methods used to precision-shape materials through the use of material-removal techniques. The present designations most commonly used for many of the newer types of manufacturing methods include in their trade designations the words "machining" or "grinding"—for example, electrical discharge *machining*, electrochemical *machining*, electrochemical *grinding*, chemical *machining*, and electron beam *machining*. Even more specific designations are also in common usage, such as electrical discharge *drilling*, electrochemical *reaming*, chemical *milling*, and electrical discharge *tapping*.

In practice, the term "machining" for the most part has been applied to those operations capable of achieving high removal rates, for holding tolerances of about ±0.001 in., and for achieving workpiece finishes as fine as 32 R_a with finishing cuts. When tolerances less than ±0.001 in. are required (i.e., of the order of ±0.0001 in.) and finishes of the order of 2 to 32 R_a, it is necessary to specify processes such as surface, cylindrical and centerless grinding, lapping, polishing, and superfinishing. Grinding processes, so-called, generally do not remove much stock or take it off at a high rate. Many exceptions exist, however, in the application of the above criteria. For example, abrasive machining is not machining per se but a grinding operation designed to provide heavy stock removal, the identical objective of so-called machining processes. Ultrasonic machining (impact grinding) does not accomplish heavy stock removal at high rates as do most machining processes, but it does provide high accuracy and high finish typical of those produced by grinding and does use abrasive grain.

The number of commercial machining and grinding processes is large. As has been shown, the naming of these various processes, including the newer methods, lacks precise terminology. It is therefore recommended that all methods of material removal, other than torch cutting and similar processes, be classified under the heading of "machining" and that each individual process be designated by the most commonly used name in the trade.

A list of the different types of machining methods plus definitions follows.

Abrasive belt grinding. Removal of material from a workpiece, roughing and/or finishing, with a power-driven abrasive-coated belt.

Abrasive belt polishing. Finishing a workpiece with a power-driven abrasive-coated belt in order to develop a very good finish.

Abrasive cutoff. Severing a workpiece by means of a thin abrasive wheel.

Abrasive flow machining. Removal of material by a viscous, abrasive media flowing under pressure through or across a workpiece.

Abrasive jet machining. Material removal from a workpiece, by impingement of fine abrasive particles which are entrained in a focused, high-velocity gas stream. Examples:
1. Abrasive jet abrading: For overall material removal, cleaning, etching, and frosting
2. Abrasive jet deburring
3. Abrasive jet drilling
4. Abrasive jet slotting

Abrasive machining. Used to accomplish heavy stock removal at high rates by use of a free-cutting grinding wheel.

Boring. Enlarging a hole by removing metal with a single or occasionally a multiple point cutting tool moving parallel to the axis of rotation of the work or tool.
1. Single-point boring: Cutting with a single-point tool
2. Precision boring: Cutting to tolerances held within narrow limits
3. Gun boring: Cutting of deep holes
4. Jig boring: Cutting of high-precision and accurate-location holes

5. Groove boring: Cutting accurate recesses in hole walls

Broaching. Cutting with a tool which consists of a bar having a single edge or a series of cutting edges (i.e., teeth) on its surface. The cutting edges of multiple-tooth, or successive single-tooth, broaches increase in size and/or change in shape. The broach cuts in a straight line or axial direction when relative motion is produced in relation to the workpiece, which may also be rotating. The entire cut is made in single or multiple passes over the workpiece to shape the required surface contour.
1. Pull broaching: Tool pulled through or over workpiece
2. Push broaching: Tool pushed over or through workpiece
3. Chain broaching: A continuous high production surface broach
4. Tunnel broaching: Work travels through an enclosed area containing broach inserts

Buffing. A two-stage operation: 1) cutting down and 2) coloring. Cutting down removes scratch marks from rough polishing, stretch marks from forming, die marks, or other surface imperfections. It makes a relatively smooth surface smoother. Coloring refines the cut-down surface and brings out maximum luster.

Burnishing. Finish sizing and smooth finishing of surfaces (previously machined or ground) by displacement, rather than removal, of minute surface irregularities with smooth point or line-contact, fixed or rotating tools.

Chemical machining. Controlled dissolution of material by contact with chemical reagents varying in type and strength depending upon the particular alloy being machined. Sometimes current is used at low current densities as an assist (electrochemical milling).

Cold form tapping. Producing internal threads by displacing material rather than removing it as either the tap or the workpiece is rotated. The thread form is produced by a tool, which has neither flutes nor cutting edges, that resembles a simple screw when viewed from the side but the end view shows that both the major and minor diameters have irregular contours for displacing the work material.

Counterboring. Removal of material to enlarge a hole for part of its depth with a rotary, pilot guided, end cutting tool having two or more cutting lips and usually having straight or helical flutes for the passage of chips and the admission of a cutting fluid.

Countersinking. Beveling or tapering the work material around the periphery of a hole creating a concentric surface at an angle less than 90° with the centerline of the hole for the purpose of chamfering holes or recessing screw and rivet heads.

Deburring. Removal of burrs, sharp edges or fins from parts by filing, grinding or rolling the work in a barrel with abrasives suspended in a suitable liquid medium. Sometimes called "burring."

Drilling. Hole making with a rotary, end-cutting tool having one or more cutting lips and one or more helical or straight flutes or tubes for the ejection of chips and the passage of a cutting fluid.
1. Center drilling: Drilling a conical hole in the end of a workpiece
2. Core drilling: Enlarging a hole with a chamfer-edged, multiple-flute drill
3. Spade drilling: Drilling with a flat blade drill tip
4. Step drilling: Using a multiple-diameter drill
5. Gun drilling: Using special straight flute drills with a single lip and cutting fluid at high pressures for deep hole drilling
6. Oil hole or pressurized coolant drilling: Using a drill with one or more continuous holes through its body and shank to permit the passage of a high pressure cutting fluid which emerges at the drill point and ejects chips

Electrical discharge grinding. Grinding by spark discharges between a negative electrode grinding wheel and a positive workpiece separated by a small gap containing a dielectric fluid such as petroleum oil.

Electrical discharge machining. Metal removed by a rapid spark discharge between different polarity electrodes, one the workpiece and the other the tool separated by a gap distance of 0.0005 in. to 0.035 in. The gap is filled with dielectric fluid and metal particles which are melted, in part vaporized and expelled from the gap.

Electrical discharge wire cutting. A special form of electrical discharge machining wherein the electrode is a continuous moving conductive wire.

Electrochemical discharge machining. Metal removal by a combination of the processes of electrochemical machining and electrical discharge machining. Most of the metal removal occurs via anodic dissolution (i.e., ECM action). Oxide films which form as a result of electrolytic action through an electrolytic fluid are removed by intermittent spark discharges (i.e., EDM action). Hence the combination of the two actions.

Electrochemical grinding. Metal is removed by deplating. The workpiece is the anode; the cathode is a conductive aluminum oxide–copper or metal-bonded diamond grinding wheel with abrasive particles. Most of the metal is removed by deplating; 0.05 to 10% is removed by abrasive cutting.

Electrochemical machining. Controlled metal removal by anodic dissolution. Direct current passes through flowing film of conductive solution which separates the workpiece from electrode-tool. The workpiece is the anode, and the tool is the cathode.

Electron beam machining. Material removal accomplished by a high velocity focused stream of electrons which melt and vaporize a workpiece at the point of impingement.

Gear cutting. Producing tooth profiles of equal spacing on the periphery, internal surface, or face of a workpiece by means of an alternate shear gear-form cutter or a gear generator.

Gear hobbing. Gear cutting by use of a tool resembling a worm gear in appearance, having helically-spaced cutting teeth. In a single-thread hob, the rows of teeth advance exactly one pitch as the hob makes one revolution. With only one hob, it is possible to cut interchangeable gears of a given pitch of any number of teeth within the range of the hobbing machine.

Gear milling. Gear cutting with a milling cutter that has been formed to the shape of the tooth space to be cut. The tooth spaces are machined one at a time.

Gear shaping. Gear cutting with a reciprocating gear-shaped cutter rotating in mesh with the work blank.

Gear shaving. A finishing operation performed with a serrated rack or gear like cutter in mesh with the gear, but with their axis skewed.

Grinding. Material removal by use of abrasive grains held by a binder.
1. Surface grinding: Producing a flat surface with a rotating grinding wheel as the workpiece passes under the wheel
2. Cylindrical grinding: Grinding the outside diameters of cylindrical workpieces held between centers
3. Internal grinding: Grinding the inside of a rotating workpiece by use of a wheel spindle which rotates and reciprocates through the length of depth of the hole being ground
4. Centerless grinding: Grinding cylindrical surfaces without use of fixed centers to rotate the work. The work is supported and rotates between three fundamental machine components: the grinding wheel, the regulating wheel, and the work guide blade.
5. Gear grinding: Removal of material to obtain correct gear tooth form by grinding. This is one of the more exact methods of finishing gears.
6. Thread grinding: Thread cutting by use of suitably formed grinding wheel

Hole sawing. The use of a cylindrical saw having end teeth which cut a circular slot through the workpiece leaving a core.

Hollow milling. Using a special end-cutting mill so designed to leave a core after feeding into or through the workpiece.

Honing. A finishing operation using fine grit abrasive stones to produce accurate dimensions and excellent finish.

Hot machining. Machining in which the workpiece shear zone is heated by auxiliary means to reduce the shear strength and increase the machinability of the material.

Hydrodynamic machining. Removal of material by the impingement of a high-velocity fluid against a workpiece.

Lapping. A finishing operation using fine abrasive grits loaded into a lapping material such as cast iron. Lapping provides major refinements in the workpiece including: 1) extreme accuracy of dimension, 2) correction of minor imperfections of shape, 3) refinement of surface finish, and 4) close fit between mating surfaces.

Laser beam machining. Use of a highly focused monofrequency collimated beam of light to melt or sublime material at the point of impingement on a workpiece.

Metal slitting. An operation using a thin circular saw blade to produce a narrow slit in the workpiece. The workpiece is fed into the saw usually on a setup similar to peripheral milling.

Milling. Using a rotary tool with one or more teeth which engage the workpiece and remove material as the workpiece moves past the rotating cutter.
1. Face milling: Milling a surface perpendicular to the axis of the cutter. Peripheral cutting edges remove the bulk of the material while the face cutting edges provide the finish of the surface being generated.
2. End milling: Milling accomplished with a tool having cutting edges on its cylindrical surfaces as well as on its end. In end milling—peripheral, the peripheral cutting edges on the cylindrical surface are used; while in end milling—slotting, both end and peripheral cutting edges remove metal.
3. Side and slot milling: Milling of the side or slot of a workpiece using a peripheral cutter
4. Slab milling: Milling of a surface parallel to the axis of a helical, multiple-toothed cutter mounted on an arbor
5. Straddle milling: Peripheral milling a workpiece on both sides at once using two cutters spaced as required

Planing. Producing flat surfaces by linear reciprocal motion of the work and the table to which it is attached relative to a stationary single-point cutting tool.

Polishing. Removal of metal by the action of abrasive grains carried to the work by a flexible support, generally either a wheel or a coated abrasive belt.

Reaming. An operation in which a previously formed hole is sized and contoured accurately by using a rotary cutting tool (reamer) with one or more cutting elements (teeth). The principal support for the reamer during the cutting action is obtained from the workpiece.
1. Form reaming: Reaming to a contour shape
2. Taper reaming: Using a special reamer for taper pins
3. Hand reaming: Using a long lead reamer which permits reaming by hand
4. Pressure coolant reaming (or gun reaming): Using a multiple-lip, end cutting tool through which coolant is forced at high pressure to flush chips ahead of the tool or back through the flutes for finishing of deep holes

Rotary filing and burring. Machining or smoothing surfaces with contour-fitting rotary tools where only a minimum amount of material is to be removed.

Routing. Cutting out and contouring edges of various shapes in a relatively thin material using a small diameter rotating cutter which is operated at fairly high speeds.

Sawing. Using a toothed blade or disc to sever parts or cut contours.
1. Circular sawing: Using a circular saw fed into the work by motion of either the workpiece or the blade
2. Power band sawing: Using a long, multiple-tooth continuous band resulting in a uniform cutting action as the workpiece is fed into the saw
3. Power hack sawing: Sawing in which a reciprocating saw blade is fed into the workpiece

Shaping. Using single-point tools fixed to a ram reciprocated in a linear motion past the work.
1. Form shaping: Shaping with a tool ground to provide a specified shape
2. Contour shaping: Shaping of an irregular surface, usually with the aid of a tracing mechanism
3. Internal shaping: Shaping of internal forms such as keyways and guides

Skiving. Generating cylindrical forms by moving a form tool laterally through a rotating workpiece.

Snagging. Heavy stock removal of superfluous material from a workpiece by using a portable or swing grinder mounted with a coarse grain abrasive wheel.

Spotfacing. Using a rotary, hole-piloted end-facing tool to produce a flat surface normal to the axis of rotation of the tool on or slightly below the workpiece surface.

Subzero machining. Using refrigerant or other means for cooling the workpiece during, or before, machining.

Superfinishing. An abrasive process utilizing either a

bonded stick for a cylindrical workpiece or a cup wheel for flat and spherical work. A large contact area, 30% approximately, exists between workpiece and abrasive. The object of superfinishing is to remove surface fragmentation and to correct inequalities in geometry, such as grinding feed marks and chatter marks.

Tapping. Producing internal threads with a cylindrical cutting tool having two or more peripheral cutting elements shaped to cut threads of the desired size and form. By a combination of rotary and axial motion, the leading end of the tap cuts the thread while the tap is supported mainly by the thread it produces.

Thermochemical machining. Removal of workpiece material—usually only burrs and fins—by exposure to hot corrosive gases which are formed by detonating an explosive mixture.

Threading. Producing external threads on a cylindrical surface.
1. Die threading: A process for cutting external threads on cylindrical or tapered surfaces by the use of solid or self-opening dies
2. Single-point threading: Turning threads on a lathe
3. Thread grinding: See definition under *Grinding*
4. Thread milling: A method of cutting screw threads with a milling cutter

Trepanning. Cutting with a boring tool so designed as to leave an unmachined core when the operation is completed.

Tumble grinding. Various surfacing operations ranging from deburring and polishing to honing and microfinishing metallic parts before and after plating.

Turning. Generating cylindrical forms by removing metal with a single-point cutting tool moving parallel to the axis of rotation of the work.
1. Single-point turning: Using a tool with one cutting edge
2. Face turning: Turning a surface perpendicular to the axis of the workpiece
3. Form turning: Using a tool with a special shape
4. Turning cutoff: Severing the workpiece with a special lathe tool
5. Box tool turning: Turning the end of a workpiece with one or more cutters mounted in a boxlike frame, primarily for finish cuts

Ultrasonic impact grinding. Material removal by means of an ultrasonic-vibrating tool usually working in an abrasive slurry in close contact with a workpiece or having diamond or carbide cutting particles on its end.

Elements of the Machining Process

The Mechanics of Chip Formation

THE BASIC METAL-CUTTING OPERATION

ALTHOUGH the variety of metal-cutting processes is very large, it is common to use the following characterization: a surface layer of constant thickness is removed by the relative movement between a cutting tool and workpiece (Fig. 1). The cutting speed is defined as the velocity of the relative movement between the cutting tool and workpiece. The thickness, t, often varies as cutting proceeds; however, to simplify the analysis, it is usually considered to be constant. The cutting tool is characterized by two angles: γ, the rake angle, and α, the relief angle (or clearance angle). The relief angle provides clearance to prevent rubbing between the cutting-tool tip and the machined surface; its value is dependent on the strength and wear resistance of the tool material and, to a lesser degree, on the work material. The rake angle is very important in chip formation and is shown positive in Fig. 1. It can be seen from Fig. 1 that the sum of the rake, relief, and wedge angles is equal to 90°; the wedge angle is the included angle between the rake and flank surfaces of the cutting tool.

Two possible top views of Fig. 1 are presented in Fig. 2. In the sketch at left in Fig. 2, the tool's cutting edge is perpendicular to the direction of relative motion. In this case, the cutting is referred to as orthogonal cutting and is easily analyzed compared to the other case where the cutting action is oblique to the cutting edge (sketch at right in Fig. 2). Oblique cutting (drilling, milling, etc.) is far more common in practice than orthogonal, which is more commonly used in theoretical and experimental work.

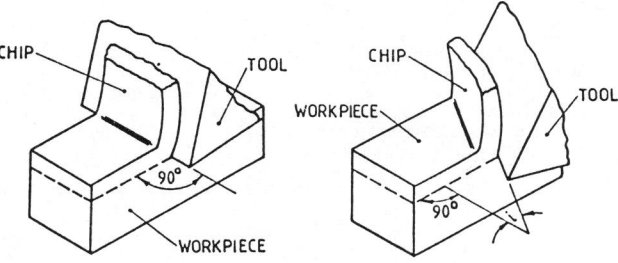

Fig. 2. (Left) Orthogonal and (right) oblique cutting

CHIP FORMATION

Traditionally, three types of chips, as numbered by Ernst (Ref 1), are classified with corresponding theories developed: Fig. 3(a)—Type 1, discontinuous or segmented chip formation; Fig. 3(b)—Type 2, continuous chip formation; and Fig. 3(c)—Type 3, chip formation with a built-up edge (BUE). Of these three, the second type, continuous chip formation, is the most commonly analyzed.

The basic mechanism of chip formation is that of shear deformation. In the simplest analysis, the material is assumed to shear along a shear plane extending from the tool tip to the free surface (sketch at left in Fig. 4). The mechanism is similar to a sliding deck of cards, an analogy suggested by Piispanen (Ref 2); see sketch at right in Fig. 4. Although the region of plastic deformation in metal cutting is referred to as the shear plane, the actual deformation at cutting speeds used in practice occurs across a narrow width or band referred to as the shear zone. The consideration of a shear zone is necessary because if the material instantly went from its undeformed condition (to the left of the shear plane) to its deformed condition (to the right of the shear plane) the transition would occur in zero time requiring infinite force, or alternatively infinite acceleration of the material. Since this is impossible, the deformation must take place in a zone of finite thickness.

REFERENCES

1. Ernst, H. Physics of metal-cutting. Contribution to the special volume *Machining of metals* published by the American Society for Metals, Cleveland, Ohio, 1938.
2. Piispanen, V. Lastunmuodostumisen Teoriaa. *Teknillinen Aikakauslehti,* Vol. 27 (1937).

Tool Wear in Metal Cutting

THE FAILURE of cutting tools can be classified into two broad categories according to the processes by which failure occurs. The categories are:

1. Failure mechanisms that bring the life of the cutting tool to an abrupt, premature end
2. Gradual tool wear that progressively develops on the tool flank surface (flank wear) or on the tool rake face (crater wear).

The first of these categories can be further subdivided into failure modes based on either excessive temperatures or excessive stresses. The progressive or gradual wear of a cutting tool occurs in the two regions illustrated in Fig. 5.

TOOL FAILURE DUE TO EXCESSIVE TEMPERATURES

The elevated temperatures that occur in the tool-chip-workpiece contact zones may cause an initially sharp cutting tool to lose some of its strength and therefore flow plastically under the pressures developed by the cutting force. The flow of the tool material along the flank surface causes the cutting tool to assume a configuration resembling that shown in Fig. 6. In this illustration, both the depression of the tool edge and the bulging on the flank surface of the cutting tool are illustrated. The clearance angle of the cutting tool is reduced to zero for a portion along the flank and, for some period of time, the contact area between the tool and the workpiece increases. During this period, layers of the tool material in contact with the workpiece gradually detach. For a short period of time, the tool may continue to cut

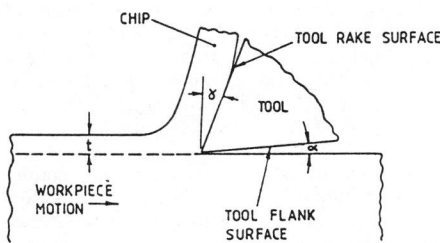

Fig. 1. Idealized view of metal cutting

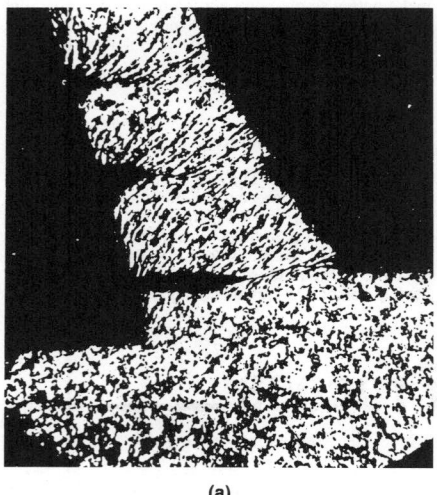 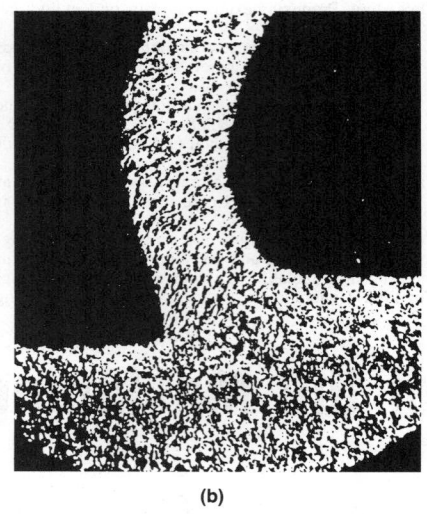 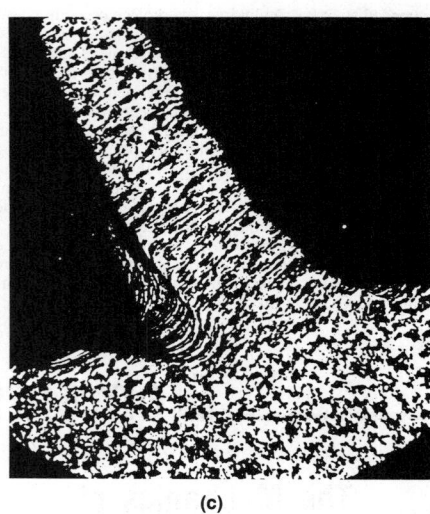

(a)   (b)   (c)

(a) TYPE 1: Discontinuous chip (segmental chip). Initially compressed layer passes off with each chip segment. This type of chip is most easily disposed of. Finish of workpiece is good when pitch of segments is small.

(b) TYPE 2: Continuous chip, with continuously escaping compressed layer adjacent to tool face. Ideal chip from standpoint of quality of finish on workpiece, temperature of tool point, and power consumption.

(c) TYPE 3: Continuous chip, with built-up edge adjacent to tool face. This type of chip is commonly encountered in ductile materials. Finish is rough because of fragments of built-up edge escaping with workpiece.

Fig. 3. Three basic types of chips produced in metal cutting

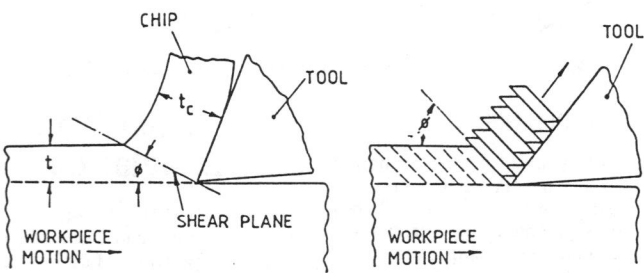

Fig. 4. Mechanism of chip formation. (Left) The shear plane. (Right) Piispanen's "deck of cards" analogy.

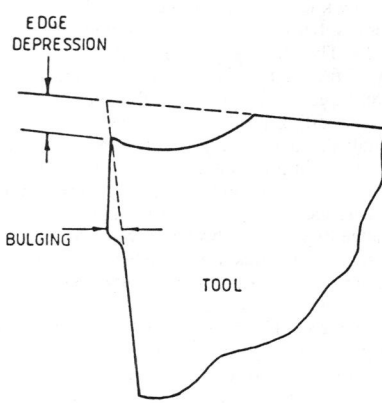

Fig. 6. Edge depression and bulging on a cutting tool due to plastic deformation

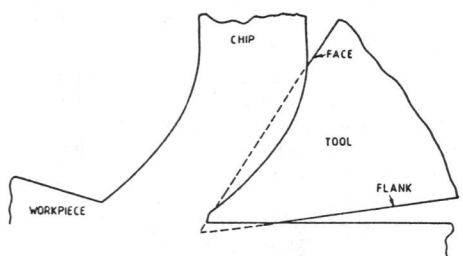

Fig. 5. Tool-wear regions in metal cutting

with this shape and is said to have "form stability." The large area of intimate contact results in substantial friction between the tool and the workpiece, causing the temperature to rise rapidly. The cutting tool then loses its form stability and fails rapidly because of the further softening of the tool material in the cutting region due to the increased level of temperature.

This type of tool failure is not limited to high-speed-steel cutting tools; for, even though cemented-carbide cutting tools are relatively brittle, they possess a certain amount of ductility under the high compressive loads and elevated temperature present during cutting.

TOOL FAILURE DUE TO EXCESSIVE STRESS

An excessive force acting on a cutting tool may cause immediate failure of the cutting edge due to a lack of tool strength. Alternatively, the mechanical failure of the cutting tool may result from a fatigue type of failure. The chipping of a tool and the development of cracks along its cutting edge can be attributed to faulty tool design, material selection, and reconditioning techniques, and to machining conditions conducive to chatter. Methods employed to minimize this type of failure include the use of small or negative tool rake angles on brittle tool materials, employing large side cutting edge angles to protect the tool tip, and honing a narrow chamfer along the cutting edge.

Cutting-tool failure by fracture is not restricted to cemented-carbide tools; fracture of high-speed-steel tools, especially drills, end mills, and form cutters, is common. Many of these tools have intricate geometries, and the correct balance between high strength and high toughness is not always easy to attain. However, the intrinsic brittleness of cemented-carbide tools makes them as a class more susceptible to fracture. Sudden loads caused by dropping the tool or the rapid engagement into a large depth of cut may cause fracture. Additionally, carbide tools are sensitive to

transient thermal stresses that occur during milling and other intermittent cutting operations. Inadequate application of coolant can result in localized cooling, increasing the level of thermal stresses. Thermal stresses can also be introduced by improper brazing and grinding techniques.

Sudden tool failure of cemented-carbide cutting tools can also be attributed to the built-up-edge phenomenon. During sudden tool disengagement, a portion of the built-up edge may break out, taking a portion of the tool material with it. In addition, during the cooling that occurs at the end of a cut, the difference between the linear coefficient of expansion of the built-up-edge material and that of the cemented carbide can introduce cracks in the carbide tool material.

The geometry of the cemented-carbide cutting tool is an important factor affecting tool performance. Larger rake angles lead to improved cutting efficiency but weaken the tool mechanically. The optimum tool geometry depends not only on the specific tool and work material combination

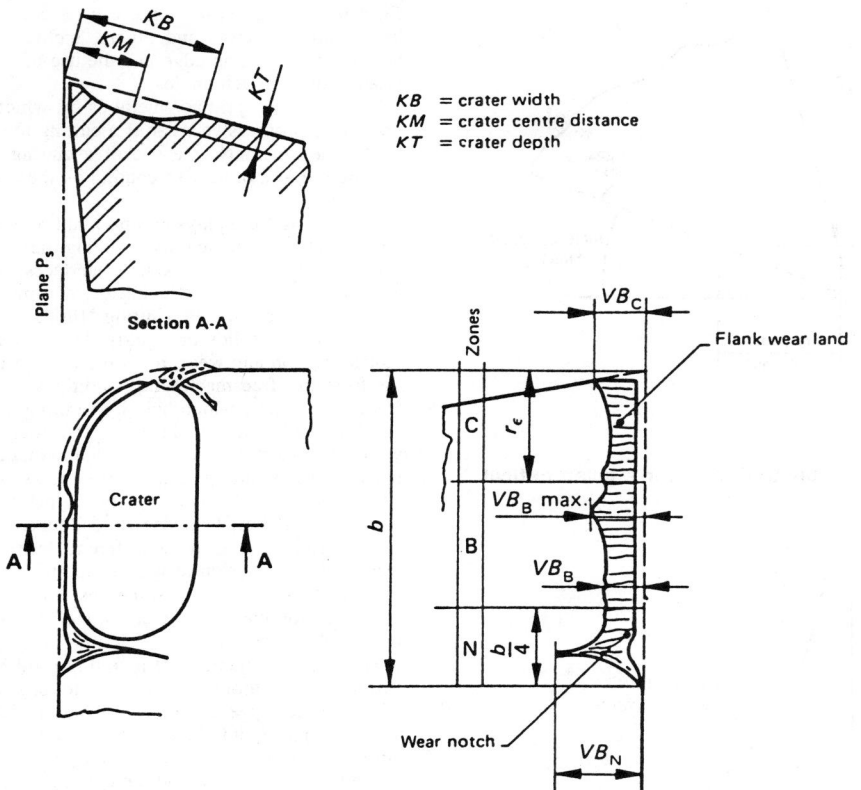

KB = crater width
KM = crater centre distance
KT = crater depth

Fig. 7. Features of single-point-tool wear in turning (ISO Standard)

acterize the maximum level of flank wear that may occur.

In Fig. 8, a typical curve for tool flank wear versus cutting time is illustrated diagrammatically. Generally, the tool-flank-wear curve consists of three stages with different slopes. The three stages can be classified as the initial wear stage, the stationary wear stage, and the final wear stage. In the first stage (Stage I) there is rapid breakdown of the initially sharp cutting edge and a finite wear land is developed. The breakdown of the sharp tool edge results from plastic deformation and the elevated temperatures. The second stage (Stage II) of flank wear is one in which the wear is characterized by a nearly uniform wear rate. Finally, in Stage III, tool flank wear accelerates rapidly and the tool fails soon after reaching this stage. In this stage the presence of a large wear land drastically increases the tool temperature, causing rapid deterioration of the tool edge.

It should be clear from Fig. 8 that at low values of flank wear, for example h_1, predictions cannot be made as to the actual wear after a longer machining period, such as when the wear is h_2. This is due to the complex phenomena associated with tool wear, complicating both the theoretical analysis and practical predictions. In practice, cutting tools should be reground prior to the time the tool wear reaches the third stage where the rapid deterioration of the tool occurs. The ISO standard recommends the criteria for effective tool life given in Table 1.

The influence of the clearance angle on the degree of permissible tool-flank wear for a given critical wear land is illustrated in Fig. 9. With a high clearance angle more flank wear is permissible before the critical wear land is reached; however, excessive clearance weakens the cutting tool.

but also upon the various cutting operations. Research on tool wear has, in the past, focused to a large extent on the progressive wear that occurs on the flanks and rake surfaces of cutting tools. However, knowledge of the causes of and method for avoiding premature failure of cutting tools is extremely important in practice. In recent years, there has been a growing awareness of the importance of tool entrance and exit conditions on the performance of carbide tools in the face milling operation (Ref 1).

GRADUAL WEAR ON THE TOOL FLANK SURFACE

The gradual or progressive wear that develops on the flank surface of a cutting tool is called flank wear. It occurs as a result of friction between the progressively increasing contact area on the tool flank and the newly generated workpiece surface. The width of the wear land on the tool, which developed parallel to the resultant cutting direction because of the rigidity of the tool-workpiece system, is commonly taken as an indication of the amount of wear. In practice, it is customary to measure the wear land with a toolmaker's microscope and use the resulting data as the measure of the amount of tool wear.

The International Organization for Standardization (ISO) standard for tool-life testing (Ref 2) describes the important features of tool wear in single-point turning. Figure 7, taken from the standard, illustrates the main features. It should be noted that the wear on the flank face is seldom uniform along the active cutting edge; it is usually greater at the ends than in the center. As a consequence, the locations and the degree of flank wear should be specified when deciding on the

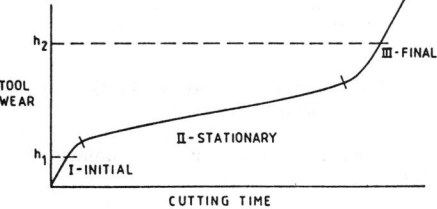

Fig. 8. Tool-flank wear versus cutting time

amount of allowable wear prior to tool resharpening. Flank wear in zone N (see Fig. 7) often contains a notch or groove that forms as a result of contact between the cutting tool and a work-hardened layer of the work material that developed either from a previous processing operation or a preceding motion of the tool itself. In this zone, the width of the wear land is denoted as VB_N.

At the tool corner, zone C in Fig. 7, the cutting action is more complicated because of the flow of the chip around the tool corner. Here the wear land is denoted as VB_C. In the central zone (zone B), the flank wear is more nearly uniform and two designations for the wear in this zone are used: VB_B to represent the average value of the wear land in this region and VB_B max to char-

GRADUAL WEAR ON THE TOOL RAKE SURFACE

The gradual or progressive wear that develops on the rake surface of a cutting tool is called *crater wear*. It occurs as a result of the friction developed as the chip flows over the rake surface of the cutting tool. The crater that is formed conforms to the chip shape and occurs in the region of contact between chip and tool (Fig. 5). Crater wear is largely a temperature-dependent phenomenon. At high cutting speeds the temperatures at the chip-tool interface are on the order of magnitude of 1000 °C; at this temperature level high speed steel tools wear very rapidly. On the other hand, cemented carbide cutting tools retain their hardness at these temperature levels although rapid crater wear can occur by solid-state diffusion. At these conditions of high speed and temperature, the development of the crater on the tool rake surface takes place at the location where the temperature is the greatest.

At the very high speed cutting conditions frequently employed for cemented carbide cutting tools, the life of the tool is often determined by the crater wear. With time, the crater enlarges to

Table 1. Tool-life criteria

High-speed-steel tools	Cemented-carbide tools
Catastrophic failure	$VB_B = 0.3$ mm (uniform wear)
$VB_B = 0.3$ mm (uniform wear)	VB_B max = 0.6 mm (irregular wear)
VB_B max = 0.6 mm (irregular wear)	$KT = 0.06 + 0.3$ f (f = feed)

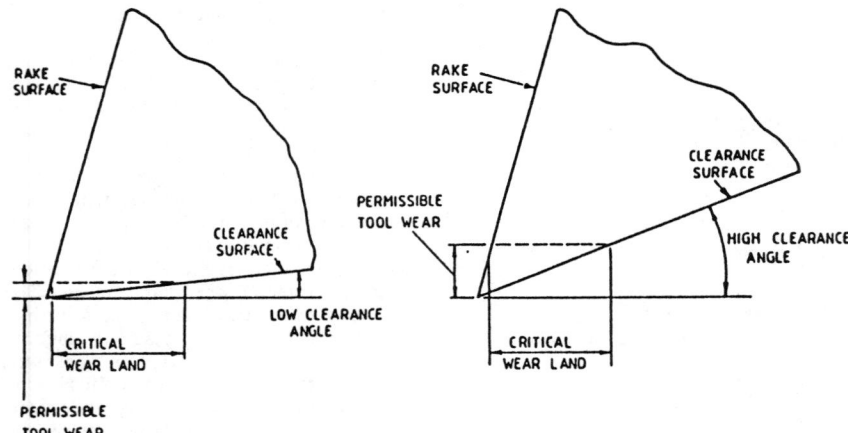

Fig. 9. Influence of clearance angle on permissible tool wear for a given critical wear land

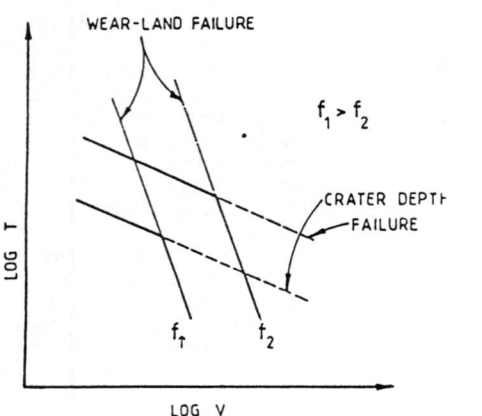

Fig. 10. Tool-life curves for combined flank- and crater-wear failure

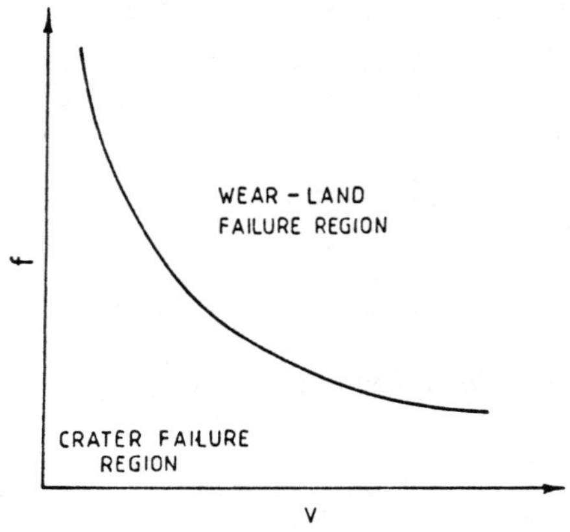

Fig. 11. Regions where flank- and crater-wear criteria apply

such a degree that the edge of the cutting tool is weakened and finally breaks. At this point the tool tip is effectively destroyed and must be replaced. The depth of the crater increases with time much in the same manner as the development of flank wear with time as illustrated in Fig. 8.

Combined flank wear and crater wear tool life criteria can be represented on individual plots of cutting speed and tool life (Fig. 10) and cutting speed and feed rate (Fig. 11). From both illustrations it can be seen that crater wear is more rapid at the higher cutting speeds, where the cutting temperature plays the important role.

While the experimental measurement of cutting-tool wear occurring on the tool nose or tool flank surfaces does not present any serious problems, the measurement of crater wear has been handicapped by a lack of an efficient measuring technique. Consequently, only the simple parameters, such as maximum depth, have been employed, and although these parameters partly serve the purpose, much information regarding the crater shape is not obtained. Experimental techniques for measuring crater wear can be categorized into two groups; namely, optically based methods and nonoptical methods. The nonoptical techniques include the tool-weighing, stylus (most commonly used), lapping, and radioisotope methods. Optical techniques include the crater-plot technique and the Schmaltz light-section

technique. These techniques are summarized in Ref 3.

RETARDATION OF TOOL WEAR

As noted earlier, tool wear is an unavoidable consequence of the metal-cutting operation. Hence, the retardation of tool wear is a subject that has received considerable attention. The optimization of the metal-cutting process involves a balance of resources, and the cutting tool is an important factor in the analysis. Consequently, tool wear and its retardation have a direct relationship with the attaining of machining optimization criteria such as minimum cost, maximum production rate, and maximum profit.

The control of tool wear can be accomplished in a wide variety of ways; here, only a few of the more important methods are summarized. Procedures to enhance tool life in situations where the life of the tool is brought to an abrupt end, such as by fracture and edge chipping, are relatively well known even though they may be somewhat difficult to attain in practice. The selection of the proper grade of tool material, the choice of an appropriate tool geometry, the use of a rigid cutting-tool–workpiece–machine-tool system, and the utilization of reasonable cutting conditions of feed rate, depth of cut, and cutting speed are all important factors to be considered for minimizing the occurrence of tool fracture.

Chipping of the cutting edge can be controlled by the above factors and others, including the honing of the cutting edge and the use of appropriate tool-approach angles.

There are two primary methods by which the gradual or progressive wear of a cutting tool can be retarded; namely, the use of a cutting fluid and the use of a protective coating on the cutting tool itself.

The cutting fluid plays an important role in retarding cutting-tool wear by reducing the friction between the tool-chip-workpiece interfaces, which also has the effect of reducing the cutting force and power requirements. Cutting "fluids" can be in the form of solids or liquids. Solids used to reduce friction and wear in cutting are mostly in the form of "free machining" additives such as lead, sulfur, etc., which help by forming a low-shear-strength film or layer at the tool-chip-workpiece interface, resulting in the reduction of friction during cutting. Liquids used to assist the cutting process are mainly of two types, water based and oil based. These fluids, mixed with various additives, provide different desired effects such as corrosion resistance, lubrication and cooling. In some cutting situations where flood cooling is not practical, the use of a mist or spray is a viable alternative.

The primary functions of a cutting fluid in retarding cutting-tool wear are (a) to provide a cooling effect by removing heat from the cutting tool, workpiece and chips, and (b) to produce a lubricating effect by reducing the friction between the flowing chip and the cutting tool, thus reducing the heat generated in cutting. In both cases, the temperature in the cutting zone is lowered, reducing the tendency for tool softening and wear thus enhancing the life of the cutting tool.

Tool wear can also be retarded by means of a protective coating on the cutting tool. While both high speed steel and cemented carbide cutting tools have been coated, most recent developments have been directed toward the carbides. Research and development into new coating materials and methods for applying these coatings is moving at a very rapid pace. Here, only a brief introduction to this important subject is made; Reference 4 covers this topic in more detail.

Tool wear is retarded by the coating of conventional cemented carbide tool tips with a very thin layer of a hard substance by chemical vapor deposition. The earliest coating materials to be commercially produced were titanium carbide; later, titanium nitride coated tools were developed. The titanium nitride coated tools have a distinctive golden color. The coatings are made of a layer of pure carbide and/or nitride that are metallurgically bonded to the cemented carbide tool tip. The thickness of the layer is on the order of 5 μm (0.0002 in.) and the layers are continuous, nearly free of cracks, and of a very fine grain size. The grain size is around 0.1 μm, which is an order of magnitude smaller than the carbide grains in the tool.

The very thin layers of the coated material remain crack-free, strongly adherent, and effective in wear resistance in many practical situations. It is common to obtain a severalfold increase in the tool life of the coated tools compared with that of the uncoated tools in the machining of steel and cast iron at high speeds. For a given tool life, the coated tools provide an increased metal-removal rate because higher cutting speeds can be used. In a sense, the coated tools combine the advantages of titanium carbide based tools in

LINEAR

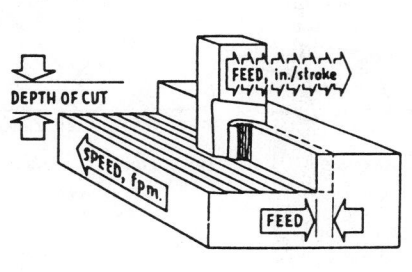

Speed = relative motion between tool and work, stated in surface feet per minute (fpm). Reciprocating motions may be crank-powered, hydraulically, or direct-reversible, electrically driven. There will always be accelerations for a portion of the stroke at least, but cutting speed is assumed to be the average speed. Feed with single-point tools is the amount the tool or work table is indexed. It is measured parallel to the major machined surface and perpendicular to tooth path. Depth of cut is the perpendicular height of the transient surface. Metal removed is feed times depth of cut times length of cut in the direction of tool travel, expressed in cubic inches.

ROTATIONAL

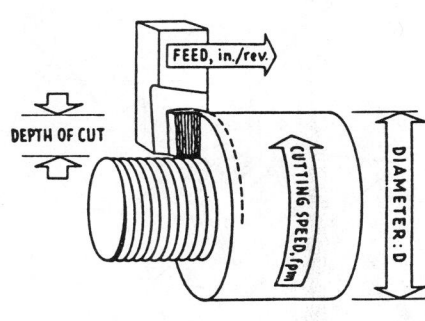

Speed, stated in surface feet per minute, (fpm) is the peripheral speed at the cutting edge. To convert rpm into fpm, use the following:

fpm = πD/12 x rpm. (converting D to ft.) This applies to milling, drilling, turning, and all rotary operations. There are many charts and nomographs available for rapid conversion. Feed per revolution in turning and drilling is a geared feed driven from the main spindle. Divide by number of teeth for feed per tooth. Milling machine feeds are independent of spindle speeds and indicate table travel only.

$$\text{Feed per tooth in milling} = \frac{\text{table feed in inches per minute}}{\text{number of teeth x spindle rpm}}$$

Fig. 12. Work-tool motion

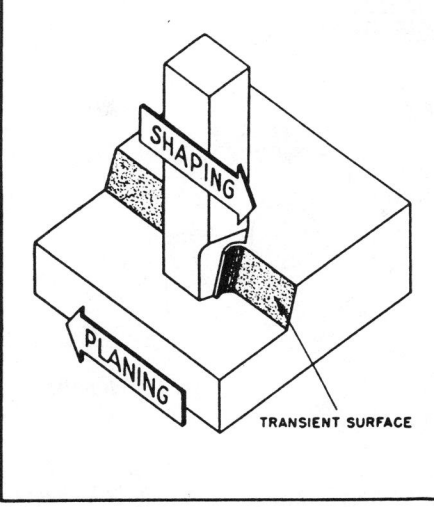

SHAPING, PLANING are similar operations. In a shaper the tool is moved by a ram while the work is fixed. In a planer the work reciprocates on a table while the tool remains stationary. Feed in both operations occurs by quick indexing at the end of every stroke.

MILLING is combined rotation and linear motion perpendicular to rotation axis. Each tooth of a milling cutter (e.g. a face mill) is like a planer tool in terms of the cutting edge and transient surface. Milling variations depend on the arrangement of the cutter and work.

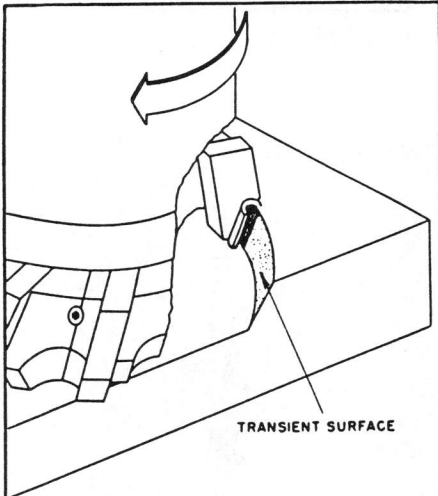

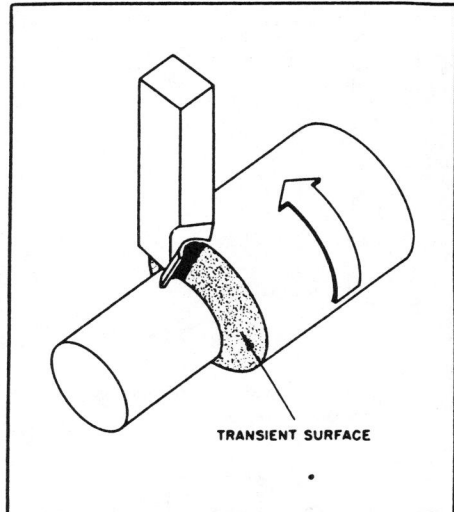

TURNING comprises a rotating workpiece and a stationary tool which feeds slowly but continuously, generating an endless transient surface. Variations of turning include facing, boring, thread chasing and tapping, and machines like vertical turret lathes or boring mills.

DRILLING is a highly-specialized version of turning. Either work or tool may rotate, (the operation can be performed in a lathe) and feed is along drill axis. Drills may have one, two, or more flutes, but two is conventional. The transient surface is continuous helicoid.

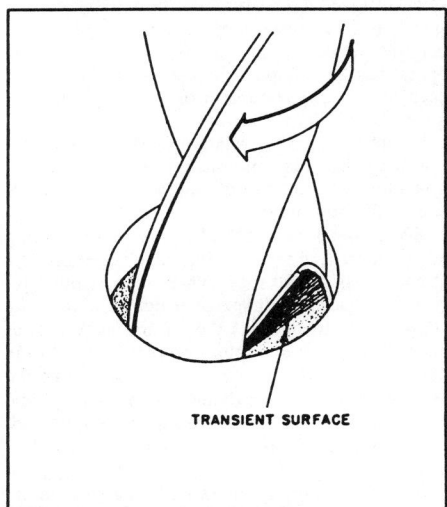

Fig. 13. Generation of the transient surface (from U.S.A.F. Machinability Report—1950, Curtiss-Wright Corp., Wood-Ridge, NJ)

terms of higher metal-removal rates with the toughness of conventional tungsten carbide–cobalt tools when machining steels and cast irons at high speeds.

The key to the success of coated tools is that the coatings act as a barrier to the diffusion that occurs during adhesion wear and at the high speed–high temperature range where diffusion occurs across the tool-chip interface. The coated tools are less successful where the cutting-tool wear is primarily by the abrasion mechanism. Obviously, the coated tools are not suitable in situations where the cutting tool must be reground and in applications involving intermittent

cutting where the coating may fracture and flake away.

REFERENCES

1. Pekelharing, A. J. The exit failure in interrupted cutting. *Annals of the CIRP*, Vol. 27, No. 1 (1978).
2. Tool-life testing with single-point turning tools. ISO 3685, International Organization for Standardization, Geneva, Switzerland, 1977.
3. DeVries, M. F. Measuring carbide tool wear. Technical Paper IQ71-921, Society of Manufacturing Engineers, Dearborn, MI, 1971.
4. Trent, E. M. *Metal cutting*. London: Butterworth, 1977.

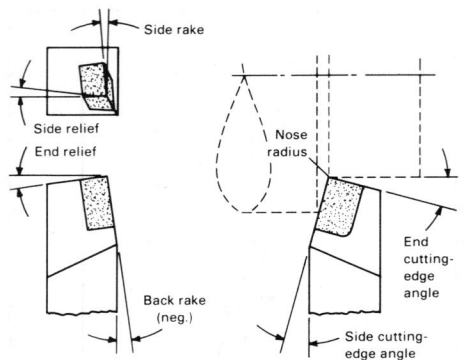

Fig. 14. Identification of angles for single-point tools

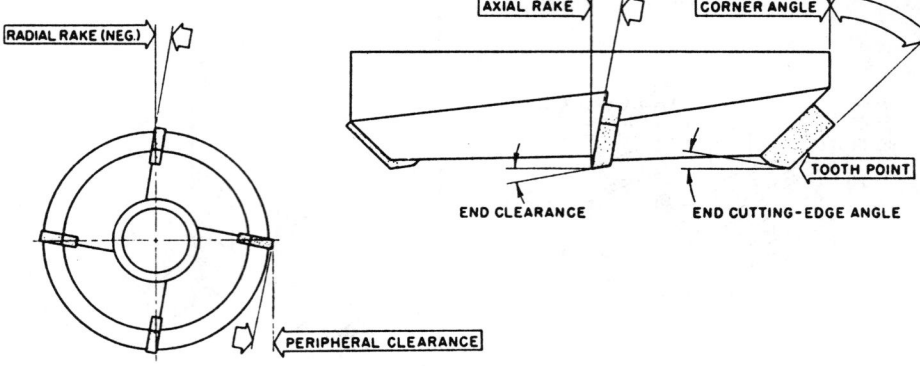

Fig. 15. Identification of angles for a face mill

Machining Parameters

CUTTING SPEED, feed, and depth of cut are terms which apply to all machining operations. Speed is always stated in linear terms, because cutter and work diameters vary and rpm is meaningless without peripheral speeds. Feed is reduced to feed per tooth, because feeding mechanisms in various operations are so different and cutting tools may have any number of teeth. Depth of cut, measured perpendicular to feed increment, permits calculation of chip-section area, regardless of slope of the cutting edge or transient surface. This area, multiplied by cutting speed, gives the metal-removal rate in cubic inches per minute per tooth. In discontinuous cutting operations like shaping or planing, the length of each stroke and the number of strokes per minute must also be known.

Figure 12 indicates relative work and tool motion for linear and rotational types of machining operations. Feed, speed, and depth of cut are shown for both modes.

Every metal-cutting tool has a major cutting edge. In planing, the cutting edge takes one cut at each pass of the tool, whereupon it must reciprocate back and index over before it can take another cut. In milling, the cutting edge rotates clear of the work and must complete a revolution before it can take another cut (at which time the work has meanwhile advanced). In turning, the cutting edge never leaves the work but cuts continuously due to combined simultaneous rotation and axial feed between work and tool.

In most machining operations, the surface cut by the major cutting edge is a temporary or transient surface, because the next tooth or the next index of feed or the next revolution removes it and generates another one. Eventually, all the transient surfaces successively are removed and the "finished" surface is usually produced by a secondary cutting edge.

In terms of the transient surface and the major cutting edge which produces it, all machining operations are fundamentally alike.

Tool angles which locate the major cutting edge and the slope of the tool face also locate the transient surface. Feed and speed settings determine the rate at which the successive transient surfaces are generated and removed, and this whole relationship forms the basis for metal-cutting theory and the analysis of the machining process and its many variations (see Fig. 13).

Chip disposal, surface finish, and tool life are influenced by direction of chip flow. Approximate effects of changing inclination and cutting edge angles are sketched below. The horizontally curling spiral, which may be hard to dispose of and which can damage the machined surface, is successively lifted and turned so it helixes out of tool-work area.

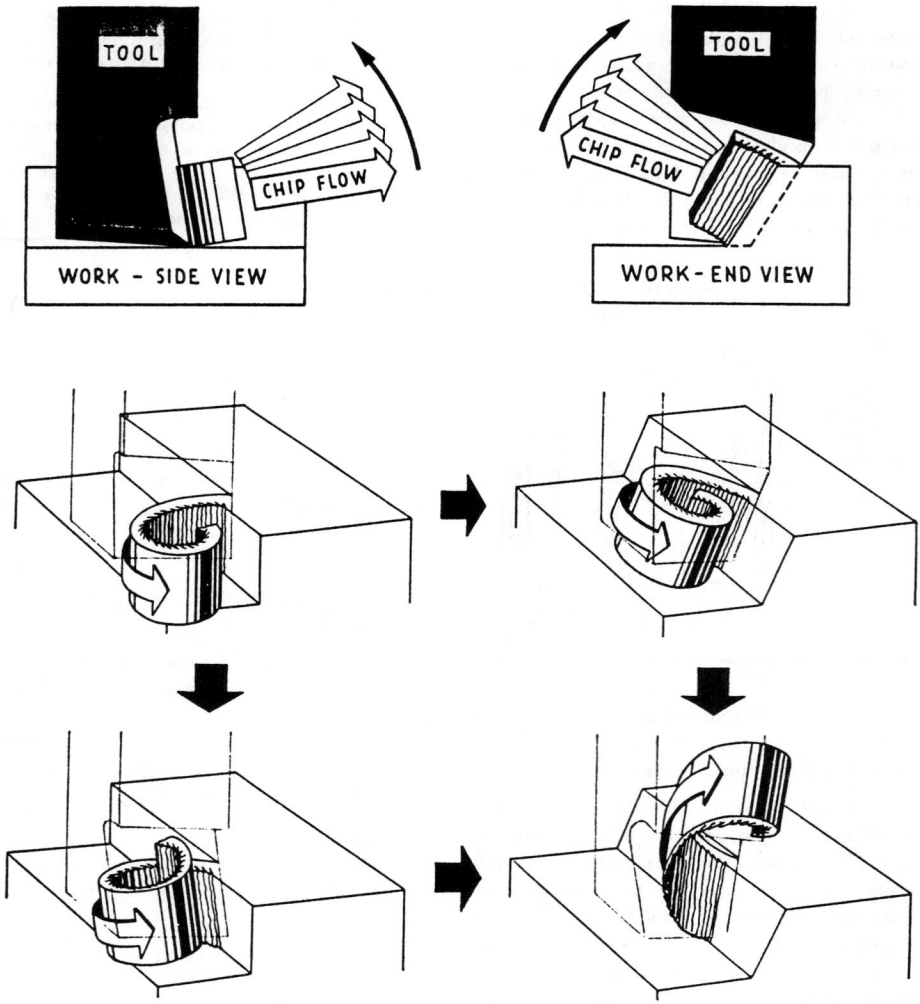

AS INCLINATION INCREASES . . . cutting-edge becomes more inclined in a direction similar to the back rake, tilting the chip plane spiral upward, away from the machined surface.

AS SIDE CUTTING-EDGE ANGLE INCREASES . . . transient surface slopes more, lifting the resultant rake plane, and the chip spiral, causing chip to flow away from the surface.

Fig. 16. Chip flow in machining (from same source as Fig. 13)

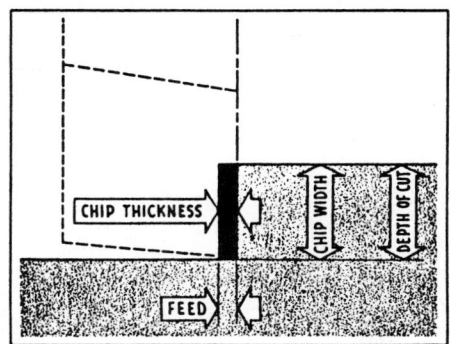

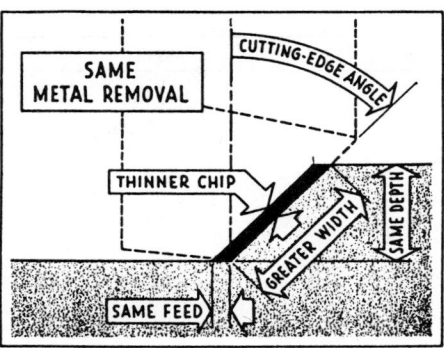

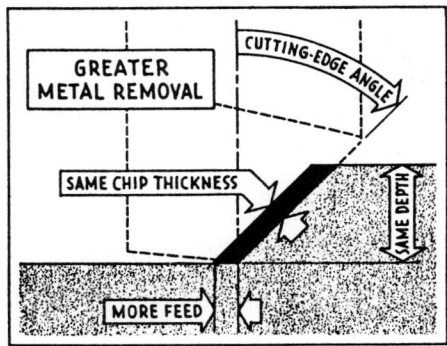

Fig. 18. Tool forces in turning

Fig. 17. Variation of chip thickness with cutting-edge angle (from same source as Fig. 13 and 16)

Chips vary with cutting-edge angle

Chip thickness can be decreased without altering the feed or depth of cut, and therefore without reducing the metal removal rate, by giving the tool a cutting-edge angle. On the other hand, when the optimum chip thickness is known, the feed and hence the removal rate can be increased at a fixed depth of cut by giving the tool the same side cutting-edge angle. In effect, this change makes use of more of the cutting-edge length, and requires more power.

In the conduct of machining processes, cutting-tool angles are carefully specified. Figure 14 shows a single-point tool such as may be used in turning and identifies important angles. Figure 15 shows similar important angles referenced in describing face-milling cutters. Tool angles control chip flow as illustrated in Fig. 16. Chips vary with the cutting-edge angle as shown in Fig. 17.

The control of chip flow has been given more careful consideration recently because of current interest in unmanned machining operations. Provisions must be made for developing chip-flow characteristics which will permit easy disposal without damaging the workpiece or the machine tool, including fixturing.

Forces and Power

THE FORCES in machining can be determined with a tool dynamometer. In turning, the

Table 2. Turning forces

Material and hardness	Tool rake angle	Feed (ipr)	Side cutting edge angle 0° F_t	F_r	F_f	Side cutting edge angle 30° F_t	F_r	F_f
Steels, 135-200 BHN	Pos., +5° to +10°	0.005	1853	541	1116	1928	588	1244
		0.010	3062	725	1324	3196	788	1476
		0.020	5060	970	1570	5265	1055	1750
		0.030	6790	1150	1736	7064	1250	1934
Steels, 200-325 BHN	Pos., +5° to +10°	0.005	1942	497	1183	2015	529	1313
		0.010	3256	738	1429	3373	785	1587
		0.020	5460	1095	1727	5662	1165	1918
		0.030	7396	1380	1930	7662	1467	2142
	Neg., −5° to −10°	0.005	2192	1040	1842	2285	1117	2038
		0.010	3592	1478	2268	3748	1587	2503
		0.020	5885	2100	2778	6134	2254	3075
		0.030	7857	2578	3134	3188	2767	3468
Cast iron, 200-260 BHN	Pos., +5° to +10°	0.005	1636	474	1097	1692	486	1180
		0.010	2790	833	1546	2976	856	1662
		0.020	4727	1466	2180	4880	1506	2344
		0.030	6447	2040	2665	6663	2096	2866

(a) Approximate values. Multiply values given by the depth of cut in inches to obtain the force. To convert to Newtons, multiply force by 4.48. To convert feed (ipr) to feed (mm/r), multiply ipr by 25.4. F_t = tangential or cutting force; F_r = radial force; F_f = feed force.

Table 3. Shop formulas for turning, milling, drilling and broaching—English units (metric units)

Parameter	Turning	Milling	Drilling	Broaching
Cutting speed, fpm	$V_c = .262 \times D_t \times rpm$	$V_c = .262 \times D_m \times rpm$	$V_c = .262 \times D_d \times rpm$	V_c
(m/min)	$\left(V_c = \dfrac{\pi}{1000} \times D_t \times rpm\right)$	$\left(V_c = \dfrac{\pi}{1000} \times D_m \times rpm\right)$	$\left(V_c = \dfrac{\pi}{1000} \times D_d \times rpm\right)$	(V_c)
Revolutions per minute	$rpm = 3.82 \times \dfrac{V_c}{D_t}$	$rpm = 3.82 \times \dfrac{V_c}{D_m}$	$rpm = 3.82 \times \dfrac{V_c}{D_d}$	—
	$\left(rpm = \dfrac{1000}{\pi} \times \dfrac{V_c}{D_t}\right)$	$\left(rpm = \dfrac{1000}{\pi} \times \dfrac{V_c}{D_m}\right)$	$\left(rpm = \dfrac{1000}{\pi} \times \dfrac{V_c}{D_d}\right)$	—
Feed rate, in./min	$f_m = f_r \times rpm$	$f_m = f_t \times n \times rpm$	$f_m = f_r \times rpm$	—
(mm/min)	$(f_m = f_r \times rpm)$	$(f_m = f_t \times n \times rpm)$	$(f_m = f_r \times rpm)$	—

(continued)

Table 3. (continued)

Parameter	Turning	Milling	Drilling	Broaching
Feed per tooth, in.	—	$f_t = \dfrac{f_m}{n \times rpm}$	—	f_t
(mm)	—	$\left(f_t = \dfrac{f_m}{n \times rpm}\right)$	—	(f_t)
Cutting time, min	$t = \dfrac{L}{f_m}$ $\left(t = \dfrac{L}{f_m}\right)$	$t = \dfrac{L}{f_m}$ $\left(t = \dfrac{L}{f_m}\right)$	$t = \dfrac{L}{f_m}$ $\left(t = \dfrac{L}{f_m}\right)$	$t = \dfrac{L}{12\,V_c}$ $\left(t = \dfrac{L}{1000\,V_c}\right)$
Rate of metal removal, in.^{3}/min	$Q = 12 \times d \times f_r \times V_c$	$Q = w \times d \times f_m$	$Q = \dfrac{\pi D_d^2}{4} \times f_m$	$Q = 12 \times w \times d \times V_c$
(cm^3/min)	$(Q = d \times f_r \times V_c)$	$\left(Q = \dfrac{w \times d \times f_m}{1000}\right)$	$\left(Q = \dfrac{\pi D_d^2}{4000} \times f_m\right)$	$(Q = w \times d_t \times V_c)$
Horsepower required at spindle*	$hp_s = Q \times P$	$hp_s = Q \times P$	$hp_s = Q \times P$	—
(Power required at spindle*)	$(kW_s = Q \times P)$	$(kW_s = Q \times P)$	$(kW_s = Q \times P)$	—
Horsepower required at motor*	$hp_m = \dfrac{Q \times P}{E}$	$hp_m = \dfrac{Q \times P}{E}$	$hp_m = \dfrac{Q \times P}{E}$	$hp_m = \dfrac{Q \times P}{E}$
(Power required at motor*)	$\left(kW_m = \dfrac{Q \times P}{E}\right)$	$\left(kW_m = \dfrac{Q \times P}{E}\right)$	$\left(kW_m = \dfrac{Q \times P}{E}\right)$	$\left(kW_m = \dfrac{Q \times P}{E}\right)$
Torque at spindle	$T_s = \dfrac{63030\,hp_s}{rpm}$ $\left(T_s = \dfrac{9549kW_s}{rpm}\right)$	$T_s = \dfrac{63030\,hp_s}{rpm}$ $\left(T_s = \dfrac{9549kW_s}{rpm}\right)$	$T_s = \dfrac{63030\,hp_s}{rpm}$ $\left(T_s = \dfrac{9549kW_s}{rpm}\right)$	—

D_t	=	Diameter of workpiece in turning, in. (mm)	kW_m	=	Power at motor, kW
D_m	=	Diameter of milling cutter, in. (mm)	kW_s	=	Power at spindle, kW
D_d	=	Diameter of drill, in. (mm)	L	=	Length of cut, in. (mm)
d	=	Depth of cut, in. (mm)	n	=	Number of teeth in cutter
d_t	=	Total depth per stroke in broaching, in. (mm)	P	=	Unit power hp/in.^{3}/min (kW/cm^3/min)
E	=	Efficiency of spindle drive	Q	=	Rate of metal removed, in.^{3}/min (cm^3/min)
f_m	=	Feed rate, in./min (mm/min)	rpm	=	Revolutions per minute of work or cutter
f_r	=	Feed, in. (mm) per revolution	T_s	=	Torque at spindle, in.·lb (N·m)
f_t	=	Feed, in. (mm) per tooth	t	=	Cutting time, minutes
hp_m	=	Horsepower at motor	V_c	=	Cutting speed, ft/min (m/min)
hp_s	=	Horsepower at spindle	w	=	Width of cut, in. (mm)

*Unit power data are given in Table 4 for turning, milling and drilling.

Table 4. Average unit power requirements for turning, drilling and milling—English units (metric units)

Material	Hardness, Bhn	Turning (P_t), HSS and carbide tools Feed, 0.005-0.020 ipr (0.12-0.50 mm/r) Sharp tool		Dull tool		Drilling (P_d), HSS drills Feed, 0.002-0.008 ipr (0.05-0.20 mm/r) Sharp tool		Dull tool		Milling (P_m), HSS and carbide tools Feed, 0.005-0.012 ipt (0.12-0.30 mm/t) Sharp tool		Dull tool	
Steels, wrought and cast	85-200	1.1	(0.050)	1.4	(0.064)	1.0	(0.046)	1.3	(0.059)	1.1	(0.050)	1.4	(0.064)
(plain carbon, alloy and	35-40 R_c	1.4	(0.064)	1.7	(0.077)	1.4	(0.064)	1.7	(0.077)	1.5	(0.068)	1.9	(0.086)
tool steels)	40-50 R_c	1.5	(0.068)	1.9	(0.086)	1.7	(0.077)	2.1	(0.096)	1.8	(0.082)	2.2	(0.100)
	50-55 R_c	2.0	(0.091)	2.5	(0.114)	2.1	(0.096)	2.6	(0.118)	2.1	(0.096)	2.6	(0.118)
	55-58 R_c	3.4	(0.155)	4.2	(0.191)	2.6	(0.118)	3.2(b)	(0.146)(b)	2.6	(0.118)	3.2	(0.146)
Cast irons	110-190	0.7	(0.032)	0.9	(0.041)	1.0	(0.046)	1.2	(0.055)	0.6	(0.027)	0.8	(0.036)
(gray, ductile, malleable)	190-320	1.4	(0.064)	1.7	(0.077)	1.6	(0.073)	2.0	(0.091)	1.1	(0.050)	1.4	(0.064)
Stainless steels, wrought and cast	135-275	1.3	(0.059)	1.6	(0.073)	1.1	(0.050)	1.4	(0.064)	1.4	(0.064)	1.7	(0.077)
(ferritic, austenitic, martensitic)	30-45 R_c	1.4	(0.064)	1.7	(0.077)	1.2	(0.055)	1.5	(0.068)	1.5	(0.068)	1.9	(0.086)
PH stainless steels	150-450	1.4	(0.064)	1.7	(0.077)	1.2	(0.055)	1.5	(0.068)	1.5	(0.068)	1.9	(0.086)
Titanium	250-375	1.2	(0.055)	1.5	(0.068)	1.1	(0.050)	1.4	(0.064)	1.1	(0.050)	1.4	(0.064)
Hi-temp alloys (Ni- and Co-base)	200-360	2.5	(0.114)	3.1	(0.141)	2.0	(0.091)	2.5	(0.114)	2.0	(0.091)	2.5	(0.114)
(Iron-base)	180-320	1.6	(0.073)	2.0	(0.091)	1.2	(0.055)	1.5	(0.068)	1.6	(0.073)	2.0	(0.091)
Refractory alloys (Tungsten)	321	2.8	(0.127)	3.5	(0.159)	2.6	(0.118)	3.3(b)	(0.150)(b)	2.9	(0.132)	3.6	(0.164)
(Molybdenum)	229	2.0	(0.091)	2.5	(0.114)	1.6	(0.073)	2.0	(0.091)	1.6	(0.073)	2.0	(0.091)
(Columbium)	217	1.7	(0.077)	2.1	(0.096)	1.4	(0.064)	1.7	(0.077)	1.5	(0.068)	1.9	(0.086)
(Tantalum)	210	2.8	(0.127)	3.5	(0.159)	2.1	(0.096)	2.6	(0.118)	2.0	(0.091)	2.5	(0.114)
Nickel alloys	80-360	2.0	(0.091)	2.5	(0.114)	1.8	(0.082)	2.2	(0.100)	1.9	(0.086)	2.4	(0.109)
Aluminum alloys	30-150(c)	0.25	(0.011)	0.3	(0.014)	0.16	(0.007)	0.2	(0.009)	0.32	(0.015)	0.4	(0.018)
Magnesium alloys	40-90(c)	0.16	(0.007)	0.2	(0.009)	0.16	(0.007)	0.2	(0.009)	0.16	(0.007)	0.2	(0.009)
Copper	80 R_B	1.0	(0.046)	1.2	(0.055)	0.9	(0.041)	1.1	(0.050)	1.0	(0.046)	1.2	(0.055)
Copper alloys	10-80 R_B	0.64	(0.029)	0.8	(0.036)	0.48	(0.022)	0.6	(0.027)	0.64	(0.029)	0.8	(0.036)
	80-100 R_B	1.0	(0.046)	1.2	(0.055)	0.8	(0.036)	1.0	(0.046)	1.0	(0.046)	1.2	(0.055)

(a) Power requirements at spindle drive motor, corrected for 80% spindle drive efficiency. (b) Carbide. (c) 500 kg.

tool dynamometer usually measures three components: the tangential or cutting forces, the thrust or separating forces, and the feed forces (Fig. 18). The cutting force is important since, when multiplied by the cutting velocity, it determines the power requirements in machining. The thrust force or separating force determines the accuracy produced on the part. Typical tables of the force components for several work materials are shown in Table 2.

The forces in milling can be determined again by mounting a dynamometer to the surface of the milling machine table. In the drilling and tapping operations, the thrust and torque can also be determined with a suitable dynamometer.

For general approximations, the power requirements in turning and milling can be obtained by measuring the horsepower input to the drive motor on the machine tool. The power required at the spindle can be more accurately determined by calibrating the horsepower output versus horsepower input characteristics of the machine tool. In calibration, one can absorb the power at the spindle with a Prony brake. Equations providing power and torque for various machining conditions are given in Tables 3 and 4.

A good approximation of the horsepower required in most machining operations can be predicted from the unit power requirements. The horsepower at the spindle can be calculated by multiplying the unit horsepower by the rate of metal-removal equations that are presented in Tables 3 and 4.

Cutting-Tool Materials

GENERAL GUIDELINES FOR SELECTION OF TOOL MATERIALS

Selection of the optimum tool material is a major factor in realizing the full potential of a particular machine tool. General guidelines for selection of tool materials for machining various work materials at different hardness ranges are shown in Fig. 1. The following supplemental guidelines serve as an approach to logical selection of the tool materials best suited for machining a specific work material.

1. High speed steel tools are generally used for the following:
   - High volume, low cutting speed operations (for example, in screw machines)
   - Complex tool forms such as form tools, drills, cutoff tools, etc.
   - All sizes of end mills, drills, reamers, taps and gear cutters
   - Certain machining operations on problem materials, such as nickel base high temperature alloys
   - High positive rake requirements
2. Cast alloy cutting tool materials are selected as an intermediate between high speed steel and carbide tool materials. The high cobalt high speed steels also serve as intermediates, and there appears to be a trend for them to supersede the cast alloy tools.
3. Carbide tools are generally applicable when one or more of the following conditions exist:
   - Rigidity of the machine tool, tooling, and workpiece is acceptable.
   - Machine tool power is adequate for higher metal removal rates.
   - Workpiece configuration and machining operation permit higher cutting speeds.
   - High production rates are required.
4. Ceramic tools, high strength carbides, diamond tools, and the cast alloy tools referred to previously have rather specific application in contrast with the wide usage of high speed steel and carbide tools.

HIGH SPEED STEEL TOOLS

High speed steels can be classified into three general types as follows:

1. Tungsten high speed steels
2. Molybdenum high speed steels
3. High speed steels containing cobalt.

Chemical compositions of the AISI high speed steels are listed in Section 18 of this volume.

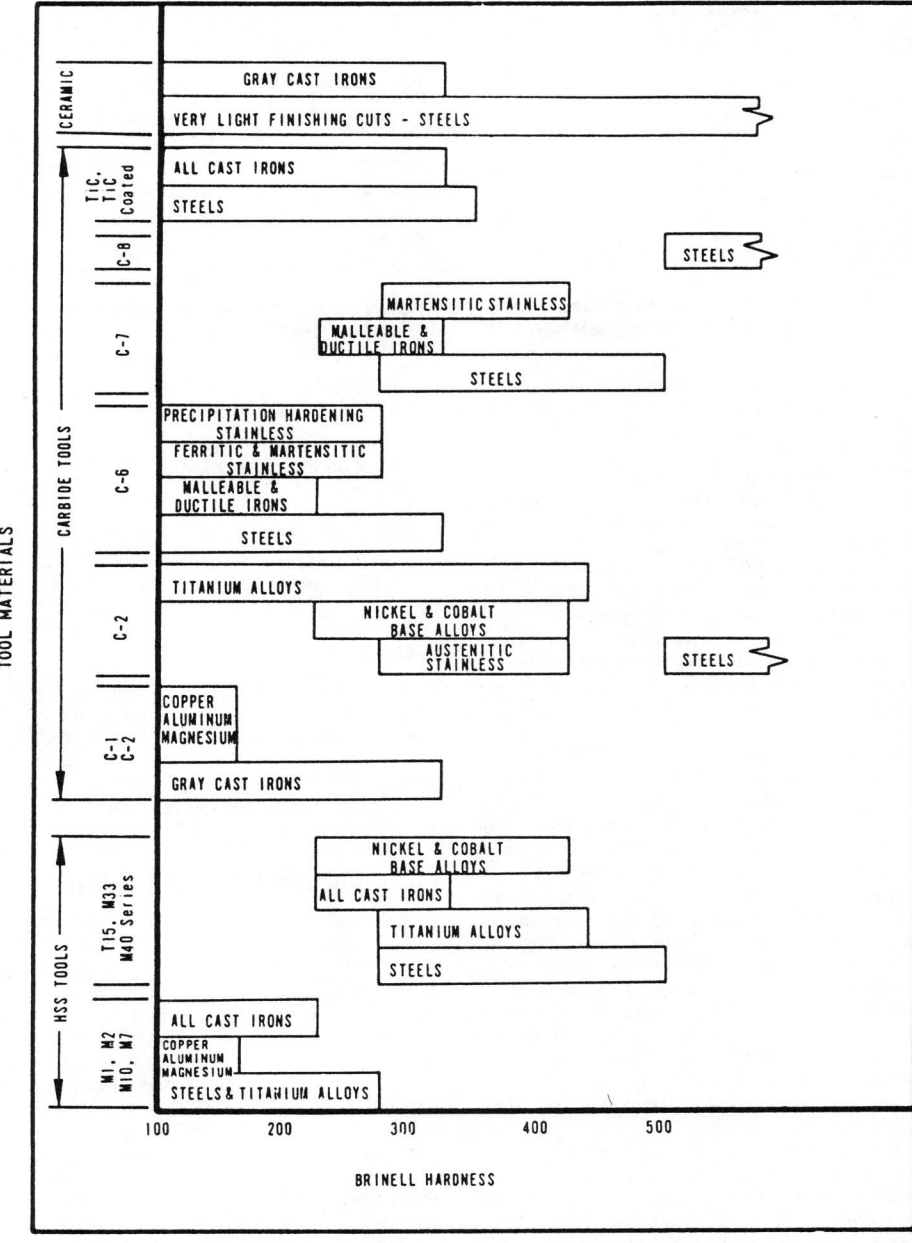

Fig. 1. General guidelines for selection of tool materials for machining of various work materials at different hardness ranges

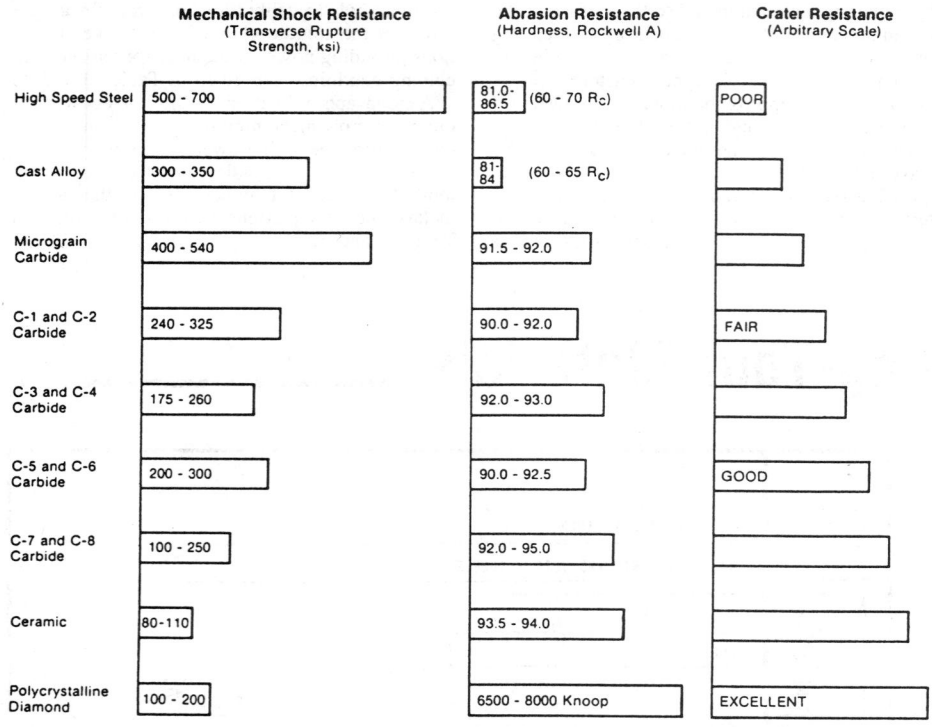

Fig. 2. Comparative properties of tool materials derived from published data for available grades. Crater resistance refers to ferrous metals.

As a general rule, the tungsten grades such as T1, T2, etc., are not quite as tough as the molybdenum grades but are much simpler to heat treat. The molybdenum high speed steels such as M1, M2, M3, M7 and M10 are much more widely used than the tungsten grades. Both the tungsten and the molybdenum high speed steels can be hardened to 64 to 66 R_c and are recommended for the machining of easy-to-machine materials. These include steels having a hardness up to 350 Bhn. While high speed steel types 1 and 2 can be used in machining steels having a higher hardness, the type 3 high speed steels containing cobalt are recommended. It should be noted that several of the high speed steels such as M3 and M4 contain more vanadium, thus providing increasing wear resistance, but at the same time making the grinding of the tools more difficult.

The high speed steels with cobalt, type 3, contain 5 to 12 percent cobalt. The addition of cobalt provides greater hot hardness and wear resistance but results in a somewhat lower toughness. In general, it has been found that these grades are not particularly advantageous in the machining of the readily machinable materials. They are, however, most beneficial for machining steels having a hardness level above 350 Bhn and for the more difficult-to-machine metals, such as titanium and nickel base high temperature alloys. The T15 grade also has a high percentage of vanadium; consequently, it is more difficult and more costly to grind than the other high speed steels containing cobalt.

In general, the high speed steels containing cobalt can be heat treated to a hardness level of 65 to 67 R_c. While the M40 series can be hardened to a level of 70 R_c, the hardness range usually recommended is 66 to 68 R_c. Above 68 R_c, high speed steel tools tend to be too brittle for most applications.

High speed steel cutters are widely used for the following machining operations:

Operations using form tools
Screw machine operations
End milling
Drilling
Reaming (carbide-tipped reamers are also widely used)
Tapping (nitrided taps are often used for difficult-to-machine alloys)
Broaching
Gear cutting

A recent development in high speed steels is the availability of indexable inserts made by the powder metal process. The sizes and shapes of the inserts correspond to those of some of the more popular carbide inserts. The current grade selection is limited, but in the future, selection is likely to expand.

MICROGRAIN CARBIDE TOOLS

A comparatively new group of high strength carbides intended for machining applications is now available. These tungsten carbide–cobalt base materials are characterized by strength levels previously found only in impact and rock bit grades, coupled with much higher levels of hardness, as shown in Fig. 2. Their properties are achieved by various proprietary processes and depend in part on achieving and maintaining a very fine carbide grain size.

Manufacturers recommend these materials for improving metal removal rates and tool life over those possible with high speed steels for conditions where carbides with normal strength levels would chip or break. The major use of micrograin carbides at present is in cutoff and form tools, replacing high speed steel in low-surface-

speed, high-production operations on screw machines and similar equipment. Some application is also being found for micrograin carbide tools in the machining of problem materials such as the high temperature alloys and for making punch and die sets intended for extended use.

The high edge-strength of micrograin carbides allows the use of high speed steel tool geometry and, when necessary, speeds as low as those used with high speed steels. Conversely, operation at normal carbide speeds may not be practical because the high cobalt binder level in these materials significantly reduces their resistance to cratering.

COATED CARBIDE TOOLS

Perhaps the most important advance in cutting tool technology during the past decade has been the development of coated carbide inserts. The impact of this improvement can be measured in terms of the major benefits of increased metal removal rate and tool life that have been found in turning and face milling of cast irons and steels. These two general types of work materials still represent the bulk of the work material being machined in the United States. The application of these tool materials overlaps much of the entire C-1 through C-8 range. The coatings are usually, but not necessarily, deposited on conventional carbide inserts. Special substrates are being developed which definitely are not suitable as metal cutting tools without the coatings.

Titanium carbide coatings are produced by reactive deposition from the gaseous phase, where titanium tetrachloride vapor is converted to fine grain titanium crystals. The process continues until the coating is about 0.0002 inch (5 μm) thick. Other coatings such as TiN, HfN, HfC, and Al_2O_3 are produced using the same technology with other gaseous components.

The improved machining performance of coated carbide inserts results from reduced friction, increased surface hardness and, most important of all, chemical inertness. The thin layer of coating provides a cobalt-free diffusion barrier between the workpiece and the insert. The reduction or elimination of diffusion results in an appropriate reduction of cratering and flank wear. For many applications, this results in either a longer tool life or a higher metal removal rate when compared to uncoated carbides. An exception to this advantage would probably occur in highly abrasive, lower cutting speed operations. Newer grades are being developed to successfully perform heavy roughing and interrupted cuts.

CERAMIC TOOLS

There are two principal types of ceramic cutting tools, the pure alumina tools and the alloyed cermets. The cermets are also aluminum-oxide-base materials containing various amounts of titanium carbide or other alloying ingredients. The alumina tools are white in appearance. Although these tools may be either hot pressed or cold pressed, they are usually cold pressed. The alumina inserts are normally applied in lighter-duty cuts, such as noninterrupted turning and boring.

The cermet tools are hot pressed. These tools are much tougher than the alumina inserts and are therefore applicable for roughing, for interrupted cuts, for some face milling and for machining very hard (60 to 68 R_c) materials.

Presently, the main uses for ceramic tool materials are in finish and semifinish turning. Typical applications are boring and facing of cast irons for the automotive industry, turning of steels for ordnance applications, and finish turning and facing of chilled iron and of hardened steel rolls. Where high rigidity has been designed into the machine tool and the tooling, roughing of cast iron and steel may be performed with ceramic materials.

In addition to the introduction of hot-pressed cermets, three recent developments in the preparation of ceramic inserts have expanded the usage of these tools. Those developments are as follows:

1. The grinding of K-lands on the cutting edges
2. The use of thicker inserts
3. The use of larger nose radii

The primary failure mode of ceramic inserts has always been either edge chipping or insert breakage. These three improvements have all helped reduce those problems.

A "K-land" is a precision-ground flat that removes the sharp cutting edge where the top and side faces of the insert intersect. The width of the flat varies from 0.002 to 0.012 inch (0.05 to 0.30 mm). The angle of the bevel may vary from 30 degrees to 45 degrees with respect to the top face.

A thicker insert will be less likely to break under the increased cutting forces produced by a chipped cutting edge. This increases insert life by preserving unused corners which would normally be lost if the insert cracks.

A larger nose radius will naturally present a stronger corner than will a smaller radius. This also helps to reduce edge chipping and nose breakage.

Even with the improved technology of adding K-lands and using thicker inserts, more care must be exercised in applying ceramic tools than is normally used in applying high speed steel or carbide inserts. In particular, it is necessary to avoid mechanical shock where possible and thermal shock at all times. Ceramic tools are rarely, if ever, used with cutting fluids. Their application in turning, boring or milling is always dry. Negative rake inserts are used almost exclusively.

Table 1. Commonly used industry code for classification of carbide tool materials

Cast iron, nonferrous and nonmetallic materials:
C-1   Roughing
C-2   General purpose
C-3   Finishing
C-4   Precision finishing

Steel and steel alloys:
C-5   Roughing
C-6   General purpose
C-7   Finishing
C-8   Precision finishing

DIAMOND TOOLS

Diamond tools are most generally employed for machining nonferrous alloys and abrasive materials, such as presintered carbides and ceramics, graphite, fiberglass, rubber and high-silicon aluminum alloys. Man-made polycrystalline diamonds that are sintered under very high temperatures and pressures are now available as cutting tool inserts of various sizes and shapes. Polycrystalline diamond tools have proved to be, in most cases, greatly superior to natural mined diamonds in normal machining operations on the work materials listed above. However, even though the polycrystalline diamonds are much tougher and able to withstand a great deal more abuse than natural mined diamonds, it is still good practice to observe as much care in handling the tools as possible. If natural mined diamonds are desired for use as the cutting tool material, then the following guidelines should be observed:

Diamond tools are purchased from suppliers who are expert in the art of cutting and orienting diamonds and the tools should be returned to the suppliers for resharpening. In-house grinding or reshaping should not be attempted. The diamond tool should be resharpened as soon as it becomes dull to minimize breakage, thereby increasing the number of possible resharpenings.

In general, diamond tools should be treated with the same care as would be used with any very hard, brittle tool material. However, even light interrupted cuts are possible with the stronger polycrystalline diamonds. Diamond tools are

generally used with positive-rake tool geometry in order to reduce the forces on the tool, but there are other applications of negative-rake geometries that have proved successful. The new strength and reduced cost of polycrystalline diamond have opened up many new fields of machining, and new applications continue to be found.

The use of a cutting fluid will usually increase the tool life of the diamond, but where this is not practical, a stream of air can be used to cool the tool and keep the cutting edge free from chips.

CARBIDE TOOLS

Each manufacturer of carbide tools has a wide variety of carbide grades so that a specific grade can be selected for a given material and operation. Although most of the carbide grades are listed under the industry code (C-1, C-2, etc.) in Table 1, this does not necessarily imply that the various manufacturers' grades under a specific code are equivalent.

The C-1 through C-4 grades are straight tungsten carbide bonded with cobalt and vary chiefly in cobalt content and grain size. They are widely used for machining cast irons, high temperature alloys, work hardening stainless steels, abrasive nonmetallic materials and nonferrous metals including titanium. Toughness decreases and hardness increases proceeding from C-1 to C-4. The C-2 grades provide a good compromise in properties and may be categorized as "general purpose" for the aforementioned materials.

The C-5 through C-8 grades include those which contain various combinations of tungsten carbide, tantalum carbide and titanium carbide bonded with cobalt. The C-8 category now also includes tools of straight titanium carbide bonded with molybdenum (or molybdenum carbide) and nickel. Other straight titanium carbide grades with properties in the C-5 through C-7 range are also available. Grades C-5 through C-8 are generally recommended for machining steels because they provide better crater resistance than the C-1 to C-4 grades. Toughness decreases and the hardness increases in going from C-5 to C-8. The C-6 grades are for general-purpose use.

For work materials such as highly alloyed cast irons, alloy steels over 50 R_c, ferritic or martensitic stainless steels and some high temperature

Table 2. ISO 513-1975 (E) classification of carbides according to use for machining

Designation(a)	Material to be machined	Use and working conditions
P 01	Steel, steel castings	Finish turning and boring; high cutting speeds, small chip section, accuracy of dimensions and fine finish, vibration-free operation
P 10	Steel, steel castings	Turning, copying, threading and milling; high cutting speeds, small or medium chip sections
P 20	Steel, steel castings Malleable cast iron with long chips	Turning, copying, milling, medium cutting speeds and chip sections; planing with small chip sections
P 30	Steel, steel castings Malleable cast iron with long chips	Turning, milling, planing, medium or low cutting speeds, medium or large chip sections, and machining in unfavorable conditions(b)
P 40	Steel Steel castings with sand inclusion and cavities	Turning, planing, slotting, low cutting speeds, large chip sections with the possibility of large cutting angles for machining in unfavorable conditions(b) and work on automatic machines
P 50	Steel Steel castings of medium or low tensile strength, with sand inclusion and cavities	For operations demanding very tough carbide: turning, planing, slotting, low cutting speeds, large chip sections, with the possibility of large cutting angles for machining in unfavorable conditions(b) and work on automatic machines
M 10	Steel, steel castings, manganese steel Gray cast iron, alloy cast iron	Turning, medium or high cutting speeds. Small or medium chip sections

(continued)

Table 2. (continued)

Designation(a)	Material to be machined — Groups of application	Use and working conditions
M 20	Steel, steel castings, austenitic or manganese steel, gray cast iron	Turning, milling. Medium cutting speeds and chip sections
M 30	Steel, steel castings, austenitic steel, gray cast iron, high temperature resistant alloys	Turning, milling, planing. Medium cutting speeds, medium or large chip sections
M 40	Mild free cutting steel, low tensile steel Nonferrous metals and light alloys	Turning, parting off, particularly on automatic machines
K 01	Very hard gray cast iron, chilled castings of over 85 scleroscope hardness, high silicon aluminum alloys, hardened steel, highly abrasive plastics, hard cardboard, ceramics	Turning, finish turning, boring, milling, scraping
K 10	Gray cast iron over 220 HB malleable cast iron with short chips, hardened steel, silicon aluminum alloys, copper alloys, plastics, glass, hard rubber, hard cardboard, porcelain, stone	Turning, milling, drilling, boring, broaching, scraping
K 20	Gray cast iron up to 220 HB nonferrous metals: copper, brass, aluminum	Turning, milling, planing, boring, broaching, demanding very tough carbide
K 30	Low hardness gray cast iron, low tensile steel, compressed wood	Turning, milling, planing, slotting, for machining in unfavorable conditions(b) and with the possibility of large cutting angles
K 40	Soft wood or hard wood Nonferrous metals	Turning, milling, planing, slotting, for machining in unfavorable conditions(b) and with the possibility of large cutting angles

(a) In each letter category, low designation numbers are for high speeds and light feeds, higher numbers for slower speeds and/or heavier feeds. Also, increasing designation numbers imply increasing toughness and decreasing wear resistance of the cemented carbide materials. (b) Unfavorable conditions include: shapes that are awkward to machine; material having a casting or forging skin; material having variable hardness; and machining that involves variable depth of cut, interrupted cut or moderate to severe vibrations.

alloys, the preferred carbide grades may come from either the C-1 to C-4 or the C-5 to C-8 group.

Table 2 has been used in classifying carbides according to use as specified by the International Organizaton for Standardization (ISO).

Additional information on carbide tools can be found in Section 18 of this volume.

CAST ALLOY TOOLS

Cast alloys have been available for many years. These materials are generally cobalt-chromium-tungsten alloys with carbon and other alloy additions. They are not heat treatable, and the maximum hardness (55 to 65 R_c) occurs near the cast surface. As a result, cast alloy tools must be used as-cast with as little grinding as possible.

Cast alloy materials are not widely used. However, they find limited use as a compromise since they perform well at higher surface speeds than conventional high speed steels and are more resistant to chipping than standard carbide grades.

Cutting Fluids

WHEN PROPERLY APPLIED, cutting fluids can increase productivity and reduce costs by making possible the use of higher cutting speeds, higher feed rates and greater depths of cut. Effective application of cutting fluids can also lengthen tool life, decrease surface roughness, increase dimensional accuracy and decrease the amount of power that is consumed when cutting dry. Knowledge of cutting fluid functions, types, physical limitations and compositions plays an important role in the selection and application of the proper fluid for a specific machining situation.

FUNCTIONS OF CUTTING FLUIDS

Depending upon the machining operation being performed, a cutting fluid has one or more of the following functions:

- Cooling the tool, workpiece, and chip
- Lubricating (reducing friction and minimizing erosion on the tool)
- Controlling built-up edge (BUE) on the tool
- Flushing away chips
- Protecting the workpiece from corrosion.

The relative importance of each function depends upon the work material, the cutting tool, the cutting conditions and the finish required on the part.

TYPES OF CUTTING AND GRINDING FLUIDS

Many cutting fluids are available today to satisfy the requirements of modern machine tools.

Although there is no all-purpose cutting fluid, some fluids offer considerable versatility while others are tailored for specific applications. For economic reasons, most metalworking firms try to use as few different fluids as possible. Preferred fluids have long life and do not require constant changing or modifying.

Each of the basic types of fluids has distinctive features, advantages and limitations, although the dividing line is not always clearly identifiable. The four basic types of cutting fluids are as follows:

Cutting oils
- straight and compounded mineral oil (plus additives)

Water-miscible (water-soluble) fluids
- emulsifiable oils (soluble oils)
- chemical (synthetic) fluids
- semichemical (semisynthetic) fluids

Gases

Paste and solid lubricants.

The cutting oils and the water-miscible fluids are the types most commonly used. Cooling is best accomplished by water-miscible fluids, while oil-base fluids provide better lubrication.

CUTTING OILS

Cutting oils have a mineral oil base and may be used straight (uncompounded) or compounded (combined with polar additives and/or chemically active or inactive additives or compounds). They have excellent lubrication properties, good rust control and long life, but they do not cool as well as water-miscible fluids. Oils may be classified as inactive or active.

Inactive Cutting Oils

Inactive cutting oils (compounded with chemically inactive additives), have relatively high lubricity but low or no anti-welding properties. They have nonstaining properties which prevent discoloration of chemically sensitive materials.

Straight mineral oils do not have as good lubrication properties as do the compounded types, but they are lower in cost. Straight mineral oils are generally restricted to light-duty operations on metals that are easier to machine, such as aluminum, magnesium, brass and sulfurized or leaded free-machining steels, where the lubrication and cooling requirements are not severe. The oils are noncorrosive and stable and, if kept clean, can be used almost indefinitely. They lubricate all exposed moving parts, and minor leakage into or from gear boxes, bearings and hydraulic systems does not upset a machine's performance.

Fatty oils were once widely used as cutting fluids. The most common types are lard and rapeseed oil. Their use has declined, partly because they are now difficult to obtain and are expensive, but mainly because modern additives blended with mineral oil are much more effective. Fatty oils are very polar and have high "oiliness"—antifriction performance—but poor anti-weld characteristics. They oxidize readily and display a tendency to fume and to emit unpleasant odors.

Compounded cutting oils are made by blending polar additives and/or chemically active additives with mineral oil. Polar additives, such as certain fats, oils, waxes and synthetic materials, increase the load-carrying and cutting capability of mineral oil. Common polar additives are lard oil and castor oil. The function of any polar additive is to wet and to penetrate the chip/tool interface by reducing the interfacial tension between the carrier mineral oil and the metal. Such additives have been refined to minimize previous objections to the formation of disagreeable odors and the tendency to gum.

Fatty-mineral oils are combinations of one or more fatty oils blended in straight mineral oil. Lard oil is frequently used for this purpose. Fatty oils may comprise up to 40 percent of the blend depending on the application and are quite effective for many operations. The advantages are not great and are confined primarily to improvement of finish in the machining of mild steel, brass, copper and aluminum. Such blends are particularly suitable for machining the harder types of brass and copper, where straight mineral oil may not give the finish required and where the use of more active oils would cause staining. These oils are widely used in automatic screw machines where the operations are not exceptionally severe.

Extreme pressure (EP) additives are added to fluids used for machining operations where cutting forces are particularly high, such as tapping and broaching, or for operations performed with heavy feeds. Chemical or EP additives provide a tougher, more stable form of lubrication at the chip-tool interface. These additives include sulfur, chlorine or phosphorus compounds that react at high temperatures in the cutting zones to form metallic sulfides, chlorides and phosphides. In addition to providing extreme pressure lubrication, these additives provide a film on the tool surface with anti-weld properties that minimize the built-up edge. Sulfurized fatty-mineral oil blends have sulfur added in a strongly bonded, inactive form which may be totally nonstaining.

Active Cutting Oils

Active cutting oils (compounded with chemically active additives) include sulfurized or phosphorized mineral oils, sulfo-chlorinated mineral oils, and sulfo-chlorinated fatty-mineral oil blends. These fluids have good anti-weld and extreme pressure lubrication qualities which improve tool life in high-temperature and high-pressure applications. They may, however, cause discoloration or staining of certain metals.

Sulfur can be added to a cutting oil in the form of sulfurized mineral oil or sulfurized fat. The sulfurized mineral oil is more active at lower temperatures and tends to severely stain aluminum, copper, brass, bronze and magnesium alloys. In comparison, the sulfurized fatty oil (which is sulfurized at a higher temperature) will not release sulfur as readily and, therefore, has less tendency to stain nonferrous metals or steel. Oils containing sulfur will form metallic sulfide films which will act as solid lubricants at temperatures up to about 1300 °F (700 °C).

Chlorine reacts and functions in essentially the same manner as sulfur. Inhibiting ingredients are added to prevent corrosion of ferrous surfaces, since chlorine is more reactive than sulfur. Chemical reaction is thus restricted to the chip-tool interface, where the temperatures are high. Chlorinated oils form an iron chloride film when reacting with ferrous work materials or high speed steel tools. This film has a low shear strength and provides low friction up to about 750 °F (400 °C), beyond which it decomposes. Chlorinated oils usually will not stain nonferrous alloys.

When both chlorine and sulfur are added to cutting oils, anti-weld or EP characteristics are effective over a wider temperature range. Sulfo-chlorinated mineral oil and sulfochlorinated fatty oils are examples of such oils which are suitable for a wide range of applications.

Phosphorus performs as a mild EP lubricant or anti-friction additive when added to a cutting oil. The phosphide film breaks down at lower temperatures than do sulfide or chloride films and, therefore, is not as effective an anti-weld agent. Phosphorus is most effective in reducing friction and wear. These oils are nonstaining to most ferrous and nonferrous alloys.

These basic types of cutting oils can have various additive levels and may be used singly or in combination and at various viscosities to suit various applications. Some recent developments include oils in which the production of fume and oil-mist have been significantly reduced.

Applications of all compounded cutting oils (both active and inactive) are generally limited to machining operations on difficult-to-machine metals or form grinding from the solid. The high cost, the possible danger from smoke and fire, and the operator health problems often restrict the use of these oils to operations where other fluids do not provide satisfactory performance.

WATER-MISCIBLE (WATER-SOLUBLE) FLUIDS

The water-miscible (water-soluble) cutting fluids are primarily used for high speed machining operations because they have better cooling capabilities. These fluids are also best for cooling machined parts to minimize thermal distortion.

Water-miscible cutting fluids are mixed with water at different ratios depending on the machining operation. For high-speed chip-making operations, they are normally mixed 1 part concentrate to 20 to 30 parts water. For many grinding operations where it is desirable to have a lighter fluid with more cooling action, the ratio is 1:40 or 1:50. Water-miscible fluids form mixtures ranging from emulsions to solutions when mixed with water. Because water has a high specific heat, high thermal conductivity and high heat of vaporization, it is one of the most effective cooling media known. Blended with water, the water-miscible fluids provide the combined cooling and moderate lubrication required by metal removal operations conducted at high speeds and lower pressures.

The water-miscible fluids can be classified as emulsifiable oils (soluble oils), chemical (synthetic) fluids or semichemical (semisynthetic) fluids. Fluids within these classes are available for light-, medium-, and heavy-duty performance.

Emulsifiable Oils

Emulsifiable oils are commonly called soluble oils, emulsions, or emulsifiable cutting fluids. An emulsion is a suspension of oil droplets in water made by blending the oil with emulsifying agents and other materials. These emulsifiers (soap or soap-like materials) break the oil into minute particles and keep the particles dispersed in water for long periods of time. Bactericides—usually nonphenolic organic compounds specifically approved by the Environmental Protection Agency (EPA)—are added to control the growth of micro-organisms such as bacteria, algae and fungi. If disposal is of no concern, phenolics may be used. The soaps, wetting agents, and couplers used as emulsifiers in water-miscible fluids reduce surface tension significantly. As a result, the liquid has a greater tendency to foam when subjected to shear and turbulence. For this reason, water-miscible fluids sometimes cause foaming problems in operations such as gundrilling or flat-bed and double-disk grinding. With the use of special wetting agents and foam depressants, however, water-miscible fluids can be rendered sufficiently nonfoaming to be effective in almost all operations.

Emulsifiable oils combine the lubricating and rust-prevention properties of oil with water's excellent cooling properties. Emulsions, with their cooling-lubricating properties, are most effectively used for metalcutting operations with high cutting speeds and low cutting pressures accompanied by considerable heat generation.

Advantages of emulsifiable oils over straight or compounded cutting oils include greater reduction of heat, cleaner working conditions, economy resulting from dilution with water, better operator acceptance and improved health and safety benefits. They can be used for practically all light- and moderate-duty cutting operations, as well as for most heavy-duty applications except those involving extremely difficult-to-machine materials. Emulsifiable oils can be used for practically all grinding operations with the exception of severe grinding operations, such as form, thread and plunge grinding where wheel form is a critical factor. Extreme-pressure, compounded emulsifiable oils do not suffer from this limitation.

Cutting fluid manufacturers supply emulsifiable oils as concentrates that the user prepares by mixing with water. Mixtures range from 1 part oil in 100 parts water to a 1:5 oil-water ratio. The leaner emulsions are used for grinding or light-duty machining operations where cooling is the essential requirement. Lubricating properties and rust prevention increase with higher concentrations of oil. The four types of emulsifiable oils are summarized in Table 1.

General-purpose soluble oils are milky fluids with mineral oil droplets of 0.0002-in. to 0.008-in. (0.005-mm to 0.2-mm) diameter. They are commonly used at dilutions of 1:10 to 1:40 for general purpose machining.

Clear-type (or translucent) soluble oils contain less oil (with higher proportions of corrosion inhibitors) and considerably more emulsifier than do milky emulsions. The clear type, therefore, consists of oil dispersions with smaller oil droplets which are more widely distributed. Since there is less dispersion of transmitted light, the fluid is less opaque, and the result is a translucent liquid. The translucency is not permanent, though, because oftentimes the tiny oil droplets tend to coalesce and form larger droplets. These oils are generally used for grinding or for light-duty machining.

Fatty soluble oils have animal or vegetable fats or oils or other esters added to the mineral oil content to provide a range of fluids with enhanced lubricating properties.

EP soluble oils contain sulfur, chlorine or phosphorus additives to improve load-carrying performance. Since the EP concentrate is diluted 5 to 20 times when the emulsion is prepared, the lubricating capability is reduced. Where the lubricating capabilities of soluble-oil emulsions and the cooling properties of cutting oils are inadequate, EP soluble oils can satisfy both requirements in many cases. These fluids, commonly known as heavy-duty soluble oils, have in some

Table 1. Water-miscible cutting fluids

Class	Type	General characteristics
Emulsifiable oils	(1) General-purpose soluble oils	Used at dilutions between 1:10 and 1:40 to give a milky emulsion. Used for general-purpose machining.
	(2) Clear-type soluble oils	Used at dilutions between 1:50 and 1:100. Their high emulsifier content results in emulsions which vary from translucent to clear. Used for grinding or light-duty machining.
	(3) Fatty soluble oils	Used at similar concentrations to (1) and of similar appearance. Their fat content makes them particularly good for general machining operations on nonferrous metals.
	(4) EP soluble oils	Generally contain sulfurized or chlorinated EP additives. Used at dilutions between 1:5 and 1:20 where a higher performance than that given by (1), (2) or (3) is required.
Chemical (synthetic) fluids	(1) True solutions	Essentially solutions of chemical rust inhibitors in water. Used at dilutions between 1:50 and 1:100 for grinding operations on iron and steel.
	(2) Surface-active chemical fluids	Contain mainly water-soluble rust inhibitors and surface-active load-carrying additives. Used at dilutions between 1:10 and 1:40 for cutting and at higher dilutions for grinding. Most are suitable for both ferrous and nonferrous metals.
	(3) EP surface-active chemical fluids	Similar in characteristics to (2) but containing EP additives to give higher machining performance when used with ferrous metals. Used at dilutions between 1:5 and 1:30.
Semichemical (semisynthetic) fluids	——	Essentially a combination of a chemical fluid and a small amount of emulsifiable oil in water forming a translucent, stable emulsion of a small droplet size. EP additives are usually included permitting their use for moderate- and heavy-duty machining and grinding applications.

cases replaced cutting oils for broaching, gear hobbing, gear shaping, and gear shaving.

Chemical Fluids

Chemical (synthetic) fluids are chemical solutions consisting of inorganic and/or other materials dissolved in water and containing no mineral oil. All of these fluids are coolants; some are also lubricants. Chemical agents that go into these fluids include: amines and nitrites* for corrosion inhibitors; nitrates, for nitrite stabilization; phosphates and borates, for water softening; soaps and wetting agents, for lubrication and reduction of surface tension; phosphorus, chlorine and sulfur compounds, for chemical lubrication; glycols, as blending agents and humectants; and germicides, to control the growth of bacteria.

In general, the advantages of chemical fluids include economy, rapid heat dissipation, good size control, detergent properties which help keep the machine surfaces and the coolant systems clean, excellent workpiece visibility, easy mixing with little agitation, and high resistance to rancidity and rust. The increased cost of oils, plus OSHA and EPA requirements, may result in increased use of these and other water-miscible fluids.

A possible disadvantage of chemical fluids that may be encountered in some severe operations is insufficient lubricity, which may cause sticking and/or wear of moving machine tool parts. The dry, powdery residue or film left by chemical fluids is easy to remove. This residue can interfere with component movements if allowed to remain on the machine surfaces. Most of the stickiness is caused by the minerals in the water used. The mineral content in water can quadruple in six weeks as water is added to replace that lost by evaporation. Deionized water should be used to avoid this problem. Improved lubricating or EP properties, however, can be provided. Other limitations may include foaming problems in high agitation applications, and high detergency and alkalinity characteristics which may irritate sensitive hands when the concentration of the mix is not controlled.

Chemical fluids are usually classified into two general groups: true solutions and surface-active types (see Table 1).

True-solution fluids (without wetting agents), also called chemical solutions or chemical grinding fluids, primarily contain rust inhibitors (inorganic and organic nitrites*), sequestering agents, amines, phosphates, borates, glycols or ethylene or propylene oxide condensates. Some of these fluids contain highly developed corrosion inhibitors such as sodium nitrite* (for cast iron), triethanolamine (for both cast iron and steel) and sodium mercaptobenzothiazole (for reducing corrosion on brass, zinc and aluminum). True solutions, used at 1:50 or 1:100 ratios, are clear in appearance but are often colored with dyes to indicate their presence in water. These fluids are restricted to grinding operations where they prevent rust and permit rapid heat removal. True solutions have high surface tension (about equivalent to water). They have a tendency to leave a residue of hard or crystalline deposits formed by water evaporating. These chemicals can also be added to emulsifiable oils or to other chemical fluids to enhance their corrosion-inhibiting properties.

Surface-active chemical fluids are extremely fine colloidal solutions composed of inorganic and organic materials dissolved in water with the addition of wetting agents (surface active agents). The wetting agents improve the wetting action of the water and provide greater uniformity of both heat dissipation and anti-rust action. This type of fluid may include anti-foaming agents, humectants, mild lubricants (organic or inorganic) and water softeners. The lubricating (anti-wear) properties may be provided by a viscous polyglycol compound. Corrosion protection may be provided by a mixture of triethanolamine and caprylic acid, and the formulation may include a deactivator for copper alloys.

The surface-active type of chemical fluid has fair lubricity, low surface tension, and good rust-inhibiting properties and usually leaves a dry, hard or powdery residue that is easily removed. The slight tendency of these fluids to foam is usually not a serious problem in most operations. They are usually used at dilutions of 1 part concentrate in 10 to 40 parts water.

EP surface-active chemical fluids are similar to the plain (general-purpose) surface-active type but have chlorine, sulfur or phosphorus additives to provide extreme-pressure lubrication effects. These fluids are diluted at one part concentrate to between 5 and 30 parts water for tougher machining operations.

Semichemical Fluids

Semichemical fluids or semisynthetic fluids are essentially a combination of chemical fluids and emulsifiable oils in water. These fluids are actually preformed chemical emulsions that contain only a small amount of emulsified mineral oil, about 5 to 30 percent of the base fluid, which has been added to form a translucent, stable emulsion of small droplet size. Since the usual EP additives can be incorporated (often more readily in the oil content than in the synthetic base), the lubricating performance can be varied to permit using such fluids for moderate- and heavy-duty machining and grinding applications.

Semichemical fluids combine some of the best qualities of chemical fluids and emulsifiable oils. The advantages and limitations are similar to those described for chemical fluids, except that semichemical fluids have better lubricating properties than do chemical fluids. They are also cleaner, with better rust and rancidity control than emulsifiable oils.

GASES

Air is the most common gaseous fluid. It is present under atmospheric pressure for dry operations and also present when fluids are used. Air is sometimes compressed to provide better cooling, with a stream directed at the cutting zone to remove heat by forced convection. This also can be used to blow chips away, but safety precautions must be observed. Gases such as argon, helium and nitrogen are sometimes used to prevent oxidation of the workpiece and the chips, but the high cost of these gases generally makes them uneconomical for production applications.

Gases such as Freon™ or CO_2, with boiling points below room temperature, can be com-

*CAUTION: The use of fluids containing nitrites may present a hazard and is presently under review by the National Institute of Occupational Safety and Health (NIOSH). Nitrites can react with amines to form nitrosamines which are carcinogenic. NIOSH may ban or control the use of nitrites in cutting fluids after completion of their review.

pressed and sprayed into the cutting zone to provide evaporative cooling. Use of liquid argon or nitrogen allows cooling to several hundred degrees below zero. Care is necessary, however, to prevent part warpage caused by large temperature differentials.

Advantages of inert gases include good cooling ability, increased tool life, a clear view of the operation, elimination of mist, and no contamination of the workpiece, chips or machine lubricants. The cost of some of these gases, however, can be extremely high.

PASTE AND SOLID LUBRICANTS

There are also paste and solid lubricants that are usually applied manually by brush or by oil-can to the tool or workpiece in operations such as tapping and hand reaming. Grinding wheels are sometimes impregnated with solids possessing lubricating qualities. In special cases, such as knife grinding, wheels are treated with sulfur to produce a cooler action in wet grinding. Also, external application of grease sticks to grinding wheels can provide some lubrication, but this requires performing the operation dry. Solid waxes in stick form can be used on grinding wheels, sanding disks or belts, taps, and band- or circular-saw blades for lubrication to improve finish and tool life or to reduce burring or metal welding. Other solids most often used as heavy-duty lubricants include graphite, molybdenum disulfide, pastes, soaps and waxes.

SELECTION OF A CUTTING FLUID

The proper choice of a cutting fluid depends on many complex interrelated factors. Of primary concern are machinability (or grindability) of the material, compatibility (metallurgical, chemical and human) and acceptability (fluid properties, reliability and stability).

Machinability

The selection of the type of cutting fluid for use should be based on:

- Type of machining operation
- Material being machined
- Tool material
- Machining conditions—cutting speed, feed, and depth of cut.

One of the most important factors in selecting a cutting fluid is the nature of the cutting operation itself. The various machining processes naturally differ in metal removal characteristics. The more difficult operations will place greater demands on a cutting fluid. Selection is, therefore, a matter of assessing the severity of the machining operation and marrying it to the appropriate cutting fluid. The approximate ratings for the machining processes, in order of decreasing severity, are as follows:

1. Internal broaching
2. External broaching
3. Tapping
4. Threading
5. Gear cutting
6. Deep hole drilling
7. Boring
8. Screw machining with form tools
9. High-speed, light-feed screw machining
10. Reaming
11. Milling
12. Drilling
13. Planing and shaping
14. Turning, single-point tools
15. Sawing
16. Grinding.

These ratings cannot be regarded as absolute, since variations in the tool geometry and workpiece material will change the severity of the machining operation.

For heavy-duty machining operations (such as tapping or broaching), medium- or heavy-duty cutting oils are generally used. Horizontal broaching of steel usually requires a heavier-bodied or more-chemically-active oil than does vertical surface broaching under comparable conditions. The heavier oil clings to the horizontal broach better than a water-miscible fluid, and chemical activity aids in efficient cutting. For vertical surface broaching of mild steels, emulsions or solutions may be used, but the usual choice is an oil.

Cutting oils can be ranked in order of increasing load-carrying capacity as follows:

1. Straight mineral oil
2. Mineral oil with fatty additives
3. Mineral oil with chlorinated additives
4. Mineral oil with sulfurized fatty additives
5. Mineral oil with free sulfur and sulfurized or chlorinated compounds.

Low-speed operations require the lubricating properties of cutting oils. As a general rule, cutting oils should be used at cutting speeds below 100 feet per minute (30 m/min). Cutting oils with EP lubricants are effective in a wide variety of machining operations on many materials up to speeds of 200 feet per minute (60 m/min). At speeds of about 200 feet per minute (60 m/min), chemical action is much less effective than at the lower speeds. Chemical action diminishes quickly and becomes virtually nil as the cutting speed is increased to 400 feet per minute (120 m/min).

Both tapping and threading involve many small cutting edges in continuous contact with the work throughout the cut. The design of the tools and the nature of these operations shield the edges of the tools from the flow and the cooling effect of the cutting fluid, particularly in tapping.

Drilling can be a difficult operation when done on difficult-to-machine materials. Drilling speeds are generally slower than those used for other operations because the cutting edge is in continuous contact with the metal when cutting and the cutting edges are shielded from the flow and beneficial cooling action of the cutting fluid. The preferred fluids for conventional drilling operations are emulsifiable oils and sulfurized or chlorinated mineral oils. These fluids provide some lubricity to prevent chatter and friction-generated heat while carrying away the heat generated by chip formation. Oil-hole drills should be used wherever possible.

There are some operations for which oils are specially formulated. For example, honing requires the use of a thinner, paraffinic-base oil.

For many high-speed operations, such as grinding and turning or milling with carbides, the most important benefit is provided by the superior cooling characteristics of the water-miscible fluids. In such instances, lubrication and anti-weld properties are less important. At high speeds, time for reaction between fluid additives and the workpiece is reduced to a minimum. Further, the higher relative velocities of the work and the tool

virtually preclude the fluid's reaching the work zone. Consequently, gross cooling is required to prevent catastrophic tool failure from crater wear, to prevent distortion in the workpiece from heat buildup and to control workpiece size.

Some emulsifiable oils and chemical fluids are formulated specifically for grinding operations and are used in concentrations of 1 part concentrate in 25 to 60 parts water. An increase in the richness of emulsifiable oil mixtures from 2.5 percent to 10 percent can improve the grinding ratio and the workpiece finish and reduce horsepower requirements. Note that the grinding ratio is a measure of the volume of material removed per unit volume of wheel wear.

Severe grinding operations, such as form, thread, and plunge grinding where wheel form is a critical factor, require the use of cutting oils or EP-compounded emulsifiable oils. It should be noted that for the low stress grinding technique, the use of an oil-base fluid results in less surface distortion. Grinding oils reduce friction, thus reducing the heat generated. They permit heavier cuts to be taken, produce smoother finishes, reduce wheel breakdown, and, hence, allow better form control. Oils suffer from the disadvantages of allowing heat buildup in the workpiece, tending to hold chips in suspension, and smoking or burning. The messiness of oil vapors is also a nuisance.

In summary, from a machinability viewpoint, cutting and grinding fluids should be selected to provide the following:

Greater Lubrication
- at relatively low speeds
- on the more difficult-to-machine materials
- for more difficult operations
- for better surface finish.

Greater Cooling
- at relatively high speeds
- on the easier-to-machine materials
- for easier operations
- where heat buildup in the part is a problem.

Other Considerations

A number of other important factors must be considered when selecting cutting and grinding fluids. These include compatibility of the fluid with the material being machined, the quality of the water in which it is mixed, tramp oil contamination and the machine tool. The acceptability of the fluid, in terms of its properties, reliability and stability, is also important. Other factors to be considered include the quantity of fluid used and the method of application, facilities for storing, handling, cleaning or recycling the fluid, the quality of water used in the aqueous emulsions or solutions, and the rust and rancidity control necessary. Economic considerations include any waste treatment facilities that may be required prior to disposal. The cutting fluid's effect on the total machining cost per piece must be considered, since the more expensive fluids may sometimes be more economical in the long run.

APPLICATION OF CUTTING FLUIDS

The way in which a cutting fluid is applied has a considerable influence on tool life and on the machining operation in general. Although there are many extremely effective devices and systems for supplying fluids to the cutting area, special equipment is not generally necessary for good results.

Even the best fluid, however, cannot perform its function unless it is effectively delivered to the cutting zone. Thus, a fluid chosen for its lubricating qualities must be directed so that it can form a film at the sliding surfaces. Likewise, a fluid used for cooling must gain reasonable access to the cutting edge of the tool. These conditions usually require that the fluid be forced into the cutting zone so that the heat can be removed as it is generated. Continuous application of the cutting fluid is preferable to intermittent application. Sporadic fluid application causes thermal cycling, which leads to the formation and propagation of microcracks in hard and brittle tool materials, such as carbides. Besides shortened tool life, intermittent fluid application can lead to irregular surface finishes.

A secondary advantage of proper fluid application is the efficient removal of chips. This also aids in prolonging tool life, since properly placed fluid nozzles can prevent blockage or packing of chips in flutes of milling cutters and drill bits.

Manual Application

Paste and solid lubricants are applied manually by brush or oilcan to the tool and to the workpiece, mainly in tapping. More recent developments include pressurized aerosol dispersants and foams which cling to the tool and workpiece.

Manual application is an effective method where a small number of holes are to be drilled or tapped on a machine that is not equipped with a coolant system. When two different operations are performed on the same machine, manual application may be used in conjunction with the flood cooling system of the machine. The flood cooling may be used for a drilling operation with a moderately active fluid, whereas a highly active cutting oil can be applied manually for a tapping operation that follows.

Flood Application

The most common method of applying fluids is to flood the tool and workpiece. A low-pressure pump delivers the cutting fluid through piping and valves to nozzles situated near the cutting zone. After flooding the cutting area, the fluid drains down over various parts of the machine into the chip pan where it flows to the sump of the coolant pump. The volume of the tank must be sufficient to allow time for cooling and for the settling of fine swarf and may require from five to fifty gallons or more depending on the type of machine. A coarse strainer on top of the collecting pan prevents larger chips from entering the tank and the fine strainer at the pump section. Important exceptions are grinding, honing, lapping and deep-hole boring machines where high-quality work depends upon removing the finer swarf and abrasive particles. Occasions arise when the inclusion of filtration equipment on other types of machines can avoid the gross contamination and overloading of the coolant with the metallic particles and help keep the fluid clean, thus prolonging its useful life.

Flood application of cutting fluids permits a continuous flow of fluid to the cutting zone and is efficient in flushing away chips. A copious stream of fluid should be applied so that the cutting tool edge and the work are completely enveloped. In addition to supplying an adequate amount of fluid to the cutting zone, a copious flow provides cooling action that prevents undue temperature rise. Proper cutting fluid application should not be neglected because of inadequate

Table 2. Typical coolant flow requirements

Operation	Flow at work	Remarks
Turning	5 gal/min (19 L/min)/tool	
Screw machining:		
1 in. (25 mm) diam	35 gal/min (132 L/min)	
2 in. (50 mm) diam	45 gal/min (170 L/min)	
3 in. (75 mm) diam	60 gal/min (227 L/min)	
Milling:		
Small cutters	5 gal/min (19 L/min)/tool	
Large cutters	Up to 60 gal/min (227 L/min)/tool	
Drilling, reaming, 1 in. (25 mm) diam	2-3 gal/min (7.6-11 L/min)	
Drilling, large	2-3 gal/min × diam, in. (0.3-0.43 L/min × diam, mm)	
Gundrilling:		
External chip removal type:		Fine filtration required.
0.18-0.37 in. (4.6-9.4 mm) diam	2-6 gal/min (7.6-23 L/min)	
0.37-0.75 in. (9.4-19 mm) diam	5-17 gal/min (19-64 L/min)	
0.75-1.25 in. (19-32 mm) diam	10-40 gal/min (38-151 L/min)	Use higher flow rates for deeper holes and the largest diameters in each range.
1.25-1.50 in. (32-38 mm) diam	17-50 gal/min (64-189 L/min)	
Internal chip removal type:		
0.31-0.37 in. (7.9-9.4 mm) diam	5-8 gal/min (19-30 L/min)	
0.37-0.75 in. (9.4-19 mm) diam	8-26 gal/min (30-98 L/min)	
0.75-1.18 in. (19-30 mm) diam	26-66 gal/min (98-250 L/min)	
1.18-2.38 in. (30-60 mm) diam	66-130 gal/min (250-492 L/min)	
Trepanning:		
External chip removal heads:		Fine filtration required.
2-3.5 in. (51-89 mm) diam	8-48 gal/min (30-182 L/min)	
3.5-6 in. (89-152 mm) diam	16-80 gal/min (61-303 L/min)	
6-8 in. (152-203 mm) diam	32-104 gal/min (121-394 L/min)	Use higher flow rates for deeper holes and the largest diameters in each range. If an emulsion is used instead of oil, increase the flow rate.
Internal chip removal heads:		
2.37-6 in. (60-152 mm) diam	110-215 gal/min (416-814 L/min)	
6-12 in. (152-305 mm) diam	215-340 gal/min (814-1287 L/min)	
12-18 in. (305-457 mm) diam	340-460 gal/min (1287-1741 L/min)	
18-24 in. (457-610 mm) diam	460-570 gal/min (1741-2158 L/min)	
Honing:		
Small	3 gal/min (11 L/min)/hole	Very fine filtration required.
Large	5 gal/min (19 L/min)/hole	
Broaching:		
Small	10 gal (38 L)/stroke	
Large	3 gal/stroke × length of cut, in. (0.45 L/stroke × length of cut, mm)	
Centerless grinding:		
Small	20 gal/min (76 L/min)	Fine filtration required.
Large	40 gal/min (151 L/min)	
Other grinding	5 gal/min/in. of wheel width (0.75 L/min/mm of wheel width)	

splash guards. Typical coolant flow requirements are given in Table 2.

The geometry of flood application directly influences the effectiveness of cutting fluids. Nozzles should direct fluid flow so that fluid is not thrown off the workpiece or tool by centrifugal force. Two or more nozzles should be used — one for directing fluid into the cutting zone and the other for auxiliary cooling and flushing away chips.

High-Pressure Application

For some operations, such as gundrilling and trepanning, high-pressure fluid systems are normally used with fluids being applied at pressures ranging from 100 to 2000 psi (690 to 13 790 kPa). A gundrilling tool is essentially a single-point end cutter similar to a boring tool, except that it has an internal passage for fluid. Trepanning is a holemaking operation that cuts a cylindrical path into the metal, leaving a solid core. This core passes through the hollow cylindrical cutting head as a tool feeds into the metal. The cutting fluid is pumped around the outside of the tool under pressure, forcing chips back through the center. Cutting fluids for trepanning must have good EP and anti-weld properties, must be low enough in viscosity to flow freely around the tool, and must have good oiliness.

Deep-hole drilling presents the problem of maintaining a sufficient flow of cutting fluid to the cutting edges. One solution is the use of oil-groove, oil-hole, or oil-tube drills, which utilize drill-flute space for cutting fluid passages. The fluid, under 50 to 100 psi (345 to 690 kPa) pressure, is transferred to the drill by a rotating gland and is forced directly into the cutting zone. The fluid flowing from the hole assists in chip removal. Oil-hole drills have become very popular in recent years, particularly for deep holes. Their use represents a significant improvement over flooding as the method of getting the cutting fluid to the drill lips. Significant increases in tool life and productivity can also be achieved (that is, oil-hole cooling gives better tool life at higher speeds than flood cooling does).

High-pressure systems are sometimes used for other operations. The high pressure facilitates the fluid's reaching the chip tool interface. In grinding, a high-pressure jet also serves to clean the wheel.

Mist Systems

Cutting fluids may also be applied in the form of an air-carried mist. Small jet equipment is used to disperse soluble oil or synthetic water-miscible cutting fluids as very fine droplets in a carrier such as air, at 10 to 80 psi (69 to 552 kPa), or occasionally as an aerosol. Water-miscible fluids are preferred over oil because oil presents possible health hazards and tends to clog. Mist application is best suited to operations where the cutting speed is high and the areas of cut are low, as in end milling. Cutting fluids normally chosen primarily for their cooling ability are used for mist application. The very small droplets come into contact with the hot tool, workpiece, or chip and evaporate and rapidly remove heat by vaporization. Mist cooling does not require the splash guards, the chip pans and the return hoses that are required for flood cooling. In addition, only small amounts of fluids are used, and these generally dry on the part or can be easily wiped away.

Mist systems are often advantageous because they:

• Provide better tool life than cutting dry
• Provide coolant when a flood system is not available or practical
• Apply fluids to otherwise inaccessible areas
• Provide higher fluid velocities at tool-workpiece interface than possible by flood cooling
• Reduce costs in some cases
• Give better visibility of the workpiece in cut.

Disadvantages of mist systems include limited cooling capability and the need for venting.

Two types of mist generators are normally used—the aspirator type and the direct-pressure type. For the aspirator, a stream of air is blown over the open end of a tube which is immersed in the fluid. A partial vacuum is created and the fluid is drawn up the tube where it becomes entrained in the air stream. In the pressurized mist generator, either pressurized gas bottles or shop air may be used to force the fluid into the air stream.

Special Application Methods

Chilled cutting fluids and highly pressurized gases have been shown effective in increasing tool life. These methods are more exotic than the conventional application methods. In special cases they may be economically justifiable. They may prove worthwhile to try in cases where nothing else works.

MAINTENANCE OF CUTTING FLUIDS

Cutting fluids, like any other fluids that are used over and over again, must be cared for properly. There are several precautions that should be observed.

Cutting Oils

Oil-type fluids perform satisfactorily if applied at full flow to the tools and the work and if sufficient volume is maintained in the system to hold the oil at around 70 to 75 °F (21 to 24 °C).

Cutting and grinding oils become contaminated rapidly during use. Extraneous materials, chips, dirt, etc., should be removed continuously or at periodic intervals by filters, strainers, centrifuges, or settling tanks. The mechanical edge-type filter, incorporating metal strips or disks as the filter element, acts principally as a strainer. The absorbent-type filter uses paper disks, cotton waste, or cloth bags as the filtering element. Magnetic filters are suitable for separation of ferrous particles. Centrifuging is used for removing heavy contaminants and particles from oil. Centrifuging, together with a heating unit and settling tanks, is often used for extracting oil from chips.

All cutting oil systems should be drained at intervals, manually cleaned, flushed and replenished with filtered or new cutting oil. Frequency of cleaning will vary, depending upon individual conditions.

Emulsifiable Oils

Emulsions generally require more maintenance and care than do cutting oils. When preparing emulsions, always add the oil to at least two or three times as much water. This initial mix should be agitated thoroughly while the oil is being added; otherwise, soap forms which combines with the water causing the mineral oil to separate out. If not enough water is used, or if the water is added to the oil, an invert emulsion will result (water particles are dispersed in the oil phase) which is undesirable for metalcutting. Premixed fluids, instead of plain water, should be used for preparing or for altering the mix concentration.

Water used in preparing emulsions is very important. Hard water containing various minerals and salts often hinders or impedes emulsification. It is not uncommon for emulsions made with hard water to "break" readily, that is, to separate into a stratified condition with a layer of oil or creamy emulsion floating on the surface. Such separation is detrimental. On the other hand, water that is too soft will cause foaming. Use of a specially formulated hard-water soluble oil with soft water may lead to the formation of a bluish-black stain on freshly ground ferrous parts.

Water used for making cutting fluid mixtures should be as pure as possible for the most economical and trouble-free use. Boiler water condensate (when available) or deionzied water (mineral free) should be used. The deionizer removes all minerals by chemical absorption so that the effluent is equivalent to distilled water. Consequently, no residues are left by evaporation of the water, and corrosion effects from minerals are eliminated.

Pretreatment of hard water is sometimes necessary. Water conditioning agents consisting usually of polyphosphate combinations are readily available, as is trisodium phosphate. The general rule for using these materials is to add about 1.5 ounces per 100 gallons (0.11 grams per liter) of water per grain of hardness. Chemical treatment of extremely hard water is far more economical than the purchase of a specially formulated soluble oil. Excessive use of polyphosphate water conditioners, however, will tend to increase bacteria, mold and fungi growth. Where water makeup rates are high, the use of deionized water is preferred.

Micro-organisms in the water shorten the service life of a soluble-oil emulsion. Micro-organisms of three types—bacteria, algae, and fungi—are often encountered in soluble oils, and all three have detrimental effects on emulsion stability. Many soluble oils are compounded with a bactericide, but the amount that can be added is limited by its solubility in the oil. When the emulsion is made, the bactericide is further diluted, reducing its effectiveness.

Rancidity, the term applied whenever a cutting fluid gives off a bad odor, is usually caused by bacterial growth. The rotten-egg stench emanating from the sump of a machine that has been shut down over a weekend is caused by bacteria that attack inorganic sulfates found in all natural waters. A quality cutting fluid and regular use of biocides where needed are the best insurance against rancidity.

Emulsion concentration is not always given the attention it deserves. In heavy-duty cutting operations, heat at the tool causes water to evaporate at a rate faster than the carryoff of oil on the machined parts. This results in an increased oil-to-water ratio, which if carried too far can cause an invert emulsion. The opposite applies to grinding operations where oil carryoff is higher and the emulsion becomes increasingly dilute with use. This may cause rusting unless the concentration is frequently checked and controlled.

The amount of time that an emulsion is kept in service varies widely—anywhere from one week to six months. Cooling an emulsion by mechanical circulation or refrigeration is useful in extending its service life and in producing better finish. Aeration, although an effective means of cooling an emulsion, will usually increase microorganism growth, which results in reduced emulsion life. In use, emulsions should be held at a temperature between 55 and 70 °F (13 and 21 °C).

Before putting an emulsion into use, wash and flush the coolant system thoroughly. Deposits of all kinds must be removed. Emulsions are extremely susceptible to contamination. If there is any reason to suspect that bacteria are present in the system, flush it with a germicidal solution before putting in the new emulsion.

Occupational Dermatitis

Dermatitis is frequently, though mistakenly, associated with the handling of petroleum products, particularly in the machine shop where the operator's hands or forearms are in prolonged contact with soluble oils and cutting oils. Combined with dirt, these fluids form grimy compounds that may become embedded in the skin, often blocking the pores and hair follicles. In many cases, these areas become infected and dermatitis sets in. This condition may occur with almost any material that is allowed to remain on the skin for a long period of time; it is not limited to petroleum products. A contributing factor may sometimes be the solvent action of the cutting fluid. Left on the skin, the fluid can dissolve the natural skin oils, inflaming the skin and causing it to crack.

The bactericides used in most cutting fluids generally have no effect on the incidence of dermatitis; however, formaldehyde-releasing bactericides may sensitize the skin of some people. People with thin skin are more susceptible to dermatitis. Fair, blond-complexioned people are usually more susceptible than are dark, oilier-skinned people. For hypersensitive workers, the only answer is a switch to another job—one in which they will not be exposed to cutting fluids.

Maintaining personal cleanliness, keeping the fluid clean, using commercially available hand creams, barrier creams or protective clothing, and installing splash guards on the machines—all can help to control dermatitis in the shop.

Machine Cleaning Practice

For mineral-oil-base fluids, machine cleaning can generally be accomplished by periodic removal of chips, metal fines and sludge, followed by flushing with clean cutting oil.

No matter how effective a coolant clarifying system is on water-miscible fluids, machine cleaning must be performed eventually. Proper machine cleaning can extend coolant life 4 to 6 times, compared to simply removing the old coolant and replacing it with fresh coolant.

For machines not in operation, the proper cleaning procedure is as follows:

1. Pump out old coolant.
2. Clean out chips and oil residue.
3. Add cleaner mixed with water at a ratio of 1:50 to machine tool (fill reservoir).
4. Circulate cleaner for at least three hours until machine is clean and apply cleaning solution directly to all machine surfaces that are not in contact with circulating system.
5. When the machine is clean, pump out cleaner and remove all accumulated sediment from the sump.
6. Fill with enough plain water to circulate through all coolant lines and circulate for at least 15 minutes while rinsing surfaces previously cleaned in step 4.
7. Drain and refill with plain water, circulate, rinse and drain again.
8. Add new cutting fluid immediately to cover all exposed metal surfaces to prevent rusting.

By thoroughly cleaning a machine each time the coolant needs changing, all bacteria are removed and, consequently, are not present to immediately begin degrading the fresh coolant. Although a thorough cleaning takes considerably more time compared to pumping out and recharging, the coolant will last considerably longer; consequently, machines can be cleaned on a predetermined schedule instead of on a "catch-as-catch-can" program. The result is controllable machine downtime, which is always less expensive than emergency downtime.

Disposal of Cutting Fluids

Disposal problems with straight oil products are minimized with proper batch-type recovery equipment. Straight oil fluids can be sterilized and water contaminants removed by heat. Settling and addition of base concentrates usually restore adequate quality for continued use. When necessary, oil-base fluids should be disposed of by burning (possibly as a fuel).

In order to meet federal, state, and local water-pollution-control laws, all water-miscible cutting fluids should undergo some sort of treatment before disposal into a lake, stream, or municipal sewer system. The chemicals considered as pollutants in water-miscible products are oil, nitrites, phenols, phosphates, PCB, and heavy metals. The oil content can be broken out of emulsion by an acid or aluminum sulfate treatment. The nitrite content of the effluent can then be destroyed by treatment with sulfamic acid.

In some states effluent containing more than 2 parts per billion of phenol or phenolic derivatives is prohibited. Where regulations are less stringent, phenols can be removed by the use of a carbon filter or slurry.

Recycling Cutting Fluids

Recycling of cutting fluids can solve water-disposal problems, reduce costs and ease pollution problems. A closed-loop system for cutting fluid recycling will have some loss, but 90 percent recovery is not unusual. Even lower rates of recovery can justify the expense of a closed-loop system.

One basic part of a recycling system is some kind of filter to remove metal chips and grinding swarf. Other elements are needed to remove tramp oils, to provide makeup fluid, and, unless the recycling system is part of a central fluid system, to haul, store, and pump fluid in and out of machine sumps.

A recycling system can be used for separate machines or used in a central fluid system as described in the following discussion.

CENTRAL FLUID SYSTEM

Wherever possible, consideration should be given to the possible use of central systems for handling of recycled fluids from groups of machines. This is only practical if the machines are using the same cutting fluid throughout. A group of grinding machines may be linked together to handle the swarf by means of an integrated conveying system; however, all the machines must be grinding similar materials since mixed metals lose their scrap value. The central collection of swarf wetted by cutting fluids also reduces physical handling and improves working conditions.

ECONOMIC CONSIDERATIONS

The cost per piece produced is more important than the initial price of a cutting fluid because in many cases more expensive fluids may actually be more economical to use. Because the cutting fluid is part of a manufacturing system, an economic analysis of cutting fluids must take into consideration both the costs associated with the fluid itself and the costs affected by the fluid. These two cost divisions can be further subdivided into direct and indirect costs. In those cases where the influence of specific costs is great, it is possible to make a rough estimate of the savings without a complete cost analysis. In any event, the economic justification of cutting and grinding fluids should include not only the cost of the fluid but also its effect on the cost of tools, wheels, downtime, etc.

Surface Texture and Integrity

Surface Texture

THE SURFACES produced by machining generally are irregular and complex. Despite this, the majority of machined parts can satisfactorily perform their functions with general, uncomplicated surface texture specifications. Many machining processes can meet these general surface texture requirements with ordinary process control and a minimum of quality control. Dimensional checks and visual examination for macro imperfections usually are sufficient. When surface roughness requirements exceed about 63 micro-in. R_a (1.6 μm), most companies will use a visual check rather than a measurement of the roughness profile. Often the visual check is aided by using sets of sample roughness specimens. Any more exacting specifications for most surfaces would serve no practical purpose and would result in needless expense.

Parts that are highly engineered, heavily stressed or subjected to unusual environments usually have more specific and detailed surface quality requirements. Tolerance requirements for these parts may also need to be closely allied to the roughness allowed. For some of these parts, history or testing has demonstrated a direct relationship between performance and surface texture. Quality

control to measure surface texture is, therefore, necessary for these parts. The following are examples of parts that require enhanced surface texture specifications in order to improve fatigue strength, corrosion resistance, cleanliness, appearance, coatability, sealing or product performance:

1. Antifriction bearings and airfoils
2. Objects operating in corrosive environments
3. Food preparation devices
4. Telescope lenses, plug gages and rolling mill rolls
5. Journal bearings
6. Painted or coated surfaces
7. Sealing surfaces or friction clamped assemblies

Parts that are critically stressed, product-life limiting or safety controlling, or those specific areas on such parts, should have additional specifications that include surface integrity considerations.

SURFACE TEXTURE DEFINED

Surface texture is the repetitive or random deviations from the nominal surface which form the three-dimensional surface topography. A variety of mechanical, electronic and optical devices are available to measure these deviations by sampling the profile of the workpiece. Figure 1 shows a comparison between the measured and the nominal profiles for a given surface.

American National Standard ANSI B46.1–1978 describes, standardizes and calls out acceptable measuring instrumentation for surface texture.* (Internationally, ISO R468 compares in content.) According to the standard, surface texture includes four elements—roughness, waviness, lay and flaws. Figure 2 shows these characteristics in relation to a unidirectional lay surface.

Definitions of the four elements of surface texture are as follows:

Roughness consists of the finer irregularities which generally result from the inherent action of the production process. These include transverse feed marks and other irregularities within the limits of the sampling length.

Waviness includes all irregularities whose spacing is greater than the roughness sampling length and less than the waviness sampling

*ANSI B46.1–1978, "Surface Texture" (which includes ANSI Y14.36–1978 "Surface Texture Symbols") should be available to and be read by everyone concerned with specifying or measuring surface texture. Copies are available from the American Society of Mechanical Engineers, United Engineering Center, 345 East 47th Street, New York, NY 10017.

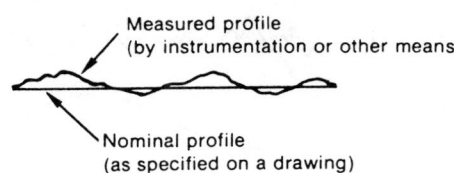

Fig. 1. Surface texture profile. (Based on ANSI B46.1 – 1978)

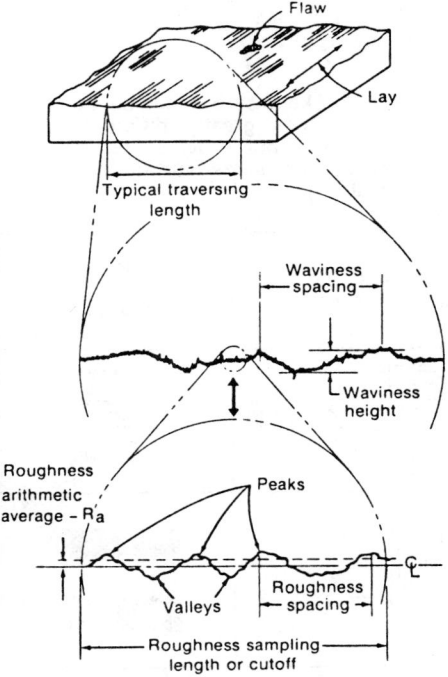

Fig. 2. Pictorial display of unidirectional lay surface characteristics. (Based on ANSI B46.1 – 1978)

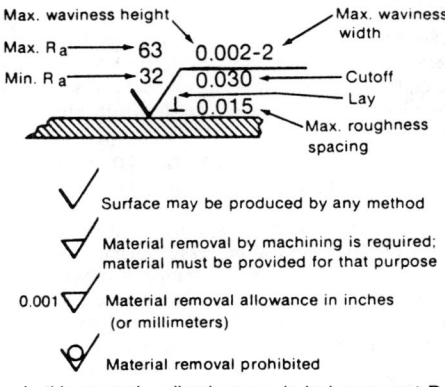

In this example, all values are in inches except R_a values, which are in micro-inches (millionths of an inch). Metric values (millimeters and micrometers, respectively) are used on metric drawings.

Fig. 3. Surface texture symbols used for drawings or specifications. (Based on ANSI Y14.36 – 1978)

length. Waviness may result from machine or work deflections, chatter, vibration, heat treatment or cutting tool runout. Roughness may be considered superimposed on a "wavy" surface.

Lay is the direction of the predominant surface pattern, ordinarily determined by the production method used.

Flaws are unintentional irregularities which occur at one place or at relatively infrequent or widely varying intervals on the surface. Flaws include cracks, blow holes, inclusions, checks, ridges, scratches, etc. Unless otherwise specified, the effect of flaws shall not be included in the roughness average measurements. Where flaws are to be restricted or controlled, a special note as to the method of inspection should be included on the drawing or in the specifications.

Surface finish is a colloquial term widely used to denote the general quality of a surface. Surface finish is not specifically tied to the texture or characteristic pattern of the surface, nor is it tied to specific roughness values; however, a "good" finish implies low roughness values and vice versa. The term surface finish is not as precisely defined as are the terminologies used in the American National Standard, nor is it necessarily expressed numerically.

SURFACE TEXTURE SYMBOLS

The description and specification of surface texture is accomplished by means of symbols and conventions as presented in the ANSI Standard Y14.36 – 1978 (almost in complete agreement with ISO 1302). The symbol used to designate surface roughness is the check mark with a horizontal extension as shown in Fig. 3. Roughness, which is the most commonly used surface parameter, is specified by placing the height rating in micro-inches (one micro-inch = 0.000001 inch) or micrometers (one micrometer = 0.000001 meter) to the left of the check mark. If there is a maximum and a minimum rating, the two numbers are placed one above the other. Roughness is defined as the arithmetic average (AA) deviation of the surface expressed in micro-inches from a mean line or centerline (see Fig. 2 and 4). R_a is a symbol for roughness that has been adopted internationally. Surface roughness is still sometimes displayed with the symbols AA, or CLA or c.l.a. Many instruments still in use employ an average deviation from the roughness centerline which is

R_a R_q

R_a is the universally recognized parameter of roughness. It is the arithmetic mean of the departures y of the profile from the mean line. It is normally determined as the mean results of several consecutive sampling lengths L.

R_q is the corresponding rms parameter.

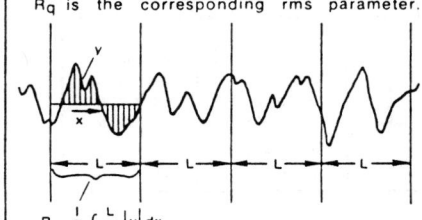

$$R_a = \frac{1}{L}\int_0^L |y|\,dx$$

$$R_q = \sqrt{\frac{1}{L}\int_0^L y^2\,(x)\,dx}$$

R_z

Ten-Point Height is the average distance between the five highest peaks and the five deepest valleys within the sampling length and measured perpendicular to it.

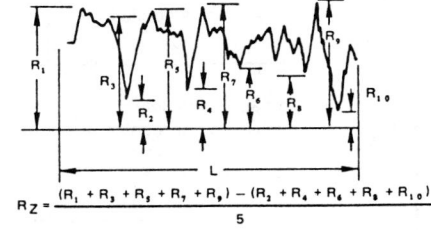

$$R_z = \frac{(R_1 + R_3 + R_5 + R_7 + R_9) - (R_2 + R_4 + R_6 + R_8 + R_{10})}{5}$$

R_t R_{max} R_{tm}

R_t is the maximum peak-to-valley height within the assessment length. R_{max} is the maximum peak-to-valley height within a sampling length L. But because the value can be greatly affected by a spurious scratch or particle of dirt on the surface, it is more usual to use the average (R_{tm}) of five consecutive sampling lengths.

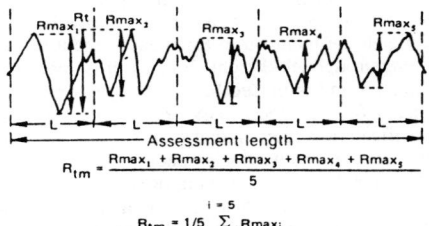

$$R_{tm} = \frac{R_{max_1} + R_{max_2} + R_{max_3} + R_{max_4} + R_{max_5}}{5}$$

$$R_{tm} = 1/5 \sum_{i=1}^{i=5} R_{max_i}$$

R_p R_{pm}

R_p is the maximum profile height from the mean line within the sampling length. R_{pm} is the mean value of R_p determined over 5 sampling lengths.

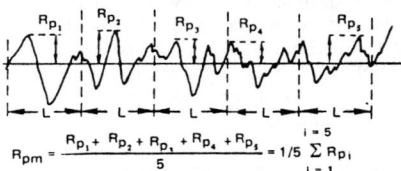

$$R_{pm} = \frac{R_{p_1} + R_{p_2} + R_{p_3} + R_{p_4} + R_{p_5}}{5} = 1/5 \sum_{i=1}^{i=5} R_{p_i}$$

P C

Peak count is the number of peak/valley pairs per inch projecting through a band of width b centered about the mean line.

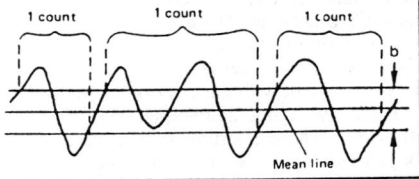

Fig. 4. Some commonly used surface texture symbols and their definitions. (Based on ANSI B46.1 – 1978)

Table 1. Ratio of root mean square roughness to arithmetic average roughness values

Root mean square roughness:	R_q
Arithmetic average roughness:	R_a
Theoretical ratio of sine waves, R_q/R_a	1.11
Actual ratios of R_q/R_a for various processes:	
turning	1.17 to 1.26
milling	1.16 to 1.40
surface grinding	1.22 to 1.27
plunge grinding	1.26 to 1.28
soft honing	1.29 to 1.48
hard honing	1.50 to 2.10
electrical discharge machining	1.24 to 1.27
shot peening	1.24 to 1.28
Practical first approximation of R_q/R_a:	
for most processes	1.25
for honing	1.45

SOURCE: J. Peters, P. Vanherck, and M. Sastrodinoto, *Assessment of surface typology analysis techniques, Annals of the CIRP* 28/2, 1979.

the root mean square (rms), also expressed in micro-inches (R_q). While still used frequently, rms actually has been obsolete since about 1950. Roughness-measuring instruments calibrated in rms read approximately 25 percent higher on a given surface than those instruments calibrated for R_a (see Table 1). The difference is usually much less than the point-to-point variations on any given machined surface.

The roughness width (or sampling) cutoff is the greatest spacing of repetitive surface irregularities to be included in the measurement of the average roughness height (Fig. 2). It is specified in inches and is placed below the horizontal extension of the check mark. A waviness height specification is rated in inches as the peak-to-valley distance; and waviness spacing or width, also rated in inches, is the spacing of successive peaks or valleys (Fig. 2). Waviness height and waviness width values are placed above the horizontal extension of the check mark. The symbol designating the lay is placed under the extension of the check mark. This symbol indicates the direction of the lay relative to the nominal surface. Figure 5 explains the various symbols used to designate lay. The pitted nondirectional or protuberant designation of lay, P, is useful to describe surfaces produced by some of the nontraditional machining operations. The surface texture symbol also can designate the extent of material removal desired—from none to any amount, as shown in Fig. 3.

SURFACE TEXTURE MEASUREMENT

The most prevalent measuring technique for surface texture employs a mechanical-electronic device whose readout indicates the roughness of the surface profile taken during the passage of a small radius stylus over a short straight line path on the surface. The most common diamond stylus has a 0.0004-in. (10-μm) radius and usually is used with a 0.030-in. (0.8-mm) cutoff width. The total stylus travel is usually 20 to 60 times the cutoff width with the electronic circuitry continuously averaging the readings over the set cutoff width. These instruments can read average roughness, R_a, peak count or other roughness designations depending on the particular instrument design.

It should be kept in mind that electronic sur-

face measuring devices generally indicate the roughness but do not indicate the physical character of the surface. Several surfaces can, in effect, be quite different in appearance and still yield similar roughness values, as shown in Fig. 6.

SURFACE TEXTURE PRODUCED BY THE MACHINING PROCESS

To a large degree, the surface texture produced by a material-removal process is characteristic for that particular process. The range of roughnesses typically obtained for a variety of manufacturing processes is shown in Fig. 7. This chart also indicates that it is possible to exceed the usual range under unusual or specially controlled conditions.

The selection of surface texture values involves more than merely designating a particular process. The ability of a processing operation to produce a specific surface roughness depends on

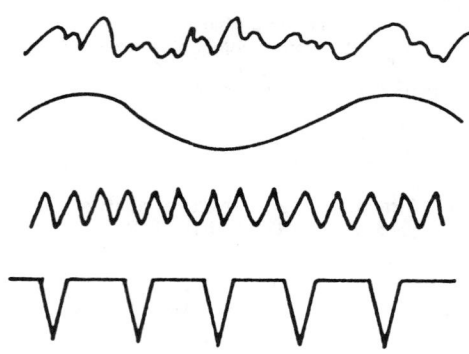

Fig. 6. Sketches of cross sections of surfaces greatly different in character but having approximately the same surface finish level

Lay Symbol	Meaning	Example Showing Direction of Tool Marks	Photographs of Examples
—	Lay approximately parallel to the line representing the surface to which the symbol is applied.	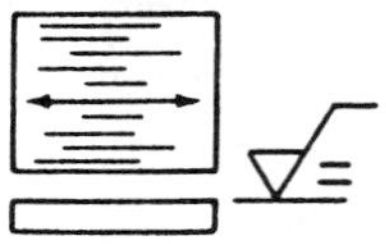	
⊥	Lay approximately perpendicular to the line representing the surface to which the symbol is applied.	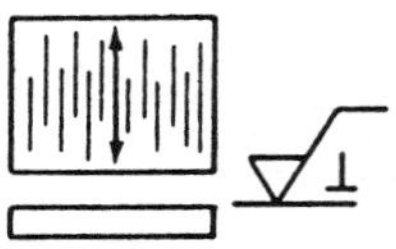	
X	Lay angular in both directions to line representing the surface to which the symbol is applied.	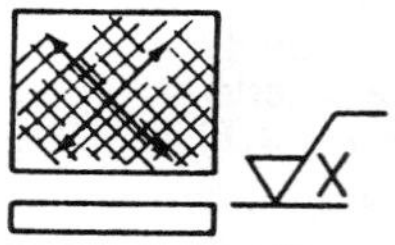	
M	Lay multidirectional.	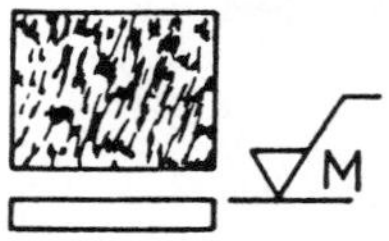	
C	Lay approximately circular relative to the center of the surface to which the symbol is applied.	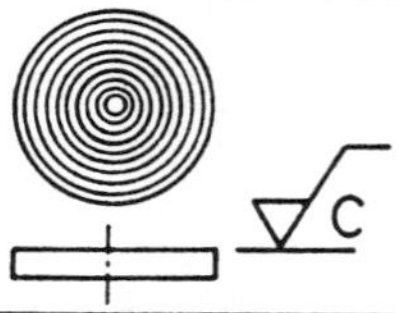	
R	Lay approximately radial relative to the center of the surface to which the symbol is applied.	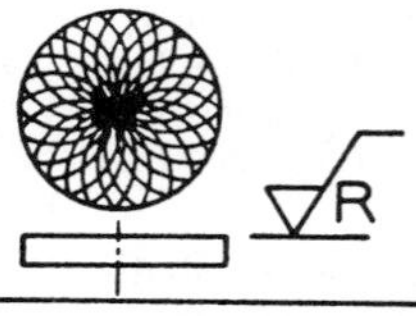	
P	Lay particulate, non-directional, or protuberant.	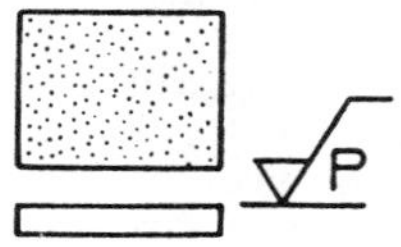	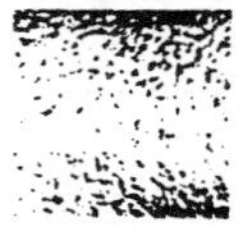

Fig. 5. Lay symbols for surface texture designation. (Based on ANSI Y14.36–1978)

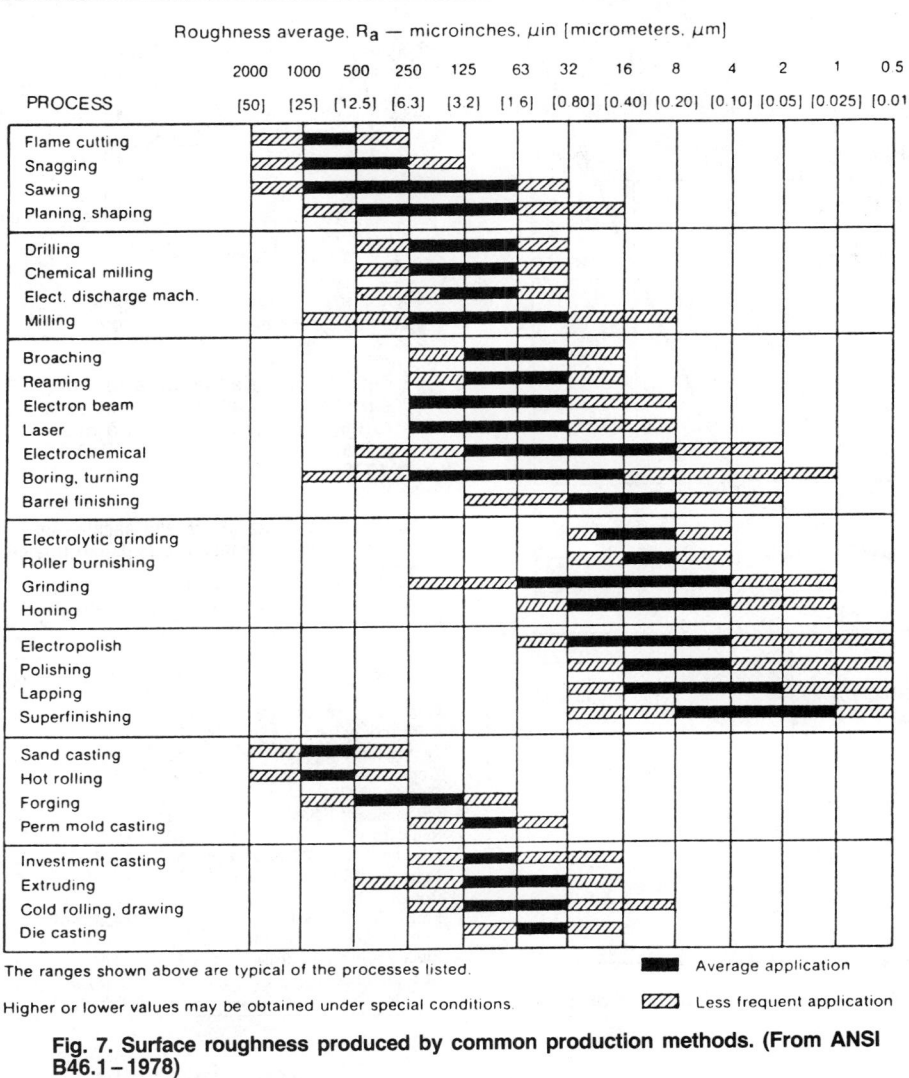

Roughness average, R_a — microinches, μin [micrometers, μm]

Fig. 7. Surface roughness produced by common production methods. (From ANSI B46.1–1978)

The ranges shown above are typical of the processes listed.

Higher or lower values may be obtained under special conditions.

■ Average application

▨ Less frequent application

Table 2. Guide to surface roughness values for close-tolerance machine work

Dimensional tolerances, in. (mm)	Surface roughness, μin. (μm)
Below 0.0002 (Below 0.005)	Below 8 (Below 0.2)
0.0002-0.0005 (0.05-0.012)	8-16 (0.2-0.4)
0.0005-0.0010 (0.012-0.025)	16-32 (0.4-0.8)
0.0010-0.0020 (0.025-0.05)	32-63 (0.8-1.6)
0.0020-0.0100 (0.05-0.25)	63-250 (1.6-6.3)

Table 3. Preferred values for arithmetic average roughness (R_a), cutoff length, and maximum waviness height
Source: ANSI Y14.36–1978

Arithmetic average roughness (R_a), μin. (μm)	Standard roughness sampling length (cutoff)(a), in. (mm)		Maximum waviness height, in. (mm)
0.5 (0.012)	0.003	(0.08)	0.00002 (0.0005)
1 (0.025)(b)	0.010	(0.25)	0.00003 (0.0008)
2 (0.050)(b)	0.030	(0.80)	0.00005 (0.0012)
3 (0.075)	0.1	(2.5)	0.00008 (0.0020)
4 (0.10)(b)-	0.3	(8.0)	0.0001 (0.0025)
5 (0.125)	1.0	(25.0)	0.0002 (0.005)
6 (0.15)			0.0003 (0.008)
8 (0.20)(b)			0.0005 (0.012)
10 (0.25)			0.0008 (0.020)
13 (0.32)			0.001 (0.025)
16 (0.40)(b)			0.002 (0.05)
20 (0.50)			0.003 (0.08)
25 (0.63)			0.005 (0.12)
32 (0.80)(b)			0.008 (0.20)
40 (1.00)			0.010 (0.25)
50 (1.25)			0.015 (0.38)
63 (1.60)(b)			0.020 (0.50)
80 (2.0)			0.030 (0.80)
100 (2.5)			
125 (3.2)(b)			
160 (4.0)			
200 (5.0)			
250 (6.3)(b)			
320 (8.0)			
400 (10.0)			
500 (12.5)(b)			
600 (15)			
800 (20)			
1000 (25)(b)			

(a) When no value is specified, the value of 0.030 in. (0.8 mm) applies. (b) Recommended.

many factors. In turning, for example, the surface roughness is geometrically related to the nose radius of the tool and the feed per revolution. For surface grinding, the final surface depends on the type of grinding wheel, the method of wheel dressing, the wheel speed, the table speed, cross feed and down feed, and the grinding fluid. For electrical discharge machining (EDM), the roughness level is related directly to the individual spark discharge energy level. A change in any of the process operating parameters may have a significant effect on the final surface produced.

MACHINING COST AND SURFACE TEXTURE

The cost of producing a machined surface increases with increasing requirements for finer finishes. Certain machining operations, such as rough turning and milling, are necessary to shape a component to its required dimensions. Additional operations to refine the surface are needed only to permit the surface to perform functions which it could not otherwise perform. A surface roughness of 63 micro-in. (1.6 μm) or coarser can be obtained at a reasonable cost by general roughing and semifinishing operations. The relationship of surface texture to the cost of machining is shown in Fig. 8.

DIMENSIONAL TOLERANCE VERSUS SURFACE TEXTURE

There is a direct relationship between the dimensional tolerance of a part and its permissible surface roughness, since the roughness measurement involves the average linear deviation of the actual surface from the nominal surface defined by the dimension. If the deviations induced by the surface roughness exceed those permitted by the dimensional tolerance, the dimension will be subject to an uncertainty beyond the tolerance, as shown in Fig. 9. On most surfaces the total profile height is approximately four times the measured (arithmetic average) roughness. When measurements are made on a diameter of a part, this value would be doubled. It follows that the roughness value on a diameter should not exceed one-eighth the dimensional tolerance on the diameter if useful dimensional controls are to be maintained.

Each application must be evaluated on its own merit. Table 2, if used with discretion, may serve as a guide to the surface roughness values that may be necessary where machine work must be held to close tolerance for other reasons than merely surface texture. It must be further realized that the practical control of tolerance is also influenced by the size of the part, the overall size of the surface being cut, and the material-removal operations involved.

SURFACE TEXTURE AND QUALITY ASSURANCE

Quality assurance for the surfaces produced by material-removal processes should include assessment of the surface roughness and all other surface texture factors. This does not imply that the effect of scratches, tool marks, sharp corners and other geometric considerations can be overlooked. It is well established that all of these elements can produce stress concentrations that can lead to premature fatigue failures. Historically, surface roughness has been the prime criterion for surface quality and a guide to acceptable fa-

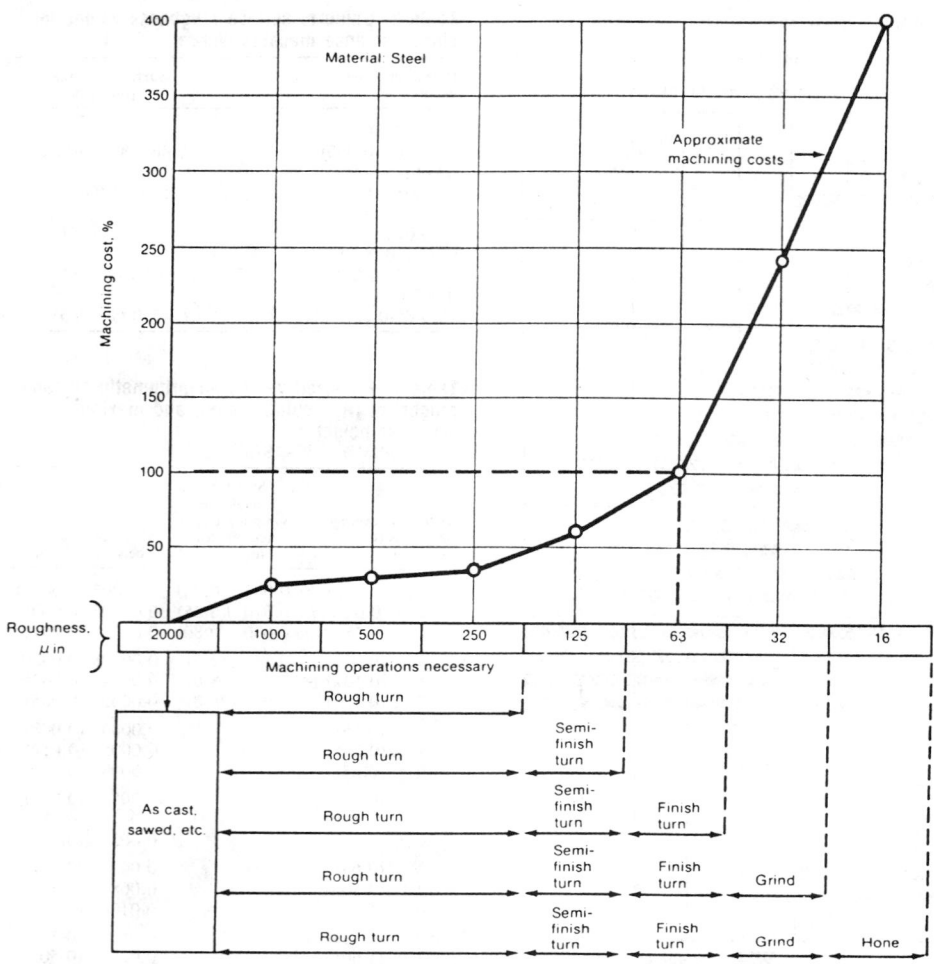

Fig. 8. Relative machining costs and surface roughness for steel parts. (Courtesy of General Electric Co.)

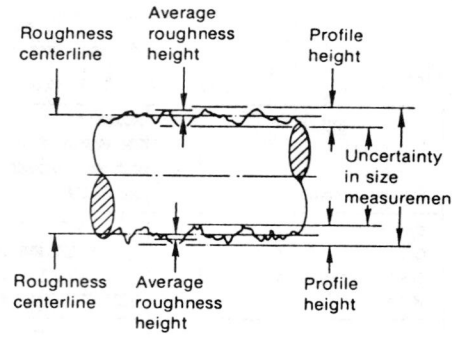

Fig. 9. Uncertainty in size measurement in relation to surface roughness parameters for a cylindrical surface. (From ANSI B46.1–1978)

Table 4. Effect of machining and peening on fatigue strength of Inconel 718 (solution treated and aged 44 R_c)

Operation	Fatigue strength(a) ksi	(MPa)	Percent of gentle grind
Gentle surface grinding	60	(414)	100
Conventional surface grinding	24	(165)	40
Gentle turning	60	(414)	100
Abusive turning	60	(414)	100
Standard ECM	39	(269)	65
Off-standard ECM	39	(269)	65
Standard ECM plus Peen . . .	78	(538)	130
Off-standard ECM plus Peen	67	(462)	112
Finish EDM	22	(152)	37
Rough EDM	22	(152)	37
Finish EDM plus Peen	66	(455)	110
Rough EDM plus Peen	75	(517)	125
Electropolishing (ELP)	42	(290)	70
ELP plus Peen	78	(538)	130

(a) Room temperature, 10^7 cycles full-reverse bending

tigue strength. Some recent data indicate that for some alloys surface roughness is not the critical criterion for high cycle fatigue strength.

PREFERRED ROUGHNESS VALUES

With all the factors that can influence the generation of a surface roughness value, with the dearth of data linking a specific roughness value to the function performed by a specific surface,

and with the variability introduced by the minute sample used in most measuring techniques, it is impractical to place undue emphasis on achieving a specific roughness number. To minimize the variety of drawing callouts, the American National Standards Institute in the standard, ANSI Y14.36–1978, promotes the use of preferred values for average roughness (R_a), roughness cutoff length and maximum waviness height. These values are listed in Table 3.

——Surface Integrity——

SURFACE INTEGRITY is the subject that covers the description and control of the many possible alterations produced in the surface layer of materials during manufacturing and their effect on material properties as well as on service performance. Surface integrity does not include the considerations that relate to surface texture, such as surface roughness.

Gentle conditions
Surface roughness: 45 μin R_a

Conventional conditions
Surface roughness: 40 μin R_a

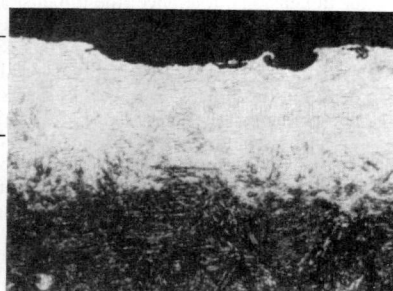

Abusive conditions
Surface roughness: 50 μin R_a

Gentle grinding produced no visible surface alterations. Conventional grinding shows evidence of spotty surface rehardening and underlying overtempering or softening. Abusive grinding produced a rehardened surface layer averaging 0.001 inch deep and an underlying overtempered zone approximately 0.004 inch deep.   1000X

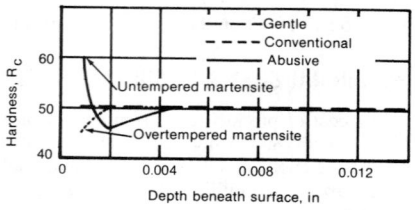

Fig. 10. Surface characteristics of 4340 steel (quenched and tempered, 50 R_c) produced by grinding

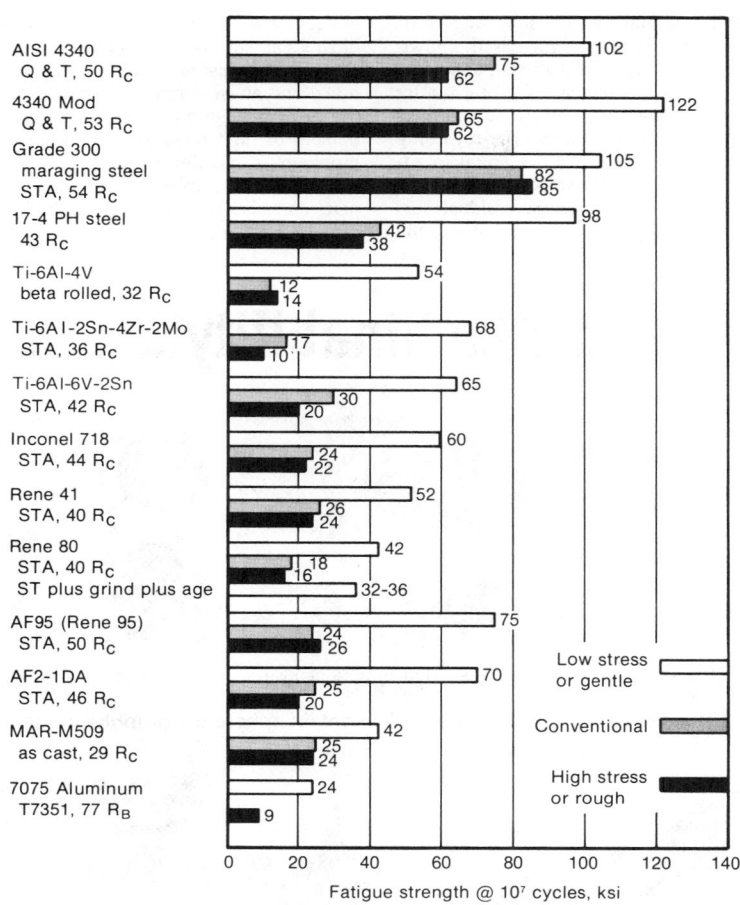

Fig. 11. High-cycle fatigue response of various materials with low-stress, conventional and high-stress grinding

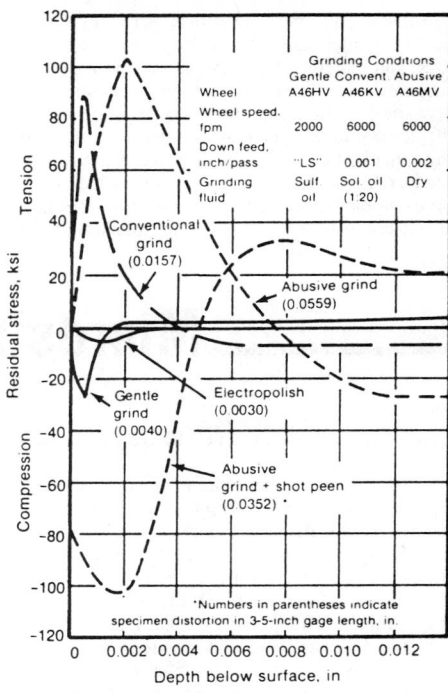

Fig. 12. Residual surface stress in 4340 steel (quenched and tempered, 50 R$_C$) produced by surface grinding. (M. Field, J. F. Kahles and J. T. Cammett)

Figure 10 illustrates an alteration that can take place as a result of a machining process and that is often very damaging to material properties. Steels such as AISI 4340 when ground will show an altered surface layer consisting of rehardened and tempered martensite because of the elevated temperature developed during grinding. Milling, drilling, reaming and tapping can cause severe plastic deformation leading to tearing and cracking. Electrical discharge machining invariably leaves a recast layer that is unacceptable for many highly stressed components because it reduces fatigue strength in great measure.

Figure 11 shows a set of bar charts for a variety of high-strength steels, titanium alloys, and high-temperature alloys. These alloys were ground conventionally, abusively, and gently (low stress) and then tested in reverse bending fatigue. The results show, by far, that the low-stress process is superior. Also, it can be concluded that conventional grinding is essentially the same as abusive grinding. For these alloys, conventional grinding should not be used for the manufacture of highly stressed or critical components. Table 4 shows fatigue data for Inconel 718 which has been machined by various processes. Notable are the reductions in fatigue strength experienced by the EDM and ECM test specimens and the possibility of recovering strength by shot peening.

Table 5. Low-stress grinding procedures

Grinding parameters	Steels and nickel-base high-temperature alloys		Titanium	
	English	Metric	English	Metric
Surface grinding				
Wheel	A46HV	A46HV	C60HV	C60HV
Wheel speed	2500 to 3000 fpm(a)	13 to 15 m/s(a)	2000 to 3000 fpm(a)	10 to 15 m/s(a)
Downfeed per pass	0.0002 to 0.0005 in.(b)	0.005 to 0.013 mm(b)	0.0002 to 0.0005 in.(b)	0.005 to 0.013 mm(b)
Table speed	40 to 100 fpm(c)	12 to 30 m/min(c)	40 to 100 fpm(c)	12 to 30 m/min(c)
Crossfeed per pass	0.040 to 0.050 in.	1 to 1.25 mm	0.040 to 0.050 in.	1 to 1.25 mm
Grinding fluid	Highly sulfurized oil	Highly sulfurized oil	(d)	(d)
Traverse cylindrical grinding				
Wheel	A60IV	A60IV	C60HV	C60HV
Wheel speed	2500 to 3000 fpm(a)	13 to 15 m/s(a)	2000 to 3000 fpm(a)	10 to 15 m/s(a)
Infeed per pass	0.0002 to 0.0005 in.(b)	0.005 to 0.13 mm(b)	0.0002 to 0.0005 in.(b)	0.005 to 0.013 mm(b)
Work speed	70 to 100 fpm(c)	20 to 30 m/min(c)	70 to 100 fpm(c)	20 to 30 m/min(c)
Grinding fluid	Highly sulfurized oil	Highly sulfurized oil	(d)	(d)

NOTE: For a wide variety of metals (including high-strength steels, high-temperature alloys, titanium and refractory alloys), low-stress grinding practices develop very low residual tensile stresses. In some materials the residual stress produced near the surface is actually in compression instead of tension.

(a) Low-stress grinding requires wheel speeds lower than the conventional 6000 feet per minute (30 meters per second). In order to apply low-stress grinding, it would be preferable to have a variable-speed grinder. Since most grinding machines do not have wheel speed control, it is necessary to add a variable speed drive or make pulley modifications.

(b) Downfeeds or infeeds in the range of 0.0002 to 0.0005 in. per pass (0.005 to 0.013 mm/pass) have been found satisfactory for steels, nickel-base high-temperature alloys, and titanium alloys. A typical feed schedule calls for removing the last 0.010 in. (0.254 mm) of stock as follows: remove 0.008 in. (0.2 mm) at 0.0005 in. per pass (0.013 mm/pass) and remove the last 0.002 in. (0.05 mm) at 0.0002 in. per pass (0.005 mm/pass). (c) Increased work speeds even above those indicated are considered to be advantageous toward improving surface integrity. (d) See Section 16.3, Codes 81 and 83, in Machining Data Handbook, Third Edition, Volume Two.

Table 5 outlines low-stress grinding procedures for steels, nickel-base high-temperature alloys, and titanium alloys. An examination of these recommendations shows that the wheel speed for low-stress grinding is approximately one-half of the conventional speed. Additionally, soft grinding wheels are used as well as very low downfeed or infeed rates. Reducing the wheel speed poses problems because most grinders do not have speed control. Also, the use of soft wheels increases wheel cost and the low feed rates seriously decrease productivity. Increased knowledge concerning the effect of various parameters has made it possible to increase feed rates after changes such as decreasing wheel speed and wheel hardness were made. Until experience regarding allowables is accumulated, it is essential to adhere to the use of conservative parameters.

During the machining process, residual surface stresses are produced. Figure 12 shows the effect of grinding and polishing on the magnitude of stress at various depths below the surface. The residual stresses near the surface for conventional and abusive grinding are tensile, whereas the low-stress grind is in compression. Development of high residual tension stresses causes problems in meeting dimensional specifications as a result of the accompanying distortion of thin parts and also increases susceptibility to stress-corrosion failure.

Microstructure and Machinability

Cast Irons

GRAY, ductile and malleable cast irons are typical of materials for which hardness testing is used to assess their machinability. A preferred method is to combine hardness testing with a microstructure evaluation, because some microconstituents that affect machinability adversely have only a minor influence on the hardness.

Figure 1 illustrates the microstructures of typical gray cast irons and a white cast iron. As the percentage of ferrite decreases and the pearlite increases, the tool life obtained in machining operations such as turning, face milling and drilling decreases. Additionally, the finer the pearlite spacing, the lower the tool life. Carbides in quantities as little as 3 to 5 percent seriously decrease tool life. Table 1 lists the hardness of the more common microconstituents of cast iron. Steadite, which is a eutectic high in phosphorus, in spite of its relatively high hardness, is not detrimental to tool life in the amounts usually found in cast irons in the United States—5 percent or less. Essentially, the tool life experienced in the machining of cast irons is a function of the matrix structure and not of the condition or amount of the graphite.

Figures 2 and 3 illustrate typical microstructures of ductile and malleable irons. The annealed structures of both types have a matrix of

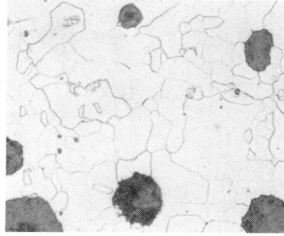 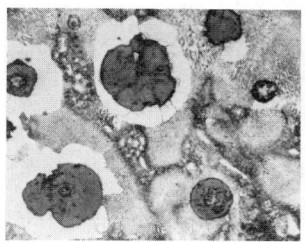 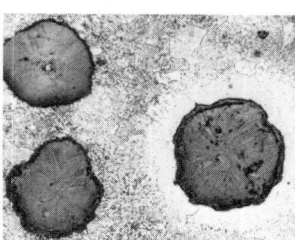

(a) 100% ferrite
170 BHN

(b) 50% ferrite, 50% pearlite
207 BHN

(c) Spheroidite
265 BHN

Fig. 2. Microstructures of ductile irons. All contain spheroidal graphite. Nital etch; 500×.

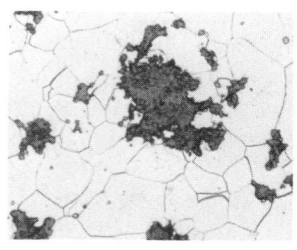 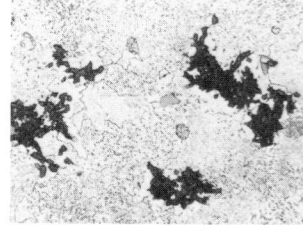 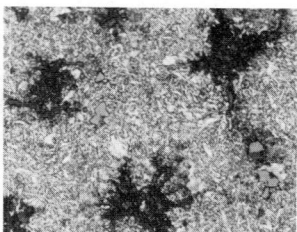

(a) Ferritic malleable
ASTM 32510
Ferritic structure

(b) Pearlitic malleable
ASTM 48004
Spheroidized structure

(c) Pearlitic malleable
ASTM 80002
Spheroidized structure

Fig. 3. Microstructure of malleable irons. All contain nodular graphite or temper carbon. Nital etch; 500×.

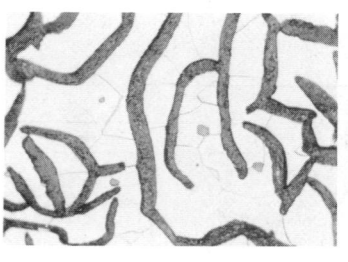 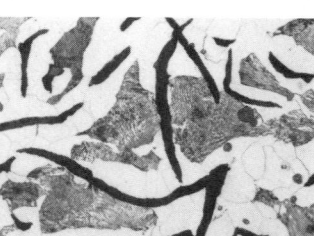 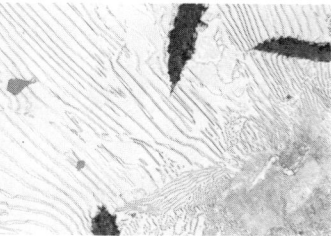 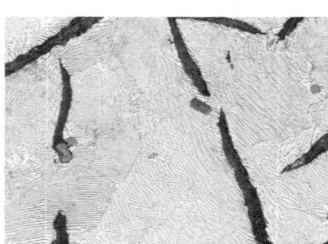

(a) 100% ferrite
120 BHN

(b) 50% pearlite, 50% ferrite
150 BHN

(c) Coarse pearlite
195 BHN

(d) Fine pearlite
215 BHN

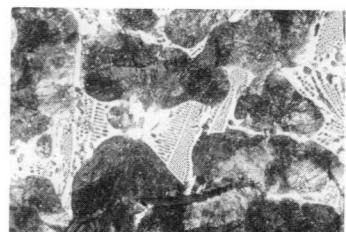 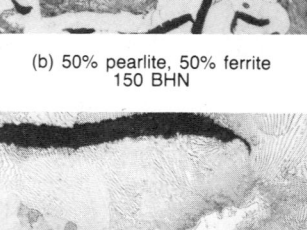 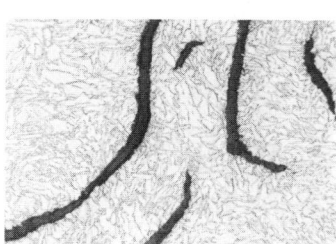 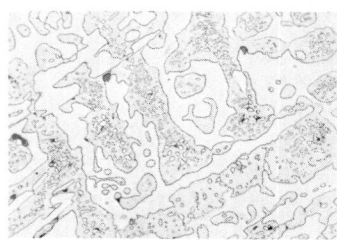

(e) Pearlite + steadite
200 BHN

(f) Pearlite + carbide
240 BHN

(g) Acicular
263 BHN

(h) White iron. Pearlite + carbide. 550 BHN

Fig. 1. Microstructure of gray irons. All except white iron contain flake graphite. Nital etch; 100×.

Table 1. Hardness of cast iron microconstituents

Microconstituent	Knoop hardness, 100-g load
Graphite	15 to 40
Ferrite	215 to 270
Pearlite	300 to 390
Steadite	600 to 1200
Carbide	1000 to 2300

Table 2. Turning cast irons with C-2 carbide
Cutting speed, ft/min, for 30-min tool life = V_{30}.

Material	Matrix microstructure	Type of graphite	BHN	V_{30}	UTS, psi	YP, psi	Elong., %
Gray Iron	100% ferrite	Flake	100	880	15 700	...	...
	Coarse pearlite	Flake	195	360	35 000	...	...
	Fine pearlite	Flake	225	340	45 000	...	...
	Acicular	Flake	263	200	59 000	...	...
Ductile Iron	100% ferrite	Spheroidal	170	810	70 000	56 000	22
	97% ferrite, 3% pearlite	Spheroidal	183	570	77 000	62 000	20
	60% ferrite, 40% pearlite	Spheroidal	207	430	84 700	69 800	17
	60% ferrite, 40% pearlite	Spheroidal	215	360	93 000	72 000	4
	20% ferrite, 80% pearlite	Spheroidal	265	240	97 250	79 000	2
Ferritic malleable, ASTM 32510	100% ferrite	Temper carbon	109	950	50 000	32 500	10
Pearlitic malleable: ASTM 48004	Spheroidite	Temper carbon	179	450	70 000	48 000	4
ASTM 60003	Spheroidite	Temper carbon	230	280	80 000	60 000	3
ASTM 80002	Spheroidite	Temper carbon	250	260	100 000	80 000	2

practically all ferrite and can be machined at relatively high speeds comparable to those used for machining of gray iron.

Table 2 lists the cutting speeds which result in a 30-minute tool life in turning for various microstructures of gray, ductile, and malleable cast irons. It is interesting to note that even though the 100% ferrite structure of ductile iron is 170 HB, considerably harder than the ferritic gray iron at 100 HB, the cutting speeds for a 30-minute tool life are comparable.

As indicated previously, carbides in cast irons are detrimental to tool life. Figure 4 shows a comparison of tool life for conventional face milling of two ferritic grades of malleable iron— 32510 and 35018. The 35018 alloy contained carbides, which were responsible for reducing the tool life from a level of 720 in. (18 m) of work travel to 90 in. (2.3 m) at a cutting speed of about 1000 ft/min (305 m/min). The use of molybdenum as an alloying element in high-strength gray irons helps avoid the presence of a carbide phase. Figure 5 compares tool life in milling of a chromium-nickel-copper gray iron with a molybdenum-copper gray iron. The content of intercellular carbides in the chromium-nickel-copper alloy was less than 5%. Based upon a tool life of 60 min, the cutting speed could be increased from 300 ft/min (91 m/min) to 420 ft/min (128 m/min) by using the carbide-free iron. On the other hand, if a tool life of 60 min was anticipated at 420 ft/min (128 m/min) and then a batch of iron with carbides was placed into the machining line, the tool life would drop to 20 min.

——— Steels ———

MICROSTRUCTURES typical of wrought steels are illustrated in Figure 6. As in the case of cast irons, microstructure plays an important role in influencing the machining characteristics of many different types of steel alloys. Tool life is increased as the proportion of ferrite increases in annealed low- and medium-carbon steels. The addition of free-machining additives such as sulfur and lead (Fig. 6a), which appear as grayish manganese sulfide inclusions, improves machinability, usually at the sacrifice of transverse mechanical properties. The effectiveness of such additives is a function of many processing variables, including the silicon and oxygen content and the sequence of mechanical working procedures. As the inclusion count increases, machinability is improved. Further, it is preferable to have inclusions with low length-to-width ratios (football shapes rather than stringers). Uniformity in the distribution of additives is extremely important.

Steels such as AISI 8640, 4140 and 4340 in the annealed condition consist of approximately 50% ferrite and 50% pearlite. Variations in annealing practices can result in increases in pearlite to levels of 75 to 90% with a resultant decrease in tool life. Normalized (air-cooled) structures of the 0.40% carbon steels have microstructures consisting of fine pearlite in amounts up to 100%.

Several other types of microconstituents that enter the shop environment are the spheroidized form (Fig. 6f) and the Widmanstätten structure (Fig. 6g). In medium-carbon steels, the spheroidized form provides excellent tool life but tends to provide poorer surface finish. The Widmanstätten structure is often present in the thinner sections of steel forgings, steel castings, and steel weldments that have not been subjected to post-heat-treatments such as annealing or normalizing. The machinability and the toughness and ductility of the Widmanstätten structure are poor, and therefore annealing or normalizing treatments are required.

The machining of forgings and castings that have not been subjected to post-heat-treatments is complicated by the fact that thin sections may have a Widmanstätten structure while heavier sections are ferritic and pearlitic, each with different machining response.

Quenched-and-tempered steels at hardness levels of 300 to 400 HB (Fig. 6h) are more difficult to machine than are the lower-hardness annealed or normalized forms. However, because the microstructure is a "single" microconstituent (namely, tempered martensite), the correlation of tool life with hardness is predictable. Figure 7 is a plot of hardness versus cutting speed for a 30-min tool life for 12 different wrought steels at various hardness levels. Testing included the use of high speed steel tools as well as carbide tools. The scatter in the hardness range around 200 HB

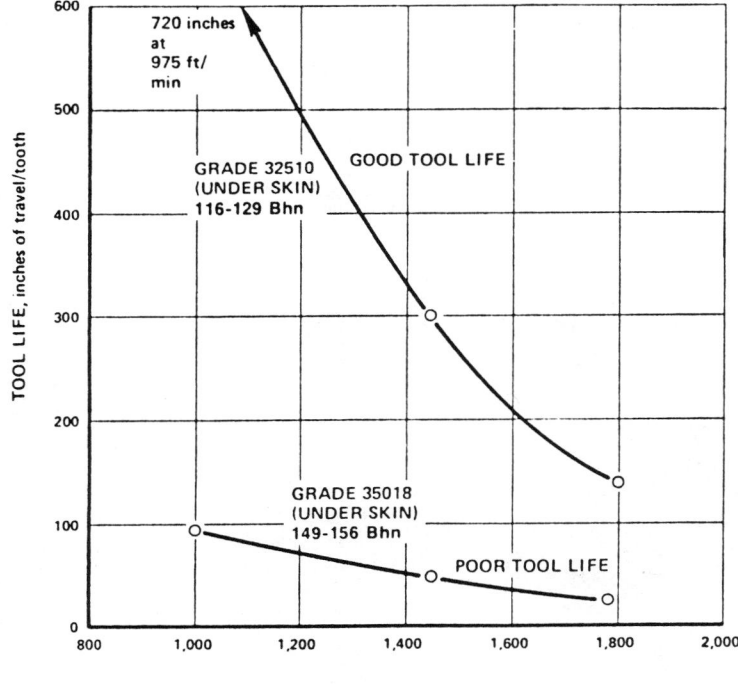

Tool: C-6 carbide single-tooth fly cutter (axial rake, 7° neg; radial rake, 7° neg; end cutting-edge angle, 5°; relief angle, 7°; corner angle, 45°). Feed: 0.010 in./rev. Depth of cut, 0.100 in. Width of cut: 4³/₈ in. Cutting fluid, none. Tool-life end point: 0.015-in. wear land.

Fig. 4. Comparison of tool life for conventional face milling of two ferritic grades of malleable iron

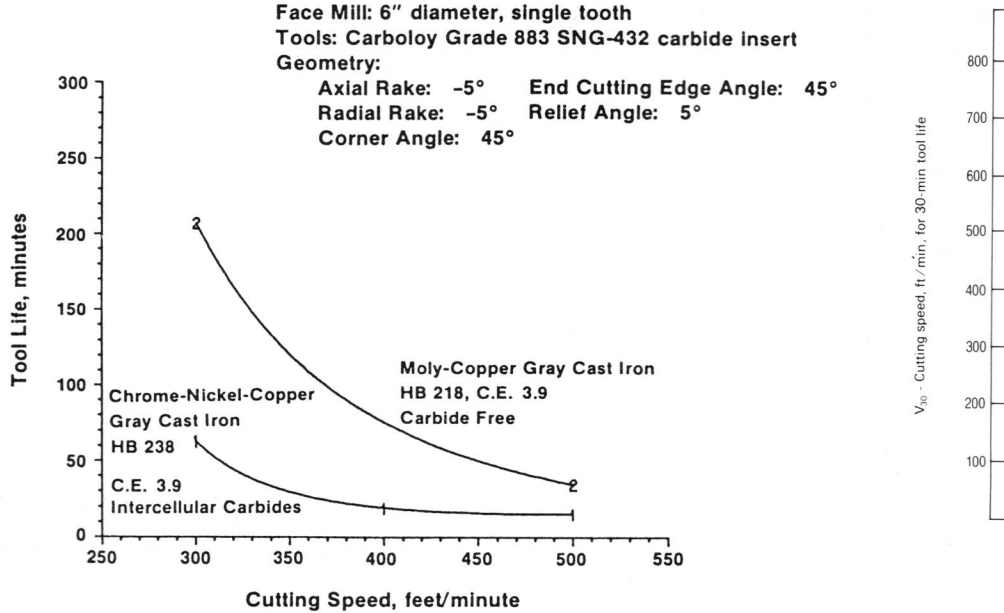

Face Mill: 6" diameter, single tooth
Tools: Carboloy Grade 883 SNG-432 carbide insert
Geometry:
Axial Rake: -5° End Cutting Edge Angle: 45°
Radial Rake: -5° Relief Angle: 5°
Corner Angle: 45°
Tool Life, minutes
300
250
200
150
100
50
0
Chrome-Nickel-Copper
Gray Cast Iron
HB 238
C.E. 3.9
Intercellular Carbides
Moly-Copper Gray Cast Iron
HB 218, C.E. 3.9
Carbide Free
250
300
350
400
450
500
550
Cutting Speed, feet/minute

Fig. 5. Effect of cutting speed and work material in face milling of class 45C gray irons

V30 - Cutting speed, ft/min, for 30-min tool life
800
700
600
500
400
300
200
100
C-6 carbide
T1 HSS
100
200
300
400
500
560
Brinell hardness number
13.2
32.4
42.3
51.3
56
Rc hardness

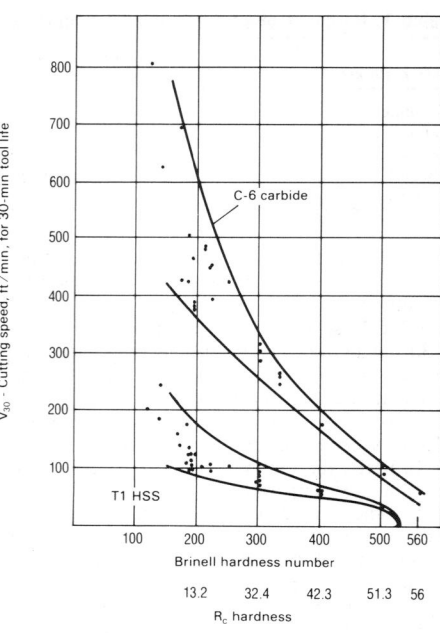

Fig. 7. V_{30} vs hardness in turning wrought steels

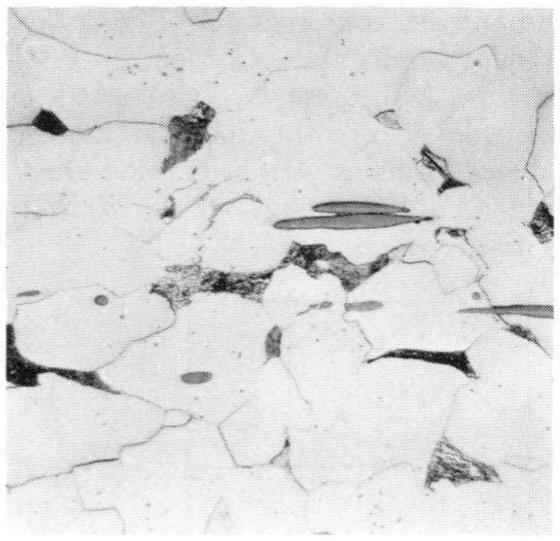

(a) AISI 1112
10 pearlite, 90 ferrite + sulfides
135 BHN 500×

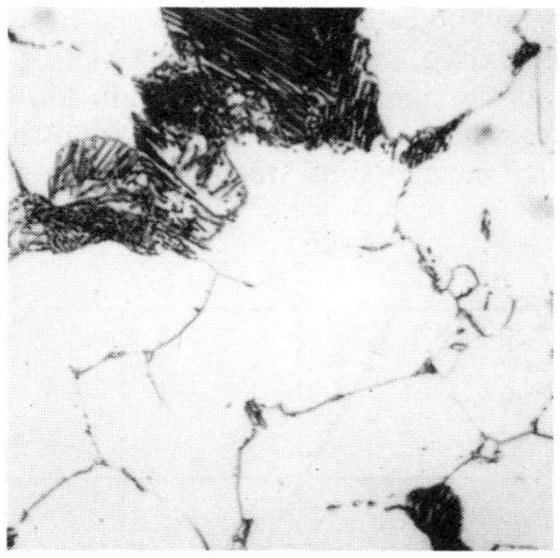

(b) AISI 8620
30 pearlite, 70 ferrite
135 BHN 2000×

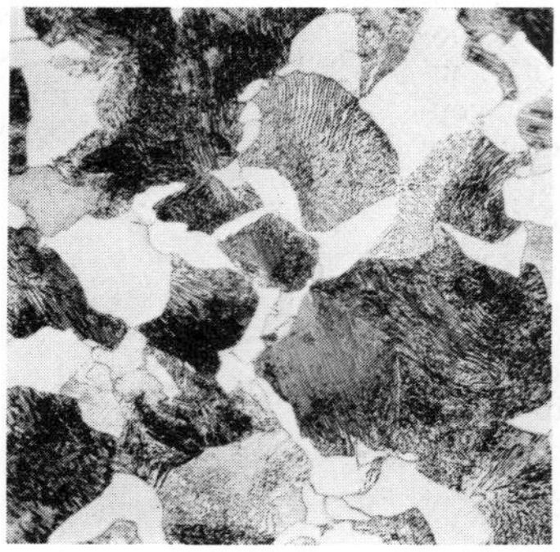

(c) AISI 8640
50 pearlite, 50 ferrite
170 BHN 500×

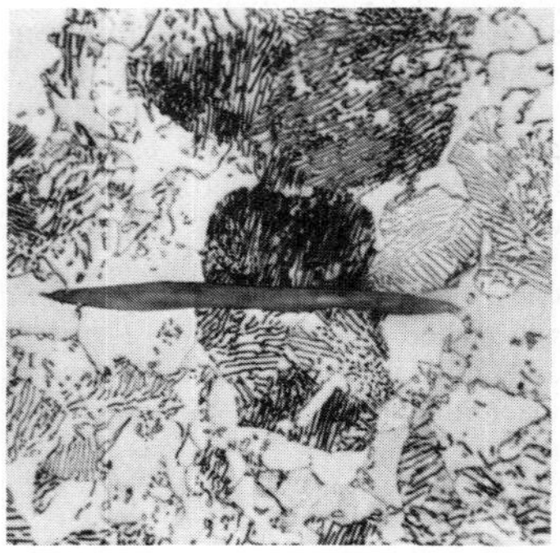

(d) AISI 8640 resulfurized
65 pearlite, 35 ferrite + sulfides
185 BHN 2000×

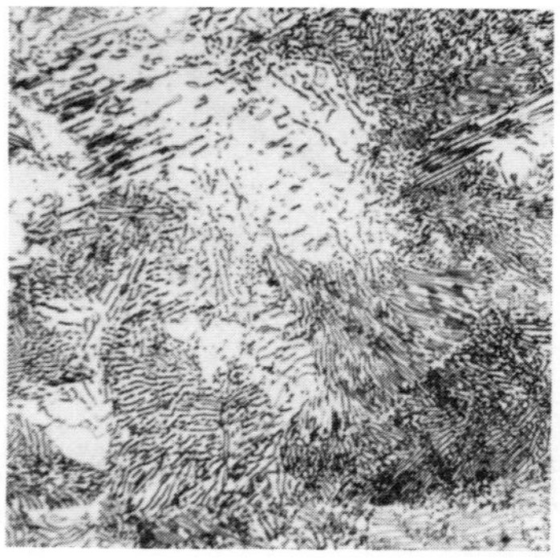

(e) AISI 4340
100% pearlite
221 BHN 2000×

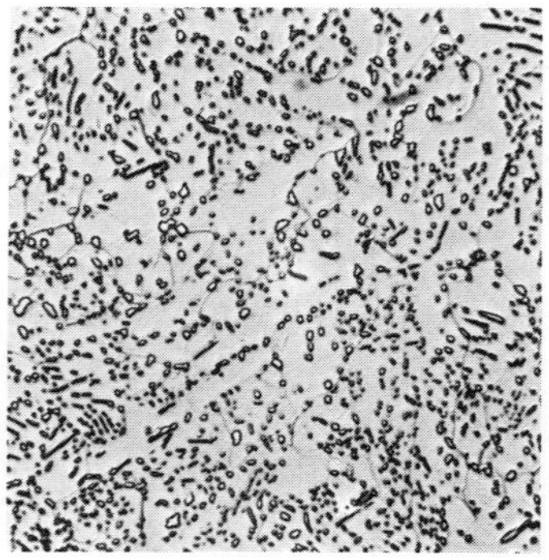

(f) AISI 8640
Spheroidite
180 BHN 2000×

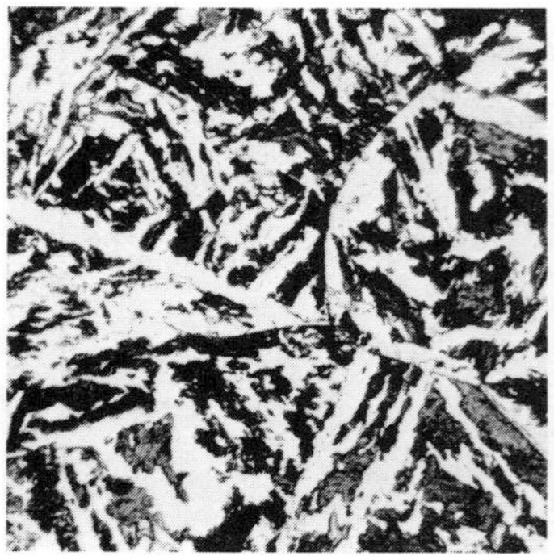

(g) AISI 8640
Widmanstätten
250 BHN 2000×

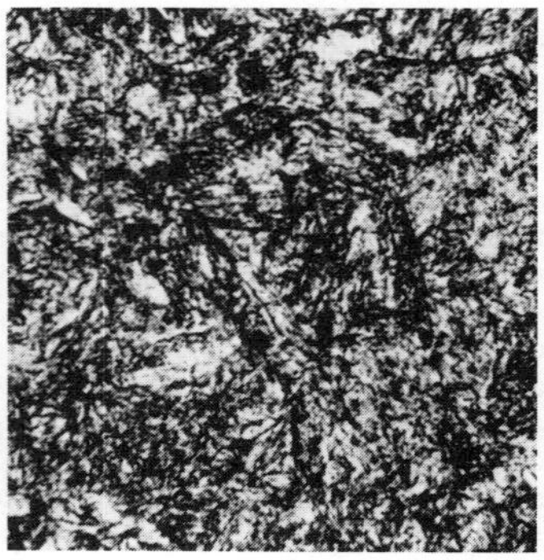

(h) AISI 8640
Tempered martensite
300 BHN 2000×

Fig. 6. Microstructures of wrought steels. Nital etch.

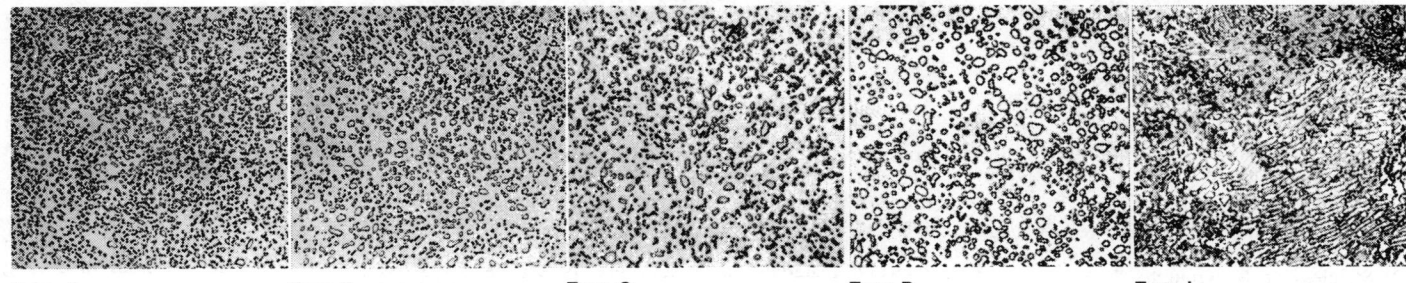

Fig. 8. Typical carbide structures. 1500×. (From J. V. Lyons and M. J. Hudson, Proceedings of the Conference on Machinability, Iron and Steel Institute, London, 1965, p 113)

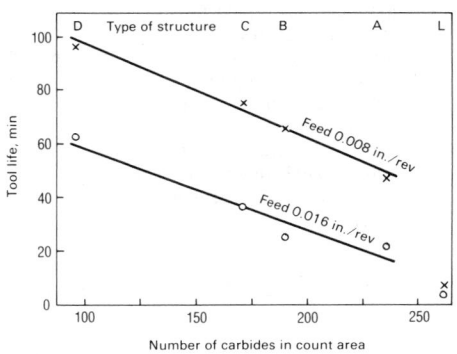

From J. V. Lyons and M. J. Hudson, Proceedings of the Conference on Machinability, Iron and Steel Institute, London, 1965, p 114.

Fig. 9. Relationship of machinability and carbide size

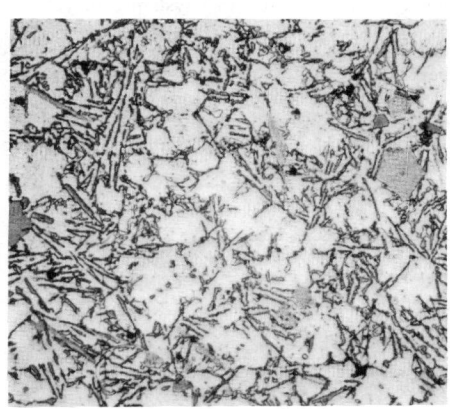

From C. A. Queener and W. L. Mitchell, AFS Transactions, Vol 73 (1965), p 18.

Fig. 10. Structure of a commercial aluminum alloy 380 die casting. Etched in 0.5% HF; 500×.

Table 3. Tool-life tests on different carbide structures

Carbide structure	Tool life, min, at 110 ft/min cutting speed, 0.05-in. depth of cut			
	Feed 0.008 in./rev	Average	Feed 0.016 in./rev	Average
Type L	9, 1, 13, 6	7	40 s, 20 s, 30 s, 9 s	25 s
Type A	49, 47, 60, 42, 37, 51	48	17, 28, 20, 21, 17, 29	22
Type B	62, 68, 64, 70, 64, 70	66	19, 27, 28, 26, 18, 32	25
Type C	71, 60, 76, 77, 95, 74	75	56, 50, 23, 28, 36, 32	37
Type D	87, 110, 88, 113, 92, 93	97	44, 92, 73, 51, 62, 56	63

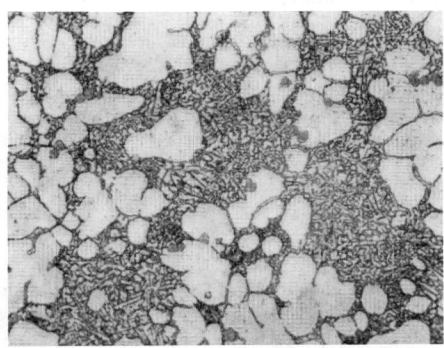

Fig. 11. Structure of a sodium-modified aluminum alloy 380 containing 0.85% Fe. Etched in 0.5% HF; 500×. (From same source as Fig. 10)

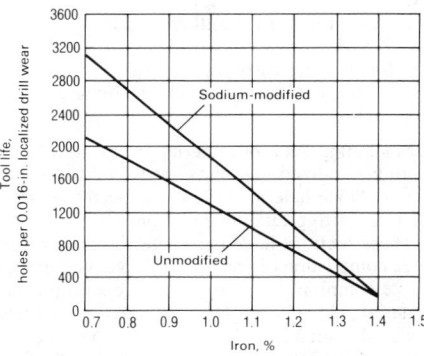

Drilling conditions: 260 ft/min, 0.0032 in./rev; water-soluble oil 1:20; carbon steel drill 7/64-in. diam, 118° point angle, 7° clearance angle; 1/4-in.-depth through holes.

Fig. 12. Tool life in drilling tests on sodium-modified and unmodified aluminum alloy 380. (From same source as Fig. 10)

no circumstances should reliance be placed upon using hardness as an index of machinability in going from one alloy system to another — for example, going from steels to high-temperature nickel-base alloys or to titanium alloys.

The microstructure of bearing steels, such as the 1% C-1% Cr type, controls the machinability of these types of materials. Usually, these steels are machined in the spheroidized condition. Tool life is a function of the carbide size and number per unit area. For a given chemical composition, as the size increases (and number decreases) tool life is improved appreciably. If any of the carbides tend to be lamellar (platelike), tool life is seriously decreased. Figure 8 shows typical carbide structures, and Fig. 9 and Table 3 illustrate the influence on tool life.

____ Aluminum Alloys ____

MOST ALUMINUM ALLOYS may be machined at very high speeds, because tool life is

is very large, thereby indicating the difficulty encountered in using hardness for correlating tool life in machining. However, about 350 HB — the hardness range of quenched-and-tempered steels — the scatter is reduced markedly and hardness may be a reasonable criterion for planning and for quality control of work materials. As shown, there are limitations in using hardness for even a given class of work materials — namely, steels. Under

not an important consideration or limitation. However, the high-silicon alloys such as 380 (8.5% Si) and 390 (17% Si) are exceptions. Figure 10 shows the microstructure of a commercial 380 die casting. The needle-like constituent is silicon, which is abrasive and detrimental to tool life. Figure 11 shows the 380 alloy with iron controlled at 0.85% and sodium treated in order to break up the silicon into a very fine dispersion. In processing the high-silicon alloys, microstructure can serve to provide a quality check for machining.

Figure 12 depicts the result of drilling tests for sodium-modified and unmodified die castings of various iron contents. A 0.85% iron sodium-modified die-casting alloy will permit drilling more than 2400 holes before resharpening — an increase of 700 holes over the unmodified 380 aluminum alloy.

Machining Data Recommendations

SPEEDS AND FEEDS

TABLES 1 to 42 provide starting recommendations in English units for the turning, face milling, end milling (peripheral), drilling, reaming and tapping of:

1. Free-machining low-carbon steels, wrought
2. Medium-carbon alloy steels, wrought
3. Stainless steels, wrought
4. Cast irons—gray, ductile and malleable
5. Aluminum alloys, wrought and cast
6. Titanium alloys, wrought
7. Nickel-base high-temperature alloys, wrought and cast.

In addition to the seven material classes noted above, machining data for the following material groups are included in Tables 43 to 54:

8. Maraging steels, wrought
9. Tool steels, wrought or cast
10. Structural steels, wrought
11. Austenitic manganese steels, cast
12. Magnesium alloys, wrought or cast
13. Copper alloys, wrought or cast
14. Nickel alloys, wrought or cast
15. Lead, tin and zinc alloys, cast
16. Zirconium alloys, wrought
17. Machinable carbides, ferritic
18. Controlled-expansion alloys (Invar and Kovar).

Machining data for material groups 8 to 18 are more generalized than those provided for groups 1 to 7. Nonetheless, these are expected to serve well as starting recommendations.

These sections have been selected from a very large compilation of machining data covering most of the commonly used machining operations and materials (*Machining Data Handbook*, 3rd Edition, Machinability Data Center, Metcut Research Associates Inc., Cincinnati, 1980).

Typically, these specific sets of data show the range of speeds and feeds and other parameters for materials with widely different machining properties. Steels represented here by the free-machining low-carbon types, the medium-carbon alloy steels, and the stainless steels machine reasonably well. In contrast are alloys such as the nickel-base high-temperature alloys and titanium alloys. Especially the nickel-base high-temperature alloys must be machined at very low cutting speeds. Aluminum alloys, in wrought form, are typical of materials that do not cause appreciable wear of cutting tools, even at high speeds. The 380-type and 390-type high-silicon aluminum die-casting alloys are exceptions, as an examination of the recommended conditions will indicate.

The speeds and feeds for the various operations, based on the data obtained from many sources, represent a tool life of approximately one to two hours of cutting time for most of the common alloys when using high speed steel tools or brazed carbide tools. A tool life of 30 to 60 minutes is applicable for indexable-insert carbide tools. In actual shop operations, the time between tool changes might be three to four times these values because of the loading and unloading times associated with most machining operations. Generally, tool life in excess of two hours of actual cutting time would indicate that the speeds and feeds are too low and could be increased to achieve more nearly optimum production at minimum cost.

In addition to the parameters presented in Tables 1 to 54, Tables 55 to 61 should be consulted for selection of tool geometry, and Tables 62 and 63 for selection and identification of cutting fluids.

TOOL GEOMETRY

Recommendations in Tables 55 to 61 are made for tool geometry in turning, face milling, end milling (peripheral), drilling, reaming and tapping, and are to be used in conjunction with the data tables suggesting speeds and feeds for various material groups (see Tables 1 to 54 in this article).

CUTTING FLUIDS

Cutting fluid recommendations for typical steels, cast irons, and aluminum, titanium and high-temperature alloys are shown in Table 62. The codes, 0 through 9, designate the type of fluid as listed in Table 63.

It is important to note that the codes 1 through 9 cover very broad general classes of cutting fluids. For example, under cutting-fluid code 3 (Oils—Heavy Duty) one may purchase the following subtypes:

Sulfurized mineral oil
Compounded sulfurized mineral oil
Chlorinated mineral oil
Sulfurized mineral oil + sulfur, chlorine and phosphorus compounding
Sulfochlorinated mineral oil
Sulfurized mineral oil + fatty oil
Chlorinated mineral oil + fatty oil
Sulfochlorinated mineral oil + fatty oil
Compounded sulfochlorinated mineral oil + fatty oil
Sulfochlorinated fatty oil
Sulfurized fatty oil
Mineral oil + added fat, sulfur, chlorine and phosphorus compounding.

At present, there are no standards to clearly indicate which subtypes are preferable for specific machining operations and materials. The subtypes as designated reflect the many varied descriptions of products of cutting fluid suppliers. In order to select a specific product, it is necessary to consider company policies regarding use of fluids (central systems, recycling), availability, preference for a particular supplier, economics and other conditions. The code numbers 0 through 9 as shown in Table 63 provide good starting recommendations based upon general practices in the machining industry. For additional information, see the article on Cutting Fluids in this section.

Table 1. Turning, free-machining low-carbon steels, wrought

Material	Hardness, Bhn	Condition	Depth of cut, in.(a)	High speed steel tool Speed, fpm	High speed steel tool Feed, ipr	Tool material, AISI	Carbide tool Uncoated Speed, fpm Brazed	Carbide tool Uncoated Speed, fpm Indexable	Carbide tool Uncoated Feed, ipr	Tool material grade	Carbide tool Coated Speed, fpm	Carbide tool Coated Feed, ipr	Tool material grade
Resulfurized steels													
1116, 1117, 1118, 1119, 1211, 1212	100 to 150	Hot rolled or annealed	.040	200	.007	M2, M3	670	790	.007	C-7	1200	.007	CC-7
			.150	150	.015	M2, M3	510	600	.020	C-6	775	.015	CC-6
			.300	120	.020	M2, M3	400	475	.030	C-6	625	.020	CC-6
			.625	90	.030	M2, M3	320	370	.040	C-6	—	—	—
	150 to 200	Cold drawn	.040	210	.007	M2, M3	680	820	.007	C-7	1225	.007	CC-7
			.150	160	.015	M2, M3	520	625	.020	C-6	800	.015	CC-6
			.300	125	.020	M2, M3	410	495	.030	C-6	650	.020	CC-6
			.625	100	.030	M2, M3	330	385	.040	C-6	—	—	—
1213, 1215	100 to 150	Hot rolled or annealed	.040	295	.008	M2, M3	725	860	.007	C-7	1300	.007	CC-7
			.150	225	.015	M2, M3	550	650	.020	C-6	850	.015	CC-6
			.300	175	.020	M2, M3	415	510	.030	C-6	675	.020	CC-6
			.625	140	.030	M2, M3	340	400	.040	C-6	—	—	—
	150 to 200	Cold drawn	.040	300	.008	M2, M3	790	900	.007	C-7	1350	.007	CC-7
			.150	230	.015	M2, M3	600	700	.020	C-6	900	.015	CC-6
			.300	180	.020	M2, M3	475	550	.030	C-6	725	.020	CC-6
			.625	140	.030	M2, M3	370	415	.040	C-6	—	—	—
1108, 1109, 1110, 1115 ...	100 to 150	Hot rolled or annealed	.040	180	.008	M2, M3	600	700	.007	C-7	1050	.007	CC-7
			.150	135	.015	M2, M3	450	525	.020	C-6	700	.015	CC-6
			.300	110	.020	M2, M3	360	420	.030	C-6	550	.020	CC-6
			.625	85	.030	M2, M3	280	330	.040	C-6	—	—	—

(continued)

Table 1. (continued)

Material	Hardness, Bhn	Condition	Depth of cut, in.(a)	High speed steel tool Speed, fpm	Feed, ipr	Tool material, AISI	Carbide tool Uncoated Speed, fpm Brazed	Indexable	Feed, ipr	Tool material grade	Coated Speed, fpm	Feed, ipr	Tool material grade
Resulfurized steels (continued)													
1108, 1109, 1110, 1115 ...	150 to 200	Cold drawn	.040	190	.008	M2, M3	615	725	.007	C-7	1100	.007	CC-7
			.150	145	.015	M2, M3	475	550	.020	C-6	700	.015	CC-6
			.300	110	.020	M2, M3	375	430	.030	C-6	575	.020	CC-6
			.625	90	.030	M2, M3	290	340	.040	C-6	—	—	—
Leaded steels													
12L13, 12L14, 12L15	100 to 150	Hot rolled, normalized, annealed or cold drawn	.040	340	.009	M2, M3	950	1000	.007	C-7	1550	.007	CC-7
			.150	260	.015	M2, M3	725	800	.020	C-6	1025	.015	CC-6
			.300	210	.020	M2, M3	575	620	.030	C-6	825	.020	CC-6
			.625	160	.030	M2, M3	450	490	.040	C-6	—	—	—
	150 to 200	Hot rolled, normalized, annealed or cold drawn	.040	350	.009	M2, M3	975	1050	.007	C-7	1500	.007	CC-7
			.150	270	.015	M2, M3	750	840	.020	C-6	1000	.015	CC-6
			.300	210	.020	M2, M3	590	640	.030	C-6	800	.020	CC-6
			.625	170	.030	M2, M3	460	475	.040	C-6	—	—	—

NOTE: For recommended tool geometry, see Tables 55 to 61; for recommended cutting fluids, see Table 62.

(a) Caution: Check horsepower requirements on heavier depths of cut.

Table 2. Face milling, free-machining low-carbon steels, wrought

Material	Hardness, Bhn	Condition	Depth of cut, in.(a)	High speed steel tool Speed, fpm	Feed per tooth, in.	Tool material, AISI	Carbide tool Uncoated Speed, fpm Brazed	Indexable	Feed per tooth, in.	Tool material grade	Coated Speed, fpm	Feed per tooth, in.	Tool material grade
Resulfurized steels													
1116, 1117, 1118, 1119, 1211, 1212	100 to 150	Hot rolled or annealed	.040	260	.008	M2, M7	730	800	.008	C-6	1200	.008	CC-6
			.150	200	.012	M2, M7	550	605	.012	C-6	785	.012	CC-6
			.300	155	.016	M2, M7	385	470	.016	C-5	610	.016	CC-5
	150 to 200	Cold drawn	.040	250	.008	M2, M7	665	730	.008	C-6	1100	.008	CC-6
			.150	190	.012	M2, M7	500	550	.012	C-6	715	.012	CC-6
			.300	150	.016	M2, M7	350	430	.016	C-5	560	.016	CC-5
1213, 1215	100 to 150	Hot rolled or annealed	.040	375	.008	M2, M7	800	880	.008	C-6	1325	.008	CC-6
			.150	290	.012	M2, M7	605	665	.012	C-6	865	.012	CC-6
			.300	225	.016	M2, M7	420	515	.016	C-5	670	.016	CC-5
	150 to 200	Cold drawn	.040	360	.008	M2, M7	730	800	.008	C-6	1200	.008	CC-6
			.150	275	.012	M2, M7	575	615	.012	C-6	800	.012	CC-6
			.300	215	.016	M2, M7	390	480	.016	C-5	625	.016	CC-5
1108, 1109, 1110, 1115 ...	100 to 150	Hot rolled or annealed	.040	235	.008	M2, M7	655	700	.008	C-6	1050	.008	CC-6
			.150	180	.012	M2, M7	495	545	.012	C-6	710	.012	CC-6
			.300	140	.016	M2, M7	345	425	.016	C-5	550	.016	CC-5
	150 to 200	Cold drawn	.040	225	.008	M2, M7	600	640	.008	C-6	950	.008	CC-6
			.150	170	.012	M2, M7	450	495	.012	C-6	645	.012	CC-6
			.300	130	.016	M2, M7	315	385	.016	C-5	500	.016	CC-5
Leaded steels													
12L13, 12L14, 12L15	100 to 150	Hot rolled, normalized, annealed or cold drawn	.040	325	.008	M2, M7	840	900	.008	C-6	1350	.008	CC-6
			.150	260	.012	M2, M7	700	725	.012	C-6	940	.012	CC-6
			.300	200	.016	M2, M7	460	565	.016	C-5	735	.016	CC-5
	150 to 200	Hot rolled, normalized, annealed or cold drawn	.040	290	.008	M2, M7	750	800	.008	C-6	1200	.008	CC-6
			.150	235	.012	M2, M7	600	650	.012	C-6	845	.012	CC-6
			.300	180	.016	M2, M7	400	500	.016	C-5	650	.016	CC-5

NOTE: For recommended tool geometry, see Tables 55 to 61; for recommended cutting fluids, see Table 62.

(a) Depth of cut is measured parallel to axis of cutter.

Table 3. End milling (peripheral), free-machining low-carbon steels, wrought

Material	Hardness, Bhn	Condition	Radial depth of cut, in.(a)	High speed steel tool Speed, fpm	Feed, in./tooth Cutter diam, in. 3/8	1/2	3/4	1-2	Tool material, AISI	Carbide tool Speed, fpm	Feed, in./tooth Cutter diam, in. 3/8	1/2	3/4	1-2	Tool material grade
Resulfurized steels															
1116, 1117, 1118, 1119, 1211, 1212	100 to 150	Hot rolled or annealed	.020	235	.001	.002	.004	.005	M2,	620	.002	.003	.005	.007	C-5
			.060	180	.002	.003	.005	.006	M3,	475	.003	.004	.006	.008	
			dia/4	160	.001	.002	.004	.005	M7	420	.002	.003	.005	.006	
			dia/2	140	.0007	.001	.003	.004		360	.0015	.002	.004	.005	

(continued)

Table 3. (continued)

Material	Hardness, Bhn	Condition	Radial depth of cut, in.(a)	High speed steel tool Speed, fpm	Feed, in./tooth Cutter diam, in. 3/8	1/2	3/4	1-2	Tool material, AISI	Carbide tool Speed, fpm	Feed, in./tooth Cutter diam, in. 3/8	1/2	3/4	1-2	Tool material grade
Resulfurized steels (continued)															
1116, 1117, 1118, 1119, 1211, 1212	150 to 200	Cold drawn	.020	200	.001	.002	.004	.005	M2,	585	.002	.003	.005	.007	C-5
			.060	150	.002	.003	.005	.006	M3,	450	.003	.004	.006	.008	
			dia/4	130	.001	.002	.004	.005	M7	390	.002	.003	.005	.006	
			dia/2	115	.0007	.001	.003	.004		360	.0015	.002	.004	.005	
1213, 1215	100 to 150	Hot rolled or annealed	.020	260	.001	.002	.004	.006	M2,	680	.002	.004	.006	.008	C-5
			.060	200	.002	.003	.005	.007	M3,	520	.003	.005	.007	.009	
			dia/4	175	.001	.002	.004	.005	M7	440	.002	.003	.006	.007	
			dia/2	150	.0007	.001	.003	.004		410	.0015	.002	.005	.006	
	150 to 200	Cold drawn	.020	230	.001	.002	.004	.006	M2,	620	.002	.004	.006	.007	C-5
			.060	180	.002	.003	.005	.007	M3,	475	.003	.005	.007	.008	
			dia/4	160	.001	.002	.004	.005	M7	400	.002	.003	.006	.007	
			dia/2	140	.0007	.001	.003	.004		375	.0015	.002	.005	.006	
1108, 1109, 1110, 1115	100 to 150	Hot rolled or annealed	.020	225	.001	.002	.004	.005	M2,	600	.002	.003	.005	.007	C-5
			.060	175	.002	.003	.005	.006	M3,	450	.003	.004	.006	.008	
			dia/4	155	.001	.002	.004	.005	M7	390	.002	.003	.005	.007	
			dia/2	135	.0007	.001	.003	.004		340	.0015	.002	.004	.006	
	150 to 200	Cold drawn	.020	190	.001	.002	.004	.005	M2,	570	.002	.003	.005	.007	C-5
			.060	145	.002	.003	.005	.006	M3,	440	.003	.004	.006	.008	
			dia/4	130	.001	.002	.004	.005	M7	385	.002	.003	.005	.007	
			dia/2	115	.0007	.001	.003	.004		340	.0015	.002	.004	.006	
Leaded steels															
12L13, 12L14, 12L15	100 to 150	Hot rolled, normalized, annealed or cold drawn	.020	270	.001	.002	.004	.006	M2,	710	.002	.004	.006	.008	C-5
			.060	200	.002	.003	.005	.007	M3,	540	.003	.005	.007	.009	
			dia/4	180	.001	.002	.004	.005	M7	460	.002	.004	.006	.008	
			dia/2	160	.0007	.001	.003	.004		430	.0015	.003	.005	.007	
	150 to 200	Hot rolled, normalized, annealed or cold drawn	.020	240	.001	.002	.004	.006	M2,	625	.002	.004	.006	.008	C-5
			.060	185	.002	.003	.005	.007	M3,	480	.003	.005	.007	.009	
			dia/4	165	.001	.002	.004	.005	M7	400	.002	.004	.006	.008	
			dia/2	145	.0007	.001	.003	.004		375	.0015	.003	.005	.007	

NOTE: For recommended tool geometry, see Tables 55 to 61; for recommended cutting fluids, see Table 62.

(a) For standard length end mills, maximum axial depth can be up to 1.5 × the cutter diameter.

Table 4. Drilling, free-machining low-carbon steels, wrought

Material	Hardness, Bhn	Condition	Speed, fpm(a)	Feed, ipr(a) Nominal hole diameter, in. 1/16	1/8	1/4	1/2	3/4	1	1 1/2	2	Tool material grade, AISI
Resulfurized steels												
1116, 1117, 1118, 1119, 1211, 1212	100 to 150	Hot rolled or annealed	70 105	.001 —	.003	.005	.012	.018	.022	.025	.030	M10, M7, M1
	150 to 200	Cold drawn	70 110	.001 —	.003	.005	.012	.018	.022	.025	.030	M10, M7, M1
1213, 1215	100 to 150	Hot rolled or annealed	90 125	.001 —	.003	.005	.012	.018	.022	.025	.030	M10, M7, M1
	150 to 200	Cold drawn	100 130	.001 —	.003	.005	.012	.018	.022	.025	.030	M10, M7, M1
1108, 1109, 1110, 1115	100 to 150	Hot rolled or annealed	70 95	.001 —	.003	.005	.012	.018	.022	.025	.030	M10, M7, M1
	150 to 200	Cold drawn	70 100	.001 —	.003	.005	.012	.018	.022	.025	.030	M10, M7, M1
Leaded steels:												
12L13, 12L14, 12L15	100 to 150	Hot rolled, normalized, annealed or cold drawn	80 150	.001 —	.003	.005	.012	.018	.022	.025	.030	M10, M7, M1
	150 to 200	Hot rolled, normalized, annealed or cold drawn	80 140	.001 —	.003	.005	.012	.018	.022	.025	.030	M10, M7, M1

NOTE: For recommended tool geometry, see Tables 55 to 61; for recommended cutting fluids, see Table 62.

(a) For drilling deep holes with twist drills, reduce speed and feed as follows: If hole depth is 3 drill diameters, reduce speed by 10%, feed by 10%; 4 diameters, speed by 20%, feed by 10%; 5 diameters, speed by 30%, feed by 20%; 6 diameters, speed by 35%; feed by 20%; 8 diameters, speed by 40%, feed by 20%.

Table 5. Reaming, free-machining low-carbon steels, wrought

Material	Hardness, Bhn	Condition	Roughing Speed, fpm	Feed, ipr(a) — Reamer diameter, in. 1/8	1/4	1/2	1	1½	2	Tool material grade, AISI or C	Finishing Speed, fpm	Feed, ipr(a) — Reamer diameter, in. 1/8	1/4	1/2	1	1½	2	Tool material grade, AISI or C
Resulfurized steels																		
1116, 1117, 1118, 1119, 1211, 1212	100 to 150	Hot rolled or annealed	120	.005	.008	.012	.020	.025	.030	M1, M2, M7	60	.006	.010	.015	.025	.030	.035	M1, M2, M7
			140	.006	.010	.015	.025	.030	.035	C-2	70	.006	.010	.015	.025	.030	.035	C-2
	150 to 200	Cold drawn	130	.005	.008	.012	.020	.025	.030	M1, M2, M7	65	.006	.010	.015	.025	.030	.035	M1, M2, M7
			150	.006	.010	.015	.025	.030	.035	C-2	70	.006	.010	.015	.025	.030	.035	C-2
1213, 1215	100 to 150	Hot rolled or annealed	170	.006	.010	.015	.025	.030	.035	M1, M2, M7	55	.008	.012	.018	.030	.035	.040	M1, M2, M7
			200	.006	.010	.015	.025	.030	.035	C-2	70	.008	.012	.018	.030	.035	.040	C-2
	150 to 200	Cold drawn	180	.006	.010	.015	.025	.030	.035	M1, M2, M7	60	.008	.012	.018	.030	.035	.040	M1, M2, M7
			220	.006	.010	.015	.025	.030	.035	C-2	70	.008	.012	.018	.030	.035	.040	C-2
1108, 1109, 1110, 1115	100 to 150	Hot rolled or annealed	115	.005	.008	.012	.020	.025	.030	M1, M2, M7	60	.006	.010	.015	.025	.030	.035	M1, M2, M7
			130	.006	.010	.015	.025	.030	.035	C-2	65	.006	.010	.015	.025	.030	.035	C-2
	150 to 200	Cold drawn	120	.005	.008	.012	.020	.025	.030	M1, M2, M7	65	.006	.010	.015	.025	.030	.035	M1, M2, M7
			140	.006	.010	.015	.025	.030	.035	C-2	70	.006	.010	.015	.025	.030	.035	C-2
Leaded steels																		
12L13, 12L14, 12L15	100 to 150	Hot rolled, normalized, annealed or cold drawn	225	.006	.010	.015	.025	.030	.035	M1, M2, M7	70	.008	.012	.018	.030	.035	.040	M1, M2, M7
			260	.006	.010	.015	.025	.030	.035	C-2	75	.008	.012	.018	.030	.035	.040	C-2
	150 to 200	Hot rolled, normalized, annealed or cold drawn	215	.006	.010	.015	.025	.030	.035	M1, M2, M7	65	.008	.012	.018	.030	.035	.040	M1, M2, M7
			250	.006	.010	.015	.025	.030	.035	C-2	70	.008	.012	.018	.030	.035	.040	C-2

NOTE: For recommended tool geometry, see Tables 55 to 61; for recommended cutting fluids, see Table 62.

(a) Based on 4 flutes for 1/8- and 1/4-inch reamers, 6 flutes for 1/2-inch reamers, and 8 flutes for 1-inch and larger reamers.

Table 6. Tapping, free-machining low-carbon steels, wrought

Material	Hardness, Bhn	Condition	Speed, fpm(a) — Threads per inch 7 or less	8 to 15	16 to 24	Over 24	HSS tool material, AISI
Resulfurized steels							
1116, 1117, 1118, 1119, 1211, 1212	100 to 150	Hot rolled or annealed	25	50	65	70	M10, M7, M1
	150 to 200	Cold drawn	30	55	70	75	M10, M7, M1
1213, 1215	100 to 150	Hot rolled or annealed	30	55	70	75	M10, M7, M1
	150 to 200	Cold drawn	35	60	75	80	M10, M7, M1
1108, 1109, 1110, 1115	100 to 150	Hot rolled or annealed	25	50	65	70	M10, M7, M1
	150 to 200	Cold drawn	30	55	70	75	M10, M7, M1
Leaded steels							
12L13, 12L14, 12L15	100 to 150	Hot rolled, normalized, annealed or cold drawn	30	60	80	85	M10, M7, M1
	150 to 200	Hot rolled, normalized, annealed or cold drawn	25	50	70	75	M10, M7, M1
	200 to 250	Hot rolled, normalized, annealed or cold drawn	20	45	60	65	M10, M7, M1

NOTE: For recommended tool geometry, see Tables 55 to 61; for recommended cutting fluids, see Table 62.

(a) These speeds are for tapping 65% to 75% threads in shallow through-holes. Reduce the speed when tapping deep holes, blind holes or higher percentage of thread.

Table 7. Turning, medium-carbon alloy steels, wrought

Material	Hardness, Bhn	Condition	Depth of cut, in.(a)	High speed steel tool			Carbide tool — Uncoated				Coated		
				Speed, fpm	Feed, ipr	Tool material, AISI	Speed, fpm Brazed	Speed, fpm Indexable	Feed, ipr	Tool material grade	Speed, fpm	Feed, ipr	Tool material grade
1340, 1345, 4042, 4047, 4140, 4142, 4145, 4147, 4340, 50B40, 50B44, 5046, 50B46, 5140, 5145, 5147, 81B45, 8640, 8642, 8645, 86B45, 8740, 8742	175-225	Hot rolled, annealed or cold drawn	.040	135	.007	M2, M3	375	500	.007	C-7	650	.007	CC-7
			.150	105	.015	M2, M3	300	400	.020	C-6	525	.015	CC-6
			.300	80	.020	M2, M3	240	315	.030	C-6	400	.020	CC-6
			.625	65	.030	M2, M3	190	250	.040	—	—	—	—
	225-275	Annealed, normalized, cold drawn or quenched & tempered	.040	115	.007	M2, M3	350	465	.007	C-7	600	.007	CC-7
			.150	90	.015	M2, M3	280	365	.020	C-6	475	.015	CC-6
			.300	70	.020	M2, M3	220	285	.030	C-6	375	.020	CC-6
			.625	55	.030	M2, M3	170	225	.040	—	—	—	—
	275-325	Normalized or quenched & tempered	.040	90	.007	T15, M42(b)	330	440	.007	C-7	575	.007	CC-7
			.150	70	.015	T15, M42(b)	260	340	.015	C-6	450	.015	CC-6
			.300	55	.020	T15, M42(b)	200	270	.020	C-6	350	.020	CC-6
			.625	—	—								
	325-375	Normalized or quenched & tempered	.040	70	.005	T15, M42(b)	275	380	.007	C-7	500	.007	CC-7
			.150	55	.010	T15, M42(b)	215	300	.015	C-6	400	.015	CC-6
			.300	40	.015	T15, M42(b)	170	235	.020	C-6	300	.020	CC-6
			.625	—	—								
	375-425	Quenched & tempered	.040	60	.005	T15, M42(b)	225	300	.007	C-7	400	.007	CC-7
			.150	45	.010	T15, M42(b)	175	240	.015	C-6	300	.015	CC-6
			.300	35	.015	T15, M42(b)	140	190	.020	C-6	250	.020	CC-6
			.625	—	—								
	45-48 R$_C$	Quenched & tempered	.040	45	.005	T15, M42(b)	200	275	.005	C-8	—	—	—
			.150	35	.010	T15, M42(b)	160	210	.010	C-7	—	—	—
			.300	25	.015	T15, M42(b)	125	165	.015	C-6	—	—	—
			.625	—	—								
	48-50 R$_C$	Quenched & tempered	.040	40	.005	T15, M42(b)	165	200	.005	C-8	—	—	—
			.150	30	.010	T15, M42(b)	130	160	.007	C-8	—	—	—
			.300	25	.015	T15, M42(b)	100	125	.010	C-6	—	—	—
			.625	—	—								
	50-52 R$_C$	Quenched & tempered	.040	35	.003	T15, M42(b)	125	175	.005	C-8	—	—	—
			.150	25	.007	T15, M42(b)	100	140	.007	C-8	—	—	—
			.300	—	—	—	80	110	.010	C-6	—	—	—
			.625	—	—								
	52-54 R$_C$	Quenched & tempered	.040	—	—	—	100	115	.005	C-8	—	—	—
			.150	—	—	—	80	90	.007	C-8	—	—	—
			.300	—	—	—	—	—					
			.625	—	—								

NOTE: For recommended tool geometry, see Tables 55 to 61; for recommended cutting fluids, see Table 62.

(a) Caution: Check horsepower requirements on heavier depths of cut. (b) Any premium HSS (T15, M33, M41-M47).

Table 8. Face milling, medium-carbon alloy steels, wrought

Material	Hardness, Bhn	Condition	Depth of cut, in.(a)	High speed steel tool			Carbide tool — Uncoated				Coated		
				Speed, fpm	Feed per tooth, in.	Tool material, AISI	Speed, fpm Brazed	Speed, fpm Indexable	Feed per tooth, in.	Tool material grade	Speed, fpm	Feed per tooth, in.	Tool material grade
1340, 1345, 4042, 4047, 4140, 4142, 4145, 4147, 4340, 50B40, 50B44, 5046, 50B46, 5140, 5145, 5147, 81B45, 8640, 8642, 8645, 86B45, 8740, 8742	175-225	Hot rolled, annealed or cold drawn	.040	175	.008	M2, M3	460	560	.008	C-6	825	.008	CC-6
			.150	135	.012	M2, M3	380	440	.012	C-6	570	.012	CC-6
			.300	115	.016	M2, M3	280	345	.016	C-5	450	.016	CC-5
	225-275	Annealed, normalized, cold drawn or quenched and tempered	.040	150	.006	M2, M3	415	510	.007	C-6	765	.007	CC-6
			.150	110	.010	M2, M3	330	400	.010	C-6	520	.010	CC-6
			.300	85	.014	M2, M3	260	315	.014	C-5	410	.014	CC-5
	275-325	Normalized or quenched and tempered	.040	105	.006	T15, M42(b)	400	485	.006	C-6	725	.005	CC-6
			.150	85	.009	T15, M42(b)	305	375	.008	C-6	485	.007	CC-6
			.300	65	.012	T15, M42(b)	235	290	.010	C-5	375	.009	CC-5
	325-375	Normalized or quenched and tempered	.040	80	.005	T15, M42(b)	345	420	.005	C-6	630	.004	CC-6
			.150	60	.008	T15, M42(b)	265	325	.007	C-6	420	.006	CC-6
			.300	50	.010	T15, M42(b)	205	250	.009	C-5	325	.008	CC-5
	375-425	Quenched and tempered	.040	60	.004	T15, M42(b)	265	325	.004	C-6	485	.003	CC-6
			.150	50	.006	T15, M42(b)	215	265	.006	C-6	345	.005	CC-6
			.300	40	.008	T15, M42(b)	170	205	.008	C-6	265	.007	CC-6
	45-48 R$_C$	Quenched and tempered	.040	55	.003	T15, M42(b)	245	300	.002	C-6	—	—	—
			.150	40	.005	T15, M42(b)	190	230	.004	C-6	—	—	—
			.300	35	.007	T15, M42(b)	155	180	.006	C-6	—	—	—
	48-50 R$_C$	Quenched and tempered	.040	40	.002	T15, M42(b)	185	225	.002	C-6	—	—	—
			.150	30	.004	T15, M42(b)	145	175	.003	C-6	—	—	—
			.300	25	.006	T15, M42(b)	110	135	.004	C-6	—	—	—
	50-52 R$_C$	Quenched and tempered	.040	35	.002	T15, M42(b)	155	190	.002	C-2	—	—	—
			.150	25	.003	T15, M42(b)	125	155	.003	C-2	—	—	—
			.300	20	.004	T15, M42(b)	100	120	.004	C-2	—	—	—
	52-54 R$_C$	Quenched and tempered	.040	—	—	—	100	125	.002	C-2	—	—	—
			.150	—	—	—	80	100	.002	C-2	—	—	—
			.300	—	—	—	65	80	.003	C-2	—	—	—

NOTE: For recommended tool geometry, see Tables 55 to 61; for recommended cutting fluids, see Table 62.

(a) Depth of cut is measured parallel to axis of cutter. (b) Any premium HSS (T15, M33, M41-M47).

Table 9. End milling (peripheral), medium-carbon alloy steels, wrought

Material	Hardness, Bhn	Condition	Radial depth of cut, in.(a)	High speed steel tool						Carbide tool						Tool material grade
				Speed, fpm	Feed, in./tooth — Cutter diam, in.				Tool material, AISI	Speed, fpm	Feed, in./tooth — Cutter diam, in.				
					$3/8$	$1/2$	$3/4$	$1-2$			$3/8$	$1/2$	$3/4$	$1-2$	
1340, 1345, 4042, 4047, 4140, 4142, 4145, 4147, 4340, 50B40, 50B44, 5046, 50B46, 5140, 5145, 5147, 81B45, 8640, 8642, 8645, 86B45, 8740, 8742	175-225	Hot rolled, annealed or cold drawn	.020	120	.001	.002	.003	.004	M2,	455	.0015	.003	.005	.006	C-5
			.060	90	.002	.003	.004	.005	M3,	350	.0025	.004	.006	.007	
			dia/4	80	.0015	.002	.003	.004	M7	295	.002	.003	.005	.006	
			dia/2	70	.001	.0015	.002	.003		275	.0015	.002	.004	.005	
	225-275	Annealed, normalized, cold drawn or quenched and tempered	.020	105	.001	.002	.003	.004	M2,	390	.001	.002	.004	.005	C-5
			.060	80	.002	.003	.004	.005	M3,	300	.002	.003	.005	.007	
			dia/4	70	.0015	.002	.003	.004	M7	255	.0015	.002	.004	.005	
			dia/2	60	.001	.0015	.002	.003		235	.001	.0015	.003	.004	
	275-325	Normalized or quenched and tempered	.020	85	.0007	.0015	.003	.004	M2,	310	.001	.002	.004	.005	C-5
			.060	65	.001	.002	.004	.005	M3,	235	.002	.003	.005	.006	
			dia/4	55	.0007	.0015	.003	.004	M7	205	.0015	.002	.004	.005	
			dia/2	50	.0005	.001	.002	.003		190	.001	.0015	.003	.004	
	325-375	Normalized or quenched and tempered	.020	65	.0005	.0015	.003	.004	M2,	260	.001	.0015	.003	.005	C-5
			.060	55	.0005	.0015	.004	.005	M3,	200	.0015	.003	.005	.006	
			dia/4	50	.0005	.0015	.003	.004	M7	170	.0015	.002	.004	.005	
			dia/2	40	.0005	.001	.002	.003		160	.001	.002	.003	.004	
	375-425	Quenched and tempered	.020	55	.0005	.0007	.001	.002	T15(b)	210	.001	.001	.002	.003	C-5
			.060	45	.0005	.0001	.002	.003		160	.0015	.002	.003	.004	
			dia/4	40	.0005	.0005	.002	.0025		135	.0015	.0015	.0025	.0035	
			dia/2	35	—	—	.0015	.002		125	.001	.001	.002	.003	
	45-48 R_C	Quenched and tempered	.020	35	.0005	.0005	.0007	.001	T15(b)	150	.0005	.0007	.002	.0025	C-5
			.060	30	—	.0007	.001	.002		110	.0007	.001	.003	.0035	
			dia/4	25	—	.0005	.001	.0015		95	.0005	.0007	.002	.003	
			dia/2	20	—	—	.001	.0015		90	—	—	.0015	.0025	
	48-50 R_C	Quenched and tempered	.020	30	.0005	.0005	.0005	.001	T15(b)	100	.0005	.0005	.0015	.002	C-5
			.060	25	—	.0007	.001	.0015		75	.0005	.001	.002	.0035	
			dia/4	20	—	.0005	.001	.0015		65	—	.0007	.0015	.003	
			dia/2	15	—	—	.0005	.001		60	—	—	.001	.0025	
	50-52 R_C	Quenched and tempered	.020	25	.0005	.0005	.0005	.001	T15(b)	80	.0005	.0005	.0015	.002	C-5
			.060	20	—	.0005	.001	.0015		60	—	.001	.002	.003	
			dia/4	15	—	—	.0005	.001		50	—	—	.002	.003	
			dia/2	—	—	—	—	—		45	—	—	.0015	.002	
	52-54 R_C	Quenched and tempered	.020	—	—	—	—	—		65	.0005	.0005	.0015	.002	C-5
			.060	—	—	—	—	—		45	—	.001	.002	.003	
			dia/4	—	—	—	—	—		40	—	—	.002	.003	
			dia/2	—	—	—	—	—		35	—	—	.0015	.002	

NOTE: For recommended tool geometry, see Tables 55 to 61; for recommended cutting fluids, see Table 62.

(a) For standard length end mills, maximum axial depth can be up to 1.5 × the cutter diameter. (b) Any premium HSS (T15, M33, M41-M47).

Table 10. Drilling, medium-carbon alloy steels, wrought

Material	Hardness, Bhn	Condition	Speed, fpm(a)	Feed, ipr(a) — Nominal hole diameter, in.								Tool material grade, AISI or C
				$1/16$	$1/8$	$1/4$	$1/2$	$3/4$	1	$1 1/2$	2	
1340, 1345, 4042, 4047, 4140, 4142, 4145, 4147, 4340, 50B40, 50B44, 5046, 50B46, 5140, 5145, 5147, 81B45, 8640, 8642, 8645, 86B45, 8740, 8742	175-225	Hot rolled, annealed or cold drawn	65	.001	.003	.006	.010	.014	.016	.020	.025	M10, M7, M1
	225-275	Annealed, normalized, cold drawn or quenched and tempered	55	.001	.003	.004	.007	.010	.012	.015	.020	M10, M7, M1
	275-325	Normalized or quenched and tempered	45	—	.002	.004	.006	.008	.009	.012	.015	M10, M7, M1
	325-375	Normalized or quenched and tempered	35	—	.002	.003	.005	.008	.009	.010	.011	M10, M7, M1
	375-425	Quenched and tempered	25	—	.002	.003	.004	.006	.008	.008	.010	T15, M42(b)
	45-48 R_C	Quenched and tempered	20	—	.001	.002	.003	.003	.004	.004	.004	T15, M42(b)
	48-50 R_C	Quenched and tempered	15	—	.0005	.001	.001	.002	.003	.004	.004	T15, M42(b)
	50-52 R_C	Quenched and tempered	10	—	.0005	.001	.001	.002	.003	.003	.004	T15, M42(b)
	52-54 R_C	Quenched and tempered	75	—	—	.001	.001	.0015	—	—	—	C-2

NOTE: For recommended tool geometry, see Tables 55 to 61; for recommended cutting fluids, see Table 62.

(a) For drilling deep holes with twist drills, reduce speed and feed as follows: If hole depth is 3 drill diameters, reduce speed by 10%, feed by 10%; 4 diameters, speed by 20%, feed by 10%; 5 diameters, speed by 30%, feed by 20%; 6 diameters, speed by 35%, feed by 20%; 8 diameters, speed by 40%, feed by 20%. (b) Any premium HSS (T15, M33, M41-M47).

Table 11. Reaming, medium-carbon alloy steels, wrought

Material	Hardness, Bhn	Condition	Roughing								Tool material grade, AISI or C	Finishing								Tool material grade, AISI or C
			Speed, fpm	Feed, ipr(a) — Reamer diameter, in.								Speed, fpm	Feed, ipr(a) — Reamer diameter, in.							
				1/8	1/4	1/2	1	1 1/2	2				1/8	1/4	1/2	1	1 1/2	2	
1340, 1345, 4042, 4047, 4140, 4142, 4145, 4147, 4340, 50B40, 50B44, 5046, 50B46, 5140, 5145, 5147, 81B45, 8640, 8642, 8645, 86B45, 8740, 8742	175-225	..Hot rolled, annealed or cold drawn	70	.004	.007	.012	.020	.025	.030	M1, M2, M7	35	.005	.008	.012	.020	.025	.030	M1, M2, M7
			85	.004	.007	.012	.020	.025	.030	C-2	45	.005	.008	.012	.020	.025	.030	C-2
	225-275	..Annealed, normalized, cold drawn or quenched and tempered	60	.003	.006	.010	.015	.020	.025	M1, M2, M7	30	.004	.007	.010	.015	.020	.025	M1, M2, M7
			75	.004	.006	.010	.015	.020	.025	C-2	40	.004	.007	.010	.015	.020	.025	C-2
	275-325	..Normalized or quenched and tempered	50	.003	.005	.008	.012	.015	.020	M1, M2, M7	25	.003	.006	.008	.012	.015	.020	M1, M2, M7
			65	.004	.006	.008	.012	.015	.020	C-2	35	.004	.006	.008	.012	.015	.020	C-2
	325-375	..Normalized or quenched and tempered	35	.002	.004	.005	.008	.010	.012	M1, M2, M7	25	.002	.005	.006	.010	.012	.015	M1, M2, M7
			50	.004	.006	.008	.010	.012	.015	C-2	35	.004	.006	.008	.010	.012	.015	C-2
	375-425	..Quenched and tempered	25	.002	.004	.005	.006	.007	.008	T15, M42(b)	20	.002	.004	.005	.006	.007	.008	T15, M42(b)
			40	.004	.006	.008	.010	.011	.012	C-2	30	.004	.006	.008	.010	.011	.012	C-2
	45-48 R_C	Quenched and tempered	25	.0015	.002	.003	.005	.007	.008	T15, M42(b)	15	.001	.0015	.002	.003	.004	.005	T15, M42(b)
			35	.003	.005	.006	.008	.009	.010	C-2	25	.003	.005	.006	.008	.009	.010	C-2
	48-50 R_C	Quenched and tempered	20	.001	.0015	.0015	.002	.002	.002	T15, M42(b)	15	.001	.0015	.0015	.002	.002	.002	T15, M42(b)
			30	.003	.004	.005	.006	.007	.008	C-2	25	.003	.004	.005	.006	.007	.008	C-2
	50-52 R_C	Quenched and tempered	15	.001	.001	.0015	.0015	.0015	.0015	T15, M42(b)	10	.001	.001	.0015	.0015	.0015	.0015	T15, M42(b)
			25	.002	.003	.004	.005	.006	.006	C-2	20	.002	.003	.004	.005	.006	.006	C-2
	52-54 R_C	Quenched and tempered	—	—	—	—	—	—	—	—	—	—	—	—	—	—	—	—
			25	.002	.002	.003	.004	.004	.004	C-2	20	.002	.002	.003	.004	.004	.004	C-2

NOTE: For recommended tool geometry, see Tables 55 to 61; for recommended cutting fluids, see Table 62.

(a) Based on 4 flutes for 1/8- and 1/4-inch reamers, 6 flutes for 1/2-inch reamers, and 8 flutes for 1-inch and larger reamers. (b) Any premium HSS (T15, M33, M41-M47).

Table 12. Tapping, medium-carbon alloy steels, wrought

Material	Hardness, Bhn	Condition	Speed, fpm(a) — Threads per inch				HSS tool material, AISI
			7 or less	8 to 15	16 to 24	Over 24	
1340, 1345, 4042, 4047, 4140, 4142, 4145, 4147, 4340, 50B40, 50B44, 5046, 50B46, 5140, 5145, 5147, 81B45, 8640, 8642, 8645, 86B45, 8740, 8742	175-225	.....Hot rolled, annealed or cold drawn	15	30	40	45	M10, M7, M1
	225-275	.....Annealed, normalized, cold drawn or quenched and tempered	10	25	30	35	M10, M7, M1
	275-325	.....Normalized or quenched and tempered	7	15	20	25	M10, M7, M1
	325-375	.....Normalized or quenched and tempered	7	15	18	20	Nitrided M10, M7, M1
	375-425	.....Quenched and tempered	5	10	14	15	Nitrided M10, M7, M1
	45-48 R_C	....Quenched and tempered	3	7	9	10	Nitrided M10, M7, M1
	48-50 R_C	....Quenched and tempered	2	6	7	8	Nitrided M10, M7, M1
	50-52 R_C	....Quenched and tempered	2	3	4	5	Nitrided M10, M7, M1

NOTE: For recommended tool geometry, see Tables 55 to 61; for recommended cutting fluids, see Table 62.

(a) These speeds are for tapping 65% to 75% threads in shallow through-holes. Reduce the speed when tapping deep holes, blind holes or higher percentage of thread.

Table 13. Turning, wrought stainless steels

Material	Hardness, Bhn	Condition	Depth of cut, in.(a)	High speed steel tool			Carbide tool — Uncoated				Carbide tool — Coated		
				Speed, fpm	Feed, ipr	Tool material, AISI	Speed, fpm Brazed	Speed, fpm Indexable	Feed, ipr	Tool material grade	Speed, fpm	Feed, ipr	Tool material grade
Ferritic steels													
405, 409, 429, 430, 434, 436, 442, 446 135	135	Annealed	.040	150	.007	M2, M3	575	650	.007	C-7	850	.007	CC-7
	to		.150	120	.015	M2, M3	450	500	.015	C-6	650	.015	CC-6
	185		.300	95	.020	M2, M3	350	400	.030	C-6	525	.020	CC-6
			.625	75	.030	M2, M3	275	310	.040	C-6	—	—	—
Austenitic steels													
201, 202, 301, 302, 302B, 304, 304L, 305, 308, 309, 309S, 310, 310S, 314, 316, 316L, 317, 321, 330, 347, 348, 384, 385 135	135	Annealed	.040	95	.007	M2, M3	325	375	.007	C-3	500	.007	CC-3
	to		.150	75	.015	M2, M3	300	325	.015	C-3	425	.015	CC-3
	185		.300	60	.020	M2, M3	225	250	.020	C-2	325	.020	CC-2
			.625	45	.030	M2, M3	175	200	.030	C-2	—	—	—
	225	Cold drawn	.040	80	.007	T15, M42(b)	300	325	.007	C-3	425	.007	CC-3
	to		.150	65	.015	T15, M42(b)	250	275	.015	C-3	350	.015	CC-3
	275		.300	50	.020	T15, M42(b)	190	215	.020	C-2	275	.020	CC-2
			.625	40	.030	T15, M42(b)	140	165	.030	C-2	—	—	—
Martensitic steels													
403, 410, 420, 422, 501, 502 135	135	Annealed	.040	155	.007	M2, M3	475	620	.007	C-7	800	.007	CC-7
	to		.150	125	.015	M2, M3	400	480	.015	C-6	625	.015	CC-6
	175		.300	100	.020	M2, M3	320	380	.030	C-6	500	.020	CC-6
			.625	80	.030	M2, M3	240	300	.040	C-6	—	—	—
	175	Annealed	.040	145	.007	M2, M3	460	570	.007	C-7	850	.007	CC-7
	to		.150	115	.015	M2, M3	385	450	.015	C-6	550	.015	CC-6
	225		.300	90	.020	M2, M3	300	350	.030	C-6	450	.020	CC-6
			.625	70	.030	M2, M3	235	265	.040	C-6	—	—	—
	275	Quenched	.040	95	.007	T15, M42(b)	360	465	.007	C-7	700	.007	CC-7
	to	and	.150	75	.015	T15, M42(b)	280	360	.015	C-6	450	.015	CC-6
	325	tempered	.300	60	.020	T15, M42(b)	225	280	.020	C-6	375	.020	CC-6
			.625	—									
	375	Quenched	.040	65	.007	T15, M42(b)	290	320	.007	C-7	475	.007	CC-7
	to	and	.150	50	.015	T15, M42(b)	225	250	.015	C-6	300	.015	CC-6
	425	tempered	.300	40	.020	T15, M42(b)	180	200	.020	C-6	250	.020	CC-6
			.625	—									

NOTE: For recommended tool geometry, see Tables 55 to 61; for recommended cutting fluids, see Table 62.

(a) Caution: Check horsepower requirements on heavier depths of cut. (b) Any premium HSS (T15, M33, M41-M47).

Table 14. Face milling, wrought stainless steels

Material	Hardness, Bhn	Condition	Depth of cut, in.(a)	High speed steel tool			Carbide tool — Uncoated				Carbide tool — Coated		
				Speed, fpm	Feed per tooth, in.	Tool material, AISI	Speed, fpm Brazed	Speed, fpm Indexable	Feed per tooth, in.	Tool material grade	Speed, fpm	Feed per tooth, in.	Tool material grade
Ferritic steels													
405, 409, 429, 430, 434, 436, 442, 446 135-185	135-185	Annealed	.040	190	.008	M2, M7	565	620	.008	C-6	925	.008	CC-6
			.150	145	.012	M2, M7	425	470	.012	C-6	600	.012	CC-6
			.300	115	.016	M2, M7	300	365	.016	C-5	475	.016	CC-5
Austenitic steels													
201, 202, 301, 302, 302B, 304, 304L, 305, 308, 309, 309S, 310, 310S, 314, 316, 316L, 317, 321, 330, 347, 348, 384, 385 135-185	135-185	Annealed	.040	115	.008	M2, M7	400	440	.008	C-2	650	.008	CC-2
			.150	90	.012	M2, M7	300	330	.012	C-2	425	.012	CC-2
			.300	70	.016	M2, M7	200	250	.016	C-2	325	.016	CC-2
	225-275	Cold drawn	.040	100	.006	M2, M7	290	350	.007	C-2	525	.007	CC-2
			.150	75	.010	M2, M7	245	300	.010	C-2	400	.010	CC-2
			.300	55	.014	M2, M7	190	235	.014	C-2	300	.014	CC-2

(continued)

Table 14. (continued)

Material	Hardness, Bhn	Condition	Depth of cut, in.(a)	High speed steel tool Speed, fpm	Feed per tooth, in.	Tool material, AISI	Carbide tool — Uncoated Speed, fpm Brazed	Speed, fpm Indexable	Feed per tooth, in.	Tool material grade	Coated Speed, fpm	Feed per tooth, in.	Tool material grade
Martensitic steels 403, 410, 420, 422, 501, 502	135-175	Annealed	.040	160	.008	M2, M7	500	540	.008	C-6	800	.008	CC-6
			.150	125	.012	M2, M7	390	480	.012	C-6	525	.012	CC-6
			.300	100	.016	M2, M7	215	260	.016	C-5	325	.016	CC-5
	175-225	Annealed	.040	145	.008	M2, M7	480	510	.008	C-6	765	.008	CC-6
			.150	115	.012	M2, M7	365	380	.012	C-6	500	.012	CC-6
			.300	90	.016	M2, M7	240	295	.016	C-5	375	.016	CC-5
	275-325	Quenched and tempered	.040	100	.006	T15, M42(b)	370	410	.006	C-6	625	.005	CC-6
			.150	75	.009	T15, M42(b)	280	310	.008	C-6	400	.007	CC-6
			.300	60	.012	T15, M42(b)	195	240	.010	C-5	315	.009	CC-5
	375-425	Quenched and tempered	.040	60	.004	T15, M42(b)	180	200	.004	C-6	300	.003	CC-6
			.150	45	.006	T15, M42(b)	135	150	.006	C-6	200	.005	CC-6
			.300	35	.008	T15, M42(b)	95	115	.008	C-6	150	.007	CC-6

NOTE: For recommended tool geometry, see Tables 55 to 61; for recommended cutting fluids, see Table 62.

(a) Depth of cut is measured parallel to axis of cutter. (b) Any premium HSS (T15, M33, M41-M47).

Table 15. End milling (peripheral), wrought stainless steels

Material	Hardness, Bhn	Condition	Radial depth of cut, in.(a)	High speed steel tool Speed, fpm	Feed, in./tooth Cutter diam, in. 3/8	1/2	3/4	1-2	Tool material, AISI	Carbide tool Speed, fpm	Feed, in./tooth Cutter diam, in. 3/8	1/2	3/4	1-2	Tool material grade
Ferritic steels 405, 409, 429, 430, 434, 436, 442, 446	135 to 185	Annealed	.020	145	.001	.002	.004	.005	M2,	455	.0005	.001	.003	.005	C-5
			.060	110	.002	.003	.005	.006	M3,	350	.001	.002	.004	.006	
			dia/4	95	.001	.002	.004	.005	M7	295	.001	.0015	.003	.004	
			dia/2	85	.001	.0015	.003	.004		275	.0007	.001	.002	.003	
Austenitic steels 201, 202, 301, 302, 302B, 304, 304L, 305, 308, 309, 309S, 310, 310S, 314, 316, 316L, 317, 321, 330, 347, 348, 384, 385	135 to 185	Annealed	.020	100	.001	.002	.004	.005	M2,	340	.0005	.001	.002	.004	C-2
			.060	75	.002	.003	.005	.006	M3,	260	.001	.002	.003	.005	
			dia/4	65	.001	.002	.004	.005	M7	225	.001	.0015	.0025	.004	
			dia/2	55	.001	.001	.003	.004		210	.0005	.001	.002	.003	
	225 to 275	Cold drawn	.020	90	.001	.002	.004	.005	M2,	300	.0005	.001	.002	.003	C-2
			.060	65	.002	.003	.005	.006	M3,	230	.001	.002	.003	.004	
			dia/4	55	.001	.002	.004	.005	M7	200	.001	.0015	.0025	.003	
			dia/2	50	.001	.001	.003	.004		180	.0005	.001	.002	.0025	
Martensitic steels 403, 410, 420, 422, 501, 502	135 to 175	Annealed	.020	145	.001	.002	.004	.005	M2,	455	.0005	.001	.003	.005	C-5
			.060	110	.002	.003	.005	.006	M3,	350	.001	.002	.004	.006	
			dia/4	95	.0015	.002	.004	.005	M7	295	.001	.0015	.0025	.004	
			dia/2	85	.001	.0015	.003	.004		275	.0005	.001	.002	.003	
	175 to 225	Annealed	.020	130	.001	.002	.003	.004	M2,	390	.0005	.001	.003	.005	C-5
			.060	100	.002	.003	.004	.005	M3,	300	.001	.002	.004	.006	
			dia/4	90	.001	.002	.003	.004	M7	255	.001	.0015	.003	.004	
			dia/2	75	.001	.001	.002	.003		235	.0005	.001	.002	.003	
	275 to 325	Quenched and tempered	.020	80	.0007	.001	.002	.003	M2,	300	.0005	.001	.003	.005	C-5
			.060	60	.001	.002	.003	.004	M3,	230	.001	.002	.004	.006	
			dia/4	50	.0007	.001	.002	.003	M7	200	.001	.0015	.003	.004	
			dia/2	45	.0005	.0007	.0015	.0025		180	.0005	.001	.002	.003	
	375 to 425	Quenched and tempered	.020	65	.0005	.0007	.001	.002	T15(b)	250	.0005	.001	.002	.003	C-5
			.060	55	.0005	.001	.002	.003		190	.001	.002	.003	.004	
			dia/4	50	.0005	.0005	.0015	.0025		165	.001	.0015	.0025	.003	
			dia/2	40	—	—	.001	.002		150	.0005	.001	.002	.0025	

NOTE: For recommended tool geometry, see Tables 55 to 61; for recommended cutting fluids, see Table 62.

(a) For standard length end mills, maximum axial depth can be up to 1.5 × the cutter diameter. (b) Any premium HSS (T15, M33, M41-M47).

Table 16. Drilling, wrought stainless steels

Material	Hardness, Bhn	Condition	Speed, fpm(a)	1/16	1/8	1/4	1/2	3/4	1	1½	2	Tool material grade, AISI
Ferritic steels												
405, 409, 429, 430, 434, 436, 442, 446	135 to 185	Annealed	65	.001	.002	.004	.007	.010	.012	.015	.018	M10, M7, M1
Austenitic steels												
201, 202, 301, 302, 302B, 304, 304L, 305, 308, 309, 309S, 310, 310S, 314, 316, 316L, 317, 321, 330, 347, 348, 384, 385	135 to 185	Annealed	55	.001	.002	.004	.007	.010	.012	.015	.018	M10, M7, M1
	225 to 275	Cold drawn	50	.001	.002	.004	.007	.010	.012	.015	.018	M10, M7, M1
Martensitic steels												
403, 410, 420, 422, 501, 502	135 to 175	Annealed	55 / 75	.001 / —	— / .003	— / .006	— / .010	— / .013	— / .016	— / .021	— / .025	M10, M7, M1 / M10, M7, M1
	175 to 225	Annealed	65	.001	.003	.006	.010	.013	.016	.021	.025	M10, M7, M1
	275 to 325	Quenched and tempered	55	—	.003	.004	.007	.010	.012	.015	.018	M10, M7, M1
	375 to 425	Quenched and tempered	45	—	.001	.002	.004	.004	.004	.004	.004	T15, M42(b)

NOTE: For recommended tool geometry, see Tables 55 to 61; for recommended cutting fluids, see Table 62.

(a) For drilling deep holes with twist drills, reduce speed and feed as follows: If hole depth is 3 drill diameters, reduce speed by 10%, feed by 10%; 4 diameters, speed by 20%, feed by 10%; 5 diameters, speed by 30%, feed by 20%; 6 diameters, speed by 35%, feed by 20%; 8 diameters, speed by 40%, feed by 20%. (b) Any premium HSS (T15, M33, M41-M47).

Table 17. Reaming, wrought stainless steels

Material	Hardness, Bhn	Condition	Roughing Speed, fpm	1/8	1/4	1/2	1	1½	2	Roughing Tool material grade, AISI or C	Finishing Speed, fpm	1/8	1/4	1/2	1	1½	2	Finishing Tool material grade, AISI or C
Ferritic steels																		
405, 409, 429, 430, 434, 436, 442, 446	135 to 185	Annealed	75	.003	.004	.005	.008	.010	.012	M1, M2, M7	35	.004	.004	.006	.008	.009	.010	M1, M2, M7
			90	.004	.008	.012	.016	.020	.024	C-2	50	.004	.006	.008	.010	.011	.012	C-2
Austenitic steels																		
201, 202, 301, 302, 302B, 304, 304L, 305, 308, 309, 309S, 310, 310S, 314, 316, 316L, 317, 321, 330, 347, 348, 384, 385	135 to 185	Annealed	70	.003	.005	.008	.010	.012	.015	M1, M2, M7	35	.003	.003	.004	.006	.007	.008	M1, M2, M7
			85	.004	.008	.012	.016	.020	.024	C-2	50	.003	.004	.006	.008	.009	.010	C-2
	225 to 275	Cold drawn	60	.003	.005	.008	.010	.012	.015	M1, M2, M7	35	.003	.003	.004	.006	.007	.008	M1, M2, M7
			75	.004	.008	.012	.016	.020	.024	C-2	50	.003	.004	.006	.008	.009	.010	C-2
Martensitic steels																		
403, 410, 420, 422, 501, 502	135 to 175	Annealed	80	.003	.004	.005	.008	.010	.012	M1, M2, M7	40	.004	.004	.006	.008	.009	.010	M1, M2, M7
			95	.004	.008	.012	.016	.020	.024	C-2	60	.004	.006	.008	.010	.011	.012	C-2
	175 to 225	Annealed	75	.003	.004	.005	.008	.010	.012	M1, M2, M7	40	.004	.004	.006	.008	.009	.010	M1, M2 M7
			90	.004	.008	.012	.016	.020	.024	C-2	60	.004	.005	.006	.008	.009	.010	C-2
	275 to 325	Quenched and tempered	60	.003	.004	.005	.008	.010	.012	M1, M2, M7	35	.003	.003	.004	.006	.007	.008	M1, M2, M7
			75	.004	.006	.009	.015	.018	.020	C-2	50	.003	.004	.005	.006	.007	.008	C-2
	375 to 425	Quenched and tempered	45	.002	.004	.005	.008	.010	.012	T15, M42(b)	30	.003	.003	.004	.006	.007	.008	T15, M42-(b)
			60	.003	.005	.006	.008	.010	.012	C-2	40	.003	.004	.005	.006	.007	.008	C-2

NOTE: For recommended tool geometry, see Tables 55 to 61; for recommended cutting fluids, see Table 62.

(a) Based on 4 flutes for 1/8- and 1/4-inch reamers, 6 flutes for 1/2-inch reamers, and 8 flutes for 1-inch and larger reamers. (b) Any premium HSS (T15, M33, M41-M47).

Table 18. Tapping, wrought stainless steels

Material	Hardness, Bhn	Condition	Speed, fpm(a) Threads per inch 7 or less	8 to 15	16 to 24	Over 24	HSS tool material, AISI
Ferritic steels							
405, 409, 429, 430, 434, 436, 442, 446	135 to 185	Annealed	15	25	35	40	M10, M7, M1
Austenitic steels							
201, 202, 301, 302, 302B, 304, 304L, 305, 308, 309, 309S, 310, 310S, 314, 316, 316L, 317, 321, 330, 347, 348, 384, 385	135 to 185	Annealed	12	18	22	25	M10, M7, M1
	225 to 275	Cold drawn	10	12	16	20	M10, M7, M1
Martensitic steels							
403, 410, 420, 422, 501, 502	135 to 175	Annealed	15	25	35	40	M10, M7, M1
	175 to 225	Annealed	12	18	25	30	M10, M7, M1
	275 to 325	Quenched and tempered	12	18	22	25	M10, M7, M1
	375 to 425	Quenched and tempered	8	10	14	18	Nitrided M10, M7, M1

NOTE: For recommended tool geometry, see Tables 55 to 61; for recommended cutting fluids, see Table 62.

(a) These speeds are for tapping 65% to 75% threads in shallow through-holes. Reduce the speed when tapping deep holes, blind holes or higher percentage of thread.

Table 19. Turning, cast irons

Material	Hardness, Bhn	Condition	Depth of cut, in.(a)	High speed steel tool Speed, fpm	Feed, ipr	Tool material, AISI	Carbide tool Uncoated Speed, fpm Brazed	Indexable	Feed, ipr	Tool material grade	Coated Speed, fpm	Feed, ipr	Tool material grade
Gray irons, ferritic													
ASTM A48, class 20; SAE J431c, grade G1800	120 to 150	Annealed	.040	185	.007	M2, M3	650	725	.010	C-3	950	.007	CC-3
			.150	145	.015	M2, M3	500	550	.020	C-2	600	.015	CC-2
			.300	120	.020	M2, M3	400	450	.030	C-2	575	.020	CC-2
			.625	100	.030	M2, M3	325	360	.040	C-2	—	—	—
Gray irons, pearlitic													
ASTM A48, classes 30, 35, 40; SAE J431c, grade G3000	190 to 220	As cast	.040	120	.007	T15, M42(b)	370	410	.007	C-3	525	.007	CC-3
			.150	80	.015	T15, M42(b)	300	340	.015	C-2	450	.015	CC-2
			.300	65	.020	T15, M42(b)	250	275	.030	C-2	350	.020	CC-2
			.625	55	.030	T15, M42(b)	200	225	.040	C-2	—	—	—
Ductile irons, ferritic													
ASTM A536, grades 60-40-18, 65-45-12; SAE J434c, grades D4018, D4512	140 to 190	Annealed	.040	200	.007	M2, M3	700	775	.010	C-7	950	.010	CC-7
			.150	150	.015	M2, M3	550	600	.020	C-7	775	.020	CC-7
			.300	125	.020	M2, M3	450	500	.030	C-6	650	.030	CC-6
			.625	100	.030	M2, M3	360	400	.040	C-6	—	—	—
Ductile irons, ferritic-pearlitic													
ASTM A536, grade 80-55-06; SAE J434c, grade D5506	190 to 225	As cast	.040	140	.007	M2, M3	480	540	.010	C-7	700	.010	CC-7
			.150	110	.015	M2, M3	375	425	.020	C-7	550	.020	CC-7
			.300	85	.020	M2, M3	310	350	.030	C-6	450	.030	CC-6
			.625	70	.030	M2, M3	250	275	.040	C-6	—	—	—
Malleable irons, ferritic													
ASTM A47, grades 32510, 35018; ASTM A602, grade M3210; SAE J158, grade M3210	110 to 160	Malleablized	.040	190	.007	M2, M3	750	800	.010	C-7	1000	.007	CC-7
			.150	145	.015	M2, M3	575	625	.020	C-7	800	.015	CC-7
			.300	120	.020	M2, M3	475	500	.030	C-6	650	.020	CC-6
			.625	95	.030	M2, M3	375	400	.040	C-6	—	—	—

(continued)

Table 19. (continued)

Material	Hard-ness, Bhn	Condition	Depth of cut, in.(a)	High speed steel tool Speed, fpm	Feed, ipr	Tool material, AISI	Carbide tool Uncoated Speed, fpm Brazed	Index-able	Feed, ipr	Tool mate-rial grade	Coated Speed, fpm	Feed, ipr	Tool mate-rial grade
Malleable irons, pearlitic ASTM A220, grades 40010, 45006, 45008, 50005; ASTM A602, grades M4504, M5003; SAE J158, grades M4504, M5003	160 to 200	Malleablized and heat treated	.040 .150 .300 .625	130 100 80 65	.007 .015 .020 .030	M2, M3 M2, M3 M2, M3 M2, M3	520 400 325 260	600 450 375 300	.010 .020 .030 .040	C-7 C-7 C-6 C-6	775 600 475 —	.010 .015 .020 —	CC-7 CC-7 CC-6 —
	200 to 240	Malleablized and heat treated	.040 .150 .300 .625	100 75 60 50	.007 .015 .020 .030	M2, M3 M2, M3 M2, M3 M2, M3	400 300 250 200	450 350 280 230	.010 .020 .030 .040	C-7 C-7 C-6 C-6	600 450 350 —	.010 .015 .020 —	CC-7 CC-7 CC-6 —

NOTE: For recommended tool geometry, see Tables 55 to 61; for recommended cutting fluids, see Table 62.

(a) Caution: Check horsepower requirements on heavier depths of cut. (b) Any premium HSS (T15, M33, M41-M47).

Table 20. Face milling, cast irons

Material	Hard-ness, Bhn	Condition	Depth of cut, in.(a)	High speed steel tool Speed, fpm	Feed per tooth, in.	Tool material, AISI	Carbide tool Uncoated Speed, fpm Brazed	Index-able	Feed per tooth, in.	Tool mate-rial grade	Coated Speed, fpm	Feed per tooth, in.	Tool mate-rial grade
Gray irons, ferritic ASTM A48, class 20; SAE J431c, grade G1800	120 to 150	Annealed	.040 .150 .300	235 185 145	.010 .014 .018	M2, M7 M2, M7 M2, M7	630 475 335	695 525 410	.010 .015 .020	C-2 C-2 C-2	1000 675 525	.010 .014 .018	CC-2 CC-2 CC-2
Gray irons, pearlitic ASTM A48, classes 30, 35, 40; SAE J431c, grade G3000	190 to 220	As cast	.040 .150 .300	120 90 70	.008 .012 .016	M2, M7 M2, M7 M2, M7	465 350 245	510 385 300	.008 .012 .016	C-2 C-2 C-2	775 500 400	.007 .010 .014	CC-2 CC-2 CC-2
Ductile irons, ferritic ASTM A536, grades 60-40-18, 65-45-12; SAE J434c, grades D4018, D4512	140 to 190	Annealed	.040 .150 .300	195 150 115	.010 .014 .018	M2, M7 M2, M7 M2, M7	665 500 350	730 550 430	.010 .015 .020	C-6 C-6 C-6	1100 715 560	.008 .012 .016	CC-6 CC-6 CC-6
Ductile irons, ferritic-pearlitic ASTM A536, grade 80-55-06; SAE J434c, grade D5506	190 to 225	As cast	.040 .150 .300	145 110 85	.008 .012 .016	M2, M7 M2, M7 M2, M7	465 350 245	510 385 300	.008 .012 .016	C-6 C-6 C-6	765 500 400	.008 .012 .016	CC-6 CC-6 CC-6
Malleable irons, ferritic ASTM A47, grades 32510, 35018; ASTM A602, grade M3210; SAE J158, grade M3210	110 to 160	Malleablized	.040 .150 .300	330 250 200	.010 .014 .016	M2, M7 M2, M7 M2, M7	800 600 425	880 660 520	.010 .015 .020	C-6 C-6 C-6	1300 855 675	.008 .012 .016	CC-6 CC-6 CC-6
Malleable irons, pearlitic ASTM A220, grades 40010, 45006, 45008, 50005; ASTM A602, grades M4504, M5003; SAE J158, grades M4504, M5003	160 to 200	Malleablized and heat treated	.040 .150 .300	175 135 105	.008 .012 .016	M2, M7 M2, M7 M2, M7	500 375 265	550 415 325	.008 .012 .016	C-6 C-6 C-6	825 540 425	.008 .012 .016	CC-6 CC-6 CC-6
	200 to 240	Malleablized and heat treated	.040 .150 .300	120 90 70	.007 .011 .016	M2, M7 M2, M7 M2, M7	430 350 245	475 400 300	.007 .010 .014	C-6 C-6 C-6	700 525 400	.007 .010 .014	CC-6 CC-6 CC-6

NOTE: For recommended tool geometry, see Tables 55 to 61; for recommended cutting fluids, see Table 62.

(a) Depth of cut is measured parallel to axis of cutter.

Table 21. End milling (peripheral), cast irons

Material	Hardness, Bhn	Condition	Radial depth of cut, in.(a)	High speed steel tool Speed, fpm	Feed, in./tooth Cutter diam, in. 3/8	1/2	3/4	1-2	Tool material, AISI	Carbide tool Speed, fpm	Feed, in./tooth Cutter diam, in. 3/8	1/2	3/4	1-2	Tool material grade
Gray irons, ferritic ASTM A48, class 20; SAE J431c, grade G1800	120 to 150	Annealed	.020 .060 dia/4 dia/2	180 145 120 100	.001 .002 .0015 .001	.002 .004 .003 .002	.005 .006 .005 .004	.007 .008 .006 .005	M2, M3, M7	625 480 400 375	.001 .002 .0015 .001	.003 .005 .003 .002	.007 .009 .006 .005	.009 .010 .007 .006	C-2
Gray irons, pearlitic ASTM A48, classes 30, 35, 40; SAE J431c, grade G3000	190 to 220	As cast	.020 .060 dia/4 dia/2	125 95 85 75	.001 .002 .0015 .001	.002 .003 .002 .0015	.003 .004 .003 .002	.004 .005 .004 .003	M2, M3, M7	390 300 255 235	.001 .002 .0015 .001	.003 .004 .003 .002	.005 .006 .005 .004	.007 .008 .006 .005	C-2
Ductile irons, ferritic ASTM A536, grades 60-40-18, 65-45-12; SAE J434c, grades D4018, D4512	140 to 190	Annealed	.020 .060 dia/4 dia/2	125 105 85 75	.001 .002 .001 .0007	.002 .003 .002 .001	.005 .006 .005 .004	.007 .008 .006 .005	M2, M3, M7	450 345 295 275	.001 .002 .0015 .001	.003 .005 .003 .002	.007 .009 .006 .005	.009 .010 .008 .006	C-5
Ductile irons, ferritic-pearlitic ASTM A536, grade 80-55-06; SAE J434c, grade D5506	190 to 225	As cast	.020 .060 dia/4 dia/2	105 90 70 60	.001 .002 .001 .0007	.002 .003 .002 .001	.004 .005 .004 .003	.006 .007 .005 .004	M2, M3, M7	375 280 245 225	.001 .002 .0015 .001	.002 .004 .003 .002	.006 .007 .005 .004	.007 .008 .006 .005	C-5
Malleable irons, ferritic ASTM A47, grades 32510, 35018; ASTM A602, grade M3210; SAE J158, grade M3210	110 to 160	Malleablized	.020 .060 dia/4 dia/2	180 145 120 100	.001 .002 .0015 .001	.002 .004 .003 .0015	.005 .006 .004 .003	.007 .008 .006 .005	M2, M3, M7	615 470 400 375	.001 .002 .0015 .001	.003 .005 .003 .002	.007 .009 .006 .005	.009 .010 .008 .006	C-5
Malleable irons, pearlitic ASTM A220, grades 40010, 45006, 45008, 50005; ASTM A602, grades M4504, M5003; SAE J158, grades M4504, M5003	160 to 200	Malleablized and heat treated	.020 .060 dia/4 dia/2	140 110 90 80	.001 .002 .001 .0007	.002 .003 .002 .001	.004 .005 .004 .003	.006 .007 .005 .004	M2, M3, M7	425 325 280 260	.001 .002 .0015 .001	.003 .004 .003 .002	.005 .007 .004 .003	.007 .008 .006 .005	C-5
	200 to 240	Malleablized and heat treated	.020 .060 dia/4 dia/2	110 90 70 60	.001 .002 .001 .0007	.002 .003 .002 .001	.003 .004 .003 .002	.004 .005 .004 .003	M2, M3, M7	375 280 245 225	.001 .002 .0015 .001	.002 .003 .0025 .002	.004 .005 .004 .003	.006 .007 .005 .004	C-5

NOTE: For recommended tool geometry, see Tables 55 to 61; for recommended cutting fluids, see Table 62.

(a) For standard length end mills, maximum axial depth can be up to 1.5 × the cutter diameter.

Table 22. Drilling, cast irons

Material	Hardness, Bhn	Condition	Speed, fpm(a)	Feed, ipr(a) Nominal hole diameter, in. 1/16	1/8	1/4	1/2	3/4	1	1½	2	Tool material grade, AISI or C
Gray irons, ferritic ASTM A48, class 20; SAE J431c, grade G1800	120 to 150	Annealed	90 160 350	.001 — —	— .003 .003	— .006 .006	— .012 .012	— .018 .018	— .022 .022	— .025 .025	— .030 .030	M10, M7, M1 M10, M7, M1 C-2
Gray irons, pearlitic ASTM A48, classes 30, 35, 40; SAE J431c, grade G3000	190 to 220	As cast	75 95 225	.001 — —	— .003 .003	— .005 .005	— .012 .012	— .018 .018	— .022 .022	— .025 .025	— .030 .030	M10, M7, M1 M10, M7, M1 C-2
Ductile irons, ferritic ASTM A536, grades 60-40-18, 65-45-12; SAE J434c, grades D4018, D4512	140 to 190	Annealed	85 115	.001 —	— .003	— .006	— .010	— .013	— .016	— .021	— .025	M10, M7, M1 M10, M7, M1

(continued)

Table 22. (continued)

Material	Hardness, Bhn	Condition	Speed, fpm(a)	1/16	1/8	1/4	1/2	3/4	1	1 1/2	2	Tool material grade, AISI or C
Ductile irons, ferritic-pearlitic ASTM A536, grade 80-55-06; SAE J434c, grade D5506	190 to 225	As cast	70	.001	.003	.006	.010	.013	.016	.021	.025	M10, M7, M1
Malleable irons, ferritic ASTM A47, grades 32510, 35018; ASTM A602, grade M3210; SAE J158, grade M3210	110 to 160	Malleablized	130	.001	.003	.006	.010	.013	.016	.021	.025	M10, M7, M1
Malleable irons, pearlitic ASTM A220, grades 40010, 45006, 45008, 50005; ASTM A602, grades M4504, M5003; SAE J158, grades M4504, M5003	160 to 200	Malleablized and heat treated	100	.001	.003	.006	.010	.013	.016	.021	.025	M10, M7, M1
	200 to 240	Malleablized and heat treated	85	—	.002	.004	.007	.010	.012	.015	.017	M10, M7, M1

Feed, ipr(a) — Nominal hole diameter, in.

NOTE: For recommended tool geometry, see Tables 55 to 61; for recommended cutting fluids, see Table 62.

(a) For drilling deep holes with twist drills, reduce speed and feed as follows: If hole depth is 3 drill diameters, reduce speed by 10%, feed by 10%; 4 diameters, speed by 20%, feed by 10%; 5 diameters, speed by 30%, feed by 20%; 6 diameters, speed by 35%, feed by 20%; 8 diameters, speed by 40%, feed by 20%.

Table 23. Reaming, cast irons

Material	Hardness, Bhn	Condition	Roughing Speed, fpm	1/8	1/4	1/2	1	1 1/2	2	Tool material grade, AISI or C	Finishing Speed, fpm	1/8	1/4	1/2	1	1 1/2	2	Tool material grade, AISI or C
Gray irons, ferritic ASTM A48, class 20; SAE J431c, grade G1800	120 to 150	Annealed	120	.006	.008	.012	.020	.025	.030	M1, M2, M7	65	.006	.010	.015	.025	.030	.035	M1, M2, M7
			200	.006	.008	.012	.020	.025	.030	C-2	90	.006	.008	.010	.020	.025	.030	C-2
Gray irons, pearlitic ASTM A48, classes 30, 35, 40; SAE J431c, grade G3000	190 to 220	As cast	90	.005	.006	.010	.015	.020	.025	M1, M2, M7	40	.005	.008	.012	.020	.025	.030	M1, M2, M7
			120	.005	.006	.010	.015	.020	.025	C-2	65	.005	.008	.012	.020	.025	.030	C-2
Ductile irons, ferritic ASTM A536, grades 60-40-18, 65-45-12; SAE J434c, grades D4018, D4512	140 to 190	Annealed	70	.004	.006	.010	.015	.020	.025	M1, M2, M7	45	.005	.008	.012	.020	.025	.030	M1, M2, M7
			120	.004	.006	.010	.015	.020	.025	C-2	60	.005	.008	.012	.020	.025	.030	C-2
Ductile irons, ferritic-pearlitic ASTM A536, grade 80-55-06; SAE J434c, grade D5506	190 to 225	As cast	65	.004	.006	.010	.015	.020	.025	M1, M2, M7	40	.005	.008	.012	.020	.025	.030	M1, M2, M7
			110	.004	.006	.010	.015	.020	.025	C-2	50	.004	.006	.010	.015	.020	.025	C-2
Malleable irons, ferritic ASTM A47, grades 32510, 35018; ASTM A602, grade M3210; SAE J158, grade M3210	110 to 160	Malleablized	85	.006	.008	.012	.020	.025	.030	M1, M2, M7	55	.006	.010	.015	.025	.030	.035	M1, M2, M7
			140	.006	.008	.012	.020	.025	.030	C-2	80	.005	.008	.012	.020	.025	.030	C-2

Feed, ipr(a) — Reamer diameter, in.

(continued)

Table 23. (continued)

Material	Hardness, Bhn	Condition	Roughing Speed, fpm	Feed, ipr(a) Reamer diameter, in. 1/8	1/4	1/2	1	1 1/2	2	Tool material grade, AISI or C	Finishing Speed, fpm	Feed, ipr(a) Reamer diameter, in. 1/8	1/4	1/2	1	1 1/2	2	Tool material grade, AISI or C
Malleable irons, pearlitic ASTM A220, grades 40010, 45006, 45008, 50005; ASTM A602, grades M4504, M5003; SAE J158, grades M4504, M5003	160 to 200	Malleablized and heat treated	80	.004	.006	.010	.015	.020	.025	M1, M2, M7	45	.005	.008	.012	.020	.025	.030	M1, M2, M7
			135	.004	.006	.010	.015	.020	.025	C-2	70	.005	.008	.012	.020	.025	.030	C-2
	200 to 240	Malleablized and heat treated	60	.004	.006	.008	.012	.016	.020	M1, M2, M7	35	.004	.007	.010	.015	.020	.025	M1, M2, M7
			100	.004	.006	.008	.012	.016	.020	C-2	60	.004	.006	.008	.012	.016	.020	C-2

NOTE: For recommended tool geometry, see Tables 55 to 61; for recommended cutting fluids, see Table 62.

(a) Based on 4 flutes for 1/8- and 1/4-inch reamers, 6 flutes for 1/2-inch reamers, and 8 flutes for 1-inch and larger reamers.

Table 24. Tapping, cast irons

Material	Hardness, Bhn	Condition	Speed, fpm(a) Threads per inch 7 or less	8 to 15	16 to 24	Over 24	HSS tool material, AISI
Gray irons, ferritic ASTM A48, class 20; SAE J431c, grade G1800	120 to 150	Annealed	35	60	75	80	M10, M7, M1
Gray irons, pearlitic ASTM A48, classes 30, 35, 40; SAE J431c, grade G3000	190 to 220	As cast	15	30	45	50	M10, M7, M1
Ductile irons, ferritic ASTM A536, grades 60-40-18, 65-45-12; SAE J434c, grades D4018, D4512	140 to 190	Annealed	20	40	55	60	M10, M7, M1
Ductile irons, ferritic-pearlitic ASTM A536, grade 80-55-06; SAE J434c, grade D5506	190 to 225	As cast	15	30	40	45	M10, M7, M1
Malleable irons, ferritic ASTM A47, grades 32510, 35018; ASTM A602, grade M3210; SAE J158, grade M3210	110 to 160	Malleablized	25	40	55	60	M10, M7, M1
Malleable irons, pearlitic ASTM A220, grades 40010, 45006, 45008, 50005; ASTM A602, grades M4504, M5003; SAE J158, grades M4504, M5003	160 to 200	Malleablized and heat treated	20	30	45	50	M10, M7, M1
	200 to 240	Malleablized and heat treated	15	25	35	40	M10, M7, M1

NOTE: For recommended tool geometry, see Tables 55 to 61; for recommended cutting fluids, see Table 62.

(a) These speeds are for tapping 65% to 75% threads in shallow through-holes. Reduce the speed when tapping deep holes, blind holes or higher percentage of thread.

Table 25. Turning, aluminum alloys

Material	Hardness, Bhn (500 kg)	Condition	Depth of cut, in.(a)	High speed steel tool Speed, fpm	Feed, ipr	Tool material, AISI	Carbide tool (uncoated) Speed, fpm Brazed	Index-able	Feed, ipr	Tool material grade
Wrought alloys All (e.g., 1100, 2024, 6061, 7075, etc.)	30 to 150	Cold drawn or solution treated and aged	.040	1000	.007	M2, M3	2000	Max.	.010	C-3
			.150	900	.015	M2, M3	1800	Max.	.020	C-2
			.300	800	.030	M2, M3	1200	Max.	.040	C-2
			.625	500	.040	M2, M3	1000	Max.	.080	C-2
Cast alloys, sand and permanent mold All (e.g., 201, 308, 355, 520, 850, etc.)	40 to 100	As cast	.040	1000	.007	M2, M3	2000	Max.	.010	C-3
			.150	900	.015	M2, M3	1800	Max.	.020	C-2
			.300	800	.030	M2, M3	1200	Max.	.040	C-2
			.625	500	.040	M2, M3	1000	Max.	.080	C-2

(continued)

Table 25. (continued)

Material	Hardness, Bhn (500 kg)	Condition	Depth of cut, in.(a)	High speed steel tool			Carbide tool (uncoated)			Tool material grade
				Speed, fpm	Feed, ipr	Tool material, AISI	Speed, fpm Brazed	Index-able	Feed per tooth, in.	

Cast alloys, sand and permanent mold (continued)

Material	Hardness	Condition	Depth	Speed	Feed	AISI	Brazed	Index	Feed	Grade
All (e.g., 201, 308, 355, 520, 850, etc.)	70 to 125	Solution treated and aged	.040	800	.007	M2, M3	1800	Max.	.010	C-3
			.150	700	.015	M2, M3	1400	Max.	.020	C-2
			.300	600	.030	M2, M3	1000	Max.	.040	C-2
			.625	400	.040	M2, M3	800	Max.	.080	C-2

Cast alloys, die castings

Material	Hardness	Condition	Depth	Speed	Feed	AISI	Brazed	Index	Feed	Grade
360.0, A360.0, 380.0, A380.0, C443.0, 518.0	40 to 100	As cast	.040	1000	.007	M2, M3	2000	Max.	.010	C-3
			.150	900	.015	M2, M3	1800	Max.	.020	C-2
			.300	—	—	—	—	—	—	—
			.625	—	—	—	—	—	—	—
	70 to 125	Solution treated and aged	.040	800	.007	M2, M3	1800	Max.	.010	C-3
			.150	700	.015	M2, M3	1400	Max.	.020	C-2
			.300	—	—	—	—	—	—	—
			.625	—	—	—	—	—	—	—
383.0, A384.0, 413.0, A413.0	40 to 100 500kg	As cast	.040	800	.007	M2, M3	1800	Max.	.010	C-3
			.150	700	.015	M2, M3	1600	Max.	.020	C-2
			.300	—	—	—	—	—	—	—
			.625	—	—	—	—	—	—	—
	70 to 125	Solution treated and aged	.040	700	.007	M2, M3	1600	Max.	.010	C-3
			.150	600	.015	M2, M3	1200	Max.	.020	C-2
			.300	—	—	—	—	—	—	—
			.625	—	—	—	—	—	—	—
390.0, 392.0	40 to 100	As cast	.040	150	.007	M2, M3	525	550	.010	C-2
			.150	135	.015	M2, M3	475	500	.015	C-2
			.300	—	—	—	—	—	—	—
			.625	—	—	—	—	—	—	—
	70 to 125	Solution treated and aged	.040	140	.007	M2, M3	450	475	.010	C-2
			.150	125	.015	M2, M3	425	450	.015	C-2
			.300	—	—	—	—	—	—	—
			.625	—	—	—	—	—	—	—

NOTE: For recommended tool geometry, see Tables 55 to 61; for recommended cutting fluids, see Table 62.

(a) Caution: Check horsepower requirements on heavier depths of cut.

Table 26. Face milling, aluminum alloys

Material	Hardness, Bhn (500 kg)	Condition	Depth of cut, in.(a)	High speed steel tool			Carbide tool (uncoated)			Tool material grade
				Speed, fpm	Feed per tooth, in.	Tool material, AISI	Speed, fpm Brazed	Index-able	Feed, ipr	

Wrought alloys

Material	Hardness	Condition	Depth	Speed	Feed/tooth	AISI	Brazed	Index	Feed	Grade
All (e.g., 1100, 2024, 6061, 7075, etc.)	30 to 150	Cold drawn or solution treated and aged	.040	1200	.010	M2, M7	2000	Max.	.010	C-2
			.150	800	.015	M2, M7	1800	Max.	.020	C-2
			.300	650	.020	M2, M7	1200	Max.	.025	C-2

Cast alloys, sand and permanent mold

Material	Hardness	Condition	Depth	Speed	Feed/tooth	AISI	Brazed	Index	Feed	Grade
All (e.g., 201, 308, 355, 520, 850, etc.)	40 to 100	As cast	.040	1200	.008	M2, M7	2000	2500	.010	C-2
			.150	800	.012	M2, M7	1800	2000	.015	C-2
			.300	650	.016	M2, M7	1200	1500	.020	C-2
	70 to 125	Solution treated and aged	.040	1000	.008	M2, M7	1800	2500	.010	C-2
			.150	700	.012	M2, M7	1400	1800	.015	C-2
			.300	550	.016	M2, M7	1000	1300	.020	C-2

Cast alloys, die castings

Material	Hardness	Condition	Depth	Speed	Feed/tooth	AISI	Brazed	Index	Feed	Grade
360.0, A360.0, 380.0, A380.0, C443.0, 518.0	40 to 100	As cast	.040	1200	.008	M2, M7	1400	1800	.005	C-2
			.150	800	.012	M2, M7	1150	1500	.010	C-2
			.300	—	—	—	—	—	—	—
	70 to 125	Solution treated and aged	.040	1000	.008	M2, M7	1250	1600	.005	C-2
			.150	700	.012	M2, M7	1000	1300	.010	C-2
			.300	—	—	—	—	—	—	—
383.0, A384.0, 413.0, A413.0	40 to 100	As cast	.040	1000	.008	M2, M7	1150	1500	.005	C-2
			.150	700	.012	M2, M7	950	1200	.010	C-2
			.300	—	—	—	—	—	—	—
	70 to 125	Solution treated and aged	.040	700	.006	M2, M7	1150	1400	.005	C-2
			.150	600	.010	M2, M7	900	1100	.010	C-2
			.300	—	—	—	—	—	—	—
390.0, 392.0	40 to 100	As cast	.040	165	.006	M2, M7	425	500	.005	C-2
			.150	140	.009	M2, M7	370	450	.010	C-2
			.300	—	—	—	—	—	—	—
	70 to 125	Solution treated and aged	.040	155	.006	M2, M7	375	475	.005	C-2
			.150	130	.009	M2, M7	350	425	.010	C-2
			.300	—	—	—	—	—	—	—

NOTE: For recommended tool geometry, see Tables 55 to 61; for recommended cutting fluids, see Table 62.

(a) Depth of cut is measured parallel to axis of cutter.

Table 27. End milling (peripheral), aluminum alloys

Material	Hardness, Bhn (500 kg)	Condition	Radial depth of cut, in.(a)	High speed steel tool Speed, fpm	Feed, in./tooth Cutter diam, in. 3/8	1/2	3/4	1-2	Tool material, AISI	Carbide tool Speed, fpm	Feed, in./tooth Cutter diam, in. 3/8	1/2	3/4	1-2	Tool material grade
Wrought alloys All (e.g., 1100, 2024, 6061, 7075, etc.)	30 to 150	Cold drawn or solution treated and aged	.020 .060 dia/4 dia/2	800 600 500 400	.003 .004 .003 .002	.004 .006 .004 .003	.005 .008 .006 .005	.007 .010 .008 .006	M2, M3, M7	1300 1000 900 800	.003 .004 .003 .002	.004 .006 .005 .004	.005 .008 .006 .005	.007 .010 .008 .006	C-2
Cast alloys, sand and permanent mold All (e.g., 201, 308, 355, 520, 850, etc.)	40 to 100	As cast	.020 .060 dia/4 dia/2	1000 800 600 500	.003 .004 .003 .002	.004 .006 .004 .003	.005 .008 .006 .005	.007 .010 .008 .006	M2, M3, M7	1300 1000 900 800	.003 .004 .003 .002	.004 .006 .005 .004	.005 .008 .006 .005	.007 .010 .008 .006	C-2
	70 to 125	Solution treated and aged	.020 .060 dia/4 dia/2	800 600 500 400	.003 .004 .003 .002	.004 .006 .004 .003	.005 .008 .006 .005	.007 .010 .008 .006	M2, M3, M7	1300 1000 900 800	.003 .004 .003 .002	.004 .006 .005 .004	.005 .008 .006 .005	.007 .010 .008 .006	C-2
Cast alloys, die castings 360.0, A360.0, 380.0, A380.0, C443.0, 518.0	40 to 100	As cast	.020 .060 dia/4 dia/2	1000 800 600 500	.003 .004 .003 .002	.004 .006 .004 .003	.005 .008 .006 .005	.007 .010 .008 .006	M2, M3, M7	1300 1000 900 800	.003 .004 .003 .002	.004 .006 .005 .004	.005 .008 .006 .005	.007 .010 .008 .006	C-2
	70 to 125	Solution treated and aged	.020 .060 dia/4 dia/2	900 700 500 400	.003 .004 .003 .002	.004 .006 .004 .003	.005 .008 .006 .005	.007 .010 .008 .006	M2, M3, M7	1300 1000 900 800	.003 .004 .003 .002	.004 .006 .005 .004	.005 .008 .006 .005	.007 .010 .008 .006	C-2
383.0, A384.0, 413.0, A413.0	40 to 100	As cast	.020 .060 dia/4 dia/2	900 700 500 400	.003 .004 .003 .002	.004 .006 .004 .003	.005 .008 .006 .005	.007 .010 .008 .006	M2, M3, M7	1300 1000 900 800	.003 .004 .003 .002	.004 .006 .005 .004	.005 .008 .006 .005	.007 .010 .008 .006	C-2
	70 to 125	Solution treated and aged	.020 .060 dia/4 dia/2	800 600 400 300	.003 .004 .003 .002	.004 .006 .004 .003	.005 .008 .006 .005	.007 .010 .008 .006	M2, M3, M7	1000 900 800 700	.003 .004 .003 .002	.004 .006 .005 .004	.005 .008 .006 .005	.007 .010 .008 .008	C-2
390.0, 392.0	40 to 100	As cast	.020 .060 dia/4 dia/2	300 200 175 150	.003 .004 .003 .002	.004 .006 .004 .003	.005 .008 .005 .004	.007 .010 .007 .005	M2, M3, M7	600 500 400 300	.003 .004 .003 .002	.004 .006 .005 .004	.005 .008 .006 .005	.007 .010 .008 .006	C-2
	70 to 125	Solution treated and aged	.020 .060 dia/4 dia/2	250 200 150 125	.003 .004 .003 .002	.004 .006 .004 .003	.005 .008 .005 .004	.007 .010 .007 .005	M2, M3, M7	500 400 300 200	.003 .004 .003 .002	.004 .006 .005 .004	.005 .008 .006 .005	.007 .010 .008 .006	C-2

NOTE: For recommended tool geometry, see Tables 55 to 61; for recommended cutting fluids, see Table 62.

(a) For standard length end mills, maximum axial depth can be up to 1.5 × cutter diameter.

Table 28. Drilling, aluminum alloys

Material	Hardness, Bhn (500 kg)	Condition	Speed, fpm	Feed, ipr(a) Nominal hole diameter, in. 1/16	1/8	1/4	1/2	3/4	1	1 1/2	2	Tool material grade, AISI
Wrought alloys All (e.g., 1100, 2024, 6061, 7075, etc.)	30 to 150	Cold drawn or solution treated and aged	140 315	.001 —	— .003	— .007	— .012	— .016	— .019	— .025	— .030	M10, M7, M1
Cast alloys, sand and permanent mold All (e.g., 201, 308, 355, 520, 850, etc.)	40 to 100	As cast	140 350	.001 —	— .003	— .007	— .012	— .016	— .019	— .025	— .030	M10, M7, M1
	70 to 125	Solution treated and aged	140 275	.001 —	— .003	— .007	— .012	— .016	— .019	— .025	— .030	M10, M7, M1

(continued)

Table 28. (continued)

Material	Hardness, Bhn (500 kg)	Condition	Speed, fpm	Feed, ipr(a) — Nominal hole diameter, in.								Tool material grade, AISI
				1/16	1/8	1/4	1/2	3/4	1	1 1/2	2	
Cast alloys, die castings												
360.0, A360.0, 380.0, A380.0, C443.0, 518.0	40 to 100	As cast	150	.001	—	—	—	—	—	—	—	M10, M7, M1
			375	—	.003	.007	.012	.016	.019	.025	.030	
	70 to 125	Solution treated and aged	150	.001	—	—	—	—	—	—	—	M10, M7, M1
			300	—	.003	.007	.012	.016	.019	.025	.030	
383.0, A384.0, 413.0, A413.0	40 to 100	As cast	100	.001	—	—	—	—	—	—	—	M10, M7, M1
			140	—	.003	.007	.012	.016	.019	.025	.030	
	70 to 125	Solution treated and aged	90	.001	—	—	—	—	—	—	—	M10, M7, M1
			120	—	.003	.007	.012	.016	.019	.025	.030	
390.0, 392.0	40 to 100	As cast	25	.001	—	—	—	—	—	—	—	M10, M7, M1
			50	—	.003	.007	.012	.016	.019	.025	.030	
	70 to 125	Solution treated and aged	25	.001	—	—	—	—	—	—	—	M10, M7, M1
			45	—	.003	.007	.012	.016	.019	.025	.030	

NOTE: For recommended tool geometry, see Tables 55 to 61; for recommended cutting fluids, see Table 62.

(a) For drilling deep holes with twist drills, reduce speed and feed as follows: If hole depth is 3 drill diameters, reduce speed by 10%, feed by 10%; 4 diameters, speed by 20%, feed by 10%; 5 diameters, speed by 30%, feed by 20%; 6 diameters, speed by 35%, feed by 20%; 8 diameters, speed by 40%, feed by 20%.

Table 29. Reaming, aluminum alloys

Material	Hardness, Bhn (500 kg)	Condition	Roughing Speed, fpm	Feed, ipr(a) — Reamer diameter, in.						Tool material grade, AISI or C	Finishing Speed, fpm	Feed, ipr(a) — Reamer diameter, in.						Tool material grade, AISI or C
				1/8	1/4	1/2	1	1 1/2	2			1/8	1/4	1/2	1	1 1/2	2	
Wrought alloys																		
All (e.g., 1100, 2024, 6061, 7075, etc.)	30 to 150	Cold drawn or solution treated and aged	500	.007	.010	.015	.020	.025	.030	M1, M2, M7	180	.004	.006	.010	.015	.018	.020	M1, M2, M7
			1000	.005	.007	.010	.015	.020	.025	C-2	250	.004	.006	.010	.015	.018	.020	C-2
Cast alloys, sand and permanent mold																		
All (e.g., 201, 308, 355, 520, 850, etc.)	40 to 100	As cast	500	.007	.010	.015	.020	.025	.030	M1, M2, M7	160	.004	.006	.010	.015	.018	.020	M1, M2, M7
			1000	.005	.007	.010	.015	.020	.025	C-2	250	.004	.006	.010	.015	.018	.020	C-2
	70 to 125	Solution treated and aged	400	.007	.010	.015	.020	.025	.030	M1, M2, M7	160	.004	.006	.010	.015	.018	.020	M1, M2, M7
			850	.005	.007	.010	.015	.020	.025	C-2	250	.004	.006	.010	.015	.018	.020	C-2
Cast alloys, die castings																		
360.0, A360.0, 380.0, A380.0, C443.0, 518.0	40 to 100	As cast	500	.007	.010	.015	.020	.025	.030	M1, M2, M7	120	.004	.006	.010	.015	.018	.020	M1, M2, M7
			1000	.005	.007	.010	.015	.020	.025	C-2	200	.004	.006	.010	.015	.018	.020	C-2
	70 to 125	Solution treated and aged	400	.007	.010	.015	.020	.025	.030	M1, M2, M7	120	.004	.006	.010	.015	.018	.020	M1, M2, M7
			850	.005	.007	.010	.015	.020	.025	C-2	200	.004	.006	.010	.015	.018	.020	C-2
383.0, A384.0, 413.0, A413.0	40 to 100	As cast	350	.007	.010	.015	.020	.025	.030	M1, M2, M7	100	.004	.006	.010	.015	.018	.020	M1, M2, M7
			700	.005	.007	.010	.015	.020	.025	C-2	175	.004	.006	.010	.015	.018	.020	C-2
	70 to 125	Solution treated and aged	350	.007	.010	.015	.020	.025	.030	M1, M2, M7	100	.004	.006	.010	.015	.018	.020	M1, M2, M7
			700	.005	.007	.010	.015	.020	.025	C-2	175	.004	.006	.010	.015	.018	.020	C-2
390.0, 392.0	40 to 100	As cast	100	.007	.010	.015	.020	.025	.030	M1, M2, M7	80	.005	.008	.012	.018	.020	.025	M1, M2, M7
			200	.007	.010	.015	.020	.025	.030	C-2	100	.005	.008	.012	.018	.020	.025	C-2
	70 to 125	Solution treated and aged	100	.007	.010	.015	.020	.025	.030	M1, M2, M7	80	.005	.008	.012	.018	.020	.025	M1, M2, M7
			200	.007	.010	.015	.020	.025	.030	C-2	100	.005	.008	.012	.018	.020	.025	C-2

NOTE: For recommended tool geometry, see Tables 55 to 61; for recommended cutting fluids, see Table 62.

(a) Based on 4 flutes for 1/8- and 1/4-inch reamers, 6 flutes for 1/2-inch reamers, and 8 flutes for 1-inch and larger reamers.

Table 30. Tapping, aluminum alloys

Material	Hardness, Bhn (500 kg)	Condition	Speed, fpm(a) Threads per inch 7 or less	8 to 15	16 to 24	Over 24	HSS tool material, AISI
Wrought alloys							
All (e.g., 1100, 2024, 6061, 7075, etc.) .... 30 to 150		Cold drawn or solution treated and aged	50	85	105	115	M10, M7, M1
Cast alloys, sand and permanent mold							
All (e.g., 201, 308, 355, 520, 850, etc.)	40 to 100	As cast	50	85	110	115	M10, M7, M1
	70 to 125	Solution treated and aged	40	65	85	90	M10, M7, M1
Cast alloys, die castings							
360.0, A360.0, 380.0, A380.0, C443.0, 518.0	40 to 100	As cast	65	110	140	150	M10, M7, M1
	70 to 125	Solution treated and aged	45	75	95	100	M10, M7, M1
383.0, A384.0, 413.0, A413.0	40 to 100	As cast	65	110	140	150	M10, M7, M1
	70 to 125	Solution treated and aged	45	75	95	100	M10, M7, M1
390.0, 392.0	40 to 100	As cast	55	100	110	120	M10, M7, M1
	70 to 125	Solution treated and aged	40	70	90	95	M10, M7, M1

NOTE: For recommended tool geometry, see Tables 55 to 61; for recommended cutting fluids, see Table 62.

(a) These speeds are for tapping 65% to 75% threads in shallow through-holes. Reduce the speed when tapping deep holes, blind holes or higher percentage of thread.

Table 31. Turning, wrought titanium alloys

Material	Hardness, Bhn	Condition	Depth of cut, in.(a)	High speed steel tool Speed, fpm	Feed, ipr	Tool material, AISI	Carbide tool (uncoated) Speed, fpm Brazed	Index-able	Feed, ipr	Tool material grade
Commercially pure Ti (99.0) .... 110 to 170		Annealed	.040	250	.005	T15, M42(b)	525	565	.005	C-3
			.150	220	.010	T15, M42(b)	450	485	.010	C-2
			.300	175	.015	T15, M42(b)	340	360	.015	C-2
			.625	—	—	—	170	180	.020	C-2
	140 to 200	Annealed	.040	190	.005	T15, M42(b)	450	500	.005	C-3
			.150	170	.010	T15, M42(b)	390	425	.010	C-2
			.300	150	.015	T15, M42(b)	290	320	.015	C-2
			.625	—	—	—	145	160	.020	C-2
	200 to 275	Annealed	.040	115	.005	T15, M42(b)	290	370	.005	C-3
			.150	105	.010	T15, M42(b)	250	320	.008	C-2
			.300	95	.015	T15, M42(b)	190	240	.015	C-2
			.625	—	—	—	95	120	.020	C-2
Alpha alloys .... 300 to 340 (Ti-5Al-2.5Sn, Ti-5Al-2.5Sn-ELI, Ti-6Al-2Cb-1Ta-0.80Mo)		Annealed	.040	80	.005	T15, M42(b)	215	250	.005	C-3
			.150	70	.010	T15, M42(b)	185	215	.008	C-2
			.300	60	.015	T15, M42(b)	140	160	.010	C-2
			.625	—	—	—	70	80	.015	C-2
Alpha-beta alloys .... 310 to 350 (Ti-6Al-4V, Ti-6Al-4V-ELI, Ti-6Al-2Sn-4Zr-2Mo, Ti-6Al-2Sn-4Zr-2Mo-0.25Si, Ti-6Al-2Sn-4Zr-6Mo)		Annealed	.040	70	.005	T15, M42(b)	170	225	.005	C-3
			.150	60	.010	T15, M42(b)	145	195	.008	C-2
			.300	50	.015	T15, M42(b)	110	145	.010	C-2
			.625	—	—	—	55	70	.015	C-2
	320 to 380	Solution treated and aged	.040	65	.005	T15, M42(b)	160	190	.005	C-3
			.150	55	.010	T15, M42(b)	135	165	.008	C-2
			.300	45	.015	T15, M42(b)	85	120	.010	C-2
			.625	—	—	—	50	60	.015	C-2
Beta alloys .... 275 to 350 (Ti-3Al-8V-6Cr-4Mo-4Zr, Ti-8Mo-8V-2Fe-3Al, Ti-11.5Mo-6Zr-4.5Sn, Ti-10V-2Fe-3Al, Ti-13V-11Cr-3Al)		Annealed or solution treated	.040	40	.005	T15, M42(b)	125	160	.005	C-3
			.150	30	.010	T15, M42(b)	105	135	.008	C-2
			.300	25	.015	T15, M42(b)	80	85	.010	C-2
			.625	—	—	—	40	50	.015	C-2
	350 to 440	Solution treated and aged	.040	35	.005	T15, M42(b)	110	125	.005	C-3
			.150	25	.010	T15, M42(b)	90	105	.008	C-2
			.300	—	—	—	70	80	.010	C-2
			.625	—	—	—	35	40	.015	C-2

NOTE: For recommended tool geometry, see Tables 55 to 61; for recommended cutting fluids, see Table 62.

(a) Caution: Check horsepower requirements on heavier depths of cut. (b) Any premium HSS (T15, M33, M41-M47).

Table 32. Face milling, wrought titanium alloys

Material	Hardness, Bhn	Condition	Depth of cut, in.(a)	High speed steel tool Speed, fpm	Feed per tooth, in.	Tool material, AISI	Carbide tool (uncoated) Speed, fpm Brazed	Speed, fpm Indexable	Feed per tooth, in.	Tool material grade
Commercially pure Ti (99.0)	110 to 170	Annealed	.040	175	.006	T15, M42(b)	530	585	.005	C-2
			.150	135	.009	T15, M42(b)	400	440	.010	C-2
			.300	105	.012	T15, M42(b)	280	345	.015	C-2
	140 to 200	Annealed	.040	145	.004	M2, M7	400	440	.004	C-2
			.150	110	.006	M2, M7	300	330	.006	C-2
			.300	85	.008	M2, M7	200	250	.008	C-2
	200 to 275	Annealed	.040	105	.004	M2, M7	325	350	.004	C-2
			.150	80	.006	M2, M7	275	300	.006	C-2
			.300	60	.008	M2, M7	190	235	.008	C-2
Alpha alloys (Ti-5Al-2.5Sn, Ti-5Al-2.5Sn-ELI, Ti-6Al-2Cb-1Ta-0.80Mo)	300 to 340	Annealed	.040	70	.004	T15, M42(b)	260	290	.004	C-2
			.150	55	.006	T15, M42(b)	225	240	.006	C-2
			.300	40	.008	T15, M42(b)	150	185	.008	C-2
Alpha-beta alloys (Ti-6Al-4V, Ti-6Al-4V-ELI, Ti-6Al-2Sn-4Zr-2Mo, Ti-6Al-2Sn-4Zr-2Mo-0.25Si, Ti-6Al-2Sn-4Zr-6Mo)	310 to 350	Annealed	.040	55	.004	T15, M42(b)	170	185	.004	C-2
			.150	45	.006	T15, M42(b)	130	145	.006	C-2
			.300	35	.008	T15, M42(b)	95	115	.008	C-2
	320 to 380	Solution treated and aged	.040	55	.003	T15, M42(b)	145	160	.004	C-2
			.150	50	.005	T15, M42(b)	110	120	.006	C-2
			.300	40	.007	T15, M42(b)	80	95	.008	C-2
Beta alloys (Ti-3Al-8V-6Cr-4Mo-4Zr, Ti-8Mo-8V-2Fe-3Al, Ti-11.5Mo-6Zr-4.5Sn, Ti-10V-2Fe-3Al, Ti-13V-11Cr-3Al)	275 to 350	Annealed or solution treated	.040	40	.003	T15, M42(b)	130	145	.004	C-2
			.150	30	.005	T15, M42(b)	100	110	.006	C-2
			.300	20	.007	T15, M42(b)	70	85	.008	C-2
	350 to 440	Solution treated and aged	.040	30	.002	T15, M42(b)	80	90	.004	C-2
			.150	25	.004	T15, M42(b)	60	65	.006	C-2
			.300	20	.006	T15, M42(b)	40	50	.008	C-2

NOTE: For recommended tool geometry, see Tables 55 to 61; for recommended cutting fluids, see Table 62.

(a) Depth of cut is measured parallel to axis of cutter. (b) Any premium HSS (T15, M33, M41-M47).

Table 33. End milling (peripheral), wrought titanium alloys

Material	Hardness, Bhn	Condition	Radial depth of cut, in.(a)	HSS Speed, fpm	HSS Feed 3/8	HSS Feed 1/2	HSS Feed 3/4	HSS Feed 1-2	Tool material, AISI	Carbide Speed, fpm	Carb Feed 3/8	Carb Feed 1/2	Carb Feed 3/4	Carb Feed 1-2	Tool material grade
Commercially pure Ti (99.0)	110 to 170	Annealed	.020	175	.002	.003	.005	.006	M2, M3, M7	425	.001	.003	.005	.007	C-2
			.060	160	.003	.004	.006	.007		390	.002	.004	.006	.008	
			dia/4	85	.0015	.002	.003	.004		240	.0015	.003	.005	.007	
			dia/2	60	.001	.0015	.002	.003		180	.001	.002	.004	.006	
	140 to 200	Annealed	.020	170	.0015	.003	.005	.006	M2, M3, M7	400	.001	.003	.005	.007	C-2
			.060	150	.002	.004	.006	.007		370	.002	.004	.006	.008	
			dia/4	85	.0015	.002	.003	.004		230	.0015	.003	.005	.007	
			dia/2	60	.001	.0015	.002	.003		175	.001	.002	.004	.006	
	200 to 275	Annealed	.020	150	.001	.002	.004	.005	M2, M3, M7	350	.001	.002	.005	.007	C-2
			.060	130	.002	.003	.005	.006		325	.002	.003	.006	.008	
			dia/4	75	.0015	.002	.003	.004		200	.0015	.002	.005	.007	
			dia/2	50	.001	.0015	.002	.003		150	.001	.001	.004	.006	
Alpha alloys (Ti-5Al-2.5Sn, Ti-5Al-2.5Sn-ELI, Ti-6Al-2Cb-1Ta-0.80Mo)	300 to 340	Annealed	.020	110	.001	.002	.004	.005	T15(b)	300	.001	.002	.005	.007	C-2
			.060	100	.002	.003	.005	.006		275	.002	.003	.006	.008	
			dia/4	55	.001	.0015	.002	.003		170	.0015	.002	.005	.006	
			dia/2	40	.001	.001	.0015	.002		130	.001	.001	.004	.005	
Alpha-beta alloys (Ti-6Al-4V, Ti-6Al-4V-ELI, Ti-6Al-2Sn-4Zr-2Mo, Ti-6Al-2Sn-4Zr-2Mo-0.25Si, Ti-6Al-2Sn-4Zr-6Mo)	310 to 350	Annealed	.020	100	.001	.002	.004	.005	T15(b)	290	.001	.002	.005	.007	C-2
			.060	90	.002	.003	.005	.006		260	.002	.003	.006	.008	
			dia/4	50	.001	.0015	.002	.003		160	.0015	.002	.005	.006	
			dia/2	35	.001	.001	.0015	.002		125	.001	.001	.004	.005	
	320 to 380	Solution treated and aged	.020	85	.001	.002	.003	.005	T15(b)	225	.001	.002	.004	.006	C-2
			.060	75	.002	.003	.004	.006		200	.002	.003	.005	.007	
			dia/4	40	.001	.0015	.002	.003		125	.0015	.002	.005	.006	
			dia/2	30	.001	.001	.0015	.002		100	.001	.001	.004	.005	
Beta alloys (Ti-3Al-8V-6Cr-4Mo-4Zr, Ti-8Mo-8V-2Fe-3Al, Ti-11.5Mo-6Zr-4.5Sn, Ti-10V-2Fe-3Al, Ti-13V-11Cr-3Al)	275 to 350	Annealed or solution treated	.020	50	.001	.002	.004	.005	T15(b)	150	.001	.002	.005	.007	C-2
			.060	45	.0015	.003	.005	.006		130	.0015	.003	.006	.008	
			dia/4	25	.001	.002	.003	.004		75	.001	.002	.005	.006	
			dia/2	20	.0007	.0015	.002	.003		50	.0007	.001	.004	.005	
	350 to 440	Solution treated and aged	.020	40	.0007	.0015	.002	.004	T15(b)	125	.0007	.0015	.003	.005	C-2
			.060	35	.001	.002	.003	.005		110	.001	.002	.004	.006	
			dia/4	20	.0007	.0015	.002	.003		60	.0007	.0015	.003	.005	
			dia/2	15	.0005	.001	.0015	.002		45	.0005	.001	.002	.004	

NOTE: For recommended tool geometry, see Tables 55 to 61; for recommended cutting fluids, see Table 62.

(a) For standard length end mills, maximum axial depth can be up to 1.5 × the cutter diameter. (b) Any premium HSS (T15, M33, M41-M47).

Table 34. Drilling, wrought titanium alloys

Material	Hardness, Bhn	Condition	Speed, fpm(a)	Feed, ipr(a) — Nominal hole diameter, in. 1/16	1/8	1/4	1/2	3/4	1	1½	2	Tool material grade, AISI
Commercially pure Ti (99.0)	110 to 170	Annealed	80 to 110	.0005	.002	.005	.008	.010	.012	.015	.017	M10, M7, M1
	140 to 200	Annealed	65 to 90	.0005	.002	.005	.008	.010	.012	.015	.017	M10, M7, M1
	200 to 275	Annealed	40 to 55	.001	.002	.005	.008	.010	.012	.015	.017	M10, M7, M1
Alpha alloys (Ti-5Al-2.5Sn, Ti-5Al-2.5Sn-ELI, Ti-6Al-2Cb-1Ta-0.80Mo)	300 to 340	Annealed	45	—	.002	.005	.007	.008	.010	.012	.015	T15, M42(b)
Alpha-beta alloys (Ti-6Al-4V, Ti-6Al-4V-ELI, Ti-6Al-2Sn-4Zr-2Mo, Ti-6Al-2Sn-4Zr-2Mo-0.25Si, Ti-6Al-2Sn-4Zr-6Mo)	310 to 350	Annealed	35	—	.002	.004	.006	.007	.008	.010	.012	T15, M42(b)
	320 to 380	Solution treated and aged	30	—	.002	.003	.005	.006	.007	.009	.010	T15, M42(b)
Beta alloys (Ti-3Al-8V-6Cr-4Mo-4Zr, Ti-8Mo-8V-2Fe-3Al, Ti-11.5Mo-6Zr-4.5Sn, Ti-10V-2Fe-3Al, Ti-13V-11Cr-3Al)	275 to 350	Annealed or solution treated	25	—	.001	.003	.004	.005	.006	.007	.008	T15, M42(b)
	350 to 440	Solution treated and aged	20	—	.001	.002	.003	.004	.004	.005	.006	T15, M42(b)

NOTE: For recommended tool geometry, see Tables 55 to 61; for recommended cutting fluids, see Table 62.

(a) For drilling deep holes with twist drills, reduce speed and feed as follows: If hole depth is 3 drill diameters, reduce speed by 10%, feed by 10%; 4 diameters, speed by 20%, feed by 10%; 5 diameters, speed by 30%, feed by 20%; 6 diameters, speed by 35%, feed by 20%; 8 diameters, speed by 40%, feed by 20%. (b) Any premium HSS (T15, M33, M41-M47).

Table 35. Reaming, wrought titanium alloys

Material	Hardness, Bhn	Condition	Roughing Speed, fpm	Feed, ipr(a) — Reamer diameter, in. 1/8	1/4	1/2	1	1½	2	Tool material grade, AISI or C	Finishing Speed, fpm	1/8	1/4	1/2	1	1½	2	Tool material grade, AISI or C
Commercially pure Ti (99.0)	110 to 170	Annealed	175	.004	.008	.012	.018	.022	.025	M1, M2, M7	60	.004	.006	.010	.012	.015	.020	M1, M2, M7
			375	.004	.008	.012	.018	.022	.025	C-2	75	.004	.006	.010	.012	.015	.020	C-2
	140 to 200	Annealed	140	.004	.008	.012	.018	.022	.025	M1, M2, M7	50	.004	.006	.010	.012	.014	.016	M1, M2, M7
			375	.004	.008	.012	.018	.022	.025	C-2	65	.004	.006	.010	.012	.014	.016	C-2
	200 to 275	Annealed	120	.004	.008	.012	.018	.022	.025	M1, M2, M7	45	.004	.006	.010	.012	.014	.016	M1, M2, M7
			300	.004	.008	.012	.018	.022	.025	C-2	60	.004	.006	.010	.012	.014	.016	C-2
Alpha alloys (Ti-5Al-2.5Sn, Ti-5Al-2.5Sn-ELI, Ti-6Al-2Cb-1Ta-0.80Mo)	300 to 340	Annealed	70	.003	.007	.009	.012	.015	.017	T15, M42(b)	20	.004	.006	.010	.012	.014	.016	T15, M42(b)
			250	.003	.008	.012	.016	.020	.023	C-2	35	.004	.006	.010	.012	.014	.016	C-2
Alpha-beta alloys (Ti-6Al-4V, Ti-6Al-4V-ELI, Ti-6Al-2Sn-4Zr-2Mo, Ti-6Al-2Sn-4Zr-2Mo-0.25Si, Ti-6Al-2Sn-4Zr-6Mo)	310 to 350	Annealed	65	.003	.006	.010	.012	.014	.016	T15, M42(b)	20	.004	.006	.010	.012	.014	.016	T15, M42(b)
			200	.003	.006	.010	.012	.014	.016	C-2	35	.004	.006	.010	.012	.014	.016	C-2
	320 to 380	Solution treated and aged	50	.003	.007	.010	.012	.014	.016	T15, M42(b)	20	.003	.005	.008	.010	.012	.014	T15, M42(b)
			160	.003	.007	.010	.012	.014	.016	C-2	35	.003	.005	.008	.010	.012	.014	C-2
Beta alloys (Ti-3Al-8V-6Cr-4Mo-4Zr, Ti-8Mo-8V-2Fe-3Al, Ti-11.5Mo-6Zr-4.5Sn, Ti-10V-2Fe-3Al, Ti-13V-11Cr-3Al)	275 to 350	Annealed or solution treated	30	.002	.005	.007	.010	.012	.014	T15, M42(b)	15	.003	.005	.008	.010	.012	.014	T15, M42(b)
			75	.002	.005	.007	.010	.012	.014	C-2	30	.003	.005	.008	.010	.012	.014	C-2
	350 to 440	Solution treated and aged	20	.002	.004	.006	.008	.010	.012	T15, M42(b)	10	.003	.005	.008	.010	.012	.014	T15, M42(b)
			50	.002	.004	.006	.008	.010	.012	C-2	25	.003	.005	.008	.010	.012	.014	C-2

NOTE: For recommended tool geometry, see Tables 55 to 61; for recommended cutting fluids, see Table 62.

(a) Based on 4 flutes for 1/8- and 1/4-inch reamers, 6 flutes for 1/2-inch reamers, and 8 flutes for 1-inch and larger reamers. (b) Any premium HSS (T15, M33, M41-M47).

Table 36. Tapping, wrought titanium alloys

Material	Hardness, Bhn	Condition	Speed, fpm(a) — Threads per inch				HSS tool material, AISI
			7 or less	8 to 15	16 to 24	Over 24	
Commercially pure Ti (99.0)	110 to 170	Annealed	20	40	55	60	Nitrided M10, M7, M1
	140 to 200	Annealed	15	30	45	50	Nitrided M10, M7, M1
	200 to 275	Annealed	12	25	35	40	Nitrided M10, M7, M1
Alpha alloys .... (Ti-5Al-2.5Sn, Ti-5Al-2.5Sn-ELI, Ti-6Al-2Cb-1Ta-0.80Mo)	300 to 340	Annealed	10	15	20	25	Nitrided M10, M7, M1
Alpha-beta alloys .... (Ti-6Al-4V, Ti-6Al-4V-ELI, Ti-6Al-2Sn-4Zr-2Mo, Ti-6Al-2Sn-4Zr-2Mo-0.25Si, Ti-6Al-2Sn-4Zr-6Mo)	310 to 350	Annealed	7	15	18	20	Nitrided M10, M7, M1
	320 to 380	Solution treated and aged	3	7	9	10	Nitrided M10, M7, M1
Beta alloys .... (Ti-3Al-8V-6Cr-4Mo-4Zr, Ti-8Mo-8V-2Fe-3Al, Ti-11.5Mo-6Zr-4.5Sn, Ti-10V-2Fe-3Al, Ti-13V-11Cr-3Al)	275 to 350	Annealed or solution treated	5	10	14	15	Nitrided M10, M7, M1
	350 to 440	Solution treated and aged	2	3	4	5	Nitrided M10, M7, M1

NOTE: For recommended tool geometry, see Tables 55 to 61; for recommended cutting fluids, see Table 62.

(a) These speeds are for tapping 65% to 75% threads in shallow through-holes. Reduce the speed when tapping deep holes, blind holes or higher percentage of thread.

Table 37. Turning, nickel-base and cobalt-base high-temperature alloys

Material	Hardness, Bhn	Condition	Depth of cut, in.(a)	High speed steel tool			Carbide tool — Uncoated				Coated		
				Speed, fpm	Feed, ipr	Tool material, AISI	Speed, fpm Brazed	Index-able	Feed, ipr	Tool material grade	Speed, fpm	Feed, ipr	Tool material grade
Nickel-base alloys, wrought .... Haynes alloy 263; Incoloy 901, 903; Inconel 617, 625, 702, 706, 718, 721, 722, X-750, 751; M252; Nimonic 75, 80; Waspaloy	200 to 300	Annealed or solution treated	.030 .100 .200 —	25 20 — —	.005 .007 — —	T15, M42(b) T15, M42(b) — —	90 70 50 —	100 80 60 —	.005 .007 .015 —	C-3 C-2 C-2 —	— — — —	— — — —	— — — —
	300 to 400	Solution treated and aged	.030 .100 .200	25 15 —	.005 .007 —	T15, M42(b) T15, M42(b) —	80 65 40	95 75 50	.005 .007 .015	C-3 C-2 C-2	600 500 450	.003 .005 .005	Borazon Borazon Borazon
Nickel-base alloys, cast .. B-1900; GMR-235, -235D; Hastelloy alloy D; IN-100 (René 100);	250 to 320	As cast or cast and aged	.030 .100 .200	15 12 —	.005 .005 —	T15, M42(b) T15, M42(b) —	45 35 30	60 45 35	.005 .007 .010	C-3 C-2 C-2	— — —	— — —	— — —
IN-738, -792; Inconel 713C, 718; M252; MAR-M200, -M246, -M421, -M432; René 80, 125; SEL; SEL 15; TRW VI A; Udimet 500, 700	320 to 425	As cast or cast and aged	.030 .100 .200	12 10 —	.005 .005 —	T15, M42(b) T15, M42(b) —	40 30 25	50 35 30	.005 .007 .010	C-3 C-2 C-2	600 450 400	.003 .005 .005	Borazon Borazon Borazon
Cobalt alloys, wrought .. AiResist 213; Haynes alloy 25 (L605); Haynes alloy 188;	180 to 230	Solution treated	.030 .100 .200 —	25 20 — —	.005 .007 — —	T15, M42(b) T15, M42(b) — —	80 60 45 —	90 70 55 —	.005 .007 .010 —	C-3 C-2 C-2 —	— — — —	— — — —	— — — —
J-1570; MAR-M905, -M918; S-816; V-36	270 to 320	Solution treated and aged	.030 .100 .200	20 15 —	.005 .007 —	T15, M42(b) T15, M42(b) —	70 55 40	80 65 50	.005 .007 .010	C-3 C-2 C-2	— — —	— — —	— — —
Cobalt alloys, cast ...... AiResist 13, 215; FSX-414; HS-6, -21, -31 (X-40);	220 to 290	As cast or cast and aged	.030 .100 .200	15 12 —	.005 .007 —	T15, M42(b) T15, M42(b) —	50 40 —	60 50 —	.005 .007 —	C-3 C-2 —	— — —	— — —	— — —
HOWMET #3; MAR-M302, -M322, -M509; NASA Co-W-Re; WI-52; X-45	290 to 425	As cast or cast and aged	.030 .100 .200	12 10 —	.005 .005 —	T15, M42(b) T15, M42(b) —	45 35 —	55 40 —	.005 .005 —	C-3 C-2 —	600 450 400	.003 .005 .005	Borazon Borazon Borazon

NOTE: For recommended tool geometry, see Tables 55 to 61; for recommended cutting fluids, see Table 62.

(a) Caution: Check horsepower requirements on heavier depths of cut. (b) Any premium HSS (T15, M33, M41-M47).

Table 38. Face milling, nickel-base and cobalt-base high-temperature alloys

Material	Hardness, Bhn	Condition	Depth of cut, in.(a)	High speed steel tool Speed, fpm	High speed steel tool Feed per tooth, in.	High speed steel tool Tool material, AISI	Carbide tool (uncoated) Speed, fpm Brazed	Carbide tool (uncoated) Speed, fpm Indexable	Carbide tool (uncoated) Feed per tooth, in.	Carbide tool (uncoated) Tool material grade
Nickel-base alloys, wrought .... Haynes alloy 263; Incoloy 901, 903; Inconel 617, 625, 702, 706, 718, 721, 722, X-750, 751; M252; Nimonic 75, 80; Waspaloy	200 to 300	Annealed or solution treated	.040 .150 .300	25 20 —	.004 .006 —	T15, M42(b) T15, M42(b) —	— — —	— — —	— — —	— — —
	300 to 400	Solution treated and aged	.040 .150 .300	20 15 —	.003 .005 —	T15, M42(b) T15, M42(b) —	— — —	— — —	— — —	— — —
Nickel-base alloys, cast .... B-1900; GMR-235, -235D; Hastelloy alloy D; IN-100 (René 100); IN-738, -792; Inconel 713C, 718; M252; MAR-M200, -M246, -M421, -M432; René 80, 125; SEL; SEL 15; TRW VI A; Udimet 500, 700	250 to 320	As cast or cast and aged	.040 .150 .300	20 12 9	.002 .002 .003	T15, M42(b) T15, M42(b) T15, M42(b)	— — —	— — —	— — —	— — —
	320 to 425	As cast or cast and aged	.040 .150 .300	15 10 7	.002 .002 .003	T15, M42(b) T15, M42(b) T15, M42(b)	— — —	— — —	— — —	— — —
Cobalt alloys, wrought .... AiResist 213; Haynes alloy 25 (L605); Haynes alloy 188; J-1570; MAR-M905, -M918; S-816; V-36	180 to 230	Solution treated	.040 .150 .300	30 25 20	.002 .003 .004	T15, M42(b) T15, M42(b) T15, M42(b)	65 60 —	70 65 —	.005 .005 —	C-2 C-2 —
	270 to 320	Solution treated and aged	.040 .150 .300	15 10 7	.002 .003 .004	T15, M42(b) T15, M42(b) T15, M42(b)	50 45 —	60 55 —	.005 .006 —	C-2 C-2 —
Cobalt alloys, cast .... AiResist 13, 215; FSX-414; HS-6, -21, -31 (X-40); HOWMET #3; MAR-M302, -M322, -M509; NASA Co-W-Re; WI-52; X-45	220 to 290	As cast or cast and aged	.040 .150 .300	15 10 7	.002 .003 .004	T15, M42(b) T15, M42(b) T15, M42(b)	45 40 —	50 45 —	.005 .006 —	C-2 C-2 —
	290 to 425	As cast or cast and aged	.040 .150 .300	12 8 5	.002 .002 .003	T15, M42(b) T15, M42(b) T15, M42(b)	30 20 —	35 25 —	.005 .006 —	C-2 C-2 —

NOTE: For recommended tool geometry, see Tables 55 to 61; for recommended cutting fluids, see Table 62.

(a) Depth of cut is measured parallel to axis of cutter. (b) Any premium HSS (T15, M33, M41-M47).

Table 39. End milling (peripheral), nickel-base and cobalt-base high-temperature alloys

Material	Hardness, Bhn	Condition	Radial depth of cut, in.(a)	HSS Speed, fpm	HSS Feed, in./tooth 3/8	HSS 1/2	HSS 3/4	HSS 1-2	Tool material, AISI	Carbide Speed, fpm	Carbide Feed, in./tooth 3/8	Carbide 1/2	Carbide 3/4	Carbide 1-2	Tool material grade
Nickel-base alloys, wrought Haynes alloy 263; Incoloy 901, 903; Inconel 617, 625, 702, 706, 718, 721, 722, X-750, 751; M252; Nimonic 75, 80; Waspaloy	200 to 300	Annealed or solution treated	.020 .060 dia/4 dia/2	20 15 12 10	.001 .002 .0015 .001	.001 .002 .0015 .001	.002 .003 .0025 .002	.002 .004 .003 .002	T15(b)	80 60 50 45	.001 .0015 .001 —	.001 .002 .0015 —	.002 .003 .002 .0015	.002 .004 .0025 .002	C-2
	300 to 400	Solution treated and aged	.020 .060 dia/4 dia/2	15 12 10 8	.001 .0015 .001 .0007	.001 .0015 .001 .0007	.0015 .002 .0015 .001	.002 .003 .002 .0015	T15(b)	60 45 40 35	.001 .0015 .001 —	.001 .0015 .0015 —	.0015 .002 .002 .001	.002 .003 .003 .0015	C-2
Nickel-base alloys, cast .... B-1900; GMR-235, -235D; Hastelloy alloy D; IN-100 (René 100); IN-738, -792; Inconel 713C, 718; M252; MAR-M200, -M246, -M421, -M432; René 80, 125; SEL; SEL 15; TRW VI A; Udimet 500, 700	250 to 320	As cast or cast and aged	.020 .060 dia/4 dia/2	20 15 12 10	.001 .002 .001 .0007	.001 .002 .001 .0007	.002 .003 .002 .0015	.002 .004 .003 .002	T15(b)	75 55 50 45	.001 .001 .001 —	.001 .002 .001 —	.002 .003 .0015 .001	.002 .004 .002 .0015	C-2
	320 to 425	As cast or cast and aged	.020 .060 dia/4 dia/2	15 10 8 5	.001 .0015 .001 .0005	.001 .0015 .001 .0007	.0015 .002 .0015 .001	.002 .003 .0025 .002	T15(b)	45 35 30 25	.001 .0015 .001 —	.001 .0015 .001 —	.0015 .002 .001 .001	.002 .003 .0015 .001	C-2
Cobalt alloys, wrought ..... AiResist 213; Haynes alloy 25 (L605); Haynes alloy 188; J-1570; MAR-M905, -M918; S-816; V-36	180 to 230	Solution treated	.020 .060 dia/4 dia/2	15 12 10 8	.001 .002 .001 .0007	.001 .002 .001 .0007	.002 .003 .002 .001	.002 .004 .003 .002	T15(b)	70 50 45 40	.001 .001 .001 —	.001 .002 .001 —	.002 .003 .0015 .001	.002 .004 .002 .0015	C-2
	270 to 320	Solution treated and aged	.020 .060 dia/4 dia/2	12 10 8 5	.001 .0015 .001 .0007	.001 .0015 .001 .0007	.0015 .002 .0015 .001	.002 .003 .002 .0015	T15(b)	60 45 40 35	.001 .0015 .001 —	.001 .0015 .001 —	.0015 .002 .0015 .001	.002 .003 .0015 .001	C-2
Cobalt alloys, cast .... AiResist 13, 215; FSX-414; HS-6, -21, -31 (X-40); HOWMET #3; MAR-M302, -M322, -M509; NASA Co-W-Re; WI-52; X-45	220 to 290	As cast or cast and aged	.020 .060 dia/4 dia/2	12 10 8 5	.001 .002 .001 .0007	.001 .002 .001 .0007	.002 .0025 .002 .001	.002 .003 .002 .0015	T15(b)	60 45 40 35	.001 .001 .001 —	.001 .002 .001 —	.002 .003 .0015 .001	.002 .004 .002 .0015	C-2
	290 to 425	As cast or cast and aged	.020 .060 dia/4 dia/2	10 8 7 5	.001 .0015 .001 .0007	.001 .0015 .001 .0007	.0015 .002 .0015 .001	.002 .003 .002 .0015	T15(b)	40 30 25 20	.001 .0015 .001 —	.001 .0015 .001 —	.0015 .002 .0015 .001	.002 .003 .0015 .001	C-2

NOTE: For recommended tool geometry, see Tables 55 to 61; for recommended cutting fluids, see Table 62.

(a) For standard length end mills, maximum axial depth can be up to 1.5 × the cutter diameter. (b) Any premium HSS (T15, M33, M41-M47).

Table 40. Drilling, nickel-base and cobalt-base high-temperature alloys

Material	Hard-ness, Bhn	Condition	Speed, fpm(a)	Feed, ipr(a) — Nominal hole diameter, in.								Tool material grade, AISI or C
				1/16	1/8	1/4	1/2	3/4	1	1 1/2	2	
Nickel-base alloys, wrought .... Haynes alloy 263; Incoloy 901, 903; Inconel 617, 625, 702, 706, 718, 721, 722, X-750, 751; M252; Nimonic 75, 80; Waspaloy	200 to 300	Annealed or solution treated	20	—	.002	.003	.003	.004	—	—	—	T15, M42(b)
	300 to 400	Solution treated and aged	15	—	.002	.003	.003	.004	—	—	—	T15, M42(b)
Nickel-base alloys, cast .... B-1900; GMR-235, -235D; Hastelloy alloy D; IN-100 (René 100); IN-738, -792; Inconel 713C, 718; M252; MAR-M200, -M246, -M421, -M432; René 80, 125; SEL; SEL 15; TRW VI A; Udimet 500, 700	250 to 320	As cast or cast and aged	7	—	.002	.003	.003	.004	—	—	—	T15, M42(b)
	320 to 425	As cast or cast and aged	15	—	.001	.002	.003	.004	—	—	—	C-2
Cobalt alloys, wrought .... AiResist 213; Haynes alloy 25 (L605); Haynes alloy 188; J-1570; MAR-M905, -M918; S-816; V-36	180 to 230	Solution treated	20	—	.002	.003	.003	.004	—	—	—	T15, M42(b)
	270 to 320	Solution treated and aged	15	—	.002	.003	.003	.004	—	—	—	T15, M42(b)
Cobalt alloys, cast .... AiResist 13, 215; FSX-414; HS-6, -21, -31 (X-40); HOWMET #3; MAR-M302, -M322, -M509; NASA Co-W-Re; WI-52; X-45	220 to 290	As cast or cast and aged	7	—	.002	.003	.003	.004	.006	—	—	T15, M42(b)
	290 to 425	As cast or cast and aged	15	—	.001	.002	.003	.004	—	—	—	C-2

NOTE: For recommended tool geometry, see Tables 55 to 61; for recommended cutting fluids, see Table 62.

(a) For drilling deep holes with twist drills, reduce speed and feed as follows: If hole depth is 3 drill diameters, reduce speed by 10%, feed by 10%; 4 diameters, speed by 20%, feed by 10%; 5 diameters, speed by 30%, feed by 20%; 6 diameters, speed by 35%, feed by 20%; 8 diameters, speed by 40%, feed by 20%. (b) Any premium HSS (T15, M33, M41-M47).

Table 41. Reaming, nickel-base and cobalt-base high-temperature alloys

Material	Hard-ness, Bhn	Condition	Roughing Speed, fpm	Feed, ipr(a) — Reamer diameter, in.						Tool material grade, AISI or C	Finishing Speed, fpm	Feed, ipr(a) — Reamer diameter, in.						Tool material grade, AISI or C
				1/8	1/4	1/2	1	1 1/2	2			1/8	1/4	1/2	1	1 1/2	2	
Nickel-base alloys, wrought ... Haynes alloy 263; Incoloy 901, 903; Inconel 617, 625, 702, 706, 718, 721, 722, X-750, 751; M252; Nimonic 75, 80; Waspaloy	200 to 300	Annealed or solution treated	20	.002	.006	.008	.010	.012	.015	T15, M42(b)	15	.003	.004	.006	.008	.010	.012	T15, M42(b)
			40	.002	.006	.008	.010	.012	.015	C-2	25	.003	.004	.006	.008	.010	.012	C-2
	300 to 400	Solution treated and aged	15	.002	.006	.008	.010	.012	.015	T15, M42(b)	12	.003	.004	.006	.008	.010	.012	T15, M42(b)
			30	.002	.006	.008	.010	.012	.015	C-2	15	.003	.004	.006	.008	.010	.012	C-2
Nickel-base alloys, cast ...... B-1900; GMR-235, -235D; Hastelloy alloy D; IN-100 (René 100); IN-738, -792; Inconel 713C, 718; M252; MAR-M200, -M246, -M421, -M432; René 80, 125; SEL; SEL 15; TRW VI A; Udimet 500, 700	250 to 320	As cast or cast and aged	— 30	— .002	— .004	— .006	— .008	— .010	— .012	C-2	— 10	— .002	— .004	— .006	— .008	— .010	— .012	C-2
	320 to 425	As cast or cast and aged	— 10	— .002	— .004	— .006	— .008	— .010	— .012	C-2	— 5	— .002	— .004	— .006	— .008	— .010	— .012	C-2
Cobalt alloys, wrought .... AiResist 213; Haynes alloy 25 (L605); Haynes alloy 188; J-1570; MAR-M905, -M918; S-816; V-36	180 to 230	Solution treated	15	.002	.006	.008	.010	.012	.015	T15, M42(b)	12	.003	.004	.006	.008	.010	.012	T15, M42(b)
			30	.002	.006	.008	.010	.012	.015	C-2	20	.003	.004	.006	.008	.010	.012	C-2
	270 to 320	Solution treated and aged	12	.002	.006	.008	.010	.012	.015	T15, M42(b)	10	.003	.004	.006	.008	.010	.012	T15, M42(b)
			25	.002	.006	.008	.010	.012	.015	C-2	15	.003	.004	.006	.008	.010	.012	C-2
Cobalt alloys, cast .... AiResist 13, 215; FSX-414; HS-6, -21, -31 (X-40); HOWMET #3; MAR-M302, -M322, -M509; NASA Co-W-Re; WI-52; X-45	220 to 290	As cast or cast and aged	— 30	— .002	— .004	— .006	— .008	— .010	— .012	C-2	— 10	— .003	— .004	— .006	— .008	— .010	— .012	C-2
	290 to 425	As cast or cast and aged	— 10	— .002	— .004	— .006	— .008	— .010	— .012	C-2	— 5	— .002	— .004	— .006	— .008	— .010	— .012	C-2

NOTE: For recommended tool geometry, see Tables 55 to 61; for recommended cutting fluids, see Table 62.

(a) Based on 4 flutes for 1/8- and 1/4-inch reamers, 6 flutes for 1/2-inch reamers, and 8 flutes for 1-inch and larger reamers. (b) Any premium HSS (T15, M33, M41-M47).

Table 42. Tapping, nickel-base and cobalt-base high-temperature alloys

Material	Hardness, Bhn	Condition	Speed, fpm(a) Threads per inch				HSS tool material, AISI
			7 or less	8 to 15	16 to 24	Over 24	
Nickel-base alloys, wrought .... Haynes alloy 263; Incoloy 901, 903; Inconel 617, 625, 702, 706, 718, 721, 722, X-750, 751; M252; Nimonic 75, 80; Waspaloy	200 to 300	Annealed or solution treated	3	7	9	10	Nitrided M7, M41-M44
	300 to 400	Solution treated and aged	2	4	6	7	Nitrided M7, M41-M44
Nickel-base alloys, cast .... B-1900; GMR-235, -235D; Hastelloy alloy D; IN-100 (René 100); IN-738, -792; Inconel 713C, 718; M252; MAR-M200, -M246, -M421, -M432; René 80, 125; SEL; SEL 15; TRW VI A; Udimet 500, 700	250 to 320	As cast or cast and aged	2	3	4	5	Nitrided M7, M41-M44
Cobalt alloys, wrought .... AiResist 213; Haynes alloy 25 (L605); Haynes alloy 188; J-1570; MAR-M905, -M918; S-816; V-36	180 to 230	Solution treated	3	5	7	8	Nitrided M7, M41-M44
	270 to 320	Solution treated and aged	2	3	4	5	Nitrided M7, M41-M44
Cobalt alloys, cast .... AiResist 13, 215; FSX-414; HS-6, -21, -31 (X-40); HOWMET #3; MAR-M302, -M322, -M509; NASA Co-W-Re; WI-52; X-45	220 to 290	As cast or cast and aged	2	3	4	5	Nitrided M7, M41-M44

NOTE: For recommended tool geometry, see Tables 55 to 61; for recommended cutting fluids, see Table 62.

(a) These speeds are for tapping 65% to 75% threads in shallow through-holes. Reduce the speed when tapping deep holes, blind holes or higher percentage of thread.

Table 43. Machining recommendations for wrought maraging steels (275-325 HB, annealed)

Machining operation	Tool material High speed steel	Carbide	Cutting speed, fpm	Feed ipr	ipt	Cutting fluid(a)
Turning						
Roughing	M42	—	60	.015	—	2, 5, 7
	—	C-6	325	.020	—	0, 4, 6
Finishing	M42	—	100	.005	—	2, 5, 7
	—	C-7	525	.007	—	0, 4, 6
Face milling						
Roughing	—	C-5	235	—	.010	0, 4, 6
Finishing	—	C-6	325	—	.006	0, 4, 6
End milling (peripheral)						
Cutter diam 1/2 in.:						
Roughing	M2, M3, M7	—	95	—	.001	2, 5, 7
Finishing	M2, M3, M7	—	150	—	.001	2, 5, 7
Cutter diam 1 in.:						
Roughing	M2, M3, M7	—	95	—	.003	2, 5, 7
Finishing	M2, M3, M7	—	150	—	.003	2, 5, 7
Drilling						
Nominal drill diameter:						
1/8 in.	M10, M7, M1	—	60	.003	—	2, 5, 7
1/4 in.	M10, M7, M1	—	60	.005	—	2, 5, 7
1/2 in.	M10, M7, M1	—	60	.008	—	2, 5, 7
1 in.	M10, M7, M1	—	60	.012	—	2, 5, 7
2 in.	M10, M7, M1	—	60	.016	—	2, 5, 7
Reaming — finishing						
Reamer diameter:						
1/8 in.	—	C-2	35	.004	—	1, 4, 6
1/4 in.	—	C-2	35	.006	—	1, 4, 6
1/2 in.	—	C-2	35	.008	—	1, 4, 6
1 in.	—	C-2	35	.012	—	1, 4, 6
2 in.	—	C-2	35	.020	—	1, 4, 6
Tapping						
Threads per inch:						
7 or less	M10, M7, M1	—	10	N/A	N/A	3, 5, 7
8 through 15	M10, M7, M1	—	25	N/A	N/A	3, 5, 7
16 through 24	M10, M7, M1	—	30	N/A	N/A	3, 5, 7
Over 24	M10, M7, M1	—	35	N/A	N/A	3, 5, 7

NOTE: For recommended tool geometry, see Tables 55 to 61; for recommended cutting fluids, see Table 62.

(a) See Table 63 for identification of numbered cutting fluids.

Table 44. Machining recommendations for wrought or cast tool steels (150-200 HB, annealed)

Machining operation	Tool material High speed steel	Carbide	Cutting speed, fpm	Feed ipr	ipt	Cutting fluid(a)
Turning						
Roughing	M2, M3	—	70	.020	—	1, 4, 6
	—	C-6	260	.020	—	0, 4, 6
Finishing	M2, M3	—	110	.007	—	1, 4, 6
	—	C-7	425	.007	—	0, 4, 6

(continued)

Table 44. (continued)

Machining operation	Tool material — High speed steel	Carbide	Cutting speed, fpm	Feed — ipr	ipt	Cutting fluid(a)
Face milling						
Roughing	—	C-5	275	—	.016	0, 4, 6
Finishing	—	C-6	450	—	.008	0, 4, 6
End milling (peripheral)						
Cutter diam 1/2 in.:						
Roughing	M2, M3, M7	—	65	—	.001	2, 5, 7
Finishing	M2, M3, M7	—	100	—	.001	2, 5, 7
Cutter diam 1 in.:						
Roughing	M2, M3, M7	—	65	—	.003	2, 5, 7
Finishing	M2, M3, M7	—	100	—	.003	2, 5, 7
Drilling						
Nominal drill diameter:						
1/8 in.	M10, M7, M1	—	60	.002	—	2, 5, 7
1/4 in.	M10, M7, M1	—	60	.003	—	2, 5, 7
1/2 in.	M10, M7, M1	—	60	.006	—	2, 5, 7
1 in.	M10, M7, M1	—	60	.011	—	2, 5, 7
2 in.	M10, M7, M1	—	60	.016	—	2, 5, 7
Reaming — finishing						
Reamer diameter:						
1/8 in.	—	C-2	45	.005	—	2, 4, 6
1/4 in.	—	C-2	45	.008	—	2, 4, 6
1/2 in.	—	C-2	45	.012	—	2, 4, 6
1 in.	—	C-2	45	.020	—	2, 4, 6
2 in.	—	C-2	45	.030	—	2, 4, 6
Tapping						
Threads per inch:						
7 or less	M10, M7, M1	—	8	N/A	N/A	3
8 through 15	M10, M7, M1	—	18	N/A	N/A	3
16 through 24	M10, M7, M1	—	25	N/A	N/A	3
Over 24	M10, M7, M1	—	30	N/A	N/A	3

NOTE: For recommended tool geometry, see Tables 55 to 61; for recommended cutting fluids, see Table 62.

(a) See Table 63 for identification of numbered cutting fluids.

Table 45. Machining recommendations for wrought or cast tool steels (225-275 HB, annealed)

Machining operation	Tool material — High speed steel	Carbide	Cutting speed, fpm	Feed — ipr	ipt	Cutting fluid(a)
Turning						
Roughing	M2, M3	—	45	.015	—	2, 5, 7
	—	C-6	225	.020	—	0, 4, 6
Finishing	M2, M3	—	75	.007	—	2, 5, 7
	—	C-7	370	.007	—	0, 4, 6
Face milling						
Roughing	—	C-5	240	—	.010	0, 4, 6
Finishing	—	C-6	320	—	.007	0, 4, 6
End milling (peripheral)						
Cutter diam 1/2 in.:						
Roughing	M42	—	55	—	.002	2, 5, 7
Finishing	M42	—	75	—	.001	2, 5, 7
Cutter diam 1 in.:						
Roughing	M42	—	55	—	.004	2, 5, 7
Finishing	M42	—	75	—	.003	2, 5, 7
Drilling						
Nominal drill diameter:						
1/8 in.	M10, M7, M1	—	35	.002	—	2, 5, 7
1/4 in.	M10, M7, M1	—	35	.003	—	2, 5, 7
1/2 in.	M10, M7, M1	—	35	.005	—	2, 5, 7
1 in.	M10, M7, M1	—	35	.008	—	2, 5, 7
2 in.	M10, M7, M1	—	35	.013	—	2, 5, 7
Reaming — finishing						
Reamer diameter:						
1/8 in.	—	C-2	50	.004	—	2, 4, 6
1/4 in.	—	C-2	50	.006	—	2, 4, 6
1/2 in.	—	C-2	50	.008	—	2, 4, 6
1 in.	—	C-2	50	.012	—	2, 4, 6
2 in.	—	C-2	50	.020	—	2, 4, 6
Tapping						
Threads per inch:						
7 or less	M10, M7, M1	—	7	N/A	N/A	3
8 through 15	M10, M7, M1	—	15	N/A	N/A	3
16 through 24	M10, M7, M1	—	18	N/A	N/A	3
Over 24	M10, M7, M1	—	20	N/A	N/A	3

NOTE: For recommended tool geometry, see Tables 55 to 61; for recommended cutting fluids, see Table 62.

(a) See Table 63 for identification of numbered cutting fluids.

Table 46. Machining recommendations for wrought structural steels (200-250 HB, hot rolled, normalized, stress relieved, or quenched and tempered)

Machining operation	Tool material High speed steel	Carbide	Cutting speed, fpm	Feed ipr	ipt	Cutting fluid(a)
Turning						
Roughing	M2, M3	—	60	.020	—	1, 4, 6
	—	C-6	260	.030	—	0, 4, 6
Finishing	M2, M3	—	105	.007	—	1, 4, 6
	—	C-7	420	.007	—	0, 4, 6
Face milling						
Roughing	—	C-5	240	—	.014	0, 4, 6
Finishing	—	C-6	410	—	.007	0, 4, 6
End milling (peripheral)						
Cutter diam 1/2 in.:						
Roughing	M2, M3, M7	—	60	—	.002	1, 4, 6
Finishing	M2, M3, M7	—	95	—	.002	1, 4, 6
Cutter diam 1 in.:						
Roughing	M2, M3, M7	—	60	—	.004	1, 4, 6
Finishing	M2, M3, M7	—	95	—	.004	1, 4, 6
Drilling						
Nominal drill diameter:						
1/8 in.	M10, M7, M1	—	45	.003	—	1, 4, 6
1/4 in.	M10, M7, M1	—	45	.004	—	1, 4, 6
1/2 in.	M10, M7, M1	—	45	.007	—	1, 4, 6
1 in.	M10, M7, M1	—	45	.012	—	1, 4, 6
2 in.	M10, M7, M1	—	45	.018	—	1, 4, 6
Reaming—finishing						
Reamer diameter:						
1/8 in.	—	C-2	40	.004	—	1, 4, 6
1/4 in.	—	C-2	40	.008	—	1, 4, 6
1/2 in.	—	C-2	40	.012	—	1, 4, 6
1 in.	—	C-2	40	.020	—	1, 4, 6
2 in.	—	C-2	40	.030	—	1, 4, 6
Tapping						
Threads per inch:						
7 or less	M10, M7, M1	—	8	N/A	N/A	2, 5, 7
8 through 15	M10, M7, M1	—	18	N/A	N/A	2, 5, 7
16 through 24	M10, M7, M1	—	25	N/A	N/A	2, 5, 7
Over 24	M10, M7, M1	—	30	N/A	N/A	2, 5, 7

NOTE: For recommended tool geometry, see Tables 55 to 61; for recommended cutting fluids, see Table 62.

(a) See Table 63 for identification of numbered cutting fluids.

Table 47. Machining recommendations for cast austenitic manganese steels (150-220 HB, annealed)(a)

Machining operation	Tool material High speed steel	Carbide	Cutting speed, fpm	Feed ipr	ipt	Cutting fluid(b)
Turning						
Roughing	—	—	—	—	—	—
	—	C-7	85	.015	—	3, 5
Finishing	—	—	—	—	—	—
	—	C-7	125	.010	—	3, 5
Face milling						
Roughing	—	C-6	90	—	.010	3, 5
Finishing	—	C-6	135	—	.007	3, 5
Drilling(c)						
Nominal drill diameter:						
1/4 in.	M42	(c)	20(c)	.002(c)	—	3, 5
1/2 in.	M42	(c)	20(c)	.003(c)	—	3, 5
1 in.	M42	(c)	20(c)	.004(c)	—	3, 5

NOTE: For recommended tool geometry, see Tables 55 to 61; for recommended cutting fluids, see Table 62.

(a) Austenitic manganese steels machine with difficulty; cast to finished size when possible. (b) See Table 63 for identification of numbered cutting fluids. (c) Carbide drilling is suggested; increase speed to 70 fpm and reduce feed by .001 ipr at each diameter.

Table 48. Machining recommendations for wrought or cast magnesium alloys
(50-90 HB, 500-kg load, annealed, cold drawn, or solution treated and aged)

Machining operation	Tool material High speed steel	Carbide	Cutting speed, fpm	Feed ipr	ipt	Cutting fluid(a)
Turning						
Roughing	M2, M3	—	600	.030	—	1, 88
	—	C-2	1500	.040	—	1, 88
Finishing	M2, M3	—	1000	.007	—	1, 88
	—	C-3	2500	.010	—	1, 88
Face milling						
Roughing	—	C-2	1500	—	.020	1, 88
Finishing	—	C-2	2500	—	.010	1, 88

(continued)

Table 48. (continued)

Machining operation	Tool material — High speed steel	Carbide	Cutting speed, fpm	Feed — ipr	ipt	Cutting fluid(a)
End milling (peripheral)						
Cutter diam ¹/₂ in.:						
Roughing	M2, M3, M7	—	600	—	.004	1, 88
Finishing	M2, M3, M7	—	1000	—	.004	1, 88
Cutter diam 1 in.:						
Roughing	M2, M3, M7	—	600	—	.009	1, 88
Finishing	M2, M3, M7	—	1000	—	.008	1, 88
Drilling						
Nominal drill diameter:						
¹/₈ in.	M10, M7, M1	—	330	.003	—	1, 88
¹/₄ in.	M10, M7, M1	—	330	.007	—	1, 88
¹/₂ in.	M10, M7, M1	—	330	.012	—	1, 88
1 in.	M10, M7, M1	—	330	.019	—	1, 88
2 in.	M10, M7, M1	—	330	.030	—	1, 88
Reaming — finishing						
Reamer diameter:						
¹/₈ in.	—	C-2	225	.005	—	1, 88
¹/₄ in.	—	C-2	225	.007	—	1, 88
¹/₂ in.	—	C-2	225	.010	—	1, 88
1 in.	—	C-2	225	.015	—	1, 88
2 in.	—	C-2	225	.025	—	1, 88
Tapping						
Threads per inch:						
7 or less	M10, M7, M1	—	65	N/A	N/A	1, 88
8 through 15	M10, M7, M1	—	110	N/A	N/A	1, 88
16 through 24	M10, M7, M1	—	140	N/A	N/A	1, 88
Over 24	M10, M7, M1	—	150	N/A	N/A	1, 88

NOTE: For recommended tool geometry, see Tables 55 to 61; for recommended cutting fluids, see Table 62.

(a) See Table 63 for identification of numbered cutting fluids. Cutting fluid 88 is a special cutting fluid for magnesium. Consult supplier or manufacturer to be sure fluids do not contain water or other chemicals that may support combustion of magnesium.

Table 49. Machining recommendations for wrought or cast copper alloys (10-70 HB, cold drawn or as-cast)

Machining operation	Tool material — High speed steel	Carbide	Cutting speed, fpm	Feed — ipr	ipt	Cutting fluid(a)
Turning						
Roughing	M2, M3	—	200	.020	—	4, 6, 87
	—	C-2	400	.030	—	4, 6
Finishing	M2, M3	—	300	.007	—	4, 6, 87
	—	C-3	500	.007	—	4, 6
Face milling						
Roughing	—	C-2	500	—	.018	4, 6
Finishing	—	C-2	750	—	.008	4, 6
End milling (peripheral)						
Cutter diam ¹/₂ in.:						
Roughing	M2, M3, M7	—	200	—	.003	4, 6, 87
Finishing	M2, M3, M7	—	325	—	.003	4, 6, 87
Cutter diam 1 in.:						
Roughing	M2, M3, M7	—	200	—	.005	4, 6, 87
Finishing	M2, M3, M7	—	325	—	.006	4, 6, 87
Drilling						
Nominal drill diameter:						
¹/₈ in.	M10, M7, M1	—	125	.003	—	4, 6, 87
¹/₄ in.	M10, M7, M1	—	125	.006	—	4, 6, 87
¹/₂ in.	M10, M7, M1	—	125	.010	—	4, 6, 87
1 in.	M10, M7, M1	—	125	.016	—	4, 6, 87
2 in.	M10, M7, M1	—	125	.025	—	4, 6, 87
Reaming — finishing						
Reamer diameter:						
¹/₈ in.	—	C-2	75	.005	—	1, 87
¹/₄ in.	—	C-2	75	.007	—	1, 87
¹/₂ in.	—	C-2	75	.012	—	1, 87
1 in.	—	C-2	75	.015	—	1, 87
2 in.	—	C-2	75	.025	—	1, 87
Tapping						
Threads per inch:						
7 or less	M10, M7, M1	—	20	N/A	N/A	1, 87
8 through 15	M10, M7, M1	—	40	N/A	N/A	1, 87
16 through 24	M10, M7, M1	—	55	N/A	N/A	1, 87
Over 24	M10, M7, M1	—	60	N/A	N/A	1, 87

NOTE: For recommended tool geometry, see Tables 55 to 61; for recommended cutting fluids, see Table 62.

(a) See Table 63 for identification of numbered cutting fluids. Cutting fluid 87 is a special cutting fluid for copper; contact suppliers.

Table 50. Machining recommendations for wrought or cast nickel alloys, 99% Ni, Monels (100-200 HB, annealed, cold drawn, or as-cast)

Machining operation	Tool material — High speed steel	Carbide	Cutting speed, fpm	Feed — ipr	ipt	Cutting fluid(a)
Turning						
Roughing	M42	—	55	.030	—	3, 5, 7
	—	C-6	165	.020	—	0, 5, 7
Finishing	M42	—	100	.007	—	3, 5, 7
	—	C-7	340	.007	—	0, 5, 7
Face milling						
Roughing	—	C-6	100	—	.010	0, 5, 7
Finishing	—	C-6	175	—	.006	0, 5, 7
End milling (peripheral)						
Cutter diam 1/2 in.:						
Roughing	M2, M3, M7	—	50	—	.002	3, 5, 7
Finishing	M2, M3, M7	—	80	—	.002	3, 5, 7
Cutter diam 1 in.:						
Roughing	M2, M3, M7	—	50	—	.004	3, 5, 7
Finishing	M2, M3, M7	—	80	—	.004	3, 5, 7
Drilling						
Nominal drill diameter:						
1/8 in.	M10, M7, M1	—	50	.003	—	3, 5, 7
1/4 in.	M10, M7, M1	—	50	.005	—	3, 5, 7
1/2 in.	M10, M7, M1	—	50	.009	—	3, 5, 7
1 in.	M10, M7, M1	—	50	.014	—	3, 5, 7
2 in.	M10, M7, M1	—	50	.021	—	3, 5, 7
Reaming — finishing						
Reamer diameter:						
1/8 in.	—	C-2	45	.006	—	5, 7, 2
1/4 in.	—	C-2	45	.008	—	5, 7, 2
1/2 in.	—	C-2	45	.010	—	5, 7, 2
1 in.	—	C-2	45	.015	—	5, 7, 2
2 in.	—	C-2	45	.020	—	5, 7, 2
Tapping						
Threads per inch:						
7 or less	M10, M7, M1	—	7	N/A	N/A	2, 3
8 through 15	M10, M7, M1	—	15	N/A	N/A	2, 3
16 through 24	M10, M7, M1	—	18	N/A	N/A	2, 3
Over 24	M10, M7, M1	—	20	N/A	N/A	2, 3

NOTE: For recommended tool geometry, see Tables 55 to 61; for recommended cutting fluids, see Table 62.

(a) See Table 63 for identification of numbered cutting fluids.

Table 51. Machining recommendations for cast lead, tin and zinc alloys (5-30 HB, 500-kg load, as-cast)

Machining operation	Tool material — High speed steel	Carbide	Cutting speed, fpm	Feed — ipr	ipt	Cutting fluid(a)
Turning						
Roughing	M2, M3	—	250	.015	—	0, 4
	—	C-2	500	.015	—	0
Finishing	M2, M3	—	500	.007	—	0, 4
	—	C-2	1000	.007	—	0
Face milling						
Roughing	—	C-2	250	—	.012	0
Finishing	—	C-2	500	—	.006	0
End milling (peripheral)						
Cutter diam 1/2 in.:						
Roughing	M2, M3, M7	—	350	—	.004	0, 1, 4
Finishing	M2, M3, M7	—	450	—	.005	0, 1, 4
Cutter diam 1 in.:						
Roughing	M2, M3, M7	—	350	—	.007	0, 1, 4
Finishing	M2, M3, M7	—	450	—	.007	0, 1, 4
Drilling						
Nominal drill diameter:						
1/8 in.	M10, M7, M1	—	300	.004	—	0, 1, 4
1/4 in.	M10, M7, M1	—	300	.007	—	0, 1, 4
1/2 in.	M10, M7, M1	—	300	.012	—	0, 1, 4
1 in.	M10, M7, M1	—	300	.017	—	0, 1, 4
2 in.	M10, M7, M1	—	300	.025	—	0, 1, 4
Reaming — finishing						
Reamer diameter:						
1/8 in.	—	C-2	150	.005	—	0, 4
1/4 in.	—	C-2	150	.010	—	0, 4
1/2 in.	—	C-2	150	.015	—	0, 4
1 in.	—	C-2	150	.020	—	0, 4
2 in.	—	C-2	150	.030	—	0, 4

(continued)

Table 51. (continued)

Machining operation	Tool material — High speed steel	Carbide	Cutting speed, fpm	Feed — ipr	ipt	Cutting fluid(a)
Tapping						
Threads per inch:						
7 or less	M10, M7, M1	—	45	N/A	N/A	1, 4
8 through 15	M10, M7, M1	—	75	N/A	N/A	1, 4
16 through 24	M10, M7, M1	—	95	N/A	N/A	1, 4
Over 24	M10, M7, M1	—	100	N/A	N/A	1, 4

NOTE: For recommended tool geometry, see Tables 55 to 61; for recommended cutting fluids, see Table 62.

(a) See Table 63 for identification of numbered cutting fluids.

Table 52. Machining recommendations for wrought zirconium alloys (140-280 HB, rolled, extruded or forged)(a)

Machining operation	Tool material — High speed steel	Carbide	Cutting speed, fpm	Feed — ipr	ipt	Cutting fluid(b)
Turning						
Roughing	M2, M3	—	80	.020	—	4, 6
	—	C-2	200	.020	—	4, 6
Finishing	M2, M3	—	150	.007	—	4, 6
	—	C-2	325	.007	—	4, 6
Face milling						
Roughing	—	C-2	225	—	.014	4, 6
Finishing	—	C-2	350	—	.006	4, 6
End milling (peripheral)						
Cutter diam 1/2 in.:						
Roughing	M42	—	100	—	.002	5, 7
Finishing	M42	—	150	—	.002	5, 7
Cutter diam 1 in.:						
Roughing	M42	—	100	—	.004	5, 7
Finishing	M42	—	150	—	.004	5, 7
Drilling						
Nominal drill diameter:						
1/8 in.	M10, M7, M1	—	55	.003	—	5, 7
1/4 in.	M10, M7, M1	—	55	.004	—	5, 7
1/2 in.	M10, M7, M1	—	55	.006	—	5, 7
1 in.	M10, M7, M1	—	55	.010	—	5, 7
2 in.	M10, M7, M1	—	55	.015	—	5, 7
Reaming — roughing(c)						
Reamer diameter:						
1/8 in.	—	C-2	175	.004	—	5, 7
1/4 in.	—	C-2	175	.007	—	5, 7
1/2 in.	—	C-2	175	.010	—	5, 7
1 in.	—	C-2	175	.016	—	5, 7
2 in.	—	C-2	175	.025	—	5, 7
Tapping						
Threads per inch:						
7 or less	M10, M7, M1	—	8	N/A	N/A	5, 7
8 through 15	M10, M7, M1	—	18	N/A	N/A	5, 7
16 through 24	M10, M7, M1	—	25	N/A	N/A	5, 7
Over 24	M10, M7, M1	—	30	N/A	N/A	5, 7

NOTE: For recommended tool geometry, see Tables 55 to 61; for recommended cutting fluids, see Table 62.

(a) Chips of zirconium may be pyrophoric. Avoid machining conditions that generate fine chips and high temperatures (e.g., avoid dry cutting). (b) See Table 63 for identification of numbered cutting fluids. (c) These cutting conditions are for roughing in order to avoid producing fine chips.

Table 53. Machining recommendations for machinable carbides — Ferro-Tic (40-50 HRC, annealed)

Machining operation	Tool material — High speed steel	Carbide	Cutting speed, fpm	Feed — ipr	ipt	Cutting fluid
Turning						
Roughing	M42	—	20	.010	—	—
	—	C-2	20	.015	—	—
Finishing	M42	—	25	.005	—	—
	—	C-3	35	.005	—	—
Face milling						
Roughing	—	C-2	30	—	.010	—
Finishing	—	C-2	40	—	.005	—
End milling (peripheral)						
Cutter diam 1/2 in.:						
Roughing	M42	—	25	—	.002	—
Finishing	M42	—	35	—	.002	—
Cutter diam 1 in.:						
Roughing	M42	—	25	—	.004	—
Finishing	M42	—	35	—	.004	—

(continued)

Table 53. (continued)

Machining operation	Tool material High speed steel	Carbide	Cutting speed, fpm	Feed ipr	ipt	Cutting fluid
Drilling						
Nominal drill diameter:						
1/8 in.	M42	—	15	.002	—	—
1/4 in.	M42	—	15	.003	—	—
1/2 in.	M42	—	15	.004	—	—
1 in.	M42	—	15	.006	—	—
2 in.	M42	—	15	.008	—	—
Reaming—finishing						
Reamer diameter:						
1/8 in.	—	C-2	25	.002	—	—
1/4 in.	—	C-2	25	.003	—	—
1/2 in.	—	C-2	25	.004	—	—
1 in.	—	C-2	25	.005	—	—
2 in.	—	C-2	25	.006	—	—
Tapping						
Threads per inch:						
7 or less	M10, M7, M1(a)	—	2	N/A	N/A	—
8 through 15	M10, M7, M1(a)	—	3	N/A	N/A	—
16 through 24	M10, M7, M1(a)	—	4	N/A	N/A	—
Over 24	M10, M7, M1(a)	—	5	N/A	N/A	—

NOTE: For recommended tool geometry, see Tables 55 to 61; for recommended cutting fluids, see Table 62.

(a) Nitrided steels.

Table 54. Machining recommendations for controlled-expansion alloys—Invar(a) and Kovar (125-250 HB, annealed or cold drawn)

Machining operation	Tool material High speed steel	Carbide	Cutting speed, fpm	Feed ipr	ipt	Cutting fluid(b)
Turning						
Roughing	M42	—	30	.003	—	1, 5, 7
Finishing	M42	—	25	.010	—	1, 5, 7
Face milling						
Roughing	M42	—	30	—	.007	1, 5, 7
Finishing	M42	—	25	—	.003	1, 5, 7
End milling (peripheral)						
Cutter diam 1/2 in.:						
Roughing	M2, M3, M7	—	25	—	.002	1, 5, 7
Finishing	M2, M3, M7	—	35	—	.002	1, 5, 7
Cutter diam 1 in.:						
Roughing	M2, M3, M7	—	25	—	.004	1, 5, 7
Finishing	M2, M3, M7	—	35	—	.004	1, 5, 7
Drilling						
Nominal drill diameter:						
1/8 in.	M42	—	30	.003	—	1, 5, 7
1/4 in.	M42	—	30	.004	—	1, 5, 7
1/2 in.	M42	—	30	.008	—	1, 5, 7
1 in.	M42	—	30	.012	—	1, 5, 7
2 in.	M42	—	30	.018	—	1, 5, 7
Reaming—finishing						
Reamer diameter:						
1/8 in.	—	C-2	50	.004	—	2, 5, 7
1/4 in.	—	C-2	50	.006	—	2, 5, 7
1/2 in.	—	C-2	50	.010	—	2, 5, 7
1 in.	—	C-2	50	.015	—	2, 5, 7
2 in.	—	C-2	50	.020	—	2, 5, 7
Tapping						
Threads per inch:						
7 or less	M10, M7, M1	—	10	N/A	N/A	2, 3
8 through 15	M10, M7, M1	—	12	N/A	N/A	2, 3
16 through 24	M10, M7, M1	—	16	N/A	N/A	2, 3
Over 24	M10, M7, M1	—	20	N/A	N/A	2, 3

NOTE: For recommended tool geometry, see Tables 55 to 61; for recommended cutting fluids, see Table 62.

(a) For free-machining controlled-expansion alloy (Invar 36), triple speeds shown here for turning, face milling, end milling (peripheral) and drilling. (b) See Table 63 for identification of numbered cutting fluids.

Table 55. Tool geometry for turning and boring tools, single point

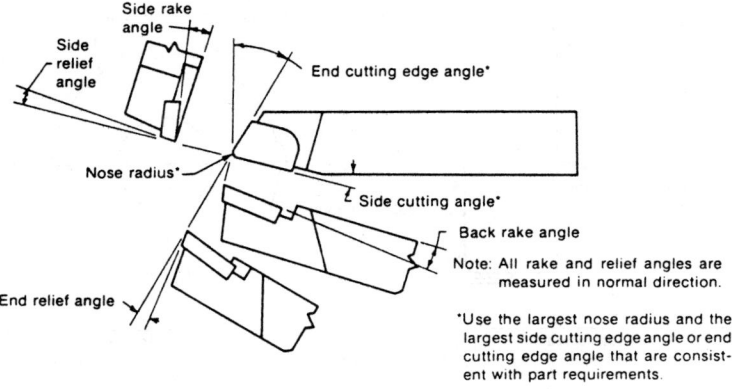

Material	Hardness, Bhn	High speed steel				Carbide					
						Brazed			Indexable		
		Back rake angle, deg	Side rake angle, deg	End relief angle, deg	Side relief angle, deg	Back rake angle, deg	Side rake angle, deg	Relief angle, deg	Back rake angle, deg	Side rake angle, deg	Relief angle, deg
Free-machining carbon steels-wrought; carbon steels-wrought and cast; free-machining alloy steels-wrought; alloy steels-wrought and cast; high-strength steels-wrought; maraging steels-wrought; tool steels-wrought; nitriding steels-wrought; armor plate-wrought; structural steels-wrought	85-225	10	12	5	5	0	6	7	0	5	5
	225-325	8	10	5	5	0	6	7	0	5	5
	325-52R$_C$	0	10	5	5	0	6	7	−5	−5	5
	52R$_C$-58R$_C$	—	—	—	—	—	—	—	−5	−5	5
Free-machining stainless steels-wrought	135-275	5	8	5	5	0	6	7	−5	−5	5
	275-425	0	10	5	5	0	6	7	−5	−5	5
Stainless steels, ferritic-wrought and cast	135-185	5	8	5	5	0	6	7	0	5	5
Stainless steels, austenitic-wrought and cast	135-275	0	10	5	5	0	6	7	0	5	5
Stainless steels, martensitic-wrought and cast	135-325	0	10	5	5	0	6	7	0	5	5
	325-425 / 48R$_C$-52R$_C$	0	10	5	5	0	6	7	−5	−5	5
Precipitation-hardening stainless steels-wrought and cast	150-450	0	10	5	5	0	6	7	−5	−5	5
Gray, ductile and malleable cast irons; compacted graphite cast irons; white cast irons	100-200	5	10	5	5	0	6	7	−5	−5	5
	200-300	5	8	5	5	0	6	7	−5	−5	5
	300-400	5	5	5	5	−5	−5	7	−5	−5	5
Aluminum alloys-wrought and cast	30-150 (500 kg)	20	15	12	10	3	15	7	0	5	5
Magnesium alloys-wrought and cast	40-90 (500 kg)	20	15	12	10	3	15	7	0	5	5
Titanium alloys-wrought and cast	110-440	0	5	5	5	0	6	7	−5	−5	5
Copper alloys-wrought and cast	40-200 (500 kg)	5	10	8	8	0	8	7	0	5	5
Nickel alloys-wrought and cast; chrome-nickel alloys; beryllium-nickel alloys	80-360	8	10	12	12	0	6	7	−5	−5	5
Nitinol alloys-wrought	210-340 / 48R$_C$-52R$_C$	—	—	—	—	0	5	7	0	5	5
High-temperature alloys-wrought and cast	140-475	0	10	5	5	0	6	7	5	0	5
Columbium alloys-wrought, cast, P/M; molybdenum alloys-wrought, cast, P/M; tantalum alloys-wrought, cast, P/M	170-290	0	20	5	5	0	20	7	—	—	5
Tungsten alloys-wrought, cast, P/M	180-320	—	—	—	—	−15	0	7	—	—	—
Zinc alloys-cast	80-100	10	10	12	4	5	5	7	0	5	5
Uranium alloys-wrought	190-210	—	—	—	—	0	0	7	−5	0	7
Zirconium alloys-wrought	140-280	15	10	10	10	5	5	7	5	5	6
Magnetic alloys, nickel- and cobalt-base; controlled expansion alloys	125-250	10	8	8	8	—	—	—	—	—	—
Powder metal alloys: Copper	All	10	8	8	8	6	12	7	—	—	—
Iron	All	0	0	8	8	6	16	7	0	0	5
Magnetic core iron	185-240	15	30	8	8	20	0	7	5	5	5
Carbon and graphite	All	0	0	20	20	0	0	20	0	0	15
Machinable carbide (Ferro-Tic)	40R$_C$-51R$_C$	−5	−5	5	5	−5	−5	7	−5	−5	5

Table 56. Tool geometry for face mills

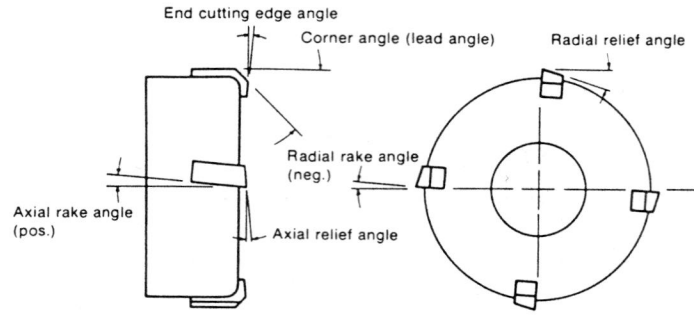

Material	Hardness, Bhn	High speed steel		Indexable carbide		Brazed carbide		Corner angle, deg	End cutting-edge angle, deg	Axial relief angle, deg	Radial relief angle, deg
		Axial rake angle, deg	Radial rake angle, deg	Axial rake angle, deg	Radial rake angle, deg	Axial rake angle, deg	Radial rake angle, deg				
Free-machining carbon steels; carbon steels-wrought and cast; free-machining alloy steels; alloy steels-wrought and cast; maraging steels-wrought; tool steels-wrought; structural steels-wrought	85-270	10 to 15	10 to 15	5 to 7	−5 to −14	0 to −7	0 to −7	30	5 to 10	5 to 7	3 to 7
	270-325	10 to 15	10 to 15	−4 to −8	−3 to −11	0 to −7	0 to −7	30	5 to 10	5 to 7	3 to 7
	325-425	10 to 12	10 to 12	−4 to −8	−3 to −11	0 to −10	0 to −10	30	5 to 10	5 to 7	3 to 7
	43R_C-50R_C	5 to 10	5 to 10	−4 to −8	−3 to −11	−5 to −15	−5 to −15	45	4 to 7	5 to 7	3 to 7
	50R_C-56R_C	—	—	−4 to −8	−3 to −11	−5 to −15	−5 to −15	45	4 to 7	8	8
High-strength steels-wrought	225-425	5 to 10	0 to 10	−4 to −8	−3 to −11	−5 to −15	−5 to −15	45	4 to 7	5 to 7	3 to 7
	45R_C-58R_C	—	—	−4 to −8	−3 to −11	−5 to −15	−5 to −15	45	4 to 7	8	8
Nitriding steels-wrought	200-350	5 to 10	5 to 10	−4 to −8	−3 to −11	0 to −10	−5 to −15	45	5 to 10	5 to 7	3 to 5
Armor plate-wrought	250-320	0 to 5	0 to 5	−4 to −8	−3 to −11	0 to −10	−5 to −15	45	4 to 7	5 to 7	3 to 5
Free-machining stainless steels-wrought	135-275	10 to 15	10 to 12	5 to 11	−5 to −11	0	0 to 5	45	5	8 to 10	8 to 10
	275-425	5 to 10	5 to 10	5 to 11	−5 to −11	0	0 to −5	45	5	8 to 10	8 to 10
Stainless steels, ferritic and austenitic-wrought and cast	135-275	10 to 15	10 to 12	5 to 11	−5 to −11	0 to 5	0 to −5	45	5	8 to 10	8 to 10
Stainless steels, martensitic-wrought and cast	135-425	5 to 10	5 to 10	5 to 11	−5 to −11	0	0 to −5	45	5	8 to 10	8 to 10
Precipitation-hardening stainless steels-wrought and cast	150-450	5 to 10	5 to 10	5 to 7	0 to 5	0	0	45	5	8 to 10	8 to 10
Gray, ductile and malleable cast irons; compacted graphite cast irons; white cast irons	100-400	20 to 30	−5 to −10	5 to 11	−5 to −11	5 to 10	5 to -10	45	5 to 10	4 to 7	4 to 7
Aluminum alloys-wrought and cast	30-150 (500 kg)	20 to 35	20 to 35	5 to 7	0 to 5	10 to 20	10 to 20	45	7 to 12	3 to 5	10 to 12
Magnesium alloys-wrought and cast	40-90 (500 kg)	20 to 35	20 to 35	5 to 7	0 to 5	10 to 20	10 to 20	45	7 to 12	3 to 5	10 to 12
Titanium alloys-wrought and cast	110-440	5	5	0 to −5	0 to −5	0 to −5	−10	45	6 to 12	10 to 12	10 to 12
Copper alloys-wrought and cast	40-200 (500 kg)	12 to 25	10 to 12	5 to 7	0 to 5	3 to 10	3 to 10	45	7 to 12	3 to 5	5 to 10
Nickel alloys, chrome-nickel and beryllium-nickel alloys-wrought and cast	80-360	7	15	5 to 11	−5 to −14	5 to 10	0 to −5	45	5	7 to 9	7 to 9
Nitinol alloys-wrought	210-340 / 48R_C-60R_C	—	—	5 to 7	0 to 5	0	0	45	10	12	12
High-temperature alloys-wrought and cast	200-475	5 to 10	5 to 10	0 to 5	0 to −5	0 to 5	0 to −5	45	5	7 to 10	7 to 10
Columbium alloys-wrought, cast, P/M	170-225	0	20	5 to 7	0 to 5	0	10	45	5 to 10	10	10
Molybdenum alloys-wrought, cast, P/M	220-290	0	20	5 to 7	0 to 5	0	0	45	5 to 10	10	10
Tantalum alloys-wrought, cast, P/M	200-250	0	20	5 to 7	0 to 5	0	0	45	5 to 10	10	10
Tungsten alloys-wrought, cast, P/M	180-320	—	—	−4 to −8	−3 to −11	−15	0	45	5 to 10	15	15
Zinc alloys-cast	80-100	10 to 15	10 to 15	5 to 7	0 to 5	10 to 12	10 to 12	45	7 to 12	10	10 to 12

Table 57. Tool geometry for end mills (peripheral), high speed steel

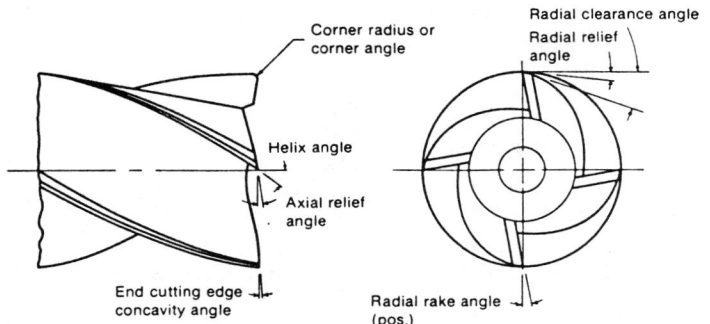

Nominal cutter diameter, in.	General-purpose—30° to 35° helix (steels, cast irons, copper alloys, titanium alloys, nickel alloys, high-temperature alloys and zinc alloys)			35° to 45° helix Aluminum and magnesium alloys		
	Radial primary relief angle, deg	Primary land width, in.	Radial secondary clearance angle, deg	Radial primary relief angle, deg	Primary land width, in.	Radial secondary clearance angle, deg
$1/16$	20 to 21	0.007-0.010	30 to 35	20 to 22	0.007-0.010	30 to 35
$1/8$	12 to 13	0.010-0.015	22 to 28	14 to 18	0.010-0.015	25 to 30
$3/16$	12 to 13	0.010-0.020	20 to 25	14 to 18	0.010-0.020	25 to 30
$1/4$	10 to 11	0.010-0.020	20 to 25	12 to 15	0.010-0.020	22 to 28
$5/16$	10 to 11	0.015-0.025	20 to 25	12 to 14	0.015-0.025	21 to 28
$3/8$	10 to 11	0.015-0.025	17 to 20	12 to 14	0.015-0.025	19 to 26
$7/16$	9 to 10	0.020-0.030	17 to 20	11 to 13	0.020-0.030	18 to 25
$1/2$	9 to 10	0.020-0.030	17 to 20	11 to 13	0.020-0.030	18 to 25
$5/8$	9 to 10	0.025-0.035	17 to 20	11 to 13	0.025-0.035	18 to 25
$3/4$	8 to 9	0.030-0.040	15 to 18	10 to 12	0.030-0.040	17 to 24
$7/8$	8 to 9	0.030-0.040	15 to 18	10 to 12	0.030-0.040	17 to 24
1	8 to 9	0.035-0.050	15 to 18	10 to 12	0.035-0.050	16 to 23
$1\text{-}1/4$	7 to 8	0.040-0.060	13 to 18	9 to 11	0.040-0.060	14 to 22
$1\text{-}1/2$	7 to 8	0.040-0.060	11 to 17	9 to 11	0.040-0.060	13 to 21
$1\text{-}3/4$	7 to 8	0.040-0.060	10 to 16	8 to 10	0.040-0.060	12 to 20
2	6 to 7	0.040-0.060	9 to 15	8 to 10	0.040-0.060	12 to 20

Table 58. Tool geometry for drills, high speed steel twist

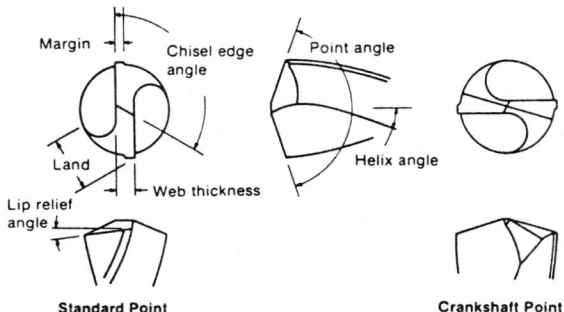

Lip relief angles at periphery	Drill size							
	#80 to #61	#60 to #41	#40 to #31	$1/8$ to $1/4$ in.	$1/4$ to $3/8$ in.	$3/8$ to $1/2$ in.	$1/2$ to $3/4$ in.	1 in. and up
A	24°	21°	18°	16°	14°	12°	10°	8°
B	20°	18°	16°	14°	12°	10°	8°	7°
C	26°	24°	22°	20°	18°	16°	14°	12°

Material	Hardness, Bhn	Drill type	Point angle, deg(a)	Lip relief angle, deg(b)	Helix angle	Point grind
Free-machining carbon steels-wrought; carbon steels-wrought, cast and P/M; free-machining alloy steels-wrought; alloy steels-wrought, cast and P/M; maraging steels-wrought; tool steels-wrought, cast and P/M; nitriding steels-wrought; armor plate-wrought; structural steels-wrought	85-225	General purpose	118	A	Standard	Standard
	225-325	General purpose	118	A	Standard	Standard
	325-425	General purpose	118-135	B	Standard	Crankshaft
	$45R_C\text{-}52R_C$	Heavy web	118-135	B	Standard	Crankshaft
High-strength steels-wrought	175-325	Heavy web	118	A	Standard	Crankshaft
	$325\text{-}52R_C$	Heavy web	118-135	B	Standard	Crankshaft
Austenitic manganese steels-cast	150-220	Rail drill	135	B	Low	Split
Free-machining stainless steels-wrought	135-425	General purpose	118	A	Standard	Crankshaft
Stainless steels, ferritic-wrought and cast; stainless steels, austenitic-wrought, cast and P/M; stainless steels, martensitic-wrought, cast and P/M; precipitation-hardening stainless steels-wrought and cast	135-200	General purpose	118-135	A	Standard	Standard
	200-325	General purpose	118-135	A	Standard	Crankshaft
	325-425	Heavy web	118-135	B	Standard	Crankshaft
	$48R_C\text{-}52R_C$	Heavy web	118-135	B	Standard	Crankshaft

(continued)

Table 58. (continued)

Material	Hardness, Bhn	Drill type	Point angle, deg(a)	Lip relief angle, deg(b)	Helix angle	Point grind
Gray, ductile and malleable cast irons; compacted graphite cast iron; white cast iron	110-220	General purpose	118	A	Standard	Standard
	220-400	Heavy web	118	A	Standard	Standard
Aluminum alloys-wrought and cast	30-150 (500 kg)	Polished flutes	90-118	C	High	Standard
Magnesium alloys-wrought and cast	40-90 (500 kg)	Polished flutes	70-118	C	High	Standard
Titanium alloys-wrought and cast	110-275	General purpose	118-135	B	Standard	Crankshaft
	275-440	Heavy web	118-135	B	Standard	Crankshaft
Copper alloys-wrought, cast, P/M	40-200 (500 kg)	Polished flutes	118	C	Low	Standard
Nickel alloys-wrought, cast, P/M; chromium-nickel alloys-cast	80-360	General purpose	118	B	Standard	Crankshaft
Nitinol alloys-wrought	210-360	General purpose	118	B	Standard	Crankshaft
	48R$_C$-52R$_C$	General purpose	118	B	Standard	Crankshaft
High-temperature alloys-wrought and cast	140-475	Heavy web	118-135	B	Standard	Crankshaft
Columbium, molybdenum and tantalum alloys-wrought, cast, P/M	170-290	General purpose	118	B	Standard	Standard
Tungsten alloys (Anviloy)(c)	290-320	General purpose	118	B	Standard	Standard
Zinc alloys-cast	80-100	General purpose	118	C	Standard	Standard
Uranium alloys-wrought(d)	190-210	Special carbide	118	5-8	20°	−5° land on drill lip
Zirconium alloys-wrought	140-280	General purpose	118	A	Standard	Crankshaft
Magnetic core iron	185-240	General purpose	100-118	A	High	Standard
Controlled-expansion alloys	125-250	General purpose	118	A	Standard	Crankshaft
Magnetic alloys (Hi Perm 49, HyMu 80)	185-240	General purpose	118	A	Standard	Crankshaft
Carbon and graphite	8-100 Shore	General purpose	90-118	C	Standard	Crankshaft
Machinable carbide (Ferro-Tic)	40R$_C$-51R$_C$	General purpose	118	B	Standard	Crankshaft

NOTE: Use stub-length drills whenever possible on high-strength materials.

(a) Chisel-edge angle, 115° to 135°. (b) See tabulation next to illustration of tool geometry at the head of this table. (c) For both high speed steel and carbide drills. (d) For carbide drills.

Table 59. Tool geometry for reamers, high speed steel

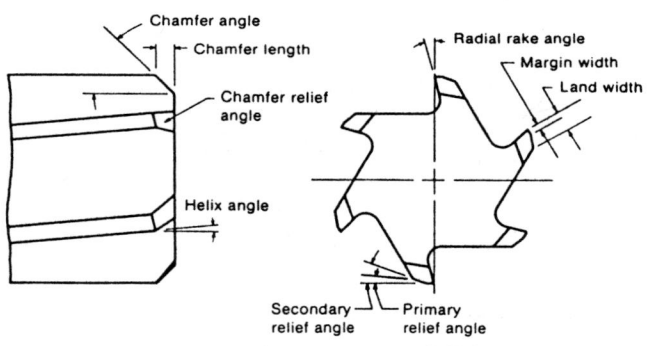

Diameter, in.	Margin width, in.	Primary radial relief angle, deg	Chamfer angle, deg	Chamfer relief angle, deg
Under 1/8	0.004-0.006	20 to 25	45	7 to 12
1/8 to 1/4	0.006-0.008	15 to 20	45	7 to 12
1/4 to 1/2	0.008-0.010	11 to 14	45	7 to 12
1/2 to 3/4	0.010-0.015	8 to 12	45	7 to 12
3/4 to 1	0.012-0.017	7 to 10	45	7 to 12
1 to 1-1/2	0.014-0.018	5 to 8	45	7 to 12
1-1/2 to 2	0.016-0.022	5 to 8	45	7 to 12
Over 2	0.018-0.025	5 to 8	45	7 to 12

Reamer Selection Guidelines

Straight flute	Right hand helix	Left hand helix
• For general purpose use.	• Freer cutting. • Improves finish. • Reduces chatter. • Requires rigid setups with no backlash.	• Reduces chatter on nonrigid setups. • Use 30° to 45° helix for spines and keyways. • Requires more thrust.

Table 60. Tool geometry for reamers, carbide

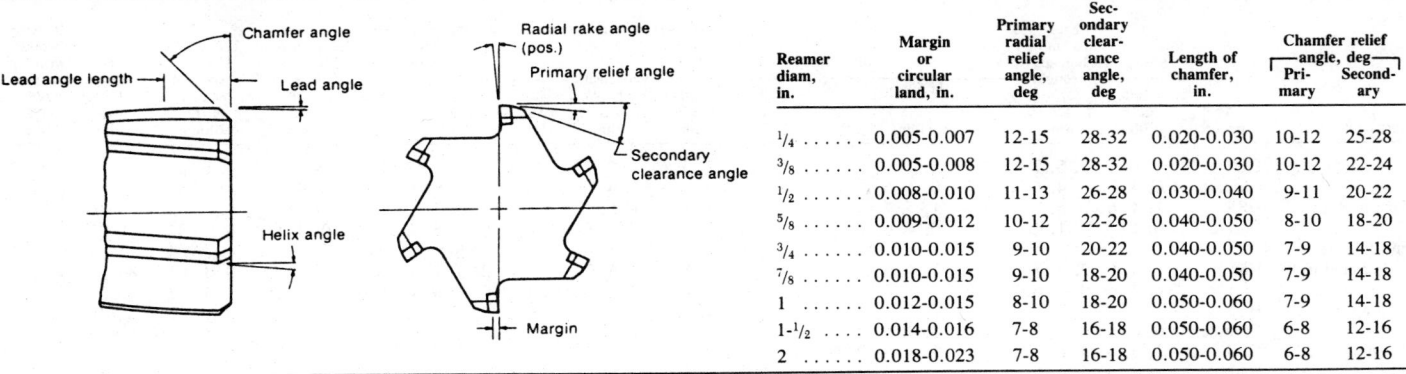

Reamer diam, in.	Margin or circular land, in.	Primary radial relief angle, deg	Secondary clearance angle, deg	Length of chamfer, in.	Chamfer relief angle, deg — Primary	Secondary
1/4	0.005-0.007	12-15	28-32	0.020-0.030	10-12	25-28
3/8	0.005-0.008	12-15	28-32	0.020-0.030	10-12	22-24
1/2	0.008-0.010	11-13	26-28	0.030-0.040	9-11	20-22
5/8	0.009-0.012	10-12	22-26	0.040-0.050	8-10	18-20
3/4	0.010-0.015	9-10	20-22	0.040-0.050	7-9	14-18
7/8	0.010-0.015	9-10	18-20	0.040-0.050	7-9	14-18
1	0.012-0.015	8-10	18-20	0.050-0.060	7-9	14-18
1-1/2	0.014-0.016	7-8	16-18	0.050-0.060	6-8	12-16
2	0.018-0.023	7-8	16-18	0.050-0.060	6-8	12-16

Table 61. Tool geometry for taps, high speed steel

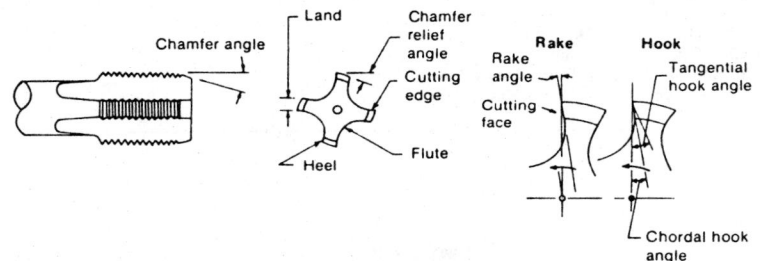

Material	Hardness, Bhn	Hook or rake angle, deg	Chamfer relief angle, deg	Type of tap — Through hole	Blind hole
Steels-wrought, cast, P/M	85-200	7 to 10	8	Spiral point	Fast spiral flute
	200-300	0 to 8	8	Spiral point	Fast spiral flute
	300-375	0	6	Modified 4-flute hand tap	
	375-425	−3 to −6	6	Modified 4-flute hand tap	
	48R$_C$-52R$_C$	−5 to −10	4 to 6	Modified 4-flute hand tap	
Stainless steels, ferritic, martensitic, precipitation-hardening-wrought, cast, P/M	135-275	8 to 12	8	Heavy-duty spiral point	Heavy-duty spiral flute
	275-325	0 to 5	8	Heavy-duty spiral point	Heavy-duty spiral flute
	325-425	0	6 to 8	Modified 4-flute hand tap	
Stainless steels, austenitic-wrought, cast, P/M	135-275	15 to 20	10	Heavy-duty spiral point	Heavy-duty spiral flute
Gray, ductile and malleable cast irons	120-260	5 to 8	6	4-flute hand tap	
	260-330	0 to 3	6	Modified 4-flute hand tap	
Aluminum and magnesium alloys-wrought, cast, P/M	All	10 to 20	12	Spiral point high helix	Fast spiral flute
Titanium alloys-wrought and cast	110-275	10 to 15	12	Modified spiral point	Modified 4-flute hand tap
	275-440	6 to 10	12	Modified spiral point	Modified 4-flute hand tap
Copper alloys-wrought, cast, P/M(a) Groups 1 and 2	All	0 to 8	10	Spiral point	Spiral flute
Group 3	All	9 to 18	12	Spiral point	Fast spiral flute
Nickel alloys, magnetic alloys, controlled-expansion alloys-wrought, cast, P/M	80-170	9 to 12	6 to 8	Spiral point	Spiral flute
High-temperature alloys-wrought and cast	140-425	0 to 10	4 to 6	2-flute spiral point(b)	3-flute hand tap(b), interrupted thread
Columbium, molybdenum and tantalum alloys-wrought, cast, P/M	170-290	10 to 12	6 to 8	2-flute spiral point(b)	4-flute hand tap(b)
Magnetic core iron	185-240	12 to 15	6 to 8	2-flute spiral point	Spiral flute
Zinc alloys-die cast	80-100	12 to 15	12	Spiral point high hook	Fast spiral flute

(a) Group 1 alloys: 314, 330, 332, 335, 340, 342, 353, 356, 360, 365, 366, 367, 368, 370, 377, 385, 485, 544, 834, 836, 838, 842, 844, 848, 852, 854, 855, 857, 858, 864, 867, 879, 928, 932, 934, 935, 937, 938, 989, 943, 944, 945, 953, 954, 956, 973, 974, 976, 978. Group 2 alloys: 226, 230, 240, 260, 268, 270, 280, 442, 443, 444, 445, 464, 465, 466, 467, 651, 655, 675, 687, 770, 817, 821, 833, 853, 861, 862, 865, 868, 872, 874, 875, 876, 878, 903, 905, 915, 922, 923, 925, 926, 927, 947, 948, 952, 955, 957, 958. Group 3 alloys: 102, 110, 113, 114, 115, 116, 122, 170, 172, 175, 210, 220, 505, 510, 521, 524, 614, 706, 715, 745, 752, 754, 757, 801, 803, 805, 807, 809, 811, 813, 814, 815, 818, 820, 822, 824, 825, 826, 827, 828, 863, 902, 907, 909, 910, 911, 913, 916, 917, 962, 963, 964, 966. (b) Special taps for these alloys are offered by some tap manufacturers.

Table 62. Cutting fluid recommendations (see Table 63 for identification of cutting fluid code numbers)

Operation	Tool material	Free-machining carbon steels, wrought 100-275 Bhn	275-425 Bhn	48-65 Rc	Alloy steels, wrought 125-275 Bhn	275-425 Bhn	45-65 Rc	Stainless steels, wrought Martensitic 135-275 Bhn	275-425 Bhn	48-56 Rc	Ferritic and austenitic 135-275 Bhn	275-375 Bhn	Gray, ductile, malleable cast irons, 110-400 Bhn	Al alloys, wrought & cast, 30-125 Bhn(a)	Ti alloys, 110-440 Bhn	Hi-temp alloys, wrought & cast, 140-475 Bhn
Turning(b)	HSS	1, 4, 6	2, 5, 7	3, 5, 7	1, 4, 6	2, 5, 7	3, 5, 7	2, 5, 6	2, 5, 7	3, 5, 7	2, 5, 6	2, 5, 7	4, 6	4, 6, 82	83	3, 5, 7
	Carbide	0, 4, 6	0, 4, 6	0, 4, 6	0, 4, 6	0, 4, 6	0, 4, 6	0, 4, 6	0, 4, 6	0, 4, 6	0, 4, 6	0, 4, 6	0, 4, 6	0, 4, 6	0, 83	0, 4, 6
Face milling	HSS	1, 4, 6	2, 5, 7	3	1, 5, 7	2, 5, 7	3	2, 5, 7	1, 5, 7	3	2, 5, 7	2, 5, 7	4, 6	0, 1, 82	83, 91	2, 5
	Carbide	0, 1, 4	0, 1, 4	0, 2	0, 4, 6	0, 4, 6	0, 2	0, 4, 6	0, 4, 6	0, 5, 7	0, 4, 6	0, 4, 6	0, 4, 6	0, 1, 82	0, 83	0, 3
End milling (peripheral)	HSS	1, 4, 6	2, 5, 7	3	1, 4, 6	2, 5, 7	3	2, 5, 7	2, 5, 7	3	2, 5, 7	2, 5, 7	4, 6	0, 1, 82	83, 91	3
	Carbide	0, 4, 6	0, 4, 6	3	0, 4, 6	0, 4, 6	3	0, 4, 6	0, 4, 6	3, 5, 7	0, 4, 6	0, 4, 6	0, 4, 6	0, 1, 82	83	3
Drilling	HSS	1, 4, 6	2, 5, 7	3	1, 4, 6	2, 5, 7	3	2, 5, 7	2	3	2, 5, 7	2, 5, 7	4, 6	0, 1, 82	83, 91	3
	Carbide	0, 4, 6	0, 4, 6	3	0, 4, 6	1, 5, 7	3	1, 4, 6	1, 5, 7	3, 5, 7	1, 4, 6	1, 5, 7	0, 4, 6	0, 1, 82	83	3
Reaming	HSS	1, 5, 6	2, 6, 7	3	1, 4, 6	2, 5, 7	3	3	3	3	3	3	4, 6	0, 1, 82	83, 91	3
	Carbide	1, 4, 6	1, 4, 6	2, 3	1, 4, 6	1, 4, 6	2, 3	1, 5, 7	2, 5, 7	3, 5, 7	2, 5, 7	2, 5, 7	4, 6	0, 1, 82	83, 91	3
Tapping	HSS	1, 5, 7	2, 5, 7	3	2, 5, 7	3, 5, 7	3	3	2	3	3	3	4, 6	0, 1, 82	1, 83	3
Surface grinding(c)	—	4, 6	4, 6	3	1, 4, 6	2, 4, 6	3	5, 7	1, 4, 6	5, 7	5, 7	5, 7	4, 6(d)	1, 82	81, 83	3

(a) 500-kg load. (b) Single-point and box tools. (c) Horizontal spindle, reciprocating table. (d) For cast irons over 400 Bhn, fluids 5 and 7.

Table 63. Identification codes for cutting fluid types

Code No.	Type	Code No.	Type
0	Dry	8	Specials—Light Duty
1	Oils—Light Duty (General Purpose)	80	Honing oils
2	Oils—Medium Duty	81	Chemical grinding fluids
3	Oils—Heavy Duty	82	Cutting fluids for aluminum
4	Emulsifiable Oils—Light Duty (General Purpose)	83	Cutting fluids for titanium
5	Emulsifiable Oils—Heavy Duty	9	Specials—Heavy Duty
6	Chemicals and Synthetics—Light Duty (General Purpose)	90	Honing oils
7	Chemicals and Synthetics—Heavy Duty	91	Cutting fluids for titanium
		92	Biodegradable soluble oils

Grinding

THE MOST widely used material-removal process for achieving high finish and high workpiece accuracy is grinding. Types of grinding processes commonly used are denoted as surface grinding, cylindrical grinding, internal grinding, centerless grinding, thread grinding, and abrasive belt grinding. Care must be used in the selection of the wheels and abrasive belts in order to meet finish and tolerance requirements without damaging the workpiece. Table 1 shows the standard marking system for identifying grinding wheels and other bonded abrasives (per ANSI B74.13-1977 and ISO 525-1975E).

Abrasive Types. The two most common abrasives used in industrial applications are aluminum oxide and silicon carbide. Aluminum oxide is a synthetic material produced in an electric furnace. Aluminum oxide abrasives have a wide variety of properties, such as toughness and friability, which can be tailored to specific job requirements. Tough aluminum oxide is used for rough grinding operations. Semifriable aluminum oxide is a general-purpose type. White friable aluminum oxide is used for tool grinding and for heat-sensitive materials; pink friable aluminum oxide is used for difficult-to-grind alloys.

Silicon carbide, also a product of the electric furnace, is harder than aluminum oxide and is used primarily for grinding nonmetallic, nonferrous and low-tensile-strength materials. Green friable silicon carbide is used for general-purpose grinding of tungsten carbide. Various combinations of aluminum oxide and silicon carbide abrasives are available only in organic bonds—resinoid, rubber and shellac.

Grinding is also commonly performed using cubic boron nitride and diamond as abrasive materials.

Grain size influences surface roughness, stock removal rate, chip size and corner holding ability. Surface roughness depends on a number of factors, such as wheel hardness, structure and dressing methods, but finer-grain abrasives tend to produce surfaces with lower roughness values.

Wheel Grade. The grade or hardness of a wheel is a measure of the strength of the bond holding the grains in the wheel. The major factor affecting bond strength is the amount of bond holding an individual grain. Wheel grades range from A to Z in order of increasing bond content. Wheel grade requirements vary inversely with the workpiece hardness. Harder materials require softer-grade wheels, while softer materials require harder-grade wheels.

Wheel Structure. The wheel structure number indicates the porosity or density of the wheel. Low numbers indicate a dense structure, while high numbers indicate an open structure. Wheels having an open structure are recommended for surface grinding, while wheels having a dense structure are recommended for cylindrical, centerless and form grinding. Dense wheel structures are recommended when low surface-roughness values are required.

Bond. Of the several bond types listed in Table 1, resinoid, rubber and vitrified are the most widely used types. Vitrified bonds are more brittle than resinoid bonds but are superior for holding form. Normal speeds for vitrified wheels are 6500 or 8500 ft/min (33 or 43 m/s). Higher speeds of 10 000 to 16 500 ft/min (51 to 64 m/s) require special wheels, tested and approved for the particular speed and machine. Normal speeds for resinoid-bond wheels range from 6500 to 9500 ft/min (33 to 48 m/s), depending on the actual grading. In some cases, cutoff wheels and snagging wheels can be run at speeds up to 16 500 ft/min (64 m/s). In any case, the speed marked on the blotter is the maximum permissible speed. Resinoid-bond wheels are used for wet grinding and for cutoff operations.

Truing and Dressing

Before truing a grinding wheel, the spindle should be run until it is up to operating temperature. Then the diamond truing tool should be placed at the center of the wheel and fed out to the wheel edge, and then from edge to edge. A copious flow of coolant should be directed at the wheel-tool interface.

The face of the wheel must be prepared for the job it is expected to do by dressing. When dressing wheels for finishing operations, the diamond

Table 1. Markings for identifying grinding wheels and other bonded abrasives
(Standard marking system chart per ANSI B74.13-1977 and ISO 525-1975E)

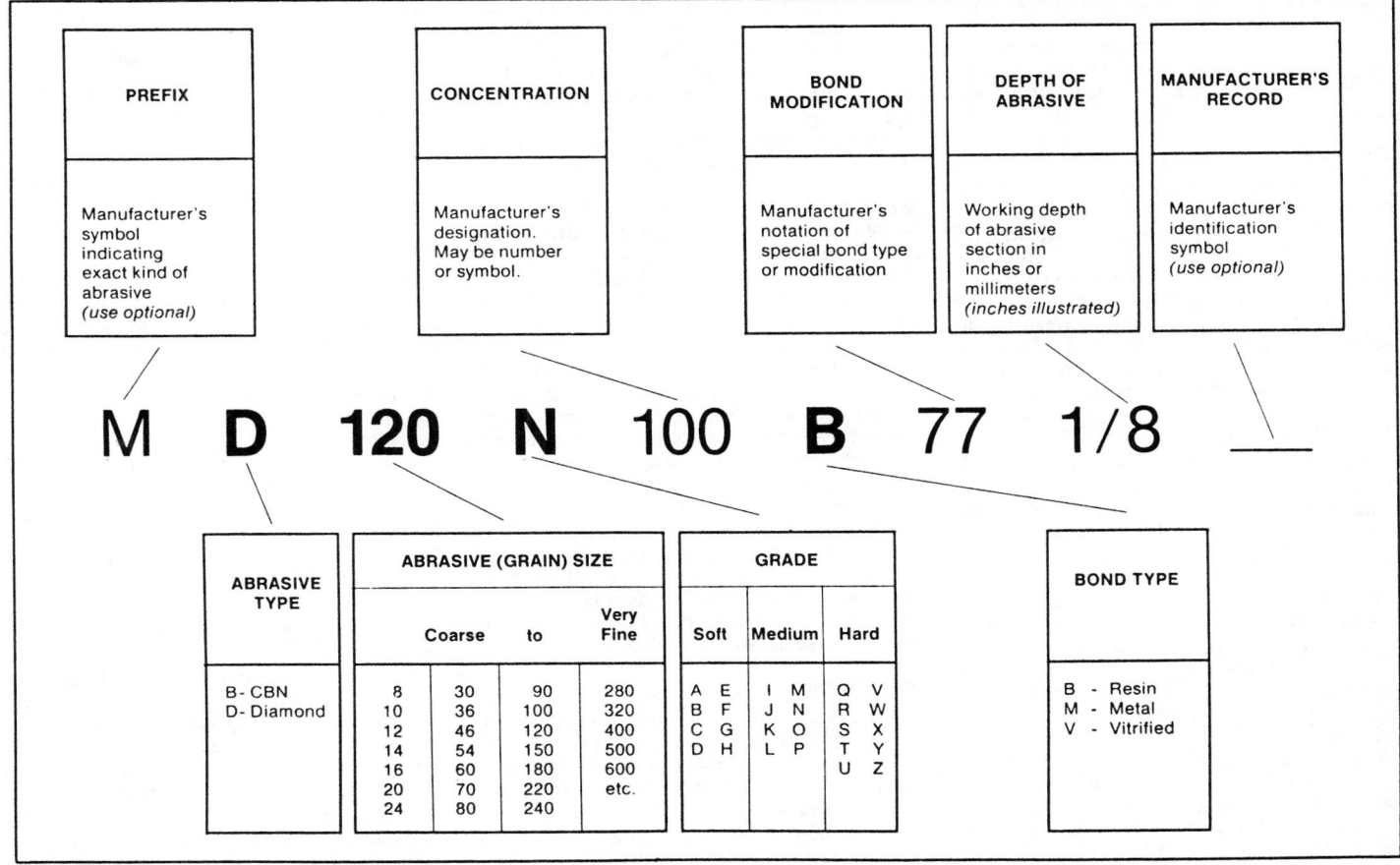

		ABRASIVE (GRAIN) SIZE				GRADE			STRUCTURE		BOND TYPE*	MANUFACTURER'S RECORD
PREFIX	**ABRASIVE TYPE**	Coarse	Medium	Fine	Very Fine	Soft	Medium	Hard	Dense to Open			
Manufacturer's symbol indicating exact kind of abrasive *(use optional)*	A - Aluminum Oxide C - Silicon Carbide	8 10 12 14 16 20 24	30 36 46 54 60	70 80 90 100 120 150 180	220 240 280 320 400 500 600	A E B F C G D H	I M J N K O L P	Q V R W S X T Y U Z	1 9 2 10 3 11 4 12 5 13 6 14 7 15 8 16 etc. *(use optional)*	B - Resinoid BF - Resinoid Reinforced E - Shellac O - Oxy-chloride R - Rubber RF - Rubber Reinforced S - Silicate V - Vitrified Mg - Magnesia	Manufacturer's private marking to identify wheel *(use optional)*

*ISO 525-1975(E) specifies a Mg-Magnesia bond not contained in ANSI B74.13-1977 and does not include bond types O-Oxychloride and S-Silicate.

Table 2. Markings for identifying diamond and cubic boron nitride (CBN) grinding wheels and other bonded abrasives
(Standard marking system chart per ANSI B74.13-1977)

PREFIX	CONCENTRATION	BOND MODIFICATION	DEPTH OF ABRASIVE	MANUFACTURER'S RECORD
Manufacturer's symbol indicating exact kind of abrasive *(use optional)*	Manufacturer's designation. May be number or symbol.	Manufacturer's notation of special bond type or modification	Working depth of abrasive section in inches or millimeters *(inches illustrated)*	Manufacturer's identification symbol *(use optional)*

M   **D**   **120**   **N**   100   **B**   77   1/8   ___

ABRASIVE TYPE	ABRASIVE (GRAIN) SIZE				GRADE			BOND TYPE
	Coarse	to		Very Fine	Soft	Medium	Hard	
B- CBN D- Diamond	8 10 12 14 16 20 24	30 36 46 54 60 70 80	90 100 120 150 180 220 240	280 320 400 500 600 etc.	A E B F C G D H	I M J N K O L P	Q V R W S X T Y U Z	B - Resin M - Metal V - Vitrified

tool is fed slowly across the face; when dressing wheels for roughing operations, the diamond tool is fed more rapidly.

CUBIC BORON NITRIDE (CBN) GRINDING WHEELS

Applications

Cubic boron nitride (CBN) grinding wheels are recommended for alloys that are difficult to grind with conventional abrasives; that is, steels and cast irons with hardnesses above 50 R_C and nickel and cobalt high-temperature alloys with hardnesses above 35 R_C.

Wheel Identification

Table 2 shows the standard marking system for diamond and cubic boron nitride abrasives (per ANSI B74.13-1977).

Bond Types. CBN grinding wheels are available with resinoid, vitrified and metal bonds. Resinoid-bond wheels produce a good finish at high metal removal rates and provide fairly good wheel life. Vitrified-bond wheels are better for holding form and provide longer wheel life. Metal-bond wheels are free cutting and give high metal removal rates.

Metal-bond, electroplated wheels contain a single layer of CBN grains and cannot be trued or dressed. Therefore, it is imperative that the spindle be running true and that all mounting surfaces be clean and free of burrs. When these wheels are mounted, the runout should not exceed 0.0005 in. (0.013 mm) TIR in either the radial or the lateral directions.

Grit Size. Because of the sharpness of CBN, it is necessary to use finer grain sizes than are used with conventional abrasives to obtain the same surface roughness level. Extremely fine finishes are difficult to obtain with CBN abrasives because there is little or no burnishing of the surface since CBN abrasives wear so little.

Grinding Fluids

Hardened steels and cast irons can be ground dry with CBN wheels; however, using a grinding fluid will improve both the surface roughness and the wheel life. A light-duty soluble oil will suffice for these materials. When grinding the superalloys, a heavy-duty, active grinding fluid, that is, sulfochlorinated oil or sulfochlorinated soluble oil, is a must. These heavy-duty grinding fluids will also improve grinding ratios when grinding hardened steels with CBN wheels.

Truing and Dressing

It is of utmost importance that a CBN wheel be properly conditioned before attempting to grind with it. Both resinoid-bond and vitrified-bond wheels can be trued using a rotary diamond dresser, a brake-controlled truing device or a metal-bonded, 100 to 180 mesh size diamond truing tool. A single-point or cluster diamond should not be used. The wheel should be flooded with a copious supply of coolant while truing.

After truing, resinoid-bond CBN wheels must be dressed to open up the glazed surface left by the truing process. A conditioning stick containing approximately 200-mesh aluminum oxide or silicon carbide abrasives in a soft vitrified bond can be used. It is best to jog the wheel and press the conditioning stick into the wheel while it is slowing down. This removes the bond, exposing the sharp cutting edges. When the conditioning

stick starts to wear rapidly, the wheel is properly dressed. Vitrified-bond CBN wheels usually do not need to be dressed or conditioned. They can be used immediately after truing.

Machine Requirements

To realize maximum benefits from using CBN wheels, the machine must have sufficient rigidity, horsepower and spindle speed. Machine vibrations that can be tolerated with conventional abrasives because of their more rapid wheel breakdown will cause CBN wheels to chatter.

Surface Integrity

Because CBN wheels remain sharp longer, they produce less heat and deformation in the workpiece, thus less residual stress. This permits the use of conventional speeds and feeds on critical parts that would otherwise require "low stress" grinding procedures if conventional abrasives were used.

DIAMOND GRINDING

Diamond grinding technology has developed considerably during the past 30 years with respect to both grinding wheels with natural grit and, more recently, grinding wheels with synthetic grit. Today, diamond wheels are available in a large variety of mesh sizes, diamond concentrations, and bond types. Table 2 lists the standard markings used for diamond wheels. Grain types have been developed for specific grinding operations and materials. More recent developments have opened up new applications for diamond wheels, especially in the field of steel and carbide grinding utilizing metal-clad, diamond-grain wheels. Diamond wheels are available in several bonds: metal, vitrified and resinoid. Each bond type provides different results, depending on the type of grinding operation and the material being ground.

Metal-bond wheels are extremely strong, long-life wheels having diamonds securely held in a metal matrix. Chief uses are cutoff operations with thin wheels, offhand grinding, chip-breaker grinding and electrolytically assisted grinding. Metal-bond wheels should always be used with a coolant.

Vitrified-bond wheels are diamonds bonded together with ceramic materials. This bond type provides long wheel life at good grinding rates. The principal applications for vitrified-bond diamond wheels are surface grinding, cylindrical grinding, offhand grinding and internal grinding.

Resinoid-bond wheels are the most popular type because they provide excellent grinding speed and relatively long life. A resinoid diamond wheel is ideally suited for grinding carbide cutting tools where maximum protection from heating and cracking are a consideration. This type of bond is best for machine grinding operations, such as cutter grinding, surface and cylindrical grinding. Resinoid bond has excellent form holding qualities, is easily dressed, and is most suitable for diamond wheels with formed faces.

Wet Grinding

Wherever possible, diamond wheels should be used wet. This promotes longer wheel life, decreases workpiece surface roughness and reduces wheel dressing frequency. For machines not equipped with flood coolant systems or where the operator must observe the work at all times, the mist or wick method can be used. The primary

reason for recommending wet grinding with diamond wheels is that the diamond itself has a relatively low thermal threshold and will "burn" at relatively low (1100 to 1300 °F [600 to 700 °C]) temperatures. The heat at the crystal-work interface can easily reach the low temperature range in dry grinding.

Wet grinding with diamond wheels eliminates some of the common problems associated with dry grinding. Since workpiece temperatures are kept at a low level, the potential for surface cracks or possible catastrophic cracking of the workpiece is greatly reduced. Dust problems also are eliminated. The flushing action of a copious supply of coolant helps to keep the diamond wheel cutting free, thus preventing loading of the wheel.

Dry Grinding

If dry grinding is absolutely necessary, wheel speed must be kept below 4500 ft/min (23 m/s), with 3500 to 4000 ft/min (18 to 20 m/s) preferable. With care and practice, an experienced operator can become proficient at dry grinding of hard materials such as carbide and ceramics. One problem associated with dry grinding is dust; therefore, adequate dust-collecting systems need to be employed. Wheel wear will be greater than with wet grinding because softer wheels must be used to help keep the grinding temperatures as low as possible. Although resinoid-bond wheels are recommended for dry diamond grinding of hard, brittle materials such as carbide, glass and ceramics, high grinding temperatures can burn or crack the grinding wheel. Metal- and vitrified-bond diamond wheels, on the other hand, if run dry on hard, brittle materials may cause cracking of the workpiece material.

Machines for Grinding With Diamonds

Diamond grinding requires a higher level of machine tool integrity than might be required for grinding with a silicon carbide or an aluminum oxide wheel. Spindles must run true, bearings should be in good condition and good maintenance procedures should be followed.

Wheel Mounting

When mounting a diamond wheel, extra care should be taken to be sure that back plates, flanges and spindles are clean, free of burrs and running true. Straight wheels should be tightened gently between the flanges and then lightly tapped into position so that they run within 0.0005 in. (0.013 mm) TIR. The flanges should then be completely tightened.

If the grinding machine has a tapered spindle nose, it is common practice to mount each diamond wheel on a separate collet or adaptor. If left as a unit, the mounted wheel and adaptor can then be removed and replaced as needed, and the time and abrasive lost in truing can be avoided.

GRINDING RECOMMENDATIONS

The grinding conditions listed in Table 3 are provided as starting recommendations. These starting conditions may have to be adjusted to reflect (1) workpiece characteristics, (2) requirements for the ground surface, (3) economic objectives and (4) operational conditions. Obviously, these grinding recommendations represent only an approximation of the optimum conditions achievable by testing and experimenting.

Table 3. Conditions for surface grinding—horizontal spindle, reciprocating table

Material	Hardness, Bhn	Condition(a)	Wheel speed, fpm	Table speed, fpm	Downfeed, in./pass (R = Rough, F = Finish)	Crossfeed, in./pass	Wheel identi-fication, ANSI(b)
Free-machining low-carbon steels, wrought(c)	50 R$_C$ max	HR, N, A, CD or Q&T	5500-6500	50-100	R: .003; F: .001 max	.050-.500 (max, $^1/_4$ of wheel width)	A46JV
	Over 50 R$_C$	C and/or Q&T	5500-6500	50-100	R: .002; F: .0005 max	.025-.250 (max, $^1/_{10}$ of wheel width)	A46IV
Medium-carbon alloy steels, wrought	50 R$_C$ max	HR, N, A, CD or Q&T	5500-6500	50-100	R: .003; F: .0005 max	.050-.500 (max, $^1/_4$ of wheel width)	A46JV
	Over 50 R$_C$	C and/or Q&T	3000-4000	50-100	R: .002; F: .0005 max	.025-.250 (max $^1/_{10}$ of wheel width)	A46IV
Gray, ductile, malleable cast irons	45 R$_C$ max	A-C, A or Q&T	5500-6500	50-100	R: .003; F: .001 max	.050-.500 (max, $^1/_3$ of wheel width)	C36IV or A46JV
	45-52 R$_C$	A-C, A or Q&T	5500-6500	50-100	R: .002; F: .0005 max	.050-.500 (max, $^1/_5$ of wheel width)	C36JV or A46IV
	48-60 R$_C$	F or IH	5500-6500	50-100	R: .002; F: .0005 max	.025-.250 (max, $^1/_{10}$ of wheel width)	C46HV or A46HV
Stainless steels, wrought:							
Ferritic(d)	135-185	A	5500-6500	50-100	R: .002; F: .0005 max	.050-.500 (max, $^1/_4$ of wheel width)	A46IV
Austenitic(e)	135-275	A or CD	5500-6500	50-100	R: .002; F: .0005 max	.050-.500 (max, $^1/_4$ of wheel width)	C46JV
Martensitic(f)	135-275	A	5500-6500	50-100	R: .002; F: .0005 max	.050-.500 (max, $^1/_4$ of wheel width)	A46IV
	Over 275	Q&T	5500-6500	50-100	R: .001; F: .0003 max	.025-.250 (max, $^1/_{10}$ of wheel width)	A46HV
Aluminum alloys, wrought and cast	40-125(g)	A-C or ST&A	4000-5000	50-100	R: .003; F: .001 max	.050-.500 (max, $^1/_3$ of wheel width)	C46JV(h)
Titanium alloys, wrought(i)	300-380	A	3000-4000	60	R: .001; F: .0005 max	.025-.250 (max, $^1/_{10}$ of wheel width)	C54JV(j)
	320-440	ST&A	3000-4000	60	R: .001; F: .0005 max	.025-.250 (max, $^1/_{10}$ of wheel width)	C54JV(j)
High-temperature alloys, wrought and cast	200-390	A or ST	3000-3500	50-100	R: .001; F: .0005 max	.020-.200 (max, $^1/_{12}$ of wheel width)	A46HV
	300-475	ST&A	3000-3500	50-100	R: .001; F: .0005 max	.020-.200 (max, $^1/_{12}$ of wheel width)	A46HV

NOTE: See Table 62 in the preceding article ("Machining Data Recommendations") for cutting fluid recommendations.

(a) A = annealed; A-C = as-cast; C = carburized; CD = cold drawn; F or IH = flame or induction hardened; HR = hot rolled; N = normalized; Q&T = quenched and tempered; ST = solution treated; ST&A = solution treated and aged. (b) Wheel recommendations are for wet grinding. For dry grinding, use a softer-grade wheel. See Table 1 for wheel markings. (c) Resulfurized steels: 1108, 1109, 1110, 1115, 1116, 1117, 1118, 1119, 1211, 1212, 1213, 1215. Leaded steels: 10L18, 11L17, 12L13, 12L14, 12L15. (d) Types 405, 409, 429, 430, 434, 436, 442, 446. (e) Types 201, 202, 301, 302, 302B, 304, 304L, 305, 308, 309, 309S, 310, 310S, 314, 316, 316L, 317, 321, 330, 347, 348, 384, 385. (f) Types 403, 410, 414, 420, 422, 431, 440A, 440B, 440C, 501, 502. Greek Ascoloy. (g) 500-kg load. (h) Wax filled. (i) CAUTION: Potential fire hazard. Exercise caution in grinding and disposing of swarf. (j) Use friable (green grit) silicon carbide.

These recommendations should handle most grinding situations. In a job shop, it is often necessary to avoid wheel changes; a few general-purpose wheels usually are used to grind almost every type of material, even if at less than optimum efficiency. Where large production quantities are involved, testing and modifications may be necessary to obtain optimum results.

Despite the use of standard grinding wheel markings (Table 1), wheels with identical markings made by various manufacturers may vary in actual performance. Individual wheel makers have their own modifications of both abrasive and bond which cannot be included in the data tables. Likewise, the structure number is omitted from the wheel identification because its significance differs among various makes of wheels. The structure provides chip clearance and is also a result of grain size and proportion of bond. The best or standard structure is usually derived for most grain size and grade combinations as a result of experience and testing. The manufacturers of grinding wheels publish their own tables of wheel recommendations which include grain and bond modifications as well as structure numbers.

Recommendations in Table 3 are predicated on average shop conditions. Deviations from such basic conditions will generally affect the manner in which grinding wheels perform. Some of the variables discussed later have opposing effects and may balance each other when present concurrently. Other variables can have similar effects which can be expected to work additively. An understanding of the directions in which process variables interact with grinding wheel per-

formance permits modification of the controllable variables to improve over-all operating conditions or economics.

MODIFYING WHEEL RECOMMENDATIONS

The following lists of guidelines are provided for use in modifying the grinding wheel recommendations provided in Table 3 to improve production or to meet the specific requirements of an application. No application, however, should be so finely tuned that normal variations would affect the output rate or the quality. The best approach is to accommodate any variations likely to occur. The best wheel for any application is the one that compromises the ability to cut rapidly with the ability to hold form, maintain surface roughness requirements and last longer.

Wheels recommended in Table 3 are suitable for both rough and finish grinding in one setup. Where two setups are used, use a wheel one grade harder for roughing and a wheel one grade softer for finishing.

General Guidelines

• To remove a substantial amount of stock—wet grinding is recommended over dry grinding.
• To remove stock faster—use a coarser grain wheel with a more open structure and a less friable abrasive.
• To produce a smoother finish—use a finer-grain wheel with a denser structure and a less friable abrasive. See Table 4.
• To generate form—use a finer-grain wheel with a more dense structure. See Table 5.

Table 4. Grinding wheel grain sizes for producing various surface roughnesses

Grain size	Surface roughness	
	μin.	μm
46	32	0.80
54	20 to 32	0.50 to 0.80
60	15 to 20	0.38 to 0.50
80	10 to 15	0.25 to 0.38
120	8 to 10	0.20 to 0.25

Table 5. Grinding wheel grain sizes for the radius of forms or fillets

Grain size	Minimum radius	
	inch	mm
80	0.010	0.254
120	0.007	0.178
180	0.005	0.127
220	0.004	0.102
280	0.003	0.076
320	0.002	0.051
500	0.001	0.025

• To grind large areas—use a softer grade and coarser-grain wheel.
• To grind small areas—use a harder grade and a finer-grain wheel.
• To grind soft metals—use a harder grade and a coarser-grain wheel.
• To grind hard metals—use a softer grade and a finer-grain wheel.
• To improve workpiece finish—dress the wheel to a fine finish.

- To minimize heat, warpage and surface damage in the workpiece — maintain wheel sharpness.
- If the grinding wheel breaks down too fast — use a wheel with a less friable abrasive, a harder grade and a denser structure.
- If the grinding wheel glazes and burns — use a wheel with a more friable abrasive, a softer grade and a more open structure.

Guidelines for Surface Grinding

- To minimize heat and warpage in the workpiece — use fast table speeds and light downfeeds, and dress the wheel before final size.
- To improve workpiece finish — dress the wheel to a fine finish.
- When using wheels larger than 14-in. (356 mm) in diameter — use one grade softer bond.

Guidelines for Surface Grinding, Vertical Spindle, Rotary Table

- To produce smoother workpiece finish and closer tolerances — use faster table speed, lighter downfeed and proper sparkout.
- To increase material removal rate — increase table speed and downfeed rate.

Guidelines for Cylindrical Grinding

- When using wheels larger than 14-in. (356 mm) in diameter — use one grade softer bond.
- For heavier stock removal — use a faster traverse speed and a slower work speed and/or increase the depth of cut.
- To improve workpiece finish — use a slower traverse speed and a faster work speed and/or decrease the depth of cut.
- When using wider wheels — use a softer grade wheel (infeed grinding).
- When using narrower wheels — use a harder grade wheel (infeed grinding).

Guidelines for Centerless Grinding

- When using wheels smaller than 20-in. (500 mm) in diameter — use a one grade harder wheel.
- To improve workpiece finish — use a smaller angle of draw and a faster regulating wheel speed.
- For heavier stock removal — use a larger angle of draw and a slower regulating wheel speed and grind closer to center.
- To produce roundness — grind as high above center as possible.
- To grind long bars — grind below center on a flat blade.

Guidelines for Internal Grinding

- For long bores — use softer-grade wheels and wider wheels, if possible.
- For low-powered machines — use softer-grade wheels.
- For light-spindle machines — use softer-grade wheels.

MODIFYING OPERATING CONDITIONS TO CHANGE WHEEL GRADE ACTION

Most grinding problems (other than those related to machine condition) arise from the action of the wheel grade (or hardness) which is a direct function of the wheel sharpness or wear rate. Indicators of too little wear (that is, the wheel is

acting hard) are: the wheel glazing, loading or not cutting freely; workpieces with burning, heat checking (grinding cracks) or out of roundness; finishes getting progressively better (due to glazing); and finely spaced chatter marks or squealing. Indicators of too much wear (that is, the wheel is acting soft) are: the wheel breaking down too fast; workpieces with finishes getting progressively worse; widely spaced chatter marks; and scratches, fishtails, taper or lack of accuracy.

The action of the wheel grade can be altered by adjusting the other grinding conditions to achieve proper grinding action when it is not desirable or possible to change the wheel. Table 6 contains guidelines for modifying wheel action.

Table 6. Guidelines for modifying wheel action

Operation	To make wheel act HARDER	To make wheel act SOFTER
Internal grinding	Decrease work rpm	Increase work rpm
	Decrease wheel reciprocation	Increase wheel reciprocation
	Increase wheel rpm	Decrease wheel rpm
	Dress wheel at slower rate	Dress wheel at faster rate
Surface grinding	Decrease table speed	Increase table speed
	Decrease crossfeed	Increase crossfeed
	Increase wheel rpm	Decrease wheel rpm
	Dress wheel at slower rate	Dress wheel at faster rate
Cylindrical grinding (traverse)	Decrease traverse speed	Increase traverse speed
	Increase wheel rpm	Decrease wheel rpm
	Dress wheel at slower rate	Dress wheel at faster rate
Centerless grinding	Decrease feed wheel rpm	Increase feed wheel rpm
	Dress grinding wheel at slower rate	Dress grinding wheel at faster rate

Table 7. Production and precision grinding tolerances

Grinding operation	Tolerances (plus or minus) — inch — Production	Precision	Tolerances (plus or minus) — mm — Production	Precision
Cylindrical grinding				
Diameters	0.00025	0.00001	0.0064	0.00025
Shoulders:				
Shoulder to shoulder	0.00025	0.0005	0.0064	0.0127
Traverse grinding to a shoulder	0.002	0.001	0.050	0.025
Corners and radii:				
External corners	Sharp	Sharp		
Internal corner radii	0.005	0.0025	0.13	0.063
Spherical sections (oscillating grinders):				
Diameters	0.00015		0.0038	
Location of centers	0.001		0.025	
Centerless grinding				
Diameters and Parallelism	0.0001	0.000025	0.0025	0.00064
Roundness	0.000012		0.0003	
Concentricity of stepped diameters	0.00025	0.0001	0.0064	0.0025
Thread grinding				
Lead error (inch per inch)	0.00025	0.00001	0.0064	0.00025
Pitch diameter	0.0005	0.0002	0.0127	0.0050
Roundness	0.00025		0.0064	
Concentricity (thread form with OD)	0.0005		0.0127	
Grooves (width)	0.001		0.025	
Surface grinding				
Reciprocating table grinder:				
Flatness	0.0002	0.00015	0.0050	0.0038
Thickness	0.0003	0.00015	0.0076	0.0038
Rotary table grinder				
Flatness	0.0002	0.0001	0.0050	0.0025
Parallelism	0.0002	0.00005	0.0050	0.0013
Thickness	0.001	0.0002	0.025	0.0050
Internal grinding				
Holes (using automatic sizing devices)	0.00025	0.00005	0.0064	0.0013
Face runout (squareness of shoulder to bore)	0.00025	0.00005	0.0064	0.0013

SOURCE: Adapted from H. E. Trucks, *Designing for economical production*, Dearborn, MI: Society of Manufacturing Engineers, 1974, p. 34.

PRECAUTIONS

Safety. Follow all requirements for safe wheel use. Do not exceed the maximum speed marked on the wheel or shown in ANSI B7.1-1978, *Safety Requirements for the Use, Care, and Protection of Abrasive Wheels*.

Cutting Fluids. Use the proper cutting fluid, use the proper flow rates, and direct the fluid so that it reaches the cutting zone. See the article on Cutting Fluids, in this section.

Surface Integrity. To avoid damage to the ground surface in the form of grinding burn, heat checks (grinding cracks), warpage, residual stress or other surface alterations, see the article on Surface Integrity, on page 27•24 in this section, for proper grinding procedures.

SURFACE ROUGHNESS

The grain size of a grinding wheel will determine the approximate surface roughness that can be obtained on the workpiece, as given in Table 4. The structure of the wheel and the dressing procedure will also affect the surface roughness to some degree. A finer-grain-size wheel will usually produce surfaces with lower roughness values at some sacrifice of stock removal capability.

Obtaining surface roughness values of less than 10 to 15 micro-in. R_a (0.25 to 0.38 μm) requires special attention to work speeds and crossfeed or traverse rates. Very smooth surfaces require abrasive grain sizes of 220 and finer.

TOLERANCES

The tolerances achievable with various grinding operations are given in Table 7. The production tolerances can be held without difficulty; however, larger tolerances, in applications where they are acceptable, will be more economical. The precision tolerances can be held with care but will be more costly.

Form Requirements. Grinding of forms and fillets usually requires the use of 80 grain size and finer wheels, as indicated in Table 5.

Thread Grinding Requirements. In conventional grinding, the standard practice is to use the coarsest practical grain size for fastest stock removal. In thread grinding, however, the coarsest practical grain size is limited by the maximum

Table 8. Coarsest allowable grain sizes for thread grinding wheels

American Standard Unified threads		Coarsest allowable grain size		American Standard Unified threads		Coarsest allowable grain size	
Threads per in.	Root width, in.	Vitrified wheels	Resinoid wheels	Threads per in.	Root width, in.	Vitrified wheels	Resinoid wheels
80	0.0016	240	220	16	0.0078	120	100
72	0.0017	240	220	14	0.0089	100	100
64	0.0019	220	180	13	0.0096	100	90
56	0.0022	220	180	12	0.0104	90	90
48	0.0026	220	180	11	0.0113	90	90
44	0.0028	220	180	10	0.0125	90	90
40	0.0031	220	180	9	0.0139	90	80
36	0.0034	180	180	8	0.0156	80	80
32	0.0039	180	150	7	0.0178	80	80
28	0.0045	150	120	6	0.0208	80	80
24	0.0052	150	120	5	0.0250	80	70
20	0.0062	120	120	4.5	0.0278	80	70
18	0.0069	120	100	4	0.0312	70	70

SOURCE: Adapted from Thread Grinding, 18th Edition, 1963, published by Norton Company.

allowable radius at the bottom of the thread. This radius is usually given in terms of root width, which is the distance between the points of tangency of the arc and the sides of the teeth. The root width varies inversely with the number of threads per inch (or metric pitch); the greater the number of threads per inch, the smaller the maximum allowable root width.

To alter the wheel recommendations in Table 3, Table 8 may be used as a guide for selecting the coarsest allowable grain size for a thread grinding wheel. If a particular standard calls for a narrower root width, the grain size must be chosen for the maximum allowable root width. Another factor to be considered is the quality of the finish. Where a very low surface roughness value is required on threads of relatively coarse pitch, it is advisable to use a finer grain wheel than would be necessary to hold the proper root width.

Nontraditional Machining Processes

THE FOLLOWING SUMMARY charts provide data and information which are helpful in considering and selecting applicable processes alternative to the more conventional material-removal processes. Usually the processing parameters are somewhat more difficult to specify than those for turning, milling, drilling, and other better known machining methods.

The Machining Data Handbook, Third Edition (published by Machinability Data Center, Metcut Research Associates, Inc., Cincinnati, 1980), provides more detailed summarized data for the following operations as well as information on other nontraditional processes which are not included here.

Process summary chart for ECM and EDM

	Electrochemical machining (ECM)	Electrical discharge machining (EDM)
Principle	Controlled metal removal by anodic dissolution. DC current passes through flowing film of conductive solution which separates workpiece from electrode-tool. Workpiece is anode, tool the cathode.	Metal is removed by rapid spark discharges between an electrode and a conductive workpiece separated by a 0.0005-0.035 in. gap filled with a dielectric fluid and metal particulates. The workpiece material is melted, vaporized in part, and expelled from the gap.
Equipment	Machine tool must be rigid to withstand high fluid separating forces; must protect mechanical and electrical systems from corrosive electrolytes, and have provisions for venting of work areas; dc power source; electrolyte system, including pumps, filters, storage tanks, and heat exchanger; electrolyte clarifier may be required; servomechanism for process control optional.	Rigid machine tool for close control of spark gap; EDM power supply; servomechanism to control electrode movement; dielectric fluid pressure or vacuum system and filter. CNC and tool change capability available.
Typical applications ...	High strength, high hardness materials; high temperature alloy forgings; odd shaped holes and cavities; jet engine blade airfoils; small deep holes; jet engine blade cooling holes; deburring; face turning of discs; tungsten carbide machining; etching of numbers and letters in hard steels; selective machining of one material from another, enlarging I.D.'s.	Manufacture of: Dies (stamping, cold heading, forging, injection molding); carbide forming tools; tungsten parts; burrfree parts; odd shaped holes and cavities; small diameter deep holes; high strength and high hardness materials; narrow slots (0.002-0.012 in. wide); honeycomb cores and assemblies; and other fragile parts.

(continued)

Process summary chart for ECM and EDM (continued)

	Electrochemical machining (ECM)	Electrical discharge machining (EDM)
Operating parameters and typical values	Gap voltage: 4-24 Amperage: Commercial units 50-40,000 Gap spacing: End 0.001 in.-0.020 in. Side 0.005 in.-0.020 in. Electrolyte flow: 20-200 ft./sec. Electrolyte pressure: 10-400 psi Electrolyte temperature: 75-150 °F Electrolyte downfeed: 0.020-0.750 in./min. at 100-2000 amp/sq. in. Electrodes: Brass, copper, stainless steel, titanium, platinum, sintered copper-tungsten, aluminum, graphite. Electrolyte: Proprietary mixtures, sodium chloride solution up to $2\frac{1}{2}$ lb./gal.; sodium nitrate solution up to 5 lb./gal.; sulfuric acid 30%, sodium nitrate and sodium chloride mix. Workpiece: Must be electrically conductive.	Voltage: Up to 300 Amperage: Commercial units 0.1-500 Frequency: 200-500,000 Hz Dielectric fluid pressure: 0-70 psi; sometimes vacuum (28 in. Hg) Electrode materials: Graphite, copper, brass, copper-tungsten, silver-tungsten, tungsten, and tungsten carbide Fluids: Hydrocarbon (petroleum) oils Workpiece: Must be electrically conductive. Power source rating: Single Units up to 500 amp.
Tolerances	Practical: ±0.005 in. Possible: ±0.0005 in.	Practical: ±0.002-0.005 in. Possible: ±0.0001-0.0005 in.
Surface	4 to 50 R_a can be attained. No heat affected surface or burrs created. Guard against selective etching in remote areas exposed to electrolyte by shields or dams.	Finish is affected by removal rate: 0.015 in³/hr-30 R_a 0.5 in³/hr-200 R_a 3.0 in³/hr-400 R_a Heat affected zone: 0.0001-0.005 in.
Practical removal rates	Approximately 0.1 cu. in./min/1000 amp or 1 cu. in./min. for a 10,000 amp unit.	0.01-25 cu. in./hr. depending upon tolerances and finish requirements and EDM paramaters.
Machining characteristics	No electrode wear; process may produce overcut, taper, and corner radiuses depending upon tool design; observed values; taper 0.001 in. over entire depth, overcut 0.005 in., corner radius 0.015 in.; tools subject to damage by arcing if process malfunctions; no machining stresses introduced; burr free machining.	Produces taper and overcut (finished cavity minus original electrode size) in workpiece, also corner radiuses. Electrode wear ratio (WR) measured and expressed as: $$\text{Wear Ratio (WR)} = \frac{\text{Volume of work removed}}{\text{Volume of electrode consumed}}$$ or $$\text{Wear Ratio (WR)} = \frac{\text{Depth of cut}}{\text{Length of electrode consumed}}$$ Typical values: Taper, 0.0005-0.005 in./in./side; overcut 0.0002-0.035 in./side; minimum corner radius 0.0005 or equal to overcut. Typical electrode wear ratios: Metallic electrodes 3:1; carbon 3:1 to 100:1.

Process summary chart for CHM and AJM

	Chemical machining (CHM)	Abrasive jet machining (AJM)
Principle	Metal removed by chemical or electrochemical attack of preferentially exposed surfaces. Essential steps: cleaning part; masking with tapes or resistant paints, or printing, using photoengraving techniques; etching; demasking; cleaning. Two processes involved: chemical milling, chemical blanking.	Fine abrasive particles carried in a high velocity gas stream are used to machine and grind materials.
Equipment	Chemical milling: Large or small, thick parts: Masking facilities, corrosion resistant processing tanks and fixtures, vented tanks or rooms. Chemical blanking: Large or small parts depending upon limitation of etching facilities (including thin sheets): Tooling (artwork and photographic negatives, silk screens for masking with etch resistant lacquer), layout tables, photoengraving equipment, including manual or automatic and continuous spray etching machines.	Mixing chamber containing abrasive particles into which gas (generally air) is introduced to entrain the abrasives and carry them to a handpiece containing a nozzle. Appropriate control switches and accessory equipment, such as a dust collector, exhaust chamber, air compressor, and air filter, are required.
Typical applications	Chemical milling: Shallow cavities or pockets; over-all weight reduction; tapered sheets, plates, or extrusions for airframes. Chemical blanking: Printed circuit etching, decorative panels, thin stampings. Applicable to aluminum, magnesium, iron, copper, nickel and cobalt base alloys, refractory alloys such as tungsten, columbium, molybdenum.	Drilling, slotting, or sawing of hard, brittle materials, such as glass, germanium, and silicon; deburring; surface finishing on glass and other materials; scribing; resistor trimming.
Operating parameters and typical values	Masking with tapes or paints common in chemical milling. Printing using photoresists common in chemical blanking. Etchant (function of metal and often proprietary): For aluminum, solutions of sodium hydroxide. For steel, solutions of hydrochloric plus nitric acids. For copper, solutions of iron chloride plus nitric acid. Etchants can be conductors. Temperature of etchant: Nominally 70 to 210 °F. Demasking solution: Various solvents. Cleaning or desmutting solution: water, chromic acid, hydrochloric acid.	Abrasive Type: Al_2O_3 and SiC Particle Sizes: 10, 27, and 50 microns (powders must be sized closely and free of silica. Quantity Used: 10 to 20 g./min. nonrecirculating for cutting—3 to 5 g./min. for fine work such as resistor trimming. Abrasive Carrier or Medium: Air, CO_2, Nitrogen at 30 to 120 psi with nozzle to provide velocities 500 to 1000 ft./sec. Quantity of Gas Used: approximately $\frac{1}{3}$ cu. ft./min. DO NOT USE OXYGEN. Nozzle Type: Tungsten carbide or sapphire. Work distance from nozzle, 0.010-3 in. Typical Sizes—Round: 0.007 to 0.032 in. dia.; Rectangular: 0.003 × 0.020 in. to 0.026 × 0.026 in. to 0.007 × 0.150 in.

(continued)

Process summary chart for CHM and AJM (continued)

	Chemical machining (CHM)	Abrasive jet machining (AJM)
Tolerances	For chemical milling and chemical blanking. Metal removal or thickness of sheet: 0.002 in. — tolerance, ±0.001-0.002 in.; 0.020 in. — tolerance, ±0.004-0.010 in.; 0.060 in. — tolerance, ±0.006-0.012 in. Tolerances are function of masking or printing technique, configuration, time, temperature, and condition of the bath.	Practical: ±0.005 in. Possible: ±0.002 in.
Surface	Average values: Aluminum, 90 R_a; magnesium, 50 R_a; steel, 60 R_a; titanium, 25 R_a; tungsten, 50 R_a.	AJM produces a matte finish. 27 micron aluminum oxide or silicon carbide — 14-20 R_a 50 micron aluminum oxide or silicon carbide — 38-55 R_a 10 micron aluminum oxide — 6-8 R_a.
Practical removal rates	Penetration: 0.0005-0.003 in./min. Note: Optimum rates for best control vary for different materials, e.g., 0.001 in./min. is excellent for aluminum.	0.001 cu. in./min. which is equivalent to making a slot 0.020 in. wide × 0.010 in. deep and 5 in. long in one minute.
Machining characteristics	In etching, any masked area is undercut. Etch factor in chemical milling is undercut divided by depth of cut. Etch factor in chemical blanking is ratio of depth of cut to undercut. Undercut limits minimum hole diameter or slot width. In chemical blanking, small etch factors desired. In chemical milling, large etch factors sometimes desired for smooth transition to minimize stress concentration. Chemical blanking, minimum hole diameter or slot width: Copper alloys, 0.7 times sheet thickness; aluminum, 1.4 times sheet thickness; steel, 1.0 times sheet thickness; stainless, 1.4 times sheet thickness. Inside corner radiuses: Equal to thickness of sheet being chemically machined for most alloys. Outside corner radiuses: equal to $\frac{1}{3}$ thickness of sheet; $\frac{1}{16}$ in. thickness is maximum practical limit for blanking of sheets.	Nozzle life: approximately 10 hrs. for tungsten carbide; approximately 300 hrs. for sapphire. Slots or holes made are subject to taper which increases as distance between nozzle and workpiece (nozzle tip distance — NTD) increases.

Process summary chart for LBM and EBM

	Laser beam machining (LBM)	Electron beam machining (EBM)
Principle	Material removal is accomplished by converting stored electrical energy into a single-wavelength, collimated beam of light in or near the visible spectrum and focusing it on a workpiece. This beam is capable of vaporizing and melting all materials and can be utilized for machining and welding. Note: Welding, an important application for lasers, is not covered in this process summary chart.	High velocity electrons focus on workpiece and vaporize material.
Equipment	Power supply, Excitation source (the "pump"), Lasing material, Focusing lens, Work handling fixtures CNC control	Electron beam cutter with workpiece in vacuum of 10^{-4} mm of mercury or better.
Typical applications	Small holes in all types of material. Thick and thin film resistor trimming. Scribing of silicon and ceramic substrates. Removal of small increments of material for balancing and frequency tuning. Cutting of titanium, cardboard, wood, and fabric.	Drilling holes as small as 0.0005 in. practically instantaneously in all materials, including ceramics; cutting closely spaced thin slots — example: $\frac{1}{2}$ in. long slots 0.005 in. wide, spaced 0.010 in. apart in 0.025 in. thick alumina; scribing of thin films; removing broken taps of small diameter.
Operating parameters and typical values	Wavelength (Lasing Material): Ruby (0.03-0.07% Cr) — .694 microns Nd-Glass (Neodymium Glass, 2-6% Nd) — 1.06 microns YAG (Yttrium aluminum garnet, 1% Nd) — 1.06 microns CO_2 (CO_2 — He — N_2 gas laser) — 10.6 microns Excitation Source: Ruby — Pulsed xenon flash lamp Nd-Glass — Pulsed xenon flash lamp YAG — Pulsed xenon flash lamp — Continuous wave (CW) tungsten or krypton lamp CO_2 — CW or pulsed electrical discharge Practical Pulse Repetition Rate: Ruby — 1-2 pulses/second Nd-Glass — 1-2 pulses/second YAG — 1-10,000 pulses/second CO_2 — 1-7000 pulses/second Peak Output Power (calculated): Ruby — 1500-19,000 watts Nd-Glass — 1000-75,000 watts YAG — 5000-17,000 watts pulsed — 300 watts CW CO_2 — 1000 watts CW	Accelerating voltage: 50,000-150,000 Beam current: 0-1000 microamp Pulse width: 4-64,000 microseconds Pulse frequency: 0.1-16,000 Hz Capacity: 1150 watts, continuous

(continued)

Process summary chart for LBM and EBM (continued)

	Laser beam machining (LBM)	Electron beam machining (EBM)
Tolerances	Same size hole reproducible within 5% of diameter. In materials more than 0.020 in. thick, taper becomes noticeable but less in plastics.	On 0.125 in. holes, ±0.001 in. On 0.0005 in. holes, ±0.00005 in.
Surface	Thin heat affected layers consisting mainly of vaporized and redeposited metal are present. Cratering occurs and is dependent upon factors such as energy, quality of optics, nature and thickness of material.	Finish data not available. Incident surface slightly cratered and walls of 0.125 in. holes contain refuse readily removed mechanically. Heat affected zone is practically nonexistent.
Practical removal rates	Holes: 0.0002 in. to 0.040 in. dia. in thicknesses ranging from 0.002 in. to 0.050 in. in metals and ceramics and to 1 in. in plastics at rates of 1-200/second. Cutting: Titanium alloys (0.020-0.400 in. thick) CO$_2$ Laser — oxygen assist 100-500 in./min. Cardboard (0.020-0.125 in. thick) CO$_2$ Laser 100-500 in./min. Plastics (0.030-0.125 in. thick) CO$_2$ Laser 100-300 in./min. Scribing: Glass — up to 500 in./min.	0.0005 in. slots 10-24 in./min. in 0.010 in. material. Holes in sheet materials of all kinds 0.001-0.025 in. thick; 0.0125 in. diameter holes, less than 2 seconds; 0.0005 in. diameter holes less than 0.1 second.
Machining characteristics	The laser has capability for rapid removal of all materials without contact of laser with workpiece and special capability for micro-machining and for machining at difficult angles and in inaccessible areas. Lasers have limited applicability for materials which have high thermal conductivity and high reflectivity (poor spectral absorptance). Spectral absorptance of materials also varies with the wavelength of the laser type being used, condition of the surface, temperature, etc. Generally, Ruby, Nd-Glass, and YAG lasers are more applicable to metals, while CO$_2$ lasers are preferred for nonmetallic materials (e.g., plastics, cardboard, and wood).	Beam is about 0.0005 in. in diameter. For holes larger than 0.001 in., spot is deflected or rotated. Holes up to $^3/_8$ in. diameter or $^3/_8$ in. square can be trepanned. Holes in 0.025 in. thick sheet have hour-glass figure, with minimum diameter at neck, largest on side of material incident to beam. Holes in 0.005 in. material practically parallel sided.

Process summary chart for USIG

	Ultrasonic impact grinding (USIG)		Ultrasonic impact grinding (USIG)
Principle	Tool vibrated around 20,000 cycles per second by magnetostrictive transducer. Fine abrasive particles in a water slurry between tool and workpiece reach high velocities, dislodge material from workpiece. Cavity produced assumes shape of tool.	Tolerances	Practical: ±0.001 in. Possible: 0.0005 in. (total).
Equipment	Machine tool equipped with transducer, generator power supply; specially shaped toolholder and tool; abrasive powder; pump for abrasive powder-water mix.	Surface	Roughing: 25 R$_a$ Finishing: 10 R$_a$ No heat affected surface produced.
Typical applications	Machining of nonmetallic, brittle, or hard materials, such as semiconductors (silicon, germanium), ceramics, glass silicon carbide, tungsten carbide; production of accurate and odd shapes in nonmetallic, brittle, or hard materials — generally applied to materials harder than 64R$_c$ ($^1/_{16}$ in. thickness max. at 64R$_c$). Minimum practical hardness for steel is 45R$_c$.	Practical removal rates	Feed rate: (in./min.), tungsten carbide, 0.005; silicon carbide, 0.010; ceramics, 0.050; silicon (pure), 0.070. Average volume removal rates: (cu. in./min.), hardened tool steel, 0.0001; boron carbide, 0.0002; hardened stainless steel, 0.0008; silicon, 0.005; germanium, 0.006; carbon, 0.015. Rate diminishes as abrasive breaks down; periodic additions and replacement desirable. Rate increased by forced application of abrasive.
Operating parameters and typical values	Frequency: 19,000-25,000 Hz Transducer feed: 1-2 in. Length of tool: Designed for maximum linear tool motion (0.0005-0.0025 in. at end of tool) at 30 ft./sec. Toolholder material: Monel or stainless steel. Tool material: Cold rolled steel or stainless steel. Power ranges: 50-2400 watts Abrasive: Usually boron carbide or silicon carbide; also aluminum oxide; diamond dust has been used. Abrasive grain size: Roughing 200-400 mesh, finishing 800-1000 mesh. Abrasive carrier: Usually water mixed with about 50% of abrasive, by volume	Machining characteristics	Finish size of cavities is function of abrasive grit size. Size of cavity equals size of tool, plus twice the particle size. Ratio of stock removed to tool wear averages about 10:1. Design and manufacture of toolholder and tools are critical. Monel excellent for toolholder. Tools are mild steel rods or tubes, annealed drill blanks, and stainless such as Type 303. In making holes, use thin wall tubes to increase rate.

Process summary chart for HDM and TCM

	Hydrodynamic machining (HDM)	Thermochemical machining (TCM)
Principle	Hydrodynamic machining (HDM) removes material by the impingement of a high-velocity fluid against the workpiece. The coherent jet of water or water with additives (to aid coherence or prevent freezing) is propelled at speeds up to Mach 3 thereby cutting or shearing the workpiece. A synthetic-sapphire nozzle controls the jet.	Thermochemical machining (TCM) removes workpiece material — usually only burrs and fins — by exposure of the workpiece to hot, corrosive gases. The process is sometimes called combustion machining, thermal deburring, or thermal energy method (TEM). The workpiece is exposed for a very short time to extremely hot gases, which are formed by detonating an explosive mixture. The ignition of the explosive — usually hydrogen or natural gas and oxygen — creates a transient thermal wave that vaporizes the burrs and fins. The main body of the workpiece remains unaffected and relatively cool because of its low surface-to-mass ratio and the brevity of the exposure to high temperatures.
Equipment	Machine and hydraulic components are available for HDM; however, a general-purpose machine tool is not regularly available. Each application is engineered to meet the requirements as found by sample test cuts made to determine the exact values for the key operating parameters. Equipment generating pressures up to 60,000 pounds per square inch [415 MPa] is commercially available with controls, filters and hydraulic seals of reasonable durability.	Automatic equipment is commercially available with chambers up to 11 inches [280 mm] diameter and 9 inches [229 mm] high. A cycle time of 15 seconds is possible with this equipment. Fixturing is primarily oriented toward simplifying the movement of workpieces into and out of the processing chamber. Delicate parts may require holding fixtures to withstand buffeting during the thermal shock wave. Bulk loading baskets are used for handling small parts in lots up to hundreds per load.
Typical applications	The ability of HDM to cut very thin, soft metals or nonmetallic materials in any position with a very narrow kerf leads to many form-cutting applications, from paper, wood, cloth, leather, plastics, composites, fiberglass, and other similar type materials.	TCM will remove burrs or fins from a wide range of materials, but it is particularly effective with materials of low thermal conductivity. It will deburr thermosetting plastics — but not thermoplastic materials. Any modest size workpiece requiring manual deburring or flash removal should be considered a candidate for thermal deburring. Die castings, gears, valves, rifle bolts and similar small parts are deburred readily, including blind, internal and intersecting holes in inaccessible locations. Carburetor parts are processed in automated equipment.

Operating parameters and typical values	Fluid
	  type: Water or water plus additives — well filtered
	  additives: Glycerine, polyethylene oxide, long-chain polymers
	  pressure: 10 to 60 ksi [69 to 415 MPa]
	  jet velocity: 1,000 to 3,000 ft/s [305 to 915 m/min]
	  flow: Up to 2 gal/min [7.5 L/min]
	  jet force on workpiece: 1 to 30 lb [4.5 to 134 N]
	Power: Up to 50 hp [38 kW] available
	Nozzle material: Synthetic sapphire most common, also hardened steels, 17-4 stainless
	  diameter: 0.003 to 0.015 inch [0.075 to 0.38 mm]
	  angle: Perpendicular to positive rake angles of 30°
	Kerf width: 0.003 to 0.016 inch [0.075 to 0.41 mm]
	Standoff distance: 0.1 to several inches [2.5 to 50 mm]; $1/8$ inch [3 mm] is typical
		Gas
		  type: Natural gas and oxygen or hydrogen and oxygen
		  mixture: 4:1 to 9:1 (oxygen:gas)
		  pressure (initial): 10 to 370 psi [69 to 255 kPa]
		Thermal wave temperature: Estimated 6,000 °F [3,315 °C]
		  time: 1 to 2 microseconds: Mach 8

Tolerances	The kerf is about 0.001-inch [0.025 mm] larger than the orifice utilized. For composite materials, the best combination is a high-pressure jet from a small nozzle operating at a slight rake angle and close to the workpiece. Taper can be a factor if too large of a standoff distance is used.	Thermal distortion is not a problem unless extremely close tolerances are involved. Thin sections of fragile parts or those made from highly oxidation-resistant materials may be difficult for TCM. As a rule of thumb, the maximum burr thickness should be less than one-fifteenth of the thinnest feature on the workpiece. Uniformity of results and greater quality assurance over hand deburring is a special advantage of TCM.
Surface	Edge quality depends upon how easily the workpiece fractures. Soft materials cut smoothly and crushable materials can be slit with high quality edges. Test cuts are recommended before selecting the final operating parameters. The high energy density of the jet stream causes a slight rise in temperature which may melt some types of plastics at low traverse rates, but the temperature is not high enough to be a concern for paper-type materials. Tendency to delaminate layers of composite materials during cutting can be eliminated by reducing the cutting rate.	Other than the deburring or radiusing action, there is little effect of TCM on the surface of the workpiece. Surface stains may result if the workpiece is not clean, dry and free of oil before TCM. There can be a thin recast layer below thick burrs or fins. A thin oxide film is deposited on most parts. These films are easily removed with commercial solvents or washes.
Practical removal rates	Cutting rates depend on the work material and vary directly with the horsepower applied and inversely with the material thickness. Rates up to 6,000 feet per minute [1,830 m/min] have been attained on paper products. Rates of 100 to 1000 feet per minute [3 to 305 m/min] are commonly used.	Removal is accomplished in 1 to 2 microseconds. Cycle time depends upon the degree of automation of equipment.
Machining characteristics	The "chips" mix in the fluid so dust is eliminated; therefore, explosion and fire hazards are reduced. Air entrapped in the jet stream can create considerable noise with some work materials. The noise level increases with standoff distance but can be reduced by using certain fluid additives or certain rake angles. The noise levels are generally below the OSHA requirements. Periodic preventive maintenance is very desirable for equipment operating at these pressures.	The equipment has a stout chamber to contain the detonation. Preplanned handling of the pressurized gases is needed. Noise can range from a gentle "ping" to the sharp report like a .22-caliber rifle when the higher pressure settings are used. Even at full load, the average sound emission level generally remains below 85 decibels. Cleaning of the operating chamber is required at regular intervals.

Process summary chart for AFM and EDWC

	Abrasive flow machining (AFM)	Electrical discharge wire cutting (EDWC)
Principle	Abrasive flow machining (AFM) is the removal of material by a viscous, abrasive media flowing, under pressure, through or across a workpiece.	Electrical discharge wire cutting (EDWC) is a special form of electrical discharge machining wherein the electrode is a continuously moving conductive wire. EDWC is often called traveling-wire EDM. A small-diameter, tensioned wire is guided to produce a straight, narrow-kerf cut. The slowly moving wire brings a fresh, constant-diameter electrode to the cutting gap, thereby enhancing kerf size control. Usually, a programmed or numerically controlled motion guides the cutting, while the width of the kerf is maintained by the wire size and discharge controls. The dielectric is oil or deionized water carried into the gap by motion of the wire. The wire is inexpensive enough to be used only once.
Equipment	Typical equipment consists of a two-chamber unit in which the viscous, abrasive media is pushed from one chamber through the workpiece by a hydraulically operated piston into a receiving chamber (100 to 3000 psi) [700 to 20,000 kPa]. The process is then reversed and then is repeated from 1 to 100 flow reversals per fixture load.	Several manufacturers regularly build EDWC equipment with NC, tracer controls and all programming accessories. Table motions as much as 2 by 3 feet [0.6 by 1 m] have been made, with 1 by 1 foot [0.3 by 0.3 m] being more common. Workpiece thickness 3 to 6 inches [75 to 150 mm] can be accommodated.
Typical applications	Edge finishing, radiusing, deburring, polishing, and minor surface material removal are accomplished with AFM. It is not a mass material-removal process, but it is particularly useful for polishing or deburring inaccessible internal passages. Materials from soft aluminum to tough nickel alloys are being processed with AFM. Removal of undesired layers produced by thermal processes such as EDM, LBM, or nitriding is also achieved with AFM.	Punches, dies and stripper plates can be cut in any of the hardened, conductive tool materials. Stacking of sheets for multiple cutting is possible. Conventional carbide dies can frequently be cut from a solid plate using EDWC, without the conventional segmenting and fitting.

Operating parameters and typical values	Media
	  viscosity: stiff to fluid
	  grit size: #8 to #1000
	  starting temp.: 90° to 125 °F [32° to 52 °C]
	  grit types: aluminum oxide, boron carbide, silicon carbide, and diamond
	Flow
	  pressure: 100 to 3000 psi [700 to 20,000 kPa]
	  volume: 3 to 100 fluid ounces [100 to 3000 ml]
	  rate: 2 to 60 gal/min [7 to 225 L/min]
	Strokes: 1 to 100
		Power supply
		  type: 55 to 60 V (open circuit volts to 300)
		  frequency: Pulse time controlled 1 to 100 μs in time or 180 to 300 kHz with 3 kHz most frequent
		  current: 1 to 32 A
		Electrode wire
		  types: Brass, copper, tungsten, molybdenum
		  diameter: 0.003 to 0.012 inch [0.076 to 0.30 mm]; most frequently used size is 0.008 inch [0.2 mm]
		  speed: 0.1 to 6 in/s [2.5 to 150 mm/s]
		
		Dielectric: Deionized water, oil, or rarely, air, gas or plain water
		Overcut
		  (working gap): 0.0008 to 0.0020 inch [0.02 to 0.05 mm]; usually 0.001 inch [0.025 mm]

Tolerances	0.0002 in. [0.005 mm]	Workpiece accuracy usually is ±0.0005 inch [±0.013 mm], with special instances of ±0.0002 inch [±0.005 mm] and ±0.0001 inch [±0.0025 mm] in unique applications.
Surface	R_a = 2 to 10 microinches [0.050 to 0.25 micrometer] Surfaces are typically smoothed to $^1/_{10}$ of the prior roughness in terms of R_a	Surface roughness is typically in the 30- to 50-microinch R_a [0.8 to 1.3 μm] range, and the surface has a matte or velvetlike texture. Special care and slower cutting can produce a surface roughness of 15 microinches R_a [0.4 μm]. The recast and the heat-affected zone are very small and uniform with the low spark energy levels that are typically used. These layers should be removed or modified on critical or fatigue-sensitive surfaces.
Practical removal rates	Typically 0.005 to 0.020 in/min [0.0013 to 0.5 mm] depending upon workpiece configuration, grit type and size, pressure, and temperature.	Cutting of 0.001 to 6 inch [0.25 to 150 mm] thick materials can be done at a rate of approximately 4 square inches per hour [26 cm²/hr], which on thin parts can yield cutting at 40 inches per minute [1 m/min]
Machining characteristics	More uniform deburring than can be accomplished with hand tools is a significant quality advantage for AFM. Careful post-operation cleaning is recommended. This cleaning can readily be accomplished by an air blast or immersion in an agitated solvent. Machining action is gentle and continuous, and the burrs or "chips" are retained in the media. The media can tolerate as much as 10 percent of its volume in "chips" or foreign material. The media life is limited at the point at which a substantial number of the grains become dull. Reconditioning of the media can be accomplished by measured additions of abrasive grit and lubricant.	The machine motions are similar to those for a band or a wire saw. Use of cutter offset control permits several sizes to be cut from the same NC tape. The cutting accuracy is high because the spark erosion emanates from a fresh, constant-size section of wire.

28 HEAT TREATING

Edited by Ross B. Shingledecker, Ladish Co.

Review Committee: Fred Bartkowski, Nelson Marshall & Associates, Inc.; Robert Blumenthal, Marquette University; Alfons De Ridder, Ladish Co.; David P. Dixon, Ampco Metal; Robert W. Foreman, Park Chemical Co.; R. E. Haimbaugh, Induction Heat Treating Corp.; J. A. Hildebrandt, Clark Equipment Corp. (retired); W. James Laird, Jr., Upton Industries, Inc.; James T. Staley, Aluminum Company of America; Dennis M. Wagen, W B Combustion, Inc., Daniel S. Zamborsky, Bendix Corp.

This section was condensed from Metals Handbook, Ninth Edition, Volume 4, Heat Treating. For more detailed information, the reader is referred to the larger work.

Physical Metallurgy and the Heat Treatment of Steel

By George Krauss, Colorado School of Mines

THE HEAT TREATMENT of steel is based on the physical metallurgical principles which relate processing, properties and structure. In heat treatment, the processing is most often entirely thermal and modifies only structure. Thermomechanical treatments, which modify component shape and structure, and thermochemical treatments, which modify surface chemistry and structure, are also important processing approaches which fall into the domain of heat treat-

ment. Scientific principles link the processing parameters to structure and properties, and are increasingly necessary for proper application of the equipment and instrumentation now available for control of heat treatment processes. Examples of scientific efforts which directly support the technology of heat treatment include characterization of mechanisms of phase transformations which produce desired structures and properties of heat treated parts; determination of

phase transformation and annealing kinetics which establish processing times, temperatures and cooling rates for heat treatments; and evaluation of mechanisms of deformation and fracture of the structures produced by heat treatment.

In view of the importance of structure and its formation to heat treatment, the purpose of this article is to describe the various microstructures which form in steels, the various factors which determine the formation of microstructures during heat treatment processing of steel, and some of the characteristic properties of each of the microstructures. Structure-sensitive properties such as strength, ductility and toughness establish the ease of manufacturing, service performance, and limitations to service conditions of heat treated steels.

The descriptions of the microstructures and principles presented here should be considered only introductory, and the references listed at the end of this article should be consulted for more information. The details of the various heat treatments for many grades of steel are presented in the subsequent articles of this section.

THE IRON-CARBON PHASE DIAGRAM

The microstructures which result from heat treatment of steel are composed of one or more phases in which the atoms of iron, carbon and other elements in steel are associated. Figure 1 shows a portion of the iron-carbon phase diagram from pure iron through the carbon concentration of cementite, 6.67 wt %. The temperature and composition ranges in which the various phases exist are shown on the diagram. Alloys containing up to 2 wt % carbon are classified as steels; alloys containing more than 2 wt % carbon are classified as cast irons. The solid lines represent conditions where carbon, when it exceeds its solubility in ferrite and austenite, is present in the form of cementite (Fig. 1). This is invariably the case in steels. The dashed lines represent the conditions where carbon is present as graphite rather than as cementite, a situation much more common in cast irons than in steels.

In steels, the temperatures which are the boundaries of the various phase fields are frequently referred to as critical temperatures. Since the critical temperatures are often identified by changes in slope or thermal *arrests* in heating and cooling curves, they are given the designation "A." If equilibrium conditions are applicable, the designations Ae_1, Ae_3 and Ae_{cm}, or simply A_1, A_3 and A_{cm}, are used as shown in Fig. 1. If heating conditions (which raise the critical temperatures relative to equilibrium) apply, Ac_1, Ac_2 and Ac_{cm} are used, the subscript "c" being derived from the French word *chauffant*. If cooling conditions (which lower the critical temperature relative to equilibrium) apply, the designations Ar_1, Ar_3 and Ar_{cm} are used, the subscript "r" being derived from the French word *refroidissant*. There is hysteresis in the transformation temperatures because continuous heating and cooling leave insufficient time to accomplish the diffusion-controlled phase transformations at the true equilibrium temperatures.

Steels and cast irons contain, in addition to iron and carbon, many other elements which shift the boundaries of the phase fields in the Fe-C diagram. Some alloying elements such as Mn and Ni are austenite stabilizers and extend the temperature range over which austenite is stable. Elements such as Cr and Mo are ferrite stabilizers and restrict the ranges of austenite stability. Therefore, care must be taken in the direct use of the Fe-C diagram to predict phase relationships in commercial alloys which contain elements in addition to Fe and C. Nevertheless, the iron-carbon diagram is the most important reference for understanding the relationships between structure and heat treatment of steels, and, subject to the above limitations, will be used in

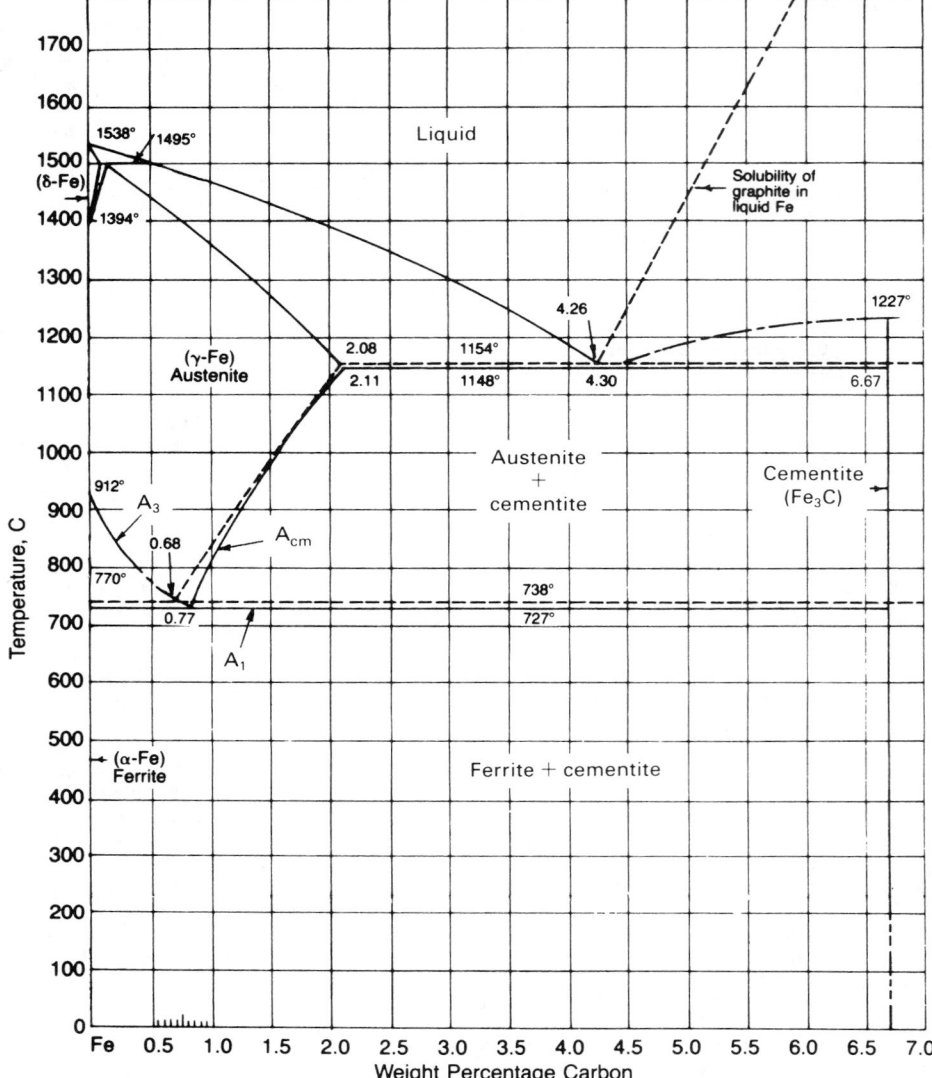

Fig. 1. The Fe-C equilibrium diagram up to 6.67% C. Solid lines indicate Fe-Fe₃C diagram; dashed lines indicate Fe-graphite diagram. (Ref 1)

this article to illustrate the basis for microstructural formation in steels as well as Fe-C alloys.

The phase diagram shown in Fig. 1 assumes equilibrium—i.e., that the carbon and iron have had sufficient time to distribute themselves in the various phases as shown. Sometimes, equilibrium is difficult to achieve, especially in steels which contain elements which diffuse only sluggishly, and, in fact, certain heat treatments such as hardening are designed to prevent formation of equilibrium structures. Thus the fact that equilibrium may not be achieved, together with the shift of the phase-field boundaries by alloying elements, place limitations on the direct use of the Fe-C phase diagram.

Austenite, also referred to as γ-iron, is the face-centered cubic crystal form or phase of iron which is stable at high temperatures. Figure 1 shows that carbon in Fe-C alloys is soluble in austenite up to just over 2 wt %, and that the single-phase austenite field dominates the Fe-C diagram at high temperatures. In all low-alloy steels, therefore, it is possible to produce a single-phase austenite microstructure. This characteristic is perhaps the

most important feature of steels in that it enables steels to be hot worked or wrought. Also, cooling from the single-phase austenite field makes possible a wide variety of heat treatments based on transformation of the austenite.

The single-phase austenite, without the obstacles which second phases present to dislocation motion and without the sites for fracture initiation which second-phase particles offer, deforms and recrystallizes readily so that substantial reductions in section size by hot rolling or forging may be accomplished. Traditionally, hot deformation is performed in the upper temperature range of the austenite field. Hot deformation of austenite at lower temperatures or even in the two-phase ferrite-austenite field (controlled rolling), and the addition of small amounts of alloying elements (microalloying) such as niobium and vanadium, which precipitate as fine alloy carbonitrides at low temperatures, are new approaches to processing of steels (Ref 2 and 3). The low-temperature deformation and/or precipitation retard or prevent austenite recrystallization and grain growth, and therefore produce finer

Fig. 2. Scanning electron micrograph showing cleavage of a body-centered cubic microstructure. Courtesy of F. Zia-Ebrahimi.

austenite grains and subsequently fine austenite transformation products during cooling after hot deformation.

Ferrite, also referred to as α-iron, is the body-centered cubic form or phase of iron which is stable at low temperatures. Microstructures in low-carbon steels which consist largely of polycrystalline ferrite are highly formable at room temperature; dislocations move readily on the many slip systems of the body-centered cubic structure (Ref 4). However, at low temperatures, dislocation motion in the body-centered cubic structure is severely restricted (Ref 5 and 6). As a result, ferrite grains fracture in a brittle manner with little plastic deformation at low temperatures. Figure 2 shows an example of the brittle fracture which develops in ferrite stressed at low temperatures and/or high strain rates. The fracture is termed cleavage because it occurs by cleaving or separation across {100} planes of the body-centered cubic structure. Thus cleavage is a direct reflection of the crystal structure of ferrite and presents a major limitation to the use of steels under certain service conditions.

Carbon, because of its small atomic size, is dissolved in the octahedral interstitial sites between iron atoms in ferrite and austenite (Ref 7). When the solubility of carbon in either austenite or ferrite is exceeded, the phase cementite, also referred to as Fe_3C or θ-carbide, forms. The compound cementite has higher strength and lower ductility than either ferrite or austenite and, depending on its morphology and distribution, contributes in a variety of ways to the strengthening, deformation and fracture of steels.

The interstitial sites for carbon in ferrite are much smaller than those in austenite, and therefore the solubility of carbon in ferrite is significantly lower than that in austenite. Figure 3 shows an expanded portion of the iron-rich side of the Fe-C diagram. The maximum solubility of carbon in ferrite is only about 0.02 wt % and with decreasing temperature becomes almost negligible. As a result of the decreasing solid solubility with decreasing temperature, on slow cooling, cementite forms on ferrite grain boundaries. If, for some reason, cooling is too rapid for cementite formation, the carbon is trapped in the interstitial sites and contributes to various aging phenomena unique to ferrite steels (Ref 5 and 8). The one process is associated with segregation of carbon atoms to dislocations and grain boundaries, and is referred to as strain aging. The other process is associated with precipitation of fine carbide particles either on dislocations or in the ferrite

matrix and is referred to as quench aging. Figure 4 shows an example of fine dendritic cementite particles which have formed by quench aging on dislocations in the ferrite of a low-carbon steel. Both strain aging and quench aging effectively pin dislocations, and are responsible for the discontinuous yielding of low-carbon steels with largely ferritic microstructures.

PEARLITE AND BAINITE

Figure 1 shows that the austenite in an iron-carbon alloy containing 0.77 wt % carbon must transform to ferrite and cementite at 727 °C. A solid-state reaction in which one phase transforms to two other phases is referred to as a eutectoid reaction. In Fe-C alloys and steels, a unique parallel array of ferrite and cementite lamellae termed pearlite develops as a result of the eutectoid reaction. Figure 5 shows pearlite which has formed in a eutectoid steel; here, the cementite appears white, and the ferrite gray.

Pearlite in a eutectoid steel is nucleated at austenite grain boundaries, and grows as spherical-shaped colonies or nodules into the austenite. Carbon must diffuse to the growing cementite lamellae of the pearlite. Also, iron atoms must rearrange themselves by short-range diffusion from the face-centered cubic structure of austenite to their arrangements in the crystal structures of ferrite and cementite at the interface of the growing pearlite colonies. The rate of transport of carbon and iron atoms is temperature dependent and in-

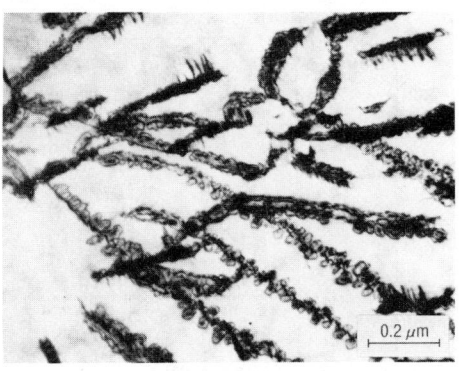

Fig. 4. Transmission electron micrograph showing cementite precipitated on dislocations in an 0.08C-0.63Mn steel aged 115 h at 97 °C. Courtesy of J. E. Indacochea (Ref 9).

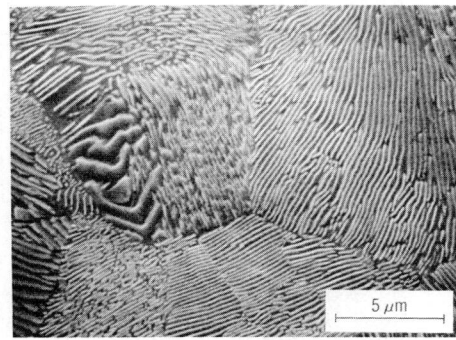

Fig. 5. Scanning electron micrograph showing pearlite in a eutectoid rail steel. Courtesy of F. Zia-Ebrahimi.

creases exponentially with increasing temperature.

At temperatures just below the eutectoid temperature, 727 °C in the Fe-C system, the thermodynamic driving force for the eutectoid reaction (the decrease in free energy per unit volume when austenite is replaced by pearlite) available to offset the increase in energy associated with pearlite colony–austenite interfaces and the ferrite-cementite interfaces within the pearlite colonies is low. As a result, the nucleation rate of colonies is low and the spacing of cementite lamellae within the colonies is large. The coarse interlamellar spacing increases the diffusion distance for carbon, and causes a low rate of growth for those colonies which manage to nucleate. Thus, pearlite transformation at temperatures close to the eutectoid temperatures is sluggish and the pearlite microstructure which forms is relatively coarse. With increased undercooling, the thermodynamic driving force increases, the nucleation rate of pearlite colonies increases, interlamellar spacings decrease, and the growth rate of colonies increases. As a result of the latter changes, the transformation of austenite to pearlite accelerates with decreasing temperature.

Figure 6 shows an isothermal transformation diagram for a eutectoid steel. The diagram shows the beginning and end of the eutectoid transformation of austenite to pearlite for specimens cooled from the single-phase austenite field and held isothermally at temperatures between A_1 and

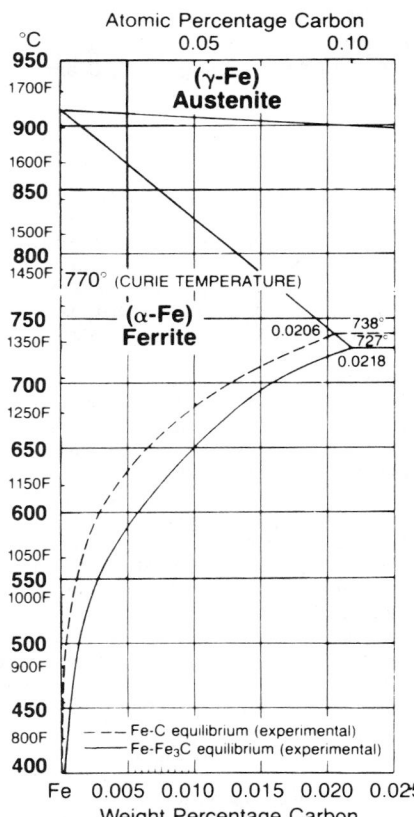

Fig. 3. Fe-rich side of Fe-C diagram, showing extent of ferrite phase field and decrease of carbon solubility with decreasing temperature (Ref 1)

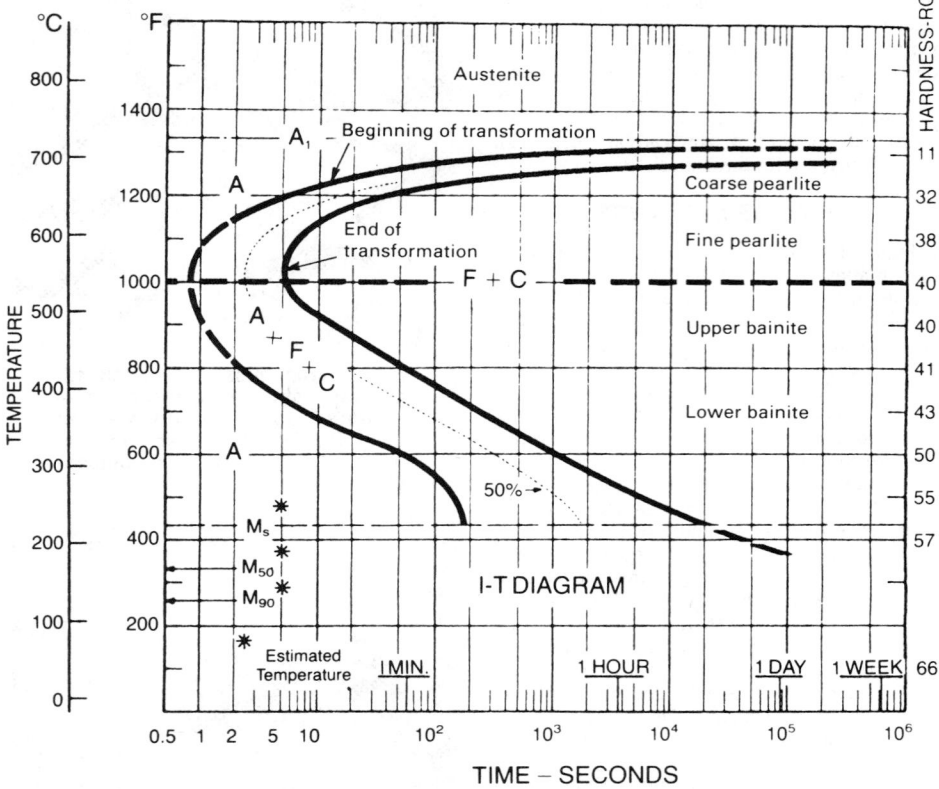

Specimens were austenitized at 900 °C and had an austenite grain size of ASTM No. 6.

Fig. 6. Isothermal transformation diagram for 1080 steel containing 0.79% C and 0.76% Mn (Ref 10)

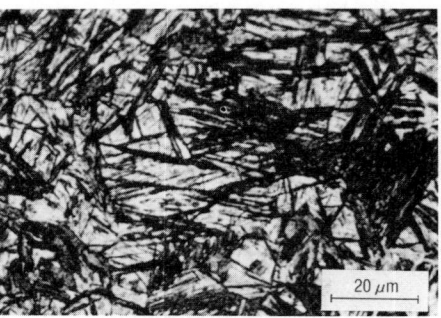

Fig. 8. Light micrograph (nital etch) showing lower bainite (dark plates) formed in 4150 steel. Courtesy of F. A. Jacobs (Ref 12).

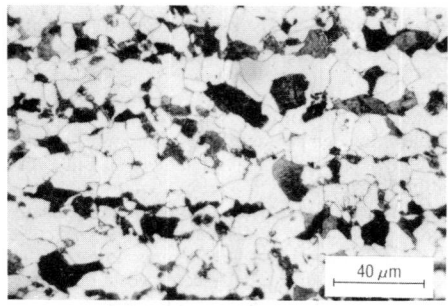

Fig. 9. Light micrograph (nital etch) showing microstructure of proeutectoid ferrite (white) and pearlite (dark) in a 0.17C-1.20Mn-0.19Si steel

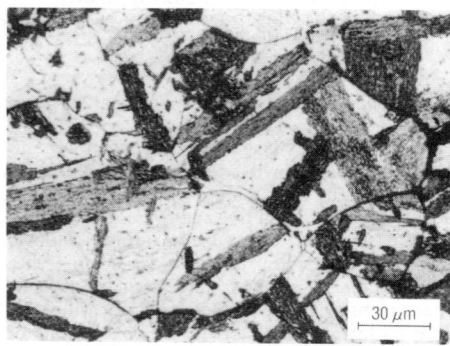

Etched in saturated picric acid plus a wetting agent to reveal austenite grain boundaries, then etched in nital to reveal bainite patches.

Fig. 7. Light micrograph showing patches of upper bainite formed in 4150 steel partially transformed at 460 °C. Courtesy of F. A. Jacobs (Ref 12).

540 °C. The acceleration of the transformation with decreasing temperature is apparent.

At temperatures below 540 °C, the diffusion of iron atoms is reduced to the extent that they can no longer be readily transferred even the very short distance across the pearlite/austenite interface. Therefore, the mechanism for the change in crystal structure from austenite to ferrite changes from diffusion to shear. Instead of an atom-by-atom transfer across an interface, large numbers of iron atoms shear or move cooperatively to form lath- or plate-shaped crystals of ferrite. Carbon

diffusion and cementite formation must still occur because of the low solubility of carbon in the body-centered cubic ferrite, but the cementite forms as separate particles rather than as continuous lamellae as in pearlite. The microstructure produced by both shear and diffusion is termed bainite, after Edgar C. Bain, who did much pioneering work in the characterization of austenite transformation and hardenability of steels (Ref 11).

Two forms of bainite develop in steels. One is termed upper bainite because it forms at relatively high temperatures, just below the range of pearlite formation. Upper bainite forms in patches containing many parallel laths of ferrite. Carbon is rejected from the ferrite and concentrates to form relatively coarse cementite particles between the ferrite laths. Figure 7 shows patches of upper bainite formed by partial transformation of the austenite at 460 °C. The austenite which did not transform at 460 °C formed martensite (light background phase) on quenching to room temperature. The general morphology of upper bainite is shown in Fig. 7, but the ferrite laths and cementite particles are too fine to be resolvable in the light micrograph.

The other type of bainite is termed lower bainite because it forms at lower temperatures than does upper bainite. The ferrite takes a plate morphology and the cementite is present as very fine particles within the ferrite plates. Figure 8 shows lower bainite which has formed in a 4150 steel. The bainite plates are at angles with respect to each other, giving an acicular or needle-like appearance to the microstructure rather than the

blocky or feathery appearance of upper bainite. Again the very fine carbide particles in the bainite plates are not resolvable in the light micrograph.

PROEUTECTOID FERRITE AND CEMENTITE

Figure 1 shows that alloys which contain either less carbon (hypoeutectoid steels) or more carbon (hypereutectoid steels) than the eutectoid composition must first form either ferrite or cementite when slowly cooled from the single-phase austenite field. The ferrite or cementite formed before the eutectoid reaction are termed proeutectoid ferrite or proeutectoid cementite.

Figure 9 shows the microstructure of a low-carbon steel after air cooling from austenite. The white grains are proeutectoid ferrite which have nucleated on and grown from the austenite grain boundaries. As the ferrite grains grew, carbon was rejected into the austenite grain boundaries. Eventually, the carbon concentration was sufficient for pearlite formation, and the balance of the microstructure transformed to pearlite. Most of the pearlite colonies appear uniformly black because the light is scattered by the lamellar structures which are too closely spaced to be resolvable in the light micrograph.

The growth of proeutectoid ferrite is dependent on the rejection of carbon atoms into the austenite and the transfer of iron atoms across the ferrite/austenite interface from the face-centered cubic to the body-centered cubic structure. The latter process is dependent on the degree of coherency or disorder in atom arrangement at the interface. Also under some conditions, substitu-

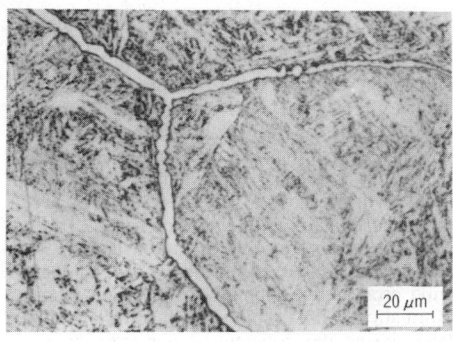

(a)

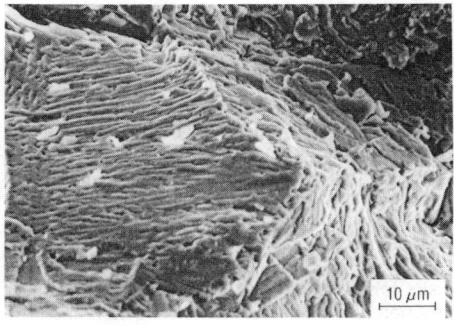

(b)

Fig. 10. (a) Light micrograph (nital etch) showing cementite network on prior austenite grain boundaries in an Fe-1.12C-1.5Cr alloy. (b) Scanning electron micrograph showing cementite interface fracture morphology in same alloy. Courtesy of T. Ando.

tional alloying elements must be incorporated into the ferrite structure if they are ferrite stabilizers or rejected from the ferrite if they are austenite stabilizers. Recent experimental and theoretical work on the effects of alloy element partitioning and interface structure on the formation of proeutectoid ferrite is reviewed in Ref 13.

Generally under conditions of slow cooling, the proeutectoid ferrite grows uniformly into austenite and an equiaxed ferrite grain structure develops as shown in Fig. 9. However, if the austenite in hypereutectoid steels is rapidly cooled, the transfer of iron atoms across ferrite/austenite interfaces is restricted, and the diffusion-controlled growth of ferrite is replaced by a shear mechanism. As a result a plate-shaped morphology of ferrite, frequently referred to as acicular or Widmanstätten ferrite, develops in rapidly cooled low-carbon steels. Substitutional alloying elements such as manganese tend to retard the formation of equiaxed ferrite grains and promote acicular ferrite formation.

In hypereutectoid steels, proeutectoid cementite nucleates and grows on austenite grain boundaries during cooling from the austenite phase field. Figure 10(a) shows a continuous network of proeutectoid cementite which has formed on austenite grain boundaries of an Fe-Cr-C alloy. The balance of the microstructure is martensite which formed on quenching after the cementite network had developed. The interface between

the proeutectoid cementite and martensite is quite brittle. Figure 10(b) shows fracture that has followed proeutectoid cementite interfaces which are characterized by many flat facets with intervening ledges.

Initial proeutectoid cementite growth appears to depend only on diffusion of carbon and therefore proceeds very rapidly. Later stages of cementite growth require partitioning of substitutional alloying elements such as chromium and therefore are very sluggish (Ref 14). The very rapid initial growth of proeutectoid cementite may occur even during oil quenching for hardening and is associated with the intergranular fracture often observed in high-carbon steel quenched from temperatures above A$_{cm}$. Figure 11 shows an example of 52100 steel quenched from 1000 °C (Ref 15). The intergranular fracture facets are quite smooth, in contrast to Fig. 10(b), and no cementite is visible in this scanning electron micrograph. The presence of small amounts of cementite is, however, established by Auger electron spectroscopy, an analytical technique capable of determining chemical compositions of very thin layers (Ref 16).

In view of the brittleness which continuous networks of proeutectoid cementite impart, hypereutectoid steels are reheated intercritically into the austenite/cementite two-phase field for annealing (if maximum ductility and machinability are desired) or for hardening (if wear and fatigue

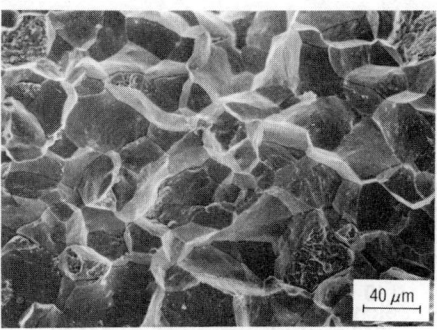

Fig. 11. Scanning electron micrograph showing intergranular fracture in 52100 steel oil quenched from 1000 °C (Ref 15)

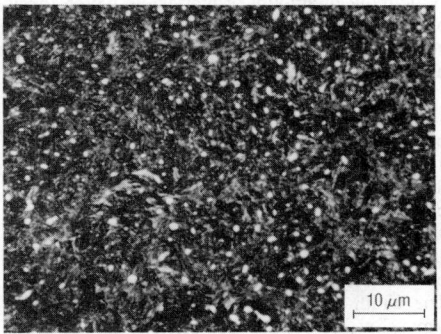

Fig. 12. Light micrograph (nital etch) of a high-carbon bearing steel, showing spheroidized cementite particles in a matrix of martensite produced by intercritical austenitizing and quenching. Courtesy of J. Bruce Kelley.

Fig. 13. Scanning electron micrograph of fracture surface of 52100 steel intercritically austenitized at 800 °C and oil quenched. Arrows point to fine spherical carbide particles. (Ref 15)

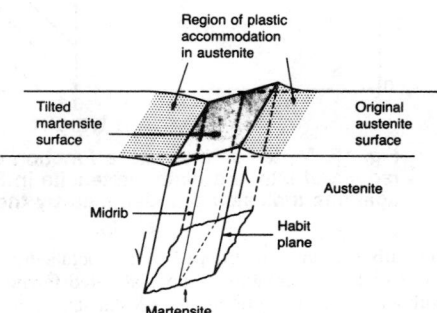

Fig. 14. Schematic illustration of shear and surface tilt associated with formation of a martensite plate. Courtesy of M. D. Geib (adapted from Ref 19).

resistance are required). During the intercritical heating, proeutectoid cementite networks as well as the lamellae of cementite in pearlite partially dissolve and spheroidize. Figure 12 shows the microstructure of an intercritically austenitized and hardened bearing steel. The spheroidized cementite particles (white) are dispersed in a matrix of martensite (dark). When a hardened steel with a microstructure similar to that shown in Fig. 12 is fractured, a transgranular fracture morphology (see Fig. 13) develops rather than intergranular fracture (Fig. 11). The fracture is initiated at the fine spherical carbide particles and the toughness is related to spacing of the particles (Ref 17).

MARTENSITE

Martensite is the phase formed in steels by a diffusionless, shear transformation of austenite, and is the base structure for hardened steels. Martensite is not shown on the Fe-C diagram because it does not form under equilibrium conditions; generally rapid cooling to temperatures well below A$_1$ is required to form martensite. As expected from the Fe-C diagram, martensite eventually decomposes to a mixture of ferrite and cementite if heated below A$_1$.

Shear or the displacive, cooperative movement of many atoms has already been mentioned as a mechanism by which bainite and acicular proeutectoid ferrite form. The formation of the latter structures, however, occurs under conditions such that carbon diffusion accompanies the formation of body-centered cubic ferrite. When martensite forms, even the carbon atoms cannot diffuse. Thus

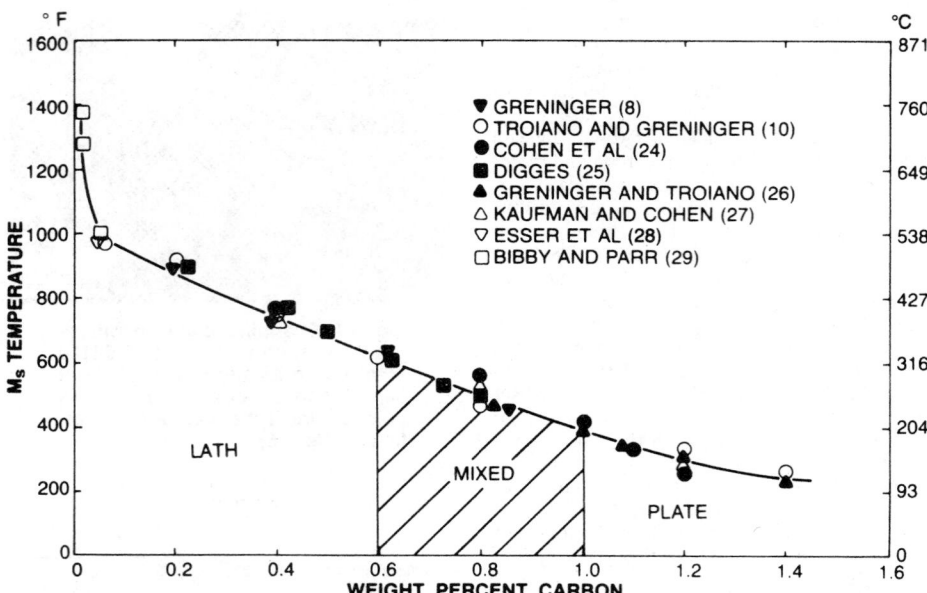

Fig. 15. M$_s$ temperature as a function of carbon content in steels. Composition ranges of lath and plate martensite in Fe-C alloys are also shown. (Ref 2; investigations indicated are identified by their numbers in that reference.)

the carbon atoms are trapped in the octahedral interstitial sites, creating a supersaturated ferrite with a body-centered tetragonal crystal structure. The higher the concentration of carbon atoms, the greater the tetragonality (Ref 18).

Figure 14 shows schematically a martensite plate which has formed in austenite adjacent to a free surface. The martensite surface is tilted by the shear transformation, and the austenite plane along which the martensite forms is termed the habit plane. In order to accomplish the shape deformation shown, not only must the face-centered cubic austenite lattice transform to the body-centered tetragonal lattice of martensite, but the martensite crystal once formed must accommodate itself to the constraints of the surrounding bulk austenite and the restrictions imposed by the plane-strain deformation parallel to the habit plane (Ref 19). This accommodation is accomplished by slip or twinning of the martensite plate, and as a result martensite in steels contains a high residual density of dislocations and/or fine twins.

The martensitic transformation is characterized by athermal kinetics—i.e., the amount of martensite formed is independent of time and is a function only of the amount of undercooling below the M$_s$ temperature, the temperature at which martensite starts to form on cooling in a given steel. The following equation has been developed (Ref 20) for estimating the volume fraction of martensite, f, formed by quenching to any temperature, T$_q$:

$$f = 1 - \exp - [0.011 \, (M_s - T_q)]$$

Thus, if the M$_s$ of a given steel is known, the amount of martensite formed on quenching to any temperature below M$_s$ can be established.

The M$_s$ temperature is a function of the carbon and alloying-element content of a steel, and a number of relationships have been developed to relate M$_s$ to composition (Ref 7). Figure 15 shows M$_s$ as a function of carbon content. The decrease in M$_s$ with increasing carbon content is related to the increased shear resistance produced by in-

Fig. 16. Light micrograph (nital etch) showing lath martensite in 4340 steel oil quenched from 940 °C and tempered at 350 °C (Ref 23)

creasing amounts of carbon in solid solution in the austenite. An important consequence of low M$_s$ temperature, according to the above equation, is the reduced amount of martensite which forms on cooling to room temperature. Therefore, large volume fractions of austenite may be retained in high-carbon steels.

Figure 15 indicates that two types of martensite form in carbon steels. The two categories are based on morphology and microstructural characteristics of the martensite (Ref 7 and 21). The lath morphology forms in low- and medium-carbon steels, and consists of regions or packets where many fine laths or board-shaped crystals are arranged parallel to one another. The habit plane of the laths is close to but not exactly {111}. The width of most of the laths is less than 0.5 μm—i.e., below the resolution of the light microscope, and therefore the microstructure appears very uniform, with only the largest laths resolvable. Figure 16 demonstrates the above characteristics of lath martensite in a 4340 steel. Electron microscopy is required to show that the fine structure of lath martensite consists of a high density of tangled dislocations and that retained

austenite is present as thin films between the martensite laths (Ref 22).

The plate morphology of martensite forms in high-carbon steels and consists of martensite plates which form at angles with respect to each other on either {225}$_γ$ or {259}$_γ$ habit planes. Figure 17 shows a plate martensite microstructure in an Fe-1.39C alloy cooled to room temperature. Consistent with the low M$_s$ of this alloy, a large amount of retained austenite is present.

The fine structure of plate martensite consists of thin twins, about 10 nm thick, and/or dislocation arrays typical of low-temperature plastic deformation. The impingement of nonparallel plates during development of a martensite microstructure sometimes causes microcracks to form in the martensite (Ref 25). Examples of microcracks are shown in the large plate of Fig. 17. The density of microcracks in plate martensite is reduced by formation of martensite in fine-grained austenite, by lowering the carbon concentration of the austenite by intercritical austenitizing (thereby developing a more parallel martensite morphology and less impingement), and by tempering.

The carbon range in which a mixed morphology of lath and plate martensite forms is sensitive to alloy content and is not well known. Even in the range of carbon contents where lath martensite forms, there is a gradual decrease in the definition of packets with increasing carbon content (Ref 26).

TEMPERED MARTENSITE

As-quenched martensite is supersaturated with carbon, has a very high interfacial energy per unit volume associated with the fine laths or plates of the martensitic microstructure, contains a high density of dislocations which store considerable strain energy, and may coexist with retained austenite. As a result of these characteristics, martensitic microstructures are quite unstable, and decompose when heated. A practical benefit of the decomposition is increased toughness, and for this reason almost all hardened steels are heated to some temperature below Ac$_1$, a heat treatment process which is referred to as tempering.

A wide range of microstructures may be produced by tempering of martensite. Carbon atoms rearrange themselves into various configurations

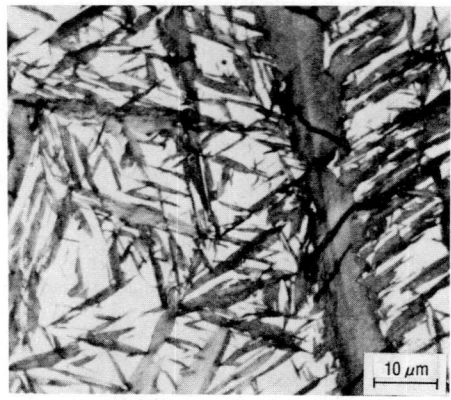

Fig. 17. Light micrograph (aqueous 10% sodium bisulfide etch) showing plate martensite and retained austenite in an Fe-1.39C alloy (Ref 24)

Table 1. Tempering reactions in steel

Temperature range, °C	Reaction and symbol (if designated)	Comments
−40 to 100	Clustering of 2 to 4 carbon atoms on octahedral sites of martensite (A1); segregation of carbon atoms to dislocations and boundaries	Clustering is associated with diffuse spikes around fundamental electron diffraction spots of martensite
20 to 100	Modulated clusters of carbon atoms on (102) martensite planes (A2)	Identified by satellite spots around electron diffraction spots of martensite
60 to 80	Long period ordered phase with ordered carbon atoms (A3)	Identified by superstructure spots in electron diffraction patterns
100 to 200	Precipitation of transition carbide as aligned 2-nm-diam particles (T1)	Recent work identifies carbides as eta (orthorhombic, Fe_2C); earlier studies identified the carbides as epsilon (hexagonal, $Fe_{2.4}C$)
200 to 350	Transformation of retained austenite to ferrite and cementite (T2)	Associated with tempered-martensite embrittlement in low- and medium-carbon steels
250 to 700	Formation of ferrite and cementite; eventual development of well-spheroidized carbides in a matrix of equiaxed ferrite grains (T3)	This stage now appears to be initiated by chi-carbide formation in high-carbon Fe-C alloys
500 to 700	Formation of alloy carbides in Cr-, Mo-, V- and W-containing steels. The mix and composition of the carbides may change significantly with time (T4)	The alloy carbides produce secondary hardening and pronounced retardation of softening during tempering or long-time service exposure around 500 °C
350 to 550	Segregation and cosegregation of impurity and substitutional alloying elements	Responsible for temper embrittlement

and structures within the martensite crystals even at temperatures well below 100 °C (Ref 27). Tempering between 100 °C and Ac_1 produces various types of carbide-particle dispersions as well as major changes in the matrix martensite. The reactions which produce the carbides have long been recognized and are classified as stages of tempering: T_1, T_2, etc. The reactions which depend on very short-range rearrangement of carbon atoms in the as-quenched martensite prior to carbide formation have only recently been studied, and to distinguish those reactions from the carbide-forming reactions it has been suggested that they be classified as aging reactions: A_1, A_2, etc. (Ref 28 and 29).

Table 1 lists the various reactions and microstructural changes which may be developed by tempering steel (Ref 29). The aging and tempering classifications serve primarily to mark microstructures which form on the way to equilibrium, ultimately a microstructure which consists of spheroidized carbide particles dispersed in a matrix of equiaxed ferrite grains. Many of the reactions or microstructural states require further characterization, some occur concurrently, and others may yet be discovered. The reactions are controlled by diffusion of carbon, iron, and/or alloying elements, and therefore steel composition, time and temperature determine where a given tempering treatment stops in the sequence of structural changes indicated in Table 1.

Significant increases in toughness are achieved by tempering at temperatures above 150 °C. In general, subject to the development of various embrittlement phenomena, as tempering temperature increases, toughness increases and hardness decreases. Therefore, in applications where high hardness must be retained, tempering is performed at relatively low temperatures, usually between 150 and 200 °C. Figure 18 shows the fine structure which develops in martensite as a result of low-temperature tempering. A portion of a single plate of martensite in an Fe-1.22C alloy tempered at 150 °C is shown. The fine dark streaks mark positions of rows of very fine carbide particles, each about 2 nm in diameter, which have precipitated from the supersaturated martensite (Ref 30). The particles themselves are masked by the strain which accompanies the precipitation and which causes the dark contrast in the transmission electron micrograph. The carbide is not the equilibrium Fe_3C, but a transition carbide, first designated as epsilon-carbide with a hexagonal structure as identified by x-ray dif-

Fig. 18. Transmission electron micrograph showing fine structure in a plate of martensite in an Fe-1.22C alloy (Ref 30)

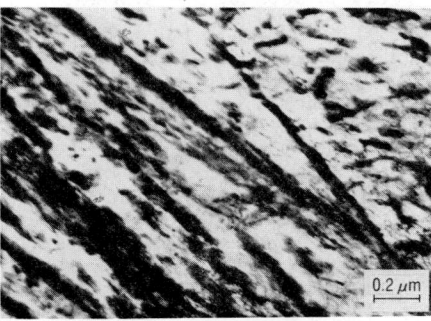

Fig. 19. Transmission electron micrograph showing chi-carbide in the martensitic microstructure of an Fe-1.22C alloy tempered at 350 °C (Ref 33)

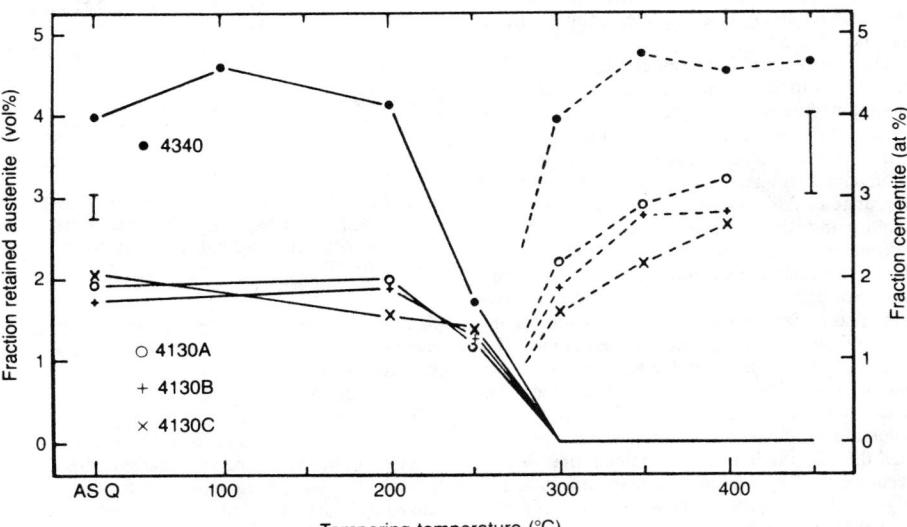

Fig. 20. Retained austenite and cementite as a function of tempering temperature in several medium-carbon steels (Ref 34)

fraction (Ref 31), but more recently designated eta-carbide with an orthorhombic structure as identified by electron diffraction (Ref 32). Both the epsilon-carbide and eta-carbide have carbon contents substantially higher than that of cementite.

Steels tempered to develop the fine transition carbides show a modest but significant increase in toughness. The hardness, however, remains high because of the extremely fine carbide dispersion and the retention of much of the dislocation substructure introduced by the martensitic transformation.

In steels tempered between 200 and 350 °C, the transition carbide is replaced by cementite or χ-carbide, and retained austenite transforms to

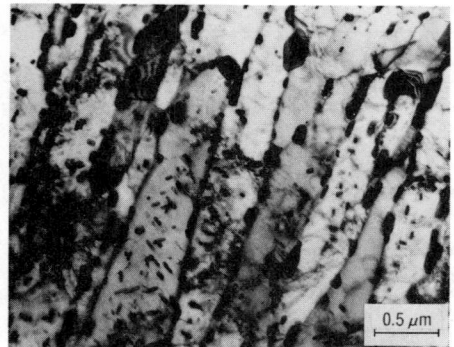

Fig. 21. Transmission electron micrograph showing microstructure of 4130 steel water quenched from 900 °C and tempered 500 min at 650 °C. Courtesy of F. Woldow.

ferrite and cementite. The χ-carbide is a complex carbide with a monoclinic structure which forms in tempered high-carbon martensites and is eventually replaced by cementite. Figure 19 shows a dense distribution of carbide particles, identified as χ-carbide, in an Fe-1.22C alloy tempered at 350 °C (Ref 33). The carbide particles are considerably coarser than the transition carbides in Fig. 19, and are present at the interfaces of the martensite plates as well as within the plates.

Figure 20 shows that small amounts of austenite are present even in medium-carbon steels, that the austenite is stable throughout the tempering-temperature range in which the transition carbide forms, and that the austenite begins to transform at temperatures above 200 °C. Austenite in medium-carbon steels is retained between martensite laths and, when it transforms on tempering, produces relatively coarse plates of interlath cementite (Ref 22).

The coarse carbides produced by replacement of the transition carbides and transformation of the retained austenite, together with a limited recovery of the dislocation substructure of the martensite, reduce impact toughness. This decrease in impact toughness produced by tempering in the range of 250 to 400 °C is referred to as tempered martensite embrittlement.

Tempering at temperatures above 400 °C produces substantial coarsening of the microstructure. Not only do the cementite particles coarsen and spheroidize, but also the martensitic matrix is significantly altered. The laths are almost dislocation-free and are now ferrite because all carbon has completely precipitated as carbides. The reduction in dislocation density is driven by the reduction of the strain energy which accompanies the elimination of the dislocations, and is accomplished by various recovery mechanisms. Figure 21 shows the structure of a 4130 steel quenched to form martensite and tempered at 650 °C. Carbide particles are present within and between the remanent martensite laths. The lath morphology, although coarsened, persists because of the pinning of the lath boundaries by carbides. In alloy steels such as 4130, which contains nominally 1% Cr and 0.2% Mo, various alloy carbides, in addition to cementite, form during high-temperature tempering. The intralath carbides in Fig. 21 have a specific habit plane and have been identified as Mo_2C (Ref 35). A number of other carbides of chromium, molyb-

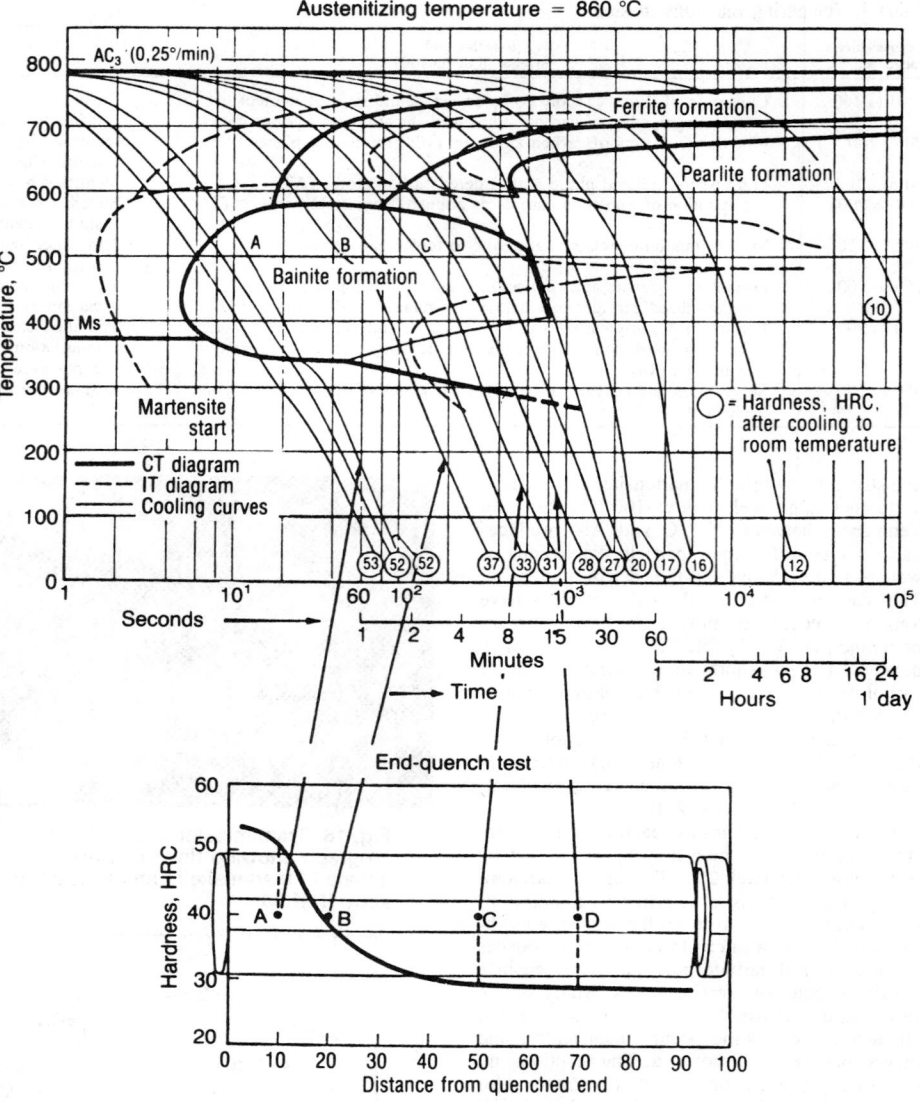

Fig. 22. Continuous transformation (solid lines) and isothermal transformation (dashed lines) diagrams for steel containing nominally 0.4 C, 1.0 Cr and 0.2 Mo. Several cooling rates are related to positions on a Jominy end-quench specimen. (Ref 38)

denum, vanadium and other carbide-forming elements may also develop depending on the composition of the steel.

As tempering temperature increases above 400 °C, hardness and strength drop rapidly and toughness improves significantly. In alloy steels, the development of fine alloy carbide dispersions offsets the softening which accompanies the changing dislocation substructure and coarsening of the lath and cementite structure. In fact, if the alloy carbide dispersions are sufficiently fine and dense, an increase in hardness may develop. This increase in hardness due to alloy carbide precipitation high in the tempering-temperature range is referred to as secondary hardening.

As noted, toughness increases significantly with increasing tempering temperature. However, if impurities such as phosphorus, antimony and tin are present in a steel, these elements may segregate to grain boundaries and/or carbide-matrix interfaces and cause large reductions in impact

toughness (Ref 36). This phenomenon develops during tempering in, or slow cooling through, the temperature range 350 to 550 °C, and is referred to as temper embrittlement. The impurity atom segregation may be accompanied by the co-segregation of the substitutional alloying elements present in steels (Ref 37).

TRANSFORMATION DIAGRAMS

The previous sections have shown that the transformation of austenite produces a wide variety of microstructures in response to such factors as steel composition, temperature of transformation, and cooling rate. In order to characterize the conditions which produce the various microstructures, two types of transformation diagrams have been developed. Isothermal transformation (IT) diagrams are based on the austenite decomposition at constant temperatures, while continuous transformation (CT)

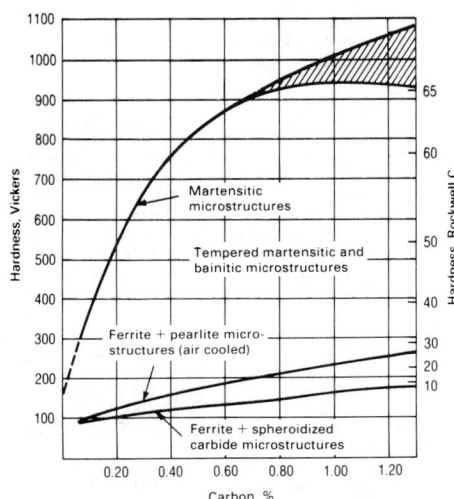

Fig. 23. Hardness as a function of carbon content for various microstructures in steels. Cross-hatched area shows effect of retained austenite. (After Ref 39)

diagrams follow microstructural development as a function of cooling rate. Most heat treatments are performed by continuous cooling, and therefore CT diagrams more accurately than IT diagrams represent the conditions encountered in commercial practice.

An example of an IT diagram for a eutectoid steel has already been shown in Fig. 6. Figure 22 shows CT (solid lines) and IT (dashed lines) diagrams for a medium-carbon alloy steel, 4140, containing nominally 0.4% C, 1% Cr and 0.2% Mo. Superimposed on the diagram are various cooling rates, some of which are related to positions on a Jominy end-quench specimen. The more rapid cooling rates produce microstructures of higher hardness as indicated.

The microstructures in the medium-carbon steel (Fig. 22) are more varied than in the eutectoid steel (Fig. 6) in that proeutectoid ferrite forms. Also the alloying elements in the 4140 steel significantly retard formation of ferrite and pearlite and thereby increase the range of cooling rates which form martensite and bainite. Comparison of the IT and CT diagrams in Fig. 22 shows that all of the phase transformations are shifted to lower temperatures and longer times by continuous cooling.

The CT diagrams provide the basis of hardenability, the technology which is concerned with estimating the depth and distribution of martensite in hardened components as a function of cooling rate and composition. Figure 22 shows that an important effect of alloying is to retard the formation of microstructures of low hardness. Therefore, martensite formation may be accomplished by less-severe quenching with the advantages of lower residual surface tensile stresses, reduced distortion and/or prevention of quench cracking. For a given quench, alloying increases depth of martensite formation in a part.

SUMMARY: CARBON CONTENT, MICROSTRUCTURE AND PROPERTIES

Figure 23 shows hardness as a function of carbon content for the various types of microstruc-

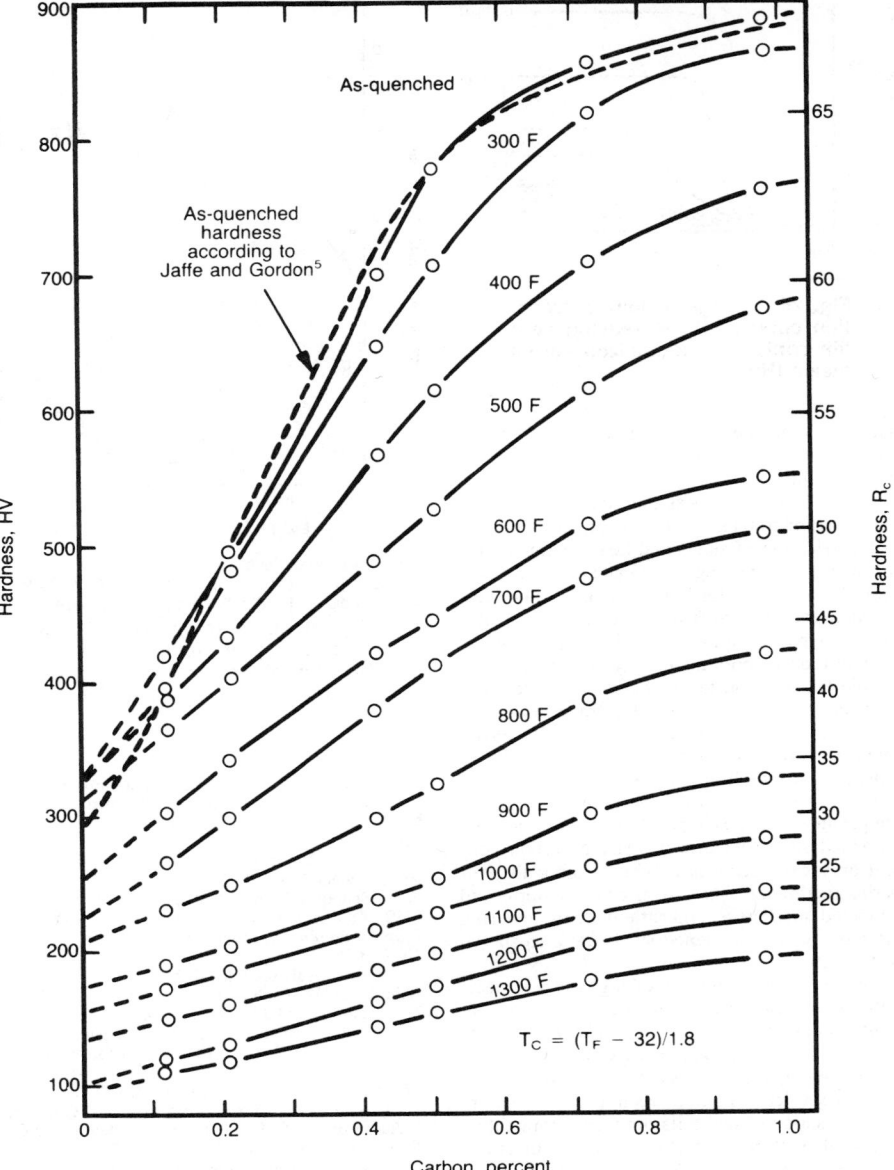

Fig. 24. Hardness as a function of carbon content of martensite in Fe-C alloys tempered at various temperatures (Ref 40)

$$T_C = (T_F - 32)/1.8$$

tures which may be formed by heat treatment of steel. More detail regarding the effect of tempering is shown in Fig. 24. Hardness is a readily measured property which in general is directly proportional to strength and inversely proportional to ductility.

Figures 23 and 24 demonstrate the great versatility of steels and the major effect of carbon content on establishing mechanical properties. Not all of the microstructures may be readily formed in all steels. For example, in low-carbon steels which have very low hardenability, it may be impossible to produce fully martensitic structures in all but the very thinnest sections. Low-carbon steels are therefore invariably used with ferrite-pearlite microstructures where the high ductility of ferrite is beneficial for cold working and formability. At the other extreme, medium- and high-carbon steels alloyed with chromium, nickel and/

or molybdenum may have such high hardenability for martensite or bainite formation that it is very difficult to form ferrite-pearlite microstructures except under conditions of very slow cooling or in very heavy sections.

Figures 23 and 24 show that the higher the carbon content the higher the hardness of a given microstructure. The microstructures with the highest hardness are formed by transformations which involve shear—i.e., the martensite or bainite transformations or the tempering of the shear-produced martensite microstructures. The shear transformations produce fine crystals (laths or plates), supersaturate the structure with carbon or create very fine carbide dispersions, and introduce a high dislocation density into the product phases. The lower-strength microstructures are produced by diffusion-controlled transformations or microstructural changes which pro-

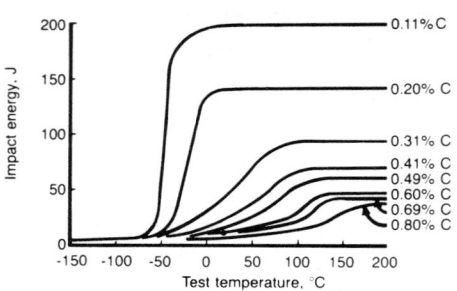

Fig. 25. Change in impact transition curves with increasing pearlite content in normalized carbon steels (Ref 41)

duce coarse ferrite-carbide microstructures without developing extensive dislocation substructures in the ferrite.

In addition to hardness and strength, fracture resistance or toughness is a major consideration in the selection of steels and heat treatments for severe applications. Figures 25 and 26 show CVN impact toughness for ferrite-pearlite and tempered martensitic microstructures, respectively. For the ferrite-pearlite steels, increasing carbon content reduces both the energy absorbed during ductile fracture and raises the transition temperature at which brittle, cleavage fracture occurs. Thus very-low-carbon steels with largely ferrite microstructures are best suited for applications which require high toughness.

Figure 26 shows that increasing tempering temperature increases impact toughness for a given hardened steel, but that increasing carbon content drastically reduces toughness after all tempering treatments. The effects of phosphorus and tempered martensite embrittlement in lowering the toughness of hardened steels are also shown in Fig. 26.

The very low toughness of high-carbon steels with either pearlite or tempered martensitic microstructures limits their use to applications where high hardness is of benefit to wear and fatigue resistance but where impact, tensile loading is not a major service condition. Examples of such applications are railroad rails produced from eutectoid steels with fully pearlitic microstructures and bearings produced from 52100 steel that has been heat treated by intercritical austenitizing, oil quenching and low-temperature tempering to produce tempered martensitic microstructures of high hardness.

REFERENCES

1. *Metals Handbook*, 8th Ed., Vol 8, ASM, Metals Park, OH, 1973
2. *Thermomechanical Processing of Microalloyed Austenite*, edited by A. J. DeArdo, G. A. Ratz and P. J. Wray: TMS-AIME, Warrendale, PA, 1982
3. *Deformation, Processing, and Structure*, edited by G. Krauss: ASM, Metals Park, OH, 1984
4. *Theory of Dislocations*, by J. P. Hirth and J. Loth: McGraw-Hill, New York, 1968
5. *The Physical Metallurgy of Steels*, by W. C. Leslie: McGraw-Hill, New York, 1981.
6. *Mechanical Properties of BCC Metals*, edited by M. Meshii: TMS-AIME, Warrendale, PA, 1982
7. *Principles of Heat Treatment of Steel*, by G. Krauss: ASM, Metals Park, OH, 1980

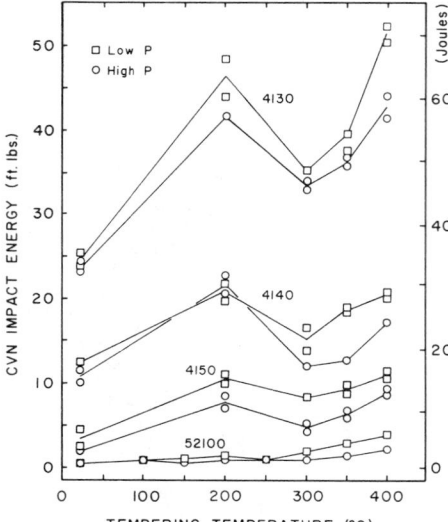

High phosphorus levels are approximately 0.02%. Low phosphorus levels are approximately 0.002% for the 41xx steels and 0.009% for the 52100 steel.

Fig. 26. CVN impact energy absorbed at room temperature as a function of tempering for medium- and high-carbon steels (Ref 42 and 43)

8. The Quench-Aging of Low-Carbon Iron and Iron-Manganese Alloys: An Electron Transmission Study, by W. C. Leslie: *Acta Metallurgica*, Vol 9, 1961, p 1004-1022
9. "Dual Phase Behavior and Aging of a Renitrogenized Steel," by J. E. Indacochea: M.S. Thesis, Colorado School of Mines, Golden, CO, 1978
10. *Atlas of Isothermal Transformation and Cooling Transformation Diagrams*: ASM, Metals Park, OH, 1977, p 28
11. Historical Account of the Contribution of E. C. Bain, by H. W. Paxton and J. B. Austin: *Metallurgical Transactions*, Vol 13, 1972, p 1035-1042
12. "The Combined Effects of Phosphorus and Carbon on Hardenability and Phase Transformation Kinetics in 41XX Steels," by F. A. Jacobs: M.S. Thesis, Colorado School of Mines, Golden, CO, 1982
13. *Solid-Solid Phase Transformations*, edited by H. I. Aaronson, D. E. Laughlin, R. F. Sekerko and C. M. Wayman: TMS-AIME, Warrendale, PA, 1982
14. Development and Application of Growth Models for Grain Boundary Allotriomorphs of a Stoichiometric Compound in Ternary Systems, by T. Ando and G. Krauss: *Metallurgical Transactions*, Vol 14A, 1983, p 1261-1269
15. Microstructure and Fracture of 52100 Steel, by K. Nakazawa and G. Krauss: *Metallurgical Transactions A*, Vol 9A, 1978, p 681-689
16. The Effect of Phosphorus Content on Grain Boundary Cementite Formation in AISI 52100 Steel, by T. Ando and G. Krauss: *Metallurgical Transactions*, Vol 12A, 1981, p 1283-1290
17. The Relationship of Microstructure to Fracture Morphology and Toughness of Hardened Hypereutectoid Steels, by G. Krauss: in *Microstructure and Residual Stress Effects on the Properties of Case Hardened Steels*, TMS-AIME, Warrendale, PA, 1984
18. Effect of Carbon on the Volume Fractions and Lattice Parameters of Retained Austenite and Martensite, by C. S. Roberts: *Transactions of AIME*, Vol 197, 1953, p 203-204
19. The Crystallography of Martensite Transformations, by B. A. Bibby and J. W. Christian: *JISI*, Vol 197, 1961, p 122-131
20. A General Equation Prescribing the Extent of the

Austenite-Martensite Transformation in Pure Iron-Carbon Alloys and Plain Carbon Steels, by D. P. Koistinen and R. E. Marburger: *Acta Metallurgica*, Vol 7, 1959, p 59-60
21. The Morphology of Martensite in Iron-Carbon Alloys, by A. R. Marder and G. Krauss: *Transactions of ASM*, Vol 60, 1967, p 651-660
22. Retained Austenite and Tempered Martensite Embrittlement, by G. Thomas: *Metallurgical Transactions A*, Vol 9A, 1978, p 439-450
23. Tempered Martensite Embrittlement in SAE 4340 Steel, by J. P. Materkowski and G. Krauss: *Metallurgical Transactions A*, Vol 10A, 1979, p 1643-1651
24. Microcracking Sensitivity in Fe-C Plate Martensite, by A. R. Marder, A. O. Benscoter and G. Krauss: *Metallurgical Transactions*, Vol 1, 1970, p 1545-1549
25. Microcracking in Fe-C Acicular Martensite, by A. R. Marder and A. O. Benscoter: *Transactions of ASM*, Vol 61, 1968, p 293-299
26. The Morphology of Microstructure Composed of Lath Martensite in Steels, by T. Maki, K. Tsuzaki and I. Tamura: *Transactions of the Iron and Steel Institute of Japan*, Vol 20, 1980, p 207-214
27. Winchell Symposium on Tempering of Steel: *Metallurgical Transactions*, Vol 14A, 1983, p 985-1146
28. Early Stages of Aging and Tempering of Ferrous Martensites, by G. B. Olson and M. Cohen: *Metallurgical Transactions A*, Vol 14A, 1983, p 1057-1065
29. Tempering and Structural Change in Ferrous Martensites, by G. Krauss: in *Phase Transformations in Ferrous Alloys*, TMS-AIME, Warrendale, PA, 1984
30. A Study of the Early Stages of Tempering in an Fe-1.22%C Alloy, by D. L. Williamson, K. Nakazawa and G. Krauss: *Metallurgical Transactions A*, Vol 10A, 1979, p 1351-1363
31. Structural Transformations in the Tempering of High Carbon Martensitic Steel, by K. H. Jack: *JISI*, Vol 169, 1951, p 26-36
32. Crystal Structure and Morphology of the Carbide Precipitated for Martensitic High Carbon Steel During the First Stage of Tempering, by Y. Hirotsu and S. Nagakura: *Acta Metallurgica*, Vol 20, 1972, p 645-655
33. Chi-Carbide in Tempered High Carbon Martensite, by C.-B. Ma, T. Ando, D. L. Williamson and G. Krauss: *Metallurgical Transactions A*, Vol 14A, 1983, p 1033-1045
34. Determination of Small Amounts of Austenite and Carbide in a Hardened Medium Carbon Steel by Mössbauer Spectroscopy, by D. L. Williamson, R. G. Schupmann, J. P. Materkowski and G. Krauss: *Metallurgical Transactions A*, Vol 10A, 1979, p 379-382
35. *Steels Microstructure and Properties*, by R. W. K. Honeycombe: Edward Arnold Publishers, Ltd., London, and ASM, Metals Park, OH, 1982
36. Temper Brittleness—An Interpretive Review, by C. J. McMahon, Jr.: in *Temper Embrittlement in Steel*, STP 407, ASTM, 1968, p 127-167
37. The Thermodynamics of Interactive Co-Segregation of Phosphorus and Alloying Elements in Iron and Temper-Brittle Steels, by M. Guttman, Ph. Dumonlin and M. Wayman: *Metallurgical Transactions A*, Vol 13A, 1982, p 1693-1711
38. *Atlas zur Wärmebehandlung der Stähle*, Vol 1-4: Max-Planck-Institut für Eisenforschung, in cooperation with the Verein Dentscher Eisenhüttenlente, Verlag Stahleisen, M. B. H., Düsseldorf, 1954-1976
39. *Alloying Elements in Steel*, 2nd Ed., by E. C. Bain and H. W. Paxton: ASM, Metals Park, OH, 1961
40. Hardness of Tempered Martensite in Carbide and Low Alloy Steels, by R. A. Grange, C. R. Hribal and L. F. Porter: *Metallurgical Transactions*, Vol 8A, 1977, p 1775-1785
41. The Optimization of Microstructures in Steel and Their Relationship to Mechanical Properties, by F. B. Pickering: in *Hardenability Concepts With Applications to Steel*, edited by D. V. Doane and J. S. Kirkaldy, TMS-AIME, Warrendale, PA, 1978, p 179-228
42. "A Study of Mechanisms of Tempered Martensite Embrittlement in Low-Alloy Medium-Carbon Steels," by F. Zia-Ebrahimi: Ph.D. Thesis, Colorado School of Mines, Golden, CO, 1982
43. "The Effects of Phosphorus and Tempering on the Fracture of AISI 52100 Steel," by D. L. Yaney: M.S. Thesis, Colorado School of Mines, Golden, CO, 1981

Stress-Relief Heat Treating of Steel

STRESS-RELIEF HEAT TREATING is used to relieve stresses that remain locked in a structure as a consequence of a manufacturing sequence.

Stress-relief heat treating is the uniform heating of a structure or portion thereof to a suitable temperature below the transformation range (Ac_1 for ferritic steels), holding at this temperature for a predetermined period of time, followed by uniform cooling. Approximate critical temperatures for selected carbon and low-alloy steels are given in Table 1.

The relief of residual stresses is a time-temperature related phenomenon (Fig. 1), parametrically correlated by the Larson-Miller equation:

$$\text{Thermal effect} = T(\log t + 20)(10^{-3})$$

where T is temperature (Rankin) and t is hours. It is evident in Fig. 1 that similar relief of residual stresses can be achieved by holding a component for longer periods of time at a lower temperature. For example, holding a piece at 595 °C (1100 °F) for 6 h provides the same relief of residual stress as heating at 650 °C (1200 °F) for 1 h.

Relief of residual stresses represents typical stress-relaxation behavior, in which the material undergoes microscopic (sometimes even macroscopic) creep at the stress-relief temperature. Creep-resistant materials, such as the chromium-bearing low-alloy steels and the chromium-rich high-alloy steels, normally require higher stress-relief heat treating temperatures than conventional low-alloy steels. Typical stress-relief temperatures for low-alloy ferritic steels are between 595 and 675 °C (1100 and 1250 °F).

Normalizing of Steel

NORMALIZING OF STEEL is a heat treating process that is often considered from both thermal and microstructural standpoints. In the thermal sense, normalizing is an austenitizing heating cycle followed by cooling in still or slightly agitated air. Typically, the work is heated to a temperature about 55 °C (100 °F) above the upper critical line of the iron–iron carbide phase diagram, as shown in Fig. 2—that is, above Ac_3 for hypoeutectoid steels and above A_{cm} for hypereutectoid steels.

Uses. A broad range of ferrous products can be normalized. All of the standard low-carbon, medium-carbon and high-carbon wrought steels can be normalized, as well as many castings. Austenitic steels, stainless steels and maraging steels either cannot be normalized or usually are not normalized.

The purpose of normalizing varies considerably. Normalization may increase or decrease the strength and hardness of a given steel in a given product form depending on the thermal and mechanical history of the product. Actually, the functions of normalizing may overlap with or be confused with those of annealing, hardening and

Table 1. Approximate critical temperatures for selected carbon and low-alloy steels

| | Critical temperatures on heating at 28 °C/h (50 °F/h) | | | | Critical temperatures on cooling at 28 °C/h (50 °F/h) | | | |
| | Ac_1 | | Ac_3 | | Ar_3 | | Ar_1 | |
Steel	°C	°F	°C	°F	°C	°F	°C	°F
1010	724	1335	877	1610	849	1560	682	1260
1020	724	1335	846	1555	816	1500	682	1260
1030	727	1340	813	1495	788	1450	677	1250
1040	727	1340	793	1460	757	1395	671	1240
1050	727	1340	768	1415	741	1365	682	1260
1060	727	1340	746	1375	727	1340	685	1265
1070	727	1340	732	1350	710	1310	691	1275
1080	729	1345	735	1355	699	1290	693	1280
1340	716	1320	777	1430	721	1330	621	1150
3140	735	1355	766	1410	721	1330	660	1220
4027	727	1340	807	1485	760	1400	671	1240
4042	727	1340	793	1460	732	1350	654	1210
4130	757	1395	810	1490	754	1390	693	1280
4140	732	1350	804	1480	743	1370	679	1255
4150	743	1370	766	1410	729	1345	671	1240
4340	724	1335	774	1425	710	1310	654	1210
4615	727	1340	810	1490	760	1400	649	1200
5046	716	1320	771	1420	732	1350	682	1260
5120	766	1410	838	1540	799	1470	699	1290
5140	738	1360	788	1450	727	1340	693	1280
5160	710	1310	766	1410	716	1320	677	1250
52100	727	1340	768	1415	716	1320	688	1270
6150	749	1380	788	1450	743	1370	693	1280
8115	721	1330	838	1540	788	1450	671	1240
8620	732	1350	829	1525	768	1415	660	1220
8640	732	1350	779	1435	727	1340	666	1230
9260	743	1370	816	1500	749	1380	713	1315

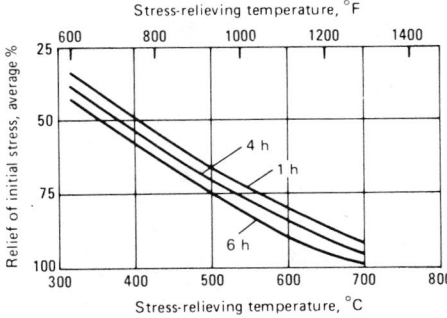

Fig. 1. Illustration of the relationship between time and temperature in the relief of residual stresses in steel

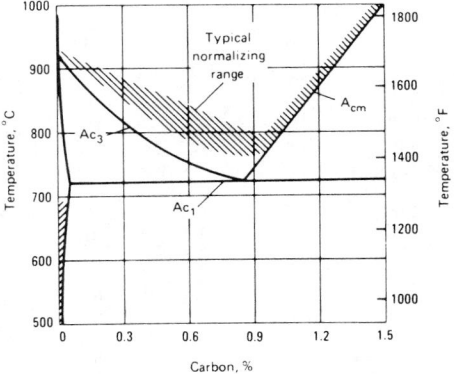

Fig. 2. Partial iron–iron carbide phase diagram, showing typical normalizing range for plain carbon steels

stress relieving. Improved machinability, grain-structure refinement, homogenization, and modification of residual stresses are among the reasons for which normalizing is done. Homogenization of castings by normalizing may be done in order to break up or refine the dendritic structure and facilitate a more even response to subsequent hardening. Similarly, for wrought products, normalization can obliterate banded grain structure due to hot rolling, as well as large grain size or mixed large and small grain size due to forging practice.

Carbon and Alloy Steels. Table 2 lists typical normalizing temperatures for some standard grades of carbon and alloy steels. These temperatures

Table 2. Typical normalizing temperatures for standard carbon and alloy steels

Based on production experience, normalizing temperature may vary from as much as 27 °C (50 °F) below to as much as 55 °C (100 °F) above indicated temperature. The steel should be cooled in still air from indicated temperature.

Grade	°C	°F	Grade	°C	°F
Plain carbon steels			**Standard alloy steels**		
1015	915	1675	4817	925	1700
1020	915	1675	4820	925	1700
1022	915	1675	5046	870	1600
1025	900	1650	5120	925	1700
1030	900	1650	5130	900	1650
1035	885	1625	5132	900	1650
1040	860	1575	5135	870	1600
1045	860	1575	5140	870	1600
1050	860	1575	5145	870	1600
1060	830	1525	5147	870	1600
1080	830	1525	5150	870	1600
1090	830	1525	5155	870	1600
1095	845	1550	5160	870	1600
1117	900	1650	6118	925	1700
1137	885	1625	6120	925	1700
1141	860	1575	6150	900	1650
1144	860	1575	8617	925	1700
			8620	925	1700
Standard alloy steels			8622	925	1700
1330	900	1650	8625	900	1650
1335	870	1600	8627	900	1650
1340	870	1600	8630	900	1650
3135	870	1600	8637	870	1600
3140	870	1600	8640	870	1600
3310	925	1700	8642	870	1600
4027	900	1650	8645	870	1600
4028	900	1650	8650	870	1600
4032	900	1650	8655	870	1600
4037	870	1600	8660	870	1600
4042	870	1600	8720	925	1700
4047	870	1600	8740	925	1700
4063	870	1600	8742	870	1600
4118	925	1700	8822	925	1700
4130	900	1650	9255	900	1650
4135	870	1600	9260	900	1650
4137	870	1600	9262	900	1650
4140	870	1600	9310	925	1700
4142	870	1600	9840	870	1600
4145	870	1600	9850	870	1600
4147	870	1600	50B40	870	1600
4150	870	1600	50B44	870	1600
4320	925	1700	50B46	870	1600
4337	870	1600	50B50	870	1600
4340	870	1600	60B60	870	1600
4520	925	1700	81B45	870	1600
4620	925	1700	86B45	870	1600
4621	925	1700	94B15	925	1700
4718	925	1700	94B17	925	1700
4720	925	1700	94B30	900	1650
4815	925	1700	94B40	900	1650

can be interpolated to obtain values for carbon contents not listed.

——Annealing of Steel——

ANNEALING is a generic term denoting a treatment that consists of heating to and holding at a suitable temperature followed by cooling at an appropriate rate, primarily for softening of metallic materials. It also is applied to produce desired changes in other properties or in microstructure. Steels may be annealed to facilitate cold working or machining, to improve mechanical or electrical properties, or to promote dimensional stability.

Annealing Cycles. In practice, specific thermal cycles of an almost infinite variety are used to achieve the various goals of annealing. These cycles fall into several broad categories that can

Table 3. Recommended temperatures and cooling cycles for full annealing of small carbon steel forgings

Data are for forgings up to 75 mm (3 in.) in section thickness. Time at temperature usually is a minimum of 1 h for sections up to 25 mm (1 in.) thick; ¹/₂ h is added for each additional 25 mm (1 in.) of thickness.

Steel	Annealing temperature °C	°F	Cooling cycle(a) °C From	To	°F From	To	Hardness range, HB
1018	855-900	1575-1650	855	705	1575	1300	111-149
1020	855-900	1575-1650	855	700	1575	1290	111-149
1022	855-900	1575-1650	855	700	1575	1290	111-149
1025	855-900	1575-1650	855	700	1575	1290	111-187
1030	845-885	1550-1625	845	650	1550	1200	126-197
1035	845-885	1550-1625	845	650	1550	1200	137-207
1040	790-870	1450-1600	790	650	1450	1200	137-207
1045	790-870	1450-1600	790	650	1450	1200	156-217
1050	790-870	1450-1600	790	650	1450	1200	156-217
1060	790-845	1450-1550	790	650	1450	1200	156-217
1070	790-845	1450-1550	790	650	1450	1200	167-229
1080	790-845	1450-1550	790	650	1450	1200	167-229
1090	790-830	1450-1525	790	650	1450	1200	167-229
1095	790-830	1450-1525	790	655	1450	1215	167-229

(a) Furnace cooling at 28 °C/h (50 °F/h).

Table 4. Recommended temperatures and time cycles for annealing of alloy steels

Steel	Austenitizing temperature °C	°F	Temperature °C From	To	°F From	To	Conventional cooling(a) Cooling rate °C/h	°F/h	Time, h	Isothermal method(b) Cool to °C	°F	Hold, h	Hardness (approx), HB
To obtain a predominantly pearlitic structure(c)													
1340	830	1525	735	610	1350	1130	11	20	11	620	1150	4.5	183
2340	800	1475	655	555	1210	1030	8.3	15	12	595	1100	6	201
2345	800	1475	655	550	1210	1020	9.1	15	12.7	595	1100	6	201
3120(d)	885	1625	...	...	...	...	...	...	...	650	1200	4	179
3140	830	1525	735	650	1350	1200	11	20	7.5	660	1225	6	187
3150	830	1525	705	645	1300	1190	11	20	5.5	660	1225	6	201
3310(e)	870	1600	...	...	...	...	...	...	...	595	1100	14	187
4042	830	1525	745	640	1370	1180	11	20	9.5	660	1225	4.5	197
4047	830	1525	735	630	1350	1170	11	20	9	660	1225	5	207
4062	830	1525	695	630	1280	1170	8.3	15	7.3	660	1225	6	223
4130	855	1575	765	665	1410	1230	20	35	5	675	1250	4	174
4140	845	1550	755	665	1390	1230	14	25	6.4	675	1250	5	197
4150	830	1525	745	670	1370	1240	8.4	15	8.6	675	1250	6	212
4320(d)	885	1625	...	...	...	...	...	...	...	660	1225	6	197
4340	830	1525	705	565	1300	1050	8.3	15	16.5	650	1200	8	223
4620(d)	885	1625	...	...	...	...	...	...	...	650	1200	6	187
4640	830	1525	715	600	1320	1110	7.6	15	15	620	1150	8	197
4820(d)	...	...	...	...	...	...	...	...	...	605	1125	4	192
5045	830	1525	755	665	1390	1230	11	20	8	660	1225	4.5	192
5120(d)	885	1625	...	...	...	...	...	...	...	690	1275	4	179
5132	845	1550	755	670	1390	1240	11	20	7.5	675	1250	6	183
5140	830	1525	740	670	1360	1240	11	20	6	675	1250	6	187
5150	830	1525	705	650	1300	1200	11	20	5	675	1250	6	201
52100(f)	...	...	...	...	...	...	...	...	...	...	...	...	...
6150	830	1525	760	675	1400	1250	8.4	15	10	675	1250	6	201
8620(d)	885	1625	...	...	...	...	...	...	...	660	1225	4	187
8630	845	1550	735	640	1350	1180	11	20	8.5	660	1225	6	192
8640	830	1525	725	640	1340	1180	11	20	8	660	1225	6	197
8650	830	1525	710	650	1310	1200	8.4	15	7.2	650	1200	8	212
8660	830	1525	700	655	1290	1210	8.4	15	8	650	1200	8	229
8720(d)	885	1625	...	...	...	...	...	...	...	660	1225	4	187
8740	830	1525	725	645	1340	1190	11	20	7.5	660	1225	7	201
8750	830	1525	720	630	1330	1170	8.4	15	10.7	660	1225	7	217
9260	860	1575	760	705	1400	1300	8.4	15	6.7	660	1225	6	229
9310(e)	870	1600	...	...	...	...	...	...	...	595	1100	14	187
9840	830	1525	695	640	1280	1180	8.4	15	6.6	650	1200	6	207
9850	830	1525	700	645	1290	1190	8.4	15	6.7	650	1200	8	223
To obtain a predominantly ferritic and spheroidized carbide structure													
1320(d)	805	1480	...	...	...	...	...	...	...	650	1200	8	170
1340	750	1380	735	610	1350	1130	5.5	10	22	640	1180	8	174
2340	715	1320	655	555	1210	1030	5.5	10	18	605	1125	10	192
2345	715	1320	655	550	1210	1020	5.5	10	19	605	1125	10	192
3120(d)	790	1450	...	...	...	...	...	...	...	650	1200	8	163
3140	745	1370	735	650	1350	1200	5.5	10	15	660	1225	10	174
3150	750	1380	705	645	1300	1190	5.5	10	11	660	1225	10	187
9840	745	1370	695	640	1280	1180	5.5	10	11	650	1200	10	192
9850	745	1370	700	645	1290	1190	5.5	10	11	650	1200	12	207

(a) The steel is cooled in the furnace at the indicated rate through the temperature range shown. (b) The steel is cooled rapidly to the temperature indicated and is held at that temperature for the time specified. (c) In isothermal annealing to obtain pearlitic structure, steels may be austenitized at temperatures up to 70 °C (125 °F) higher than temperatures listed. (d) Seldom annealed. Structures of better machinability are developed by normalizing or by transforming isothermally after rolling or forging. (e) Annealing is impractical by the conventional process of continuous slow cooling. The lower transformation temperature is markedly depressed, and excessively long cooling cycles are required to obtain transformation to pearlite. (f) Predominantly pearlitic structures are seldom desired in this steel.

Table 5. Austenitizing temperatures for direct-hardening carbon and alloy steels (SAE)

Steel	Temperature °C	°F	Steel	Temperature °C	°F	Steel	Temperature °C	°F
Carbon steels			1146	800-845	1475-1550	50B50	800-845	1475-1550
1025	855-900	1575-1650	1151	800-845	1475-1550	50B60	800-845	1475-1550
1030	845-870	1550-1600	1536	815-845	1500-1550	5130	830-855	1525-1575
1035	830-855	1525-1575	1541	815-845	1500-1550	5132	830-855	1525-1575
1037	830-855	1525-1575	1548	815-845	1500-1550	5135	815-845	1500-1550
1038(a)	830-855	1525-1575	1552	815-845	1500-1550	5140	815-845	1500-1550
1039(a)	830-855	1525-1575	1566	855-885	1575-1625	5145	815-845	1500-1550
1040(a)	830-855	1525-1575				5147	800-845	1475-1550
1042	800-845	1475-1550				5150	800-845	1475-1550
1043(a)	800-845	1475-1550	**Alloy steels**			5155	800-845	1475-1550
1045(a)	800-845	1475-1550	1330	830-855	1525-1575	5160	800-845	1475-1550
1046(a)	800-845	1475-1550	1335	815-845	1500-1550	51B60	800-845	1475-1550
1050(a)	800-845	1475-1550	1340	815-845	1500-1550	50100	775-800(c)	1425-1475(c)
1055	800-845	1475-1550	1345	815-845	1500-1550	51100	775-800(c)	1425-1475(c)
1060	800-845	1475-1550	3140	815-845	1500-1550	52100	775-800(c)	1425-1475(c)
1065	800-845	1475-1550	4037	830-855	1525-1575	6150	845-885	1550-1625
1070	800-845	1475-1550	4042	830-855	1525-1575	81B45	815-855	1500-1575
1074	800-845	1475-1550	4047	815-855	1500-1575	8630	830-870	1525-1600
1078	790-815	1450-1500	4063	800-845	1475-1550	8637	830-855	1525-1575
1080	790-815	1450-1500	4130	815-870	1500-1600	8640	830-855	1525-1575
1084	790-815	1450-1500	4135	845-870	1550-1600	8642	815-855	1500-1575
1085	790-815	1450-1500	4137	845-870	1550-1600	8645	815-855	1500-1575
1086	790-815	1450-1500	4140	845-870	1550-1600	86B45	815-855	1500-1575
1090	790-815	1450-1500	4142	845-870	1550-1600	8650	815-855	1500-1575
1095	790-815(a)	1450-1500(b)	4145	815-845	1500-1550	8655	800-845	1475-1550
			4147	815-845	1500-1550	8660	800-845	1475-1550
			4150	815-845	1500-1550	8740	830-855	1525-1575
Free-cutting carbon steels			4161	815-845	1500-1550	8742	830-855	1525-1575
1137	830-855	1525-1575	4337	815-845	1500-1550	9254	815-900	1500-1650
1138	815-845	1500-1550	4340	815-845	1500-1550	9255	815-900	1500-1650
1140	815-845	1500-1550	50B40	815-845	1500-1550	9260	815-900	1500-1650
1141	800-845	1475-1550	50B44	815-845	1500-1550	94B30	845-885	1550-1625
1144	800-845	1475-1550	5046	815-845	1500-1550	94B40	845-885	1550-1625
1145	800-845	1475-1550	50B46	815-845	1500-1550	9840	830-855	1525-1575

(a) Commonly used on parts where induction hardening is employed. All steels from SAE 1030 up may have induction hardening applications. (b) This temperature range may be employed for 1095 steel that is to be quenched in water, brine or oil. For oil quenching, 1095 steel may alternatively be austenitized in the range 815 to 870 °C (1500 to 1600 °F). (c) This range is recommended for steel that is to be water quenched. For oil quenching, steel should be austenitized in the range 815 to 870 °C (1500 to 1600 °F).

be classified according to the temperature to which the steel is heated and the method of cooling used. The maximum temperature may be below the lower critical temperature, A_1 (subcritical annealing); above A_1 but below the upper critical temperature, A_3 in hypoeutectoid steels or A_{cm} in hypereutectoid steels (intercritical annealing); or above A_3 (full annealing).

Temperatures, cooling cycles and associated Brinell hardness ranges for simple annealing of carbon steels are given in Table 3, and similar data for annealing of alloy steels to produce two different structures are presented in Table 4.

SPHEROIDIZING

Steels may be spheroidized—that is, heated and cooled to produce a structure of globular carbides in a ferritic matrix—by the following methods:

1. Prolonged holding at a temperature just below Ae_1
2. Heating and cooling alternately between temperatures that are just above Ac_1 and just below Ar_1
3. Heating to a temperature above Ac_1, and then either cooling very slowly in the furnace or holding at a temperature just below Ar_1
4. Cooling at a suitable rate from the minimum temperature at which all carbide is dissolved, to prevent reformation of a carbide network, and then reheating in accordance with method 1 or 2 above (applicable to hypereutectoid steel containing a carbide network).

PROCESS ANNEALING

As the hardness of steel increases during cold working, ductility decreases and additional cold reduction becomes so difficult that the material must be annealed to restore its ductility. Such annealing between processing steps is referred to as "in-process" or simply "process" annealing.

Process annealing usually consists of heating to a temperature below Ae_1, soaking for an appropriate time and then cooling, usually in air. In most instances, heating to a temperature between 11 and 22 °C (20 and 40 °F) below Ae_1 produces the best combination of microstructure, hardness and mechanical properties.

Austenitizing Temperatures for Hardening Carbon and Low-Alloy Steels

TEMPERATURES recommended for austenitizing carbon and low-alloy steels prior to hardening are given in Table 5 (for direct-hardening grades) and Table 6 (for carburized steels). Table 6 is applicable to carburized steels that have been cooled slowly from the carburizing temperature and are to be furnace hardened in a subsequent operation.

For most applications, the rate of heating to the austenitizing temperature is less important than

other factors in the hardening process, such as maximum temperature attained throughout the section, temperature uniformity, time at temperature and rate of cooling. The thermal conductivity of the steel, the nature of the furnace atmosphere (scaling or nonscaling), thickness of section, method of loading (spaced or stacked), and the degree of circulation of the furnace atmosphere all influence the rate of heating of the steel part to the required temperature selected from Tables 5 and 6.

The difference in temperature rise within thick and thin sections of articles of varying cross section is a major problem in practical heat treating operations. When temperature uniformity is the ultimate objective of the heating cycle, this is more safely attained by slow heating than by rapid heating. Furthermore, the maximum temperature in the austenite range should not exceed that required to achieve the necessary extent of solution of carbide. The temperatures listed in Tables 5 and 6 conform with this requirement. When heating with significant cross-section variations, provisions should be made for slower heating to minimize thermal stresses and distortions.

Quenching of Steel

QUENCHING OF STEEL is the rapid cooling of steel from a suitable elevated temperature. This generally is accomplished by immersion in water, oil, polymer solution or salt, although forced air is sometimes used. As a result of quenching,

Table 6. Reheating (austenitizing) temperatures for hardening of carburized(a) carbon and alloy steels (SAE)

Steel	Temperature °C	°F	Steel	Temperature °C	°F	Steel	Temperature °C	°F
Carbon steels			1527 760-790		1400-1450	4626 815-845		1500-1550
1010 760-790		1400-1450				4718 815-845		1500-1550
1012 760-790		1400-1450	**Free-cutting carbon steels**			4720 815-845		1500-1550
1015 760-790		1400-1450	1109 760-790		1400-1450	4815 800-830		1475-1525
1016 760-790		1400-1450	1115 760-790		1400-1450	4817 800-830		1475-1525
1017 760-790		1400-1450	1117 760-790		1400-1450	4820 800-830		1475-1525
1018 760-790		1400-1450	1118 760-790		1400-1450	8115 845-870		1550-1600
1019 760-790		1400-1450				8615 845-870		1550-1600
1020 760-790		1400-1450	**Alloy steels**			8617 845-870		1550-1600
1022 760-790		1400-1450	3310 790-830		1450-1525	8620 845-870		1550-1600
1513 760-790		1400-1450	4320 830-845		1525-1550	8622 845-870		1550-1600
1518 760-790		1400-1450	4615 815-845		1500-1550	8625 845-870		1550-1600
1522 760-790		1400-1450	4617 815-845		1500-1550	8627 845-870		1550-1600
1524 760-790		1400-1450	4620 815-845		1500-1550	8720 845-870		1550-1600
1525 760-790		1400-1450	4621 815-845		1500-1550	8822 845-870		1550-1600
1526 760-790		1400-1450				9310 790-830		1450-1525

(a) Carburizing is commonly carried out at 900 to 925 °C (1650 to 1700 °F); slow cooled and reheated to given austenizing temperature.

production parts must develop an acceptable as-quenched microstructure and, in critical areas, mechanical properties that will meet minimum specifications after the parts are tempered.

The effectiveness of quenching depends on the cooling characteristics of the quenching medium as related to the ability of the steel to harden. Thus, results may be varied by changing the steel composition or the agitation, temperature and type of quenching medium. The design of the quenching system and the thoroughness with which the system is maintained contribute to the success of the process. The design of the part also contributes to the mechanical properties and the distortion that will result from a particular quench.

MECHANISM OF QUENCHING

Several factors are involved in the mechanism of quenching: (a) internal conditions of the workpiece that affect the supply of heat to the surface, (b) surface and other external conditions that affect the removal of heat, (c) the heat-extracting potential of the quenching fluid in the quiescent state at normal fluid temperatures and pressures ("standard" conditions), and (d) changes in the heat-extracting potential of the fluid brought about by "nonstandard" conditions of agitation, temperature or pressure.

The typical surface and center cooling curves shown in Fig. 3 graphically describe the four stages of heat transfer from a hot solid to a cold liquid.

Stage A' in Fig. 3 illustrates the first effects of immersion. Sometimes called the "initial liquid contact stage," this stage is characterized by the formation of vapor bubbles that precedes the establishment of an enveloping vapor blanket.

Stage A, called the "vapor blanket cooling stage," is characterized by the Leidenfrost phenomenon—namely, the formation of an unbroken vapor blanket that surrounds the test piece. It occurs when the supply of heat from the surface of the test piece exceeds the amount of heat needed to form the maximum vapor per unit area of the piece. This stage is one of slow cooling, because the vapor envelope acts as an insulator and cooling occurs principally by radiation through the vapor film.

Stage B, the "vapor transport cooling stage," which produces the highest rate of heat transfer, begins when the temperature of the surface metal has been reduced somewhat and the continuous vapor film collapses; violent boiling of the

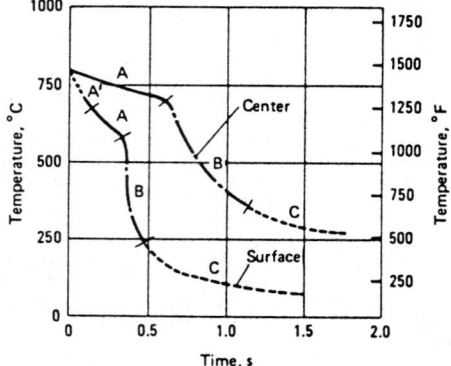

Fig. 3. Typical surface and center cooling curves, indicating the stages of heat transfer from a hot solid to a cold liquid

quenching liquid then occurs and heat is removed from the metal at a very rapid rate, largely as heat of vaporization. The boiling point of the quenchant determines the conclusion of this stage. Size and shape of the vapor bubbles are important in controlling the duration of stage B, as well as the cooling rate developed within it.

Stage C is called the "liquid cooling stage"; the cooling rate in this stage is slower than that developed in stage B. Stage C begins when the temperature of the metal surface is reduced to the boiling point (or boiling range) of the quenching liquid. Below this temperature, boiling stops and slow cooling takes place thereafter by conduction and convection.

Agitation is externally produced movement of the quenching liquid relative to the part, either by stirring the liquid or moving the part, or both in combination. This activity has an extremely important influence on the heat-transfer characteristics of the quenching liquid. It causes an earlier mechanical disruption of the vapor blanket in stage A and produces smaller, more frequently detached vapor bubbles during the vapor transport cooling stage (stage B). It mechanically disrupts or dislodges gels and solids, whether they are on the surface of the test piece or suspended at the edge of the vapor blanket, thus producing faster heat transfer in liquid cooling (stage C). In addition to producing the above effects, agitation also brings cool liquid to replace heat-laden liquid (see Table 7).

Temperature of Quenchant. The temperature of the liquid may markedly affect its ability to extract heat. As an example, water temperature is very important because it loses its cooling power as it approaches its boiling point. In oil, this effect is not as pronounced because oil becomes less viscous as the temperature is increased. This "thinning" of the oil offsets the temperature rise by a substantial amount.

QUENCHING MEDIA

Many different media have been used for quenching. Most are included in the list that follows, but some of these are used only to a very limited extent.

- Water
- Brine solutions (aqueous)
- Caustic solutions
- Oils
- Polymer solutions
- Molten salts
- Molten metals
- Gases, including still or moving
- Fog quenching
- Dry dies, commonly water cooled.

Water. As a quenching medium, plain water approaches the maximum cooling rate attainable in a liquid. Other advantages are that it is inexpensive and readily available and is easily disposed of without attendant problems of pollution or health hazard.

One disadvantage of plain water as a quenchant is that its rapid cooling rate persists throughout the lower temperature range, in which distortion or cracking is likely to occur. Consequently, water usually is restricted to the quenching of simple, symmetrical parts made of the shallower-hardening grades of steel (plain carbon

Table 7. Effect of agitation on the effectiveness of quenching

Circulation or agitation	H-value or quenching power		
	Oil	Water	Caustic soda or brine
None	0.25 to 0.30	0.9 to 1.0	2
Mild	0.30 to 0.35	1.0 to 1.1	2 to 2.2
Moderate	0.35 to 0.40	1.2 to 1.3	...
Good	0.4 to 0.5	1.4 to 1.5	...
Strong	0.5 to 0.8	1.6 to 2.0	...
Violent	0.8 to 1.1	4	5

or low-alloy). Another disadvantage of using plain water is that its vapor blanket stage may be prolonged. This prolongation, which varies with the degree to which the complexity of the part being quenched encourages vapor entrapment and with the temperature of the quench water, results in uneven hardness.

Brine Solutions. The term "brine," as applied to quenching, refers to aqueous solutions containing various percentages of salt (such as sodium chloride or calcium chloride).

Brine offers the following advantages compared with plain water or oil, for quenching:

- Cooling rate is faster than that of water for the same degree of agitation, or less agitation is required for a given cooling rate.
- Temperature is less critical than for water, thus requiring less control.
- Possibility of soft spots from steam pockets is less than in water quenching.

A principal disadvantage of brine is that its corrosive nature requires that, for reasonable service life, the quench tank, pumps, conveyors and other equipment in constant contact with the brine solution either be protected from corrosion by coating, plating or sheathing or be made of corrosion-resistant metals (such as copper-bearing or stainless steels).

Caustic Solutions. Aqueous solutions of 5 to 10% sodium hydroxide (NaOH) often are used for quenching. The performance of such solutions is similar to brine solutions — sharing about the same advantages as a quenchant. Its main shortcoming is its high alkalinity, which is harmful to human skin.

OIL QUENCHING

Quenching oils can be divided into several distinct groups. Based on their composition, quenching effect and use temperature, quenching oils are categorized as conventional, fast, martempering, or hot quenching.

Conventional quenching oils are mineral oils, sometimes containing antioxidants; however, they are free from additives that alter their quenching effects. The typical viscosity of conventional quenching oils is in the range of 100 to 110 SUS at 40 °C (100 °F), but it can reach about 200 SUS at 40 °C (100 °F).

Fast quenching oils are mineral oil blends, usually with a viscosity between 50 and 110 SUS at 40 °C (100 °F), but for the most part between 85 and 105 SUS at 40 °C (100 °F). They contain specially developed proprietary additives that provide faster quenching effects; in addition they can be compounded further with antioxidants, wetting agents and other additives.

Martempering or hot quenching oils are solvent-refined paraffin-type mineral oils of very good thermal and oxidation stability. They are used at temperatures between about 95 °C (200 °F) and 230 °C (450 °F) for modified and actual martempering of ferrous metals. Martempering oils may contain antioxidants to improve their aging stability. They are also available with comparatively high quenching effects, even at high use temperatures, because of the additions of very effective speed improvers.

Emulsions of soluble oils, usually employed as coolants for metal-working such as grinding, cutting and sometimes forming, are used for quenching with concentrations of 3 to about 15%. Such quenching emulsions are similar in char-

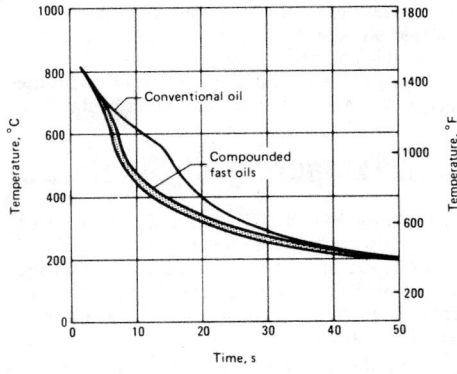

Center cooling curves for austenitic stainless steel quenched in conventional oil and compounded oils with fast quenching capacities. All oils at 52 °C (125 °F). Specimens were 13 mm (0.5 in.) in diam by 64 mm (2.5 in.) long.

Fig. 4. Center cooling curves in oil

acteristics to water-based quenchants, particularly when compared to synthetic polymer solutions.

Cooling Characteristics. The ideal quenchant would have a high initial quenching effect during vapor-phase and boiling-range periods but would cool slowly through the final convection range (liquid-cooling phase). Cold water, and especially the aqueous solutions of inorganic salts, show the highest initial quenching speeds, but they also quench very fast at the end of the quenching process — that is, during the convection phase. Thus, they are restricted to quenching simple shapes and steels of comparatively low hardenability. For other workpieces, they would cause either intolerable degrees of distortion, or warpage and high quench crack rate. All quenching oils have considerably lower quenching effects than water or aqueous inorganic salt solutions. However, the heat extraction is more uniform and particularly slow at the end of the cooling cycle — that is, during the convection range. As a result, the dangers of distortion or cracking are diminished decisively.

Conventional quenching oils exhibit a comparatively long vapor-phase period, during which period the quenching speed is very low. The rate of cooling becomes higher during the boiling range, followed again by very slow cooling in the convection range. Thus, the quenching power of conventional quenching oils is far less than that of water and often inadequate for steel of lower hardenability.

Fast quenching oils show a high initial quenching speed, in some situations approaching the initial speed of water, followed by fast cooling during the boiling range. The cooling rate in the convection range is usually about the same for fast and conventional quenching oils.

Figure 4 shows the spread of cooling curves of eight different compounded (speed-improving additive-containing) fast quenching oils as a shaded band, and gives a curve for one typical conventional additive-free 100 SUS at 40 °C (100 °F) mineral quenching oil. These cooling curves, which show clearly the difference in quenching effects of conventional and fast quenching oils, were recorded without agitation and using a cylindrical austenitic stainless steel test probe 13 mm (0.5 in.) in diameter by 64 mm (2.5 in.) long, containing a thermocouple at its geometrical center. All oils were at a temperature of 50 °C (125 °F).

POLYMER SOLUTIONS

Cooling Characteristics. Polyvinyl alcohol (PVA) was introduced in the mid-1950's as an additive to water to modify its cooling rate. As the curves in Fig. 5 indicate, only slight variations in solution concentration are needed to produce changes in cooling characteristics of PVA solutions. At concentrations of less than 0.01%, the cooling characteristics at room temperature are only modestly different from those of water alone. With such small concentration variations, close control of PVA solutions is necessary. Control is complicated by the fact that quenched parts can become coated with an insoluble layer of resin, thus reducing the bath concentration. Maintaining an "effective" concentration requires specific control measures.

Polyalkylene Glycols (PAG). Proper selection of the polymer composition, and its molecular weight, provides a PAG product that is completely soluble in water at room temperature. However, the selected PAG molecules exhibit the unique behavior of inverse solubility in water — that is, water insolubility at elevated temperatures. This

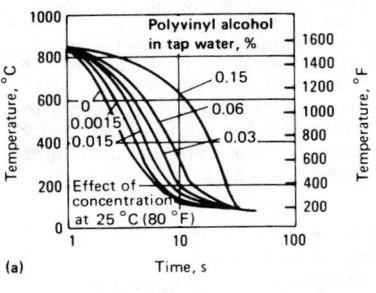

(a)

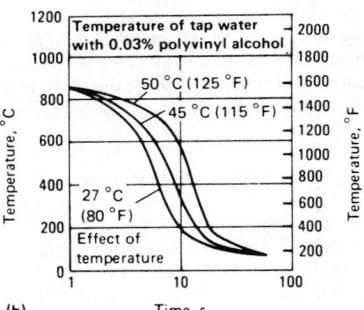

(b)

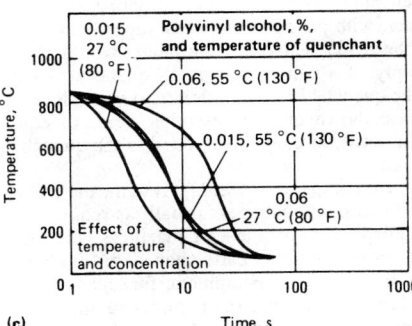

(c)

Center cooling curves for type 304 stainless steel specimens 13 mm (¹/₂ in.) diam by 100 mm (4 in.) long quenched in still tap water at 25 °C (80 °F) (or at other temperatures shown) containing various concentrations of polyvinyl alcohol. Thermocouple was placed in geometric center of each specimen. Water hardness, 130 ppm.

Fig. 5. Cooling curves for polyvinyl alcohol solutions

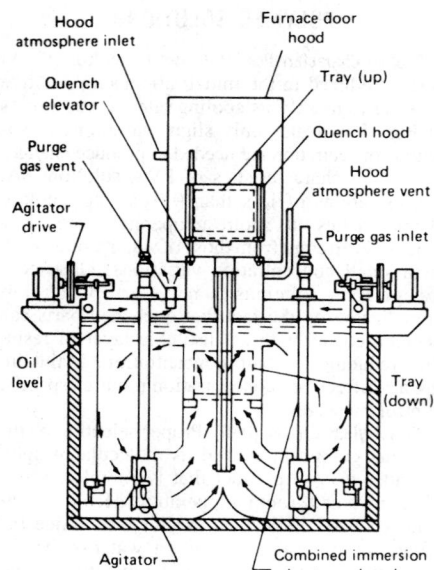

Directional vanes in the oil stream distribute the oil flow uniformly. Unit contains combined heating and cooling elements and provision for blanketing the surface of the oil with an inert gas atmosphere. Radiant tube is used for heating and cooling.

Fig. 6. Schematic of a typical installation for high-volume batch quenching of carburized or hardened parts on trays

phenomenon provides the unique mechanism for cooling hot metal by surrounding the metal piece with a polymer-rich coating that serves to govern the rate of heat extraction into the surrounding aqueous solution. As the metal part approaches in temperature the temperature of the quenchant itself, the PAG polymer coating dissolves to again provide a uniform concentration in the quenchant bath.

QUENCHING SYSTEMS

Equipment requirements for quenching may vary widely. A small plant making machine parts might each day require the hardening of only a few simple carbon steel parts weighing about 1.4 kg (3 lb) apiece. For such an application, the quenching "system" would comprise a barrel of water, with piping to the water supply and a drain to the sewer. Handling equipment would consist simply of a pair of tongs. As quantity of work to be quenched and complexity of workpieces increase, however, various other items of equipment must, of course, be added to the quenching system.

For a complete quenching system, the following functional equipment usually is required and installed: (*a*) work tank or machines, (*b*) facilities for handling the parts quenched, (*c*) quenching medium, (*d*) equipment for agitation, (*e*) coolers, (*f*) heaters, (*g*) pumps and strainers or filters, (*h*) quenchant supply tank, (*i*) equipment for ventilation and for protection against hazards, and (*j*) equipment for automatic removal of scale from tanks.

A typical batch-quenching tank that can be used for high-volume production of work on trays is illustrated in Fig. 6. Atmosphere protection, agitation, temperature control and other features are included in this arrangement. The pressure of the

atmosphere gas must be greater than that of the outside air; otherwise, air will enter the chain-drive unit opposite the agitator motors, and the revolving chain will carry air into the oil and cause oxidation, especially when hot oil is being used.

Tempering of Steel

TEMPERING OF STEEL is a process in which previously hardened or normalized steel is heated to a temperature below the transformation range and cooled at a suitable rate, primarily to increase ductility and toughness. Steels are tempered by reheating after hardening to obtain specific values of mechanical properties and to relieve quenching stresses and ensure dimensional stability. Tempering usually follows quenching from above the critical temperature.

PRINCIPAL VARIABLES

Tempering Temperature. Several empirical relationships have been made between the tensile

strength and hardness of tempered steels such that measurement of hardness is commonly used to evaluate the response of a steel to tempering. For example, Fig. 7 shows the effect of tempering temperature on the hardness of a plain carbon steel (AISI 1050) held at temperature for 1 h. Note that both room-temperature hardness and strength decrease as the tempering temperature is increased. Ductility at ambient temperatures, as measured by either elongation or reduction in area, increases with tempering temperature.

Tempering Time. The diffusion of carbon and alloying elements necessary for the formation of carbides is temperature and time dependent. The effect of tempering time on the hardness of an 0.82% carbon steel tempered at various temperatures is shown in Fig. 8. The changes in hardness are approximately linear over a large portion of the time range when the time is presented on a logarithmic scale. Rapid changes in room-temperature hardness occur at the start of tempering in times less than 10 s. Less rapid, but still large, changes in hardness occur in times from 1 to 10 min, and smaller changes occur in times from 1

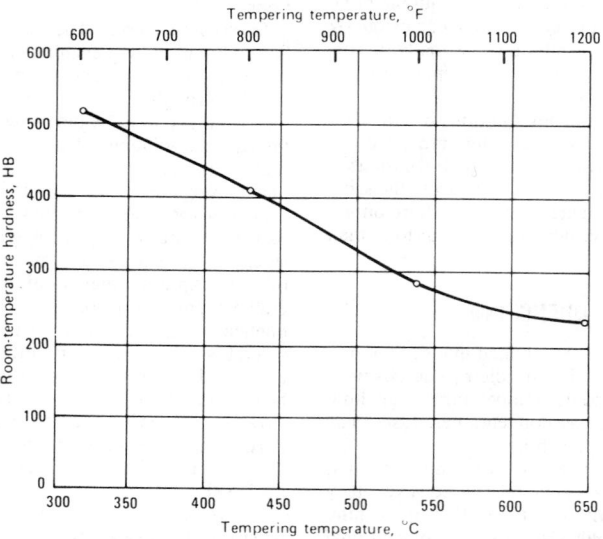

Fig. 7. Effect of tempering temperature on room-temperature hardness of 1050 steel

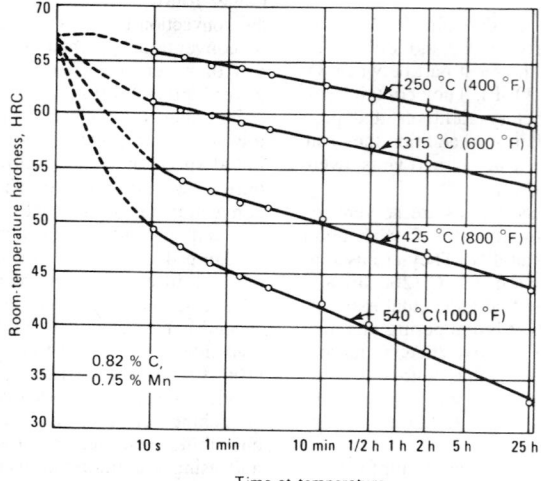

Note nearly straight lines on logarithmic time scale.

Fig. 8. Effect of time at four tempering temperatures on room-temperature hardness of quenched 0.82% C steel

to 2 h. For consistency and less dependency on variations in time, components generally are tempered for 1 to 2 h. The levels of hardness produced by very short tempering cycles, such as in induction tempering, would be quite sensitive to both the temperature achieved and the time at temperature.

Cooling Rate. Another factor that can affect the properties of a steel is the cooling rate from the tempering temperature. Although tensile properties are not affected by cooling rate, toughness (as measured by notched-bar impact testing) can be decreased if the steel is cooled slowly through the temperature range from 375 to 575 °C (705 to 1065 °F), especially in steels that contain carbide-forming elements. Elongation and reduction in area may be affected also.

EQUIPMENT FOR TEMPERING

Steel usually is tempered in either an air (convection) furnace or a salt bath. Molten metal baths, oil baths, and flame or induction heating units are used also.

Convection Furnaces. The most commonly employed tempering method utilizes the recirculating or forced-air convection furnace. Forced recirculating air is the most common and efficient method of tempering because it lends itself to a wide selection of furnace designs to accommodate a variety of products and capacities. Moreover, the metallurgical results are very good in terms of price per unit weight of yield.

Convection furnaces generally are designed for tempering temperatures of 150 to 750 °C (300 to 1380 °F). For temperatures up to 550 °C (1020 °F), recirculated hot air is supplied to the product from a chamber separate from the work-holding area, to avoid uneven heating by radiation. For temperatures of 550 to 750 °C, either forced convection or radiant heating is used depending on the metallurgical requirements of the product. To obtain closer control of metallurgical properties, the recirculated forced hot air is employed; but for greater efficiency, radiant heating is used, because transfer of radiant heat is greater as the temperature approaches 750 °C (1380 °F).

Salt bath furnaces may be employed for tempering at 160 °C (320 °F) and above. Good heat transfer and natural convection in the bath promote uniformity of workpiece temperature.

All moisture must be removed from parts before they are immersed in the molten salt, because hot salt reacts violently with moisture. If dirty or oily parts are immersed in the bath, the salt will become contaminated and require more frequent rectification.

All parts tempered in salt must be cleaned soon after being removed from the bath, because the salt that clings to them is hygroscopic and may cause severe corrosion. Parts with small or blind holes from which salt is difficult to clean should not be tempered in salt.

TEMPERING PROCEDURES

Tempering can be accomplished by soaking entire parts in the furnace for periods of time sufficient to bring the tempering mechanism to the desired point of completion or by selective heating of certain portions of the part to achieve toughness or plasticity in those areas.

Bulk processing may be done in convection furnaces or in molten salt, hot oil or molten metal baths. Selection of the type of furnace depends primarily on number and size of parts and on desired temperature. Table 8 gives temperature ranges, most likely reasons for use, and fundamental problems of these four types of equipment.

Selective tempering techniques are used to soften specific areas of fully hardened parts or to temper areas that were selectively hardened previously. The purpose of this treatment is to improve the machinability, the toughness or the resistance to quench cracking in the selected zone.

Induction and flame tempering are the most commonly utilized selective techniques because of their controllable local heating capabilities. Immersion of selected areas in molten salt or molten metal can be accomplished, but with somewhat less control.

Martempering of Steel

MARTEMPERING is a term used to describe an elevated-temperature quenching procedure aimed at reducing cracks, distortion or residual stresses. It is not a tempering procedure, as the name implies, and is more properly termed "marquenching."

Martempering of steel (and of cast iron) consists of: (*a*) quenching from the austenitizing temperature into a hot fluid medium (hot oil, molten salt, molten metal or a fluidized particle bed) at a temperature usually above the martensitic range (M$_s$ point); (*b*) holding in the quenching medium until the temperature throughout the steel is substantially uniform; and then (*c*) cooling (usually in air) at a moderate rate, to prevent large differences in temperature between the outside and the center of the section. Formation of martensite occurs fairly uniformly throughout the workpiece during cooling to room temperature, thereby avoiding formation of excessive amounts of residual stress. The microstructure after martempering is essentially primary martensite, which is untempered and brittle for most applications. After being air cooled to room temperature, martempered parts are tempered in the same manner as if they had been conventionally quenched. Figure 9 shows the significant difference between conventional quenching (*a*) and martempering (*b*).

Any steel part or grade of steel responding to oil quenching can be martempered to provide similar physical properties. The quenching severity of molten salt is greatly enhanced by agitation and by water additions to the nitrate-salt bath. Both techniques are particularly beneficial in heat treating of carbon steels having limited hardenability. Table 9 compares the properties obtained in 1095 steel by martempering and tempering with those obtained by conventional quenching and tempering.

Table 8. Temperature ranges and general conditions of use for four types of tempering equipment

Type of equipment	Temperature range °C	°F	Service conditions
Convection furnace	50 to 750	120 to 1380	For large volumes of nearly common parts; variable loads make control of temperature more difficult
Salt bath	160 to 750	320 to 1380	Rapid, uniform heating; low to medium volume; should not be used for parts whose configurations make them hard to clean
Oil bath	Up to 250	Up to 480	Good if long exposure is desired; special ventilation and fire control are required
Molten metal bath	Above 390	Above 735	Very rapid heating; special fixturing is required (high density)

Table 9. Mechanical properties of 1095 steel heat treated by two methods

Heat treatment	Hardness, HRC	Impact energy J	ft·lb	Elongation(a), %
Water quench and temper:				
Specimen 1	53.0	16	12	0
Specimen 2	52.5	19	14	0
Martemper and temper:				
Specimen 3	53.0	38	28	0
Specimen 4	52.8	33	24	0
(a) In 25 mm or 1 in.				

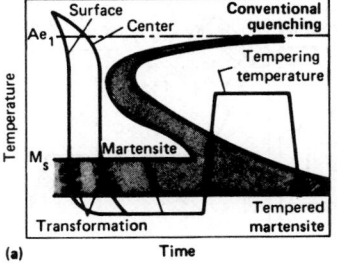
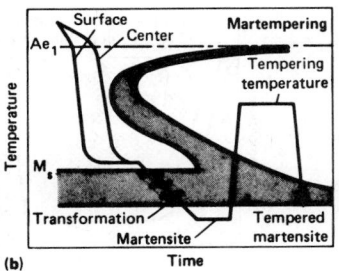
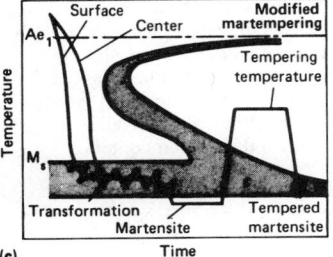

(a) Conventional process. (b) Martempering. (c) Modified martempering.

Fig. 9. Time-temperature transformation diagrams with superimposed cooling curves showing quenching and tempering

MODIFIED MARTEMPERING

Modified martempering differs from "standard" martempering only in that the temperature of the quenching bath is below the M_s point (Fig. 9c). The lower temperature increases the severity of quenching. This is important for steels of lower hardenability that require faster cooling in order to harden to sufficient depth, or when the M_s is high and some bainite is detrimental to the finished part. Thus, modified martempering is applicable to a greater range of steel compositions than is the standard process.

Although hot oil is invariably the quenchant employed for modified martempering at 175 °C (350 °F) and lower, molten nitrate-nitrite salts (with water addition and agitation) are effective at temperatures as low as 175 °C. Due to their higher heat-transfer coefficients, molten salts offer some metallurgical and operational advantages.

MARTEMPERING MEDIA

Molten salt and hot oil are both widely used for martempering. Several factors must be considered when choosing between salt and oil. Operating temperature is the most common deciding factor. Oils are widely used for martempering at up to 205 °C (400 °F), and sometimes at temperatures as high as 230 °C (450 °F). Molten salt is used for martempering in the range from 160 to 400 °C (320 to 750 °F).

SUITABILITY OF STEELS FOR MARTEMPERING

Alloy steels generally are more adaptable than carbon steels to martempering. In general, any steel that is normally quenched in oil can be martempered. Some carbon steels that are normally water quenched can be martempered at 205 °C (400 °F) in sections thinner than 5 mm ($^3/_{16}$ in.), using vigorous agitation of the martempering medium. In addition, thousands of gray cast iron parts are martempered on a routine basis.

The grades of steel that are commonly martempered to full hardness include 1090, 4130, 4140, 4150, 4340, 4640, 5140, 6150, 8630, 8640, 8740, 8745, SAE 1141 and SAE 52100. Carburizing grades such as 3312, 4620, 5120, 8620 and 9310 also are commonly martempered after carburizing. Occasionally, higher-alloy steels such as type 410 stainless are martempered, but this is not a common practice.

Success in martempering is based on a knowledge of the transformation characteristics (TTT curves) of the steel being considered. The temperature range in which martensite forms is especially important. Figure 10 shows the martensite temperature ranges for 14 carbon and low-alloy steels. Two trends may be observed in these data: (*a*) as carbon content increases, the martensite range widens and the martensite transformation temperature is lowered; and (*b*) the martensite range of a triple-alloy (Ni-Cr-Mo) steel is usually lower than that of either a single-alloy or a double-alloy steel of similar carbon content.

Any steel that is to be martempered successfully must contain sufficient carbon or alloying additions to move the nose of the TTT curve to the right, thus permitting sufficient time for quenching past the nose of the TTT curve without transformation in this region. Figure 11 shows the TTT curve for 1090 steel, which is the sim-

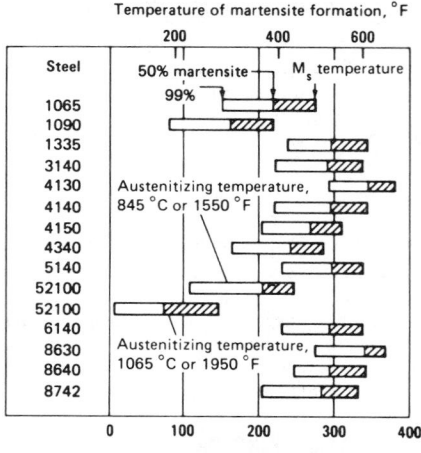

Fig. 10. Temperature ranges of martensite formation in 14 carbon and low-alloy steels

plest form of transformation diagram, because no proeutectoid constituents (free ferrite or free carbide) are involved in the transformation at temperatures above that corresponding to the nose of the curve. The speed of transformation at the nose is related to the hardenability of the steel; when the nose of the TTT curve is far to the left on the diagram, the steel has lower hardenability; when the nose is to the right, the steel has higher hardenability. To achieve full hardening during quenching, the cooling curve of the steel must pass to the left of the curve farthest to the left on the diagram. In production, some loss in as-quenched hardness is usually accepted in order to achieve minimum distortion.

Austempering of Steel

AUSTEMPERING is the isothermal transformation of a ferrous alloy at a temperature below that of pearlite formation and above that of martensite formation. Steel is austempered by being:

- Heated to a temperature within the austenitizing range, usually 790 to 870 °C (1450 to 1600 °F)
- Quenched in a bath maintained at a constant temperature, usually in the range of 260 to 400 °C (500 to 750 °F)
- Allowed to transform isothermally to bainite in this bath
- Cooled to room temperature, usually in still air.

The fundamental difference between austempering and conventional quenching and tempering is shown schematically in Fig. 12. Austempering of steel and hardenable grades of cast iron offers several potential advantages:

- Increased ductility or notch toughness at a given hardness
- Reduced distortion, which lessens subsequent machining time, stock removal and cost
- The shortest over-all time cycle to through-harden within the hardness range of 35 to 55 HRC, with resulting savings in energy and capital investment.

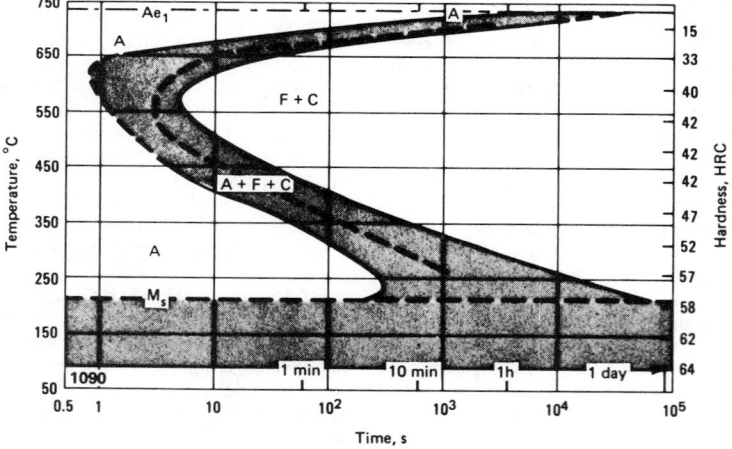

The 1090 steel was austenitized at 885 °C (1625 °F) and had a grain size of 4 to 5.
Fig. 11. Time-temperature transformation diagram for 1090 steel

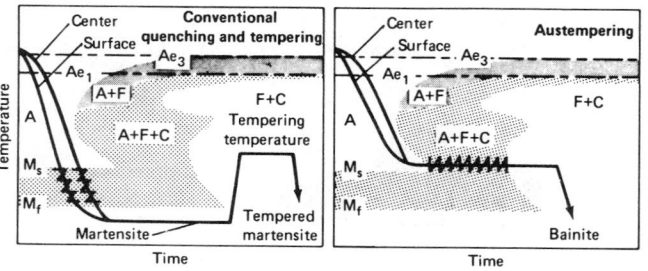

Fig. 12. Schematic comparison of time-temperature transformation cycles for conventional quenching and tempering and for austempering

Table 10. Compositions and characteristics of salts used for austempering

Constituent or characteristic	High range	Wide range
Sodium nitrate	45-55%	15-25%
Potassium nitrate	45-55%	45-55%
Sodium nitrate	. . .	25-35%
Melting point (approx)	220 °C (430 °F)	150-165 °C (300-330 °F)
Working temperature range	260-595 °C (500-1100 °F)	175-540 °C (345-1000 °F)

QUENCHING MEDIA FOR AUSTEMPERING

Molten salt is the quenching medium most commonly used in austempering, because: (a) it transfers heat rapidly; (b) it virtually eliminates the problem of a vapor-phase barrier during the initial stage of quenching; (c) its viscosity is uniform over a wide range of temperatures; (d) its viscosity is low at austempering temperatures (near that of water at room temperature), thus minimizing dragout losses; (e) it remains stable at operating temperatures and is completely soluble in water, thus facilitating subsequent cleaning operations; and (f) the salt can be easily recovered from wash waters so that there is no discharge to drain.

Formulations and characteristics of two typical salt quenching baths are given in Table 10. The high-range salt is suitable for austempering only, whereas the wide-range salt may be used for austempering, martempering, and modifications thereof.

STEELS FOR AUSTEMPERING

The selection of steel for austempering must be based on transformation characteristics as indicated in time-temperature transformation (TTT) diagrams. Three important considerations are (a) the location of the nose of the TTT curve and the time available for bypassing it, (b) the time required for complete transformation of austenite to bainite at the austempering temperature, and (c) the location of the M_s point. Suggested steels are:

- Plain carbon steels containing 0.50 to 1.00% C and a minimum of 0.60% Mn
- High-carbon steels containing more than 0.90% C and, possibly, a little less than 0.60% Mn
- Certain carbon steels (such as 1041) with carbon contents less than 0.50% but with manganese contents in the range from 1.00 to 1.65%
- Certain low-alloy steels (such as series 5100 steels) containing more than 0.30% C; series 1300 to 4000 steels with carbon contents in excess of 0.40%; and other steels, such as 4140, 6145 and 9440.

APPLICATIONS

Austempering usually is substituted for conventional quenching and tempering for either or both of two reasons: (a) to obtain improved mechanical properties (particularly higher ductility or notch toughness at a given high hardness), and (b) to decrease the likelihood of cracking and distortion. In some applications, austempering is less expensive than conventional quenching and tempering. This is most likely when small parts are treated in an automated setup wherein conventional quenching and tempering comprise a three-step operation—that is, austenitizing, quenching and tempering. Austempering requires only two processing steps: austenitizing and isothermal transformation in an austempering bath.

MODIFIED AUSTEMPERING

As mentioned previously, modifications of austempering practice that give rise to mixed structures of pearlite and bainite are quite common in industrial practice. The amounts of pearlite and bainite may vary widely in different modifications of processing.

Patenting, a treatment used in the wire industry, is a significant and useful form of modified austempering, in which austenitized wire or rod is continuously quenched into a bath maintained at 510 to 540 °C (950 to 1000 °F) and held in the bath for periods ranging from 10 s (for small wire) to 90 s (for rod). Patenting provides a combination of moderately high strength and high ductility. As indicated in Fig. 13 by the line designated "Modified practice," the process varies from true austempering in that the quenching rate, instead of being rapid enough to avoid the nose of the TTT curve, is sufficiently slow to intersect the nose, which results in the formation of fine pearlite.

Similar practice is usefully employed in applications involving plain carbon steels when a hardness between about 30 and 42 HRC is desirable or acceptable. The hardness of plain carbon steel quenched at a rate that intersects the nose of the TTT curve will vary with carbon content.

Modified practices can be applied to parts having sections thicker than those normally considered practicable for austempering.

Cold Treating of Steel

WHENEVER hardening is to be done during heat treating, complete transformation from austenite to martensite generally is desired prior to tempering. From a practical standpoint, however, conditions vary widely and 100% transformation rarely, if ever, occurs. Cold treating may be useful, in many instances, for improving the percentage of transformation and thus enhancing properties.

During hardening, martensite develops as a continuous process from start (M_s) to finish (M_f) through the martensite-formation range. Except in a few highly alloyed steels, martensite starts to form at well above room temperature. In many instances, transformation is essentially complete at room temperature. Retained austenite tends to be present in varying amounts, however, and, when considered excessive for a particular application, must be transformed to martensite and then tempered.

Cold Treating vs Tempering. Immediate cold treating without delays at room temperature or at other temperatures during quenching offers the best opportunity for maximum transformation to martensite. In some instances, however, there is a risk that this will cause cracking of parts. Therefore, it is important to ensure that the grade of steel and the product design will tolerate immediate cold treating rather than immediate tempering. Some steels must be transferred to a tempering furnace when still warm to the touch, to minimize the likelihood of cracking. Design features such as sharp corners and abrupt changes in section create stress concentrations and promote cracking.

In most instances, cold treating is not done before tempering. In some types of industrial applications, tempering is followed by deep freezing and retempering without delay. For example, such parts as gages, machineways, arbors, mandrels, cylinders, pistons, and ball and roller bearings are treated in this manner for dimensional stability. Several freeze-draw cycles are used for critical applications.

ADVANTAGES OF COLD TREATING

Unlike heat treating, which requires that temperature be precisely controlled to avoid reversal, successful transformation through cold treating depends only on attainment of the minimum low temperature and is not affected by lower temperatures. As long as the material is chilled to −84 °C (−120 °F), transformation will occur; additional chilling will not cause reversal.

Time at Temperature. After thorough chilling, additional exposure has no adverse effect. When heat is used, holding time and temperature are critical. In cold treatment, materials of different compositions and of different configurations may be chilled at the same time even though each may have a different high-temperature transformation point. Moreover, the warm-up rate of a chilled

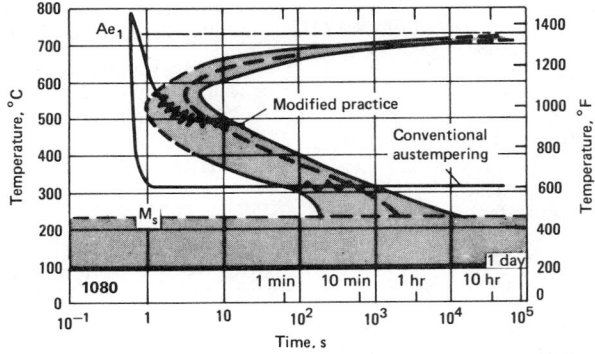

When applied to wire, the modification shown is known as "patenting."

Fig. 13. Time-temperature transformation diagram for 1080 steel, showing difference between conventional and modified austempering

material is not critical as long as uniformity is maintained and gross temperature-gradient variations are avoided.

Heat Treating of Ultrahigh- Strength Steels

ULTRAHIGH-STRENGTH STEELS are heat treated by use of equipment and techniques similar to those employed for heat treating constructional alloy steels. The ultrahigh-strength steels ordinarily are quenched and tempered to specified hardnesses, but for critical applications it may be necessary to pull tensile specimens to ensure that a required combination of strength and ductility has been achieved. In still other instances, it may be necessary to conduct impact tests or fracture-toughness tests in order to ensure that a required level of resistance to brittle fracture has been attained.

MEDIUM-CARBON LOW-ALLOY STEELS

Medium-carbon low-alloy steels are readily hot forged. To avoid stress cracks resulting from air hardening, the forged part should be slowly cooled in a furnace or in an insulating medium. Prior to machining, usual practice is to normalize at 870 to 925 °C (1600 to 1700 °F) and temper at 650 to 675 °C (1200 to 1250 °F) or, if the steel is a deep air-hardening grade, to anneal by furnace cooling from 815 to 845 °C (1500 to 1550 °F) to about 540 °C (1000 °F). These treatments impart moderately hard microstructures suitable for machining. A very soft spheroidized structure can be obtained by full annealing. Such a structure is less well suited for machining than the normalized-and-tempered structure. However, for severe cold forming operations such as spinning, deep drawing and wiredrawing, the soft and ductile spheroidized structure is preferred. If blanks for parts are produced by flame cutting, they are annealed before being formed or machined. Welded parts, especially if complex, are stress relieved, or hardened and tempered, immediately after welding.

Heat treatments and properties of type 4340, the most popular medium-carbon low-alloy steel, are discussed below. For corresponding information on other steels of this class — types 4130, 4140, 300M, D-6a and D-6ac (300M, D-6a and D-6ac are modifications of 4340), 6150 and 8640 — as well as on medium-alloy air-hardening steels and high-alloy hardenable (9Ni-4Co) steels, refer to Metals Handbook, 9th Edition, Volume 4, pages 119 to 129.

4340

Type 4340 is a deep-hardening steel. In thin sections, this steel is air hardening, although in practice it is usually oil quenched. When 4340 is heat treated to tensile strengths greater than about 1400 MPa (about 200 ksi), it is subject to hydrogen embrittlement. Parts exposed to hydrogen, such as in pickling or electroplating, should be baked at 185 to 195 °C (365 to 385 °F) for at least 8 h, and for 23 h if thicker than 38 mm (1½ in.), as soon a possible after the pickling or plating operation.

Heat Treatments. The following standard heat treatments apply to 4340 steel:

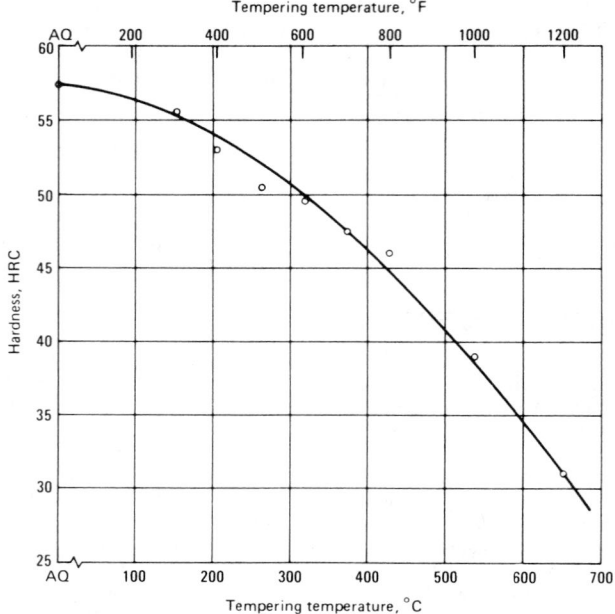

All specimens oil quenched from 845 °C (1550 °F) and tempered 2 h at temperature.
Fig. 14. Variation of hardness with tempering temperature for 4340 steel

- *Normalize:* Heat to 845 to 900 °C (1550 to 1650 °F) and hold for a period that depends on section thickness; air cool.
- *Anneal:* Heat to 830 to 860 °C (1525 to 1575 °F) and hold for a period that depends on section thickness or furnace load; furnace cool.
- *Harden:* Heat to 800 to 845 °C (1475 to 1550 °F) and hold 15 min for each 25 mm (1 in.) of thickness (minimum, 15 min); oil quench to below 65 °C (150 °F); or quench into fused salt at 200 to 210 °C (390 to 410 °F), hold 10 min, then air cool to below 65 °C (150 °F).
- *Temper:* At least ½ h at 200 to 650 °C (400 to 1200 °F); air cool. Temperature and time at temperature depend mainly on desired strength or hardness.
- *Spheroidize:* Preheat to 690 °C (1275 °F) and hold 2 h, increase temperature to 750 °C (1375 °F) and hold 2 h, cool to 650 °C (1200 °F) and hold 6 h, furnace cool to about 600 °C (1100 °F), and finally air cool to room temperature. An alternative schedule is to heat to 730 to 750 °C (1350 to 1375 °F) and hold several hours, then furnace cool to room temperature.
- *Stress relieve:* After straightening, forming or machining, parts may be stress relieved at 650 to 675 °C (1200 to 1250 °F).

Properties. Through hardening of 4340 steel can be achieved by oil quenching round sections up to 75 mm (3 in.) in diameter, and by water quenching larger sections (up to the limit of hardenability).

Hardness of type 4340 as a function of tempering temperature is plotted in Fig. 14. Typical mechanical properties of oil-quenched 4340 are given in Table 11.

Heat Treating of Maraging Steels

MARAGING STEELS are ultrahigh-strength steels that differ from conventional steels in that they are not hardened by carbon content. Instead, these steels are strengthened by precipitation of intermetallic compounds produced by age hardening a matrix of very-low-carbon martensite. Carbon, in fact, is an impurity in maraging steels and is kept at the lowest possible concentration.

Grades of maraging steels have been developed that provide specific levels of yield strength ranging from 1030 to 3450 MPa (150 to 500 ksi).

The absence of carbon and the use of intermetallic precipitation to achieve hardening produce several unique characteristics of maraging steels that set them apart from conventional steels.

HEAT TREATING

Conventional heat treatments for the standard grades of maraging steel are given in Table 12.

Table 11. Typical mechanical properties of 4340 steel
Oil quenched from 845 °C (1550 °F) and tempered at various temperatures.

Tempering temperature °C	°F	Tensile strength MPa	ksi	Yield strength MPa	ksi	Elongation in 50 mm or 2 in., %	Reduction in area, %	Hardness HB	HRC	Izod impact energy J	ft·lb
205	400	1980	287	1860	270	11	39	520	53	20	15
315	600	1760	255	1620	235	12	44	490	49.5	14	10
425	800	1500	217	1365	198	14	48	440	46	16	12
540	1000	1240	180	1160	168	17	53	360	39	47	35
650	1200	1020	148	860	125	20	60	290	31	100	74
705	1300	860	125	740	108	23	63	250	24	102	75

Table 12. Heat treatments and typical mechanical properties of standard 18Ni maraging steels

Grade	Heat treatment (a)	Tensile strength MPa	ksi	Yield strength MPa	ksi	Elongation in 50 mm or 2 in., %	Reduction in area, %	Fracture toughness MPa$\sqrt{m}$	ksi$\sqrt{in.}$
18Ni(200)	A	1500	218	1400	203	10	60	155-200	140-220
18Ni(250)	A	1800	260	1700	247	8	55	120	110
18Ni(300)	A	2050	297	2000	290	7	40	80	73
18Ni(350)	B	2450	355	2400	348	6	25	35-50	32-45
18Ni(Cast)	C	1750	255	1650	240	8	35	105	95

(a) Treatment A: solution treat 1 h at 820 °C (1500 °F), then age 3 h at 480 °C (900 °F). Treatment B: solution treat 1 h at 820 °C (1500 °F), then age 12 h at 480 °C (900 °F). Treatment C: anneal 1 h at 1150 °C (2100 °F), age 1 h at 595 °C (1100 °F), solution treat 1 h at 820 °C (1500 °F) and age 3 h at 480 °C (900 °F).

Alloys with higher titanium contents are susceptible to formation of TiC films at austenite grain boundaries after holding at temperatures on the order of 900 to 1100 °C (1650 to 2000 °F). These films can severely embrittle the alloy when it is subsequently age hardened, leading to low-energy fractures along prior austenite grain boundaries. Prolonged annealing in this temperature range should be avoided for all compositions.

Solution Treatment. Maraging steels normally are solution annealed (austenitized) 1 h for each 25 mm (1 in.) of section size. Atmosphere control may be necessary to minimize surface damage. Ordinarily, dry hydrogen or dissociated ammonia atmospheres are used. The cooling rate after annealing is of little consequence because it has no effect on either microstructure or properties. It is essential, however, that the steel be cooled to room temperature before it is age hardened. If this is not done, the steel may contain untransformed (retained) austenite and may be much softer than expected.

Age hardening normally is done at 455 to 510 °C (850 to 950 °F) for 3 to 12 h. In typical treatments at 480 °C (900 °F), grades 18Ni(200), 18Ni(250) and 18Ni(300) are held 3 to 6 h, and grade 18Ni(350) is held 6 to 12 h. The 350 grade also is aged for 3 to 6 h at 495 to 510 °C (925 to 950 °F). For applications such as die casting tooling, aging at temperatures on the order of 530 °C (985 °F) is employed.

The standard age-hardening heat treatments listed in Table 12 produce 0.04% contraction in length in the 18Ni(200) grade, 0.06% contraction in the 18Ni(250) grade and 0.08% contraction in both the 18Ni(300) and 18Ni(350) grades. These very small dimensional changes during hardening allow many maraging steel components to be finish machined in the annealed condition. The finished parts then can be hardened without further machining. When greater dimensional accuracy is required, an allowance for contraction is readily made.

Case Hardening of Steel

Gas Carburizing

GAS CARBURIZING, in current commercial practice, uses carbon from hydrocarbon gases and easily vaporized hydrocarbon liquids to produce a hard surface layer on steel parts. Gas carburizing is often referred to as "case carburizing." The main function of gas carburizing is to provide an adequate supply of carbon for absorption and diffusion into the steel.

CARBON SOURCES

Gaseous Sources. Gases most commonly used are natural gas, "manufactured" gas and certain propanes (Table 1). Butane is used infrequently. Where the demand for carbon is low, as in carbon restoration, endothermic generator gas is a suitable source of carbon.

Natural gas and propane are the preferred sources when available in high-purity forms. The most desirable form of propane is derived from natural gas, rather than from petroleum. Propane obtained as a by-product of oil refining frequently contains excessive amounts of ethylene, propylene and other unsaturated hydrocarbons that break down rapidly to oily soot or coke.

For uniform carburizing, circulation of furnace gases is necessary. Because hydrocarbon gases provide large quantities of available carbon, relatively small flows of gas are required. The circulation resulting from gas flow alone is not always sufficient to produce uniform carburizing. It is usually essential that the furnace have high-volume forced circulation of the gases to all parts of the work load.

A common commercial practice is to use an endothermic gas or a purified exothermic gas as a carrier that is enriched with one of the hydrocarbon gases. The ratio of carrier gas to hydrocarbon gas varies widely in industry but is usually in the range from 8-to-1 to 30-to-1. The ratio used depends on the types of carrier and hydrocarbon gases, furnace size and condition, amount of circulation, and the work surface.

Although the use of carrier gases results in improved circulation because of the larger volume used, the additional forced circulation provided by a fan is usually needed to obtain maximum uniformity of carburizing and furnace temperature, particularly with batch furnaces, in which the load is densely packed.

Liquid Hydrocarbon Sources. Liquids also are used as sources of carburizing gas. These liquids are usually proprietary compounds and range in composition from pure hydrocarbons such as terpenes, dipentene or benzene to oxygenated hydrocarbons such as alcohols, glycols or ketones. When a liquid is used, it normally is fed in droplet form to a target plate in the furnace, where it volatilizes almost instantaneously. The vapors dissociate thermally to provide a carburizing atmosphere containing carbon monoxide, carbon dioxide, methane and water vapor.

CARRIER GASES

Most gas carburizing furnaces employ one of several carrier gases to dilute and react with the hydrocarbon gas used as the principal source of carbon. The three common carrier gases, classes 201, 202 and 302, are listed in Table 2. Class 302 gas is most commonly used today, because of its ease of use.

Class 202 gas offers a moderate range of control, some carbon availability for carburizing, and continuous operation with dew points of −40 °C (−40 °F) or lower.

Class 302 (endothermic) gas is generally the preferred carrier gas for use in gas carburizing furnaces and is the most widely used. It offers a broad range of carbon control, a moderate amount of carbon availability for carburizing and, when operated with dew points of −1 °C (+30 °F) and above, continuous operation without weekend shutdowns for burnout. Lower temperatures may be used, down to −7 °C (+20 °F), but periodic burnout may be necessary.

Nitrogen gas is available from liquefaction plants, as a by-product from oxygen plants, or from modified exothermic generators as a class 201 atmosphere. Nitrogen alone is neutral, but in commercial form it usually contains one or more minor impurities that may be oxidizing or decarburizing.

The inertness of pure nitrogen presents a problem of carbon control when this gas is used as a carrier for hydrocarbon gas. Because of the great capacity of hydrocarbons to supply carbon to steel, the ratio of nitrogen carrier gas to hydrocarbon gas must be large—from about 50-to-1 to 100-

Table 1. Principal gases used as sources of carbon in gas carburizing

Constituent	Composition, vol %					
	Natural gas			Coke oven gas	Commercial (normal) butane	Commercial propane
	Pittsburgh	Kansas City	Indianapolis			
CH_4 (methane)	83.4	84.1	87.2	32.1	...	...
C_2H_6 (ethane)	15.8	6.7	6.0	...	...	...
C_2H_4 (ethylene)	...	...	...	3.5	...	2.5
C_6H_6 (benzene)	...	...	...	0.5	...	...
C_4H_{10} (butane)	...	...	...	...	93.0	...
C_3H_8 (propane)	...	...	...	...	7.0	96.0
C_4H_{10} (isobutane)	...	...	...	...	...	1.5

Table 2. Principal carrier gases used in gas carburizing

Class	Method of preparation	Air-to-gas ratio(a)	N₂	Composition(a), vol % CO(b)	CO₂	H₂(b)	CH₄	Dew point °C	°F	Fuel gas required(c) m³	ft³	Nature of atmosphere
201	Prepared nitrogen base with lean mixture	9.0	97.1	1.7	...	1.2	...	−40	−40	3.8	135	Noncombustible; inert
202	Prepared nitrogen base with rich mixture	6.0	75.3	11.0	...	13.2	0.5	−40	−40	4.5	160	Combustible; toxic; medium reducing
302	Endothermic base (completely reacted and cooled to eliminate breakdown of CO into C + CO₂)	2.5	39.8	20.7	...	38.7	0.8	−4 to −21(d)	+25 to −5(d)	5.7(e)	200(e)	Combustible; toxic; very reducing

(a) Analyses based on 1055 kJ (1000 Btu) natural gas requiring 9.6 volumes of air for complete combustion. For high-H_2 artificial gas, multiply the quoted ratio of air to gas by 0.5; for medium-H_2, high-CO artificial gas, by 0.4; for propane, by 2.5; for butane, by 3.2. (b) If made with artificial gas, the CO will be slightly lower and H_2 somewhat higher. With propane and butane, the reverse will be true. (c) Cubic metres (cubic feet) of 1055 kJ (1000 Btu) natural gas required to make 28 m³ (1000 ft³) of atmosphere. For high-H_2 artificial gas, multiply value shown by 2.0; for propane, by 0.4; for butane, by 0.3. (d) Dew point is varied by changing the ratio of air to gas going to the generator. Most carburizing is done with dew point (at the generator) from −12 to −4 °C (+10 to +25 °F). (e) Plus 7 m³ (250 ft³) of fuel gas per 28 m³ (1000 ft³) of prepared atmosphere, for heating retort

to-1. Even more important, the hydrocarbon gas must be controlled within limits of about 0.1% to maintain carburizing potential within the close ranges of ±0.05% carbon.

EQUIPMENT

Gas carburizing furnaces vary widely as to physical construction, but they can be divided into two major categories: batch and continuous. The two types differ largely in their method of handling the work. In a batch-type furnace, the work is charged and discharged as a single unit or batch; in the continuous furnace, workpieces enter and leave the furnace as units in a continuing stream. The continuous furnace generally is favored for large-volume production of similar parts with total case-depth requirements of 0.4 to 3 mm (0.015 to 0.120 in.).

Pit-Type Batch Furnaces. Pit-type carburizing furnaces consist essentially of two parts: the furnace, which is placed in a pit and extends to floor level or slightly above, and a cover or lid, which extends upward from floor level. These furnaces should be as gastight as possible; a positive pressure should be maintained within the furnace. It must be equipped with a fan to circulate gases.

Horizontal batch furnaces are ideal for gas carburizing small parts requiring light case depths and direct quenching. These furnaces are well suited to handling batches of less than 900 kg (2000 lb) and, when equipped with an integral oil-quenching system, can be made part of a continuous production line. These furnaces may also be equipped to cool loads slowly from the carburizing temperature. They can economically handle parts in quantities that are too small to justify the use of continuous equipment.

Rotary-retort batch furnaces are designed to handle loads of 45 to 680 kg (100 to 1500 lb). They are best suited for carburizing relatively small parts that may be indiscriminately loaded.

Rotary-hearth furnaces also may be used as batch-type units for uniform and controlled carburizing of parts such as spiral bevel gears, rings, sleeves, races and disks, which are individually loaded and die quenched. However, use of these furnaces is recommended for low-volume production only.

Continuous Furnaces. Continuous carburizing furnaces are generally preferred for production loads exceeding 180 kg/h (400 lb/h) and requiring the same case depth, or for loads of sufficient size that require 24-h continuous operation with a minimum number of changes in required case depth. Continuous furnaces permit individual quenching of larger parts and batch quenching of smaller parts. Some types are equipped to provide cooling under cover of protective atmospheres.

Shaker-hearth furnaces use a reciprocating shaker motion to move the work along the hearth; this motion may be regulated to control the time cycle and case depth. Heating is efficient and confined mainly to the work load. Parts may be fed into the furnace by hand or by means of automatic metering and may be individually quenched. Use of this type of furnace is generally limited to lightweight parts that are to be carburized to case depths of 0.3 mm (0.010 in.) or less.

Rotary-retort continuous furnaces are used to carburize the same types of small parts that can be handled in a rotary-retort batch furnace. The advantage of the continuous rotary furnace is that it can be automatically loaded and unloaded, thus eliminating the need for removing and replacing the head.

Pusher-type continuous furnaces are widely used continuous carburizing units. Construction usually consists of a gastight welded shell with radiant tubes for heating. The work is pushed through on trays with or without fixtures and, after completion of the carburizing cycle, may be quenched or cooled slowly. Circulating fans are almost always used for more uniform temperature and carburization. Most pusher-type furnaces are built with purging vestibules at the charge and discharge ends to reduce contamination of the atmosphere by air. In many instances, washing and tempering equipment is incorporated to provide a fully automated heat treating line.

Continuous pusher-type furnaces are designed for high-volume production.

PREPARATION AND HANDLING OF PARTS

While parts are being transferred from the machine shop to the carburizing furnace, they should be handled carefully to avoid nicks and surface damage, which are costly to correct after hardening is completed.

Cleaning. All parts should be thoroughly cleaned before they are charged into the furnace. Trays should be degreased by washing, although some users burn off organic matter in a furnace.

Loading Methods. To obtain the maximum net load that can be carburized and quenched uniformly, the method of supporting the work in the furnace by trays, baskets, screens, spacers or other fixtures should be worked out carefully. Contact between the work and trays or baskets, contact between parts, and overly dense loads all result in uneven case depth and quenching, and should therefore be minimized or avoided. The size and shape of parts will determine the method of loading for proper gas circulation and, for parts that are quenched directly on fixtures, will determine uniformity of hardening in the quench.

PROCESS VARIABLES

Successful operation of the gas carburizing process depends on control of three principal variables: temperature, time and atmosphere composition.

Effect of Temperature. The maximum rate at which carbon can be added to steel is limited by the rate of diffusion of carbon in austenite. This diffusion rate increases greatly with temperature; the rate of carbon addition at 925 °C (1700 °F) is about 40% greater than at 870 °C (1600 °F).

The temperature most commonly used for carburizing is 925 °C (1700 °F). This temperature permits a reasonably rapid carburizing rate without excessive deterioration of furnace equipment, particularly of heat-resistant alloys. The carburizing temperature is sometimes raised to 955 or 980 °C (1750 or 1800 °F) for certain deep-case requirements. For shallow-case carburizing in which case depth must be kept within a specified narrow range, lower temperatures are frequently used, because case depth can be more accurately controlled with the slower rates of carburizing obtained with lower temperatures.

For consistent results in carburizing, the temperature must be uniform. Uniformity among various locations throughout the work load depends on furnace design, load density, recirculation and heating rate. With high density of load, batch furnaces should have effective recirculating fans.

For best control in batch furnaces, the thermocouple should be placed so that it reaches control temperature before any part of the charge. In continuous furnaces, the thermocouple should be as close as possible to the work without interfering with the flow of work through the furnace. For easy access in checking procedures, the thermocouple and protection tube usually are located through the sidewall of the continuous furnace. Because the first zone of a continuous furnace is a heating zone, the temperature-control thermocouple of this zone should be placed near the last part of the zone, to ensure against overheating the work. The thermocouple in the carburizing zone should be approximately in the middle of the zone.

Effect of Time. F. E. Harris developed a formula for the effect of time and temperature on case depth for normal carburizing (*Metal Progress*, Aug 1943), which can be shown in English units as:

$$\text{Case depth} = \frac{31.6\sqrt{t}}{10^{(6700/T)}}$$

where case depth is in inches; t is time at temperature, in hours; and T is the absolute temperature, in degrees Rankine (Fahrenheit + 460).

The equivalent metric formula, using SI units, is:

$$\text{Case depth} = 660 \cdot e^{-8287/T} \cdot \sqrt{t}$$

where case depth is in millimetres, t is time in hours, and T is temperature in Kelvins (Celsius + 273).

For a specific carburizing temperature, the relationship becomes simply:

$$\text{Case depth} = K\sqrt{t}$$

Case depth, mm = $0.635\sqrt{t}$ for 925 °C

(Case depth, in. = $0.025\sqrt{t}$ for 1700 °F)

Case depth, mm = $0.533\sqrt{t}$ for 900 °C

(Case depth, in. = $0.021\sqrt{t}$ for 1650 °F)

Case depth, mm = $0.457\sqrt{t}$ for 870 °C

(Case depth, in. = $0.018\sqrt{t}$ for 1600 °F)

Values of case depth calculated for times of 2 to 36 h at three common carburizing temperatures are given in Table 3.

When carburizing is purposely controlled to produce surface carbon concentrations somewhat less than saturated austenite, case depth will be slightly less than the equation shows.

In addition to the time at carburizing temperature, several hours may be required for bringing large workpieces to operating temperature. For work quenched directly from the carburizer, the cycle may be further lengthened to allow time for the work to cool from the carburizing temperature to a quenching temperature of about 845 °C (1550 °F). Although some diffusion of carbon from case to core occurs during this time, diffusion is slower than it would be at the carburizing temperature. This period may be used deliberately as a moderate diffusion period, to lower the carbon concentration at the surface by maintaining an atmosphere of low carbon potential in contact with the work during this time.

Harris also developed a method for calculating the carburizing time and diffusion time to produce a carburized case of predetermined depth and carbon concentration at the surface, which can be shown as:

$$\text{Carburizing time} = \text{total time}\left(\frac{C - C_i^2}{C_0 - C_i}\right)$$

and

Diffusion time = total − carburizing time

where total time, in hours, is calculated from the equation in Table 3; C is the final desired surface carbon concentration; C_0 is the surface carbon concentration at the end of the carburizing cycle; and C_i is the concentration of carbon at the core.

Gas-Carburizing Atmospheres. The furnace atmosphere used for gas carburizing consists of an endothermic carrier gas enriched with methane or propane. The main constituents of the atmosphere are CO, N_2, H_2, CO_2, H_2O and CH_4. Of these constituents, N_2 is inert, acting only as a diluent. The amounts of CO, CO_2, H_2O and H_2 present are very nearly the proportions expected at equilibrium from the reversible reaction:

$$CO + H_2O \rightleftarrows CO_2 + H_2 \qquad (\text{Eq 1})$$

given the particular ratios of C, O and H in the atmosphere. Methane is invariably present in amounts well in excess of the amount expected if all the gaseous constituents were in equilibrium.

Although the sequence of reactions involved in carburizing is not known in detail, it is known that carbon can be added or removed rapidly from steel by the over-all reversible reactions:

$$2\,CO \rightleftarrows C\,(\text{in Fe}) + CO_2 \qquad (\text{Eq 2})$$

and

$$CO + H_2 \rightleftarrows C\,(\text{in Fe}) + H_2O \qquad (\text{Eq 3})$$

Methane or propane enrichment of endothermic gas provides the primary source of carbon for carburizing by slow reactions such as:

$$CH_4 + CO_2 \rightarrow 2\,CO + 2\,H_2 \qquad (\text{Eq 4})$$

and

$$CH_4 + H_2O \rightarrow CO + 3\,H_2 \qquad (\text{Eq 5})$$

which reduce the concentrations of CO_2 and H_2O, respectively. These reactions regenerate CO and H_2, thereby directing the reactions of Eq 2 and 3 to the right. Reactions in Eq 4 and 5 do not approach equilibrium. With methane contents typical of carburizing atmospheres, thermodynamic calculations show that, if the reactions did approach equilibrium, atmosphere CO_2 contents and dew points would be much lower than the values customarily observed. The sum of reactions in Eq 2 and 4 and in Eq 3 and 5 is reduced to:

$$CH_4 \rightarrow C\,(\text{in Fe}) + 2\,H_2 \qquad (\text{Eq 6})$$

Thus, with constant CO_2 content and constant dew point, the net atmosphere composition change during carburizing is a reduction in methane content and an increase in hydrogen content. In most commercial operations, atmosphere flow rates are high enough, and the rate of methane decomposition is low enough, to prevent significant hydrogen buildup during a carburizing cycle. Carburization rates measured in CH_4-H_2 atmospheres have shown that, if the furnace-atmosphere methane content is high enough—above 1% at

Table 3. Values of case depth calculated by the Harris equation

| Time(t), h | Case depth(a), after carburizing at: | | | | | |
| | 870 °C (1600 °F) | | 900 °C (1650 °F) | | 925 °C (1700 °F) | |
	mm	in.	mm	in.	mm	in.
2	0.64	0.025	0.76	0.030	0.89	0.035
4	0.89	0.035	1.07	0.042	1.27	0.050
8	1.27	0.050	1.52	0.060	1.80	0.071
12	1.55	0.061	1.85	0.073	2.21	0.087
16	1.80	0.071	2.13	0.084	2.54	0.100
20	2.01	0.079	2.39	0.094	2.84	0.112
24	2.18	0.086	2.62	0.103	3.10	0.122
30	2.46	0.097	2.95	0.116	3.48	0.137
36	2.74	0.108	3.20	0.126	3.81	0.150

(a) Case depth, mm = $0.635\sqrt{t}$ (case depth, in. = $0.025\sqrt{t}$) for 925 °C (1700 °F); $0.533\sqrt{t}$ ($0.021\sqrt{t}$) for 900 °C (1650 °F); $0.457\sqrt{t}$ ($0.018\sqrt{t}$) for 870 °C (1600 °F). For normal carburizing (saturated austenite at the steel surface while at temperature).

925 °C (1700 °F), for example—some carburizing can occur by direct decomposition of methane on the steel surface, according to the reaction in Eq 6.

Carbon-potential control during carburizing is achieved by varying the flow rate of the hydrocarbon enrichment gas and maintaining a steady flow of endothermic carrier gas. As a basis for regulating the enrichment gas, the concentration of some constituent of the furnace atmosphere is monitored; for example, water vapor content by dew-point measurement, carbon dioxide content by infrared gas analysis or oxygen potential by zirconia oxygen sensor. The first two quantities provide measures of carbon potential according to the reactions of Eq 2 and 3. Oxygen potential is related to carbon potential by the reaction:

$$C\,(\text{in Fe}) + \tfrac{1}{2}\,O_2 \rightleftarrows CO \qquad (\text{Eq 7})$$

When the carbon dioxide content of the atmosphere remains relatively constant, both carbon dioxide and oxygen potential provide good measures of the carbon potential.

CARBON CONCENTRATION GRADIENTS AND SURFACE CARBON CONTENT

The carbon concentration gradient of carburized parts is influenced by carburizing temperature and time, type of cycle (various combinations of carburizing and diffusion times), carbon potential of the furnace atmosphere and the original composition of the steel.

In carburizing, carbon is transferred from the carburizing atmosphere surrounding the steel part to the part surface. The carbon then diffuses slowly into the bulk of the part, establishing a carbon concentration gradient below the surface. The driving force for the carburizing reaction is called the carbon potential. Within the steel part, the high carbon surface has a higher carbon potential then the low-carbon interior; thus, carbon tends to move from the surface toward the center. Similarly, the carburizing atmosphere has a higher carbon potential than does the surface of the steel. If during processing the atmosphere carbon potential should fall below the carbon potential at the steel surface, carbon will be removed from the steel.

Normal Carbon Gradients. A normal carbon gradient is one produced by maintaining saturated austenite at the surface of the steel during the entire cycle. Figure 1 illustrates the influence of carburizing temperature on the carbon gradient for normal carburizing of a 1020 steel in a batch furnace. Comparable data also are given for 8620 steel carburized for $7\frac{1}{2}$ h in an atmosphere containing 12% methane. Carburizing temperatures for both steels were 870, 900 and 925 °C (1600, 1650 and 1700 °F).

—————— Carbonitriding ——————

CARBONITRIDING is a modified form of gas carburizing, rather than a form of nitriding. The modification consists of introducing ammonia into the gas carburizing atmosphere to add nitrogen to the carburized case as it is being produced. Nascent nitrogen forms at the work surface by the dissociation of ammonia in the furnace atmosphere; the nitrogen diffuses into the steel simultaneously with carbon. Typically, carbonitriding is carried out at a lower temperature and

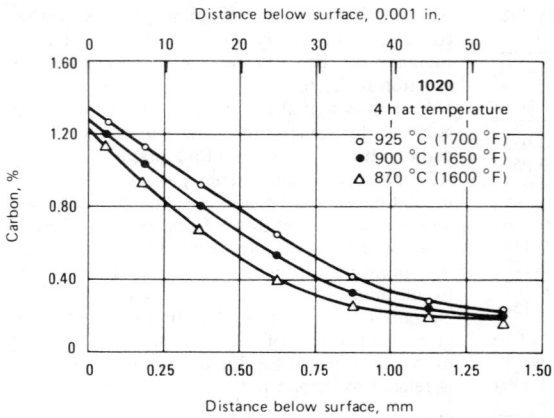
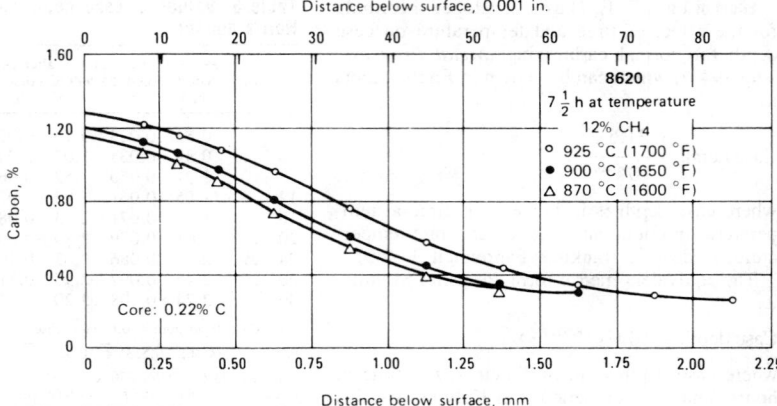

Carburized at three temperatures. The 1020 steel was carburized in a batch furnace; the 8620 steel, in a recirculating pit furnace.

Fig. 1. Carbon gradients for 1020 and 8620 steels

for a shorter time than gas carburizing, producing a shallower case than is usual in production carburizing.

In its effects on steel, carbonitriding is similar to liquid cyaniding. Because of problems in disposing of cyanide-bearing wastes, carbonitriding is often preferred over liquid cyaniding. In terms of case characteristics, carbonitriding differs from carburizing and nitriding in that (a) carburized cases normally do not contain nitrogen, and (b) nitrided cases contain nitrogen primarily, whereas carbonitrided cases contain both.

APPLICATIONS

Although carbonitriding is a modified carburizing process, its applications are more restricted than those of carburizing. As has been stated previously, carbonitriding is largely limited to case depths of about 0.75 mm (0.03 in.) or less, while no such limitation applies to carburizing. Two reasons for this are: (a) carbonitriding is generally done at temperatures of 870 °C (1600 °F) and below, whereas, because of the time factor involved, deeper cases are produced by processing at higher temperatures; and (b) the nitrogen addition is less readily controlled than is the carbon addition, a condition that can lead to an excess of nitrogen and, consequently, to high levels of retained austenite and case porosity when processing times are too long.

HARDNESS GRADIENTS

Hardness at various levels in the case depends on the microstructure. Hardness gradients associated with the microstructures of 1117 steel are presented in Fig. 2. When the carbonitriding atmosphere was relatively high in ammonia (11% NH₃), the nitrogen content of the case was high, and enough austenite was retained after quenching to lower the hardness to 510 HK (48 HRC), 500-g load, at a depth of 0.025 mm (0.001 in.) below the surface. The amount of retained austenite was decreased, and hardness consequently increased, either by lowering the ammonia flow rate from 0.57 to 0.14 m³/h (20 to 5 ft³/h), which reduced the ammonia content of the furnace atmosphere from 11 to 3%, or by introducing a 15-min diffusion period at the end of the carbonitriding operation. Either treatment increased the hardness to meet or exceed a specified minimum

value of 630 HK (55 HRC), 500-g load, at 0.025 mm below the surface.

HARDENABILITY OF CASE

One major advantage of carbonitriding is that the nitrogen absorbed during processing lowers the critical cooling rate of the steel. That is, the hardenability of the case is significantly greater when nitrogen is added by carbonitriding than when the same steel is carburized only, as is shown in Fig. 3.

Because of the hardenability effect of nitrogen, carbonitriding makes it possible to oil quench steels such as 1010, 1020 and 1113 to obtain martensitic case structures. Because of lower processing temperatures and/or the use of less severe quenches, carbonitriding may produce less part distortion and better control of dimensions than carburizing, and thus may eliminate the need for straightening or final grinding operations.

EFFECT OF TIME AND TEMPERATURE ON DEPTH OF CASE

Based on a survey of industrial practice, Fig. 4 shows case depths for different combinations of total furnace treating time and temperature.

Figure 5(a) shows the effects of total furnace time on case depth for 1020 steel. Specimens were heated to 705, 760, 815 and 870 °C (1300, 1400, 1500 and 1600 °F) for periods of 15, 30 and 45 min. Figure 5(b) indicates the total case depths that can be obtained on an 1112 steel held for 15 min at various temperatures between about 750 and 900 °C (1380 and 1650 °F). All data in Fig. 5 were obtained in a single plant.

FURNACES

Almost any furnace suitable for gas carburizing can be adapted to carbonitriding. If dense loads are to be processed, the furnace must be equipped

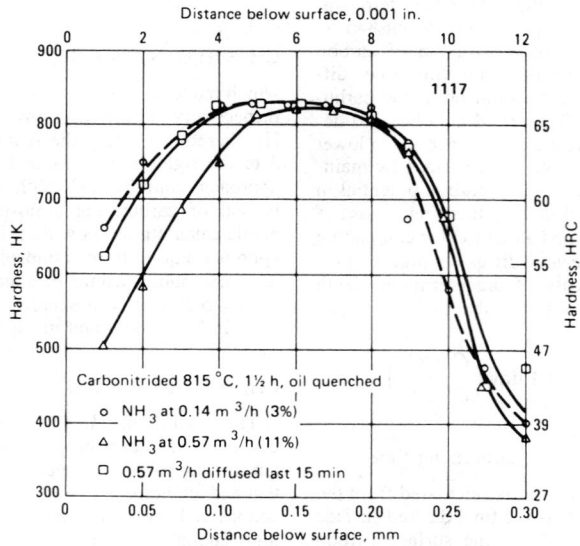

Required minimum hardness of 630 HK at 0.025 mm below surface was met by reducing the percentage and flow rate of ammonia or by adding a diffusion period after carbonitriding, as indicated. Atmosphere consisted of endothermic carrier gas (dew point, −1 °C) at 4.25 m³/h (150 ft³/h), natural gas at 0.17 m³/h (6 ft³/h), and ammonia in the amounts indicated.

Fig. 2. Hardness gradients in 1117 steel carbonitrided at 815 °C (1500 °F) for 1½ h and quenched in oil

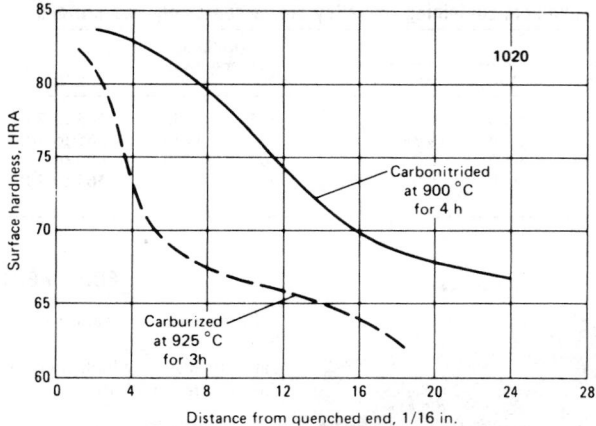

Hardness was measured along the surface of the as-quenched hardenability specimen. Ammonia and methane contents of the inlet carbonitriding atmosphere were 5%; remainder, carrier gas.

Fig. 3. End-quench hardenability curve for 1020 steel carbonitrided at 900 °C (1650 °F) compared with curve for the same steel carburized at 925 °C (1700 °F)

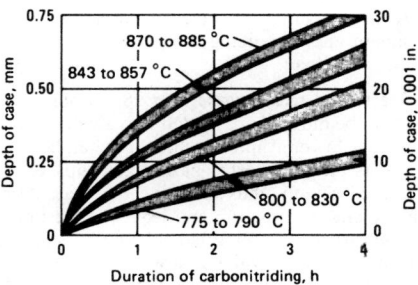

Fig. 4. Results of a survey of industrial practice regarding effects of time and temperature on depth of carbonitrided cases

with a fan to circulate the atmosphere. With shallow or openly spaced work loads, fan circulation of the atmosphere may not be required. For work that is to be clean and bright after quenching, the furnace must be equipped with protective-atmosphere vestibules to the quench area.

ATMOSPHERE CONSTITUENTS

The atmospheres used in carbonitriding generally comprise a mixture of carrier gas, enriching gas and ammonia. Basically, the atmospheres used in carbonitriding are produced by adding from about 2 to 12% ammonia to a standard gas carburizing atmosphere.

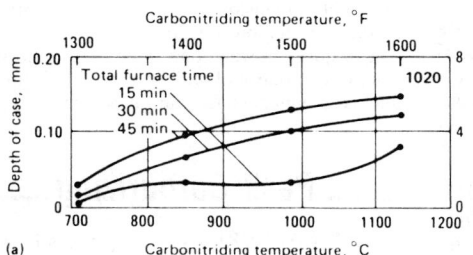

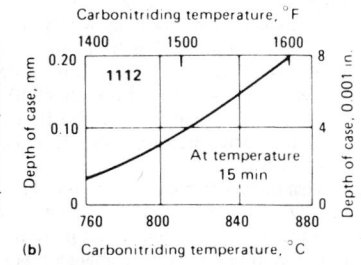

Both sets of data were obtained in the same plant. (a) 1020 steel at total furnace time. (b) 1112 steel is for 15 min at temperature.

Fig. 5. Effects of temperature and of duration of carbonitriding on depth of case

Gas Nitriding

GAS NITRIDING is a case hardening process whereby nitrogen is introduced into the surface of a solid ferrous alloy by holding the metal at a suitable temperature (below Ac_1, for ferritic steels) in contact with a nitrogenous gas, usually ammonia. Quenching is not required for the production of a hard case. The nitriding temperature for all steels is between 495 and 565 °C (925 and 1050 °F).

APPLICATION FACTORS

Principal reasons for nitriding are:

- To obtain high surface hardness
- To increase wear resistance and antigalling properties
- To improve fatigue life
- To improve corrosion resistance (except for stainless steels)
- To obtain a surface that is resistant to the softening effect of heat at temperatures up to the nitriding temperature.

Because of the absence of a quenching requirement, with attendant volume changes, and the comparatively low temperatures employed in this process, nitriding produces less distortion and deformation than either carburizing or conventional hardening. Some growth does occur as a result of nitriding, but volumetric changes are relatively small.

Nitridable Steels. Of the alloying elements commonly used in commercial steels, aluminum, chromium, vanadium, tungsten and molybdenum are beneficial in nitriding, because they form nitrides that are stable at nitriding temperatures. Molybdenum, in addition to its contribution as a nitride-former, also reduces the risk of embrittlement at nitriding temperatures. Other alloying elements, such as nickel, copper, silicon and manganese, have little, if any, effect on nitriding characteristics.

The following steels can be gas nitrided for specific applications:

- Aluminum-containing low-alloy steels (Table 4)
- Medium-carbon, chromium-containing low-alloy steels of the 4100, 4300, 5100, 6100, 8600, 8700, 9300 and 9800 series
- Hot work die steels containing 5% chromium, such as H11, H12 and H13
- Ferritic and martensitic stainless steels of the 400 series
- Austenitic stainless steels of the 300 series
- Precipitation-hardening stainless steels, such as 17-4 PH, 17-7 PH and A-286.

Aluminum-containing steels produce a nitrided case of very high hardness and excellent wear resistance. However, the nitrided case also has low ductility, and this limitation should be carefully considered in the selection of aluminum-containing steels.

SINGLE-STAGE AND DOUBLE-STAGE NITRIDING

Either a single- or a double-stage process may be employed when nitriding with anhydrous ammonia. In the single-stage process, a temperature in the range of about 495 to 525 °C (925 to 975 °F) is used, and the dissociation rate ranges from 15 to 30%. This process produces a brittle, nitrogen-rich layer, known as the "white nitride layer," at the surface of the nitrided case.

The double-stage process, known also as the Floe process (U.S. Patent 2 437 249), has the advantage of reducing the thickness of the white nitrided layer.

The first stage of the double-stage process is, except for time, a duplication of the single-stage process. The second stage may proceed at the nitriding temperature employed for the first stage, or the temperature may be increased to from 550 to 565 °C (1025 to 1050 °F); however, at either temperature, the rate of dissociation in the second stage is increased to from 65 to 85% (preferably, 80 to 85%). Generally, an external ammonia dissociator is necessary for obtaining the required higher second-stage dissociation.

The principal purpose of double-stage nitriding is to reduce the depth of the white layer produced on the surface of the case (see Fig. 6).

OPERATING PROCEDURES

Surface Preparation of Parts To Be Nitrided

After hardening and tempering, and before nitriding, parts should be thoroughly cleaned. Most parts can be successfully nitrided immediately after vapor degreasing. However, some machine finishing processes such as buffing, finish grinding, lapping and burnishing may produce surfaces that retard nitriding and result in uneven case depth and distortion. There are two methods by which

Table 4. Nominal compositions and preliminary heat treating cycles for aluminum-containing low-alloy steels commonly gas nitrided

SAE	Steel AMS	Nitralloy	C	Mn	Si	Composition, % Cr	Ni	Mo	Al	Se	Austenitizing temperature(a) °C	°F	Tempering temperature(a) °C	°F
...	...	G	0.35	0.55	0.30	1.2	...	0.20	1.0	...	955	1750	565 to 705	1050 to 1300
7140	6470	135M	0.42	0.55	0.30	1.6	...	0.38	1.0	...	955	1750	565 to 705	1050 to 1300
...	6475	N	0.24	0.55	0.30	1.15	3.5	0.25	1.0	...	900	1650	650 to 675	1200 to 1250
...	...	EZ	0.35	0.80	0.30	1.25	...	0.20	1.0	0.20	955	1750	565 to 705	1050 to 1300

(a) Sections up to 50 mm (2 in.) in diameter are quenched in oil; larger sections may be water quenched.

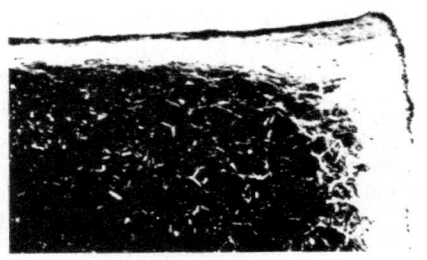 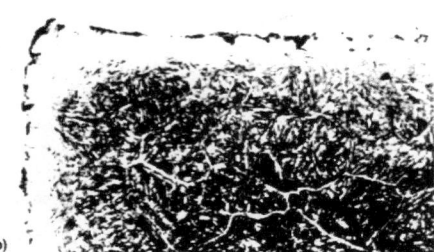

(a) 0.03-mm (0.0013-in.) white layer formed after single-stage nitriding at 525 °C (975 °F) for 60 h with 28% dissociation. (b) 0.02-mm (0.0008-in.) white layer formed after double-stage nitriding at 525 °C (975 °F) for 9¹⁄₂ h with 25 to 28% dissociation, then at 550 °C (1025 °F) for 50¹⁄₂ h and 80 to 84% dissociation. Buildup of white layer at corners during single-stage nitriding was 0.08 mm (0.0033 in.); during double-stage nitriding, 0.05 mm (0.0020 in.). Etched in 2% nital. Magnification, 150×

Fig. 6. Micrographs of white nitride layers developed on vacuum-melted AMS 6470 steel

the surfaces of parts finished by such methods may be successfully conditioned before nitriding.

One method consists of vapor degreasing parts and then abrasive cleaning them by aluminum oxide grit blasting immediately prior to nitriding. Any residual grit must be brushed off before parts are loaded into the furnace. Parts should be handled with clean gloves.

The second method is to apply a light phosphate coating.

Furnace Purging

After loading and sealing the furnace at the start of the nitriding cycle, it is necessary to purge the air from the retort before the furnace is heated to a temperature above 150 °C (300 °F). This prevents oxidation of parts and furnace components and, when ammonia is used as the purging atmosphere, *avoids production of a potentially explosive mixture*. Nitrogen is preferred in place of ammonia for purging, but the same precautions should be taken to avoid oxidation of parts.

Dissociation Rates. The nitriding process is based on the affinity of nascent nitrogen for iron and certain other metallic elements. Nascent nitrogen is produced by the dissociation of gaseous ammonia when it contacts hot steel parts. Although various rates of dissociation can be used successfully in nitriding, it is important that the nitriding cycle begin with a dissociation rate of about 15 to 30% and that this rate be maintained for 4 to 10 h, depending on the duration of the total cycle; temperature should be maintained at about 525 °C (975 °F). This initial cycle develops a shallow white layer from which diffusion of nitrogen into the main case structure proceeds.

Furnace Cooling

Most nitriding furnaces are equipped with a heat exchanger that will accelerate cooling of the furnace and work load at the conclusion of the nitriding cycle. When an external water-cooled heat exchanger is used, the furnace heating elements

are turned off when the nitriding cycle is completed, and the furnace temperature is allowed to drop approximately 55 °C (100 °F). At this point, the ammonia flow is approximately doubled, and the cooling water is turned on in the heat-exchanger.

Extreme care must be exercised to ensure a positive gas flow through the furnace as evidenced by the exit gas bubbles. When gas flow through the furnace has been stabilized, the flow may be reduced to the minimum required for positive pressure. After cooling to 150 °C (300 °F) or below, the furnace may be opened.

CONTROL OF CASE DEPTH

Case depth and case hardness, the two criteria most commonly referred to in the control of case properties, vary not only with the duration and other conditions of nitriding, but also with steel composition, prior structure and core hardness.

Aluminum-Containing Steels. Of the aluminum-containing nitriding steels, the most widely used is SAE 7140 (AMS 6470). Figure 7 indicates the hardness gradients and case depths obtained with this steel, as a function of cycle time and nitriding conditions. Results were obtained in single-stage nitriding for various lengths of time up to 800 h and at temperatures ranging from 510 to 540 °C (950 to 1000 °F); several different dissociation rates are represented.

Chromium-Containing Low-Alloy Steels. Of these steels, 4140 exhibits the best nitriding characteristics because of its higher chromium content and nickel-free composition. Although 4340 develops a heavier case than 8640 in the first 24 h of nitriding, this difference begins to decrease at the end of a 48-h cycle.

Chromium-containing tool steels, such as H11, H12, H13 and D2, provide high core strength with high case hardness, an excellent combination for applications involving severe impact or very high unit loading. Use of these steels is limited primarily by high cost and fabricating difficulties.

EQUIPMENT

Furnaces of several designs are in common use in gas nitriding installations. Most of these are batch furnaces, which incorporate certain essential features, including:

- A means of sealing the charge, to exclude air and other contaminants while containing the controlled atmosphere
- An inlet line for introducing atmosphere and an outlet line for exhausting used atmosphere
- A means of heating and appropriate temperature controls
- A means, such as a fan, for circulating atmosphere and equalizing temperature throughout the work load.

The vertical retort furnace (Fig. 8) is stationary; parts to be nitrided are loaded into a work basket, which is lowered into the heating chamber. The lid rests on an asbestos gasket and dips into an oil-filled trough, thus effecting the seal. Atmosphere enters at the top and leaves at the bottom of the furnace. Cooling is achieved by starting a fan and opening a valve in a water-jacketed cooling manifold. Furnaces of similar design, but without the water-jacketed manifold, are used when rapid cooling as a means of increasing furnace output is not required; the quality of nitriding achieved is equivalent to that of manifolded furnaces.

ION NITRIDING

Since the mid-1960's, nitriding equipment utilizing the glow-discharge phenomenon has been commercially available. Initially termed "glow-discharge" nitriding, the process is now generally known as ion nitriding. The term "plasma nitriding" is gaining acceptance.

Ion nitriding is an extension of conventional nitriding processes using plasma-discharge physics. In vacuum, high-voltage electrical energy is used to form a plasma, through which nitrogen ions are accelerated to impinge on the workpiece. This ion bombardment heats the workpiece, cleans the surface and provides active nitrogen.

Metallurgically versatile, the process provides excellent dimensional control and retention of surface finish. Ion nitriding can be conducted at temperatures lower than those conventionally employed. Control of white-layer composition and thickness enhances fatigue properties.

Pack Carburizing

PACK CARBURIZING is a process in which carbon monoxide derived from a solid compound decomposes at the metal surface into nascent carbon and carbon dioxide. The nascent carbon is

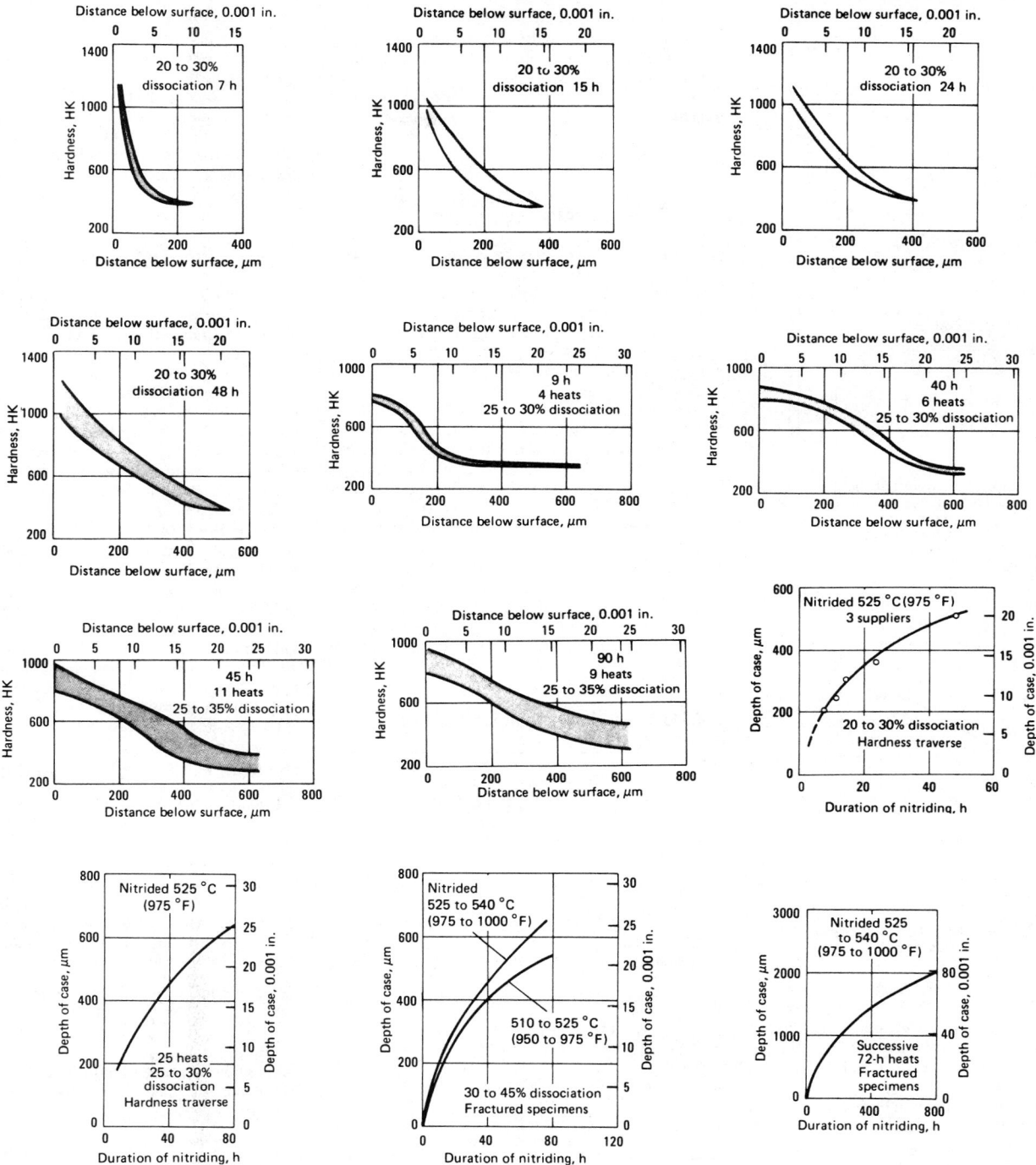

Fig. 7. Hardness gradients and case-depth relations for single-stage nitrided aluminum-containing SAE 7140 steel

absorbed into the metal, and the carbon dioxide immediately reacts with carbonaceous material present in the solid carburizing compound to produce fresh carbon monoxide. The formation of carbon monoxide is enhanced by energizers or catalysts, such as barium carbonate ($BaCO_3$), calcium carbonate ($CaCO_3$) and sodium carbonate (Na_2CO_3), that are present in the carburizing compound. These energizers facilitate the reduction of carbon dioxide with carbon to form carbon monoxide. Thus, in a closed system, the amount of energizer does not change. Carburiz-

ing continues as long as enough carbon is present to react with the excess carbon dioxide.

CARBURIZING COMPOUNDS

The common commercial carburizing compounds are reusable and contain 10 to 20% alkali or alkaline earth metal carbonates bound to hardwood charcoal or to coke by oil, tar or molasses. Barium carbonate is the principal energizer, usually comprising about 50 to 70% of the total carbonate content. The remainder of the energizer

usually is made up of calcium carbonate, although sodium carbonate also may be used.

PROCESS CONTROL

In pack carburizing, as in other carburization processes, the carbon-concentration gradient obtained is a function of carbon potential, carburizing temperature and time.

Temperature. Pack carburizing normally is performed at temperatures from 815 to 955 °C (1500 to 1750 °F). In recent years, the upper limits have

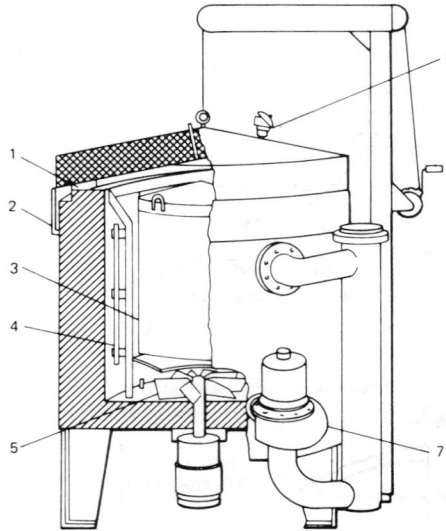

1, gasket; 2, oil seal; 3, work basket; 4, heating elements; 5, circulating fan; 6, thermocouple; and 7, cooling assembly. At end of cycle, a valve is opened and a fan (not shown) incorporated in the external cooler circulates atmosphere through the water-jacketed cooling manifold.

Fig. 8. Vertical retort nitriding furnace

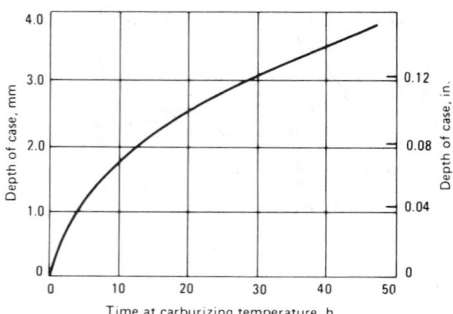

Fig. 9. Effect of time on case depth at 925 °C (1700 °F)

been steadily raised, and carburizing temperatures as high as 1095 °C (2000 °F) have been used. Steelmaking processes have improved to the extent that fine grain size is maintained at temperatures approaching or exceeding 1040 °C (1900 °F).

Time. The rate of change in case depth at a particular carburizing temperature is proportional to the square root of time. The rate of carburization is thus highest at the beginning of the cycle and gradually diminishes as the cycle is extended (see Fig. 9).

Liquid Carburizing and Cyaniding

LIQUID CARBURIZING is a process used for case hardening of steel or iron parts. The parts are held at a temperature above Ac_1 in a molten salt that will introduce carbon and nitrogen, or carbon alone, into the metal. Diffusion of the carbon from the surface toward the interior produces a case that can be hardened, usually by fast quenching from the bath. Carbon diffuses from

the bath into the metal and produces a case comparable with one resulting from gas carburizing in an atmosphere containing some ammonia. However, because liquid carburizing involves faster heat-up (due to the superior heat-transfer characteristics of salt bath solutions), cycle times for liquid carburizing are shorter than those for gas carburizing.

CYANIDE-CONTAINING LIQUID CARBURIZING BATHS

"Light case" and "deep case" are arbitrary terms that have been associated with liquid carburizing in baths containing cyanide. There is necessarily some overlapping of bath compositions for the two types of case. In general, the two types are distinguished more by operating temperature than by bath composition. Hence, the terms "low temperature" and "high temperature" are preferred.

Low-temperature cyanide-type baths (light-case baths) are those usually operated in the temperature range from 845 to 900 °C (1550 to 1650 °F), although for certain specific effects this range is sometimes extended to 790 to 925 °C (1450 to 1700 °F). Low-temperature baths are best suited to formation of shallower cases. Low-temperature baths are generally of the accelerated cyanogen type containing various combinations and amounts of the constituents listed in Table 5 and differ from cyaniding baths in that the case produced by a low-temperature bath consists predominantly of carbon.

Table 5. Operating compositions of liquid carburizing baths

Constituent	Composition of bath, %	
	Light case, low temperature(a), 845-900 °C (1550-1650 °F)	Deep case, high temperature(b), 900-955 °C (1650-1750 °F)
Sodium cyanide	10-23	6-16
Barium chloride	...	30-55(c)
Salts of other alkaline earth metals(d) ...	0-10	0-10
Potassium chloride ..	0-25	0-20
Sodium chloride	20-40	0-20
Sodium carbonate ...	30 max	30 max
Accelerators other than those involving compounds of alkaline earth metals(e)	0-5	0-2
Sodium cyanate	1.0 max	0.5 max

(a) Density of molten salt, 1760 kg/m³ at 900 °C (110 lb/ft³ at 1650 °F). (b) Density of molten salt, 2000 kg/m³ at 925 °C (125 lb/ft³ at 1700 °F). (c) Proprietary barium chloride–free deep-case baths are available. (d) Calcium and strontium chlorides have been employed. Calcium chloride is more effective, but its hygroscopic nature has limited its use. (e) Among these accelerators are manganese dioxide, boron oxide, sodium fluoride and sodium pyrophosphate.

High-temperature cyanide-type baths (deep-case baths) are usually operated in the temperature range from 900 to 955 °C (1650 to 1750 °F).

High-temperature baths are used for producing cases 0.5 to 3.0 mm (0.020 to 0.120 in.) deep. In some instances, even deeper cases are produced (up to about 6 mm, or 0.250 in.), but the most important use of these baths is for the rapid development of cases 1 to 2 mm (0.040 to 0.080 in.) deep. These baths consist of cyanide and a major proportion of barium chloride (Table 5).

CYANIDING (LIQUID CARBONITRIDING)

Cyaniding, or salt bath carbonitriding, is a heat treating process that produces a file-hard, wear-resistant surface on ferrous parts. When steel is heated above Ac_1 in a suitable bath containing alkali cyanides and cyanates, the surface of the steel absorbs both carbon and nitrogen from the molten bath. When quenched in mineral oil, paraffin-base oil, water or brine, the steel develops a hard surface layer, or case, that contains less carbon and more nitrogen than the case developed in activated liquid carburizing baths.

Bath Composition. A sodium cyanide mixture such as grade 30 in Table 6, containing 30% NaCN, 40% Na₂CO₃ and 30% NaCl, is generally used for cyaniding on a production basis. This mixture is preferable to any of the other compositions given in Table 6. The inert salts sodium chloride and sodium carbonate are added to cyanide to provide fluidity and to control the melting points of all mixtures. The 30% NaCN mixture, as well as those containing 45, 75 and 97% NaCN, may be added to the operating bath to maintain a desired cyanide concentration for a specific application.

NONCYANIDE LIQUID CARBURIZING

Liquid carburizing can be accomplished in a bath containing a special grade of carbon instead of cyanide as the source of carbon. In this bath, carbon particles are dispersed in the molten salt by mechanical agitation, which is achieved by means of one or more simple propeller stirrers that occupy a small fraction of the bath. The basic composition and major operational features of this bath are as follows:

Composition .....Alkali carbonates/ chlorides; proprietary carbon cover
Decomposition ....None
Contaminants .....Metallics (Fe), SiO₂/Al₂O₃
Maintenance .....Cover replenishment
Control ....Physical: C1012 test-piece hardness. Chemical: H₂O insolubles (% C)
Correction ....Add carbon; adjust agitation; sludge.

Table 6. Compositions and properties of sodium cyanide mixtures

Constituent or property	Grade			
	96-98(a)	75(b)	45(b)	30(b)
Composition, %:				
Sodium cyanide	97	75	45.3	30.0
Sodium carbonate	2.3	3.5	37.0	40.0
Sodium chloride	Trace	21.5	17.7	30.0
Melting point, °C (°F)	560 (1040)	590 (1095)	570 (1060)	625 (1155)
Specific gravity:				
At 25 °C (75 °F)	1.50	1.60	1.80	2.09
At 860 °C (1580 °F)	1.10	1.25	1.40	1.54

(a) Appearance: white crystalline solid. This grade also contains 0.5% sodium cyanate (NaNCO) and 0.2% sodium hydroxide (NaOH); sodium sulfide (Na₂S) content, nil. (b) Appearance: white granular mixture.

The chemical reaction involved is not fully understood, but is thought to involve adsorption of carbon monoxide on carbon particles. The carbon monoxide is generated by reaction between the carbon and carbonates, which are major ingredients of the molten salt. The adsorbed carbon monoxide is presumed to react with steel surfaces much as in gas or pack carburizing.

Operating temperatures for this type of bath are generally higher than those for cyanide-type baths. A range of about 900 to 955 °C (1650 to 1750 °F) is most commonly employed. Temperatures below about 870 °C (1600 °F) are not recommended, and may even lead to decarburization of the steel. The case depths and carbon gradients produced are in the same range as for high-temperature cyanide-type baths (see Fig. 10), but there is no nitrogen in the case.

Temperatures above 955 °C (1750 °F) produce more rapid carbon penetration and do not adversely affect noncyanide baths, because no cyanide is present to break down and cause carbon scum or frothing. Equipment deterioration is the chief factor that limits operating temperature.

CARBON GRADIENTS

Figure 10 shows carbon gradients produced by liquid carburizing 1020 steel bars at 845, 870 and 955 °C (1550, 1600 and 1750 °F) for various lengths of time at carburizing temperature. Carbon-gradient data for two wrought alloy steels (3312 and 8620) and one cast alloy steel (4615 mod) are also shown. After carburizing, the 8620 steel parts were austenitized at 840 °C (1540 °F) and quenched in oil at 55 °C (130 °F). The 4615 mod steel parts were austenitized at 790 °C (1450 °F), quenched in salt at 190 °C (375 °F) for 3 min, and cooled in air.

HARDNESS GRADIENTS

The indentation hardness data presented in Fig. 11 for five different steels indicate the effects of normal variations in practice on the hardness gradient. The shaded bands represent the scatter in results obtained from multiple tests of each steel. Although similar surface hardnesses are obtained with all five steels, depth of hardness varies with the alloy content of the steel. A comparison among the hardnesses of these five steels at a depth of 1 mm (0.040 in.) illustrates this variation.

Although a minimum case hardness of 60 HRC cannot be maintained to a depth of 1 mm (0.040 in.) with 1020 (0.30 to 0.60% Mn) steel, it can sometimes be achieved with 1113 (0.70 to 1.00% Mn) steel and can almost always be achieved with 1117 (1.00 to 1.30% Mn), 4615 and 8620 steels.

FURNACES AND EQUIPMENT

Liquid carburizing is carried out in a salt bath furnace that may be heated either externally or internally. In an externally heated furnace, heat is introduced into an annular space between the salt pot and the surrounding insulation, which usually is made of firebrick. In an internally heated furnace, heat is introduced directly into the salt. Both internally and externally heated furnaces generally have insulated lids that slide to open the bath and allow workpieces and fixtures to be positioned, usually with an overhead crane or with similar mechanized lifting equipment. For more detailed information, see the article in this section entitled "Salt Bath Equipment."

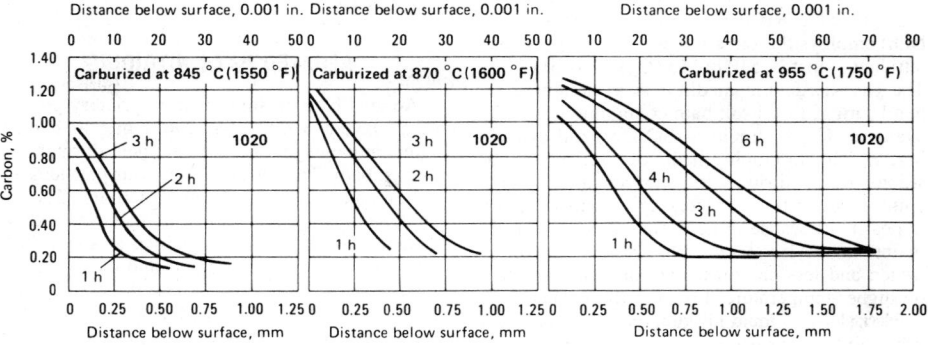

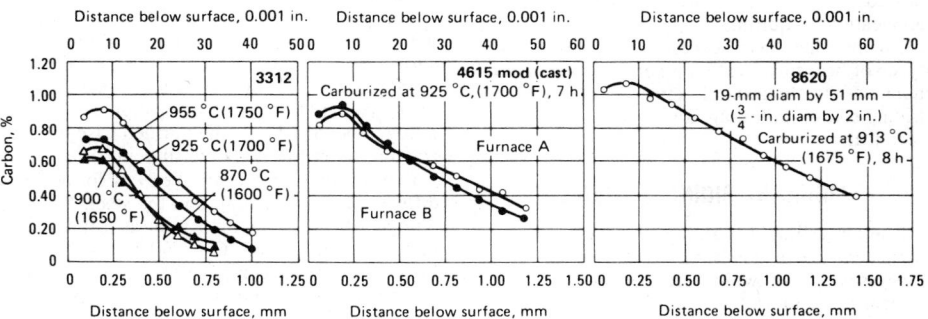

Carbon gradients produced by liquid carburizing carbon and alloy steels in low-temperature and high-temperature baths. The 1020 carbon steel bars were carburized at 845, 870 and 955 °C (1550, 1600 and 1750 °F) for the periods shown. The data on 3312 alloy steel show the effect of four different carburizing temperatures on carbon gradient (time constant at 2 h). The data on modified 4615 steel castings indicate the slight differences in gradients obtained in two furnaces employing the same carburizing conditions (7 h at 925 °C, or 1700 °F). These data and the data on 8620 steel parts show a decrease in carbon content near the surface caused by diffusion of carbon during reheating to austenitizing temperature.

Fig. 10. Carbon gradients produced by liquid carburizing of carbon and alloy steels

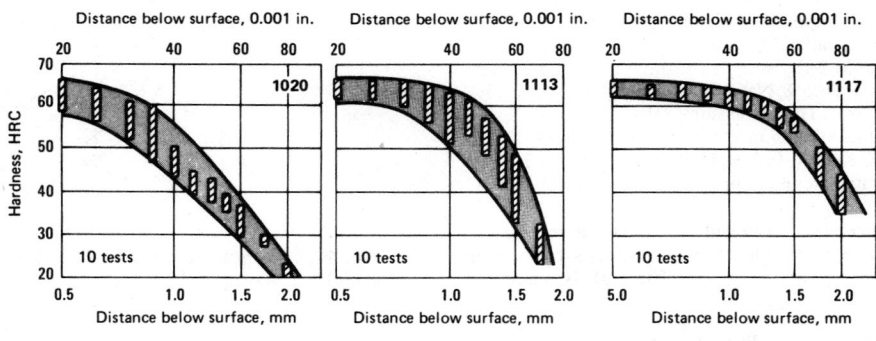

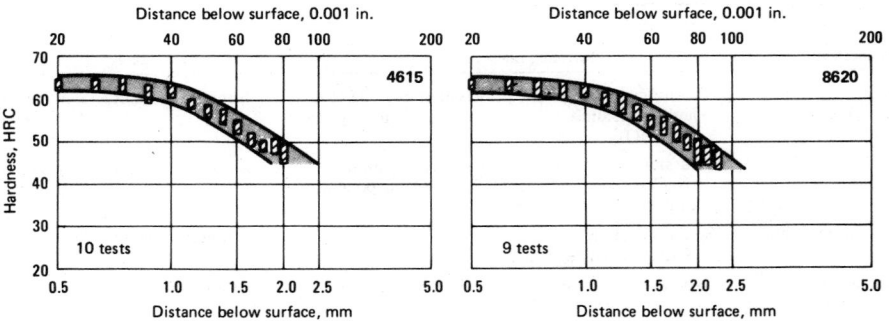

Fig. 11. Case-hardness gradients, showing scatter resulting from normal variations

Liquid Nitriding

LIQUID NITRIDING (nitriding in a molten salt bath) employs the same temperature range as gas nitriding—that is, 510 to 565 °C (950 to 1050 °F). The case-hardening medium is a molten nitrogen-bearing, fused-salt bath containing primarily cyanates. Unlike liquid carburizing and cyaniding, which can employ baths of similar compositions, liquid nitriding is a subcritical case-hardening process; thus, processing of finished parts is possible because dimensional stability can be maintained. Also, liquid nitriding adds more nitrogen and less carbon to ferrous materials than do higher-temperature diffusion treatments.

Principal Uses. Liquid nitriding processes are used primarily to improve wear resistance of surfaces and to increase the endurance limit in fatigue. For many steels, resistance to corrosion is improved. These processes are not suitable for many applications requiring deep cases and hardened cores, but they have successfully replaced other types of heat treatment on a performance or economic basis. In general, the uses of liquid nitriding and gas nitriding are similar, and at times identical.

LIQUID NITRIDING SYSTEMS

The term "liquid nitriding" has become a generic term for a number of different fused salt processes, all of which are performed at subcritical temperature. Operating at these temperatures, the treatments are essentially chemical diffusion operations, and they influence metallurgical structures primarily through absorption and reaction of nitrogen rather than through the minor amount of carbon that is assimilated. Although the different processes are represented by a number of commercial trade names, the basic subclassifications of liquid nitriding are those presented in Table 7.

One type of commercial bath for liquid nitriding is composed of a mixture of sodium and potassium salts. The sodium salts, which comprise 60 to 70% (by weight) of the total mixture, consist of 96.5% NaCN, 2.5% Na_2CO_3 and 0.5% NaCNO. The potassium salts, 30 to 40% (by weight) of the mixture, consist of 96% KCN, 0.6% K_2CO_3, 0.75% KCNO and 0.5% KCl. The operating temperature of this salt bath is 565 °C (1050 °F).

Recent developments eliminate or reduce cyanide content. One mixture consists of cyanate ions and carbonate ions. The bath is stabilized by lithium. Sulfur additions keep minimal levels of cyanide (formed *in situ*) within the range from 0.3 to 0.5%.

AERATED BATH NITRIDING

Aerated bath nitriding is a proprietary process (U.S. Patent 3 022 204; 1962) in which measured amounts of air are pumped through the molten bath. The introduction of air provides agitation and stimulates chemical activity. The cyanide content of this bath, calculated as sodium cyanide, is preferably maintained at about 50 to 60% of the total bath content, and the cyanate is maintained at 32 to 38%. The potassium content of the fused bath, calculated as elemental potassium, is between 10 and 30%, preferably about 18%. The potassium may be present as the cyanate or the cyanide, or both. The remainder of the bath is sodium carbonate.

This process produces a nitrogen-diffused case 0.3 mm (0.012 in.) deep on plain carbon or low-alloy steels in a $1\frac{1}{2}$-h cycle. The surface layer (0.005 to 0.01 mm or 0.0002 to 0.0005 in. deep) of the case is composed of epsilon Fe_3N and a nitrogen-bearing Fe_3C; the nitrided case does not contain the brittle Fe_2N constituent.

Beneath the outer layer, Fe_3N is formed in a diffusion zone extending into the steel. Depth of nitrogen diffusion in 1015 steel as a function of nitriding time at 565 °C (1050 °F) is shown in Fig. 12. The outer compound layer provides wear resistance, while the transition zone improves fatigue strength.

Aerated Noncyanide Nitriding. Environmental concerns have led to the development of cyanide-free processes for liquid nitriding. In one proprietary process, the base salt is supplied as a cyanide-free mixture of potassium cyanate and a combination of sodium carbonate and potassium carbonate or sodium chloride and potassium chloride. Cyanide is produced as a by-product.

In another process, sulfur reacts with the measured air to produce thiosulfate and sulfite, which reduce the cyanide formed to a controlled level with no adverse metallurgical effects. Lithium carbonate retards the decline of cyanate ions to produce a more stable bath. The controlled sulfur addition allows the user to adjust the proportion of porous to compact layers in the compound zone. The nitrogen diffusion zone is typically 0.5 mm (0.020 in.), depending on alloy and treatment time.

CASE DEPTH AND CASE HARDNESS

Data indicating depth of case obtained in liquid nitriding various steels in a conventional bath at 525 °C (975 °F) for up to 70 h are shown in Fig. 13. The steels include three chromium-containing low-alloy steels (4140, 4340 and 6150), two aluminum-containing nitriding steels (SAE 7140 and AMS 6475), and four tool steels (H11, H12, M50 and D2). All were nitrided in a salt bath with an effective cyanide content of 30 to 35% and a cyanate content of 15 to 20%. Case depths were measured visually on metallographically prepared samples that were etched in 3% nital. Before being nitrided, samples were tempered to the core hardnesses indicated.

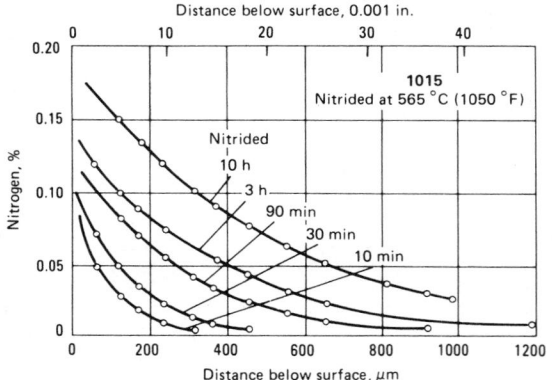

Fig. 12. Nitrogen gradients in 1015 steel as a function of time of nitriding at 565 °C (1050 °F), using the aerated bath process

Table 7. Liquid nitriding processes

Process identification	Operating range composition	Chemical nature	Suggested post treatment	Operating temperature °C	Operating temperature °F	U.S. patent number
Aerated cyanide-cyanate	Sodium cyanide (NaCN), potassium cyanide (KCN) and potassium cyanate (KCNO), sodium cyanate (NaCNO)	Strongly reducing	Water or oil quench; nitrogen cool	570	1060	3 208 885
Casing salt	Potassium cyanide (KCN) or sodium cyanide (NaCN), sodium cyanate (NaCNO) or potassium cyanate (KCNO) or mixtures	Strongly reducing	Water or oil quench	510 to 650	950 to 1200	. . .
Pressure nitriding	Sodium cyanide (NaCN), Sodium cyanate (NaCNO)	Strongly reducing	Air cool	525 to 565	975 to 1050	. . .
Regenerated cyanate-carbonate	Type A—Potassium cyanate (KCNO), potassium carbonate (K_2CO_3)	Mildly oxidizing	Water, oil, or salt quench	580	1075	4 019 928
	Type B—Potassium cyanate (KCNO), potassium carbonate (K_2CO_3), 1-10 ppm, sulfur (S)	Mildly oxidizing	Water or oil quench, or salt quench	540 to 575	1000 to 1070	4 006 043

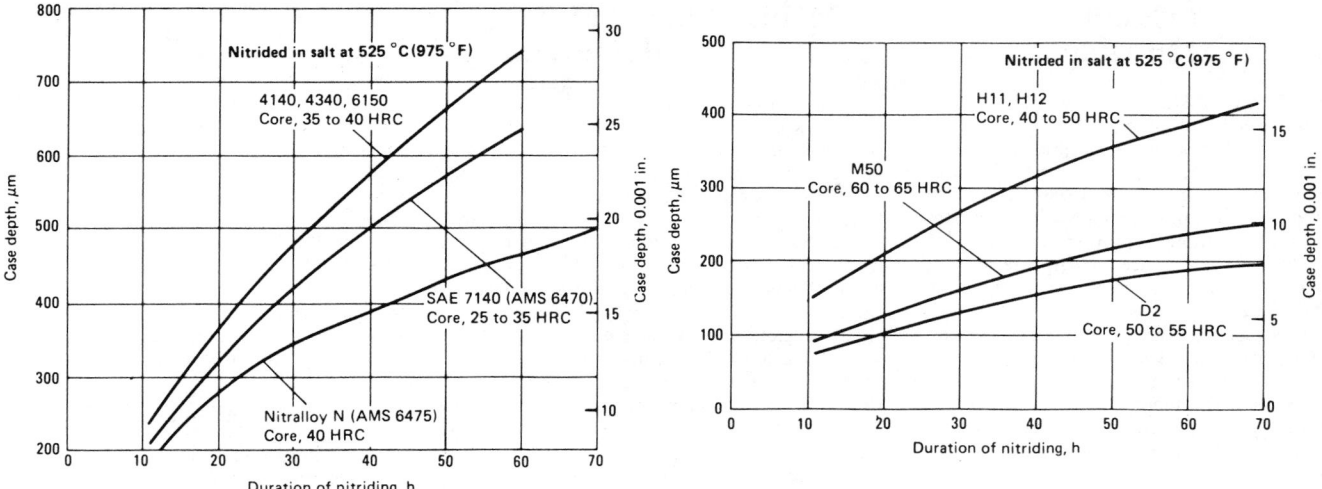

Nitrided in a conventional salt bath at 525 °C (975 °F) for up to 70 h.

Fig. 13. Depth of case for several chromium-containing low-alloy steels, aluminum-containing steels and tool steels after liquid nitriding

"BLUEING"/CORROSION RESISTANCE

For many years, old types of cyanide-base nitriding baths have employed subsequent methods of blueing nitrided steel, usually nitrate-base salt baths used at 425 to 455 °C (800 to 850 °F). Proprietary mixes used alkalis and/or strong oxidizing agents added to the nitrate salt. The new nitriding processes permit quenching directly into oxidizing salt baths. Polishing the treated steel, followed by further blueing, results in an appealing lustre with remarkable improvements in corrosion resistance.

EQUIPMENT

Salt bath furnaces used for nitriding may be heated by gas, oil or electricity, and are essentially similar in design to salt bath furnaces used for other processes. Although batch installations are most common, semicontinuous and continuous operations are feasible. Generally, the same furnace equipment can be used for other heat treating applications by merely changing the salt. (Further details on specific types of furnaces may be found in the article on salt bath equipment, in this volume.)

Gaseous Ferritic Nitrocarburizing

FERRITIC NITROCARBURIZING processes are thermochemical treatments that involve diffusional addition of both nitrogen and carbon to the surface of ferrous materials at temperatures completely within the ferrite-phase field. Cycle times are usually less than 3 h; these processes are termed "short-cycle" nitriding. The primary objective of such treatments is usually to improve antiscuffing characteristics of ferrous engineering components by providing the surface with a compound layer—really, a surface zone—exhibiting good wear/friction-resistant properties. In addition, fatigue characteristics can be considerably improved, particularly when nitrogen is retained in solid solution in the "diffusion zone" beneath the compound layer. This retention normally is achieved by quenching in oil or water from the treatment temperature. Some distinction should be drawn between two different processes with similar names—ferritic nitrocarburizing and carbonitriding. Carbonitriding is performed with the steel in an austenitic phase, above 760 °C (1400 °F). Ferritic nitrocarburizing is performed in the ferritic range, below 675 °C (1250 °F).

GASEOUS NITROCARBURIZING

Gaseous nitrocarburizing commonly employs sealed-quench batch furnaces of the same design used for carburizing and carbonitriding. Furnace operating temperatures are low enough to maintain steels in the ferritic condition. The atmosphere employed consists of ammonia diluted with a carrier gas. In one process, the atmosphere is formed from equal amounts of ammonia and endothermic gas, American Gas Association (AGA) type 302. In another process, a typical atmosphere consists of 35% ammonia and 65% refined exothermic gas (AGA type 201, nominally 97% nitrogen), which may be enriched with a hydrocarbon gas. High-purity nitrogen is used as a diluent in a variant of this process. Gaseous nitrocarburizing is performed near 570 °C (1060 °F), a temperature just below the austenite range for the Fe-N system. Treatment times generally range from 1 to 5 h.

Testing Results. The compound layer formed on AISI 1015 steel by gaseous nitrocarburizing in an ammonia/endothermic gas mixture is shown in Fig. 14. The layer is somewhat more dense than that formed by salt bath nitrocarburizing. X-ray diffraction analysis has confirmed the predominance of epsilon carbonitride phase in the compound layer.

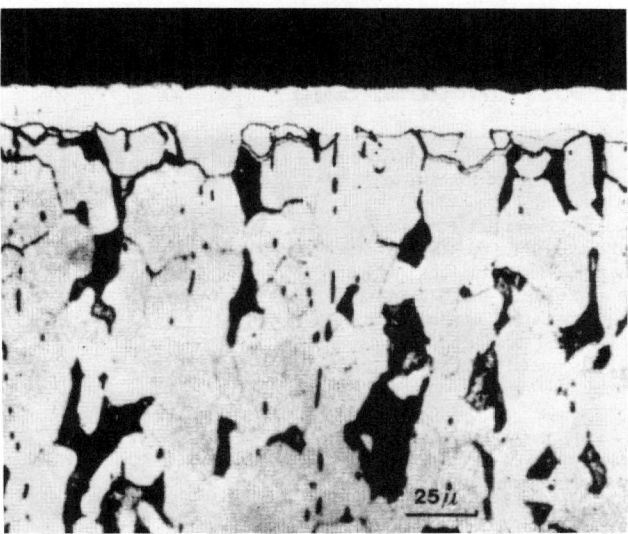

Fig. 14. AISI 1015 steel after 3 h of gaseous nitrocarburizing in an ammonia/endothermic gas mixture at 570 °C (1060 °F) followed by oil quenching

Vacuum Carburizing

VACUUM CARBURIZING is a high-temperature gas carburizing process that is carried out at pressures below atmospheric pressure (below 100 kPa, or 760 torr). Vacuum carburizing temperatures typically range from 980 to 1050 °C (1800 to 1925 °F); in some cases, however, the range is extended from about 900 to 1095 °C (1650 to 2000 °F). The furnace atmosphere usually consists solely of an enriching gas such as natural gas, pure methane or propane. Nitrogen is sometimes used as a carrier gas. During the carburizing portion of the cycle, furnace pressures are maintained in the range of about 7 to 55 kPa (50 to 400 torr). The atmosphere may flow through the furnace constantly while a constant pressure is maintained; alternatively, the furnace may be repeatedly backfilled and evacuated.

Vacuum carburizing proceeds by dissociation of hydrocarbon gas at the steel surfaces and by direct absorption of carbon; hydrogen gas is liberated. The reaction with methane is:

$$CH_4 + Fe \leftrightarrows Fe(C) + 2H_2$$

The carburizing step is carried out at 7 to 55 kPa (50 to 400 torr), but most commonly at 300 torr. The carburizing gas is used to maintain the pressure. The diffusion step is carried out at normal vacuum pressures of 100 μm Hg or less.

PROCESS CYCLES

Process cycles usually start with loading of workpieces that are relatively free of surface contaminants onto fixtures. The loads are then transferred to the cold furnace. The furnace is evacuated, usually with a mechanical pump, and the parts are heated to carburizing temperature. The combination of vacuum and heat removes most surface contaminants, leaving clean surfaces that can readily absorb carbon. Natural gas is admitted to raise the furnace pressure to about 15 to 55 kPa (100 to 400 torr). After the natural gas atmosphere is circulated for a predetermined and closely controlled length of time, the furnace is evacuated for the diffusion cycle, which is carried out under soft vacuum conditions.

After the carburizing gas has been evacuated, the diffusion cycle is timed to produce the desired surface carbon content and case depth. In certain instances, the carburizing and diffusion steps are alternated one or more times to obtain the desired carbon gradient.

Following diffusion, the carburized parts normally are treated by backfilling the furnace with nitrogen and cooling the work with fans to about 425 °C (800 °F) to refine the grains. The furnace then is immediately evacuated, and the work is reheated to about 845 °C (1550 °F) for austenitizing. The load is subsequently transferred to an internal cooling zone and lowered into an integral oil quench.

When the entire sequence of heat processing is done under vacuum or in an inert, nonoxidizing atmosphere, the work comes out clean and bright. A typical single carburizing-diffusion cycle, with no repetition of the two steps, is illustrated in Fig. 15. Figure 15(a) shows the variation of temperature with time and Fig. 15(b) the variation of furnace pressure with time.

Methods of Measuring Case Depth

CASE HARDENING may be defined as a process by which a ferrous material is hardened in such a manner that the surface layer, known as the case, becomes substantially harder than the remaining material, known as the core. Case hardening processes include carburizing, nitriding, carbonitriding, cyaniding, and induction and flame hardening. In every instance, case hardening affects chemical composition or mechanical properties, or both.

An accurate and repeatable method of measuring case depth is essential for quality control of the case hardening process and for evaluation of workpieces for conformance with specifications, such as might be done during a failure analysis.

Because measurements made by the various methods are not necessarily taken at the same location in a case, confusion and misunderstanding can result if the method of measurement is not specified.

Effective case depth is the perpendicular distance from the surface of a hardened case to the deepest point at which a specified level of hardness is maintained. The hardness criterion, except when otherwise specified, is 50 HRC.

Total case depth may be defined as the perpendicular distance from the surface of a hardened or unhardened case to the point at which differences in chemical or physical properties of the case and core can no longer be distinguished. Total case depth sometimes is considered to be the distance from the surface to the deepest point at which the carbon content is 0.04% higher than the carbon content of the core.

CHEMICAL METHOD

The chemical method of measuring case depth generally is used only for carburized cases but may be used for cyanided or carbonitrided cases

as well. This method consists of determining the carbon content (and, when applicable, the nitrogen content) by chemical analysis at incremental depths below the surface. The chemical method is considered to be the most accurate method of measuring total case depth. One of two common methods is used to analyze for carbon content: combustion analysis or spectrographic analysis. Combustion carbon analysis currently is the most widely employed.

Spectrographic Analysis. Carbon content may be determined accurately by spectrographic analysis. This method makes use of a vacuum spectrometer, which permits measurement of spectral lines in the ultraviolet region where air would ordinarily absorb much of the emitted radiation.

MECHANICAL METHOD

In the mechanical method of measuring case depth, hardness traverses are taken on the case and core of a specimen that has been prepared by one of three procedures. It is considered the most accurate method of measuring effective case depth (depth to 50 HRC). This method also is preferred for measuring total depth of thin cases (0.25 mm or less).

For measurement of effective case depth, read to point of specified hardness, which is 50 HRC (or approved equivalent) except for selectively hardened cases, for which the following values are recommended:

Carbon content, %	Case hardness, HRC
0.28 to 0.32	35
0.33 to 0.42	40
0.43 to 0.52	45
0.53 and over	50

Hardness testers that produce small, shallow impressions should be used for both of the following procedures, so that the hardness values obtained will be representative of the surface or

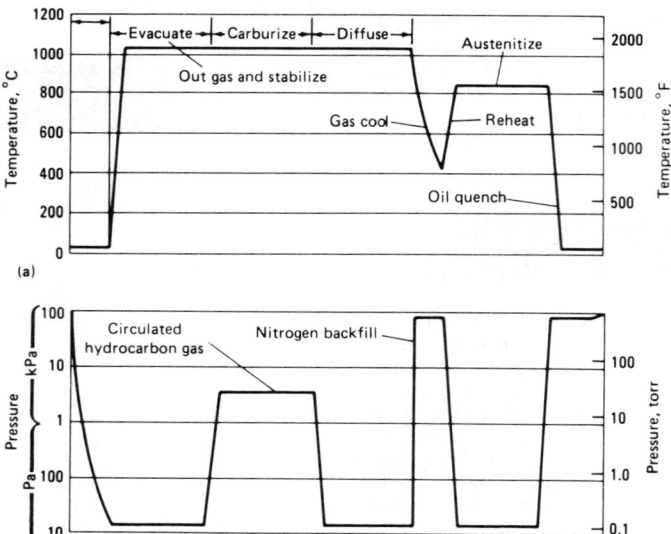

(a) Variation of temperature. (b) Variation of pressure.

Fig. 15. Typical single vacuum carburizing cycle (with no repetition of carburizing and diffusion steps)

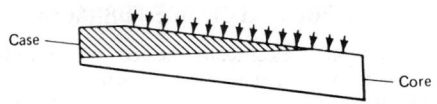

Arrows show locations of hardness-indenter impressions.

Fig. 16. Taper-ground specimen for hardness-traverse method of measuring depth of light and medium cases

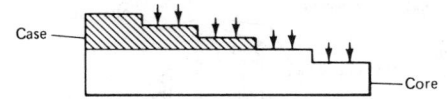

Arrows show locations of hardness-indenter impressions.

Fig. 17. Step-ground specimen for hardness-traverse method of measuring depth of medium and heavy cases

area being tested. Testers that produce Vickers or Knoop microhardness numbers with loads of at least 0.5 kg are recommended, although testers using heavier loads (such as Rockwell superficial) can be used in some instances.

Taper-Grind Procedure. This procedure, illustrated in Fig. 16, sometimes is used for measurement of light and medium cases.

A shallow taper is ground through the case, and hardness measurements are made along the surface thus prepared. The angle is chosen so that readings spaced equal distances apart will represent the hardnesses at the desired increments below the surface of the case.

Step-Grind Procedure. This procedure, illustrated in Fig. 17, is recommended for measurement of medium and heavy cases. It is essentially the same as the taper-grind procedure, with the exception that hardness readings are made on steps that are known distances below the surface.

Heat Processing Equipment

Types of Heat Treating Furnaces

FURNACES commonly used in heat treating are classified in two broad categories, batch furnaces and continuous furnaces. In batch furnaces, workpieces normally are loaded and unloaded manually. A continuous furnace has an automatic conveying system that provides a constant workload, in kilograms (pounds) per hour, to the unit.

Brief discussions of batch and continuous furnaces are presented below. For more detailed information on heat treating furnaces, see Metals Handbook, 9th Edition, Volume 4, pages 285 to 292.

BATCH FURNACES

The basic batch furnace normally consists of an insulated chamber with an external reinforced steel shell, a heating system for the chamber, and one or more access doors to the heated chamber. Standard batch furnaces such as box, car-bottom and pit types are most commonly used when a wide variety of heat-hold-cool temperature cycles are required.

With the addition of powered workhandling systems—integral quench tanks, slow-cool chambers and some automatic controls—the basic box-type batch furnace is upgraded to a semicontinuous batch furnace, which is a commonly used piece of heat treating equipment. The design of a semicontinuous batch furnace is shown in Fig. 1.

CONTINUOUS FURNACES

Continuous furnaces consist of the same basic components as batch furnaces: an insulated chamber, a heating system and access doors. Figure 2 shows one configuration of a continuous furnace.

Common types of continuous furnaces include pusher, belt- or chain-conveyor, roller-hearth, shaker-hearth, rotary-hearth and walking-beam furnaces. In pusher and roller-hearth furnaces, the total workload, in kilograms (pounds) per hour, consists of the work, trays, and any fixtures required. In belt- or chain-conveyor fur-

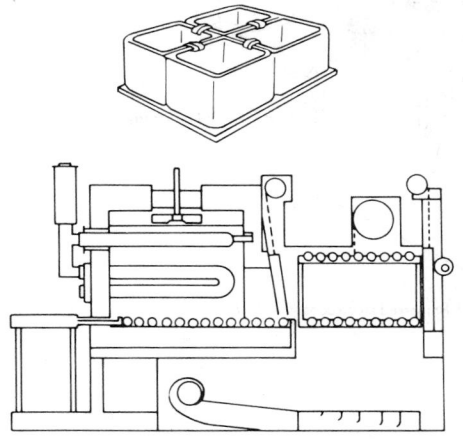

Fig. 1. Semicontinuous batch furnace

naces, both the conveyor and the work must be heated unless the belt or chain is returned inside the furnace. The work generally is placed directly on the belt or chain, and in some cases the belt or chain may be equipped with some type of attachment for holding and positioning the work. In shaker-hearth and rotary-hearth furnaces, the work generally can be placed directly on the hearth. In walking beam-furnaces, the work can be placed directly on the hearth but sometimes may be placed on a tray.

Salt Bath Equipment

SALT BATHS are used in a wide variety of commercial heat treating operations, including cyaniding, liquid carburizing, liquid nitriding, austempering, martempering and tempering applications. Salt bath equipment is well adapted to heat treatment of tool steels, as well as to treatment of nonferrous alloys. Advantages of using salt bath equipment include thermal control and rapid heating rates.

EXTERNALLY HEATED FURNACES

Externally heated salt bath furnaces may be fired by gas or oil, or heated by means of electrical-resistance elements. A typical gas- or oil-fired furnace that is commonly used in liquid carburizing applications is shown in Fig. 3(a). Electrical-resistance furnaces for neutral heating or liquid carburizing (see Fig. 3b) are less widely used than furnaces fired by gas or oil.

IMMERSED-ELECTRODE FURNACES

Introduction of the immersed-electrode furnace greatly extended the useful range and capacity of molten carburizing baths. The electrodes can be removed and replaced without bailing the furnace. This design is also suitable for neutral heating, as well as for cyanide and noncyanide carburizing processes. The molten salt is contained in a steel or ceramic pot surrounded by suitable insulating materials, which separate it from an exterior casing or framework of heavy-gage steel. The salt is heated by passing alternating current through it with immersed electrodes. As a result of the resistance built up to passage of current through salt, heat is generated within the salt itself. This heat is quickly dissipated by a downward stirring action created by the electrodes.

A typical immersed-electrode furnace is shown in Fig. 3(c).

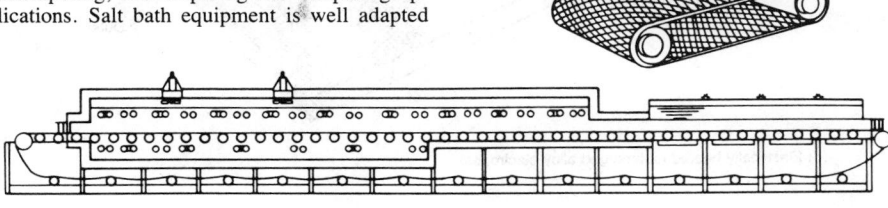

Fig. 2. Continuous furnace

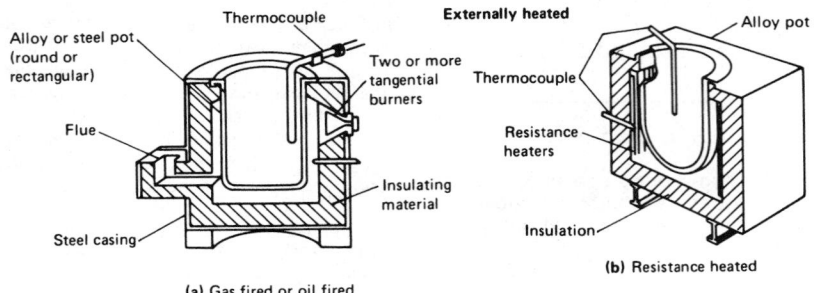

(a) Gas fired or oil fired

(b) Resistance heated

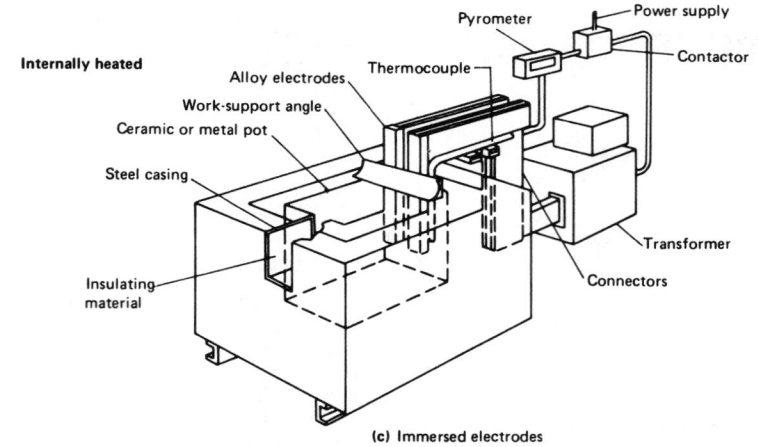

(c) Immersed electrodes

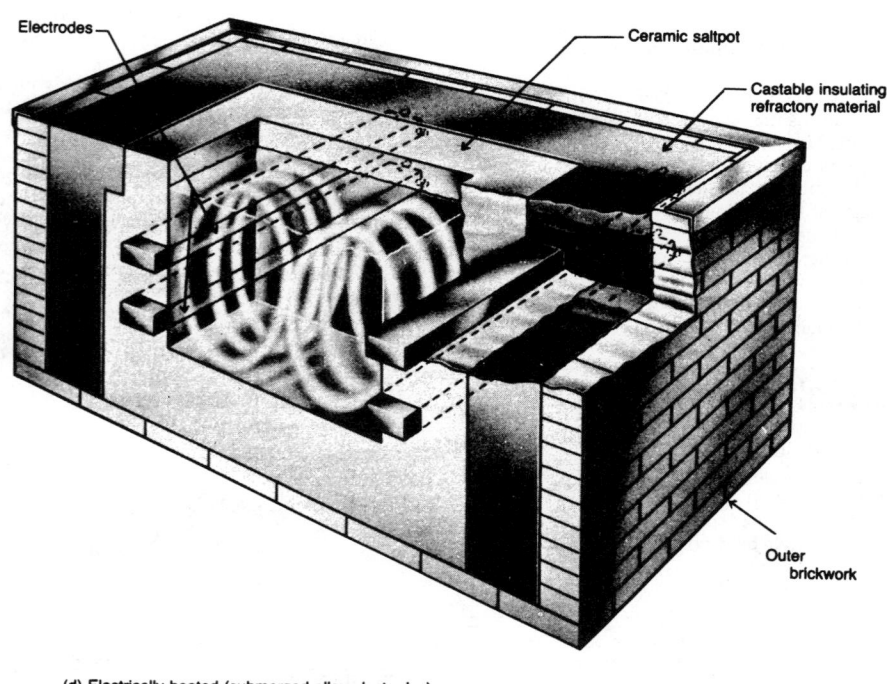

(d) Electrically heated (submerged alloy electrodes)

Fig. 3. Principal types of externally and internally heated salt bath furnaces used for liquid carburizing

SUBMERGED-ELECTRODE FURNACES

In a typical electrically heated submerged-electrode furnace, the frame is made of heavy angle iron with a steel plate at the base beneath the brickwork. The outer brickwork consists of hollow ceramic tile or common building brick. The salt pot is made of burned alumina firebrick. Castable insulating refractory fills the space between the sidewalls and the ceramic pot. An electrically heated submerged-electrode furnace is shown in Fig. 3(d).

Fluidized-Bed Equipment

PRINCIPLES OF FLUIDIZED-BED HEAT TREATING

In fluidization, a bed of dry, finely divided particles, typically aluminum oxide in the heat treating context, is made to behave like a liquid by a moving gas fed upward through the bed. A gas-fluidized bed is considered a dense-phase fluidized bed when it exhibits a clearly defined upper limit or surface. At a sufficiently high fluid-flow rate, however, the terminal velocity of the solids is exceeded, the bed goes into motion, and the upper surface of the bed disappears. This state constitutes a disperse, dilute or lean-phase fluidized bed with pneumatic transport of solids. The general types of fluidized beds are shown in Fig. 4. The majority of beds used for heat treatment are of the aggregative or bubbling type.

Although the properties of solid and fluid alone determine the quality of fluidization (that is, whether smooth or bubbling fluidization occurs), many factors influence the rate of solid mixing,

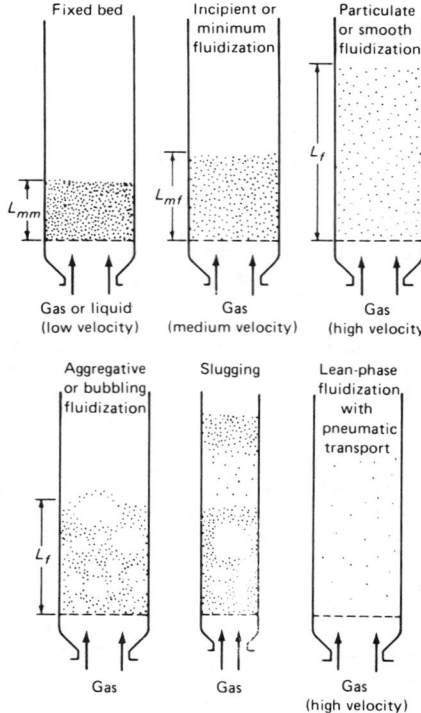

Fig. 4. Various types of contacting in fluidized beds

the size of bubbles and the extent of heterogeneity in the bed. These factors include bed geometry; gas-flow rate; type of gas distributor; and internal-vessel features such as screens, baffles and heat exchangers.

Internal-Resistance-Heated Fluidized Beds. In this type of unit, the gas and particles are heated by suitably sheathed internal-resistance-heated elements. For high-temperature operation between 500 and 1000 °C (930 and 1830 °F), silicon carbide elements can be used, but they must be sheathed to prevent reactions between the elements and the bed material. At lower temperatures, a mineral-insulated heater with an integral metal sheath can be used.

External-Resistance-Heated Fluidized Beds. A fluidized bed contained in a heat-resisting pot can be heated by external resistance elements, as shown in Fig. 5. Waste-heat recovery can be used to increase thermal efficiency, and the fluidizing gas can be maintained at any desired composition. Although this appears to be a good method of applying heat to the fluidized bed, there is a severe limitation on the rate of heat input that can be achieved through the wall of the pot. Heat-up rates from cold to operating temperatures of 700 to 800 °C (1290 to 1470 °F) can be as long as 5 to 6 h.

APPLICATIONS OF FLUIDIZED-BED FURNACES

Applications of fluidized-bed furnaces to heat treatment of metals include continuous units for all types of wire and strip processing (patenting, austenitizing, annealing, tempering, quenching, etc.); continuous rotary types for fasteners, bearings and other small parts; and all configurations of batch-type units for general heat treating applications.

Heat Treating in Vacuum Furnaces and Auxiliary Equipment

VACUUM HEAT TREATING is a relatively new development in metallurgical processing. Vacuum heat processing consists of thermally treating metals in heated enclosures that are evacuated to partial pressures compatible with the particular metals and processes. Vacuum is substituted for the more commonly used protective gas atmospheres during either part or all of the heat treatment. Furnace equipment used in vacuum heat treating differs widely in size, shape, construction and method of loading.

Although originally developed for processing of electron-tube materials and refractory metals for aerospace applications, vacuum furnaces are now employed in brazing, sintering, heat treating and diffusion bonding of metals. Vacuum furnaces also are used for annealing, carburizing, heating and quenching, tempering, and stress relieving. Furnaces for vacuum heat treating are equipped for workloads ranging from several pounds up to 100 tons, and heated working chambers range in size from 0.03 m³ (1 ft³) up to hundreds of cubic feet. Although most vacuum furnaces are batch-type installations, continuous vacuum furnaces with multiple zones for

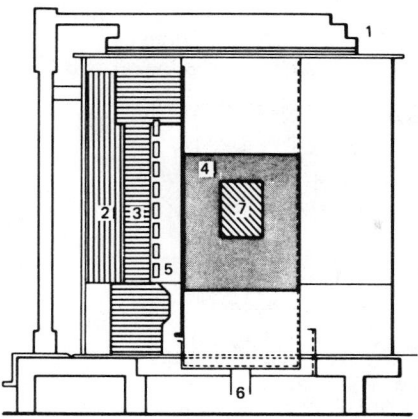

(1) Pivoting cover. (2) Insulating lagging. (3) Refractory material. (4) Fluidized bed. (5) Resistance elements. (6) Intake for fluidizing gas (air or nitrogen). (7) Parts to be treated.

Fig. 5. Fluidized-bed furnace with external heating by electrical-resistance elements

purging, preheating, high-temperature processing and cooling by gas or liquid quenching also are used.

VACUUM MEASUREMENTS

A theoretical or ideal vacuum is an empty space that does not contain either vapors, particles, gases or other matter and consequently has no atmospheric pressure. Because this condition does not exist, even in outer space, an ideal vacuum cannot be achieved. Therefore, a manufactured vacuum is expressed in relative terms of pressure compared with the standard atmospheric pressure surrounding the earth. Standard atmospheric pressure at sea level, 45° latitude and 0° C has the following values: 760 mm Hg, 760 000 μ or μm Hg, 29.921 in. Hg, or 14.696 psi.

FURNACE EQUIPMENT

Although conventional atmosphere furnaces can be adapted for vacuum heat treating by adding a vacuum-tight retort connected to a suitable

pumping system, furnace equipment developed especially for vacuum heat treating is generally used.

COLD WALL FURNACES

Cold wall furnace units consist of a water-cooled vacuum vessel maintained near ambient temperature during high-temperature operations. Consequently, because the operating temperature does not affect the strength of the vessel material, large units can be constructed for use at high operating temperatures.

In the cold wall design, the water-cooled vacuum vessel contains and supports the internal insulation, the electrical heating elements and the hearth on which the workload rests. The vacuum acts as (a) a substitute for the normal heat treating atmosphere to protect the workload; (b) an insulating medium in the furnace, because the thermal conductivity of a vacuum is essentially zero; and (c) an effective protective coating around the heating elements, the heat shields and the supporting hearth.

Bottom-Loading Furnaces. As shown in Fig. 6, bottom-loading furnaces are stationary and elevated well above floor level. The bottom descends to floor level for ease of loading. The work is loaded on trays that are placed on the hearth by a fork lift when the bottom is in the lowered position. Such furnaces are built to handle large, heavy loads and are cooled rapidly by a high-velocity internal or external circulating gas system.

Horizontal-Loading Furnaces. A box-type horizontal-loading furnace consists of a gastight cylindrical shell with circular convex end plates. In some designs, both of the end plates are hinged to permit easy access to the furnace interior.

Cross-sectional views of a three-chamber oil-quench furnace are shown in Fig. 7. The front chamber is equipped with internal cooling coils and a circulating fan for accelerated gas cooling. The center chamber is the heating chamber, which can be sealed at both ends during the heating cycle by internal moving heat shields and doors equipped with O-rings. The third chamber contains the oil quench and the vertical transport system required to immerse the work in the circulated quenching oil.

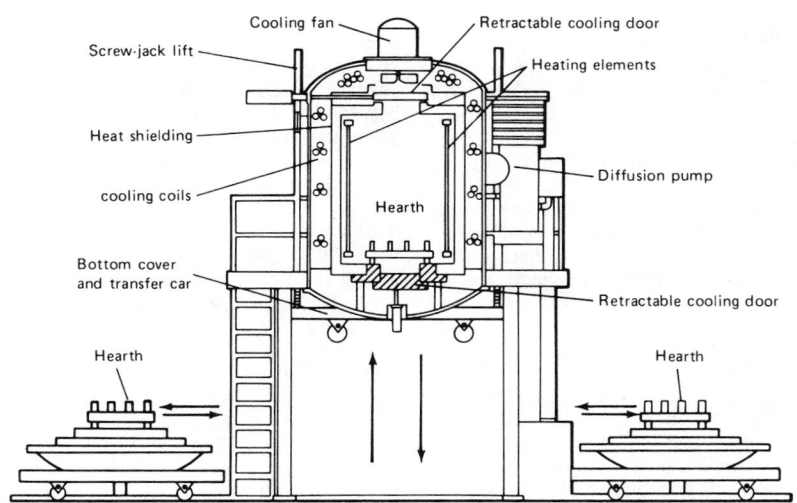

Fig. 6. Bottom-loading cold wall vacuum furnace

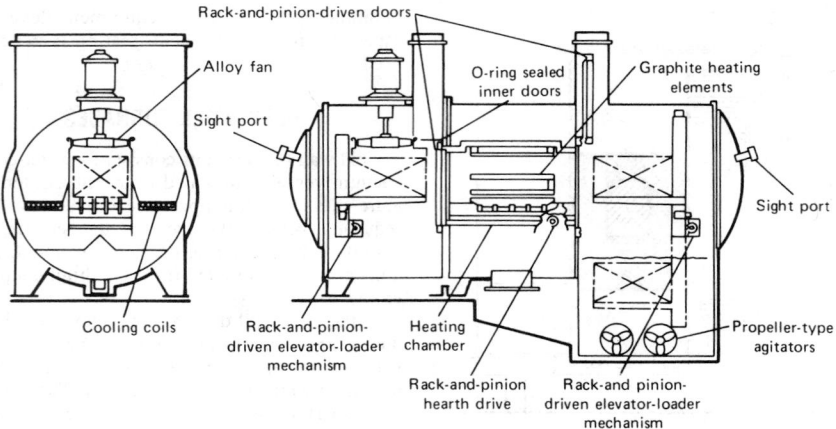

Fig. 7. Three-chamber vacuum oil quench furnace

HEATING ELEMENTS

Resistance heating and induction heating are the two most common methods of heating within cold wall furnaces. When vacuum furnaces are heated inductively, a graphite cylinder is used as a susceptor; the graphite is heated by induction and radiates the heat to the work inside the cylinder. When heating is provided by the more common resistance elements, the heat transfer is also completed by radiation; therefore, the active heating surface should be large enough to effect a rapid transfer of heat.

Resistance heating elements operating in a vacuum do not require oxidation-resistant properties equal to those required in oxidizing atmospheres. Because, to improve operating efficiency, resistance heating elements are heated to higher temperatures than are the elements used in conventional furnaces, resistance heating elements require low vapor pressures to ensure long life. Materials meeting these requirements are:

- Refractory metals, such as tungsten, molybdenum and tantalum
- Pure solid graphite in the form of bar, rod or tube
- Pure graphite cloth woven from fine filaments of pyrolyzed graphite
- Chromium-nickel elements for operating temperatures below 980 °C (1800 °F).

Refractory Metals

Tungsten is capable of withstanding higher operating temperatures than the other refractory metals. As a heating-element material, it is used as a thin sheet or as sections of woven wire screen.

Molybdenum, in the form of solid rod, strip or thin sheet, is the most widely used metallic heating-element material. Molybdenum in sheet form is normally preferred because its electrical power density (watts per square inch of radiating surface) is low compared with that of cylindrical rod, resulting in lower operating temperatures and thus longer service life.

Solid Graphite Heaters

All metals lose some strength when heated, whereas crystalline carbon in the form of graphite increases in strength as the temperature increases. Pure graphite in the form of flat bar and rod is less expensive than other high-temperature metallic resistors. Graphite also has a much lower

heat expansion coefficient and is more resistant to thermal shock than most metallic materials, and has a high melting point and a low vapor pressure; thus, it is an excellent choice for a vacuum furnace heating-element material.

Graphite Cloth Heaters

A third type of material used for vacuum heating elements is a cloth composed of fine graphite fibers. This material is made from rayon cloth pyrolyzed at high temperature to convert the carbon in the rayon to crystalline graphite. The cloth is strong and very flexible. It can be cut with ordinary scissors to the desired size and shape. Because the cloth is flexible, the supporting system can be simplified considerably. The ends usually are clamped in graphite electrodes.

PUMPING SYSTEMS

Vacuum vessels are evacuated by various types of pumping systems that depend, to a great extent, on the pressure range needed for processing. An adequate vacuum pumping system must attain the specified pressure and must have sufficient capacity to handle the processing gas load, not only at the ultimate pressure but at all intermediate pressures during the pumpdown cycle. Pumping systems are usually divided into two subsystems: the roughing pump and the high-vacuum pump. For certain requirements, a single pumping system is sufficient for the entire range and cycle. Pumps usually are classified as mechanical pumps or diffusion pumps.

Mechanical pumps operate on the fluid-flow principle and are primarily positive-displacement pumps with suitable seals to permit operation at low pressures. Piston pumps or rotary blowers in various pumping-speed ratings are available. Vacuum-system levels down to 25 μm Hg can be obtained with oil-sealed rotary mechanical pumps.

Diffusion Pumps. For pumping at a vacuum-system level below 10^{-3} torr, a vapor-diffusion pump generally is used. Pumping action is directed by a high-velocity stream of heavy molecules in the form of a pump fluid, usually oil.

A schematic diagram of a three-nozzle vapor-diffusion pump is shown in Fig. 8. Vapor from a liquid held in a closed boiler heated at the bottom is forced upward inside the boiler. The vapor passes quickly through a narrow circumfer-

ential opening in the nozzles at a downward angle. Molecules of gas that stray from the vacuum chamber above the pump toward the vapor jet streaming from the nozzles encounter the downward-directed stream of heavy molecules. The over-all effect is to compress the gas molecules and force them downward to a point where they can be removed by the mechanical forepump.

Heat-Resistant Alloys for Furnace Parts, Trays and Fixtures

TRAYS AND FIXTURES made of heat-resistant alloys are among the many parts used in industrial heat treating furnaces that operate at temperatures from 540 to 980 °C (1000 to 1800 °F).

BASIC METALLURGY AND PRODUCT FORMS

A partial list of typical products can be divided into two categories: parts that go through the furnace, and parts that remain in the furnace. Parts that go through the furnace and are, therefore, subjected to thermal and/or mechanical shock include trays, fixtures, conveyor chains and belts, and quenching fixtures. Parts that remain in the furnace, and thus undergo less thermal or mechanical shock, include support beams, hearth plates, combustion tubes, roller and skid rails, conveyor rolls, walking beams, rotary retorts, pit-type retorts, muffles, and drive and idler drums.

The heat-resistant alloys used for these parts are supplied in either wrought or cast form and, in some situations, may be a combination of the two. The properties and costs of the two forms vary, even though their chemical compositions are similar. Because there are many foundries and fabricators with experience in the design and application of these products, it is important to seek their advice when purchasing high-alloy parts.

Five types of heat-resistant alloys are used for furnace parts, trays and fixtures:

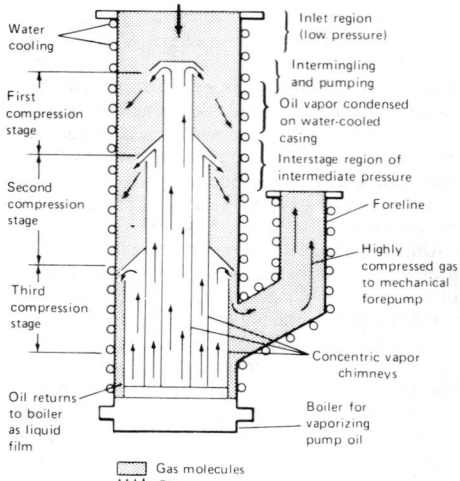

Fig. 8. Oil vapor-diffusion pump

Table 1. Recommended materials for furnace parts and fixtures for hardening, annealing, normalizing, brazing and stress relieving
Where more than one material is recommended for a specific part and operating temperature, each has proved satisfactory in service. Multiple choices are listed in order of increasing alloy content.

Retorts, muffles(a) Wrought	Cast	Radiant tubes(a) Wrought	Cast	Mesh belts, wrought	Chain link Wrought	Cast	Sprockets, rolls, guides, trays Wrought	Cast
595 to 675 °C (1100 to 1250 °F)								
430	HF	430	HF	430	430	HF	430	HF
304		304		304	304		446	
							304	
675 to 760 °C (1250 to 1400 °F)								
304	HF	347	HF	309	309	HF	304	HF
347	HH	309	HH			HH	316	HH
309(b)							309	
760 to 925 °C (1400 to 1700 °F)								
310	HH	310(c)	HH	314	314	HH	310	HH
35-18(d)	HT(e)	35-18(d)	HK	35-18(d)	35-18(d)	HL	35-18(d)	HK
Inconel	HW(e)(f)	Inconel	HL			HT		HL, HT
925 to 1010 °C (1700 to 1850 °F)								
35-18(d)	HK	Inconel	HK	314	314	HL	310	HL
Inconel	HT		HL	35-18(d)	35-18(d)	HT	35-18(d)	HT
	HW		HT	Inconel	Inconel		Inconel	
1010 to 1095 °C (1850 to 2000 °F)								
35-18(d)	HK	Inconel	HL	35-18(d)	35-18(d)	HL	35-18(d)	HL
Inconel	HL		HX	80-20	80-20	HT	Inconel	HX
	HW							
	HX							
	NA22H							
1095 to 1205 °C (2000 to 2200 °F)								
Hastelloy X	HL	Inconel	HL	35-18(d)	35-18(d)	HX	Inconel	HL
Inconel	HU		HX	80-20	80-20			HX
	HX							

(a) Temperature gradients of 40 to 95 °C (100 to 200 °F) are assumed between heat-source side and work-zone side of retorts, muffles and radiant tubes. (b) The stabilized grade 309S is recommended for applications involving mechanical or thermal shock. (c) Recommended for vertical mounting only. (d) A series of alloys generally of the 35Ni-15Cr type or modifications that contain from 30 to 40% Ni and 15 to 23% Cr and include RA-330, 35-19, Incoloy and other proprietary alloys. (e) HK or HL is recommended where greater strength is needed. (f) Recommended for applications requiring shock resistance, such as shaker hearths.

Table 2. Recommended materials for parts and fixtures for carburizing and carbonitriding furnaces
Where more than one material is recommended for a specific part and product form (wrought or cast), each has proved satisfactory in service. Multiple choices are listed in order of increasing alloy content.

Type of part	Recommended materials(a) Wrought	Cast
Retorts(b)	35-18(c), Inconel	HK, HT
Muffles(b)	35-18(c), Inconel	HT
Radiant tubes(b)	35-18(c), Inconel	HT, HU, HX
Structural parts	35-18(c), Inconel	HT
Pier caps, rails	35-18(c), Inconel	HT
Tube supports	35-18(c), Inconel	HT
Trays, baskets, fixtures: Not quenched	35-18(c), Inconel	HT, HT(Nb), HU, HU(Nb)
Oil quenched	35-18(c), Inconel	HT, HT(Nb), HU, HU(Nb), HW

(a) Operating temperature, 815 to 1010 °C (1500 to 1850 °F). (b) Temperature gradients of 40 to 95 °C (100 to 200 °F) difference in temperature between heat-source side and work-zone side of retorts, muffles and radiant tubes. (c) A series of alloys generally of the 35Ni-15Cr type or modifications that contain from 30 to 40% Ni and 15 to 23% Cr and that include RA-330, 35-19, Incoloy and other proprietary alloys.

- Fe-Cr alloys
- Fe-Cr-Ni alloys
- Fe-Ni-Cr alloys
- Nickel-base alloys
- Cobalt-base alloys.

The great majority of heat treating furnaces use only the second and third types, because the Fe-Cr alloys do not have sufficient high-temperature strength to be useful, and the nickel- and cobalt-base types (except Inconel) are generally too expensive except for very special applications.

SPECIFIC APPLICATIONS

Recommended applications for alloys for parts and fixtures for various types of heat treating furnaces, based on atmosphere and temperature, are summarized in Tables 1, 2 and 3. Where more than one alloy is recommended, each has proved adequate, although service life varies in different installations because of differences in exposure conditions.

Trays and Grids. Many parts to be heat treated are irregular in shape and as such must be conveyed through the continuous heat treating furnaces or loaded and unloaded from the batch furnaces on grids or trays. These trays or grids must withstand exposure to the same furnace conditions as the product and as such are subject to heating and cooling, to detrimental and beneficial atmospheres, as well as to compression and tensile loading. Heat-resistant alloys are used extensively for these parts, although there are instances where dispensable carbon or low-alloy steel fabricated trays are employed. In this instance, the choice is based on the economics of the particular situation, taking into account the cost of materials as well as the service life expected.

Baskets and Fixtures. In many situations, parts being heat treated are of a size that does not permit them to be loaded directly on a furnace hearth, tray or grid. They require some type of container, such as a basket; design of baskets varies because each product is developed for a specific application and loading and unloading must function with a specific type of furnace equipment.

Pots. Furnace design is the most important consideration in the selection of material for pots holding molten lead or salt. Externally heated pots act as a muffle or barrier between the heating and work zones. This type of service is severe because of the great difference between outside and inside temperatures, especially while the furnace is being heated to the operating temperature, when the outside of the pot is subjected to maximum heat input and the lead or salt it contains is still solid.

Table 3. Recommended materials for parts and fixtures for salt baths
Where more than one material is recommended for a specific part and operating temperature, each has proved satisfactory in service. Multiple choices are listed in order of increasing alloy content (except ceramic parts).

Process and temperature range	Electrodes	Pots	Thermocouple protection tubes
Salt quenching, 205 to 400 °C (400 to 750 °F)	Low-carbon steel	Low-carbon steel	Low-carbon steel, 446
Tempering, 400 to 675 °C (750 to 1250 °F)	Low-carbon steel, 446, 35-18(a)	Aluminized low-carbon steel, 309	Aluminized low-carbon steel, 446
Neutral hardening, 675 to 870 °C (1250 to 1600 °F)	446, 35-18(a)	35-18(a), HT, HU, ceramic Inconel	446, 35-18(a)
Carburizing, 870 to 940 °C (1600 to 1720 °F)	446, 35-18(a)	Low-carbon steel(b), 35-18(a), HT	446, 35-18(a)
Tool steel hardening, 1010 to 1315 °C (1850 to 2400 °F)	Low-carbon steel(c), 446	Ceramic	446, 35-18(a), ceramic

(a) A series of alloys generally of the 35Ni-15Cr type or modifications that contain from 30 to 40% Ni and 15 to 23% Cr and include RA-330, 35-19, Incoloy and other proprietary alloys. (b) Immersed-electrode furnaces only. (c) Low-carbon steel is recommended for completely submerged electrodes only.

Furnace Control Instrumentation

Temperature Control

TEMPERATURE INSTRUMENTATION AND CONTROL SYSTEMS used in heat treating include temperature sensors, controllers, final control elements, measurement instruments and set-point programmers. A basic control loop includes a temperature sensor, a controller and a final control element.

BASIC CONTROL LOOP AND AUXILIARY DEVICES

The basic control loop is illustrated schematically in Fig. 1. Auxiliary devices used with this basic control loop include measurement instruments and set-point programmers (Fig. 2). The measurement instrument monitors the same temperature sensor as that used by the controller. The set-point programmer automatically varies the controller set point to provide a temperature cycle or temperature program in accordance with an established plan.

TEMPERATURE SENSORS

Thermocouples and resistance temperature detectors (also known as resistance thermometers) are the most important contact-type electrical temperature sensors used in the metals industry. Well over 90% of the sensors used in this industry are estimated to be thermocouples. A thermocouple is rugged, inexpensive and accurate; covers wide temperature ranges; and is fast in response. A resistance thermometer is more accurate and stable than a thermocouple. However, the resistance thermometer is more expensive, slower in response, and limited to lower temperatures, typically 540 °C (1000 °F).

Thermocouples

Thermocouples consist of two dissimilar wires that are metallurgically homogeneous. They are joined at one end, called the measuring or hot junction. The other end, which is connected to the copper wire of the measuring instrument circuitry, is called the reference or cold junction. The electrical signal output in millivolts is proportional to the difference in temperature between the measuring junction (hot) and the reference junction (cold). The different types of thermocouples, classified by their metallurgical compositions, have different output signal calibrations.

Thermocouples most commonly used in heat processing applications are listed in Table 1.

Resistance Temperature Detectors

Resistance temperature detectors are contact-type sensors. Their electrical resistance is proportional to temperature. Typical detector materials are platinum, copper and nickel. They are more stable and accurate than thermocouples, but even the platinum detectors have an upper temperature limit of approximately 540 °C (1000 °F), which reduces their usage in the metals industry.

Resistance temperature detectors are normally larger in size and slower in response than thermocouples. However, the new thin-film deposited detectors minimize this disadvantage, which characterizes conventional wire-wound detectors.

Noncontact Sensors

Radiation sensors are noncontact-type temperature sensors used with radiation pyrometers. One type of radiation pyrometer is the optical pyrometer.

Radiation sensors respond to radiant energy. They are classified as total-radiation (wide-band) or narrow-band types, depending on the width of the radiation wavelength band to which they respond. Total radiation sensors use thermal detectors, and narrow-band sensors use photoelectric detectors. Radiant energy from metals is characterized at lower temperatures as red hot (with longer wavelengths). At higher tempera-

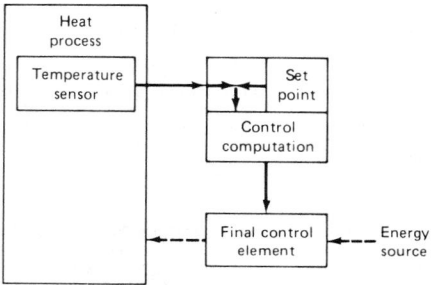

Fig. 1. Basic control loop

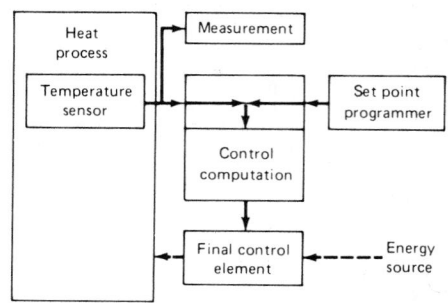

Fig. 2. Basic control loop with auxiliary devices

Table 1. Comparison of thermocouple types

Type	Usable temperature range °C	°F	Advantages	Restrictions
J (iron-constantan)	−185 to 870	−300 to 1600	Comparatively inexpensive; suitable for continuous service to 870 °C (1600 °F) in neutral or reducing atmospheres	Maximum upper limit in oxidizing atmosphere is 760 °C (1400 °F), due to the oxidation of the iron; protection tubes should be used above 480 °C (900 °F); protection tubes should always be used in a contaminating medium
K (nickel, chromium-nickel, aluminum)	−20 to 1370	0 to 2500	Suitable for oxidizing atmospheres; in higher temperature ranges, provides a more mechanically and thermally rugged unit than platinum or rhodium-platinum, and longer life than iron-constantan	Especially vulnerable to reducing atmospheres, requiring substantial protection when used
T (copper-constantan)	−185 to 370	−300 to 700	Resists atmosphere corrosion; applicable in reducing or oxidizing atmospheres below 315 °C (600 °F); its stability makes it useful at subzero temperatures; has high conformity to published calibration data	Copper oxidizes above 315 °C (600 °F)
E (nickel, chromium-constantan)	−185 to 870	−300 to 1600	Has high thermoelectric power; both elements are highly corrosion-resistant, permitting use in oxidizing atmospheres; does not corrode at subzero temperatures	Stability is unsatisfactory in reducing atmospheres
S (platinum, 10% rhodium-platinum) R (platinum, 13% rhodium-platinum)	−20 to 1480	0 to 2700	Usable in oxidizing atmospheres; provides a higher usable range than type K; frequently more practical than noncontact pyrometers; has high conformity to published calibration data	Easily contaminated in other than oxidizing atmospheres
B (platinum, 30% rhodium-platinum, 6% rhodium)	870 to 1650	1600 to 3000	Better stability than types S or R; increased mechanical strength; usable to higher temperatures than types S and R; reference-junction compensation is not required if junction temperature does not exceed 65 °C (150 °F)	Available in standard grade only; high temperature limit requires the use of alumina insulators and protection tubes; easily contaminated in other than oxidizing atmospheres

tures, it is characterized as white hot (with shorter wavelengths).

Radiation sensors are used typically to control annealing furnaces as well as brazing and forge furnaces. They also are used for blast-furnace stoves, salt pots, checker bricks, rolling and strip mills, and induction-heating processes. One or more of the following conditions justify the use of noncontact sensors instead of contact sensors.

- Temperatures are too high for contact sensors.
- Work is moving too fast for detection by contact sensors.
- Required response rate is too fast for contact sensors.

MEASUREMENT AND CONTROL INSTRUMENTS

Measurement instruments measure the output signal of the temperature sensor and convert it to a temperature indication or recording in engineering units. Transmitters are used in some measurement systems to amplify and condition the temperature signal. The accuracy of the measurement depends greatly on the accuracy of the temperature sensor and the connecting leadwire. The accuracy of the measurement instrument is defined in its specifications under referenced conditions for its power supply, ambient conditions (temperature and humidity), electrical noise rejection and maximum source impedance. The accuracy of the transmitter has similar qualifications.

Measurement instruments are classified by their displays, analog or digital, and whether they are recording or nonrecording types. Analog displays include meters and motor-driven pointers. Analog strip-chart or round-chart recorders include analog temperature indication. Digital displays are available with or without digital printers. In general, digital equipment is more accurate than analog equipment, but specifications must be checked in either case.

——Atmosphere Control——

ALL METHODS of atmosphere control can effectively be divided into two groups: those involving control of the atmosphere-generating system, and those involving control of the atmosphere once it is inside the furnace. Both are important in maintaining a controlled condition throughout the heating process.

Most operations performed in the heat processing industry can be done under endothermic or exothermic protective atmospheres. Effective control of the generators that produce such atmospheres may be accomplished by means of certain types of controllers—namely, combustibles controllers, infrared CO_2 controllers and oxygen probes.

ANALYSIS AND CONTROL OF ENDOTHERMIC AND EXOTHERMIC ATMOSPHERES

Endothermic gas is used in heat treating furnaces as a protective atmosphere for hardening, stress relieving, carbon restoration and carburizing. The most recognized guides to endothermic generator operation are CO_2 and dew point. The ultimate guide is the relationship be-

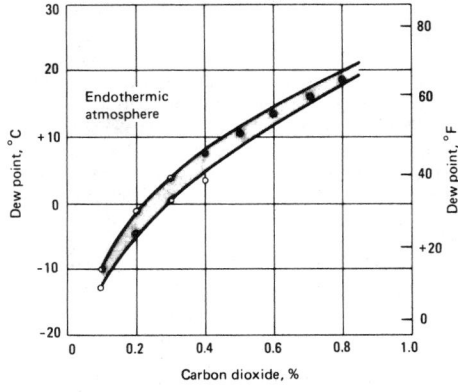

The generator in each plant was operated at a different temperature, all within the range from 1005 to 1095 °C (1840 to 2000 °F).

Fig. 3. Variation in the relation between dew point and carbon dioxide in generation of an endothermic atmosphere, as obtained from four plants

tween CO_2 and dew point. Any deviation from this relationship indicates a generator problem, such as a leak or a carbon-coated catalyst. Figure 3 shows the proper carbon dioxide/dew point relationship.

Exothermic-type atmospheres are used where inert atmospheres are required. Applications are common in the metal treating industry for bright annealing of copper, annealing of motor laminations, processing of aluminum, and annealing of coils of wire and steel sheet.

OXYGEN PROBES

The oxygen probe is based in theory on a hot ceramic electrochemical cell. The probe will respond to oxygen, hydrogen, carbon monoxide, water and carbon dioxide and thus can determine the oxidization potential of a gas. The output of the oxygen probe is a direct measurement of the oxidation potential of the atmosphere at the process temperature of the furnace. Therefore, when the probe temperature is close to the furnace temperature, the response of the probe is a direct indication of whether the atmosphere will oxidize or reduce steel, provided that the composition of the atmosphere with regard to the proportions of carbon gases and hydrogen is known. Under such conditions, the probe will give a reliable indication of the oxidation/reduction situation for all furnace temperatures.

——Furnace Safety——

HEAT TREATING FURNACES require safety procedures common to all industrial installations, but in addition, they have requirements specific to the use of high-temperature energy sources and potentially explosive gases and liquids used as aids to chemical processing.

Because these heat treating processes require careful control for optimum technical results as well as for safety, proper training of operating personnel is a primary consideration. Proper equipment design is also critical.

The information presented here is not intended to be interpreted as a safety standard but is offered only as a set of guidelines. Safety standards for furnaces are maintained by the National Fire Protection Association, by the U.S. Occupational Safety and Health Administration and by insurance underwriters.

All equipment should be installed and operated with awareness of the potentials for fire and explosion and the hazards to operators and equipment. Equipment designs should ensure reliable, safe operation over the expected maximum life of the equipment.

FUEL-FIRED FURNACES

Electrical Power for Fuel-Fired Furnaces

The safe use of electrical energy employed in heat treating control processes requires adherence to National Electrical Codes and to local requirements of states and communities. Good practice dictates that a circuit breaker be positioned within view of the operator.

Control Circuits for Fuel-Fired Furnaces

Combustion-Air Blower Control. Combustion-air blowers must be interlocked with the combustion-limit circuits to shut down the process in the event of failure. The motor starter should be wired so that it will disconnect when any phase is interrupted or when the motor malfunctions. A pressure switch in the air line for direct sensing is a must.

Gas-Pressure Control. Fuel must arrive at the burner in the correct quantity and at the correct time for safe combustion. Fuel pressure thus must be proven within an allowable range. Gas-pressure switches for both high and low gas limits are installed in the main gas lines.

Pressure Regulators. Good, safe design normally requires one regulator for pilot fuel and another regulator for main-burner fuel. The regulators should be vented to a safe location outside the plant. Positive lockup regulators are recommended to prevent downstream pressure buildup during shutdown periods.

Purging of Fuel-Fired Furnaces

The furnace must be purged of any possible combustible materials. This is best accomplished by opening the furnace doors, which should be equipped with a limit switch to ensure that they are open. The combustion blower or exhaust fans can be timed to allow for a minimum of four changes of air in the combustion chamber.

Pilot Control

Pilot assemblies can be of either the atmospheric or the blast type. The atmospheric type is similar to an atmospheric burner, in which the air is inspirated from the atmosphere by the gas stream. In the blast type, air and gas are brought to a mixer under pressure.

Flame Detection

The types of flame-detection devices and their modes of operation include:

- *Thermocouple.* Limited to atmospheric burners; responds to heat of flame
- *Flame Electrode (Flame Rod).* For gas-fired burners firing lean or on ratio; transmits dc current generated by flame impingement on the rod
- *Ultraviolet Scanner.* Used for gas or oil flames; transmits current resulting from sighting of ultraviolet light in the flame proper.

A flame electrode or an ultraviolet scanner is connected to a flame relay that is electrically tied in to the main fuel valve; a thermocouple is directly connected to a gas valve that operates mechanically.

Burner Operation

The main fuel supply for fuel-fired heat treating furnaces normally is natural gas, propane-air, propane, butane or one of the fuel oils.

The main gas valve may be of the manual-reset type, requiring an operator, or may be fully automatic. For furnaces with capacities greater than 422 MJ/h (400 000 Btu/h), it is recommended that a second blocking valve be inserted into the main gas line.

Burner Control. The gas-air ratio ordinarily is controlled to about 10 parts air to 1 part natural gas for good combustion efficiency. There are several devices involved in control of this ratio, and these devices fall into two broad categories: the diaphragm type, or proportionator, which uses the pulse line to keep air and gas at a specified ratio; and the mechanical-linkage type. Both are effective and common, but the diaphragm type is the more positive, because there are no linkages that can slip and require adjustment. Also, if air lines should become dirty, resulting in a lessening of air pressure, the gas pressure will follow, maintaining the correct ratio.

Temperature Control

Temperature-control devices fall into two categories: primary controls and process limiting devices. Safe operation, especially when furnace practices require long cycles and little operator attention, dictates that limits be placed on the process to ensure adequate alarm and perhaps to shut down the operation to prevent destruction of the product, the furnace or the plant itself. Whether an analog device, a strip-chart recorder, a digital readout or a printout is used is a matter of operator preference and depends on the nature of the product.

The typical temperature sensor is either a thermocouple or a resistive temperature device (RTD). The thermocouple is most common. Several types of thermocouple junctions are available, with the choice depending on such factors as temperature range and furnace atmosphere. They are comparatively inexpensive and can be easily protected from atmospheres with protective "wells," which are immersion tubes that project into the furnace zone to be controlled. RTDs, although more accurate than thermocouples by factors ranging from 10 to 1 up to 50 to 1, are expensive and less rugged. For most purposes, thermocouples are satisfactory. Some firms are using heat-flow sensing to remotely ascertain interior temperatures and to provide an element of redundancy for protection of furnaces and their contents. Good temperature-sensing devices will detect failure of a thermocouple or RTD, cause the process firing rate to be reduced to its minimum rate, and perhaps provide an alarm.

Furnace temperature can be regulated by one of two very common procedures. Simple high and low firing rates are used when temperature can be allowed to vary within a fairly large range. More common, in heat treating, is the use of proportional control, wherein the temperature is held nearly constant through the use of a bridge circuit. This circuit balances the signal between the controller and the butterfly valve and holds the latter at the proper opening to maintain the desired temperature. The latter scheme, although more costly, is required for close control.

Supervisory Gas-Cock System

A supervisory gas-cock system is used to ensure a safe "lightoff" procedure on a manually ignited, multiburner furnace that does not have flame-safety equipment and a programmed sequence of piloting the main burners. Supervisory gas-cock systems are used on radiant-tube furnaces and other furnaces where flame-safety systems are difficult to apply.

ELECTRIC FURNACES

Electric-furnace installations are made up of various electrical and mechanical components, many of which are water cooled and equipped with protective devices.

Furnace manufacturers generally issue instructions concerning safe practices, and these instructions should never be ignored. Potential hazards can be avoided by ensuring that operating personnel are trained thoroughly and that installations conform to safety practices and local codes.

Original equipment usually contains devices for preventing overloads and short circuits. In addition, ground detectors and surge detectors protect motor-generator units from faulty coil or transformer installations at heating stations and from breakdown of insulation in the generator windings.

Protective devices commonly used with induction-heating radio-frequency generators are as follows:

- Door interlocks
- Grounding devices to ground high-voltage circuit when furnace doors are open
- Warning lights
- Warning signs
- Circuit breaker for entire unit
- Overload relays
- Water-flow switches
- Water-temperature switches
- Time-delay relays (tube warmup)
- Grid overload relays
- Control-circuit overload relays
- Arc gaps on blocking and tank capacitors
- Surge protection
- Electronic crowbar

Operators should become familiar with these safety devices and should inspect them periodically to ensure that they are in good working condition.

Electrical Power. Although motor generators account for the largest total power output of installed induction-heating equipment, vacuum-tube oscillators probably occur in the greatest numbers of units. Many small vacuum-tube oscillators are required to account for as many kilowatts as one 1250-kW, 3-kilocycle motor-generator set. Although many vacuum-tube oscillators for induction heating are made in small sizes, 25-kW and 50-kW outputs are also common ratings.

Furnace Atmospheres and Carbon Control

Furnace Atmospheres

CONTROL OF FURNACE ATMOSPHERES has become increasingly critical to successful heat treating with more precise metallurgical specifications. The prevention of surface oxidation or scaling when metals are exposed to elevated temperatures remains an important task of the furnace atmosphere.

Properly applied and controlled, furnace atmospheres provide the source of elements in some heat treating processes, surface cleansing of parts being treated in other processes, and a protective environment to guard against adverse effects of air when metals are exposed to elevated temperatures in still other processes.

CLASSIFICATIONS OF PREPARED ATMOSPHERES

Most prepared atmospheres are commonly referred to in the field by their generic names or, in some instances, by trade names. The American Gas Association has classified the commercially important prepared atmospheres into groups based on the method of preparation or the original constituents employed. These groups are designated and defined as follows (see also Table 1):

- *Class 100—Exothermic Base.* Formed by partial or complete combustion of a gas-air mixture; water vapor may be removed to produce a desired dew point
- *Class 200—Prepared Nitrogen Base.* An exothermic base with carbon dioxide and water vapor removed
- *Class 300—Endothermic Base.* Formed by partial reaction of a mixture of fuel gas and air in an externally heated catalyst-filled chamber
- *Class 400—Charcoal Base.* Formed by passing air through a bed of incandescent charcoal
- *Class 500—Exothermic-Endothermic Base.* Formed by complete combustion of a mixture of fuel gas and air, removing water vapor, and re-forming the carbon dioxide to carbon monoxide by means of reaction with fuel gas in an externally heated catalyst-filled chamber
- *Class 600—Ammonia Base.* This can consist of raw ammonia, dissociated ammonia or partially or completely combusted dissociated ammonia with dew point regulated

These broad areas of classification are subclassified and numerically designated to indicate variations in the method by which they are pre-

Table 1. Classification, applications and compositions of principal furnace atmospheres

Class	Description	Common application	Nominal composition, vol % N₂	CO	CO₂	H₂	CH₄
101	Lean exothermic	Oxide coating of steel	86.8	1.5	10.5	1.2	...
102	Rich exothermic	Bright annealing; copper brazing; sintering	71.5	10.5	5.0	12.5	0.5
201	Lean prepared nitrogen	Neutral heating	97.1	1.7	...	1.2	...
202	Rich prepared nitrogen	Annealing, brazing stainless steel	75.3	11.0	...	13.2	0.5
301	Lean endothermic	Clean hardening	45.1	19.6	0.4	34.6	0.3
302	Rich endothermic	Gas carburizing	39.8	20.7	...	38.7	0.8
402	Charcoal	Carburizing	64.1	34.7	...	1.2	...
501	Lean exothermic-endothermic	Clean hardening	63.0	17.0	...	20.0	...
502	Rich exothermic-endothermic	Gas carburizing	60.0	19.0	...	21.0	...
601	Dissociated ammonia	Brazing, sintering	25.0	...	...	75.0	...
621	Lean combusted ammonia	Neutral heating	99.0	...	...	1.0	...
622	Rich combusted ammonia	Sintering stainless powders	80.0	...	...	20.0	...

pared. This subclassification is differentiated by replacing the two zeros of the six basic designators by the following:

- **01** indicates the use of a lean air-to-gas mixture.
- **02** indicates the use of a rich air-to-gas mixture.
- **03** and **04** indicate that preparation of the gas was completed within the furnace itself without the use of a separate machine or generator.
- **05** and **06** indicate that the original base gas was passed through incandescent charcoal before admission to the work chamber.
- **07** and **08** indicate addition of a raw hydrocarbon fuel gas to the base gas before admission to the work chamber.
- **09** and **10** indicate addition of a raw hydrocarbon fuel gas and raw dry anhydrous ammonia to the base gas before admission to the work chamber.
- **11** and **12** indicate addition of a combusted mixture of chlorine, hydrocarbon fuel gas and air to the base gas before admission to the work chamber.
- **13** and **14** indicate that the base gas has had all sulfur or all sulfur and odors removed before admission to the work chamber.
- **15, 16, 17** and **18** indicate addition of lithium vapor to the base gas before admission to the work chamber.
- **19** and **20** indicate that preparation of the gas was completed within the furnace itself with the addition of lithium vapor.
- **21** and **22** indicate that some additional special treatment was given to the base gas before admission to the work chamber.
- **23** and **24** indicate addition of steam and air in conjunction with a catalyst within the generator to convert CO to CO₂, which is then removed.
- **25** and **26** indicate addition of steam in conjunction with a catalyst within the generator to convert CH₄ to H₂ and CO₂, which is then removed.

This classification system provides for a large number of possibilities.

ENDOTHERMIC-BASE ATMOSPHERES

Endothermic-base atmospheres are produced in generators that use air and a hydrocarbon gas as fuel. These two gases are mixed in a controlled ratio, slightly compressed, and then passed into a chamber filled with a nickel-bearing catalyst. This chamber has been heated externally to approximately 1040 °C (1900 °F). The gases react in this chamber to form endothermic gas.

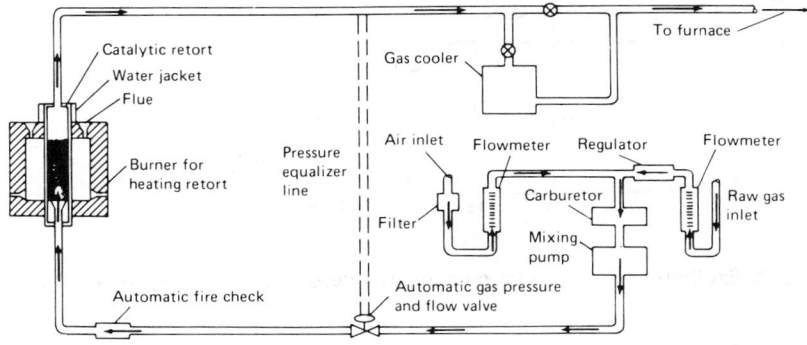

Fig. 1. Schematic flow diagram of an endothermic gas generator

Endothermic atmospheres thus produced must be cooled rapidly to ensure the integrity of their chemical compositions. Figure 1 is a schematic diagram of an endothermic gas generator.

Endothermic gas produced from natural gas, which is primarily methane, has a typical analysis as follows: 40.4% hydrogen, 39.0% nitrogen, 19.8% carbon monoxide, 0.5% methane, 0.2% water vapor, and 0.1% carbon dioxide.

Common Applications. Endothermic atmospheres can be used in virtually all furnace processes that operate about 760 °C (1400 °F) and require strong reducing conditions. The most common use is as carrier gases in gas carburizing and carbonitriding applications.

EXOTHERMIC-BASE ATMOSPHERES

Exothermic gases (class 100) have been used extensively for many years as lower-cost prepared furnace atmospheres. Exothermic atmospheres are divided into two basic classes, rich and lean. Rich exothermic atmospheres (class 102) have moderate reducing capabilities of 10 to 21% combined carbon monoxide and hydrogen, and lean exothermic atmospheres (class 101), usually with 1 to 4% combined carbon monoxide and hydrogen, have minimal reducing qualities.

Rich Exothermic Atmospheres. The principal uses of rich exothermic furnace atmospheres include clean heat treating of certain ferrous and nonferrous materials, such as annealing and tempering of steel, brazing of copper and silver, and sintering of powdered metals.

Reducing properties of rich exothermic atmospheres may be varied to make them suitable for specific processes. Figure 2 indicates the usual operating range of the gas generator and reflects changes (by dry volumetric measurement) in the

following constituents of the product gas at any particular setting: carbon dioxide, carbon monoxide, hydrogen and unburned methane.

Lean exothermic atmospheres generally have limited use in most heat treating applications, particularly for ferrous materials, except when these atmospheres are used as intentional surface oxidizing agents or for specialized low-temperature operations. Lean atmospheres are used to some extent for processes such as copper annealing and are employed more widely when the primary processing aim is to exclude oxygen or to provide purging and blanket gas.

PREPARED NITROGEN-BASE ATMOSPHERES

Prepared nitrogen-base atmospheres are exothermic atmospheres (produced by combustion of a mixture of air and fuel gas) from which almost all of the carbon dioxide and water vapor has been removed. The combination of very low dew point, approximately −40 °C (−40 °F), and the virtual absence of carbon dioxide, accounts for the marked difference between the properties and applications of prepared nitrogen-base atmospheres and those designated exothermic base.

Advantages and Disadvantages. The principal advantage of prepared nitrogen-base atmospheres is their applicability to a variety of heat treating operations for low-carbon, medium-carbon and high-carbon steels and for some other metals. Because of their low dew point and the virtual absence of carbon dioxide, these atmospheres (in the absence of oxygen-bearing contaminants introduced as a result of furnace operations) are neither oxidizing nor decarburizing.

The main disadvantages of these atmospheres lie in the high initial cost of equipment, the large space requirements, and the need for more exacting maintenance and control of the generators.

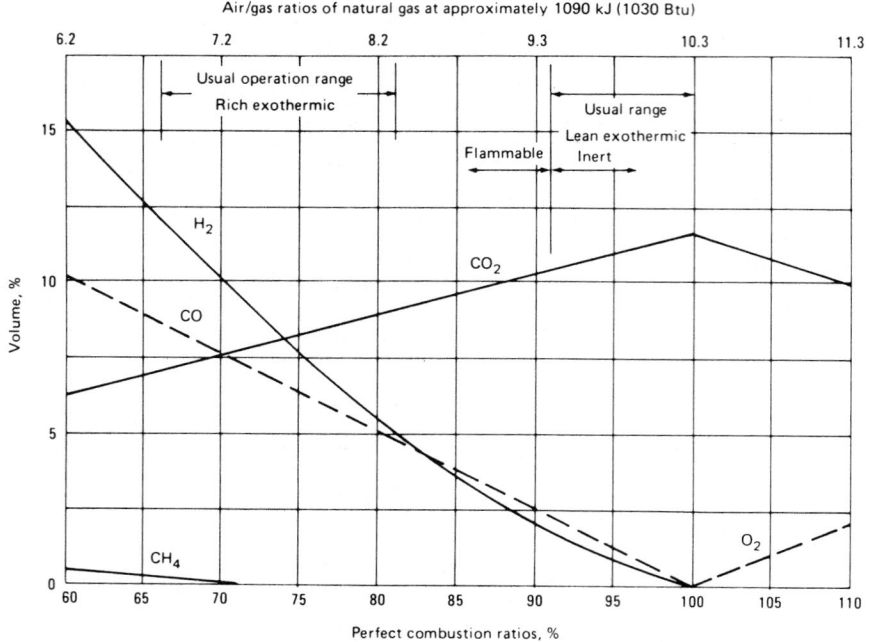

Fig. 2. Exothermic atmosphere composition versus air-fuel ratio (natural gas)

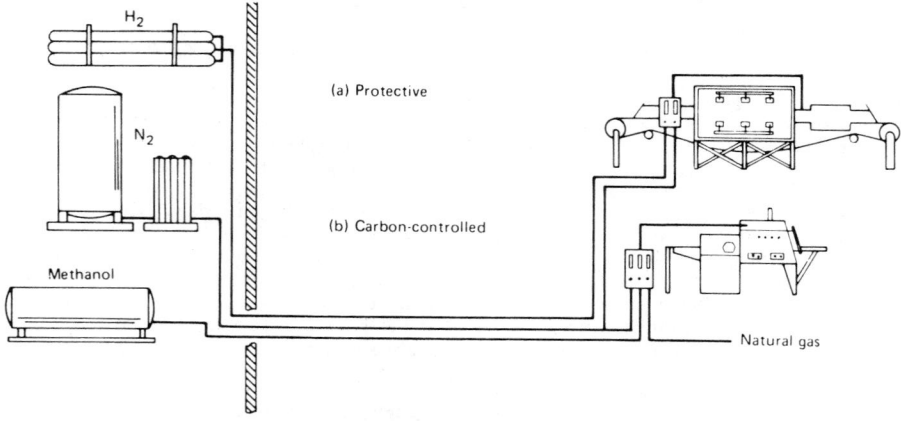

Fig. 3. Industrial gas atmosphere systems

COMMERCIAL NITROGEN-BASE ATMOSPHERES

Commercial nitrogen-base atmosphere systems employed by the metalworking and heat treating industry use gases and equipment that are common among all applications. In most instances, the major atmosphere component is industrial gas nitrogen, which is supplied to the furnace from a system consisting of a storage tank, a vaporizer, and a station controlling pressure and flow rate. The basic components of industrial gas atmosphere systems are illustrated in Fig. 3.

DISSOCIATED AMMONIA ATMOSPHERES

Dissociated ammonia (class 601) is a medium-cost prepared furnace atmosphere providing a dry, carbon-free source of reducing gas. Typical composition is 75% hydrogen, 25% nitrogen, less than 300 ppm residual ammonia, and less than −50 °C (−60 °F) dew point.

Principal uses of dissociated-ammonia furnace atmospheres include bright copper and silver brazing; bright heat treating of selected nickel alloys, copper alloys and carbon steels; bright annealing of electrical components; and as a carrier mixed gas for certain nitriding processes, including the "Floe" system of nitriding, which is a method of controlling white layer.

HYDROGEN ATMOSPHERES

Commercially available hydrogen is 98 to 99.9% pure. All cylinder hydrogen contains traces of water vapor and oxygen. Methane, nitrogen, carbon monoxide and carbon dioxide may be present as impurities in very small amounts, depending on the method of manufacture.

Applications. Dry hydrogen is used in annealing of stainless and low-carbon steels, electrical steels and several nonferrous metals. It is used also in sintering of refractory materials such as tungsten carbide and tantalum carbide, in nickel

brazing of stainless steel and heat-resisting alloys, and in copper brazing.

STEAM ATMOSPHERES

Steam may be used as an atmosphere for scale-free tempering and stress relieving of ferrous metals in the temperature range of 345 to 650 °C (650 to 1200 °F). The steam causes a thin, hard and tenacious blue-black oxide to form on the metal surface. This oxide film, which is about 0.00127 to 0.008 mm (0.00005 to 0.0003 in.) thick, improves certain properties of various metal parts.

Control of Surface Carbon Content in Heat Treating of Steel

MOST HEAT TREATING ATMOSPHERES are gaseous mixtures containing carbon monoxide, carbon dioxide, methane, nitrogen, hydrogen and water vapor. The relative amounts of these gases depend on the type of generator gas used, on the processing temperature and on the amount of gas added during processing. For example, endothermic generator gas produced by catalytic reaction of natural gas with air results in the following composition (approximate percentages by volume): 20 CO, 40 H_2, 40 N_2, 0.1 to 0.5 CO_2, 0.2 to 1.2 H_2O, 0.2 to 0.8 CH_4. In gas carburizing, a common commercial practice is to use an endothermic gas as a carrier and to enrich it with natural or propane gas additions.

SAMPLING OF ATMOSPHERES FOR ANALYSIS

In carburizing of steel, three properties of the finished case are important: surface concentration of carbon, case depth and carbon gradient. Case depth depends primarily on carburizing time and temperature; it often can be presumed to depend only on time, because the temperature used in a given plant is often standardized. The other two case characteristics depend strongly on the carbon potential of the atmosphere and on accurate control of carbon potential. Carbon control is difficult unless the amounts of both carburizing and decarburizing constituents present in the atmosphere can be analyzed and controlled. The first requisite of analysis is that a representative sample of the furnace atmosphere be obtained. The sample of gas should be taken from a point in the furnace chamber as close as possible to the work being treated.

Probe Materials and Design. The probe should be made of a heat-resisting alloy that does not react with the gas sample. Probes made of iron-chromium alloys are preferred to those made of high-nickel alloys, because nickel catalyzes the breakdown of carbon monoxide into carbon dioxide and soot.

INFRARED ANALYZERS

Infrared analyzers are based on the principle that any compound present in the mixture will absorb infrared energy in proportion to its con-

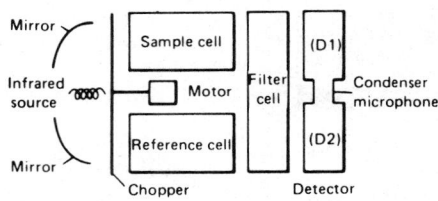

Fig. 4. Elements of a positive-filtering infrared analyzer for measuring CO, CO$_2$ and methane contents of an atmosphere

centration in the mixture. The wavelengths absorbed are different for each compound. Elemental gases such as hydrogen and oxygen do not absorb infrared radiation and hence cannot be analyzed by this method.

Infrared analyzers normally are used to measure carbon monoxide, carbon dioxide and methane. For automatic control of carburizing processes where the carrier gas and enriching gas are reasonably uniform in composition, infrared analysis of carbon dioxide is considered to be the most accurate method and is the most widely used.

Positive-Filtering Analyzer. In this type of analyzer (illustrated schematically in Fig. 4), an electrically heated helix of nickel-chromium alloy wire is the source of infrared radiation. Radiation from this source is split into two beams by mirrors. Both beams are simultaneously interrupted by a motor-driven chopper. The resulting pulses of radiation cause alternate heating and cooling of gas in the two sides of the detector. A condenser microphone consisting of a movable metal diaphragm and a fixed metal plate is mounted between the two sides of the detector. A measurement of differential pressure can thus be obtained.

DEW-POINT ANALYZERS

Dew-point analyzers measure the partial pressure of water vapor in the furnace atmosphere. Dew point is defined as the exact temperature (at a given pressure) at which a mixture of gases will begin to precipitate its moisture content. When air and gas are mixed in consistent, fixed proportions and the mixture is heated to allow chemical reactions to reach equilibrium, the dew point will reflect the chemical balance of the various components comprising the reacted products.

The use of dew point for controlling carbon potential is a fast, inexpensive and relatively simple procedure. Control of endothermic-base atmospheres by the dew-point method is widely accepted in industry.

Chilled Mirror. One of the first methods devised for automatic control of dew point involves use of the chilled-mirror instrument, a schematic representation of which is presented in Fig. 5. This method utilizes refrigeration and heating to condense and evaporate moisture from an illuminated mirror while the temperature of the mirror is being monitored. A photoelectric cell is used to detect the intensity of the light that the mirror reflects from the source of illumination. The intensity of this reflection depends on the amount of moisture present on the mirror. When the photoelectric cell registers a reflection that represents a deviation from a desired dew point, it generates an electric signal that in turn actuates appropriate auxiliary equipment at the furnace or

atmosphere generator, which restores the dew point to the desired level.

OXYGEN PROBES

One of the most recently developed methods of measuring carbon potential in a furnace atmosphere is the oxygen probe, or oxygen meter. In an endothermic-base or exothermic-base carburizing atmosphere, one of the reversible reactions is

$$CO + {}^1/_2O_2 \rightleftarrows CO_2$$

The carbon potential of such an atmosphere is inversely related to the square root of the partial pressure of oxygen. Thus, by monitoring the concentration of oxygen, carbon potential can be defined without considering the concentrations of hydrogen, water vapor or carbon dioxide. The only atmosphere constituent that directly influ-

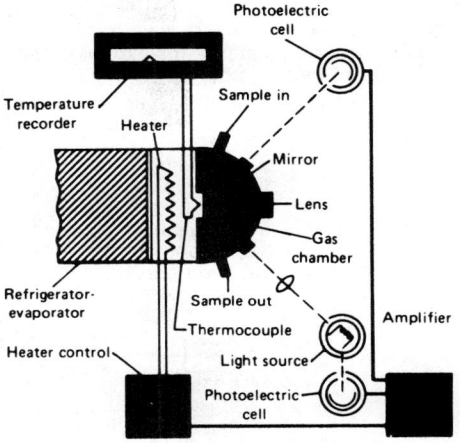

Fig. 5. Elements of chilled-mirror apparatus for measuring dew point

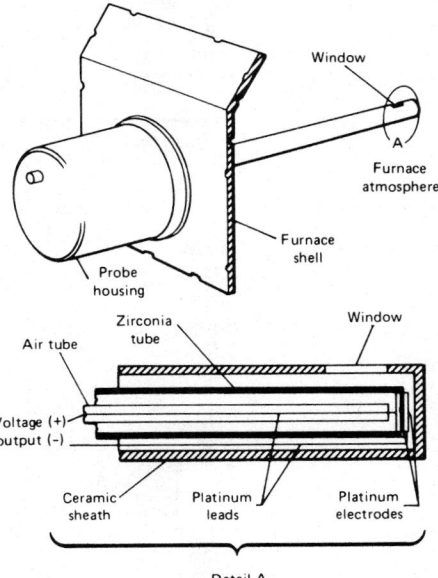

Fig. 6. Elements of a typical oxygen probe for controlling carburizing atmospheres

ences the relationship is carbon monoxide. As long as the carbon monoxide content is reasonably constant, carbon activity can be controlled by controlling oxygen content. This method of carbon-potential control has the advantage that it is less sensitive to changes in CO and/or H$_2$ content of the carburizing atmosphere.

An oxygen probe usually consists of two platinum electrodes separated by a solid electrolyte in the form of a gastight zirconia tube closed at one end (see Fig. 6). The probe, which usually is enclosed in a ceramic sheath, is inserted into the furnace. The furnace atmosphere enters the probe through a window in the sheath and contacts the outer electrode. The other electrode, inside the zirconia tube, is in contact with air, which serves as a reference gas of constant oxygen content. The difference between the partial pressure of oxygen in the furnace atmosphere and that in air induces an electromotive force (voltage), or emf, across the electrodes. The partial pressure of oxygen in the furnace atmosphere is determined by the voltage output (emf) of the sensor. Thus, carbon potential can be controlled by controlling the temperature in the furnace and the voltage output of the sensor.

Evaluation of Carbon Control in Processed Parts

WHEN FABRICATED STEEL PARTS are heated in a carbonaceous atmosphere that will either carburize or decarburize the surfaces of the parts, evaluation of the precise effect of the atmosphere on the parts usually is desirable. This is particularly important when the carbon content at the surface, and at significant depths below the surface, is to be controlled by adjusting the composition of the furnace atmosphere. Several methods of evaluating carbon control of processed parts are considered below.

HARDNESS TESTS

Hardness tests for evaluating carbon control should be used with caution. The type of test selected should be one in which the depth of metal affected is properly related to the depth to which carbon control is desired. When hardness tests are used, one of the following methods should be adopted only after all possible sources of error in the specific application have been thoroughly investigated:

- Surface hardness measurements taken under at least two conditions of loading (Rockwell C and superficial Rockwell 15N, for example)
- Superficial Rockwell 15N tests on steps ground below the surface to significant depths
- Microhardness measurements, either on steps or, preferably, on a cross section through the carbon-control zone.

In instances where hardness measurements do not give adequate information, one or more of the following techniques should also be used:

- Microscopic examination
- Analysis of consecutive cuts
- Analysis of shim stock
- Spectrographic analysis
- Electromagnetic testing.

MICROSCOPIC EXAMINATION

Microscopic examination is the only method of determining surface carbon variations that shows the effects of such variations on microstructure. Because of this, it will indicate which corrective action must be taken to alter surface carbon concentration. Microscopic examination is also useful in determining whether or not an improper surface carbon content is detrimental to the part.

ANALYSIS OF CONSECUTIVE CUTS

The consecutive-cuts method of analysis can be used for accurate evaluation of carbon control at any significant depth below the surfaces of parts. Because of the extremely accurate machining operations required to obtain reliable information, this type of evaluation usually is performed on cylindrical test bars treated with the work. After a test bar has undergone the heat treatment to be evaluated, consecutive cuts of 0.03 to 0.3 mm (0.001 to 0.020 in.) or more are turned from the surface of the test bar, and the turnings from each cut are analyzed for carbon concentration.

The test bar should be made of the same grade of steel as that of the workpieces. It should be accurately machined on centers to true cylindrical shape. The diameter should be such that the section of the test bar is representative of the critical section of the workpieces to be evaluated. The length of the test bar need only be sufficient to allow enough turnings for a carbon analysis and a check analysis.

ANALYSIS OF SHIM STOCK

Thin-gage shim stock can be used effectively for determination of the carbon potential of an atmosphere. The success of this method is due to rapid through carburizing of the thin test strip, which eliminates carbon diffusion as a variable in the atmosphere evaluation. The accuracy of this test is adequate for application to carbon control of commercial carburizing, carbon restoration and carbonitriding.

In practice, a strip of annealed 1010 steel, 32 by 75 by 0.1 to 0.15 mm ($1\frac{1}{4}$ by 3 by 0.004 to 0.006 in.), is weighed before and after 1 h of exposure to the furnace atmosphere. The gain in weight, as determined on an analytical balance, is used to calculate the carbon potential of the atmosphere as follows:

Carbon potential =

$$\left(\frac{\text{Gain in weight} \times 100}{\text{Final weight}} \right) +$$

% original carbon

Total test time varies from $\frac{1}{2}$ to $1\frac{1}{2}$ h.

Figure 7 illustrates the carbon-potential rod employed in the shim-stock test and procedures for exposure of shim stock to the furnace atmosphere.

SPECTROGRAPHIC ANALYSIS

Carbon content can be determined accurately by spectrographic analysis. This method makes use of a vacuum spectrometer, which permits measurement of spectral lines in the ultraviolet

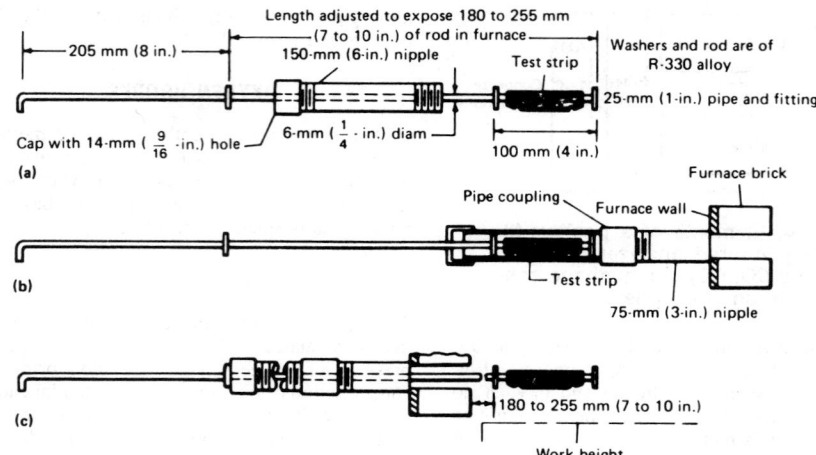

(a) Assembly of test strip for determination of carbon potential. (b) Assembly test strip on rod attached to furnace wall preparatory to insertion into furnace (same position used for cooling of test strip). (c) Test strip inserted into furnace above work in unobstructed stream of furnace atmosphere.

Fig. 7. Arrangements for exposing shim stock to furnace atmosphere for evaluation of carbon potential

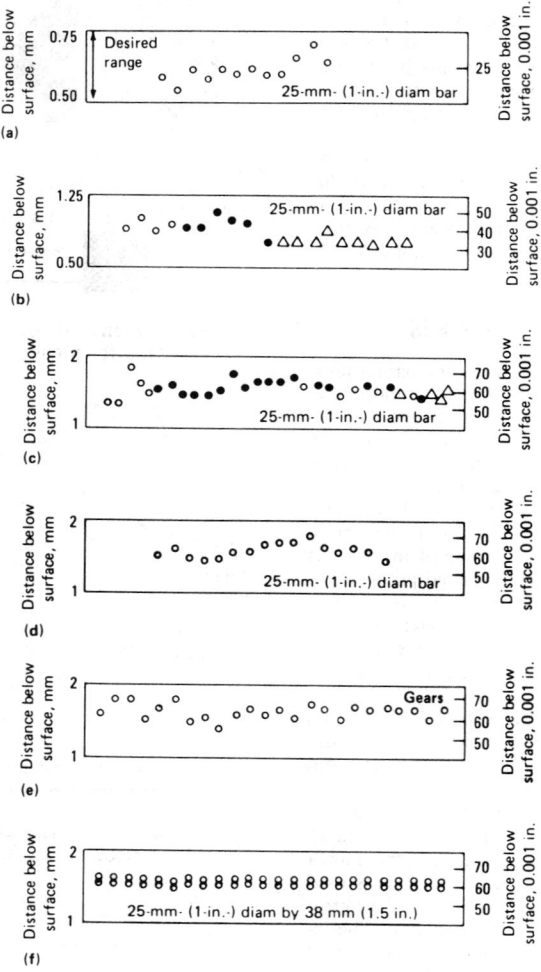

Data points represent individual heats, and are plotted in chronological order (left to right). Data were collected in studies in two different plants. Legend for parts (b) and (c): ○ Furnace A; ● Furnace B; △ Furnace C.

Fig. 8. Reproducibility of case depth in 8620 steel, using the criterion of depth to 0.40% carbon

Table 2. Carbon content of shim stock and workpiece surfaces as determined by spectrographic and combustion analyses

Shim stock and workpieces were heat treated, in the same load, in a 915-mm (36-in.) continuous belt furnace with an endothermic-base atmosphere (class 301; dew point, −9 to −1 °C or +15 to +30 °F).

Sample No.	Amount of carbon present, %		Workpiece surface (spectrographic analysis)
	Spectrographic analysis	Combustion analysis	
1	0.36	0.36	0.38
2	0.24	0.27	0.25
3	0.22	0.24	0.225
4	0.35	0.35	0.34
5	0.30	0.30	0.305

region where air would ordinarily absorb much of the emitted radiation.

Spectrographic analysis normally is performed on flat test specimens that can be taper ground, step ground or reground incrementally after each carbon determination. A very small amount of material is ground from the surface (to remove oxides). Successive cuts are made, and analyses are performed after each cut.

Whereas carbon determination by the combustion method provides an average carbon content for the amount of material removed by machining, spectrographic analysis determines the local carbon content of the specimen to a depth of 0.03 mm (0.001 in.) below the surface. Table 2 compares analyses performed on five samples of shim stock by the spectrographic and combustion methods.

ELECTROMAGNETIC TESTING

In one type of electromagnetic testing, known as magnetic-comparator testing, the part to be tested is placed in an induction coil. A reference part of known electromagnetic response is placed in a second coil. Both parts are simultaneously subjected to identical electromagnetic fields, and their responses to these fields are compared by an electronic balancing circuit. Any imbalance between responses is indicated by a meter.

Electromagnetic (eddy-current) testing can be used only as a comparison test; its accuracy and usefulness depend on proper development of standards and test procedures.

CASE-DEPTH VARIATION

In commercial practice, case depth is controlled within certain tolerances, which depend on the intended service of the part, the type and condition of the carburizing furnace, the cycles employed, and the limitations of control equipment. The end use of the carburized part is the principal determinant of how wide or narrow a range of case depths is acceptable.

The data shown in Fig. 8 provide a summary of the reproducibility of case depths within prescribed limits for test specimens and parts made of 8620 steel.

Localized Heat Treating

Induction Hardening and Tempering

ELECTROMAGNETIC INDUCTION is one method of generating heat within a part for hardening or tempering a steel or cast iron part. Any electrical conductor can be heated by electromagnetic induction. As alternating current from the converter flows through the inductor, or work coil, a highly concentrated, rapidly alternating magnetic field is established within the coil. The strength of this field depends primarily on the magnitude of the current flowing in the coil. The magnetic field thus established induces an electric potential in the part to be heated, and because the part represents a closed circuit, the induced voltage causes the flow of current. The resistance of the part to the flow of the induced current causes heating.

The pattern of heating obtained by induction is determined by the (*a*) shape of the induction coil producing the magnetic field, (*b*) number of turns in the coil, (*c*) operating frequency, (*d*) alternating-current power input, and (*e*) nature of the workpiece. Four examples of magnetic fields and induced currents produced by induction coils are shown in Fig. 1.

The rate of heating obtained with induction coils depends on the strength of the magnetic field to which the part is exposed. In the workpiece, this becomes a function of the induced currents and of the resistance to their flow.

The depth of current penetration depends on workpiece permeability, resistivity, and the alternating-current frequency. Since the first two factors vary comparatively little, the greatest variable is frequency. Depth of current penetration decreases as frequency increases. High-frequency current is generally used when shallow heating (thin case) is desired; intermediate and low frequencies are used in applications requiring deeper heating.

Most induction surface-hardening applications require comparatively high power densities and short heating cycles in order to restrict heating to the surface area. The principal metallurgical advantages that may be obtained by surface hardening with induction include increased wear resistance and improved fatigue strength.

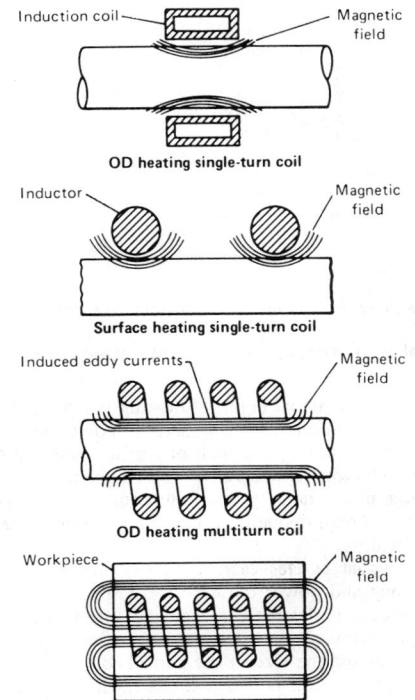

Fig. 1 Magnetic fields and induced currents produced by various induction coils

CHARACTERISTICS OF COMMERCIAL EQUIPMENT

Induction hardening is generally done at frequencies of 1000 Hz or higher.

Medium-frequency motor-generator units consist of a high-frequency generator driven by a motor. Induction motors, which may be mounted integrally with the generator or separately on a common base, are used with 1-, 3- and 10-kHz generators. The integrally mounted units can be either vertical or horizontal. These units generally are extremely reliable. The major maintenance requirement is infrequent bearing replacement.

Medium-frequency solid-state units are used to produce frequencies from 180 Hz to 50 kHz. While several different types of power circuits are used, all convert the 60-Hz main-line frequency into single-phase high frequency at relatively high efficiency. Both fixed-frequency and variable-frequency output units are available.

Solid-state units can be made to provide the same heating results as motor-generator units. Selection of frequency, power and duration of power should be made independent of whether a motor generator of solid-state induction heater is to be used.

Solid-state units have the advantage of a conversion efficiency greater than 90%, compared with 75 to 80% for a motor generator. Solid-state units also do not have the starting in-rush of a motor generator.

Radio frequency (RF) arc vacuum tube units consist of a power supply section and an oscillator section. The power section provides the high voltage for the oscillator tube after rectification to a pulsating direct current, usually by solid-state diodes. The oscillator tube and a tank circuit consisting of a matched inductor coil and capacitor comprise the oscillator section. These units operate at frequencies ranging from 200 kHz to 5 MHz. Output transformers usually are desir-

Table 1. Selection of power source and frequency for various applications of induction hardening

Depth of hardening mm	in.	Section size mm	in.	Power lines, 50 or 60 Hz	Frequency converter, 180 Hz	Solid state or motor generator 1000 Hz	3000 Hz	10 000 Hz	Vacuum tube, over 200 kHz
Surface hardening									
0.38-1.27	0.015-0.050	6.35-25.4	1/4-1	...	...	...	...	...	Good
1.29-2.54	0.051-0.100	11.11-15.88	7/16-5/8	...	...	...	...	Fair	Good
		15.88-25.4	5/8-1	...	...	...	...	Good	Good
		25.4-50.8	1-2	...	...	...	...	Good	Fair
		Over 50.8	Over 2	...	...	Fair	Good	Good	Poor
2.56-5.08	0.101-0.200	19.05-50.8	3/4-2	...	...	...	Good	Good	Poor
		50.8-101.6	2-4	...	...	Good	Good	Fair	...
		Over 101.6	Over 4	...	...	Good	Fair	Poor	...
Through hardening									
		1.59-6.35	1/16-1/4	...	...	...	...	...	Good
		6.35-12.7	1/4-1/2	...	...	...	...	Fair	Good
		12.7-25.4	1/2-1	...	...	...	Fair	Good	Fair
		25.4-50.8	1-2	...	...	Fair	Good	Fair	...
		50.8-76.2	2-3	...	...	Good	Good	Poor	...
		76.2-152.4	3-6	Fair	Good	Good	Poor	Poor	...
		Over 152.4	Over 6	Fair	Good	Poor	Poor	Poor	...

Good indicates frequency that will most efficiently heat the material to austenitizing temperature for the specified depth. For through hardening, **Good** indicates the amount of material approximately equal to the mg/J shown in Fig. 3. **Fair** indicates a frequency that is lower than optimum but high enough to heat the material to austenitizing temperature for the specified depth. With this frequency, the current penetration relative to the section size causes current cancellation and lowered efficiency. For through hardening, **Fair** indicates an amount of material less than the pounds per kilowatt-hour shown in Fig. 3. **Fair** may also indicate a frequency higher than optimum that can overheat the surface at high-energy inputs. Converters cost more per kilowatt-hour than the converters of optimum frequency. With some equipment, the efficiency may be lower. **Poor** indicates a frequency that will overheat the surface unless low-energy input is used. Efficiency and production are low and capital cost of converters per kilowatt-hour is high.

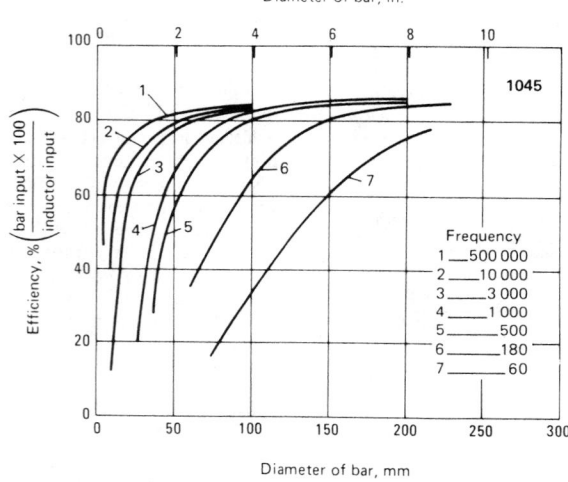

Bars are of various diameters heated to 1095 °C (2000 °F) with inside diameters of inductors 28.6 mm (1.125 in.) larger than outside diameters of bars.

Fig. 2. Efficiency of energy transfer at several frequencies to type 1045 steel bars

able where it is necessary to couple the power into small, concise areas. Options such as SCR power control are available and usually are desirable.

SELECTION OF FREQUENCY, POWER AND DURATION OF HEATING

The distribution of induced current in a part is maximum on the surface and decreases rapidly within the part; the effective penetration of current increases with a decrease in the frequency. The distribution of induced current is influenced also by the magnetic and electrical characteristics of the part being heated; and, since these properties change with temperature, the current distribution will change as the work is heated.

Because the heat rapidly progresses to the interior by conduction as soon as the surface is heated, the actual depth of heating is determined by the duration of heating and the power density (kilowatts per square inch of surface exposed to the inductor), as well as by the frequency. Maximum power density, minimum duration of heating, and high frequency produce a minimum depth of heating.

Selection of Frequency. In analyzing the frequency and power required for a specific application, it is desirable to consider the frequency first. Primary considerations are the depth of heating and the size of the part. Table 1 lists the frequencies and power sources most commonly used in induction hardening. As shown in this tabulation, the lower frequencies are more suitable as the size of the part and the case depth increase. Use of the wrong frequency will result in a decrease in electrical efficiency; sometimes in failure to maintain a minimum case depth where

shallow cases are required; or in failure to heat uniformly throughout the piece where through hardening is required. Figure 2 illustrates the decrease in transfer of energy or heating efficiency that could result if the incorrect frequency is selected.

Selection of Power. The size of the converter or the power required should be determined on the basis of power density, section size, heating method and production requirements.

In surface hardening, the area heated at one time, multiplied by power density, indicates the total power input (kilowatts). This area is obtained by multiplying the perimeter of the part by the length of the inductor. To calculate the power required for through heating, divide the desired production load by the values for pounds per kilowatt-hour given in Fig. 3.

Selection of Duration of Heating. When the frequency and power density have been selected, the duration of the heating cycle becomes a fixed value for a specific set of conditions. To calculate duration of heating for surface hardening by the static method, divide the value for kilowatt seconds per square inch by power density (kilowatts per square inch). The value of kilowatt seconds per square inch is affected by case-depth requirements, type of steel, and prior structure, and may be derived by experiment or be based on previous experience. To calculate heating time for surface hardening by the scanning method, divide kilowatt seconds per square inch by power density and inductor length.

SELECTION OF COIL DESIGN

Coil design is influenced by a number of factors, including the dimensions and configuration of the part to be heated, the heat pattern desired, whether the part is heated throughout its length at the same time or progressively, the number of parts to be heated, and the frequency and power of the induction heater.

Basic Designs. Five basic designs of work coils for use with high-frequency (over 200 kHz) units and the heat patterns developed by each are shown in Fig. 4(a) through (e). These basic shapes are: (a) a simple solenoid for external heating; (b) a coil to be used internally for heating bores; (c) a pie-plate type of coil designed to provide high current densities in a narrow band for scanning applications; (d) a single-turn coil for scanning a rotating surface, provided with a contoured half-turn that will aid in heating the fillet; and (e) a pancake coil for spot heating. Solenoid coils for external heating are most efficient and should be used whenever possible.

The same designs are used for lower frequencies, although the higher powers may require milled copper coil construction. This type of coil construction involves milling or drilling out of holes, followed by brazing in of inserts, to form the cooling passages.

Ferrite concentrators can be used on coils to increase coil efficiency. Laminated iron concentrators can be used at 1 to 10 kHz to increase coil efficiency.

It usually is important to keep coil lead lengths as short as possible. If the lead lengths provide excessive power drops, they should be made wider or brought closer together (or both). The number of turns in a coil depends on the requirements of the area to be heated and on the ability to match the impedance of the power supply.

Commercial copper tubing may be used for

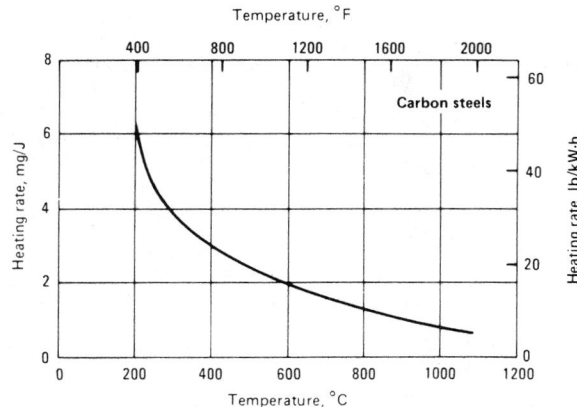

For converted frequencies, the total power transmitted by the inductor to the work is less than the power input to the machine, because of converter losses.

Fig. 3. Heating rate for through heating of carbon steels by induction

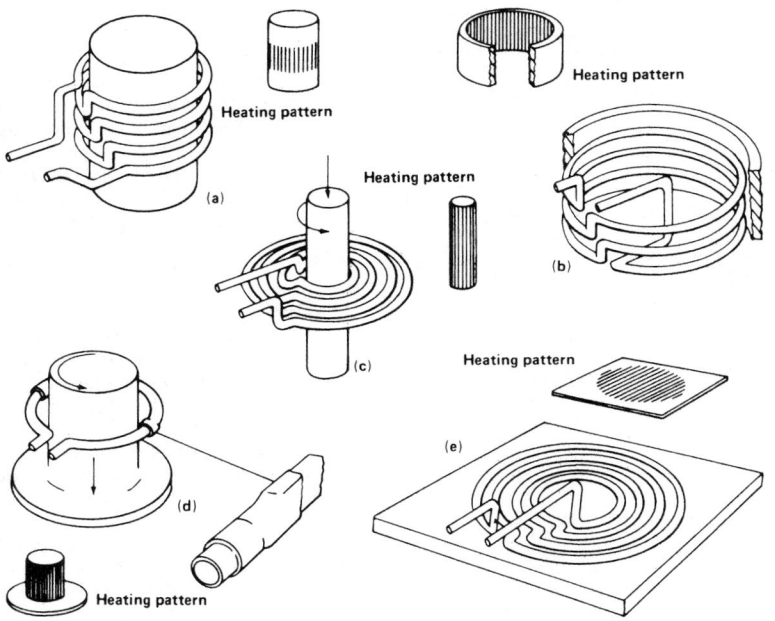

Fig. 4. Typical work coils for high-frequency units

coils. The tubing must be large enough to permit an adequate flow of water for cooling.

Coil Coolants. Water is commonly used for cooling inductors, although in some applications oil, modified water, or a plastic quench may be employed to serve the dual purpose of cooling the inductor and quenching the workpiece in a continuous heating and quenching operation. Generally, the water should have a hardness of less than 10 grains/gal. If the water-cooling passages are small relative to the current load carried by the inductor, it may be necessary to use distilled or deionized water to avoid a deposit buildup that could eventually stop circulation. Preferably, the water should be filtered to remove foreign particles that might clog small passageways, especially when intricately designed inductors are being used. The water should have an inlet temperature below 35 °C (95 °F), and flow should be sufficient to prevent the outlet temperature from rising above 66 °C (150 °F).

QUENCHING

The type of quench used will depend on metallurgical considerations. A great many induction hardening applications employ water as the quenching medium. Other media, such as conventional quenching oil, water modified by organic polymer, and compressed air, are occasionally used. Water is easiest to handle, simple to install and maintain, and generally less hazardous than other media. Oil quenching produces the least distortion and provides the smallest tendency toward cracking. The modified-water compounds are compounds with organic polymers that are soluble in water. The temperature and concentration determine the quenching rate. Compressed air is used in shallow-case applications where the air and the massive heat sink of the workpiece are used to produce the required cooling rate.

Basic Systems for Quenching. Eleven basic ar-

rangements for quenching induction hardened parts are shown schematically in Fig. 5(a) through (k). In correlation with the lettering there, these arrangements are briefly described as follows:

(a) Heat in coil; manually lift part out of coil; submerge part in tank of agitated quench medium. Used where limited production does not warrant the cost of an automated quench.

(b) Heat and quench in one position; quench by means of integral quench chamber in inductor. Called single-shot method.

(c) Heat in coil with part stationary; quench ring moves in place. Single-shot adaptation of scanning method.

(d) Part is hydraulically lowered into quench tank after single-shot heating. Quench medium is agitated by submerged spray ring or propeller.

(e) Vertical or horizontal scanning with integral spray quench. Single-turn inductor. Used for shallow hardening.

(f) Vertical or horizontal scanning with multiturn coil and separate multirow quench ring. Used for deep-case or through hardening.

(g) Coil scans and heats workpiece; self-quench or compressed air quench. Used in special applications with high-hardenability steels.

(h) Horizontal cam-fed parts are pushed through coil, then dropped onto submerged quench conveyor.

(i) Vertical scanning with single-turn inductor in combination with integral dual quench: one quench ring for scan hardening; the second for stationary quenching when the scanning travel stops. Used for parts having a diameter or a flange section too large to travel through the inductor, wherein it is desired to harden up to the shoulder or flange.

(j) Vertical scanning with single-turn inductor with integral spray quench and submerged quench in tank.

(k) Split inductor and integral split quench ring. Used for hardening crankshaft bearing surfaces.

INDUCTION TEMPERING

Extensive production experience with thousands of tons of steel per month has demonstrated the commercial success of induction tempering for many applications. Induction tempering has proved particularly adaptable to automation in production lines.

Applications. At present, two principal areas of application exist for induction tempering:

- Selective tempering
- Progressive tempering of bar stock previously hardened by scanning.

Flame Hardening

FLAME HARDENING employs direct impingement of a high-temperature flame or of high-velocity combustion-product gases; the part is then cooled at a rate that will produce the desired levels of hardness and other properties. The high-temperature flame is obtained by combustion of a mixture of fuel gas with oxygen or air; flame heads are used for burning the mixture. Depths of hardening from about 0.8 to 6.4 mm (^{1}/$_{32}$ to ^{1}/$_4$ in.) or more can be obtained, depending on the fuels used, the design of the flame head, the duration of heating, the hardenability of the work material, and the quenching medium and method of quenching used. The process can be used for through hardening of work 75 mm (3 in.) or less in cross section, depending on the hardenability of the steel.

SCOPE AND APPLICATION

Flame hardening is applied to a wide diversity of workpieces and ferrous materials (when required mechanical properties can be provided by selective or localized hardening) for one or more of the following reasons:

- Because parts are so large as to make conventional furnace heating and quenching impracticable or uneconomical. Typical examples include large gears, machineways, and large dies and rolls.
- Because only a small segment, section or area of a part requires heat treatment, or because heat treating all over would be detrimental to the function of the part.
- Because dimensional accuracy of a part is impracticable or difficult to attain or control by furnace heating and quenching.
- Because the use of flame hardening permits a part to be made from a less costly material.

METHODS OF FLAME HARDENING

The versatility of flame-hardening equipment and the wide range of heating conditions obtainable with gas burners often permit flame hardening to be done by a variety of methods, of which the principal ones are:

- Spot, or stationary
- Progressive
- Spinning
- Combination progressive-spinning.

Selection of the appropriate method depends on the shape, size and composition of the workpiece; the area to be hardened; the depth of case required; and the number of pieces to be hardened.

The spot (stationary) method, illustrated in Fig. 6(a), consists of locally heating selected areas with a suitable flame head and subsequently quenching. The heating head may be of either single-orifice or multiple-orifice design, depending on the extent of the area to be hardened. The heat input must be balanced to obtain a uniform temperature over the entire selected area. After being heated, the parts usually are immersion quenched; however, in some mechanized operations, a spray quench may be used.

Basically, the spot method requires no elaborate equipment (except, perhaps, fixtures and timing devices to ensure uniform processing of each piece). However, the operation may be automated by indexing the heated parts into either a spray quench or a suitable quench bath.

The progressive method, illustrated in Fig. 6(b), is used to harden large areas that are beyond the scope of the spot method. The size and shape of the workpiece, as well as the volume of oxygen and fuel gas required to heat the specified area, are factors in selection of this method. In progressive hardening, the flame head is usually of the multiple-orifice type, and quenching facilities may be either integrated with the flame head or separate from it. The flame head progressively heats a narrow band that is subsequently quenched as the head and quench traverse the workpiece.

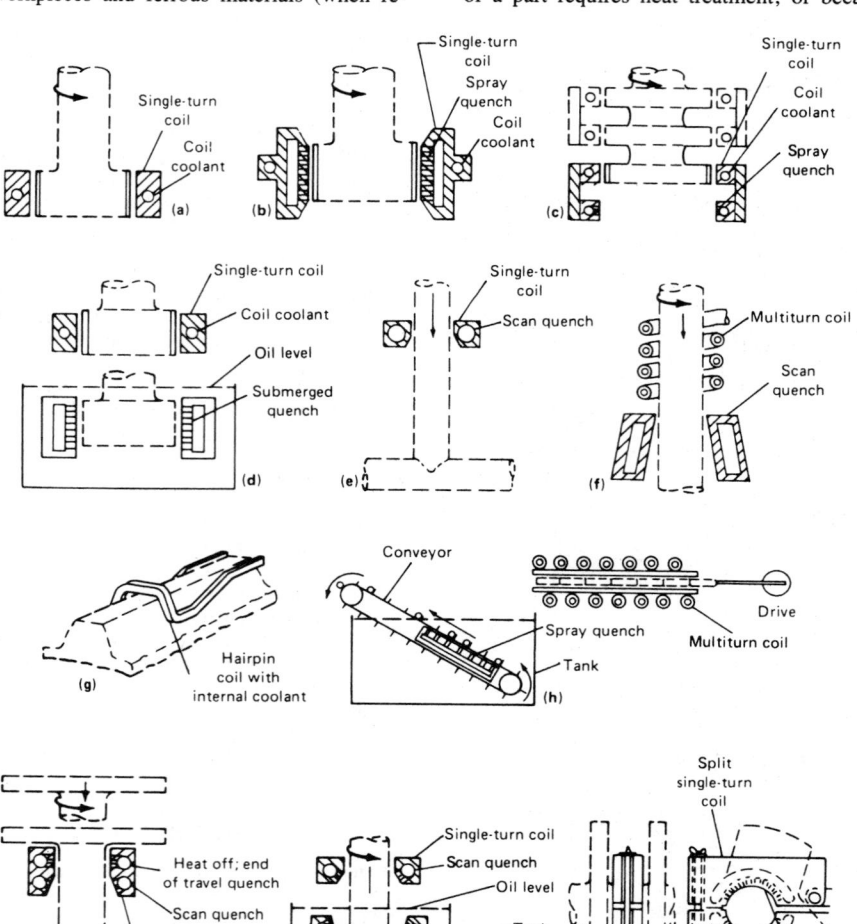

Fig. 5. Basic arrangements for quenching induction heated parts

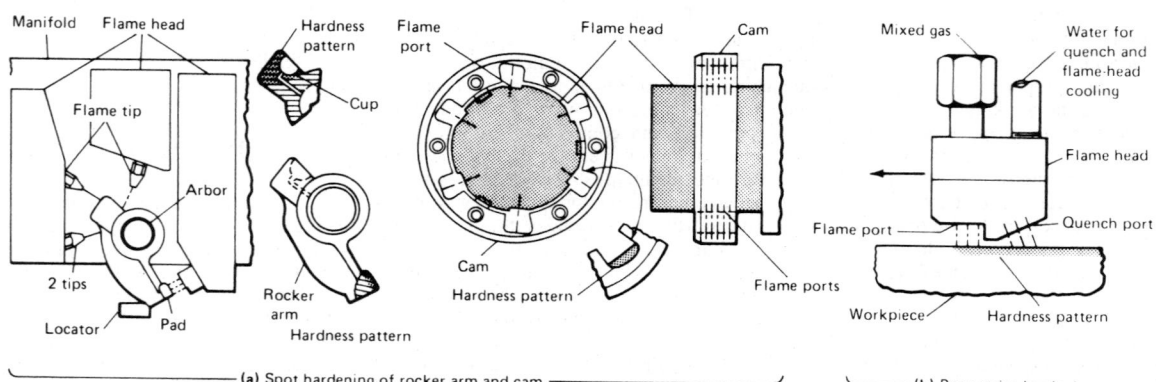

(a) Spot (stationary) method of flame hardening a rocker arm and the internal lobes of a cam; quench not shown. (b) Progressive method.

Fig. 6. Spot (stationary) and progressive methods of flame hardening

The equipment needed for flame hardening by the progressive method consists of one or more flame heads and a quenching means mounted on a movable carriage that runs on a track at a regulated speed (flame-cutting machines are adaptable to this type of flame hardening).

The spinning method (Fig. 7) is applied to round or semiround parts such as wheels, cams or gears. In its simplest form, the method employs a mechanism for rotating or spinning the workpiece, in either a horizontal or a vertical plane, while the surface is being heated by the flame head. One or more water-cooled heating heads equal in width to the surface to be heated are employed. The speed of rotation is relatively unimportant, provided that uniform heating is obtained. After the surface has been heated to the desired temperature, the flame is extinguished or withdrawn and the work is quenched by immersion or spray, or a combination of both.

The spinning method is particularly adaptable to extensive mechanization and automation; this makes it possible, for example, for all the cams on a camshaft to be hardened at the same time.

The combination progressive-spinning method (Fig. 8), as the name implies, combines the progressive and spinning methods for hardening long parts such as shafts and rolls. The workpiece is rotated as in the spinning method; but in addition, the heating heads traverse the roll or shaft from one end to the other. Only a narrow circumferential band is heated progressively as the flame head moves from one end of the work to the other. The quench follows immediately behind the heating head, either as an integral part of the head or as a separate quench ring.

FUEL GASES

Several different fuel gases are used in flame hardening. In selecting a fuel gas for a given application, the required rate of heating and the cost of the gas must be considered along with the initial cost of equipment and maintenance.

Flame hardening does not alter the composition of the base metal if done properly. Carburizing, neutral and oxidizing flames can be used. Oxidizing flames have high oxygen ratios and can be detrimental because they produce extremely hot temperatures that can cause decarburization and overheating. A carburizing flame can prevent some decarburization but also can introduce unwanted carbon into the surface. For best results, neutral or slightly carburizing flames should be used.

A comparison of the rates of heating of fuel gases can be made when certain fundamental properties of usable mixtures with oxygen are known. A parameter that correlates well with actual heating speed is "combustion intensity," or "specific flame output." This is the product of the normal velocity of burning multiplied by the net heating value of the mixture of oxygen and

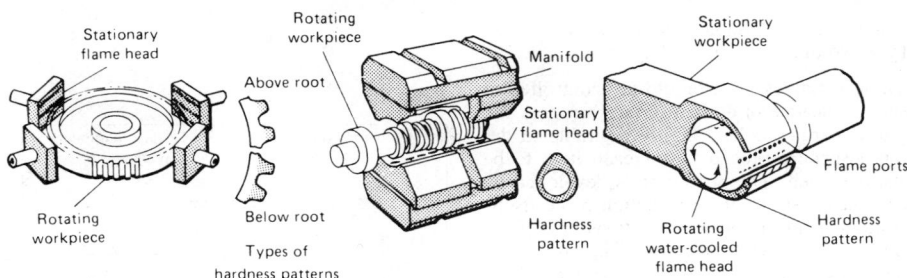

In methods illustrated at left and at center, the part rotates; in method at right, the flame head rotates. Quench not shown.

Fig. 7. Spinning methods of flame hardening

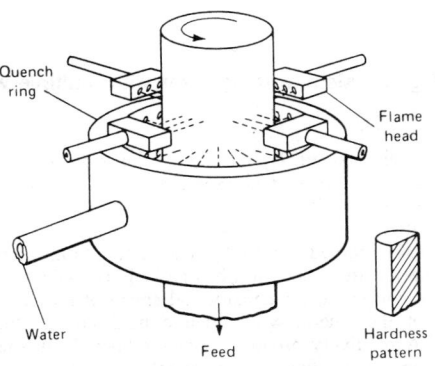

Fig. 8. Combination progressive-spinning method of flame hardening

fuel gas. Thus, a knowledge of these two properties often permits selection of the most suitable fuel gas for a specific hardening speed and depth of case. The fuels of greatest commercial interest are ranked by combustion intensity (at metallurgically suitable ratios of mixture with oxygen) in the following order: acetylene, propane, methane. Values of normal burning velocity and the heating values of metallurgically suitable mixtures are listed in Table 2.

The time required for heat penetration is another good criterion for judging the heating qualities of a fuel provided that all other variables remain constant. Figure 9 shows comparative heating times for stabilized MAPP (methylacetylene propadiene), acetylene and propane using an efficient coupling distance for each fuel. These curves show that a greater depth of hardness can be obtained with MAPP in a shorter length of time.

Depth of Heating. Shallow hardness patterns (less than 3.2 mm or 0.125 in. deep) can be attained only with oxy-gas fuels. The high-temperature flames obtained with oxy-gas fuels provide the fast heat transfer necessary for effective localization of the heat pattern. Deeper hardness patterns permit the use of either oxy-gas fuels or air-

gas fuels. Oxy-gas fuels will localize the heat, but care is required in their application to avoid overheating the surface during development of the deeper-seated heat. Air-gas fuels, with their slower rates of heat transfer (lower flame temperatures), minimize or eliminate surface overheating but generally extend the heat pattern beyond the desired hardness pattern.

OPERATING PROCEDURES AND CONTROL

The success of many flame-hardening applications depends largely on the skill of the operator. This is especially true when the volume of work is so small or so varied that the cost of automatic-control equipment is not justified.

The principal operating variables are:

- Distance from inner cone of oxy-fuel gas flames, or from air-fuel gas burner, to work surface
- Flame velocities and oxygen-to-fuel ratios

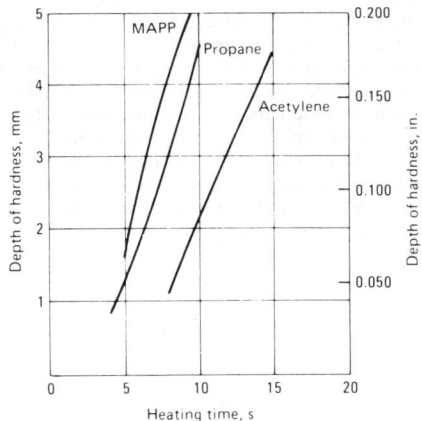

Flame velocity, 170 m/s (550 ft/s); port size, No. 69 drill (0.74 mm, or 0.0292 in.); coupling distance, 9.5 mm (³⁄₈ in.); material, 1036 steel. Oxygen-to-fuel ratios: MAPP, 5.0; acetylene, 1.33; propane, 4.5.

Fig. 9. Comparison of heating times for MAPP, acetylene and propane

Table 2. Fuel gases used for flame hardening

Gas	Heating value MJ/m³	Heating value Btu/ft³	Flame temperature With oxygen °C	With oxygen °F	With air °C	With air °F	Usual ratio of oxygen to fuel gas	Heating value of oxy-fuel gas mixture MJ/m³	Heating value of oxy-fuel gas mixture Btu/ft³	Normal velocity of burning mm/s	Normal velocity of burning in./s	Combustion intensity(a) mm/s × MJ/m³	Combustion intensity(a) in./s × Btu/ft³	Usual ratio of air to fuel gas
Acetylene	53.4	1433	3105	5620	2325	4215	1.0	26.7	716	535	21	14 284	15 036	...
City gas	11.2-33.5	300-900	2540	4600	1985	3605	(b)	(b)	(b)	(b)	(b)	(b)	(b)	(b)
Natural gas (methane)	37.3	1000	2705	4900	1875	3405	1.75	13.6	364	280	11	3 808	4 004	9.0
Propane	93.9	2520	2635	4775	1925	3495	4.0	18.8	504	305	12	5 734	6 048	25.0

(a) Product of normal velocity of burning multiplied by heating value of oxy-fuel gas mixture. (b) Varies with heating value and composition.

• Rate of travel of flame head or work
• Type, volume and angle of quench.

These variables must be closely controlled to ensure duplication of desired surface hardness and depth of hardness. It is highly desirable to develop a specific procedure for each item to be flame hardened. The procedure is developed by preliminary tests on the production piece itself, if warranted, or on mock-up sections of approximately the same cross section as the production piece. After the desired contour and depth of hardened zone have been developed, the procedure is applied to production pieces and, when established, is made a part of the heat treating specification.

Laser Surface Transformation Hardening

A LASER can generate very intense energy fluxes at the workpiece surface, and the resulting temperature profiles in the workpiece usually can be made steep enough to negate the need for external quench media. The laser beam is a beam of light, which is essentially independent of the workpiece, is easily controlled, requires no vacuum, and generates no combustion products. It is ideally suited, therefore, for this purpose.

FUNDAMENTALS OF LASER SURFACE HARDENING

When a laser beam impinges on a surface, part of its energy is absorbed as heat at the surface. If the power density of the laser beam (usually given in watts per square centimetre) is sufficiently high, heat will be generated at the surface faster than heat conduction to the interior can remove it, and the temperature in the surface layer will increase rapidly. In a very short time, a thin surface layer will have reached austenitizing temperatures, whereas the interior of the workpiece is still cool. Even with a relatively moderate power density of 500 W/cm^2 (3300 W/in.^{2}), temperature gradients of 500 °C/mm (25 °F/mil)

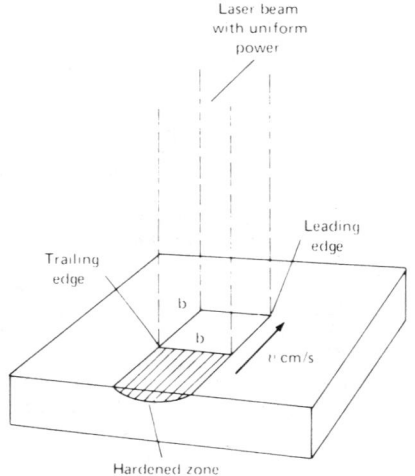

Fig. 10. Square laser beam with uniform power density on a flat plate

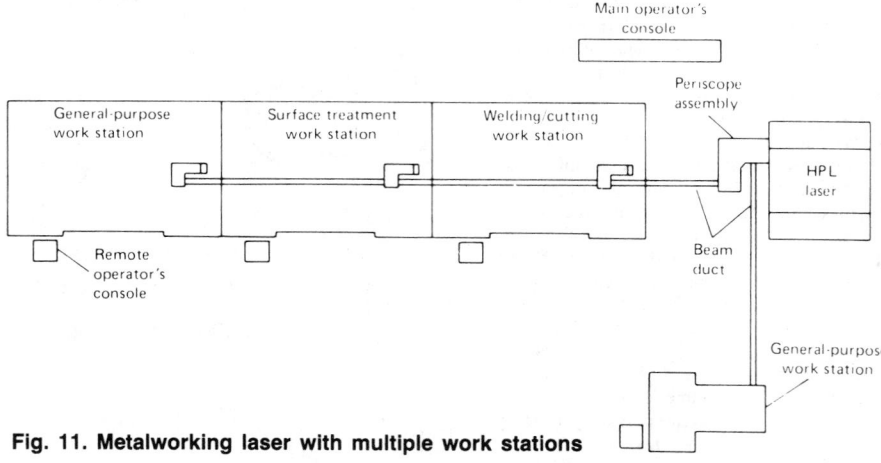

Fig. 11. Metalworking laser with multiple work stations

can be obtained. By moving the laser beam over the workpiece surface (see Fig. 10), a point on the surface within the path of the beam is rapidly heated as the beam passes. This area is subsequently cooled rapidly by heat conduction to the interior after the beam has passed. By selecting the correct power density and speed of the laser spot, the material will harden to the desired depth.

A relatively broad area beam, usually in the shape of a square or a rectangle, is used in the laser hardening process. The power density of a focused laser beam used for hardening is much lower than the power density of the small, intense focused spots used for welding and cutting. The power density is typically in the 1000 to 2000 W/cm^2 (6400 to 13 000 W/in.^{2}) range, occasionally as high as 5000 or as low as 500 W/cm^2 (32 000 or 3200 W/in.^{2}).

The resulting depth of case will depend on the hardening response of the material, but it will rarely be more than 2.5 mm (0.1 in.). For steel with low hardenability, such as plain carbon steel, the depth of case obtainable is much smaller, varying from perhaps 0.25 mm (0.01 in.) in mild steels to 1.3 mm (0.05 in.) in a medium-carbon steel.

The major advantages of laser surface hardening include: close control of the power input with modern metalworking lasers; the high power density provided by the laser, which in turn minimizes the total energy input and, thereby, dimensional distortion; and the ability of the laser to reach normally inaccessible areas on the workpiece surface. Because no vacuum or protective atmosphere enclosure is needed, and because the distance from the workpiece to the last optical element of the laser system can be quite long, it is possible to process very large or irregular-shaped workpieces.

On the negative side, the depth of case obtainable is limited to about 2.5 mm (0.1 in.) and is usually less than half of this, and the capital cost of the equipment may be high. Also, the surface to be heated must be nonreflective, and thus it is usually necessary to paint black the surfaces to be heated or to black oxide the parts. Therefore, careful analysis of a potential application for laser hardening is needed to ascertain the cost-effectiveness of the process.

METALWORKING LASERS

Several models of metalworking lasers of both domestic and foreign manufacture are commer-

cially available. The majority of these are of either the neodymium, YAG solid-state type or the carbon dioxide gas type. These lasers may have pulsed or continuous output power. Both types, whether pulsed or continuous wave, can be used for transformation surface hardening.

The primary output beam from the laser rarely is used in metalworking applications. Instead, the output beam is directed and shaped by optical systems to generate a laser spot of the desired size and shape on the workpiece surface. Such an arrangement allows substantial flexibility in the use of a laser system. Because the coherent radiation from a laser has low loss of power with distance, the laser itself can be situated at a considerable distance from the work area. Furthermore, different types of metalworking applications can be performed with the same laser by changing the optical system.

Figure 11 shows a typical laser system arrangement. In this instance, each work station has its own control console, allowing manual or automatic control of beam power, duration of power delivery to the workpiece, rate of increase of power at the start of the run, rate of decrease at the end (ramp-up and ramp-down) and manipulation of the workpiece fixture protective gas flow. The laser power is automatically maintained at the desired level by a feedback control device in the laser. Positive feedback from the temperature in the beam/workpiece interaction zone also can be used. The entire operation can be controlled by microprocessors. By directing the laser beam to the individual work stations in sequence and by utilizing the time when the power is used at another station for workpiece manipulation, maximum usage of the laser can be achieved.

Electron-Beam Heat Treating

ELECTRON-BEAM HEAT TREATING is a selective hardening process in which the surface of a hardenable ferrous alloy is heated rapidly above the transformation temperature of the alloy by direct bombardment or impingement of an accelerated stream of electrons. At the end of a heating cycle of 0.5 to 2.5 s, the flow of electrons is stopped abruptly to allow the part or workpiece being processed to self-quench and to form a martensitic structure with a compressive stress on the surface of the hardened area. The

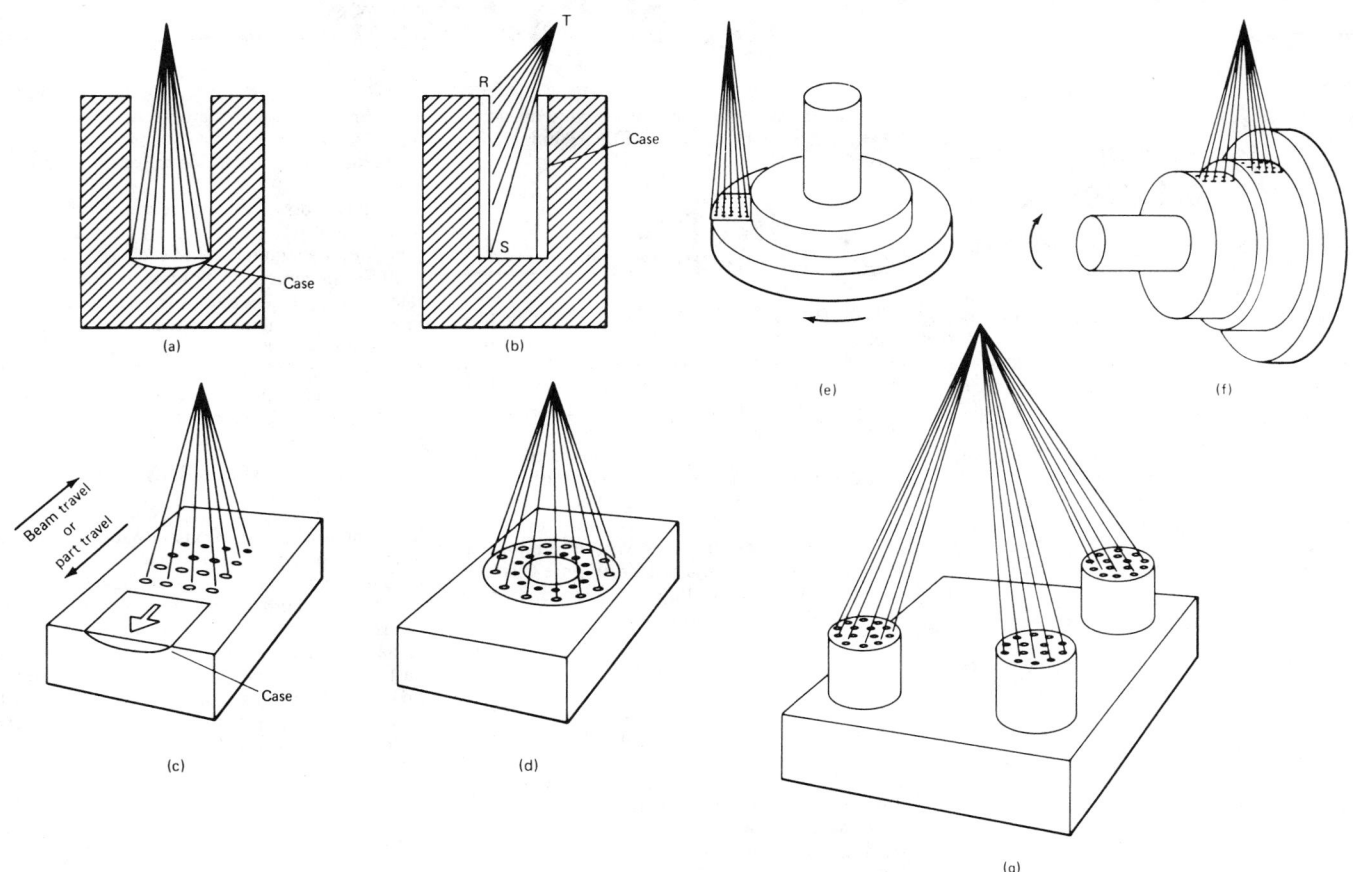

(a) Display static pattern within cavity in workpiece. (b) Maintain angle of workpiece rotation, RST, at 25° minimum. (c) Display static pattern and move the pattern or the workpiece to heat treat large areas. (d) Display static pattern; this annular pattern has well-defined inside and outside diameters. (e) Display static pattern and rotate workpiece. (f) Display more than one pattern and rotate workpiece. (g) Display multiple patterns on one workpiece or on a small group of workpieces for simultaneous hardening; patterns may be similar or dissimilar in geometric shape.

Fig. 12. Workpiece configurations and heating patterns for electron-beam heat treating

electron-beam hardening process normally is applied to finish-machined or ground surfaces. Because the buildup of energy is rapid and well controlled, postheat treatment operations such as grinding or straightening usually are not needed.

APPLICATION CRITERIA

A part or workpiece to be heat treated is considered a suitable candidate for electron-beam heat treating if it meets the following criteria:

* The material must contain adequate carbon to produce satisfactory case hardness.
* The stream of electrons must have line-of-sight access to the area requiring heat treatment and a beam-impingement angle of at least 25°. Guidelines for acceptable part configurations are shown in Fig. 12.
* The component being heat treated may be processed in a vacuum envelope or chamber, or at pressures up to 1 atm in air or inert gas. Vacuum chambers in high-production heat treating systems typically have interior volumes of 0.02 to 0.11 m³ (0.8 to 4.0 ft³).
* The surface to be heat treated should be machined or ground to final dimensions. If grinding is required after heat treating, removal of 0.05 to 0.25 mm (0.002 to 0.010 in.) of stock normally is sufficient.
* To prevent magnetic interaction or unintentional deflection of the electron beam, the component being heat treated must be demagne-

tized prior to hardening. Demagnetization normally is required if the part has been fixtured with magnetic clamps or chucks in operations prior to heat treating.

* The mass of the part must be sufficient to self-quench the heat treated area. The ability of a part to self-quench is determined by its composition as well as its configuration. Steels of high hardenability (such as AISI 4150) can be through hardened in many instances. For plain carbon steels, however, five to eight units of mass beneath the heated area are required for each unit of mass of hardened case.

ELECTRON-BEAM EQUIPMENT

In electron-beam heat treating, a highly concentrated beam of high-velocity electrons is used to heat selective surface areas. These electrons are accelerated and collimated into a dense, extremely energetic beam by the accelerating potential between the cathode and the anode. The high-energy beam thus formed passes through a small-diameter hole in the anode. Because of the mutual repulsion among neighboring electrons, the beam requires further collimation below the anode. This additional collimation is controlled with a focus coil that allows variation of the distance from the gun to the workpiece. A deflection coil deflects the reconverging beam to a designated location on the workpiece.

A high vacuum is needed in the region where

the electrons are emitted and accelerated, both to protect the emitter from oxidation and to prevent interference with the electrons while they are still at low velocity. Therefore, the electron-gun housing is pumped and maintained at a vacuum of 10^{-5} torr. The workpieces are contained in an enclosure under a vacuum of approximately 5×10^{-2} torr. An intermediate vacuum level provides short evacuation times and higher production rates. Treating at one atmosphere does not require any evacuation time.

In electron-beam heat treating, the energy exchange is simply a matter of the electrons in the beam transferring their kinetic energy to the atomic structure of the target material in the form of heat. The electron beam, when sharply focused for welding, is capable of impingement power densities on the order of 10 MW/cm² (65 MW/in.²). Because this powerful concentration of energy is easily controllable in power magnitude, power density and beam position, it is well suited for surface hardening as well. These power densities are much too high for nondestructive heat treating, however. Destructive heat treating in this context refers to controlled remelting of ferrous and nonferrous materials.

An energy concentration of 3.1 kW/cm² (20 kW/in.²) is more suitable for selective heat treating. To reduce the beam energy to this level, a single electron beam is programmed through a group of discrete beam positions referred to as a raster pattern.

Heat Treating of Cast Irons

Introduction to Heat Treating of Cast Irons

CAST IRONS may be compared with steels in their reactions to hardening. However, because cast irons (except white iron) contain graphite and substantially higher percentages of silicon, they require higher austenitizing temperatures. The graphitizing effect of silicon is so powerful that an unalloyed gray iron may become completely graphitized below the A_1 temperature during heating for austenitizing. Thus, some high-silicon irons require longer intervals at the austenitizing temperature for reabsorption of the desired carbon content in austenite. This interval is extended because silicon retards the absorption of carbon in austenite. The lower-silicon irons respond best to heat treatment.

HARDNESS MEASUREMENTS

Conventional hardness measurements on cast irons always indicate lower values than the true hardness of the metal matrix. This discrepancy, which is more pronounced in gray iron than in ductile and malleable irons, occurs because conventional hardness readings are composite values that reflect the hardnesses of both the matrix material and soft graphite.

Figure 1(a) shows the relation between observed HRC readings and those converted from microhardness values for five gray irons of different carbon equivalents. Hardness measurements were taken at two laboratories after quenching and after tempering of each iron. The data in Fig. 1(a) show why the observed values obtained by conventional hardness testing may be misleading, and help to explain the good wear resistance of gray irons with apparently low hardness. Note that there is a correlation with carbon equivalent for all five irons tested and that the discrepancy between observed and converted hardness values diminishes at the lower hardness level.

Another comparison between observed and converted HRC values for gray and ductile irons

is shown in Fig. 1(b). These irons were quenched in water from 900 °C (1650 °F) and tempered at 425 °C (800 °F) for 2 h.

Heat Treating of Gray Irons

THE HEAT TREATMENT most frequently applied to gray iron, with the possible exception of stress relieving, is annealing. Annealing of gray iron consists of heating the iron to a temperature high enough to soften it, and/or to minimize or eliminate massive eutectic carbides, thus improving its machinability. This heat treatment reduces mechanical properties substantially, however. It will reduce the grade level approximately to the next lower grade; for example, the properties of a class 40 gray iron will be diminished to those of a class 30 gray iron. Figure 2 shows the effect of full annealing on a tensile strength of class 30 gray iron arbitration bars.

ANNEALING

Gray iron commonly is subjected to one of three annealing treatments, each of which involves heating to a different range of temperature. These treatments are ferritizing annealing, medium (or "full") annealing and graphitizing annealing.

Ferritizing Annealing. For most gray irons, a ferritizing annealing temperature between 705 and 760 °C (1300 and 1400 °F) is recommended. The furnace temperature profile must be such that castings are sure to reach the set temperatures. Precise temperatures within this range depend on the exact composition of the iron.

Medium ("full") annealing is usually performed at temperatures between 790 and 900 °C (1450 and 1650 °F). This treatment is used when a ferritizing anneal would be ineffective because of the high alloy content of a particular iron. It is recommended, however, that the efficacy of temperatures at or below 760 °C (1400 °F) be tested before a higher annealing temperature is adopted as part of a standard procedure.

Graphitizing Annealing. If the microstructure of gray iron contains massive carbide particles, higher

annealing temperatures are necessary. Graphitizing annealing may have the purpose simply of converting massive carbide to pearlite and graphite.

To break down massive carbide with reasonable speed, temperatures of at least 870 °C (1600 °F) are required. With each additional 55 °C (100 °F) increment in holding temperature, the rate of carbide decomposition doubles; consequently, it is general practice to employ holding temperatures of 900 to 955 °C (1650 to 1750 °F). However, at 925 °C (1700 °F) and above, the phosphide eutectic present in irons containing 0.10% P or more may melt.

NORMALIZING

Gray iron is normalized by being heated to a temperature above the transformation range, held at this temperature for a period of about 1 h per inch of maximum section thickness, and cooled in still air to room temperature. Normalizing may be used to enhance mechanical properties, such as hardness and tensile strength, or to restore as-cast properties that have been modified by another heating process, such as graphitizing or the preheating and postheating associated with repair welding.

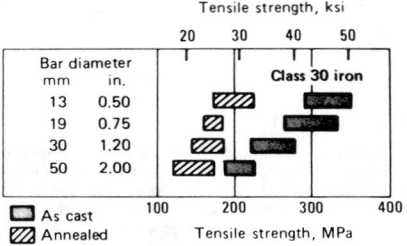

Specimens were arbitration bars from 31 heats. Bars were annealed at 925 °C (1700 °F) for 2 h, plus 1 h per 25 mm (1 in.) of section over 25 mm, and cooled at a maximum rate of 160 °C/h (285 °F/h) from 925 to 565 °C (1700 to 1050 °F). Cooling continued from 565 °C at a maximum rate of 130 °C/h (230 °F/h) to 200 °C (390 °F); bars were then air cooled to room temperature.

Fig. 2. Effect of annealing on tensile strength of class 30 gray iron

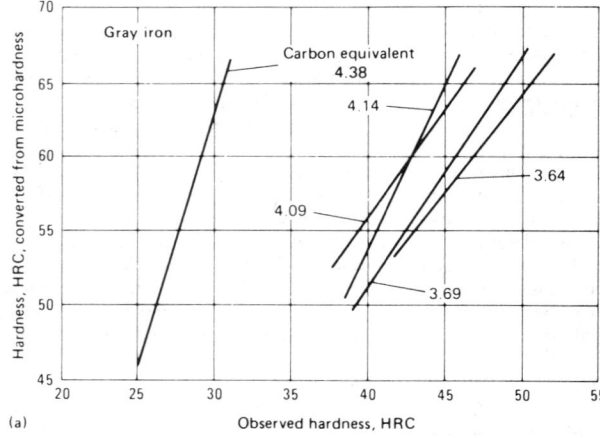

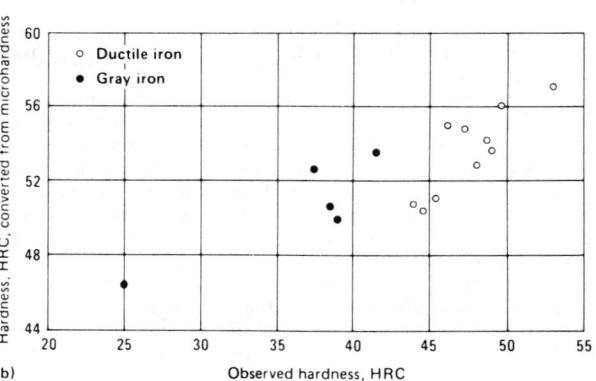

(a) Relation, as influenced by carbon equivalent, for gray iron containing type 3 graphite. (b) Relation for gray and ductile irons quenched in water from 900 °C (1650 °F) and tempered 2 h at 425 °C (800 °F).

Fig. 1. Relations between observed and converted hardness values for gray and ductile irons

Table 1. Effect of air cooling from various temperatures on typical properties of gray irons

Condition(a)		Unalloyed iron(b)			Alloyed iron(c)			
		Hardness, HB	Tensile strength MPa	ksi	Hardness, HB	Tensile strength MPa	ksi	Combined carbon, %
As cast		207	265	38.1	212	265	38.7	0.84
Air cooled from:								
°C	°F							
540	1000	202	210	30.4	212	275	39.7	0.82
595	1100	190	255	36.8	210	275	40.0	0.86
650	1200	138	195	28.2	202	265	38.6	0.80
705	1300	125	180	26.4	187	265	38.4	0.81
760	1400	131	190	27.4	170	235	34.2	0.60
815	1500	152	205	29.7	212	295	42.6	0.76
870	1600	152	205	29.7	217	305	44.2	0.81
925	1700	152	205	29.6	223	290	42.4	0.82
980	1800	152	210	30.1	255	340	49.0	0.80

(a) Specimens 30 mm in diameter by 180 mm (1.2 in. in diameter by 7 in.) were held at temperature for 1 h before being cooled in still air to room temperature. (b) As-cast composition: 3.15 total C, 0.54 combined C, 2.59 Si, 0.09 P, 0.135 S, 0.88 Mn, 0.01 Cr, 0.10 Ni. (c) As-cast composition: 3.33 total C, 0.84 combined C, 2.27 Si, 0.076 P, 0.122 S, 0.72 Mn, 0.44 Cr, 0.36 Ni, 0.28 Mo.

The temperature range for normalizing gray iron is approximately 885 to 925 °C (1625 to 1700 °F). Heating temperature has a marked effect on microstructure and on mechanical properties such as hardness and tensile strength. This is demonstrated in Table 1.

HARDENING AND TEMPERING

Gray irons are hardened and tempered to improve their mechanical properties, particularly strength and wear resistance.

Austenitizing. In hardening gray iron, the casting is heated to a temperature high enough to promote formation of austenite, held at that temperature until the desired amount of carbon has been dissolved, and then quenched at a suitable rate. Heating for austenitizing may be accomplished in a salt bath or in an electrically heated, gas-fired or oil-fired furnace.

The temperature to which the casting must be heated is determined by the transformation range of the particular gray iron of which it is made. The transformation range can extend more than 55 °C (100 °F) above the A_1 (transformation-start) temperature. A formula for determining the approximate A_1 transformation temperature of unalloyed gray iron is:

°C: 730 + 28.0 (% Si) − 25.0 (% Mn)

°F: 1345 + 50.4 (% Si) − 45.0 (% Mn)

Chromium raises the transformation range of gray iron. In high-nickel, high-silicon irons, for example, each percent of chromium raises the transformation range by about 40 °C (72 °F). Nickel, on the other hand, lowers the critical range. In a gray iron containing from 4 to 5% Ni, the upper limit of the transformation range is about 710 °C (1310 °F).

Provided that recommended limits are not exceeded, the higher the casting is heated above the transformation range, the greater will be the amount of carbon dissolved in the austenite (Fig. 3) and the higher will be the hardness of the casting after quenching.

Quenching. Oil is the quenching medium most frequently used for gray iron. Water generally is not a satisfactory quenching medium for furnace-heated gray iron; it extracts heat so rapidly that distortion and cracking are likely in all except small parts of simple design. Recently developed water-soluble polymer quenches can provide the convenience of water quenching along with lower cooling rates, which can minimize thermal shock.

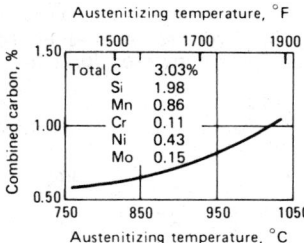

Specimens were furnace heated and water quenched. Combined carbon by difference.

Fig. 3. Increase in combined carbon with increase in austenitizing temperature for gray iron

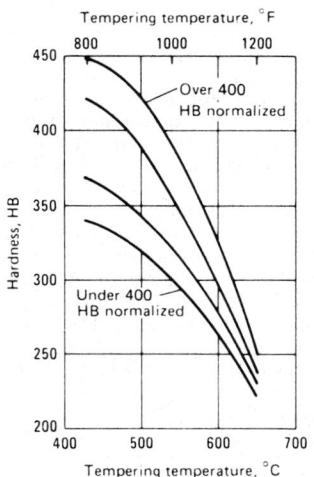

Fig. 4. Hardness of normalized ductile iron tempered at various temperatures

The least severe quenching medium is air. Unalloyed or low-alloy gray iron castings usually cannot be air quenched, because the cooling rate is not high enough to form martensite. However, for irons of high alloy content, forced-air quenching is frequently the most desirable cooling method.

Tempering. After quenching, castings usually are tempered at temperatures well below the transformation range for about 1 h per inch of thickest section. As the quenched iron is tempered, its hardness decreases, whereas it usually gains in strength and toughness.

FLAME HARDENING

Flame hardening is the method of surface hardening most commonly applied to gray iron. Both unalloyed and alloyed gray irons can be successfully flame hardened. However, some compositions yield much better results than others. One of the most important aspects of composition is the combined carbon content, which should be in the range of 0.50 to 0.70%.

INDUCTION HARDENING

Gray iron castings can be surface hardened by the induction method when the number of castings to be processed is large enough to warrant the relatively high equipment cost and the need for special induction coils.

Heat Treating of Ductile Irons

WHEN MAXIMUM DUCTILITY and good machinability are desired and high strength is not required, ductile iron castings are generally given a full ferritizing anneal. The microstructure is thus converted to ferrite and spheroidal graphite.

Two different annealing cycles may be used satisfactorily. Selection of one or the other will depend on the type of heat treating equipment that is available. These two cycles are:

• Hold at 900 to 955 °C (1650 to 1750 °F) for 1 h plus 1 h or more per inch of section thickness. For thin-section castings containing 2.20 to 2.70% Si, holding at 955 °C (1750 °F) for 1 to 3 h is sufficient. In heavy-section castings where chill has formed on corners, holding at 955 °C (1750 °F) for 3 to 8 h may be required. Cool to 690 °C (1275 °F) in any convenient manner (but uniformly, if residual stress is to be avoided), and hold at 690 °C (1275 °F) for 5 h plus 1 h per inch of casting section.

• Hold at 900 to 955 °C (1650 to 1750 °F) as above, but furnace cool to 650 °C (1200 °F) so that the cooling rate between 790 and 650 °C (1450 and 1200 °F) does not exceed 20 °C/h (35 °F/h).

A shorter, subcritical annealing cycle can be used when carbides can be tolerated and maximum impact properties are not required.

NORMALIZING OF DUCTILE IRON

Normalizing can result in a considerable improvement in tensile properties and may be used in production of ductile iron of types 100-70-03 and 120-90-02. The following temperatures and minimum holding times are recommended for normalizing unalloyed ductile iron (see Fig. 4):

Section			Time,
mm	in.	Temperature	h
Under 13	Under 1/2	870 °C (1600 °F) min	1
13 to 25	1/2 to 1	940 °C (1725 °F)	1
Over 25	Over 1	940 °C (1725 °F)	2

HARDENING AND TEMPERING OF DUCTILE IRON

A temperature of 845 to 925 °C (1550 to 1700 °F) is normally used for austenitizing commercial castings and produces the highest as-quenched hardness. Oil is preferred as a quenching medium, to minimize stresses, but water or brine may be used for simple shapes. Complicated castings may have to be quenched in oil at 80 to 100 °C (180 to 210 °F) to avoid cracks.

To relieve quenching stresses, castings should be tempered immediately after quenching. Tempered hardness depends on as-quenched hardness level, alloy content and tempering time, as well as on temperature. Precise data on ductile irons are not available for use in drawing precise tempering curves such as those for steels. Figure 5 can be used as a first approximation by the heat treater. More accurate control of hardness can be attained by close control of material composition and heat treating cycle.

SURFACE HARDENING OF DUCTILE IRON

Ductile iron responds readily to surface hardening by flame or induction processes. Because of the short heating cycle in these processes, the pearlitic types of ductile iron, 80-60-03 and 100-70-03, are preferred. Irons without free ferrite in their microstructures respond almost instantly to flame or induction heating and require very little holding time at the austenitizing temperature in order to be fully hardened.

Heat Treating of Malleable Irons

FERRITIC AND PEARLITIC malleable irons are both produced by annealing white iron of controlled composition. Thus, annealing is an essential part of the manufacturing process for these irons.

The annealing treatment involves three important steps. The first causes nucleation of graphite and is initiated during heating to a high holding temperature and occurs very early during the holding period.

The second step consists of holding at 900 to 970 °C (1650 to 1780 °F); this step is called first-stage graphitization (FSG). During FSG, massive carbides are eliminated from the iron structure. At this point, the iron is rapidly cooled to 725 to 740 °C (1340 to 1360 °F) prior to entering second-stage graphitization.

The third step in annealing consists of slow cooling through the allotropic transformation range of the iron; this step is called second-stage graphitization (SSG). During SSG, a completely ferritic matrix free of pearlite and carbides is obtained when the cooling rate is 2 to 17 °C/h (3 to 30 °F/h). This cooling rate, which depends on the silicon content of the iron and the temper carbon nodule count, may be increased to 85 °C/min (150 °F/min) to form a pearlitic matrix. Oil quenching from the FSG temperature following completion of that step will produce a martensitic matrix.

HARDENING AND TEMPERING OF PEARLITIC MALLEABLE IRON

A typical procedure for producing a hardened pearlitic malleable iron consists of (a) air quenching castings after first-stage annealing, which results in retention of about 0.75% combined carbon in the matrix; (b) reheating and holding for 1 h at 845 to 870 °C (1550 to 1600 °F)

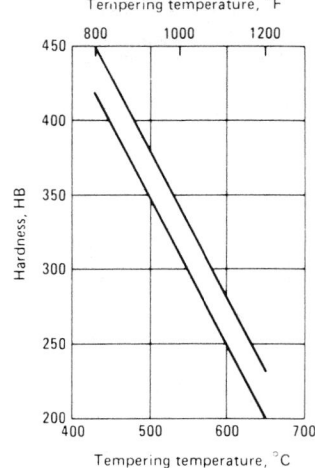

As-quenched hardness more than 500 HB.
Fig. 5. Hardness of oil-quenched ductile iron tempered for 1 h

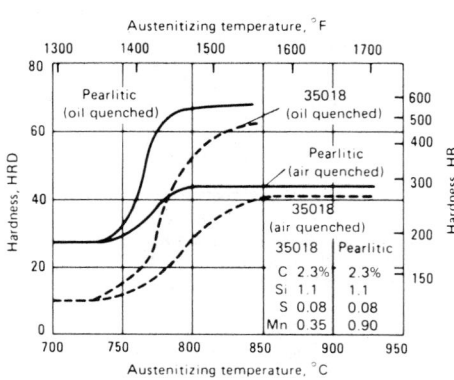

Fig. 6. Effects of austenitizing temperature, quenching medium and manganese content on hardness of as-quenched malleable iron

to reaustenitize the matrix; and then (c) quenching in heated (80 to 105 °C; 180 to 220 °F) and agitated oil, thus developing a matrix consisting of martensite and bainite with a hardness of 555 to 627 HB. Figure 6 shows the effects of austenitizing temperature, quenching medium and manganese content on the hardness of ferritic and pearlitic malleable iron both before and after heat treating. If direct oil quenching is used, caution must be exercised to prevent cracking due to high combined carbon.

Tempering treatments consist of cycles of no less than 2 h at temperature to ensure uniformity

of product. Tempering times must also be adjusted for section thickness and quenched microstructures. Fine pearlite and bainite require longer tempering times than that for martensite. In general, final hardness is controlled with process controls approximately the same as those encountered in heat treatment of medium-carbon and higher-carbon steels.

Heat Treating of Austenitic Irons

FLAKE-GRAPHITE corrosion-resistant austenitic cast irons are susceptible to work hardening during machining and require careful cooling from the casting operation and/or subsequent heat treating operations. Compositions of these irons are given in Table 2.

Stress Relieving. For most applications, it is recommended that austenitic cast irons be stress relieved at 620 to 675 °C (1150 to 1250 °F), for 1 h per inch of section, to remove residual stresses resulting from casting or machining, or both. Stress relieving should follow rough machining, particularly for castings that must conform to close dimensional tolerances, that have been extensively welded, or that are to be exposed to high stresses in service.

Holding of castings at 480 °C (900 °F) for 1 h per inch of thickness will remove about 60% of the stress; stress relieving at 675 °C (1250 °F) will remove almost 95%. It is usually acceptable to cool castings in air at a rate of 1 to 2 h per inch of section thickness, although furnace cooling produces maximum stress relief. Stress relieving does not affect tensile strength, hardness or ductility.

Spheroidize Annealing. Castings with hardnesses above 190 HB may be softened by heating to 980 to 1040 °C (1800 to 1900 °F) for 1/2 to 5 h, except those alloys containing 4% or more chromium. Excessive carbides cause this high hardness and may occur in rapidly cooled castings and thin sections. Annealing dissolves or spheroidizes carbides. Although it lowers hardness, spheroidize annealing does not adversely affect strength.

High-Temperature Stabilization. Except for castings of alloy type 1, which are not recommended for service above 430 °C (800 °F), castings used for either static or cyclic service at 480 °C (900 °F) or above should be given a stabilization heat treatment. This stabilization treatment consists of holding at 760 °C (1400 °F) for 4 h minimum or at 870 °C (1600 °F) for 2 h minimum, furnace cooling to 540 °C (1000 °F), and then cooling in air.

Table 2. Compositions of flake-graphite corrosion-resistant austenitic cast irons

| Type | Composition, % | | | | | |
	TC(a)	Si	Mn	Ni	Cu	Cr
1(b)	3.00 max	1.00-2.80	0.50-1.50	13.50-17.50	5.50-7.50	1.50-2.50
1b	3.00 max	1.00-2.80	0.50-1.50	13.50-17.50	5.50-7.50	2.50-3.50
2(c)	3.00 max	1.00-2.80	0.50-1.50	18.00-22.00	0.50 max	1.50-2.50
2b	3.00 max	1.00-2.80	0.50-1.50	18.00-22.00	0.50 max	3.00-6.00(d)
3	2.60 max	1.00-2.00	0.50-1.50	28.00-32.00	0.50 max	2.50-3.50
4	2.60 max	5.00-6.00	0.50-1.50	29.00-32.00	0.50 max	4.50-5.50
5	2.40 max	1.00-2.00	0.50-1.50	34.00-36.00	0.50 max	0.10 max(e)
6(f)	3.00 max	1.50-2.50	0.50-1.50	18.00-22.00	3.50-5.50	1.00-2.00

(a) Total carbon. (b) Type 1 is recommended for applications in which the presence of copper offers corrosion-resistance advantages. (c) Type 2 is recommended for applications in which copper contamination cannot be tolerated, such as handling of foods or caustics. (d) Where some machining is required, 3.0 to 4.0 Cr is recommended. (e) Where increased hardness, strength and heat resistance are desired, and where increased expansivity can be tolerated, Cr may be increased to 2.5 to 3.0%. (f) Type 6 also contains 1.0% Mo.

Heat Treating of Tool Steels

FOR classifications and compositions of various tool steels, refer to Section 18 of this volume.

WATER-HARDENING TOOL STEELS

Water-hardening tool steels containing 0.90 to 1.00% carbon are the most widely used. Carbon content affects heat treating temperatures as indicated in Table 1, which outlines recommended heat treating practices for these steels.

As a class, water-hardening tool steels are relatively low in hardenability, although they are arbitrarily classified and available as shallow-hardening, medium-hardening and deep-hardening types. Their low hardenability is frequently an advantage, because it allows tough core properties in combination with high surface hardness.

Normalizing. Except in special instances where experience has proved it beneficial, normalizing is not recommended for water-hardening tool steels as received from the supplier.

Annealing. Tool steels of the W types are received from the supplier in the annealed condition. Thus, annealing by the user is usually unnecessary. Annealing is applied to forged or cold worked carbon tool steel to soften it for easier machining.

Austenitizing temperatures for water-hardening tool steels normally vary from 760 to 845 °C (1400 to 1550 °F), as indicated in Table 1. Higher temperatures are sometimes used for special purposes. Hardenability increases as austenitizing temperature increases. The optimum time at austenitizing temperature is from 10 to 30 min. Preheating is unusual except for very large tools or those with intricate cross sections.

Quenching. To produce maximum depth of hardness in water-hardening tool steels, it is essential that they be quenched as rapidly as possible. In most instances, water or a brine solution consisting of 10% NaCl (by weight) in water is used. Occasionally, for an even faster quench, an iced brine solution is employed. Cooling rate is a function of workpiece size as well as of quenching medium; for this reason, small pieces can be quenched in oil.

Tempering. Water-hardening tool steels should be tempered immediately after hardening, preferably before they reach room temperature; about 50 °C (120 °F) is optimum. Salt baths, oil baths and air furnaces are all satisfactory for tempering.

All parts made of these steels should be tempered at temperatures not lower than 175 °C (350 °F). One hour at temperature is usually adequate; additional soaking time will further lower hardness. Figure 1 shows the effect of tempering temperature on hardness of water-hardening tool steels austenitized at 790, 815 and 845 °C (1450, 1500 and 1550 °F) and quenched in brine.

SHOCK-RESISTING TOOL STEELS

Recommended heat treating practices for shock-resisting tool steels are outlined in Table 2.

Normalizing is not recommended for shock-resisting tool steels.

Annealing. The high-silicon types (S2, S4, S5 and S6) are susceptible to graphitization and decarburization. Annealing these types at temperatures higher than those indicated in Table 2 may produce a softer structure, but it will also increase the danger of graphitization. The silicon types should not be soaked at temperature.

Austenitizing temperatures for shock-resisting tool steels vary from 815 to 955 °C (1500 to 1750 °F). Preheating is not mandatory, but it is sometimes desirable for large tools, to minimize distortion, shorten time at the austenitizing temperature and speed up production.

Tempering. Both the tungsten and the silicon types of shock-resisting tool steel resist softening from tempering to a greater degree than carbon tool steels. Secondary hardening does not occur in these steels, except to a minimal degree in some compositions of the tungsten type.

The effect of tempering temperature on hardness for six different compositions of type S1 tool steel is shown in Fig. 2.

OIL-HARDENING COLD WORK TOOL STEELS

Recommended heat treating practice for oil-hardening cold work tool steels is summarized in Table 3.

Normalizing is desirable and sometimes necessary for parts that have been forged or heated previously to temperatures much higher than the proper austenitizing temperature, because it produces a more uniformly refined grain structure.

Annealing. Finished or semifinished tools made from oil-hardening cold work steels should be protected from decarburization or carburization during annealing. This can be accomplished by the use of dry exothermic furnace atmospheres.

Table 1. Recommended heat treating practice for water-hardening tool steels

°C	Temperature °F	Carbon content	Hardness after treatment	Procedure
Normalizing				
815	1500	0.60 to 0.75	. . .	Heat through uniformly; hold for
790	1450	0.75 to 0.90	. . .	15 min (light sections) to 1 h
870	1600	0.90 to 1.10	. . .	(heavy sections), then air cool
870 to 925	1600 to 1700	1.10 to 1.40	. . .	
Annealing				
740 to 760	1360 to 1400	0.60 to 0.90	156 to 201 HB	Heat through uniformly; hold for 1
760 to 790	1400 to 1450	0.90 to 1.40	156 to 201 HB	to 4 h(a); furnace cool to 510 °C (950 °F) at 20 °C/h (40 °F/h), then air cool
Hardening(b)				
790 to 845	1450 to 1550	0.60 to 0.80	65 to 68 HRC	Hold at austenitizing temperature
775 to 845	1425 to 1550	0.85 to 1.05	65 to 68 HRC	for 10 to 30 min; quench in
760 to 830	1400 to 1525	1.10 to 1.40	65 to 68 HRC	water or brine (very small pieces may be oil quenched)

(a) Holding times vary from about 1 h, for light sections and small furnace charges, to about 4 h, for heavy sections and large furnace charges. (b) For large tools and tools with intricate sections, preheating at 565 to 650 °C (1050 to 1200 °F) is recommended.

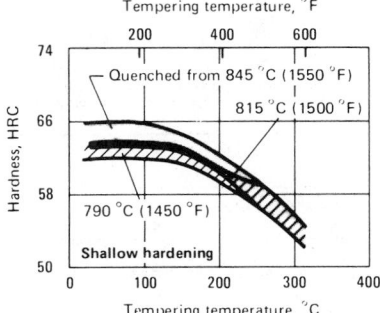
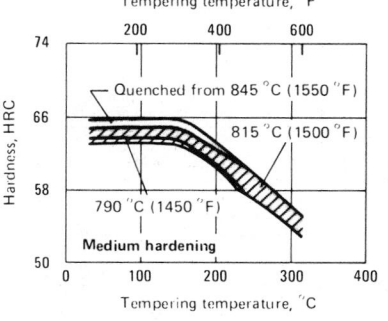
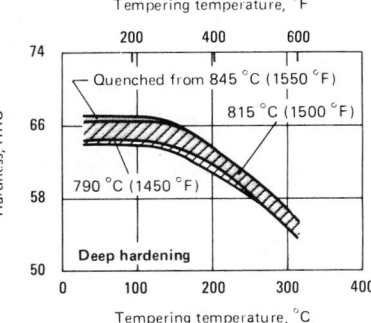

Specimens held for 1 h at the tempering temperature in a recirculating-air furnace. Cooled in air to room temperature. Data represent 20 25-mm (1-in.) diam specimens for each steel. Compositions of steels: shallow hardening, 0.90 to 1.00 C, 0.18 to 0.22 Mn, 0.20 to 0.22 Si, 0.18 to 0.22 V; medium hardening, 0.90 to 1.00 C, 0.25 Mn, 0.25 Si, no alloying elements; deep hardening, 0.90 to 1.00 C, 0.30 to 0.35 Mn, 0.20 to 0.25 Si, 0.23 to 0.27 Cr.

Fig. 1. Effect of tempering temperature on surface hardness of water-hardening tool steels austenitized at three different temperatures and quenched in brine

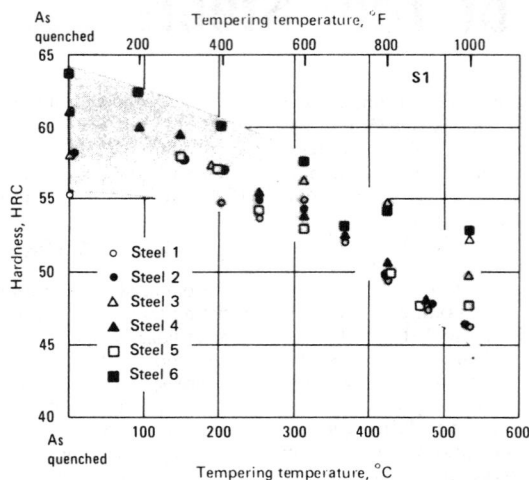

No.	Steel Type	C	Si	Composition, % W	Cr	V	Quenching Temperature °C	°F	Medium
1	S1	0.43	...	2.00	1.30	0.25	955	1750	...
2	S1	0.53	...	2.00	1.65	0.25	900	1650	...
3	S1	0.50	...	2.75	1.25	0.20	925	1700	...
4	S1	0.55	...	2.50	1.50	0.35	925	1700	...
5	S1	0.50	0.75	2.50	1.15	0.20	955	1750	Oil
6	S1	0.58	0.95	2.25	1.25	0.25	925	1700	Oil

Fig. 2. Effect of tempering temperature on hardness of type S1 shock-resisting tool steels

More often, however, it is accomplished by pack annealing, wherein work to be annealed is packed in a box and surrounded with inert protective material, such as clean cast iron chips or 6-to-8-mesh spent pitch coke.

Preheating of the O steels will minimize distortion during subsequent hardening. It is almost always required for tools that are to be austenitized in liquid baths. Recommended preheating temperatures are listed in Table 3.

Austenitizing. Recommended austenitizing temperatures for oil-hardening tool steels are given in Table 3.

Quenching. The optimum temperature range for quenching baths consisting of conventional oils is 40 to 60 °C (100 to 140 °F); agitation is recommended.

Tempering. The O steels should be tempered immediately after quenching (preferably before they quite reach room temperature). These steels usually are not tempered below 120 °C (250 °F) or above 540 °C (1000 °F); the most commonly used temperature range is from 175 to 205 °C (350 to 400 °F). Tempering times vary with section size. Often, a time at temperature of 1 h per inch of thickness (minimum dimension of heaviest sec-

tion) or per inch of diameter, with a minimum of 1 h, is used. Typical hardness values obtained after tempering are shown in Fig. 3.

MEDIUM-ALLOY AIR-HARDENING, AND HIGH-CARBON HIGH-CHROMIUM, COLD WORK TOOL STEELS

Recommended heat treating practice for medium-alloy air-hardening cold work tool steels (group A) and high-carbon high-chromium cold work tool steels (group D) is summarized in Table 4.

Normalizing. Except for type A10 (see Table 4), normalizing is not recommended for any of the steels in groups A and D.

Annealing. These steels are usually supplied in the annealed condition by the manufacturer. However, they should be annealed after forging and prior to rehardening. Annealing is required also for previously hardened or welded tools that are to be reworked. Recommended annealing temperatures for the various types are given in Table 4.

Preheating. Steels of the A and D groups are usually preheated before being austenitized for hardening. Preheating reduces subsequent distortion in the hardened parts by minimizing nonuniform dimensional changes during austenitizing. Preheating of simpler tools made of grades A4, A5, A6 and A10 can often be eliminated if they are austenitized in a furnace instead of a liquid bath, because these steels are austenitized at lower temperatures.

Recommended preheating temperatures are listed in Table 4.

Austenitizing. Steels of groups A and D can be austenitized in molten salt baths or in various types of furnaces using gaseous atmospheres. Because of their lower austenitizing temperatures, types A4, A5, A6 and A10 may also be austenitized in molten lead, or in open furnaces with oxidizing atmospheres. However, the latter methods are not satisfactory for the other A steels or for the D steels, because of their higher austenitizing temperatures (see Table 4).

Table 2. Recommended heat treating practice for shock-resisting tool steels

Steel	Normalizing	Annealing Temperature(a) °C	°F	Cooling rate(b) °C/h	°F/h	Annealed hardness, HB	Hardening Preheat °C	°F	Temperature Austenitize °C	°F	Holding time, min	Quenching medium(c)	Quenched hardness, HRC
S1	Not rec	790-815	1450-1500	22	40	183-229	...	...	900-955	1650-1750	15-45	O	57-59
S2	Not rec	760-790	1400-1450	22	40	192-217	650(d)	1200(d)	845-900	1550-1650	5-20	B, W	60-62
S4	Not rec	760-790	1400-1450	22	40	192-229	650	1200	870-925	1600-1700	5-20	B, W	61-63
									900-925	1650-1700	5-20	O	57-59
S5	Not rec	775-800	1425-1475	14	25	192-229	760	1400	870-925	1600-1700	5-20	O	58-61
S6	Not rec	800-830	1475-1525	14	25	192-229	760	1400	915-955	1675-1750	10-30	O	56-60
S7	Not rec	815-845	1500-1550	14	25	187-223	650-705	1200-1300	925-955	1700-1750	15-45(e)	A, O	60-61

(a) Lower limit of range should be used for small sections, upper limit for large sections. Holding time varies from about 1 h, for light sections and small furnace charges, to about 4 h, for heavy sections and large charges; for pack annealing, hold for 1 h per inch of pack cross section. (b) Maximum. Rate is not critical after work is cooled to about 510 °C (950 °F). (c) O, oil; B, brine; W, water; A, air. (d) Preferable for large tools, to minimize decarburization. (e) For open furnace heat treatment. For pack hardening, hold for 1/2 h per inch of pack cross section.

Table 3. Recommended heat treating practice for oil-hardening cold work tool steels

Steel	Normalizing temperature(a) °C	°F	Annealing Temperature(b) °C	°F	Cooling rate(c) °C/h	°F/h	Annealed hardness, HB	Hardening Preheat °C	°F	Temperature Austenitize °C	°F	Holding time, min	Quenching medium(d)	Quenched hardness, HRC
O1	870	1600	760-790	1400-1450	20	40	183-212	650	1200	790-815	1450-1500	10-30	O	63-65
O2	845	1550	745-775	1375-1425	20	40	183-212	650	1200	760-800	1400-1475	5-20	O	63-65
O6	870	1600	765-790	1410-1450	10	20	183-217	...	...	790-815	1450-1500	2-5	O	63-65
O7	900	1650	790-815	1450-1500	20	40	192-217	650	1200	790-830	1450-1525	10-30	W	64-66
										845-885	1550-1625	10-30	O	64-66(e)

(a) Holding time, after uniform through heating, varies from about 15 min, for small sections, to about 1 h, for large sections. Work is cooled from temperature in still air. (b) Lower limit of range should be used for small sections, upper limit for large sections. Holding time varies from about 1 h, for light sections and small furnace charges, to about 4 h, for heavy sections and large charges; for pack annealing, hold for 1 h per inch of pack cross section. (c) Maximum. Rate is not critical after cooling to below 540 °C (1000 °F). (d) O, oil; W, water. (e) Sections larger than 38 mm (1 1/2 in.) will be softer.

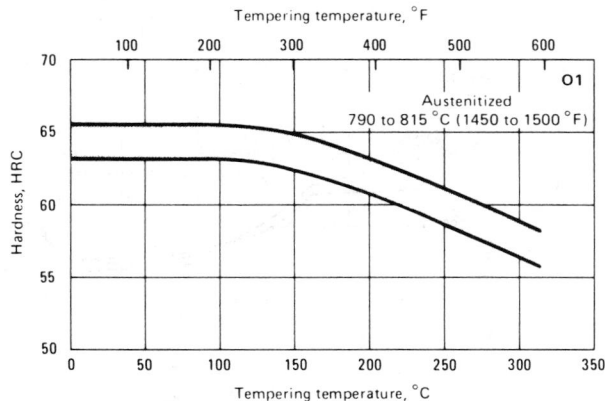

Fig. 3. Effect of tempering temperature on as-quenched hardness of type O1 oil-hardening tool steel

Quenching. Steels of groups A and D, except D3, will attain maximum hardness by cooling in still air, unless sections are extremely large.

Tempering practices for A and D steels parallel those described for O steels in the preceding section. Tempering is usually begun when the work reaches a temperature of about 50 to 65 °C (120 to 150 °F). However, these steels retain some austenite at this temperature range.

Multiple tempering is effective in decreasing the amount of austenite retained in A and D steels and is a common practice in heat treating them. The general precautions and tempering practices outlined for O steels in the preceding section are followed for the A and D steels. However, because most of the steels in groups A and D (except A4, A5 and A6) soften less rapidly than group O steels with an increase in tempering temperature (Fig. 4), higher tempering temperatures can be used for the A and D steels.

HOT WORK TOOL STEELS

Nominal compositions of chromium, tungsten and molybdenum types of hot work tool steels are given in Section 18. Table 5 summarizes the heat treating practices commonly employed for this composite group of tool steels.

Normalizing. Because these steels as a group are either partially or completely air-hardening, normalizing is not recommended.

Annealing. To minimize scaling and decarburization, small parts are usually pack annealed, while large and heavy die blocks are more commonly annealed in controlled-atmosphere furnaces (see Table 5).

Preheating prior to austenitizing is nearly always recommended for all hot work steels.

Austenitizing temperatures recommended for hardening of hot work tool steels are given in Table 5. Rapid heating from the preheating temperature to the austenitizing temperature is preferred for types H16 through H43.

Tempering practice is essentially the same as that recommended for A and D grades. All of the H grades show some secondary hardening characteristics; therefore, double tempering is recommended (see representative examples in Fig. 5 and 6).

HIGH SPEED TOOL STEELS

Nominal compositions of high speed tool steels are given in Section 18. Recommended heat treating practice is summarized for two standard groups of high speed tool steels and one intermediate group in Table 6; note that normalizing of high speed tool steels is not recommended.

Annealing. High speed steel must be fully annealed after forging or when rehardening is required. To minimize decarburization, pack annealing in tightly closed containers is recommended. The packing material can be dry sand or lime to which a small amount of charcoal has been added; burned cast iron chips also are satisfactory. Because the packing material acts to insulate the container and thereby slow down heating, the container should be filled in such a way with the steel to be annealed that a minimum amount of packing material is required.

After the steel has reached the annealing temperature range (Table 6), it should be held at temperature for 1 h per inch of thickness of the container and should then be slowly cooled in the furnace (at a rate not exceeding 20 °C or 40 °F per h) until it reaches a temperature of 650 °C (1200 °F), when a faster rate of cooling is permissible.

Preheating. Austenite begins to form at about 760 °C (1400 °F), and preheating for hardening to slightly above this temperature will minimize stresses that might be set up because of the transformation.

Double preheating—in one furnace at 540 to 650 °C (1000 to 1200 °F) and in another at 845 to 870 °C (1550 to 1600 °F)—is often recommended, to minimize thermal shock.

Austenitizing. High speed steels depend on the solution of various complex alloy carbides during austenitizing to develop their heat-resisting qualities and cutting ability. These carbides do not dissolve to an appreciable extent unless the steel is heated to temperatures near the melting point (Table 6). Therefore, exceedingly accurate temperature control is required in austenitizing high speed steels.

Quenching. High speed steels can be quenched in air, oil or molten salt. However, except for thin tools, which are air quenched between plates to keep them straight, it is customary to quench in oil from muffle or semimuffle furnaces and in molten salt from a high-temperature salt bath.

Table 4. Recommended heat treating practice for medium-alloy air-hardening, and high-carbon high-chromium, cold work tool steels

Steel	Normalizing temperature(a) °C (°F)	Annealing Temperature(b) °C	°F	Cooling rate(c) °C/h	°F/h	Annealed hardness, HB	Hardening Preheat °C	°F	Temperature Austenitize °C	°F	Holding time, min	Quenching medium(d)	Quenched hardness, HRC
Medium-alloy air hardening cold work tool steels													
A2	Not rec	845-870	1550-1600	22(e)	40(e)	201-229	790	1450	925-980	1700-1800	20-45(f)	A	62-65(g)
A3	Not rec	845-870	1550-1600	22	20	207-229	790	1450	955-1010	1750-1850	25-60(f)	A	...
A4	Not rec	740-760	1360-1400	14(h)	25(h)	200-241	675	1250	815-870	1500-1600	15-90	A	61-64(g)
A5	Not rec	740-760(j)	1360-1400(j)	14	25	229-255	595	1100	790-845	1450-1550	15-45	A	62-63(g)
A6	Not rec	730-745	1350-1375	14	25	217-248	650	1200	830-870	1525-1600	20-45	A	59-63(g)
A7	Not rec	870-900	1600-1650	14(e)	25(e)	235-262	815	1500	955-980	1750-1800	30-60(f)	A	64-67(g)
A8	Not rec	845-870	1550-1600	22	40	192-223	790	1450	980-1010	1800-1850	20-45(f)	A	60-62(g)
A9	Not rec	845-870	1550-1600	14	25	212-248	790	1450	980-1025	1800-1875	20-45(f)	A	56-58(g)
A10	790 (1450)	765-795	1410-1460	8	15	235-269	650	1200	790-815	1450-1500	30-60	A	62-64(g)
High-carbon high-chromium cold work tool steels													
D1	Not rec	870-900	1600-1650	22	40	207-248	815	1500	970-1010	1775-1850	15-45(f)	A	61
D2	Not rec	870-900	1600-1650	22	40	217-255	815	1500	980-1025	1800-1875	15-45(f)	A	64
D3	Not rec	870-900	1600-1650	22	40	217-255	815	1500	925-980	1700-1800	15-45(f)	O	64
D4	Not rec	870-900	1600-1650	22	40	217-255	815	1500	970-1010	1775-1850	15-45(f)	A	64
D5	Not rec	870-900	1600-1650	22	40	223-255	815	1500	980-1025	1800-1875	15-45(f)	A	64
D7	Not rec	870-900	1600-1650	22	40	235-262	815	1500	1010-1065	1850-1950	30-60(f)	A	65

(a) Holding time, after uniform through heating, varies from about 15 min, for small sections, to about 1 h, for large sections. Work is cooled from temperature in still air. (b) Lower limit of range should be used for small sections, upper limit for large sections. Holding time varies from about 1 h, for light sections and small furnace charges, to about 4 h, for heavy sections and large charges; for pack annealing, hold for 1 h per inch of pack cross section. (c) Maximum rate, to 540 °C (1000 °F) unless footnoted to indicate otherwise. (d) A, air; O, oil. (e) To 705 °C (1300 °F). (f) For open furnace heat treatment. For pack hardening, hold for ¹/₂ h per inch of pack cross section. (g) Hardness varies with austenitizing temperature. (h) To 650 °C (1200 °F). (j) One manufacturer recommends cooling from 760 to 540 °C (1400 to 1000 °F), then reheating to 730 °C (1350 °F) and cooling.

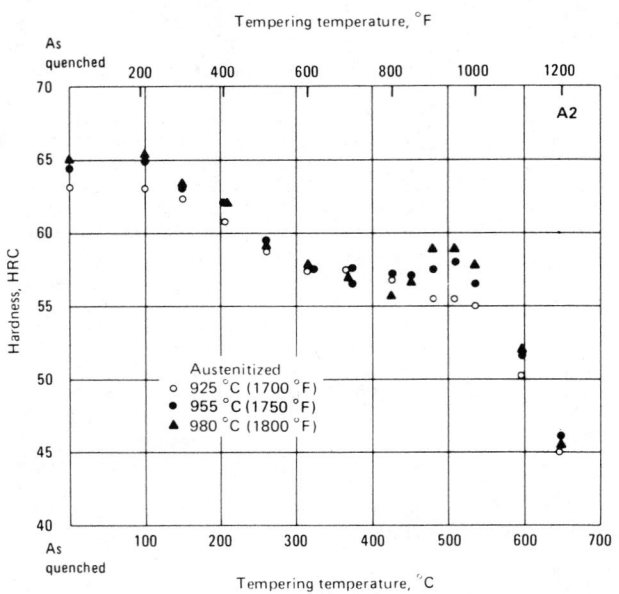
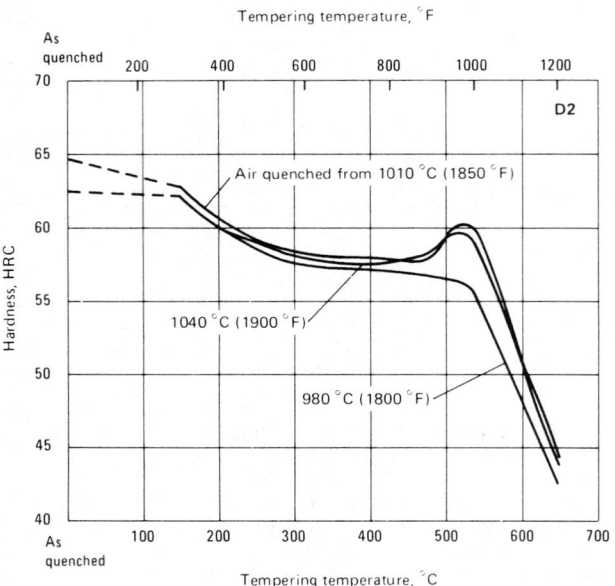

Fig. 4. Effect of tempering temperature on as-quenched hardness of types A2 and D2 cold work tool steels

Table 5. Recommended heat treating practice for hot work tool steels

Steel	Normalizing	Annealing Temperature(a) °C	Annealing Temperature(a) °F	Cooling rate(b) °C/h	Cooling rate(b) °F/h	Annealed hardness, HB	Hardening Preheat °C	Hardening Preheat °F	Hardening Austenitize °C	Hardening Austenitize °F	Holding time, min	Quenching medium(c)	Quenched hardness, HRC
Chromium hot work tool steels													
H10	Not recommended	845-900	1550-1650	22	40	192-229	815	1500	1010-1040	1850-1900	15-40(d)	A	56-59
H11	Not recommended	845-900	1550-1650	22	40	192-229	815	1500	995-1025	1825-1875	15-40(d)	A	53-55
H12	Not recommended	845-900	1550-1650	22	40	192-229	815	1500	995-1025	1825-1875	15-40(d)	A	52-55
H13	Not recommended	845-900	1550-1650	22	40	192-229	815	1500	995-1040	1825-1900	15-40(d)	A	49-53
H14	Not recommended	870-900	1600-1650	22	40	207-235	815	1500	1010-1065	1850-1950	15-40(d)	A	55-56
H16	Not recommended	870-900	1600-1650	22	40	212-241	815	1500	1120-1175	2050-2150	2-5	A, O	55-58
H19	Not recommended	870-900	1600-1650	22	40	207-241	815	1500	1095-1205	2000-2200	2-5	A, O	52-55
Tungsten hot work tool steels													
H20	Not recommended	870-900	1600-1650	22	40	207-235	815	1500	1095-1205	2000-2200	2-5	A, O	53-55
H21	Not recommended	870-900	1600-1650	22	40	207-235	815	1500	1095-1205	2000-2200	2-5	A, O	43-52
H22	Not recommended	870-900	1600-1650	22	40	207-235	815	1500	1095-1205	2000-2200	2-5	A, O	48-57
H23	Not recommended	870-900	1600-1650	22	40	212-255	815	1500	1205-1260	2200-2300	2-5	O	33-35(e)
H24	Not recommended	870-900	1600-1650	22	40	217-241	815	1500	1095-1230	2000-2250	2-5	A, O	44-55
H25	Not recommended	870-900	1600-1650	22	40	207-235	815	1500	1150-1260	2100-2300	2-5	A, O	46-53
H26	Not recommended	870-900	1600-1650	22	40	217-241	870	1600	1175-1260	2150-2300	2-5	A, O, S	63-64
Molybdenum hot work tool steels													
H41	Not recommended	815-870	1500-1600	22(f)	40(f)	207-235	730-845	1350-1550	1095-1190	2000-2175	2-5	A, O, S	64-66
H42	Not recommended	845-900	1550-1650	22	40	207-235	730-845	1350-1550	1120-1220	2050-2225	2-5	A, O, S	54-62
H43	Not recommended	815-870	1500-1600	22(g)	40(g)	207-235	730-845	1350-1550	1095-1190	2000-2175	2-5	A, O, S	54-58

(a) Lower limit of range should be used for small sections, upper limit for large sections. Holding time varies from about 1 h, for light sections and small furnace charges, to about 4 h, for heavy sections and large charges; for pack annealing, hold for 1 h per inch of pack cross section. (b) Maximum rate to 425 °C (800 °F) unless footnoted to indicate otherwise. (c) A, air; O, oil; S, salt. (d) For open-furnace heat treatment. For pack hardening, hold for $^1/_2$ h per inch of pack cross section. (e) Temper to precipitation harden. (f) To 540 °C (1000 °F). (g) to 480 °C (900 °F).

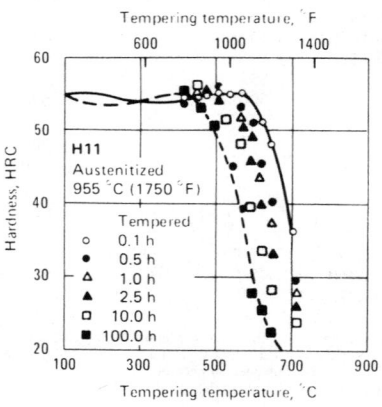

Fig. 5. Effect of tempering time and temperature on hardness of air-cooled type H11 hot work tool steel

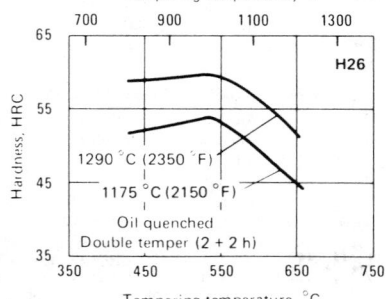

Fig. 6. Effect of tempering temperature on hardness of type H26 hot work tool steel for two different austenitizing temperatures

Table 6. Recommended heat treating practice for high speed tool steels

Steel	Normalizing	Annealing Temperature(a), °C	Annealing Cooling rate(b) °C/h	Annealed hardness, HB	Temperature Preheat, °C	Temperature Austenitize(c), °C	Hardening Holding time, min	Quenching medium(d)	Quenched hardness, HRC
Tungsten high speed tool steels, standard group									
T1	Not recommended	870-900	20	217-255	815-870	1260-1300	2-5	O, A, S	63-65
T2	Not recommended	870-900	20	223-255	815-870	1260-1300	2-5	O, A, S	64-66
T4	Not recommended	870-900	20	229-269	815-870	1260-1300	2-5	O, A, S	64-66
T5	Not recommended	870-900	20	235-285	815-870	1275-1300	2-5	O, A, S	64-66
T6	Not recommended	870-900	20	248-302	815-870	1275-1300	2-5	O, A, S	64-66
T8	Not recommended	870-900	20	229-255	815-870	1260-1300	2-5	O, A, S	64-66
T15	Not recommended	870-900	20	241-277	815-870	1205-1260	2-5	O, A, S	65-67
Molybdenum high speed tool steels, standard group									
M1	Not recommended	815-870	20	207-235	730-845	1175-1220	2-5	O, A, S	64-66
M2	Not recommended	870-900	20	212-241	730-845	1190-1230	2-5	O, A, S	65-66
M3	Not recommended	870-900	20	223-255	730-835	1205-1230	2-5	O, A, S	64-66
M4	Not recommended	870-900	20	223-255	730-845	1205-1230	2-5	O, A, S	64-66
M6	Not recommended	870	20	248-277	790	1175-1205	2-5	O, A, S	63-66
M7	Not recommended	815-870	20	217-255	730-845	1175-1220	2-5	O, A, S	64-65
M10	Not recommended	815-870	20	207-255	730-845	1175-1220	2-5	O, A, S	64-66
M30	Not recommended	870-900	20	235-269	730-845	1205-1230	2-5	O, A, S	64-66
M33	Not recommended	870-900	20	235-269	730-845	1205-1230	2-5	O, A, S	64-66
M34	Not recommended	870-900	20	235-269	730-845	1205-1230	2-5	O, A, S	64-66
M36	Not recommended	870-900	20	235-269	730-845	1220-1245	2-5	O, A, S	64-66
M41	Not recommended	870-900	20	235-269	730-845	1190-1215	2-5	O, A, S	63-66
M42	Not recommended	870-900	20	235-269	730-845	1165-1190	2-5	O, A, S	63-66
M43	Not recommended	870-900	20	248-269	730-845	1150-1175	2-5	O, A, S	63-66
M44	Not recommended	870-900	20	248-285	730-845	1200-1225	2-5	O, A, S	63-66
M46	Not recommended	870-900	20	235-269	730-845	1190-1220	2-5	O, A, S	63-66
M47	Not recommended	870-900	20	235-269	730-845	1175-1205	2-5	O, A, S	63-66
High speed tool steels, intermediate group									
M50	Not recommended	830-845	20	197-235	730-845	1095-1120	2-5	O, A, S	63-65
M52	Not recommended	830-845	20	197-235	730-845	1120-1175	2-5	O, A, S	63-65

(a) Pack annealing is recommended, for minimum decarburization. Steels should be held at temperature for 1 h per inch of thickness of the container. (b) Maximum. Rate is not critical after work (in pack, if employed) has been furnace cooled to 650 °C (1200 °F). (c) If steels are austenitized in a salt bath, austenitizing temperatures should be 14 °C (25 °F) lower than those in the ranges given. (d) O, oil; A, air; S, salt.

Table 7. Recommended heat treating practice for low-alloy special-purpose tool steels

Steel	Normalizing temperature(a) °C	Normalizing temperature(a) °F	Annealing Temperature(b) °C	Annealing Temperature(b) °F	Cooling rate(c) °C/h	Cooling rate(c) °F/h	Annealed hardness, HB	Austenitizing temperature(d) °C	Austenitizing temperature(d) °F	Hardening Holding time, min	Quenching medium(e)	Quenched hardness, HRC(f)
L1	900	1650	775-800	1425-1475	22	40	179-207	790-845	1450-1550	10-30	O, W	64
L2	870-900	1600-1650	760-790	1400-1450	22	40	163-197	790-845	1450-1550	10-30	W	63
L3	900	1650	790-815	1450-1500	22	40	174-201	845-925	1550-1700	10-30	O	63
								775-815	1425-1500	10-30	W	64
								815-870	1500-1600	10-30	O	64
L6	870	1600	760-790	1400-1450	22	40	183-212	790-845	1450-1550	10-30	O	62
L7	900	1650	790-815	1450-1500	22	40	183-212	815-870	1500-1600	10-30	O	64

(a) Holding time, after uniform through heating, varies from about 15 min, for small sections, to about 1 h, for large sections. Work is cooled from temperature in still air. (b) Lower limit of range should be used for small sections, upper limit for large sections. Holding time varies from about 1 h, for light sections and small furnace charges, to about 4 h, for heavy sections and large charges; for pack annealing, hold for 1 h per inch of pack cross section. (c) Maximum. Rate is not critical after cooling to below 540 °C (1000 °F). (d) These steels are seldom preheated. (e) O, oil; W, water. (f) Typical average values; subject to variations depending on austenitizing temperature and quenching medium.

Table 8. Recommended heat treating practice for mold steels

Steel	Normalizing temperature(a) °C	Annealing Temperature(b) °C	Annealing Cooling rate(c) °C/h	Annealed hardness, HB	Carburizing temperature °C	Hardening (after carburizing) Austenitizing temperature, °C	Hardening (after carburizing) Holding time, min	Quenching medium(d)	Quenched hardness, HRC
P1	Not required	730-900	20	81-101	900-925	790-800	15	W, B	62-64
P2	Not required	730-815	20	103-123	900-925	830-845	15	O	62-65
P3	Not required	730-815	20	109-137	900-925	800-830	15	O	62-64
P4	Not recommended	870-900	15	116-128	970-995	970-995	15	A	62-65
P5	Not required	845-870	20	105-116	900-925	845-870	15	O, W	62-65
P6	Not required	845	8	183-217	900-925	790-815	15	A, O	60-62
P20	900	760-790	20	149-179	870-900(e)	815-870	15	O	58-64
P21	900	Not recommended				Hardened by solution treating and aging(f)			

(a) Holding time, after uniform through heating, varies from about 15 min, for small sections, to about 1 h, for large sections. Work is cooled from temperature in still air. (b) Lower limit of range should be used for small sections, upper limit for large sections. Holding time varies from about 1 h, for light sections and small furnace charges, to about 4 h, for heavy sections and large charges; for pack annealing, hold for 1 h per inch of pack cross section. (c) Maximum. Rate is not critical after cooling to below 540 °C (1000 °F). (d) W, water; B, brine; O, oil; A, air. (e) When applicable. (f) Solution treatment: Hold at 705 to 730 °C (1300 to 1350 °F) for 1 to 3 h, quench in air or oil; approximate solution treated hardness, 24 to 28 HRC. Aging treatment: Reheat to 510 to 550 °C (950 to 1025 °F); approximate aged hardness, 40 to 30 HRC.

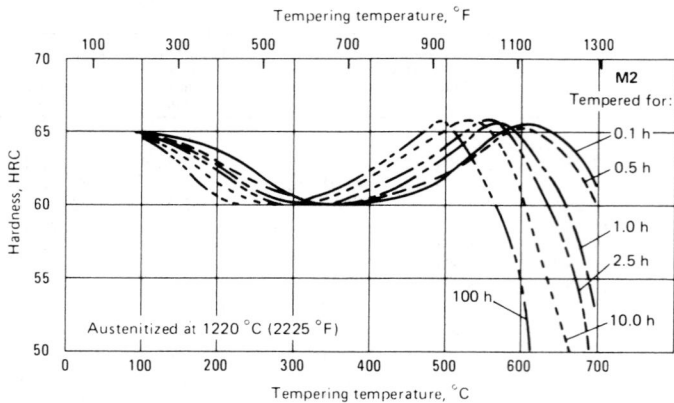

Fig. 7. Effect of tempering temperature and time on hardness of type M2 high speed steel

After its temperature has been equalized in the salt quench, the tool is air cooled. For large cutters heated in a furnace, an interrupted oil quench is often used to minimize quenching strains and prevent cracking. This consists of cooling the cutters in the oil only until they lose color (about 540 °C or 1000 °F) and then cooling them in air.

Tempering. As shown in Fig. 7 for an M2 steel austenitized at 1220 °C (2225 °F), the hardness of high speed steel is directly affected by tempering temperature and time. From the slope of the curves in Fig. 7, it can be seen that M2 undergoes secondary hardening at temperatures above approximately 370 °C (700 °F) and that secondary hardening proceeds at higher temper-

atures up to about 595 °C (1100 °F), depending on time at temperature. These temperatures approximate the practical limits for most tempering operations; lower temperatures do not evoke the secondary hardening response, and higher temperatures produce hardnesses considerably lower than those usually desired.

High speed steels normally are subjected to a minimum of two separate tempering treatments within the range of 540 to 595 °C (1000 to 1100 °F). The duration of each treatment is usually 2 h or more at temperature. This process ensures attainment of consistent martensitic structures, because the amount of retained austenite in the as-quenched condition will vary signifi-

cantly because of variations in heat chemistry, prior thermal history, hardening temperature and quenching conditions.

LOW-ALLOY SPECIAL-PURPOSE TOOL STEELS

Nominal compositions of low-alloy special-purpose tool steels are given in Section 18.

Because of their relatively low austenitizing temperatures, L steels are easily heat treated. Recommended heat treating practice is summarized in Table 7.

Tempering. Tools made of L steels should be quenched only to a temperature at which they can be handled with bare hands, about 50 °C (125 °F), and should be tempered immediately thereafter; otherwise, cracking is likely to occur. The tempering characteristics of these steels are similar to those of O grades (see Fig. 3).

MOLD STEELS

The principal use of these type P steels is for plastic molds. However, some steels, such as P4, P20 and P21, are used also for die-casting dies. The several types vary widely in composition. When molds are carburized or nitrided, the same procedures are used as for production steels.

Heat treating practice for mold steels is summarized in Table 8. P21 is a special type heat treated by the manufacturer and delivered ready for the user to machine and place in operation without further treatment. As noted in Table 8, this steel is hardened by solution treating and aging.

Heat Treating of Stainless Steels and Heat-Resisting Alloys

Heat Treating of Stainless Steels

AUSTENITIC STAINLESS STEELS

Austenitic stainless steels may be divided into five groups: (*a*) conventional austenitics, such as types 301, 302, 303, 304, 305, 308, 309, 310, 316 and 317; (*b*) stabilized compositions, primarily types 321, 347 and 348; (*c*) low-carbon grades, such as types 304L, 316L and 317L; (*d*) high-nitrogen grades, such as AISI types 201, 202, 304N and 316N, and the Nitronic series of alloys; and (*e*) highly alloyed austenitics.

Conventional austenitics cannot be hardened by heat treatment but will harden as a result of cold working. These steels are usually purchased in an annealed or cold worked state. Following welding or thermal processing, a subsequent re-anneal may be required for optimum corrosion resistance, softness and ductility. During annealing, chromium carbides, which markedly decrease resistance to intergranular corrosion, are dissolved. Annealing temperatures, which vary somewhat with the composition of the steel, are given in Table 1.

Because carbide precipitation can occur at temperatures between 425 and 900 °C (800 and 1650 °F), it obviously is desirable that the annealing temperature should be safely above this limit. Moreover, because all carbides should be in solution before cooling begins, and because the chromium carbide dissolves slowly, the highest practical temperature consistent with limited grain growth is selected. This temperature is in the vicinity of 1095 °C (2000 °F).

Cooling from the annealing temperature must be rapid, but it must also be consistent with limitations of distortion. Whenever considerations of distortion permit, water quenching is used, thus ensuring that dissolved carbides remain in solution (because it precipitates carbides more rapidly, type 310 invariably requires water quenching).

FERRITIC STAINLESS STEELS

Ferritic stainless steels may be divided into two groups: (*a*) conventional ferritics, such as types 409, 430, 434 and 446; and (*b*) low-interstitial ferritics, such as types 439, 444, E-BRITE, SEA-CURE, AL 29-4C and AL 29-4-2. Ferritic stainless steels are not hardened by quenching but

rather develop minimum hardness and maximum ductility, toughness and corrosion resistance in the annealed and quenched condition. Therefore, the only heat treatment applied to ferritics is annealing. This treatment relieves stresses developed during welding or cold working and provides a more homogeneous structure by dissolving transformation products formed during welding. Postweld heat treatment of low-interstitial ferritic stainless steels is generally unnecessary and is frequently undesirable. Table 2 summarizes current annealing practices for ferritic grades.

MARTENSITIC STAINLESS STEELS

Heat treating of martensitic stainless steel is essentially the same as for plain carbon or low-alloy steels, in that maximum strength and hardness depend chiefly on carbon content. The principal metallurgical difference is that the high alloy content of the stainless grades causes the transformation to be so sluggish, and the hardenability to be so high, that maximum hardness is produced by air cooling in the center of sections up to approximately 30.5 cm (12 in.) thick.

Annealing. Temperatures and resulting hardnesses for process (subcritical) annealing, full

Table 1. Recommended annealing temperatures for austenitic stainless steels

UNS No.	Designation	Temperature(a) °C	Temperature(a) °F
Conventional grades			
S30100, S30200, S30215	....301, 302, 302B	1010 to 1120	1850 to 2050
S30300, S30323	..........303, 303Se	1010 to 1120	1850 to 2050
S30400, S30500, S30800	....304, 305, 308	1010 to 1120	1850 to 2050
S30900, S30908	.........309, 309S	1040 to 1120	1900 to 2050
S31000, S31008	......310, 310S	1040 to 1065	1900 to 1950
S31600	316	1040 to 1120	1900 to 2050
S31700	317	1065 to 1120	1950 to 2050
Stabilized grades			
S32100	321	955 to 1065	1750 to 1950
S34700, S34800	.......347, 348	980 to 1065	1800 to 1950
N08020	.................Carpenter 20Cb-3	925 to 955	1700 to 1750
Low-carbon grades			
S30403	.........304L, 304LN	1010 to 1120	1850 to 2050
S31603, S31703	.........316L, 316LN, 317L	1040 to 1110	1900 to 2025
High-nitrogen grades			
S20100, S20200	...........201, 202	1010 to 1120	1850 to 2050
S30451	................304N	1010 to 1120	1850 to 2050
S31651	................316N	1010 to 1120	1850 to 2050
S24100	...........Nitronic 32, Carpenter 18Cr-2Ni-12Mn	1010 to 1065	1850 to 1950
S24000	...........Nitronic 33	1040 to 1095	1900 to 2000
S21904	...........Nitronic 40, Carpenter 21Cr-6Ni-9Mn	980 to 1175	1800 to 2150
S20910	...........Nitronic 50, Carpenter 22Cr-13Ni-5Mn	1065 to 1120	1950 to 2050
S21800	...........Nitronic 60	1040 to 1095	1900 to 2000
S28200	.................Carpenter 18-18 PLUS	1040 to 1095	1900 to 2000
Highly alloyed grades			
....	317LM, 317LX, 317L PLUS, 317LMO, 7L4	1120 to 1150	2050 to 2100
....	JS700, JS777	1065 to 1150	1950 to 2100
N08904	.........904L, AL-4X, 2RK65	1075 to 1125	1965 to 2055
N08028	.................Sanicro 28	...	...
N08366	.................AL-6X	1205 to 1230	2200 to 2250
S31254	.................254 SMO	1150 to 1205	2100 to 2200

(a) Temperatures given are for annealing a composite structure. Time at temperature and method of cooling depend on thickness. Light sections may be held at temperature for 3 to 5 min per 2.5 mm (0.10 in.) of thickness, followed by rapid air cooling. Thicker sections are water quenched. For many of these grades, a postweld heat treatment is not necessary. For proprietary alloys, alloy producers may be consulted for details. Although cooling from the annealing temperature must be rapid, it must also be consistent with limitations of distortion.

Table 2. Recommended annealing treatments for ferritic stainless steels

UNS No.	Designation	Treatment temperature °C	Treatment temperature °F
Conventional ferritic grades			
S40500	405	650-815	1200-1500
S40900	409	870-900	1600-1650
S43000	430	705-790	1300-1450
S43020	430F	705-790	1300-1450
S43400	434	705-790	1300-1450
S44600	446	760-830	1400-1525
Low-interstitial ferritic grades			
S43035	439	870-925	1600-1700
S44400	444	955-1010	1750-1850
S44626	E-BRITE	760-955	1400-1750
S44660	SEA-CURE, SC-1	1010-1065	1850-1950
...	AL 29-4C	1010-1065	1850-1950
S44800	Al 29-4-2	1010-1065	1850-1950
S44635	MONIT	1010-1065	1850-1950

Note: Postweld heat treating of low-interstitial ferritic stainless steels is generally unnecessary and frequently undesirable. Any annealing of these grades should be followed by water quenching or very rapid cooling.

Table 3. Annealing temperatures and procedures for wrought martensitic stainless steels

	Process (subcritical) annealing Temperature(a), °C	Hardness	Full annealing Temperature(b)(c), °C	Hardness	Isothermal annealing(c) Procedure(d)	Hardness
Type						
403, 410 ...	650-760	82-92 HRB	830-885	75-85 HRB	Heat to 830 to 885 °C; hold 6 h at 705 °C	85 HRB
414	650-730	99 HRB-24 HRC	Not recommended		Not recommended	
416, 416(Se) ..	650-760	86-92 HRB	830-885	75-85 HRB	Heat to 830 to 885 °C; hold 2 h at 720 °C	85 HRB
420	675-760	94-97 HRB	830-885	86-95 HRB	Heat to 830 to 885 °C; hold 2 h at 705 °C	95 HRB
431	620-705	99 HRB-30 HRC	Not recommended		Not recommended	
440A	675-760	90 HRB-22 HRC	845-900	94-98 HRB	Heat to 845 to 900 °C; hold 4 h at 690 °C	98 HRB
440B 440C,	675-760	98 HRB-23 HRC	845-900	95 HRB-20 HRC	Same as 440A	20 HRC
440F ...	675-760	98 HRB-23 HRC	845-900	98 HRB-25 HRC	Same as 440A	25 HRC

(a) Air cool from temperature; maximum softness is obtained by heating to temperature at high end of range. (b) Soak thoroughly at temperature within range indicated; furnace cool to 790 °C; continue cooling at 15 to 25 °C/h to 595 °C; air cool to room temperature. (c) Recommended for applications in which full advantage may be taken of the rapid cooling to the transformation temperature and from it to room temperature. (d) Preheating to a temperature within the process annealing range is recommended for thin-gage parts, heavy sections, previously hardened parts, parts with extreme variations in section or with sharp re-entrant angles, and parts that have been straightened or heavily ground or machined to avoid cracking and minimize distortion, particularly for types 420 and 431, and 440A, B, C and F.

annealing and isothermal annealing are given in Table 3. Full annealing is an expensive and time-consuming treatment; it should be used only when required for subsequent severe forming. Types 414 and 431 do not respond to full or isothermal annealing procedures within a reasonable soaking period.

Austenitizing temperatures, soaking times, quenching media and tempering temperatures are summarized in Table 4. The 440 grades require all the extra precautions that are taken, to prevent quench cracking, during quenching of high-hardenability steels. When maximum corrosion resistance and strength are desired, the steel should

be austenitized at the high end of the temperature range. For alloys that are to be tempered above 565 °C (1050 °F), the low side of the austenitizing range is recommended, because it enhances ductility and impact properties.

PRECIPITATION-HARDENING STAINLESS STEELS

Recommended procedures for homogenization, austenite conditioning, transformation cooling and precipitation hardening (age-tempering) of a semiaustenitic precipitation-hardening stainless steel are given in Table 5. For procedures pertaining to other precipitation-hardening grades, see Metals Handbook, 9th Edition, Volume 4.

Heat Treating of Heat-Resisting Alloys

STRESS RELIEVING

Stress relieving of heat-resisting alloys and refractory metals frequently entails a compromise; the desirability of maximum relief of residual stress must be weighed against possible effects deleterious to high-temperature properties and corrosion resistance.

True stress relieving of wrought material usually is confined to alloys that are not age-hardenable. Thus, the time and temperature cycles may vary considerably, depending on the metallurgical characteristics of the alloy and on the type and magnitude of residual stresses developed by previous fabricating processes.

Stress-relieving temperatures are usually below the annealing or recrystallization temperatures. Typical cycles for wrought alloys are listed in Table 6; temperatures at least 25 °C (50 °F) higher or lower than those listed are usually satisfactory.

ANNEALING

When applied to heat-resisting alloys, annealing implies full annealing—that is, complete re-

Table 4. Procedures for hardening and tempering wrought martensitic stainless steels to specific strength and hardness levels

Type	Austenitizing(a) Temperature(b) °C	Austenitizing(a) Temperature(b) °F	Quenching medium(c)	Tempering temperature(d) °C min	Tempering temperature(d) °C max	Tempering temperature(d) °F min	Tempering temperature(d) °F max	Tensile strength MPa	Tensile strength ksi	Hardness, HRC
403, 410	925 to 1010	1700 to 1850	Air or oil	565	605	1050	1125	760 to 965	110 to 140	25 to 31
				205	370	400	700	1105 to 1515	160 to 220	38 to 47
414	925 to 1050	1700 to 1925	Air or oil	595	650	1100	1200	760 to 965	110 to 140	25 to 31
				230	370	450	700	1105 to 1515	160 to 220	38 to 49
416, 416(Se)	925 to 1010	1700 to 1850	Oil	565	605	1050	1125	760 to 965	110 to 140	25 to 31
				230	370	450	700	1105 to 1515	160 to 220	35 to 45
420	985 to 1065	1800 to 1950	Air or oil(e)	205	370	400	700	1550 to 1930	225 to 280	48 to 56
431	985 to 1065	1800 to 1950	Air or oil(e)	565	605	1050	1125	860 to 1035	125 to 150	26 to 34
				230	370	450	700	1210 to 1515	175 to 220	40 to 47
440A	1010 to 1065	1850 to 1950	Air or oil(e)	150	370	300	700	...	...	49 to 57
440B	1010 to 1065	1850 to 1950	Air or oil(e)	150	370	300	700	...	...	53 to 59
440C, 440F	1010 to 1065	1850 to 1950	Air or oil(e)	...	160	...	325	...	...	60 min
				...	190	...	375	...	...	58 min
				...	230	...	450	...	...	57 min
				...	355	...	675	...	...	52 to 56

(a) Preheating to a temperature within the process annealing range is recommended for thin-gage parts, heavy sections, previously hardened parts, parts with extreme variations in section or with sharp re-entrant angles, and parts that have been straightened or heavily ground or machined, to avoid cracking and minimize distortion, particularly for types 420, 431, and 440A, B, C and F. (b) Usual time at temperature ranges from 30 to 90 min. The low side of the austenitizing range is recommended for all types subsequently tempered to 25 to 31 HRC; generally, however, corrosion resistance is enhanced by quenching from the upper limit of the austenitizing range. (c) Where air or oil is indicated, oil quenching should be used for parts more than 6.4 mm ($^1/_4$ in.) thick; martempering baths at 150 to 400 °C (300 to 750 °F) may be substituted for an oil quench. (d) Generally, the low end of the tempering range of 150 to 370 °C (300 to 700 °F) is recommended for maximum hardness, the middle for maximum toughness, and the high end for maximum yield strength. Tempering in the range of 370 to 565 °C (700 to 1050 °F) is not recommended, because it results in low and erratic impact properties and poor resistance to corrosion and stress corrosion. (e) For minimum retained austenite and maximum dimensional stability, a subzero treatment −75 °C ± 10 °C (−100 °F ± 20 °F) is recommended; this should incorporate continuous cooling from the austenitizing temperature to the cold transformation temperature.

Table 5. Recommended heat treating procedures for a semiaustenitic precipitation-hardening stainless steel (UNS S17400)

Homogenization. 1175 ± 15 °C (2150 ± 25 °F), 2 h + 30 min per 25 mm (1 in.)(a)

Austenite conditioning (solution treatment). 1040 ± 15 °C (1900 ± 25 °F), 30 min + 30 min per 25 mm (1 in.)(a)

Transformation cooling. To below +30 °C (+90 °F)

Precipitation hardening. To obtain minimum tensile strengths shown, use the following treatments for wrought alloys(b):

Tensile strength MPa	ksi	Treatment
1310	190	1 h at 480 ± 5 °C (900 ± 10 °F)
1170	170	4 h at 495 ± 5 °C (925 ± 10 °F)
1070	155	4 h at 550 ± 5 °C (1030 ± 10 °F)
1030	150	4 h at 565 ± 5 °C (1050 ± 10 °F)
1000	145	4 h at 580 ± 5 °C (1075 ± 10 °F)
930	135	4 h at 620 ± 5 °C (1150 ± 10 °F)

(a) To prevent cracking and ensure uniform properties, cool as follows: 75 mm (3 in.) and less, oil quench or air cool; 75 to 150 mm (3 to 6 in.), air cool; 150 mm (6 in.) and over, air cool under cover. *All parts must be cooled to below +30 °C (+90 °F)* prior to the precipitation-hardening cycle. (b) If hardness exceeds maximum specified, reheat treat at a slightly higher temperature for a minimum of 30 min.

crystallization and the attainment of maximum softness. The practice is usually applied to wrought alloys of the nonhardening type. For a majority of the hardenable alloys, annealing cycles are the same as those used for solution treating. However, the two treatments serve different purposes. Annealing is used mainly to increase ductility (and reduce hardness) to facilitate forming or machining, prepare for welding, relieve stresses after welding, produce specific microstructures, or soften age-hardened structures by re-solution of second phases. Solution treating is intended to dissolve second phases to produce maximum corrosion resistance or to prepare for aging. Additionally, it will homogenize microstructure prior to aging.

Annealing practices vary considerably among different plants. Representative annealing temperatures, holding times, and cooling procedures are given in Table 6.

Table 6. Typical stress-relieving and annealing cycles for wrought heat-resisting alloys

Alloy	Stress relieving Temperature °C	Stress relieving Temperature °F	Stress relieving Holding time per inch of section, h	Annealing(a) Temperature °C	Annealing(a) Temperature °F	Annealing(a) Holding time per inch of section, h
Iron-base and iron-nickel-chromium alloys						
RA-330	900	1650	1(b)	1110(c)	2025(c)	$^1/_4$(d)
19-9 DL	675(e)	1250(e)	4	980	1800	1
A-286	(f)	(f)	...	980	1800	1
Discaloy	(f)	(f)	...	1035	1900	1
Nickel-base alloys						
Astroloy	(f)	(f)	...	1135	2075	4
Hastelloy B	(f)	(f)	...	1175	2150	1
Hastelloy C	(f)	(f)	...	1215	2225	1
Hastelloy W	(f)	(f)	...	1175	2150	1
Hastelloy X	(f)	(f)	...	1175	2150	1
Incoloy 800	870	1600	$1^1/_2$	980	1800	$^1/_4$
Incoloy 800H	...	...	...	1175	2150	...
Incoloy 825	...	...	...	980	1800	...
Incoloy 901	(f)	(f)	...	1095	2000	2
Inconel 600	900	1650	1	1010	1850	$^1/_4$(d)
Inconel 601	...	...	...	980	1800	...
Inconel 625	870	1600	1	980	1800	1
Inconel 690	...	...	...	1040	1900	$^1/_2$
Inconel 718	(f)	(f)	...	955	1750	1
Inconel X-750	880(g)	1625(g)	...	1035	1900	$^1/_2$
Nimonic 80A	(f)	(f)	...	1080	1975	2
Nimonic 90	(f)	(f)	...	1080	1975	2
René 41	(f)	(f)	...	1080	1975	2
Udimet 500	(f)	(f)	...	1080	1975	4
Udimet 700	(f)	(f)	...	1135	2075	4
Waspaloy	(f)	(f)	...	1010	1850	4
Cobalt-chromium-nickel-base alloys						
L-605 (HS-25)	(h)	(h)	...	1230	2250	1
N-155 (HS-95)	(h)	(h)	...	1175	2150	...
S-816	(h)	(h)	...	1205	2200	1
Refractory metals(j)						
Ta-10W	1205(k)	2200(k)	1	1425(k)	2600(k)	1
FS-80	1095(k)	2000(k)	1	1315(k)	2400(k)	1
FS-82	1095(k)	2000(k)	1	1315(k)	2400(k)	1
Mo-0.5 Ti	1095(m)	2000(m)	$^1/_2$	1315(m)(n)	2400(m)(n)	1
TZM	1205(m)	2200(m)	1	1425(m)(n)(p)	2600(m)(n)(p)	1

(a) Minimum hardness is achieved by cooling rapidly from the annealing temperature, to prevent precipitation of hardening phases. Water quenching is preferred, and is usually necessary for heavy sections; air cooling is preferred for heavy sections of Waspaloy, Udimet 500, Udimet 700 and Inconel X-750, because water quenching causes cracking. However, for complex shapes subject to excessive distortion, oil quenching is often adequate and more practical. Rapid air cooling usually is adequate for parts formed from strip or sheet. Rapid cooling from the annealing or solution treating temperature does not suppress the aging reaction of some alloys, such as Astroloy; these alloys become harder and stronger. (b) Time given is minimum; some plants use as long as 3 h per inch. (c) Nominal temperature; 1035 to 1175 °C (1900 to 2150 °F) is commonly used. (d) Short time is required for prevention of grain coarsening. (e) Nominal temperature; 650 to 705 °C (1200 to 1300 °F) is permissible. (f) Full annealing is recommended, because intermediate temperatures cause aging. (g) Used only for stress equalizing of warm worked grades. (h) Full annealing is recommended, if further fabrication is performed; otherwise, material can be stress relieved at approximately 55 °C (100 °F) below annealing temperature. (j) Annealing temperatures depend on prior plastic deformation, degree of cold work, alloy content and interstitial purity. Annealing temperatures given are those most frequently used for cold worked sheet or plate; in many instances, more precise determination of the recrystallization temperature is necessary for a specific application. (k) Heat and cool in vacuum or inert-gas atmosphere. (m) Heat and cool in hydrogen or vacuum. (n) Seldom used as finished product in annealed condition, because recrystallization raises the ductile-brittle transition temperature, resulting in brittleness at low temperatures. (p) For vacuum-arc-cast material with a minimum of 50% cold work.

Table 7. Typical solution-treating and aging cycles for wrought heat-resisting alloys

Alloy	Solution treating Temperature °C	°F	Time, h	Cooling procedure	Aging Temperature °C	°F	Time, h	Cooling procedure
Iron-base alloys								
A-286	980	1800	1	Oil quench	720	1325	16	Air cool
Discaloy	1010	1850	2	Oil quench	730	1350	20	Air cool
					650	1200	20	Air cool
N-155	1175	2150	1	Water quench	815	1500	4	Air cool
Nickel-base alloys								
Astroloy	1175	2150	4	Air cool	845	1550	24	Air cool
	1080	1975	4	Air cool	760	1400	16	Air cool
Hastelloy B	1175	2150	1/2	(a)	(b)	(b)	...	...
Hastelloy B-2	1065	1950	1/2	Rapid quench	...	...	...	...
Hastelloy C-4	1065	1950	1/2	Rapid quench	...	...	...	...
Hastelloy C-276	1120	2050	1/2	Rapid quench	...	...	...	...
Hastelloy N	1175	2150	1/2	Rapid quench	...	...	...	...
Hastelloy S	1065	1950	1/2	Rapid quench	...	...	...	...
Hastelloy C	1220	2225	1	(a)	(b)	(b)	...	...
Hastelloy W	1175	2150	1	(a)	(b)	(b)	...	...
Hastelloy X	1175	2150	1	(a)	...	...	...	...
Inconel 901	1095	2000	2	Water quench	790	1450	2	Air cool
					720	1325	24	Air cool
Inconel 600	1120	2050	2	Air cool	...	...	...	...
Inconel 601	1150	2100	1	Air cool	...	...	...	...
Inconel 617	1175	2150	2	(a)	...	...	...	...
Inconel 625	1150	2100	2	(a)	...	...	...	...
Inconel 706	925-1010	1700-1850	...	...	845	1550	3	Air cool
					720	1325	8	Furnace cool
					620	1150	8	Air cool
	925-1010	1700-1850	...	...	730	1350	8	Furnace cool
					620	1150	8	Air cool
Inconel 718	980	1800	1	Air cool	720	1325	8	Furnace cool
					620	1150	8	Air cool
Inconel X-750 (AMS 5667)	855	1625	24	Air cool	705	1300	20	Air cool
Inconel X-750 (AMS 5668)	1150	2100	2	Air cool	845	1550	24	Air cool
					705	1300	20	Air cool
Nimonic 80A	1080	1975	8	Air cool	705	1300	16	Air cool
Nimonic 90	1080	1975	8	Air cool	705	1300	16	Air cool
René 41	1065	1950	1/2	Air cool	760	1400	16	Air cool
Udimet 500	1080	1975	4	Air cool	845	1550	24	Air cool
					760	1400	16	Air cool
Udimet 700	1175	2150	4	Air cool	845	1550	24	Air cool
	1080	1975	4	Air cool	760	1400	16	Air cool
Waspaloy	1080	1975	4	Air cool	845	1550	24	Air cool
					760	1400	16	Air cool
Cobalt-base alloys								
Haynes 25; L-605	1230	2250	1	Rapid air cool	(b)	(b)	...	...
Haynes 188	1175	2150	1/2	Rapid air cool	...	...	...	...
Haynes 556	1175	2150	1/2	Rapid air cool	...	...	...	...
S-816	1175	2150	1	(a)	760	1400	12	Air cool
Stellite 6B	1230	2250	1	Air cool	...	...	...	...

Note: Alternate treatments may be used to improve specific properties. (a) To provide an adequate quench after solution treating, it is necessary to cool below about 540 °C (1000 °F) rapidly enough to prevent precipitation in the intermediate temperature range. For sheet metal parts of most alloys, rapid air cooling will suffice. Oil or water quenching is frequently required for heavier sections that are not subject to cracking. (b) Aging occurs in service at elevated temperatures.

SOLUTION TREATING AND AGING

Solution treating and aging practices for iron, nickel- and cobalt-base heat-resisting alloys are summarized in Table 7.

PROTECTIVE ATMOSPHERES

Protective atmospheres are used in annealing or solution treating if heavy oxidation cannot be tolerated. If oxidation can be tolerated, because of subsequent stock removal, heat-resisting alloys can be solution treated in air or in the normal mixture of air and combustion products found in gas-fired furnaces. However, refractory metals must always be heat treated in a vacuum or in an inert-gas atmosphere (argon, helium, or an ArHe mixture) or hydrogen. In some cases, ceramic coatings are used to prevent surface attack.

Exothermic Atmosphere. A lean and dilute exothermic atmosphere is relatively safe and economical. The surface scale formed in such an at-mosphere can be removed by pickling or by salt bath descaling and pickling. Such an atmosphere, formed by burning fuel gas with air, contains about 85% nitrogen, 10% carbon dioxide, 1.5% carbon monoxide, 1.5% hydrogen, and 2% water vapor. This atmosphere will produce a scale rich in chromium oxides.

Endothermic atmospheres prepared by reacting fuel gas with air in the presence of a catalyst are not recommended, because of their carburizing potential. Similarly, the endothermic mixture of nitrogen and hydrogen formed by dissociating ammonia is not used, because of the probability of nitriding.

Dry hydrogen (dew point, −50 °C [−60 °F] or lower) is used in preference to dissociated ammonia for bright annealing of heat-resisting alloys. Hydrogen is not recommended for bright annealing of alloys containing significant amounts of elements (such as aluminum or titanium) that form stable oxides not reducible at normal heat treating temperatures and dew points. Hydrogen is not recommended for annealing or solution treating alloys that contain boron, because of the danger of deboronization through formation of boron hydrides. Nor can hydrogen be used for heat treating niobium and tantalum, because of its embrittling effect.

Dry argon (dew point, −50 °C [−60 °F] or lower) should be used if no oxidation can be tolerated. It is mandatory that this type of atmosphere be used in a sealed retort or sealed furnace chamber. A purge of at least ten times the volume of the retort is recommended before the retort is placed in the furnace. The argon must be kept flowing continually during and after the treatment until the workpieces have cooled nearly to room temperature, to prevent the formation of an oxide film.

Heat-resisting alloys containing stable-oxide formers such as aluminum and titanium, with or without boron, must be bright annealed in a vacuum or in a chemically inert gas such as argon. If used, argon must be pure and dry—dew point, −50 °C (−60 °F) or lower.

Principles of Heat Treatment of Nonferrous Alloys

By Charlie R. Brooks, University of Tennessee

THE PRINCIPLES which govern heat treatment of metals and alloys are applicable, of course, to both ferrous and nonferrous alloys. However, in practice there are sufficient differences to make it convenient to emphasize as separate topics the peculiarities of the alloys of each class in their response to heat treatment. For example, in nonferrous alloys, eutectoid transformations, which play such a prominent role in steels, are seldom encountered, so that the principles associated with time-temperature-transformation diagrams and with martensite formation are not emphasized in this review (they are covered in another section of this handbook). On the other hand, the principles associated with chemical homogenization of cast structures are applicable to many alloys in both classes.

Examination of the heat treatments used for nonferrous alloys reveals that a wide variety of processes are employed. However, because the process of diffusion underlies nearly all heat treatments, the concepts of diffusion are summarized first in this article. Annealing after cold working is a very important heat treatment for nonferrous alloys, and this topic is discussed next. Then the subject of homogenization annealing is reviewed, because it is an important heat treatment for as-cast structures. The process of precipitation, and the hardening that accompanies it, are described next, because these phenomena are especially important in Al-base alloys (and also in some Mg-, Cu- and Ni-base alloys). Then, to illustrate the formation of structures in which two phases are present in comparable quantities (e.g., Ti-base alloys, some Cu brasses, etc.), the heat treatments of a specific type of Cu-Zn alloy are examined. Finally, references are listed which provide additional information on the principles of heat treatment of nonferrous alloys.

DIFFUSION IN METALS AND ALLOYS

In heat treatment of metals and alloys, the rate of structural changes is usually controlled by the rate at which the atoms in the lattice change position. Thus, when cold worked copper is annealed and softens, or an aluminum-base alloy is aged, we are interested in how the atoms move relative to each other so as to bring about the observed changes in properties. The movement of the atoms involved here is called diffusion, and it is this process of diffusion which is examined in this section.

Diffusion in Pure Metals (Self-Diffusion). Atoms in a lattice at finite temperature are not static, but are vibrating in three dimensions around the normal atom position, usually the lattice site. Thus, consideration arises as to whether these atoms, by some mechanism, can exchange positions with each other and thereby move through the lattice. Such movement of the atoms of a pure metal is termed self-diffusion, and it is usually detected by experiments in which a thin layer of a radioactive atom is placed on the surface (e.g., by plating) of the same metal which is not radioactive and then the sample is given an annealing treatment at sufficient temperature and for sufficient time to allow diffusion. Because the difference between the radioactive and nonradioactive atoms is in the nuclear structure, and not in the valence electrons which are related to bonding, it is assumed that the radioactive atoms move through the lattice by the same mechanism and at the same rate as do the nonradioactive atoms. Thus, the movement of the radioactive atoms, which can be followed by a suitable radioactivity detector, reflects the type of movement the atoms in the metal undergo.

Such an experiment is illustrated schematically in Fig. 1. The radioactive layer is depicted as only two atoms thick, whereas it will really be much thicker (e.g., 1 mm). The sequence of time from 0 to t_3 shows increasing amounts of radioactive atoms (closed circles) moving into the lattice of the nonradioactive atoms (open circles), and simultaneously the lattice sites of the radioactive atoms are occupied by the nonradioactive atoms. The amount of radioactivity is measured as a function of depth into the sample from the surface, giving the profiles shown at the bottom of the figure.

Vacancies. The movement of atoms in the lattice, as depicted in Fig. 1, can be conceived to occur by several mechanisms. For example, at any instant of time, it is possible that the nearest two neighboring atoms have vibrated in directions so that space is left around the two atoms, allowing them to exchange positions simultaneously. Such an event is depicted in Fig. 2(a). It is clear that the two atoms which exchange positions must move, to some extent, the neighboring atoms in order to pass each other during the exchange process. It may also be possible for four atoms to vibrate at some instant so that they move cooperatively in a ring, allowing all four to move simultaneously to new neighboring positions, as depicted in Fig. 2(b).

Although mechanisms such as those just suggested probably occur in some alloys, in most

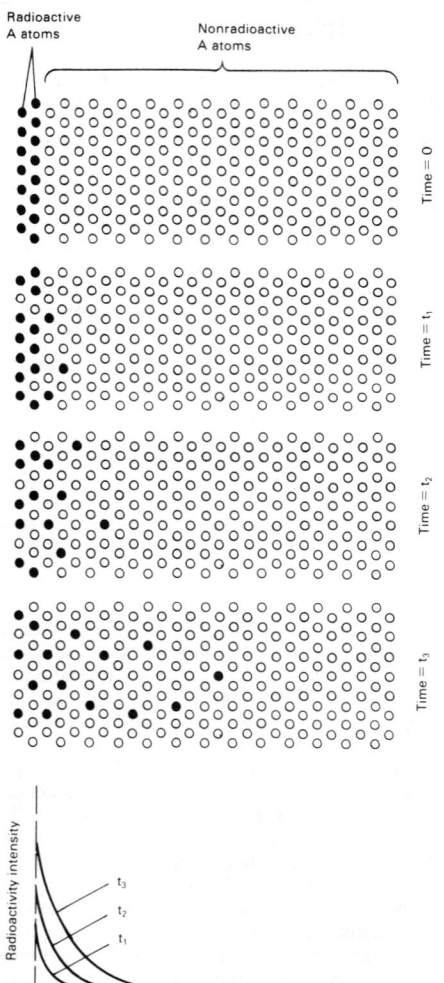

Fig. 1. Schematic diagram showing self-diffusion in a pure metal (radioactive atoms represented by filled circles)

Radioactive A atoms — Nonradioactive A atoms

Time = 0
Time = t_1
Time = t_2
Time = t_3

Radioactivity intensity

t_3
t_2
t_1

Distance from surface

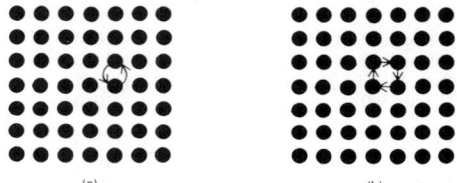

(a)   (b)

In (a), two atoms move simultaneously to exchange positions. In (b), four atoms move cooperatively to rotate simultaneously to move to new positions.

Fig. 2. Schematic representation of two possible diffusion mechanisms

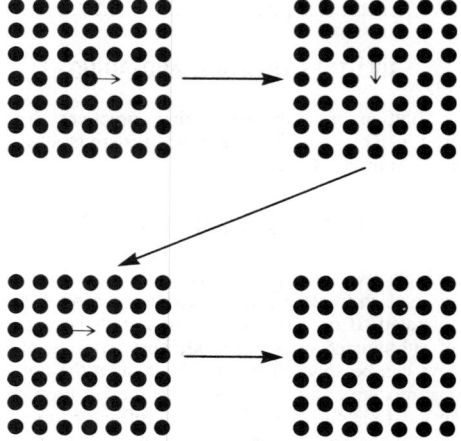

The vacancy moves to the new positions with time as shown by the small arrows. The large arrows show the changes with time.

Fig. 3. Schematic depiction of diffusion by vacancy movement

metals and alloys diffusion occurs by vacancy movement. An unoccupied normal atom position in the crystal structure (usually a lattice site) is a vacancy. The presence of vacancies in a lattice at equilibrium is a consequence of a balance between the energy required to form the vacancies ΔH and the entropy ΔS created by their presence. Thus, there is an equilibrium concentration which minimizes the free energy change ($\Delta G = \Delta H - T\Delta S$).

If a vacancy exists in a lattice, then it requires much less energy for an atom to change positions than in the mechanisms depicted in Fig. 2. An atom has only to move into the vacancy, with much less energy. Such movement is shown in Fig. 3. It is to be noted that the diffusion occurs by rather random movement of the vacancies throughout the lattice.

Diffusion in Alloys (Chemical Diffusion). When two metals (or alloys) are placed in contact, atoms will begin to migrate across the contacting interface. Such diffusion of unlike species is called chemical diffusion, and is illustrated schematically in Fig. 4. (For the process to occur as shown in Fig. 4, the metals have to be soluble in each other; otherwise, when sufficient amounts of one metal diffuse into the other to reach a concentration corresponding to the solubility limit, precipitation of a second phase occurs.) The chemical diffusion depicted in Fig. 4 actually occurs by vacancy diffusion.

Fick's Laws of Diffusion. The mathematical relation that connects the concentration of the diffusing species with distance is Fick's law, a phenomenological equation which fits well most diffusion data. Fick's first law states that the diffusion flux, J (in one-dimensional diffusion), is given by

$$J = -D(dC/dx)$$

where C is concentration and x is distance. D is a constant at a given temperature, but may be concentration-dependent; it is called the diffusivity or diffusion coefficient. Figure 5 illustrates the relation between these terms and the concentration profile associated with chemical diffusion, such as illustrated in Fig. 4. Figure 6 shows data typical of those obtained by machining thin layers from a diffusion couple and analyzing each for the amount of the metals present. The diffusion flux (if concentration is put in proper units) is defined as the number of atoms of the diffusing species which pass through a plane of unit area, which is normal to the diffusion direction, per unit time. Thus, the flux may be given in terms of number of atoms per square centimetre per second.

The effect of time t on the flux is incorporated in Fick's second law (again, for one-dimensional diffusion):

$$dC/dt = D\,d/dx(dC/dx) = D\frac{d^2C}{dx^2}$$

If the diffusion couple consists of two pure metals A and B that are completely soluble in each other, the solution to this equation is

$$C_A = 1/2\,[1 - \phi\,(x/2\,Dt)]$$

where ϕ is the Gauss error function and C_A is the concentration of A at distance x from the original interface. (Similar expressions are obtained for different starting conditions—e.g., an alloy coupled against a pure metal, etc.) To extract D,

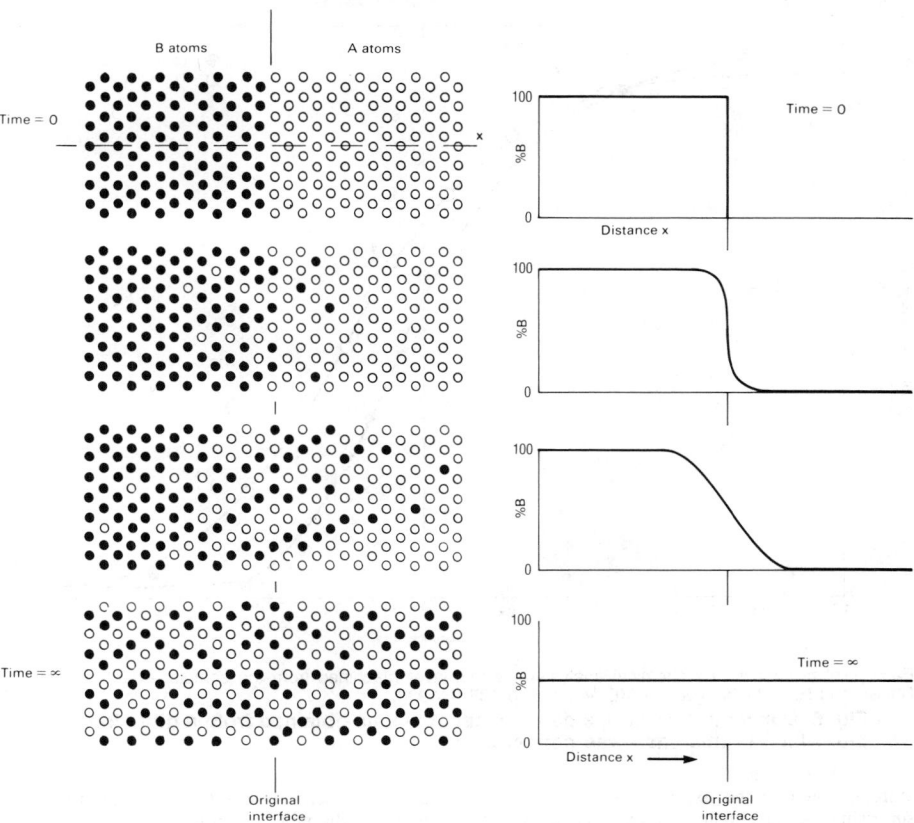

The diffusion couple is made up of pure B (filled circles) and pure A (open circles). As time progresses, mixing on the two sides occurs. At infinite time, complete mixing has been achieved, with the chemical composition being identical on both sides.

Fig. 4. Schematic illustration of chemical diffusion involving two different metals

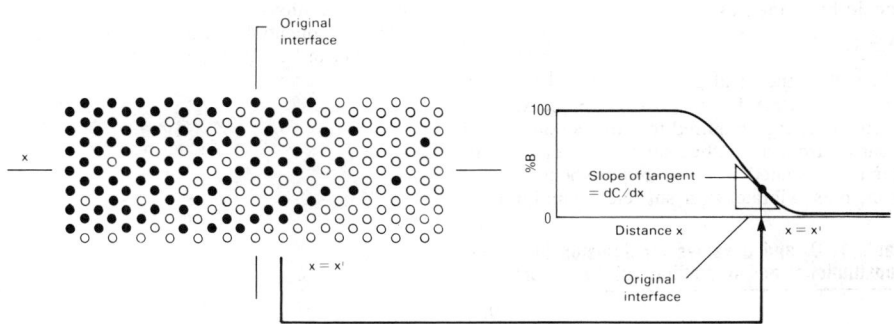

Flux of atoms across the plane at x = x′ is the number of atoms crossing a plane 1 cm square per unit time (s) and is proportional to the gradient dC/dx at that location (x = x′):
$$J = -D(dC/dx)$$
The proportionality constant is the diffusivity or diffusion coefficient. The negative sign is required to make the flux positive to be physically realistic, as the gradient dC/dx is negative.

Fig. 5. Illustration of the meaning of the terms in Fick's first law of diffusion

then, for a given diffusion time t at a given distance x, the value of C_A is obtained (e.g., read from Fig. 6). This allows a value of ϕ to be obtained. Then, error function tables are used to determine the argument of ϕ—that is, to determine a value for (x/2 Dt). Then D is obtained.

Such a procedure should yield the same value of D no matter what value of x is chosen. However, it is found that D will usually vary, meaning that it is a function of composition. In this case, the equation to use is

$$dC/dt = d/dx(D\,dc/dx)$$

The solution is more complicated, but allows determination of the diffusion coefficient as a function of composition.

An important practical relation evolves from the solution to Fick's second law—namely, that the time-distance relation for a given concentration C is $x^2 \cong Dt$. This means, for example, that during a homogenization treatment designed to remove the effects of dendritic segregation (coring), the time is proportional to x^2, where x is approximately the dendritic arm spacing.

Temperature Dependence of the Rate of Diffusion. The dependency of the rate of diffusion on temper-

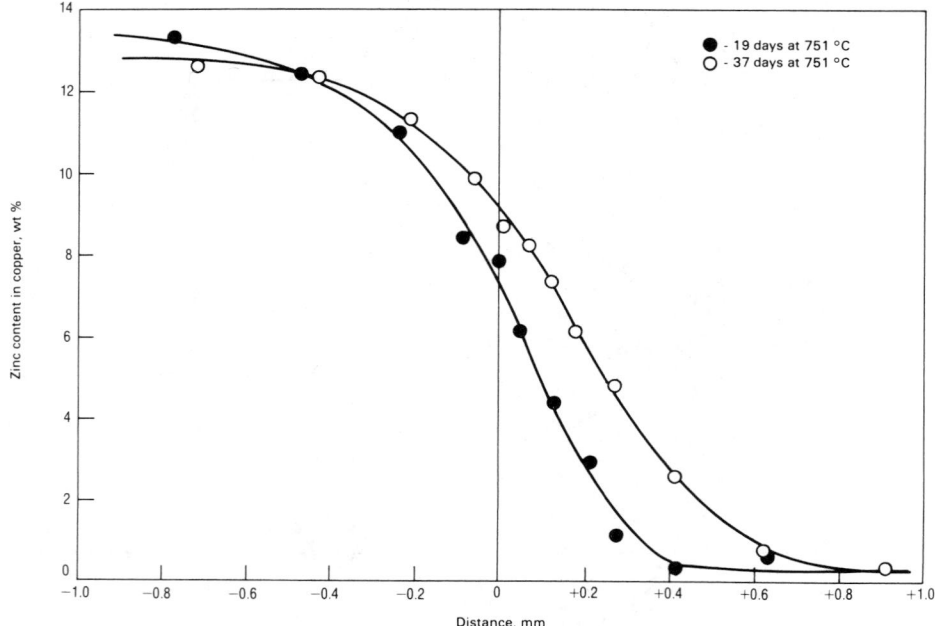

Each point represents the chemical analysis of a thin layer machined from the sample. Adapted from F. N. Rhines and R. F. Mehl, *Trans AIME*, Vol 128, p 185 ff, 1938.

Fig. 6. Concentration profile data typical of metals obtained from a diffusion couple, which in this case was copper:zinc

ature is found to be exponential, which is not surprising, because many rate reactions obey such a dependency. Thus, D is given by

$$D = D_0 \, e^{-B/T}$$

where D_0 and B are constants, and T is absolute temperature. Theoretical treatments show that this should be written as

$$D = D_0 e^{-Q/RT}$$

where R is the ideal gas constant and Q is the activation energy for the diffusion process. Q reflects the energy required to move an atom over a barrier from one lattice site to another; the barrier is associated with the requirement that the atom must vibrate with sufficient amplitude to

break the nearest neighboring bonds in order to move to the new locations.

The values of D_0 and Q shown in Table 1 typify those found in metals. The equation above for the temperature dependency of D predicts that log D plotted versus $1/T$ should be a straight line, and Fig. 7 shows some typical linear results for metals and alloys.

The exponential temperature dependence is important in heat treating. It shows that the rate

of change in processes which are diffusion-controlled will increase greatly with an increase in temperature. Thus, an increase in temperature of 10 K will approximately double the rate of the process.

Intrinsic Diffusion Coefficients. If the original interface of the diffusion couple is identifiable, then experiments show that the location where half of the diffusing species will have moved from one side to the other does not coincide with the original interface. This is sometimes referred to as the *Kirkendall effect*, and is taken as strong experimental evidence of the vacancy mechanism of diffusion in metals. Darken showed that the relation between the measured diffusion coefficient (as described above) and the intrinsic diffusion diffusivities of the individual atom species (for a binary system of atoms A and B) is

$$D = C_A D_A + C_B D_B$$

Here C_A and C_B are the mole fractions of A and B, respectively, and D_A and D_B are the intrinsic diffusivities of A and B, respectively. D_A and D_B are concentration-dependent.

Interstitial Diffusion. If the solute atom is sufficiently small, it will locate in an interstice between the larger solvent atoms, forming an interstitial solid solution. Diffusion of interstitial atoms occurs, not by a vacancy mechanism, but by the atoms jumping from one interstitial site to another. (Fick's laws still apply.) As the interstitial solute atom increases in size, the activation energy increases (Table 1), showing that it becomes more difficult for the atom to move between the solute solvent atoms to a neighboring interstitial site. In general, the activation energy for interstitial diffusion is less than that for substitutional diffusion.

Grain-boundary diffusion. Experimental studies have shown that diffusion along grain boundaries, along the core of dislocations and on free surfaces is considerably more rapid than diffusion through

Table 1. D_0 and Q values for diffusion in various substitutional and interstitial solid solutions

Solute	Solvent (host structure) D_0, cm²/s	Q, calories per mole
Substitutional diffusion		
Cu	Cu 0.78	50 500
Cu	Sn 0.11	45 000
Cu	Ni 1.92	68 000
Ni	Cu 1.1	53 800
Cu	Al 0.647	32 270
Zn	Cu 0.73	47 500
Pb	Pb 0.887	25 500
Ti	Ti 0.000358	31 200
Al (4%)	Cu 0.0455	39 500
Zn (24-29%)	Cu 0.095	35 000
Interstitial diffusion		
H	Cu ..... 10^{-2}	10 000
O	Cu ..... 10^{-3}	46 000
C	Ti 0.00302	20 000
O	Ti 1	40 000
H	Ta	6 000
C	Ta 0.0061	38 520
N	Ta 0.0056	37 840
O	Ta 0.0044	25 450

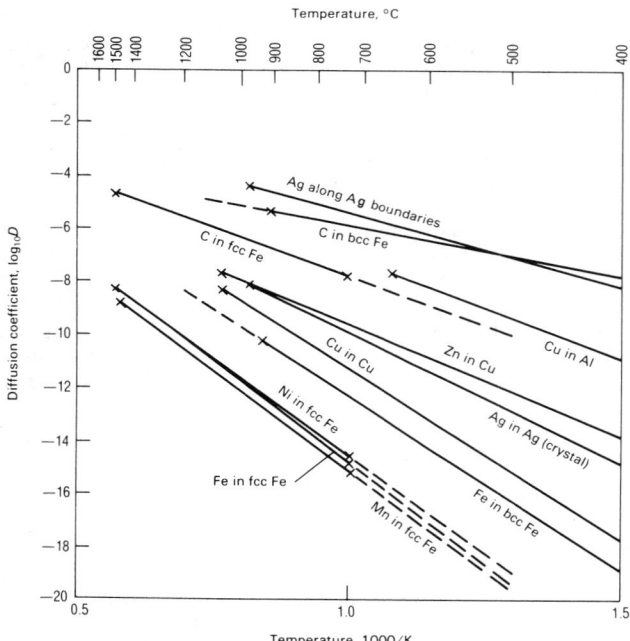

The straight lines are prominent and commonly found. From L. H. Van Vlack, *Elements of Materials Science*, 2nd Ed., Addison-Wesley Publishing Co., Reading, MA, 1964.

Fig. 7. Plots of log D versus $1/T$ for several metals

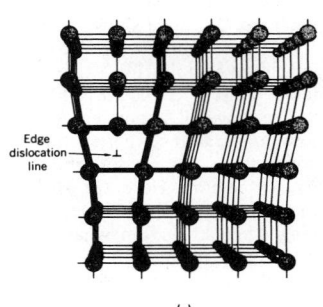 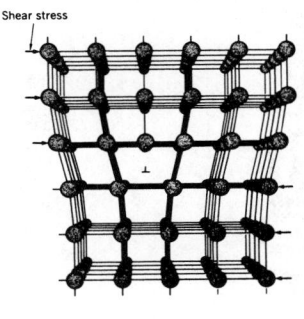 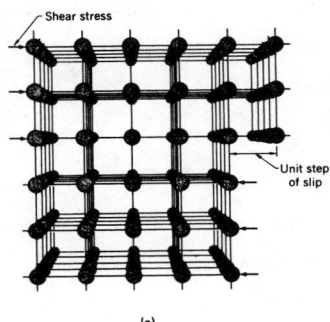

(a) An edge dislocation in a crystal. (b) The dislocation has moved one lattice spacing due to the shearing force. (c) The dislocation has reached the edge of the crystal and produced unit slip. Adapted from A. G. Guy, *Essentials of Materials Science*, McGraw-Hill, New York, 1976.

Fig. 8. The motion of an edge dislocation and the production of a unit step of slip at the surface of the crystal

the interior of a crystal. Of particular interest here is grain-boundary diffusion, which influences precipitation and phase changes at the boundary. The data in Fig. 7 for self-diffusion in silver show that the grain-boundary diffusivity is several orders of magnitude greater than bulk diffusion. Also, as temperature decreases, bulk diffusion becomes slower and grain-boundary diffusion becomes more important.

ANNEALING OF COLD WORKED METALS

Dislocations. Plastic deformation in metals and alloys occurs primarily by relative movement or slip of blocks of material on specific crystallographic planes (slip planes) and in certain directions (slip directions). (Plastic deformation in metals can also occur by twinning. However, in the brief treatment here, this mechanism will not be discussed.) This occurs not by movement of regions of the crystal as a whole, but by movement of successive dislocations. A dislocation is a lattice defect (either edge or screw) which is present in even well-annealed metals as a consequence of prior processing. Dislocations play a central role in plastic deformation because less energy is required to produce slip by movement of the dislocations than by movement of entire regions of a crystal past each other. This process is illustrated in Fig. 8 for an edge dislocation.

Obviously, millions of dislocations must repeat this process in order to generate visually obvious shape changes. This is possible, however, because the dislocations, which are present in the metal prior to plastic deformation, create other dislocations by a multiplication mechanism during plastic deformation.

In hexagonal close-packed crystals, the prominent slip plane is the close-packed (001) plane, and the slip directions in this plane are the close-packed directions, of which there are three nonparallel, identical choices. Thus, this crystal structure exhibits three slip systems. In the face-centered cubic structure, the slip plane is also the close-packed plane {111}. However, in this system there are four types of nonparallel {111} planes. In each plane there are three possible slip directions (⟨110⟩ type), and hence 12 slip systems. In the body-centered cubic structure, the slip plane is of the {110} type (also the most closely packed plane in this system), and the slip directions are of the ⟨111⟩ type, of which there are three in each plane. Thus, the body-centered cubic structure also has 12 slip systems. The types of slip plane and slip direction are sensitive to temperature, and in some alloys other slip systems are activated when temperature changes.

Effect of Cold Working on Properties and Microstructure. The multiplication of dislocations on several slip systems upon plastic deformation leads to their interaction with each other, and this restricts their movement, so that further deformation requires an increase in external load. Thus, the material work (or strain) hardens. This effect is illustrated in Fig. 9, which shows the strengthening induced by deformation in rolling of pure copper, and of copper-zinc solid-solution alloys, at 25 °C. Plastic deformation such that strengthening or hardening occurs is called *cold working;* plastic deformation such that work hardening does not occur is called *hot working.* (Alternative definitions are given below, under "Hot Working.") Note that these definitions have no particular attachment to room temperature.

Cold working increases hardness, yield strength and tensile strength, and lowers ductility. It also increases electrical resistivity because the increasing density of dislocations scatters the electrons. Fig. 10 illustrates the effects of cold working on several properties.

Cold working of a metal causes distortion of grains, and the specific nature of this distortion depends on the type of deformation (e.g., rolling, swaging, etc.). If the plane of observation is parallel to the rolling direction, the grains will appear elongated in the rolling direction. Also

observed in the microstructure are parallel striations within the grains, the density of which increases with the amount of deformation. These striations are actually rows of etch pits, or etched grooves, where the etchant has removed metal preferentially at surface locations at which the dislocations emerge. Such striations are sometimes called *deformation bands.* In metals and alloys which show annealing twins (mainly face-centered cubic metals, such as copper and brass), the twins, originally appearing as straight lines crossing (or nearly crossing) the grains, become bent, distorted and fragmented. All of these microstructural features of cold worked metals are illustrated in Fig. 11.

Recovery, Recrystallization and Grain Growth. In shaping of metals and alloys by cold working, there is a limit to the amount of plastic deformation attainable without fracture. However, proper heat treatment prior to reaching this limit restores the metal or alloy to a structural condition similar to that prior to deformation, and then

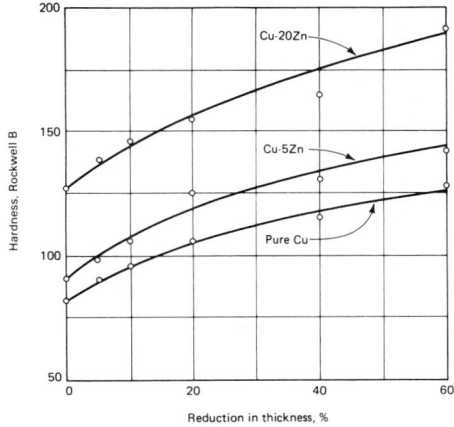

From Charlie R. Brooks, *Heat Treatment, Structure and Properties of Nonferrous Alloys*, American Society for Metals, Metals Park, OH, 1982.

Fig. 9. The effect of plastic deformation (by rolling at 25 °C) on hardness of pure copper and two Cu-Zn solid-solution alloys

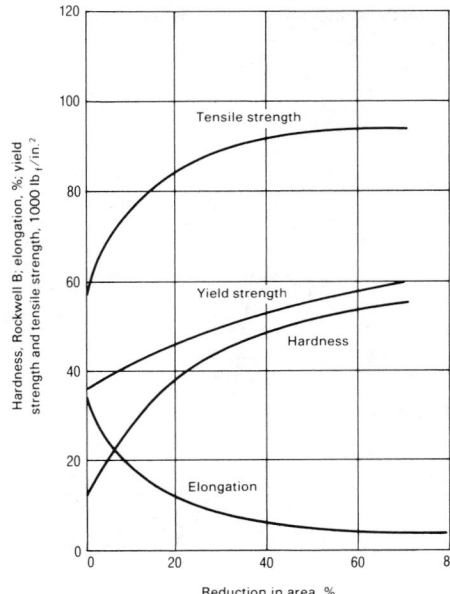

Adapted from R. A. Wilkins and E. S. Bunn, *Copper and Copper Base Alloys*, McGraw-Hill, New York, 1943.

Fig. 10. The effect of cold working (by rolling at 25 °C) on the tensile mechanical properties and hardness of oxygen-free, high-conductivity (OFHC) copper

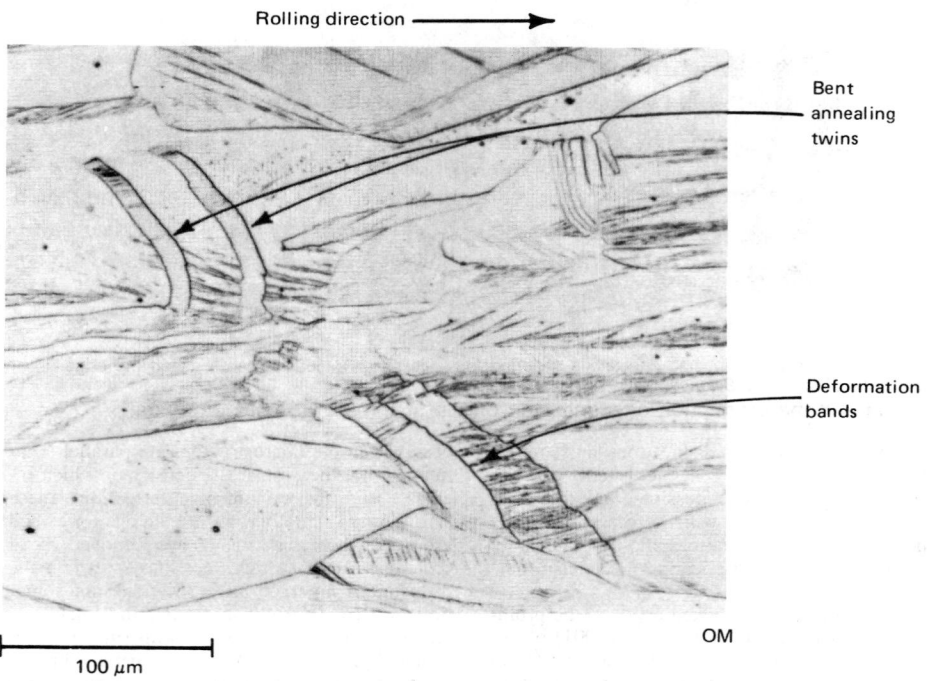

Rolling direction ——————→

Bent annealing twins

Deformation bands

OM

100 μm

From same source as Fig. 9.

Fig. 11. The microstructure of a Cu-15Zn alloy cold rolled at 25 °C to a 40% reduction in thickness, showing deformation bands and bent annealing twins revealed by etching the polished surface

additional cold working can be conducted. This type of heat treatment is called *annealing,* and in this section some of the principles involved and the effects which occur are summarized.

Because cold working produces an increasing concentration of lattice defects (e.g., dislocations), the energy of the crystals is increased. Thus, there is a thermodynamic driving force for the metal to undergo changes which will return

it to the original, low-energy condition. The rates of these changes depend on the mechanisms involved, and are sensitive functions of temperature and alloy.

The changes in strength that occur during annealing are illustrated by the hardness data in Fig. 12(a). The hardness (and the yield and tensile strengths) initially remains approximately constant (or increases slightly), then shows an abrupt

decrease, followed by a continued, but gradual, decrease. The data shown in Fig. 12(a) are for a fixed temperature. A similar result is obtained by annealing samples for a fixed time at increasing temperatures, as shown in Fig. 12(b).

The stage of annealing for short times or at low temperatures wherein the hardness remains constant, or increases slightly, is called the *recovery* region. Here the dislocations undergo movement by thermal activation, being rearranged into arrays somewhat more stable and more difficult to move than in the cold worked, unannealed condition, and hence cause a slight increase in hardness. In this period, such rearrangement allows some properties to attain their values prior to cold working, and hence is referred to as recovery. One such property is electrical resistivity, as illustrated in Fig. 13. The cellular arrangement of the dislocations, compared with that of the cold worked condition, in-

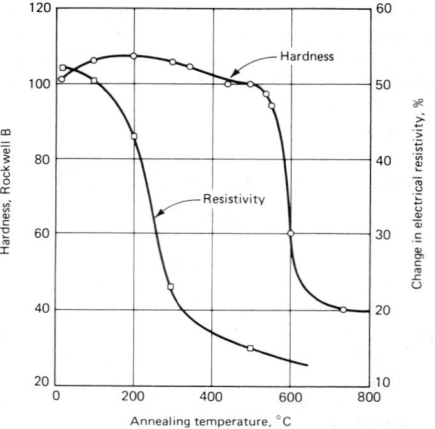

The metal had been cold worked at 25 °C almost to fracture. Annealing time, 1 h. Adapted from J. E. Wilson and L. Thomassen, *Trans ASM*, Vol 22, p 769, 1934.

Fig. 13. Effect of annealing temperature on hardness and electrical resistivity of nickel

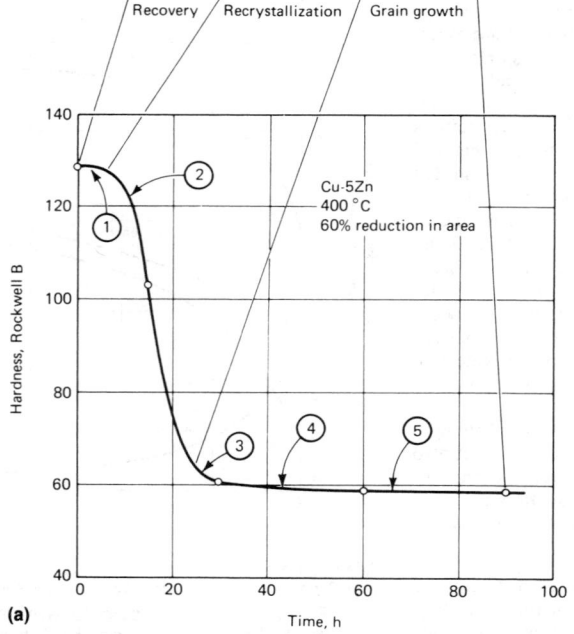

(a)

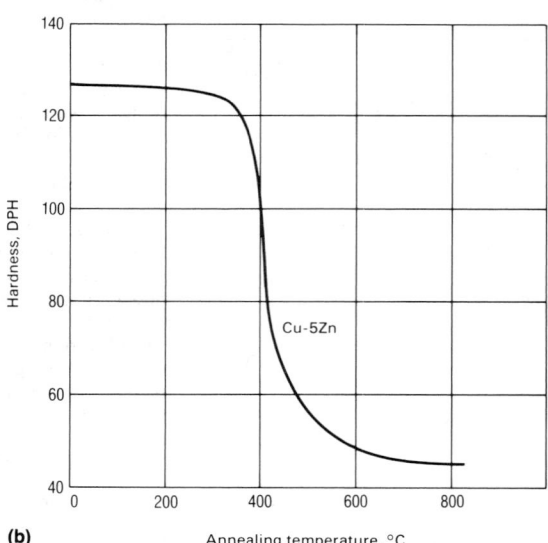

(b)

Part (a) from same source as Fig. 9.

Fig. 12. (a) Effect of annealing time at fixed temperature (400 °C) on hardness of a Cu-5Zn solid-solution alloy cold worked 60%. (b) Effect of annealing temperature at fixed time (15 min) on hardness of a Cu-5Zn solid-solution alloy cold worked 60%.

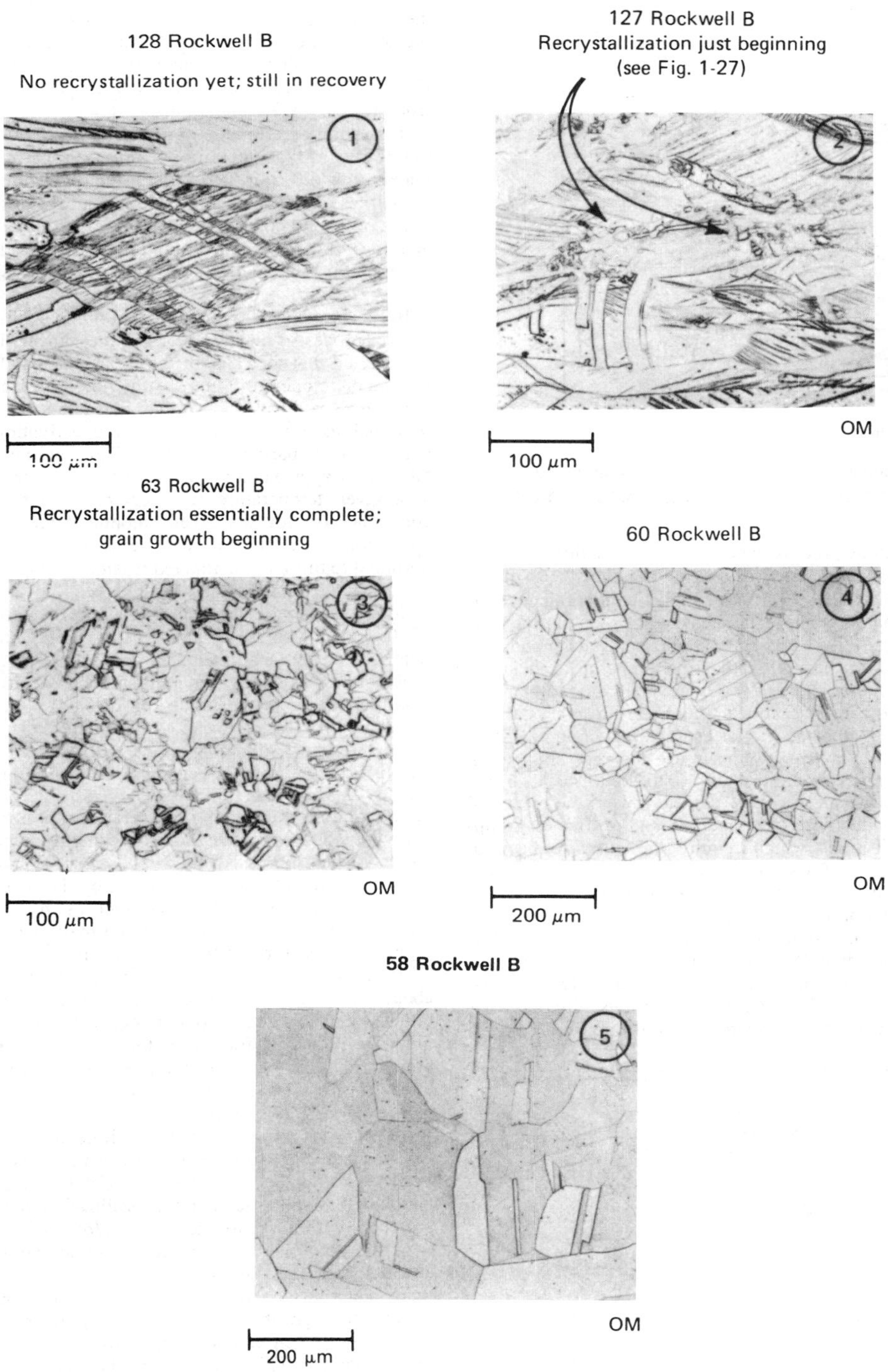

128 Rockwell B

No recrystallization yet; still in recovery

100 μm

127 Rockwell B
Recrystallization just beginning
(see Fig. 1-27)

OM

100 μm

63 Rockwell B
Recrystallization essentially complete;
grain growth beginning

OM

100 μm

60 Rockwell B

OM

200 μm

58 Rockwell B

OM

200 μm

From same source as Fig. 9.

Fig. 14. Microstructure of a Cu-5Zn alloy, cold rolled to 60%, then annealed for different times at 400 °C

strength decreases as grain size increases, during this period the hardness decreases, although only gradually (Fig. 13).

The microstructural changes which occur during annealing are illustrated in Fig. 14. During recovery, there is a decrease in the density of deformation bands, although this effect is not prominent. When recrystallization commences, small, equiaxed grains begin to appear (see micrograph 2 in Fig. 14, and Fig. 15) in the structure. These continue to form and grow until the cold worked matrix is consumed, which marks the end of the recrystallization period and the beginning of grain growth. Further annealing causes only an increase in grain size (see micrographs 3, 4 and 5 in Fig. 14).

Factors Affecting Recrystallization. Because annealing of cold worked metals is usually carried out to soften the material, the temperature and time required to complete recrystallization must be known in order to determine the proper heat treatment. It is common to refer to the *recrystallization temperature* as an indicator of the temperature at which the metal must be annealed for softening. (This temperature can be taken to be that which gives any specified amount of recrystallization.)

Several factors affect the value of the recrystallization temperature. Two of the most important are annealing time and amount of prior cold work. Figure 16 illustrates the effect of annealing time. The longer the time at a given temperature, the farther the metal progresses in the annealing process. Thus, if a metal just commences recrystallization at 200 °C in 15 min, then it may be completely recrystallized in 30 min.

The effect of the amount of prior cold work is illustrated in Fig. 17. Increasing amounts of plastic deformation increase the concentration of lattice defects (e.g., dislocations) and make the metal more thermodynamically unstable. Hence, recrystallization occurs at lower temperatures, or in shorter times, the greater the amount of cold work. Although this is the main effect, it is to be noted that the type of deformation, the rate of deformation and the deformation temperature also affect the rate of recrystallization.

Chemical composition affects the recrystalli-

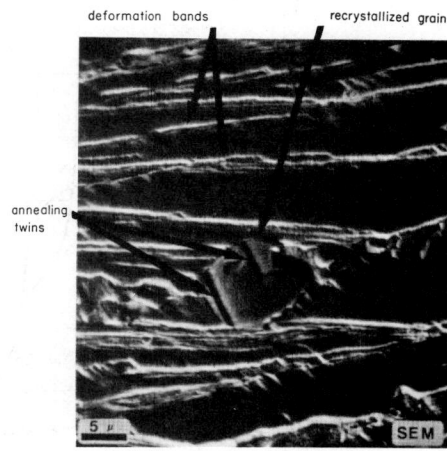

Cu–5% Zn alloy, cold worked by rolling at 20 °C to reduction in thickness of 60%; annealed 60 min at 350 °C. From same source as Fig. 9.

Fig. 15. High-magnification scanning electron micrograph showing a small recrystallized nucleus

creases the mean free path of the electrons and lowers the resistance.

After longer times or at higher temperatures, the structure undergoes a more radical change. Small crystals appear which contain a low dislocation density (of magnitude similar to that prior to cold working) and hence are relatively soft. These crystals nucleate in regions of high dislocation density, and thus in the microstructure appear at or near deformation bands. With time, these nuclei grow, and more nuclei form in the

remaining cold worked matrix. Eventually, these grains contact each other (at that time the original worked material has disappeared). The formation of these grains is referred to as *recrystallization*. During this recrystallization period, strength decreases drastically (Fig. 12 and 13).

Following recrystallization, the energy of the alloy is reduced further by a decrease in the grain-boundary area by grain growth. Thus, the long-time or high-temperature region of the annealing curve is referred to as *grain growth*. Because

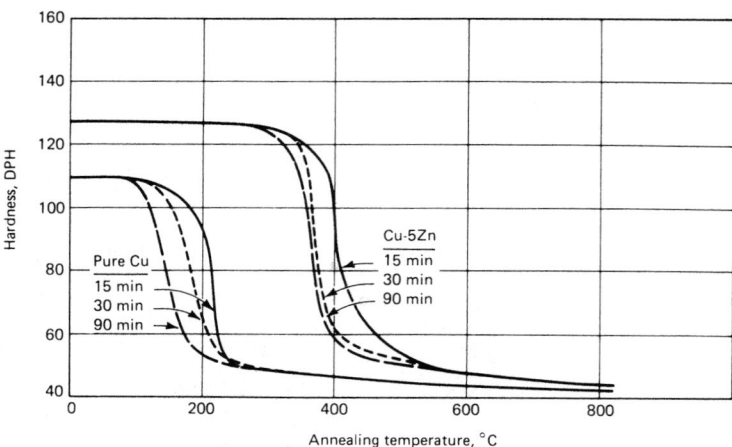

Both materials were originally cold rolled at 25 °C to 60% reduction in thickness. From same source as Fig. 9.

Fig. 16. Illustration of effect of annealing time on the annealing process in pure Cu and a Cu-5Zn alloy

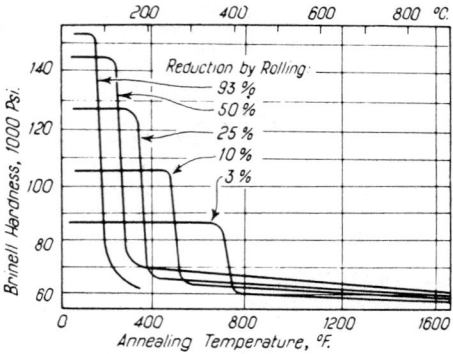

From G. Sachs and K. R. Van Horn, *Practical Metallurgy,* American Society for Metals, Metals Park, OH, 1951.

Fig. 17. Illustration of effect of amount of cold working on the annealing process for pure copper

zation process, and here a distinction must be made between solid-solution alloys and multiphase alloys. In many alloys containing second-phase particles, the presence of such particles favors formation of recrystallization nuclei and thus lowers the recrystallization temperature. In solid-solution alloys, even quite small amounts of solute can have potent effects on the recrystallization temperature. For example, addition of 0.05% Ag to copper will increase the recrystallization temperature from about 140 to about 340 °C. Thus, because silver only slightly lowers the electrical conductivity of copper, this alloy is used in applications which require the alloy to be cold worked for strength but in which slight heating may occur, and stress relaxation and recrystallization must be prevented.

If the solubility is sufficiently high to allow considerable solute concentration, the recrystallization temperature may decrease. This is illustrated in Fig. 18 for Cu-Zn alloys. This effect is expected to be related to the influence of zinc on

the atom mobility in Cu-Zn alloys, and indeed the activation energy Q for diffusion in these alloys increases slightly with additions of up to 10% Zn, then decreases considerably with additions from 10 to 20% Zn.

It is useful here to note a rule of thumb—that the recrystallization temperature is approximately 0.3 to 0.6 of the absolute melting point. In the case of Cu-Zn solid-solution alloys, addition of zinc to copper lowers the melting point, and thus the recrystallization temperature will decrease for high zinc contents (e.g., 20 to 30%) (see Fig. 18).

Because recovery, recrystallization and grain-growth processes all involve atom movement, it is expected that the rates of these processes will depend on temperature in the same functional relation as does diffusion—that is, the rate is proportional to $e^{-Q/RT}$, where Q is the activation energy for the particular process. Thus, we may take as an approximation that the time required at a given temperature for recrystallization to commence (or for any given amount of recrystallization to be attained) will be inversely proportional to this exponential expression. Using as typical activation energies those given for diffusion, it is found that a decrease in temperature of 10 °C may increase by a factor of two the time required for recrystallization to commence.

Abnormal Grain Growth. The recrystallization process referred to in the preceding discussions is sometimes called *primary recrystallization,* to distinguish from other situations which lead to unusually large grains on annealing. Under conditions of very high amounts of plastic deformation and high annealing temperatures, abnormally large grains can develop following primary recrystallization: this is called *secondary recrystallization.* Such behavior is favored by the presence of grain-growth inhibitors, such as insoluble particles (e.g., inclusions). Abnormally large grains can also form if the metal has received a critical, but small, amount of deformation (e.g., about 10% or less) prior to annealing. In this case, primary recrystallization does not occur, but a few grains with less deformation than neighboring grains grow relatively rapidly at the expense of the cold worked grains. This effect is also called *germinative grain growth.*

Hot Working. Alternative definitions of cold working and hot working to those given previously can now be presented. *Cold working* is plastic deformation such that recrystallization does not occur within a reasonable time. *Hot working* is plastic deformation at or above a temperature at which recrystallization occurs in a rather short time. Thus, if the deformation temperature is sufficiently high, the metal cannot be cooled rapidly enough, even in a short time, to prevent recrystallization. This rather "spontaneous" recrystallization is depicted in Fig. 19.

HOMOGENIZATION OF CASTINGS

One of the most important commercial heat treatments is homogenization of castings. Such a treatment is used prior to mechanical processing of the cast ingot, and it is often used even when an object is cast into essentially the final shape. The temperatures and times used depend on the diffusion rate and the starting structure (the latter dictates the concentration gradients and the diffusion path). To understand how this enters into the situation, it is important to know how solidification occurs in alloys, and especially how

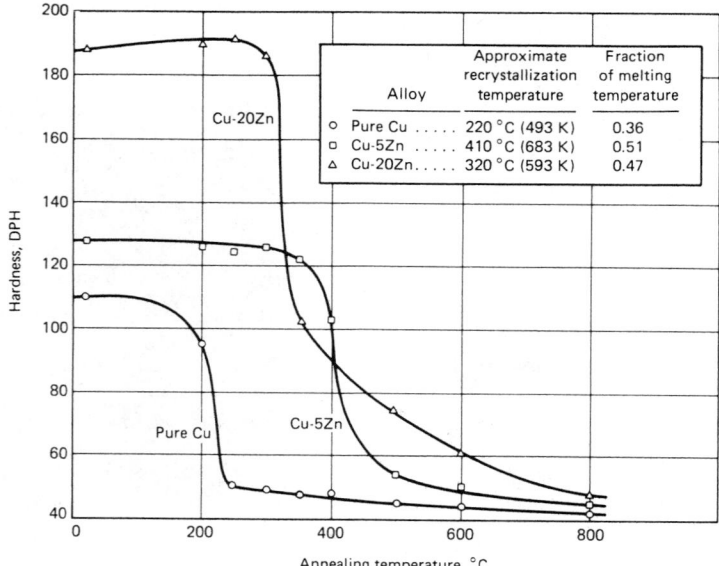

Alloy	Approximate recrystallization temperature	Fraction of melting temperature
○ Pure Cu	220 °C (493 K)	0.36
□ Cu-5Zn	410 °C (683 K)	0.51
△ Cu-20Zn	320 °C (593 K)	0.47

The alloys were originally cold rolled at 25 °C to 60% reduction in thickness. The recrystallization temperatures listed are based on the inflection point of each curve. From same source as Fig. 9.

Fig. 18. Illustration of effect of Zn content of Cu-Zn solid-solution alloys on the annealing process

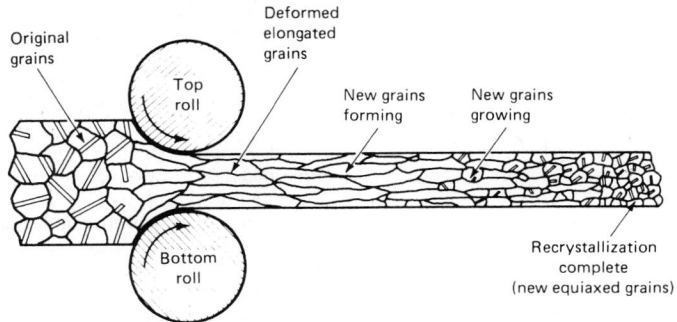

From R. A. Grange, in *Fundamentals of Deformation Processing*, ed. E. A. Backofen, J. J. Burke, L. F. Coffin, N. T. Reed and V. Weiss, Syracuse Univ. Press, Syracuse, N.Y., 1964, as adapted from J. M. Camp and C. B. Francis, *The Making, Shaping and Treating of Steel*, 5th Ed., U.S. Steel Corp., Pittsburgh, 1940.

Fig. 19. Schematic illustration of the change in grain structure on hot rolling

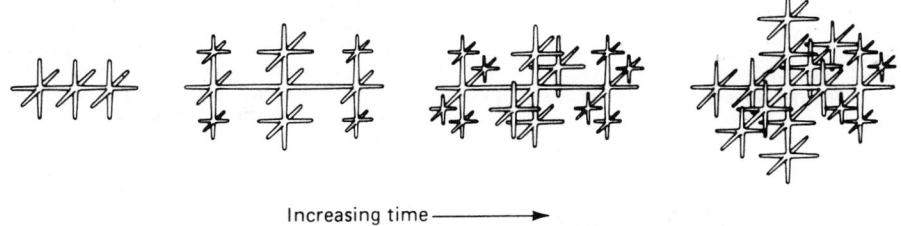

Increasing time ⟶

Adapted from P. S. Hurd, *Metallic Materials*, Holt, Rinehart and Winston, New York, 1968.

Fig. 20. Schematic illustration of a dendritic crystal forming in a liquid

chemical segregation develops during solidification.

Dendrite Formation. In metals and alloys, the crystals which form in the liquid during freezing generally have a configuration consisting of a main branch with many appendages. A crystal of such a morphology is called a *dendrite* ("fern-like"), and its formation is illustrated schematically in Fig. 20. During freezing, many crystals form, usually on the cold sidewalls of the mold, but also in the center of the casting. As these dendritic crystals grow, they eventually become large enough so that impingement occurs. Then the remaining liquid freezes, with a boundary formed between the differently oriented grains. The original dendritic pattern may not be apparent by observation of only the geometry of the grain boundaries outlining the grains.

Coring. In solidification of most alloys, chemical segregation intrinsically accompanies dendrite formation. To see how this develops, consider a hypothetical alloy whose phase diagram is that shown in Fig. 21. On slow cooling of a liquid alloy containing 30% B, crystallization commences at temperature T_0. The chemical composition of this crystal will be 10% B. As cooling continues, the crystal grows in size (as a dendrite). The phase diagram shows that the equilibrium composition of the crystal must follow the solidus line (line abc in Fig. 21). Thus, the crystal continuously changes its chemical composition, approaching 30% B as the temperature approaches that of completion of freezing, T_2. At T_2, the metal consists of crystals each containing uniformly 30% B. Note that the center of each crystal corresponds to the original nucleus, which had only 10% B when freezing commenced. Thus, on cooling, as the dendrites increased in size, from each layer frozen onto the crystal some B atoms must move throughout the crystal, including some to the center, to maintain the chemical composition uniformly at the value

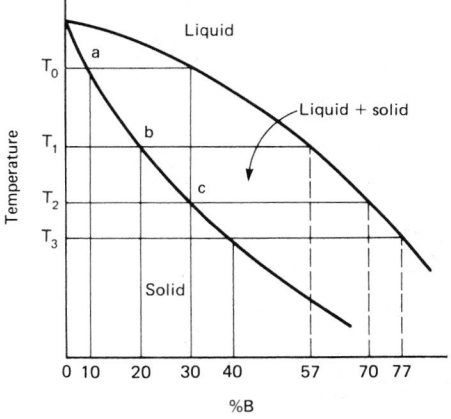

Adapted from same source as Fig. 9.

Fig. 21. Hypothetical phase diagram of system A-B, showing the composition of the solid as a 30% B alloy freezes

given by the solidus at any given temperature.

Clearly, such atom movements require finite time, and the question naturally arises as to what deviations from equilibrium will occur if the alloy is cooled rapidly from the liquid. A simplified picture of what occurs is as follows. On rapid cooling, the first crystals to form have a composition of 10% B. As these grow, the interface between the liquid and the solid crystals maintains the chemical composition given in the phase diagram. Thus, when the crystal has grown as the temperature has decreased from T_0 to T_1, the outside of the crystal will have a composition of 20% B. However, due to the rapid cooling, the center of the crystal will still be 10% B. Between the center and the outside, the composition varies

smoothly between 10 and 20% B. The rapid cooling has not allowed sufficient time for significant diffusion to occur in this composition gradient. On slow cooling, freezing would be complete when the temperature reached T_2. However, at this temperature the outside of the crystal has a composition of 30% B, but the center only 10% B. Thus the average composition of the crystal is somewhere between 10 and 30% B. Freezing cannot be complete until the average composition reaches 30% B (since this is the composition of the alloy), and hence undercooling occurs. Layers continue to add to the dendrite, until the sidearms impinge, and finally all of the dendrites impinge on each other, and freezing is complete. In the example used here, the last layer to freeze, when the sidearms make contact, contains 40% B (Fig. 21).

The frozen structure consists of dendrites in which the central regions of the main branch and of the sidearms contain about 10% B, and the regions where the sidearms met on completion of freezing contain about 40% B. If an etchant is used for which the rate of attack of the metal is sensitive to this compositional difference, then certain regions will be dissolved or attacked more readily than others. The surface then will consist of low and high regions, which reflect light differently, causing contrast in the appearance of the microstructure. An example is shown in Fig. 22 for a Ni-Cu alloy containing 30% Cu. Note that at low magnification the uneven etching has revealed the dendritic structure of the crystals.

This chemically segregated, dendritic structure is referred to as *cored*, and the process of its formation is called *coring*.

Chemical Homogenization Annealing. The chemical gradients in a dendritically cored structure can be reduced to an acceptable level by annealing at a sufficiently high temperature for a sufficient time. The rate of diffusion is given by an appropriate solution to Fick's law. As an approximation, the required time is $x^2 \cong Dt$, where x is the distance between the regions of low and of high concentration in the dendrite cell, which is one-half of the cell size. As an example, in Fig. 22(d) the cell size is approximately 40 μm, so x = 20 μm. Taking $D = 2 \times 10^{-10}$ cm²/s at 1000 °C for a Ni-30Cu alloy, then the required homogenization time is about 6 h. At 1100 °C, $D = 10^{-9}$, and the required time is 1 h. Obviously, higher temperatures lower considerably the required time, but other factors, such as excessive oxidation, must be considered.

If an ingot with a cored, cast structure, such as that shown in Fig. 22, is reduced in thickness 50% by rolling, then the dendritic cells will (on the average) be elongated in the rolling direction but reduced in thickness 50% in the through-thickness direction of the rolled plate. Thus the effective diffusion distance x becomes about 10 μm. Then, at 1000 °C, the required homogenization annealing time becomes about 1 h, instead of the 6 h for the as-cast structure. This points out the advantage in processing of coupling a homogenization anneal with plastic deformation to remove coring present in the as-cast structure.

In many alloy ingots, there also occurs gross, or ingot, segregation, where the chemical composition of the outside of the ingot may be different from that along its centerline. Here the final liquid freezes, and rejection of solute elements (frequently impurities) from the advancing front of the freezing crystals, in which they have a lower solubility, results in a region rich in these ele-

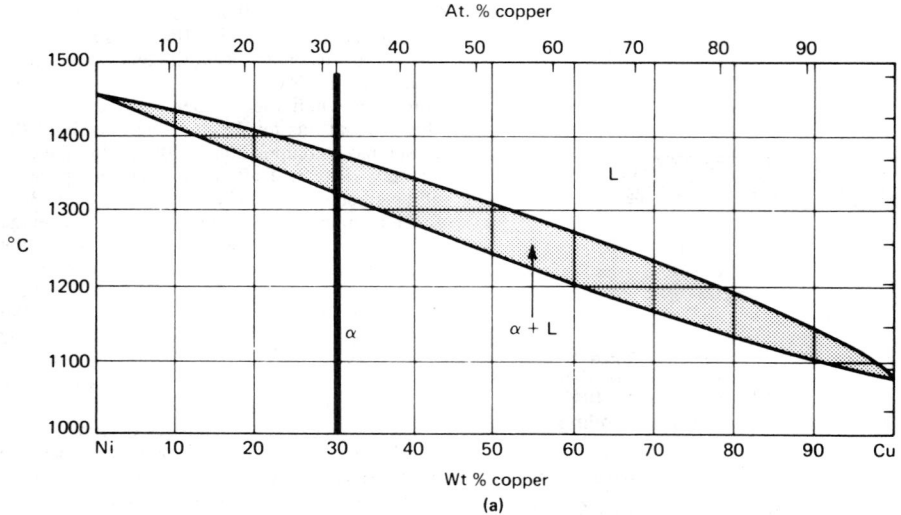

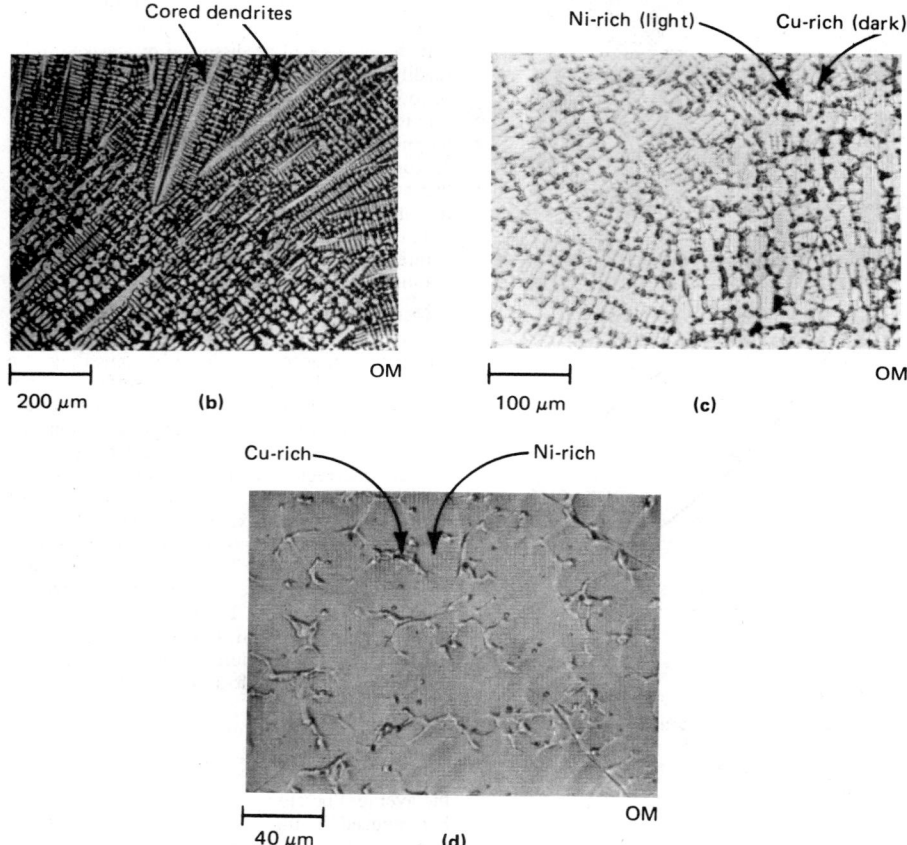

(a) Ni-Cu phase diagram. (b, c and d) The microstructure at increasingly higher magnifications. Note that the dendrite cells are approximately 40 μm across (d). From same source as Fig. 9.

Fig. 22. The Ni-Cu phase diagram, and the microstructure of a Ni-30Cu alloy that has been cooled rapidly from the liquid, developing a nonequilibrium cored structure

ments near the center. However, a calculation similar to that above shows that the diffusion distance in this type of chemical inhomogeneity is much too great to be reduced appreciably by homogenization annealing.

In many commercial nonferrous alloys, the as-cast structure will not only be cored, but also will contain nonequilibrium, second-phase particles. In such systems, on slow cooling, when freezing

is complete a single-phase solid will be present (as described above). However, on rapid cooling, in which coring occurs, the liquid composition may increase to the value of the eutectic before freezing is completed. Then this liquid freezes to a solid eutectic structure. The microstructure then consists of a dendritically cored matrix containing small regions of multiphase, eutectic solid. These regions will dissolve on

proper solution heat treatment, and thus will be removed along with the coring.

PRECIPITATION HARDENING HEAT TREATMENTS

In designing alloys for strength, an approach often taken is to develop an alloy in which the structure consists of particles which impede dislocation motion dispersed in a ductile matrix. The finer the dispersion, for the same amount of particles, the stronger the material.

Such a dispersion can be obtained by choosing an alloy which, at elevated temperature, is single phase, but which on cooling will precipitate another phase in the matrix. A heat treatment is then developed to give the desired distribution of the precipitate in the matrix. If hardening occurs from this structure, then the process is called *precipitation hardening*. It is to be noted that not all alloys in which such a dispersion can be developed will harden. However, in this section attention is placed on systems which do harden if the precipitation process is properly controlled.

Solution Heat Treatment. A prerequisite to precipitation hardening is the ability to heat the alloy to a temperature range wherein all of the solute is dissolved, so that a single-phase structure is attained. This is shown schematically in Fig. 23 for a 10% B alloy in a hypothetical system A-B. Heating above the solvus temperature T_2 for this alloy, and holding in the α range for sufficient time, will form the single phase α. This is the required *solution heat treatment*. This structure is then retained at ambient temperature by cooling rapidly (e.g., water quenching) from the α range to prevent the precipitate from forming. The structure is supersaturated with respect to the solute, and hence is unstable.

The Process of Precipitation. After quenching from the α region (Fig. 23), precipitation is achieved by reheating the alloy below the solvus (T_2 in Fig. 23) at a suitable temperature for a suitable time. During this time, at localized regions (e.g., grain boundaries), the precipitates nucleate. Because these precipitates have a higher solute content than the matrix, the region in the matrix surrounding them is reduced in solute content. This forms a concentration gradient such that the solute atoms diffuse from the adjacent matrix toward the particles, allowing the precipitates to continue to grow. The rate of growth is diffu-

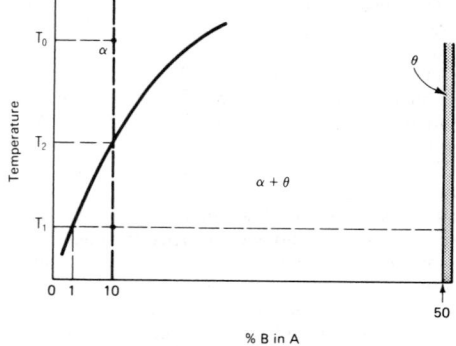

The decreasing solubility of B in α with decreasing temperature allows an alloy containing 10% B to be single-phase at high temperature (i.e., above T_2) but two-phase at low temperature (T_1). Adapted from same source as Fig. 9.

Fig. 23. Hypothetical phase diagram of system A-B

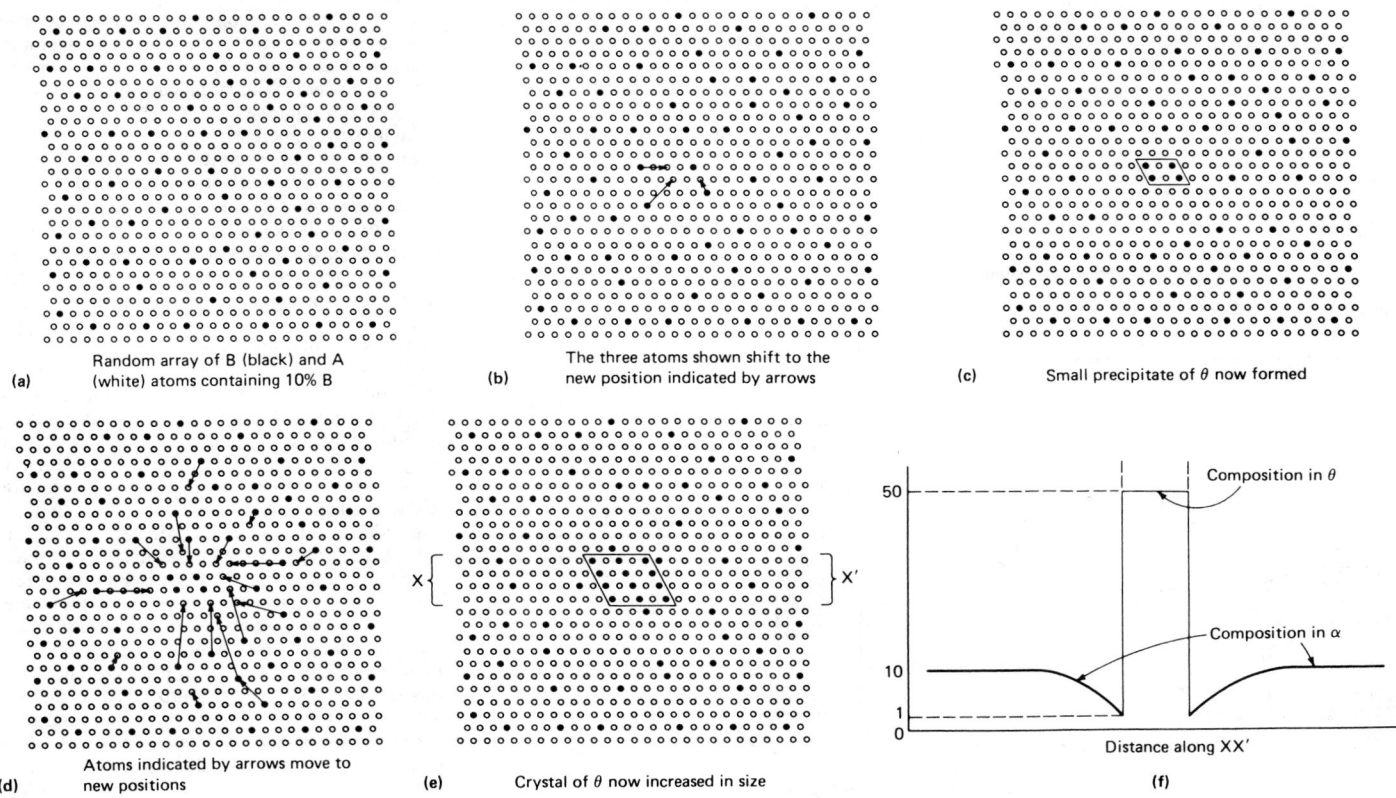

(a) Random array of B (black) and A (white) atoms containing 10% B

(b) The three atoms shown shift to the new position indicated by arrows

(c) Small precipitate of θ now formed

(d) Atoms indicated by arrows move to new positions

(e) Crystal of θ now increased in size

(f) [graph: Composition in θ / Composition in α / Distance along XX′ / 50, 10, 1, 0]

Time is increasing from (a) to (e), but at (e) equilibrium is not yet attained. In (f) is shown the concentration profile through the precipitate in (e). From same source as Fig. 9.

Fig. 24. Schematic illustration of formation of a precipitate in a supersaturated matrix

sion-controlled and is given by an appropriate solution to Fick's law. The precipitation process is depicted schematically in Fig. 24. Here the precipitate contains 50% B (see Fig. 23).

The maximum amount of precipitate which can form is given by the equilibrium amount, which can be calculated from a mass balance (lever rule). Once this equilibrium amount of precipitate has been attained, then further change in the precipitates is caused by the tendency for the system to reduce the precipitate/matrix interfacial area. Thus, with time at a given aging temperature, the smaller precipitates dissolve, with the solute diffusing through the matrix to contribute to the growth of the larger particles. This results in a microstructure containing larger, but fewer, particles. An equivalent effect is obtained by using a high aging temperature for a given time. These changes are depicted schematically in Fig. 25.

Control of Precipitation Through Heat Treatment. The precipitation heat treatment for the desired properties is determined empirically. Higher precipitation temperatures usually are associated with a lower nucleation rate and thus a coarser precipitate distribution. Also, as the precipitation temperature used approaches the solvus, the amount of precipitate decreases (vanishing at the solvus).

The microstructural effects will be demonstrated by referring to the aging of an Al-5Cu alloy. Figure 26 shows that this alloy must be solution heat treated at temperatures between 500 °C (solvus) and 575 °C (solidus). If this alloy is quenched from 545 °C (after 1 h), then aged for 12 h at 400 °C, the structure obtained will be that shown in Fig. 27(a). The precipitates are fine

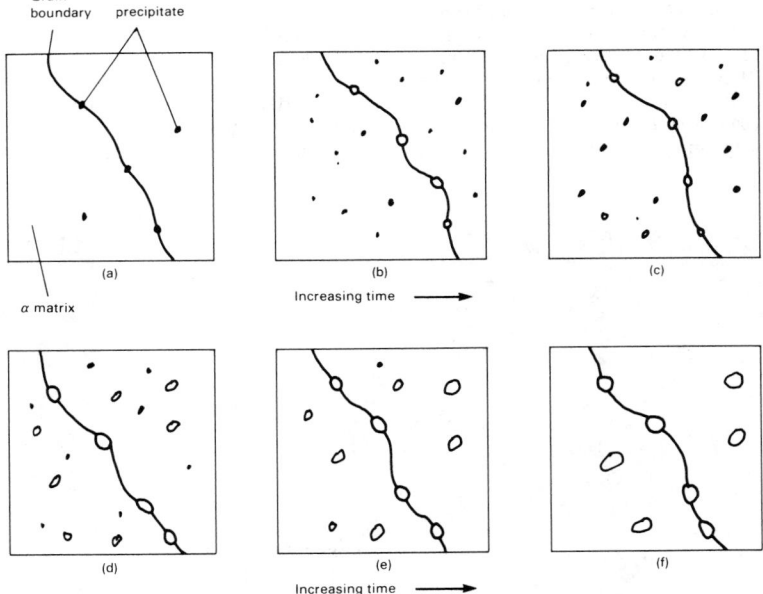

Grain boundary / θ precipitate / α matrix / Increasing time / Increasing time / (a) (b) (c) (d) (e) (f)

Fig. 25. Schematic illustration of formation of Θ precipitates in the α matrix (a and b) and their coarsening (c through f)

and evenly distributed, and are about 1 μm in size. If an aging temperature of 300 °C is used (for 12 h), then the structure in Fig. 27(b) is obtained. It can be seen that this higher aging temperature produced a somewhat coarser distribution of Θ than that at 300 °C.

In the Al-5Cu alloy (and in most other precipitation-hardenable alloys), the precipitation process is not as simple as that depicted schematically in Fig. 24. Instead, formation of the equilibrium precipitate (Θ in the Al-Cu alloy) is preceded by formation of one or more nonequi-

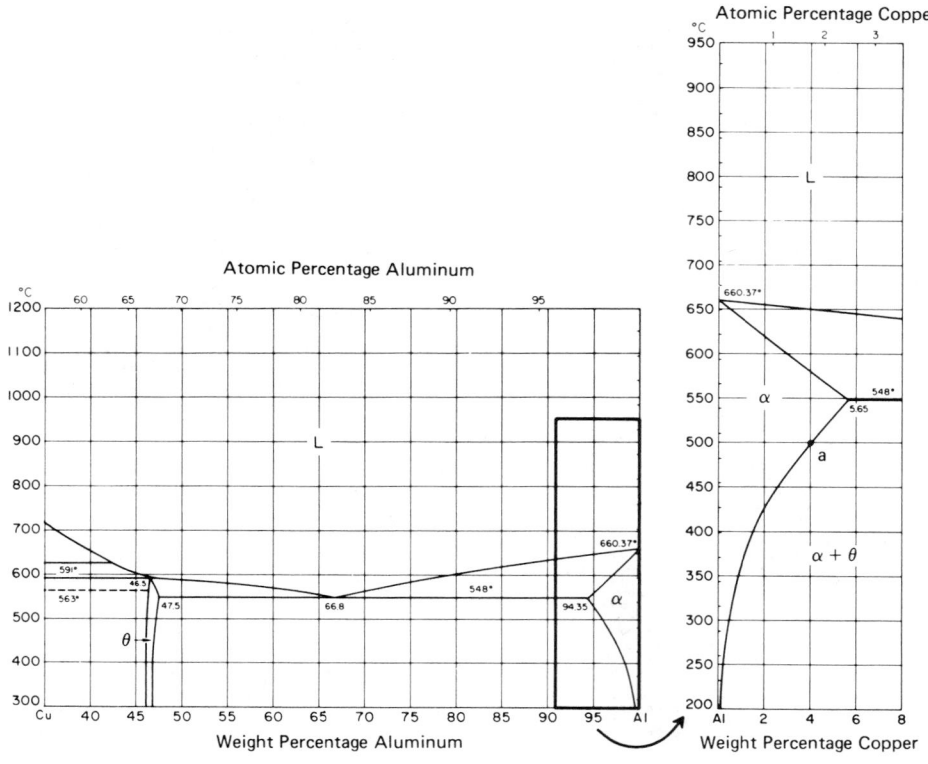

Atomic Percentage Aluminum

Atomic Percentage Copper

Adapted from same source as Fig. 9.

Fig. 26. The Al-rich end of the Al-Cu phase diagram, showing the 5% Cu line

ture, the lower the maximum hardness, because less precipitate forms as the solvus temperature is approached. However, the higher the temperature, the higher the rate of precipitation, and hence the maximum hardness is attained in less time.

In most commercial precipitation-hardenable alloys, the rate of precipitation is low at ambient temperature, although sufficiently rapid to bring about measurable hardness changes in a reasonable time, as shown in Fig. 30 for aging at 30 °C. If hardening occurs at or near ambient temperature, it is termed *age hardening;* aging at other temperatures is called *precipitation hardening.*

Commercial alloys usually contain multiple elements, so that the required heat treating temperatures cannot always be deduced from an examination of related binary phase diagrams. In many alloys which contain mainly two alloying additions, the ternary phase diagram can be used as a guide for establishing the required heat treatments. For example, consider the Al-base alloy 2024. It contains approximately 4% Cu and 1% Mg, with lesser amounts of Mn, Si, Fe, Cr and Zn. Considering the alloy to be an Al-Cu-Mg ternary alloy, and using 4% Cu and 1% Mg to represent the average concentrations of these elements, then the Al-rich end of the ternary phase diagram can be used to illustrate the required heat treatments. This is shown in Fig. 31. The liquidus is about 650 °C, the solidus 570 °C and the solvus 500 °C. Thus, the solution annealing temperature must be between 500 and 570 °C. When consideration is given to the allowable range of the amounts of Cu and Mg in the 2024 alloy, then the solution annealing range is narrowed. Avoiding heating above the liquidus is of particular importance, because this will allow formation of small regions of liquid in the structure, which on cooling form compounds, and can lead to problems in achieving desired properties. The specification for the solution annealing temperature for alloy 2024 is 488 to 499 °C, only an 11 °C spread. Thus, if 494 °C is used, only a deviation of about ±5 °C is allowed.

Aging of alloy 2024 must be carried out below the solvus, about 500 °C. The response to aging for this alloy is typified by the data in Fig. 32.

librium configurations or precipitates. For an Al-4.6Cu alloy, this is shown in Fig. 28. In the earliest stage, Cu-rich zones form (called *Guinier-Preston Zones*), followed by two metastable precipitates (Θ'' and Θ'), before the equilibrium Θ appears. Note how fine these metastable phases are. In Fig. 28, the Θ'' particles are approximately 0.01 μm in size, corresponding to particles about 50 atoms in size.

Precipitation hardening. The strengthening which occurs during aging of an Al-4Cu alloy is illus-

trated in Fig. 29. Note that the maximum hardness is about double that in the as-quenched (supersaturated α) condition. Also note that the maximum hardness does not correspond to formation of the equilibrium Θ phase, but to the metastable, transition phases, which form in a considerably finer distribution than does Θ (compare Fig. 28 and 27).

The effect of temperature and time on aging is illustrated by the data in Fig. 30. As pointed out previously, the higher the precipitation tempera-

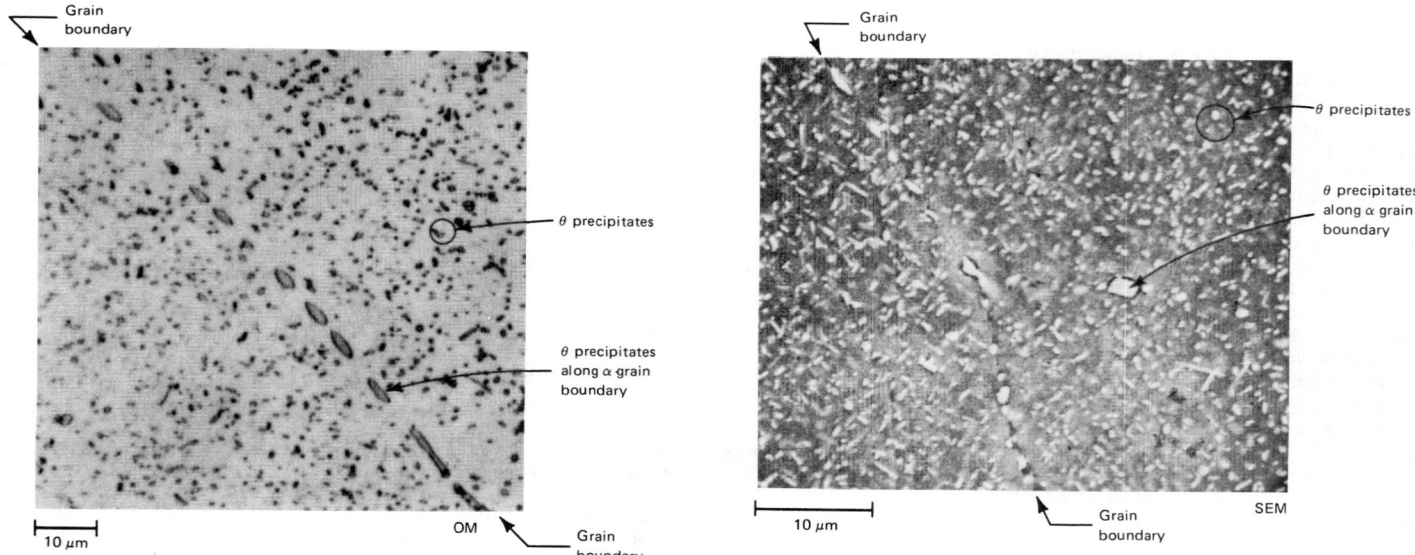

(a) is an optical micrograph, and (b) is a scanning electron micrograph. From same source as Fig. 9.

Fig. 27. Microstructure of Al-5Cu alloy heated for 1 week at 545 °C, cooled rapidly to 25 °C, then held 12 h at (a) 400 °C and (b) 300 °C

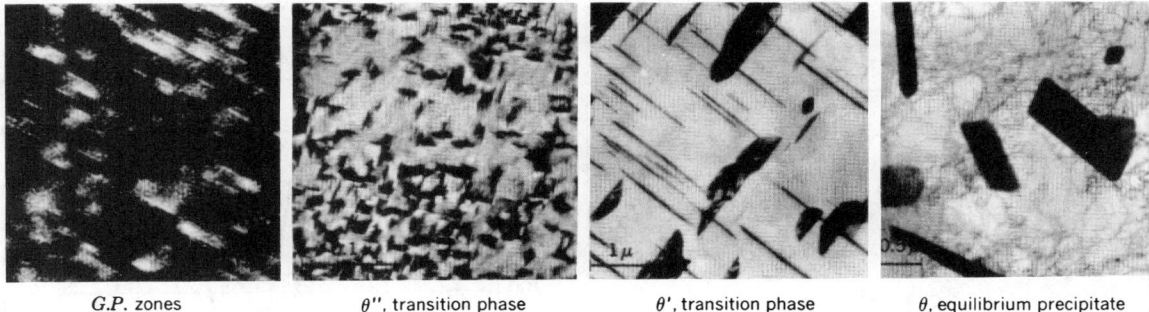

| G.P. zones | θ'', transition phase | θ', transition phase | θ, equilibrium precipitate |

The micrograph at far right shows θ precipitates similar in size to those shown in Fig. 27(b). Adapted from *Introduction to Materials Science*, by A. G. Guy, McGraw-Hill, New York, 1979.

Fig. 28. Transmission electron micrographs of precipitates formed in an Al-4.6Cu alloy with increasing aging time (left to right)

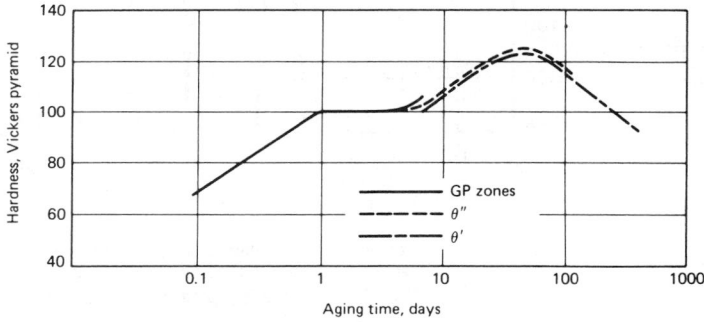

Compare this curve to the structures shown in Fig. 28. Adapted from J. M. Silcock, T. J. Heal, and M. K. Hardy, *J Inst Metals*, Vol 83, 1953, p 239.

Fig. 29. Hardness curve for an Al-4Cu alloy showing the relationship between the various precipitates formed and the hardness on aging at 130 °C

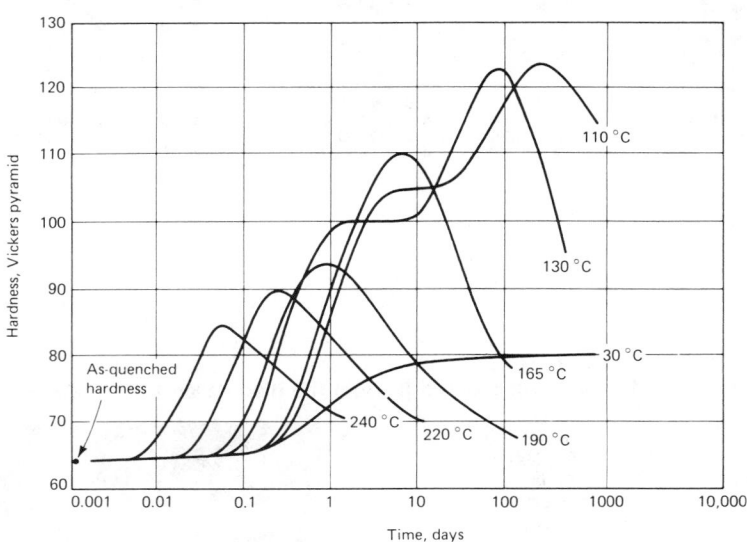

The alloy was solution annealed for at least 48 h at 520 °C, then cooled quickly (water quenched) to 25 °C. Adapted from H. K. hardy, *J Inst Metals*, Vol 79, 1951, p 321.

Fig. 30. Hardness as a function of aging time for an Al-4Cu alloy

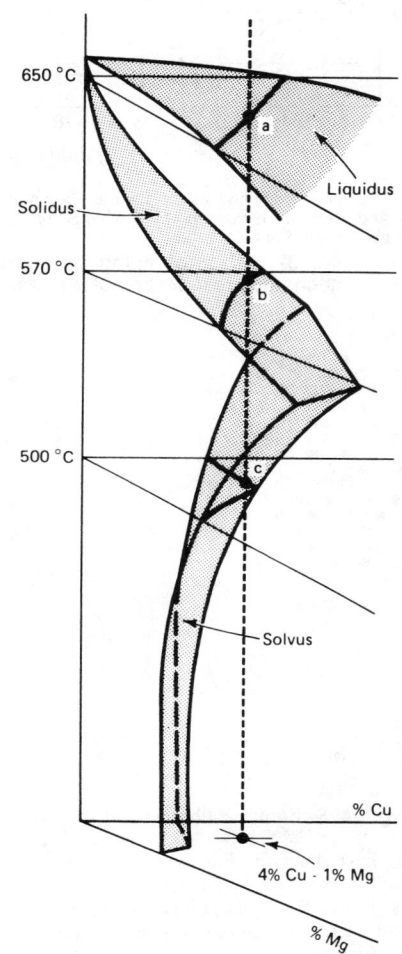

These temperatures are based on experimental observations. Adapted from same source as Fig. 9.

Fig. 31. Schematic illustration of the Al-rich end of the Al-Cu-Mg phase diagram, with the liquidus, solidus and solvus shown for a 4Cu-1Mg alloy

Note that there is a range of combinations of temperature and time which will give about the same optimum mechanical properties.

DEVELOPMENT OF TWO-PHASE STRUCTURES

In some nonferrous alloys (e.g., Ti-base alloys and high-zinc Cu-Zn alloys), the desired structure consists of a mixture of two phases of comparable quantity (unlike the two-phase structures developed in precipitation hardening, where the precipitate is in the minority). The morphology and amount of each are varied by control of the high temperature used and the cooling rate from that temperature. The preferred microstructure can be quite complex, and the required treatment differs considerably for different systems, so that a systematic treatment of the principles involved is difficult. Instead, in this section a specific alloy will be used to illustrate the types of treatments involved.

In the Cu-Zn system, alloys containing about 40% Zn serve as the basis for some commercial alloys (e.g., Muntz metal and naval brass). The

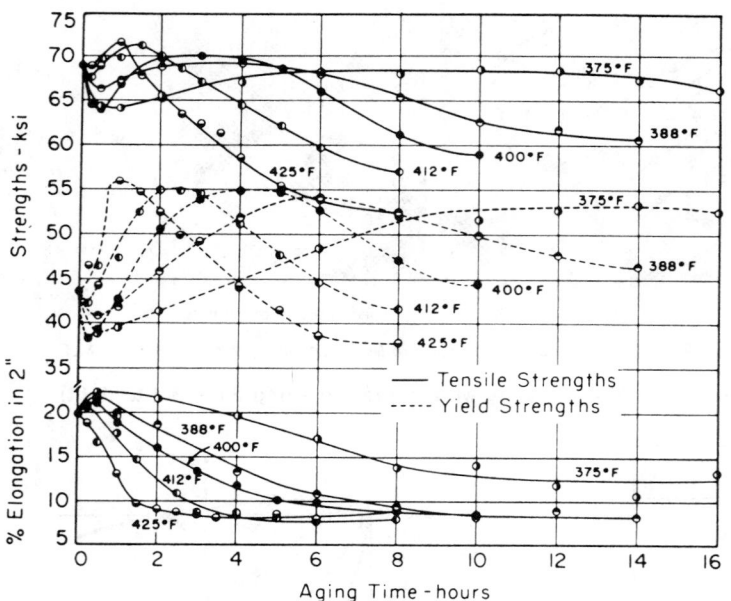

The initial condition was the natural aged state (temper T4). From W. A. Anderson, in *Precipitation from Solid Solution,* American Society for Metals, Metals Park, OH, 1958.

Fig. 32. Effect of aging time and temperature on mechanical properties of Al-base alloy 2024

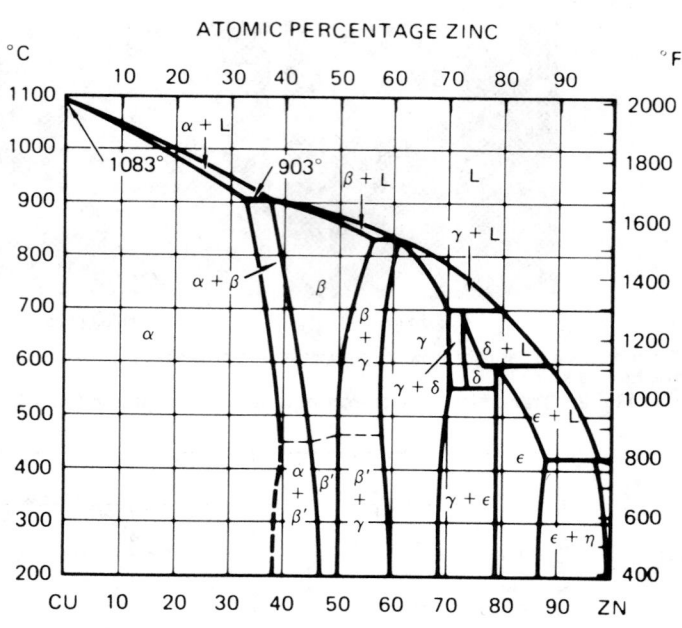

The β phase is body-centered cubic; the β' phase is an ordered structure based on this arrangement.

Fig. 33. The Cu-Zn phase diagram

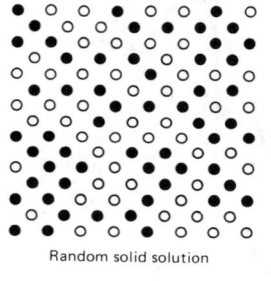

Random solid solution

(a)

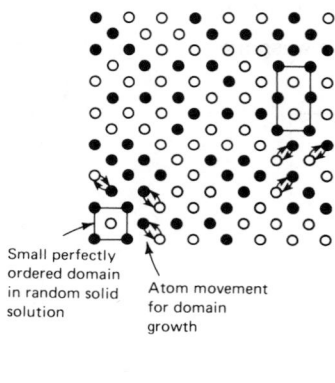

Small perfectly ordered domain in random solid solution

Atom movement for domain growth

(b)

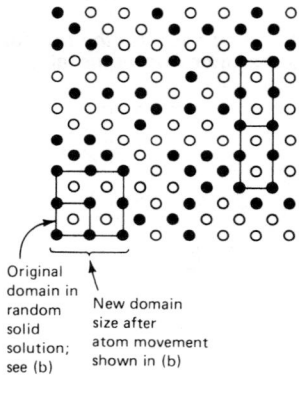

Original domain in random solid solution; see (b)

New domain size after atom movement shown in (b)

(c)

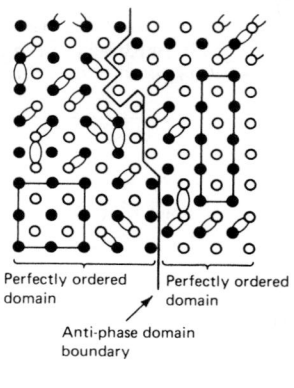

Perfectly ordered domain

Perfectly ordered domain

Anti-phase domain boundary

(d)

Adapted from same source as Fig. 9.

Fig. 34. Schematic illustration of a possible mechanism for the formation of ordered β' from the disordered β in Cu-Zn alloys

Cu-Zn phase diagram (Fig. 33) shows that the alloys of interest are in the region of α and β phase stability. The β phase is body-centered cubic, with the Cu and Zn atoms located at random on the lattice sites.

On cooling to temperatures below the dashed line (about 450 °C), the Cu and Zn atoms take specific relative positions on the sites, forming an *ordered structure,* or a *superlattice.* This phase is denoted β' in Fig. 33. If the composition is exactly 50 at.% Zn, then the ordered structure is based on a body-centered cubic cell with Zn atoms at the center and Cu atoms on the corners (or vice versa).

The formation of an ordered structure from a disordered matrix of the same basic lattice involves the localized exchange of atom positions (via the vacancy mechanism) to the desired structure. This process is depicted schematically in Fig. 34. It can be seen that an ordered region grows by atoms at the β/β' interface, taking on

the arrangement of the ordered β' region. When two interfaces from neighboring regions meet, the arrangement of atoms may be out of sequence (out of phase). Such an interface is called an *antiphase boundary,* and the enclosed regions are called *domains.* The properties of the β' ordered structure depends on the degree of perfection (correctness of relative atom location) within the domains and on the domain size, both of which depend on the temperature and time involved in forming β' from β.

These alloys in the β' form are not suitable for commercial use, because this structure is brittle. However, alloys in which the β' phase coexists with the ductile α phase are useful. The Cu-40Zn alloy can be heat treated at high temperature so that it is all β. The structure developed at lower temperatures depends on the heat treatment, because this controls precipitation and formation of the α phase. If the alloy is cooled slowly from 800 °C, the phase diagram (Fig. 33) shows that

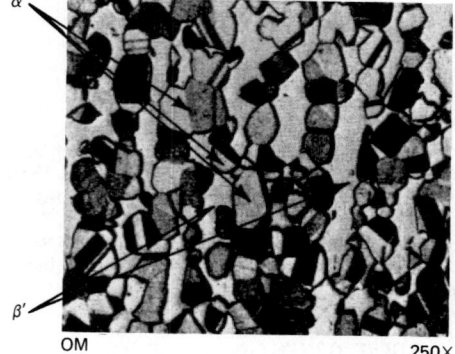

The clear, white regions are the β', and the dark and gray regions showing annealing twins are α. Adapted from D. K. Crampton, *Metal Progress,* Vol 46, 1944, p 276.

Fig. 35. Typical microstructure of annealed Muntz metal (Cu-40Zn)

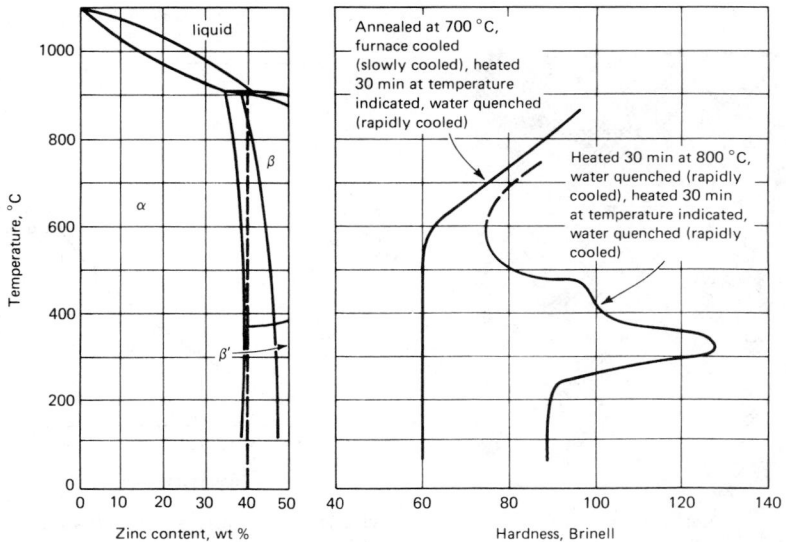

Adapted from T. Matsuda, *J Inst Metals*, Vol 39, 1928, p 67.

Fig. 36. Influence of heat treatment on the hardness at 25 °C of a Cu-40Zn alloy

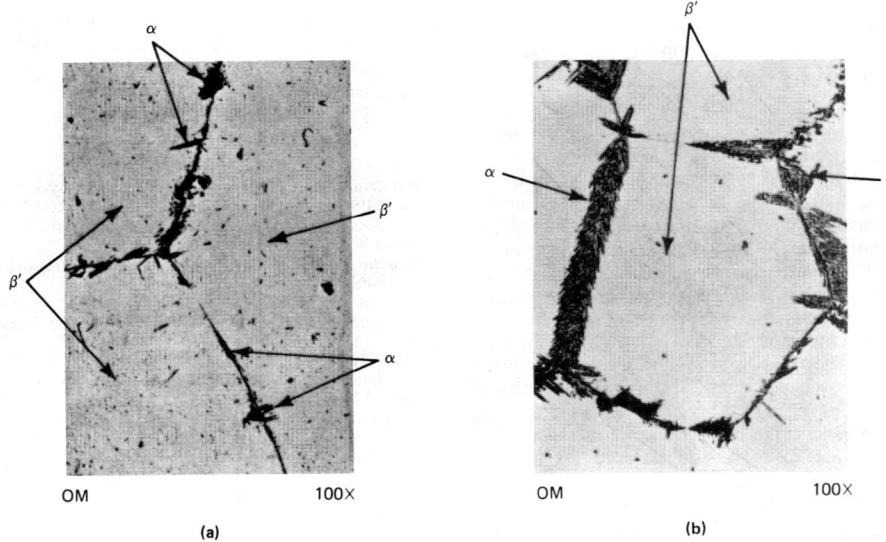

Even rapid cooling has not prevented some α from forming. (a) Cu-40Zn alloy quenched into ice water from 825 °C. Adapted from *Engineering Physical Metallurgy*, by R. H. Heyer, Van Nostrand Reinhold, 1939, used with permission of Brooks/Cole Publishing Co. (b) Quenched Muntz metal. From *Metals Handbook*, 8th Ed., Vol 7, American Society for Metals, Metals Park, OH, 1972.

Fig. 37. Microstructures typical of Cu-40Zn alloys cooled rapidly from the β region to 25 °C

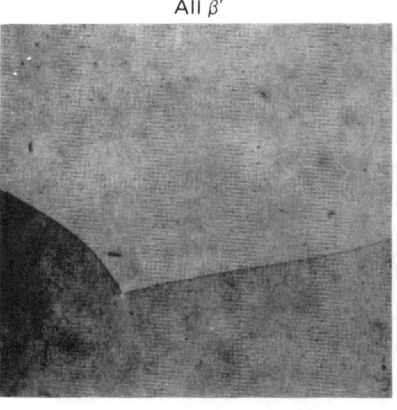

All β′

OM   100×

Quenched from 800 °C

(a)

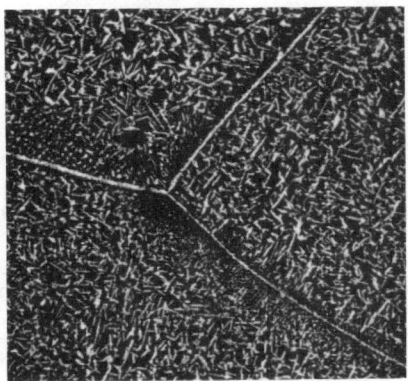

White α in β′

OM   100×

Quenched from 800 °C, reheated for 30 min at 400 °C

(b)

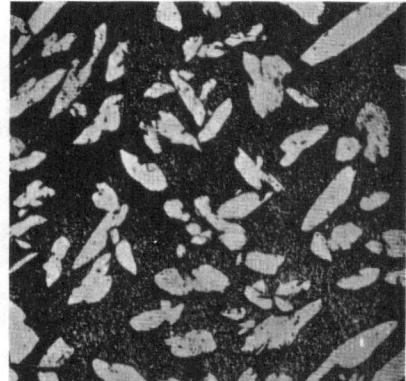

White α in β′

OM   100×

Quenched from 800 °C, reheated for 30 min at 600 °C

(c)

The higher reheating temperature gives a coarser structure, and hence a softer material. Adapted from C. H. Samans, *Metallic Materials in Engineering*, MacMillan Co., New York, 1963.

Fig. 38. Microstructures of Cu-42Zn alloy quenched from the β region, then reheated to develop an α precipitate structure

at 25 °C the alloy should consist of approximately equal amounts of α and β′. Figure 35 shows a typical microstructure.

One of the curves in Fig. 36 shows that the amount of β′ influences hardness. The alloy was cooled slowly from 700 °C, where it was mostly β, to 25 °C, then reheated to temperature for 30 min, followed by rapid cooling. On heating at 800 °C, the structure is all β, and on rapid cooling little α forms. However, the β orders to β′, giving a hardness around 90 HB. Reheating for 30 min in the lower temperature range (25 to 500 °C) is not sufficient to affect significantly the originally slowly cooled structure, and the hardness remains constant. In this temperature range, the structure consists of approximately equal

amounts of α and β′. However, as the temperature increases from 500 °C, 30 min is sufficient time to allow the equilibrium amounts and α and β to form. Thus, as the temperature increases, increasing amounts of β and decreasing amounts of α are present at temperature, giving increasing amounts of β′ on cooling rapidly to 25 °C, and hence a rise in hardness.

If the Cu-40Zn alloy is cooled rapidly to 25 °C after sufficient holding (e.g., 30 min) above about 750 °C, a structure of essentially all β′ is obtained. Often some α is observed to have formed in the β grain boundaries, and the morphology will vary somewhat depending on the exact cooling rate. Usually, the α is present as "needles" emanating from the boundaries, with a clear

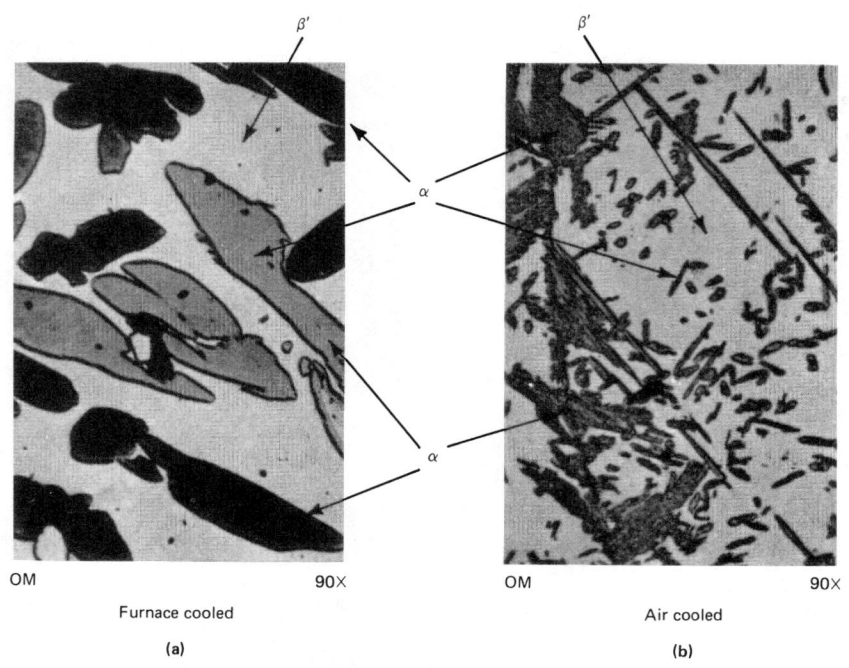

OM   90×

Furnace cooled

(a)

OM   90×

Air cooled

(b)

Adapted from R. F. Mehl and G. T. Marzke, *Trans AIME*, Vol 93, 1931, p 123.

Fig. 39. Microstructures of a Cu-43Zn alloy after cooling from 700 °C, the β region, showing effect of cooling rate on structure of α crystals

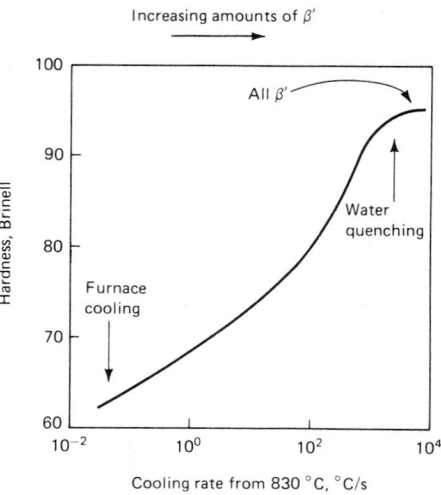

Adapted from T. Matsuda, *J Inst Metals*, Vol 39, 1928, p 67.

Fig. 40. Effect of cooling rate from the β region on hardness of a Cu-40Zn alloy

crystallographic relation between the α and the β′ in which it has formed. Figure 37 shows two examples.

On reheating β′ in the intermediate temperature range, the morphology of the α formed will vary depending on the exact heat treatment. Also, reheating will influence the change in the ordered structure. Both changes affect properties, and the hardness can be increased considerably by judicious treatment. In Fig. 36 are shown hardness data for a Cu-40Zn alloy after reheating for 30 min following an initial treatment of quenching from 800 °C. Supposedly the maximum hardness obtained by treatment around 300 °C is caused by formation of a fine α precipitate and some changes in the ordered β′ phase. The types of microstructures obtained by such heat treatments are illustrated in Fig. 38 for a Cu-42Zn alloy. In this alloy the zinc content is sufficiently high to

completely suppress any α formation on rapid cooling from β, giving at 25 °C only β′ (Fig. 38a). Reheating for 30 min at 400 °C gives a fine α precipitate on the β′ grain boundaries, and a fine intercrystalline precipitate of α (Fig. 38b). Reheating for 30 min at a higher temperature, 600 °C, gives a coarser α structure (Fig. 38c).

If the rate of cooling from the α region is quite low (several hours to 25 °C), then α nucleates at a high temperature at which the nucleation rate is low, and the α crystals grow relatively large as few crystals nucleate. This gives a rather coarse structure, typified by Fig. 39(a). As the cooling rate increases, the nucleation rate increases, but the individual α crystals do not have time to grow large before the temperature becomes too low for significant growth to continue. This gives a finer structure (see Fig. 39b) and increases strength. Eventually, the cooling rate becomes sufficient

to suppress formation of α altogether, giving a structure entirely of highly unstable β′ at 25 °C. Figure 40 illustrates the influence of cooling rate from β on hardness.

SELECTED REFERENCES

Physical Metallurgy Principles, 2nd Ed., by R. E. Reed-Hill: Van Nostrand Reinhold, New York, 1973

Fundamentals of Physical Metallurgy, by J. D. Verhoeven: Wiley, New York, 1975

Structure and Properties of Engineering Alloys, by W. F. Smith: McGraw-Hill, New York, 1981

Structure and Properties of Alloys, by R. M. Brick, R. B. Gordon and A. Phillips: McGraw-Hill, New York, 1965

Heat Treatment, Structure and Properties of Nonferrous Alloys, by C. R. Brooks: American Society for Metals, Metals Park, OH, 1982

An Introduction to the Solidification of Metals, by W. C. Winegard: Institute of Metals, London, 1964

Principles of Solidification, by B. Chalmers: Wiley, New York, 1964

Precipitation Hardening, by J. W. Martin: Pergamon, New York, 1968

Aluminum Alloys: Structure and Properties, by L. F. Mondolfo: Butterworths, Boston, 1976

Heat Treating of Nonferrous Alloys and Powder Metallurgy Parts

Heat Treating of Aluminum Alloys

ONE essential attribute of a precipitation-hardening alloy system is a temperature-dependent equilibrium solid solubility characterized by increasing solubility with increasing temperature. Although this condition is met by most of the binary aluminum alloy systems, many exhibit very little precipitation hardening, and these alloys ordinarily are not considered heat treatable. Alloys

of the binary Al-Si and Al-Mn systems, for example, exhibit relatively insignificant changes in mechanical properties as a result of heat treatments that produce considerable precipitation.

The solubility/temperature relationship required for precipitation hardening is illustrated by the Al-Cu system (see Fig. 1). The equilibrium solid solubility of copper in aluminum increases as temperature increases—from about 0.20% at 250 °C (480 °F) to a maximum of 5.65% at the eutectic melting temperature of 548 °C (1018 °F). (It is considerably lower than 0.20%

at temperatures below 250 °C.) For Al-Cu alloys containing from 0.2 to 5.6% Cu, two distinct equilibrium solid states are possible. At temperatures above the lower curve in Fig. 1 (solvus), the copper is completely soluble, and when the alloy is held at such temperatures for sufficient time to permit needed diffusion, the copper will be taken completely into solid solution. At temperatures below the solvus, the equilibrium state consists of two solid phases: solid solution, α, plus an intermetallic-compound phase, θ (Al₂Cu). When such an alloy is converted to all solid solution by holding above the solvus temperature and then the temperature is decreased to below the solvus, the solid solution becomes supersaturated and the alloy seeks the equilibrium two-phase condition; the second phase tends to form by solid-state precipitation.

Commercial alloys whose strength and hardness can be significantly increased by heat treatment include 2*xxx*, 6*xxx* and 7*xxx* series wrought alloys (except 7072) and 2*xx*.0, 3*xx*.0 and 7*xx*.0 series casting alloys. Some of these contain only copper, or copper and silicon, as the primary strengthening alloy addition(s). Most of the heat treatable alloys, however, contain combinations of magnesium with one or more of the elements copper, silicon and zinc.

Precipitation heat treatments generally are low-temperature, long-term processes. Temperatures range from 115 to 190 °C (240 to 375 °F); times vary from 5 to 48 h.

Choice of time/temperature cycles for precipitation heat treatment should receive careful consideration. Larger particles of precipitate result from longer times and higher temperatures; however, the larger particles must, of necessity, be fewer in number with greater distances between them. The objective is to select the cycle that produces optimum precipitate size and distribution pattern. Consequently, the cycles used represent compromises that provide the best combinations of properties.

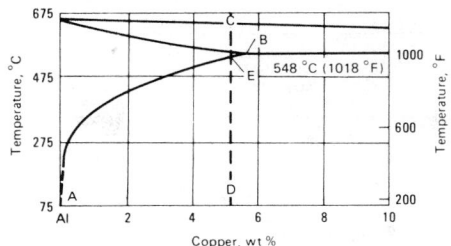

Line AB represents the increase in solubility of copper in solid aluminum with increasing temperature. See text for discussion.

Fig. 1. Aluminum-rich end of the aluminum-copper equilibrium diagram

To recap, heat treatment to increase strength of aluminum alloys is a three-step process:

1. Solution heat treatment: dissolution of soluble phases
2. Quenching: development of supersaturation
3. Aging: precipitation of solute atoms either at room temperature (natural aging) or elevated temperature (artificial aging or precipitation heat treatment).

Typical solution and precipitation heat treatments for wrought and cast aluminum alloy products are given in Tables 1 and 2.

SOLUTION HEAT TREATING

To take advantage of the precipitation-hardening reaction, it is necessary first to produce a solid solution. The process by which this is accomplished is called solution heat treating, and its objective is to take into solid solution the maximum practical amounts of the soluble hardening elements in the alloy. The process consists of soaking the alloy at a temperature sufficiently high and for a time long enough to achieve a nearly homogeneous solid solution.

Solution Treating Temperature. Nominal commercial solution heat treating temperature is determined by the composition limits of the alloy and an allowance for unintentional temperature variations. Although ranges normally listed allow variations of ±6 °C (±10 °F) from the nominal, some highly alloyed, controlled-toughness, high-strength alloys require that temperature be controlled within more restrictive limits (see Tables 1 and 2).

Solution Treating Time. The time at the nominal solution heat treating temperature ("soak time") required to effect a satisfactory degree of solution of the undissolved or precipitated soluble phase constituents and to achieve good homogeneity of the solid solution is a function of microstructure before heat treatment. This time requirement can vary from less than a minute for thin sheet to as much as 20 h for large sand or

Table 1. Typical solution and precipitation heat treatments for aluminum alloy mill products

These times and temperatures are typical for various forms, sizes and methods of manufacture and may not exactly describe optimum treatments for specific items.

Alloy	Product form	Solution heat treatment(a) Metal temperature(b) °C	°F	Temper designation	Precipitation heat treatment Metal temperature(b) °C	°F	Time(c), h	Temper designation
2011	Rolled or cold finished rod and bar	525	975	T3(d)	160	320	14	T8(d)
				T4	...	...	...	...
				T451(e)	...	...	...	...
2014(f)	Flat sheet	500	935	T3(d)	160	320	18	T6
				T42	160	320	18	T62
	Coiled sheet	500	935	T4	160	320	18	T6
				T42	160	320	18	T62
	Plate	500	935	T42	160	320	18	T62
				T451(e)	160	320	18	T651(e)
	Die forgings	500(g)	935(g)	T4	170	340	10	T6
	Hand forgings and rolled rings	500(g)	935(g)	T4	170	340	10	T6
				T452(h)	170	340	10	T652(h)
2017	Rolled or cold finished wire, rod and bar	500	935	T4	...	...	...	...
				T42	...	...	...	...
				T451(e)	...	...	...	...
2018	Die forgings	510(j)	950(j)	T4	170	340	10	T61
2024(f)	Flat sheet	495	920	T3(d)	190	375	12	T81(d)
				T361(d)	190	375	8	T861(d)
				T42	190	375	9	T62
					190	375	16	T72
2025	Die forgings	515	960	T4	170	340	10	T6
2036	Sheet	500	930	T4	...	...	...	...
2117	Rolled or cold finished wire and rod	500	935	T4	...	...	...	...
				T42	...	...	...	...
2218	Die forgings	510(j)	950(j)	T4	170	340	10	T61
		510(k)	950(k)	T41	240	460	6	T72
2219(f)	Flat sheet	535	995	T31(d)	175	350	18	T81(d)
2618	Forgings and rolled rings	530	985	T4	200	390	20	T61
4032	Die forgings	510(g)	950(g)	T4	170	340	10	T6
6005	Extruded rod, bar, shapes and tube	530(m)	985(m)	T1	175	350	8	T5
6009(n)	Coiled sheet	555	1030	T4	175	350	8	T6
6010(n)	Coiled sheet	565	1050	T4	175	350	8	T6
6053	Die forgings	520	970	T4	170	340	10	T6
6061(f)	Sheet	530	985	T4	160	320	18	T6
6063	Extruded rod, bar, shapes and tube	(m)	(m)	T1	205(p)	400(p)	1	T5
		520(m)	970(m)	T4	175(q)	350(q)	8	T6
		520	970	T42	175(q)	350(q)	8	T62
	Drawn tube	520	970	T4	175	350	8	T6

(continued)

Table 1. (continued)

Alloy	Product form	Solution heat treatment(a) Metal temperature(b) °C	°F	Temper designation	Precipitation heat treatment Metal temperature(b) °C	°F	Time(c), h	Temper designation
6070	Extruded rod, bar, shapes and tube	545(m)	1015(m)	T4	160	320	18	T6
				T42	160	320	18	T62
6151	Die forgings	515	960	T4	170	340	10	T6
	Rolled rings	515	960	T4	170	340	10	T6
				T452(h)	170	340	10	T652(h)
6262	Rolled or cold finished wire, rod and bar	540	1000	T4	170	340	8	T6
					170	340	12	T9(r)
				T451	170	340	8	T651(e)
				T42	170	340	8	T62
6463	Extruded rod, bar, shapes and tube	(m)	(m)	T1	205(p)	400(p)	1	T5
		520(m)	970(m)	T4	175(q)	350(q)	8	T6
		520	970	T42	175(q)	350(q)	8	T62
7001	Extruded rod, bar, shapes and tube	465	870	W	120	250	24	T6
					120	250	24	T62
				W510(e)	120	250	24	T6510(e)
				W511(e)	120	250	24	T6511(e)
7005	Extruded rod, bar and shapes	...	...	...	...	...	...	T53(s)
7050	Plate	475	890	W51(e)	(t)	(t)	(t)	T7651(u)
7075(f)	Rolled or cold finished wire, rod and bar	490	915	W	120	250	24	T6
					120	250	24	T62
					(v)(w)	(v)(w)	(v)(w)	T73(u)
				W51(e)	120	250	24	T651(e)
					(v)(w)	(v)(w)	(v)(w)	T7351(e)(u)
	Extruded rod, bar, shapes and tube	465	870	W	120(x)	250(x)	24	T6
7475	Sheet	515(y)	960(y)	W	120 plus 155	250 315	3 3	T61(y)

(a) Material should be quenched from the solution-treating temperature as rapidly as possible and with minimum delay after removal from the furnace. When material is quenched by total immersion in water, unless otherwise indicated, the water should be at room temperature, and should be suitably cooled so that it remains below 38 °C (100 °F) during the quenching cycle. Use of high-velocity, high-volume jets of cold water also is effective for some materials. (b) The nominal temperatures listed should be attained as rapidly as possible and maintained within ±6 °C (±10 °F) of nominal during the time at temperature. (c) Approximate time at temperature. The specific time will depend on the time required for the load to reach temperature. The times shown are based on rapid heating, with soak time measured from the time the load reaches a temperature within 6 °C (10 °F) of the applicable temperature. (d) Cold working subsequent to solution heat treatment and prior to any precipitation heat treatment is necessary to attain the specified properties for this temper. (e) Stress relieved by stretching to produce a specified amount of permanent set subsequent to solution heat treatment and prior to any precipitation heat treatment. (f) These heat treatments also apply to alclad sheet and plate in these alloys. (g) Solution heat treatment is followed by quenching in water 60 to 82 °C (140 to 180 °F). (h) Stress relieved by 1 to 5% cold reduction subsequent to solution heat treatment and prior to precipitation heat treatment. (j) Solution heat treatment is followed by quenching in water at 100 °C (212 °F). (k) Solution heat treatment is followed by quenching in room-temperature air blast. (m) By suitable control of extrusion temperature, product may be quenched directly from extrusion press to provide specified properties for this temper. Some products may be adequately quenched in room-temperature air blast. (n) See U.S. Patent 4 082 578. (p) An alternative treatment of 3 h at 182 °C (360 °F) also may be used. (q) An alternative treatment of 6 h at 182 °C (360 °F) also may be used. (r) Cold working subsequent to precipitation heat treatment is necessary to attain the specified properties for this temper. (s) No solution heat treatment; 72 h at room temperature following press quench, followed by two-stage precipitation heat treatment comprised of 8 h at 107 °C (225 °F) plus 16 h at 149 °C (300 °F). (t) Aging practice varies with product, size, nature of equipment, loading procedures and furnace-control capabilities. The optimum practice for a specific item can be ascertained only by actual trial treatment of the item under specific conditions. Typical procedures involve a two-stage treatment comprised of 3 to 30 h at 121 °C (250 °F) followed by 15 to 18 h at 163 °C (325 °F) for plate; 8 h at 99 °C (210 °F) followed by 24 to 28 h at 163 °C (325 °F) also may be used. (u) Aging of aluminum alloys 7050, 7075, 7175 and 7475 from any temper to the T73 or T76 temper series requires closer-than-normal controls on aging variables such as time, temperature, heatup rate, etc., for any given item. In addition, when material in a T6-type temper is reaged to a T73- or T76-type temper, the specific condition of the T6 temper (such as property levels and other effects of processing variables) is extremely important and will affect the capability of the reaged material to conform to the requirements specified for the applicable T73- or T76-type temper. (v) Two-stage treatment comprised of 6 to 8 h at 107 °C (225 °F) followed by: 24 to 30 h at 163 °C (325 °F) for sheet and plate; 8 to 10 h at 177 °C (350 °F) for rolled or cold finished rod and bar; 6 to 8 h at 177 °C (350 °F) for extrusions and tube; 8 to 10 h at 177 °C (350 °F) for forgings in the T73 temper and 6 to 8 h at 177 °C (350 °F) for forgings in the T7352 temper. (w) For sheet, plate, tube and extrusions, an alternative two-stage treatment comprised of 6 to 8 h at 107 °C (225 °F) followed by 14 to 18 h at 168 °C (335 °F) may be used, provided that a heatup rate of approximately 14 °C/h (25 °F/h) is employed. For rolled or cold finished rod and bar, the alternative treatment is 10 h at 177 °C (350 °F). (x) An alternative three-stage treatment comprised of 5 h at 99 °C (210 °F), 4 h at 121 °C (240 °F) and then 4 h at 149 °C (300 °F) also may be used. (y) Must be preceded by soak at 466 to 477 °C (870 to 890 °F). See U.S. Patent 3 791 880.

plaster-mold castings. Guideline information for soak times required for wrought products of various section thicknesses is given in Table 1. Similar guidelines for castings are presented in Table 2. The time required to heat a load to the treatment temperature in furnace heat treatment also increases with section thickness and furnace loading, and thus total cycle time increases with these factors.

Nonequilibrium Melting. When high heating rates are employed, the phenomenon of nonequilibrium melting must be considered. This phenomenon can also be explained with the help of the Al-Cu phase diagram (Fig. 1). The room-temperature microstructure of an F-temper product containing 4% Cu consists of a solid solution of copper in aluminum and particles of Al_2Cu. When this product is heated slowly, the Al_2Cu begins to dissolve, and, if heating is slow enough, all of the Al_2Cu is dissolved when temperatures above the solvus (500 °C, or 932 °F) are reached. When the heating rate is high, however, much of the Al_2Cu remains undissolved. If a material with this microstructure is heated at or above the eutectic temperature of 548 °C (1018 °F), melting will begin at the interface between the Al_2Cu and the matrix. With sufficient time above the eutectic

temperature, this metastable liquid will dissolve to form a solid solution and will leave no trace provided that hydrogen gas has not condensed at the interface to form a void. If the product is quenched before the liquid has time to equilibrate, however, it will solidify and form fine eutectic rosettes. This nonequilibrium melting should not be confused with true equilibrium melting, which would occur in an alloy containing more than 5.65% Cu. In such an alloy, eutectic melting is equilibrium melting. No matter how long such an alloy is held above the eutectic temperature, the liquid will never solidify.

Underheating. When the temperatures attained by the parts or pieces being heat treated are appreciably below the normal range, solution is incomplete, and strength somewhat lower than normal is expected. The shallow slope of the solvus at its intersection with the composition line (typified by point E on line AB in Fig. 1) indicates that a slight decrease in temperature below point E will result in a large reduction in the concentration of the solid solution and a correspondingly significant decrease in final strength. The effect of solution-treating temperature on the strength of two aluminum alloys is illustrated by the following data:

Solution treating temperature °C	°F	Tensile strength MPa	ksi	Yield strength MPa	ksi
6061-T6 sheet 1.6 mm (0.064 in.) thick					
493	920	 301	43.7	272	39.4
504	940	 316	45.8	288	41.7
516	960	 333	48.3	305	44.3
527	980	 348	50.5	315	45.7
2024-T4 sheet 0.8 mm (0.032 in.) thick					
488	910	 419	60.8	255	37.0
491	915	 422	61.2	259	37.5
493	920	 433	62.8	269	39.0
496	925	 441	63.9	271	39.3

In the tabulation above, note especially the effects of small increments of temperature, within the normal range, on the properties of 0.8-mm (0.032-in.) 2024-T4 sheet.

The soak times for wrought alloys take into account the normal thermal lag between furnace and part and the difference between surface and center temperatures for commercial equipment qualified to the standards of MIL-H-6088. The rapid heating rates of salt baths permit all immersion time to be counted as soak time unless the bath temperature drops below the minimum

Table 2. Typical heat treatments of aluminum alloy sand and permanent mold castings

Alloy	Temper	Type of casting(a)	Solution heat treatment(b) Temperature(c) °C	°F	Time, h	Aging treatment Temperature(c) °C	°F	Time, h
201.0	T6	S	510-515; 525-530	950-960; 980-990	2 14-20	155	310	20
	T7	S	510-515; 525-530	950-960; 980-990	2 14-20	190	370	5
204.0	T4	S or P	520	970	10	...	...	...
208.0	T55	S	...	...	...	155	310	16
222.0	O(d)	S	...	...	...	315	600	3
	T61	S	510	950	12	155	310	11
	T551	P	...	...	...	170	340	16-22
295.0	T4	S	515	960	12	...	...	...
	T6	S	515	960	12	155	310	3-6
	T62	S	515	960	12	155	310	12-24
	T7	S	515	960	12	260	500	4-6
296.0	T4	P	510	950	8	...	...	...
	T6	P	510	950	8	155	310	1-8
	T7	P	510	950	8	260	500	4-6
319.0	T5	S	...	...	...	205	400	8
	T6	S	505	940	12	155	310	2-5
355.0	T51	S or P	...	...	...	225	440	7-9
	T6	S	525	980	12	155	310	3-5
		P	525	980	4-12	155	310	2-5
	T62	P	525	980	4-12	170	340	14-18
	T7	S	525	980	12	225	440	3-5
		P	525	980	4-12	225	440	3-9
	T71	S	525	980	12	245	475	4-6
		P	525	980	4-12	245	475	3-6
C355.0	T6	S	525	980	12	155	310	3-5
	T61	P	525	980	6-12	Room temperature 155	310	8 (min) 10-12
356.0	T51	S or P	...	...	...	225	440	7-9
	T6	S	540	1000	12	155	310	3-5
		P	540	1000	4-12	155	310	2-5
	T7	S	540	1000	12	205	400	3-5
		P	540	1000	4-12	225	440	7-9
	T71	S	540	1000	10-12	245	475	3
		P	540	1000	4-12	245	475	3-6
A356.0	T6	S	540	1000	12	155	310	3-5
	T61	P	540	1000	6-12	Room temperature 155	310	8 (min) 6-12
357.0	T6	P	540	1000	8	175	350	6
	T61	S	540	1000	10-12	155	310	10-12
A357.0	...	(e)	540	1000	8-12	(f)	(f)	(f)
359.0	...	(e)	540	1000	10-14	(f)	(f)	(f)
A444.0	T4	P	540	1000	8-12	...	...	...
520.0	T4	S	430	810	18(g)	...	...	...
535.0	T5(d)	S	400	750	5	...	...	...
705.0	T5	S	...	...	...	Room temperature		21 days
710.0	T5	S	...	...	...	Room temperature		21 days
711.0	T1	P	...	...	...	Room temperature		21 days
712.0	T5	S	...	...	...	Room temperature 155	315	21 days 6-8
713.0	T5	S or P	...	...	...	Room temperature		21 days
850.0	T5	S or P	...	...	...	220	430	7-9
851.0	T5	S or P	...	...	...	220	430	7-9
	T6	P	480	900	6	220	430	4
852.0	T5	S or P	...	...	...	220	430	7-9

(a) S, sand; P, permanent mold. (b) Unless otherwise indicated, solution treating is followed by quenching in water at 65 to 100 °C (150 to 212 °F). (c) Except where ranges are given, listed temperatures are ±6 °C or ±10 °F. (d) Stress relieve for dimensional stability as follows: hold 5 h at 413 ± 14 °C (775 ± 25 °F); furnace cool to 345 °C (650 °F) over a period of 2 h or more; furnace cool to 230 °C (450 °F) over a period of not more than 1/2 h; furnace cool to 120 °C (250 °F) over a period of approximately 2 h; cool to room temperature in still air outside the furnace. (e) Casting process varies (sand, permanent mold or composite) depending on desired mechanical properties. (f) Solution heat treat as indicated, then artificially age by heating uniformly at the temperature and for the time necessary to develop the desired mechanical properties. (g) Quench in water at 65 to 100 °C (150 to 212 °F) for 10 to 20 s only.

of the range. Even then, soak time begins as soon as the bath temperature returns to the minimum. In air furnaces, soak time does not begin until all furnace instruments return to their original set temperature—that is, the temperature reading before insertion of the load.

In air furnaces, thermocouples may also be attached to, or buried in, parts located in the load in such a manner as to represent the hottest and coldest temperatures in each zone. In this way, it is possible to ensure that adequate soaking is obtained.

QUENCHING

In most instances, to avoid those types of precipitation that are detrimental to mechanical properties or to corrosion resistance, the solid solution formed during solution heat treatment must be quenched rapidly enough (and without interruption) to produce a supersaturated solution at room temperature—the optimum condition for precipitation hardening. The resistance to stress-corrosion cracking of certain Cu-free Al-Zn-Mg alloys, however, is improved by slow quench-

ing. Most frequently, parts are quenched by immersion in cold water or, in continuous heat treating of sheet, plate or extrusions in primary fabricating mills, by progressive flooding or high-velocity spraying with cold water. However, parts of complex shape, often with both thin and thick sections (such as die forgings, most castings, impact extrusions, and components formed from sheet) are commonly quenched in a medium that provides somewhat slower cooling. This medium may be water at 65 to 80 °C (150 to 180 °F), boiling water, an aqueous solution of polyalkaline glycol, or some other fluid medium such as forced air or mist.

If appreciable precipitation during cooling is to be avoided, two requirements must be satisfied. First, the time required for transfer of the load from the furnace to the quenching medium must be short enough to preclude slow precooling into the temperature range where very rapid precipitation takes place. For alloy 7075, this range was determined to be 400 to 290 °C (750 to 550 °F). The second requirement for avoidance of appreciable precipitation during quenching is that the volume, heat-absorption capacity and rate of flow of the quenching medium be such that little or no precipitation occurs during cooling. Any interruption of the quench that might allow reheating into a temperature range where rapid precipitation can occur must be prohibited.

TREATMENTS THAT PRECEDE PRECIPITATION HEAT TREATING

Immediately after being quenched, most aluminum alloys are nearly as ductile as they are in the annealed condition. Consequently, it is often advantageous to form or straighten parts in this temper. Moreover, at the mill level, controlled mechanical deformation is the most common method of reducing residual quenching stresses. Because precipitation hardening will occur at room temperature, forming or straightening usually follows as soon after quenching as possible. In addition, maximum effectiveness in stress relief is obtained by working the metal immediately after quenching.

In some alloys, notably those of the 2xxx series, cold working of freshly quenched material greatly increases its response to later precipitation heat treatment. Mills take advantage of this phenomenon by applying a controlled amount of rolling (sheet and plate) or stretching (extrusion, bar and plate) to produce higher mechanical properties. However, if the higher properties are used in design, reheat treatment must be avoided.

Forming and Straightening. These operations vary in degree from minor corrections of warpage to complete forming of complex parts from solution-treated flat blanks. Particular value is gained when enough forming can be done at this stage of processing to eliminate the distortion caused by quenching. However, production operations must be adjusted so that most of the plastic deformation is accomplished before an appreciable amount of precipitation hardening takes place.

Although the most severe forming operations may have to be arranged to avoid natural aging, it often is desirable to allow some natural aging to occur and thus avoid formation of Lüders lines. This condition of nonuniform deformation is most likely to occur shortly after quenching and diminishes significantly after a few hours of natural aging. Complete freedom from Lüders lines, however, may require one or two days of natural aging prior to forming. Thus, the forming operation may have to be timed so as to obtain the most appropriate trade-off of these characteristics for the specific parts involved. Lüders lines also can be reduced by employing low strain rates or by forming at temperatures of 150 to 175 °C (300 to 350 °F).

Residual stresses in sheet-metal parts formed in the quenched condition are higher than those in parts formed in the annealed condition. Consequently, forming in the quenched condition

should be selected judiciously for parts that are critical in fatigue or stress corrosion.

Re-solution heat treatment of parts formed after quenching often causes excessive grain growth in critically strained regions and thus is not recommended.

Natural Aging. The more highly alloyed members of the 6xxx wrought series, the copper-containing alloys of the 7xxx group, and all of the 2xxx alloys are almost always solution heat treated and quenched. For some of these alloys—particularly the 2xxx alloys—the precipitation hardening that results from natural aging alone produces useful tempers (T3 and T4 types) that are characterized by high ratios of tensile to yield strength and high fracture toughness and resistance to fatigue. For the alloys that are used in these tempers, the relatively high supersaturation of atoms and vacancies retained by rapid quenching causes rapid formation of GP zones, and strength increases rapidly, attaining nearly maximum stable values in four or five days. Tensile-property specifications for products in T3- and T4-type tempers are based on a nominal natural aging time of four days. In alloys for which T3- or T4-type tempers are standard, the changes that occur on further natural aging are of relatively minor magnitude, and products of these combinations of alloy and temper are regarded as essentially stable after about one week.

In contrast to the relatively stable condition reached in a few days by 2xxx alloys that are used in T3- or T4-type tempers, the 6xxx alloys and to an even greater degree the 7xxx alloys are considerably less stable at room temperature and continue to exhibit significant changes in mechanical properties for many years. Because of the relative instability of the 7xxx alloys, the naturally aged temper (after solution heat treatment and quenching) is designated by the suffix letter W. For specific description of this condition, the time of natural aging should be included (example: 7075-W, 1 month).

Aging characteristics vary from alloy to alloy with respect to both time to initial change in mechanical properties and rate of change, but aging effects always are lessened by reductions in aging temperature. With some alloys, aging can be suppressed or delayed for several days by holding at a temperature of −18 °C (0 °F) or lower. It is usual practice to complete forming and straightening before aging changes mechanical properties appreciably. When scheduling makes this impractical, aging may be avoided in some alloys by refrigerating prior to forming. It is conventional practice to refrigerate alloy 2024-T4 rivets to maintain good driving characteristics. Full-size wing plates for current-generation jet aircraft have been solution heat treated and quenched at the primary fabricating mill, packed in dry ice in specially designed insulated containers and transported by rail about 2000 miles to the aircraft manufacturer's plant for forming.

Unanticipated difficulties may arise as a result of failure to control refrigerator or part temperature closely enough. If opening of the cold box to insert or remove parts is done too frequently, the cooling capacity of the refrigerator may be exceeded. At times, the rate at which heavy-gage parts can be cooled in a still-air cold box has been found to be insufficient. This problem has been solved in one plant by immersing parts in a solvent at −40 °C (−40 °F) before placing them in the refrigerator.

Cold Work Strain Hardening. The T3-type tempers are distinguished from T4-type tempers by significant mechanical-property differences resulting from cold work strain hardening associated with certain mechanical operations performed after quenching. Roller or stretcher leveling to achieve flatness or straightness introduces modest strains (on the order of 1 to 4%) that cause changes in mechanical properties (primarily, increases in strength). Further increases in strength can be obtained by cold rolling, additional stretching, combinations of these operations or, for products such as hand forgings, compressive deformation. The tempers produced by these operations followed by natural aging alone (no precipitation heat treatment) are classified as T3-type tempers, and an additional digit is used to indicate a variation in strain hardening that results in significant changes in properties. In the most recently introduced 2xxx aircraft alloy, 2324, high strength is achieved by cold rolling plate to a T39 temper.

Precipitation Heat Treatment. Production of material in T5- through T10-type tempers necessitates precipitation heat treating at elevated temperatures (artificial aging). Although the hardening precipitate developed by this operation is submicroscopic, structures before and after precipitation heat treatment often can be distinguished by etching metallographic specimens. In aluminum alloys in the solution heat treated and quenched condition, coloration contrast between grains of differing orientation is relatively high, particularly in 2xxx series wrought alloys and 2xx.0 series casting alloys. This contrast is noticeably decreased by precipitation heat treatment.

Differences in type, volume fraction, size and distribution of the precipitated particles govern properties as well as the changes observed with time and temperature, and these are all affected by the initial state of the structure. The initial structure may vary in wrought products from unrecrystallized to recrystallized and may exhibit only modest strain from quenching or additional strain from cold working after solution heat treatment. These conditions, as well as the time and temperature of precipitation heat treatment, affect the final structure and the resulting mechanical properties.

Because mechanical properties and other characteristics change continuously with time and with temperature, treatment to produce a combination of properties corresponding to a specific alloy/temper combination requires one or more rather specific and coordinated combinations of time and temperature, with both parameters being subject to practical limitations. Recommended commercial treatments often are compromises between (a) time and cost factors and (b) the probability of obtaining the intended properties, with consideration of allowances for variables such as composition within specified range and temperature variations within the furnace and load. Use of higher temperatures may reduce treatment time; but if the temperature is too high, characteristic features of the precipitation-hardening process reduce the probability of obtaining the required properties.

T6 and T7 Tempers. Precipitation heat treatment following solution heat treatment and quenching produces T6- and T7-type tempers. Alloys in T6-type tempers generally have the highest strengths practical without sacrifice of the other properties and characteristics found by experience to be satisfactory and useful for engineering applications.

Alloys in T7 tempers are "overaged," which means that some degree of strength has been sacrificed or "traded off" to improve one or more other characteristics. Strength may be sacrificed to improve dimensional stability, particularly in products intended for service at elevated temperatures, or to lower residual stresses in order to reduce warpage or distortion in machining. T7-type tempers frequently are specified for cast or forged engine parts. Precipitation heat treating temperatures used to produce these tempers generally are higher than those used to produce T6-type tempers in the same alloys.

Two important groups of T7-type tempers — the T73 and T76 types — have been developed for the wrought alloys of the 7xxx series, which contain more than about 1.25% copper. These tempers are intended to improve resistance to exfoliation corrosion and stress-corrosion cracking, but as a result of over-aging, they also increase fracture toughness and, under some conditions, reduce rates of fatigue-crack propagation. The T73-type temper has greatly minimized stress-corrosion cracking of large and complex machined parts made of these alloys, which occasionally occurred with T6-type tempers. The precipitation heat treatments used to produce the T73- and T76-type tempers consist either of a two-stage isothermal precipitation heat treatment or of heating at a controlled rate to a single treatment temperature. The microstructural/electrochemical relationships that are required in order to achieve the desired corrosion-resisting characteristics can be developed by using only a single-stage precipitation heat treatment above about 150 °C (300 °F), but higher strength is obtained by preceding this with a lower-temperature stage or with a slow, controlled heatup. Extended natural aging can provide the same results, but the times required at room temperature are impractical. Either during the preliminary stage or during slow heatup, a fine, high-density dispersion of GP zones is nucleated. Either the time and temperature of the first step or the rate of heating must be controlled to produce GP zones that will not dissolve but will transform to the η′ precipitate when heated to the aging temperature above 150 °C (300 °F). The aging practice that produces the results in the shortest time depends on the GP-zone solvus temperature. This temperature, in turn, depends on vacancy concentration, a factor influenced by solution heat treating temperature and quench rate, and on composition. If first-step aging time is too short, if first-step aging temperature is too far below the GP-zone solvus, or if heating rates are too high, the GP zones will dissolve above 150 °C (300 °F), and the resultant coarse and widely distributed precipitate will provide lower strength. The T76-type treatments have the same operational sequence but employ second-stage heating only long enough to develop a resistance to exfoliation corrosion higher than that provided by the T6-type tempers. Materials in the T73-type temper also have high resistance to exfoliation corrosion.

T8-Type Tempers. In some alloys — particularly certain alloys of the 2xxx series — strain introduced by cold working after solution heat treatment and quenching nucleates a finer, denser precipitate dispersion that considerably increases strength. This effect is the basis for the higher-strength T8-type tempers of alloys 2011, 2024, 2124, 2219 and 2419, which are produced by applying controlled amounts of cold rolling, stretching, or combinations of these operations.

Alloys 2024, 2124 and 2219 in T8-type tempers are particularly well suited for supersonic and military aircraft; alloy 2219 in such tempers, and alloy 2014-T65, were the principal materials for the fuel and oxidizer tanks (which also served as the primary structure) of the Saturn V space vehicles. Re-solution heat treatment of mill products supplied in these tempers can result in grain growth and in substantially lower strength than is normal for the original temper. Such reheat treatment is not recommended.

Alloys of the 7xxx series do not respond favorably to the sequence of operations used to produce T8-type tempers, and no such tempers are standard for these alloys. The strains associated with stretching or compressive stress relief of 7xxx alloys have relatively little effect on the mechanical properties of material precipitation heat treated to T6-type tempers. On the other hand, these operations have measurable detrimental effects on final strength when T73-, T736- or T76-type tempers are produced, particularly in the direction opposite the direction of cold work. Accordingly, specification properties are somewhat lower for the stress-relieved versions of these tempers. Decreasing the over-aging time to compensate for the loss in strength is not advisable, because this would impair development of the desired corrosion characteristics.

Temperature control and uniformity present essentially the same problems in precipitation heat treating as they do in solution heat treating.

Good temperature control and uniformity throughout the furnace and load is required for all precipitation heat treating. Recommended temperatures are generally those that are least critical and that can be used with practical time cycles. Except for 7xxx alloys in T7x tempers, these temperatures generally allow some latitude and should have a high probability of meeting property specification requirements. Furnace radiation effects seldom are troublesome except in those few furnaces that are used for both solution and precipitation heat treating. Generally, such situations should be avoided, because the high heat capacity needed for the higher temperatures may be difficult to control at normal aging temperatures.

Soak time in precipitation heat treating is not difficult to control; the specified times carry rather broad tolerances. Heavier loads with parts racked closer together, and even nested, are not abnormal. The principal hazard is undersoaking due to gross excesses in loading practices. Some regions of the load may reach soak temperature long after soak time has been called. Placement of load thermocouples is critical, and limiting the size and spacing of a load may be necessary for aging to the T73 and T76 tempers. As discussed above, soak time is not as critical for peak aged (T6 and T8) tempers.

Heat Treating of Copper Alloys

HEAT TREATING PROCESSES that are applied to copper and copper alloys include homogenizing, annealing, stress relieving, solution treating, precipitation (age) hardening, and quench hardening and tempering.

HOMOGENIZING

Homogenization is required most frequently for alloys with wide freezing ranges, such as tin (phosphor) bronzes, copper nickels and silicon bronzes. Although coring occurs to some extent in alpha brasses, alpha aluminum bronzes and copper-beryllium alloys, these alloys survive primary mill processing and become homogenized during normal process working and annealing. There is rarely a necessity to apply homogenization to finished or semifinished copper alloy mill products.

The time and temperature required for this process varies with the alloy, the cast grain size and the desired degree of homogenization. Typical soak times vary from 3 to over 10 h. Temperatures normally are above the upper annealing range, to within 50 °C (90 °F) of the solidus temperature.

ANNEALING

Wrought Products

Annealing of cold worked metal is accomplished by heating to a temperature that produces recrystallization and, if desirable, by heating beyond the recrystallization temperature to produce grain growth. Temperatures commonly used for annealing cold worked coppers and copper alloys are given in Table 3.

Annealing is primarily a function of metal temperature and time at temperature. Except for multiphase alloys and certain precipitation-hardening alloys, or alloys susceptible to fire cracking, rates of heating and cooling are relatively unimportant.

Because of the multiplicity of influential variables, it is difficult to tabulate a definite annealing schedule that will result in completely recrystallized metal of a specific grain size.

In commercial mill practice, copper alloys usually are annealed at successively lower temperatures as the material approaches the final anneal, with intermediate cold reductions of at least 35% and as high as 50 to 60% in single or multiple passes wherever practicable. The higher temperatures initially used accelerate homogenization, and the resulting large grains permit more economical reduction during the early working operation.

During subsequent anneals, the grain size should be dereased gradually to approximate the final grain size required. This point usually is reached one or two anneals before the final anneal. With such a sequence and with sufficiently severe intermediate reductions, it is then possible to produce a uniform final grain size within a lot and from lot to lot.

Annealing to Specific Properties

Although specific properties are most frequently produced by controlled cold working of annealed material, there are occasions in which annealing to temper is necessary or advantageous. In hot rolling of copper alloy plate — particularly large patterns — the finishing temperature may not be consistent or controllable, and varying degrees of work hardening may occur. Also, small quantities and/or odd sizes of required drawn or roll tempered materials may not be readily available, while appropriate stocks of harder material are.

Table 3. Annealing temperatures for cold worked coppers and copper alloys

Alloy	Common name	Annealing temperature °C	°F
Wrought coppers			
C10200	Oxygen-free copper	425-650	800-1200
C11000	Electrolytic tough pitch copper	250-650	500-1200
C11300, C11400,			
C11500, C11600	Silver-bearing tough pitch copper	400-475	750-900
C12000	Phosphorus-deoxidized copper, low residual phosphorus	325-650	600-1200
C12200	Phosphorus-deoxidized copper, high residual phosphorus	375-650	700-1200
C14500	Phosphorus-deoxidized, tellurium-bearing copper	425-650	800-1200
Wrought copper alloys			
C17000, C17200,			
C17500	Beryllium copper	775-925(a)	1425-1700(a)
C21000	Gilding metal	425-800	800-1450
C22000	Commercial bronze	425-800	800-1450
C22600	Jewelry bronze	425-750	800-1400
C23000	Red brass	425-725	800-1350
C24000	Low brass	425-700	800-1300
C26000	Cartridge brass	425-750	800-1400
C26800, C27000,			
C27400	Yellow brass	425-700	800-1300
C28000	Muntz metal	425-600	800-1100
C31400	Leaded commercial bronze	425-650	800-1200
C33000, C33500	Low-leaded brass	425-650	800-1200
C33200, C34200,			
C35300	High-leaded brass	425-650	800-1200
C34000, C35000	Medium-leaded brass	425-650	800-1200
C35600	Extra-high-leaded brass	425-650	800-1200
C36000	Free-cutting brass	425-600	800-1100
C36500, C36600,			
C36700, C36800	Leaded Muntz metal	425-600	800-1100
C37000	Free-cutting Muntz metal	425-650	800-1200
C37700	Forging brass	425-600	800-1100
C38500	Architectural bronze	425-600	800-1100
C44300, C44400,			
C44500	Inhibited admiralty	425-600	800-1100
C46200	Naval brass	425-600	800-1100
C48200, C48500	Leaded naval brass	425-600	800-1100
C50500	Phosphor bronze	475-650	900-1200
C51000, C52100,			
C54200	Phosphor bronze	475-675	900-1250
C53200, C53400,			
C54400	Free-cutting phosphor bronze	475-675	900-1250
C60600, C60800	Aluminum bronze	550-650	1000-1200
C61000	Aluminum bronze	600-675	1100-1250
C61300, C61400	Aluminum bronze	750-875	1400-1600
C61800, C61900,			
C62400	Aluminum bronze	600-650(b)	1100-1200(b)
C63000	Aluminum bronze	650-700(c)	1200-1300(c)
C63200	Aluminum bronze	675-725(c)	1250-1350(c)
C64200	Aluminum bronze	Above 650	Above 1200
C65100	Low-silicon bronze	475-675	900-1250
C65500	High-silicon bronze	475-700	900-1300
C67000, C67500	Manganese bronze	425-600	800-1100
C68700	Aluminum brass	425-600	800-1100
C70600	Copper nickel, 10%	600-825	1100-1500
C71500	Copper nickel, 30%	650-825	1200-1500
C75200, C75700,		600-825	1100-1500
C77000	Nickel silver		

(a) Solution-treating temperature. (b) Cool rapidly (cooling method important in determining result of annealing. (c) Air cool (cooling method important in determining result of annealing).

Tensile strengths and hardness levels similar to those of $1/8$, $1/4$ and $1/2$ hard cold worked tempers can be produced by annealing hard worked brasses, nickel silvers and phosphor bronzes. While the yield strength for a given final hardness tends to be lower for alloys annealed to temper than for those cold worked to temper, the fatigue resistance of some phosphor bronze spring materials in annealed $1/2$ hard tempers appears to be superior to that of cold worked material.

General Precautions

For best results in annealing copper and copper alloys, the following precautions should be observed.

Sampling and Testing. Test specimens must represent the extreme conditions of the furnace load. For copper alloys that do not contain grain-growth inhibitors, the best and most accurate test for the extent of annealing is the size of the average grain. Grain size is usually the basis for acceptance or rejection of the material.

Hydrogen Embrittlement. When copper that contains oxygen is to be annealed, the hydrogen in the furnace atmosphere must be kept to a minimum. This reduces the embrittlement caused by the combination of the hydrogen in the atmosphere with the oxygen in the copper, forming water vapor under pressure and resulting in minute porosity in the metal.

Impurities. Occasionally, it is difficult to obtain proper grain growth by annealing under standard conditions that previously have resulted in the desired grain size. This difficulty sometimes may be traced to impurities in the alloy.

Fire cracking occurs when some alloys that contain residual stresses are heated too rapidly. Leaded alloys are particularly susceptible to fire cracking. The remedy is to heat slowly until the stresses are relieved.

Thermal shock or fatigue occurs when rapid and extreme changes in temperature occur. Stresses that result in thermal shock are influenced by thermal expansion, thermal conductivity, strength, toughness, the rate of change of the temperature and the condition of the material.

Castings

Annealing is applied to castings of some duplex alloys, such as manganese bronzes and aluminum bronzes, where it is intended to correct the effects of mold cooling. The extremely slow cooling of sand and plaster castings, or the rapid cooling of permanent mold or die castings, can produce microstructures resulting in high hardness and/or low ductility and occasionally inferior corrosion resistance. Typical annealing treatments for castings are in the range from 580 to 700 °C (1075 to 1300 °F) for 1 h at temperature. For aluminum bronzes, rapid cooling by water quenching or high-velocity air is advisable.

STRESS RELIEVING

During processing or fabrication of copper alloys by cold working, strength and hardness increase due to plastic strain. Because plastic strain is accompanied by elastic strain, residual stresses remain in the resultant product. If allowed to remain in sufficient magnitude, residual surface tensile stresses can result in stress-corrosion cracking of material in storage or service, unpredictable distortion of material during cutting or machining, and hot cracking of materials during processing, brazing or welding.

Stress-relief heat treatments are carried out at temperatures below those normally used for annealing. Typical process stress-relieving temperatures for 19 wrought copper alloys are given in Table 4. Temperatures for treatment of cold formed or welded structures are generally 50 to 110 °C (90 to 200 °F) higher.

Table 4. Typical stress-relieving temperatures for 19 wrought copper alloys

Alloy	Common name	Stress-relieving temperature(a) °C	°F
C21000	Gilding metal	190	375
C22000	Commercial bronze	205	400
C23000	Red brass	230	450
C24000	Low brass	260	500
C26000	Cartridge brass	260	500
C27000	Yellow brass	260	500
C28000	Muntz metal	205	400
C36000	Free-cutting brass	245	475
C44300, C44400,			
C44500	Inhibited admiralty	290	550
C51000, C52100	Phosphor bronze	205	400
C61300, C61400	Aluminum bronze	345	650
C65500	High-silicon bronze	345	650
C70600, C71500	Copper nickel	260	500
C75200	Nickel silver	260	500

(a) Time at temperature, 1h.

HARDENING

Copper alloys that are hardened through heat treatment are of two general types: those that are softened by high-temperature quenching and hardened by lower-temperature treatments, and those that are hardened by quenching from high temperatures through martensitic-type reactions. Alloys that harden during low-to-intermediate-temperature treatments following solution quenching include precipitation-hardening, spinodal-hardening and order-hardening types. Quench-hardening alloys comprise aluminum bronzes, nickel-aluminum bronzes, and a few copper-zinc alloys. Quench-hardened alloys normally are tempered to improve toughness and ductility and reduce hardness in a manner similar to that for alloy steels.

Low-Temperature-Hardening Alloys

For purposes of comparison, Table 5 lists examples of the various types of low-temperature-hardening alloys, as well as typical heat treatments and attainable property levels for these alloys. Additional details are given in the three subsections that follow.

Precipitation-Hardening Alloys. Most copper alloys of the precipitation-hardening type find use in electrical- and heat-conduction applications. The heat treatment must therefore be designed to develop the necessary mechanical strength and electrical conductivity. The resulting hardness and strength depend on both the effectiveness of the solution quench and the control of the precipitation (aging) treatment. Note: "age hardening" and "aging" are used in heat treating practice as substitutes for the terms "precipitation" and "spinodal hardening." Copper alloys harden by elevated-temperature treatment rather than ambient-temperature (natural) aging as in the case of some aluminum alloys. As solute atoms proceed through the coagulation, coherency and precipitation cycle in the quenched alloy lattice, the hardness increases, reaches a peak, and then decreases with time. Electrical conductivity increases continuously with time until some maximum is reached, normally in the fully precipitated condition. The optimum condition generally preferred results from a precipitation treatment of temperature and duration just beyond those that correspond to the hardness aging peak. Cold working prior to precipitation aging tends to improve the heat treated hardness. In the case of lower-strength wrought alloys such as C18200 (Cu-Cr) and C15000 (Cu-Zr), some heat treated hardness may be sacrificed to attain increased conductivity, with final hardness and strength being enhanced by cold working. Two precipitation treatments are necessary in order to develop maximum electrical conductivity and hardness in alloy C18000 (Cu-Ni-Si-Cr) because of two distinct precipitation mechanisms.

Spinodal-Hardening Alloys. Alloys that harden by spinodal decomposition are hardened by a treatment similar to that used for precipitation-hardening alloys. The soft and ductile spinodal structure is generated by a high-temperature solution treatment followed by quenching. The material can be cold worked or formed in this condition. A lower-temperature spinodal-decomposition treatment, commonly referred to as aging, is then used to increase the hardness and strength of the alloy. Spinodal-hardening alloys are basically copper-nickel alloys with chromium or tin additions. The hardening mechanism is related to a miscibility gap in the solid solution and does not result in precipitation. Since no crystallographic changes take place, spinodal-hardening alloys retain excellent dimensional stability during hardening.

Order-Hardening Alloys. Certain alloys, generally those that are nearly saturated with an alloying element dissolved in the alpha phase, will undergo an ordering reaction when highly cold worked material is annealed at a relatively low temperature. Alloys C61500, C63800, C68800 and C69000 are examples of copper alloys that exhibit this behavior. Strengthening is attributed to short-range ordering of the solute atoms within the copper matrix, which greatly impedes the motion of dislocations through the crystals.

Quench Hardening and Tempering

Quench hardening and tempering (also referred to as "quench-and-temper hardening") is used primarily for aluminum bronze and nickel aluminum bronze alloys, and occasionally for some cast manganese bronze alloys with zinc equivalents of 37 to 41%. Aluminum bronzes with 9 to 11.5% Al, and nickel aluminum bronzes with 8.5 to 11.5% Al, respond in a practical way to quench hardening by a martensitic-type reaction. Alloys higher in aluminum content generally are too susceptible to quench cracking, whereas those with lower aluminum contents do not contain enough high-temperature beta phase to respond to quench treatments.

COPPER-ALUMINUM (ALUMINUM BRONZE) ALLOYS

The microstructures and consequent heat treatabilities of aluminum bronzes vary with aluminum content much the same as these characteristics vary with carbon content in steels. Unlike steels, aluminum bronzes are tempered above the normal transformation temperature, typically in the range from 565 to 675 °C (1050 to 1250 °F). In selection of tempering temperatures, consideration must be given both to required properties and to hardness obtained on quenching. Normal tempering time is 2 h at temperature. Moreover, heavy or complex sections should be heated slowly to avoid cracking. After the tempering cycle has been completed, it is important that aluminum bronzes be cooled rapidly, using water quenching, spray cooling or fan cooling. Slow cooling through the range from 565 to 275 °C (1050 to 530 °F) can cause the residual tempered martensitic beta phase to decompose, forming the embrittling alpha-gamma eutectoid. The presence of appreciable amounts of this eutectoid structure can result in low tensile elongation, low energy of rupture, severely reduced impact values, and reduced corrosion resistance in some media. For adequate protection against detrimental eutectoid transformation, cooling after tempering should bring the alloy to a temperature below 370 °C (700 °F) within about 5 min, and to a temperature below 275 °C (530 °F) within 15 min. Normally, the danger of eutectoid transformation is much smaller in nickel aluminum bronzes, and these alloys can be air cooled after tempering.

Alpha aluminum bronzes are those aluminum bronzes that contain less than 9% aluminum, or less than 8.5% aluminum with up to 3% iron. They are essentially single-phase alloys, except for fine iron-rich particles in those alloys that contain iron. For alpha aluminum bronzes, effective strengthening can be attained only by cold work, and annealing and/or stress relieving are the only heat treatments of practical use.

Annealing of alpha aluminum bronzes is carried out at temperatures from about 540 to about 870 °C (1000 to 1600 °F), with the iron-containing alloys requiring temperatures nearer the high end of this range.

Complex (alpha-beta) aluminum bronzes are those aluminum bronzes whose normal microstructures contain more than one phase to the extent that beneficial quench-and-temper treatments are possible. These copper aluminum alloys, with and without iron, are heat treated by procedures somewhat similar to those used for heat treatment of steel, and have isothermal transformation diagrams that resemble those of carbon steels. For these alloys, the quench-hardening treatment is essentially a high-temperature soak intended to dissolve all of the alpha phase into the beta phase. Quenching results in a hard room-temperature beta martensite structure; and subsequent tempering reprecipitates fine alpha needles in the structure, forming a tempered beta martensite.

Table 6 gives typical tensile properties and hardnesses of alpha-beta aluminum bronzes after various stages of heat treatment.

COPPER-BERYLLIUM ALLOYS

Wrought copper-beryllium alloy mill products generally are supplied solution treated or solution

Table 5. Typical heat treatments and resulting properties for several low-temperature-hardening alloys

Alloy	Solution-treating temperature(a) °C	°F	Aging treatment Temperature °C	°F	Time, h	Hardness	Electrical conductivity, % IACS
Precipitation hardening							
C15000	980	1795	500-550	930-1025	3	30 HRB	87-95
C17000, C17200, C17300	760-800	1400-1475	300-350	575-660	1-3	35-44 HRC	22
C17500, C17600	900-950	1650-1740	455-490	850-915	1-4	95-98 HRB	48
C18000(b), C81540	900-930	1650-1705	425-540	800-1000	2-3	92-96 HRB	42-48
C18200, C18400, C18500, C81500	980-1000	1795-1830	425-500	800-930	2-4	68 HRB	80
C94700	775-800	1425-1475	305-325	580-620	5	180 HB	15
C99400	885	1625	482	900	1	170 HB	17
Spinodal hardening							
C71900	900-950	1650-1740	425-760	800-1400	1-2	86 HRC	4-4
C72800	815-845	1500-1550	350-360	660-680	4	32 HRC	...

(a) Solution treating is followed by water quenching. (b) Alloy C18000 (81540) must be double aged—typically, 3 h at 540 °C (1000 °F) followed by 3 h at 425 °C (800 °F) (U.S. Patent No. 4 191 601)—to develop the higher levels of electrical conductivity and hardness.

Table 6. Typical heat treatments and resulting properties for complex (alpha-beta) aluminum bronzes

Alloy	Typical conditions	Tensile strength MPa	ksi	Yield strength(b) MPa	ksi	Elongation(c), %	Hardness, HB
C62400	As forged or extruded	620-690	90-100	240-260	35-38	14-16	163-183
	Solution treated at 870 °C (1600 °F) and quenched, tempered 2 h at 620 °C (1150 °F)	675-725	98-105	345-385	50-56	8-14	187-202
C63000	As forged or extruded	730	106	365	53	13	187
	Solution treated at 855 °C (1575 °F) and quenched, tempered 2 h at 650 °C (1200 °F)	760	110	425	62	13	212
C95300	As cast	495-530	72-77	185-205	27-30	27-30	137-140
	Solution treated at 855 °C (1575 °F) and quenched, tempered 2 h at 620 °C (1150 °F)	585	85	290	42	14-16	159-179
C95400	As cast	585-690	85-100	240-260	35-38	14-18	156-179
	Solution treated at 870 °C (1600 °F) and quenched, tempered 2 h at 620 °C (1150 °F)	655-725	95-105	330-370	48-54	8-14	187-202
C95500	As cast	640-710	93-103	290-310	42-45	10-14	183-192
	Solution treated at 855 °C (1575 °F) and quenched, tempered 2 h at 650 °C (1200 °F)	775-800	112-116	440-470	64-68	10-14	217-234

(a) As-cast condition is typical for moderate sections shaken out at temperatures above 540 °C (1000 °F) and fan cooled; or mold cooled, annealed at 620 °C (1150 °F) and fan (rapid) cooled. (b) At 0.5% extension under load. (c) In 50 mm or 2 in.

treated and cold worked. Material in these conditions can be fabricated without further heat treatment. Thus, solution treating typically is not a part of the fabricating process unless it is necessitated by a special requirement such as softening of the material for additional forming or is employed as a salvage operation for parts that have been incorrectly treated for precipitation hardening.

Solution treating must be carefully controlled to produce desired grain size, dimensional tolerances and mechanical properties, and to prevent surface oxidation.

In cast products, the as-cast structure usually contains a large amount of microsegregation within the dendritic pattern. Therefore, castings must be heated for a time sufficient to homogenize the structure. A minimum of 3 h at temperature is recommended for this purpose.

Quenching is a critical phase of the solution-treating process. Successful treatment requires that the material be quenched immediately, and at the highest possible rate, after being removed from the furnace. Any time lapse during transfer from the furnace to the quenching medium will permit some cooling and will cause precipitation. Precipitation is rapid at elevated temperatures, and its occurrence significantly affects the properties obtained during subsequent precipitation hardening.

PRECIPITATION HARDENING

Cold working of solution-treated copper-beryllium alloys influences the strength attainable through subsequent aging; the greatest response to aging occurs in material in the cold rolled hard temper. In general, work hardening offers no advantages beyond the hard temper, because formability is poor and control of the precipitation-hardening treatment for maximum strength is critical. For some applications, however, wire is drawn to higher levels of cold work prior to precipitation hardening.

Close control of temperature is critical in conventional aging of copper-beryllium alloys. A change in temperature affects the time required for development of maximum properties. Also, the higher temperatures can result in lower property values. Normal commercial control of ±6 °C (±10 °F) is adequate for temperatures in the range from 315 to 370 °C (600 to 700 °F).

COPPER-CHROMIUM ALLOYS

Copper-chromium alloys containing 0.5 to 1.0% Cr are solution treated, in molten salt or in controlled-atmosphere furnaces to avoid scaling, at 980 to 1010 °C (1800 to 1850 °F) and rapidly quenched. Solution-treated chromium copper is soft and ductile; therefore, it can be cold worked in a manner similar to that used for unalloyed copper.

After being solution treated, the material may be aged for several hours at 400 to 500 °C (750 to 930 °F) to produce special mechanical and physical properties. A typical aging cycle is 4 h or more at 455 °C (850 °F).

COPPER-ZIRCONIUM ALLOYS

Solution treatment of zirconium copper (99.7 min Cu, 0.13 to 0.30 Zr) consists of heating to 900 to 980 °C (1650 to 1795 °F) and quenching in water. The material may then be precipitation hardened for 1 to 4 h at 500 to 550 °C (930 to 1020 °F). If cold working is done prior to aging, the aging temperature is reduced to 370 to 480 °C (700 to 900 °F) for 1 to 4 h.

Time at the solution-treating temperature should be minimized, to limit grain growth and possible internal oxidation by reaction of the zirconium with the furnace atmosphere. Because solution and diffusion of the zirconium occur rapidly at the solution-treating temperature, holding at temperature is not required.

Maximum mechanical properties and resistance to softening are developed with maximum solution of zirconium. If material containing 0.15% Zr or more is heated above 980 °C (1795 °F), the Cu_3Zr phase will begin to melt. A slight amount of melting will not affect mechanical properties; however, if an excessive amount of melting should occur, ductility of the alloy will decrease.

Normally, as the solution temperature is increased from 900 to 980 °C (1650 to 1795 °F), the aging temperatures should also be increased to maintain high electrical conductivity.

The increase in strength of zirconium copper depends primarily on cold work. Although aging results in some increase in strength, its chief effect is to increase electrical conductivity.

Heat Treating of Lead and Lead Alloys

LEAD normally is considered to be unresponsive to heat treatment. Yet, some means of strengthening lead and lead alloys may be required for some applications.

SOLUTION TREATING AND AGING

Useful strengthening of lead can be attained by adding sufficient quantities of antimony to produce hypoeutectic lead-antimony alloys. Small amounts of arsenic have particularly strong effects on the age-hardening response of such alloys, and these effects are enhanced by solution treating followed by rapid quenching prior to aging.

Application. To reduce the antimony contents of the positive plates in lead-acid storage batteries, because of antimony's detrimental effect on charge retention, there has been a trend toward replacing eutectic alloys with a Pb-6Sb-0.15As alloy. Battery grids made of this arsenical alloy will age harden slowly after casting and air cooling. However, storing grids for several days constitutes unproductive use of floor space and results in undesirable interruptions in manufacturing sequences.

DISPERSION HARDENING

Another mechanism for strengthening of lead alloys involves elements that have low solubilities in solid lead, such as copper and nickel. Alloys that contain these elements can be processed so that no homogenization results; most of the strengthening that occurs is developed through dispersion hardening, with some solid-solution hardening taking place as a secondary effect. The resulting structure is more stable than those developed by other hardening processes. Dispersion strengthening also has been achieved through powder metallurgy methods in which lead oxide, alumina or similar materials are dispersed in pure lead.

FABRICATION

Although alloy selection is important, care must be taken in fabrication as well. Castings should be cooled rapidly to a temperature below that at which the structure breaks down, or a coarse structure will be obtained.

Heat Treating of Magnesium Alloys

MAGNESIUM ALLOYS usually are heat treated either to improve mechanical properties or as a means of conditioning for specific fabricating operations. The type of heat treatment selected depends on alloy composition and form (cast or wrought), and on anticipated service conditions.

Solution heat treatment improves strength and results in maximum toughness and shock resistance. Artificial aging (precipitation heat treatment) subsequent to solution treatment gives maximum hardness and yield strength, but with some sacrifice of toughness. As applied to castings, artificial aging without prior solution treatment or annealing is a stress-relieving treatment that also somewhat increases tensile properties.

The basic temper designations for magnesium alloys are given in Table 7.

In most wrought alloys, maximum mechanical properties are developed through strain hardening, and these alloys generally are either used without subsequent heat treatment or merely aged to a T5 temper. Occasionally, however, solution treatment, or a combination of solution treatment with strain hardening and artificial aging, will substantially improve mechanical properties. Wrought alloys that can be strengthened by heat treatment are grouped into four general classes according to composition:

Table 7. Basic temper designations

F	As fabricated
O	Annealed, recrystallized (wrought products only)
H	Strain hardened (wrought products only):
H1	Strain hardened only
H2	Strain hardened and partially annealed
H3	Strain hardened and stabilized
W	Solution heat treated; unstable temper
T	Heat treated to produce stable tempers other than F, O, or H:
T2	Annealed (cast products only)
T3	Solution heat treated and cold worked
T4	Solution heat treated
T5	Artificially aged only
T6	Solution heat treated and artificially aged
T7	Solution heat treated and stabilized
T8	Solution heat treated, cold worked and artificially aged
T9	Solution heat treated, artificially aged and cold worked
T10	Artificially aged and cold worked

- Magnesium-aluminum-zinc (example: AZ80A)
- Magnesium-thorium-zirconium (example: HK31A)
- Magnesium-thorium-manganese (examples: HM21A, HM31A)
- Magnesium-zinc-zirconium (example: ZK60A).

Solution Treating and Aging. Schedules for solution treating and aging of magnesium alloy castings are summarized in Table 8. In solution treating of magnesium-aluminum-zinc alloys, parts should be loaded into the furnace at approxi-

mately 260 °C (500 °F) and then raised to the appropriate solution-treating temperature slowly, to avoid fusion of eutectic compounds and resultant formation of voids. The time required to bring the load from 260 °C to the solution-treating temperature is determined by the size of the load and by the composition, size, weight and section thickness of the parts, but 2 h is a typical time. All other heat treatable magnesium alloys can be loaded into the furnace at the solution-treating temperature. For alloy HK31A, it is important to bring the load to temperature as rapidly as possible, to avoid grain coarsening.

During aging, magnesium alloy parts should be loaded into the furnace at the treatment temperature, held for the appropriate period of time, and then cooled in still air. As indicated in Table 8, there is a choice of artificial aging treatments for some alloys; results are closely similar for the alternative treatments given.

Heat Treating of Nickel and Nickel Alloys

NICKEL and nickel alloys may be subjected to one or more of five principal types of heat treatment, depending on chemical composition, fabrication requirements and intended service. These methods include:

Annealing—A heat treatment designed to produce a recrystallized grain structure and softening in work-

Table 8. Recommended solution-treating-and-aging schedules for magnesium alloy castings
For castings up to 51 mm (2 in.) in section thickness; heavier sections may require longer times at temperature.

Alloy	Final temper	Aging(a) Temperature °C, ±6(b)	Aging(a) Temperature °F, ±10(b)	Aging(a) Time, h	Solution treating(c) Temperature °C, ±6(b)	Solution treating(c) Temperature °F, ±10(b)	Solution treating(c) Time, h	Maximum temperature °C	Maximum temperature °F	Aging after solution treating Temperature °C, ±6(b)	Aging after solution treating Temperature °F, ±10(b)	Aging after solution treating Time, h
Magnesium-aluminum-zinc alloys(d)												
AM100A	T5	232	450	5	...	...	...	...	...	...	...	...
	T4	...	...	...	424(e)	795(e)	16-24(e)	432	810	...	...	...
	T6	...	...	...	424(e)	795(e)	16-24(e)	432	810	232	450	5
	T61	...	...	...	424(e)	795(e)	16-24(e)	432	810	218	425	25
AZ63A	T5	260(f)	500(f)	4(f)	...	...	...	...	...	...	...	...
	T4	...	...	...	385	725	10-14	391	735	...	...	...
	T6	...	...	...	385	725	10-14	391	735	218(f)	425(f)	5(f)
AZ81A	T4	...	...	...	413(e)	775(e)	16-24(e)	418	785	...	...	...
AZ91C	T5	168(g)	335(g)	16(g)	...	...	...	...	...	...	...	...
	T4	...	...	...	413(e)	775(e)	16-24(e)	418	785	...	...	...
	T6	...	...	...	413(e)	775(e)	16-24(e)	418	785	168(h)	335(h)	16(h)
AZ92A	T5	260	500	4	...	...	...	...	...	...	...	...
	T4	...	...	...	407(j)	765(j)	16-24(j)	413	775	...	...	...
	T6	...	...	...	407(j)	765(j)	16-24(j)	413	775	218	425	5
Magnesium-zirconium alloys												
EZ33A	T5	216(k)	420(k)	5(k)	...	...	...	...	...	...	...	...
HK31A(m)	T6	...	...	...	566	1050	2	571	1060	204	400	16
HZ32A	T5	316	600	16	...	...	...	...	...	...	...	...
QE22A(n)	T6	...	...	...	527	980	4-8	538	1000	204	400	8
QH21A(n)	T6	...	...	...	527	980	4-8	538	1000	204	400	8
ZE41A	T5	329(p)	625(p)	2(p)	...	...	...	...	...	...	...	...
ZE63A(q)	T6	...	...	...	480	895	10-72	491	915	141	285	48
ZH62A	T5	329	625	2	...	...	...	...	...	...	...	...
	plus:	177	350	16	...	...	...	...	...	...	...	...
ZK51A	T5	177(r)	350(r)	12(r)	...	...	...	...	...	...	...	...
ZK61A	T5	149	300	48	...	...	...	...	...	...	...	...
	T6	...	...	...	499(s)	930(s)	2(s)	502	935	129	265	48

(a) Aging of castings to the T5 temper is done from the as-cast condition. (b) Except where quoted differently. (c) After solution treatment and before subsequent aging, castings are cooled to room temperature by fast fan cooling, except where otherwise indicated. Use carbon dioxide or sulfur dioxide atmosphere above 400 °C (750 °F). (d) For solution treating, Mg-Al-Zn alloys are loaded into the furnace at 260 °C (500 °F) and brought to temperature over a 2-h period at a uniform rate of temperature increase. (e) Alternative treatment, to prevent germination (excessive grain growth): 6 h at 413 ± 6 °C (775 ± 10 °F), 2 h at 352 ± 6 °C (665 ± 10 °F), 10 h at 413 ± 6 °C (775 ± 10 °F). (f) Alternative treatment: 5 h at 232 ± 6 °C (450 ± 10 °F). (g) Alternative treatment: 4 h at 216 ± 6 °C (420 ± 10 °F). (h) Alternative treatment: 5-6 h at 216 ± 6 °C (420 ± 10 °F). (j) Alternative treatment, to prevent germination (excessive grain growth): 6 h at 407 ± 6 °C (765 ± 10 °F), 2 h at 352 ± 6 °C (665 ± 10 °F), 10 h at 407 ± 6 °C (765 ± 10 °F). (k) Alternative treatment, which can be used where maximum resistance to creep at elevated temperature is not of prime importance: 2 h at 343 ± 6 °C (650 ± 10 °F). (m) Alloy HK31A castings must be loaded into the furnace already at temperature and brought back to temperature as quickly as possible. (n) Quench from solution-treating temperature either in water at 65 °C (150 °F) or in other suitable quenching medium. (p) This treatment is adequate for development of satisfactory properties; it may be followed by 16 h at 177 ± 6 °C (350 ± 10 °F), to provide very slight improvements in mechanical properties. (q) Alloy ZE63A must be solution treated in a special hydrogen atmosphere, because its mechanical properties are developed through hydriding of some of its alloying elements. Hydriding time depends on section thickness; as a guide, 6.4-mm (¹/₄-in.) sections require approximately 10 h, and 19-mm (³/₄-in.) sections require about 72 h. Following solution treatment, ZE63A should be quenched in oil, water spray or air blast. (r) Alternative treatment: 8 h at 218 ± 6 °C (425 ± 10 °F). (s) Alternative treatment: 10 h at 482 ± 6 °C (900 ± 10 °F).

Table 9. Annealing, stress-relieving and stress-equalizing schedules for nickels and nickel alloys

	Open annealing				Soft annealing Closed annealing			
	Temperature		Time,	Cooling	Temperature		Time,	Cooling
Material	°C	°F	min	method(a)	°C	°F	h	method(a)
Nickel 200	815 to 925	1500 to 1700	¹/₂ to 5	AC or WQ	705 to 760	1300 to 1400	2 to 6	AC
Nickel 201	760 to 870	1400 to 1600	¹/₂ to 5	AC or WQ	705 to 760	1300 to 1400	2 to 6	AC
Monel 400	870 to 980	1600 to 1800	2 to 10	AC or WQ	760 to 815	1400 to 1500	1 to 3	AC
Monel R-405	870 to 980	1600 to 1800	2 to 10	AC or WQ	760 to 815	1400 to 1500	1 to 3	AC
Monel K-500	870 to 1040	1600 to 1900	5 to 15	WQ	—	Not applicable	—	—
Inconel 600	925 to 1040	1700 to 1900	15 to 30	AC or WQ	925 to 980	1700 to 1800	1 to 3	AC
Inconel 601	1095 to 1175	2000 to 2150	15 to 30	AC or WQ	1095 to 1175	2000 to 2150	1 to 3	AC
Inconel 617	1120 to 1175	2050 to 2150	15 to 30	AC or WQ	1120 to 1175	2050 to 2150	1 to 3	AC
Inconel 625	980 to 1150	1800 to 2100	15 to 30	AC or WQ	980 to 1150	1800 to 2100	1 to 3	AC
Inconel 718	955 to 980	1750 to 1800	15 to 30	AC	—	Not applicable	—	—
Inconel X-750	1095 to 1150	2000 to 2100	15 to 30	AC	—	Not applicable	—	—
Hastelloy B	1095 to 1185	2000 to 2165	5	AC or WQ	...	...	...	...
Hastelloy C	1215	2220	5	WQ	...	...	...	...
Hastelloy X	1175	2150	5 to 15	AC or WQ	1175	2150	1	AC or WQ

(a) AC, air cool; WQ, water quench.

	Stress relieving				Stress equalizing			
	Temperature		Time,	Cooling	Temperature		Time,	Cooling
Material	°C	°F	min	method(a)	°C	°F	h	method(a)
Nickel 200	480 to 705	900 to 1300	¹/₂ to 3	AC	260 to 480	500 to 900	1 to 2	AC
Nickel 201	480 to 705	900 to 1300	¹/₂ to 3	AC	260 to 480	500 to 900	1 to 2	AC
Monel 400	540 to 565	1000 to 1050	1 to 2	AC	230 to 315	450 to 600	1 to 3	AC
Monel R-405	...	...	...	...	...	...	...	...
Monel K-500	...	...	...	...	...	...	...	...
Inconel 600	760 to 870	1400 to 1600	1 to 2	AC	...	...	...	...
Inconel 601	...	...	...	...	...	...	...	...
Inconel 617	...	...	...	...	...	...	...	...
Inconel 625	...	...	...	...	...	...	...	...
Inconel 718	...	...	...	...	...	...	...	...
Inconel X-750	...	...	...	...	...	...	...	...
Hastelloy B	1095 to 1185	2000 to 2165	¹/₁₂	AC or WQ	—	Not applicable	—	—
Hastelloy C	1215	2220	¹/₁₂	WQ	—	Not applicable	—	—
Hastelloy X	...	...	...	...	—	Not applicable	—	—

(a) AC, air cool; WQ, water quench.

Table 10. Solution-treating and age-hardening schedules for nickel alloys

	Solution treating				
	Temperature		Time,	Cooling	
Alloy	°C	°F	h	method(a)	Age hardening
Monel K-500	980	1800	¹/₂ to 1	WQ	Heat to 595 °C (1100 °F), hold 16 h; furnace cool to 540 °C (1000 °F), hold 6 h; furnace cool to 480 °C (900 °F), hold 8 h; air cool
Inconel 718	980	1800	1	AC	Heat to 720 °C (1325 °F), hold 8 h; furnace cool to 620 °C (1150 °F), hold until furnace time for entire age-hardening cycle equals 18 h; air cool
Inconel X-750	1150	2100	2 to 4	AC	Heat to 845 °C (1550 °F), hold 24 h; air cool; reheat to 705 °C (1300 °F), hold 20 h; air cool
	980	1800	1	AC	Heat to 730 °C (1350 °F), hold 8 h; furnace cool to 620 °C (1150 °F), hold until furnace time for entire age-hardening cycle equals 18 h; air cool
Hastelloy X	1175	2150	1	AC	Heat to 760 °C (1400 °F), hold 3 h; air cool; reheat to 595 °C (1100 °F), hold 3 h; air cool

(a) WQ, water quench; AC, air cool.

hardened alloys. Annealing usually requires temperatures between 705 and 1205 °C (1300 and 2200 °F), depending on alloy composition and degree of work hardening.

Stress relieving—A heat treatment used to remove or reduce stresses in work-hardened non-age-hardenable alloys without producing a recrystallized grain structure. Stress-relieving temperatures for nickel and nickel alloys range from 425 to 870 °C (800 to 1600 °F), depending on alloy composition and degree of work hardening.

Stress equalizing—A low-temperature heat treatment used to balance stresses in cold worked material without an appreciable decrease in the mechanical strength produced by cold working.

Solution treating—A high-temperature heat treatment designed to put age-hardening constituents and carbides into solid solution. Normally applied to age-hardenable materials before the aging treatment.

Age hardening (precipitation hardening)—A treatment performed at intermediate temperatures (425 to 870 °C; 800 to 1600 °F) on certain alloys in order to develop maximum strength by precipitation of a dispersed phase throughout the matrix.

ANNEALING PRACTICE

The differences in chemical composition among nickel and nickel alloys necessitate modifications in annealing temperatures (Table 9), as well as in furnace atmospheres. The precipitation-hardening alloys must be cooled rapidly after annealing if maximum softness is desired.

AGE HARDENING

Addition of magnesium, aluminum, silicon, titanium and certain other alloying elements to nickel and nickel alloys, separately or in combinations, produces an appreciable response to age hardening. The effect is dependent on both chemical composition and aging temperature; it is caused by precipitation of submicroscopic particles throughout the grains, which results in a marked increase in hardness and strength.

Prior Solution Treating. Unlike precipitation-hardening stainless steels and aluminum-base alloys, nickel alloys normally do not require solution treating in the upper annealing temperature range prior to age hardening. However, solution treating may be employed to enhance special properties (Table 10). For example, Inconel X-750 may be solution treated for 2 to 4 h at 1150 °C (2100 °F) and air cooled prior to a double (high and low temperature) aging cycle to develop maximum creep, relaxation and rupture strength at temperatures above about 600 °C (1100 °F). This combination of heat treatments is considered essential for high-temperature springs and turbine blades made of Inconel X-750.

Age-hardening practices for several nickel alloys are summarized in Table 10. In general, nickel alloys are soft when quenched from temperatures ranging from 790 to 1220 °C (1450 to 2225 °F);

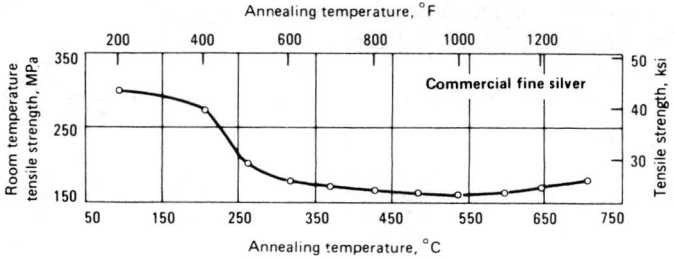

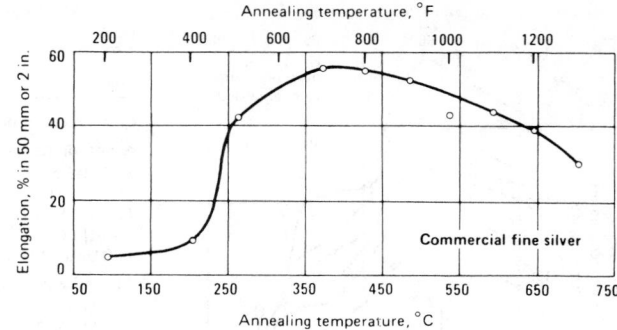

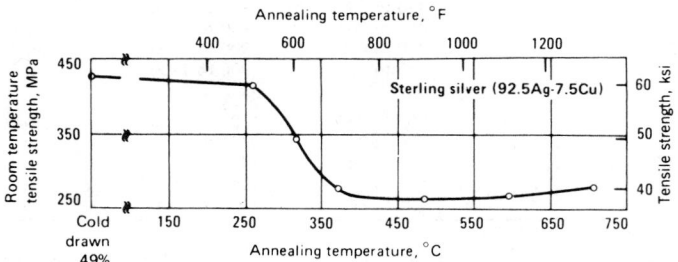

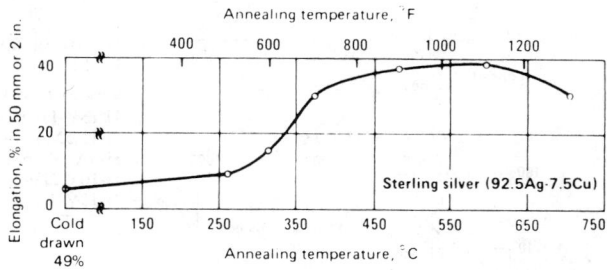

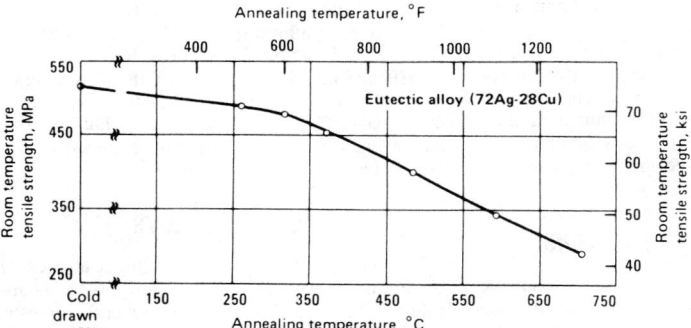

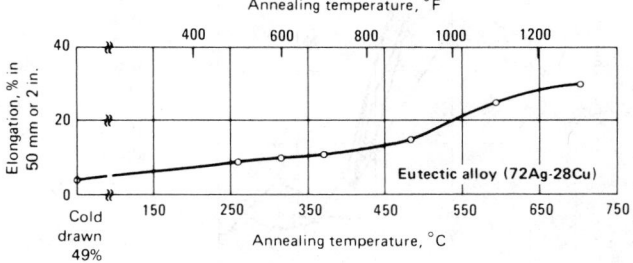

Fig. 2. Effects of annealing on strength and elongation of cold drawn 2.3-mm- (0.091-in.-) diam silver and silver alloy wire

however, they may be hardened by holding at 480 to 870 °C (900 to 1600 °F) or above and then furnace or air cooling.

Annealing of Precious Metals

ANNEALING BEHAVIOR of commercial fine silver and silver alloys, of commercial rhodium, and of commercial platinum and palladium is given in Fig. 2, 3 and 4.

Consolidated polycrystalline ruthenium usually is hot worked at high temperatures. For thicknesses of less than 0.5 mm (0.020 in.), previously hot worked material can be cold worked with very small reductions to thicknesses of about 0.25 mm (0.010 in.) using intermediate anneals at 1050 to 1250 °C (1920 to 2280 °F). At 20 °C (68 °F), the hardness of an annealed bar of ruthenium would be 200 to 350 HV.

Iridium, like tungsten, is initially hot worked. Subsequent fabrication, which is done warm (that is, below the recrystallization temperature), results in a fibrous structure. Recrystallization of warm worked iridium occurs at a temperature of

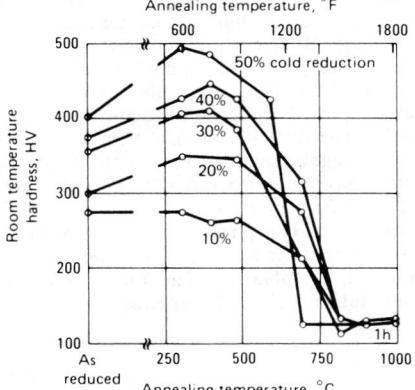

Fig. 3. Room-temperature hardness of commercial rhodium after annealing at various temperatures

1000 °C (1830 °F) or higher. The hardness of a warm drawn 0.5-mm- (0.020-in.-) diam iridium wire, annealed at 1000 °C, would be 200 to 240 HV; as-warm-drawn, iridium wire would have a hardness of 600 to 700 HV. Recrystallized irid-

ium, like recrystallized tungsten, is relatively brittle at room temperature.

Heat Treating of Titanium and Titanium Alloys

TITANIUM AND TITANIUM ALLOYS are heat treated for the following purposes:

- To reduce residual stresses developed during fabrication (stress relieving)
- To produce an optimum combination of ductility, machinability, and dimensional and structural stability (annealing)
- To increase strength (solution treating and aging)
- To optimize special properties such as fracture toughness, fatigue strength and high-temperature creep strength.

Various types of annealing treatments (single, duplex, beta and recrystallization annealing, for example), and solution-treating-and-aging treatments, are imposed to achieve selected mechan-

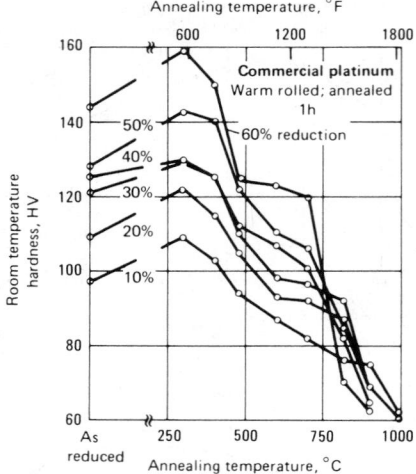

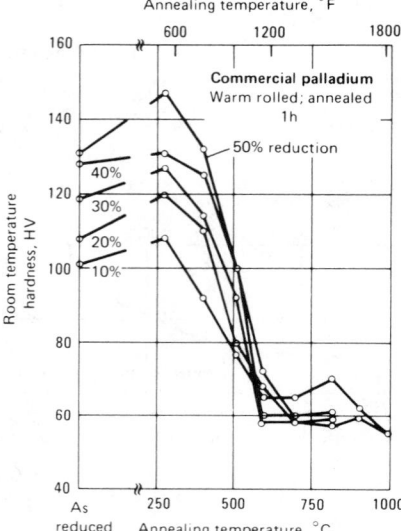

Fig. 4. Room-temperature hardness of commercial grades (99.9% +) of platinum and palladium after annealing

ical properties. Stress relieving and annealing may be employed to prevent preferential chemical attack in some corrosive environments, to prevent distortion (a stabilization treatment) and to condition the metal for subsequent forming and fabricating operations.

STRESS RELIEVING

Titanium and titanium alloys can be stress relieved without adversely affecting strength or ductility. Stress-relieving treatments decrease the undesirable residual stresses that result from (a) nonuniform hot forging deformation from cold forming and straightening, (b) asymmetric machining of plate (hogouts) or forgings, and (c) welding and cooling of castings. Removal of such stresses helps maintain shape stability and eliminates unfavorable conditions, such as the loss of compressive yield strength commonly known as the Bauschinger effect.

Table 11 presents combinations of time and temperature that are used for stress relieving titanium and titanium alloys.

Table 11. Recommended stress-relief treatments for titanium and titanium alloys
Parts can be cooled from stress relief by either air cooling or slow cooling.

Alloy	Temperature °C	°F	Time, h
Commercially pure Ti (all grades)	480 to 595	900 to 1100	$\frac{1}{4}$ to 4
Alpha or near-alpha titanium alloys			
Ti-5Al-2.5Sn	540 to 650	1000 to 1200	$\frac{1}{4}$ to 4
Ti-8Al-1Mo-1V	595 to 705	1100 to 1300	$\frac{1}{4}$ to 4
Ti-6Al-2Sn-4Zr-2Mo	595 to 705	1100 to 1300	$\frac{1}{4}$ to 4
Ti-6Al-2Cb-1Ta-0.8Mo	595 to 650	1100 to 1200	$\frac{1}{4}$ to 2
Ti-0.3Mo-0.8Ni (Ti Code 12)	480 to 595	900 to 1100	$\frac{1}{4}$ to 4
Alpha-beta titanium alloys			
Ti-6Al-4V	480 to 650	900 to 1200	1 to 4
Ti-6Al-6V-2Sn (Cu + Fe)	480 to 650	900 to 1200	1 to 4
Ti-3Al-2.5V	540 to 650	1000 to 1200	$\frac{1}{2}$ to 2
Ti-6Al-2Sn-4Zr-6Mo	595 to 705	1100 to 1300	$\frac{1}{4}$ to 4
Ti-5Al-2Sn-4Mo-2Zr-4Cr (Ti-17)	480 to 650	900 to 1200	1 to 4
Ti-7Al-4Mo	480 to 705	900 to 1300	1 to 8
Ti-6Al-2Sn-2Zr-2Mo-2Cr-0.25Si	480 to 650	900 to 1200	1 to 4
Ti-8Mn	480 to 595	900 to 1100	$\frac{1}{4}$ to 2
Beta or near-beta titanium alloys			
Ti-13V-11Cr-3Al	705 to 730	1300 to 1350	$\frac{1}{12}$ to $\frac{1}{4}$
Ti-11.5Mo-6Zr-4.5Sn (Beta III)	720 to 730	1325 to 1350	$\frac{1}{12}$ to $\frac{1}{4}$
Ti-3Al-8V-6Cr-4Zr-4Mo (Beta C)	705 to 760	1300 to 1400	$\frac{1}{6}$ to $\frac{1}{2}$
Ti-10V-2Fe-3Al	675 to 705	1250 to 1300	$\frac{1}{2}$ to 2
Ti-15V-3Al-3Cr-3Sn	790 to 815	1450 to 1500	$\frac{1}{12}$ to $\frac{1}{4}$

ANNEALING

Recommended annealing treatments for several alloys are given in Table 12. Mill annealing is a general-purpose treatment given to all mill products. It is not a full anneal, and may leave traces of cold or warm working in the microstructures of heavily worked products (particularly sheet). Duplex and triplex annealing alter the shapes, sizes and distributions of phases to those required for improved creep resistance or fracture toughness.

SOLUTION TREATING AND AGING

Time/temperature combinations for solution treating are given in Table 13. A load may be charged directly into a furnace operating at the solution-treating temperature. Although preheating is not essential, it may be used to minimize distortion of complex parts.

Quenching. The rate of cooling from the solution-treating temperature has an important effect on strength. If the rate is too low, appreciable diffusion may occur during cooling, and decomposition of the altered beta phase during aging may not provide effective strengthening.

Aging. The final step in heat treating titanium alloys to high strength consists of reheating to an aging temperature between 425 and 650 °C (800 and 1200 °F). Aging causes decomposition of the supersaturated beta phase retained on quenching. A summary of aging times and temperatures is presented in Table 13. The time/temperature combination selected depends on required strength.

Heat Treating of Tin-Rich Alloys

BINARY ALLOYS

Tin-antimony, tin-bismuth, tin-lead and tin-silver alloys can be temper hardened by solution treatment and aging. However, only the tin-antimony alloys can be permanently strengthened by heat treatment; all other tin-rich binary alloys will gradually soften at room temperature. The greatest improvement obtainable in binary tin-antimony alloys occurs in the alloy that contains 9% antimony; a hardness of 21 HB and a tensile strength of 51 MPa (7.4 ksi) can be increased to 26 HB and 65 MPa (9.4 ksi). This alloy is tempered for 48 h at 100 °C (212 °F) after being quenched from 225 °C (435 °F). During this tempering treatment, elongation decreases from 20 to 10% (in 50 mm or 2 in.).

TERNARY ALLOYS

Permanent effects of heat treatment also carry over into ternary alloys of tin, antimony and cadmium. This was discovered in an early investigation of the strength and hardness of ternary alloys containing up to 43% cadmium and 14% antimony using chill cast specimens. It was found that the strengthening effect of cadmium in the terminal solution tin phase alpha is much greater than that of antimony. The presence of the sigma phase as primary cuboids has no effect on strength or hardness, but the presence of primary epsilon destroys the useful mechanical properties.

Optimum properties (tensile strength: 92 MPa, or 13.4 ksi) were obtained in a Sn-9Sb-1.5Cd alloy quenched from 220 °C (430 °F) and then aged for 1000 h at 140 °C (285 °F). This alloy consists of finely divided sigma and epsilon phases in a matrix of alpha.

PEWTER

The hardness values of spun pewterware, or of other articles that have been manufactured by mechanically working the metal, can be restored by heat treatment at temperatures from 110 to 150 °C (230 to 300 °F). The time required varies from 3 h at the lower temperature to a few minutes at the higher temperature. A tin alloy containing 6% antimony and 2% copper will harden to 90% of the hardness of the as-cast material after annealing for 1 h at 200 °C (390 °F). Longer annealing times at lower temperatures have smaller but similar effects on the recovery from work softening.

Table 12. Recommended annealing treatments for titanium and titanium alloys

Alloy	Temperature °C	°F	Time, h	Cooling method
Commercially pure Ti (all grades)	650 to 760	1200 to 1400	$1/10$ to 2	Air
Alpha or near-alpha titanium alloys				
Ti-5Al-2.5Sn	720 to 845	1325 to 1550	$1/6$ to 4	Air
Ti-8Al-1Mo-1V	790(a)	1450(a)	1 to 8	Air or furnace
Ti-6Al-2Sn-4Zr-2Mo	900(b)	1650(b)	$1/2$ to 1	Air
Ti-6Al-2Cb-1Ta-0.8Mo	790 to 900	1450 to 1650	1 to 4	Air
Alpha-beta titanium alloys				
Ti-6Al-4V	705 to 790	1300 to 1450	1 to 4	Air or furnace
Ti-6Al-6V-2Sn (Cu + Fe)	705 to 815	1300 to 1500	$3/4$ to 4	Air or furnace
Ti-3Al-2.5V	650 to 760	1200 to 1400	$1/2$ to 2	Air
Ti-6Al-2Sn-4Zr-6Mo	(c)	(c)	...	...
Ti-5Al-2Sn-4Mo-2Zr-4Cr (Ti-17)	(c)	(c)	...	...
Ti-7Al-4Mo	705 to 790	1300 to 1450	1 to 8	Air
Ti-6Al-2Sn-2Zr-2Mo-2Cr-0.25Si	705 to 815	1300 to 1500	1 to 2	Air
Ti-8Mn	650 to 760	1200 to 1400	$1/2$ to 1	(d)
Beta or near-beta titanium alloys				
Ti-13V-11Cr-3Al	705 to 790	1300 to 1450	$1/6$ to 1	Air or water
Ti-11.5Mo-6Zr-4.5Sn (Beta III)	690 to 760	1275 to 1400	$1/6$ to 1	Air or water
Ti-3Al-8V-6Cr-4Zr-4Mo (Beta C)	790 to 815	1450 to 1500	$1/4$ to 1	Air or water
Ti-10V-2Fe-3Al	(c)	(c)	...	...
Ti-15V-3Al-3Cr-3Sn	790 to 815	1450 to 1500	$1/12$ to $1/4$	Air

(a) For sheet and plate, follow by $1/4$ h at 790 °C (1450 °F), then air cool. (b) For sheet, follow by $1/4$ h at 790 °C (1450 °F), then air cool (plus 2 h at 595 °C or 1100 °F, then air cool, in certain applications). For plate, follow by 8 h at 595 °C (1100 °F), then air cool. (c) Not normally supplied or used in annealed condition (see Table 13). (d) Furnace or slow cool to 540 °C (1000 °F), then air cool.

Table 13. Recommended solution-treating-and-aging (stabilizing) treatments for titanium alloys

Alloy	Solution temperature °C	°F	Solution time, h	Cooling rate	Aging temperature °C	°F	Aging time, h
Alpha or near-alpha alloys							
Ti-8Al-1Mo-1V	980 to 1010(a)	1800 to 1850(a)	1	Oil or water	565 to 595	1050 to 1100	...
Ti-6Al-2Sn-4Zr-2Mo	955 to 980	1750 to 1800	1	Air	595	1100	8
Alpha-beta alloys							
Ti-6Al-4V	955 to 970(b)(c)	1750 to 1775(b)(c)	1	Water	480 to 595	900 to 1100	4 to 8
	955 to 970	1750 to 1775	1	Water	705 to 760	1300 to 1400	2 to 4
Ti-6Al-6V-2Sn (Cu + Fe)	885 to 910	1625 to 1675	1	Water	480 to 595	900 to 1100	4 to 8
Ti-6Al-2Sn-4Zr-6Mo	845 to 890	1550 to 1650	1	Air	580 to 605	1075 to 1125	4 to 8
Ti-5Al-2Sn-2Zr-4Mo-4Cr	845 to 870	1550 to 1600	1	Air	580 to 605	1075 to 1125	4 to 8
Ti-6Al-2Sn-2Zr-2Mo-2Cr-0.25Si	870 to 925	1600 to 1700	1	Water	480 to 595	900 to 1100	4 to 8
Beta or near-beta alloys							
Ti-13V-11Cr-3Al	775 to 800	1425 to 1475	$1/4$ to 1	Air or water	425 to 480	800 to 900	4 to 100
Ti-11.5Mo-6Zr-4.5Sn (Beta III)	690 to 790	1275 to 1450	$1/8$ to 1	Air or water	480 to 595	900 to 1100	8 to 32
Ti-3Al-8V-6Cr-4Mo-4Zr (Beta C)	815 to 925	1500 to 1700	1	Water	455 to 540	850 to 1000	8 to 24
Ti-10V-2Fe-3Al	760 to 780	1400 to 1435	1	Water	495 to 525	925 to 975	8
Ti-15V-3Al-3Cr-3Sn	790 to 815	1450 to 1500	$1/4$	Air	510 to 595	950 to 1100	8 to 24

(a) For certain products, use solution temperature of 890 °C (1650 °F) for 1 h, then air cool or faster. (b) For thin plate or sheet, solution temperature can be used down to 890 °C (1650 °F) for 6 to 30 min, then water quench. (c) This treatment is used to develop maximum tensile properties in this alloy.

Heat Treating of Special-Purpose Alloys

HEAT TREATING procedures for several special-purpose alloys used in military and aerospace applications, and in other products where special properties are required, are discussed in this article. Included are procedures for treating depleted uranium (DU), tantalum, niobium and alloys of these metals.

ANNEALING OF DEPLETED URANIUM AND ITS ALLOYS

Annealing of cold worked DU is similar to annealing of other metals. The first stage is recovery, in which there is a slight decrease in hardness, a small decrease in electrical resistivity and a pronounced sharpening of x-ray line shape. Recovery is followed by recrystallization. The variation of recrystallization temperature as a function of cold work is shown in Fig. 5 for an

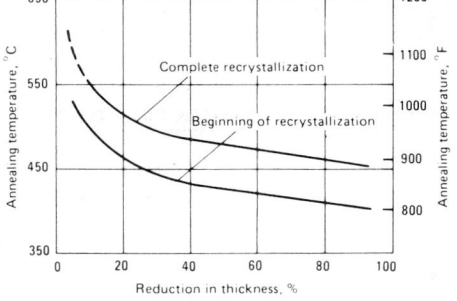

Annealing time, $1^{1}/_{2}$ h.

Fig. 5. Recrystallization temperature as a function of cold work for rolled depleted uranium of moderate purity

annealing time of $1^{1}/_{2}$ h. Recrystallization begins at 400 °C (750 °F) and is complete at 450 °C (840 °F) in material cold worked 90 to 94%. Light cold working (about 4%) causes recrystallization to begin at 525 °C (975 °F), but recrystal-lization is not complete after $1^{1}/_{2}$ h at 600 °C (1110 °F).

Solution Treating and Aging. Heat treating for improved hardness and mechanical properties in dilute DU alloys consists of solution treating in the gamma-phase temperature range of 800 to 850 °C (1470 to 1560 °F), quenching to room temperature, and aging in the alpha temperature range. Alternatively, interrupted quenching in a molten metal or salt bath held at the appropriate temperature can be used.

An important consideration in the selection of conditions for gamma solution treatment is the hydrogen level required in the final product. Hydrogen is detrimental to ductility in U-0.75Ti and must be maintained at 1 ppm or less to ensure high ductility in heat treated parts. These levels have been achieved consistently in extruded 36-mm (1.4-in.)-diam rod by vacuum outgassing for $2^{1}/_{2}$ h at 850 °C (1560 °F) at 10^{-5} torr. Unalloyed uranium and uranium alloys are sensitive to hydrogen and for maximum material properties require extensive outgassing. The literature should be consulted before selecting conditions for these alloys.

Table 14. Cooling rates for DU-0.75Ti in various quench media

Media	Quench rate °C/s	Quench rate °F/s
Flowing argon	3.8	6.8
Conventional or soluble oil	38-40	68-72
0.05% PVA(a)	80	145
Water	98	175
10% brine	190	340
(a) Polyvinyl alcohol.		

Quenching. Table 14 gives the rates at which DU-0.75Ti cools when quenched in various media. The test slugs used for measuring these rates were 22 mm (0.875 in.) in diameter by 21 mm (0.845 in.) long. Cooling rates for other DU alloys should be similar. Except for DU-0.75Ti, which is cooled by very slow argon gas quenching, response to subsequent aging at 350 °C (660 °F), as determined by hardness measurements, was independent of quench rates above 40 °C/s (72 °F/s). Because DU-Ti alloys are shallow hardening, higher quench rates are needed to achieve uniform hardening response in larger-diameter bars or thicker plates.

ANNEALING OF TANTALUM AND NIOBIUM AND THEIR ALLOYS

Tantalum and niobium and their alloys are most often used in the fully recrystallized condition to achieve the best fabrication response, although some applications require stress relieved or cold worked properties. The recrystallization temperature is so highly dependent on purity, amount of cold work and prior history, that current practice is to anneal pilot samples to ensure that the correct temperatures are used. Time at temperature is typically 1 h.

Table 15 can be used as a guide for choosing pilot temperatures. Materials given heavy fabrication reductions will recrystallize to finer grain sizes at lower temperatures than will those given lighter fabrication reductions. The recrystallization annealing temperature is also somewhat dependent on interstitial purity. For example, pure tantalum containing 200 ppm oxygen requires a higher recrystallization annealing temperature than that for pure tantalum containing less than 50 ppm oxygen.

───── Sintering ─────

SINTERING is the process by which loose or compressed powders are bonded by heating at temperatures below the melting points of the major constituents. Densification may or may not occur. If powders of two or more different metals are heated together to a sufficiently high temperature, alloying may take place simultaneously with sintering. Sometimes a liquid phase forms and assists in consolidation, or a compact may be sintered for a short time and then infiltrated with a molten metal of lower melting point.

SINTERING FURNACES

The burn-off chamber of a sintering furnace used for sintering ferrous preforms is usually controlled to heat the preforms to temperatures from 500 to 800 °C (930 to 1470 °F). It is important that all lubricants, including zinc stea-

rate, stearic acid and waxes, be volatilized and expelled from the furnace before the preforms enter the high-temperature section, and both the flow of gas and the time of heating should be sufficient to ensure that this is done. If lubricants pass into the sintering section, they will be decomposed, and the liberated products may adversely affect the sintering process, the parts and the furnace.

The high-temperature heating section must be long enough to allow sufficient time for the preforms to heat up to temperature and enough soak time at temperature for adequate sintering. Multiple-control zones are used to obtain suitable temperature gradients.

The cooling section often begins with a short, insulated zone in which the preforms cool slowly enough to avoid thermal shock and to allow for carbon restoration. The cooling section follows.

SINTERING ATMOSPHERES

Protective atmospheres are used in powder metallurgy (a) to prevent oxidation and reduce oxides, (b) to control carbon contents of iron and iron alloy preforms, and (c) to flush volatilized lubricants from the furnace.

The atmospheres most commonly used for sintering are: hydrogen, dissociated ammonia, nitrogen-base exothermic gas, purified rich exothermic gas, endothermic gas, and vacuum.

Vacuum is used mainly for sintering preforms of stainless steels and tool steels; soft magnetic materials; and refractory metals such as tantalum, titanium, zirconium and uranium — all of which react with most of the usual protective gases, including hydrogen. Vacuum is also being used to an increasing extent for sintering of conventional ferrous materials at high temperatures.

Heat Treating of Ferrous Powder ──── Metallurgy Parts ────

HEAT TREATING procedures used for powder metallurgy parts are theoretically the same as those used for wrought or cast parts of the same composition; however, some practical limitations must be considered in dealing with iron-base alloy (steel) powder metallurgy parts. If a powder compact has the same composition as its wrought

or cast counterpart, it should likewise respond in the same manner to heat treatment. The principles are the same — that is, the maximum hardness that can be achieved is controlled by the content of carbon that is in solution at the austenitizing temperature, not necessarily by the total carbon content. In powder metallurgy, because alloys are produced by mixing various powders and then sintering the mixture, some of the carbon can be incorporated in a graphitic state and can remain so at the austenitizing temperature. Because graphitic carbon does not contribute to hardening, combined carbon must be known before an optimum heat treating procedure can be selected.

Furthermore, as with wrought or cast parts, the hardenabilities of powder metallurgy parts are controlled by the amounts of alloying elements they contain. It is therefore essential to know the complete composition before attempting to perform any heat treatment.

For example, if a specific powder metallurgy part is composed essentially of iron with a combined carbon content of 0.5% and if section thickness does not exceed about 4.8 mm ($^3/_{16}$ in.), full hardness should be obtained by austenitizing at about 870 °C (1600 °F) and quenching in oil.

For sections greater than about 4.8 mm, an aqueous (water-brine or synthetic polymer) quench will probably be necessary to attain full hardness. This quenching treatment may induce cracking, because powder metallurgy parts always contain some voids and thus are more likely to crack than their wrought or cast counterparts. Therefore, oil quenching is always preferred for powder metallurgy parts. Thicker sections should contain sufficient alloying elements such as manganese and/or chromium to provide the required hardenability.

Heating Media. Powder metallurgy parts for through hardening normally should be heated in a gaseous atmosphere, such as the type provided by an endothermic generator, using a carbon potential equal to that of the combined carbon in the compact. Heating of powder metallurgy parts in a molten salt bath is not recommended even for those having maximum density, because they are extremely difficult to wash free of salt. If the salt remains in the voids, the parts will corrode.

Tempering. Austenitized-and-quenched powder metallurgy parts should be tempered just as are their wrought counterparts. Tempering temperature should be 150 °C (300 °F) or higher if some decrease in hardness can be tolerated.

Table 15. Annealing temperatures for tantalum and niobium and their commercial alloys

Alloy designation	Nominal alloy additions, %	Annealing temperature Stress relief °C	Annealing temperature Stress relief °F	Annealing temperature Recrystallization °C	Annealing temperature Recrystallization °F
Tantalum alloys					
Ta	None	850	1560	1000-1250	1830-2280
Ta	None(a)	1000	1830	1200-1350	2190-2460
FS63	2.5W, 0.15Nb	1000	1830	1200-1300	2190-2370
FS61	7.5W(a)	...	...	1400-1550	2550-2820
FS60	10W	1100	2010	1300-1600	2370-2910
T111	8W, 2Hf	1100	2010	1400-1650	2550-3000
T222	9W, 2.4Hf, 0.01C	1100	2010	1400-1650	2550-3000
Niobium alloys					
Nb	None	800	1470	900-1200	1650-2190
FS80	1Zr	875-1150	1610-2100	1150-1250	2100-2280
SNb 291	10Ta, 10W	1000	1830	1150-1200	2100-2190
Nb 752	10W, 2.5Zr	...	...	1300-1400	2370-2550
C 129Y	10W, 10Hf, 0.1Y	900	1650	1150-1250	2100-2280
FS85	28Ta, 11W, 0.8Zr	1150	2100	1300-1400	2370-2550
C103	10Hf, 1Ti, 0.7Zr	...	...	1250-1375	2280-2510
(a) Powder metallurgy; all other compositions are vacuum melted.					

29 SURFACE TECHNOLOGY

Edited by William G. Wood, Kolene Corp.

This section was condensed from Metals Handbook, Ninth Edition, Volume 5, Surface Cleaning, Finishing, and Coating. For more detailed information on the topics covered herein, the reader is referred to the larger work.

Metal Cleaning

Edited by George A. Shepard, Republic Steel Research Center

Selection of Cleaning Process

IN SELECTING a metal cleaning process, many factors must be considered, including: (*a*) identification and characterization of the soil to be removed; (*b*) identification of the substrate to be cleaned and the importance of the condition of the surface or structure to the ultimate use of the part; (*c*) degree of cleanness required, which depends on subsequent operations such as phosphating, plating and painting; (*d*) capabilities of the available facilities; (*e*) impact of the process on the environment; and (*f*) over-all cost of the process.

Types of soil may be broadly classified into six groups: (*a*) pigmented drawing compounds, (*b*) unpigmented oil and grease, (*c*) chips and cutting fluids, (*d*) polishing and buffing compounds, (*e*) rust and scale, and (*f*) miscellaneous surface contaminants, such as lapping compounds and residue from magnetic particle inspection.

Dried or oxidized greases, oils and drawing compounds are very difficult to remove, and thus every effort should be made to clean the parts as soon after processing as possible.

REMOVAL OF PIGMENTED DRAWING COMPOUNDS

All pigmented drawing lubricants are difficult to remove from metal parts. Consequently, many plants review all aspects of press forming operations to avoid the use of pigmented compounds.

Table 1 indicates cleaning processes typically selected for removing pigmented compounds from drawn and stamped parts.

REMOVAL OF UNPIGMENTED OIL AND GREASE

Common shop oils and greases, such as unpigmented drawing lubricants, rust-preventive oils, and quenching and lubricating oils, can be effectively removed by several different cleaners.

Table 1 lists cleaning methods frequently used for removing oils and greases.

REMOVAL OF CHIPS AND CUTTING FLUIDS FROM STEEL PARTS

Cutting and grinding fluids used for machining may be classified into three groups, as follows:

- Plain or sulfurized mineral and fatty oils (or combinations of the two), chlorinated mineral oils, and sulfurized chlorinated mineral oils
- Conventional or heavy-duty soluble oils with

Table 1. Metal cleaning processes
Processes are listed in order of decreasing preference.

Type of production	In-process cleaning	Preparation for painting	Preparation for phosphating	Preparation for plating
Removal of pigmented drawing compounds(a)				
Occasional or intermittent	Hot emulsion hand slush, spray emulsion in single stage, vapor slush degrease(b)	Boiling alkaline, blow off, hand wipe; Vapor slush degrease, hand wipe; Acid clean(c)	Hot emulsion hand slush, spray emulsion in single stage, hot rinse, hand wipe	Hot alkaline soak, hot rinse (hand wipe, if possible), electrolytic alkaline, cold water rinse
Continuous high production	Conveyorized spray emulsion washer	Alkaline soak, hot rinse, alkaline spray, hot rinse	Alkaline or acid(d) soak, hot rinse, alkaline or acid(d) spray, hot rinse	Hot emulsion or alkaline soak, hot rinse, electrolytic alkaline, hot rinse
Removal of unpigmented oils and greases				
Occasional or intermittent	Solvent wipe; Emulsion dip or spray; Vapor degrease; Cold solvent dip; Alkaline dip, rinse, dry (or dip in rust preventive)	Solvent wipe; Vapor degrease or phosphoric acid clean(d)	Solvent wipe; Emulsion dip or spray, rinse; Vapor degrease	Solvent wipe; Emulsion soak, barrel rinse, electrolytic alkaline rinse, hydrochloric acid dip, rinse
Continuous high production	Automatic vapor degrease; Emulsion, tumble, spray, rinse, dry	Automatic vapor degrease	Emulsion power spray, rinse; Vapor degrease; Acid clean(c)	Automatic vapor degrease, electrolytic alkaline rinse, hydrochloric acid dip, rinse(e)
Removal of chips and cutting fluids				
Occasional or intermittent	Solvent wipe; Alkaline dip and emulsion surfactant; Stoddard solvent or trichlorethylene; Steam	Solvent wipe; Alkaline dip and emulsion surfactant; Solvent or vapor	Solvent wipe; Alkaline dip and emulsion surfactant(f); Solvent or vapor	Solvent wipe; Alkaline dip, rinse, electrolytic alkaline(g), rinse, acid dip, rinse(h)
Continuous high production	Alkaline (dip or spray) and emulsion surfactant	Alkaline (dip or spray) and emulsion surfactant	Alkaline (dip or spray) and emulsion surfactant	Alkaline soak, rinse, electrolytic alkaline(g), rinse, acid dip and rinse(h)
Removal of polishing and buffing compounds				
Occasional or intermittent	Seldom required	Solvent wipe; Surfactant alkaline (agitated soak), rinse; Emulsion soak, rinse	Solvent wipe; Surfactant alkaline (agitated soak), rinse; Emulsion soak, rinse	Solvent wipe; Surfactant alkaline (agitated soak), rinse, electroclean(j)
Continuous high production	Seldom required	Surfactant alkaline spray, spray rinse; Agitated soak or spray, rinse(k)	Surfactant alkaline spray, spray rinse; Emulsion spray, rinse	Surfactant alkaline soak and spray, alkaline soak, spray and rinse, electrolytic alkaline(j), rinse, mild acid pickle, rinse

(a) For complete removal of pigment, parts should be cleaned immediately after the forming operation, and all rinses should be by spraying where practical. (b) Used only when pigment residue can be tolerated in subsequent operations. (c) Phosphoric acid cleaner-coaters are often sprayed on the parts to clean the surface and leave a thin phosphate coating. (d) Phosphoric acid for cleaning and iron phosphating. Proprietary products for high- and low-temperature application are available. (e) Some plating processes may require additional cleaning dips. (f) Neutral emulsion or solvent should be used before manganese phosphating. (g) Reverse-current cleaning may be necessary to remove chips from parts having deep recesses. (h) For cyanide plating, acid dip and water rinse are followed by alkaline and water rinses. (j) Other preferences: stable or diphase emulsion spray or soak, rinse, alkaline spray or soak, rinse, electroclean; or solvent presoak, alkaline soak or spray, electroclean. (k) Third preference: emulsion spray rinse.

sulfur or other compounds added and soluble grinding oils with wetting agents
- Chemical cutting fluids, which are water-soluble and generally act as cleaners. They contain soaps, amines, sodium salts of sulfonated fatty alcohols, alkyl aromatic sodium salts of sulfonates, or other types of the soluble addition agents.

Usually, all three types of fluids are easily removed, and the chips fall away during cleaning, unless the chips or the part become magnetic. Plain boiling water is often suitable for removing these soils, and in some plants, mild detergents are added to the water to increase its effectiveness. Steam is widely used for in-process cleaning, especially for large components. Table 1 indicates cleaning processes typically used for removing cutting fluids to meet specific production requirements.

REMOVAL OF POLISHING AND BUFFING COMPOUNDS

Table 1 lists preferred and alternate methods for removing polishing and buffing compounds from sheet metal parts. However, some modification may be required for complete removal of all classes of these soils.

REMOVAL OF RUST AND SCALE

The seven basic methods used for removing rust and scale from ferrous mill products, forgings, castings and fabricated metal parts are:

- Abrasive blasting (dry or wet)
- Tumbling (dry or wet)
- Brushing
- Acid pickling
- Salt bath descaling
- Alkaline descaling
- Acid cleaning.

The most important considerations in selecting one of the above methods are:

- Thickness of rust or scale
- Composition of metal
- Condition of metal (product form or heat treatment)
- Allowable metal loss
- Surface finish tolerances
- Shape and size of workpieces
- Production requirements
- Available equipment
- Cost
- Freedom from hydrogen embrittlement.

Combinations of two or more of the available processes are frequently used to advantage.

____ Alkaline Cleaning ____

By Donald P. Murphy, Occidental Chemicals, Parker Surface Treatments

ALKALINE CLEANING is used to remove soils from the surface of metals. Soils removed through alkaline cleaning include oil, grease, waxy solids, metallic particles, dust, carbon particles and silica. Alkaline cleaners are applied by immersion or spray, and the metals are cleaned by emulsification, dispersion, saponification, or combinations of these mechanisms. The cleaning step is usually followed by a water rinse and a subsequent operation such as conversion coating or electroplating.

METHOD OF APPLICATION

Immersion Cleaning. When an alkaline cleaner is used in immersion processes, the parts to be cleaned are placed in the cleaning solution, which allows the cleaning solution to come in contact with the entire surface of the part. After the alkaline cleaner has affected the soil on the part, the soil is removed from the metal surface by convection currents in the solution which are created by heating units or through some mechanical means of solution movement.

Electrocleaning is a specialized form of alkaline cleaning in which electrodes are placed in a cleaning solution. Direct current is passed through the solution, and the part to be cleaned is made the anode and the electrode is the cathode. The cleaning process is enhanced by the scrubbing action of the oxygen which evolves at the anode. Additional information on this process can be found in the next article, "Electrolytic Cleaning."

Relatively expensive electrolytic cleaners known as derusters are used extensively, especially by electroplaters. Derusters are used with periodic reverse cycles, in which both anodic and cathodic cleaning are used in the operation, as well as for nonelectrolytic immersion cleaning.

Other variations of immersion cleaning include (a) barrel cleaning, in which small parts are agitated inside a barrel which rotates in the cleaner; (b) immersion cleaning, where parts are moved by either a rotary screw conveyor or a moving conveyor chain; (c) immersion cleaning, where solution is agitated by either a recirculating pump, a mechanical mixer or ultrasonic sound waves; and (d) cleaning with the aid of external forces, such as brushes or squeegees.

Spray cleaning is accomplished by pumping the cleaning solution from a reservoir through a large pipe, or header, to a series of smaller pipes off of the header, called risers, and finally out of spray nozzles on the riser onto the part to be cleaned (Fig. 1). Spray pressure can vary from as low as 14 kPa (2 psi) to as much as 1380 kPa (200 psi) or more. Pressures usually range from 70 to 210 kPa (10 to 30 psi). In general, the higher the spray pressure, the more mechanical help is provided in removing soil from the metal surface. This is especially true with small electri-

cally charged particles such as dust, carbon smut, and silica. Spray cleaners are prepared with low-foaming detergents, which are not usually as effective as those found in immersion cleaners. Thus, the impingement of a spray cleaner plays a vital role in the removal of soil.

CLEANING MECHANISMS

Cleaning is accomplished by one or more of three major mechanisms: (a) saponification, (b) emulsification, and (c) dispersion. The three mechanisms can operate independently or in combination with each other. The saponification mechanism is limited to soils containing fats or other compounds that will chemically react with alkaline salts. The emulsification and dispersion mechanisms are effective on almost any liquid organic soil that is insoluble in water. The following is a brief description of these mechanisms.

Saponification. In saponification, fatty compounds, both animal and vegetable, react with the alkaline salts in an alkaline cleaner to form a water-soluble soap. Fatty soils are removed by dissolving them in the cleaning solution.

Emulsification joins together two mutually insoluble liquids such as oil and water. This is made possible by the detergent, more properly called a surface-active agent or surfactant, which has a water-soluble chemical grouping on one end and an oil-soluble chemical grouping on the other end. The surfactant acts as a connector to keep the two insoluble liquids together as though they were one unit. In a cleaner dissolved in water, the water-soluble end of the detergent has been solubilized in the water with the oil-soluble end still needing a media where it will be soluble. Upon encountering the oil on a part, the oil-soluble end is immediately solubilized, and because there is more water than oil, the water phase completely surrounds the oil. Most emulsions are cloudy or milky in appearance, although it is possible to form what appears to be a clear emulsion.

Environmental limitations in many areas, especially fish breeding areas, minimize the use of emulsifiers. Emulsifiers tie up oils and minute amounts of metals in the oils and prevent their removal during the treatment of solutions and rinse waters before they are discharged into rivers and streams.

Dispersion. In dispersion, the surfactant acts to lower the surface tension of the cleaner at the

metal surface, allowing the cleaner to cover the metal uniformly. The interfacial tension of the cleaner is lowered, permitting the cleaner to penetrate the oil film and break it into smaller units. As a result, the oil is dispersed into small droplets, which are undercut by the film of cleaner spreading across the metal surface. These droplets lose their attraction to the metal, are released, and, because they are lighter than the cleaner solution, float to the cleaner surface where they reassemble into a semicontinuous film.

RINSING

A good water rinse is essential for good cleaning. The water rinse may be either hot, warm or cold, but it should be kept relatively clean. In order to ensure good rinsing, the water rinse should contain no more than 3% of the concentration of the cleaner solution. If the cleaner is at 30 g/L (4 oz/gal), the rinse water should contain no more than 0.9 g/L (0.12 oz/gal).

ALKALINE CLEANER COMPOSITION

Alkaline cleaners are comprised basically of three major types of components: (a) the builders, which make up the largest portion of the cleaner; (b) organic or inorganic additives, which promote better cleaning or act to affect the metal surface in some way; and (c) the surfactants.

Builders are the alkaline salts in an alkaline cleaner. They are usually blends selected from the following groups: alkali metal orthophosphates, alkali metal condensed phosphates, alkali metal hydroxides, alkali metal silicates, alkali metal carbonates, alkali metal bicarbonates, and alkali metal borates. A blend of two or more of these builders is used for cleaning performance, physical properties of the dry blend, and economics (Table 2). The alkali metal is usually sodium.

Phosphates serve a multiple function in the cleaner. They act to soften the water, eliminating the flocculent precipitate caused by calcium, magnesium and iron. They act as a soil dispersant, as a source of alkalinity, and as a buffer, which prevents large changes in the level of alkalinity. Some common phosphates used are trisodium phosphate, disodium phosphate, tetrasodium pyrophosphate, and sodium tripolyphosphate, known as tripoly. Phosphates also have a moderate detersive value. Like emulsifiers, phosphates are subject to environmental regulations which limit the amount of phosphorus allowed in discharged water.

Silicates are also multifunctional. They provide alkalinity, keep soil suspended, provide some detergency, and act as inhibitors protecting metals such as aluminum and zinc from attack by other alkaline salts. Commonly used silicates include sodium metasilicate and sodium orthosilicate. Silicates are difficult to rinse and may cause trouble in subsequent plating operations if they are not completely removed during rinsing.

Carbonates are a cheap source of alkalinity. They act as buffers and as absorbing media for the liquid components of the cleaner. Hydroxides are another relatively inexpensive source of strong alkalinity. Borates are somewhat like silicates in that they provide some detergency, act as buffers and provide some metal protection.

Additives are either organic or inorganic compounds that provide additional cleaning or surface modification. Chemical compounds such as

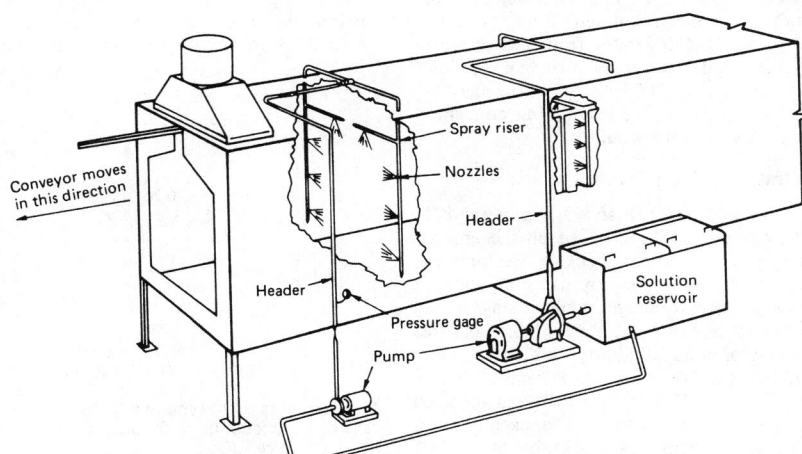

Fig. 1. Equipment for spray cleaning operation

Table 2. Alkaline cleaning formulas for various metals

Constituent	Aluminum Immersion	Aluminum Spray	Steel Immersion	Steel Spray	Steel Electrocleaning	Zinc Immersion	Zinc Spray	Zinc Electrocleaning
Sodium hydroxide	...	...	38	50	55	...	...	16
Sodium carbonate(a)	55	18	36	17	18	10	20	50
Sodium metasilicate(b)	37	...	12	...	10	15	10	...
Sodium metasilicate(c)	...	60	...	...	10	...	...	32
Tetrasodium pyrophosphate	...	20	9	20	6	20	65	...
Sodium tripolyphosphate	...	...	...	...	...	50	...	...
Trisodium phosphate	...	...	...	10	...	...	...	...
Fatty acid esters	1	...	3	0.6	...	...	...	1
Ethoxylated alkylphenol	...	...	2	0.2	...	...	...	...
Ethoxylated alcohol	...	2	...	2	...	...	5	...
Sodium alkylbenzene sulfonate	5	...	...	...	...	5	...	1
Naphthalene sulfonate	2	...	...	0.2	1	...	...	...

(a) Dense. (b) Anhydrous. (c) Hydrate.

glycols, glycol ethers, chelating agents and polyvalent metal salts could be considered additives. For environmental reasons, chelating agents are replacing phosphates in some cleaning formulas. These additives can soften water and complex or tie up metal ions. Some widely used chelating agents include sodium gluconate, sodium citrate, tetrasodium ethylenediamine tetraacetate (EDTA), trisodium nitrilotriacetate (NTA) and triethanolamine.

Surfactants are organic compounds which provide detergency, emulsification and wetting in an alkaline cleaner. There are four major types of surfactants: (*a*) anionic, (*b*) cationic, (*c*) nonionic and (*d*) amphoteric. An anionic surfactant is one in which the largest part of the molecule is the anion, or the negatively charged ion. Conversely, in the cationic type, the cation, or positively charged ion, is the larger molecule. The amphoteric type is pH dependent, being anionic in an alkaline medium and cationic in an acidic medium. In all three, the surfactant ionizes in water, separating into negative ions and positive ions which function independently of each other. A nonionic surfactant, as its name implies, does not ionize, but rather remains as an intact molecule. Examples of the four types of surfactants are listed below:

- Anionic: sodium alkylbenzene sulfonate
- Nonionic: ethoxylated long-chain alcohol
- Cationic: quaternary ammonium chloride
- Amphoteric: alkyl substituted imidazoline

In alkaline spray cleaners, the nonionic type of surfactant is used almost exclusively because, in general, it is the only type that can be obtained in a low-foaming form and still provide good detergency. The immersion cleaners can use any of the four types because foam is not a problem. However, the surfactants used are usually of the anionic or nonionic type.

The amphoteric type of surfactant is seldom used because it becomes anionic in an alkaline medium and cationic in an acidic medium. Since the anionic and cationic types are already available at a lower cost, there is generally no advantage in using an amphoteric. The cationic type is good for certain special applications as an emulsifier, but generally is not as effective a detergent as the anionic or nonionic types. Additionally, the cationics are frequently substantive to metal—that is, they form a film on the metal, thereby defeating the intended purpose of providing a clean metal surface.

TESTING AND CONTROL OF CLEANERS

An alkaline cleaner loses strength through use and through dilution caused by replacement of lost cleaning solution with water. For this reason, a method of determining not only what the strength of the cleaner is at any given time, but also what it should be to provide the best performance, is necessary. A chemical procedure known as acid-base titration is most commonly used. In this procedure, a known amount of alkaline cleaner is placed in a container, and an acid of precise concentration (titrating solution) is measured into the cleaner until a certain pH is achieved. Certain organic compounds known as indicators change color at specific pH levels, and this color change is used to identify when the desired pH level is achieved in the titration. By determining the amount of acid titrating solution required to achieve the desired pH level in an alkaline cleaner of known volume and concentration, a factor can be developed which will allow calculation of the concentration of any solution of the same cleaner by multiplying the factor by the number of millilitres of titrating solution required to achieve the desired pH.

The following example demonstrates the use of acid-base titration for testing alkaline cleaner. A 5-mL sample of cleaner is tested with 1N acid used for titrating. (1N refers to a specific concentration of acid.) The indicator is phenolphthalein, which changes from red to colorless at a pH of about 8.7. When the 5-mL sample of a precisely measured 10 g/L (1.3 oz/gal) concentration of cleaner is titrated, 5 mL of acid is needed to turn the indicator from red to colorless. 5 mL equals 10 g/L (1.3 oz/gal), or each 1 mL equals 2 g/L (0.27 oz/gal). The factor is 2.0, and the equation for calculating concentration in grams per litre would be:

Concentration = 2 × millilitres of acid

The procedure outlined above is a method for determining concentration based on the amount of free alkalinity. Another check on the condition of the cleaner can be done by using another indicator called methyl orange, which changes from yellow to orange at about 3.9 pH. This measures the amount of total alkalinity, which in many cleaners is in the ratio of 1.2 to 1, total alkalinity to free alkalinity. This ratio is different for each cleaner and should be determined for each cleaner. A rule of thumb sometimes used is that the cleaner should be dumped, and a new solution prepared, when the ratio doubles. This comes about because the total alkalinity constantly increases with additions of cleaner while the free alkalinity is held fairly constant. As a result of this, total-alkalinity titration is a direct measure of the amount of contamination in the cleaner because the ratio of total alkalinity to free alkalinity increases only when cleaner is added to replace free alkalinity lost through reaction with the soil. Dragout of solution on parts does not cause an increase in the ratio because both free and total alkalinity are lost in proportion to their concentration in the solution.

Because of the blend of builders used, some alkaline cleaners do not have any free alkalinity, and they must be controlled by total-alkalinity titrations. These cleaners are usually specialized, and the pH of the cleaner solution is intentionally maintained below 9 to maintain stability of one or more of the cleaner's components, to keep the cleaner from attacking the metal substrate, or for some other special purpose. Total-alkalinity titration is performed exactly like free-alkalinity titration, except that the indicator is methyl orange (3.9 pH) instead of phenolphthalein (8.7 pH). The only drawback to this method of control is the inability to measure the degree of soil contamination of the cleaner by a titration procedure.

Electrolytic Cleaning

ELECTROLYTIC CLEANING, or electrocleaning, is the process by which a workpiece is made anodic or cathodic in a specially formulated alkaline cleaning solution. Direct currents of 3 to 12 V are applied to yield current densities ranging from 1 to 15 A/dm^2 (10 to 150 A/ft^2) of work area. An electrocleaning operation usually follows an alkaline soak and serves two purposes: (*a*) it removes any residual soils that may have been left behind by the soak cleaner, and (*b*) the electrocleaner activates the metal surface—that is, it eliminates the passive condition but only when cleaned in the cathodic mode.

ANODIC ELECTROCLEANING

Also referred to as reverse-current cleaning, anodic electrocleaning is most commonly used on ferrous metals. Figure 2 illustrates a workpiece being cleaned anodically and gives the accompanying reaction. Because the workpiece is the anode (positive), free electrons are discharged by the hydroxyl ions to the metal, re-

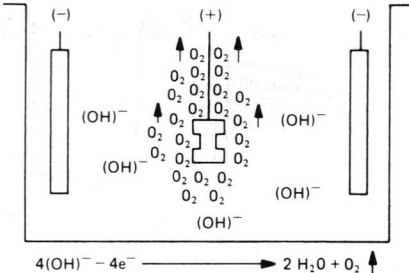

$$4(OH)^- - 4e^- \longrightarrow 2H_2O + O_2 \uparrow$$

Four electrons are discharged by four hydroxyl (OH)$^-$ ions at the anode, or workpiece, to liberate one molecule of oxygen (O$_2$).

Fig. 2. Anodic electrocleaning

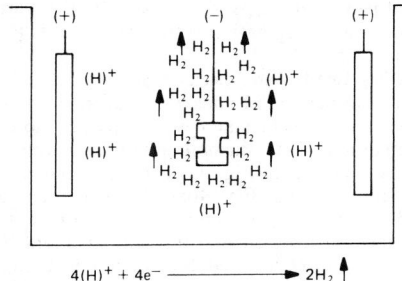

$$4(H)^+ + 4e^- \longrightarrow 2H_2 \uparrow$$

Reaction of electrons with positively charged hydrogen ions results in liberation of hydrogen gas.

Fig. 3. Cathodic electrocleaning

sulting in liberation of gaseous oxygen. The oxygen bubbles liberated at the workpiece create a scrubbing action, which blasts dirt particles off the workpiece. These bubbles rise to the top of the cleaner, increasing solution agitation.

Anodic electrocleaning eliminates the possibility of any metal contaminants (in the cleaner) being plated onto the work being cleaned.

CATHODIC ELECTROCLEANING

Cathodic electrocleaning, also called direct-current electrocleaning, employs a negative charge on the part. Insoluble steel or nickelplated steel anodes are used. The cleaning action obtained is somewhat better than that of anodic cleaning because hydrogen gas is evolved at the cathode (workpiece), and, at a given current density, twice the volume of hydrogen is liberated at the cathode than oxygen at the anode. The added mechanical scrubbing action and solution agitation, coupled with the fact that the negatively charged workpiece repels negatively charged particles of soil, are the major advantages of cathodic electrocleaning. This method of cleaning should not be used for parts susceptible to hydrogen embrittlement. An illustration of cathodic electrocleaning is presented in Fig. 3.

——Emulsion Cleaning——

EMULSION CLEANING is a process for removing heavy soils from the surfaces of metals and nonmetals by using organic solvents dispersed in an aqueous medium aided by an emulsifying agent. Depending on the solvent used, cleaning is done at temperatures from 10 to 82 °C (50 to 180 °F).

An emulsion system contains two mutually insoluble or nearly insoluble phases, one of which is dispersed in the other in the form of globules. One phase is usually a hydrocarbon and the other is water. The dispersed phase is distributed as globules in the liquid, continuous phase.

An unstable single-phase emulsion has a uniformly dispersed phase that tends to separate and form a solvent layer. Solvents with specific gravities less than 1.0 form top layers and those with specific gravities greater than 1.0 form bottom layers. These cleaners require moderate to considerable agitation to maintain complete dispersion.

A diphase, multiphase or floating-layer emulsion cleaner forms two layers in the cleaning tank and is used in this separated condition. Work is immersed through the solvent-rich surface layer

into the water-rich lower layer, permitting both cleaning phases to come in contact with the surfaces to be cleaned. When used in a spray system, a diphase cleaner resembles an unstable single-phase cleaner, because the solvent and water phases are mixed by the pumping action.

COMPOSITION OF EMULSION CLEANERS

Stable, unstable, diphase and other emulsion cleaners cover a wide range of solvent and emulsifier compositions. The solvent is generally of petroleum origin and may be heterocyclic (M-pyrol), naphthenic, aromatic, and of hydrocarbon nature (kerosine). Solvents are available with boiling points of 60 to 260 °C (140 to 500 °F) and flash points ranging from room temperature to above 93 °C (200 °F).

Emulsifiers include: (a) nonionic polyethers and high-molecular-weight sodium or amine soaps of hydrocarbon sulfonates, (b) amine salts of alkyl aryl sulfonates (anionic), (c) fatty acid esters of polyglycerides, (d) glycerols and (e) polyalcohols. Cationic ethoxylated long-chain amines and their salts are also used in emulsions.

Emulsifiers must have some solubility in the solvent phase. When solubility is low, it can be increased by adding a coupling agent (hydrotrope), such as a higher-molecular-weight alcohol, ester or ether. These additives are soluble in oil and water.

Concentration ranges of emulsion cleaners are 2 to 5% concentrate for spray applications and 4 to 10% for soak applications. Floating-layer diphase systems usually require about a 50-mm (2-in.) layer of solvent over a sufficient depth of water to permit the workpiece to be submerged. Water-in-solvent emulsions are operated at higher concentrations, ranging from 15 to 25%.

The compositions of emulsion cleaners for spray and soak operations are similar; however, soaps or wetting agents used in spray operations must have low-foaming characteristics. Compositions and operating temperature ranges for emulsion concentrates are given in Table 3.

——Solvent Cleaning——

SOLVENT CLEANING is a surface-preparation process that is especially well-suited for removal of organic compounds such as grease and oil from metal surfaces. Organic compounds are easily solubilized by solvent and then removed from the workpieces. In some cases, solvent

cleaning can precede other surface preparations and can extend the life of cleaning operations and reduce costs. In other cases, solvent cleaning is used to prepare workpieces for the next operation, such as assembly, painting, inspection, further machining, or packaging. Prior to plating, solvent cleaning is followed by an alkaline wash or other similar process which provides an oil-free surface. Solvent cleaning can also be used to remove water from electroplated parts, a common procedure in the jewelry industry.

Solvent cleaning can be accomplished in room-temperature baths or by use of vapor decreasing techniques. Room-temperature solvent cleaning is referred to as cold cleaning. Vapor degreasing is the process in which parts are solvent cleaned by condensing solvent vapors of a nonflammable solvent on the workpieces.

COLD CLEANING

Cold cleaning is a process for removing oil, grease, loose metal chips and other contaminants from the surfaces of metal parts. Common organic solvents, such as aliphatic petroleums, chlorinated hydrocarbons, chlorofluorocarbons, and blends of these various solvents, are used. Cleaning is usually performed at, or slightly above, room temperature. Parts are cleaned by being immersed and soaked in the solvent, with or without agitation. Parts that are too large to be immersed are sprayed or wiped with the solvent. Ultrasonic cleaning is sometimes used in conjunction with solvent cleaning to loosen and remove soils, such as abrasive compounds, from deep recesses or other difficult-to-reach areas. This reduces the time required for solvent cleaning of complex shapes.

Solvents

Table 4 lists aliphatic petroleums, chlorinated hydrocarbons, chlorofluorocarbons, alcohols and other solvents commonly used in cold cleaning. Stoddard solvent, mineral spirits and VM & P naphtha are widely used, because of their low cost and relatively high flash points.

VAPOR DEGREASING

Vapor degreasing is a generic term applied to a cleaning process that uses the hot vapors of a chlorinated or fluorinated solvent to remove soils, particularly oils, greases and waxes. A vapor degreasing unit consists of an open steel tank with a heated solvent reservoir, or sump, at the bot-

Table 3. Compositions and operating temperatures for emulsion concentrates
Maximum safe temperature depends on the flash point of the hydrocarbon (petroleum) solvent used as the major component.

Component	Composition, parts by volume		
	Stable(a)	Unstable(b)	Diphase(c)
Petroleum solvent(d)	250-300	350-400	250-300
Soaps(e)	10-15	15-25	None
Petroleum (or mahogany) sulfonates(f)	10-15	None	1-5
Nonionic surface-active agents(g)	5-10	None	1-5
Glycols, glycol ethers(h)	1-5	1-5	1-5
Aromatics(j)	5-10	25-50	5-10
Water(k)	5-10	None	None

(a) Operating temperature range: 4 to 66 °C (40 to 150 °F). (b) Operating temperature range: 4 to 66 °C (40 to 150 °F). (c) Operating temperature range: 10 to 82 °C (50 to 180 °F). (d) Two frequently used solvents are deodorized kerosine and mineral seal oil. (e) Most soaps are based on rosin or other short-chain fatty acids, saponified with organic amines or potassium hydroxide. (f) Low-molecular-weight petroleum sulfonates (mahogany sulfonates) are used for good emulsification plus some rust protection. High-molecular-weight sulfonates, with or without alkaline-earth sulfonates, offer good rust inhibition and fair emulsification. (g) Increased content improves stability in hard water, but increases cost. (h) Glycols and glycol ethers are used in amounts necessary to act as couplers in stable and unstable emulsions. These agents are frequently used with diphase and detergent cleaners to provide special cosolvency of unique or unusual soils. (j) Aromatic solvents are used to provide cosolvency for special or unique soils. They also serve to inhibit odor-causing or rancidifying bacteria. (k) Water or fatty acids, or both, are used to adjust the clarity and the stability of emulsion concentrates, particularly those which are stable or unstable.

Table 4. Properties of cold cleaning solvents

Solvent	Flash point(a) °C	Flash point(a) °F	OSHA TWA ppm(b)
Aliphatic petroleums			
Kerosine	63	145	...
Naphtha, hi-flash	43	110	...
Mineral spirits	14	57	500
Naphtha, VM & P	9	48	500
Stoddard solvent	41	105	100
Chlorinated hydrocarbons(c)			
Methylene chloride	None	None	500
Perchlorethylene	None	None	100
Trichloroethane (1,1,1)	None	None	350
Trichlorethylene	None	None	100
Trichlorotrifluoro- ethane	None	None	1000
Alcohols			
Ethanol, SD	14	57	1000
Isopropanol	10	50	400
Methanol	12	54	200
Other solvents			
Acetone	−18	0	750
Benzol	−11	12	10
Cellosolve(d)	40	104	50
Toluol	4	40	100

(a) Tag closed up. (b) OSHA exposure values expressed as parts of vapor or gas per million parts of air by volume at 25 °C (77 °F) and a pressure of 760 mm Hg. These values should not be regarded as precise boundaries between safe and dangerous concentrations. They represent conditions under which it is believed that nearly all workers may be repeatedly exposed, day after day, without adverse effect. The values refer to time-weighted average concentrations for a normal workday. (c) Also used for vapor degreasing. (d) 2-ethoxy-ethanol.

tom and a cooling zone near the top. Sufficient heat is introduced into the sump to boil the solvent and generate hot solvent vapor. Because the hot vapor is heavier than air, it displaces the air and fills the tank up to the cooling zone. The hot vapor is condensed when it reaches the cooling zone, thus maintaining a fixed vapor level and creating a thermal balance. The temperature differential between the hot vapor and the cool workpiece causes the vapor to condense on the workpiece and dissolve the soil.

Solvents

Only halogenated solvents are used in vapor degreasing and have all or most of the following characteristics:
- High solvency for oil, grease and other contaminants to be removed
- Low heat of vaporization and low specific heat to maximize the amount of solvent that condenses on a given weight of metal and to minimize heat requirements
- Boiling point high enough so that sufficient

solvent vapor is condensed on the work to ensure adequate final rinsing in clean vapor
- Boiling point low enough to permit the solvent to be separated easily from oil, grease or other contaminants by simple distillation
- High vapor density, in comparison with air, and low rate of diffusion into air, to minimize loss of solvent to the atmosphere
- Chemical stability in the process
- Noncorrosiveness to metals used in workpieces and in construction of equipment for the process
- Nonflammability, nonexplosiveness and controllability with respect to health hazards under operating conditions.

Table 5 lists pertinent properties of halogenated solvents used for vapor degreasing.

Degreasing Systems and Procedures

The four principal systems used for vapor degreasing are illustrated schematically in Fig. 4. Regardless of the system used, the distinctive features of vapor degreasing are a final rinse in pure vapors and a dry final product.

Vapor Phase Only. The simplest form of degreasing system uses the vapor phase only (Fig. 4a). The work to be cleaned is lowered into the vapor zone, where the relative coolness of the work causes the vapor to condense on its surface. The condensate dissolves the soil and removes it from the surface of the work by dripping back into the boiling solvent. When the work reaches the temperature of the hot vapor, condensation and cleaning action cease. Workpieces are dry when removed from the tank.

Vapor-Spray-Vapor. If the workpiece contains blind holes or recesses that are not accessible to the vapor, or if the soil cannot be removed by the vapor, a spray stage may be added. The system then consists of vapor, spray, vapor (Fig. 4b).

Warm Liquid–Vapor. Small parts with thin sections may attain temperature equalization before the work is clean. For these parts, and for other

small parts that are packed in baskets, the warm liquid–vapor degreasing system (Fig. 4c) is recommended.

Boiling Liquid–Warm Liquid–Vapor. For cleaning parts with particularly heavy or adherent soil or small workpieces that are nested or packed closely together in baskets, the boiling liquid–warm liquid–vapor system (Fig. 4d) is recommended. In the unit shown in Fig. 4(d), the work may be held in the vapor zone until condensation ceases and then be lowered into the boiling liquid, or the work may be lowered directly into the boiling liquid. In the boiling liquid, the violent boiling action scrubs off most of the heavy deposit as well as metal chips and insolubles. Next, the work is transferred to the warm liquid, which removes any remaining dirty solvent and lowers the work temperature. Finally, the work is transferred to the vapor zone, where condensation provides a final rinse.

Ultrasonic Degreasing. Ultrasonic transducers, which convert electrical energy into ultrasonic vibrations, can be used in conjunction with the vapor degreasing process. Cleaning efficiency in the liquid phase of a vapor degreasing cycle can be considerably augmented by application of ultrasonic energy.

Acid Cleaning of Iron and Steel

ACID CLEANING is a process in which a solution of a mineral acid, organic acid, or acid salt, in combination with a wetting agent and detergent, is used to remove oxide, shop soil, oil, grease and other contaminants from metal surfaces, with or without application of heat. The distinction between acid cleaning and acid pickling is a matter of degree, and some overlapping in the use of these terms occurs. Acid pickling is a more severe treatment for removal of scale from semifinished mill products, forgings or

Table 5. Vapor degreasing solvents

Solvent	Flash point	TLV, ppm(a)	Solvency	Photochemical reactivity	Vapor density (air = 1.0)	Volume of condensate L	Volume of condensate gal	Stabilization	Boiling point °C	Boiling point °F	Molecular weight
Trichlorethylene	None	100	Strong	Yes	4.5	3.8	1.00	Yes	88	190	131
1,1,1-trichloroethane	None	350	Moderate	No	4.6	3.3	0.86	Yes	74	165	133
Perchlorethylene	None	100	Moderate	Yes	5.7	6.0	1.60	Yes	121	250	166
Trichlorotrifluoroethane fluorocarbon 113 ..	None	1000	Mild	No	6.5	2.0	0.54	No	49	120	187
Methylene chloride ..	None	100	Strong	No	2.9	0.72	0.19	Yes	41	105	85

(a) Adopted by the American Conference of Governmental Industrial Hygienists, 1981.

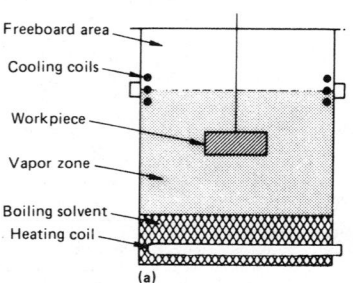 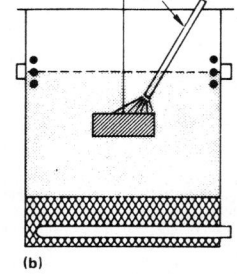 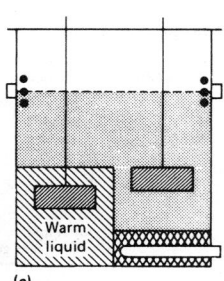 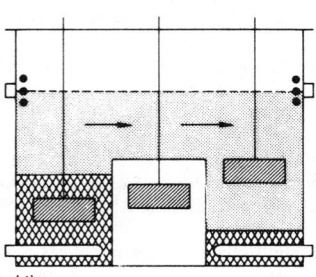

(a) Vapor phase only. (b) Vapor-spray-vapor. (c) Warm liquid–vapor. (d) Boiling liquid–warm liquid–vapor.

Fig. 4. Principal systems for vapor degreasing

Table 6. Typical compositions of acid cleaners for cleaning ferrous metals
The amounts of each constituent are given in weight percent.

Constituent	Immersion		Spray		Barrel	Wipe	Electro-lytic
Phosphoric acid	70	...	70	...	...	15-25	...
Sodium acid pyrophosphate	...	16.5	...	16.5	16.5	...	...
Sodium bisulfate	...	80	...	80	80	...	...
Sulfuric acid	...	...	...	...	...	...	55-70
Nonionic wetting agent(a)	5	...	5	...	...	7-20	...
Anionic wetting agent	...	3	...	3	3	...	...
Other additives	(b)	(b)	(b)(c)	(b)(c)	(b)(c)	(b)(d)	(b)
Water	25(e)	...	25(e)	...	...	rem	rem

(a) Ethylene glycol monobutyl ether is used. (b) Inhibitors up to 1% concentration may be used to minimize attack on metal. (c) An antifoaming agent is usually required when the cleaner is used in a spray or barrel system. (d) A small addition of sodium nitrate is often used as an accelerator in cleaning rolled steel; nickel nitrate is used in cleaning galvanized steel. (e) Before dilution.

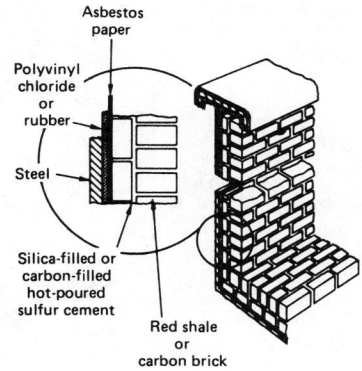

Fig. 5. Section of an acid cleaning tank

castings, whereas acid cleaning generally refers to the use of acid solutions for final or near-final preparation of metal surfaces before plating, painting or storage.

CLEANER COMPOSITION

A variety of mineral acids and solutions of acid salts can be used, either with or without surfactants (wetting agents), inhibitors and solvents. The large number of compositions that are used may be classified as:

• Inorganic (mineral) acid solutions
• Acid-solvent mixtures
• Solutions of acid salts.

Many acid cleaners are available as proprietary compounds, either as liquid concentrates or powders to be mixed with water. Compositions of several solutions used for cleaning ferrous metals are given in Table 6. Formulas suitable for use with nonferrous metals can be found in the article on Cleaning and Finishing of Nonferrous Metals, in this section. Table 7 presents conditions for acid cleaners used in cleaning ferrous metals.

METHODS OF APPLICATION

Wipe on/wipe off, spray, immersion and rotating-barrel methods are all used extensively for acid cleaning. Although heating greatly increases efficiency, cleaning is frequently done at room temperature for superior process control and economy of operation. When heat is used, the temperature range of the cleaner is usually 60 to 82 °C (140 to 180 °F) with temperatures up to 98 °C (200 °F) being used occasionally. Time cycles for acid cleaning are short compared with those for acid pickling, especially when stronger acids are being used. Selection of method depends on the nature of soil being removed, the size and shape of the workpiece, quantity of similar pieces to be cleaned, and type of acid cleaner used.

EQUIPMENT

Wipe on/wipe off cleaning requires only the simplest equipment. Acid-resistant pails and protective clothing, and common mops, brushes and wiping cloths are all that is needed.
Immersion systems require equipment varying from earthen crocks for hand dipping at room temperature to fully automated systems using heat and ultrasonic or electrolytic assistance. The construction of an acid tank is shown in Fig. 5.

Table 7. Operating conditions for acid cleaners for ferrous metals

Type of acid cleaner	Concentration		Temperature	
	g/L	oz/gal	°C	°F
Immersion	120	16	71	160
	60-120	8-16	60	140
Spray	60	8	60	140
	15-30	2-4	60	140
Barrel	15-60	2-8	Room	
Wipe	...	...	Room	
Electrolytic(a)	...	...	21	70

(a) Current density, 10 A/dm² (100 A/ft²).

Tanks for sulfuric acid may be lined with natural rubber and acid-resistant red shale or carbon brick joined with silica-filled hot poured sulfur cement.

Pickling of Iron and Steel

PICKLING is the chemical removal of surface oxides (scale) and other contaminants such as dirt from metal by immersion in an aqueous acid solution. Wide variations are possible in the type, strength and temperature of the acid solutions used. Because of its economy and adaptability to continuous operations, pickling is the most efficient method of scale removal for large-tonnage

Inner lining of brick acts only as a thermal shield and as a protection against mechanical damage to the corrosion-resistant polyvinyl chloride or rubber membrane.

products such as merchant bar, blooms, billets, sheet, strip, wire and tubing. Pickling is also applicable to many types of forgings and castings.

PICKLING SOLUTIONS

Sulfuric acid, despite certain drawbacks, is still the most common pickling liquor. It produces satisfactory results when used (a) for batch descaling of carbon steel rod and wire (to 0.60% carbon) and (b) for continuous cleaning, provided that the iron concentration in the bath is less than 8 wt %. Table 8 lists the types of carbon and alloy steel products that are pickled in sulfuric acid and also lists the ranges of acid concentrations and temperatures used.

In comparison with hydrochloric acid, sulfuric acid offers the advantages of lower cost, less fumes, and less volume of acid to handle.

Hydrochloric acid is preferred for batch pickling of hot rolled or heat treated high carbon steel rod and wire. This acid produces a uniform light gray surface and decreases the possibility of overpickling. When used in large quantities, the spent acid can be regenerated (recycled), which eliminates an acid-disposal problem.

Table 8. Solution concentrations and operating temperatures used for pickling carbon and alloy steel products

Product	Sulfuric acid concentration, wt%		Bath temperature			
			min		max	
	min	max	°C	°F	°C	°F
Bar, low carbon	7	15	68	155	85	185
Bar, alloy	9	12	66	150	77	170
Billet, low-carbon	7	12	74	165	82	180
Billet, alloy	9	12	82	180	93	200
Pipe for galvanizing	7	15	71	160	88	190
Sheet for galvanizing	4	12	66	150	77	170
Sheet, tin plate (white pickle)	9	12	66	150	85	185
Strip, soft	6	12	77	170	88	190
Strip, alloy and high-carbon	7	12	66	150	77	170
Strip, continuous pickling	23	38	77	170	100	212
Tubing, low-carbon seamless	7	12	77	170	88	190
Tubing, high-carbon and alloy structural	9	12	71	160	93	200
Tubing (over 0.40% carbon)	9	12	60	140	71	160
Wire, soft	4	11	77	170	88	190
Wire, alloy and high-carbon	3	7	63	145	74	165
Fabricated parts (for tinning):						
Initial pickle	5	10	66	150	88	190
Final dip	(a)	(a)	...	...	38	100

(a) Concentrated hydrochloric acid, 1.14 to 1.16 sp gr.

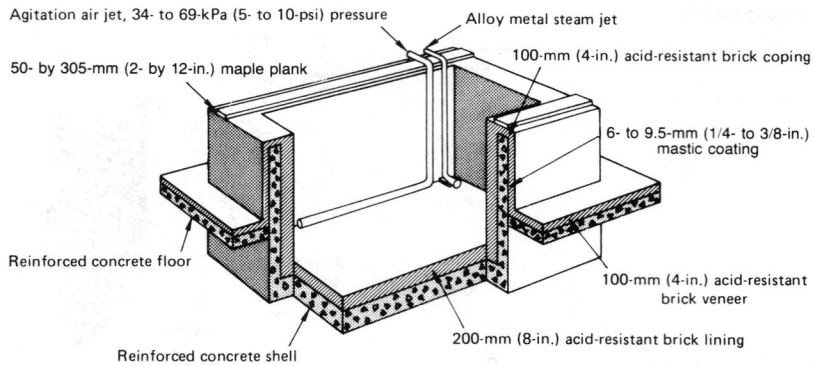

Fig. 6. Materials used in construction of 3.6- to 4.5-t (4- to 5-ton) capacity tank for pickling coils of steel

Operating conditions for batch and continuous pickling in hydrochloric acid solutions are as follows:

Operating variable	Batch	Continuous
Hydrochloric acid, wt %	8-12	6-14
Temperature:		
°C	38-41	77-93
°F	100-105	170-200
Immersion time	5-15 min	1-20 s
Iron concentration(a), wt %	13	13
(a) Maximum allowable in bath.		

INHIBITORS

Inhibitors are added to acid pickling solutions primarily to protect the steel being cleaned by retarding the chemical action of the acid on the base metal. Inhibitors (a) minimize the loss of iron, (b) reduce the extent of hydrogen embrittlement, (c) protect the metal against pitting (caused by overpickling) and poor surface quality, (d) reduce acid fumes resulting from excessive reaction between the acid and base metal, and (e) reduce acid consumption. In production operations, inhibitors do not appreciably affect the rate of scale or rust removal.

PRECLEANING

Alkaline precleaning before acid pickling is beneficial for removing soils that do not readily react with acid. These soils include greases, oils, soaps, lubricants and carrier coatings. If such materials are carbonized by exposure to heat, they become more difficult to remove and usually contribute to the formation of smut on the surface of the metal during the pickling operation.

EQUIPMENT

Construction materials for pickling tanks include wood, concrete, brick, plastic and steel. Acid-resistant linings provide protection for the outer shell of the tank. These are commonly of natural, pure gum or synthetic rubber. Acid-resistant brick linings line the sides and floor of the tank. Figure 6 shows the materials used in construction of a tank of 3.6- to 4.5-t (4- to 5-ton) capacity used for pickling coiled steel.

Abrasive Blast Cleaning

ABRASIVE BLAST CLEANING entails the forceful direction of abrasive particles, either dry or suspended in a liquid, against the surfaces of parts or products. Abrasive blast cleaning removes contaminants and conditions the surfaces for subsequent finishing. Typical uses include:

- Removing rust, scale, dry soils, mold sand or paint
- Roughening surfaces in preparation for bonding, painting or other coating
- Removing burrs
- Developing a matte surface finish
- Removing flash from molding operations
- Carving in glass or porcelain.

ABRASIVES FOR DRY BLAST CLEANING

The materials used in dry abrasive blast cleaning may be categorized as metallic grit, metallic shot, sand, glass and miscellaneous. Hardness, density, size and shape are important considerations in choosing an abrasive for a specific application (see Table 9).

PROPELLING ABRASIVE MEDIA

Three basic methods are used to propel the abrasive medium against the surfaces of the workpieces: (a) airless abrasive blast blade or vane-type wheels, (b) pressure blast nozzle systems and (c) suction (induction) blast nozzle systems.

Wheels. Airless abrasive blast wheels are generally of the slider blade type as shown in Fig. 7.

Pressure blast nozzle systems generally rely on a 690-kPa (100-psig) air supply to propel the abrasive through a special nozzle. A typical intermittent pressure tank has dimensions of 610 by 610 mm (24 by 24 in.) and an abrasive discharge capacity of 0.12 m³ (4.2 ft³). This capacity is adequate to operate one 6-mm (¼-in.) diam blast nozzle for 30 to 60 min. This type of tank is refilled through the filling valve by gravity when the air supply is shut off. Without air pressure in the tank, the filling valve is pushed down and open by the weight of the abrasive.

Suction blast systems are generally considered the simplest form of abrasive blast equipment. The suction blast cabinets may be used manually or may have fixed or oscillating nozzles. Figure 8 illustrates a cabinet measuring 1220 by 915 by 840 mm (48 by 36 by 33 in.). The pressure tank and filling valve may be vertically doubled with a timer and proper valving to provide a continuous automatic pressure tank.

Figure 9 illustrates a suction blast nozzle assembly. The nozzle in the suction cabinet is an induction nozzle which creates a blasting mixture by the siphon effect of the air that is discharged through the nozzle body. This effect pulls abrasive through the abrasive hose from the cabinet hopper, and the blast mixture is formed within the nozzle body.

Salt Bath Descaling

SALT BATH DESCALING processes are classified as oxidizing, electrolytic and reducing. Complete removal of scale requires the use of an acid dip or acid pickling treatment after salt bath

Table 9. Abrasives, equipment and cycles used for dry blasting

Material or product	Reason for blasting	Abrasive Type	Size No.	Type	Horse-power	Nozzle diameter mm	Nozzle diameter in.	Blasting cycle
Ferrous metals								
Cast iron	Prepare for zinc impregnation	Iron grit	G80	Air, table(a)	...	6	¼	1 h
	Remove molding sand	Steel shot	S230	Wheel, barrel	15	...	...	10 min
Cold rolled steel	Remove graphite for painting	Iron grit	G80	Wheel, barrel	15	...	...	10 min
				Air, table(a)	...	6	¼	40 min
Gray iron exhaust manifolds, bearing caps	Clean for machining	Malleable iron shot	S460	Wheel, tumble(b)	80	...	...	1500 pieces/h
Gray iron motor blocks and heads	Remove sand and scale after heat treatment	Malleable iron shot	S550	Wheel, blast cabinet(c)	150	...	...	0.19-0.27 min/piece

Table 9 (continued)

Material or product	Reason for blasting	Abrasive Type	Size No.	Type	Horse-power	Nozzle diameter mm	Nozzle diameter in.	Blasting cycle
Hardened steel screws	Remove heat treat scale	Iron grit	G80	Wheel, barrel	10	...	...	5 min
Hot rolled steel	Prepare for painting	Iron grit	G80	Air, table(a)	...	6	1/4	1 h
Malleable iron castings	Prepare for galvanizing	Steel grit	G50	Wheel, barrel	40	...	...	15 min
Pole-line hardware	Prepare for galvanizing	Steel grit	G50	Wheel, barrel	40	...	...	15-20 min
Round steel bar	Etch for adhesive coating	Iron grit	G80	Air, blast room	...	6	1/4	2 min
Soil pipe fittings	Remove molding sand	Steel shot	S330	Wheel, barrel	30	...	...	181 kg (400 lb) in 5 min
Steel drums	Prepare for painting	Iron grit	G80	Air, blast room	...	6	1/4	4 min
Steel rod	Clean for wire-drawing	Steel grit	G40	Wheel, continuous(d)	80	...	...	0.2-1.5 m/s (40-300 ft/min)
Steel screws	Prepare for plating	Iron grit	G80	Air, barrel(a)	...	8	5/16	20 min
Structural steel	Prepare for painting	Steel grit	G40	Wheel, continuous(d)	80	...	...	0.15 m/s (30 ft/min)
Weldments (steel)	Remove scale, welding flux and splatter for painting	Steel grit	G25	Wheel, barrel	30	...	...	136-272 kg (300-600 lb) in 7 min
Engine parts for rebuilding	Remove paint, scale and carbon deposits	Glass beads	60-100 mesh	Air	...	6	1/4	5-20 min
Nonferrous metals								
Aluminum	Produce frosted surface	Sand	50	Air, barrel	...	6	1/4	20 min
	Prepare for painting	Iron grit	G80	Wheel, barrel	15	...	...	5 min
Bronze	Produce frosted surface	Sand	50	Air, barrel	...	6	1/4	20 min
Aluminum and bronze	Prepare and condition surface	Glass beads	20-400	Air	...	6	1/4	5-20 min
Nonmetallic materials								
Clear plastic parts	Produce frosted surface	Sand	50	Air, barrel	...	6	1/4	15 min
Hard rubber	Improve appearance	Sand	50	Air, barrel	...	6	1/4	20 min
Molded plastic parts	Remove flash	Walnut shells	...	Wheel, barrel	10	...	...	8 min
Phenolic fiber	Produce frosted surface	Sand	50	Air, barrel	...	6	1/4	30 min
	Prepare for painting	Sand	50	Air, barrel	...	6	1/4	20 min

(a) Four air nozzles. (b) Two wheels, 40 hp each. (c) Six wheels, 25 hp each. (d) Four wheels, 20 hp each.

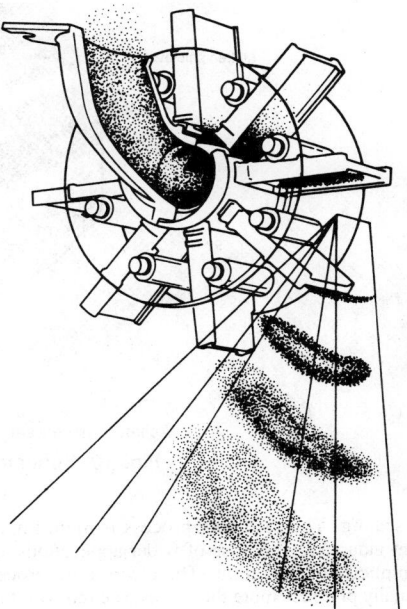

Fig. 7. Slider blade airless abrasive blast wheel

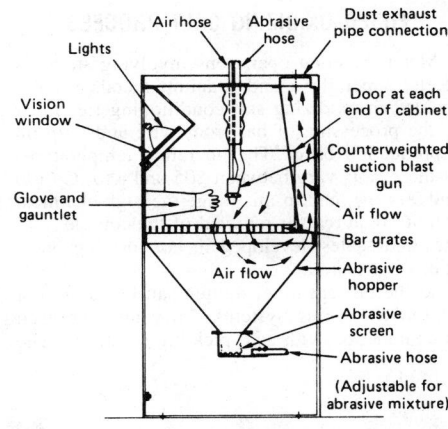

Fig. 8. Suction blast cabinet

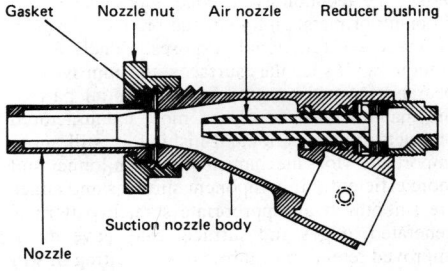

Fig. 9. Suction blast nozzle assembly

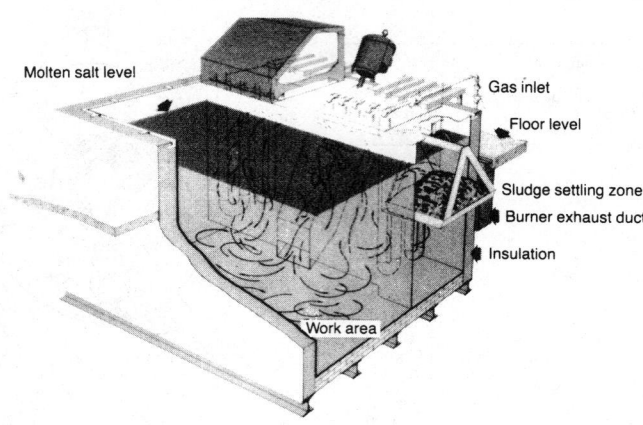

Agitated molten salt bath with sludge settling zone.
Fig. 10. Schematic of salt bath furnace

descaling. The oxidizing process is more important industrially because of wider applications and simplicity of operation. The electrolytic process usually provides more thorough scale removal than the reducing process, and it also has the capability of functioning in both an oxidizing and a reducing mode. Because the reducing process employs a bath that is operated at a lower temperature, it is sometimes advantageous for descaling metals that undergo changes in properties at higher temperatures.

FUSED OXIDIZING SALT PROCESS

Metal finishing operations involving stainless steels, superalloys and titanium metals usually require an oxidizing salt conditioning treatment in the processing of bar, rod, wire and strip on a production basis. The operating temperatures of these salts vary between 205 and 480 °C (400 and 900 °F), and in this range the high chemical activity required for removal of the complex oxides and scales developed in hot forming operations is ensured.

At the present time, molten salt baths are not complete cleaning systems. They must be used in conjunction with acid pickling solutions. The required concentrations and temperatures of pickling acids can be considerably reduced by the conditioning action of the fused salt.

ELECTROLYTIC PROCESS

Fused salts that are neither chemically oxidizing nor chemically reducing can be activated to produce either of these conditions by the input of electrical energy. The electrical system involved is rather simple. It employs a direct current source, a reversing switch, a positive and a negative pole that can be either the furnace wall or the work load, and a conducting medium that is the molten salt bath.

Electrolytic fused salt systems are used primarily for removal of sand from iron castings, especially where designs permit only limited access to internal surfaces and where dislodged particles of retained sand can cause substantial damage under operating conditions. For this purpose, current reversal is not required, and the electrolytic salt bath is operated with the work at a negative or reducing polarity. Cycles are usually developed through experience and may vary from 15 to 30 min, followed by water rinsing and drying.

THE REDUCING PROCESS

The reducing or sodium hydride descaling process is performed in a fused caustic bath in which sodium hydride (NaH) is generated *in situ* from metallic sodium and dissociated ammonia. The reaction takes place in generators immersed in the molten sodium hydroxide, and the recommended concentration of 1.5 to 2% sodium hydride is adjusted and maintained by controlling the feed of sodium and gaseous hydrogen. This control of the sodium hydride concentration of the bath is necessary for efficient descaling and is achieved by regular analyses. The uniformity of sodium hydride throughout the bath is dependent on convection currents and the periodic immersion of products to be descaled.

Heat process scale can be removed from all metallic alloys that do not react with the required caustic base and that are not metallurgically affected by the temperature of operation of between 370 and 400 °C (700 and 750 °F).

FUSED SALT EQUIPMENT

The internal structure of a salt bath unit which could have any conceivable dimensions is shown in Fig. 10. Basically, it is imperative that the equipment be properly designed and engineered to fit the exact requirements if maximum efficiency is to be obtained from a fused salt cleaning system.

Selected References

• *Metal Finishing Guidebook and Directory* (anonymous): 1978, p 71-183
• *Principles of Electroplating and Electroforming*, 3rd Ed., by Blum and Hogaboom: McGraw-Hill, p 200-219
• *Modern Electroplating*, 2nd Ed., by Lowenheim: John Wiley, p 545-555
• *Synthetic Detergents*, by McCutcheon: MacNair-Dorland, 1950, p 255-336
• *Encyclopedia of Chemical Technology*, 3rd Ed., Vol 15, by Kirk-Othmer: John Wiley, p 296-312
• *Finishing Metal Products*, 2nd Ed., by Simonds and Bregman: McGraw-Hill, p 47-110
• *Protective Coatings for Metals*, 3rd Ed., by Burns-Bradley: Reinhold, p 27-54

Mechanical Finishing

By J. Bernard Hignett, The Harper Company

THE TERM "mechanical finishing" encompasses the technology of edge and surface conditioning of metal and nonmetal products for both cosmetic and functional purposes. Generation of smooth and specular surfaces to improve appearance of components is an essential part of the manufacturing cycle of most manufactured products, but functional finishing is still more important. Most mechanisms will run longer and more efficiently if component surfaces and edges are smooth. If an appropriate scratch pattern is generated, edges and surfaces may have much improved retention of lubricants, resulting in still smoother operation. Removal of stress raisers at sharp corners and generation of controlled radii on edges can substantially improve thermal and mechanical fatigue strength of highly stressed components. Removal of tensile stresses by improved surface integrity will reduce or eliminate their contribution to service failures. Moreover, generation of high compressive stresses, which can be achieved by several mechanical finishing processes, can significantly increase resistance to fatigue stresses and thus increase the service life of highly stressed parts. Improved edge and surface condition in passages through which gases or fluids flow reduces "drag" and thus increases flow rates.

Mechanical finishing is an essential part of the manufacturing cycle for most products. A high standard of mechanical finish will normally result in a better product that is also better looking. Proper attention to the technology of mechanical finishing will result in improved productivity as well as an improved product. In view of the fact that the cost of mechanical finishing of most manufactured products is generally more than 10% of total manufacturing cost, industry should invest appropriately in engineering personnel and capital equipment. Mechanical finishing has been neglected by many manufacturing organizations. Now that excellent mechanical finishing processes are generally available, this neglect is no longer justifiable.

Mechanical finishing processes may be classified as follows:

• Manual filing, scraping, etc.
• Machining
• Polishing, buffing and brushing
• Abrasive and nonabrasive blasting
• Mass finishing
• Chemically and electrochemically related processes

• Other processes, frequently classified as "nontraditional mechanical finishing." Major processes in this group are thermal-energy and abrasive-flow finishing.

Polishing, Buffing and Brushing

POLISHING is the mechanical finishing of a product using abrasives which are firmly adhered to a flexible backing, such as with an abrasive belt or with abrasives bonded to a flexible wheel or a flap wheel (while grinding is the use of abrasives firmly bonded to a rigid backing, as is the case with a grinding wheel).

Buffing refers to the finishing of a product by means of abrasive loosely adhering to a flexible backing, typically liquid or bar compounds applied to cloth wheels.

Brushing is the use of filament wheels for edge or surface conditioning of a product. Filaments are normally nonabrasive fibers or metal wires, but can be abrasive-laden filaments. Brushing may be done either wet or dry, and may employ compounds loosely applied.

Polishing processes are primarily abrasive processes, whereas in buffing and brushing a substantial part of the action may be plastic deformation of surfaces and edges.

Traditionally, polishing and buffing have been associated with decorative surface finishing operations and surface finishing prior to plating or painting. Modern automated polishing and buffing are more frequently used for improving functional edge and surface condition and are often used to improve product shape and tolerances (replacing some machining operations). Brushing is most often used for cleaning (removing scale, oxide films, rust and old paint), although it also finds many applications in deburring and in edge and surface conditioning.

More than one-third of the polishing and buffing machines currently in use in the U.S. are more than 20 years old. Less than one-third of the equipment has been installed during the past 10 years, and only a fairly small proportion of these more modern machines are used for functional finishing. It is clear that there are tremendous opportunities for the metalworking industry to improve operations by paying closer attention to automated polishing, buffing and brushing processes—after coating; for deflashing, deburring and descaling; to improve lubricity of surfaces and the cutting edges of parts; to improve air flow around edges and corners; to improve resistance to fatigue failure; and to satisfy safety and environmental requirements.

POLISHING, BUFFING AND BRUSHING EQUIPMENT

Manual polishing and buffing in standard polishing lathes are operations that are in general use in industry today. Semiautomatic equipment, such as simple indexing spindles for use with manual lathes, attachments for centerless buffing machines, and simple irregular polishing attachments, all facilitate mechanical finishing of parts, but all have the same inherent disadvantages: high labor content, high cost, variable quality and lack of control. Automated finishing equipment is as important in manufacturing operations as is automated equipment for machining and forming.

The basic types of automated polishing, buffing and brushing machines are as follows:

• Indexing rotary tables
• Continuous rotary tables
• Strip and sheet finishing machines
• Centerless tube and rod finishing machines
• In-line continuous finishing systems
• Rectangular conveyor finishing equipment.

Continuous Rotary-Table Polishing and Buffing Machines

Continuous rotary-table equipment consists of a circular table rotating in a horizontal plane. Components are fixtured on spindles mounted around the periphery of the table, and the spindles may be fixed, rotating or indexing as they pass under a series of heads mounted around the table. Polishing and buffing heads are situated for finishing of all significant surfaces; an initial coarse operation may be followed by finer polishing which is followed in turn by final color buffing, thus completing several operations in a single rotation of the table.

Continuous rotary-table equipment is best suited for fairly small and symmetrically shaped components.

Lengthy contact time against the wheel or belt cannot be achieved with this type of equipment. It is easy to fully automate loading and unloading operations. Modern equipment normally includes automated addition of compound with automatic monitoring to ensure that precise quantities are applied. Automatic control of wheel pressure against components compensates for buffing-wheel wear and ensures uniform results. Control of buffing-wheel, table and spindle speeds is also readily available to ensure consistent results from this type of equipment.

Indexing Rotary-Table Polishing, Buffing and Brushing Machines

Standard indexing rotary-table equipment comprises a limited number of work spindles mounted on a table. The table indexes to locate parts on each spindle at a polishing or buffing head. For example, a six-spindle machine will normally be used where there are four polishing, buffing or brushing operations, one loading station and one unloading station. This type of equipment is more versatile than continuous rotary tables, permitting longer periods of contact and enabling larger parts to be processed. Spindles are normally rotated at each station, but can be cammed to permit finishing of irregularly shaped parts. When equipment is to be used for handling a variety of parts, N/C control is frequently justified.

Flat-Part Buffing and Polishing Machines

There is an immense variety of flat-part polishing machines for strip, sheet, small parts and large parts. There is the capability for polishing one or both faces in a single setting with single or multiple heads. All equipment is based on a table or conveyor feeding work under polishing belts or wheels, buffs or brushes. Modern equipment permits precise control of wheel pressure, quick adjustment for different thicknesses of material, and usually easy replacement of wheels and belts.

Straight-Line and Conveyor-Type Equipment

It is the developments in conveyor polishing and buffing that have been the most important improvements of equipment during recent years, giving greater versatility and improved productivity, and facilitating total automation. With straight-line finishing, "mush buffing" can be utilized for greatest flexibility, versatility and economy.

Success of conveyor buffing equipment depends on a rigid conveyor for precise location of components to be finished. Conveyors are normally built to suit specific applications for equipment, and spindles on which parts will be fixtured and mounted along the conveyor. The distance between these spindles depends on the size of the parts being buffed. Spindles can normally be rotated in either direction, or they may index while passing under the polishing and buffing heads.

Polishing and buffing heads are placed around the conveyor, mounted on individual stands or suspended from overhead framework. The latter normally offers greater versatility for positioning of heads and for maintaining cleanness of the equipment.

"Mush buffing" entails the use of wide wheels, along the conveyor so that parts may be rotated or indexed across the face of the wheel. This results in very high efficiency with considerable contact time of buff against component. This is the most effective means of handling most parts of complex shape. The buffing wheel can run at comparatively low speeds so that the wheel remains very flexible and will follow the contours of parts being finished.

Modern fully automated conveyor buff, brush and polish machines can be built to handle a variety of different parts, frequently simultaneously. Where considerable versatility is needed and for large systems then N/C or CNC is frequently justifiable.

BRUSHING AND POLISHING MATERIALS

Development of machine wheels and belts, buffs, bar, liquid compound and the newer impregnated fiber materials have kept pace with development of equipment and needs for better and more consistent products. Some of the important developments include:

• Modern polishing belts which maintain a uniform cut throughout their useful life. Modern belt polishing equipment depends on use of such products; inferior belts result in inconsistent results and the need for additional heads to compensate for reduced abrasion.
• "Air flow" buffs with the buff material cut at a bias, which give consistent performance as the buff wears because the quantity of material exposed remains uniform throughout the life of the buff. Air cooling at the face between buff and component is better and produces more consistent performance.
• Modern liquid compounds can be used for aggressive cut operations as well as superfine finishing. These compounds have good storage life and consistent performance with excellent adhesion to the buff.
• "Three-dimensional" and nonwoven fiber buffs and wheels extend capabilities for achieving new surface conditions and cleaner results.
• Modern abrasive flap wheels can frequently perform to achieve results which hitherto could only be obtained by use of grinding wheels or very coarse abrasive belts. These wheels are capable of maintaining consistent performance and of providing the flexibility needed for automated polishing.

Abrasive and Nonabrasive Blasting

By R. Steven Marcus, The Markee Corp.

BLASTING processes, both abrasive and nonabrasive, are based on the principle of propelling a series of particles against a surface for the purpose of modifying the condition of that surface. In the vast majority of cases, the particles are propelled in a matrix of air or water, but in some rare applications the liquid or gas itself is used to transfer the energy to the work surface. The basic means of propelling the blasting material is with either air pressure or a centrifugal wheel.

APPLICATIONS

Blast finishing processes are used for removal of contaminants, deflashing and trimming, effecting changes in surface condition, deburring, and peening.

Removal of Contaminants. Major applications in this group are removal of sand from castings and removal of scale from forgings. Blasting is also the standard means of removing all scale from steel products and of cleaning paint, rust, oil and other undesirable coatings from part surfaces.

Deflashing and Trimming. Deflashing of nonferrous die castings and thermoset plastic parts is normally accomplished by use of blasting techniques. Trimming of excess insulation from electrical wires, and deflashing of circuit boards, are accomplished by blasting.

Changes in Surface Condition. Blasting can be used to improve the mechanical bonding of most coatings—paints, galvanized coatings, elastomer bonds and enamel coatings.

Deburring. Blasting may be used to break off small burrs on some products; if the material is ductile, blasting may blunt burrs and sharp edges. The blasting process is not suitable for generating radii on edges and corners.

Peening. Blasting, particularly with nonabrasive materials such as steel shot, can generate very high compressive stresses in the surfaces of parts. This results in improvement of fatigue life in highly stressed parts such as springs, gears, cutting tools, and aircraft engine and airframe components.

BLASTING EQUIPMENT

A blasting system generally comprises five basic components: a propelling device, a blast cabinet, a separator, a system for return of the blasting medium, and a dust collector.

Propelling Device. There are two basic types of propelling devices: centrifugal wheels or slingers, and air blast equipment. Centrifugal wheels comprise the most efficient method of propelling large volumes of blasting material over fairly large work areas. Wheels may be powered by motors rated at 5 to 100 hp, and may be capable of throwing from 50 to 900 kg (100 to 2000 lb) of blasting material per minute. In air-blast equipment, an abrasive or nonabrasive blasting medium is propelled in an air stream, generally by use of a blast nozzle containing a venturi chamber. Compressed air passing through the nozzle creates suction, which is then used to draw the medium from an unpressurized chamber and feed it into the air stream. The medium is then accelerated in the air stream through the nozzle.

Alternatively, a pressurized container may be used for the blasting medium, which is transferred by compressed air through a hose to the nozzle. Either system is capable of accelerating the medium to supersonic velocities—that is, to speeds greater than those achievable by wheel blasting. Pressure blast systems may utilize a liquid, generally water, and such systems are well suited to removal of oily or wet contaminants from work surfaces. Chemicals are normally added to the water to prevent corrosion of equipment and workpieces and to provide additional cleaning capability. In some specialized deburring applications, blasting is done with water (no abrasive) at very high pressures.

Blast Cabinet. The simplest form of blast cabinet is a small cabinet into which the operator places his hands, protected by rubber gauntlets. Such units are popular in tool rooms, and for small-volume applications where one piece can be blasted at a time. Where parts are to be handled in bulk, the most efficient method is to tumble them inside a blast zone, typically an endless-belt tumbling system.

Rotary table systems of both continuous and indexing types are well established for blasting of more fragile parts. The parts may be conveyorized through a blast zone or roll-through, or may be hung on hooks attached to overhead conveyors. When automatic blasting is required, blast cabinets usually are designed to meet specific requirements. Room-size cabinets are used for blasting very large parts, and for still larger items (such as complete ships) portable cabinets, sealed to prevent escape of the blasting medium, are used for section-by-section blasting.

Separator. Used blasting material normally is contained within the blast cabinet and returned to the propelling device for reuse. Generally it is desirable to remove fines, scale, sand or other contaminants from the blasting medium before it is reused. For some applications it is necessary to reclassify the blasting medium.

The most popular separation system is an air-wash system in which the used medium is spread into a thin curtain and air is drawn through the curtain to cause lighter particles to be deflected and hence removed. Another means of separation is by centrifugal (or cyclone) separators suited for small-volume applications and lighter media. On occasion, screening separators may be suitable where contaminants are of a much smaller size than that of the medium.

Medium-Return System. In most wheel-type blasting machines the spent abrasive is collected at the bottom of the blast cabinet by screw conveyors that carry it to the side of the machine and deposit it at the bottom of a vertical belt conveyor, which in turn transfers it to the top of an air-wash separator. At some point there is normally a separating screen that removes large contaminant particles or workpieces which have been spilled into the medium.

Air blast machines generally employ vacuum tubes to return the abrasive to the separator, and even media such as steel shot can be handled in this manner.

Dust Collector. A dust collector is usually necessary in a blast system as a means of removing the finest contaminants and of maintaining clean working conditions.

BLASTING MEDIA

For wheel-type blasting equipment the most common media are ferrous shot and grit, and the most popular sizes are between 0.5 and 1.4 mm (0.020 and 0.055 in.). In general, the smallest size of shot or grit that can provide satisfactory results should be selected. Also used in wheel-type blasting are nonferrous abrasives such as plastic pellets, corn cob, nutshells, aluminum nuggets and cut wire.

Air blast applications generally employ a wider variety of blasting media. Glass beads, which are noncontaminating and are capable of light deburring, peening, cleaning, and generation of decorative surface finishes, are widely used. Abrasive grains are in general use, and very fine sizes may be used in air blasting to generate finer finishes than those obtainable by wheel blasting.

Types, sizes and applications of various commonly used blasting media are given in Table 1.

Mass Finishing

MASS FINISHING normally involves loading of components to be finished into a container together with some abrasive medium, water, and compound. Action is applied to the container to cause the medium to rub against the surfaces,

Table 1. Media commonly used in blasting

Type of medium	Sizes normally available	Applications
Glass beads	8 to 10 sizes from 30- to 440-mesh; also many special gradations	Decorative blending; light deburring; peening; general cleaning; texturing; noncontaminating
Aluminum oxide	10 to 12 sizes from 24- to 325-mesh	Fast cutting; matte finishes; descaling and cleaning of coarse and sharp textures
Garnet	6 to 8 sizes (wide-band screening) from 16- to 325-mesh	Noncritical cleaning and cutting; texturing; noncontaminating for brazing steel and stainless steel
Crushed glass	5 sizes (wide-band screening) from 30- to 400-mesh	Fast cutting; low cost; short life; abrasive; noncontaminating
Steel shot	12 or more sizes (close gradation) from 8- to 200-mesh	General-purpose rough cleaning (foundry operations, etc.); peening
Steel grit	12 or more sizes (close gradation) from 10- to 325-mesh	Rough cleaning; coarse textures; foundry welding applications; some texturing
Cut plastic	3 sizes (fine, medium, coarse); definite-size particles	Deflashing of thermoset plastics; cleaning; light deburring
Crushed nutshells	6 sizes (wide-band screening)	Deflashing of plastics; cleaning; very light deburring; fragile parts

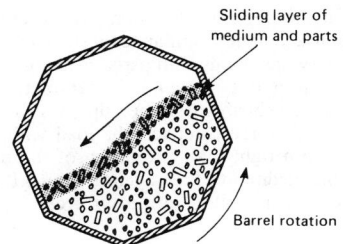

Fig. 1. Action of medium and parts within a rotating barrel

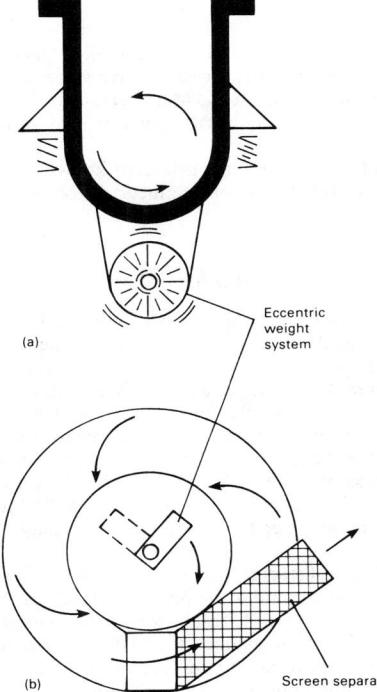

(a) Tub vibrator. (b) Bowl vibrator.
Fig. 2. Vibratory finishing machines

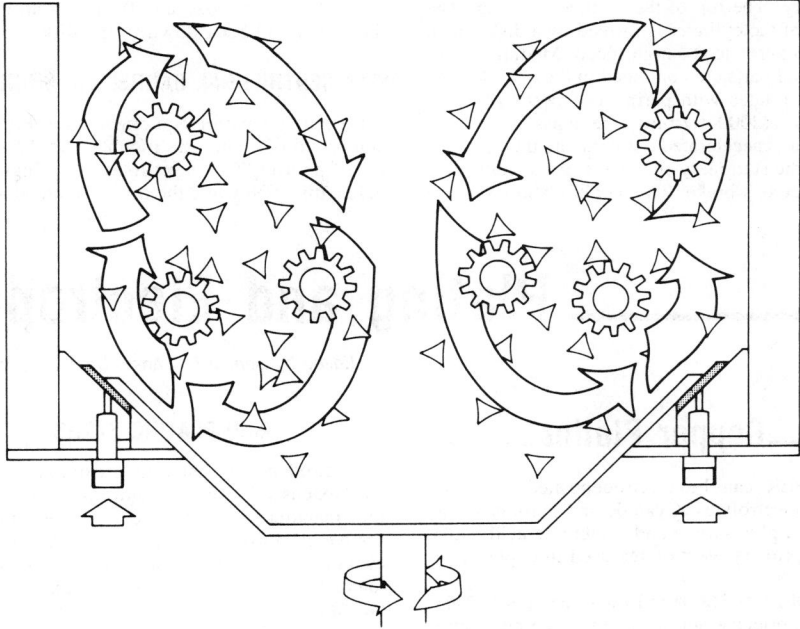

Fig. 3. Centrifugal disk machine

edges and corners of the components, or to cause the components to rub against each other, or both. This action may deburr, generate edge and corner radii, clean the parts by removing rust and scale, and modify the surface stress. The basic mass finishing processes include:

• Barrel finishing
• Vibratory finishing
• Centrifugal disk finishing
• Centrifugal barrel finishing
• Spindle finishing.

BARREL FINISHING

The rotary barrel, or tumbling barrel, utilizes the sliding movement of an upper layer of workload in the barrel, as shown in Fig. 1. The barrel is normally loaded about 60% full with a mixture of parts, medium, compound, and water. As the barrel rotates, the load moves upward to a turnover point; then the force of gravity overcomes the tendency of the mass to stick together, and the top layer slides toward the lower area of the barrel.

VIBRATORY FINISHING

A vibratory finishing machine is an open-topped tub or bowl mounted on springs, usually lined with polyurethane. Parts and medium are loaded in a fashion similar to that of a tumbling barrel. With a vibratory machine, the container can be almost completely filled. Vibratory action is created by a vibratory motor attached to the bottom of the container, by a shaft or shafts with eccentric loads driven by a standard motor, or by a system of electromagnets operating at 50 or 60 Hz. The action of the medium against the components takes place throughout the load, so that process cycles are substantially shorter than those of conventional tumbling in barrels.

Tub Vibrators. A tub vibrator consists of an open container, mounted on springs, whose cross section is in the shape of a "U" (see Fig. 2a) or a round inverted keyhole, or is a modification of one of these shapes.

Bowl Vibrators. A bowl vibrator (see Fig. 2b) is comprised of a round bowl-shape or doughnut-shape container. As in a tub vibrator, the container is mounted on springs. Vibratory action is imparted to the bowl by eccentric weights mounted on a vertical shaft at the center of the bowl. Eccentric weights at either end of the shaft are adjustable in their relationship to one another.

SPINDLE FINISHING MACHINES

Spindle finishing is categorized as a mass finishing process, although parts to be deburred or finished are mounted on fixtures. The process uses fine abrasive media for finishing. The spindle machine consists of a circular rotating tub which holds the abrasive medium, and a rotating or oscillating spindle to which the part is fixed. The workpiece mounted on the spindle is immersed into the rapidly moving abrasive slurry, causing the abrasive to flow swiftly over rough edges and over part surfaces. In some designs, the medium container is stationary, and the fixtures move the parts rapidly through the medium.

CENTRIFUGAL DISK FINISHING

The centrifugal disk process is a high-energy mass finishing process. The basic design (see Fig. 3) is a vertical cylinder with side walls which are

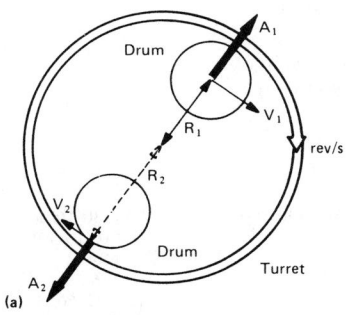
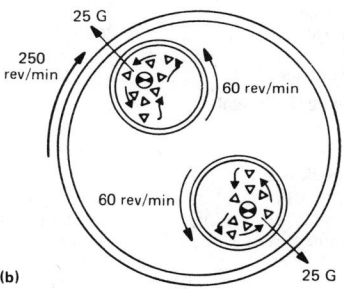

(a) Principle of centrifugal barrel process. (b) Action within a centrifugal barrel machine.
Fig. 4. Centrifugal barrel machines

stationary. The top of the cylinder is open. The bottom of the cylinder is formed by a disk which is driven to rotate at a high speed. Medium, compound and parts are contained in the cylinder. As the disk rotates with peripheral speeds of up to 10.2 m/s (2000 ft/min), the mass within the container is accelerated outward and then upward against the stationary side walls of the container, which act as a brake. The medium and parts rise to the top of the load and then flow in towards the center and back down to the disk.

CENTRIFUGAL BARREL FINISHING

Centrifugal barrel equipment (see Fig. 4) is comprised of containers mounted on the periphery of a turret. The turret rotates at a high speed in one direction while the drums rotate at a lower speed in the opposite direction. The drums are loaded in a manner similar to normal tumbling or vibratory operations with parts, medium, water, and some form of compound. Turret rotation creates a high centrifugal force, up to 100 times gravity. This force compacts the load within the drums into a tight mass. Rotation of the drums causes the medium to slide against the work load, to remove burrs and to refine surfaces.

Plating and Electropolishing

Edited by Donald F. Gagas, Centralab, Inc.

Copper Plating

COPPER can be electrodeposited from numerous electrolytes. Cyanide and pyrophosphate alkalines, plus sulfate and fluoborate acid baths, are the primary electrolytes used in copper plating.

High-Efficiency Sodium and Potassium Cyanide Baths. With proprietary additives, the high-concentration baths are used to produce deposits of various degrees of brightness and leveling, in thicknesses ranging from 8 to 50 μm (0.3 to 2.0 mils). Thick deposits that are ductile and bright can be produced in routine operations. Under most plating conditions, the high throwing power of the electrolyte produces adequate coverage in recessed areas. Antipitting additives are generally used in these baths to promote pore-free (nonpitted) deposits.

Current interruption is used frequently for operating high-efficiency electrolytes to produce greater leveling and uniform distribution of copper on complex shapes and to reduce plating time and the amount of metal required for plating complex shapes to a specified minimum thickness. Periodic reversal may be used to provide even higher leveling and better metal distribution than can be obtained with current interruption. Periodic reversal also improves the pore-filling characteristics of the high-efficiency electrolytes. Compositions and operating conditions of cyanide copper plating baths are given in Table 1.

ACID PLATING BATHS

Electrodeposition of copper from acid baths is used extensively for electroforming, electrorefining, manufacturing of copper powder, and decorative electroplating. Acid copper plating baths contain copper in the bivalent form and are more tolerant of ionic impurities than alkaline baths, but have less macrothrowing power and poorer metal distribution. Acid baths have excellent microthrowing power, which can be effective in sealing porous die castings.

Copper Sulfate Bath. The copper sulfate bath is the most frequently used of the acid copper electrolytes and is used primarily in electroforming.

Copper Fluoborate Bath. The copper fluoborate bath produces high-speed plating and dense deposits to any required thickness, usually up to 500 μm (20 mils). This bath is simple to prepare, stable, and easy to control. Concentration limits and operating conditions of acid copper plating baths are given in Table 2.

Hard Chromium Plating

HARD CHROMIUM PLATING is produced by electrodeposition from a solution containing chromic acid (CrO_3) and a catalytic anion in proper proportion. The metal so produced is extremely hard and corrosion resistant.

The process is used for rebuilding mismachined or worn parts, for automotive valve stems, piston rings, shock rods, MacPherson struts, the bores of diesel and aircraft cylinders, and for hydraulic shafts.

Hard chromium plating is also known as industrial, functional or engineering chromium plating.

PRINCIPAL USES

Wear Resistance. Extensive performance data indicate the effectiveness of chromium plate in reducing wear of piston rings due to scuffing and abrasion. The average life of a chromium-plated ring is approximately five times that of an unplated ring made of the same basis metal. Piston rings for most engines have a chromium plate thickness of 100 to 175 μm (4 to 7 mils) on the wearing face, although thicknesses up to 250 μm (10 mils) are specified for some heavy-duty engines.

Tooling Applications. Various types of tools are plated with chromium for one or more of the following reasons: (a) to minimize wear, (b) to prevent seizing and galling, (c) to reduce friction and (d) to prevent or minimize corrosion. Steel or beryllium copper dies for molding of plastics are usually plated with chromium.

Worn gages can be salvaged by being built up with hard chromium plate. Also, chromium plate provides steel gages with good protection against

Table 1. Compositions and operating conditions of cyanide copper plating baths

Constituent or condition	Dilute cyanide	Standard barrel	Rochelle cyanide — Low concentration(a)	High concentration(a)	High-efficiency — Sodium cyanide(b)	Potassium cyanide(b)
Bath composition, g/L (oz/gal)						
Copper cyanide	22 (3)	45 (6)	26 (4)	60 (8)	80 (11)	80 (11)
Sodium cyanide	33 (4)	68 (9)	35 (5)	80 (11)	105 (14)	105 (14)
Sodium carbonate	15 (2)	. . .	30 (4)	30 (4)	. . .	. . .
Sodium hydroxide	To pH	. . .	To pH	To pH	30 (4)	. . .
Rochelle salt	. . .	45-75 (6-10)	45 (6)	90 (12)	. . .	. . .
Potassium hydroxide	. . .	8-15 (1-2)	. . .	. . .	. . .	35 (5)
Bath analysis, g/L (oz/gal)						
Copper	16 (2)	32 (4)	18 (2)	43 (6)	56 (7)	56 (7)
Free cyanide	9 (1)	18 (2)	7 (0.8)	15 (2)	18 (2)	18 (2)
Operating conditions						
Temperature, °C (°F)	30-50 (86-120)	55-70 (130-160)	55-70 (130-160)(c)	60-75 (140-170)	60-75 (140-170)	60-75 (140-170)
Current density, A/dm² (A/ft²)	1.0-1.5 (10-15)	. . .	1.0-4.0 (10-40)	2.0-5.0 (20-50)	2.0-6.0 (20-60)	2.0-6.0 (20-60)
Cathode efficiency, %	30-50	. . .	40-60	60-90	70-100	70-100
Voltage, V	6	6(d)	6	6	6	6
pH	12.0-12.6	. . .	12.0-12.6(c)	13	>13	>13
Anodes	Copper, steel	Copper	Copper	Copper	Copper	Copper

(a) Low concentration typical for strike; high concentration typical for plating. (b) Used with addition agents, as proprietary or patented processes. (c) For zinc-base die castings, maintain temperature at 60 to 71 °C (140 to 160 °F) and pH between 11.6 and 12.3. (d) At 6 V, the bath draws approximately 0.3 A/L (2 A/gal) through the solution. At 12 V, the bath draws 0.4 A/L (3 A/gal).

Table 2. Compositions and operating conditions of acid copper plating baths

Constituent or condition	Copper sulfate bath		Copper fluoborate bath	
	General	Printed circuit through-hole	Low copper	High copper
Bath composition, g/L (oz/gal)				
Copper sulfate, $CuSO_4 \cdot 5H_2O$	200-240 (27-32)	60-110 (8-15)	...	...
Sulfuric acid, H_2SO_4	45-75 (6-10)	180-260 (24-35)	...	...
Copper fluoborate, $Cu(BF_4)_2$	...	...	225 (30)	450 (60)
Fluoboric acid, HBF_4	...	...	To pH	40 (5)
Bath analysis, g/L (oz/gal)				
Copper	50-60 (7-8)	15-28 (2-4)	8 (1)	16 (2)
Sulfuric acid	45-75 (6-10)	180-260 (24-35)	...	...
Specific gravity at 25 °C (77 °F)	...	...	1.17-1.18	1.35-1.37
Operating conditions				
Temperature, °C (°F)	20-50 (68-120)	20-40 (68-105)	20-70 (68-160)	20-70 (68-160)
Current density, A/dm^2 (A/ft^2)	2.0-10.0 (20-100)	0.1-6.0 (1-60)	7.0-13.0 (70-130)	12-35 (120-350)
Cathode efficiency, %	95-100	95-100	95-100	95-100
Voltage, V	6	6	6	6-12
pH	...	...	0.8-1.7	<0.6
Anodes	Copper(a)	Copper(a)	Copper(b)	Copper(b)

(a) Phosphorized copper is recommended. (b) High-purity, oxygen-free, nonphosphorized copper is recommended.

Table 3. Sulfate baths for hard chromium plating

Type of bath	Chromic acid(a)		Sulfate(a)		Current density		Bath temperature	
	g/L	oz/gal	g/L	oz/gal	A/dm^2	$A/in.^2$	°C	°F
Low concentration	250	33	2.5	0.33	31-62	2-4	52-63	125-145
High concentration	400	53	4.0	0.53	16-54	1-3.5	43-63	110-145

(a) Concentration usually can deviate ±10% without creating problems. It is recommended that adjustments be such that the concentrations listed above lie in the middle of the range permitted. For example, chromic acid can fluctuate by ±23 g/L (±3 oz/gal); therefore the concentration range should be 225 to 270 g/L (30 to 36 oz/gal), rather than 205 to 250 g/L (27 to 33 oz/gal).

Table 4. Rates of deposition of hard chromium from low-concentration baths

Thickness of plate		Plating time, h:min, at current density of:		
µmm	mils	31 A/dm^2 (2.0 $A/in.^2$)	47 (3.0)	62 (4.0)
Conventional sulfate bath(a)				
25	1	1:05	0:40	0:25
50	2	2:05	1:20	0:55
125	5	5:20	3:20	2:20
Mixed catalyst bath(b)				
25	1	0:50	0:30	0:20
50	2	1:40	1:00	0:40
125	5	4:05	2:25	1:45

(a) Bath containing 250 g/L (33 oz/gal) of chromic acid and with 100-to-1 ratio of chromic acid to sulfate, operated at 54 °C (130 °F). (b) Proprietary bath containing 250 g/L (33 oz/gal) of chromic acid, operated at 54 °C (130 °F).

rusting in normal exposure and handling.

Deep drawing tools often are plated with chromium, in thicknesses up to 100 µm (4 mils), for improvement of tool performance or building up of worn areas, or for both reasons.

PLATING BATHS

Chromic acid is the source of metal in hard chromium-plating baths. However, a chromic acid solution does not deposit chromium unless a definite amount of catalyst is present. If there is either too much or too little catalyst, no chromium metal is deposited. Catalysts that have proved successful are acid anions, the first of which to be used was sulfate.

CONVENTIONAL SULFATE BATHS

Composition of conventional chromic acid baths catalyzed by sulfate can vary widely, provided that the ratio by weight of chromic acid to sulfate radical is within a range between 75 to 1 and 120 to 1. Throwing power, or distribution of plate, is optimum at ratios between 90 to 1 and 110 to 1; however, in the range from 75 to 1 to 90 to 1, brighter deposits are obtained, less burning occurs and a higher current density can be used.

Solutions containing chromic acid in concentrations as low as 50 g/L (7 oz/gal) have been reported but are not practical for production for several reasons: (a) plating range is too limited, (b) such solutions are more sensitive to contamination, (c) they have higher electrical resistance and (d) they require higher voltage for operation. Compositions and operating conditions for two chromic acid–sulfate baths (low and high concentrations) for hard chromium plating are given in Table 3.

PROCESS CONTROL

In addition to bath composition, the principal variables that must be controlled for satisfactory hard chromium plating are (a) anodes, (b) current density and (c) bath temperature.

Anodes. In contrast to other plating baths, which use soluble anodes to supply the bath with a large part of the metal ion being plated, chromium-plating baths are operated with insoluble lead alloy anodes. Therefore, additions of chromic acid must be made as required to keep the chromium-plating bath supplied with chromium metal ions.

Current Density and Efficiency. Cathode current efficiency varies with current density and temperature of the plating bath. Efficiency increases with increasing current density and decreasing temperature. These two variables have a definite effect on appearance and hardness of the deposit. A high bath temperature results in a milky, dull, softer deposit at lower current efficiencies, unless the current density is increased substantially. Raising current density causes the deposit to change successively at specific temperatures. Because tank time is an important economic factor, the highest rates of deposition that are produced by the highest available currents may determine which plating bath temperature is most useful.

Deposition Rates. Times required for plating hard chromium deposits of various thicknesses are shown as a function of current density in Table 4 (for low-concentration baths) and Table 5 (for high-concentration baths).

Bath temperature affects both the conductivity and the current required. If limited power is available, satisfactory hard chromium plating can be obtained at lower temperatures (43 to 49 °C, or 110 to 120 °F); but if power supply is adequate, it is advantageous to work at higher temperatures up to 66 °C (150 °F), because of the faster deposition rate and the improved durability of the deposit.

Tanks and Linings. Figure 1 illustrates a hard chromium plating tank arrangement. Most tanks for chromium plating are made of steel and lined with an acid-resisting material. Because of their excellent resistance to corrosion by chromic acid, lead alloys containing antimony or tin may be used as tank linings.

Table 5. Rates of deposition of hard chromium from high-concentration baths

Thickness of plate		Plating time, h:min, at current density of:				
µm	mils	23 A/dm^2 (1.5 $A/in.^2$)	31 A/dm^2 (2.0 $A/in.^2$)	39 A/dm^2 (2.5 $A/in.^2$)	47 A/dm^2 (3.0 $A/in.^2$)	54 A/dm^2 (3.5 $A/in.^2$)
Conventional sulfate bath(a)						
25	1	2:20	1:35	1:15	0:55	0:45
50	2	4:35	3:10	2:30	1:55	1:30
125	5	11:30	8:00	6:15	4:40	3:50
255	10	23:00	16:00	12:30	9:25	7:35
380	15	34:30	24:00	18:45	14:05	11:25
510	20	46:00	32:00	25:00	18:50	15:10
Mixed catalyst bath(b)						
25	1	1:25	0:55	0:45	0:35	0:25
50	2	2:50	1:50	1:25	1:05	0:50
125	5	7:00	4:40	3:35	2:45	2:10
255	10	14:00	9:20	7:10	5:25	4:20
380	15	21:00	14:00	10:45	8:10	6:25
510	20	8:00	18:40	14:20	10:55	8:35

(a) Bath containing 400 g/L (53 oz/gal) of chromic acid with 100-to-1 ratio of chromic acid to sulfate, operated at 54 °C (130 °F). (b) Bath containing 400 g/L (53 oz/gal) of chromic acid, 1.5 g/L (0.20 oz/gal) of sulfate, and sufficient fluoride catalyst to give 100-to-1 ratio results; operating temperature, 54 °C (130 °F).

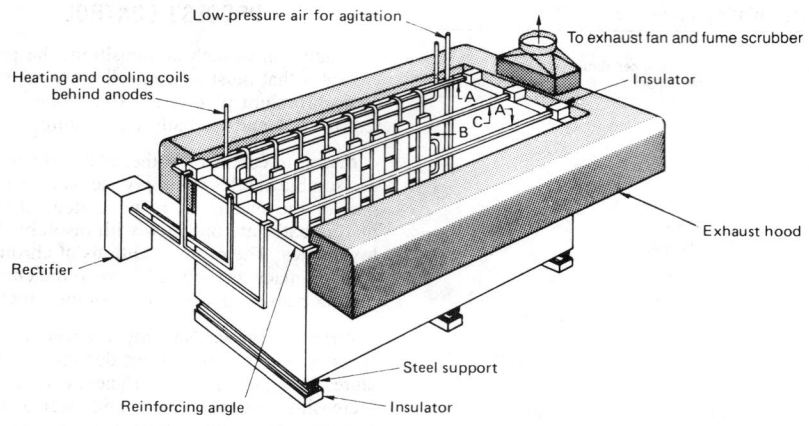

A: anode rods; B: lead or lead-tin anodes; C: cathode rod.
Fig. 1. Tank and accessory equipment used for hard chromium plating

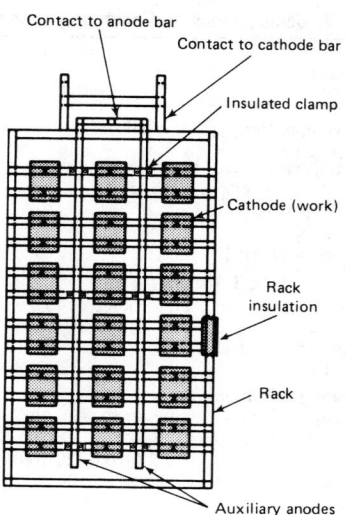

Shows position of cathodes and auxiliary anodes.
Fig. 2. Rack assembly for decorative chromium plating

Decorative Chromium Plating

DECORATIVE CHROMIUM PLATING is differentiated from hard chromium plating by thickness and by the type of undercoating used. Decorative chromium coatings are very thin, usually not exceeding an average thickness of 1.25 μm (50 μin.). Decorative chromium is applied over undercoatings, such as nickel or copper plus nickel, which impart a bright, semibright or satin cosmetic appearance to the chromium. The choice of undercoatings, as well as the type of chromium applied, can also provide corrosion protection. Currently, most decorative chromium coatings are applied from hexavalent chromium processes, based on chromic anhydride (CrO_3).

CHROMIUM BATH COMPOSITION

Sometimes referred to as the ordinary or conventional bath, the oldest chromium plating bath used for decorative plating consists of an aqueous solution of chromic anhydride (CrO_3) that also contains a small amount of soluble sulfate ($SO_4^=$), referred to as a catalyst, added as sulfuric acid or as a soluble sulfate salt such as sodium sulfate. When dissolved in water, the chromic anhydride forms chromic acid, which is believed to exist in the following equilibrium:

$$H_2Cr_2O_7 + H_2O \rightleftarrows 2H_2CrO_4$$

Most chromium is deposited within the following operating limits:

Chromic anhydride .... 200 to 400 g/L
(27 to 54 oz/gal)
Chromic anhydride-
to-sulfate ratio .... 80:1 to 125:1
Cathode current
density .... 7.5 to 17.5 A/dm²
(75 to 175 A/ft²)

With the development of duplex, microcracked and crack-free applications, specialized bath compositions and operating conditions have come into use. Many of these, however, are either proprietary or are not subjects of general agreement. Table 6 provides information for a general decorative chromium-plating bath and a bright crack-free bath.

TEMPERATURE OF CHROMIUM BATHS

Chromium plating is performed within the range from 38 to 60 °C (100 to 140 °F); 46 to 52 °C (115 to 125 °F) is the most common operating range. At room temperature, the bright plating range is impractically narrow.

CURRENT DENSITY IN CHROMIUM PLATING

The standard sulfate bath is usually operated in the range from 10 to 16 A/dm² (100 to 160 A/ft²). A current density of about 10 A/dm² (100 A/ft²) is used for solutions maintained at 38 °C (100 °F). A higher current density, sometimes as high as 30 A/dm² (300 A/ft²), is required for solutions at 55 °C (130 °F). The choice for a specific use depends on such variables as the complexity of the article being plated, and the equipment available. After the current density has been established, close control must be maintained.

ANODES FOR CHROMIUM PLATING

In chromium plating, insoluble lead or lead-alloy anodes are almost always used. Chromium metal is supplied by chromic acid in the electrolyte.

Anodes with round cross sections are most commonly used. When maximum anode area is desired, corrugated, ribbed, ridged and multi-edged anodes are used. The round anode is preferred because its surface is active on its entire circumference, enabling it to carry higher amperage at lower voltage.

Auxiliary anodes are mounted on the plating rack, insulated from the cathode current-carrying members and provided with means of direct connection to the anodic side of the electrical circuit (Fig. 2).

EQUIPMENT FOR CHROMIUM PLATING

Tanks for chromium plating may be constructed of steel and lined with one of the following materials:

• Flexible plastic-type materials, such as fiber glass or polyvinyl chloride, either in sheet form or sprayed on
• Lead alloy (6% antimony)
• Acid-resistant, high-temperature baked brick or tile, set in a silica cement.

Heating. Chromium plating tanks may be heated internally or externally. Internal heating, by steam coils or electric immersion heaters, is usually used for small tanks; external heating, by heat exchangers, is used for large tanks.

Nickel Plating

ELECTRODEPOSITS of nickel exhibit a wide variety of properties, depending on composition of the plating bath and operating conditions. They may be classified according to application or appearance as general-purpose, special-purpose, black, and bright.

General-purpose nickel deposits, produced by Watts, sulfamate and fluoborate baths, are essentially sulfur-free. They are used primarily to protect alloys based on iron, copper or zinc against corrosive attack in rural, marine and industrial atmospheres.

GENERAL-PURPOSE PLATING BATHS

Table 7 lists ranges of compositions and operating conditions for three general-purpose nickel plating baths (Watts, sulfamate and fluoborate) and indicates mechanical properties of deposits produced by these solutions. These baths are sometimes referred to as gray nickel baths.

The Watts bath and its higher nickel chloride modification (Table 7) are the standard general-purpose nickel plating solutions. The major por-

Table 6. Compositions and operating conditions for two chromium plating baths

Constituent or condition	General decorative	Bright crack-free
Chromic acid content	250 g/L (33 oz/gal)	260-300 g/L (35-40 oz/gal)
Ratio of chromic acid to sulfate	100:1 to 125:1	150:1
Operating temperature	38-49 °C (100-120 °F)	52-54 °C (125-130 °F)
Cathode current density	7.5-17.5 A/dm² (75-175 A/ft²)	25-30 A/dm² (250-300 A/ft²)

Table 7. General-purpose nickel plating baths
Compositions, operating conditions and typical mechanical properties of deposits.

Constituent condition or property	Watts bath(a)	Sulfamate bath	Fluoborate bath
Composition			
Nickel sulfate, $NiSO_4 \cdot 6H_2O$	225-410 g/L (30-55 oz/gal)	...	...
Nickel chloride, $NiCl_2 \cdot 6H_2O$	30-60 g/L (4-8 oz/gal)(a)	0-30 g;/L (0-4 oz/gal)	0-15 g/L (0-2 oz/gal)
Nickel sulfamate, $Ni(SO_3NH_2)_2$	...	263-450 g/L (35-60 oz/gal)	...
Nickel fluoborate, $Ni(BF_4)_2$	...	...	225-300 g/L (30-40 oz/gal)
Total nickel as metal	55-105 g/L (7.7-14.2 oz/gal)	62-113 g/L (8.2-15 oz/gal)	55-80 g/L (7.6-10.5 oz/gal)
Boric acid, H_3BO_3	30-45 g/L (4-6 oz/gal)	30-45 g/L (4-6 oz/gal)	15-30 g/L (2-4 oz/gal)
Antipitting additives	(b)	(b)	(b)
Operating conditions			
pH	15-5.2	3-5	2.5-4
Temperature	46-71 °C (115-160 °F)	38-60 °C (100-140 °F)	38-71 °C (100-160 °F)
Current density	1-10 A/dm² (10-100 A/ft²)	2.5-30 A/dm² (25-300 A/ft²)	2.5-30 A/dm² (25-300 A/ft²)
Mechanical properties of deposits(c)			
Tensile strength	345-690 MPa (50-100 ksi)	380-1070 MPa (55-155 ksi)	380-830 MPa (55-120 ksi)
Vickers hardness	100-250 HV	130-600 HV	125-300 HV
Elongation in 50 mm (2 in.)	10-35%	3-30%	5-30%
Stress	105-205 MPa (15-30 ksi)	3-110 MPa (0.5-16 ksi)	90-205 MPa (13-30 ksi)
References			
(See list following footnotes)	6, 7	1, 2, 3, 6	4, 5, 6, 7

(a) A high-chloride modification of the Watts bath may be made by increasing content of nickel chloride to 113 g/L (15 oz/gal). (b) In many instances, antipitting additives are not required if solution is pure. When required, either of the following may be added: 0.03 to 0.075 mL/L (0.1 to 0.25 mL/gal) of solution (not to be added to solutions containing wetting agents or organic stress-reducers), or a surfactant and wetting agent of approved type to give 50 to 32 dynes/cm. (c) Wide ranges of values reflect variations in composition and degree of purity of plating baths.
References: (1) R. Barrett, *Plating*, Vol 41, 1954, p 1027. (2) M. Diggin, *Metal Progress*, Vol 66, Oct 1954, p 132. (3) D. Fanner and R. Hammond, *Trans Inst Met Fin*, Vol 36, 1958-1959, p 32. (4) C. Struyk and A. Carlson, *Plating*, Vol 37, 1950, p 1242. (5) E. Roehl and W. Wesley, *Plating*, Vol 37, 1950, p 142. (6) A. K. Graham, *Electroplating Engineering Handbook*, Reinhold Publishing Corp., New York, 1962. (7) AES Research Report No. 20.

tion of the nickel-ion content is contributed by the relatively inexpensive nickel sulfate; a high nickel sulfate concentration is used when high current densities are required.

The sulfamate bath (Table 7) is a general-purpose bath that yields deposits of low stress, has a wide operating range, and is easy to control. Because of the very high solubility of nickel sulfamate, maintaining a higher nickel-metal concentration in this solution than in other nickel baths is possible, permitting the use of lower operating temperatures and higher plating rates. A small amount of nickel chloride is usually added to the bath to minimize anode passivity, especially when higher plating rates are used.

The fluoborate bath (Table 7) can be operated over wide ranges of nickel concentration, temperature and current density. However, because the bath is well buffered, the maximum pH is limited, even for high-pH treatments. The reactivity of the fluoborate ion with some materials of construction requires consideration. Silica filter aids cannot be used on a continuous basis, although cellulose filters are satisfactory. Lead, titanium, and high-silicon cast iron are readily attacked. Stainless steels containing 20% Cr, 25 to 30% Ni and 2 to 3% Mo are suitably resistant.

BLACK NICKEL PLATING BATHS

Black nickel deposits are used primarily for decorative effect and to provide nonreflecting surfaces. Uses include (a) typewriter and camera parts, (b) military instruments, (c) clothes fasteners and (d) costume jewelry. Deposits are brittle and readily chip or flake on bending or impact; thus, deposits usually are not permitted to exceed 1.0 to 1.5 μm (0.04 to 0.06 mil) in thickness.

There are several successful compositions for producing black nickel deposits; all incorporate zinc (Zn^{++}) and thiocyanate (CNS^-) ions. Table 8 gives the compositions and operating conditions for a sulfate and a chloride black nickel plating bath. The sulfate bath is more commonly used, but the chloride formula is said to permit higher current densities and a wider pH range, enhancing adhesion by permitting the use of an acid–nickel chloride strike pretreatment without rinsing.

BRIGHT PLATING BATHS

Bright nickel plating baths are modifications of the Watts nickel solution and contain organic or combined organic and inorganic brightening agents; these additions serve to produce a high degree of brightness, leveling reflectivity, and hardness. A great variety of additives is used, usually in highly specific combinations. Their function is to produce as brilliant and ductile a deposit as possible over a wide range of current densities and operating conditions. Some of these additives are consumed very slowly during electrolysis; others are consumed more rapidly.

ANODES

A chill cast anode containing 99% nickel came into acceptance with the widespread use of the Watts bath; subsequently, a rolled, depolarized anode was developed that contained small amounts of nickel oxide but retained a 99% nickel content. Rolled depolarized anodes are used in high-pH baths, although they may be used throughout the entire pH range of nickel plating solutions. To avoid the formation of rough deposits, these anodes usually are covered with cotton or synthetic fiber anode bags while in use.

Cast or rolled carbon anodes (99% nickel anodes containing about 0.2% carbon) may be used in baths with a pH of 4.5 or less. Although they are capable of forming an adherent carbon-silica film that retains loose anode particles, they are covered that with anode bags during use to prevent the formation of rough deposits.

Virtually all nickel anodes made today are of high purity and contain a minimum of 99% nickel.

BATH TEMPERATURE

Variations in the operating temperature of a nickel plating bath can have a marked effect on the properties of the electrodeposited nickel. To obtain consistent results, the temperature of a nickel plating bath should be maintained within ±2 °C (±4 °F) of the recommended temperature for a given application. In general, most industrial nickel electroplating baths are operated in the range from 38 to 60 °C (100 to 140 °F).

CURRENT DENSITY

Without agitation, Watts nickel plating solutions generally are used at average current densities in the range from 0.5 to 3.5 A/dm² (5 to 35 A/ft²). With mechanical agitation this range can be increased significantly depending on the rate of agitation. When these baths are operated at room temperature, plating must be done at very low current densities—usually less than 1.0 A/

Table 8. Compositions and operating conditions of two black nickel plating baths

Constituent or condition	Sulfate bath	Chloride bath
Composition		
Nickel sulfate, $NiSO_4 \cdot 6H_2O$	75 g/L (10 oz/gal)	...
Nickel chloride, $NiCl_2 \cdot 6H_2O$	...	75 g/L (10 oz/gal)
Zinc sulfate, $ZnSO_4 \cdot 7H_2O$	30 g/L (4 oz/gal)	...
Zinc chloride, $ZnCl_2$	...	30 g/L (4 oz/gal)
Ammonium sulfate, $(NH_4)_2SO_4$	35 g/L (5 oz/gal)	...
Ammonium chloride, NH_4Cl	...	30 g/L (4 oz/gal)
Sodium thiocyanate, NaSCN	15 g/L (2 oz/gal)	15 g/L (2 oz/gal)
Operating conditions		
pH	5.6	5.0
Temperature	21-24 °C (70-75 °F)	21-24 °C (70-75 °F)
Current density	0.15 A/dm² (1.5 A/ft²)	0.15-0.6 A/dm² (1.5-6 A/ft²)

dm² (10 A/ft²). The pH can have a significant influence on the current densities that can be used. From pH 4.8 to 5.4, midrange current densities must be used. Low-pH solutions may be used at current densities from the middle to high end of the range.

PLATING EQUIPMENT

Nickel can be deposited in still tanks, plating barrels, and a variety of automatic equipment. The design of such equipment need not vary significantly from that of the conventional plating equipment used in depositing other widely used metals, including copper, cadmium and zinc. Significant differences, where they occur, are likely to be found in the materials used and the means provided for filtering, agitating and heating the plating solutions.

Tank Linings. Most tanks used for nickel plating are made of steel and are lined with hard rubber or polyvinyl chloride. The lining protects the steel from attack by the plating bath, minimizing contamination of the solution, and prevents some bipolar effects. Plastisol-coated steel grids or heavy rubber mats are sometimes placed in the bottom of the tanks to prevent puncture of the linings by workpieces, racks, anodes or other objects inadvertently dropped into the tank. Brick bottoms, with the bricks placed loosely, are also used.

Plating Barrels. Many material systems have been used for the construction of nickel plating barrels, including rubber-covered steel. Recent developments in plastic materials have made available several compositions with excellent mechanical properties over a wide temperature range, and resistance to many corrosive chemicals. Of the more common barrel materials, hard rubber is acceptable for resistance to heat and chemical corrosion.

Electroless Nickel Plating

ELECTROLESS NICKEL PLATING is used to deposit nickel without the use of an electric current. The coating is deposited by an autocatalytic chemical reduction of nickel ions by hypophosphite, aminoborane or borohydride compounds.

Electroless nickel is an engineering coating, normally used because of excellent corrosion and wear resistance. Electroless nickel coatings are also frequently applied on aluminum to provide a solderable surface and are used with molds and dies to improve lubricity and part release. Because of these properties, electroless nickel coatings have found many applications, including those in petroleum, chemicals, plastics, optics, printing, mining, aerospace, nuclear, automotive, electronics, computers, textiles, paper, and food machinery. Some advantages and limitations of electroless nickel coatings include:

Advantages

- Good resistance to corrosion and wear
- Excellent uniformity
- Solderability and brazability
- Low labor costs

Limitations

- High chemical cost
- Brittleness

- Poor welding characteristics of nickel phosphorus deposits
- Need to copper strike plate alloys containing significant amounts of lead, tin, cadmium and zinc before electroless nickel can be applied
- Slower plating rate, as compared with electrolytic methods.

BATH COMPOSITION AND CHARACTERISTICS

Electroless nickel coatings are produced by the controlled chemical reduction of nickel ions onto a catalytic surface. The deposit itself is catalytic to reduction, and the reaction continues as long as the surface remains in contact with the electroless nickel solution. Because the deposit is applied without an electric current, its thickness is uniform on all areas of an article in contact with fresh solution.

Early electroless nickel formulations were ammoniacal and operated at high pH. Later acid solutions were found to have several advantages over alkaline solutions. Among these are (a) higher plating rate, (b) better stability, (c) greater ease of control and (d) improved deposit corrosion resistance. Accordingly, most hypophosphite reduced electroless nickel solutions are operated between 4 and 5.5 pH. Compositions of alkaline and acid plating solutions are listed in Table 9.

Aminoborane Baths. The use of aminoboranes in commercial electroless nickel plating solutions

has generally been limited to two compounds: (a) N-dimethylamine borane (DMAB)—(CH₃)₂ NHBH₃, and (b) N-diethylamine borane (DEAB)—(C₂H₅)₂ NHBH₃. Compositions and operating conditions for aminoborane baths are listed in Table 10.

Sodium Borohydride Baths. The borohydride ion is the most powerful reducing agent available for electroless nickel plating. Any water-soluble borohydride may be used, although sodium borohydride is preferred. Compositions of borohydride-reduced electroless nickel baths are also shown in Table 10.

Energy

The amount of energy or heat present in an electroless nickel solution is one of the most important variables affecting coating deposition. In a plating bath, temperature is a measure of its energy content.

Temperature has a strong effect on the deposition rate of acid hypophosphite-reduced solutions. The rate of deposition is usually very low at temperatures below 65 °C (150 °F), but increases rapidly with increased temperature. This is illustrated in Fig. 3, which gives the results of tests conducted using bath 3 in Table 9.

Complexing Agents

To avoid spontaneous decomposition of electroless nickel solutions and to control the reaction so that it occurs only on the catalytic sur-

Table 9. Hypophosphite-reduced electroless nickel plating solutions

Constituent or condition	Alkaline			Acid		
	Bath 1	Bath 2	Bath 3	Bath 4	Bath 5	Bath 6
Composition						
Nickel chloride, g/L (oz/gal)	45 (6)	30 (4)	30 (4)	...	...	...
Nickel sulfate, g/L (oz/gal)	...	...	...	21 (2.8)	34 (4.5)	45 (6)
Sodium hypophosphite, g/L (oz/gal)	11 (1.5)	10 (1.3)	10 (1.3)	24 (3.2)	35 (4.7)	10 (1.3)
Ammonium chloride, g/L (oz/gal)	50 (6.7)	50 (6.7)	...	...	...	...
Sodium citrate, g/L (oz/gal)	100 (13.3)	...	...	...	...	...
Ammonium citrate, g/L (oz/gal)	...	65 (8.6)	...	...	...	...
Ammonium hydroxide,	To pH	To pH	...	...	...	...
Lactic acid, g/L (oz/gal)	...	...	...	28 (3.7)	...	...
Malic acid, g/L (oz/gal)	...	...	...	...	35 (4.7)	...
Amino-acetic acid, g/L (oz/gal)	...	...	...	...	...	40 (5.3)
Sodium hydroxy-acetate, g/L (oz/gal)	...	...	10 (1.3)	...	...	...
Propionic acid, g/L (oz/gal)	...	...	...	2.2 (0.3)	...	...
Acetic acid, g/L (oz/gal)	...	...	...	...	...	10 (1.3)
Succinic acid, g/L (oz/gal)	...	...	...	...	10 (1.3)	...
Lead, ppm	...	...	...	1	...	...
Thiourea, ppm	...	...	...	...	1	...
Operating conditions						
pH	8.5-10	8-10	4-6	4.3-4.6	4.5-5.5	4.5-5.5
Temperature, °C (°F)	90-95 (195-205)	90-95 (195-205)	88-95 (190-205)	88-95 (190-205)	88-95 (190-205)	88-95 (190-205)
Plating rate, μm/h (mil/h)	10 (0.4)	8 (0.3)	10 (0.4)	25 (1)	25 (1)	25 (1)

Table 10. Aminoborane and borohydride-reduced electroless nickel plating solutions

Constituent or condition	Aminoborane		Borohydride	
	Bath 7	Bath 8	Bath 9	Bath 10
Composition				
Nickel chloride, g/L (oz/gal)	30 (4)	24-48 (3.2-6.4)	...	20 (2.7)
Nickel sulfate, g/L (oz/gal)	...	...	50 (6.7)	...
DMAB, g/L (oz/gal)	...	3-4.8 (0.4-0.64)	3 (0.4)	...
DEAB, g/L (oz/gal)	3 (0.4)	...	...	...
Isopropanol, mL (fluid oz)	50 (1.7)	...	...	...
Sodium citrate, g/L (oz/gal)	10 (1.3)	...	...	...
Sodium succinate, g/L (oz/gal)	20 (2.7)	...	...	...
Potassium acetate, g/L (oz/gal)	...	18-37 (2.4-4.9)	...	...
Sodium pyrophosphate, g/L (oz/gal)	...	...	100 (13.3)	...
Sodium borohydride, g/L (oz/gal)	...	...	...	0.4 (0.05)
Sodium hydroxide, g/L (oz/gal)	...	...	...	90 (12)
Ethylene diamine, 98%, g/L (oz/gal)	...	...	...	90 (12)
Thallium sulfate, g/L (oz/gal)	...	...	...	0.4 (0.05)
Operating conditions				
pH	5-7	5.5	10	14
Temperature, °C (°F)	65 (150)	70 (160)	25 (77)	95 (205)
Plating rate, μm/h (mil/h)	7-12 (0.5)	7-12 (0.5)	...	15-20 (0.6-0.8)

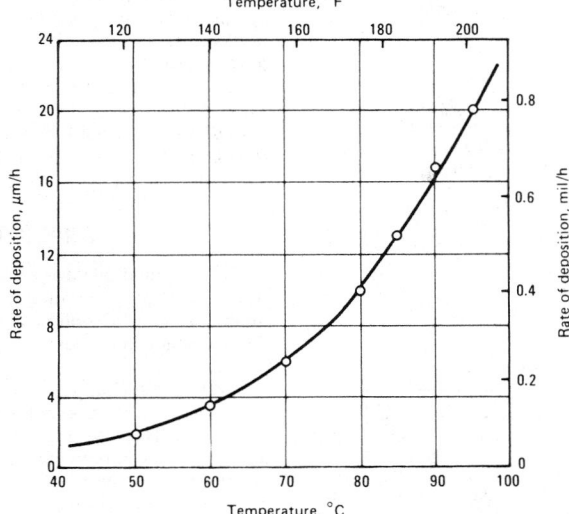

Tests conducted on bath 3 in Table 9 at 5 pH.
Fig. 3. Effect of solution temperature on rate of deposition

face, complexing agents are added. Complexing agents are organic acids or their salts, added to control the amount of free nickel available for reaction. They act to stabilize the solution and to retard precipitation of nickel phosphite.

Inhibitors

The amount of inhibitor used is critical. The presence of only about 1 mg/L (4 mg/gal) of HS⁻ ion completely stops deposition, whereas at a concentration of 0.01 mg/L (0.04 mg/gal), this ion is an effective inhibitor. Excess inhibitor absorbs preferentially at sharp edges and corners, resulting in incomplete coverage (edge pullback) and porosity.

EQUIPMENT FOR ELECTROLESS NICKEL PLATING

Because electroless nickel is applied by a chemical reaction rather than by electrolytic deposition, unique tanks and auxiliary equipment are required to ensure trouble-free operation and quality coatings.

Plating Tanks

Cylindrical or bell-shaped tanks have been used for electroless nickel plating, although rectangular tanks have been found to be the most convenient to build and operate. Rectangular tanks have been constructed from various materials in many different sizes.

Physical Dimensions. The following factors should be considered when selecting the size of an electroless nickel plating tank:

- Size of the part to be plated
- Number of parts to be plated each day
- Plating thickness required
- Plating rate of the solution (most conventional electroless nickel solutions deposit between 12 and 25 μm/h, or 0.5 and 1 mil/h)
- Type of rack, barrel or basket used to support parts
- Size of supporting rack, barrel or basket
- Number of production hours available each day to process parts
- Maximum recommended work load of 1.2 dm²/L (0.5 ft²/gal) of working solution

Construction Materials. The most widely used materials for tank construction have been polypropylene, stainless steel, and steel or aluminum with a 635 μm (25 mil) thick polyvinyl chloride bag liner. Although all of these materials have been used successfully, a 6 to 12 mm (0.25 to 0.5 in.) thick polypropylene liner, installed in a steel or fiberglass support tank, has proven to be the most trouble-free material and has gained the widest acceptance. Polypropylene is relatively inexpensive and is very resistant to plate-out. The smooth surface of polypropylene also reduces the possibility of deposit nucleation.

Zinc Plating

ZINC is anodic to iron and steel and therefore offers more protection when applied in thin films of 7 to 15 μm (0.3 to 0.5 mil) than similar thicknesses of nickel and other cathodic coatings, except in marine environments where it is surpassed by cadmium. Because it is relatively inexpensive and readily applied in barrel, tank or continuous plating facilities, zinc is often preferred for coating iron and steel parts when protection from either atmospheric or indoor corrosion is the primary objective.

CYANIDE ZINC BATHS

Bright cyanide zinc baths may be divided into four broad classifications based on their cyanide content: (a) regular cyanide zinc baths, (b) midcyanide or half-strength cyanide baths, (c) low-cyanide baths and (d) microcyanide baths. Table 11 gives the general compositions and operating conditions for these various systems.

Standard cyanide zinc baths provide excellent throwing and covering power. The ability of the standard cyanide zinc bath to cover at very low current densities is greater than that of any other zinc plating system. This capability depends on the bath composition, temperature, base metal, and proprietary additives used, but it is generally superior to the acid chloride systems.

Midcyanide Zinc Bath. In an effort to reduce cyanide waste as well as treatment and operating costs, most cyanide zinc baths are currently at the so-called midcyanide, half-strength, or dilute cyanide bath concentration, indicated in Table 11. Plating characteristics of midcyanide baths and regular cyanide baths are practically identical. The only drawback of the midcyanide bath when compared with the standard bath is a somewhat lower tolerance for impurities and poor preplate cleaning.

Low-cyanide zinc baths, which are generally defined as those baths operating at approximately 6 to 12 g/L (0.68 to 1.36 oz/gal) sodium cyanide and zinc metal, are substantially different in plating characteristics from the midcyanide and standard cyanide baths. The plating additives normally used in regular and midstrength cyanide baths do not function well with low metal and cyanide contents. Special low-cyanide brighteners have been developed for these baths.

Low-cyanide zinc baths are more sensitive to extremes of operating temperature than either regular or midcyanide baths. The efficiency of the bath may be similar to that of a regular cyanide bath initially, but tends to drop off more rapidly (especially at higher current densities) as the bath ages. These baths are used extensively for rack plating of wire goods. Unlike the other

Table 11. Compositions and operating conditions of cyanide zinc baths

Cathode current density: limiting, 0.002 to 25 A/dm² (0.02 to 250 A/ft²); average barrel, 0.6 A/dm² (6 A/ft²); average rack, 2.0 to 5 A/dm² (20 to 50 A/ft²). Bath voltage: 3 to 6 V, rack; 12 to 25 V, barrel.

Constituent	Standard cyanide bath(a) Optimum g/L	oz/gal	Range g/L	oz/gal	Midcyanide or half-strength cyanide bath(b) Optimum g/L	oz/gal	Range g/L	oz/gal
Preparation								
Zinc cyanide	61	8.1	54-86	7.2-11.5	30	4.0	27-34	3.6-4.5
Sodium cyanide	42	5.6	30-41	4.0-5.5	20	2.7	15-28	2.0-3.7
Sodium hydroxide	79	10.5	68-105	9.0-14.0	75	10.0	60-90	8.0-12.0
Sodium carbonate	15	2.0	15-60	2.0-8.0	15	2.0	15-60	2.0-8.0
Sodium polysulfide	2	0.3	2-3	0.3-0.4	2	0.3	2-3	0.3-0.4
Brightener	(g)	(g)	1-4	0.1-0.5	(g)	(g)	1-4	0.1-0.5
Analysis								
Zinc metal	34	4.5	30-48	4.0-6.4	17	2.3	15-19	2.0-2.5
Total sodium cyanide	93	12.4	75-113	10.0-15.1	45	6.0	38-57	5.0-7.6
Sodium hydroxide	79	10.5	68-105	9.0-14.0	75	10.0	60-90	8.0-12.0
Ratio: NaCN to Zn	2.75	0.37	2.0-3.0	0.3-0.4	2.6	0.3	2.0-3.0	0.2-0.4

Constituent	Low-cyanide bath(c) Optimum g/L	oz/gal	Range g/L	oz/gal	Microcyanide bath(d) Optimum g/L	oz/gal	Range g/L	oz/gal
Preparation								
Zinc cyanide	9.4(b)	1.3(e)	7.5-14(b)	1.0-1.9	(f)	(f)	(f)	(f)
Sodium cyanide	7.5	1.0	6.0-15.0	0.8-2.0	1.0	0.1	0.75-1.0	0.4-0.13
Sodium hydroxide	65	8.7	52-75	6.9-10.0	75	10.0	60-75	8-10
Sodium carbonate	15	2.0	15-60	2.0-8.0	...	...	...	...
Sodium polysulfide	...	...	...	...	...	...	...	...
Brightener	(g)	(g)	1-4	0.1-0.5	(g)	(g)	1-5	0.1-0.7
Analysis								
Zinc metal	7.5	1.0	...	0.8-1.5	7.5	1.0	6.0-11.3	0.8-1.5
Total sodium cyanide	7.5	1.0	6.0-15.0	0.8-2.0	1.0	0.1	0.75-1.0	0.1-0.13
Sodium hydroxide	75	10	60-75	8.0-10.0	75	10.0	60-75	8-10
Ratio: NaCN to Zn	1.0	0.1	1.0	0.1	...	...	...	...

(a) Operating temperature: 29 °C (84 °F) optimum; range of 21 to 40 °C (69 to 105 °F). (b) Operating temperature: 29 °C (84 °F) optimum; range of 21 to 40 °C (69 to 105 °F). (c) Operating temperature: 27 °C (79 °F) optimum; range of 21 to 35 °C (69 to 94 °F). (d) Operating temperature: 27 °C (79 °F) optimum; range of 21 to 35 °C (69 to 94 °F). (e) Zinc oxide. (f) Dissolve zinc anodes in solution until desired concentration of zinc metal is obtained. (g) As specified.

Table 12A. Compositions of cadmium plating cyanide solutions

Solution No.	Ratio of total sodium cyanide to cadmium metal	Cadmium oxide g/L	oz/gal	Cadmium metal g/L	oz/gal	Composition(a) Sodium cyanide g/L	oz/gal	Sodium hydroxide(b) g/L	oz/gal	Sodium carbonate(c) g/L	oz/gal
1	4:1	23	3	19.6	2.62	78	10.4	14.2	1.90	30-75	4-10
2	7:1	23	3	19.6	2.62	138	18.4	14.2	1.90	30-45	4-6
3	5:1	26	3.5	22.9	3.06	115	15.3	16.4	2.19	30-60	4-8
4	4.5:1	40	5.5	36.1	4.82	163	21.7	25.8	3.44	30-45	4-6

(a) Metal-organic agents are added to cyanide solutions to produce fine-grained deposits. The addition of excessive quantities of these agents should be avoided, because this will cause deposits to be of inferior quality and to have poor resistance to corrosion. The addition of these agents to solutions used for plating cast iron is not recommended. (b) Sodium hydroxide produced by the cadmium oxide used. In barrel plating, 7.5 g/L (1 oz/gal) is added for conductivity. (c) Sodium carbonate produced by decomposition of sodium cyanide and absorption of carbon dioxide, and by poor anode efficiency. Excess sodium carbonate causes anode polarization, rough coatings, and lower efficiency. Excess sodium carbonate may be reduced by freezing, or by treatment with calcium sulfate.

Table 12B. Operating conditions for cadmium plating cyanide solutions

Solution No.	Current density(a) Range A/dm²	A/ft²	Average A/dm²	A/ft²	Operating temperature °C	°F	Remarks
1	0.5-6	5-60	2.5	25	27-32	80-90	For use in still tanks. Good efficiency, fair throwing power. Also used in bright barrel plating
2	1-8	10-80	2.5	25	27-32	80-90	For use in still tanks and automatic plating. High throwing power, uniform deposits, fair efficiency. Not for use in barrel plating
3	0.5-9	5-90	3.5	35	24-29	75-85	Primarily for use in still tanks, but can be used in automatic plating and barrel plating. High efficiency and good throwing power
4	0.5-15	5-150	5.0	50	27-32	80-90	Used for plating cast iron. High speed and high efficiency(b)

(a) For uniform deposits from cyanide solutions, the use of a current density of at least 2 A/dm² (20 A/ft²) is recommended. Agitation and cooling of solution are required at high current densities. (b) Agitation and cooling are required when current density is high (above 2 A/dm², or 20 A/ft²).

cyanide systems, low-cyanide baths are quite sensitive to sulfide treatments to reduce impurities. Regular sulfide additions may be deleterious to the plating brightness produced (see Table 11).

Microcyanide zinc baths are essentially a retrogression from the alkaline noncyanide zinc process. Since the alkaline bath is often difficult to operate within its somewhat limited parameters, many platers use an absolute minimal amount of cyanide in these baths—1.0 g/L (0.13 oz/gal), for example. This acts essentially as an additive, increasing the over-all bright range of the baths and simplifying operation. It negates the purpose of the alkaline noncyanide bath, which is to totally eliminate cyanide, and should be viewed as an expedient crutch rather than a normally recommended zinc plating system (see Table 11).

Cadmium Plating

ELECTRODEPOSITS OF CADMIUM are used extensively to protect steel and cast iron against corrosion. Because cadmium is anodic to iron, the underlying ferrous metal is protected at the expense of the cadmium plate even if the cadmium becomes scratched or nicked, exposing the substrate.

Cadmium is usually applied as a thin coating (less than 25 μm or 1 mil thick) intended to withstand atmospheric corrosion.

PLATING BATHS

Most cadmium plating is done in cyanide baths, which generally are made by dissolving cadmium oxide in a sodium cyanide solution. Sodium cyanide provides conductivity and makes the corrosion of the cadmium anodes possible.

Cyanide Baths. Compositions and operating conditions of four cyanide baths are given in Tables 12A and 12B. Note that for each of these baths a ratio of total sodium cyanide to cadmium metal is indicated; maintenance of the recommended ratio is important to the operating characteristics of the bath.

For still tank or automatic plating of steel, selection of a bath on the basis of cyanide-to-metal ratio depends on the type of work being plated and the results desired:

- For parts with no recesses and when protection of the basis metal is the sole requirement, solution 1 in Table 12A (ratio, 4 to 1) is recommended
- For plating parts with deep recesses and when a bright, uniform finish is required, solution 2 in Table 12A (ratio, 7 to 1) is recommended
- For all-purpose bright plating of various shapes, solution 3 in Table 12A (ratio, 5 to 1) is recommended
- For high-speed, high-efficiency plating, solution 4 in Table 12A (ratio, 4.5 to 1) is recommended.

Although the use of brighteners produces maximum improvement in uniformity and throwing power in solution 3 in Table 12A, brighteners also improve these properties in solutions 1 and 2.

Normally, the sodium hydroxide content of cyanide baths is not critical. Usual limits are 22 g/L (3 oz/gal); the preferred concentration for best results is 15 ± 4 g/L (2 ± 0.5 oz/gal). Sodium hydroxide contributes to conductivity and, in excess, affects the current-density range for obtaining bright plate.

Brighteners. The most used, and probably the safest, brightening agents for cyanide baths are organics such as:

- Aldehydes
- Ketones
- Alcohols
- Furfural
- Dextrin
- Gelatin
- Milk sugar
- Molasses
- Piperonal
- Some sulfonic acids.

These materials form complexes with the electrolyte in cyanide baths and influence the orientation and growth of electrodeposited crystals, resulting in the formation of fine longitudinal crystals and hence a bright deposit.

ANODES

The anode system for cadmium plating from a cyanide solution consists of ball-shaped cadmium anodes in a spiral cage of bare steel (Fig. 4). The spherical shape provides a large surface area in relation to weight, without a large investment in cadmium. Ball anodes also make it possible to maintain an approximately constant anode area, and little or no anode scrap is produced. Cadmium balls are usually 50 mm (2 in.) in diameter and weigh 0.57 kg ($1^{1}/_{4}$ lb) per ball.

CURRENT DENSITY

Cyanide cadmium baths may be operated over a wide range of cathode current densities, as indicated in Table 12B. In a properly formulated bath operated within its intended current-density range, the cathode efficiency is 90% ± about 5%. Thus, to apply a 25-μm (1-mil) deposit of cadmium requires 1.1 A · h/dm² (11 A · h/ft²).

The ranges of current density given in Table 12B are suggested limiting values. Choice of current density is governed mainly by the type of work being plated; for example, low current densities are suitable for small lightweight parts, current densities up to 4 A/dm² (40 A/ft²) for medium-weight parts of fairly uniform shape and high current densities for uniform heavy parts like cylinders and shafts.

Tin Plating

TIN ELECTRODEPOSITS are very useful, because a thin coating can provide the desirable properties of tin, such as excellent solderability, ductility, softness, and corrosion or tarnish resistance. Thus, stronger materials, required for their engineering properties, may exhibit the desirable properties of tin on their surfaces. A tin deposit provides sacrificial protection to copper, nickel, and many nonferrous metals and alloys. It does not provide the same properties to steel in a normal atmosphere. Thick, nonporous coatings provide long-term protection in almost any application. The required coating thickness is established by the application.

TYPES OF ELECTROLYTES

Tin may be deposited from alkaline or acid solutions. Electrolyte compositions and process operating details are readily available in published references (Ref 1, 2, and 3). Basic details of electrolyte composition and operating conditions

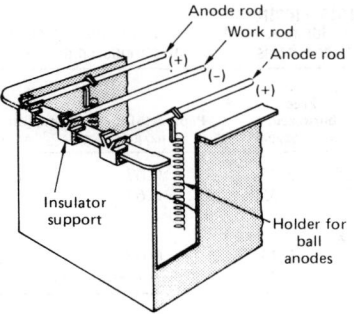

Fig. 4. Still plating tank with spiral steel holders for cadmium ball anodes

are shown in Table 13 for alkaline solutions and in Tables 14 and 15 for acid solutions. Tin ions in the alkaline electrolytes have a valence of +4. Those in the acid electrolytes have a valence of +2. Consequently, the alkaline systems require the passage of twice as much current to deposit one gram molecule of tin at the cathode.

Alkaline electrolytes usually contain only an alkaline stannate and the applicable hydroxide to obtain satisfactory coatings. Unlined steel equipment is acceptable, but the electrolyte must be heated. Factors such as operating temperature, solution constituent concentration, and operating current density affect the efficiency and plating rate of the system and must be properly balanced and controlled.

Acid Electrolytes. Several acid tin plating electrolytes are available. Two of these can be considered as general systems adaptable to almost any application, and electrolytes such as Halogen (chloride-fluoride based) or Ferrostan (a special sulfate-based system) have been developed for highly specialized use. The acid electrolytes differ from the alkaline in many respects. The solution of a stannous salt in a water solution of the applicable acid does not produce a smooth, adherent deposit on a cathode, and a grain-refining addition agent (gelatin or peptone) must be used. Usually such materials are not directly soluble in water solution and a wetting-agent-type material (such as β-naphthol) is also necessary.

Acidic electrolytes require lined or plastic tanks. The throwing power of acid tin electrolytes is inferior to that of the alkaline tin solutions, but they are sufficiently superior to other acid processes so that special anode systems are not usually required.

REFERENCES

1. *Modern Electroplating*, 3rd Ed., edited by F. A. Lowenheim: Wiley-Interscience, Electrochemical Society Series, New York, 1974

2. *Metal Finishing Guidebook and Directory*, annual issue: Metals and Plastics Publications, Inc., Hackensack, NJ, 1982

3. Publications of the International Tin Research Institute, Greenford, England (with offices in Palo Alto, CA, and Columbus, OH)

Lead Plating

LEAD has been deposited from a variety of electrolytes among which are fluoborates and fluosilicates. Fluoborate baths are the most widely used because of the availability of lead fluoborate and the simplicity of bath preparation, operation and stability. Fluoborate baths provide finer-grained, denser lead deposits. Fluosilicate baths, although less costly to use for large operations, are difficult to prepare for small-scale plating. They are not suitable for plating directly on steel and are subject to decomposition, which produces silica and lead fluoride. (See Tables 16 and 17 for compositions and operating conditions for lead fluoborate and lead fluosilicate baths.)

EQUIPMENT

Anodes should be bagged in dynel or polypropylene cloth to prevent sludge from entering the plating bath. Bags should be leached in hot water to remove any sizing agents used in their manufacture before use in the plating bath. Nylon and cotton materials deteriorate rapidly and should not be used in any of the baths. Fluoborate and fluosilicate baths attack equipment made of titanium, neoprene, glass or other silicated materials, and such materials should not be used in these solutions. Anode hooks should be made of Monel metal.

Tanks or tank linings should be made of rubber, polypropylene or other plastic materials in-

Table 14. Composition and operating conditions for sulfate electrolyte

Temperatures for sulfate electrolytes are 21 to 38 °C (70 to 100 °F); they do not require heating; cooling may be considered if temperature rises, to reduce adverse effects of temperature on the electrolyte constituents; cathode current density is 1 to 10 A/dm² (10 to 100 A/ft²).

Constituent	Amount g/L	oz/gal	Operating limits of g/L	oz/gal
Stannous sulfate	80	10.6	60-100	8-13
Tin metal, as sulfate	40	5.3	30-50	4-6.5
Free sulfuric acid	50	6.7	40-70	5.3-9.3
Phenolsulfuric acid(a)	40	5.3	30-60	4-8
β-naphthol	1	0.125	1	0.125
Gelatine	2	0.25	2	0.25

(a) Phenolsulfuric acid is most often used. Cresolsulfonic acid performs equally well and is a constituent of some proprietary solutions.

Table 13. Composition and operating conditions for stannate plating electrolytes

Values of composition are for electrolyte start-up; operating limits for the electrolyte composition are approximately −10 to +10% of start-up values.

Bath	Potassium stannate g/L	oz/gal	Sodium stannate g/L	oz/gal	Potassium hydroxide g/L	oz/gal	Sodium hydroxide g/L	oz/gal	Tin metal(a) g/L	oz/gal	Temperature °C	°F	Cathode current density A/dm²	A/ft²
A	105	14	...	...	15(b)	2(b)	...	...	40	5.3	66-88	150-190	3-10	30-100
B	210	28	...	...	22	3	...	...	80	10.6	77-88	170-190	to 16	to 160
C	420	56	...	...	22	3	...	...	160	21.2	77-88	170-190	to 40	to 400
D	...	...	105(c)	14	...	...	10(b)	1.3(b)	42	5.6	60-82	140-180	0.5-3	5-30

(a) As stannate. (b) Free alkali may need to be higher for barrel plating. (c) Na$_2$SnO$_3$·3H$_2$O; solubility in water is 61.3 g/L (8.2 oz/gal) at 16 °C (60 °F) and 50 g/L (6.6 oz/gal) at 100 °C (212 °F).

Table 15. Composition and operating conditions for fluoborate electrolyte

Three electrolyte compositions are given; the standard is generally used for rack or still plating; the high speed for applications like wire plating; the high throwing power for barrel plating or applications where a great variance exists in cathode current density as a result of cathode configuration.

Electrolyte	Stannous fluoborate g/L	oz/gal	Tin metal(a) g/L	oz/gal	Free fluoboric acid g/L	oz/gal	Free boric acid g/L	oz/gal	Paptone(b) g/L	oz/gal	β-naphthol g/L	oz/gal	Hydroquinone g/L	oz/gal	Temperature °C	°F	Cathode current density A/dm²	A/ft²
Standard	200	26.8	80	10.8	100	13.4	25	3.35	5	0.67	1	0.13	1	0.13	16-38(c)	60-100(c)	2-20	20-200
High speed . .	300	39.7	120	16.1	200	26.8	25	3.35	5	0.67	1	0.13	1	0.13	16-38	60-100	2-20	20-200
High throwing power	75	9.9	30	4.0	300	40.2	25	3.35	5	0.67	1	0.13	1	0.13	16-38	60-100	2-20	20-200

(a) As fluoborate. (b) Dry basis. (c) Electrolytes do not require heating. Cooling may be considered if temperature rises, to reduce adverse effects of temperature on the electrolyte constituents.

Table 16. Compositions and operating conditions for lead fluoborate baths

Bath	Lead g/L	oz/gal	Fluoboric acid (min) g/L	oz/gal	Peptone solution vol%	Free boric acid g/L	oz/gal	Temperature °C	°F	Cathode current density(a) A/dm²	A/ft²	Anode composition	Anode/cathode ratio
High-speed	225	30	100	13.4	1.7	1 to saturation	0.13 to saturation	20-41	68-105	5	50	Pure lead	2:1
High throwing power	15	2	400	54	1.7	. . .	. . .	24-71	75-160	1	10	Pure lead	2:1

(a) Values given are minimums. Current density should be increased as high as possible without burning the deposit; this is influenced by the degree of agitation.

Table 17. Compositions and operating conditions for lead fluosilicate baths

Bath	Lead, g/L (oz/gal)	Animal glue, g/L (oz/gal)	Peptone equivalent, g/L (oz/gal)	Total fluosilicate, g/L (oz/gal)	Temperature °C	°F	Cathode current density A/dm²	A/ft²	Anode current density A/dm²	A/ft²	Anode composition
1	10 (1.3)	0.19 (0.025)	5 (0.67)	150 (20)	35-41	95-105	0.5-8	5-80	0.5-3	5-30	Pure lead
2	180 (24)	5.6 (0.75)	150 (20.1)	140 (18.75)	35-41	95-105	0.5-8	5-80	0.5-3	5-30	Pure lead

ert to the solution. Pumps and filters of type 316 stainless steel or Hastalloy C are satisfactory for intermittent use; for continuous use, however, equipment should be made from or lined with graphite, rubber, polypropylene or other inert plastic.

Tin-Lead Plating

ELECTRODEPOSITION (plating) of tin-lead alloys is used to protect steel against corrosion, serve as an etch resistant, and facilitate soldering. Tin-lead plating is a relatively simple process because the standard electrode potentials of tin and lead differ by only 10 mV. Tin-lead alloys have been deposited by electrolytes such as sulfamates, fluosilicates, pyrophosphates, chlorides and fluoborates. Of these, fluoborate is available commercially and is generally used to plate tin-lead alloys.

Bath Components. Because concentrated solutions of stannous and lead fluoborates and fluoboric acid are available commercially, alloy plating baths are made by mixing and diluting concentrates. Compositions of concentrates are given in Table 18.

Anodes. Tin-lead alloy anodes of at least 99.9% purity must be used. The most objectionable anode impurities are arsenic, silver, bismuth, antimony, copper, iron, sulfur, nickel and zinc.

Silver Plating

ELECTRODEPOSITION of silver onto base-metal alloys has increased significantly in volume as a result of the increasing popularity of stainless steel and pewter designs for flatware and

Table 18. Compositions of alloy plating bath concentrates

Constituent	wt %	Amount g/L	oz/gal
Lead fluoborate			
Lead fluoborate, Pb(BF₄)₂	51.0	893	119
Lead, Pb(a)	27.7	485	65
Fluoboric acid, free HBF₄	0.6	10.5	1.4
Boric acid, free H₃BO₃	1.0	18	2.4
Stannous (tin) fluoborate			
Stannous fluoborate, Sn(BF₄)₂	51.0	816	109.0
Tin, Sn⁺²(a)	20.7	331	44.3
Fluoboric acid, free HBF₄	1.8	29	3.9
Boric acid, free H₃BO₃	1.0	16	2.1
Fluoboric acid			
Fluoboric acid, HBF₄	49	671	89.9
Boric acid, free H₃BO₃	0.6	8.3	1.1
Hydrofluoric acid, free HF	None	. . .	. . .

(a) Equivalent.

tableware. This decorative application constitutes the largest usage of this white metal. However, with rapid escalation of the price of gold, the electronics industry has taken to increased usage of silver in areas where its tendency to migrate, tarnish or destabilize electrically poses no performance problems.

The unique corrosion-resistant characteristics of silver make it applicable as a coating for storage containers for specific chemicals, such as tank cars for bulk transport of specific chemicals. In medical-instrument applications, silver plating has been used frequently on electrocardiogram probes. More recently, electrodeposition of alloys identified as low-karat golds have been applied to jewelry, watch cases and similar items. Although these alloys are referred to as gold alloys, the silver content is substantially greater than the gold content.

SOLUTION FORMULATIONS

For well over 100 years, silver has been plated out of alkaline cyanide electrolytes produced by dissolving silver cyanide in a solution of sodium cyanide. Currently, a double cyanide salt (potassium silver cyanide containing 54.2% metallic silver) is used for bath makeup and replenishment. Typical compositions and operating temperatures and current densities for conventional and high-speed electroplating solutions are listed in Table 19.

FUNCTIONS OF BATH INGREDIENTS

In cyanide formulations, free cyanide ensures a stable metal-ion concentration by providing good anode corrosion. Electrolyte conductivity is also increased by cyanide, carbonate and hydroxide. Hydroxide reduces the rate of cyanide decomposition. Without addition of a brightener such as ammonium thiosulfate, selenium and/or antimony, the deposit will have a matte appearance.

ANODES

The anodes for silver plating should be composed of silver of the highest purity.

Table 19. Suggested plating solutions for silver plating

Solution type	Potassium silver cyanide, KAg(CN)₂ g/L	oz/gal	Potassium cyanide, KCN g/L	oz/gal	Potassium carbonate, K₂CO₃ g/L	oz/gal	Temperature °C	°F	Current density A/dm²	A/ft²
Conventional	45-60	6-8	30-45	4-6	30-90	4-12	20-25	68-77	0.55-1.6	5.5-16
High speed ..	150-225	20-30	110-150	15-20	15-75	2-10	20-30	68-86	5.4-10.8	54-108

Table 20. Formulas for hot cyanide baths

Formula No.	Gold(a) troy g/L	oz/gal	Potassium cyanide g/L	oz/gal	Potassium carbonate g/L	oz/gal	Dipotassium phosphate g/L	oz/gal	Temperature °C	°F	Current density A/m²	A/ft²
1	1-8	0.1-1.0	30	4.0	...	...	...	...	48-66	120-150	11-54	1-5
2	4-12	0.5-1.5	30	4.0	30	4.0	30	4.0	48-66	120-150	11-54	1-5
3	10	1.2	...	...	...	...	120	16	48-71	120-160	11-54	1-5

(a) As potassium gold cyanide.

Gold Plating

GOLD PLATING research has received great impetus in the period since the early 1950's, encouraged by the needs of the electronics-related industries. These industries required thick, low-porosity coatings with greater hardness, wear resistance and corrosion resistance. The need for gold deposits with these improved properties has led to coating processes that also offer high purity, silicon eutectic-forming ability, low electrical contact resistance, good solderability and weldability, and high infrared emissivity. In the decorative field as well, processes have been developed which can plate thick, ductile and wear-resistant deposits in a variety of colors and karats.

PLATING SOLUTIONS

Hot Cyanide Baths. Typical formulations of hot gold cyanide baths can be found in Table 20. Gold metal is added as potassium gold cyanide, and the solutions are made more conductive by adding potassium cyanide, potassium carbonate or dipotassium phosphate, either singly or in combination. These baths are used to plate 24-karat pure soft gold deposits.

Potassium cyanide is added frequently in formulas 1 and 2 of Table 20, although it is not required in formula 3. Dipotassium phosphate, however, helps to brighten the deposit and is replenished periodically.

The pH of gold cyanide baths is regularly maintained at less than 11.8. As the pH exceeds 12, the bright plating range of any gold bath is reduced, and the bath tends to plate red or smutty deposits are eventually produced. At a pH below 10.5, solutions containing more than 6 g/L (0.7 oz/gal) free cyanide darken on electrolysis and eventually form a black precipitate consisting of cyanogen derivatives, while solutions containing 5 g/L (0.6 oz/gal) or less free cyanide slowly liberate hydrogen cyanide without changing color.

For barrel plating, a gold concentration of 1 to 4 g/L (0.1 to 0.5 troy oz/gal) and a current density of 11 to 32 A/m² (1 to 3 A/ft²) are preferred. Current efficiency during barrel plating is greatly affected by gold concentration, current density, size of the load, size and shape of the work, ease of tumbling, rotation rate of the barrel, and number and size of the perforations in the barrel.

Brass Plating

COMMERCIAL BRASS PLATING SOLUTIONS are cyanide based. Noncyanide solutions have enjoyed limited utilization because they suffer from lack of stability, difficulties with alloy control, or wide ranges of current density.

The basic ingredients of a cyanide brass plating solution are sodium (or infrequently potassium) cyanide, copper cyanide and zinc cyanide. Also present in the bath make-up are ammonia (added intentionally or by decomposition of cyanide) and carbonate. Sodium carbonate is necessary in a new plating solution to provide buffering action without which the solution will not produce a consistent color. A mixture of sodium carbonate and bicarbonate is used to produce buffering action in the proper range of approximately 10 pH. Table 21 provides compositions and operating conditions for various brass plating solutions.

Current densities of over 2.2 A/dm² (22 A/ft²) are easily possible with the low current densities remaining with the same color below 0.1 A/dm² (1 A/ft²). Poor performance at higher current densities is an early indication that solution stability requires checking.

Anodes for brass plating should be of the approximate alloy being plated. For yellow brass, 70% copper and 30% zinc are recommended. Anodes may be of ball, slug, or other shapes used in steel or titanium baskets. Bar anodes are also used. Purity is essential with lead content below 0.01% and all foreign metals below 0.02%.

Equipment is of standard design, with steel being suitable for tanks, coils and filters. Preference should be given to rubber- or plastic-lined tanks with stainless or titanium coils, because iron forms ferrocyanides that precipitate as zinc ferrocyanide, which causes a grayish sludge. Filters also may be of plastic construction.

Bronze Plating

BRONZE PLATING SOLUTIONS have copper as the cyanide complex and tin as the stannate complex, with excess cyanide and hydroxide. Both sodium and potassium salt formulations give higher efficiencies and permit lower concentrations and temperatures.

While solutions of materials other than cyanide and stannate complexes have been explored, none have been commercially successful. Table 22 presents compositions and operating conditions for two typical bronze plating solutions.

Temperature of the solution is an important plating variable. Temperatures below 41 °C (105 °F) produce poor deposits almost always higher in copper. Higher temperatures create higher efficiencies and greater permissible current densities. Normal temperatures are from 60 to 80 °C (140 to 175 °F).

Anodes are usually made of pure copper. Bronze anodes dissolve poorly; the use of separate tin and copper anodes has been tried, but this arrangement is almost impossible to control. Thus,

Table 21. Composition and operating conditions for typical brass plating solutions

Type	Sodium cyanide, g/L (oz/gal)	Potassium cyanide, g/L (oz/gal)	Copper cyanide, g/L (oz/gal)	Copper(a), g/L (oz/gal)	Zinc cyanide, g/L (oz/gal)	Zinc(b), g/L (oz/gal)	Sodium carbonate, g/L (oz/gal)	Sodium hydroxide, g/L (oz/gal)	Potassium hydroxide, g/L (oz/gal)	Ammonia(c), %	Addition agents, %	Temperature, °C (°F)	pH	Current density, A/dm² (A/ft²)
Conventional brass for flash or rack or still plating	36 (4.8)	...	26 (3.5)	18 (2.4)	11 (1.5)	6 (0.8)	10 (1.3)	...	...	0.5	...	32 (90)	10.0	0-3 (0-30)
Conventional brass for bulk (barrel) plating	60 (8)	...	36 (4.9)	26 (3.5)	14 (1.7)	8 (1.1)	10 (1.3)	...	...	0.5	...	32 (90)	10.0	0-2 (0-20)
Conventional brass for high-speed plating	60 (8)	...	36 (4.9)	25 (3.3)	14 (1.7)	8 (1.1)	10 (1.3)	...	...	...	1	70 (158)	10.0	0-7 (0-70)

(a) As Cu. (b) As Zn. (c) Aqua.

Table 22. Compositions and operating conditions for two typical bronze plating baths

Solution	Sodium cyanide, g/L (oz/gal)	Potassium cyanide, g/L (oz/gal)	Copper cyanide, g/L (oz/gal)	Copper(a), g/L (oz/gal)	Sodium stannate, g/L (oz/gal)	Potassium stannate, g/L (oz/gal)	Tin(a), g/L (oz/gal)	Sodium hydroxide, g/L (oz/gal)	Potassium hydroxide, g/L (oz/gal)	Rochelle salt(b), %	Addition agent, %	Temperature, °C (°F)	Current density, A/dm² (A/ft²)
Standard bronze plating solution, 15% tin	70 (9.3)	...	40 (5.3)	28 (3.7)	20 (2.7)	...	9 (1.2)	19 (2.5)	...	...	...	60 (140)	0.1-8 (1-80)
Bright bronze plating solution, 12% tin	...	90 (12)	40 (5.3)	28 (3.7)	...	60 (8)	23 (3.1)	...	7.5 (1.0)	5	0.25	60-71 (140-160)	to 11 (to 110)

(a) As metal. (b) Or equivalent.

the normal practice is to use copper anodes and add the necessary tin as sodium or potassium stannate.

Equipment used for bronze plating may be made of steel, but rubber- or plastic-lined tanks are preferable. Heating coils may be of steel or stainless steel. Barrels for bulk plating should be of any standard design that is suitable for the elevated temperature to be encountered.

Rhodium Plating

RHODIUM PLATING was not used to any appreciable extent until 1930, when the superior properties of rhodium as a noble, white finish for jewelry were first recognized. Since that time, rhodium plate has been used for decorative and, especially since 1945, engineering applications.

Decorative Applications. Rhodium is used to provide whiter finishes on platinum goods and to impart nontarnishing coatings over sterling silver products. For such applications, rhodium is deposited from phosphate and sulfate baths.

Engineering Applications. Two properties of rhodium make it suitable as a coating in many electronic applications: (a) low electrical resistivity and (b) high hardness. Thicker deposits from conventional jewelry electrolytes tend to stress crack, causing coatings on silver to tarnish the substrate at exposed areas. This is controlled by a thin layer of nickel (1.25 μm or 0.05 mil) or a thinner layer of palladium (0.1 μm or 0.004 mil) on the silver and under the rhodium. For electronic applications where undercoatings of nickel are undesirable, low-stress compositions have been developed. One electrolyte contains selenic acid and another contains magnesium sulfamate. Deposit thicknesses obtained from these solutions range from 25 μm (1 mil) to 200 μm (8 mils), respectively.

Rhodium solutions for engineering applications are summarized in Table 23. The low-stress sulfamate process is used to electroplate rhodium on small electronic parts through the use of plating barrels.

Table 23. Low-stress engineering rhodium solutions

Solution	Selenic acid process	Magnesium sulfamate process
Rhodium (sulfate complex)	10 g/L (1.3 oz/gal)	2-10 g/L (0.3-1.3 oz/gal)
Sulfuric acid (concentrated)	15-200 mL/L (2-26 fluid oz/gal)	5-50 mL/L (0.7-7 fluid oz/gal)
Selenic acid	0.1-1.0 g/L (0.01-0.1 oz/gal)	...
Magnesium sulfamate	...	10-100 g/L (1.3-13 oz/gal)
Magnesium sulfate	...	0-50 g/L) (0-7 oz/gal)
Current density	1-2 A/dm² (10-20 A/ft²)	0.4-2 A/dm² (4-22 A/ft²)
Temperature	50-75 °C (120-165 °F)	20-50 °C (68-120 °F)

Selective Plating

SELECTIVE PLATING, also referred to as electrochemical metallizing, is a method of depositing metal from a concentrated electrolyte solution without using immersion tanks.

SELECTIVE PLATING PROCESS

Selective plating is a process similar to a combination of arc welding and electroplating. A flexible cable, or lead, connects the positive output terminal of the power pack, a highly specialized rectifier, with a hand-held, or fixtured, insulated handle called the stylus. The end of the stylus is connected to a shaped graphite block, called the anode. Another flexible cable, or lead, connects the negative output terminal of the power pack with a clamp that is attached to the metal part on which the buildup is to occur. The metal part, or workpiece, becomes the cathode.

The anode is wrapped with an absorbent material, such as cotton batting or polyester sleeving, and dipped into a liquid solution of the metal to be deposited. When this saturated anode wrap is brought into contact with the cathodic workpiece, electrical current permits the flow of metal ions from the solution in the wrap onto the surface of the workpiece, creating a buildup of metal on the workpiece surface. A schematic diagram of this operation is shown in Fig. 5. This technique allows the metal buildup to be limited only to the area of contact between wrapped anode and workpiece. Little or no masking of workpieces is required, and the thickness of buildups can be controlled very accurately.

EQUIPMENT

The equipment required for selective plating, in addition to its very specialized solutions, consists of the following: (a) a specially designed power pack; (b) an assortment of working tools, called styli, which include shaped graphite anodes; and (c) accessories and auxiliary equipment which provide flow of solution and/or motion of the anode or part.

Mechanical Coating

MECHANICAL COATING is a method which utilizes kinetic energy to deposit metallic coatings onto parts. It is also known as mechanical plating or peen plating when the coating applied is less than 25 μm (1 mil) thick.

MECHANICAL PLATING

Mechanical plating (plating less than 25 μm or 1 mil thick) is widely used in the automotive industry on hardened (greater than 32 HRC) steel

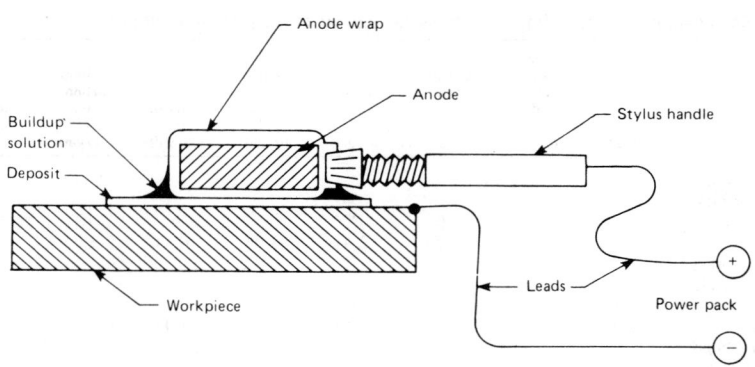

Courtesy of Selectrons Ltd., Waterbury, CT.

Fig. 5. Manual operation of selective plating process

parts as a replacement for electroplating because of hydrogen embrittlement problems.

MECHANICAL GALVANIZING

Where coatings 25 μm (1 mil) or greater in thickness are required for extended outdoor corrosion protection, mechanical galvanizing has gained wide acceptance as an alternative to hot dip galvanizing of fasteners and other small components. The primary benefits are the uniformity of the coating and the elimination of stickers (parts that are welded together). When small parts are coated in bulk, uniform thickness is difficult to control using hot dip methods. Coatings of zinc up to 75 μm (3 mils) thick can be mechanically deposited.

PROCESS CYCLE

Parts which have been degreased, descaled and copper flashed are tumbled in rubber-lined barrels with water, glass bead impact media, promoter chemicals, and a finely divided powder of the metal to be mechanically plated. The promoter chemical cleans the metal powder and controls the size of the metal powder agglomerates that form. It also acts as a catalyst. The mechanical energy generated from the rotation of the barrel is transmitted through the glass impact medium and causes the clean metal powder to be cold welded to the clean metal parts, that is, joined below their melting points, providing an adherent metallic coating. Chemicals used in this process are available commercially.

EQUIPMENT

A mechanical coating system is comprised of a series of operations which include (a) weighing, (b) loading, (c) cleaning, (d) galvanizing or plating, (e) rinsing, (f) surging, (g) separation, (h) handling of glass impact media, (i) post-treatments, (j) rinsing and (k) curing or drying.

Electropolishing

ELECTROPOLISHING is an electrochemical process for removing metal. Etching, deburring, smoothing, coloring and machining are typical electropolishing processes. The removal of metal is done anodically in acid and alkaline solutions.

During the process, products of anodic metal dissolution react with the electrolyte to form a film at the metal surface. Two types of films have been observed: (a) a viscous liquid that is nearly saturated, or is supersaturated, with the dissolution products; and (b) anodically discharged gas, usually oxygen. Both types of films exist simultaneously in most commercial electropolishing solutions. The gas appears to be a blanket on the outside of the viscous film. Which type of film predominates depends on (a) the kind of metal, (b) the nature of the electrolyte and (c) the surface condition prior to electropolishing (i.e., surface contamination, grain size, inclusions).

ELECTROPOLISHING SOLUTIONS AND OPERATING CONDITIONS

The most widely used electropolishing solutions contain one or more of the concentrated inorganic acids—sulfuric, phosphoric and chromic.

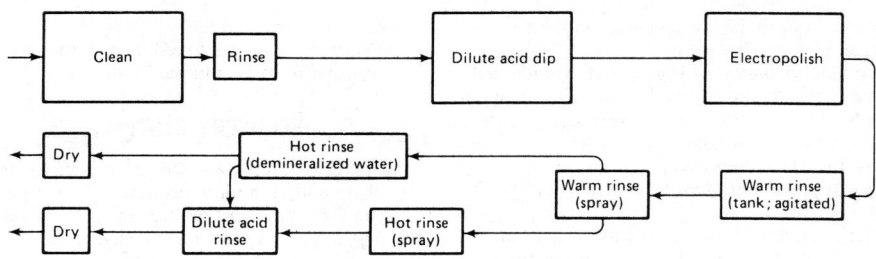

Fig. 6. Process cycles typically employed in electropolishing

Sometimes, an inorganic acid, such as hydrofluoric or hydrochloric acid, or an organic acid, such as acetic, citric, tartaric or glycolic acid, may be used with one or more of the concentrated inorganic acids mentioned above. Figure 6 shows a typical processing sequence used in electropolishing.

EQUIPMENT AND PROCESSING PROCEDURES

Equipment and layout for electropolishing generally resemble those of an electroplating installation.

Racks. Items to be electropolished are placed with firm electrical contact on a rack to be positioned on the work bar. The parts are placed so that current distribution will be uniform when they are immersed between cathodes in the electropolishing solution. Usually, racks are made of copper or titanium and have spring clips or fingers of bronze, steel or titanium to hold the workpieces.

Tanks are made of stainless steel, low-carbon steel, plastic or fiberglass. Tanks may require linings of rubber, polyvinyl chloride, polypropylene or lead, depending on the electrolyte used. The use of plastic-lined tanks is limited to electropolishing in which the operating temperature is below that at which the lining can soften.

Cathodes are made of copper, lead, stainless steel or carbon. The material to be used depends on the temperature and the kind of electropolishing bath. Cathodes are individual rods, flat strips, or expanded metal in strips, which can be suspended from the cathode bar in a pattern for good current distribution.

Treatment of Plating Wastes

TREATMENT OF PLATING WASTES involves one of two possible approaches—destruction or recovery. Destructive-type techniques convert toxic or environmentally harmful wastes into legally acceptable forms for subsequent disposal. Recovery techniques recycle plating wastes for further use. Either approach must be used to comply with government regulations, which determine the extent and quality of the waste treatment operation.

CONVENTIONAL WASTE-WATER TREATMENT

The wastewater treatment flowsheet shown in Fig. 7 is typical for the majority of plating shops where there are various types of heavy-metal wastes. It consists of the following unit operations:

- Chromium reduction of segregated chromium waste streams to chemically reduce the chro-

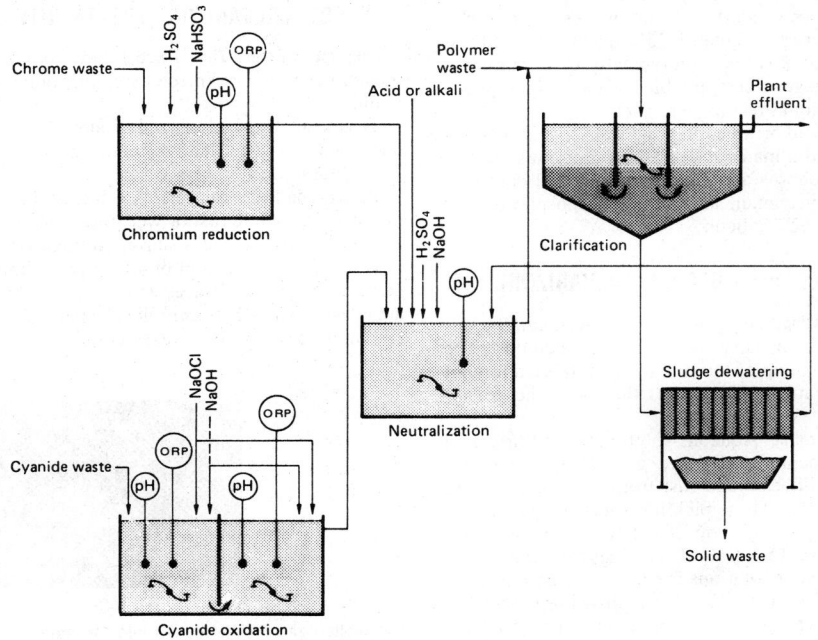

Fig. 7. Electroplating treatment flowsheet

mium from its hexavalent state to the trivalent state
• Cyanide oxidation of segregated cyanide-bearing waste streams to oxidize the toxic cyanides to harmless carbon and nitrogen compounds
• Neutralization of combined wastewaters to adjust the pH within acceptable discharge limits and to precipitate dissolved metals as hydroxides
• Clarification of neutralized wastewater streams to separate decontaminated wastewater from the metal hydroxide sludge
• Dewatering of sludge to reduce volume of solid waste that must be ultimately disposed.

RECOVERY SYSTEMS

Recovery systems are attractive because they can abate pollution more economically than conventional destruction-type treatment techniques. Savings in purchases of recovered chemicals and water, plus substantial reductions in solid waste disposal costs, make recovery worth considering wherever technically feasible. Installation of a recovery system can often reduce hydraulic and chemical loading on an existing waste treatment system and consequently improve its operation. In some instances, installation of a properly designed, efficient recovery system may obviate the need of installing a conventional waste treatment plant. Typical recovery systems are (*a*) evaporation, (*b*) reverse osmosis (RO), (*c*) electrodialysis, (*d*) electrolytic and (*e*) ion exchange.

Metallic Coating Processes Other Than Plating

Hot Dip Galvanized Coatings

Edited by Daryl E. Tonini, American Hot Dip Galvanizers Association, Inc., Serge Belisle, Noranda Sales Corp., and David C. Pearce, ASARCO, Inc.

HOT DIP GALVANIZING is a process in which an adherent, protective coating of zinc and zinc compounds is developed on the surfaces of iron and steel products by immersing them in a bath of molten zinc. The protective coating usually consists of several layers (see Fig. 1). Those closest to the basis metal are composed of iron-zinc compounds; these, in turn, are covered by an outer layer consisting almost entirely of zinc.

IRON AND STEEL SUBSTRATES

The chemical compositions of irons and steels, and even the forms in which certain elements such as carbon and silicon are present, determine the suitability of ferrous metals for hot dip galvanizing and may markedly influence the appearance and properties of the coating. Steels that contain less than 0.25% carbon, less than 0.05% phosphorus, less than 1.35% manganese and less than 0.05% silicon, individually or in combination, are generally suitable for galvanizing using conventional techniques.

To avoid brittleness of the iron-zinc alloy layer in cast iron materials, substrate iron must be low in phosphorus and silicon; a preferred composition may contain about 0.01% phosphorus and about 0.12% silicon.

CLEANING BEFORE GALVANIZING

Degreasing. Organic contaminants can be removed from the work by several methods. The most common of these in the post fabrication hot dip galvanizing process is the use of heated alkaline cleaning baths.

Acid Pickling. Aqueous solutions of sulfuric acid or hydrochloric acid are generally used to remove mill scale and rust from steel parts before galvanizing. These pickling solutions may either be sulfuric acid, 3 to 10 wt %, or hydrochloric acid, 5 to 15 wt %. To increase effectiveness, sulfuric acid solutions are always used hot at 60 to 79 °C (140 to 175 °F); hydrochloric acid solutions are usually used at about room temperature, 24 to 38 °C (75 to 100 °F), to avoid excessive fuming. To avoid overpickling, inhibitors are often used with both sulfuric and hydrochloric acid solutions.

Fluxing is accomplished by immersing the workpiece in an aqueous flux solution or by maintaining a flux blanket on the surface of the molten zinc bath. A combination of these two techniques is often used. The fluxing agent is usually a zinc–ammonium chloride salt.

GALVANIZING BATH

The molten zinc bath is operated at temperatures usually in the range from 445 to 465 °C (830 to 870 °F). At 480 °C (900 °F) and above, the dissolution rate of iron and steel in zinc is extremely rapid, and the effects of these temperatures on both workpiece and galvanizing tank are generally harmful.

The thickness of the coating is primarily controlled by immersion time, withdrawal rate and the composition of the iron or steel substrate material. Immersion time usually is from 1 to 5 min, and withdrawal rate for most articles is about 1.5 m/min (5 ft/min).

POSTGALVANIZING TREATMENTS

Removal of Excess Zinc. Excess zinc may be removed from the workpiece by centrifuging or by wiping.

Quenching. The galvanized material is often quenched in a water bath to inhibit growth of the intermetallic alloy layers.

Wet Storage Stain Inhibitors. A white stain, commonly called white rust or wet storage stain, may appear on zinc surfaces during storage or shipment. The stain is found on material with newly galvanized, bright surfaces and especially in such areas as crevices between closely packed sheets and angle bars if the surfaces come into contact with condensate or rainwater and the moisture does not dry quickly. Zinc surfaces that have developed a normal protective layer of corrosion products are seldom attacked.

Wet storage stain is best avoided by preventing newly galvanized surfaces from coming into contact with rain or condensate water during storage and transport. Materials stored outdoors should be arranged so that water can easily run off the surfaces and so that all surfaces are well ventilated.

BATCH GALVANIZING EQUIPMENT

Because the galvanizing kettle is the most important piece of equipment used in galvanizing, its selection should be based on the careful evaluation of several major variables, among which are size, shape, wall thickness, tank material, source of heat, and requirements for auxiliary equipment.

Aluminum Coating of Steel

ALUMINUM-COATED STEEL PRODUCTS may be classified with respect to their intended service, their properties, their behavior, and the economy of their nonrusting surfaces. Properties of aluminum-coated steel are beneficial for products which require the advantages of good corrosion resistance, bright metallic appearance, receptiveness to finishes, high reflectivity, and good electrical conductivity. For some types of products, the behavior of the aluminum-iron interfacial compound is relied upon for resistance to oxidation, scale formation and abrasion, and for high hardness.

Eta (100%Zn)
Zeta (94%Zn 6%Fe)
Delta (90%ZN 10%Fe)
Gamma (75%Zn 25%Fe)
Steel

The molten zinc is interlocked into the steel by the alloy reaction which forms zinc iron layers and creates a metallurgical bond. Magnification, 250×.

Fig. 1. Photomicrograph of a typical hot dip galvanized coating

Table 1. Effect of silicon and diffusion treatment on thickness of coating

	Thickness			
	Intermetallic layer		Total coating	
Condition	μm	mils	μm	mils
Pure aluminum coating(a)				
As coated	23	0.9	51	2.0
Diffused for:				
1100 h at 480 °C (900 °F)	33	1.3	51	2.0
1100 h at 540 °C (1000 °F)	36	1.4	51	2.0
1000 h at 595 °C (1100 °F)	43	1.7	51	2.0
456 h at 675 °C (1250 °F)	66	2.6	66	2.6
120 h at 760 °C (1400 °F)	66	2.6	66	2.6
360 h at 845 °C (1550 °F)	71	2.8	71	2.8
3 min at 1090 °C (2000 °F)	71	2.8	71	2.8
Aluminum-silicon alloy coating(b)				
As coated	8	0.3	28	1.1
Diffused for:				
1100 h at 480 °C (900 °F)	28	1.1	30	1.2
1100 h at 540 °C (1000 °F)	33	1.3	33	1.3
1000 h at 595 °C (1100 °F)	38	1.5	38	1.5
456 h at 675 °C (1250 °F)	56	2.2	56	2.2
24 h at 790 °C (1450 °F)	69	2.7	69	2.7
5 min at 1120 °C (2050 °F)	76	3.0	76	3.0
3 min at 1260 °C (2300 °F)	114	4.5	114	4.5

(a) Coating weight, 0.3 kg/m² (1 oz/ft²). (b) Coating weight, 0.15 kg/m² (0.5 oz/ft²); alloy contained 9% silicon.

CONTINUOUS HOT DIP COATING OF MILL PRODUCTS

Aluminum hot dip coating of steel strip and wire is performed by a number of patented processes. The Sendzimir method, which is the most widely used for sheet, consists of oxidizing the surface of the steel, reducing the oxidized surface in a reducing atmosphere, and immersing the steel in molten aluminum. This procedure expedites wetting, or formation of the alloy between the aluminum and the steel.

The use of a nonoxidizing, direct-fired furnace in place of an oxidizing furnace offers substantial benefits in increased production, reduction of hearth roll pickup, and ability to maintain a positive pressure throughout the furnace line, preventing air leaks into the furnace atmosphere. Combustion products of a direct-fired furnace are maintained with a slight excess of combustibles, ensuring fast removal of oil and smut and reduction of surface oxides on the incoming strip.

Other processes are also based on the use of a reducing atmosphere, but without preliminary oxidation of the strip or wire. In a broad sense, the reducing atmosphere may be considered as a gaseous flux.

The Lundin process uses aqueous fluxes that are applied to the wire before immersion in molten aluminum. This eliminates the need for a reducing atmosphere. In this process, adequate ducts are necessary to remove corrosive salt fumes from the coating surface. Fumes are carried through wash systems to the outside surface.

Procedures and Control. Most commercial hot dipped aluminum-coated steel strip is produced on continuous, anneal in-line equipment similar to that used for galvanizing. The process consists essentially of three operations: surface preparation, heat treatment of the steel base, and aluminum coating.

Surface preparation is a two-phase operation. First, all soil is removed from the surface by oxidation at elevated temperature or by chemical cleaning. Then, the surface oxides are reduced in a suitable atmosphere to prepare the strip for coating.

Because the reaction between aluminum and steel is extremely rapid, the immersion time, the temperature of the molten aluminum, and the temperature of the strip before and after coating must all be controlled to prevent formation of an excess of iron-aluminum interfacial alloy (Table 1). Unless a void layer separates the alloyed coating from the base metal, the amount of iron in the alloyed coating increases with time as the aluminum continues to diffuse into the base metal.

The amount of brittle interfacial alloy layer can be altered also by addition of silicon to the coating bath (Table 1). This increases the apparent ductility of the coating, permitting more severe fabrication of the sheet without peeling of the coating. As shown in Fig. 2, there is a rapid decrease in the thickness of the interfacial layer as the silicon content increases to about 2.5%. A smaller decrease occurs as the silicon content is further increased.

Equipment for a continuous line for hot dip aluminum coating consists of a feeding section, a furnace section and a delivery section. In the feeding section, equipment uncoils incoming strip and feeds it into the coating line at a designated constant speed under specified tension. The furnace section contains the preheating or oxidizing furnace, the annealing furnace, the cooling furnace and the coating pot. If chemical cleaning is used, alkaline cleaner and water rinse tanks are substituted for the preheating furnace. The cooling furnace is connected directly with the annealing furnace and extends to the coating bath with its end sealed by means of a snout extending into the molten aluminum bath. A dry reducing atmosphere of hydrogen and nitrogen is maintained within the annealing and cooling furnaces. The delivery section is equipped to provide rapid cooling and sufficient time for setting the coating before the strip contacts the support roll over the coating bath. Drive rolls and equipment for looping, roller leveling, coiling and shearing, as well as stretch leveling and surface conditioning, are all contained in the delivery section.

Effect of Coating on Strength and Fabricability. For different reasons, aluminum coating of steel strip and wire results in measurable decreases in their strength. The strength of strip decreases because it is normally annealed prior to coating in order to improve its fabricability. Wire strength decreases as a result of the high temperature, above 650 °C (1200 °F), of the hot dip coating bath. An example is aluminum conductor, steel-reinforced ACSR wire, which is either aluminum coated or galvanized. Its tensile strengths range from 1140 to 1450 MPa (165 to 210 ksi) depending on the type and class of coating specified. The data in Table 2 illustrate this effect for cold drawn or air patented steel wire of various carbon contents.

Table 2. Effect of hot dip aluminum coating on tensile strength of steel wire

	Tensile strength			
	Before coating		After coating	
Steel type and condition(a)	MPa	ksi	MPa	ksi
0.10% C, cold drawn	896	130	551	80
0.45% C, cold drawn	1241	180	896	130
0.75% C, cold drawn	1896	275	1379	200
0.45% C, air patented	745	108	724	105
0.75% C, air patented	1131	164	1117	162

(a) 1.62-mm (0.064-in.) diam specimens.

Steel sheet coated with aluminum-silicon alloy withstands moderate forming, drawing and spinning operations without flaking or peeling of the coating. Steel sheet coated with commercially pure aluminum withstands moderate brake and roll forming operations and can be spun or embossed, but is not suitable for drawing. Sheet with either type of coating can be given a 180° bend around a diameter equal to twice the thickness of the material; however, in any forming operation, it is advisable to allow liberal radii to prevent crazing of the coating.

Because sheared edges are susceptible to corrosion, the use of aluminum-coated sheet for fabricated assemblies may be limited for appearance considerations. Corrosion protection is not impaired, because there is no undercutting of the coating.

Coated sheet should be fabricated before the coating is diffused. Diffusion converts the coating to an iron-aluminum compound, which is very brittle.

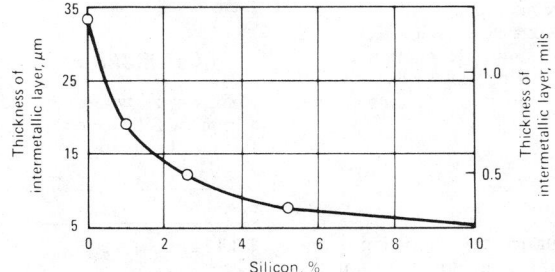

Immersed 15 s at 700 °C (1290 °F).
Fig. 2. Effect of silicon on formation of iron-aluminum interfacial layer

Typical Applications. A wide variety of industrial, farm and consumer products are fabricated from steel sheet aluminum coated at the mill. The following products require resistance to oxidation and corrosion at temperatures from 95 to 680 °C (200 to 1250 °F): combustion-chamber and outer casings, agricultural crop dryers, automotive mufflers, space heaters, furnace flues, oven interiors, barbecue grills, and wrappers for water-heater elements. Fabrication of these parts requires moderate drawing, forming, punching and spot welding.

PACK DIFFUSION PROCESSES

Pack diffusion processes are analogous to pack carburizing and can be referred to as cementation or impregnation processes. Alloys of iron, nickel, cobalt and copper are commonly coated by these methods. Because of the high temperature of the substrate, the aluminum being deposited alloys immediately with the basis metal; thus, a pure aluminum overlay never forms. If required, a separate diffusion treatment may be used after the cementation process has been completed.

Calorizing, by definition, is a term relating to all pack diffusion processes for coating metal with aluminum. It is also a trade name for a particular two-stage pack diffusion process, in which (a) the parts are coated with a high concentration of aluminum, which penetrates the material to a depth of up to 150 μm (6 mils), and (b) the aluminum coating is diffused into the material to a depth of 1000 μm (40 mils) to form an alloy with the basis metal.

In current practice, the process consists of packing the parts with halide salts and aluminum powder in a closed container. The container is held at 820 to 980 °C (1500 to 1800 °F) for 6 to 24 h, after which the parts are removed from the container and heated in air to diffuse the aluminum. The concentration of aluminum in the surface alloy layer is about 50 to 60% after coating and about 25% after the diffusion cycle.

55% Aluminum-Zinc Alloy Coated Steel Sheet and Wire

ALUMINUM-ZINC alloy coating of steel sheet and wire is performed on continuous hot dip coating lines with inline gas cleaning and heat treating of the steel substrate. With a nominal composition of 55% aluminum, 43.4% zinc and 1.6% silicon, the coating provides the durability and high temperature resistance of aluminum coatings along with the sacrificial protection characteristics of zinc coatings.

A 55% aluminum-zinc coating line for either sheet or wire is very similar to continuous annealed hot dip galvanizing lines. The coating process involves the following functions:

- Cleaning
- Heat treating
- Coating
- Accelerated cooling.

Cleaning the steel surface before coating is important to avoid surface imperfections on the coated product. Because residual surface contaminants differ significantly for sheet and wire-feed stock, removal techniques vary.

Heat Treating. A critical phase of the 55% aluminum-zinc coating process is the dual purpose heat treating of the steel before coating. The elevated temperature serves to (a) promote gas-metal reactions between the protective atmosphere and the steel surface, and (b) achieve metallurgical conditioning of the steel in the form of recrystallization or stress relieving.

Coating. After heat treating, the hot steel enters the coating bath which is nominally 55% aluminum, 1.6% silicon, remainder zinc. The bath temperature is maintained at approximately 600 °C (1110 °F). An induction-heated, ceramic-lined pot is used to contain the spelter. A similar but smaller pot is used as a premelt container to ensure that the melt has been homogenized before being added to the coating pot. Coating weight is controlled to the following limits by wiping the excess coating with gas or jet wipers:

Application	Average coating weight, min	
	g/m²	mg/in.²
Conventional sheet products	150	97
Culvert sheet	215	140
Class 30 wire	98	63
Class 45 wire	144	93

Cooling. To obtain maximum corrosion resistance, the coated sheet is forced-air cooled from the coating temperature to about 370 °C (700 °F), at a minimum rate of 11 °C (20 °F) per second, after exiting the coating bath.

Hot Dip Tin Coating of Steel and Cast Iron

HOT DIP tinning is accomplished by applying a thin coating of molten tin to a metallic object. Such coatings are applied to iron and steel to (a) provide a nontoxic, protective or decorative coating for food-handling, packaging or dairy equipment; (b) facilitate soldering of a variety of components used in electronic and electrical equipment; and (c) assist in bonding another metal to the basis metal, as in tinning of cast iron bearing shells prior to lining with lead-base or tin-base alloy. The usual thickness range of hot dip tin coatings is 3.8 to 18 μm (0.15 to 0.7 mils).

STEELS FOR HOT DIP TINNING

Low-carbon steels, containing less than 0.2% carbon, are intrinsically well suited for hot dip tinning. Medium- to high-carbon steels (0.3 to 1.0% C) may require greater care in pickling, but ordinarily there is little difficulty in processing them.

CAST IRONS FOR HOT DIP TINNING

Cast irons whose chemical compositions fall within the following ranges are generally suitable for hot dip tinning:

Element	Composition, %
Total carbon	3.2-3.5
Silicon	1.7-2.7
Manganese	0.5-0.8
Sulfur	0.05-0.12
Phosphorus	Up to 1.3

Annealed irons containing subsurface oxides of silicon may be less suited to tinning than as-cast irons.

CLEANING BEFORE HOT DIP TINNING

Degreasing. Oil, grease, soap and other lubricants used in machining, drawing and forming can be removed by one or more of several methods, including vapor degreasing, solvent cleaning, alkaline cleaning and emulsion cleaning.

FLUXING

Fluxing facilitates and speeds the reaction of molten tin with iron or steel, promoting the formation of a continuous thin layer of tin-iron or other intermetallic phases on which the liquid tin coating can spread in an even, smooth, continuous film. The material compositions of two aqueous flux solutions are given in Table 3.

Flux Covers. A cover of molten flux should be maintained on the surface of the first tin dipping bath. Compositions of two effective flux covers are given in Table 4. The salt components of the flux solutions given in Table 3 form suitable flux covers.

SINGLE-POT TINNING

Single-pot tinning is used to provide a preliminary coating for bonding or soldering, or to coat workpieces that do not require the highest-quality finish. The process involves a single immersion of fluxed workpieces in a molten tin bath heated to 280 to 325 °C (535 to 615 °F). The average operating temperature of the bath is maintained at about 300 °C (575 °F). When the workpiece is withdrawn from the bath, the surface of the work may have spots of flux which must be removed by suitable washing.

TWO-POT TINNING

In two-pot tinning, the work is dipped first in a tin bath with a flux cover and then in a tin bath covered with oil or molten grease. The process develops thick, high-quality coatings and offers the following advantages over single-pot tinning.

- Because no reactive tinning involving iron takes place during the second dip, metal in the second tinning pot can be kept lower in iron content. This results in final coatings that are low in contaminants.
- Flux residues entrained on the work from the first dip are absorbed into the oil cover of the second dip.
- The finished work retains a thin film of oil that protects the coating during shipment and storage.

Babbitting

BABBITTING is a process by which softer metals (basically tin/lead combinations) are bonded chemically or mechanically to a shell or stiffener, which supports the weight and torsion of a rotating, oscillating or sliding shaft.

MECHANICAL BONDING

Mechanical bonding of a babbit to a shell is a simple fastening process. Mechanical methods

Table 3. Compositions of flux solutions used in hot dip tinning

Solution	Zinc chloride kg	Zinc chloride lb	Ammonium chloride kg	Ammonium chloride lb	Sodium chloride kg	Sodium chloride lb	Hydrochloric acid(a) cm³	Hydrochloric acid(a) oz	Water L	Water gal
A	11	25	0.7	1.5	...	...	296-591	10-20	38	10
B	11	24	1.4	3.0	3	6	296-591	10-20	45	12

(a) Commercial grade, 28%.

Table 4. Compositions of flux covers for hot dip tinning baths

Mixture	Melting point °C	Melting point °F	Constituent	Composition, wt %
A	260	500	Zinc chloride	78
			Sodium chloride	22
B	260	500	Zinc chloride	73
			Sodium chloride	18
			Ammonium chloride	9

used in babbitting shops include anchor-groove dovetails and drilled and tapped holes into which molten babbitt flows, locking the babbitt in place. Copper or brass screws inserted into threaded holes also help hold the babbitt to the shell.

STATIC BABBITTING

Static babbitting requires a babbitting mandrel to form the babbitt in the bearing shell. Mandrels can be designed for use in either the vertical or the horizontal position. Plates are placed at each end when the mandrel is designed for use in the horizontal position. The mandrel is heated to 300 ± 10 °C (570 ± 18 °F). If a riser is used at the top of the vertical mandrel to hold extra babbitt, the riser must be heated to the same temperature. The tinned bearing shells are heated to 325 ± 25 °C (615 ± 45 °F), preferably by submersion in a large tinning pot.

CENTRIFUGAL BABBITTING

Centrifugal babbitting requires a machine expressly designed and built or modified for this purpose. No mandrel is used in horizontal applications. Submersion of the bearing shell in a 40Sn-60Pb alloy is recommended for preheating because this provides closer control of temperature. Fit plates are used to center the bearing in the machine.

Hot Dip Lead Alloy Coating of Steel

HOT DIPPING with lead alloys containing 2 to 25% tin requires close control of the surface preparation of the steel base, the composition of the chloride flux, the composition and temperature of the molten lead-tin alloy coating bath, the time of immersion in the bath, and regulation of the amount of coating left on the steel.

Because lead alone does not alloy with iron, tin or other elements such as antimony must be added to form on the surface of the steel an iron-tin alloy to which the lead will adhere. The lead-tin alloy, which is usually called a terne coating, also provides excellent solderability and good corrosion resistance in various environments and corrosive media. A range of coating thicknesses from 60 to over 250 g/m² (0.28 to over 1.17 oz/ft²) may be specified, as shown in Table 5.

In hot dipping of formed and fabricated articles, the steel is first cleaned of oils and greases by use of solvents or detergents. Then, the steel is pickled in a 6 to 12 vol % sulfuric acid solution at 71 °C (160 °F), or in an inhibited 5 to 10% hydrochloric acid solution at 49 to 65 °C (120 to 150 °F). The iron content of these solutions should not exceed 5%. Next, the steel is immersed for 5 to 20 s in either a zinc chloride or zinc–ammonium chloride flux.

The lead bath is maintained in the range from 325 to 390 °C (620 to 735 °F). The bath temperature and duration of immersion of the work is increased as the mass of the article being coated increases. When withdrawn from the bath, the coated item is centrifuged or shaken to remove excess coating metal and flux. The workpiece may be quenched in water to solidify the coating before it is contacted by other articles. The thickness of the terne coating on formed articles normally varies from 5 to 15 μm (0.2 to 0.6 mil) depending on the application. The thickness may be controlled by variations in (a) the temperature of the bath, (b) the duration of immersion and (c) the amount of shaking or centrifuging.

Thermal Spray Coatings

THERMAL SPRAY is a generic term for a group of commonly used processes for depositing metallic and nonmetallic coatings. These processes, sometimes known as "metallizing," comprise the plasma-arc spray, electric-arc spray and flame spray processes. Coatings can be sprayed from rod or wire stock, or from powdered material.

PLASMA-ARC SPRAY PROCESS

Figure 3 shows the modules constituting a basic plasma-arc spray system and the variables that must be controlled. The power level, the pressure and flow of the arc gases, and the rate of flow of powder and carrier gas are controlled at the console of the system. The spray-gun position and gun-to-work distance are usually preset. The movement of the workpiece is controlled by using automated or semiautomated tooling. Substrate temperatures can be controlled by pre-

heating and by limiting the temperature increase during processing by interrupted spraying.

ELECTRIC-ARC SPRAY PROCESS

The electric-arc spray process utilizes metal in wire form. This process differs from the other thermal spray processes in that there is no external heat source such as gas flame or electrically induced plasma. Heating and melting occur when two electrically opposed charged wires, comprising the spray material, are fed together in such a manner that a controlled arc occurs at the intersection. The molten metal is atomized and propelled onto a prepared substrate by a stream of compressed air or gas (Fig. 4).

FLAME SPRAY PROCESS

The flame spray process utilizes combustible gas as a heat source to melt the coating material. Flame spray guns are available to spray materials in rod, wire or powder form. Most flame spray guns can be adapted for use with several combinations of gases to balance operating cost and coating properties. Acetylene, propane, Mapp gas and oxygen-hydrogen are commonly used flame spray gases. In general, changing the nozzle and/or air cap is all that is required to adapt the gun. A typical wire or rod flame spray gun is shown in cross section in Fig. 5.

Oxidation Protective Coatings for Superalloys and Refractory Metals

OXIDATION protective coatings for superalloys and refractory metals serve the exclusive purpose of preventing corrosive attack of the substrate for the maximum possible time with the maximum degree of reliability.

TYPES OF COATINGS

Diffusion Coatings. In the diffusion coating application process, aluminum is made to react at the surface of the substrate, forming a layer of monoaluminide (known generically as MAl). For coatings applied over nickel-base superalloys, nickel aluminide (NiAl) is the resulting species, and over cobalt, it is cobalt aluminide (CoAl).

Overlay coatings do not rely on reaction with the substrate for their formation, although some moderate interdiffusion usually occurs during service. Rather, the material applied over the substrate during the coating process and the specific method of processing determine the com-

Table 5. Coating designations and minimum coating test limits

Coating designation	Corresponds to obsolete class g/m²	Corresponds to obsolete class oz/ft²	Minimum coating test limits(a) Triple spot test g/m²	Minimum coating test limits(a) Triple spot test oz/ft²	Minimum coating test limits(a) Single spot test g/m²	Minimum coating test limits(a) Single spot test oz/ft²
LT01	Commercial		No test		No test	
LT25	75	0.35	76	0.25	61	0.20
LT35	105	0.45	107	0.35	76	0.25
LT40	120	0.55	122	0.40	92	0.30
LT55	170	0.75	168	0.55	122	0.40
LT85	260	1.10	259	0.85	214	0.70
LT110	340	1.45	336	1.10	275	0.90

(a) Total of both sides.

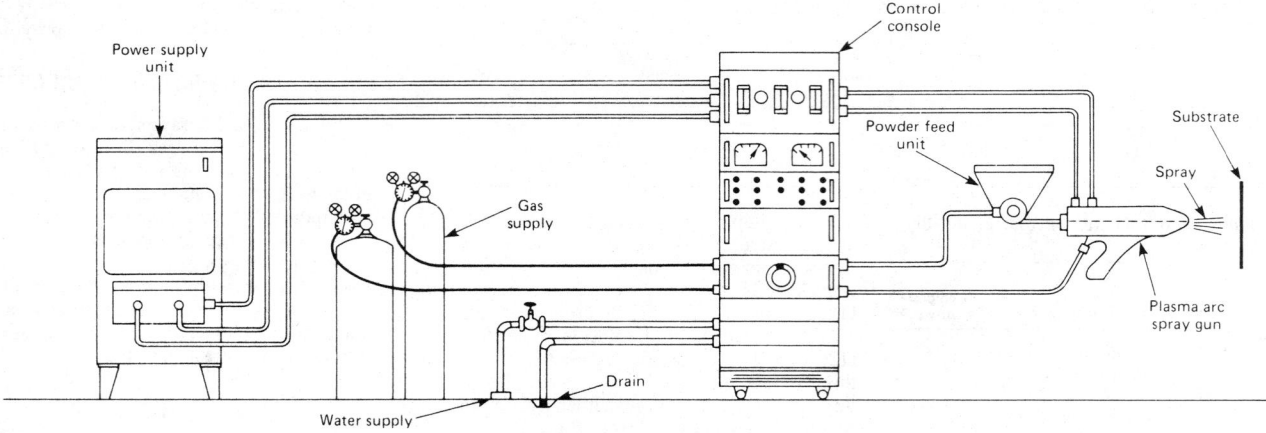

Fig. 3. A complete plasma-arc spray system

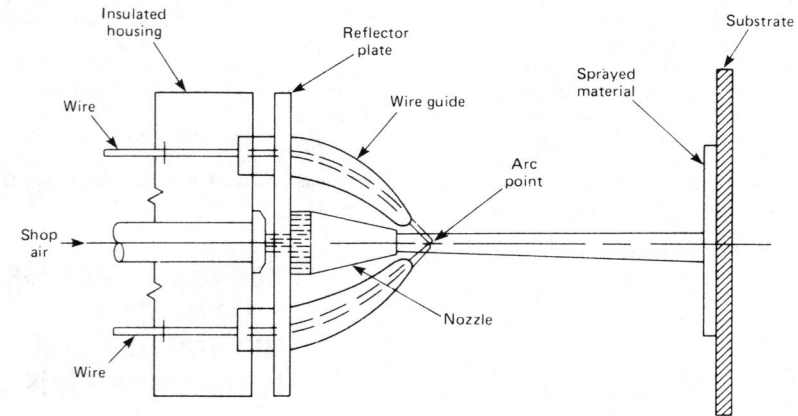

Fig. 4. Typical electric-arc spray device

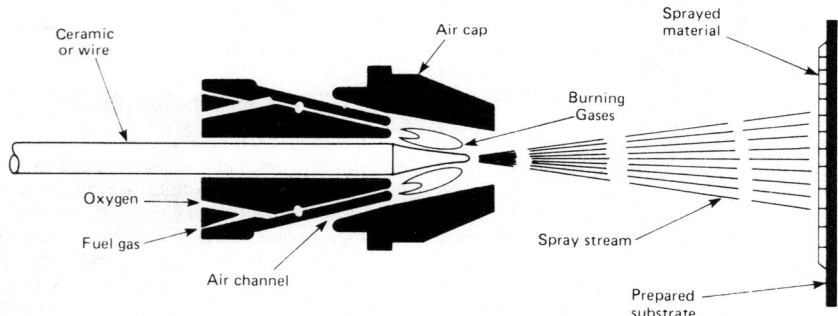

Fig. 5. Cross section of typical wire or rod flame spray gun

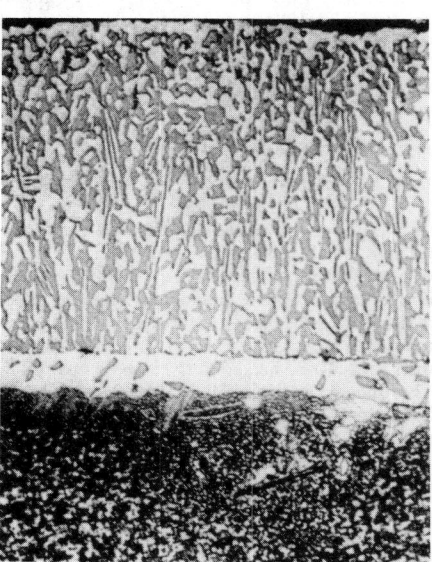

Fig. 6. Microstructure of two-phase cobalt-chromium-aluminum-yttrium (CoCrAlY) overlay coating

position and microstructure of the coating (see Fig. 6).

METHODS OF APPLYING DIFFUSION COATINGS

Two basic methods are used for the application of diffusion coatings, but many variations of these methods exist. One basic operation is the application and diffusion of the coating in essentially one operation, although a postapplication heat treatment is sometimes used to complete the diffusion process. The other comprises first applying a green coating, which may be applied by mechanical means before diffusion and which contains unreacted coating species, and then heat treating to accomplish final formation of the coating by fusion/diffusion processes. For some coatings, both methods are combined.

Chemical
—— Vapor Deposition ——

CHEMICAL VAPOR DEPOSITION (CVD) is a process which in some ways is like the gas carburizing and the carbonitriding processes. In the CVD process, a reactant atmosphere gas is fed into the processing chamber where it decomposes at the surface of the workpiece, liberating one material for either absorption by or accumulation on the workpiece. A second material is liberated in gas form and is removed from the processing chamber, along with excess atmosphere gas, as a mixture referred to as off-gas.

Reactant atmospheres used in CVD include chlorides, fluorides, bromides and iodides, as well as carbonyls, organometallic compounds, hydrides and hydrocarbons. Hydrogen is often included as a reducing agent. The reactant atmosphere must be reasonably stable until it reaches the substrate, where reaction must occur with reasonably efficient conversion of the reactant. It is sometimes necessary to heat the reactant to produce the gaseous atmosphere.

A few reactions for deposition occur at substrate temperatures below 200 °C (390 °F). Some organometallic compounds deposit at temperatures below 600 °C (1110 °F). Most reactions and reaction products require temperatures above 800 °C (1470 °F).

COMMON CVD COATINGS

Nickel. CVD nickel is generally separated from a nickel carbonyl, $Ni(Co)_4$, atmosphere. The properties of the deposited nickel are equivalent to those of sulfamate nickel deposited electrolytically.

Tungsten may be deposited by thermal decomposition of tungsten carbonyl at 300 to 600 °C (570 to 1110 °F), or it may be deposited by hydrogen reduction of tungsten hexachloride at 700 to 900 °C (1290 to 1650 °F). The most convenient and most widely used reaction is the hydrogen reduction of tungsten hexafluoride.

Chromium. Coating of steel and other alloys with chromium may be done by pack cementation, a process similar to pack carburizing, or by a dynamic, flow-through CVD process (open-tube CVD). Pack cementation may be considered CVD, because coating occurs by thermal decomposition of a gaseous chromium carrier.

Titanium Carbide. Wear-resistant coatings of titanium carbide are formed by the hydrogen reduction of titanium tetrachloride in the presence of methane or some other hydrocarbon. The substrate temperature ranges from 900 to 1010 °C (1650 to 1850 °F), depending on the substrate.

PROCESS CONDITIONS

Surface preparation for CVD coating by degreasing or grit blasting is similar to methods used for other dry coating processes such as vacuum evaporation and sputtering. In addition, a CVD precoating treatment may be given. For substrates whose superficial oxides may be reduced by hydrogen, a soak in a hydrogen atmosphere at an appropriate temperature is used.

Deposition Temperature. With few exceptions, the rate of deposition from CVD reactions increases with temperature in a manner specific to each reaction. Deposition at the highest possible rate is preferable; however, there are limitations which require a processing compromise. As temperature is increased, the deposition goes through two and sometimes three stages.

───── Vacuum Coating ─────

VACUUM COATING is the process of depositing metals and metal compounds from a source in a high-vacuum environment onto a substrate. Three principal techniques used to accomplish the deposition process are evaporation, ion plating, and sputtering. In each technique, the transport of vapor is carried out in an evacuated, controlled environment chamber at a residual air pressure of 1 to 10^{-5} Pa (10^{-2} to 10^{-7} torr).

In the evaporation process, vapor is generated by heating the source material to a temperature such that the vapor pressure significantly exceeds the ambient chamber pressure and produces sufficient vapor for practical deposition.

EVAPORATION

Evaporation is a surface phenomenon, and does not necessarily constitute boiling. Surface evaporation occurs from conduction, not from formation of subsurface vapor bubbles at a depth below the surface, if the liquid has sufficient thermal conductivity such as that possessed by metals. The rate of evaporation, assuming that none of the vaporized material returns to the surface, is given by the Langmuir equation:

$$w = 0.0585 \, P \left(\frac{M}{T} \right)^{1/2} \text{ g/cm}^2 \text{ s}$$

where P is the vapor pressure of the evaporant

Table 6. Temperatures at which vapor pressure is 1 Pa (10^{-2} torr) for several common elements

Element	Temperature	
	°C	°F
Aluminum	1150	2100
Beryllium	1245	2270
Bismuth	680	1260
Cadmium	265	510
Carbon	2460	4460
Chromium	1400	2550
Cobalt	1520	2770
Copper	1260	2290
Germanium	1400	2550
Gold	1400	2550
Indium	945	1730
Lead	715	1320
Magnesium	440	825
Manganese	940	1720
Molybdenum	2350	4260
Nickel	1530	2780
Osmium	2910	5270
Platinum	2090	3790
Rhenium	3065	5550
Silicon	1470	2680
Silver	1630	2970
Tantalum	3060	5540
Tin	1250	2280
Titanium	1740	3160
Tungsten	3230	5840
Uranium	1930	3510
Vanadium	1850	3360
Yttrium	1630	2970
Zinc	345	650
Zirconium	2400	4350

in torr, M is the molecular weight of the evaporating material, T is absolute temperature, K, and w is the amount of vapor material evaporated per unit time.

The evaporating metal atom (molecule) leaves the surface in a straight line. Collision of the evaporating atoms with residual gas molecules randomly moving about the vacuum chamber lowers the energy of the evaporant and also changes its direction. A sufficient number of collisions results in too great a loss in energy and too great a randomization of the flow from the source to produce an adherent deposit. Therefore, proper positioning of the substrate with respect to the source and the quality of the vacuum in the chamber must be considered for each application. Highest quality coatings (electronic and optical applications) are deposited when the source-to-substrate distances are less than the mean path distance between collisions of a gas molecule and the equipment. Deposits acceptable for decorative purposes can be deposited at distances of several mean free paths. At a chamber pressure of 10^{-1} Pa (10^{-3} torr), the source-to-substrate distance should be less than 500 mm (20 in.). At 10^{-2} Pa (10^{-4} torr), the source-to-substrate distance can be increased to over 4000 mm (160 in.).

To coat the entire surface of a substrate, it must be rotated and translated over the vapor source. Deposits made on substrates positioned at low angles to the vapor source result in fibrous, poorly bonded structures. Deposits resulting from excessive gas scattering are poorly adherent, amorphous, and generally dark in color. Thus, if only one surface is to be coated, other surfaces should be masked to prevent coating deposition. The highest-quality deposits are made on surfaces nearly normal to the vapor flux. Such deposits faithfully reproduce the substrate surface texture. Highly polished substrates produce lustrous deposits, and the bulk properties of the deposits are maximized for the given deposition conditions.

For practical deposition rates, source material should be heated to a temperature so that its vapor pressure is at least 1 Pa (10^{-2} torr) or higher. Temperatures at which the vapor pressure is 1 Pa (10^{-2} torr) for several common elements are shown in Table 6. Deposition rates for evaporating bulk vacuum coatings can be very high. Commercial coating equipment can deposit up to 500 000 Å/min (2000 μin./min) using large ingot material sources and high-powered electron beam heating techniques.

As indicated previously, the directionality of evaporating atoms from a vapor source generally requires the substrate to be articulated (positioned in a specific fashion) within the vapor cloud. To obtain a specific film distribution on a substrate, the shape of the object, the arrangement of the vapor source relative to the component surfaces, and the nature of the evaporation source must be accounted for. The distribution of the evaporant leaving the surface of the source is given by the cosine law, as illustrated in Fig. 7 for a small surface source of area dA, emitting a vapor flux (m). The material passing through a solid angle ψ at a direction of angle ϕ with the normal to the surface dA, per unit time, is given by:

$$dm = \frac{m}{\pi} \cos \phi \, d\psi \qquad \text{(Eq 1)}$$

Material arriving at a small substrate area (dA_2), inclined at an angle θ to the vapor stream direction, at a distance r from the source can be expressed in the following manner. The projection of solid angle $d\psi$ onto area dA_2 is:

$$d\psi = \frac{\cos \theta}{r^2} \, dA_2 \qquad \text{(Eq 2)}$$

Combining Eq 1 and 2 gives:

$$dm = \frac{m}{\pi r^2} \cos \phi \cos \theta \, dA_2 \qquad \text{(Eq 3)}$$

These relationships have held true at relatively low rates of evaporation.

At high rates of evaporation, such as those experienced in the practical production of thick films using high-power electron beam sources, the mass distribution is much steeper than predicted by Eq 3. This has been attributed to: (*a*) evaporant interactions above the source, (*b*) formation of a depression in the surface by the impact of energetic electrons from the heating source, and (*c*)

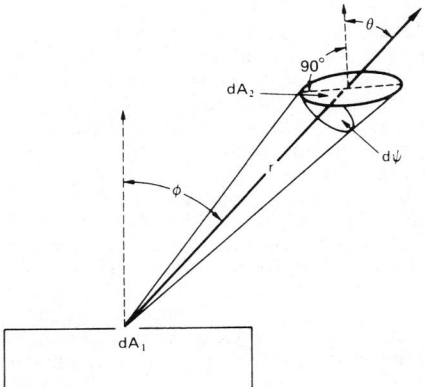

Fig. 7. Surface element (dA_2) receiving deposit from a small-area source (dA_1)

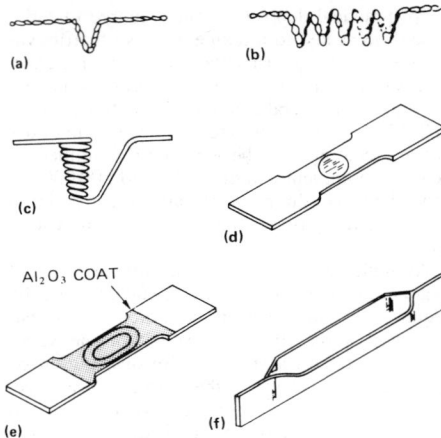

(a) Hairpin source. (b) Wire helix. (c) Wire basket. (d) Dimpled foil. (e) Dimpled foil with alumina coating. (f) Canoe-type source.

Fig. 8. Wire and metal-foil sources

formation of waves on the surface of the molten source.

Evaporation Sources

Most elemental metals, semiconductors, compounds, and many alloys can be directly evaporated in vacuum.

Resistance Sources. The simplest sources are resistance wires and metal foils of various types (Fig. 8). They are generally constructed of refractory metals such as tungsten, molybdenum and tantalum. The filaments serve the dual function of heating and holding the material for evaporation.

Sublimation Sources. Several elements, such as chromium, palladium, molybdenum, vanadium, iron and silicon, can be evaporated directly from the solid phase.

Crucible sources comprise the greatest applications in high-volume production for evaporating refractory metals and compounds. The crucible materials are usually refractory metals, oxides and nitrides, and carbon. Crucibles offer the capability of evaporating intermediate quantities of materials.

Heating can be accomplished by radiation from a secondary refractory heating element (Fig. 9), by a combination of radiation and conduction, and by radio frequency induction heating.

Evaporation Methods

Electron beam heating provides a flexible heating method that can concentrate heat on the evaporant. Portions of the evaporant next to the container can be kept at lower temperatures, thus minimizing interaction. Two principal electron guns in use are the linear focusing gun, which uses magnetic and electrostatic focusing methods (Fig. 10a), and the bent-beam magnetically focused gun (Fig. 10b). The bent-beam gun has the advantage of compactness; a large filament area allows operation below 10 kV without the sacrifice in power experienced with the linear system. The disadvantages of electron beam heating methods are (a) interference with evaporation-rate monitors as a result of charge accumulation, and (b) electrical charging of substrates by stray electrons; interference of electronic rate monitors can be minimized by placing a grounded grid in front of the monitor. Substrates may be allowed to float

or be maintained at the filament voltage to resist charging. The guns must be operated in a vacuum environment of 10^{-2} Pa (10^{-4} torr) or less to preserve filament life and to prevent gas scattering of the emitted electrons.

Continuous Feed. High-rate evaporation of alloys to form film thicknesses of 100 to 150 μm (4 to 6 mil) requires electron beam heating sources and large quantities of evaporation source material. Electron beams of 45 kW or higher are used to melt evaporants in water-cooled copper hearths up to 150 by 450 mm (6 by 18 in.) in cross section.

APPLICATIONS

Coatings applied by vacuum deposition may be broadly classified as decorative and functional. Decorative coatings are widely used in the automotive, home appliance, hardware, costume jewelry, and novelty fields. Functional coatings have numerous applications such as reflection, antireflection, filter and beam-splitter coatings on optical instruments; current-carrying, dielectric and semiconductor coatings on electronic components; hot corrosion-resistant coatings on aircraft and missile parts; and oxidation-resistant coatings for gas turbine blades and vanes.

REQUIREMENTS OF THE SUBSTRATE MATERIAL

The primary requirement of the material to be coated is that it be stable in vacuum. It must not evolve gas or vapor when exposed to the metal vapor. Gas evolution may result from:

- Release of gas adsorbed on the surface
- Release of gas trapped in the pores of a porous substrate
- Evolution of a material such as plasticizers used in plastics
- Actual vaporization of an ingredient in the substrate material.

PROCESSING TECHNIQUES AND LIMITATIONS

Vacuum deposition is used to accomplish (a) bulk coating of powders or small parts (encap-

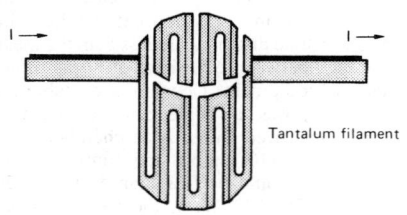

Fig. 9. Molybdenum crucible source with tantalum sheet filament

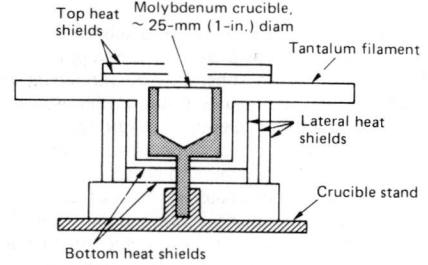

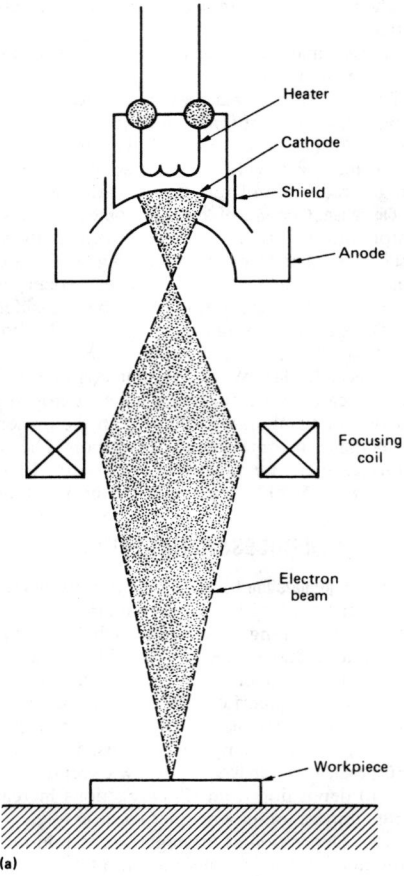

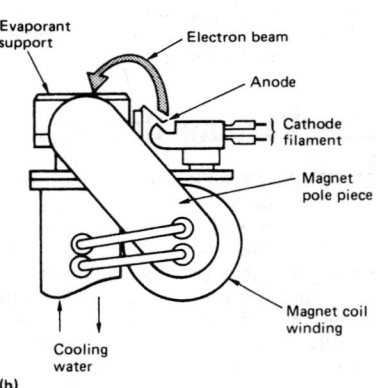

(a) Linear focusing gun. (b) Bent-beam electron gun with water-cooled evaporant support.

Fig. 10. Two principal electron guns for electron beam heating

sulation), (b) batch coating of individual parts, or (c) continuous or semicontinuous coating of rolled materials.

Encapsulation. Tiny parts or finely divided powders are transferred through a coating zone while held in a monolayer on a supporting oscillating tray. The material to be coated must be free-flowing and not excessively agglomerated. The material to be coated also should preferably be spherical, although irregular crystals of powder can be coated, provided reentrant angles are not present at the edges of the crystal. Sharp edges are undesirable but are not impossible to coat.

The minimum particle size of powder that can be encapsulated is approximately 10 μm (0.4 mil) in diameter. Particles of smaller size resist rotation during the coating operation, and therefore only one surface is exposed.

Batch Coating. Irregularly shaped finished parts are coated for either decorative or functional applications by batch coating. These parts are individually fixtured and are rotated during the coating operation to expose all surfaces.

Continuous and Semicontinuous Coating. Rolls of material that vary from monofilaments a few μm (mils) in diameter to sheets 1.8 m (6 ft) in width may be continuously coated. For coating these products, materials of evaporation that are at least as volatile as gold should be used. With metals of this type, the beam of metal vapor may be directed downward if necessary, so that complex surfaces can be more readily coated from multiple sources.

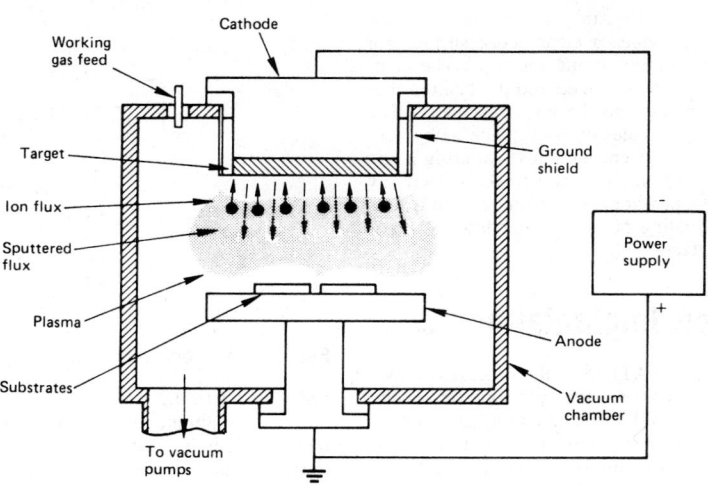

Fig. 11. Planar electrode system used for sputtering

Sputtering

SPUTTERING is a process wherein material is ejected from the surface of a solid or liquid because of the momentum exchange associated with bombardment by energetic particles. The bombarding species are generally ions of a heavy inert gas. Argon is most commonly used. The source of ions may be an ion beam or a plasma discharge into which the material to be bombarded is immersed.

In the plasma-discharge sputter coating process, a source of coating material called a target is placed into a vacuum chamber which is evacuated and then backfilled with a working gas, such as argon, to a pressure adequate to sustain a plasma discharge. A negative bias is then applied to the target so that it is bombarded by positive ions from the plasma.

Sputter coating chambers are typically evacuated to pressures ranging from 10^{-3} to 10^{-5} Pa before backfilling with argon to pressures of 0.1 to 10 Pa. The intensity of the plasma discharge, and thus the ion flux and sputtering rate that can be achieved, depends on the shape of the cathode electrode, and on the effective use of a magnetic field to confine the plasma electrons.

The deposition rate in sputtering depends on the target sputtering rate and the apparatus geometry — for example, the target shape and the position of the substrates relative to the target or targets. It also depends on the working gas pressure, since high pressures limit the passage of sputtered flux to the substrates.

For many years, most sputtering has been done using simple planar electrode systems of the type shown in Fig. 11, but the development of a class of sputtering sources with magnetic plasma confinement, called magnetrons, has greatly enhanced the capabilities of the sputtering process.

Ion Plating

ION PLATING is a generic term applied to atomistic film deposition processes in which the substrate surface and/or the depositing film is subjected to a flux of high-energy particles (usually gas ions) sufficient to cause changes in the interfacial region or film properties. Such changes may be in film adhesion to the substrate, film morphology, film density, film stress, or surface

coverage by the depositing film material. The above definition of ion plating refers only to processes that affect the film and/or substrate and does not define either the source of the depositing material or the origin of the bombarding species.

Ion plating is typically done in an inert-gas discharge system similar to that used in sputter deposition (see previous article), except that the substrate is the sputtering cathode and the bombarded surface often has a complex geometry.

A schematic of a system typical of many ion-plating operations is shown in Fig. 12. Basically, the ion-plating apparatus is comprised of a vacuum chamber and a pumping system, which is typical of any conventional vacuum deposition unit. There is also a film atom vapor source and an inert-gas inlet. For a conductive sample, the workpiece is the high-voltage electrode, which is insulated from the surrounding system. In the more generalized situation, a workpiece holder is the

high-voltage electrode and either conductive or nonconductive materials for plating are attached to it. Once the specimen to be plated is attached to the high-voltage electrode or holder and the boat or filament vaporization source is loaded with the coating material, the system is closed and the chamber is pumped down to a pressure in the range of 10^{-3} to 10^{-4} Pa (10^{-5} to 10^{-6} torr). When a desirable vacuum has been achieved, the chamber is backfilled with argon to a pressure of approximately 1 to 0.1 Pa (10^{-2} to 10^{-3} torr). A potential of -3 to -5 kV is then introduced across the high-voltage electrode (specimen or specimen holder) and the ground for the system. A glow discharge occurs between the electrodes which results in the specimen being bombarded by the high-energy argon ions produced in the discharge, which is equivalent to direct-current sputtering. The argon-ion bombardment effectively cleans the specimen surfaces by removing any adsorbed layers or surface contamination. The

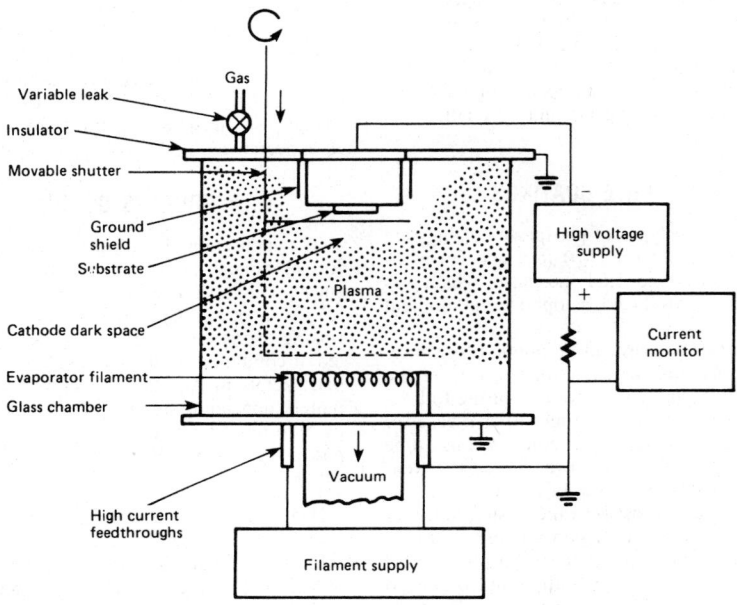

This apparatus employs a dc gas discharge and an evaporator filament.
Fig. 12. Simple ion-plating apparatus

cleaning step usually can be completed in a few minutes and provides a clean work surface for receipt of the plating atoms and ions. The coating source is then energized and the coating material is vaporized into the glow discharge. The above process provides the workpiece with a uniform ion bombardment, and gas scattering along with ion deflection gives rise to the effectively high throwing power of this technique and results in uniform coating of even rather intricate variations in surface contour.

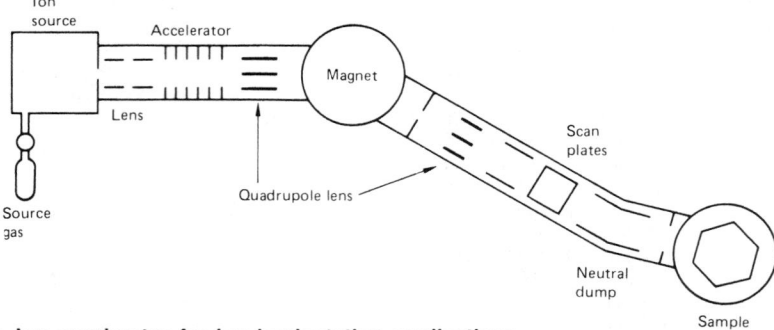

Fig. 13. Ion accelerator for ion implantation applications

Ion Implantation

ION IMPLANTATION is the process of modifying the physical or chemical properties of the near surface of a solid by embedding appropriate atoms into it from a beam of ionized particles. The properties to be modified may be electrical, optical or mechanical, and they may relate to the semiconducting behavior of the material or its corrosion behavior. The solid may be crystalline, polycrystalline or amorphous and need not be homogeneous.

The use of accelerated electron beams can produce penetrations into a substrate surface on the order of 0.1 to 0.2 μm (1000 to 2000 Å) at 100 kV, and with higher accelerating voltages the potential for depth of penetration is increased. Surfaces may be treated by ion implantation to produce an effective alloyed surface layer where the composition varies as a function of depth. An ion beam implanter, which can provide a range of ion energies, can produce remarkable variations in characteristics at the surface and to limited depths below the surface. In addition, unique alloys not possible through normal alloying can be produced through this technique. Theoretically, two or more metals, completely insoluble in each other, can be alloyed in this manner. An ion beam accelerator is illustrated schematically in Fig. 13.

Although relatively thin, an implanted ion layer can alter mechanical and chemical properties significantly. For example, the optimum surface-layer thickness for wear resistance under high-load conditions can be quite thin—only 2 μm (0.08 mil) for a molybdenum disulfide compound applied in a conventional manner.

Nonmetallic Coating Processes

Edited by Thomas A. Taylor, Union Carbide Coatings Service

Phosphate Coating

PHOSPHATE COATING is the treatment of iron, steel, galvanized steel, or aluminum with a dilute solution of phosphoric acid and other chemicals in which the surface of the metal, reacting chemically with the phosphoric acid medium, is converted to an integral, mildly protective layer of insoluble crystalline phosphate.

Phosphate coatings range in thickness from less than 3 to 50 μm (0.1 to 2 mils). Coating weight (grams per square metre of coated area), rather than coating thickness, has been adopted as the basis for expressing the amount of coating deposited.

PHOSPHATE COATINGS

Three principal types of phosphate coatings are in general use: (a) zinc, (b) iron and (c) manganese. A fourth type, lead phosphate, which was more recently introduced, is operated at ambient temperatures.

Zinc phosphate coatings can be applied by spraying, by immersion, or by a combination of the two. Coatings can be used for any of the following applications of phosphating: (a) as a base for paint or oil; (b) as an aid to cold forming, tube drawing or wiredrawing; (c) for increasing wear resistance; or (d) for rustproofing.

Iron phosphate coatings were the first to be used commercially. Early iron phosphating solutions consisted of ferrous phosphate/phosphoric acid used at temperatures near boiling and produced dark gray coatings with coarse crystals. The term "iron phosphate coatings" refers to coatings resulting from alkali-metal phosphate solutions operated at pH values ranging from 4.0 to 5.0, which produce exceedingly fine crystals.

Although iron phosphate coatings are applied to steel to provide receptive surfaces for bonding of fabrics, woods and other materials, they are used chiefly as base coatings for subsequent painting.

Manganese phosphate coatings are applied to ferrous parts (bearings, gears, and internal-combustion engine parts, for example) for break-in and for prevention of galling. These coatings are usually dark gray. However, because almost all manganese phosphate coatings are used as an oil base and because the oil intensifies the coloring, manganese phosphate coatings are usually black in appearance.

PROCESS DETAILS

The application of a phosphate coating as a base for subsequent painting normally comprises five successive operations: (a) cleaning, (b) rinsing, (c) phosphating, (d) rinsing and (e) chromic acid rinsing. Some of these operations may be omitted or combined, such as cleaning and coating in one operation. Additional operations may be required, depending on the surface condition of the parts to be phosphated or on the function of the phosphate coating. Parts exemplifying these exceptions are:

- Heavily scaled parts, which may require pickling before cleaning
- Parts with extremely heavy coatings of oil or drawing compounds, which may require rough cleaning before the normal cleaning operation
- Parts that are tempered in a controlled atmosphere before being phosphated, which may not require cleaning and rinsing before phosphating
- Parts that are phosphated and later oiled for antifriction purposes, which may have the chromic acid rinse omitted, because corrosion resistance is not required. Some rust-preventive oils negate the need for a chromic rinse while still providing excellent corrosion resistance
- Automotive parts when electrodeposition of a primer is involved. A deionized water rinse is required after the chromic acid rinse.

Phosphating Methods

Phosphate coatings may be applied to a surface by either immersion or spraying, or by a combination of immersion and spraying.

Immersion. All three types of phosphate coatings (zinc, iron and manganese) can be deposited by immersion. Immersion is applicable to coating of racked parts, barrel coating of small parts, and continuous coating of strip. In general, smaller parts are more economically coated by immersion than by spraying.

Spraying. Zinc and iron phosphate coatings are applied by the spray method, although manganese phosphate is not. The spray method is used to apply a phosphate coating to racked parts, such as panels for household appliances, or to a continuous strip. Occasionally, baskets of parts are passed through a spray system, but this is not a preferred method. Spray phosphating, because of the equipment required, is usually most applicable to high-volume coating of parts.

Phosphating Time

In general, the spray method produces a given coating weight at a faster rate than the immersion

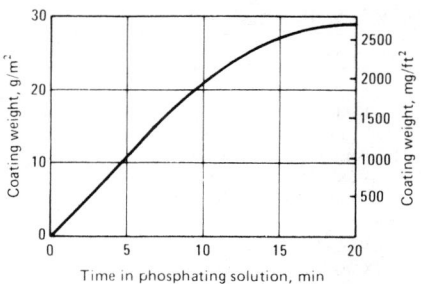

Fig. 1. Weight of manganese phosphate coatings as a function of time of exposure of steel surface to phosphating solution

Table 1. Operating temperature ranges for phosphating solutions in phosphating applications

Phosphate coating/method	Metal treated	Reason for treatment	Operating temperature °C	°F
Medium iron, immersion	Steel	Paint bonding	60-82	140-180
Heavy zinc, immersion	Steel	Corrosion resistance	88-96	190-205
Medium zinc, immersion	Steel	Paint bonding	32-82	90-180
Medium zinc, immersion	Steel plated with zinc or cadmium	Paint bonding	60-82	140-180
Medium zinc, spray	Steel sheet	Paint bonding	38-60	100-140
Medium zinc, spray	Steel	Cold drawing	60-74	140-165
Medium zinc, spray	Galvanized steel	Paint bonding	49-60	120-140
Manganese, immersion	Steel	Wear resistance	93-99	200-210

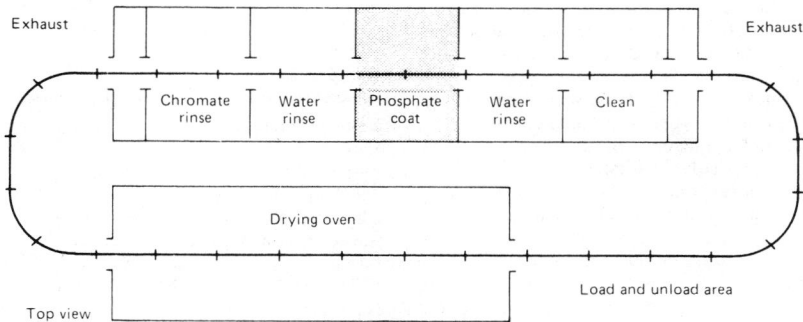

Fig. 3. Continuous conveyorized line for spray phosphating

method. In spray zinc phosphating, a coating of 1.6 to 2.1 g/m² (150 to 200 mg/ft²) normally can be obtained in 1 min or less, whereas obtaining a coating of this weight by the immersion method may require as much as 2 to 5 min.

The weight of manganese phosphate coatings on steel surfaces is a function of immersion time, as indicated by the curve in Fig. 1.

Operating Temperature

Although operating temperatures of different phosphating solutions may range from 32 to 99 °C, or 90 to 210 °F (Table 1), individual solutions are compounded to operate at maximum efficiency within specific temperature limits.

EQUIPMENT FOR IMMERSION SYSTEMS

An immersion phosphating system for all types of coatings (zinc, manganese and iron) should include (a) the required number of tanks, (b) temperature and solution-level controls, (c) overflow and drainage systems, (d) vapor-exhaust systems and (e) materials-handling devices. When drums are used to contain the parts, devices are required at each tank to rotate the drums at approximately 4 rpm while the drums and the parts within them are submerged.

Phosphating tanks are usually made from low-carbon steel plate about 6 mm (¼ in.) thick. A tank and drum for immersion phosphating of small parts are shown in Fig. 2.

Tank accessories, including (a) steam coils or other heating media, (b) piping, (c) screens, (d) drum trunnions and (e) drum-rotating mechanisms, may be made of low-carbon steel or stainless steel. Electropolished stainless steel steam coils permit less sludge buildup on the coils.

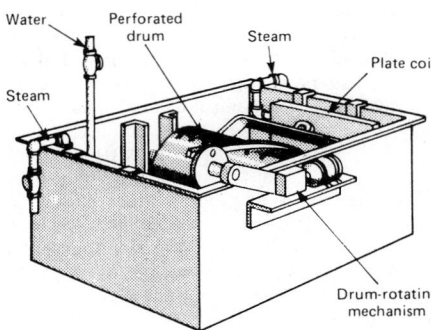

Drum into which parts are loaded is shown in immersion position.

Fig. 2. Immersion phosphating tank for batch coating of small parts

Tank Design. Tanks should have sufficient capacity to stabilize solution temperature and solution concentration, and to prevent rapid buildup of solution contamination. Tanks for the phosphating stage should have a sloping bottom, with at least 0.46 m (1.5 ft) of space below the lowest work level to accommodate sludge buildup.

Rinse Tanks. Water rinse tanks and associated equipment, including (a) steam coils or other heating media, (b) piping and (c) screens, may be constructed of low-carbon steel. Rinse tanks for certain parts sometimes require drum-rotating devices.

Drying equipment for immersion phosphating systems can be of several types. For small parts, such as washers, a centrifuge may be used to spin off moisture. If parts are hot enough, no additional heated air is required. However, if parts are cold, heated air may be introduced into the centrifuge.

EQUIPMENT FOR SPRAY SYSTEMS

Spray systems usually are completely enclosed in a continuous, chambered tunnel or cabinet for better control of the process and cleanness of the operation. Parts or panels to be processed are hung on racks or hooks, or are placed in baskets, and are automatically carried through the various stages of the spray phosphating line (Fig. 3).

Chromate Conversion Coating

CHROMATE CONVERSION COATINGS are formed on metal surfaces as a result of the chemical attack that occurs when a metal is immersed in or sprayed with an aqueous solution of chromic acid, chromium salts such as sodium or potassium chromate or dichromate, hydrofluoric acid or hydrofluoric acid salts, phosphoric acid, or other mineral acids. The chemical attack causes dissolution of some surface metal and formation of a protective film containing complex chromium compounds.

Chromate treatments are of two types: (a) those that are complete in themselves and deposit substantial chromate films on the substrate metal, and (b) those that are used to seal or supplement oxide, phosphate or other types of nonmetallic protective coatings. Two chromate treatments for aluminum are complete in themselves: chromium chromate and chromium phosphate. To form the chromium phosphate coating, phosphoric acid or phosphoric acid salts are required in addition to chromic and hydrofluoric acids or their corresponding salts.

For a chromate film to be deposited, the passivity that develops on a metal in a solution of strictly chromate anions must be broken down in solution in a controlled way. This is achieved by adding other anions, such as sulfate, nitrate, chloride or fluoride, as activators to attack the metal.

Painting

PAINTING is a generic term for the application of a thin organic coating to the surface of a material for decorative, protective or functional purposes.

TYPES OF PAINT

The general terms "paint" and "organic coating" are essentially interchangeable and are used to designate certain coatings having an organic base. Most organic coatings are based on a film former or binder which is dissolved or dispersed in a solvent or in water. This film-forming liquid constitutes the vehicle in which pigments are dispersed to give color, opacity and other properties to the dried film. Many other ingredients may be added to the vehicle to achieve specific film properties. These would include such things as driers to aid curing, plasticizers to impart flexibility and other properties, and stabilizers to lessen the deleterious effects of heat or sunlight. A wide variety of film-forming materials is available and includes oils, varnishes, synthetic resins and polymers such as cellulose, vinyl, epoxy and polyester. In general, major performance characteristics depend on the binder used.

Enamels are topcoats characterized by their ability to form a smooth surface that is typically of high gloss, but may also include lower degrees of gloss such as flat enamels. Enamels may air dry or bake. Air-dry enamels are cured essentially by a combination of solvent evaporation and oxidation.

Lacquers are compositions based on synthetic thermoplastic film-forming materials dissolved in organic solvent. These dry primarily by solvent evaporation.

Water-borne paints are dilutable with water. There are three principal types: solutions, colloidal dispersions, and emulsions. Solution coatings are based on water-soluble binders. Many conventional binders (alkyds, acrylics and epoxies) can be made water soluble by chemically attaching polar groups such as carboxyl, hydroxyl and amide, which are strongly hydrophilic.

Electrophoretic paints are special water-reducible paints. Resin and pigment materials are shipped and stored as concentrates to be added to the production tank as needed. Electrophoretic films are always deposited from a dip tank. The operating bath consists of resin concentrate and pigment concentrate mixed with deionized water and small amounts of solubilizers and defoamers. The concentration of nonvolatile solids in the bath varies from about 10 to 20%, depending on type and composition.

Autophoretic paints are water-reducible paints deposited on metal surfaces by the catalytic action of the metal on the paint materials in the bath. Currently, only ferrous surfaces activate the autophoretic paints available commercially. Tubular automotive frames are coated by this method, because the entire length of the tubing can be coated inside and outside with equal ease.

High-solids paints contain 50% or more solids by volume. One method of obtaining high-solids paints is to use lower-molecular-weight polymers, which require less solvent to attain the desired application viscosity. Another method of reducing viscosity of high-solids paints is by heating the paint material to a temperature of about 32 to 52 °C (90 to 125 °F). Many two-component systems use a catalyst to increase the rate of the curing reaction. Fast-reacting two-component systems are usually applied with special spray guns that mix the two components at the spray nozzle.

SELECTION OF A PAINT SYSTEM

Although a wide variety of coatings is available, the ideal coating system — one with all desired performance properties, simple application, and low cost — is difficult to find. Factors such as (a) regulations, (b) service environment, (c) substrate and service conditions, (d) basic function, (e) application limitations and (f) cost usually must be compromised.

Service Environment. In selection of a paint system, many factors concerning service environment must be considered. Needs are different for interior and exterior coatings. The specific properties needed, such as resistance to heat, cold, sunlight and weathering, must be determined. If

Table 2. Organic coatings selected for corrosion resistance in various environments

Coatings	Applications
Outdoor exposure	
Oil paints	Buildings, vehicles, bridges; maintenance
Alkyds	Trim paints, metal finishes, product finishes
Amino resin-modified alkyds	Automotive, metal awnings, aluminum siding
Nitrocellulose lacquers	Product finishes, aerosol lacquers
Acrylics	Automotive finishes
Marine atmosphere	
Alkyds, chlorinated rubber, phenolics, vinyls, vinyl-alkyds	Superstructures and shore installations
Urethanes	Clear marine varnishes
Water immersion	
Phenolics	Ship bottoms
Vinyls	Ship bottoms, locks
Chlorinated rubber	Ship bottoms, swimming pools
Urethanes	Clear marine varnishes
Chemical fumes	
Epoxies, chlorinated rubber, vinyls, urethanes	Chemical-processing equipment
Extreme sunlight	
Vinyls	Metal awnings
Acrylics	Automotive finishes
Silicone alkyds	Petroleum-industry processing equipment
High humidity	
Amino resin-modified alkyds	Refrigerators, washing machines
Epoxies	Air conditioners
Catalyzed epoxies, chlorinated rubber, phenolics	Maintenance; chemical and paper plants
High temperature	
Epoxies	Motors, piping, 120 °C (250 °F) max
Modified silicones	Stove parts, roasters, 205 °C (400 °F) max
Silicones	Stove parts, roasters, 290 °C (550 °F) max; aluminum-pigmented paints, 650 °C (1200 °F) max

Table 3. Coating recommendations for environmental protection of parts made from steel sheet
Each of these parts should receive an appropriate phosphate coating before being painted.

Part No.	Priming	Finishing	Application
For 3-yr exposure to marine atmosphere			
1	Epoxy-ester	Modified acrylic	Electrostatic spray
2	Epoxy-ester	Epoxy-ester	Dip or spray
3	Zinc chromate	Phenolic or vinyl	Dip, spray, or flow coat
For 10-yr (minimum) indoor exposure			
4	Alkyd or modified epoxy	Urea-alkyd or modified acrylic	Spray or flow coat

the coating needs to be resistant to chemicals, the specific chemical, such as acid, alkali, solvent, or water immersion, should be specified. Mechanical properties required, including hardness, flexibility, impact resistance and abrasion resistance, should also be determined.

Corrosion of steel and cast iron occurs in all environments. The rate and extent of corrosion vary from mild attack in dry, clean environments to highly accelerated attack in marine or industrial areas where corrosive fumes are present in the air. Table 2 lists paints selected for service in a wide range of corrosive conditions.

Table 3 lists the paints and methods of application used for prime and finish coats on three steel parts required to withstand three years of outdoor exposure in a marine environment and on a steel part required to withstand ten years of indoor exposure. All four parts would be phosphate coated before being painted.

Paint films also may be required to resist acids and alkalis, solvents, staining, heat, impact, marring and abrasion. Some coatings must be able to withstand flexing without cracking or flaking. Table 4 lists paints that have proven successful in withstanding mechanical and chemical action.

SURFACE PREPARATION

The importance of proper surface preparation to the durability of any coating system cannot be overemphasized. Without proper surface preparation, the finest paint, applied with the greatest of skill, will fall short of its maximum performance or may even fail miserably. A coating can perform its function only so long as it remains intact and firmly bonded to the substrate.

Cleaning. Before being painted, metals usually are exposed to one or more fabricating processes, such as rolling, stamping, forming, forging, ma-

Table 4. Paints selected for resistance to mechanical or chemical action

Action	Paint
Abrasion	Vinyls; plastisols; polyurethanes
Impact	Epoxies; vinyls; polyurethanes
Marring	Thermosetting acrylics; vinyls
Flexing	Epoxies; vinyls
Acids	Chlorinated rubber; vinyls; epoxies
Solvents	Epoxies; phenolics
Detergents	Thermosetting acrylics; epoxies
Staining	Thermosetting acrylics
Gasoline	Alkyds; lacquers
Alkalis	Phenolics
Heat	Alkyd-amines; silicone resins

chining and heat treating. In these processes, the metal surfaces pick up various contaminants that can either interfere with the adhesion of the paint film or allow corrosion to progress beneath the paint film and cause it to fail prematurely.

The principal surface contaminants that adversely affect the performance of paint films include oils, greases, dirt, rust, mill scale, water, and salts such as chlorides and sulfides. These contaminants must be removed from the surface before paint is applied.

Selection of cleaning process is governed by the soil or contaminant to be removed, the degree of cleanness required, the type of paint to be applied, and the size, shape, material and end use of the part.

A high degree of cleanness can be obtained by combining mechanical and chemical cleaning methods. Before being painted, steel surfaces of tank cars used to carry corrosive materials were alkaline or solvent cleaned and then abrasive blasted. This removed all contaminants. A corrosion-inhibiting primer was applied, followed by a chemical-resistant topcoat. This paint film, controlled to a minimum thickness of 75 μm (3 mils), has a life expectancy of 4 to 6 years.

The influences of part size, shape and required paint-film durability on the method of surface preparation are shown in Table 5.

SPRAYING

Spraying is adaptable to either large-volume or low-volume production. Applications may be limited because of solvent emissions, possible fire hazards, or potential damage from overspray. Spraying methods include (a) conventional air spraying, in which the paint is atomized and propelled against the work by means of compressed air; (b) hot spraying; (c) hydraulic airless spraying; and (d) air and airless electrostatic spraying.

Hot spraying is a method in which air-atomized spraying equipment is used in conjunction with a heat exchanger to heat the paint to a predetermined temperature. The temperature range for hot spraying is usually from 60 to 82 °C (140 to 180 °F). In air-atomized spraying, viscosities suitable for application are obtained with the use of solvents. The hot spraying method uses heat to lower the viscosity to the optimum range for spraying, allowing application of paint with higher solids content.

Airless spraying, with either heated or unheated paint, uses hydraulic (pumping) pressure to propel the paint through the hoses and atomizing nozzle. The main advantages of airless spray painting stem from the elimination of air as the force for atomizing and propelling the paint. Airless spraying requires the release of considerably less energy at the nozzle, and overspray is minimized.

Electrostatic Spraying. In electrostatic spray painting, electrically charged atomized particles of paint are attracted to the grounded part. Paint for electrostatic application can be atomized by conventional air, airless or rotational techniques.

One of the chief advantages of electrostatic spraying is the small loss of paint from overspray. Whereas in conventional spray painting as much as 70% of the paint sprayed may be lost because of overspray, as little as 10% may be lost in electrostatic spraying. Another advantage is the ability of this method to produce a consistent paint film over a long production run.

The main disadvantage of electrostatic spraying is that the electrostatic attraction of the part for the paint draws the paint to the nearest edge or surface of the part, and it is difficult or impossible to get paint into deep recesses, corners and shielded areas.

DIP PAINTING

Dip painting consists of submerging a part in paint contained in a tank, withdrawing the part, and permitting the part to drain. Parts with complex surfaces may be coated efficiently by dipping. Larger parts, produced in quantity, are racked or hung on conveyors, which carry the parts to the paint tank, automatically immerse and withdraw the parts, and carry them over drip troughs into which the excess paint runs off.

Dip painting is seldom used where uniformity of paint thickness is required. This procedure may be unsatisfactory for painting parts having machined holes or surfaces where masking is impractical. Usually, paint films applied by dipping are heavier at the bottom than at the top, because of paint runoff, and are thin at sharp edges. In addition, bubbles and bumps may be found at the bottom edges of painted pieces. Parts may be dip painted to obtain complete coverage and then spray painted to obtain a required surface appearance.

The blower wheel and wire fan guard shown in Fig. 4(a) and (b) are examples of parts that require coating of all surfaces, but for which a variation in the coating thickness is acceptable. Dip painting, followed by spinning to remove excess paint, is the preferred method for coating such parts.

FLOW COATING

In flow coating, paint is pumped from a storage tank through properly positioned nozzles, onto all surfaces of parts, as they are conveyed. Excess paint drains back into the storage tank for recirculation. Paint films applied by flow coating are wedge shaped, and are thinner at the top and thicker at the bottom of the painted part. In flow coating, utilization of paint approaches 95% as opposed to about 50% for atomized air spraying and 70 to 80% for dipping. Properly designed flow coating machines with vapor chamber flow-out reduce solvent losses and eliminate tears, sags and curtains.

Flow coating is used extensively to paint panels for home appliances. Flow coating is also used (a) to coat parts with recesses inaccessible to spraying, (b) to coat parts (such as bedsprings) for which good appearance is desirable but is secondary to complete coverage and economical application, and (c) to coat intricate parts that are too open in design to permit efficient spray painting and too large to be practical for dip painting.

Figure 5 shows a large assembly which, be-

Table 5. Surface preparation and painting procedures for providing three steel parts with two degrees of durability in industrial atmospheres

Part No.	Size	Procedures for 1-yr durability Surface preparation	Prime painting	Finish painting	Procedures for 3-yr durability Surface preparation	Prime painting	Finish painting
1	Under 0.76 m (2.5 ft) long	(a)(b)(c)(d)	None	(h)(j)	(f)(b)(g)(b)(c)(d)	(h)(j)	(n)(m)
	Over 0.76 m (2.5 ft) long	(a)(b)(c)(d)	None	(k)(m)	(f)(b)(g)(b)(c)(d)	(k)(m)	(h)(j)
2	100 mm (4 in.) long	(a)(b)(d)	None	(n)(m)	(a)(b)(c)(d)	(n)(m)	(r)(p)
	0.6 m (2 ft) long	(a)(b)(d)	None	(h)(p)	(a)(b)(c)(d)	(h)(j)	(h)(p)
	3.0 m (10 ft) long	(e)(a)(b)(d)	None	(k)(m)	(e)(g)(b)(c)(d)	(k)(m)	(k)(m)
3	100-mm (4-in.) diam	(e)(a)(b)(d)	None	(n)(m)	(e)(a)(b)(c)(d)	(n)(m)	(n)(m)
	0.3-m (1-ft) diam	(e)(a)(b)(d)	None	(h)(q)	(e)(a)(b)(c)(d)	(h)(p)	(h)(q)
	1-m (4-ft) diam	(e)(a)(b)(d)	None	(h)(q)	(e)(a)(b)(c)(d)	(h)(p)	(h)(q)

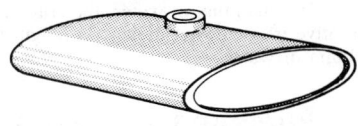

Part 1

Part 2

Part 3

(a) Phosphoric acid clean. (b) Water rinse. (c) Chromic acid rinse. (d) Forced-air dry. (e) Abrasive blast clean. (f) Alkaline clean. (g) Zinc phosphate coat. (h) Flow coat with baking alkyd. (j) Bake at 150 °C (300 °F) for 15 min. (k) Hand spray with air drying alkyd. (m) Air dry. (n) Dip in air-drying alkyd. (p) Bake at 150 °C (300 °F) for 20 min. (q) Bake at 150 °C (300 °F) for 25 min. (r) Hand spray with baking alkyd.

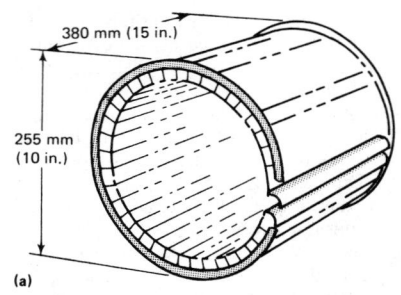 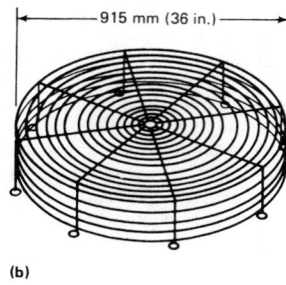

If considerably larger, parts like these could be painted more efficiently by the flow coating process. (a) Blower wheel. (b) Wire fan guard.

Fig. 4. Parts that can be efficiently coated by dip painting

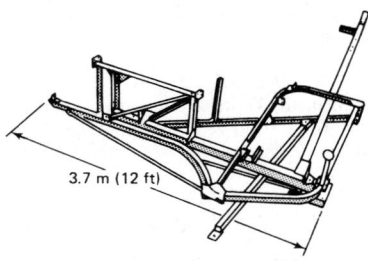

Fig. 5. Assembly for which flow coating is an efficient method

cause of its size and construction, is an example of a part for which flow coating is the most efficient method of painting. Spraying would be inefficient because of overspray, and dip coating would require a large quantity of paint to fill the dip tank. However, flow coating would be impractical for a similar assembly twice as large, and spray painting would be preferred.

Rust-Preventive Compounds

RUST-PREVENTIVE COMPOUNDS are removable coatings used to protect iron and steel surfaces against corrosive environments during fabrication, storage or use.

TYPES OF COMPOUNDS

Rust-preventive compounds dmay be divided into seven general categories:

1. *Petrolatum compounds* have a greaselike consistency. These compounds provide protection by the inclusion of effective corrosion inhibitors and by the continuous physical barrier presented to corrodents by the film.
2. *Oil compounds* are similar to lubricating oils but contain active rust inhibitors and corrosion inhibitors, in amounts that afford effective protection under various conditions of exposure.
3. *Hard dry-film compounds* establish a film either by evaporation of a solvent diluent or by chemical reaction after application. The coatings generally are then but fairly hard, being similar in appearance to varnish.
4. *Solvent-cutback petroleum-base compounds* are deposited as residual coatings through evaporation of the solvent. This class encompasses materials ranging from those that leave

a thin, transparent film to those forming a relatively heavy asphaltic or bitumastic film that hardens sufficiently to form an impervious barrier.
5. *Emulsion compounds* rely on the polarity and preferential attraction of emulsifiers or inhibitors to metal surfaces for protection of the surfaces after the water phase has evaporated.
6. *Water-displacing polar compounds* contain preferential wetting agents that have an affinity for metal surfaces.
7. *Fingerprint removers and neutralizers* are low-viscosity compounds containing suitable solvents to dissolve, suppress and neutralize acids, salts and residues from handling and from other sources of surface contamination. They are used for in-process or other short-term protection, or as pretreatments for longer-term protection.

SELECTION OF MATERIAL

The occurrence of rust on the surfaces of steel or other iron-base alloys depends on the contact of the surfaces with moisture and oxygen. The choice of rust-preventive compounds is influenced by:

- Environment, which includes climatic conditions, geographic location, and type of storage facilities
- Anticipated time in storage
- Material or object to be protected
- Necessity for removing compound

The severity of climatic, geographic and storage environments varies considerably. Mild conditions are encountered in a rural inland area with a daily temperature that ranges from −1 to 16 °C (30 to 60 °F) and a range of relative humidity of 25 to 70%. Severe conditions are those of a seacoast area where the temperature differential may exceed 33 °C (60 °F) and where 100% relative humidity is reached once or twice each day. When such conditions result in the dew point being reached, the result is moisture condensation. Therefore, the requirements for a successful rust preventive for any specific surface stored for six months inside a completely enclosed building would range from a very light oil for the rural area to a heavy-film petrolatum for the seacoast location.

Outside storage in either location would require the use of a material designed to cope with the erosive effect of rainfall. In an industrial inland area, the corrosion-preventive requirement is increased to provide resistance to the increased reactivity of the fumes present in the industrial atmosphere; for example, a very light oil might

need to be replaced by a soft petrolatum. Careful determination of the environmental factors against which the rust preventive must perform is of major importance in selecting a proper rust-preventive material.

The requirements of a rust preventive can be affected significantly by the degree of auxiliary protection afforded by the type of enclosure used for storing coated articles. Maximum auxiliary protection is provided by storage in a fully enclosed, humidity-controlled area in a permanent building, or in a container sealed against moisture and holding desiccants to absorb any moisture originally present. There are four general categories of parts or assemblies to which rust-preventive materials are applied for storage:

1. Assembled machinery or equipment in standby storage
2. Finished parts in stock, or spare parts for replacement
3. Tools, such as drills, taps, dies and gages
4. Mill products such as sheet, strip, rod and bar.

METHODS OF APPLICATION

Rust preventives are applied by spraying or fogging, dipping, flowing or slushing, and brushing or wiping. The method used is determined by the type of preventive and by the quantity, size, complexity and surface finish of the articles to be coated. The equipment and methods of application used are similar to those used in painting.

Petrolatum compounds may be applied either hot or cold. Cold application generally is restricted to parts too large and bulky for practical tank immersion, or to parts on which only localized protection is desired, such as the ways for a lathe bed.

Oil compounds can be applied by dipping, spraying, flowing, brushing or wiping. For spray application, oil rust preventives usually require no dilution; they are applied as received. Moderate air pressure is used to avoid misting and overspraying of the material. A wetting spray is usually sufficient. It is more difficult to control coating weight and uniformity by spray application than by dipping. To ensure an adequate coat, the material should be applied until it just begins to flow.

Emulsion compounds, most widely used for small items, are oil-in-water emulsions containing 8 to 12% solids. These compounds are available as concentrates, and are diluted with water at ratios of 1 part concentrate to from 4 to 10 parts water, as indicated by specification MIL-C-40084.

Water-displacing compounds are most effective when dip-applied, although spray application also may be used. These materials actually remove films or droplets of water from the metal surfaces by preferential wetting.

A schematic representation of a suitable dip tank for the automatic removal of water from a water-displacing preservative system is shown in Fig. 6. Because the specific gravity of the preservative is less than that of water, the column of preservative is somewhat greater in height than the column of water.

Porcelain Enameling

PORCELAIN ENAMELS are glass coatings applied primarily to products made of steel sheet, cast iron, or aluminum to improve appearance

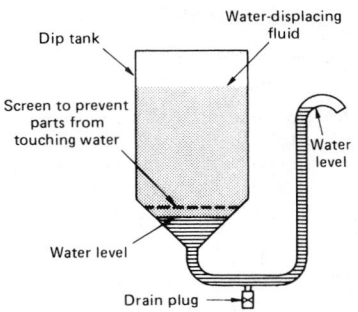

Fig. 6. Tank for dip application of water-displacing rust-preventive compounds

and protect the metal surface. Porcelain enamels are distinguished from other ceramic coatings by their predominantly vitreous nature and the types of applications for which they are used, and from paint by their inorganic composition and the fusion of the coating matrix to the substrate metal. Porcelain enamels of all compositions are matured at 425 °C (800 °F) or above.

TYPES OF PORCELAIN ENAMELS

Porcelain enamels for steel sheet and cast iron are classified as either ground-coat or cover-coat enamels. Ground-coat enamels contain oxides that promote adherence of the enamel to the metal substrate. Cover-coat enamels are applied over ground coats to improve the appearance and properties of the coating.

The basic material of the porcelain enamel coating is called a frit; it is a special glass of small friable particles produced by quenching a molten glassy mixture. Because porcelain enamels are usually designed for specific applications, the compositions of the frits from which they are made vary widely.

Enamel Frits for Steel Sheet. All the frits for which compositions are given in Table 6 are classified as alkali borosilicates for use as ground coats on steel sheet. Their compositions differ depending on the application environment of the enameled product.

Cover coats for steel sheet are applied over ground coats or directly to properly prepared decarburized steel. Compositions of frits for cover-

Table 6. Melted-oxide compositions of frits for ground-coat enamels for steel sheet

Constituent	Composition, wt %			
	Regular blue-black enamel	Alkali-resistant enamel	Acid-resistant enamel	Water-resistant enamel
SiO$_2$	33.74	36.34	56.44	48.00
B$_2$O$_3$	20.16	19.41	14.90	12.82
Na$_2$O	16.74	14.99	16.59	18.48
K$_2$O	0.90	1.47	0.51	...
Li$_2$O	...	0.89	0.72	1.14
CaO	8.48	4.08	3.06	2.90
BaO	9.24	8.59	...	...
ZnO	...	2.29	...	...
Al$_2$O$_3$	4.11	3.69	0.27	...
ZrO$_2$	...	2.29	...	8.52
TiO$_2$	...	...	3.10	3.46
CuO	...	...	...	0.39
MnO$_2$ ..	1.43	1.49	1.12	0.52
NiO	1.25	1.14	0.03	1.21
Co$_3$O$_4$	0.59	1.00	1.24	0.81
P$_2$O$_5$	1.04	...	...	0.20
F$_2$	2.32	2.33	1.63	1.94

Table 7. Melted-oxide compositions of frits for cover-coat enamels for steel sheet

Constituent	Composition, wt %		
	Titania white enamel		Weather-resistant blue enamel
	Fused at 815 °C (1500 °F)	Alkali resistant	
SiO$_2$	41.55	43.10	43.97
B$_2$O$_3$	12.85	13.81	6.51
Na$_2$O	7.18	5.99	13.83
K$_2$O	7.96	10.12	0.21
Li$_2$O	0.59	0.57	2.37
CaO	...	...	2.68
PbO	...	...	14.96
ZnO	1.13	...	...
Al$_2$O$_3$	...	...	0.43
ZrO$_2$	...	2.05	...
TiO$_2$	21.30	19.39	5.86
P$_2$O$_5$	3.03	0.54	...
Co$_3$O$_4$	...	...	3.72
F$_2$	4.41	4.43	5.46

Table 8. Melted-oxide compositions of frits for enamels for cast iron

Constituent		Composition, wt %		
		Cover coats		
	Ground coat(a)	Zirconium-opacified enamel(a)	Antimony-opacified enamel(b)	Acid-resistant enamel (a)(b)
SiO$_2$	77.7	28.0	22.9	37.0
B$_2$O$_3$	6.8	8.8	11.2	4.9
Na$_2$O	4.3	10.0	12.3	16.8
K$_2$O	...	4.1	6.0	1.7
PbO	4.0	17.8	9.8	8.8
CaO	...	8.7	8.0	2.0
ZnO	...	6.1	7.5	5.9
Al$_2$O$_3$	7.2	4.5	6.4	1.9
Sb$_2$O$_3$	...	...	13.9	13.1
ZrO$_2$	...	6.1	...	...
TiO$_2$	...	...	...	7.9
F$_2$	...	5.9	2.0	...

(a) For dry process. (b) For wet process.

Table 9. Melted-oxide compositions of frits for enamels for aluminum

Constituent	Composition, wt %		
	Lead-base enamel	Barium enamel	Phosphate enamel
PbO	14-45	...	...
SiO$_2$	30-40	25	...
Na$_2$O	14-20	20	20
K$_2$O	7-12	25	...
Li$_2$O	2-4	...	4
B$_2$O$_3$	1-2	15	8
Al$_2$O$_3$	...	3	23
BaO	2-6	12	...
P$_2$O$_5$	2-4	...	40
F$_2$	...	...	5
TiO$_2$	15-20	(a)	(a)

(a) TiO$_2$, 7 to 9 wt %, added to frit during mill preparation of the enamel slip.

coat enamels are given in Table 7. Electrostatic dry-powder cover coats may be applied over an electrostatic dry-powder ground coat and the entire two-coat/one-fire system matured in a single firing.

Enamel Frits for Cast Iron. Compositions of frits for enamels for cast iron vary depending on whether the frit is applied by the dry process or the wet process (Table 8). Dry-process enamels are commonly used for large cast iron fixtures because of their brilliance and ability to cover small surface irregularities.

Enamel frits for aluminum are usually based on lead silicate and on cadmium silicate, but may be based on phosphate or barium. Table 9 gives the compositions of some frits for aluminum.

The high-lead enamels for aluminum have a high gloss, good acid and weather resistance, and good mechanical properties.

PREPARATION OF FRITS

Porcelain enamel is usually applied as a suspension of finely milled frit in water; however, it may also be applied as a dry powder by electrostatically spraying on steel sheet or by dredging on cast iron. The wet-process frit is reduced to a fine powder in a ball mill. For milling, the ball charge should occupy 50 to 55% of the mill volume. After loading of the frit charge and mill additions such as clay, bentonite, electrolytes and coloring oxides, the water is added. Frits for dry electrostatic application are ground without water by the frit supplier and furnished to the porcelain enameler in a ready-to-use form.

STEELS FOR PORCELAIN ENAMELING

Typical compositions of the various grades of low-carbon steel sheet that are commercially available for porcelain enameling are listed in Table 10.

PREPARATION OF STEEL FOR PORCELAIN ENAMELING

When chemical treatments are used, mechanized equipment is usually used in production operations. The parts are placed on corrosion-resistant racks and dipped in or sprayed with a series of solutions. The sequence of processing steps and the solutions used in conventional production operations are indicated in Fig. 7. After drying at 93 to 150 °C (200 to 300 °F), the parts have a light straw color.

THE PORCELAIN ENAMELING PROCESS

Several basic methods are used to apply the porcelain enamel to the base metal. Included are

Table 10. Compositions of low-carbon iron and steel sheet for porcelain enameling

Enameled metal	Element, wt %						
	Carbon	Manganese	Phosphorus	Sulfur	Aluminum	Titanium	Niobium
Enameling iron	0.03	0.05(a)	0.01	0.02	(b)	...	...
Decarburized enameling steel	0.005	0.20-0.30	0.01	0.02	(b)	...	...
Titanium-stabilized enameling steel	0.05	0.30	0.01	0.02	0.05	0.30	...
Interstitial-free enameling steel	0.005	0.20	0.01	0.02	...	0.04	0.09
Cold rolled steel	0.06	0.35	0.01	0.02	(b)	...	...

(a) Some enameling iron may have manganese contents of 0.20 wt %. (b) None specified, but these materials may be supplied as aluminum-killed products.

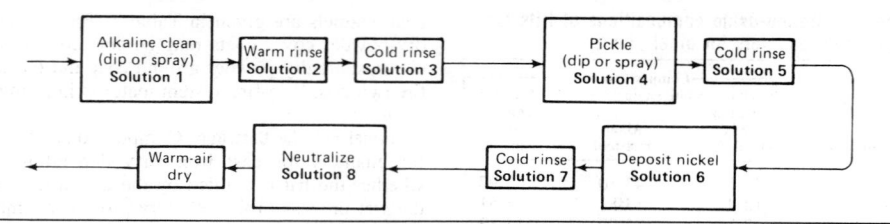

No.	Solution	Composition	°C	°F	Dip	Spray
				Temperature		
					Cycle time, min	
1	Alkaline cleaner(a)	Cleaner, 15-60 g/L (2-8 oz/gal)(b)	Ambient to 100(c)	Ambient to 212(c)	6-12	1-3
2	Warm rinse	Water	49-60	120-140	1/2-4	1/2-1
3	Cold rinse	Water	Ambient		2-4	1/2-1
4	Pickle(d)	H_2SO_4, 6-8%	66-71	150-160	5-10	3-5
5	Cold Rinse	Water, H_2SO_4(e)	Ambient		1/2-4	1/2-1
6	Nickel-deposition(f)	$NiSO_4 6H_2O$, 5.6-7.5 g/L (0.75-1.0 oz/gal)(e)	60-82	140-180	5-10	4-6
7	Cold rinse	Water, H_2SO_4(e)	Ambient		1/2-4	1/2-1
8	Neutralize	$^2/_3$ Na_2CO_3 and $^1/_3$ borax, 0.60-2.10 g/L (0.008-0.28 oz/gal) as Na_2O	49-71	120-160	1-6	1-2

For dip or spray application

(a) For spray cleaning, use a two-stage process. (b) For spray cleaning, use 3.8 to 15 g/L (0.5 to 2.0 oz/gal). (c) 60 to 82 °C (140 to 180 °F) for spray cleaner. (d) Weight loss of metal is 3 to 5 g/m^2 (0.3 to 0.5 g/ft^2). (e) Solution pH, 3 to 3.5, to prevent formation of ferric iron. (f) Nickel deposit should be 0.2 to 0.6 g/m^2 (0.02 to 0.06 g/ft^2). Continuous filtration is commonly used to remove Fe(OH)$_3$.

Fig. 7. Process for preparing steel surfaces for ground-coat porcelain enameling

dipping, low-coating, electrodeposition, manual spray, electrostatic spray and dry-powder spray. The best method of application for a particular part is determined by quantity and quality requirements, the type of material being applied, units produced per hour, capital investment, labor cost and ultimately, part cost.

Dipping is widely used as a method for applying the porcelain enamel, particularly when both sides of the parts require coverage. Dipping can be used for both ground-coat application and cover-coat application. It is performed by immersing the part in the prepared porcelain enamel slip, then withdrawing it and allowing the excess material to drain from the part.

Flow Coating. In flow coating, the porcelain enamel slip is flowed onto the surface of the part. The process is applicable to high-volume continuous operations for parts requiring the same porcelain enamel. In automatic flow coating, the parts are placed on hangers at the correct angle for draining and are carried by conveyor through the flow coating chamber. The porcelain enamel slip is pumped at a high volume, 570 L/min (150 gal/min), and low pressure, 70 to 105 kPa (10 to 15 psi), through a series of nozzles that are directed at various areas of the part to ensure complete coverage.

Spraying of the porcelain enamel slip is done primarily for one-side coverage. It is also used for reinforcing enamel bisque and for making repairs on enameled surfaces. Spraying is ideal for parts that are too large for hand or mechanical manipulation, particularly where service and appearance requirements permit no drain lines, beading, or buildup of the porcelain enamel.

ENAMELING FURNACES

Firing is accomplished in continuous, intermittent or batch furnaces heated by oil, natural gas, propane gas or electricity. With oil heating, a muffle furnace is used to prevent the products of combustion from contaminating the enamel coating. Gas-fired furnaces are either muffle, radiant-tube or luminous-wall types.

Firing of porcelain enamel involves the flow and consolidation of a viscous liquid and the escape of gases through the coating during its formation. Within limits, firing time and temperature are varied in a compensating manner. For example, similar properties and appearance develop when liners for household refrigerators are fired at 805 °C (1480 °F) for $2^1/_2$ min or at 790 °C (1450 °F) for 4 min. In all instances, there is a minimum practical temperature for the attainment of complete fusion, acceptable adherence and desired appearance. Most ground-coat enamels for high-production steel parts exhibit acceptable properties over a firing range of 55 °C (100 °F) at an optimum firing time. However, control within 11 °C (20 °F) is ordinarily maintained to produce uniform appearance and allow interchangeability of parts. Cycles for continuous-furnace firing of ground coats and cover coats on a number of different types of steel parts are given in Table 11.

Ceramic Coating

CERAMIC COATINGS include the super porcelains, which are based on silicates and oxides. Ceramic coatings also refer to high-temperature coatings based on oxides, carbides, silicides, borides, nitrides, cermets and other inorganic materials. Ceramic coatings are applied to metals to protect them against wear, oxidation and corrosion at room and elevated temperatures.

SELECTION FACTORS

Several factors must be considered in selecting a ceramic coating. These include:

- Service environment to be encountered by the coated metal
- Mechanisms by which the coatings provide protection at elevated temperature
- Compatibility of the coating with the substrate metal
- Method of applying the coating
- Quality control of the coating
- Ability of coating to be repaired.

COATING MATERIALS

Silicates. Coatings prepared from silicate powders (frits), with or without mill-added refractories, find the greatest industrial usage of all ceramic coatings. Silicate-frit coatings are used for such long-duration elevated-temperature applications as aircraft combustion chambers, turbines and exhaust manifolds, and heat exchangers. Variations in composition of the silicate frits are virtually unlimited. Frits range from alkali-alumina borosilicate glasses, which are relatively soft, low melting and highly fluxed, to barium crown glasses. Compositions of unmelted frits for silicate-base coatings for high-temperature service are indicated in Table 12.

Silicate coatings can be applied by spraying or air brushing (for which the material is atomized and carried by compressed air), dipping and draining (which may be followed by spraying), slushing and draining, filling and draining, and flow coating. Under certain conditions, electrostatic spraying also can be used.

Oxides. Coatings based on oxide materials provide underlying metals, except refractory metals, with protection against oxidation at elevated temperature and with a high degree of thermal insulation. Flame-sprayed oxide coatings do not provide refractory metals with the necessary protection against oxygen because of their inherent porosity. Oxide coatings can be readily applied in thicknesses up to 6.4 mm (0.25 in.), but their resistance to thermal shock decreases with increasing thickness.

Carbides are used as ceramic coatings principally for wear and seal applications, in which the high hardness of carbides is an advantage. These applications include jet-engine seals, rubber-skiving knives, paper-machine knives, and plug gages. Carbide coatings for wear resistance are applied by flame spraying or detonation-gun techniques.

Silicides are the most important coating materials for protecting refractory metals against oxidation. Silicide-base coatings protect by means of a thin coating of silica that forms on the coating surface when heated in an oxygen-containing atmosphere. To improve the self-healing, emit-

Table 11. Cycles for firing ground-coated and cover-coated steel sheet parts in a continuous furnace

Type of part	Gage of steel	°C	°F	Firing time, min(b)
		Operating temperature(a)		
Architectural panels	16-22	805	1480	4-5
Home laundry equipment	18-22	805	1480	4-5
Water-heater tanks	7-16	870	1600	8-12
Range ware	18-24	805	1480	3
Refrigerator liners	20-22	805	1480	$2^1/_2$-3
Sanitary ware	14-18	815	1500	5-8
Signs	16-22	805	1480	3-5

(a) Temperature varies with composition of frit. (b) Time in hot zone of furnace.

Table 12. Compositions of unmelted frit batches for high-temperature-service silicate-base coatings

Constituent	Parts by weight for specific frits(a)						
	UI-32	UI-285	UI-346	UI-418	NBS-11	NBS-331	NBS-332
Quartz	29.3	21.2	18.3	31.2	18.0	38.0	37.5
Feldspar	42.0	30.2	47.4	...	31.0	...	...
Hydrated borax	28.9	21.0	17.9	...	37.1	...	...
Sodium carbonate	7.7	5.3	6.1	...	5.9	...	...
Sodium nitrate	5.0	4.0	4.4	...	3.8	...	...
Fluorspar	4.5	3.2	2.8	...	3.0	...	...
Tricobalt tetroxide	0.6	...	0.4	...	0.5	...	...
Nickel oxide	0.6	...	0.4	...	0.6	...	...
Manganese dioxide	1.8	...	1.1	...	1.1	...	...
Barium carbonate	...	...	...	26.3	...	56.6	56.6
Zinc oxide	...	...	...	4.2	...	5.0	5.0
Whiting	...	...	...	7.5	...	7.1	6.3
Vanadium pentoxide	...	...	1.3	...	...	...	...
Aluminum hydrate	...	15.1	...	...	...	...	1.5
Boric acid	...	...	...	12.0	...	11.5	11.5
Cerium oxide	...	...	...	4.2	...	...	...
Titania	...	...	...	4.2	...	...	...
Bismuth nitrate	...	...	...	4.2	...	...	...
Bismuth oxide	...	...	...	6.2	...	...	...
Beryllia	...	...	...	...	...	2.5	...
Zirconia	...	...	...	...	...	...	2.5

(a) UI numbers designate frit compositions developed at the University of Illinois; NBS numbers, frits developed at the National Bureau of Standards.

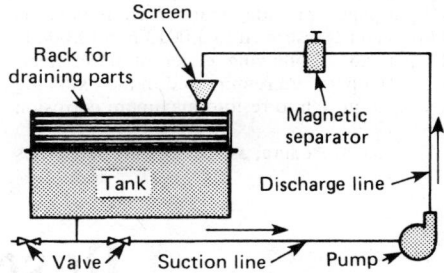

Fig. 8. Recirculating dip tank for application of ceramic coatings

spraying, (b) plasma-arc flame spraying and (c) detonation gun spraying. The first two methods use coating materials in powder or rod form. Detonation gun spraying uses only powder materials.

Development of Corrosion and Wear Resistance by Oxidation of Liquid Nitrided Surfaces

By William G. Wood, Kolene Corp.

AERATED LIQUID NITRIDING, a widely accepted heat treatment which provides ferrous surfaces with superior wear resistance, can be modified by controlled oxidation to provide corrosion resistance comparable to that of chromium plating or copper-nickel-chromium plating. The surface developed by aerated liquid nitriding is composed of epsilon iron nitride, F_3N. This compound, which has corrosion properties equivalent to those developed by flash nickel plating or flash chromium plating of iron, extends to a depth of 0.01 to 0.015 mm (0.0004 to 0.0006 in.) in ferrous surfaces liquid nitrided for 60 to 180 min under controlled conditions. In moderately corrosive environments, ferrous parts so treated can be substituted for stainless steel parts or chromium-plated ferrous parts.

The corrosion-resistance characteristic of the epsilon iron surface is appreciably enhanced by a fused salt oxidation process while hardness patterns and wear characteristics are relatively unchanged. The salt quench was included because of environmental requirements for elimination of cyanide and cyanate compounds in effluents from hardening processes. Laboratory testing indicates that this improvement is not the result of initial oxidation but rather of a complex subsurface combination of diffused nitrogen, oxygen and iron. Auger profiles have verified the observation that the epsilon iron nitride is combined with diffused oxygen and that there is a relationship among quench-salt dwell time, oxygen content of the compound zone, and corrosion protection.

The aerated liquid nitriding treatment combined with the oxidizing quench is a viable process for development of wear and corrosion resistance under a variety of circumstances, but the resulting surface lacks compatibility with nonmetallic seals and bushings because of surface roughness. To overcome this difficulty, liquid-nitrided-and-quenched parts can be polished by vibratory finishing or by fixtured finishing with

tance, chemical stability, or adherence of this thin silica coating, other elements, such as chromium, niobium, boron or aluminum, are added to the coating formula.

Phosphate-Bonded Coatings. Phosphates for metal-protective coating systems are formed by the chemical reaction of phosphoric acid and a metal oxide such as aluminum oxide, chromium oxide, hafnium oxide, zinc oxide or zirconium oxide. The phosphate-bonded materials are used to protect metals against heat and to act as binders in thin ceramic paint films. Phosphate-bonded composites, depending on composition, can withstand temperatures up to 2425 °C (4400 °F) and have been applied in thicknesses up to 50 mm (2 in.).

COATING METHODS

Ceramic coatings may be applied by brushing, spraying, dipping, flow coating, combustion flame spraying, plasma-arc flame spraying, detonation gun spraying, pack cementation, fluidized-bed deposition, vapor streaming, troweling, and electrophoresis. Most of these methods have been used for coating of production parts.

SPRAYING AND DIPPING

Spraying and dipping are two methods of applying ceramic coatings in a slip or slurry form. Spraying and dipping methods are used to apply silicates and other coatings onto engine exhaust ducts, space heaters, radiators and other high-production parts.

Spraying can be used when the shape of the work permits direct access to all surface areas to be coated. This method is usually used for applying a closely controlled thickness of coating to exterior surfaces only.

Dipping can be used for almost all parts. This includes riveted or spot welded assemblies, ex-

cept those in which faying surfaces would be inadequately covered by the slurry. For a uniform coating thickness, a handling cycle must be established for each part to produce drainage of each surface at the proper angle.

Surface Preparation. Parts must be thoroughly cleaned before spraying or dipping. Oily spots prevent adherence of the coating and cause blistering or spalling during firing. When sand blasting is used, the abrasive must be free of contaminants. Sharp workpiece edges should be rounded, because they are difficult to coat. If sharp edges are coated without being rounded off, the coating will often spall after firing.

Equipment for spraying ceramic coatings is available commercially. The spray gun should have a nozzle with an orifice diameter of 1.30 to 2.80 mm (0.050 to 0.110 in.). Efficient nozzles have a spraying capacity of 260 000 to 330 000 mm³ (16 to 20 in.³) of coating material per minute using an air pressure of 345 kPa (50 psi) to propel the coating material to the work surface. The compressed-air supply should be filtered to remove dirt, rust, oil and moisture. A reliable air-pressure regulator should be used to permit accurate adjustment of pressures, particularly in the range from 205 to 550 kPa (30 to 80 psi).

For most dipping applications, the equipment consists of a tank such as that shown in Fig. 8. The tank should be large enough to permit complete submersion of the part into the slip. An easel or rack is required for draining.

FLAME SPRAYING

Most ceramic coating materials used currently can be applied by flame spraying. Silicates, silicides, oxides, carbides, borides and nitrides are among the principal materials deposited by this process. Three methods of heating and propelling the particles in the plastic condition to the substrate surface include: (a) combustion flame

320-grit paper, provided that surface removal is uniform and is restricted to 0.0013 mm (0.00005 in.) per side. Following either of these treatments, the parts are reimmersed in the oxidizing fused salt quench to restore maximum corrosion resistance.

The final procedure, as specified for numerous engine and chassis parts subject to wear and corrosive conditions, is as follows:

1. Aerated liquid nitride in a cyanide-free fused salt for 60 to 120 min at 575 °C (1070 °F).
2. Quench in a controlled-rate oxidizing fused salt at 400 °C (750 °F).
3. Water rinse.
4. Abrasive polish to desired surface finish with maximum stock removal of 0.0013 mm (0.00005 in.).
5. Reimmerse in the controlled-rate oxidizing fused salt for a dwell time of 20 min.
6. Water rinse and oil.

Descaling and Cleaning of Stainless Steels and Heat-Resisting Alloys

By Robert R. Gaugh, Armco, Inc.

STAINLESS steels and heat-resisting alloys can be descaled in several ways, with the choice depending on the amount and type of oxidation (scale) present and on the available equipment.

ABRASIVE BLAST CLEANING

Sand blasting is effective for rapid removal of heavy or tightly adhering scale. The types of work for which this technique is frequently used include heavily scaled plate sections, forgings and castings, and parts made of straight chromium steel that have developed tightly adhering scale during annealing. Sand blasting is fast and economical, but must be followed by an acid pickling treatment for removal of embedded scale particles.

Shot Blasting. Carbon steel shot, steel wire and iron grit are not recommended for use as blasting media because they may cause particles of iron to become embedded in stainless steel surfaces and thus severely detract from their corrosion resistance. Unless these contaminants are completely removed by acid pickling, they can rust and provide sites for initiation of pitting.

Wet blasting is adaptable for use with stainless steels and heat-resisting alloys. Various abrasives conveyed in liquid carriers are discharged at the work by compressed air. A variety of finishes can be obtained through selection of abrasives and adjustment of pressures. Wet blasting is used primarily for removing light amounts of oxidation such as the heat tinting that results from precipitation-hardening treatments.

ACID PICKLING

An alternative method of removing scale from stainless steels and heat-resisting alloys is by immersion in an aqueous solution containing about 10 vol % nitric acid and 1 to 2 vol % hydrofluoric acid. This solution should be used at a maximum temperature of 60 °C (140 °F). The time required for pickling depends on the alloy, the thickness of the scale and the temperature of the solution. For removal of heat tinting or for pickling following abrasive blast cleaning, the time need be only several minutes. Pickling with these acids should be used with care for martensitic alloys in the fully hardened condition because of the danger of hydrogen-embrittlement cracking.

SALT BATH DESCALING

The removal of heavy scale may be accelerated by using baths of molten sodium hydroxide to which certain reagents are added. These baths can be used with virtually all stainless and heat-resisting steels, but no one type of molten salt is equally effective with all alloys. Salt bath descaling does not completely remove all of the scale, but must be followed by a short pickling treatment. Salt bath descaling has several advantages: it acts only on the scale and does not result in metal loss or etching; and it does not preferentially attack areas in which intergranular carbides are present. Use of molten salts is not recommended for those alloys that precipitation harden at the operating temperature of the bath.

PASSIVATION OF STAINLESS STEELS

During handling and processing operations such as forming, machining, tumbling and lapping, particles of iron or tool steel may be embedded in or smeared on the surfaces of stainless steel components. If allowed to remain, these particles may corrode and produce rust spots on the stainless steel. To prevent this condition, semifinished or finished parts are given a passivation treatment. This treatment, which consists of immersing stainless steel parts in a solution of nitric acid, or of nitric acid plus oxidizing salts, dissolves the embedded or smeared iron and restores the original corrosion-resistant surface. Passivation should be the last operation given to fabricated parts made of stainless steel. This treatment will not remove particles of scale or rusty areas from workpiece surfaces.

Table 1. Compositions and operating conditions for two solutions commonly used for passivation of stainless steels and heat-resisting alloys

Constituent or condition	Solution A(a)	Solution B(b)
70% nitric acid (HNO_3)	20 to 40 vol %	20 vol %
Sodium dichromate ($Na_2Cr_2O_7 \cdot 2H_2O$)	None	4 to 6 wt %
Water	Remainder	Remainder
Operating temperature	Room to 60 °C (140 °F)	Room to 50 °C (120 °F)
Exposure time	30 to 60 min	30 min

(a) For series 200, 300 and 400 grades containing 17% Cr or more (except type 440) and for precipitation-hardening alloys. (b) For all free-machining grades, all grades with highly polished surfaces, type 440, and series 400 grades containing less than 17% Cr.

Table 1 gives compositions of two solutions commonly used for passivation of various stainless steels and heat-resisting alloys, together with their operating temperatures and exposure times. Parts should be thoroughly degreased prior to immersion in order to allow the solution to contact the metal surfaces completely. These solutions also will remove lead, copper, cadmium or zinc applied to stainless steel wire for cold heading, wiredrawing or spring winding. However, it is preferred that these contaminants be removed prior to the passivation treatment so as to prevent contamination of the solution.

Control of passivating solutions consists mainly of replenishing the nitric acid, sodium dichromate (if used) and water that are lost as a result of dragout.

REMOVAL OF METALLIC CONTAMINANTS FROM NICKEL-BASE AND COBALT-BASE ALLOYS

Chemical methods are usually used for removing metallic contaminants from Ni- and Co-base alloys. A typical procedure for removing iron, zinc, and thin films of lead is as follows:

- Vapor degrease or alkaline clean.
- Immerse in a 1-to-1 solution (by volume) of nitric acid (1.41 sp gr) and water for 15 to 30 min at about 35 °C (95 °F).
- Water rinse and dry.

Another procedure, which has proved successful for removing brass, lead, zinc, bismuth and tin from Ni- and Co-base alloys, is as follows:

- Vapor degrease or alkaline clean.
- Soak at room temperature, for 20 min to 4 h (depending on severity of contamination), in a solution containing 54 g/L (7.22 oz/gal) nitric acid, 150 to 375 g/L (20 to 50 oz/gal) acetic acid and 19 to 64 g/L (2.8 to 8.5 oz/gal) hydrogen peroxide.

SELECTED REFERENCES

- Standard Recommended Practice for Cleaning and Descaling Stainless Steel Parts, Equipment, and Systems: ASTM A380-78, American Society for Testing and Materials, Philadelphia, 1983
- *Cleaning and Descaling Stainless Steels:* from *Designers' Handbook Series,* American Iron and Steel Institute, Washington, May 1982
- Cleaning Stainless Steel: ASTM STP 538, American Society for Testing and Materials, Philadelphia, 1973
- Update on Cleaning Stainless Steels: *Metal Progress,* June 1973, p 38–60

Cleaning and Finishing of Nonferrous Metals

Edited by R. Terrence Webster, Teledyne Wah-Chang Albany

Cleaning and Finishing of Aluminum and Aluminum Alloys

ALUMINUM and aluminum alloy products are cleaned and finished by various techniques in order to enhance their appearance or improve their functional surface properties, or both.

ABRASIVE BLAST CLEANING

Abrasive blasting is most efficient in removing scale, sand, and mold residues from castings. It is readily adaptable to cleaning of castings, because they are usually thick enough so that no distortion results from blasting.

Blast cleaning of parts with relatively thin sections is not recommended, because such parts are readily warped by the compressive stresses that are set up in the surface by blasting. Blasting of thin sections with coarse abrasive is not recommended because the coarse abrasive can wear through the aluminum. Typical conditions for dry blasting with silica abrasive are given in Table 1.

In wet blasting, a fine abrasive is mixed with water to form a slurry that is forced through nozzles and directed at the part. Abrasive grits from 100 to 5000 mesh may be used. Wet blasting is generally used when a fine-grain matte finish is desired for decorative purposes.

Typical wet blasting procedures are given in Table 2. Wet blasting is used also for preparing surfaces for organic or electroplated coatings. Ultrafine glass bead blasting can also be used in place of wet blasting.

BARREL BURNISHING

Barrel burnishing is used to produce a smooth, mirror like texture on aluminum parts. Bright dipping immediately prior to burnishing will aid in producing a better finish. Other preliminary treatments also are helpful in specific instances, particularly for cast aluminum parts. One of these pretreatments entails etching the castings for 20 s in an alkaline solution at 82 °C (180 °F) and then dipping them for 2 to 3 s in a solution consisting of (by volume) 3 parts nitric acid (36° Bé) and 1 part hydrofluoric acid at 21 to 24 °C (70 to 75 °F).

The principle of barrel burnishing is to cause surface metal to flow, rather than to remove metal from the surface. Burnishing compounds must have lubricating qualities. Soaps made especially for burnishing are usually used. They are readily obtainable, and many of them have a pH of about 8, although more acidic materials can be used.

POLISHING AND BUFFING

Because aluminum is more easily worked than many other metals, few aluminum parts require polishing prior to buffing for final finish. In some instances, polishing may be required for removal of burrs, flash or surface imperfections. Usually, buffing with a sisal wheel prior to final buffing is sufficient.

Polishing. Most polishing operations can be performed using either belts or setup wheels. Setup wheels may be superior to belts for rough cutting down when canvas wheels in a relatively crude setup can be used. For fine polishing work, a specially contoured wheel may be more satisfactory than a belt.

The conditions for wheel polishing of die cast aluminum soleplates for steam irons are as follows:

Type of polishing wheel .... Felt
Setup time .... 10 min
Wheel speed .... 1800-2000 rev/min
Lubricant .... Tallow grease stick

The medium-hard felt polishing wheel is 350 to 400 mm (14 to 16 in.) in diameter, with a 125-mm (5-in.) face. The surface of the wheel is double coated with 240-mesh alumina abrasive bonded with hide glue.

Table 3 gives the conditions and sequence of operations for belt polishing of die cast steam-iron soleplates made of aluminum alloy 380.0. Ten polishing heads are used to produce a bright finish on the sides and bottoms of the soleplates.

Buffing. Selection of procedure for buffing depends mainly on cost, because it is usually possible to obtain the desired results by any one of several different procedures. For example, in hand buffing, combinations of the various influences might call for the use of equipment ranging from simple, light-duty machines to heavy-duty, variable-speed, double-control units.

Die cast aluminum soleplates for steam irons (Table 4) were buffed to a bright finish on an automatic machine with eight buffing heads. The soleplates were made of alloy 380.0 and were prepolished with 320-mesh grit. A liquid buffing compound was applied by one gun per wheel for the first four heads and by two guns per wheel for the last four heads. The guns were on for 0.12 s and off for 13 s. Service life was 72 000 pieces for each buff of the first four heads, and 24 000 pieces for each buff of the last four heads.

CHEMICAL CLEANING

The degree and nature of cleanness required are governed by the subsequent finishing operations. For example, cleaning requirements for plating or for application of chromate or other mild-reaction conversion coatings are somewhat more stringent than those for anodizing.

Solvent Cleaning. The primary function of solvent cleaners is the removal of oil and grease compounds. Organic solvents alone rarely provide sufficient cleaning to permit final finishing operations; solvents usually are used to remove large amounts of organic contaminants to minimize overloading of the subsequently used alkaline cleaners.

Emulsifiable solvents also are used to clean aluminum. These are organic solvents, such as kerosine, Stoddard solvent and mineral spirits, to which small amounts of emulsifiers and surfactants are added. In use, this type of cleaner emul-

Table 1. Conditions for abrasive blast cleaning with silica

Grit size	Mesh	Nozzle diameter mm	Nozzle diameter in.	Nozzle to work(a) mm	Nozzle to work(a) in.	Air pressure kPa	Air pressure psi
20-60 ....Coarse		10-13	$^3/_8$-$^1/_2$	300-500	12-20	205-620	30-90
40-80 ....Medium		10-13	$^3/_8$-$^1/_2$	200-350	8-14	205-620	30-90
100-200 ......Fine		6-13	$^1/_4$-$^1/_2$	200-350	8-14	205-515	30-75
Over 200 ....Very fine		13	$^1/_2$	200-300	8-12	310	45

(a) Nozzle approximately 90° to work.

Table 2. Conditions for wet blasting of aluminum-base materials

Nozzle-to-work distance, 75 to 100 mm (3 to 4 in.); operating pressure, 550 kPa (80 psi).

Operation	Abrasive Type	Abrasive Mesh size
Deburr and clean ....Alumina		220
Blend and grind .....Silica flour		325
Lap and hone ....Glass		1000
	Diatomite	625-5000

Table 3. Conditions of belt polishing for bright finishing of die cast soleplates

Operation	Area polished	Polishing head, No.	Contact wheel Type	Contact wheel Size mm	Contact wheel Size in.	Hardness, durometer	Belt(a) Size mm	Belt(a) Size in.	Abrasive mesh size	Life, pieces
1 ....Side		1,2	Plain face	50 by 380	2 by 15	60	50 by 3050	2 by 120	280(b)	600
2 ....Side		3,4	Plain face	50 by 380	2 by 15	60	50 by 3050	2 by 120	320(b)	600
3 ....Bottom		5	Serrated(c)	150 by 380	6 by 15	45	150 by 3050	6 by 120	120(b)	1200
4 ....Bottom		6	Serrated(c)	150 by 380	6 by 15	45	150 by 3050	6 by 120	150(b)	2000
5 ....Bottom		7	Serrated(c)	150 by 380	6 by 15	45	150 by 3050	6 by 120	220(b)	2000
6 ....Bottom		8	Serrated(c)	150 by 380	6 by 15	45	150 by 3050	6 by 120	280(b)	2000
7 ....Bottom		9	Plain face	150 by 380	6 by 15	60	150 by 3050	6 by 120	320(b)	2000
8 ....Bottom		10	Plain face	150 by 380	6 by 15	60	150 by 3050	6 by 120	320(d)	600

(a) Belt speed for all operations was 35 m/s (6900 sfm). All belts were cloth; bond, resin over glue. (b) Aluminum oxide abrasive. (c) 45° serration, 13-mm ($^1/_2$-in.) land, 10-mm ($^3/_8$-in.) groove. (d) Silicon carbide abrasive.

Table 4. Automatic bright-finish buffing of aluminum soleplates

Operation	Area buffed	Buffing head No.	Type	Size	°	Overall mm	in.	Arbor hole mm	in.	No. of Sections	Speed m/s	sfm	Life, pieces	No. of guns	g per shot	oz per shot
1	Side	1,2	Sisal	10-mm (³/₈-in.) spiral sewed		410	16	40	1¹/₄	2	37	7350	72 000	1	0.5	0.02
2	Side	3	Bias	16-ply, 20-spoke sewed		430	17	45	1³/₄	2	40	7800	72 000	1	0.5	0.02
3	Side	4	Bias	16-ply, 20-spoke sewed		430	17	45	1³/₄	2	40	7800	72 000	1	0.5	0.02
4	Top	5	Sisal	10-mm (³/₈-in.) spiral sewed		410	16	45	1³/₄	15	37	7350	24 000	2	3.0	0.1
5	Top	6	Sisal	10-mm (³/₈-in.) spiral sewed		410	16	45	1³/₄	15	37	7350	24 000	2	3.0	0.1
6	Top	7,8	Bias	16-ply, 20-spoke sewed		430	17	45	1³/₄	10	40	7800	24 000	2	3.0	0.1

(a) All wheels had 180-mm (7-in.) diam centers. (b) Proprietary liquid compound was used. Cycle time: 0.12 s on, 13.0 s off.

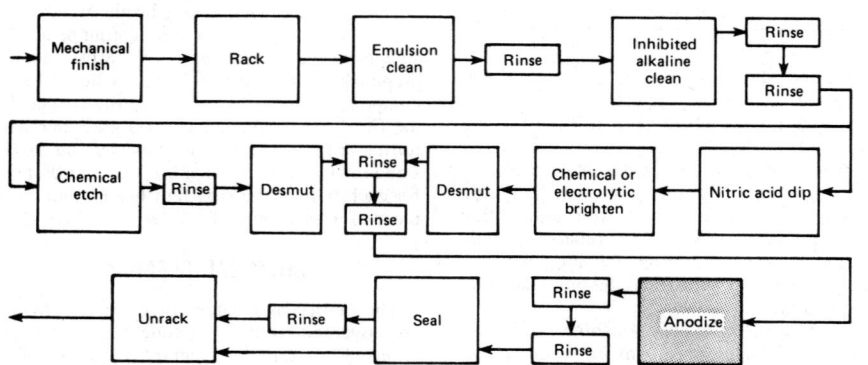

Fig. 1. Anodizing operations

sifies the oil or grease on the surface. The soil and cleaner are removed with water, preferably applied by spraying.

Alkaline cleaning is the most widely used method for cleaning aluminum and aluminum alloys. This method is easy to apply in production operations, and equipment costs are low. Aluminum is readily attacked by alkaline solutions. Most solutions are maintained at a pH between 9 and 11, and they are often inhibited to some degree to minimize or prevent attack on the metal. The most frequently used cleaner is the mildly inhibited type.

Acid cleaning may be used alone or in conjunction with other acid, alkaline or solvent cleaning systems. Vapor degreasing and alkaline cleaning may be required for removal of heavy oils and grease from workpieces before they are immersed in an acid bath. One of the main functions of an acid cleaner is the removal of surface oxides prior to resistance welding, painting, conversion coating, bright dipping, etching or anodizing.

A mixture of chromic and sulfuric acids is commonly used to remove surface oxides, burnt-in oil, water stains or other films, such as the iridescent or colored films formed during heat treating. This acid mixture cleans and imparts a slightly etched appearance to the surface, preparing it for painting, caustic etching, conversion coating or anodizing. Nonpolluting, proprietary products free of chromic acid are available for acid cleaning and deoxidizing.

ANODIZING PROCESSES

The basic reaction in all anodizing processes is conversion of the aluminum surface to aluminum oxide while the part is the anode in an electrolytic cell.

The three principal types of anodizing processes are (a) chromic, in which the active agent is chromic acid; (b) sulfuric, in which the active agent is sulfuric acid; and (c) hard processes that use sulfuric acid alone or with additives. Other processes, used less frequently or for special purposes, use sulfuric-oxalic, phosphoric, oxalic, boric, sulfosalicylic or sulfophthalic acid solutions. Except for those produced by hard anodizing processes, most anodic coatings range in thickness from 5 to 18 μm (0.2 to 0.7 mils).

The succession of operations typically employed in anodizing is illustrated in Fig. 1.

Surface Preparation. A chemically clean surface free of oxides is a basic requirement for successful anodizing. The cleaning method is selected on the basis of the type of soils or contaminants that must be removed. Usually, the cleaning cycle consists of removing the major organic contaminants by vapor degreasing or solvent cleaning, then making the surface chemically clean so that acid pickling or etching solutions can react uniformly over the entire surface.

Chromic Acid Process. The sequence of operations used in the chromic acid process depends on the type of part, the alloy to be anodized, and the principal objective for anodizing. This process is used where it is difficult to remove all of the electrolyte. It will color the finish from yellow to dark olive depending on thickness. Table 5 gives a typical sequence of operations that meets the requirements of military specification MIL-A-8625.

Chromic acid anodizing solutions contain from 3 to 10 wt % CrO_3. A solution is made up by filling the tank about half full of water, dissolving the acid in the water, and then adding water to adjust to the desired operating level.

A chromic acid anodizing solution should not be used unless:

- pH is between 0.5 and 1.0
- Concentration of chlorides (as sodium chloride) is less than 0.02%
- Concentration of sulfates (as H_2SO_4) is less than 0.05%
- Total chromic acid content, as determined by pH and Baumé readings, is less than 10%. When this percentage is exceeded, part of the bath is withdrawn and is replaced with fresh solution.

When the anodizing process is started, the voltage is controlled so that it will increase from 0 to 40 V within 5 to 8 min. The voltage is regulated to produce a current density of not less than 0.1 A/dm² (1.0 A/ft²), and anodizing is continued for the specified time period (usually, 30 to 40 min). At the end of the cycle the current is gradually reduced to zero, and the part is removed from the bath within 15 s, rinsed, and sealed. Weight of coating after sealing should be 200 mg/m² (19 mg/ft²) minimum.

Table 5. Sequence of operations for chromic acid anodizing

Operation	Solution	Solution temperature °C	°F	Treatment time, min
Vapor degrease	Suitable solvent	...	...	...
Alkaline clean	Alkaline cleaner	(a)	(a)	(a)
Rinse(b)	Water	Ambient	Ambient	1
Desmut(c)	HNO₃, 10-25 vol %	Ambient	Ambient	As required
Rinse(b)	Water	Ambient	Ambient	1
Anodize	CrO₃, 46 g/L (5¹/₄ oz/gal)(d)	32-35	90-95	30(e)
Rinse(b)	Water	Ambient	Ambient	1
Seal(f)	Water(g)	90-100	190-210	10-15
Air dry	...	105 max(h)	225 max(h)	As required

(a) According to individual specifications. (b) Running water or spray. (c) Generally used in conjunction with alkaline-etch type of cleaning. (d) pH 0.5. (e) Approximate; time may be increased to produce maximum coating weight desired. (f) Optional. (g) Water may be slightly acidulated with chromic acid, to a pH of 4 to 6. (h) Drying at elevated temperature is optional.

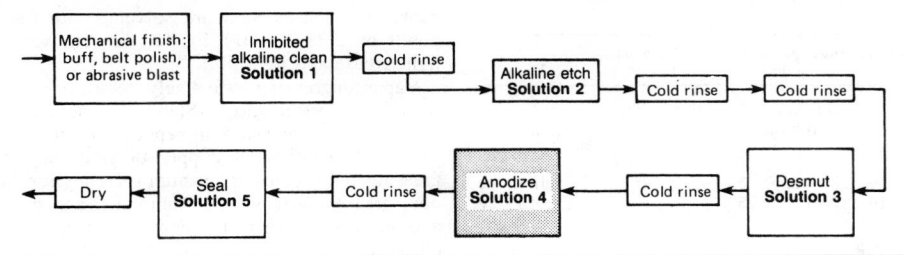

Fig. 2. Operations sequence in sulfuric acid anodizing of architectural parts

Solution No.	Type of solution	Composition	Operating temperature °C	°F	Cycle time, min
1	Alkaline cleaning	Alkali, inhibited	60-71	140-160	2-4
2	Alkaline etching	NaOH, 5 wt %	50-71	120-160	2-20
3	Desmutting	HNO_3, 25-35 vol %	Room	Room	2
4	Anodizing	H_2SO_4, 15 wt %	21-25	70-75	5-60
5	Sealing	Water (pH 5.5-6.5)	100	212	5-20

Sulfuric Acid Process. The basic operations for this process are the same as for the chromic acid process. Parts or assemblies that contain joints or recesses that could entrap the electrolyte should not be anodized in the sulfuric acid bath. The concentration of sulfuric acid (1.84 sp gr) in the anodizing solution is 12 to 20 wt %. A solution containing 36 L (9.5 gal) of H_2SO_4 per 380 L (100 gal) of solution is capable of producing an anodic coating that, when sealed in a boiling dichromate solution, meets the requirements of MIL-A-8625.

A sulfuric acid anodizing solution should not be used unless:

- The concentration of chlorides (as sodium chloride) is less than 0.02%
- Aluminum concentration is less than 20 g/L (2.7 oz/gal)
- Sulfuric acid content is between 165 and 200 g/L (22 to 27 oz/gal)

At the start of the anodizing operation, the voltage is adjusted to produce a current density of 0.9 to 1.2 A/dm² (9 to 12 A/ft²). The voltage will increase slightly as the aluminum content of the bath increases. The approximate voltages required for anodizing various wrought aluminum alloys in a sulfuric acid bath at 1.2 A/dm² (12 A/ft²) are as follows:

Alloy	Volts
1100	15.0
2011	20.0
2014	21.0
2017	21.0
2024	21.0
2117	16.5
3003	16.0
3004	15.0
5005	15.0
5050	15.0
5052	14.5
5056	16.0
5357	15.0
6053	15.5
6061	15.0
6063	15.0
6151	15.0
7075	16.0

A flow chart and a table of operating conditions for operations typically used in anodizing architectural parts by the sulfuric acid process are presented in Fig. 2.

Hard Anodizing. The primary differences between the sulfuric acid and hard anodizing processes are the operating temperature and the current density at which anodizing is accomplished. Hard anodizing produces a considerably heavier coating than conventional anodizing in a given length of time.

The hard anodizing process uses a sulfuric acid bath containing 10 to 15 wt % acid, with or without additions. The operating temperature of the bath ranges from 0 to 10 °C (32 to 50 °F), and current density is between 2 and 3.6 A/dm² (20 and 36 A/ft²). High temperatures cause the formation of soft and more porous outer layers of the anodic coating. This change in coating characteristics reduces wear resistance significantly and tends to limit coating thickness. Excessive operating temperatures result in dissolution of coating and can burn and damage the work.

Cleaning and Finishing of Copper and Copper Alloys

COATINGS on copper alloys vary according to treatment, appearance and corrosion resistance. The applications range from simple, low-cost chemical treatments that provide a uniform surface appearance to expensive electroplates that provide maximum corrosion resistance. Prior to the application of a protective or decorative coating, the metal surface must be prepared by suitable cleaning procedures.

PICKLING AND BRIGHT DIPPING

Pickling in solutions containing 4 to 15 vol % sulfuric acid or 40 to 90 vol % hydrochloric acid is used for removal of oxides formed on the surfaces of copper-base materials during mill processing and fabricating operations. The sulfuric acid solution is used to remove (a) black copper oxide scale on brass extrusions, forgings and machined parts; (b) oxide on copper tubing, forgings and machined parts; and (c) light annealing scale or tarnish. The hydrochloric acid solution is primarily used for finishing, but is also used to remove scale and tarnish from brass forgings and machined parts and oxide from copper forgings and machined parts. Conditions for pickling copper-base metals with sulfuric acid and hydrochloric acid are shown in Table 6. Sometimes no additional surface preparation is necessary to produce the uniformity of appearance required for further finishing of copper; however, heavily scaled material may need a bright dip or color dip after pickling.

Except for bright annealed material, copper alloys must be (a) pickled after each annealing treatment, (b) completely descaled, and (c) bright dipped to produce a natural surface color and luster suitable for other finishing treatments such as electroplating or painting. Scale dip and bright dip solutions are given in Table 7.

ABRASIVE BLAST CLEANING

Abrasive cleaning is used to remove molding, core sand and investment material from the exterior and interior surfaces of copper-base castings. Selection of the proper kind and particle size of grit determines the type and color of the finish. The coarser grits clean faster but give a rougher finish.

Dry abrasive cleaning of berryllium copper is usually confined to castings. Steel shot is used for general cleaning to remove sand and slight surface imperfections from the casting after mold shakeout. Sands are used to blend in surface areas, to remove heat treat scale and to produce a uni-

Table 6. Pickling conditions for copper-base materials

Acid condition	Amount, vol %	35% hydrogen peroxide	Water	Temperature of solution	Immersion time, min
Sulfuric acid(a)	15-20	3-5 vol %	Rem	Room temperature to 60 °C (140 °F)	¼-5
Hydrochloric acid(b)	40-90	...	Rem	Room temperature	1-3

(a) 1.83 sp gr. The bath needs additives to stabilize peroxide, and accelerators to maintain etch rate. Proprietary products are available from metal finishing suppliers. (b) 1.16 sp gr.

Table 7. Scale dip and bright dip conditions for copper-base metals
These solutions remove scale that is not removed by sulfuric or hydrochloric acid solutions; lower concentrations of nitric acid and higher concentrations of sulfuric acid produce a bright lustrous finish; these solutions can remove 0.0255 mm (0.001 in.) of metal and should not be used when close dimensional tolerances must be maintained.

Solution	Sulfuric acid, vol %(a)	Nitric acid, vol %(b)	Hydrochloric acid(c) g/L	oz/gal	Water, vol%	Temperature of solution	Immersion time, s
Scale dip:							
Solution A	0	50	4	½	50	Room temperature	15-60
Solution B	25-35	35-50	4	½	35-40	Room temperature	15-60
Bright dip	50-60	15-25	4	½	rem	Room temperature	5-45

(a) 1.83 sp gr. (b) 1.41 sp gr. (c) 1.16 sp gr, excess hydrochloric acid spots brass. Wood soot and activated charcoal are added to the solution to prevent this condition.

Table 8. Operating conditions for mass finishing

Material	Medium	Size of abrasive particles mm	in.	Tumbling time, h	Finish
Heavy cutting					
Brass or bronze castings	Aluminum oxide	6.4-19	0.25-0.75	6-16	Matte
Moderate cutting					
Brass stampings	Aluminum oxide	6.4-19	0.25-0.75	1-6	Light matte
Brass screw-machine parts	Aluminum oxide or granite	6.4-19	0.25-0.75	1/2-6	Light matte or bright
Light cutting(a)					
Brass stampings or screws(b)	Limestone	3.2-13	0.13-0.50	2-6	Bright

(a) Submerged tumbling is used for fragile and precision parts. (b) Screw-machine parts.

Table 9. Offhand belt polishing and wheel buffing operations for sand cast red brass parts
Sand cast lavatory fittings made of red brass are finished in a sequence of six operations in preparation for decorative chromium plating; sequential finishing of spout with flat surfaces.

Operation	Type of contact wheel	Wheel speed, rev/min	Pieces per hour	Type of abrasive belt	Belt life, pieces	Polishing lubricant or buffing compound
Rough polishing	. . . Cloth(a)	2100	23	80-grit silicon carbide	29	None or light application of grease stick
Final polishing	. . . . . Cloth(a)	2100	30	220-grit Al_2O_3	49	Grease stick
Spot polishing	. . . . . Cloth(a)	2100	46	220-grit Al_2O_3	77	Grease stick
General buffing	. . . . Spiral-sewn, treated cloth sections with intermediate airway(b)	2400	32	. . .	. . .	Tripoli
Spot buffing	. . . . . . Spiral-sewn, treated cloth sections with intermediate airway(b)	2400	115	. . .	. . .	Tripoli
Color buffing	. . . . . Spiral-sewn, treated cloth sections with intermediate airway(b)	1700	75	. . .	. . .	Silica compound

(a) 355-mm (14-in.) diam, 45-mm (1¾-in.) width, 90 density. (b) 355-mm (14-in.) diam, 60.3-mm (2³/₈-in.) width, 18 ply, 86/93.

form surface texture. Graded bronze chips, together with the regular commercial abrasives, are used in some applications to impart a better color and finish.

MASS FINISHING

Mass finishing is best suited for stamped, formed or machined parts. Castings with remnants of gates and parting lines, forgings with heavy scale, flash lines or die marks, and heavily burred, pitted or dented parts are not well suited for mass finishing. Light burrs can sometimes be removed by a prior bright dip, after which tumbling may be used for radius blending, polishing and burnishing. High thin burrs of soft alloys are likely to peen over. Mass finishing of soft alloys at excessive speeds with insufficient amounts of solution can result in roughened and indented surfaces. Dry tumbling is generally restricted to small parts of simple shape and of maximum dimensions less than 50 mm (2 in.).

Abrasives. Aluminum oxide, silicon carbide, limestone and flintstone are the abrasive materials most often used in mass finishing of copper and copper alloys. Combinations of these abrasives may be used for specific applications.

Surface Finishes. Although mass finishing produces the final finishes for many parts, it is used more extensively for cleaning prior to plating and painting or for deburring and polishing before a final finish is applied. Examples of mass finishing applications, together with operating conditions, are given in Table 8.

POLISHING AND BUFFING

Copper alloy parts are polished after scale removal and dressing or rough cutting, but before final finishing operations, which include buffing, burnishing and honing. Rough castings normally require two polishing operations before buffing. Forgings and stampings require one polishing operation before buffing. Pipe, tubing and some stampings can be buffed without previous polishing. Buffing is not required when a brushed or satin final finish is desired.

Because copper-base materials are softer than steel, fewer stages of successively finer polishing are required to achieve a uniformly fine surface finish. When flawless chromium-plated surfaces are required, it is necessary both to buff and to color buff the polished copper alloy surfaces before plating. Chromium reproduces all imperfections in the underlying plating or base metal and, because chromium is hard and has a high melting point, it is more resistant to flow and is not readily buffed by normal methods. A good chromium-plated surface can be obtained without the color buff operation, by only polishing and cutdown buffing. Examples of offhand belt polishing and wheel buffing operations are given in Table 9.

CHEMICAL AND ELECTROCHEMICAL CLEANING

Solvent and Vapor Degreasing. Solvent cleaning of copper alloys involves immersion in special

naphthas, such as Stoddard solvent, with flash points over 38 °C (100 °F), for the removal of light grease and light oil.

Vapor degreasing effectively removes many soils from copper alloys. Stabilized trichlorethylene is used extensively in vapor degreasing because it does not attack copper alloys during degreasing and because it has high solvency for the oils, greases, waxes, tars, lubricants and coolants in general use in the copper and brass industry.

Emulsion and Alkaline Cleaning. Parts with heavy soils such as machine oils, greases and buffing compounds are treated first with emulsion cleaners to remove most of the soil. After the parts have been rinsed, the remaining soil is removed by alkaline soak or electrolytic cleaning. Precleaning reduces the contamination of the alkaline solution, extending the life of the solution.

Electrolytic alkaline cleaning is the most reliable method for cleaning parts for plating. The work is the cathode, and steel electrodes are the anodes. Reverse-current anodic cleaning cannot be used for more than a few seconds because copper dissolves in the solution. Copper alloys will tarnish readily during exposure to the oxygen that is released at the anode, but this may be minimized by the addition of inhibitors.

Cleaning and Finishing of Magnesium Alloys

FINISHING systems are applied to magnesium alloys primarily for corrosion resistance, although solderability, wear resistance and other properties sometimes are also desired.

The selection of a suitable finishing system depends on the service environment, particularly with respect to oxygen (air), moisture, chlorides and temperature. The following environments are listed in the order of increasing corrosiveness: (a) oil immersion, (b) indoor, (c) rural, (d) industrial and (e) marine. Magnesium alloys are afforded increasing protection in the following order: (a) bare, (b) pickled, (c) chromate conversion coated, (d) anodized, (e) electroplated and (f) coated with organic finishes. Anodic coatings are porous and provide no corrosion protection unless sealed with a paint.

MECHANICAL CLEANING

Mechanical cleaning of magnesium alloy products is accomplished by grinding and rough polishing, dry or wet abrasive blast cleaning, and wire brushing.

Grinding and Rough Polishing. Grinding with belts, wheels and rotary files is used for cleaning sand castings. Belt grinding, however, usually is used as a finishing operation for the removal of flash and surface imperfections from die castings, or of die marks and scratches from significant surfaces of extrusions. No great danger of surface contamination exists, and no special abrasive or belt backings are necessary.

Dry Abrasive Blast Cleaning. Sand blasting is the method of dry abrasive blast cleaning most frequently used on magnesium alloys. Many foundries use flint silica sand with a fineness of 25 or 35 AFS. Occasionally, however, steel grit is used. Usually, castings are blasted immediately after shakeout to reveal any major surface defects.

Table 10. Acid pickling treatments for magnesium alloys

Treatment	Principal applications	Metal removed μm	mils	Constituents	Amount g/L	oz/gal	Operating temperature °C	°F	Immersion time, min	Tank material or lining
For cast or wrought alloys										
Chromic acid	Remove oxide, flux, corrosion products	None		CrO_3	180	24	21-100 (a)	70-212 (a)	1-15	Stainless steel, 1100 aluminum, lead
Ferric nitrate(b)	Bright finish; maximum corrosion resistance of bare metal; finishing of die castings	8	0.3	CrO_3 $Fe(NO_3)_3 \cdot 9H_2O$ NaF	180 40.0 3.5	24 5.3 0.47	16-38	60-100	¼-3	Type 316 stainless steel, vinyl, polyethylene
Hydrofluoric acid	Active surface for chemical treatment	3	0.1	50% HF	230	31	21-32	70-90	½-5	Type 316 stainless steel, lead, rubber
Nitric acid	Prepickle for ferric nitrate treatment(c)	13-25	0.5-1.0	70% HNO_3	50	6.7	21-32	70-90	⅕-½	Stainless steel
For wrought alloys only										
Acetic-nitrate	Remove mill scale; improve corrosion resistance of bare metal	13-25	0.5-1.0	CH_3COOH $NaNO_3$	192 50.0	25.6 6.7	21-27	70-80	½-1	3003 aluminum, ceramic, lead
Glycolic-nitrate	Remove mill scale or surface oxides; improve corrosion resistance(d)	12-25	0.5-1.0	70% $CH_2OHCOOH$ 70% HNO_3 $NaNO_3$	230 40 40	31 5.3 5.3	16-49	60-120	½-1	Rubber
Chromic-nitrate	Remove mill scale, burned-on graphite; preclean for welding	13	0.5	CrO_3 $NaNO_3$	180 30	24 4	21-32	70-90	3	Stainless steel, lead, rubber, vinyl
Chromic sulfuric	Preclean for spot welding	8	0.3	CrO_3 96% H_2SO_4	180 0.4	24 0.05	21-32	70-90	3	Stainless steel, 1100 aluminum, ceramic, rubber
For cast alloys only										
Nitric sulfuric	Remove effects of blasting from sand castings	50	2.0	70% HNO_3 96% H_2SO_4	77.0 20	10.3 2.7	21-32	70-90	⅙-¼	Ceramic, rubber, glass
Phosphoric acid	Remove surface segregation from die castings; maximum corrosion resistance of bare metal	13	0.5	85% H_3PO_4	866	116	21-27	70-80	½-1	Lead, glass, ceramic, rubber
Sulfuric acid	Remove effects of blasting from sand castings	50	2.0	96% H_2SO_4	30	4	21-32	70-90	⅙-¼	Ceramic, rubber, lead, glass

(a) For removal of flux, solution must be at 88 to 100 °C (190 to 212 °F). (b) For most uniform appearance, die castings must be mechanically finished before being pickled, because the ferric nitrate solution accentuates flow marks and segregation on die-cast surfaces. (c) Use of nitric acid prepickle increases solution life and decreases treatment time in ferric nitrate pickling. (d) Nonvolatile glycolic acid reduces costs compared with acetic acid.

Wet abrasive blast cleaning of magnesium alloys is used for (a) final finishing before electroplating, (b) producing a matte surface before chemical treatments, (c) removing carbonaceous matter or heavy corrosion products and (d) removing residual paint after stripping operations.

Wire Brushing. Magnesium alloy sheet is wire brushed for in-process cleaning and for removal of oxides before arc or resistance welding.

CHEMICAL CLEANING

Solvent cleaning and vapor degreasing are used to remove oils, forming lubricants, waxes, quenching oils, corrosion-protective oils, polishing and buffing compounds, and other soluble soils and contaminants.

Emulsion cleaning may be used for removal of oils and buffing compounds. The emulsion cleaner should be neutral or alkaline, with a pH of 7.0 or above, so as not to etch magnesium surfaces. Emulsion cleaners incorporating water with the solvent should be tested before use to avoid possible attack or pitting of the metal.

Alkaline cleaning is the most frequently used method of cleaning magnesium alloys preparatory to painting, chemical treatments or plating. Alkaline cleaners are also used to remove chromate films from magnesium.

Most magnesium alloys are not attacked by common alkalis except pyrophosphates and some polyphosphates, and even these alkalis do not appreciably attack magnesium above a pH of 12.0. Nearly any heavy-duty alkaline cleaner suitable for low-carbon steel performs satisfactorily on magnesium alloys. The pH of alkaline cleaners for magnesium alloys should be 11.0 or higher.

Acid pickling is required for removal of contamination that is tightly bound to the surface or insoluble in solvents or alkalis. These contaminants include natural oxide tarnish, embedded sand or iron, chromate coatings, welding residues and burned-on lubricants.

In selecting an acid pickling treatment, consideration should be given to the type of surface contamination to be removed, the type of magnesium alloy to be treated and the dimensional loss allowable, as well as the desired surface appearance. Table 10 gives details of acid pickling treatments used for magnesium alloys.

ANODIC TREATMENTS

The chemical conversion coatings produced by anodic treatments provide protection against corrosion, an adhesive base for paint, and a decorative finish. One such treatment is called Chemical Treatment No. 9, a galvanic anodizing treatment for which a source of electric power is not required. Chemical Treatment No. 9 is ap-

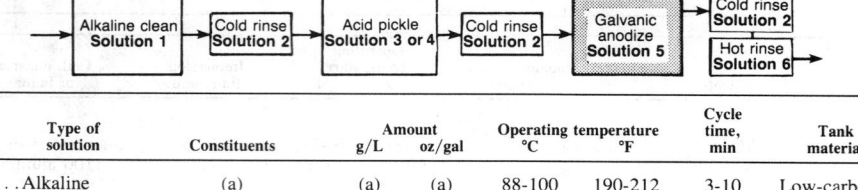

Solution No.	Type of solution	Constituents	Amount g/L	Amount oz/gal	Operating temperature °C	Operating temperature °F	Cycle time, min	Tank material
1	Alkaline cleaner	(a)	(a)	(a)	88-100	190-212	3-10	Low-carbon steel
2	Cold rinse	Water	...	...	Ambient	Ambient	(b)	Low-carbon steel
3	Acid pickle(c)	60%HF Water(d)	180 Rem	24 Rem	21-32	70-90	5(e)	Low-carbon steel(f)
4	Acid pickle(g)	NaHF$_2$, KHF$_2$ or NH$_4$HF$_2$ Water(d)	50 Rem	6.6 Rem	21-32	70-90	5	Low-carbon steel(f)
5	Galvanic anodize(h)	(NH$_4$)$_2$SO$_4$ Na$_2$Cr$_2$O$_7$ · 2H$_2$O NH$_4$OH (sp gr 0.880) Water(d)	30 30 2.2 Rem	4 4 0.3 Rem	49-60	120-140	10-30(j)	(k)
6	Hot rinse	Water	...	...	71-82	160-180	(m)	Low-carbon steel

Note: Racks and baskets for use with all tank materials may be of stainless steel, Monel, or phosphor bronze.

(a) Type and strength of solution governed by degree of surface contamination. (b) Rinse thoroughly; agitate. (c) May be used for all alloys in all forms; must be used for castings that have not been pickled after being sand blasted. (d) Water from steam condensate or water treated by ion exchange should be used, when available, instead of well or hard tap water. (e) For AZ31A and B, ¹/₂ to 1 min. (f) Lined with lead or with natural or synthetic rubber. (g) An alternative pickle for wrought products and for castings that have been pickled after sand blasting; preferred for AZ31B and C. (h) Current density should not exceed 1 A/dm² (10 A/ft²); at least 753 A · min/m² (70 A · min/ft²) is required for uniform coating. Maintain solution at pH 5.6 to 6.0 by adding solution containing 5% CrO₃ and concentrated H₂SO₄. (j) Treat parts until a uniform black coating is obtained. (k) If made of low-carbon steel, the tank acts as a cathode. If tank is made of or lined with nonmetallic material, use steel cathode plates. (m) Immerse long enough to sufficiently heat parts to facilitate rapid drying; keep rinse clean with adequate flow of fresh water.

Fig. 3. Chemical Treatment No. 9 (MIL-M-3171A) for galvanic anodizing

plied to all forms and alloys of magnesium to produce a protective black coating with good paint-base characteristics. Parts with attachments of other metals may also be treated. Because this process does not result in appreciable dimensional change, the parts are machined to close tolerances before treatment.

Proper galvanic action requires the use of racks made of stainless steel, Monel or phosphor bronze. When the workpieces are immersed in the anodizing solution, they are made the anodes, and the tank, if made of low-carbon steel, acts as the cathode. If the tank is equipped with a nonmetallic lining, separate steel cathodes must be used.

A processing diagram, and details of solution compositions and operating conditions for Chemical Treatment No. 9, are given in Fig. 3.

Cleaning and Finishing of Nickel and Nickel Alloys

NICKEL ALLOYS do not require special techniques or precautions for removing shop soils such as soap, drawing compound, oil, grease, cutting fluid and polishing compound. Oxide, scale, tarnish or discoloration can be removed from nickel and nickel alloys by mechanical methods such as grinding or abrasive blasting or by chemical methods such as pickling. Conventional methods of cleaning with alkaline compounds, emulsions or solvents, or by vapor degreasing, may be employed.

PICKLING

Pickling is a standard method for producing bright, clean surfaces on nickel alloys, either as an intermediate step during fabrication or as a last step on finished parts. Procedures used for pickling of nickel alloys are governed by both material composition and prior thermal treatment. The necessity of pickling can be avoided by using bright-heating practices—that is, by using inert furnace atmospheres to prevent oxidation. Pickling should not be used to overhaul material by dissolving away appreciable amounts of metal. Table 11 provides selected formulas for pickling nickel alloys. To aid in preparing these solutions, the acids, and their respective specific gravities and concentrations, are given below:

Acid	°Baumé	Specific gravity	Concentration, wt %
HNO$_3$	42	1.41	67
H$_2$SO$_4$	66	1.84	93
HCl	20	1.16	32
HF	30	1.26	70

FINISHING

Nickel alloys can be ground, polished, buffed or brushed by all methods commonly used for other metals.

For high-nickel alloys, a series of operations is required to produce a satisfactory finish. The number and type of operations required depend on the initial finish of the material, the desired final finish, and the type of equipment used. The pressures and speeds of the finishing equipment must be closely controlled. The high-nickel alloys, particularly nickel-chromium and nickel-iron-chromium alloys, do not conduct heat away as rapidly as copper and aluminum. Excessive heat will destroy the true color of the metal and may warp flat, thin articles.

Some general recommendations for finishing operations are given in Table 12.

Abrasive blasting or grinding, followed by flash pickling, is usually the best method for the removal of heavy scale. Abrasive blasting requires low capital investment and eliminates the use and disposal of acids.

Cleaning and Finishing of Zinc Alloys

ALLOYS used in the manufacture of zinc alloy die castings are made with high-grade zinc with about 4% aluminum, 0.04% magnesium, and either 0.25% (maximum) or 1.0% copper. Die castings are usually dense and fine grained, but do not always have smooth surfaces. Defects sometimes found in the surface layers include cracks, cold shut crevices, skin blisters and hemispherical pores. Burrs are usually left at parting lines where fins and gates are removed by die trimming. The normal sequence of preparation steps prior to plating includes:

- Smoothing of parting lines
- Smoothing of rough or defective surfaces, if necessary
- Buffing, if necessary
- Precleaning and rinsing
- Alkaline electrocleaning and rinsing
- Acid dipping and rinsing
- Copper striking

Polishing of parting lines is the initial finishing operation on zinc die castings after fins, gates and risers have been removed. This operation may be carried out by using setup wheels or abrasive belts, tumbling with abrasive media, or vibrating with abrasives. Abrasive sizes from 180 to 220-mesh are commonly used.

Polishing Other Surfaces. Although many die castings are smooth enough to require only buffing, spot or over-all polishing may be required. A 220-mesh abrasive is used on setup wheels or abrasive belts, although finer sizes may suffice at times.

Preparation for Plating. Zinc-base die castings may be prepared for plating using the procedures shown in the flow chart in Fig. 4. Duration of immersion in the alkaline cleaning solution must be kept to a minimum, or the cleaner will attack the zinc. Anodic cleaning is preferred, because any film of alkali remaining on the parts after anodic cleaning is removed more easily in the subsequent acid dip than is alkaline film from cathodic cleaning.

Cleaning and Finishing of Reactive and Refractory Metals and Alloys

CLEANING AND FINISHING processes for titanium, tungsten, molybdenum, tantalum, niobium (columbium), zirconium, and hafnium, and their alloys, are similar to those for other metals.

CLEANING AND FINISHING OF TITANIUM AND TITANIUM ALLOYS

The metallurgical and chemical properties of titanium create a number of very special cleaning problems. These include:

- Affinity of titanium to common gases
- Galvanic effects caused by discontinuities in scaled surfaces
- Metallurgical restrictions on the temperature of the descaling medium
- Variety of scales encountered in titanium descaling

Table 11. Formulas for pickling nickel alloys

Formula No.	Reagents	Weight, %	Amount	Temperature °C	°F
1	Nitric acid (HNO_3), 1.41 sp gr	20	300 mL 10 oz	70	160
	Water		1000 mL 34 oz		
2(a)	Nitric acid (HNO_3), 1.41 sp gr	10	133 mL 4 oz	75	170
	Sodium chloride (NaCl)	5	63 g 2 oz		
	Water		1000 mL 34 oz		
3	Nitric acid (HNO_3), 1.41 sp gr	20	315 mL 11 oz	50	125
	Hydrofluoric acid (HF), 1.26 sp gr	2	34 mL 1 oz		
	Water		1000 mL 34 oz		
4	Sulfuric acid (H_2SO_4), 1.84 sp gr	25	200 mL 8 oz	80	180
	Water		1000 mL 34 oz		
5	Sulfuric acid (H_2SO_4), 1.84 sp gr	15	111 mL 4 oz	80	180
	Sodium chloride (NaCl)	5	63 mL 2 oz		
	Water		1000 mL 34 oz		
6	Sulfuric acid (H_2SO_4), 1.84 sp gr	15	119 mL 4 oz	20-40	70-100
	Sodium dichromate ($Na_2Cr_2O_7$)	10	135 g 5 oz		
	Water		1000 mL 34 oz		
7	Sulfuric acid (H_2SO_4), 1.84 sp gr	12	82 mL 3 oz	Ambient	
	Sodium fluoride (NaF)	2	23 g 1 oz		
	Water		1000 mL 34 oz		
8	Sulfuric acid (H_2SO_4), 1.84 sp gr	20	171 mL 6 oz	80	180
	Sodium chloride (NaCl)	5	73 g 3 oz		
	Sodium nitrate ($NaNO_3$)	5	73 g 3 oz		
	Water		1000 mL 34 oz		
9	Sulfuric acid (H_2SO_4), 1.84 sp gr	35	1200 mL 41 oz	20-40	70-100
	Nitric acid (HNO_3), 1.41 sp gr	30	1860 mL 63 oz		
	Sodium chloride (NaCl)	0.5	30 g 1 oz		
	Water		1000 mL 34 oz		
10	Hydrochloric acid (HCl), 1.16 sp gr	6	200 mL 7 oz	60	140
	Water		1000 mL 34 oz		
11	Hydrochloric acid (HCl), 1.16 sp gr	12	535 mL 18 oz	80	180
	Cupric chloride ($CuCl_2$)	2	33 g 1 oz		
	Water		1000 mL 34 oz		
12	Hydrochloric acid (HCl), 1.16 sp gr	1	30 mL 1 oz	Ambient	
	Ferric chloride ($FeCl_3$)	1	11 g 0.3 oz		
	Water		1000 mL 34 oz		
13	Sodium hydroxide (NaOH)	15	188 g 7 oz	80	180
	Potassium permanganate ($KMnO_4$)	5	63 g 2 oz		
	Water		1000 mL 34 oz		
14	Ammonium hydroxide (NH_4OH)	2(b)	20 mL 0.5 oz	Ambient	
	Water		1000 mL 34 oz		
15	Alkaline cleaner		60-75 g/L 7-9 oz/gal	80	180
	Water		1000 mL 34 oz		
16	Agar-agar	1	10 g 0.3 oz	20-65	70-150
	Potassium ferricyanide ($K_3Fe(CN)_6$)	0.1	1 g 0.03 oz		
	Sodium chloride (NaCl)	0.1	1 g 0.03 oz		
	Water		1000 mL 34 oz		

(a) An addition of at least 40 g of nickel per litre to formula 2 will prevent overpickling of chromium-bearing alloys. (b) Volume %.

• Protective coatings used in titanium manufacturing.

Removal of Scale

Scale is removed from titanium products by several mechanical methods. Abrasive methods, such as grinding and grit blasting, are preferred for removing heavy scale from large sections. Centerless grinding is used for finishing round bars, and wide-belt grinding is used for finishing sheet and strip. Grinding is usually most efficient when it is performed at low wheel and belt speeds.

The belt-grinding sequence is usually begun with an 80-grit belt, when it is necessary to remove more than 0.07 mm (0.003 in.) of stock from the surface of the sheet. Descaling and pickling of the sheet before grinding prolong belt life. A flow chart for belt grinding of alloy Ti-6Al-4V sheet is shown in Fig. 5.

Molten Salt Descaling Baths

Molten salt descaling baths are primarily used for descaling bar, sheet products and tubing.

A primary producer of titanium sheet uses an oxidizing salt bath for removing the hot work scale in the following sequence of operations:

• Immerse in oxidizing salt for 5 to 20 min at 400 to 480 °C (750 to 895 °F).
• Quench with water for 1 min.
• Immerse in sulfuric acid, 10 to 40 vol %, for 2 to 5 min at 50 to 60 °C (120 to 140 °F).
• Rinse with water for 1 min.
• Recycle if necessary.
• Pickle in nitric-hydrofluoric acid solution (15 to 40% nitric, 1 to 3% hydrofluoric) for 1 to 5 min.

Removal of Greases and Other Soils

Greases and other soils usually are removed by vapor degreasing, emulsion and solvent cleaning, or alkaline cleaning. Vapor degreasing normally employs either trichlorethylene or perchlorethylene. Methyl ethyl ketone is used in situations where chlorinated solvents are not desirable.

Chemical Conversion Coatings

Chemical conversion coatings are applied by immersing the material in a tank containing the coating solution. Spraying and brushing are alternative methods of application. One coating bath consists of an aqueous solution of sodium orthophosphate, potassium fluoride and hydrofluoric acid, and can be used with various constituent amounts, immersion times and bath temperatures. The resultant coatings are composed primarily of titanium and potassium fluorides and phosphates. Several solutions, together with immersion times, are listed in Table 13.

CLEANING AND FINISHING OF TUNGSTEN AND MOLYBDENUM

The processes and equipment used for cleaning and finishing tungsten, molybdenum and their alloys are similar to those for steel and heat-resistant metals with some exceptions and modifications.

Abrasive Blasting

The oxides of tungsten and molybdenum become volatile at relatively low temperatures. Oxides formed on the metal surface during hot working are porous. To remove these oxides, abrasive blasting is not generally required. When this process is used, special caution must be taken. When a plate of these metals is subjected to abrasive blasting on one side for an extended length of time without turning over, the thermal stress caused by the temperature difference across the thickness of the plate may result in cracking of the metal.

Molten Caustic Process

To remove the heavy oxide scale from tungsten, molybdenum and their alloys, the molten caustic process is used. The caustic may be straight sodium hydroxide, or with an addition of 10% sodium nitrate or sodium nitrite, or with 0.5 to 2.5% sodium hydroxide. The operating temperature ranges from 340 to 400 °C (645 to 750 °F), and the immersion time is from 5 to 20 min or until the bubbling reaction stops. Caution should be taken to prevent water from getting into the bath, or a violent reaction will occur. When the workpiece is removed from the caustic, it is rinsed immediately with a jet of hot water in order to blast off the dissolved material and attached salt.

Acid Cleaning and Pickling

When tungsten and molybdenum are slightly oxidized on the surface or after the heavily oxidized workpiece is cleaned with molten caustic, acid cleaning is used. The acid solution consists of 50 to 70 vol% concentrated nitric acid, 10 to 20% concentrated hydrofluoric acid, remainder water. The cleaning solution is best when maintained at temperatures of 50 to 65 °C (120 to 150 °F).

Electrolytic Cleaning

Electrolytic etching may be used for preparation and activation of the metal before electroplating. The electrolytes may be either acid or alkaline solutions such as:

• *Acid electrolyte:* aqueous solution of 5 to 50 vol % concentrated hydrofluoric acid, using 5-V, 60-Hz alternating current
• *Alkaline electrolyte:* aqueous solution of 2 to 10% sodium hydroxide or potassium hydroxide, by weight, with a nickel cathode, using a current density of 230 A/dm² (2300 A/ft²).

Anodizing

For anodizing of tungsten and molybdenum, an acetic-base electrolyte for vanadium may be used. It consists of (*a*) acetic acid, (*b*) 0.02 *M*

Table 12. Recommended procedures for finishing of nickel and nickel alloys

Operation	Wheel	Grit No.	Compound	m/s	sfpm
Grinding	Rubber bond	24 or 36	None	40-45	8000-9000
Grinding	Vitrified bond	24 or 36	None	25-30	5000-6000
Roughing	Cotton fabric, sewn sections	60 or 80	None	30-40	6000-7500
Dry fining	Cotton fabric, sewn sections	100 or 120	None	30-40	6000-7500
Greasing	64-68 unbleached sheeting, spirally sewn sections	150 or 180	Polishing tallow or No. 180 emery grease cake	30-40	6000-7500
Grease coloring	88-88 unbleached sheeting, spirally sewn or loose disk; or quilted sheepskin	200 or 220	Polishing tallow or "F" emery grease cake	30-40	6000-7500
Bobbing and sanding	Leather wheel for two bobbing operations, second with medium-density felt wheel	...	Grout	25	5000
Cutting down	88 unbleached sheeting, loose spirally sewn sections or loose-disk wheel	...	Tripoli	40-45	8000-9000
Coloring (bright finish)	88-88 unbleached sheeting, loose spirally sewn sections or loose-disk wheel	...	White aluminum oxide	50	10 000
Coloring (mirror finish)	Loose-disk, 88-88 unbleached sheeting or Canton flannel	...	Green chromium oxide	50	10 000
Brushing	Tampico	...	"F" emery grease cake or grout	5-15	1200-3000

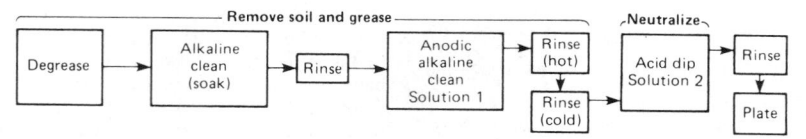

Solution No.	Type of solution	Composition	Operating temperature	Cycle time, s
1	Anodic alkaline cleaner	Alkali(a)	(b)	(b)
2	Acid dip	H$_2$SO$_4$(c)	Room temperature	30-60

(a) Special formula for zinc-base die castings. (b) Temperature, time and current density are adjusted to minimize attack on castings. (c) 0.25 to 0.75%.

Fig. 4. Surface preparation of zinc-base die castings for electroplating

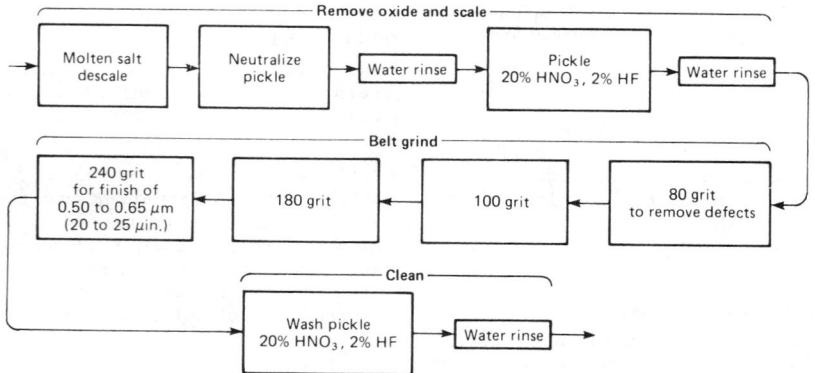

Fig. 5. Cleaning and belt-grinding sequences for alloy Ti-6Al-4V sheet

sodium tetraborate·decahydrate and (c) 1.0 M additional water. When the water content is less than 1.0 M, the conductivity is reduced to an inconvenient extent. When the water content is more than 2 M, the film formed on molybdenum becomes excessively unstable. In this process, a platinum cathode is preferred, and the temperature is maintained around 25 °C (77 °F), with the presence of air.

The anodic film formed on molybdenum is so unstable that it will change interference colors and eventually disappear when exposed to air. It can be stabilized either by dipping in glacial acetic acid, followed by drying with filter paper, or by using a jet of compressed air immediately after withdrawal from the electrolyte.

CLEANING AND FINISHING OF ZIRCONIUM AND HAFNIUM ALLOYS

Zirconium and hafnium surfaces may require cleaning and finishing for purposes such as preparation for joining, heat treatment, plating, forming, and producing final surface finishes. Special surface preparation and cleaning generally are not required for corrosion resistance because the naturally formed surface oxide protects the metal regardless of surface condition.

Removal of Surface Soil

Grease, oil and lubricants used in machine forming and other fabricating operations may be removed by a number of techniques. Alkaline or emulsion cleaners used in simple soak tanks or in ultrasonic units; acetone or trichlorethylene solvent washing; and vapor degreasing and detergent cleaning are all widely used. Hand wiping with a solvent such as acetone, alcohol or trichlorethylene is used for light soil removal. Electrolytic alkaline cleaning is also used. In the electrolytic system, the work can be of either anodic or cathodic polarity provided that the voltage and current can be controlled to avoid anodizing or spark discharge, and subsequent pitting. Removal of these soils is essential before acid etching to provide uniform acid attack. The soils must be removed before heat treatment and joining to prevent contamination and consequent loss of ductility.

Blast Cleaning

Mechanical descaling methods such as sandblasting, shot blasting and vapor blasting are used to remove hot work scales and hard lubricants from zirconium and hafnium surfaces. Aluminum oxide, silicon carbide, silica sand and steel grit are satisfactory media for mechanical descaling. Periodic replacement of used medium may be required to avoid excessive working of the surfaces by dull particulates. Roughening of exposed surface areas results from grit or shot impingement, depending on the grit size used. Any abrasive or shot blast cleaning may induce residual compressive stresses and warpage in the surface of the material, particularly thin sheet. Warpage also may occur in sections which are subsequently chemical milled or contour machined.

Blast cleaning is not intended to eliminate pickling procedures. Abrasive blasting does not remove surface layers contaminated with interstitial elements such as carbon, oxygen and nitrogen. Generally, blast cleaning is followed by a pickling step to ensure complete removal of

Table 13. Conversion coating baths for titanium alloys

Bath No.	Bath solution	Composition	Amount g/L	Amount oz/gal	Temperature °C	Temperature °F	pH	Immersion time, min
1	Degreasing solution	$Na_3PO_4 \cdot 12H_2O$	50	6.5	85	185	5.1-5.2	10
		$KF \cdot 2H_2O$	20	2.6				
		HF solution(a)	11.5	1.5				
2	Pickling solution	$Na_3PO_4 \cdot 12H_2O$	50	6.5	27	81	<1.0	1-2
		$KF \cdot 2H_2O$	20	2.6				
		HF solution(a)	26	3.4				
3	Chemical immersion solution	$Na_2B_4O_7 \cdot 10H_2O$	40	5.2	85	185	6.3-6.6	20
		$KF \cdot 2H_2O$	18	2.3				
		HF solution(a)	16	2.1				

(a) Hydrofluoric acid, 50.3% by weight.

surface contamination and cold worked layers and to produce a smooth bright finish.

Chemical Descaling

Some scale, as well as forming lubricants, can be removed by proprietary water solutions of strong caustic compounds, or by the use of molten alkaline-base salt baths. The salt baths operate at temperatures of 650 to 705 °C (1200 to 1300 °F) and must be used carefully according to the manufacturer's instructions. Salt bath descaling is done by a series of cycles through (a) the salt, (b) a water rinse and (c) a sulfuric acid bath to remove scale before final pickling.

Pickling or Etching

Metal removal by a chemical bath of nitric-hydrofluoric acid is used most commonly, although other baths have been used. The usual bath for zirconium, Zircaloys and hafnium is composed of (a) 25 to 50% nitric acid, 70 vol %; (b) 2 to 5% hydrofluoric acid, 49 vol %; and (c) remainder water. The acid bath for zirconium-niobium alloys consists of (a) 28 to 32% sulfuric acid, sp gr 1.84; (b) 28 to 32% nitric acid; (c) 5 to 10% hydrofluoric acid; and (d) remainder water.

The hydrofluoric acid attacks the zirconium and hafnium, and the nitric acid oxidizes the hydrogen formed by the reaction and prevents its absorption by the metal. The ratio of nitric to hydrofluoric acid should not be less than 10 to 1. Except for zirconium-niobium alloys, the rate of metal removal is linear with hydrofluoric acid concentration and doubles as the bath temperature rises from 43 to 71 °C (110 to 160 °F).

Material etched in the nitric-hydrofluoric acid bath must be rinsed quickly and completely with flowing water, to prevent an insoluble fluoride surface stain from forming. This stain is extremely detrimental to corrosion resistance in hot water and steam environments of nuclear reactors. Extreme precautions to ensure rapid and effective rinsing are required for this service.

CLEANING AND FINISHING OF TANTALUM AND NIOBIUM

The processes and equipment used for cleaning and finishing of tantalum, niobium and their alloys are essentially similar to those used for steel and heat-resistant alloys with some exceptions and modifications.

Mechanical Grinding and Finishing

Although commercially pure tantalum and niobium are worked at room temperature, their alloys are usually worked at elevated temperatures. Heavy oxide scale is formed. To remove such heavy oxide scale, mechanical grinding is the most effective method. For coarse grinding, a vitrified-bond grinding wheel with 46 to 60-mesh aluminum oxide is used. For finishing, a finer-grit, vitrified- or resinoid-bond grinding wheel with 60 to 120-mesh silicon carbide is used. For cutting of tantalum or niobium, an abrasive wheel is also preferred. The grinding operation usually is performed dry, but the abrasive cutting is done with water cooling.

Abrasive Blasting

To remove intermediate thicknesses of oxide scale from tantalum or niobium, abrasive blasting is applied. Abrasives used are usually silicon carbide in grit sizes from 36 to 120 mesh.

Alkaline Cleaning Process

When oxide scale is combined with grease, graphite, molybdenum disulfide and other lubricants on workpieces of tantalum and niobium, an alkaline cleaning process is usually used. The starting product for the solution is a solid alkaline material that usually consists of 50 to 80% sodium hydroxide, with the remainder being sodium metasilicate and sodium carbonate. The solids are dissolved in water at a concentration of 0.6 to 1.2 kg/L (5 to 10 lb/gal), and the solution is kept at 65 to 80 °C (150 to 180 °F). The

soaking time for the workpiece ranges from a few minutes to a few hours. After soaking, the workpiece should be immediately rinsed with a jet of water in order to blast off the loosened scale and attached salt.

Acid Cleaning and Pickling

After mechanical grinding, abrasive blasting or alkaline cleaning, tantalum and niobium are cleaned further with an acid solution. This consists of 40 to 60 vol % concentrated nitric acid, 10 to 30% concentrated hydrofluoric acid, remainder water. This cleaning solution is best when maintained at temperatures of 50 to 65 °C (120 to 150 °F). After acid pickling, the workpiece should be washed with water or rinsed thoroughly with a jet of water to remove any traces of acids.

Good ventilation and drainage systems should be installed in the acid pickling room. A recycling system to remove the residues and to refresh the acid is preferred for both economical and ecological reasons. For small acid pickling operations, without recycling, the acid solution should be neutralized and diluted with a large volume of water before being poured into the drainage system.

Electrolytic Cleaning

Electrolytic etching may be used for preparation and activation of tantalum and niobium before electroplating. The electrolytes usually used are as follows:

• For niobium, a 49% solution of concentrated hydrofluoric acid, with alternating current at 1 to 5 V and a current density of 22 to 108 A/dm² (220 to 1080 A/ft²) is employed for 1 to 3 min at room temperature.
• For tantalum, a solution consisting of 90% concentrated sulfuric acid and 10% concentrated hydrofluoric acid, used in the 25 to 40 °C (75 to 105 °F) range with a platinum cathode at a current density of 10 to 50 A/dm² (100 to 500 A/ft²), gives excellent results.

Solvent Cleaning

To remove oil, grease and other contaminants from the surface of tantalum or niobium, a common organic solvent, such as trichlorethylene, acetone or isopropanol, may be used. Cleaning is performed at room temperature by immersing workpieces in the solvent. Ultrasonic vibration is sometimes used to loosen soils from deep recesses. For large parts, spraying or wiping with the solvent may serve the same purpose. A proper ventilation system should be installed in the room for solvent cleaning.

Corrosion Theory

By Dean M. Berger, Gilbert/Commonwealth Companies

CORROSION of metal is a chemical or electrochemical process in which surface atoms of a solid metal react with a substance in contact with the exposed surface. The corroding medium is usually a liquid, but can be a gas or a solid.

All structural metals corrode to some extent in natural environments. Bronze, brass, most stainless steels, zinc and pure aluminum corrode so slowly in service conditions that long service life is expected without protective coatings. Corrosion of structural grades of iron and steel, the 400 series stainless steels, and some aluminum alloys, however, proceeds rapidly unless the metal is protected against corrosion. Corrosion of iron and steel is of particular concern because annual losses attributed to corrosion of steel have been estimated at nearly 10 billion dollars. Corrosion of metal is commonly identified by appearance, the method by which it takes place or the method by which it is accelerated, as shown in the following list (Ref 1):

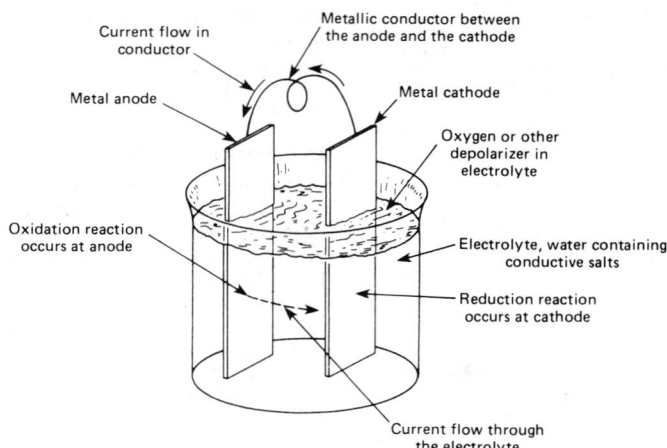

Fig. 1. Simple cell showing components necessary for corrosion (Ref 3)

- *Uniform*: uniform surface effect as opposed to localized pitting
- *Electrochemical*: corrosion occurring by chemical dissolution
- *Galvanic*: corrosion accelerated by a difference in potential between metals
- *Concentration cell*: corrosion accelerated by a difference in concentration of an ion or other dissolved substance
- *Erosion corrosion*: corrosion accelerated by flow of liquid or gas
- *Embrittlement*: corrosion which causes a ductile material to fail in a manner such that no significant localized shearing or yielding occurs
- *Stress corrosion*: corrosion accelerated or activated by stress
- *Filaform*: a form of corrosion producing porosity resembling worm holes
- *Corrosion fatigue*: fatigue accelerated by the presence of a corrosive environment
- *Intergranular*: corrosion which proceeds along grain boundaries
- *Fretting*: metal-to-metal wear which is accelerated by corrosion
- *Impingement*: corrosion accelerated by fluid impingement
- *Dezincification*: corrosion of brass in which zinc is preferentially leached out of a copper matrix
- *Chemical reaction*: corrosion which takes place by dissolution of the metal or by reaction of the metal and the corrosive medium.

Detailed discussions of these different forms of corrosion can be found in the articles "Corrosion Failures," "Stress-Corrosion Cracking," "Hydrogen-Damage Failures" and "Corrosion-Fatigue Failures" in *Metals Handbook*, 8th Ed., Vol 10.

ELECTROCHEMICAL CORROSION BASICS

Electrochemical corrosion in metals in a natural environment, whether in the atmosphere, in water or underground, is caused by a flow of electricity from one metal to another, or from one part of a metal surface to another part of the same surface where conditions permit the flow of electricity. For the flow of energy to take place, either a moist conductor or an electrolyte must be present. An electrolyte is an electricity-conducting solution containing ions which are atomic particles or radicals bearing an electrical charge.

Charged ions are present in solutions of acids, alkalis and salts. The presence of an electrolyte is necessary for corrosion to occur. Water, especially salt water, is an excellent electrolyte.

Electricity passes from a negative area to a positive area through the electrolyte. For corrosion to occur in metals, there must be (a) an electrolyte, (b) an area or region on a metallic surface with a negative charge, (c) a second area with a positive charge and (d) an electrically conductive path between (b) and (c) (Ref 2). These components are arranged to form a closed electrical circuit. In the simplest case, the anode would be one metal, such as iron, the cathode would be another, perhaps copper, and the electrolyte might or might not have the same composition at both anode and cathode. The anode and cathode could be of the same metal under conditions described later in this article.

The cell shown in Fig. 1 illustrates the corrosion process in its simplest form. This cell includes the following essential components: (a) a metal anode, (b) a metal cathode, (c) a metallic conductor between the anode and the cathode and (d) an electrolyte in contact with the anode and the cathode. If the cell were constructed and allowed to function, an electrical current would flow through the metallic conductor and the electrolyte, and if the conductor were replaced by a voltmeter, a potential difference between the anode and the cathode could be measured. The anode would corrode. Chemically, this is an oxidation reaction. The formation of hydrated red iron rust by electrochemical reactions may be expressed as follows:

$$4Fe \rightarrow 4Fe^{++} + 8\,e^-$$

$$4Fe + 3O_2 + H_2O \rightarrow$$

$$2\,Fe_2O_3 \cdot H_2O \qquad (Eq\ 1)$$

$$4Fe + 2O_2 + 4H_2O \rightarrow 4Fe\,(OH)_2$$

$$4Fe\,(OH)_2 + O_2 \rightarrow$$

$$2Fe_2O_3 \cdot H_2O + 2H_2O \qquad (Eq\ 2)$$

During metallic corrosion, the rate of oxidation equals the rate of reduction. Thus, a nondestructive chemical reaction, reduction, would proceed simultaneously at the cathode. In most cases, hydrogen gas is produced on the cathode. When the gas layer insulates the cathode from the electrolyte, current flow stops, and the cell is polarized. However, oxygen or some other de-

polarizing agent is usually present to react with the hydrogen, which reduces this effect and allows the cell to continue to function.

Contact between dissimilar metallic conductors or differences in the concentration of the solution cause the difference in potential that results in electrical current. Any lack of homogeneity on the metal surface or in its environment may initiate attack by causing a difference in potential, and this results in localized corrosion. The metal undergoing electrochemical corrosion need not be immersed in a liquid, but may be in contact with moist soil, or may have moist areas on the metal surface.

CORROSIVE CONDITIONS

If oxygen and water are both present, corrosion will normally occur on iron and steel. Rapid corrosion may take place in water, the rate of corrosion being accelerated by several factors such as: (a) the velocity or the acidity of the water, (b) the motion of the metal, (c) an increase in temperature or aeration and (d) the presence of certain bacteria. Corrosion can be retarded by protective layers or films consisting of corrosion products or adsorbed oxygen. High alkalinity of the water also retards the rate of corrosion on steel surfaces. Water and oxygen remain the essential factors, however, and the amount of corrosion is generally controlled by one or the other. For example, corrosion of steel does not occur in dry air and is negligible when the relative humidity of the air is below 30% at normal or lower temperatures. This is the basis for prevention of corrosion by dehumidification (Ref 4).

Water can readily dissolve a small amount of oxygen from the atmosphere, thus becoming highly corrosive. When the free oxygen dissolved in water is removed, the water becomes practically noncorrosive unless it becomes acidic, or anaerobic bacteria incite corrosion. If oxygen-free water is maintained at a neutral pH or at slight alkalinity, it is practically noncorrosive to structural steel. Steam boilers and water-supply systems are effectively protected by deaerating the water. Additional information can be obtained in the articles "Corrosion in Fresh Water" and "Corrosion in Seawater" in *Metals Handbook*, 9th Ed., Vol 1.

Soils. Dispersed metallic particles or bacteria pockets can provide a natural electrical pathway for buried metal. If an electrolyte is present, and the soil has a negative charge in relation to the metal, an electrical path from the metal to the soil will occur, resulting in corrosion. Differences in soil conditions, such as moisture content and resistivity, are commonly responsible for creating anodic and cathodic areas (Fig. 2). Where a difference exists in the concentration of oxygen in the water or in moist soils in contact with metal at different areas, cathodes develop at points of relatively high oxygen concentration and anodes at points of low concentration. Further information is available under "Soil Corrosion" in *Metals Handbook*, 9th Ed., Vol 1.

Chemicals. In an acid environment, even without the presence of oxygen, the metal at the anode is attacked at a rapid rate. At the cathode, atomic hydrogen is released continuously, to become hydrogen gas. Corrosion by an acid can result in the formation of a salt, which slows the reaction because the salt formation on the surface is then attacked.

Corrosion by direct chemical attack is the single most destructive force against steel surfaces.

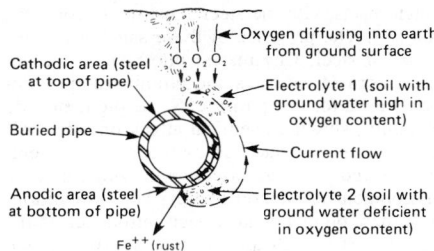

A difference in oxygen content at different levels in the electrolyte will produce a difference in potential. Anodic and cathodic areas will develop, and a corrosion cell, called a concentration cell, will form.

Fig. 2. A metal pipe buried in moist soil forming a corrosion cell

Substances having chlorine or other halogens in their compositions are particularly aggressive. Galvanized roofing has been known to corrode completely within six months of construction, the building being downwind of an aluminum ingot plant where fluorides were always present in the atmosphere. Consequently, galvanized steel should not have been specified. Selection of materials and evaluation of service conditions are extremely important in combating corrosion.

Atmospheric corrosion differs from the corrosive action that occurs in water or underground because sufficient oxygen is always present. In atmospheric corrosion, the formation of insoluble films and the presence of moisture and deposits from the atmosphere control the rate of corrosion. Contaminants such as sulfur compounds and salt particles can accelerate the corrosion rate. Nevertheless, atmospheric corrosion occurs primarily through electrochemical means and is not directly caused by chemical attack. The anodic and cathodic areas are usually quite small and close together so that corrosion appears uniform, rather than in the form of severe pitting which can occur in water or soil. A more detailed discussion can be found in the article "Atmospheric Corrosion" in *Metals Handbook*, 9th Ed., Vol 1.

GALVANIC CORROSION

The potential available to promote the electrochemical corrosion reaction between dissimilar metals is suggested by the galvanic series which lists a number of common metals and alloys arranged according to their tendency to corrode when in galvanic contact (Table 1). Metals close to one another on the table generally do not have a strong effect on each other, but the farther apart any two metals are separated, the stronger the corroding effect on the one higher in the list. It is possible for certain metals to reverse their positions in some environments, but the order given in Table 1 is maintained in natural waters and the atmosphere. The galvanic series should not be confused with the similar electromotive force series which shows exact potentials based on highly standardized conditions which rarely exist in nature (Ref 3).

The three-layer iron oxide scale formed on steel during rolling varies with the operation performed and the rolling temperature. The dissimilarity of the metal and the scale can cause corrosion to occur, with the steel acting as the anode in this instance. Unfortunately, mill scale is cathodic to steel, and an electric current can easily be produced between the steel and the mill scale. This electrochemical action will corrode the steel without affecting the mill scale (Fig. 3).

Table 1. Galvanic series
These metals are arranged by their tendency to corrode galvanically; consult text for further details.

Corroded end (anodic)
  Magnesium
  Magnesium alloys
  Zinc
  Aluminum 1100
  Cadmium
  Aluminum 2017
Steel or iron
Cast iron
Chromium-iron (active)
Ni-Resist
18-8 chromium-nickel-iron (active)
18-8-3 chromium-nickel-molybdenum-iron (active)
Lead-tin solders
Lead
Tin
Nickel (active)
Inconel (active)
Hastelloy C (active)
Brass
Copper
Bronzes
Copper-nickel alloys
Monel
Silver solder
Nickel (passive)
Inconel (passive)
Chromium-iron (passive)
18-8 chromium-nickel-iron (passive)
18-8-3 chromium-nickel-molybdenum-iron (passive)
Hastelloy C (passive)
Silver
Graphite
Gold
Platinum
Protected end (cathodic)

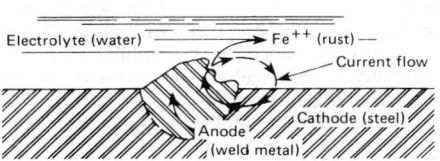

Fig. 3. Mill scale forming a corrosion cell on steel

A galvanic couple may be the cause of premature failure in metal components of water-related structures or may be advantageously exploited. Galvanizing iron sheet is an example of useful application of galvanic action or cathodic protection. Iron is the cathode and is protected against corrosion at the expense of the sacrificial zinc anode. Alternatively, a zinc or magnesium anode may be located in the electrolyte close to the structure and may be connected electrically to the iron or steel. This method is referred to as cathodic protection of the structure. Iron or steel can become the anode when in contact with copper, brass or bronze; however, they corrode rapidly while protecting these metals. Also, weld metal may be anodic to the basis metal, creating a corrosion cell when immersed (Fig. 4).

While the galvanic series (Table 1) represents the potential available to promote a corrosive reaction, the actual corrosion is difficult to predict. Electrolytes may be poor conductors, or long distances may introduce large resistance into the corrosion-cell circuit. More frequently, scale formation forms a partially insulating layer over the anode. A cathode having a layer of adsorbed gas bubbles, as a consequence of the corrosion-cell reaction, is polarized. The effect of such conditions is to reduce the theoretical consumption

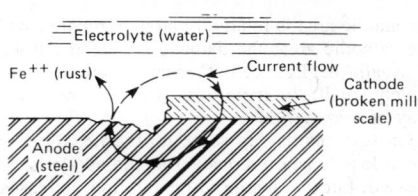

Weld metal may be anodic to steel, creating a corrosion cell when immersed.

Fig. 4. Weld metal forming a corrosion cell on steel

of metal by corrosion. The area relationship between the anode and cathode may also strongly affect the corrosion rate; a high ratio of cathode area to anode area produces more rapid corrosion. In the reverse case, the cathode polarizes, and the corrosion rate soon drops to a negligible level.

The passivity of stainless steels is attributed to either the presence of a corrosion-resistant oxide film or an oxygen-caused polarizing effect, durable only as long as there is sufficient oxygen to maintain the effect, over the surfaces. In most natural environments, stainless steels will remain in a passive state and thus tend to be cathodic to ordinary iron and steel. Change to an active state usually occurs only where chloride concentrations are high, as in seawater or reducing solutions. Oxygen starvation also produces a change to an active state. This occurs where the oxygen supply is limited, as in crevices and beneath contamination on partially fouled surfaces.

PITTING

Pitting is a type of localized cell corrosion. It is predominantly responsible for the functional failure of iron and steel water-related installations. Pitting may result in the perforation of water pipe, rendering it unserviceable, even though less than 5% of the total metal has been lost through rusting. Where confinement of water is not a factor, pitting causes structural failure from localized weakening while considerable sound metal still remains.

Pitting develops when the anodic or corroding area is small in relation to the cathodic or protected area. For example, pitting can occur where large areas of the surface are covered by mill scale, applied coatings, or deposits of various kinds, and breaks exist in the continuity of the protective coating. Pitting may also develop on bare, clean metal surfaces because of irregularities in the physical or chemical structure of the metal. Localized, dissimilar soil conditions at the surface of steel can also create conditions that promote pitting (Ref 3).

Electrical contact between dissimilar materials or concentration cells (areas of the same metal where oxygen or conductive salt concentrations in water differ) accelerates the rate of pitting. In closed-vessel structures, these couples cause a difference of potential which results in the flow of an electric current through the water or across the moist steel from the metallic anode to a nearby cathode. The cathode may be copper, brass, mill scale, or any portion of a metal surface that is cathodic to the more active metal areas. In practice, mill scale is cathodic to steel and is found to be a common cause of pitting. The difference of potential generated between steel and mill scale often amounts to 0.2 to 0.3 V. This couple is nearly as powerful a generator of corrosion currents as is the copper-steel couple. However, when the anodic area is relatively large compared with the cathodic area, the damage is spread out and is usually negligible, but when the anode is relatively small, the metal loss is concentrated and may be very serious.

On surfaces having some mill scale, the total metal loss is nearly constant as the anode is decreased, but the degree of penetration increases. Figure 3 shows how a pit forms where a break occurs in mill scale. When contact between dissimilar materials is unavoidable and the surface is painted, it is preferred to paint both materials. If only one surface is painted, it should be the cathode. If only the anode is coated, any weak points such as pinholes or holidays in the coating will probably result in intense pitting (Ref 5).

Severe pitting is often caused by concentration differences in the electrolyte, especially dissolved oxygen. When part of the metal is in contact with water relatively low in dissolved oxygen, it is anodic to adjoining areas in contact with water higher in dissolved oxygen. The lack of oxygen may be caused by exhaustion of dissolved oxygen as in a crevice (Fig. 5). This figure illustrates another type of concentration cell. This cell, at the mouth of a crevice, is caused by a difference in concentration of the metal in solution. These two effects sometimes blend together as, for example, in a re-entrant angle in a riveted seam.

As a pit, perhaps at a break in mill scale, becomes deeper, an oxygen concentration cell is started by depletion of oxygen in the pit. The rate of penetration by such pits is accelerated proportionately as the bottom of the pit becomes more anodic. Fabrication operations may crack mill scale and result in accelerated corrosion.

Stray Currents. Accelerated corrosion of steel and iron can be produced by stray currents. Direct currents in the soil or water associated with nearby cathodic protection systems, industrial activities or direct-current electric railways can be intercepted and carried for considerable distances by buried steel structures. Corrosion takes place where the stray currents are discharged from the steel to the environment, and damage to the structure can occur very rapidly (Ref 3).

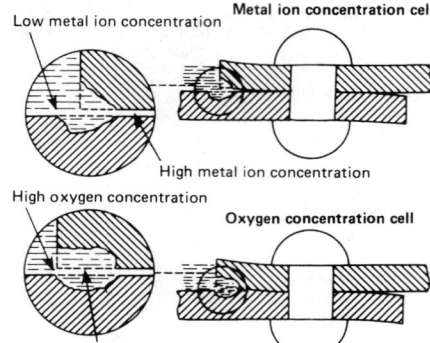

Attacks may occur simultaneously.

Fig. 5. Corrosion caused at crevices by concentration cells

COATINGS AND CORROSION PREVENTION

The problems of corrosion should be approached in the design stage, and the selection of a protective coating is important. Paint systems and lining materials exist which slow the corrosion rate of carbon steel surfaces. High-performance organic coatings such as epoxy, polyesters, polyurethanes, vinyl, or chlorinated rubber help to satisfy the need for corrosion prevention. Special primers are used to provide passivation, galvanic protection, corrosion inhibition, or mechanical or electrical barriers to corrosive action (Ref 6).

Corrosion Inhibitors. A water-soluble corrosion inhibitor reduces galvanic action by making the metal passive or by providing an insulating film on the anode or the cathode, or both. A very small amount of chromate, polyphosphate or silicate added to water creates a water-soluble inhibitor. A slightly soluble inhibitor incorporated into the prime coat of paint may also have a considerable protective influence. Inhibitive pigments in paint primers are successful inhibitors except when they dissolve sufficiently to leave holes in the paint film. Most paint primers contain a partially soluble inhibitive pigment such as zinc chromate which reacts with the steel substrate to form the iron salt. The presence of these salts slows corrosion of steel. Chromates, phosphates, molybdates, borates, silicates and plumbates are commonly used for this purpose. Some pigments add alkalinity, slowing chemical attack on steel. Alkaline pigments, such as metaborates, cement, lime, or red lead, are effective provided that the environment is not too aggressive. In addition, many new pigments have been introduced to the paint industry such as zinc phosphosilicate and zinc flake.

Sacrificial Coatings. Zinc-rich primers are applied at 75 μm (3.0 mils) dry film thickness to provide galvanic protection. These primers are very effective, even in chemical environments, because the zinc is affected before the steel is attacked. Adequate high-performance topcoats are recommended to prolong life of this coating system.

Barrier Coatings. Protective coatings are the most widely used and recognized forms of barrier materials in engineered and remedial construction. Barrier coatings may vary in thickness from thin paint films of only a few mils to heavy mastic coatings applied from 6 to 13 mm (0.25 to 0.50 in.) thicknesses to acid proof brick linings where they may be several inches thick. Barrier coatings are effective because they keep moisture, oxygen and corrosive chemicals away from the structure. Protective barrier coatings vary considerably in composition, performance and applied cost. Various rubberlike materials, plastics, tars and waxes are used.

REFERENCES

1. Corrosion Principles Can Never Be Forgotten in Organic Finishing, by D. M. Berger: *Metal Finishing,* Nov 1974
2. "Introduction to Corrosion": Carboline Co., 1968
3. "Paint Manual": Bureau of Reclamation, Denver, 1976
4. *Theory of Corrosion,* Vol I: Steel Structures Painting Council, 1968
5. *Corrosion Causes and Prevention,* by F. N. Speller: McGraw-Hill, New York, 1951
6. Designing To Prevent Corrosion in the Process Industry, by F. L. Whitney: 59-SA-58, American Society of Mechanical Engineers, May 1959

Selected References on Surface Technology

Compiled by James A. Snide, University of Dayton

Metals Handbook, 9th Ed., Vol 5, *Surface Cleaning, Finishing, and Coating:* American Society for Metals, Metals Park, 1982
Specialized Cleaning, Finishing and Coating Processes: American Society for Metals, Metals Park, 1981
Industrial Abrasive Materials and Compositions, edited by M. J. Collier: Noyes, New York, 1981
Pickling of Metals, by M. Straschill: Portcullio Press, New York, 1981
Surface Finishing System, by G. J. Rudzki: American Society for Metals and Finishing Publication Ltd., Metals Park, 1983
An Introduction to Metallic Corrosion, 3rd Ed., by U. R. Evans: Edward Arnold Ltd. and American Society for Metals, Metals Park, 1981
Corrosion Engineering, by M. G. Fontana and N. G. Greene: Edward Arnold, New York, 1963
Electroplating: Fundamentals of Surface Finishing, by F. A. Lowenheim: McGraw-Hill, New York, 1977

Metal Surface Treatment: Chemical and Electrochemical Surface Conversions, edited by M. H. Gutcho: Noyes, New York, 1982
Electrodeposition Processes, Equipment and Compositions, edited by J. I. Duffy: Noyes, New York, 1982
Electroplating Engineering Handbook, 3rd Ed., by A. K. Graham: Van Nostrand Reinhold, New York, 1971
Sputtering by Particle Bombardment I: Physics and Applications, edited by R. Behrisch: Springer-Verlag, New York, 1973
Ion Implantation, by G. Dearnaley: Elsevier, New York, 1973
Electroless and Other Nonelectrolytic Plating Techniques—Recent Developments, edited by J. I. Duffy: Noyes, New York, 1981
Diffusion Cladding of Metals, by G. U. Somsonov: Plenum Press, New York, 1967
The Technique of Enamelling, by G. Clark: Van Nos-trand Reinhold, New York, 1977
Surfaces and Interfaces in Ceramic-Metals Systems, edited by J. Pask and A. Evans: Plenum Press, New York, 1981
Electropainting, 2nd Ed., by R. L. Yeates: Portcullio Press, New York, 1981
Coatings of High-Temperature Materials, by H. H. Hauser: Plenum Press, New York, 1966
The Surface Treatment and Finishing of Aluminum and Its Alloys, 4th Ed., by S. Wernick and R. Pinner: Robert Draper Ltd., Teddington, England, 1972
Finishing of Aluminum, edited by G. H. Kissin: Reinhold, New York, 1963
Plating on Less Common Metals, by J. G. Beach and C. L. Faust: in *Modern Electroplating,* edited by F. A. Lowenheim, John Wiley & Sons, New York, 1974, p 618–635
High Temperature Inorganic Coatings, edited by J. Huminik: Reinhold, New York, 1963

30 JOINING

Edited by Ernest F. Nippes, Rensselaer Polytechnic Institute

This section was condensed from Metals Handbook, Ninth Edition, Volume 6, Welding, Brazing, and Soldering. For more detailed information on the topics covered herein, the reader is referred to the larger work. Additional information on joining can be located in this volume by consulting the index.

Introduction and Overview

Principles of Joining

By Jack H. Devletian and William E. Wood, Oregon Graduate Center

WELDING is the joining of two or more pieces of metal by applying heat or pressure, or both, with or without the addition of filler metal, to produce a localized union through fusion or re-crystallization across the interface. Ideally, welding a particular alloy with filler metal that matches exactly provides several advantages:

• Uniform composition throughout the weld joint
• Excellent match of physical properties such as color, density, and electrical and thermal conductivities
• Uniform mechanical properties throughout the weld joint and base metal after postweld heat treatment.

In commercial arc welding practice, however, a steel plate of one composition, such as A242, A441, A588 or API-5LX, is most likely to be welded with a steel electrode of a different chemical composition, such as E7018 or ER70S-3.

Similarly, nonferrous metals, including the aluminum alloys 3004, 5005, 6061, 6070 and A357.0, are all ordinarily welded with ER4043 filler metal for general-purpose gas metal-arc and gas tungsten-arc applications. The majority of filler metal selection recommended by the American Welding Society (AWS), American Society of Mechanical Engineers (ASME), American Petroleum Institute (API), and military welding codes is based on providing crack-free welds and closely

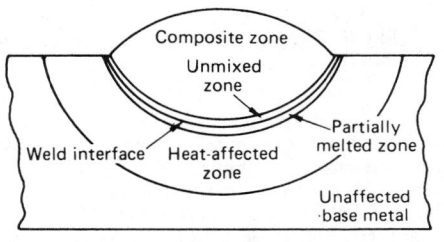

Source: Ref 1.

Fig. 1. Metallurgical zones developed in a typical weld

matching the tensile properties of the as-deposited filler metal with those of the base metal. The composition match, although important, is the secondary consideration.

As a result of the nonmatching filler metal and heat-distribution characteristics, the weld joint is usually a chemically heterogeneous composite consisting of as many as six metallurgically distinct regions. Based on the work by Savage, Nippes and Szekeres (Ref 1), a typical weld consists of (1) the composite zone, (2) the unmixed zone, (3) the weld interface, (4) the partially melted zone, (5) the heat-affected zone (HAZ) and (6) the unaffected base metal. These zones are illustrated in Fig. 1.

Composite Zone. The admixture of filler metal and melted base metal comprises a completely melted and homogeneous weld fusion zone in the composite zone or region. For example, when gray cast iron is welded with a nickel electrode, the composite region contains a homogeneous molten pool of nickel filler metal diluted with melted gray iron base metal. Similarly, when E10018 electrodes are used to weld HY-80 steel, the chemical composition of the composite zone is the weighted average of the elements (i.e., carbon, nickel or manganese) from both the filler metal and the melted base metal. Even totally dissimilar metals such as copper and nickel can be welded autogenously (no filler metal) to each other by gas tungsten-arc welding (GTAW), and the bulk composition throughout the composite zone will be surprisingly uniform. Thorough mixing is promoted by forced convection in the molten pool combined with the substantial reduction of free energy contributed by the great increase in mixing entropy.

Unmixed Zone. The narrow region surrounding the bulk composite zone is the unmixed zone (Fig. 1), which consists of a boundary layer of melted base metal that froze before undergoing any mixing in the molten composite zone. This layer at the extremities of the weld pool is characterized by a composition essentially identical to the base metal with a typical thickness of about 0.05 to 0.10 in., depending on welding process and weld cooling rate. Although the unmixed zone is present in all fusion welds, it is readily visible only in those welds utilizing a filler-metal alloy of substantially different composition than the base metal.

Consider, for example, shielded metal-arc welding (SMAW) of class 30 gray cast iron with nickel-rich filler metal. The unmixed zone is clearly visible (Fig. 2) because the molten gray iron solidifies as a white iron (with eutectic of Fe_3C plus γ) structure, while the composite zone, containing a majority of nickel filler metal, solidifies as austenite. Conversely, consider gas tungsten-arc welding of pure nickel deposited with a nickel filler metal. In this instance (see Fig. 3), no unmixed zone is visible, because the composition and cooling conditions of the composite-zone liquid are identical to those of the liquid in the unmixed zone.

Weld Interface. The third region defined in a weldment is the weld interface. This surface clearly delineates the boundary between the unmelted base metal on one side and the solidified weld metal on the other side. Often in pure metals or very dilute alloys using matching filler metal, the transition from base metal to weld metal is difficult to observe metallographically, but can be revealed through alloy-sensitive etching of the solidification substructure. Generally, as the alloy content and the solidification range between liquidus and solidus of a given weld increase, the solidification structure is more easily revealed by etching.

Partially Melted Zone. In the base metal immediately adjacent to the weld interface, where some localized melting may occur, the partially melted zone can be observed. In many alloys that contain low-melting inclusions and impurity or alloy segregation at grain boundaries, liquation of these low-melting microscopic regions may occur and extend from the weld interface into the partially melted zone. The depth to which a liquated region penetrates into the base metal depends on the solidus temperature of the liquated matter. In steels, the classic example of constitutional liquation in the partially melted zone occurs in HY-80 steel weldments. Typically, the liquation of manganese sulfide inclusions results in a hot crack or microfissure which extends from the unmixed zone, across the weld interface, and into the partially melted zone.

Heat-Affected Zone. The true HAZ (Fig. 1) is the portion of the weld joint which has been subjected to peak temperatures high enough to produce solid-state microstructural changes but too

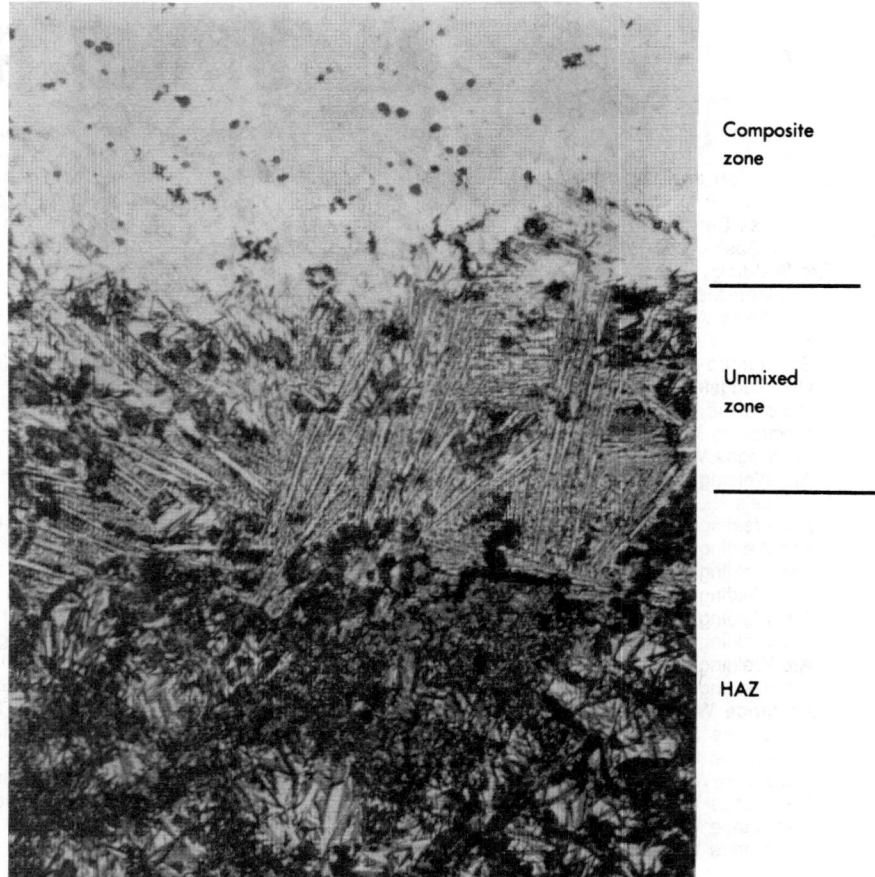

Composite zone

Unmixed zone

HAZ

Composite zone is austenite. Unmixed zone is white iron. HAZ is martensite and undissolved graphite. Etched in 2% nital; magnification, 160×.

Fig. 2. Weld deposited on gray iron with nickel filler metal

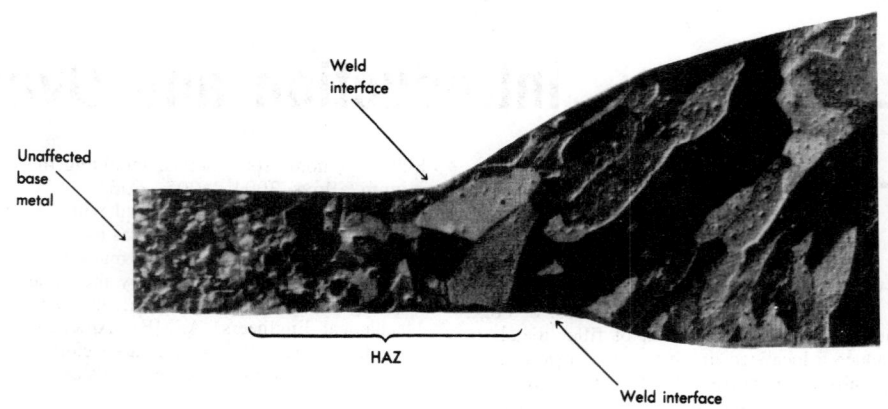

Weld interface

Unaffected base metal

HAZ

Weld interface

Etched in aqua regia; magnification, 20×.

Fig. 3. Gas tungsten-arc weld of pure nickel deposited with matching nickel filler metal

low to cause any melting. For example, this zone in single-phase wrought alloy is characterized by a steadily increasing grain size from the outer extremity of the HAZ to a maximum grain size at the weld interface, as shown in Fig. 3.

Unaffected Base Metal. Finally, that part of the workpiece that has not undergone any metallurgical change is the unaffected base metal. Although metallurgically unchanged, the unaffected base metal, as well as the entire weld joint, is likely to be in a state of high residual transverse and longitudinal shrinkage stress, depending on the degree of restraint imposed on the weld.

SOLIDIFICATION OF WELDS

Epitaxial Growth. Fundamental solidification mechanics developed primarily for cast metals have been successfully applied to the solidification of welds. The outstanding difference between the solidification of a casting and that of a weld (aside from the relative sizes and cooling rates) is the phenomenon of epitaxial growth in welds. In castings, formation of solid crystals from the melt requires heterogeneous nucleation of solid particles, principally on the mold walls, followed by grain growth. In contrast, the nucleation event in welds is eliminated during the initial stages of solidification because of the mechanism of epitaxial growth wherein atoms from the molten weld pool are rapidly deposited on pre-existing lattice sites in the adjacent solid base metal. As a result, the structure and crystallographic orientation of the HAZ grains at the weld interface continue into the weld fusion zone as shown in the pure nickel weld in Fig. 3. In fact, the exact location of the weld interface is very difficult to determine in any weld deposited on pure metals using matching filler metal. Even microstructural features, such as annealing twins located in the HAZ weld joints, will continue to grow epitaxially into the weld during solidification. Similarly, nonmatching filler metals will also solidify epitaxially, particularly if the filler metal and base metal have the same crystal structure upon solidification, e.g., welding Monel (fcc) with nickel (fcc) filler metal.

Weld-Pool Shape. Because it controls the grain structure of the weld, weld-pool shape is an important factor in welding. For example, if a single-phase metal is gas tungsten-arc welded at a low velocity, the weld pool is elliptical (nearly circular), as shown in Fig. 4(a). The columnar grains grow in the direction of the thermal gradient produced by the moving heat source (arc). The grains grow epitaxially from the base metal toward the arc. Because the direction of maximum temperature gradient is constantly changing from approximately 90° to the weld interface at position A to nearly parallel to the weld axis at position B, the grains must grow from position A and continuously turn toward the position of the moving arc. The process of "competitive growth" provides a means whereby grains less favorably oriented for growth are pinched off or crowded out by grains better oriented for continued growth. The $\langle 001 \rangle$ and $\langle 10\bar{1}0 \rangle$ are the generally favored directions for crystal growth in cubic (fcc and bcc) and hexagonal (hcp) metals, respectively. In fcc metals, for example, the $\langle 001 \rangle$ most favored direction leads each solidifying grain because the four close-packed {111} planes symmetrically located around the $\langle 001 \rangle$ axes require the greatest time to solidify and, therefore, serve both to drag and guide the growth of solidifying grains.

The shape of the weld pool tends to become more elongated with increasing welding speed. In Fig. 4(b), the direction of maximum temperature gradient is perpendicular to the weld interface at positions A and B, but, because the weld pool is trailing a greater distance behind the arc, the temperature gradient at position B is no longer strongly directed toward the electrode. Therefore, the columnar grains do not turn as much as in the case of a nearly circular weld pool.

Finally, the weld takes on a teardrop shape at the high welding speeds that are usually encountered in commercial welding practice. The weld pool is elongated so far behind the welding arc that the directions of the maximum temperature gradient at position A and B in Fig. 4(c) have changed only slightly. As a result, the grains grow from the base metal and converge abruptly at the centerline of the weld with little change in direction. Welds that solidify in a teardrop shape have the poorest resistance to centerline hot

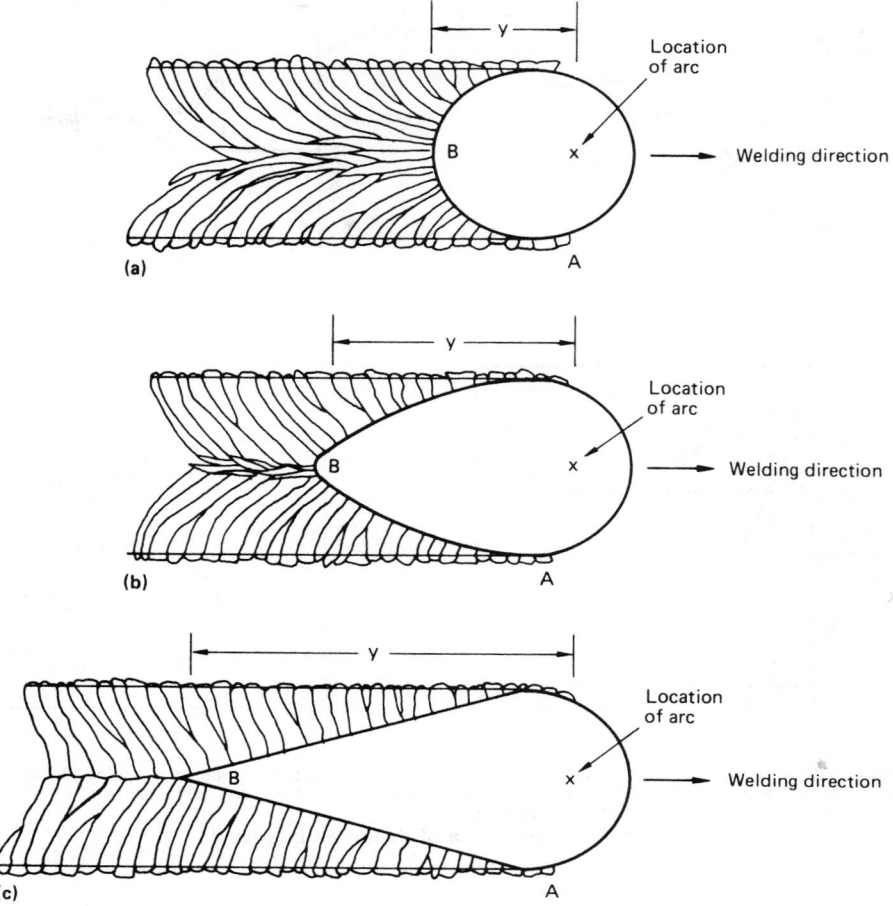

Travel speeds: (a) low, (b) intermediate, (c) high.

Fig. 4. Comparison of weld-pool shapes

cracking because low-melting impurities and other low-melting constituents tend to segregate at the centerline. Unfortunately, this solidification geometry occurs most frequently in commercial welding applications, because high heat input and high travel speeds produce the most cost-effective method of welding.

Cells, Dendrites and Microsegregation. Each columnar weld-metal grain may contain a solidification substructure. No substructures are visible metallographically in the pure nickel weld shown in Fig. 3, but the weld in Fig. 5 shows a definite substructure within each grain. Although the bulk weld-metal composition is homogeneous, cells or cellular dendrites represent a commonly observed pattern of microsegregation that is developed during nonequilibrium solidification of a weld or casting. Microsegregation is characterized by a compositional difference between the cores and peripheries of individual cells and cellular dendrites. Cells are microscopic pencil-shape protrusions of solid metal that freeze ahead of the solid/liquid interface in the weld. Cellular dendrites are more developed than cells and appear to have a "treelike" shape; the main stalk is called the "primary dendrite arm," and the orthogonal branches are called the "secondary dendrite arms." The cores of the cells and dendrite arms have a higher solidus temperature and contain less solute than the intercellular and interdendritic regions. In actual welding practice, cellular or dendritic microsegregation is virtually impossible to

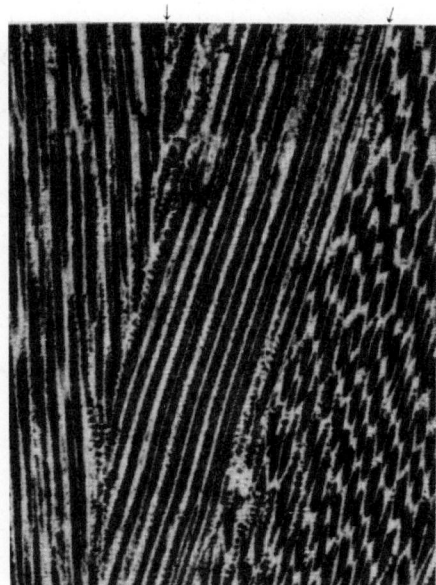

Matching filler metal. Arrows indicate grain boundaries, and dark-etching regions are cores of cellular dendrites. Etched in Stead's reagent; magnification, 65×.

Fig. 5. Cellular dendritic substructure in Hadfield steel weld metal deposited by the GTAW process

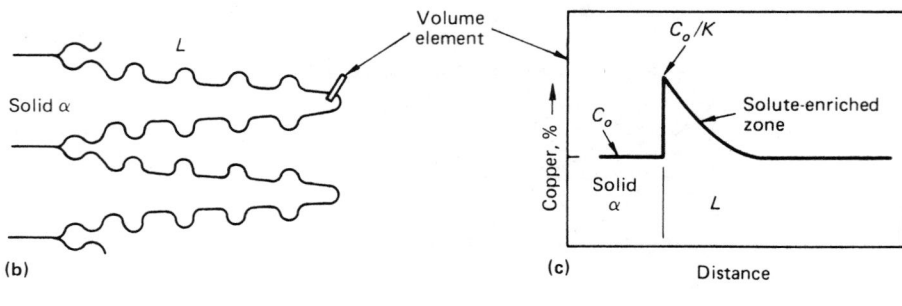

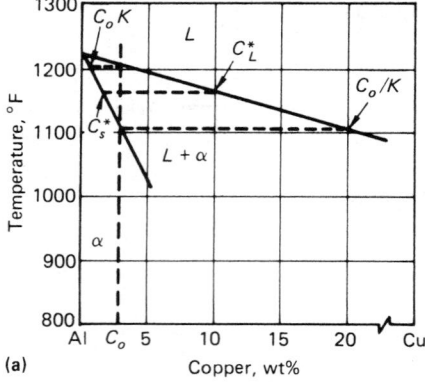

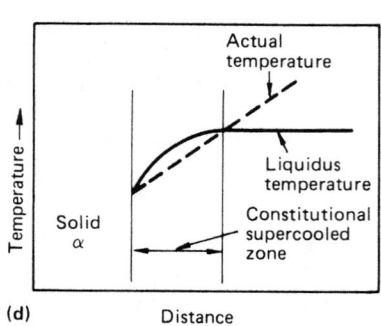

(a) Solidification of 3%Cu-Al alloy by (b) growth of dendrites. (c) Solute redistribution occurring ahead of the solid/liquid interface. (d) Constitutional supercooling develops when the actual temperature of liquid in the copper-rich zone is greater than the liquidus temperature.

Fig. 6. Solidification of dendrites in a weld

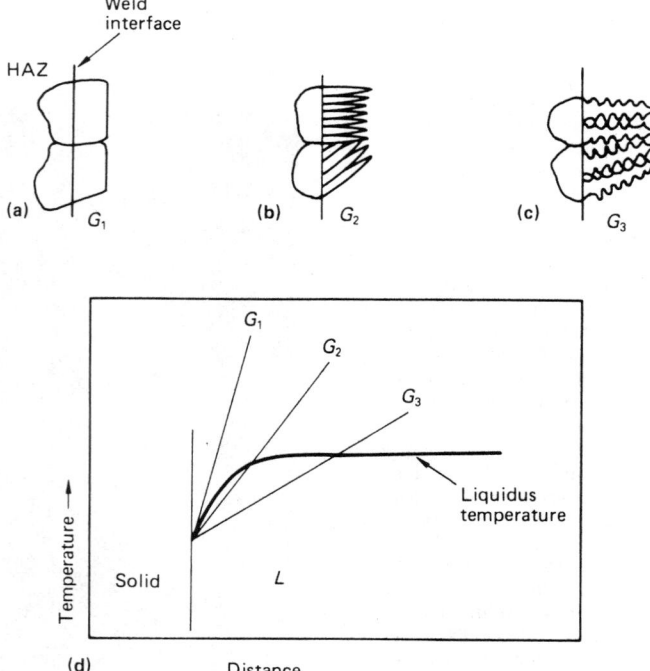

(a) Steep G_1 planar growth. (b) Intermediate G_2 cellular growth. (c) Small G_3 cellular dendritic growth. (d) Solidification of the weld.

Fig. 7. Effect of thermal gradient on mode of solidification in welds for constant growth rate

avoid unless the metal being welded is a pure element.

Generally, the important parameters controlling the cellular or cellular dendritic substructures in welds are (1) the equilibrium partition ratio, K, which is an index of the segregation potential of an alloy:

$$K = C_s^*/C_L^* \qquad \text{(Eq 1)}$$

where C_s^* is the solute content of the solid at the solid/liquid interface and C_L^* is the solute content of the liquid at the solid/liquid interface; (2) the alloy composition itself, C_o; (3) the temperature gradient, G, in the liquid at the weld interface in °F/in.; and (4) the growth rate, R, or velocity of the interface in in./s.

For example, consider an alloy of composition C_o equal to 3%Cu-97%Al. Thus in Fig. 6(a):

$$K = C_s^*/C_L^* = 1.7\%/10\% = 0.17$$

The first metal to solidify will contain only:

$$C_o K = (3)0.17 = 0.51\% \text{ Cu}$$

while the last liquid to solidify between cells or cellular dendrites is rich in copper:

$$C_o/K = 3/0.17 = 17.6\% \text{ Cu}$$

These values represent the short transients at the start and finish of solidification of a cell or cellular dendrite. As the cell or dendrite grows in the weld, a dynamic equilibrium between the newly forming solid of composition $C_o = 3\%$ Cu and the copper-rich liquid containing a maximum of $C_o/K = 17.6\%$ Cu is achieved at the solid/liquid interface as shown in Fig. 6(b) and (c). If the actual temperature distribution ahead of the solid/liquid interface is less than the liquidus temperature, constitutional supercooling occurs (Fig. 6d). Supercooling means that the solute-enriched liquid ahead of the solid/liquid interface has been cooled below its equilibrium freezing temperature, and constitutional indicates that the supercooling originated from an enrichment in composition rather than temperature.

Microsegregation results when the copper-rich liquid at the solid/liquid interface solidifies between the cellular dendrites. The interdendritic regions are so segregated with copper (solute) that a small amount of eutectic $(\alpha + \Theta)$ is frequently observed. Eutectic structures can occur only when the composition of solidifying metal exceeds the maximum solid solubility of 5.65% Cu in α.

Whether or not a planar, cellular or dendritic substructure occurs upon solidification is largely determined by G and R, which control the amount of constitutional supercooling. If a weld is deposited at a constant travel speed, R becomes fixed. By inducing an extremely steep temperature gradient G_1 (Fig. 7a), no constitutional supercooling occurs and the solidified weld-metal grain structure is planar. For example, the epitaxially grown columnar grains in the pure nickel weld (Fig. 3) contain no solidification substructure. Therefore, the location of the weld/base metal interface is extremely difficult to distinguish.

When the gradient is decreased slightly to G_2 (Fig. 7b), any protuberance of solid metal on the interface will grow faster than the remaining flat interface because the solid is growing into supercooled liquid; that is, the solid protuberance exists at a temperature below the liquidus for that alloy. As a result, a cellular substructure devel-

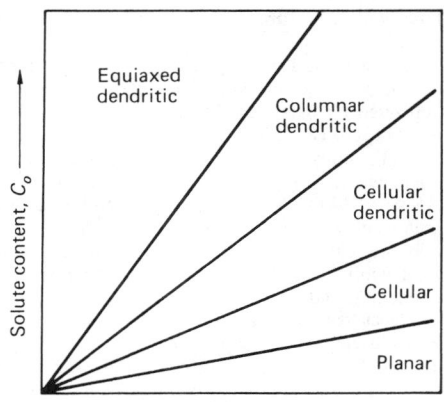

Fig. 8. Dependence of mode of solidification on G/R parameter for different solute concentrations (C_o)

ops in each epitaxially grown grain. The liquid ahead and alongside of each cell contains greater solute content than the cell core.

If the value of the temperature gradient is further decreased to G_3 (Fig. 7c), constitutional supercooling becomes so extensive that secondary arms form and cellular dendritic growth is observed. The greatest degree of microsegregation occurs during columnar dendritic solidification, while no measurable segregation is encountered in planar growth. Whether planar, cellular, or cellular dendritic, growth is always anisotropic.

Investigators have found that these solidification substructures can be characterized by the combined parameter G/R. Figure 8 shows that a large value of G/R combined with a very dilute alloy will result in a planar solidification structure, while a low G/R and high solute concentration will produce a heavily segregated columnar dendritic structure. Both columnar dendritic and equiaxed dendritic structures, although common in large castings, are infrequently encountered in welds. In practice, cellular and cellular dendritic substructures are most frequently observed in welds. The difference between the cellular dendritic and columnar dendritic structures is related to the length of the constitutionally supercooled zone ahead of the solid/liquid interface. This zone is typically much smaller for cellular dendritic than for columnar dendritic solidification. Therefore, each grain will contain many cellular dendrites (Fig. 5), whereas only one columnar dendrite occupies one grain. Unfortunately, it is very difficult to control G and R independently in welding practice. As a general rule, a high welding speed (R) will produce a steep G. The relative values of G and R, however, determine the solidification morphology for a given alloy of fixed C_o and K.

Solidification Rate. While G/R controls the mode of solidification, the weld cooling rate, in terms of the parameter GR (solidification rate in units of °F/s), determines both the size and spacing of cells and dendrites. Flemings and others have demonstrated that the effect of solidification rate on dendrite-arm spacing (d) is:

$$d = a(GR)^{-n} \qquad \text{(Eq 2)}$$

where a is a constant and n is approximately $1/2$ for primary arms and between $1/3$ and $1/2$ for sec-

ondary arms. The dendrite-arm spacing of stainless steel in an electroslag weld is often several hundred times greater than that found in a rapidly cooled laser weld.

Solute Banding. The phenomenon of solute banding occurs to some degree in all alloy welds. The formation of ripples on the weld surface and solute banding within the weld are both caused by the discontinuous nature of weld-metal solidification and occurs in manual as well as in automatic welds where the travel speed is mechanically constant. During weld-metal solidification,

however, R fluctuates cyclically above and below a mean value of growth rate that is determined by the weld travel speed. Fluctuations in R result in not only ripple formation but also solute banding. Because an abrupt increase in R causes a reduction in the amount of solute that can be held in the solute-enriched liquid (Fig. 6c), excess solute is dumped and appears as a solute-rich band. Similarly, a sudden decrease in R produces a solute-poor band. Solute banding lines (Fig. 9) are very helpful in welding research because they always outline the weld-pool shape at

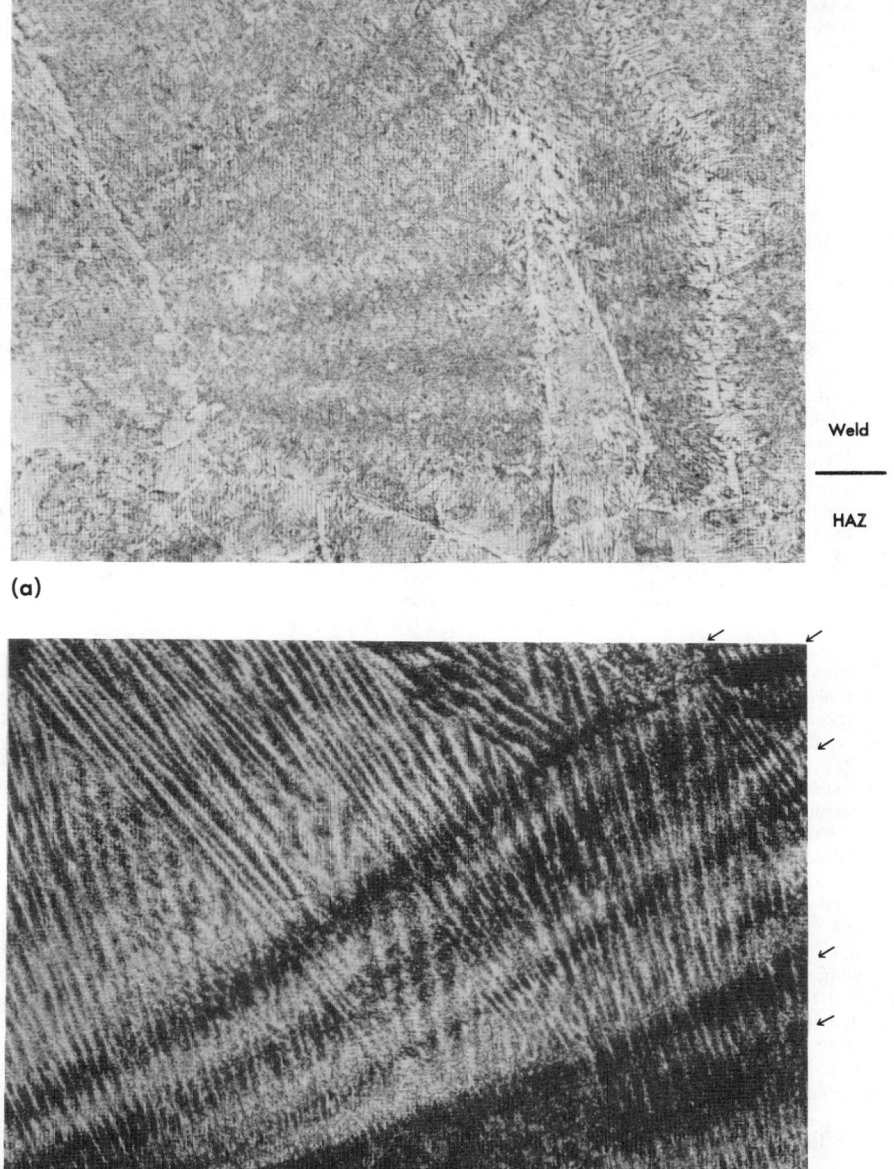

(a) Etched in nital. (b) Etched in solute-sensitive Stead's reagent; solute banding lines are indicated by arrows. Magnification, 25×.

Fig. 9. Solute banding lines in electroslag weld metal of 6%Ni steel

a given instant during solidification. For example, the form factor (ratio of width to depth of weld pool), which is so important in electroslag welding (ESW), can be easily measured metallographically using solute banding lines.

PREHEATING

Preheating of joints to be welded is an extremely effective method ordinarily used to reduce (1) the cooling rates of the weld and HAZ, (2) the magnitude of distortion and residual shrinkage stress, and (3) the arc energy input required to deposit a given weld. The first two factors are essential to prevent cracking in hardenable steels. The third is often necessary to weld thick sections of highly conductive metals, such as copper or aluminum.

From the Adams cooling rate and peak HAZ temperature equations (Ref 2), it is clear that preheating of the workpiece can significantly reduce weld cooling rate and increase the width of the HAZ. The accompanying changes in microstructure and hardness of the weld joint can be substantial. As an example, adequate preheating of a high-carbon 1080 steel will produce a crack-free pearlitic microstructure in the HAZ, while a similar weld without preheating will result in a brittle martensitic structure in the HAZ.

Calculation of Preheat Temperatures. Methods used to determine the proper preheating temperature for welding a given plain or low-alloy steel are (1) consultation of the ASME Boiler Code (Section IX), AWS D1.1 Structural Welding Code, API or American Association of State Highway and Transportation Officials (AASHTO) codes, or other recognized welding codes; (2) carbon-equivalent calculations; (3) reference to published literature; and (4) use of published CCT curves. When welding to a code, preheating temperatures are mandated for a particular grade of steel and thickness to be welded. For example, the AWS D1.1 code specifies that 2-in.-thick A588 steel must be submerged-arc welded with a minimum preheating temperature of 150 °F.

Reduction of Distortion and Residual Stress. The second purpose of preheating both ferrous and nonferrous metals is to reduce distortion and residual stress. As a weld cools through the austenite range in steels and through an elevated temperature range in nonferrous alloys, the metal has little strength and good plasticity. Therefore, the weld metal and HAZ deform plastically to accommodate the change in dimensions imposed by the shrinking weld. Upon cooling to room temperature, residual stresses build up because of continued shrinkage, but not to the extent experienced in a similar weld without preheating. The degree of reduction of distortion and residual stress is difficult to predict in practical welding applications, because it depends on many variables, including magnitude of restraint, preheating temperature, groove preparations and heat input.

POSTWELD HEAT TREATMENT

A wide variety of metallurgical objectives may be accomplished through postweld heat treatments, including stress relief, dimensional stability, resistance to stress corrosion and, occasionally, improved toughness and mechanical properties. The most common postweld heat treatments for steels are subcritical stress relieving, normalizing, and quenching and tempering.

Typical treatments for nonferrous metals, such as aluminum alloys, include postweld stress relieving, full solution heat treatment and aging, aging only, and annealing.

Stress relieving is probably the treatment most frequently used to reduce the residual welding stresses in welds that are heavily restrained or susceptible to cracking. The dominant mechanisms of stress relief are relaxation of stress and tempering of martensite or overaging of precipitation-hardening alloys. In steels, the stress-relief temperature ranges from 895 to 1240 °F, which is below the eutectoid transformation, for a minimum of 1 h per inch of thickness. Most often, a weld deposited on a high-hardenability steel, such as 4130, is put into the stress-relieving furnace before having a chance to cool below the preheating or interpass temperature. Thus, the microstructure does not contain any martensite, because any austenite remaining after welding is transformed to bainite during stress relieving in accordance with the TTT diagram for 4130 steel. If the steel weld does form martensite after welding because of insufficient preheating, the stress-relieving operation tempers the martensite to a lower value of hardness, but improves toughness and ductility.

Postweld stress relieving can virtually eliminate caustic stress-corrosion cracking occurring in the HAZ of ASTM A516, grade 70, steel, which is ordinarily used in the pulp and paper industry. The combined reduction of residual stress and glavanic differences among the weld metal, the HAZ and the base metal undoubtedly contribute significantly to the improved stress-corrosion resistance of these weldments.

Postweld normalizing applies primarily to steels. This treatment has generally the same function in welding as in casting—promoting toughness and eliminating coarse grain structure. Therefore, the ESW process especially benefits from normalizing. Because of the high heat input, 2000 kJ/in. for a 2-in.-thick weld, the large HAZ and weld-metal grain sizes result in a severe loss of Charpy V-notch toughness. Several welding codes permit electroslag welds in primary tension members only if the welds are normalized. Unfortunately, normalizing is a far more expensive operation than stress relieving. Although stress relieving can be conducted in the field, normalizing requires such high temperatures, 1600 to 1730 °F for 1 h per inch of thickness, that the entire welded assembly must be transported to a large furnace.

Benefits of postweld normalizing include the elimination of both the coarse columnar grain structure in the weld and the large grain size in the HAZ and substantial improvement in toughness at the weld centerline and HAZ. The microstructure of typical normalized weld joints contains a fine-grain mixture of pearlite and polygonal ferrite. For example, the microstructural changes induced by normalizing electroslag welds deposited on A588 steel will raise the CVN toughness of the weld centerline from approximately 7 ft·lb at 0 °F to more than 50 ft·lb.

Normalizing is generally not as beneficial for gas tungsten-arc, shielded metal-arc or gas metal-arc welds as for electroslag or submerged-arc welds, because grain structures in the weld and HAZ are already refined by virtue of their comparatively low heat input and small bead size. Nonetheless, normalizing can be applied to any weld and will replace the weld and HAZ struc-

tures with a uniform structure of polygonal ferrite plus pearlite.

Quenching and tempering of welded joints is an expensive operation and is reserved only for welds deposited on heat treatable steels such as 4130, 4140, 4340, H-11, and other steels used for high-strength, high-hardness applications. Unlike the quenched and tempered steels, which include A514, A517 (T-1) and A508 (HY-80), heat treatable steel welds contain high carbon levels and alloying elements to match the hardness and hardenability of the base metal after postweld quenching and tempering.

Solution treating and aging are postweld heat treatments used with precipitation-hardening alloys for uniform strength of the weld, HAZ and base metal. If the weld cooling is rapid, as in electron beam welding, aging alone can yield significant strengthening. In the latter case, the weld and the large-grain portion of the HAZ are considered after welding to be essentially solution treated.

REFERENCES

1. A Study of Weld Interface Phenomena in a Low Alloy Steel, by W. F. Savage, E. F. Nippes and E. S. Szekeres: *Welding Journal*, Sept 1976, p 260s-268s
2. Cooling Rates and Peak Temperatures in Fusion Welding, by C. M. Adams, Jr.: *Welding Journal*, Vol 37, No. 5, 1958, p 210s-215s

ADDITIONAL READING

Atlas of Continuous Cooling Transformation Diagrams for Engineering Steels, by M. Atkins: American Society for Metals and British Steel Corp., 1980
Metals Handbook, 8th Ed., Vol 8, *Metallography, Structures and Phase Diagrams*, edited by T. Lyman: American Society for Metals, 1973
Metals Handbook, 8th Ed., Vol 1, *Properties and Selection: Irons and Steels*, edited by T. Lyman: American Society for Metals, 1961, p 46, 52
Solidification Processing, by M. C. Flemings: McGraw-Hill, New York, 1974
Fundamentals of Welding, by C. Weisman: in *Welding Handbook*, 7th Ed., Vol 1, American Welding Society, Miami, 1976
The Influence of Cooling Rate and Composition on Weld Metal Microstructure in a C/Mn and HSLA Steel, by A. G. Glover *et al*: *Welding Journal*, Vol 56, No. 9, 1977, p 267s-273s
Sources of Weld Metal Oxygen Contamination During Submerged Arc Welding, by T. W. Eager: *Welding Journal*, Vol 57, No. 3, 1978, p 76s-80s
Improvement of Weld Fusion Zone Toughness by Fine TiN, by S. Kanazawa *et al*: *Transactions of Iron and Steel Institute of Japan*, Vol 16, 1976, p 487-495
Factors Controlling HAZ and Weld Metal Toughness in C-Mn Steels, by R. E. Dolby: *Proceedings of First National Conference on Fracture*, Johannesburg, South Africa, 1979, in *Engineering Applications of Fracture Analysis*, by G. G. Garrett and D. L. Marriott, Pergamon Press, p 117-134
The Effect of Heat Input on the Microstructure and Properties of C-Mn All-Weld-Metal Deposits, by G. M. Evans: *Welding Journal*, Vol 61, No. 4, 1982, p 125s-132s
F. R. Coe: *Welding in the World*, Vol 14 (1-2), 1976, p 1-7
J. C. Barland: *British Welding Journal*, Vol 7, 1960, p 508

Process Selection

SELECTION OF A JOINING PROCESS requires a basic knowledge of the various processes and of their relationships to such variables as joint design, base-metal properties, equipment cost and welder skill.

For increased economy, automatic welding modes should be considered in process selection. Most basic welding processes generally can be automated through the use of robotics, mechanical indexing, and positioning systems. The success of any automated mode depends on the adaptability of the item to be welded and the time devoted to the proper design of equipment.

Welding processes that employ an electric arc are the most widely used in industry. The arc may be established between an electrode and the base material, as in shielded metal-arc welding (SMAW) and gas tungsten-arc welding (GTAW), or the arc may occur within the welding heat source, as in plasma-arc welding (PAW). Furthermore, the arc and molten metal may be protected by an inert gas, granulated flux, or gaseous slag products of a consumable electrode.

Other welding processes include (1) oxyfuel gas welding (OFW), in which a combustible gas is burned with additions of oxygen to produce a high-temperature flame; (2) resistance welding in which high current density is introduced to create a high metal temperature and pressure is applied to produce a weld; (3) flash welding, in which an arc is created and followed by instantaneous force to bring the parts being welded together; (4) diffusion welding (DFW), in which clean metallic parts are brought together with high force to create bonding through diffusion; (5) friction welding (FRW), in which two parts to be welded are brought together with force and movement at high speed to create high temperature and bonding; (6) electron beam welding (EBW), in which a focused stream of electrons produces melting and joining; (7) laser beam welding (LBW), in which a coherent light beam is focused on the workpiece to create melting for welding or cutting; (8) ultrasonic welding (USW), in which a concentrated beam of sound waves is used; and (9) explosion welding (EXW), in which a high-energy explosive is used to create very high forces between two workpieces, thus bonding them together.

The OFW process is the oldest in use, but it is usually limited to manual techniques for braze welding or as a heat source for silver brazing and soldering. Fusion welding is possible with OFW, but the high heat input may create metallurgical problems.

The SMAW Process (stick) is widely used because of its versatility, portability and low cost, which make it useful for field fabrication and installation. Although individual weld joint costs may be high compared with those for automated processes, initial equipment cost and portability often are deciding factors in selection.

The shielded GTAW process (heliarc) is a strong competitor with SMAW because of its adaptability to certain materials, such as titanium, zirconium, stainless steel and aluminum, as well as its capability to produce high-quality welds. Shielded GTAW is well suited to automated welding modes.

Submerged-arc welding (SAW) is a high-production process that can be used for shop, field and semiautomated applications. Submerged-arc welding has certain limitations for weld-position requirements.

Plasma-arc welding is a high-energy-heat-source application that is particularly adaptable to automated welding techniques. It has been used advantageously for hard facing with special metal alloys for wear and abrasion applications.

The LBW, EBW, DFW, EXW, FRW, USW and flash welding processes are somewhat specialized and limited in application. Resistance welding is a low-cost, high-production process for use in industrial applications. It is an excellent substitute for riveted construction of thin metal members.

For a detailed list of joining processes and recommended joining thicknesses for several alloys, see Table 1.

INDUSTRIAL USAGE

The industrial usage of a welding process depends to a great extent on the following considerations:

- The material and its weldability
- Production requirements
- Design specifications and intended service
- Size and complexity of weldment
- Fabrication site — shop or field
- Cost of welding equipment
- Welder skill and training required.

Welding processes enjoying the greatest industrial usage are the manual welding processes — SMAW, OFW, GTAW, GMAW and FCAW. The mechanized processes used most frequently in industry are GTAW, GMAW, FCAW and SAW.

Industrial usage of a welding process depends, for the most part, on the material to be welded. Carbon steel, the most widely used material, can be welded by most manual or automatic welding processes. Aluminum, on the other hand, frequently is welded with an inert-gas process, such as GTAW or GMAW.

At the other end of the spectrum are the more sophisticated processes such as EBW and LBW. These processes generally are used for more specialized applications.

APPLICATIONS

The choice of welding process used on any construction assignment depends, to a great extent, on the type of job involved. Applicable codes and standards and the location where the welding is to be carried out have a direct bearing on the choice of process. Although welding can be applied to a broad spectrum of manufacturing and construction activities, there are certain applications that encompass the major volume of welding in industry. These are discussed below, with specific attention to the welding processes involved.

For structural welding, there are two categories of welding activity — buildings and bridges — that normally are recognized. Although there are different design considerations and requirements for these two categories of structures, they both require the use of AWS D1.1, "Structural Welding Code." This code covers qualification of procedures and welders for building and structural bridge fabrication and erection. One of the options suggested in AWS D1.1 is the use of prequalified procedures involving four processes (SMAW, SAW, GMAW and FCAW). Three other processes (ESW, EGW and stud welding) are permitted but must be qualified for the specific application. It is the responsibility of the designer to specify the type of joint required; the fabricator then must select the welding process to meet joint-design requirements. Because the selection of base metals is limited to those permitted in AWS D1.1, certain variables are decided upon before a fabricator is contacted.

Piping, pressure-vessel, boiler and storage-tank construction involves a large portion of current welding applications encompassing the petroleum, petrochemical, chemical, power-generation (utility) and transmission pipeline industries. Codes and standards pertaining to these applications include the following:

- ASME "Boiler and Pressure Vessel Code"
- ANSI B31.1, "Pressure Piping Code"
- API standard 1104
- API standard 620
- API standard 650
- American Water Works Association (AWWA) standard D100 (AWS D5.2).

Power-piping, pressure-vessel, and boiler welding construction normally is governed by Section IX of the ASME "Boiler and Pressure Vessel Code" for procedure and operator qualification. The manufacturer or fabricator is responsible for qualifiying the process used with the procedures to be followed in construction. The only constraints are processes that may not be permitted by governing codes or standards of job specifications. Processes that are impractical for use at the location of the construction are also eliminated. The most widely used processes in field erection/construction are SMAW and GTAW, with limited use of SAW, GMAW and FCAW. For shop fabrication, the only limiting factors to process use are design, equipment cost, and governing codes and standards.

Transmission-piping construction normally is controlled through API standard 1104. Welding processes generally are limited by field conditions. Although SMAW has dominated transmission-piping applications for many years, automation has led to the widespread use of the GMAW and FCAW processes.

Storage-tank construction generally is governed by API standard 620, API standard 650 or AWWA standard D100 (AWS D5.2), depending on the type of storage tank. A considerable amount of storage tank construction is performed by the SMAW process, but automated SAW also is used to a great extent. For storage tanks constructed from aluminum-base metal, GTAW and GMAW are used extensively.

Shipbuilding construction is governed by American Bureau of Shipping (ABS) requirements. The major welding process utilized in shipbuilding is SMAW, but there is increased development in the use of SAW, GMAW and FCAW. Some applications are well suited to the use of ESW for joining thick plate.

Aircraft and aerospace welded construction is governed by military specifications, and welding processes are governed by material, production and quality considerations. Gas tungsten-arc welding continues to be the dominant process used, but SMAW, GMAW, EBW, PAW, resistance welding, stud welding and brazing also are used for aircraft and aerospace construction because of the flexibility of these processes.

Automotive and railroad industries use almost every welding process available because of the many types of materials and applications encountered. Resistance welding is the dominant process employed on automotive assembly lines. Other processes used in the automotive industry include FRW, GMAW, EBW, brazing, soldering and

Table 1. Recommended joining processes for various metal groups

Joining process	Recommended thickness (in.) for:			
	Carbon steels	Low-alloy steels	Stainless steels	Cast irons
Arc welding				
AHW	Up to 1/4	Up to 1/4	Up to 1/8	1/8 and up
BMAW	Up to 1/4	...	...	...
CAW	Up to 1/4	Up to 1/4	...	1/8 to 3/4
CAW-G	Up to 1/4	Up to 1/4	...	1/8 to 3/4
CAW-S	Up to 1/4	Up to 1/4	...	1/8 to 3/4
CAW-T	Up to 1/4	Up to 1/4	...	1/8 to 3/4
EGW	1/4 and up	1/4 and up	1/4 and up(a)	...
FCAW	1/8 and up	1/8 and up	1/8 and up	1/8 to 3/4
GMAW	1/8 and up	1/8 and up	1/8 and up	1/8 to 3/4
GMAW-P	All thicknesses	All thicknesses	All thicknesses	1/8 and up
GMAW-S	Up to 1/4	Up to 1/4	Up to 1/4	...
GTAW	Up to 1/4	Up to 1/4	Up to 1/4	...
GTAW-P	Up to 1/4	Up to 1/4	Up to 1/4	...
PAW	...	...	Up to 3/4	...
SAW	All thicknesses	All thicknesses	All thicknesses	1/4 and up
SAW-S	1/4 and up	1/4 and up	1/4 and up	...
SMAW	All thicknesses	All thicknesses	All thicknesses	All thicknesses
SW	All thicknesses	All thicknesses	All thicknesses	...
Resistance welding				
FW	All thicknesses	All thicknesses	All thicknesses	...
HFRW	Up to 1/4	Up to 1/4	Up to 1/4	...
PEW	Up to 1/4	Up to 1/4	Up to 1/4	...
RPW	Up to 1/4	Up to 1/4	Up to 1/4	...
RSEW	Up to 1/4	Up to 1/4	Up to 1/4	...
RSW	Up to 1/4	Up to 1/4	Up to 1/4	...
UW	Up to 1/4	Up to 1/4	Up to 1/4	...
Solid-state welding				
CW	1/4 and up	...	...	...
DFW	...	All thicknesses	All thicknesses	...
EXW	Up to 3/4	Up to 3/4	Up to 3/4	...
FOW	All thicknesses	...	...	...
FRW	1/8 and up	1/8 and up	1/8 and up	...
HPW	1/8 and up	1/8 and up	1/8 and up	...
USW	Up to 1/8	Up to 1/8	Up to 1/8	...
Oxyfuel gas welding				
AAW	Up to 1/8	Up to 1/8	Up to 1/8	...
OAW	Up to 3/4	Up to 1/8	Up to 1/8	All thicknesses
OHW	Up to 1/4	Up to 1/8	Up to 1/8	Up to 1/4
Other welding processes				
EBW	All thicknesses	All thicknesses	All thicknesses	...
ESW	3/4 and up	3/4 and up	3/4 and up	...
IW	Up to 1/8	...	...	...
LBW	Up to 3/4	Up to 3/4	Up to 3/4	...
Brazing				
AB	Up to 1/4	Up to 1/4	Up to 1/4	Up to 3/4
DFB	All thicknesses	All thicknesses	All thicknesses	All thicknesses
DB	Up to 1/4	Up to 1/8	Up to 1/8	...
FB	All thicknesses	All thicknesses	All thicknesses	All thicknesses
IB	Up to 3/4	Up to 3/4	Up to 3/4	Up to 1/4
IRB	Up to 1/4	Up to 1/8	Up to 1/8	...
LB	Up to 1/8	Up to 1/8	Up to 1/8	...
RB	Up to 1/4	Up to 1/8	Up to 1/8	...
TB	Up to 3/4	Up to 3/4	Up to 3/4	Up to 1/4
TCAB	Up to 1/8	Up to 1/8	Up to 1/8	Up to 1/8
Soldering				
DS	Up to 1/8	Up to 1/8	Up to 1/8	...
FS	Up to 1/8	Up to 1/8	Up to 1/8	...
IS	Up to 1/8	Up to 1/8	Up to 1/8	...
IRS	Up to 1/8	Up to 1/8	Up to 1/8	...
INS	Up to 1/8	Up to 1/8	Up to 1/8	...
RS	Up to 1/8	Up to 1/8	Up to 1/8	...
TS	Up to 1/8	Up to 1/8	Up to 1/8	...
WS	Up to 1/8	Up to 1/8	Up to 1/8	...

(continued)

(a) Applicable to EGW using solid electrode wire

Table 1 (continued)

Joining process	Recommended thickness (in.) for:					
	Nickel and nickel alloys	Aluminum and aluminum alloys	Titanium and titanium alloys	Copper and copper alloys	Magnesium and magnesium alloys	Refractory metals and alloys
Arc welding						
AHW	Up to 1/8	Up to 1/8	...	Up to 1/8	Up to 1/8	...
BMAW	...	...	...	...	...	...
CAW	...	...	...	...	...	...
CAW-G	...	...	...	...	...	...
CAW-S	...	...	...	...	...	...
CAW-T	...	...	...	...	...	...
EGW	1/4 and up (a)	...	...	...	...	...
FCAW	Up to 3/4	...	...	...	...	...
GMAW	All thicknesses	Up to 3/4	Up to 3/4	Up to 3/4	Up to 3/4	1/8 to 1/4
GMAW-P	All thicknesses	Up to 1/4	All thicknesses	Up to 1/4	All thicknesses	1/8 to 3/4
GMAW-S	Up to 1/4	...	...	...	...	...
GTAW	Up to 1/4	Up to 3/4	Up to 3/4	Up to 1/8	Up to 1/4	Up to 1/8
GTAW-P	Up to 1/4	Up to 3/4	Up to 3/4	Up to 1/8	Up to 1/4	Up to 1/8
PAW	Up to 3/4	Up to 1/8	Up to 3/4	Up to 1/4	...	Up to 1/4
SAW	1/4 and up	...	...	...	...	...
SAW-S	3/4 and up	...	...	...	...	...
SMAW	All thicknesses	...	...	...	...	...
SW	...	All thicknesses	...	...	All thicknesses	...
Resistance welding						
FW	All thicknesses	All thicknesses	All thicknesses	All thicknesses	1/8 and up	1/8 and up
HFRW	Up to 1/8	Up to 1/4	Up to 1/4	Up to 1/4	Up to 1/4	...
PEW	Up to 1/4	Up to 1/4	Up to 1/4	Up to 1/4	Up to 1/4	...
RPW	Up to 1/4	Up to 1/4	Up to 1/4	Up to 1/4	Up to 1/4	...
RSEW	Up to 1/4	Up to 1/4	Up to 1/4	Up to 1/4	Up to 1/4	...
RSW	Up to 1/4	Up to 1/4	Up to 1/4	Up to 1/4	Up to 1/4	...
UW	Up to 1/4	Up to 1/4	Up to 1/4	Up to 1/4	Up to 1/4	...
Solid-state welding						
CW	...	Up to 3/4	...	Up to 1/4	Up to 1/4	...
DFW	...	Up to 1/4	All thicknesses	...	...	...
EXW	Up to 3/4	All thicknesses	All thicknesses	All thicknesses	All thicknesses	Up to 3/4
FOW	...	...	...	...	...	1/8 and up
FRW	1/8 and up	1/8 and up	1/8 and up	1/8 and up	1/8 and up	...
HPW	1/8 and up	...	...	1/8 and up	...	...
USW	Up to 1/8	Up to 1/4	Up to 1/8	Up to 1/8	Up to 1/8	Up to 1/8
Oxyfuel gas welding						
AAW	...	...	...	...	...	...
OAW	Up to 1/8	Up to 1/8	...	...	...	...
OHW	Up to 1/8	Up to 1/8	...	...	...	...
Other welding processes						
EBW	All thicknesses	All thicknesses	All thicknesses	All thicknesses	All thicknesses	Up to 1/4
ESW	3/4 and up	...	...	...	...	...
IW	...	...	...	...	...	...
LBW	Up to 3/4	Up to 1/4	Up to 3/4	...	Up to 3/4	...
Brazing						
AB	Up to 1/4	Up to 3/4	Up to 1/8	Up to 3/4	Up to 3/4	Up to 1/8
DFB	All thicknesses	All thicknesses	All thicknesses	All thicknesses	Up to 3/4	Up to 1/4
DB	Up to 1/8	Up to 3/4	...	...	Up to 1/4	...
FB	All thicknesses	Up to 3/4	All thicknesses	All thicknesses	Up to 3/4	Up to 1/4
IB	Up to 1/4	Up to 1/8	Up to 1/8	Up to 1/8	...	Up to 1/8
IRB	Up to 1/8	Up to 1/8	Up to 1/8	...	...	Up to 1/8
LB	Up to 1/8	Up to 1/8	Up to 1/8	Up to 1/8	...	Up to 1/8
RB	Up to 1/8	...	...	Up to 1/4	...	...
TB	Up to 3/4	Up to 3/4	...	Up to 3/4	Up to 1/4	Up to 1/4
TCAB	Up to 1/8	...	...	Up to 1/8	...	...
Soldering						
DS	Up to 1/8	Up to 1/8	...	Up to 1/8	...	...
FS	Up to 1/8	Up to 1/8	...	Up to 1/8	...	...
IS	Up to 1/8	Up to 1/8	...	Up to 1/8	...	...
IRS	Up to 1/8	Up to 1/8	...	Up to 1/8	...	...
INS	Up to 1/8	Up to 1/8	...	Up to 1/8	...	...
RS	Up to 1/8	Up to 1/8	...	Up to 1/8	...	...
TS	Up to 1/8	Up to 1/8	...	Up to 1/8	...	...
WS	Up to 1/8	Up to 1/8	...	Up to 1/8	...	...

(a) Applicable to EGW using solid electrode wire

flash welding. The railroad industry utilizes flash welding, SMAW, GMAW, FCAW, SAW and thermit welding. Although there are no established codes or standards that specifically cover welding procedures and operator qualifications for these industries, guidelines set forth in ASME Section IX and AWS D1.1 should be followed extensively.

SAFETY

The utilization of welding in manufacturing, construction and maintenance activities involves many considerations. Of greatest importance, however, is the need to be attentive to safety. Factors involved in personnel safety include protection against eye and ear damage, burns, radiation, respiratory damage, and crushed or broken limbs. Loss of and damage to equipment and buildings are generally the result of fire or explosion.

The following publications and documents may help the reader gain an understanding of the precautions that should be taken to meet minimum levels of safety:

- American National Standards Institute (ANSI) Standard Z49.1, "Safety in Welding and Cutting"
- ANSI/NFPA No. 70, "National Electric Code"
- Occupational Safety and Health Administration (OSHA), various publications and documents
- AWS 6.3, "Recommended Safe Practices for Plasma Arc Cutting"
- AWS F2.1, "Recommended Safe Practices for Electron Beam Welding and Cutting"
- ANSI Z136.1, "Safe Use of Lasers"
- Compressed Gas Association (CGA) P-1, "Safe Handling of Compressed Gases"
- ANSI Z87.1, "Practice for Occupational and Educational Eye and Face Protection"
- ANSI Z88.2, "Practices for Respiratory Protection"
- ANSI Z89.1, "Safety Requirements for Industrial Head Protection."

Each welding process and its application determines the requirements for protection of personnel. General requirements for protection of personnel involved in welding are covered by ANSI Z49.1.

Joint Design and Preparation

A WELD JOINT serves to transfer the stresses between the joined members and throughout the welded assembly. Forces and loads, which are introduced at different points, are transmitted to different areas throughout the weldment. The type of loading and service of the weldment influence the selection of joint design. The names of joint types—butt, T, corner, lap and edge—describe how the members meet.

The weld selected may or may not require preparation in the form of a groove to permit proper access to the root of the joint. Several methods—machining, chipping, shearing, grinding, gas cutting, gas gouging, and air carbon arc gouging—may be used to create single- or double-bevel, V-, J- or U-grooves.

No preparation is needed for a square groove weld. The members can be abutted in the same

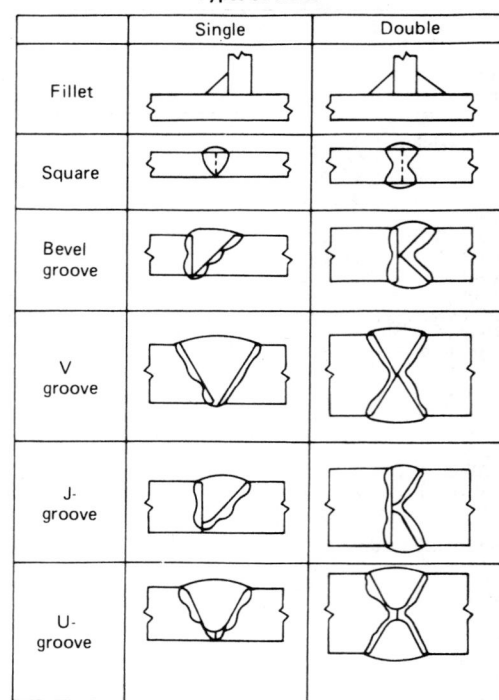

Fig. 10. Types of joints and welds

plane. If there is zero root opening, the weld arc essentially fuses the members together. Usually, very little or no weld metal bridges the weld members, although some weld metal or reinforcement may be left on the top and bottom surfaces. Similarly, fillet welds, which join pieces perpendicular to each other, can be used on T-joints without preparation. The welds and the fusion into both members join the two pieces.

Groove preparations are often necessary, however, for successful welding of various corner-, T- and butt-joint applications. Frequently required with greater metal thicknesses, these weld preparations ensure that the welding heat and weld metal reach and fuse the root of the joint.

NOMENCLATURE

Terminology describing various joints and welds is often confused. Types of welds and joints recognized by the American Welding Society (AWS) are illustrated in Fig. 10. Note that joints are the junctions where the members join and welds are the mechanisms (groove or preparation) which complete the joints.

Although precise nomenclature to differentiate between welds and joints is desirable, engineers will probably continue to use the terms interchangeably. Thus, in this article, as well as in other practical applications, the terms "joint" and "weld" are used interchangeably to describe how a joint is constructed.

WELD-JOINT DESIGN

The first consideration in the design of a weld joint is its ability to transfer load; the second is cost. The ideal weld joint is one that can handle the loads imposed, usually with a substantial safety margin, and still be produced at minimal cost.

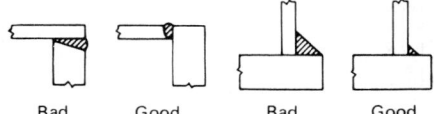

Fig. 11. Determination of weld size

Therefore, once the type of joint has been selected primarily on the basis of load requirements, the choice of weld to complete the joint should be determined by the effects of the structural design and layout on weld-metal, accessibility and preparation requirements—variables that directly influence the cost of the weld joint.

Weld Metal. In general, joint and weld types specified should require the least amount of filler metal to avoid unnecessary expense. The size of the weld should always be designed with reference to the size of the thinner member. The joint cannot be made any stronger by matching the weld size to the thicker member, which would require a greater amount of weld metal (Fig. 11).

In the design stage, joints that create extremely deep grooves should be avoided, because these require more filler metal. Deep-penetration welding processes, such as submerged-arc welding, especially automatic welding, reduce the volume of weld metal needed. Because greater thicknesses of plate can be welded, use of these methods may make the additional cost of groove preparations unnecessary.

If groove welds are required, minimum root openings and included angles should be used to reduce the amount of filler metal required. The use of double-groove instead of single-groove welds on thick plate where welding from both sides is feasible decreases the total filler metal.

To prevent waste of weld metal, welding spec-

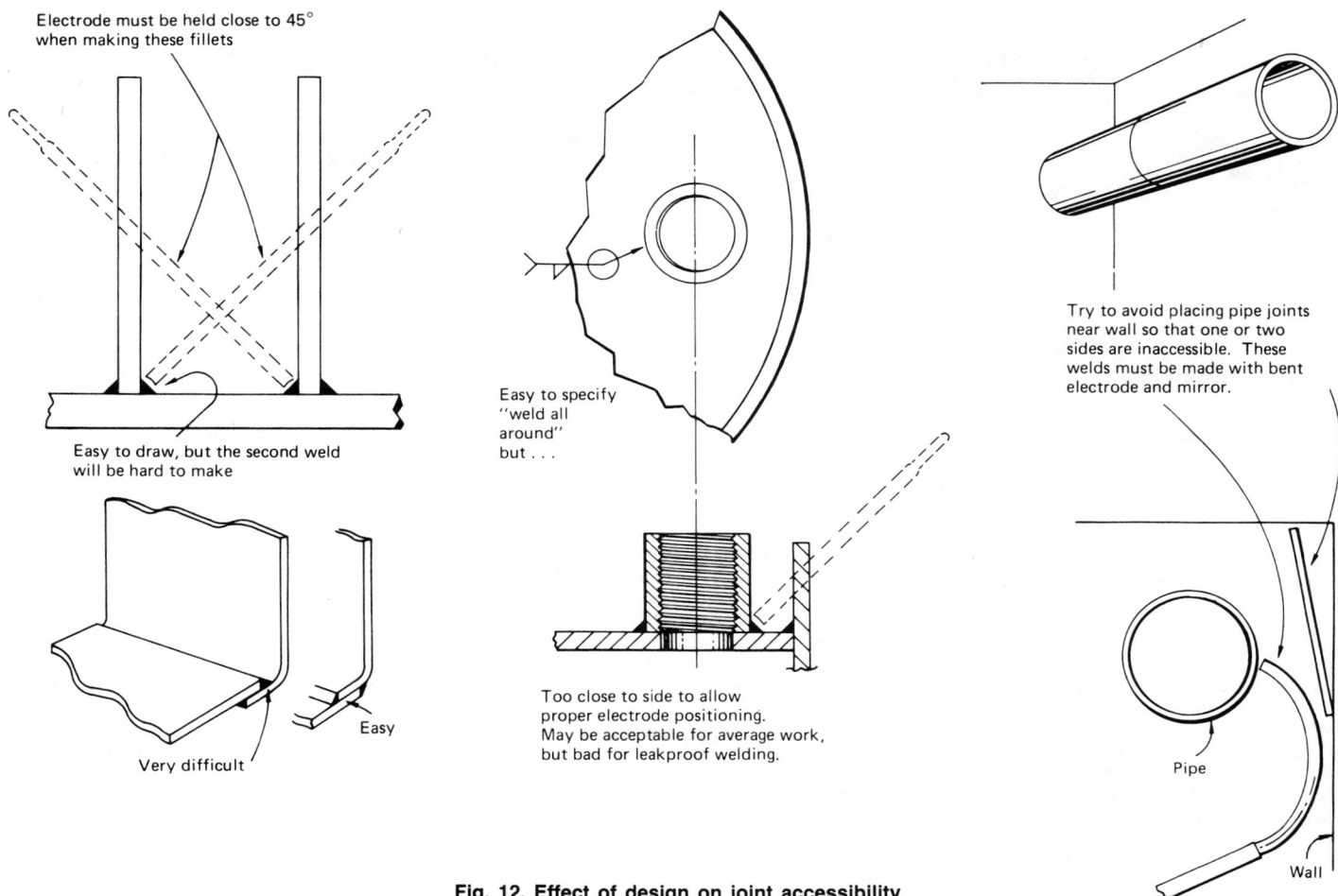

Electrode must be held close to 45°
when making these fillets

Easy to draw, but the second weld
will be hard to make

Very difficult

Easy

Easy to specify
"weld all
around"
but . . .

Too close to side to allow
proper electrode positioning.
May be acceptable for average work,
but bad for leakproof welding.

Try to avoid placing pipe joints
near wall so that one or two
sides are inaccessible. These
welds must be made with bent
electrode and mirror.

Pipe

Wall

Fig. 12. Effect of design on joint accessibility

ifications should avoid unnecessary use of the all-around welding symbol (Fig. 12). Welding of everything that touches has no engineering advantages, yet increases the amount of weld metal used.

When joining high-strength materials, cost can be minimized by specifying the use of high-strength weld metal only where required for primary load-carrying welds.

Accessibility. An important factor in the design of a weld joint is the accessibility of the members to be welded. The welder needs space to manip-ulate the electrode when making the weld. Frequently, what appears straightforward on the drawing board may be impractical in the shop or very costly to produce. Figure 12 illustrates examples of joint placement that are often difficult to weld.

Arc Welding

Shielded Metal-Arc Welding

SHIELDED METAL-ARC WELDING (SMAW) is a manual arc welding process in which the heat for welding is generated by an arc established between a flux-covered consumable electrode and a workpiece. The electrode tip, molten weld pool, arc, and adjacent areas of the workpiece are protected from atmospheric contamination by a gaseous shield obtained from the combustion and decomposition of the electrode covering. Additional shielding is provided for the molten metal in the weld pool by a covering of molten flux or slag. Filler metal is supplied by the core of the consumable electrode and from metal powder mixed with the electrode coverings of certain electrodes. Shielded metal-arc welding is often referred to as arc welding with stick electrodes, manual metal-arc welding, and stick welding.

PROCESS CAPABILITIES

Shielded metal-arc welding is the most widely used welding process for joining metal parts because of its versatility and its less complex, more portable and less costly equipment.

Versatility. Shielded metal-arc welding can be done indoors or outdoors. Joints in any position that can be reached with an electrode (i.e., overhead joints and vertical joints) can be welded. By the use of bent electrodes, joints in blind areas can be welded, including the back sides of pipes in restricted areas, which are inaccessible for most welding processes.

The power-supply leads can be extended for relatively long distances, and no hoses are required for shielding gas or water cooling.

PRINCIPLES OF OPERATION

An adequate power supply is required for SMAW. Suitable cables are used to attach one terminal of the power supply to the electrode holder and the other terminal to a ground clamp (Fig. 1).

To start welding, an arc is struck by briefly touching the workpiece with the tip of the electrode. The welder guides the electrode by hand

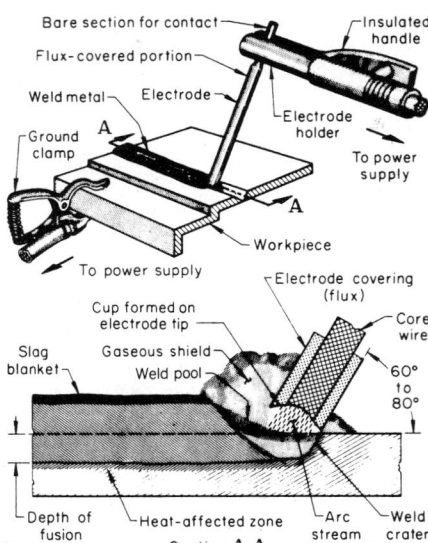

Fig. 1. Setup and fundamentals of operation for shielded metal-arc welding

in welding a joint, and controls its position, direction, travel speed and arc length (the distance between the end of the electrode and the work surface). In many applications in which electrodes with heavy coverings are used, the welder actually drags the electrode in the joint or on the work and uses the electrode angle to control arc length. Electrodes must be discarded at a length of about 2 in.

A power supply with a drooping volt-ampere characteristic is required. With such a power

Table 1. Output ratings of power supplies used in SMAW

	Output current, A, at load, V	
Rated (60% duty cycle)(a)		**Maximum (35% duty cycle)(b)**
Constant-current motor-generators (dc)		
300 at 32	....	375 at 35
400 at 36	....	500 at 40
500 at 40	....	625 at 44
600 at 44	....	750 at 44
Transformer-rectifiers (dc)		
400 at 36	....	500 at 40
500 at 40	....	625 at 44
600 at 44	....	750 at 44
800 at 44	....	1000 at 44
Transformers (ac)		
400 at 36	....	500 at 40
500 at 40	....	625 at 44
600 at 44	....	750 at 44
(c)		(b)
750 at 44	....	925 at 44
1000 at 44	....	1250 at 44
1500 at 44	....	1875 at 44

(a) Rated current can be delivered continuously for 6 min out of every 10 min. (b) Maximum current can be delivered continuously for 3½ min out of every 10 min. (c) One-hour duty rating. Rated current can be delivered continuously for 1 h, then for 45 min of every hour for the next 3 h.

supply, the current decreases as the arc becomes longer and increases as the arc becomes shorter.

POWER SUPPLIES

Many types and sizes of power supply are used for SMAW. When alternating current (ac) is used, high-voltage input power is transformed or stepped down through a transformer to a voltage low enough for safe use. When direct current (dc) is used, a motor-generator or a transformer-rectifier unit is used. A transformer-rectifier consists of a step-down voltage transformer with means to rectify alternating current to direct current.

Both alternating current and direct current produce acceptable results in welding low-carbon steel. Although each has distinct advantages, the choice usually depends on cost, availability of equipment, and the electrode used.

Combination ac/dc power supplies are widely used and are versatile for general-purpose applications. They consist of a transformer and a rectifier in combination and are capable of supplying either alternating current or direct current. When direct current is used, either straight or reverse polarity is available. Output ratings for various power supplies are given in Table 1.

Direct current electrode negative can be used for SMAW of all steels. Melting and deposition rates are higher than with DCEP, while penetration is shallower and narrower (Fig. 2).

ELECTRODES

Electrodes used in SMAW have many different compositions of core wire and a wide variety of types and weights of flux covering. Standard electrode diameters (diameters of the core wire) range from $\frac{1}{16}$ to $\frac{5}{16}$ in. Length is usually 9 to 18 in., although electrodes up to 36 in. long have been made for special applications. A bare (uncoated) end of the electrode, standardized at length of $\frac{3}{4}$ to $1\frac{1}{2}$ in., is provided for making electrical contact with the electrode holder.

Classification of low-carbon steel covered electrodes according to the system devised by the American Welding Society (AWS) is generally used throughout industry. In this system, designations consist of the letter E (for electrode) and four digits (five digits for weld-metal strength of 100 ksi or more). For electrodes producing under 100 ksi tensile strength, the first two digits indicate minimum tensile strength, in ksi, of deposited weld metal in the as-welded condition. The third digit indicates welding positions for which the electrode can be used successfully. For example, Exx1x indicates all positions; Exx2x, flat welds and horizontal fillet welds only; and Exx4x, specific design for vertical-down welds. The fourth digit indicates the type of covering and suitable current characteristics (Table 2). For example, E6011 is an electrode that deposits weld metal with a minimum tensile strength of 60 ksi (first two digits); can be used for welds in all positions (third digit); has a high-cellulose, potassium covering and can be used either with alternating current or with DCEP (fourth digit). A suffix added to the classification number indicates chemical composition of the deposited metal as specified in AWS specification A5.5

Electrode Coverings. The composition of the electrode covering largely determines the performance of the electrode and the soundness of the weld. Table 3 lists materials used in making electrode coverings. More than 18 can be used;

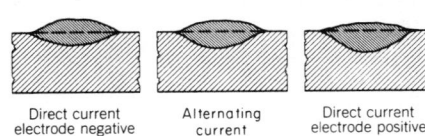

Fig. 2. Relative depths of penetration for different current characteristics

12 or more may be included in a specific covering. Each material is often used for more than one purpose. Table 3 gives compositions of electrode coverings for classes of electrodes used in welding low-carbon steel and lists the primary and secondary functions of the constituents of the coverings. The thickness of the covering varies from 10 to 55% of the total diameter of the covered electrode, depending on the type of covering.

EFFECT OF MOISTURE IN ELECTRODE COVERINGS

Moisture is not as harmful in the coverings of some low-carbon steel electrodes as has been assumed. Precautions should be taken to store electrodes in dry places. However, redrying, which is often done after prolonged storage, can impair both quality and operation of electrodes with cellulose coverings, especially E6010 and E6011 electrodes.

Table 4 gives recommended moisture contents of coverings and storage and redrying conditions for different classes of electrodes. Some brands of E6010 and E6011 electrodes operate satisfactorily and produce satisfactory weld deposits when the moisture content of the covering is above the recommended range. All other electrodes usually operate best when the moisture content is lowest.

SELECTION OF ELECTRODE CLASS

Mechanical Properties. Carbon steel electrodes classified in the type E60xx series may be used

Table 2. Coverings and currents indicated by fourth digit in AWS classifications of low-carbon steel covered arc welding electrodes

Fourth digit	Covering	Current(a)
0	High cellulose, sodium(b)	DCEP(b)
	High iron oxide(c)	ac or dc(c)(d)
1	High cellulose, potassium	ac or DCEP
2	High titania, sodium(e)	ac or dc(f)
3	High titania, potassium	ac or dc(f)
4	Iron powder, titania	ac or dc(f)
5	Low hydrogen, sodium	DCEP
6	Low hydrogen, potassium	ac or DCEP
7	Iron powder, iron oxide	ac or dc(d)
8	Iron powder, low hydrogen	ac or DCEP

(a) ac, alternating current; dc, direct current; DCEP, direct current electrode positive. (b) When third digit is 1. (c) When third digit is 2. (d) Either polarity for flat welds; DCEN for horizontal fillet welds. (e) Can also imply high iron oxide, ac or dc, either polarity. (f) Either polarity

Table 3. Functions and composition ranges of constituents of coverings on low-carbon steel arc welding electrodes

Constituent of covering	Function of constituent — Primary	Secondary	E6010, E6011	E6012, E6013	E6020, E6022	E6027	E7014	E7016	E7018, E7048	E7024	E7028
Cellulose ....Shielding gas		...	25-40	2-12	1-5	0-5	2-6	...	...	1-5	...
Calcium carbonate ....Shielding gas		Fluxing agent	...	0-5	0-5	0-5	0-5	15-30	15-30	0-5	0-5
Fluorspar ....Slag former		Fluxing agent	...	...	...	...	...	15-30	15-30	...	5-10
Dolomite ....Shielding gas		Fluxing agent	...	...	...	...	...	...	...	...	5-10
Titanium dioxide (rutile) ....Slag former		Arc stabilizer	10-20	30-55	0-5	0-5	20-35	15-30	0-5	20-35	10-20
Potassium titanate ....Arc stabilizer		Slag former	(a)	(a)	...	...	...	...	0-5	...	0-5
Feldspar ....Slag former		Stabilizer	...	0-20	5-20	0-5	0-5	0-5	0-5	...	0-5
Mica ....Extrusion		Stabilizer	...	0-15	0-10	...	0-5	...	...	0-5	...
Clay ....Extrusion		Slag former	...	0-10	0-5	0-5	0-5	...	...	...	...
Silica ....Slag former		...	...	...	5-20	...	...	...	...	...	...
Manganese oxide ....Slag former		Alloying	...	...	0-20	0-15	...	...	...	...	...
Iron oxide ....Slag former		...	...	...	15-45	5-20	...	...	...	...	...
Iron powder ....Deposition rate		Contact welding	...	...	...	40-55	25-40	...	25-40	40-55	40-55
Ferrosilicon ....Deoxidizer		...	...	...	...	0-5	0-10	0-5	5-10	5-10	2-6
Ferromanganese ....Alloying		Deoxidizer	5-10	5-10	5-20	5-15	5-10	2-6	2-6	5-10	2-6
Sodium silicate ....Binder		Fluxing agent	20-30	20-30	5-15	5-10	0-10	0-5	0-5	0-10	0-5
Potassium silicate ....Arc stabilizer		Binder	(a)	5-15(a)	0-5	0-5	5-10	5-10	5-10	0-10	0-5

(a) Used (in place of constituent on line above) in E6011 and E6013 electrodes to permit welding with alternating current.

Table 4. Recommended moisture content of coverings, and storage and redrying conditions, for low-carbon steel covered arc welding electrodes

Electrode class	Recommended moisture content of covering, %	Relative humidity, %(a)	Temperature of holding oven, °F	Redrying temperature, °F(b)
E6010	3.0-5.0	20-60	(c)	(c)
E6011	2.0-4.0	20-60	(c)	(c)
E6012, E6013, E6020, E6022	Less than 1	60 max	100-120	275 ± 25
E6027, E7014, E7024	Less than 0.5	60 max	100-120	275 ± 25
E7015, E7016	Less than 0.4	50 max	130-330	650 ± 50
E7018, E7028, E7048	Less than 0.4	50 max	130-330	750 ± 50

(a) For storage at normal temperature of 80 °F ± 20 °F. (b) 1 h at temperature. (c) Follow manufacturer's recommendation.

for welding lower-carbon-content steels. Where higher strengths are required, electrodes possessing higher weld-metal deposit strength, i.e., type E70xx series, are used.

Material Composition. Carbon steels which contain less than 0.30% C (low-carbon steel) are readily weldable with any class of low-carbon steel electrodes. Steels which have carbon contents from 0.30 to 0.60% (medium-carbon steel) have restrictions in the applied welding procedure and electrode selection. In this range, low-hydrogen electrodes are necessary.

Quality. Covered electrodes for SMAW are rated for radiographic soundness characteristics in three categories: grade 1, grade 2, and not required (AWS A5.1).

Gas Metal-Arc Welding (MIG Welding)

GAS METAL-ARC WELDING (GMAW), which often is called MIG (metal inert gas) welding, is an arc welding process in which the heat for welding is generated by an arc between a consumable electrode and the work metal. The electrode, a bare solid wire that is continuously fed to the weld area, becomes the filler metal as it is consumed. The electrode, weld pool, arc, and adjacent areas of the base metal are protected from atmospheric contamination by a gaseous shield provided by a stream of gas, or mixture of gases, fed through the welding gun. The gas shield must provide full protection, because even a small amount of entrained air can contaminate the weld deposit.

Gas metal-arc welding overcomes the restriction of using an electrode of limited length, as in shielded metal-arc welding (SMAW), and overcomes the inability to weld in various positions, which is a limitation of submerged-arc welding (SAW).

Gas metal-arc welding is widely used in semiautomatic, machine and automatic modes. In semiautomatic welding, the most popular method of applying this process, the welder guides the gun along the joint and adjusts the welding conditions. The wire feeder continuously feeds the filler-wire electrode, and the arc length is maintained by the power source. In automatic GMAW, the machinery controls the welding parameters, arc length, joint guidance and wire feed, observed by the operator. Machine GMAW has only limited popularity and is characterized by machine control of arc length, wire feed and joint guidance. The operator adjusts the welding parameters.

Metals Welded. The GMAW process was first applied to welding of magnesium and aluminum alloys and stainless steels, because it was often the only method by which satisfactory welds could be produced at an economical rate. The nature of the process permits its use for welding most metals and alloys.

All ferrous and most nonferrous materials can be joined using GMAW. A wide range of material thicknesses can be joined in all positions by selecting the right mode of metal transfer from among those possible with GMAW.

Polarity. Most GMAW applications require the use of direct current electrode positive (DCEP), which is also referred to as reverse polarity. This type of electrical connection provides a stable arc, smooth metal transfer, relatively low spatter loss, and good weld-bead characteristics for the entire range of welding currents used.

Direct current electrode negative (DCEN), which is also referred to as straight polarity, is seldom used, because the arc can become very unstable and erratic even though the electrode melting rate is higher than that achieved with DCEP. Penetration is lower with DCEN than with DCEP.

Alternating current is not normally used with GMAW for two reasons: (1) the arc is extinguished during each half cycle as the current reduces to zero, and it may not reignite if the cathode cools sufficiently; and (2) rectification of the reverse-polarity cycle promotes erratic arc operation.

PRINCIPLES OF OPERATION

Essential requirements for GMAW are: (1) a power supply that provides sufficient voltage to push the current across the gap to make the arc, and sufficient current to melt the electrode to make the weld deposit; (2) a wire feeder that continuously advances the electrode as it melts; (3) a smooth flow of shielding gas; and (4) a welding gun that carries the current, electrode wire, shielding gas and, depending on the gun design, cooling water.

These essentials are illustrated schematically in Fig. 3. The wire-feed system is of the constant-speed, push type, by means of which a specific rate of wire feed can be obtained. The power supply is a constant-voltage type that will maintain any desired voltage output within its capability, regardless of current flow. Under these conditions, only enough current will flow to melt the electrode wire at a rate equal to that of the wire feed.

METAL TRANSFER

The type of arc obtainable in GMAW is identified by the mode of metal transfer. These modes of transfer are commonly referred to as spray,

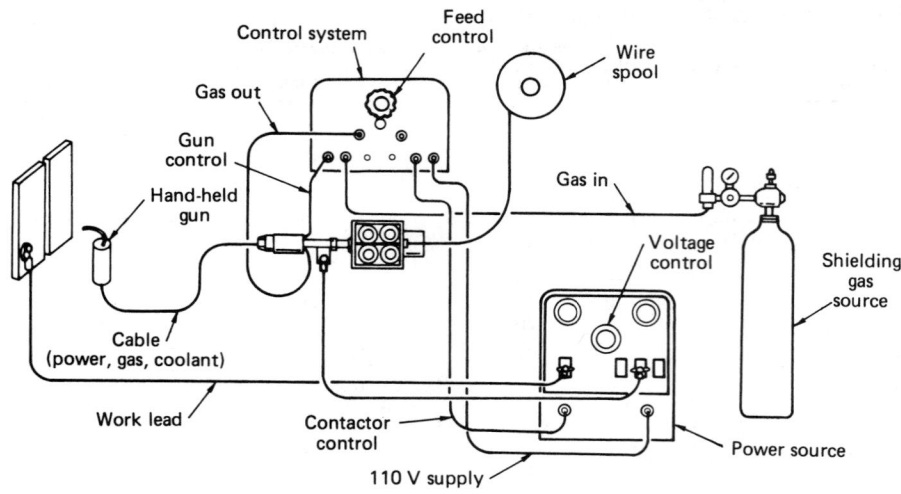

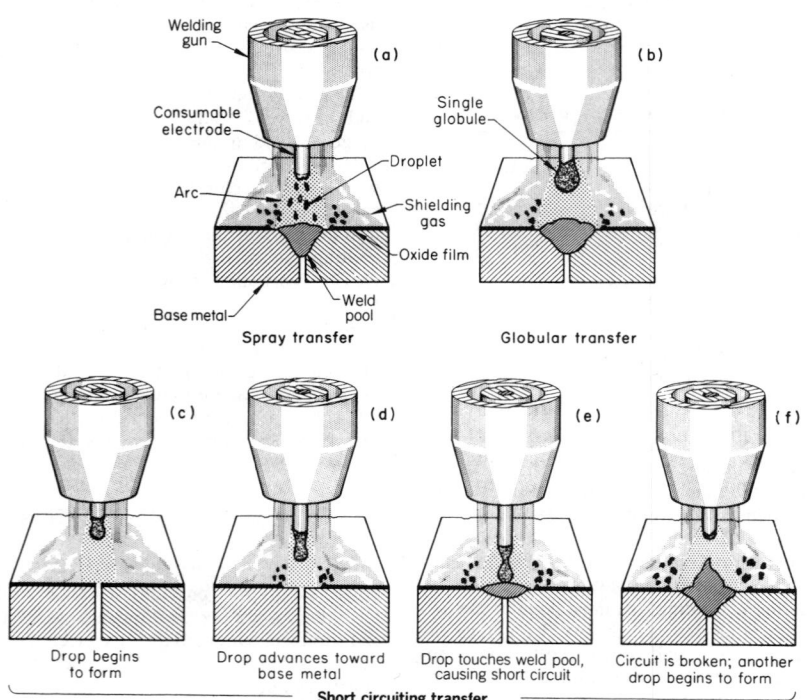

(a) Spray transfer. (b) Globular transfer. (c), (d), (e) and (f) Steps in short-circuiting transfer.
Fig. 4. Modes of metal transfer in GMAW

tions of GMAW. It is especially well adapted to welding of thin sections, because heat input is low; it is less often used for welding thick sections. This mode of transfer permits welding in any position and occurs with carbon dioxide, argon–carbon dioxide mixtures, and helium-base shielding gases.

Steps that occur with the short-circuiting mode of transfer are shown in Fig. 4. At the start of the short-circuiting arc cycle, the end of the electrode wire melts into a small globule of liquid metal (Fig. 4c).

Pulsed-current transfer is a spray-type transfer that occurs in pulses at regularly spaced intervals rather than at random intervals. In the time interval between pulses, the welding current is reduced and no metal transfer occurs.

The pulsing action is obtained by combining the outputs of two power supplies working at two current levels. One acts as a background current to preheat and precondition the advancing continuously fed electrode; the other power supply furnishes a peak current for forcing the drop from the electrode to the joint being welded. The peaking current is a half-wave direct current; because it is tied into line frequency, drops will be transferred from the electrode to the joint 60 to 120 times per second.

POWER SOURCES

Alternating current is seldom used in GMAW. Direct current electrode positive is used for most applications, although DCEN is sometimes used when penetration must be minimal. Figure 5 compares the depths of penetration obtained in welding with DCEN and DCEP under otherwise identical conditions.

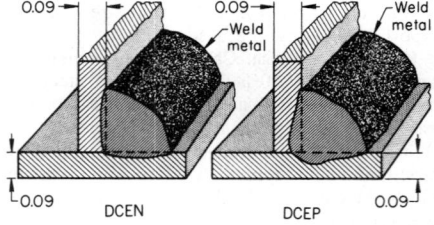

Low-carbon steel; low-carbon steel filler metal

Joint type	T
Weld type	Fillet
Welding position	Horizontal
Electrode wire	$1/32$-in.-diam low-carbon steel
Shielding gas	Carbon dioxide
Current	120 A (dc)
Voltage	19 V
Melting rate	280 in./min

Fig. 5. Depths of penetration obtained in GMAW with DCEP and DCEN under otherwise identical conditions

The GMAW process uses power sources similar to those used with other continuous electrode-feed welding processes, such as flux-cored arc welding (FCAW) and SAW. Many types of direct-current power sources may be used, including rotating (generator) or static (single- or three-phase transformer-rectifier) welding machines. Any of these types of machines are available to produce constant-current or constant-voltage output, or both.

globular, short-circuiting and pulsed-current transfer.

Spray Transfer. In this mode, metal is transferred from the end of the electrode wire to the pool in an axial stream of fine droplets. This condition is illustrated in Fig. 4(a). These small droplets emanate from the tapered end of the electrode; one droplet follows another, but they are not connected. The size of the droplets may vary, but in a spray arc the maximum diameter is less than that of the electrode wire.

The spray arc occurs at high current density, generally with argon or an argon-rich shielding gas. A true spray arc cannot be obtained with a shielding gas composed of more than 10 to 15%

carbon dioxide. The spray transfer mode gives high heat input, maximum penetration and a high deposition rate. In welding of steel, it generally is limited to welding in the flat position and the horizontal fillet position. This mode also produces the least amount of spatter.

Globular transfer occurs at lower current densities and is characterized by the formation of a relatively large drop of molten metal at the end of the tapered electrode wire (Fig. 4b). The drop forms at the end of the electrode wire until the force of gravity overcomes the surface tension of the molten drop, at which time the drop falls into the weld pool.

Short-circuiting transfer is used in many applica-

WELDING GUNS

The welding gun used in GMAW transmits welding current to the electrode. Because the wire is fed continuously, a sliding electrical contact is used. The welding current is passed to the electrode through a copper alloy contact tube. The contact tubes have various hole sizes, corresponding to the diameter of the electrode wire. The gun also has a gas-supply connection and a nozzle to direct the shielding gas around the arc and weld pool. To prevent overheating of the welding gun, cooling is required to remove the heat generated. Shielding gas or water circulating in the gun, or both, are used for cooling. Some guns are also air cooled.

Semiautomatic Guns. Handheld semiautomatic guns usually have a curved neck, which makes them applicable to all welding positions. A semiautomatic gun is shown in Fig. 6. The gun is attached to the service lines, which include the power cable, water hose, gas hose, and wire conduit or liner. The guns have metal nozzles that have orifice diameters from $^3/_8$ to $^7/_8$ in. to direct the shielding gas to the arc and weld pool, depending on the welding requirements.

ELECTRODES

The electrodes for GMAW are usually quite similar or identical in composition to those used for welding by most other bare electrode processes. In many cases, the electrode wires are chosen to match the chemical composition of the base metal as closely as possible. In some cases, electrodes with a somewhat different chemical composition are used to obtain maximum mechanical properties or better weldability.

Composition of electrode wire has a significant effect on results. Because of the importance of electrode composition, many large users of electrode wire have established their own specifications, which they have developed from their own experience. Table 5 lists AWS classifications and composition limits for different types of electrodes that are used for GMAW of low-carbon steels. For compositions of electrode wire used for welding other metals, see the articles in this volume that deal with the welding of specific metals.

SHIELDING GASES

The primary purpose of shielding gas is to protect the molten weld metal and the heat-affected zone (HAZ) from oxidation and other contamination. Reactive metals such as titanium require protection over a much greater area in the weld vicinity.

Originally, only the inert gases (argon and helium) were used for shielding, but carbon dioxide is now used extensively, and oxygen and carbon dioxide are often mixed with the inert gases. Chemical behavior and welding applications of the gases and mixtures of shielding gases commonly used in GMAW are presented in Table 6.

Argon. Welding-grade argon is 99.995% pure. It is a monatomic gas (one atom per molecule), it is inert, and it is insoluble in molten metal. Argon is 38% heavier than air, which is advantageous for welding in the flat position and the horizontal fillet position. As shown in Table 6, pure argon can be used as a shielding gas for virtually all metals, but it is not ordinarily used in welding of steels, for which argon-base mix-

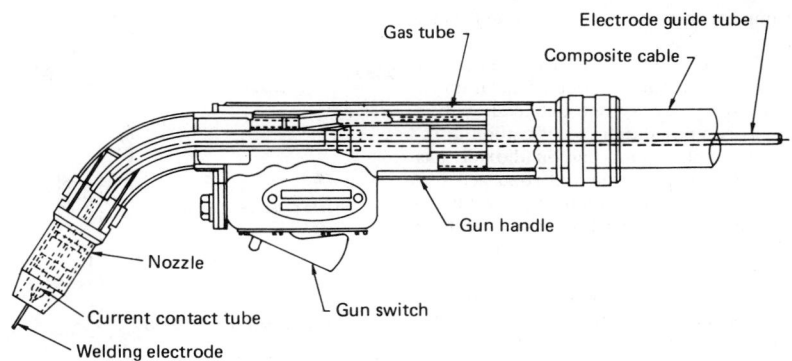

Fig. 6. Typical semiautomatic gas-cooled, curved-neck GMAW gun

Table 5. AWS classifications and composition limits for carbon steel electrode wires for GMAW (AWS A5.18)

AWS classification(a)	Composition, %						
	Carbon	Manganese	Silicon	Phosphor	Sulfur	Copper	Other
ER70S-2	0.07	0.90-1.40	0.40-0.70	0.025	0.035	0.50	Ti, Zr, Al
ER70S-3	0.06-0.15	0.90-1.40	0.45-0.70	0.025	0.035	0.50	···
ER70S-4	0.07-0.15	1.00-1.50	0.65-0.85	0.025	0.035	0.50	···
ER70S-5	0.07-0.19	0.90-1.40	0.30-0.60	0.025	0.035	0.50	Al
ER70S-6	0.07-0.15	1.40-1.85	0.80-1.15	0.025	0.035	0.50	···
ER70S-7	0.07-0.15	1.50-2.00	0.50-0.80	0.025	0.035	0.50	···

(a) ER70S-G, which is not shown in this table, has no chemical requirements.

Table 6. Uses of different shielding gases for GMAW

Type of gas	Typical mixtures, %	Primary uses
Argon	···	Nonferrous metals
Helium	···	Aluminum, magnesium, and copper alloys
Carbon dioxide	···	Low-carbon and low-alloy steels
Argon-helium	20-50A-50-80He	Aluminum, magnesium, copper and nickel alloys
Argon-oxygen	1-2O_2	Stainless steels
	3-5O_2	Low-carbon and low-alloy steels
Argon-carbon dioxide	20-50CO_2	Low-carbon and low-alloy steels
Helium-argon-carbon dioxide	90He-7$^1/_2$A-2$^1/_2$$CO_2$	Stainless steels
	60-70He-25-35Ar-5CO_2	Low-alloy steels
Nitrogen	···	Copper alloys

Table 7. Selection of shielding gases for GMAW of carbon steels

Metal	Shielding gas	Advantages
Short circuiting transfer		
Carbon steel	Argon-20-25% carbon dioxide	Less than $^1/_8$ in. thick: high welding speeds without melt-through; minimum distortion and spatter; good penetration
	Argon-50% carbon dioxide	Greater than $^1/_8$ in. thick: minimum spatter; clean weld appearances; good weld pool control in vertical and overhead positions
	Carbon dioxide	Deeper penetration; faster welding speeds; minimum cost
Spray transfer		
Carbon steel	Argon-3-5% oxygen	Good arc stability; produces a more fluid and controllable weld pool; good coalescence and bead contour; minimizes undercutting; permits higher speeds, compared with argon
	Carbon dioxide	High-speed mechanized welding; low-cost manual welding

tures are preferred (see the "typical mixtures" column in Table 6).

Helium shielding gas is chemically inert and is used primarily on aluminum, magnesium and copper alloys. Helium is a light gas that is obtained by separation from natural gas. It may be distributed as a liquid, but it is more often used as compressed gas in cylinders.

Carbon Dioxide. Most reactive gases cannot be used alone for shielding. Carbon dioxide is the outstanding exception; it is extensively used both by itself and as a component of gas mixtures, to which it imparts improved arc action and metal transfer.

Carbon dioxide is widely used in welding of steel by the short-circuiting mode of metal transfer. A true spray arc is not obtained with a shielding gas composed entirely of carbon dioxide. In many applications, carbon dioxide has provided good welding speed and good penetration, and in many applications the welds so shielded have proved to be less expensive than argon-shielded welds.

Carbon dioxide decomposes to carbon monoxide and oxygen at arc temperatures, producing an oxidizing effect approximately equal to that obtained by the use of an inert gas with 8 to 10% oxygen. In spite of this oxidizing effect, sound weld deposits, free of porosity, can be consistently obtained with carbon dioxide shielding when a deoxidizing electrode wire is used with the globular mode of metal transfer.

The major disadvantage of carbon dioxide shielding is the rather "harsh" arc and excessive spatter that it produces. Weld spatter can be minimized by maintaining a short, uniform arc length.

Selection of a shielding gas for a given application depends on the type and thickness of the base metal, cost and effectiveness of the different gases, joint design, position of welding, technique to be employed, fixturing, speed and required quality. Information on selection of shielding gases for GMAW of carbon steels, using the short-circuiting and spray-arc modes of transfer, is given in Table 7.

Flux-Cored Arc Welding

FLUX-CORED ARC WELDING (FCAW) is a process in which the heat for welding is produced by an arc between a tubular consumable electrode wire and the work metal, with shielding provided by gas evolved during combustion and decomposition of a flux contained within the tubular electrode wire, or by the flux gas plus an auxiliary shielding gas. Thus, there are two major versions of the process: one that uses both an auxiliary shielding gas (usually carbon dioxide or a mixture of argon and carbon dioxide) and shielding obtained from the flux core of the electrode; and the self-shielding method, which depends on combustion and decomposition of flux-core compounds for shielding.

Both methods of FCAW are closely related to other arc welding processes. The method that uses an auxiliary gas shield is similar to gas metal-arc welding (GMAW), which employs a solid consumable electrode and depends on an externally applied gas shield for protecting the arc and molten metal from contamination by the atmosphere.

FUNDAMENTALS OF THE PROCESS

The flux-cored wire is the main difference between GMAW and FCAW with an auxiliary gas shield (Fig. 7a). The flux core provides a molten slag that covers the weld metal and a gas that assists in shielding of the arc. The necessary equipment, including the electrode holder, is essentially the same for both of these processes.

Aside from the use or nonuse of auxiliary shielding gas, the self-shielding and auxiliary-gas-shielded methods differ mainly in the type of electrode holder used and in the length of elec-

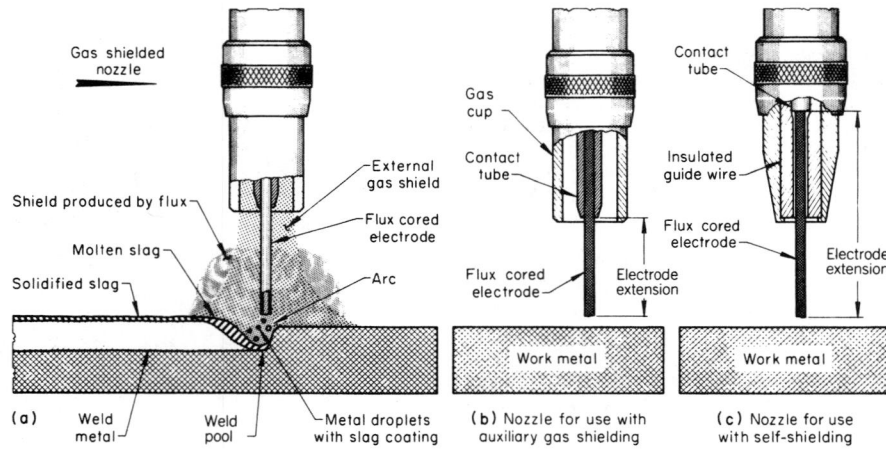

Note differences in length of electrode extension.

Fig. 7. Operating principles for FCAW with auxiliary gas shielding, and nozzles for auxiliary-gas-shielded and self-shielding methods

trode extension. As illustrated in Fig. 7(b), with the type of electrode holder used with the auxiliary-gas-shielded process, the contact tube extends nearly to the end of the gas cup.

For the self-shielding method, a much greater electrode extension is used ($2^1/_2$ in. or more), as indicated in Fig. 7(c).

POWER SUPPLY

Alternating current is seldom used for welding with flux-cored electrode wires. Direct current supplied by a rectifier, motor-generator or engine-driven generator and operated with reverse polarity (direct current electrode positive, or DCEP) is generally used; straight polarity (direct current electrode negative, or DCEN) is also used, although only to a limited extent.

ELECTRODE HOLDERS

Electrode holders for auxiliary-gas-shielded FCAW are similar to those used for GMAW. Various sizes, ratings and styles, suitable for both semiautomatic and automatic welding, are available. Both air-cooled and water-cooled electrode holders are made.

The same electrode holder can be used for self-shielding FCAW as is used for the auxiliary-gas-shielded method. Ordinarily, the electrode holder requires minor modification for use without auxiliary gas shielding — for example, a longer electrode extension is used (Fig. 7).

WIRE-FEED SYSTEMS

A four-roll feeder is shown in Fig. 8(a). In this feeder, all four rolls are driven. Push-type feeding systems vary considerably among manufacturers. Some have only two rolls — one driven roll and one pressure roll. The design must ensure that the roll speed and pressure can be varied and that the rolls can be changed quickly.

FLUX-CORED ELECTRODE WIRES

Flux-cored electrode wire consists of a low-carbon steel sheath surrounding a core of fluxing and alloying material.

Manufacture of flux-cored electrode wire is a specialized and precise operation. Most flux-

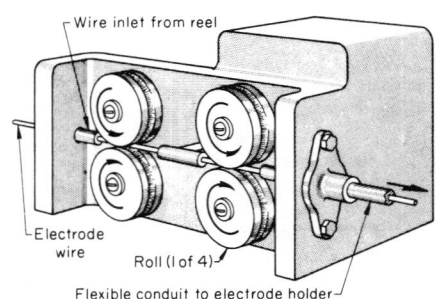

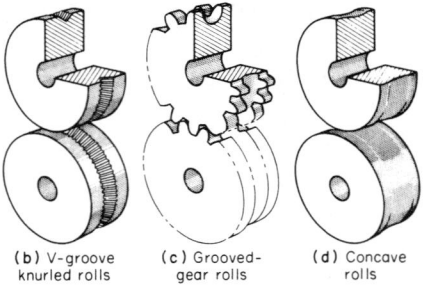

(a) Four-roll push-type feeder, in which all rolls are driven. (b) V-groove knurled feed rolls, used for electrode wire of medium to large diameter. (c) Grooved-gear feed rolls, used for soft flux-cored electrode wire. (d) Concave, smooth-face rolls, used for small-diameter wire.

Fig. 8. Electrode wire-feed system and three designs of feed rolls

cored electrode wire is made by passing low-carbon steel strip through contour-forming rolls that bend the strip into a U-shape cross section. The U-shape product is then filled with a measured amount of granular core material (flux) by passing it through a filling device. Next, the flux-filled U-shape strip passes through closing rolls that form it into a tube and tightly compress the core materials. The tube is then pulled through drawing dies that reduce the diameter of the tube and further compress the core materials. The drawing operation secures the core materials inside the tube.

Functions of the compounds contained in the core are similar to those of the compounds in the cov-

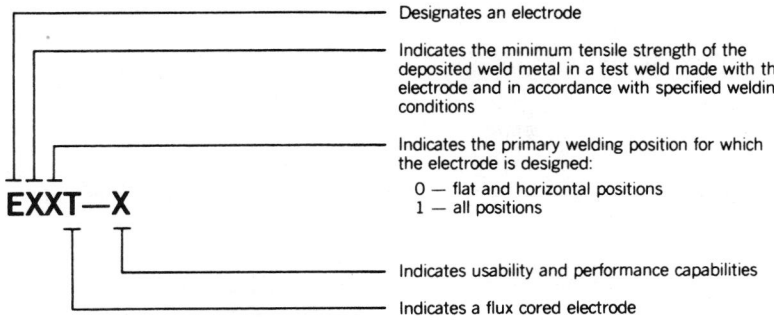

EXXT—X

- Designates an electrode
- Indicates the minimum tensile strength of the deposited weld metal in a test weld made with the electrode and in accordance with specified welding conditions
- Indicates the primary welding position for which the electrode is designed:
  - 0 — flat and horizontal positions
  - 1 — all positions
- Indicates usability and performance capabilities
- Indicates a flux cored electrode

The letter X as used in this figure and in electrode classification designations in this specification substitutes for specific designations indicated by this figure.

Fig. 9. Classification system for carbon steel flux-cored electrodes (AWS A5.20-79)

Table 8. Shielding gases and current types for carbon steel flux-cored electrodes (AWS A5.20-79)

AWS classification	Shielding gas	Current and polarity
EXXT-1	CO_2	DCEP
EXXT-2	CO_2	DCEP
EXXT-3	None	DCEP
EXXT-4	None	DCEP
EXXT-5	CO_2	DCEP
EXXT-6	None	DCEP
EXXT-7	None	DCEN
EXXT-8	None	DCEN
EXXT-10	None	DCEN
EXXT-11	None	DCEN
EXXT-G	Not defined	Not defined
EXXT-GS	Not defined	Not defined

erings on the stick electrodes used for SMAW, which are to:

- Act as deoxidizers or scavengers to help purify the weld metal and produce a sound deposit
- Form slag to float on the molten weld metal and protect it from the atmosphere during solidification
- Act as arc stabilizers to produce a smooth welding arc and reduce weld spatter
- Add alloying elements to the weld metal to increase weld strength and to provide other required weld-metal properties
- Provide shielding gas.

Classification of Electrodes. The various types of flux-cored electrodes are classified by the American Welding Society (AWS). The classification for flux-cored electrodes follows the standard pattern used in other AWS specifications. The AWS specification A5.20-79 covers carbon steel flux-cored electrodes for welding carbon and low-alloy steels. Figure 9 illustrates this classification system, which can be explained further by considering a typical designation such as E70T-1:

- The prefix "E" indicates an electrode.
- The first digit, 7, indicates a minimum tensile strength of 70 ksi.
- The second digit, 0, indicates that the electrode can be welded in the flat and horizontal positions.
- The "T" indicates a tubular or flux-cored electrode.

Hydrogen. Most flux-cored electrodes are considered to be low-hydrogen electrodes, because the materials used in the cores do not contain appreciable levels of hydrogen. Sometimes, however, certain materials are hygroscopic and may absorb moisture when exposed to a high-humidity atmosphere. Therefore, after removal from their original container, flux-cored electrodes should be treated in the same manner as low-hydrogen flux-covered electrodes are treated after removal from their original container.

SHIELDING GASES

Gases used for auxiliary-gas-shielding FCAW can be the same as those used for GMAW, which include carbon dioxide, 98% argon with 2% oxygen, and 75% argon with 25% carbon dioxide. The chemical reactions of carbon dioxide with the carbon steel base metal and with the cored electrodes make it desirable as a shielding gas. In addition, carbon dioxide is less costly than the mixtures containing argon, based on the cost of

gas per pound of weld metal deposited. Additional information on shielding gases and types of current is given in Table 8.

DEPOSITION RATE

As is true for all arc welding processes, the deposition rate for FCAW depends on the welding current and the electrode diameter. Typical relations between current and deposition rate for four different electrode-wire diameters are given in Fig. 10, which indicates for each electrode-wire diameter the current range that is normally appropriate. Deposition efficiency in FCAW is generally high, 70 to 85%, but occasionally as high as 92%.

Submerged-Arc Welding

SUBMERGED-ARC WELDING (SAW) is an arc welding process in which the heat for welding is supplied by an arc (or arcs) developed between a bare metal (or flux-cored) consumable electrode (or electrodes) and a workpiece. The arc is shielded by a layer of granular and fusible flux, which blankets the molten weld metal and the base metal near the joint and protects the molten weld metal from atmospheric contamination.

PRINCIPLES OF OPERATION

In all types of equipment, mechanically powered drive rolls continuously feed the consumable electrode wire through a contact tube (nozzle) and through the flux blanket to the joint being welded. The electrode wire is coiled on a reel or in a drum. The electrode wire melts off at the weld zone and is deposited along the joint. Granular flux is deposited ahead of the arc, and after the weld metal solidifies, unfused flux is removed by a vacuum pickup system to be screened and reused. In automatic welding, flux recovery may be an integral function of the equipment, as a flux-recovery tube follows directly behind the contact tube.

Submerged-arc welding is adaptable to both semiautomatic and fully automatic operation, although the latter, because of inherent advantages, is more widely used. In semiautomatic welding, the welder manually controls the rate of travel by guiding a welding gun that feeds the flux and the electrode to the joint. In fully au-

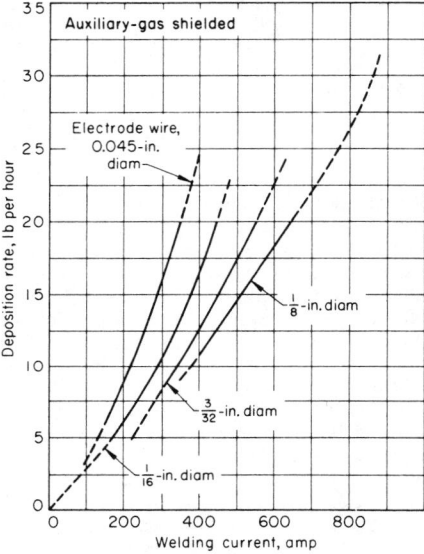

The solid part of each curve indicates the normal current range for that size of electrode.

Fig. 10. Effect of welding current and electrode-wire diameter on deposition rate

tomatic welding, the equipment automatically feeds and guides the electrode and the flux along the joint and controls the rate of deposition. A typical machine for automatic SAW is shown in Fig. 11.

ADVANTAGES AND LIMITATIONS

Submerged-arc welding, either semiautomatic or fully automatic, offers the following advantages over some other welding processes:

- Joints can be prepared with a shallow V-groove, resulting in less filler metal being used. In some applications, a groove is not required.
- The arc operates under the flux cover, thus eliminating weld spatter and arc flash.
- The process can be used at high welding speeds and deposition rates to weld flat plate or the surfaces of cylindrical shapes of virtually any size or thickness. It also can be used for hard facing or weld-overlay applications.
- The flux acts as a scavenger and deoxidizer to remove undesirable contaminants from the molten weld pool and to produce sound welds with good mechanical properties. The flux may,

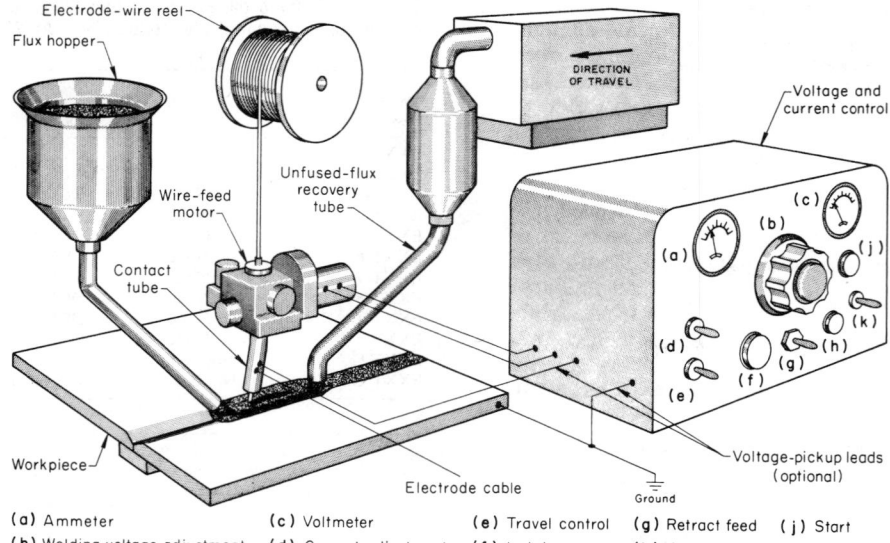

(a) Ammeter   (c) Voltmeter   (e) Travel control   (g) Retract feed   (j) Start
(b) Welding-voltage adjustment   (d) Current adjustment   (f) Inch button   (h) Weld stop   (k) Contactor

Fig. 11. Typical automatic SAW unit

if required, supply alloying elements to the weld metal.

- For welding unalloyed low-carbon steels, inexpensive electrode wires can be used — usually carbon steel wire, either bare or flash plated with copper to improve electrical contact and protect against rusting. Flux-cored electrodes also may be used, generally with neutral fluxes.
- The SAW process can be used for welding in exposed areas with relatively high winds; the granular flux shielding provides protection superior to that obtained from the electrode covering in shielded metal-arc welding (SMAW) or the gas shielding in gas metal-arc welding (GMAW).
- Low-hydrogen weld metal can be produced.

Limitations of SAW include the following:

- Flux, flux-handling equipment and workholding fixtures are required. Many joints also require the use of backing plates, strips or rings.
- Flux is subject to contamination that may cause weld discontinuities.

- To obtain welds of good quality, the base metal must be homogeneous and essentially free of scale, rust, oil and other contaminants.
- Slag (solidified residue from the fused flux) must be removed from the weld bead; this is sometimes difficult. In multiple-pass welding, slag must be removed after each pass to avoid entrapment in the weld metal.
- The process usually is unsuited for use on metal less than $3/16$ in. thick because melt-through is likely, unless backup methods are provided.
- Except for special applications, welding is largely restricted to the flat and horizontal positions in order to avoid runoff of flux.

WELDABILITY OF STEELS

Steels that are suitable for welding by other major welding processes are equally well suited to welding by the SAW process. Submerged-arc welding is used most widely in production welding of unalloyed (plain) low-carbon steels containing less than 0.30% C and 0.05% S.

SUBMERGED-ARC WELDING CONSUMABLES

The filler metal and flux used during SAW are classified as consumables. The specific chemical and physical properties of the wire, base metal and flux, as well as process parameters, control weld-metal composition, microstructure and properties. Because of the complex interactions of these factors, specific fluxes and wires must be combined to optimize weld-metal properties.

Electrode Wire

Solid electrode wire for SAW of steel is available commercially in sizes from $1/16$ to $1/4$ in. in diameter and $5/64$ to 0.120 in. in tubular form. Electrode wire $3/8$ in. in diameter is seldom used. Electrodes are produced in various ferrous alloy compositions, ranging from unalloyed low-carbon steel to high-alloy steel, and also in a variety of nonferrous compositions.

Composition. The AWS classifications and composition limits for the different types of electrode wires that are widely used for SAW of low-carbon and low-alloy steels are listed in Tables 9 and 10, respectively. These electrodes may be used in suitable combination with AWS classes of flux to produce weld metal with specified mechanical properties. Many carbon and alloy steel electrodes of compositions different from those shown in Tables 9 and 10 are available.

Wire Surface. Most non-stainless steel electrode wire is lightly coated with copper during manufacture. The copper coating provides some protection from rust and ensures good electrical contact between the electrode and the contact tube nozzle of the welding head. Good electrical contact is essential for maintaining satisfactory arc characteristics.

Selection of wire size (diameter) depends on equipment capabilities and application. Small-diameter electrode wire ($1/16$ to $3/32$ in.) is used almost exclusively with semiautomatic welding equipment. The $3/32$-in.-diam electrode wire is used with either semiautomatic or fully automatic equipment. Larger sizes ($1/8$ in. diam and above) are used only with fully automatic equipment.

Current Range. In SAW, an electrode of specific diameter can operate within a wide current range, as shown in Table 11. The overlap of current ranges makes it possible to use any of several wire sizes at a particular welding-current setting. Changing to a smaller-diameter electrode wire at a given current may serve to increase depth of fusion and reduce the width of the weld bead.

Fluxes

Fluxes used in SAW are granular, fusible mineral materials containing oxides of manganese, silicon, titanium, aluminum, calcium, zirconium and magnesium, as well as other compounds such as calcium fluoride. They are melted by the welding arc and, in the molten condition, blanket the weld metal and shield it from the atmosphere.

Fluxes are classified by AWS on the basis of the mechanical properties of a weld deposit. Figure 12 illustrates the classification system for fluxes used for SAW, which can be further explained by considering a typical designation such as F7A6-EM12K. Following the guidelines, illustrated in Fig. 12, this designation refers to a flux that will produce weld metal which, in the as-welded condition, will have a tensile strength no lower than 70 ksi and Charpy V-notch impact strength of at least 20 ft·lb at −60 °F when de-

Table 9. AWS classifications and composition limits for electrodes for SAW of low-carbon steels (AWS A5.17-80)

Electrode classification	Composition(a), wt%					
	C	Mn	Si	S	P	Cu(b)
Low-manganese steel electrodes						
EL8 . . . . 0.10	0.25-0.60	0.07	0.035	0.035	0.35	
EL8K 0.10	0.25-0.60	0.10-0.25	0.035	0.035	0.35	
EL12 . . . . 0.05-0.15	0.25-0.60	0.07	0.035	0.035	0.35	
Medium-manganese steel electrodes						
EM12 . . . . 0.06-0.15	0.80-1.25	0.10	0.035	0.035	0.35	
EM12K 0.05-0.15	0.80-1.25	0.10-0.35	0.035	0.035	0.35	
EM13K 0.07-0.19	0.90-1.40	0.35-0.75	0.035	0.035	0.35	
EM15K 0.10-0.20	0.80-1.25	0.10-0.35	0.035	0.035	0.35	
High-manganese steel electrodes						
EH14 . . . . 0.10-0.20	1.70-2.20	0.10	0.035	0.035	0.35	

(a) Single values are maximums. Electrodes shall be analyzed for those elements for which specific values are shown. Elements other than those shown, which are intentionally added (except iron), shall also be reported. The total of these latter elements and all other elements not intentionally added shall not exceed 0.50%. (b) The copper limit includes any copper coating that may be applied to the electrode.

Table 10. AWS classifications and composition limits for electrodes for SAW of low-alloy steels (AWS 5.23-80)

Electrode classification	C	Mn	Si	S	P	Cr	Ni	Mo	Cu(b)	V	Al	Ti	Zr
Carbon-molybdenum steels													
EA1	0.07-0.17	0.65-1.00	0.20	0.035	0.025	···	···	0.45-0.65	0.35	···	···	···	···
EA2	0.07-0.17	0.95-1.35	0.20	0.035	0.025	···	···	0.45-0.65	0.35	···	···	···	···
EA3	0.10-0.18	1.65-2.15	0.20	0.035	0.025	···	···	0.45-0.65	0.35	···	···	···	···
EA4	0.10-0.18	1.25-1.65	0.25	0.035	0.025	···	···	0.45-0.65	0.35	···	···	···	···
Chromium-molybdenum steels													
EB2	0.07-0.15	0.45-0.80	0.05-0.30	0.030	0.025	1.00-1.75	···	0.45-0.65	0.35	···	····	···	···
EB2H	0.28-0.33	0.45-0.65	0.55-0.75	0.015	0.015	1.00-1.50	···	0.40-0.65	0.30	0.20-0.30			
EB3	0.07-0.15	0.45-0.80	0.05-0.30	0.030	0.025	2.25-3.00	···	0.90-1.10	0.35	···	···	···	···
EB5	0.18-0.23	0.40-0.70	0.40-0.60	0.025	0.025	0.45-0.65	···	0.90-1.10	0.30	···	···	···	···
EB6(c)	0.10	0.40-0.65	0.20-0.50	0.025	0.025	4.50-6.00	···	0.45-0.65	0.35	···	···	···	···
EB6H	0.25-0.40	0.75-1.00	0.25-0.50	0.030	0.025	4.80-6.00	···	0.45-0.65	0.35	···	···	···	···
Nickel steel													
ENi1	0.10	0.75-1.25	0.05-0.25	0.010	0.010	0.15	0.80-1.20	0.30	0.35	···	···	···	···
ENi2	0.10	0.75-1.25	0.05-0.25	0.010	0.010	···	2.10-2.90	···	0.35	···	···	···	···
ENi3	0.13	0.60-1.20	0.05-0.25	0.012	0.012	0.15	3.10-3.80	···	0.35	···	···	···	·
ENi4	0.12-0.19	0.60-1.00	0.10-0.30	0.020	0.015	···	1.60-2.10	0.10-0.30	0.35	···	···	···	···
Other low-alloy steels													
EF1	0.07-0.15	0.90-1.70	0.15-0.35	0.025	0.025	···	0.95-1.60	0.25-0.55	0.35	···	···	···	···
EF2	0.10-0.18	1.70-2.40	0.20	0.025	0.025	···	0.40-0.80	0.40-0.65	0.35	···	···	···	···
EF3	0.10-0.18	1.70-2.40	0.20	0.025	0.025	···	0.70-1.10	0.45-0.65	0.35	···	···	···	···
EF4	0.16-0.23	0.60-0.90	0.15-0.35	0.035	0.025	0.40-0.60	0.40-0.80	0.15-0.30	0.35	···	···	···	···
EF5	0.10-0.17	1.70-2.20	0.20	0.010	0.010	0.25-0.50	2.30-2.80	0.45-0.65	0.50	···	···	···	···
EF6	0.07-0.15	1.45-1.90	0.10-0.30	0.015	0.015	0.20-0.55	1.75-2.25	0.40-0.65	0.35	···	···	···	···
EM2(d)	0.10	1.25-1.80	0.20-0.60	0.010	0.010	0.30	1.40-2.10	0.25-0.55	0.25	0.05	0.10	0.10	0.10
EM3(d)	0.10	1.40-1.80	0.20-0.60	0.010	0.010	0.55	1.90-2.60	0.25-0.65	0.25	0.04	0.10	0.10	0.10
EM4(d)	0.10	1.40-1.80	0.20-0.60	0.010	0.010	0.60	2.00-2.80	0.30-0.65	0.25	0.03	0.10	0.10	0.10
EW	0.12	0.35-0.65	0.20-0.35	0.040	0.030	0.50-0.80	0.40-0.80	···	0.30-0.80	···	···	···	···
EG						No requirements specified							

(a) Single values are maximums. Electrodes shall be analyzed for those elements for which specific values are shown. Elements other than those shown, which are intentionally added (except iron), shall also be reported. The total of these latter elements and all other elements not intentionally added shall not exceed 0.50%. The letter "N" as a suffix to a classification indicates that the electrode is intended for welds in the core belt region of nuclear reactor vessels, as described in paragraph A2.2 of the Appendix to this specification. This suffix changes the limits on the phosphorus, vanadium, and copper, as follows: P, 0.012% max; V, 0.05% max; Cu, 0.08% max. "N" electrodes shall not be coated with copper or any material containing copper. The "EF5" and "EW" electrodes shall not be designated as "N" electrodes. (b) The copper limit includes any copper coating which may be applied to the electrode. (c) The EB6 classification is similar to, but not identical with, the ER502 classification in A5.9-77, "Specification for Corrosion-Resisting Chromium and Chromium-Nickel Steel Bare and Composite Metal Cored and Stranded Arc Welding Electrodes and Welding Rods." (d) The composition ranges of classifications with the "EM" prefix are intended to conform to the ranges for similar electrodes in the military specifications.

Table 11. Current ranges for electrode wires used in SAW

Wire diameter, in.	Current range(a), A	Wire diameter, in.	Current range(a), A
1/16	115-500	5/32	340-1100
5/64	125-600	3/16	400-1300
3/32	150-700	7/32	500-1400
1/8	220-1000	1/4	600-1600

(a) Upper and lower limits of ranges are extremes and are rarely used.

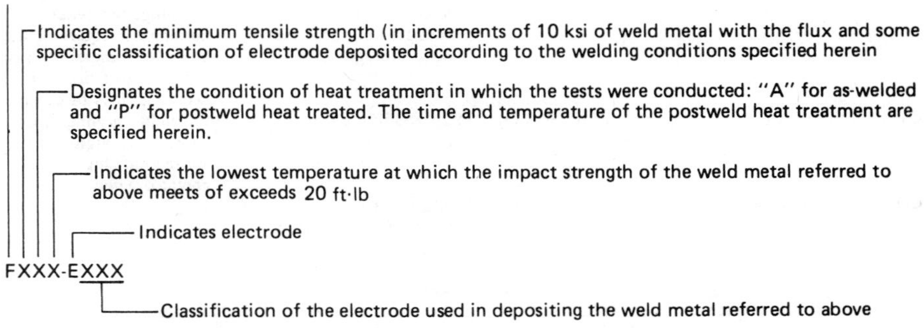

Fig. 12. AWS classification system for SAW fluxes

posited with an EM12K electrode. Submerged-arc welding fluxes are produced in three forms: prefused, bonded and agglomerated.

Prefused Fluxes. In the production of a prefused flux, the ingredients are dry mixed and then melted. Typical melting and pouring temperatures are between 2700 and 3100 °F. The molten flux may either be water shotted or poured on chill plates, then crushed and sized. With proper chilling, a glassy product is obtained. The product is passed over a series of screens that set upper and lower limits on the particle size—for example, through 12 mesh and on 200 mesh.

Bonded Fluxes. In the production of bonded flux, the raw materials are ground to approximately 100 mesh; they are dry mixed and then bonded with an addition of potassium silicate or sodium silicate. The resulting mixture is pelletized, dried

at a relatively low temperature, broken up by mechanical means, and screened. Advantages of bonded fluxes are:

- Because of the low temperatures involved in the bonding process, metallic deoxidizers and ferroalloys can be included in the flux.
- The density of the flux is lower, which permits use of a thicker layer of flux in the weld zone.
- The solidified slag is readily detachable after welding.

One disadvantage of bonded fluxes is that fines cannot be removed without some alteration in the flux composition. Another disadvantage is that bonded fluxes are likely to absorb moisture, which can cause weld-metal porosity or hydrogen-induced cracking.

Agglomerated fluxes are similar to bonded fluxes except that a ceramic binder is used. The problem of moisture pick up by agglomerated or bonded fluxes can affect the operability of these particular formulations. For example, some manufacturers recommend use of controlled temperature/humidity storage environments. Also, careful flux-handling practices may be used, such as flux preheating prior to use and disposal of

Table 12. Typical compositions of manganese silicate fluxes for SAW

Substance	Prefused flux	Bonded flux
MnO	42.0%	36.5%
MnO$_2$	···	5.2(a)
SiO$_2$	45.0	38.0
CaF$_2$	6.9	3.9
CaO	1.2	0.8
MgO	0.3	2.7
BaO	0.1	0.3
Al$_2$O$_3$	2.0	1.1
FeO	1.5	···
Fe$_2$O$_3$	···	2.7
TiO$_2$	0.1	0.1
K$_2$O	0.4	···
Na$_2$O	0.4	1.5
PbO	0.1	0.1
FeSi (50%)	···	7.1(b)
Ratio MnO/SiO$_2$	$\dfrac{42.0}{45.0} = 0.93$	$\dfrac{40.7}{45.6} = 0.89$

(a) At welding temperatures, MnO$_2$ reacts with silicon in ferrosilicon to yield additional MnO equal to 4.2% of the total flux (0.815 × 5.2% = 4.2%). (b) In reaction with MnO$_2$ at welding temperatures, silicon in ferrosilicon forms additional SiO$_2$ equal to 7.6% of the total flux (2.14 × 7.1% ÷ 2 = 7.6%).

flux that has been exposed to high temperature/humidity environments for long periods. Redrying of large volumes of flux after exposure to moisture-absorbing conditions is difficult.

Composition. In the development of the SAW process, prefused fluxes consisting of complex silicates were used. Formulations were chiefly alumina silicates of manganese, calcium and magnesium. Manganese silicate compositions corresponding to the typical analyses shown in Table 12 have been extensively used. Flux formulations of this type generally have excellent operating characteristics, such as good current/voltage stability, and produce weld deposits with excellent bead shape.

The demand for submerged-arc deposits having lower oxygen contents (in the range of 250 to 450 ppm of oxygen) and optimum toughness has been satisfied by the use of fluxes containing substantial contents of CaO, Al$_2$O$_3$, MgO and CaF$_2$. Typical low-oxygen-potential flux compositions (sometimes referred to as basic) are presented in Table 13. The various types of flux are classified as shown in Table 14.

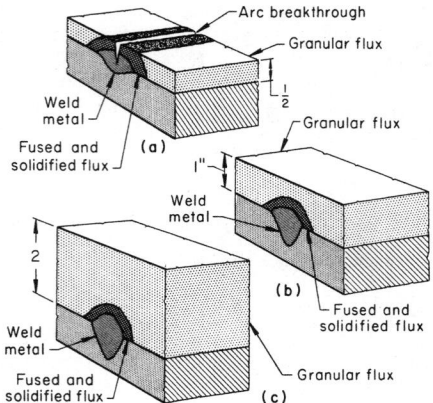

(a) Flux layer too shallow, resulting in arc breakthrough (from loss of shielding), shallow penetration, and weld porosity or cracking. (b) Flux layer at correct depth for good weld-bead shape and penetration. (c) Flux layer too deep, resulting in peaked weld bead with above-average penetration.

Fig. 13. Effect of depth of flux layer on shape and penetration of submerged-arc surface welds made at 800 A

Table 13. Typical low-oxygen-potential fluxes

Flux	Al$_2$O$_3$	SiO$_2$	TiO$_2$	MgO	CaF$_2$	CaO	MnO	Na$_2$O	K$_2$O	Basicity index (BI)(a)
A	49.9	13.7	10.1	2.9	5.7	···	15.1	1.6	0.2	0.4
B	45.7	12.9	10.7	0.1	15.6	···	10.9	0.2	1.7	0.6
C	24.9	18.4	0.2	28.9	24.2	···	1.8	2.1	0.07	1.8
D	23.2	14.5	0.9	30.8	17.7	4.2	2.3	2.3	0.1	2.1
E	19.3	16.3	0.8	27.2	23.6	9.8	0.08	0.9	1.1	2.4
F	18.5	15.2	1.0	31.8	29.8	1.8	1.5	1.0	1.05	2.7
G	14.6	13.5	0.3	31.2	25.7	···	3.5	1.3	1.9	3.0
H	18.1	13.2	0.5	28.2	31.8	4.5	0.1	0.9	0.9	3.0
I	17.0	12.2	0.7	36.8	29.2	0.7	8.9	1.6	0.1	3.5

(a) BI < 1.0, acidic; BI 1.0 to 1.5, neutral; BI > 1.5, basic.

Type of Welding Current. Fluxes usually are designed to operate with either alternating current or direct current. The type of welding current used has an effect on all weld properties. Joint penetration (and resulting dilution) are greatly affected by the type of welding current used. Direct current can be used in the electrode positive (DCEP), also known as reverse polarity, or the electrode negative (DCEN), also known as straight polarity, mode. The joint penetration will change

Table 14. Fluxes for SAW

Flux type	Chemical constituents	Advantages	Limitations	Basicity	Flux form	Comments
Manganese silicate	MnO + SiO$_2$ > 50%	Moderate strength, tolerant to rust, fast welding speeds, high heat input, good storage	Limited use for multipass welding, use where no toughness requirement, high weld-metal oxygen, increase in silicon on welding, loss in carbon	Acid	Fused	Associated manganese gain; maximum current, ≅1100 A; higher welding speeds
Calcium-high silica	CaO + MgO + SiO$_2$ > 60%	High welding current, tolerant to rust	Poor weld toughness, use where no toughness requirement, high-weld metal oxygen	Acid	Agglomerated fused	Differ in silicon gain, some capable of 2500 A, wires with high manganese
Calcium silicate-neutral	CaO + MgO + SiO$_2$ > 60%	Moderate strength and toughness, all current types, tolerant to rust, single- or multiple-pass weld		Neutral	Agglomerated fused	
Calcium silicate-low silica	CaO + MgO + SiO$_2$ > 60%	Good toughness with medium strength; fast welding speeds, less change in composition and lower oxygen	Not tolerant to rust, not used for multiwire welding	Basic	Agglomerated fused	
Aluminate basic	Al$_2$O$_3$ + CaO + MgO > 45% Al$_2$O$_3$ > 20%	Good strength and toughness in multipass welds No change in carbon; loss of sulfur and silicon	Not tolerant to rust, limited to DCEP welding, poor slag detachability	Basic	Agglomerated	Usually manganese gain; maximum current, ≅1200 A; good mechanical properties
Alumina	Bauxite base	Less change in weld composition and lower oxygen than for acid type, moderate to fast welding speeds		Neutral	Agglomerated fused	
Basic fluoride	CaO + MgO + MnO + CaF$_2$ > 50% SiO$_2$ ≤ 22%, CaF$_2$ ≤ 15%	Very low oxygen, moderate to good low-temperature toughness	May present problems of slag detachability May present problem of moisture pickup	Basic	Agglomerated fused	Can be used with all wires, preferably direct current welding, very good weld properties

with type of current used; however, that change will be influenced by the current range and flux type. Most submerged-arc fluxes are designed for alternating current or DCEN.

Melting Rate. The amount of flux fused per minute in SAW depends on the welding current and voltage. For a given current, the amount of flux fused per minute increases with voltage.

Depth of flux layer affects the shape and penetration of welds, as shown in Fig. 13. When the flux layer is too shallow (Fig. 13a), the arc is exposed and a cracked or porous weld results. When the flux is too deep (Fig. 13c), the result is peaked weld beads with above-average joint penetration. When the flux is neither too shallow nor too deep, very faint flashes appear around the interface between the electrode wire and the flux, and the weld bead appears as shown in Fig. 13(b).

Gas Tungsten-Arc Welding (TIG Welding)

GAS TUNGSTEN-ARC WELDING (GTAW), often called TIG (tungsten inert gas) welding, is an arc welding process in which the heat is produced between a nonconsumable electrode and the work metal. The electrode, weld pool, arc, and adjacent heated areas of the workpiece are protected from atmospheric contamination by a gaseous shield. This shield is provided by a stream of gas (usually an inert gas), or a mixture of gases. The gas shield must provide full protection; even a small amount of entrained air can contaminate the weld.

The arc and weld pool are visible to the welder in GTAW. Slag that may be entrapped in the weld is not produced, and filler wire is not transferred across the arc, thus eliminating weld spatter. Because the electrode is nonconsumable, a weld can be made by fusion of the base metal without the addition of filler metal. A filler metal may be used, however, depending on the requirements that have been established for the particular joint.

Gas tungsten-arc welding is an all-position welding process and is especially well adapted to welding of thin metal—often as thin as 0.005 in. The process can be applied by the manual, semiautomatic, machine and automatic methods. Manual GTAW uses a hand-held torch. Filler metal, if used, is added by hand. Figure 14 illustrates the equipment used for manual welding. Manual welding is used for most GTAW applications.

Metals That Can Be Welded. The nature of GTAW permits its use for welding of most metals and alloys. Metals that are gas tungsten-arc welded include carbon and alloy steels, stainless steels, heat-resistant alloys, refractory metals, aluminum alloys, beryllium alloys, copper alloys, magnesium alloys, nickel alloys, titanium alloys and zirconium alloys.

Base-Metal Thickness. Gas tungsten-arc welding is applicable to a wide range of base-metal thicknesses. The process is well adapted to welding of sections $1/8$ in. thick or less, because of the intense, concentrated heat produced by the arc, which results in high welding speeds. Multiple-pass welding with the addition of filler metal can be done.

For base metal more than $1/4$ in. thick, other welding processes are generally used.

FUNDAMENTALS OF THE PROCESS

The electric arc is produced by the passage of current through the ionized inert shielding gas. The ionized atoms lose electrons and are left with a positive charge. The positive gas ions flow from the positive to the negative pole of the arc. The electrons travel from the negative to the positive pole. The power expended in an arc is the product of the current passing through the arc and the voltage drop across the arc.

Arc Initiation. Some preliminary means of initiating the emission of electrons and ionization of the gas is generally used for initiating (striking) the arc. Energy for this emission and ionization can be obtained by (1) touching the energized electrode to the work and quickly withdrawing it to the desired arc length, (2) using a pilot arc that provides an ionized path for the main arc or (3) using auxiliary apparatus that produces a high-frequency spark between the electrode and the work.

Electrode and filler-metal positions in manual GTAW are shown in Fig. 15. Once the arc is started, the torch is held so that the electrode is positioned at an angle of about 75° to the surface of the workpiece and points in the direction of welding, as shown in all views in Fig. 15. To start welding, the arc usually is moved in a circular fashion until enough base metal melts to produce a weld pool of suitable size (Fig. 15a). As adequate fusion is achieved, a weld is made by gradually moving the electrode along the adjoining edges of the parts to be welded, so as to progressively fuse the parts together. Filler metal, when added manually, is often held at an angle of about 15° to the surface of the work and is slowly fed into the weld pool (Fig. 15c). Filler metal must be fed carefully to avoid disturbing the gas shield or touching the electrode and thereby causing oxidation at the end of the filler rod or contamination of the electrode. The filler metal may be added continuously from a rod, or the rod may be dipped in and out.

Filler metal can be added continuously by holding the filler rod in line with the weld (as is often done in multiple-pass welding of V-joints), or by oscillating the rod and the torch from side to side, with the filler rod feeding into the weld pool. The latter technique is one often used in surfacing.

To stop welding, first the filler metal is withdrawn from the pool (Fig. 15d), but is momentarily kept under the gas shield to prevent oxidation of the filler metal; then the torch is moved to the leading edge of the pool (Fig. 15e) before the arc is extinguished. The arc can be extinguished by raising the torch just enough to extinguish the arc, but not enough to cause contamination of the weld crater and the electrode.

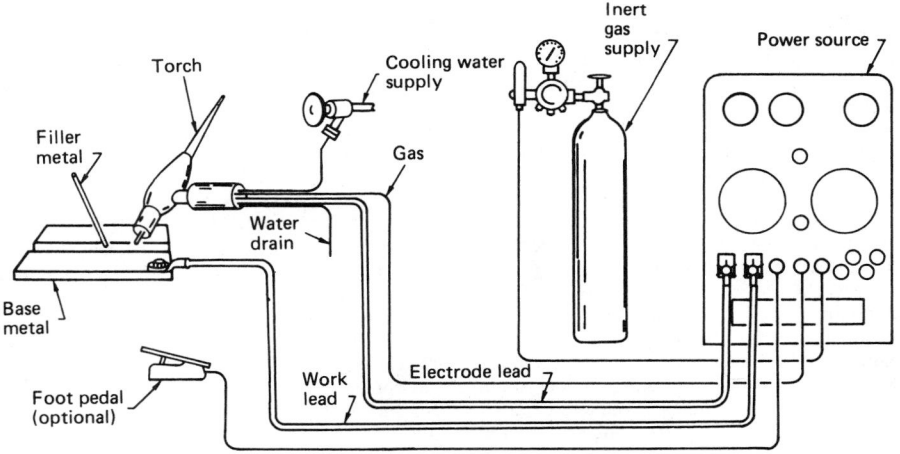

Fig. 14. Manual GTAW equipment

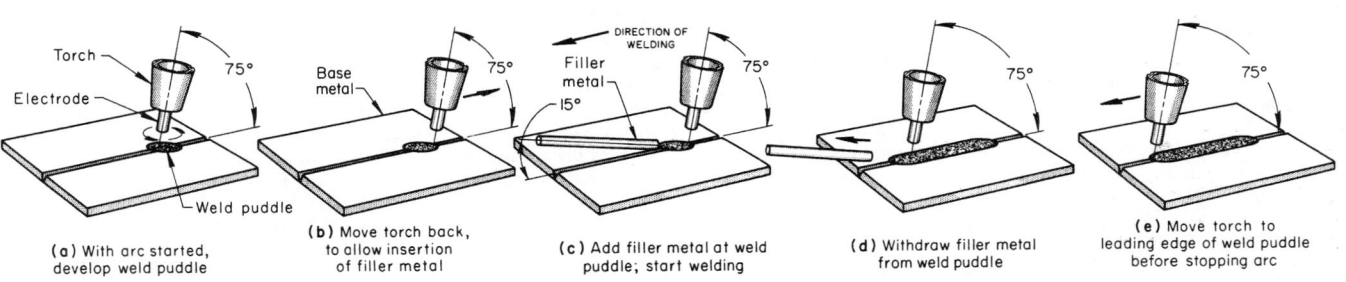

See text for discussion.
Fig. 15. Positions of torch and filler metal in manual GTAW

Table 15. Suitability of types of current for GTAW of various metals

Metal welded	Alternating current(a)	DCEN	DCEP
Low-carbon steel:			
0.015 to 0.030 in.(a)	G(b)	E	NR
0.030 to 0.125 in.	NR	E	NR
High-carbon steel	G(b)	E	NR
Cast iron	G(b)	E	NR
Stainless steel	G(b)	E	NR
Heat-resistant alloys	G(b)	E	NR
Refractory metals	NR	E	NR
Aluminum alloys:			
Up to 0.025 in.	E	NR(c)	G
Over 0.025 in.	E	NR(c)	NR
Castings	E	NR(c)	NR
Beryllium	G(b)	E	NR
Copper and alloys:			
Brass	G(b)	E	NR
Deoxidized copper	NR	E	NR
Silicon bronze	NR	E	NR
Magnesium alloys:			
Up to 1/8 in.	E	NR(c)	G
Over 3/16 in.	E	NR(c)	NR
Castings	E	NR(c)	NR
Silver	G(b)	E	NR
Titanium alloys	NR	E	NR

Note: E, excellent; G, good; NR, not recommended
(a) Stabilized. Do not use alternating current on tightly jigged assemblies. (b) Amperage should be about 25% higher than when DCEN is used. (c) Unless work is mechanically or chemically cleaned in the areas to be welded

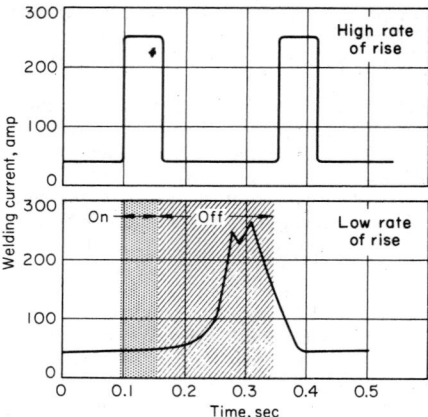

Fig. 17. Representative shapes of current pulses with high and low rates of rise

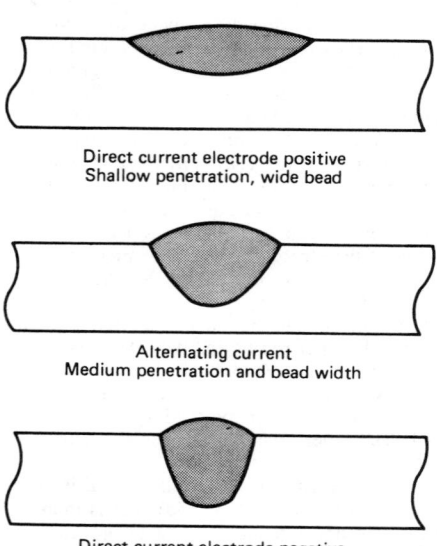

Direct current electrode positive
Shallow penetration, wide bead

Alternating current
Medium penetration and bead width

Direct current electrode negative
Deep penetration, narrow bead

Fig. 16. Effects of welding current on weld shape

WELDING CURRENT

Current is one of the most important operating conditions to control in any welding operation, because it is related to the depth of penetration, welding speed, deposition rate, and quality of the weld.

Fundamentally, there are but three choices of welding current: (1) direct current electrode negative (DCEN), (2) direct current electrode positive (DCEP) and (3) alternating current. Certain desirable effects can be obtained by superimposing high-frequency current on all three. A guide to selection of the type of current for welding various metals is presented in Table 15.

Oxide Removal by Direct Current Electrode Positive. When the electrode is positive, argon or helium ions travel to the surface of the base metal. Positively charged gas ions are produced through action of the arc on the surrounding inert-gas atmosphere. The gas ions have considerable mass and hence acquire large amounts of kinetic energy while speeding to the surface of the base metal. When these ions collide with the surface, they clean it by tearing away particles of oxide in a manner somewhat analogous to grit blasting.

Alternating current gives good penetration and surface-oxide reduction, and the form of the gas tungsten-arc weld bead it produces when filler metal is added is more nearly that of a satisfactory sheided metal-arc deposit but with slightly less reinforcement. The bead produced in GTAW with alternating current is wider and shallower than a DCEN bead, but narrower and deeper than a DCEP bead, and it has more reinforcement than either a DCEN or a DCEP bead. Figure 16 illustrates the effects of the three different types of welding current on weld-bead geometry. Alternating current is therefore preferred for the welding of aluminum, magnesium and beryllium copper because of its ability to remove oxides.

Pulsed-Current Welding. Pulsed-current GTAW employing a high rate of current rise and decay and a high pulse-repetition rate is widely used in joining of precision parts.

Pulsed-current GTAW has several advantages over constant-current welding of thin materials. This mode of welding is best suited to poor joint fit-up. Also, improved control over distortion is possible. Conventional fixturing can be used with thinner materials. High pulse provides the high current levels needed to complete penetration in open root welding. Low pulse cools the pool, thus preventing melt-through at the root of the joint. Pulsing also reduces the heat input to the base metal, which is desirable for welding thin sheet metal. Figure 17 illustrates representative shapes of current pulses with high and low rates of rise, correlated with a scale of time in seconds.

EQUIPMENT

The equipment used for GTAW includes a power supply, a welding torch, a nonconsumable electrode and a gas shielding system. Optional equipment also may include a water circulator, a foot-control pedal, voltage-control devices, motion devices, oscillators and wire feeders.

Power Supplies

Power-supply units for GTAW include (1) power-driven generators, either electric-motor driven or engine driven; (2) transformer-rectifier welding machines; (3) three-phase rectifier welding machines; and (4) transformer welding machines.

Welding Torches

A torch for manual GTAW should be compact, lightweight and fully insulated. It must provide a handle for holding it, a means for conveying the shielding gas to the arc area, and a collet, chuck or other means for securing the tungsten electrode and conducting welding current to it. The torch assembly normally includes various cables, hoses and adapters for connecting the torch to sources of power, gas and water (if water is used for cooling). A sectional view of a water-cooled manual torch is shown in Fig. 18.

The major distinction among torches for GTAW is whether they are air cooled or water cooled. Air-cooled torches might more correctly be called gas-cooled torches, because much of the cooling is achieved by the flow of shielding gas; the only true air cooling is by radiation into the surrounding air. On the other hand, for water-cooled torches, some cooling is provided by the flow of shielding gas, but it is supplemented by water that is circulated through the torch (Fig. 18a).

Electrodes

The use of a nonconsumable electrode—an electrode that does not supply filler metal—constitutes the major difference between GTAW and other metal-arc welding processes. Tungsten, which has the highest melting temperature of all metals (6170 °F), has proved to be the best material for nonconsumable electrodes. In addition to having an extremely high melting point, tungsten is a strong emitter of electrons, which stream across the arc path, ionize it, and thus facilitate the maintenance of a stable arc.

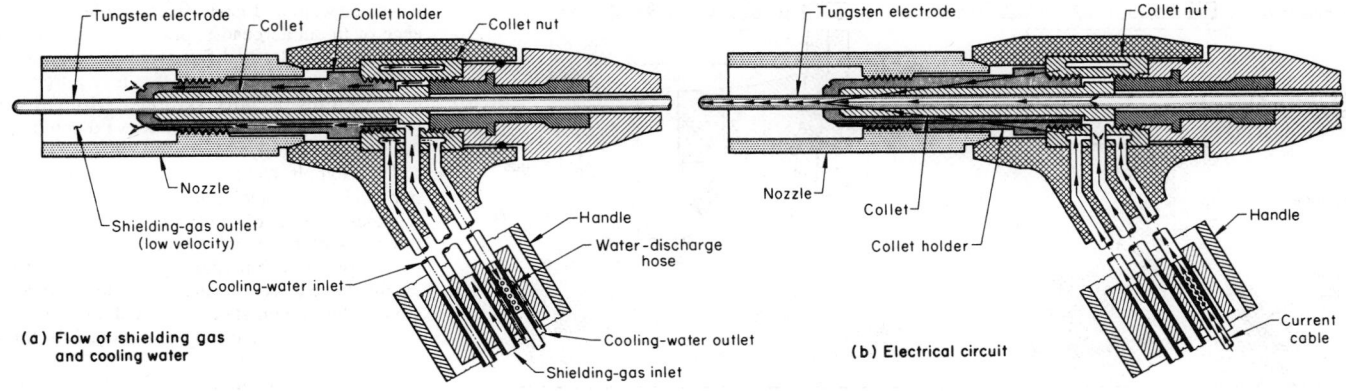

Fig. 18. Sectional views of a typical torch assembly for GTAW

Tungsten of commercial purity (99.5% W) and tungsten alloyed with either thoria or zirconia are the electrode materials used in virtually all applications of GTAW. Pure tungsten electrodes cost about 25 to 35% less than the thoriated types, depending on finish.

Table 16 gives AWS classifications and compositions for tungsten and tungsten alloy electrodes. Pure tungsten electrodes (EWP) are the least costly type of electrode, and they are used on less critical applications with alternating current. Pure tungsten electrodes have low current-carrying capacities and low contamination resistance. These electrodes are designated with green markings.

Tungsten electrodes with 1 to 2% thoria (EWTh-1 and EWTh-2) are superior to pure tungsten electrodes. They have improved electron emissivity and current-carrying capacity, longer life, and greater contamination resistance. Arc starting is easier, and the arc is more stable (see Table 17).

SHIELDING GASES

The main requirement of a shielding gas is that it exclude air from the weld pool, the electrode, and the heated end of the filler rod (if used), to avoid contamination of the weld deposit. Shielding gas does not directly add heat to the weld, but it does affect heat input. The gases ordinarily used in GTAW are argon, helium, argon-helium mixtures and argon-hydrogen mixtures.

The choice of shielding gas can significantly affect weld quality as well as welding speed. Argon, helium and argon-helium mixtures do not react with tungsten or tungsten alloy electrodes and have no adverse effect on the quality of the weld metal.

Argon Versus Helium. There is considerable difference of opinion about the relative merits of argon and helium for welding purposes. Each gas possesses characteristics that make it more suitable than the other for certain applications (see Table 17).

FILLER METALS

The selection of the proper filler metal is based primarily on the composition of the base metal being welded. Filler metals usually are matched as closely as possible to the base-metal composition.

Closer control of composition, purity and quality is exercised for filler metals than for base metals.

Choice of a filler metal depends on the proposed application. Tensile properties and impact toughness, as well as electrical conductivity, thermal conductivity, corrosion resistance and weld appearance, are important considerations in the choice of a filler metal.

Deoxidizers may be added to improve weld soundness. In GTAW, loss of deoxidizers is minimal; the filler metal is not transferred across the arc. Generally, a solid bare wire (electrode) manufactured for GMAW is suitable for GTAW. Further modifications may be made to some filler-metal compositions to improve postweld heat treatment response.

Plasma-Arc Welding

PLASMA-ARC WELDING (PAW) is an arc welding process in which heat is produced by a constricted arc between an electrode and a workpiece (transferred arc), or between a nonconsumable tungsten electrode and a constricting orifice (nontransferred arc). Shielding is generally obtained from the hot, ionized gas issuing from the orifice of the constricting nozzle, which may be supplemented by an auxiliary source of shielding gas. Shielding gas may be an inert gas or a mixture of gases.

Plasma-arc welding is closely related to gas tungsten-arc welding (GTAW). Plasma is present in all arcs. If a constricting orifice (nozzle) is placed around the arc, the amount of ionization, or plasma, is increased. This results in a higher arc temperature and a more concentrated heat pattern than exists in GTAW. For more information, see the article "Gas Tungsten-Arc Welding" in this Volume.

For plasma-arc welding, constriction of the arc is produced by the design of the welding torch. Figure 19 shows the heat patterns and arc temperatures for a nonconstricted arc, used in GTAW, and a constricted arc, used in PAW.

METALS WELDED

Plasma-arc welding is used to join most of the metals commonly welded by GTAW. These met-

Table 16. AWS classifications and composition limits for GTAW electrodes (AWS A5.12)

AWS classification	Tungsten (min)(a), %	Thoria, %	Zirconia, %	Other (max)(b), %
EWP	99.5	...	...	0.5
EWTh-1	98.5	0.8-1.2	...	0.5
EWTh-2	97.5	1.7-2.2	...	0.5
EWTh-3(c)	98.95	0.35-0.55	...	0.5
EWZr	99.2	...	0.15-0.40	0.5

(a) By difference. (b) Total. (c) EWTh-3 is a tungsten electrode with an integral lateral segment throughout its length that contains 1.0 to 2.0% thoria; average thoria content of the electrode is as shown in this table.

Table 17. Recommended tungsten electrodes and shielding gases for welding different metals

Type of metal	Thickness	Type of current	Electrode	Shielding gas
Aluminum	All	Alternating current	Pure or zirconiated	Argon or argon-helium
	Thick only	DCEN	Thoriated	Argon-helium or argon
	Thin only	DCEP	Thoriated or zirconiated	Argon
Copper, copper alloys	All	DCEN	Thoriated	Argon or argon-helium
	Thin only	Alternating current	Pure or zirconiated	Argon
Magnesium alloys	All	Alternating current	Pure or zirconiated	Argon
	Thin only	DCEP	Zirconiated or thoriated	Argon
Nickel, nickel alloys	All	DCEN	Thoriated	Argon
Plain carbon, low-alloy steels	All	DCEN	Thoriated	Argon or argon-helium
	Thin only	Alternating current	Pure or zirconiated	Argon
Stainless steel	All	DCEN	Thoriated	Argon or argon-helium
	Thin only	Alternating current	Pure or zirconiated	Argon
Titanium	All	DCEN	Thoriated	Argon

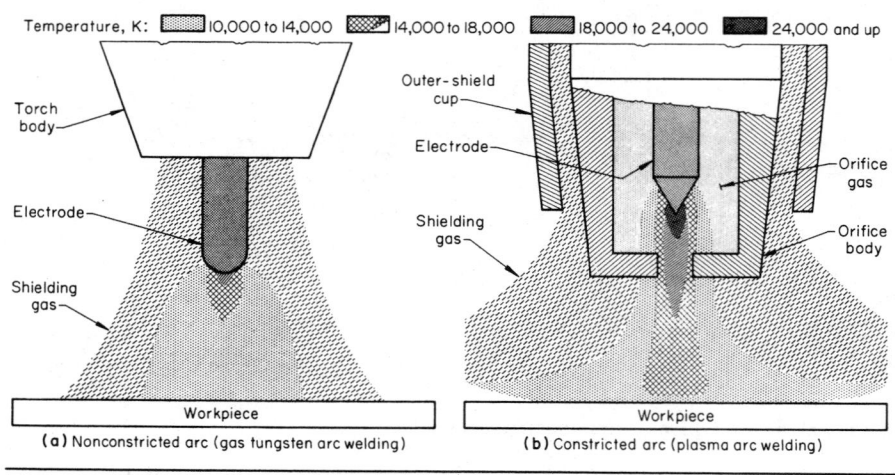

Temperature, K: ▢ 10,000 to 14,000 ▨ 14,000 to 18,000 ▧ 18,000 to 24,000 ■ 24,000 and up

(a) Nonconstricted arc (gas tungsten arc welding)

(b) Constricted arc (plasma arc welding)

Nonconstricted arc	**Constricted arc (³/₁₆-in.-diam orifice)**
Shielding gas .... Argon, at 40 ft³/h	Shielding gas .... Argon, at 40 ft³/h
Current .... 200 A	Current .... 200 A
Voltage .... 15 V	Voltage .... 30 V

This illustration shows the effect of constriction on temperature and heat pattern.

Fig. 19. Comparison of a nonconstricted arc used for GTAW and a constricted arc used for PAW

als include carbon and low-alloy steels, stainless steels, copper alloys, nickel- and cobalt-base alloys, and titanium alloys. Thicknesses ranging from 0.001 to 0.25 in. can be welded in one pass. They are welded with a transferred arc using direct current electrode negative (DCEN).

ARC MODES

Two modes of operation are the nontransferred arc and the transferred arc. In the nontransferred mode, the current flow is from the electrode inside the torch to the nozzle containing the orifice and back to the power supply. The nontransferred mode normally is used for plasma spraying or for generating heat in nonmetals. In the transferred arc, current is transferred from the tungsten electrode inside the welding torch, through the orifice to the workpiece, and back to the power supply. The difference between these two modes of operation is shown in Fig. 20.

POWER SOURCES

Direct current electrode negative from a constant-current power source is used for most PAW applications; however, alternating current or DCEP power sources can be used.

Rectifiers having an open-circuit voltage of 65 to 80 V are most commonly used as the basic unit. Power sources for GTAW can be used for PAW. Also, power sources that have appropriate adjustments are made especially for PAW.

WELDING TORCHES

Torches for PAW are more complex than those for GTAW, because separate passages are required for the orifice gas and the shielding gas, and because the orifice body must be protected by a water-cooled jacket.

A torch for manual PAW is shown in Fig. 21. It is provided with a handle for holding, a means for securing the tungsten electrode in position and

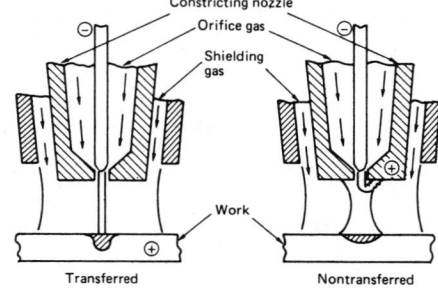

Fig. 20. Transferred and nontransferred plasma-arc modes

conducting current to it, separate passages for the orifice and shielding gases, a water-cooled orifice body (copper) and an outer shield cup (usually of ceramic material). Manual PAW torches are available for operation on DCEN at currents up to 225 A. Controls for gas and welding current usually are separate from the torch and are operated either by a foot control or automatically.

KEYHOLE WELDING

Because of the intense heat and mechanical force of the plasma arc, the keyhole technique can be used with PAW. A hole is produced at the leading edge of the weld pool by the force of the plasma arc displacing the molten metal, allowing the arc to pass completely through the workpiece. As welding progresses, surface tension causes the molten metal to flow in behind the hole to form the weld bead.

The major advantages of the keyhole technique are the ability to penetrate rapidly through relatively thick root sections and to produce a uniform underbead without mechanical backing. Also, the ratio of the depth of penetration to the width of the weld is much higher, resulting in a

narrower weld and heat-affected zone. The presence of the underbead is proof of complete joint penetration and simplifies weld inspection. Because of the precise control that is required, keyhole welding usually is done automatically. However, a successful keyhole weld can be produced manually by a skilled welder.

The keyhole technique can be applied to carbon and alloy steel and stainless steel in the thickness range of 0.090 to 0.325 in. — total thickness of plate or thickness of root face for thicker plates with prepared joint edges. For metals that have lower density or greater surface tension in the molten state, such as titanium alloys, the keyhole technique can be employed for thicker sections, often up to 0.600 in.

Aluminum and aluminum alloys can also be welded by the keyhole technique using a variable-polarity squarewave direct-current power supply.

Backing Requirements. The weld pool of a plasma-arc keyhole weld is supported by the surface tension of the molten metal. Close-fitting backing bars, which affect chill and ability to hold assembly tolerances, are not necessary. However, as in GTAW, shielding gas generally is required on the back side of the weld to protect the molten underbead from atmospheric contamination. A backing such as that shown in Fig. 22 provides a duct for flow of the shielding gas at the weld root. This type of backing supports and aligns the workpiece, contains backing gas, and provides a vent space for the plasma jet. The groove generally is about ³/₄ in. wide and 1 in. deep.

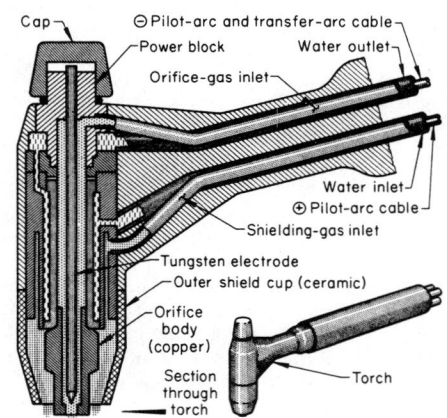

Fig. 21. Torch for manual PAW

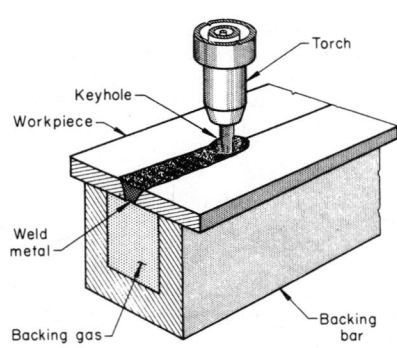

Fig. 22. Backing used with the keyhole technique

Electroslag Welding

ELECTROSLAG WELDING (ESW) is a process that uses the heat generated by passing an electrical current through a pool of molten slag (flux) to melt the edges of the joint (base metal) and a filler wire (electrode). The electrical resistivity of the molten slag continuously produces the heat necessary to continue the welding process. The molten slag pool also acts as a protective cover over the liquid metal weld pool. The slag pool and weld pool are contained in a cavity formed between the edges of the parts being joined and copper shoes used as dams. The copper shoes are normally water cooled, but this is not mandatory if the necessary amount of heat can be rejected. Typical electroslag weld-joint cross sections are shown in Fig. 23.

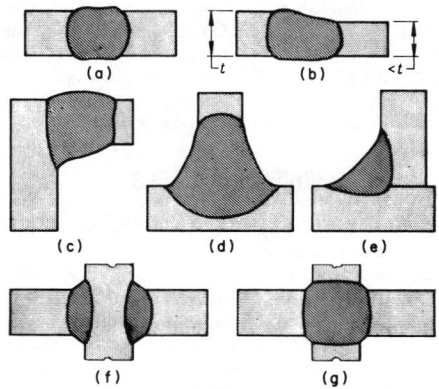

(a) Butt; square-groove weld. (b) Butt; square-groove weld, transition between two plates of different thicknesses. (c) Corner; square-groove weld. (d) T; square-groove weld. (e) Corner; fillet weld. (f) Double-T; two square-groove welds. (g) Modified butt; square-groove weld.

Fig. 23. Seven weld-joint combinations made by ESW

PRINCIPLES OF OPERATION

Conventional Electroslag Welding. Equipment for welding and a detailed view of a joint being welded are presented in Fig. 24. The sectional view shows typical depths of molten weld metal and slag. In Fig. 24, two guide tubes and two sets of electrode wire-feed rolls are shown. One, two or three guide tubes and sets of feed rolls are commonly used, depending on base-metal thickness, although more may sometimes be required. The electrode wire guides are held a short distance of $1/2$ to 2 in. above the molten slag. This distance is controlled by moving the entire welding machine upward at a pre-established rate that is consistent with the deposition rate. Usually, the welding machine is moved on a vertical track or rail. The mechanism for achieving vertical movement of the machine is not shown in Fig. 24.

Consumable-Guide Electroslag Welding. With the consumable-guide system, the mode of metal transfer is also arcless. The principal difference from the conventional system is that none of the welding-machine components moves upward during the welding operation; the only vertical movement is that of the electrode wire feeding down through the feed rolls and guide tube. As shown in Fig. 25, the main components of the

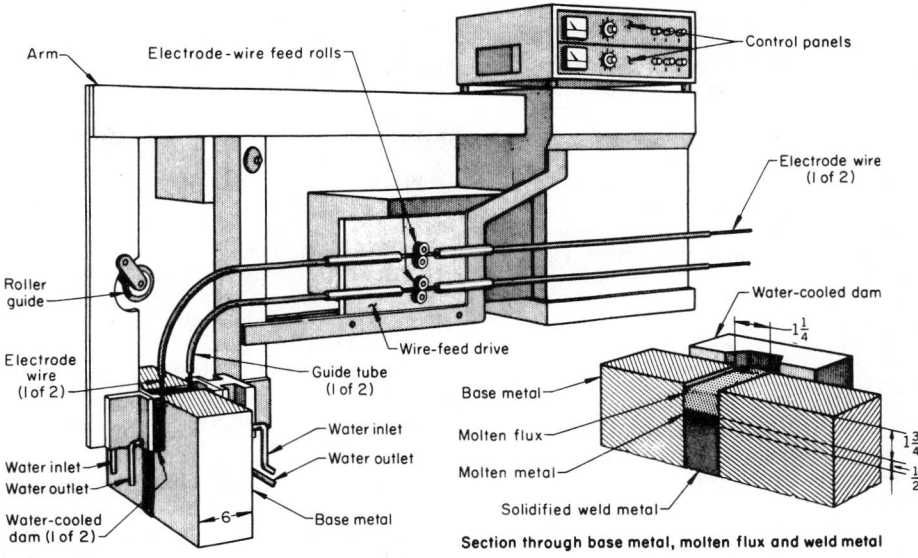

The mechanism for providing vertical movement of the welding machine is not shown.

Fig. 24. Setup for conventional ESW

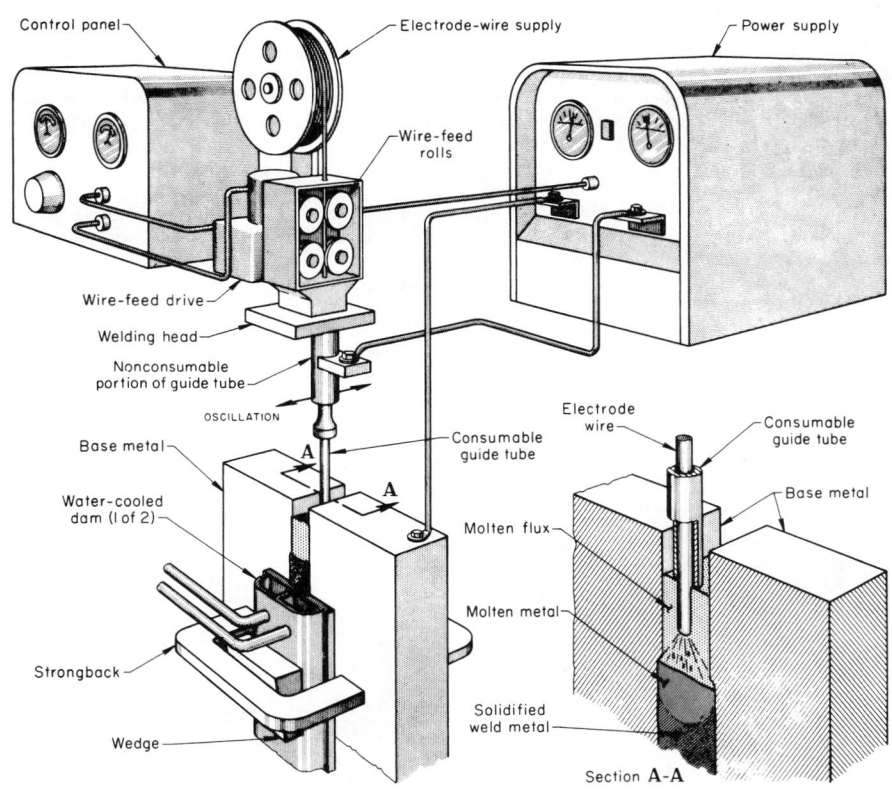

The ground cable is normally connected to the sump, not the top of the plate as shown above. It can also be connected to a clamp-on starting sump made from copper.

Fig. 25. Setup for consumable-guide ESW

consumable guide system are the power supply, control panel, wire-feed drive, consumable guide, and water-cooled dams. The wire is fed vertically down through the consumable guide to the molten slag pool. Prior to welding, the consumable guide is lowered to within approximately $1^1/2$ in. of the bottom of the joint; it melts off as the level of the weld rises. Welding begins by arcing between the electrode wire and the base metal, or steel wool, as in the conventional system.

Power Source. Direct-current, constant-voltage power supplies are recommended for ESW. For continuous-duty welding, power supplies with adequate current and voltage ratings should be used. Usually, the power rating required is 750 A, 50 V, at 100% duty cycle. A power source of this rating is ncessary for each wire in multiwire applications. Constant-current power supplies may be used for ESW; however, they are not recommended for most applications.

ELECTRODE WIRES

Electrode wires for ESW are available in two types: solid wire and metal-cored wire. Solid wire electrodes similar to those used for gas metal-arc welding (GMAW) or SAW are most frequently used. Solid wires used for ESW of low-carbon steel and high-strength low-alloy (HSLA) steel conform to American Welding Society (AWS) A5.25.

FLUXES

The molten flux is the slag that gives the electroslag process its name. Fluxes are marketed as proprietary materials; there are no standard specifications other than the classification in AWS A5.25, where fluxes are classified on the basis of the mechanical properties of the weld deposit made with a particular electrode.

Electrogas Welding

ELECTROGAS WELDING (EGW) is an automatic method of gas metal-arc vertical butt welding using a solid wire consumable electrode or a flux-cored tubular wire consumable electrode. With solid wire electrodes, an external shielding gas is required; with flux-cored tubular wire electrodes, the composition of the electrode core supplies all or part of the shielding. An essential feature of the EGW process is the use of copper dams to confine the molten weld metal. The dams are usually water cooled, although air

cooling is also used. Often referred to as "molding shoes," the dams shape the weld. Although the axis of the weld is vertical, the process is actually flat-position welding with vertical travel. The consumable electrode is fed down into a cavity formed by the opposing faces of the components to be welded and the two water-cooled dams (Fig. 26).

In its mechanical aspects, and its application to welding practice, EGW resembles conventional electroslag welding (ESW), from which it was developed (see the article "Electroslag Welding" in this volume). Electrically, EGW differs from ESW primarily in that the heat is produced by an electric arc and not by electrical resistance of a slag.

APPLICATIONS

Electrogas welding is most often used for joining relatively thick plates, such as those required in the construction of ships, bridges, storage tanks and pressure vessels. Large-diameter thick-wall pipes and longitudinal seams of pressure vessels also can be butt welded by EGW.

Metals Welded. The electrogas process has been generally restricted to welding of low-carbon and medium-carbon steels. These groups include structural steels such as ASTM A36 and carbon-manganese-silicon steels. It has also been successfully used for welding of alloy steels and the austenitic grades of stainless steel. Experimental work has also been carried out on aluminum.

EQUIPMENT

The essential components of equipment for EGW are a power supply, an electrode-wire guide, water-cooled dams, a system for feeding the electrode wire, a mechanism for oscillating the

electrode-wire guide, and methods for supplying shielding gas to the area immediately above the weld pool. Except for the power supply, the major components of the equipment are incorporated in an assembly that moves as an integral unit as welding proceeds. A typical unit for EGW is shown in Fig. 26.

Power Supplies. Electrogas welding is done with direct current electrode positive (DCEP), normally supplied by a transformer-rectifier. Motor-driven and engine-driven generators are sometimes used in field construction sites. The power supply may be of either the constant-current or constant-voltage type; the constant-current type is used for welding units in which vertical travel is controlled by changes in arc voltage.

ELECTRODE WIRES

Either solid or flux-cored electrode wire (filler metal) can be used in EGW. American Welding Society (AWS) Specification A5.26 covers the requirements of both types of electrodes for welding carbon and high-strength low-alloy (HSLA) steels (non-heat-treatable types).

SHIELDING GASES

A mixture of approximately 80% argon and 20% carbon dioxide is widely used, and is generally preferred, as a shielding gas for most EGW applications. This mixture is well suited for use with both solid and flux-cored electrode wire. Carbon dioxide alone is also used and is particularly satisfactory when employed with flux-cored wire. Self-shielded flux-cored electrodes contain core materials which generate gases that shield the molten metal from atmospheric contamination.

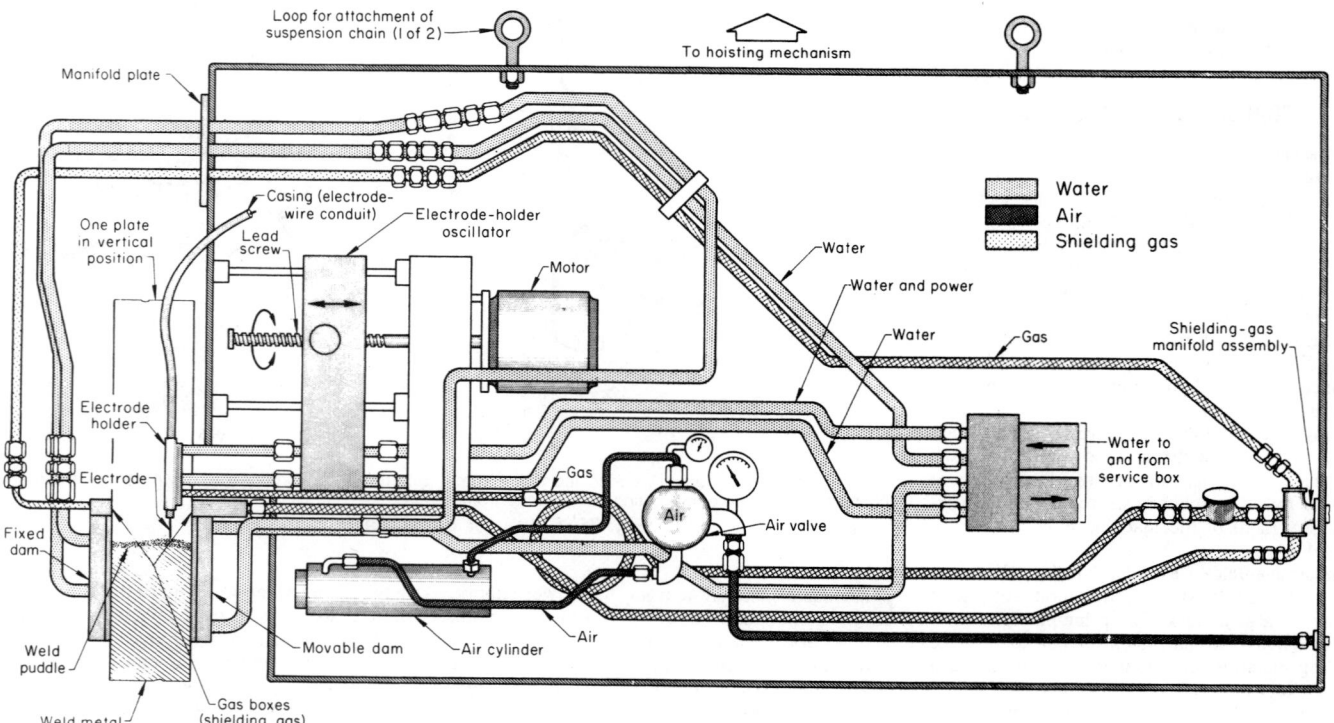

Fig. 26. Electrogas welding unit

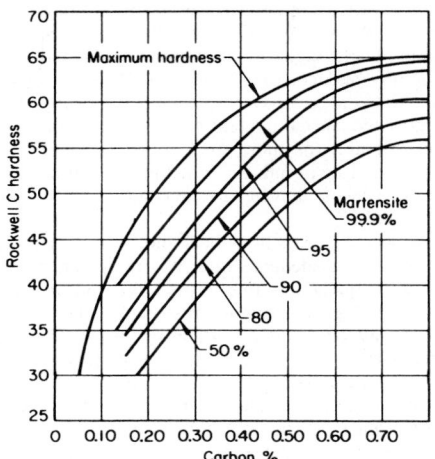

Cooling produced specific percentages of martensite and (top line) the maximum hardness obtainable in severe water quenching of small specimens of carbon steel.

Fig. 27. Effect of carbon content on the hardness of carbon steel cooled rapidly

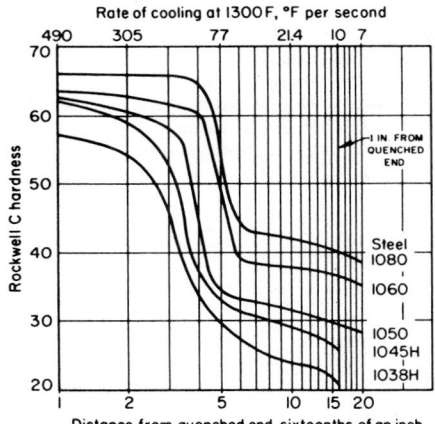

Fig. 28. Maximum hardenability of carbon steels, determined by the standard end-quench test

Arc Welding of Hardenable Carbon and Alloy Steels

HARDENABLE CARBON AND ALLOY STEELS share certain metallurgical characteristics that govern some of the guidelines used in arc welding. These steels may form martensite and/or bainite as a result of welding. These constituents can cause cracking as a result of welding and have a marked effect on the mechanical properties of the weldment. The presence of martensite and/or bainite and the resulting mechanical properties depend on the chemical composition of the steel, the cooling rate following welding, and any postweld heat treatment.

HARDNESS AND HARDENABILITY

Weldability of hardenable carbon and alloy steels differs from that of plain carbon steels because of their greater tendency to form harder regions in the heat-affected zone (HAZ). Hardenability determines the depth and distribution of hardness for a given cooling rate. The thermal cycle of the weld metal and HAZ in a typical arc weld involves a short-time, high-temperature austenitizing treatment, followed by rapid cooling. Hardenability data that have been used in the development of steels can be used as a guide to the hardness levels expected in the HAZ of a weld. Generally, a higher carbon and/or alloy content leads to the production of a more crack-sensitive microstructure in the HAZ. Higher carbon content not only produces more martensite, but also increases the crack sensitivity of the martensite. In a carbon steel without any significant alloy content, only very rapid cooling produces 100% martensite and the maximum obtainable hardness. The effect of carbon content and of the amount of martensite formed on the hardness of carbon steel is shown in Fig. 27.

In most arc welding applications involving unalloyed carbon steel, the cooling rate of the weld metal and the HAZ is too low to develop the maximum hardness that the steel of a specific carbon content can attain, because the hardenability of the steel is low. Nevertheless, an undesirable amount of hardening can occur. In

welding of alloy steels, maximum hardness often is developed in the heat-affected base metal even when the cooling rate is low, because of the high hardenability of alloy steels.

As the carbon content of plain carbon steel is increased, the hardenability (as well as the hardness) increases. This is shown by the end-quench maximum hardenability curves for five plain carbon steels plotted in Fig. 28; when these five steels are compared at 1 in. from the quenched end of the hardenability specimen (which corresponds to a cooling rate of 10 °F/s at 1300 °F), it can be seen that the hardness increases from 21 HRC for 1038H steel to 40 HRC for 1080 steel. Cooling rates of 10 °F/s are not uncommon in arc welding. If the hardness shown for a particular steel at the 1-in. position in Fig. 28 is unacceptable (or is associated with cracking), measures must be taken to avoid the development of this hardness, or to decrease it.

Carbon and low-alloy steels that have the same carbon content will have the same maximum hardness when cooled rapidly enough to achieve maximum martensite in the microstructure. In Fig. 29(a), the maximum hardness is shown at the $^1/_{16}$-in. end-quench distance for five alloy steels that have the same nominal carbon content of 0.40% (maximum of the specification range is 0.44% for each steel). Despite the major differences in alloy content and hardenability among these five steels, each steel has the same maximum hardness, 60 HRC. The two steels in Fig. 29(a) with slightly lower maximum carbon contents (4037H and 1038H) have slightly lower maximum hardnesses, as shown.

Ten 41xxH steels are compared in Fig. 29(b). Each of these steels has essentially the same alloy content (nominally 1% Cr and 0.20% Mo, with 0.80% Mn; there are slight variations through the series), but the steels range in maximum carbon content from 0.23% for 4118H to 0.65% for 4161H. As shown in Fig. 29(b), in this series of ten chromium-molybdenum steels, the maximum hardness (at the $^1/_{16}$-in. end-quench distance) increases from 48 to 65 HRC as maximum carbon content increases from 0.23 to 0.65%.

In welding of high-strength quenched-and-tempered steels that contain not more than 0.25% C, welding procedures are deliberately chosen so that martensitic structures are obtained. These

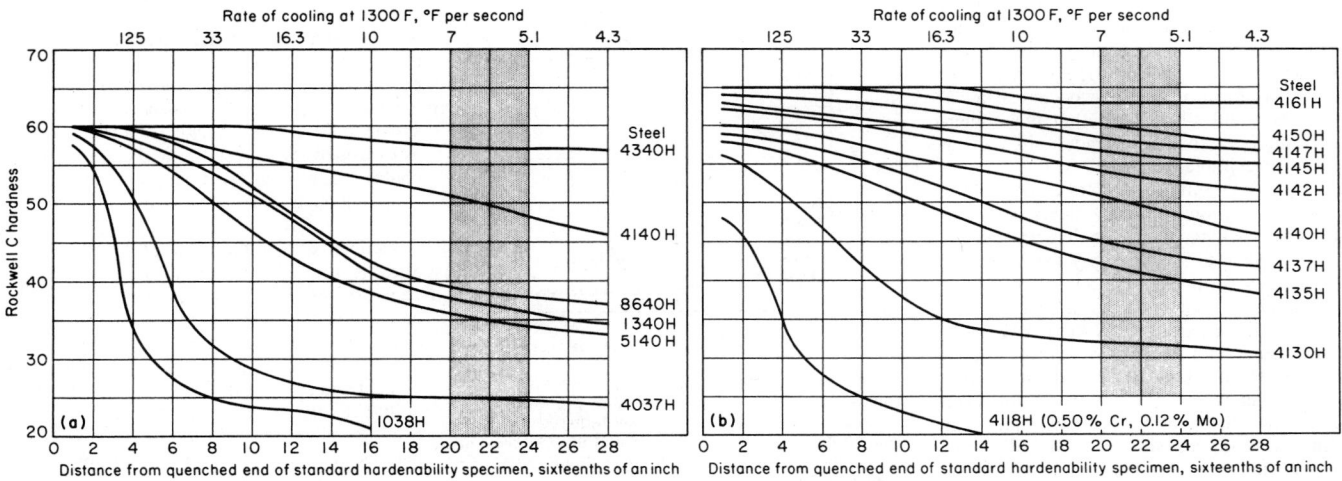

(a) Effect of various amounts and combinations of alloying elements in steel with a nominal carbon content of 0.40%, compared with carbon steel 1038H. (b) Effect of carbon content in 41xxH alloy steels (nominal 1% Cr and 0.2% Mo, except for 4118H). All data from SAE Handbook, 1970. See text for discussion.

Fig. 29. Maximum hardenability of alloy steels and 1038H carbon steel

steels are intended to be welded in the quenched-and-tempered condition. The low-carbon martensite formed during postweld cooling has desirable strength, and the as-welded joints have adequate toughness. However, as carbon content is increased, the higher-carbon martensites that are formed are harder and less ductile in the untempered condition, and are a major contributor to cold cracking of welds.

As the alloy content or the carbon content of a steel is increased, the hardenability is increased. Figure 29(a) shows the maximum hardenability of 1038H carbon steel and six widely used alloy steels of 0.40% nominal carbon content. The maximum hardenability of carbon steel 1038H (bottom curve) is low; that is, hardness decreases rapidly as cooling rate (top scale) decreases. As various amounts and combinations of alloying elements (chromium, nickel, molybdenum, manganese) are added to steel containing 0.40% C, the maximum hardenability increases to that shown for 4340H. Figure 29(b) shows the large effect of increasing carbon content on maximum hardenability in 41*xx*H steels. The effect on hardenability of increasing carbon from 0.23% (the maximum in 4118H) to 0.65% (the maximum in 4161H) in these alloy steels is somewhat greater than the effect, shown in Fig. 29(a), of increasing total alloy content (manganese, chromium, molybdenum, nickel) from 1% (maximum in 1038H) to 4.15% (maximum in 4340H).

WELDABILITY OF HARDENABLE CARBON AND ALLOY STEELS

In general, weldability of steel decreases as hardenability increases, because higher hardenability promotes the formation of microstructures which are more sensitive to cold cracking.

In welding, it is seldom possible to achieve cooling rates lower than 5 or 6 °F/s. This cooling rate corresponds to positions of 20 to 24 sixteenths of an inch from the quenched end of the standard end-quench hardenability specimen (Fig. 29a and b). Thus, if the hardness corresponding to this cooling rate (end-quench distance) for a particular steel is too high to be acceptable in the HAZ of a weldment, the weldment, or at least the zone of excessive hardness, must receive a tempering treatment after welding.

Solidification cracking, or hot cracking, is usually not as serious a problem with these steels as with the higher-alloy austenitic steels. Hot cracking can occur with the right combination of high sulfur, carbon and nickel together with high restraint. This form of cracking occurs in hardenable carbon and alloy steels, but the higher-strength and/or high-nickel grades are considered more susceptible.

By far the most common form of cold cracking in hardenable carbon and alloy steels is hydrogen-induced cracking. Occasionally, weldability of a steel is taken to mean simply the resistance of steel to this form of cracking. Plain carbon steels also experience hydrogen-induced cracking, but the higher hardenability and HAZ hardness associated with the higher-carbon and/or higher-alloy steels frequently require more precautions during welding to avoid hydrogen-induced cracking. Steels of high carbon content can produce martensite of very poor ductility that cannot withstand the shrinkage strains in an arc weld, even in the absence of appreciable hydrogen interaction. In these cases, steels are susceptible to cracking from inadequate ductility.

Hydrogen-induced cracking is usually more prevalent in welding of hardenable carbon and alloy steels than in welding of plain carbon steels. When other factors such as hydrogen, restraint and thermal cycle are equal, a steel with higher carbon and/or alloy contents has a greater tendency to form a harder microstructure, which is more susceptible to hydrogen-induced cracking.

Failures associated with hydrogen-induced cracking usually occur in the HAZ, but may also occur in the weld metal if four conditions are present simultaneously. These conditions are (1) a critical concentration of hydrogen, (2) a stress intensity of significant magnitude, (3) a susceptible microstructure and (4) a temperature between −150 and 400 °F. Hydrogen-induced cracks are generally transgranular and initiate either immediately after welding or after a delayed period.

Hydrogen in the welding-arc atmosphere is converted to the atomic state and readily dissolves in the weld pool. Because the solubility of hydrogen in steel decreases with decreasing temperature, hydrogen is strongly driven out of solution in the HAZ and weld metal during cooling. To escape, atomic hydrogen must diffuse to some interface, collect, and then re-form as molecular hydrogen. Atomic hydrogen may, however, interact with dislocations and diffuse to triaxially stressed regions where it acts as an embrittling agent. Because it is difficult for atomic hydrogen to escape from lattice imperfections by diffusion, extremely high internal stresses may develop and cracking may occur.

The amount of hydrogen absorbed by a weldment depends on several factors, including the cooling rate, size of the weld bead, and initial concentration of hydrogen in the arc atmosphere. Generally, the risk of cracking increases with increasing hydrogen concentration. Control of the hydrogen level may be achieved by minimizing the available hydrogen and providing sufficient time for hydrogen to diffuse from the weldment. The major sources of hydrogen are hydrogenous compounds and moisture in fluxes and electrode coatings, contamination of shielding gas, contamination of bare filler wires, and surface contamination of the workpiece.

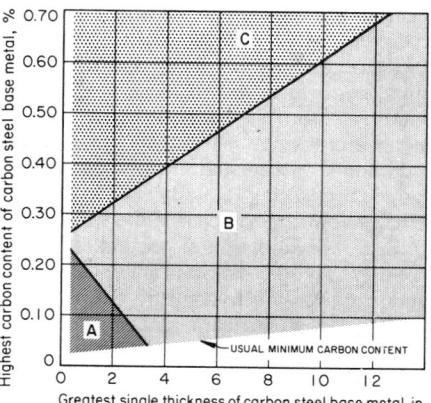

Expressed in terms of the need for preheating and postweld stress relieving. (A) Neither preheating nor postweld stress relieving is usually required. (B) Preheating is usually required; postweld stress relieving is not usually required. (C) Both preheating and postweld stress relieving are usually required.

Fig. 30. Combined influence of base-metal thickness and carbon content on weldability

Microstructure is an important factor in determining the susceptibility of a steel to hydrogen-induced cracking. Any microstructure that has low ductility and contains internal stresses is sensitive to hydrogen-induced cracking. Generally, the higher the carbon content and the harder the microstructure, the more susceptible the steel will be to hydrogen-induced cracking. High-carbon martensite is the most crack-sensitive microstructure.

Proper preheat, high heat input, and maintenance of an adequate interpass temperature reduce the quenching rate in the HAZ and provide a softer, less-sensitive microstructure. The HAZ also may be softened either by postweld heat treatment or by the tempering effect of subsequent weld passes. Where the procedure allows some flexibility in altering these variables, it is important to recognize these effects and to use caution when conditions increase the tendency for cracking.

Austenitic filler materials sometimes are used to reduce the amount of hydrogen available to sensitive HAZ microstructures and thus reduce the tendency for hydrogen-induced cracking. Austenitic weld metals have a higher solubility for hydrogen than the ferritic weld metal or HAZ; hydrogen in the HAZ escapes more quickly than it is replenished by hydrogen in the austenitic weld deposit.

Carbon Equivalent. Several formulas have been developed to assist in evaluating the weldability of hardenable carbon and alloy steels. These formulas reduce the significant composition variables to a single number, known as the carbon equivalent (*CE*). For example:

$$CE = \%C + \frac{\%Mn}{6}$$
$$+ \frac{\%Cr + \%Mo + \%V}{5}$$
$$+ \frac{\%Si + \%Ni + \%Cu}{15}$$

Steels having carbon equivalents of less than 0.35% (using this formula) usually require no preheating or postheating. Steels with carbon equivalents between 0.35 and 0.55% usually require preheating, and steels with carbon equivalents greater than 0.55% may require both preheating and postheating. Because the carbon equivalent is calculated from the base-metal composition and includes no other variables, it is only an approximate measure of weldability or susceptibility to weld cracking. Other factors, such as hydrogen level, restraint and cooling rate, that contribute to weld cracking must be considered simultaneously, in relation to a specific application.

As noted above, the value of the carbon equivalent is limited because it only considers the composition of the base material. Section thickness and weldment restraint are of equal or greater importance than the carbon equivalent. Figure 30 demonstrates the relationship between carbon content and section thickness as they affect weldability, expressed in terms of the need for preheating and postweld stress relieving. Combinations of carbon content and section thickness in area A of Fig. 30 are easily welded. Combinations in area B of Fig. 30 usually require preheating; those in area C usually require both preheating and postweld stress relieving. The use of

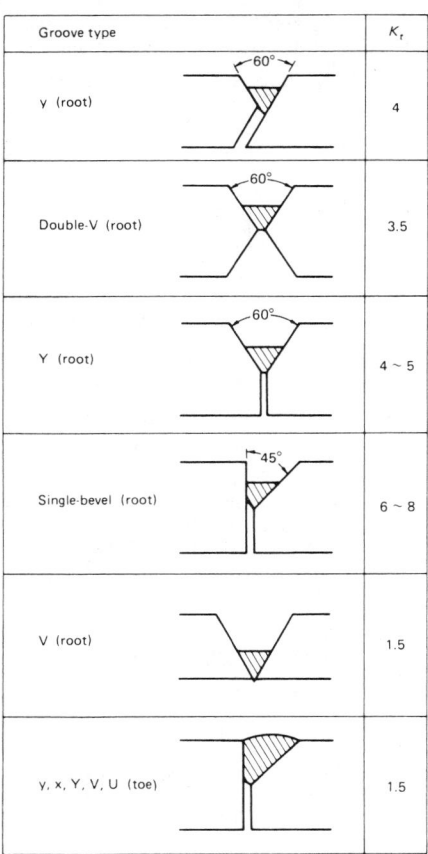

Groove type		K_t
y (root)		4
Double-V (root)		3.5
Y (root)		4 ~ 5
Single-bevel (root)		6 ~ 8
V (root)		1.5
y, x, Y, V, U (toe)		1.5

Fig. 31. Stress-concentration factors at root and toe weld positions

Table 18. Recommended minimum preheat and interpass temperatures for arc welding of typical ASTM quenched-and-tempered steels

Thickness range, in.	Minimum preheat and interpass temperatures, °F(a)				
	A514/A517	A533	A537	A543	A678
Up to 0.50	50	50	50	100	50
0.56 to 0.75	50	100	50	125	100
0.81 to 1.00	125	100	50	150	100
1.1 to 1.5	125	200	100	200	150
1.6 to 2.0	175	200	150	200	150
2.1 to 2.5	175	300	150	300	150
Over 2.5	225	300	225	300	...

(a) With low-hydrogen welding practices. Maximum temperature should not exceed the given value by more than 150 °F.

Table 19. Recommended minimum preheat and interpass temperatures for several AISI low-alloy steels

AISI steel	Thickness range, in.	Minimum preheat and interpass temperatures, °F(a)
4027	Up to 0.5	50
	0.6-1.0	150
	1.1-2.0	250
4037	Up to 0.5	100
	0.6-1.0	200
	1.1-2.0	300
4130, 5140	Up to 0.5	300
	0.6-1.0	400
	1.1-2.0	450
4135, 4140	Up to 0.5	350
	0.6-1.0	450
	1.1-2.0	500
4320, 5130	Up to 0.5	200
	0.6-1.0	300
	1.1-2.0	400
4340	Up to 2.0	550
8630	Up to 0.5	200
	0.6-1.0	250
	1.1-2.0	300
8640	Up to 0.5	200
	0.6-1.0	300
	1.1-2.0	350
8740	Up to 1.0	300
	1.1-2.0	400

(a) Low-hydrogen welding processes only

Table 20. Recommended minimum preheat and interpass temperatures for ASTM high-strength low-alloy structural steels using low-hydrogen welding procedures

ASTM steel	Thickness, in.(a)	Minimum temperature, °F
A242; A441; A572, grades 42, 50; A588; A633, grades A, B, C, D	Up to 0.75	32
	0.81 to 1.50	50
	1.56 to 2.50	150
	Over 2.50	225
A572, grades 60, 65; A633, grade E	Up to 0.75	50
	0.81 to 1.50	150
	1.56 to 2.50	225
	Over 2.50	300

(a) Thickness of thicker section at the joint

low-hydrogen processes or filler materials can offset, to some extent, the requirement for preheating, by shifting the lines that separate the three zones upward.

PREHEATING, INTERMEDIATE HEAT TREATMENTS, AND POSTHEATING

The most critical and quite often least understood part of any welding procedure involves the specification of a suitable thermal treatment. Control of the temperature of a weld can prevent problems related directly to welding, or can reduce the severity of some effects. In any welded joint, the following problems may arise:

• Hydrogen-induced cold cracking may occur in the HAZ, or possibly the weld metal, of partially completed welds.
• Residual stresses may locally exceed the yield strength of the materials.
• Restraint, or rigidity, of the over-all structure may contribute to the generation of reaction stresses transverse to the direction of welding.
• Heat of welding may produce unacceptably low toughness or mechanical properties in the weld area compared with the base plate.

Preheating. The minimum preheat and interpass temperatures are chosen primarily so as to prevent hydrogen-induced cracking. They are chosen with considerations of the chemical compositions of the base material and electrode, the hydrogen content of the deposited weld metal, and the stresses imposed across the weld. Local stress concentrations at the weld root and toe increase the possibility of cracking (Fig. 31).

The structural welding codes in Canada (CSA W59) and the United States (AWS D1.1) outline the minimum preheat and interpass requirements. Tables 18, 19 and 20 give recommended minimum preheat and interpass temperatures for various alloy steel classifications. The temperatures depend upon the welding process used and are chosen for the more susceptible steel or plate thickness. Welding is not permitted when the ambient temperature surrounding the weld area is lower than 0 °F. When no preheat is specified and the plate temperature is below 32 °F, the base metal should be preheated to at least 50 °F, and this temperature should be maintained during welding. When the base-metal temperature is below the specified minimum in Table 18, 19 or 20, preheating should be done such that the surfaces of the parts being welded are at or above the specified minimum temperature for a distance equal to the thickness of the part being welded, but not less than 3 in., both laterally and in advance of welding.

The temperatures in Tables 18, 19 and 20 are guidelines based on average conditions. The codes recommend higher temperatures for (1) highly restrained welds, (2) certain combinations of steel thickness and energy input, (3) high-strength weld metal and (4) joints where transfer of tensile stress occurs in the through-thickness direction of the weld metal. Although the need for higher temperatures is recognized, the increase is left entirely up to the discretion of the fabricator.

ARC WELDING OF HARDENABLE CARBON STEELS

Carbon steels are defined as steels that contain up to 2% C, 1.65% Mn, 0.6% Si and 0.6% Cu. Phosphorus and sulfur may also be present in these steels in concentrations up to 0.04 and 0.05%, respectively. Carbon content and the steelmaking process used (acid or basic) are frequently used as guidelines for classification of carbon steels. Carbon content, however, because of its well-defined effect on the properties of steel, is considered the most common criterion for classification.

Low-carbon steels, those containing less than 0.25% C, are generally easy to join by any arc welding process. Welds of acceptable quality usually can be produced without the need for preheating, postheating or special welding techniques, provided that the sections being welded are less than 1 in. thick and that severe joint restraint does not exist. Filler-metal selection is seldom critical for welding low-carbon steel and is based mainly on tensile-strength requirements.

Medium-carbon steels, those containing 0.25 to 0.50% C, can also be satisfactorily welded by all of the arc welding processes. Because of the formation of greater amounts of martensite in the weld zone and the higher hardness of the martensite, prehating or postheating, or both, are often necessary.

For joint designs and welding processes and procedures that induce high weld cooling rates, preheating is used to inhibit martensite formation and allow upper transformation products to form. Postweld heat treatment also is used to temper martensite and restore toughness in the HAZ. Modifications in welding procedure—for example, the use of large V-grooves or multiple passes—also decrease the cooling rate and the probability of weld cracking. In multiple-pass welding, the final weld bead should be deposited in such a manner that it is surrounded on both sides by weld metal from previous passes. When this is done, the HAZ that results from the deposition of the previous-pass weld beads is tempered by the heat from the final-pass bead.

Selection of filler materials for arc welding becomes more critical as the carbon content of the steel increases. As discussed earlier in this article, steels with higher carbon contents are more susceptible to hydrogen-induced cracking; therefore, low-hydrogen electrodes and processes ordinarily are used. As the carbon content of the steel being welded approaches 0.50% low-hydrogen conditions become mandatory.

High-carbon steels, with more than 0.50% C, are difficult to weld because of their susceptibility to cracking. Excessive hardness and brittleness often occur.

For best results in SMAW, the use of low-hydrogen electrodes is recommended. Similarly, for other arc welding processes, low-hydrogen practice is mandatory. Both preheating and postweld stress relieving or tempering usually are required. Austenitic stainless steel electrodes sometimes are used for welding high-carbon steel to obtain greater notch toughness in the joint. However, the HAZ may still be hard and brittle, and preheating and postweld stress relieving may be necessary.

ARC WELDING OF HIGH-STRENGTH STRUCTURAL STEELS AND HSLA STEELS

The steels included in these groups have higher strengths and better toughness than plain carbon steels; some grades also have improved corrosion resistance in comparison with carbon steels. This classification includes as-rolled pearlitic structural steels with minimum specified yield strengths of 40 to 50 ksi; normalized carbon steels with minimum yield strengths of 42 to 100 ksi and, if required, specified toughness and impact strengths; and microalloyed HSLA steels with properties that result from a combination of alloy additions and controlled rolling procedures. These steels are frequently used in the as-rolled condition and do not require subsequent heat treatments after forming or welding. They are widely used for structural applications and components for transportation, materials handling, agriculture and construction.

Welding Processes. The HSLA steels can be welded with any of the arc welding processes using suitable welding procedures. The selection of a process is dictated by thickness, position of the joint, and the physical location of the part to be welded. Most of the applications are structural uses such as bridges and buildings; in these cases, GTAW and plasma-arc welding (PAW) are uneconomical and thus are not used. Electroslag welding (ESW) and electrogas welding (EGW) have been used only to a limited extent because

their high heat inputs may cause excessive grain growth and precipitation of microalloying elements, both of which lead to reduced mechanical properties. The most commonly used processes are SMAW, FCAW, GMAW and SAW.

Filler metals designated for SMAW of high-strength structural and HSLA steels are listed in Table 21. The electrode is selected primarily by strength level and to a lesser extent by chemical composition. Alloying elements are added to the electrodes to achieve the necessary mechanical properties or chemical compositions.

Low-hydrogen potassium electrodes or low-hydrogen iron-powder electrodes are recommended for these grades. These steels can also be welded with cellulosic electrodes, provided that necessary precautions are taken to prevent hydrogen-induced cracking. Low-hydrogen electrodes often can be used without preheating. Preheating is advised with cellulosic electrodes. The use of adequate preheating and/or postheating can prevent hydrogen cracking in the HAZ.

ARC WELDING OF HIGH-STRENGTH QUENCHED-AND-TEMPERED ALLOY STEELS

The steels discussed in this section are the quenched-and-tempered weldable alloy steels containing not more than 0.25% C, and with a total content of alloying elements (not including manganese and silicon) ranging from 0.85% to about 16%. These steels are welded in the quenched-and-tempered condition and have yield strengths of 50 to 180 ksi, depending on alloy content, section thickness and heat treatment. They have high strength in combination with good ductility. Various combinations of notch toughness, fatigue strength and corrosion resistance can be developed to meet the requirements of different applications, such as structures and pressure vessels for use at atmospheric, cryogenic or elevated temperatures.

Although high-strength alloy steels with up to 0.25% C cannot be successfully welded with simple procedures and minimal control, they are less difficult to weld than are the higher-carbon alloy steels such as 4140. The quenched-and-tempered alloy steels with up to 0.25% C were designed to be welded with moderate or no preheating, and to be used in most applications in the as-welded condition. Knowing the correct procedures that should be used for welding these steels, and rigorously following them, are fundamental to welding these steels successfully.

Composition, Properties and Microstructure. The composition ranges (ladle analysis) for representative high-strength quenched-and-tempered alloy steels are given in Table 22.

Many of the quenched-and-tempered steels in Table 22 usually are produced with sulfur contents of less than 0.025%, and some, such as HY-130 and HP 9-4-20, are produced with sulfur contents not exceeding 0.010%. The manganese-to-sulfur ratio is generally greater than 30 to 1, so that with a carbon content of about 0.20% or less, the susceptibility to hot cracking is negligible.

The A533 steel, with a somewhat higher carbon content, has negligible susceptibility to hot cracking because it has a high manganese-to-sulfur ratio, usually about 50 to 1. The A543 steel, with a low manganese content, is susceptible to cracking when the carbon content is at the maximum unless sulfur content is extremely low.

Table 21. Covered electrodes commonly used for SMAW of high-strength structural and HSLA steels

Steel grade	SMAW electrode
A225	
Grade C	E11018-M
Grade D	E9018-M
A242	E7016, E7018
A299	E8016-C3, C8018-C3
A302	
Grade A	E7016-A1, E7018-A1
Grade B	E8016-B2, E8018-B2
Grades C, D	E10016-D2, E10018-D2
A441	E7016, E8016-C3, E8018-C3
A537	
Class 1	E7018-A1
Class 2	E8016-C1, C11018-M
A572	
Grades 42, 50	E7016, E7018
Grades 60, 65	E8016-C3, E8018-C3
A588	
Grades A, B, C, D, E	E7016, E7018
Grade F	E8016-B1, E8018-B1
Grade G	E8016-C1, E8018-C1
Grade H	E8015-G, E8018-G
A606	E7016, E7018, E7028
A607	
Grades 45, 50, 55	E7016, E7018, E7028
Grades 60, 65, 70	E9018-M
A618	E7018
A633	E7016, E7018, E7028
A656	E10018-D2
A662	E7016, E7018, E7028
A678	
Grade A	E7016, E7018, E7028
Grade B	E9018-M
Grade C	E10018-M
A709	
Grade 36T	E6012, E6013, E7014, E7016, E7018, E7028
Grade 50T, 50WT	E7016, E7018, E7028
Grade 100T, 100WT	E11018-M
A737	
Grade B	E7016, E7018, E7028
Grade C	E9018-M

Susceptibility to cold cracking under conditions of high restraint decreases with increased M_s temperature. This effect has been attributed to the self-tempering of the martensite that forms at high temperatures within the martensite transformation range. Furthermore, cold cracking is directly proportional to the hydrogen content in the welding atmosphere.

All of the steels in Table 22 have low susceptibility to cold cracking if suitable care is taken to limit the amount of hydrogen in the welding atmosphere.

Selection of Welding Process. Shielded metal-arc, submerged-arc, flux-cored arc and gas metal-arc welding are most commonly used for joining quenched-and-tempered alloy steels with carbon contents up to 0.25%. These four processes can be used effectively for welding steels having yield strengths up to approximately 150 ksi. Gas tungsten-arc or electron beam welding (EBW) must be used for steels with yield strengths over 150 ksi, including the HP 9-4-20 steel.

Weld cooling rates for arc welding processes are usually so high that the mechanical properties of the HAZ in the high-strength quenched-and-tempered steels approach those of the steel in the quench-hardened condition. Thus, postweld heat

Table 22. Compositions of representative high-strength alloy steels (quenched and tempered)

ASTM designation	Composition type	C	Mn	Si	Ni or Cu	Cr	Mo	Other
					Composition, %			
A533, grade B(a) ...... Mn-Mo-Ni		0.25 max	1.15-1.50	0.15-0.30	0.40-0.70 Ni	···	0.45-0.60	···
A517(b):								
Grade A .... Mn-Si-Cr-Mo-Zr-B		0.15-0.21	0.80-1.10	0.40-0.80	···	0.50-0.80	0.18-0.28	0.05-0.15 Zr; 0.0025 max B
Grade B .... Mn-Cr-Mo-V-B		0.15-0.21	0.70-1.00	0.20-0.35	···	0.40-0.65	0.15-0.25	0.01-0.03 Ti; 0.0005-0.005 B(c)
Grade C .... Mn-Mo-B		0.10-0.20	1.10-1.50	0.15-0.30	···	···	0.20-0.30	0.001-0.005 B
Grade D .... Cr-Mo-Cu-Ti-B		0.13-0.20	0.40-0.70	0.20-0.35	0.20-0.40 Cu	0.85-1.20	0.15-0.25	0.04-0.10 Ti; 0.0015-0.005 B
Grade E .... Cr-Mo-Cu-Ti-B		0.12-0.20	0.40-0.70	0.20-0.35	0.20-0.40 Cu	1.40-2.00	0.40-0.60	0.04-0.10 Ti; 0.0015-0.005 B
Grade F .... Mn-Ni-Cr-Mo-Cu-V-B		0.10-0.20	0.60-1.00	0.15-0.35	0.70-1.00 Ni(d)	0.40-0.65	0.40-0.60	0.03-0.08 V; 0.002-0.006 B
Grade G .... Mn-Si-Cr-Mo-Zr-B		0.15-0.21	0.80-1.10	0.50-0.90	···	0.50-0.90	0.40-0.60	0.05-0.15 Zr; 0.0025 max B
Grade H .... Mn-Ni-Cr-Mo-V-B		0.12-0.21	0.95-1.30	0.20-0.35	0.30-0.70 Ni	0.40-0.65	0.20-0.30	0.03-0.08 V; 0.0005 min B
Grade J .... Mn-Mo-B		0.12-0.21	0.45-0.70	0.20-0.35	···	···	0.50-0.65	0.001-0.005 B
Grade K .... Mn-Mo-B		0.10-0.20	1.10-1.50	0.15-0.30	···	···	0.45-0.55	0.001-0.005 B
Grade L .... Cr-Mo-Cu-Ti-B		0.13-0.20	0.40-0.70	0.20-0.35	0.20-0.40 Cu	1.15-1.65	0.25-0.40	0.04-0.10 Ti; 0.0015-0.005 B
Grade M .... Mn-Ni-Mo-B		0.12-0.21	0.45-0.70	0.20-0.35	1.20-1.50 Ni	···	0.45-0.60	0.001-0.005 B
Grade P .... Mn-Ni-Cr-Mo-B		0.12-0.21	0.45-0.70	0.20-0.35	1.20-1.50 Ni	0.85-1.20	0.45-0.60	0.001-0.005 B
Grade Q .... Mn-Ni-Cr-Mo-V		0.14-0.21	0.95-1.30	0.15-0.35	1.20-1.50 Ni	1.00-1.50	0.40-0.60	0.03-0.08 V
A542(e) 2¹/₄Cr-1Mo		0.15 max	0.30-0.60	0.15-0.30	···	2.00-2.50	0.90-1.10	···
A543(f) 3Ni-Cr-Mo		0.23 max	0.40 max	0.20-0.35	2.60-4.00 Ni	1.50-2.00	0.45-0.60	0.03 max V
HY-130(g) 5Ni-Cr-Mo-V		0.12 max	0.60-0.90	0.20-0.35	4.75-5.25 Ni	0.40-0.70	0.30-0.65	0.05-0.10 V
HP 9-4-20(h) 9Ni-4Co-Cr-Mo-V		0.17-0.23	0.20-0.30	0.10 max	8.50-9.50 Ni	0.65-0.85	0.90-1.10	4.25-4.75 Co; 0.06-0.10 V
Mod A203, grade D 3¹/₂Ni		0.17 max	0.70 max	0.15-0.30	3.25-3.75 Ni	···	···	···
A553, grade A 9Ni		0.13 max	0.90 max	0.15-0.30	8.50-9.50 Ni	···	···	···
A553, grade B 8Ni		0.13 max	0.90 max	0.15-0.30	7.50-8.50 Ni	···	···	···
A333, grade 3 .... 3¹/₂Ni		0.19 max	0.31-0.64	0.18-0.37	3.18-3.82 Ni	···	···	···
A514, type F ..... Mn-Ni-Cr-Mo-Cu-V-B		0.10-0.20	0.60-1.00	0.15-0.35	0.70-1.00 Ni(d)	0.40-0.65	0.40-0.60	0.03-0.08 V; 0.002-0.006 B

Note: Phosphorus content is 0.035 max for all steels except HY-130 and HP 9-4-20, which contain 0.010 max P; and A333, grade 3, which contains 0.05 max P. Sulfur content is 0.040 max for all steels except HY-130 and HP 9-4-20, which contain 0.010 max S; A542, which contains 0.035 max S; and A333, grade 3, which contains 0.05 max S.
(a) See A541, class 3, and A508, class 3, for forging steels. (b) Pressure-vessel quality; see A514 for structural quality. (c) Vanadium, 0.03 to 0.08. (d) Copper, 0.15 to 0.50. (e) See A541, class 6, for forging steel. (f) See A541, class 7, and A508, class 4, for forging steels. (g) See A579, grade 12, for forging steel. (h) See A579, grade 81, for forging steel.

Table 23. Typical hardening media for tool steels

Group	Type	Hardening medium	Typical grades
Water hardening	Plain carbon	W, B	W1
Shock resisting	Medium carbon, low-alloy	O	S1, S5, S6
Cold work	High-carbon, low-alloy	O	O1, O2, O6, O7
	High-carbon, medium-alloy	A	A2, A6, A7
	High-carbon, high-chromium	A	D2, D4, D7
Hot work	Chromium	A	H11, H12, H13
	Tungsten	A	H21
	Molybdenum	A	H42
High speed	Tungsten	A	T1, T4, T15
	Molybdenum	A	M1, M2, M3
Mold steels	Low-carbon, low-alloy	O	P1, P20
Special-purpose	Low-alloy	O	L2, L6

Note: A, air; B, brine; O, oil; W, water.

treatment, such as quenching and tempering, is unnecessary unless stress corrosion is a factor. Electroslag welding, which subjects the base metal to prolonged heating and consequently lower cooling rates, generally requires quenching and tempering of these steels after welding.

Welding of structures in the field often requires welding conditions similar to those used in the shop. To minimize the problem of duplicating shop practice in the field, large sections are often fabricated in the shop and transported to the field for final assembly by welding.

ARC WELDING OF TOOL STEELS

The compositions of tool steels range from that of the plain low-carbon mold steel P1 to that of the high-alloy high speed steels, some of which have a total alloy content that exceeds 25%. It follows that their weldability also varies over a broad range. Commercial tool steels are classified into seven major AISI groups. These groups are further classified as to type (hot work, high speed, etc.). Hardening media for these seven groups of tool steels are listed in Table 23. Virtually all of these steels are weldable with vary-

ing degrees of difficulty. Generally, weldability varies with hardenability.

Steels such as P1, like other plain low-carbon steels, can be welded without special procedures such as preheating and postheating. However, most tool steels have high carbon contents (some as high as 2.50%) and relatively high contents of alloying elements such as manganese, silicon, chromium, molybdenum, tungsten, vanadium and cobalt. Therefore, most tool steels require the use of carefully controlled preheating and postheating, and in most applications a considerable amount of welder skill is required.

Arc Welding of Cast Irons

CAST IRONS include a large family of ferrous alloys covering a wide range of chemical compositions and metallurgical microstructures. Some of these materials are readily welded, while others require great care to produce a sound weldment. Some cast irons are considered nonweldable.

WELDING METALLURGY OF CAST IRONS

Cast irons have carbon contents in excess of 2% and silicon contents in excess of $1/2$%. Carbon can be present in the form of (1) eutectic graphite flakes, (2) graphite nodules caused by modifications of eutectic graphite during solidification, (3) pearlitic iron carbide, (4) eutectic iron carbide or (5) carbon retained in a solid phase such as martensite. Carbon in the form of iron carbide is in a metastable state and can transform to graphite if the kinetics are suitable. Depending on alloy content, melting practice and thermal treatment, cast irons represent a wide range of alloys, including the following categories: white, malleable, gray, ductile, and compacted graphite irons. Mechanical properties and weldability are dependent on microstructure, which is directly related to partitioning of carbon during solidification and subsequent cooling. White iron is generally considered to be nonweldable because of its extreme hardness and brittleness. Ductile iron is easier to weld than gray iron, partly due to the lower levels of sulfur and phosphorus in ductile iron. It offers superior base-metal properties, but weaknesses not critical in a gray iron weldment are unacceptable in a ductile iron one.

MICROSTRUCTURE

Cast irons have various microstructures and physical properties, resulting in marked differences in weldability. Variations in thermal gradients across the weldment result in variations in microstructure and properties. The various microstructures are classified into different zones and regions, as shown in Fig. 32. The nature and size of these zones in cast iron weldments are determined by the thermal weld cycle, composition of the base metal, and welding consumables. To develop welding procedures that minimize the deleterious effects of these zones, the influences of welding variables on mechanical properties must be considered.

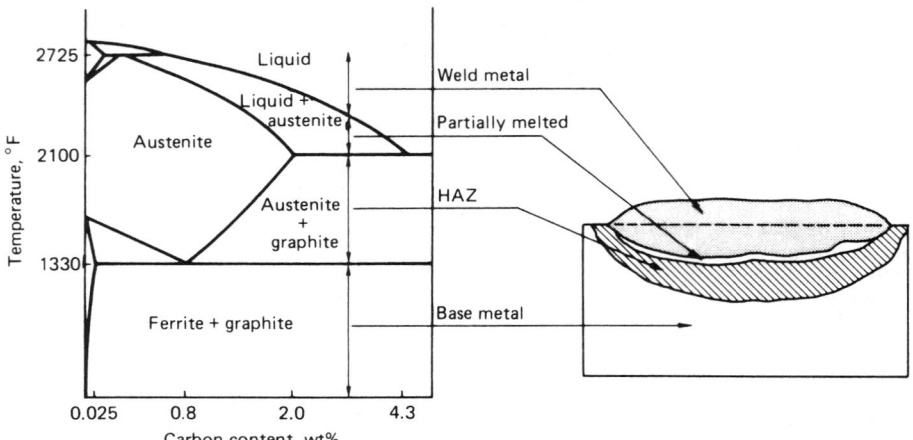

Fig. 32. Schematic representation of the zones in a typical cast iron weldment

HEAT-AFFECTED ZONE

Welding of cast irons is characterized by rapid cooling as compared with cooling rates during casting. Consequently, properties of the weld and the sections of the casting exposed to elevated temperatures (the heat-affected zone, or HAZ) differ from those in the remainder of the casting. Portions of the cast iron HAZ reach temperatures during welding which cause the carbon to diffuse into the austenite. On cooling, this austenite transforms into hard eutectoid decomposition products such as martensite. The amount of martensite formed depends on the cast iron composition and thermal treatment. Ferritic cast irons contain most of their carbon in the form of graphite, which dissolves slowly, thus producing less martensite. The greatest percentage of carbon in pearlitic cast irons is finely divided into the pearlitic structure. This carbon dissolves readily, producing a large amount of martensite. The brittle martensite may be tempered to a lower-strength, more ductile structure through (1) preheating and interpass temperature control, (2) multiple-pass welding or (3) postweld heat treatments such as stress-relief annealing.

WELDING PROCEDURES AND PROCESSES

More than 90% of all industrial welding of cast iron is done by arc welding processes. Arc welding has a lower heat input than oxyfuel gas welding because of a higher welding speed and a higher deposition rate. The welding operation can be automated to varying degrees, and distortion due to welding heat is more readily controlled. Arc welding achieves temperatures (approximately 5430 °F) in excess of those required for fusion of the base metal. The intensity of the heat source allows the necessary fusion while heating only a small portion of the weldment. This may cause high cooling rates and may result in large thermal expansion and contraction stresses. However, arc welding processes can produce welds of good quality with proper selection of the welding process, consumables and procedures.

CONSUMABLES

Filler Metals for Shielded Metal-Arc Welding of Cast Irons. A variety of covered electrodes are used for SMAW of iron castings. Economic considera-tions and weld requirements determine the appropriate product for each application. Electrodes designed specifically for welding of iron castings are described in American Welding Society (AWS) Specification A5.15, "Welding Rods and Covered Electrodes for Welding Cast Iron."

Covered electrodes utilizing a cast iron rod as the core are classified by AWS as ECI. They are of comparatively low cost and produce a weldment with chemical and mechanical properties similar to those of the base metal.

Austenitic stainless steel electrodes, such as AWS Specification A5.4 ("Specification for Corrosion-Resisting Chromium-Nickel Steel Covered Welding Electrodes") classes E308, E309, E310 and E312, have found some application in welding of iron castings.

Electrodes using steel as a core wire but depositing a cast iron type of deposit are similar in usage to the ECI electrodes.

Nickel-base electrodes have been widely accepted for welding of iron castings. Unlike iron, nickel has a low solubility for carbon in the solid state. Thus, as the weld pool solidifies and cools, carbon is rejected from the solution and precipitates as graphite. This reaction increases the volume of the weld deposit, thereby offsetting shrinkage stresses and lessening the likelihood of fusion-zone cracking.

Electrodes classified as ENi-CI utilize a commercially pure nickel-core wire and thus produce a deposit of high nickel content. Even when highly diluted by the base metal, the deposit remains ductile and machinable. ENiFe-CI electrodes produce a nickel-iron deposit and, as a result, have four distinct advantages over ENi-CI electrodes:

- The deposits are stronger and more ductile; this increased strength level makes the product suitable for welding higher-strength gray and ductile irons, as well as for many dissimilar metal joint applications.
- Nickel-iron deposits are more tolerant of phosphorus than are nickel deposits; thus, ENiFe-CI electrodes are preferred for welding gray iron castings with higher phosphorus contents.
- The coefficient of expansion of the nickel-iron deposit is somewhat less than that of the nickel deposit; thus, nickel-iron products may be used to weld heavier sections and yet avoid fusion-line cracking due to expansion differences.
- ENiFe-CI electrodes are generally lower in cost than ENi-CI electrodes.

PREHEATING

In many cases, preheating is not needed to produce an acceptable weld, but can be used to reduce the thermal gradient and conductivity and to decrease the rate of cooling. Preheating is also useful in reducing differential mass effects, such as those encountered when welds are made between light and heavy sections. A thorough analysis and judicious use of preheating for each shape enable the welder to obtain a uniform cooling rate and thus reduce cooling stresses (Fig. 33). A fundamental preheating rule is illustrated in Fig. 33; the preheat is applied in a manner such that, on cooling, the weld is under compressive stress. Preheating in the areas indicated in Fig. 33(b) expands the metal. Thus, the crack is expanded. After welding, the weld is in compression as the part cools and the metal shrinks (Fig. 33d).

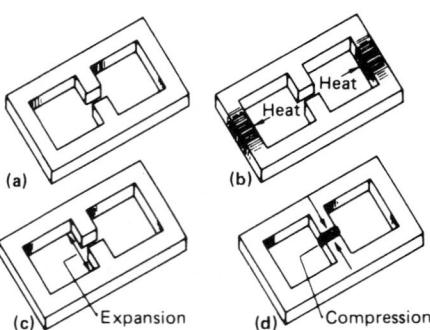

Specific part should be heated in such a manner that the weld is under compressive stress on cooling. See text for discussion.
Fig. 33. Preheating of castings

WELDING PRACTICES

The following general practices have been found to be useful in the welding process:

- When no preheat is used, the interpass temperature should not exceed 200 °F.
- When preheat is used, the interpass temperature should not exceed the preheat temperature by more than 100 °F.
- To minimize welding stresses, a backstep sequence should be used with stringer beads no more than 2 to 3 in. in length; allow each deposit to cool to approximately 100 °F before making subsequent deposits.
- Avoiding melting more of the casting than is necessary.
- Whenever possible, deposit two or more layers for best machinability.
- Always strike the arc in the weld groove, never on the casting.
- The arc length should be kept as short as possible, typically $^1/_8$ to $^3/_{16}$ in.

Arc Welding of Stainless Steels

MOST STAINLESS STEELS that do not contain more than 0.03% sulfur, phosphorus or selenium are considered to be weldable. All weldable stainless steels can be joined by the various arc welding processes. However, variations in composition and physical and mechanical properties affect their relative weldability, as do fabricating conditions and service requirements.

Shielded metal-arc (SMAW), submerged-arc (SAW), gas metal-arc (GMAW), gas tungsten-arc (GTAW), and plasma-arc welding (PAW) are used extensively for joining stainless steels. Flux-cored arc welding (FCAW) is also used, but to a lesser extent. This article discusses the weldability of the various grades and the suitability of arc welding processes for specific conditions and requirements.

AUSTENITIC STAINLESS STEELS

Differences in composition among the standard austenitic stainless steels affect weldability and performance in service. For example, types 302, 304 and 304L differ primarily in carbon content, and consequently there is a difference in the amount of carbide precipitation that can occur in the heat-affected zone (HAZ) after the heating and cooling cycle encountered in welding. Types 303 and 303Se contain 0.20% P maximum plus 0.15% Se or S for free machining. These elements are detrimental to weldability and can cause severe hot cracking in the weld metal. Types 316 and 317 contain molybdenum for increased corrosion resistance and higher creep strength at elevated temperatures. However, unless controlled by extra-low carbon content, as in type 316L, carbide precipitation occurs in the HAZ during welding. Types 318, 321, 347 and 348 are stabilized with titanium, or niobium-plus-tantalum, to prevent intergranular precipitation of chromium carbides when the steels are heated to a temperature in the sensitizing range, as during welding.

Welding Characteristics. The austenitic stainless steels, except for the free-machining grades, are the easiest to weld and produce welded joints that are characterized by a high degree of toughness, even in the as-welded condition.

The high electrical resistivity of stainless steel makes it suitable for welding with low heat inputs. With reduced heat, good penetration and fusion result because low thermal conductivity retains heat in the weld area.

Effect of Carbide Precipitation on Corrosion Resistance of Welded Joints. The precipitation of intergranular chromium carbides is accelerated by an increase in temperature within the sensitizing range and by an increase in time at temperature. When intergranular chromium carbides are precipitated at welded joints, resistance to intergranular corrosion and stress corrosion markedly decreases. The decrease in corrosion resistance is attributed to the presence of the chromium-rich carbides at the grain boundaries and the depletion of chromium in the adjacent matrix material. Although intergranular carbide precipitation generally occurs between 800 and 1600 °F, sensitization is restricted to a narrower range by the fairly rapid heating and cooling that usually occur in welding. The narrower range varies with time at temperature and steel composition, but is approximately 1200 to 1600 °F.

Solution annealing puts carbides back into solution and restores normal corrosion resistance, but is generally inconvenient. The solution annealing temperature range is very high, 1900 °F minimum, and unless stainless steel is protected from air at these temperatures, it oxidizes rapidly, forming adherent oxide scale. Thin sections, unless adequately supported, may sag or be severely distorted at these temperatures or during

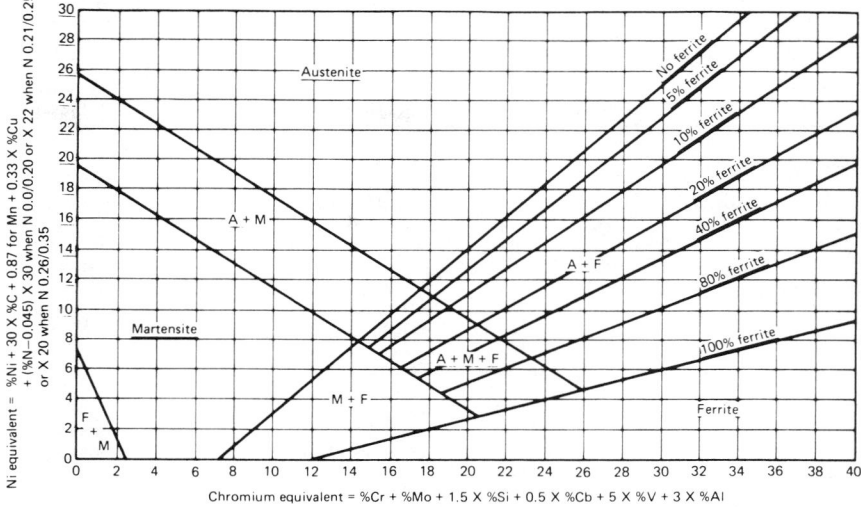

Vanadium, copper and aluminum added. (Source: *Welding Journal*, May 1982, p 152s. Adapted by R. Harry Espy)

Fig. 34. Schaeffler constitution diagram for stainless steel weld metal modified for manganese with nitrogen

rapid cooling from them. Rapid cooling in solution annealing may present other problems. Water quenching, although effective, is seldom feasible except for small workpieces of simple shape. Unless adequate safeguards are available, water quenching of large workpieces from the solution annealing temperature is hazardous. Often solution annealing is impractical because the workpiece is too large for available furnace and cooling facilities.

Extra-low-carbon stainless steels (types 304L and 316L) are resistant to carbide precipitation in the 800 to 1600 °F range and can thus undergo normal welding without reduction in corrosion resistance. Carbides precipitate in significant quantities when extra-low-carbon steels are heated and held in the sensitizing temperature range for an extended period, as in service. These steels are generally recommended for use below 800 °F.

Stabilized Steels. Compared with the extra-low-carbon steels, the stabilized steels exhibit higher strength at elevated temperature. For service in a corrosive environment in the sensitizing temperature range of 800 to 1600 °F, an austenitic steel stabilized with niobium-plus-tantalum or with titanium is needed. The filler metal used for welding should also be of a stabilized composition.

Microfissuring in Welded Joints. Interdendritic cracking in the weld area that occurs before the weld cools to room temperature is known as hot cracking or microfissuring. The occurrence of microfissuring is related to the:

- Microstructure of the weld metal as solidified
- Composition of the weld metal, especially the content of certain residual or trace elements
- Amount of stress developed in the weld as it cools
- Ductility of the weld metal at high temperatures
- Presence of notches.

Microfissuring can be prevented or minimized by proper control of ferrite in the weld metal. Wide use has been made of the modified Schaeffler diagram (Fig. 34) to determine the approximate amount of ferrite that will be obtained in

the austenitic weld metal of a given composition.

Selection of Filler Metals. Electrodes and welding rods suitable for use as filler metal in welding of austenitic stainless steels are shown in Table 24. These filler metals, with AWS standard composition specifications, are for GMAW, SAW and SMAW. The notes in this table should be carefully studied because selection of filler metals for welding austenitic stainless steels requires consideration of the microstructural constituents of the as-deposited weld metal. Ultimately, these microstructural constituents determine the mechanical properties, crack sensitivity and corrosion resistance of the weld. The constituents of principal concern are austenite, delta ferrite and precipitated carbides.

Some filler metals, such as types 310, 310Cb, 310Mo and 330, invariably deposit a fully austenitic weld metal. In these alloys, the ratio of ferrite formers to austenite formers cannot, within permissible limits, be raised high enough to produce any delta ferrite in the austenite. Consequently, when these filler metals are applied to restrained joints or to base metals containing additions of elements such as phosphorus, sulfur, selenium and silicon, only procedures proved suitable by experience should be used.

The compositions of most filler metals are adjusted by the manufacturers to produce weld deposits that have ferrite-containing microstructures. Thus, ferrite-forming elements, such as chromium and molybdenum, are maintained at the high side of their allowable ranges, and austenite-forming elements, such as nickel, are kept low. The amount of ferrite in the structure of the weld metal depends on the ratio or balance of these elements. At least 3 or 4 FN delta ferrite is needed in the as-deposited weld metal for effective suppression of hot cracking.

Preheating. In general, no benefit is derived from preheating of austenitic stainless steels. In some applications, preheating can increase carbide precipitation, cause shape distortion of the workpiece, or increase hot-cracking tendencies.

Postweld Stress Relieving. Although the effects of residual stress from welding on the properties of austenitic stainless steels are limited in compar-

Table 24. Electrodes or welding rods used as filler metals in arc welding of stainless steels

Type of steel welded	Condition of weldment for service(a)	Electrode or welding rod(b)	Type of steel welded	Condition of weldment for service(a)	Electrode or welding rod(b)
Austenitic steels			**Martensitic steels**		
301, 302, 304, 305,			403, 410, 416,..		
308(c) ....1 or 2		308	416Se(k) ....2 or 3		410
302B(d)...................1		309	403, 410(m)................1		308, 309, 310
304L ....1 or 4		347, 308L	416, 416Se(m)..............1		308, 309, 312
303, 303Se(e) ....1 or 2		312	420(n)...................2 or 3		420
309, 309S 1		309	431(n)...................2 or 3		410
310, 310S 1		310	431(p) 1		308, 309, 310
316(f) ....1 or 2		316			
316L(f)...............1 or 4		318, 316L	**Ferritic steels**		
317(f) ....1 or 2		317			
317L(f)................1 or 4		317Cb	405(q) 2		405Cb, 430
318, 316Cb(f)...........1 or 4		318	405, 430(m)................1		308, 309, 310
321(g)................1 or 5		347	430F, 430FSe(m)...........1		308, 309, 312
347(h)................1 or 5		347	430, 430F, 430FSe(r) 2		430
348(j)................1 or 5		347	446 2		446
			446(s)...................1		308, 309, 310

(a) 1, as welded; 2, annealed; 3, hardened and stress relieved; 4, stress relieved; 5, stabilized and stress relieved. (b) Prefix E or ER omitted. (c) Type 308 weld metal is also referred to as 18-8 and 19-9 composition. Actual weld analysis requirements are: 0.08% C max, 19.0% Cr min, and 9.0% Ni min. (d) Type 310 (1.50% Si max) may be used as filler metal, but the pickup of silicon from the base metal may result in weld hot cracking. (e) Free-machining base metal will increase the probability of hot cracking of the weld metal. Type 312 filler metal provides weld deposits that contain a large amount of ferrite to prevent hot cracking. (f) Welds made with types 316, 316L, 317, and 317Cb electrodes or welding rods may occasionally display poor corrosion resistance in the as-welded condition. In such cases, corrosion resistance of the weld metal may be restored by the following heat treatments: for types 316 and 317 base metal, full anneal at 1950 to 2050 °F; for types 316L and 317L base metal, 1600 °F stress relief; for type 318 base metal, 1600 to 1650 °F stabilizing treatment. Where postweld heat treatment is not possible, other filler metals may be specially selected to meet the requirements of the application for corrosion resistance. (g) Type 321 covered electrodes are not regularly manufactured, because titanium is not readily recovered during deposition. (h) Caution is needed in welding thick sections, because of cracking problems in heat-affected zones. (j) In base metal and weld metal, for nuclear service, tantalum is restricted to 0.10% max and cobalt to 0.20% max. (k) Annealing softens and imparts ductility to heat-affected zones and weld. Weld metal responds to heat treatment in a manner similar to the base metal. (m) These austenitic weld metals are soft and ductile in as-welded condition, but the heat-affected zone will have limited ductility. (n) Requires careful preheating and postweld heat treatment to avoid cracking. (p) Requires careful preheating. Service in as-welded condition requires consideration of hardened heat-affected zones. (q) Annealing increases ductility of heat-affected zones and weld metal. Type 405 weld metal contains niobium, rather than aluminum, to reduce hardening. (r) Annealing is employed to increase ductility of the welded joint. (s) Type 308 filler metal will not display scaling resistance equal to that of the base metal. Consideration must be given to differences in the coefficients of thermal expansion of the base metal and the weld metal.
Source: George E. Linnert, Welding Characteristics of Stainless Steels, *Metals Engineering Quarterly*, Nov 1967

ison with the effects of cold working, residual stress may significantly affect mechanical properties. Because the effective yield strength varies from point to point, the application of further stresses at later stages of fabrication can cause excessive distortion and even premature failure. Nonuniform heating, which relieves some local residual stress, may also contribute to distortion. For these reasons, stress relieving may be required to ensure dimensional stability.

Stress relieving can be performed over a wide range of temperatures, depending on the amount of relaxation required. Time at temperature ranges from about 1 h per inch of section thickness at temperatures above 1200 °F to 4 h per inch of section thickness at temperatures below 1200 °F. Because of the high coefficient of expansion and the low thermal conductivity of austenitic stainless steels, cooling from the stress-relieving temperature must be slow. The stress-relieving temperature selected must be compatible with the extent of carbide precipitation acceptable and with the corrosion resistance desired. Nonstabilized stainless steels cannot be stress relieved in the sensitizing temperature range without sacrifice of corrosion resistance. Extra-low-carbon stainless steels are affected much less, because carbide precipitation in these steels is sluggish. Stabilized stainless steels exhibit minimal chromium carbide precipitation tendencies.

Shielded Metal-Arc Welding

Electrodes are available in diameters ranging from $1/16$ to $1/4$ in. If operating amperages are not given by the manufacturer, the suggested values in Table 25 should be used. Electrodes of the

Table 25. Suggested operating parameters when electrode guidelines are not available
Electrode designations, all classifications; suffixes –15 and –16.

Electrode size	Average arc current, A	Maximum arc voltage, V
$1/16$	35-45	24
$5/64$	45-55	24
$3/32$	65-80	24
$1/8$	90-110	25
$5/32$	120-140	26
$3/16$	160-180	27
$1/4$	220-240	28

Source: AWS A5.4-81

–15 and –16 types up to $5/32$ in. in diameter can be used in all welding positions; electrodes $3/16$ in. in diameter and larger should be used in the flat and horizontal fillet positions only.

For welding special grades of stainless steel, composite electrodes of special composition may be necessary. For SMAW, a composite electrode consists of a flux-covered carbon steel tube containing the alloying elements, in powder form, in the core. These electrodes also are available for depositing the standard compositions of stainless steel.

Welding Procedure. The most significant difference in the procedure used for welding stainless steel as opposed to plain low-carbon steel is that less welding heat is required. The workpieces should be carefully prepared and fitted. Thin-gage stainless steel, in particular, should be properly clamped and held in alignment to reduce buckling. Large electrodes of more than $1/4$ in. diam

and excessive arc length contribute to loss of chromium in the weld deposit. Excessive weaving of electrodes of any size should be avoided. Maximum width of weave should be limited to four times the core wire diameter. The stringer-bead technique is generally recommended for depositing weld metal.

Special care is required between weld passes to remove all slag from the deposited bead. Only stainless steel wire brushes and tools should be used for this purpose.

Austenitic stainless steels should be welded in the annealed condition. The stabilized grades, such as types 321 and 347, should be in the stabilized annealed condition, while the nonstabilized grades should be in the solution annealed condition. Because austenitic steels are not hardenable by heat temperature, preheating is not used. Their mechanical properties are not substantially changed by welding, although stainless steel that has been purposely work hardened to increase its strength softens in the HAZ.

Gas Metal-Arc Welding

Shielding Gases. Although the range of choice of shielding gas for welding stainless steels is considerably narrower than for carbon and low-alloy steels, several gas mixtures have proved satisfactory. The mode of metal transfer influences the choice of shielding gas. For instance, with spray-arc or pulsed-arc transfer, a shielding-gas mixture containing 99% argon and 1% oxygen has been widely used and generally recommended. In some plants, 98% argon and 2% oxygen is used with success. For short-circuiting transfer, a mixture of 90% helium, 7.5% argon and 2.5% carbon dioxide has been extensively used for shielding, but helium is gradually losing favor because of its high cost. However, a mixture in which the above proportions of helium and argon have been reversed (90% argon, 7.5% helium, 2.5% carbon dioxide) has proved successful for short-circuiting transfer. Regardless of other variations, the shielding gas for GMAW of stainless steel should contain at least 97.5% inert gas (argon or helium, or a mixture of the two). When carbon dioxide is used, the maximum is usually 2.5% to retain weld quality and corrosion resistance.

NITROGEN-STRENGTHENED AUSTENITIC STAINLESS STEELS

This family of stainless steels offers two specific advantages over the conventional austenitic stainless steels: (1) increased strength at all temperatures, cryogenic through elevated; and (2) improved resistance to pitting corrosion. They differ from the conventional austenitic stainless steels in that manganese has been substituted for all or part of the nickel, thus allowing greater amounts of nitrogen to be dissolved in the matrix of the alloy. The nitrogen acts as a solid solution strengthener and increases the annealed yield strength to approximately twice that of the conventional austenitic stainless steels.

Welding characteristics of the conventional austenitic stainless steels also apply to the nitrogen-strengthened austenitic stainless steels. The measures taken to prevent unwanted changes in composition in the conventional austenitic stainless steels should be taken with the nitrogen-strengthened steels. Specific attention should be paid to the control of nitrogen.

The composition of the weld deposit must be

maintained through proper welding procedures to ensure that the delta ferrite content is constant and weld hot cracking is avoided.

In Fig. 34, for average stainless steel welds ($^5/_{32}$-in.-diam electrode, SMAW process), including the modifications (up to 15% Mn and 0.35% N), the percent of ferrite (up to ≈30%) in a matrix of austenite, martensite, or austenite and martensite can be predicted within ≈4%.

FERRITIC STAINLESS STEELS

The ferritic stainless steels are generally less weldable than the austenitic stainless steels and produce welded joints having lower toughness because of grain coarsening that occurs at the high welding temperatures. The standard ferritic stainless steels are: (1) type 446 (25% Cr); (2) types 430, 430F and 430F-Se (17% Cr); and (3) types 405 and 409 (13% Cr). Type 409 is ferritic because it has a low carbon content (0.08% maximum) and a minimum titanium content equal to six times the carbon content. Type 405, which also contains only 0.08% maximum C, contains an average of 0.20% Al, which promotes ferrite formation.

Effect of Welding Heat on Ductility and Grain Size. Although most ferritic stainless steels have compositions that ensure a ductile ferritic structure at room temperature, variations in composition within the standard composition limits can result in the formation of small amounts of austenite during heating to elevated temperature. On cooling, the austenite transforms to martensite, resulting in a duplex structure of ferrite and a small amount of martensite. The martensite reduces both ductility and toughness of the steel. Annealing transforms the martensite and restores normal ferritic properties, but annealing increases cost and can result in an excessive amount of distortion, particularly in parts that were previously formed by a cold working process.

Preheating. The recommended preheating temperature range for ferritic stainless steels is 300 to 450 °F. The need for preheating is determined largely by the composition, mechanical properties and thickness of the steel being welded. Steels less than $^1/_4$ in. thick are much less likely to crack during welding than those more than $^1/_4$ in. thick. The type of joint, joint location, restraints imposed by clamping and jigging, the welding process, and the rate of cooling from the welding temperature can also affect weld cracking.

Postweld Annealing. The temperature range for postheating or postweld annealing of ferritic stainless steels is 1450 to 1550 °F, which is safely below the temperatures for austenite formation and grain coarsening. Annealing transforms a mixed structure into a wholly ferritic structure and restores the mechanical properties and corrosion resistance that may have been adversely encountered in welding. Thus, except for its inability to refine coarsened ferrite grains, annealing is generally beneficial.

Selection of Filler Metal. As shown in Table 24, both ferritic and austenitic stainless steel filler metals are used in arc welding of ferritic stainless steel. Ferritic stainless steel filler metals offer the advantages of having the same color and appearance, the same coefficient of thermal expansion, and essentially the same corrosion resistance as the base metal. However, austenitic stainless steel filler metals are often used to obtain more ductile weld metal in the as-welded condition. Designations for ferritic stainless steel

filler metals are covered in AWS specification A5.9-81 and are listed in Table 26.

MARTENSITIC STAINLESS STEELS

The standard martensitic stainless steels are types 403, 410, 414, 416, 416Se, 420, 431, 440A, 440B and 440C. These steels derive their corrosion resistance from chromium, which they contain in proportions ranging from 11.5 to 18%. Martensitic stainless steels are the most difficult stainless steels to weld because they are chemically balanced to become harder, stronger and less ductile through thermal treatment. These same metallurgical changes occur from the heat of welding. As a result of welding, these changes are restricted to the weld area and are not uniform over the entire section. The nonuniform metallurgical condition of the part makes it susceptible to cracking when subjected to the high stresses from welding. Increasing carbon content in martensitic stainless steels generally results in increased hardness and reduced ductility. Thus, the three type 440 stainless steels are seldom considered for applications that require welding, and filler metals of the type 440 compositions are not readily available.

Preheating and Postweld Heat Treating. The usual preheating temperature range of martensitic steels is 400 to 600 °F. Carbon content of the steel is the most important factor in determining whether or not preheating is necessary. On the basis of carbon content alone, a steel containing not more than 0.10% C seldom requires preheating, and one with more than 0.10% C requires preheating to prevent cracking. Other factors that determine the need for preheating are the mass of the joint, degree of restraint, presence of a notch effect, and the composition of the filler metal. The following can be used to correlate preheating and postweld heat treating practice with carbon contents and welding characteristics of martensitic stainless steels.

- *Carbon below 0.10%*: Neither preheating nor postweld annealing generally is required; steels with carbon contents this low are not standard.
- *Carbon 0.10 to 0.20%*: Preheat to 500 °F; weld at this temperature; cool slowly.
- *Carbon 0.20 to 0.50%*: Preheat to 500 °F; weld at this temperature; anneal.
- *Carbon over 0.50%*: Preheat to 500 °F; weld with high heat input; anneal.

If the weldment is to be hardened and tempered immediately after welding, annealing may be omitted. Otherwise, the weldment should be annealed immediately after welding, without cooling to room temperature.

Martensitic Precipitation-Hardening Steels

These steels have a predominantly austenitic structure at the solution-annealing temperature of approximately 1900 to 1950 °F, but they undergo an austenite-to-martensite transformation when cooled to room temperature.

These steels can be readily welded. The welding procedures resemble those ordinarily used for the 300 series stainless steels, despite differences in composition and structure between the two classes. The formation of martensite, which occurs during cooling from elevated temperatures as in welding, does not result in full hardening. These steels are not sensitive to cracking and do not require preheating.

Semiaustenitic Precipitation-Hardening Steels

Unlike martensitic PH steels, semiaustenitic PH steels are soft enough in the annealed condition to permit cold working. When cooled rapidly from the annealing temperature to room temperature, they retain their austenitic structure, which displays good toughness and ductility in cold forming operations. The M_s temperatures for these steels are well below room temperature, but they vary depending on composition and annealing temperature.

The semiaustenitic PH steels are normally welded in the annealed condition. The tough austenitic structure imparts welding characteristics similar to those of 300 series stainless steels. The semiaustenitic PH steels are not susceptible to cracking when welded, even when welded after transformation to martensite, because the low-carbon martensite developed is not of high hardness or low ductility. Also, cold cracking does not occur in the base metal adjacent to the weld because the HAZ is austenitized during welding and remains substantially austenitic as the joint cools to room temperature.

Austenitic Precipitation-Hardening Steels

The alloy content of these steels is high enough to maintain an austenitic structure after annealing and after any aging or hardening treatment. The precipitation-hardening phase is soluble at the annealing temperature of 2000 to 2050 °F, and it remains in solution during rapid cooling from the annealing temperature. When these steels are reheated to about 1200 to 1400 °F, precipitation occurs, and the hardness and strength of the austenitic structure increase. The hardness attained is lower than that of the martensitic or semiaustenitic PH steels, but the nonmagnetic properties are retained. Although the austenitic PH steels remain austenitic during all phases of forming, welding and heat treatment, some contain alloying elements (for precipitation-hardening purposes) that greatly affect behavior in welding.

Welding Techniques for Precipitation-Hardening Stainless Steels. The precipitation-hardening stainless steels can be welded using the arc welding techniques described under the section on austenitic stainless steels in this article. The major difference is that these steels are usually heat treated after welding to achieve the required mechanical properties, which is usually unnecessary with austenitic stainless steels. Precipitation-hardening stainless steels may be welded with matched or dissimilar filler metals, or without filler metals, as is the case with most stainless steels. There

Table 26. Chemical compositions for ferritic stainless steel electrodes for SAW and GMAW
All are maximum percentages, unless otherwise noted.

AWS classification	C	Cr	Ni	Mo	Mn	Si	P	S	N	Cu
ER430	0.10	15.5-17.0	0.6	0.75	0.6	0.50	0.03	0.03	⋯	0.75
ER26-1	0.01	25.0-27.5	(a)	0.75-1.50	0.40	0.40	0.02	0.02	0.015	0.20(a)

(a) Nickel, max = 0.5% − Cu
Source: AWS A5.9-81

is a wide variety of hardenable filler metals available for these PH grades through the consumable manufacturer.

Arc Welding of Heat-Resistant Alloys

HEAT-RESISTANT ALLOYS considered in this article include nickel-base, iron-base and cobalt-base alloys. The procedures used in welding heat-resistant alloys depend to some extent on the mechanism by which they are strengthened for high-temperature service — primarily solid-solution strengthening or precipitation hardening.

WELDING PROCESSES

Heat-resistant alloys can be welded by all arc welding processes. Gas tungsten-arc welding (GTAW) is widely used, especially for joining thin sections. In general, shielded metal-arc welding (SMAW) and gas metal-arc welding (GMAW) are used in joining sections more than $1/4$ in. thick, where the heat input does not adversely affect the weld metal or the base metal. Submerged-arc welding (SAW) generally is used in high-volume production welding of sections more than 1 in. thick.

The data in Table 27 are intended to serve as starting points for the establishment of machine settings for GTAW. These conditions are, in general, suitable for welding nickel, iron-nickel-chromium, iron-chromium-nickel and cobalt-base heat-resistant alloys, when making butt, corner or T-joints, using an appropriate groove design based on stock thickness and application. An increase in welding current of 10 to 20 A may be needed for melt-through T-joints. Generally, the interpass temperature should range from 200 to 350 °F, depending on the alloy. Oscillation of the welding torch may help to prevent cracking by changing the solidification pattern. This may also improve the appearance of the weld.

CLEANING OF WORKPIECES

The weldability of heat-resistant alloys is markedly affected by the cleanness of the base metal and the filler metal. Shop dirt, paint, grease, oil, machine lubricants, processing chemicals, temperature-indicating sticks, marking crayons, oxide films, and scale are the main surface contaminants. Sulfur and lead in foreign material on the workpiece surface can diffuse into the base metal when it is heated and result in severe cracking.

NICKEL-BASE HEAT-RESISTANT ALLOYS

The nickel-base alloys are commonly welded by GTAW, SMAW and GMAW. Frequently, a root pass is made by GTAW and the subsequent passes by GMAW. Submerged-arc welding can be used on certain alloys, but the welding flux must be carefully selected to obtain adequate protection and provide correct elemental additions to the weld pool. The welding conditions chosen must avoid excessive heat input. In welding of metal more than 3 in. thick, shrinkage stresses decrease ductility slightly, and a postweld stress-relieving treatment may be necessary. The manufacturer of the alloy should be consulted for specific details.

Table 27. Conditions for GTAW of heat-resistant alloys(a)

Base-metal thickness, in.	Diameter of filler metal, in.(b)	Electrode diameter, in.(c)	Shielding gas		Welding current, A(d)
			Gas	Flow rate, f³/h	
0.010	0.020	0.040-0.060	Ar	12-15	10-15
0.020	0.030	0.060	Ar	12-15	15-25
0.030	0.030; 0.045	0.060	Ar	12-15	25-35
0.045	0.045	0.060	Ar	12-15	40-50
0.050	0.045	0.060	Ar	12-15	45-55
0.060	0.045	0.060	Ar	12-15	55-65
0.080	0.060	0.060	Ar	12-15	75-85
0.100	0.060; 0.090	0.093	Ar or He	12-20	95-105
0.125	0.060; 0.090	0.093	Ar or He	12-20	110-135
0.250	0.060; 0.090	0.093	Ar or He	12-20	130-200

(a) The data in this table are intended to serve as starting points for the establishment of optimum machine settings for welding workpieces on which previous experience is lacking. The data are subject to adjustment as necessary to meet the special requirements of individual applications. Torch nozzle diameter was $7/16$ in.; nozzle had a gas lens. (b) Minimum wire diameters were applicable. (c) EWTh-2 electrodes. (d) DCEN with high-frequency arc starting. An increase of 10 to 20 A may be needed for melt-through T-joints.

Precipitation-hardenable alloys are susceptible to cracking in the weld metal or in the heat-affected zone (HAZ) unless they are properly heat treated before and after welding. These alloys are normally welded in the solution-annealed condition. If they are welded in the precipitation-hardened condition, a solution anneal is required prior to high-temperature service. In solution annealing, a rapid heating rate should be used to avoid parent-metal strain-age cracking. This usually can be accomplished by charging into a hot furnace.

Preweld and Postweld Heat and Mechanical Treatments. The solid-solution (non-age-hardenable) alloys are welded in both the annealed and moderately cold worked conditions. The precipitation-hardenable alloys usually are welded in the solution-treated condition.

Weldments made of solid-solution alloys can be used as welded or after stress relieving, depending on the alloy and application. Stress relieving in the range from 800 to 1600 °F, depending on the alloy and its condition, can be used to reduce or remove stresses in work-hardened solid-solution alloys without producing a recrystallized grain structure. A low-temperature stress-equalizing heat treatment of 600 to 800 °F can be used to redistribute stresses without appreciably decreasing the mechanical strength produced by the previous cold working.

Precipitation-hardenable alloys are given a solution treatment after welding to relieve residual stresses, and then they are hardened by an aging treatment.

Filler metals used with nickel-base heat-resistant alloys usually have the same general composition as the alloy being welded. Because of high arc currents and high welding temperatures, compositions of filler metals are often modified to resist porosity and hot cracking of the weld metal. Tack welding and root-pass welding without filler metal are permissible for some alloys. However, care must be taken to avoid centerline splitting and crater cracking when no filler metal is used. To minimize cracking, concave welds should be avoided. Table 28 gives the compositions of filler metals commonly used in GTAW; several of these filler metals are used for welding metals other than nickel-base alloys.

For welding the precipitation-hardenable nickel-base alloys, either a precipitation-hardenable or a solid-solution filler metal may be used, depending on service requirements.

IRON-NICKEL-CHROMIUM AND IRON-CHROMIUM-NICKEL HEAT-RESISTANT ALLOYS

Iron-base heat-resistant alloys include strain-hardenable, solid-solution-strengthened and precipitation-hardenable types. All contain appreciable amounts of nickel and chromium, with either one or the other of these elements constituting the principal alloying addition. Other alloying elements generally are added to increase high-temperature strength (molybdenum, tungsten and cobalt), to act as stabilizers (niobium and tantalum), or to promote strengthening (aluminum, titanium, copper and boron).

The usual range of service temperature for these alloys, 1200 to 1400 °F, limits the selection of filler metals and preheat and postheat treatments for welding.

The 16-25-6 and 19-9 DL alloys are easily joined by arc welding. Weld deposits can be made with an austenitic stainless steel filler metal, with a nickel-base alloy filler metal, or with a filler metal of the same composition as the base metal. Generally, preheating and postheating are used.

The solid-solution-strengthened alloys, such as N-155, are weldable by SMAW, GMAW and GTAW; however, the heat input should be kept low, and welds should be cooled rapidly to maintain ductility.

Some precipitation-hardenable alloys, such as A-286, are considerably more difficult to weld. These alloys are extremely sensitive to hot cracking in the weld metal and in the HAZ. Cracking is most likely to occur when aged metal or highly restrained parts are joined. Cracks in root passes or crater cracks can be minimized by using suitable welding procedures and techniques to control heat input during welding.

COBALT-BASE HEAT-RESISTANT ALLOYS

Cobalt-base heat-resistant alloys are available in both cast and wrought forms. Generally, cast alloys are more difficult to weld than wrought alloys. Where the application requires very high-reliability welds, only GTAW and GMAW are recommended; otherwise, SMAW is used.

Joint design and weld grooves for cobalt-base alloys are essentially the same as for nickel-base alloys. A square-groove butt joint is used for sheet metal up to about $7/64$ or $1/8$ in. thick, a V-groove for plate up to $3/8$ in. thick, a double-V-groove or a double-U-groove for thicknesses of $3/8$ to $5/8$ in., and a double-U-groove for thicknesses over $5/8$ in. Where T-joints are used, the same groove

Table 28. Compositions of filler metals and electrode wires for arc welding of heat-resistant alloys

AWS classification or trade name	C	Mn	Fe	S	Si	Cu	Ni(a)	Composition, % Co	Al	Ti	Cr	Nb + Ta	Mo	Other
Nickel-based bare electrodes for GTAW and GMAW														
ERNiCr-3	0.10	2.5-3.5	3.0	0.015	0.50	0.50	67 min	(b)	···	0.75	18.0-22.0	2.0-3.0(c)	···	0.50
ERNiCrFe-5	0.08	1.0	6.0-10.0	0.015	0.35	0.50	67 min	···	···	···	14.0-17.0	1.5-3.0	···	1.0
ERNiCrFe-6	0.08	2.0-2.7	10.0	0.015	0.35	0.50	67 min	···	···	2.5-3.5	14.0-17.0	···	···	0.50
ERNiCrFe-7	0.08	1.0	5.0-9.0	0.01	0.50	0.50	70 min	···	0.40-1.00	2.00-2.75	14.0-17.0	0.70-1.20	···	0.50
ERNiCrMo-3	0.10	0.5	5.0	0.015	0.5	···	rem	1.0	0.4	0.4	20.0-23.0	3.15-4.15	8.0-10.0	···
GMR-235	0.16	0.25	9.0-11.0	0.03	0.6	···	rem	2.5	1.75-2.25	2.25-2.75	14.0-17.0	···	4.5-6.5	0.009 B
ERNiCrMo-2	0.05-0.15	1.0	17.0-20.0	0.03	1.0	···	rem	0.5-2.5	···	···	20.5-23.0	···	8.0-10.0	0.2-1.0 W
Hastelloy S	0.01	0.2	1.0	0.005	0.20	···	67	···	0.2	···	15.5	···	15.5	0.009 B, 0.02 La
ERNiCrMo-7	0.007	0.50	1.5	0.005	0.04	···	65	1.0	···	···	16	···	15.5	···
Haynes 556	0.10	1.5	···	0.005	0.40	···	20	20	0.3	···	22	0.1	3	0.9 Ta, 0.2 N, 2.5 W
ERNiCrMo-4	0.01	0.5	5.5	0.005	0.04	···	62	1.2	···	···	16	···	16	3.5 W, 0.35 V
Inconel 601	0.05	0.5	14.1	0.007	0.25	0.25	60.5	···	1.35	···	23.0	···	···	···
Inconel 617	0.07	0.02	0.4	0.005	0.14	···	54	12.5	1.0	0.24	22	···	9	···
Inconel 718	0.08	0.35	rem	0.015	0.35	0.3	50-55	1.0	0.2-0.8	0.65-1.15	17.0-21.0	4.75-5.5	2.8-5.5	(d)
René 41 (AMS 5800)	0.12	0.1	5.0	0.015	0.5	···	rem	10.0-12.0	1.4-1.6	3.0-3.3	18.0-20.0	···	9.0-10.5	(e)
Waspaloy (AMS 5828C)	0.07	0.10	0.75	···	0.1	···	rem	13.5	1.4	3.0	19.75	···	4.45	(f)
Nickel-based covered electrodes for SMAW														
ENiCrFe-1	0.08	1.5	11.0	0.015	0.75	0.50	68 min(a)	···	···	···	13.0-17.0	1.5-4.0	···	0.50
ENiCrFe-2	0.10	1.0-3.5	6.0-12.0	0.020	0.75	0.50	rem	···	···	···	13.0-17.0	0.5-3.0	0.5-2.5	0.50
ENiCrFe-3	0.10	5.0-9.5	6.0-10.0	0.015	1.0	0.50	rem	(g)	···	1.0	13.0-17.0	1.0-2.5(h)	···	0.50
EniMo-1	0.12	1.0	4.0-7.0	0.030	1.0	···	rem	2.5	···	···	1.0	···	26.0-30.0	(j)
ENiMo-3	0.12	1.0	4.0-7.0	0.030	1.0	···	rem	2.5	···	···	2.5-5.5	···	23.0-27.0	(j)
ENiCrMo-3	0.10	0.5	5.0	0.015	0.50	···	rem	1.0(a)	0.40	0.40	20.0-23.0	3.15-4.15	8.0-10.0	···
ENiCrMo-2	0.10	0.5	18.5	0.005	0.5	···	47	1.5	···	···	22	···	9	0.005 B
ENiCrMo-7	0.007	0.5	1.5	0.005	0.10	···	65	1.0	···	···	16	···	15.5	···
ENiCrMo-4	0.01	0.5	5.5	0.005	0.04	···	62	1.2	···	···	16	···	16	3.5 W, 0.35 V
Inconel 117	0.01	0.6-1.4	1.7	0.008	0.50	0.20	52	12.0	0.2	···	23.5	0-0.5	9.0	···
Iron-nickel-chromium, iron-chromium-nickel, and cobalt-based heat-resistant alloy filler metals														
19-9 W (AMS 5782)	0.07-0.13	1.00-2.00	rem	0.030	1.00	0.50	8.00-9.50	···	···	0.10-0.30	19.0-22.0	1.00-1.30	0.35-0.65	(k)
Multimet (N-155) (AMS 5794)	0.1	1.00-2.00	rem	0.030	1.00	···	19.00-21.00	18.5-21.0	···	···	20.0-22.5	0.75-1.25	2.5-3.5	(m)
A-286 (AMS 5804)	0.04-0.05	1.25-1.35	rem	0.008	0.70	···	25	···	0.24-0.32	2.2	15	0.10-0.12	1.25	(n)
HS-25 or L-605 (AMS 5796)	0.10	1.5	3 max	···	10 max	···	10	rem	···	···	20	···	···	15 W
Haynes 188	0.10	0.6	1.5	0.005	0.35	···	22	39	···	···	22	···	···	14.5 W, 0.04 La

(a) Contains incidental cobalt. (b) Cobalt, 0.10% max, when specified. (c) Tantalum, 0.30% max, when specified. (d) Phosphorus, 0.015%; boron, 0.006%. (e) Boron, 0.01%; total of other elements, 0.003%. (f) Boron, 0.005%; zinc, 0.04%. (g) Cobalt, 0.12% max, when specified. (h) Tantalum, 0.30% max, when specified. (j) Vanadium, 0.60%; phosphorus, 0.04%; total of other elements, 0.50%. (k) Phosphorus, 0.04% max; tungsten, 1.25 to 1.75%. (m) Phosphorus, 0.040% max; tungsten, 2.00 to 3.00%. (n) Phosphorus, 0.02% max; boron, 0.0015 to 0.0022%

limitations apply as for butt joints. Corner joint welds should be backed by fillet welds if possible. This type of joint should be avoided where high stresses are likely to occur. V-grooves should have a 60° groove angle for GTAW.

Arc Welding of Aluminum Alloys

GAS METAL-ARC WELDING (GMAW) and gas tungsten-arc welding (GTAW) have almost entirely replaced other arc welding processes for joining aluminum alloys. These gas shielded-arc welding processes result in optimum weld quality and minimum distortion, and they require no flux.

BASE METALS

Most aluminum alloys can be joined by either GMAW or GTAW, and the weldabilities of aluminum alloys are essentially the same for both processes. The most common alloys are grouped by weldability rating as follows:

Readily weldable

- *Wrought alloys*: Pure aluminum, 1350, 1060, 1100, 2219, 3003, 3004, 5005, 5050, 5052, 5083, 5086, 5154, 5254, 5454, 5456, 5652, 6010, 6061, 6063, 6101, 6151, 7005, 7039
- *Casting alloys*: 356.0, 443.0, 413.0, 514.0, A514.0

Weldable in most applications

- *Wrought alloys*: 2014, 2036, 2038, 4032
- *Casting alloys*: 208.0, 308.0, 319.0, 333.0, 355.0, C355.0, 511.0, 512.0, 710.0, 711.0, 712.0

Limited weldability

- *Wrought alloys*: 2024
- *Casting alloys*: 222.0, 238.0, 295.0, 296.0, 520.0

Welding not recommended

- *Wrought alloys*: 7021, 7029, 7050, 7075, 7079, 7129, 7150, 7178, 7475
- *Casting alloys*: 242.0

FILLER METALS

Classifications and compositions of filler metals for GMAW and GTAW of aluminum alloys are given in Table 29. In addition, filler metals having the same composition as the base-metal alloy are often used for repairing casting defects.

Selection of Filler Metal. Common criteria to be considered in selecting a filler metal are ease of welding, strength, ductility, corrosion resistance of the filler metal/base metal combination, color match with the base metal after anodizing, and service at elevated temperature. The filler metals listed in Table 29 have been developed to satisfy these requirements.

JOINT DESIGN AND EDGE PREPARATION

In general, joint design for aluminum alloys is similar to that for steel. Some recommended butt-joint designs for GMAW by direct current electrode positive (DCEP; reverse polarity) and GTAW with alternating current are shown in Fig. 35. When direct current electrode negative (DCEN; straight polarity) GTAW is used, the root

Table 29. Chemical composition requirements of filler metals for GMAW and GTAW of aluminum alloys

AWS classification	Composition(a), %								Other elements(b)		
	Silicon	Iron	Copper	Manganese	Magnesium	Chromium	Zinc	Titanium	Each(c)	Total	Aluminum
ER1100 . . . .(d)	(d)	0.05-0.20	0.05	···	···	0.10	···	0.05	0.15	99.00 min(c)
ER2319(e) 0.20	0.30	5.6-6.8	0.20-0.40	0.02	···	0.10	0.10-0.20	0.05	0.15	rem
ER4043 . . . . .4.5-6.0	0.8	0.30	0.05	0.05	···	0.10	0.20	0.05	0.15	rem
ER4047 . . . . .11.0-13.0	0.8	0.30	0.15	0.10	···	0.20	···	0.05	0.15	rem
ER4145 . . . . .9.3-10.7	0.8	3.3-4.7	0.15	0.15	0.15	0.20	···	0.05	0.15	rem
ER5183 . . . . .0.40	0.40	0.10	0.50-1.0	4.3-5.2	0.05-0.25	0.25	0.15	0.05	0.15	rem
ER5356 . . . . .0.25	0.40	0.10	0.05-0.20	4.5-5.5	0.05-0.20	0.10	0.06-0.20	0.05	0.15	rem
ER5554 . . . . .0.25	0.40	0.10	0.50-1.0	2.4-3.0	0.05-0.20	0.25	0.05-0.20	0.05	0.15	rem
ER5556 . . . . .0.25	0.40	0.10	0.50-1.0	4.7-5.5	0.05-0.20	0.25	0.05-0.20	0.05	0.15	rem
ER5654 . . . .(f)	(f)	0.05	0.01	3.1-3.9	0.15-0.35	0.20	0.05-0.15	0.05	0.15	rem
R242.0(g)(h) 0.7	1.0	3.5-4.5	0.35	1.2-1.8	0.25	0.35	0.25	0.05	0.15	rem
R295.0(g) 0.7-1.5	1.0	4.0-5.0	0.35	0.03	···	0.35	0.25	0.05	0.15	rem
R355.0(g) 4.5-5.5	0.6(j)	1.0-1.5	0.50(j)	0.40-0.6	0.25	0.35	0.25	0.05	0.15	rem
R356.0(g) 6.5-7.5	0.6	0.25	0.35	0.20-0.40	···	0.35	0.25	0.05	0.15	rem

(a) Single values shown are maximum percentages, except where a minimum is specified. Analysis shall be made for the elements for which specific limits are shown. If, however, the presence of other elements is suspected or indicated in the course of routine analysis, further analysis shall be made to determine that these other elements are not in excess of the limits specified for "other elements." (b) Beryllium shall not exceed 0.0008%. (c) The aluminum content is the difference between 100.00% and the sum of all other metallic elements present in amounts of 0.010% or more each, expressed to the second decimal before determining the sum. (d) Silicon plus iron shall not exceed 0.95%. (e) Vanadium content shall be 0.05 to 0.15%. Zirconium content shall be 0.10-0.25%. (f) Silicon plus iron shall not exceed 0.45%. (g) For repair of castings. (h) Nickel content shall be 1.7 to 2.3%. (j) If iron exceeds 0.45%, manganese content shall not be less than half the iron content.
Source: AWS A5.10

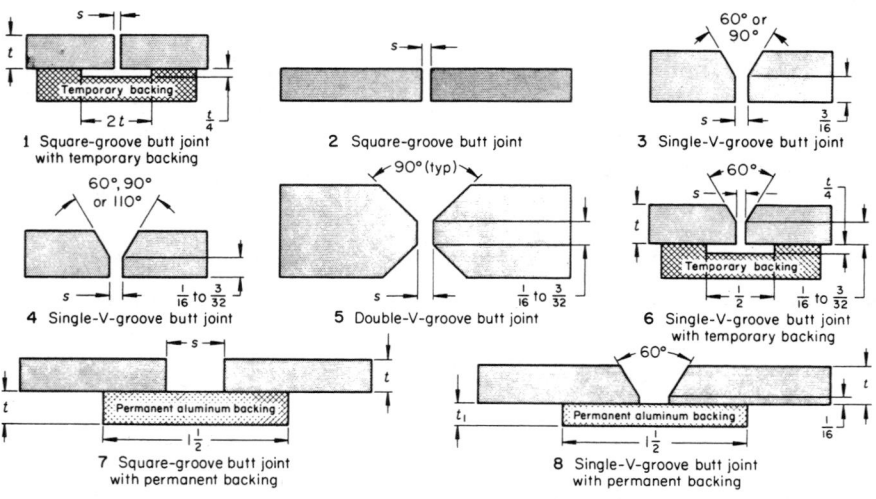

face can be thicker and the grooves narrower.

Lap joints are used more often for aluminum alloys than for most other metals. The efficiency of lap joints is 60 to 80%, depending on the alloy and temper. Lap joints offer the advantages of ease of fit-up and the fact that no edge preparation is required.

Edge Preparation and Assembly. Materials up to about ³/₈ in. thick can be sheared to a reasonably square edge that can be cleaned readily. Dull or improperly designed tools result in lapping of material on prepared edges that can trap lubricant, which can cause weld porosity.

PREWELD CLEANING

Preweld cleaning of aluminum is essential for optimum weld quality. Precleaning requirements are especially stringent prior to direct current electrode negative GTAW, because under such conditions, the arc exerts no cleaning action.

First, the work-metal surface should be cleaned of contaminants. The following manual cleaning methods can be used for small production runs. Dirt can be removed easily by washing and scrubbing with a detergent solution; drying is necessary to ensure that no moisture is present on the surfaces to be welded. Removal of grease and oil can be accomplished by swabbing with cloths soaked in an approved nontoxic solvent.

Next, thick oxide layers should be mechanically removed with a wire brush, steel wool, mill file, portable milling tool, or scraper. The use of abrasive paper or grinding disks alone is not recommended, because particles of the abrasive may become embedded in the aluminum and, unless subsequently removed, can cause inclusions in the weld. Wire-brush bristles preferably should be 0.012 to 0.016 in. in diameter and made of stainless steel to minimize iron oxide pickup.

Preweld cleaning also can be done by chemical removal of oxides.

PREHEATING

In gas shielded-arc welding of aluminum alloys, preheating of parts to be welded is normally done only when the temperature of the parts is below 32 °F or when the mass of the parts is such that the heat is conducted away from the joint faster than the welding process can supply

Metal thickness, *t*, in.	Semiautomatic GMAW			Manual GTAW		
	Welding position(a)	Joint design(b)	Root opening, *s*, in.	Welding position(a)	Joint design	Root opening, *s*, in.
¹/₁₆	F	1, 7	0-³/₃₂	F,H,V,O	2	0
³/₃₂	F	1	0	F,H,V	2	0
	F,H,V,O	7	¹/₈	O	2	0
¹/₈	F,H,V	1	0-³/₃₂	F	2	0
	F,H,V,O	7	³/₁₆	H,V,O	2	0
³/₁₆	F,H,V	2	0	F	4-60°	0-¹/₈
	F,H,V,O	6	0-¹/₁₆	H	4-90°	0-³/₃₂
	F,V	8	³/₃₂-³/₁₆	V	4-60°	0-³/₃₂
	H,O	8	³/₁₆	O	4-110°	0-³/₃₂
¹/₄	F	2	0	F	4-60°	0-¹/₈
	F,H,V,O	6	0-³/₃₂	H	4-90°	0-³/₃₂
	F,V	8	¹/₈-¹/₄	V	4-60°	0-³/₃₂
	H,O	8	¹/₄	O	4-110°	0-³/₃₂
³/₈	F	3-90°	0-³/₃₂	F	4-60°	0-¹/₈
	F	6	0-³/₃₂	F	5	0-³/₃₂
	H,V,O	6	0-³/₃₂	V	4-60°	0-³/₃₂
	F,V	8	¹/₄-³/₈	H,V,O	5	0-³/₃₂
	H	8	³/₈	H	4-90°	0-³/₃₂
	O	8	³/₈	O	4-110°	0-³/₃₂
³/₄	F	3-60°	0-³/₃₂	···	···	···
	F	8	0-¹/₈	···	···	···
	H,V,O	8	0-¹/₁₆	···	···	···
	F,H,V,O	8	0-¹/₁₆	···	···	···

(a) F = flat; H = horizontal; V = vertical; O = overhead. (b) For design 8, $t_1 = t$ for *t* less than ³/₈ in., and $t_1 = $ ³/₈ in. for *t* greater than ³/₈ in.

Joints 2, 3, 4, 5 and 6 should be back gouged to solid weld metal before applying a pass on the root side.
Fig. 35. Recommended butt-joint designs for direct current electrode positive GMAW and alternating current GTAW of aluminum alloys

it. Preheating may be advantageous for GTAW with alternating current of parts thicker than about $^3/_{16}$ in. and for GMAW of parts thicker than about 1 in. Gas tungsten-arc welding with DCEP is limited to thin material, and preheating is not necessary with this process. Thick parts also should not be preheated for GTAW using DCEN, because of the high heat input provided to the work. Preheating can also reduce production costs because the joint area reaches welding temperature faster, thus permitting higher welding speeds.

Various methods can be used to preheat the entire part or assembly to be welded, or only the area adjacent to the weld can be heated by use of a gas torch. In mechanized welding, local preheating and drying can be done by gas or tungsten-arc torches installed ahead of the welding electrode.

The preheating temperature depends on the job. Often 200 °F is sufficient to ensure adequate penetration on weld starts, without readjustment of the current as welding progresses. Preheating temperature for wrought aluminum alloys seldom exceeds 300 °F, because the desirable properties of certain aluminum alloys and tempers may be adversely affected at higher temperatures. Aluminum-magnesium alloys containing 4.0 to 5.5% Mg (5083, 5086 and 5456) should not be preheated to more than 200 °F, because their resistance to stress-corrosion cracking is reduced.

GAS METAL-ARC WELDING

The ability of GMAW to deposit large quantities of weld metal in a short period of time has played a large part in the increased use of aluminum since the late 1940's. Typical welding schedules for GMAW of aluminum alloys are given in Table 30.

Power Supply and Equipment. Only direct current electrode positive, which gives good penetration and a cathodic cleaning action at the work surface, is used in GMAW of aluminum alloys. The constant and pulsed direct-current power supplies, and the wire-feed systems, electrode holders and control systems used for GMAW of aluminum alloys, are the same as those used for GMAW of other metals.

Argon-Helium Mixtures. To take advantage of the higher arc heat of helium without the disadvantages associated with using the pure gas, mixtures of argon and helium are usually used for joining thick metal. Although users have individual preferences, mixtures ranging between 50 and 75% helium are used. A helium-rich mixture, such as 75% helium and 25% argon, is frequently used for welding workpieces more than 2 in. thick. For workpieces more than 3 in. thick, helium-rich mixtures maximize weld penetration and minimize porosity. For welding workpieces 1 to 3 in. thick in the flat position, increasing the current or voltage, or both, allows the helium content to be decreased.

Flow Rates. Typical shielding-gas flow rates for GMAW of aluminum and aluminum alloys using $^1/_{16}$-in.-diam electrode wire are:

Shielding gas	Flow rate(a), ft³/h
100% argon	30-70
75% helium, 25% argon	50-110
100% helium	60-140

(a) The lower rates are more suitable for indoor work and moderate welding current. The higher rates are more suitable for high current, maximum speed, and outdoor welding.

Table 30. Typical welding schedules for GMAW of aluminum alloys

Material thickness or fillet size, in.	Type of weld fillet or groove	Electrode diameter, in.	Welding power(a) Current, A (DCEP)	Welding power(a) Voltage, V	Wire-feed speed, in./min	Shielding gas flow, ft³/h	No. of passes	Travel speed (per pass), in./min
$^3/_{64}$	Square groove and fillet	0.030	50	12-14	268-308	30	1	17-25
$^5/_{64}$	Square groove and fillet	0.030	55-60	12-14	295-320	30	1	17-25
$^5/_{64}$	Square groove and fillet	$^3/_{64}$	110-125	19-21	175-185	30	1	20-27
$^3/_{32}$	Square groove and fillet	0.030	90-100	14-18	330-370	30	1	24-36
$^1/_8$	Fillet	0.030	110-125	19-22	410-460	30	1	20-24
$^1/_8$	Square grove	$^3/_{64}$	110-125	20-24	175-190	40	1	20-24
$^3/_{16}$	Square groove and fillet	$^3/_{64}$	160-195	20-24	215-225	40	1	20-25
$^1/_4$	Fillet	$^3/_{64}$	160-195	20-24	215-225	40	1	20-25
$^1/_4$	V-groove	$^1/_{16}$	175-225	22-26	150-195	40	3	20-25
$^3/_8$	V-groove and fillet	$^1/_{16}$	200-300	22-26	170-275	40	2-5	25-30
$^1/_2$	V-groove and fillet	$^1/_{16}$	220-230	22-27	195-205	40	3-8	12-18
$^1/_2$	Double-V-groove	$^3/_{32}$	320-340	22-29	140-150	45	2-5	15-17
$^3/_4$	Double-V-groove	$^1/_{16}$	255-275	22-27	230-250	50	4-10	8-18
$^3/_4$	Double-V-groove	$^3/_{32}$	355-375	22-29	155-160	50	4-10	4-16
1	Double-V-groove	$^1/_{16}$	255-290	22-27	230-265	50	4-14	6-18
1	Double-V-groove	$^3/_{32}$	405-425	22-27	175-180	50	4-8	8-12

Notes: (1) For groove and fillet welds, material thickness also indicates fillet weld size. Use V-groove for $^3/_{16}$ in. and thicker. (2) Use argon for thin and medium material; use 50% argon and 50% helium for thick material. Increase gas flow rate 10% for overhead position. (3) Increase amperage 10 to 20% when backup is used. (4) Decrease amperage 10 to 20% when welding out of position.
(a) Direct current electrode positive (DCEP)
Source: Cary, H. B., *Modern Welding Technology*, Prentice-Hall, New York, 1979, p 440

Table 31. Welding schedules for alternating-current GTAW of aluminum

Material thickness or fillet size, in.	Type of weld fillet or groove	Tungsten electrode diameter, in.	Filler rod diameter, in.	Nozzle size (inside diameter), in.	Shielding gas flow, ft³/h	Welding current (ac), A	No. of passes	Travel speed (per pass), in./min
$^3/_{64}$	Square groove and fillet	$^1/_{16}$	$^1/_{16}$	$^1/_4$-$^3/_8$	20	40-60	1	14-18
$^1/_{16}$	Square groove and fillet	$^3/_{32}$	$^3/_{32}$	$^5/_{16}$-$^3/_8$	20	70-90	1	8-12
$^3/_{32}$	Square groove and fillet	$^3/_{32}$	$^3/_{32}$	$^5/_{16}$-$^3/_8$	20	95-115	1	10-12
$^1/_8$	Square groove and fillet	$^1/_8$	$^1/_8$	$^3/_8$	20	120-140	1	9-12
$^3/_{16}$	Fillet	$^5/_{32}$	$^5/_{32}$	$^7/_{16}$-$^1/_2$	25	160-200	1	9-12
$^3/_{16}$	V-groove	$^5/_{32}$	$^5/_{32}$	$^7/_{16}$-$^1/_2$	25	160-180	2	10-12
$^1/_4$	Fillet	$^3/_{16}$	$^3/_{16}$	$^7/_{16}$-$^1/_2$	30	230-250	1	8-11
$^1/_4$	V-groove	$^3/_{16}$	$^3/_{16}$	$^7/_{16}$-$^1/_2$	30	200-220	2	8-11
$^3/_8$	V-groove	$^3/_{16}$	$^3/_{16}$	$^1/_2$	35	250-310	2-3	9-11
$^1/_2$	V- or U-groove	$^1/_4$	$^1/_4$	$^5/_8$	35	400-470	3-4	6

Notes: (1) Increase amperage when backup is used. (2) Data are for all welding positions. Use low side of range for out-of-position welding. (3) For tungsten electrodes: 1st choice—pure tungsten EWP; 2nd choice—zirconiated EWZr. (4) Normally, argon is used for shielding; however, mixtures of 10% or more helium with argon are sometimes used for increased penetration in aluminum $^1/_4$ in. thick or more. The gas flow should be increased when helium is added. A mixture of 75% helium plus 25% argon is popular. When 100% helium is used, gas flow rates are about twice those used for argon.
Source: Cary, H.B., *Modern Welding Technology*, Prentice-Hall, New York, 1979, p. 339

GAS TUNGSTEN-ARC WELDING

The advantages of welding aluminum alloys by GTAW are the same as for other metals, as discussed in the article "Gas Tungsten-Arc Welding" in this Volume. Welding can be done with or without filler-metal additions, depending on the aluminum alloys to be alclad and on the joint configuration.

Thicknesses of aluminum alloys commonly welded by GTAW range from 0.040 to $^3/_8$ in. for manual welding and from 0.010 to 1 in. for automatic welding. Gas tungsten-arc welding is especially suitable for automatic welding of thin workpieces that require the utmost in quality of finish, because of the precise heat control possible and the ability to weld with or without filler metal. Metal thicker than $^3/_8$ in. can be manually welded; however, either GMAW or automatic GTAW is preferred.

Power Supply and Equipment. For joining aluminum alloys, GTAW uses either alternating or direct current. Both negative and positive electrode polarities are used with direct-current welding. The same power supplies, arc-stabilization accessories, torches and control systems that are used for other metals are used for the GTAW process on aluminum.

ELECTRODES FOR GAS TUNGSTEN-ARC WELDING

For alternating-current GTAW, unalloyed tungsten and tungsten-zirconia electrodes are recommended. Zirconiated electrodes are less likely to be contaminated by aluminum and have a slightly higher current rating.

SHIELDING GASES FOR GAS TUNGSTEN-ARC WELDING

Argon, helium, and mixtures of argon and helium are used as shielding gases in GTAW of aluminum alloys. The selection of shielding gas is somewhat dependent on the type of current used. Welds made with alternating current show little difference in soundness or strength whether made with argon or helium shielding.

Helium. With helium shielding, penetration is deeper, but a higher flow rate is required; hence, helium is sometimes used with higher speeds and for thicker sections.

Argon is used as a shielding gas more often than helium, because it:

• Is more readily available and costs less than helium

Table 32. Nominal compositions, melting points, relative thermal conductivities, and weldabilities of wrought coppers and copper alloys that are commonly arc welded

UNS No.	Alloy name	Nominal composition, %	Melting point (liquidus), °F	Relative thermal conductivity(a)	Weldability(b) GTAW	GMAW	SMAW
OFC and ETP coppers							
C10200	Oxygen-free copper (OFC)	99.95 Cu	1981	100	G	G	NR
C11000	Electrolytic tough pitch copper (ETP)	99.90 Cu, 0.04 O_2	1981	100	F	F	NR
Deoxidized coppers							
C12000	Phosphorus-deoxidized copper, low-P (DLP)	99.9 Cu, 0.008 P	1981	99	E	E	NR
C12000	Phosphorus-deoxidized copper, high-P (DHP)	99.9 Cu, 0.02 P	1981	87	E	E	NR
Beryllium coppers							
C17500	High-conductivity beryllium copper, 0.6%	96.9 Cu, 0.6 Be, 2.5 Co	1955	53-66(c)	F	F	F
C17000	High-strength beryllium copper, 1.7%	98.3 Cu, 1.7 Be	1800	27-33(c)	G	G	G
C17200	High-strength beryllium copper, 1.9%	98.1 Cu, 1.9 Be	1800	27-33(c)	G	G	G
Low-zinc brasses							
C21000	Gilding, 95%	95 Cu, 5 Zn	1950	60	G	G	NR
C22000	Commercial bronze, 90%	90 Cu, 10 Zn	1910	48	G	G	NR
C23000	Red brass, 85%	85 Cu, 15 Zn	1880	41	G	G	NR
C24000	Low brass, 80%	80 Cu, 20 Zn	1830	36	G	G	NR
High-zinc brasses							
C26000	Cartridge brass, 70%	70 Cu, 30 Zn	1750	31	F	F	NR
C26800	Yellow brass, 66%	65 Cu, 35 Zn	1710	30	F	F	NR
C28000	Muntz metal, 60%	60 Cu, 40 Zn	1660	31	F	F	NR
Tin brasses							
C44300	Admiralty brass	71 Cu, 28 Zn, 1 Sn(d)	1720	28	F	F	NR
C46400	Naval brass	60 Cu, 39.25 Zn, 0.75 Sn(d)	1650	30	F	F	NR
Special brasses							
C67500	Manganese bronze A	58.5 Cu, 39 Zn, 1.4 Fe, 1 Sn, 0.1 Mn	1630	27	F	F	NR
C68700	Aluminum brass, arsenical	77.5 Cu, 20.5 Zn, 2 Al (0.06 As)	1780	26	F	F	NR
Nickel silvers							
C74500	Nickel silver, 65-10	65 Cu, 25 Zn, 10 Ni	1870	12	F	F	NR
C75200	Nickel silver, 65-18	65 Cu, 17 Zn, 18 Ni	2030	8	F	F	NR
C75400	Nickel silver, 65-15	65 Cu, 20 Zn, 15 Ni	1970	9	F	F	NR
C75700	Nickel silver, 65-12	65 Cu, 23 Zn, 12 Ni	1900	10	F	F	NR
C77000	Nickel silver, 55-18	55 Cu, 27 Zn, 18 Ni	1930	8	F	F	NR
Phosphor bronzes							
C50500	Phosphor bronze, 1.25% E	98.7 Cu, 1.3 Sn (0.2 P)	1970	53	G	G	F
C51000	Phosphor bronze, 5% A	95 Cu, 5 Sn (0.2 P)	1920	18	G	G	F
C52100	Phosphor bronze, 8% C	92 Cu, 8 Sn (0.2 P)	1880	16	G	G	F
C52400	Phosphor bronze, 10% D	90 Cu, 10 Sn (0.2 P)	1830	13	G	G	F
Aluminum bronzes							
C61300	Aluminum bronze D, Sn-stabilized	89 Cu, 7 Al, 3.5 Fe (0.35 Sn)	1915	14	G	E	G
C61400	Aluminum bronze D	91 Cu, 6-8 Al, 1.5-3.5 Fe, 1 max Mn	1915	17	G	E	G
C63000	Aluminum bronze E	82 Cu, 10 Al, 5 Ni, 3 Fe	1930	10	G	G	G
Silicon bronzes							
C65100	Low-silicon bronze B	98.5 Cu, 1.5 Si	1940	15	E	E	F
C65500	High-silicon bronze A	97 Cu, 3 Si	1880	9	E	E	F
Copper nickels							
C70600	Copper nickel, 10%	88.6 Cu, 9-11 Ni, 1.4 Fe, 1.0 Mn	2100	12	E	E	G
C71500	Copper nickel, 30%	70 Cu, 30 Ni	2260	8	E	E	E

(a) Based on the thermal conductivity of alloy C10200 (226 Btu/ft^2 in feet per hour at 68 °F) as 100. For comparison, carbon steel has a thermal conductivity of 30 Btu/ft^2 in feet per hour at 68 °F, which is 13 on this scale. (b) E = excellent, G = good, F = fair, NR = not recommended. (c) In the precipitation-hardened condition. (d) Alloys C44300 and C46500 contain a nominal 0.06% As; alloys C44400 and C46600, a nominal 0.06% Sb; alloys C44500 and C46700, a nominal 0.06% P.

- Affords better control of the weld pool
- Gives a smoother, quieter arc, greater arc cleaning action, and easier arc starting
- Requires less gas for specific applications
- Has better cross-draft resistance than helium
- Causes less clouding, and the metal stays brighter; the operator can thus see the weld pool more easily.

FILLER METALS FOR GAS TUNGSTEN-ARC WELDING

Gas tungsten-arc welding can be done with or without filler metal. Sometimes the joint design permits the base metal to provide the weld metal. In some square-groove butt joints, the weld metal comes from the straight sides of the groove, or extra metal may be provided on a corner or flange that is melted to form the weld. If restraint is likely to cause cracking, best results are achieved by the addition of separate filler metal.

Usually, filler metal is added in the form of a bare rod for manual welding or as a coil of wire for automatic feeding.

ALTERNATING-CURRENT GAS TUNGSTEN-ARC WELDING

Usually, alternating-current welding provides the optimum combination of current-carrying capacity, arc controllability and arc cleaning action for welding of aluminum alloys. Typical welding schedules for manual alternating-current GTAW are given in Table 31.

Arc Welding of Copper and Copper Alloys

COPPER AND COPPER ALLOYS offer a unique combination of properties, among which the most important are conductivity, strength and corrosion resistance. Other useful attributes include spark resistance, wear resistance, nonmagnetic or low-permeability properties, and color. In manufacturing, copper is often joined by welding, and arc welding is the most important of the processes employed. Arc welding can be achieved by shielded metal-arc welding (SMAW); gas tungsten-arc welding (GTAW), including the pulsed-current mode; gas metal-arc welding (GMAW), including the pulsed-current and fine-wire modes; plasma-arc welding (PAW); and submerged-arc welding (SAW). In all processes, the dominant factors in establishing weldability are thermal conductivity and the alloy type with reference to solidification range and low-melting-point constituents.

The copper and copper alloys that are most frequently arc welded are listed in Table 32, along with other data pertinent to welding.

FACTORS THAT AFFECT WELDABILITY

Other than the elements that comprise a specific alloy, the principal factors that influence weldability are thermal conductivity of the alloy being welded, shielding gas, type of current, joint design, welding position, and surface condition (cleanness). Effect of type of current is discussed under the sections on individual processes and alloys in this article.

Effect of Thermal Conductivity. The welding behavior of copper and copper alloys is strongly influenced by thermal conductivity, which varies greatly among these alloys. Table 32 gives relative thermal conductivities that are based on the conductivity of alloy C10200 (oxygen-free copper), which is 226 Btu/ft^2 in feet per hour at 68 °F, as 100. The range shown in Table 32 is from 100, for alloys C10200 and C11000, to lows of 8 to 12 for nickel silvers and copper nickels, and 9 for alloy C65500 (high-silicon bronze A). For comparison, the thermal conductivity of carbon steel, 30 Btu/ft^2 in feet per hour at 68 °F, is 13 on this scale.

In welding of commercial coppers and lightly alloyed copper materials having high thermal conductivity, the type of current and shielding gas must be selected for maximum heat input to counteract the rapid heat dissipation from the weld region.

Even the less-conductive copper alloys may require preheating (depending on section thickness), in spite of the concentrated heat input of arc welding processes.

Shielding gas for gas shielded-arc welding is usually argon, or argon plus 25 to 75% helium. Because helium is more expensive than argon, it is advantageous to develop welding procedures compatible with the use of argon or high-argon mixtures. Argon or argon-helium mixtures produce more uniform welds than helium, give a more stable arc and cause less weld spatter.

Helium alone or in mixtures with argon is preferred where high heat input is needed, as in welding of highly conductive coppers or copper alloys and aluminum bronze. Helium gives about one-third greater heat input than does argon at equal welding current.

Joint design for arc welding of copper and copper alloys does not differ greatly from that used in arc welding of steel. Sections up to $1/8$ in. thick can be joined by use of square-groove welds without root openings. Thicker sections ordinarily are joined using either single-V-groove or double-V-groove welds with root faces not more than $1/8$ in. wide.

Precipitation-Hardenable Alloys. The most important precipitation-hardening reactions in copper alloys are obtained with beryllium, chronium, boron, nickel-silicon and zirconium. Care must be taken in welding precipitation-hardenable copper alloys to avoid oxidation and incomplete fusion. Wherever possible, the components should be welded in the annealed condition, and then the weldment should be given a precipitation-hardening heat treatment.

GAS TUNGSTEN-ARC WELDING

Gas tungsten-arc welding is well suited for copper and copper alloys because of its intense arc, which produces an extremly high temperature at the joint and a narrow heat-affected zone (HAZ). In welding copper and the more heat-conductive copper alloys, the intensity of the arc is important in completing fusion with minimum heating of the surrounding, highly conductive base metal. In welding copper alloys that have been precipitation hardened, a narrow HAZ is particularly desirable.

Filler metals most frequently used in GTAW of copper and copper alloys are listed in Table 33. Frequently, the filler-metal composition is matched closely to the base-metal composition, but a filler metal of composition different from that of the base-metal may be selected. The reasons for this are dealt with in the sections on GTAW of various copper alloys in this article.

Type of Current. Gas tungsten-arc welding is done on most copper and copper alloys with direct current electrode negative (DCEN) to permit use of an electrode of minimum size for a given welding current and to provide maximum penetration. Alternating current stabilized by high frequency is used on beryllium coppers and aluminum bronzes to prevent buildup of tenacious oxide films on these base metals.

Electrodes. Any of the standard tungsten or alloyed tungsten electrodes described in the article "Gas Tungsten-Arc Welding" in this volume can be used in GTAW of copper and copper alloys.

Coppers

Although the weld quality of commercial coppers joined by GTAW differs depending on the cuprous oxide content of the copper, the nominal welding conditions for coppers of a given thickness and joint design are approximately the same.

High-Strength Beryllium Coppers

High-strength beryllium coppers, alloys C17000 (1.7% Be) and C17200 (1.9% Be), are more easily welded than the higher-melting and less-fluid high-conductivity beryllium coppers. Factors governing the suitability of GTAW for high-strength beryllium copper are as described for high-conductivity beryllium copper. However, contrary to practice with the high-conductivity alloys, GTAW can be used on thicknesses greater than $1/2$ in. when it is not practical to weld by the preferred GMAW process.

Cadmium and Chromium Coppers

Generally, the procedures recommended for GTAW of copper are good bases for determining welding parameters for cadmium and chromium coppers. These alloys have lower thermal and electrical conductivities than copper and can be welded with lower preheats and heat inputs than those required for copper. In addition to GTAW, cadmium and chromium coppers can be welded by the other gas-shielded processes.

Nickel Silvers

Nickel silvers, which are alloys composed of copper (65%), zinc (17 to 27%) and nickel (10 to 18%), can be joined by the GTAW process, although welding of these alloys is not widely

Table 33. Filler metals most frequently used in GTAW of copper and copper alloys(a)

Filler metal	AWS classification	Principal constituents(b)
Copper	ERCu	98.0 min Cu + Ag, 1.0 Sn, 0.5 Mn, 0.50 Si, 0.15 P
Phosphor bronze	ERCuSn-A	93.5 min Cu + Ag, 4.0-6.0 Sn, 0.10-0.35 P
Aluminum bronze	RCuAl-A2	1.5 Fe, 9.0-11.0 Al, rem Cu + Ag
Aluminum bronze	ERCuAl-B	3.0-4.25 Fe, 11.0-12.0 Al, rem Cu + Ag
Silicon bronze	ERCuSi-A	94.0 min Cu + Ag, 2.8-4.0 Si, 1.5 Zn, 1.5 Sn, 1.5 Mn, 0.5 Fe
Copper nickel	ERCuNi	1.00 Mn, 0.40-0.70 Fe, 29.0-32.0 Ni + Co, 0.20-0.50 Ti, rem Cu + Ag

(a) Based on AWS A5.27, A5.7, and A5.6; see current editions of those specifications for complete compositions and qualifications.
(b) Single percentages are maximums unless otherwise stated. Optional elements and impurities have been omitted.

Table 34. Filler metals for GMAW of copper and copper alloys

Bare wire(a)	Common name	Base-metal applications
ERCu	Copper	Coppers
ERCuSi-A	Silicon bronze	Silicon bronzes, brasses
ERCuSn-A	Phosphor bronze	Phosphor bronzes, brasses
ERCuNi	Copper nickel	Copper-nickel alloys
ERCuAl-A2	Aluminum bronze	Aluminum bronzes, brasses, silicon bronzes, manganese bronzes
ERCuAl-A3	Aluminum bronze	Aluminum bronzes
ERCuNiAl	...	Nickel-aluminum bronzes
ERCuMnNiAl	...	Manganese-nickel-aluminum bronzes
RBCuZn-A	Naval brass	Brasses, copper
RCuZn-B	Low-fuming brass	Brasses, manganese bronzes
RCuZn-C	Low-fuming brass	Brasses, manganese bronzes

(a) See AWS A5.7-77, "Specification for Copper and Copper Alloy Bare Welding Rods and Electrodes" or AWS A5.27-78, "Specification for Copper and Copper Alloy Gas Welding Rods."

practiced. From a welding standpoint, nickel silvers are similar to brasses of comparable zinc content. These alloys are frequently used for decorative purposes where color match is important. However, there are no zinc-free filler metals that are suitable for arc welding that give good color match.

Copper-Zinc (Brass) Alloys

Of the copper-zinc alloys rated as to weldability in Table 32, the low-zinc brasses are shown to have good weldability by GTAW. High-zinc brasses, tin brasses, special brasses, and nickel silvers are shown to have only fair weldability, either because of high zinc content or because of moderate zinc content in combination with other elements, such as oxide-forming aluminum or nickel.

Gas tungsten-arc welding, because of its ability to weld rapidly with a highly localized heat input, is sometimes used for welding copper-zinc alloys (20% Zn or less) that contain up to 1% Pb, even though leaded copper alloys are generally not recommended for arc welding.

Phosphor Bronzes

Gas tungsten-arc welding is used to join sheet and other forms of wrought phosphor bronze up to about $1/2$ in. thick. This process is also used to join or repair phosphor bronze castings. Copper-tin alloys solidify with large, weak dendritic grain structures. Such structures in the weld metal have a tendency to crack. Hot peening at each layer of multiple-pass welds reduces cracking stresses and, therefore, the likelihood of weld-metal cracking.

Aluminum Bronzes

Aluminum bronzes up to about $3/8$ in. thick are readily joined by gas tungsten-arc welding, although welding conditions differ somewhat from those for most copper alloys. Porosity is minimized by the presence of iron, manganese or nickel in the filler metal or base metal, or in both. Aluminum bronze castings also are repair welded by GTAW.

Silicon Bronzes

Gas tungsten-arc welding is used on thin to moderately thick nonleaded silicon bronzes, which are the most weldable of the copper alloys. Characteristics of these bronzes that contribute to weldability are their low thermal conductivity, good deoxidation of the weld metal by silicon, and the protection offered by the resulting slag. Silicon bronzes have a relatively narrow hot-short

range just below the solidus, and they must be rapidly cooled through this range to avoid weld cracking.

Copper Nickels

Gas tungsten-arc welding is the preferred process for joining copper nickels in thicknesses up to about $1/16$ in. and may be used for greater thicknesses. Manual welding is normally used for sheet and plate up to $1/4$ in. thick.

GAS METAL-ARC WELDING

Gas metal-arc welding is used to join all of the coppers and copper alloys listed in Table 32. It is preferred for joining the aluminum bronzes, silicon bronzes and copper nickels in section thicknesses greater than about $1/8$ in. Gas tungsten-arc welding is preferred for thicknesses less than about $1/8$ in.

The major application of GMAW to copper alloys is in joining material from $1/8$ to $1/2$ in. thick, and the process is almost invariably selected for arc welding sections of copper alloys thicker than about $1/2$ in., where its high deposition rate is a major advantage over GTAW or SMAW. The greater rate of heat input to the weld, compared with that for GTAW, is a disadvantage in some applications because of the wider HAZ.

Direct current electrode positive (DCEP) is used exclusively for GMAW of copper alloys. Argon is normally used for shielding. Helium or mixtures of argon and helium are used where hotter arcs are needed than are possible at given current levels with pure argon.

Electrode wires used for GMAW and their basemetal applications are given in Table 34.

Arc Welding of Magnesium Alloys

MOST MAGNESIUM ALLOYS can be joined by gas tungsten-arc welding (GTAW) and gas metal-arc welding (GMAW).

WELDABILITY

Table 35 lists some magnesium alloys that are weldable, along with their respective weldability ratings based on a scale of A (excellent) to D (limited). This rating is based largely on freedom from susceptibility to cracking, and to some extent on joint efficiency. Under optimum welding conditions, including favorable joint design, joint

efficiencies of 60 to 100% can be obtained for virtually all of the magnesium alloys. Alloys rated A in Table 35 are likely to have high joint-efficiency ratings.

FILLER METALS

Compositions of the four most commonly used electrode wires for GMAW and filler metals (when used) for GTAW are given in Table 36. The choice of electrode wire or filler metal is governed by the composition of the base metal.

Electrode wire or filler metals having compositions conforming to ER AZ61A or ER AZ92A (Mg-Al-Zn) are considered satisfactory for welding wrought alloys AZ10A, AZ31B, AZ31C, AZCOML, AZ61A, AZ80A, ZE10A and ZK21A to themselves or to each other. ER AZ61A is usually preferred for welding wrought products because of its tendency to resist crack sensitivity. The ER AZ92A filler metal shows less crack sensitivity for welding the cast Mg-Al-Zn and Mg-Al alloys. The same electrode wires or filler metals are used for joining any one of the above alloys to high-temperature alloys HK31A, HM21A and HM31A. However, when the high-temperature alloys are joined to each other, ER EZ33A is recommended.

SHIELDING GASES

Only the inert gases are used for shielding in arc welding of magnesium alloys. Argon is the most widely used. Helium and various mixtures of argon and helium have also proved satisfactory, but because of the higher cost per unit volume of helium, and because two to three times

Table 35. Relative arc weldability of magnesium alloys

Alloy	Rating
Casting alloys	
AM100A	B+
AZ63A	C
AZ81A	B+
AZ91C	B+
AZ92A	B
EK30A	B
EK41A	B
EZ33A	A
HK31A	B+
HZ32A	B
K1A	A
QE22A	B
ZE41A	B−
QH21A	B
ZH62A	C−
ZK51A	D
ZK61A	D
Wrought alloys	
AZCOML	A
AZ10A	A
AZ31B,C	A
AZ61A	B
AZ80A	B
HK31A	A
HM21A	A
HM31A	A
ZE10A	A
ZK21A	B
ZK60A	D

Note: A, excellent; B, good; C, fair; D, limited weldability

Table 36. Compositions of electrodes and filler metals used in GMAW and GTAW of magnesium alloys (AWS A5.19)

Element	ER AZ61A	ER AZ101A	ER AZ92A	ER EZ33A
Aluminum	5.8-7.2	9.5-10.5	8.3-9.7	...
Beryllium	0.0002-0.0008	0.0002-0.0008	0.0002-0.0008	...
Manganese	0.15 min	0.13 min	0.15 min	...
Zinc	0.40-1.5	0.75-1.25	1.7-2.3	2.0-3.1
Zirconium	...	...	...	0.45-1.0
Rare earth	...	...	...	2.5-4.0
Copper	0.05 max	0.05 max	0.05 max	...
Iron	0.005 max	0.005 max	0.005 max	...
Nickel	0.005 max	0.005 max	0.005 max	...
Silicon	0.05 max	0.05 max	0.05 max	...
Others (total)	0.30 max	0.30 max	0.30 max	0.30 max
Magnesium	rem	rem	rem	rem

more helium than argon is required for the same degree of shielding, the use of pure helium has gradually decreased.

PREHEATING

The need for preheating of castings is determined largely by section thickness and amount of restraint. Thick sections, particularly if the magnitude of joint restraint is small, seldom need preheating. Thin sections and highly restrained joints often require preheating to prevent weld cracking, particularly in the high-zinc alloys.

GAS TUNGSTEN-ARC WELDING

Gas tungsten-arc welding is used more extensively than GMAW for joining magnesium alloys. It is well suited for welding thin sections. Control of heat input and the molten weld pool is better with GTAW than with GMAW. Alternating-current machines with a high-frequency current superimposed on the normal welding current for arc stabilization, and direct-current ma-

chines with continuous amperage control, are used for gas tungsten-arc welding.

Arc Welding of Nickel Alloys

NICKEL ALLOYS can be joined by arc welding. The wrought nickel alloys listed in Table 37 can be arc welded under conditions similar to those used in arc welding of austenitic stainless steel. Cast nickel alloys, particularly those of high silicon content, present difficulties in welding.

The most widely employed processes for welding the non-age-hardenable (solid-solution-strengthened) wrought nickel alloys are gas tungsten-arc welding (GTAW), gas metal-arc welding (GMAW) and shielded metal-arc welding (SMAW). Submerged-arc welding (SAW) has limited applicability, as does plasma-arc welding (PAW). The GTAW process is preferred for welding the precipitation-hardenable alloys, although GMAW and SMAW are also used.

Preweld Heating and Heat Treating. Preweld heating of wrought alloys is not required unless the base metal is below 60 °F, in which case a path 10 to 12 in. wide on both sides of the joint should be warmed to 60 to 70 °F to avoid condensation of moisture that may cause porosity in the weld metal.

Nickel alloys are usually welded in the solution-treated condition. Precipitation-hardenable alloys should be annealed before welding if they have undergone any operations that introduce high residual stresses.

Postweld Treatment. No postweld treatment, either thermal or chemical, is needed or recommended to maintain or restore corrosion resistance. Heat treatment may be necessary to meet specification requirements, such as stress relief of a fabricated structure to avoid age hardening or stress-corrosion cracking of the weldment in hydrofluoric acid vapor or caustic soda. If welding induces moderate to high residual stresses, the precipitation-hardenable alloys require a stress-relief anneal after welding and before aging.

GAS TUNGSTEN-ARC WELDING

Nickel alloys, both cast and wrought, either solid-solution-strengthened or precipitation-hardenable, can be welded by the GTAW process. The addition of filler metal is usually recommended. Direct current electrode negative (DCEN) is recommended for both manual and machine welding.

Shielding Gas. Either argon or helium, or a mixture of the two, is used as a shielding gas for welding nickel and nickel alloys. Additions of oxygen, carbon dioxide or nitrogen to argon gas will usually cause porosity or erosion of the electrode. Argon with small quantities of hydrogen (about 5%) can be used for single-pass welding and may help to avoid porosity in pure nickel.

Table 37. Nominal compositions of weldable wrought nickel and nickel alloys

Alloy designation	Ni	C	Mn	Fe	S	Si	Cu	Cr	Al	Ti	Nb	Other
Nickel 200	99.5	0.08	0.18	0.2	0.005	0.18	0.13	...	...	...	...	...
Nickel 201	99.5	0.01	0.18	0.2	0.005	0.18	0.13	...	...	...	...	...
Nickel 205	99.5	0.08	0.18	0.10	0.004	0.08	0.08	...	...	0.03	...	0.05 Mg
Nickel 211	95.0	0.10	4.75	0.38	0.008	0.08	0.13	...	...	...	...	...
Nickel 220	99.5	0.04	0.10	0.05	0.004	0.03	0.05	...	...	0.03	...	0.05 Mg
Nickel 230	99.5	0.05	0.08	0.05	0.004	0.02	0.05	...	...	0.003	...	0.06 Mg
Nickel 270	99.98	0.01	<0.001	0.003	<0.001	<0.001	<0.001	<0.001	...	<0.001	...	Mg <0.001, Co <0.001
Monel 400	66.5	0.15	1.0	1.25	0.012	0.25	31.5	...	...	...	...	...
Monel 401	42.5	0.05	1.6	0.38	0.008	0.13	Bal.	...	...	...	...	...
Monel 404	54.5	0.05	0.08	0.25	0.012	0.05	44.0	...	0.03	...	...	...
Monel R-405	66.5	0.15	1.0	1.25	0.043	0.25	31.5	...	...	...	...	...
Monel K-500	66.5	0.13	0.75	1.00	0.005	0.25	29.5	...	2.73	0.60	...	...
Monel 502	66.5	0.05	0.75	1.00	0.005	0.25	28.0	...	3.00	0.25	...	...
Inconel 600	76.0	0.08	0.5	8.0	0.008	0.25	0.25	15.5	...	...	...	...
Inconel 601	60.5	0.05	0.5	14.1	0.007	0.25	0.50	23.0	1.35	...	...	...
Inconel 617	54.0	0.07	...	...	...	...	...	22.0	1.0	...	...	12.5 Co, 9.0 Mo
Inconel 625	61.0	0.05	0.25	2.5	0.008	0.25	...	21.5	0.2	0.2	3.65	9.0 Mo
Inconel 671	Bal.	0.05	...	...	...	...	...	48.0	...	0.35	...	...
Inconel 702	79.5	0.05	0.50	1.0	0.005	0.35	0.25	15.5	3.25	0.63	...	...
Inconel 706	41.5	0.03	0.18	40.0	0.008	0.18	0.15	16.0	0.20	1.75	2.9	...
Inconel 718	52.5	0.04	0.18	18.5	0.008	0.18	0.15	19.0	0.50	0.90	5.13	3.05 Mo
Inconel 721	71.0	0.04	2.25	6.5	0.005	0.08	0.10	16.0	...	3.05	...	...
Inconel 722	75.0	0.04	0.50	7.0	0.005	0.35	0.25	15.5	0.70	2.38	...	...
Inconel X-750	73.0	0.04	0.50	7.0	0.005	0.25	0.25	15.5	0.70	2.50	0.95	...
Inconel 751	72.5	0.05	0.5	7.0	0.005	0.25	0.25	15.5	1.20	2.30	0.95	...
Incoloy 800	32.5	0.05	0.75	46.0	0.008	0.50	0.38	21.0	0.38	0.38	...	...
Incoloy 801	32.0	0.05	0.75	44.5	0.008	0.50	0.25	20.5	...	1.13	...	...
Incoloy 802	32.5	0.35	0.75	46.0	0.008	0.38	...	21.0	0.58	0.75	...	...
Incoloy 804	41.0	0.05	0.75	25.4	0.008	0.38	0.25	29.5	0.30	0.60	...	...
Incoloy 825	42.0	0.03	0.50	30.0	0.015	0.25	2.25	21.5	0.10	0.90	...	3.0 Mo
Ni-span-C 902	42.25	0.03	0.40	48.5	0.02	0.50	0.05	5.33	0.55	2.58	...	...

Table 38. Chemical compositions of titanium and titanium alloy filler metals(a) AWS A5.16

AWS classification	C	O	H	N	Al	V	Sn	Cr	Fe	Mo	Nb	Ta	Pd	Ti
ERTi-1(b)	0.03	0.10	0.005	0.012	...	...	...	...	0.10	...	...	...	...	rem
ERTi-2	0.05	0.10	0.008	0.020	...	...	...	...	0.20	...	...	...	...	rem
ERTi-3	0.05	0.10-0.15	0.008	0.020	...	...	...	...	0.20	...	...	...	...	rem
ERTi-4	0.05	0.15-0.25	0.008	0.020	...	...	...	...	0.30	...	...	...	...	rem
ERTi-0.2Pd	0.05	0.15	0.008	0.020	...	...	...	...	0.25	...	...	...	0.15-0.25	rem
ERTi-3Al-2.5V	0.05	0.12	0.008	0.020	2.5-3.5	2.0-3.0	...	...	0.25	...	...	...	...	rem
ERTi-3Al-2.5V-1(b)	0.04	0.10	0.005	0.012	2.5-3.5	2.0-3.0	...	...	0.25	...	...	...	...	rem
ERTi-5Al-2.5Sn	0.05	0.12	0.008	0.030	4.7-5.6	...	2.0-3.0	...	0.40	...	...	...	...	rem
ERTi-5Al-2.5Sn-1(b)	0.04	0.10	0.005	0.012	4.7-5.6	...	2.0-3.0	...	0.25	...	...	...	...	rem
ERTi-6Al-2Nb-1Ta-1Mo	0.04	0.10	0.005	0.012	5.5-6.5	...	...	...	0.15	0.5-1.5	1.5-2.5	0.5-1.5	...	rem
ERTi-6Al-4V	0.05	0.15	0.008	0.020	5.5-6.75	3.5-4.5	...	...	0.25	...	...	...	...	rem
ERTi-6Al-4V-1(b)	0.04	0.10	0.005	0.012	5.5-6.75	3.5-4.5	...	...	0.15	...	...	...	...	rem
ERTi-8Al-1Mo-1V	0.05	0.12	0.008	0.03	7.35-8.35	0.75-1.25	...	...	0.25	0.75-1.25	...	...	...	rem
ERTi-13V-11Cr-3Al	0.05	0.12	0.008	0.03	2.5-3.5	12.5-14.5	...	10.0-12.0	0.25	...	...	...	...	rem

(a) Single values are maximum. (b) Extra-low interstitials for welding similar base metals

Table 39. Dimensions of typical joints for welding titanium and titanium alloys

t is base-metal thickness.

Base-metal thickness, in.	Root opening, in.	Groove angle,°	Weld-bead width, in.
Square-groove butt joint			
0.010-0.090	0	...	...
0.031-0.125	0-0.10t	...	...
Single-V-groove butt joint			
0.062-0.125	0-0.10t	30-60	0.10-0.25t
0.090-0.125	(a)	90	...
0.125-0.250	0-0.10t	30-60	0.10-0.25t
Double-V-groove butt joint			
0.250-0.500	0-0.20t	30-120	0.10-0.25t
Single-U-groove butt joint			
0.250-0.750	0-0.10t	15-30	0.10-0.25t
Double-U-groove butt joint			
0.750-1.500	0-0.10t	15-30	0.10-0.25t
Fillet weld			
0.031-0.125	0-0.10t	0-45	0-0.25t
0.125-0.500	0-0.10t	30-45	0.10-0.25t

(a) Root face, 0.030 in.
Source: J.J. Vagi. *et al.*, "Welding Procedures for Titanium and Titanium Alloys," NASA TMX 53432, 1965

GAS METAL-ARC WELDING

The high-nickel and nickel-copper alloys can be joined by GMAW. With special procedures, precipitation-hardenable alloys, such as Monel K-500, can be gas metal-arc welded as well.

Spray, globular and short-circuiting metal transfer are suitable. Varying the power input produces the different types of metal transfer. The pulsed-arc process is also used. In these methods, the filler metal is a current-carrying consumable and is typically 0.035, 0.045 or 0.062 in. in diameeter.

Shielding gas used for GMAW of nickel alloys by spray or globular transfer may be argon, which yields good results. The addition of 15 to 20% helium is beneficial in welding nickel alloys with short-circuiting-arc or pulsed-arc transfer. As the helium content is increased from 0 to 20%, the weld beads become progressively wider and flatter, and penetration decreases. The addition of oxygen or carbon dioxide to argon stabilizes the arc, but results in heavily oxidized and irregular bead surfaces.

PLASMA-ARC WELDING

Plasma-arc welding (PAW), using the keyholing mode, can produce acceptable welds in nickel alloys up to about 0.3 in. thick. Argon-hydrogen mixtures are used as orifice and shielding gas, 5 to 8% H_2 being optimum. Current needed for keyholing decreases as hydrogen content is increased, up to about 7% H_2, above which (or when helium is used) torch starting is more difficult.

SHIELDED METAL-ARC WELDING

The SMAW process can be used for welding nickel and nickel alloys. Although the minimum metal thickness is usually about 0.050 in., thinner metal can be welded when appropriate fixtures are provided. The types of joints used and bead and groove dimensions are given in the article "Arc Welding of Heat-Resistance Alloys" in this volume.

SUBMERGED-ARC WELDING

Submerged-arc welding can be used for joining solid-solution nickel alloys. Monel 400 is the alloy most frequently welded by this process. Joints in metal up to 3 in. thick have met American Society of Mechanical Engineers (ASME) codes and other specification requirements. The SAW process cannot be used for welding the precipitation-hardenable nickel alloys.

Arc Welding of Titanium and Titanium Alloys

TITANIUM and most titanium alloys can be welded by the gas tungsten-arc, plasma-arc and gas metal-arc processes. Procedures and equipment are generally similar to those for welding austenitic stainless steel or aluminum. Because titanium and titanium alloys are extremely reactive above 1000 °F, however, additional precautions, exceeding those required during welding of austenitic stainless steel or aluminum alloys, must be taken to shield the weld and hot root side of the joint from air.

WELDABILITY

Unalloyed titanium and all alpha titanium alloys are weldable. Although the alpha-beta alloy Ti-6Al-4V and other weakly beta-stabilized alloys are also weldable, strongly beta-stabilized alpha-beta alloys are embrittled by welding. Most beta alloys can be successfully welded, but because aged welds in beta alloys can be quite brittle, heat treatment to strengthen the weld by age hardening should be used with caution.

Unalloyed titanium is generally available in several grades ranging in purity from 98.5 to 99.5% Ti. These grades are increased in strength by variations in oxygen, nitrogen, carbon and iron. Strengthening by cold working is possible but is seldom used. All grades are usually welded in the annealed condition. Welding of cold worked alloys anneals the heat-affected zone (HAZ) and negates the strength produced by cold working.

Alpha alloys Ti-5Al-2.5Sn, Ti-6Al-2Sn-4Zr-2Mo, Ti-5Al-5Sn-2Zr-2Mo, Ti-6Al-2Cb-1Ta-1Mo and Ti-8Al-1Mo-1V are always welded in the annealed condition.

Alpha-beta alloys, such as Ti-6Al-4V, can be welded in the annealed condition or in the solution-treated-and-partially-aged condition, with aging completed during postweld stress relieving. In contrast to unalloyed titanium and the alpha alloys, which can be strengthened only by cold work, the alpha-beta and beta alloys can be strengthened by heat treatment.

Metastable beta alloys Ti-3Al-13V-11Cr, Ti-11.5Mo-6Zr-4.5Sn, Ti-8Mo-8V-2Fe-3Al, Ti-15V-3Cr-3Al-3Sn and Ti-3Al-8V-6Cr-4Zr-4Mo are weldable in the annealed or solution heat treated condition. In the as-welded condition, welds are low in strength but ductile. Beta alloy weldments are sometimes used in the as-welded condition. Welds in the Ti-3Al-13V-11Cr alloy embrittle more severely when age hardened.

WELDING PROCESSES

Gas tungsten-arc welding (GTAW) is the most widely used process for joining titanium and titanium alloys, except in large thicknesses. Square-groove butt joints can be welded without filler metal in base metal up to 0.10 in. thick. For thicker base metal, the joint should be grooved and filler metal is required. Where possible, welding should be done in the flat position. Hot wire GTAW can be used for welding titanium alloys more than 1/4 in. thick.

Gas metal-arc welding (GMAW) is employed to join titanium and titanium alloys more than 1/8 in. thick. It is applied using pulsed current or the spray mode and is less costly than GTAW, especially when base-metal thickness is greater than $1/2$ in.

Plasma-arc welding (PAW) is also applicable to welding of titanium and titanium alloys. It is faster than GTAW and can be used on thicker sections, such as one-pass welding of titanium alloy plate up to $1/2$ in. thick, using square-groove butt joints and the keyhole technique. Titanium and titanium alloys are also welded by the electron beam process.

FILLER METALS

For welding titanium thicker than about 0.10 in. by the GTAW process, a filler metal must be used. For PAW, a filler metal may or may not be used for welding metal less than $1/2$ in. thick.

Fourteen titanium and titanium alloy filler-metal (or electrode) classifications are given in AWS A5.16. Five of these are essentially unalloyed titanium and the remainder are titanium alloy filler metals. Maximums are set on carbon, oxygen, hydrogen and nitrogen contents. Compositions for titanium and titanium alloy filler metals are given in Table 38.

Filler-metal composition is usually matched to the grade of titanium being welded. For improved joint ductility in welding the higher-strength grades of unalloyed titanium, filler metal of yield strength lower than that of the base metal is occasionally used. Because of the dilution that occurs during welding, the weld deposit acquires the required strength. Unalloyed filler metal is sometimes used to weld Ti-5Al-2.5Sn and Ti-6Al-4V for improved joint ductility. The use of unalloyed filler metals lowers the beta content of the weldment, thereby reducing the extent of the transformation which occurs and improving ductility.

SHIELDING GASES

In welding titanium and titanium alloys, only argon and helium, and occasionally a mixture of these two gases, are used for shielding. Because

it is more readily available and less costly, argon is more widely used.

Because of high purity (99.985% min) and low moisture content, liquid argon is often preferred. The argon gas shuld have a dew point of -75 °F or lower. The hose used for the shielding gas should be clean, nonporous and flexible, made of Tygon or vinyl plastic. Because rubber hose absorbs air, it should not be used. Excessive gas flow rates that cause turbulence should be avoided, and flowmeters are usually employed for all gas shields. Pressure (psi) gages may be employed for trailing and backup shields.

JOINT PREPARATION

If welding is done outside a controlled-atmosphere welding chamber, joints must be carefully

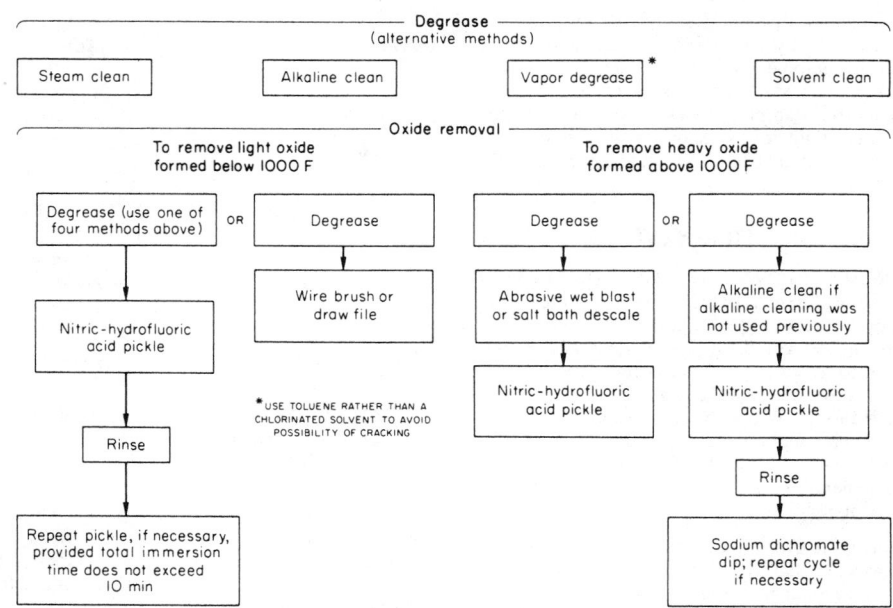

Fig. 36. Flow chart of procedures for cleaning titanium alloys prior to welding

designed so that both the top and the underside of the weld can be shielded. Dimensions of typical joints are given in Table 39. For welding titanium alloys, joint fit-up should be better than for welding other metals, because of the possibility of entrapping air in the joint. The joint should be clamped to prevent separation during welding.

CLEANING

To obtain a good weld, the joint and the surfaces of the workpieces at least 2 in. beyond the width of the gas trailing shield on each side of the weld groove must be meticulously cleaned. As shown in Fig. 36, the cleaning procedure depends on whether the oxide layer in the joint area is light or heavy.

Resistance Welding

Resistance Spot Welding

RESISTANCE SPOT WELDING (RSW) is a process in which faying surfaces are joined in one or more spots by the heat generated by resistance to the flow of electric current through workpieces that are held together under force by electrodes. The contacting surfaces in the region of current concentration are heated by a short-time pulse of low-voltage, high-amperage current to form a fused nugget of weld metal. When the flow of current ceases, the electrode force is maintained while the weld metal rapidly cools and solidifies. The electrodes are retracted after each weld, which usually is completed in a fraction of a second.

The size and shape of the individually formed welds are limited primarily by the size and con-

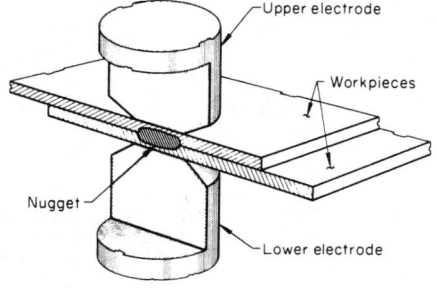

Sectional view showing shape of nugget and position of nugget relative to inner and outer surfaces of workpieces.

Fig. 1. Setup for resistance spot welding

tour of the electrode faces. The weld nugget forms at the faying surfaces (Fig. 1), but does not extend completely to the outer surfaces. In a cross section, the nugget in a properly formed spot weld is obround or oval in shape; in plan view, it has the same shape as the electrode face (which usually is round) and approximately the same size. The spots should be at a sufficient distance from the edge of the workpiece (edge distance) so that there is enough base metal to withstand the electrode force and to ensure that the local distortion during welding does not allow expulsion of metal from the weld.

APPLICATIONS

Spot welded lap joints are widely used in joining sheet steel up to about $1/8$ in. thick and are used occasionally in joining steel $1/4$ in. or more in thickness. Thicknesses of 1 in. or more have been joined by spot welding, but this requires special equipment and would not ordinarily be economical. Many assemblies of two or more sheet metal stampings that do not require gastight or liquid-tight joints can be more economically joined by high-speed RSW than by mechanical methods.

EQUIPMENT

Nearly all RSW of low-carbon steel is direct-energy welding, in which single- or three-phase 60-cycle alternating current, drawn ordinarily from 220- or 440-V inplant plower lines and stepped down to about 2 to 20 V, is fed directly to the electrodes as each weld is made. The equipment needed for RSW may be simple and inexpensive, or complex and costly, depending on the degree of automation. Machines for direct-energy welding generally are composed of these principal elements:

- *Electrical circuit*: Consists of a welding transformer, a tap switch and a secondary circuit. The secondary circuit includes the electrodes that conduct the welding current to the workpieces.
- *Control equipment*: Initiates and times the duration of current flow and may also be used instead of (or in addition to) the transformer tap switch to regulate the welding current. Controls may also sequence, time and regulate the overall operation of the machine, including initiation, automatic adjustment, and termination of welding force and current.
- *Mechanical system*: Consists of the frame, fixtures and other devices that hold and clamp the workpieces and apply the welding force.

CONTROLS FOR DIRECT-ENERGY MACHINES

Electrical controls for direct-energy resistance welding machines perform three principal functions: (1) initiating and terminating the flow of current to the welding transformer, (2) controlling the magnitude of the current and (3) timing and controlling the mechanical operations of the welding machine. The controls fall into three groups: welding contactors, timing and sequencing controls, and other current controls and regulators.

Welding contactors are devices for making and breaking an electric power circuit. On resistance welding machines, contactors and other controls are applied to the primary circuit of the welding transformer. A welding contactor should be large enough to handle the maximum input from the electric power line to the machine with the tap switch at the highest position.

Electronic contactors use ignitron tubes, thyratron tubes or silicon-controlled rectifiers to control the flow of current to the primary winding of the welding transformer. Ignitron tubes are used for applications requiring an extremely high welding machine current or a very large number of welding operations per minute. Thyratron tubes or silicon-controlled rectifiers are used to control currents that are too low for ignitron tubes to handle (less than 40 A). Electronic contactors, either synchronous or nonsynchronous, open the circuit when the current wave passes through zero.

An ignitron contactor consists of two ignitron tubes connected in inverse parallel so that one

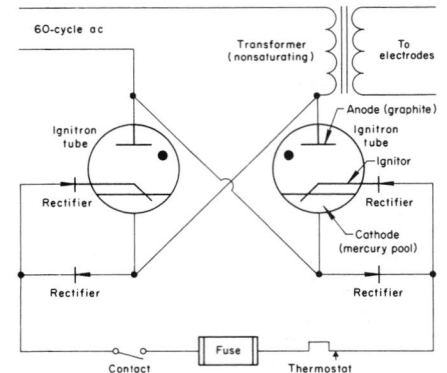

Fig. 2. Power-supply circuit and control circuit of an electronic contactor

tube carries the positive half cycle of the welding current and the other tube carries the negative half cycle. Figure 2 shows the power-supply circuit and control circuit of an electronic contactor with two ignitron tubes.

Time and Sequence Controls. The duration of current flow to the welding machine is controlled by a welding timer.

A nonsynchronous timer, by NEMA standards, may start and stop the flow of welding current at random points with respect to the line-current wave form. Variations of timing and of current input to the machine result from closing and opening the welding contactor at random points on the wave form. The time variable is at least plus or minus one half cycle and sometimes more. Ordinarily, nonsynchronous timing is sufficiently accurate for weld times of 20 cycles or longer, because the percentage variation is low and can usually be neglected. The nature and quality of the work being done determine the type of timing required for weld times of 10 to 20 cycles. Nonsynchronous timing is not recommended for weld times shorter than 10 cycles.

A synchronous timer provides a more accurate timing period and closes the primary circuit of the welding transformer at the same point (electrical angle) with respect to the power circuit voltage in making each weld. Thus, the current wave form is consistent, and the energy delivered to the welding transformer is the same for consecutive operations. Besides providing accuracy and reproducibility in these respects, a synchronous timer also eliminates variation in initial current caused by load transients.

Heat Control. The tap switch on the primary circuit of the welding transformer is used to change the ratio of transformer turns for major adjustment of the welding current. When an interme-

diate setting is needed, or for fine adjustment of the welding current, electronic heat control is used. Electronic heat control is standard equipment when synchronous timing controls are used, and it can be added to nonsynchronous controls.

For electronic heat control, the semiconductor-type rectifiers in an ignitron contactor (see Fig. 2) are replaced as firing devices by thyratrons or silicon-controlled rectifiers.

EQUIPMENT FOR DIRECT-ENERGY MACHINES

The electrical system of a direct-energy resistance welding machine is shown in Fig. 3.

Welding Transformer. The transformer used in a direct-energy resistance welding machine changes the alternating current from high voltage, low-amperage current in the primary winding, or coil, to low-voltage, high-amperage current in the secondary winding. The primary winding is connected to the power supply and is made from edge-bent strip copper, insulated between turns with the entire coil also thoroughly insulated.

There are three principal arrangements of transformer windings: multistep, series-parallel, and a combination consisting of multistep and series-parellel.

Transformer ratings for resistance welding machines are expressed in kV · A for a specified duty cycle. Standard practice is to rate resistance welding transformers on a 50% duty cycle. This duty-cycle rating is a thermal rating and states the amount of power the transformer can deliver for a stated percentage of a time period, usually 1 min, without exceeding a specified temperature rise. Thus, a welding machine rated at 100 kV · A for a 50% duty cycle can deliver 100 kV · A for 30 s of each minute without the transformer components reaching a temperature greater than that for which they were designed.

For a repetitive load, the integrating period (the sum of one "on" time and one "off" time) must not be in excess of 1 min. The kV · A demand rating of a welding transformer is the secondary open-circuit voltage multiplied by the secondary (or welding) current, divided by 1000.

The maximum permissible kV · A demand for a transformer used at a duty cycle other than the one for which the transformer is rated can be calculated from the factors listed in Table 1.

DIRECT AND SERIES (INDIRECT) WELDING

Direct Single-Spot Welding. Single-spot welds are usually made by direct welding. Figure 4 shows schematically three arrangements used for making this type of weld; these arrangements may be modified to meet special requirements. In all of the arrangements shown, one transformer sec-

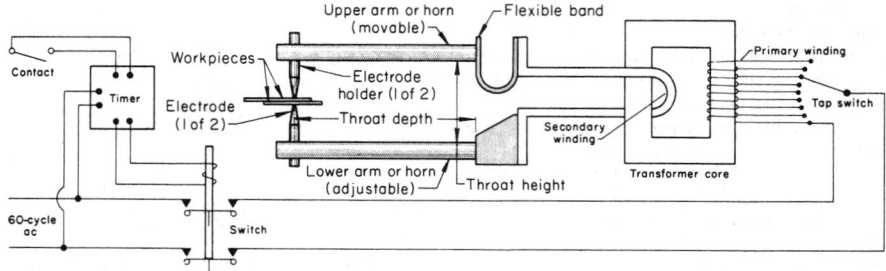

Fig. 3. Electrical system of direct-energy resistance welding machine

Table 1. Rating factors for resistance welding machines (RWMA)

Duty cycle, %	Rating factor	Duty cycle, %	Rating factor
1	7.08	30	1.29
2	5.00	35	1.195
3	4.07	40	1.115
5	3.15	50	1.000
7.5	2.57	60	0.912
10	2.23	70	0.843
15	1.82	80	0.787
20	1.58	90	0.745
25	1.41	100	0.707

ondary circuit makes one spot weld.

The simplest and most common arrangement, in which two workpieces are sandwiched between opposing upper and lower electrodes, is shown in Fig. 4(a). In Fig. 4(b), a conductive plate or mandrel having a large contacting surface is used as the lower electrode; this reduces marking on the lower workpiece and conducts heat away from the weld more rapidly and may be necessary because of the shape of the workpiece. In the arrangement in Fig. 4(c), a conductive plate or mandrel beneath the lower workpiece is used for the same purposes but in conjunction with a second upper electrode.

Direct Multiple-Spot Welding. Three arrangements of the secondary circuit for making two or more spot welds simultaneously by direct welding are shown in Fig. 5(a), (b) and (c). One transformer secondary circuit can be arranged as shown in Fig. 5(a) to make two spot welds, joining two upper workpieces to one lower workpiece. In this application, the plate or mandrel need not be an electrical conductor.

In direct multiple-spot welding (as well as in series multiple-spot welding, described below), tip contour and surface condition must be the same for each electrode. Also, the force exerted by all the electrodes on the workpieces must be equal, regardless of inequalities in work-metal thickness. The force can be equalized by using a spring-loaded electrode holder or a hydraulic equalizing system. The use of a conductive plate or mandrel, as in Fig. 5(c), minimizes weld marks on the lower workpiece.

Series Multiple-Spot Welding. Two arrangements for making a number of spot welds simultaneously by series welding are shown in Fig. 5(d) and (e). In Fig. 5(d), each of the two transformer secondary circuits makes two spot welds. A portion of the current bypasses the weld nuggets through the upper workpiece.

Resistance Seam Welding

RESISTANCE SEAM WELDING (RSEW) is a process in which heat caused by resistance to the flow of electric current in the work metal is combined with pressure to produce a welded seam.

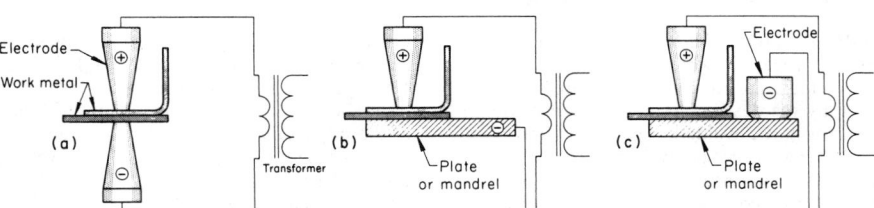

Fig. 4. Setup of work metal and electrodes for making single-spot welds

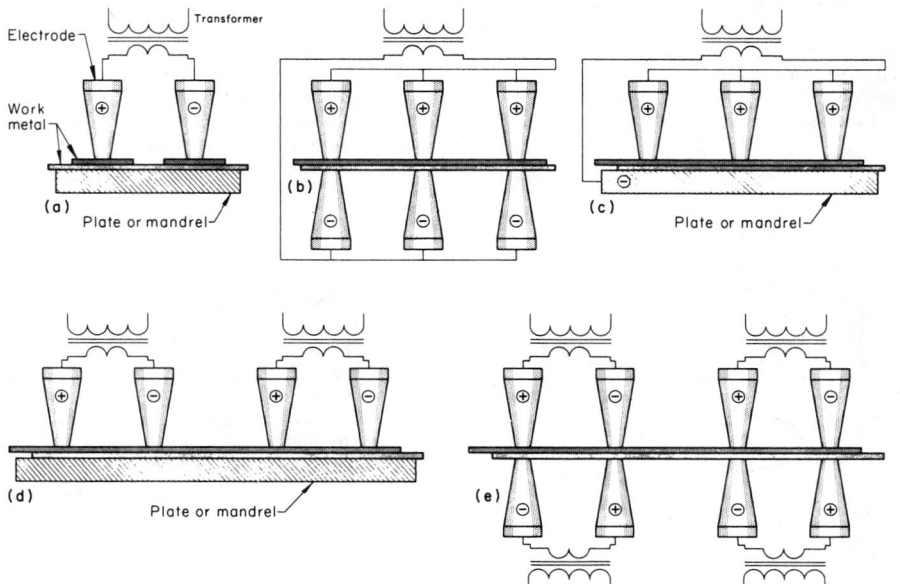

Fig. 5. Setup of work metal and electrodes for making multiple-spot welds using direct and series welding

This seam, consisting of a series of overalpping spot welds, is normally gastight or liquid-tight. Two rotating, circular electrodes (electrode wheels), or one circular and one bar-type electrode, are used for transmitting the current to the work metal. When two electrode wheels are used, one or both wheels are driven either by means of a gear-driven shaft or by a knurl or friction drive that contacts the peripheral surface of the electrode wheel.

The series of spot welds is made without retracting the electrode wheels or releasing the electrode force between spots, although the electrode wheels may advance either continuously or intermittently.

APPLICATIONS

Resistance seam welding can be applied to a variety of workpiece shapes. Girth welds can be made in round, square or rectangular parts by using electrode wheels of suitable diameter. Longitudinal welds can also be made.

Advantages of RSEW, as compared with resistance spot and projection welding, are:

- Gastight or liquid-tight joints can be produced.
- Overlap can be less than for resistance spot welds or projection welds, and seam width can be less than the diameter of spot or projection welds.

Limitations of RSEW, apart from those it shares with spot and projection welding, are:

- The weld ordinarily must proceed in a straight or uniformly curved line.
- Obstructions along with the path of the electrode wheel must be avoided or be compensated for in the design of the wheel.
- Sharp corner radii or abrupt changes in contour along the path of the electrode wheel must be avoided.
- Fatigue life of resistance seam welds is usually shorter than that of welds made by other seam welding methods.

Metals Welded. Low-carbon, high-carbon, low-alloy, high-strength low-alloy (HSLA), stainless, and many coated steels can be resistance seam welded satisfactorily. Alloys with carbon levels above about 0.15% may tend to form areas of hard martensite upon cooling.

SEAM WELDING MACHINES

A seam welding machine is similar in construction to a spot welding machine, except that one or two electrode wheels are substituted for the spot welding electrodes. Generally, seam welding is done in a press-type resistance welding machine.

There are four basic types of RSEW machines:

- *Circular*: Axis of rotation of the electrode wheels is at right angles to the front of the machine (Fig. 6a).
- *Longitudinal*: Axis of rotation of the electrode wheels is parallel to the front of the machine (Fig. 6b), and throat depth is typically 12 to 36 in.
- *Universal*: A swivel-type head and interchangeable lower arms allow the axis of rotation of the electrode wheels to be set either at right angles or parallel to the front of the machine.
- *Portable*: Work is clamped in a fixture, and a

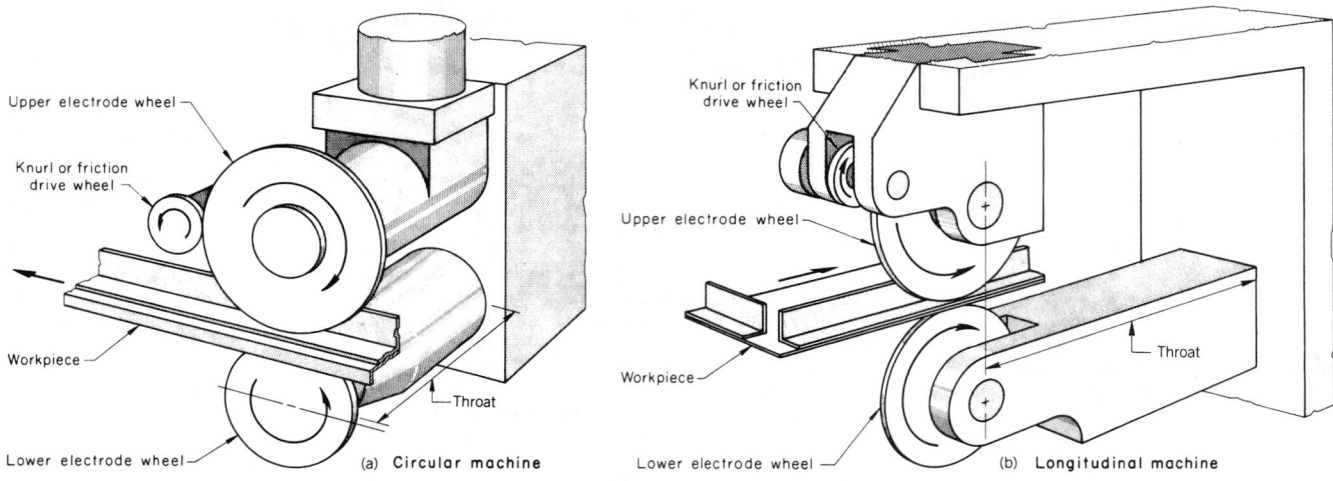

Fig. 6. Position of electrode wheels on RSEW machines

portable welding head is moved over the seam. This type of machine is used for workpieces that are too bulky to be handled by regular machines.

ELECTRODES

Electrode wheels made of Resistance Welder Manufacturers Association (RWMA) class 1 material have been used for seam welding of aluminum and magnesium alloys, galvanized steel and tin-plated steel.

Electrode wheels range in diameter from 2 to 24 in. Table 2 shows the diameters and body widths of electrodes most often used, in relation to the size of the welding machine. Narrower wheels are used in machines with knurl or fric-

Table 2. Common sizes of electrode wheels used for RSEW

Machine size	Wheel diameter, in.	Wheel width(a), in.
Small	7	$^3/_8$
Medium	8	$^3/_8$-$^1/_2$
Large	10-12	$^3/_8$-$^3/_4$

(a) Data are for body width

tion drive: wider wheels, in gear-drive and idler machines.

A small-diameter electrode wheel (Fig. 7a), or a large-diameter wheel mounted on a canted axis (Fig. 7b), can be used to avoid interference with a sidewall when a narrow flange is being welded on an inside radius or on a re-entrant curve.

Projection Welding

PROJECTION WELDING is a resistance welding process in which current flow and heating are localized at a point or points predetermined by the design or shape of one or both of two parts to be welded. The process is closely related to resistance spot welding (RSW), in which current flow and heating are localized by one or both electrode contact faces, which determine the location, size and shape of the weld produced.

Projections may be of any practical shape that can properly concentrate the welding current. In cross-wire projection welding, the curved surfaces of two intersecting wires perform the function of a projection. The shapes of parts may also take the place of conventional projections in projection welding of special types of joints.

Formation of Weld. The formation of a projection

weld nugget, which depends on the design of the projection, the selection of welding conditions and the adequacy of the resistance welding equipment, is shown in Fig. 8. In this application, 0.092-in.-thick low-carbon steel is projection welded using embossed spherical projections and a weld time of 20 cycles ($^1/_3$ s).

In the first stage of fabrication (Fig. 8a), the workpieces are brought together under pressure without applicaton of welding current, and the projection may be slightly compressed and indented into the surface of the mating workpiece. Figure 8(b), (c) and (d) show the stages of formation of a typical weld at 20, 60 to 70, and 100% of the weld time (or heat time).

At about 20% of weld time (Fig. 8b), collapse of the projection is nearly complete, and a pressure weld is formed. A nugget of fused weld metal does not begin to form until about 50% of the weld time has elapsed. At about 60 to 70% of weld time (Fig. 8c), fusion has progressed a sufficient distance from the interface to produce a well-defined weld nugget that has about half its final thickness (penetration) and diameter. Sheet separation adjacent to the weld nugget has been reduced to zero, and the softened metal above the nugget has been flattened against the face of the upper electrode (compare with Fig. 8b). Nugget diameter, penetration and shear strength continue to increase as weld time progresses. Figure 8(d) shows the fully developed weld nugget.

APPLICABILITY

The principal application of projection welding is the joining of stamped low-carbon, low-alloy, and high-strength low-alloy (HSLA) steel parts. During stamping (punching, drawing or forming), one of the parts must have a projection which has been formed during the stamping operation to enable projection welding. Projection welding is also used for joining screw-machine parts to stamped parts; the projection is machined or cold formed on the end of the screw-machine part.

Projection welding is most successful in workpieces 0.022 to 0.125 in. thick. Stock 0.010 in. thick has been projection welded; projection design is critical, however, and machines with low-inertia heads and fast follow-up are needed. Sec-

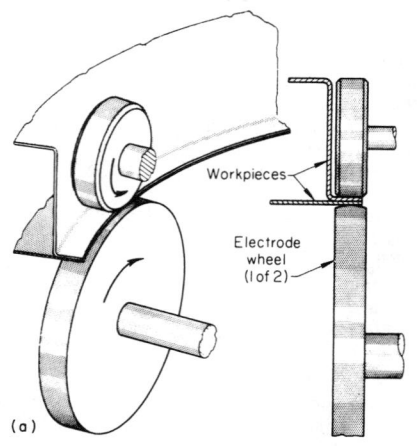

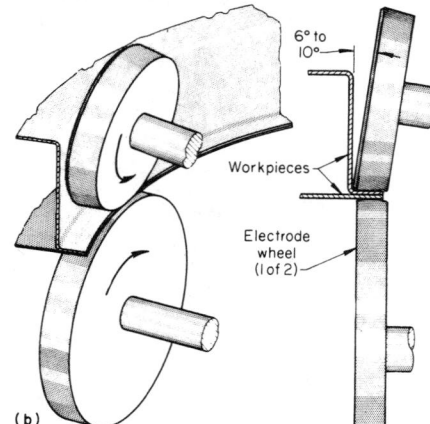

(a) Small-diameter wheel. (b) Canted, large-diameter wheel.
Fig. 7. Upper electrode wheels used to avoid interference with a sidewall

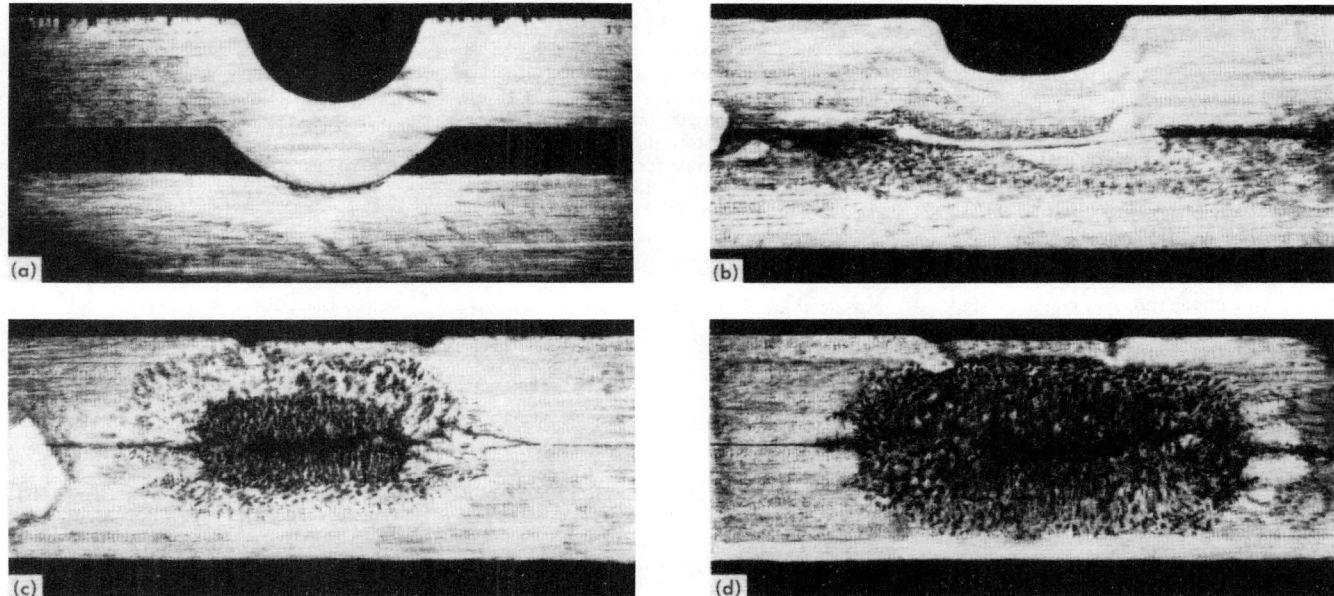

Fig. 8. Development of weld nugget during projection welding of embossed spherical projections

tions less than 0.010 in. thick are more adaptable to spot welding.

Advantages. The principal advantages of projection welding are:

- The number of welds that can be made simultaneously with one operation of the welding machine is limited only by the ability of the controls to regulate current and force.
- Because of greater current concentration at the weld, and thus less chance of shunting, narrower flanges can be welded, and welds can be spaced closer together by projection welding than by spot welding.
- Electrodes used in projection welding have faces larger than the projection or pattern of projections and larger than the faces of electrodes used for making spot welds of comparable nugget diameter. Consequently, because of lower cur-

rent density, electrodes require less maintenance than do spot welding electrodes.

- Projection welds can be made in metal that is too thick to be joined by RSW.
- Flexibility in selection of projection size and location allows welding of workpieces in thickness ratios of 6 (or more) to 1. Workpieces in thickness ratios greater than about 3 to 1 sometimes are difficult to spot weld.

Limitations of projection welding include:

- Forming of one or more projections on one of the workpieces may require extra operations unless the parts are press formed to design shape.
- When several welds are made at once with the same electrode, alignment of the work and dimensions (particularly height) of the projections must be held to close tolerances to obtain consistent weld quality.

WELDING MACHINES

Press-type machines, with either single-phase or three-phase transformers, usually are used for projection welding The welding head in these machines is guided by bearings or ways and moves in a straight line. Platens with T-slots or tapped holes are used for mounting the welding dies or electrodes. Rocker-arm machines generally are not used for projection welding because the electrode moves in an arc that can cause slippage between the components as the projection collapses.

At the start of the welding stroke (Fig. 9a), equal air pressure is applied to the top of the piston and to the diaphragm. By the time the piston has completed its travel, the diaphragm has been compressed by the retraction of the internal shaft. Retraction of the internal shaft simultaneously

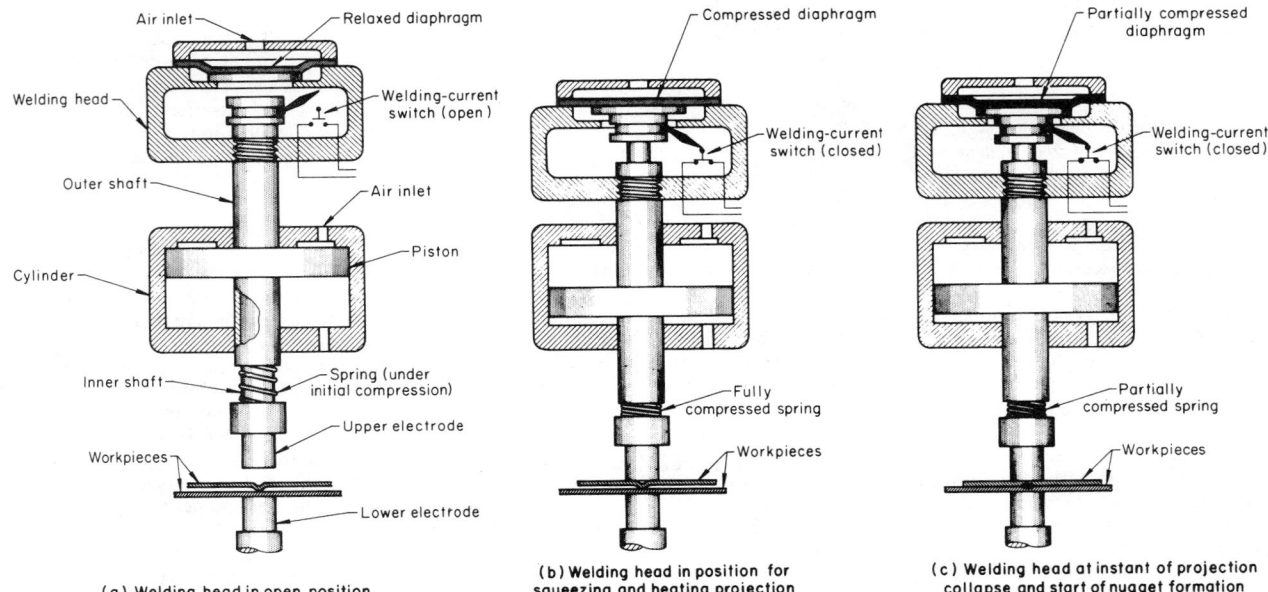

(a) Welding head in open position

(b) Welding head in position for squeezing and heating projection

(c) Welding head at instant of projection collapse and start of nugget formation

Fig. 9. Low-inertia welding head for projection welding setup

compresses the diaphragm and closes the welding-current switch; the control lever for the switch rides in the shaft collar.

When the welding current is initiated (Fig. 9b), the only remaining mass to be moved is the internal shaft and its attached electrode. Air pressure acting on the diaphragm and the force of the compressed spring between the inner and outer shafts easily overcome the low inertia of the system and move the upper electrode downward as the projection collapses. Thus, the workpieces are kept in close contact under partial electrode force until the welding head follows and forges the weld region with the full electrode force (Fig. 9c).

PROCESS VARIABLES

The major variables that affect projection welding are welding current, electrode force and weld time.

Schedules for projection welding of low-carbon steel from 0.014 to 0.125 in. thick are given in Table 3. These data are intended to serve as starting points and should be adjusted to suit the specific application and the equipment used.

Table 3. Conditions for projection welding of 1010 steel 0.014 to 0.125 in. thick using RWMA class 2 electrodes

Thickness of thinnest outside piece(a), in.	Electrode face diam (min)(b), in.	Net electrode force, lb	Weld time, cycles (60 cycles/s)	Hold time, cycles (60 cycles/s)	Welding current(c), A
0.014	$^1/_8$	175	7	15	5 000
0.021	$^5/_{32}$	300	10	15	6 000
0.031	$^3/_{16}$	400	15	15	7 000
0.044	$^1/_4$	400	20	15	7 000
0.062	$^5/_{16}$	700	25	15	9 500
0.078	$^3/_8$	1200	30	30	13 000
0.094	$^7/_{16}$	1200	30	30	14 500
0.109	$^1/_2$	1700	30	45	16 000
0.125	$^9/_{16}$	1700	30	45	17 000

Note: Steel to be welded should be free from scale, oxides, paint, grease and oil.
(a) Data based on thickness of thinner sheet and for two thicknesses only. Maximum ratio between two thicknesses, 3 to 1. (b) Face diameter equals twice the diameter of the projection. (c) Approximate current at electrodes, using 60-cycle ac
Source: "Recommended Practices for Resistance Welding," AWS C1.1

Welding current required for projection welding, although slightly less per weld than that needed for spot welding, must be high enough to cause fusion before the projection is completely flattened. The recommended current is the highest current that, when used with the correct electrode pressure, does not cause excessive expulsion of metal.

Electrode force used in projection welding depends on the work metal, the size and design of the projection, the number of projections in the joint, and the welding machine. Excessive force causes the projection to collapse before the weld area has reached the proper temperature, resulting in the formation of ring welds, in which fusion occurs around the periphery of the projection but is incomplete at the center. For best weld appearance, the electrode force should be such that the projection is flattened completely after the metal has reached welding temperature.

Weld time for a given type and thickness of work metal depends on welding current and rigidity of the projection. Weld time is less important than electrode force in projection welding of low-carbon, low-alloy and HSLA steel, provided that the time is sufficient to produce a nugget of adequate size at the chosen welding current. A short weld time creates higher production efficiency and less discoloration and distortion of the workpiece. After the proper electrode force and welding current are determined, the weld time is adjusted to make the desired weld.

ELECTRODES

An electrode designed for RSW can be used for projection welding if the electrode face is large enough to cover the projection being welded, or the pattern of projections being welded simultaneously. To minimize marking and indentation of workpieces, the recommended electrode-face diameter for making a single projection weld usually is two or more times the diameter of the projection. In multiple-projection welding, the electrode face should be large enough to extend beyond the boundaries of the pattern of projections by approximately the diameter of one projection. RWMA class 2 electrode materials generally are preferred because they provide the best compromise among electrical and conductivity, strength, hardness and temperature resistance.

Resistance Welding of Stainless Steels

STAINLESS STEELS are readily resistance welded by spot, seam and projection methods. Generally, the weld time and welding current are less than those used for welding carbon steel, but the electrode force is usually greater. Austenitic stainless steels of the 300 series are resistance welded more often than any other metal except low-carbon steel. Ideally, stainless steel to be resistance welded should contain a maximum of 0.08% C (as in types 304, 316 and 347), although steels of higher carbon content (such as types 301, 302, 309 and 310) can be successfully resistance welded.

The martensitic and ferritic types of stainless steel can be welded satisfactorily. The martensitic types are less frequently resistance welded, because joints made in them are hard and brittle in the as-welded condition.

EQUIPMENT

Resistance welding of stainless steel requires less transformer capacity than for equivalent thicknesses of any other metal, under the same conditions. The welding machines use a single-phase alternating-current power supply, a three-phase rectifier, or a frequency-converter power supply. A weld made with a three-phase power supply usually requires slightly higher current and a longer weld time than a weld made with a single-phase supply.

FACTORS AFFECTING RESISTANCE WELDING OF STAINLESS STEELS

The characteristics of stainless steel that affect resistance welding include electrical resistivity, thermal conductivity, melting temperature, strength at elevated temperatures, coefficient of thermal expansion, and contact resistance.

Electrical Resistivity. Stainless steels have much higher electrical resistance than carbon steels. As a result, more heat is generated in a stainless steel with the same current. Therefore, resistance welding of a stainless steel requires lower currents or shorter weld times than welding of a carbon steel.

Thermal Conductivity. Stainless steel has a lower thermal conductivity than carbon steel and, therefore, heat is conducted away from the weld zone more slowly. This, in conjunction with the electrical resistivity, means that less heat has to be applied to reach the melting temperature.

Melting temperatures of stainless steels have an effect on the amount of heat required to produce fusion for welding. The austenitic stainless steels melt in various ranges between 2500 and 2650 °F, and the martensitic and ferritic alloys melt in ranges between 2550 and 2790 °F. Plain low-carbon steels melt at temperatures between 2700 and 2800 °F.

Coefficient of Thermal Expansion. Austenitic and nitrogen-strengthened austenitic stainless steel expands and contracts with changing temperature almost 50% more than does plain carbon steel. These dimensional changes and the slower heat diffusion in austenitic grades result in greater thermal stress, which leads to warping. The ferritic, martensitic and precipitation-hardening grades of stainless steel have coefficients of thermal expansion from 6 to 11% lower than that of plain carbon steel.

High strength at room and elevated temperatures of nitrogen-strengthened austenitic precipitation-hardening stainless steels and, to a lesser extent, of straight-chromium grades makes it necessary to use greater electrode force than is required for carbon steel to bring the work-metal surfaces together for the required intimate contact at points of welding.

Contact resistance of stainless steel is higher than that of carbon steel and, therefore, greater electrode pressure is needed to make good resistance welds.

WELDING CHARACTERISTICS OF STAINLESS STEELS

Austenitic Stainless Steels. All of the austenitic stainless steels in the solution-annealed condition contain carbon in solution. When the steel is heated to the temperature range of 800 to 1500 °F, as in the weld heat-affected zone (HAZ), carbon combines with chromium, resulting in chromium carbide precipitation at the grain boundaries. Precipitation of chromium carbide (sensitization) makes the material susceptible to intergranular corrosion. Sensitization to intergranular corrosion is influenced in austenitic stainless steel by (1) the carbon content, (2) the solubility of the carbon, (3) the presence of one or more stabilizing elements, (4) the proximity of the actual metal temperature to 1200 °F and (5) the time period during which the metal is held in that temperature range. The first three factors are controllable only through selection of composition; the latter two are controllable by adjustment of welding conditions such as heat input, size of spot, production rate and provisions for cooling.

Martensitic stainless steels most often welded are types 403, 410, 414 and 431. These steels can be resistance welded in the annealed, hardened or hardened-and-tempered condition. Regardless of prior condition, welding produces a hardened martensitic zone adjacent to the weld. The hardness of this zone depends mainly on carbon content, although it can be controlled to a degree by the welding procedure. As the hardness of the metal in the HAZ increases, its susceptibility to cracking increases and its toughness decreases. Steels having a maximum carbon content of 0.15%, such as types 403 and 410, often produce satisfactory welds without postweld heat treatment. Steels with higher carbon content, such as types 420 and 440A, generally require postweld heat treatment.

Ferritic grades of stainless steel most often welded are types 405, 430, 442 and 446. Resistance welds in these steels can exhibit reduced ductility at room temperature because of grain growth in the HAZ. Ferritic stainless steels can also become embrittled if heated to approximately 885 °F. This phenomenon, known as "885 °F embrittlement," may be encountered in the HAZ of a ferritic stainless steel weld. If properties in the as-welded condition are unsuitable, annealing must follow welding. Postweld annealing, in addition to improving ductility, helps to restore normal corrosion resistance.

Precipitation-hardening grades of stainless steel can be resistance welded in the annealed or aged (hardened) condition. If they are welded in the aged condition, the strength of the weld area is significantly reduced because the heat of welding anneals the material subjected to it. Generally, most of the strength may be restored by simply re-aging the weldment. When maximum strength is required in the weld area, it is best to resistance weld these grades in the annealed condition and then age the weldment.

Resistance Welding of Aluminum Alloys

ALUMINUM ALLOYS, both the non-heat-treatable and heat treatable types, either wrought or cast, can be resistance welded. Some of these alloys are welded more readily than others. Characteristics of aluminum alloys include comparatively high thermal and electrical conductivity, a relatively narrow plastic range (about 200 to 400 °F temperature differential between softening and melting), considerable shrinkage during cooling, a troublesome surface oxide, and an affinity for copper electrode materials.

BASE-METAL CHARACTERISTICS

Although all aluminum alloys can be resistance spot and seam welded, some alloys or combinations of alloys have higher as-welded properties than others. Table 4 gives melting ranges, electrical and thermal conductivities, and resistance weld-ability of some wrought alloys and casting alloys.

Alclad Alloys. Resistance welding is also done on alclad products made by roll cladding some of the alloys listed in Table 4 with a thin layer of aluminum or an aluminum alloy. Because this layer is anodic to the core alloy, it provides electrochemical protection for exposed areas of the core. Alclad alloys 2219, 3003, 3004, 6061 and 7075 have a cladding of alloy 7072, which contains 1% Zn; alclad alloy 2014 has a cladding of alloy 6003 or sometimes alloy 6053, both of which contain about 1.2% Mg; and alclad alloy 2024 has a cladding of alloy 1230, which contains a minimum of 99.3% Al.

Table 4. Melting ranges, electrical and thermal conductivities, and resistance weldability of common aluminum alloys
Wrought and casting alloys are identified by Aluminum Association designations.

Alloy and temper	Melting range, °F	Electrical conductivity, % IACS(a)	Relative thermal conductivity(b), %	Resistance weldability(c)
Non-heat-treatable wrought aluminum alloys				
1350-H19	1195-1215	62	60	ST
1060-H18	1195-1215	61	57	ST
1100-H18	1190-1215	57	55	RW
3003-H18	1190-1210	40	39	RW
3004-H38	1165-1205	42	42	RW
5005-H38	1170-1205	52	51	RW
5050-H38	1160-1205	50	49	RW
5052-H38	1100-1200	35	35	RW
5083-H321	1065-1180	29	30	RW
5086-H38	1084-1184	31	32	RW
5154-H38	1100-1190	32	32	RW
5182-O	1065-1185	31	31	RW
5454-H34	1115-1195	34	34	RW
5456-H321	1060-1180	29	30	RW
Heat treatable wrought aluminum alloys				
2014-T6	950-1180	40	39	ST
2024-T361	935-1180	30	31	ST
2036-T4	1030-1200	41	40	RW
2219-T37	1010-1190	28	29	ST
6009-T4	1040-1200	44	43	RW
6010-T4	1085-1200	39	38	RW
6061-T6	1100-1200	43	43	RW
6063-T6	1140-1210	53	51	RW
6101-T6	1140-1205	57	55	RW
7075-T6	890-1180	33	33	ST
Aluminum casting alloys				
413.0-F	1065-1080	31	32	LW
443.0-F	1065-1170	37	37	RW
308.0-F	970-1135	37	37	ST
238.0-F	945-1110	25	26	LW
513.0-F	1075-1180	34	34	ST
520.0-T4	840-1120	21	22	NR
333.0-T6	960-1085	29	30	ST
C355.0-T61	1015-1150	39	38	ST
356.0-T6	1035-1135	39	38	ST
712-F	1120-1190	40	39	RW

(a) International Annealed Copper Standard, volume basis at 68 °F. For comparison, copper alloy 102 (oxygen-free copper) is 101% and low-carbon (1010) steel about 14%. (b) Based on copper alloy 102 as 100%, which has a thermal conductivity of 226 Btu/ft·h·°F at 68 °F. Low-carbon steel has a thermal conductivity of about 13% on this relative scale. (c) RW, readily weldable; ST, weldable in most applications but may require special techniques for specific applications; LW, limited weldability and usually requires special techniques; NR, welding not recommended

Effects on Weldability. The hardness of an alloy is one variable influencing weldability. Any alloy in the annealed condition (O temper) is more difficult to weld than the same alloy in a harder temper. In general, alloys in the softer tempers are much more susceptible to excessive indentation and sheet separation and to low or inconsistent weld strength. Greater deformation under the welding force causes an increase in the contact area and variations in the distribution of current and pressure. Therefore, welding of aluminum alloys in the annealed condition or in the softer tempers is not recommended without special electromechanical or electronic controls.

High-strength alloys such as 2024 and 7075 are easy to resistance weld, but may require application of a forge pressure, because they are more

susceptible to cracking and porosity than the lower-strength alloys.

RESISTANCE WELDING MACHINES

Aluminum alloys can be resistance welded with single-phase direct-energy, three-phase direct-energy, and stored-energy machines. Best results are obtained by using a machine that has these features:

- Ability to handle high welding currents for short weld times
- Synchronous electronic controls for weld time and welding current
- Low-inertia welding head for rapid follow-up of electrode force
- Slope control (for single-phase welding machines)
- Multiple-electrode-force system to permit proper forging of the weld nugget and re-dressing of electrodes.

ELECTRODES AND ELECTRODE HOLDERS

Selection of electrode material and face shape, maintenance of the face, and cooling of the electrode are important in producing consistent spot and seam welds in aluminum alloys.

Copper alloy electrodes, Resistance Welder Manufacturers Association (RWMA) classes 1, 2 and 3 are used for welding aluminum alloys. These electrode materials have high electrical and thermal conductivities, which combined with adequate cooling help keep the temperature of the electrode below the temperature at which aluminum will alloy with copper and cause electrode pickup.

Design of spot welding electrodes suitable for spot welding of aluminum includes both straight and offset electrodes. Construction details of each are shown in Fig. 10. Straight electrodes should be used whenever possible, because deflection and skidding may occur with offset electrodes under similar welding conditions. If offset electrodes are used, the amount of offset should be the minimum permitted by the shape of the assembly being welded.

Only electrodes that have the cooling-water hole within 3/8 in. of the face surface should be used.

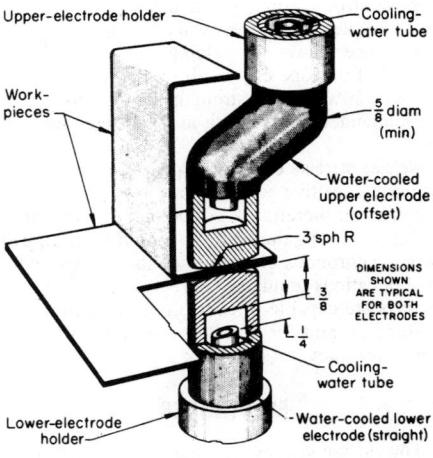

Fig. 10. Construction details of straight and offset radius-face electrodes used in resistance spot welding of aluminum alloys

A design utilizing fluted cooling-water holes provides more cooling surface than one specifying round holes.

PREWELD SURFACE PREPARATION

Although welds for some purposes can be made satisfactorily without any preweld surface preparation, welds that are free of cracks, porosity and sheet separation, and that have the most uniform strength and symmetry, are obtained only with correct procedures for cleaning and reduction or removal of oxide film. In addition, adequate surface preparation reduces electrode contamination.

Resistance Welding of Copper and Copper Alloys

RESISTANCE SPOT WELDING is widely used for joining copper and copper alloys. Principal applications include welding sections up to about 0.060 in. thick, particularly those alloys with low electrical conductivities. Many copper alloys with low conductivities can be seam welded easily. Coppers are difficult to seam weld. Projection welding is not recommended for copper or for most brasses. Bronzes can be projection welded with satisfactory results in many applications.

WELDING CHARACTERISTICS

The resistance weldability of any copper alloy is inversely proportional to its electrical and thermal conductivities. Generally, alloys with lower conductivities are easier to weld. Compared with steel, most copper alloys require shorter weld time, lower electrode force, higher current and different electrode materials that are compatible with the alloy being welded.

WELDING EQUIPMENT

Single-phase and three-phase direct-energy and electrostatic stored-energy (capacitor-discharge) welding machines are used for resistance welding of copper and copper alloys. The addition of slope control to single-phase direct-energy welding machines is not necessary for spot welding most copper alloys. In welding high-zinc brasses, the use of upslope can result in an increase of as much as 20% in weld strength. Downslope is not recommended for welding any of the copper alloys.

Welding Machine Controls. Copper alloys are particularly sensitive to variations in welding conditions, and therefore all direct-energy machines used for welding these alloys should be equipped with synchronous electronic controls, especially in applications requiring short weld times. These devices are capable of controlling weld time and welding current for repeated operations with extreme accuracy.

ELECTRODES

The current used for resistance welding of copper alloys is much higher than that used for welding low-carbon steel, and therefore, the electrode must have high electrical conductivity to minimize heat buildup.

Electrode Materials. The Resistance Welder Manufacturers Association (RWMA) class 1 electrode materials, containing copper and cadmium, are sometimes used for welding copper and high-conductivity brass and bronze. Class 2 materials, containing copper and chromium, are used on low-conductivity brass and bronze and the copper-nickel alloys. Class 3 materials are used in electrodes for seam welding.

Electrodes must be efficiently water cooled to minimize sticking to the work metal and to prolong their life. Face contours must be carefully prepared and the electrodes must be properly aligned for welding.

SELECTION OF PROCESS

Weldability of the work metal often determines which process should be used for a given application. Some of the coppers and copper alloys can be spot welded, but not seam welded because of high conductivity, and not projection welded because of low compressive strength of the projections at elevated temperature.

COPPERS

Coppers and copper alloys having electrical conductivity higher than about 30% IACS are the least well suited for resistance spot, projection or seam welding, mainly because of severe electrode pickup.

Thin copper stock can be welded using electrodes faced with RWMA class 13 (tungsten) or class 14 (molybdenum), but surface appearance is poor and frequent electrode maintenance is required. A tinned coating on wire or sheet is helpful in welding copper.

BERYLLIUM COPPER

Beryllium copper alloys can be resistance welded most successfully in thin gages. Spot welding produces satisfactory welds; seam welding is less successful. Projection welding is satisfactory, provided that the projections can be formed with the work metal in the annealed condition and without cracking the work metal around the projection. Close control of welding conditions is required for consistent weld size and joint strength.

Oxide films produced by heat treating must be removed to ensure low and consistent contact resistance. Work metals that have not been heated after rolling frequently need only degreasing before welding.

Low electrical conductivity (22% IACS for alloys C17000 and C17200) contributes to the weldability of beryllium copper alloys. However, they are more difficult to resistance weld than low-carbon steel. Alloy C17500 has an electrical conductivity of 45% IACS and is more difficult to resistance weld than higher-strength, lower-conductivity beryllium copper.

LOW- AND HIGH-ZINC BRASSES

The low-zinc brasses are difficult to weld, although easier than copper, and are subject to electrode pickup. Welds made in these brasses may lack strength, principally because of comparatively high electrical conductivity (32 to 56% IACS).

The high-zinc brasses have an electrical conductivity of 27 to 28% IACS and can be both spot and projection welded over a wide range of conditions. Electrode pickup can be a problem unless weld time, welding current and electrode force are properly selected.

Excessive electrode pickup and blowthrough of the weld may occur when long weld times, high energy input and low electrode forces are used. Yellow brasses (alloys C26800 and C27000) are less susceptible to electrode pickup than cartridge brass, except when long weld times and high energy input are used. Electrode force should be sufficient to prevent arcing or expulsion of molten metal, to which these alloys are subject because of their 30 to 40% content of zinc, which boils at about 1665 °F. As shown in Table 5, the recommended electrode force, when using electrodes having a face diameter of $^3/_{16}$ in., is approximately 400 lb.

COPPER NICKELS

The copper-nickel alloys have electrical conductivities of 4.6 to 11% IACS, are readily spot and seam welded with relatively low welding current, and generally do not alloy with the electrode material and cause electrode pickup.

NICKEL SILVERS

Nickel silvers, which have about the same conductivities (6 to 10.9% IACS) as copper nickels, are spot welded as readily as copper nickels but are more difficult to seam weld. Surface contaminants such as lead and bismuth (which form low-melting eutectics with copper and nickel) or sulfur (which may be introduced in forming) must be removed before resistance welding.

Table 5. Conditions for RSW of various copper alloys(a)

UNS No.	Alloy name	Weld time, cycles	Electrode force, lb	Welding current, A
C23000	Red brass	6	400	25 000
C24000	Low brass	6	400	24 000
C26000	Cartridge brass	4	400	25 000
C26800-C27000	Yellow brass	4	400	24 000
C28000	Muntz metal	4	400	21 000
C51000-C52400	Phosphor bronze	6	510	19 500
C62800	Aluminum bronze	4	510	21 000
C65100-C65500	Silicon bronze	6	400	16 500
C66700	Manganese brass	6	400	22 000
C68700	Aluminum brass	4	400	24 000
C69200	Silicon brass	6	510	22 000

(a) For spot welding 0.036-in.-thick sheet using RWMA type E electrodes with $^3/_{16}$-in.-diam face and 30° chamfer and made of RWMA class 1 material

Table 6. Recommended spot spacing, contacting overlap and approximate shear load of joint for RSW of high-zinc brasses

Thickness of thinnest sheet, in.	Spot spacing, in., min	Minimum contacting overlap(a), in.,	Shear load of joint, lb
0.032	5/8	1/2	330
0.050	5/8	5/8	512
0.064	3/4	3/4	680
0.094	1	1	1168
0.125	1 1/2	1 1/4	1872

(a) Minimum edge distance is equal to one half the contacting overlap.

BRONZES

The phosphor bronzes, except alloy C50500, which is not recommended for resistance spot and seam welding because of its high electrical conductivity (48% IACS), have relatively low electrical conductivity (11 to 20% IACS) and are readily spot and seam welded using low welding currents. Electrode pickup can be reduced by use of a type F (radius) electrode face and frequent redressing to keep the face clean and smooth. Hot shortness can be minimized by supporting the workpieces to prevent strain during welding and by using a greater minimum overlap than recommended by the data in Table 6.

Flash Welding

FLASH WELDING commonly is used to join sections of metals and alloys in production quantities. It is a resistance/forge welding process in which the items to be welded are securely clamped to electric current-carrying dies, heated by the electric current, and upset. Clamping ensures good electrical contact between the current-carrying dies and the workpiece and prevents the parts from slipping during the upsetting action. Flash welding equipment must be durable to withstand high clamping force and upset pressures without deflecting. If deflection occurs, misalignment of workpieces may occur during welding. Flash welding is rapid and economical, and when properly executed, welds of uniform high quality are produced. A typical flash welding machine comprises a horizontal press-transformer combination with conducting work-clamping dies mounted on the press platens.

APPLICATIONS

Flash welding can be used for joining many ferrous and nonferrous alloys and combinations of dissimilar metals. In addition to low-carbon steels, metals that are flash welded on a production basis include low-alloy steels, tool steels, stainless steels, aluminum alloys, magnesium alloys, nickel alloys and copper alloys.

WELDING PARAMETERS

Fundamentally, flash welding involves heating the ends of the pieces to be welded and subsequently forging them together. During the heating phase, a thermal distribution pattern is established along the axial length of the pieces being joined, which is characterized by a steep temperature gradient.

The major difference between the temperature pattern developed in flash welding and that developed in resistance welding is that flash welding produces a much steeper thermal gradient. This steep thermal gradient, combined with the resulting characteristic upset pattern, enables flash welding to accommodate a much greater variety of materials and shapes than can be welded by resistance welding.

Once the proper temperature-distribution pattern has been established, the abutting surfaces are rapidly forced together. The parts must be securely held together during the forging process to prevent slipping. Three distinct peaks are characteristic of flash welds (Fig. 11). The two

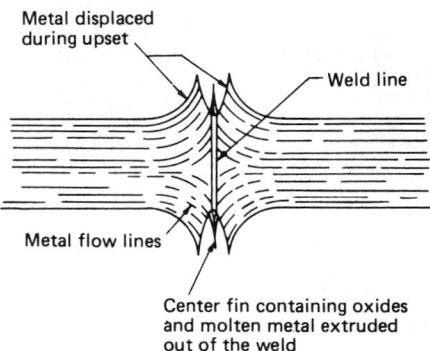

Fig. 11. Typical peaks and flow lines in a flash weld

peaks on either side of the weld line represent the material displaced by the upsetting action; the center peak is the molten metal extruded out of the weld, including oxides or contaminants formed during heating.

HEAT (ENERGY) SOURCES

One of the major considerations for flash welding is the electrical power service. Flashing is a term used to describe the major heating process during flash welding. When the ends of the workpiece are brought together under light pressure, an electrical short circuit is established through the material. Because the abutting surfaces are not perfectly matched, the short-circuit current flows across the joint only at a few small contact areas. The large amount of current flowing through a relatively small area causes very rapid heating to the melting point. Heating is so rapid and intense that molten metal is expelled explosively from the joint area. Following this explosion, a brief period of arcing occurs. Arcing is not sustained due to the low voltages employed. Studies have shown that stable flashing can be maintained with as low as 2 1/2 to 3 V, but typically the flashing voltage for alternating-current machines is from 5 to 10 V.

Following the expulsion of molten metal and subsequent arcing, small craters are formed on the ends of the abutting surfaces. The pieces are steadily advanced toward one another, and other short circuits are formed and additional molten metal is expelled. This process continues as random melting, arcing and expulsion occur over the entire cross-sectional surface. During flashing, many active areas are in various stages of this sequence (Fig. 12). Flashing surfaces act as heat sources, and the steep thermal profile is established primarily from these heat sources. Temperatures of the flashes are at or above the melting point of the material and are progressively lower as distance progresses from the flashing surface toward the clamp.

FORCE

Parts to be welded must be clamped or fixtured securely to provide good electrical contact with current-carrying dies and to transmit the upset forces. Also, a reliable source of force for the upsetting action is required. Systems for generating these forces vary greatly and are determined by the cross sections to be welded. The simplest welding machines derive their power for clamping and upsetting from the operator, i.e., the parts are clamped to the dies via screw, lever, cam or toggle force-multiplying linkages. Upset forces are generated in a similar manner. In larger machines, these forces are generated by pneumatic systems, oil hydraulic systems, combination air/oil hydraulic systems, or motor-driven cam systems.

MACHINES

Machines used for flash welding generally consist of a mainframe; a low-impedance welding transformer; a stationary platen; a movable platen on which clamping dies, electrodes and other tools needed to position and hold the workpieces are mounted; flashing and upsetting mechanisms; and the necessary electrical, air, or hydraulic controls (Fig. 13).

Flash welding machines may be manual, semiautomatic or fully automatic. With manual operation, the operator controls the speed of the platen from the time flashing is initiated until the upset is completed. In semiautomatic operation, the operator manually initiates flashing and then actuates an automatic cycle that completes the weld. In fully automatic operation, after the operator initiates the welding sequence, a fully automatic cycle can be used through the use of position-indicating devices and timers. Values for the various welding parameters are preselected by the operator. Automatic feedback control is

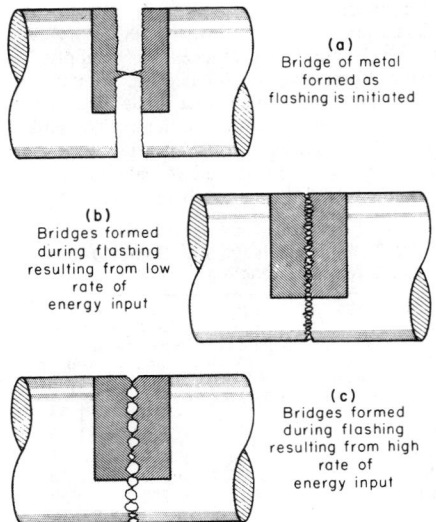

Fig. 12. Effect of energy input on bridges produced during flash welding

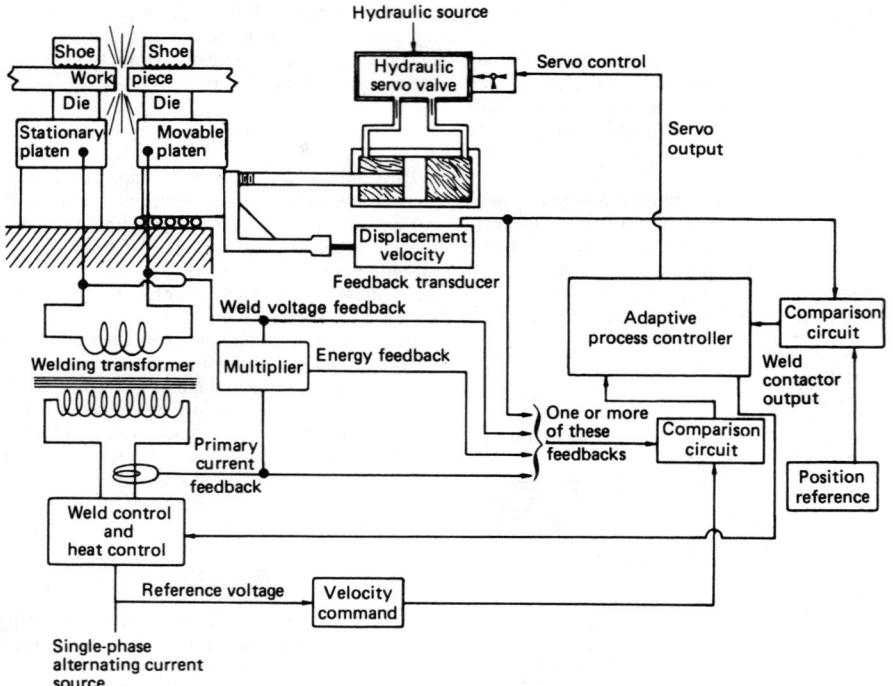

Fig. 13. Adaptive controls for flash welding

used in some applications. These fully adaptive circuits vary from current to voltage feedback circuits, and data are obtained to control the welding sequences. Figure 13 shows various adaptive controls used by different flash welding unit manufacturers.

Transformers are used to supply electrical energy for the welding operation. The transformer tap switch frequently is a rotary, eight-step, knife-type, fully enclosed locking switch, which provides convenient adjustment of the welding voltage to suit work requirements. Voltages from 4 to 16 V are common.

CLAMPING DIES AND FIXTURES

Workpieces must be accurately clamped to maintain alignment, to allow the secondary current to pass into the workpieces, and to apply the upsetting force properly. Generally, the parts of the clamping mechanism that actually grip the workpiece are the electrodes, often called clamping dies. The surface of the material in contact with the dies must be cleaned of scale, rust, paint or contaminants to ensure proper current flow.

The materials commonly used for clamping dies are RWMA class 3 and hardened tool steels such as H11, L6 and O1. Bronze and other copper-base electrode materials can be used in some applications. The die half (upper or lower) that conducts the current usually is made of a copper-base material. The other half often is made of the same copper-base material or of hardened steel.

Cooling of Electrodes. The need for water cooling of flash welding electrodes depends on the size of the electrode, die or fixture, the magnitude of the flashing current, and the production rate. When the mass of the electrodes is large compared with that of the workpiece, the heat sink generally is large enough to dissipate the heat generated by the resistance to current flow and the heat absorbed from the workpiece. With high flashing currents and high production rates, electrodes generally must be cooled.

Shape and Size of Clamping Dies. Generally, the shape of the clamping-die surface is such that the die encompasses almost the entire workpiece surface. The required area of clamping-die contact depends on the current needed for heating the

workpiece and the pressure needed for holding it. Semicircular dies are used where the line contact provided by V-shape dies gives insufficient surface for the current to flow without burning the workpiece, or for holding the workpiece without marking it.

Table 7 gives the minimum lengths of clamping dies for welding rounds, tubing and other sections of various diameters or minimum dimensions; these data are for welding steels of low or medium forging strength.

PREWELD PROCESSING

Parts that are flash welded come in a variety of forms, including forgings; rolled or extruded bar; sheet, strip or plate; ring preforms; and castings. Each of the above parts or assemblies requires preweld preparation.

Cleaning

As a minimum precaution, the ends of the workpiece that is clamped on the current-conducting die must be free from dirt, scale, surface oxidation, and grease. In addition, the ends of the workpiece that extend into the weld zone must be free from any contamination that could react with the base metal at the high temperatures developed during welding. Cleaning is needed because of the high current densities developed in the workpiece at the current-conducting dies. If insufficient electrical contact is made, weld quality is poor and localized hot spots can develop between the current-conducting dies and the workpiece. These localized hot spots are called "die burns." Many of the alloys welded have tightly adherent, highly resistive oxides that must be removed prior to welding.

In addition, poor fit-up, loose scale, and grease may cause the parts to slip during upsetting. Common cleaning techniques include (1) abrading the surfaces by grinding, grit blasting or wire brushing; (2) pickling or chemical descaling; and (3) vapor degreasing.

End Preparation

Die burns, upset slippage and inferior welds may be caused by poor fit-up between the current-conducting dies and the workpiece. In some cases, the ends of the workpiece must be machined or ground to fit the dies, particularly in welding of rough forgings or castings. Also, a chamfer may be required on the ends of the workpiece to initiate flashing action.

Alignment of Workpiece

After the workpieces are clamped in the welding unit, the alignment of the workpieces with

Table 7. Minimum lengths of clamping dies, with and without backup, for flash welding various diameters of workpieces made from steels of low or medium forging strength(a)

Workpiece diameter, in.(a)	Minimum length of clamping die, in.		Workpiece diameter, in.(a)	Minimum length of clamping die with backup, in.	Workpiece diameter, in.(a)	Minimum length of clamping die with backup, in.
	With backup	Without backup(b)				
0.250, 0.312	0.375	1.00	2.00	1.25	6.00	3.25
0.375	0.375	1.50	2.50	1.75	6.50	3.50
0.500	0.375	1.75	3.00	2.00	7.00	3.75
0.750	0.500	2.00	3.50	2.25	7.50	4.00
1.000	0.750	2.50	4.00	2.50	8.00	4.25
1.50	1.000	3.00	4.50	2.75	8.50	4.50
			5.00	2.75	9.00	4.75
			5.50	3.00	9.50	5.00

(a) Diameter of rounds or tubing, or minimum dimension of other sections. (b) Backup is recommended for all workpiece diameters or minimum dimensions over 1.50 in.

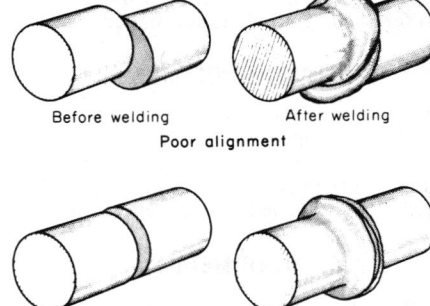

Before welding   After welding
Poor alignment

Before welding   After welding
Good alignment

Fig. 14. Effect of workpiece alignment on joint quality and weld upset

Table 8. Lengths of joints commonly flash welded in various thicknesses of flat sheet

Sheet thickness, in.	Joint length, in.	Sheet thickness, in.	Joint length, in.
0.010	1	0.060	25
0.020	5	0.080	35
0.030	10	0.100	45
0.040	15	0.125	57
0.050	20	0.187	88

Note: Length of joints in stock thicknesses of 0.050 in. and greater can be up to 100 in., depending on material extension, platen travel, die alignment, and clamping.

Table 9. Maximum flash-weldable diameters of tubing of various wall thicknesses

Wall thickness, in.	Maximum diameter, in.	Wall thickness, in.	Maximum diameter, in.
0.020	$^1/_2$	0.125	4
0.030	$^3/_4$	0.187	6
0.050	$1^1/_4$	0.250	9
0.062	$1^1/_2$	0.375	15
0.080	2	0.500	20
0.100	3		

respect to each other along the axis of upset must be maintained. After the flashing process and during upsetting, any misalignment may cause the parts to overlap one another or cause a skewed weld. If this occurs, insufficient metal will be extruded out of the weld zone during upsetting and a poor-quality weld will be formed (Fig. 14).

Dimensional Tolerances. Some combinations of joint length and stock thickness are difficult or even impossible to weld. Table 8 gives relationships between joint length (sheet width) and sheet thickness that have proven successful in a large number of production jobs. Relationships of tube diameter to wall thickness that have been successfully welded are listed in Table 9.

Burnoff

After the parts are clamped in the welding machine and the welding power is activated, the abutting surfaces are brought together for a brief period of violent flashing. This phase of the welding sequence is called "burnoff." It serves the function of squaring off the abutting surfaces and compensates for inconsistencies in end preparation.

Preheating

Preheating is a resistive heating phase of the welding process in which the heat is generated by the electrical resistance of the workpieces. It is accomplished by bringing the workpieces together under light pressure, creating a short circuit. The pressure must be great enough to prevent flashing, but not so great that workpieces are prematurely welded.

In practice, preheating is performed in a cyclic manner; workpieces are brought together briefly, then separated for a brief period to allow the heat generated to diffuse into them. This sequence is repeated several times. Fully automatic machines perform this function, thus eliminating operator variables.

Flashing

The primary purpose of flashing is to generate enough heat to produce a plastic zone that per-

mits adequate upsetting. The rate of energy input must be in proper proportion to the travel of the platen or movable die, so that constant flashing is maintained until the appropriate amount of metal is flashed off and the required plastic zone is obtained.

Upsetting (Forging)

Bonding takes place during the upsetting action, and some metal must be extruded from the weld zone to remove slag and other inclusions not expelled during flashing. The extruded metal must extend beyond the cross-sectional boundaries of the workpiece to ensure that maximum amounts of slag and inclusions are removed when the weld upset is removed during subsequent trimming. When the weld upset is removed, no evidence of the weld should remain. Porosity near the outer surface on an etched section of the weld, or a crevice around the workpiece after the weld upset has been removed, indicates incomplete bonding because of either insufficient upsetting force or insufficient plasticity during upsetting.

A large upset produced by an excessive upsetting force can be detrimental to weld quality, can waste metal, and indicates that most of the plastic metal has been extruded, requiring bonding to take place in metal where plasticity may not have been sufficient to ensure a good weld.

Other Welding Processes

Oxyfuel Gas Welding Processes and Their Application to Steel

OXYFUEL GAS WELDING (OFW) is a manual process in which the metal surfaces to be joined are melted progressively by heat from a gas flame, with or without filler metal, and are caused to flow together and solidify without the application of pressure to the parts being joined. The most important source of heat for OFW is the oxyacetylene welding torch.

MAJOR APPLICATIONS

Oxyfuel gas welding can be used to join thin carbon steel sheet and carbon steel tube and pipe. The advantages of OFW include: the ability to control heat input, bridge large gaps, avoid melt-through and clearly view the weld pool. Carbon steel sheet, formed in a variety of shapes, can often be welded more economically by OFW than by other processes. Oxyfuel gas welding is capable of joining small-diameter carbon steel pipe (up to about 3 in. in diameter) with resulting weld quality equal to competitive processes and often with greater economy. Pipe in wall thicknesses up to $^3/_8$ in. can be welded in a single pass.

GASES

Oxygen and acetylene are the principal gases used in OFW. Oxygen supports combustion of the fuel gases. Acetylene supplies both the heat intensity and the atmosphere needed to weld steel. Hydrogen, natural gas, propane and proprietary gases are used only to a limited extent in oxyfuel gas welding or brazing of metals with low melting temperatures. In welding of carbon steel, the gas flame shields the weld adequately, and no flux is required. Adjustment for correct flame atmosphere is important, but the absence of flux results in one less variable to control.

Oxygen. Only by burning selected fuel gases with high-purity oxygen in a high-velocity flame can the high heat-transfer intensity required in OFW be obtained. Oxygen is supplied for oxyfuel gas welding and cutting at a purity of 99.5% or higher, because small percentages of contaminants have a noticeable effect on combustion efficiency.

When the consumption requirement is relatively small, the oxygen is supplied and stored as a compressed gas in a standard steel cylinder under an initial pressure of up to 2.6 ksi. The most frequently used cylinder (Fig. 1) has a capacity of 244 scf.

Acetylene is a hydrocarbon gas with the chemical formula C_2H_2. When under pressure of 29.4 psi and above, acetylene is unstable, and a slight shock can cause it to explode, even in the absence of oxygen or air. Safety rules for use of acetylene and handling of acetylene equipment are extremely important.

Acetylene should not be used at pressures greater than 15 psi. Acetylene generators for on-site gas production are constructed so that the gas is not given off at pressures much greater than 15 psi. Commercially supplied portable cylinders are specially constructed (Fig. 1) to store acetylene under high pressure.

TORCHES

Welding torches control the operating characteristics of the welding flame and enable the flame to be manipulated during welding. The choice of torch size and style depends on the work to be performed. Aircraft welding torches, for example, are small and light to permit ease of

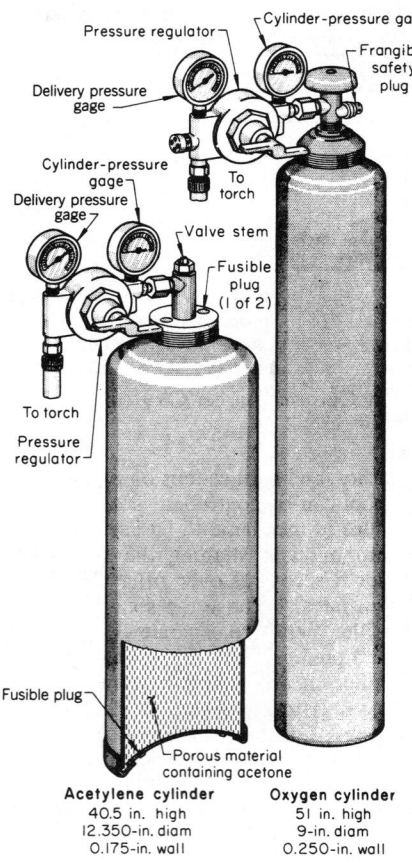

Fig. 1. Gas cylinders and regulators used in OFW

Acetylene cylinder
40.5 in. high
12.350-in. diam
0.175-in. wall

Oxygen cylinder
51 in. high
9-in. diam
0.250-in. wall

handling. Most torch styles permit one of several sizes of welding tips or a cutting attachment to be added.

The general construction of an OFW torch is shown schematically in Fig. 2. The principal operating parts are inlet valves, rear body, handle and head.

FLAME ADJUSTMENT

Different welding atmospheres and flame temperatures can be produced by varying the relative amounts of oxygen and fuel gas in the gas flowing to the tip of the torch. Usually, a welder makes the appropriate adjustments in gas flow based on the appearance of the flame. This is not true for oxyhydrogen welding, however. For all fuels, the oxyfuel flame can be classified as reducing, neutral or oxidizing.

WELDING RODS

Filler metal for OFW of low-carbon steel is available in the form of cold drawn steel rods 36 in. long and $1/16$ to $1/4$ in. in diameter. Welding rods for OFW of other metals are supplied in various lengths, depending on whether they are wrought or cast.

Oxyacetylene Braze Welding of Steel and Cast Irons

OXYACETYLENE BRAZE WELDING is a method of oxyfuel gas welding (OFW) capable of joining many base metals, but it is used primarily on steel, cast iron and malleable iron with a copper alloy filler metal (rod) and a flux. Braze welding is similar to torch brazing with a filler rod, except that joint openings are wider and distribution of filler metal takes place by deposition rather than by capillary melting action. Equipment and some filler metals used in braze welding are the same as those used in torch brazing (see the article "Torch Brazing of Steels" in this volume).

In braze welding of ferrous metals, the base metal is not melted. The filler-to-base-metal bond is the same as in torch brazing. Flux is applied to the joint surfaces, which (together with the surrounding area) are preheated to the point where the filler metal wets or "tins" these surfaces.

APPLICABILITY

Braze welding is used for making groove, fillet, plug or slot welds in metal ranging from thin sheet to heavy castings. Weld layers can be built up, as in OFW. The process is often used as a low-temperature substitute for OFW. Braze welding resembles brazing in that nonferrous filler metals are used, and bonding is achieved without melting the base metal. Braze welding resembles welding because it can be used for filling grooves and for building up fillets as required.

FLAME ADJUSTMENT

Oxyacetylene flame for braze welding is adjusted to the neutral condition for carbon steels. For cast irons, the flame is adjusted to a slightly oxidizing condition by increasing oxygen flow, which reduces the telltale acetylene "feather."

FILLER METALS

Compositions of four common copper-zinc or copper-zinc-nickel filler metals often used for braze welding are given in Table 1. These four compositions are available as welding rods in the standard diameters from $1/16$ to $1/4$ in.; length is usually 36 in. The order in which the four filler metals are listed in Table 1 is roughly that of increasing tensile strength. Weld-metal minimum tensile strengths for the compositions shown range from approximately 40 to 60 ksi. Melting temperatures of these alloys range from about 1600 to 1750 °F.

FLUXES

There are no standard specifications for braze welding fluxes. The few existing government specifications are based on composition, and the fluxes so specified are not considered as effective as the proprietary flux formulations that are available commercially.

Electron Beam Welding

ELECTRON BEAM WELDING (EBW) is a high-energy-density fusion process that is accomplished by bombarding the joint to be welded with an intense (strongly focused) beam of electrons that have been accelerated up to velocities 0.3 to 0.7 times the speed of light at 25 to 200 kV, respectively. The instantaneous conversion of the kinetic energy of these electrons into thermal energy as they impact and penetrate into the workpiece on which they are impinging causes the weld-seam interface surfaces to melt and produces the weld-joint coalescence desired. Electron beam welding is used to weld any metal that can be arc welded; weld quality in most metals

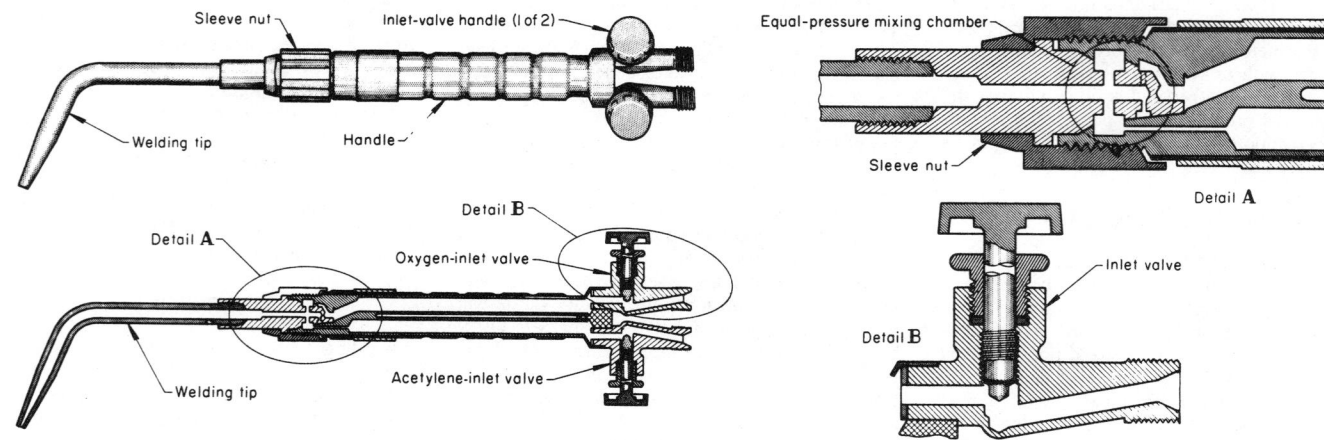

Fig. 2. Oxyfuel gas welding torch

Table 1. Approximate compositions of four common filler metals used for braze welding

| AWS classification(a) | Cu | Chemical composition, % | | | | | Minimum tensile strength, ksi | Liquidus temperature, °F |
		Zn	Sn	Fe	Ni		
RBCuZn-A 60		39	1	...	...	40	1650
RBCuZn-B 60		37.5	1	1	0.5	50	1630
RBCuZn-C 60		38	1	1	...	50	1630
RBCuZn-D 50		40	...	...	10	60	1715

(a) See AWS Specifications A5.7 and A5.8 for additional data. Additional filler metals are available. User should follow manufacturer's recommendations.

is equal to or superior to that produced by gas tungsten-arc welding (GTAW).

Because the total kinetic energy of the electrons can be concentrated onto a small area on the workpiece, power densities as high as 10^6 W/cm² can be achieved. That is higher than is possible with any other known continuous beam, including laser beams. The high power density plus the extremely small intrinsic penetration of electrons in a solid workpiece result in almost instantaneous local melting and vaporization of the workpiece material. That characteristic distinguishes EBW from other welding methods in which the rate of melting is limited by thermal conduction.

PRINCIPLES OF OPERATION

Basically, the electron beam is formed (under high-vacuum conditions) by employing a triode-style electron gun consisting of a cathode — a heated source (emitter) of electrons that is maintained at some high negative potential; a grid cup — a specially shaped electrode that can be negatively biased with respect to the hot cathode emitter (filament); and an anode — a ground potential electrode through which the electron flow passes in the form of a collimated beam.

ADVANTAGES

One of the prime advantages of EBW is the ability to make welds that are deeper and narrower than arc welds, with a total heat input that is much lower than that required in arc welding. This ability to achieve a high weld depth-to-width ratio eliminates the need for multiple-pass welds, as is required in arc welding. The lower heat input results in a narrow workpiece heat-affected zone (HAZ) and noticeably less thermal effects on the workpiece.

Laser Beam Welding

LASER BEAM WELDING (LBW) is a joining process that produces coalescence of materials with the heat obtained from the application of a concentrated coherent light beam impinging upon the surfaces to be welded. The word laser is an acronym for "light amplification by stimulated emission of radiation." The laser can be considered, for metal joining applications, as a unique source of thermal energy, precisely controllable in intensity and position. For welding, the laser beam must be focused to a small spot size to produce a high power density. This controlled power density melts the metal and, in the case of deep-penetration welds, vaporizes some of it. When solidification occurs, a fusion zone or weld joint results. The laser beam, which consists of a stream of photons, can be focused and directed by optical elements (mirrors or lenses).

The laser beam can be transmitted through the air for appreciable distances without serious power attenuation or degradation.

The two kinds of industrial laser welding processes, conduction limited and deep penetration, are normally autogenous; they use only the parent metal with no added filler. In conduction-limited LBW, the metal absorbs the laser beam at the work surface. The subsurface region is heated entirely by thermal conduction. Conduction-limited LBW uses solid-state and moderate-power carbon dioxide (CO_2) lasers and is normally performed with low average power ($\lesssim$1kW).

Deep-penetration LBW requires a high-power CO_2 laser. Thermal conduction does not limit penetration; laser beam energy is delivered to the metal through the depth of the weld, not just to the top surface. Deep-penetration LBW is similar to high-power electron beam welding (EBW) in vacuum.

APPLICATIONS

The automotive, consumer products, aerospace and electronics industries all use LBW to join a variety of materials. Among the weldable metals are lead, precious metals and alloys, copper and copper alloys, aluminum and aluminum alloys, titanium and titanium alloys, refractory metals, hot and cold rolled carbon and low-alloy steels, high-strength low-alloy (HSLA) steels, stainless steels, and heat-resistant nickel and iron-base alloys. Porosity-free welds can be attained with tensile strengths equal to or exceeding those of the base metal.

Solid-State Welding

SOLID-STATE WELDING (SSW) processes are those that produce coalescence at temperatures below the melting point of the base metal being joined. These processes involve either the use of deformation, or diffusion and limited deformation, to produce high-quality joints between both similar and dissimilar materials.

One form of SSW, called diffusion welding, is accomplished by bringing together the surfaces to be welded (faying surfaces) under moderate pressure and elevated temperature in a controlled atmosphere so that a coalescence of the interfaces or faying surfaces can occur. The other form, called deformation welding, is accomplished by subjecting the surfaces to be welded to extensive deformation. Melting or fusion is not associated with either process.

DIFFUSION WELDING

Because diffusion welding requires the application of heat and pressure, the specific equipment utilized is a function of part shape and size and the type of atmosphere and temperature re-

quired. In most cases the equipment is custom built by or for the user, and the actual welding is accomplished in a vacuum, inert-gas or reducing atmosphere.

A protective envelope is supplied by either a stationary hard shell furnace or by encapsulating the parts to be welded in a thin sheet metal container or retort made from a suitable alloy, usually low-carbon steel or 304 stainless steel. These thin containers are generally fabricated by fusion welding and subsequently evacuated through a small tube welded to the container. In some instances evacuation continues during the application of temperature and pressure, and in other cases the tube is closed by pinch welding after an initial evacuation is completed.

DEFORMATION WELDING

Cold Welding

Cold welding is accomplished at or near room temperature. The major divisions of cold welding are lap welding, butt welding and slide welding. Other deformation welding processes such as roll welding and ultrasonic welding also are often performed at room temperature.

Lap welding involves the welding of ductile sheet metal by overlapping the two pieces and applying pressure by means of suitable tools. Normally this is accomplished by means of tool steel indentors, although end lap joints have sometimes been produced employing flat anvil tools.

Butt Welding. In butt welding, the ends of pairs of wires and rods are pushed against each other with sufficient force to upset the ends. Normally this is accomplished by mechanically clamping both parts at some distance from their ends and applying pressure with the clamps, which results in the upset materials extruding between the clamps.

Thermit Welding

THERMIT* WELDING is a process that produces coalescence of metals by heating them with superheated liquid metal from an aluminothermic reaction between a metal oxide and aluminum with or without the application of pressure. Filler metal is obtained from the liquid metal. The aluminothermic reaction associated with the thermit process produces pure carbon-free heavy metals, such as chromium, manganese or vanadium from ores, oxides or chlorides. Introduction of this process constituted a major technological breakthrough; when these metals are generated by the electrothermic processes, they have high carbon contents that are unacceptable for many metallurgical applications.

PRINCIPLES OF THERMIT WELDING

The aluminothermic reaction that utilizes aluminum as a reductant takes place according to the following general equation:

Metal oxide + aluminum →

aluminum oxide + metal + heat

The intense superheat set free during the reaction generates iron and aluminum oxide in liquid form. Because each component has a differ-

*"Thermit" is a worldwide registered trademark of Th. Goldschmidt AG, Essen, West Germany.

ent density, they separate automatically within seconds, and the liquid iron can be used for different welding applications or for the production of metals and alloys. The theoretical temperature achieved by reducing iron oxide with aluminum is about 5600 °F.

PRESSURE THERMIT WELDING

Originally, pressure thermit welding required sufficient heat to be generated by the aluminothermic reactants to preheat the metal parts and to produce a butt weld by forging the pieces together. Parts are aligned so that faces to be welded are abutted tightly against one another. They must be straight, clean, and lightly clamped into position. The entire joint is surrounded by a ceramic or sand mold; an aluminothermic compound creating a slag with high melting point is used.

FUSION THERMIT WELDING

Thermit welding technology found large-scale application with the introduction of the "center-pour" weld design, where the exothermically produced iron is used as a metallurgical means of joining two parts and not merely as a heating device. When workpieces are properly aligned, with an adequate gap between the parts, a mold that is either built on the parts or premanufactured on a pattern from the parts is placed around the area to be welded.

Explosion Welding

EXPLOSION WELDING (EXW), or explosion bonding and explosion cladding, is a method in which the controlled energy of a detonating explosive is used to create a metallurgical bond between two or more similar or dissimilar metals. No intermediate filler metal, e.g., a brazing compound or soldering alloy, is needed to promote bonding, and no external heat need be applied. Diffusion does not occur during bonding.

PROCESS FUNDAMENTALS

Explosion bonding is a cold pressure welding process in which the contaminant surface films are plastically jetted off the base metals as a result of the high-pressure collision of the two metals. During the high-velocity collision of metal plates, a jet is formed between the metal plates if the collision angle and the collision velocity are in the range required for bonding. Contaminant surface films that are detrimental to the establishment of a metallurgical bond are swept away in the jet. The metal plates, cleaned of any surface films by the jet action, are joined at an internal point under the influence of the extremely high pressure that is obtained near the collision point.

Jetting Phenomenon. A layer of explosive is placed in contact with one surface of the prime metal plate, which is maintained at a constant parallel separation from the backer or base plate.

The explosive is detonated at a point or line, and as the detonation front moves across the plate, the prime metal is deflected and accelerated to plate velocity, thus establishing an angle between the two plates.

NATURE OF BONDING

Explosively welded metals that are commercially manufactured preferably exhibit a wavy bond zone interface. Aside from its technological importance, the wavy bond is remarkable because of its very regular pattern. Bond zone wave formation is analogous to fluid flowing around an obstacle. When the fluid velocity is low, the fluid flows smoothly around the obstacle; above a certain fluid velocity, the flow pattern becomes turbulent.

The obstacle in explosion bonding is the point of highest pressure in the collision region. Because the pressures in this region are many times higher than the dynamic yield strength of the metals, they flow plastically in a manner similar to fluids. The microstructure of the metals at the bond zone shows clearly that the metals did not melt but flowed plastically during the process. Electron microprobe analysis across such plastically deformed areas showed that no diffusion occurred due to extremely rapid self-quenching of the metals.

Friction Welding

FRICTION WELDING (FRW) is a process in which the heat for welding is produced by direct conversion of mechanical energy to thermal energy at the interface of the workpieces without the application of electrical energy, or heat from other sources, to the workpieces. Friction welds are made by holding a nonrotating workpiece in contact with a rotating workpiece under constant or gradually increasing pressure until the interface reaches welding temperature and then stopping rotation to complete the weld. The frictional heat developed at the interface rapidly raises the temperature of the workpieces, over a very short axial distance, to values approaching, but below, the melting range; welding occurs under the influence of pressure that is applied while the heated zone is in the plastic temperature range.

Friction welding is classified as a solid-state welding process, in which joining occurs at a temperature below the melting point of the work metal. If incipient melting does occur, there is no evidence in the finished weld because the metal is worked during the welding stage.

There are two methods of joining workpieces by FRW: continuous-drive FRW and inertia-drive FRW.

PROCESS CAPABILITIES

Many ferrous and nonferrous alloys can be friction welded. Friction welding also can be used to join metals of widely differing thermal and mechanical properties. Often combinations that can be friction welded cannot be joined by other welding processes because of the formation of brittle phases that would make such joints unserviceable. The submelting temperatures and short weld times of FRW allow many combinations of work metals to be joined.

End preparation of workpieces, other than that necessary to ensure reasonably good axial alignment and to produce the required length tolerance for a specific set of welding conditions, is not critical. Frictional wear removes irregularities from the joint surfaces and leaves clean, smooth surfaces heated to welding temperature.

Sections Welded. In FRW, the joint face of at least one of the workpieces must be essentially round. The rotating workpiece should be somewhat concentric in shape because it revolves at a relatively high speed. Workpieces that are not round, such as hexagon-shape workpieces, have been friction welded successfully, but the resulting weld upset is rough, asymmetrical, and difficult to remove without damaging the welded assembly. For special applications, welding machines have been modified so that the spindle stops at the same place each time, thus making it possible for workpieces to be oriented to each other.

Stud Welding

STUD WELDING is an arc welding process in which the contact surfaces of a stud, or similar fastener, and a workpiece are heated and melted by an arc drawn between them. The stud is then plunged rapidly onto the workpiece to form a weld. Arc initiation, arc time, and plunging are controlled automatically.

The two basic methods of stud welding are known as stud arc welding and capacitor-discharge stud welding. Both methods involve direct current and arcing. For stud arc welding, a motor-generator, a transformer-rectifier or a storage battery provides the power supply. The power supply for capacitor-discharge stud welding is a low-voltage electrostatic storage system, and the arc is produced by a rapid discharge of stored electrical energy. A welding-current controller and a welding tool or gun (stud gun) complete the necessary welding equipment for both methods of stud welding. In the arc method, a ceramic arc shield known as a ferrule is generally used to shield the arc and retain the molten weld metal.

In both methods, the stud, or fastener, serves as the electrode; the gun is the electrode holder. Flux is generally used for stud arc welding of ferrous alloys and is an integral part of the stud. Flux provides cleaning action, arc stability and a protective atmosphere. The arc time for capacitor-discharge welding is so short that flux is not needed. In welding of aluminum alloys by the stud arc welding method, a shielding gas is introduced, but this is not required with the capacitor-discharge method.

PROCESS CAPABILITIES AND LIMITATIONS

Stud welding is a rapid process. Welding time, which depends on the method and on the diameter of the stud, varies from 1 to 6 ms for the capacitor-discharge method and from 0.10 s to slightly more than 1 s for stud arc welding. In both methods, melt-through and distortion are minimal.

Studs can be welded in places that are not readily accessible, and the welded area need not be in view of the operator. Stud welding does not require access to the back of the workpiece. Stud welded fasteners can often replace fasteners normally secured by riveting, drilling and tapping; manual arc welding; resistance welding; or brazing.

The shank of a stud or other weld fastener can be of any size, shape or type that can be gripped in the stud holder. Usually, the weld base, which is the end of the stud or other fastener that is to be welded, is round for both processes. However, square- and rectangular-shape weld-base studs can be welded by the arc method.

Percussion Welding

PERCUSSION WELDING (PEW) is an arc welding process in which the heat is obtained from an arc produced by a rapid discharge of electrical energy and force is percussively applied during or immediately after the electrical discharge. A shallow layer of metal on the contact surfaces of the workpieces is melted by the heat of the arc produced between them, and one of the workpieces is impacted against the other, extinguishing the arc, expelling molten metal and completing the weld.

Arc initiation, arc time and welding force are controlled and synchronized automatically. The power supply usually is a welding transformer or a capacitor (or bank of capacitors). The welding force (forging force) is applied by electromagnetic devices, electromechanical devices, a cam-actuated direct drive, springs, or gravity.

RELATION OF PERCUSSION WELDING TO STUD WELDING

Percussion welding and stud welding are similar in three important respects:

- Welding heat is obtained from an arc.
- Force is applied percussively.
- Arc-starting methods used in the several variations of the two processes are similar.

Percussion welding and stud welding differ in certain aspects of equipment, technique, and process variables. Although a clear-cut distinction exists between PEW and stud arc welding, there are close similarities between the capacitor-discharge methods of percussion and stud welding.

Percussion welding is used principally for making electrical connections and electrical contact devices; stud welding is used chiefly for joining studs or similar shapes to larger parts for fastening other components.

APPLICATIONS

Percussion welding can be employed to join like and unlike metals that cannot usually be flash or stud welded. It is used for welding fine wire leads to filaments in lamps and terminals of electrical and electronic components where reliable joints are needed to withstand shock, vibration, and extended service at elevated temperature. Additionally, PEW is used in making telephone equipment and other electronic and electrical devices and for attaching large-area contacts to switch components. Because the total heat input is small and can be highly localized, these welds can be made a few thousandths of an inch away from glass seals or other heat-sensitive materials without damaging them.

Ultrasonic Welding

ULTRASONIC WELDING (USW) is used effectively for joining both similar and dissimilar metals with lap-joint welds. High-frequency vibrations, introduced into the areas to be joined, disrupt the metal atoms at the interface of the weld components and produce an interlocking of these atoms to achieve a mechanical joint. No significant heating is involved; the maximum temperature at the weld interface is usually in the range of 35 to 50% of the absolute melting point of the metal. The base metal does not melt and subsequently solidify with a brittle cast structure, as with high-temperature joining processes. Moderate pressure is applied during joining to maintain intimate contact between the parts, but the pressure does not cause significant deformation in the weld zone—seldom more than about 10%. Preweld cleaning requirements are minimal, while postweld cleaning or heat treatment is not necessary. Fluxes or filler metals also are not used.

Ultrasonic energy is produced through a transducer, which converts high-frequency electrical vibrations to mechanical vibrations at the same frequency, usually above 15 kHz (above the audible range). Mechanical vibrations are transmitted through a coupling system to the welding tip and into the workpieces. The tip vibrates laterally, essentially parallel to the weld interface, while static force is applied perpendicular to the interface.

Spot welding and continuous seam welding can be accomplished by this method. In spot welding, the duration of the ultrasonic pulse (usually 1 s or less) is selected according to the thickness and hardness of the materials being joined. For seam welding, the tip is a roller disk that rotates across the parts to be joined.

Most industrial metals can be joined to themselves or to other metals by ultrasonic welding.

WELDABILITY OF METALS

Ductile metals, aluminum and copper alloys, as well as the precious metals (gold, silver, platinum and palladium), are among the easiest to weld. Both aluminum alloys and precious metals can be joined to semiconductor materials such as germanium and silicon. Various types of iron and steel are somewhat more difficult to join. The most troublesome are refractory metals and some of the higher-strength metals, including titanium, nickel and zirconium, and their alloys, which should be welded only in thin gages.

High-Frequency Welding

HIGH-FREQUENCY WELDING includes a number of processes in which metal-to-metal bonding is accomplished by using heat caused by the flow of high-frequency current at the faying surfaces, with upsetting forces perpendicular to the interface added in most cases. Although similar in many respects, two separate high-frequency welding processes can be identified: high-frequency resistance welding (HFRW) and high-frequency induction welding (HFIW), sometimes referred to as induction resistance welding. In HFRW, heating current enters the work through electrical contacts on the surface. In HFIW, heating current is induced in the work through an external induction coil, and no physical or electrical contact between the workpiece and the power supply is needed.

In more conventional resistance welding processes, heating is accomplished at the joint interface by the flow of current across the interface as it passes between two electrodes pressed against the work. The current is normally direct current (dc) or low frequency alternating current (ac), from 60 to 360 Hz. In some cases, the current may be the result of a capacitor discharge. Very high currents are required to heat the metal; large, well-cooled electrical contacts must be placed as close as possible to the desired weld, and high contact pressures are normally required.

In high-frequency welding, current flows in work surfaces parallel to the joint. For HFRW, the location of the current flow in the work surfaces is determined by the location of the electrical contacts and external electrical conductors, and for HFIW, by the design and location of the induction coil. The shape and relative position of the workpiece surfaces as they are brought together immediately before welding have a major influence in determining the current flow and concentration in continuous seam welding by both processes.

The depth to which current flows depends on the frequency and the resistive and magnetic properties of the workpiece. At higher frequencies, current flow is shallower and more concentrated near the work surface. The range of frequencies used most often in this process is from 300 to 450 kHz, although frequencies as low as 10 kHz may be used in some instances. This concentration of current at the surface permits welding temperatures to be achieved with power consumption at much lower levels than with conventional resistance welding. The efficiency of welding is greatly increased, and relatively small contacts may be used.

Several factors must be considered in making successful high-frequency welds. Because of the advantages of concentrated heating, the process is inherently fast. Excessive conduction of heat away from the faying surfaces can diminish weld quality; thus, seam welding is performed at relatively high throughput speeds.

Hard Facing

HARD FACING is the application of a hard, wear-resistant material to the surface of a component by welding, spraying or allied welding processes to reduce wear or loss of material by abrasion, impact, erosion, galling and cavitation. The stipulation that the surface be modified by welding, spraying or allied welding processes excludes the use of heat treatment or surface-modification processes such as flame hardening, nitriding or ion implantation as a hard facing process. The stipulation that the surface be applied for the main purpose of reducing wear excludes the application of materials primarily used for prevention or control of corrosion or high-temperature scaling. Corrosion and/or high-temperature scaling may, however, have a major effect on the wear rate and, hence, may become a significant factor in selection of proper materials for hard facing.

Hard facing applications for wear control vary widely, ranging from very severe abrasive wear service, such as rock crushing and pulverizing, to applications that require minimization of metal-to-metal wear, such as control valves where a few thousandths of an inch of wear is intolerable. Hard facing is used for controlling abrasive wear, such as encountered by mill hammers, digging tools, extrusion screws, cutting shears, parts of earthmoving equipment, ball mills, and crusher parts. It is also used to control wear of unlubricated or poorly lubricated metal-to-metal sliding contacts such as control valves, undercarriage parts of tractors and shovels, and high-performance bearings.

HARD FACING MATERIALS

Hard facing materials include a wide variety of alloys, ceramics, and combinations of these materials. Conventional hard facing materials are normally classified as steels or low-alloy ferrous materials, chromium white irons or high-alloy ferrous materials, carbides, nickel-base alloys, or cobalt-base alloys, A few copper-base alloys are sometimes used for hard facing applications, but for the most part, hard facing alloys are either iron-, nickel- or cobalt-base. Microstructurally, hard facing alloys generally consist of hard-phase precipitates such as borides, carbides, or intermetallics bound in a softer iron-, nickel- or cobalt-base alloy matrix.

Cobalt-Base Alloys

The alloys listed in Table 2 that contain 2.5% C have more than 30 vol% total carbides, which results in extremely high abrasion resistance. The microstructure of the Co-30Cr-12W-2.5C alloy, sometimes referred to as Alloy No. 1, reveals a large volume fraction of carbides. As the carbon content is increased, the volume fraction of the matrix is decreased, and the impact resistance, weldability and machinability also decrease. Thus, the gain in abrasive wear resistance is obtained at the expense of other properties that may be more desirable.

Nickel-Base Alloys

Most commercially available nickel-base hard facing alloys can be divided into three groups: boride-containing alloys, carbide-containing alloys, and Laves phase−containing alloys. The compositions of some typical nickel-base hard facing alloys are listed in Table 3.

The boride-containing nickel-base alloys were first commercially produced as spray-and-fuse powders. These alloys are currently available from most manufacturers of hard facing products under various trade names and in a variety of forms such as bare cast rod, tube wires, and powders for plasma weld and manual torch. This group of alloys is primarily composed of Ni-Cr-B-Si-C. Usually, the boron content ranges from 1.5 to 3.5%, depending on chromium content, which varies from 0 to 15%. The higher-chromium alloys generally contain a large amount of boron, which forms very hard chromium borides with hardnesses of approximately 1800 DPH (kg/mm²). Other borides high in nickel and with lower melting points are also present to facilitate fusing.

The abrasion resistance of these alloys is a function of the amount of hard borides present. Alloys containing large amounts of boron such as Ni-14Cr-4Si-3.4B-0.75C are extremely abrasion resistant but have poor impact resistance. Because most of the boride-containing nickel-base alloys contain only small amounts of solid-solution strengtheners, considerable loss of room-temperature hardness occurs at elevated temperatures.

Iron-Base Alloys

Iron-base hard facing alloys are more widely used than cobalt- and/or nickel-base hard facing alloys and constitute the largest volume use of hard facing alloys. Iron-base hard facing alloys offer low cost and a broad range of desirable properties. Most equipment that undergoes severe wear, such as crushing and grinding equipment and earthmoving equipment, is usually very large and rugged, and often is subject to contamination. Parts subjected to wear usually require downtime for repair. For this reason, there is a general temptation to hard face them with the lowest-cost and most readily available materials. As a result, literally hundreds of iron-base hard facing alloys are in use today.

Due to the great number of alloys involved, iron-base hard facing alloys are best classified by their suitability for different types of wear and their general microstructures rather than by chemical composition. Most iron-base hard facing alloys can be divided into the following classes (see Table 4):

Table 2. Composition and hardness of selected cobalt-base hard facing alloys

AWS designation or tradename	Nominal composition	Nominal macrohardness DPH	Nominal macrohardness HRC	Matrix, DPH	Hard particles Type	Hard particles DPH
Alloy 21	Co-27Cr-5Mo-2.8Ni-0.2C	255	24-27	250	Eutectic	900
RCoCrA	Co-28Cr-4W-1.1C	424	39-42	370	Eutectic	900(a)
RCoCrB	Co-29Cr-8W-1.35C	471	40-48	420	Eutectic	900(a)
RCoCrC	Co-30Cr-12W-2.5C	577	52-54	510	M_7C_3 M_6C	900(a) 1540 1700
Alloy 20	Co-32Cr-17W-2.5C	653	53-55	540	M_7C_3 M_6C	900
Tribaloy T-800	Co-28Mo-17Cr-3Si	653	54-64	800(b)	Laves phase	1100

(a)Matrix and M_7C_3 eutectic. (b) Matrix and Laves phase eutectic

Table 3. Composition and hardness of selected nickel-base hard facing alloys

AWS designation or tradename	Nominal composition	Nominal macrohardness DPH	Nominal macrohardness HRC	Matrix, DPH	Hard particles Type	Hard particles DPH
RNiCr-C	Ni-15Cr-4Si-3.5B-0.75C	633	57	420	Primary boride Secondary boride Eutectic Carbide (M_7C_3)	2300 950 750 1700
RNiCr-B	Ni-12Cr-3.5Si-2.5B-0.35C	530	51	410	Primary boride Secondary boride Eutectic phase	2300 950 750
Hastelloy C	Ni-17Cr-17Mo-0.12C	200	HRB 95	180	M_6C	1700
Haynes 716	Ni-11Co-26Cr-29Fe-3.5W-3Mo-1.1C-0.5B	315	32	215	M_7C_3	1500
Tribaloy T-700	Ni-32Mo-15Cr-3Si	470	45	800(a)	Laves phase	...

(a) Matrix and Laves phase eutectic

Table 4. Composition and hardness of selected iron-base hard facing alloys

Nominal composition	Nominal hardness DPH	Nominal hardness HRC	Unlubricated sliding wear(a), mm³	Abrasive wear(b), mm³	Density, lb/in.³
Pearlitic steels					
Fe-2Cr-1Mn-0.2C	318	32	0.5	55	0.28
Fe-1.7Cr-1.8Mn-0.1C	372	38	0.6	67	0.27
Austenitic steels					
Fe-14Mn-2Ni-2.5Cr-0.6C	188 RHB	88 RHB	0.4	86	0.28
Fe-15Cr-15Mn-1.5Ni-0.2C	230	18	0.3	113	0.28
Martensitic steels					
Fe-5.4Cr-3Mn-0.4C	544	52	0.4	54	0.27
Fe-12Cr-2Mn-0.3C	577	54	0.3	60	0.27
High-alloy irons					
Fe-16Cr-4C	595	55	0.3	13	0.27
Fe-30Cr-4.6C	560	53	0.2	15	0.26
Fe-36Cr-5.7C	633	57	0.1	12	0.27

(a) Wear measured from tests conducted on Dow-Corning LFW-1 against 4620 steel ring at 80 rpm for 2000 revolutions varying the applied loads. (b) Wear measured from dry sand rubber wheel abrasion tests. Tested for 2000 revolutions at a load of 30 lb using a 9-in.-diam rubber wheel and AFS test sand.

Table 5. Approximate hardness of selected materials

Material	DPH	Hardness HK	Mohs
Diamond	· · ·	8000	10
SiC	3200	2750	9.2
W₂C	3000	2550	+9
VC	2800		+9
TiC	2800	2750	+9
Cr₃C₂	2700		
Alumina	· · ·	2100	9
WC	2400	1980	+9
Cr₇C₃	2100	· · ·	· · ·
Cr₂₃C₆	1650	· · ·	· · ·
Mo₂C	1570	· · ·	8
Zircon	· · ·	1340	· · ·
Fe₃C	1300	Cementite	· · ·
Quartz	1000	800	7
Lime	· · ·	560	· · ·
Glass	· · ·	500-600	· · ·

Table 6. Hard facing processing

Process	Heat source	Mode of application	Hardfacing alloy form
Oxyfuel gas welding	Oxyfuel gas	Manual or automatic	Bare cast rods or powder
Shielded metal arc welding	Electric arc	Manual	Flux coated rods
Open arc welding	Electric arc	Semiautomatic	Flux cored tube wire
Gas tungsten arc welding	Inert gas shielded electric arc	Manual or automatic	Bare rods or wire
Submerged arc welding	Flux covered electric arc	Semiautomatic	Bare solid or tubular wire
Plasma transferred welding	Inert gas shielded plasma arc	Automatic	Powder, hot wire
Plasma arc welding	Inert gas shielded plasma arc	Manual or automatic	Same as GTAW
Spray and fuse	Oxyfuel gas	Manual	Powder
Plasma spray	Plasma arc	Manual or automatic	Powder
Detonation gun	Oxyacetylene detonation	Automatic	Powder

- Pearlitic steels
- Austenitic steels
- Martensitic steels
- High-alloy irons.

Carbides

The amount of carbides used for hard facing applications is small compared with iron-base hard facing alloys, but carbides are extremely important for severe abrasion and cutting applications. Historically, tungsten-base carbides were used exclusively for hard facing applications. Recently, however, carbides of other elements such as titanium, molybdenum, tantalum, vanadium and chromium have proved useful in many hard facing applications.

The widespread use of carbides for hard facing is primarily based on the general belief that all carbides, due to their high hardness, resist abrasion. In reality, wear resistance of carbide composites is a function of the abrasion resistance of the matrix, as well as the resistance of the carbides to fracture and fragmentation, especially under high-stress applications. While the hardness values of various carbides are readily available, crushing strengths unfortunately are not. As a result, a general tendency to select carbides based solely on their hardness is widespread. Table 5 lists the hardnesses of various carbides and selected other materials of comparison.

Copper-Base Alloys

The copper-base hard facing alloys are similar to bronzes and are used in applications where copper-base bearing materials normally are employed as homogeneous parts. It is often more economical to apply copper-base hard facing alloys as overlays in less-expensive base metals such as low-carbon steels.

The properties of copper-base hard facing alloys are similar to the properties of corresponding bronzes. Copper-base hard facing alloys are used for applications where resistance to corrosion, cavitation erosion and metal-to-metal wear is desired as in bearing materials. Copper-base hard facing alloys have poor resistance to corrosion by sulfur compounds, abrasive wear and elevated-temperature creep, are not as hard as all the classes of alloys previously discussed, and are not easily welded.

HARD FACING ALLOY SELECTION

Hard facing alloy selection is guided primarily by wear and cost considerations. However, other manufacturing and environmental factors must also be considered, such as base metal, deposition process, and impact, corrosion, oxidation and thermal requirements. The factors affecting hard facing process selection are discussed in other sections of this article. Usually, the hard facing process dictates the hard facing or filler-metal product form.

Hard facing alloys usually are available as bare rod, flux-coated rod, long-length solid wires, long-length tube wires (with and without flux), or powders. Table 6 lists various welding processes, heat sources, and the proper forms of consumables for each process. In general, the impact resistance of hard facing alloys decreases as the carbide content increases. As a result, in situations where a combination of impact and abrasion resistance is desired, a compromise between the two must be made. Where impact resistance is extremely important, austenitic manganese steels are used to build up worn parts.

HARD FACING PROCESS SELECTION

Hard facing process selection may be as important as hard facing alloy selection, depending on the engineering application. Service performance requirements not only dictate hard facing alloy selection, but have a strong influence on hard facing process selection as well. Other technical factors involved in hard facing process selection include (but are not limited to) hard facing property and quality requirements, physical characteristics of the workpiece, metallurgical properties of the base metal, form and composition of the hard facing alloy, and welder skill. Ultimately, economic considerations predominate, and cost is the determining factor in the final process selection.

For years, hard facing has been limited, by definition, to welding processes. The definition adopted in this article has been expanded to include thermal spraying (THSP) as a hard facing process. The first consideration in hard facing process selection is frequently to determine if welding processes or THSP processes are preferred or required. As a rule, welding processes are preferred for hard facing applications requiring dense, relatively thick coatings with high bond strengths between the hard facing and the workpiece. Thermal spraying processes, on the other hand, are preferred for hard facing applications requiring thin, hard coatings applied with minimal thermal distortion of the workpiece.

Brazing

Furnace Brazing of Steels

FURNACE BRAZING is a mass-production process for joining the components of small assemblies by a metallurgical bond, using a nonferrous filler metal as the bonding material and a furnace as the heat source. Furnace brazing is only practical if the filler metal can be placed on the joint before brazing and retained in position during brazing.

Furnace brazing requires the use of a suitable atmosphere to protect the steel assemblies against oxidation, or oxidation and decarburization, during brazing and during cooling, which is accomplished in chambers adjacent to the brazing furnace. The proper brazing atmosphere facilitates proper wetting of the joint surfaces by the molten copper filler metal, usually without use of a brazing flux.

Although filler metals other than copper can be used in furnace brazing of carbon and low-alloy steels, copper generally is preferred because of its low cost and the high strength of the joints produced. The high brazing temperature necessary when copper filler metals are used (2000

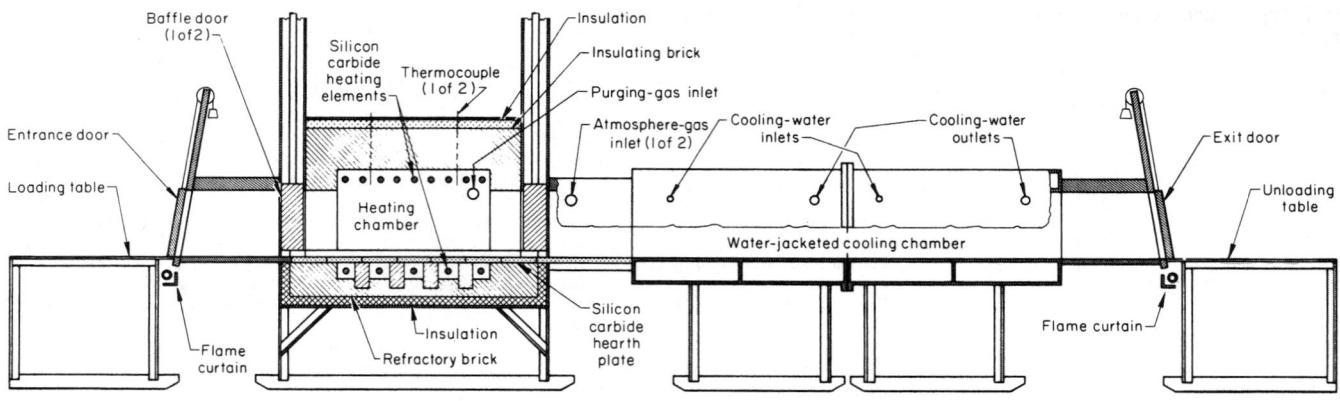

Fig. 1. Batch box-type brazing furnace

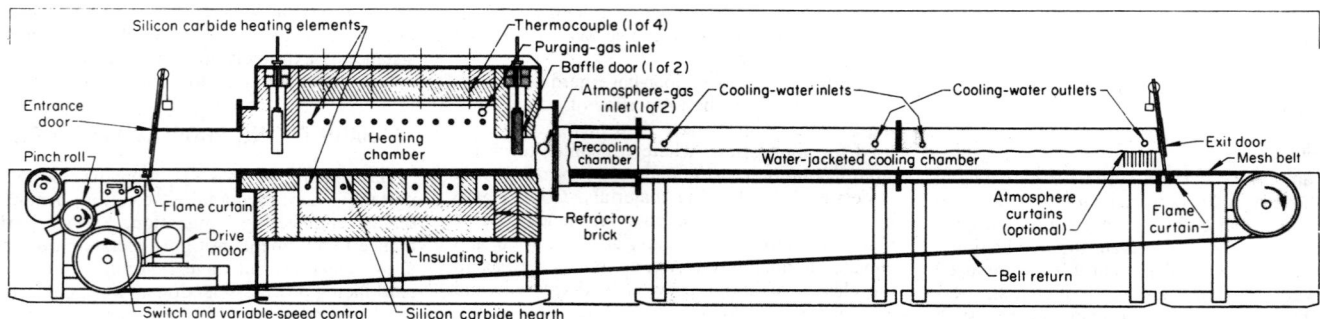

Fig. 2. Mesh-belt conveyor brazing furnace utilizing a water-jacketed cooling chamber

to 2100 °F) is also advantageous when steel assemblies are to be heat treated after brazing.

APPLICABILITY

Generally, steel assemblies that are brazed most efficiently and economically are small and weigh less than 5 lb. Much larger assemblies can be brazed in specially built furnaces; the size of assemblies is limited by the heat required to bring them to the brazing temperature.

The brazing temperature, which is considerably higher than those employed in heat treatment of steel, imposes limitations on furnace design and operation, including the maximum feasible size of the heating chamber, the degree of tightness and temperature uniformity that can be maintained, the time required to heat the workpieces to the brazing temperature, and the weight of loads that can be supported at 2000 °F without sagging of furnace fixtures.

SEQUENCE OF OPERATIONS

Furnace brazing entails four processing operations: cleaning, assembling and fixturing, brazing, and cooling.

Cleaning generally is limited to removal of oils used in machining operations. The preferred cleaning methods are alkaline cleaning, solvent cleaning and vapor degreasing. When alkaline cleaning is used, it is important that all alkaline compounds be removed from workpieces before they enter the brazing furnace. Pigmented drawing compounds containing lead are generally removed by mechanical cleaning methods, such as dry grit blasting or wet blasting with an abrasive slurry. If they are not completely removed,

drawing compounds containing lead are extremely detrimental to the quality of the brazed joint and to the life of furnace components.

Assembling and Fixturing. Components to be furnace brazed are generally designed for assembly by press fitting, expanding, swaging, or other means that eliminate the need for fixtures.

Brazing. The assemblies are moved into the brazing chamber of the furnace, where they are heated under a suitable protective atmosphere. When the assembly reaches a temperature higher than the melting point of the filler metal, the filler metal wets and flows over the steel surfaces and is drawn into the joints by capillary action. In making the bond, the filler metal forms a solid solution with, but does not melt, the steel surface. Heating time for furnace brazing of most steel assemblies is from 10 to 15 min.

Cooling. The assemblies are moved to the cooling chamber of the furnace, where they are cooled under a protective atmosphere (usually the same atmosphere as was used in the brazing chamber). They remain in the cooling chamber until they have cooled enough so that they will not discolor when exposed to air, usually to about 300 °F.

BRAZING FURNACES

Furnaces used for brazing are classified into four groups: (1) batch type, with either air or controlled atmospheres, in which workpieces are loaded and unloaded manually; (2) continuous type, with either air or controlled atmospheres, which feature an automatic conveying system; (3) retort type, with controlled atmospheres; and (4) vacuum type. The batch- and continuous-type furnaces are used most frequently for brazing of carbon and low-alloy steel assemblies. The method

of heating varies with the application. Some furnaces are heated by gas or oil, but most are heated by electrical resistance.

Batch-type furnaces, which heat each workload separately, normally consist of an insulated chamber with an external reinforced steel shell, a heating system for the chamber, and one or more access doors to the heated chamber. Standard batch furnaces may be box type (Fig. 1), top loading (pit type), side loading or bottom loading. Gas- or oil-fired batch furnaces without retorts require that flux be used on the parts for brazing.

Typically, such a furnace accommodates four trays at a time—one in the heating chamber and three in the cooling chamber. As soon as the tray in the heating chamber reaches brazing temperature, the operator pulls the end tray out of the cooling chamber and pulls the other two trays closer to the end. The operator then pushes the hot tray of brazed assemblies into the empty space in the cooling chamber and pushes a new tray of unbrazed assemblies into the heating chamber.

Continuous-type furnaces receive a steady flow of incoming assemblies. The two common furnace types used for copper brazing of steels are the mesh-belt and roller-hearth conveyor furnaces. Mesh-belt conveyor furnaces offer the advantages of continuous operation at high capacity and accurate, automatic cycle timing in both the heating and cooling chambers. An electrically heated mesh-belt conveyor furnace, incorporating several special features, is shown in Fig. 2.

PROTECTIVE FURNACE ATMOSPHERES

The gas atmospheres used in furnace brazing serve primarily to protect the steel assemblies from

Table 1. Protective atmospheres commonly used in furnace brazing

| Description | AGA class | Nominal composition, vol% | | | | | Dew point, °F | Fuel required(a), ft³ | Air-gas ratio(b) |
		N₂	CO	CO₂	H₂	CH₄			
Rich exothermic-based	102	71.5	10.5	5.0	12.5	0.5	(c)	155	6.0
Products of combustion of hydrocarbon gas passed through incandescent charcoal	402	rem	30.0	···	16.0	···	−15	80	6.0
Rich prepared nitrogen-based	202	75.3	11.0	···	13.2	0.5	−40	160	6.0
Lean endothermic-based	301	45.1	19.6	0.4	34.6	0.3	+20 to 50	190(d)	2.6
Rich endothermic-based	302	39.8	20.7	···	38.7	0.8	+25 to −5	200(d)	2.5
Dissociated ammonia	601	25.0	···	···	75.0	···	−60	(e)	···
Hydrogen purified	···	···	···	···	100.0	···	−75	···	···
Commercial nitrogen-hydrogen(f)	···	95.0	···	···	5.0	···	−35	···	···
Commercial nitrogen-methanol(f)	···	79.0	7.0	···	14.0	···	+10 to +55	···	···

(a) Per 1000 ft³ of atmosphere; based on use of natural gas rated at 1000 Btu/ft³. For other fuel gases, multiply by: 2.0, for high-hydrogen artificial gas; 2.5, for medium-hydrogen, high-CO artificial gas; 0.4, for propane; and 0.3, for butane. (b) Values indicate number of parts of air to one part of gas (based on use of natural gas at 1000 Btu/ft³). (c) Dew point is about 10 °F above temperature of cooling water; dew point may be reduced to +40 °F by refrigeration, or to −50 °F by adsorbent-tower dehydration. (d) Plus 250 ft³/1000 ft³ for heating gas. (e) 23.5 lb of ammonia per 1000 ft³ of atmosphere. (f) Percentage of atmosphere components can significantly vary depending on the requirements of the base metal. Manufacturer's recommendations should be followed.

oxidation or scaling and to assist the flow of filler metal by promoting wetting of steel surfaces. Both functions require a gas atmosphere that is reducing. When required, the atmosphere may also serve to maintain the carbon content of the steel by preventing carburization or decarburization at elevated temperatures. To satisfy all requirements, the atmosphere must provide complete protection to assemblies in both the heating and cooling chamber of the brazing furnace.

In theory, almost any reducing atmosphere can be used in furnace brazing of low-carbon steel with a copper filler metal. In practice, a rich exothermic atmosphere (AGA class 102) is usually selected, because it (1) is the least expensive of the generated atmospheres, (2) is adequately reducing, (3) has relatively low sooting potential compared with drier atmospheres containing more carbon monoxide and (4) requires a minimum of generator maintenance (see atmosphere compositions in Table 1).

VACUUM BRAZING

There are two types of vacuum brazing—high vacuum, and medium or partial vacuum. High vacuum is well suited for brazing of base metals containing hard-to-dissociate oxides such as nickel-base superalloys. Partial vacuums are used when base or filler metal volatilizes at its brazing temperature under high-vacuum conditions.

BRAZING WITH COPPER FILLER METALS

Copper is the preferred filler metal for furnace brazing of carbon and low-alloy steel assemblies without flux in reducing protective atmospheres. Significant amounts of two trace elements, arsenic and phosphorus, should be avoided because they form brittle compounds in the brazed joint. The copper should be essentially arsenic-free, and if it was deoxidized with phosphorus, the residual phosphorus content should be low.

Filler Metals. There are three standard copper brazing filler metals, bearing the American Welding Society (AWS) designations BCu-1, BCu-1a and BCu-2 (Table 2).

Table 2. Copper filler metals commonly used in furnace brazing (AWS A5.8)

AWS classification	Minimum copper, %	Brazing temperature, °F
BCu-1	99.90	2000-2100
BCu-1a	99.0	2000-2100
BCu-2	86.5	2000-2100

Torch Brazing of Steels

TORCH BRAZING is a brazing process in which the heat is obtained from a gas flame or flames impinging on or near the joint to be brazed. Torches used in this process may be of the hand-held type or may consist of fixed burners with one or many flames. Several types of fuel gas are available for combustion with oxygen or air. Torch brazing can be performed as a completely manual, partly mechanized or a completely automatic process.

Joint Design. In torch brazing of steel, the frequently used filler metals need a joint clearance (at brazing temperature) of 0.001 to 0.005 in. for capillary flow. Where thermal expansion is significant, an allowance is made on room-temperature measurements. Lap joints designed for shear loads are generally preferred to butt joints designed for tensile loads. For maximum joint efficiency, the length of the overlap should measure at least three times the thickness of the thinnest member to be joined.

FUEL GASES

Acetylene, natural gas, propane and proprietary gas mixtures are the types of fuel gas most often used in torch brazing of steel. Hydrogen, butane and producer (city) gas are seldom employed. In manual torch brazing, pure oxygen is chiefly used as the combustion agent because of its high heating rate. As a cheaper source of lower-grade oxygen, compressed air or a high-volume low-pressure blower is also suitable.

EQUIPMENT FOR MANUAL TORCH BRAZING

Manual brazing torches designed for oxyfuel gas consumption have three principal components: the torch body, which serves as the handle and is equipped with needle valves to control the flow of oxygen and fuel gas; a mixing head; and a set of torch tips, which may be supplied with or without extension tubes. Essential details of these parts are shown in Fig. 3.

FILLER METALS

Silver and copper-zinc brazing alloys are the filler metals used in torch brazing of low-carbon and low-alloy steels. The product forms, nominal compositions, and melting and brazing tem-

perature ranges of the filler metals most frequently used are given in Table 3.

Silver-alloy filler metals BAg-1 through 7, 20, 27 and 28 are used for torch brazing most types of steel to themselves or to other metals, except aluminum and magnesium, and are available in several product forms. Alloys BAg-8 through 19 and BAg-21 through 26 are used chiefly in furnace or induction brazing, and are, for the most part, used in joining base metals other than low-carbon and low-alloy steels.

The most frequently used alloys of the group shown in Table 3 are BAg-1, 1a and 3; the first two are outstanding for high fluidity, low melting temperature and narrow melting range.

Copper-zinc filler metals are used extensively in manual torch brazing and braze welding of low-carbon and low-alloy steels. They can also be used to join nickel-base and copper-nickel alloys to themselves or to steel, where corrosion resistance is not required. Table 3 gives the nominal compositions and melting characteristics of four copper-zinc alloys used as filler metals for torch brazing of steels. The two RB types are classified for braze welding and brazing; the two R types are for braze welding only. The R types are included in Table 3 to clarify a distinction among the four standard copper-zinc filler metals that is often overlooked; only the two RB types are used in torch brazing.

FLUXES

Surface oxide films inhibit the wetting of the base metal by the filler metal and, therefore, the capillary flow of the filler metal in the joint. They also prevent the formation of a true metal-to-metal braze bond. Fluxes must have sufficient chemical and physical activity to reduce or dissolve the thin surface oxide films without attacking the base metal severely. They are not made to dissolve heavy oxides, greases or dirt. For these reasons, fluxes are not intended to serve as substitutes for prebraze cleaning or for removal of heavy oxide films.

Flux Types. For torch brazing of low-carbon and low-alloy steels, fluxes can be grouped roughly into three general types, as shown in Table 4. The flux constituents are formulated to ensure that useful temperature ranges meet the applicable filler-metal temperature requirements. Because of their stability, boron compounds are present in all three types. For compatibility with the silver-alloy filler metals, fluorine compounds are added to lower the useful temperature range and the viscosity.

Table 3. Filler metals for torch brazing low-carbon and low-alloy steels(a)

AWS classification	Product form	Ag	Cu	Zn	Cd	Ni	Sn	Fe	Mn	Si	P	Solidus	Liquidus	Brazing
Silver alloys														
BAg-1	Strip, wire, powder	45	15	16	24	...	...	...	...	...	...	1125	1145	1145-1400
BAg-1a	Strip, wire, powder	50	15.5	16.5	18	...	...	...	...	...	...	1160	1175	1175-1400
BAg-2	Strip, wire, powder	35	26	21	18	...	...	...	...	...	...	1125	1295	1295-1400
BAg-2a	Strip, wire, powder	30	27	23	20	...	...	...	...	...	...	1125	1310	1310-1400
BAg-3	Strip, wire, powder	50	15.5	15.5	16	3.0	...	...	...	...	...	1170	1270	1270-1400
BAg-4	Strip, wire, powder	40	30	28	...	2.0	...	...	...	...	...	1220	1435	1435-1650
BAg-5	Strip, wire, powder	45	30	25	...	...	...	...	...	...	...	1225	1370	1370-1550
BAg-6	Strip, wire, powder	50	34	16	...	...	...	...	...	...	...	1250	1425	1425-1600
BAg-7	Strip, wire, powder	56	22	17	...	...	5.0	...	...	...	...	1145	1205	1205-1400
BAg-20	Strip, wire, powder	30	38	32	...	...	...	...	...	...	...	1250	1410	1410-1600
BAg-27	Strip, wire, powder	25	35	26.5	13.5	...	...	...	...	...	...	1125	1375	1300-1400
BAg-28	Strip, wire, powder	40	30	23	...	...	2	...	...	...	...	1200	1310	1310-1500
Copper-zinc alloys														
RBCuZn-A(a)	Strip, rod, wire, powder	...	59	40	...	...	0.6	...	...	...	...	1630	1650	1670-1750
RBCuZn-D(a)	Strip, rod, wire, powder	...	48	41	...	10.0	...	...	...	0.15	0.25	1690	1715	1720-1800
RCuZn-B(b)	Rod	...	58	38	...	0.5	0.95	0.7	0.25	0.08	...	1590	1620	...
RCuZn-C(b)	Rod	...	58	39	...	...	0.95	0.7	0.25	0.08	...	1595	1620	...

(a) Classified for braze welding and brazing. (b) Classified for braze welding.
Source: Abstracted from the mandatory and nonmandatory sections of AWS A5.7, AWS A5.8, and other sources

Table 4. Types of flux ordinarily used in torch brazing of low-carbon and low-alloy steels

AWS type	Useful temperature range, °F	Principal constituents	Available forms	Applicable filler metals
3A	1050-1600	Boric acid, borates, fluorides, fluoborates, wetting agent	Paste, liquid, slurry, powder	BAg
3C	1050-1800	Boric acid, borates, boron, fluorides, fluoborates, wetting agent	Paste, slurry, powder	BAg, RBCuZn
3D	1400-2200	Boric acid, borates, borax, fluorides, fluoborates, wetting agent	Paste, slurry, powder	RBCuZn

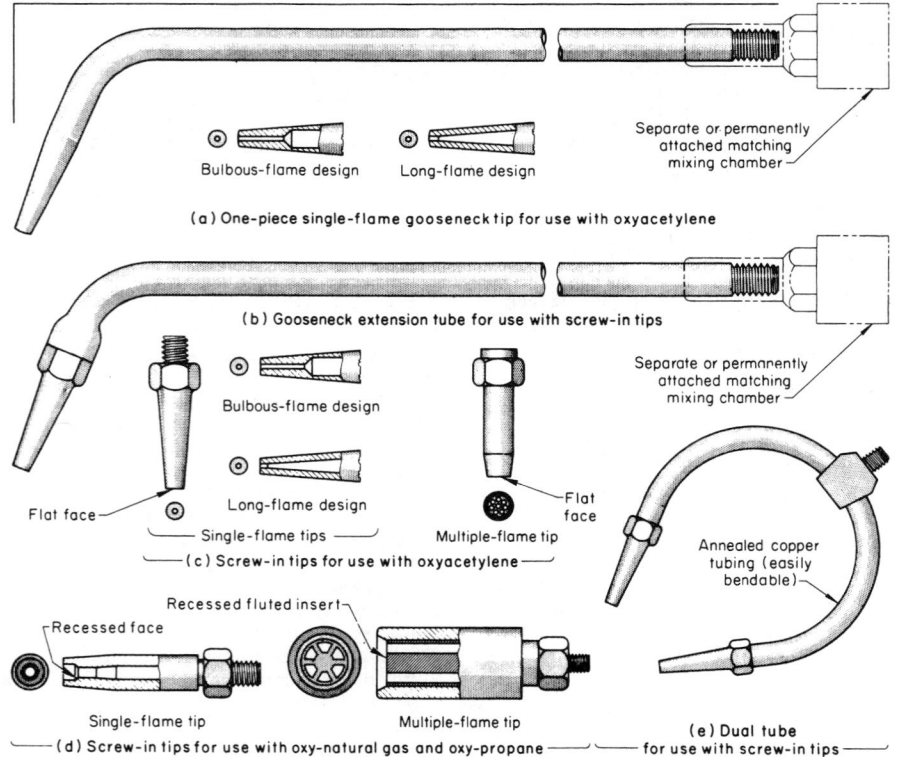

(a) One-piece single-flame gooseneck tip for use with oxyacetylene

(b) Gooseneck extension tube for use with screw-in tips

(c) Screw-in tips for use with oxyacetylene

(d) Screw-in tips for use with oxy-natural gas and oxy-propane

(e) Dual tube for use with screw-in tips

(a) One-piece tip and tube extension. (b) Two-piece tip and tube assembly that permits quick replacement of tips. (c) and (d) Basic screw-in tips. (e) Dual tip holder. Designs are available in various sizes, styles and capacities, used with complementary torch systems and gas settings.

Fig. 3. Tip designs used in manual torch brazing

Induction Brazing of Steels

INDUCTION BRAZING is a process in which the surfaces of components to be joined are selectively heated to brazing temperature by electrical energy transmitted to the workpiece by induction, rather than by a direct electrical connection, using an inductor or work coil. Heating is the result of eddy current or I^2R losses in the workpiece; because of the electrical resistivity of the workpiece and the flow of induced alternating current through it, heat is generated. When the work metal being heated is ferromagnetic, as most steels are, some additional heating results from hysteresis.

The depth of heating by induction depends primarily on the frequency of the alternating current. As the frequency is increased, both the theoretical depth of current penetration and the depth of the heated zone in the workpiece decrease. For example, the theoretical depth of current penetration is about 0.035 in. at a frequency of 3 kHz, but decreases to about 0.003 in. at 500 kHz.

PROCESS CAPABILITIES

The primary advantage of induction brazing over other brazing processes is high-speed localized heating that minimizes oxidation and thus reduces cleaning after brazing. Because the heating is localized, warpage is often less than when the entire assembly is heated, and the nature and extent of metallurgical changes, such as softening of cold worked or heat treated metal, are also minimized.

Metals Brazed. With the exception of aluminum and magnesium, most of the common metals and alloys that can be joined by other brazing processes can be brazed satisfactorily by induction in air.

Size Limitations. Induction brazing is applied most conveniently to small- and medium-size assemblies. Brazing of large assemblies, such as cylindrical bodies several feet in diameter, entails major design and installation problems, even with an adequate power supply.

Shape Limitations. Assemblies of almost any shape

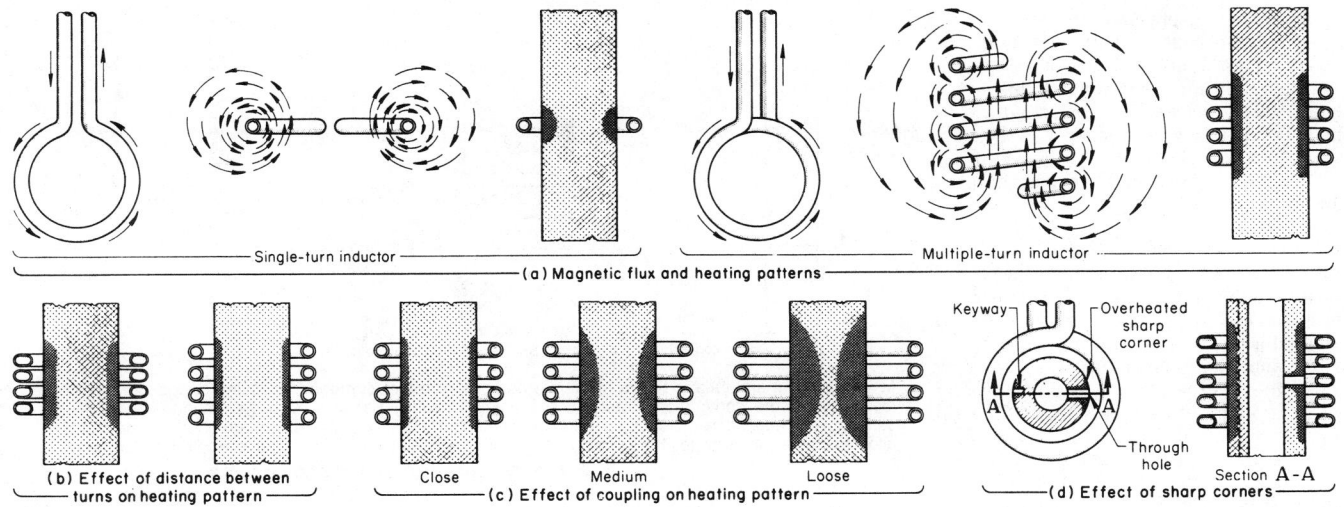

Fig. 4. Magnetic fields and heating patterns produced by various inductors

can be heated for brazing by induction, depending primarily on limitations imposed by construction of a suitable inductor, matching the impedance of the inductor and setup with output characteristics of the power supply, efficiency in heating, and cost.

POWER SUPPLY

Induction brazing is usually done at frequencies of 10 kHz and higher. A variety of types of commercial power supplies are available, ranging in rate from about $1/2$ to several hundred kilowatts, thus providing a wide selection that can be used for single or multiple inductors.

INDUCTORS

Variables that affect the pattern of heating obtained by induction and that are therefore pertinent to induction brazing include: (1) the shape of the inductor that produces the magnetic field, (2) the number of turns in the inductor, (3) the spacing between turns of the inductor, (4) the distance (air gap) between the turns of the inductor and the workpiece, (5) the presence of sharp corners on that portion of the workpiece within the magnetic field, (6) the presence of metallic shields within or near the inductor, (7) the operating frequency and (8) the alternating-current power input.

Magnetic Fields and Heating Patterns. Figure 4 shows examples of magnetic fields and heating patterns produced by induction. The patterns of magnetic flux for a single-turn and a multiple-turn inductor and the heating patterns developed by these inductors are shown in Fig. 4(a). Figure 4(b), (c) and (d), respectively, illustrate heating patterns of inductor pitch, or the distance between turns in a multiple-turn inductor, showing that finer pitch windings develop a deeper heat pattern than loose windings; coupling, or the air gap between inductor and workpiece, showing that deepest heat patterns occur with loose couplings; and sharp corners in a multiple-turn inductor, such as corners in a keyway.

FILLER METALS

Requirements of a filler metal for induction brazing are:

- Melting temperature lower than temperatures at which the metals being brazed are adversely affected
- Ability to wet the metals being brazed
- Narrow melting range (difference between solidus and liquidus temperatures)
- Sufficient fluidity at the brazing temperature to enable the filler metal to flow rapidly through the joint by capillary action
- Composition chemically compatible with the base metal
- Ability to form joints that have the required mechanical properties.

Selection of Alloy. Compositions, solidus and liquidus temperatures, and brazing-temperature ranges for the alloys that are most widely used as filler metals in induction brazing of steel are given in Table 5. Of these alloys, BAg-1 is used much more extensively than the others.

FLUXES

Flux or another means of oxygen exclusion is required for induction brazing. The flux used should decompose oxides without corroding the base metal or the filler metal, should be extremely active because of the short brazing times employed, and should be easy to remove after brazing.

Type 3A flux meets these requirements and is used for an estimated 95% of the induction brazing applications that involve steel. This flux is composed of a wetting agent and one or more of the following: boric acid, borates, fluorides, and fluoborates. It is effective within the temperature range of 1050 to 1600 °F.

Table 5. Compositions, solidus and liquidus temperatures, and brazing-temperature ranges for filler metals frequently used in induction brazing

AWS classification	Composition, %					Temperature, °F		
	Ag	Cu	Zn	Cd	Ni	Solidus	Liquidus	Brazing
BAg-1	44-46	14-16	14-18	23-25	...	1125	1145	1145-1400
BAg-2	34-36	25-27	19-23	17-19	...	1125	1295	1295-1550
BAg-3	49-51	14.5-16.5	13.5-17.5	15-17	2.5-3.5	1170	1270	1270-1500
BAg-1a	49-51	14.5-16.5	15.5-17.5	17-19	1160-1175	...	...	1145-1400
BAg-5	44-46	29-31	24-26	...	1225-1370	...	...	1370-1550
BAg-7(a)	55-57	21-23	16-18	...	1145-1205	...	...	1205-1400

(a) Contains 5.0% Sn

Resistance Brazing

RESISTANCE BRAZING is a resistance joining process in which the workpieces are heated locally and filler metal that is preplaced between the workpieces is melted by the heat obtained from resistance to the flow of electric current through the electrodes and the work. In the usual application of resistance brazing, the heating current is passed through the joint itself. Equipment is the same as that used for resistance welding, and the pressure needed for establishing electrical contact across the joint is ordinarily applied through the electrodes. The electrode pressure also is the usual means for providing the tight fit needed for capillary behavior in the joint. The heat for resistance brazing can be generated mainly in the workpieces themselves, in the electrodes, or in both, depending on the electrical resistivity and dimensions.

APPLICABILITY

Parts of many different shapes can be resistance brazed, provided that the surfaces to be joined are either flat or conform over a sufficient contact area and that they can be held together under pressure to permit the heating current to flow through the joint and the filler metal to be distributed throughout the joint by capillary action. Workpieces that can be joined by resistance brazing range from 0.001-in.-diam wire to assemblies with joint areas of about 10 to 15 in.². Joint area in most high-production resistance brazing is small, usually not more than 0.1 to 0.6 in.².

Metals Joined. The work metal most frequently

joined by resistance brazing is copper. Resistance brazing with high-resistivity electrodes or electrode facings is an efficient method of providing localized heating at the joint in this highly conductive metal, but avoiding fusion of the copper base metal. In addition, copper is the only frequently used metal that can be brazed in air with self-fluxing filler metals (copper-phosphorus alloys, BCuP type) and, thus, without the use of a flux.

In plants where copper and copper alloy assemblies are resistance brazed, the process occasionally is applied to assemblies made of steel or other metals. Typical resistance brazed low-carbon steel assemblies are transformer brackets made by joining hat-shape strips $1/8$ in. thick by 1 in. wide to flat strips $1/8$ by 1 by 6 in., making a 1-in.-square joint at each end of the flat strip using preplaced foil of BAg-1a filler metal. Fins made of steel or other metals are resistance brazed to low-carbon steel tubing for heat exchangers.

Stainless steel, nickel alloys, and aluminum are resistance brazed to a limited extent. For example, stainless steel internal baffle plates are joined by this process to the inner walls of 1020 steel tubes in heat-exchanger applications. Additional metals, not mentioned above, are also occasionally resistance brazed.

EQUIPMENT

Resistance brazing ordinarily is done with conventional resistance welding equipment, as described in the article "Resistance Spot Welding" (RSW) in this volume. This equipment can be used for some resistance brazing applications without modification. Generally, heating and cooling times are longer and electrode force is lower for resistance brazing than for RSW.

Resistance spot welding machines may be modified to provide ranges of operating conditions suitable for resistance brazing, or machines may be designed especially for resistance brazing. Other changes often needed to adapt RSW equipment for resistance brazing are in eletrode holders and electrodes.

METAL ELECTRODES

The materials used most frequently in resistance brazing are RWMA class 2 (chromium copper), RWMA class 14 (molybdenum), and various grades of carbon-graphite and graphite. Other standard and special electrode materials are sometimes used for special applications.

CARBON ELECTRODES

Ordinarily, two general types of carbon electrodes are used in resistance brazing: carbon-graphite and electrographite (artificial graphite).

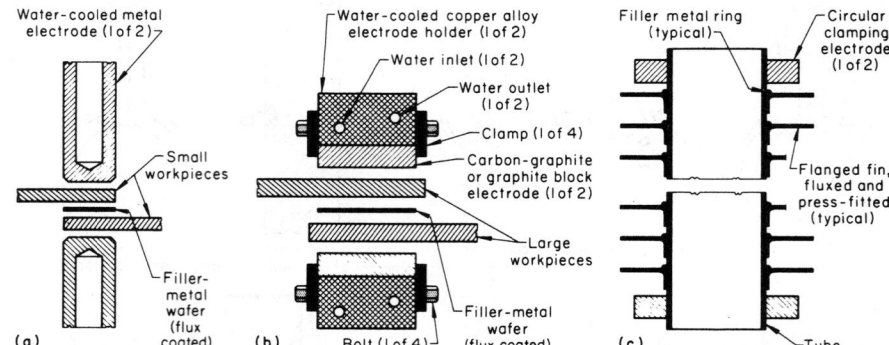

(a) For brazing small flat parts or small flat portions of larger components, using opposed water-cooled metal electrodes of the conventional resistance welding type. (b) For brazing large flat parts, typically of a highly conductive metal such as copper, using opposed carbon block electrodes attached to water-cooled copper alloy electrode holders. (c) For brazing flanged fins to a tube, using circular clamping electrodes.
Fig. 5. Arrangements for resistance brazing

These electrode materials are made by simultaneously heating and blending the finely divided raw materials with coal tar pitch, which serves as a binder.

Compositions and properties of the commercial carbon electrode materials vary; no generally accepted industry standards and terminology exist. The properties of five grades that are generally typical of the materials used in carbon electrodes for resistance brazing are given in Table 6.

ARRANGEMENT OF ELECTRODES

In most resistance brazing, the electrodes apply the brazing force and are arranged in line with the workpieces between them. This arrangement of electrodes, workpieces and filler metal is shown in Fig. 5. In making some butt joints or where space limitations or work configurations do not permit the use of opposed electrodes, the electrodes are merely connected on either side of the joint to provide the brazing current; other means are used to apply the brazing force to the joint.

FILLER METALS

Of the large number of filler metals available, only a few are used extensively in resistance brazing. Selection of filler metal for resistance brazing is similar to that for other brazing processes (see "Torch Brazing of Steels") and is discussed in the articles on those processes and on brazing of cast irons, stainless steels, aluminum alloys and copper alloys in this volume.

More attention is given in resistance brazing to selecting compatible filler metals having the lowest brazing temperature, because, in resistance brazing, the maximum local temperature reached by the work must be kept as low as pos-

sible while providing uniform heating of the abutting joint surfaces and the filler metal. Fluidity of the filler metal is not critical in most resistance brazing, because the filler metal is usually preplaced and the bond area is relatively large. The general types of filler metal usually selected for resistance brazing various classes of work metals are:

Work metal	Filler-metal alloys
Steel, stainless steel, heat-resistant alloys, copper, copper alloys, nickel alloys	Silver (BAg type)
Aluminum alloys	Al-Si
Copper and copper alloys	Cu-P

These types of filler metal all have relatively low brazing temperatures.

FLUXES AND CLEANING

A flux is used in almost all resistance brazing. It serves the same purposes in resistance brazing as in other brazing processes: providing a coating to prevent or minimize oxidation of the work metal during heating; dissolving oxides that are present or that may form during heating; and assisting the molten filler metal in wetting the work metal to promote capillary flow. The flux in resistance brazing, however, has the additional function of serving as an electrical conductor to permit passage of the brazing current through the joint; most dry fluxes are nonconductors and must be mixed with water in order to conduct current.

Application. The flux is usually applied as a dilute water-base paste shortly before the parts and filler metal are assembled for brazing. Arcing and an explosion may occur if the paste is not a thin, uniform layer and free from lumps. If the flux should dry before brazing is started, it may be possible to restore electrical conductivity by moistening it, but results are not always consistent. Once melted, the flux remains conductive. If the filler metal is in powder form, flux can be combined with it in a fine-particle paste.

Selection. The same fluxes are used for resistance brazing as for other brazing processes on the same work metals. Selection and properties of fluxes are described in this volume in the article "Torch Brazing of Steels" and articles describing brazing of specific types of alloys. Type 3A fluxes are general-purpose fluxes suitable for most metals that are commonly resistance brazed.

Table 6. Properties of five carbon electrode materials used in resistance brazing

Electrode material	Electrical resistivity, $\Omega \cdot$ in.	Scleroscope hardness	Flexural strength, psi (min)	Apparent density, g/cm³
Carbon-graphite, hard(a)	0.00080	70	3500	1.74
Carbon-graphite, hard, oxidation resistant(b)	0.00080	70	3500	1.75
Carbon-graphite, soft	0.00075	40	2400	1.57
Electrographite(a)	0.00042	50	2500	1.73
Electrographite, oxidation resistant(b)	0.00042	50	2500	1.75

(a) This type of carbon electrode material is also frequently used in air carbon arc cutting. (b) Similar to the electrode material listed immediately above, but impregnated with a small percentage of an oxidation retardant, usually an inorganic compound containing boron or phosphorus, for longer life

Dip Brazing of Steels in Molten Salt

DIP BRAZING in molten salt is also referred to as salt-bath dip brazing and molten chemical-bath dip brazing. In this process, the assembly to be brazed is immersed in a bath of molten salt, which provides the heat and may supply the fluxing action for brazing as well. The bath temperature is maintained above the liquidus of the filler metal, but below the melting range of the base metal.

FURNACES

A salt-bath furnace consists essentially of a metal or ceramic (refractory) pot that serves as a container for the molten salt. Some salt-bath furnaces are externally heated by gas, oil or electrical resistance; this type of furnace lends itself more readily to intermittent operation and is not widely used for high-volume production. On the other hand, furnaces that are internally heated by immersed or submerged electrodes are not well suited to intermittent operation; therefore, they are used for high-volume production (see other articles in this volume which deal with salt-bath furnaces).

SALTS

The types of salts used in dip brazing of carbon and low-alloy steels are neutral chloride salts, neutral chloride salts plus a fluxing agent such as borax or cryolite, and carburizing and cyaniding salts, which are also fluxing types of salts. Types and compositions of brazing salts and temperatures used for brazing of carbon and low-alloy steels with various filler metals are given in Table 7.

Fluxing agents such as borax and cryolite are added to neutral chloride salts to produce a fluxing environment in the bath. When these fluxing agents are used with silver alloy or copper-zinc filler metals, periodic flux additions are required to maintain the fluxing potential of the bath. Above 1200 °F, the fluxing potential can decrease rapidly because of oxidation from contact with air or the parts being brazed; therefore, the fluxing agent must be replenished more frequently.

Carburizing and cyaniding salts provide their own fluxing action. In addition, they supply carbon or carbon and nitrogen to the surface of the steel assembly as it is being brazed. Although silver brazing alloys have been used successfully, RBCuZn-A filler metal is generally preferred.

FLUXES

An adequate fluxing environment is needed to ensure good flow and penetration of the brazing alloy in salt-bath dip brazing. When brazing is done in a neutral chloride salt bath, a flux is usually applied to the assemblies before brazing. Generally, the application of flux to the assembly is not necessary when using a cyanide bath or other fluxing bath.

Flux can be applied by brushing, dipping or spraying the parts to be brazed before, during or after assembly. After flux application, if any moisture is present, the assemblies must be preheated to dry them before immersion in the salt bath. Typical fluxes employed for prefluxing of carbon steels and low-alloy steels that are to be brazed in a salt bath are American Welding Society (AWS) types 3A and 3B.

FILLER METALS

The brazing filler metals shown in Table 7 are the most widely used for salt-bath dip brazing of carbon and low-alloy steels. Although silver alloys BAg-13 and 13A (not shown in Table 7) can be used for brazing in a salt bath, they have been supplanted in most applications by copper-zinc alloys, which are less costly and have similar brazing temperature ranges. The rapid heating rate and nonoxidizing environment in a salt bath minimize dezincification of copper-zinc alloys, thereby facilitating the use of these alloys.

JOINT DESIGN

The filler metal providing the bond in brazed joints is drawn by capillary action between closely adjacent, matching surfaces. A diametral clearance of 0.001 to 0.003 in. is considered necessary for good flow and penetration of silver or copper-zinc filler metals in most joints. For copper brazing, joint clearance can range from a slight interference fit to a positive diametral clearance of about 0.002 in. When brazing dissimilar metals that have differing coefficients of thermal expansion or dissimilar masses of the same metal, the designer must be aware of the differing rates of expansion to ensure that the required joint clearance is obtained between the components at the brazing temperature.

Brazing of Cast Irons

BRAZING of gray, ductile and malleable cast irons differs from brazing of steel in two principal respects: special precleaning methods are necessary to remove graphite from the surface of the iron, and the brazing temperature is kept as low as feasible to avoid reduction in the hardness and strength of the iron.

The processes used for brazing cast irons are the same as those used for brazing steel—furnace, torch, induction and dip brazing. As with other metals, selection of the brazing process depends largely on the size and shape of the assembly, the quantity of assemblies to be brazed, and the equipment available.

FILLER METAL AND FLUX

Because most cast irons are brazed at relatively low temperatures, the filler metals used are almost exclusively silver brazing alloys. Compositions and other information concerning the more common silver alloy filler metals are listed in the articles "Furnace Brazing of Steels" and "Brazing of Stainless Steels" in this volume. Of these silver alloys, BAg-1 is most often used for brazing of cast iron, principally because it has the lowest brazing-temperature range. A fluoride-type flux such as American Welding Society (AWS) type 3A is usually used with BAg-1 filler metal.

BRAZEABILITY

Relatively high silicon content and sand inclusions on as-cast surfaces have some adverse effects on the brazeability of cast iron. These effects, however, are less significant than the adverse effect of graphite, which is present in all gray, ductile and malleable cast irons. Graphite has essentially the same effect on machined joint surfaces as on as-cast surfaces. Malleable iron is generally considered to be the most brazeable, and gray iron the most difficult to braze, of the three types of cast irons.

PREPARATION OF CASTINGS FOR BRAZING

Preferred joint designs for brazing cast irons are generally the same as for steel. Best results are obtained using diametral clearances in the range of 0.002 to 0.005 in. Diametral clearances up to 0.010 in. may be used, but this much clearance will result in lower joint strength and added filler-metal cost.

Methods of Surface Preparation. A number of methods have been tried for preparing cast iron surfaces for brazing; most of them have been only partly successful. Abrasive blasting with steel shot or grit has proved reasonably successful for preparing the surfaces of ductile and malleable iron castings, but is seldom suitable for preparing surfaces of gray iron castings. Electrolytic treatment in a molten salt bath, alternately reducing and

Table 7. Typical salts used for salt-bath dip brazing of carbon and low-alloy steels with various filler metals

Filler metal(a)	Type of salt	Nominal composition, %	Brazing temperature range(b), °F
BAg-1 through BAg-8, and BAg-18	Neutral	55 BaCl$_2$, 25 NaCl, 20 KCl	1150-1600
	Cyaniding-fluxing	20-30 Na$_2$CO$_3$, 20-30 KCl, 30-40 NaCN	1200-1600
	Neutral	50 NaCl, 50 KCl	1350-1600
RBCuZn-A	Neutral	80 BaCl$_2$, 20 NaCl	1675-1725
	Fluxing	79 BaCl$_2$, 20 NaCl, 1 borax	1675-1725
	Carburizing-fluxing (water soluble)	30 NaCl, 30 KCl, 20 carbonate, 15-20 NaCN, activator (proprietary)	1675-1725
	Carburizing and self-fluxing	50 carbonate, 50 chloride with graphite addition(c)	1500-1700
RBCuZn-D	Neutral	90 BaCl$_2$, 10 NaCl	1900-1925
BCu-1 and 1a	Neutral	95 BaCl$_2$, 5 NaCl	2000-2100
	Neutral	100 BaCl$_2$	2000-2100

(a) Nominal compositions and brazing temperature ranges are given for silver alloys and copper-zinc alloys in the article "Torch Brazing of Steels" in this Volume and for copper filler metals in the article "Furnace Brazing of Steels" in this Volume. (b) Temperatures shown are those of the salt bath. (c) Used with mechanical agitation

oxidizing, has been the most successful method for surface preparation and is applicable to all graphitic cast irons.

FUSED SALT CLEANING

Alkaline-base fused salts operating at 750 to 900 °F are extremely effective in removing surface oxides and sand from iron castings. At these temperatures, salt baths exhibit the required high chemical activity. This action is further enhanced in the cleaning operations by the introduction of electrical energy.

The excellent metallurgical interface obtainable with this cycle is shown in Fig. 6. Bonds of this quality have permitted direct babbitting of cast iron bearings and silver brazing of fittings to cast iron surfaces. With proper cleaning and preparation of cast iron, it is possible to produce a brazed bond between metal surfaces that exceeds the strength of the parent metal.

Brazing of Stainless Steels

STAINLESS STEELS, as a class, are no more difficult to braze than carbon and low-alloy steels. The high quantities of chromium present in stainless steels cause the chromium oxide films on the

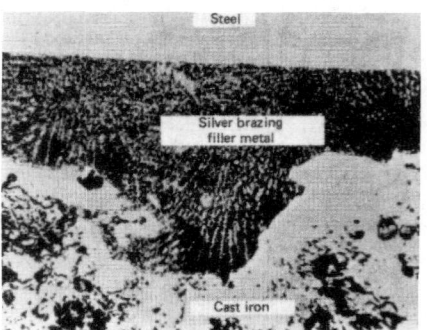

Fig. 6. Interface of brazed cast iron and steel after fused salt cleaning

surfaces of all stainless steels, as well as films of titanium oxide that form on the surfaces of titanium-stabilized stainless steels such as 321. These oxides, which are refractory and strongly adherent, prevent wetting of the base metal by the molten filler metal.

The formation of chromium oxide is accelerated when stainless steels are heated in air. Therefore, although the oxide may have been removed from the surface by chemical cleaning at room temperature, a new oxide layer that seriously interferes with wetting forms rapidly when the steel is heated in air to the brazing temperature.

Brazing Processes. Stainless steels can be brazed by all conventional brazing processes, including furnace, torch, induction, resistance, and salt-bath dip brazing. Furnace brazing is most widely used, because applications generally require brazing in a prepared atmosphere or vacuum.

FILLER METALS

Most stainless steels can be brazed with any one of several different filler metals, including silver alloys, nickel alloys, copper, and gold alloys. In most applications, filler metals are selected for mechanical properties, corrosion resistance, service temperature and compatibility, rather than for brazeability. Table 8 lists composition requirements for the filler metals most often used in brazing of stainless steels.

Silver Alloys. The most widely used filler metals for brazing stainless steels are the silver alloys (the BAg group). Alloy BAg-3, which contains 3% Ni, is probably the silver alloy selected most frequently, although several other silver alloys can also be used successfully.

Silver brazed joints cannot be used for high-temperature service; the recommended maximum service temperature is 700 °F (BAg-13). Recommended allowances on joint fit for silver brazing are relatively loose—generally, 0.002- to 0.004-in. diametral clearance.

Of the silver alloy filler metals shown in Table 8, all except BAg-19, and possibly BAg-13, are

Table 8. Typical compositions and properties of standard brazing filler metals for brazing stainless steels

Filler metal	Ag	Cu	Zn	Cd	Ni	Sn	Li	Mn	Other elements total	Solidus temperature, °F	Liquidus temperature, °F	Brazing temperature range, °F
Silver alloys												
BAg-1	44.0-46.0	14.0-16.0	14.0-18.0	23.0-25.0	···	···	···	···	0.15	1125	1145	1145-1400
BAg-1a	49.0-51.0	14.5-16.5	14.5-18.5	17.0-19.0	···	···	···	···	0.15	1160	1175	1175-1400
BAg-2	34.0-36.0	25.0-27.0	19.0-23.0	17.0-19.0	···	···	···	···	0.15	1125	1295	1295-1550
BAg-2a	29.0-31.0	26.0-28.0	21.0-25.0	19.0-21.0	···	···	···	···	0.15	1125	1310	1310-1550
BAg-3	49.0-51.0	14.5-16.5	13.5-17.5	15.0-17.0	2.5-3.5	···	···	···	0.15	1170	1270	1270-1500
BAg-4	39.0-41.0	29.0-31.0	26.0-30.0	···	1.5-2.5	···	···	···	0.15	1240	1435	1435-1650
BAg-5	44.0-46.0	29.0-31.0	23.0-27.0	···	···	···	···	···	0.15	1250	1370	1370-1550
BAg-6	49.0-51.0	33.0-35.0	14.0-18.0	···	···	···	···	···	0.15	1270	1425	1425-1600
BAg-7	55.0-57.0	21.0-23.0	15.0-19.0	···	···	4.5-5.5	···	···	0.15	1145	1205	1205-1400
BAg-8	71.0-73.0	rem	···	···	···	···	···	···	0.15	1435	1435	1435-1650
BAg-8a	71.0-73.0	rem	···	···	···	···	0.25-0.50	···	0.15	1410	1410	1410-1600
BAg-9	64.0-66.0	19.0-21.0	13.0-17.0	···	···	···	···	···	0.15	1240	1325	1325-1550
BAg-10	69.0-71.0	19.0-21.0	8.0-12.0	···	···	···	···	···	0.15	1275	1360	1360-1550
BAg-13	53.0-55.0	rem	4.0-6.0	···	0.5-1.5	···	···	···	0.15	1325	1575	1575-1775
BAg-13a	55.0-57.0	rem	···	···	1.5-2.5	···	···	···	0.15	1420	1640	1600-1800
BAg-18	59.0-61.0	rem	···	···	···	9.5-10.5	···	···	0.15	1115	1325	1325-1550
BAg-19	92.0-93.0	rem	···	···	···	···	0.15-0.30	···	0.15	1400	1635	1610-1800
BAg-20	29.0-31.0	37.0-39.0	30.0-34.0	···	···	···	···	···	0.15	1250	1410	1410-1600
BAg-21	62.0-64.0	27.5-29.5	···	···	2.0-3.0	5.0-7.0	···	···	0.15	1275	1475	1475-1650
BAg-22	48.0-50.0	15.0-17.0	21.0-25.0	···	4.0-5.0	···	···	7.0-8.0	0.15	1260	1290	1290-1525
BAg-23	84.0-86.0	···	···	···	···	···	···	rem	0.15	1760	1780	1780-1900
BAg-24	49.0-51.0	19.0-21.0	26.0-30.0	···	1.5-2.5	···	···	···	0.15	1220	1305	1305-1550
BAg-25	19.0-21.0	39.0-41.0	33.0-37.0	···	···	···	···	4.5-5.5	0.15	1360	1455	1455-1555
BAg-26	24.0-26.0	37.0-39.0	31.0-35.0	···	1.5-2.5	···	···	1.5-2.5	0.15	1305	1475	1475-1600
BAg-27	24.0-26.0	34.0-36.0	24.5-28.5	12.5-14.5	···	···	···	···	0.15	1125	1375	1375-1575
BAg-28	39.0-41.0	29.0-31.0	26.0-30.0	···	···	1.5-2.5	···	···	0.15	1200	1310	1310-1550

Filler metal	Cu	Zn	Sn	Fe	Mn	Ni	P	Pb	Al	Si	Other elements total	Solidus temperature, °F	Liquidus temperature, °F	Brazing temperature range, °F
Copper alloys														
BCu-1	99.90 min	···	···	···	···	···	0.075	0.02	0.01	···	0.10	1981	1981	2000-2100
BCu-1a	99.0 min	···	···	···	···	···	···	···	···	···	0.30	1981	1981	2000-2100
BCu-2	86.5 min	···	···	···	···	···	···	···	···	···	0.50	1981	1981	2000-2100

(continued)

used at brazing temperatures that fall within the effective range of sensitizing temperatures (1000 to 1600 °F) for austenitic stainless steels. Chromium carbide precipitation occurs in the sensitizing temperature range, resulting in impairment of the corrosion resistance of the base metal. Carbide precipitation, however, depends on time as well as temperature, and exposure to the sensitizing-temperature range for only a few minutes is unlikely to result in a significant amount of precipitate.

Nickel Alloys. After the silver alloys, nickel alloys usually rank next in frequency of use as brazing filler metals for stainless steels. Nickel alloy filler metals provide joints that have excellent corrosion resistance and high-temperature strength. These filler metals alloy with stainless steel, however, and form phases with two undesirable characteristics: the phases are considerably less ductile than either the base metal or the filler metal, even at elevated temperatures, and for this reason they are a potential source of rupture; and the alloys that are formed with stainless steel are higher-melting alloys that are likely to freeze and block further flow into the joint during brazing.

To achieve flow in deep joints, diametral clearances of as much as 0.004 to 0.008 in. are necessary.

Copper Filler Metals. The high brazing temperature and the need for a protective atmosphere generally restrict the use of copper filler metals to furnace brazing. These filler metals (the BCu group) melt at about 1980 °F and flow freely at 2050 °F.

Copper is not recommended for exposure to certain corrosive substances, such as the sulfur in jet fuel and in sulfur-bearing atmospheres. Furthermore, copper filler metals exhibit poor oxidation resistance at elevated temperatures and should not be exposed to service temperatures above 800 °F. When a copper filler metal is used, recommended diametral allowances on joint fit range from 0.004-in. clearance to 0.002-in. interference.

Gold alloys (the BAu group) are sometimes used for brazing stainless steel, although their high cost restricts the use of these filler metals to specialized applications, such as fabrication of aerospace equipment. When a gold alloy is used, alloying with the stainless steel base metal is minimized, and, as a result, joints exhibit good ductility.

Cobalt Alloys. Cobalt-base filler metals are very rarely used for brazing stainless steel. The alloy included in Table 8, however, is available for that purpose.

FLUXES

For furnace brazing in strongly reducing or inert atmospheres, flux usually is not required. In some furnace brazing applications, however, a flux is necessary. A flux is always required for torch brazing and is usually required for induction and resistance brazing, unless atmospheric protection is provided.

Either of two American Welding Society (AWS) types of flux (3A and 3B) is suitable for all stainless steel brazing applications where a flux is needed. Both types are available in powder, paste and liquid forms. Type 3A flux contains boric acid, borates, fluorides, fluoborates and a wetting agent, and has an effective temperature range of 1050 to 1600 °F. This flux is suitable for use with silver alloy filler metals.

FURNACE ATMOSPHERES

Almost all furnace brazing of stainless steel is done in a protective atmosphere, or vacuum. The protective atmospheres most often used are dry hydrogen and dissociated ammonia.

Brazing of Heat-Resistant Alloys

HEAT-RESISTANT ALLOYS are frequently referred to as superalloys because of their strength, oxidation resistance and corrosion resistance at elevated temperatures (1200 to 2200 °F).

BRAZING FILLER METALS

The American Welding Society (AWS) has classified several gold-, nickel- and cobalt-base brazing filler metals which can be used for elevated-temperature service (see Table 8 in preceding article, "Brazing of Stainless Steels"). In

Table 8 (continued)

Filler metal	Cr	B	Si	Fe	C	P	S	Al	Ti	Mn	Cu	Zr	Ni	Other elements total	Solidus temperature, °F	Liquidus temperature, °F	Brazing temperature range, °F
Nickel alloys																	
BNi-1	13.0-15.0	2.75-3.50	4.0-5.0	4.0-5.0	0.6-0.9	0.02	0.02	0.05	0.05	...	...	0.05	rem	0.50	1790	1900	1950-2200
BNi-1a	13.0-15.0	2.75-3.50	4.0-5.0	4.0-5.0	0.06	0.02	0.02	0.05	0.05	...	...	0.05	rem	0.50	1790	1970	1970-2200
BNi-2	6.0-8.0	2.75-3.50	4.0-5.0	2.5-3.5	0.06	0.02	0.02	0.05	0.05	...	...	0.05	rem	0.50	1780	1830	1850-2150
BNi-3	...	2.75-3.50	4.0-5.0	0.5	0.06	0.02	0.02	0.05	0.05	...	...	0.05	rem	0.50	1800	1900	1850-2150
BNi-4	...	1.5-2.2	3.0-4.0	1.5	0.06	0.02	0.02	0.05	0.05	...	...	0.05	rem	0.50	1800	1950	1850-2150
BNi-5	18.5-19.5	0.03	9.75-10.50	...	0.10	0.02	0.02	0.05	0.05	...	...	0.05	rem	0.50	1975	2075	2100-2200
BNi-6	...	...	...	...	0.10	10.0-12.0	0.02	0.05	0.05	...	...	0.05	rem	0.50	1610	1610	1700-2000
BNi-7	13.0-15.0	0.01	0.10	0.2	0.08	9.7-10.5	0.02	0.05	0.05	0.04	...	0.05	rem	0.50	1630	1630	1700-2000
BNi-8	...	...	6.0-8.0	...	0.10	0.02	0.02	0.05	0.05	21.5-24.5	4.0-5.0	0.05	rem	0.50	1800	1850	1850-2000

Filler metal	Au	Cu	Pd	Ni	Other elements total	Solidus temperature, °F	Liquidus temperature, °F	Brazing temperature range, °F
Precious metal alloys								
BAu-1	37.0-38.0	rem	...	...	0.15	1815	1860	1860-2000
BAu-2	79.5-80.5	rem	...	...	0.15	1635	1635	1635-1850
BAu-3	34.5-35.5	rem	...	2.5-3.5	0.15	1785	1885	1885-1995
BAu-4	81.5-82.5	...	...	rem	0.15	1740	1740	1740-1840
BAu-5	29.5-30.5	...	33.5-34.5	35.5-36.5	0.15	2075	2130	2130-2250
BAu-6	69.5-70.5	...	7.5-8.5	21.5-22.5	0.15	1845	1915	1915-2050

Filler metal	Cr	Ni	Si	W	Fe	B	C	P	S	Al	Ti	Zr	Co	Other elements total	Solidus temperature, °F	Liquidus temperature, °F	Brazing temperature range, °F
Cobalt alloys																	
BCo-1	18.0-20.0	16.0-18.0	7.5-8.5	3.5-4.5	1.0	0.7-0.9	0.35-0.45	0.02	0.02	0.05	0.05	0.05	rem	0.50	2050	2100	2100-2250

(a) Single values are maximum percentages, unless otherwise indicated.
Source: AWS A5.8-81, "Specification for Brazing Filler Metals"

addition to these brazing filler metals, there are many that are not classified by AWS. The AWS-classified brazing filler metals are suitable for high-temperature service. It should be noted that for lower service temperatures, copper (BCu) and silver (BAg) brazing filler metals have been used for many successful applications.

Product Forms. Available forms of AWS-classified and proprietary brazing filler metals include wire, foil, tape, paste and powder. The form used frequently is dictated by the application. If the filler metal required for a specific application is only available as a dry powder, then brazing aids such as cements and pastes are available to help position the brazing filler metal.

SURFACE CLEANING AND PREPARATION

Cleaning of all surfaces that are involved in the formation of the desired brazed joint is necessary to achieve successful and repeatable brazed joints.

Chemical cleaning methods are most widely used. As part of any chemical cleaning procedure for preprocessing assemblies for brazing, solvent degreasing to remove all oils and greases should be the first operation. This is necessary to ensure wettability of the chemicals used for cleaning.

Mechanical cleaning usually is confined to those metals with heavy tenacious oxide films or to repair brazing on components exposed to service. Mechanical methods are standard machining processes—abrasive grinding, grit blasting, filing, or wire brushing (stainless steel bristles must be used). These are used not only to remove surface contaminants, but also to slightly roughen or fray the surfaces to be brazed.

CONTROLLED ATMOSPHERES

Controlled atmospheres (including vacuum) are used to prevent the formation of oxides during brazing and to reduce the oxides present so that the brazing filler metal can wet and flow on clean base metal. Controlled-atmosphere brazing is widely used for production of high-quality joints. Large tonnages of assemblies of a wide variety of base metals are mass produced by this process.

Controlled atmospheres are not intended to perform the primary cleaning operation for the removal of oxides, coatings, grease, oil, dirt or other foreign materials from the parts to be brazed. All parts for brazing must be subjected to appropriate prebraze cleaning operations as dictated by the particular metals. Controlled atmospheres commonly are employed in furnace brazing; however, they may also be used with induction, resistance, infrared, laser and electron beam brazing. In applications where a controlled atmosphere is used, postbraze cleaning is generally not necessary. In special cases, flux may be used with a controlled atmosphere (1) to prevent the formation of oxides of titanium and aluminum during brazing in a gaseous atmosphere, (2) to extend the useful life of the flux and (3) to minimize postbraze cleaning. Fluxes should not be used in a vacuum environment.

Pure dry hydrogen is used as a protective atmosphere because it dissociates the oxides of many elements. Hydrogen with a dew point of −60 °F dissociates the oxides of most elements found in heat-resistant alloys.

Inert gases, such as helium and argon, do not form compounds with metals. In equipment designed for brazing at ambient pressure, inert gases reduce the evaporation rate of volatile elements, in contrast to brazing in a vacuum. Inert gases permit the use of weaker retorts than required for vacuum brazing. Elements such as zinc and cadmium, however, vaporize in pure dry inert atmospheres.

Vacuum. An increasing amount of brazing of heat-resistant alloys, particularly precipitation-hardenable alloys that contain titanium and aluminum, is done in vacuum. Vacuum brazing in the range of 10^{-4} torr has proved adequate for brazing most of the superalloys.

NICKEL-BASE ALLOYS

In the selection of a brazing process for a nickel-base alloy, the characteristics of the alloy must be carefully considered. The nickel-base alloy family includes alloys that differ significantly in physical metallurgy, such as precipitation strengthened versus solid-solution strengthened, and in process history, cast versus wrought. These characteristics can have a profound effect on brazeability.

Precipitation-hardenable alloys present several difficulties not normally encountered with solid-solution alloys. Precipitation-hardenable alloys often contain appreciable (greater than 1%) quantities of aluminum and titanium. The oxides of these elements are almost impossible to reduce in a controlled atmosphere (vacuum, hydrogen). Therefore, nickel plating or the use of a flux is necessary to obtain a surface that allows wetting by the filler metal.

Because these alloys are hardened at temperatures of 1000 to 1500 °F, brazing at or above these temperatures may alter the alloy properties. This frequently occurs when using silver-copper (BAg) filler metals, which occasionally are used on heat-resistant alloys.

Liquid-metal embrittlement is another difficulty encountered in brazing of precipitation-hardenable alloys. Many nickel-, iron- and cobalt-base alloys crack when subjected to tensile stresses in the presence of molten metals.

COBALT-BASE ALLOYS

Brazing of cobalt-base alloys is readily accomplished by the same techniques used for nickel-base alloys. Because most of the popular cobalt-base alloys do not contain appreciable amounts of aluminum or titanium, brazing atmosphere requirements are less stringent.

Cobalt alloys, much like nickel alloys, can be subject to liquid-metal embrittlement or stress-corrosion cracking when brazed under residual or dynamic stresses. This frequently is observed when using silver or silver-copper (BAg) filler metals. Liquid-metal embrittlement of cobalt-base alloys by copper (BCu) filler metals occurs with or without the application of stress; therefore, BCu filler metals should be avoided in brazing of cobalt.

Brazing of Aluminum Alloys

BRAZING of aluminum alloys was made possible by the development of fluxes that disrupt the oxide film on aluminum without harming the underlying metal and filler metals (aluminum alloys) that have suitable melting ranges and other

Table 9. Melting ranges and brazeabilities of some common aluminum alloys

Alloy	Melting range, °F	Brazeability(a)
Non-heat-treatable wrought alloys		
1350	1195-1215	A
1100	1190-1215	A
3003(b)	1190-1210	A
3004	1165-1205	B
5005	1170-1205	B
5050	1160-1205	B
5052	1100-1200	C
Heat treatable wrought alloys		
6053	1100-1205	A
6061	1100-1200	A
6063	1140-1210	A
6951(c)	1140-1210	A
7005	1125-1200	B
Casting alloys(d)		
443.0	1065-1170	B
356.0	1035-1135	B
710.0	1105-1195	B
711.0	1120-1190	A

(a) A, generally brazeable by all commercial procedures; B, brazeable with special techniques or in specific applications that justify preliminary trials or testing to develop the procedure and to check the performance of brazed joints; C, limited brazeability. (b) Used both plain and as the core of brazing sheet. (c) Used only as the core of brazing sheet. (d) Sand and permanent mold castings only

desirable properties. The aluminum-base filler metals used for brazing aluminum alloys have liquidus temperatures much closer to the solidus temperature of the base metal than those for brazing most other metals. For this reason, close temperature control is required in brazing of aluminum. The brazing temperature should be approximately 70 °F below the solidus temperature of the base metal, but if temperature is accurately controlled and the brazing cycle is short, it can be as close as 10 °F. Aluminum alloys, depending on composition, can be brazed with commercial filler metals from 1020 to 1180 °F. Most brazing is done at temperatures between 1040 and 1140 °F.

Much of the equipment and many of the techniques used to prepare, braze and inspect aluminum alloys are the same as those used for other metals.

BASE METALS

The non-heat-treatable wrought alloys that have been brazed most successfully are the 1*xxx* and 3*xxx* series and low-magnesium members of the 5*xxx* series. The alloys with higher magnesium contents are more difficult to braze by the usual flux methods, because of poor wetting and excessive penetration by the filler metal.

Some common wrought and cast aluminum alloys are listed in Table 9 with their melting temperature ranges and brazeability ratings. Brazing of aluminum is generally limited to parts more than 0.015 in. thick, but dip brazing and fluxless vacuum brazing have been done successfully on aluminum fin stock as thin as 0.005 in.

FILLER METALS

Commercial filler metals for brazing of aluminum are aluminum-silicon alloys containing 7

Table 10. Compositions and solidus, liquidus and brazing-temperature ranges of brazing filler metals for use on aluminum alloys

AWS classification	Composition(a), %						Temperature, °F		
	Si	Cu	Mg	Zn	Mn	Fe	Solidus	Liquidus	Brazing
BAlSi-2	6.8-8.2	0.25	...	0.20	0.10	0.8	1070	1135	1110-1150
BAlSi-3(b)	9.3-10.7	3.3-4.7	0.15	0.20	0.15	0.8	970	1085	1060-1120
BAlSi-4	11.0-13.0	0.30	0.10	0.20	0.15	0.8	1070	1080	1080-1120
BAlSi-5(c)	9.0-11.0	0.30	0.05	0.10	0.05	0.8	1070	1095	1090-1120
BAlSi-6(d)	6.8-8.2	0.25	2.0-3.0	0.20	0.10	0.8	1038	1125	1110-1150
BAlSi-7(d)	9.0-11.0	0.25	1.0-2.0	0.20	0.10	0.8	1038	1105	1090-1120
BAlSi-8(d)	11.0-13.0	0.25	1.0-2.0	0.20	0.10	0.8	1038	1075	1080-1120
BAlSi-9(d)	11.0-13.0	0.25	0.10-0.5	0.20	0.10	0.8	1044	1080	1080-1120
BAlSi-10(d)	10.0-12.0	0.25	2.0-3.0	0.20	0.10	0.8	1038	1080	1080-1120
BAlSi-11(d,e)	9.0-11.0	0.25	1.0-2.0	0.20	0.10	0.8	1038	1105	1080-1120

(a) Principal alloying elements. (b) Contains 0.15% Cr. (c) Contains 0.20% Ti. (d) Solidus and liquidus temperature ranges vary when used in vacuum. (e) Contains 0.02-0.20% Bi.

Table 11. Physical properties of typical fluxes for brazing aluminum alloys

Property of flux	Dip brazing (flux 33)	Torch and furnace brazing (flux 34)
Solidus temperature, °F	900	915
Liquidus temperature, °F	1035	1115
Density at 1100 °F, lb/ft³	104	107
Specific heat, Btu/lb °F (approx)	0.2	(a)
Heat of fusion, Btu/lb (approx)	168	(a)
Heat requirement, Btu to heat 1 lb of flux from solid at 70 °F to liquid at 1150 °F (approx)	385	(a)
Resistivity, Ω · cm at:		
1080 °F	0.43	(a)
1130 °F	0.36	(a)
1150 °F	0.33	(a)
1180 °F	0.29	(a)

(a) These properties are not pertinent to torch and furnace brazing.

to 12% Si. Lower melting points are attained, with some sacrifice in resistance to corrosion, by adding copper and zinc. Filler metals for vacuum brazing of aluminum usually contain magnesium. The compositions and the solidus, liquidus and brazing temperatures of the most frequently used brazing filler metals for aluminum are given in Table 10.

The optimum brazing-temperature range for an aluminum-base filler metal is determined not only by the melting range of the filler metal and the amount of molten filler metal needed to fill the joint, but also is limited by the mutual solubility between the filler metal and the base metal being brazed.

Filler metals for separate application from the base metal to be brazed are available as wire and sheet (thin-gage shim stock). The manufacture of filler metal in sheet and wire forms becomes more difficult as silicon content increases. Only filler metals BAlSi-2 (alloy 4343), BAlSi-4 (alloy 4047) and alloy 4004 are available as sheet.

FLUXES

Conventional brazing, performed in air or other oxygen-containing atmospheres, requires the use of a chemical flux. Fluxes, which become active before brazing temperature is reached and are molten over the entire brazing range, penetrate the film of oxide, exclude air and promote wetting of the base metal by the filler metal. A satisfactory flux must: (1) begin to melt at a tem-

perature low enough to minimize oxidation of the parts, (2) be essentially molten at the time the filler metal melts, (3) flow over the joint and the filler metal to shield them from oxidizing gases, (4) penetrate the oxide films, (5) lower the surface tension between the solid and liquid metals to encourage wetting, (6) remain liquid until the filler metal has solidified and (7) be relatively easy to remove after brazing is complete.

Less-active fluxes are required for dip brazing than for torch or furnace brazing, because the parts are totally immersed in flux during dip brazing and oxygen cannot reach the surfaces of the parts to re-form oxide. Physical properties of typical fluxes are given in Table 11.

Brazing of Copper and Copper Alloys

MOST COPPERS AND COPPER ALLOYS can be brazed satisfactorily, using one or more of the conventional brazing processes. These processes include furnace, torch, induction, resistance and dip brazing.

BRAZEABILITY

Brazeability is generally rated from good to excellent. With some alloys, however, difficul-

ties may be encountered. For example, some lead-containing alloys can form a dross that interferes with wetting, and tin-containing alloys, if not stress relieved before brazing, may crack when subjected to rapid localized heating.

FILLER METALS

Table 12 presents the nominal compositions, solidus and liquidus temperatures, and electrical conductivities of some filler metals used in brazing of copper and copper alloys, as well as the joint clearance used with each. The filler metals listed in Table 12 represent four series: (1) copper-zinc (RBCuZn) alloys, (2) copper-phosphorus and copper-silver-phosphorus (BCuP) alloys, (3) silver (BAg) alloys and (4) gold (BAu) alloys. The copper (BCu) filler metals are omitted. Because of their high liquidus temperature (1980 °F), these filler metals are restricted to use with the copper-nickel alloys only. Of the filler metals listed in Table 12, the BCuP and BAg alloys are by far the most widely used in brazing copper and its alloys.

Gold alloy (BAu) filler metals, such as BAu-4 in Table 12, are high-cost compositions that are generally restricted to highly specialized applications such as joining vacuum-tube components that are hermetically sealed. In this application, the low vapor pressure of gold is advantageous. The high liquidus temperatures of gold alloy filler metals further limit their use to brazing of coppers and a few high-melting-temperature copper-nickel alloys.

BRAZING FLUXES

The types of fluxes used in brazing of coppers and copper alloys are listed in Table 13. All are marketed as proprietary compositions with no standard composition ranges.

Type 3A is a general-purpose, low-temperature flux suitable for use with all copper and copper alloy base metals except those containing substantial amounts of aluminum. The filler metals that are compatible with this flux include most of the copper-phosphorus alloys (with the exception of BCuP-1) and most of the silver alloys

Table 12. Nominal compositions, solidus and liquidus temperatures, and electrical conductivities of filler metals used in brazing of copper and copper alloys, and joint clearances used with these filler metals

AWS filler metal	Nominal composition, %							Solidus temperature, °F	Liquidus temperature, °F	Conductivity(a), % IACS	Typical diametral joint clearance, in.
	Ag	Cu	P	Zn	Cd	Ni	Other				
RBCuZn-A	...	59.25	...	40	...	...	0.75Sn	1630	1650	26	0.002-0.005
RBCuZn-D	...	48	...	42	...	10	...	1690	1715	...	...
BCuP-1	...	95	5	...	...	...	...	1310	1650	...	0.002-0.005
BCuP-2	...	92.75	7.25	...	...	...	...	1310	1460	...	0.001-0.003
BCuP-4	6	86.75	7.25	...	...	...	...	1190	1335	...	0.001-0.003
BCuP-5	15	80	5	...	...	...	...	1190	1475	10	0.001-0.005
BAg-1	45	15	...	16	24	...	...	1125	1145	28	0.002-0.005
BAg-1a	50	15.5	...	16.5	18	...	...	1160	1175	24	0.002-0.005
BAg-2	35	26	...	21	18	...	...	1125	1295	29	0.002-0.005
BAg-3	50	15.5	...	15.5	16	3	...	1170	1270	18	0.002-0.005
BAg-5	45	30	...	25	...	...	...	1250	1370	19	0.002-0.005
(b)	75	22	...	3	...	...	...	...	1365	...	0.002-0.005
BAg-8	77	23	...	...	...	...	...	1435	1435	...	0.002-0.005
BAg-8a	72	27.8	...	...	...	...	0.2Li	1410	1410	89(c)	0.002-0.005
BAg-19	92.5	7.3	...	...	...	...	0.2Li	1435	1635	88(c)	0.002-0.005
BAu-4	...	...	...	...	...	18	82Au	1740	1740	6	0.002-0.005

(a) Ratio of the resistivity of the material at 68 °F to that of IACS, expressed as a percentage and calculated on a volume basis. (b) Special filler metal used in brazing nickel silver knife handles. (c) Conductivity of filler metal after volatilization of lithium in brazing

Table 13. Fluxes used in brazing of copper and copper alloys

AWS flux type	Working temperature, °F	Constituents	Available forms	Base metals	Filler metals
3A	1050-1600	Boric acid Borates Fluorides Fluoborates Wetting agent	Powder Paste Liquid	All coppers and copper alloys except aluminum bronzes	BCuP and BAg series, except for BCuP-1 and BAg-19
3B	1350-2100	Boric acid Borates Fluorides Fluoborates Wetting agent	Powder Paste Liquid	All coppers and copper alloys except aluminum bronzes	All listed in Table 1
4	150-1600	Chlorides Fluorides Borates Wetting agent	Powder Paste	Aluminum bronzes	BAg series (principally)
5	1400-2200	Borax Boric acid Borates Wetting agent	Powder Paste Liquid	All coppers and copper alloys except aluminum bronzes	RBCuZn series (principally)

Table 14. Recommended protective atmospheres for furnace brazing of copper and copper alloys

Base metal	Suitable atmosphere	Maximum dew point, °F
Coppers, phosphor bronzes, and copper nickels	Lean or rich exothermic	20
	Reacted endothermic	20
	Dissociated ammonia	20
Red brasses(a)	Purified lean exothermic	10
	Reacted endothermic	10
	Dissociated ammonia	20
Yellow brasses(b), leaded brasses, tin brasses(b), and nickel silvers	Purified lean exothermic	−40
	Reacted endothermic	−20
	Dissociated ammonia	20
Silicon and aluminum bronzes	Purified lean exothermic	−40
	Dissociated ammonia	−40

(a) Low zinc. (b) High zinc

(including all those with liquidus temperatures below 1600 °F).

Type 3B flux is a modification of type 3A for use at higher temperatures. The active temperature range of type 3B flux is 1350 to 2100 °F. This flux can be used for brazing with any of the filler metals listed in Table 12, provided that the brazing temperature is above 1350 °F.

Type 4 flux is specifically prepared for brazing the aluminum-bearing copper alloys, and has the same working temperature range as type 3A flux, 1050 to 1600 °F. This flux is generally used with the silver alloy filler metals.

Type 5 flux, which has a working temperature range of 1400 to 2200 °F, is used with the copper-zinc filler metals. It is less active than type 3B flux, but it remains active longer (as required by the longer heating cycles used to braze heavy components) and costs less.

JOINT CLEARANCE

Joint clearance is a principal factor in determining the mechanical strength of brazed joints. It is also a factor in minimizing voids in the joint area and establishing the capillary force required to fill a joint.

Typical diametral joint clearances used with the filler metals commonly used in joining copper and copper alloys are given in Table 12. These are clearances at room temperature and are applicable to brazing of components of about the same mass made from the same copper or copper alloy. Adjustments may be required for brazing dissimilar metals to compensate for different coefficients of thermal expansion.

SELECTION OF BRAZING PROCESS

Often the size and shape of an assembly suggest a preferred brazing process to the exclusion of other processes. When two or more brazing processes have approximately equal suitability for brazing a given assembly, the quantity of assemblies to be brazed, because of the direct bearing on cost, is likely to be the deciding factor in process selection.

FURNACE BRAZING

Furnace brazing is a mass-production process. Its primary advantage is that it can be used to

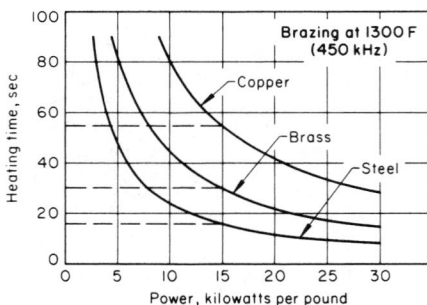

Power is expressed as kilowatts required to heat 1 lb of metal at the joint to 1300 °F. (Source: Wilkinson, W. D., *British Welding Journal*, Oct 1965)

Fig. 7. Power input and heating time required for high-frequency induction brazing

process a large number of assemblies on a batch or continuous basis at low unit cost. Furnace brazing can be used to braze a number of joints on the same assembly simultaneously or to braze a variety of different assemblies simultaneously. Furnace brazing also provides an enclosed container for atmospheres that can protect assemblies against surface oxidation and other undesirable effects encountered during heating in air (see Table 14).

TORCH BRAZING

Torch brazing of copper and copper alloys follows the same basic principles and uses the same equipment as for torch brazing of steel (see the article "Torch Brazing of Steels" in this volume).

INDUCTION BRAZING

The efficiency of heating by induction varies directly with the electrical resistivity of the alloy. Brass, because it has higher electrical resistivity, can be heated more efficiently than copper; steel, which has even higher resistivity, can be heated more efficiently than brass. In terms of the high-frequency power input and the time required to heat 1 lb of metal in a joint assembly to a brazing temperature of about 1300 °F, a power input of 15 kW (at 450 kHz) can heat a steel joint to this

temperature in about 16 s, whereas brass requires about 30 s and copper about 55 s (Fig. 7).

The general advantages of induction brazing are applicable to copper and its alloys. One of the general limitations of induction brazing is the cost of induction heating equipment, which far exceeds the cost of torch brazing equipment and usually exceeds the cost of equipment for resistance brazing or dip brazing in molten salt. Because efficiency in heating copper and copper alloys by induction is generally lower than that for heating steel, power requirements and consequently the cost of the equipment for achieving a given production rate are proportionately higher.

RESISTANCE BRAZING

Resistance brazing is often used for joining of copper conductors, terminals and other parts in lap joints for electrical connections where heating must be localized and closely controlled during brazing and where the brazed joint must have low electrical resistance. Generation of heat in the filler metal and nearly complete filling with a thin layer of the filler metal in the joint help in meeting both of these objectives.

The filler metal most frequently used is BCuP-5, which is used without a flux for brazing copper in most applications. In spite of the comparatively low electrical conductivity of BCuP-5 (approximately 10% IACS, Table 12), producing brazed joints having a low voltage drop acceptable for nearly all applications is not difficult, because the layer of filler metal in the brazed joint is very thin and the joints are designed to provide a conducting area larger than the cross section of the smaller member.

Soldering

SOLDERING is defined by the American Welding Society (AWS) as metal coalescence below 800 °F. Soldering facilitates joining of parts without heat damage and provides a system for rapid multiple-part joining of products such as printed circuits.

Principal soldering alloys are combinations or alloys of tin and lead. The tin component in the alloy reacts with the metals to be joined to form a metallurgical bond. Variations in tin and lead content and the addition of various alloying elements result in different melting ranges and joining characteristics of the alloy. Figure 1 compares soldering-temperature ranges with some base-metal melting points. Solders may contain antimony, silver, zinc, indium and bismuth. These alloys allow specific melting resistance of the completed joints.

PRINCIPLES OF SOLDERING

Soldering involves metallurgy, physics and chemistry in the interaction of elements, the constitution of fluxes, the thermal chemistry involved in heating both fluxes and metals to the molten state, and the underlying thermodynamics and fluid dynamics in promoting formation of a soldered joint. When the selected solder is heated with the appropriate flux to make a joint

with a particular base metal, the liquid materials formed flow over the surfaces to be joined. Relative surface tensions of the materials dictate these flow characteristics and define the capability for work adhesion in making the soldered joint.

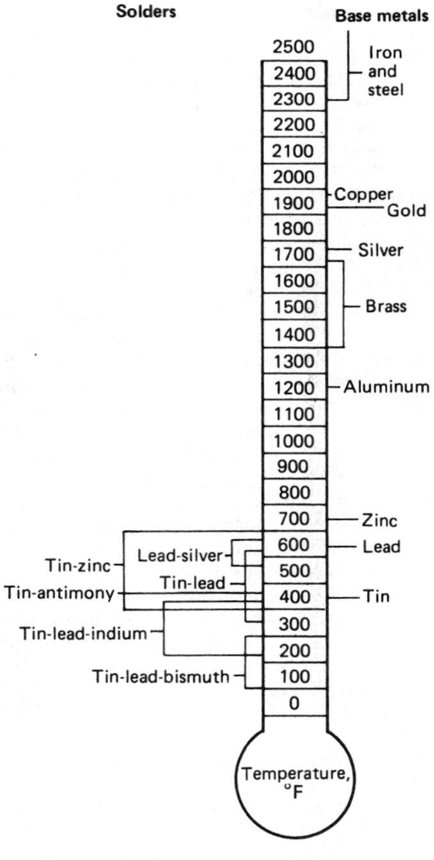

The cleanness and chemical composition of the surfaces to be joined are critical to the process. One function of the flux is to ensure that the base metal is sufficiently clean to provide adequate spread and flow of the soldering alloy to promote joint formation. Under most soldering conditions, a very low contact angle at the edge of the solder and good wetting are necessary for joint formation. Under certain conditions, the contact angle required is higher, and the capability to bridge or fill a gap between two metal surfaces is equally important.

At the soldering temperature, the liquid metal displaces the flux from the joint surfaces and is in intimate contact with the base metal to be joined, making possible a metallurgical reaction between the liquid solder and the base metal. This reaction varies in direct relation to the proximity and compatibility of the liquid solder and the solid base metal. In many soldering systems, an intermetallic compound is developed at the interface between the solder and the base metal, producing an essentially complete metallurgical joint. Reference to the constitutional diagrams of the metals involved is essential to understanding the soldering process. During cooling of the soldered joints, reaction products modify the metallurgical properties of the soldering alloy, which is reflected in its melting range and solidification pattern. The intermetallic compound formed at the interface continues to grow at a substantial rate until solidification takes place. When the joint is completely solidified, diffusion between the base metal and soldered joint continues until the completed item is cooled to room temperature. Mechanical properties of soldered joints, therefore, are generally related to, but not equivalent to, the mechanical properties of the soldering alloy.

Most soldering operations are carried out in air, with the flux acting as a barrier to surface oxidation and interaction with the atmosphere. In some cases, however, a protective atmosphere can enhance the formation of the desired soldered joint and also serve to reduce the amount of flux used in particular products.

TIN-LEAD SOLDERS

Care should be taken in specifying the correct solder for the job, because each alloy is unique with regard to its composition and properties. When tin-lead solders are referred to, the tin content is customarily given first—for example, 40%Sn-60%Pb. Table 1 presents information on analysis and use of tin-lead solders.

Solders in the tin-lead system are the most widely used of all joining materials. Industrial soldering alloys are in use that contain combinations of materials from 100% Pb to 100% Sn, as demanded by the particular application. The utility of the tin-lead combination is highlighted by examination of the constitution diagram between these two materials, shown in Fig. 2. Soldering alloys can be obtained with melting temperatures as low as 360 °F and as high as 600 °F within this system. Except for the pure metals and the eutectic solder at 63%Sn-37%Pb, all soldering alloys melt within a temperature range that varies according to the alloy composition. Each alloy has unique characteristics. In general, properties are influenced by the melting characteristics of the alloys, which in some measure are related to their load-carrying and temperature capabilities.

Applications. Soldering alloys containing less than 5% Sn are used for joining tin-plated containers and for automobile radiator manufacture. For automobiles, a small additional amount of silver is usually added to provide extra joint strength at automobile radiator operating temperatures. Soldering alloys of 10%Sn-90%Pb and 20%Sn-80%Pb are also used in radiator joints. With

Fig. 1. Soldering-temperature ranges compared with base-metal melting points

Source: Lead Industries Association, Inc.

Table 1. Tin-lead solders

Composition, %			Temperature, °F			Uses
Tin	Lead	Solidus	Liquidus	Pasty range		
2	98	518	594	76		Side seams for can manufacturing
5	95	518	594	76		Coating and joining metals
10	90	514	570	56		
15	85	440	550	110		
20	80	361	531	170		Coating and joining metals, or filling dents or seams in automobile bodies
25	75	361	511	150		Machine and torch soldering
30	70	361	491	130		
35	65	361	477	116		General purpose and wiping solder
40	60	361	460	99		Wiping solder for joining lead pipes and cable sheaths. For automobile radiator cores and heating units
45	55	361	441	80		Automobile radiator cores and roofing seams
50	50	361	421	60		General purpose. Most popular of all
60	40	361	374	13		Primarily used in electronic soldering applications where low soldering temperatures are required
63	37	361	361	0		Lowest melting (eutectic) solder for electronic applications

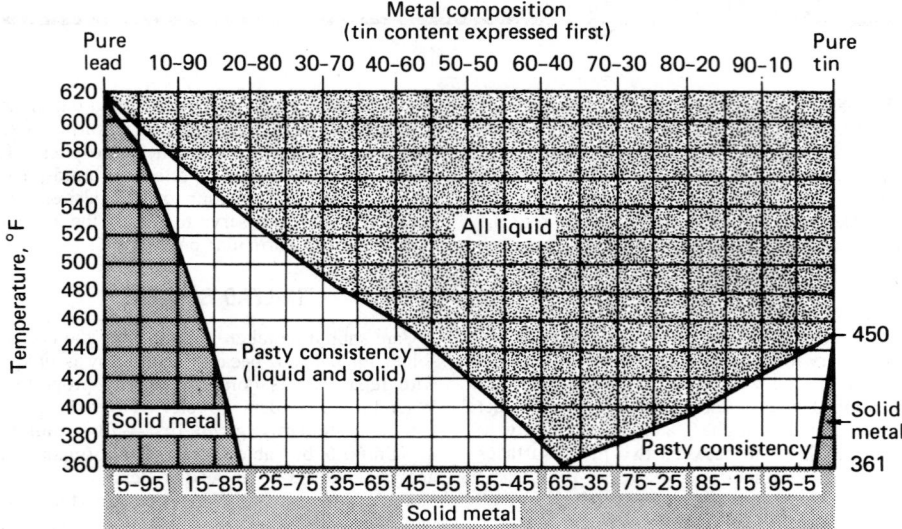

Source: Lead Industries Association, Inc.

Fig. 2. Tin-lead phase diagram

compositions between 10%Sn-90%Pb and 25%Sn-75%Pb, care must be taken to avoid any kind of movement during the solidification phase to prevent hot tearing in solders with a wide freezing range, as indicated by the constitution diagram (Fig. 2).

Higher-tin-content solders at the 25%Sn-75%Pb and 30%Sn-70%Pb compositions have lower liquidus temperatures and can be used for joining materials with sensitivity to high temperature, or where the wetting characteristics of the tin are important to providing sound soldered joints. Soldering alloys in the composition range described above usually are applicable to industrial products and generally are used in conjunction with inorganic fluxing materials.

The widely used general-purpose soldering alloys contain 40 to 50% Sn. These solders are used for plumbing applications, electrical connections and general soldering of domestic items. The 60%Sn-40%Pb and 63%Sn-37%Pb alloys are used most extensively in the electronic industries.

Impurities in Tin-Lead Solders

Impurities in solders can affect their performance and must be kept to a minimum. American Society for Testing and Materials (ASTM) standards for soldering alloys set maximum tolerable impurities in alloys as provided by the supplier or refinery. Impurities can be inadvertently picked up during normal usage of the alloys, especially when solder pots with recirculation systems and passage of components through the molten materials are used. The purity of solders supplied by reputable manufacturers usually is adequate for most applications. Particular soldering operations may require the use of superpurity materials that can be supplied upon request. Impurities present in sufficient quantities can affect wetting properties, flow within the joint, melting temperature of the solder, strength capabilities of joints, and oxidation characteristics of the soldering alloys. The most common impurity elements are listed below with their principal levels and effects.

Aluminum. Traces of aluminum in a tin-lead solder bath can seriously affect soldering qualities.

More than 0.005% of this metal can cause grittiness, lack of adhesion, and surface oxidation of the soldering alloy. A deterioration in surface brightness of a molten bath sometimes is an indication of the presence of aluminum.

Antimony is often used in solders as a deliberate addition. Some specifications even require a minumum antimony content. As an impurity, antimony tends to reduce the effective spread of a soldering alloy. High-lead solder specifications usually require a 0.5% Sb maximum limit. The general rule is that antimony should not exceed 6% of the tin content, although in some applications this rule can be invalid.

Arsenic. A progressive deterioration in the quality of the solder is observed with increases in arsenic content. As little as 0.005% As induces some dewetting, which becomes more severe as the percentage of arsenic is increased to 0.02%. Arsenic levels should be kept within this range.

Bismuth. Low levels of bismuth in the soldering alloy generally do not cause any difficulties, although some discoloration of soldered surfaces occurs at levels above 0.5%.

Cadmium. A progressive decrease in wetting capability occurs with additions of cadmium to tin-lead solders. While there is no significant change in the molten solder appearance, small amounts of cadmium can increase the risk of bridging and icicle formation in printed circuits.

Copper. The role of copper as a contaminant in solder appears to be variable and related to the particular product. A molten tin-lead solder bath is capable of dissolving copper at a high rate, easily reaching 0.3% Cu. Copper in liquid solder does not appear to have any deleterious effect upon wetting rate or joint formation. Excess copper settles to the bottom of a solder bath as an intermetallic compound sludge. New soldering alloy allows a maximum copper content of 0.08%.

Iron and nickel are not naturally present in soldering alloys. The presence of iron-tin compounds in tin-lead solders can be identified by a grittiness. Generally, iron is limited to 0.02% maximum in new solder. There are no specification limits for nickel, but levels as low as 0.02% can somewhat reduce wetting characteristics.

Phosphorus and Sulfur. Phosphorus at a 0.01% level is capable of producing dewetting and some grittiness. At higher levels, surface oxidation occurs, and some identifiable problems such as grittiness and dewetting become readily discernible. Sulfur causes grittiness in solders at a very low level, and should be held to 0.001%. Discrete particles of tin sulfide can be formed. Both of these elements are detrimental to good soldering.

Zinc. The ASTM new soldering alloy specification states that zinc content must be kept to 0.005% maximum in tin-lead solders. At this maximum limit, even with new solders in a molten bath, some surface oxidation can be observed and oxide skins may form, encouraging icicles and bridging. Up to 0.01% Zn has been identified as the cause of dewetting on copper surfaces.

The combined effects of the above impurity elements can be significant. Excessive contamination in a solder bath or dip pot generally can be identified through surface oxidation, changes in the product quality, and the appearance of grittiness or frostiness in joints made in this bath. A general sluggishness of the solder also is observed. In addition to analysis, experience with solder-bath operation is helpful in determining the point at which the material should be renewed for good solder joint production.

OTHER SOLDERING ALLOYS

A wide range of alternate alloys are available. At low temperatures, the ternary gallium-indium-tin eutectic (62.5%Ga-16%Sn-21.5%In), with a melting point of 50 °F, is useful. Bismuth-base fusible alloys with melting points ranging from 110 to 485 °F are manufactured. Alloys based on indium with lead, tin and silver additions are available to cover the temperature range from 200 to 600 °F. Solders available in the temperature range from 600 to 750 °F are limited, but combinations available include cadmiun-zinc and cadmium-silver alloys (liquidus temperatures from 510 to 750 °F), the zinc-aluminum eutectic (liquidus 720 °F), and three gold-base solders—the gold-tin eutectic at 80%Au-20%Sn, with a melting point of 535 °F; the gold-germanium eutectic at 88%Au-12%Ge, with a melting point of 675 °F; and the gold-silicon eutectic at 96.4%Au-3.6%Si, with a melting point of 700 °F. The temperature range from 750 to 840 °F is limited to the only available alloy, the aluminum-germanium eutectic (45%Al-55%Ge), with a melting point of 795 °F. The gold-indium composition (82%Au-18%In), with a liquidus of 905 °F and a solidus of 845 °F, is occasionally used.

Tin-Antimony Solder. The tin-antimony solder listed in Table 2 has excellent soldering and strength characteristics. It is used where a slightly higher temperature range is needed and in applications for joining stainless steels where lead contamination must be avoided. This alloy also is used in plumbing and refrigeration because of its good creep strength and fatigue resistance.

Tin-Silver Solders. Because of their comparatively high cost, tin-silver solders, shown in Table 3, are used in fine instrument work and in food applications. They have good wetting characteristics and superior joint strength compared with conventional solders.

Tin-zinc solders were developed primarily for joining aluminum. These alloys resist galvanic corrosion of soldered joints in aluminum. Soldering of aluminum is more difficult than, and differs in important ways from, soldering of other commonly used materials. Aluminum forms a tenacious surface oxide that is very difficult to remove, necessitating the use of very active flux. The wetting properties of solders on aluminum surfaces are generally not as good as on other common metals, and this has resulted in the development of special processes such as rub soldering and ultrasonic soldering to alleviate these difficulties.

Typical tin-zinc compositions are:

Composition, %		Temperature, °F	
Tin	Zinc	Solidus	Liquidus
91	9	390	390
80	20	390	520
70	30	390	590
60	40	390	645
30	70	390	710

Lead-Silver and Lead-Silver-Tin Solders. The lead-silver solders listed in Table 4 have solidus temperatures high enough to make then useful where strength at moderately elevated temperatures is required. Flow characteristics are, however, not satisfactory. The addition of 1% Sn improves wetting and flow characteristics and reduces susceptibility to humid-atmosphere corrosion. The listed alloys with tin contents over 1% are used primarily in electronic applications where high wettability is required. The 97.5%Pb-1%Sn-1.5%Ag alloy is useful in cryogenic duty, where the high lead content is advantageous, and also for soldering fine copper wire, where it reduces the tendency of the wire to dissolve in the solder.

Cadmium-silver solders are used primarily in applications where high service temperatures are required. High-tensile-strength joints can be produced that retain a substantial portion of their strength at service temperatures up to 425 °F. A 95%Cd-5%Ag solder has a solidus temperature of 640 °F and a liquidus temperature of 740 °F.

Cadmium-zinc solders are useful for soldering aluminum. Joints of intermediate strength and corrosion resistance are attainable with proper techniques. Several cadmium-zinc solders are:

Composition, %		Temperature, °F	
Cadmium	Zinc	Solidus	Liquidus
82.5	17.5	509	509
40	60	509	635
10	90	509	750

Zinc-Aluminum Solder. The 95%Zn-5%Al solder was developed specifically for use on aluminum. It develops joints with good strength and corrosion resistance where the soldering temperature is high. It has been used in certain specialized electronic applications where the high solidus and liquidus temperature (720 °F) is advantageous.

Indium-base solders, listed in Table 5, are generally specialty solders. They possess properties that make them valuable for certain applications. The 52%In-48%Sn and 97%In-3%Ag alloys will wet glass, quartz and many ceramics, making them useful in glass-to-metal seals. Because of their low vapor pressure, they also are useful for seals in vacuum systems. The indium-lead and indium-lead-silver solders have improved resistance to thermal fatigue compared with the conventional lead-tin solders. This, coupled with their marked reduction in scavenging and leaching of gold surfaces, has led to their use in electronic assemblies.

Table 2. Tin-antimony solder

ASTM alloy grade	Federal specification QQS-571	Composition, %		Temperature, °F	
		Tin	Antimony	Solidus	Liquidus
95TA	Sb 5	95	5	452	464

Table 3. Tin-silver solders

ASTM alloy grade	Federal specification QQS-571	Composition, %		Temperature, °F	
		Tin	Silver	Solidus	Liquidus
96.5 TS	Sn 96	96.5	3.5	430	430
(a)	···	97.5	2.5	430	438
(a)	···	95.0	5.0	430	465
(a)	···	90.0	10.0	295	563

(a) No ASTM or federal designation is available.

Table 4. Lead-silver and lead-silver-tin solders

ASTM alloy grade	Federal specification QQS-571	Composition, %			Temperature, °F	
		Lead	Silver	Tin	Solidus	Liquidus
2.5S	Ag 2.5	97.5	2.5	···	579	579
5.5S	Ag 5.5	94.5	5.5	···	579	579
1.5S	Ag 1.5	97.5	1.5	1.0	588	588
(a)	···	92.5	2.5	5.0	549	565
(a)	···	88	10.0	2.0	514	576
(a)	···	36	1.5	62.5	354	372

(a) No ASTM or federal designation is available.

Table 5. Indium-base solders

Composition, %				Temperature, °F	
Indium	Lead	Tin	Other	Solidus	Liquidus
44	···	42	(a)	200	200
52	···	48		244	244
70	9.6	15	(b)	257	(c)
97	···	···	(d)	290	290
80	15	···	(e)	301	(c)
12	18	70		324	(c)
25	37.5	37.5	···	274	358
50	50	···	···	356	408
90	···		(f)	290	446
75	25	···		264	508
5	92.5	···	(g)	572	(c)
5	90	···	(h)	554	590
5	95	···		558	598

(a) 14% Cd. (b) 5.4% Cd. (c) Melting point only determined. (d) 3% Ag. (e) 5% Ag. (f) 10% Ag. (g) 2.5% Ag. (h) 5% Ag

Table 6. Fusible solders

Composition, %					Temperature, °F	
Bismuth	Lead	Tin	Cadmium	Other	Solidus	Liquidus
44.7	22.6	8.3	5.3	(a)	117	117
49	18	12	···	(b)	136	136
48	25.63	12.77	9.60	(c)	142	149
50	26.7	13.3	10.0	···	158	158
42.5	37.7	11.3	8.5	···	160	190
55.5	44.5	···	···	···	255	255
58	···	42	···	···	281	281
40	···	60	···	···	281	338
48	28.5	14.5	···	(d)	217	440

(a) 19.1% In. (b) 21% In. (c) 4.0% In. (d) 9.0% Sb

Fusible Solders. The bismuth-base solders, often called fusible solders, are useful for soldering operations where temperatures below 360 °F are required. A representative group is listed in Table 6.

Alloys rich in bismuth generally are not good solders. They do not wet base metals readily, and either corrosive fluxes or pretinning is required.

Precious-Metal Solders. Gold-base solders find application primarily in semiconductor-device assembly and package sealing. Although the base cost of these solders is high, their corrosion resistance, good wettability and strength, and compatibility with silicon justify their use. Compositions of several precious-metal solders are:

Composition, %		Temperature, °F	
Gold	Other	Solidus	Liquidus
80	(a)	536	536
88	(b)	673	673
96.4	(c)	698	698
82	(d)	843	905

(a) 20% Sn. (b) 12% Ge. (c) 3.6% Si. (d) 18% In.

FLUXES

Flux technology is multifaceted and complex. Published data generally do not suggest any relationship among flux composition, strength properties of joints, and optimized processing, but experimentation has clearly proved that these factors are inextricably intertwined.

Corrosive general-purpose fluxes are effective on low-carbon steel, copper, brass and bronze. Applications are in the production of auto radiators, air conditioning and refrigerating equipment, and sheet-metal assembly. Compositions of these fluxes include:

1 Zinc chloride .... 40 oz
  Ammonium chloride .... 4 oz
  Water to make .... 1 gal
2 Zinc chloride .... 36 oz
  Sodium chloride .... 10 oz
  Ammonium chloride .... 1/2 oz
  Hydrochloric acid .... 1 oz
  Water to make .... 1 gal
3 Zinc chloride .... 21 oz
  Sodium chloride .... 6 oz
4 Zinc chloride .... 25 oz
  Ammonium chloride .... 3 1/2 oz
  Petroleum jelly .... 65 oz
  Water .... 6 1/2 oz

Intermediate fluxes contain organic compounds that decompose at soldering temperatures. When properly used, the mildly corrosive elements in the flux volatilize, leaving a residue relatively inert and easily removed with water. They are effective on all materials that are solderable with mild fluxes. Typical compositions are as follows:

1 Glutamic acid hydrochloride .... 19 oz
  Urea .... 11 oz
  Water .... 1 gal
  Wetting agent .... 0.2 wt%
2 Hydrazine monohydrobromide .... 10 oz
  Water .... 90 oz
  Nonionic wetting agent .... 1/20 oz
3 Lactic acid (85%) .... 9 oz
  Water .... 42 oz
  Wetting agent .... 1/10 oz

Noncorrosive fluxes are the rosin-base fluxes—nonactivated, mildly activated, and activated. For all electronic and critical soldering applications, water-white rosin dissolved in an organic solvent (item 1 below) is the safest known flux. Activators added to the rosin increase activity, but the flux residue from these fluxes should pass tests for noncorrosivity and nonconductance when used on electronic applications. These fluxes are effective on clean copper, brass, bronze, tinplate, terneplate, electrodeposited tin, cadmium, nickel and silver. Compositions of these fluxes are as follows:

1 Water-white rosin .... 10 to 25 wt%
  Alcohol, turpentine or petroleum .... rem
2 Water-white rosin .... 40 wt%
  Glutamic acid hydrochloride .... 2 wt%
  Alcohol .... rem
3 Water-white rosin .... 40 wt%

Cetyl pyridinium bromide .... 4 wt%
  Alcohol .... rem
4 Water-white rosin .... 40 wt%
  Stearine .... 4 wt%
  Alcohol .... rem
5 Water-white rosin .... 40 wt%
  Hydrazine hydrobromide .... 2 wt%
  Alcohol .... rem

PRECLEANING AND SURFACE PREPARATION

Oil, film, grease, tarnish, paint, pencil markings, cutting lubricants and general atmospheric dirt interfere with the soldering process. A clean surface is imperative to ensure a sound and uniform-quality soldered joint. Fluxing alone cannot substitute for adequate precleaning. Therefore, a variety of techniques are used to clean and prepare the surfaces of metals to be soldered. The importance of cleanness and surface preparation cannot be overemphasized. These steps help ensure sound soldered joints, as well as a rapid production rate. Precleaning can also greatly reduce repair work due to defective soldered joints. Two general methods of cleaning are chemical and mechanical. The most common of these are degreasing, acid cleaning, mechanical cleaning with abrasives, and chemical etching.

SOLDER-JOINT DESIGN

Soldering alloys generally have lower-strength properties than the materials to which they will be joined. Over-all design of a product involving soldered joints must therefore be evaluated to ensure that the joints are capable of carrying the supplied loads for the expected life of the product. Stress-rupture and creep properties are therefore important to solder joints under load in service. Care must be taken also not to use bulk solder properties for this evaluation because these do not take into account the effect of joint formation, interfacial solder-joint reactions, and stress-transfer capabilities across soldered joints. In designing a product, several solder/base metal selections can be made that will adequately perform the task. In addition to design aspects, over-all costs of materials and of manufacturing the product are usually taken into account. The lap joint provides a capability for conservative design by allowing larger areas of joints to be utilized at lower unit stress (see various joint designs in Fig. 3).

Most data available in the literature on joint strengths are not directly applicable to the design of a soldered joint. It is often necessary to fabricate sample parts and test the joints to ensure their producibility.

HEATING METHODS

In addition to surface preparation, solder selection and fluxing, another important part of the soldering process is the choice of heating method. Several methods are available:

- Soldering iron or bit
- Flame or torch soldering
- Hot dip soldering
- Induction soldering
- Resistance soldering
- Furnace soldering
- Infrared soldering
- Ultrasonic soldering
- Wave soldering.

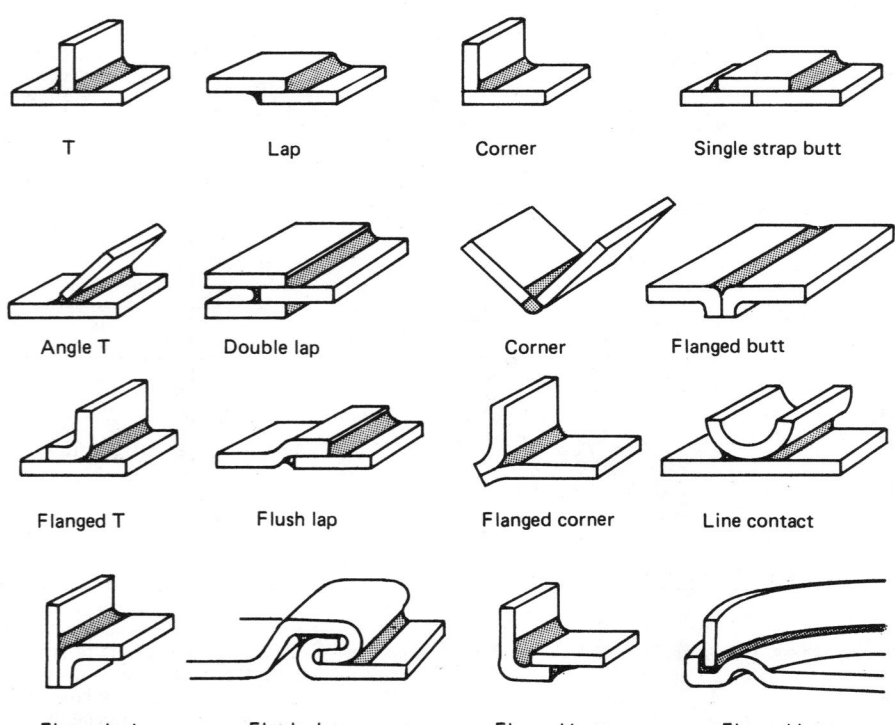

T Lap Corner Single strap butt

Angle T Double lap Corner Flanged butt

Flanged T Flush lap Flanged corner Line contact

Flanged edge Flat lock seam Flanged bottom Flanged bottom

Solder-joint terminology has not been standardized. (Source: American Welding Society Soldering Manual)

Fig. 3. Joint designs frequently used in soldering

31 RECYCLING OF METALS AND ALLOYS

By Howard E. Boyer

Introduction and Overview

THE TERM "scrap" can have a number of different meanings, depending on the backgrounds of the people involved. Unfortunately, to the average layman the term "scrap" seems to imply something that is inferior, for which reason this term was probably poorly selected for the area it now covers. However, it would not be practical at this time to change the terminology. In certain industries (tool steel, for example), metal for recycling often is termed "recovery metal" just to avoid the term "scrap." Within the scope of this article, "scrap" and "recovery metal" mean the same and will be used interchangeably as deemed appropriate.

In the machine shop the term "scrap" is used principally to denote generated metal—turnings and borings. In the press shop, metal scrap consists mainly of punchings and skeletons. Furthermore, in a forge shop, trimmings from the various types of die forgings comprise most of the total generated scrap.

In the foundry, sprues, risers and gates are commonly called "return scrap." Actually the foundries were the first segment of the metalworking industry to deal with scrap or recovery-metal problems. Here it was essential because in many foundries (notably in steel foundries) as much as 50% of the total tonnage melted becomes return scrap. Thus, in any foundry it is merely a matter of keeping the return scrap sorted as to composition and then cutting it up as required for remelting.

The types of scrap discussed above comprise one of the two principal categories of scrap—generally identified as "processing scrap." The second principal source, or category, of scrap consists of that recovered from used (usually worn-out) products. Salvaging this metal becomes more complex in terms of sorting so as to keep the various types of metals separated. However, metals recovered from machinery or other manufactured articles can be of the highest quality, and, furthermore, the cost of such recovery can be less than that of producing virgin metal.

It is often stated that there are two types of "mines" from which metals are produced: (*a*) those mines below the ground that yield ores for metal production; and (*b*) another form of "mine" above the ground—that is, a vast source of scrap metal for recovery.

Table 1. Preliminary identification of metals and alloys by color

Color	Metal or alloy
Red or reddish	Copper
Light brown or tan	90/10 cupronickel
Dark yellow	Bronzes, gold
Light yellow	Brasses
Bluish or dark gray	Lead, zinc, zinc alloys
White or light gray	Nearly all others

Table 2. Preliminary identification of metals by weight

Weight	Metals	Specific gravity
Very heavy . . .	Gold, platinum group, tungsten	19.3-21.4
Heavy	Lead, silver, molybdenum	10.2-11.3
Light	Magnesium, aluminum, titanium	1.7-4.5
Intermediate . .	Nearly all others	6-9

Recovery of metal from worn-out or obsolete products is not new. In fact, back at the time when enough automobiles had been produced to create scrap piles of worn-out automobiles, many foundries began to replace their melting charges of pig iron with cast iron engine blocks—an excellent source of high-grade cast iron. Steel railroad rails which have been worn out of dimensional tolerance represent another notable example of easy-to-recover scrap for remelting in cupolas as well as in steelmaking furnaces.

Organized Procedures for Recycling

In any plant where recycled scrap is to become all or at least a part of the melting stock, it is essential to establish a raw-material team, just as holds true for achieving other objectives such as energy conservation.

As a rule, depending to some extent on the size of the plant and how it is organized, the "scrap team" is comprised of melt-shop managers, plant metallurgists and some purchasing personnel. This combination should be able not only to maintain close control of the internal scrap regardless of how it is generated, but also to maintain high-

quality performance from the scrap dealers within practical distances. The objective in all cases is to achieve and sustain the least-cost, satisfactory melting charge, commonly termed the LCC.

The above paragraphs are intended primarily to emphasize that a successful scrap-recovery program is in no way a one-person operation and requires input from several departments to achieve LCC.

A major problem in any scrap program is control of composition, which is usually the responsibility of the plant metallurgists because they inherit the job of scrap identity.

SORTING

Identification of the various metals is not an easy task. Lack of proper identification is very often a source of metal contamination in the melt. Some of the well-recognized procedures for identification are discussed briefly in the paragraphs which follow. More elaborate discussions of methods for identification are provided in Vol. 11 of the 8th Edition of Metals Handbook (also see Section 33 in this Desk Edition).

Identification Methods

There are many different methods of identifying metal compositions, varying from the simplest spot tests, or just a visual examination, to tests made with highly sophisticated instruments. A few of the more common methods for identification are indicated in the paragraphs below.

Color. Many metals and alloys have a characteristic color which can be utilized for initial separation. A preliminary sorting based on color can be carried out according to Table 1.

Weight. Various metals may vary greatly in specific gravity—for example, from 1.74 for magnesium to 21.4 for platinum. Thus, it may be possible to make an initial separation, classifying them into categories as shown in Table 2.

Magnetic testing is very often a valuable tool in sorting, but it must be used with much discretion. In many instances, condition (usually referring to the amount of cold work), as well as composition, determines whether a metal is attracted to a magnet.

There are three ferromagnetic metals: iron, nickel and cobalt. Among alloys, iron-base al-

Table 3. Preliminary identification of metals and alloys by magnetic response

Response	Metal or alloy
Strongly magnetic	Cast irons, steels, 400 stainless steels, nickel, cobalt
Slightly magnetic	Monel (not K or S Monel), aluminum bronze, manganese bronze, silicon bronze
Nonmagnetic	Nearly all others

loys are most likely to be ferromagnetic, although a few nickel alloys also are magnetic. The 300 series stainless steels are nonmagnetic in their annealed condition, but they become magnetic when cold worked. Certain iron-containing copper alloys also are slightly magnetic. Table 3 summarizes identification by magnetic response.

Spark testing is based on the property of some metals (principally ferrous metals) to exhibit specific spark patterns when subjected to a high-speed grinding wheel in relatively subdued light. A fairly coarse wheel at a speed of about 7500 fpm is generally best. With a little experience, and by the use of known standards, it is easily possible to separate carbon steels on the basis of carbon content. Other elements also can be detected depending largely on the experience of the operator.

Proficiency in spark testing requires practice in identifying the sparks and reproducing spark results, so that a given material will always show the same spark patterns. Use of the same wheel, the same pressure and constant lighting conditions are important factors for reducing the variables encountered in spark testing. In the descriptions of spark trails, a number of terms are used to describe parts of the trail. These are listed and shown schematically in Fig. 1.

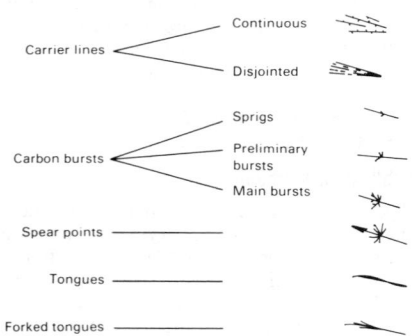

Fig. 1. Schematic representation of spark-testing terminology

Descriptions and diagrammatic representations of the sparks obtained are readily available. Of the more common alloys, those containing iron or nickel give off characteristic sparks. Alloys of cobalt, tungsten, molybdenum and titanium also give off characteristic sparks. Abbreviated descriptions of sparks obtained during scrap testing are given in Tables 4 and 5.

In making any spark test the requirement for known standards cannot be overemphasized. These standards must be quickly available to aid the sorter in matching knowns and unknowns.

Chemical spot testing for sorting or final identification of materials range from simple tests to show attack or lack of attack by specific media, to more involved spot tests which determine the presence or absence of various alloying elements. Such tests are generally qualitative, although some degree of quantitative results often can be obtained by experienced operators aided by simultaneous tests made on known alloys (standards).

Spot tests usually are made using one of the following techniques:

1. By bringing together one drop each of the test solution and reagent on porous or nonporous supporting surfaces, such as paper, glass or porcelain
2. By placing a drop of the test solution on a medium impregnated with appropriate reagents
3. By placing a drop of the reagent solution on a small quantity of the solid material.

The electrographic method represents a more sophisticated approach to chemical spot testing. By this method a sample of the material to be tested is obtained by simultaneously dissolving the material electrolytically and transferring it to a filter paper. The filter paper, wetted with an appropriate solution that acts as an electrolyte, is placed on a clean area of the material to be tested. The material then is connected into an electrical circuit as the anode. The cathode is connected to the wet area of the paper, allowing a direct current to flow through the electrolyte into the paper. The anode material is dissolved electrolytically and the cations produced are transferred to the surface of the paper in contact with the material (Fig. 2).

In addition to the section on nondestructive testing in this book and Vol. 11, 8th Edition of the Metals Handbook, chemical spot testing procedures are described in detail by ASTM, by INCO (International Nickel Co.) and by NASA.

Chemical analysis, either by the age-old wet chemistry methods or by more modern means that involve the use of spectroscopy, offers methods of positive identification. However, extensive use of analysis is expensive so that, as a rule, it is

Table 5. Spark-stream colors of metals and alloys

Material	Spark-test color
Nickel	Coarse red
D nickel	Coarse red
Z nickel	Coarse red
Monel	Coarse red
K Monel	Coarse red
S Monel	Coarse red
Cupronickel	Coarse red
Nickel silver	None
Inconels	Very dark red
Nimonics	Very dark red
Nichrome	Fine orange-red
330 stainless (31-15)	Coarse orange-red
310 stainless (25-20)	Fine orange-red turning white
309 stainless (25-12)	Coarse light orange turning white
300 stainless (18-8)	Light and diffused
400 stainless	Very light and diffused
Cobalt	Coarse red
Tungsten	Short yellow-white
Tungsten carbide	Short yellow-white
Molybdenum	Short yellow-white
Titanium	Brilliant white
Muntz metal (75% Ni, 6% Cu, 2% Cr, rem Fe)	Coarse red

not economically practical to make analyses of all scrap. More often, complete analyses on a limited basis are used in conjunction with the simpler but less positive methods outlined in the foregoing.

Summary on Identification. Because of the many different situations involved in recovering and using scrap, there cannot be any hard and fast rules or fixed procedures for ascertaining identity.

As discussed briefly in the foregoing, scrap may be identified by object recognition (usually color), apparent density, spark patterns, chemical spot tests, and the more time-consuming and expensive analytical procedures. Certain commercial devices also are available, including fluorescent x-ray, spectrographic analyzers, portable optical emission devices, and thermoelectric sorters.

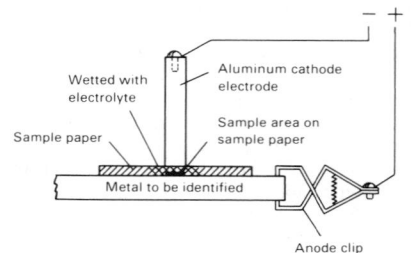

Fig. 2. Arrangement for electrographic sampling

Table 4. Spark-stream characteristics of metals and alloys

Material	Description of spark system
Normal carbon steel	Heavy dense sparks 18 to 24 in. long that travel completely around the grinding wheel. Sparks are white to straw colored with main bursts throughout.
400 series chromium stainless steel	Sparks are not as heavy or dense as in normal carbon steel. Sparks are 14 to 18 in. long, travel completely around the grinding wheel, and are orange to straw colored, ending with a forked tongue. Preliminary bursts and few main bursts.
300 series 18-8 stainless steel	Sparks are not as heavy or as dense as those of normal carbon steel. Sparks are 12 to 18 in. long, travel completely around the grinding wheel, and are orange to straw colored, ending in a straight line with few, if any, bursts.
310 series 25-20 stainless steel	The spark stream is thin and from 4 to 6 in. long. Sparks are orange to red in color, do not travel around the grinding wheel, and there are no bursts.
Nickel and cobalt high-temperature alloys	The spark stream is thin and about 2 in. long. The sparks are dark red in color, do not travel around the grinding wheel, and there are no bursts.

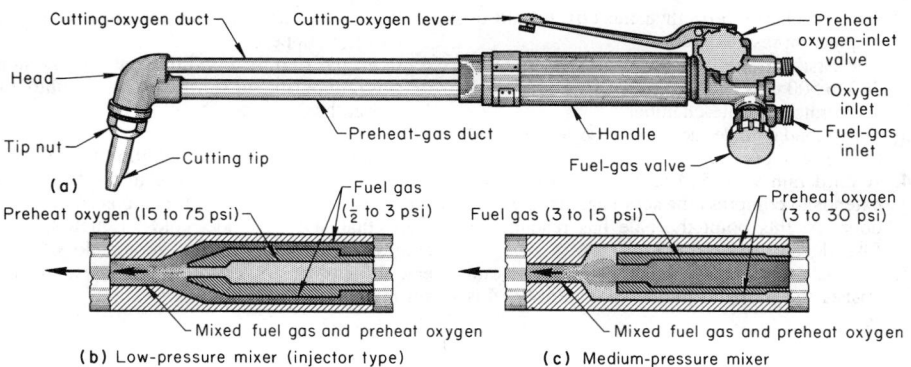

Fig. 3. (a) Typical manual cutting torch in which preheat gases are mixed before entering torch head. (b) and (c) Sections through preheat-gas duct showing two types of mixers commonly used with the torch shown. Injector mixer (b) is used for low-pressure fuel gases.

In all instances the potential scrap user must determine the answers to at least two questions:

1. How strict must identification be?
2. How much can be spent on scrap identification?

Answers to these questions indicate the approach which must be taken to identify, and thereby control composition of, the scrap.

SCRAP PROCESSING

Since the pieces of scrap being processed may range from an aluminum beverage can to a large machinery component weighing many tons, it follows that a wide range of equipment in terms of type and size will be needed.

Further, equipment and procedures vary greatly depending on the size and scope of the operation. For example, the procedures and equipment requirements for a relatively small operation involving mostly "in-house" scrap and perhaps supplemented with a variety of locally purchased scrap would be one consideration. On the other hand, a large scrapping operation where a business is made of purchasing, processing and selling scrap metal is quite another consideration. Therefore, these conditions will be used as examples in the brief discussion of procedures and equipment which follows.

Equipment and Procedures for In-House Operation

For relatively small-scale operation a gas cutting torch such as shown in Fig. 3 has proved to be a very versatile tool for cutting large forgings, castings, railroad rails, car wheels, heavy plates and other odd-shaped and/or heavy articles into pieces that are of appropriate size for the melting units involved.

Shears of various types also are extremely useful tools for processing small to large quantities of scrap. For small shops the alligator shear is probably the most widely used because it is capable of handling so many different sizes and shapes.

The two common types of alligator shears (low-knife and high-knife) are shown in Fig. 4. As indicated in the illustration, the shearing action is similar to that of a pair of scissors. A crankshaft transmits power to the shearing arms, and the leverage produces the force for shearing.

The weight of alligator shears ranges from about 2500 to 43 000 lb. The smaller shears can be wheel or skid-mounted, thus increasing their versatility. Speeds of alligator shears may range from about 50 strokes per minute for the smaller machines down to about 18 strokes per minute for the largest shears.

Alligator shears can be used for cutting up a wide variety of product forms, including all kinds of flat stock, bars, some types of forgings and castings, and structurals.

Other types of shears are used in small operations, but the alligator shear has been popular for a long time because of its versatility.

Disposal of Processed Scrap. In a small scrapping operation involving the use of such tools as described above, the cut pieces usually are placed in tote boxes and then are transferred to storage bins where identity is preserved pending further transfer to the melting unit.

Equipment and Procedures for Large-Scale Operation

Handling equipment is primary in nearly every scrap-processing plant. Although it does not process scrap, at least one crane is needed (most scrap plants have several). The traditional cable cranes are available on crawler, truck, pedestal, gantry, rail, and overhead mountings.

At the end of each crane is a grapple, a clam bucket or a magnet. For ferrous scrap, magnets that may weigh up to 12 tons are commonly used. Such a magnet can lift more than 4000 lb of heavy melting scrap.

Conveyors are widely used to move scrap from processing equipment to railroad cars or trucks, or to stockpiles for later processing.

Shredders. The giant of the scrap industry (in terms of size, output and cost) is the shredder (Fig. 5).

Although the predominant raw material for the shredder is automobile hulks, consumer goods such as refrigerators and washers, and other sheet steel products, are also processed in shredders.

Shredders are available which can shred 60 to 90 tons per hour. Such a shredder can completely shred an automobile hulk in a matter of seconds. Maintenance of shredders is high. For example, the hammers generally require replacement after processing about 6000 automobile hulks.

Guillotine shears are the real "workhorses" of the scrap industry. These machines are capable of slicing heavy pieces of steel such as I-beams, ship plate, pipe, and railroad-car sides. Guillotine shears may vary in capacity from 300 to more than 2000 tons of shear force. The shear knives, made of an alloy steel, have cutting edges on four sides, and are rotated as required to expose sharp cutting edges.

Baling Presses. The hydraulic baling press is the

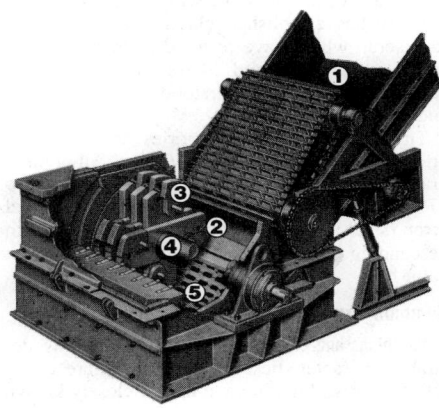

Automobile hulks are fed continuously into the entrance to the shredder (1). At the mouth of the shredder (2), the hulks meet alloy steel hammers (3) that are driven by a massive rotor (4). The size of the shredded scrap is determined by the size of the grate openings (5). After falling through the grates, the iron and steel scrap, nonferrous metals and nonmetallic materials are removed from the shredder by conveyor.

Fig. 5. Internal view of a metal shredder

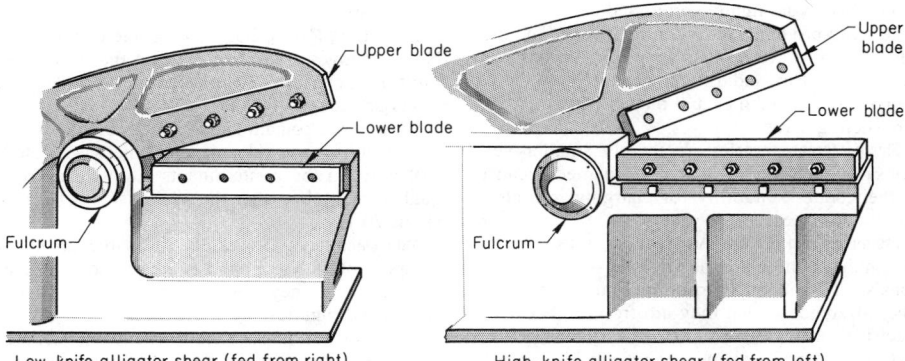

Low-knife alligator shear (fed from right)   High-knife alligator shear (fed from left)

Fig. 4. Two types of alligator shears. See text for discussion.

most widely used piece of equipment in the scrap industry. The larger balers are double- and triple-stroke compression units with up to 600 horsepower. The largest baler will take three flattened automobiles (less the engines) and, in less than two minutes, produce a 2200-lb bale having dimensions of 36 by 24 by 24 inches. At maximum efficiency this machine is capable of processing over 40 tons per hour.

A large baler operates in six distinct steps as described in the following:

1. Scrap is loaded into a hopper and then dumped into the charge box; then the box lid closes to precompress the scrap.
2. The first ram applies a compressive force of 200 to 500 tons as it extends to force the loose scrap into the bale chamber.
3. The second cylinder descends with 500 to 1500 tons of compression.
4. A third ram with 500 to 1500 tons of force extends to compress the scrap against the bale door. At this point the bale has reached its final density and size.
5. The bale door then opens and the third ram pushes the bale out of the chamber. By this time the charge for the next bale is loaded into the hopper.
6. In a final step the box lid opens and the hopper dumps the next load of scrap into the charge box.

Miscellaneous Equipment. In addition to the major pieces of equipment discussed in the foregoing, scrap processors employ various pieces of equipment that may include cold briquetting machines, crushers for turnings, motor-block breakers, alligator shears, gas cutting torches, and arc-cutting equipment.

Recycling of Specific Metals and Alloys

RECYCLING OF IRON AND STEEL SCRAP

As indicated in the foregoing, scrap is prepared for recycling in a variety of ways and by means of various equipment, depending largely on size and condition of the scrap metal. The scrap industry has established classifications for ferrous scrap which has resulted in more than 20 categories for steel and 10 for cast iron. These categories should be understood by the processor and the potential user.

The more common categories are discussed briefly in the paragraphs which follow. All of these grades have a relatively high yield — often 93 to 97% when melted in a cupola. The grades of steel scrap (steel as well as cast iron) are discussed in descending order of quality as well as of market value.

Common Grades of Steel Scrap

Punchings and plate scrap have the following desirable characteristics: they generally are clear, free from rust, high in yield and of closely known composition. These products of stamping plants generally are low in carbon, manganese, phosphorus and sulfur.

No. 1 busheling scrap is similar to punchings and plate and often is marketed as a mixture. For the most part this grade contains more steel clippings (up to 12 in. in length).

Shredded clippings represent another good factory grade of scrap; actually they may consist of one of the two grades mentioned above which has been run through a shredding machine. Thus they resemble shredded automobile scrap and are similar in density.

No. 1 bundles are tightly compacted bales of light-gage scrap produced in hydraulic balers. The initial material is from press shops and consists of clean sheet, strip, and trimmings which could have been sold as punchings and plate scrap or as No. 1 busheling scrap.

Electric furnace bundles are merely smaller versions of No. 1 bundles and are so named because of their better suitability for charging into electric-arc furnaces.

Prompt Silicon Grades. As to form, these grades are similar to those described above — that is, they consist of punchings, trimmings and skeletons. The difference is that they are from high-silicon electrical sheet. These represent very pure grades in terms of tramp elements, but are high in silicon, which must be considered in their use.

Flashings. This grade, also called "forging scrap," consists mainly of croppings, and very often may include defective forgings. This grade of scrap exhibits the effects of forging temperature, mainly in the form of scale.

Heavy Home Scrap from Steel Mills. "Home scrap" includes items such as ingot butts, billets, blooms, slab crops, forge crops and rail crops.

Railroad wheels and track materials represent a class of scrap where impurities are generally low. However, in using this grade of scrap it must be considered that railroad materials are being alloyed to an increasing extent with chromium and molybdenum, which may be present in the scrap. Further, it must be remembered that most of the earlier cast iron wheels have been replaced with medium-carbon steel forgings.

Cut plate and structural scrap are sold under several different codes, depending largely on the size of the pieces: one code permits lengths up to 5 ft, widths up to 24 in. and thicknesses up to $^1/_4$ in. These grades are more commonly used for charging large electric furnaces than for cupolas.

No. 1 Heavy Melting Steel. This grade is characterized by a higher percentage of impurities than that found in cut plate and structural scrap, as well as higher alloy content because it generally is composed partly of HSLA steels. It is usually available in lengths under 5 ft for charging into BOF's and large arc furnaces.

Shredded scrap generally is similar to shredded clippings, discussed above, except that shredded scrap is likely to contain more shreddings from auto bodies, and may contain more plastics, aluminum and other contaminants.

Foundry steel scrap may be up to 2 in. in cross section and is comprised mainly of auto frames, axles and steering systems — that is, the heavier auto scrap.

No. 2 Heavy Melting Steel. This grade differs from No. 1 (above) mainly in the fact that the lower limit of thickness is $^1/_8$ in., and that more coated steels are added.

No. 2 bundles contain significant amounts of steel sheet that has been galvanized or otherwise coated. Not only do the contaminants result in a poor-quality melt, but also the yield is poor — only about 70%.

Auto slabs have essentially the same chemical composition as No. 2 bundles but provide a greater yield because they contain less trash compared with baled bundles.

Briquetted steel turnings are classified in accordance with several codes which include a number of types (and grades of purity) that largely depend on whether or not the turnings are mixed with cast iron borings.

Steelmaking slag scrap generally is considered to be a low-grade melting material and is priced accordingly. This material consists of irregular nuggets of steel which have been separated magnetically out of crushed slag. Melting yield usually ranges from 70 to 80%.

Common Grades of Cast Iron Scrap

The grades of cast iron scrap are substantially fewer than those of steel scrap. The most common grades are discussed briefly in the paragraphs which follow, generally in order of descending quality.

Broken Ingot Molds and Stools. This grade of scrap is derived from worn-out ingot molds; since this type of scrap is not well suited to BOF steelmaking, most of it goes to the foundries. This grade is comprised of fairly high-quality cast iron that generally is low in tramp metals.

Drop-Broken Machinery-Cast (also called No. 1 machinery-cast). By definition this grade is machinery scrap and must show evidence of having been machined, thus designating it as a good grade of relatively soft iron.

Auto Cast. This grade is generally comprised of the power trains from trucks and buses. For the most part steel components are "stripped out" and represent profitable scrap items for other categories.

Cupola cast is a designated grade that can be virtually any grade of iron except stove plate, burnt iron or brake shoes. It is often known as "mixed cast."

Briquetted cast iron borings have become a prominent grade of cast iron scrap. However, the scrap user should be thoroughly familiar with the various available grades of borings.

Stove plate consists of thin sections which were purposely cast from high-fluidity iron, and thus runs very high in silicon and phosphorus.

Burnt iron consists of burnt stove plate, grates, furnace parts, fire pots and annealing boxes and tubes. It represents the lowest grade of cast iron scrap.

Cost Variations Among Different Grades

In the foregoing an attempt has been made to show the potential user of recycled metal that a great deal of consideration must be given to selection because many grades are available which very often overlap in quality and price.

Delivered prices of the common grades of iron and steel scrap are published every day by the American Metal Market. These data show how the prices of the various grades of scrap reflect

Table 1. Partial chemical compositions of typical stainless steelmaking waste products

Waste material	Composition, %						
	Cr	Ni	Mo	Fe	Mn	Pb	Zn
Electric furnace dust	9.3	2.2	1.1	27.8	3.6	0.8	4.9
AOD vessel dust	11.1	3.8	0.7	40.5	5.5	0.6	0.8
Grinding swarf	11.7	6.8	1.2	61.6	1.0	0.1	<0.1
Mill scale	8.6	3.9	0.5	54.8	0.8	0.1	<0.1

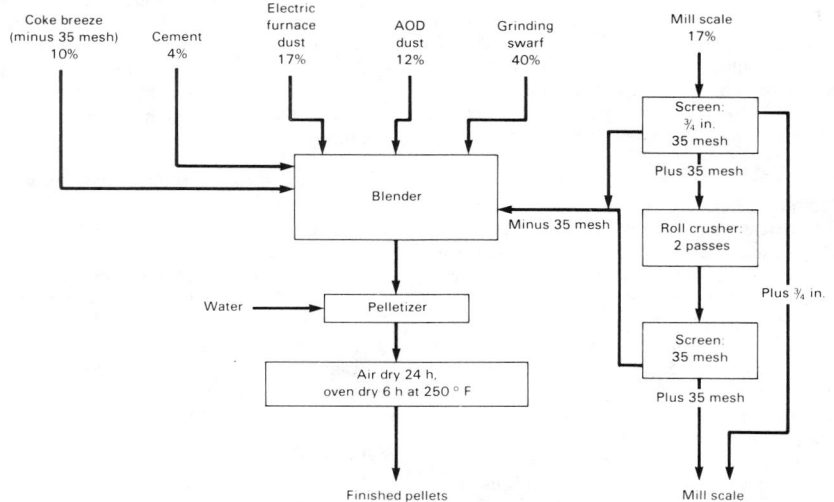

Fig. 1. Flow diagram for laboratory agglomeration of four stainless steelmaking wastes by pelletizing with coke breeze reductant and cement binder

the quality levels of the various grades. It often is observed that a given grade will vary in price from one region to another. Further, one may sometimes note that there is little difference in price between two grades of different quality. All such factors are interrelated with supply and demand.

RECYCLING OF STAINLESS STEEL SCRAP

In general, the procedures for salvaging stainless steel scrap from generated materials or from used and worn-out components are the same as those for carbon and alloy steels discussed above. However, because of the greater value per unit weight of stainless steels, it is appropriate to recover additional waste materials. Such efforts are proving successful. Typical waste materials and their partial chemical compositions are given in Table 1.

Pelletizing. As an example, the four stainless steel wastes shown in Table 1 were mixed with coke breeze reductant and cement binder and blended into a composite mixture. Pellets $^3/_8$ to $^3/_4$ in. in diameter were then produced in a drum-type pelletizer according to the flow diagram in Fig. 1. The pellets were then charged into the furnace for processing.

RECYCLING OF NONFERROUS METALS

Producing nonferrous metals from scrap consumes much less energy than producing them from ores, as indicated in Table 2. The majority of the scrap processed is termed "new scrap" or "in-process scrap" and is derived mainly from processor's waste. "Old scrap," sometimes referred to as "post-consumer scrap," largely originates from discarded components. Much of this type of scrap is highly contaminated and costly to process.

Aluminum

Because of the highly reactive characteristics of aluminum, only magnesium, calcium and sodium may be removed efficiently by current refining practice. The "secondary" aluminum industry is, therefore, concerned mainly with specification alloys, most of which are used in foundries. This results in major emphasis being placed on sorting, identification and blending of the scrap to achieve the required melt composition.

Aluminum sheet, obsolete components, and castings are first pulverized in a hammer mill. Ferrous metals are removed magnetically. Swarf and borings are degreased and dried. If their lengths are short they can generally be added directly to the melt. Bushy swarf may have to be shredded and baled.

Melting of Secondary Alloys. Bulky aluminum scrap is usually melted in a reverberatory-type furnace under low-melting-point fluxes such as NaCl plus KCl, along with fluorides and CaCl to improve fluidity and promote separation of the impurities.

Magnesium is removed from the aluminum by bubbling chlorine through the melt via a graphite tube. This forms $MgCl_2$, which rises to the top and separates to the slag.

Table 2. Unit energy for production of primary and secondary metals

Metal	Energy, GJ/t		
	Primary from ore	Secondary from scrap	Energy savings
Magnesium	372	10	362
Aluminum	353	13	340
Nickel	150	16	134
Copper	116	19	97
Zinc	68	19	49
Steel	33	14	19
Lead	28	10	18

Alloying, as required, then can be achieved by adding the element, or in some cases an alloy. Lump silicon is added to produce the aluminum-silicon grades.

It is not possible at present to extract pure aluminum commercially from secondary sources.

Copper

Both copper and copper alloys are produced in large quantities from a wide variety of waste materials.

Medium- to high-grade scrap (greater than 40% Cu) is usually melted and refined in a reverberatory furnace. Most of the impurities (Fe, Zn, Ni, Sn, Pb, Si, Al, Mn and Mg) can be oxidized preferentially by blowing air through the unit.

Since tin is a valuable by-product of secondary copper production, efforts are often made to recover it—commonly by use of a silicate slag; thus, the slag will contain up to 25% Sn.

Waste materials having low copper content (as low as 5%) which are highly contaminated with other metals—for example, brass components with soldered joints, and materials that vary widely in shape and size—are better suited to melting in a blast furnace.

To obtain high-purity copper (purity greater than 99.9%), electrorefining can be carried out on impure copper (greater than 95%) which has been cast into anodes.

To reclaim scrap which has a higher copper content, other systems may be used, depending greatly on the amount of copper present. As an example, for scrap which contains approximately 20% copper, a processing cycle such as that shown at left in Fig. 2 is appropriate. Here it is seen

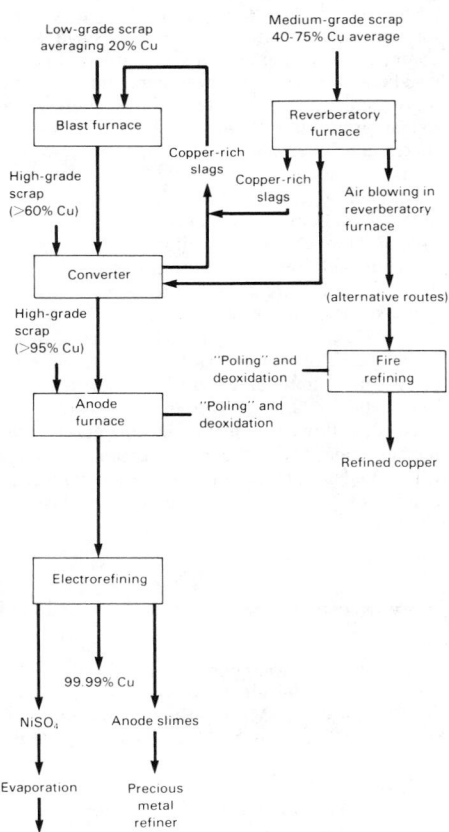

Fig. 2. Flow diagram for recycling of copper scrap

that the scrap is processed first in a blast furnace, then in a converter, and next in an anode furnace where a high-grade scrap (95% Cu) is produced. Final electrolytic processing results in 99.99% pure copper, as shown.

For processing of scrap having a higher copper content (40 to 75%), an alternative and simpler method is appropriate, as indicated in the right-hand portion of the flow diagram in Fig. 2.

Zinc

A large potential recovery source of zinc is galvanized scrap, although this source remains essentially untapped because of the problems involved in separating zinc coatings from steel.

About 50% of the recovered zinc comes from dross, zinc dust and zinc oxide produced during melting of foundry alloys. Because of the demand for zinc chemical products (such as zinc oxide), only about 10% of the waste zinc is recycled as refined metal.

Recovery from Process Residues. During galvanizing, an intermetallic compound is precipitated with entrained zinc and sinks to the bottom of the bath, which accounts for about 11% loss of zinc during galvanizing. At the same time, zinc forms an oxide ash on the surface of the bath, with a zinc content of up to 70%. However, the majority of these products are recycled to form various zinc chemicals rather than to produce pure zinc for zinc alloys.

Lead

High-grade lead scrap can be melted satisfactorily and refined to final composition in a simple kettle-type furnace. However, medium- to low-grade residues, and lead-rich slurries, after sintering, usually are refined in a blast, reverberatory or rotary-hearth-type furnace. Lead scrap commonly is contaminated with Cu, Sb, Sn, Al, As, Fe and S. Most of these contaminants form more stable oxides than lead. Thus, preferential oxidation by air blowing or by use of oxidizing fluxes can be employed to remove a great deal of the contaminating products.

Another approach to the refining of lead scrap involves addition of elements which form high-melting-point compounds with the contaminating elements and use of slag-forming materials. For example, zinc may be added to remove silver.

Recovery from Storage Batteries. Due to the fact that some 47% of the lead in the U.S. is used in the manufacture of storage batteries, these batteries offer a most important source of lead recovery. The batteries are first drained of acid, then crushed, followed by magnetic separation of the ferrous materials and density separation of the lead. The lead subsequently is smelted with coke and fluxes in a rotary, reverberatory or blast fur-

nace in a one- or two-stage process. The crude lead then is subjected to preferential oxidation.

Tin

Tin, like zinc, is used mainly as a coating for other metals, and therefore both electrochemical methods and selective melting are used in recovering tin. Secondary tin sources include dusts from smelting operations and drosses from melting, as well as solder and tin-bearing bronzes.

Tin presently is being recovered from new tin-plate scrap by a hydrometallurgical process involving dissolution in hot caustic containing sodium nitrate. The tin then is recovered from the stannate solution by electrolysis. At present, tin is not being recovered from "tin cans."

Precious Metals

Precious metals are recovered from solutions and electrochemical wastes, from smeltable materials and slags, from electronic scrap, and from richer sources such as brazing alloys, old jewelry, dental wastes and catalytic reagents.

Silver. Photographic films and papers are important secondary sources of silver. These materials are incinerated before being mixed with other constituents for melting.

Richer silver scrap is melted in a reverberatory furnace, resulting in separation of the charge into slag, matte and bullion. Fines are first nodulized to minimize loss in melting. The bullion penetrates the hearth refractories, which are removed periodically, crushed, and returned to the melting unit for salvage of the silver.

Gold. In the majority of applications, recovery of gold is complicated by the fact that platinum, ruthenium, gold and/or silver are alloyed with each other. Residues containing these elements are first processed by chemical treatment, wherein a recovery of 99.9% is achieved.

Anode slimes from electrorefining of metals contain significant quantities of precious metals. Other precious metals are removed preferentially by chemical treatment. The "gold sand" left after leaching is treated electrolytically to reclaim the gold.

Gold also is being recovered by an ion-exchange technique on a commercial basis from rinse waters containing 3 to 5 ppm gold, which are used in gold plating.

Titanium

Refining of contaminated titanium scrap has not been feasible because of titanium's high degree of activity with oxygen. Thus, it is essential to segregate the scrap at the source. Common practice is to melt the uncontaminated scrap under vacuum in an electron beam furnace.

Most titanium recovery (about 22% of total in-

got production) is from unalloyed, uncontaminated scrap which is reprocessed by means of vacuum melting techniques.

Apart from the technique described above, reclamation of titanium is still on a small-scale experimental basis.

Refractory Metals

Reclamation of refractory alloy wastes offers a source of refractory metals, but it has not been developed sufficiently to be of commercial value.

Used tool bits offer one source for recovery of refractory metals, although the cost is high and the amount of recovered metal from carbides is low. The three main methods of recovering refractory metals from their carbides are:

1. Leaching in nitric acid to dissolve the cobalt binder
2. Fused salt dissolution in which the carbide scrap is added to a bath of molten $NaNO_3$-$NaNO_2$ at 930 °F (500 °C), producing Na_2CO_3 and Na_2WO_4 which is recycled to ammonium paratungstate
3. Mechanical disintegration in which the carbide scrap is fired at a screen; the resulting powder retains its original cobalt content and can be used without further additions.

Superalloys

Because of the high cost and/or periodic scarcity of superalloys, recycling of scrap is used extensively for their recovery.

A survey of the industry in 1976 showed that the superalloys produced in the United States were melted using an average of 42% "home" (in-house) scrap, 17% purchased scrap and 41% primary metal.

Traditionally, the superalloy industry has required that the highest-quality raw materials be used to begin the melt because the furnaces used have had limited refining capacity. However, the innovation of the argon-oxygen decarburization (AOD) process has provided a means of removing many impurities, thus permitting use of a much lower grade of scrap. Other new systems, including the vacuum-oxygen decarburization (VOD) process and plasma-arc refining, offer additional refining capabilities for producing high-grade superalloys from lower grades of scrap.

Progress also is being made in segregating and identifying superalloy turnings. "Pedigree" turnings are routinely degreased, fragmented and compressed for remelting.

Considerable attention now is being directed to the recovery of superalloys from grindings, although they are generally lower in purity than turnings and thus offer more problems in reclamation.

══ Selected References on Recycling ══

A Review of Methods for Identifying Scrap Metals, by R. Newell, R. E. Brown, D. M. Soboroff and V. H. Maker: Bureau of Mines Information Circular 8902, U.S. Department of the Interior, 1982

System Aspects of Materials Recycling, by M. B. Bever: in *Conservation of Recycling*, Vol 2, Pergamon Press, Ltd., Oxford, 1978, p 1-17

In-Plant Recycling of Stainless and Other Specialty Steelmaking Wastes, by L. W. Higley, Jr., R. L. Crosby and L. A. Neumeier: Bureau of

Mines Report of Investigation 3721, U.S. Department of the Interior, 1982

Joint Symposium on Technologies for Scrap Processing and Steel Making in the 80's: Institute of Scrap Iron and Steel, Inc., and American Iron and Steel Institute, Pittsburgh, Nov 17-18, 1982, 82 pages

Scrap — A Nuisance or an Asset, by Keith Gall: *Maintenance Engineering*, London, May 1977

The Use of Steel Mill Waste Solids in Iron and Steelmaking: AISI Conference, New York, May

24, 1978 (retroactive coverage)

Expected Increase of Scrap Rate in Charge and Its Effect on Structure and Character of Steelmaking Processes, by I. Zadny: *Steel Times*, Vol 209, No. 8, Aug 1981, p 409-413

Processing of Complex Alloy Scrap for Remelting, by W. D. Bentall and G. Horn: *Metallurgia*, Vol 48, No. 1, Jan 1981, p. 22-26

Recycling of Non-Ferrous Metals, by J. J. Moore: *International Metals Reviews*, No. 5, 1978, p 241

Part IV
TESTING AND INSPECTION

32 FAILURE ANALYSIS

Additional and more detailed information on failure analysis can be found in Metals Handbook, Eighth Edition, Volume 10: Failure Analysis and Prevention. For additional information within this volume, the reader should consult the index.

Engineering Aspects of Failure and Failure Analysis

General Practice in Failure Analysis

By Robert Clark Anderson, Anderson & Associates, Inc.

A FAILURE INVESTIGATION may have various objectives. An investigation may be made to affix blame for the failure, and may involve product-liability litigation. It may involve pinpointing a lax supplier or an individual or department performing improperly. Another type of failure analysis involves simply finding the cause so the problem won't occur again. Occasionally a failure analysis is made purely for academic reasons. In any case, the importance of the procedure cannot be overemphasized. Many failures involve loss of life, limb, and millions of dollars in property damage. Most metal failures can be adequately analyzed by the metallurgist. Others are more complex and require experts in other engineering disciplines.

SEQUENCE OF ANALYSIS

The old adage of the carpenter, "measure twice and cut once," could well hold true for the failure analyst. Organize your thinking and procedures carefully before starting. Failure analysis is like playing detective with inanimate objects. Due to complexity, the following sequence is merely a suggestion and a guideline to follow:

1. Background information:
 • Learn all you can about what happened prior to and at the time of the failure.
 • Learn all you can about the procedure used for manufacturing the failed part.
2. Visual examination:
 • Visit the scene, make sketches, and take measurements, notes and photographs.
 • Select parts to be removed to the investigative laboratory.
 • Very carefully examine all parts in the laboratory. Sketch, measure, take notes, and photograph. Start by using the naked eye and proceed to any necessary magnification required for understanding.
3. Perform nondestructive testing (testing which in no way alters or changes the part). Document the results.
4. Perform destructive testing procedures. (If other investigators are involved, allow them to proceed through step 3 before destructive tests are performed.)
5. Examine all data and draw conclusions if possible.
6. Prepare a documented presentation of the findings, including:
 • Background information
 • Descriptions of specimens examined
 • Descriptions of test procedures and results
 • A discussion of why tests were performed and of the significance of the results
 • Positive conclusions.

This list may seem less than specific. In failure analysis, each step dictates the next procedure. It is a waste of time and money to routinely perform many tests that have no significance.

BACKGROUND INFORMATION

Background information falls into three categories: hearsay information, documentable information gained at the scene, and information gained about the manufacture of the equipment and the process involved in its use at the time of the accident (all obtained away from the scene of the accident).

Hearsay information is a scenario of the events as described by eyewitnesses. Any such information furnished should be weighed carefully since it often involves bias. It has been shown that various witnesses to the same event tell widely differing stories as to what they observed. Witnesses tend to embellish the event. They often relate what they assume you want to hear. They think in terms of protecting themselves and their fellow workers from blame.

Documentable information is much more reliable. This includes such factual data as date and time of the failure, weather conditions, written reports, charts, and photographs taken at the scene.

The following is a list of some of the background data which may be gathered.

• Material specifications
• Design operating conditions
• Length of service
• Temperatures and pressures
• Static and dynamic loading conditions
• Corrosive and erosive conditions
• Vibration and cyclic loading
• Testing, inspection and maintenance schedules
• Any unusual occurrences.

Any information about the product that failed is helpful. Generally this falls into the "how was it made" and "how was it used" categories. This is obtained either from the manufacturer or from available literature.

Ask questions, and get answers in writing — signed if possible. Be sure you understand all of the background information before proceeding. Then write it down, starting generally, stating what happened and then becoming more specific. Failure-analysis reports have a habit of being kept for long periods of time. When referred to later the background information is all that is available on what happened.

INSPECTION KIT

It is wise to prepare an inspection kit to take to the accident site. Such a kit should include measuring equipment, markers of various types, lighting (flashlight), inspection tools, tools for removing and storing small samples, and such other things as you think you might need. A second kit is necessary for photographic and other recording equipment.

VISUAL EXAMINATION

Visual examination is the first and the most important step in failure analysis. In most cases, as is true with the medical profession, the diagnosis is made by visual examination, and testing is performed to either confirm or deny it.

If the failure is complex, as in the case of an explosion or an airplane crash, it is necessary to start by examining and perhaps mapping the over-all scene of the failure. More commonly, visual examination can start much closer to the known failure site or part. A common error in visual examination is jumping to the conclusion that the first broken part you find is the over-all cause of the failure; it may be just a result.

Begin with the naked eye, then increase magnification using a simple magnifying device. Progress eventually to a stereoscopic microscope and then perhaps to an electron microscope.

Try to account for all markings you observe. Record all identification marks. Look for operating instructions, warnings and limitations. You may find mill heat numbers that identify a particular batch of material. Look for evidence of abuse. Did it occur before or as the result of the failure? How did fracturing take place? Did the part deform, indicating an overload, or is the fracture a brittle, abnormal type of break? What was the direction of loading that produced the deformation? If the fracture is brittle, is that really abnormal for the material? Look for stress concentrations, corrosion, abrasion, heat discoloration, etc. If corrosion is observed, determine the type and remove a corrosion-product sample for later analysis.

After completing a general examination, concentrate on the fracture. Look for evidence of a fracture origin. Is there more than one point of fracture origin? Particularly observe the fracture mechanism at the origin point; it may be much different than in the areas of propagation.

Measurements, sketches and notes go hand-in-hand. Sketch everything involved. Sketching, regardless of quality, makes you see things you wouldn't otherwise see. Add dimensional measurements to the sketches. Notes can be made on the sketches as well as separately. A pocket tape recorder is helpful in making field notes. Review and rewrite your field notes as soon as possible while they are fresh in your mind.

The final step in the field examination is selection and removal of samples. It is good to remove as much of the failure as possible to the laboratory for observation under ideal conditions. Many times parts are large and must be cut into smaller sections for removal. Mark the areas where cuts are to be made and photograph prior to cutting. It is desirable to cover fracture surfaces for protection before beginning any cutting procedure. Avoid cutting procedures that might affect the structure of the specimen. If abrasive or torch cutting methods are used, be sure to cut sufficiently far back so that any heat generated does not affect areas to be examined later. Never clean surfaces by abrasive means; this will severely damage them for future fractographic examination. Do not acid clean for the same reason. Field cleaning should be confined to a soft bristle brush. Even solvent cleaning may contaminate trace corrosion products. Carefully label all specimens. If parts cannot be removed, techniques exist for making replicas of fracture origins which can be examined in the laboratory.

Parts can also be prepared for metallographic examination and examined in the field.

Once samples have been removed to the laboratory they can be examined under more ideal conditions. After specimens have been examined in the as-received condition and after any surface materials have been noted and analyzed, the specimens can be cleaned, if necessary, using a mild detergent and a soft brush. Ultrasonic cleaning baths can also be used. There are some special electrolytic cleaning methods that are excellent. Parts or specimens cleaned with water should be rinsed with alcohol and dried, to prevent rusting. Small parts can be kept sealed with desiccants to preserve them in temporary storage.

One of the principal goals of laboratory visual examination is to confirm the mechanism of fracture origin. This often requires higher-magnification study. Use of an electron microscope, especially a scanning electron microscope (SEM), gives excellent fractographic characterization.

The basic types and causes of fractures to look for are summarized as follows:

Ductile types	Brittle types	Brittle causes
Tension	Intergranular	Manufacturing
Torsion	Transgranular	defects
Compression	Fatigue	Hydrogen and
Shear	Porosity	caustic
Creep		embrittlement
		Inherent
		brittleness
		Corrosion
		Stress corrosion
		Heat treating
		Welding
		Triaxiality
		High transition
		temperature
		Precipitation and
		phase
		phenomena

NONDESTRUCTIVE TESTS

Nondestructive examination covers a wide variety of tests, but generally is distinguishable by the fact that the parts are not effectively altered in any way. The only alteration that may occur is superficial cleaning. Sometimes only nondestructive tests are allowed so as to preserve the parts intact for evidentiary purposes. Even superficial cleaning may not be allowed for the same reason. Nondestructive testing includes such tests (other than visual examination and measurements which have already been discussed) as radiographic examination (x-ray), magnetic-particle testing, dye-penetrant examination, ultrasonic testing and eddy-current testing, among many others.

It is important to be very knowledgeable in the performance of the test prior to using it. It is also important that all test equipment be properly calibrated. For other nondestructive tests, see the section on nondestructive testing, which follows. Most nondestructive tests can be recorded photographically. For best photographic documentation of nondestructive tests, special techniques may be desired for enhancement.

DESTRUCTIVE TESTING

Destructive testing is most widely known and used in failure analysis for confirmation of prop-

erties. There are many types of destructive tests, but a few are very popular and are used in most failure-analysis procedures. Tensile testing gives strength properties, tells when the material starts to deform plastically, and gives information on how much deformation takes place prior to fracture (how brittle it is). Chemical analysis tells what kind of material it is and if any of the chemical elements that make up the material are out of specification range. It also shows if unwanted trace elements (unspecified components) are present and in what quantities. In addition to identifying the composition of the material, this test can also analyze contaminants, such as corrosion products. This is accomplished either by the so-called wet methods or electronically. Tests can be quantitative, semiquantitative or qualitative. Hardness testing gives information similar to tensile testing (an approximation of strength), but is less destructive and can be used to survey a considerable area for property variation. Metallographic examination is one of the most revealing destructive tests. This test reveals cleanness, soundness, cracking, heat treatment, structural variations, heat effects, surface conditions and many other aspects. Other popular destructive tests are impact testing, to determine the response of the material to rapidly applied loads; fatigue testing, to determine reaction to repeated or cyclic loading; and macroetching, to ascertain gross soundness and structural uniformity. Many other confirming-type destructive tests exist for special problems.

CONCLUDING REMARKS

When examination and testing are completed, the pieces should fall in place like a jigsaw puzzle. Carefully examine all of the results and documentation. Be sure you have not accepted only tests that confirmed your diagnosis and discarded other results. This is a common error of the failure analyst. If tests are questionable, retest. Reexamine all data and documentation. If answers are not readily apparent, don't give up easily. Review all tests and results, and, if necessary, perform additional, confirmatory tests. When you are convinced that your conclusions are sound, put them in writing. Do this in such a manner that nontechnical people will understand, because such people often are called upon to make decisions based on your report.

Fracture Mechanics—Fatigue and Fracture

By J. M. Barsom, U.S. Steel Corp.

FRACTURE-MECHANICS TECHNOLOGY has significantly improved the ability to design safe and reliable structures. The application of fracture-mechanics concepts has identified and quantified the primary parameters that affect structural integrity. These parameters include the magnitude and range of the applied stresses; the size, shape orientation, and rate of propagation of the existing crack; and the fracture toughness of the material.

Linear-elastic fracture-mechanics technology is based on an analytical procedure that relates the stress-field magnitude and distribution in the vicinity of a crack tip to the nominal stress applied to the structure; to the size, shape and orientation of the crack or cracklike imperfection; and to the material properties. The stress-field equations show that the magnitude of the elastic stress field can be described by a single parameter, K, which is designated the "stress-intensity factor." Relations between the stress-intensity factor and various body configurations; crack sizes, shapes and orientations; and loading conditions are available in the published literature (Ref 1 and 2).

The stress-intensity factor, K_I, for a crack tip in any body that is subjected to tensile stresses, σ, perpendicular to the plane of the crack (mode I deformation) is given by the relationship

$$K_I = \sigma \sqrt{\pi a}\, f(g) \qquad \text{(Eq 1)}$$

where a is crack length and f(g) is a function that accounts for crack geometry and structural configuration. This general relationship makes it possible to translate laboratory results into practical design information without the need for extensive service experience or correlations.

Failure of structural and equipment components usually is caused by initiation, subcritical propagation and unstable extension of cracks. Fracture-mechanics methodology has been used to identify and investigate the parameters that govern the behavior of each of these stages in the life of components. Because of space limitations, the following sections present a very brief discussion of fatigue in the absence of an aggressive environment, and fracture in the absence of significant crack-tip plastic deformation. For further information and for application of fracture-mechanics technology to corrosion-fatigue crack initiation and propagation, stress-corrosion cracking and elastic-plastic fracture, the reader should consult other published literature, such as Ref 3 and 4.

FATIGUE

Fracture of structural and equipment components as a result of cyclic loading has long been a major design problem and the subject of numerous investigations. Although a considerable amount of fatigue data are available, the majority have been concerned with the nominal stress required to cause failure in a given number of cycles—namely, S-N curves. Usually, such data are obtained by testing smooth specimens which, although of some qualitative use for guiding material selection, are subject to limitations caused primarily by the failure to adequately distinguish between fatigue-crack-initiation life and fatigue-crack-propagation life. The existence of surface irregularities and cracklike imperfections reduces and may eliminate the crack-initiation portion of the fatigue life of the component. Fracture-mechanics methodology offers considerable promise in improving our understanding of the initiation and propagation of fatigue cracks and in solving the problem of designing to prevent failures by fatigue.

FATIGUE-CRACK INITIATION

Initiation of fatigue cracks in structural and equipment components occurs in regions of stress concentrations, such as notches, as a result of stress fluctuation. The material element at the tip of a notch in a cyclically loaded component is

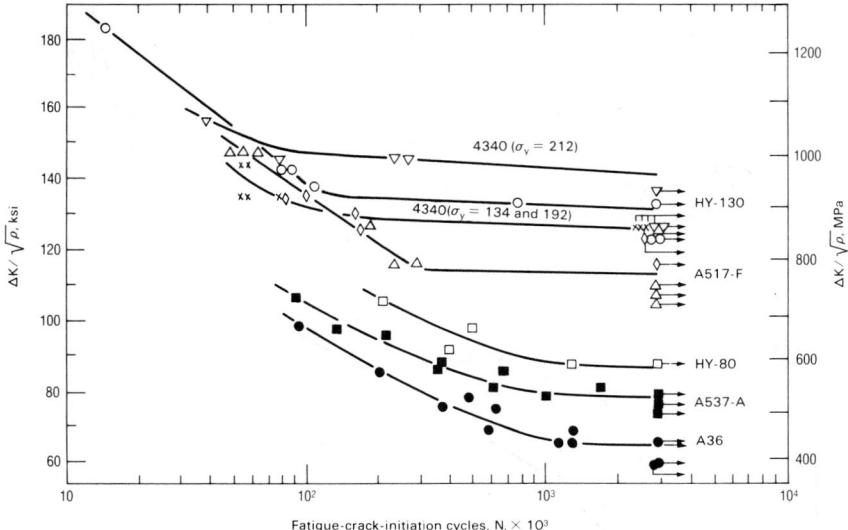

Fig. 1. Fatigue-crack-initiation behavior of various steels at a stress ratio of +0.1 (Ref 3)

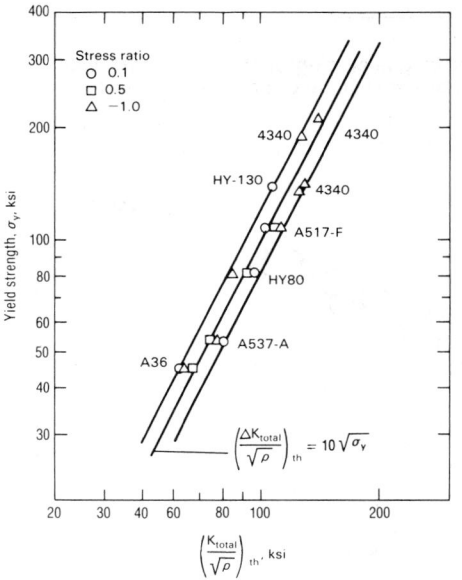

Fig. 2. Dependence of fatigue-crack-initiation threshold on yield strength

subjected to the maximum stress range, $\Delta\sigma_{max}$. Consequently, this material element is most susceptible to fatigue damage and is, in general, the origin of fatigue-crack initiation. It can be shown that, for sharp notches, the maximum-stress range on this element can be related to the stress-intensity-factor range, ΔK_I, as follows (Ref 3):

$$\Delta\sigma_{max} = \frac{2}{\sqrt{\pi}} \frac{\Delta K_I}{\sqrt{\rho}} = \Delta\sigma(k_t) \qquad \text{(Eq 2)}$$

where ρ is the notch-tip radius, Δσ is the range of applied nominal stress and k_t is the stress-concentration factor.

Fatigue-crack-initiation behavior of various steels is presented in Fig. 1 (Ref 3) for specimens subjected to zero-to-tension bending stress and containing a smooth notch that resulted in a stress-

concentration factor of about 2.5. The data show that $\Delta K_I/\sqrt{\rho}$, and therefore $\Delta\sigma_{max}$, is the primary parameter that governs fatigue-crack-initiation behavior in regions of stress concentration for a given steel tested in a benign environment. The data also indicate the existence of a fatigue-crack-initiation threshold, $\Delta K_I/\sqrt{\rho})_{th}$, below which fatigue cracks would not initiate at the roots of the tested notches. The value of this threshold is characteristic of the steel and increases with increasing yield or tensile strength of the steel. The data show that the fatigue-crack-initiation life of a component subjected to a given nominal-stress range increases with increasing strength. However, this difference in fatigue-crack-initiation life among various steels decreases with increasing stress-concentration factor (Ref 3).

Finally, fatigue-crack-initiation data for various steels subjected to stress ratios (ratio of nominal minimum applied stress to nominal maximum applied stress) ranging from −1.0 to +0.5 indicate that fatigue-crack-initiation life is governed by the total maximum stress (tension plus compression) range at the tip of the notch (Ref 5). The data presented in Fig. 2 (Ref 6) indicate that the fatigue-crack-initiation threshold, $\Delta K_I/\sqrt{\rho})_{th}$, for various steels subjected to stress ratios ranging from −1.0 to +0.5 can be estimated from

$$\frac{\Delta K_{total}}{\sqrt{\rho}} = 10\sqrt{\sigma_{ys}} \qquad \text{(Eq 3)}$$

where ΔK_{total} is the stress-intensity-factor range calculated by using the tension-plus-compression stress range, and σ_{ys} is the yield strength of the material.

FATIGUE-CRACK PROPAGATION

Extensive data have shown that the fatigue-crack-propagation behavior of metals is controlled primarily by the stress-intensity-factor range, ΔK_I. The fatigue-crack-propagation behavior of metals can be divided into three regions, as shown in Fig. 3 (Ref 7). The behavior in region 1 exhibits a fatigue-crack-propagation

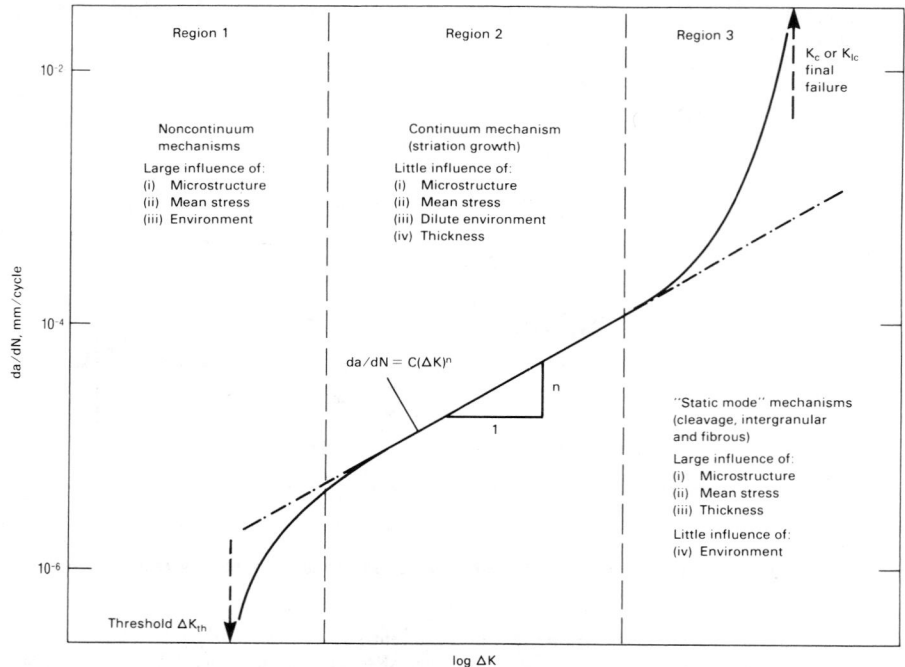

Fig. 3. Schematic illustration of variation of fatigue-crack-growth rate, da/dN, with alternating stress intensity, ΔK, in steels, showing regions of primary crack-growth mechanisms (Ref 7)

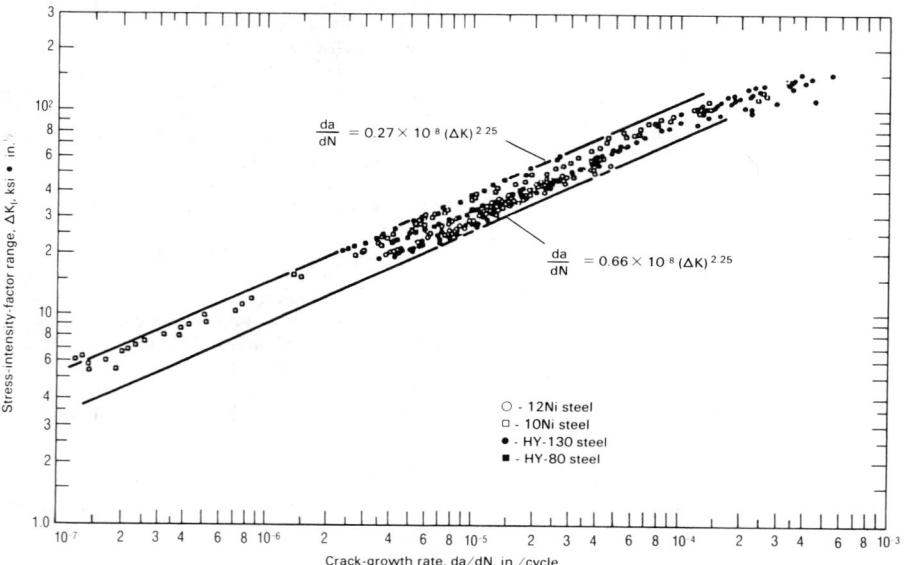

Fig. 4. Summary of fatigue-crack-growth data for martensitic steels (Ref 3)

threshold, ΔK_{th}, which corresponds to the stress-intensity-factor range below which cracks do not propagate under cyclic-stress fluctuations. An analysis of experimental results published on nonpropagating fatigue cracks shows that conservative estimates of ΔK_{th} for various steels subjected to different stress ratios, R, can be predicted (Ref 3) from

$$\Delta K_{th} = 6.4 (1 - 0.85R)$$
$$\text{for } R \geq +0.1 \qquad \text{(Eq 4a)}$$

and

$$\Delta K_{th} = 5.5 \text{ for } R < +0.1 \qquad \text{(Eq 4b)}$$

where ΔK_{th} is in ksi · in.$^{1/2}$.

Equation 4 indicates that the fatigue-crack-propagation threshold for steels is primarily a function of the stress ratio and is essentially independent of chemical or mechanical properties. The behavior in region 2 (Fig. 3) represents the fatigue-crack-propagation behavior above ΔK_{th}, which can be represented by the power-law relationship

$$\frac{da}{dN} = A(\Delta K_I)^n \qquad \text{(Eq 5)}$$

where a is crack length, N is number of cycles, and A and n are constants.

Extensive fatigue-crack-growth-rate data for various steels show that the primary parameter

affecting growth rate in region 2 is the stress-intensity-factor range, and that the mechanical and metallurgical properties of these steels have negligible effects on the fatigue-crack-growth rate in a room-temperature air environment. The data for martensitic steels fall within a single band, as shown in Fig. 4 (Ref 3), and the upper bound of scatter can be obtained (Ref 3) from

$$\frac{da}{dN} = 0.66 \times 10^{-8}(\Delta K_I)^{2.25} \qquad \text{(Eq 6)}$$

where a is in inches and ΔK_I is in ksi · in.$^{1/2}$. Similarly, as shown in Fig. 5 (Ref 3), data for ferrite-pearlite steels fall within a single band (different from the band for martensitic steels), and the upper bound of scatter can be calculated from

$$\frac{da}{dN} = 3.6 \times 10^{-10}(\Delta K_I)^{3.0} \qquad \text{(Eq 7)}$$

where a is in inches and ΔK_I is in ksi · in.$^{1/2}$.

The stress ratio and mean stress have negligible effects on the rate of crack growth in region 2. Also, the frequency of cyclic loading and the wave form (sinusoidal, triangular, square, trapezoidal) do not affect the rate of crack propagation per cycle of load for steels in benign environments (Ref 3).

The acceleration of fatigue-crack-growth rates that determines the transition from region 2 to region 3 appears to be caused by the superposition of a brittle or a ductile-tearing mechanism onto the mechanism of cyclic subcritical crack extension, which leaves fatigue striations on the fracture surface. These mechanisms occur when the strain at the tip of the crack reaches a critical value (Ref 3). Thus, the fatigue-rate transition from region 2 to region 3 depends on the maximum stress-intensity factor, on the stress ratio, and on the fracture properties of the material (Ref 3).

FRACTURE TOUGHNESS

The stress-intensity value for a given applied stress increases with increasing crack length, and for a given crack length increases with increasing applied stress. One of the underlying principles of fracture mechanics is that unstable fracture occurs when the stress-intensity factor at the crack tip reaches a critical value, K_c. For mode I deformation and for small crack-tip plastic deformation (plane-strain conditions), the critical stress-intensity factor for fracture instability, K_{Ic}, represents the inherent ability of a material to resist progressive tensile crack extension. However, this fracture-toughness property varies with constraint, and like other material properties such as yield strength, varies with temperature and loading rate as follows:

K_c = critical stress-intensity factor for static loading and plane-stress conditions of variable constraint. Thus, this value depends on specimen thickness and geometry, as well as on crack size.

K_{Ic} = critical stress-intensity factor for static loading and plane-strain conditions of maximum constraint. Thus, this value is a minimum value for thick plates.

K_{Id} = critical stress-intensity factor for dynamic (impact) loading and plane-strain conditions of maximum constraint

$K_c, K_{Ic} \text{ or } K_{Id} = C\sigma\sqrt{a}$

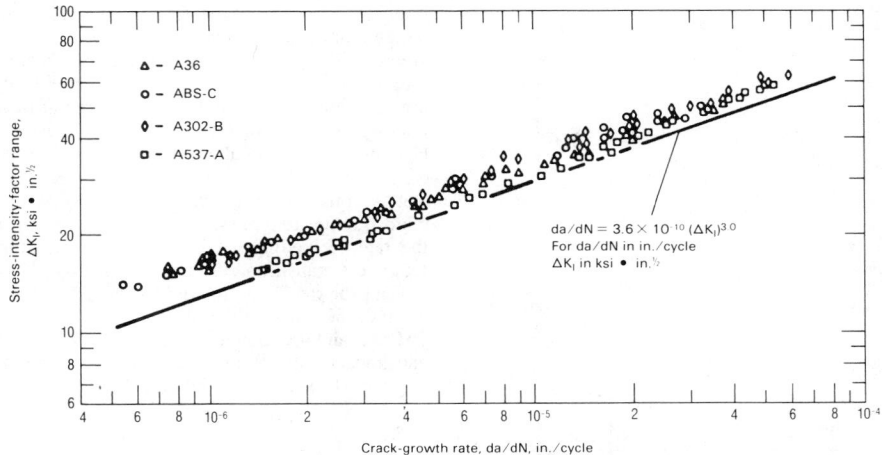

Fig. 5. Summary of fatigue-crack-growth data for ferrite-pearlite steels (Ref 3)

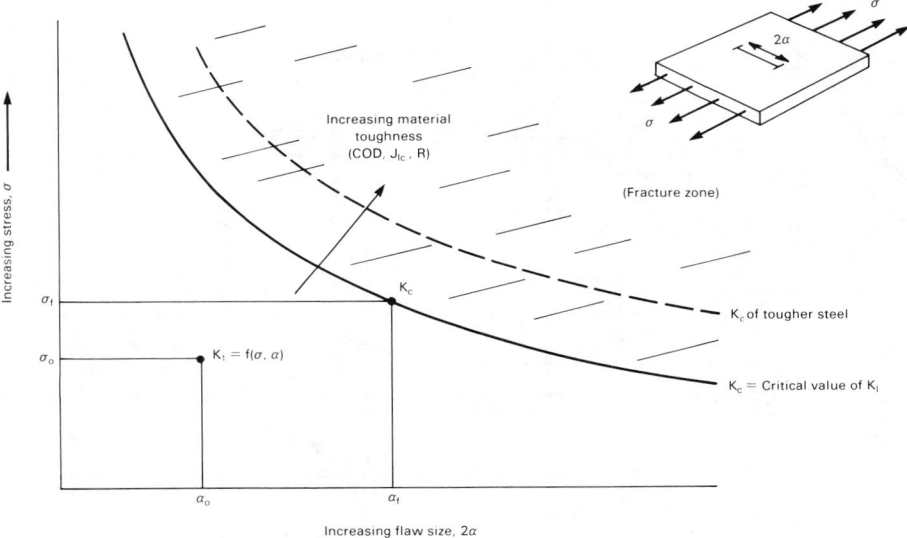

Fig. 6. Schematic illustration of relationship among stress, flaw size and material toughness (Ref 3)

where C is a constant that is a function of specimen and crack geometry, σ is nominal stress in ksi, and a is flaw size in inches.

Each of these values (K_c, K_{Ic} and K_{Id}) is also a function of temperature, particularly for those structural materials exhibiting a transition from brittle to ductile behavior.

By knowing the critical value of K_I at failure (K_c, K_{Ic} or K_{Id}) for a given material of a particular thickness and at a specific temperature and loading rate, the designer can determine flaw sizes that can be tolerated in structural members for a given design stress level. Conversely, he can determine the design stress level that can be safely used for an existing crack that may be present in a structure. In general, the relationship among fracture toughness (K_c), stress (σ) and crack size (a) is shown schematically in Fig. 6 (Ref 3) for a through-thickness crack in a plate. The figure shows that there are many combinations of stress and crack size (e.g., σ_f and a_f) that may cause fracture, and many combinations (e.g., σ_0 and a_0) that will not cause fracture, of the particular material under slow loading and at test temperature.

The effects of temperature and rate of loading on the fracture toughness of steels are discussed in the article that follows.

REFERENCES

1. *Stress Analysis of Cracks Handbook,* edited by H. Tada, P. C. Paris and G. R. Irwin: Del Research Corp., Hellertown, PA, 1973
2. *Handbook of Stress-Intensity Factors for Researchers and Engineers,* by G. C. Sih: Institute of Fracture and Solid Mechanics, Lehigh University, Bethlehem, PA, 1973
3. *Fracture and Fatigue Control in Structures—Applications of Fracture Mechanics,* by S. T. Rolfe and J. M. Barsom: Prentice-Hall, Englewood Cliffs, NJ, 1977
4. *Application of Fracture Mechanics for Selection of Metallic Structural Materials,* edited by J. E. Campbell, W. W. Gerbrich and J. H. Underwood: American Society for Metals, Metals Park, OH, 1982
5. "Effect of Cyclic Frequency on the Corrosion-Fatigue Crack-Initiation Behavior of ASTM A517 Grade F Steel," by M. E. Taylor and J. M. Barsom: in *Fracture Mechanics: Thirteenth Conference,* ASTM STP 743, American Society for Testing and Materials, 1981
6. "Fracture Mechanics for Bridge Design," by R. Roberts, J. M. Barsom, J. W. Fisher and S. T. Rolfe: FHWA-RD-78-69, Federal Highway Administration,

Office of Research and Development, July 1977
7. Near-Threshold Fatigue-Crack Propagation in Steels, by R. O. Ritchie: *International Metals Reviews,* Vol 24, No. 5 and 6, 1979

Fracture Toughness

By J. M. Barsom, U.S. Steel Corp.

TOUGHNESS is defined as the ability of a material to absorb energy. It is usually characterized by the area under a stress-strain curve for a smooth (unnotched) tension specimen loaded slowly to fracture. Notch toughness represents the ability of a material to absorb energy usually determined under impact loading in the presence of a notch. Notch toughness is measured by using a variety of specimens such as the Charpy V-notch impact specimen, the dynamic-tear specimen, and plane-strain fracture-toughness specimens under static loading (K_{Ic}) and under impact loading (K_{Id}) (see preceding article), as well as others.

DUCTILE-TO-BRITTLE FRACTURE TRANSITION

Traditionally, the notch-toughness characteristics of low- and intermediate-strength steels have been described in terms of the transition from ductile to brittle behavior as test temperature increases. Most structural steels can fail in either a ductile or a brittle manner depending on several conditions such as temperature, loading rate and constraint.

The most widely used specimen for characterizing the ductile-to-brittle transition behavior of steels has been the Charpy V-notch impact specimen, which is described in ASTM Standards A23 and A370. These specimens may be tested at different temperatures and the impact notch toughness at each test temperature may be determined from the energy absorbed during fracture, the percent shear (fibrous) fracture on the fracture surface, or the change in the width of the specimen (lateral expansion). An example of the ductile-to-brittle transition with temperature for each of these parameters is presented in Fig. 7. The actual values for each parameter and the locations of the curves along the temperature axis are usually different for different steels and even for a given steel composition.

The rate of change from ductile to brittle behavior depends on many parameters, including strength and composition of the material. Because the transition occurs over a range of temperatures, it has been customary to define a single temperature within the transition range that reflects the behavior of the steel under consideration. Several equally useful definitions are in use, including the 15 ft·lb temperature, the 15 mil temperature and the 50% shear temperature.

FRACTURE TOUGHNESS

Fracture-toughness behavior can be established best by using fracture-mechanics concepts. The toughness values are obtained by testing fatigue-cracked specimens at a given test temperature and rate of loading. Standard specimen geometries and test procedures used to obtain such data for structures under low rates of loading have been developed by the American Society for Testing and Materials (Ref 1). These

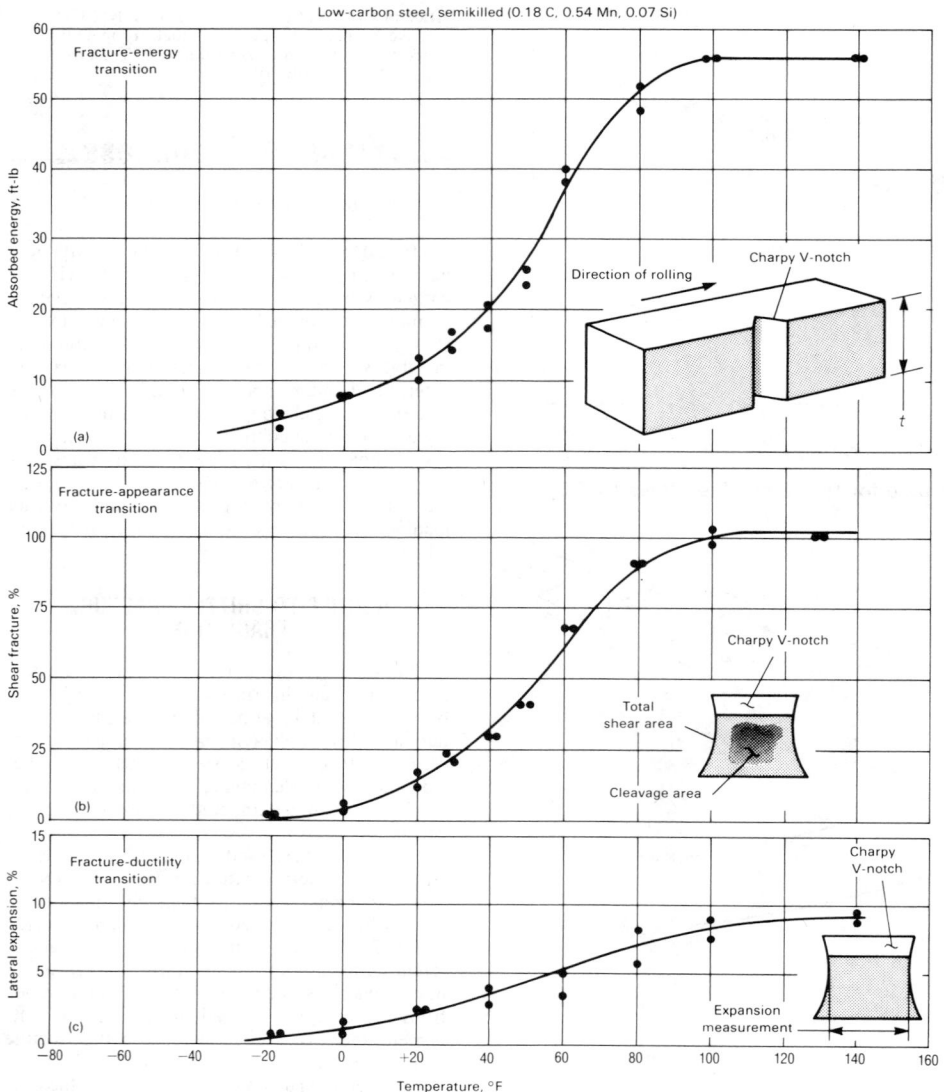

Low-carbon steel, semikilled (0.18 C, 0.54 Mn, 0.07 Si)

The drawings at lower right in the graphs indicate: (a) orientation of the specimen notch with plate thickness, *t*, and direction of rolling; (b) location of the total shear area on the fracture surface; and (c) location of the expansion measurement in this series of tests—all illustrated for a Charpy V-notch specimen. Percentage of shear fracture and lateral expansion were based on the original dimensions of the specimen.

Fig. 7. Characteristics of the transition-temperature range for Charpy V-notch testing of low-carbon steel plate, as determined by (a) fracture energy, (b) fracture appearance and (c) fracture ductility

tests establish the fracture-toughness behavior in terms of a critical stress-intensity factor, K_c, which is a function of the critical fracture stress and the size of the fatigue crack. Thus, the results of a fracture-mechanics characterization for a specified application (geometry, temperature and rate of loading) would yield the specific combinations of stress, crack size and crack shape that would cause fracture.

EFFECTS OF CONSTRAINT, TEMPERATURE AND LOADING RATE

Fracture toughness, K_c, varies with the degree of localized constraint to plastic flow along the tip of the fatigue crack. Thus, cracks in very thick members are subjected to higher constraints than are cracks in thinner members. The maximum constraint, as defined in ASTM E399, occurs under plane-strain conditions and results in the low-

est value of fracture toughness, K_{Ic}. Under identical test conditions, the K_c values for thinner plates are usually higher than those observed under plane-strain conditions (i.e., $K_c > K_{Ic}$).

Fracture toughness, K_{Ic}, of constructional steels under a constant rate of loading increases with increasing temperature (Ref 2 and 3). The rate of increase of K_{Ic} with temperature does not remain constant, but increases markedly above a given test temperature. An example of this behavior is shown in Fig. 8 (Ref 2 and 4) for A36 steel plate tested at three different loading rates. This transition in plane-strain fracture toughness is related to a change in the microscopic mode of crack initiation at the crack tip from cleavage to increasing amounts of ductile tearing.

An analysis of plane-strain fracture-toughness data that were obtained for constructional steels and that were valid according to ASTM standard procedures shows that the fracture-toughness

transition curve is translated (shifted) to higher temperature values as the rate of loading is increased. Thus, at a given temperature, fracture-toughness values measured at high loading rates are generally lower than those measured at lower loading rates. Also, the fracture-toughness values for constructional steels decrease with decreasing test temperature to a minimum K_{Ic} value that is equal to about 25 ksi·in.$^{1/2}$. This minimum fracture-toughness value is independent of the rate of loading used to obtain the fracture-toughness transition curve.

Data for steels having yield strengths between 36 and 250 ksi, such as those presented in Fig. 9 (Ref 2 and 4), show that the shift between static and impact plane-strain fracture-toughness curves is given (Ref 2) by the relationship

$$T_{shift} = 215 - 1.5\sigma_{ys}$$

$$\text{for } 28 \text{ ksi} < \sigma_{ys} \leq 130 \text{ ksi} \qquad \text{(Eq 1a)}$$

and

$$T_{shift} = 0 \text{ for } \sigma_{ys} > 130 \text{ ksi} \qquad \text{(Eq 1b)}$$

where T is temperature in °F and σ_{ys} is room-temperature yield strength. The temperature shift

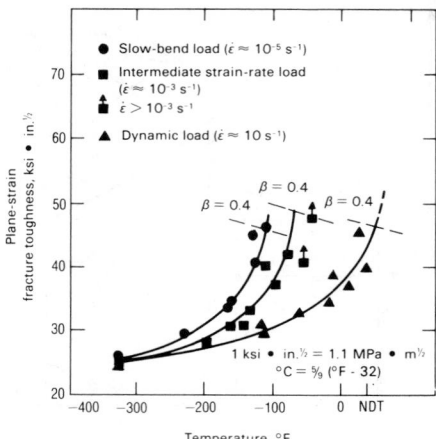

Fig. 8. Effect of temperature and strain rate on plane-strain fracture-toughness behavior of ASTM type A36 steel

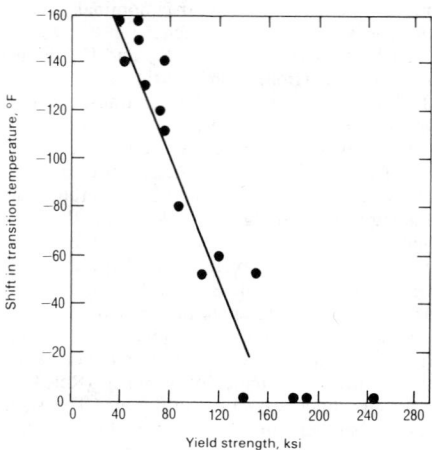

Fig. 9. Effect of yield strength on shift in transition temperature between impact and static plane-strain fracture-toughness curves

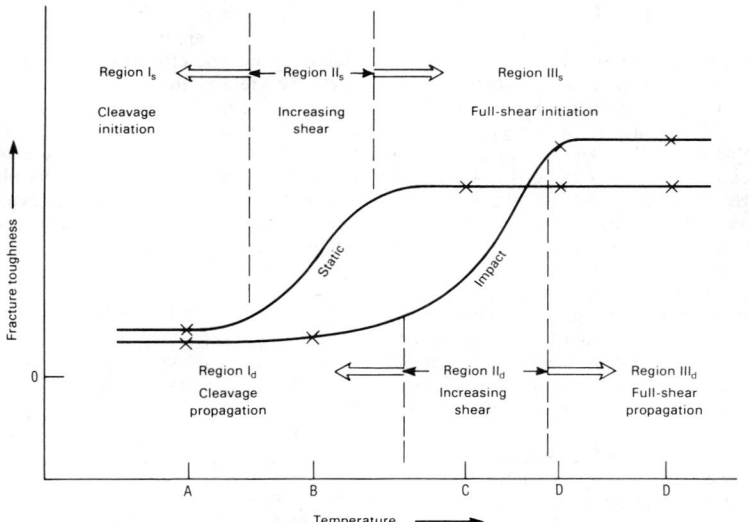

Fig. 10. Fracture-toughness transition behavior of steel under static and impact loading

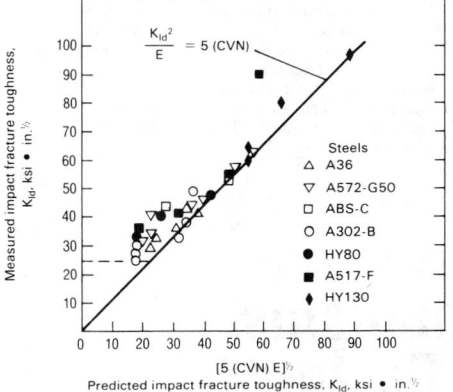

Fig. 11. Correlation of plane-strain impact fracture toughness and impact Charpy V-notch energy absorption for various grades of steel

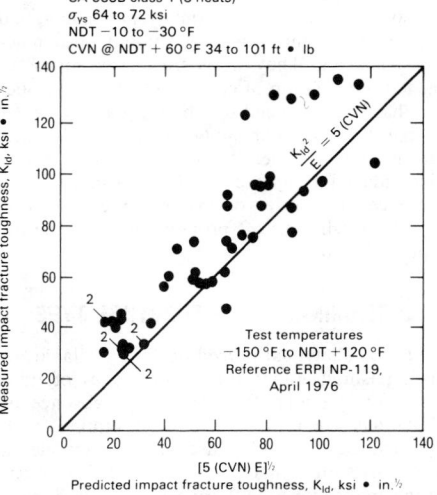

Fig. 12. Correlation of plane-strain impact fracture toughness and impact Charpy V-notch energy absorption for SA 533B, class 1, steel

where t is the loading time for the test and E is the elastic modulus for the material.

A proper use of fracture-mechanics methodology for fracture control of structures necessitates the determination of fracture toughness for the material at the temperature and loading rate representative of the intended application.

MORPHOLOGY OF FRACTURE SURFACES

The morphology of fracture surfaces for steel can be understood by considering the fracture-toughness transition behavior under static and impact loading (Fig. 10). The static fracture-toughness transition curve depicts the mode of crack initiation at the crack tip. The dynamic fracture-toughness transition curve depicts the mode of crack propagation.

The fracture-toughness curve for either static or dynamic loading can be divided into three regions as shown in Fig. 10. In region I_s for the static curve, the crack initiates in a cleavage mode from the tip of the fatigue crack. In region II_s, the fracture toughness to initiate unstable crack propagation increases with increasing temperature. This increase in crack-initiation toughness corresponds to an increase in the size of the plastic zone and in the zone of ductile tearing (shear) at the tip of the crack prior to unstable crack extension. In this region, the ductile-tearing zone is usually very small and is difficult to delineate by visual examination. In region III_s, the static fracture toughness is quite large and somewhat difficult to define, but the fracture initiates by ductile tearing (shear).

Once a crack has initiated under a static load, the morphology (cleavage or shear) of the fracture surface for the propagating crack is determined by the dynamic behavior and degree of

between static and any intermediate or impact plane-strain fracture-toughness curves is given (Ref 5) by the equation

$$T_{shift} = (150 - \sigma_{ys})\dot{\epsilon}^{0.17} \qquad (Eq\ 2)$$

where T is temperature in °F, σ_{ys} is room-temperature yield strength in ksi, and $\dot{\epsilon}$ is strain rate in s^{-1}. The strain rate is calculated for a point on the elastic-plastic boundary (Ref 6) according to the equation

$$\dot{\epsilon} = \frac{2\sigma_{ys}}{tE} \qquad (Eq\ 3)$$

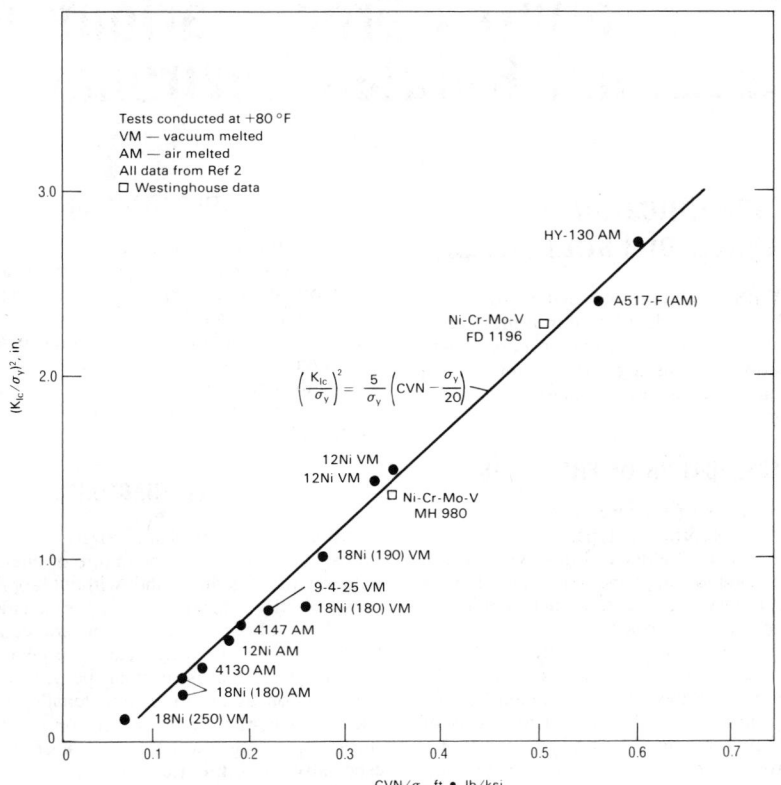

Fig. 13. Relation between K_{Ic} and CVN values in the upper-shelf region (Ref 2)

plane strain at the temperature. Regions I_d, II_d and III_d in Fig. 10 correspond to cleavage, increasing ductile tearing (shear) and full-shear crack propagation, respectively. Thus, at temperature A, the crack initiates and propagates in cleavage. At temperatures B and C, the crack exhibits ductile initiation but propagates in cleavage. The only difference between the behaviors at temperatures B and C is that the ductile-tearing zone for crack initiation is larger at temperature C than at temperature B. At temperatures D, cracks initiate and propagate in full shear. Consequently, full-shear fracture initiation and propagation occur only at temperatures for which the static and dynamic (impact) fracture behaviors are on the upper shelf.

CORRELATIONS OF K_{Id}, K_{Ic} AND CHARPY V-NOTCH IMPACT ENERGY ABSORPTION

The Charpy V-notch impact specimen is the most widely used specimen for material development, specifications and quality control. Moreover, because the Charpy V-notch impact energy absorption curve for constructional steels undergoes a transition in the same temperature zone as the impact plane-strain fracture toughness (K_{Id}), a correlation between these test results has been developed for the transition region and is given (Ref 2 and 4) by the equation

$$\frac{(K_{Id})^2}{E} = 5(CVN) \qquad \text{(Eq 4)}$$

where K_{Id} is in $ksi \cdot in.^{1/2}$, E is in ksi and CVN is in $ft \cdot lb$. The validity of this correlation is apparent from the data presented in Fig. 11 for various grades of steel ranging in yield strength from about 36 to about 140 ksi and in Fig. 12 for eight heats of SA 533B, class 1, steel. Consequently, a given value of CVN impact energy absorption corresponds to a given K_{Id} value (Eq 4), which in turn corresponds to a given toughness behavior at lower rates of loading. The behavior for rates of loading less than impact are established by shifting the K_{Id} value to lower temperatures by using Eq 1 or 2. Conversely, for a desired behavior at the minimum operating temperature and maximum in-service rate of loading, the corresponding behavior under impact loading can be established by using Eq 1 or 2, and the equivalent CVN impact value can be established by using Eq 4.

Barsom and Rolfe (Ref 2) suggested a relationship between K_{Ic} and upper-shelf Charpy V-notch impact energy absorption. This upper-shelf correlation, shown in Fig. 13, was developed empirically for steels having room-temperature yield strength, σ_{ys}, higher than about 110 ksi and is given by the equation

$$\left(\frac{K_{Ic}}{\sigma_{ys}}\right)^2 = \frac{5}{\sigma_{ys}}\left(CVN - \frac{\sigma_{ys}}{20}\right) \qquad \text{(Eq 5)}$$

where K_{Ic} is in $ksi \cdot in.^{1/2}$; σ_{ys} is in ksi; and CVN is energy absorption, in $ft \cdot lb$, for a Charpy V-notch impact specimen tested in the upper-shelf (100% shear fracture) region.

At the upper shelf, the effects of loading rate and notch acuity are not as critical as in the transition region. The effect of loading rate is to elevate the yield strength by about 25 ksi. Thus, Eq 5 may be used to calculate K_{Id} values by replacing σ_{ys} with the dynamic yield strength, σ_{yd}, where $\sigma_{yd} \approx \sigma_{ys} + 25$ ksi. This use of Eq 5 to calculate K_{Id} is consistent with the observation that, in the upper-shelf region, the dynamic fracture toughness of steels is higher than the static fracture toughness.

REFERENCES

1. Standard Method of Test for Plane-Strain Fracture Toughness of Metallic Materials: ASTM Standards, Part 10 (E399-78), American Society for Testing and Materials, 1978
2. *Fracture and Fatigue Control in Structures—Applications of Fracture Mechanics,* by S. T. Rolfe and J. M. Barsom: Prentice-Hall, Englewood Cliffs, NJ, 1977
3. K_{Ic} Transition-Temperature Behavior of A517-F Steel, by J. M. Barsom and S. T. Rolfe: *Engineering Fracture Mechanics,* Vol 2, No. 4, June 1971
4. Development of the AASHTO Fracture-Toughness Requirements for Bridge Steels, by J. M. Barsom: *Engineering Fracture Mechanics,* Vol 7, No. 3, Sept 1975
5. Effect of Temperature and Rate of Loading on the Fracture Behavior of Various Steels, by J. M. Barsom: in *Dynamic Fracture Toughness,* The Welding Institute, Abington, Cambridge, 1976
6. The Static and Dynamic Low-Temperature Crack-Toughness Performance of Seven Structural Steels, by A. K. Shoemaker and S. T. Rolfe: *Engineering Fracture Mechanics,* Vol 2, No. 4, June 1971

Failures From Various Mechanisms and Related Environmental Factors

Identification of Types of Failures

ANALYSIS of a failure of a metal structure or part usually requires identification of the type of failure. Failure can occur by one or more of several mechanisms, including surface damage such as corrosion or wear, elastic or plastic distortion, and fracture.

CLASSIFICATION OF FRACTURES

Many elements of fracture have been used to describe and categorize the types of fractures encountered in the laboratory and in service. These elements include loading conditions, rate of crack growth, and macroscopic and microscopic appearance of fracture surfaces.

Failure analysts often find it useful to classify fractures on a macroscopic scale as ductile fractures, brittle fractures, fatigue fractures, and fractures resulting from the combined effects of stress and environment. The last group includes stress-corrosion cracking and liquid-metal embrittlement.

DUCTILE FRACTURES

Ductile fractures are characterized by tearing of metal accompanied by appreciable gross plastic deformation and expenditure of considerable energy. Ductile tensile fractures in most materials have a gray, fibrous appearance and are classified on a macroscopic scale as either flat (perpendicular to the maximum tensile stress) or shear (at a 45° slant to the maximum tensile stress) fractures.

BRITTLE FRACTURES

Brittle fractures are characterized by rapid crack propagation with less expenditure of energy than with ductile fractures and without appreciable gross plastic deformation. Brittle tensile fractures have a bright, granular appearance and exhibit little or no necking. They are generally of the flat type—that is, normal (perpendicular) to the direction of the maximum tensile stress. A chevron pattern may be present on the fracture surface, pointing toward the origin of the crack, especially in brittle fractures in flat, platelike components.

FATIGUE FRACTURES

Fatigue fractures result from cyclic loading, and appear brittle and generally smooth on a macroscopic scale. They are characterized by incremental propagation of cracks until the cross section has been reduced to where it can no longer support the maximum applied load, and fast fracture ensues. Frequently, the progress of a service-induced fatigue crack is indicated by the presence of a series of macroscopic crescents, called "beach marks," progressing from the origin of the crack.

DETERMINATION OF FRACTURE TYPE

Several analytical procedures are available for distinguishing among the various types of fracture. For example, the presence or absence of plastic macrodeformation can be determined with the unaided eye, or by use of a steel scale, a machinist's micrometer or a machinist's or measuring microscope. Differences in some dimensional attribute of parts (such as width or thickness) at and well away from the fracture can serve to define macrodeformation after assurance that

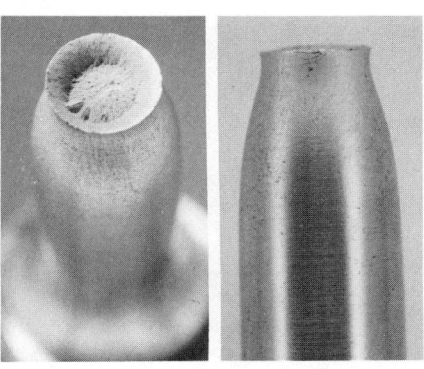 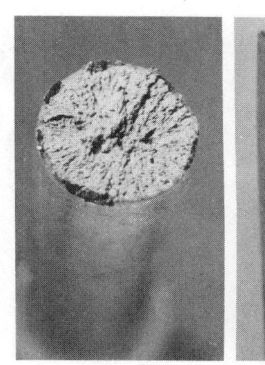 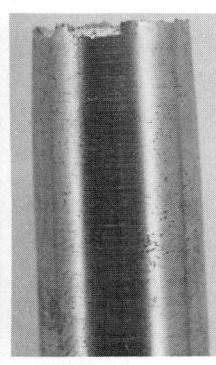

Source: Ductile and Brittle Fractures, by G. F. Vander Voort, *Metals Engineering Quarterly*, Vol 16, No. 3, 1976, p 33

Fig. 1. Appearance of ductile (left) and brittle (right) tensile fractures

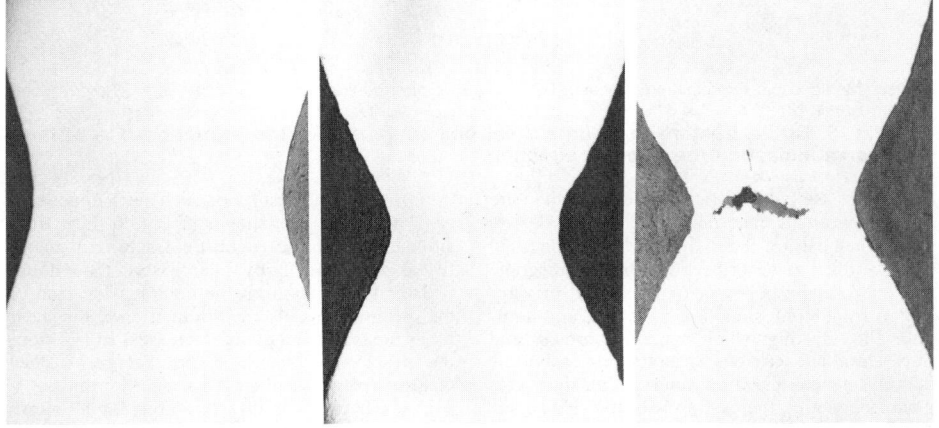

Fig. 2. Sections of a tensile specimen at various stages of deformation during development of a cup-and-cone fracture, showing that the fracture begins internally. Magnification, 10×.

both points of measurement had the same dimension before fracture.

Fracture-surface matching is also used to determine the presence or absence of plastic deformation. It is very important, however, to resist the temptation to fit the matching fracture surfaces together, because this almost always destroys (smears) microscopic features. The fracture surfaces should never actually touch during fracture-surface matching.

Ductile and Brittle Fractures

By Gordon W. Powell, The Ohio State University, and George F. Vander Voort, Carpenter Technology Corp.

MACHINE and structural components often fail in service as a result of either ductile or brittle fracture. The latter occurs far more frequently than the former, and, consequently, greater emphasis is placed on brittle fracture in the following discussion. The objectives of this review are to discuss the metallurgical and environmental factors and the service conditions which promote ductile or brittle fracture and to cite some of the structural features of fracture surfaces which provide information relative to the manner and direction of crack propagation and to service conditions. In addition, some examples of fractures

which have occurred in service will be presented. Most of the discussion pertains to fracture of low-strength ferritic steels.

FRACTURE INDUCED BY UNIAXIAL TENSION

The terms "ductile" and "brittle" as used in this discussion indicate the extent of macroscopic plastic deformation which precedes fracture. However, it must be pointed out that these terms also can be applied, and are applied, to fracture on a microscopic level. Ductile fractures are those which occur by microvoid formation and coales-

cence, whereas brittle fractures are those characterized by the propagation of a crack by quasicleavage. In high-strength aluminum alloys or high-strength steels, brittle (low-energy) fracture may occur by formation of large, shallow dimples. Consequently, the terms "ductile" and "brittle" also relate to the relative toughness of materials. The use of these terms at both the macroscopic and microscopic levels of observation can lead to some confusion, as will be pointed out below. But, as stated above, the usage herein is restricted to the macroscopic level.

Clearly, the cup-and-cone fracture shown at left in Fig. 1 has occurred as a result of appreciable plastic deformation and thus is a ductile fracture, whereas the other fracture in Fig. 1 is a brittle fracture. The sequence of events which culminates in a cup-and-cone fracture is illustrated vividly in Fig. 2, which shows the development of voids within the necked region (triaxial tensile stresses) of a tensile specimen and the coalescence of the voids to produce an internal crack by normal rupture. Final separation of the cross section occurs by shear rupture, which produces the wall of the cup. SEM fractographs of the bottom and the sidewall of the cup are presented in Fig. 3. On the microscopic level, a crack is formed by coalescence of microvoids which form as a result of particle-matrix decohesion or cracking of second-phase particles; the microvoids and the associated particles are shown at high magnification in Fig. 4. The process of microvoid for-

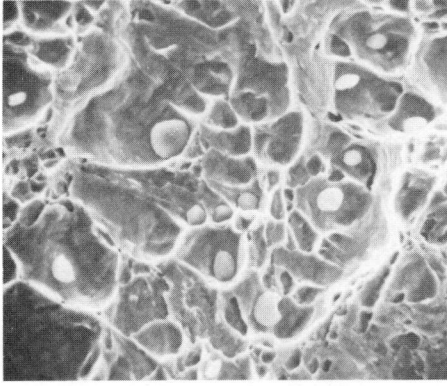

Courtesy of R. D. Buchheit, Battelle-Columbus

Fig. 4. SEM fractograph showing large and small sulfide inclusions in a ductile-dimple fracture. Magnification, 5000×.

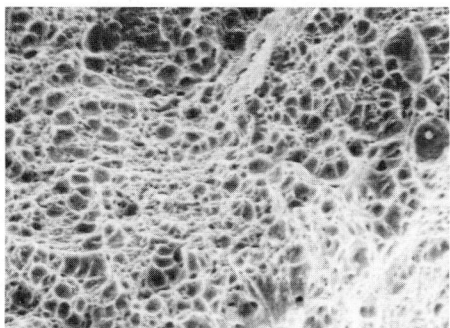 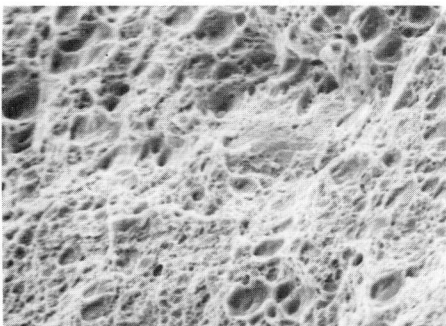

Fig. 3. SEM fractographs of the bottom (left) and sidewall (right) of a ductile cup-and-cone fracture surface. Magnification, 1000×.

mation and coalescence involves considerable localized plastic deformation and requires the expenditure of a large amount of energy, which is the basis of selection of a material with good fracture toughness. The reduction in area of ultrahigh-purity aluminum and copper approaches 100% because of the absence from these materials of void-nucleating particles. In terms of their visual appearance, ductile fractures have a matte or silky texture.

With regard to the brittle fracture in Fig. 1, it will be noted that the fracture surface is characterized by a pattern of radial ridges which emanate from the center of the fracture surface. The ridges run parallel to the direction of crack propagation, and a ridge is produced when two cracks which are not coplanar become connected by tearing of the intermediate material. The cracks, which propagate predominantly by quasicleavage, move rapidly toward the periphery of the specimen cross section and, as shown in Fig. 1, penetrate the external surface of the specimen by shear rupture along a relatively small shear lip. The shear lip develops as a result of the change in the state of stress from one of triaxial tension to one of plane stress. The extent or width of the shear lip is dependent on the temperature at which fracture occurs, formation of a shear lip being favored by higher temperatures.

The radial pattern and the shear lip provide useful information relative to the origin and the direction of propagation of a crack. For example, the radial pattern on the fracture surface of the machine component shown in Fig. 5 indicates that fracture initiated on the upper edge of the fracture surface, the actual initiation site being designated by the arrow. The ridge pattern shown in Fig. 5 is commonly referred to as a river pattern and is formed by the convergence and joining of tear ridges into a single tear ridge as the crack front advances. The extent to which a distinctive ridge pattern is developed is dependent on the temperature at which fracture occurs, higher temperatures (relative to the nil-ductility transition temperature) promoting the formation of a readily visible river pattern (this matter is considered again below). Hence, the absence or presence of a ridge pattern on the fracture surface of a brittle fracture may be used to provide a qualitative estimate of the fracture temperature relative to the nil-ductility transition temperature of the steel.

As noted above, the brittle fracture shown in Fig. 1 terminates with a very small shear lip. This fact can prove helpful when one is faced with the problem of determining the origin of a brittle fracture. The origin of the fracture is invariably characterized by the absence of a shear lip whereas a shear lip is expected to be present along the periphery of the fracture surface where the crack emerges from the interior of the material. Consequently, the periphery of a fracture surface should be examined with these facts in mind. If fracture occurs at a low temperature, then a shear lip may not be formed.

The preceding discussion has been confined implicitly to wrought materials. If one considers the fracture of castings which may contain porosity and microshrinkage cavities, then the development of a readily visible ridge pattern and shear lip is highly improbable because the topography of the fracture surface is determined largely by the distribution of the porosity and the microshrinkage cavities; the fracture surfaces of castings are highly irregular in comparison with those of wrought materials. Therefore, fractures

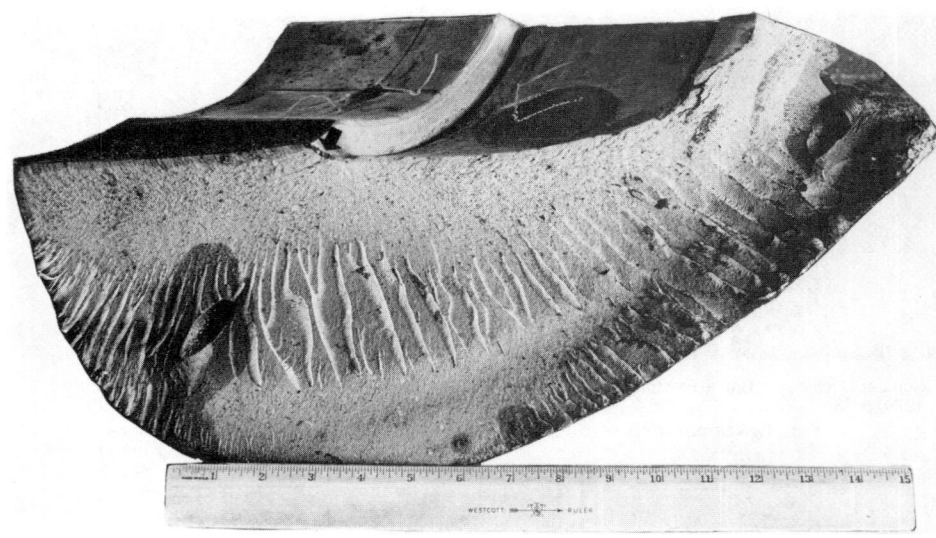

Source: Macroscopic Examination Procedures, in *Metallography in Failure Analysis*, by G. F. Vander Voort, Plenum Press, New York, 1978

Fig. 5. Brittle fracture in a large steel component. Note the pronounced radial marks indicating the fracture direction.

of castings are more difficult to analyze than are those of wrought materials.

As noted earlier, the use of the terms "ductile" and "brittle" to describe both the macroscopic and microscopic behavior of a material may result in some confusion. For example, an aluminum alloy casting which contains porosity and microshrinkage cavities exhibits relatively low ductility when tested in uniaxial tension and, macroscopically, the casting is brittle. However, the ligaments between the pores and microshrinkage cavities in the weakest cross section of the specimen behave as very small tensile specimens with a correspondingly small gage length and they fracture in a ductile manner but with very little extension. Thus, the casting is brittle on a macroscopic scale but fractures in a ductile manner on a microscopic scale; the toughness of the casting is low. Another example is provided by a cast-and-extruded ferritic Fe–8 wt % Al alloy. This alloy is ductile at ambient temperature as its reduction in area is at least 40%, but fracture occurs by quasicleavage. In this case, the alloy is ductile on a macroscopic level but fractures in a brittle manner on a microscopic scale. Thus, there are cases in which fractures that are ductile or brittle macroscopically exhibit the opposite behavior on a microscopic scale.

DUCTILE-TO-BRITTLE TRANSITION

Body-centered-cubic metals, such as ferritic steels, and also some hexagonal-close-packed metals, undergo a ductile-to-brittle transition in their fracture behavior. Fractures which arise from this phenomenon have an especially insidious character because they may occur under static loading at stresses well below the yield strength and without warning. Brittle fracture is induced by low temperatures, high strain rates and a state of triaxial tensile stresses, such as produced by a notch. Only two of the three need be present to initiate brittle fracture. In many cases (ships, storage tanks), brittle fracture is induced by low temperature and a state of triaxial tensile stresses at a critical location in the structure.

The ductile-to-brittle transition can be demonstrated quite readily by the Charpy V-notch

impact test. The results of such tests conducted on a mild steel are shown in Fig. 6. The transition from ductile to brittle fracture is manifested geometrically by a decrease in the amount of lateral expansion on the compression side of the specimen and also by elimination of the shear lip as the test temperature decreases. In addition, the usual appearance of the fracture surface changes from that of a rough, ductile fracture to a smooth, planar brittle fracture. On a microscopic level, the ductile fracture is characterized by microvoid coalescence, dimples of various orientations, and transgranular cracking, whereas the low-temperature brittle fracture consists of transgranular quasicleavage. At temperatures within the transition range, crack propagation occurs by both microvoid coalescence and quasicleavage, with the latter being confined predominantly to the central region of the fracture surface. Face-centered-cubic alloys, such as AISI type 316 stainless steel, do not exhibit a ductile-to-brittle transition, as is demonstrated in Fig. 7 by SEM fractographs of fractured Charpy specimens. These fracture surfaces were produced by microvoid coalescence and contain dimples of various orientations. The existence of a ductile-to-brittle transition in mild steel is related to the structure and mobility of dislocations in the body-centered-cubic ferrite; the strong temperature dependence of the flow stress and dislocation–interstitial atom interactions are specific aspects of this phenomenon.

Another example of the ductile-to-brittle transition in mild steel is shown in Fig. 8, which is a photograph of a series of notched ship-plate samples broken over a range of temperatures; the notched end of each sample is located at the top of the photograph. At −25 and −35 °F, the material is ductile, the fracture surface being characterized by prominent shear lips and a central region of fibrous fracture. The −50 °F fracture surface has a distinct chevron (or herringbone) pattern which is produced by tear ridges running perpendicular to the crack front; the crack front is more advanced at the center of the plate (plane-strain condition) and lags behind at the sides of the plate (plane-stress condition) where there is a shear lip. The chevron pattern becomes less

Temperature, °F	75	150	200
Energy, ft·lb	25	99	112
Lateral expansion, in.	0.032	0.073	0.073
% fibrous	65	95	100

Temperature, °F	0	25	50
Energy, ft·lb	4	10	17
Lateral expansion, in.	0.006	0.014	0.021
% fibrous	15	20	40

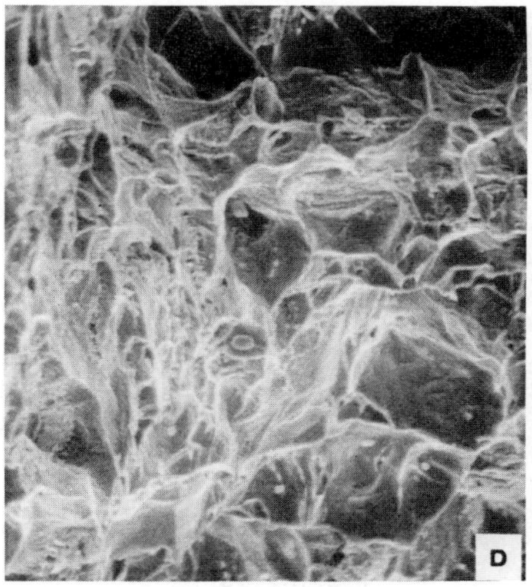

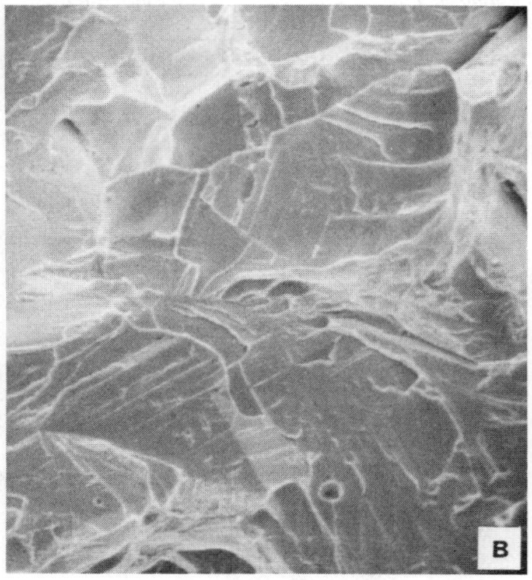

Source: Ductile and Brittle Fractures, by G. F. Vander Voort, *Metals Engineering Quarterly*, Vol 16, No. 3, 1976, p 32

Fig. 6. SEM fractographs (both at 500×) of ductile (D) and brittle (B) fractures in Charpy V-notch impact specimens shown at top

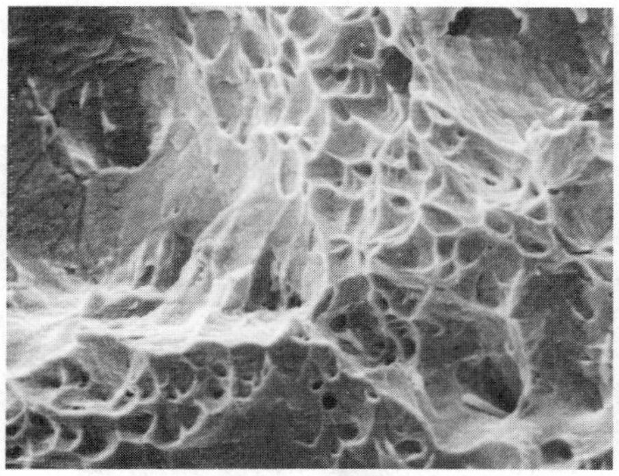

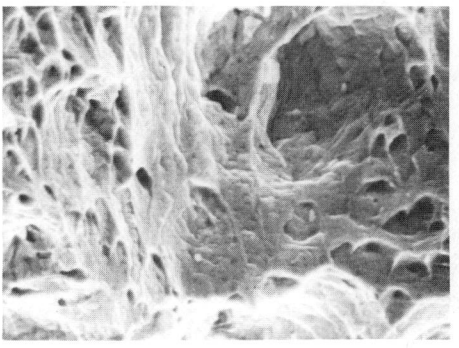

Fig. 7. SEM fractographs (all at 1000×) of Charpy specimens of AISI type 316 stainless steel broken at −300 °F (a), −200 °F (b), −100 °F (c) and 0 °F (d)

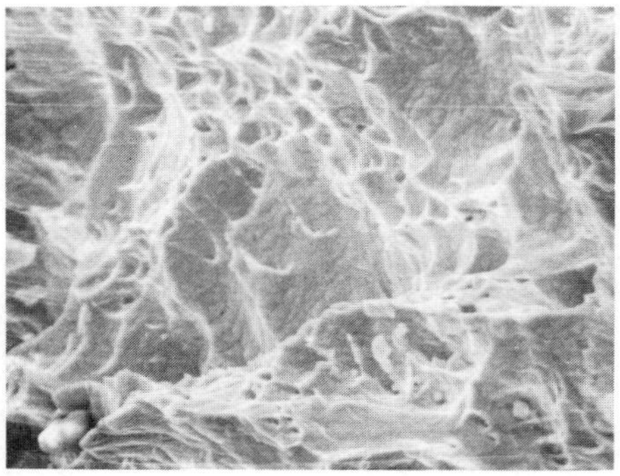

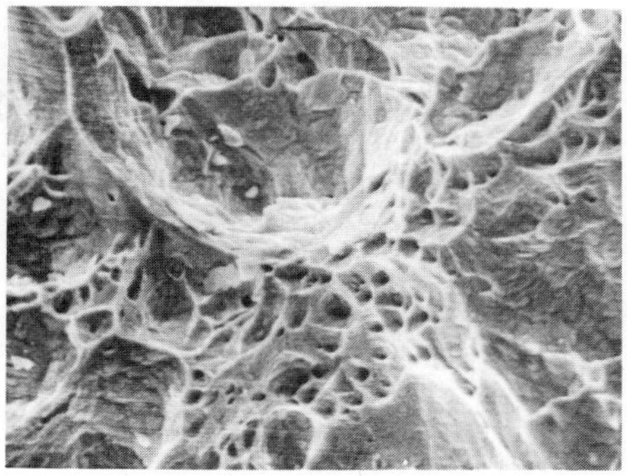

distinct at lower temperatures because the steel is less ductile and the crack front is not curved. Recently, a crude-oil storage tank split open and the fracture surface of the steel plate displayed essentially the same surface features shown in Fig. 8. The ambient temperature had been about −15 °F for 2 or 3 days and the tank was empty during this period. At 7 a.m. on one of these cold days, pumping of crude oil at a temperature of 50 °F into the tank was initiated. Approximately 10 h later the tank split open. The crack started at the base of the tank at the toe of a fillet weld and propagated vertically through the wall of the tank. Within 2 ft of the fracture-initiation site, the fracture surface was featureless and a shear lip was not in evidence. Approximately 10 ft above the initiation site, where the wall of the tank had been heated by the warm oil, there was a distinct chevron pattern and a shear lip was present along each edge of the fracture surface. When present, the chevron pattern provides useful information to the failure analyst because the chevrons point

in the direction from which the crack originated. Further up the wall of the tank, the steel plate failed by slant shear primarily because the steel plate was thinner ($^5/_{16}$ in.) at the top of the tank.

Brittle fractures may occur by either transgranular or intergranular cracking. Intergranular fractures are specific to certain conditions which induce the embrittlement — temper embrittlement, hydrogen embrittlement, stress-corrosion cracking, liquid-metal embrittlement, etc.

HIGH-TEMPERATURE TENSILE FRACTURE

The results of Gleeble tests performed at relatively high strain rates and temperatures on AISI type 1040 steel are shown in Fig. 9. The ductility (% reduction in area) is lower at the lower strain rate (0.08 in./s) because of extensive grain-boundary decohesion which occurs at the lower strain rate. Intergranular fracture becomes more prevalent at low strain rates and high temperatures because the grain boundary can undergo

Source: Macroscopic Examination Procedures, in *Metallography in Failure Analysis,* by G. F. Vander Voort, Plenum Press, New York, 1978

Fig. 8. Chevron patterns in mild steel ship-plate samples broken over a range of temperatures. Each fracture began at the notch (top). Mating fracture halves are shown for each temperature.

Tensile strength, ksi	Reduction in area, %					Tensile strength, ksi	Reduction in area, %

Crosshead speed
0.08 in./s — 5 in./s

		SLOW	FAST			
6.1	96		2000°F		15.2	99.9
3.7	81		2200°F		11.2	98.6
2.6	60		2400°F		7.4	93

2400 °F; 0.08 in./s   2400 °F; 5 in./s

Nital   200×   Nital   200×

Fig. 9. Macrographs (at 2.5×) and micrographs of AISI type 1040 carbon steel Gleeble test samples

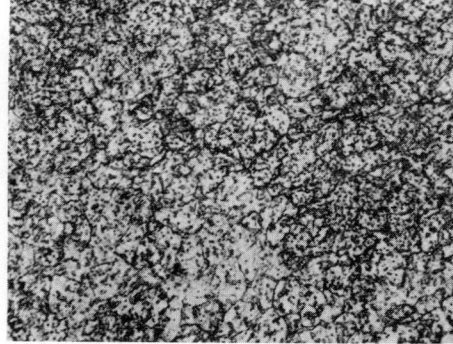

3% nital   700×

Source: Ductile and Brittle Fractures, by G. F. Vander Voort, *Metals Engineering Quarterly,* Vol 16, No. 3, 1976, p 62

Fig. 10. "Super-duty" 12-in.-long axle shaft made of AISI type S7 tool steel that failed in torsion (note torsion marks on axle shaft) as a result of improper hardening

shear and subsequent decohesion under the action of an applied stress. Structural components such as heat-exchanger tubes which have fractured as a result of creep will often display readily visible surface cracks at locations away from the main fracture.

SERVICE FAILURES

Service failures which occur solely by ductile fracture are relatively infrequent, but, when they do occur, the structure or machine component is often stated to have failed as a result of an overload. Overload failures may be the result of the part having been underdesigned (a term which includes the selection and heat treatment of the materials) for a specific set of service conditions, improperly fabricated or fabricated from defective materials, or of the part having been abused — i.e., subjected to conditions of load and environment which exceeded those of the intended use. A simple example of ductile fracture of a machine component is shown in Fig. 10. A "super-duty" rear axle which had been fabricated from an AISI type S7 tool steel failed as a result of a torsional overload; note that the smaller-diameter section of the shaft has undergone considerable twisting. The axle was intended to have a quenched-and-tempered microstructure, but it was actually underaustenitized (or not austenitized at all) during heat treatment as indicated by its spheroidized microstructure and hardness (22 to 27 HRC).

Brittle fractures which occur in service are initiated invariably by defects which are present initially in the manufactured product or fabricated structure, or by defects which develop during service. The defects are essentially stress concentrators and may take any one of the following forms: large inclusions, laps or seams introduced into the material during rolling, forging, etc.; sharp re-entrant corners or notches; machining, quenching, fatigue or stress-corrosion cracks; cracks which evolve as a result of hydrogen embrittlement (hydrogen flakes). This list of defects, although not complete, includes many of the common initiators of brittle fracture. The single most prevalent initiator of brittle fracture is the fatigue crack, which conservatively accounts for at least 50% of all brittle fractures in manufactured products. Although not defects in a geometric sense, residual stresses which arise from working, heat treatment or welding of a material can be an important factor in initiation of brittle fractures.

Fig. 11. Circular spall that began at a large subsurface inclusion in a hardened steel roll

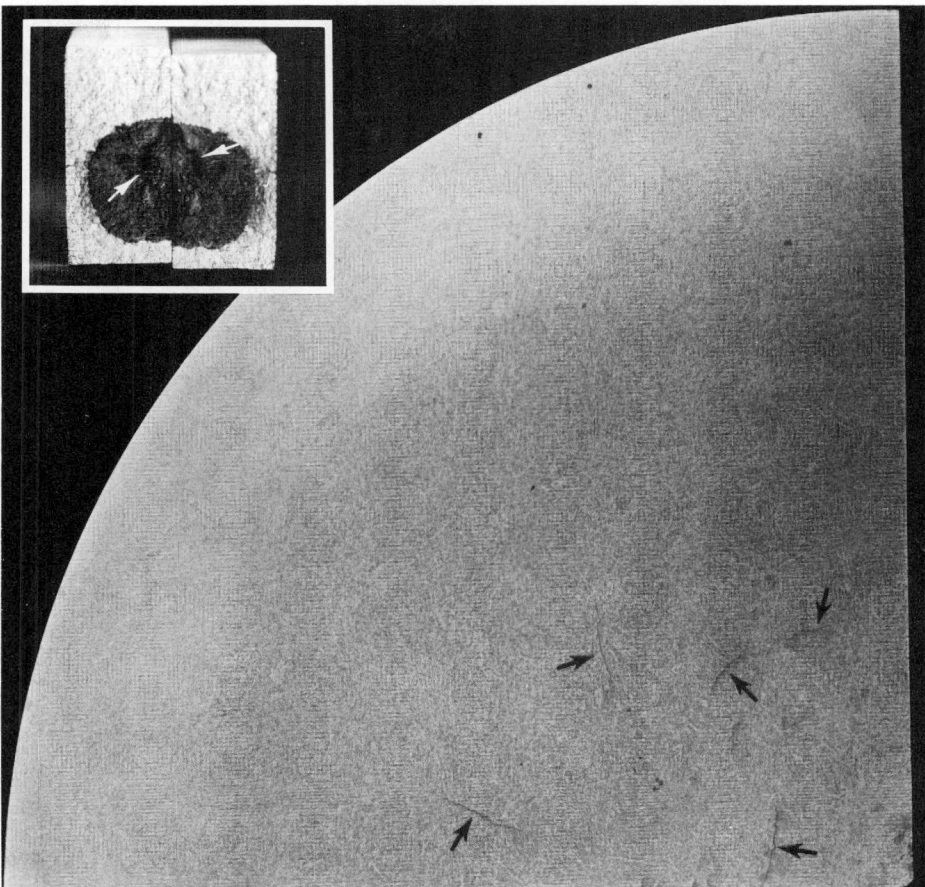

Source: Ductile and Brittle Fractures, by G. F. Vander Voort, *Metals Engineering Quarterly*, Vol 16, No. 3, 1976, p 57

Fig. 12. Hydrogen flakes (arrows) in a 16-in.-diam forging. Hairline crack that opened after cold etching (see inset) shows characteristic shape of hydrogen flake. Some staining (arrows in inset) resulted from cold etching.

Examples of defects arising from manufacturing operations are shown in Fig. 11, 12 and 13. In Fig. 11, a subsurface inclusion initiated a circular fatigue crack which ultimately produced brittle fracture; note the well-defined river pattern radiating from the periphery of the defect. The hydrogen flakes shown in Fig. 12 resulted from an excessive amount of hydrogen being dissolved in the steel. Hydrogen can be picked up by the steel during melting, teeming, welding, pickling or electroplating. Figure 13 is a display of some mechanical notches which were produced by improper welding procedures.

As mentioned above, residual stresses may be an important element in brittle fracture. Figure 14 shows a brittle fracture which originated from a hole punched in the steel plate; note the chevron pattern. The surface of the punched hole is plastically deformed and the strain gradient produces residual stresses which in conjunction with the applied stresses initiated a small crack at the surface of the hole. In a situation such as this one, moisture may have had access to the hole and contributed to formation of the initial crack.

Finally, an example of a brittle fracture which was initiated by a fatigue crack is shown in Fig. 15. The surface of a fatigue crack is relatively smooth but it may contain the well-known beach or clamshell markings. In addition, the periphery of a fatigue crack is a relatively smooth curve.

Fatigue Failures

By Campbell Laird, University of Pennsylvania

FATIGUE is the progressive localized permanent structural change that occurs in a material subjected to repeated or fluctuating strains at stresses having a maximum value less than the tensile strength of the material. Fatigue may culminate in cracks or fracture after a sufficient number of fluctuations.

Fatigue fractures are caused by the simultaneous action of cyclic stress, tensile stress and plastic strain. If any one of these three is not present, fatigue cracking will not initiate and propagate. The cyclic stress starts the crack; the tensile stress produces crack growth (propagation). Although compressive stress will not cause fatigue, compression loads may do so.

The process of fatigue may be considered as consisting of three stages:

1. Initial fatigue damage leading to crack initiation

2. Crack propagation until the remaining uncracked cross section of a part becomes too weak to carry the loads imposed

3. Final, sudden fracture of the remaining cross section.

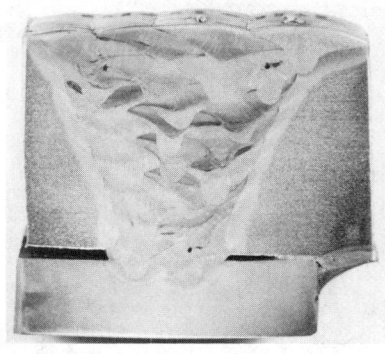

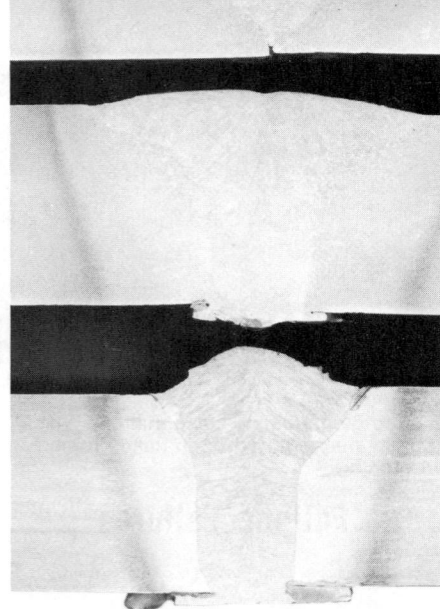

Source: Ductile and Brittle Fractures, by G. F. Vander Voort, *Metals Engineering Quarterly,* Vol 16, No. 3, 1976, p 36

Fig. 13. Examples of mechanical notches caused by improper welding procedures. Magnifications: top left, 1×; bottom left and right, 2×.

Fig. 14. Brittle fracture that began at a punched hole (shown at right) in steel plate. Mating fracture halves show chevrons pointing back toward the punched hole.

Fatigue cracking normally results from cyclic stresses that are well below the static yield strength of the material. (In low-cycle fatigue, however, or if the material has an appreciable work-hardening rate, the stresses also may be above the static yield strength.) Generally, a fatigue crack is initiated in a highly stressed region of a component subjected to cyclic stresses of sufficient magnitude. The crack propagates under the applied stress through the material until complete fracture results. On the microscopic scale, the most important feature of the fatigue process is nu-cleation of one or more cracks under the influence of reversed stresses that exceed the flow stress, followed by development of cracks at persistent slip bands or at grain boundaries.

PREDICTION OF FATIGUE LIFE

In practice, except for a few relatively brittle materials, prediction of the fatigue life of a material is complicated because fatigue life is very sensitive to small changes in loading conditions, local stresses and local characteristics of the material. Because it is difficult to account for these minor changes in either the dynamic stress-prediction techniques or in fatigue-failure criteria, there is a large uncertainty inherent in analytical predictions of fatigue life. Thus, the designer also is required to rely on experience with similar parts and eventually on qualification testing of prototypes or production parts. Although laboratory fatigue tests performed on small specimens are not sufficient for precisely establishing the fatigue life of a part, it is useful to examine these data because laboratory tests (*a*) are the major source of fatigue-failure criteria, (*b*) isolate the loading variables involved in fatigue, (*c*) are useful in rating materials in terms of their relative resistance to fatigue and (*d*) can be used to establish the relative influences of such items as fabrication method, surface finish, heat treatment, assembly technique and environment on fatigue life.

In general, fatigue life can be expected to depend on the following:

1. Type of loading (uniaxial, bending, torsional)
2. Shape of loading curve
3. Frequency of load cycling
4. Loading pattern (periodic loading at constant or variable amplitude, programmed loading or random loading)
5. Magnitude of stresses
6. Part size
7. Fabrication method and surface roughness
8. Operating temperature
9. Operating atmosphere.

Traditionally, fatigue life has been expressed as the total number of stress cycles required for a fatigue crack to form and grow right through a structure. In this article, fatigue data are expressed in terms of total life.

Fatigue data also can be expressed in terms of crack-growth rate. Depending on the material, and the structure, cracks can form early in the fatigue life, and grow continuously until catastrophic failure occurs. However, the rate of growth is usually slow at the beginning of fatigue life, unless the stress is high.

CONCEPTS RELATED TO FATIGUE

Most laboratory fatigue testing is done either with uniform axial loading or in uniform bending, thus producing only tensile and compressive stresses. The stress is usually cycled either between a maximum and a minimum tensile stress or between a maximum tensile stress and a maximum compressive stress. The latter is given an algebraic minus sign, and therefore is called the minimum stress.

Stress Ratio. The algebraic ratio of two specified stress values in a stress cycle is called the stress ratio. Two commonly used stress ratios are: the ratio, A, of the alternating stress amplitude to the mean stress ($A = S_a/S_m$); and the ratio, R, of the minimum stress to the maximum stress

Source: Macroscopic Examination Procedures, in *Metallography in Failure Analysis*, by G. F. Vander Voort, Plenum Press, New York, 1978, p 58

Fig. 15. Fracture surface of an ASTM type A36 steel member. This failure began by fatigue (arrows) but progressed only a short distance before brittle fracture occurred.

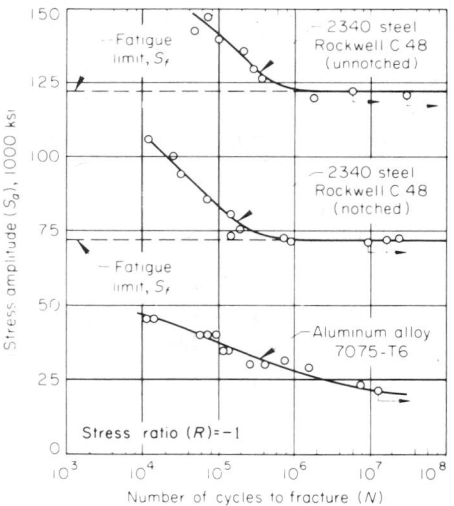

Fig. 16. Typical S-N curves for constant stress amplitude, and sinusoidal loading

$(R = S_{min}/S_{max})$. For example, if the stresses are fully reversed, the stress ratio R becomes -1; if the stresses are partially reversed, R becomes a negative number less than 1. If the stress is cycled between a maximum stress and no load, the stress ratio R becomes zero.

Applied Stresses. Three descriptions of the applied stress are sometimes given. The mean stress, S_m, is the algebraic average of the maximum and minimum stresses in one cycle, $S_m = (S_{max} +$

$S_{min})/2$. In the completely reversed test, the mean stress is zero. The range of stress, S_r, is the algebraic difference between the maximum and minimum stresses in one cycle, $S_r = S_{max} - S_{min}$. The stress amplitude, S_a, is one-half the range of stress, $S_a = S_r/2 = (S_{max} - S_{min})/2$.

During a fatigue test the stress cycle is usually maintained constant, so that the applied stress conditions can be written $S_m \pm S_a$, where S_m is

the static or mean stress and S_a is the alternating stress, equal to half the stress range. When $S_m = 0$, the maximum tensile stress is equal to the maximum compressive stress; this is called an alternating stress, or a completely reversed stress. When $S_m = S_a$, the minimum stress of the cycle is zero; this is called a pulsating or repeated tensile (or compressive) stress. Any other combination is known as a fluctuating stress.

Stress Intensity. Because of the importance of crack growth in fatigue, modern workers have conceived of a parameter called the stress-intensity factor, K, which takes account of the local stresses acting to drive a crack, independent of the structure which contains it. The stress-intensity factor is derived from the solution of the stress field near the tip of a crack in linear elastic material and is given by $K = F \cdot S_{max} \cdot (\pi a)^{1/2}$, where a is the length of the crack, and F is a factor which depends on the type of loading, whether the crack is being stressed in tension or in shear, and on the geometry of the structure. Numerous compilations of this factor are available for different conditions.

S-N Curves. The results of fatigue tests are often plotted as maximum stress or stress amplitude to number of cycles, N, to fracture using a logarithmic scale. The resulting curve of data points is called an S-N curve. Three typical S-N curves are shown in Fig. 16. The two curves for 2340 steel are typical for steels — a fairly straight slanting portion at low cycles straightening into a horizontal line at higher cycles, with a sharp transition between the two.

With the onset of the electronic revolution, techniques for precise measurement of strain have greatly improved. Fatigue tests therefore are conducted in strain control and the results are often plotted as ϵ/N curves where ϵ is either total strain or plastic strain. When such results are presented in log-log plots, linear behavior is usually observed (see Fig. 17), and this is referred to as the Coffin-Manson Law. The slope of the straight line is usually between -0.5 and -0.6, but for extreme materials it may reach -0.9.

Fatigue Properties. The horizontal portion of an S-N curve represents the maximum stress that the metal can withstand for an infinitely large number of cycles with 50% probability of failure and

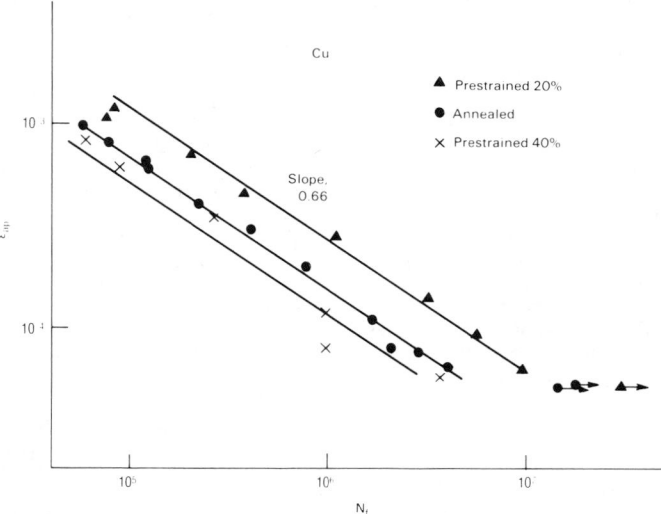

Courtesy of P. Lukas and M. Klesnil. Source: *Materials Science and Engineering*, Vol 11, 1973, p 345
Fig. 17. Typical Coffin-Manson curves for tests at constant strain amplitude, for annealed and cold worked copper

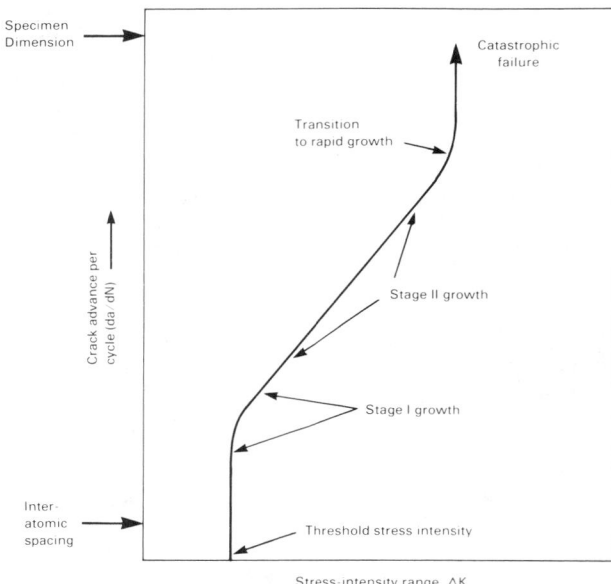

Fig. 18. Schematic representation of crack-growth rate as a function of stress-intensity range

is called the fatigue (endurance) limit, S_f. Nonferrous metals are traditionally regarded as not possessing a fatigue limit, because their S/N curves differ in shape from those of ferrous materials (see the curve for aluminum alloy 7075-T6 in Fig. 16). However, modern research shows that there is a stress/strain limit below which damaging slip bands do not form in nonferrous metals. However, for these metals, instead of reporting the fatigue limit, it is customary to report the fatigue strength, which is the stress to which the metal can be subjected for a specified number of cycles. There is no standard number of cycles, so it is necessary that each table of fatigue strengths specify the number of cycles for which the strengths are reported.

Contemporary workers have approached fatigue analysis by defining more elaborate fatigue properties analogous to the yield stress, ultimate strength, etc., familiar from monotonic deformation and fracture. Cyclic deformation is approached by the cyclic stress-strain curve—that is, a plot of the saturated stress attained by a specimen cycled in plastic strain control against the value of the applied strain. The properties are derived from the parameters of this curve, as well as those of the S/N curve and the Coffin-Manson curve. Likewise, the properties of a material in fatigue-crack propagation are obtained from a plot of crack-propagation rate against the range of the stress-intensity factor, ΔK (see the schematic plot shown in Fig. 18). It will be noted that the propagation curve typically consists of three parts: steeply sloping parts at high and low values of ΔK, and a more gently sloping part at intermediate values of ΔK. The exponent for this last part and the threshold stress intensity below which the crack-growth rate is insignificant are important properties.

Stress-Concentration Factor. Stress is concentrated in a metal by structural discontinuities, such as notches, holes or scratches, which act as stress raisers. The stress-concentration factor, K_t, is the ratio of the greatest stress in the region of the notch (or other stress concentrator) to the corresponding nominal stress. For determination of K_t, the greatest stress in the region of the notch

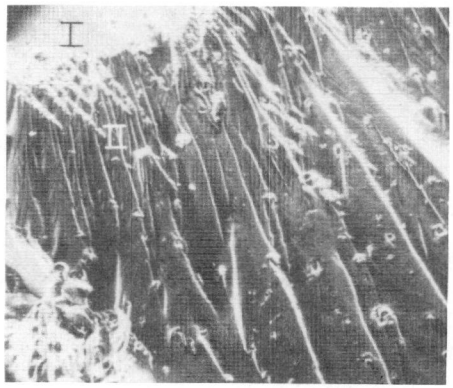

Fig. 19. Transition from stage I to stage II of a fatigue fracture in a coarse-grain specimen of aluminum alloy 2024-T3

is calculated from the theory of elasticity, or equivalent values are derived experimentally.

STAGES OF FATIGUE FRACTURE

The fracture surface that results from fatigue failure has a characteristic appearance that can be divided into three zones or progressive stages of fracture.

Stage I is the initiation of cracks and their propagation by slip-plane fracture, extending inward from the surface at approximately 45° to the stress axis. A stage I fracture never extends over more than about two to five grains around the origin. In each grain the fracture surface is along a well-defined crystallographic plane, which should not be confused with a cleavage plane although it has the same brittle appearance. There are usually no visible fatigue striations associated with a stage I fracture surface. In some instances (depending on the material, environment and stress level), a stage I fracture may not be discernible.

Stage II. The transition from stage I to stage II fatigue fracture is the change of orientation of the main fracture plane in each grain from one or

two shear planes to many parallel plateaus separated by longitudinal ridges. The plateaus are usually normal to the direction of maximum tensile stress.

A transition from stage I to stage II in a coarse-grain specimen of aluminum alloy 2024-T3 is shown in Fig. 19. The presence of inclusions rich in iron and silicon did not affect the fracture path markedly. The inclusions, which were fractured, ranged from 5 to 25 μm in diameter. In Fig. 19 the stage II area shows a large number of approximately parallel fatigue patches containing very fine fatigue striations that are not resolved at the magnification used. Fine striations are typical in stage II, but frequently are seen only under high magnification. The rates of growth associated with stages I and II are shown in Fig. 18.

Stage III occurs during the last stress cycle when the cross section is unable to sustain the applied load. The final fracture, which is the result of a single overload, can be brittle or ductile, or a combination of the two.

FRACTURE CHARACTERISTICS REVEALED BY MACROSCOPY

Beach Marks. The most characteristic feature usually found on fatigue-fracture surfaces is beach marks, which are centered around a common point that corresponds to the fatigue-crack origin. Also called clamshell, conchoidal and arrest marks, beach marks are perhaps the most important characteristic feature in identifying fatigue failures. In tough materials, with thick or round sections, the final-fracture zone will consist of a fracture by two distinct modes: (*a*) tensile fracture (plane-strain mode) extending from the fatigue zone and in the same plane, and (*b*) shear fracture (plane-stress mode) at 45° to the surface of the part bordering the tensile fracture. These two modes are illustrated in the surface of a fatigue fracture through a thick section shown in Fig. 20. In Fig. 20, two features in the final-fracture zone that aid in determining the origin of fracture are: (*a*) fatigue usually originates at the surface, and therefore the fatigue origin is not included in the shear-lip fracture; and (*b*) the presence of characteristic chevron marks in the tensile fracture that point back to the origin of fracture.

FRACTURE CHARACTERISTICS REVEALED BY MICROSCOPY

Examination of fatigue fractures is routinely carried out in the scanning electron microscope because of the high resolution, high depth of field and convenience of this instrument.

Metallographic examinations of cross sections through suspected fatigue fractures typically show that the crack path was transgranular. However, this might be misleading, and secure results on whether crack paths were transgranular or intergranular are best obtained by studying the specimen surface, fracture surface and metallographic section in combination.

Striations. In electron-microscope examination of fatigue-fracture surfaces, the most prominent features found are patches of finely spaced parallel marks, called fatigue striations. The fatigue striations are oriented perpendicular to the microscopic direction of crack propagation and, with uniform loading, generally increase in spacing as they progress from the origin of fatigue, because

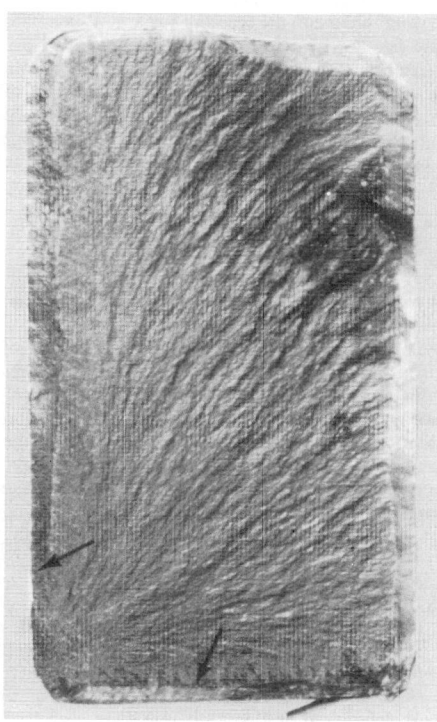

Chevron marks point to origin of fatigue in lower left-hand corner. Shear rupture along the periphery is indicated by arrows.

Fig. 20. Surface of a fatigue fracture in a 4330V steel part

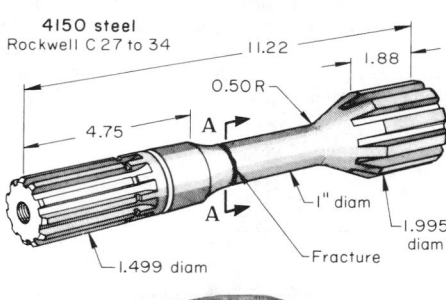

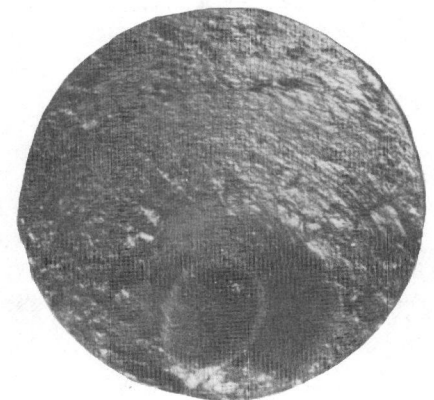

View A-A

The shaft, made of 4150 steel, was used in a piston-type pump. View A-A shows beach marks over a large area of a fracture surface; oval region near bottom center is the final-fracture area. Dimensions are in inches.

Fig. 21. Pump shaft that fractured in service as a result of reversed-bending and torsional fatigue

the stress intensity increases. It is firmly established that each striation is caused by one stress cycle.

Rotational Bending. A machine component that is commonly subjected to a bending load is a rotating round shaft. A unique feature of rotational-bending loading is that during one revolution of the shaft both maximum and minimum loading are exerted around the entire circumference of the shaft in the region of maximum bending moment. Because the loading is axially symmetrical, a fatigue crack can be initiated at any point, or at several points, around the periphery of the shaft (see Fig. 21).

Distortion Failures

DISTORTION FAILURE occurs when a structure or component is deformed so that it (a) no longer can support the load it was intended to carry, (b) is incapable of performing its intended function, or (c) interferes with the operation of another component. Distortion failures can be either plastic or elastic, and may or may not be accompanied by fracture. There are two main types of distortion: *size distortion,* which refers to a change in volume (growth or shrinkage); and *shape distortion* (bending or warping), which refers to a change in geometrical form.

Distortion failures ordinarily are considered to be self-evident—for example, damage of a car body in a collision or bending of a nail being driven into hard wood. However, failure analysts often are faced with more subtle situations. For example, the immediate cause of distortion (bending) of an automobile-engine valve stem is contact of the valve head with the piston, but a failure analyst must go beyond this immediate cause in order to recommend proper corrective measures. The valve may have stuck open because of faulty lubrication; the valve spring may have broken because corrosion had weakened it; the spring may have had insufficient strength and taken a set, allowing the valve to drop into the path of the piston; or the engine may have been over-revved many times, causing coil clash and subsequent fatigue fracture of the spring. Without careful consideration of all the evidence, a failure analyst may miss the true cause of a distortion failure.

OVERLOADING

Every structure has a load limit beyond which it is considered unsafe or unreliable. Applied loads that exceed this limit are known as overloads and sometimes result in distortion or fracture of one or more structural members. Estimation of load limits is one of the most important aspects of design and is commonly computed by one of two methods—classical design or limit analysis.

Classical design keeps allowable stresses entirely within the elastic region and is used routinely in design of parts. Generally, allowable stresses for static service are set at one-half the yield strength for ductile materials and one-sixth for brittle materials, although other fractions may be more suitable for specific applications.

Limit Analysis. The upper limit in design is defined as the load at which a structure will break or collapse under a single application of force.

Limit analysis assumes in idealized material—one that behaves elastically up to a certain yield strength and then does not work harden but

undergoes an indefinite amount of plastic deformation with no change in stress. The inherent safety of a structure is more realistically estimated by limit analysis in those instances when the structure will tolerate some plastic deformation before it collapses.

Figure 22 illustrates the relative stress-strain behavior of a low-carbon steel, a strain-hardening material and an idealized material—all with the same yield strength (the upper yield point for the low-carbon steel, and the stress at 0.2% offset for the strain-hardening material). Load limits for parts made of materials that strain harden significantly when stressed in the plastic region can be estimated by limit analysis, as can those for parts made of other materials whose stress-strain behavior differs from that of the idealized material. In these situations, the designer bases his design calculations on an assumed strength that actually may lie well within the plastic region for the material.

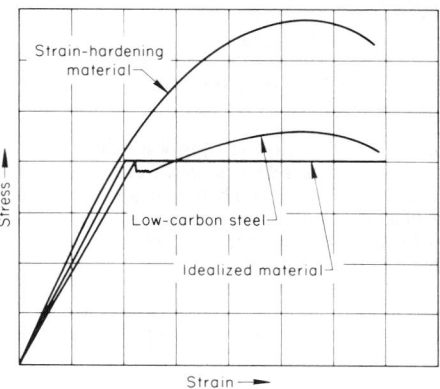

Fig. 22. Schematic comparison of the conventional stress-strain behavior of a low-carbon steel, a strain-hardening material, and the idealized material assumed in limit analysis, all having the same yield strength

Amount of Distortion. When designing structures using limit analysis, designers do not always consider the amount of distortion that will be encountered. A rough illustration of the distortion that resulted from overloading of small cantilever beams is given in Fig. 23. Known loads were applied to rectangular-section beams of low-carbon steel and of stainless steel, and the permanent deflection at the loading point was measured. Maximum fiber stresses were calculated from the applied load and original specimen dimensions.

Effect of Temperature. Distortion failures caused by overload can occur at any temperature at which the flow strength of the material is less than the fracture strength. In this discussion, flow strength is defined as the average true stress required to produce detectable plastic deformation caused by a relatively slow, continuously increasing application of load; fracture strength is the average true stress at fracture caused by a relatively slow, continuously increasing application of load. The flow strength and fracture strength of a material are temperature dependent, as is the elastic modulus (Young's modulus, bulk modulus or shear modulus). Figure 24 illustrates this temperature dependence schematically for polycrystalline materials that do not undergo a solid-state trans-

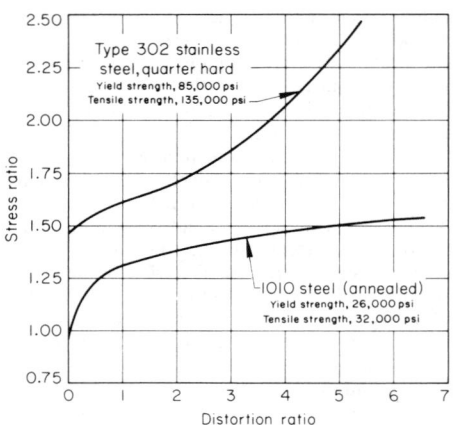

Distortion ratio is permanent deflection, measured at a distance from the support ten times the beam thickness, divided by beam thickness. Stress ratio is maximum stress, calculated from applied load and original beam dimensions, divided by yield strength.

Fig. 23. Relation of distortion ratio to stress ratio for two steel cantilever beams of rectangular cross section

formation. Two flow strengths are shown — one for a material that does not have a ductile-to-brittle transition in fracture behavior, such as a metal with a face-centered-cubic (fcc) crystal structure, and one for a body-centered-cubic (bcc) material that exhibits a ductile-to-brittle transition.

As shown in Fig. 24, the flow strength, fracture strength and elastic modulus of a material generally decrease as temperature increases. If a structure can carry a certain load at 20 °C (70 °F), it can carry the same load without deforming at lower temperatures.

——Wear Failures——

Edited by David Rigney, The Ohio State University

WEAR is a surface phenomenon that occurs by displacement and detachment of material. Because wear usually implies a progressive loss of weight and alteration of dimensions over a period of time, wear problems generally differ from those entailing outright breakage. However, wear may progress far enough to cause catastrophic failure. For example, fatigue failure may initiate in a worn area.

All mechanical components that undergo sliding or rolling contact are subject to some degree of wear. Typical of such components are bearings, gears, seals, guides, piston rings, splines, brakes and clutches. Wear of these components may range from mild, polishing-type attrition to rapid and severe removal of material with accompanying surface roughening. Whether or not wear constitutes failure of a component depends on whether the wear deleteriously affects the ability of the component to function. For example, even mild, polishing-type wear of a close-fitting spool in a hydraulic valve may cause excessive leakage and thus constitute failure, even though the surface of the spool is smooth and apparently undamaged. On the other hand, a hammer in a rock crusher, for example, can continue to function satisfactorily in spite of severe denting, gouging and the removal of as much as several inches of surface metal.

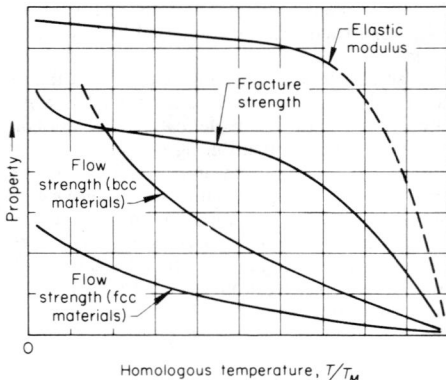

T is the instantaneous absolute temperature, and T_M is the absolute melting temperature of the material.

Fig. 24. Schematic diagram of the temperature dependence of elastic, plastic and fracture behavior of polycrystalline materials that do not exhibit a solid-state transformation

Lubrication implies the intentional use of a substance that reduces friction between contacting surfaces.

TYPES OF WEAR

In general, wear may be defined as damage to a solid surface caused by the removal or displacement of material by the mechanical action of a contacting solid, liquid or gas. Wear is usually detrimental, but in mild form (such as "breaking in"), wear may be beneficial. When a failure is caused predominantly by one type of wear, analysis may be relatively simple. However, many wear failures result from a combination of types or modes of wear. In addition, as wear progresses there may be a change in the predominant wear mode. Under these conditions, analysis is more complex.

"Sliding wear," commonly known as adhesive wear, occurs when two surfaces slide against each other under pressure. The process involves adhesion, plastic deformation and fracture (see Fig. 25). The following terms, in order of increasing severity, are often used to describe the damage associated with sliding wear: scoring/scuffing, galling, and seizing. Wear debris may include flakes, cylinders, spheres and irregular particles, and may even include some microcutting chips, as in abrasive wear.

Abrasive wear is displacement of material from a surface by contact with hard projections on a mating surface, or with hard particles, that are

moving relative to the wearing surface. When hard particles are involved, they may be trapped between two sliding surfaces and abrade one or both of them, or they may be embedded in either of the surfaces and abrade the opposing surface (see Fig. 26). Abrasive wear may occur in the dry state or in the presence of a liquid. Microcutting chips generally are present in abrasive wear debris.

Surface fatigue is a special type of surface damage whereby particles of metal are detached from a surface under high cyclic stresses, causing pitting or spalling. Surface fatigue is an important failure mechanism in rolling-contact systems.

Erosive wear is abrasive wear involving loss of surface material by contact with a fluid that contains particles. Relative motion between the surface and the fluid is essential to this process, and the force on the particles that actually inflict the damage is applied kinetically. Although erosive wear most often involves solid particles, one type — *liquid-impingement erosion* — is caused by liquid droplets carried in a rapidly moving stream of gas. Erosion in which the relative motion of solid particles is nearly parallel to the eroded surface is called *abrasive erosion*, whereas erosion in which the relative motion of particles is nearly normal to the eroded surface is called *impingement erosion*.

Chemical or corrosive wear is a type of wear in which chemical or electrochemical reaction with the environment significantly contributes to the wear rate. In some instances, chemical reaction takes place first and is followed by the removal of corrosion products by mechanical action. However, mechanical action may precede chemical action and result in formation of very small particles of debris, which subsequently react with the environment. Chemical reaction, even when mild, and mechanical action may be mutually enhancing (see Fig. 27).

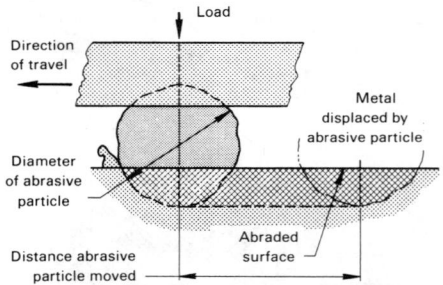

Fig. 26. Idealized representation of abrasive wear resulting from mechanical application of force to an abrasive particle

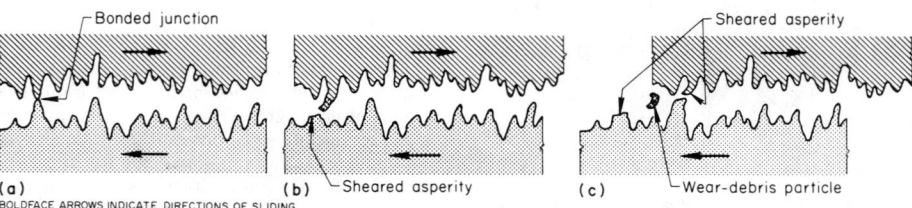

As the surfaces slide across one another, a bonded junction (a) is torn from one asperity (b), then is sheared off by an adjacent asperity to form the particle of wear debris (c). Alternatively, the sheared-off material may transfer to the opposing surface, becoming part of the wear debris at a later time. Note that the vertical dimension has been exaggerated in these drawings.

Fig. 25. Schematic illustration of one process by which a particle of wear debris may be detached during sliding wear

Fig. 27. Worn type 440A stainless steel drive-roller sleeve that was used in a belt conveyor under mildly corrosive conditions

"**Fretting**" is a special form of wear which occurs between two contacting surfaces subjected to repeated, small-amplitude relative sliding, such as from vibration, in the presence of oxygen.

Erosion-corrosion is a type of wear in which there is relative movement between a surface and a corrosive fluid (which also may carry abrasive particles), the wear rate being directly related to the rate of relative movement. When abrasive particles are present, material removal is effected mainly by contact with the particles (erosive wear).

A special form of erosion-corrosion called *cavitation erosion* can occur on a surface in contact with a liquid that does not contain particulate matter. In cavitation erosion, the repeated formation and collapse of vapor bubbles at the surface imposes large repetitive contact stresses that can cause pitting or spalling.

LUBRICATED WEAR

On lubricated surfaces, the wear process is mild and generates fine debris of a particle size as small as one or two microns. Abrasive wear predominates under lubricated conditions. Electron-microscope examination of worn surfaces from lubricated assemblies usually reveals a multitude of fine scratches oriented in the direction of relative motion. The fine debris that is generated by abrasion becomes suspended in the oil or grease. In devices using circulating-oil lubrication, advantage has been taken of the fact that wear debris can be analyzed by spectroscopy, and that deterioration of the device by wear can be diagnosed from these results.

Modes of Lubrication. There are several basic modes of lubrication. In all modes, contact surfaces are separated by a lubricating medium, which may be a solid, a semisolid or a pressurized liquid or gaseous film. *Hydrodynamic lubrication* is a system in which the shape and relative motion of the sliding surfaces cause the formation of a fluid film having sufficient pressure to separate the surfaces (see Fig. 28). *Hy-*

drostatic lubrication is a system in which the lubricant is supplied under sufficient external pressure to separate the opposing surfaces by a fluid film. *Elastohydrodynamic lubrication* is a system in which the friction and film thickness between the two bodies in relative motion are determined by the elastic properties of the bodies, in combination with the viscous properties of the lubricant at the prevailing pressure, temperature and rate of shear. *Dry-film (solid-film) lubrication* is a system in which a coating of solid lubricant separates the opposing surfaces and the lubricant itself wears away. *Boundary lubrication* and *thin-film lubrication* are two modes in which friction and wear are affected by properties of the contacting surfaces as well as by properties of the lubricant. In boundary lubrication, each surface is covered by a chemically bonded fluid or semisolid film, which may or may not serve to separate opposing surfaces.

LUBRICANTS

Many kinds of surface films can act as lubricants, preventing cold welding of asperities on opposing surfaces or allowing opposing surfaces to slide across one another at a lower frictional force than would prevail if a film were not present. Lubricants may be either liquid or solid. (On some occasions, gas films may act as lubricants.) One of the functions of a lubricant is to carry away heat generated by two surfaces sliding under contact pressure. Liquid lubricants can dissipate heat better than solid or semifluid lubricants, but in all types the shear properties of the lubricant are critical to its performance.

Lubricant Failures Leading to Wear. In devices that depend on lubricants to combat friction and avoid deterioration by wear, failure of the lubricant can be disastrous. Most lubricant failures occur by (*a*) chemical decomposition, (*b*) contamination, (*c*) changes in properties caused by excessive heat, or (*d*) outright loss from, or inadequate flow of a pressurized fluid into, lubricated areas. Lubricating oils and greases can fail by any one of the foregoing processes alone. However, in most situations, chemical decomposition, contamination and temperature are all involved and are interrelated.

SELECTED REFERENCES

Books:

Sourcebook on Wear Control Technology, ASM, 1978
Wear Control Handbook, ASME, 1980
Metals Handbook, 8th Ed., Vol 10, *Failure Analysis and Prevention: Wear Failures,* p 134-153; Fretting Failures, p 154-160; Liquid-Erosion Failures, p 160-167, ASM, 1975

Metals Handbook, 9th Ed., Vol 1, *Properties and Selection: Irons and Steels:* Wear Resistance, p 597-638, ASM, 1978
Fundamentals of Friction and Wear of Materials, ASM, 1981
Proceedings, Wear of Materials, St. Louis (1977), Dearborn (1979), San Francisco (1981), Reston (1983), Vancouver (1985), published by ASME, New York (co-sponsored by ASM)
Treatise on Materials Science and Technology, Vol 13, *Wear,* edited by D. Scott, Academic Press, New York, San Francisco, London, 1979
Tribology in Metalworking: Friction, Lubrication, and Wear, by J. A. Schey, ASM, 1983
Principles of Tribology, edited by J. Halling, Macmillan, London, 1975
Tribology, by H. Czichos, Vol 1 in Tribology Series (also other books in Tribology Series), Elsevier, 1978

Journals:

Wear
Tribology International
Journal of Tribology (formerly, *Journal of Lubrication Technology, Trans. ASME*)
Note: Tribology is the science and technology of interacting surfaces in relative motion.

Fretting Failures

FRETTING is a wear phenomenon that occurs between two mating surfaces; it is adhesive in nature, and vibration is its essential causative factor. Usually, fretting is accompanied by corrosion. In general, fretting occurs between two tight-fitting surfaces that are subjected to a cyclic, relative motion of extremely small amplitude.

Fretting is also referred to as fretting corrosion, false brinelling, friction oxidation, chafing fatigue, molecular attrition and wear oxidation.

FRETTING CHARACTERISTICS

The difference between fretting and ordinary wear is that fretting generally occurs at contacting surfaces that are intended to be fixed in relation to each other, but that actually undergo minute alternating relative motion, called "slip," that is usually produced by vibration.

Common sites for fretting are in joints that are bolted, keyed, pinned, press fitted and riveted; in oscillating bearings, splines, couplings, clutches, spindles and seals; in press fits on shafts; and in universal joints, baseplates, shackles and prosthetic devices. One problem with fretting is that it may initiate fatigue cracks—which, in shafts and other highly stressed components, often result in fatigue fracture. Localized wear and removal of material by fretting usually are not sufficient to cause serious problems, although material removal as deep as $1/16$ in. has been observed.

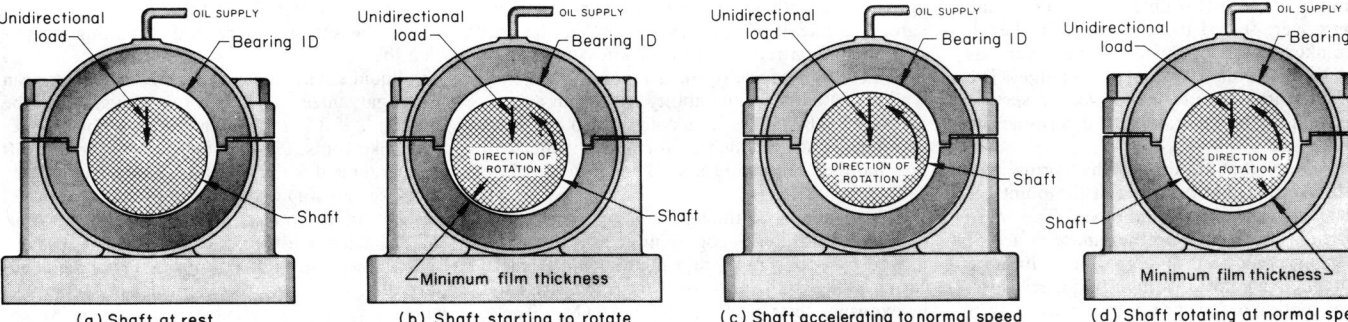

(a) Shaft at rest   (b) Shaft starting to rotate   (c) Shaft accelerating to normal speed   (d) Shaft rotating at normal speed

Fig. 28. Step-by-step development of a hydrodynamic lubricating film in a unidirectionally loaded journal bearing

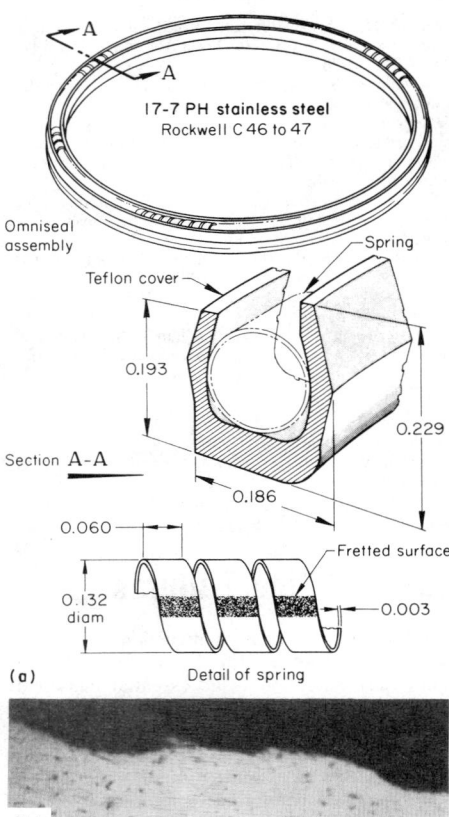

17-7 PH stainless steel
Rockwell C 46 to 47

Omniseal assembly

Spring

Teflon cover

0.193

0.229

Section A-A

0.186

0.060

Fretted surface

0.132 diam

0.003

(a)

Detail of spring

(b)

Fig. 29. (a) Omniseal assembly in which a 17-7 PH stainless steel spring in a Teflon cover failed by fretting. Dimensions are in inches. (b) Micrograph of a section taken through the fretted area

RECOGNITION OF FRETTING

Fretting of ferrous metals in air produces a characteristic reddish-brown debris of ferric oxide which, when mixed with oil or grease, produces a debris that is often called "blood," "cocoa" or "red mud" debris. In components that are lubricated so that ordinary corrosion is not likely to occur, presence of the reddish-brown debris is indicative of fretting.

EXAMPLE OF FRETTING FAILURE

The omniseal assembly shown in Fig. 29(a) served as a hydraulic seal around the cylinder liners in a diesel-engine block. The assembly, consisting of a 17-7 PH stainless steel spring and a Teflon cover, sealed the coolant from the oil in the crankcase. The open side of the cover was exposed to the coolant side of the engine. The spring failed after about six months of service and was sent to the laboratory to determine the cause of failure.

It was found that failure of the spring was caused by fretting that occurred at the points of contact between the spring and cover. The small black particles of residue on the inner surface of the cover promoted fretting corrosion. Vibration of the diesel engine imparted high-frequency, low-amplitude motion to the spring, thus causing fretting between the contacting surfaces of the spring and the cover.

Corrosion Failures

By Seymour Coburn, Corrosion Consultants, Inc.

CORROSION is the inevitable result of permitting unprotected plain carbon steel, zinc, copper and aluminum, among the common structural metals, to be exposed to specific environments that can aggressively attack them. Such environments involve a humid atmosphere (70 to 80% relative humidity), contaminated principally with chlorides and sulfates, and a nocturnal temperature drop of 8 to 11 °C (15 to 20 °F).

FACTORS CONTRIBUTING TO CORROSION FAILURES

The principal factor responsible for atmospheric corrosion of metals is condensation of dew containing such water-soluble ionic species as chlorides and sulfates that facilitate the electrochemical reaction between the metal and the condensed film of moisture.

The next most important factor is the absence from the specification, or the unawareness of the design engineer, of the fundamental chemical reactions of the metal in the presence of dew condensate which can have a pH as low as 4.5 and/or contain chlorides or sulfates. Examples include rusting of plain carbon steel exposed to dew condensate; slow deterioration of galvanized steel in industrial environments; the blue-green patina developed by copper, brass and bronze items in both chloride and sulfate atmospheres generated by deicing salts and sulfur oxides; and tarnishing of copper and silver household items by hydrogen sulfide generated by boiling of eggs.

Factors that contribute to initiation, maintenance and acceleration of corrosion involve the prevailing range of temperatures; the washing action of rainwater; the degree of exposure to wind and sun; the geometric configuration of the structure; and the presence of crevices, faying surfaces, joints, improperly made welds, vibrations, design stresses and the like. In addition, processing and fabricating operations such as surface grinding, forming, drilling, welding and heat treatment, among others, produce conditions that influence the susceptibility of the metal to attack by condensed moisture films. Immersion in liquids, such as in process equipment for pipe, tanks, mixers, filter presses, etc., can result in misapplication and in slow or rapid attack, depending on the aggressiveness of the fluid medium.

CORROSION RATES

Corrosion researchers have studied rates of corrosion as both a safety and an economic measure. The rate at which metal is lost can be crucial, because a reduction in thickness can result in either (a) perforation of a pipeline or a tank wall or (b) a loss in the ability to sustain a compressive load (such as for a column in a building) or a tensile stress (such as for a sheet pile wall in a harbor, supporting a soil/rock backfill in a cell).

The reduction in thickness of test specimens exposed in different geographic locations can be judged by exposure of thin plates. (Plates measuring 100 by 150 mm, or 4 by 6 in., are used in exposure tests performed by ASTM.) These specimens are cleaned, weighed, exposed for various intervals of time, and then cleaned and reweighed to determine the loss due to corrosion. In instances where the corrosion was uniform rather than of a pitting nature, the loss in weight can be converted to a reduction in thickness. The loss in thickness or weight can be expressed in inches or mils per year, micrometres per year, milligrams per square decimetre per day, or any such variation that fits the specific order of magnitude of loss.

TYPES OF CORROSION

The various types of corrosion have been classified into eight basic categories: uniform corrosion, pitting corrosion, selective leaching, intergranular corrosion, crevice corrosion, galvanic corrosion, erosion-corrosion and stress corrosion.

Uniform Corrosion

The metals most commonly attacked by uniform, or general, corrosion are plain carbon steels, high-strength low-alloy weathering-type steels, zinc and galvanized steel, cadmium-coated steel and, to a more limited extent, copper and copper alloys. This form of attack occurs most frequently during outdoor exposure to the atmosphere and during total immersion in liquids that constitute uniform environments from the standpoint of temperature gradients and suspended abrasive solids.

Uniform corrosion of steels takes place as the result of electrochemical action between closely spaced microanodes and microcathodes on the metal surface. In the case of zinc and galvanized steel, the acid pH of the uniform film of condensed dew exhausts itself by reacting with and being neutralized by the basic zinc carbonate film that is present on the surface of the zinc. This repetitious action that occurs nightly is like removing a sheet of paper from a pad each night. The same approach applies to the formation of a patina on copper and its alloys. However, in this instance, the corrosion products are insoluble toward rainwater and seal the surface against further attack. The same phenomenon occurs in weathering-type steels, where a tightly adherent nonporous oxide film develops that protects the substrate from further atmospheric attack. In contrast, the rust or corrosion product on plain carbon steel is not protective and merely slows the rate of attack.

The corrosion products that form on zinc and cadmium in industrial environments are not completely protective. They result in limited service life, because they form water-soluble cadmium and zinc sulfates. However, in a rural environment the basic zinc carbonate film that forms offers a long service life. In a marine atmosphere, basic zinc chloride films form, offering protection to the substrate zinc and extending its service life.

A liquid environment such as potable water can affect galvanized piping in two ways. It can be protective if it is a hard water and deposits limestone-like films. On the other hand, if it is a soft water saturated with carbon dioxide, it can quickly attack the zinc and soon begin to attack the steel surface. In chemical environments, some unusual behaviors are noted for the inorganic acids — hydrochloric, sulfuric and nitric. These acids behave differently toward plain carbon steel piping when they are at different strength levels. An appreciation and an understanding of these possibilities enable a specification engineer to make

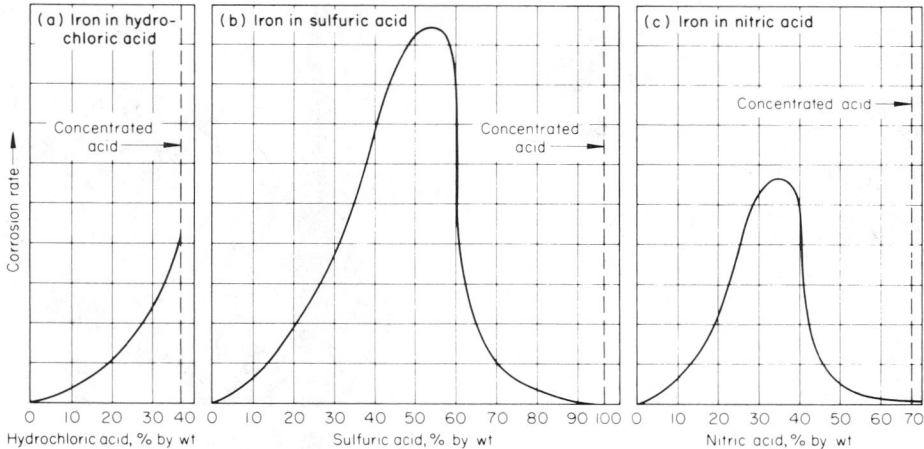

It should be noted that the scales for corrosion rate are not the same for all three charts. As discussed in text, the corrosion rate of iron (and steel) in nitric acid in concentrations of 70% or higher, although low compared to the maximum rate, is still sufficient to make it unsafe to ship or store nitric acid in these metals. (Source of charts: M. Henthorne, "Corrosion Causes and Control," Carpenter Technology Corp., Reading, PA, 1972, p 30)

Fig. 30. Effect of acid concentration on the corrosion rate of iron completely immersed in aqueous solutions of three inorganic acids at room temperature

the best economic choice of materials. This variation in behavior with concentration is illustrated in Fig. 30.

Pitting Corrosion

One of the most damaging types of attack is the formation of pits in a pipeline, in the wall of a vessel or in the bottom of a storage tank. Pitting can occur in carbon steels as well as in stainless steels. No metal seems to be immune from this form of attack, and even the most highly alloyed composition is vulnerable under certain conditions. In the case of stainless steel, the presence of chloride ions in a slightly acidic solution can result in widely scattered pit formation. It is said that the chloride ion is responsible for disrupting the passive or protective films on stainless steel and aluminum. In a carbon steel oil storage tank from which the mill scale has not been removed, pits can form at cracks in the scale and perforate the tank by virtue of what is termed a small anode–large cathode relationship. Grains of sand lying on a damp stainless steel surface can stimulate pit formation beneath the grains by shutting off access to oxygen. This leads to the formation of a minute differential oxidation or aeration cell.

Pits require a long incubation period before they are made evident by sudden failure (see Fig. 31). Pits grow downward and rarely originate on vertical walls except in the case of cracked mill scale or a break in a protective coating under conditions of total immersion. Pit growth is unique in that it is autocatalytic or self-stimulating and self-propagating when once started. Each pit, as a result of hydrolysis, produces within its confines more acid conditions than exist outside of the pit and thus promotes its continued life. The direct action of the pit is localized and, in a sense, protects the surrounding metal cathodically.

Corrosion inhibitors, which operate in a somewhat mysterious fashion, can stimulate attack if present in lower-than-optimum concentrations in aqueous solutions. Other stimuli for pitting are the bubbling of air against a submerged surface, residual stresses along a line of bending in fabrication, and certain types of inclusions inherent in the steelmaking process.

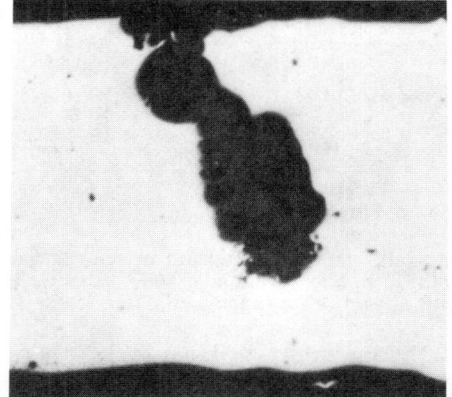

Fig. 31. Unetched section, at 95×, through the bottom of a type 321 stainless steel aircraft fresh-water storage tank that failed in service as a result of pitting, showing subsurface enlargement of one of the pits

Selective Leaching

Selective leaching is the removal of an element from an alloy by corrosion. The most common example is dezincification, the selective removal of zinc in brasses. Many alloys are susceptible to selective leaching under certain conditions. The elements that are more resistant to the environment remain behind, provided that they have a sufficiently continuous structure to prevent their breaking away in small particles.

Mechanisms. Two mechanisms have been described for selective leaching: (a) two metals in an alloy are dissolved, and one redeposits on the surface; and (b) one metal is selectively dissolved, leaving the other metals behind. Dezincification of brasses occurs by the first mechanism; the loss of molybdenum from nickel alloys in molten sodium hydroxide occurs by the second.

Dezincification occurs in brasses containing less than 85% copper. Zinc corrodes preferentially, leaving a porous residue of copper and corrosion products. Alpha brass containing 70% copper and 30% zinc (copper alloy 260) is particularly susceptible to dezincification when exposed in an aqueous electrolyte at elevated temperatures.

Dezincification proceeds as follows: (a) the brass dissolves, (b) the zinc ions stay in solution and (c) the copper plates back on. Dezincification can proceed in the absence of oxygen, as evidenced by the fact that zinc corrodes slowly in pure water. However, oxygen increases the rate of attack when it is present. Analyses of dezincified areas usually show 90 to 95% copper, with some of it present as copper oxide.

Dezincification may be either uniform or of the "plug" type (see Fig. 32). High zinc content in a brass favors uniform attack; relatively low zinc content favors plug-type attack. The composition of the liquid in contact with the metal has a greater effect on the type of dezincification, but the pattern of behavior is neither completely consistent nor fully understood. Slightly acidic water, low in salt content and at room temperature, is likely to produce uniform attack, whereas neutral or alkaline water, high in salt content and above room temperature, often produces plug-type attack.

Graphitic Corrosion. Perhaps the second most frequently observed type of selective leaching is graphitic corrosion of gray iron, which occurs in relatively mild aqueous environments and on buried pipe.

The graphite in gray iron is cathodic to iron and remains behind as a porous mass when iron is leached out. Graphitic corrosion usually occurs at a low rate. The graphite mass is porous and very weak, and graphitic corrosion produces little or no change in metal thickness. A corroded surface usually does not appear different from that of uncorroded gray iron.

Graphitic corrosion does not occur in ductile iron or malleable iron, because no graphite network is present to hold together the residue. White iron has essentially no free carbon and is not subject to graphitic corrosion.

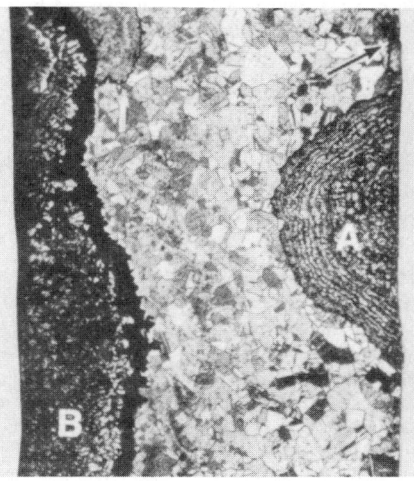

Area A shows plug-type attack on the nickel-chromium-plated outside surface of the brass pipe that initiated below a break in the plating (at arrow). Area B shows uniform attack on the bare inside surface of the pipe. (Etched in NH_4OH-H_2O_2; 85×)

Fig. 32. Micrograph showing difference in dezincification of inside and outside surfaces of a plated copper alloy 260 (cartridge brass, 70%) pipe for domestic water supply

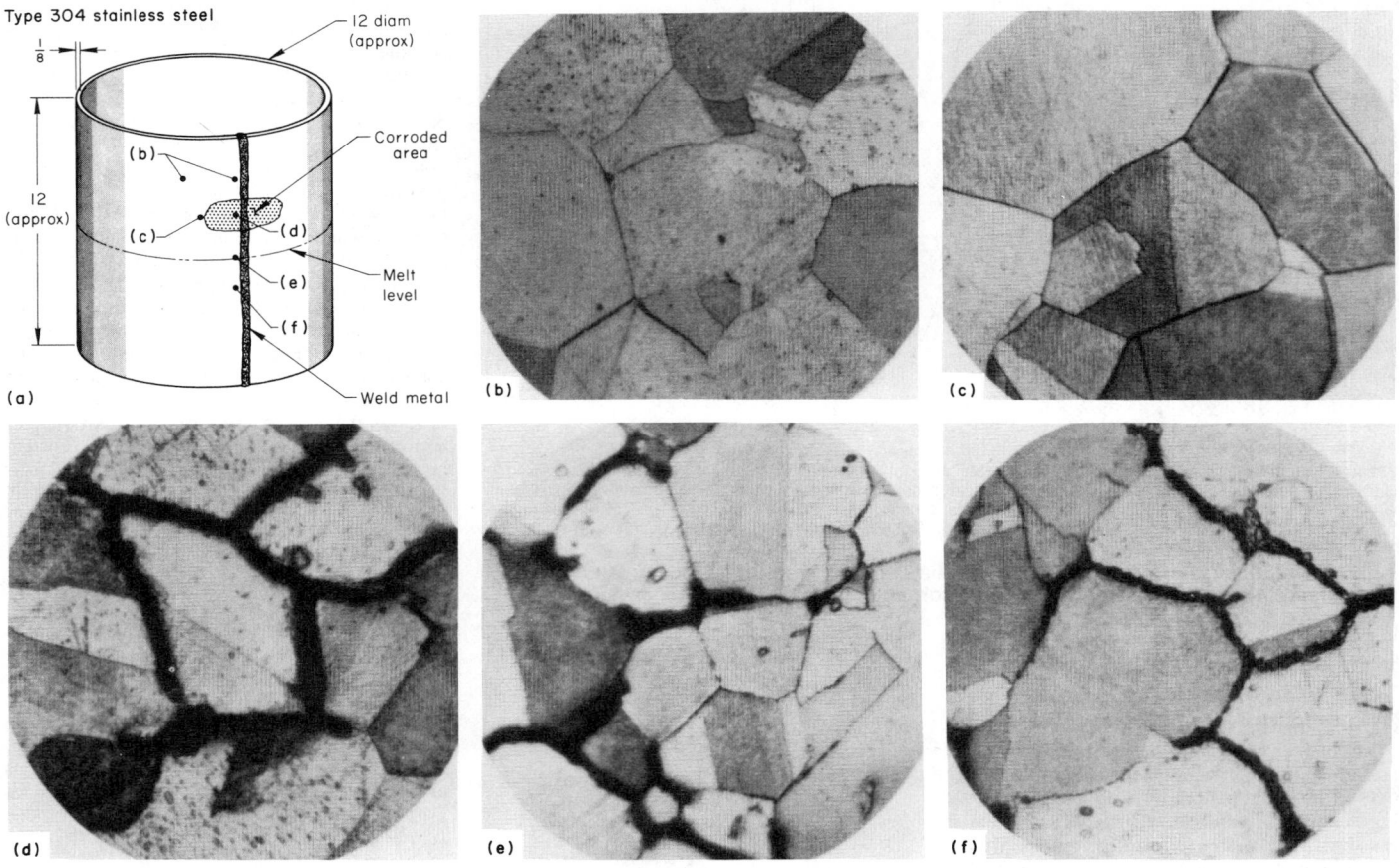

Fig. 33. (a) Schematic illustration of a fused-salt, electrolytic-cell pot of type 304 stainless steel that failed by intergranular corrosion as a result of metal sensitization. (b) to (f) Micrographs, at 500×, of corroded and uncorroded specimens taken from the correspondingly lettered areas on the pot shown in (a); specimens were etched in CuCl₂.

Detection of Selective Leaching. On many alloys selective leaching is not readily detected by visual examination. A copper flash is usually visible on dezincified copper alloys, but this is not positive evidence of dezincification because a copper flash can deposit from even small amounts of copper salts in aqueous solution without the occurrence of dezincification.

An area where selective leaching has occurred will sound dull when struck. However, severe intergranular corrosion may produce the same effect.

Intergranular Corrosion

Intergranular corrosion of stainless steels is encountered most frequently in welded assemblies. The contact areas between crystals, commonly termed grain boundaries, become especially susceptible to corrosion because of the temperatures reached during welding. This phenomenon is called "heat sensitization."

For austenitic stainless steels the sensitizing temperature range is between 850 and 400 °C (1550 and 750 °F). The extent of damage is based on the length of time the metal is held at either the top or the bottom of the sensitizing temperature range. Slow cooling through this range results in a vulnerable situation. Rapid cooling through this range avoids any serious damage. The susceptible area extends just a few millimetres on either side of the weld area. This distance is termed the "heat-affected zone." An example of intergranular attack is shown in Fig. 33.

The degree of sensitization is a function of the carbon content of the metal. An 18-8 stainless steel containing 0.1% C or more will be extensively sensitized after only a five-minute exposure at 580 °C (1080 °F). A steel containing 0.03% C sustains very little damage. Although the physical properties of a steel exposed to such conditions are hardly affected, exposure of the steel to an aggressive environment, such as seawater spray, will cause extensive damage. Under such conditions the carbon diffuses to the grain boundary at the elevated temperature and there combines with the chromium and some iron to form a metallic carbide represented as $M_{23}C_6$. Under such conditions the boundary area is severely depleted of its chromium to the point where the chromium content can drop below the 12% necessary for maintenance of a passive state. Thus, a strong active-passive cell forms between a large cathodic grain and a small anodic grain boundary, with the corrosion occurring along the anodic grain boundaries, permitting crystals to fall out and leading to fracture of the metal following the path of the grain boundaries.

Three measures can be taken to avoid heat sensitization. If welding is to be done, the carbon content can be decreased to below 0.03%. Another method is to incorporate titanium or niobium in the composition, in order to tie up carbon preferentially. A third option is to heat the system to about 1100 °C (2000 °F) to dissolve the carbides, then cool rapidly through the sensitizing temperature range.

Crevice Corrosion

The electrochemical theory of corrosion is best illustrated by the cell or battery concept. The familiar dry cell represents the two-metal or two-electrode type of cell. With the two electrodes being of the same metal, variations in the character of the medium can be responsible for creation of a cell leading to localized attack of the metal. One of the most common, and one that is responsible for a large economic loss, is the differential oxygen or aeration cell. Accumulations of leaves and twigs in the corners of bridge girders, and of clods of mud beneath automobile fenders, result in relatively rapid attack of the metal just beneath the damp accumulation of debris. This is caused by the inability of air to penetrate the clod of debris by comparison with the ease of penetration at the thin edges of the debris. The disproportionate access of air results in formation of a differential aeration cell. The deposit of mud can be considered to be a type of crevice. More recognizable crevices occur at back-to-back angles on transmission towers, bolted joints of steel and aluminum, the faying surfaces in bolted joints, and accidental or design arrangements involving contact of metal with wood, rubber, plastic, glass, wax, asbestos, concrete, etc. The contact area is generally of a capillary nature and thus draws in a certain amount of liquid from which the air is quickly exhausted, in comparison with the liquid film outside the capillary, which is constantly saturated with air.

Table 1. Standard electromotive-force (emf) series of metals

Metal	Electrode potential vs hydrogen electrode
Potassium	−2.29
Sodium	−2.71
Magnesium	−2.36
Zinc	−1.66
Chromium	−0.76
Iron	−0.44
Cadmium	−0.40
Cobalt	−0.28
Nickel	−0.25
Tin	−0.13
Lead	−0.12
Hydrogen	0.00
Copper	+0.34
Mercury	+0.79
Silver	+0.80
Palladium	+0.99
Platinum	+1.2
Gold	+1.5

Aluminum and stainless steels are notoriously susceptible to attack in oxygen-deficient situations. One of the simplest yet most effective means of demonstrating this phenomenon is to place a rubber band around a thin plate of stainless steel and immerse it in a dilute salt water solution. The rubber band will be literally cutting the plate of stainless steel because of the differential aeration cell created.

Bolted joints, unless they are checked for tightness or involve the use of high-strength bolts, can undergo such attack beneath the bolt head or beneath the washer as well as within the faying surfaces of the joint.

It is the opinion of some investigators that conditions within the crevice can become far more severe than those that exist outside the crevice. For example, the production of additional soluble iron can increase the level of acidity within the crevice compared with that in the moisture film outside the crevice much the same as is found within the depths of pits in stainless steel.

Galvanic Corrosion

When dissimilar metals are in electrical contact in an electrolyte, the less noble metal (anode) is attacked to a greater degree than if it were exposed alone, and the more noble metal (cathode) is attacked to a lesser degree than if it were exposed alone. This behavior, which is known as galvanic corrosion, can often be recognized by the fact that the corrosion is more severe near the junction of the two metals than elsewhere on the metal surfaces. Galvanic corrosion is usually the result of poor design and selection of materials, or the plating-out of a more noble metal from solution on a less noble metal.

The greater the difference in potential between the two metals, the more rapid will be the galvanic attack. The textbook electromotive-force (emf) series ranks the metals according to their chemical reactivity, but applies only to the laboratory conditions under which the reactivity was determined (see Table 1). In practice, the solution potentials of metals are affected by the presence of passive or other protective films on some metals, polarization effects, degree of aeration, complexing agents, and temperature.

Galvanic Series in Seawater. A galvanic series based on immersion in seawater is more generally applicable than the electromotive-force series as an

indication of the rate of corrosion between different metals or alloys when they are in contact in an electrolyte. In most electrolytes the metal close to the active end of the galvanic-series chart will behave as an anode, and the metal closer to the noble end will act as a cathode. The amount of separation between two metals in the chart is a rough measure of the difference in potential that can be expected and is usually related to the rate of galvanic corrosion between the two metals in a given electrolyte.

This galvanic series, which includes most of the industrially important metals, is given in Table 2. In most cases, metals from one group can be coupled with other metals from the same group without causing a substantial increase in the corrosion rate of the more active metal.

Erosion-Corrosion

The erosion-corrosion phenomena show up in a variety of ways. One of the most common is that occurring in pipes through which a variety of corrosive and noncorrosive fluids flow at differing velocities. The pipes may develop protective films of a passive nature such as the characteristic thin oxide films on aluminum and stainless steel, compared with the mixed lead oxide – lead sulfate film that forms when lead pipe is used to convey sulfuric acid.

Generally simple lamellar or straight-line flow causes no problem until an obstruction appears and turbulent flow develops. Such turbulence has the ability to physically disrupt and tear away the various types of protective films that different types of pipe metals can develop, exposing bare metal for subsequent corrosion. The cycle repeats itself so that thinning and perforation occur at an accelerated rate. Disruption of a surface film can result in a form of galvanic corrosion in which the small exposed surface acts as an anode surrounded by a large (coated) cathode. Accelerated attack occurs at the anode.

Turbulent conditions are created by changes in diameter in piping systems. Bends and elbows likewise create conditions conducive to changes in velocity that create turbulent conditions, sometimes called cavitation, wherein bubbles of air contained in the fluid can collapse and exert an abrasive action against any protective films formed on the pipe surface. Water flowing at 12 m/s (39 ft/s) can be aggressive at pH 3 to 5 and 7 to 9. In contrast, no damage is done at pH 6 and 10 to 13. Corrosion inhibitors can minimize the corrosive effects of turbulent fluid action. Use of smooth interiors with few bends and protective coatings are means for mitigating the effects of fluid flow.

SELECTED REFERENCES

Corrosion: Causes and Prevention, F. N. Speller, McGraw-Hill, New York, 1951
Protection Against Atmospheric Corrosion: Theories and Methods, K. Barton, John Wiley, New York, 1976
Corrosion Resistance of Metals and Alloys, R. J. McKay and R. Worthington, Reinhold, New York, 1936
Metallic Corrosion Passivity and Protection, U. R. Evans, Edward Arnold, London, 1948
The Corrosion and Oxidation of Metals, U. R. Evans, St. Martin's Press, New York, 1963
The Corrosion and Oxidation of Metals, 1st Supplemental Volume, St. Martin's, 1968, 2d, 1976
Corrosion and Corrosion Control, H. H. Uhlig, 2d ed., John Wiley, New York, 1971
Corrosion, Volumes I and II, L. L. Shreir, ed., 2d ed., Newnes-Butterworths, London, 1976
An Introduction to Corrosion and Protection of Metals, G. Wranglen, Stockholm, 1972

Table 2. Galvanic series in seawater

Corroded end (anodic, or least noble)
Magnesium Magnesium alloys
Zinc Galvanized steel or galvanized wrought iron
Aluminum alloys 5052, 3004, 3003, 1100, 6053, in this order
Cadmium
Aluminum alloys 2117, 2017, 2024, in this order
High-strength low-alloy steel Plain carbon steel Cast iron
Ni-Resist (high-nickel cast iron)
Type 410 stainless steel (active)
50-50 lead-tin solder
Type 304 stainless steel (active) Type 316 stainless steel (active)
Lead Tin
Copper alloy 280 (Muntz metal, 60%) Copper alloy 675 (manganese bronze A) Copper alloys 464, 465, 466, 467 (naval brass)
Nickel 200 (active) Inconel alloy 600 (active)
Hastelloy B Chlorimet 2
Copper alloy 270 (yellow brass, 65%) Copper alloys 443, 444, 445 (admiralty brass) Copper alloys 608, 614 (aluminum bronze) Copper alloy 230 (red brass, 85%) Copper 110 (ETP copper) Copper alloys 651, 655 (silicon bronze) Copper alloy 715 (copper nickel, 30%) Copper alloy 923, cast (leaded tin bronze G) Copper alloy 922, cast (leaded tin bronze M)
Nickel 200 (passive) Inconel alloy 600 (passive)
Monel alloy 400
Type 410 stainless steel (passive) Type 304 stainless steel (passive) Type 316 stainless steel (passive) Incoloy alloy 825
Inconel alloy 625 Hastelloy C Chlorimet 3
Silver
Titanium
Graphite
Gold
Platinum
Protected end (cathodic, or most noble)

Corrosion Engineering, M. G. Fontana and N. D. Greene, McGraw-Hill, New York, 1967
Theory of Corrosion: Protection of Metals, N. D. Tomashov, MacMillan, New York, 1966
Fundamentals of Corrosion, J. C. Scully, Pergamon Press, New York, 1966
Corrosion: A Compilation, M. G. Fontana, The Press of Hollenbeck, Columbus, Ohio, 1957
Corrosion Testing Procedures, F. A. Champion, Chapman and Hall, London, 1964

Corrosion Testing, F. L. La Que, Marburg Lecture, Proceedings of the American Society for Testing and Materials, 1951

Design and Corrosion Control, V. R. Pludek, John Wiley, New York, 1977

Marine Corrosion: Causes and Prevention, F. L. La Que, John Wiley, New York, 1975

Metals Handbook, Properties and Selection: Irons and Steels, Vol. 1, 9th ed., American Society for Metals, 1978

Corrosion Handbook, H. H. Uhlig, ed., John Wiley, New York, 1948

Handbook on Corrosion Testing and Evaluation, W. H. Ailor, ed., John Wiley, New York, 1971

Atmospheric Factors Affecting the Corrosion of Engineering Metals, STP 646, S. K. Coburn, ed., American Society for Testing and Materials, 1978

Metal Corrosion in the Atmosphere, STP 435, American Society for Testing and Materials, 1968

Underground Corrosion, M. Romanoff, Circular 579, U.S. Govt., Printing Office, Washington, D.C., 1957

Corrosion Inhibitors, I. L. Rozenfeld, McGraw-Hill, New York, 1981

Corrosion Inhibitors, C. C. Nathan, ed., National Association of Corrosion Engineers, 1973

Protective Coatings for Metals, R. M. Burns and W. W. Bradley, Reinhold, 3d ed., New York, 1967

Coatings for Corrosion Protection, E. W. Cochran and D. Tonini, eds., American Society for Metals, 1979

Corrosion Control by Coatings, H. Leidheiser, Jr., ed., Science Press, Princeton, N.J., 1979

The Corrosion of Light Metals, H. P. Godard, ed., John Wiley, New York, 1967

Corrosion of Stainless Steels, A. J. Sedriks, John Wiley, 1979

Zinc: Its Corrosion Resistance, C. J. Slunder and W. K. Boyd, Zinc Institute, New York, 1971

The Corrosion of Copper, Tin and Their Alloys, H. Leidheiser, Jr., John Wiley, New York, 1971

Corrosion Data Survey—Metals and Nonmetals, N. E. Hamner, compiler, National Association of Corrosion Engineers, 5th ed., 1974

Stress Corrosion Cracking and Embrittlement, W. D. Robertson, John Wiley, New York, 1956

The Theory of Stress Corrosion Cracking in Alloys, J. C. Scully, North Atlantic Treaty Organization (NATO), Brussels, 1971

Handbook of Corrosion Protection for Steel Pile Structures in Marine Environments, T. D. Dismuke, S. K. Coburn and C. M. Hirsch, eds., American Iron and Steel Institute, Washington, D.C., 1981

Corrosion Chemistry, G. R. Brubaker and P. B. Phipps, eds., American Chemical Society, Washington, D.C., 1979

Corrosion-Erosion Behavior of Materials, K. Natesan, ed., The Metallurgical Society of AIME, Warrendale, Pa., 1980

Good Painting Practice, Volumes I and II, Steel Structures Painting Council, J. Bigos, ed., Pittsburgh, 2nd ed., 1983

Corrosion of Building Materials, D. Knofel, Van Nostrand Reinhold, New York, 1975

The Chemistry of Building Materials, R. M. E. Diamont, Business Books, London, 1970

Journals

Corrosion, National Association of Corrosion Engineers, Houston

Materials Performance, National Association of Corrosion Engineers, Houston

British Corrosion Journal, The Metals Society, London

Corrosion Science, Pergamon Press, Oxford, United Kingdom

Anti-Corrosion: Methods and Materials, Sawell Publications, Ltd., London

Stress-Corrosion Cracking

By W. R. Warke and S. W. Ciaraldi, Standard Oil Company (Indiana)

STRESS-CORROSION CRACKING (SCC) is a mechanical-environmental process in which sustained tensile stresses and chemical influence combine to initiate and propagate cracks in ma-

terials. Stress-corrosion cracking is produced by the synergistic action of tensile stresses, a specific environment and a susceptible alloy, and results in a failure where one would not necessarily occur if only two of these conditions were met. The process of stress-corrosion cracking generally involves crack initiation, subcritical crack growth, and final failure when the stress-corrosion crack reaches a critical size such that the tensile strength or fracture toughness of the remaining material is exceeded. It is the subcritical crack-growth phase of this process that is actually stress-corrosion cracking.

Although the physical manifestation of stress-corrosion cracking is obvious (i.e., a cracked or broken sample or part), the mechanisms and atomic-level processes involved are many and complex. As a result, the effects of environmental and metallurgical variables are difficult if not impossible to predict. Despite many years of research and experience, stress-corrosion cracking remains the predominant cause of unexpected failures in many segments of industry, such as the petrochemical and chemical process industries, and is a continuing concern to the aerospace industry.

CHARACTERISTICS OF STRESS-CORROSION CRACKING

In 1972, B. F. Brown published the following list of characteristics of stress-corrosion cracking (in *Stress-Corrosion Cracking of Metals—A State of the Art*, edited by H. L. Craig, STP 518, American Society for Testing and Materials, 1972):

1. Tensile stress is required. This stress may be supplied by service loads, cold work, mismatch in fit-up, heat treatment, and by the wedging action of corrosion products.
2. Only alloys are susceptible (no pure metals), although there may be a few exceptions to this rule.
3. Generally only a few chemical species in the environment are effective in causing SCC of a given alloy.
4. The species responsible for SCC in general need not be present either in large quantities or in high concentrations.
5. With some alloy/corrodent combinations, temperatures substantially above room temperature may be required to activate some process essential to SCC.
6. An alloy is usually almost inert to the environment which causes SCC.
7. Stress-corrosion cracks are always macroscopically brittle in appearance, even in alloys which are very tough in purely mechanical fracture tests.
8. Microscopically, the fracture mode for SCC is usually different from the fracture mode for plane-strain fractures in the same alloy.
9. There appears to be a threshold stress below which SCC does not occur, at least in some systems.

These guidelines are still generally considered to be valid observations. The following three comments can be added to those above:

1. Cathodic protection has been successful, in some cases, in preventing initiation and/or propagation of stress-corrosion cracking.
2. Addition of soluble salts containing certain specific anions can inhibit the crack-produc-

ing effect of a given environment on a given alloy.
3. Certain aspects of the metallurgical structure of an alloy (such as grain size, crystal structure and number of phases) influence the susceptibility of the alloy to stress-corrosion cracking in a specific environment.

CRACK INITIATION AND PROPAGATION

Initiation of stress-corrosion cracking can occur by a variety of mechanisms. In some cases, it is due to the presence of unstable protective surface films. This is often manifested by crack initiation from a corrosion pit or trench. On other occasions, cracking begins directly from a severe stress concentration, a pre-existing flaw, or a crack which was formed previously by a different mechanism such as metal fatigue. In those cases where stress-corrosion cracking is associated with hydrogen-stress cracking, the initiation phase can involve generation of hydrogen by corrosion reactions, and atomic hydrogen entry and concentration in the metal, followed by crack initiation.

A number of theories have been proposed to account for the mechanism of stress-corrosion-crack propagation:

1. Successive formation and rupture of a passive layer at the crack tip, with resultant brief periods of localized anodic dissolution.
2. Preferential dissolution of one microstructural constituent (such as a grain boundary region) which is anodic with respect to the main body of the metal.
3. Adsorption of damaging ions which weakens atomic bonding at the crack tip.
4. When fresh metal is exposed by slip at the crack tip there is a competition between dissolution, which would blunt the crack tip, and passivation, which would prevent the environment from reaching the metal. Between these two extremes there exists a range where the slip step is attacked but the crack walls are passivated.
5. Hydrogen is generated at local cathodes as a result of corrosion reactions. It is then able to enter the metal, diffuse to the crack tip and cause crack propagation by hydrogen-stress cracking.

Although these theories are all in accord with some of the observed facts, no single theory has been proposed to date that completely explains all of the items listed above under "Characteristics of Stress-Corrosion Cracking."

On a macroscopic scale, stress-corrosion cracks propagate due to tensile stresses, generally on a plane perpendicular to the maximum principal stress. On a microscopic scale, however, the cracks frequently undergo extensive branching and will, on the scale of the microstructure, be at a variety of angles to the stress axes. Cracking may be intergranular, transgranular or mixed-mode depending on the alloy, its microstructure and the particular environment involved. For example, as a general rule, chloride stress-corrosion cracking of an austenitic stainless steel is transgranular (Fig. 34), whereas polythionic acid cracking of the same material is intergranular. Caustic cracking of carbon steels is generally intergranular (Fig. 35). In copper-zinc alloys, the path of cracking can be made transgranular or intergranular by adjusting alloy composition or the pH of aqueous solutions in which these alloys are immersed (Fig. 36).

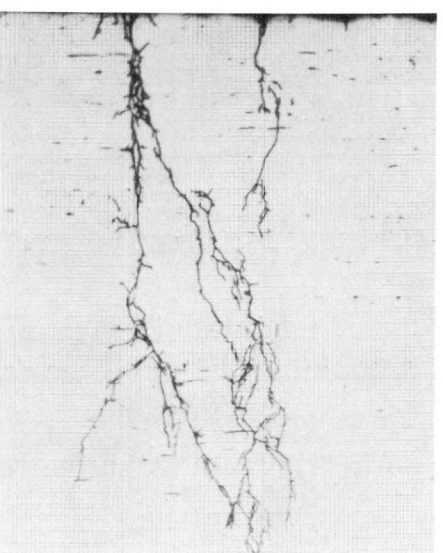

Fig. 34. As-polished cross section, at 100×, through a stress-corrosion-cracked type 304 stainless steel part, showing branching of cracks as they proceed downward from the surface (top of micrograph)

STRESSES

The stresses producing stress-corrosion cracking are sustained, but not necessarily constant, tensile stresses. These stresses may result from a variety of causes and may be residual stresses or service stresses, or both. Some examples of sources of stress which could lead to stress-corrosion cracking are:

1. *Residual:*
   • Differential strains in metalworking (e.g., cold drawn wire)
   • Welding shrinkage stresses
   • Press straightening of shafts
   • Assembly operations such as bolt pretorquing and shrink fitting
   • Bending and expansion-rolling of boiler and heat-exchanger tubes
   • Shrinkage stresses from heat treating.
2. *Service:*
   • Normal service loads
   • Differential thermal expansion due to thermal gradients or dissimilar metals in an assembly
   • Wedging of corrosion products in crevices.

The time required for a part or specimen to fail by stress-corrosion cracking generally increases with decreasing stress, down to a stress, called the threshold stress, below which stress corrosion occurs at such a low rate that it does not affect service life. Figure 37 illustrates this behavior for two 18-8 stainless steels (types 304 and 304L) and two high-alloy stainless steels (types 310 and 314). This threshold stress, however, may be as low as 10% of the yield stress and usually is not a practical design stress.

Similarly, when a fracture-mechanics approach is used, there is a stress-intensity threshold, K_{Iscc}, below which the crack-growth velocity is immeasurably small. Above K_{Iscc}, crack-growth velocity often increases sharply to a plateau where velocity is not a function of stress

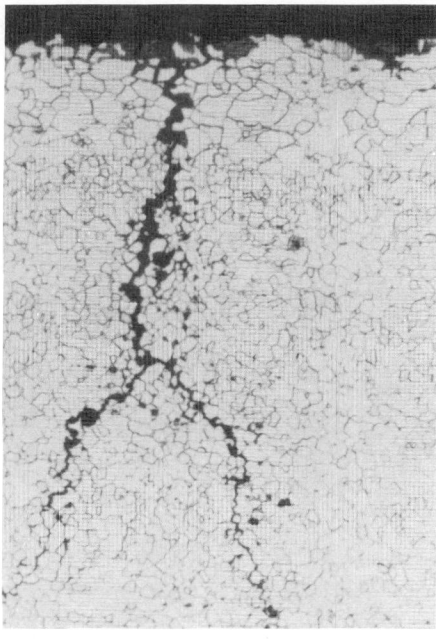

Fig. 35. Micrograph, at 100×, of a nital-etched specimen of ASTM A245 carbon steel, showing stress-corrosion cracking that occurred in a concentrated solution of ammonium nitrate

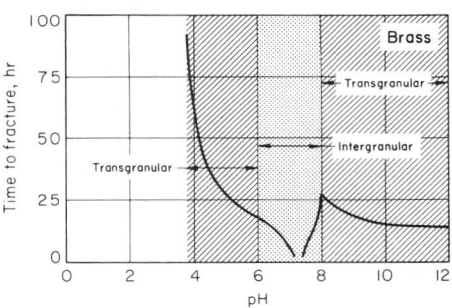

Fig. 36. Effect of pH on time to fracture by stress-corrosion cracking of brass in ammoniacal copper sulfate solution at room temperature

intensity. In this stress-intensity range, stress-corrosion-crack velocity is believed to be controlled by environmental and/or mechanistic factors such as rate of access of the solution to the crack tip, the rate of a chemical reaction, kinetics of passive-film rupture and repair, or hydrogen diffusivity.

ENVIRONMENTAL FACTORS

On the basis of decades of field experience and laboratory research, it has been found that stress-corrosion cracking of various alloy groups is promoted by certain environments. The dependence of stress-corrosion cracking on the presence of one or more unique species sometimes is called the "specific-ion effect." The presence of ions known to induce cracking in a given alloy type is widely used as a criterion for predicting whether or not cracking will occur. For example, the presence of aqueous chlorides in contact with

austenitic stainless steels at elevated temperatures is always regarded as potentially hazardous. Although this is a reasonable criterion for predicting stress-corrosion cracking in some environments, it is known that ions are not always required and that environments as simple as pure water, dry hydrogen gas, anhydrous ammonia and other pure substances can contribute to stress-corrosion cracking. Table 3 is a limited compilation of environments known to cause stress-corrosion cracking of the indicated alloy types. The amount of the damaging ion needed to promote cracking may be extremely small, as is the case for both polythionic acid and sodium thiosulfate cracking of austenitic stainless steel. Also, determining the source of a harmful ionic species may be quite difficult. For example, chloride cracking can occur due to perspiration drops which fell on the steel during construction in a system otherwise thought to be essentially chloride-free. Other less obvious chloride sources include leaching from insulation by rainwater and concentration to significant levels by alternate wetting and drying.

In addition to the presence and concentration of damaging ions, many other environmental variables influence the occurrence and severity of stress-corrosion cracking. Some of these variables are temperature, pH, electrochemical potential, aeration, and the presence of other ions which may either promote or inhibit cracking. A rough rule of thumb for chloride stress-corrosion cracking of austenitic stainless steels is that cracking does not occur below about 50 °C (120 °F). Figure 36 depicts the effect of pH on the time-to-fracture of brass in ammoniacal copper sulfate. As mentioned above, cathodic protection has been employed to avoid stress-corrosion cracking.

METALLURGICAL VARIABLES

As a general but not universal rule, stress-corrosion cracking occurs primarily in alloys rather than pure metals. Susceptibility to cracking is often a function of the content of a major alloying element, such as nickel in iron-chromium-nickel alloys (austenitic stainless steels). This behavior is depicted by the well-known Copson curve of failure time in boiling 42% $MgCl_2$ as a function of nickel content, which shows a pronounced minimum at 5 to 10% nickel. (See also Fig. 37.)

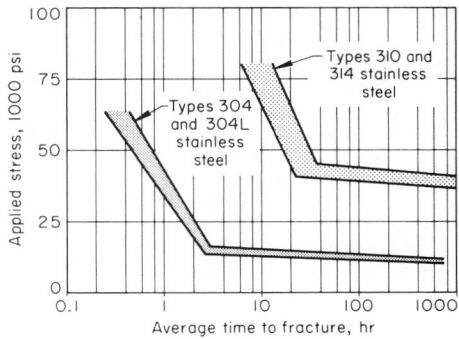

Effect is shown by relation of applied stress to average time to fracture for two 18-8 stainless steels (types 304 and 304L) and two high-alloy stainless steels (types 310 and 314) in boiling 42% magnesium chloride solution.

Fig. 37. Effect of alloy composition on threshold stress

Table 3. Specific ions and substances that have been known to cause stress-corrosion cracking in various alloys when present at low concentrations and as impurities

Damaging specific ions and substances	Alloys susceptible to stress-corrosion cracking	Temperature
Halogen group		
Fluoride ions	Sensitized austenitic stainless steel	Room
Gaseous chlorine	High-strength low-alloy steel	Room
Fused chloride salt	Zirconium alloys and titanium alloys	Above melting point of fused salts
Gaseous iodine	Zirconium alloys	300 °C (570 °F)
Gaseous HCl and HBr	High-strength low-alloy steels (rapid crack growth)	Room
Halides in aqueous solutions	High-strength aluminum alloys	Room
	High-strength steels	Room
	Austenitic stainless steels	Hot
Oxygen group (H_2O-O_2-H_2 systems)		
O_2 dissolved in liquid H_2O	Sensitized stainless steels	300 °C (570 °F)
Gaseous hydrogen at ambient pressure	High-strength low-alloy steels	Room
Gaseous hydrogen at high temperature and pressure	Low-strength and medium-strength steels	>200 °C (>390 °F)
Gaseous H_2O	High-strength aluminum alloys	Room
Gaseous H_2O-O_2-H_2	High-strength uranium alloys	. . .
Hydroxides (LiOH, NaOH, KOH)	Carbon steels; Fe-Cr-Ni alloys (caustic cracking)	>100 °C (>210 °F)
Oxygen group (S, Se, Te systems)		
Polythionic acids ($H_2S_nO_6$)	Sensitized stainless steels, sensitized Inconel 600	Room
H_2S gas	High-strength low-alloy steels	Room
Sulfide impurities in aqueous solutions	Medium-strength to high-strength steels (accelerated hydrogen-induced cracking)	Room
MnS and MnSe inclusions	High-strength steels (initiation sites for cracking)	Room
SO_2 gas with moisture	Copper alloys	Room
Nitrogen group		
N_2O_4 liquid	High-strength titanium alloys	50 °C (120 °F)
Fuming nitric acid	Pure titanium; high-strength aluminum alloys	>100 °C (>210 °F)
Nitrates in aqueous solution	Carbon steels	>100 °C (>210 °F)
Nitrogen oxides with moisture	Copper alloys	Room
Aerated aqueous NH_3 and ammonium salts in aqueous solution	Copper alloys	Room
Nitrogen, phosphorus, arsenic, antimony and bismuth, as alloying species in metal	Stainless steels (in presence of Cl−) and copper alloys (in presence of aerated aqueous NH_3); accelerated cracking	Room
Arsenic, antimony and bismuth, as ions in aqueous solutions	High-strength steels; accelerated hydrogen entry and hydrogen-induced cracking	Room
Carbon group (C, Si, Ge, Sn, Pb)		
Carbonate ions in aqueous solutions . . .	Carbon steel	100 °C (210 °F)
CO-CO_2-H_2O gas	Carbon steel	. . .
Lead ions in aqueous solutions	High-nickel alloys	. . .

Increasing zinc content decreases the life of Cu-Zn alloys in ammoniacal solutions. Composition can also have an effect by changing the relative amounts of various phases in the microstructure. Titanium alloys become more resistant to cracking in seawater as the proportion of beta phase increases as influenced by alloying. Nominally austenitic stainless steel castings and weld deposits, and dual-phase stainless steels, have varying resistances to stress-corrosion cracking, depending largely on the relative amounts of ferrite and austenite in the structure.

Next to alloy composition, the metallurgical variable having the greatest influence is probably strength level, with susceptibility to cracking generally increasing with increasing strength. This behavior is the basis for the 22 HRC maximum on heat treated alloy steel bolts to be used in areas where aqueous hydrogen sulfide solution might be encountered (ASTM A193, grade B7M). The above effect of strength level has also been observed in quenched-and-tempered carbon and alloy steels, in martensitic stainless steels, and in

hardenable alloys based on iron, nickel, titanium and aluminum. Other significant metallurgical parameters include cold work, grain size, inclusion content (pitting nucleation sites), banding, grain-flow directionality and, in austenitic alloys, sensitization, which is a prerequisite for polythionic acid cracking of stainless steel. In addition to introducing residual stresses, welding can, in some alloys, produce metallurgical conditions in the weld metal or heat-affected zone which render them more susceptible to stress-corrosion cracking than the base metal.

SELECTED REFERENCES

Stress-Corrosion Cracking and Hydrogen Embrittlement of Iron Base Alloys (conference held at Unieux-Firminy, France, June 12-16, 1973), edited by J. Hochmann, J. Slater and R. W. Staehle: National Association of Corrosion Engineers, Houston (in publication by NACE as of 1975)

Stress-Corrosion Cracking in High-Strength Steels and in Titanium and Aluminum Alloys, edited by B. F. Brown: Naval Research Laboratory, Washington, 1972

Stress-Corrosion Cracking of Metals—A State of the Art, edited by H. L. Craig: STP 518, American Society for Testing and Materials, 1972

The Theory of Stress-Corrosion Cracking in Alloys, by J. C. Scully: North Atlantic Treaty Organization, Brussels, 1971

Fundamental Aspects of Stress-Corrosion Cracking (conference held at the Ohio State University, Columbus, OH, Sept 11-15, 1967), edited by R. W. Staehle, A. J. Forty and D. van Rooyen: National Association of Corrosion Engineers, Houston, 1969

The Stress Corrosion of Metals, by H. L. Logan: John Wiley & Sons, New York, 1966

Liquid-Metal Embrittlement

By William Rostoker, University of Illinois at Chicago

LIQUID-METAL EMBRITTLEMENT results in a loss in ductility of a solid metal, or its fracture below the normal yield stress, under the circumstances that its surface is wetted by some lower-melting liquid metal. Thus, for example, a 70-30 brass wetted by mercury will fracture at a stress near but below that for yielding under simple tensile- or bend-testing conditions. Although the ductility of 70-30 brass is very high and although tensile rupture of this material exhibits substantial necking, the fracture associated with mercury embrittlement is flat, intercrystalline cleavage under SEM examination, and the gage length shows no measurable elongation.

Fracture can be initiated on a smooth surface without the assistance of any observable stress concentration. Fracture above the critical stress threshold initiates immediately upon the establishment of wetting by the liquid. Crack initiation and resultant fast fracture may be delayed by slow development of wetting. Many potential failure events do not occur because the conditions for wetting do not exist. Delayed fast fracture can also occur because of the existence of subcritical stress levels which become fracture-functional only when grain-boundary grooving or penetration by the liquid, regulated by diffusion, reaches a depth dictated by the reduced K_c of the wetted metal. Precracked test specimens under static loading show continuing crack growth proportional to K as long as liquid metal can follow the crack-root advance. The fatigue strengths of vulnerable alloys exposed to the appropriate liquid metals are significantly reduced. This does not represent penetration of existing films by mechanical flexure — it occurs on specimens which have been carefully wetted prior to the cyclic stress exposure. Alloys strengthened by transformations such as age hardening, and tempered martensites, seem to be much more vulnerable to extreme embrittlement. In such instances, brittle fracture can occur at stress levels far below the engineering yield stress.

SUSCEPTIBILITY OF VARIOUS METALS

Susceptibility to liquid-metal embrittlement is unique to specific metals. Liquid mercury embrittles copper and aluminum alloys but not simple carbon or alloy-carbon steels. Molten lithium, zinc, indium and cadmium, and alloys containing these elements, can embrittle carbon steels but have smaller effects, or no effects at all, on aluminum and copper alloys. Molten bismuth embrittles copper alloys but has no effect on aluminum or on carbon steels. There is a specificity in the identities of liquid metals which exert large, small or no embrittling effects on any given solid alloy under conditions of tensile stress and wetting. Liquid alloying complicates the definition of specificity. Liquid lead, for example, has no embrittling effect on carbon steels, but small percentages of dissolved tin, cadmium and nickel create a potential for cracking under tensile stresses which are either simple stresses or components of complex stress systems. Commercially pure metals are usually not, in any practical sense, subject to liquid-metal embrittlement. Two exceptions are zinc, which is embrittled by mercury, and copper, which is embrittled by molten bismuth. Thus copper and aluminum are not embrittled by liquid mercury whereas their alloys are very much in hazard. There is a transition from brittle to ductile behavior with increasing temperature. For example, the temperature range for embrittlement of carbon steels by liquid lead alloys is bracketed on the low side by the melting temperature of the lead alloy and by an upper limit of temperature which is only about 300 °C higher. Above this upper limit, the steel has undiminished ductility.

ENGINEERING CONTEXT OF LIQUID-METAL CRACKING

The engineering context of liquid-metal cracking includes many different situations. Perhaps the most obvious is brazing. If residual stresses are not eliminated ahead of time, or if thermal stresses are generated by high rates of heating or cooling, and the brazing alloy is one potentially capable of causing liquid-metal embrittlement of the metal being brazed, then the brazed assembly can exhibit unexpected cracks in the base metal. Hot dipping for coating purposes with the same scenario can produce the same effect. Nondeliberate overheating should be recognized as creating a potential cracking situation. A case in point is the cracking of steel shafting when the bronze bearing overheats to the point of even superficial melting. Rupture of mercury indicator or control devices constitutes a potential hazard to aluminum structures.

"Hot shortness" should be regarded as another term for liquid-metal embrittlement. Thus, for example, very small levels of contamination of copper and brass ingots by bismuth have long been correlated with cracking in hot rolling or forging. The same is true of leaded tool and die steels. In both of these cases, exposure of microalloy-free metals to the embrittling liquid as a droplet wetted to the surface produces the same cracking. In both cases, the very small volumes of liquid in relation to the extent of cracking is to be recognized. The distinction between liquid-metal embrittlement and hot shortness is often obscured by the differences in origin of the liquids. Thus, sodium is not regarded as being likely to embrittle structural aluminum alloys, but as a contaminant in melting operations it is not to be ignored. Yet aluminum alloys contaminated by very small amounts of sodium are hot short, and the same alloys which are sodium-free can be caused to crack under stress when only a tiny droplet of sodium is wetted to the surface.

MECHANISMS OF LIQUID-METAL EMBRITTLEMENT

Mechanisms of liquid-metal embrittlement are credible but are not able to account for the circumstances of cracking—most particularly in defining the specificity between the identity of the liquid metal which will probably embrittle some engineering alloy material. It should be kept in mind that the number of critical combinations of liquid and solid to which cracking potential can be ascribed is small compared with the number for which there is no experience with such events.

Theoretical models for cleavage fracture seem to apply. The grain-boundary local stresses generated by impinging slip bands can be identified with the initiation of microcracks. The atomic model has been associated with reduction of cohesive forces between the atoms of the solid metal by interaction with atoms of the liquid metal at the actual crack base. Whether the mobile atoms reach that point by surface diffusion or by vapor transport is a continuing subject of discussion. Fracture energies measured by fracture-mechanics experimental procedures show levels of K_c which are consonant with brittle behavior. In an equiaxed, polycrystalline structure the fracture path is invariably intercrystalline. Deviations from this are associated with highly anisotropic grain shapes. Single crystals are not normally subject to embrittlement by surface wetting with a liquid metal that is active on the same alloy in a polycrystalline form.

SELECTED REFERENCES

Embrittlement by Liquid Metals, by W. Rostoker, J. M. McCaughey and H. Markus: Reinhold, New York, 1960
Adsorption-Induced Brittle Fracture in Liquid-Metal Environments, by A. R. C. Westwood, C. M. Preece and M. H. Kamdar: Chapter 10 in *Fracture,* Vol III, edited by H. Liebowitz, Academic Press, New York, 1971

Deleterious Effects of Hydrogen

By R. A. Oriani, University of Minnesota

HYDROGEN exerts deleterious effects on virtually all metals and alloys. The severity of these effects differs from one family of alloys to another, and differs within any one family depending on strength level, chemical composition, impurity content and microstructure. Furthermore, for any one alloy having specific values of all of the above parameters, the harmful effects vary according to the temperature, the thermodynamic potency of the hydrogen and its kinetic availability, the state of stress of the alloy, the rate of deformation of the alloy, and probably still other factors. The large number of relevant parameters, many of which cannot be independently varied, complicates the understanding of the mechanisms underlying these effects.

TYPES OF DELETERIOUS HYDROGEN EFFECTS

A phenomenological classification of harmful hydrogen effects on metallic alloys is as follows:

1. Decrease of ductility as measured by elongation at fracture or reduction in area at fracture
2. Enhancement of the rate of propagation of a pre-existing crack
3. Delayed failure under a static load
4. Reduced fatigue strength
5. Some forms of stress-corrosion cracking
6. Blistering and "staircase" cracking
7. Hydrogen attack (methane formation).

All of these effects ultimately lead to lowering of the amount of externally applied energy or work necessary to produce material failure, although in the cases of effects 6 and 7 sufficient harm to the metal part or to the device or equipment may be produced by the excessive plastic deformation caused in that part.

Which of the above-enumerated effects is manifested depends on alloy properties, the manner of application of external mechanical loading, the temperature, and the source of the hydrogen. If the source of hydrogen is the gas and the metal does not undergo a change of temperature, the potency of the hydrogen—that is, its potential for causing deleterious effects—may be measured simply by the pressure (more accurately, the fugacity) of the hydrogen gas. In these circumstances, pressures of less than one atmosphere can cause effects 1 and 2 in steels with yield strengths above 1380 MPa (200 ksi) if the hydrogen can enter the metal. Entry of the gaseous hydrogen is strongly impeded by the naturally occurring surface oxide. Gaseous hydrogen decreases the ductility (effect No. 1) of lower-strength steels, and the effect increases with increasing hydrogen gas pressure. Gaseous hydrogen under isothermal conditions is capable of lowering the intrinsic fatigue resistance of steels and nickel alloys.

The potency of gaseous hydrogen under conditions of constant temperature cannot be greater than its pressure unless the temperature is sufficiently high so that a chemical reaction involving the hydrogen can take place within the metal. A steel at 200 to 300 °C (390 to 570 °F) will absorb hydrogen from gas streams containing hydrogen or hydrogen-bearing compounds. The metal-dissolved hydrogen will then react with dissolved carbon and cementite in the metal to produce methane within pre-existing microvoids. The pressure of the methane can reach sufficiently large values such that the microvoids, usually at grain boundaries, are enlarged and link up to produce failure. A less calamitous result of high-temperature absorption of hydrogen is the change of mechanical properties due to decarburization.

If hydrogen at moderate pressures is absorbed by a low-strength steel at an elevated temperature, subsequent lowering of the temperature will generate a much higher gas pressure within microvoids because the hydrogen solubility in the iron lattice decreases with decreasing temperature. The atomically dissolved hydrogen precipitates at internal surfaces of microvoids and recombines to form molecular hydrogen in the void volumes. If the gas pressure thereby generated is greater than the flow stress of the surrounding metal, the microvoids will be expanded. Adjoining microvoids can then coalesce by pressure-induced plastic deformation to form large voids which manifest themselves as blisters if they occur near an external surface.

A more ubiquitous source of high-potency hydrogen in steels is corrosion. Whenever the cathodic reaction during corrosion is the reduction of hydrogen ions, some of the hydrogen generated is taken up by the corroding steel. Depending on many factors, the thermodynamic potency of such hydrogen can attain values equivalent to many hundreds of atmospheres of gaseous hydrogen, and the kinetic availability for harmful effects is much greater than that of gaseous hydrogen. Effects 1 through 4 can be the results of corrosion-generated hydrogen. Because the absorption of such hydrogen is greatly enhanced by sulfide ions in water, corrosion of steels by sulfide-bearing waters is particularly conducive to hydrogen embrittlement. This is sometimes called "sulfide stress cracking." The "staircase" cracking of line-pipe steels is an example of such damage. Pickling of steels is an example of corrosion applied purposefully to achieve descaling. The absorbed hydrogen, however, is high-potency hydrogen which can generate large internal molecular gas pressures capable of propagating preexisting microfissures and expanding microvoids.

Clearly, any factor that increases the cathodic generation of hydrogen on a steel surface produces the risk of increasing the potency of the cathodically deposited hydrogen. For example, physical contact between a steel and a more electrochemically active metal such as zinc, magnesium or aluminum can result in enhanced absorption of hydrogen by the steel, leading to various forms of damage. Cathodic protection of steel structures by use of sacrificial anodes or impressed currents can be particularly damaging, because hydrogen potencies of millions of atmospheres can be generated by excessive applied cathodic currents. Plating of metals is often accompanied by unwanted deposition of hydrogen. For this reason, elevated-temperature baking after plating is required, but it should be remembered that electroplated coatings often act as barriers to the egress of the absorbed hydrogen. As a final example of corrosive attack we mention stress-corrosion cracking, characterized by the rapid propagation of transgranular or intergranular cracks. Although there is uncertainty about the mechanism of stress-corrosion cracking, it appears that in some cases the cathodically generated hydrogen is responsible for the damage.

Corrosion-Fatigue Failures

Edited by Charles R. Morin, Packer Engineering

CORROSION FATIGUE is the combined action of repeated or fluctuating stress and a corrosive environment to produce progressive cracking. Usually, environmental effects are deleterious to fatigue life, producing cracks in fewer cycles than would be required in a more inert environment. Once fatigue cracks have formed, the corrosive aspect also may accelerate the rate of crack growth.

In corrosion fatigue, the magnitude of cyclic stress and the number of times it is applied are not the only critical loading parameters. Time-dependent environmental effects also are of prime importance. When failure occurs by corrosion fatigue, stress-cycle frequency, stress-wave shape and stress ratio all affect the cracking processes.

EFFECT OF FREQUENCY

In nonaggressive environments, cyclic frequency generally has little effect on fatigue behavior. On the other hand, in aggressive environments fatigue strength is strongly dependent on frequency. Corrosion-fatigue strength (endur-

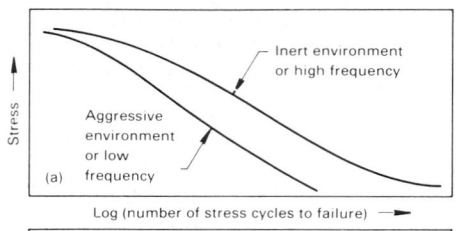

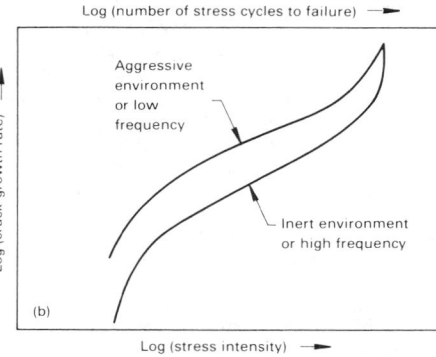

Fig. 38. Typical fatigue behavior in aggressive environment or at low frequency compared with fatigue behavior in inert environment or at high frequency, shown (a) as an S-N curve and (b) as variation of crack-growth rate with stress intensity

ance limit at a prescribed number of cycles) will generally decrease as the cyclic frequency is decreased. This effect is most important at frequencies of less than 10 Hz.

The frequency dependence of corrosion fatigue is thought to result from the fact that interaction of a material and its environment is essentially a rate-controlled process. Low frequencies, especially at low strain amplitudes or when there is substantial elapsed time between changes in stress levels, allow time for interaction between material and environment; high frequencies do not, particularly when high strain amplitude is involved also. At very high frequencies, or in the plastic-strain range, localized heating may seriously affect the properties of the part. Such effects normally are not considered to be related to a corrosion-fatigue phenomenon.

When environments have a deleterious effect on fatigue behavior, there may exist a critical range of frequencies of loading in which the mechanical/environmental interaction is significant. Above this range the effect usually disappears, while below this range the effect may diminish.

EFFECT OF STRESS AMPLITUDE

In general, a low amplitude of cyclic stress favors relatively long fatigue life, permitting greater opportunity for involvement of the environment in the fatigue process. Where stresses are sufficiently high to cause significant macroscopic plastic deformation, environmental interaction may be insignificant unless the strain rate is in a critical range for stress-corrosion cracking in certain alloy/environment systems.

Stress amplitude must be considered together with mean stress and frequency. Low stress levels may allow more time for environmental interaction, but if the frequency is high, the crack tip may not be exposed to the environment for a

time sufficient for the corrosion processes to do significant damage. Typical fatigue behavior, illustrating the effect of frequency and/or stress, as discussed above, is illustrated in Fig. 38.

EFFECTS OF ENVIRONMENT

Nucleation and propagation of corrosion-fatigue cracks in service are influenced by corrosive environments—mainly, bulk aqueous solutions or environments produced by continuous or periodic condensation of vapor.

Effect on Fatigue Strength. For any given material, the fatigue strength, or fatigue life at a given value of maximum stress, generally decreases in the presence of an aggressive environment. This effect varies widely, depending primarily on the characteristics of the material/environment combination. The environment affects crack-growth rate or probability of fatigue-crack initiation, or both. For many materials, the stress range required to cause fatigue failure diminishes progressively with time and with number of cycles. This effect is illustrated in Fig. 38(a) as a progressively wider separation between fatigue curves for inert and aggressive environments with increasing life or decreasing stress.

Corrosion-fatigue tests on smooth specimens of high-strength steel indicate that very large reductions in fatigue strength or fatigue life can occur in salt water. For instance, the fatigue strength at ten million cycles could be reduced to as little as 10% of that in dry air. In these tests, the main role of the environment was corrosive attack of

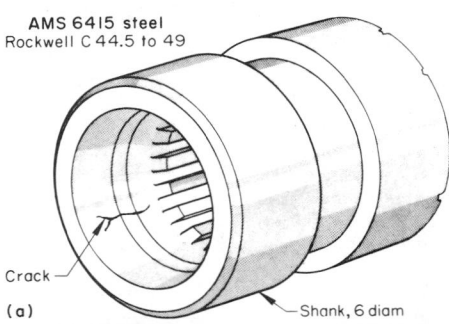

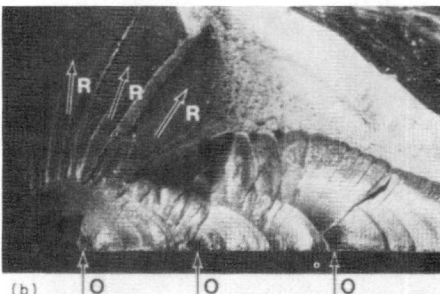

(a) View of a portion of the shaft, showing the location of the corrosion-fatigue crack on the tapered interior of the shank. (b) Light fractograph, at about 4½×, showing fracture origins at corrosion pits (arrows O) and direction of fast fracture (arrows R). Region where crack was opened in the laboratory, by sawing and then breaking by hand, is shown at upper right corner.

Fig. 39. Hollow, splined alloy steel shaft that failed by corrosion-induced fatigue in aircraft service because of exposure to hydraulic oil that was contaminated with water

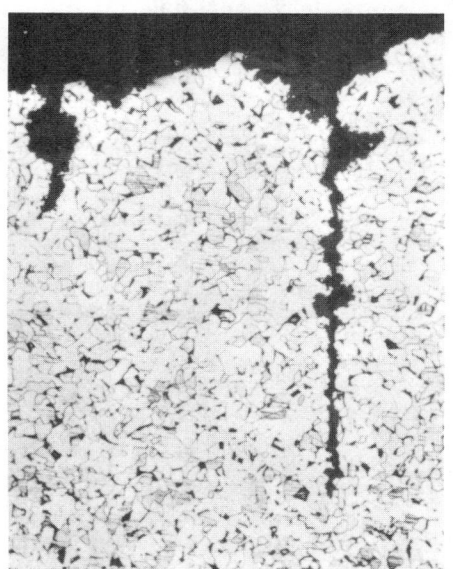

Micrograph, at 250×, of a nital-etched section through corrosion-fatigue cracks that originated at corrosion pits in a carbon steel boiler tube. Corrosion products are present along the entire length of the cracks.

Fig. 40. Corrosion-fatigue cracks in carbon steel

the polished surface, creating local stress raisers that initiated fatigue cracks. Salt water also increases crack-growth rate in steels.

As an example, corrosion by water-contaminated hydraulic oil, which reduced fatigue strength to about one-fourth of that expected in the absence of corrosion, initiated fatigue cracking of an alloy steel aircraft shaft (Fig. 39).

Effect on Crack Initiation. Surface features at origins of corrosion-fatigue cracks vary with the alloy and with specific environmental conditions. In carbon steels, cracks often originate at corrosion pits and often contain significant amounts of corrosion products (see Fig. 40). The cracks are predominantly transgranular and may exhibit a slight amount of branching.

Effect on Crack Propagation. Once fatigue cracks have formed, the rate of enlargement of the crack (crack-propagation rate) is often accelerated by corrosive environments. For some stress wave forms (such as ramp, hold, reverse) the enhancement of crack-growth rate can be very large, as much as several orders of magnitude. Such effects usually are evaluated by fracture-mechanics methods relating crack growth per stress cycle, da/dN, to cyclic stress intensity, ΔK (Fig. 38b).

SELECTED REFERENCES

Corrosion Fatigue: Chemistry, Mechanics, and Microstructure (proceedings of conference held at the University of Connecticut, June 14-18, 1971), edited by O. Devereux, A. J. McEvily and R. W. Staehle: National Association of Corrosion Engineers, Houston

Corrosion-Fatigue Technology (proceedings of symposium held at Denver, Nov 14-19, 1976), edited by H. L. Craig, T. W. Crooker and P. W. Hoeppner: American Society for Testing and Materials, Philadelphia

Pulp and Paper Industry Corrosion Problems (proceedings of international symposia), Vol I, II and III: National Association of Corrosion Engineers, Houston

Marine Corrosion, Causes and Prevention, by F. L. LaQue: John Wiley & Sons, New York, 1975

Corrosion and Corrosion Control, 2nd Ed., by H. H. Uhlig: John Wiley & Sons, New York, 1971

Elevated-Temperature Failures

By R. J. Fields, National Bureau of Standards, and T. Weerasooriya, University of Dayton

IN SERVICE at elevated temperature, the life of a metal component subjected to either sustained or oscillatory loads is limited. Stress imposed at elevated temperature produces a continuous strain in the component and results in creep. Creep, by definition, is time-dependent deformation occurring under stress. After a period of time, creep terminates in creep fracture (also called stress rupture). The variety of creep mechanisms at elevated temperature is demonstrated in the deformation-mechanism maps shown in Fig. 41 for thoria-dispersed nickel and type 316 stainless steel. Against axes of applied stress and temperature, contours of material response (i.e., creep strain rate) are plotted. Superimposed on this diagram are regions of deformation mechanisms. Diagrams such as these have been constructed for a wide variety of materials and are available in Ref 1. A similar variety of fracture mechanisms is observed at elevated temperatures, as demonstrated in the fracture-mechanism maps shown in Fig. 42 for $2^1/_4$Cr-1Mo steel and type 316 stainless steel. These maps are similar to deformation-mechanism maps except that the contours are of time-to-fracture and the regions are of fracture mechanisms. Maps such as these are available in the literature (Ref 2 to 4) and summarize an enormous body of data, particularly elevated-temperature failure data.

From Fig. 41 and 42, it can be seen that the conditions of temperature, stress and time under which creep and creep-fracture failures occur depend on the metal or alloy and on the service environment. In general, elevated-temperature behavior begins at about one-third of the absolute melting temperature for metals. This is an oversimplification for alloys. The temperature at which the mechanical strength of an alloy becomes limited by creep rather than by yield strength must be determined individually for alloys on the basis of behavior. Elevated-temper-

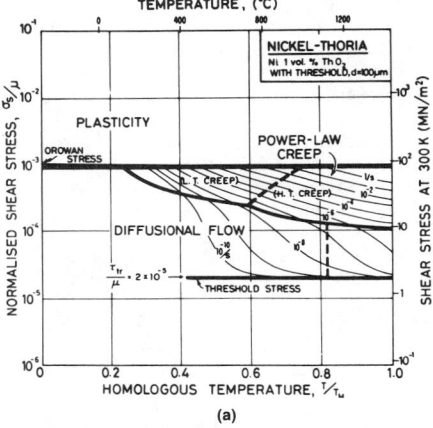

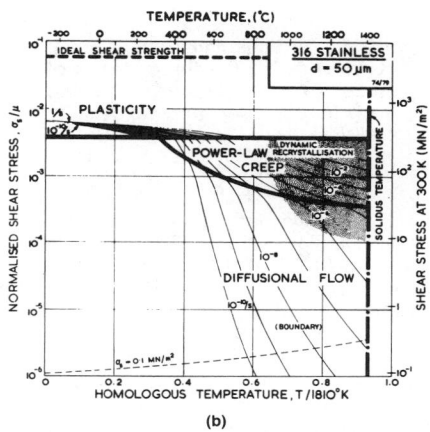

These diagrams show how these alloys deform at various stress levels and elevated temperatures. Diffusional flow is a type of creep that occurs at very high temperatures and very low stresses. Detailed explanations of these maps may be found in Ref 1. Courtesy of H. J. Frost and M. F. Ashby, Pergamon Press, 1982.

Fig. 41. Deformation-mechanism maps for (a) thoria-dispersed nickel and (b) type 316 stainless steel

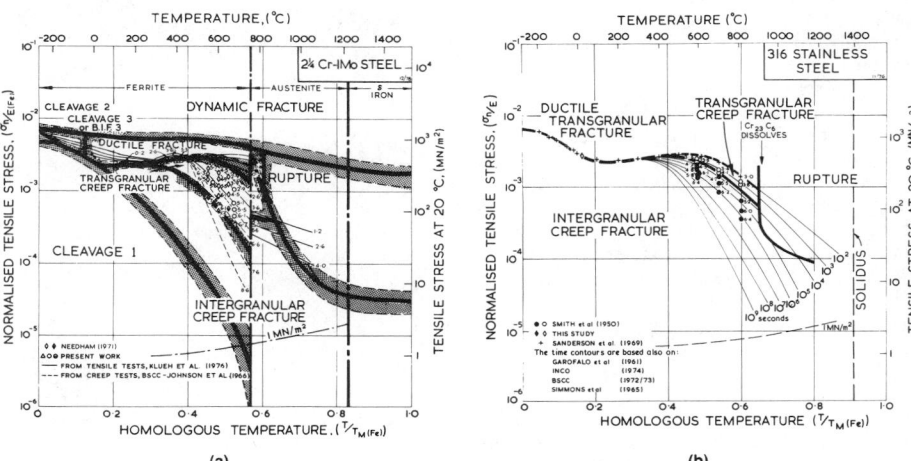

These diagrams show the conditions of stress and elevated temperature under which ductile fracture, transgranular creep fracture, intergranular creep fracture and rupture occur. Detailed explanations of these maps may be found in Ref 2, 3 and 4.

Fig. 42. Fracture-mechanism maps for (a) $2^1/_4$Cr-1Mo steel and (b) type 316 stainless steel

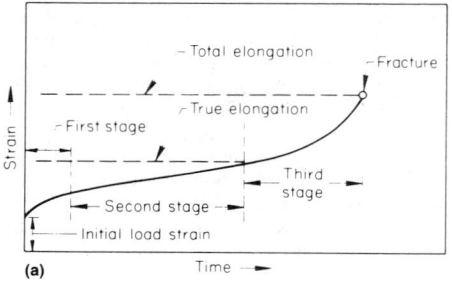

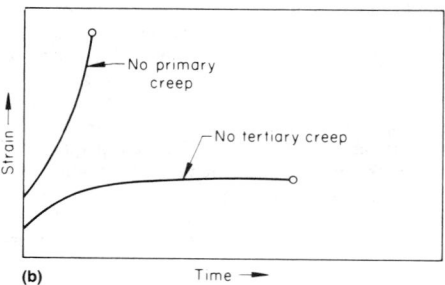

Fig. 43. (a) Schematic tensile-creep curve showing the three stages of creep. (b) Creep curves showing no primary creep and no tertiary creep.

ature behavior begins at approximately 205 °C (400 °F) for aluminum alloys, 315 °C (600 °F) for titanium alloys, 370 °C (700 °F) for low-alloy steels, 540 °C (1000 °F) for austenitic, iron-base high-temperature alloys, 650 °C (1200 °F) for nickel-base and cobalt-base high-temperature alloys, and 980 to 1540 °C (1800 to 2800 °F) for refractory metals and alloys. New alloys are continually being developed that raise the temperature at which the elevated-temperature region begins. Such is the case for mechanically alloyed superalloys and rapidly solidified aluminum alloys.

The principal types of elevated-temperature failure are excessive creep deformation and creep fracture, fatigue or creep-fatigue interaction, thermal fatigue, tension overload, and combinations of these, as modified by environment. Generally, the type of failure is established by examination of fracture surfaces, metallographic sections extracted from regions adjacent to the fracture, and comparisons of component operating conditions with available data on creep, creep-fracture, tensile, creep-fatigue and thermal-fatigue behavior. Such analysis is usually sufficient for most failure investigations, but a more thorough analysis may be required when stress, time, temperature and environment have acted to change the metallurgical structure of the alloy.

CREEP DEFORMATION

Most creep deformation consists of three distinct, chronological stages (Ref 5); these stages are shown schematically in Fig. 43(a). Following an initial elastic strain resulting from the immediate effects of the load, there is a region of increasing plastic strain at a decreasing strain rate (called first-stage, primary or transient creep). Following first-stage creep is a region of nominally minimum-rate plastic straining (second-stage or secondary creep). Under a constant stress, as opposed to a constant load, a constant creep rate

may be obtained in secondary creep. This is referred to as steady-state creep. Finally, there is a region of rapidly increasing strain rate with rapid extension to fracture. This is third-stage or tertiary creep.

Primary creep has no distinct end point and tertiary creep has no distinct beginning. The accelerating strain observed in tertiary creep may be due to a reduction in cross-sectional area resulting from cracking or necking. Environmental effects (such as oxidation) that reduce cross-sectional area may increase the tertiary creep rate. In many commercial creep-resistant alloys, tertiary creep is caused by inherent deformation processes or by changes in microstructure such as precipitate coarsening and consequential loss of creep strength.

In designing components for service at elevated temperatures, data pertaining to the elapsed time and extension that precede tertiary creep are of the utmost importance; design for creep resistance is based on such data. However, the magnitude of the tertiary creep strain is also important, because excessive deformation may allow detection of a failing component prior to catastrophic fracture.

Under certain conditions, some metals may not exhibit all three stages of creep. For example, at high stresses or temperatures, the absence of primary creep is not uncommon, with secondary creep or, in extreme cases, tertiary creep following immediately upon loading. At the other extreme, notably in cast alloys, no tertiary creep may be observed and fracture may occur with only a minimum of extension. Both of these phenomena are illustrated by the creep curves in Fig. 43(b). Oscillating creep rates may occur if dynamic recrystallization takes place during creep.

CREEP FRACTURE

A component under creep loading will eventually fracture (rupture) provided that the strain occurring during creep does not relieve the stress. Examples of ruptured superheater tubes are shown in Fig. 44(a). These tubes had supported an internal steam pressure of 12.8 MPa at 538 °C (1850 psi at 1000 °F) for more than 20 years. Depending on the alloy and the service conditions, creep fracture can be either macroscopically brittle or ductile (see Fig. 42). Ductile fracture is transgranular and, typically, is accompanied by discernible elongation and necking similar to room-temperature ductile fractures. Brittle fracture is intergranular and occurs with little or no elongation or necking. Intergranular cracking may not be readily discernible on the surface of a part; however, if the oxide scale developed during elevated-temperature service in air is removed or the part is sectioned and polished metallographically, such cracking will usually be visible as shown in Fig. 44(b). These creep cracks originate as tiny cavities (Fig. 44c) which grow by diffusion and link together to form the larger cracks (Ref 6 and 7).

Creep-fracture behavior is portrayed graphically by plotting the logarithm of the rupture stress versus the logarithm of the rupture life. These curves are known as stress-rupture curves and are used for design purposes and for improving metallurgical knowledge of the failure process. They provide an indication of metallurgical instabilities or changes in creep-fracture mechanism. The fracture-mechanism map for type 316 stainless steel (Fig. 42b) has been replotted as a stress-rupture curve in Fig. 45, indicating the regions in which failure is transgranular or intergranular.

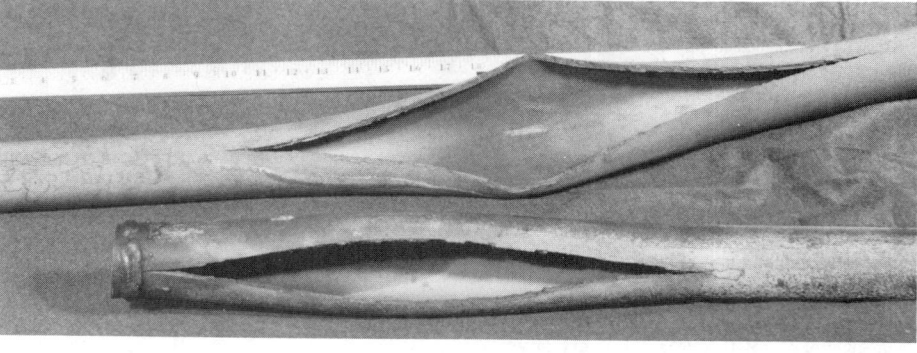

(a)

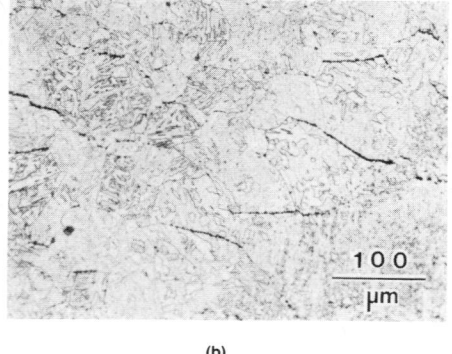

(b)

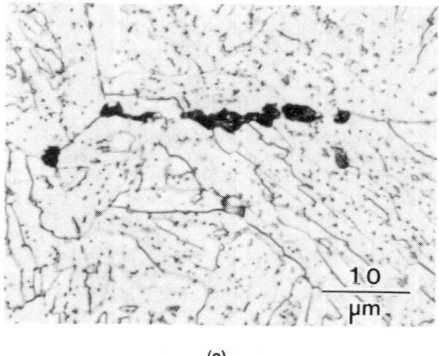

(c)

Fig. 44. (a) Ruptured superheater tubes. Magnification, 0.2×. (b) Creep cracks found in metallographic section of 1¼Cr-½Mo steel steam pipe. Magnification, 160×; nital etch. (c) Creep cavities that are linking to form cracks visible in (b). Magnification, 1500×; nital etch.

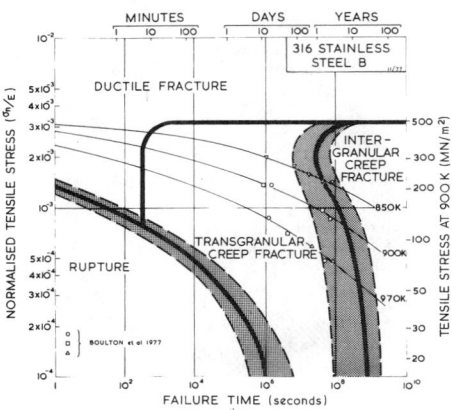

This diagram shows the dependence of fracture life on stress and temperature. The mechanism by which failure occurred is also indicated.

Fig. 45. Stress-rupture curve for type 316 stainless steel

This mechanism information is important for extrapolation of data in time or in multiaxial loading situations.

CREEP DUCTILITY

The two common measures of ductility are elongation and reduction in area. In the standard tensile-creep curve, as shown in Fig. 43(a), there are actually two measures of elongation that are of interest. First, there is true elongation, which is defined as the elongation at the end of second-stage creep. Second, there is total elongation, which is the elongation at fracture. In some instances, the difference between these two measures of elongation is due to extension caused by crack separation or necking. Ductility data from creep-fracture data are generally erratic, even in replicate tests; they are more erratic for castings than for wrought products. In some metals and alloys, the values of total elongation follow a smooth curve that either increases or decreases with increasing time-to-fracture and temperature. However, many alloys exhibit a series of maximums and minimums in ductility with time or temperature. True elongation nominally consists of extension resulting from nonlocalized creep and some extension due to intergranular void formation. As shown by the solid circles in Fig. 46, the variation of true elongation with rupture life for a Co-Cr-Ni alloy (S-590) follows a smooth curve, with true elongation decreasing with increasing rupture life. On the other hand, total elongation (open circles in Fig. 46) does not exhibit any clear-cut correlation with rupture life.

Creep ductility is an important factor in alloy selection. In conventionally cast nickel-base superalloys, for example, a creep ductility at fracture of about 1% is common at 760 °C (1400 °F), compared with values above 5% for the strongest wrought superalloys. Because component designs are frequently limited by a 1% creep strain, a high creep ductility may preclude use of an alloy to its full strength potential. On the other hand, as shown in Fig. 47, a higher stress-rupture ductility for the same load and temperature conditions means a higher safety margin. Because premature failures have resulted from lack of ductility during creep, fail-safe systems or retirement-for-cause programs must often trade creep strength for creep ductility. Creep-fatigue and thermal-cycle fatigue resistance are also related to creep ductility. Generally, superalloys with the highest

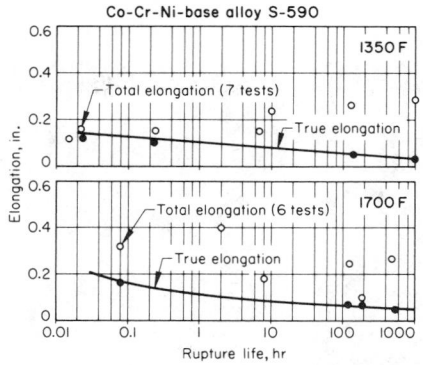

Fig. 46. Relation of elongation and rupture life for Co-Cr-Ni-base alloy S-590 tested at two temperatures

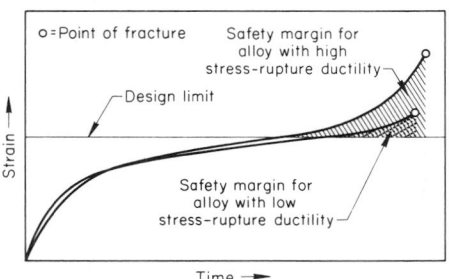

Fig. 47. Schematic creep curves for alloys having low and high stress-rupture ductilities, showing the increased safety margin provided by the alloy with high stress-rupture ductility

ductilities for a given strength level show the greatest resistance to creep-fatigue crack growth and thermal-cycle fatigue.

CREEP-FATIGUE INTERACTION

Although creep and creep-fracture tests are performed under static loading on unnotched

specimens, components in practice have cracks, notches and other stress raisers and experience cyclic loading. In many critical applications, particularly in aircraft engines, the growth of a creep or fatigue crack to a predetermined length will be cause for retirement of a component. Hence, failure in these applications is not defined as excessive strain or fracture, but as the growth of a crack. Alloys used in these designs must be crack-growth-resistant and flaw-tolerant. A great deal of research is focused on the interaction of creep and cyclic loading (creep-fatigue) to produce crack growth.

In general, creep-fatigue crack-growth rates and growth mechanisms are functions of the maximum stress-intensity factor, the minimum/maximum stress-intensity ratio (r-ratio), the frequency, the temperature, the hold or sustained-load time, the loading wave shape, and the material. If crack growth is studied as a function of frequency, keeping the other parameters constant, a curve such as that shown in Fig. 48 is obtained (Ref 8). These results are for IN718, a nickel-base superalloy which is used in aircraft engine disks. In Fig. 48, there are three distinct regions of crack growth as the frequency changes. If crack growth is studied as a function of temperature, keeping the other parameters constant, three regions similar to those above are again observed. Figure 49(a), (b) and (c) are fractographs characterizing these three regions.

At high frequencies and/or lower (but still elevated) temperatures, crack growth is independent of frequency or temperature. The material just ahead of the crack tip does not undergo any time-dependent processes, such as oxidation or creep relaxation. In this region, the micromechanism of crack growth is predominantly striation formation, as shown in Fig. 49(a). This fatigue mechanism is essentially the same as that which occurs at room temperature. At low frequencies and/or higher temperatures, crack growth is a fully time-dependent process. In this case, crack growth can be predicted by integrating the sustained-load or creep crack-growth behavior over the applied stress history (Ref 9). As shown in Fig. 49(c), crack growth in this region is fully

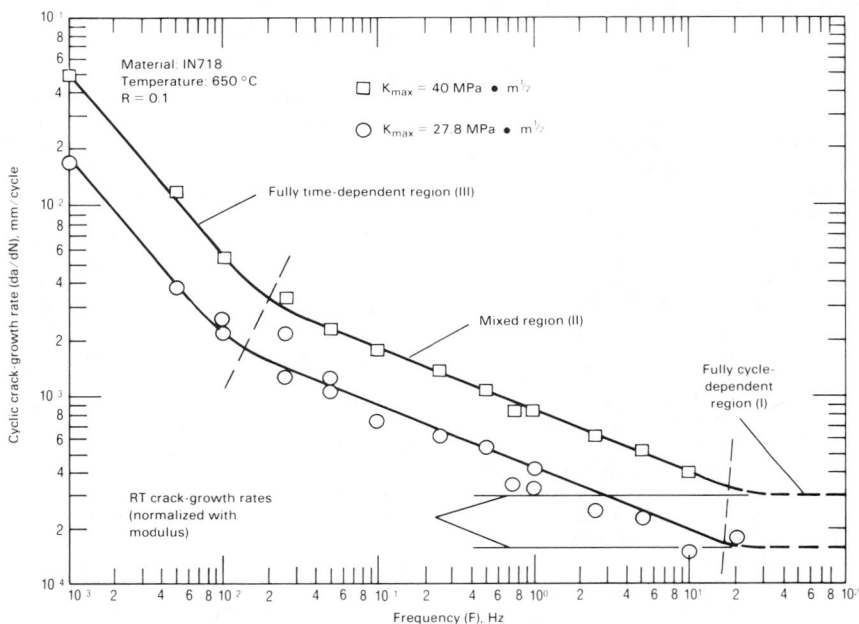

Fig. 48. Cyclic growth rate of a crack in alloy IN718 at 650 °C (1200 °F) as a function of frequency. Note the three regions or slopes of the curves.

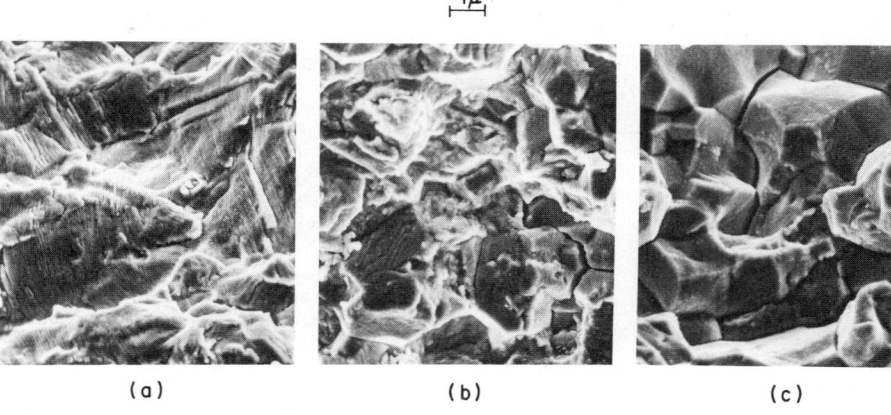

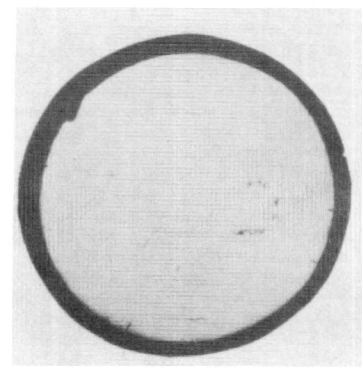

Fig. 49. Scanning electron fractographs of typical fracture surfaces for (a) fully cycle-dependent (10-Hz) region, (b) mixed (0.5-Hz) region and (c) fully time-dependent (0.001-Hz) region. These tests were conducted on IN718 at 650 °C (1200 °F), r = 0.1 and K_{max} = 40 MPa·m$^{1/2}$.

intergranular and has been shown to be due to both time-dependent degradation (e.g., oxidation or hot stress corrosion) and creep cavitation along the grain boundaries. In the mixed region or the creep-fatigue interaction region (Fig. 49b), one observes a mixture of intergranular and transgranular crack-growth surfaces. It is important to determine the mechanism region in which a component is operating because the appropriate crack-growth predictive models that have been developed differ for the three regions.

THERMAL FATIGUE

Mechanical loading is not the only damaging force applied to components at elevated temperatures. Transient thermal gradients within a component can induce plastic strains, and if these gradients are applied repeatedly, the resulting cyclic strain can cause component failure. This process is known as thermal-cycle fatigue or, simply, thermal fatigue. The effects of the strains that are induced by thermal transients on the airfoils of gas-turbine engines are illustrated schematically in Fig. 50. The thermal strains are generated in the airfoils because the outer surfaces change temperature more rapidly than the metal within. The effects of start-up and cool-down can often account for more reduction in life expectancy than creep and mechanical fatigue at elevated temperatures. Thermal-fatigue cracks initiate along the surface and progress inward. They are oriented normal to the surface and may occur singly or in multiples. Because the crack initiates externally, the amount of corrosion or oxidation along the surface of a thermal-fatigue crack is inversely proportional to the depth of the crack. Thermal-fatigue cracks can progress intergranularly or transgranularly, just as creep-fatigue cracks can. Whereas creep tends to produce numerous subsurface cracks (which contain no oxide layer), thermal fatigue often produces only a few surface cracks.

One way to reduce thermal fatigue is to reduce section size. The effect of section size on resistance to thermal fatigue is illustrated by the macrographs in Fig. 51. These photos show two RA 330 alloy bars — one 9.5 mm ($^3/_8$ in.) in diameter and the other 12.7 mm ($^1/_2$ in.) in diameter — taken from the same furnace basket after many service cycles of heating and quenching. The 9.5-mm-diam bar exhibited better resistance to thermal fatigue.

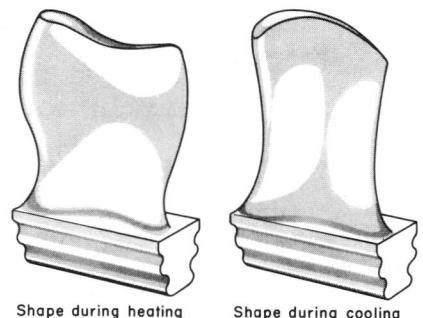

Shape during heating   Shape during cooling

Failure by fracture in thermal fatigue is caused by these cyclic thermal stresses.

Fig. 50. Schematic portrayal of the expansion, contraction and distortions in shape that occur in the airfoils of turbine blades as a result of cyclic heating and cooling or of uneven heating

ENVIRONMENTALLY INDUCED FAILURE

A critical factor in the performance of metals in elevated-temperature service is the environment and the resulting surface/environment interactions. Control of environment or protection of materials by coatings or self-protective oxide films is essential to most elevated-temperature applications.

General oxidation can lead to premature failure by reducing the cross-sectional area; grain-boundary oxidation may produce a notch effect that can also limit life. Some environments may be more harmful than others. Attack of fire-side surfaces of steam-boiler tubes by ash from vanadium-bearing fuel oils can be quite severe. Salt-containing atmospheres exert a deleterious effect on steels and high-temperature alloys, particularly titanium alloys. A vacuum may be a more deleterious environment than air because the chromium content of the alloy can be reduced by evaporation in vacuum. One of the most aggressive environments is found in the high-temperature incinerators of naval vessels and in coal conversion plants. In some cases, no metal alloy can resist the combined effects of stress, temperature and corrosion. It is then necessary to resort to the use of a structural ceramic such as silicon carbide or alumina. Great care must be

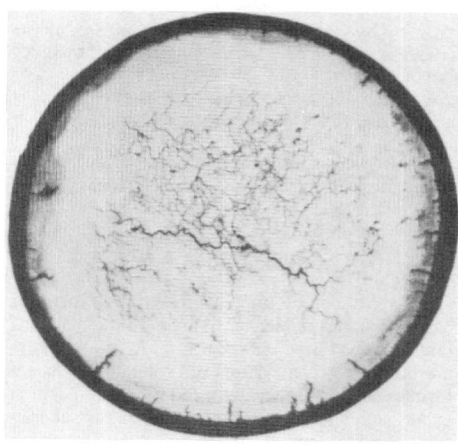

These macrographs of alloy RA 330 bars, 9.5 and 12.7 mm ($^3/_8$ and $^1/_2$ in.) in diameter, that were taken from the same furnace basket, show evidence of carburization, but only the 12.7-mm-diam bar exhibits severe cracking due to thermal fatigue. Magnification, about 4.5×; etched in mixed acids.

Fig. 51. Effect of section size on resistance to thermal fatigue

exercised in the use of these materials, because they are extremely brittle.

REFERENCES

1. *Deformation Mechanism Maps: The Plasticity and Creep of Metals and Ceramics*, by H. J. Frost and M. F. Ashby: Pergamon Press, Oxford, 1982
2. Fracture-Mechanism Maps and Their Construction for F.C.C. Metals and Alloys, by M. F. Ashby, C. Gandhi and D. M. R. Taplin: *Acta Met.*, Vol 27, 1979, p 699-729
3. Fracture-Mechanism Maps for Materials Which Cleave, by C. Gandhi and M. F. Ashby: *Acta Met.*, Vol 27, 1979, p 1565-1602
4. Fracture Mechanisms in Pure Iron, Two Austenitic Steels, and One Ferritic Steel, by R. J. Fields, M. F. Ashby and T. Weerasooriya: *Met. Trans.*, Vol 11A, 1980, p 333-342
5. *Fundamentals of Creep and Creep-Rupture in Metals*, by F. E. C. Garofalo: MacMillan, New York, 1965
6. Introduction to the Viewpoint Set on Creep Cavitation, by W. D. Nix: *Scripta Met.*, Vol 17, 1983, p 1-4
7. Intergranular Cavitation in Creeping Alloys, by A. S. Argon: *Scripta Met.*, Vol 17, 1983, p 5-12
8. Effect of Frequency on Fatigue Crack Growth Rate at High Temperature, by T. Weerasooriya: in *Proceedings of 16th National Symposium on Fracture Mechanics*, Columbus, OH, 1983, published by ASTM
9. A Model for Creep-Fatigue Interactions in Alloy 718, by T. Nicholas, T. Weerasooriya and N. E. Ashbaugh: in *Proceedings of 16th National Symposium on Fracture Mechanics*, Columbus, OH, 1983, published by ASTM

33 Nondestructive Testing

Edited by Ron Mason, Multispec

This section was condensed from Metals Handbook, Eighth Edition, Volume 11, Nondestructive Inspection and Quality Control, pages 20 to 286. For more detailed information on the topics covered herein, the reader is referred to the larger work.

Liquid-Penetrant Inspection

Edited by Ronald H. Selner, Universal Technology Corp.

LIQUID-PENETRANT INSPECTION is a nondestructive method for finding discontinuities that are open to the surface of solid and essentially nonporous materials. Indications of flaws can be found regardless of the size, configuration, internal structure or chemical composition of the workpiece being inspected and regardless of flaw orientation. Liquid penetrants can seep into (and be drawn into) various types of minute surface openings (reportedly, as fine as 0.1 μm, or 4 μin., in width) by capillary action. Because of this, the process is well suited for detection of all types of surface cracks, laps, porosity, shrinkage areas, laminations and similar discontinuities. It is used extensively for inspection of wrought and cast products of both ferrous and nonferrous metals, powder metallurgy parts, ceramics, plastics, and glass objects.

Advantages. In practice, the liquid-penetrant process is relatively simple (no electronic systems are involved). Equipment generally is simpler and less costly than that for most other nondestructive-inspection methods. Establishment of procedures and standards for inspection of specific parts or products is usually less difficult than for the more highly sophisticated inspection methods.

The liquid-penetrant method does not depend on ferromagnetism, and the arrangement of the discontinuities is not a factor. The penetrant method is good not only for detecting surface flaws in nonmagnetic metals, but also for revealing surface flaws in a variety of other nonmagnetic materials. Penetrant inspection also is used for inspecting items made from ferromagnetic steels; in some instances, its sensitivity is greater than that of magnetic-particle inspection.

Limitations. The major limitation of liquid-penetrant inspection is that it can detect only imperfections that are open to the surface; some other method must be used for detecting subsurface defects or discontinuities. Another factor that may inhibit the effectiveness of liquid-penetrant in-spection is the surface roughness of the object being inspected. Extremely rough or porous surfaces are likely to produce false indications.

Although the liquid-penetrant method often is used to inspect some types of powder metallurgy parts, the process generally is not well suited to inspection of low-density powder metallurgy parts or other porous materials, because the penetrant enters the pores and thus registers each pore as a defect.

PHYSICAL PRINCIPLES

Liquid-penetrant inspection depends mainly on a liquid's effectively wetting the surface of a solid workpiece or specimen, flowing over that surface to form a continuous and reasonably uniform coating, and then migrating into cavities that are open to the surface. The cavities of interest usually are exceedingly small, often invisible to the unaided eye. The ability of a given liquid to flow over a surface and enter surface cavities depends principally on the following:

- Cleanness of the surface
- Configuration of the cavity
- Size of the cavity
- Surface tension of the liquid
- Ability of the liquid to wet the surface.

The cohesive forces between molecules of a liquid cause surface tension. An example of the influence of surface tension is the tendency of free liquid, such as a droplet of water, to contract into a sphere. In such a droplet, surface tension is counterbalanced by the internal hydrostatic pressure of the liquid. When the liquid comes into contact with a solid surface, the cohesive force responsible for surface tension competes with the adhesive force between the molecules of the liquid and the solid surface. These forces jointly determine the contact angle between the liquid and the surface. If the angle is less than 90°, the liquid has good wetting ability.

DESCRIPTION OF THE PROCESS

Regardless of the type of penetrant used, and regardless of other variations in the basic process, liquid-penetrant inspection requires at least five essential steps:

1. **Surface Preparation.** All surfaces of a workpiece must be thoroughly cleaned and completely dried before it is subjected to liquid-penetrant inspection. Discontinuities exposed to the surface must be free from oil, water or other contaminants for at least 25 mm (1 in.) beyond the area being inspected if they are to be detected.

2. **Penetration.** After the workpiece has been cleaned, liquid penetrant is applied in a suitable manner so as to form a film of the penetrant over the surface for at least 13 mm ($^1/_2$ in.) beyond the area being inspected. This film should remain long enough to allow maximum penetration of the penetrant into any surface openings that are present.

3. **Removal of Excess Penetrant.** Next, excess penetrant should be removed from the surface. The cleaning method is determined by the type of penetrant used. Some can be simply wiped off or washed away with water; others require the use of solvents. Uniform removal of excess penetrant is necessary for effective inspection, but overcleaning must be avoided.

4. **Development.** A developing agent is applied so that it forms a film over the surface. The developer acts as a blotter to assist the natural seepage of the penetrant out of surface openings and to spread it at the edges, so as to greatly magnify the apparent width of the flaw. The developer also provides a uniform background to assist visual inspection.

5. **Inspection.** After being sufficiently developed, the surface is visually examined for indications of penetrant bleedback from surface

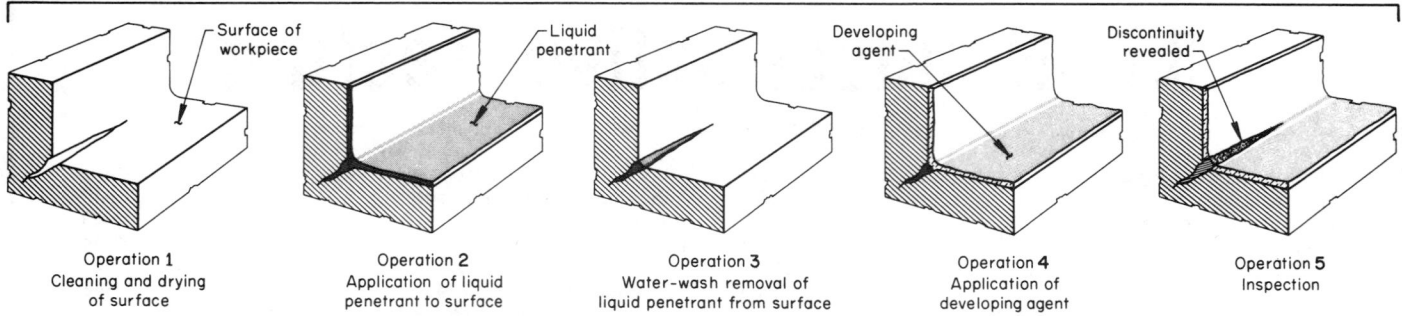

Fig. 1. Five essential operations for liquid-penetrant inspection using the water-washable system

openings. This examination must be performed in a suitable inspection environment. Visible-penetrant inspection is performed in good white light. When fluorescent penetrant is used, inspection is performed in a suitably darkened area using black (ultraviolet) light, which causes the penetrant to emit visible light.

These five essential operations are shown schematically for the water-washable system in Fig. 1. The operations are similar for the other liquid-penetrant systems.

PENETRANT SYSTEMS

Because of the vast differences among applications for liquid-penetrant inspection, it has been necessary to develop three basic penetrant systems. These three systems are broadly classified as (*a*) the water-washable system, (*b*) the post-emulsifiable system and (*c*) the solvent-removable system. The three systems are discussed below.

The water-washable penetrant system is designed so that the penetrant is directly water-washable from the surface of the workpiece; it does not require a separate emulsification step as do post-emulsifiable penetrant systems. It can be used to process workpieces quickly and efficiently. It is important, however, that the washing operation be carefully controlled, because water-washable penetrants are susceptible to overwashing. The degree and speed of removal depend on such processing conditions as spray-nozzle characteristics, water pressure and temperature, duration of rinse cycle, surface condition of the workpiece, and inherent removal characteristics of the penetrant employed. The essential operations entailed in this specific system are illustrated schematically in Fig. 1.

The Post-Emulsifiable System. To ensure detection of minute discontinuities in some materials, high-sensitivity penetrants that are not water-washable

are employed. Because they are not water-washable, the danger of washing the penetrant out of the flaws is reduced. These penetrants have an oil base and require an additional operation in the inspection process: application of an emulsifier after the penetrant has been applied and has been allowed the proper penetration (dwell) time. The emulsifier makes the penetrant soluble in water so that the excess penetrant can be removed by water rinsing. Therefore, the emulsification time must be carefully controlled so that the surface penetrant becomes water-soluble but penetrant in the flaws does not. Operations involved in the post-emulsifiable system (in addition to precleaning) are illustrated schematically in Fig. 2.

The Solvent-Removable System. Occasionally, it is necessary to inspect only a small area of a workpiece or to inspect a workpiece on the site rather than at a regular inspection station. For such situations, solvent-removable penetrants are available. Normally, the same type of solvent is used both for precleaning and for removal of excess penetrant. This penetrant process is convenient and broadens the range of applications of penetrant inspection. The operations for this process are illustrated schematically in Fig. 3.

The solvent-removable penetrants have an oil base. Optimum solvent removal is accomplished by wiping off as much of the excess penetrant as possible with a paper towel or a lint-free cloth, then slightly dampening a clean cloth with solvent and wiping off what remains. Final wiping with a dry paper towel or clean cloth is required.

The penetrant may also be removed by flooding the surface with solvent, in the same manner as for water-washable penetrants. The flooding technique is particularly useful for large workpieces, but it must be very carefully used to prevent removal of the penetrant from the flaws.

The solvent-removable system is used mainly for special applications; because it involves too much labor, it is not practical for production applications.

LIQUID-PENETRANT MATERIALS

There are two basic types of liquid penetrants: fluorescent and visible. Each type is obtainable for any one of the three systems (water-washable, post-emulsifiable or solvent-removable).

Fluorescent-penetrant inspection utilizes penetrants that fluoresce brilliantly under ultraviolet light. The sensitivity of a fluorescent penetrant depends on its ability to form indications that appear as small sources of light in an otherwise dark area. There are three basic sensitivity levels: regular or normal sensitivity, high sensitivity, and ultrahigh sensitivity.

Visible-penetrant inspection employs a penetrant that usually is red in color and that produces vivid red indications in contrast with the light background of the applied developer under visible light. The visible-penetrant process does not require the use of ultraviolet light. Visible-penetrant indications must be viewed, however, under adequate white light. Although the sensitivity of visible penetrants is not as great as that of fluorescent penetrants, it is adequate for many applications.

Water-washable penetrants are designed for removal by water rinsing after a suitable penetration (dwell) time. The emulsifier is "built in" to the water-washable penetrant. When this type of penetrant is used, it is extremely important that removal of excess surface penetrant be properly controlled to prevent overwashing, which may cause the penetrant to be washed out of the discontinuities.

Post-emulsifiable penetrants are insoluble in water and thus are not removable by water rinsing alone. They are designed to be selectively removed from the surface of a part by the use of a separate emulsifier. The emulsifier, properly applied, and left for a suitable emulsification time, combines with the excess surface penetrant to form a water-washable mixture that can be rinsed from the surface of the part. The penetrant that re-

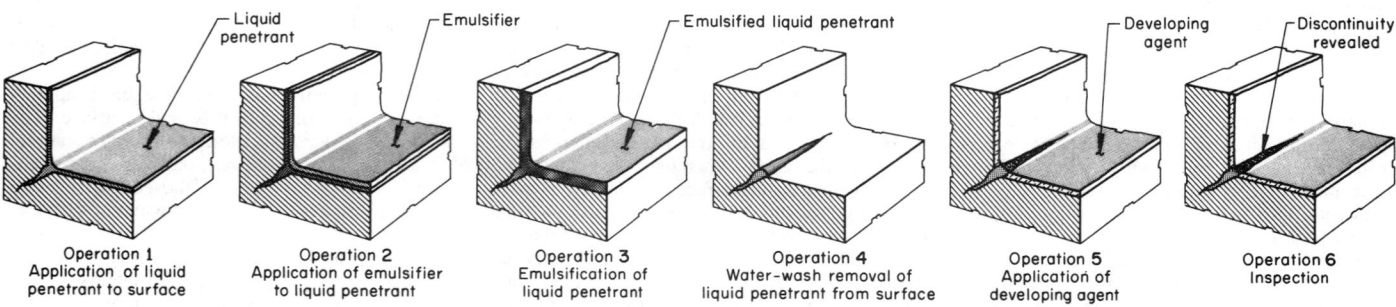

Fig. 2. Operations (in addition to precleaning) for the post-emulsifiable liquid-penetrant system

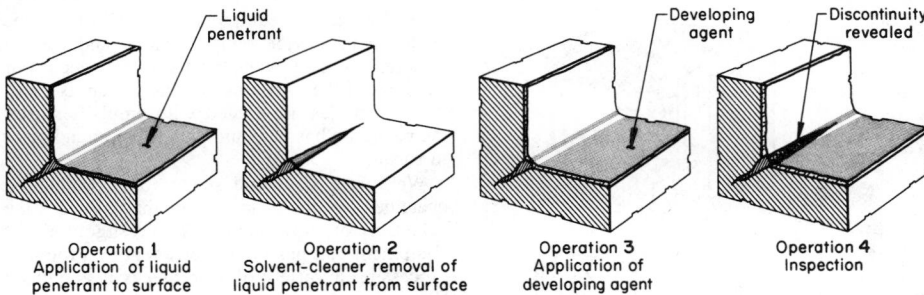

Fig. 3. Operations (in addition to precleaning) for the solvent-removable liquid-penetrant system

mains within the discontinuity is not subject to overwashing if the emulsifier is confined to the surface and if the discontinuity is tight (no mechanical rinsing).

Physical and Chemical Characteristics. Both fluorescent and visible penetrants, whether water-washable, post-emulsifiable or solvent-removable, must have certain chemical and physical characteristics if they are to perform their intended functions. Principal requirements of penetrants are as follows:

- Chemical stability and uniform physical consistency
- A flash point not lower than 60 °C (140 °F); penetrants that have lower flash points constitute a potential fire hazard.
- A high degree of wettability
- Low viscosity, to permit better coverage and minimum dragout
- Ability to penetrate discontinuities quickly and completely
- Sufficient brightness and permanence of color
- Chemical inertness with materials being inspected and with containers
- Low toxicity, to protect personnel
- Slow drying characteristics
- Ease of removal
- An inoffensive odor
- Low cost.

EMULSIFIERS

Emulsifiers are liquids used to render excess oily penetrant on the surface of a workpiece water-washable. There are two types of emulsifiers: oil-base and water-base.

Oil-base emulsifiers function by diffusion. The emulsifier film diffuses into the penetrant film and renders it spontaneously emulsifiable in water. The rate at which it diffuses into the oily penetrant establishes its emulsification time.

Water-base emulsifiers (hydrophilic removers) usually are supplied as liquid concentrates that are diluted in water to concentrations of 5 to 50% for dip-tank applications and of 0.05 to 5% for spray applications. Water-base emulsifiers function by displacing excess surface penetrant from the surface of the part by detergent action. The force of the water spray or air agitation of open dip tanks provides the scrubbing action while the detergent displaces the excess surface penetrant.

SOLVENT REMOVERS (CLEANERS)

Solvent removers, sometimes referred to as cleaners, differ from emulsifiers in that they remove excess surface penetrant through direct solvent action. The penetrant is dissolved by the

solvent remover. There are two basic types of solvent removers: flammable and nonflammable. Flammable cleaners are free of halogens, but are potential fire hazards. Nonflammable cleaners usually contain halogenated solvents, which render them unsuitable for some applications—usually because of their high toxicity or because they have undesirable effects on some materials.

DEVELOPERS

The amount of penetrant that emerges from a small surface opening is minute, and thus the visible evidence of its presence must be enhanced. Penetrants are made as brilliant and intense in color as possible. In addition, developers are used to spread the penetrant available at the defect, thus increasing the amount of light emitted, or the amount of contrast, that makes the defect visible to the unaided eye.

Developers act in several different ways, all of which serve to increase visibility. Dry developers consist of fine powder that is applied over the surface of the workpiece after the penetrant has been applied and the excess removed. The developer powder then performs the following functions:

- It has a blotting action, which serves to draw more penetrant from the surface opening.
- It provides a reflective base over which the penetrant can spread and disperse, thus increasing the amount of penetrant-covered surface exposed to the eye.
- It acts to cover up confusing background, and in some forms provides a complete background layer for contrast. Also, it reduces glare.
- It acts to remove from flaws penetrant that may contaminate the workpiece during further processing (for example, penetrant allowed to remain in a weld crack may cause trouble in rewelding).

There are four types of developers: dry, water-suspendible, water-soluble and solvent-suspendible.

Dry Developers

Dry powders were the first developers to be used with fluorescent penetrants. Ideally, dry-powder developers should be light and fluffy and should cling to dry metallic surfaces in a fine film. However, adherence of powder should not be excessive, because the amount of penetrant at fine flaws is so small that it cannot work through a thick coating of powder. Also, the powder should not float and fill the air with dust. Unfortunately, the powders that make the best dry developers do float to the extent that dusty air at the developer station is unavoidable.

Application and Removal of Dry Powders. Hand processing equipment usually includes a developer station, which for use with dry developers is an open tank. Workpieces are dipped into the powder; or powder is picked up with a scoop, or with the hands, and dropped onto the workpiece. Excess powder is removed by shaking and tapping the workpiece. Some powders are so light and fluffy that parts are dipped into them as easily as into a liquid.

Other effective methods of application make use of rubber spray bulbs or air-operated spray guns. An electrostatic-charged powder gun that can apply an extremely even and adherent coating of dry powder on metal parts also is used. For simple application—especially when only a portion of the surface of a large part is being inspected—a very soft bristle brush often is adequate.

Wet Developers

Wet developers are of three types: suspensions of developer powder in water (the most widely used), aqueous solutions of suitable salts, and suspensions of powder in volatile solvents.

Water-suspendible developers permit high-speed application of developer in mass inspection of small to medium-size workpieces by the fluorescent method. A basket of small, irregularly shaped workpieces that has gone through the steps of penetrant application, penetrant dwell and washing can be coated with developer in one quick dip in a water suspension. This method not only is quick, but also provides thorough and complete coverage of all surfaces of the pieces being inspected. No dry-powder application method has all these advantages to the same degree.

Wet developer is applied just after excess penetrant is washed away and immediately before drying. After drying, the surfaces are thus uniformly coated with a thin film of developer. Developing time is decreased because the heat from the drier helps to bring penetrant back out of surface openings and because, with the developer film already in place, the developing action proceeds at once. Workpieces are ready for inspection in a shorter period of time, before excessive bleedout from large openings takes place, and thus better definition of flaw indications often is obtained.

The material for water-suspendible developer is furnished as a dry powder, which is added to water in recommended proportions—usually from 0.04 to 0.12 kg/L ($^1/_3$ to 1 lb/gal).

Water-Soluble Developers. By using a material that is soluble in water, many of the problems inherent in suspension-type wet developers can be avoided. Some organizations, including the U.S. Air Force, have changed from water-suspended to water-soluble developers. Water-soluble developers are not, however, recommended for use with water-washable or visible penetrants.

Solvent-Suspendible Developers. The forerunner of modern solvent-suspendible developers (also called nonaqueous developers) was the whiting-alcohol mixture of the old kerosine-and-whiting method. The solvent technique is a very effective means of applying a smooth coating of developer to the workpiece surface. Because the solvents used are moderately quick-drying, there is very little running of developer, even on vertical surfaces, and uniform coating is not difficult to obtain. The solvent may or may not also be a solvent for the penetrant. It is sometimes a partial solvent, at least, for the dye in the penetrant.

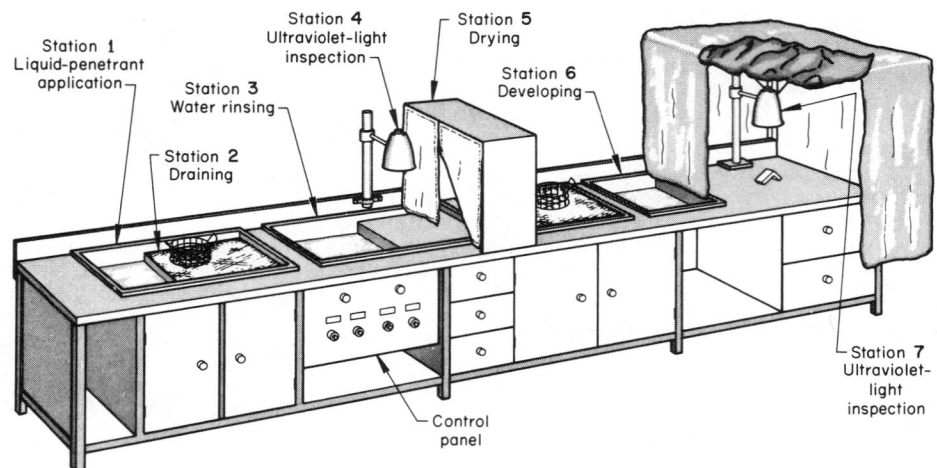

Fig. 4. Typical seven-station "package" equipment unit for inspecting workpieces by the water-washable fluorescent-penetrant system

With fluorescent penetrants, this type of developer is used primarily in portable kits, with spray cans, but is seldom used for inspection of large lots. It is almost universally used with color-contrast (visible) penetrants.

On rough surfaces, solvent developers sometimes react unfavorably with very brilliant fluorescent penetrants. They draw out small traces of fluorescent material remaining in rough spots and cause undesirable over-all background glow.

Chlorinated solvents have been used extensively for this type of developer. They have the advantage of being nonflammable, but the disadvantage of being toxic; thus, their use is often prohibited.

The powder in a solvent-suspendible developer must possess the usual properties of a developer. In addition, in the color-contrast system, it must form the uniform, dead-white layer that serves as an opaque background for the colored-dye indications. Chalk, and mixtures of chalk with other white powders, are commonly used to improve the color and texture of the background.

Solvent developers are almost always premixed by the manufacturer to the optimum concentration. The exact ratio of powder to solvent is not extremely critical, but cans of mixed developer must be kept tightly closed to prevent evaporation of solvent.

Solvent developers are sometimes applied with a paintbrush, but this is likely to result in smeared indications; application by a pressure spray can is a preferred method.

Selection of Developer

Because developers play such an important role in penetrant inspection, it is important that the right one be used for a given job. It has been found, for instance, that on very smooth or polished surfaces, dry powder does not adhere satisfactorily, and wet developers do a better job. On the other hand, on very rough surfaces dry powder is far more effective.

Following are a few general rules regarding choice of developers:

- Use wet developer in preference to dry on very smooth surfaces.
- Use dry developer in preference to wet on very rough surfaces.
- Wet developers are better suited to high-production inspection of small workpieces, because of their greater ease and speed of application.
- Wet developers cannot be used reliably where sharp fillets unavoidably accumulate developer so as to mask indications of flaws.
- Solvent developers are effective for revealing fine, deep cracks, but are not satisfactory for finding wide, shallow flaws.
- Cleaning and reinspecting a rough surface is difficult if a wet developer was used for a prior inspection.

It should always be remembered that the developer does not produce indications but simply absorbs the penetrant already present in or at the flaw and makes it more visible.

EQUIPMENT REQUIREMENTS

With the exception of a source of ultraviolet ("black") light for use with fluorescent penetrants, there is no special equipment that is absolutely essential for liquid-penetrant inspection. Reasonably effective inspection operations have been performed with a minimum of simple and relatively crude equipment.

From a more practical standpoint, however, the above approach should be considered only when: (a) no more than a few workpieces are involved; (b) specific portions of very large workpieces are being inspected; (c) maximum sensitivity is not required; or (d) inspection must be performed in the field. Therefore, most liquid-penetrant inspection is done with equipment designed specifically for the purpose.

A variety of equipment is available. "Package units" that incorporate all the necessary stations and controls are widely used, especially where relatively small workpieces in a variety of sizes and shapes are being inspected. A typical package unit for inspection by the water-washable fluorescent-penetrant system is shown in Fig. 4. This system is designed to process a steady flow of workpieces, which move through seven stations: application of penetrant, draining of excess penetrant, water rinsing, inspection under ultraviolet light to check thoroughness of rinsing, drying, application of developer, and final ultraviolet-light inspection for flaws. This unit does not include stations for preliminary cleaning or postcleaning; these operations are often done in

another area. The equipment shown in Fig. 4 is available in a wide range of sizes and can be modified in many ways to fit specific needs. For example, if the post-emulsifiable system is used, the workpieces are coated with emulsifier after the penetrant has been allowed to drain and prior to rinsing.

Workpiece sizes and shapes, and production quantities, are the major factors that influence the selection of equipment. An arrangement of equipment that has proved efficient in a foundry processing a variety of workpieces is shown in Fig. 5.

Facilities for Precleaning. Regardless of size, shape or composition, workpieces must be thoroughly cleaned and dried prior to liquid-penetrant inspection. Cleaning equipment must be selected to fit both the workpiece and the type of contaminants to be removed. A variety of such equipment is available commercially. Examples are vapor degreasers, sand and grit blasters, water-detergent washing machines, solvent or chemical tanks, and sprays.

Penetrant Station. The principal requirement of a penetrant station is that it provide a means for coating workpieces with penetrant—either all over, for small workpieces, or over small areas of large workpieces when only local inspection is required. In addition, a means should be provided for draining excess penetrant back into the penetrant reservoir, unless the expendable technique is being used. Draining racks usually serve the additional purpose of providing a storage place for parts during the time required for penetration (dwell time).

Emulsifier Station. The emulsifier liquid is contained in a tank of sufficient size and depth to permit immersion of the workpieces, either individually or in batches. Usually provided are covers to reduce evaporation, and drain valves for cleanout when the bath has to be renewed. Suitable drain racks are also a part of this station, to permit excess emulsifier to drain back into the tank.

If large workpieces must be coated with emulsifier, methods must be devised to achieve the fastest possible coverage. Multiple spraying or copious flowing of emulsifier from troughs or perforated pipes can be used on some types of automatic equipment. For local coating of large workpieces, spraying often is satisfactory, using the expendable technique described for application of penetrant.

Rinse Station. Water rinsing (washing) of small workpieces is frequently done by hand, either individually or in batches in wire baskets. The workpieces are held in the wash tank and cleaned with a hand-held spray using water at tap pressure and temperature (water temperature should not, however, be below 10 °C, or 50 °F).

Drying Station. The recirculating hot-air drier is one of the most important equipment components. The drier must be large enough to handle easily the type and number of workpieces being inspected. Heat input, air flow, and rate of movement of workpieces through the drier, as well as temperature control, are all factors that must be balanced. The drier may be of the cabinet type illustrated in Fig. 4, or it may be designed so that the workpieces pass through on a conveyor (Fig. 5). If conveyor operation is used, the speed must be coordinated with the required drying cycle.

Developer Station. The type and location of developer station depend on whether dry or wet

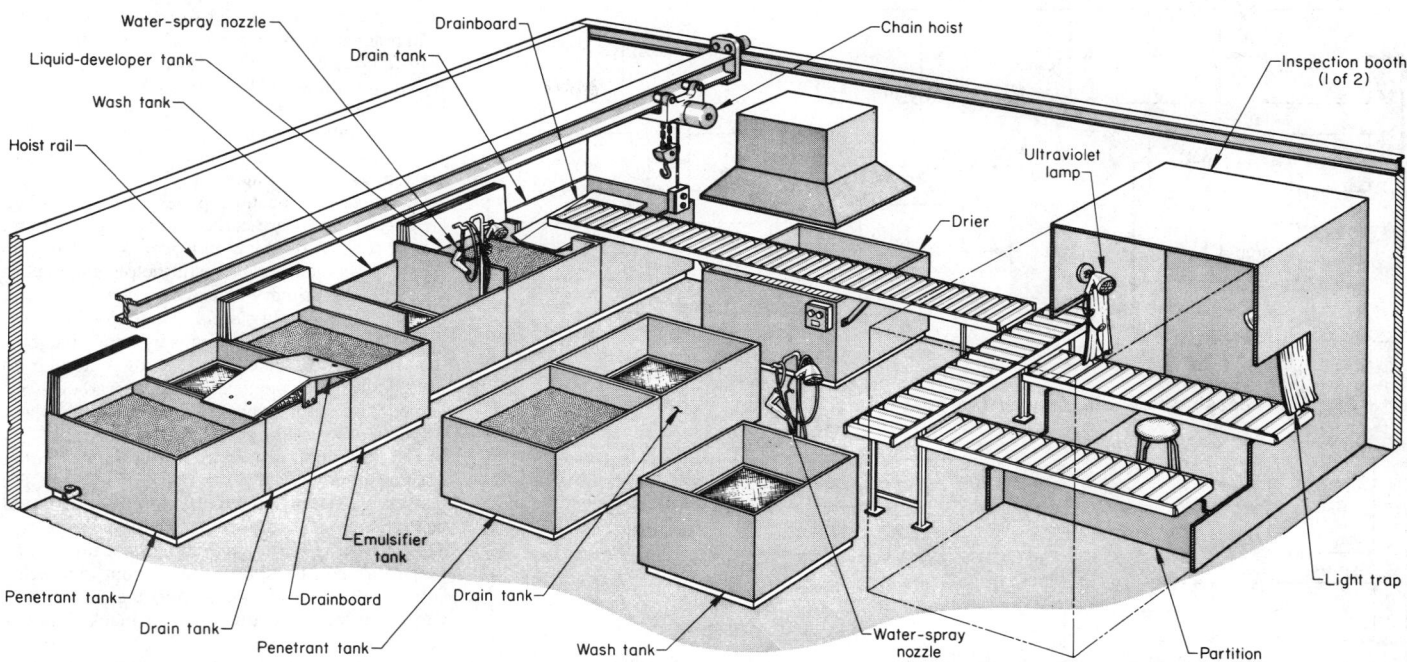

Fig. 5. Arrangement of equipment used in one foundry for liquid-penetrant inspection of a large variety of castings to rigid specifications. Many of the castings require handling by crane or roller conveyor.

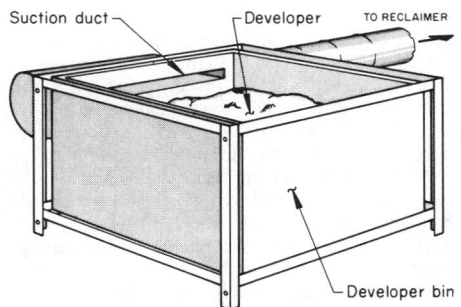

Fig. 6. Dry-developer bin equipped with dust-control and reclaimer system

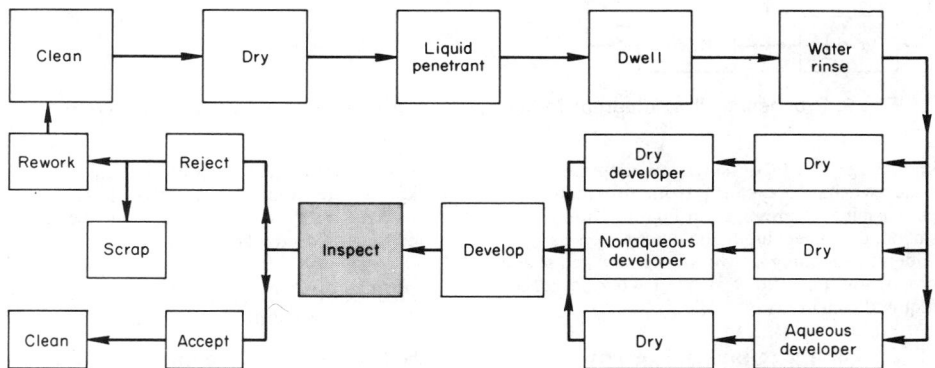

Fig. 7. Processing flow diagram for the water-washable liquid-penetrant system

developer is to be used. For dry developer, the developer station is downstream from the drier, whereas for wet developer it immediately precedes the drier, following the rinse station.

The dry-developer station usually consists of a simple bin containing the powder. Dried workpieces are dipped into the powder and the excess powder is shaken off. Larger workpieces may not be so easily immersed in the powder, so a scoop is usually provided for throwing powder over the surfaces, after which the excess is shaken off. The developer bin should be equipped with an easily removable cover to protect the developer from dust and dirt when not in use.

Dust-control systems are sometimes needed when dry developer is used (Fig. 6).

Inspection Station. Essentially, the inspection station is simply a worktable on which workpieces can be handled under proper lighting. For fluorescent methods, the table is usually surrounded by a curtain or hood to exclude most of the white light from the area (see Fig. 4). For visible penetrants, a hood is not necessary.

Black (ultraviolet) lights may be batteries of 100- or 400-watt lamps for area lighting, or, in small stations, may be one or two 100-watt spot lamps mounted on brackets from which they can be lifted and moved about by hand. Because of the heat given off by black lights, good air circulation is essential in black-light booths.

For automatic inspection, workpieces are moved through booths equipped with split curtains, either by hand or by conveyor (see Fig. 5).

Postcleaning Station. Postinspection cleaning often is necessary to remove all traces of penetrant and developer. Equipment for postcleaning is available commercially, and special designs are seldom required.

SELECTION OF PENETRANT SYSTEM

Size, shape and weight of workpieces, as well as number of similar workpieces to be inspected, often influence the selection of a penetrant system.

Sensitivity and Cost. The desired degree of sensitivity, and cost, usually are the most important factors in selecting a system. The methods ca-

pable of the greatest sensitivity are also the most costly. There are many inspection operations that require the ultimate in sensitivity, but there are also many where extreme sensitivity not only is not required but may produce misleading results and thus is undesirable.

On a practical basis, the three major penetrant systems are broken down into six systems or variations of systems. Following is a listing of these six systems, in order of decreasing sensitivity and decreasing cost.

1. Post-emulsifiable fluorescent
2. Solvent-removable fluorescent
3. Water-washable fluorescent
4. Post-emulsifiable visible
5. Solvent-removable visible
6. Water-washable visible.

PROCESSING CYCLES FOR WATER-WASHABLE SYSTEMS

After the workpieces have been precleaned, processing for penetrant inspection should begin

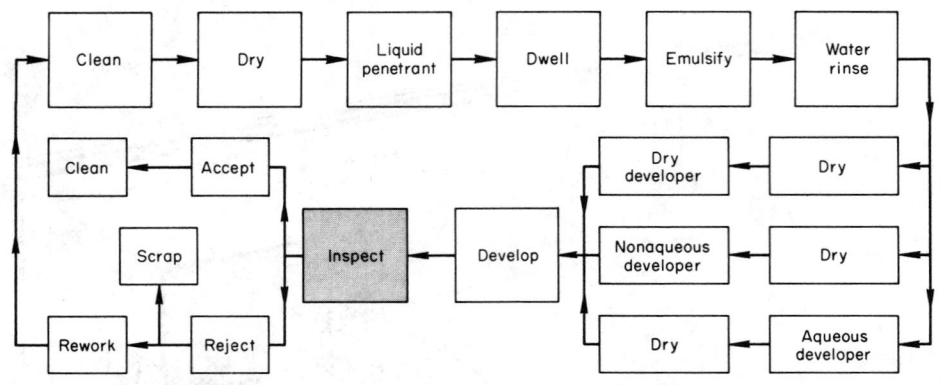

Fig. 8. Processing flow diagram for the post-emulsifiable liquid-penetrant system

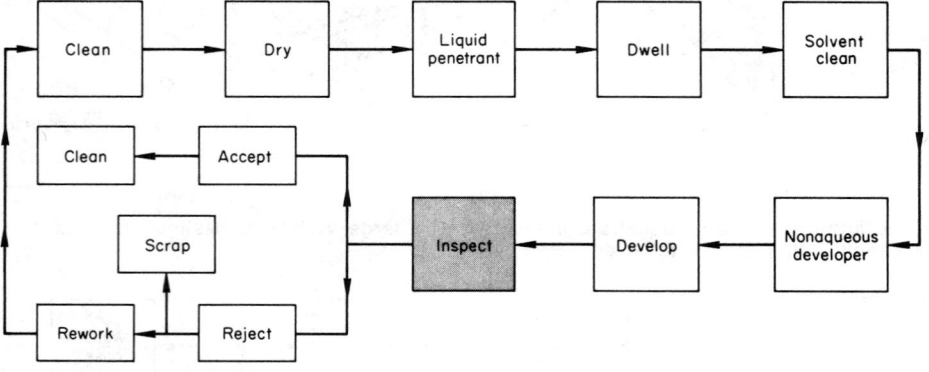

Fig. 9. Processing flow diagram for the solvent-removable liquid-penetrant system

immediately. A processing flow diagram for the water-washable system, from precleaning to postcleaning, is presented in Fig. 7. Time in each station, equipment used, and other factors can vary widely, depending on workpiece size and shape, production quantities of similar workpieces, and required sensitivity.

PROCESSING CYCLES FOR POST-EMULSIFIABLE SYSTEMS

As indicated by the flow diagram in Fig. 8, the processing cycle for the post-emulsifiable system is the same as for the water-washable sys-

tem (see preceding description) except that the workpieces are dipped or otherwise treated with the emulsifier immediately following the dwell period that succeeds the penetrant treatment.

PROCESSING CYCLES FOR SOLVENT-REMOVABLE SYSTEMS

The basic sequence of operations for the solvent-removable liquid-penetrant system is generally similar to that followed for the other systems. A typical sequence is shown by the flow diagram in Fig. 9. A notable difference is that with the solvent-removable system, excess pen-

etrant is very often wiped from the workpiece.

In practice, however, the techniques of the solvent-removable system are likely to be quite different from those of the other systems—largely because of differences in application. The solvent-removable system finds its greatest use when inspection must be done in areas where it is not convenient to use water.

For instance, the tube plates used in nuclear steam generators are carbon steel forgings more than 0.6 m (2 ft) thick and over 3 m (10 ft) in diameter. Because one side of the plate is in contact with the primary-side water, that side must be clad with stainless steel. One way in which such cladding has been done is by explosive joining. Because the outer edges of the cladding do not always bond completely with the carbon steel, and because of the critical role played by the plate, inspection of bonding integrity is essential.

The inspection procedure follows a machining cut that removes about 25 mm (1 in.) from the periphery. The machined circumference is tested using the solvent-removable visible-dye system. This system is the logical choice because it can be performed in a short time without the necessity of removing the plate from the machine or darkening the inspection area for black-light examination.

TECHNIQUES USING PORTABLE EQUIPMENT

Liquid-penetrant inspection is one of the inspection processes in which the testing materials can be brought to the work site whenever the workpiece cannot be brought to the inspection area. Portable penetrant inspection, by the solvent-removable system, is used extensively to inspect welded joints where only limited areas are being examined and where total immersion of the weldment would be impractical. In general, the methods and procedures used with portable equipment are essentially the same as those used with fixed, immersion-type inspection equipment.

Portable kits are available for both the fluorescent and the visible solvent-removable systems. The three components—cleaner, penetrant and developer—are packaged both in bulk containers and in aerosol spray cans. The choice between spray and swabbing depends on the type and size of defect anticipated, the size of the area to be tested, and the degree of precleaning and postcleaning to be performed.

Magnetic-Particle Inspection

Edited by Ron Mason, Multispec

MAGNETIC-PARTICLE INSPECTION is a method for locating surface and subsurface discontinuities in ferromagnetic materials. It depends for its operation on the fact that when the material or part under test is magnetized, discontinuities that lie in a direction generally transverse to the direction of the magnetic field will cause a leakage field to be formed at and above the surface of the part. The presence of this leakage field, and therefore the presence of the discontinuity, is detected by the use of finely divided ferromagnetic particles applied over the surface, some of the particles being gathered and

held by the leakage field. This magnetically held collection of particles forms an outline of the discontinuity and generally indicates its location, size, shape and extent. Magnetic particles are applied over a surface as dry particles, or as wet particles in a liquid carrier such as water or oil.

Nonferromagnetic materials cannot be inspected by this method. Such materials include aluminum alloys, magnesium alloys, copper and copper alloys, lead, titanium and titanium alloys, and austenitic stainless steels.

Applications. The principal industrial uses of magnetic-particle inspection are final inspection,

receiving inspection, in-process inspection and quality control, maintenance and overhaul in the transportation industries, plant and machinery maintenance, and inspection of large components.

Although in-process magnetic-particle inspection is used to detect discontinuities and imperfections in material and parts as early as possible in the sequence of operations, final inspection is needed to ensure that rejectable discontinuities or imperfections detrimental to the use or function of the part have not developed during processing.

Advantages. The magnetic-particle method is a

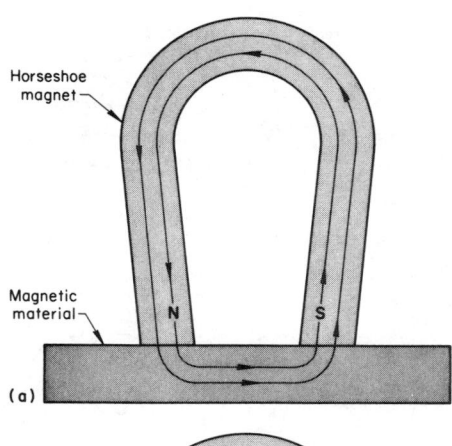

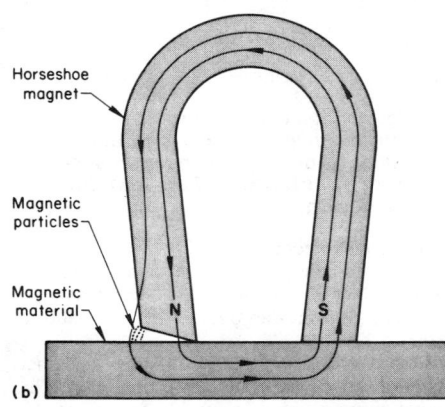

Fig. 1. (a) Horseshoe magnet with a bar of magnetic material across poles, forming a closed, ringlike assembly, which will not attract magnetic particles. (b) Ringlike magnet assembly with an air gap, to which magnetic particles are attracted.

sensitive means of locating small and shallow surface cracks in ferromagnetic materials. Indications may be produced at cracks that are large enough to be seen by the naked eye, but exceedingly wide cracks will not produce a particle pattern if the surface opening is too wide for the particles to bridge.

Discontinuities that do not actually break through the surface also are indicated in many instances by this method, although certain limitations must be recognized and understood. If a discontinuity is fine and sharp and close to the surface, such as a long stringer of nonmetallic inclusions, a sharp indication can be produced.

If the discontinuity lies deeper, the indication is less distinct.

Limitations. Magnetic-particle inspection has certain limitations that the operator must be aware of; for instance, thin coatings of paint and other nonmagnetic coverings, such as plating, adversely affect sensitivity of magnetic-particle inspection. Other limitations are:

- This method will work only on ferromagnetic materials.
- For best results, the magnetic field must be in a direction that will intercept the principal plane of the discontinuity at right angles. Sometimes this requires two or more sequential inspections with different magnetizations.
- Demagnetization following inspection is often necessary.
- Postcleaning to remove remnants of the magnetic particles or carrying solutions clinging to the surface may sometimes be required after testing and demagnetization.
- Exceedingly large currents sometimes are required for very large parts.
- Care is necessary to avoid local heating and burning of finished parts or surfaces at the points of electrical contact.
- Although magnetic-particle indications are easily seen, experience and skill in interpreting their significance sometimes are needed.

DESCRIPTION OF MAGNETIC FIELDS

Magnetized Ring. When a magnetic material is placed across the poles of a horseshoe magnet having square ends, forming a closed or ringlike assembly, the lines of force flow from the north pole through the magnetic material to the south pole (see Fig. 1a). (Magnetic lines of force flow preferentially through magnetic material rather than through nonmagnetic material or air.) The magnetic lines of force will be enclosed within the ringlike assembly because no external poles exist, and iron filings or magnetic particles dusted over the assembly are not attracted to the magnet even though there are lines of magnetic force flowing through it.

If one end of the magnet is not square and an air gap exists between that end of the magnet and the magnetic material, the poles will still attract magnetic materials. Magnetic particles will cling to the poles and bridge the gap between them, as shown in Fig. 1(b). Any radial crack in a circularly magnetized piece will create a north and a south magnetic pole at the edges of a crack. Magnetic particles will be attracted to the poles created by such a crack, forming an indication of the discontinuity in the piece.

The fields set up at cracks or other physical or magnetic discontinuities in the surface are called leakage fields. The strength of a leakage field determines the number of magnetic particles that will gather to form indications: strong indications are formed at strong fields, weak indications at weak fields. The density of the magnetic field determines its strength and is partly governed by the shape, size and material of the part being inspected.

Magnetized Bar. A straight piece of magnetized material (bar magnet) has a pole at each end. Magnetic lines of force flow through the bar from the south pole to the north pole. Because the magnetic lines of force within the bar magnet run the length of the bar, it is said to be longitudinally magnetized or to contain a longitudinal field.

If a bar magnet is broken into two pieces, a leakage field with north and south poles is created between the pieces, as shown in Fig. 2(a). This field exists even if the fracture surfaces are brought together (see Fig. 2b). If the magnet is cracked but not broken completely in two, a somewhat similar result occurs. A north and a south pole form at opposite edges of the crack, just as though the break were complete (see Fig. 2c). It is this field that attracts the iron particles that outline the crack. The strength of these poles will be different from that of the fully broken pieces, and will be a function of the depth of the crack and the width of the air gap at the surface.

Circular Magnetization. Electric current passing through any straight conductor such as a wire or bar creates a circular magnetic field around the conductor. When the conductor of electric current is a ferromagnetic material, the passage of current induces a magnetic field in the conductor as well as in the surrounding space. A part magnetized in this manner is said to have a circular field or to be circularly magnetized, as shown in Fig. 3(a).

Longitudinal Magnetization. Electric current also can be used to create a longitudinal magnetic field in magnetic materials. When electric current is passed through a coil of one or more turns, a magnetic field is established lengthwise or longitudinally within the coil, as shown in Fig. 3(b). The nature and direction of the field around the conductor that forms the turns of the coil produce longitudinal magnetization.

Effect of Flux Direction. To form an indication, the magnetic field must approach a discontinuity at an angle great enough to cause the magnetic lines of force to leave the part and return after bridging the discontinuity. For best results, an intersection approaching 90° is desirable. For this reason, the direction, size and shape of the discontinuity are important. The direction of the magnetic field also is important for optimum results, and so is the strength of the field in the area of the discontinuity.

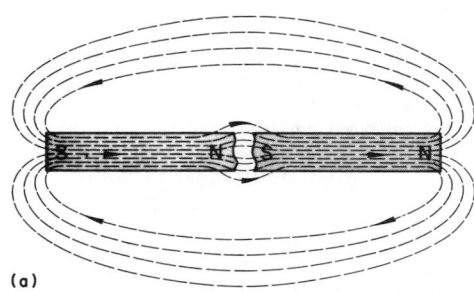

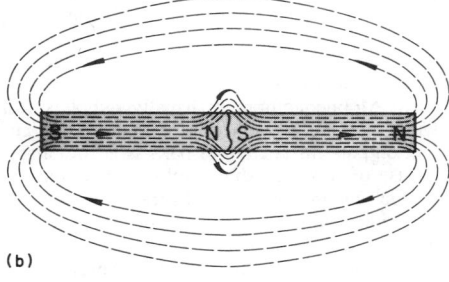

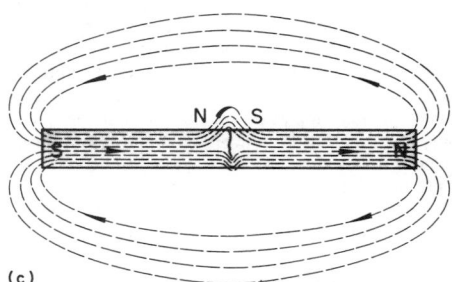

Fig. 2. Leakage fields between two pieces of a broken bar magnet: (a) with magnet pieces apart, and (b) with magnet pieces together (which would simulate a flaw). (c) Leakage field at a crack in a bar magnet.

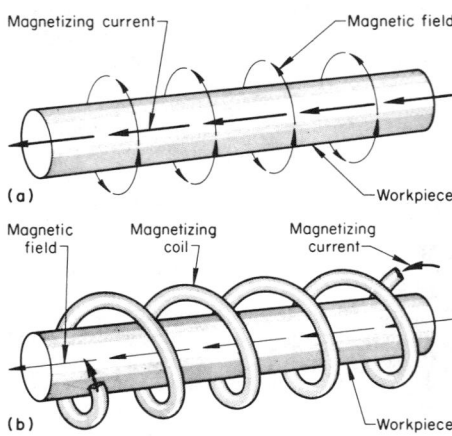

Fig. 3. Magnetized bars showing directions of magnetic field: (a) circular and (b) longitudinal

Figure 4(a) illustrates a condition where the current is passed through the part, causing formation of a circular field around the part. Under normal circumstances, a discontinuity such as that shown as A in Fig. 4(a) would give no indication of its presence, because it is regular in shape and lies in a direction parallel to that of the magnetic field. If the discontinuity had an irregular shape but was predominantly parallel to the magnetic field, such as at B, there is a good possibility that a weak indication would form. Where the predominant direction of the discontinuity is at a 45° angle to the magnetic field, such as at C, D and E, the conditions are more favorable for detection regardless of the shape of the discontinuity. Discontinuities whose predominant directions, regardless of shape, are at a 90° angle to the magnetic field produce the most pronounced indications (see discontinuities F, G and H).

A bar that has been longitudinally magnetized is shown in Fig. 4(b). Discontinuities L, M and N, which are at about 45° to the magnetic field, would produce detectable indications, as they would with a circular field. Discontinuities J and K would display pronounced indications, but discontinuities P, Q and R would probably not be detected.

Magnetiziation Methods. In magnetic-particle inspection, the magnetic particles may be applied to the part while the magnetizing current is flowing or after the current has ceased to flow, depending largely on the retentivity of the part. The first technique is known as the "continuous method"; the second, as the "residual method."

If the magnetism remaining in the part after the current has been turned off for a period of time (residual magnetism) does not provide a leakage field strong enough to produce readable indications when magnetic particles are applied to the surface, the part must be continuously magnetized during application of the particles. Consequently, the residual method can be used only on materials having sufficient retentivity; usually the harder the material, the higher the retentivity. The continuous method is the only method used on low-carbon steels or iron having little or no retentivity.

MAGNETIZING CURRENT

Both direct current and alternating current are suitable for magnetizing parts for magnetic-par-

ticle inspection. the strength, direction and distribution of magnetic fields are greatly affected by the type of current used for magnetization.

The fields produced by direct and alternating current differ in many characteristics. The difference that is of importance in magnetic-particle inspection is that fields produced by direct current generally penetrate the cross section of the part, whereas fields produced by alternating current are confined to the metal at or near the surface of the part, which commonly is known as the skin effect. Therefore, alternating current should not be used in searching for subsurface discontinuities.

Direct Current. The most satisfactory source of direct current is the rectification of alternating current. Both single-phase and three-phase alternating current are furnished commercially. By the use of rectifiers, the reversing alternating current can be converted to unidirectional current, and when three-phase alternating current is so rectified, the delivered direct current is entirely the equivalent of straight direct current for purposes of magnetic-particle inspection. The only difference between rectified three-phase alternating current and straight direct current is a slight ripple in the value of the rectified current, amounting to only about 5% of the maximum current value.

Alternating Current, which must be single-phase when used directly for magnetizing purposes, is taken from commercial power lines and usually has a frequency of 50 or 60 Hz. When used for magnetizing, line voltage is stepped down, by means of transformers, to the low voltages required. At these low voltages, magnetizing currents of several thousand amperes often are used.

One problem encountered when alternating current is used is that the resultant residual magnetism in the part may not be at a level as high as that of the magnetism generated by the peak current of the alternating-current cycle. This is because the level of residual magnetism depends on where in the cycle the current was discontinued.

POWER SOURCES

At present, portable, mobile and stationary equipment is available; selection from among these types depends on the nature and location of testing.

Portable equipment is available in lightweight (16- to 41-kg, or 35- to 90-lb) power-source units that can be readily taken to the inspection site. Generally, these portable units are designed to use 115-, 230- or 460-V alternating current and supply magnetizing-current outputs of 750 to 1500 A in half-wave or alternating current. Small, lightweight pulsed dc units also are available which can produce up to 7000 A of magnetizing current.

Mobile units generally are mounted on wheels to facilitate their being moved to various inspection sites. Mobile equipment usually supplies full-wave, half-wave or alternating magnetizing-current outputs. Inspection of parts is accomplished by use of flexible cables, yokes, prod contacts, contact clamps and coils. Instruments and controls are mounted on the front of the unit. Magnetizing current usually is controlled by a remote-control switch connected to the unit by an electric cord. Quick-coupling connectors for connecting magnetizing cables are on the front of the unit.

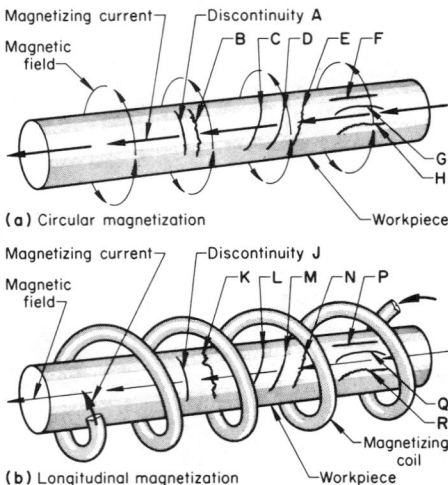

(a) Circular magnetization

(b) Longitudinal magnetization

Fig. 4. Effect of direction of magnetic field or flux flow on detectability of discontinuities with various orientations. See text for discussion.

Mobile equipment usually is powered by single-phase or three-phase, 60-Hz alternating current (230 or 460 V), depending on current requirements, and has an output range of 1500 to 10 000 A.

Stationary equipment may be obtained as either general-purpose or special-purpose units. The general-purpose unit is primarily for use in the wet method, and has a built-in tank that contains the bath pump, which continuously agitates the bath and forces the fluid through hoses onto the part being inspected. Pneumatically operated contact heads, together with a rigid-type coil, provide capabilities for both circular and longitudinal magnetization. Self-contained ac or dc power supplies are available in amperage ratings from 1000 to 10 000 A.

Stationary power packs serve as sources of high-amperage magnetizing current to be used in conjunction with special fixtures, or with cable-wrap or clamp-and-contact techniques. Rated output varies from a customary 4000 to 6000 A to as high as 30 000 A. The higher-amperage units are used for over-all magnetization of large forgings or castings that otherwise would require systematic prod inspection at much lower current levels. Some units feature three output circuits that are systematically energized in rapid sequence, either electrically or mechanically, for effectively magnetizing a part in several directions at virtually the same time. This permits disclosure of discontinuities lying in any direction after only a single processing step.

Special-purpose stationary units are designed for handling and inspecting large quantities of similar items. Generally, conveyors, automatic markers, and alarm systems are included in such units to expedite handling and disposition of parts.

METHODS OF GENERATING MAGNETIC FIELDS

One of the basic requirements of magnetic-particle inspection is that the part undergoing inspection be properly magnetized so that leakage fields created by discontinuities will attract the magnetic particles. Permanent magnets serve some useful purpose in this respect, but generally mag-

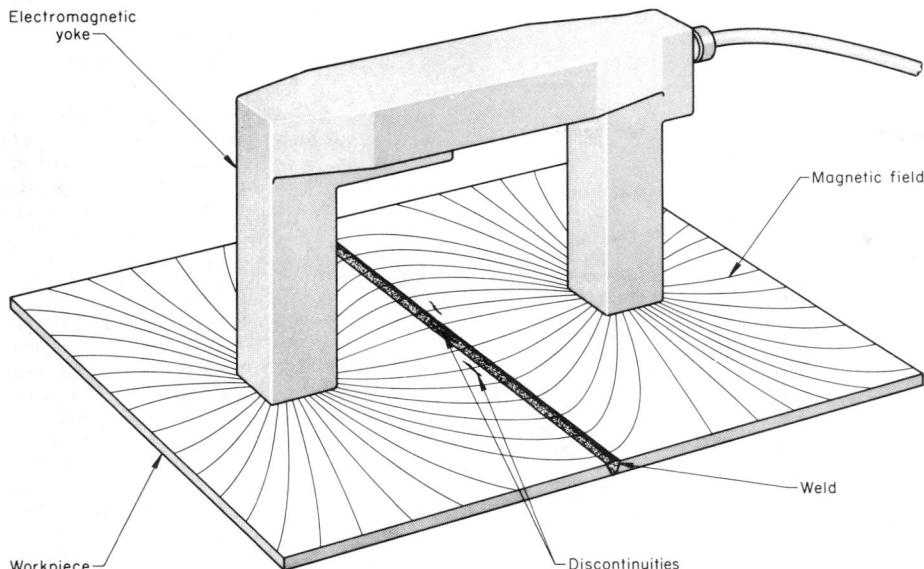

Fig. 5. Electromagnetic yoke, showing position and magnetic field for detection of discontinuities parallel to a weld bead. Discontinuities across a weld bead may be detected by placing the contact surfaces of the yoke next to and on either side of the bead (rotating yoke about 90° from position shown here).

netization is produced by electromagnets or the magnetic field associated with the flow of electric current. Basically, magnetization is derived from the circular magnetic field generated when an electric current flows through a conductor. The direction of this field is dependent on the direction of current flow.

Yokes. There are two basic types of yokes that are commonly used for magnetizing purposes: permanent-magnet and electromagnetic yokes. Both are hand held, and thus are quite mobile.

Electromagnetic yokes (see Fig. 5) consist of a coil wound around a U-shape core of soft iron. The legs of the yoke can be either fixed or adjustable. Adjustable legs permit changing the contact spacing and the relative angle of contact to accommodate irregular-shape parts. Unlike a permanent-magnet yoke, an electromagnetic yoke can readily be switched on or off. This feature makes it convenient to apply and remove the yoke from the test piece.

The design of an electromagnetic yoke can be based on the use of either direct or alternating current, or both. The flux density of the magnetic field produced by the direct-current type can be changed by varying the amount of current in the coil. The direct-current type of yoke has greater penetration whereas the alternating-current type concentrates the magnetic field at the surface of the test piece, providing good sensitivity for the disclosure of surface discontinuities over a relatively narrow area. Yokes utilizing alternating current for magnetization have numerous applications, and can be used for demagnetization as well. In general, discontinuities to be disclosed should be centrally located in the area between pole pieces and oriented perpendicular to an imaginary line connecting them (see Fig. 5).

In operation, the part completes the magnetic path for the flow of magnetic flux. The yoke is a source of magnetic flux, and the part becomes the preferential path completing the magnetic circuit between the poles. (In Fig. 5, only those portions of the flux lines near the poles are shown.)

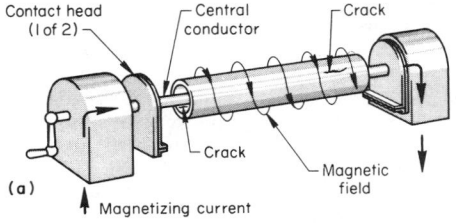

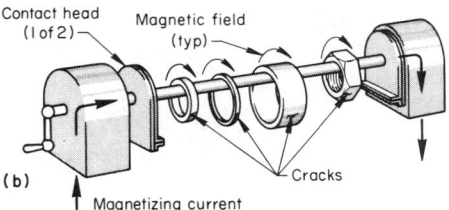

Fig. 6. Use of central conductors for circular magnetization of (a) long hollow cylindrical parts and (b) short hollow cylindrical or ringlike parts, for detection of discontinuities on inner and outer surfaces

Coils. Single-loop and multiple-loop coils (conductors) are used for longitudinal magnetization of components (see Fig. 3b and 4b). The field within the coil has a definite direction, corresponding to the direction of the lines of force running through it. The flux density passing through the interior of the coil is proportional to the product of the current, I, in amperes, and the number of turns in the coil, N. Thus, the magnetizing force of such a coil can be varied by varying either the current or the number of turns in the coil.

For large parts, a coil may be produced by winding several turns of a flexible cable around the part; however, care must be taken to ensure that no indications are concealed beneath the cable.

Portable magnetizing coils are available that can be plugged into an electrical outlet. These coils can be used for in-place inspection of shaft-like parts in railroad shops, aircraft-maintenance shops, and shops for automobile, truck and tractor repair. Transverse cracks in spindles and shafts are easily detected with use of such coils.

Most coils used for magnetizing are short, especially those wound on fixed frames. The relation of the length of the part being inspected to the width of the coil must be considered. For a simple part, the effective over-all distance that can be inspected is 150 to 225 mm (6 to 9 in.) on either side of the coil. Thus, a part 300 to 450 mm (12 to 18 in.) long can be inspected using a normal coil approximately 25 mm (1 in.) thick. In testing longer parts, either the part must be moved at regular intervals through the coil, or the coil must be moved along the part.

The ease with which a part can be longitudinally magnetized in a coil is significantly related to the length-to-diameter (L/D) ratio of the part. This is due to the demagnetizing effect of the magnetic poles that are set up at the ends of the part. This demagnetizing effect is pronounced for L/D ratios of less than 10 to 1, and very significant for ratios of less than 3 to 1. Where the L/D ratio is extremely unfavorable, pole pieces of similar cross-sectional area may be introduced to effectively increase the length of the part and consequently improve the L/D ratio.

Central Conductors. For many tubular or ring-shape parts, it is advantageous to use a separate conductor to carry the magnetizing current, rather than the part itself. Such a conductor, commonly referred to as a central conductor, is threaded through the inside of the part (see Fig. 6) and is a convenient means of circularly magnetizing a part without the need for making direct contact to the part itself. Central conductors are made of solid and tubular nonmagnetic and ferromagnetic materials and are good conductors of electricity.

The basic rules regarding magnetic fields around a circular conductor carrying direct current are:

- The magnetic field outside a conductor of uniform cross section is uniform along the length of the conductor.
- The magnetic field is 90° to the path of the current through the conductor.
- The flux density outside the conductor varies inversely with the radial distance from the center of the conductor.

Direct-Contact Method. For small parts having no openings through the interior, circular magnetic fields are produced by direct contact to the part. This is done by clamping the parts between contact heads (head shot), generally on a bench unit (see Fig. 7) that incorporates the source of current. A similar unit can be used for supplying the magnetizing current to a central conductor (see Fig. 6).

The contact heads must be so constructed that the surfaces of the part are not damaged—either physically by pressure, or structurally by heat from arcing or from high resistance at the points of contact. Such heat can be especially damaging to hardened surfaces such as bearing races.

For complete inspection of a complex part, it may be necessary to attach clamps at several points on the part or to wrap cables around the part to get fields in the proper directions at all points on the surface. This often requires several magnetizations. Multiple magnetizations may be minimized by use of the "over-all" magnetization

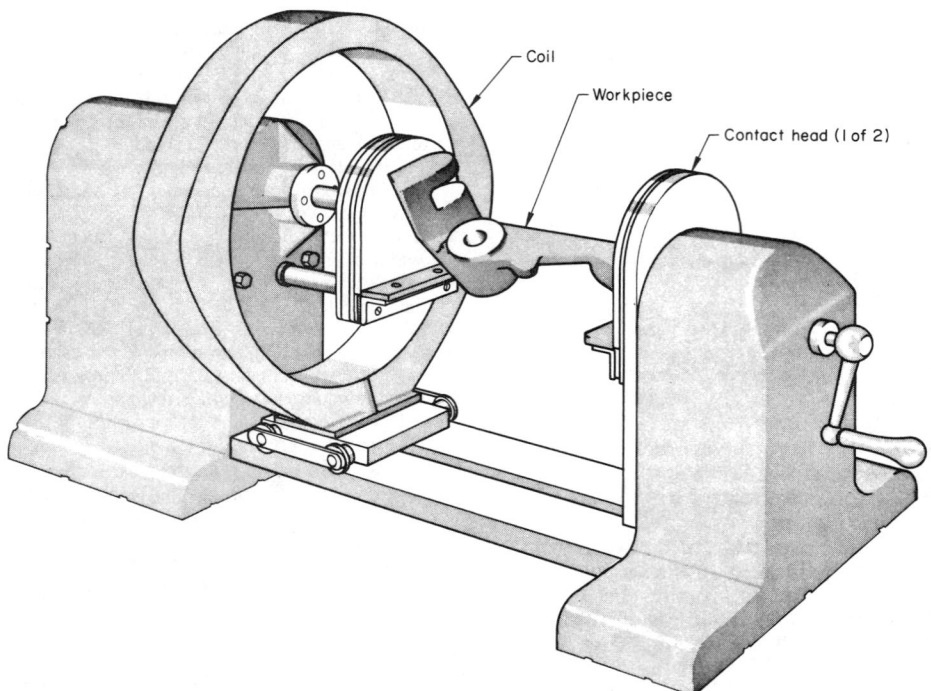

Fig. 7. Bench unit for circular magnetization of workpieces that are clamped between contact heads (direct-contact, head-shot method). The coil on the unit can be used for longitudinal magnetization.

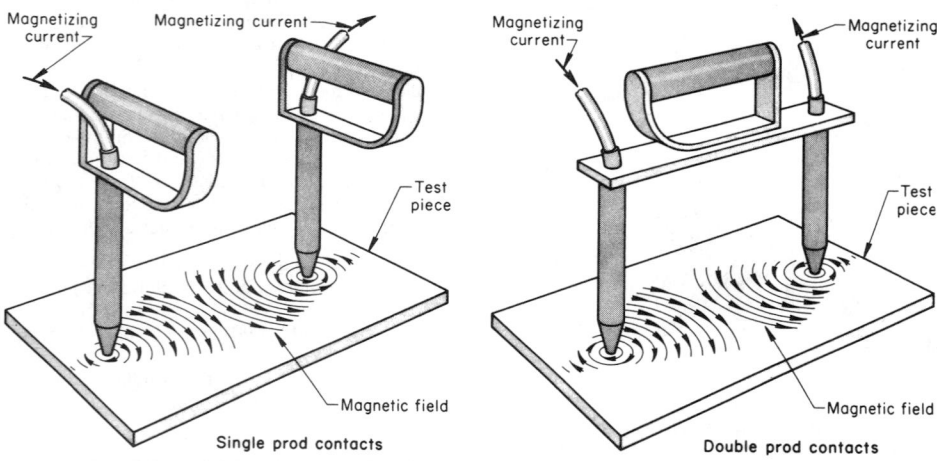

Fig. 8. Single and double prod contacts. Discontinuities are detected by magnetic field generated between the prods.

method, by use of multidirectional magnetization, or by use of induced-current magnetization.

Prod Contacts. For inspection of large and massive parts too bulky to be put into a unit having clamping contact heads, magnetization is often done by using prod contacts (see Fig. 8) to pass the current directly through the part or through a local portion of it. Such local contacts do not always produce true circular fields, but are very convenient and practical for many purposes; prod contacts are often used in magnetic-particle inspection of large castings and weldments.

Prod contacts are widely used and have many advantages. Easy portability makes them most convenient to use for field inspection of large tanks and welded structures. Sensitivity to defects lying wholly below the surface is greater with this

method of magnetization than with any other, especially when half-wave current is used in conjunction with dry powder and the continuous method of magnetization.

The use of prod contacts has disadvantages of which the operator should be aware:

• Suitable magnetic fields exist only between and near the prod contact points. These points are seldom more than 300 mm (12 in.) apart, and usually much less; therefore, it is sometimes necessary to relocate the prods so that the entire surface of a part can be inspected.
• Interference of the external field that exists between the prods sometimes makes observation of pertinent indications difficult. The strength of the current that can be used is limited by this effect.

• Great care must be used to avoid burning of the part under the contact points. Burning may be caused by dirty contacts, insufficient contact pressure, or excessive currents.

Induced current provides a convenient method of generating circumferential magnetizing current in ring-shape parts without making electrical contact. This is accomplished by properly orienting the ring within a magnetizing coil such that it links or encloses lines of magnetic flux (flux linkage), as shown in Fig. 9(a). As the level of magnetic flux changes (increases or decreases), a current flows around the ring in a direction opposing the change in flux level. The magnitude of this current depends on the total flux linkages, rate of flux-linkage changes, and the electrical impedance associated with the current path within the ring. Increasing the flux linkages and the rate of change increases the magnitude of current induced in the ring. The circular field associated with this current takes the form of a toroidal magnetic field that encompasses all surface areas on the ring and that is conducive to the disclosure of circumferential types of discontinuities. This is shown schematically in Fig. 9(b).

The choice of magnetizing current for the induced-current method depends on magnetic properties of the part to be inspected. In instances where the residual method is applicable, such as for most bearing races or similar parts having high magnetic retentivity, direct current is used for magnetizing. The rapid interruption of this current, by quick-break circuitry, results in a rapid collapse of the magnetic flux and the generation of a high-amperage, circumferentially

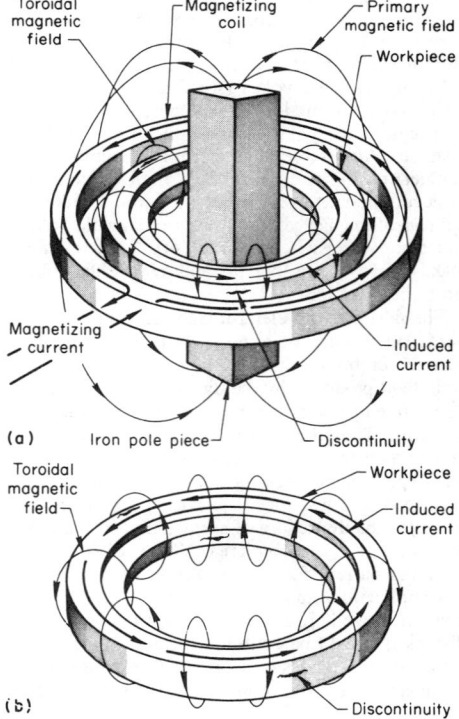

(a) Ring being magnetized by induced current. Current direction corresponds to decreasing magnetizing current. (b) Resulting induced current and toroidal magnetic field in a ring.

Fig. 9. Induced-current method of magnetizing a ring-shape part

directed single pulse of current in the part. Thus, the part is residually magnetized with a toroidal field, and subsequent application of magnetic particles will produce indications of circumferentially oriented discontinuities. A type of current that is similar, but of opposite polarity and lower amplitude, is associated with the increasing flux due to the rapidly rising current (make), but in this case only that current generated by the sudden breaking of the direct current serves a useful purpose. (With direct current there is no change in the flux level once the current is "on" and a steady-state condition exists. Hence, an eddy current or induced current is caused to flow when the magnetizing current is increasing from zero to its steady-state value. An eddy current of opposite polarity is caused to flow as the magnetizing current is reduced from its steady-state value to zero.)

Passing an alternating current through a conductor will set up a fluctuating magnetic field as the level of magnetic flux rapidly changes from a maximum value in one direction to an equal value in the opposite direction. This is similar to the current that would flow in a single-shorted-turn secondary of a transformer. The alternating induced current in conjunction with the continuous method renders the method applicable for processing magnetically soft, or less retentive, parts.

The induced-current method, in addition to eliminating the possibility of damaging the part, also is capable of magnetizing in one operation parts that otherwise would require more than one head shot. Two examples of this type of part are illustrated in Fig. 10 and 11. These parts cannot be completely processed by one head shot to disclose circumferential defects, because regions at the contact points are not properly magnetized. Thus, a two-step inspection process would be required for full coverage, with the part rotated approximately 90° prior to the second step. On the other hand, the induced-current method provides full coverage in one processing step. The disk-shape part shown in Fig. 11 presents an additional problem when the contact method is employed to disclose circumferential defects in the vicinity of the rim. Even when a two-step process is employed, as with the ring in Fig. 10, the primary current path through the part may not develop a circular field of ample magnitude in the rim area. The induced current can be selectively concentrated in the rim area by proper pole-piece selection to provide full coverage (rim area)

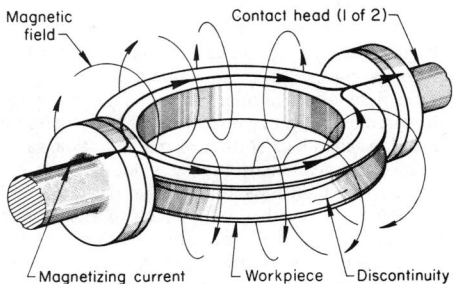

Because regions at contact points are not magnetized, two operations are required for full coverage. With use of the induced-current method, parts of this shape can be completely magnetized in one operation.

Fig. 10. Current and magnetic-field distribution in a ring being magnetized with a head shot

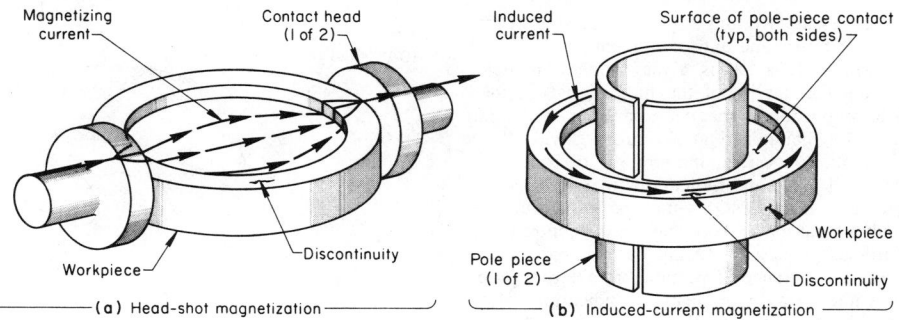

Fig. 11. Current paths in a rimmed disk-shape part that has been magnetized by a head shot and by induced current

in a single processing step. The pole pieces depicted in Fig. 11(b) are hollow and cylindrical, with one on each side of the disk. These pole pieces direct the magnetic flux through the disk such that the rim is the only portion constituting a totally enclosing current path.

Pole pieces used in conjunction with this method are preferably constructed of laminated ferromagnetic material to minimize the flow of eddy currents within the pole pieces, which detract from the induced (eddy) current developed within the part being processed. Pole pieces also can be made of rods, of wire-filled nonconductive tubes, or of thick-wall pipe saw cut to break up the eddy-current path.

MAGNETIC PARTICLES AND SUSPENDING LIQUIDS

Magnetic particles are classified according to the vehicle by which they are carried to the part: by air (dry-particle method) or by a liquid (wet-particle method). Magnetic particles can be made of any low-retentivity ferromagnetic material, finely subdivided. The characteristics of this material, including magnetic properties, size, shape, density, mobility, and degree of visibility and contrast, vary over wide ranges for different applications.

Magnetic Properties. Particles used for magnetic-particle inspection should have high magnetic permeability so that they can be readily magnetized by the low-level leakage fields that occur around discontinuities and can be drawn by these fields to the discontinuities themselves to form readable indications. (Sometimes the fields at very fine discontinuities are extremely weak.)

Low coercive force and low retentivity are desirable for magnetic particles.

Effect of Particle Size. Large, heavy particles are not likely to be arrested and held by weak fields when such particles are moving over a part surface, but fine particles will be held by very weak fields. However, extremely fine particles also may adhere to surface areas where there are no discontinuities (especially if the surface is rough) and form confusing backgrounds. Coarse dry particles fall too fast and are likely to bounce off the part surface without being attracted by the weak leakage fields at imperfections. Finer particles can adhere to fingerprints, rough surfaces, and soiled or damp areas, thereby obscuring indications.

Effect of Particle Shape. Long, slender particles develop stronger polarity than globular particles. Because of the attraction exhibited by opposite poles, these tiny slender particles, which have pronounced north and south poles, arrange themselves into strings more readily than do globular particles. The ability of dry particles to flow freely and to form uniformly dispersed clouds of powder that will spread evenly over a surface is a necessary characteristic for rapid and effective dry-powder testing.

Visibility and contrast are promoted by choosing particles with colors that make them easy to see against the color of the surface of the part being inspected. The natural color of the metallic powders used in the dry method is silver-gray, but pigments are used to color them. The colors of particles for the wet method are limited to the black and red of the iron oxides commonly used as the base for wet particles.

For increased visibility, particles are coated with fluorescent pigment by the manufacturer. The search for indications is conducted in total or partial darkness, using ultraviolet light to activate the fluorescent dyes. Fluorescent magnetic particles are available for both the wet and dry methods. However, the wet method is the one with which they are more commonly used.

Dry particles are available in a variety of colors, some of them fluorescent. Dry particles are most sensitive for use on very rough surfaces and for detecting flaws beneath the surface. They are ordinarily used with portable equipment. The reclaiming and reuse of dry particles is not recommended.

Wet particles are best suited for detection of fine discontinuities such as fatigue cracks. Wet particles commonly are used in stationary equipment where the bath can remain in use until contaminated or until the properties of the particles are exhausted. They are also used in field operations with portable equipment, but care must be taken to agitate the bath constantly.

Oil Suspending Liquid. The oil used as a suspending liquid for magnetic particles should be an odorless, well-refined light petroleum distillate of low viscosity and a high flash point. The viscosity of the oil should not exceed 0.03 cm^2/s (3 cSt) when tested at 38 °C (100 °F), and must not exceed 0.05 cm^2/s (5 cSt) when tested at the temperature prevailing at the point on the test piece being inspected. Above 0.05 cm^2/s, the movement of magnetic particles in the bath is sufficiently retarded to have a definite effect in reducing buildup, and therefore visibility, of an indication of a small discontinuity. Parts should be precleaned to remove oil and grease, because oil from the surface builds up in the bath and increases its viscosity.

Water Suspending Liquid. The use of water instead of oil for magnetic-particle wet-method baths reduces costs and eliminates bath flammability. Water-suspendible particle concentrates include the necessary wetting agents, dispersing agents, rust inhibitors and antifoam agents.

Strength of the bath is a major factor in determining the quality of the indications obtained. The proportion of magnetic particles in the bath must be maintained at a uniform level. If the concentration varies, the strength of indications also will vary, and indications may be misinterpreted. Fine indications may be missed entirely with a weak bath. Too heavy a concentration of particles gives a confusing background and excessive adherence of particles at external poles, thus interfering with clean-cut indications of extremely fine discontinuities.

The best method for ensuring optimum bath concentration for any given combination of equipment, bath application, type of part and discontinuities sought is to test the bath using parts with known discontinuities. Bath strength can be adjusted until satisfactory indications are obtained. This bath concentration can then be adopted as standard for those conditions.

Concentration of the bath can be measured reasonably accurately by the settling test. In this test, 100 mL (0.03 gal) of well-agitated bath is placed in a pear-shape centrifuge tube. The volume of solid material that settles out after a predetermined interval (usually 30 min) is read on the graduated cylindrical part of the tube. Dirt in the bath also will settle and usually shows as a separate layer on top of the oxide. The layer of dirt usually is easily distinguishable, because it is different in color from the magnetic particles.

ULTRAVIOLET LIGHT

A mercury-vapor lamp is a convenient source of ultraviolet light. This type of lamp emits light whose spectrum has several itensity peaks within a wide band of wavelengths. When used for a specific purpose, emitted light is passed through a suitable filter so that only a relatively narrow band of ultraviolet wavelengths is available. For instance, a band in the long-wave ultraviolet spectrum is used for fluorescent liquid-penetrant or magnetic-particle inspection.

Fluorescence is the quality of an element or combination of elements to absorb the energy of light at one frequency and emit light of a different frequency. Fluorescent materials used in liquid-penetrant and magnetic-particle inspection are combinations of elements chosen to absorb light in the peak energy band of the mercury-vapor lamp fitted with a Kopp glass filter. This peak occurs at about 365 nm (14.4 μin.). The ability of fluorescent materials to emit light in the greenish-yellow wavelengths of the visible spectrum depends on the intensity of ultraviolet light at the workpiece surface.

DETECTABLE DISCONTINUITIES

The usefulness of magnetic-particle inspection in the search for discontinuities or imperfections depends on exactly what types of discontinuities the method is capable of finding. Of importance are the size, shape, orientation and location of the discontinuity, with respect to its ability to produce leakage fields.

Surface Discontinuities. The largest and most important type of discontinuities consists of those that are exposed to the surface. Surface cracks or discontinuities are effectively located with magnetic particles. Surface cracks, such as those shown in Fig. 12, also are more detrimental to

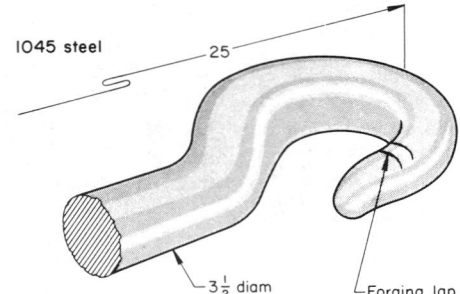

Fig. 12. A 1045 steel crane hook showing indications of forging laps of the type revealed by magnetic-particle inspection. Dimensions are in inches.

service life of a component than are subsurface discontinuities, and therefore they are more frequently the object of inspection.

Magnetic-particle inspection is capable of locating seams, laps, quenching and grinding cracks, and surface ruptures occurring in castings, forgings and weldments. The method also will detect surface fatigue cracks developed during service. Magnetizing and particle-application methods may be critical in certain instances, but in most applications the requirements are relatively easily met, because leakage fields usually are strong and highly localized.

For successful detection of a discontinuity, there must be a field of sufficient strength and in a generally favorable direction to produce strong leakage fields. For maximum detectability, the field set up in the part should be at right angles to the length of a suspected discontinuity (see Fig. 4). This is especially true if the discontinuity is small and fine.

Subsurface discontinuities comprise those voids or nonmetallic inclusions that lie just beneath the surface. Nonmetallic inclusions are present in all steel products to some degree. They occur as scattered individual inclusions, or they may be aligned in long stringers. These discontinuities usually are very small and cannot be detected unless they lie very close to to the surface, because they produce highly localized but rather weak fields.

NONRELEVANT INDICATIONS

Nonrelevant indications are true patterns caused by leakage fields that do not result from the presence of flaws. The term "false indications" is sometimes used to describe this type of indication, because the indication falsely implies the presence of a flaw, even though the particle buildup actually results from a leakage field. There are several possible causes of nonrelevant indications; these require evaluation but should not be interpreted as flaws.

DEMAGNETIZATION AFTER INSPECTION

All ferromagnetic materials, after having been magnetized, will retain a residual magnetic field to some degree. This field may be negligible in magnetically soft metals, but in harder metals it may be comparable to the intense fields associated with the special alloys used for permanent magnets.

It is not always necessary to demagnetize parts after magnetic-particle inspection. However, al-
though demagnetization involves time and expense, it is essential in many instances. Demagnetization may be easy or difficult, depending on the type of metal. Metals having high coercive force are the most difficult to demagnetize. High retentivity is not necessarily related directly to high coercive force, so that the strength of the retained magnetic field is not always an accurate indicator of the ease of demagnetizing.

Reasons for Demagnetizing. There are many reasons for demagnetizing a part after magnetic-particle inspection (or, for that matter, after magnetization for any other reason). Demagnetization may be necessary if:

- The part will be used in an area where a residual magnetic field will interfere with the operation of instruments that are sensitive to magnetic fields or may affect the accuracy of instrumentation incorporated in an assembly that contains the magnetized part
- During subsequent machining, chips may adhere to the surface being machined and adversely affect surface finish, dimensions and tool life
- During cleaning operations, chips may adhere to the surface and interfere with subsequent operations such as painting or plating
- Abrasive particles may be attracted to magnetized parts such as bearing surfaces, bearing raceways or gear teeth, resulting in abrasion or galling, or may obstruct oil holes and grooves
- During some arc-welding operations, strong residual magnetic fields may deflect the arc away from the point at which it should be applied
- A residual magnetic field in a part may interfere with remagnetization of the part at a field intensity too low to overcome the remanent field in the part.

Reasons for Not Demagnetizing. Demagnetization may not be necessary if:

- Parts are made of magnetically soft steel having low retentivity; such parts usually will become demagnetized as soon as they are removed from the magnetizing source
- The parts are subsequently heated above their Curie point and consequently lose their magnetic properties
- The magnetic field is such that it will not affect the function of the part in service
- The part is to be remagnetized for further magnetic-particle inspection or for some secondary operation in which a magnetic plate or chuck may be used to hold the part.

This last reason may appear to conflict with the last item in the list under "Reasons for Demagnetizing". The establishment of a longitudinal field after circular magnetization negates the circular field, because two fields in different directions cannot exist in the same part at the same time. If the magnetizing force is not of sufficient strength to establish the longitudinal field it should be increased, or other steps taken to ensure that the longitudinal field actually has been established. The same is true in changing from longitudinal to circular magnetization. If the two fields (longitudinal and circular) are applied simultaneously, a field will be established that is a vector combination of the two in both strength and direction. However, if the fields are impressed successively, the last field applied, if strong enough to establish itself in the part, will destroy the remanent field from the previous magnetization.

Eddy-Current Inspection

Edited by Allen E. Wehrmeister, Babcock & Wilcox Co.

EDDY-CURRENT INSPECTION is based on the principles of electromagnetic induction and is used to identify or differentiate a wide variety of physical, structural and metallurgical conditions in electrically conductive ferromagnetic and nonferromagnetic metals and metal parts. Eddy-current inspection can be used:

- To measure or identify conditions and properties related to electrical conductivity, magnetic permeability and physical dimensions (primary factors affecting eddy-current response)
- To detect seams, laps, cracks, voids and inclusions
- To sort dissimilar metals and detect differences in their composition, microstructure and other properties (such as grain size, heat treatment and hardness)
- To measure the thickness of a nonconductive coating on a conductive metal, or the thickness of a nonmagnetic metal coating on a magnetic metal.

Because eddy-current inspection is an electromagnetic-induction technique, it does not require direct electrical contact with the part being inspected. The eddy-current method is adaptable to high-speed inspection, and because it is nondestructive, it can be used to inspect an entire production output if desired. The method is based on indirect measurement, and the correlation between the instrument readings and the structural characteristics and serviceability of the parts being inspected must be carefully and repeatedly established.

Eddy-current inspection is extremely versatile, which is both an advantage and a disadvantage. The advantage is that the method can be applied to many inspection problems provided that the physical requirements of the material are compatible with the inspection method. In many applications, however, the sensitivity of the method to the many properties and characteristics inherent within a material can be a disadvantage; some variables in a material that are not important in terms of material or part serviceability may cause instrument signals that mask critical variables or are mistakenly interpreted to be caused by critical variables.

Eddy-Current vs Magnetic Inspection Methods. In eddy-current inspection, the eddy currents create their own electromagnetic field, which may be sensed either through the effects of the field on the primary exciting coil or by means of an independent sensor. In nonferromagnetic materials, the secondary electromagnetic field is derived exclusively from eddy currents. However, with ferromagnetic materials, additional magnetic effects occur that usually are of sufficient magnitude to overshadow the basic eddy-current effects from electrical conductivity only. These magnetic effects result from the magnetic permeability of the material being inspected, and may be virtually eliminated by magnetizing the material to saturation in a static (direct-current) magnetic field. When the permeability effect is not eliminated, the inspection method is more correctly categorized as electromagnetic or magnetoinductive inspection.

PRINCIPLES OF OPERATION

Functions of a Basic System. The part to be inspected is placed within or adjacent to an electric coil in which an alternating current is flowing. As shown in Fig. 1, this alternating current, called the exciting current, causes eddy currents to flow in the part as a result of electromagnetic induction. These currents flow within closed loops in the part, and their magnitude and timing (or phase) depend on (*a*) the original or primary field established by the exciting currents, (*b*) the electrical properties of the part and (*c*) the electromagnetic fields established by currents flowing within the part.

The electromagnetic field in the region in the part and surrounding the part depends on both the exciting current from the coil and the eddy currents flowing in the part. The flow of eddy currents in the part depends on the electrical characteristics of the part, the presence or absence of flaws or other discontinuities in the part, and the total electromagnetic field within the part.

The change in flow of eddy currents caused by the presence of a crack in a pipe is shown in Fig. 2. The pipe travels along the length of the inspection coil, as shown. In section A-A in Fig. 2, no crack is present and the eddy-current flow is symmetrical. In section B-B, where a crack is present, the eddy-current flow is impeded and changed in direction, causing significant changes in the associated electromagnetic field. Finally, the condition of the part can be monitored by observing the effect of the resulting field on the electrical characteristics of the exciting coil, such as its electrical impedance, induced voltage or induced currents. Alternatively, the effect of the electromagnetic field can be monitored by observing the induced voltage in one or more other coils placed within the field near the part being monitored.

Each and all of these changes can have an effect on the exciting coil or other coil or coils used for sensing the electromagnetic field adjacent to a part. The effects most often used to monitor the condition of the part being inspected are the electrical impedance of the coil and the induced voltage of either the exciting coil or other adjacent coil or coils.

Eddy-current systems vary in complexity depending on individual inspection requirements. However, most systems must provide for the following functions:

- Excitation of the inspection coil, with one or more frequencies
- Modulation of the inspection-coil output signal by the part being inspected
- Processing of the inspection-coil signal prior to amplification
- Amplification of the inspection-coil signals
- Detection or demodulation of the inspection-coil signal, usually accompanied by some analysis or discrimination of signals which may be performed by a computer
- Display of signals on a meter, an oscilloscope, an oscillograph or a strip-chart recorder; or recording of signals on paper punch tape or magnetic tape
- Handling of the part being inspected and support of inspection-coil assembly.

Elements of a typical inspection system are shown schematically in Fig. 3. The particular elements in Fig. 3 are for a system developed to inspect bar or tubing. The generator supplies excitation current to the inspection coil and a synchronizing signal to the phase shifter, which provides switching signals for the detector. The loading of the inspection coil by the part being inspected modulates the electromagnetic field of the coil. This causes changes in the amplitude and phase of the inspection-coil voltage output.

The output of the inspection coil is fed to the amplifier and detected or demodulated by the detector. The demodulated output signal, after some further filtering and analyzing, is then displayed on an oscilloscope or a chart recorder. The displayed signals, having been detected or demodu-

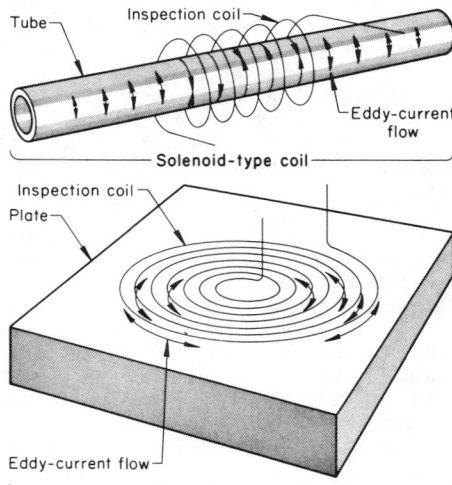

Fig. 1. Two common types of inspection coils, and the patterns of eddy-current flow generated by the exciting current in the coils. Solenoid-type coil is applied to cylindrical or tubular parts; pancake-type coil, to a flat surface.

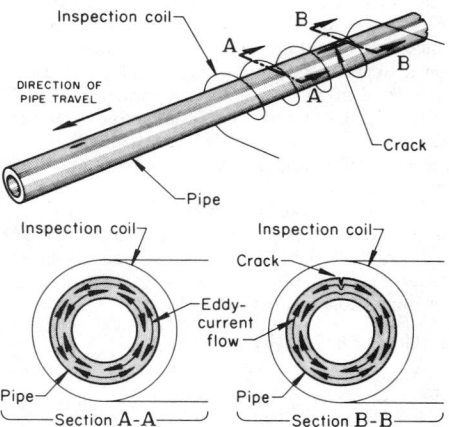

Fig. 2. Effect of a crack on the pattern of eddy-current flow in a pipe

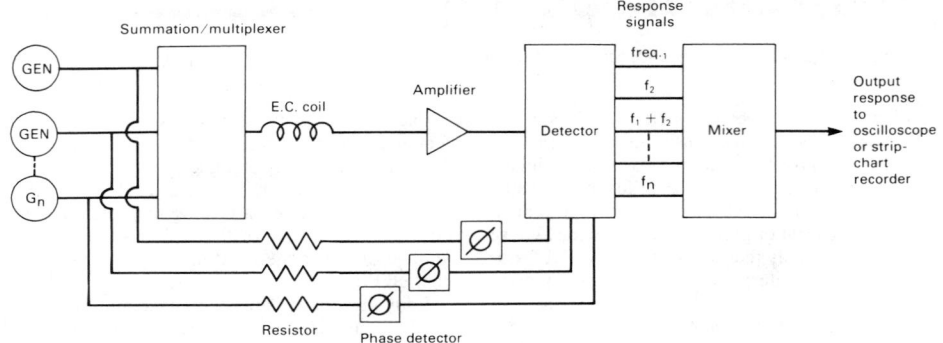

Fig. 3. Principal elements of a typical system for eddy-current inspection of bar or tubing. See description in text.

Table 1. Electrical resistivity and conductivity of several common metals and alloys

Metal or alloy	Resistivity $\Omega \cdot cm \times 10^{-6}$	Conductivity, % IACS
Silver	1.63	105
Copper, annealed	1.72	100
Gold	2.44	70
Aluminum	2.82	61
Aluminum alloys:		
6061-T6	4.1	42
7075-T6	5.3	32
2024-T4	5.2	30
Magnesium	4.6	37
70-30 brass	6.2	28
Phosphor bronzes	16	11
Monel	48.2	3.6
Zirconium	50	3.4
Zircaloy-2	72	2.4
Titanium	54.8	3.1
Ti-6A1-4V alloy	172	1.0
304 stainless steel	70	2.5
Inconel 600	98	1.7
Hastelloy X	115	1.5
Waspaloy	123	1.4

lated, vary at a much slower rate, depending on (*a*) the rate of changing the inspection probe from one part being inspected to another, (*b*) the speed at which the part is fed through an inspection coil or (*c*) the speed with which the inspection coil is caused to scan past the part being inspected.

OPERATING VARIABLES

The principal operating variables encountered in eddy-current inspection include coil impedance, electrical conductivity, magnetic permeability, lift-off and fill factors, edge effect and skin effect. Each of these variables is considered in this section.

Coil Impedance

When direct current is flowing in a coil, the magnetic field reaches a constant level and the electrical resistance of the wire is the only limitation to the flow of current. However, when alternating current is flowing in a coil, two limitations are imposed: the alternating-current resistance of the wire and a quantity known as inductive reactance (X_L).

Impedance is usually plotted on an impedance-plane diagram. In such a diagram, resistance is plotted along one axis and inductive reactance (or inductance) along the other axis. Because each specific condition in the material being inspected may result in a specific coil impedance, each condition may correspond to a particular point on the impedance-plane diagram. For instance, if a coil were placed sequentially on a series of thick pieces of metal, each with a different resistivity, each piece would cause a different coil impedance and would correspond to a different point on a locus in the impedance plane. The curve generated might resemble that shown in Fig. 4, which is based on IACS conductivity ratings. Other curves would be generated for other material variables, such as section thickness and types of surface flaws.

By use of more than one test frequency, the impedance planes can be manipulated to accept a desirable variable (in flaws) and reduce the effects of undesirable variables—i.e., lift-off and/or dimensional effects (see Fig. 3).

Electrical Conductivity

All materials have a characteristic resistance to the flow of electricity. Those with the highest resistivity are classified as insulators; those having intermediate resistivity are classified as semiconductors; and those having low resistivity are classified as conductors. The conductors, which

include most metals, are of greatest interest in eddy-current inspection. The relative conductivities of the common metals and alloys vary over a wide range.

Capacity for conducting current may be measured in terms of either conductivity or resistivity. In eddy-current inspection, frequent use is made of measurement based on the International Annealed Copper Standard (IACS). In this system, the conductivity of annealed, unalloyed copper is arbitrarily rated at 100%, and the conductivities of other metals and alloys are expressed as percentages of this standard. Thus, the conductivity of unalloyed aluminum is rated 61% IACS, or 61% that of unalloyed copper. The resistivity and IACS conductivity ratings of several common metals and alloys are given in Table 1.

Magnetic Permeability

Ferromagnetic metals and alloys, including iron, nickel, cobalt and some of their alloys, act to

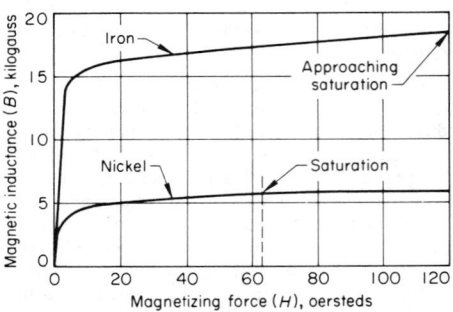

Fig. 5. Magnetization curves for annealed commercially pure iron and nickel

concentrate the flux of a magnetic field. They are strongly attracted to a magnet or an electromagnet, have exceedingly high and variable susceptibilities, and have very high and variable permeabilities.

Magnetic permeability is not a constant for a given material but depends on the strength of the magnetic field acting on it. For instance, consider a sample of steel that has been completely demagnetized and then placed in a solenoid coil. As current in the coil is increased, the magnetic field associated with the current will increase. The magnetic flux within the steel, however, will increase rapidly at first and then level off so that an additionally large increase in the strength of the magnetic field will result in only a small increase in flux within the steel. The steel sample will then have achieved a condition known as magnetic saturation. The curve showing the relation between magnetic-field intensity and the magnetic flux within the steel is known as a magnetization curve. Magnetization curves for annealed commercially pure iron and nickel are shown in Fig. 5. The magnetic permeability of a material is the ratio between the strength of the magnetic field and the amount of magnetic flux within the material. As shown in Fig. 5 at saturation (where there is no appreciable change in induced flux in the material for a change in field strength) the permeability is nearly constant for small changes in field strength.

Magnetic permeability of the material being

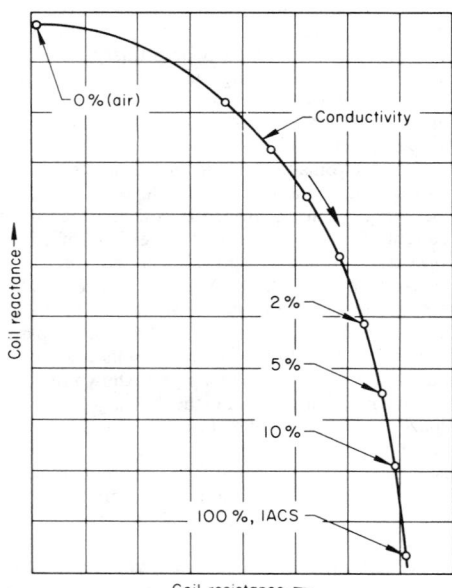

Fig. 4. Typical impedance-plane diagram derived by placing an inspection coil sequentially on a series of thick pieces of metal, each with a different IACS electrical resistance or conductivity rating. The inspection frequency was 100 kHz.

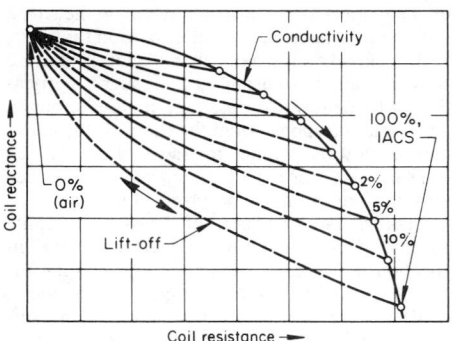

Fig. 6. Impedance-plane diagram showing curves for electrical conductivity and lift-off. Inspection frequency was 100 kHz.

inspected strongly influences the eddy-current response. Consequently, the techniques and conditions used for inspecting magnetic materials differ from those used for inspecting nonmagnetic materials.

"Lift-off" Factor

When a probe inspection coil, attached to a suitable inspection instrument, is energized in air, it will give some indication even if there is no conductive material in the vicinity of the coil. The initial indication will begin to change as the coil is moved closer to a conductor. Because the field of the coil is strongest close to the coil, the indicated change on the instrument will continue to increase until the coil is directly on the conductor. These changes in indication with changes in spacing between the coil and the conductor, or part being inspected, are called "lift-off." The lift-off effect is so pronounced that small variations in spacing can mask many indications resulting from the condition or conditions of primary interest. Consequently, it is usually necessary to maintain a constant relationship between the size and shape of the coil and the size and shape of the part being inspected.

The change of coil impedance with lift-off can be derived from the impedance-plane diagram shown in Fig. 6. When the coil is suspended in air away from the conductor, impedance is at a point at the upper end of the curve at far left in Fig. 6. As the coil approaches the conductor, the impedance moves in the direction indicated by the dashed lines until the coil is in contact with the conductor. When contact occurs, the impedance is at a point corresponding to the impedance of the part being inspected, which in this instance represents its conductivity. The fact that the lift-off curves approach the conductivity curve at an angle can be utilized in some instruments to separate lift-off signals from those resulting from variations in conductivity or some other parameter of interest.

Although lift-off can be troublesome in many applications, it can also be useful. For instance, by utilizing the lift-off effect, eddy current instruments are excellent for measuring the thickness of nonconductive coatings, such as paint and anodized coatings, on metals.

Fill Factor

In an encircling coil, a condition comparable to lift-off is known as "fill factor." It is a measure of how well the part being inspected fills the coil. As with lift-off, changes in fill factor resulting from such factors as variations in outside diameter must be controlled because small changes can give large indications. The lift-off curves shown in Fig. 6 are very similar to those for changes in fill factor. For a given lift-off or fill factor, the conductivity curve will shift to a new position, as indicated in Fig. 6. Fill factor can sometimes be used as a rapid method for checking variations in outside-diameter measurements in rods and bars.

Edge Effect

When an inspection coil approaches the end or edge of a part being inspected, the eddy currents are distorted, because they are unable to flow beyond the edge of a part. The distortion of eddy currents results in an indication known as "edge effect." Because the magnitude of the effect is very large, it limits inspection near edges. Unlike lift-off, little can be done to eliminate edge effect. A reduction in coil size will lessen the effect somewhat, but there are practical limits that dictate the sizes of coils for given applications. In general, it is not advisable to inspect any closer than 3.2 mm ($^1/_8$ in.) from the edge of a part.

One alternative for inspection near an edge with minimal edge effect is to scan in a line parallel to the edge. Inspection can be carried out by maintaining a constant probe-to-edge relationship, but each new scan-line position will require adjustment of the instrument. Fixturing of the probe is recommended.

Skin Effect

Eddy currents are not uniformly distributed throughout a part being inspected; rather, they are densest at the surface immediately beneath the coil and become progressively less dense with increasing distance below the surface. The concentration of eddy currents at the surface of a part is known as "skin effect." At some distance below the surface of a thick part there will be essentially no currents flowing. The depth of eddy-current penetration should be considered for thickness measurements and for detection of subsurface flaws.

Figure 7 shows how the eddy current varies as a function of depth below the surface. The depth at which the density of the eddy current is reduced to about 37% of the density at the surface is defined as the standard depth of penetration. This depth depends on the electrical conductivity and magnetic permeability of the material and on the frequency of the magnetizing current. Depth of penetration decreases with increases in conductivity, permeability or inspection frequency. The standard depth of penetration can be calculated from the equation:

$$S = 1980 \sqrt{\rho/\mu f}$$

where S is standard depth of penetration, in inches; ρ is resistivity, in ohm-centimeters; μ is magnetic permeability (1 for nonmagnetic materials); and f is inspection frequency, in hertz. The standard depth of penetration, as a function of inspection frequency, is shown for several metals of various electrical conductivities in Fig. 8.

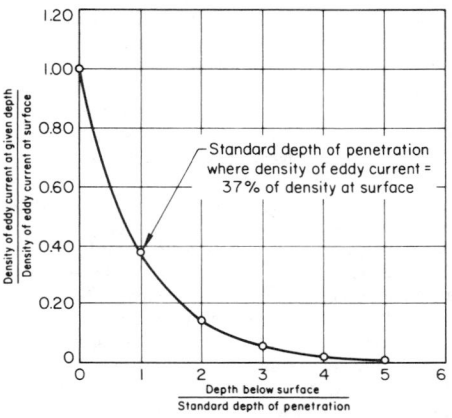

Fig. 7. Variation in density of eddy current as a function of depth below the surface of a conductor — a variation commonly known as "skin effect"

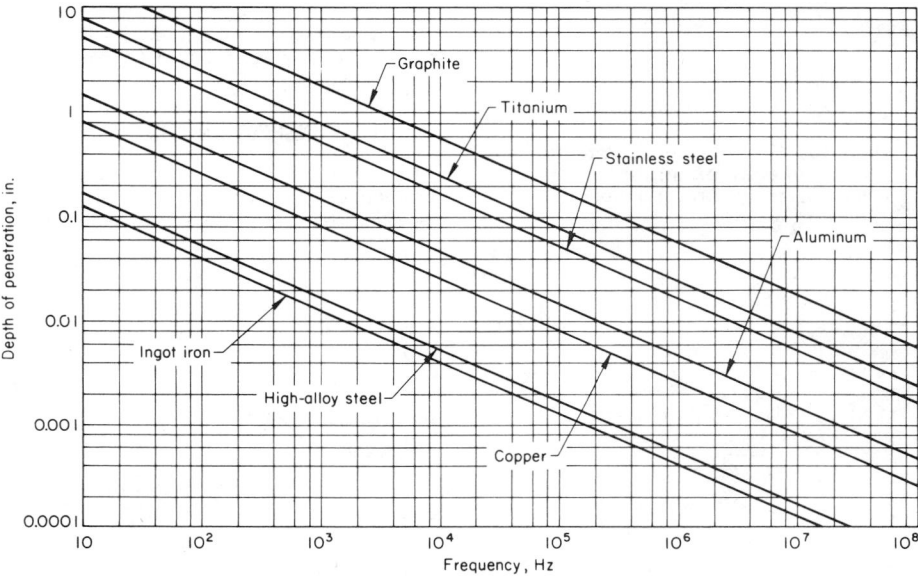

Fig. 8. Standard depths of penetration as a function of frequencies used in eddy-current inspection, for several metals of various electrical conductivities

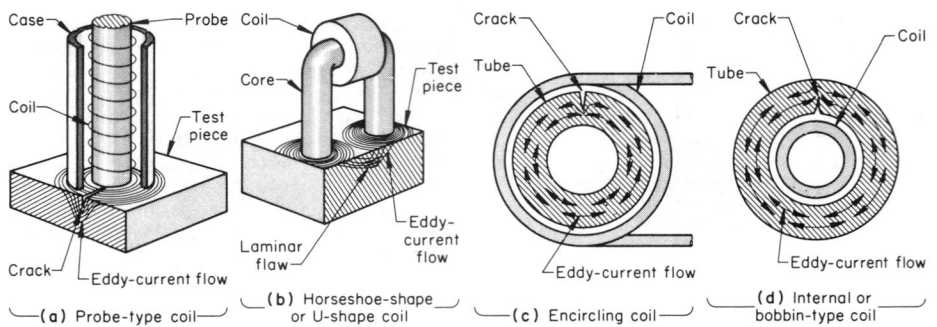

(a) Probe-type coil applied to a flat plate, for detection of a crack. (b) Horseshoe-shape or U-shape coil applied to a flat plate, for detection of a laminar flaw. (c) Encircling coil applied to a tube. (d) Internal or bobbin-type coil applied to a tube.

Fig. 9. Types and applications of coils used in eddy-current inspection

INSPECTION FREQUENCIES

The inspection frequencies used in eddy-current inspection range from about 60 Hz to 6 MHz or more. Most inspection of nonmagnetic materials is performed at a few kilohertz. In general, the lower frequencies are used for inspecting magnetic materials. However, the actual frequency used in any specific eddy-current inspection will depend on the thickness of the material being inspected, the desired depth of penetration, the degree of sensitivity or resolution required, and the purpose of the inspection.

Selection of inspection frequency is normally a compromise. For instance, penetration should be sufficient to reach any subsurface flaws that must be detected or to determine material parameters (such as case hardness). Although penetration is greater at lower frequencies, it does not follow that the lowest possible frequency should be used. Unfortunately, as the frequency is lowered, the sensitivity to flaws decreases somewhat and the speed of inspection may be curtailed. Normally, therefore, the highest possible inspection frequency that is still compatible with the penetration depth required is selected. The choice is relatively simple when only surface flaws must be detected, in which instance frequencies up to several megahertz may be used. However, when flaws at some considerable depth below the surface must be detected, or when flaw depth or flaw size are to be determined, low frequencies must be used and sensitivity may be sacrificed.

In inspection of ferromagnetic materials, relatively low frequencies are normally used because of the low penetration in these materials. Higher frequencies can be used when it is necessary to inspect for surface conditions only. However, even the higher frequencies used in these applications are still considerably lower than those used to inspect nonmagnetic materials for similar conditions.

INSPECTION COILS

The inspection coil is an essential part of every eddy-current inspection system. The shape of the inspection coil depends to a considerable extent on the purpose of the inspection and on the shape of the part being inspected. In inspection for flaws, such as cracks or seams, it is essential that the flow of the eddy currents be as nearly perpendicular to the flaws as possible, in order to obtain a maximum response from the flaws. If the eddy-current flow is parallel to flaws, there will be little or no distortion of the currents and hence very little reaction on the inspection coil.

Probe and Encircling Coils. Of the almost infinite variety of coils employed in eddy-current inspection, probe coils and encircling coils are the most commonly used. Normally, in inspection of a flat surface for cracks at an angle to the surface, a probe-type coil would be used, because this type of coil induces currents that flow parallel to the surface, and therefore across a crack, as shown in Fig. 9(a). On the other hand, a probe-type coil would not be suitable for detecting a laminar type of flaw. For such a discontinuity, a U-shape or horseshoe-shape coil, such as the one shown in Fig. 9(b), would be satisfactory.

To inspect tubing or bar, an encircling coil (Fig. 9c) is generally used, both because of complementary configuration and because of the testing speeds that can be obtained with this type of coil. However, an encircling coil is sensitive only to discontinuities that are parallel to the axis of the tube or bar. The coil is satisfactory for this particular application because, as a result of the manufacturing process, most discontinuities in tubing and bar are parallel to the major axis. If it is necessary to locate discontinuities that are not parallel to the axis, a probe coil must be used and either the coil or the part must be rotated during scanning. To detect discontinuities on the inside surface of a tube, an internal or bobbin-type coil (Fig. 9d) may be used. An alternative is to use an encircling coil with a depth of penetration sufficient to detect flaws on the inside surface. The bobbin-type coil, in common with the encircling coil, is sensitive to discontinuities that are parallel to the axis of the tube or bar.

Multiple Coils. In many setups for eddy-current inspection, two coils are used. The two coils are normally connected in a series-opposing arrangement so that when their impedances are the same, there is no output from the pair. Pairs of coils can be used in either an absolute or a differential arrangement (see Fig. 10). In the absolute arrangement (Fig. 10a), a sample of acceptable material is placed in one coil and the other coil is used for inspection. In this manner the coils are comparing an unknown against a standard, the differences between the two (if any) being indicated by a suitable instrument. Arrangements of this type are commonly employed in sorting applications. Fixtures are commonly used to

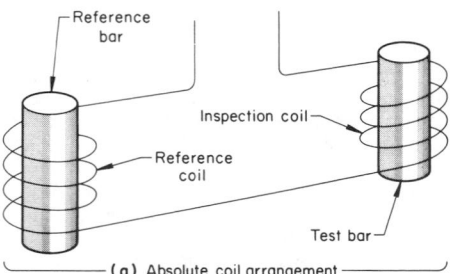

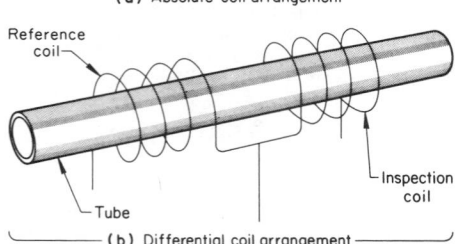

Fig. 10. Absolute and differential arrangements of multiple coils used in eddy-current inspection. See text.

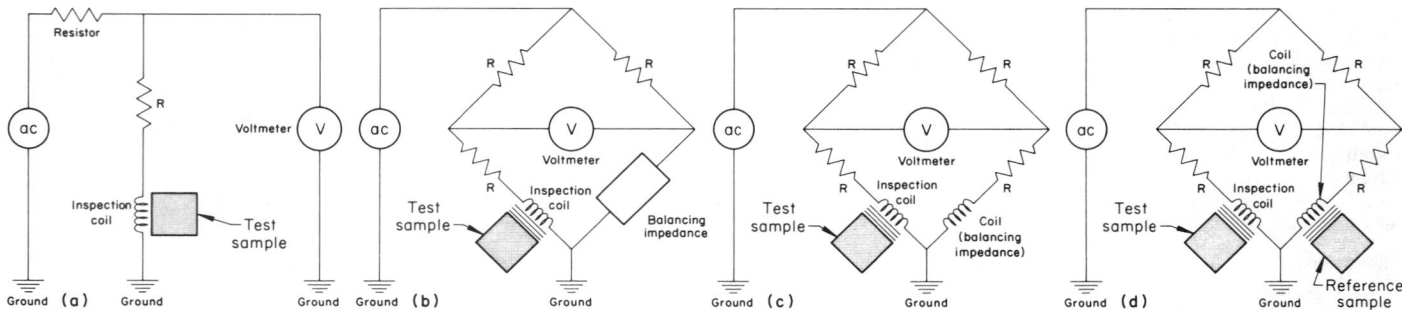

Fig. 11. Four types of eddy-current instruments: (a) a simple arrangement, in which voltage across the coil is monitored; (b) typical impedance bridge; (c) impedance bridge with dual coils; and (d) impedance bridge with dual coils and a reference sample in the second coil

maintain a constant geometrical relationship between coil and part.

In many applications, an absolute coil arrangement is undesirable. For instance, in inspection of tubing, an absolute arrangement will indicate dimensional variations in both outside diameter and wall thickness even though such variations may be well within allowable limits. To avoid this problem, a differential coil arrangement such as that shown in Fig. 10(b) can be used. Here, the two coils compare one section of the tube with an adjacent section. When the two sections are the same, there is no output from the pair of coils and hence no indication on the eddy-current instrument. Gradual dimensional variations within the tube or gross variations between individual tubes are not indicated, whereas discontinuities—which normally occur abruptly—are very apparent. In this way, it is possible to have an inspection system that is sensitive to flaws and relatively insensitive to changes that normally are not of interest.

Sizes and Shapes. Inspection coils are made in a variety of sizes and shapes. Selection of a coil for a particular application depends on the type of discontinuity. For instance, when an encircling coil is used to inspect tubing or bar for short discontinuities, best resolution is obtained with a short coil. Alternatively, a short coil has the disadvantage of being sensitive to the position of the part in the coil. Longer coils are not as sensitive to position of the part, but are not as effective in detecting very small discontinuities. Small-diameter probe coils have greater resolution than larger ones but are more difficult to manipulate and are more sensitive to lift-off variations.

EDDY-CURRENT INSTRUMENTS

A simple eddy-current instrument, in which the voltage across an inspection coil is monitored, is shown in Fig. 11(a). This circuit is adequate for measurement of large lift-off variations, if accuracy is not of great importance. A circuit designed for greater accuracy is shown in Fig. 11(b). This instrument consists of a signal source, an impedance bridge with dropping resistors, an inspection coil in one leg and a balancing impedance in the other leg. The differences in voltage between the two legs of the bridge are measured by an alternating-current voltmeter. Alternatively, the balancing impedance in the leg opposite the inspection coil may be a coil identical to the inspection coil, as shown in Fig. 11(c), or it may have a reference sample in the coil, as shown in Fig. 11(d). In the latter, if all the other components in the bridge were identical, a signal would occur only when the inspection-coil impedance deviated from that of the reference sample.

There are other methods of achieving bridge balance, such as varying the values of resistance of the resistor in the upper leg of the bridge and one in series with the balancing impedance. The most accurate bridges can measure absolute impedance to within 0.01%. However, in eddy-current inspection, it is not how an impedance bridge is balanced that is important, but rather how it is unbalanced by the effects of a flaw.

Another type of bridge system is an induction bridge, in which the power signal is transformer-coupled into an inspection coil and a reference coil. In addition, the entire inductance-balance system is placed in the probe, as shown in Fig. 12. The probe consists of a large transmitter (or

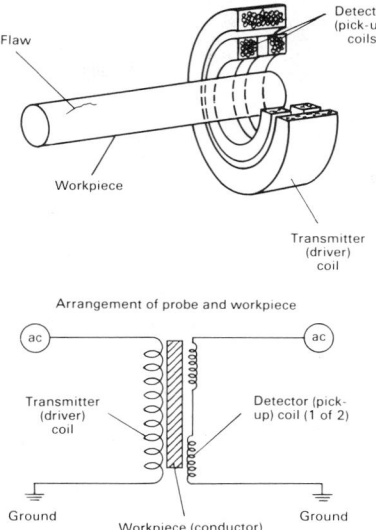

Arrangement of probe and workpiece

Wiring schematic for probe and workpiece

Fig. 12. Induction-bridge probe, in place at the surface of a workpiece. Schematic shows how power signal is transformer-coupled from a transmitter coil into two detector coils — an inspection coil (at bottom) and a reference coil (at top).

driver) coil and two small detector (or pickup) coils wound in opposite directions as mirror images of each other. An alternating current is supplied to the large transmitter coil to generate a magnetic field. If the transmitter coil is not in the vicinity of a conductor, the two detector coils detect the same field, and since they are wound in opposition to each other the net signal is zero. If, however, one end of the probe is placed near a metal surface, the field is different at the two ends of the probe, and a net voltage appears across the two coils. The resultant field is the sum of a "transmitted" signal, which is present all the time, and a "reflected" signal due to the presence of a conductor (the metal surface). This coil arrangement can be used as a probe or as an encircling coil (see Fig. 12).

Readout Instrumentation. An important part of an eddy-current inspection system is the instrument used for a readout. The readout device may be an integral part of the system, an interchangeable plug-in module, or a solitary unit connected by cable. The readout instrument should be of adequate speed, accuracy and range to meet the inspection requirements of the system. Frequently, several readout devices are employed in a single inspection system. The more common types of readout, listed in order of increasing cost and complexity, are:

Alarm lights alert the operator that a test-parameter limit has been exceeded.

Sound alarms serve the same purpose as alarm lights but free the attention of the operator so that he can manipulate the probe in manual scanning.

Kick-out relays activate a mechanism that automatically rejects or marks a part when a test parameter has been exceeded.

Analog meters give a continuous reading over an extended range. They are fairly rapid (with a frequency of about 1 Hz), and the scales can be calibrated to read parameters directly. The accuracy of these devices is limited to about 1% of

full scale. They can be used to set the limits on alarm lights, sound alarms and kick-out relays.

Digital meters are easier to read and can have greater ranges than analog meters. Numerical values are easily read without extrapolation, but fast trends of changing readings are more difficult to interpret. Although many digital meters have binary coded decimal (bcd) output, they are relatively slow.

X-Y plotters can be used to display impedance-plane plots of the eddy-current response. They are very helpful in designing and setting up eddy-current bridge-unbalance inspections and discriminating against undesirable variables. They are also useful in sorting out the results of inspections. They are fairly accurate and provide a permanent copy.

X-Y storage oscilloscopes are very similar to X-Y plotters but can acquire signals at high speed. However, the signals have to be processed manually, and the screen can quickly become cluttered with signals. In some instruments, high-speed X-Y gates can be displayed and set on the screen.

Strip-chart recorders furnish a fairly accurate (about 1% of full scale) recording at reasonably high speed (about 200 Hz). However, once on the chart, the data must be read by an operator. Several channels can be recorded simultaneously, and the record is permanent.

Magnetic-tape recorders are fairly accurate and capable of recording at very high speed (10 MHz). Moreover, the data can be processed by automated techniques.

Computers. The data from several channels can be fed directly to a high-speed computer, either analog or digital, for on-line processing. The computer can separate parameters and calculate the variable of interest and significance, catalog the data, print summaries of the result and store all data on tape for reference in future scans.

DISCONTINUITIES DETECTABLE BY EDDY-CURRENT INSPECTION

Basically, any discontinuity that appreciably alters the normal flow of eddy currents can be detected by eddy-current inspection. With encircling-coil inspection of either solid cylinders or tubes, surface discontinuities having a combination of predominantly longitudinal and radial dimensional components are readily detected. When discontinuities of the same size are located beneath the surface of the part being inspected at progressively greater depths, they become increasingly difficult to detect, and can be detected at depths greater than 13 mm ($^1/_2$ in.) only with special equipment designed for this purpose.

On the other hand, laminar discontinuities such as may be found in welded tubes may not alter the flow of the eddy currents enough to be detected unless the discontinuity breaks either the outside or inside surfaces, or unless it produces a discontinuity in the weld from upturned fibers caused by extrusion during welding. A similar difficulty could arise for the detection of a thin planar discontinuity that is oriented substantially perpendicular to the axis of the cylinder.

Regardless of the limitations, a majority of objectionable discontinuities can be detected by eddy-current inspection at high speed and at low cost. Some of the discontinuities that are readily detected are seams, laps, cracks, slivers, scabs, pits, slugs, open welds, miswelds, misaligned welds, black or gray oxide weld penetrators, pinholes, hook cracks and surface cracks.

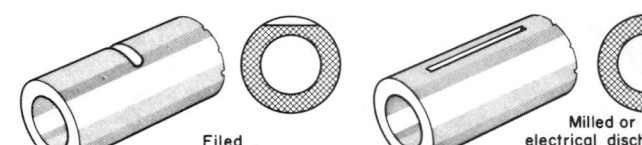
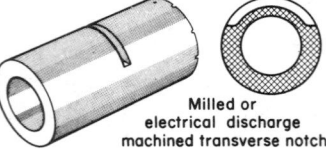
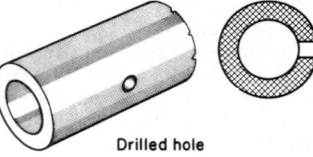

Filed transverse notch

Milled or electrical discharge machined longitudinal notch

Milled or electrical discharge machined transverse notch

Drilled hole

ASTM standards for eddy-current testing include E215 (aluminum alloy tube), E376 (measurement of coating thickness), E243 (copper and copper alloy tube), E566 (ferrous metal sorting), E571 (nickel and nickel alloy tube), E690 (nonmagnetic heat-exchanger tubes), E426 (stainless steel tube) and E309 (steel tube).

Fig. 13. Several fabricated discontinuities used as reference standards in eddy-current inspection

Reference Samples. A basic requirement for eddy-current inspection is a reliable and consistent means for setting the sensitivity of the tester to the proper level each time it is used. A standard reference sample must be provided for this purpose. Without this capability, eddy-current inspection would be of little value. In selecting a standard reference sample, the usual procedure is to select a sample of product that can be run through the inspection system without producing appreciable indications from the tester. Several samples may have to be run before a suitable one is found; the suitable one then has reference discontinuities fabricated into it.

The type of reference discontinuities that must be used for a particular application are specified (for instance, by ASTM and API). In selecting reference discontinuities some of the major considerations are: (*a*) they must meet the required specification, (*b*) they should be easy to fabricate, (*c*) they should be reproducible, (*d*) they should be producible in precisely graduated sizes and (*e*) they should produce an indication on the eddy-current tester that closely resembles those produced by the natural discontinuities.

Several discontinuities that have been used for reference standards are shown in Fig. 13; these include a filed transverse notch, milled or electrical discharge machined longitudinal and transverse notches, and drilled holes.

Electromagnetic Sorting and Testing

Edited by Peter J. Suhr, Magnetic Analysis Corp.

MAGNETIC testing and inspection broadly comprise all methods in which magnetic fields play an essential role. Partly for convenience, several methods that employ magnetic fields are accorded independent status. One basis for division is the frequency range within which the units that generate the magnetic fields operate. These ranges of frequency are to some extent arbitrary, but they can be differentiated on an approximate basis as being static, low, intermediate and high. A second basis for division is the absence or presence of electric current, which serves to differentiate magnetic and electromagnetic methods. Finally, in classifying methods in which the essential magnetic field involved arises from an electric current, the methods of transmitting current can be differentiated as being either conductive or inductive.

In this article, the electromagnetic method is set apart from certain other magnetic-field methods that employ electric current—specifically, the magnetic-particle and eddy-current methods, both of which are dealt with in separate articles in this volume. The electromagnetic method, also called the magnetoinductive method, employs induced electric current as the source of the magnetic field and can be applied to ferromagnetic materials only. The method employs current at low frequencies, generally not exceeding about 400 Hz and more often in the range of 60 to 100 Hz. Trends in equipment design indicate a preference for the use of frequencies in the range of 10 to 60 Hz. (Under certain circumstances, eddy-current instruments can be used for sorting, particularly when properties of surface layers are involved and when differences between test pieces are clear-cut.) The magnetic permeability (μ) of the metal to which the electromagnetic method is applied is a dominant variable, as are other factors that influence magnetic hysteresis, such as chemical composition, variations in processing, metallurgical structure and heat treatment of the metal.

The electromagnetic method is a principal means for sorting or differentiating ferromagnetic materials on the basis of chemical, physical, mechanical and processing variables, including chemical composition, metallurgical structure, hardness and residual stresses. The method serves, therefore, as a rapid, convenient and economical substitute for conventional chemical analysis and for tests such as hardness and tensile tests, provided that reference standards of acceptability have been established and related to indications or measurements derived from the electromagnetic test instrument.

MAGNETISM

The ferromagnetic materials of industrial importance, notably the irons and steels, have high and variable magnetic permeabilities. These materials are strongly attracted to a magnet or an electromagnet; they respond positively to a magnetizing force.

Magnetic Hysteresis. It is a familiar phenomenon that some ferromagnetic materials, when magnetized by an external field, do not return to a completely unmagnetized state when removed from that field. In fact, these materials must be subjected to a reversed field of a certain strength to demagnetize them (discounting heating the material to a characteristic temperature, called the Curie point, above which ferromagnetic ordering of atomic moments is thermally destroyed, or mechanically working the material to reduce the magnetization). If an external field that can be varied in a controlled manner is applied to a completely demagnetized ("virgin") specimen, and if instrumentation for measuring the magnetic induction within the specimen is at hand, the magnetization curve of the material may be determined. A representative magnetization (hysteresis) curve for a ferromagnetic material is shown in Fig. 1.

As shown in Fig. 1(a), starting at the origin

(*O*) with the specimen in the unmagnetized condition and increasing the magnetizing force (*H*) in small increments, the flux density (*B*) in the material increases quite rapidly at first, then it increases more slowly until it reaches a point beyond which any increase in the magnetizing force does not increase the flux density. This is shown by the dashed curve *Oa* in Fig. 1(a). In this condition, the specimen is said to be magnetically saturated.

When the magnetizing force is gradually reduced to zero, the curve *ab* results (Fig. 1b). The amount of magnetism that the steel retains at point *b* is called residual magnetism or remanence (B_r).

When the magnetizing current is reversed and gradually increased in value, the flux continues to diminish. The flux does not become zero until point *c* is reached, at which time the magnetizing force is represented by *Oc* (see Fig. 1c), which graphically designates the coercive force (H_c) in the material.

As the reversed field is increased beyond *c*, point *d* is reached (Fig. 1d). At this point the specimen is again saturated, but in the opposite polarity. The magnetizing force is now decreased to zero, and the *de* line is formed and retains reversed-polarity residual magnetism (B_r) in the specimen. Again increasing the magnetizing force in the original direction completes the curve *efa*. Now the cycle is complete, and the curve (*abcdefa*) is called a hysteresis loop. In alternating-current applications, the ferromagnetic material goes through this cycle for every reversal in current—60 times per second on a 60-Hz power line.

The definite lag throughout the cycle between the magnetization force and the flux is called hysteresis. If the hysteresis loop is slender (Fig. 1e), the indication is that the material has low retentivity (low residual field) and is easy to magnetize (has low reluctance). A wide loop (Fig. 1f) indicates that the material has high reluctance and is difficult to magnetize.

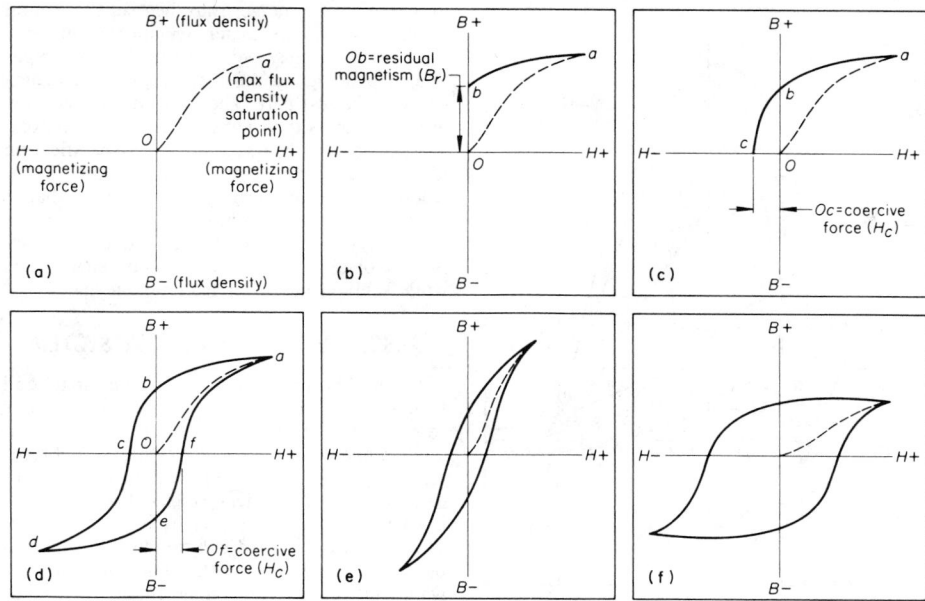

Fig. 1. Representative magnetization (hysteresis) curve for a ferromagnetic material. See text for discussion.

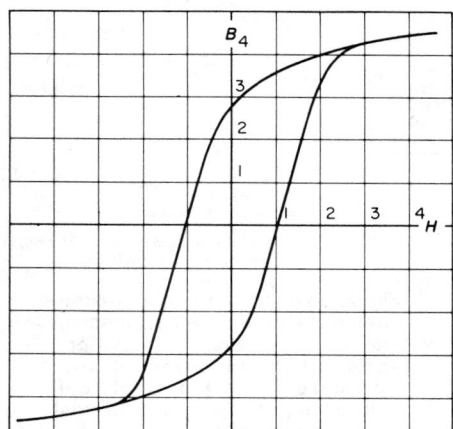

Fig. 2. Distortion-free hysteresis loop as portrayed on the screen of an instrument with calibrated B and H scales.

HYSTERESIS-LOOP TESTS

The variables portrayed in the hysteresis loop—saturation, remanence, coercive force and magnetic permeability—are influenced to some extent by basic characteristics and properties of ferromagnetic materials, including their chemical composition, metallurgical structure and heat treatment. Magnetic saturation, for instance, is closely related to chemical composition and crystal structure. Recognition of these relationships led to the design and development of testing and inspection equipment capable of measuring and portraying magnetic variables that could be related directly to metallurgical properties and other characteristics of interest, such as carbon content, hardness and case depth. In one series of instruments, the hysteresis loop for a given magnetized specimen is portrayed on the screen of a cathode-ray oscilloscope. A hysteresis loop as

portrayed on the screen of an instrument with calibrated B and H scales is shown in Fig. 2.

COMPARATOR-BRIDGE TESTS

The concept of hysteresis-loop analysis in nondestructive testing of steel was introduced in 1939 and is still used in some comparator-bridge equipment. However, for the purpose of sorting, the hysteresis loop is not the best representation of magnetic-material values. An improved representation of the characteristic values of magnetic and electrical material was introduced with a magnetic comparator test instrument. With this instrument, depending on the selection made, the magnetization variations, the magnetic-permeability variations, or variations in the curvature of the hysteresis loop can be presented pictorially. The instrument can be used to sort or evaluate on the basis of chemical composition, metallurgical structure, hardness, tensile strength, depth of case, or surface decarburization, or combinations of these variables.

Operating Procedures. The procedures that would be followed in sorting two steels of different chemical compositions, A and B, using a magnetic comparator instrument, are outlined in Fig. 3. The principal objective, in this instance, is to obtain the greatest possible difference in amplitude and shape between the patterns that are displayed on the screen for the two different steels. This will provide clear-cut differentiation that can be used as a basis for sorting. In Fig. 3, the shaded symbols represent the controls that are adjusted during each step of the setup procedure.

As shown in Fig. 3(a), a sample of steel A is placed in the test coil, causing a pattern to appear on the screen. The phase shifter and coarse sensitivity are adjusted until the entire pattern is displayed and centered as shown in Fig. 3(a). A fine sensitivity adjustment is used to bring the curve to a specific indication height called the "absolute value." This permits all sensitivity positions on the sensitivity selector to be expressed as percentages of absolute value. If a second sample

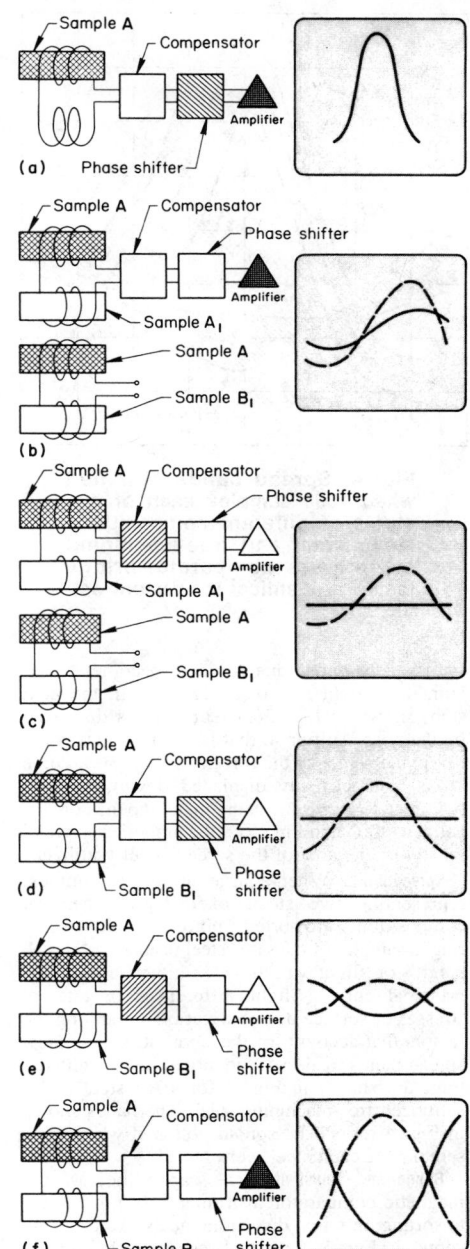

Fig. 3. Operating procedures followed, and screen pictures obtained, in sorting two steels of different compositions, A and B, using a magnetic comparator instrument. See text for description. In (b) through (f), solid line on display is for sample A_1 and broken line is for sample B_1.

of steel A, noted as A_1 in Fig. 3(b), is placed in the unoccupied difference coil and the sensitivity control is adjusted to a higher level, a difference between the magnetic properties of samples A and A_1 will appear on the screen as indicated by the solid line on the display of Fig. 3(b). If A_1 is then replaced by sample B_1, as shown at lower left in Fig. 3(b), a curve for this sample might be displayed as indicated by the broken line on the display. Next, sample A_1 is reinserted in the difference coil, replacing B_1, and the compen-

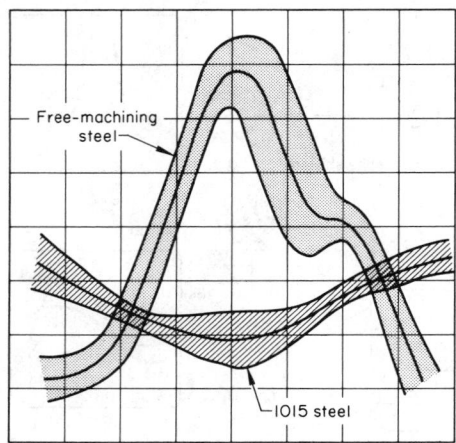

Fig. 4. Spread bands obtained when 1000 samples each of two steels of different compositions (1015 steel and free-machining steel) were comparator-bridge tested at identical instrument settings

sator is adjusted so that curve A_1 becomes a horizontal, straight line (see Fig. 3c). In the next step, shown in Fig. 3(d), curve B_1 is displaced by the phase shifter so that its peak or maximum lies in the center of the screen. In Fig. 3(e), curve A_1 is symmetrically displaced downward with reference to curve B_1, using the compensator. Finally, as shown in Fig. 3(f), the amplifier is turned up so that the area of the screen is fully utilized.

Spread Bands. When a large number of samples, representing two steels of different chemical compositions, are sorted, minor variables occurring in samples of the same steel result in a "spread band," or slight variations to either side of the principal curve. Slight differences in internal stresses or surface decarburization are among the factors that account for these variations. Typical spread bands made with identical instrument settings are shown in Fig. 4 for 1015 steel and a sulfurized free-machining steel. The bands shown in Fig. 4 reflect the spread obtained when 1000 samples of each steel were tested.

Range and Sensitivity. The capabilities of the magnetic comparator instrument are not limited to sorting only two dissimilar steels at a time. As shown in Fig. 5, four steels can be differentiated using only one instrument setting; in this instance, hot rolled billets of 1075, 5132, 5140 and 6150 steels were being sorted.

Comparator With Meter Display. Figure 6 shows the control panel of a comparator-bridge instrument with meter display and three-way gating that permits sorting into three categories. If desired, a cathode-ray-tube display unit can be connected to the instrument to provide visual display. Although such a unit may be helpful as a setup aid, it is not essential to the operation of the instrument. The instrument operates at a test frequency of 60 Hz and will accommodate standard test coils ranging from 2 to 500 mm (0.08 to 20 in.) in diameter. The instrument can be used at a maximum production rate of 10 000 parts per hour.

Setup Procedure for Sorting. In the following procedure, it is assumed that test parts are to be sorted into three categories—namely, within specified requirements, above specified requirements and below specified requirements. Furthermore, both

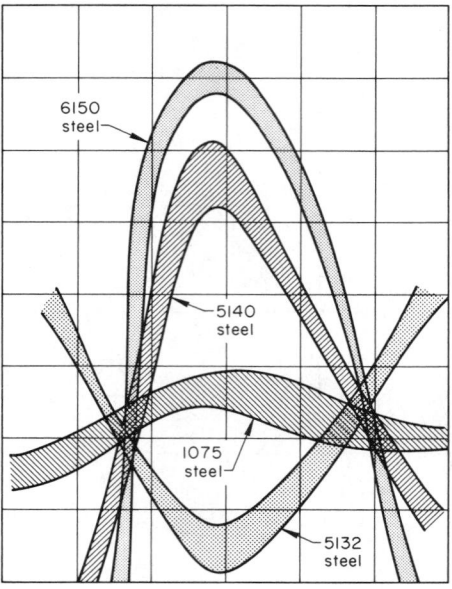

Fig. 5. Screen patterns and spread bands for four steels of different compositions. Patterns were obtained at identical instrument settings.

test parts used in the balancing procedure are to be within specified requirements.

The setup procedure begins with establishment of test limits. With the unit balanced and with acceptable test parts in the reference and test coils, sensitivity is reduced to position 12 and minor readjustments in balancing are made. The acceptable (within specification) test part is then removed from the test coil and is replaced with an out-of-specification test part representative of an unknown number of parts to be sorted. This test part is placed in the coil at a position that results in a meter reading that is farthest from -100. If the meter does not deviate from -100 or moves full scale in the positive direction, sensitivity should be further reduced and, with the test part in position, the phase control should be rotated

to within a $\pm60°$ range from the "start" position to obtain maximum meter deviation from zero. This deviation may be either positive or negative, depending on the metallurgical characteristics of the test part. The first out-of-specification test part is then removed from the test coil and replaced with a second out-of-specification test part with the objective of producing a meter reading that is significantly different from that of the first. From this point on, minor adjustments in sensitivity, coil current and phase controls are made to obtain maximum spread between test parts while keeping the meter readings on scale.

ABSOLUTE SINGLE-COIL TEST SYSTEM

The basic circuitry of an absolute single-coil system designed to operate at low frequencies is shown in Fig. 7. A power source providing constant-amplitude sinusoidal alternating current feeds a primary magnetizing coil, within which is located a secondary or pickup coil so that both coils, in an enclosed assembly, encircle the test object. The coils are designed to provide a field distribution in which the field strength increases toward both ends of the magnetizing coil, thus compensating for the demagnetization effect caused by the ends of the test object. (Conven-

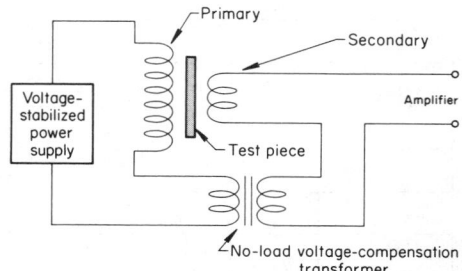

Fig. 7. Basic circuitry of a no-load-compensated single-coil electromagnetic test system. A no-load voltage-compensation transformer removes the empty-coil voltage signal from output test signal to amplifier.

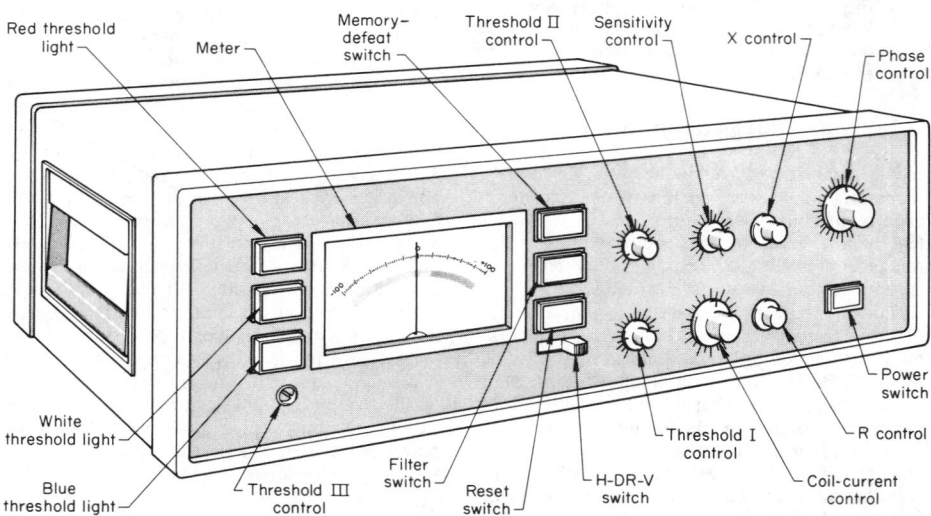

Fig. 6. Control panel of a comparator-bridge instrument with meter display and three-way gating that permits sorting into three categories

tional solenoid coils are wound uniformly from end to end.) Also, the secondary coil is located in the minimum-field-strength area of the magnetizing coil, thereby increasing pickup sensitivity.

Each single-coil instrument contains a single fixed-frequency oscillator, designed to operate at either 50 or 10 Hz. These low frequencies have been found to be highly effective in tests related to changes in characteristics of magnetic-hysteresis loss and to variations in the shapes of hysteresis loops, rather than to changes in eddy-current characteristics. They also ensure greater penetration depths of the exciting magnetic field.

In the absence of the test object, the alternating electromagnetic field of the primary magnetizing coil induces an empty-coil, no-load voltage in the secondary coil. When a test object is placed in the coil assembly, an induction voltage signal due to the presence of the test object is superimposed on the empty-coil voltage signal picked up by the secondary coil. Because the empty-coil, no-load voltage provides no useful test information, it is completely suppressed by an equal and opposite voltage derived from the secondary winding of the no-load voltage-compensation transformer, also shown in Fig. 7. In the test-coil assemblies, which cover a range of dimensions, transmitters of no-load voltage-compensation signals are cast integrally within the test-coil housings.

Instrumentation and Method of Operation. Details of the circuitry of the single-coil test system are shown in Fig. 8. The power supply provides alternating current to the test coil. This current is provided at a stabilized alternating voltage having a frequency that is fixed at 10 or 50 Hz, depending on the instrument being used. A stabilized reference voltage from the power supply is sent to the rectifier and the precision potentiometer. The test signal is derived from the secondary winding of the test coil, and is corrected for a "no load" test condition by a voltage from the secondary winding of the transmitter. This test signal passes through the phase-controlled rectifier, filter and amplifier unit, which provides three separate output signal channels for V, R and H. The signal desired for any specific sorting application is selected by the selector switch. This signal is then compared with the reference signal from the potentiometer in the precision differential amplifier, whose sensitivity can be adjusted by the precision sensitivity switch. Test-signal amplitudes are displayed on the panel meter, which provides three sorting ranges. Indicator lamps indicate the sorting range into which each test object falls.

Operating the Single-Coil Unit. For quantitative sorting and measurement applications, a suitable test coil or probe is connected by cable to a socket on the panel of the unit. When encircling test coils are used, the test-coil diameter selected should permit the test object to fill the coil opening as completely as possible. This ensures maximum coil-fill factor and test sensitivity. With the coil or probe connected, the power-supply on-off switch is pressed to energize the system.

By pressing one of the three pushbuttons, the test signal (V, R or H) is selected to provide optimum characteristics for specific sorting problems or measurement applications. For instance, if the voltage-vector signal V is selected, the panel meter will indicate the magnitude of this signal.

To carry out a series of measurements with conventional electromagnetic test instruments, it

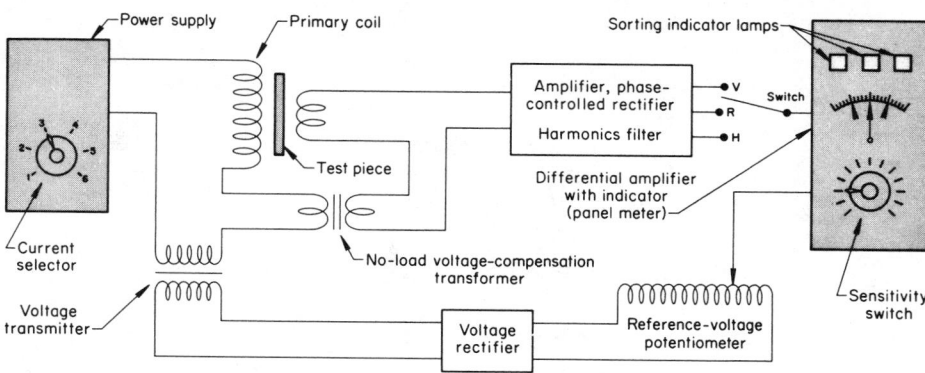

Fig. 8. Functional components and basic circuitry of a single-coil absolute test system

is usually necessary to make many nulling adjustments and to vary the test sensitivity in order to obtain signal amplitudes that can be displayed on the test instrument. With the single-coil test instrument, the measurement series can be obtained immediately without resorting to trial runs. All significant test data for each specific test problem can be taken from the values measured in the first test run.

APPLICATIONS

Selection of Standards. Modern electromagnetic test instrumentation offers a high degree of sensitivity to all variables that affect magnetic characteristics. This is an outstanding advantage in detecting relatively subtle differences in essentially similar materials. However, it imposes rather rigorous requirements on the selection of test parts as standards for sorting or analysis, as well as on their manipulation or positioning before and during testing. Among the conditions that should be considered in the selection and use of standards are:

* Standards must be of the same size and shape as the test parts.
* The condition of heat treatment of the standards must be the same as that specified for the test parts.
* Surface finishes of the standards and of the test parts must be the same. If the test parts are ground, plated or coated, the standards must be finished in like manner.
* The magnitude of residual stresses resulting from forming, bending and other cold working operations should be the same in both the standards and the test parts.
* Two or three standards of each of the categories of parts that are to be sorted should be available for use in the event that a standard is lost or mislaid.
* Standards will warm to a moderate temperature as a result of exposure to induced current and will stabilize at this temperature. This rise in temperature will affect readout and require rebalancing of the test instrument.
* Care must be exercised to avoid the use of standards or test parts that have been magnetized and not fully demagnetized. Residual magnetism will result in false readings.
* The metallurgical characteristics of standards should be established by thorough metallographic examination and by hardness tests and other appropriate tests as required.

Interpretation of Readout. The success of all electromagnetic sorting and analysis depends at the outset on the interpretation of readout—the nature of readout varying with the type of readout instrumentation used. Trace patterns obtained on the screen of an oscilloscope and readings obtained on a precision voltmeter are among the most common readout media. In all cases, initial readouts are obtained with known standards; these readouts, in turn, are compared with the readouts obtained from test samples. For practical reasons, the differences that exist between reference standards must be known at the outset and must be carefully defined in most applications.

Much effort has been expended in correlating various complicated oscilloscope patterns with differences in metallurgical properties and structure. A significant increase in hardness, for instance, appears to have a striking effect on the shape of oscilloscope pattern—namely, an increase in angularity. This effect is shown by the three patterns in Fig. 9, which differentiate three

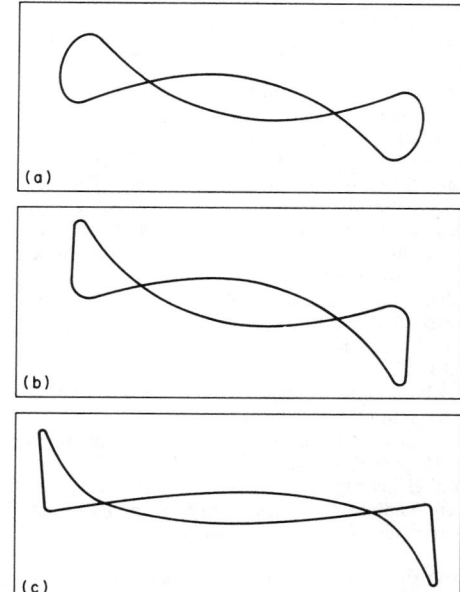

(a)

(b)

(c)

Fig. 9. Oscilloscope lissajous patterns for a steel of the same composition in three conditions: annealed (a), and hardened and tempered to medium hardness (b) and high hardness (c)

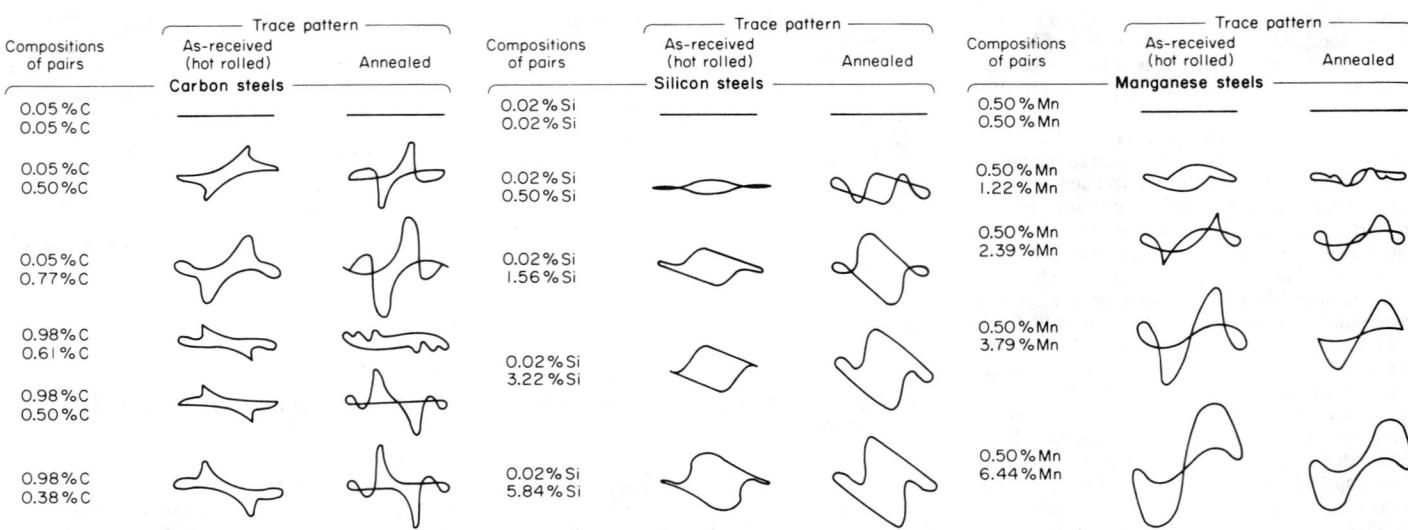

Fig. 10. Oscilloscope lissajous patterns obtained on a comparator-bridge instrument for three series of steels: a carbon, a silicon and a manganese series. See text.

conditions of heat treatment of a steel of the same chemical composition—annealed (Fig. 9a), hardened and tempered to medium hardness (Fig. 9b), and hardened and tempered to high hardness (Fig. 9c). The increase in angularity of the oscilloscope patterns is attributed to increased internal residual stress.

Sorting by Chemical Composition. Three series of steels—a carbon series, a silicon series and a manganese series—were prepared; in each of the series, the element named was made the major variable. These steels were prepared in an effort to obtain patterns on the oscilloscope of a comparator-bridge instrument that would (a) differentiate each of the three elements and (b) differentiate varying amounts of each of the elements. The oscilloscope patterns obtained, together with the amounts of each of the three elements contained in the steels, are given in Fig. 10. Also given are the patterns obtained when each of the steels was annealed from the as-received (hot rolled) condition.

Sorting Based on Hardness. Low-frequency comparator-bridge instruments are commonly used to sort steel parts on the basis of hardness. The reference and test coils are initially balanced with sample parts of known hardness. In subsequent tests, parts of unknown hardness are substituted for one of the reference coils. The degree of unbalance that results is then correlated with an increase or decrease in hardness.

Following heat treatment of a batch of steel bolts, 13 mm ($\frac{1}{2}$ in.) in diameter by 75 mm (3 in.) long, it was determined by hardness testing that some of the bolts were considerably softer than others. The difference in hardness amounted to 15 points on the Rockwell C (HRC) scale. Further investigation revealed that the soft bolts were made of 1020 steel, whereas the hard bolts were made of 1035 steel—the material specified for the application. Because the soft bolts did not meet strength requirements, it was essential that they be sorted from the hard bolts.

A soft bolt was placed in the test coil of a comparator-bridge instrument, and the current level and harmonic phase were adjusted to give the oscilloscope pattern shown in Fig. 11(a). Replacing the soft bolt with a hard bolt made of 1035 steel produced the pattern shown in Fig. 11(b).

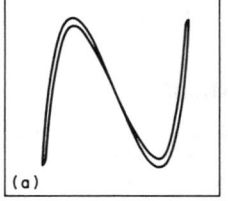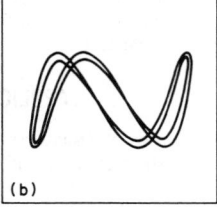

Fig. 11. Oscilloscope lissajous patterns for (a) soft 1020 steel bolts and (b) hard 1035 steel bolts. Both current level and harmonic phase were adjusted to provide the necessary contrast used in sorting.

After testing several hard and soft bolts, it was determined that these patterns did not overlap. Using the patterns as standards, the mixed lot of bolts was sorted at the rate of one bolt per second.

Luminous point patterns, representing three ranges of hardness, are shown as they appeared on the screen of an oscilloscope in Fig. 12. These patterns were developed in connection with the sorting of mixed lots of tapered studs, using a comparator-bridge instrument. The required hardness range established for the steel studs, following hardening and tempering, was Rockwell C 34 to 40. However, some of the studs exceeded this hardness range because they were not tempered, whereas others fell short of it because of surface decarburization. The mixed lots were successfully sorted into three hardness ranges based on the luminous point patterns shown in Fig. 12.

Case Hardening. For many steel components, it is desirable to combine a hard, wear-resistant surface with a tough and relatively ductile core material. This combination is attained by case hardening by any one of several processes, including carburizing, carbonitriding, cyaniding, nitriding, induction hardening and flame hardening.

Depending on the consistency of the magnetic properties of case and core, it is usually possible to measure case depth or to sort on the basis of case depth using electromagnetic sorting equip-

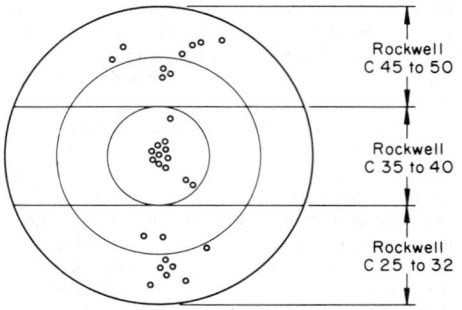

Fig. 12. Luminous point patterns, representing three ranges of hardness, as they appeared on an oscilloscope. The patterns were used in sorting mixed lots of tapered studs on the basis of hardness.

ment. In general, these measurements are more readily obtainable on parts that have been case hardened by induction hardening or by flame hardening than on parts case hardened by other processes. This is because induction-hardened and flame-hardened parts undergo changes in metallurgical structure without alteration of chemical composition.

When changes in both metallurgical structure and chemical composition are involved, the problem of obtaining accurate measurements becomes more complicated. Carburized cases are difficult to measure with accuracy using electromagnetic methods, and no satisfactory measurements of case depth have been obtained with carbonitrided or nitrided cases. Comparative or qualitative analyses are achieved more readily than are quantitative measurements.

When samples from a production lot of 700 000 case-hardened steel pins were sectioned for hardness testing, it was determined that the core hardness of some pins did not meet the minimum hardness requirement of Rockwell C 32. However, the case hardness of all pins, as determined by surface-hardness measurements, was satisfactory. Therefore, the unsatisfactory pins could not be sorted on the basis of hardness tests without destroying all the pins.

Radiography

Edited by Harold Berger, Industrial Quality, Inc.

Radiographic Inspection

RADIOGRAPHY is a nondestructive-inspection method that is based on differential absorption of penetrating radiation — either electromagnetic radiation of very short wavelength or particulate radiation — by the part or test piece (object) being inspected. Because of differences in density and variations in thickness of the part, or differences in absorption characteristics caused by variations in composition, different portions of a test piece absorb different amounts of penetrating radiation. Unabsorbed radiation passing through the part can be recorded on film or photosensitive paper, viewed on a fluorescent screen, or monitored by various types of radiation detectors. The term "radiography" usually implies a radiographic process that produces a permanent image on film (conventional radiography) or paper (paper radiography or xeroradiography), although in a broad sense it refers to all forms of radiographic inspection. When inspection involves viewing of a real-time image on a fluorescent screen or image-intensifier, the radiographic process is termed "real-time inspection." When electronic, nonimaging instruments are used to measure the intensity of radiation, the process is termed "radiation gaging." Tomography, a radiation inspection method adapted from the medical computerized axial tomography CAT scanner, provides a cross-sectional view of an inspection object. All the above terms are used mainly in connection with inspection that involves penetrating electromagnetic radiation in the form of x-rays or gamma rays. "Neutron radiography" refers to radiographic inspection using neutrons rather than electromagnetic radiation. This article discusses radiography using x-rays and gamma rays; neutron radiography is discussed in the article which follows.

In conventional radiography, an object is placed in a beam of x-rays and the portion of the radiation that is not absorbed by the object impinges on a detector such as film. The unabsorbed radiation exposes the film emulsion, similar to the way that light exposes film in photography. Development of the film produces an image that is a two-dimensional "shadow picture" of the object. Variations in density, thickness and composition of the object being inspected cause variations in the intensity of the unabsorbed radiation and appear as variations in photographic density (shades of gray) in the developed film. Evaluation of the radiograph is based on a comparison of the differences in photographic density with known characteristics of the object itself or with standards derived from radiographs of similar objects of acceptable quality.

USES OF RADIOGRAPHY

Radiography is used to detect features of a component or assembly that exhibit differences in thickness or physical density compared with surrounding material. Large differences are more easily detected than small ones. In general, radiography can detect only those features that have a reasonable thickness or radiation path length in

a direction parallel to the radiation beam. This means that the ability of the process to detect planar discontinuities such as cracks depends on proper orientation of the test piece during inspection. Discontinuities such as voids and inclusions, which have measurable thickness in all directions, can be detected as long as they are not too small in relation to section thickness. In general, features that exhibit differences in absorption of a few percent compared with the surrounding material can be detected.

Applicability. Radiographic inspection is used extensively on castings and weldments, particularly where there is a critical need to ensure freedom from internal flaws. For instance, radiography often is specified for inspection of thick-wall castings and weldments for steam-power equipment (boiler and turbine components and assemblies) and other high-pressure systems. Radiography also can be used on forgings and mechanical assemblies. When used with mechanical assemblies, radiography provides a unique NDT capability of inspecting for condition and proper placement of components. Certain special devices are more satisfactorily inspected by radiography than by other methods. For instance, radiography is well suited to the inspection of semiconductor devices for cracks, broken wires, unsoldered connections, foreign material and misplaced components, whereas other methods are limited in ability to inspect semiconductor devices.

Sensitivity of x-radiography, real-time x-ray methods and gamma-ray radiography to various types of flaws depends on many factors, including type of material, type of flaw, and product form. (Type of material in this context is usually expressed in terms of atomic number — for instance, metals having low atomic numbers are classified as light metals and those having high atomic numbers as heavy metals.) Table 1 indicates the general degrees of suitability of the

Table 1. Comparison of suitabilities of three radiographic methods for inspection of light and heavy metals

Inspection application	Suitability for light metals(a)			Suitability for heavy metals(a)		
	X-ray	Real-time radiography(b)	Gamma ray	X-ray	Real-time radiography(b)	Gamma ray
General:						
Surface cracks(c)	F(d)	F(d)	F(d)	F(d)	F(d)	F(d)
Internal cracks	F(d)	F(d)	F(d)	F(d)	F(d)	F(d)
Voids	G	G	G	G	G	G
Thickness	F	F	F	F	F	F
Metallurgical variations	F	F	F	F	F	F
Sheet and plate:						
Thickness	G(e)	G(e)	G(e)	G(e)	U	G(e)
Laminations	U	U	U	U	U	U
Voids	G	G	G	G	G	G
Bar and tube:						
Seams	P	P	P	P	P	P
Pipe	G	G	G	G	F	F
Cupping	G	G	G	G	F	F
Inclusions	F	F	F	F	F	F
Castings:						
Cold shuts	G	G	G	G	G	G
Surface cracks	F(d)	F(d)	F(d)	F(d)	F(d)	F(d)
Internal shrinkage	G	G	G	G	G	G
Voids, pores	G	G	G	G	G	G
Core shift	G	G	G	G	G	G
Forgings:						
Laps	P(d)	P(d)	P(d)	P(d)	U	U
Inclusions	F	F	F	F	F	U
Internal bursts	G	G	G	F	F	G
Internal flakes	P(d)	P(d)	U	P(d)	P(d)	U
Cracks and tears	F(d)	F(d)	F(d)	F(d)	F(d)	F(d)
Welds:						
Shrinkage cracks	G(d)	G(d)	G(d)	G(d)	G(d)	G(d)
Slag inclusions	G	G	G	G	G	G
Incomplete fusion	G	G	G	G	G	G
Pores	G	G	G	G	F	G
Incomplete penetration	G	G	G	G	G	G
Processing:						
Heat treating cracks	U	F	U	P	P	U
Grinding cracks	U	F	U	U	U	U
Service:						
Fatigue and heat cracks	F(d)	F(d)	P(d)	P	P	P
Stress corrosion	F	F	P	F	F	P
Blistering	P	P	P	P	P	P
Thinning	F	F	F	F	F	F
Corrosion pits	F	F	P	G	G	P

(a) G = good, F = fair, P = poor, U = unsatisfactory. (b) Real-time radiography offers the advantage that the part can be manipulated to present the best view — for example, to align a crack. Also, when microfocus, magnification methods are used, real-time radiography presents excellent resolution and contrast. (c) Includes only visible cracks. Minute surface cracks normally are undetectable by radiographic inspection methods. (d) Radiation beam must be parallel to the cracks, laps or flakes. (e) When calibrated using special thickness gages.

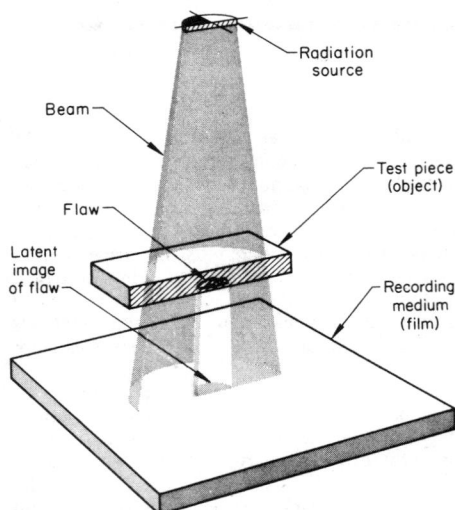

Fig. 1. Diagram of the basic elements of a radiographic system, showing method of detecting and recording an internal flaw in a plate of uniform thickness

three main radiographic methods for detection of discontinuities in various product forms and applications. In some instances, radiography cannot be used even though it appears suitable from Table 1, because the part is accessible from one side only. Both sides must be accessible for radiography.

Radiography can be used to inspect most types of solid material, with the possible exception of assemblies containing materials of very high or very low density. (Neutron radiography, however, often can be used in such instances, as discussed in the article that follows.) Both ferrous and nonferrous alloys can be radiographed, as can nonmetallic materials and composites.

Limitations. Compared with other nondestructive methods of inspection, radiography is expensive. Relatively large capital costs and space allocations are required for a radiographic laboratory or a real-time inspection station. On the other hand, when portable x-ray or gamma-ray sources are used, capital costs can be relatively low. Operating costs can be high; large percentages of the total inspection time is spent in setting up for radiography. With real-time radiography, operating costs usually are much lower,

because setup times are shorter and there are no extra costs for x-ray film and processing.

Field inspection of thick sections is a time-consuming process. Portable x-ray sources generally emit relatively low-energy radiation, up to about 400 keV, and also are limited as to the intensity of radiation output. These characteristics of portable sources combine to limit x-radiography in the field to sections having absorption equivalent to that of about 75 mm (3 in.) of steel. Radioactive sources also are limited in the thickness that can be inspected, primarily because high-activity sources require heavy shielding for protection of personnel. This limits field usage to sources of lower activity that can be transported in relatively lightweight containers. Because portable x-ray and gamma-ray sources are limited in effective radiation output, exposure times usually are long for thick sections. Recent developments, such as a portable linear accelerator, can speed up and increase the penetrating power of field radiographic methods.

Certain types of flaws are difficult to detect by radiography. Laminar defects such as cracks present problems unless they are essentially parallel to the radiation beam. Tight, meandering cracks in thick sections usually cannot be detected even when properly oriented. Minute discontinuities such as inclusions in wrought material, flakes, microporosity and microfissures cannot be detected unless they are sufficiently segregated to yield a detectable gross effect. Laminations normally are not detectable by radiography because of their unfavorable orientation—usually parallel to the surface. Laminations seldom yield differences in absorption that enable laminated areas to be distinguished from lamination-free areas.

PRINCIPLES OF RADIOGRAPHY

Three basic elements—a radiation source or probing medium, the test piece or object being evaluated, and a recording medium (usually film)—combine to produce a radiograph. These elements are shown schematically in Fig. 1. The test piece in Fig. 1 is a plate of uniform thickness containing an internal flaw that has absorption characteristics different from those of the surrounding material. Radiation from the source is absorbed by the test piece as the radiation passes through it; the flaw and surrounding material absorb different amounts of radiation. Thus, the amount of radiation that reaches the film in the area beneath the flaw is different from the amount

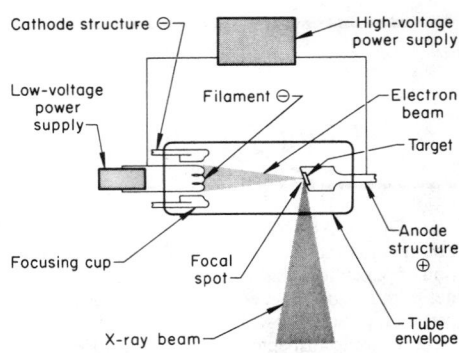

Fig. 3. Schematic diagram of the principal components of an x-ray unit

that impinges on adjacent areas. This produces on the film a latent image of the flaw that, when the film is developed, can be seen as a "shadow" of different photographic density from that of the image of the surrounding material.

Geometric Factors in Radiography. Because a radiograph is a two-dimensional representation of a three-dimensional object, the radiographic images of most test pieces are somewhat distorted in size and shape.

In conventional radiography, the position of a flaw within the volume of a test piece cannot be determined exactly with a single radiograph; depth in the direction of the radiation beam cannot be determined exactly. Conclusions regarding depth sometimes can be drawn from the sharpness of the flaw image. Images of flaws close to the detector tend to appear sharper than images of flaws near the source side of the object. However, techniques such as stereoradiography, tomography, triangulation, or simply making two or more exposures (with the radiation beam being directed at the test piece from a different angle for each exposure) can be used to locate flaws more exactly within the test-piece volume.

SOURCES OF RADIATION

Two types of radiation are used for nondestructive inspection—electromagnetic radiation and particulate radiation. Figure 2 shows a portion of the electromagnetic spectrum that includes several types of electromagnetic radiation. Only x-rays and gamma rays, which have relatively short wavelengths, have the capability of penetrating opaque materials to image internal flaws. X-rays and gamma rays differ from other types of electromagnetic radiation (including visible light, microwaves and radio waves) only in their wavelengths and, as shown in Fig. 2, there is not always a distinct transition from one type of electromagnetic radiation to another.

X-rays and gamma rays are physically indistinguishable; they differ only in the manner in which they are produced. X-rays result from the interaction between a rapidly moving stream of electrons and atoms in a solid target material or from electron transitions in the atomic shells, whereas gamma rays are emitted during the radioactive decay of unstable atomic nuclei.

X-ray tubes are electronic devices that convert electrical energy into x-rays. Typically, an x-ray tube consists of a cathode structure containing a filament and an anode structure containing a target—all within an evacuated chamber or envelope (see Fig. 3).

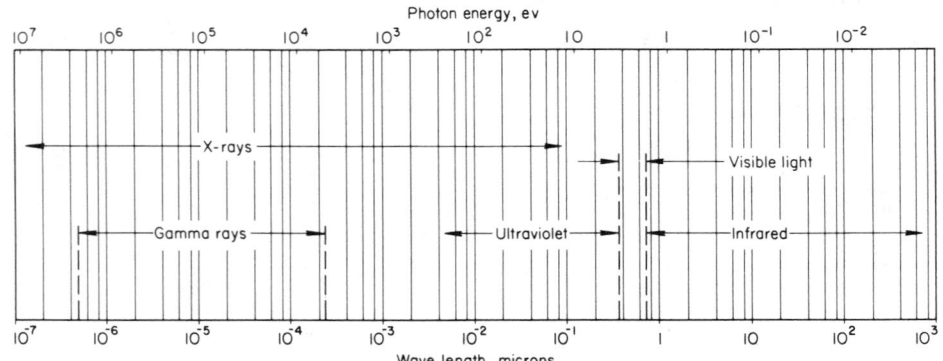

Fig. 2. Schematic representation of the portion of the electromagnetic spectrum that includes x-rays, gamma rays, ultraviolet and visible light, and infrared radiation, showing their relationship with wave length and photon energy

The cathode structure incorporates a filament and a focusing cup, which surrounds the filament. The focusing cup functions as an electrostatic lens whose purpose is to direct the electron beam toward the anode. The filament, usually a coil of tungsten wire, is heated to incandescence by an electric current produced by a relatively low voltage, similar to the operation of an ordinary incandescent light bulb. The heated filament emits electrons, which are accelerated across the evacuated space between the cathode and the anode. The driving force for acceleration is a high electrical potential (voltage) between anode and cathode, which is applied during exposure.

The anode usually consists of a button of the target material embedded in a mass of copper that absorbs much of the heat generated by electron collisions with the target. Tungsten is the preferred target material for x-ray tubes used in radiography because tungsten is (a) an efficient emitter of x-rays because of its high atomic number, and (b) withstands high temperatures of operation because of its high melting point. Gold and platinum also are used in x-ray tubes for radiography, but targets made of these metals must be more effectively cooled than targets made of tungsten. Most industrial x-ray tubes have forced liquid cooling to dissipate the large amounts of anode heat generated during operation.

Production of X-Rays. A low-voltage power supply, usually controlled by a rheostat, generates the electric current that heats the filament to incandescence. This creates an electron cloud around the filament. When a high voltage is applied between the cathode and the anode, electrons are accelerated across the evacuated space between anode and cathode and strike the target. The focusing cup is shaped so that the beam of electrons is concentrated on a small area of the target, called the focal spot. In the portion of the target immediately beneath the focal spot, electrons are slowed and absorbed, and both bremsstrahlung (a broad spectrum of x-radiation whose minimum wavelength is determined by the accelerating voltage in the x-ray tube: literally, braking radiation) and characteristic x-rays are produced.

Most of the energy in the impinging electron beam is transformed into heat, which must be dissipated. Severe restrictions are imposed on the design and selection of materials for the anode and target to ensure that structural damage from overheating does not prematurely destroy the target. Anode heating also limits the size of the focal spot. Because smaller focal spots produce sharper radiographic images, the design of the anode and target represents a compromise between maximum radiographic definition and maximum target life. Usually, the electron spot on the target is made to appear as a smaller focal spot by inclining the anode face at a small angle (usually about 20°) to the centerline of the x-ray beam, as shown in Fig. 4.

There are three important electrical characteristics of x-ray tubes; (a) the filament current, which controls the filament temperature and in turn the quantity of electrons emitted; (b) the tube voltage, or anode-to-cathode potential, which controls the energy of impinging electrons and thus the energy, or penetrating power, of the x-ray beam; and (c) the tube current, which is directly related to filament temperature and is usually referred to as the milliamperage of the tube. The intensity of a radiation beam of given energy is approximately proportional to milliamperage,

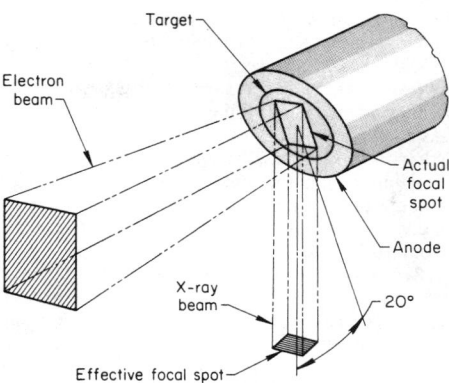

Fig. 4. Schematic diagram of the actual and effective focal spots of an anode that is inclined at 20° to the centerline of the x-ray beam

which is used as one of the variables in exposure calculations. Intensity of x-rays or gamma rays is actually measured in roentgens, one roentgen being the amount of radiation that produces ions carrying one electrostatic unit of electricity in 1.293 mg of air. The roentgen is a measure of the energy absorbed in a given volume of air; it corresponds to 83.3 ergs/g. A unit called the rad is the amount of radiation that leads to the absorption of 100 ergs/g. The SI unit for absorbed dose is the gray (Gy); 1 Gy = 100 rads = 1 J/kg.

Table 2 gives penetrating capabilities of x-ray beams of various energy levels expressed as the range of steel thickness that can be satisfactorily inspected. The maximum values in this table represent thicknesses of steel that can be routinely inspected using exposures of several minutes' duration and with medium-speed film. Thicker sections can be inspected for each x-ray energy value by using faster films and long exposure times, but for routine work the use of higher-energy x-rays is more practical. Sections thinner

Table 2. Penetrating capabilities of conventional x-ray tubes and high-energy sources

Maximum accelerating potential	Penetration range for steel	
	mm	in.
X-ray tubes		
150 kV	Up to 16	Up to ⁵/₈
250 kV	Up to 38	Up to 1¹/₂
400 kV	Up to 64	Up to 2¹/₂
1000 kV (1 MV)	6.4 to 89	¹/₄ to 3¹/₂
High-energy sources		
2.0 MeV	6.4 to 250	¹/₄ to 10
4.5 MeV	25 to 305	1 to 12
7.5 MeV	57 to 460	2¹/₄ to 18
20.0 MeV	75 to 610	3 to 24

than minimum thicknesses shown in Table 2 can easily be penetrated, but radiographic contrast may not be optimum.

Gamma rays are high-energy electromagnetic waves of relatively short wavelength that are emitted during the radioactive decay of both naturally occurring and artificially produced unstable isotopes. In all respects other than origin, gamma rays and x-rays are identical. Unlike the broad-spectrum radiation produced by an x-ray tube, gamma-ray sources emit one or more discrete wavelengths of radiation, each having its own characteristic photon energy.

Many of the elements in the periodic table either have naturally occurring radioactive isotopes or have isotopes that can be made radioactive by irradiation usually with neutrons in the core of a nuclear reactor. However, only a few radioisotopes — thulium-170, iridium-192, cesium-137, cobalt-60 and radium — have been used extensively for radiography.

Radiation Characteristics of Gamma-Ray Sources. Because gamma radiation is produced by the radioactive decay of unstable atomic nuclei, there is a continuous reduction in the intensity of emitted radiation with time as more and more unstable nuclei transform to stable nuclei. This reduction follows a logarithmic law, and each radioactive isotope has a characteristic "half-life," or amount of time that it takes for the intensity of emitted radiation to be reduced by one half. The term "half-life" should not be misinterpreted as meaning that the intensity of emitted radiation will be zero at the end of a second half-life. Rather, the intensity remaining at the end of a second half-life will be one-half of the intensity at the end of the first half-life, or one-fourth of the initial intensity. The intensity at the end of the third half-life will be one-eighth of the initial intensity, and so on.

The more important characteristics of gamma-ray sources are summarized in Table 3. Source strength and specific activity are not listed, because these characteristics vary according to the physical size of the source, the material and design of its encapsulation, and the degree of concentration at the time the source was originally produced. At present, the well-used radioisotope sources, iridium-192 and cobalt-60, are available with very high specific activities. Therefore, these sources can provide short exposure times and good image sharpness in field applications.

ATTENUATION OF ELECTROMAGNETIC RADIATION

X-rays and gamma rays interact with any substance, even gases such as air, as the rays pass through the substance. It is this interaction that enables parts to be inspected by differential attenuation of radiation, and that enables differences in the intensity of radiation to be detected

Table 3. Characteristics of gamma-ray sources used in industrial radiography

Gamma-ray source	Half-life	Photon energy, MeV	Radiation output, rhm/ci(a)	Penetrating power for steel	
				mm	in.
Thulium-170	128 days	0.054 and 0.084(b)	0.003	13	¹/₂
Iridium-192	74 days	12 rays from 0.21 to 0.61	0.55	75	3
Cesium-137	33 years	0.66	0.39	75	3
Cobalt-60	5.3 years	1.17 and 1.33	1.35	230	9
Radium	1620 years	11 principal rays from 0.24 to 2.20(c)	0.84	125	5

(a) Output for typical unshielded, encapsulated sources; rhm/ci = roentgens per hour at one metre per curie. (b) Against strong background of higher-MeV radiation. (c) Derived primarily from radioactive decay of daughter products.

and recorded. Both these effects are essential to the radiographic process. Attenuation characteristics of materials vary with type, intensity and energy of the radiation, and with density and atomic structure of the material.

The attenuation of electromagnetic radiation is a complex process. The intensity of radiation varies exponentially with the thickness of homogeneous material through which it passes. This behavior is expressed as

$$I = I_0 \exp(-\mu t)$$

where I is the intensity of the emergent radiation, I_0 is the initial intensity, t is the thickness of homogeneous material and μ is a characteristic of the material known as the linear-absorption coefficient. The coefficient μ is constant for a given situation, but varies with the material and with the photon energy of the radiation. The units of μ are reciprocal length (for instance, cm^{-1}). The absorption coefficient of a material is sometimes expressed as a mass-absorption coefficient (μ/ρ) where ρ is the density of the material.

There are three primary attenuation processes: the photoelectric effect, Compton scattering and pair production.

Photoelectric effect is an interaction with orbital electrons in which a photon of electromagnetic radiation is consumed in breaking the bond between an orbital electron and its atom. Energy in excess of the bond strength imparts kinetic energy to the electron.

The photoelectric effect generally decreases with increasing photon energy (E) as $E^{-3.5}$. For elements of low atomic number, the photoelectric effect is negligible at photon energies exceeding about 100 keV. However, the photoelectric effect varies with the fourth to fifth power of atomic number; thus, for elements of high atomic number, the effect accounts for an appreciable portion of total absorption at photon energies up to about 2 MeV.

Compton scattering is a form of direct interaction between an incident photon and an orbital electron in which the electron is ejected from the atom and only a portion of the kinetic energy of the photon is consumed. The photon is scattered incoherently, emerging in a direction that is different from the direction of incident radiation and emerging with reduced energy. The relationship of the intensity of the scattered beam to the intensity of the incident beam, scattering angle and photon energy in the incident beam is complex, yet is amenable to theoretical evaluation. Compton scattering varies directly with atomic number of the scattering element, and approximately inversely with photon energy in the energy range that is of major interest.

Pair production is an absorption process that creates two 0.5-MeV photons of scattered radiation for each photon of high-energy incident radiation consumed; a small amount of scattered radiation of lower energy also accompanies pair production. Pair production is more important for heavier elements; the effect varies with atomic number, Z, approximately as $Z(Z + 1)$. The effect also varies approximately logarithmically with photon energy.

In pair production, a photon of incident electromagnetic radiation is consumed in creating an electron-positron pair that then is ejected from an atom. This effect is possible only at photon energies exceeding 1.02 MeV, because, according to the theory of relativity, 0.51 MeV is consumed in the creation of the mass of each particle, elec-

tron or positron. Any energy of the incident photon exceeding 1.02 MeV imparts kinetic energy to the pair of particles.

Total absorption is the sum of the absorption or scattering effects of all the attenuation processes. As an example, Fig. 5 shows the variation of the mass-absorption coefficient for uranium as a function of photon energy, and indicates the amount of total absorption that is attributable to the various atomic processes over the entire range from 10 keV to 100 MeV.

Radiographic Equivalence. The absorption of x-rays and gamma rays by various materials becomes less dependent on composition as radiation energy increases. For instance, at 150 kV, 25 mm (1 in.) of lead is equivalent to 350 mm (14 in.) of steel, but at 1000 kV, 25 mm of lead is equivalent to only 125 mm (5 in.) of steel. Approximate radiographic absorption equivalence factors for several metals are given in Table 4. When exposure charts are available only for certain common materials (such as steel or aluminum), exposure times for other materials can be estimated by determining the exposure time for an equal thickness of a common material from the chart, then multiplying by the radiographic equivalence factor.

PRINCIPLES OF SHADOW FORMATION

The image formed on a radiograph is similar to the shadow cast on a screen by an opaque object placed in a beam of light. Although radiation used in radiography penetrates an opaque object whereas light does not, the geometric laws of shadow formation are basically the same. X-rays, gamma rays and light all travel in straight lines. Straight-line propagation is the chief characteristic of radiation that permits formation of a sharply discernible shadow. The geometric relationships of source, object and screen to each other determine the three main characteristics of the shadow—the degrees of enlargement, distortion and unsharpness (see Fig. 6).

Enlargement. The shadow of the object (test piece) is always farther from the source than the object itself. Thus, as illustrated for a point source in Fig. 6(a), dimensions of the shadow are always greater than corresponding dimensions of the object. Mathematically, the size of the image or degree of enlargement may be calculated from the relationship:

$$M = S_i/S_o = L_i/L_o$$

where M is the degree of enlargement (magnification), S_i is the size of the image, S_o is the size

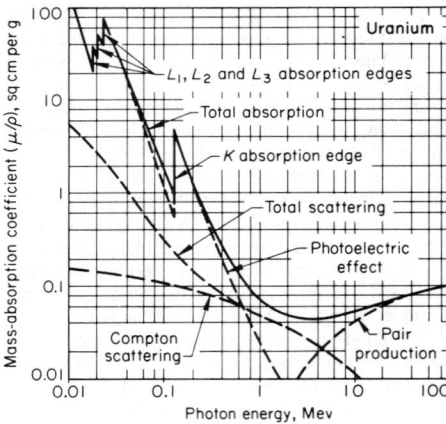

Fig. 5. Calculated mass-absorption coefficient for uranium as a function of photon energy (solid line) and contributions of various atomic processes (dashed lines).

of the object, L_i is the source-to-image distance, and L_o is the source-to-object distance.

With very small focal spots, large values of geometric magnification can be used effectively. Values of 6 to 20 are common; magnification values as high as 100 can be used. Focal spots in microfocal x-ray equipment range from 5 to 20 μm (0.0002 to 0.0008 in.). In addition to increased image size, magnification systems also offer improved contrast because radiation scattered in the object does not reach the detector.

Distortion. As long as the plane of a two-dimensional object and the plane of the recording surface are parallel to each other, the image of that object plane will be undistorted regardless of the angle at which the beam of radiation impinges on the object. Also, the degree of enlargement for different points in a given object plane is constant because the ratio L_i/L_o is invariant. However, as shown in Fig. 6(b), if the plane of the object and the plane of the recording surface are not parallel, the image will be distorted. For objects of appreciable thickness, the magnification for different object planes will vary because L_o varies.

Geometric Unsharpness. In reality, most radiation sources are too large to be approximated by a point. Most conventional x-ray tubes have focal spots several millimetres in size. Even high-energy sources have focal spots of appreciable size, although seldom exceeding 2 mm (0.08 in.) in

Table 4. Approximate radiographic absorption equivalence for various metals

Material	X-rays, kV 50	100	150	220	400	X-rays, MeV 1	2	4 to 25	Gamma rays Ir-192	Cs-137	Co-60	Ra
Magnesium	0.6	0.6	0.05	0.08	...	...	...	...	...	...	...	...
Aluminum	1.0	1.0	0.12	0.18	...	...	...	...	0.35	0.35	0.35	0.40
Aluminum alloy												
2024	2.2	1.6	0.16	0.22	...	...	...	...	0.35	0.35	0.35	...
Titanium	...	...	0.45	0.35	...	...	...	...	...	...	...	...
Steel	12.0	1.0	1.0	1.0	1.0	1.0	1.0	1.0	1.0	1.0	1.0	1.0
18-8 stainless steel	12.0	1.0	1.0	1.0	1.0	1.0	1.0	1.0	1.0	1.0	1.0	1.0
Copper	18.0	1.6	1.4	1.4	1.4	...	...	1.3	1.1	1.1	1.1	1.1
Zinc	...	1.4	1.3	1.3	1.3	...	...	1.2	1.1	1.0	1.0	1.0
Brass(a)	...	1.4	1.3	1.3	1.3	1.2	1.2	1.2	1.1	1.1	1.1	1.1
Inconel alloys	...	16.0	1.4	1.3	1.3	1.3	1.3	1.3	1.3	1.3	1.3	1.3
Zirconium	...	...	2.3	2.0	...	1.0	...	...	...	...	...	...
Lead	...	...	14.0	12.0	...	5.0	2.5	3.0	4.0	3.2	2.3	2.0
Uranium	...	...	...	25.0	...	...	...	3.9	12.6	5.6	3.4	...

(a) Containing no tin or lead; absorption equivalence is greater than these values when either element is present.

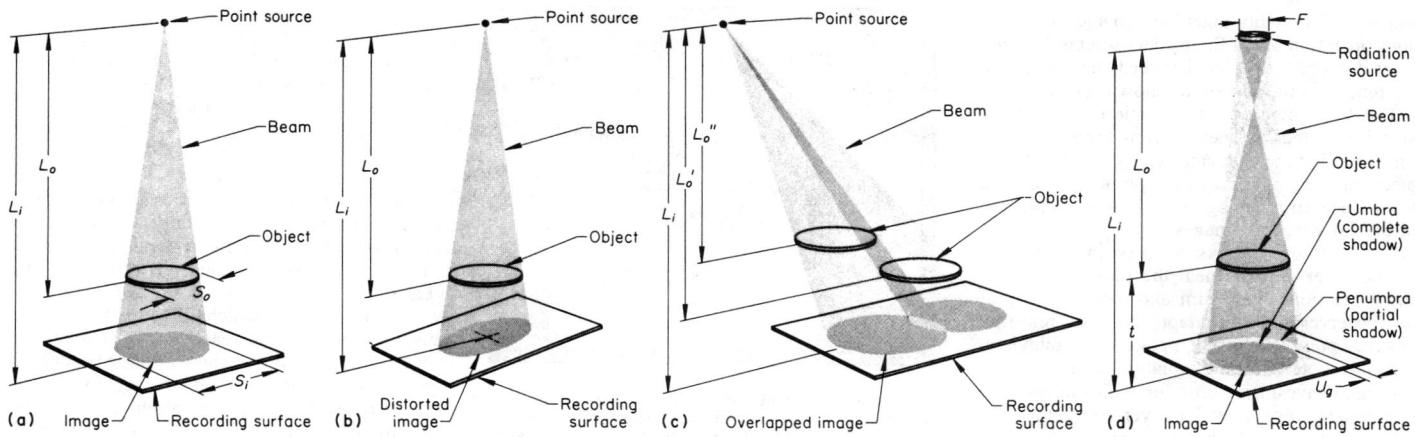

L_o = Source-to-object distance; L_i = source-to-image distance; S_o = size of object; S_i = size of image; U_g = geometric unsharpness; F = size of focal spot; t = object-to-image distance.

Fig. 6. Schematic representation of the effect of geometric relationships on radiographic image from point sources and actual radiation source. (a) Image size, (b) image distortion and (c) image overlap for point sources of radiation; (d) degree of image unsharpness from an actual radiation source. See text for discussion.

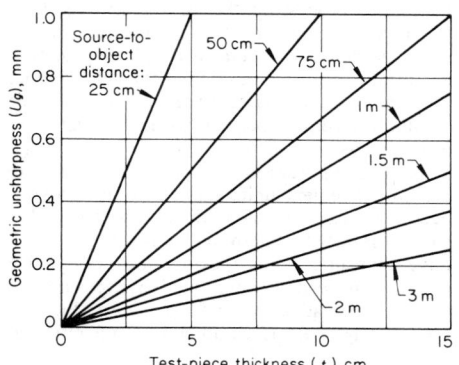

Fig. 7. Relation of geometric unsharpness to test-piece thickness for various source-to-object distances when the source is 5 mm in diameter

diameter. Gamma-ray sources vary widely in size, depending on source strength and specific activity, but seldom are less than about 2.5 mm (0.1 in.) in diameter.

Geometric unsharpness is one of several unsharpness factors, and, at low and medium x-ray energies, is usually the largest contributor to maximum unsharpness. Neglecting the distance between the actual surface of the recording medium and the adjacent (facing) surface of the test piece, which usually is quite small in relation to test-piece thickness, the geometric unsharpness can be calculated for any source size, and can be expressed as a series of straight-line plots relating geometric unsharpness, U_g, to test-piece thickness, t, for various values of source-to-object distance, L_o. A typical series is shown in Fig. 7 for a 5-mm-diam (0.2-in.-diam) source. It is helpful to prepare graphs like the one in Fig. 7 for each source size used.

RECORDING METHODS

Variations in the intensity of x-rays or gamma rays that pass through a material can be presented (a) as a visible permanent image, (b) as a visible real-time image, (c) as a meter reading or (d) graphically on a strip chart. By definition, radiography implies that the recording medium provides a visible permanent image. Permanent images are recorded on x-ray film, radiographic paper, or electrostatically sensitive paper such as is used in the xeroradiographic process, and are called radiographs. Real-time images, such as those presented on a fluorescent screen, image amplifier or television monitor, differ in appearance from those on radiographs; records of these images may be made by photography or video recording. If the information is sensed or recorded using radiation-measuring instruments and does not appear as an image, the recording process is termed "radiation gaging." X-ray film is used more extensively than all other recording media combined.

X-ray film is constructed of a thin, transparent plastic support called a film base, which usually is coated on both sides (but occasionally on one side only) with an emulsion consisting mainly of grains of silver salts that are embedded in gelatin (see Fig. 8). These salts are very sensitive to electromagnetic radiation, especially x-rays, gamma rays and visible light. The film base, usually tinted blue, is about 0.18 mm (0.007 in.) thick. An adhesive undercoat fastens the emulsion to the film base. A very thin but tough coating of gelatin called a protective overcoat covers the emulsion to protect it against minor abrasion. The total thickness of the x-ray film is approximately 0.23 mm (0.009 in.), including film base, two emulsions, two adhesive undercoats and two protective overcoats.

Radiographic Paper. Ordinary photographic paper can be used to record x-ray images, although its characteristics are not always satisfactory. Photographic paper has a low speed, and the resulting image is low in contrast. However, photographic paper in various forms can be used effectively for some applications.

Radiographic paper can exhibit excellent sensitivity, which in many respects matches or exceeds that of fast direct-exposure x-ray films. Radiographic paper does not match the sensitivities of slow x-ray films, but because of their speed, convenience and low cost, radiographic papers are being used both for radiography of materials that do not require critical examination and for "in-process" control.

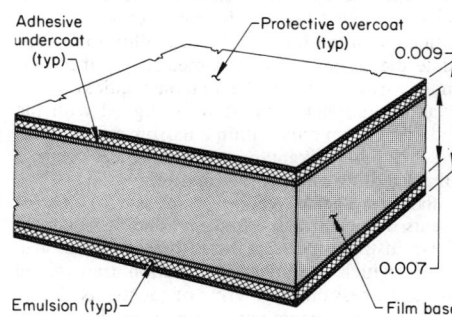

Fig. 8. Schematic cross section of a typical x-ray film

Xeroradiography (dry radiography) is a form of imaging that uses electrostatic principles for formation of a radiographic image. In film radiography, a latent image is formed in the emulsion of a film; in xeroradiography, the latent image is formed on a plate coated with a photoconductive layer of selenium. Before use, the plate is given an even charge of static electricity over the entire surface. As soon as the plate is charged, it becomes sensitive to light as well as to x-radiation and must be protected from light by a rigid holder similar to a film cassette. In practice, the holder is used for radiography as though it contained film. X-radiation will differentially discharge the plate according to the amount of radiation received by different areas. This forms an electrostatic latent image of the test piece on the plate.

Development of the exposed plate is done by subjecting the plate, in the absence of light, to a cloud of fine powder charged opposite to the electrostatic charges remaining on the plate. The charged powder is attracted to the residual charges on the plate. The visible radiographic image can be made permanent by placing a piece of specially treated paper over the plate and transferring the powder to the paper, which then is heated to fix the powder in place.

Selenium-coated plates can be easily damaged by fingerprints, dirt and abrasion. For this reason, automated equipment is used for charging, and for development and image transfer to paper.

Fluoroscopy, or real-time radiography, produces x-ray images very quickly. A simple fluoroscope consists of an x-ray machine capable of continuous operation and a fluorescent screen. The test piece (object) is placed between the source (x-ray tube) and the screen, as shown schematically in Fig. 9. There are many variations of this basic unit that protect the viewer from exposure to excessive radiation, that can move or rotate the test piece, and that also may incorporate attachments to produce still photographs or motion pictures of the fluoroscopic images.

Simple fluoroscopy is not used much because the observer must be near the radiation and because the images are dim and therefore require the observer to dark-adapt. Modern real-time systems present a bright image on a television screen remote from the radiation area.

Capital investment cost for real-time systems is comparatively high; however, once this cost has been amortized, the cost per unit inspected usually is lower than for other radiographic-inspection processes.

The chief operational advantage of real-time radiography is the ability to see the image in real time. The test piece may be moved past the screen, or it may be rotated by handling devices, or it may be moved toward the x-ray tube for magnification of the image. The capability to manipulate the inspection object means that the optimum x-ray view can be obtained quickly. For example, a crack may be well aligned with the radiation beam only within a narrow range of angles. Another advantage is convenience, because film handling and processing are avoided.

Radiation Gaging. Radiation-measuring instruments do not produce images. The output from these instruments is a meter reading or a strip chart, which records the radiation transmitted through a test piece in terms of roentgens. Many of these instruments are routinely used to check areas surrounding a radiographic-inspection site for excessive radiation.

Radiation gaging can be applied to certain automated processes, such as thickness gaging of materials or determination of liquid levels in sealed containers. In these applications, it may not be necessary to actually measure the amount of radiation passing through the material, but only to detect changes in the level of radiation—in other words, for a "go, no-go" type of inspection.

When a highly absorbing material such as thick lead or concrete must be inspected for voids and when usual radiographic techniques are impractical, radiation gaging can be used effectively. Voids can be located in these materials by noting increases in the readings of radiation-detecting instruments.

Tomography. Cross-sectional images of an inspection object may be obtained by a series of radiation attenuation measurements all around the object. Typically, a fan beam of radiation about 1 or 2 mm in height is used along with a bank of detectors on the opposite side of the object. The attenuation data permit a computer reconstruction showing density differences in the cross section of the object.

CHARACTERISTICS OF X-RAY FILM

Three general characteristics of film—speed, gradient and graininess—are primarily responsible for performance of the film during exposure and processing and for quality of the resulting image. Film speed, gradient and graininess are interrelated; that is, the faster the film, the larger the graininess and the lower the gradient—and

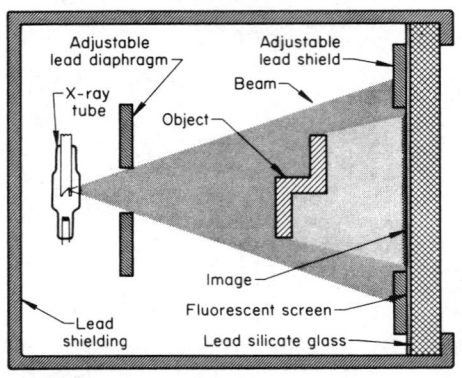

In a modern real-time system, a sensitive TV camera would detect the light or the fluorescent screen would be inside an image-intensifier tube. With either system, viewing would typically be done on a TV monitor remote from the radiation area.

Fig. 9. Diagram of the components and principles of operation of a simple fluoroscope

vice versa. Film speed and gradient are derived from the characteristic curve for a film emulsion, which is a plot of film density versus the exposure required for producing that density in the processed film. Graininess is an inherent property of the emulsion, but can be affected somewhat by conditions of exposure and development.

There are other characteristics of x-ray film that can be used to produce special effects. However, the following discussion is mainly confined to film speed, gradient and graininess, and the factors that affect them.

Density. The quantitative measure of blackening of a photographic emulsion is called density.

Density, measured directly with an instrument called a densitometer, is the logarithm of the ratio of the light intensity incident on the film to that transmitted by the film. Therefore, a film with a density of 1.0 will transmit only 10% of the light, a film with a density of 2.0 will transmit only 1/100 of the light, and so on.

There are two kinds of density: (*a*) the density associated with transparent-base radiographic film, called transmission density; and (*b*) the density associated with opaque-base imaging material such as radiographic paper, called reflection density.

Exposure is the intensity of radiation multiplied by the time during which it acts; that is, the amount of energy that reaches a particular area of the film and that is responsible for producing a particular density on the developed film.

Characteristic Curves. The relation between the exposure applied to a given type of radiographic film and the resulting density is expressed in a curve known as the characteristic curve of that particular type of film. Other names for this curve are H and D curve, D log E curve and sensitometric curve. Such curves are determined by applying a series of known exposures to the film and, after processing the film according to a standard procedure, reading the resulting densities. The curve is generated by plotting density against the logarithm of relative exposure. Figure 10(a) shows the characteristic curves of three commercial films exposed to x-radiation between lead screens.

Relative exposure is used partly because there are no convenient units suitable to all kilovoltages and scattering conditions, and partly because it is easy to determine the logarithm of relative exposure. Using the logarithm of relative exposure instead of only relative exposure has several other advantages; for instance, the otherwise long scale is compressed, and ratios of

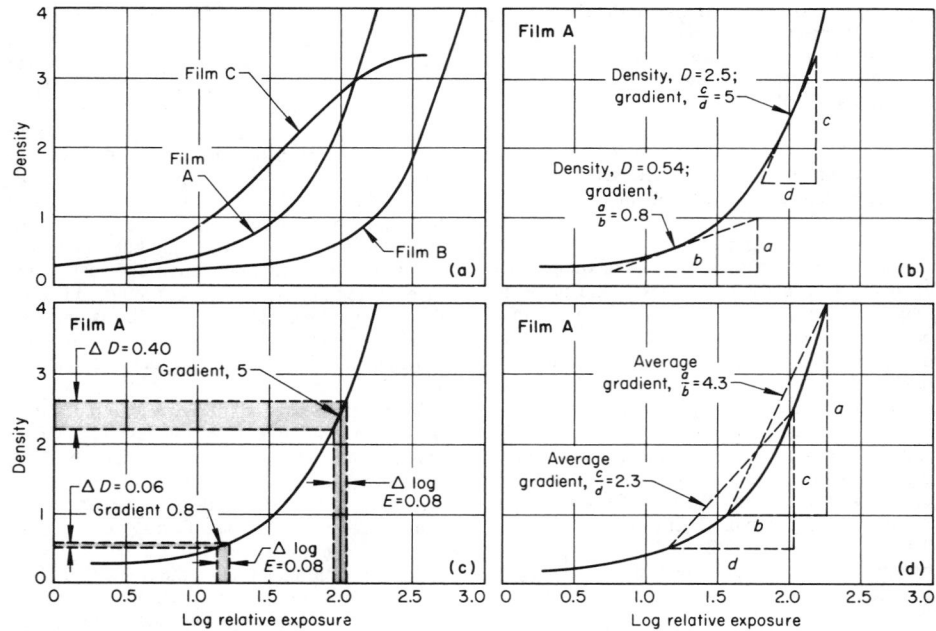

(a) Typical curves for three industrial x-ray films exposed to x-radiation between lead screens. (b) Evaluation of gradients at two points on the curve for film A in (a). (c) Density differences (ΔD) corresponding to a 20% difference in relative exposure (Δ log E = 0.08) determined for the two values of gradients evaluated in (b). (d) Average gradients for film A determined over two density ranges.

Fig. 10. Characteristic curves for x-ray film that determine type of film and film gradient, speed and density

Table 5. General characteristics of four types of radiographic film (ASTM E94)

Film type	Film characteristic Speed	Gradient	Graininess
1	Low	Very high	Very fine
2	Medium	High	Fine
3	High	Medium	Coarse
4(a)	Very high(b) Medium(d)	Very high(b) Medium(d)	(c) Medium(d)

(a) Normally used with fluorescent screens. (b) When used with fluorescent screens. (c) Graininess is mainly a characteristic of the fluorescent screens. (d) When used for direct exposure or with lead screens.

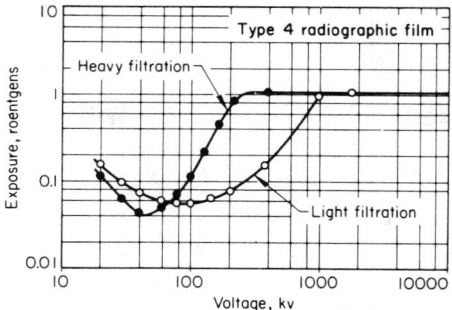

Fig. 11. Spectral-sensitivity curves for a type 4 radiographic film, showing exposure required to produce a density of 1.0

intensities or exposures (which are determined by simply subtracting logarithms) are usually more significant in radiography than actual exposures or intensities.

Characteristic curves are very useful in determining speed and gradient of the film as well as indicating the type of film. For instance, in Fig. 10(a), characteristic curves for films A and B are J-shape curves, which is typical of industrial x-ray film of types 1, 2 and 3 (as defined in Table 5). The curve for film C begins to flatten at a density between 3.0 and 3.5, giving it the S-shape typical of type 4 x-ray films used with fluorescent screens.

Film speed is inversely related to the time required for a given intensity of radiation to produce a particular density on the film—the shorter the exposure, the faster the film. In absolute units, film speed is inversely proportional to the total energy (roentgens) of a particular radiation spectrum (wavelength distribution at a given kilovoltage) that produces a given density on the film. For most practical applications, it is convenient and effective to deal with relative speeds. In using relative speeds, film speeds are expressed in terms of the speed of one particular film whose relative speed is arbitrarily assigned a value, for instance 100. Thus, if film A requires half the exposure of film B, the slower film (film B) is chosen as the standard and assigned an arbitrary speed of 100. Film A, which is twice as fast, would have a relative speed of 200.

Film gradient, also called film contrast, relates to the difference in density between adjacent areas on a radiograph. If the difference is great, the gradient (contrast) is said to be high. If the difference is slight, the gradient is said to be low. The contrast seen on a radiograph is known as radiographic contrast, and is composed of two factors: (*a*) subject contrast, which is the result of variations in the amount of radiation absorbed by the test piece, and which causes variations in radiation intensity impinging on the film; and (*b*) film gradient (or film contrast), which is a measure of the response of the emulsion to the intensity of impinging radiation and is a characteristic of a given film.

Film gradient is determined from the characteristic curve by finding the slope of the curve at a given density. The slope of the characteristic curve changes continuously over its entire length, as indicated in Fig. 10(a). The steeper the slope of the curve over a range of relative exposures, the greater the difference in density and hence the greater the resolution of detail; thus, high gradient is usually important for good radiography.

Graininess. The silver halide grains that are contained in the emulsion of x-ray film are minute and can be seen only with a high-power micro-

scope, such as an electron microscope. Even though the emulsion on each side of the film is only about 0.01 mm (about 0.0005 in.) thick, the grains are piled on top of each other in countless numbers. When the exposed-and-processed radiograph is viewed, these small individual silver grains appear grouped together in relatively large masses. This "clumping," which is visible to the unaided eye or at low magnification, produces the visual impression called graininess.

All films exhibit some degree of graininess. In general, slower films have less graininess than faster films. Thus, type 1 films have the least graininess, type 3 films exhibit the most graininess, and type 2 films are intermediate.

Spectral Sensitivity. The shape of the characteristic curve of a given x-ray film is for all prac-

tical purposes unaffected by the wavelength distribution in the x-ray or gamma-ray beam used for the exposure. However, the sensitivity of the film in terms of roentgens required to produce a given density is strongly affected by radiation energy (beam spectrum of a given kilovoltage or given gamma-ray source).

Figure 11 shows the exposure required for producing a density of 1.0 on type 4 radiographic film developed in an x-ray developer (made from powder) for 5 min at 20 °C (68 °F). The exposures were made directly, without screens. The spectral-sensitivity curves for all x-ray films have approximately the same general features as the curves shown in Fig. 11.

The classification of radiographic film is complicated; however, a relatively simple classification has been adopted by ASTM. According to the classification in ASTM E94, radiographic films are grouped into four types. The general characteristics of these four types are summarized in Table 5. The relative image quality of x-ray film may also be determined in a quantitative manner using a multihole test piece as described in ASTM E746.

Screens are often used with x-ray films during exposure. Metal screens, typically used at x-ray energies of over 150 kV, intensify the image by emission of photoelectrons and help reduce the effects of scatter by attenuating the lower-energy scattered radiation. Lead is typically used. Fluorescent screens are used in some situations to help reduce exposure times.

The quality level of an industrial radiograph is governed by the radiographic sensitivity exhibited on the radiograph itself. Radiographic sen-

Table 6. Guide to selection of radiographic film for steel, aluminum, bronze and magnesium in various thicknesses (ASTM E94)

Thickness mm	in.	Type of film(a) for use with these x-ray-tube voltages, or radioactive isotopes: 50 to 80 kV	80 to 120 kV	120 to 150 kV	150 to 250 kV	Ir-192	250 to 400 kV	1 MeV	Co-60	2 MeV	Ra	6 to 31 MeV
Steel												
0 to 6.4	0 to 1/4 ...	3	3	2	1	...	...	...	...	...	...	...
6.4 to 13	1/4 to 1/2...	4	3	2	2	...	1	...	...	...	...	...
13 to 25	1/2 to 1 ...	...	4	3	2	2	2	1	...	1	2	...
25 to 50	1 to 2	...	...	...	3	2	2	1	2	1	2	1
50 to 100	2 to 4	...	...	...	4	3	4	2	2	2	3	1
100 to 200	4 to 8	...	...	...	...	...	4	3	3	2	3	2
Over 200	Over 8 ...	...	...	...	...	...	...	...	...	3	...	2
Aluminum												
0 to 6.4	0 to 1/4 ...	1	1	...	...	...	...	...	...	...	...	...
6.4 to 13	1/4 to 1/2...	2	1	1	1	...	...	...	...	...	...	...
13 to 25	1/2 to 1 ...	2	1	1	1	...	1	...	...	...	...	...
25 to 50	1 to 2	3	2	2	1	1	1	...	...	...	...	...
50 to 100	2 to 4	4	3	2	2	1	2	...	...	...	...	...
100 to 200	4 to 8	...	4	3	3	2	3	...	...	...	...	...
Over 200	Over 8 ...	...	...	...	...	4	...	...	...	...	...	...
Bronze												
0 to 6.4	0 to 1/4 ...	4	3	2	1	1	1	1	...	...	...	...
6.4 to 13	1/4 to 1/2...	...	3	2	2	2	1	1	...	1	...	...
13 to 25	1/2 to 1 ...	...	4	4	3	2	2	1	2	1	2	...
25 to 50	1 to 2	...	...	4	4	3	3	1	2	1	2	1
50 to 100	2 to 4	...	...	...	...	3	4	2	3	2	3	1
100 to 200	4 to 8	...	...	...	...	...	...	3	3	2	...	2
Over 200	Over 8 ...	...	...	...	...	...	...	...	...	3	...	2
Magnesium												
0 to 6.4	0 to 1/4 ...	1	1	...	...	...	...	...	...	...	...	...
6.4 to 13	1/4 to 1/2...	1	1	1	...	...	...	...	...	...	...	...
13 to 25	1/2 to 1 ...	2	1	1	...	1	...	...	...	...	...	...
25 to 50	1 to 2	2	1	1	1	1	...	...	...	...	...	...
50 to 100	2 to 4	3	2	2	1	2	...	...	...	...	...	...
100 to 200	4 to 8	...	3	2	2	3	...	...	...	...	...	...
Over 200	Over 8 ...	...	...	...	4	...	...	...	...	...	...	...

(a) These recommendations represent a usually acceptable level of radiographic quality. Optimum radiographic quality will be promoted by use of the lowest-number film type that economic and technical considerations will allow. The recommendations for type 4 film are based on the use of fluorescent screens.

sitivity is determined through the use of pene-trameters or image-quality indicators (IQI), as discussed in the following section. Because sensitivity depends on the type of material and the thickness to be radiographed as well as on the photon energies of the incident radiation, specific recommendations for selection of film cannot be given. However, Table 6 may be used as a general guide for selection of film for a usually acceptable level of radiographic quality for various metals and radiation-source energies. It should be noted, however, that better radiographic quality will be promoted by the lowest film-type number that economic and technical considerations will allow.

PENETRAMETERS (IMAGE-QUALITY INDICATORS)

Penetrameters, or image-quality indicators (IQI), are of known size and shape, and have the same attenuation characteristics as the material in the test piece. They are placed on the test piece or on a block of identical material during setup and are radiographed at the same time as the test piece. Penetrameters preferably are located in regions of maximum test-piece thickness and greatest test-piece-to-film distance, and near the outer edge of the central beam of radiation. The degree to which features of the penetrameter are visible in the developed image is a measure of the quality of that image. The image of the penetrameter that appears on the finished radiograph is evaluated during interpretation to ensure that the desired sensitivity, definition and contrast have been achieved in the developed image.

Penetrameters of different designs have been developed by various standards-making organizations. Common types are plaques containing holes and a second type containing a series of wires.

Plaque-type penetrameters consist of strips of material of uniform thickness with holes drilled through them. The designs specified by ASTM and ASME comprise 13-by-38-mm ($\frac{1}{2}$-by-$1\frac{1}{2}$-in.) plaques containing three holes; the holes are T, 2T and 4T in diameter, where T is the thickness of the plaque. Sensitivity is expressed in terms of IQI or penetrameter thickness (as a percentage of test-piece thickness), and resolution is determined by the smallest hole size visible in the radiograph. For instance, an image-quality level of 2-2T indicates that the thickness of the penetrameter equals 2% of section thickness and the 2T hole is visible. If image quality of 1-1T were required, a radiograph would be acceptable if the outline of a 1% penetrameter and the 1T hole in that penetrameter were distinguishable. Alternatively, image quality can be expressed as

Table 7. Equivalent sensitivities of various standard ASTM or ASME sensitivity levels(a)

Image-quality level	Penetrameter thickness, % of test-piece thickness	Smallest visible hole size	Equivalent sensitivity, %
1-1T	1	1T	0.7
1-2T	1	2T	1.0
2-1T	2	1T	1.4
2-2T	2	2T	2.0
2-4T	2	4T	2.8
4-2T	4	2T	4.0

(a) Equivalent sensitivity is a percentage equivalent for penetrameter thickness in which 2T is the smallest distinguishable hole size. For instance, 1-1T is equivalent to 0.7-2T.

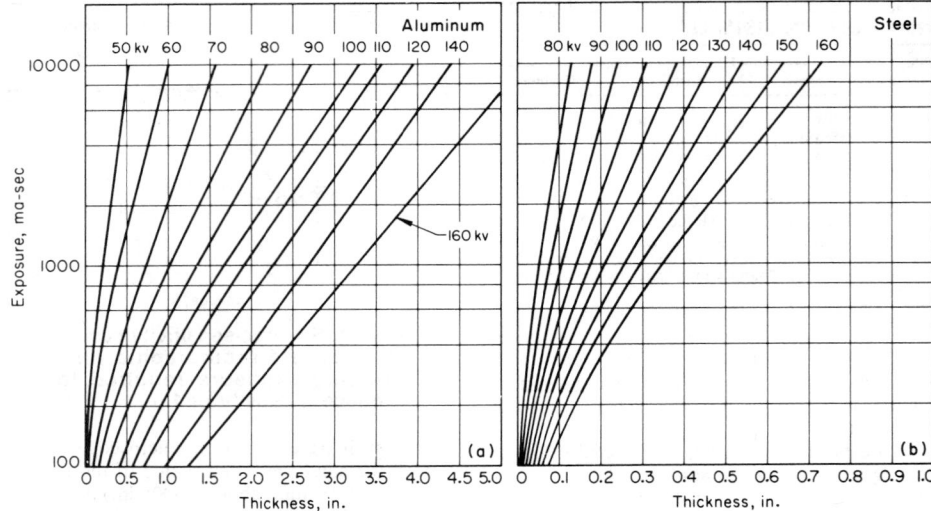

Charts for aluminum and steel were prepared specifically for an Andrex 160-kv directional x-ray machine, using a source-to-film distance of 36 in. and Industrex AA film (Eastman Kodak) developed in a manual process for 7 min in PIX developer (Picker).

Fig. 12. Typical radiographic-exposure charts for (a) aluminum and (b) steel for a film density of 2.0 without screens that relate exposure to thickness of test pieces for several values of tube voltage

a percentage only. In the ASTM or ASME systems, the equivalent sensitivity in percent is based on visibility of the 2T hole. Table 7 lists equivalent sensitivities for various standard image-quality levels. Plaque-type IQI are described in ASTM E142.

Wire-type penetrameters, widely used in Europe (see German standard DIN 54109), are also used in the U.S. and are described in ASTM standard E747. The sensitivity of a wire-type penetrameter is expressed in terms of wire diameter divided by object thickness.

EXPOSURE FACTORS

The amount of radiation that penetrates a test piece and produces a satisfactory image is governed by the intensity and spectral quality of the incident radiation, the source-to-film distance and the exposure time. In practice, the quality of the incident radiation, which depends mainly on tube voltage of an x-ray machine or on the radioactive isotope in a gamma-ray source, is chosen to be sufficiently penetrating for the type of material and thickness to be inspected. The recording medium is selected according to its sensitivity to expected variations in the intensity of transmitted radiation and its ability to record these variations in a form that will permit satisfactory interpretation. With these two factors fixed, the remaining interrelated factors—radiation intensity (determined by tube current in milliamperes for x-ray sources, or by source strength in curies for gamma-ray sources), source-to-film distance and exposure time—determine the amount of radiation impinging on the film, which is called the radiographic exposure or simply exposure.

Inverse-Square Law. Similar to visible light, x-rays or gamma rays diverge when emitted from the radiation source, and cover an increasingly larger area, with lessened intensity, as the distance from the source increases. The intensity, or amount of radiation falling on a unit area, var-

ies inversely with the square of the distance from the source. This can be expressed mathematically as:

$$IL^2 = \text{constant}$$

where I is the intensity of radiation at a given distance, L, from a source of constant R-output. More often, this so-called inverse-square law is expressed as a ratio:

$$I_1/I_2 = L_2^2/L_1^2$$

where the subscripts 1 and 2 refer to different points along a line radiating from the source. Because of this inherent characteristic of radiation, if the radiation has a certain intensity at 1 m from the source, it will have four times that intensity at 0.5 m but only one-quarter that intensity at 2 m and only one-ninth that intensity at 3 m.

Exposure charts may be made for specific materials and film types. Typical charts for aluminum and steel are presented in Fig. 12.

APPLICATIONS

Radiography offers the advantage of a picture. It represents the changes in density within the inspection object. Changes in thickness on the order of a few percent are normally detected. Therefore, flaws having some three-dimensional characteristic, such as porosity or inclusions, are normally seen. Components within an assembly can be checked. For laminar flaws, such as cracks, the radiographic view must be reasonably aligned with the flaw if a radiographic image is to appear.

As a precaution, it is well known that large doses of x-rays or gamma rays can damage skin and blood cells, can produce blindness and sterility, and in massive doses can cause severe disability or death. Protection of personnel—not only those engaged in radiographic work but also those in the vicinity of radiographic inspection—is of major importance.

Table 8. Characteristics of neutron radiography at various neutron-energy ranges

Type of neutrons	Energy range	Characteristics
Cold	Below 0.01 eV	High-absorption cross sections decrease transparency of most materials, but also increase efficiency of detection. An advantage is reduced scatter at energies below the Bragg cutoff, where neutrons can no longer undergo Bragg reflection.
Thermal	0.01 to 0.3 eV	Good discrimination between materials, and ready availability of sources.
Epithermal	0.3 eV to 10 keV	Excellent discrimination for particular materials by working at energy of resonance. Greater transmission and less scatter in samples containing materials such as hydrogen and enriched reactor fuels.
Fast	10 keV to 20 MeV	Good point sources are available. At low-energy end of spectrum, fast-neutron radiography may be able to perform many inspections done with thermal neutrons, but with a panoramic technique. Good penetration capability because of low-absorption cross sections in all materials. Poor material discrimination.

Inspection by Neutron Radiography

NEUTRON RADIOGRAPHY is a form of nondestructive inspection that uses a specific type of particulate radiation, called neutrons, to form a radiographic image of a test piece. The geometric principles of shadow formation, variation of attenuation with test-piece thickness, and many other factors that govern the exposure and processing of a neutron radiograph are similar to those for radiography using x-rays or gamma rays.

The present article deals mainly with the characteristics that differentiate neutron radiography from x-ray or gamma-ray radiography. The application of neutron radiography is described, especially in terms of its advantages for improved contrast on low-atomic-number materials, discrimination between isotopes, or inspection of radioactive specimens.

Neutrons are subatomic particles that are characterized by relatively large mass and a neutral electric charge. The attenuation of neutrons differs from the attenuation of x-rays in that the processes of attenuation are nuclear rather than ones that depend on interaction with electron shells surrounding the nucleus.

Neutrons are produced by nuclear reactors, accelerators or certain radioactive isotopes, all of which emit neutrons of relatively high energy (fast neutrons). Because most neutron radiography is performed with neutrons of lower energy (thermal neutrons), the sources are usually surrounded by a "moderator," which is a material that reduces the kinetic energy of the neutrons by scattering.

Neutron radiography differs from conventional radiography in that the attenuation of neutrons as they pass through the test piece is more related to the specific isotope present than to density or atomic number. X-rays are attenuated more by elements of high atomic number than by elements of low atomic number, and this effect varies relatively smoothly with atomic number. Also, x-rays are generally attenuated more by materials of high density than they are by materials of low density. For thermal neutrons, the attenuation tends to decrease with increasing atomic number, although the trend is by no means a smooth relationship. In addition to the high attenuation of several light elements (hydrogen, lithium and boron), certain medium to heavy elements (especially cadmium, samarium, europium, gadolinium and dysprosium) and certain specific isotopes have exceptionally high capabilities for absorbing thermal neutrons. This means that neutron radiography is capable of detecting these highly attenuating elements or isotopes when present in a structure of lower absorption capability.

Using neutrons, it is possible to radiographically detect certain isotopes—for instance, certain isotopes of hydrogen, cadmium or uranium. Some neutron-image-detection methods are insensitive to gamma rays or x-rays, and can be used to inspect radioactive materials such as reactor fuel elements. The high attenuation of hydrogen, in particular, opens many application possibilities, including inspection of assemblies for detection of adhesives, explosives, lubricants, water, hydrides, corrosion, plastic or rubber.

NEUTRON SOURCES

The excellent discrimination capabilities of neutrons generally refer to neutrons of low energy—that is, thermal neutrons. Characteristics of neutron radiography corresponding to various ranges of neutron energy are summarized in Table 8. Although any of these energy ranges can be used for radiography, this article emphasizes the thermal-neutron range, which is the most widely used for inspection.

In thermal-neutron radiography, an object (test piece) is placed in a thermal-neutron beam in front of an image detector. The neutron beam may be obtained from a nuclear reactor, a radioactive source, or an accelerator. Several characteristics of these sources are summarized in Table 9. For thermal-neutron radiography, fast neutrons emitted by these sources must first be moderated and then collimated. The radiographic intensities listed in Table 9 typically do not exceed 10^{-5} times the total fast-neutron yield of the source. Part of this loss is incurred in moderating the neutrons, and the remainder in bringing a collimated beam out of a large-volume moderator.

Collimation is necessary for thermal-neutron radiography because there are no useful point sources of low-energy neutrons. Good collimation in thermal-neutron radiography is comparable to small focal-spot size in conventional radiography; the images of thick objects will be sharper with good collimation. On the other hand, it should be noted that available neutron intensity decreases with increasing collimation.

NEUTRON DETECTION METHODS

Detection methods for neutron radiography generally make use of photographic or x-ray films. In the direct-exposure method, film is exposed

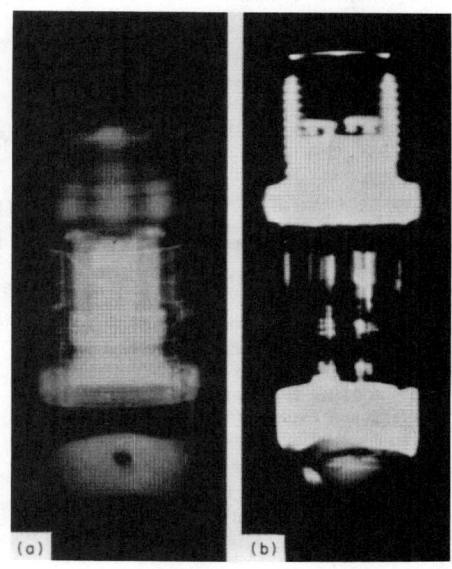

Fig. 13. Comparison of positive prints of (a) a thermal-neutron radiograph and (b) a conventional radiograph of an explosive bolt 50 mm (2 in.) long. Neutron radiograph reveals details of paper, explosive compound and plastic components not revealed by x-rays.

Table 9. Several characteristics of thermal-neutron sources

Type of source	Typical radiographic intensity(a)	Resolution	Exposure time	Characteristics
Radioisotope	10^1 to 10^4	Poor to medium	Long	Stable operation, low to medium investment cost, possibly portable
Accelerator	10^3 to 10^6	Medium	Average	On-off operation, medium cost, possibly transportable
Subcritical assembly	10^4 to 10^6	Good	Average	Stable operation, medium to high investment cost, movement difficult
Nuclear reactor	10^5 to 10^8	Excellent	Short	Medium to high investment cost, movement difficult

(a) Neutrons per sq cm per second.

directly to the neutron beam, with a conversion screen or intensifying screen providing secondary radiation that actually exposes the film. Alternatively, film can be used to record an autoradiographic image from a radioactive, image-carrying screen in a technique called the transfer method.

Direct-Exposure Method. Conversion screens of thin gadolinium foil or a scintillator have been most widely used in the direct-exposure method. When bombarded with a beam of neutrons, some of the gadolinium atoms absorb neutrons and promptly emit gamma rays and internal conversion electrons. Scintillators are fluorescent screens, often made of zinc sulfide crystals that also contain a specific isotope such as $_3Li^6$ or $_5B^{10}$. Gadolinium oxysulfide, a scintillator originally developed for x-ray radiography, has been widely used for neutron radiography.

Scintillators provide useful images with total exposures as low as 5×10^5 neutrons per sq cm. The high speed and favorable relative response make scintillators attractive for use with non-reactor neutron sources. Gadolinium screens provide greater uniformity and image sharpness (high-contrast resolution of 10 μm has been reported), but an exposure about 30 or more times that of a scintillator is required, even with fast films.

Transfer Method. In the transfer method, a thin sheet of metal, typically of indium or dysprosium, is exposed to the neutron beam transmitted through the specimen. Neutron capture induces radioactivity—indium having a half-life of 54 min and dysprosium a half-life of 2.35 h. The "radiograph" to be interpreted is made by placing the radioactive transfer screen in contact with a sheet of film. Beta-particles and gamma-rays from the transfer screen expose the film.

The transfer method is especially valuable for inspection of a radioactive specimen. Although radiation emitted by the specimen (especially gamma rays) causes heavy film fogging during x-ray radiography or direct-exposure neutron radiography, the same radiation will not induce radioactivity in a transfer screen. Thus, a clear image of the specimen can be obtained even when there is a high level of background radiation.

In comparing the two primary detection methods, the direct-exposure method offers high speed, indefinite image-integration time and the best spatial resolution. The transfer method offers insensitivity to gamma rays emitted by the specimen, and greater contrast because of lower amounts of scattered and secondary radiation.

Real-time imaging, in which light from a scintillator is observed by a television camera, also can

be used for neutron radiography. Because of low brightness, most real-time neutron radiographic images are enhanced by an image-intensifier tube, which may be separate or integral with a television camera. This method can be used for applications such as the study of fluid flow in a closed system such as a heat pipe or engine or the study of metal flow in a mold during casting.

APPLICATIONS

Various applications that are discussed in ASTM STP 586 emphasize the value of neutron radiography for inspection of ordnance, explosive, aerospace and nuclear components. The presence, absence or correct placement of explosives (see Fig. 13), adhesives, O-rings, plastic components and similar materials can be verified. The presence of fluids or corrosion can be detected. Nuclear fuel and control materials can be inspected to determine distribution of isotopes and to detect foreign or imperfect material. Hydride deposition in metals and diffusion of boron in heat treated boron-fiber composites can be observed.

The characteristics of neutron radiography complement those of conventional x-radiography; one radiation provides a capability lacking or difficult for the other.

Ultrasonic Inspection

Edited by G. J. Posakony, Battelle Pacific Northwest Laboratories

ULTRASONIC INSPECTION is a nondestructive method in which beams of high-frequency acoustic energy are introduced into the material under evaluation in order to detect surface and subsurface flaws and to measure the thickness of the material or the distance to a flaw. An ultrasonic beam will travel through a material until it strikes an interface or discontinuity such as a flaw. Interfaces and flaws interrupt the beam and reflect a portion of the incident acoustic energy. The amount of energy reflected is a function of (a) the nature and orientation of the interface or flaw and (b) the acoustic impedance of such a reflector. Energy reflected from various interfaces or flaws may be used to define the presence and locations of flaws, the thickness of the material or the depth of a flaw beneath a surface.

Most ultrasonic inspections are performed at frequencies between 1 and 25 MHz. Short shock bursts of ultrasonic energy are aimed into the material from the ultrasonic search unit of the ultrasonic flaw-detector instrument. The electrical pulse from the flaw detector is converted into ultrasonic energy by the piezoelectric transducer element in the search unit. The beam pattern from the search unit is determined by the operating frequency and size of the transducer element. Ultrasonic energy travels through the material at a specific velocity that is dependent on the physical properties of the material and on the mode of propagation of the ultrasonic wave. The amount of energy reflected from or transmitted through an interface or other type of discontinuity or reflector is dependent on the properties of the reflector. These phenomena provide the basis for establishing two of the most common measurement parameters used in ultrasonic inspection—

namely, the amplitude of the energy reflected from an interface or flaw, and the time required (from pulse initiation) for the ultrasonic beam to reach the interface or flaw.

ULTRASONIC FLAW DETECTORS

Even though the nomenclature used by different instrument manufacturers may vary, certain controls are required for the basic functions of any instrument. These functions include the power supply, clock or timer, pulser, receiver-amplifier and display. In most instances the entire electronic assembly, including the controls and display, is contained in a single instrument. A typical ultrasonic flaw detector is shown in Fig. 1. The major controls include:

- Frequency selector, for selection of the operating or test frequency
- Pulse-tuning control, for fine adjustment of the test frequency
- Pulse-repetition-rate control, which determines the number of times per second that an ultra-

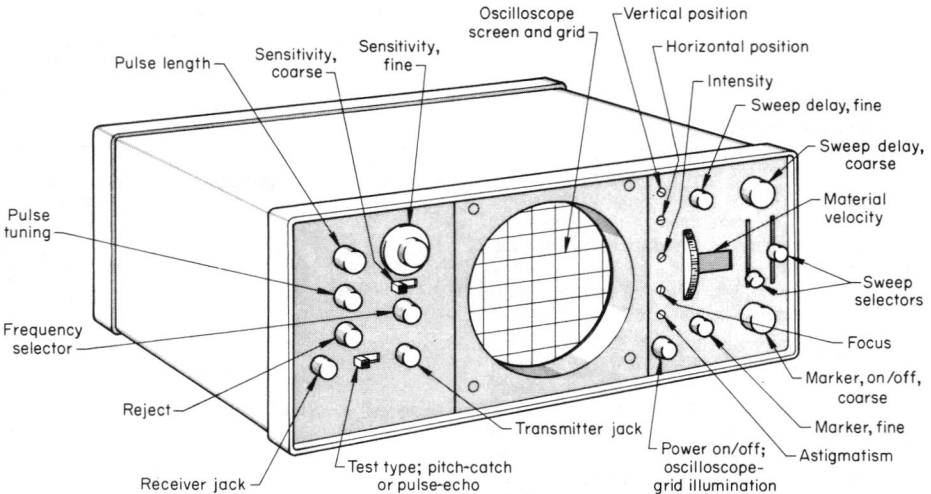

Fig. 1. Typical pulse-echo instrument, showing types and arrangement of controls

Table 1. Characteristics and applications of transducer (piezoelectric) elements (E, excellent; G, good; F, fair; P, poor)

Piezoelectric element	Characteristics of piezoelectric elements							Suitability of element in:		
	Efficiency		Coupling		Tolerance to elevated temperature	Damping ability	Undesired modes (inherent noise)	Contact inspection		Immersion inspection
	Transmit	Receive	To water	To metal				Straight-beam	Angle-beam	
Quartz	P	G	G	F	G	F	G	G	F	G
Lithium sulfate	F	E	E	P	P	E	E	P	F	E
Barium titanate	G	P	G	G	P	P	P	G	G	F
Lead zirconate titanate	E	F	F	E	E	F	P	E	E	F
Lead metaniobate	G	F	G	E	E	E	G	E	E	G

Table 2. Primary applications of four basic types of ultrasonic search units

Straight-beam contact-type units
Manufacturing-induced flaws:
  Billets—inclusions, stringers, pipe
  Forgings—inclusions, cracks, segregations, seams, flakes, pipe
  Rolled products—laminations, inclusions, tears, seams, cracks
  Castings—slag, porosity, cold shuts, tears, shrinkage cracks, inclusions
Service-induced flaws: fatigue cracks, corrosion, erosion, stress-corrosion cracks

Angle-beam contact-type units
Manufacturing-induced flaws:
  Forgings—cracks, seams, laps
  Rolled products—tears, seams, cracks, cupping
  Welds—slag inclusions, porosity, incomplete fusion, incomplete penetration, dropthrough, suckback, cracks in filler metal and base metal
  Tubing and pipe—circumferential and longitudinal cracks
Service-induced flaws: fatigue cracks, stress-corrosion cracks

Dual-element contact-type units
Manufacturing-induced flaws:
  Plate and sheet—thickness measurement, lamination detection
  Tubing and pipe—measurement of wall thickness
Service-induced flaws: wall thinning, corrosion, erosion, stress-corrosion cracks

Immersion-type units
Manufacturing-induced flaws:
  Billets—inclusions, stringers, pipe
  Forgings—inclusions, cracks, segregations, seams, flakes, pipe
  Rolled products—laminations, inclusions, tears, seams, cracks
  Welds—inclusions, porosity, incomplete fusion, incomplete penetration, dropthrough, cracks, base-metal laminations
  Adhesive-bonded, soldered or brazed products—lack of bond
  Composites—voids, resin rich, resin poor, lack of filaments
  Tubing and pipe—circumferential and longitudinal cracks
Service-induced flaws: corrosion, fatigue cracks

• Gate-position control, for isolation of that portion of the inspection zone that is to be used for additional processing
• Oscilloscope, which provides the visual display of the time and amplitude parameters used to interpret the data from the ultrasonic inspection.

ULTRASONIC TRANSDUCERS AND SEARCH UNITS

Generation and detection of ultrasonic waves for inspection is accomplished by means of a transducer element. The transducer element is contained within a device most often referred to as a search unit (or sometimes as a probe). The active element in a search unit is a piezoelectric

sonic pulse is initiated from the transducer (typically, 100 to 2000 pulses per second)
• Test-type or mode-selection switch, for adjustment of the instrument to pulse-echo or pitch-catch operation
• Sensitivity controls, for adjustment of the sensitivity or gain of the receiver-amplifier
• Sweep selector and delay, for adjustment of the time base and that portion of the inspection zone that is to be displayed

crystal. Piezoelectricity is "pressure electricity"; this property is characteristic of certain naturally occurring crystalline compounds and some man-made materials.

As the name "piezoelectric" implies, an electrical charge is developed by the crystal when pressure is applied to it. Conversely, when an electrical field is applied, the crystal mechanically deforms (changes shape). Piezoelectric crystals exhibit various deformation modes; thickness expansion is the principal mode used in transducers for ultrasonic inspection.

The most common types of piezoelectric materials used for ultrasonic search units are quartz; lithium sulfate; and polarized ceramics such as barium titanate, lead zirconate titanate and lead metaniobate. Characteristics and applications of these materials are summarized in Table 1.

Search units are of many types and shapes. Variations in search-unit construction include: transducer-element material; transducer-element thickness, surface area and shape; and type of backing material and degree of loading. Four basic types of search units are the straight-beam contact type, the angle-beam contact type, the dual-element contact type, and the immersion type, both flat and focused. Table 2 lists their primary areas of application. Sectional views of these search units, together with a special type (delay-tip contact-type search unit), are shown in Fig. 2.

COUPLANTS

Air is a poor transmitter of sound waves at megahertz frequencies. Also, the acoustic

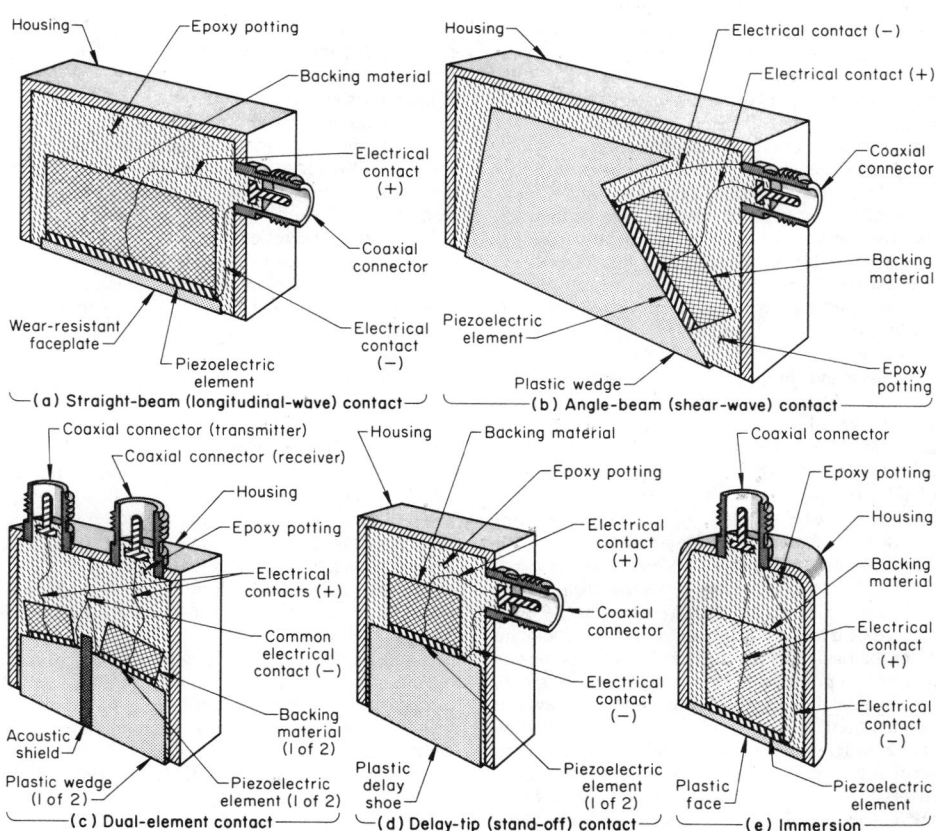

Fig. 2. Sectional views of five types of search units used in ultrasonic inspection

impedance mismatch between air and most solids is great enough that even a very thin layer of air will severely retard the transmission of sound waves from the transducer to the test piece.

To perform satisfactory contact inspection it is necessary to eliminate air between the transducer and the test piece by use of a couplant.

Couplants normally used for contact inspection include water, oils, glycerin, petroleum greases, silicone grease, cellulose gum and various commercial pastelike substances. Certain soft rubbers that transmit sound waves may be used where adequate coupling can be achieved by applying pressure to the search unit.

The following should be considered in selecting a couplant:

- Surface finish of test piece
- Temperature of test surface
- Possibility of chemical reactions between test surface and couplant
- Cleaning requirements (some couplants are difficult to remove).

Water is a suitable couplant for use on a relatively smooth surface; however, a wetting agent should be added. It is sometimes appropriate to add glycerin to increase viscosity.

Heavy oil or grease should be used on hot or vertical surfaces, or on rough surfaces where irregularities need to be filled.

Cellulose gum is especially useful on rough surfaces when good coupling is needed to minimize background noise and yield an adequate signal-to-noise ratio.

BASIC INSPECTION METHODS

In terms of applications, ultrasound can be used to measure thickness by (a) determining resonant frequencies of a test piece or (b) precise measurement of time required for an ultrasonic-wave packet (pulse) to traverse the test piece. The former uses reflected ultrasound to create standing waves in the test piece; the frequencies at which standing waves occur are used to compute thickness. In the latter method, the time that it takes for a pulse of ultrasonic energy to be transmitted through the test piece is measured; this time period may be 100 nanoseconds or less. Thickness is calculated as the product of the measured time of flight and the known acoustic-wave velocity.

Ultrasound can be used to detect flaws by measuring (a) the amplitude of the acoustic pressure wave and time of flight of reflected acoustic waves or (b) the amplitude of the acoustic pressure wave of either transmitted or reflected acoustic waves. The first of these is the pulse-echo technique, which is the most widely used of all ultrasonic techniques. Flaws are detected and their sizes estimated by comparing the amplitude of a reflected echo from an interface (either within the test piece or at the back surface) with the amplitude of an echo reflected from a reference interface of known size or from the back surface of a test piece having no flaws. The echo from the back surface (back reflection) serves as a reference point for time-of-flight measurements that enable the depth of some internal flaws to be measured. It is necessary that an internal flaw reflect at least part of the sound energy onto the receiving transducer for such depth measurements to be made. However, echoes from flaws are not essential to their detection. Merely the fact that the amplitude of an echo from back reflection of a test piece is lower than that of an

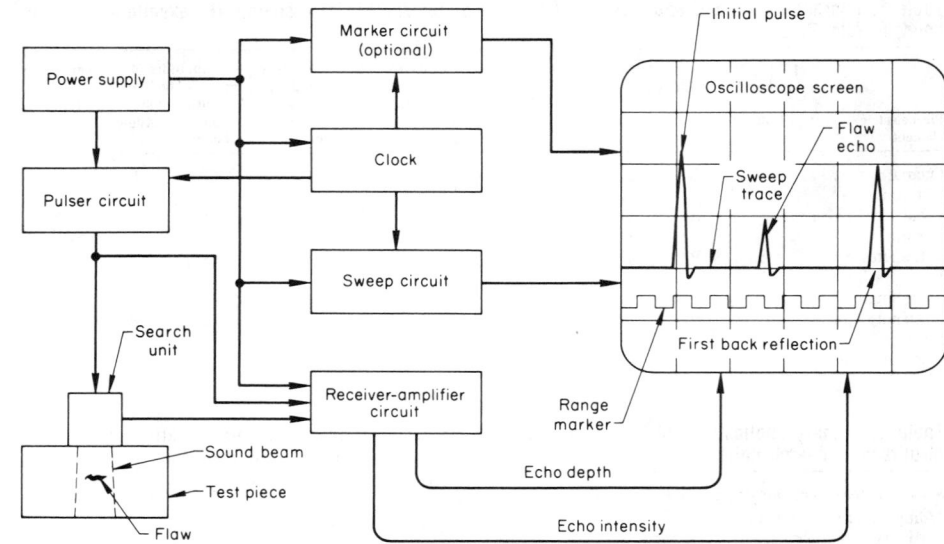

Fig. 3. Typical A-scan setup, including video-mode display, for a basic pulse-echo ultrasonic-inspection system

echo from an identical workpiece known to be free of flaws implies that the test piece contains one or more flaws. This second method of detecting the presence of flaws—that is, by sound attenuation—is used in transmission techniques as well as in the pulse-echo technique. The main disadvantage of attenuation techniques is that flaw depth cannot be measured.

PULSE-ECHO METHOD

In pulse-echo inspection, short bursts of ultrasonic energy (pulses, or wave packets) are introduced into a test piece at regular intervals of time. If the pulses encounter a reflecting surface, some or all of the energy is reflected. The proportion of energy that is reflected is highly dependent on the size of the reflecting surface in relation to the size of the incident ultrasonic beam. The direction of the reflected beam (echo) depends on the orientation of the reflecting surface with respect to the incident beam. Reflected energy is monitored; both the amount of energy reflected in a specific direction and the time delay between transmission of the initial pulse and receipt of the echo are measured.

Principles of Operation. Most pulse-echo systems consist of (a) an electronic clock; (b) an electronic signal generator, or pulser; (c) a sending transducer; (d) a receiving transducer; (e) an echo-signal amplifier; and (f) a display device. In the most widely used version of pulse-echo systems, a single transducer acts alternatively as a sending and receiving transducer. The clock and signal generator usually are combined in a signal electronic unit. Frequently, circuits that amplify and demodulate echo signals from the transducer are housed in the same unit.

A pulse-echo system with a single search unit operates as follows. At regular intervals, the electronic clock triggers the signal generator, which imposes a short burst of high-frequency alternating voltage on the transducer element. Simultaneously, the clock activates a time-measuring circuit connected to the display device. The operator preselects a constant interval between pulses by means of a "pulse-repetition-rate" control on the instrument; pulses usually are re-

peated 100 to 2000 times per second. The operator also may preselect the output frequency of the signal generator or pulser. For best results, the frequency (and sometimes the pulse-repetition rate) should be tuned to achieve the maximum response of the transducer (resonance in the vibrating element) and maximum signal-to-noise ratio (lowest amount of electronic noise) in the electronic equipment. The transducer converts the pulse of alternating voltage into a pulse of mechanical vibration having essentially the same frequency as the imposed alternating voltage. The mechanical vibration (ultrasound) is introduced into a test piece through a couplant, and travels by wave motion through the test piece at the speed of sound. When the pulse of ultrasound encounters a reflecting surface that is perpendicular to the direction of travel, ultrasonic energy is reflected and returns to the transducer. The returning pulse travels along the same path and at the same speed as the initial pulse, but in the opposite direction.

Data Presentation. Information from pulse-echo inspection can be displayed in one of three forms: (a) A-scan, which is a quantitative display of echo amplitude and time-of-flight data obtained at a single point on the surface of the test piece; (b) B-scan, which is a quantitative cross-sectional display of time-of-flight data obtained along a plane perpendicular to the surface of the test piece; or (c) C-scan, which is a semiquantitative display of echo amplitude obtained over an area of the surface of the test piece. The A-scan display, which is the most widely used form, may be analyzed in terms of the type, size and location (chiefly depth) of internal flaws.

A-scan Display. A typical A-scan setup that illustrates the essential elements in a basic system for pulse-echo inspection is shown in Fig. 3. These elements include:

- Power supply, which may run on 110-volt alternating current or on batteries
- Electronic clock, or timing circuit, for triggering of pulser and display circuits
- Pulser circuit, or rate generator, for control of frequency, amplitude and pulse-repetition rate of the voltage pulses that excite the search unit

- Receiver-amplifier circuit, for conversion of output signals from the search unit into a form suitable for oscilloscope display
- Sweep circuit, for control of (*a*) time delay between search-unit excitation and start of oscilloscope trace and (*b*) rate at which oscilloscope trace travels horizontally across the screen
- Marker circuit (optional), for production of a secondary trace, on or below the main trace, usually in the form of a square wave, which is used for precise depth measurements
- Oscilloscope screen, including separate controls for trace brightness, trace focus and illuminated measuring grid
- Flaw gate (not shown), for isolation of the echo of interest for further processing.

The search unit and the coaxial cable connecting the unit to the instrument, although not strictly part of the electronic circuitry, nevertheless must be matched to the electronics.

B-scan display is a plot of time versus distance, in which one orthogonal axis on the display corresponds to elapsed time while the other axis represents the position of the search unit along a line on the surface of the test piece relative to the position of the search unit at the start of the inspection. Echo amplitude is not measured directly as it is in A-scan inspection, but often is indicated semiquantitatively by the relative brightness of echo indications on an oscilloscope screen.

A typical B-scan system is shown in Fig. 4. The system functions are identical to those of the A-scan system except for the following basic differences:

- The display is generated on an oscilloscope screen that is composed of a long-persistence phosphor; that is, a phosphor that continues to fluoresce long after the means of excitation ceases to fall on the fluorescing area of the screen. This characteristic of the oscilloscope in a B-scan system allows the imaginary cross section to be viewed as a whole without having to resort to permanent imaging methods such as photographs. (Photographic equipment, facsimile recorders or x-y plotters can be used to record B-scan data, especially when a permanent record is desired for later reference.)
- The oscilloscope input for one axis of the display is provided by an electromechanical device that generates an electrical voltage proportional to the position of the search unit relative to a reference point on the surface of the test piece. Most B-scans are generated by scanning the search unit in a straight line across the surface of the test piece at a uniform rate. One axis of the display, usually the horizontal axis, represents the distance traveled along this line.
- Echoes are indicated by bright spots on the screen rather than by deflections of the time trace. The position of a bright spot along the axis orthogonal to the search-unit position axis, usually measured top to bottom on the screen, indicates the depth of the echo within the test piece.
- To ensure that echoes are recorded as bright spots, the echosignal from the receiver-amplifier is connected to the trace-brightness control on the oscilloscope. In some systems, the brightnesses corresponding to different values of echo amplitude may exhibit enough contrast to permit semiquantitative appraisal of echo amplitude, which is related to flaw size and shape.

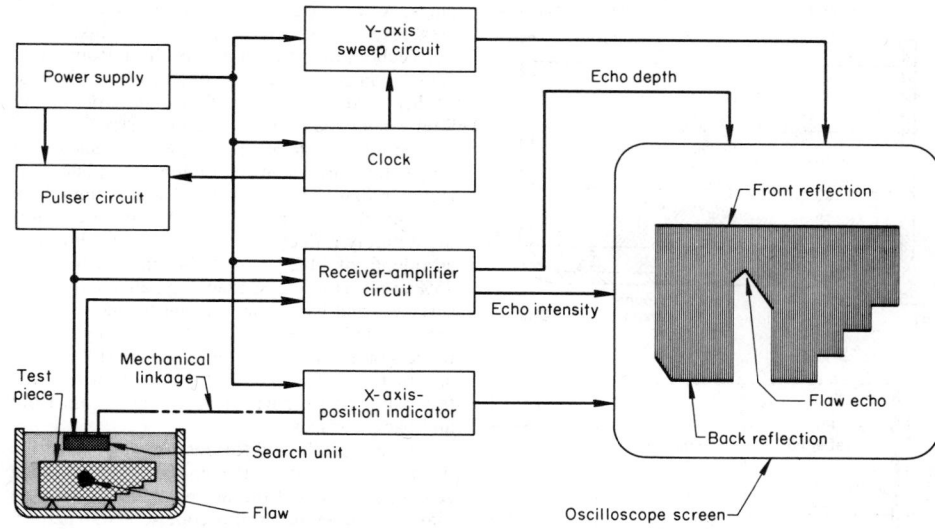

Fig. 4. Typical B-scan setup, including video-mode display, for a basic pulse-echo ultrasonic-inspection system

The oscilloscope screen in Fig. 4 illustrates the type of video-mode display that is generated by B-scan equipment. On this screen the internal flaw in the test piece shown at left in Fig. 4 is shown only as a profile view of its top reflecting surface. Portions of the test piece that are behind this large reflecting surface are in shadow.

C-scan display records echoes from internal portions of test pieces as a function of the position of each reflecting interface within an area. Flaws are shown on a readout, superimposed on a plan view of the test piece, and both flaw size (flaw area) and position within the plan view are recorded. Flaw depth normally is not recorded, although it may be measured semiquantitatively by restricting the range of depths within the test piece that is covered in a given scan.

In a basic C-scan system, shown schematically in Fig. 5, the search unit is moved over the surface of the test piece in a search pattern. The search pattern may take many forms—for instance, a series of closely spaced parallel lines, a fine zigzag pattern or a spiral pattern (polar scan). Mechanical linkage connects the search unit to x-axis and y-axis position indicators, which in turn feed position data to the x-y plotter or facsimile device. Echo-recording systems vary; some produce a shaded-line scan with echo amplitude recorded as a variation in line shading, while others indicate flaws by an absence of shading so that each flaw shows up as a blank space on the display (see Fig. 5).

INTERPRETATION OF PULSE-ECHO DATA

Interpretation of pulse-echo data is relatively straightforward for B-scan and C-scan presentations. The B-scan always records the front reflection, while internal echoes or loss of back reflection, or both, are interpreted as flaw indications. Flaw depth is measured as the distance from the front reflection to a flaw echo, the latter representing the front surface of the flaw.

In contrast to normal B-scan and C-scan dis-

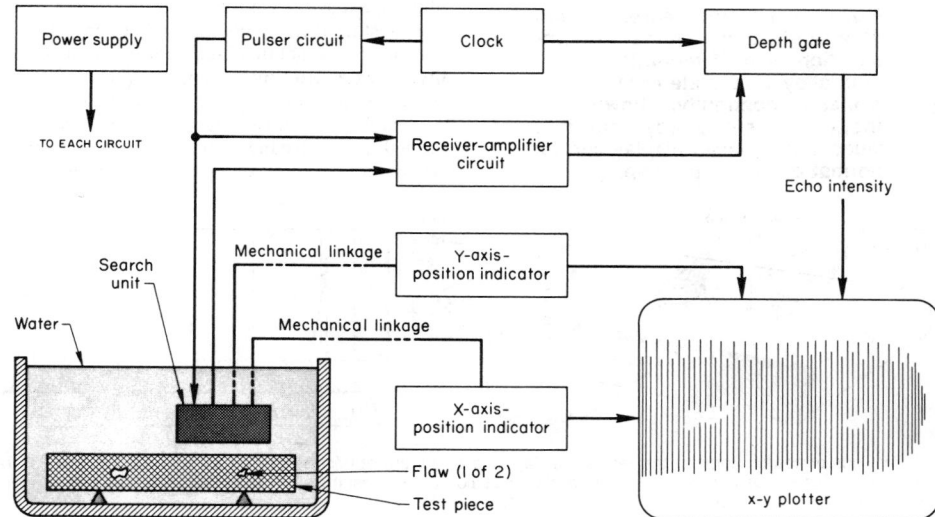

Fig. 5. Typical C-scan setup, including display, for a basic pulse-echo ultrasonic-inspection system

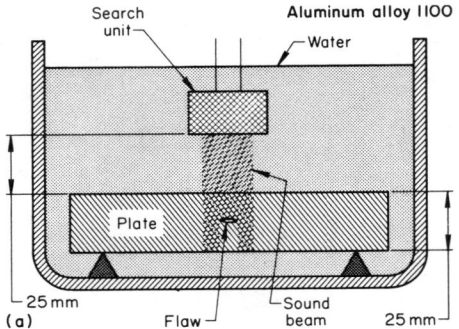

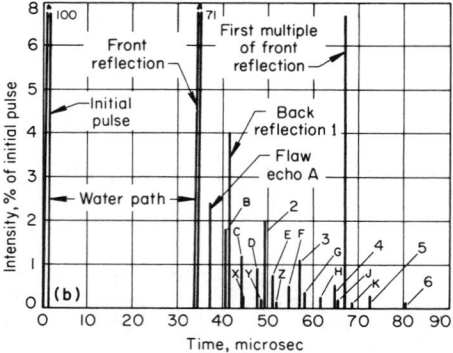

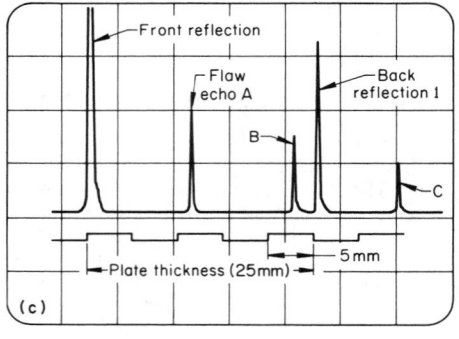

Fig. 6. Schematic representation of straight-beam immersion inspection of a 25-mm-thick aluminum alloy 1100 plate containing a planar discontinuity, showing (a) inspection setup, (b) complete video-mode A-scan display and (c) normal oscilloscope display

plays, A-scan displays are sometimes quite complex. They may contain electronic noise, spurious echoes or extra echoes resulting from mode conversion of the initial pulse, all of which must be disregarded in order to focus attention on any flaw echoes that may be present.

Basic A-scan displays are of the type shown in Fig. 6 for immersion inspection of a plate containing a flaw. The test material was 25-mm thick (1-in.-thick) aluminum alloy 1100 plate containing a purely reflecting planar flaw; the flaw was at a depth of 11.25 mm (0.443 in.), which was 45% of plate thickness, exactly parallel to the plate surfaces, and had an area equal to one-third the cross section of the ultrasonic beam; straight-beam immersion testing was done in a water-filled tank; and there were no attenuation losses within the test plate, only transmission losses across front and back surfaces.

Figure 6(a), (b) and (c), respectively, illustrate the inspection setup, the complete video-mode A-scan display and the normal video-mode display as seen on the oscilloscope screen. The normal display (Fig. 6c) represents only a portion of the complete display (Fig. 6b). The normal display is obtained by adjusting two of the oscilloscope controls (horizontal position and horizontal sweep) to display only the portion of the trace corresponding to the transit time (time of flight) required for a single pulse of ultrasound to traverse the test piece from front surface to back surface and return. Also, the gain of the receiver-amplifier is adjusted until the height of the first back reflection equals some arbitrary distance on the screen, usually a convenient number of grid lines.

Angle-Beam Technique. Most angle-beam testing is accomplished with shear waves, although refracted longitudinal waves and surface waves may be used for some applications. In contrast to straight-beam testing, only flaw indications appear on the display in an angle-beam test. Only rarely will a back surface be oriented properly to give a back-reflection indication. In most instances, ultrasonic beams are reflected from the back surface at an angle away from the search unit. These reflected pulses are capable of detecting discontinuities and are used extensively in angle-beam testing of plate and tubing.

The time base (horizontal sweep) on the oscilloscope must be carefully calibrated; in angle-beam testing there is no back-reflection echo to provide an internal depth-calibration point. Usually, an extended time base is used so that flaws located within two or more skip distances from the search unit can be detected. Skip distance is defined as the distance from the point at which the sound beam first enters the test piece to the

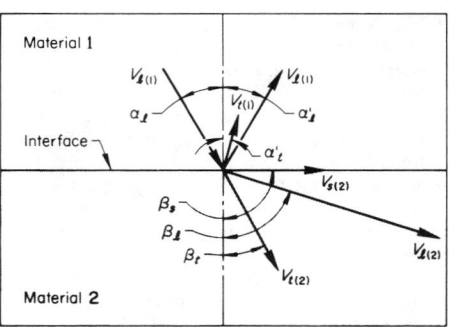

Fig. 7. Diagram showing relationship (by vectors) of all possible reflected and refracted waves to an incident longitudinal wave of velocity $V_{l(1)}$ impinging on an interface at angle α_l relative to normal to the interface

point at which the back-reflected pulse first encounters the front surface. In a plate, skip distance, S, can be calculated from:

$$S = 2t \tan \beta$$

where t is the plate thickness and β is the angle of refraction of the sound beam at the point of entry (see Fig. 7).

Sometimes, moving the search unit in an arc about the position of a suspected flaw or swiveling the search unit about a fixed position can be equally useful (see Fig. 8a). As shown in Fig. 8(b), traversing the search unit in an arc about the location of a gas hole produces little or no change in the echo; the indication on the oscilloscope screen remains constant in both height and position on the trace as the search unit is moved. On the other hand, if the search unit were to be swiveled on the same spot, the indication would abruptly disappear after the search unit had been swiveled only a few degrees.

TRANSMISSION METHODS

Regardless of whether transmission ultrasonic testing is done with direct beams or reflected beams, flaws are detected by comparing the amount of ultrasound transmitted through the test piece with the amount transmitted through a reference standard made of the same material. Transmission testing requires two search units, one to transmit the ultrasonic waves and one to receive them.

The main application of transmission methods is the inspection of plate for cracks or lamina-

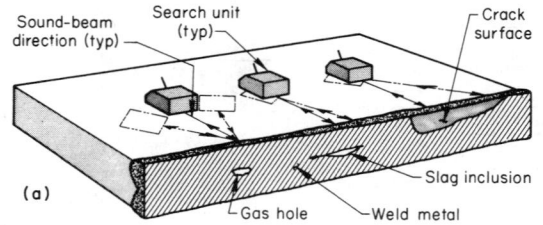
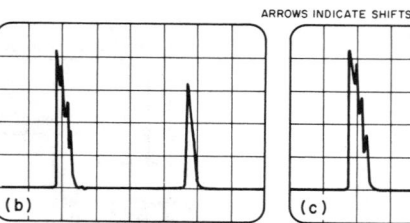
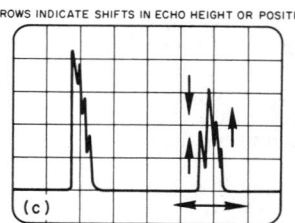
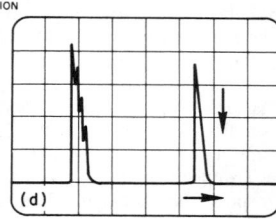

(a) Positions of search units on the test piece. (b) Display pattern obtained from a gas hole as the result of traversing the search unit in an arc about the location of the flaw. (c) Display pattern obtained from a slag inclusion as the result of swiveling the search unit on a fixed point. (d) Display pattern obtained from a crack, using the same swiveling search-unit movement as in (c).

Fig. 8. Angle-beam inspection of a weldment, showing effect of search-unit movements on oscilloscope-screen display patterns from three different types of flaws in welds. See text for discussion.

tions that have relatively large dimensions compared with the size of the search units. Immersion techniques or water-column (bubbler or squirter) techniques are most effective, because these techniques provide efficient and relatively uniform coupling between the search units and the test piece.

Displays of transmission-test data can be either oscilloscope traces, strip-chart recordings or meter readings. Oscilloscopes are used to record data mainly when pulsed sound beams are used for testing; strip charts and meters are more appropriate for continuous beams. With all three types of display, alarms or automatic sorting devices can be used to give audible warning or to shunt defective workpieces out of the normal flow of production.

Pitch-catch testing may be done either with direct beams (through-transmission testing) or with reflected beams. In both instances, pulses of ultrasonic energy pass through the material and the intensities of the pulses are measured at the point of emergence. An oscilloscope display is triggered simultaneously with the initial pulse, and the transmitted-pulse indication appears on the screen to the right of the initial-pulse indication in a manner quite similar to the back-reflection indication in pulse-echo testing. A major advantage of pitch-catch testing is that disturbances and spurious indications can be separated from the transmitted pulse by their corresponding transit times. Only the amplitude of the transmitted pulse is monitored; all other sound waves reaching the receiver are ignored. An electronic gate can be set to operate an alarm or a sorting device when the monitored amplitude of the ultrasonic wave drops below a preset value.

When reflected pulses are used, the technique is almost identical to the loss-of-back-reflection technique that is often used in ordinary pulse-echo testing.

GENERAL CHARACTERISTICS OF ULTRASONIC WAVES

In contrast to electromagnetic waves such as light or x-rays, ultrasonic waves are mechanical waves that consist of oscillations or vibrations of the atomic or molecular particles of a substance about the equilibrium position of those particles. Ultrasonic waves can propagate in elastic media, which may be solid or liquid. Ultrasonic waves in the megahertz region are severely attenuated in air and cannot propagate in a vacuum. In many respects an ultrasonic beam is similar to a light beam. Both obey general wave equations and each travels at a characteristic velocity that depends on the properties of that medium. Ultrasonic beams, like light beams, are reflected from surfaces and are refracted when they cross boundaries between two media that have different acoustic velocities. Depending on the mode of particle motion, ultrasonic waves are classified as longitudinal waves, vertically and horizontally polarized shear or transverse waves, surface waves, Lamb waves, etc. Four of these wave modes are described in the following paragraphs.

Longitudinal waves, sometimes called compression waves, are most widely used in the inspection of metals. They travel through metal as a series of alternate compressions and rarefactions in which the particles transmitting the wave vibrate back and forth in the direction of travel of the waves.

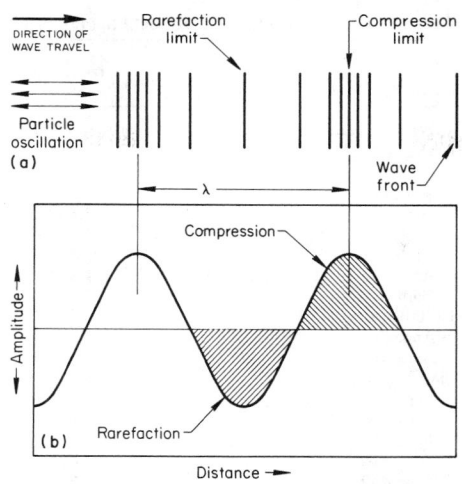

(a) Particle oscillation and resultant rarefaction and compression. (b) Amplitude of particle displacement versus distance of wave travel. The wavelength (λ) is the distance corresponding to one complete cycle.

Fig. 9. Schematic representation of longitudinal ultrasonic waves

Longitudinal ultrasonic waves and the corresponding particle oscillation and resultant rarefaction and compression are represented schematically in Fig. 9(a); a plot of amplitude of particle displacement versus distance of wave travel, together with the resultant rarefaction trough and compression crest, is shown in Fig. 9(b). The distance from one crest to the next (which equals the distance for one complete cycle of rarefaction-and-compression) is the wavelength (λ). The vertical axis in Fig. 9(b) could represent pressure instead of particle displacement; the horizontal axis could represent time instead of travel distance, because the speed of sound is constant in a given material, and this relation is used in the measurements made in ultrasonic inspection.

Longitudinal ultrasonic waves are readily propagated in liquids and elastic solids. The mean free paths of the molecules of liquids are so short that longitudinal waves can be propagated simply by the elastic collision of one molecule with the next. The velocity of longitudinal ultrasonic waves is about 6000 m/s (19 700 ft/s) in steel and about 1500 m/s (4900 ft/s) in water.

Transverse waves (shear waves) also are used extensively in ultrasonic inspection of metals. Transverse waves are visualized readily in terms of vibrations of a rope that is shaken rhythmically, in which each particle vibrates up and down in a plane perpendicular to the direction of prop-

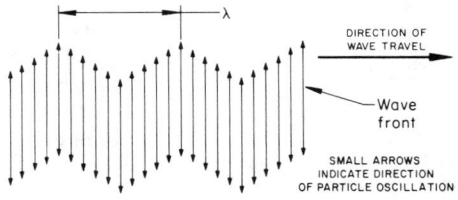

Fig. 10. Schematic representation of transverse (shear) waves. The wavelength (λ) is the distance corresponding to one complete cycle.

agation. A transverse wave is represented schematically in Fig. 10, which shows particle oscillation, wave front, direction of wave travel and the wavelength (λ) corresponding to one cycle.

Air and water will not support transverse waves. In gases, the forces of attraction between molecules are so small that shear waves cannot be transmitted. The same is true of a liquid, unless it is particularly viscous or is present as a very thin layer.

Surface waves (Rayleigh waves) are another type of ultrasonic waves used in the inspection of metals. These waves travel along the flat or curved surfaces of relatively thick solid parts. For propagation of waves of this type, the waves must be traveling along an interface bounded on one side by the strong elastic forces of a solid and on the other side by the practically negligible elastic forces between gas molecules. Surface waves, therefore, are essentially nonexistent in a solid immersed in a liquid, unless the liquid covers the solid surface only as a very thin film.

Surface waves are subject to less attenuation in a given material than are longitudinal or transverse waves. They have a velocity approximately 90% of the transverse-wave velocity in the same material. The region within which these waves propagate with effective energy is not much thicker than about one wavelength beneath the surface of the metal. At this depth, wave energy is about 4% of the wave energy at the surface, and the amplitude of oscillation decreases sharply to a negligible value at greater depths.

In Rayleigh waves, particle oscillation generally follows an elliptical orbit, as shown schematically in Fig. 11. The major axis of the ellipse is perpendicular to the surface along which the waves are traveling. The minor axis is parallel to the direction of propagation. Rayleigh waves can exist in complex forms that are variations of the simplified wave form illustrated in Fig. 11.

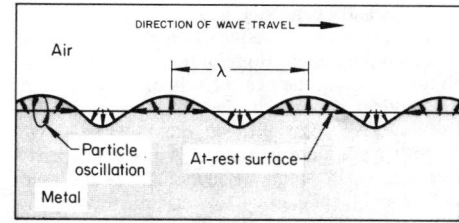

SMALL ARROWS INDICATE DIRECTIONS OF PARTICLE DISPLACEMENT

Fig. 11. Diagram of surface (Rayleigh) waves propagating at the surface of a metal along a metal-air interface. The wavelength (λ) is the distance corresponding to one complete cycle.

Lamb waves, also known as plate waves, are propagated in a mode in which the ultrasonic beam is contained within two parallel boundary surfaces (such as a plate or the wall of a tube). A Lamb wave consists of a complex vibration that occurs throughout the thickness of the material. The propagation characteristics of Lamb waves depend on the density, elastic properties and structure of the metal and are influenced by the thickness of the material.

There are two basic forms of Lamb waves: (*a*) symmetrical, or dilatational; and (*b*) asymmetrical, or bending. The form is determined by

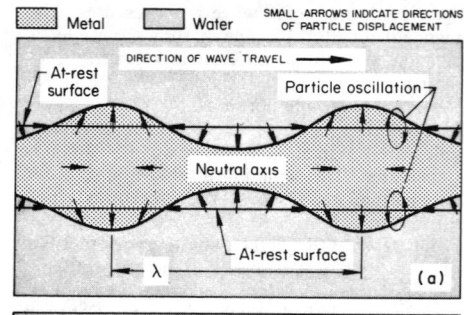

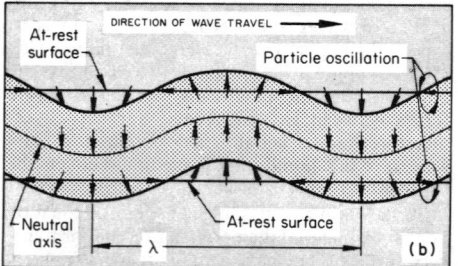

Fig. 12. Diagram of the basic patterns of (a) symmetrical (dilatational) and (b) asymmetrical (bending) Lamb waves. The wavelength (λ) is the distance corresponding to one complete cycle.

whether the particle motion is symmetrical or asymmetrical with respect to the neutral axis of the test piece. Each form is further subdivided into several modes having different velocities, which can be controlled by the angle at which the waves enter the test piece. Theoretically, there are an infinite number of specific velocities at which Lamb waves can travel in a given material. Within a given plate, the specific velocities of Lamb waves are complex functions of plate thickness and cyclic frequency.

In symmetrical (dilatational) Lamb waves, there is a compressional (longitudinal) particle displacement along the neutral axis of the plate and an elliptical particle displacement on each surface (see Fig. 12a). In asymmetrical (bending) Lamb waves, there is a shear (transverse) particle displacement along the neutral axis of the plate and an elliptical particle displacement on each surface (see Fig. 12b). The ratio of the major to minor axes of the ellipse is a function of the material in which the wave is being propagated.

FACTORS INFLUENCING ULTRASONIC INSPECTION

The parameters that must be considered in ultrasonic inspection include both the characteristics of the ultrasonic waves used and the characteristics of the parts being inspected. Equipment type and capability interact with these variables; often, different types of equipment must be selected in order to accomplish different inspection objectives.

Inspection frequency is chosen as a compromise between the ability of the ultrasonic beam to penetrate the material and the time or depth resolution desired. A high frequency generally is desired to achieve high resolution and high definition, but a lower frequency may be required to achieve the desired penetration.

Sensitivity, or the ability of an ultrasonic-in-

Table 3. Acoustic properties of several metals and nonmetals

Material	Density (ρ), g/cm³	Sonic velocities, km/s			Acoustic impedance (Z_l), g/cm³·s(d)
		V_l(a)	V_t(b)	V_s(c)	
Ferrous metals					
Carbon steel, annealed	7.85	5.94	3.24	3.0	4.66
Alloy steel:					
Annealed	7.86	5.95	3.26	3.0	4.68
Hardened	7.8	5.90	3.23	...	4.6
Cast iron	6.95-7.35	3.5-5.6	2.2-3.2	...	2.5-4.0
52100 steel:					
Annealed	7.83	5.99	3.27	...	4.69
Hardened:	7.8	5.89	3.20	...	4.6
D6 tool steel:					
Annealed	7.7	6.14	3.31	...	4.7
Hardened:	7.7	6.01	3.22	...	4.6
Stainless steels:					
Type 302	7.9	5.66	3.12	3.12	4.47
Type 304L	7.9	5.64	3.07	...	4.46
Type 347	7.91	5.74	3.10	2.8	4.54
Type 410	7.67	5.39	2.99	2.16	4.13
Type 430	7.7	6.01	3.36	...	4.63
Nonferrous metals					
Aluminum 1100-O	2.71	6.35	3.10	2.90	1.72
Aluminum alloy 2117-T4	2.80	6.25	3.10	2.79	1.75
Beryllium	1.85	12.80	8.71	7.87	2.37
Copper 110	8.9	4.70	2.26	1.93	4.18
Copper alloys:					
260 (cartridge brass, 70%)	8.53	3.83	2.05	1.86	3.27
464 to 467 (naval brass)	8.41	4.43	2.12	1.95	3.73
510 (phosphor bronze, 5% A)	8.86	3.53	2.23	2.01	3.12
752 (nickel silver 65-18)	8.75	4.62	2.32	1.69	4.04
Lead:					
Pure	11.34	2.16	0.70	0.64	2.45
Hard (94Pb-6Sb)	10.88	2.16	0.81	0.73	2.35
Magnesium alloy M1A	1.76	5.74	3.10	2.87	1.01
Mercury, liquid	13.55	1.45	...	...	1.95
Molybdenum	10.2	6.25	3.35	3.11	6.38
Nickel:					
Pure:	8.8	5.63	2.96	2.64	4.95
Inconel	8.5	5.82	3.02	2.79	4.95
Inconel X-750	8.3	5.94	3.12	...	4.93
Monel	8.83	5.35	2.72	2.46	4.72
Titanium, commercially pure	4.5	6.10	3.12	2.79	2.75
Tungsten	19.25	5.18	2.87	2.65	9.98
Nonmetals					
Air(e)	0.00129	0.331	...	...	0.00004
Ethylene glycol	1.11	1.66	...	...	0.18
Glass:					
Plate	2.5	5.77	3.43	3.14	1.44
Pyrex	2.23	5.57	3.44	3.13	1.24
Glycerin	1.26	1.92	...	...	0.24
Oil:					
Machine (SAE 20)	0.87	1.74	...	...	0.150
Transformer	0.92	1.38	...	...	0.127
Paraffin wax	0.9	2.2	...	...	0.2
Plastics:					
Methyl methacrylate (Lucite, Plexiglas)	1.18	2.67	1.12	1.13	0.32
Polyamide (nylon)	1.0-1.2	1.8-2.2	...	...	0.18-0.27
Polytetrafluoroethylene (Teflon)	2.2	1.35	...	...	0.30
Quartz, natural	2.65	5.73	...	...	1.52
Rubber, vulcanized	1.1-1.6	2.3	...	...	0.25-0.37
Tungsten carbide	10-15	6.66	3.98	...	6.7-9.9
Water:					
Liquid(f)	1.0	1.49	...	...	0.149
Ice(g)	0.9	3.98	1.99	...	0.36

(a) Longitudinal (compression) waves. (b) Transverse (shear) waves. (c) Surface waves. (d) For longitudinal waves $Z_l = \rho V_l$. (e) At standard temperature and pressure. (f) At 4 °C (39 °F). (g) At 0 °C (32 °F).

spection system to detect a very small discontinuity, is generally increased by using relatively high frequencies (short wavelengths).

Acoustic Impedance. When ultrasonic waves traveling through one medium impinge on the boundary of a second medium, a portion of the incident acoustic energy is reflected back from the boundary while the remaining energy is transmitted into the second medium. The char-

acteristic that determines the amount of reflection is the acoustic impedance of the two materials on either side of the boundary. If the impedances of the two materials are equal, there will be no reflection; if the impedances differ greatly (as between a metal and air, for instance), there will be virtually complete reflection.

This characteristic is used in ultrasonic inspection of metals to calculate the amounts of energy reflected and transmitted at impedance discontinuities and to aid in the selection of suitable materials for effective transfer of acoustic energy between components in ultrasonic-inspection systems.

The acoustic impedance for a longitudinal wave (Z_l), in grams per square centimetre – second, is defined as the product of material density (ρ), in grams per cubic centimetre, and longitudinal-wave velocity (V_l), in centimetres per second:

$$Z_l = \rho V_l$$

Acoustic properties of several metals and nonmetals are listed in Table 3. The acoustic properties of metals and alloys are influenced by variations in structure and metallurgical condition. Hence, for a given test piece the properties may differ somewhat from the values shown in Table 3.

Angle of Incidence. Only when an ultrasonic wave is incident at right angles on an interface between two materials (normal incidence — that is, angle of incidence = 0°) do transmission and reflection occur at the interface without any change in beam direction. At any other angle of incidence, the phenomena of mode conversion (a change in the nature of the wave motion) and refraction (a change in direction of wave propagation) must be considered. These phenomena may affect the entire beam or only a portion of the beam, and the sum total of the changes that occur at the interface depends on the angle of incidence and the velocity of the ultrasonic waves leaving the point of impingement on the interface. All possible ultrasonic waves leaving this point are shown for an incident longitudinal ultrasonic wave in Fig. 7. Not all the waves shown in Fig. 7 will be produced in any specific instance of oblique impingement of an ultrasonic wave on an interface between two materials. The waves that propagate in a given instance depend on the angle of incidence of the initial beam, the velocities of the wave forms in both materials and the ability of a wave form to exist in a given material.

Critical Angles. If the angle of incidence (α_l in Fig. 7) is small, sound waves propagating in a given medium undergo mode conversion at a boundary, resulting in simultaneous propagation of longitudinal and transverse (shear) waves in a second medium. If the angle is increased, the direction of the refracted longitudinal wave will approach the plane of the boundary ($\beta_l \rightarrow 90°$). At some specific value of α_l, β_l will exactly equal 90° and the refracted longitudinal wave will disappear, leaving only a refracted (mode-converted) shear wave to propagate in the second medium. This value of α_l is known as the "first critical angle." If α_l is increased beyond the first critical angle, the direction of the refracted shear wave will approach the plane of the boundary ($\beta_t \rightarrow 90°$). At a second specific value of α_l, β_t will exactly equal 90° and the refracted transverse wave will disappear. This second value of α_l is called the "second critical angle."

In ordinary angle-beam inspection, it is usually desirable to have only a shear wave propagating in the test material. Because longitudinal waves and shear waves propagate at different speeds, echo signals will be received at different times, depending on which type of wave produced the echo. When both types are present in the test material, confusing echo patterns may be displayed on the readout device, which can lead to an erroneous interpretation. Frequently, it is

Table 4. Critical angles for immersion and contact testing, and incident angle for 45° shear-wave transmission, in various metals

Metal	First critical angle, degrees(a), for:		Second critical angle, degrees(a), for:		45°-shear-wave incident angle, degrees(a), for:	
	Immersion testing(b)	Contact testing(c)	Immersion testing(b)	Contact testing(c)	Immersion testing(b)	Contact testing(c)
Steel	14.5	26.5	27.5	55	19	35.5
Cast iron	15 to 25	28 to 50	...	...	...	...
Type 302 stainless steel	15	28	29	59	19.5	37
Type 410 stainless steel	11.5	21	30	63	20.5	39
Aluminum alloy 2117-T4	13.5	25	29	59.5	20	37.5
Beryllium	6.5	12	10	18	7	12.5
Copper alloy 260 (cartridge brass, 70%)	23	44	46.5	...	31	67
Inconel	11	20	30	62	20.5	38.5
Magnesium alloy M1A	15	27.5	29	59.5	20	37.5
Monel	16.5	30	33	79	23	44
Titanium	14	26	29	59	20	37

(a) Measured from a direction normal to surface of test material. (b) In water at 4 °C (39 °F). (c) Using angle block (wedge) made of acrylic plastic.

desired to produce shear waves in a material at an angle of 45° to the surface. In most materials, incident angles for mode conversion to a 45° shear wave lie between the first and second critical angles. Typical values of α_l for all three of these — first critical angle, second critical angle, and incident angle for mode conversion to 45° shear waves — are listed in Table 4 for various metals.

Absorption of ultrasonic energy occurs mainly by conversion of mechanical energy into heat. Elastic motion within a substance as a sound wave propagates through it alternately heats the substance during compression and cools it during rarefaction. Because heat flows so much more slowly than an ultrasonic wave, thermal losses are incurred, which progressively reduces energy in the propagating wave. A related thermal loss occurs in polycrystalline materials: a thermoelastic loss arises from heat flow away from grains that have received more compression or expansion in the course of wave motion than did adjacent grains. For most polycrystalline materials this effect is most pronounced at the low end of the ultrasonic-frequency spectrum.

Scattering of an ultrasonic wave occurs because most materials are not truly homogeneous. Crystal discontinuities such as grain boundaries, twin boundaries and minute nonmetallic inclusions tend to deflect small amounts of ultrasonic energy out of the main ultrasonic beam. Also, especially in mixed microstructures or anisotropic materials, mode conversion at crystallite boundaries tends to occur because of slight differences in acoustic velocity and acoustic impedance across boundaries.

Scattering is highly dependent on the relation of crystallite size (mainly grain size) to ultrasonic wavelength. When grain size is less than 0.01 times the wavelength, scatter is negligible. Scattering effects vary approximately with the third power of grain size, and when the grain size is 0.1 times the wavelength or larger, excessive scattering may make it impossible to do valid ultrasonic inspections.

Diffraction. A sound beam propagating in a homogeneous medium is coherent — that is, all particles that lie along any given plane parallel to the wave front vibrate in identical patterns. When a wave front passes the edge of a reflecting surface, the front bends around the edge in a manner similar to that in which light bends around the edge of an opaque object. When the reflector is very small in comparison with the sound beam, as is usual for a pore or an inclusion, wave bending (forward scattering) around

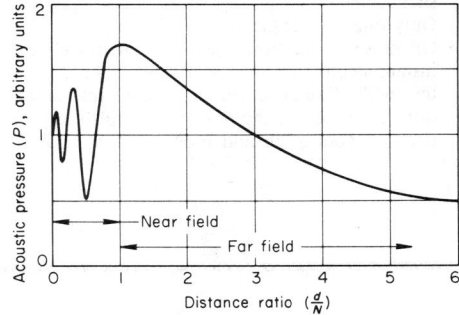

Fig. 13. Variation of acoustic pressure with distance ratio for a circular search unit. Distance ratio is distance from crystal face, d, divided by length of near field, N.

the edges of the reflector produces an interference pattern in a zone immediately behind the reflector because of phase differences among different portions of the forward-scattered beam. The interference pattern consists of alternate regions of maximum and minimum intensity that correspond to regions where interfering scattered waves are respectively in phase and out of phase.

Diffraction phenomena must be taken into account during development of ultrasonic-inspection procedures.

Near-Field and Far-Field Effects. The face of the transducer element vibrates in a complex manner that can be most easily described as a mosaic of tiny, individual crystals, each vibrating in the same direction but slightly out of phase with its neighbors. Each element in the mosaic acts like a point (Huygens) source, and radiates a spherical wave outward from the plane of the transducer face.

Along the central axis of the composite ultrasonic beam, the series of acoustic-pressure maximums and minimums become broader and more widely spaced as the distance from the transducer face, d, increases. Where d becomes equal to N (with N denoting the length of the near field), the acoustic pressure reaches a final maximum and decreases approximately exponentially with increasing distance, as shown in Fig. 13.

Beam Spreading. In the far field of an ultrasonic beam, the wave front expands with distance from a transducer. The angle of divergence from the central axis of the beam from a circular transducer is determined from ultrasonic wavelength and transducer size.

ADVANTAGES, DISADVANTAGES AND APPLICATIONS

Advantages. The principal advantages of ultrasonic inspection in comparison with other methods of nondestructive inspection of metal parts are:

- Superior penetrating power, which permits detection of flaws deep in the part. Ultrasonic inspection is done routinely to depths of several feet in many types of parts, and to depths of about 6 m (20 ft) in axial inspection of parts such as long steel shafts and rotor forgings.
- High sensitivity, permitting detection of extremely small flaws.
- Greater accuracy than other nondestructive methods in determining the positions of internal flaws, estimating their sizes, and characterizing them in terms of nature, orientation and shape.
- Only one surface need be accessible.
- Operation is electronic, which provides almost instantaneous indications of flaws. This makes the method suitable for immediate interpretation, automation, rapid scanning, in-line production monitoring and process control. With

most systems, a permanent record of inspection results can be made.
- Volumetric scanning ability, permitting inspection of a volume of metal extending from the front surface to the back surface of a part.
- Ultrasonic inspection presents no radiation hazard to operations or to nearby personnel and has no effect on equipment and materials in the vicinity.
- Portability.

Disadvantages of ultrasonic inspection include the following:

- Manual operation requires careful attention by experienced technicians.
- Technical knowledge is required for development of inspection procedures.
- Parts that are rough, irregular in shape, very small or thin, or not homogeneous are difficult to inspect.
- Discontinuities that are present in a shallow layer immediately beneath the surface may not be detectable.
- Couplants are needed to provide effective transfer of the ultrasonic beam between search units and parts being inspected.
- Reference standards are required, both for cal-

ibrating the equipment and for characterizing flaws.

Applications. Some of the major types of components that are ultrasonically inspected for the presence of flaws are:

- Mill components; rolls, shafts, drives, press columns
- Power equipment: turbine forgings, generator rotors, pressure piping, weldments, pressure vessels, nuclear fuel elements and other reactor components
- Jet-engine parts: turbine and compressor forgings, gear blanks
- Aircraft components: forging stock, frame sections, honeycomb sandwich assemblies
- Machinery materials: die blocks, tool steels, drill pipe
- Railroad parts: axles, wheels, bolted and welded rail
- Automotive parts: forgings, ductile castings, brazed and/or welded components.

Ultrasonic inspection is an effective and inexpensive method for volumetric examination of structures and components of both regular and complex shapes.

Holography

Edited by John Romanko, General Dynamics Corp.

Inspection by Optical Holography

HOLOGRAPHY is basically a two-step process for creating a whole image—that is, a three-dimensional image—of a diffusely reflecting object having some arbitrary shape. In the first step, both the amplitude and phase of any type of coherent wave motion emanating from the object are recorded by encoding this information in a suitable medium. This recording is called a hologram. At a later time, the wave motion is reconstructed from the hologram by a coherent beam in a process that results in the regeneration of an image having the true shape of the object. The utility of holography for nondestructive inspection of metal components and metal-containing structures lies in the fact that this regenerated image can then be used as a kind of three-dimensional template against which any deviations in the shape or dimensions of the object can be observed and measured.

In principle, holography can be performed with (*a*) any wave radiation encompassed in the entire electromagnetic spectrum; (*b*) any particulate radiation, such as neutrons and electrons, that possesses wave-equivalent properties; and (*c*) non-electromagnetic wave radiation, such as sound waves. The two methods currently available for practical nondestructive inspection are optical holography, using visible light waves, which is discussed in this article; and acoustical holography, using ultrasonic waves, which is discussed in the article that follows.

HOLOGRAPHIC RECORDING

When visible light waves are employed in holography, the hologram is recorded using an op-

tical system called a holocamera (see Fig. 1a). A monochromatic laser beam of phase-coherent light is divided into two beams by a variable beam splitter. One beam, the object beam, is expanded by a spatial filter into a divergent beam directed to illuminate the object uniformly. A portion of the laser light reflected from the object is intercepted by a high-resolution photographic plate, as shown in Fig. 1(a). The second beam, the reference beam, originating from the beam splitter is directed by a mirror, diverges from a second spatial filter, and is directed onto the photographic plate by a second mirror. With either object beam or reference beam absent, a uniformly exposed photographic plate will result. However, with both coherent beams falling on the plate simultaneously, an interference pattern is generated as a result of the coherent interaction of the two beams, and this pattern is recorded by the photographic emulsion.

HOLOGRAPHIC RECONSTRUCTION

In the reconstruction process, which is illustrated in Fig. 1(b), the hologram is used as a diffraction grating. When it is illuminated with the reference beam only, three beams emerge — a zero-order, or undeflected, beam and two first-order diffracted beams. The diffracted beams produce the real and virtual images of the object to complete the holographic process. The real image is pseudoscopic, or depth inverted, in appearance. Hence, the virtual image (also referred to as the true, primary or nonpseudoscopic image) is the one that is of primary interest in practical applications of holography. In Fig. 1(b), only the first-order diffracted beam that yields the virtual image is shown, the other two beams having been omitted from the figure for the sake of clarity. If the original object is three-dimensional,

the virtual image is a genuine three-dimensional replica of the object, possessing both parallax and depth of focus. However, if the configuration of the optical system or the wavelength of light used during reconstruction differs from that used during recording, distortion, aberration and changes in magnification can occur. (The holographic recording and reconstruction systems can be designed to minimize these effects.)

INTERFEROMETRIC TECHNIQUES OF INSPECTION

When optical holography is applied to inspection of parts, generation of a three-dimensional image of the object, per se, is of little value. Furthermore, for opaque materials, optical holography is strictly limited to surface observations. Hence, if optical holography is to be used for nondestructive inspection, supplementary means must be used to either stress or otherwise excite test objects so as to produce surface manifestations of the feature of interest. It is for measurement of such manifestations that the techniques of optical holography have been further developed to form the subfield of holography termed optical holographic interferometry.

As in conventional interferometry, holographic interferometric measurements can be made with great accuracy (to within a fraction of the wavelength of the light being used). Whereas conventional interferometry is usually restricted to the examination of objects having highly polished surfaces and simple shapes, holographic interferometry can be used to examine objects of arbitrary shape and surface condition. Because holographic interferometry produces a three-dimensional fringe-field image, which can be examined from many different perspectives (limited only by the size of the hologram), a single

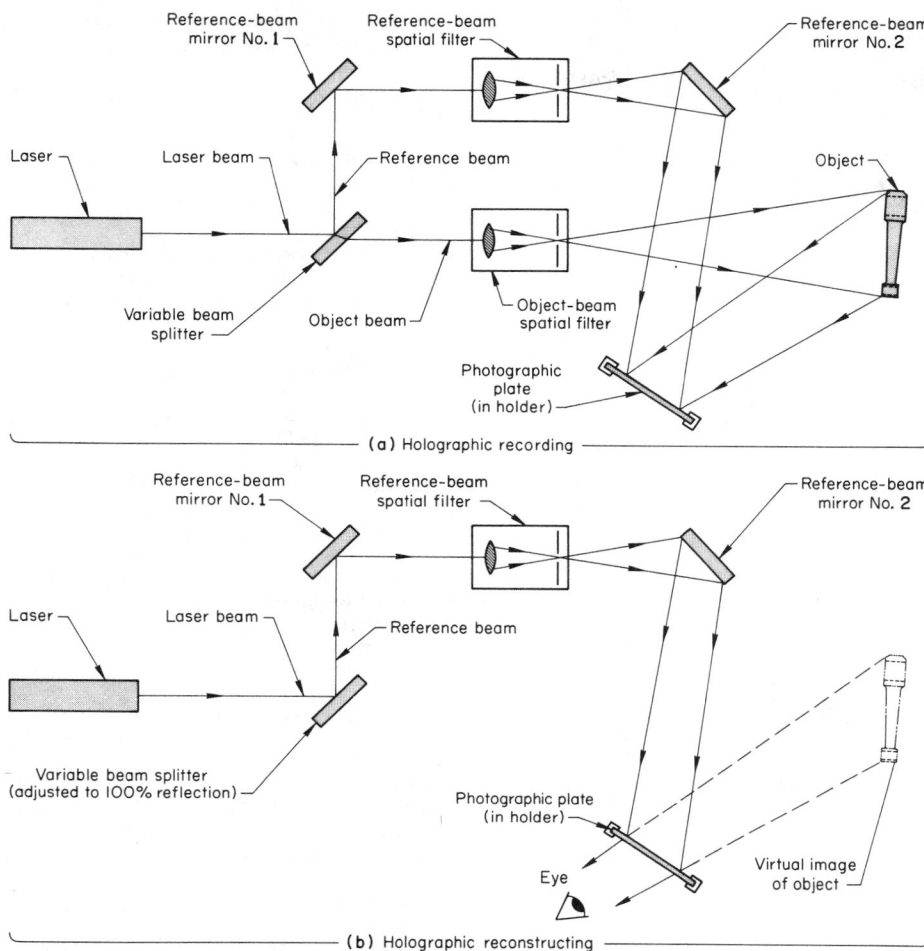

Fig. 1. Schematic diagrams of the basic optical systems used in continuous-wave holography: (a) holocamera used to record hologram of an object on a photographic plate; (b) optical system for reconstructing a virtual image of the object from the hologram on the photographic plate

holographic interferogram is equivalent to a series of conventional two-dimensional interferograms.

In contrast to conventional interferometry, which must be performed in real time, holographic interferometry can be performed either in real time or at two different times (time-lapse technique). In the time-lapse technique, advantage is taken of the fact that more than one hologram may be made on the same recording medium. Examples of this approach include the double-exposure, multiple-exposure and continuous-exposure techniques.

Inspection by Acoustical Holography

ACOUSTICAL HOLOGRAPHY is the extension of holography into the ultrasonic domain. The principles of acoustical holography are the same as those of optical holography, which are discussed in the previous article, because the laws of interference and diffraction apply to all forms of radiation obeying the wave equation. Differences arise only because the methods for record-

ing and reconstructing the hologram must accommodate the form of radiation used. This need to accommodate the form of radiation restricts the practical range of sound-wave frequency that can be used in acoustical holography.

At present, only two types of basic systems for acoustical holography are available — the liquid-surface type and the scanning type. These utilize two different detection methods, and these methods in turn dictate the application of the systems to nondestructive inspection. Neither of these two types of systems relies on the interferometric techniques of optical holographic inspection, where information on flaws at or near the surface of a test object is obtained from the pattern formed by interference between two nearly identical holographic images that are created while the object is differentially stressed. Instead, systems for acoustical holography obtain information on internal flaws directly from the image of the interior of the object.

LIQUID-SURFACE ACOUSTICAL HOLOGRAPHY

The basic system for liquid-surface acoustical holography is similar to the basic system for optical holography, except for the method of readout.

In liquid-surface systems, two separate ultrasonic transducers supply the object beam and reference beam, which are usually pulsed. The two transducers and the test object are immersed in a water-filled tank. The test object is positioned in the object plane of an acoustic (ultrasonic) lens, which also is immersed in the tank.

The practical limits for the object-beam transducer in a commercial system are a wave frequency of 1 to 10 MHz, a pulse length of 50 to 300 microseconds, and a pulse-repetition rate of 60 to 100 per second. This transducer is placed so that its beam passes through the test object. As the object beam passes through the test object, it is modified by the object. The modification is generally in both amplitude and phase. The object beam then passes through an acoustic lens, which focuses the image of the test object at the liquid surface. This image contains a wave front nearly identical to that emanating from the test object.

The reference-beam transducer is connected to the same oscillator as the object-beam transducer so that it emits a second wave front coherent with the wave front from the object-beam transducer. The reference-beam transducer is aimed at the same region of the liquid surface as the object beam, where the wave fronts interfere. The image is reconstructed from the hologram by reflecting a beam of coherent light from the ripple pattern in the isolation tank.

SCANNING ACOUSTICAL HOLOGRAPHY

The basic system for scanning acoustical holography is shown in Fig. 2. No reference-beam transducer is required in this system because electronic phase detection is used to produce the hologram: that is, the required interaction (mixing) between the piezoelectrically detected object-beam signal and the simulated reference-beam signal occurs in the electronic domain. A pulser circuit — consisting of a continuous-wave oscillator and a pulse gate that is triggered by an electronic clock — and a power amplifier feed a single focused ultrasonic transducer that is scanned over an area above the test object while alternately transmitting and receiving the ultrasonic signals. The transducer and the test object are immersed in a water-filled tank, as in Fig. 2(b), or they are coupled by a water column.

The signal is pulsed so that time gating can be used to reject undesired surface echoes. The pulse length may be set to any desired value from a few periods of the wave frequency to an upper limit of 50% of the time between successive pulses. Long pulse lengths are used to examine regions lying deep within the metal, while short pulse lengths are required for the regions near the surface so that transmitted energy is not mixed with reflected energy. The practical limits of the transducer in a commerical system are a wave frequency of 1 to 10 MHz, a pulse length of 5 to 20 microseconds, and a pulse-repetition rate of 500 to 1000 per second. The frequency band of a given transducer is relatively narrow ($\pm 5\%$ of the mean frequency).

The echo from the flaw is received by the transducer and processed by a time-interval counter or a pair of balanced mixers in quadrature. The resulting data are digitized and entered into the computer. As the transducer scans a rectangular area, a matrix of complex numbers accumulates in the computer. This matrix represents the phase and amplitude of the reflected wave taken on a set of sample points.

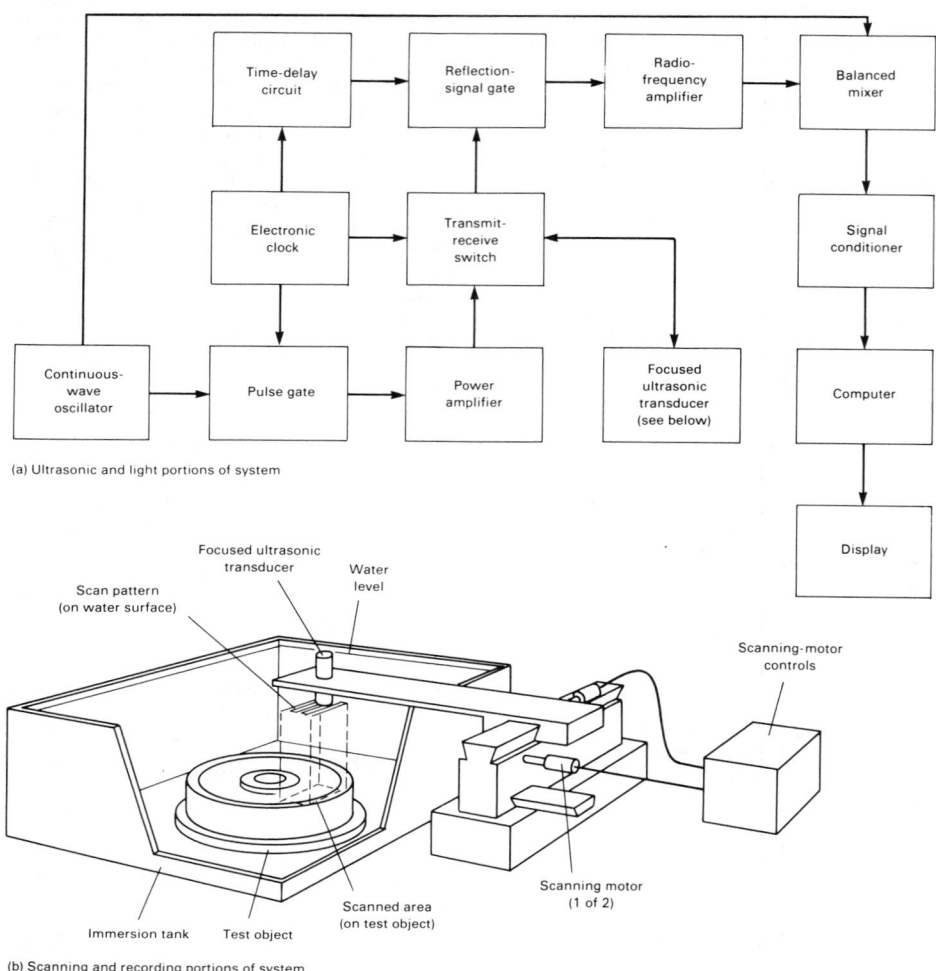

(a) Ultrasonic and light portions of system

(b) Scanning and recording portions of system

Fig. 2. Diagrams of the basic system used in nondestructive inspection by scanning acoustical holography. See text for description.

COMPARISON OF LIQUID-SURFACE AND SCANNING SYSTEMS

The outstanding feature of the liquid-surface system of acoustical holography is that it provides a real-time image, whereas the image provided by the scanning system requires reconstruction. The real-time feature makes the liquid-surface system suitable for rapid inspection of large amounts of material on a continuous basis. In contrast, inspection with the scanning system is relatively slow. Photographic or videotape records for later study may be made with either system.

The outstanding feature of the scanning system is its ability to determine accurately the position and dimensions of flaws lying deep in opaque test objects, especially when only one side of the object is accessible. (In contrast to the scanning system, the liquid-surface system is usually operated in the transmission mode, which requires access to both sides of the test object.) Although both systems offer about the same resolution, the sensitivity of the scanning system is greater by a factor of about 10^6. The excellent flaw-measuring ability of the scanning system may be used to characterize accurately the flaws detected previously by faster inspection methods, such as scanning with a conventional ultrasonic search unit.

Another important feature of the scanning system is that the commercial equipment for this system is usually transportable, whereas liquid-surface equipment is usually stationary. In addition, scanning transducers can be coupled to very large test objects by water columns, whereas inspection by the liquid-surface system usually requires that the test objects be small enough to be placed in a water-filled tank and completely immersed.

A useful advantage of the scanning system is its capability of selectively producing either longitudinal or shear waves in the volume of metal under examination, by adjustment of the angle of incidence of the incoming ultrasonic beam.

Acoustic-Emission Inspection

Edited by James R. Mitchell, Physical Acoustics Corp.

ACOUSTIC EMISSION is defined as the high-frequency stress waves generated by the rapid release of strain energy that occurs within a material during crack growth, plastic deformation, phase transformation, etc. This energy may originate from stored elastic energy as in crack propagation, or from stored chemical-free energy as in phase transformation.

Sources of acoustic emission that generate stress waves in material include local dynamic movements such as the initiation and propagation of cracks, twinning, slip, sudden reorientation of grain boundaries, bubble formation during boiling, or martensitic phase transformations. The stresses in a metallic system may be well below the elastic design limits, and yet the region near a flaw or crack tip may undergo plastic deformation and fracture from locally high stresses, ultimately resulting in premature or catastrophic failure under service conditions.

Acoustic-emission inspection detects and analyzes minute acoustic-emission signals generated by discontinuities in materials under applied stress. Proper analysis of these signals can provide information concerning the location and structural significance of the detected discontinuities.

Another important feature of acoustic emission in general is its irreversibility. If a material is loaded to a given stress level and then unloaded, usually no emission will be observed upon immediate reloading until the previous load has been exceeded.

TYPES OF ACOUSTIC EMISSIONS

Basically, there are two types of acoustic emissions: continuous and burst. The wave form of continuous-type emissions is similar to Gaussian random noise, but the amplitude varies with acoustic-emission activity. In metals and alloys, this form of emission is thought to be associated with the dislocation movements in the grains.

Burst-type emissions are of short-duration pulses (ten microseconds to a few milliseconds in length) and are associated with discrete releases of strain energy. Burst-type emissions are generated by twinning, microyielding, and the development of microcracks and macrocracks. Burst-type emissions have a greater amplitude than the continuous type.

APPLICATIONS

The generation of acoustic emission usually requires that a stress be applied to the structure being tested. The level of stress applied need not reach general yielding, nor does the stress generally need to be of a specific type. Bending stress can be applied to beamed structures, torsional stress can be applied to rotary shafts, thermal stresses are applied to materials during welding, and pressure-induced stress can be applied to pressure-containment systems.

Some of the significant applications of acoustic-emission inspection are:

- Continuous surveillance of pressure vessels and nuclear primary-pressure boundaries for the detection and location of active flaws
- Detection of incipient fatigue fracture in aircraft structures
- Monitoring of both fusion and resistance weldments during welding and during cooling
- Determination of the onset of stress-corrosion cracking and hydrogen damage in susceptible structures
- Use as a study tool for the investigation of fracture mechanisms and of behavior of materials
- Periodic inspection of tanks and aerial-device booms made of composite materials.

INSTRUMENTATION

A typical system for detecting and recording acoustic emission is shown schematically in Fig. 1. Most sensors used for acoustic-emission experiments are piezoelectric ceramic elements. Perhaps the most commonly used of these elements is a variety of lead zirconium titanate. The output of this sensor is amplified by a low-noise preamplifier with a gain of 40 or 60 db and with

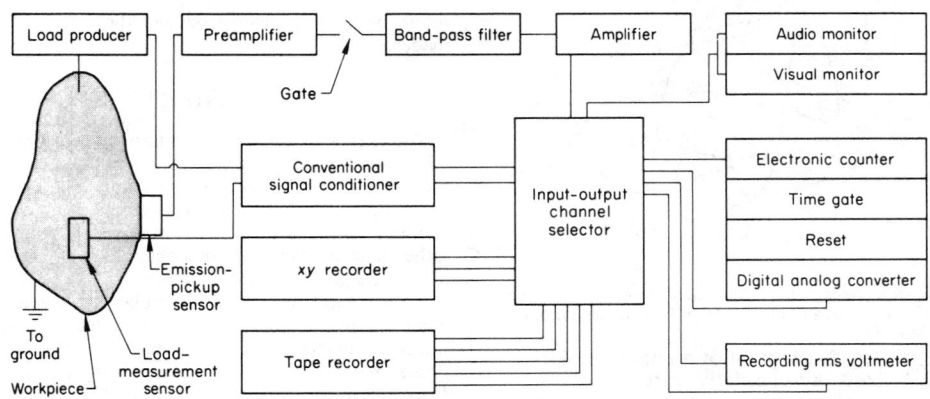

Fig. 1. Schematic representation of a typical system for detecting and recording acoustic emission

a low-impedance output capable of driving a long cable. The preamplifier typically contains a band-pass filter with sharp cutoff characteristics. The output of the filter is fed through a main amplifier that has a variable gain (maximum total gain of 100 db). Sophisticated microprocessor-based devices measure and record each feature of the acoustic-emission signal. Real-time and post-test examination of the data, as well as data manipulation and correlation, are common objectives.

The system shown in Fig. 1 employs a single emission-pickup sensor; multiple sensors can be used with triangulation techniques to locate particular sources of acoustic emission.

Microwave Inspection

Edited by William L. Rollwitz, Southwest Research Institute

MICROWAVES (or radar waves) are a form of electromagnetic radiation located in the electromagnetic spectrum at the frequencies shown in Table 1. Major subintervals of the microwave frequency band are designated by various letters; these are listed in Table 2. Solid-state devices can now generate power at frequencies up to and above 300 GHz.

Applications. The use of microwaves for evaluation of material properties and discontinuities in materials other than radomes began with the evaluation of the concentration of moisture in dielectric materials. Next, the thicknesses of thin metallic coatings on nonmetallic substrates and the thicknesses of dielectric slabs were measured. The measurement of thickness was followed by determination of voids, delaminations, macroporosity, inclusions and other flaws in plastic or ceramic materials. Microwave techniques also were used to detect flaws in bonded honeycomb structures and fiber-wound and laminar composite materials.

Lately, microwaves have been used for holographic-type images of the surfaces of conductors, insulators and semiconductors. Both active- and passive-type images have been used in medical examination of both humans and animals.

Limitations. The use of microwaves is limited by their inability to penetrate deeply into conductors or metals. This means that nonmetallic materials inside a metallic container cannot be easily inspected through the metallic container. Another limitation of microwaves is their comparatively low power of resolving localized flaws.

Advantages. Use of microwaves provides the advantages of resonance absorption (usually broadband in solids) and of coherent detection. Ad-

Table 1. Divisions of radiation, frequencies, wavelengths, and photon energies of the electromagnetic spectrum

Division of radiation	Frequency, Hz	Wavelength, m	Photon energy, eV
Radio waves (FM and TV) ...	3×10^{8}	1	10^{-6}
Microwaves	3×10^{9}	10^{-1}	10^{-5}
	3×10^{10}	10^{-2}	10^{-4}
	3×10^{11}	10^{-3}	10^{-3}
Infrared	3×10^{12}	10^{-4}	10^{-2}
	3×10^{13}	10^{-5}	10^{-1}
Visible light	3×10^{14}	10^{-6}	1
Ultraviolet light	3×10^{15}	10^{-7}	10
	3×10^{16}	10^{-8}	10^{2}
X and gamma radiation	3×10^{17}	10^{-9}	10^{3}
	3×10^{18}	10^{-10}	10^{4}
	3×10^{19}	10^{-11}	10^{5}
	3×10^{20}	10^{-12}	10^{6}
	3×10^{21}	10^{-13}	10^{7}
Cosmic radiation	3×10^{22}	10^{-14}	10^{8}

Table 2. Letter designations of major subintervals of the microwave frequency band

Letter	Frequency band, GHz	Wavelength band in vacuum, cm
L	0.390 to 1.550	76.9 to 19.3
S	1.550 to 5.200	19.3 to 5.77
X	5.200 to 10.900	5.77 to 2.75
K	10.900 to 36.000	2.75 to 0.834
Q	36.000 to 46.000	0.834 to 0.652
V	46.000 to 56.000	0.652 to 0.536
W	56.000 to 100.000	0.536 to 0.300

ditional advantages are ease of scanning and the lack of a need for surface contact.

PHYSICAL PRINCIPLES OF MICROWAVES

In free space, an electromagnetic wave is transverse; that is, the oscillating electric and magnetic fields that compose it are transverse to the direction of travel of the wave. The relative directions of these two fields and the direction of propagation of the wave are shown schematically in Fig. 1.

A particularly simple form of a propagating electromagnetic wave is the linearly polarized, sinusoidally varying, plane electromagnetic wave illustrated in Fig. 2. The velocity (v) at which a wave front travels along the z-axis is given by the relation $v = f\lambda$, where f is frequency and λ

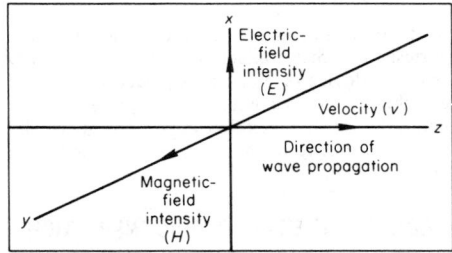

Fig. 1. Relative directions of the electric-field intensity (E), the magnetic-field intensity (H), and the direction of propagation (z) for a linearly polarized, plane electromagnetic wave

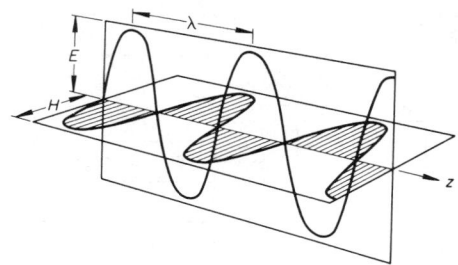

λ = wavelength; z = direction of wave propagation; E = amplitude of electric field; H = amplitude of magnetic field.

Fig. 2. Diagram of a linearly polarized, sinusoidally varying, plane electromagnetic wave propagating in empty space

is wavelength. In free space, this velocity is the speed of light, which has the value 2.998×10^8 m/s and usually is designated by the letter c.

When an electromagnetic wave meets an object or a surface, it is both reflected and transmitted. The phase angles and amplitudes of the reflected and transmitted waves are controlled by the boundary conditions, and by the differences between the permittivities and permeabilities of the two media at the boundary. For this reason, both the reflected waves and the transmitted waves have the characteristics of the materials at the interface. These properties make microwaves use-

ful in both the transmission and the reflection modes.

INSTRUMENTATION

The higher frequencies of microwaves do not allow the use of 2-GHz electronic circuits which possess discrete inductance, capacitance and resistance. Thus, microwave circuit design and description are handled from a wave standpoint, rather than on the basis of conduction of electrons through wires and components.

Hollow metal tubes (both circular and rectangular in cross section) called waveguides convey microwaves between two parts of a circuit. Sometimes the microwaves are conveyed, as a wave, along coaxial or parallel conductors and along a single wire.

Microwave energy can be conveyed through a waveguide in definite patterns called modes which consist of repetitive distributions of the electrical and magnetic fields. When the waveguide dimensions are close to the wavelength the modes are identified as either transverse electric (TE) or transverse magnetic (TM), referring to whether the electric or the magnetic field is perpendicular to the waveguide axis, or direction of propagation. One of the modes most often used is the lowest-order transverse electric mode ($TE_{1,0}$), whose waveguide pattern is shown in Fig. 3. The intensity of the electric field varies sinusoidally along and across the waveguide, with a peak intensity in the middle of the waveguide and zero

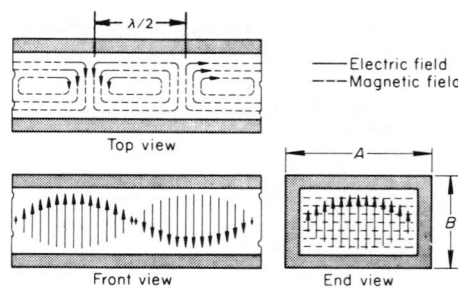

Fig. 3. Diagrams showing the lowest-order transverse electric mode ($TE_{1,0}$) in a waveguide

intensity at either side wall (see end view in Fig. 3.) A top view of the waveguide shows that the magnetic field is in the form of loops, spaced at one-half wavelength intervals.

Each mode has a cutoff frequency below which it cannot be propagated down a given size of waveguide. For the commonly used $TE_{1,0}$ mode, the cutoff frequency is equal to the free-space velocity divided by twice the A dimension. Thus, large waveguides imply low frequencies and small ones imply high frequencies.

It used to be that microwaves were generated only by vacuum devices like klystrons, magnetrons and traveling wave tubes. However, solid-state (semiconductor) types of amplifiers and generators are now available at relatively high power levels at frequencies up to 300 GHz.

Thermal Inspection

Edited by Sidney Allinikov, Wright-Patterson Air Force Base

THERMAL INSPECTION comprises inspection by all methods in which heat-sensing devices or substances are used to detect irregular temperatures. Thermal inspection of workpieces can be used to detect flaws and to detect undesirable distribution of heat during service. There are several methods of thermal inspection and many types of temperature-measuring devices and substances. This article, however, is limited mainly to the discussion of:

- Thermography, which is the mapping of isotherms, or contours of equal temperature, over a test surface
- Thermometry, which is the measurement of temperature.

These techniques are separated into two categories: (a) direct contact, in which a thermally sensitive device or material is placed in physical and thermal contact with the test piece; and (b) noncontact techniques that depend on thermally generated electromagnetic energy radiated from the test piece.

CONTACT THERMOGRAPHIC INSPECTION

Contact thermographic inspection consists of coating the surface of the test piece with a material that reacts to a change in temperature by changing color or other aspect of appearance. The reaction may be permanent or reversible. Many

coatings having reversible reactions can be recovered for reuse.

The most commonly used materials for contact thermographic inspection can be classified as heat-sensitive paints, heat-sensitive papers, thermal phosphors, liquid crystals, and other temperature-sensitive coatings.

Heat-sensitive paints that are effective from 40 to 1600 °C (about 100 to 2900 °F) are available. Some of these paints undergo several color transitions as their temperature is increased, and, under favorable conditions, an accuracy of ±5 °C (±9 °F) is attainable with them. These paints have been used effectively for monitoring isotherms in heat-affected zones during welding, for monitoring preheating prior to welding, and for inspecting castings for porosity.

An example of how photochromic paint can be used on composites of construction different from that of honeycomb-core composites is shown in Fig. 1, which illustrates a delaminated section between 20%-density stainless steel fibers brazed to Hastelloy X. Here, the white area is the delaminated section, because the workpiece was heated on the painted surface, after exposure to ultraviolet radiation.

Heat-Sensitive Papers. Several types of heat-sensitive papers have been used in nondestructive inspection. They are applied by bonding directly to the surface of the test piece, or by means of a vacuum holddown arrangement.

One type consists of a porous paper coated with an organic compound. A finely divided white pigment is applied to a highly absorptive black paper in a binder soluble in a solvent that does not dissolve the pigment. When the melting temperature of the pigment is reached, the paper absorbs the coating and the initial white color is replaced by a black appearance. These papers are used to indicate when a specific temperature has been reached.

NONCONTACT THERMOGRAPHIC INSPECTION

A large number of devices will respond to the radiant energy in the infrared bandwidth produced by an object at a temperature above absolute zero and convert it to a proportional electric signal that then may be displayed.

Infrared Detectors. There are two basic types of infrared detectors: photon-effect devices and thermal devices. The response of the photon-effect devices depends on the wavelength of the received radiation; therefore, the output signal depends on the wavelength of the infrared signal. Thermal detectors respond only to the heating caused by the incoming radiation, and the output signal is largely independent of the radiation wavelength. At certain wavelengths, the photon-effect devices will produce a much larger signal than thermal detectors.

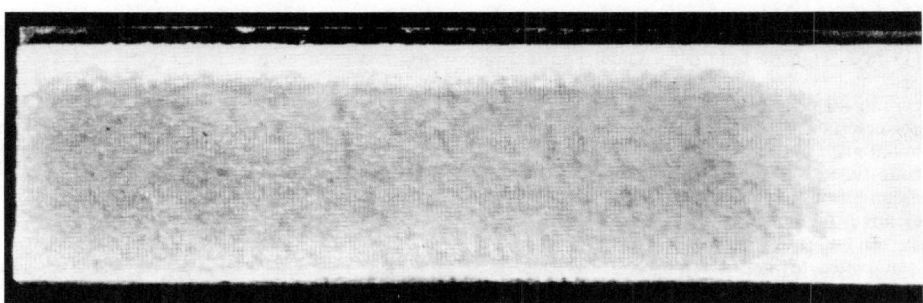

Fig. 1. Delaminated area (white) between 20%-density stainless steel fibers brazed to Hastelloy X, as revealed by photochromic paint. The delaminated area is white because the workpiece was heated on the surface that was painted.

Infrared scanning microscopes and cameras require a very rapid detector-response time because the resultant thermograph is actually nothing more than a series of many temperature measurements.

CONTACT THERMOMETRIC INSPECTION

There are several basic thermal detectors: bolometers, thermistors, thermocouples and thermopiles. For the most part, they can be used either in direct contact with the test item or as radiation detectors (noncontact applications). These devices can detect infrared radiation of both short and long wavelengths, and very low-temperature or low-radiation levels without cryogenic cooling.

Bolometers are thermal detectors that are based on the principle that the resistance of a material changes as it is heated. The bolometer allows the radiation to impinge on a very fine wire or a thin metallic film, blackened to increase absorption. The change in resistance is then a direct function of the radiation absorbed. The temperature coefficient of a bolometer is from 0.3 to 0.5%/°C.

Thermocouples. A thermocouple consists of a junction of two dissimilar metals. As the junction temperature is raised, a thermoelectric electromotive force is produced. Thermocouples are always used in pairs in a bridge circuit so that the measured temperature is a direct function of the electromotive force produced by the sensing thermocouple as subtracted from the electromotive force produced by a reference thermocouple held at a known temperature.

Thermopiles. A thermopile is merely a series of thermocouple junctions; it produces an increase in electromotive force as a direct function of the number of junctions.

Meltable Substances. Waxlike crayons that melt at temperatures in the range of 38 to 1760 °C (100 to 3200 °F) are commercially available. Such a crayon has a melting point within a nominal tolerance of ±1% of its rated temperature value. These crayons are normally used by making a mark with one or more of them on a surface before it is heated.

NONCONTACT THERMOMETRIC INSPECTION

The temperature-measuring devices used in noncontact thermometric inspection depend on response of a thermal detector to infrared radiation. They are particularly useful when it is necessary to monitor or measure surface temperatures remotely.

Radiometers are instruments used to measure incident radiation. They consist of some type of hollow cavity with an aperture in one end and a thermal detector mounted internally. The thermal detector is so located that the radiation is focused on it. Because thermal detectors have a uniform response without regard to infrared wavelength, a radiometer is often used to measure total radiation. If the radiometer has a lens system, it will be restricted to the infrared-transmission characteristics of the lenses.

Pyrometers (or infrared thermometers) are used in nondestructive testing in much the same way as are the scanning devices previously discussed under "Infrared Detectors." These instruments have less accuracy than other scanning devices, but they are simpler, more rugged, more portable and less expensive. A wide variety of this type of equipment is commercially available, with a corresponding variety of performance.

——Leak Testing——

Edited by Francis J. McElroy, Struthers Wells Corp.

LEAK TESTING is the determination of the rate at which a liquid or gas will penetrate from inside a "tight" component or assembly to the outside, or vice versa, as a result of a pressure differential between the two regions, or of permeation of a somewhat extended barrier. It has become conventional to use the term "leak" to refer to an actual discontinuity or passage through which a fluid flows or permeates.

Types of Leaks. There are two basic types of leaks—real leaks and virtual leaks. A real leak is an essentially localized leak—that is, a discrete passage through which fluid may flow (crudely, a hole). Such a leak may take the form of a tube, a crack, an orifice or the like. A system may also leak through permeation of a somewhat extended barrier; this type of real leak is called a distributed leak. A gas may flow through a solid having no holes large enough to permit more than a small fraction of the gas to flow through any one hole. This process involves diffusion through the solid and may also involve various surface phenomena such as absorption, dissociation, migration and desorption of gas molecules. Virtual leaks are leaks that involve the gradual desorption of gases from surfaces or components within a vacuum system. It is not uncommon for a vacuum system to have both real leaks and virtual leaks at the same time.

LEAK TESTING OF FLUID SYSTEMS AT PRESSURE

Leak-testing methods can be classified according to the pressure and fluid (gas or liquid) in the system. The following paragraphs describe the common fluid-system leak-testing methods in the general order shown in Table 1. Table 1 also lists methods used in leak testing of vacuum systems.

Leak detection by monitoring changes in pressure of the internal fluid often is used when leak-detection equipment is not immediately available. For the most part, detection can be accomplished with instruments that are already installed in the system.

Acoustic Methods

Turbulent flow of a pressurized gas through a leak produces sound of both sonic and ultrasonic frequencies. If the leak is large, it probably can be detected with the ear. This is an economical and fast method for finding gross leaks. Sonic emissions also are detected by such instruments as stethoscopes or microphones that have a limited ability to locate as well as determine the approximate size of a leak. Electronic transducers enhance detection sensitivity.

Bubble Testing

A simple method for leak testing of small vessels pressurized with any gas is to submerge them in a liquid and observe bubbles. If the test vessel is sealed at atmospheric pressure, a pressure differential may be obtained by pumping a partial vacuum over the liquid or by heating the liquid. The sensitivity of this test is increased by reducing (a) the pressure above the liquid, (b) the liquid density, (c) the depth of immersion in the liquid and (d) the surface tension of the liquid.

Specific-Gas Detectors

Many available types of leak detectors will react to either a specific gas or a group of gases that have some specific physical or chemical property in common.

Sulfur hexafluoride detectors operate on the principle of electron-capture detectors, used widely in the field of gas chromatography. The sensing chamber of a sulfur hexafluoride detector consists of a cylindrical cell that has a centrally mounted insulated probe. The inner wall of the

Table 1. Methods of leak-testing systems at pressure or at vacuum

Gas systems at pressure
Direct sensing by:
  Acoustic methods
  Bubble testing
  Flow detection
  Gas detection by:
    Smell
    Chemical reaction
    Halogen gas
    Sulfur hexafluoride
    Combustible gas
    Thermal-conductivity gages
    Infrared gas analyzers
    Mass spectrometry
    Radioisotope count
    Ionization gages
    Gas chromatography
Quantity-loss determination by:
  Weighing
  Gaging differential pressure

Liquid systems at pressure
Unaided visual methods	Surface wetting
	Weight loss
Aided visual methods	Water-soluble paper with aluminum foil

Vacuum systems
Manometers	Thermal-conductivity gages
Halogen gas	
Mass spectrometry	Gas chromatography
Ionization gages	

cell is coated with a radioactive element (300 millicuries of tritium). Low-energy electrons emitted by the tritium are collected on the central probe by means of a polarizing voltage maintained between the probe and the cell wall. The resulting electric current is amplified and displayed on a conventional meter. The display meter is a taut band-suspension microammeter reading 0 to 50, and response time of the instrument is 1 s. Leak-rate sensitivity is 10^{-8} ml/s.

Mass-Spectrometer Testing. A mass spectrometer is basically a device for sorting charged particles. the sample gas enters the analyzer, where its molecules are bombarded by a stream of electrons emitted by a filament. The bombarded molecules lose an electron and become positively

charged ions, which are electrostatically accelerated to a high velocity. Because the analyzer lies in a magnetic field perpendicular to the ion path, the ions travel distinct, curved paths according to their mass. The radii of these paths are determined by ion mass, the magnitude of initial acceleration and the strength of the magnetic field. With a constant magnetic field, any group of ions having the same mass can be made to travel the specific radius necessary to strike the ion collector. The positive charge of the ion is imparted to the target, or collector, and the resulting current flow is proportional to the quantity of the ions of that particular mass.

Application of Specific-Gas Detectors

The proper method for using a specific-gas detector is based on the function of the leak detector, the fluid that is leaking and the type of vessel being tested. One of the best methods of using a leak detector is discussed below.

Probing. Probes, which will react to a number of gases, may be used in either of two ways: the probe mode or the monitoring of an enclosure placed around the pressurized item.

In the first method—the probe mode—the external surface of the pressurized vessel is scanned either with a portable detector having a short probe attached, or with a long probe connected to a stationary leak detector by flexible tubing (Fig. 1). In general, the connection of a long probe to a stationary detector reduces sensitivity because of the slow release of absorbed gases in the probe tubing, which results in a high background reading. Because a correspondingly longer time is required for the gas to flow up the tube to the sensing element, it is difficult to pinpoint the location of the leak.

In the second method—monitoring of an enclosure placed around the pressurized item—the item is enclosed in a plastic bag and the accumulation of tracer gas in the bag is monitored by a leak detector connected to the bag by a short probe.

Back pressuring is a method of pressurized testing that normally is used with small, hermetically sealed electronic components such as integrated circuits, relays and transistors. In this

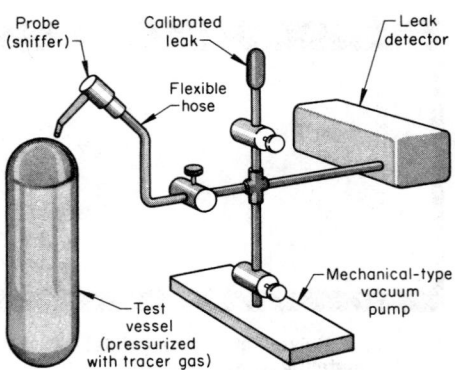

Fig. 1. Schematic showing connection of a long probe by a flexible hose to a stationary leak detector for scanning the external surface of a pressurized vessel

method, the test unit is placed in a pressurized container filled with a tracer gas and is kept there for a time to allow tracer gas to flow into the unit through any leaks that exist.

In using back pressuring, care must be taken to ensure that the leaking components contain tracer gas. Therefore, it is important that the time between back pressuring and leak testing be suitably controlled. For example, in a test specification for transistors, the transistors were subjected to helium pressure of 100 psi for 16 h, air washed for 4 min and then leak tested within 3 h. If more than 3 h elapsed before leak testing, the transistors were pressurized again.

Because leaks generally are expected to be quite small, very sensitive tracer-detector combinations must be used. The tracer gas normally used is helium or krypton 85, and detection is by use of a helium mass spectrometer or a nuclear-radiation detector, respectively. Absolute values of the size of the leak sometimes are difficult to determine with back pressuring, because of (a) adsorption and absorption of the tracer gases and (b) different detector-response times for different leak directions.

Rapid Identification of Metals and Alloys

QUALITY ASSURANCE during the fabrication of hardware, subassemblies and assemblies sometimes requires a reliable system of rapid identification of metals and alloys. Because various metals may become mixed during storage or use, and because lengths of strip, sheet, plate, billets, bars, wire and fabricated products may have lost their identifying marks, some means of sorting mixed lots is necessary. The best way of identifying such items is by quantitative chemical analysis. However, chemical or spectrographic analyses require extensive, time-consuming procedures and expensive equipment that may not be fully utilized. Also, a complete chemical analysis often may be unnecessary.

Common methods of rapid identification of metals include techniques involving the magnetic properties and weight of the metal; spark testing (of ferrous alloys); and chemical spot testing.

Sorting by Magnetic Properties and Weight

Edited by Ron Mason, Multispec

TWO of the simplest methods by which metals and alloys can be rapidly identified are methods involving magnetic properties and weight.

MAGNETIC PROPERTIES

Applying a hand magnet to a metal is the simplest method of distinguishing between ferromagnetic and nonferromagnetic metals. The attraction of the hand magnet to the metal will be strongly magnetic, slightly magnetic or nonmag-

netic. Ferromagnetic metals, particularly alloys of iron, cobalt and nickel, are attracted or repulsed strongly.

Another means for examining the magnetic properties of a metal is the use of a Tinsley thickness gage. This gage normally is used for measuring the thicknesses of nonmagnetic coatings on ferritic steel surfaces. With this instrument, a completely nonmagnetic metal is given a rating of ten and a magnetic metal such as plain carbon steel is assigned a rating of zero. Known specimens can be rated using the Tinsley thickness gage and recorded for reference.

Strongly magnetic metals include gray, ductile and malleable irons; carbon steels; alloy steels; tool steels; ferritic and martensitic stainless steels; and high-nickel alloys. Other strongly magnetic metals are the iron-silicon alloys containing 0.5 to 4.5% Si; iron-nickel alloys, particularly those containing more than 28% Ni; iron-cobalt alloys; and iron-molybdenum alloys.

Nonmagnetic Metals. The commercially pure nonferrous metals, and their alloys, except for cobalt and some nickel alloys, are nonmagnetic. Also nonmagnetic are Monel K-500, Monel 400, the Inconel alloys, and the Incoloys.

Although all austenitic steels fulfill the condition of being nonmagnetic, consideration must be given to the stability of the structure on heating, welding and cold working. Frequently, magnetic permeability is affected by cold working. Permeability of types 302 and 304 stainless steel is 1.007 to 1.010 in the annealed condition, but after 70% cold reduction, permeability increases to 36 to 40 at 100 oersteds. There is very little change in permeability of types 310 and 316 stainless steel even after 70% cold reduction.

WEIGHT

Metals can be separated into groups on the basis of their weight (specific gravity). A preliminary classification is as follows:

- Light metals and alloys (specific gravity, 1.5 to 4.75): aluminum, magnesium, beryllium and titanium
- Average-weight metals and alloys (specific gravity, 6 to 9): steels, cast irons, copper and copper alloys, nickel and nickel alloys, stainless steels, cadmium, tin, zinc and antimony
- Heavy metals (specific gravity, 9.8 to 11.3): lead, silver, molybdenum and bismuth
- Very heavy metals (specific gravity, 12 to 22): gold, tungsten, tantalum, palladium, platinum, osmium, iridium, rhodium, ruthenium and uranium (depleted).

Spark Testing of Ferrous Alloys

Edited by Ron Mason, Multispec

SPARK TESTING is a method used for the classification of ferrous alloys according to their chemical compositions, by visual examination of the spark pattern or stream that is thrown off when the alloys are held against a grinding wheel rotating at high speed. The test is a fast and economical method of separating alloys of different compositions. Experienced operators can use this method for identification of a number of ferrous alloys with reasonable accuracy.

PRINCIPLES

When a piece of steel is held in contact with an abrasive grinding wheel that is rotating at sufficient speed, a stream of sparks is produced at the point of contact. The sparks are the result of abrasion of infinitesimal particles of metal by the grinding wheel, although the heating produced by this action will barely succeed in heating the projected metal particles to a red heat. These metal particles that are torn away by the grinding wheel become incandescent in the air because oxidation is intense, the initial heating was high, and the particles are exceedingly small. As the particles are projected through the air, the trajectory (called a carrier line) is easily followed (against a dark background).

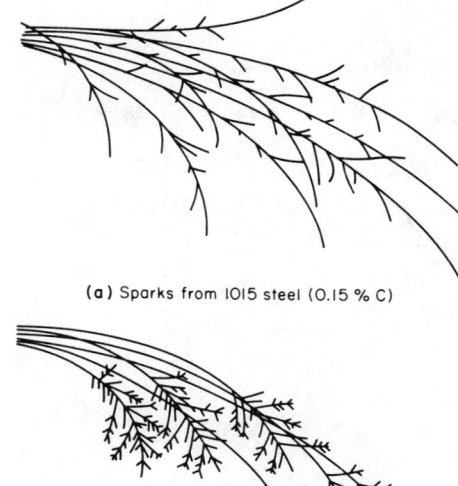

(a) Sparks from 1015 steel (0.15 % C)

(b) Sparks from 1045 steel (0.45 % C)

(c) Sparks from 1095 steel (1.0 % C)

(a) Sparks from low-carbon steel, showing slight forking effect. (b) Sparks from medium-carbon steel, showing pronounced bursts. (c) Sparks from high-carbon steel, showing the intensity of bursts that is characteristic of steel having high carbon content.

Fig. 1. Effect of carbon content of steels on the spark pattern or stream

Spark Bursts. The spark burst, or "carbon spark," is the most useful characteristic of the spark stream, because the variations in the number and intensity of the bursts indicate the changes in carbon content of the alloy (see Fig. 1). Alloys that have the same carbon content but different amounts of alloying elements are not always easily identified. Most of the alloying elements, however, have some influence on the characteristics of the spark stream; they may affect the carrier lines, the bursts, or the type of characteristic sparks. Alloying elements may retard or accelerate the carbon spark, or make the carrier line lighter or darker.

Characteristic Sparks. Figure 2 illustrates several carrier lines that exhibit the various characteristic shapes of sparks. A wavy tapering carrier line with an oxidation part that is only slightly visible is shown in Fig. 2(a). This carrier line is characteristic of several unalloyed metals. One of the most easily recognized of all the characteristic sparks is that of alloys containing molybdenum (Fig. 2b). The carrier line tapers off to a long

pointed end, and at a distance from this point there is a distinctly detached spear point.

A spark formation characteristic of aluminum-alloyed steel is shown in Fig. 2(c). The oxidation part is spherically thickened and tapers off to a short point. A few buds are evident above the thickened section. The center of the oxidation part is incandescent. The split oxidation part with two elongated points shown in Fig. 2(d) is characteristic of sparks from free-cutting steels. Slightly swollen reaction products are visible at the root of the fork.

The oxidation part of the upper carrier line in Fig. 2(e) has the shape of molten droplets of slag. Two carbon sprigs extend from these droplets. In the lower carrier line in Fig. 2(e), the short bud at the left is caused by manganese in the steel. The longer sprig is a carbon sprig. The intense formation of bursts in the carrier lines close to the oxidation parts in Fig. 2(f) indicates the presence of molten reaction products. The bursting oxidation part shown in Fig. 2(g) is characteristic of ferromanganese. Many arrows are formed, conveying the impression of a star.

Sparks typical of gray iron are shown in Fig. 2(h). Intense explosions leading to the formation of buds shoot out from the oxidation part of the carrier line.

EQUIPMENT

The minimum apparatus necessary for spark testing is either a portable or a stationary grinding machine and a suitable grinding wheel.

Grinding Machine. A stationary grinder, suitable for sharpening tools, can be used to test material that can be transported to the machine. Where it is impractical to move a part or component to the grinding machine, portable grinding equipment, with approximately 3.2 kg (7 lb) of well-distributed weight, can be used. To provide a spark stream that is satisfactory for examination, the grinder should provide a peripheral wheel speed of not less than 23 m/s (4500 ft/min), and preferably higher. Speeds of 38 to 58 m/s (7500 to 11 500 ft/min) are considered optimum in most plants.

Grinding wheels should be coarse and very hard. An aluminum oxide wheel A-46-P-V is suggested for general-purpose use, including hard alloys, and an aluminum oxide wheel A-46-S-V is suggested for soft alloys.

Spark-Test Cabinets. The examination of spark streams under nearly ideal conditions can be done in a spark-test cabinet. A cabinet about 1 m long, 900 mm wide and 900 mm deep (40 by 36 by 36 in.) is suitable for spark testing. The inside of the cabinet should be painted a flat black and should be equipped for artificial illumination. The spark-test cabinet should be located where no direct light will strike either the operator's eyes or the inside of the cabinet. The grinding machine may be mounted at the left front of the cabinet, and should throw a spark stream across the front of the cabinet at right angles to the operator's line of vision.

WORK AREA

Proper lighting conditions are necessary for satisfactory spark testing. Spark testing should be done in a shadow, and against a dark background. A semidark room is preferable. Total darkness is acceptable, but it makes handling of

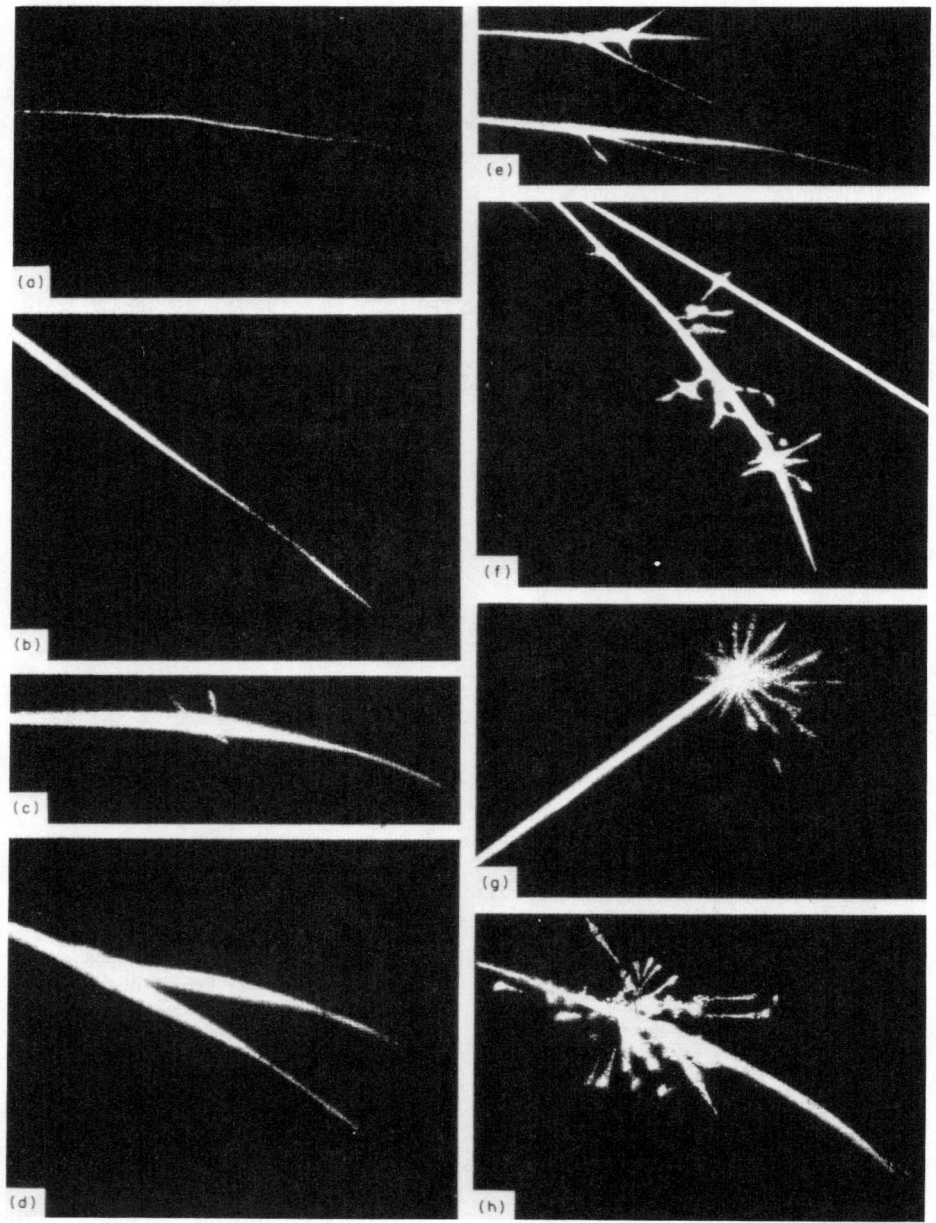

(a) Unalloyed metal. (b) to (g) Presence of: (b) molybdenum, (c) aluminum, (d) free-machining additives, (e) manganese, (f) reaction products, (g) ferromanganese. (h) Gray iron.

Fig. 2. Spark patterns that characterize an unalloyed metal and the presence of several alloying elements. See text for discussion.

test parts difficult. Bright sunlight is not satisfactory, because direct sunlight reduces the luminosity of sparks so that the distinction between a shower of sparks from high-carbon steel and their practical absence from low-carbon steel is about the only difference that can be observed.

The spark stream should be protected from heavy drafts of air, because a heavy draft will cause the tail sparks to hook, thus making correct interpretation of the spark stream difficult.

STANDARD REFERENCE SPECIMENS

Standard reference specimens can be prepared from bar stock. A convenient size of specimen is 25 mm (1 in.) in diameter by 75 mm (3 in.) long. One end of the specimen should be stamped with the identification marking of the alloy and a specimen number. Chemical compositions of the specimens should be kept for reference.

An inexperienced operator should have access to specimens of those metals that will be tested most. For example, if experience in identifying a spark stream from molybdenum steel is necessary, pieces of 1020 steel and pieces of 4615 steel, both of the same diameter and length and stamped with identification markings and specimen numbers, will provide the standard specimens necessary. A collection of specimens of varying but known compositions can be amassed

over a period of time. These specimens may be used as needed for classifying alloys of unknown composition by comparing their spark patterns.

Chemical Spot Testing

Edited by Richard L. Lessard, World Aerospace Corp.

CHEMICAL SPOT TESTING is a relatively simple, qualitative method that can be used for rapid laboratory or field identification of metals and alloys. Information on the chemical composition of the test piece is gained by observing the color change occurring during a chemical reaction taking place on one spot on the test piece, on filter paper or on a spot plate. The test method is not dependent on the use of auxiliary optical magnification.

An important part of chemical spot testing is the manipulation of drops of test or reagent solutions. The success of the test method derives from the nature of the reagents used (see Table 1 for identification of reagents mentioned in this article), together with the advantageous use of reactive conditions, so that the desired sensitivity and selectivity can be obtained with a minimum of physical and chemical operations. Tests ordinarily are performed by using one of the following techniques:

- Bringing together a few drops of the test solution and of the reagent on porous or nonporous supporting surfaces such as paper, glass or porcelain
- Placing a few drops of test solution on a medium (filter paper, asbestos or gelatin) impregnated with appropriate reagents
- Placing a drop or two of reagent on a small quantity of the solid specimen
- Subjecting a drop of reagent or a strip of reagent paper to the action of liberated gases from a drop of the test solution or from a small quantity of the solid specimen
- Adding a drop of test solution to a larger volume (0.5 to 2 ml) of reagent solution and then extracting the reaction products with organic solvents.

The successful application of spot-testing procedures is enhanced by (a) a knowledge of the chemical basis of the test used, so that every step of the procedure is understood and executed intelligently; (b) strict observance of the experimental conditions; (c) cleanness of the laboratory or test site and equipment; and (d) use of the purest reagents available. If possible, tests should be repeated to ensure reproducibility, and tests on unknown materials should be compared with tests on known materials.

In spot-testing analysis, metals almost always are detected by characteristic reactions of their mobile hydrated ions. Consequently, it must be possible to make test solutions of the material being tested. Because spot tests uniformly employ more sensitive and less disputable reactions, the presence or absence of one particular substance often can be detected with as little as one drop of the dilute test solution, even though considerable quantities of other materials are present.

Laboratory and Equipment Requirements. Although chemical spot testing should be conducted in a laboratory using laboratory equipment and con-

Table 1. Spot-testing reagents referred to in this article

All aqueous solutions are to be prepared with distilled water unless specified otherwise.

	Reagent	Preparation or description
1	a, a′-dipyridyl	Dissolve 1.2 g a, a′-dipyridyl in 60 ml 0.1M hydrochloric acid.
2	Acetic acid, 6M	Add 20.6 ml concentrated acetic acid to 39.4 ml water.
3	Acetone	Specific gravity, 0.80.
4	Alizarin, saturated	Saturated solution of alizarin in 95% methyl alcohol.
5	Alizarin S, 0.1%	Dissolve 0.06 g alizarin S in 60 ml water.
6	Ammonium acetate	Ammonium acetate crystals.
7	Ammonium hydroxide	Concentrated; specific gravity, 0.90.
8	Ammonium hydroxide, 50%	Add 30 ml concentrated ammonium hydroxide to 30 ml water.
9	Ammonium hydroxide, 3M	Add 12 ml concentrated ammonium hydroxide to 48 ml water.
10	Ammonium hydroxide, 2M	Add 8 ml concentrated ammonium hydroxide to 52 ml water.
11	Ammonium mercuric thiocyanate	(a) Dissolve 5.8 g mercuric chloride and 1.9 g ammonium thiocyanate in 60 ml water. (b) Dissolve 8 g mercuric chloride and 9 g ammonium thiocyanate in 100 ml water. Let both solutions stand several days before use.
12	Ammonium molybdate solution	Dissolve 5 g ammonium molybdate in 100 ml cold water; pour into 35 ml concentrated nitric acid.
13	Ammonium oxalate	Saturated solution of ammonium oxalate in water.
14	Ammonium persulfate	Ammonium persulfate crystals.
15	Bromine water	Saturated at 20 C (68 F).
16	Cacotheline solution	Saturated solution of cacotheline in water.
17	Cadmium chloride	Dissolve 3 g cadmium chloride and 3 g sodium chloride in 3 ml concentrated hydrochloric acid and dilute with water to 60 ml.
18	Citric acid, 30%	Dissolve 30 g citric acid in 100 ml water.
19	Cobaltous sulfate	Dissolve 0.012 g cobaltous sulfate in 2.5 ml concentrated hydrochloric acid and dilute with water to 60 ml.
20	Dimethylglyoxime. (See also No. 66)	Saturated solution of dimethylglyoxime in 95% ethyl alcohol.
21	Dimethylglyoxime (See also No. 66)	Dissolve 1 g dimethylglyoxime powder in 30 ml hot ethyl alcohol; cool and add 60 ml glacial acetic acid.
22	Dimethylglyoxime (See also No. 66)	Dissolve 1 g dimethylglyoxime in 60 ml glacial acetic acid, then mix with a solution of 10 g ammonium acetate dissolved in 30 ml ammonium hydroxide.
23	Diphenylcarbazide	Dissolve 0.6 g diphenylcarbazide in 60 ml 95% ethyl alcohol.
24	Diphenylcarbazide	Dissolve 0.1 g diphenylcarbazide in 100 ml of a solution containing 5 ml glacial acetic acid and 95 ml water.
25	Diphenylcarbazone	Dissolve 0.6 g diphenylcarbazone in 60 ml 95% ethyl alcohol.
26	EDTA sodium solution (See also No. 67)	0.003M ethylenediamine tetraacetic acid, disodium salt, 5 ml 0.0575M EDTA solution diluted to 100 ml; or dissolve 0.067 g EDTA disodium salt in 60 ml water.
27	Eriochrome black T indicator	2 ml 0.4 Eriochrome black T solution diluted to 50 ml with 25 ml triethanolamine and 23 ml reagent-grade methyl alcohol. (This solution is stable for approximately two months. Replace when it produces a grayish color rather than a bluish color with fine silver.)
28	Ferric chloride, acidified	Dissolve 6 g ferric chloride in 6 ml concentrated hydrochloric acid; dilute with water to 60 ml.
29	Glyoxal-bis (2-hydroxyanil)	Dissolve 0.6 g glyoxal-bis (2-hydroxyanil) in 60 ml 95% methyl alcohol.
30	Hydrochloric acid	Concentrated, specific gravity 1.19.
31	Hydrochloric acid, 6M	Add 30 ml concentrated hydrochloric acid to 30 ml water
32	Hydrochloric acid, 3M	Add 15 ml concentrated hydrochloric acid to 45 ml water
33	Hydrochloric acid, 2M	Add 10 ml concentrated hydrochloric acid to 50 ml water.
34	Hydrochloric acid – mercuric chloride solution	Dissolve 10 g mercuric chloride in 20 ml concentrated hydrochloric acid; add to 100 ml water.
35	Hydrochloric acid, 1M	Add 5 ml concentrated hydrochloric acid to 45 ml water.
36	Hydrogen peroxide, 6%	Add 12 ml of 30% hydrogen peroxide to 48 ml water.
37	Hydrogen peroxide, 3%	Add 6 ml 30% hydrogen peroxide to 54 ml water.
38	Iodine – sodium azide solution	Dissolve 1.3 g sodium azide in 100 ml 0.1N iodine.
39	Masking solution	Mix equal volumes of 20% sodium thiosulfate, 20% sodium tartrate solution and saturated sodium fluoride solution.
40	Mercuric chloride	Saturated solution of mercuric chloride in water.
41	Morin solution, 0.02%	Dissolve 0.012 g morin in 60 ml acetone.
42	Nitric acid	Concentrated, specific gravity 1.42.
43	Nitric acid, 50%	Add 30 ml concentrated nitric acid to 30 ml water.
44	Nitric acid, 6M	Add 22.5 ml concentrated nitric acid to 37.5 ml water.
45	Nitric acid, 3M	Add 11.3 ml concentrated nitric acid to 48.7 ml water.
46	Nitric acid, 0.2M	Add 0.75 ml concentrated nitric acid to 49.25 ml water.
47	Phosphoric acid, ortho	Concentrated, 85%.
48	Piperidine	(CH₂)₅NH liquid, specific gravity 0.86; also known as hexahydropyridine.
49	Potassium cyanide, 5%	Dissolve 3 g potassium cyanide in 60 ml water.
50	Potassium ethyl xanthogenate	Dry crystals.
51	Potassium ethyl xanthogenate, 10%	Dissolve 6 g potassium ethyl xanthogenate crystals in 60 ml water.
52	Potassium ferrocyanide, 1%	Dissolve 0.6 g potassium ferrocyanide in 60 ml water.
53	Potassium fluoride, 0.5%	Dissolve 0.3 g potassium fluoride in 60 ml water.
54	Potassium iodide, 5%	Dissolve 3 g potassium iodide in 60 ml water.
55	Potassium iodide, 50%	Dissolve 30 g potassium iodide in 60 ml water.
56	Potassium nitrite	Potassium nitrite crystals.
57	Potassium thiocyanate, 10%	Dissolve 6 g potassium thiocyanate in 60 ml water.
58	Quinalizarin	Dissolve 0.012 g quinalizarin in 60 ml 95% ethyl alcohol.
59	Quinalizarin	Dissolve 0.1 g quinalizarin in 2 ml pyridine and 20 ml acetone.
60	Rhodamine B	Dissolve 0.3 g rhodamine B in 60 ml 50% HCl.
61	Silver nitrate	Silver nitrate crystals.
62	Sodium bismuthate	Dissolve 3.5 g sodium bismuthate in 60 ml water. (Stir before using.)
63	Sodium carbonate	Saturated solution of sodium carbonate in water.
64	Sodium cyanide, 10%	Dissolve 6 g sodium cyanide in 60 ml water.
65	Sodium diethyl-dithiocarbamate, 0.1%	Dissolve 0.1 g sodium diethyldithiocarbamate in 100 ml water.
66	Sodium dimethylglyoxime, 3% (See also No. 20, 21 and 22)	Dissolve 1.8 g sodium dimethylglyoxime in 60 ml water.

(continued)

Table 1 (continued). Spot-testing reagents referred to in this article
All aqueous solutions are to be prepared with distilled water unless specified otherwise.

Reagent	Preparation or description
67	Sodium (Di) ethylenediamine tetraacetate (EDTA) . . . . . . Saturated solution of sodium (Di) ethylenediamine tetraacetate in 5 ml concentrated ammonium hydroxide
	(See also No. *26*) and 50 ml water.
68	Sodium fluoride . . . . Saturated solution of sodium fluoride in water.
69	Sodium fluoride, 0.5% . . . . Dissolve 0.3 g sodium fluoride in 60 ml water.
70	Sodium hydroxide, 10*M* . . . . Dissolve 80 g sodium hydroxide in 200 ml water.
71	Sodium hydroxide, 6*M* . . . . Add 36 ml 10*M* sodium hydroxide to 24 ml water.
72	Sodium hydroxide, 5*M* . . . . Add 30 ml 10*M* sodium hydroxide to 30 ml water.
73	Sodium hydroxide, 3*M* . . . . Add 18 ml 10*M* sodium hydroxide to 42 ml water.
74	Sodium hydroxide, 2*M* . . . . Add 12 ml 10*M* sodium hydroxide to 48 ml water.
75	Sodium hydroxide, 1*M* . . . . Add 6 ml 10*M* sodium hydroxide to 54 ml water.
76	Sodium nitrite . . . . Sodium nitrite crystals.
77	Sodium peroxide, 40% . . . . Dissolve 24 g sodium peroxide in 60 ml water.
78	Sodium peroxide, 20% . . . . Dissolve 12 g sodium peroxide in 60 ml water.
79	Sodium rhodizonate, 0.2% . . . . Dissolve 0.12 g sodium rhodizonate in 60 ml water. (Solution is unstable; make fresh daily.)
80	Sodium tartrate . . . . Saturated solution of sodium tartrate in water.
81	Sodium thiocyanate, 10% . . . . Add 6 g sodium thiocyanate to 60 ml water.
82	Sodium thiosulfate, 25% . . . . Dissolve 15 g sodium thiosulfate in 60 ml water.
83	Stannous chloride, 25% . . . . Dissolve 15 g stannous chloride in 15 ml concentrated hydrochloric acid, boil until clear, cool before use.
84	Stannous chloride, 20% . . . . Dissolve 12 g stannous chloride in 60 ml water and add 0.5 g metallic tin.
85	Stannous chloride, 5% . . . . Dissolve 3 g stannous chloride in 15 ml concentrated hydrochloric acid and boil until clear; add 45 ml
	water and 0.5 g metallic tin.
86	Stannous chloride solution . . . . Mix 50 ml stannous chloride solution, 50 ml concentrated hydrochloric acid and 100 ml water.
87	Sulfuric acid . . . . Concentrated; specific gravity, 1.84.
88	Sulfuric acid, 50% . . . . Add 30 ml concentrated sulfuric acid to 30 ml water.
89	Sulfuric acid, 3*M* . . . . Add 10 ml concentrated sulfuric acid to 50 ml water.
90	Sulfuric acid, 10% . . . . Add 6 ml concentrated sulfuric acid to 54 ml water.
91	Sulfuric acid, 5% . . . . Add 3 ml concentrated sulfuric acid to 57 ml water.
92	Tartaric acid and sodium bitartrate, pH 2.8 . . . . Dissolve 0.9 g tartaric acid and 1.4 g sodium bitartrate in 60 ml water.
93	Zinc oxide . . . . Zinc oxide powder.

ditions, portable kits can be assembled that contain the necessary chemicals and equipment for field testing.

CARBON STEELS

Carbon steels can be identified by placing one drop of 50% nitric acid (reagent *43* in Table 1) on a fresh surface. After 5 min, flush the spot with clean water; a brown spot indicates carbon steel.

The carbon content of a carbon steel can be estimated by placing two drops of 50% nitric acid (*43*) on the clean metal surface. After one minute, transfer one drop to a filter-paper disk. The intensity of the stain on the paper will vary with the carbon content. Thus it can be compared visually with identical tests made on steels of known carbon content.

ALLOY STEELS

Test for Iron. Place one drop of 50% nitric acid (*43*) on the clean metal surface and allow to react for 1 min. Add one drop of 10% potassium thiocyanate (*57*). A blood-red color identifies the presence of iron.

Test for Chromium. Place six drops of concentrated hydrochloric acid (*30*) on the clean metal surface or on several metal filings in a small test tube; add three drops of concentrated nitric acid (*42*) and three drops of water.

Add ten drops of 10*M* sodium hydroxide (*70*) and five drops of 3% hydrogen peroxide (*37*). Stir with a glass rod and allow 5 min before proceeding.

Place a small ball of absorbent cotton in the solution. Using a medicine dropper, draw off clear liquid through the cotton ball by pressing the tip firmly down in the cotton. Transfer five drops of liquid (for the chromium test) and five drops (for the molybdenum test described below) to each of two depressions in a spot plate. Add two drops of diphenylcarbazide reagent (*23*) and stir with a glass rod for 1 min; while stirring add one to two drops of 3*M* sulfuric acid (*89*). A bright-red color will appear and then disappear, leaving a faint-violet color that intensifies after 1 or 2 min, which indicates the presence of chromium in the metal.

Test for Molybdenum. To the five drops of test solution set aside from the first test for chromium described above, add a few grains of solid potassium ethyl xanthogenate (*50*), stir, and add three or four drops of 3*M* hydrochloric acid (*32*). A deep-pink color identifies the presence of molybdenum.

Test for Manganese. Place two drops of 50% nitric acid (*43*) on the surface of the metal; transfer one drop of the acid test solution to a spot-plate depression; add sodium bismuthate reagent (*62*) dropwise until a brown precipitate appears. Let stand 2 or 3 min and add one drop of concentrated nitric acid (*42*). A pink color confirms the presence of manganese.

Three Tests for Nickel. In one test for nickel, the chemicals are combined into two solutions, which are prepared and used as follows:

Test Solution No. 1: 100 cc of concentrated nitric acid (*42*), 125 cc of water and 25 cc of phosphoric acid (*47*).
Test Solution No. 2: (*a*) Dissolve 10 g of ammonium acetate crystals (*6*) in 30 ml of ammonium hydroxide (*7*); (*b*) prepare solution of dimethylglyoxime (*21*) as in Table 1; mix solutions (*a*) and (*b*).

Cut clean, white paper toweling into 50-mm (2-in.) squares and form a slight depression in the center of each square with clean fingers. Apply one drop of solution No. 1 to the surface of the metal being tested and let stand for 1 min. Place the center of the paper square over the reaction solution for 15 s, then pick off the paper without wiping. Apply two drops of solution No. 2 to the center of the paper. A pink color signifies the presence of nickel; the greater the intensity, the higher the nickel content. The nickel content can be determined to some accuracy by conducting simultaneous tests on known and unknown specimens and comparing intensities of the pink (not brown) color. If the results are not conclusive the solutions can be checked by conducting tests on specimens having widely different nickel contents, such as 4340 and 8620 steels. The intensity of the pink color should be markedly different.

A second test for nickel is as follows: Allow one or two drops of 50% nitric acid (*43*) to react to completion on the surface of the metal to be tested. Neutralize with a slight excess of zinc oxide (*93*). Add two or three drops of saturated dimethylglyoxime (*20*). A pink color identifies nickel. The minimum nickel content detectable is 0.05%.

A third test for nickel is as follows: Place two or three drops of 50% nitric acid (*43*) on the surface of the metal and allow to react to completion (about 2 min). Transfer one drop of the test reaction solution to a spot-plate depression. Add one drop of phosphoric acid (*47*) and stir. Add two drops of dimethylglyoxime reagent (*20*) and stir. Add one drop of concentrated ammonium hydroxide (*7*). A pink-to-red precipitate identifies the presence of nickel.

Test for Cobalt. Place two drops each of concentrated hydrochloric acid (*30*) and concentrated nitric acid (*42*) on the surface of the metal. Allow 2 to 3 min to react and transfer two drops of the reaction test solution to a spot-plate depression. Add two drops of sodium fluoride solution (*68*) or phosphoric acid (*47*) to mask the interference of iron. Allow chemicals to react while stirring until the solution is clear. Add two drops of 10% potassium thiocyanate (*57*) and two drops of acetone (*3*). An aqua-green to blue-green color identifies the presence of cobalt. The color is not stable and will disappear; however, the color will return on the addition of a few drops of acetone. A pink color following the aqua-green color confirms the presence of molybdenum also.

STAINLESS STEELS

Nitric Acid Test. Stainless steels are noted for their inherent resistance to attack by either concentrated or dilute solutions of nitric acid. This characteristic makes it easy to separate them from most other metals or alloys. However, the higher-carbon types (420 and 440) may be attacked slightly by nitric acid. In contrast, carbon and alloy steels are vigorously attacked by dilute nitric acid.

Copper Sulfate Test. One of the simplest methods for differentiating quickly between carbon or alloy steels and all types of stainless steel is a spot test using a solution of 5 to 10% copper sulfate (blue vitriol) in water.

Before the spot test is run, the areas to be tested should be thoroughly cleaned of grease or any foreign substances, and a mild abrasive should be used on a small area before application of the test solution, which is dropped on the clean area from a drop bottle. Ordinary carbon steel or iron will become coated with the metallic copper in a few seconds, whereas stainless steel will show no deposit or copper color.

ALUMINUM ALLOYS

Test for Silicon. Place one drop of 5M sodium hydroxide solution (72) on the thoroughly cleaned surface of the metal and allow to react for 3 to 5 min. Remove the solution by blotting with a carefully applied dry filter paper. A partly adherent grayish-brown stain, which is insoluble after the addition of a drop of 3M nitric acid (45), indicates the presence of silicon.

Test for Copper. Place one drop of 5M sodium hydroxide solution (72) on the thoroughly cleaned surface of the metal and allow to react for 5 min. Wash off the drop with water and dry the spot with an acetone (3) wash. Add two drops of 50% nitric acid (43) and two drops of 30% citric acid (18) to the stain, and mix with a pointed glass rod. Add four drops of 50% ammonium hydroxide (8), followed by one drop of sodium diethyldithiocarbamate (65). A greenish-brown to dark-brown color or precipitate indicates the presence of copper. The minimum copper content that can be detected by this technique is approximately 0.5%.

Test for Nickel. Place one drop of 5M sodium hydroxide solution (72) on the cleaned metal surface and allow to react for 10 min. Wash off the drop with water and dry with acetone (3). Dissolve 10 g of citric acid in 20 ml of water, dilute 1 ml of orthophosphoric acid (47) to 20 ml and mix these two solutions with 10 ml of 50% nitric acid (43). Place four drops of mixture on the stain and let stand for 10 min. Mix 5 ml of dimethylglyoxime (20) and 10 ml of 50% ammonium hydroxide (8). Add six drops of this solution and mix with a pointed glass rod. A red precipitate that forms quickly indicates the presence of nickel. In the absence of nickel but in the presence of iron, a reddish coloration may form that can be dispelled by the addition of more dimethylglyoxime.

Test for Zinc. Place three drops of 6M hydrochloric acid (31) on the clean metal surface and allow to react for 2 to 3 min. Add three drops of water and stir carefully with a pointed glass rod. With a medicine dropper, add three drops to a white spot plate. Add three drops of cobaltous

sulfate solution (19), three drops of ammonium mercuric thiocyanate solution (11a), and stir with the glass rod for about 15 s, rubbing the sides of the spot-plate depression with the rod. A blue precipitate indicates the presence of zinc. A violet color may occur in the presence of both copper and zinc, but copper alone produces a greenish color.

BRASSES

Test for Tin. Dissolve a few grams of the metal in 50% nitric acid (43) and bring to a boil. A finely divided white precipitate, metastannic (β-stannic) acid, indicates that tin is present in the alloy.

Another test for tin involves placing four drops of concentrated nitric acid (42) on the clean metal surface. Allow 2 min for reaction and transfer two drops of the reaction solution to a small test tube. Add ten drops of concentrated hydrochloric acid (30) and mix thoroughly. Add small particles of magnesium metal to precipitate the copper out of solution. When all the copper has precipitated, the solution will be clear and colorless. Add 1 ml of water, stir, and add four drops of mercuric chloride (40). A white-to-gray precipitate confirms the presence of tin. This test will detect amounts of tin as small as 0.25%, although for such small amounts the solution should be allowed to remain for about 1 h at 32 C (90 F).

Test for Zinc. Place two drops of concentrated nitric acid (42) on the clean metal surface. After 2 min add two drops of water and transfer three drops of the reaction solution to a spot-plate depression. Add four drops of 6M sodium hydroxide (71) to make alkaline and stir. After 1 min, add another four drops of 6M sodium hydroxide (71) and stir. (The first addition of sodium hydroxide precipitates the zinc as zinc hydroxide and the latter addition dissolves the zinc hydroxide.)

Place a ball of absorbent cotton in the solution and, using a medicine dropper, draw off the liquid using the cotton as a filter and place in a small test tube. Add four drops of 3M hydrochloric acid (32) to the filtrate, stir thoroughly, and add another four drops of 3M hydrochloric acid (32) and stir. Add three drops of cobaltous sulfate (19) and three drops of ammonium mercuric thiocyanate (11a). Stir for 15 s with a glass rod, rubbing the sides of the tube. This starts precipitation of the zinc. A blue precipitate identifies the presence of zinc.

The presence of zinc only indicates the metal to be a copper-zinc alloy. A test for lead is needed to determine whether the alloy is leaded. Presence of both zinc and tin indicates a copper-zinc-tin alloy. Chemical or spectrographic analysis is needed to determine whether the alloy is an admiralty metal or naval brass. However, naval brass usually is darker than admiralty metal.

Test for Lead. Place three to four drops of concentrated nitric acid (42) on the surface of the metal and allow to react for about 2 min. Transfer one drop to a Whatman No. 3 (thick) filter paper. After the reaction solution has soaked into the filter paper, add one drop of 0.2% sodium rhodizonate (79) to the center of the fleck on the filter paper. A blue-to-violet color develops immediately. Add one to two drops of tartaric acid–sodium bitartrate buffer solution (92). The formation of a scarlet precipitate identifies the presence of lead. This procedure will determine the

presence of lead in brass with a lead content as low as 0.15%.

The presence of zinc and lead in a specimen indicates a copper-zinc-lead alloy (alloy No. 310 to 385). The absence of lead and tin signifies a copper-zinc alloy (alloy No. 205 to 298).

Tests for Aluminum. Place six drops of concentrated nitric acid (42) on the clean metal surface and allow 2 min to react to completion. Transfer four drops of the reaction solution to a small test tube and add four drops of 6M sodium hydroxide (71) and stir. After 1 min, add another four drops of 6M sodium hydroxide (71), stir, place in a boiling water bath for 2 min and centrifuge. Place one drop of the supernatant fluid on a spot plate, add one drop of 0.1% alizarin S reagent (5), and stir. Add two drops of 6M acetic acid (2) and stir. An orange precipitate confirms the presence of aluminum.

A second test method can be used. Place two drops of the supernatant liquid from the first method on a spot plate and add two drops of 6M hydrochloric acid (31) and stir. A red precipitate confirms the presence of aluminum.

BRONZES

The chemical spot tests for tin and aluminum given in the preceding section on Brasses are applicable also to bronzes. A test for silicon follows.

Test for Silicon. Dissolve a few chips or shavings of the test material in 50% nitric acid (43) and bring to a boil. A gelatinous mass is evidence of silicon and indicates that the alloy is a silicon bronze (alloy 651 or 655). If the solution is clear, there is no silicon in the alloy.

COPPER-NICKEL AND COPPER-NICKEL-ZINC ALLOYS

The copper-nickel and copper-nickel-zinc alloys can be distinguished from nickel alloys or stainless steel by placing a drop of concentrated nitric acid (42) on a clean metal surface. If the acid reacts rapidly to a blue-green color, the alloy is a copper-nickel or a copper-nickel-zinc alloy. If the acid reacts to a green-blue color or if no reaction occurs, the metal is a nickel alloy or stainless steel.

To distinguish between copper-nickel and copper-nickel-zinc alloys, place two or three drops of concentrated nitric acid (42) on a clean metal surface. After the reaction stops, place one drop of the reaction solution on a piece of filter paper and add one drop of ammonium mercuric thiocyanate (11b). A violet or greenish-blue precipitate indicates the presence of zinc, and the metal is a copper-nickel-zinc alloy. If a yellowish precipitate appears, the metal is a copper-nickel alloy. If the test is inconclusive, use the test for copper-nickel-zinc alloys that is described below.

Weigh and place 0.09 to 0.11 g of the metal in a 500-ml volumetric flask and add 20 ml of 50% nitric acid (43). Keep the solution warm until the metal has dissolved. Dilute the solution to 500 ml with distilled water. Mix well and transfer one drop to a spot-plate depression. Add one or two drops of 50% ammonium hydroxide (8) and one drop of 3% sodium dimethylglyoxime (66). Copper-nickel-zinc alloys produce a pale-brown coloration with no precipitate within 5 min. Monels or other high-nickel alloys produce a bright-red coloration and an immediate precipitate.

Selected References on Nondestructive Testing

Magnetic-Particle Inspection

Military Specification MIL-I-83387 (USAF), "Inspection Process, Magnetic Rubber," Aug 21, 1972

"Description and Applications of Magnetic Rubber Inspection," General Dynamics, Fort Worth Div.

J. E. Halkias, W. T. Kaarlela, J. D. Reynolds and H. J. Weltman, MRI—Help for Some Difficult NDT Problems, *Materials Evaluation,* Vol 31, No. 9, Sept 1973

M. Pevar, New Magnetic Test Includes Stainless Steel, *Product Engineering,* Vol 32, No. 6, Feb 6, 1961, p 41-43

Radiographic Inspection

G. L. Clark, "Applied X-Rays," 4th Ed., McGraw-Hill, New York, 1955

R. C. McMaster (Ed.), "Nondestructive Testing Handbook, Vol I," Ronald Press, New York, 1959

H. M. Muncheryan, "Industrial Radiology and Related Phenomena," Aircraft X-Ray Laboratories, Huntington Park, Calif., 1943

R. Halmshaw (Ed.), "Physics of Industrial Radiology," American Elsevier, New York, 1966

J. C. Rockley, "An Introduction to Industrial Radiology," Butterworths, London, 1964

"Nondestructive Testing/Radiographic Testing," Classroom Training Handbook CT-6-6, Convair Division of General Dynamics, San Diego, 1967

"Nondestructive Testing Series/Radiography," U. S. Department of Defense Handbook H55

"Radiography in Modern Industry," Eastman Kodak Co., Rochester, N. Y., 1969

"Nondestructive Testing/Radiographic Testing," Programmed Instruction (Self Study) Handbook PI-4-6 (5 Volumes), Convair Division of General Dynamics, San Diego

H. Richardson, A. Bertrand and D. Shipp, "Industrial Radiography Manual," USAEC, Contract AT-(40-1)-3112, Nov 15, 1964

"Radiological Emergency Operations," Instructor's Manual TID 24918: Student's Manual TID 24919, USAEC, Washington, D. C.

"Fundamentals of Industrial X-Ray and Safety," Picker Corp., Cleveland, June 1971

"Radiation Safety Using Industrial Radioisotopes," Diano Corp., Cleveland, Sept 1973

"A Guide on Radiation Safety Considerations in the Preparation of License Applications," USAEC, Washington, D. C., Nov 1970

A. E. Zimmer, "Artifacts, and Handling and Processing Faults on X-Ray Films," Grune & Stratton, New York, 1960

Inspection by Neutron Radiography

H. Berger, "Neutron Radiography," Elsevier Publishing Co., Amsterdam, 1965

H. Berger, The Present State of Neutron Radiography and Its Potential, *Materials Evaluation,* Vol 30, 1972, p 55-65

H. Berger, Recent Progress in Neutron Imaging, *British Journal of Nondestructive Testing,* Vol 10, 1968, p 26-33

H. Berger (Ed.), "Practical Applications of Neutron Radiography and Gauging," ASTM STP 586, American Society for Testing and Materials, Philadelphia, 1976

D. C. Cutforth, Dimensioning Reactor Fuel Specimens from Thermal Neutron Radiographs, *Nuclear Technology,* Vol 18, 1973, p 67-70

Ultrasonic Inspection

P. M. Morse, "Vibration and Sound," 2nd Edition, McGraw-Hill, New York, 1948

"Nondestructive Testing," SP-5113, Technology Utilization Office, NASA, Washington, 1973

"Nondestructive Testing—Ultrasonic Testing," CT-6-4, Convair Div., General Dynamics, San Diego, 1967

"Classroom Training Handbook—Ultrasonic Testing," CR-61228, NASA, Washington, 1965

"Ultrasonics. Vol I—Basic Principles," CR-61209, NASA, Washington, 1967

"Ultrasonics. Vol II—Equipment," CR-62110, NASA, Washington, 1967

"Ultrasonics. Vol III—Applications," CR-62111, NASA, Washington, 1967

Inspection by Optical Holography

R. K. Erf *et al,* Nondestructive Holographic Techniques for Structures Inspection, U. S. Air Force Technical Report AFML-TR-72-204, Oct 1972 (AD-757 510)

L. Kersch, Advanced Concepts of Holographic Nondestructive Testing, *Materials Evaluation,* Vol 29, 1971, p 125

J. P. Waters, Object Motion Compensation by Speckle Reference Beam Holography, *Applied Optics,* Vol 11, 1972, p 630

L. Rosen, Focused-Image Holography, *Applied Physics Letters,* Vol 9, 1969, p 1421

H. J. Caulfield *et al,* Local Reference Beam Generation in Holography, *Proceedings IEEE,* Vol 55, 1967, p 1758

V. J. Corcoran *et al,* Generation of a Hologram from a Moving Target, *Applied Optics,* Vol 5, 1966, p 668

D. B. Neumann *et al,* Object Motion Compensation Using Reflection Holography, *Journal of the Optical Society of America,* Vol 62, 1972, p 1373

D. B. Neumann *et al,* Improvement of Recorded Holographic Fringes by Feedback Control, *Applied Optics,* Vol 6, 1967, p 1097

H. W. Rose *et al,* Stabilization of Holographic Fringes by FM Feedback, *Applied Optics,* Vol 7, 1968, p 87

J. C. Palais, Scanned Beam Holography, *Applied Optics,* Vol 9, 1970, p 709

R. K. Erf (Ed.), "Holographic Nondestructive Testing," Academic Press, New York, 1974

Inspection by Acoustical Holography

B. P. Hildebrand and B. B. Brenden, "An Introduction to Acoustical Holography," Plenum Press, New York, 1972

B. P. Hildebrand and H. D. Collins, Evaluation of Acoustical Holography for the Inspection of Pressure Vessel Sections, *Materials Research and Standards,* Vol 12, No. 12, Dec 1972, p 23-32

S. Tolansky, "Multiple-Beam Interferometry of Surfaces and Films," Clarendon Press, Oxford, 1948

Microwave Inspection

A. Harvey, "Microwave Engineering," Academic Press, Inc., New York, 1963

R. J. Botsco, Nondestructive Testing of Plastics with Microwaves, Parts 1 and 2, *Plastic Design and Processing* (reprinted from Nov and Dec 1968 issues, Lake Publishing Corp., Libertyville, Ill.)

M. W. Standart, A. D. Lucian, T. E. Eckert and B. L. Lamb, Development of Microwave NDT Inspection Techniques for Large Solid Propellant Rocket Motors,

NAS7-544, Final Report 1117, Aerojet-General Corp., Sacramento, Calif., June 15, 1969

A. D. Lucian and R. W. Cribbs, The Development of Microwave NDT Technology for the Inspection of Nonmetallic Materials and Composites, *Proceedings of the 6th Symposium on Nondestructive Evaluation of Aerospace and Weapons Systems Components and Materials,* 1967, p 199-232 (available from Western Periodicals Co., North Hollywood, Calif.)

L. Feinstein and R. J. Hruby, Surface Crack Detection by Microwave Methods, *Proceedings of the 6th Symposium on Nondestructive Evaluation of Aerospace and Weapons Systems Components and Materials,* 1967, p 92-106 (available from Western Periodicals Co., North Hollywood, Calif.)

Thermal Inspection

S. Allinikov, Application of Photochromic Coatings for Nondestructive Inspection, Air Force Materials Laboratory, Wright-Patterson Air Force Base, Ohio, Technical Report AFML-TR-70-246, 1970

W. E. Woodmansee, Contact Thermographic Materials Applied to Nondestructive Testing, Proceedings 7th Symposium on Nondestructive Evaluation of Components and Materials in Aerospace, Weapons Systems and Nuclear Applications, American Society for Nondestructive Testing, Inc., and Southwest Research Institute (available from Western Periodicals Co., North Hollywood, Calif.), 1969, p 67-78

W. H. Byler and F. R. Hays, Fluorescence Thermography, *Nondestructive Testing,* May-June 1961, p 177-180

C. E. Searles, Thermal Image Inspection of Adhesive Bonded Structure, "1968 Symposium on NDT of Welds and Materials Joining," American Society for Nondestructive Testing, Inc., 1968

Spark Testing

E. Pitois, "Speaking of Steel" (translated by J. D. Gat), Chemical Publishing, New York, 1929

G. Tschorn, "Spark Atlas of Steels," Macmillan, New York, 1963

Spark Characteristics of Typical Tool Steels, AISI Committee of Tool Steel Producers, *Metal Progress,* Feb 1974, p 60-61

Chemical Spot Testing

F. Feigl and U. Anger, "Spot Tests in Inorganic Analysis," 6th Edition, American Elsevier, 1972

"Nondestructive Rapid Identification of Metals and Alloys by Spot Tests," STP 550, American Society for Testing and Materials, 1974

J. Catoggio, *Chemist Analyst,* Vol 48, 1959, p 58-63, 65

"Analysis of Aluminum and Its Alloys," British Aluminium Company, Ltd., Norfolk House, St. James Square, London, 1961

F. J. Bowen, Spot Testing Aluminum Alloys, *Steel,* Vol 130, 1952, p 67, 74

R. P. Nevers, Rapid Methods for Identification of Copper Base Alloys, in "Symposium on Rapid Methods for the Identification of Metals," STP 98, American Society for Testing and Materials, 1950, p 54-57

C. Goldberg, Spot-Test Differentiation of Monel Metal and Nickel Silvers, *Chemist Analyst,* Vol 42, 1953, p 35

F. J. Welcher, "The Analytical Uses of Ethylenediamine, Tetra-Acetic Acid," Van Nostrand, 1958

34 MECHANICAL TESTING

Edited by Marc A. Meyers, New Mexico Institute of Mining and Technology

The material on torsion testing that appears in this section was reprinted from Workability Testing Techniques, edited by G. E. Dieter, American Society for Metals, 1984. The material on fatigue-crack-growth-rate test methods and fracture-toughness testing was adapted from Application of Fracture Mechanics for Selection of Metallic Structural Materials, edited by J. E. Campbell, W. W. Gerberich and J. H. Underwood, American Society for Metals, 1982. Additional material in this section is based on the Metals Engineering Institute course on Mechanical Testing of Metals, by the M.E.I. Mechanical Testing of Metals Development Committee: W. T. Becker (Chairman), University of Tennessee; J. A. Alic, Office of Technology Assessment; K. T. Bassett, Danly Machine Co.; B. Boardman, Deere & Co.; J. P. Bosscher, Calvin College; T. M. Broughton, Tinius Olsen Testing Machine Co.; J. T. Cammett III, Metcut Research Associates, Inc.; J. Devis, American Society for Metals, retired; G. E. Dieter, University of Maryland; J. L. Herron, Herron Testing Labs; P. M. Mumford, United Calibration Corp.; C. H. Philleo, General Electric Co.; D. Socie, University of Illinois at Urbana-Champaign; J. Woodruff, CMW, Inc.

Introduction to Mechanical Testing

DESIGN of structures and systems requires determination of component dimensions and is based on the appropriate mechanical properties of materials. For static loading at low temperature, design criteria may include:

1. Excessive elastic deformation
2. Initiation of plastic deformation
3. Excessive plastic deformation
4. Unstable crack propagation.

Criterion 1 states that a certain component can undergo a specific maximum elastic strain. The elastic modulus of the material is the important property, because it allows, through Hooke's Law, establishment of the maximum stress. Criterion 2 limits the strains to elastic (recoverable) strains and prohibits plastic strains; the important property of the material is its yield stress. Criterion 3 accepts a certain amount of plastic deformation; important material properties are ultimate strength, uniform elongation, and work hardening. Criterion 4 accepts the possible existence of cracks or defects, and maximum loads are based on fracture toughness.

In conjunction with static loads, most structures are subjected to cyclical loading conditions. When the loads are not constant, the possible damage due to fatigue must be considered and the appropriate fatigue properties must be considered along with the static design requirements.

At high temperature, the mechanical properties of metals are time-dependent and the properties used at low temperatures are not applicable. Under high-temperature conditions, metals tend to deform with time, and the appropriate mechanical properties of interest are creep properties.

In this section, procedures for evaluating the mechanical properties of metals—i.e., the response of a metal to a particular imposed loading condition—are reviewed.

STANDARDIZED TESTS

In the United States, most common mechanical tests have been standardized by the American Society for Testing and Materials (ASTM). In other countries, similar standards have been developed by the appropriate standardization agencies. If a mechanical test is being conducted by the manufacturer, user or testing laboratory, and if the results are used outside the organization, it is imperative that the procedures outlined in ASTM standards be followed and that this is specified in the presentation of the results. For internal work, comparative studies and research studies, specific procedures which may differ from those of ASTM can be developed on approval of all parties; in such instances, specific test conditions should be specified. For specific information on ASTM standards, test methods, definitions, specifications and recommended practices, the reader is referred to ASTM. Each year a new edition of the Book of Standards is published. Because modifications and revisions may be introduced on an annual basis, it is not wise to use obsolete editions. For reference, the appropriate ASTM specifications which are applicable to this section are listed in Table 1. In this table, those standards that have been approved as American National Standards are given. These standards (except for those under the headings "Corrosion" and "Erosion and Wear") are presented in the 1983 Annual Book of ASTM Standards. Additional information can be obtained in the Special

Technical Publications (STP's) published by ASTM, a number of which are listed under "Additional Reading" at the end of this article, together with other useful sources of information.

MECHANICAL TESTING MACHINES

Numerous machines have been designed to perform the wide variety of mechanical tests listed in Table 1. In each of the following articles, test machines which are designed for a specific test are described in the separate section of that article devoted to that particular test. In this article, the basics of "universal testing machines," which can be used for tension, compression, formability and fatigue tests, are reviewed.

A wide variety of universal testing machines, produced by several manufacturers, are available. The basic elements of a universal testing machine are shown schematically in Fig. 1. In most modern machines, load is measured with a load cell (i.e., force transducer). Most load cells consist of a deflecting member which is instrumented with strain gages and which is coupled to a signal-conditioning system. The voltage output of the signal-conditioning system varies linearly with load and is used as input to a display or recording system.

The various universal testing machines differ primarily in how the displacement-control (or force-control) system operates. Displacement is usually imposed by either a mechanical or a hydraulic system.

Hydraulic (Open-Loop) Machines. Conventional hydraulic test equipment operates largely on the basis of open-loop control. Figure 2 shows a typical system of this type that is used to position the piston rod of a hydraulic actuator. An oper-

Table 1. ASTM procedures for mechanical testing

Subject	Reference
Bend and flexure testing	
Methods for:	
Bend Testing for Metallic Flat Materials for Spring Applications	E855-81
Guided Bend Test for Ductility of Welds	E190-80
Semi-Guided Bend Test for Ductility of Metallic Materials	E290-81
Methods and Definitions for:	
Mechanical Testing of Steel Products	A370-77
Calibration of mechanical testing machines, extensometers and strain gages	
Methods of Calibration of:	
Force-Measuring Instruments for Verifying the Load Indication of Testing Machines	E74-81
Methods of Verification of:	
Extensometers, and Classification of	E83-67 (1980)
Testing Machines	E4-83
Test Methods for:	
Performance Characteristics of Bonded Resistance Strain Gages	E251-67 (1980)
Compression testing	
Methods for:	
Compression Testing of Metallic Materials at Room Temperature	E9-81
Pin-Type Bearing Test of Metallic Materials	E238-68 (1978)
Practice for:	
Compression Tests of Metallic Materials at Elevated Temperatures with Conventional or Rapid Heating Rates and Strain Rates	E209-65 (1981)
Ductility and formability testing	
Methods for:	
Bend Testing for Metallic Flat Materials for Spring Applications	E855-81
Conducting a Ball Punch Deformation Test for Metallic Sheet Material	E643-78
Ductility Testing of Metallic Foil	E796-81
Test Method for:	
Tensile Strain-Hardening Exponents (*n*-Values) of Metallic Sheet Materials	E646-78
Definitions on mechanical testing	
Definitions of Terms Relating to:	
Mechanical Testing of Steel Products	A370-77
Methods of Mechanical Testing	E6-81
Elastic properties	
Test Methods for:	
Poisson's Ratio at Room Temperature	E132-61 (1979)
Shear Modulus at Room Temperature	E143-61 (1979)
Static Determination of Young's Modulus of Metals at Low and Elevated Temperatures	E231-69 (1981)
Young's Modulus, Tangent Modulus, and Chord Modulus	E111-82
Hardness testing	
Test Methods for:	
Brinell Hardness of Metallic Materials	E10-78
Hardness Conversion Tables for Metals (Relationship Between Brinell Hardness, Vickers Hardness, Rockwell Hardness, Rockwell Superficial Hardness, and Knoop Hardness)	E140-79
Indentation Hardness of Metallic Materials by Portable Hardness Testers	E110-82
Mechanical Testing of Steel Products	A370-77
Microhardness of Materials	E384-73 (1979)
Rapid Indentation Hardness Testing of Metallic Materials	E103-61 (1979)
Rockwell Hardness and Rockwell Superficial Hardness of Metallic Materials	E18-79
Vickers Hardness of Metallic Materials	E92-82
Practice for:	
Scleroscope Hardness Testing of Metallic Materials	E448-82
Impact testing	
Methods for:	
Mechanical Testing of Steel Products	A370-77
Notched Bar Impact Testing of Metallic Materials	E23-82
Linear thermal expansion test	
Test Methods for:	
Linear Expansion of Metals	B95-39 (1979)
Linear Thermal Expansion of Rigid Solids with a Vitreous Silica Dilatometer	E228-71 (1979)

Subject	Reference
Linear Thermal Expansion of Rigid Solids with Interferometry	E289-70 (1979)
Machinability test method	
Method for:	
Machining Performance of Ferrous Metals Using an Automatic Screw/Bar Machine, Evaluating	E618-81
Shear and torsion testing	
Methods for:	
Shear Testing of Aluminum and Aluminum-Alloy Rivets and Cold-Heading Wire and Rods	B565-76 (1981)
Torsion Testing of Wire	E558-75
Residual stress measurement	
Methods for:	
Determining Residual Stresses by the Hole-Drilling Strain-Gage Method	E837-81
Verifying the Alignment of X-ray Diffraction Instrumentation for Residual Stress Measurement	E915-83
Stress-relaxation tests	
Practice for:	
Stress-Relaxation Tests for Materials and Structures	E328-78
Tension testing	
Methods for:	
Mechanical Testing of Steel Products	A370-77
Tension Testing of Metallic Foil	E345-81
Tension Testing of Metallic Materials	E8-82
Tension Testing Wrought and Cast Aluminum- and Magnesium-Alloy Products	B557-81
Tension Testing Wrought and Cast Aluminum- and Magnesium-Alloy Products [Metric]	B557M-81
Test Method for:	
Plastic Strain Ratio *r* for Sheet Metal	E517-81
Thickness measurement	
Test Method for:	
Thickness of Thin Foil and Film by Weighing	E252-78
Fatigue	
Definitions of Terms Relating to:	
Constant-Amplitude, Low-Cycle Fatigue Testing	E513-74 (1980)
Fatigue Loading	E912-83
Fatigue Testing and the Statistical Analysis for Fatigue Data	E206-72 (1979)
Fluid Aqueous and Chemical Environmentally Affected Fatigue Testing	E742-81
Practices for:	
Constant-Amplitude Axial Fatigue Tests of Metallic Materials	E466-82
Constant-Amplitude Low-Cycle Fatigue Testing	E606-80
Presentation of Constant-Amplitude Fatigue Test Results for Metallic Materials	E468-82
Statistical Analysis of Linear or Linearized Stress-Life (*S-N*) and Strain-Life (ϵ-*N*) Fatigue Data	E739-80
Verification of Constant-Amplitude Dynamic Loads in an Axial Load Fatigue Testing Machine	E467-76 (1982)
Test Method for:	
Constant-Load-Amplitude Fatigue Crack Growth Rates Above 10^{-8} m/Cycle	E647-83
Fracture Testing	
Test Methods for:	
Crack Strength of Slow-Bend Precracked Charpy Specimens of High-Strength Metallic Materials	E812-81
Drop-Weight Tear Tests of Ferritic Steels	E436-74 (1980)
Dynamic Tear Testing of Metallic Materials	E604-83
J_{Ic}, a Measure of Fracture Toughness	E813-81
Plane-Strain Fracture Toughness of Metallic Materials	E399-83
Sharp-Notch Tension Testing of High-Strength Sheet Materials	E338-81
Sharp-Notch Tension Testing with Cylindrical Specimens	E602-81
Practices for:	
Fracture Testing with Surface-Crack Tension Specimens	E740-80
R-Curve Determination	E561-81
Terminology Relating to:	
Fracture Testing	E616-82
Corrosion	
Method for:	
Preparation and Use of Bent-Beam Stress-Corrosion Specimens	G39-73

(continued)

Table 1. (continued)

Subject	Reference	Subject	Reference
Corrosion (continued)		**Effect of Temperature on Metals**	
Recommended Practices for:		Methods for:	
Alternate Immersion Stress-Corrosion Testing in 3.5% Sodium Chloride Solution	G44-75	Conducting Drop-Weight Test to Determine Nil-Ductility Transition Temperature of Ferritic Steels	E208-81
Making and Using the C-Ring Stress-Corrosion Cracking Test Specimen	G38-73	Static Determination of Young's Modulus of Metals at Low and Elevated Temperatures	E231-69 (1981)
Making and Using U-Bend Stress-Corrosion Test Specimens	G30-72	Practices for:	
Performing Stress-Corrosion Cracking Tests in a Boiling Magnesium Chloride Solution	G36-73	Compression Tests of Metallic Materials at Elevated Temperatures with Conventional or Rapid Heating Rates and Strain Rates	E209-65 (1981)
Preparation and Use of Direct Tension Stress-Corrosion Test Specimens	G49-76	Conducting Creep and Creep-Rupture Tension Tests of Metallic Materials Under Conditions of Rapid Heating and Short Times (Intent to Withdraw)	E150-64 (1981)
Susceptibility of Stainless Steels and Related Nickel-Chromium-Iron Alloys to Stress-Corrosion Cracking in Polythionic Acids, Determining	G35-73	Conducting Creep, Creep-Rupture, and Stress-Rupture Tests of Metallic Materials	E139-83
Erosion and Wear		Conducting Time-for-Rupture Notch Tension Tests of Materials	E292-78
Method of:		Elevated Temperature Tension Tests of Metallic Materials	E21-79
Vibratory Cavitation Erosion Test	G32-77	Tension Tests of Metallic Materials at Elevated Temperatures with Rapid Heating and Conventional or Rapid Strain Rates (Intent to Withdraw)	E151-64 (1981)
Test Method for:			
Abrasiveness of Ink-Impregnated Fabric Printer Ribbons	G56-77		
Terminology Relating to:			
Erosion and Wear	G40-77		

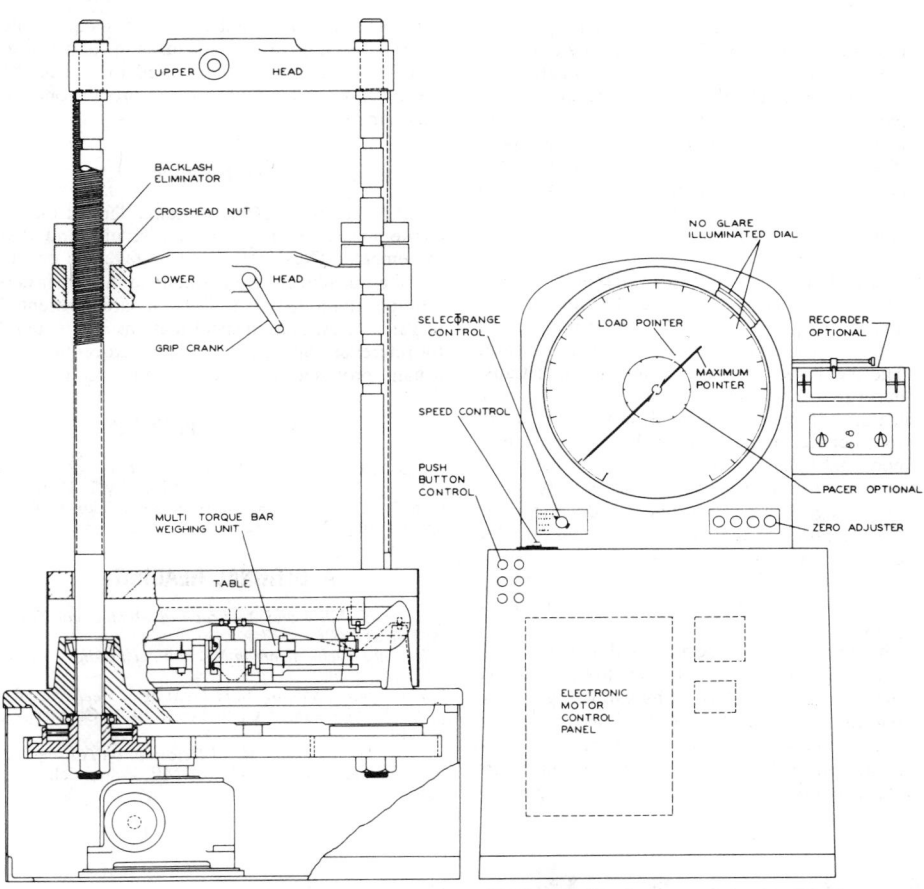

Fig. 1. Schematic drawing of universal testing machine

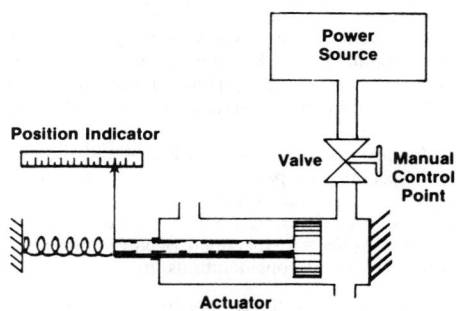

Fig. 2. Open-loop control system

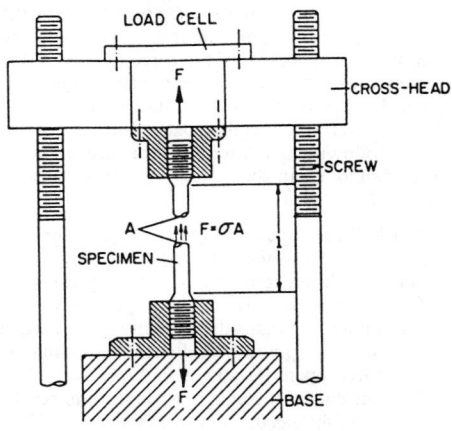

Fig. 3. Screw-driven test system

ator must observe the position indicator and constantly regulate the valve, adjusting for variations in the power source and actuator load to maintain a given position. This is the simplest form of closed-loop control, with the operator closing the loop and thus inserting his response into the loop. If the operator is provided with a suitable indicator, he can also control other parameters, such as velocity or force. If the valve is left unattended, there will be no regulation of position or any other parameter. The Baldwin and Tinius Olsen machines are examples of these machines.

Electromechanical Machines. In electromechanical machines, the cross-head usually is driven vertically by twin screws (Fig. 3). The cross-head velocity is determined by the rotational velocity of the screws. This velocity can be maintained constant throughout the test, in contrast with hydraulic machines, in which the operator continuously adjusts the velocity. These screw-driven machines are reliable, simple to operate, and perfectly suited for routine tensile and compressive testing.

Servo-Hydraulic (Closed-Loop) Systems. A typical closed-loop system that operates in position control is shown in Fig. 4. Without a human operator within the loop, a visual indication of piston position is not necessary. A transducer provides a signal that is proportional to piston position. That transducer signal is connected to a servo controller that compares it to the signal from the manually adjusted command control. The signal from the manual command control is called simple "command"; the signal from the transducer is called "feedback." If command and feedback

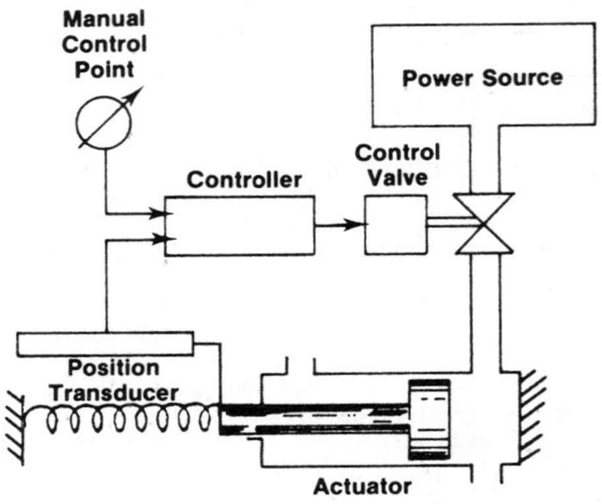

Manual Control Point

Power Source

Controller

Control Valve

Position Transducer

Actuator

Fig. 4. Closed-loop control system

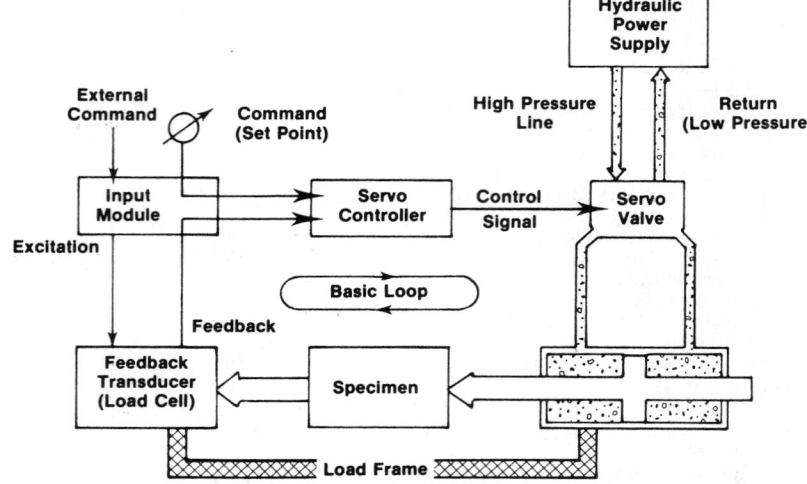

Fig. 5. Basic testing system

are not equal, the actuator is not positioned at the desired, or command, position. The servo controller reacts to relative differences between these signals (both polarity and magnitude) and applies a control signal to the control valve that causes the difference to be corrected. Thus, the system automatically maintains the piston at any given point, even when the power source or the actuator load varies. Servo-hydraulic machines also operate with the strain or load as controlling parameters.

Figure 5 is very similar to Fig. 4 and shows a test specimen as part of the closed servo loop. The force applied to the specimen is sensed by a transducer (load cell) and is controlled by the system. When the specimen is part of the loop, some important considerations are:

• Characteristics of the specimen directly affect the behavior of the system.
• Selection of the proper transducer permits close control of the parameter of interest within the specimen itself. For example, to provide the most precise control of the force applied to the transducer, a load cell (force transducer) is used, or, to control specimen strain, a strain transducer is attached to the specimen.

The following list describes some of the equipment employed in the testing system illustrated in Fig. 5:

• The hydraulic power supply is the source of power for the system. Its output is rated in gallons (litres) per minute at a certain pressure.
• The servo valve, also rated in gallons (litres) per minute, regulates the flow of fluid between the hydraulic power supply and the actuator. The valve opens to power high-pressure fluid to either end of the actuator, depending on the polarity of the electrical input of the valve. When

the servo valve is open to permit high-pressure fluid to be admitted to one end of the cylinder, the valve also provides a fluid-return path from the other end of the cylinder to the supply. When the system is in a static condition, the electrical input to the valve is zero and the valve is closed (fluid is not permitted to flow into or out of the actuator). The valve opens in proportion to the magnitude of the input signal and is fully open when the input current reaches the current input rating of the valve.
• The hydraulic actuator has a maximum static-force rating that is determined by the effective area of the piston (area of cylinder bore minus area of piston rod) and the actuating pressure.
• The feedback transducer is any suitable transducer properly coupled to the specimen to sense the quantity being controlled, in this case force. Its electrical output (feedback) is directly proportional to the mechanical input to the transducer and to the specimen. The polarity of the output signal of the transducer is dependent on the direction of the mechanical input. In the case of load cells, the output signal is positive when the input force is tensile. If the input force is compressive, the output signal has a negative polarity.
• The servo controller accepts both the command and feedback signals and reacts to their relative magnitudes and polarities by supplying an output control signal that results from any difference between command and feedback. If command and feedback are equal, the output is zero (the controlled variable is at the command level).

If command and feedback are not equal, the output has a polarity that causes the actuator to stroke in the direction required to reduce the error and a magnitude that is proportional to the amount of error.

CALIBRATION

All mechanical testing systems depend on accurate measurement of loads, strains and displacements. Thus, care must be taken to ensure that the transducers are maintained in proper calibration. In the list of ASTM standards presented in Table 1, calibration specifications are included for reference. Specific calibration procedures are usually provided by system manufacturers.

SELECTED REFERENCES

1983 Annual Book of ASTM Standards
Mechanical Measurements, 3rd Ed., by T. B. Beckwith, N. L. Buck and R. D. Marangoni: Addison-Wesley, Reading, MA, 1982

ADDITIONAL READING

Reproducibility and Accuracy of Mechanical Tests, ASTM STP 626, 1977
Recent Developments in Mechanical Testing, ASTM STP 608, 1976
The Making, Shaping, and Treating of Steel, 9th Ed., by H. E. McGannon: U. S. Steel, 1971, Chapter 49, p 1214
Novel Techniques in Metal Deformation Testing, edited by R. H. Wagoner: TMS-AIME, Warrendale, PA, 1983

Hardness Testing

HARDNESS is a term that has different meanings to different people: it is resistance to penetration to a metallurgist, resistance to wear to a lubrication engineer, a measure of flow stress to a design engineer, resistance to scratching to a mineralogist, and resistance to cutting to a machinist. Although these various definitions of hardness appear to differ significantly in character, they are all related to the plastic flow stress of the material.

Only static indentation and rebound testing are discussed in this article. These two methods account for virtually all routine hardness testing in the metalworking industry. Static indentation hardness testing is the more widely used of the two methods, although rebound testing is extensively employed, particularly for hardness measurements on large workpieces or for applications in which visible or sharp impressions in the test surface cannot be tolerated.

BRINELL HARDNESS TESTING

The Brinell hardness test is basically simple, and consists of applying a constant load, usually 500 to 3000 kg, on a hardened steel ball-type indenter, 10 mm in diameter, to the flat surface of a workpiece (Fig. 1). The 500-kg load is usually used for testing nonferrous metals such as copper and aluminum alloys, whereas the 3000-kg load is most often used for testing harder met-

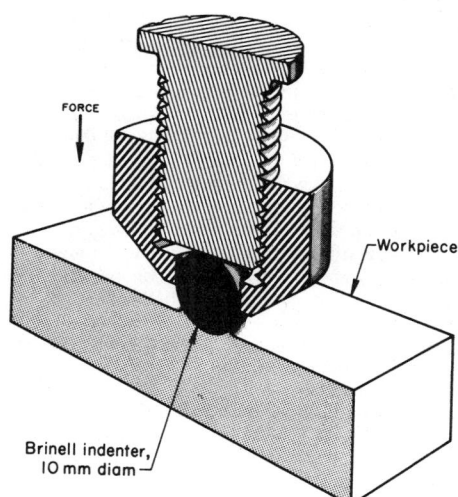

Fig. 1. Sectional view of a Brinell indenter, showing the manner in which the application of force by the indenter causes the metal of the workpiece to flow

als such as steels and cast irons. The load is held for a specified time (10 to 15 s for iron or steel and about 30 s for softer metals), after which the diameter of the recovered indentation is measured in millimetres. This time period is required to ensure that plastic flow of the work metal has stopped.

Hardness is evaluated by taking the mean diameter of the indentation (two readings at right angles to each other) and calculating the Brinell hardness number (HB) by dividing the applied load by the surface area of the indentation according to the following formula:

$$HB = L/(\pi D/2)[D - (D^2 - d^2)^{1/2}]$$

where L is the load, in kilograms; D is the diameter of the ball, in millimetres; and d is the diameter of the indentation, in millimetres.

It is not necessary, however, to make the calculation for each test. Such calculations are available in table form for all diameters of indentations in Section 1 of this Desk Edition.

Highly hardened steel (or other very hard metals) cannot be tested by a hardened steel ball by the Brinell method, because the ball will flatten during penetration and a permanent deformation will take place. This problem is recognized in specifications for the Brinell tests.

Tungsten carbide balls are recommended for Brinell testing materials of hardness from 444 HB up to about 627 HB (indentation of 2.45 mm in diameter). However, higher Brinell values will result when using carbide balls instead of steel balls, because of the difference in elastic properties.

Surface Preparation. The degree of accuracy that can be attained by the Brinell hardness test can be greatly influenced by the surface smoothness of the workpiece being tested.

The surface of the workpiece on which the Brinell indentation is to be made must be filed, ground, machined or polished with emery paper (3/0 emery paper is suitable) so that the identation diameter is clearly enough defined to permit its measurement. There should be no interference from tool marks.

Indentation Measurement. The diameter of the indentation is measured by a microscope to the

nearest 0.05 mm (0.002 in.). This microscope contains a scale, and usually a built-in light to facilitate easy reading.

The indentations produced in Brinell hardness tests may exhibit different surface characteristics. These have been carefully studied and analyzed. In some instances there is a ridge around the indentation extending above the original surface of the workpiece. In other instances the edge of the indentation is below the original surface. Sometimes there is no difference at all. The first phenomenon is called a "ridging" type of indentation and the second a "sinking" type. Cold worked metals generally have the former type of indentation, and annealed metals the latter type.

Brinell Hardness Testers. Several types of testers that exert the prescribed force on the indenter are in general use. The one most commonly used is the hydraulic, manually operated type shown in Fig. 2.

The workpiece is placed on the anvil and raised, by means of the elevating screw, to a position

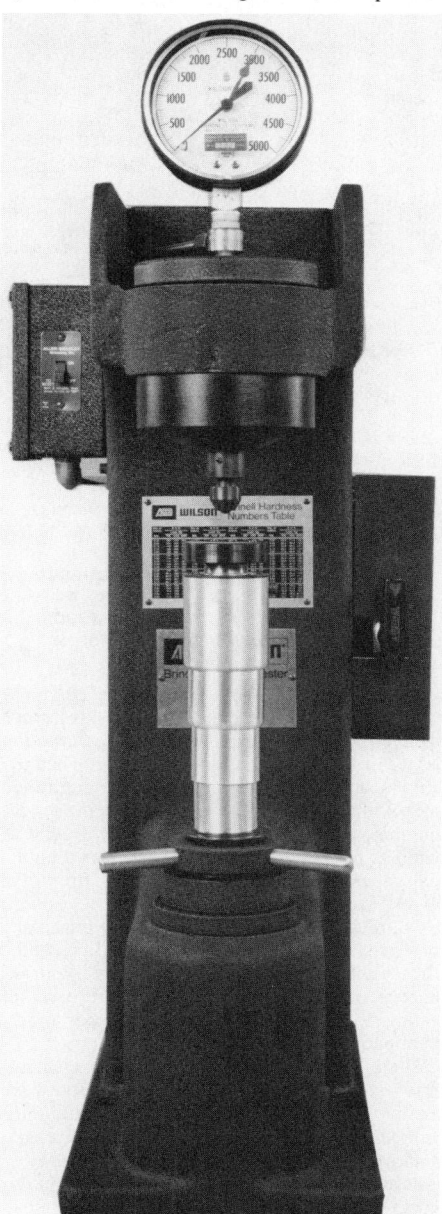

Fig. 2. Brinell hardness tester

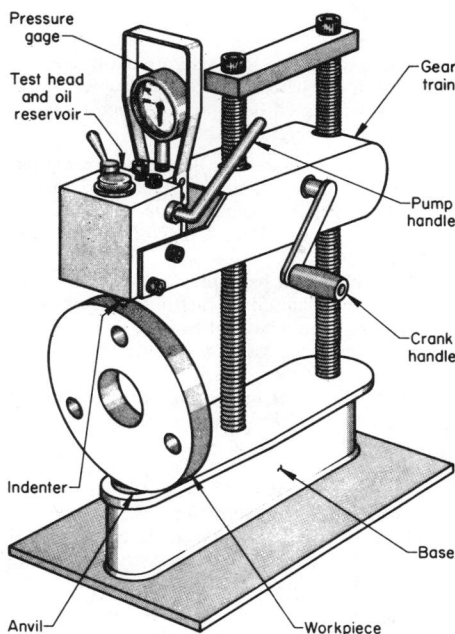

Fig. 3. Hydraulic, manually operated portable Brinell hardness tester

near the indenter. Fingertip rotation of the control knob allows a selected force (in kilograms), indicated on the gage, to be applied. This force is held for a pre-established length of time and then released. The specimen is removed and the indentation measured. The entire cycle, including indenting and measurement, requires approximately one minute.

For production testing, speed direct-reading Brinell-type testers also are available.

Portable Brinell Hardness Testers. Conventional Brinell hardness testers described above have limited use, for two reasons: (a) the workpieces to be tested must be brought to the testers, and (b) size and design of the workpieces must be such that they can be placed between the anvil and the indenter.

Some of the problems posed by the above limitations can often be solved by use of a portable hardness tester (Fig. 3). A hardness tester of the general design shown in Fig. 3 weighs no more than about 25 lb; thus, it can be easily transported to the workpieces.

Spacing of Indentations. To ensure accurate results, indentations should not be made too close to the edge of the workpiece being tested. Lack of sufficient supporting material on one side of the workpiece will cause the resulting indentation to be large and unsymmetrical. It is generally agreed that the error in a Brinell hardness number is negligible if the distance from the center of the indentation is not less than $2\frac{1}{2}$ times (and preferably 3 times) the diameter of the indentation from any edge of the workpiece.

Similarly, indentations should not be made too close to one another. If indentations are too close together, the work metal may be cold worked by the first indentation, or there may not be sufficient supporting material for the second indentation. The latter condition would produce too large an indentation, whereas the former may produce too small an indentation. To prevent this, the distance between centers of adjacent indentations should be at least three times the diameter of the indentation.

General Precautions. To avoid misapplication of Brinell hardness testing, the fundamentals and limitations of the test procedure must be clearly understood. Further, to avoid inaccuracies, some general rules should be followed. Such rules include the following:

1. Indentations should not be made on a curved surface having a radius of less than 1 in.
2. Spacing of indentations should be correct, as outlined above under "Spacing of Indentations."
3. The load should be applied steadily to avoid overloading caused by inertia of the weights.
4. The load should be applied in such a way that the direction of loading and the test surface are perpendicular to each other within 2°.
5. The thickness of the workpiece being tested should be such that no bulge or mark showing the effect of the load appears on the side of the workpiece opposite the indentation. In any event, the thickness of the specimen shall be at least ten times the depth of indentation.
6. The surface finish of the workpiece being tested should be such that the indentation diameter is clearly outlined.

Limitations. The Brinell hardness test has three principal limitations:

1. Size and shape of the workpiece must be capable of accommodating the relatively large indentations.
2. Because of the relatively large indentations, the workpiece may not be usable after testing.
3. The limit of hardness range—about 11 HB with the 500-kg load to 627 HB with the 3000-kg load—is generally considered the practical range.

ROCKWELL HARDNESS TESTING

Rockwell hardness testing is the most widely used method for determining hardness. There are several reasons for this distinction: (*a*) the Rockwell test is simple to perform and does not require highly skilled operators; (*b*) by use of different loads and indenters, Rockwell hardness testing can be used for determining hardness of most metals and alloys, ranging from the softest bearing materials to the hardest steels; (*c*) a reading can be taken in a matter of seconds with conventional manual operation and in even less time with automated setups; and (*d*) no optical measurements are required (all readings are direct).

Rockwell hardness testing differs from Brinell hardness testing in that the hardness is determined by the depth of indentation made by a constant load impressed upon an indenter. Although a number of different indenters are used for Rockwell hardness testing, the most common type is a diamond ground to a 120° cone with a spherical apex having a 0.2-mm radius, which is known as a Brale indenter (see Fig. 4).

As shown in Fig. 5, the Rockwell hardness test consists of measuring the additional depth to which an indenter is forced by a heavy (major) load (Fig. 5b) beyond the depth of a previously applied light (minor) load (Fig. 5a). Application of the minor load eliminates backlash in the load train and causes the indenter to break through slight surface roughness and to crush particles of foreign matter, thus contributing to much greater accuracy in the test. The basic principle involving minor and major loads illustrated in Fig. 5 applies to steel-ball indenters as well as to diamond indenters.

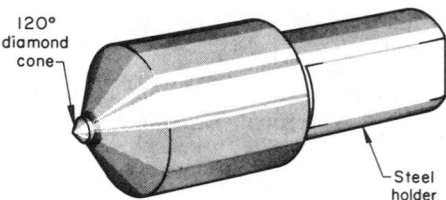

Fig. 4. Diamond-cone Brale indenter used in Rockwell hardness testing (shown at about 2×)

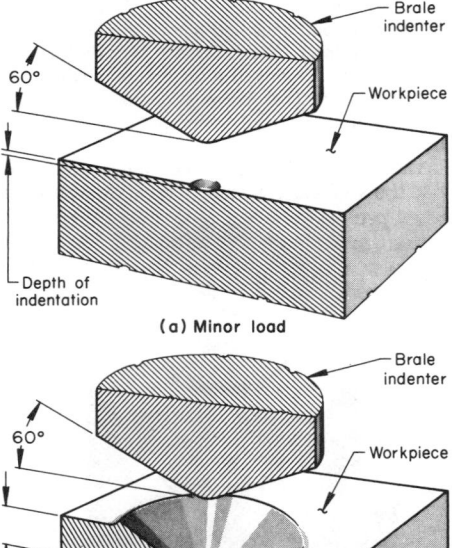

The hardness value is based on the difference in depths of indentation produced by the minor and major loads.

Fig. 5. Indentation in a workpiece made by application of (a) the minor load, and (b) the major load, on a diamond Brale indenter in Rockwell hardness testing

The minor load is applied first, and a reference or "set" position is established on the measuring device of the Rockwell hardness tester. Then the major load is applied at a prescribed, controlled rate. Without moving the workpiece being tested, the major load is removed and the Rockwell hardness number is automatically indicated on the dial gage. The entire operation takes from 5 to 10 s.

Diamond indenters are used mainly for testing materials such as hardened steels and cemented carbides. Steel-ball indenters, available with diameters of $1/16$, $1/8$, $1/4$ and $1/2$ in., are used for testing materials such as soft steel, copper alloys, aluminum alloys and bearing metals.

There are two basic types of Rockwell hardness testers—regular and superficial. Both have similar basic mechanical principles and significant components. A schematic view of a regular Rockwell hardness tester is shown in Fig. 6.

Regular Rockwell Hardness Testing. In regular Rockwell hardness testing the minor load is always 10 kg. The major load, however, can be

60, 100 or 150 kg. No Rockwell hardness value is expressed by a number alone. A letter has been assigned to each combination of load and indenter, as shown in Table 1. Each number is suffixed by first the letter H (for hardness), then the letter R (for Rockwell), and finally the letter that indicates the scale used. For example, a value of 60 on the Rockwell C scale is expressed as 60 HRC, and so on. Regardless of the scale used, the "set" position is the same; however, when the diamond Brale indenter is used, the readings are taken from the black divisions on the dial gage. When testing with any of the ball indenters, the readings are taken from the red divisions.

One Rockwell number represents an indentation of 0.002 mm (0.00008 in.). Therefore, a reading of 60 HRC indicates indentation from minor to major load of $(100 - 60) \times 0.002$ mm = 0.080 mm or 0.0032 in. A reading of 80 HRB indicates an indentation of $(130 - 80) \times 0.002$ mm = 0.100 mm or 0.004 in.

Superficial Rockwell hardness testing employs a minor load of 3 kg, but the major load can be 15, 30 or 45 kg.

Just as in regular Rockwell testing, the indenter may be either a diamond or a steel ball, depending mainly on the nature of the metal being tested. Regardless of load, the letter N designates use of the superficial Brale, and the letters T, W, X and Y designate use of steel-ball indenters. Scale and load combinations are presented in Table 1. Superficial Rockwell hardness values are always expressed with the number suffixed by a number and a letter that show the load/indenter combination. For example, if a load of 30 kg is used with a diamond indenter and a reading of 80 is obtained, the result is reported as 80 HR30N (where H means hardness, R means Rockwell, 30 means a load of 30 kg, and N indicates use of a diamond indenter).

All tests are started from the "set" position. One Rockwell superficial hardness number represents an indentation of 0.001 mm or 0.00004 in. Therefore, a reading of 80 HR30N indicates indentation from minor to major load of $(100 - 80) \times 0.001$ mm = 0.020 mm or 0.0008 in.

Dials on the superficial hardness testers contain only one set of divisions, which is used with all types of superficial indenters.

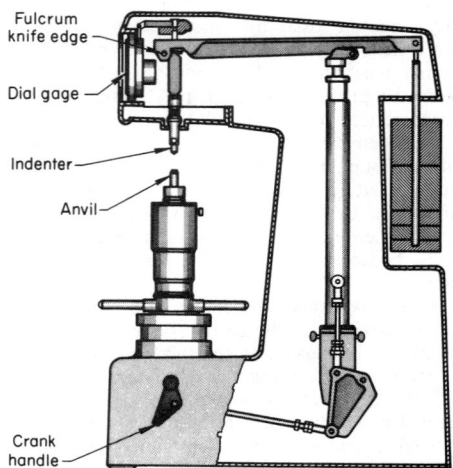

Fig. 6. Principal components of a regular (normal) Rockwell hardness tester. Superficial Rockwell testers are similarly constructed.

Table 1. Rockwell-hardness-scale designations for combinations of type of indenter and major load

Scale designation	Indenter Type	Diam, in.	Major load, kg	Dial figure	Scale designation	Indenter Type	Diam, in.	Major load, kg
Regular Rockwell tester					**Superficial Rockwell tester**			
B	Ball	1/16	100	Red	15N	N Brale	...	15
C	Brale	...	150	Black	30N	N Brale	...	30
A	Brale	...	60	Black	45N	N Brale	...	45
D	Brale	...	100	Black	15T	Ball	1/16	15
E	Ball	1/8	100	Red	30T	Ball	1/16	30
F	Ball	1/16	60	Red	45T	Ball	1/16	45
G	Ball	1/16	150	Red	15W	Ball	1/8	15
H	Ball	1/8	60	Red	30W	Ball	1/8	30
K	Ball	1/8	150	Red	45W	Ball	1/8	45
L	Ball	1/4	60	Red	15X	Ball	1/4	15
M	Ball	1/4	100	Red	30X	Ball	1/4	30
P	Ball	1/4	150	Red	45X	Ball	1/4	45
R	Ball	1/2	60	Red	15Y	Ball	1/2	15
S	Ball	1/2	100	Red	30Y	Ball	1/2	30
V	Ball	1/2	150	Red	45Y	Ball	1/2	45

Table 2. Typical applications of regular Rockwell hardness scales

Scale(a)	Typical applications
B	Copper alloys, soft steels, aluminum alloys, malleable iron
C	Steel, hard cast irons, pearlitic malleable iron, titanium, deep case-hardened steel and other materials harder than 100 HRB
A	Cemented carbides, thin steel and shallow case-hardened steel
D	Thin steel and medium case-hardened steel and pearlitic malleable iron
E	Cast iron, aluminum and magnesium alloys, bearing metals
F	Annealed copper alloys, thin soft sheet metals
G	Phosphor bronze, beryllium copper, malleable irons. Upper limit is 92 HRG, to avoid flattening of ball.
H	Aluminum, zinc, lead
K,L,M,P, R,S,V	Bearing metals and other very soft or thin materials. Use smallest ball and heaviest load that do not give anvil effect.

(a) The N scales of a superficial hardness tester are used for materials similar to those tested on the Rockwell C, A and D scales, but of thinner gage or case depth. The T scales are used for materials similar to those tested on the Rockwell B, F and G scales but of thinner gage. When minute indentations are required, a superficial hardness tester should be used. The W, X and Y scales are used for very soft materials.

Selection of Rockwell Scale

Where no specification exists or there is doubt about the suitability of a specified scale, an analysis should be made of those factors that influence the selection of the proper scale. These influencing factors are found in the following four broad categories:

1. Type of work metal
2. Thickness of work metal
3. Width of area to be tested
4. Scale limitations.

Influence of Type of Work Metal. The types of work metal normally tested using the different regular Rockwell hardness scales are given in Table 2. This information also can be helpful when one of the superficial Rockwell scales may be required. For example, note that the C, A and D scales — all with diamond indenters — are used on hard materials such as steel and tungsten carbide. Any material in this hardness category would be tested with a diamond indenter. The choice to be made is whether the C, A, D, or the 45N, 30N, or 15N scale is applicable. Whatever the choice, the number of possible scales has been reduced to six. The next step is to find a scale, either regular or superficial, that will guarantee accuracy, sensitivity and repeatability of testing.

Influence of Thickness of Work Metal. The metal immediately surrounding the indentation in a Rockwell hardness test is cold worked. The depth of material affected during testing is on the order of ten times the depth of the indentation. Therefore, unless the thickness of the metal being tested is at least ten times the depth of the indentation, an accurate Rockwell hardness test cannot be expected.

The depth of indentation for any Rockwell hardness test can easily be computed; in practice, however, computation is not necessary, because tables of minimum thicknesses are available (for example, see Table 3). The values for minimum thickness do follow the 10-to-1 ratio in some ranges, but they are actually based on experimental data accumulated on various thicknesses of low-carbon steels and of steel strip that has been hardened and tempered.

To use the values in Table 3, assume that it is necessary to check the hardness of a strip of steel 0.014 in. thick, of an approximate hardness of 63 HRC. According to Table 3, material having a hardness of 63 HRC must be approximately 0.028 in. thick for an accurate test using the C scale. Therefore, this steel strip should not be tested on the C scale. At this point, check the approximate converted hardness on the other Rockwell scales equivalent to 63 HRC. These values — taken from a conversion table — are: 83 HRA, 70 HR45N, 80 HR30N and 91 HR15N. Referring again to Table 3 for hardened 0.014-in.-thick material, there are only three Rockwell scales to choose from — 45N, 30N and 15N. The 45N scale is not suitable because the material should be at least 74 HR45N. On the 30N scale, 0.014-in.-thick material must be at least 80 HR30N, and the material at hand is 80 HR30N. On the 15N scale, the material must be at least 76 HR15N, and this material is 91.5 HR15N. Therefore, either the 30N or 15N scale may be used. After all limiting factors have been eliminated, and a choice exists between two or more scales, the scale applying the heavier load should be used. The heavier load will produce a larger indentation, covering a greater portion of the material, and a Rockwell hardness number more representative of the material as a whole will be obtained. In addition, the heavier the load, the greater the sensitivity of the scale. Checking any conversion table and comparing the 15N scale to the 30N scale will show that in the hard-steel range a difference in hardness of one point on the 30N scale represents a difference of only 0.5 point on the 15N scale. Therefore, smaller differences in hardness can be detected when using the 30N scale.

Table 3. Minimum work-metal hardness values for testing various thicknesses of metals with regular and superficial Rockwell hardness testers(a)

Metal thickness, in.	Minimum hardness for superficial hardness testing						Minimum hardness for regular hardness testing					
	Diamond Brale indenter			Ball indenter, 1/16 in.			Diamond Brale indenter			Ball indenter, 1/16 in.		
	15N (15 kg)	30N (30 kg)	45N (45 kg)	15T (15 kg)	30T (30 kg)	45T (45 kg)	A (60 kg)	D (100 kg)	C (150 kg)	F (60 kg)	B (100 kg)	G (150 kg)
0.005	...	...	...	93	...	...	...	...	...	...	...	...
0.006	92	...	...	...	...	...	...	...	...	...	...	...
0.008	90	...	...	...	...	...	...	...	...	...	...	...
0.010	88	...	...	90	87	...	...	...	...	...	...	...
0.012	83	82	77	...	...	...	...	...	...	...	...	...
0.014	76	80	74	...	...	...	...	...	...	...	...	...
0.015	...	...	...	78	77	77	...	...	...	...	...	...
0.016	68	74	72	...	...	...	86	...	...	...	...	...
0.018	(b)	66	68	...	...	...	84	...	...	...	...	...
0.020	(b)	57	63	(b)	58	62	82	77	...	100	...	...
0.022	(b)	47	58	...	...	...	78	75	69	...	...	...
0.024	(b)	(b)	51	...	...	...	76	72	67	...	...	...
0.025	...	...	...	(b)	(b)	26	...	...	...	92	92	90
0.026	(b)	(b)	37	...	...	...	71	68	65	...	...	...
0.028	(b)	(b)	20	...	...	...	67	63	62	...	...	...
0.030	(b)	(b)	(b)	(b)	(b)	(b)	60	58	57	67	68	69
0.032	(b)	(b)	(b)	...	...	...	(b)	51	52	...	...	...
0.034	(b)	(b)	(b)	...	...	...	(b)	43	45	...	...	...
0.035	...	...	...	(b)	(b)	(b)	...	...	...	(b)	44	46
0.036	(b)	(b)	(b)	...	...	...	(b)	(b)	37	...	...	...
0.038	(b)	(b)	(b)	...	...	...	(b)	(b)	28	...	...	...
0.040	(b)	(b)	(b)	(b)	(b)	(b)	(b)	(b)	20	...	20	22

(a) These values are approximate only and are intended primarily as a guide; see text for example of use. Materials thinner than shown may be tested on a Tukon microhardness tester. The thickness of the workpiece should be at least 1 1/2 times the diagonal of the indentation when using a Vickers indenter, and at least 1/2 times the long diagonal when using a Knoop indenter. (b) No minimum hardness for metal of equal or greater thickness.

The above approach would also apply in determining which scale should be used to measure the hardness of a case of known approximate depth and hardness.

Influence of Test-Area Width. In addition to the limitation of indentation depth for a workpiece of given thickness and hardness, there is a limiting factor on the minimum width of material. If the indentation is placed too close to the edge of a workpiece, the edge will deform outward and the Rockwell hardness number will be decreased accordingly.

Experience has shown that the distance from the center of the indentation to the edge of the workpiece must be at least $2\frac{1}{2}$ times the diameter of the indentation to ensure an accurate test. Therefore, the width of a narrow test area must be at least five indentation diameters when the indentation is placed in the center.

Limitations of Rockwell Scales. The potential range of each Rockwell scale can be determined readily from the dial-gage divisions on the tester: the black scale (for diamond indenter) on all regular hardness-tester dial gages is numbered from 0 to 100, with 100 corresponding to the "set" position; the red scale (for ball indenters) is numbered from 0 to 130, with 130 being the "set" position. On the superficial hardness tester, the dial gage has only one set of divisions, numbered from 0 to 100.

Use of the diamond indenter when readings fall below 20 is not recommended, since there is loss of sensitivity when indenting this far down the conical section of the indenter. Brale indenters are not calibrated below values of 20, and if used on soft materials there is no assurance that there will be the usual degree of agreement in results when replacing the indenters. Another scale should be used—for example, the B scale.

Support for Workpiece. A fundamental requirement of the Rockwell hardness test is that the surface of the workpiece being tested be approximately normal to the indenter and that the workpiece must not move or slip in the slightest degree as the major load is applied. The depth of indentation is measured by the movement of the plunger rod holding the indenter; therefore, any slipping or moving of the workpiece will be followed by the plunger rod and the motion transferred to the dial gage, causing an error to be introduced into the hardness test. As one point of hardness represents a depth of only 0.00008 in., a movement of only 0.001 in. could cause an error of over 10 Rockwell numbers. The support must be of sufficient rigidity to prevent its permanent deformation in use.

Figure 7 shows five types of anvils that, collectively, can accommodate a fairly broad range of workpiece shapes.

VICKERS HARDNESS TESTING

The Vickers hardness test follows the Brinell principle, in that an indenter of definite shape is pressed into the material to be tested, the load removed, the diagonals of the resulting indentation measured, and the hardness number calculated by dividing the load by the surface area of indentation.

The indenter is made of diamond, and is in the form of a square-base pyramid having an angle of 136° between faces (Fig. 8). This indenter thus has angle across corners, or so-called edge angle, of 148° 6' 42.5". The facets are highly polished and free from surface imperfections, and the point

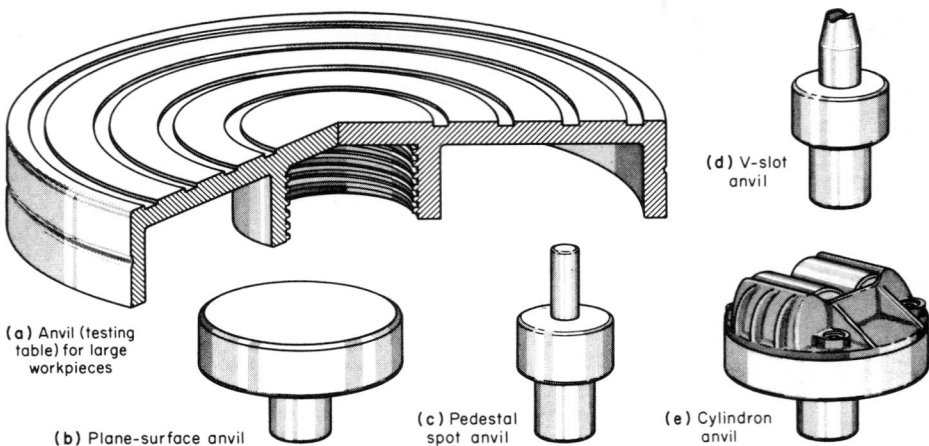

(d) V-slot anvil

(a) Anvil (testing table) for large workpieces

(b) Plane-surface anvil

(c) Pedestal spot anvil

(e) Cylindron anvil

Fig. 7. Several commonly used types of anvils that are designed to support various shapes of workpieces during hardness testing in a Rockwell tester

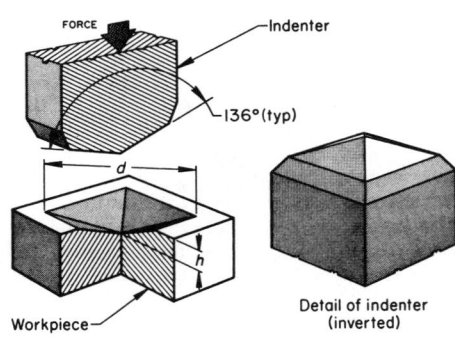

FORCE — Indenter

136° (typ)

d

h

Workpiece

Detail of indenter (inverted)

Fig. 8. Schematic representation of the square-base pyramidal diamond indenter used in a Vickers hardness tester, and of the resulting indentation in the workpiece

is sharp. The loads applied vary from 1 to 120 kg; the standard loads are 5, 10, 20, 30, 50, 100 and 120 kg. For most hardness testing, 50 kg is maximum.

With the Vickers indenter, the depth of indentation is about one-seventh of the diagonal length of the indentation. For certain types of investigation, there are advantages to such a shape. The Vickers hardness number (HV) is the ratio of the load applied to the indenter to the surface area of the indentation. By formula:

$$HV = 2P \sin(\theta/2)/d^2$$

where P is the applied load, in kilograms, d is the mean diagonal of the indentation, in millimetres; and θ is the angle between opposite faces of the diamond indenter (136°).

Equipment for determining the Vickers hardness number should be designed to apply the load without impact, and friction should be reduced to a minimum. The actual load on the indenter should be correct to less than 1% and the load should be applied slowly, because the Vickers is a static test. Some standards require that the full load be maintained for 10 to 15 s.

To obtain the greatest accuracy in hardness testing, the applied load should be as large as possible, consistent with the dimensions of the workpiece. Loads of more than 50 kg are likely to fracture the diamond, especially when used on hard materials.

The accuracy of the micrometer microscope should be checked against a stage micrometer, which consists of ruled lines, usually 0.1 mm apart, that have been checked against certified length standards. The average length of the two diagonals is used in determining the hardness value.

The corners of the indentation provide indicators of the length of the diagonals. The area must be calculated from the average of readings of both diagonals. The indentations are usually measured under vertical illumination with a magnification of about 125 diameters.

The included angle of the diamond indenter should be 136° with a tolerance of less than ±0.50°, which is readily obtainable with modern diamond-grinding equipment. This would mean an error of less than 1% in the hardness number. The indenters must be carefully controlled during manufacture so that in use the indentations produced will be symmetrical. Tables are available for converting the values of the diagonals of indentation in millimetres to the Vickers hardness number.

Several types of hardness testers have proved acceptable for making the Vickers test in accordance with the above requirements. One type is illustrated in Fig. 9. This hardness tester, which has a mainframe section that carries the stage and a starting handle having a 20-to-1 ratio, applies the load through a thrust rod to a tube, which is free to reciprocate vertically, and which carries the Vickers indenter at its lower end. Attached to the mainframe is a smaller frame that contains the control mechanism. The plunger reciprocates vertically under the influence of a rotating cam, its purpose being to apply and release the test load. The cam is mounted on a drum, and when the starting handle has been depressed, the cam is rotated by a weight attached by a flexible wire, the speed of rotation being controlled by a piston and an oil-filled dashpot. The rate of displacement of the oil is regulated by an adjustable control valve. The plunger carries a rubber pad at its upper end, which engages with a cone mounted in the beam, thereby ensuring a very slow and diminishing rate of application for the last portion of the loading cycle. Because the cam both lowers and raises the plunger, errors due to inertia and premature load removal are eliminated.

The microscope is usually mounted on a hinged bracket and may be moved to position over the indentation after the workpiece has been lowered sufficiently to clear the microscope. A knife-edge type of micrometer ocular is provided, and the indentations are read to knife-edges. The readings are taken from a digital counter mounted on the microscope. Tables for converting digital readings to Vickers hardness numbers are supplied. The micrometer ocular may be rotated through 90° so that each diagonal may be measured.

In use, the workpiece is placed on the stage, which is raised by a handwheel on the side of the hardness tester until the workpiece nearly touches the diamond indenter. The load is applied by tripping the starting handle. The time taken in the application and duration of the load may be adjusted by the oil control valve in the dashpot within a range of 10 to 30 s minimum.

If the workpiece has not been elevated sufficiently for the testing load to be applied satisfactorily, a warning is given by an automatically actuated buzzer. A foot pedal will ready the hardness tester for the next test after a test cycle is completed. The stage may be fitted with a V-block for supporting cylindrical work.

If routine hardness testing is to be carried out, a sliding table may be attached to the stage and the microscope may be mounted on an auxiliary bracket on the right-hand side of the tester so that hardness testing may be continuous without the need for winding the stage up or down.

SCLEROSCOPE HARDNESS TESTING

The Scleroscope hardness test is essentially a dynamic indentation hardness test, wherein a diamond-tipped hammer is dropped from a fixed height onto the surface of the material being tested. The height of rebound of the hammer is a measure of the hardness of the metal. The Scleroscope scale consists of units that are determined by dividing the average rebound of the hammer

from a quenched (to maximum hardness) and untempered water-hardening tool steel into 100 units. The scale is continued above 100 to permit testing of materials having hardnesses greater than that of fully hardened tool steel.

Testers. Two types of Scleroscope hardness testers are shown in Fig. 10. The Model C Scleroscope consists of a vertically disposed barrel containing a precision-bore glass tube. A base-mounted version of a Model C Scleroscope is shown in Fig. 10(a). The scale is graduated from 0 to 140. It is set behind and is visible through the glass tube. Hardness is read from the vertical scale, usually with the aid of the reading glass attached to the tester. A pneumatic actuating head, affixed to the top of the barrel, is manually operated by a rubber bulb and tube. The hammer drops and rebounds with the glass tube.

The Model D Scleroscope hardness tester (Fig. 10b) is a dial-reading tester. The tester consists of a vertically disposed barrel that contains a clutch to arrest the hammer at maximum height of rebound. This is made possible because of the short rebound height. The hammer is longer and heavier than the hammer in the Model C Scleroscope and develops the same striking energy although dropping a shorter distance.

Both models of the Scleroscope hardness tester may be mounted on various types of bases. The C-frame base, which rests on three points and is for bench use in hardness testing small workpieces, has a capacity about 3 in. high by 2$\frac{1}{2}$ in. deep. A swing arm and post is also for bench use, but has height and reach capacities of 9 and 14 in., respectively. Another type of base is used for mounting the Scleroscope hardness tester on rolls and other cylindrical objects having a minimum diameter of 2$\frac{1}{2}$ in., or on flat, horizontal surfaces having a minimum dimension of 3 by 5 in. The Model C Scleroscope hardness tester is commonly used unmounted. However, when the hardness tester is unmounted, the workpiece should have a minimum weight of 5 lb. The Model D Scleroscope hardness tester should not be used unmounted.

Workpiece Surface-Finish Requirements. As with other metallurgical hardness testers, certain surface-finish requirements on the workpiece must be met for Scleroscope hardness testing to make an accurate hardness determination. An excessively coarse surface finish will yield erratic readings. Hence, when necessary, the surface of the workpiece should be filed, machined, ground or polished to permit accurate, consistent readings to be obtained.

Limitations on Workpiece and Case Thickness. Case-hardened steels having cases as thin as 0.010 in. can be accurately hardness tested provided the core hardness is no less than 30 Scleroscope. Softer cores require a minimum case thickness of 0.015 in. for accurate results.

Thin strip or sheet may be tested, with some limitations, but only when the Scleroscope hardness tester is mounted in the clamping stand. Ideally, the sheet should be flat and without undulation. If the sheet material is bowed, the concave side should be placed up to preclude any possibility of erroneous readings due to spring effect. The minimum thicknesses of sheet in various categories that may be hardness tested are as follows:

Hardened steel	0.005 in.
Cold finished steel strip	0.010
Annealed brass strip	0.015
Half-hard brass strip	0.010

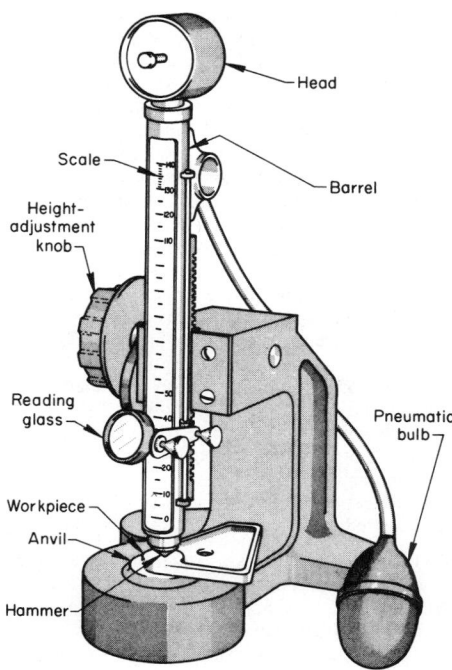

(a) Model C (vertical scale) Scleroscope hardness tester

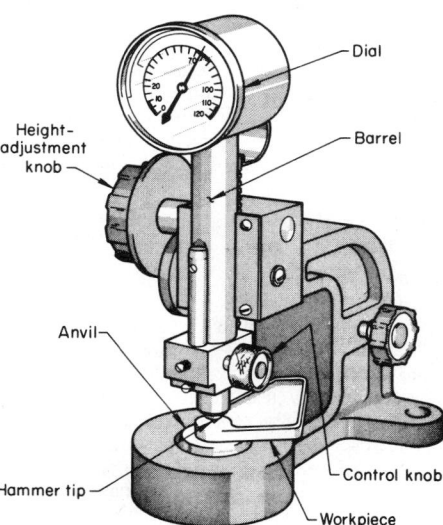

(b) Model D (dial reading) Scleroscope hardness tester

Fig. 10. Principal components of two types of base-mounted Scleroscope hardness testers

Test Procedure. To perform a hardness test with either the Model C or the Model D Scleroscope hardness tester, the tester should be held or set in a vertical position, with the bottom of the barrel in firm contact with the workpiece. The hammer is raised to the elevated position and then allowed to fall and strike the surface of the workpiece. The height of rebound is then measured, which indicates the hardness. When using the Model C Scleroscope hardness tester, the hammer is raised to the elevated position by squeezing the pneumatic bulb. The hammer is released by again squeezing the bulb. When using the Model D Scleroscope hardness tester, the hammer is raised to the elevated position by turning the knurled control knob clockwise until a definite stop is reached. The hammer is allowed to

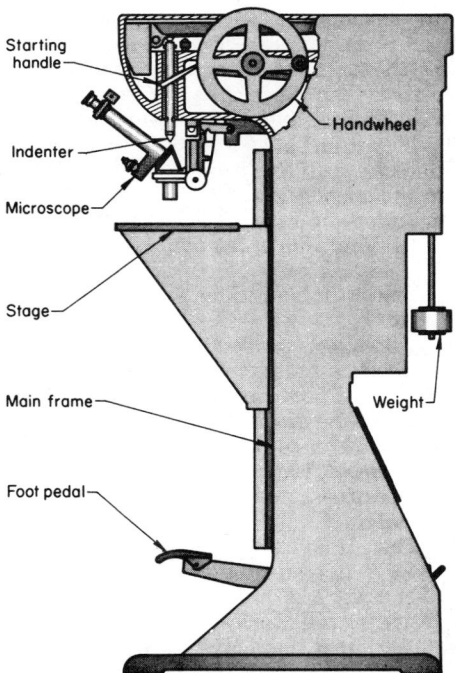

Fig. 9. Principal components of one type of Vickers hardness tester

strike the workpiece by releasing the control knob. The reading is recorded on the dial.

Spacing of Indentations. Indentations should be at least 0.50 mm (0.020 in.) apart and only one at the same spot. Flat workpieces with parallel surfaces may be hardness tested within 1/4 in. (6 mm) of the edge when properly clamped.

Taking the Readings. Experience is necessary to interpret the hardness readings accurately on a Model C Scleroscope hardness tester. Thin materials or those weighing less than 5 lb must be securely clamped to absorb the inertia of the hammer. The sound of the impact is an indication of the effectiveness of the clamping: a dull thud indicates that the workpiece has been clamped solid, whereas a hollow ringing sound indicates that the workpiece is not tightly clamped or is warped and not properly supported. Five hardness determinations should be made and their average taken as representative of the hardness of a particular workpiece.

Advantages of the Scleroscope hardness test are summarized as follows:

1. Tests can be made very rapidly—over 1000 tests per hour are possible.
2. Operation is simple, and does not require highly skilled technicians.
3. The Model C Scleroscope tester is portable and may be used unmounted for hardness testing workpieces of unlimited size—rolls, large dies and machine-tool ways.
4. The Scleroscope hardness test is a nonmarring test; no crater is left, and only in the most unusual instances would the tiny hammer mark be objectionable on a finished workpiece.
5. A single scale accommodates the entire hardness range from the softest to the hardest metals.

Limitations of the Scleroscope hardness test are summarized as follows:

1. The hardness tester must be in a vertical position or the free fall of the hammer will be impeded and result in erratic readings.
2. Scleroscope hardness tests are more sensitive to variations in surface conditions than some other hardness tests.
3. Because readings taken with the Model C Scleroscope hardness tester are those observed from the maximum rebound of the hammer on the first bounce, even the most experienced operators may disagree among themselves by one or two points in the reading.

MICROHARDNESS TESTING

The term "microhardness" usually refers to indentation hardness tests made with loads not exceeding 1 kg (1000 g). Such hardness tests have been made with a load as light as 1 g, although the majority of microhardness tests are made with loads of 100 to 500 g. In general, the term is related to the size of the indentation rather than to the load applied.

Fields of Application. Microhardness testing is capable of providing information regarding the hardness characteristics of materials that cannot be obtained with hardness tests such as the Brinell, Rockwell or Scleroscope.

Because of the required degree of precision for both equipment and operation, microhardness testing is usually, although not necessarily, performed in a laboratory. Such a laboratory, however, is often a process-control laboratory and may

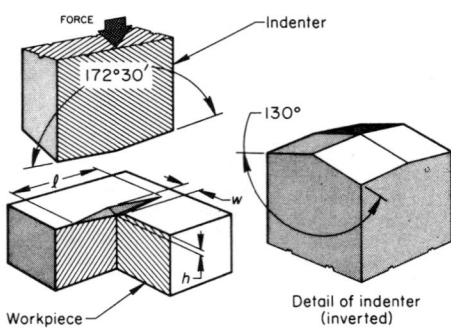

Fig. 11. Schematic representation of a pyramidal Knoop indenter, and of the resulting indentation in the workpiece

be located close to production operations. Microhardness testing is recognized as a valuable method for controlling numerous production operations in addition to its use in research applications. Specific fields of application of microhardness testing include:

1. Measuring hardness of precision workpieces that are too small to be measured by the more common hardness-testing methods
2. Measuring hardness of product forms such as foil or wire that are too thin or too small in diameter to be measured by the more convenient methods
3. Monitoring of carburizing or nitriding operations, which is usually accomplished by hardness surveys taken on cross sections of test pieces that accompanied the workpieces through production operations
4. Measuring hardness of individual microconstituents
5. Measuring hardness close to edges, thus detecting undesirable surface conditions such as grinding burn and decarburization
6. Measuring hardness of surface layers such as plating or bonded layers.

Indenters. Microhardness testing is peformed with either the Knoop or the Vickers indenter. The Knoop indenter is the more widely used in the United States; the Vickers indenter is the more widely used in Europe.

Knoop indentation testing is performed with a diamond ground to pyramidal form that produces a diamond-shape indentation having an approximate ratio between long and short diagonals of 7 to 1 (Fig. 11). The pyramidal shape employed has an included longitudinal angle of 172° 30′ and an included transverse angle of 130°. The depth of indentation is about one thirtieth of its length. Because of the shape of the indenter, indentations of accurately measurable lengths are obtained with light loads.

The Knoop hardness number (HK) is the ratio of the load applied to the indenter to the unrecovered projected area of indentation. By formula:

$$HK = P/A = P/Cl^2$$

where P is the applied load, in kilograms; A is the unrecovered projected area of indentation, in square millimetres; l is the measured length of the long diagonal, in millimetres; and C is 0.07028, a constant of the indenter relating projected area of the indentation to the square of the length of the long diagonal.

For details of the Vickers indenter, the reader is referred to Fig. 8.

Figure 12 presents a comparison of the indentations made by the Knoop and Vickers indenters. Each has some advantages over the other. For example, the Vickers indenter penetrates about twice as far into the workpiece as does the Knoop indenter, and the diagonal of the Vickers indentation is about one-third of the total length of the Knoop indentation. Therefore, the Vickers indenter is less sensitive to minute differences in surface conditions than is the Knoop indenter. However, the Vickers indentation, because of the shorter diagonal, is more sensitive to errors in measurement than is the Knoop indentation.

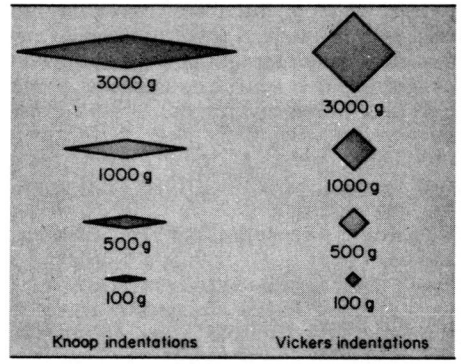

Fig. 12. Comparison of indentations made by Knoop and Vickers indenters in the same work metal and at the same loads

Microhardness Testers. Several types of microhardness testers are available. The most accurate operate through the direct application of load by dead weight, or by weights and lever.

The Tukon tester is widely used for microhardness testing. Several different designs of this microhardness tester are available; they vary mainly in load range, but all can accommodate both Knoop and Vickers indenters.

The Tukon microhardness tester shown in Fig. 13 has a load range of 1 to 1000 g. Loads are applied by dead weight. The microscope is furnished with three objective lenses having magnifications of about 150, 300 and 600 diameters.

Sources of tester error include inaccuracy in loading, vibration, rate of load application, duration of contact period, and impact. To limit the shock that can occur when the operator removes the load (this generally has an adverse effect on indentations made with loads below 500 g), an automatic test cycle is built into the Tukon microhardness tester. With this automatic test cycle, the load is applied at a constant rate, maintained in the work for 18 s, and smoothly removed. Thus, the operator does not need to touch the tester while the load is being applied and removed. The design of microhardness testers will vary from one type to another, but it is essential to remove the applied load without touching the tester if clearcut indentations are to be obtained.

A movable stage to support the workpiece is an essential component of a microhardness tester. In many applications the indentation must be in a selected area, usually limited to a few thousandths of a square millimetre. In testing with the type of Tukon microhardness tester shown in Fig. 13, first the required area is located by looking through the microscope and moving the mechanical stage until the desired location is centered within the optical field of view. The stage is then indexed under the indenter, and the automatic indentation cycle is initiated by tripping the handle. After the cycle is completed, signaled by a telltale light, the stage is again indexed back under the objective for indentation measurement.

Optical equipment used in microhardness testers for measuring the indentation must focus on both ends of the indentation at the same time, as well as be rigid and free from vibration. Lighting is also important. Complete specifications of measurement, including the mode of illumination, are necessary in microhardness-testing techniques. Polarized light, for instance, results in larger measurements than does unpolarized light. Apparently, this is caused by the reversal of the diffraction pattern; that is, the indentation appears brighter than the background. When test data are recorded, it is recommended that both the magnification and the type of illumination used be reported.

Measuring the Indentation. In measuring the indentation, the proper illumination to obtain optimum resolution is essential, and the appropriate objective lens should be selected. In operation, the ends of the indentation diagonals should be brought into sharp focus. With the Knoop indenter, one leg of the long diagonal should not be more than 20% longer than the other. If this is not apparent or if the ends of the diagonal are not in focus, the surface of the workpiece should be checked to make sure it is normal to the axis of the indenter. With the Vickers indenter, both diagonals should be measured and the average used for calculating the Vickers hardness number (HV).

Fig. 13. Principal components of a Tukon microhardness tester

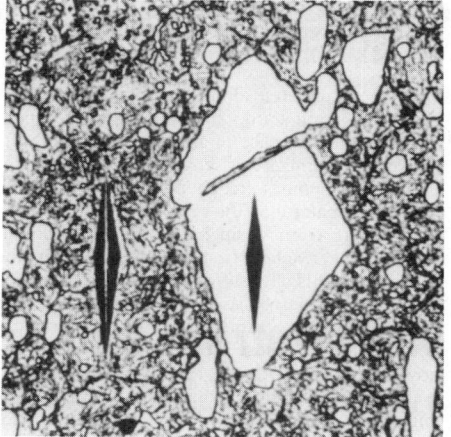

Indentation in a particle of chromium-vanadium carbide (white constituent) showed a value of 1930 HK, whereas the indentation in the matrix (darker constituent) was 801 HK. Both indentations were made with a 50-g load. (Specimen was polished, and etched in Vilella's reagent; shown at 1000×.)

Fig. 14. Comparison of Knoop indentations in two microconstituents of quenched-and-tempered D2 tool steel

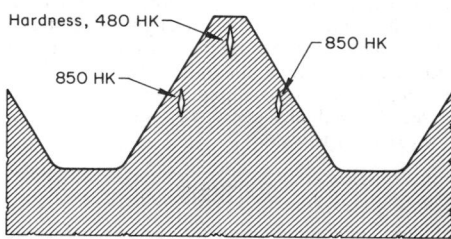

Fig. 15. Cross section of a tap tooth showing hardness variations caused by overheating during grinding

Preparing and Holding the Specimen. Regardless of whether the metal being tested for microhardness is an actual workpiece or a representative specimen, surface finish is of prime importance.

To permit accurate measurement of the length of the Knoop indentation or diagonals of the Vickers indentation, the indentation must be clearly defined. In general, as the test load decreases, the surface-finish requirements become more stringent. When the load is 100 g or less, a metallographic finish is recommended.

Specific Applications of Microhardness Testing

Microhardness testing is used extensively in research and for controlling the quality of manufactured products, as well as for solving shop problems.

Testing of small workpieces is an important use of microhardness testing. Many manufactured products, notably in the instrument and electronics industries, are too small to be tested for hardness by the more conventional methods. Many such workpieces can be tested without impairing their usefulness, generally by means of various types of holding and clamping fixtures.

Microhardness testing is also applied to product forms that cannot be tested by other means. Thin foils and small-diameter wires are typical examples.

Monitoring of Surface-Hardening Operations. Microhardness testing is the best method in present use for accurately determining case depth and certain case conditions of carburized or nitrided workpieces, using the hardness-survey procedure. In most instances this is accomplished by use of test coupons that have accompanied the actual workpieces through the heat treating operation. The coupons are then sectioned and usually mounted for testing. To ensure accurate readings close to the edge of the cross section, the 100-g load is most often used, although a 500-g load is sometimes preferred. If the 100-g load is used, a metallographic finish is essential.

Readings are taken at pre-established intervals (commonly, 0.004 or 0.005 in.), usually beginning at least 0.001 in. from the edge of the workpiece.

Measuring Hardness of Microconstituents. A great deal can be learned about metals and their potential properties (for example, their resistance to wear) by knowing the actual hardnesses of their various microconstituents. Notable examples are the highly alloyed tool steels. Figure 14 shows a micrograph of polished and etched D2 tool steel. Knoop indentations taken on the matrix (darker constituent) and on the particles of complex alloy carbide (white) show an obvious variation in size. In this instance the tests were made with a load of 50 g. The indentation on the matrix was 801 HK, whereas the carbide showed a value of 1930 HK. Actual Rockwell hardness on the C scale was 64 HRC (822 HK by conversion). Therefore, the Rockwell C scale did not register the true conditions.

Shop problems are often solved with the help of microhardness testing. Some of the most notable examples involve cutting edges of tools.

Cutting tools made from high speed steels, even though they have been correctly heat treated, are frequently damaged in grinding. Taps are among the most vulnerable, because the crests of teeth are thin and thus are likely to become overheated during grinding.

In one instance, taps were failing prematurely from dulling of the tap teeth. Hardness measurements taken at various locations (where it was possible to measure) on the taps showed consistent values of 65 HRC, which was entirely acceptable. However, when one of the taps was sectioned and a tooth area was examined with a microhardness tester, results were quite different. Measurements taken at various locations with a Knoop indenter and a 100-g load showed readings of about 850 HK (by conversion, about 65 HRC) in the center of the tooth and to within 0.003 to 0.005 in. of the edge (see Fig. 15). At the very edge of the tooth crest, however, readings were as low as 480 HK (see Fig. 15), which converts to approximately 46 HRC. Variation in size of the indentations can be clearly seen in Fig. 15. Obviously, the softened condition was a result of abusive grinding practice and rendered the tap useless.

SELECTED REFERENCES

Hardness Testing, by E. R. Petty: in *Techniques of Metals Research*, Vol 5, Part 2, edited by R. F. Bunshah, Wiley-Interscience, 1971, p 157

Hardness Measurements of Metals and Alloys, 2nd Ed., by H. O'Niel: Chapman and Hall, London, 1967

The Hardness of Metals, by B. Tabor: Oxford University Press, London, 1951

The Science of Hardness Testing and Its Research Applications, edited by J. H. Westbrook and H. Conrad: ASM, Metals Park, OH, 1973

Tension and Compression Testing

Tension Testing

By George E. Dieter, University of Maryland

THE TENSION TEST is the test most commonly used to evaluate the mechanical properties of materials (Ref 1). Its chief use is the determination of properties related to the elastic design of machines and structures. In addition, the tension test gives information on a material's plasticity and fracture. The chief advantages of the tension test are that the stress state is well established, that the test has been carefully standardized (Ref 2) and that the test is relatively easy and inexpensive to carry out. This article will not detail this test technique, because it is well covered by standard methods (Ref 2); instead, the interpretation and limitations of the test results will be discussed.

ENGINEERING STRESS-STRAIN CURVE

In the conventional engineering tension test, stress is defined by the applied load divided by the original cross-sectional area of the specimen. Engineering strain, e, is the change in length divided by the initial length:

$$e = \frac{L - L_0}{L_0} = \frac{\Delta L}{L_0} \qquad \text{(Eq 1)}$$

In the elastic region of the stress-strain curve (Fig. 1), stress is linearly related to strain, $\sigma = Ee$, where E is the elastic modulus. As long as the specimen is loaded within the elastic region, the strain is totally recoverable and the specimen will return to its original length when the load is relaxed to zero. However, when the load exceeds a value corresponding to the yield stress, the specimen undergoes gross plastic deformation and is permanently deformed when the load returns to zero. The stress to produce continued plastic deformation increases with increasing strain—the metal strain hardens. To a good engineering approximation, the volume remains constant during plastic deformation, $AL = A_0L_0$, and, as the specimen elongates, it decreases uniformly in cross-sectional area along its gage length. Initially the strain hardening more than compensates for this decrease in area, and the engineering stress continues to rise with increasing strain. Eventually a point is reached where the decrease in area is greater than the increase in deformation load arising from strain hardening. This condition will be reached first at some point in the specimen that is slightly weaker than the rest. All further plastic deformation is concentrated in this region, and the specimen begins to neck or thin down locally. Because the cross-sectional area is now decreasing far more rapidly than the deformation load is being increased by strain hardening, the engineering stress continues to decrease until fracture occurs.

The maximum in the engineering stress-strain curve corresponds to the ultimate tensile strength, σ_u. The strain at maximum load, up to which point the cross-sectional area decreases uniformly along the gage length as the specimen elongates, is the uniform elongation, e_u. For stretching-type forming operations, such as stretch forming of aircraft components or forming of automobile fenders, local necking determines the formability limit, and in such applications uniform elongation can be an important measure of ductility. In many metals, the engineering stress-strain curve is relatively flat in the vicinity of necking, and it may not be possible to establish the maximum load without ambiguity. In these cases, the method suggested by Nelson and Winlock (Ref 3) is useful.

DUCTILITY

The conventional measures of ductility that are obtained from the tension test are the engineering strain at fracture, e_f (usually expressed as a percentage elongation), and the reduction of area at fracture, RA (also usually expressed as a percentage):

$$e_f = \frac{L_f - L_0}{L_0} \qquad \text{(Eq 2)}$$

$$RA = \frac{A_0 - A_f}{A_0} \qquad \text{(Eq 3)}$$

Because an appreciable fraction of the deformation will be concentrated in the necked region of the specimen, the value of e_f will depend on the gage length L_0 over which the measurement was taken. The smaller the gage length, the greater the contribution from the neck and the higher the value of e_f.

To eliminate this difficulty and to provide a measure of ductility that correlates with forming operations in which the gage length is very short, it is possible to determine the zero-gage-length elongation, e_0. From the constancy-of-volume relationship for plastic deformation, $AL = A_0L_0$:

$$\frac{L}{L_0} = \frac{A_0}{A} = \frac{1}{1 - RA}$$

$$e_0 = \frac{L - L_0}{L_0} = \frac{A_0}{A} - 1$$

$$= \frac{1}{1 - RA} - 1 = \frac{RA}{1 - RA} \qquad \text{(Eq 4)}$$

Thus, the zero-gage-length elongation may be determined directly from the reduction of area at fracture or from the change in length of grid marks near the actual fracture. The data presented in Table 1 show how basing a comparison of the formability of aluminum alloys on the elongation in a 2-in. (50.8-mm) gage length would lead to erroneous conclusions for forming operations where local ductility determines the forming limit (Ref 4).

TRUE-STRESS/TRUE-STRAIN CURVE

The necking phenomenon which occurs in the tension test clouds the usefulness of the engineering stress-strain curve beyond the maximum load. The falloff in stress which occurs beyond P_{max} is artificial and occurs only because the stress continues to be calculated on the basis of the original cross-sectional area, A_0, when in fact the area at the necked region is now much smaller than A_0. If the *true stress*, based on the actual cross-sectional area of the specimen, is used, the stress-strain curve increases continuously up to fracture. Then, if strain is expressed as *true strain*, we have a plot called the true-stress/true-strain curve (Fig. 2).

Note that this curve continues to rise beyond necking all the way to fracture. However, once necking occurs, the constraints produced by the nondeforming region outside the neck produce a state of triaxial stress in the neck. Thus, the average stress required to cause flow from maximum load to fracture is higher than would be required if only a uniaxial stress were present. Bridgman (Ref 5) has made a mathematical analysis of the stresses in the neck that permits correction of the true-stress/true-strain curve for the

Table 1. Comparison of several measures of ductility for two aluminum alloys (Ref 4)

Alloy	e_0	$e_{2.0}$	e_u
24S-O (2024-O)	1.22	0.18	0.16
24S-T (2024-T)	0.64	0.18	0.15
75S-O (7075-O)	1.55	0.16	0.11
75S-T (7075-T)	0.44	0.11	0.09

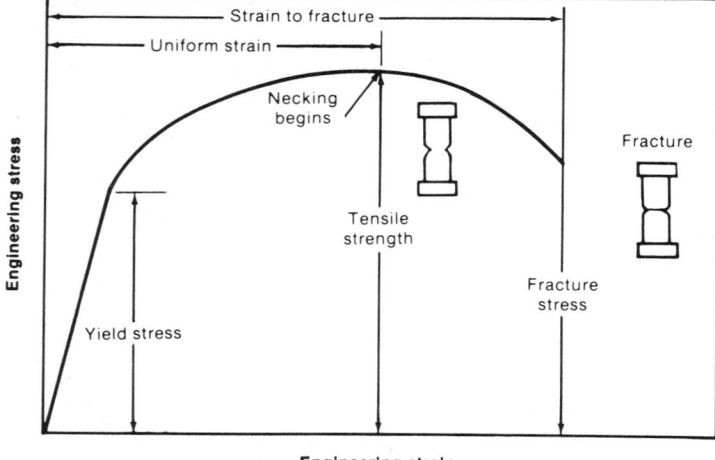

Fig. 1. Engineering stress-strain curve

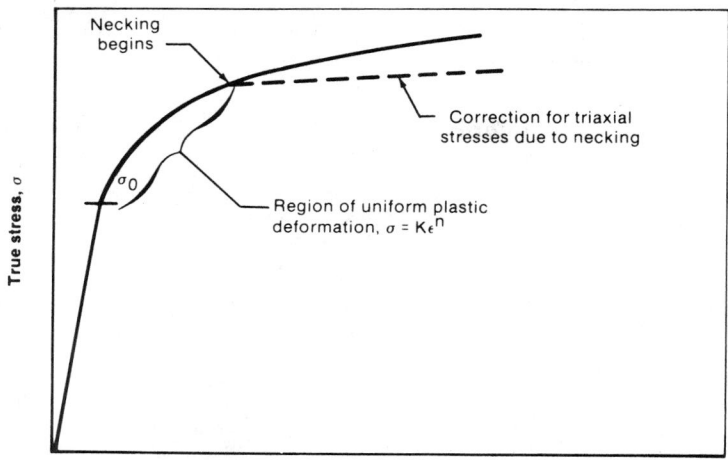

Fig. 2. True-stress/true-strain curve (flow curve)

existence of triaxial stresses. More recent studies have utilized finite-element analysis (Ref 6).

The region from yielding to necking is one of uniform plastic deformation in which the specimen gage length increases and the diameter decreases uniformly along the gage length. In this region the true-stress/true-strain curves for many ductile metals can be expressed by a power-law relation:

$$\sigma = K\epsilon^n \qquad \text{(Eq 5)}$$

where K is the strength coefficient, defined by the value of true stress at $\epsilon = 1.0$, and n is the strain-hardening exponent. Equation 5 is valid if a plot of σ versus ϵ is a straight line on log-log coordinates. The strain-hardening exponent is the slope of this line. Thus

$$n = \frac{d(\log \sigma)}{d(\log \epsilon)} = \frac{d(\ln \sigma)}{d(\ln \epsilon)} = \frac{\epsilon \, d \sigma}{\sigma \, d \epsilon} \qquad \text{(Eq 6)}$$

The strain-hardening exponent may have values from n = 0 (a perfectly plastic solid) to n = 1 (an elastic solid). Values of n for most metals are between 0.05 and 0.50.

An increase in strain rate increases the stress to produce plastic deformation (the flow stress). The effect is modest for cold working, but is quite significant for hot working. The dependence of flow stress on strain rate, at constant strain and temperature, is given by

$$\sigma = C(\dot{\epsilon})^m \big|_{\epsilon,T} \qquad \text{(Eq 7)}$$

where m is the strain-rate sensitivity. The exponent m can be evaluated from the slope of a plot of log σ versus log $\dot{\epsilon}$, or it can be obtained from rate-change tests in which the change in flow stress caused by a step change in $\dot{\epsilon}$ is measured:

$$m = \left(\frac{d \log \sigma}{d \log \dot{\epsilon}}\right) \approx \left(\frac{\Delta \log \sigma}{\Delta \log \dot{\epsilon}}\right)_{\epsilon,T}$$

$$= \frac{\log \sigma_2 - \log \sigma_1}{\log \dot{\epsilon}_2 - \log \dot{\epsilon}_1} = \frac{\log \dfrac{\sigma_2}{\sigma_1}}{\log \dfrac{\dot{\epsilon}_2}{\dot{\epsilon}_1}} \qquad \text{(Eq 8)}$$

ANALYSIS OF TENSILE INSTABILITY

The development of a necked region in a specimen loaded in uniaxial tension represents a plastic instability. Because this disturbs the simple analysis of the tension test and limits the engineering usefulness of the test, it has become the subject of considerable study (Ref 7 to 9). A practical application of the ideas presented below is given in the work of Ghosh (Ref 10), who developed a numerical analysis for predicting the shape of the engineering strain-stress curve beyond maximum load as a function of strain hardening, strain-rate hardening, and plastic anisotropy properties of the metal.

Consider a tensile specimen loaded to a value P. At any point a distance L along the specimen, the cross-sectional area is A and $P = \sigma A$. Since P does not vary along the length of the specimen, and $\sigma = f(\epsilon, \dot{\epsilon})$,

$$\frac{dP}{dL} = 0 = A\left\{\left(\frac{d\sigma}{d\epsilon}\right)_{\dot{\epsilon}} \frac{d\epsilon}{dL}\right.$$

$$\left. + \left(\frac{d\sigma}{d\dot{\epsilon}}\right)_{\epsilon} \frac{d\dot{\epsilon}}{dL}\right\} + \sigma \frac{dA}{dL} \qquad \text{(Eq 9)}$$

Because the volume of the specimen remains constant, the true strain can be written as

$$d\epsilon = \frac{dL}{L} = -\frac{dA}{A} \qquad \text{(Eq 10)}$$

and

$$\frac{d\epsilon}{dL} = -\frac{1}{A} \frac{dA}{dL} \qquad \text{(Eq 11)}$$

Also, from Eq 10, we can express the strain-rate $\dot{\epsilon}$ by

$$d\dot{\epsilon} = \frac{d\epsilon}{dt} = -\frac{1}{A} \frac{dA}{dt} = -\frac{\dot{A}}{A} \qquad \text{(Eq 12)}$$

so that

$$\frac{d\dot{\epsilon}}{dL} = -\frac{1}{A} \frac{d\dot{A}}{dL} + \frac{\dot{A}}{A^2} \frac{dA}{dL} \qquad \text{(Eq 13)}$$

The material parameters which are important to the necking process are the dimensionless work-hardening coefficient:

$$\gamma = \frac{1}{\sigma} \frac{d\sigma}{d\epsilon} \qquad \text{(Eq 14)}$$

and the strain-rate sensitivity:

$$m = \left(\frac{d \ln \sigma}{d \ln \dot{\epsilon}}\right)_\epsilon = \frac{\dot{\epsilon}}{\sigma}\left(\frac{d\sigma}{d\dot{\epsilon}}\right)_\epsilon \qquad \text{(Eq 15)}$$

When Eq 12 and 13 are substituted into Eq 9, and the definitions for γ and m are added through Eq 14 and 15, the result is

$$\frac{dA}{dL}(\sigma - m\sigma - \gamma\sigma) = -\frac{d\dot{A}}{dL} \frac{m\sigma A}{\dot{A}} \qquad \text{(Eq 16)}$$

A final rearrangement gives

$$\frac{\dfrac{1}{\dot{A}} \dfrac{d\dot{A}}{dL}}{\dfrac{1}{A} \dfrac{dA}{dL}} = \frac{\dfrac{d}{dL}(\ln \dot{A})}{\dfrac{d}{dL}(\ln A)} = \frac{m + \gamma - 1}{m} \qquad \text{(Eq 17)}$$

This equation describes the rate of change of area with length, and gives the criterion for the onset of necking.

Any real tension specimen will have variations in cross-sectional area along its length. These can arise from an intentional taper, from machining errors or from heterogeneities of structure which lead to weaker cross sections. Deformation becomes unstable when the smallest cross section of the specimen shrinks faster than the rest. This occurs when $d\dot{A}/dA > 0$. Deformation will be uniform and stable when $dA/dA < 0$. Since A/A is negative in tension, stable deformation in tension occurs when $d\dot{A}/dA \geq 0$. Therefore, from Eq 17, the condition for stable, uniform tensile deformation is

$$\gamma + m \geq 1 \qquad \text{(Eq 18)}$$

Necking is involved with the interplay between the applied stress and the flow resistance of the material. As the specimen elongates under a given load the area decreases and the stress increases. If necking is not to occur, the material's strength must increase through strain hardening (γ) and strain-rate hardening (m).

For room-temperature deformation, $m \rightarrow 0$ and the instability criterion reduces to $\gamma \geq 1$. Thus, stable tensile deformation occurs for

$$\frac{d\sigma}{d\epsilon} \geq \sigma \qquad \text{(Eq 19)}$$

If the true-stress/true-strain curve is given by $\sigma = K\epsilon^n$, then

$$\frac{d\sigma}{d\epsilon} = nK\epsilon^{n-1} = \sigma = K\epsilon^n$$

and necking occurs when

$$\epsilon = n \qquad \text{(Eq 20)}$$

Because n in tension rarely exceeds 0.5, we can see that the available uniform strain in the tension test is limited.

ELONGATION MEASUREMENTS IN TENSION TESTING

The measured elongation depends on the gage length or the dimensions of the cross section of the specimen. This is because the total extension consists of two components, the uniform extension up to the point of necking and the localized extension after necking (Fig. 3). The extent of uniform extension will depend on the metallurgical condition of the material (through n) and the

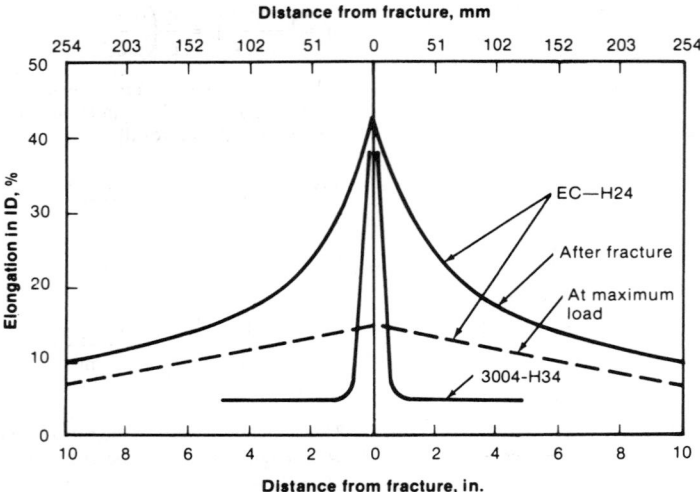

Fig. 3. Local elongation measured at positions away from fracture in tension specimens for two aluminum alloys (Ref 11)

effect of specimen size and shape on the development of a neck.

However,

$$L_f - L_0 = a + e_u L_0 \qquad (Eq\ 21)$$

where a is the local necking extension and $e_u L_0$ is the uniform extension. We then have

$$e_f = \frac{L_f - L_0}{L_0} = \frac{a}{L_0} + e_u \qquad (Eq\ 22)$$

which clearly indicates that the total elongation is a function of the gage length. Numerous attempts to rationalize the strain distribution in the tension test (Ref 12) have been made, dating back to 1850. Following Barba's law (Ref 13), which states that geometrically similar specimens develop geometrically similar necks, it is usually assumed that the local extension at the neck is proportional to the linear dimension of the cross-sectional area, $a = \beta\sqrt{A_0}$, so that the elongation equation becomes

$$e_f = \frac{\beta\sqrt{A_0}}{L_0} + e_u \qquad (Eq\ 23)$$

This equation for elongation, which is usually attributed to Unwin (Ref 14), clearly shows the rationale for the use of fixed ratios of gage length to diameter or gage length to square root of cross-sectional area in specifying tensile-specimen dimensions. It also reinforces the importance of stating the gage length over which the measurement was made when reporting elongation values. In the United States, the standard tensile specimen is 0.505 in. (12.83 mm) in diameter and 2 in. (50.8 mm) in gage length, so $1/D \simeq 4$. However, the testing standards in other countries specify different gage lengths for the measurement of elongation (Table 2).

To compare the ductilities of different metals by elongation measurements, the gage length should be adjusted as a function of the cross-sectional area of the test specimen. However, when flat specimens are cut from sheet or plate primarily to determine whether quality of individual lots meets specifications, it is usual to use fixed gage lengths, because of the lower cost of preparing and testing a large number of such specimens.

Table 2. Dimensional relationships for specimens used in different countries for measurement of elongation

Type of specimen	Dimensional relationship	United Kingdom Before 1962	United Kingdom Current	United States (ASTM)	Germany
Sheet	$L_0/\sqrt{A_0}$	4.0	5.65	4.5	11.3
Round bar	L_0/D_0	3.54	5.0	4.0	10.0

The effect of specimen geometry on total elongation is of particular concern in testing of sheet-metal specimens. Various proposals concerning the effect of ratio of width to thickness can be found in the literature (Ref 15). However, the correlation is more meaningful in terms of specimen cross-sectional area. A study of the effect of specimen geometry on elongation of sheet specimens (Ref 16) showed that, although Unwin's equation was obeyed (within considerable scatter), a simpler equation due to Templin (Ref 17) was possibly in better agreement with the results:

$$e_f = C(A)^b \qquad (Eq\ 24)$$

Analysis of the results showed that the exponent b in Templin's equation depends on both uniform strain and localized fracture strain. Thus it will vary with processing and heat-to-heat differences in the metal, and it cannot be considered to be a real constant of the metal. A general trend was shown to exist between the exponent b in Templin's equation and the logarithm of the ratio of the zero-gage-length (fracture-strain) elongations to the infinite-gage-length (uniform-strain) elongations.

Although (even after 100 years of study) opinions differ in detail concerning the effect of specimen geometry on elongation, there is general agreement concerning the validity and importance of one factor—the ratio $\sqrt{A}/L$. Even here it appears to be better validated for round bars than for rectangular specimens. However, if sufficient data on the influence of specimen size on elongation are not available, the elongation of a specimen of arbitrary size can be estimated by using the concept that a constant elongation is obtained if $\sqrt{A}/L$ is maintained constant, as suggested by Eq 23. Then, at a constant value of elongation, $\sqrt{A_1}/L_1 = \sqrt{A_2}/L_2$, where A and L are the areas and gage lengths of two different specimens, 1 and 2, of the same metal. To predict the elongation in length L_2 on a specimen with area A_2 from measurements on a specimen with area A_1, it is only necessary to adjust the gage length of specimen 1 to conform to $L_1 = L_2\sqrt{A_1/A_2}$. As an example, suppose that sheet $1/8$ in. (3.2 mm) thick is available and it is desired to predict the elongation in 2 in. (50.8 mm) in identical material 0.080 in. (2.03 mm) thick. Using sheet specimens $1/2$ in. (12.7 mm) wide, we would predict that a test specimen with L = $2\sqrt{0.125/0.080}$ = 2.5 in. (63.5 mm) from the $1/8$-in. (3.2-mm) sheet would give the same elongation as a 2-in. (50.8 mm) gage length in sheet 0.080 in. (2.0 mm) thick. The usefulness of this procedure is shown in Fig. 4, where solid lines

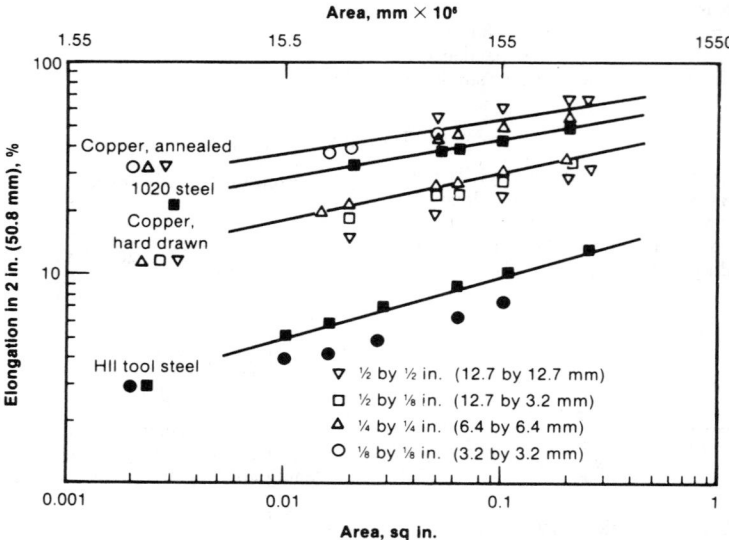

Fig. 4. Calculated variation of elongation in 2 in. (50.8 mm) with specimen cross-sectional area (Ref 16)

are experimental and points indicate predicted elongations for specimens of different areas.

The above discussion indicates that the measures of ductility that are available with the tension test leave a great deal to be desired in providing quantitative values. The main difficulty arises from the necking of the specimen. The occurrence of uniform and localized deformation makes the percentage elongation at fracture of little value as a quantitative measure of ductility, although it is usually required in metallurgical specifications. Reduction of area is a better measure of ductility, but its quantitative use is made difficult by the poorly defined triaxial stress state that is introduced by the formation of a neck.

PLANE-STRAIN TENSION TEST

A special type of tension-test specimen has been designed (Ref 18) to give maximum plastic constraint so as to emphasize the differences in fracture behavior of nominally ductile materials (Fig. 5). The deep grooves in the specimen restrict the deformation to the grooved region. The ratio B/L is large enough so that approximately plane-strain conditions are achieved in the test section (Ref 19). Thus, strain occurs in the thickness and length directions but not in the width direction. The true strain is given by

$$\epsilon = \ln h_0/h \qquad \text{(Eq 25)}$$

where h_0 is the initial thickness of the reduced section and h is the thickness at any time after deformation has begun. However, the ratio L/h is large enough so that there is no notch effect, and thus a specimen designed to the specification in Fig. 5 is an unnotched plane-strain specimen. Thus, the true stress can be determined directly from the load divided by the area (h × B).

The ability of the plane-strain tension test to distinguish between material ductilities is shown in Table 3.

The plane-strain tension specimen described in Fig. 5 has the disadvantage that it may not be practical to machine such a specimen in a thin sheet. Also, since all deformation is confined to the notch region, it may be difficult to make axial-strain measurements in those limited confines. A special clip-on fixture allows a regular sheet specimen to be converted to a plane-strain specimen (Ref 20).

This plane-strain tension specimen is 38 mm wide and 200 mm long. The specimen contains two circular edge notches with a 19-mm radius. A special fixture containing four knife edges is clamped to the surface of the specimen. The knife edges run parallel to the tensile axis of the specimen and fall just inside the reduced cross section. These knife edges prevent any deformation in the width direction. As a result, necking and failure occur perpendicular to the tensile axis.

SUMMARY

The tension test is a "benchmark test" or "reference test" that provides much basic information about the mechanical state of a material. This test provides information on the flow of a material and its ductility.

The flow resistance is evaluated on the basis of yield stress. The yield stress is variously determined by the first deviation from linear elastic behavior, or more precisely by the stress corresponding to the intersection with the stress-strain curve at an offset strain of 0.002 (the 0.2% offset

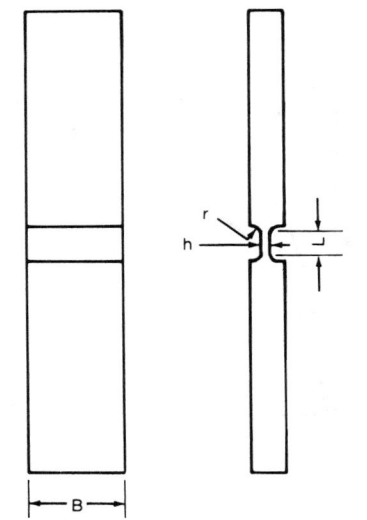

B = 1 in.; L = ¹/₄ in.; h = 0.080 in.; r = 1/16 in.

Fig. 5. Plane-strain tension specimen

yield strength). If the stress-strain curve can be expressed by Eq 5, then the yield stress corresponding to a particular cold reduction, expressed as reduction of area, RA, is given by the following equation (Ref 21):

$$\sigma_y = K \ln \left(\frac{1}{1 - RA} \right)^n \qquad \text{(Eq 26)}$$

This shows that a high strain-hardening exponent n leads to higher flow stress. In addition, with high n values the deformation is spread out and local points of weakness, which can lead to thinning or fracture, are minimized. At elevated temperatures, strain-rate sensitivity becomes important, and strain hardening becomes less important.

Elongation and reduction of area in the tension test cannot be calculated from each other because the occurrence of necking prevents the constant-volume relationship ($A_0L_0 = A_1L_1$) from being invoked over a distance containing the necked region. While elongation and reduction of area usually vary in the same way — for example, as a function of test temperature or alloy content — this is not always the case. Generally speaking, elongation and reduction of area measure different types of material behavior. Elongation measurements in the tension test are chiefly influenced by uniform elongation (except when the gage length is very short) and thus depend on the strain-hardening capacity of the material. Reduction of area is more a measure of the deformation required to produce fracture. It is the most structure-sensitive ductility parameter.

REFERENCES

1. *Mechanical Metallurgy*, 2nd Ed., by G. E. Dieter: McGraw-Hill, New York, 1976, Chapter 9
2. "Standard Methods of Tension Testing of Metallic Materials": ASTM Designation E8, Annual Book of ASTM Standards (see current index for tension tests of various product forms)
3. P. G. Nelson and W. Winlock: *ASTM Bull.*, Vol 156, Jan 1949, p 53
4. *Measurement of Ductility in Sheet Metals*, by J. R. Low and T. A. Prater: ASTM STP 87, 1948
5. P. W. Bridgman: *Trans. ASM*, Vol 32, 1944, p 553
6. A. S. Argon, J. Im and A. Needleman: *Met. Trans.*, Vol 6A, 1975, p 815

7. E. W. Hart: *Acta Met.*, Vol 15, 1976, p 351
8. *The Inhomogeneity of Plastic Deformation*, by A. S. Argon: American Society for Metals, Metals Park, OH, 1973, Chapter 7
9. J. W. Hutchinson and K. W. Neale: *Acta Met.*, Vol 25, 1977, p 839
10. A. K. Ghosh: *Met. Trans. A*, Vol 8A, 1977, p 1221
11. G. W. Stickley and D. J. Brownhill: *Proc. ASTM*, Vol 65, 1965, p 597
12. T. C. Hsu, G. S. Littlejohn and B. W. Marchbank: *Proc. ASTM*, Vol 65, 1965, p 874
13. M. J. Barba: *Mem. Soc. Ing. Civils*, Part I, 1880, p 682
14. W. C. Unwin: *Proc. Inst. Civil Eng.*, Vol 155, Aug 1903, p 170
15. "The Influence of Sheet Thickness on Tensile Properties of Sheet Metal," by A. Rudnick and R. L. Carlson: DMIC Memo 5, Battelle Memorial Institute, Columbus, OH, Jan 23 1959
16. E. G. Kula and N. N. Fahey: *Mat. Res. Std.*, Vol 1, Aug 1961, p 631
17. R. L. Templin: *Proc. ASTM*, Vol 26, 1926, p 378
18. D. P. Clausing: *Int. J. Fracture Mechanics*, Vol 6, 1970, p 71
19. J. P. Elington: *J. Mech. Phys. Solids*, Vol 6, 1958, p 276
20. H. Sang and Y. Nishikawa: *J. Metals*, Feb 1983, p 30
21. L. A. Erasmus: *Metallurgia and Metal Forming*, Apr 1975, p 94

Table 3. Fracture strains for various steels (Ref 18)

Steel	Fracture strain (ϵ_f) Tension test(a)	Plane-strain tension test
ABS-2	1.04	0.75
A302-B	0.98	0.72
HY-80	1.22	0.86
18Ni (180 ksi)	1.00	0.42
18Ni (250 ksi)	0.89	0.15
(a) Standard 0.252-in.-diam specimen.		

Compression Testing

THE CONCEPTS of uniaxial tensile and compressive loading are quite similar. In both cases, a strain is produced parallel to the applied load that has the same sign as the applied load, and two transverse strains are produced that are opposite in sign to the applied load. Below the proportional limit, the strain in the load direction, in both cases, can be calculated using a single value of Young's modulus. Similarly, the transverse strain can be calculated using a second material constant, Poisson's ratio (ν):

$$\epsilon_{transverse} = \nu\epsilon_{axial}$$

Compression testing is an extremely valuable testing procedure which is often overlooked because it is not properly understood. One of the main advantages of the compression test is that tests can be performed with a minimum of material, and thus mechanical properties can be obtained from specimens that are too small for tension testing. Compression tests are also very helpful for predicting the bulk formability of materials (behavior in forging, extrusion, rolling, etc.).

In compression testing, the material does not neck as in tension, but undergoes barreling; failure occurs by different mechanisms and therefore there is no UTS (ultimate tensile strength). In general, ductile materials do not fail in compression but tend to flow in response to the imposed loads. Brittle cylindrical specimens loaded in

Compression test of gray cast iron, grade 40, illustrating brittle fracture. The compression strength of this specimen is 110.517 ksi (762 MPa), which is approximately three times its tensile strength of 40 ksi (276 MPa).

Fig. 6. Brittle compression

compression fail in shear on a plane inclined to the load, and therefore actually break into two or more pieces (Fig. 6). In this case, an ultimate (compressive) stress can be defined.

In comparison with tension testing, several difficulties are encountered in conducting compression tests and interpretation of the experimental data. For example, maintaining complete axiality of the applied load is important. In tension testing, self-aligning grips make this relatively simple to accomplish. In compression testing, if the specimen is tall in relation to its diameter, this can present a major difficulty. Nonaxiality of the load induces a bending load in the specimen, in addition to the axial load, that potentially will cause buckling. Alignment of the loading platens to impose strict uniaxial loading is easier if the specimen contact area is large, but this in turn introduces other difficulties.

Frictional forces exist at the specimen/platen interface that tend to restrict the increase in diameter of the specimen as it decreases in height. These forces are directly related to the coefficient of friction, μ, and thus care must be given to minimizing μ. P is the compressive load. Due to these frictional forces, loading on the specimen is not uniaxial. The effect of these frictional forces is two fold; (*a*) an analysis indicates that the magnitude of the applied stress is increased over what it would be if the specimen were loaded uniaxially (i.e., the situation if μ equals 0), and (*b*) diametral expansion is hindered near the platens, but not in material well removed from the platen, so that the specimen becomes barrel shaped (Fig. 7). Because of the increased magnitude of

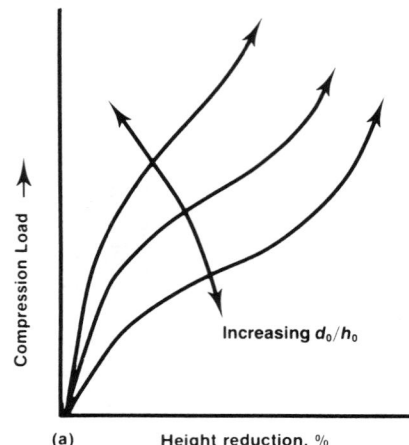

(a)   **Height reduction, %**

Increasing d_0/h_0

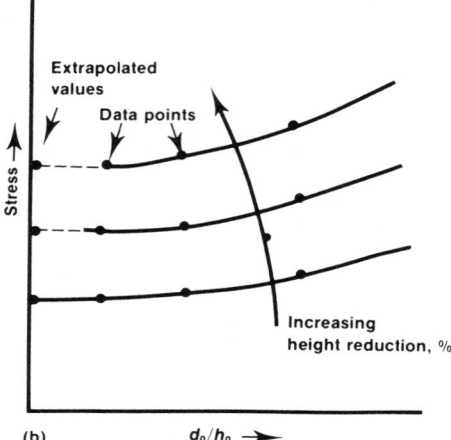

Extrapolated values

Data points

Increasing height reduction, %

(b) d_0/h_0 →

(a) Curves illustrating the relationship of compression load to height reduction for various d_0/h_0 ratios. (b) Replot of curves for extrapolation to $d_0/h_0 = 0$. d_0 is initial diameter; h_0 is initial height.

Fig. 8. Load-deflection curves

the applied stress, deformation at midheight is plastic, whereas it is still elastic near the specimen/platen interface. The ratio of the elastically strained material to the plastically strained material increases as the specimen height decreases, so that barreling increases during the course of a test. The unfortunate consequence of barreling is that specimens selected for easy axial alignment are the same specimens that show extensive barreling and for which the internal stresses in the

material have a large biaxial component. Therefore, some compromise must be made in selecting specimen dimensions. A height-to-diameter ratio of about 3 to 1 often is selected to minimize buckling that occurs due to bending loads generated by nonaxial alignment of the load.

Because of the increasing contact area and the elastically strained material near the platen, the load-deflection curve bends upward as the specimen decreases in height (Fig. 8). A dramatic increase in load occurs if the elastically strained regions of Fig. 7 overlap.

Due to the presence of frictional forces, the pressure distribution across the specimen is not uniform, as shown in Fig. 9. The pressure distribution is given by:

$$p = \sigma_0 \exp[(2\mu/h)(r - x)]$$

where p is pressure, σ_0 is yield stress, μ is coefficient of friction, r is cylinder radius, x is distance from center of cylinder to data point on x-axis, and h is cylinder height.

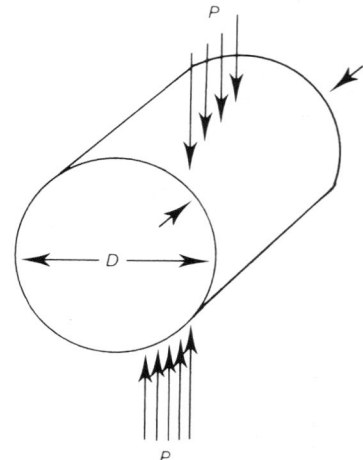

Fig. 9. Line-loaded compression disk

Figure 9 also shows how the pressure distribution changes as the height decreases and as the coefficient of friction changes. The load needed to deform the cylinder can be estimated by multiplying the average pressure on the specimen by the contact area. Figure 9 indicates that the load required to deform materials increases dramatically as the diameter-to-height ratio becomes greater and also increases with a rise in the coefficient of friction. Without overload protection on the load cell, precautions must be taken so that the large loads required to cause plastic flow do not damage the load cell by exceeding its capacity, especially at the large plastic strains characteristic of forming processes.

Coefficients of friction are kept as low as possible to minimize barreling and the development of large loads by providing lubrication at the platen/specimen interface. Two possibilities are the use of lubricating oil and oil grooves in the specimen or platen, or thin Teflon (trademark of DuPont) sheet between the platen and specimen.

An alternative procedure to minimize friction effects and the attendant barreling phenomenon is sequential loading of the specimen—that is, to load until plastic deformation just starts, unload, determine the change in dimensions and relubricate, then reload again until plastic flow initiates.

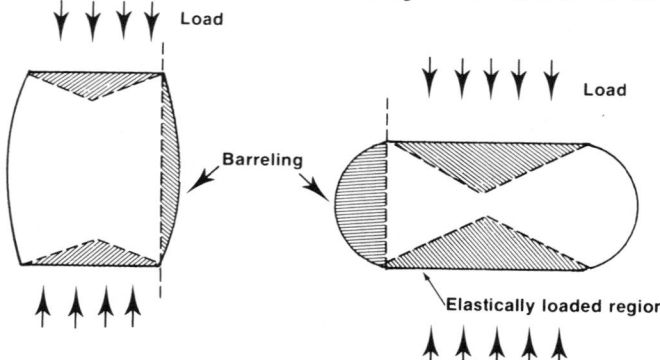

Load

Barreling

Load

Elastically loaded region

Fig. 7. Elastically loaded region and barreling in a compression specimen for two different height-to-width ratios

This locus of points is then connected to provide a constructed stress-strain curve (Fig. 8).

A tedious but accurate way to determine a true uniaxial compression stress-strain curve is to use a series of specimens having varying l/D ratios. The stress at a given height reduction is then determined for each l/D ratio, plotted as a function of l/D ratio, and extrapolated to an l/D ratio of 0. This might be done at only one value of offset (i.e., corresponding to 0.2% strain) to provide a true flow stress, or a complete curve could be constructed. This procedure is illustrated in Fig. 8 and described in detail in Cook and Clarke (Ref 1).

ASTM STANDARDIZED TESTS

ASTM Method E9 gives the standard procedures for room-temperature compression testing; for elevated-temperature testing, ASTM E209 provides the standard recommended practice. For the high-temperature tests, ASTM E209 specifies that a lubricant, such as molybdenum disulfide or graphite, should be used.

LINE-LOADED COMPRESSION TESTING

For the line-loaded test, a flat disk is loaded, as in Fig. 9. If the disk is of brittle material so that fracture occurs before plastic flow initiates, the thickness of the disk does not affect the calculated stress to cause fracture. The tensile stress at fracture for this case is calculated by:

$$s = \frac{2P}{\pi Dt}$$

where s is maximum tensile stress, P is applied load, D is specimen diameter, and t is specimen thickness. This test is known as the "Brazilian test" and is routinely applied to determine the tensile strength of rocks. More detailed information can be found in Jaeger and Cook (Ref 2).

PLANE-STRAIN COMPRESSION TESTING

The common forming operations—rolling, swaging and forging—are forming operations in which there is little or no change in dimension in one direction. For example, in rolling, a decrease in thickness is converted into an increase in length with little increase in width. Such a state of strain is then two dimensional; it is planar strain. Plane-strain deformation is referred to when the development of a local neck is discussed, and in that case the plane-strain deformation occurred due to an internal state of strain in the body. Two loading situations that develop plane-strain deformation are illustrated here; one in which the plane-strain deformation is developed due to external constraint on the flowing material, and a second, in which the constraint is developed internally in the material.

In Fig 10(a), metal can flow in the x and z directions due to the applied stress in the z direction, but cannot flow in the y direction because of the die wall. In Fig. 10(b), the load P causes flow in the z and y directions, but flow occurs in the x direction if the x dimension or the width-to-thickness ratio of the sheet is large. This is because the material not under the die has no load imposed on it and, therefore, has no tendency to spread in either the x or y direction. Therefore, this unloaded material restrains flow of the material in the x direction (but not the y

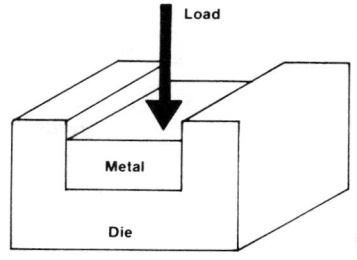

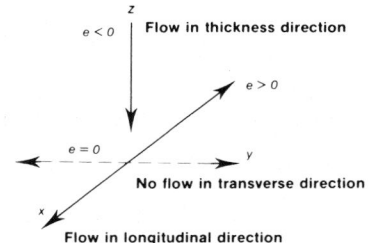

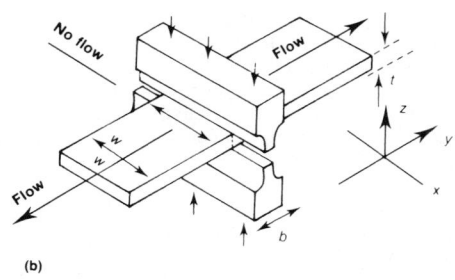

(a) Plane-strain compression of a block in a die. (b) Indenting dies for plane-strain compression testing. Source: *Mechanical Treatment of Metals*, by R. N. Parkins, American Elsevier, New York, 1968, p 22.

Fig. 10. Plane-strain compression

direction) under the die. There is net flow in the y direction, because the unloaded material is simply pushed out by the expanding material under the die. Plane-strain conditions are realized when the width-to-thickness ratio is about 10 to 1. Because of frictional forces under the die, the b/t ratio should be held to between 2 to 1 and 4 to 1. If large decreases in thickness are obtained in the test, sequential loading with a change in die dimension is necessary to maintain this ratio. In any event, it is again possible to construct a stress-strain curve.

Because the strains are large, true stresses and strains frequently are calculated rather than the nominal stresses and strains. The stress and strain developed in the plane-strain compression test are then given by:

$$\sigma_T = \frac{load}{area} = \frac{P}{(w)(b)}$$

and

$$\epsilon_T = \ln \frac{t}{t_0}$$

Although not derived here, the mean stresses and the strain in the load direction in plane-strain deformation are related to the stress and strain developed in a uniaxial compression test of a cylinder by:

$$\sigma_{plane\ strain} = 1.15\sigma_{uniaxial\ compression}$$

and

$$\epsilon_{plane\ strain} = 1.15\epsilon_{uniaxial\ compression}$$

Therefore, stresses and strains measured in the plane-strain compression test must be divided by the factor 1.15 if an equivalent curve for uniaxially loaded material is to be constructed or if the data are to be compared with data obtained in a uniaxial test.

UPSET TESTING

Bulk forming processes, such as forging, extrusion and rolling, are also evaluated for formability by upset testing. In the simplest form of this test, a short cylinder is flattened (upset) into a pancake shape. The bulging at the edges pro-

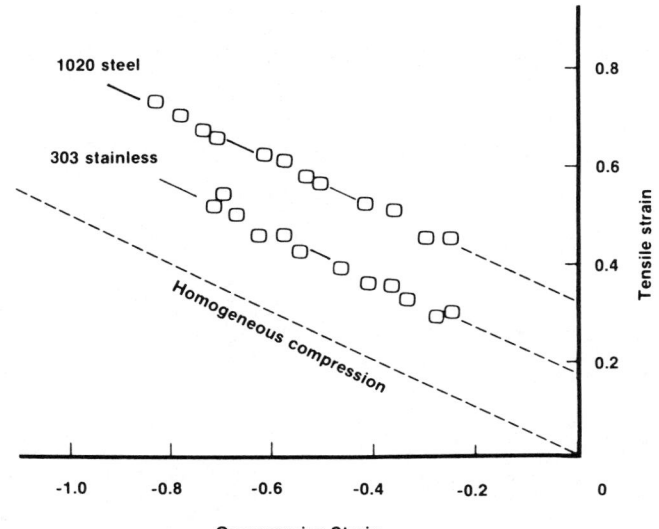

Fig. 11. Fracture-locus lines

duces tensile stresses that cause fractures. Workability is evaluated by determining the largest deformation that can be achieved without producing edge cracking. More typical, however, is use of the hot upset test over a range of temperatures to establish the working temperature that minimizes fracture.

A precision upset test has been developed to determine the three-dimensional analog of the forming-limit diagram. If grid lines are electroetched on the surfaces of small cylinders, then the compressive axial strain (ϵ_z) and the tensile

hoop strain (ϵ_θ) at which fracture occurs can be determined. By varying the length-to-diameter ratios (l/D) of the cylinders and the lubrication at the cylinder ends, a wide range of stress states can be developed. By observing the strain in the upset cylinder at which surface fracture just occurs for each of the stress states, a fracture-locus line can be established (Fig. 11). For any combination of strains below the fracture-locus line, fracture has not yet occurred and the condition is safe. When the strain path crosses the fracture line, surface fracture has occurred. This helpful procedure in predicting the bulk formability of

metals is discussed by Kuhn (Ref 3) and by Hosford and Caddell (Ref 4).

REFERENCES

1. Cook and Clarke: *J. Inst. Met.*, Vol 71, 1945, p 371
2. *Fundamentals of Rock Mechanics*, 3rd Ed., by J. C. Jaeger and N. G. W. Cook: Chapman and Hall, London, 1979, p 169
3. H. A. Kuhn: in *Formability Topics—Metallic Materials*, ASTM STP 647, 1978, p 206
4. *Metal Forming: Mechanics and Metallurgy*, by W. F. Hosford and R. M. Caddell: Prentice Hall, Englewood Cliffs, NJ, 1983, p 247

Fracture Testing

THE MOST IMPORTANT fracture tests can be grouped into two categories: impact tests and fracture-toughness tests. However, the variety of service circumstances—different types of materials, differing crack morphologies, differing environments and loading rates, effects of size—have spawned a large number of fracture tests, some highly specialized. The most common impact tests are the Charpy test and the Izod test. Both are used primarily for low- and medium-strength materials (typically steels). These materials may break at stresses either above or below yield, depending on the circumstances (temperature, size of crack, etc.). Fracture-toughness tests are intended primarily for medium- and high-strength materials that may break at below-yield stresses, if a crack or other sharp flaw—often quite small—is present. Fracture-toughness tests are based on the theoretical developments of fracture mechanics and give results that can be directly used in calculations relating the size of cracks to applied loads and stresses. In fracture-toughness tests, unlike in impact tests, loads are normally applied relatively slowly—at about the same rate as in an ordinary tension test—and temperature need not be of concern. Although fracture-toughness tests can be conducted in a tension-test machine, the specimen looks quite

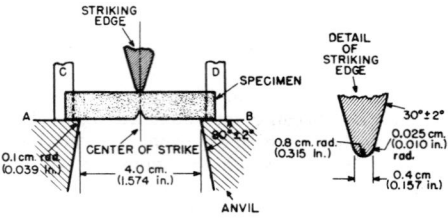

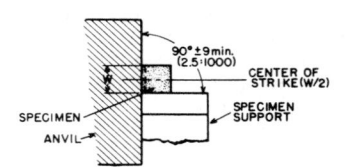

Fig. 2. Setup and specimen for Charpy impact testing

different. Generally, it is a plate containing a crack grown from a machined notch, rather than the smooth round bar used in tension testing.

Impact Testing

IMPACT TESTS feature a high but generally indeterminate rate of loading, typically generated by a swinging pendulum or falling weight. The results are not directly related to the stresses and deformations normally calculated during the course of engineering analysis and design, although the results may relate to the temperatures experienced in service. In general, impact-test results have significance based on empirical correlations with service experience, or as a means of comparing materials. The Charpy test is most commonly used to evaluate the effects of metallurgical processes on dynamic mechanical properties. Another test, the Izod impact test, employs a cantilevered specimen hit by a swinging pendulum.

CHARPY TESTS

Specimens used for Charpy tests come in several different configurations. Two examples are the V-notch test specimen, containing a shallow 45° machined notch as a stress concentrator, and the Charpy keyhole specimen, with a stress raiser

that looks something like a keyhole (Fig. 1a and b). The V-notch test is the more common of the two tests.

Charpy tests have the virtue of being simple and inexpensive. Furthermore, the specimen is small—also one of the limitations of the test—and Charpy testing machines are widely available. The principal application is for delineating the transition-temperature region in low- and medium-strength steels; in the common test sequence, a series of nominally identical specimens is broken at different temperatures. (Occasionally this test is used for materials other than low- and medium-strength steel.) Material specifications often include required levels of Charpy test performance, but the results of the test have limited fundamental significance. Results only have meaning in terms of correlations with ductile or brittle behavior under service conditions based on actual experience. However, because Charpy tests have been employed for so many years, a good deal of this experience is available in the form of general rules that can be exploited by designers.

Figure 2 illustrates the Charpy bar supported at its ends and struck on the surface opposite the notch so that the loading is three-point bending. The source of the blow is a heavy, swinging pendulum, as can be seen in Fig. 3, with a range of 25 to 240 ft · lb (35 to 325 J) of energy at impact. The weight and height of the pendulum striking head may be modified to produce various foot-pounds (joules) of energy. Testing-machine and specimen details for Charpy tests, together with test procedures, are standardized, as described in ASTM E23.

Machining of specimens should be controlled to provide specimen uniformity. The orientation of the bars, with respect to the rolling direction, often has considerable effect on impact behavior.

In its simplest form, a Charpy test is conducted by inserting a specimen into the machine, cocking the pendulum and releasing it to fracture the test bar. Typically, a series of tests is performed at different temperatures. Charpy machines should be periodically checked against standardized specimens to determine their accuracy.

A dial on the machine indicates the energy absorbed in fracturing the specimen (Fig. 3). The test is scaled so that specimens made of virtually any material will be broken by the impact. A small proportion of the initial potential energy of the pendulum is absorbed in overcoming air resis-

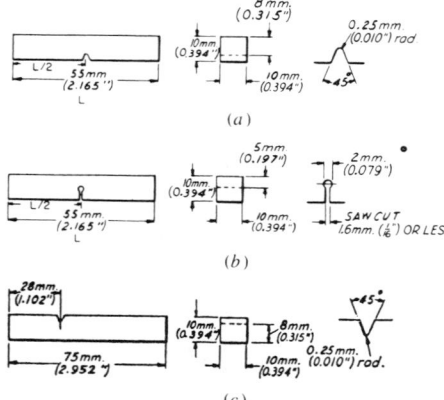

(a) Simple beam V-notch Charpy specimen. (b) Simple beam keyhole-notch Charpy specimen. (c) Cantilever beam notched Izod specimen. Source: Notched Bar Impact Testing of Metallic Materials, E23-81, ASTM, Philadelphia, 1981.

Fig. 1. Notched-bar impact-test specimens

Photographed at American Society for Metals, Metals Park, OH.

Fig. 3. Standard impact-testing machine

tance; this amount, usually $1/2$ to 1 ft·lb (0.7 to 1.5 J), can be determined by operating the machine without a specimen in place. Note that the commonly used machine is entirely manual in operation, with no instrumentation beyond the dial. The dial simply gages the height of the pendulum after it has swung through the specimen. The pendulum carries a great deal of energy even after fracturing a specimen and must be treated with respect; safety measures should always be taken. Injury may result if a guard shield is not in place during the pendulum swing. The shield is intended to prevent bodily access during test motion.

In addition to data on energy absorbed, two other quantities are commonly determined for each impact-test specimen. First, the percentage of fibrous fracture area visible on the cross section of the broken specimen is compared with the percentage of cleavage area. Second, the lateral contraction of the broken bar at the root of the notch is measured.

Figure 4 shows a series of fracture surfaces from Charpy V-notch tests at different temperatures. The fracture surface is entirely fibrous in appearance above the transition-temperature region, indicating ductile behavior; it is entirely cleavage below. In the transition region, small decreases in temperature increase the percentage of cleavage fracture. As Fig. 4 shows, the texture difference between fibrous and cleavage regions makes identification straightforward for specimens (such as annealed low-carbon steel) which fail by cleavage at low temperatures. The percentages can be estimated visually or, if more precision is needed, the fracture surfaces can be photographed and further analyzed. Note that in the transition region, cleavage takes place in the center of the specimen, with fibrous fracture near the outer surfaces of the bar. This is a result of the differing states of stress in the interior and near the surface. The inner material is con-

strained against plastic deformation and more likely to fracture by cleavage.

Notch-root contraction, measured as indicated in Fig. 5, is also a direct indication of fracture behavior—in this case, of the amount of plastic deformation accompanying the fracture. The contraction is usually easiest to measure with a micrometer caliper. The thickness of the unbroken specimen at the notch root should be measured *before* the test.

The energy absorbed and the fracture-surface appearance are the most commonly specified results of the Charpy test. For example, steels for certain applications may be required to have a certain minimum level of energy absorption—

100%	85%
70%	60%
50%	40%
30%	20%
10%	0%

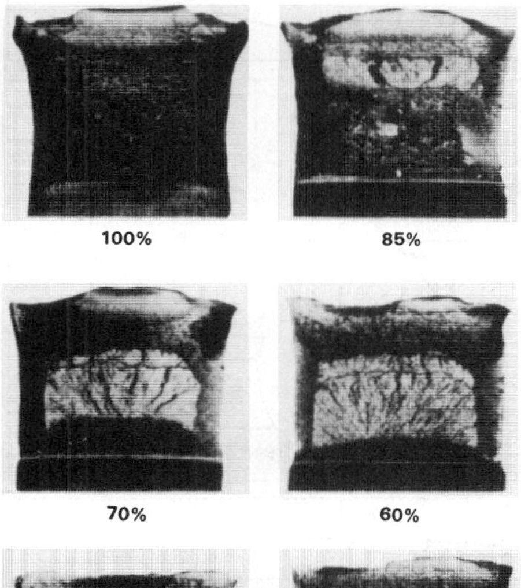

Source: Army Materials and Mechanics Research Center, Watertown Arsenal.

Fig. 4. Series of fractographs of Charpy V-notch specimens of 4340 steel tested at different temperatures, showing the change in appearance and estimated percentages of fibrous fracture

10 or 15 ft·lb (15 or 20 J) are common values—at the lowest expected service temperature. Alternately, some minimum percentage of fibrous fracture (e.g., 50%) may be required. Again, note that such requirements have no intrinsic significance and can only be defined based on correlations with service experience.

The energy absorbed in impact, percentage of fibrous fracture, and notch-root contraction can be plotted against temperature to determine the ductile-to-brittle transition temperature. However, the particular definition of the ductile-to-brittle transition temperature must be clearly specified in the data report.

The standard Charpy impact test yields direct readings only of the energy absorbed. "Instrumented" tests make use of testing machines with auxiliary sensors to acquire other data, typically load data. A specially made striking tup instrumented with strain gages can give a load-time history that allows initiation of the crack from the machined notch to be distinguished from propagation of that crack through the specimen. This is a relatively specialized test that is utilized for specific limited applications. The advantages of the basic Charpy test—small specimen size and low cost—are retained, while allowing fracture-mechanics parameters such as toughness to be estimated under specific circumstances.

THE IZOD TEST

The Izod test is a cantilever-beam test as compared to the simple-beam Charpy test. As shown in Fig. 6, the Izod specimen is held in a fixture with the V-notch facing the striking anvil of the pendulum. The center of the V-notch is in the same plane as, and parallel to, the supporting fixture. The actual fracture test is performed in the same manner as the Charpy test, and data on energy consumed are reported in foot-pounds (joules). The Izod test does not lend itself to variable-temperature testing because of the appreciable time required to place and clamp the specimen, which results in rapid temperature change due to specimen and fixture contact. Izod tests are generally specified for materials tested at room temperature and where the engineering part is designed to operate under cantilever loading.

DYNAMIC TEAR TESTS

Dynamic tear tests and drop-weight tear tests come in several varieties, two of which have been standardized by ASTM. All are similar to the Charpy test in the use of the kinetic energy of a swinging pendulum (or occasionally a falling weight) to break an artificially notched test specimen. In essence, they are larger versions of the Charpy test, with test specimens that are both thicker and wider to represent the fracture behavior of thick-section structural materials. While the Charpy test is used almost exclusively to investigate the transition-temperature behavior of low- and medium-strength steels, dynamic tear (DT) tests can be used as well for high-strength steels and for aluminum and titanium alloys. When such materials are broken in a Charpy machine, the energy absorbed is typically so low, compared to the testing machine's capacity, that little meaningful information can be obtained. The larger DT specimen allows different alloys and heat treatments to be more reliably compared.

Typical DT specimens are shown in Fig. 7. The larger specimen is used only where full-

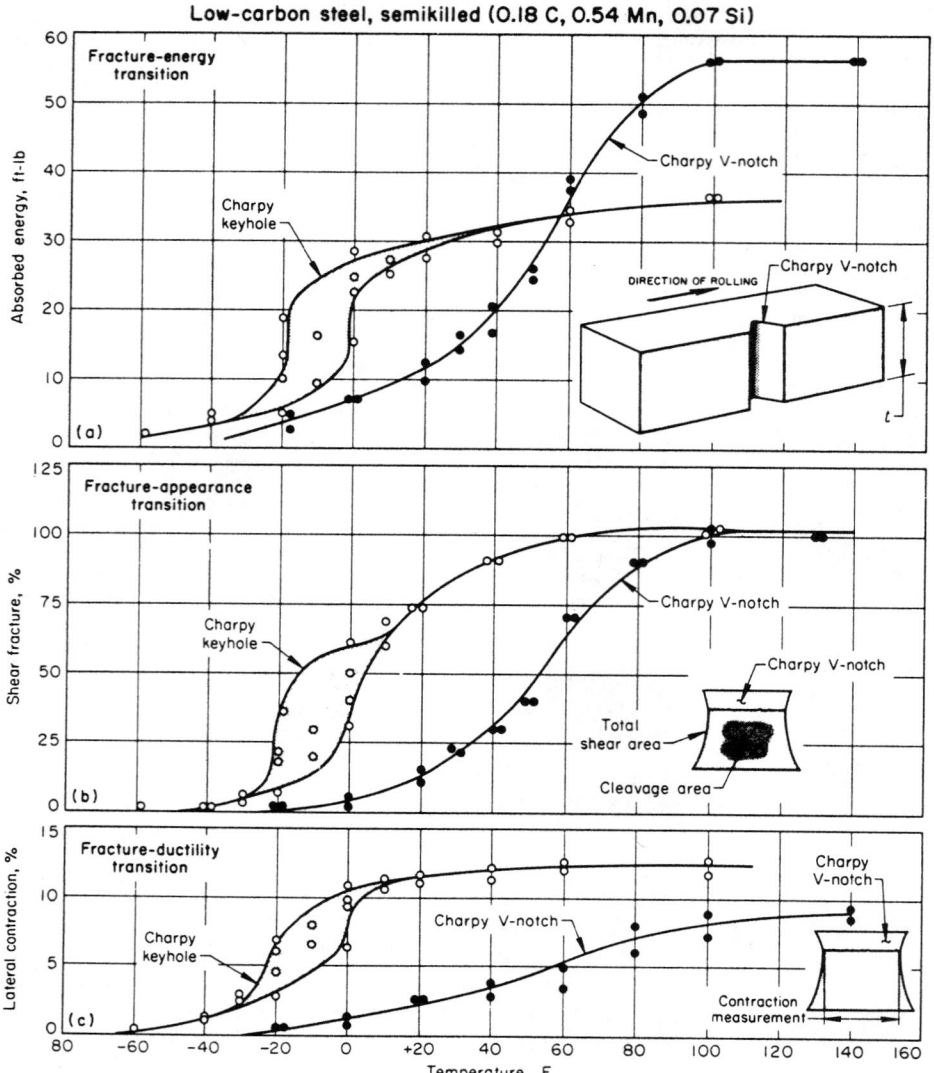

Low-carbon steel, semikilled (0.18 C, 0.54 Mn, 0.07 Si)

Tests determined by (a) fracture energy, (b) fracture appearance and (c) fracture ductility. The drawings of the Charpy V-notch specimens at lower right in the graphs indicate: (a) orientation of the specimen notch with plate thickness, t, and direction of rolling; (b) location of the total fibrous area and cleavage area on the fracture surface; and (c) location of the contraction measurement in this series of tests. Percentage of fibrous fracture and lateral contraction were based on the original dimensions of the specimen.

Fig. 5. Characteristics of the transition-temperature range of Charpy V-notch and Charpy keyhole tests of low-carbon steel plate

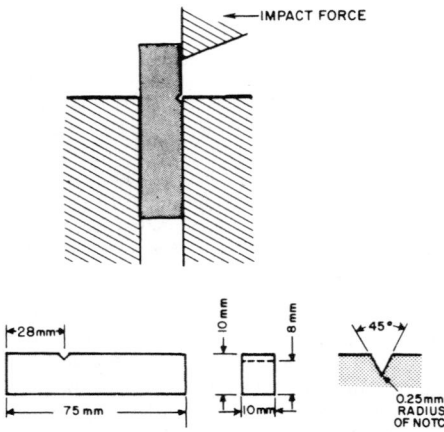

Fig. 6. Setup and specimen for Izod impact testing

thickness tests representative of actual structures must be undertaken.

As in a Charpy test, the basic information obtained from a DT test is the energy absorbed in breaking the specimen, although fracture-surface appearance and the extent of plastic deformation near the fracture can also be useful.

The two varieties of ASTM standard dynamic tear or drop-weight tests have the designations E436 (Drop-Weight Tear Tests for Ferritic Steels) and E604 (Dynamic Tear Energy of Metallic Materials). As the names imply, the purposes are quite different, even though the general features are similar.

The E436 drop-weight tear test uses a specimen, normally of the actual thickness of the steel plate being investigated, with an easily produced, pressed notch (see ASTM E436 for details). The test is used only for steels, and indicates the transition temperature between ductile and brittle behavior. To this end, specimens are

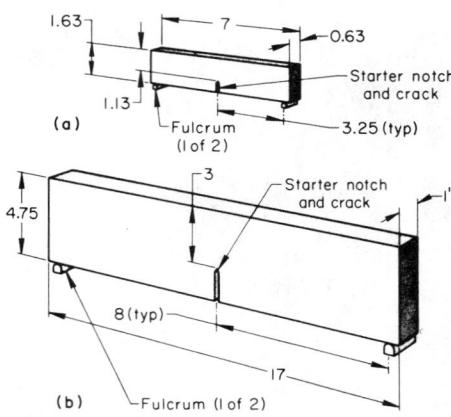

In the specimens, a crack with a sharp tip is produced by making a brittle electron beam weld or by pressing with a knife edge. With either method of providing the crack tip, and with either size of specimen, maximum-constraint conditions are attained. Dimensions are in inches.

Fig. 7. Two sizes of the standard specimen for the Naval Research Laboratory standardized dynamic tear test

tested at increasing (or decreasing) temperatures until the crack propagates as a fully developed fibrous fracture away from the notch (it makes no difference if the crack initiates by a cleavage mechanism). This is determined by visual examination of the fractured specimen; no quantitative data need be gathered. Thus, the test is simple and inexpensive. It is used primarily for quality control purposes on structural steels—notably, for pipeline applications.

The ASTM E604 test differs, as the name "Dynamic Tear Energy of Metallic Materials" implies, in two fundamental ways:

1. The energy absorbed in fracturing the specimen is measured.
2. The test can be used on materials other than steels.

The standard dimensions of the specimen are less than for either of the specimens shown in Fig. 7; the notch is machined, and the specimen is always $5/8$ in. (15.9 mm) thick. Tests can be conducted at a fixed temperature to compare several materials, or over a range of temperatures to investigate transition-temperature effects. Considerable effort has gone into correlating DT test results with plane-strain fracture-toughness values, because the more sophisticated testing procedures required for the latter are more costly and time consuming. In some cases, the DT test can be used to estimate plane-strain toughness.

DROP WEIGHT TEST

As the name implies, in a drop weight test (DWT) the specimen is broken by a falling weight rather than by a pendulum (Fig. 8). However, the essential difference between this test and those described earlier is that it provides the operational definition of an important parameter termed the nil-ductility temperature (NDT). The NDT is the highest temperature at which a particular steel is likely to fracture in brittle fashion by cleavage. Above this temperature, the fracture will be accompanied by some macroscopic plastic flow associated with fibrous fracture and a microvoid-coalescence process on the microscale. The DWT

Fig. 8. Drop weight test equipment

procedure does not, however, require interpretation of the fracture appearance, nor must any quantitative data be gathered beyond the temperature of the specimen. As the definition of the term NDT implies, the test is used only for low- and medium-strength steels that show a cleavage-fibrous transition as the temperature is increased.

The DWT also has been standardized by ASTM (E208). The specimen is a flat plate with a hard-surfacing weld bead deposited on one side, as shown in Fig. 9. The brittle weld is notched to act as a crack-starter. The fixture for DWT tests holds the specimen with the notched weld facing downward so that the falling weight will load the specimen in bending. The weld is thus on the tensile surface, and the specimen is backed up so that it can bend only a limited amount (approximately 5°). Well below the NDT, a DWT will produce complete fracture (Fig. 9). Well above the NDT, the specimen will bend elastically and plastically but will not break. The NDT, as defined by this test, is the temperature at which a crack starting at the notched hard-surfacing weld propagates to one or both edges of the tensile surface of the specimen. To conduct a DWT series, specimens are tested over a range of tem-

Fig. 9. Fracture appearance of drop weight nil-ductility transition specimens

peratures; the operationally defined NDT is the highest of these temperatures at which the specimen "breaks," according to this definition. The NDT is a widely applied parameter for structural steels—referenced, for example, in the ASME Boiler and Pressure Vessel Code.

Fracture-Toughness Testing

By J. H. Underwood, U.S. Army Armament Research and Development Command, and W. W. Gerberich, University of Minnesota

CONCEPTS OF FRACTURE MECHANICS

The concepts of fracture mechanics are basic ideas for developing methods of predicting the load-carrying capabilities of structures and components containing cracks. The concepts deal with basic quantities or parameters of fracture mechanics. These quantities can be discussed in relation to a simple example: a center crack in a plate remotely loaded by a uniform tensile stress (see Fig. 10). When the half-crack length, a, is less than 10% of the total plate width, the relationship among stress-intensity factor, K, ap-

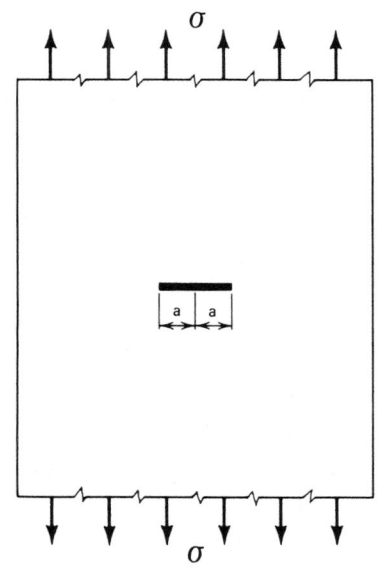

σ is remotely applied uniform tensile stress; a is half-crack length.

Fig. 10. Schematic illustration of a center crack in a wide plate

plied stress, σ, and half-crack length, a, is very close to the relationship for a crack in an infinitely wide plate, which is:

$$K = \sigma\sqrt{\pi a} \qquad \text{(Eq 1)}$$

The stress applied to the component, the length of the crack, and the stress-intensity factor in the loaded component with a crack are the basic quantities of fracture mechanics. The example in Fig. 10 also provides a simple explanation for the units of stress-intensity factor—i.e., the product of stress and square root of length. But more important is the concept that the stress-intensity factor, K, is a single parameter which includes both the effect of the stress applied to a specimen and the effect of a crack of a given size

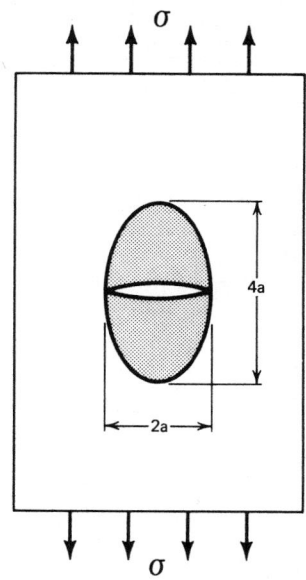

Fig. 11. Schematic illustration of the concept of energy release around a center crack in a loaded plate

in a specimen. Still using the example of Fig. 10, if the combination of σ and a in Eq 1 were to exceed a critical value of K, then the fracture strength of the plate would be exceeded and the crack would be expected to propagate by one of the several mechanisms mentioned in this article.

Energy-Release Rate

The origins of modern-day fracture mechanics may be traced to Griffith (Ref 1), who established an energy-release-rate criterion for brittle materials. Observations of the fracture strength of glass rods had shown that the longer the rod, the lower the strength. Thus the idea of a distribution of flaw sizes evolved, and it was discovered that the longer the rod, the larger the chance of finding a large natural flaw. This physical insight led to an instability criterion which involved the elastic energy released in a solid at the time a flaw grew catastrophically under an applied stress.

From the theory of elasticity comes the concept that the strain energy contained in an elastic body per unit volume is simply the area under the stress-strain curve, or:

$$U_0 = \frac{\sigma^2}{2E} \qquad \text{(Eq 2)}$$

where σ is the applied stress and E is Young's modulus. However, there is a reduction (that is, a release) of energy in an elastic body containing a flaw or a crack because of the inability of the unloaded crack surfaces to support a load. We shall assume that the volume of material whose energy is released is the area of an elliptical region around the crack (as shown in Fig. 11) times the plate thickness, B; the volume is $\pi(2a)\,(a)B$. This is based on the area of an ellipse being $\pi r_a r_b$, where r_a and r_b are the major and minor radii of the ellipse. Then, the total energy released from the body due to the crack is the energy per unit volume times the volume, which is:

$$U = \pi(2a)(a)B\,\frac{\sigma^2}{2E} = \frac{\pi\sigma^2 a^2 B}{E} \qquad \text{(Eq 3)}$$

In ideally brittle solids, the released energy can be offset only by the surface energy absorbed, which is:

$$W = (2aB)(2\gamma_s) = 4aB\gamma_s \qquad \text{(Eq 4)}$$

where 2aB is the area of the crack and $2\gamma_s$ is twice the surface energy per unit area (because there are two crack surfaces).

Griffith's energy-balance criterion, in the simplest sense, is that crack growth will occur when the amount of energy released due to an increment of crack advance is larger than the amount of energy absorbed:

$$\frac{dU}{da} \geq \frac{dW}{da} \qquad \text{(Eq 5)}$$

Performing the derivatives indicated in Eq 5 and rearranging give the Griffith criterion for crack growth:

$$\sigma\sqrt{\pi a} = \sqrt{2E\gamma_s} \qquad \text{(Eq 6)}$$

Fracture theory was built upon this criterion in the early 1940's by considering that the critical strain-energy-release rate, G_c, required for crack growth was equal to twice an effective surface energy, γ_{eff}:

$$G_c = 2\gamma_{eff} \qquad \text{(Eq 7)}$$

This γ_{eff} is predominantly the plastic energy absorption around the crack tip, with only a small part due to the surface energy of the crack surfaces. Then, with the development of complex variable and numerical techniques to define the stress fields near cracks, this energy view was supplemented by stress concepts—that is, the stress-intensity factor, K, and a critical value of K for crack growth, K_c. Replacing γ_s with γ_{eff} in Eq 6 and noting that the energy and stress concepts are essentially identical (that is, $K = \sqrt{EG}$) give:

$$K_c = \sqrt{EG_c} = \sigma\sqrt{\pi a} \qquad \text{(Eq 8)}$$

which is the crack-growth-criterion equivalent of Eq 1. Thus, K_c is the critical value of K which, when it is exceeded by a combination of applied stress and crack length, will lead to crack growth. For thick-plate plane-strain conditions, this critical value became known as the plane-strain fracture toughness, K_{Ic}, and any combination of applied stress and crack length that exceeds this value could produce unstable crack growth, as indicated schematically in Fig. 12(a) (linear-elastic).

In work with tougher, lower-strength materials, it was later noted that stable slow crack growth could occur even though accompanied by considerable plastic deformation. Such phenomena led to the nonlinear J-integral and R-curve concepts which could be used to predict the onset of stable slow crack growth and final instability under elastic-plastic conditions, as noted in Fig. 12(b). Finally, the fracture-mechanics approach was applied to characterize subcritical-crack-growth phenomena where time-dependent slow crack growth, da/dt, or cyclic crack growth, da/dN, may be induced by special environments or fatigue loading. For combinations of stress and crack length above some environmental threshold, K_{Iscc}, or fatigue threshold, ΔK_{th}, subcritical growth occurs, as indicated in Fig. 12(c).

Assumptions

Additional concepts of fracture mechanics have evolved which are related to the assumptions made in using linear-elastic fracture mechanics. Three

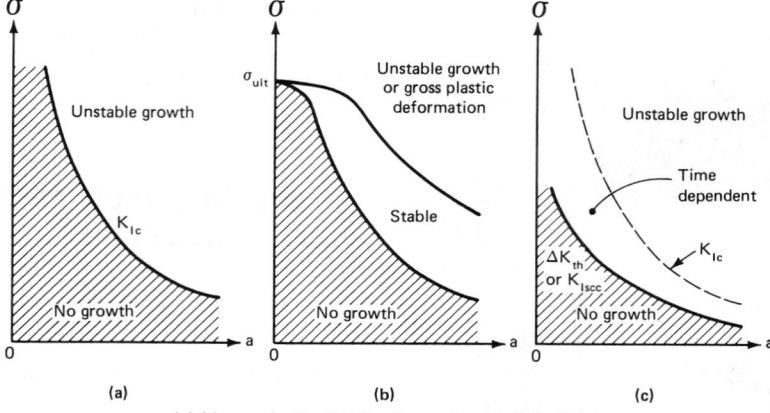

(a) Linear-elastic. (b) Elastic-plastic. (c) Subcritical.

Fig. 12. Relationships between stress and crack length, showing regions and types of crack growth

assumptions basic to fracture-mechanics analysis are as follows:

1. Cracks exist in the body to be analyzed. Indeed, cracks, or flaws with cracklike behavior, are frequently present in real structural materials.
2. A crack can be represented by a flat, free surface in a linear-elastic, homogeneous, isotropic continuum.
3. A characteristic stress field surrounds any crack in a loaded body. The magnitude of this field, the K value, at the onset of crack extension is a material property which is independent of specimen size and geometry for many conditions of loading and environment.

Stress-Intensity Factor. The second assumption given above makes possible a mathematical description of the stresses in the vicinity of the crack tip. This assumption of a crack in a linear-elastic solid at first appears contradictory to what is known about fracture of metals, because some plastic deformation is always found to accompany fracture. However, when the region of plastic deformation around the crack is small compared with the size of the crack, which often is true with large structures and with high-strength materials, this is a good assumption. Using linear-elastic theory and referring to Fig. 13, the stress at a point P near the crack tip can be expressed as:

$$\sigma_y = \frac{K_I}{\sqrt{2\pi r}} \cos\frac{\theta}{2}\left(1 + \sin\frac{\theta}{2}\sin\frac{3\theta}{2}\right) \qquad \text{(Eq 9)}$$

where σ_y is the stress perpendicular to the crack plane, r is the distance from the crack tip to point P, θ is the angle between the crack plane and the line from point P to the crack tip, and K_I is the applied opening-mode stress-intensity factor. In Eq 9, as the distance from the crack tip, r, approaches zero, the stress, σ_y, approaches infinity. This $1/\sqrt{r}$ singularity cannot actually occur, because plastic deformation relieves elastic stresses very near the crack tip. Nevertheless, in many cases the $1/\sqrt{r}$ singularity provides an adequate over-all description of the stresses near the crack tip and therefore a good description of the conditions for crack growth.

The important implications of Eq 9 are: (a) a crack in a loaded component or specimen generates its own intensified stress field near the crack

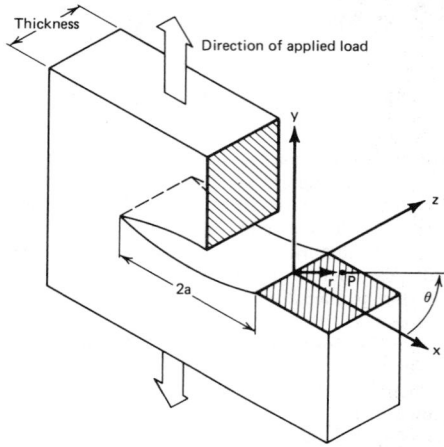

x is the direction perpendicular to the crack tip and in the plane of the crack; y is the direction perpendicular to the plane of the crack; z is the direction parallel to the crack tip; P is a point near the crack tip; r is the distance from the crack tip to point P; θ is the angle between the plane of the crack and a line from point P to the crack tip; a is half-crack length.

Fig. 13. Schematic illustration of a through-thickness crack

tip, a stress field that differs from another crack-tip stress field only by the scaling factor represented by K; and (b) the factor K expresses how much the stress intensifies at the crack tip, and thereby allows the loading and geometry factors that influence crack growth in a specimen to be described on a uniform basis using a single parameter.

The stress-intensity factor, K, can have a simple relation to applied stress and crack length, as in Eq 1. But, more often, the K relation is of greater complexity because of complex loading, various configurations of real structural components, or variations in crack shapes. The K relations for many different types of loading and specimen and crack geometries have been obtained by various experimental and analytical methods. Two handbooks that give a variety of K relations are listed as Ref 2 and 3. These handbooks give K relations for the three basic types of crack-face displacement shown in Fig. 14. Nearly all crack-related fracture processes of practical significance for metals involve mode I, opening-mode deformation, in which the dis-

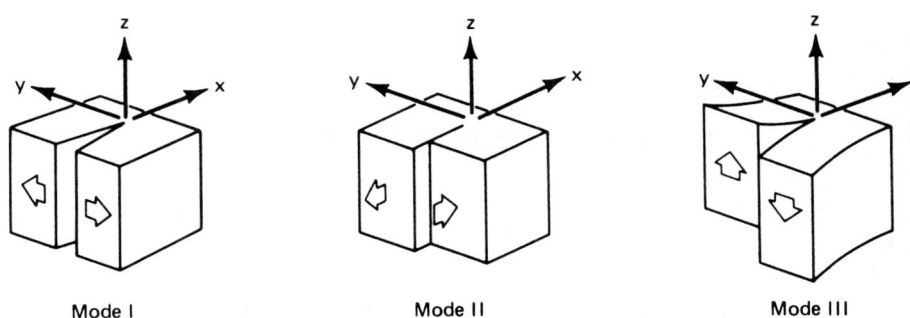

Mode I—opening mode: tension stress in y direction (perpendicular to crack surfaces). Mode II—edge-sliding mode: shear stress in x direction (perpendicular to crack tip). Mode III—screw-sliding mode: shear stress in z direction (parallel to crack tip).

Fig. 14. Modes of crack deformation

placement of the crack faces is in a direction perpendicular to the crack plane. There are cases in which shear deformation by modes II and III accompanies opening-mode deformation, but shear deformation often has little significant effect on the over-all, macroscopic fracture process.

Fracture Toughness. The third assumption given above states that for many test conditions the magnitude of the stress field, K, at the onset of crack extension is a material constant. Tests on precracked specimens of a wide variety of materials have shown that the critical K value at the onset of crack extension approaches a constant value as specimen thickness increases. Figure 15 shows this effect in tests with AISI 4340 steel specimens over a range of thickness (Ref 4). In general, when the specimen thickness and the in-plane dimensions near the crack are large enough relative to the size of the plastic zone, then the value of K at which growth begins is a constant and generally minimum value called the plane-strain fracture-toughness factor, K_{Ic}, of the material. The parameter K_{Ic} is a true material property in the same sense as is the yield strength of a material. The value of K_{Ic} determined for a given material is unaffected by specimen dimensions or type of loading, provided that the specimen dimensions are large enough relative to the plastic

zone to ensure plane-strain conditions around the crack tip (strain is zero in the through-thickness or z-direction).

Plane-strain fracture toughness, K_{Ic}, is directly related to the energy required for the onset of crack propagation by the formula:

$$K_{Ic} = \sqrt{\frac{EG_{Ic}}{1 - \nu^2}} \qquad \text{(Eq 10)}$$

where E is the elastic modulus (in MPa or psi), ν is Poisson's ratio (dimensionless), and G_{Ic} is the critical plane-strain energy-release rate for crack extension (in kJ/m^2 or $in. \cdot lb/in.^2$). In simplified concept, G_{Ic} is the critical amount of strain energy that is released from the elastic stress field of the specimen per unit area of new cracked surface for the first small increment of crack extension. The concepts of K_{Ic} and G_{Ic} are essentially interchangeable; K_{Ic} is generally preferred because it is more easily associated with the stress or load applied to a specimen. The value of K_{Ic} for a given material can be measured directly using ASTM Standard Test Method E399 (Ref 5).

Plastic Zone

As mentioned above, under "Stress-Intensity Factor," when the stresses get sufficiently high

near the crack tip, plastic deformation must occur. A description of the size of the plastic zone is essential to the understanding of how section size affects the fracture behavior of a specimen or component. The simplest concept is to imagine the point at which σ_y, according to Eq 9, reaches the yield strength of the material. This value of r represents the radius of the plastic zone. This is schematically depicted in Fig. 16; in equation form it is:

$$r_p\big|_{\theta=0°} = \frac{K_I^2}{2\pi\sigma_{ys}^2} \qquad \text{(Eq 11)}$$

where σ_{ys} is the yield strength. Of course, material at the tip of a crack does not respond simply to the σ_y stress but responds to the complex triaxial state of stress. One yield criterion which has successfully modeled complex states of stress

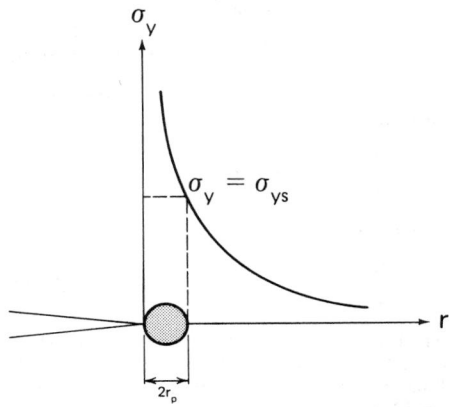

σ_y is the stress perpendicular to the crack plane; σ_{ys} is the yield strength; r is the distance from the crack tip; r_p is the radius of the plastic zone.

Fig. 16. Schematic illustration of plastic zone and stress distribution at a crack tip

in metals is the von Mises or distortional strain energy criterion. Simply stated, yielding is reached under a complex state of stress when the sum of the squares of the principal stress differences is equal to two times the uniaxial tensile yield squared, or:

$$2\sigma_{ys}^2 = (\sigma_1 - \sigma_2)^2$$
$$+ (\sigma_2 - \sigma_3)^2 + (\sigma_3 - \sigma_1)^2 \qquad \text{(Eq 12)}$$

From linear-elastic theory, all of the values of principal stress may be determined, which for mode I are:

$$\sigma_1 = \frac{K_I}{\sqrt{2\pi r}} \cos\frac{\theta}{2}\left[1 + \sin\frac{\theta}{2}\right] \qquad \text{(Eq 13a)}$$

$$\sigma_2 = \frac{K_I}{\sqrt{2\pi r}} \cos\frac{\theta}{2}\left[1 - \sin\frac{\theta}{2}\right] \qquad \text{(Eq 13b)}$$

$$\sigma_3 = 2\nu\frac{K_I}{\sqrt{2\pi r}} \cos\frac{\theta}{2} \text{ (plane strain)} \qquad \text{(Eq 13c)}$$

$$\sigma_3 = 0 \text{ (plane stress)} \qquad \text{(Eq 13d)}$$

where ν is Poisson's ratio. Combining Eq 12 and 13 gives the elastic-plastic boundary by the inverse method. Although only a first approximation, this approach illustrates the differences between thin-section and thick-section behavior. For plane stress, which would apply to relatively thin

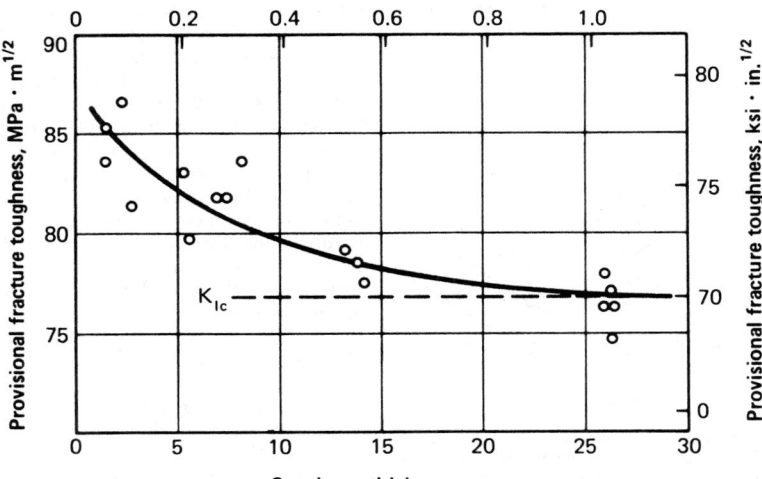

Specimen thickness, in.

The material was 4340 steel plate that had been hardened and tempered at 400 °C (750 °F) to a yield strength of 1470 MPa (213 ksi).

Fig. 15. Effect of specimen thickness on the critical K value for crack extension in steel specimens (Ref 4)

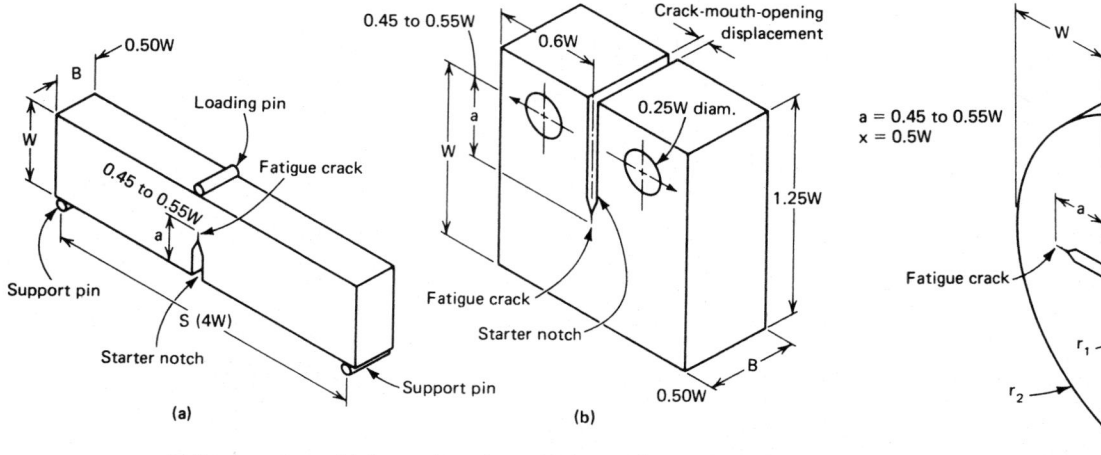

(a) Bend specimen. (b) Compact specimen. (c) Arc specimen. a is notch plus crack length; B is specimen thickness; W is specimen depth.

Fig. 17. Fracture-toughness test specimens

sections having $\sigma_3 = 0$ (stress is zero in the through-thickness direction),

$$r_p \bigg|_{\text{plane stress}} \simeq \frac{K_I^2}{2\pi\sigma_{ys}^2} \cos^2\left(\frac{\theta}{2}\right)\left\{1 + 3\sin^2\left(\frac{\theta}{2}\right)\right\}$$

$$\approx \frac{K_I^2}{2\pi\sigma_{ys}^2} \qquad \text{(Eq 14)}$$

which is equivalent to Eq 11 for $\theta = 0°$. However, for thick plates where plane-strain conditions prevail, σ_3 is as given in Eq 13, and now Eq 12 and 13 give:

$$r_p \bigg|_{\text{plane strain}} \simeq \frac{K_I^2}{2\pi\sigma_{ys}^2} \cos^2\left(\frac{\theta}{2}\right)\left\{1 + 3\sin^2\left(\frac{\theta}{2}\right)\right.$$

$$\left. - 4\nu(1 - \nu)\right\} \approx \frac{K_I^2}{6\pi^2\sigma_{ys}} \qquad \text{(Eq 15)}$$

For a Poisson's ratio of $\nu = 0.3$, the average plane-strain plastic-zone size is about $1/3$ the plane-stress value of Eq 14, considering all values of θ. The important point is that yielding in thick plates is more difficult, the plastic zones are smaller, and hence the energy absorbed around the crack is less.

FRACTURE TEST METHODS

Plane-Strain Fracture Toughness, K_{Ic}

ASTM Method E399. The first fracture-mechanics test method to become widely accepted in the United States is ASTM Method E399, for measuring plane-strain fracture toughness, K_{Ic}, of metals (Ref 5). Because of this, ASTM E399 is perhaps the most important method to consider in the application of fracture mechanics to selection of structural alloys. In addition, many other test methods use some of the same test specimens and test procedures as those in E399.

Method E399 first appeared in the 1969 ASTM Standards, and the latest revision was in 1981. As discussed previously, ASTM E399 specifies the requirements and procedures for measuring the critical value of the stress-intensity factor, K_{Ic}. The measurement corresponds to at most a 2% extension of a pre-existing fatigue crack in a

specimen large enough that plane-strain conditions predominate around the crack. Three types of test specimens which can be used with this method are shown in Fig. 17. They are, in the order in which they were developed for use with E399, the bend specimen, the compact specimen and the arc specimen. These specimens may be taken from plate and other product forms in any of six orientations for various crack-growth directions. These are shown, using the compact specimen as an example, in Fig. 18, which was taken from ASTM E399. For cylindrical products, a similar procedure is used, except that the circumferential and radial (C and R) directions are indicated instead of the long transverse and short transverse (T and S) directions.

The critical dimensions for each of the test specimens are thickness, B, width, W, and overall crack length, a, which includes the machined starter notch and the fatigue precrack. One of the important requirements of this test method is that the specimen thickness, B, and the crack length, a, be at least equal to the quantity $2.5\ (K_{Ic}/\sigma_{ys})^2$, where σ_{ys} is the yield strength of the material. Thus, it is desirable to determine the yield strength and to estimate the K_{Ic} value for any material to be tested by this method before the specimens are prepared. For some materials, often those of

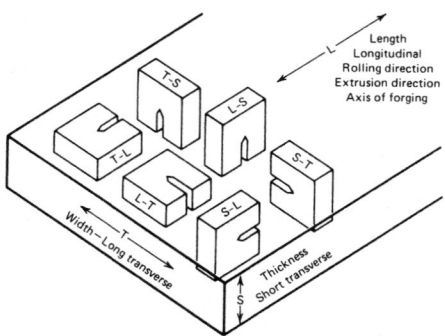

First letter designates the direction perpendicular to the crack plane; second letter designates the direction parallel to the direction of crack growth.

Fig. 18. Orientations of crack plane and direction for fracture specimens taken from rectangular product forms such as plate (from ASTM Method E399)

lower strength, the dimensions required by the relation $2.5\ (K_{Ic}/\sigma_{ys})^2$ are greater than can be obtained from the available section sizes of material. For such combinations of material and section size, measurement of K_{Ic} is not possible. In these cases, alternative methods of fracture-toughness measurement are necessary, as discussed below, under "Non-Plane-Strain Toughness Tests." When measurement of K_{Ic} *is* possible, the K_{Ic} specimen dimensions should be somewhat greater than the minimum requirements estimated from $2.5\ (K_{Ic}/\sigma_{ys})^2$. For a few materials, notably some aluminum alloys, recent testing (Ref 6) has shown that dimensions about twice as great as those normally required give more consistent K_{Ic} results.

Another important set of requirements for a valid K_{Ic} test involves the fatigue-precracking process. The test results will be valid only if the length of the fatigue crack and the straightness and flatness of the crack are within the prescribed limits. Furthermore, the maximum cyclic load used in producing the fatigue crack must be held below a designated limit to restrict the size of the plastic zone ahead of the fatigue crack.

Once the test specimen has been precracked, the procedure of the K_{Ic} test itself is quite similar to the load-versus-displacement procedure of a standard tension test. The displacement used in the K_{Ic} test is the opening displacement of the notch surfaces at the notch mouth and in the direction perpendicular to the plane of the notch and crack. This displacement is called the crack-mouth-opening displacement (see Fig. 17). A calibrated displacement gage and autographic load-versus-displacement recording equipment are needed to measure and record the test data. Special fixtures with loading pins which are free to rotate during the test are required. This ensures that the same free-rotation loading condition used for the stress and K analysis of the specimens is also present during the test. A load-versus-displacement plot for a high-strength aluminum alloy, which is generally representative of the plots from many steel and aluminum alloys, is shown in Fig. 19 (Ref 7).

Interpretation of the load-versus-displacement plot and calculation of K_{Ic} are described in detail for the various specimen types in ASTM E399. The procedures are described briefly here for the compact specimen. The provisional value of the load at which onset of crack propagation occurs,

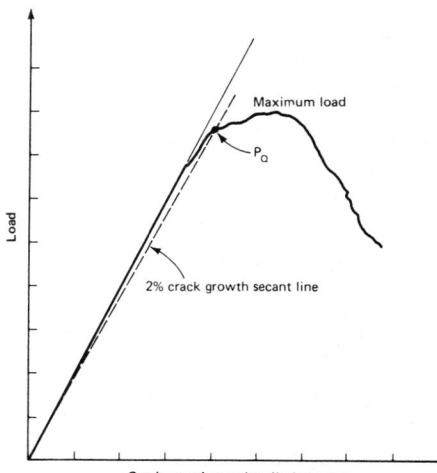

Fig. 19. Load vs crack-mouth-opening displacement typical of high-strength aluminum alloys (Ref 7)

P_Q, is obtained from the test record. The P_Q value is either the maximum load during the test or, more commonly and as shown in Fig. 19, the load corresponding to 2% crack growth as determined by the intersection of a secant line and the plot. The provisional value of the stress-intensity factor, K_Q, is determined by means of equations such as the following, for compact specimens in which K_Q is in units of $MPa \cdot m^{1/2}$ ($psi \cdot in.^{1/2}$):

$$K_Q = (P_Q/BW^{1/2}) \cdot f(a/W) \qquad (Eq\ 16)$$

where

$$f(a/W) =$$

$$\frac{[2 + a/W][0.886 + 4.64a/W - 13.32(a/W)^2 + 14.72(a/W)^3 - 5.6(a/W)^4]}{(1 - a/W)^{3/2}}$$

and where P_Q is load, in MN (lbf); B is specimen thickness, in m (in.); W is specimen width, in m (in.); and a is crack length, in m (in.). It is important to note that, for the compact specimen, the crack length is measured from the centerline of the loading-pin holes. Equation 16 is considered to be accurate within 0.5% over a wide range of crack lengths, for $0.2 \le a/W < 1$, so it can be used for a variety of fracture-mechanics tests and analyses as well as in the range of interest for K_{Ic} tests, $0.45 \le a/W \le 0.55$.

A K_Q value calculated from Eq 16 must meet several criteria in order to be a valid measurement of K_{Ic}. The first, mentioned earlier, is that the specimen thickness and over-all crack length must be equal to or greater than 2.5 $(K_{Ic}/\sigma_{ys})^2$. Also, the maximum load supported by the specimen prior to complete fracture must not be more than 10% greater than P_Q. This further ensures that P_Q corresponds to the load at which crack extension takes place rather than the load at which excessive plastic deformation occurs around the crack. The fatigue-crack portion of the fracture surface is examined and measured as designated in the test method, in regard to straightness of the crack front and other criteria.

Because of the complexity of the calculations and the requirements for validity, some of the laboratories in which these tests are run have de-

veloped computer programs which compute K_Q values and indicate their validity based on appropriate input data. The complexity of the K_{Ic} test method cannot be denied. Nevertheless, this method has been applied to a wide variety of metallic materials in the past ten years, and very similar methods have been adopted by many other standards organizations around the world. In spite of its complexity, ASTM Method E399 is the preferred standard for measurement of plane-strain fracture toughness, K_{Ic}.

New K_{Ic} Tests. A number of test specimens and methods have been proposed as new K_{Ic} procedures. Three new procedures are described here — two that are related primarily to new specimen geometries and a third that is primarily a new method with existing specimen geometries. These procedures are now being generally used to some extent and are in various stages of consideration by ASTM as K_{Ic} test methods.

A test specimen that is much like the compact specimen, but that is round rather than rectangular in shape, has been used in the United States since 1975. Recently, a round compact specimen and its associated K solution have been added to ASTM Method E399 (Ref 8) and is referred to as the disk specimen. The specimen geometry is sketched in Fig. 20. The K solution is similar both in form and in value (within ±6%) to that of the rectangular compact specimen in the K_{Ic} testing range. The advantage of the disk specimen is that it can be less costly to fabricate because turning operations are often faster than milling. In addition, when round cores are cut from structures or product forms for use in K_{Ic} tests, the disk specimen has a clear advantage.

Another test specimen which has been proposed for use in K_{Ic} measurements is the short rod specimen (Ref 9). The specimen geometry is shown in Fig. 21. The significant advantage in the proposed use of this specimen is that, due to the triangular shape of the area to be cracked, a crack initiates and grows stably during a single

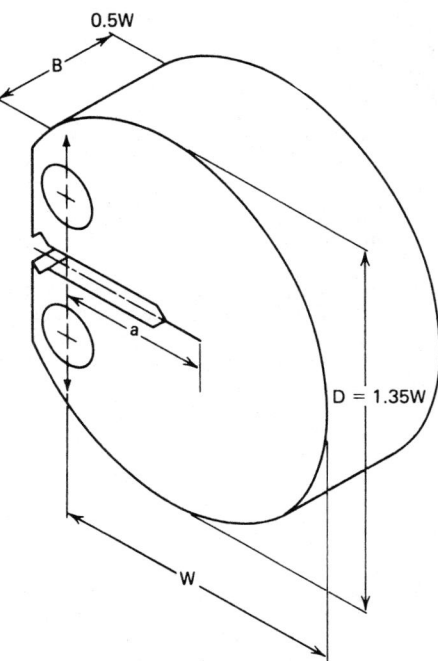

Fig. 20. A round compact K_{Ic} test specimen (the disk specimen)

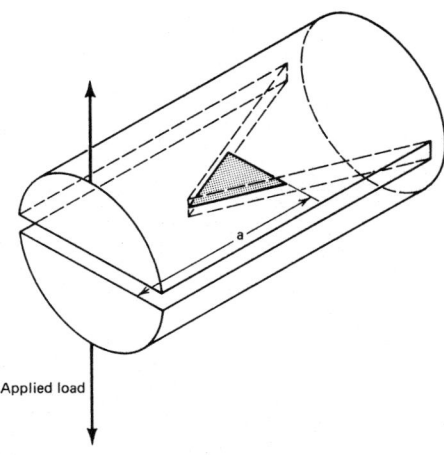

a is notch plus crack length; shaded area indicates crack growth.

Fig. 21. Proposed short rod fracture-toughness specimen

application of a relatively low load. This could make possible K_{Ic} tests without the necessity of a fatigue precrack. However, because cracks which have been produced by a single application of load may be sufficiently different from cracks produced by fatigue loading, the resulting K values may not be consistent with those obtained on fatigue-precracked specimens. Further evaluation is needed to determine whether the short rod specimens qualify for plane-strain fracture toughness testing or for screening tests.

For many years, various laboratories have performed K_{Ic}-type tests at significantly higher loading rates than those specified in E399. Reference 10 describes one example of high-loading-rate K_{Ic} tests on specimens of an alloy steel. A K_{Ic} test procedure based on E399 but at much higher loading rates is being considered by ASTM. Interlaboratory tests have been completed using a pressure-vessel steel tested in closed-loop machines at loading rates in the range of 10^4 $MPa \cdot m^{1/2}/s$; this is about a factor of 10^4 faster than the essentially static loading rate of E399. Compact specimens (Fig. 17b) and many guidelines presented in E399 are generally used in obtaining fracture-toughness data by dynamic K_{Ic} tests. The primary application of dynamic K_{Ic} tests is with the low-to-medium-strength steels, which are known to show loading-rate effects on mechanical properties, particularly yield strength. Problems with the development of the dynamic K_{Ic} test method are related to uncertainties in the yield strength of materials at high loading rates and uncertainties in load measurement at high loading rates.

Non-Plane-Strain Toughness Tests

R-Curve Determination. A consensus R-curve test method is described in ASTM Recommended Practice E561 (Ref 11), first published in 1974 and last revised in 1981. Method E561 describes the determination of the crack-growth resistance of a material under relatively thin-sheet plane-stress conditions which result in a significant amount of plastic deformation at the crack tip. The practice describes the use of center-cracked tensile specimens of the same general type shown in Fig. 10, and compact-type specimens similar to those in Fig. 17(b) except generally much thinner. The basic procedures of the R-curve

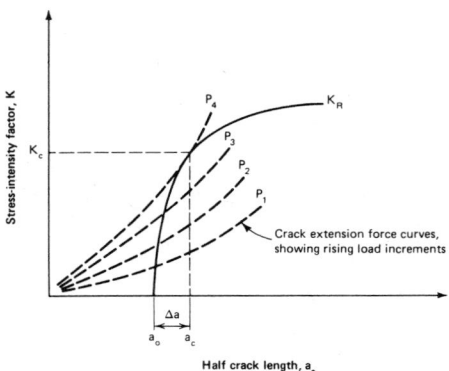

Fig. 22. Typical R-curve test results and analysis (from ASTM Practice E561)

method can be well described by referring to a sketch of typical measurements and calculations from the method using a center-cracked sheet specimen as the example. This sketch, shown in Fig. 22, has coordinates of stress-intensity factor, K, and crack length, a. The solid curve in Fig. 22, marked K_R, is a plot of K applied to the specimen versus the measured crack extension, Δa, produced by the applied K; thus the K_R curve represents the measured crack-extension *resistance* of the material for the particular specimen thickness of the test. The dashed curves are calculations of crack-extension *force* which is available to cause extension of the crack at a given load level. When the applied load is sufficiently high (curve P_4), the crack-extension force exceeds the crack-extension resistance of the material and the crack will grow. Moreover, since the slope of the crack-extension force curve is larger than that of the K_R curve, the crack will become unstable and grow to failure unless the load is decreased. The K value at this point, K_c, is the plane-stress fracture toughness, at which unstable crack growth occurs for this particular combination of material, thickness and other specimen dimensions. The K_R curve is unique to the material and specimen thickness, whereas K_c values may depend on other specimen dimensions in addition to thickness.

The measured data required to plot the K_R curve — that is, load and crack extension — can often be obtained using the same equipment and procedures as those used in K_{Ic} tests. The calculations required to plot the crack-extension force curves are based on the K solution for the specimen, such as Eq 16; the only special requirement is that an adjustment be added to the actual physical crack length, a_p, to account for the effective extension of the crack due to a crack-tip plastic zone. Thus, the effective crack size is:

$$a_e = a_p + \frac{K^2}{2\pi\sigma_Y} \qquad \text{(Eq 17)}$$

where σ_Y is the effective yield strength — that is, the average of the yield and ultimate strengths of the material. Note that the estimate of plastic-zone size from Eq 14 is used here.

Once a K_R curve is obtained for a given material and thickness, it can be used to predict the load at which unstable crack extension will occur in any other geometry, providing that the K solution is known so that the crack-extension curve can be calculated. The K_R-curve approach accounts for crack-tip plastic deformation as dis-

cussed above, with the general limitation that the specimen must remain predominantly elastic. Additional information on R-curve methods and their use is available in Ref 12 as well as in ASTM Practice E561 (Ref 11).

J_{Ic} Fracture Toughness. The development of elastic-plastic fracture-toughness tests based on the J-integral concept followed soon after the analyses and estimation of J by Rice, Paris and Merkle (Ref 13) for specimen geometries commonly used in fracture-toughness testing. ASTM has sponsored two cooperative J_{Ic} testing programs, and, based on results from these programs, a proposed J_{Ic} test method has been described in the literature (Ref 14) and has been published as an ASTM standard (Ref 15).

The J_{Ic} test method can be outlined and described in relation to the sketch of typical results in Fig. 23. Values of J are calculated from load versus load-point-displacement data obtained from bend or compact specimens using generally the same specimens, procedures and equipment as in E399. A key requirement is that the displacement must be at the load point; this is allowed but not required in a K_{Ic} test. With this requirement, the area A under the load-versus-displacement curve is a true measure of the combined elastic-and-plastic-strain-energy input to the specimen. The J value is calculated from:

$$J = \frac{2A}{B(W-a)} \qquad \text{(Eq 18)}$$

which applies directly for the bend specimen and with some modification for the compact specimen. The dimensions B, W and a are as described in Fig. 17. At least four specimens are

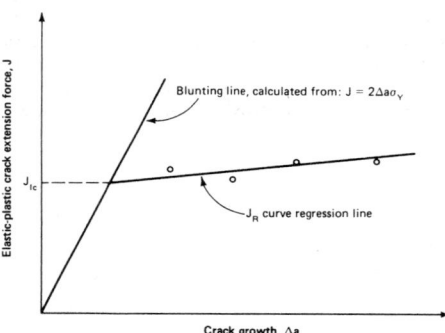

Fig. 23. Typical J_{Ic} test results and analysis (Ref 14)

loaded to produce a range of crack extension, Δa, of about 1 to 2 mm. The value of J calculated from each specimen is plotted versus Δa, and a linear-regression line is fitted to the data. This is the J_R curve as represented by a regression line (see Fig. 23). The blunting line, calculated from $J = 2\Delta a\sigma_Y$, is drawn to represent the amount of apparent crack extension associated with the crack-tip blunting which occurs before actual crack extension occurs. The value of J_{Ic} is determined at the intersection of the J_R curve regression line and the blunting line; so J_{Ic} is a measure of the fracture toughness of the material at the start of actual crack extension from the blunted crack tip.

An important advantage of the J_{Ic} test method is that it can accommodate a significant amount of crack-tip blunting and general plastic deformation in the specimen. If the amount of plastic deformation is small enough, J_{Ic} will be identical

to G_{Ic}, and thus J_{Ic} can be converted to an approximately equivalent measure of K_{Ic} (see Eq 10). For large amounts of plastic deformation, a specimen size requirement limits the size of the specimen and, indirectly, the amount of plastic deformation which can be allowed. The specimen size requirement allows a significantly smaller specimen, often ten times smaller, to be tested with the J_{Ic} procedure than with the K_{Ic} procedure. So, although the J_{Ic} test is relatively time consuming due to multiple tests, it can be used over a wider range of material properties and specimen sizes than the K_{Ic} test. In addition, single-specimen J_{Ic} test procedures, such as incremental unloading methods, can reduce both testing time and the required number of specimens in obtaining J_{Ic} test data.

COD Methods. A British test method, "Methods for Crack Opening Displacement (COD) Testing" (Ref 16), is generally similar to the ASTM K_{Ic} method. In this method, a bend specimen is used as in Fig. 17(a), and a load versus crack-mouth-opening-displacement (CMOD) plot is obtained as in Fig. 19. Then, by use of results from experiments and analyses, the displacement at the crack tip, called the crack-tip-opening displacement, or CTOD, is calculated from CMOD. Critical values of CTOD are defined including values corresponding to elastic, plane-strain conditions and values corresponding to significant amounts of plastic deformation. Methods which determine CTOD have the advantage of concentrating on the area in which the actual crack-extension process occurs; however, CTOD methods have the disadvantage of being indirect measurements of the parameter of interest — that is, the opening of the crack faces at the crack tip. Harrison *et al* (Ref 17) have described the British COD method and its application to welded structural components.

Correlation Fracture Tests

Several tests with notched or cracked specimens have been developed for specific engineering applications, in contrast to the tests described above under "Plane-Strain Fracture Toughness, K_{Ic}" and "Non-Plane-Strain Toughness Tests," which were developed to measure directly the fracture-toughness properties of materials. These engineering-application tests have been used extensively for correlation and screening in relation to fracture-toughness tests.

The most long-standing and probably the most widely used test which is used for correlation with fracture toughness is ASTM Method E23, "Notched Bar Impact Testing of Metallic Materials" (Ref 18), commonly known as the Charpy impact test. The Charpy specimen is a three-point bend-type specimen of one size, with B = W = 10 mm and with a length between support points of 40 mm (see Fig. 24). A somewhat similar test and test specimen are described in ASTM Method E604, "Dynamic Tear Energy of Metallic Materials" (Ref 19), which uses a bend specimen with B = 16 mm, W = 38 mm, and a length of 165 mm. Both of these methods measure the energy required to break the notched bend specimen by impact loading with a falling mass. The Charpy impact test, as well as the dynamic tear test, measure the total energy required to initiate and then grow a crack to complete failure of the specimen, whereas K_{Ic} is a measure of K for initial growth of a pre-existing crack. This difference, along with the differences in loading rate and specimen size between Charpy impact and

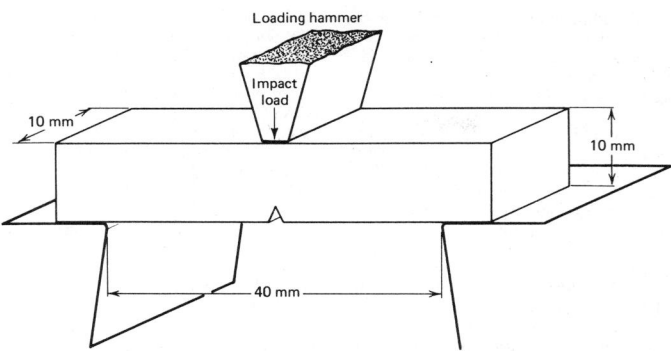

Fig. 24. Schematic illustration of Charpy impact specimen and test arrangement

K_{Ic} tests, limits the correlations between Charpy energy and K_{Ic} to certain materials and to certain ranges of Charpy energy and K_{Ic} values. However, the use of Charpy impact tests to correlate with K_{Ic} has resulted in significant time and cost savings, particularly with medium-to-high strength steels. Many structural components are purchased and inspected to a Charpy energy specification which has been shown to correspond in laboratory testing to a certain required level of K_{Ic}. However, errors could result if the correlation between Charpy energy and K_{Ic} is extended beyond the range of laboratory data for which it was established.

Two tests which are used for correlation with K_{Ic}, generally with high-strength aluminum alloys, are ASTM Method E338, "Sharp-Notch Tension Testing of High-Strength Sheet Materials" (Ref 20), and ASTM Method E602, "Sharp-Notch Tension Testing with Cylindrical Specimens" (Ref 21). Method E338 describes tests on center-notched and edge-notched sheet specimens; Method E602 describes tests on circumferentially notched cylindrical specimens. Both methods require a maximum radius of 0.018 mm (0.0007 in.) at the notch, and this effectively limits these methods to aluminum and magnesium alloys, which are readily machinable. The sharp-notch tensile-strength parameters are used extensively in the aluminum industry to screen materials in respect to their ability to meet corresponding K_{Ic} requirements. The test results are very sensitive to notch sharpness, and thus machining and inspection of the notch must be done carefully to ensure proper correlation.

Two tests that employ the Charpy specimen (Fig. 24) modified by a fatigue precrack ahead of the notch are now being considered by ASTM. They are the proposed "Method for Nominal Crack Strength of Slow-Bend Precracked Charpy Specimens of High-Strength Metallic Materials" and the proposed "Method for Impact Testing of Precracked Charpy Specimens of Metallic Materials." The slow-bend precracked Charpy method uses the maximum load during the test to calculate a nominal crack strength of the specimen which may be correlated with K_{Ic}. The impact-precracked Charpy method uses an instrumented tup on the load hammer to obtain a load-versus-time plot which, if it meets certain criteria, can be used to calculate a maximum K value which is a measure of dynamic fracture toughness. Both of these proposed precracked Charpy test methods measure a maximum stress or K value associated with crack growth from a pre-existing crack; this type of measurement is similar to that of a K_{Ic} test, so the results from these methods should correlate better with K_{Ic} than do the re-

sults of other methods. This advantage is offset by the increased difficulty in precracking the specimens and in performing and analyzing the tests.

A review of the correlation fracture tests described above, as well as others, is given in a report of the National Materials Advisory Board (Ref 22).

Sustained-Loading Crack-Growth Tests

The two general types of sustained-loading crack-growth tests are (a) determination of K_{Iscc}, the lowest threshold value of K_I at which stress-corrosion cracking occurs, and (b) determination of stress-corrosion crack-growth rate, da/dt, resulting from the combination of sustained loading and an aggressive environment. Similar terminology and testing procedures are used for threshold, K_{th}, and crack-growth rate, da/dt, involving internal or external embrittling species such as hydrogen or liquid metals. A standard method for measuring K_{Iscc} is now (1982) being prepared by ASTM Committee E24 on Fracture. The method is based on a large body of research, such as in Ref 23, as well as on the experience and results of an ASTM interlaboratory testing program. Tests were run at 17 laboratories using AISI 4340 steel specimens in 3.5% NaCl for times up to 7000 h. Additional interlaboratory tests on 7075-T7651 aluminum alloy are under way in a cooperative program of ASTM Committees E24 and G1 on Corrosion of Metals.

In determining K_{Iscc}, a group of fatigue-precracked specimens are held at different values of K_I (calculated from load and crack length) in the environment of interest, and the lowest value of K_I at which cracking occurs is K_{Iscc}. Another procedure for determining K_{Iscc} is the rising-load procedure, in which the load on a single precracked specimen is very slowly increased until crack growth occurs, and the associated value of

K_I is K_{Iscc}. Rising-load K_{Iscc} tests of an AISI 4340 steel in an H_2S atmosphere are described by Clark in Ref 24.

Measurements of stress-corrosion crack-growth rate, da/dt, are performed at constant load with crack-length measurements taken at various time increments depending on the value of da/dt. Direct measurements of crack length on the specimen surfaces are made if the environment and test plan allow it. Indirect measurement of crack length is often made by measuring specimen displacement—usually the crack-mouth-opening displacement (CMOD)—as shown in Fig. 17. The crack length can then be determined by using the known compliance relations between CMOD and crack length.

Various specimen geometries are used for stress-corrosion-cracking tests. Specimen orientation is designated as it is for fracture-toughness specimens (Fig. 18). The more commonly used specimens are described here. The most frequently used is the compact specimen; it is tension-loaded through pins as in other fracture tests (Fig. 25a), or it is self-loaded using a bolt (Fig. 25b). A double-beam specimen (often called double cantilever beam), with a configuration similar to that of the compact specimen, is used particularly for da/dt tests (see Fig. 25c). Because the crack is liable to grow out of the intended crack plane, side grooves are usually added to the double-beam specimen. Bend specimens also are used, either in three-point bending or in cantilever loading, as shown in Fig. 25(d).

Use of Fracture-Test Results

Measured values of plane-strain fracture toughness, K_{Ic}, and non-plane-strain and correlation measures of fracture toughness, are used in general to calculate the critical crack size in a loaded component which will lead to an abrupt failure. A critical-size crack may be assumed to have been present in the component as fabricated, or may have grown to critical size during service by subcritical crack growth due to fatigue loading or stress-corrosion cracking. Regardless of the cause of the crack, the measured value of toughness is used to calculate the combination of crack size and load which will cause failure.

Measured values of da/dN, K_{Iscc} and da/dt are used to determine the conditions for subcritical crack growth. These test results are used to calculate the crack sizes and loading combinations which will cause cracks to grow from small initial sizes to critical sizes.

Compilations of fracture-mechanics data for specific materials have been included in various handbooks such as the *Damage Tolerant Design Handbook* (Ref 25) and in reports of testing programs such as "Fracture Mechanics Evaluation of B-1 Materials" (Ref 26). Data from such sources may be used in a preliminary assessment

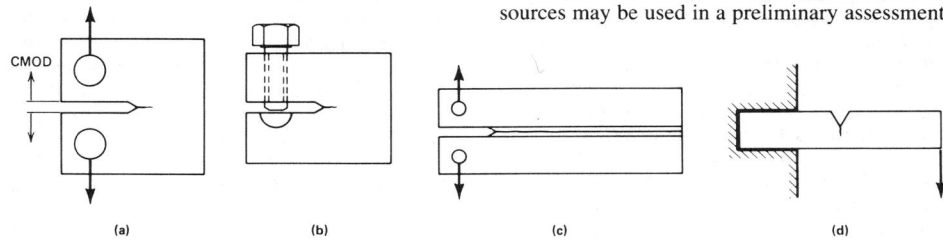

(a) Compact specimen: tension loaded; general use. (b) Modified compact specimen: bolt loaded; K_{Iscc} testing. (c) Double-beam specimen: tension loaded; K_{Iscc} and da/dt testing. (d) Bend specimen: cantilever loaded; K_{Iscc} and da/dt testing.

Fig. 25. Schematic illustration of test specimens and loading used in sustained-load crack-growth testing

of materials and their fracture properties, but when dealing with critical structures, test data should be obtained on specimens that are truly representative of the material in the structures.

REFERENCES

1. The Phenomena of Rupture and Flow in Solids, by A. A. Griffith: *Philosophical Transactions,* Vol 221, 1921, p 163-198

2. *Compendium of Stress Intensity Factors,* by D. P. Rooke and D. J. Cartwright: Hillingdon Press for Her Majesty's Stationery Office, Uxbridge, Middlesex, UK, 1976

3. *The Stress Analysis of Cracks Handbook,* by H. Tada, P. C. Paris and G. R. Irwin: Del Research Corporation, Hellertown, PA, 1973

4. The Influence of Crack Length and Thickness in Plane Strain Fracture Toughness Tests, by M. H. Jones and W. F. Brown, Jr.: in *Review of Developments in Plane Strain Fracture Toughness Testing,* STP 463, American Society for Testing and Materials, Philadelphia, 1970, p 63-101

5. Standard Test Method for Plane-Strain Fracture Toughness of Metallic Materials: E399-81, *1981 Annual Book of ASTM Standards,* Part 10, American Society for Testing and Materials, Philadelphia, 1981, p 588-618

6. Experience in Plane-Strain Fracture Toughness Testing per ASTM Method E399, by J. G. Kaufman: in *Developments in Fracture Mechanics Test Methods Standardization,* edited by W. F. Brown, Jr., and J. G. Kaufman, STP 632, American Society for Testing and Materials, Philadelphia, 1977, p 3-24

7. Fracture Toughness of Plain and Welded 3-In.-Thick Aluminum Plate, by F. G. Nelson and J. G. Kaufman: in *Progress in Flaw Growth and Fracture Toughness Testing,* STP 536, American Society for Testing and Materials, Philadelphia, 1973, p 350-376

8. A Proposed Standard Round Compact Specimen for Plane Strain Fracture Toughness Testing, by J. H. Underwood, J. C. Newman, Jr., and R. R. Seeley: *Journal of Testing and Evaluation,* Vol 8, No. 6, 1980, p 308-313

9. A Simplified Method for Measuring Plane Strain Fracture Toughness, by L. M. Barker: *Engineering Fracture Mechanics,* Vol 9, 1977, p 361-369

10. Fracture Toughness Properties of SA540 Steels for Nuclear Bolting Applications, by R. R. Seeley, W. A. Van Der Sluys and A. L. Lowe, Jr.: *Journal of Pressure Vessel Technology,* Vol 99, 1977, p 419-426

11. Standard Recommended Practice for R-Curve Determination: E561-81, *1981 Annual Book of ASTM Standards,* Part 10, American Society for Testing and Materials, Philadelphia, 1981, p 673-692

12. *Fracture Toughness Evaluation by R-Curve Methods:* STP 527, American Society for Testing and Materials, Philadelphia, 1973

13. Some Further Results of J-Integral Analysis and Estimates, by J. R. Rice, P. C. Paris and J. G. Merkle: in *Progress in Flaw Growth and Fracture Toughness Testing,* STP 536, American Society for Testing and Materials, Philadelphia, 1973, p 231-245

14. A Procedure for the Determination of Ductile Fracture Toughness Values Using J Integral Techniques, by G. A. Clarke, W. R. Andrews, J. A. Begley, J. K. Donald, G. T. Embley, J. D. Landes, D. E. McCabe and J. H. Underwood: *Journal of Testing and Evaluation,* Vol 7, Jan 1979, p 49-56

15. Standard Test for J_{Ic}, A Measure of Fracture Toughness: E813-81, *1981 Annual Book of ASTM Standards,* Part 10, American Society for Testing and Materials, Philadelphia, 1981, p 810-828

16. *Methods for Crack Opening Displacement (COD) Testing*: BS.5762, British Standards Institution, London, 1979

17. The COD Approach and Its Application to Welded Structures, by J. D. Harrison, M. G. Dawes, G. L. Archer and M. S. Kamath: in *Elastic Plastic Fracture,* STP 668, American Society for Testing and Materials, Philadelphia, 1979, p 606-631

18. Standard Methods for Notched Bar Impact Testing of Metallic Materials: E23-81, *1981 Annual Book of ASTM Standards,* Part 10, American Society for Testing and Materials, Philadelphia, 1981, p 273-296

19. Standard Test Method for Dynamic Tear Energy of Metallic Materials: E604-80, *1981 Annual Book of ASTM Standards,* Part 10, American Society for Testing and Materials, Philadelphia, 1981, p 702-710

20. Standard Method of Sharp-Notch Tension Testing of High-Strength Sheet Materials: E338-81, *1981 Annual Book of ASTM Standards,* Part 10, American Society for Testing and Materials, Philadelphia, 1981, p 553-560

21. Standard Method for Sharp-Notch Tension Testing with Cylindrical Specimens: E602-81, *1981 Annual Book of ASTM Standards,* Part 10, American Society for Testing and Materials, Philadelphia, 1981, p 693-701

22. *Rapid Inexpensive Tests for Determining Fracture Toughness:* National Materials Advisory Board, National Academy of Sciences, Washington, 1976

23. *Stress Corrosion—New Approaches,* edited by H. L. Craig, Jr.: STP 610, American Society for Testing and Materials, Philadelphia, 1976

24. Effect of Cold Working on K_{Iscc} in a 4340 Steel, by W. G. Clark, Jr.: in *Flaw Growth and Fracture,* STP 631, American Society for Testing and Materials, Philadelphia, 1977, p 331-344

25. *Damage Tolerant Design Handbook:* Report MCIC-HB-01, Metals and Ceramics Information Center, Battelle Columbus Laboratories, Columbus, OH, 1975

26. Fracture Mechanics Evaluation of B-1 Materials, by R. R. Ferguson and R. C. Berryman: Report AFML-TR-76-137, Rockwell International, B-1 Div., Los Angeles, Contract F33657-70-C-0800, Air Force Materials Laboratory, Wright-Patterson AFB, OH, Oct 1976

Fatigue Testing

A METAL subjected to repetitive loads below the yield stress may fail after a certain number of cycles. A large number of components are subjected to this type of loading (e.g., automotive engines and turbine components), and the design has to obey the constraints of the material strength under these conditions. Failure of materials under cyclic loading is called "fatigue failure," and techniques for measuring susceptibility to fatigue failure are called "fatigue testing." It has been stated (Ref 1) that fatigue accounts for at least 90% of all service failures due to mechanical causes; this shows the importance of the subject. Most fatigue failures start at the surface and progress initially in a slow manner; after the slowly growing crack has reduced the cross-sectional area sufficiently, fracture occurs suddenly. The first portion of the fracture-propagation process yields a fracture surface that usually is significantly different from the second portion. The slowly growing portion produces striations at each cycle, corresponding to each small advance of the crack front. Sometimes these striations are eliminated by subsequent "rubbing" of the surfaces. The second portion of the fracture, corresponding to the rapid propagation of the crack, shows surface features typical of tensile fracture: dimples (in the case of ductile fracture) or cleavage facets (in the case of brittle fracture).

In this article, the more traditional fatigue-testing methods, which provide S-N curves, are described first; the rotating-bending technique with the Moore machine is the best-known test. Next, the methods for analysis and presentation of these data are given. The more modern techniques, in which fatigue-crack growth is monitored as the process proceeds and in which use is made of fracture mechanics, are then presented. Fatigue life is highly susceptible to stress concentrations and to the condition of the surface.

ASTM STANDARDIZED SPECIMENS

ASTM E466 specifies specimens to be used in axial fatigue tests. For bending tests (rotating bending and peak bending), there are no ASTM specifications. The specific dimensions of specimens depend on the objective of the experimental program, on the machine to be used, and on the available material. ASTM does not specify dimensions, but details preparation techniques and reporting techniques. In reporting, a sketch of the specimen, with dimensions, should be given. The surface-roughness and out-of-flatness dimensions should be included. Specimens should not be subjected to any surface treatment. The surface preparation is extremely critical to all fatigue specimens. For axial loading, ASTM E466 states that, regardless of the machining, grinding or polishing method used, the final metal removal should be in a direction approximately parallel to the longitudinal axis of the specimen. Improper preparation methods can greatly bias the results. For instance, Fluck (Ref 2) reports that AISI 3130 steel tested under completely reversed bending at 95 000 psi has a fatigue life of 24 000 cycles when lathe-formed (surface roughness of 105 μin.) and a fatigue life of 234 000 cycles when ground and polished (surface roughness of 2 μin.). Hence, preparation techniques should be carefully developed; if a change in the preparation technique is made, it has to be demonstrated that it does not introduce any bias in the results.

TESTING MACHINES

Rotating-Bending Machines

Rotating-bending fatigue tests of the simple beam type are performed in testing machines such as that illustrated in Fig. 1, sometimes called the R. R. Moore testing machine. In operation, an electric motor rotates a cylindrical specimen, usually at 1800 rpm or higher, while a simple mechanical counter records the number of cycles. Loads are applied to the center of the specimen by a system of bearings and dead weights. A limit switch stops the test when the specimen breaks and the weights descend.

The weights produce a moment that causes the specimen to bend. A strain gage placed on the specimen shows compressive stresses on the top and tensile stresses on the bottom when the gage is rotated to the bottom. Stresses range from maximum tension to maximum compression during each revolution of the testing machine. Figure 2 shows a typical R. R. Moore machine.

Bending moments can be converted to stress

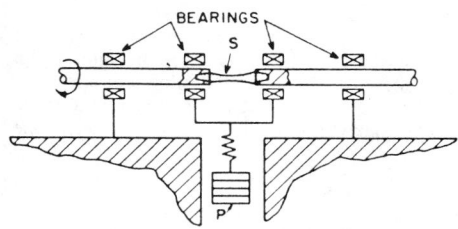

S indicates specimen. P indicates load.

Fig. 1. Loading arrangement for a rotating-beam fatigue-testing machine

by assuming that they are elastic and by employing the flexure formula:

$$\sigma = \frac{MC}{I}$$

For circular specimens, $I = \pi C^4/4$, where C is the specimen radius. The maximum stress at the outer fiber, σ, is proportional to the bending moment, M. This moment is the product of the moment arm and the force. In Fig. 2, it can be seen that the weight is equally distributed between the two sides. Hence, $M = L(P/2)$, where L is the distance from the end of support to the center of application of load.

The specimen is machined from the material to be tested, as shown in Fig. 3, and is fastened

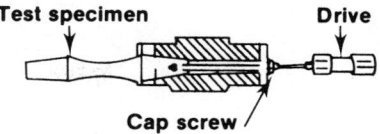
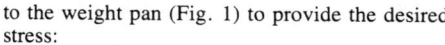

Fig. 4. Cap screw used to fasten the specimen into the bearing housing for a rotating-beam fatigue-testing machine

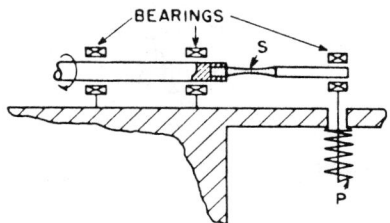

S indicates specimen. P indicates load.

Fig. 5. Loading arrangement for a cantilever-beam fatigue machine for rotating-bending testing

into the bearing housing with special cap screws, as shown in Fig. 4. The effective dead weight of the R. R. Moore machine and weighing apparatus (Fig. 2) is 10 lb (4.54 kg), which is deducted from the total weight required and added

to the weight pan (Fig. 1) to provide the desired stress:

$$40.13 \text{ lb} - 10 \text{ lb} = 30.13 \text{ lb}$$

$$(18.22 \text{ kg} - 4.54 \text{ kg} = 13.68 \text{ kg})$$

When the drive motor is actuated, a counter records the number of revolutions. If the specimen breaks, the bearing housing descends and actuates a switch that shuts off the drive motor. If the specimen does not break (carbon and low-alloy steels may achieve a million or more cycles), the stress is at or below the endurance limit. Next, the machine is shut off and another specimen is run at a higher stress level. A series of tests is performed to provide sufficient data at varying stress levels.

For cantilever-beam rotating-bending machines of the White-Souther type (Fig. 5), a different bending moment is used in stress calculation. A weight, P, is supported by fixture to a ball-bearing housing at the free end of the specimen. This produces a bending moment, M, that equals $P \times L$, which is the distance of the specimen from the center of the applied load, 3 in. (75 mm). The stress in the outer fiber is

$$S = \frac{PL}{Ic}$$

The weight added to the weight pan is the calculated weight minus the weight of the weighing apparatus.

The cantilever specimen is machined from the material to be tested, as shown in Fig. 6, and is fastened into the bearing housing at one end. The weighing apparatus is assembled, proper weights are applied, and the drive motor is actuated. A series of tests is run and data are plotted to provide an S-N graph.

Plate-Bending Machines

In rotating-bending tests, the mean or average stress is always zero. The effect of mean stress, which is very important in fatigue, is evaluated by cantilever bending machines that are used to test plate materials. In these machines, sometimes called Kraus plate-bending machines, specimens are loaded with constant deflection by means of an eccentric crank (Fig. 7). Stresses can be calculated by assuming that they are elastic. In many cases this is not a good assumption, because, when tested, some soft materials may involve small amounts of plastic stress.

Specimens are usually tapered to provide a constant-stress test area. The approximate stress, S, is given by:

$$S = \frac{ytE}{l^2}$$

where y is the specimen deflection, t is the thickness of the specimen, E is the elastic modulus, and l is the distance from load application to the back of the specimen. Constant-deflection beam-type machines are used to test both strip and plate.

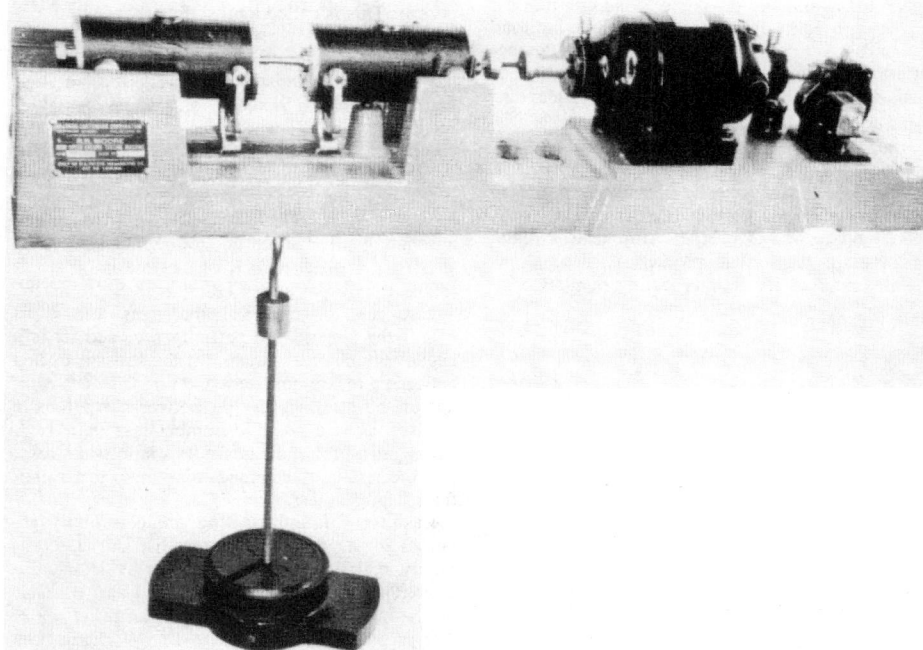

Fig. 2. Typical R. R. Moore machine for rotating-bending fatigue testing

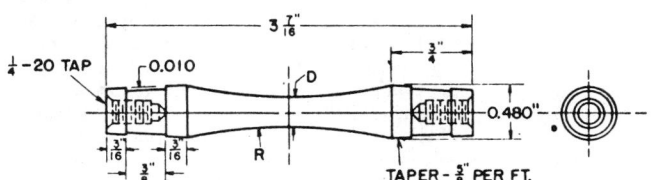

D = 0.200 to 0.400 in., selected on basis of ultimate strength of material.
R = 3.5 to 10 in.

Fig. 3. Simple rotating-beam specimen for fatigue testing

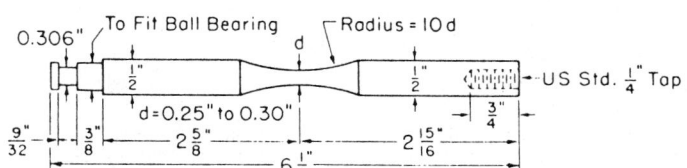

Fig. 6. Rotating cantilever-beam specimen for fatigue testing

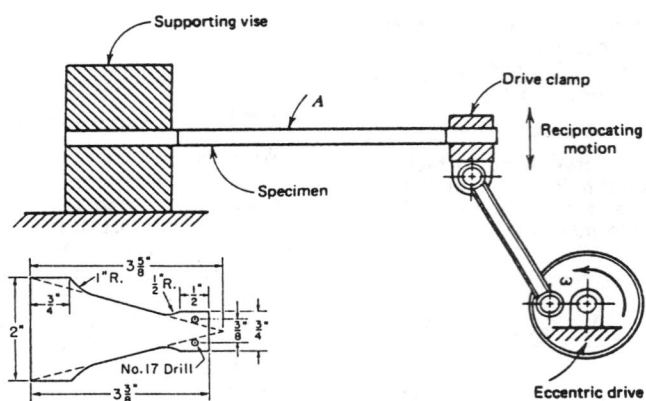

Fig. 7. Reciprocating-bending fatigue-testing machine, and typical specimen (at lower left) for testing of sheet

A typical specimen is shown at lower left in Fig. 7.

Resonant-Testing Machines

Machines for resonant testing are basically spring-mass, vibrating systems. The frame design is based on a resonant, spring-mass system that consists of two masses linked by the specimen and grip string and that oscillates as a dipole. The system is excited by an electromagnet housed in the machine base. The masses and load string are positioned in the vertical frame, which is suspended and guided on leaf springs. The eight springs are arranged in a special configuration to make a unique and compact design without the need for a heavy seismic block.

Mean load is applied by a motor located in the base of the system. The motor drives the four corner gearboxes through two shafts and applies mean loads in both tension and compression. The mean-load force is carried by the box-type structure of springs, and the level either is adjusted by a hand-held controller or is maintained at a level preset by the controller. The magnet air gap is maintained automatically by the action of the gap servomotor driving a wedge beneath the electromagnet. A linear variable displacement transformer (LVDT) constantly monitors the air gap, and control is maintained even when the mean load is changed while the machine is run-

ning. A manually operated drive located in the upper mass permits major adjustment of specimen spacing.

The electromagnet excites the dual-mass system at its natural frequency by means of pulse excitation. This feature enables a simple switch to replace the conventional power amplifier, thus providing high reliability at low cost and, in addition, eliminating the need to tune an oscillator to the natural frequency of the system. Closed-loop amplitude control is achieved by controlling the pulse power to the magnet from the error between the actual and demanded load amplitudes, thereby providing fast response to changing load demands. A strain-gaged load cell provides accurate load monitoring and digital indication of peak dynamic load, mean load, and frequency.

STRESS-LIFE DATA

ASTM Specifications for Constant Amplitude Axial Fatigue Tests of Metallic Materials, E466, and Standard 468, Constant Amplitude Fatigue Test Results for Metallic Materials, are recommended sources for fatigue-testing results.

Data Presentation

In laboratory fatigue testing, the specimen is loaded so that stress is cycled between either a maximum and a minimum tensile stress or a

maximum tensile stress and a maximum compressive stress. The latter, considered a negative tensile stress, is given an algebraic minus sign and, therefore, is called the minimum stress.

The mean stress, S_m, is the algebraic average of the maximum and minimum stresses in one cycle:

$$S_m = (S_{max} + S_{min})/2$$

In the completely reversed test, the mean stress is zero. The range of stress, ΔS, is the algebraic difference between the maximum and minimum stresses in one cycle:

$$\Delta S = S_{max} - S_{min}$$

The stress amplitude, S_a, is one half the range of stress:

$$S_a = \Delta S/2 = (S_{max} - S_{min})/2$$

During a fatigue test, the stress cycle is usually maintained constant so that applied stress conditions can be written as $S_m \pm S_a$, where S_m is the static or mean stress, and S_a is the alternating stress equal to half the stress range. The positive sign is used to denote a tensile stress; the negative sign denotes a compressive stress. Some of the possible combinations of S_m and S_a are illustrated in Fig. 8. When $S_m = 0$ (Fig. 8a), the maximum tensile stress is equal to the maximum compressive stress. This is called an alternating stress or a completely reversed stress. When $S_m = S_a$ (Fig. 8b), the minimum stress of the cycle is zero. This is called a pulsating or repeated tensile (or compressive) stress. Any other combination is known as a fluctuating stress, which may be a fluctuating tensile stress (Fig. 8c), a fluctuating compressive stress, or a stress that fluctuates between a tensile and a compressive value (Fig. 8d).

Stress ratio is the algebraic ratio of two specified stress values in a stress cycle. Two commonly used stress ratios are the ratio A of the alternating stress amplitude to the mean stress (A = S_a/S_m) and the ratio R of the minimum stress to the maximum stress (R = S_{min}/S_{max}). If the stresses are fully reversed, the stress ratio, R, is -1; if the stresses are partially reversed, R is a negative number less than 1; if the stress is cycled between a maximum stress and no load, R is zero; and if the stress is cycled between two tensile stresses, R is a positive number less than 1. A stress ratio of 1 indicates no variation in stress, and the test is a sustained-load creep test rather than a fatigue test.

Results of fatigue testing are commonly reported in the form of a stress-life curve, commonly called an S-N curve. Stress is usually plotted vertically on a linear scale and life, N, on the horizontal axis using a logarithmic scale. A typical curve is shown in Fig. 9. Each data point represents a single test. Typically a stress level is selected and the specimen is cycled until it breaks into two pieces or a predetermined number of cycles is reached. Usually, this number is between 10^6 and 10^7. If a specimen does not fail during the test it is called a "runout" and the data points are denoted with an arrow.

Fatigue limit (or endurance limit) is the value of the stress below which a material can presumably endure an infinite number of stress cycles—that is, the stress at which the S-N diagram becomes and appears to remain horizontal. The existence of a fatigue limit only occurs for carbon and low-alloy steels. It is the exception, rather than the rule.

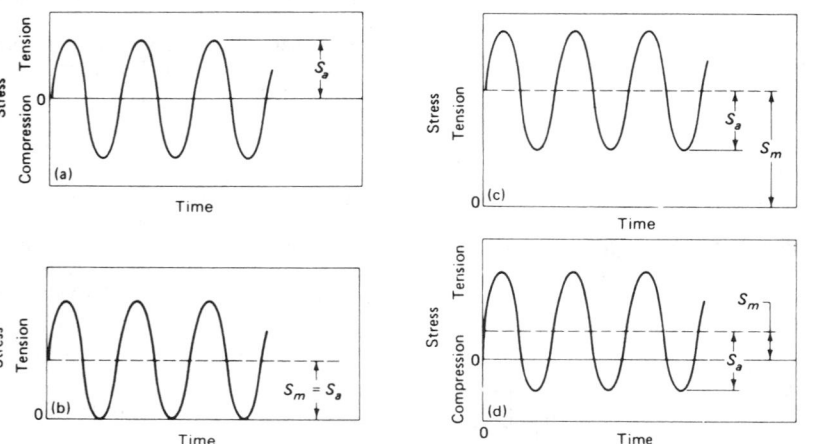

(a) Alternating stress for which $S_m = 0$ and for which R (ratio of minimum stress to maximum stress) $= -1$. (b) Pulsating tensile stress for which $S_m = S_a$, the minimum stress is zero, and R = 0. (c) Fluctuating tensile stress for which both the minimum and maximum stresses are tensile stresses, R = 1/3. (d) Fluctuating tensile-to-compressive stress for which the maximum stress is a tensile stress, R = $-1/3$.

Fig. 8. Four types of fatigue-test stresses

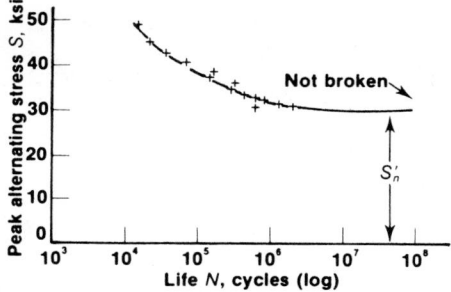

Fig. 9. S-N curve with data plotted on semilog coordinates

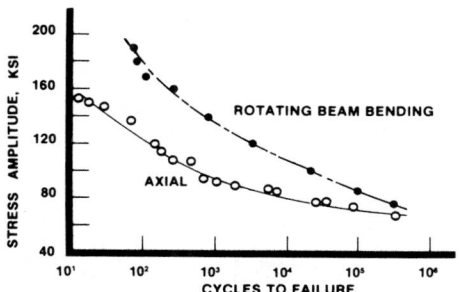

Fig. 10. Fatigue data under axial loading and in rotating bending for 4130 steel

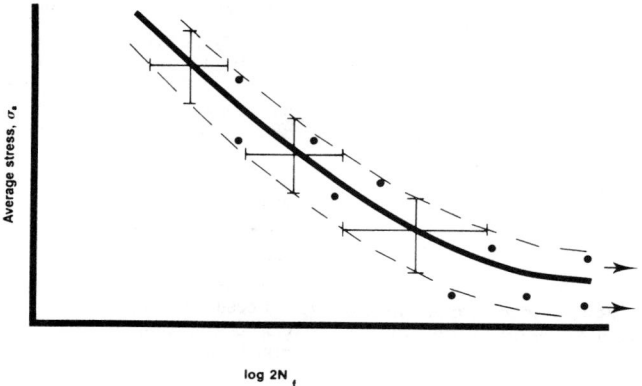

$2N_f$ is reversals to failure. σ_a is the steady-state stress amplitude.

Fig. 11. Scatter band showing S-N curves for fatigue data

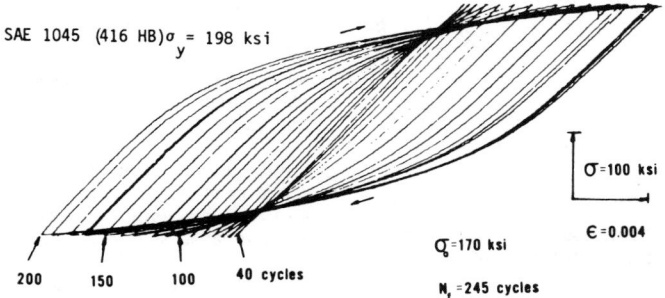

Fig. 12. Cyclic softening of a steel under controlled-stress cycling

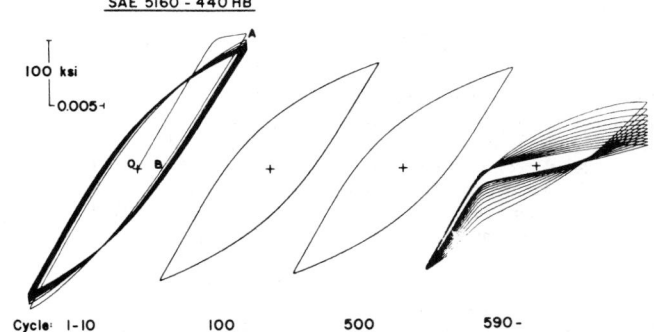

Fig. 13. Cyclic softening of a steel under controlled-strain cycling

Fatigue strength, which should not be confused with fatigue limit, is the stress to which the material can be subjected for a specified number of cycles. Fatigue strength is used for materials, such as most nonferrous metals, that do not exhibit well-defined fatigue limits.

In rotating-beam tests, only the outer fiber is under maximum stress, while in axial tests the entire cross section is under maximum stress. The increased volume of highly stressed material makes failure more likely in axial tests. As a result, data from rotating-bending tests should not be employed for estimating the fatigue lives of structures and components, but only for a rough comparison between two materials. Caution is advised when using the data for design of structures and components. Figure 10 shows S-N curves for both cases; the difference in fatigue life is significant.

Statistical Aspects

In fatigue testing—perhaps more than in other types of mechanical testing—there are always some variations in the material properties. It is not uncommon for two "identical" specimens to have fatigue lives differing by a factor of two. Distributions in fatigue life, as well as fatigue strength, are important, as shown in Fig. 11. Typical S-N curves represent the mean of all test results. Curves often are drawn to display the scatter band of data obtained with multiple tests.

Transient Behavior of Metals

Metals are metastable under application of cyclic loads, and their stress-strain response can be drastically altered when they are subjected to repeated plastic strains. Depending on its initial state (quenched and tempered, normalized, or annealed and softened) and its test condition, a metal may (a) cyclically harden, (b) cyclically soften, (c) remain cyclically stable or (d) exhibit mixed behavior (soften or harden, depending on strain amplitude).

The following discussion develops and presents equations similar to those describing the monotonic stress-strain behavior. In addition, properties more appropriate to fatigue analyses—fatigue properties—are defined.

Determination of constant-amplitude fatigue lives of specimens is customarily performed under conditions of controlled stress (as in the rotating-bending or cantilever-bending type of test) or controlled strain. Figure 12 shows the ramifications of controlling stress, providing justification for the use of controlled strain while observing the stress response. As shown, the applied stress amplitude is less than the initial or monotonic yield strength of the steel (as noted by the "linear-elastic" strain response during the first 40 cycles). Because plastic deformation occurs at a microscopic level, however, the macrolevel response of the steel is the accrual of ever-increasing amounts of plastic strain. As stress cycling proceeds beyond 40 cycles, a "runaway" process occurs as the steel undergoes cyclic softening.

When the above response is compared to the response of a steel of similar hardness (shown in Fig. 13) under conditions of controlled strain, no instability, as happened under controlled stress, is observed, although the stress limits decrease with increased cycles. These test conditions represent extremes of completely unconstrained or stress-cycling conditions and completely constrained or strain-cycling conditions. In actual engineering structures, stress-strain gradients do exist, and there is usually a certain degree of structural constraint of the material at critical locations. Because this condition is reminiscent of strain control, it is more advantageous to characterize material response under strain-controlled than stress-controlled conditions.

For these reasons, consider the cases illustrated in Fig. 14 and 15, in which total strain is controlled and the stress response is observed. As shown in Fig. 14, if the stress required to enforce the strain increases on subsequent re-

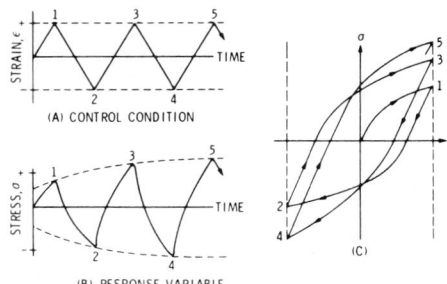

Fig. 14. Cyclic hardening under controlled-strain-amplitude cycling

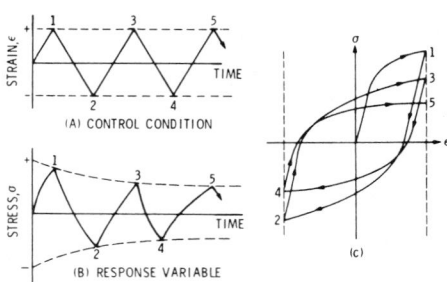

Fig. 15. Cyclic softening under controlled-strain-amplitude cycling

versals, the metal experiences cyclic hardening. The hardness, yield strength and ultimate strength increase. Such behavior is characteristic of annealed pure metals (for example, copper), many aluminum alloys, and as-quenched steels.

As illustrated in Fig. 15, the strain amplitude is controlled, but the stress required to enforce the strain decreases with subsequent reversals. This phenomenon is called cyclic softening and is characteristic of cold worked pure metals and many steels at small strain amplitudes. During cyclic softening, the flow properties (for example, hardness, yield strength and ultimate strength) decrease.

By plotting the stress amplitude versus reversals from controlled-strain test results, cyclic strain hardening and softening can be observed, as illustrated in Fig. 16. Thus, through cyclic hardening and softening, some intermediate strength level is attained that represents a steady-state condition (in which the stress required to enforce the controlled strain does not vary significantly). Some metals are cyclically stable, in which case their monotonic stress-strain behavior adequately describes their cyclic response. The steady-state condition is usually achieved in about 20 to 40% of the total fatigue life in either hardening or softening materials.

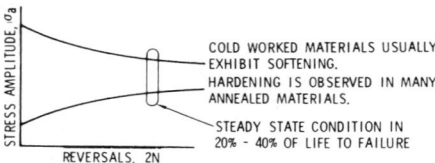

Fig. 16. Steady-state stress response for strain-controlled cycling

Cyclic Stress-Strain Behavior

The cyclic behavior of metals is best described in terms of a stress-strain hysteresis loop, as shown in Fig. 17. For completely reversed, strain-controlled conditions with zero mean strain, the total width of the loop is $\Delta\varepsilon$, or the total strain range:

$$\Delta\varepsilon = 2\varepsilon_a$$

where ε_a is strain amplitude. The total height of the loop is $\Delta\sigma$, or the total stress range:

$$\Delta\sigma = 2\sigma_a$$

where σ_a is stress amplitude.

Because the difference between the total- and elastic-strain amplitudes is the plastic-strain amplitude, the second equation below follows from the first:

$$\frac{\Delta\varepsilon}{2} = \frac{\Delta\varepsilon_e}{2} + \frac{\Delta\varepsilon_p}{2}$$

$$\frac{\Delta\varepsilon_p}{2} = \frac{\Delta\varepsilon}{2} - \frac{\Delta\varepsilon_e}{2} = \frac{\Delta\varepsilon}{2} - \frac{\Delta\sigma}{2E}$$

where E is the modulus of elasticity, ε_e is the elastic-strain amplitude and ε_p is the plastic-strain amplitude.

Changes in stress response of a metal occur rapidly during the first several percent of the total reversals to failure. The metal, under controlled-strain amplitude, eventually attains a steady-state stress response. To construct a cyclic stress-strain curve, the tips of the stabilized hysteresis loops from comparison specimen tests at several controlled-strain amplitudes can be connected.

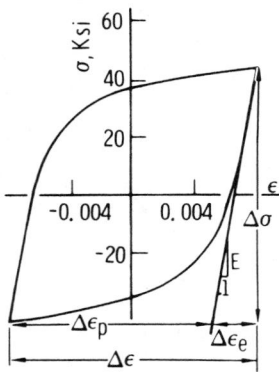

Fig. 17. Steady-state stress-strain hysteresis loop

In the particular example shown in Fig. 18, three companion specimens were tested to failure at three different controlled-strain amplitudes. The steady-state stress response, measured at approximately 50% of the life to failure, is thereby obtained. These stress values are then plotted at the appropriate strain levels to obtain the cyclic stress-strain curve. The cyclic stress-strain curve can be compared directly with the monotonic or tensile stress-strain curve to quantitatively assess cyclically induced changes in mechanical behavior (Fig. 19).

Note that in Fig. 19(a), when a material cyclically softens, the cyclic yield strength is considerably lower than the monotonic yield strength. Using monotonic properties in a cyclic application can result in prediction of fully elastic strains, when in fact considerable plastic strains are present. In T1 steels or an equivalent HSLA steel,

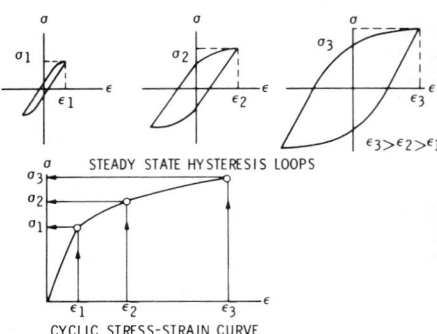

Fig. 18. Construction of cyclic stress-strain curve by joining tips of stabilized hysteresis loops

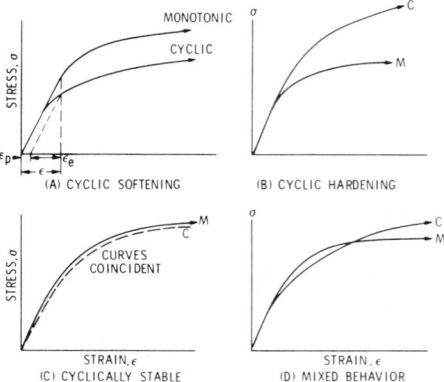

Fig. 19. Various types of cyclic stress-strain curves

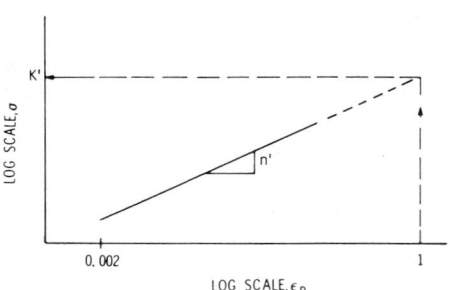

Fig. 20. True stress versus plastic strain for cyclic response (log-log coordinates)

for example, the cyclic yield strength is only about 50% of the monotonic yield strength.

Whereas the steady-state process consumes 20 to 40% of total life in constant-amplitude testing, a single large overload in actual service can produce an immediate change from the monotonic curve to the cyclic. Assembly or even driving of the completed machine "out the door" can cause an instantaneous loss of 50% of the monotonic yield strength.

Using the same approach as with the monotonic stress-strain curve, a plot of true stress versus true strain from constant-strain-amplitude test data of companion specimens on log-log paper results in a straight line (Fig. 20). Again, a power-law function between true stress and plastic strain may be represented as:

$$\sigma_a = K'(\varepsilon_p)^{n'}$$

where σ_a is the steady-state stress amplitude (measured at 50% of life to failure), ε_p is the plastic-strain amplitude, K' is the cyclic-strength coefficient and n' is the cyclic strain-hardening exponent.

The cyclic stress-strain response of a material is characterized by the following relationship:

$$\varepsilon = \frac{\sigma}{E} + \left(\frac{\sigma}{K'}\right)^{1/n}$$

The value of n' varies between 0.10 and 0.20, with an average value very close to 0.15. In general, metals with high monotonic strain-hardening exponents ($n > 0.15$) cyclically harden; those with low monotonic strain-hardening exponents ($n < 0.15$) cyclically soften.

STRAIN-LIFE RELATIONSHIPS

The $S - \log N$ plot can be linearized with full log coordinates. If true stress amplitudes are used instead of engineering stress, the entire stress-life curve may be linearized, as illustrated in Fig. 21. Thus, stress amplitude can be related to life by:

$$\sigma_a = \sigma_f'(2N_f)^b$$

where $\Delta\sigma$ is σ_a in zero-mean-constant-amplitude test, which is true stress; $2N_f$ is reversals to failure (one cycle is two reversals); σ_f' is the fatigue-strength coefficient; b is the fatigue-strength exponent (Basquin's exponent); and S_f is the fatigue-strength limit applicable to certain steels under very specific loading situations. σ_f and b are fatigue properties of the metal, with σ_f' approximately equal to σ_f for many metals and b varying between approximately -0.05 and -0.12.

Around 1955, Coffin and Manson, who were working independently on the thermal-fatigue problem, established that plastic strain$-$life data could also be linearized with log-log coordinates (Fig. 22). As with the true stress$-$life results, the plastic strain$-$life data can be related by the power-law function:

$$\frac{\Delta\varepsilon_p}{2} = \varepsilon_f'(2N_f)^c$$

where $\Delta\varepsilon_p/2$ is plastic-strain amplitude, $2N_f$ is the reversals to failure, ε_f' is the fatigue-ductility coefficient, and c is the fatigue-ductility exponent. ε_f' and c are also fatigue properties, with ε_f' approximately equal for many metals and c varying between approximately -0.5 and -0.7 for many metals.

It was mentioned previously that total strain has two components: elastic and plastic, or:

$$\varepsilon = \varepsilon_e + \varepsilon_p$$

This can also be expressed as the strain amplitudes from a constant-amplitude, zero-mean-strain controlled test:

$$\frac{\Delta\varepsilon}{2} = \frac{\Delta\varepsilon_e}{2} + \frac{\Delta\varepsilon_p}{2}$$

Therefore:

$$\sigma_a = \sigma_f'(2N_f)^b$$

$$\frac{\Delta\varepsilon_e}{2} = \frac{\sigma_a}{E}$$

Division by E, the modulus of elasticity, obtains:

$$\frac{\Delta\varepsilon_e}{2} = \frac{\sigma_f'}{E}(2N_f)^b$$

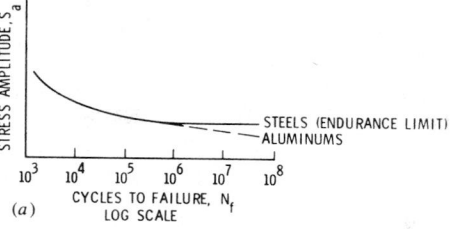

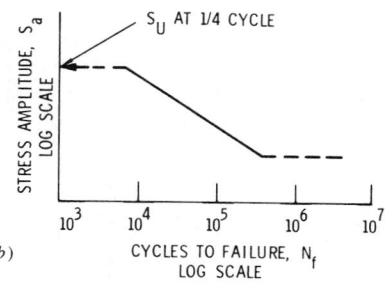

(a) Stress versus log cycles to failure. (b) Log stress versus log cycles to failure.

Fig. 21. Stress-life curves

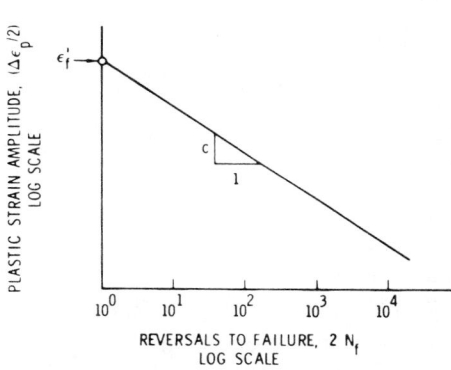

Fig. 22. Log plastic strain versus log reversals to failure

or, with elastic and plastic combined:

$$\frac{\Delta\varepsilon}{2} = \underset{\text{(Elastic)}}{\frac{\sigma_f'}{E}(2N_f)^b} + \underset{\text{(Plastic)}}{\varepsilon_f'(2N_f)^c}$$

This equation is the foundation for the strain-based approach to fatigue and is called the strain-life relationship.

The two straight lines, one for the elastic strain and one for the plastic strain, also can be plotted, as in Fig. 23. Several conclusions may be drawn from the total strain$-$life curve in Fig. 23. At short lives, less than $2N_t$ (the transition fatigue

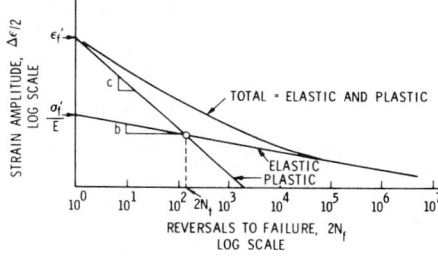

Fig. 23. Log strain versus log reversals to failure

life where $\Delta\varepsilon_p/2 = \Delta\varepsilon_e/2$), plastic strain predominates and ductility controls performance. At longer lives, greater than $2N_t$, the elastic strain is more dominant than the plastic, and strength controls performance. An "ideal material" is one with both high ductility and high strength. Unfortunately, strength and ductility are usually a trade-off; the optimum compromise must be tailored to the expected load or strain environment being considered in a real history for a fatigue analysis. By equating the elastic and plastic components of total strain, the transition fatigue life can be calculated as:

$$2N_t = \left(\frac{\varepsilon_f'E}{\sigma_f'}\right)^{1/(b-c)}$$

FATIGUE-CRACK-GROWTH-RATE TEST METHODS

By J. H. Underwood, U.S. Army Armament Research and Development Command, and W. W. Gerberich, University of Minnesota

The general nature of fatigue-crack growth and its description using fracture mechanics can be briefly summarized by the example data shown in Fig. 24. This figure, based on the work of Paris *et al* (Ref 3), shows a logarithmic plot of the crack growth per cycle, da/dN, versus the stress-intensity-factor range, ΔK, corresponding to the load cycle applied to a specimen. The da/dN-versus-ΔK plot shown is from five specimens of ASTM A533 B1 steel tested at 24 °C (75 °F). A plot of similar shape is expected with most structural alloys; the absolute values of da/dN and ΔK are dependent on the material. Results of fatigue-crack-growth-rate tests for nearly all metallic structural materials have shown that the da/dN-versus-ΔK curves have the following characteristics: (a) a region at low values of da/dN and ΔK in which fatigue cracks grow extremely slowly or not at all below a lower limit of ΔK called the threshold of ΔK, ΔK_{th}; (b) an intermediate region of power-law behavior described by the Paris equation (Ref 4):

$$\frac{da}{dN} = C(\Delta K)^n$$

where C and n are material constants; and (c) an upper region of rapid, unstable crack growth with an upper limit of ΔK which corresponds either to K_{Ic} or to gross plastic deformation of the specimen.

Testing procedures for measuring fatigue-crack-growth rates are described in ASTM Method E647. This method applies to medium-to-high crack-growth rates—that is, above 10^{-8} m/cycle (3.9×10^{-7} in./cycle). Procedures for growth rates below 10^{-8} m/cycle are under consideration by ASTM. For applications involving fatigue lives of up to about 10^6 load cycles, the procedures of E647 can be used. Fatigue lives greater than about 10^6 cycles correspond to growth rates below 10^{-8} m/cycle, and these require special testing procedures, which are related to the threshold of fatigue-crack growth illustrated in Fig. 24.

ASTM Method E647 describes the use of center-cracked specimens and compact specimens. The specimen thickness-to-width ratio, B/W, is smaller than the 0.5 value for K_{Ic} tests; the maximum B/W values for center-cracked and compact specimens are 0.125 and 0.25, respectively.

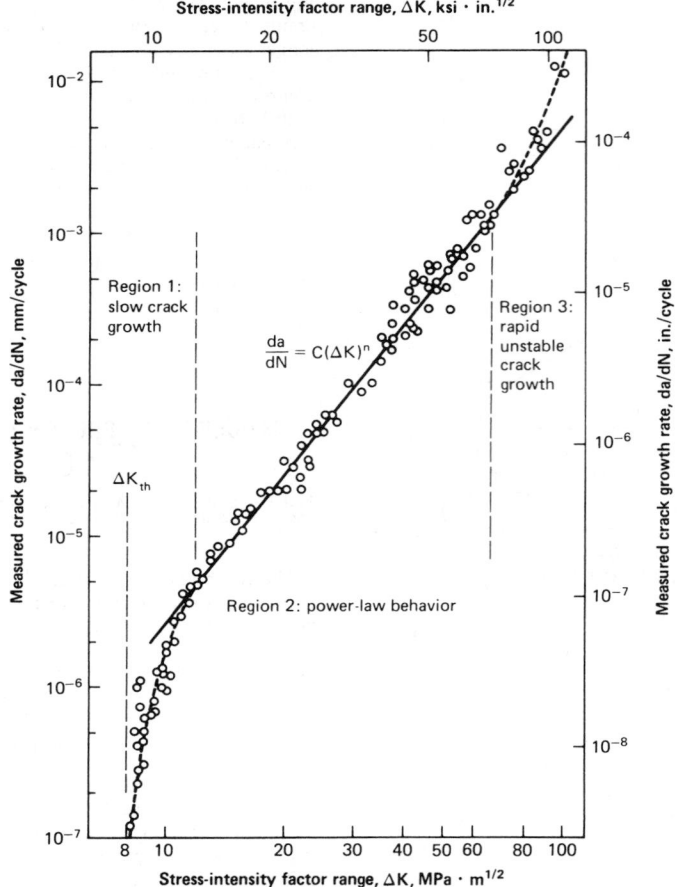

The material was ASTM A533 B1 steel, with a yield strength of 470 MPa (70 ksi). Test conditions: R = 0.10; ambient room air; 24 °C (75 °F).

Fig. 24. Fatigue-crack-growth behavior of A533 steel (Ref 3)

With the thinner specimens, it is feasible to use crack-length measurements on the sides of the specimens as representations of through-thickness crack-growth behavior. The specimens are loaded in the same general manner as for K_{Ic} testing. For tension-tension fatigue loading, the K_{Ic} loading fixtures often can be used. For this type of loading, both the maximum and minimum loads are tensile, and the load ratio, $R = P_{min}/P_{max}$, is in the range $0 < R < 1$. A ratio of $R = 0.1$ is commonly used. Tension-compression loading can be performed with the compact specimen, but it is a more complex type of loading and requires more care.

Testing normally is performed in laboratory air at room temperature; however, any gaseous or liquid environment and temperature of interest may be used in order to determine the effect of corrosion or other chemical reaction on cyclic loading. Cyclic loading may involve various wave forms for constant-amplitude loading, spectrum loading or random loading.

For constant-amplitude loading, a set of crack-length-versus-elapsed-cycle data (a versus N) is collected, with the specimen loading, P_{max} and P_{min}, generally held constant. The minimum crack-length increment, Δa, between data points is required by ASTM E647 to be larger than a certain value. This prevents the measurement of erroneous growth rates from a group of data points which are too closely spaced relative to the precision of data measurement and relative to the scatter of the data. The growth rates may be calculated by either of two methods. The secant method is simply the slope of the straight line connecting two adjacent data points. This method, although simpler, results in more scatter in measured crack-growth rate. The polynomial method fits a second-order polynomial expression (parabola) to typically 5 to 7 adjacent points, and the slope of this expression is the growth rate. The polynomial method, particularly when used with a large number of adjacent points, eliminates some of the scatter in growth rate which is inherent in fatigue testing. The measured values of growth rate typically are plotted as in Fig. 24, where ΔK is calculated from $\Delta P = P_{max} - P_{min}$ (for tension-tension loading) using a K expression such as Eq 16 in the preceding article.

The measured growth-rate data are represented by an equation of the form of the Paris equation:

$$\frac{da}{dN} = C(\Delta K)^n$$

where the material constants C and n apply only within a certain range of da/dN and ΔK values. Other relationships based on the Paris equation, such as the commonly used Forman equation (Ref 5), are used to represent the variation of da/dN with other key variables, including load ratio, R, and the critical K value, K_c, at which fast fracture of the specimen occurs. The Forman equation is:

$$\frac{da}{dN} = \frac{C(\Delta K)^n}{(1 - R)K_c - \Delta K}$$

where C and n are material constants of the same types as those in the Paris equation, but of different values. An advantage of the Forman equation is that it describes the type of accelerated da/dN behavior which is often observed at high values of ΔK and which is not described by the Paris equation. For example, for zero-to-tension loading (in which $R = P_{min}/P_{max} = 0$), as ΔK approaches K_c in the Forman equation, da/dN increases rapidly, and this is often observed in tests. In addition, the Forman equation describes the often-observed decrease in da/dN associated with an increase in R from zero toward one. So when it is necessary to describe the effect of ΔK approaching K_c or the effect of R on da/dN, the Forman equation can be used to represent the da/dN behavior. When only ΔK, the primary variable affecting da/dN, is involved, the less complex Paris equation may be used.

REFERENCES

1. *Mechanical Metallurgy*, 2nd Ed., by G. E. Dieter: McGraw-Hill, New York, 1976, p 403
2. P. G. Fluck: *ASTM Proc.*, Vol 51, 1951, p 584
3. Extensive Study of Low Fatigue Crack Growth Rates in A533 and A508 Steels, by P. C. Paris, R. J. Bucci, E. J. Wessel, W. R. Clark and T. R. Mager: in *Stress Analysis and Growth of Cracks*, Proceedings of the 1971 National Symposium on Fracture Mechanics, Part I, STP 513, American Society for Testing and Materials, Philadelphia, 1972, p 141-176
4. A Critical Analysis of Crack Propagation Laws, by P. C. Paris and F. Erdogan: *Journal of Basic Engineering, Transactions of ASME*, Vol 85, Dec 1963, p 528-534
5. Numerical Analysis of Crack Propagation in Cyclic-Loaded Structures, by R. G. Forman, V. E. Kearney and R. M. Engle: *Journal of Basic Engineering, Transactions of ASME*, Vol 89, Sept 1967, p 459-464

Creep, Shear and Torsion Testing

Creep and Creep-Rupture Testing

THE FLOW or plastic deformation of a metal held for long periods of time at stresses below the normal short-time yield strength is known as creep. Although we normally think of creep as occurring only at elevated temperatures, room temperature can be high enough for creep to occur in some metals. In lead, for example, creep at room temperature is common. In many cases, lead pipes must be supported to prevent sagging under their own weight.

The development of steam turbines and jet engines has greatly increased interest in creep because, in these, the metal parts must withstand high loads at high temperatures for long times. The high centrifugal loads tend to cause certain parts to elongate or distort. Tolerances must be kept close to be efficient; yet if the metal parts deform too much, this spacing will be eliminated and failure will occur. In most cases the parts cannot be made sufficiently heavy to prevent all creep because the weight penalty would reduce efficiency too much. Many such parts are therefore designed for a certain expected life span. For this, accurate data are needed to determine how much the metal part can be expected to deform under the conditions of stress and temperature to be encountered in service. Tests which measure the deformation of a metal as a function of time at constant load and temperature are known as creep tests.

CREEP PHENOMENA

A typical creep curve is shown in Fig. 1. The vertical (Y) axis is creep strain and the horizontal (X) axis is time plotted on logarithmic coordinate. The curve consists of three parts: primary, secondary and tertiary creep, or first-, second- and third-stage creep. The strain shown is plastic or permanent strain. When a creep specimen is loaded, there will be some elastic extension of the specimen, but this is not shown in this curve. In the primary stage, the initial creep rate shows a continuous decrease with time. In second-stage creep, the creep rate is considered essentially

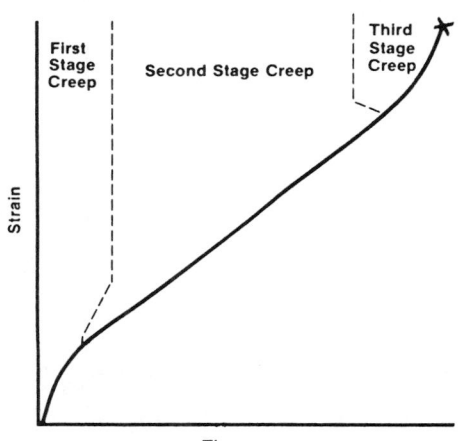

Fig. 1. Idealized creep curve

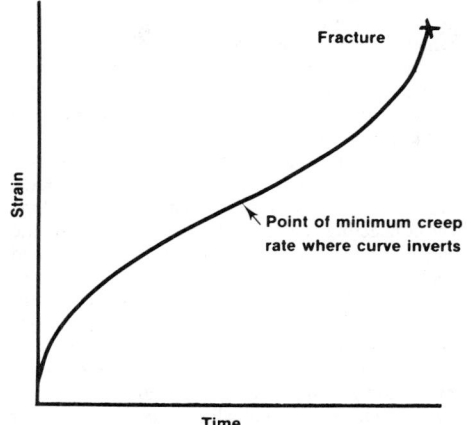

Fig. 2. Creep curve with minimum creep rate and point of inversion

constant. In the third stage, the strain rate increases rapidly to fracture. This increase in the third stage is due, in part, to the reduction in cross-sectional area and thus the increasing true stress. Measurements made of specimen cross section during third-stage creep indicate that the increase in strain rate is not due *only* to necking or reduction in cross-sectional area, however.

Although in the idealized creep curve shown in Fig. 1, the creep rate is shown as constant in the second stage, this does not occur in practice. If the test is long enough to show all stages of creep, the curve will show a continually diminishing rate of creep to a point where the curve inverts and the creep rate starts to increase again (see Fig. 2). The change in strain rate may be very slight over time; in some cases, the curve may approach a straight line.

RUPTURE TESTS

The rupture test is valuable in determining tendencies of materials that may have to break under an overload. It finds much use in selection of materials for applications where dimensioned tolerances are not critical but rupture could not be tolerated. The rupture test is similar to a creep test except that no strain measurements are made during the test. The specimen is stressed under a constant load at constant temperature as in the creep test, and the time for fracture to occur is measured. Measurements are also made of the elongation and reduction in area of the broken specimen. Stresses are higher than those used for creep tests. An example of a typical application of the rupture test would be for testing boiler pipes. This test is also called the stress-rupture test, or time-to-rupture test.

RELAXATION TESTS

The relaxation test is somewhat similar to the creep test, but the load continually decreases instead of remaining constant. This test is primarily of value in evaluating bolt materials. When a bolt is drawn up tight, a tensile load is present in the bolt and the bolt is elongated slightly. This causes a clamping load on whatever the bolt is mounted on. If the bolt creeps (extends or re-laxes) this clamping load will be reduced. If the bolt elongates sufficiently to remove all tension, it no longer fulfills its function. In a relaxation test, the load is reduced at intervals in order to maintain a constant elongation (strain). A relaxation curve thus takes the general shape shown in Fig. 3. Note that the y axis in this curve is stress or load rather than strain (elongation) as in the creep curve.

TYPICAL PROCEDURE FOR MAKING A CREEP TEST

Selection and Preparation of Specimen. The same precautions used in selecting and preparing a specimen for a short-time tension test apply to specimens for creep testing. The specimen should be selected to be truly representative of that which it is supposed to represent. Machining and grinding should follow procedures to produce a surface as nearly stress-free as possible. There should be no undercutting at the fillets, and the gage length should be uniform in cross section or very slightly smaller at the center of the gage length.

The specimen is carefully identified in as much detail as is appropriate—type of metal, heat number, vendor, etc.—and this information is recorded with the specimen's measurements. Sometimes gage marks, for measuring total extension, are made on the specimen. Such marks or scribe lines must be used with care, because the depressions or scribe lines can cause premature failure on some materials. Any operation such as stamping the ends of the specimen must be used with care to avoid any damage to the specimen.

Loading. In mounting the specimen in the adapters and load train, care is needed to avoid straining of the specimen in handling. This can occur when threading the specimen into the adapters and when handling the load train with the specimen in place, especially if the specimen is very small or brittle. The load train (specimen adapters or grips, pull rods, etc.) with the specimen in place should be carefully examined for any misalignment which will cause bending of the specimen under load. The upper load train should be suspended from the lever arm and the compensating weight adjusted so that the lever arm balances. The strain-measuring clamps and extensometer, or the platinum strips, are attached

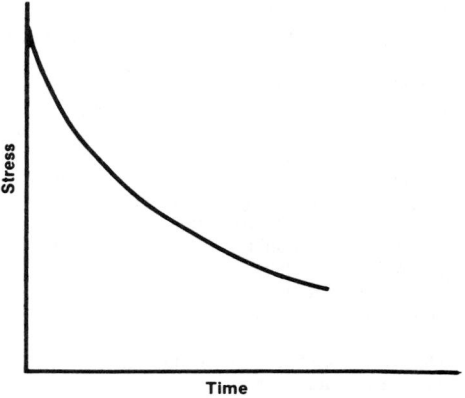

Fig. 3. Relaxation curve

to the specimen, and the load train is inserted into the furnace with the specimen centered. The specimen must be stabilized at temperature before being loaded. Also, the extensometer should be adjusted and zeroed.

Loading of the weight pan should be done smoothly and without excessive shock. If the specimen is to be step loaded, the weight is placed on the weight pan in measured increments and the strain corresponding to each step of loading is recorded. The loading curve thus obtained is used in determining the elastic modulus. If step loading is not used, a method of applying the load smoothly must be used. This can be done by having a support such as a scissors jack under the load pan during loading. When all weights are in place, the supporting jack is smoothly lowered from under the weight pan.

Data Collection. Reading of strain should be made frequently enough to define the curve well. This will necessitate much more frequent readings during the early part of the test than later. The elastic portion of the stress-strain curve can be obtained by measurement of the instantaneous contraction when the load is removed at the end of the test if the specimen has not broken.

Temperature Control. In bringing the specimen to temperature, it is important that the specimen not be overtemperatured. A common practice is to bring the specimen up to about 50 °F (about 30 °C) below the desired temperature in about 1 to 4 h and then take considerably longer in bringing the specimen to the desired temperature and adjusting for good stabilization. It should be understood that a period of time above the desired temperature is not cancelled in effect by an equal period at a temperature the same amount below the desired temperature. Any rise in temperature above the desired temperature of more than a small amount (such as defined by ASTM Recommended Practice for Conducting Creep and Time for Rupture Tension Tests of Materials) should be rejected. The limits specified in this recommendation are ±3 °F (±1.7 °C) up to 1800 °F (980 °C) and ±5 °F (±2.8 °C) above 1800 °F. At temperatures very much above 2000 °F (1095 °C), the limits are broadened somewhat. Variation of temperature along the specimen from the nominal test temperature should vary no more than these limits at these temperatures. These limits refer to indicated variations in temperature according to the temperature recorder.

Every effort should be made to ensure that the indicated temperature is as close to true temperature as possible. There is the possibility of both thermocouple error and instrument error. Thermocouples, especially base-metal thermocouples, drift in calibration with use or because of contamination. Other possible errors can result from incorrect lead wires or incorrect connection of lead wires, direct radiation on the thermocouple bead or other causes. Representative thermocouples should be calibrated from each lot of wires used for base-metal thermocouples and, except at low temperatures, base-metal thermocouples should not be re-used without slipping back to remove the wire exposed to high temperatures and rewelding. Noble-metal couples are generally more stable. However, they are also subject to error due to contamination and need to be annealed periodically. This can be done by connecting a variable transformer to the two wires and sending enough current through the wire to make it incandescent.

When the thermocouple is attached to the specimen, the junction must be kept in intimate contact with the specimen. The bead at the junction should be as small as possible, and there must be no twisting of the thermocouple elsewhere which could cause shorting. Any other metal contact across the two wires will cause shorting and erroneous readings. Many authorities recommend shielding the thermocouple junction from radiant heating.

Temperature-measuring, controlling and recording instruments must be calibrated periodically against some standard. This is usually done by connecting a precision potentiometer to the thermocouple terminals on the instrument and feeding in millivoltages corresponding to the output of the thermocouple at each of several temperatures. Tables of millivolt output for various types of thermocouples are readily available from manufacturers of precision potentiometers. Most creep and rupture machines are equipped with a switch which automatically shuts off the timer when the specimen breaks. In creep tests, the load is usually selected low enough so that rupture does not occur. The microswitch which shuts off the timer also shuts off or lowers the temperature of the furnace on many other creep-rupture units. In some furnaces the life of the heating element is severely reduced if the furnace is shut off after each test, so for some furnaces the temperature is lowered to some lower control temperature such as 1000 or 1200 °F (540 or 650 °C).

Interrupted Tests. Sometimes, because of a power failure or other problem, it becomes necessary to interrupt a test—that is, the specimen is cooled, then reheated. For many materials, this appears to have little effect on either creep properties or time-to-rupture if the times of cooling and heating are not very great. It cannot be stated, however, that such treatment will not affect any materials. Any interruption of a test should be reported.

PRESENTATION OF DATA

Creep. The usual method for presenting creep data is in the form of a curve showing percent creep strain as the vertical axis and time as the horizontal axis. The time is usually plotted on a log scale to show the early part of the curve in good detail and yet prevent the curve from being excessively long. Sometimes a whole family of

curves is plotted on the same coordinates to show the effect of different temperatures or different stresses on one material.

Other methods for plotting data include time to reach a given percent of creep versus load at a constant temperature or time to reach a given percent of creep versus temperature at constant load.

The loading curve, showing the strain versus load as the specimen is loaded, is plotted separately and is used in computing the elastic modulus of the material at temperature.

Rupture. Rupture data are presented in several types of graphs. One has stress as the vertical axis versus log of time-to-rupture (at constant temperature) on the horizontal axis. Usually, stress-rupture data are presented by means of a parameter plot—i.e., stress is plotted against a parameter value which relates it to both time and temperature. Several different parameters have been used. A widely used one, the Larson-Miller parameter, follows the formula $P = (T + 460)$ $(\log t + c)$. This means that the parameter value P equals the Rankine temperature (460 + the temperature in degrees Fahrenheit) times the log (base 10) of the time in hours plus a constant. The constant (c) has various values depending on the material but usually runs from about 17 to 23 for most materials tested. The value of c is determined by plotting log of time versus $1/[T(°F) + 460]$ using rupture data from several tests at constant stress but different temperatures on the same material. This produces a series of straight lines converging as on a single point. At this point $\log t = c$, and this constant is theoretically the best constant to use for the data involved. Figure 4 shows a parameter plot of stress-rupture data.

Shear Testing

IN GENERAL, the amount of existing shear-strength data is seriously less than the published data available for other mechanical properties. Stores of data that deal with mechanical properties such as tensile and yield strength, hardness and ductility for virtually all metals and metal alloys, and in a wide variety of conditions, are readily obtainable.

At least two reasons can be identified to explain the scarcity of shear-strength data. First, the demand is low, because the number of compo-

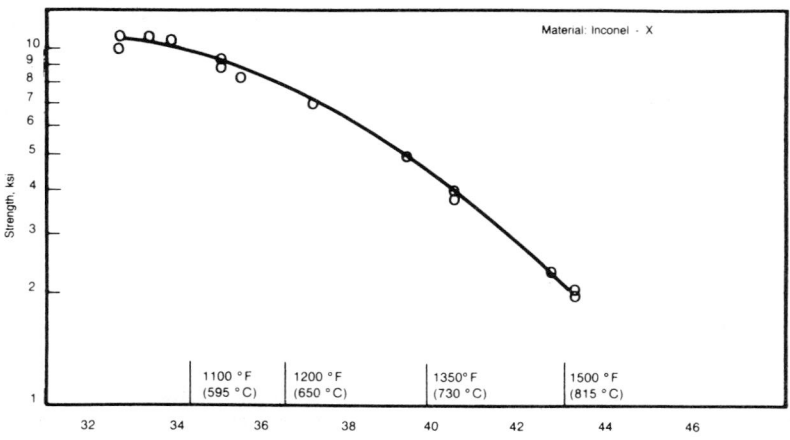

To convert values in ksi to MPa, multiply by 6.8948.

Fig. 4. Plot of stress-rupture data using the Larson-Miller parameter

Fig. 5. Shear test fixture of the compression loading type used for single or double shear test (Tinius Olsen Testing Machine Company)

nents that are loaded in shear under service conditions is far less than that of components loaded in tension, compression, bending or torsion. Probably the primary reason for the lack of published data on shear strength is the difficulty in obtaining accurate test data. Shear testing inherently involves a number of variables; thus, the tests are less reproducible than testing for properties such as tensile or yield strength. Therefore, most shear testing has been performed by means of nonstandard equipment and procedures operating on arbitrary bases, thus producing results that are empirical.

The greatest needs for shear-test data are in the designing of structures that are riveted, pinned or bolted together and where service stresses are actually in shear. Notable examples of such structures are found in the aerospace industry. The required standardization is given by ASTM Standard B565.

SINGLE- AND DOUBLE-SHEAR TESTING

In the many tests that have been devised for evaluating shear strength, both single- and double-shear testing have been used. The double-shear technique is far more accurate, however, making those results more reproducible than results for the single-shear technique.

Compression-type loading for a shear fixture is shown in Fig. 5, with a specimen being tested in double shear. This type of fixture may also be used for single-shear tests.

Procedure. The test specimen is assembled in the fixtures, as shown in Fig. 6, and loaded in tension until complete failure occurs. Crosshead speed during the test should not exceed 0.750 in./min (19.1 mm/min), and loading rate should not exceed 100 ksi/min (690 MPa/min). The maximum load in double shear is determined by the direct reading on the testing machine.

CALCULATION OF SHEAR STRENGTH

The calculation of strength in double shear is a simple matter of dividing the machine load by the area of the cross section ($\pi D^2/4$). It follows, then, that single-shear strength is one-half of this value, or:

$$\text{single-shear strength} = \frac{P/2}{\pi D^2/4}$$

where P is load in pounds (kilograms), and D is diameter in inches (millimetres).

As previously stated, shear testing is more vulnerable to the effects of variables than certain other mechanical tests, such as tests for tensile or yield strength. The reader should note from examining Fig. 6 that, even when the fixtures and test specimens meet specified tolerances, some variations are bound to exist in the test-jig assembly that will be reflected as variables in the results.

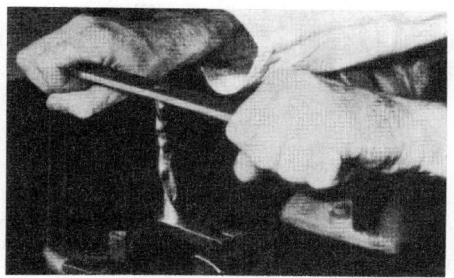

Fig. 7. Torsion testing of a brazed T-joint specimen

The presence or absence of lubricant on the surfaces of the specimens and test fixtures can be responsible for substantial variations in the results. For example, a lubricated specimen may cause a reduction in shear strength of as much as 3%. To minimize this variable, it is recommended that the test fixtures and specimens be carefully cleaned prior to testing — preferably by means of ultrasonic cleaning in a suitable solvent.

────Torsion Testing────

IN THE TORSION TEST, a specimen is subjected to twisting or torsional loads to simulate service stresses for such parts as axles, crankshafts, twist drills and spring wire. The test has not been standardized, and is rarely specified. However, the torsion test provides information such as modulus of elasticity in shear (sometimes called modulus of rigidity), the shearing yield strength and the modulus of rupture (apparent ultimate shear strength). The torsion test may also be performed as a high-temperature twist test on materials such as tool steels to determine forgeability. The test does not provide meaningful results for very brittle materials such as cast irons, because these materials would fail in diagonal tension before the shear-strength limit was reached.

GENERAL PROCEDURE

In torsion testing, the specimen is clamped in clamping heads so that the specimen remains as straight as possible during testing. The test specimen is then twisted at a slow, uniform rate until it breaks, or until a specified number of turns is obtained. The number of turns is recorded. If the number of turns falls within an acceptable range, the test specimen is considered to have passed the test. Results of the torsion test are largely comparative, and have no standardized values.

Torsion testing is frequently employed to assess the quality of brazed joints for sheet-metal products. A T-joint of sufficient length is brazed and then subjected to two full turns in torsion (Fig. 7). Visual examination is made to determine if failure has occurred in the brazed joint.

One of the only standardized applications of the torsion test applies to torsion testing of wire (ASTM E558).

An example of a torsion-testing machine is presented in Fig. 8.

DATA

Torsion data are usually presented as torque-twist curves, in which the applied torque is plotted against the angle of twist. Torsion produces a state of stress known as pure shear, and the shear stress at yielding can be calculated from the torque at yielding and the specimen dimensions. The maximum stress for a cylindrical specimen (at the surface) can be calculated from the following relation:

$$S = \frac{16T}{\pi d^3}$$

where S is maximum shear stress in psi (MPa), T is torque in lb·in. (N·m) and d is specimen diameter in inches (cm). This formula holds only

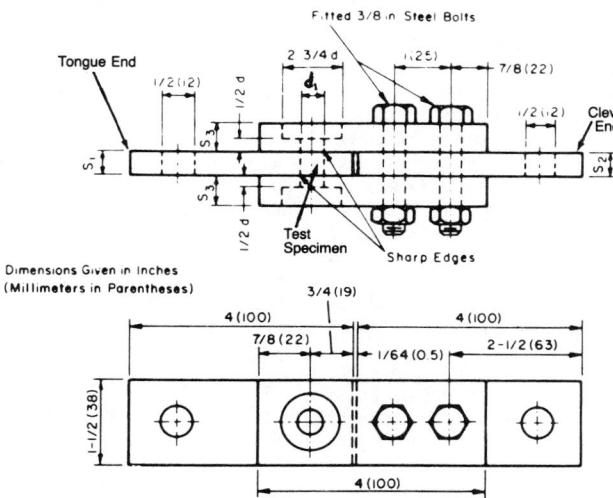

See ASTM B565 for a complete explanation of size requirements.

Fig. 6. Details of test jig used for testing specimens in double shear

Fig. 8. Close-up of a 10 000 in · lb (1100 N · m) torsion-testing machine with special tooling for Phillips screwdriver bits (Tinius Olsen Testing Machine Company)

when the strain is proportional to stress, but is commonly used for computing higher stresses and for determining modulus of rupture (apparent ultimate shear strength).

The total torsional deformation is measured as angular twist of one end of the gage length in relation to the other. In order to obtain the angular twist per inch of gage length, the total angular twist is divided by the gage length. The angular twist per inch of gage length can then be converted into shear strain, in inches per inch, by multiplying by half the diameter of the specimen.

E_s, the modulus of elasticity in shear (sometimes called the modulus of rigidity), can be calculated from the following formula:

$$E_s = \frac{SL}{r\theta}$$

where S is maximum shear stress, in psi (MPa); L is the gage length of the specimen, in inches (cm); r is the distance from the axis of the specimen to the outermost fiber (half the diameter), in inches (cm); and θ is the angle of twist, expressed in radians, in length L.

The yield strength is generally defined as the maximum stress developed by a torque producing an offset of 0.2% from the original modulus line, analogous to the method used for determining tensile yield strength.

COMPARISON OF TORSIONAL AND TENSION DATA

From the torque-twist diagram it is simple to obtain a shear stress–shear strain diagram. The great advantage of the torsion test over the tension test is that large values of strain can be obtained without complications such as necking. One problem of torsional tests is that the stress is not constant throughout the cross section. This problem can be circumvented by using tubular specimens. If the results of a tension test and a torsion test are plotted for the same low-carbon steel, the two curves will be markedly different. However, if the two curves are normalized by converting the normal stress and longitudinal strain in the uniaxial test and the shear stress and strain in the torsion test into effective stress and strain, the two curves come into close correspondence. The effective stresses and strains are determined by well-known equations (for instance, Eq 1.67 and 1.81 in *Mechanical Metallurgy: Principles and Applications*, by M. A. Meyers and K. K. Chawla, Prentice-Hall, Englewood Cliffs, NJ, 1984.) These results show that the work hardening of the material is a function of the amount of plastic strain and does not depend on the state of stress. Such is not the case for all materials, however. Differences in texture due to different constraints can be responsible for substantial differences in the effective stress-strain curve.

Formability Testing

FORMABILITY is the technical term used to describe the ease with which a metal can be shaped through plastic deformation. Usually, it is synonymous with the term "workability." The evaluation of the formability of a metal involves both measurement of resistance to deformation (strength) and determination of the extent of plastic deformation that is possible before fracture (ductility). The emphasis in most formability tests, however, is on the amount of deformation required to cause fracture.

Because of the diverse geometries of the tools and workpieces and the various ways that forces of deformation are applied, different metalworking processes produce varying stress states. These can be divided into two broad categories: (*a*) bulk-deformation processes, such as forging, extrusion and rolling, where the stress state is three-dimensional; and (*b*) sheet-forming processes, such as deep drawing and stretch forming, where the stress state is two-dimensional and lies in the plane of the sheet. The tests that simulate bulk-formability testing are given in the article on compression testing.

BEND TESTS

Bend tests are among the most frequently used tests for evaluating the ductility of a metal or welded joint by measuring its ability to resist cracking during bending. Bending is the process by which a straight length is transformed into a curved length. The fibers of the metal on the outer (convex) surface of the bend are stretched, thus inducing tensile stresses. Simultaneously, the fibers on the inner (concave) surface of the bend

are placed in compression. ASTM methods E190, E290 and E855 provide descriptions of the various procedures.

Bend Radius. For a given bending operation, the bend radius, R, cannot be made smaller than a certain value, or the metal cracks on the outer (tensile) surface. Usually, this *minimum bend radius* (R_{min}) is expressed in terms of multiples of the specimen thickness. Thus, a material with a 3t minimum bend radius can be bent without cracking through a radius equal to three times the specimen thickness. It follows, then, that a material with a minimum bend radius of 1t has greater formability; whereas a minimum bend radius of 5t indicates a less formable material.

Test Specimens. Bend-test specimens are usually in the form of a rectangular beam. Wherever possible, as with a plate or a sheet, the full thickness of the material should be used. Generally, the specimen thickness should not exceed 40 mm ($1\frac{1}{2}$ in.). When using a machined specimen of reduced thickness, the as-fabricated surface should be retained as a surface of the bend specimen. This surface should be oriented in the bend fixture as the tensile surface. For specimens cut from plate material, the width should be twice the thickness, but no less than 20 mm ($\frac{3}{4}$ in.). For thin specimens cut from sheet, the width should exceed eight times the thickness. The ratio of width to thickness affects the stress state produced in bending and, therefore, the ductility measured in the test. For this reason, bend-test results made on thin sheet should not be compared with those obtained with thicker plate to avoid erroneous conclusions about the formability of the materials.

The length of a bend-test specimen must be of some minimum that varies with thickness. Length, however, is not critical if the specimen is long enough to accomplish the bending operation. The edges of the specimen may be rounded to a radius not to exceed 1.6 mm ($\frac{1}{16}$ in.) to minimize edge cracking. Flame-cut surfaces should be machined to remove heat-affected metal. Sheared edges should be machined or smoothed on an abrasive belt to remove the sheared edge. Although bend testing usually is performed with specimens of rectangular cross section, round specimens may also be used.

Bend specimens may be cut from sheet or plate to evaluate the basic formability of the material or test the formability of an as-fabricated surface. Because most fabricated products have mechanical properties that are directional (anisotropic), directionality is an important consideration in making the test. Figure 1 shows the orientation of the bend-test specimen with the rolling direction for a longitudinal orientation and a transverse orientation. The transverse orientation generally shows lower ductility, because the tensile bending stresses are oriented perpendicular to the fiber structure developed by the rolling deformation.

The quality of welds often is evaluated by bend testing (ASTM E190). A specimen is cut from the welded assembly with the weld in the center of the specimen. The weld may be either transverse or parallel to the length of the specimen.

Free Bend Tests

A free bend test is one in which the curvature of the bend is left "free" to take its natural shape.

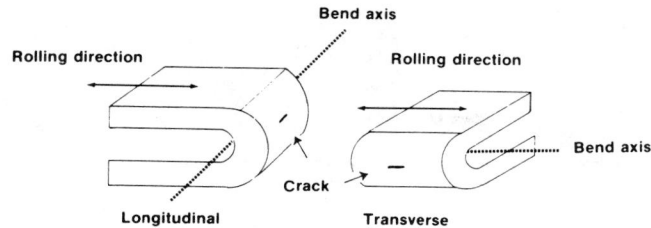

Arrows indicate direction of rolling. Source: Semi-Guided Bend Test for Ductility of Metallic Materials, ASTM E290-80, American Society for Testing and Materials, Philadelphia.

Fig. 1. Relative orientations of specimens for longitudinal and transverse bend tests

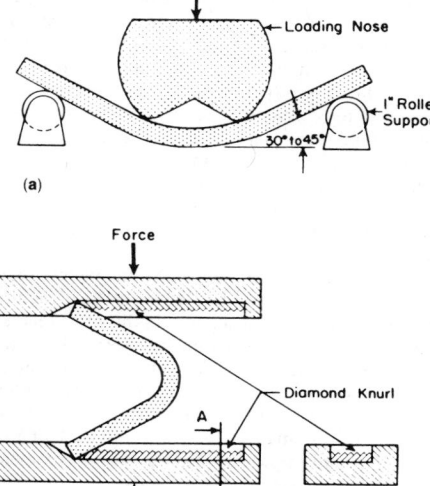

(a) A partial bend is made with the specimen in a horizontal position. (b) The specimen is positioned vertically, and the two knurled jaws are forced together until the specimen fractures or makes a 180° U-bend.

Fig. 2. Free bend tests

As shown in Fig. 2, the specimen is given a preliminary bend in a bending fixture (Fig. 2a) and then is transferred to a free bend fixture (Fig. 2b) where the bend is completed.

For moderately ductile materials, the formability is evaluated by the bend angle (α) that can be achieved before cracking occurs on the tensile face (outside surface) of the bend. For a highly ductile material that can be bent flat on itself ($\alpha = 180°$), the ductility is evaluated on the thickest specimen for which this can be done without cracking.

Restricted (Controlled) Bend Tests

A restricted bend test is one in which the test specimen is made to bend closely around a predetermined radius, R. Various examples of this test are shown in Fig. 3. The test shown in Fig. 3(a) usually is called a guided-bend test. The need for a test fixture sometimes may be eliminated by using a soft metal support to accommodate the punch, as in Fig. 3(b). For thin sheet metal, the bending force may be applied by a hand-operated lever or, alternatively, the sheet may be hammered over the bending die with a plastic or rawhide mallet (Fig. 3c). ASTM E290 describes this test in detail.

Ordinarily, a grid pattern is lightly scribed on

the tensile surface of the bend specimen before the restricted bend test. This surface is observed during the test, either with a mirror or by bending in small increments, to determine when the cracks first appear. At this point, the angle of bend is recorded, or the elongation of the tensile surface is determined from the grid network. Alternatively, the minimum bend radius that will permit bending through a fixed bend angle is determined as the measure of formability.

Bend Tests on Very Ductile Materials

Bend tests on very ductile materials are less controlled than those discussed above, but they are more severe tests. For a sheet, the basic test is to determine whether the sheet can be bent flat on itself through 180° without cracking. A further test of ductility is to cross-fold the sheet once again across the first fold (Fig. 4a). Bend tests are made on tubes by first flattening the tube, as shown in Fig. 4(b). This applies two separate transverse bends of nearly 180°. Subsequently, the flattened tube can be folded along its longitudinal axis (Fig. 4c).

SHEET-FORMABILITY TESTS

Several tests have been developed to evaluate the formability of sheet metal. Most complex sheet-forming operations can be resolved into a combination of bending plus stretching and drawing. In a pure stretch-forming operation, the edges of the blank strip are clamped, and the shape is produced by multidirectional stretching over the contours of the deforming tool or punch. Sheet-metal drawing, usually called deep drawing, utilizes the radial drawing of the sheet-metal blank into the die under the action of the punch. In deep drawing, the outer portion of the blank shrinks in diameter under circumferential compression. To prevent the blank from buckling, the blank-holder must exert sufficient pressure to prevent wrinkling, but not enough pressure to restrict the sheet from drawing into the die. Thus, in deep drawing, no deformation occurs in the central re-

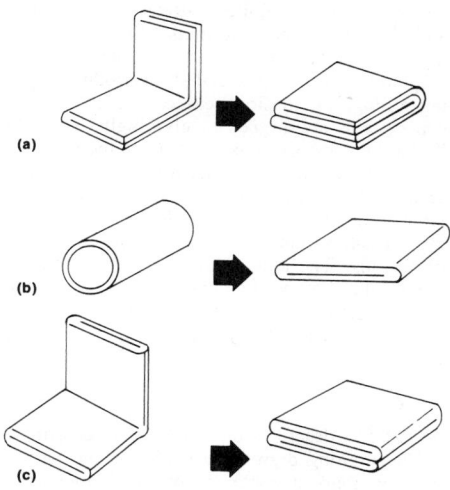

Fig. 4. Fold tests on ductile sheet or tube (see text)

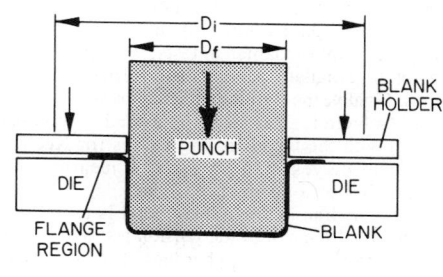

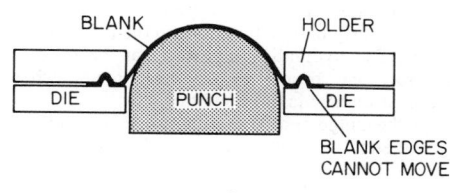

Fig. 5. Two operations that simulate stamping: (a) deep drawing; (b) stretching

gion of the punch directly under the punch; whereas in stretch forming, the maximum deformation occurs in this region. Figure 5 shows the essential differences between stretching and drawing.

The ability of the metal to undergo stretching is enhanced by a high value of strain hardening. Thus, a high value of the strain-hardening ex-

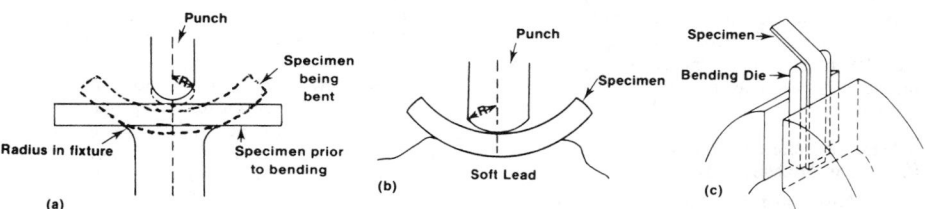

(a) Guided bend test wherein the test material is forced through a fixture of predetermined radius. (b) Modification of guided bend test using soft metal for the fixture. (c) Method of clamping the specimen while bending it over a predetermined radius. Source: same as Fig. 1.

Fig. 3. Restricted bend tests

ponent minimizes failure in stretch forming. The ability to withstand deformation in deep drawing, without failure, derives from the crystallographic texture of the metal sheet produced during rolling. The desired texture is such that the slip systems are aligned to give higher strength in the thickness of the sheet than in the plane of the sheet. As the plastic-strain ratio, r, becomes greater, the limiting draw ratio, LDR, becomes larger. The plastic-strain ratio is obtained by taking a tensile specimen and straining it to the point of necking. The longitudinal, thickness and lateral (width-direction) strains are determined and are, respectively, ϵ_l, ϵ_t and ϵ_w. The plastic-strain ratio is defined as:

$$r = \frac{\epsilon_w}{\epsilon_t}$$

ASTM 517 describes the test used to determine r. The limiting draw ratio (LDR) is the largest ratio of blank diameter to punch diameter for which the blank can be drawn into a cup of diameter D_p without tearing.

Many laboratory tests have been developed to measure and control the formability characteristics of sheet metals. Some, such as the hydraulic bulge test, are fundamental tests, while others attempt to simulate actual sheet-forming operations. Finally, the forming of actual parts on which a grid of circles has been imprinted in combination with the forming-limit curves (or Keeler-Goodwin curves) can be used to measure the formability of a given sheet metal.

In the hydraulic bulge test, metal is tested under uniaxial tension in the tension test and under local compression in the hardness test. In a typical press-forming operation, the metal is deformed under biaxial tension or biaxial tension-compression, in which the metal is strained simultaneously in two directions in the plane of the sheet. The hydraulic bulge test can be used to measure the properties of sheet metal when strained under biaxial conditions.

In the bulge test, a circular sheet is clamped at the edge and deformed by hydraulic pressure into a dome. For an isotropic sheet, essentially uniform biaxial stress and strain exist over an appreciable region at the center of the diaphragm. Failure eventually occurs in this central region.

Another sheet-formability test is the stretch bend test, which measures the ability of a sheet metal to be bent around a sharp radius under tension. It is a more severe test than the simple bend test and, in addition, can be used to measure the sensitivity of a metal to tearing from a stretched cut edge (a major problem in components with hole or stretch flanges).

In the stretch bend test, a sheared strip specimen of the material to be tested is clamped firmly between jaws and bent under tension, burr side outward, over a radiused punch. Normally, an autographic record of punch load and punch travel is obtained during the test. The punch travel — either at maximum load when cracks start to run into the material from the sheared edges, or at failure — is taken as the measure of specimen formability.

Ball Punch Deformation Test (Olsen and Erichsen Tests)

The Olsen test simulates sheet-metal performance under stretching conditions. It is a simple test in which the sheet metal is clamped rigidly in a blankholder, then stretched over a small hemispherical punch 22.2 mm ($^7/_8$ in.) in diam-

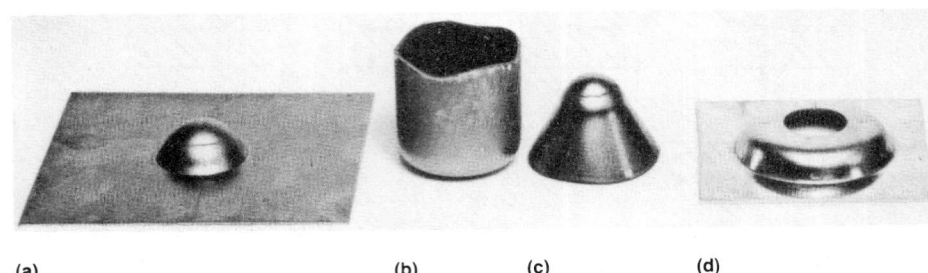

(a)   (b)   (c)   (d)

(a) Olsen and Erichsen tests. (b) Deep draw cup test. (c) Fukui test. (d) Hole-expansion test. Courtesy of Tinius Olsen Testing Machine Co.

Fig. 6. Results typical of ductility tests on sheet-metal blanks

eter. The stretchability of the sheet is then assessed by measuring the height to which the sheet can be stretched before fracture occurs. In a typical Olsen tester, both the punch travel and punch load are recorded, and the fracture point is established by noting the point at which the load suddenly decreases. Figure 6 shows sheet specimens that were subjected to four different formability tests.

The Olsen test has been replaced by the "ball punch deformation test" standardized by ASTM (ANSI/ASTM E643-78). In this test, many of the test parameters that previously were left to the discretion of the individual performing the test are normalized. The standardized test applies to specimens with thicknesses between 0.2 and 2.0 mm. The machine to which the tooling is attached should have the capability of holding down the specimen (pressure between the top and bottom die) with a force of at least 10 000 N (2200 lbf).

Because the punch surface and the sheet-metal surface are in contact during this test, the friction between the two surfaces has a large effect on the test conditions. To maintain standard friction conditions from one test to the next, the lubricant is standardized. It is commercially available petroleum jelly (vaseline) applied to the punch only. ASTM E643 also states that other lubrication systems (e.g., polyethylene sheet plus oil) may be used as agreed between supplier and user.

The speed of the penetrator shall be between 0.2 and 1 in./min (0.08 and 0.4 mm/s). The end of the test corresponds to the drop in load, which is caused by necking of the sheet. If the machine is not equipped with a load indicator, the end point will be either visible necking or fracture of the test specimen in the dome. The cup height is measured at this point and is the penetrator (punch) displacement.

The Erichsen test, which is common in Europe, where it was standardized, is similar to the Olsen test in principle — that is, the test simulates sheet-metal performance under stretching conditions. The punch diameter for the Erichsen test is slightly smaller than the punch used for the Olsen test (20 mm or 0.79 in.).

The Erichsen test may be performed with or without lubrication, but the use of lubrication introduces a new variable, as described in the above discussion of the Olsen test. A portable instrument for performing the Erichsen test is available and has been widely used for control of formability or drawability in sheet-metal working, especially for quality control of incoming material.

Limiting Dome Height Test

In the Erichsen, Olsen and bulge tests, fracture occurs at conditions that are close to equibiaxial

strain (when the strain is the same in the two perpendicular directions). In the uniaxial tension test, fracture occurs at a combination of tensile strain plus a small amount of contraction strain in the width direction. In practical press-forming operations, most fractures occur at close to plane-strain conditions, such as a tensile strain in one direction with zero strain in the other direction — which is somewhere between the conditions in the Olsen, Erichsen and bulge tests on the one hand and conditions in the tension test on the other.

The limiting dome height test has been developed to simulate more effectively the fracture conditions found in most parts. In this test, a large-diameter hemispherical punch, usually 100 mm (4 in.) in diameter, is used, and strips of sheet steel of varying widths are clamped and then stretched over the punch. The strips are marked with a grid of small circles, 2.5 mm (0.1 in.) in diameter, and the width strain at fracture is measured from the circle closest to the fracture. The width strain increases as the width of the sheet becomes greater.

The advantage of the limiting dome height test is that it more closely simulates the fracture conditions in a practical press-forming operation. It is a complex and time-consuming test, however, and the results are critically dependent on sheet thickness. In this test, lubrication is not critical; the standard practice is to perform the test dry (without lubricant).

Swift Cup Test

This test simulates the drawing operation and involves drawing of a small flat- or hemispherical-bottom, parallel-side cup. The sheet is held under a blankholder, as shown in Fig. 7, but is well lubricated with polyethylene and oil to ensure that the blank can be drawn in under the blankholder. Typical Swift cup test forming tools are available in 19-, 32- and 50-mm diameters for use with specimens ranging in thickness from 0.3 to 1.24, 0.32 to 1.30 and 0.45 to 1.86 mm, respectively. For drawing 40-mm-square cups from 80-mm-diam round specimens from 0.2 to 2 mm thick, a 40-mm-square forming tool is recommended.

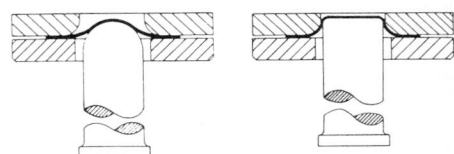

Punch diameter is 50 or 32 mm (2 or 1.3 in.).

Fig. 7. Swift cup test

The drawability of the metal is estimated by drawing a series of blanks of increasing diameter. The maximum blank size that can be drawn without fracture occurring over the punch nose is used to calculate the limiting draw ratio. For example, forming a 66-mm-diam disk using a 33-mm forming tool provides an LDR of 2.0. Because the condition of the edge of each blank can have an important effect on the test result, the blank edges usually are turned in a lathe to ensure strain-free, burr-free edges.

The results of this test correlate well with the performance of sheet metal in deep-drawn components, but, because of shape and alignment, reproducibility between laboratories is not good. The main problem with this test, however, is that it is time-consuming, and a large number of blanks of different sizes must be tested to obtain a reliable result.

Apart from measuring drawability, this test also can be used as a quality-control check to measure the tendency toward earing of the sheet metal. In this case, a blank of fixed diameter is drawn, and the height between the peaks and troughs in the cup wall are measured.

The Englehardt or draw fracture test is a variation of the Swift cup test for measuring drawability that overcomes the problems of complexity and time involved in that test. The draw fracture test involves drawing of a cup to the point of maximum drawing load, then clamping the flange and continuing the punch travel to fracture. A load-penetration curve similar to that in Fig. 8 is obtained and the Englehardt value, T, is calculated from the maximum draw and fracture loads, P_d and P_f:

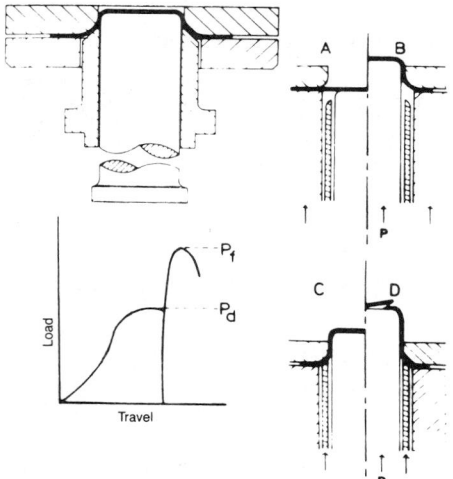

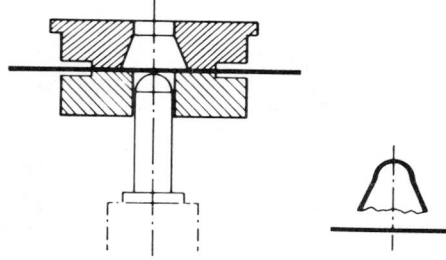

A and B: drawing. C and D: clamping and fracture.
Fig. 8. Draw fracture test

Fig. 9. Fukui conical cup test

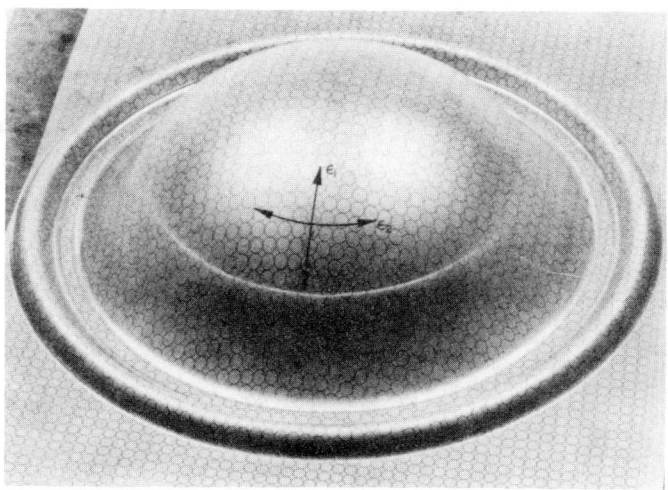

Fig. 10. Sheet specimen subjected to punch-stretch test to the point of necking (indicated by clear line). Courtesy of S. S. Hecker, Los Alamos National Laboratory.

$$T = \frac{P_f - P_d}{P_f}$$

This result depends on strip thickness and usually is corrected, using an empirical relationship, to a nominal thickness. Because of its simplicity of operation and reproducibility, the draw fracture test is the most suitable for testing of drawability on a routine basis.

Fukui Conical Cup Test

The Fukui conical cup drawing test (Fig. 9) was developed to assess the performance of a material during forming operations involving both drawing and stretching. The advantage of this test is that no holddown is necessary if the correct relationship between sheet thickness and blank diameter is maintained.

A blank of the appropriate size is laid over a 30° conical entry die and forced into the cavity by a flat-bottom or hemispherical punch. The height of the cup at failure is used as a measure of formability. The test requires various tooling for different sheet thicknesses, and the result is thickness-dependent. It has been demonstrated that the Fukui cup depth is influenced mainly by stretchability, but with some dependence on drawability. Thus, this test does not correlate as highly with uniform elongation and r-values as do other tests that are predominantly stretch or draw, which may explain why the conical cup test has not been as widely accepted as other simulative tests.

Typical tooling commercially available for the Fukui test includes a cutting ring, cutting ram and ball indenter available for specimen thicknesses from 0.5 to 1.6 mm.

FORMING-LIMIT CURVES

The poor correlation often found between results of the common "cupping" test and actual metal performance led investigators to look at some more fundamental parameters. Localized necking requires a critical combination of major and minor strains (along two perpendicular directions in the sheet plane). This concept led to the development of diagrams known as the Keeler-Goodwin or forming-limit curves (Ref 1 and

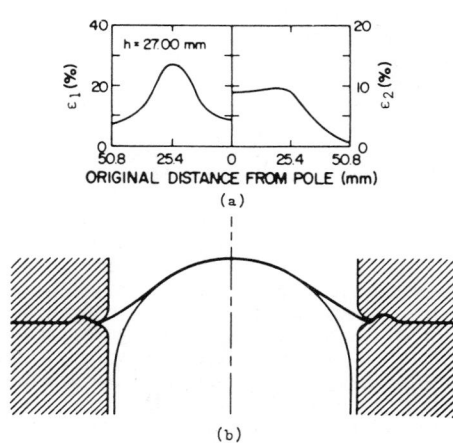

(a) Representation of strain distribution: ϵ_1, meridional strains; ϵ_2, circumferential strains; h, cup height. (b) Geometry of deformed sheet.
Fig. 11. Schematic illustration of sheet deformed by punch stretching

2). The forming-limit curve (FLC) is an important addition to the arsenal of testing techniques in formability.

Hecker (Ref 3) developed a punch-stretch apparatus and technique well suited for the determination of forming-limit curves. This apparatus consists of a punch with hemispherical head 101.6 mm (4 in.) in diameter. The die plates are mounted in a servohydraulic testing machine with the punch mounted on the actuator. The hold-down pressure on the die plates (rings) is provided by three hydraulic jacks (the hold-down load is 133 kN). The bead-and-groove arrangement in the rings eliminates any possible drawing in. The specimens are all gridded with 2.54-mm circles by a photoprinting technique. The load versus displacement is measured and recorded during the test, and the maximum load is essentially coincident with localized instability and the onset of fracture. A gridded specimen after failure is shown in Fig. 10. The circles become distorted into ellipses. The clear circumferential mark is due to necking. The strains ϵ_1 and ϵ_2 are called meridional and circumferential strains, respectively, and are measured at various points when the test is interrupted. Figure 11 shows how these strains vary with distance from the axis of symmetry of the punch, at the point where the punch has advanced a total distance of h = 27 mm. ϵ_1, the meridional strain, is highest at about 25 mm from the center ($\epsilon_1 \simeq 0.25$); ϵ_2, the circumferential strain, shows a definite plateau. By using sheets with different widths and varying lubricants between the sheet and the punch, different strain patterns are obtained. These tests are conducted to obtain different combinations of minor-major strains leading to failure. Figure 12 shows how the FLC curve is obtained. The minor strain (circumferential) is plotted on the abscissa and the major strain (meridional) is plotted on the ordinate axis. Four different specimen geometries are shown. The V-shape curve (FLC) marks the boundary of the safe-fail zone. The region above the line corresponds to failure; the region below is safe. In order to have both major and minor strains positive, a full-size specimen is used. By increasing lubrication, the major strain is increased; a polyurethane spacer is used to decrease friction. The drawings at the lower left- and right-hand corners of Fig. 12 show the deformation undergone by a circle of the grid. When both strains are positive, there is a net increase in area. Consequently, the thickness of the sheet has to decrease proportionately. On the left-hand side of the plot, negative strains are made possible by reducing the lateral dimensions of the blank. This allows free contraction in this dimension. The strains in an FLC are obtained by carefully measuring the dimensions of the ellipses adjacent to the neck-failure region.

These FLC's provide helpful guidelines for press-shop formability. Coupled with circle-grid analysis, they can serve as a guide in modifying the shapes of stampings. Circle-grid analysis consists of photoprinting a circle pattern on a blank and stamping it, determining the major and minor strains in its critical areas. This is then compared with the FLC to verify the available safety

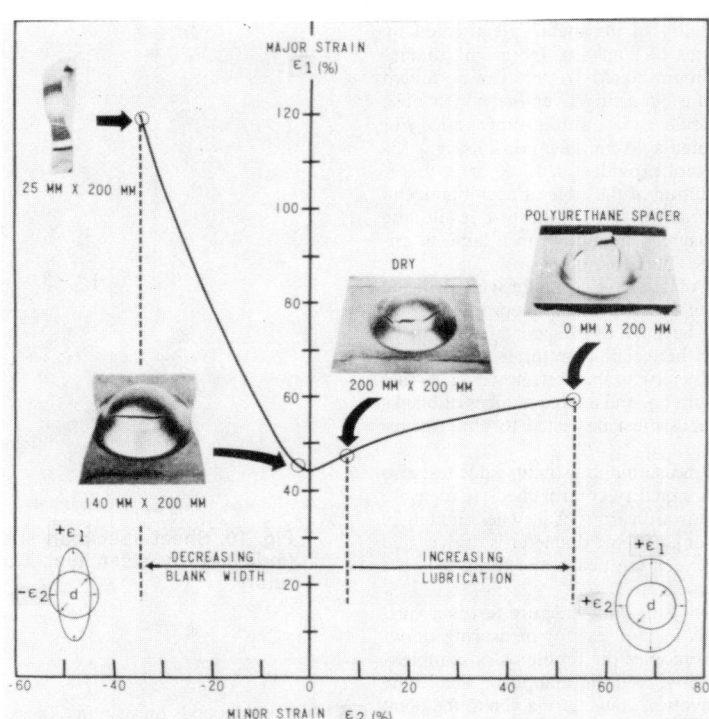

Fig. 12. Construction of a forming-limit curve (or Keeler-Goodwin diagram). Courtesy of S. S. Hecker, Los Alamos National Laboratory.

margin. The strain pattern can be monitored with changes in lubrication, hold-down pressure, and size and shape of drawbeads and the blank; this can lead to changes in experimental procedure. Circle-grid analysis also serves, in conjunction with the FLC, to indicate whether a certain alloy might be replaced by another one, possibly cheaper or lighter. During production, the use of occasional circle-grid stampings provides valuable help with respect to wear, faulty lubrication, and changes in hold-down pressure.

REFERENCES

1. S. P. Keeler and W. A. Backofen: *Trans. ASM*, Vol 56, 1963, p 25
2. "Application of Strain Analysis to Sheet Metal Forming Problems in the Press Shop," by G. M. Goodwin: SAE Automotive Engineering Congress, Detroit, Jan 1968 (SAE Paper No. 680093)
3. S. S. Hecker: *Met. Eng. Quart.*, Vol 14

Selected References on Mechanical Testing

Hardness Testing

The Hardenability of Steels, by C. A. Siebert, D. V. Doane and D. H. Breen: American Society for Metals, Metals Park, OH, 1977
The Science of Hardness Testing and Its Research Applications, edited by J. H. Westbrook and H. Conrad: American Society for Metals, Metals Park, OH, 1973
Hardness Testing Handbook, by V. E. Lysaght and A. DeBellis: American Chain and Cable Co., 1969
Metals Handbook, 8th Ed., Vol 11, *Nondestructive Testing and Quality Control:* American Society for Metals, 1976

Fracture Testing

Fracture Toughness Testing and Its Applications, STP 381: American Society for Testing and Materials, Philadelphia, 1965
Plane Strain Crack Toughness Testing of High Strength Metallic Materials, STP 410: American Society for Testing and Materials, Philadelphia, 1967
Fracture Toughness, STP 514: American Society for Testing and Materials, Philadelphia, 1972
Fracture Toughness Evaluation by R-Curve Methods, STP 536: American Society for Testing and Materials, Philadelphia, 1973

Fundamentals of Fracture Mechanics, by J. F. Knott: Butterworths, London, 1973

Fatigue Testing

Manual on Low Cycle Fatigue Testing, STP 465: American Society for Testing and Materials, Philadelphia, 1970
Metal Fatigue Damage—Mechanism, Detection, Avoidance, and Repair, STP 495: American Society for Testing and Materials, Philadelphia, 1971
Probabilistic Aspects of Fatigue, STP 511: American Society for Testing and Materials, Philadelphia, 1972
Fatigue and Fracture Toughness—Cryogenic Behavior, STP 556: American Society for Testing and Materials, Philadelphia, 1974
Handbook of Fatigue Testing, STP 566: American Society for Testing and Materials, Philadelphia, 1974
Manual on Statistical Planning and Analysis for Fatigue Experiments, STP 588: American Society for Testing and Materials, Philadelphia, 1975
Fatigue Crack Growth Under Spectrum Loads, STP 595: American Society for Testing and Materials, Philadelphia, 1976
Use of Computers in Fatigue Laboratory, STP 613: American Society for Testing and Materials, Philadelphia, 1976

Creep and Creep-Rupture Testing

Stress Rupture Parameters: Origin, Calculation, and Use, by J. B. Conway: Gordon and Breach, New York, 1969
Time-Temperature Parameters for Creep-Rupture Analysis, by S. S. Manson: Publication No. D8-100, American Society for Metals, Metals Park, OH, 1968
Source Book on Materials for Elevated-Temperature Applications, edited by E. F. Bradley: American Society for Metals, Metals Park, OH, 1979
High Temperature High Strength Nickel Base Alloys, 3rd Ed.: International Nickel Co., New York, 1972

Formability Testing

Formability: Analysis, Modeling, and Experimentation, edited by S. S. Hecker, A. K. Ghosh and H. L. Gegel: TMS-AIME, New York, 1977
Mechanics of Sheet-Metal Forming, edited by D. P. Koistinen and N. M. Wang: Plenum Press, New York, 1978
Formability Topics—Metallic Materials, STP 647: American Society for Testing and Materials, Philadelphia, 1978
Workability Testing Techniques, edited by G. E. Dieter: American Society for Metals, Metals Park, OH, 1984

35 METALLOGRAPHY

Edited by Robert J. Gray, Oak Ridge National Laboratory, and George F. Vander Voort, Carpenter Technology Corp.

Review Committee: W. A. Baeslack III, The Ohio State University; J. E. Bennett, Bell Laboratories; George Blann, Buehler Ltd.; Lee Dillinger, Leco Corp.; Rene Haig, Struers, Inc.; Richard J. Henry, University of Pittsburgh—Johnstown; James Nelson, Buehler Ltd.; M. H. Rafiee, C-E Power Systems; Donna B. Reese, Consultant; Ken Summerton, Imperial Clevite; Elisabeth Weidmann, Struers, Inc.

This section was condensed in part from Metals Handbook, Eighth Edition, Volume 8, Metallography, Structures and Phase Diagrams, pages 1 to 142. For more detailed information on the subjects covered herein, the reader is referred to the larger work.

Metallographic Practice Generally Applicable to All Metals

Metallographic Methods

THE METHODS and equipment discussed in this article are those used in the branch of metallography that employs visible light as the illumination source. This branch of metallography is called light optical metallography, to distinguish it from electron microscopy.

The preparation of metallographic specimens described in this article is applicable also to other types of studies, such as electron microscopy, microhardness testing, image analysis, and electron microprobe analysis. A well-prepared metallographic specimen is:

1. A representative sample
2. Sectioned, ground and polished so as to minimize the amount of disturbed or flowed surface metal caused by mechanical deformation, and thus to allow the true microstructure to be revealed by etching
3. Free from polishing scratches and pits and liquid staining
4. Polished so that inclusions and phases are preserved intact and not in relief
5. Flat enough to permit examination at high magnification
6. Prepared so that its edges are preserved to reveal surface details.

Preparation of metallographic specimens generally requires five major operations: (*a*) sectioning, (*b*) mounting (optional), (*c*) grinding, (*d*) polishing and (*e*) etching (optional).

SECTIONING

Many metallographic studies require more than one specimen. For example, a study of deformation in wrought metals usually requires two sections—one perpendicular to, and the other parallel to, the direction of deformation. A failed part may best be studied by selecting a specimen that intersects the origin of the failure, if the origin can be identified on the surface. Depending on the type of failure, it may be necessary to take several specimens from the area of failure and from adjacent areas.

Sampling. Bulk samples for sectioning may be removed from larger pieces or parts using methods such as core drilling, band or hacksawing, flame cutting, etc. However, when these techniques are used, precautions must be taken to avoid alteration of the microstructure in the area of interest. Laboratory controlled abrasive-wheel cutting is recommended to establish the ultimate plane of polish. In the case of relatively brittle materials, sectioning may be accomplished by fracturing the specimen at one desired location.

Abrasive-Wheel Cutting. By far the most widely used sectioning devices in metallographic laboratories are abrasive cutoff machines.

All abrasive-wheel sectioning should be done wet. An ample flow of water or water-soluble oil coolant should be directed onto the cut. Some laboratory cutoff machines provide for submerged wet cutting. Wet cutting will produce a smooth surface finish and, most importantly, will guard against excessive surface damage caused by overheating. Abrasive wheels should be selected according to the recommendations of the manufacturer. Samples must be fixtured securely during cutting and cutting pressure should be applied carefully to prevent wheel breakage.

MOUNTING OF SPECIMENS

The primary purpose of mounting metallographic specimens is for convenience in handling specimens of difficult shapes or sizes during the subsequent steps of metallographic preparation and examination. A secondary purpose is to protect and preserve extreme edges or surface defects during metallographic preparation.

The method of mounting should in no way be injurious to the microstructure of the specimen. Mechanical deformation and heat are the most likely sources of injurious effects.

Clamp Mounting. Clamps have been used for mounting metallographic cross sections in the form of thin sheets. Several specimens can be clamped conveniently in sandwich form. Although this is a rapid method, problems encountered with stains and snagging of cloths make this method less desirable than encapsulation. On the other hand, hairline separations between specimens can occur, and these separations can entrap abrasive particles or liquid solutions during metallographic preparation. These problems can be minimized by proper tightening of clamps, by use of plastic spacers between specimens or by coating specimen surfaces with epoxy before tightening.

Compression mounting, the most common mounting method, uses pressure and heat to encapsulate the specimen with a thermosetting or thermoplastic mounting material. Common thermosetting resins include phenolic (Bakelite) and diallyl phthalate, while methyl methacrylate is the most common thermoplastic mounting resin. Both thermosetting and thermoplastic materials require heat and pressure during the molding cycle, but after curing, mounts made of thermosetting materials may be ejected from the mold at the maximum molding temperature. Because some shrinkage of the resin from the specimen surface may occur, it is recommended that the temperature be lowered to about 55 °C (130 °F) before ejection. Maximum edge retention is obtained by using thermosetting epoxy resins containing hard filler particles. Thermoplastic materials remain molten at the maximum molding temperature and must be cooled under pressure before ejection.

Cold mounting materials require neither pressure nor heat, and are recommended for mounting specimens that are sensitive to heat and/or pressure.

Epoxy resins are the most widely used cold mounting materials. They adhere tenaciously to most metallurgical, mineral and ceramic specimens. They also exhibit lower volume shrinkage than either polyesters or acrylics and are very useful for impregnating porous structures or cracks by vacuum methods. For minimal shrinkage, epoxy mounts should be cured at room temperature for 8 to 12 h. Hard filler particles may be added to enhance edge retention.

Taper sectioning (mounting) generally is regarded as a special mounting technique; it enables the metallographer to examine in greater detail the immediate subsurface structure or surface topography of a specimen. Microhardness determinations and thickness measurements of thin surface coatings or diffusion zones are often more easily accomplished on taper-sectioned specimens.

Taper sectioning (Fig. 1) is accomplished by establishing a plane of polish at a small angle to the surface of the specimen.

Edge preservation can be aided by use of backup material in the mount, by application of coatings to the sides of the specimen before mounting, or by use of a filled mounting material.

Sheet-metal strips mounted on edge immediately adjacent to the specimen edge usually serve as backup material. For best results, the backup and the specimen should be of the same material, or at least should have similar grinding and polishing characteristics.

Application of surface coatings to the sides of the specimen is usually accomplished by electroplating or electroless plating with suitable metals. Iron, nickel and copper are frequently used as coatings.

GRINDING

Grinding should commence with the finest grit size that will establish an initial flat surface and remove the effects of sectioning within a few minutes. An abrasive grit size of 180 or 240 grit is coarse enough to use on specimen surfaces sectioned by an abrasive cutoff wheel. Hacksawed, band-sawed, or other rough surfaces usually require abrasive grit sizes in the range of 120 to 180 grit. The abrasive used for each succeeding grinding operation should be one or two grit sizes smaller than that used in the preceding operation. A satisfactory fine grinding sequence might involve grit sizes of 240, 320, 400 and 600 grit.

As in abrasive-wheel sectioning, all grinding should be done wet, provided that water has no adverse effects on any constituents of the microstructure. Wet grinding minimizes loading of the abrasive with metal removed from the specimen being prepared and minimizes specimen heating.

The purpose of grinding is to lessen the depth of deformed metal to the point where the last vestiges of damage can be removed by a series of polishing steps. The scratch depth and the depth of cold worked metal underneath the scratches decrease with decreasing particle size of abrasive. However, the depth of cold worked metal is roughly inversely proportional to the hardness of the specimen and may be 10 to 50 times the depth of penetration of the abrasive particles.

Grinding Media. The grinding abrasives commonly used in the preparation of metallographic specimens are silicon carbide (SiC), aluminum oxide (Al_2O_3), emery ($Al_2O_3 - Fe_3O_4$), composite ceramics and diamond. All except diamond are generally bonded to paper or cloth backing materials of various weights in the form of sheets, disks and belts of various sizes. Limited use is made of grinding wheels consisting of abrasives embedded in a bonding material. The abrasives may be used also in powder form by charging the grinding surfaces with loose abrasive particles or with abrasive in a premixed slurry or suspension.

Belt, Disk and Surface Grinders. The most common types of motor-driven grinding equipment are the belt grinder and the disk grinder. In the use of either, the metallographic specimen is held by hand against a moving, fixed-abrasive grinding material supported by a platen. Belt grinders and disk grinders may be used in either a horizontal or a vertical position. Abrasive belts are generally cloth-backed.

Lapping is an abrasive technique in which the abrasive particles roll freely on the surface of a carrier disk. During the lapping process, the disk is charged with small amounts of a hard abrasive such as diamond or silicon carbide. Lapping disks can be made of many different materials; cast iron and plastic are used most commonly. Lapping produces a flatter specimen surface than grinding, but it is a slow method and produces deeper deformation. Consequently, polishing of a lapped surface takes longer than polishing of a ground surface.

POLISHING

Polishing is the final step in producing a deformation-free surface that is flat, scratch-free, and mirrorlike in appearance. Such a surface is necessary for subsequent metallographic interpretation, both qualitative and quantitative. The polishing technique used should not introduce extraneous structures such as disturbed metal, pitting, dragging out of inclusions, "comet tailing" and staining. Polishing usually is conducted in several stages. Rough polishing generally is done with 6- or 9-μm (235- or 355-μin.) diamond abrasive charged onto napless or low-nap cloths. For hard materials such as through-hardened steels, ceramics and cemented carbides, an additional polishing step may be required. For such materials, initial rough polishing may be followed by polishing with 1-μm (40-μin.) diamond on a napless, low-nap or medium-nap cloth. A compatible lubricant should be used sparingly to prevent overheating or deformation of the surface. Intermediate polishing should be performed thoroughly so that final polishing may be of minimal duration.

Mechanical Polishing

The term "mechanical polishing" is frequently used to describe the various polishing procedures involving the use of fine abrasives or cloth. The cloth may be attached to a rotating wheel or a vibrating bowl. The specimens are held by hand, held mechanically, or merely confined within the polishing area.

Hand Polishing. Aside from the use of improved polishing cloths and abrasives, hand-polishing techniques still follow the basic practice established many years ago:

1. *Specimen Movement.* The specimen is held with one or both hands, depending on the operator's preference, and is rotated in a direction counter to the rotation of the polishing wheel. In addition, the specimen is continually moved back and forth between the center and the edge of the wheel, thereby ensuring even distribution of the abrasive and uniform wear of the polishing cloth. (Some metallographers use a small wrist rotation while moving the specimen from the center to the edge of one side of the wheel.) The main reason for rotating the specimen is to prevent formation of "comet tails." This polishing artifact, shown at left in Fig. 2, is a result of directional polishing of materials containing inclusions, fine precipitates, voids or other similar features. View at right in Fig. 2 shows the same area properly polished.

2. *Polishing Pressure.* The correct amount of applied pressure must be determined by ex-

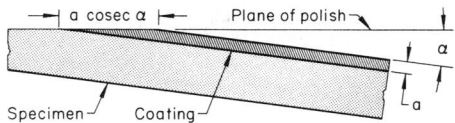

Fig. 1. Schematic illustration of taper sectioning (mounting), as applied to a coated specimen. Taper magnification equals cosecant of taper angle α.

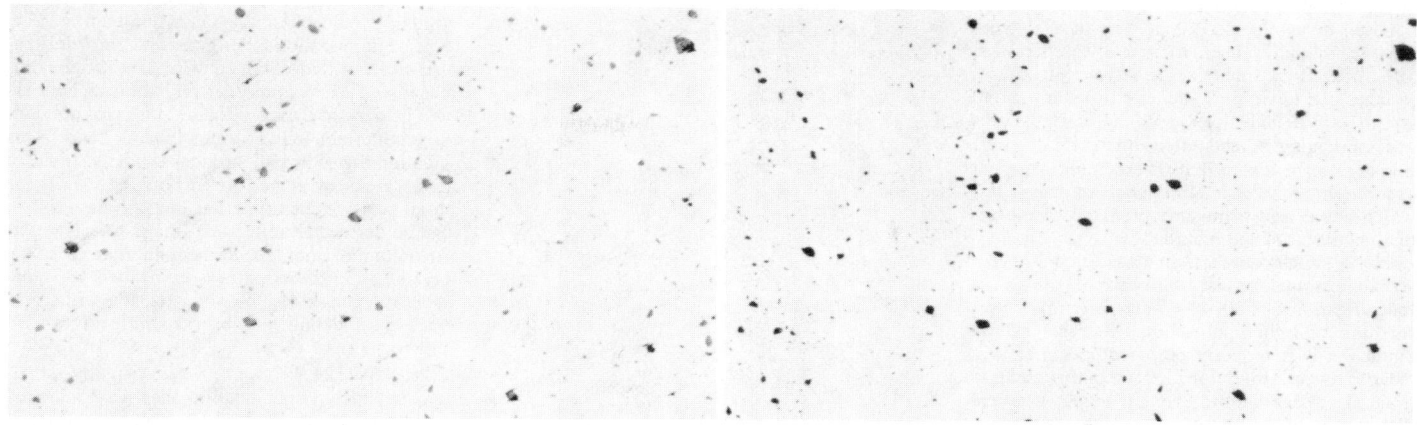

Fig. 2. As-polished (unetched) surface of a steel specimen, showing "comet tails" (left) resulting from directional polishing. View at right shows same area of specimen surface, but properly polished (note absence of "comet tails"). Magnification, 100×.

perience; in general, firm hand pressure is applied to the specimen in the initial polishing step and is proportionally decreased with successively finer polishing steps. For very soft metals, pressure other than that from the weight of the specimen itself may be eliminated entirely in the last polishing operation.

3. *Washing and Drying.* The specimen is preferably washed and swabbed in warm running water, rinsed with ethanol or any other alcohol that does not leave a residue, and dried in a stream of warm air. Alcohol usually can be used for washing when the abrasive carrier is not soluble in water.

4. *Cleanness.* The precautions for cleanness, as previously mentioned, must be strictly observed.

Final polishing is often performed using a fine (0.3- to 0.05-μm, or 12- to 2-μin.) alumina slurry on a medium-nap cloth. A wide variety of final-polishing abrasives have been employed. Prolonged application may lead to overpolishing defects such as pits, relief and edge-rounding. The final polishing step may also employ 0.25-μm (10-μin.) or finer diamond on a medium-nap cloth.

Automatic Polishing. Mechanical polishing can be automated to a high degree using a wide variety of devices ranging from relatively simple systems to rather sophisticated, minicomputer- or microprocessor-controlled devices. Units also vary in capacity from a single specimen to a dozen or more at a time. While some of these devices are used only for the final polishing step, others can be used for all grinding and polishing steps. These latter devices enable the operator to prepare a large number of specimens per day with the same degree of quality as that of hand polishing and at reduced consumable costs. Automatic polishing devices also are desirable for preparing radioactive specimens by remote control or for using corrosive-attack polishing procedures safely without hand contact.

Polishing Cloths. The requirements of a good polishing cloth include the ability to hold an abrasive, long life, absence of any foreign material that may cause scratches, and absence of any processing chemical (such as dye or sizing) that may react with the specimen. More than a hundred cloths of different fabrics, with a wide variety of naps, are manufactured specifically for metallographic polishing. Napless or low-nap cloths are recommended for rough polishing us-

ing diamond abrasive compounds. Low-, medium-, and occasionally high-nap cloths are used for final polishing, but this step should be as brief as possible to minimize relief.

Polishing Abrasives. Polishing usually involves the use of one or more of the following abrasives: diamond, aluminum oxide (Al_2O_3), magnesium oxide (MgO) and silicon dioxide (SiO_2). For certain materials, cerium oxide, chromium oxide or iron oxide may be used. With the exception of diamond, these abrasives normally are used in a distilled-water suspension, but if the metal to be polished is not compatible with water, other suspensions, such as ethylene glycol, alcohol, kerosine or glycerin, may be required. The diamond abrasive should be extended only with the carrier recommended by the manufacturer.

Electrolytic Polishing

Even with the most careful mechanical polishing, some disturbed metal, however small the amount, will remain after preparation of a metallographic specimen. This is no problem if the specimen is to be etched for structural investigation, because etching is usually sufficient to remove the slight layer of disturbed metal. If the specimen is to be examined in the as-polished condition using polarized light, or if no surface disturbance can be tolerated, either electrolytic polishing (also called "electropolishing") or chemical polishing is preferred. The simple laboratory setup shown in Fig. 3, which is sufficient for most electropolishing requirements, and the

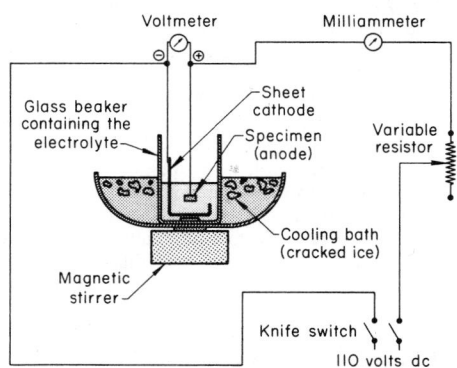

Fig. 3. Basic laboratory setup for electropolishing and electrolytic etching

more sophisticated commercial units are all based on the same principle. Direct current from an external source is applied to the electrolytic cell under specific conditions, and anodic dissolution produces leveling and brightening of the specimen surface.

Not all materials respond equally well to electrolytic polishing. Wrought solution-type alloys, such as aluminum, nickel, nickel-iron and titanium alloys, are particularly good candidates for electrolytic polishing. Electropolishing is usually reserved for single-phase alloys, because second phases and inclusions may be preferentially attacked during polishing.

Chemical Polishing

Chemical polishing involves simple immersion of a metal specimen in a suitable solution to obtain a metallographic polish. The results of chemical polishing are similar to those of electropolishing in that they vary from an etched specimen surface that has been macrosmoothed but not brightened, to a bright dipped surface that has been brightened but not macrosmoothed.

ETCHING

Metallographic etching encompasses all processes used to reveal particular structural characteristics of a metal that are not evident in the as-polished condition. Examination of a properly polished specimen before etching may reveal structural aspects such as porosity, cracks and nonmetallic inclusions. In certain nonferrous alloys, grain size can be revealed adequately only in the as-polished condition using polarized light.

Electrolytic Etching. The procedure for electrolytic etching is basically the same as for electropolishing, except that voltage and current densities are considerably lower. The specimen is made the anode, and some relatively insoluble but conductive material such as stainless steel, graphite or platinum is used for the cathode. Direct-current electrolysis is used for most electrolytic etching, and for small specimens (13-by-13-mm, or ½-by-½-in., surface to be etched), one or two standard 1½-volt flashlight batteries provide an adequate power source. A setup like that shown in Fig. 3 is usually all that is required.

Etching for Macrostructure. Macroscopic examination differs from microscopic examination in that it employs very low magnifications (up to approximately 50×) and is used for the investigation of defects and structure of a large area as opposed to a microscopic portion of that area. This technique is used to reveal solidification structure, flow lines, segregation, structural changes due to welding, general distribution and size of sulfide inclusions, porosity, ingot defects and fabricating defects. It is important that the investigator be aware that macroetching can exaggerate the size of inhomogeneities or defects, which could lead to misinterpretation of the actual condition of the material.

Etching for Microstructure. In this article, microscopic examination is limited to a maximum magnification of 2500× — the approximate useful limit of light microscopy. Microscopic examination of a properly prepared specimen will clearly reveal structural characteristics such as grain size; segregation; and the shape, size and distribution of the phases and inclusions that are present. The microstructure revealed also indicates prior mechanical and thermal treatment that the metal has received.

MICROSCOPIC EXAMINATION

Metallurgical microscopes differ from biological microscopes primarily in the manner by which the specimen is illuminated; unlike biological microscopes, metallurgical microscopes must use reflected light. Figure 4 is a simplified ray diagram of a metallurgical microscope. The prepared specimen is placed perpendicular to the optical axis of the microscope and is illuminated through the objective lens by light from the source, which is focused by the condenser into a beam that is made approximately parallel to the optical axis of the microscope by the half-silvered mirror. The light is then reflected from the surface of the specimen.

Light Sources

The amount of light lost during passage from the source through a reflecting type of microscope is appreciable because of the intricate path the light follows. For this reason, it is generally preferable that the intensity of the source be high. Several types of light sources are used, including tungsten-filament lamps, quartz-halogen lamps and xenon lamps.

Tungsten-filament lamps generally operate at low voltage and high current. They are widely used for visual examination, because of their low cost and ease of operation.

Quartz-halogen lamps provide a more intense steady illumination than tungsten-filament lamps and have a color temperature close to daylight when operated at higher voltages. They are very dependable and produce no odors.

Xenon Lamps. The intensity of xenon lamps is also very high, and their spectra make them suitable for color photomicrography. The first xenon lamps produced ozone, but modern units have overcome this problem.

Microscopic Techniques

Most microscopic studies of metals are made using bright-field illumination. In addition to this type of illumination, several special techniques (oblique illumination, dark-field illumination, opaque-stop microscopy, phase-contrast microscopy and polarized-light microscopy) have particular applications for metallographic studies.

Bright-Field Illumination. In bright-field illumination, the surface of the specimen is normal to the

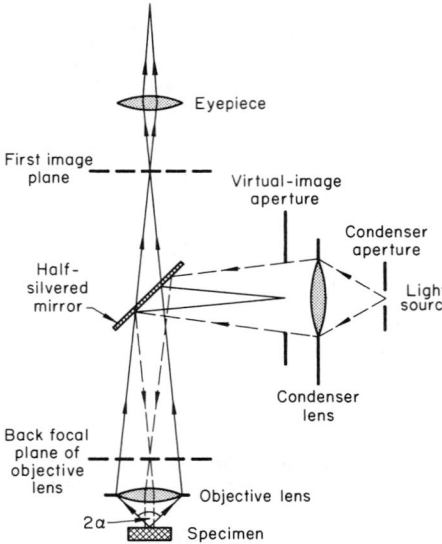

Fig. 4. Image formation in a metallurgical microscope employing bright-field illumination

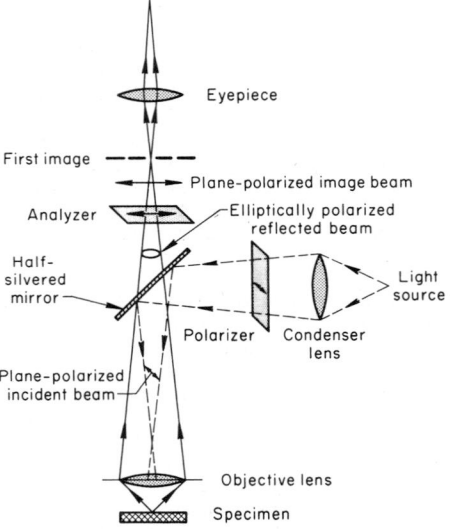

Fig. 5. Basic components of a polarizing light microscope

optical axis of the microscope, and white light is used. The ray diagram for this type of illumination in a standard type of bench microscope is illustrated in Fig. 4.

Oblique Illumination. The surface relief of a metallographic specimen can often be enhanced by using oblique illumination. This involves offsetting the condenser lens system or, as is more usually done, moving the condenser aperture to a position slightly off the optical axis. Although it should be possible to continually increase the contrast achieved by oblique illumination by moving the condenser farther and farther from the light axis, the numerical aperture of a lens is reduced when this happens, because only a portion of the lens is used. For this reason, there is a practical limit to the amount of contrast that can be achieved.

Dark-Field Illumination. Another method that often is used to distinguish features not in the plane of

the polished-and-etched surface of a metallographic specimen is dark-field illumination. This type of illumination gives contrast completely reversed from that obtained with bright-field illumination — the features that are light in bright field will be dark in dark field, and those that are dark in bright field will be light in dark field. This highlighting of angled surfaces (namely, those of pits, cracks or etched grain boundaries) allows more positive identification of their nature than can be derived from a black image under bright-field illumination. Due to the high image contrast obtained and the brightness associated with features at an angle to the optic axis, it is often possible to see details not observed with bright-field illumination.

Polarized-Light Microscopy. Because many metals and metallic and nonmetallic phases are optically anisotropic, polarized light is particularly useful in metallography.

Polarized light is obtained by placing a polarizer in front of the condenser lens of the microscope and placing an analyzer behind the eyepiece, as illustrated in Fig. 5. The polarizer produces plane-polarized light which strikes the surface and is reflected through the analyzer to the eyepieces. If an anisotropic metal is examined with the analyzer set 90° to the polarizer, the grain structure will be visible. However, viewing of an isotropic metal (cubic metals) under such conditions will produce a dark, "extinguished" condition.

Polarized light is particularly useful in metallography for revealing grain structure and twinning in anisotropic metals and alloys and for identifying anisotropic phases and inclusions.

Differential Interference Microscopy (DIC). When crossed polarized light is used along with a double quartz prism (Wollaston prism) placed between the objective and the vertical illuminator, two light beams are produced which exhibit coherent interference in the image plane. This leads to two slightly displaced (laterally) images differing in phase ($\lambda/2$), which produces height contrast. The image produced reveals topography detail somewhat similar to that produced by oblique illumination but without the loss of resolution. Images can be viewed with natural colors similar to those observed in bright field, or artificial coloring can be introduced.

High-Temperature Microscopy. Several microscopes have been developed with devices that allow simultaneous heating and examination of specimens. These have permitted direct examination of thermal effects in metals, such as grain growth, precipitation reactions, phase changes, sintering, diffusion, and certain types of surface reactions. For some limited studies of this type, it is possible simply to surround the specimen with a small furnace and microscopically examine a polished surface. An alternative means of heating the specimen is by electrical resistance. Unfortunately, these simple methods have very limited usefulness, because most metals and alloys oxidize at high temperatures and oxidation must be avoided if unobstructed observations are to be made.

Heating in a vacuum presents a problem in that certain phases exposed on the surface of the specimen can evaporate and then condense on the viewing window, thereby hindering observation. This can be partly overcome by using a double viewing window, with the one nearer the specimen being removable and the remaining one window being used for photography. Evapora-

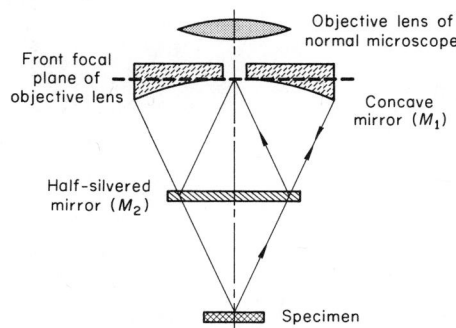

Fig. 6. Dyson-type objective for obtaining long working distance in microscopy

tion usually can be eliminated by operating in an inert-gas atmosphere, but to prevent surface oxidation or contamination, the inert gas generally must be of very high purity.

A serious problem in high-temperature microscopy is potential damage to the microscope parts, particularly the objective lens, by the high temperatures. This is partly corrected by water cooling the parts of the hot stage near the objective; however, high magnifications are still not possible, because the short working distances of most objectives do not permit examination through the viewing windows. To overcome this, long-working-distance objectives have been used. The most widely used type employs a reflecting concave mirror in conjunction with a standard objective. A simple type of long-working-distance objective is shown in Fig. 6.

Low-Temperature Microscopy. Certain reactions that occur in metals at low temperatures can be observed by microscopy. Stages have been constructed for this purpose: most are either adaptations of high-temperature stages or similar to them. Generally, a refrigerant, such as liquid nitrogen, cools the stage that holds the specimen. A thermocouple on the specimen measures the temperature, which is controlled by the supply of refrigerant. Low-temperature microscopy has found only limited use in metallography.

Straining Stages. Several other devices can be fitted to the stages of microscopes so that a specimen can be viewed while experiments are being performed on it. Stages constructed so as to allow straining of a specimen while it is being viewed have been particularly useful for studies of deformation, twinning, slip, and strain-induced transformations. In one such stage, the force from a small hydraulic cylinder is applied to the specimen through two cantilever arms. A load cell in series with the cylinder is used to monitor the applied force. The application of loads up to 900 kg (2000 lb) is possible with this type of stage.

Interferometry is probably the most sensitive and most accurate optical method of measuring the microtopography of surfaces. Two interference methods are in common use in metallography: the two-beam and the multiple-beam methods.

The principles of the two-beam method are illustrated in Fig. 7. In a two-beam interferometer, monochromatic light from the source is split into two beams. One beam travels through the microscope objective to the specimen and then is reflected back through the objective and into the eyepiece. The other beam passes through an identically matched objective onto an optically

flat reference plate and then back through the same objective, and is directed by the beam splitter to the eyepiece. The two beams meet in the eyepiece and either (*a*) reinforce each other (where the optical-path difference between them is equal to, or a multiple of, half the wavelength of the monochromatic light), or (*b*) interfere with each other (where the optical-path difference does not satisfy the above conditions). From this reinforcement or interference, contour lines are formed, with each line connecting points of the same level. The difference in height between fringes is one-half the wavelength of the light. Thallium light, which has a wavelength of 540 nm, is commonly used. It is usually possible to measure to an accuracy of $^1/_{10}$ of a displacement, which means that differences in height of about 27 nm can be measured.

Figure 8 illustrates the principles of multiple-beam interferometry. Instead of creating interference between two light beams, the multiple-beam method produces interference among many beams. This is done by placing an optically flat

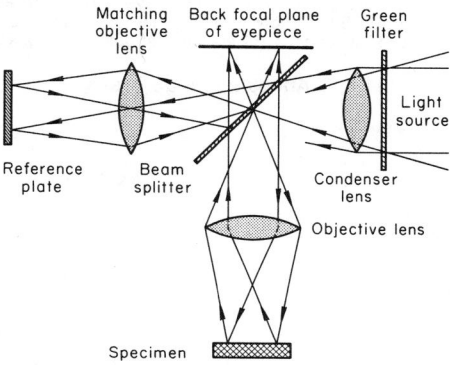

Fig. 7. Principles of two-beam interferometry

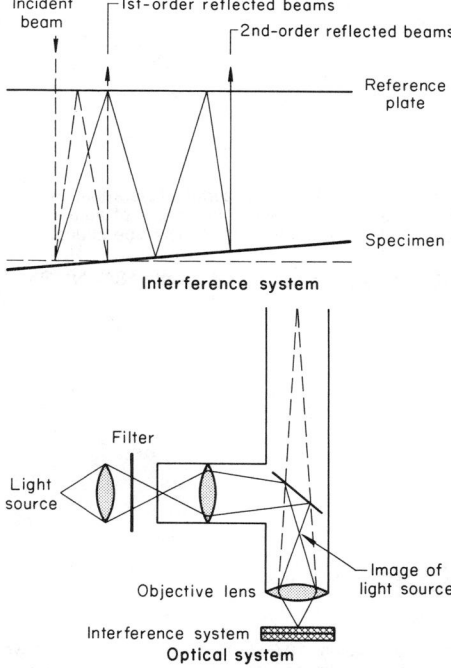

Fig. 8. Principles of multiple-beam interferometry

reference plate that is partly transmitting and partly reflecting to light onto the surface of the specimen. The plate and the specimen are positioned slightly out of parallel, but usually by not more than one or two wavelengths. Several objectives with reference plates built into them are commercially available for use on standard metallurgical microscopes. Monochromatic light is directed normal to the specimen surface through the objective. Some of this light is reflected from the surface of the reference plate and onto the eyepiece of the microscope, whereas most of the light passes through the plate and onto the specimen. A series of reflections occurs between the reference plate and the specimen such that, with each reflection, some of the light passes through the reference plate and into the eyepiece of the microscope. This light either reinforces or interferes with the light reflected from the surface of the plate, and contour lines result. If the components are correctly positioned, the multiple-beam method can produce such fine fringes that displacements as small as $^1/_{100}$ of the fringe displacement can be measured, thereby allowing measurements of differences in level as small as about 3 nm.

Principles of Technique Selection in Mechanical Polishing

THREE distinct operations are involved in determining the microstructure of metals with the use of an optical microscope: (*a*) preparation of a section surface; (*b*) development of the structure on that surface, usually by chemical etching; and (*c*) microscopic examination. Unfortunately, the effectiveness of the examination often is determined by the operation carried out least effectively, and frequently this is surface preparation.

A primary objective of a preparation procedure must be to produce a surface that fully represents the microstructure as it existed in the metal before sectioning. All structural features characteristic of the metal must be detectable, and false structures must not be introduced. This is a more demanding requirement than the mere production of a highly polished surface.

The main purpose of this article is to illustrate how objective experiments and comparisons can be used to develop preparation procedures that not only give better results but also are simpler and less laborious to use. The emphasis will be on principles that can be used as guides in the development of practical preparation procedures, rather than on the details of those procedures.

ABRASION DAMAGE AND ABRASION ARTIFACTS

The general pattern of a surface layer that has been plastically deformed is shown in Fig. 9(a) for abraded 70-30 brass, an alloy in which the effects of prior plastic deformation can readily be revealed by a range of etchants. This micrograph illustrates several characteristic features. The shallow, dark-etching, unresolved band contouring the surface scratches is known as the outer fragmented layer; it is a layer in which the strains have been very large. Beneath this extends a layer

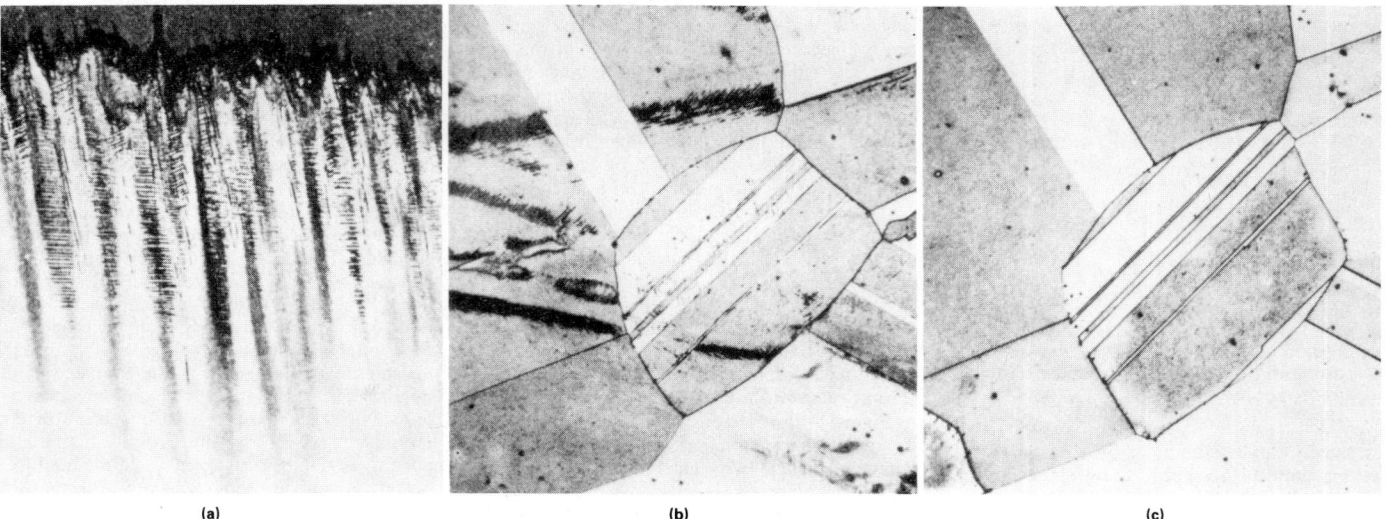

(a) (b) (c)

(a) Taper section (horizontal magnification, 600×; vertical magnification, 4920×) of surface layers that were abraded on 220-grit silicon carbide paper. The structures shown are discussed in the text. (b) and (c) Micrographs (both at 250×) showing the results of abrading on 220-grit silicon carbide paper and then polishing until about 5 μm (b) and 15 μm (c) of metal are removed. The banded markings in (b) are false structures (abrasion artifacts). The true structure is shown in (c). All three specimens etched in aqueous ferric chloride.

Fig. 9. Annealed 70-30 brass

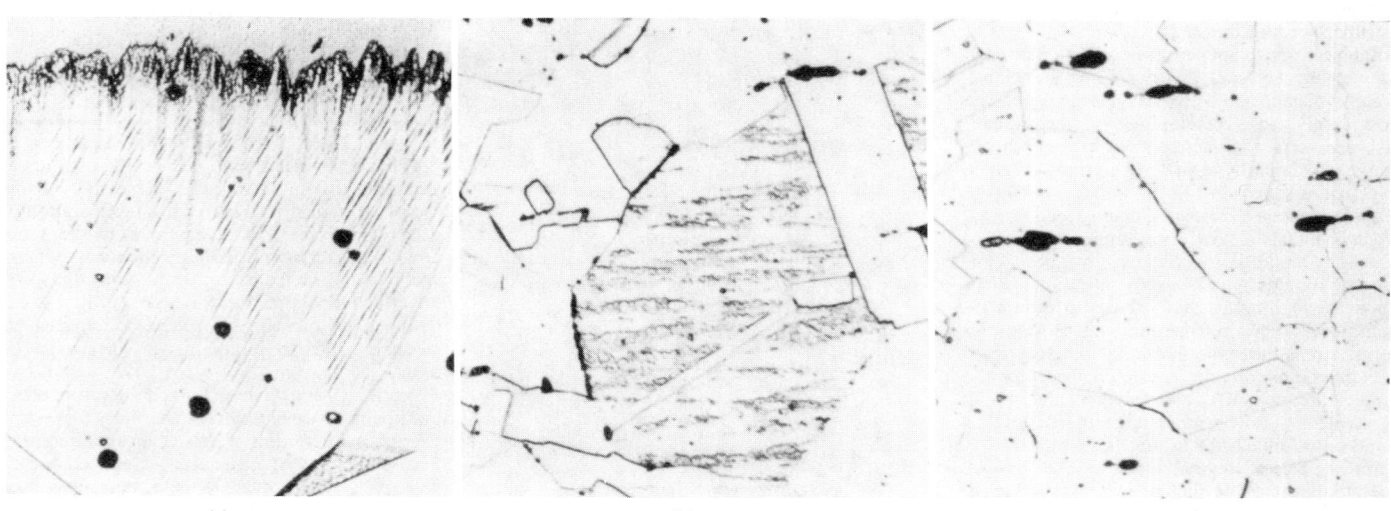

(a) (b) (c)

(a) Taper section (horizontal magnification, 600×; vertical magnification, 6060×) of surface layers that were abraded on 220-grit silicon carbide paper. The structures shown are discussed in the text. (b) and (c) Micrographs (both at 500×) showing the results of abrading on 600-grit silicon carbide paper and then polishing until about 1 μm (b) and 3 μm (c) of metal are removed. Abrasion artifacts are shown in (b). The true structure is shown in (c). All three specimens electrolytically etched in oxalic acid.

Fig. 10. Austenitic stainless steel (18% Ni, 8% Cr)

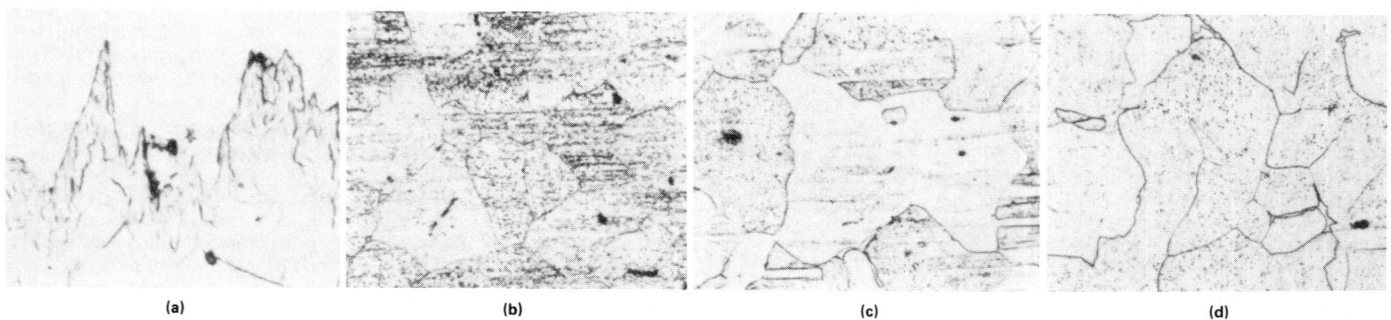

(a) (b) (c) (d)

(a) Taper section (horizontal magnification, 1000×; vertical magnification, 10 000×) of surface layers that were abraded on 220-grit silicon carbide paper. Note the outer fragmented layer. (b) Micrograph (at 250×) showing the results of abrading on 000 emery paper and then polishing only long enough to remove abrasion scratches. (c) Micrograph (at 250×) showing results of abrading on 600-grit silicon carbide paper and polishing only long enough to remove abrasion scratches. Micrographs (b) and (c) show banded markings and a generally confused structure. (d) Micrograph (at 250×) showing results of abrading on 600-grit silicon carbide paper and polishing for a longer time than for specimen shown in (c). Micrograph (d) shows the true structure. All four specimens etched in nital.

Fig. 11. Ferritic steel

in which the strains have been comparatively modest and in which they tend to concentrate in rays extending beneath individual surface scratches. This is shown by the bands of etch markings, which are known to develop at the sites of slip bands, and by the more diffuse rays, which are known to indicate the presence of kink bands. These effects extend for many times the depth of the surface scratches.

The importance of the surface damage shown in Fig. 9(a) is illustrated in Fig. 9(b) and (c). A sample of annealed 70-30 brass was abraded on 220-grit silicon carbide paper and then polished to remove a surface layer about 5 μm (200 μin.) thick. All traces of the abrasion scratches were removed and, ostensibly, a satisfactory surface was produced, but the bands of deformation etch markings shown in Fig. 9(b) appeared when the surface was etched. When layers of greater thickness were removed during polishing, these bands were gradually reduced in number and intensity, and eventually were eliminated, as can be seen in Fig. 9(c), which shows the true structure.

Abrasion Artifacts in Austenitic Steels. Austenitic steels generally are quite susceptible to abrasion artifacts, and the common etchants reveal effects due to prior deformation with considerable sensitivity. The structure of a typical abrasion-damaged layer (see Fig. 10a) is comparable to that for brass. A shallow, unresolved layer contours the surface scratches, and deep rays of deformation etch markings extend beneath the surface scratches. Bands of these deformation etch markings may appear in a final-polished surface as abrasion artifacts (see Fig. 10b). Good abrasion practice and efficient polishing processes are required for removal of the abrasion artifacts in acceptable polishing time (see Fig. 10c).

When a surface is found to contain artifacts of the type shown in Fig. 10(b), it can be assumed that a surface layer of considerable depth will have to be removed before an artifact-free surface will be obtained. Therefore, the specimen must be returned to rough polishing in order to obtain a sufficiently high cutting rate.

Abrasion Artifacts in Ferritic Steels. The deep abrasion-damage effects discussed thus far cause difficulties in only a limited range of alloys, but effects due to an outer fragmented layer are likely to be found in all metals. A section of the outer fragmented layer in a ferritic steel is shown in Fig. 11(a). The structure of the fragmented layer cannot be properly resolved by optical microscopy, but it is clear that the structure is different from that of the parent-metal ferrite grains. The types of artifacts that may be found in final-polished surfaces of ferritic steel are illustrated in Fig. 11(b) and (c). These artifacts obscure the true structure, shown in Fig. 11(d); they can be developed in virtually all metals. However, as shown in Fig. 11(a), the damaged layer is quite thin, and a polishing treatment continued for twice the time it takes to remove the abrasion scratches will eliminate the abrasion artifacts. Thus, structures with abrasion artifacts are usually the result of very inefficient preparation procedures.

Abrasion Artifacts in Pearlitic Steels. Certain distinctive artifacts caused by disturbance in the outer fragmented layer are observed in pearlitic steels. Taper sections of abraded surfaces of these steels show that the cementite plates of pearlite may simply be bent adjacent to some scratches (Fig. 12a) and may be completely fragmented adjacent to others (Fig. 12b).

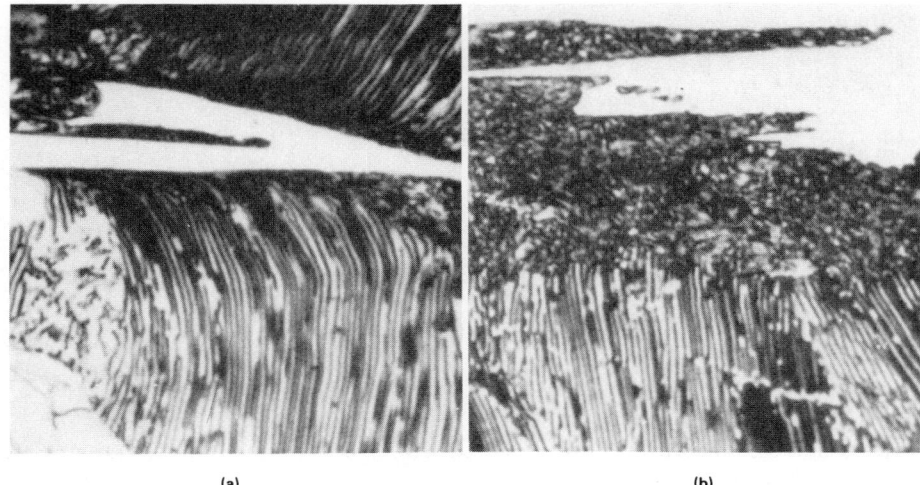

(a)   (b)

Longitudinal taper sections (horizontal magnification, 2000×; vertical magnification, 20 000×) of surface layers that were belt abraded on 100-mesh alumina, showing that cementite plates of pearlite are merely bent adjacent to some scratches (a) and are completely fragmented adjacent to others (b). Both specimens etched in picral.

Fig. 12. Pearlitic steel

Abrasion Damage in Gray Iron. Cast irons represent an important group of alloys for which a purpose of metallographic examination often is the determination of the true size and shape of the particles of free graphite that are present. The apparent size and shape of the graphite can be severely altered at several stages of the preparation sequence, thus giving rise to false structures.

The true graphite form for a particular gray iron is most closely represented in Fig. 13(c), but effects of abrasion are shown in Fig. 13(a) and (b).

Embedding of Abrasive. The points of the contacting abrasive particles of an abrasive paper fracture readily during abrasion, and these fragments may become embedded in the surface of a very soft metal, such as lead or annealed high-purity aluminum. Embedded particles are difficult to discern in the surface by optical microscopy, but a surface with a high concentration of embedded abrasive characteristically has a rough, torn appearance.

POLISHING DAMAGE

The mechanical polishing procedures commonly used in metallographic practice remove metal by mechanical cutting processes analogous to those occurring during abrasion. Mechanical polishing always produces a series of scratch grooves on the surface of the specimen; these scratches are difficult to detect by optical microscopy, particularly with bright-field illumination. Moreover, a plastically deformed, damaged layer is also introduced; the layer produced by mechanical polishing is much shallower than that produced by grinding, but its structure is similar. A layer analogous to the outer fragmented layer in ground surfaces can be recognized contouring the surface scratches, and there are occasional rays of deformed metal extending to greater depths, many times the depth of the polishing scratches. The presence of this damaged layer has important practical consequences — it affects the response of the surface to etching.

Even on well-polished surfaces, a very fine layer of damaged material will remain; this layer can

be removed by mechanical-chemical polishing ("attack" polishing). In mechanical-chemical polishing, a small amount of a dilute etchant is used together with the abrasive suspension. The etchant attacks the surface chemically while the abrasive removes the product of this chemical attack.

An excessive amount of the chemical component in a mechanical-chemical process may cause detrimental effects, such as severe etch pitting. Proper balance between the mechanical and chemical components can preserve most of the benefits provided by mechanical polishing and yet produce a damage-free surface — a most desirable combination in a final-polishing stage.

Enlargement of Polishing Scratches by Etching. It is a frequent annoyance in metallographic practice to find that a surface that appeared to be free of scratches when examined as-polished under bright-field illumination turns out to be severely scratched after etching (see examples in Fig. 14). The scratches were actually there all the time, but they were too fine to be detected when the specimen was in the unetched condition; they were enlarged, or shown in greater contrast, by etching.

Scratches are attacked preferentially during etching because of the disturbed metal, or damaged layer, associated with them. Severity of attack varies directly with the ability of the etchant to reveal deformation. The appearance of scratches also depends on the etching time. A certain minimum etching time is necessary to develop the scratches fully; thereafter, the scratches recede with increasing etching time, because etching progressively removes the damaged layer.

It may be difficult to distinguish scratches enlarged by the final polishing stages from scratch traces introduced during the previous polishing stage. The problem can be resolved by making the earlier set of scratches unidirectional and parallel to a known direction in the specimen surface. The scratch traces can then be recognized. This technique was used in preparing the specimen for the micrograph shown in Fig. 15.

Flatness. Quality in polishing practice means (a) a surface that is adequately free from confusing polishing scratches and (b) a surface that is sufficiently flat for all constituents and local regions to be examined properly.

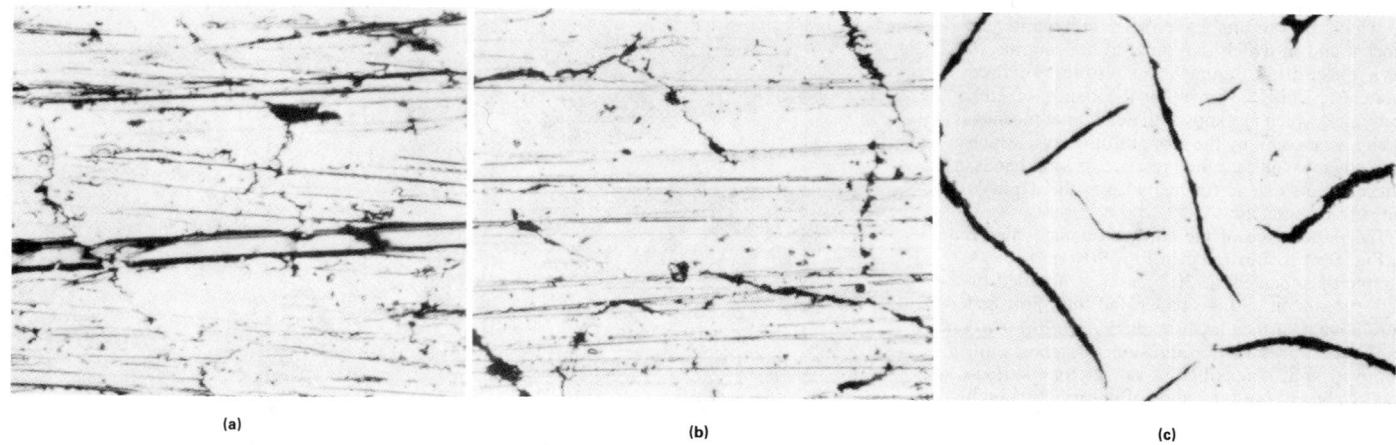

(a)   (b)   (c)

Micrographs (all at 500×) of as-polished specimens, showing results of abrading (a) on 220-grit silicon carbide paper, (b) on 600-grit silicon carbide paper and (c) on a fine fixed-abrasive lap.

Fig. 13. Effects of abrasion on flake graphite in gray iron

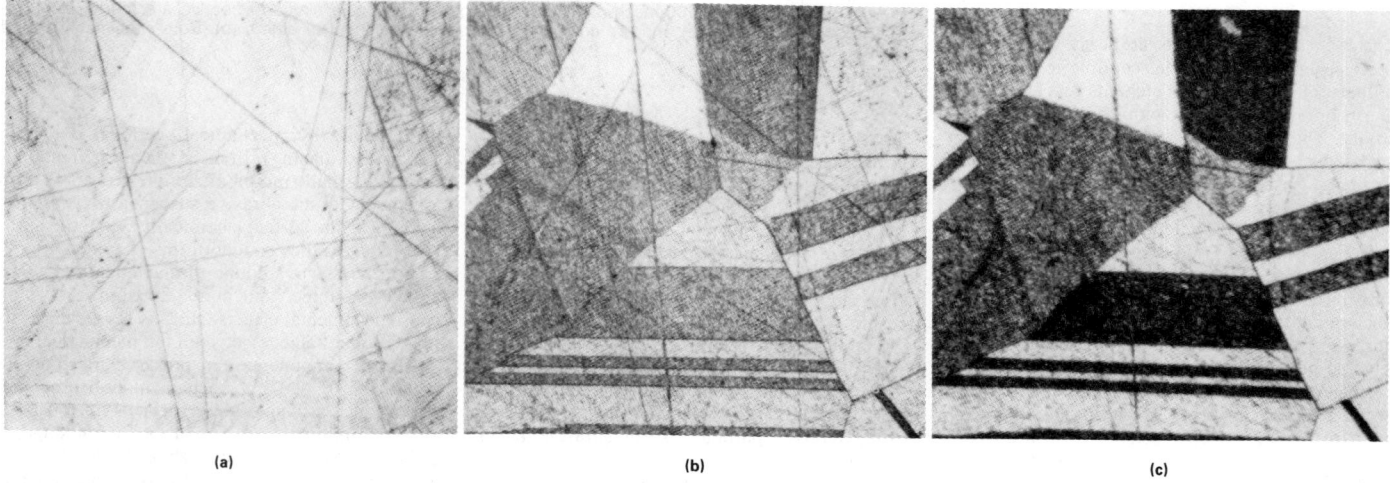

(a)   (b)   (c)

Micrographs (all at 250×) showing effect of incremental increases in etching time (etching time increases from a to b to c) on appearance and disappearance of scratches in a specimen that was polished on fine alumina. Etchant: aqueous ferric chloride. Micrograph (c) shows that a longer etching time removes most of the scratches and the damaged layer.

Fig. 14. Effect of etching time on appearance of scratches in annealed 70-30 brass

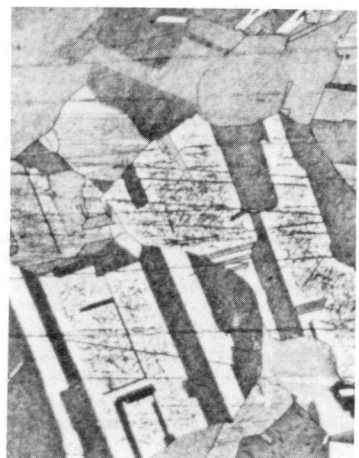

Micrograph, at 250×, showing results of fine polishing for a short time prior to etching in aqueous ferric chloride.

Fig. 15. Effect of polishing damage on response to etching for annealed 70-30 brass

Two examples of how markedly the choice of polishing abrasive and polishing cloth can affect surface flatness in specimens of duplex structure are given in Fig. 16 and 17. These micrographs show that alumina abrasive on billiard cloth produced a result inferior to diamond abrasive on synthetic suede cloth in polishing wrought iron and an aluminum alloy. The alumina on billiard cloth produced marked relief between the silicon constituent and the aluminum matrix of the aluminum alloy (Fig. 17a) and removed a portion of the silicate inclusion in the wrought iron (Fig. 16a).

Retention of Graphite in Gray Iron. Earlier in this article, it was demonstrated that although the graphite in cast iron can be damaged severely during the abrasion stage of preparation, it is possible by suitable choice of abrasion process to obtain a reasonably true representation of the structure. There remains the problem of retaining the graphite during polishing. The solution to the problem depends heavily on the length of the nap of the polishing cloth.

Graphite flakes in a gray iron invariably look much larger when long-nap cloths are used for polishing, as demonstrated in Fig. 18(a). This apparent enlargement is caused by erosion, which occurs at the interface between graphite and matrix, producing an enlarged cavity from which the flake itself eventually is removed (see Fig. 19a). With a cloth of reasonably short nap, many of the flakes are well retained, although some appear slightly larger (see Fig. 18b and 19b). Examination of sections of such a surface indicates that flakes aligned perpendicular to the surface are well sectioned (flakes at right in Fig. 19b) but that slight erosion occurs around flakes that happen to be acutely aligned to the section surface (flake at left in Fig. 19b). Correct representation of the graphite flakes is obtained after polishing with a napless cloth, as shown in Fig. 18(c).

Only a limited number of abrasives, notably diamond abrasives, produce satisfactory results on napless cloths. Even then, a fairly heavily scratched polish is obtained. If this finish is not acceptable, a finishing treatment with a fine abrasive on a napped cloth is necessary. The treatment must be kept brief to avoid enlargement of the cavities.

Elimination of Polishing Scratches. Only rarely is it required that final-polished surfaces be completely free of scratches. A more reasonable and

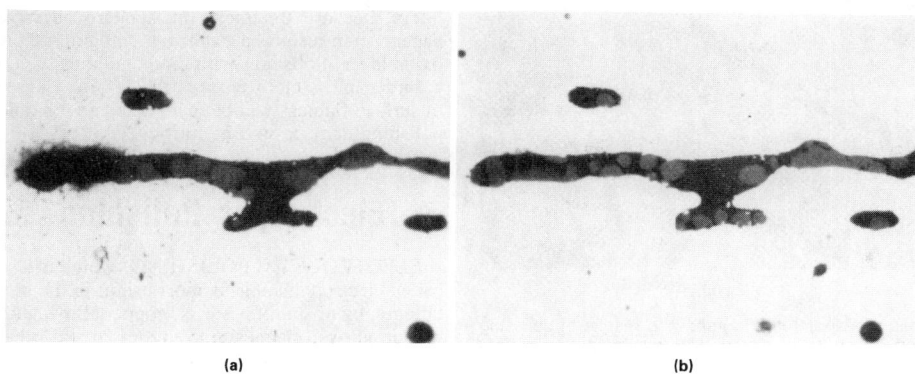

Micrographs (both at 350×) of as-polished specimen that was abraded on a fixed-abrasive lap and then polished (a) on 10-to-20-μm alumina on billiard cloth and (b) on 4-to-8-μm diamond on a synthetic suede cloth.

Fig. 16. Comparison of two polishing methods for showing inclusions in wrought iron

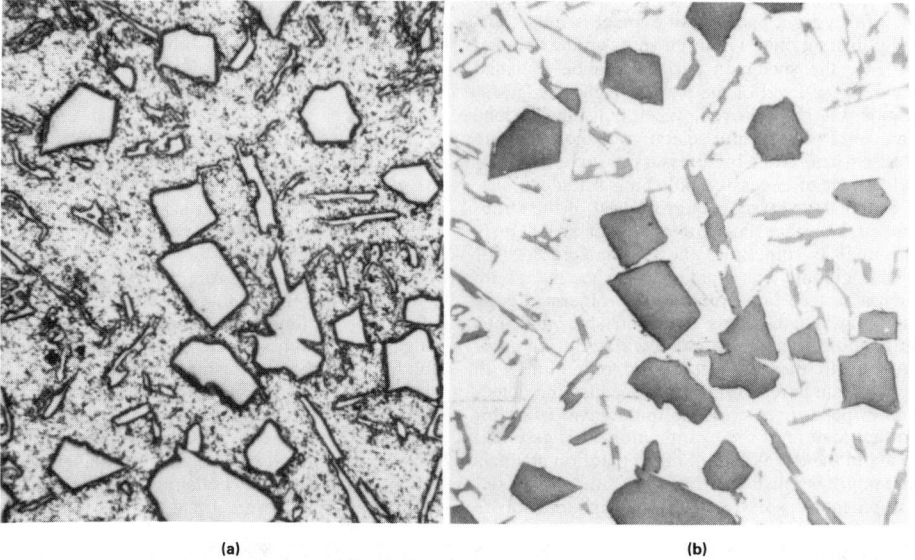

Micrographs (both at 250×) of as-polished specimen that was abraded on a fixed-abrasive lap and then polished (a) on 10-to-20-μm alumina on billiard cloth and (b) on 4-to-8-μm diamond on a synthetic suede cloth.

Fig. 17. Comparison of two polishing methods for showing phases in Al-13Si alloy

practical requirement is that no scratches should be detectable under the particular conditions of examination. Attainment of this objective will thus depend on the specimen material (more difficult with soft materials), the etching conditions (more difficult with etchants that are sensitive to deformed structures), and the optical conditions (more difficult with optical conditions that are sensitive to surface irregularities). In general, high-standard polishing processes are more laborious and require greater operator skill. A metallographer should have available a variety of final-polishing processes capable of producing increasingly higher qualities of finish from which to select the most suitable for a particular need.

Vibratory Polishing. Vibratory methods are attractive for final polishing because they operate automatically. An advantage of vibratory polishing is that it can be adapted to chemical-mechanical polishing. The important variables in vibratory polishing are the abrasive, the nature of the liquid in which the abrasive is suspended, and the load applied to the specimen.

Sometimes it is necessary to add a more aggressive etching reagent to the suspending liquid to ensure an adequate chemical component in the polishing mechanism. For example, the mechanism for an alpha-beta brass that was polished with the use of a straight glycol-water mixture had an excessive mechanical component, and final-polishing scratches became apparent as a result.

Electrochemical Differences. A further example of chemical effects arising during mechanical polishing is found in specimens containing constituents that differ considerably in their electrochemical characteristics. In galvanized steels, marked electrochemical effects arise between the zinc of the coating and the steel base metal. Severe etching of the coating occurs when the specimen is polished with an aqueous suspension or cleaned with water. This effect can be eliminated by using a suspending liquid that has a pH very close to 7.0, thus suppressing electrochemical effects. This pH can be achieved with the use of a standard buffer solution.

Edge Retention. With few exceptions, the abrasion rates of the plastics in which metallographic specimens are mounted are much greater than those of metals. The plastic abrades to a lower

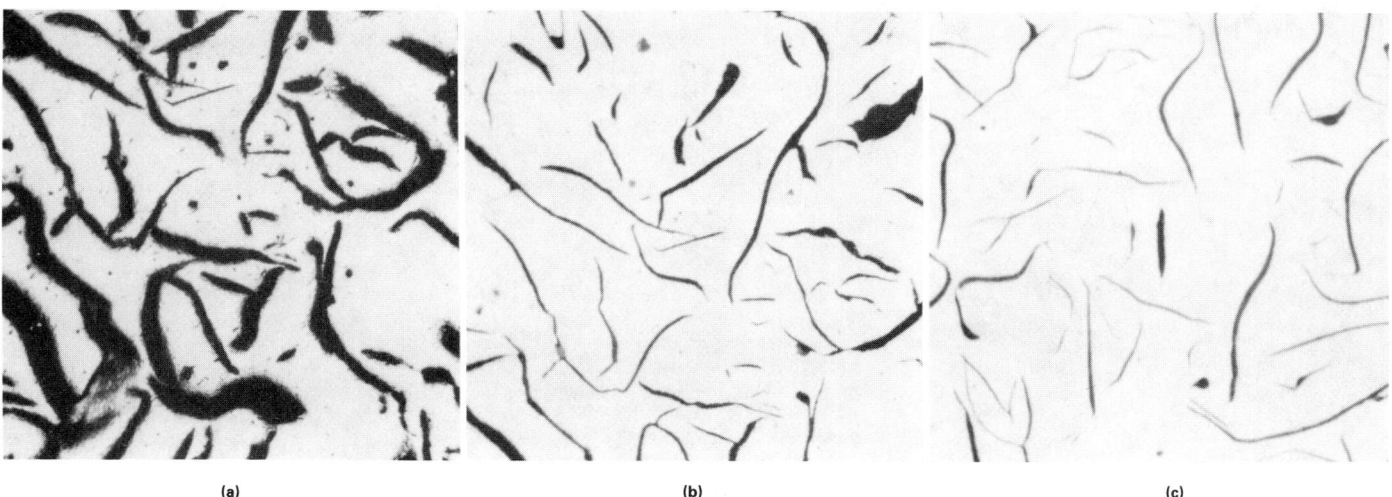

Micrographs (all at 250×) of as-polished specimens that were abraded on a fixed-abrasive lap and then polished (a) on 10-to-20-μm alumina on billiard cloth (long nap), (b) on 1-μm diamond on synthetic suede cloth (short nap) and (c) on 1-μm diamond on cotton drill (napless).

Fig. 18. Comparison of three polishing methods for retention of graphite in gray iron

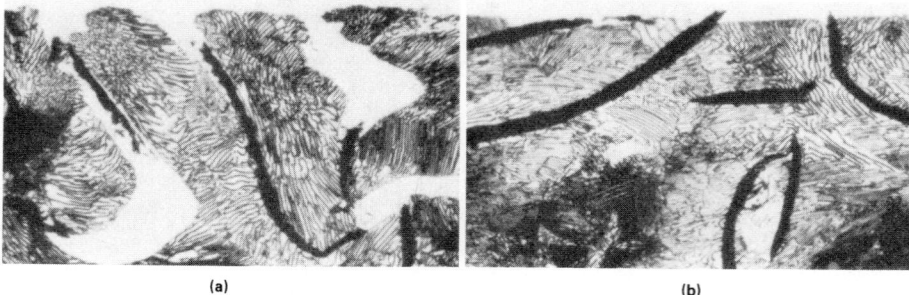

(a)   (b)

Taper sections (horizontal magnification, 1000×; vertical magnification, 10 000×) showing results of polishing (a) on 10-to-20-μm alumina on billiard cloth (long nap) and (b) on 1-μm diamond on synthetic suede cloth (short nap). Both sections etched in picral.

Fig. 19. Comparison of two polishing methods for retention of graphite in gray iron

general level than the metal, and rounding of the specimen edge occurs to adjust for differences in level. The degree of edge rounding may be increased or decreased during polishing; long-nap polishing cloths increase edge rounding.

However, the abrasion rates of different types of plastics differ significantly, and edge retention can be improved by choosing a mounting plastic that has an abrasion rate matching as closely as possible that of the specimen. For example, progressively improved edge retention is obtained, as shown in Fig. 20, with the change from a phenolic (Fig. 20a) to an allyl (Fig. 20b) to a polyvinyl formal (Fig. 20c) mounting plastic. Matching of the abrasion rates of specimen and mount is important. Metals such as chromium and tungsten, which have very low abrasion rates, show poorer edge retention than that illustrated in Fig. 20(c) even when mounted in a polyvinyl formal plastic. Metals such as copper and aluminum, which have high abrasion rates, show good edge retention even when mounted in phenolic or epoxy plastics.

Reducing the difference in abrasion rate between the specimen and mount will improve edge

retention. This may be accomplished by incorporating chips or pellets of a metal similar to the specimen in the mount face. If polishing is done on a fairly rigid pad so that contact is made during polishing only with high spots on the abraded surface, the specimen surface can be polished down to the level of the plastic. Hence, edge retention will be improved. When polishing is done on a rigid pad, careful selection of the polishing abrasive and cloth is necessary to avoid the development of excessive polishing scratches.

Surface Oxide Layers. Determination of the structure of a surface layer of oxide, or scale, on a specimen is sometimes the principal reason for metallographic examination. A specimen with such a surface layer presents a problem in edge retention. The oxide is usually friable, and thus susceptible to chipping and cracking during preparation. In these cases, an impregnation with epoxy resin is recommended. Because the detection of porosity or cracking in the layer is an important feature of the examination, it is essential to avoid the development of preparation artifacts that might be mistaken for such features. The development of such artifacts during abrasion is

likely, because treatment on standard abrasive papers often results in extensive chipping of the oxide layer. Polishing with diamond abrasive on a hard napless cloth ensures that a high degree of surface flatness will be maintained and that no polishing damage will be introduced.

Electrolytic Polishing

ELECTROLYTIC POLISHING, which is also called electropolishing, is most useful in the metallography of stainless steels, copper alloys, aluminum alloys, magnesium, zirconium and other metals that are difficult to polish by conventional mechanical methods. The electrolytic technique can completely remove all traces of worked metal remaining from the cutting, grinding and mechanical polishing operations used in preparing specimens.

When electropolishing is used in metallography, it is usually preceded by at least preliminary mechanical polishing and followed by etching.

MECHANISM

Although the mechanism of electropolishing is not understood in all its aspects, the process is generally considered to include both a leveling (or smoothing) action and a brightening action. The nature of these phenomena is discussed below.

Smoothing. According to a theory developed by P. A. Jacquet in 1936, smoothing is accomplished by preferential solution of the "hills" or ridges on a rough surface, such as those that result from mechanical finishing. When such a rough surface is made the anode of a suitable electrolytic cell, a viscous liquid layer immediately adjacent to this surface is produced by the reaction between metal and electrolyte. This layer of solution, known as the polishing film (Fig. 21), which has a greater electrical resistance than the

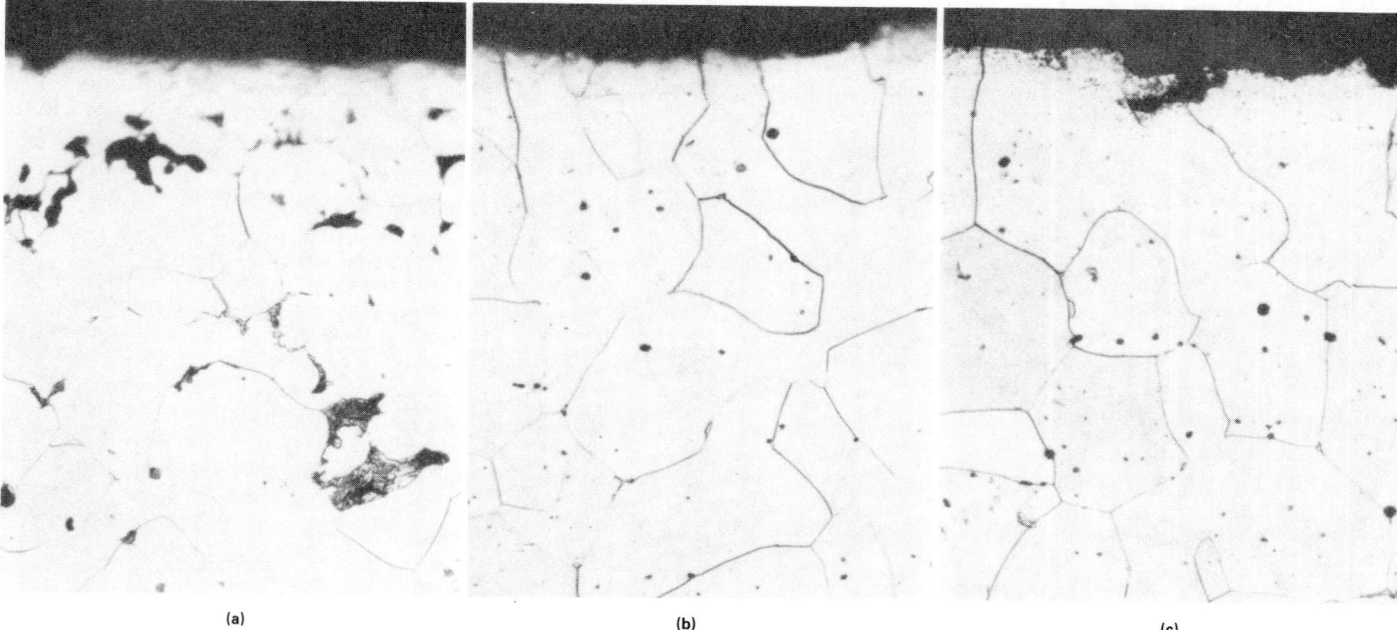

(a)   (b)   (c)

Micrographs (all at 500×) of specimens mounted (a) in a phenolic plastic (also representative of edge retention using an epoxy), (b) in an allyl plastic and (c) in a polyvinyl formal plastic (also representative of edge retention using polyvinyl chloride plastic). All three specimens etched in nital.

Fig. 20. Effect of type of mounting plastic on edge retention of steel specimens polished by standard technique

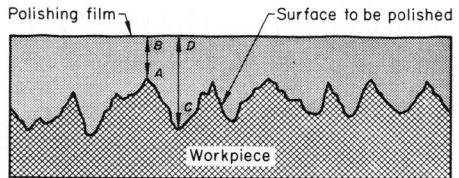

Fig. 21. Mechanism of electrolytic polishing

remainder of the solution, controls the smoothing action.

The resistance at a peak A, represented by the distance A-B, will be lower than at depression C, represented by the distance C-D, because the film is thinner at A-B. The current at A will be much higher than at C, causing metal to dissolve faster at A than at C, and producing a nearly level, gently undulating surface by removing asperities 1 μm or more in size.

More rapid ionic and molecular diffusion through the thinner polishing film at A, as well as differences in anodic polarization phenomena at A and C, may also contribute to the leveling or smoothing action.

Brightening. The brightening action is related to the elimination of irregularities as small as about 0.01 μm and to the suppression of etching on the metal surface. This behavior is generally attributed to the formation of a thin, partly passivating film directly on the surface of the metal and following its contours.

Optimum brightening conditions have been related to local differences in anodic passivation at heterogeneities and between secondary peaks and crevices, along with the effects of passivation inhibitors that influence oxide-film formation and gas evolution. Similar factors may also contribute to the primary leveling or smoothing action in electropolishing (see *Electrolytic and Chemical Polishing of Metals*, by P. V. Shchigolev: Freund, Holon, Israel, 1970).

Current-voltage relations in electropolishing vary in different electrolytes and for different metals. The simple relation wherein polishing occurs over an extensive continuous range of currents and voltages is shown in Fig. 22. At low voltages, a film forms on the surface and little or no current passes; thus, etching occurs but not polishing. At higher voltages, polishing occurs. The perchloric acid electrolytes used for aluminum conform to this relation.

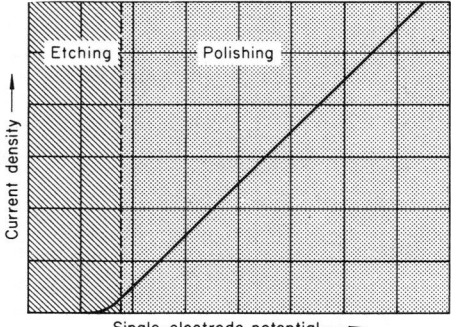

Fig. 22. Schematic relation between current density and single-electrode potential for electrolytes having a polishing action over a wide range of voltages and currents

A more complex relation, frequently encountered, is shown by the curve in Fig. 23. This curve depicts cell voltage as a function of anode current density for electropolishing of copper in an aqueous solution of orthophosphoric acid, using a potentiometric circuit.

Five distinct regions can be distinguished on the cell-voltage curve. In the region A-B, current density increases with the potential, some metal dissolves, and the surface has a dull etched appearance. The region B-C reflects an unstable condition, while region C-D indicates a stable plateau at which the polishing film, previously formed, reaches a point of equilibrium and polishing occurs; during the latter stage, current density remains constant.

Optimum polishing conditions occur along C-D near D. In the region D-E, gas bubbles evolve slowly, breaking the polishing film and causing severe pitting. Polishing with rapid evolution of gas is represented by the region E-F.

Electrolytes of the sulfuric-phosphoric acid and chromic-acetic acid types used for stainless steels also typify the complex, multistage relationship shown in Fig. 23 and discussed above.

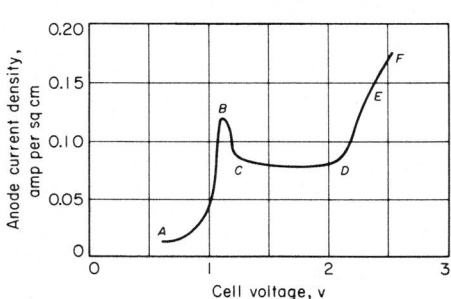

Fig. 23. Cell voltage as a function of anode current density for electropolishing of copper in orthophosphoric acid (900 grams per litre of water), using a potentiometric circuit

In establishing voltage-current curves like those in Fig. 22 and 23, the electrolysis must be allowed to proceed under fixed conditions until enough metal has dissolved to produce a "steady state" condition at the anode.

APPARATUS AND PROCEDURE

The electrical equipment used for electropolishing can vary from the simplest arrangement of dry cells to elaborate arrays of rectifiers and electronic control devices. Various types of apparatus are available commercially. The choice of equipment depends on the number and type of specimens to be treated and the versatility and control desired.

Current Source. Direct current is usually employed. The current source may consist of a battery, a direct-current generator or a rectifier. In general, a battery supply is used for low voltages only, because a bank of batteries would be needed to produce higher voltages. These three types of current source deliver a constant supply of direct current.

Electrical Circuits. Two typical circuits, one for low and one for high current densities, are shown in Fig. 24. For solutions in which a small drop in potential occurs across the cell, a potentiometric circuit, for low current densities, is more

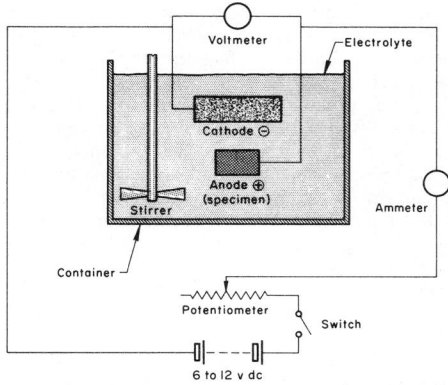

(a) Potentiometric circuit (for low current densities)

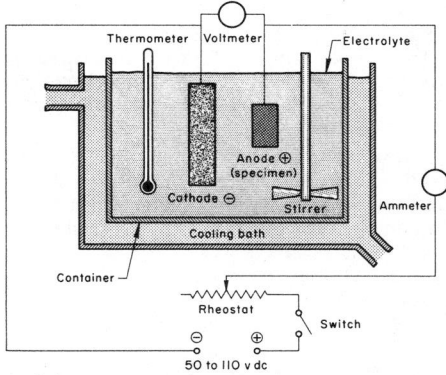

(b) Series circuit (for high current densities)

Fig. 24. Two electrical circuits, and arrangements of equipment, used for electropolishing of metallographic specimens

suitable (see Fig. 24a). Conversely, when the drop in potential across the cell is high, a series circuit, for high current densities, should be used (see Fig. 24b). Provision must be made for controlling both voltage and current.

Alternating current is used for electropolishing and electroetching the metals of the platinum group (platinum, iridium, palladium, rhodium, osmium and ruthenium), using a series circuit and schematic arrangement as shown in Fig. 24(b) with an ac source.

The electrolytic cell is simply a container for the electrolyte, in which are suspended the cathode and the anode. The cell is usually made of glass, but may be of polyethylene or polypropylene for solutions containing fluoride ions. Sometimes a stainless steel cell is used, which may serve also as the cathode. Frequently, the cell is surrounded by water or an ice bath, or is cooled in some other manner.

The specimen to be polished (anode) should be arranged for quick removal from the electrolyte. The electrical connection to the specimen should be simple and easily broken so that the specimen can be rinsed immediately after polishing (see "Arrangement of Anode and Cathode," which follows).

The cathode should be made of a metal that is inert in the electrolyte being used; stainless steel is satisfactory for most applications.

For many applications, stirring or air agitation of the electrolyte is necessary. During electropolishing under steady-state conditions, the anodic-reaction products accumulate on the surface of the polished metal. Often, natural diffusion and convection processes cannot remove these products from the anode surface into the bulk of the electrolyte rapidly enough, and excessive accumulation of reaction products interferes with the electropolishing process. Stirring or air agitation speeds up the removal of these products, prevents localized heating of the surface, maintains a uniform bath temperature and removes gas bubbles that otherwise might adhere to the surface and cause pitting. However, the use of agitation usually requires an increase in the current density in order to maintain a sufficiently thick polishing film. In some applications, vibratory motion of the specimen can be substituted for stirring. In others, both agitation of the electrolyte in the cell and control of the electropolishing temperature can be accomplished by circulating the electrolyte by means of a pump and an external cooling bath or device.

To prevent "furrowing" of the surface being electropolished, the movement of the electrolyte (and gas) across the metal surface should be gentle and nondirectional.

Arrangement of Anode and Cathode. Two ways to position the specimen (anode) and the cathode are shown in Fig. 24. In each arrangement, only the portion of the specimen to be polished is exposed to the electrolyte.

In Fig. 24(a), the surface to be polished is horizontal and facing upward, toward the cathode. This arrangement helps to maintain a stable layer near the surface being polished, and is ordinarily used when polishing occurs under a viscous layer.

In Fig. 24(b), the surface to be polished is vertical and facing toward the cathode. This arrangement is sometimes used when polishing occurs with gas evolution, because it allows easy escape of the gas bubbles. However, directional streaming can cause furrowing of the surface being polished, unless special attention is given to positioning and agitation. Reciprocating movement of the specimen is helpful in preventing furrowing.

Both pitting and furrowing are readily avoided in the cell arrangement shown in Fig. 25, in which gentle, nondirectional movement of the electrolyte at the surface being polished is provided by introducing air through perforations in a horizontal cathode at the bottom of the cell. Although the electrical circuit shown in Fig. 25 (a series circuit, same as in Fig. 24b) is ordinarily used, the potentiometric circuit shown in Fig. 24(a) can also be used.

The electrical connection to the specimen is made indirectly through a metal block and a contact wire that is spot welded to the back of the specimen and the metal block before the assembly is mounted in epoxy resin or other suitable material (see detail A in Fig. 25). After mounting, a hole is drilled through the back of the mount to the metal block to permit attachment of the electrical connector wire. The indirect connection avoids the danger of loosening the bond of specimen to mount that would exist with a direct connection through a hole drilled into the specimen itself.

The arrangement shown in Fig. 25 is especially suited for electropolishing at medium to high current densities. The mount is conveniently held in an alligator clip having stainless steel extensions welded to the jaws. The clip is attached to a hook that can be supported on a horizontal anode bar for ease of manipulation. By placing the hook on the bar, electrical contact to the specimen is made almost simultaneously with immersion in the electrolyte. Contact is broken almost simultaneously with removal from the electrolyte when the hook is lifted from the anode bar, thus allowing quick rinsing to prevent staining of the polished surface.

MOUNTING OF SPECIMENS

Only the portion of the specimen to be polished should be in contact with the electrolyte. Small specimens may be mounted by conventional molded-plastic-mounting procedures, for ease in handling for mechanical preparation and subsequent electropolishing. Electrical contact can be made through a small hole drilled through the back of the mount into the metal specimen, or by the use of an indirect connection, as shown in Fig. 25.

When specimens are mounted in plastic, the possibility of violent reaction between the plastic and some electrolytes must be considered. For example, phenol-formaldehyde mounting materials, acrylic-resin mounting materials, and cellulose-base insulating lacquers and materials should not be used in solutions containing perchloric acid, because of the danger of explosion. Polyethylene, polystyrene, epoxy resins and polyvinyl chloride can be used as mounting materials in perchloric acid solutions without danger.

Mounting of specimens in dissimilar metals is undesirable, because the metal in contact with the electrolyte is likely to interfere with polishing and also because fusible mounting alloys containing bismuth may be dangerously reactive in certain electrolytes that contain oxidizing agents. Bismuth-containing alloys may form explosive compounds in perchloric acid solutions.

In preparing an unmounted specimen for electropolishing, a suitable chemically inert, electrically insulating coating can be applied to all surfaces of the specimen (and specimen holder) except the surface to be polished. Plastic electricians' tape is also an effective stop-off, being impervious to most electrolytes and readily removable from the specimen after electropolishing.

Another arrangement that has been used for electropolishing an unmounted specimen is to clamp it against a hole in the wall of the cell.

DEVELOPMENT OF PROCEDURE FOR ELECTROPOLISHING

Metallographers often are asked to electropolish a metal or alloy that has not previously been encountered. In developing a suitable procedure, the problem should be viewed in comparison with known procedures and with information gained through previous experience.

Effect of Alloy Type. It is generally helpful to compare the position of the major component of the alloy with elements of the same general group in a periodic table, and to study the phase diagram, if available, to predict the number of phases and their characteristics. Single-phase alloys are generally easy to electropolish, whereas multiphase alloys are likely to be difficult or impossible to polish by the electrolytic method. Even minor alloying additions to a metal may profoundly affect the response of the metal to polishing in a given electrolyte.

General Principles. The possibility of polishing a metal and the conditions for polishing a metal in a given electrolyte can sometimes be ascertained by plotting current density versus electrode potential. The curve in Fig. 22 is typical of those electrolytes that polish over a very wide range or that will not polish at all. The curve in Fig. 23 is characteristic of those electrolytes that form an ionic film; polishing will occur between points *C* and *D* on this curve, and is usually best near point *D*.

In a cell designed so that the anode is clearly visible during electrolysis, the polishing plateau can be determined by observing the anode while gradually increasing the current. For stable and reproducible results, current is passed for 30 min before recording data, and the current is increased slowly.

In working with radioactive metals, the specimen is held close to a thin transparent window in a special cell, and the polishing action is observed using an external optical system that has a focal length of 4.75 mm or more, while circulating electrolyte between the specimen and the window.

After the polishing range is determined, the other constants, such as preparation, agitation and time, can be determined experimentally. The amount of preparation required depends on the nature of the specimen and on the results desired.

Specimen preparation often can be accomplished through the use of 600-grit silicon carbide paper on a rotating disk. In other cases it is necessary or desirable to prepolish the specimen with 6-μm diamond before starting to electropolish. Soft metals (for example, lead and copper) may need to be chemically etched after mechanical polishing.

The surface to be polished should be clean, to allow uniform attack by the electrolyte. To avoid contamination with oil from the fingers, the specimen should be handled with forceps or tongs after final preparation for electropolishing.

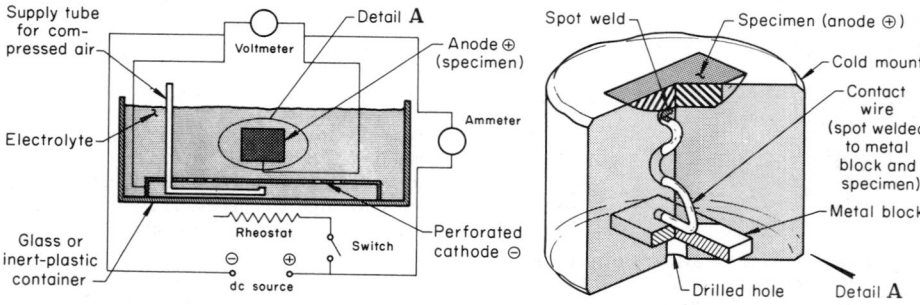

Detail at right shows an indirect electrical connection to a mounted specimen.

Fig. 25. Arrangement of equipment for electropolishing in which air agitation of electrolyte is provided through a perforated cathode

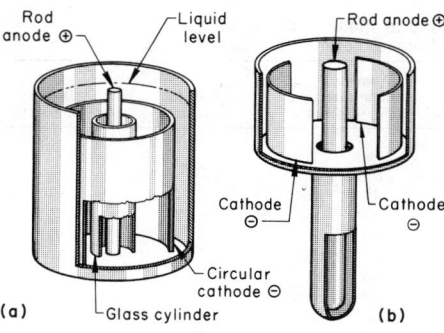

Fig. 26. Test cells proposed for use in evaluating operating conditions in electropolishing over a range of anode current densities

Use of Test Cells. A simple method for determining optimum electropolishing conditions after a suitable polishing solution has been selected was proposed by Bertein. Two test cells of his design are shown in Fig. 26.

In the cell shown in Fig. 26(a), the rod anode, the 360° glass insulating cylinder around it and the 360° circular cathode rest on the bottom of the cell, and the liquid level is maintained some distance above the upper end of the cathode and the glass cylinder. In the cell shown in Fig. 26(b), the cathode consists of two opposing circular segments, the rod anode does not extend to the bottom of the cell, and the liquid level is maintained slightly below the upper end of the anode and the two-segment cathode.

In each cell, the anode current density is greater near the liquid level and is progressively lower at greater depths. In operation, when a constant current is passed through the cell, the finish at any depth on the anode is related to the current density at that depth.

If the electrolyte is of a composition that makes it suitable for electropolishing, the optimum ranges of current density can be estimated roughly from the positions and lengths of the polished zones. Additional information can be obtained from such cells by measurements of anode potential.

The cell shown in Fig. 26(b) has the advantages of allowing accurate temperature control and observation of the anode during the passage of current. Similar results can be obtained in a Hull cell, widely used for evaluating operating conditions in electroplating.

ADVANTAGES AND LIMITATIONS OF ELECTROPOLISHING

When properly applied, electropolishing can be a useful tool for the metallographer. The principal advantages of electropolishing are:

1. For some metals, electropolishing can produce a high-quality surface finish that is equivalent to the best that can be obtained by mechanical methods.
2. Once a procedure has been established, good results can be obtained with less operator skill than that required for mechanical polishing.
3. There can be a marked saving of time if many specimens of the same material are to be polished sequentially.
4. Electropolishing is especially suited for the softer metals, which may be difficult to polish by mechanical methods.

5. No scratches are produced in electrolytic polishing. The absence of scratches is a definite advantage in viewing high-quality electropolished surfaces of optically active materials under polarized light.
6. Artifacts resulting from mechanical deformation, such as disturbed metal or mechanical twins, which are produced on the surface even by careful grinding and mechanical polishing, do not occur in electropolishing.
7. Surfaces resulting from electropolishing are completely unworked by the polishing procedure. This feature is important in low-load hardness testing or x-ray studies.
8. In some applications, etching can be accomplished by simply reducing the voltage to approximately one-tenth the potential required for polishing, then continuing electrolysis for a few seconds.
9. Electropolishing is frequently useful in electron metallography (where high resolution is often important) because it can produce clean, undistorted metal surfaces.

Metallographic preparation by electropolishing is subject to several limitations; these should be recognized to prevent misapplication of the method and disappointment in the results. The principal disadvantages include:

1. The chemicals and combinations of chemicals used in electropolishing are poisonous; many are highly flammable or potentially explosive. Only well-trained personnel thoroughly familiar with chemical laboratory procedures should be permitted to handle or mix the chemicals, or to operate the polishing baths.
2. The conditions and electrolytes required to obtain a satisfactorily polished surface differ for different alloys; hence, considerable time may be required to develop a procedure for a new alloy, if it can be developed at all. This limitation does not apply if appropriate procedures exist.
3. In multiphase alloys, the rates of polishing of different phases often are not the same. Polishing results depend heavily on whether the second or third phases are strongly cathodic or anodic with respect to the matrix. The matrix is dissolved preferentially if the other phases are relatively cathodic, thus causing the latter to stand in relief. Preferential attack may also occur at the interface between two phases. These effects are most pronounced when phases other than the matrix are virtually unattacked by the polishing bath. The effects are reversed when the matrix phase is relatively cathodic.
4. A large number of electrolytes may be needed to polish the variety of metals encountered by a given laboratory.
5. Plastic or metal mounting materials may react with the electrolyte.
6. Electropolished surfaces exhibit an undulating rather than a plane surface, and in some cases may not be suited for examination at all magnifications. Under some conditions, furrowing and pitting may be produced.
7. Edge effects limit applications involving small specimens, surface phenomena, coatings, interfaces and cracks.
8. Attack around nonmetallic particles and adjacent metal, voids, and various inhomogeneities may not be the same as that of the matrix, thus exaggerating the size of the voids and inclusions.

9. Electropolished surfaces of certain materials may be passive and difficult to etch.

ELECTROLYTES

Formulas for the eight groups of electrolytes may be found in Metals Handbook, 8th Edition, Vol 8, and in ASTM E3.

Preferred (or sometimes required) characteristics of an electrolyte are:

1. It should be somewhat viscous.
2. It must be a good solvent for the anode metal (the specimen) during electrolysis conditions.
3. It should preferably not attack the anode metal when no current is flowing.
4. It should contain one or more ions of large radii, such as $(PO_4)^{-3}$, $(ClO_4)^{-1}$, or $(SO_4)^{-2}$, and sometimes large organic molecules.
5. It should be simple to mix, stable, and safe to handle (many effective electrolytes are deficient in these respects).
6. It should function effectively at room temperature and not be sensitive to temperature changes.

Procedures and Precautions for Preparation and Handling of Etchants

IN THIS article, the term "etchants" is used in its broadest sense, to include reagents used in metallographic work for microetching, macroetching, electropolishing, chemical polishing and similar operations. Applications and compositions of these reagents are described in articles in this volume that deal with metallographic practice and with techniques for specific metals.

The formulations of etchants given elsewhere in this volume are adequate for the majority of applications, but they may occasionally require modification. Adjustments in etchant composition (as well as in etching time and technique), based on the experience and skill of the metallographer, and depending on the specific application and the magnification to be used, are frequently needed in order to obtain satisfactory results.

EXPRESSION OF COMPOSITION

Etchants are generally either aqueous or alcoholic solutions containing one or more active chemicals (acids, bases or salts). Liquids other than water or alcohol are used as solvents in some formulations.

Compositions of most etchants described in this book are expressed in terms of the amounts of the substances to be used in preparing small quantities of these reagents.

For etchants that are solutions of solid substances in liquids, the amounts of the solid substances are usually expressed in grams (g), and the amounts of liquids (or the total volumes of solution) are expressed in millilitres (mL). The liquids may be individual commercially available substances, or they may be stock solutions containing two or more substances.

To prepare large quantities, as for some macroetching, kilograms (kg) and litres (L) may be taken instead of grams and millilitres, or the amounts specified may be converted to pounds

(1 kg = 2.2 lb) and gallons (1 litre = 0.264 gal).

Other generally accepted methods for expressing composition are also used where appropriate in some articles in this book. Compositions of some etchants that are prepared by mixing together two or more liquids are given in parts by volume or percentage by volume. Compositions of some etchants consisting of solutions of solid substances in liquids are described in terms of percentage by weight.

In long-established (although nonstandard) usage in metallography, such terms as 1%, 2% and 4% have been used to describe the approximate strength of picral, and are understood to mean 1, 2 and 4 g, respectively, of picric acid per 100 mL of alcohol.

PURITY OF CHEMICALS

In the preparation of solutions for microetching and electropolishing, recommended practice is to use chemicals meeting the requirements of: NF (National Formulary), USP (U. S. Pharmacopoeia), "laboratory" or "purified" grades, or grades of still higher purity (reagent, ACS, or "certified" grades). The commercial or technical grades of certain special-purpose industrial chemicals (such as CrO_3 and synthetic methanol) are extremely pure and are equivalent to reagent, ACS or "certified" grades for use in microetching and electropolishing.

Where water is specified, distilled water is preferred, because of wide variations in the purity of tap water.

For macroetching, technical grades of chemicals are satisfactory, unless specifications indicate otherwise, and potable tap water of good quality is generally acceptable.

IDENTIFICATION OF CHEMICALS

The practices generally followed in the technical literature on metallography are used in this book to identify the chemicals used in the preparation of etchants.

Aqueous Acids. In identifications of aqueous acids, the name or formula alone, sometimes followed by "conc" or "concentrated," refers to the common commercially available concentrated laboratory grade. (See Table 1.) Where more than one concentration is commonly available, the percentage by weight of the active constituent is shown after the name or formula.

Where an acid is designated as "tech," the technical grade having the same concentration as the common laboratory grade is meant. The concentration of technical grades is sometimes expressed by suppliers in terms of specific gravity or degrees Baumé (°Bé) as shown in Table 1. (Most technical-grade chemicals are available in several different concentrations.)

Miscellaneous aqueous chemicals, such as ammonium hydroxide and hydrogen peroxide (see Table 1), which are used in various etchants, are identified similarly to aqueous acids (see above). Concentration must always be specified for hydrogen peroxide, which is available in several widely differing concentrations.

Alcohols. The alcohols most frequently used in etchants are methanol and ethanol, which are described in Table 2. It is important to use alcohol that has the desired water content (anhydrous or "95%" alcohol, whichever is specified) in etchants that contain only a small percentage of water. However, either grade can be used where the etchant is a dilute aqueous solution.

Table 1. Characteristics of aqueous liquid chemicals used in many metallographic etchants

Except for sulfuric acid, all data apply to both laboratory and technical or commercial grades of chemicals.

Name	Active constituent	Nominal composition, % by weight(a)	Specific gravity	Degrees Baumé(b)
Aqueous acids				
Acetic acid, glacial	$HC_2H_3O_2$	99.5	1.05	7.0
Fluoboric acid	HBF_4	48	1.32	35
Hydrochloric acid(c)	HCl	37	1.18	22
Hydrofluoric acid	HF	48	1.15	19
Lactic acid	$HC_3H_5O_3$	85	1.20	24
Nitric acid	HNO_3	70	1.42	43
Perchloric acid	$HClO_4$	70	1.67	58
		60	1.53	50
Phosphoric acid (ortho)	H_3PO_4	85	1.70	60
Sulfuric acid	H_2SO_4	96(d)	1.84(e)	66(e)
Miscellaneous aqueous chemicals				
Ammonium hydroxide	NH_4OH	28(f)	0.90	26
Hydrogen peroxide	H_2O_2	3(g)	1.01	1.4
		30(h)	1.11	15
		50(j)	1.20	24

(a) Nominal percentage of the active constituent; remainder is water. Reagents made by different manufacturers may differ slightly in nominal concentration and allowable range of concentration. (b) Specific gravity as indicated on the Baumé scale; sometimes used for technical grades and in laboratory measurements. (c) Technical grade is also called muriatic acid. (d) Laboratory grade. Technical grade has concentration of 93%. (e) Specific gravity and degrees Baumé are nearly constant for 93 to 100% sulfuric acid. (f) Per cent NH_3. (g) Sometimes called "10 volume." (h) Sometimes called "100 volume." (j) Sometimes called "170 volume."

Table 2. Characteristics of pure methanol and ethanol

Name	Active constituent	Nominal composition, % by volume(a)
Methanol (methyl alcohol)	CH_3OH	99.5(b)
Methanol (methyl alcohol), 95%	CH_3OH	95(c)
Ethanol (ethyl alcohol), anhydrous	C_2H_5OH	99.5(d)(e)
Ethanol (ethyl alcohol), 95%	C_2H_5OH	95(e)

(a) Nominal percentage of the active constituent; remainder is water, unless otherwise specified. (b) Synthetic methanol; the commercial grade is of high purity and is satisfactory for use in all ordinary metallographic etchants where methanol is specified (wood alcohol has not been manufactured commercially in the United States since 1969). Methanol is available only as an anhydrous (also called absolute) grade containing less than 0.1 or 0.2% water as packaged, and usually not more than about 0.5% water at time of use, depending on storage and handling. (c) Where methanol, 95%, is called for, the ordinary anhydrous grade must be diluted by the user with 5% water by volume. (d) The anhydrous (also called absolute) grade of ethanol is ordinarily used only where no significant amount of water can be tolerated. It contains less than 0.1 or 0.2% water as packaged, and usually not more than about 0.5% water at time of use, depending on storage and handling. (e) Available only with special government permit.

Practice with regard to the substitution of methanol for ethanol, or vice versa, and with regard to the use of some grades of denatured ethanol in etchants, varies greatly among metallographic laboratories.

Although many etchant formulations show the use of methanol or ethanol as alternative materials, caution should be exercised in substituting one for the other in formulations where their equivalence is not indicated. Safety considerations rule out any changes in accepted formulations for electropolishing without a thorough chemical study. Also, ethanol or higher alcohols should not be substituted for methanol in nital containing more than 5% by volume of concentrated nitric acid, or in other methanol-base etchants that contain strong oxidants and only a small percentage of water.

In a wide variety of applications for which the etchant is specified to contain ethanol (excluding electropolishing electrolytes), a number of laboratories have found it convenient and satisfactory to substitute for pure "anhydrous" or "absolute" (99.5%) ethanol, and for pure 95% ethanol

(see Table 2), either a proprietary solvent or denatured "reagent" alcohol having a suitable water content (as shown in the lower part of Table 3), which are available without permit from suppliers of laboratory chemicals. These grades have been formulated in accordance with U. S. government regulations to be suitable for general laboratory purposes and have been denatured with small percentages of volatile solvents; they may be substituted for pure ethanol having the same water content, except where pure ethanol is required for some special reason.

The specially denatured (SD) alcohols described in Table 3 are generally suitable for use in etchants. However, SD alcohol is obtainable only with special government permits and usually can be purchased only in larger quantities than the proprietary solvent and "reagent" alcohol in Table 3, and only from major suppliers of solvents.

Water of Hydration. With some exceptions, it has been common practice since the earliest days of metallography to identify solid salts and acids used in etchants only by names and abbreviated formulas, without showing the presence or absence of water of hydration (see *Metallographers' Handbook of Etching,* by T. Berglund: Pitman & Sons, Ltd., 1931, p 49-65 and Table IV, p 66).

Historically, in developing and preparing etchants, the most stable hydrate (which was the common commercial form) was ordinarily used, except for salts that do not form hydrates. Current practice varies from laboratory to laboratory.

Using the specified amount of either the anhydrous or a hydrated form of a solid salt or acid in preparing an etchant will in most cases produce essentially the same etching behavior; any difference in results will usually be small in comparison with the effects of normal differences in technique and other variables in specimen preparation.

Exceptions are the preparation of etchants that must be anhydrous or must contain only a small and fairly critical percentage of water for proper etching activity; for such etchants, the need to use specific anhydrous or hydrated forms of each component should be clearly stated.

Some salts (such as ferric nitrate) do not exist in an anhydrous form. Conversely, some nomi-

Table 3. Nominal compositions of various grades of denatured alcohol (ethanol) used in some metallographic etchants(a)

Component	Parts by volume in specially denatured alcohol(b)					
	Formula SD-1(c)		Formula SD-3A		Formula SD-30	
	Anhydrous	95%(d)	Anhydrous	95%(d)	Anhydrous	95%(d)
Ethanol, anhydrous	100	95	100	95	100	95
Water	…	5	…	5	…	5
Methanol	4	4	5	5	10	10
Methyl isobutyl ketone	1	1	…	…	…	…

Component	Parts by volume in proprietary solvent(e)		Parts by volume in "reagent" alcohol(e)	
	Anhydrous	95%(d)	Anhydrous	95%(d)
SD-1 anhydrous(c)	100	…	…	…
SD-1, 95%(c) (d)	…	100	…	…
SD-3A, anhydrous	…	…	95	…
SD-3A, 95%(d)	…	…	…	95
Methyl isobutyl ketone	1	1	…	…
Hydrocarbon solvent or gasoline	1	1	…	…
Ethyl acetate	1	1	…	…
Isopropyl alcohol	…	…	5	5

(a) See text for discussion of suitability of the various grades for use in etchants. (b) Specially denatured alcohol is available only with special government permit. (c) The formula shown here has replaced the old SD-1 formula in which wood alcohol was specified: wood alcohol has not been manufactured commercially in the United States since 1969. (d) The designation of type of denatured alcohol as "95%" means that the denatured product contains 5 parts of water for every 95 parts of anhydrous (absolute) ethanol, plus denaturants as specified. (e) Available without government permit from suppliers of laboratory chemicals, for scientific and general laboratory purposes.

nally water-free compounds contain a substantial percentage of water. One of these is picric acid, for which the 10 to 15% water content found in laboratory grades is necessary for satisfactory performance of etchants based on it (see Table 4).

Miscellaneous chemicals for which correct identification may present problems because of similarity in names of different chemicals or because of misleading or nonstandard nomenclature and trade names are described in Table 4. Also included are certain chemicals for which some aspects of composition or behavior are important.

SAFETY PRECAUTIONS

All chemicals are potentially dangerous, and persons formulating and using etchants should be thoroughly familiar with the chemicals involved, and with the proper procedures for handling and mixing them. The discussion that follows indicates many of the potential hazards that attend the use of chemicals, and describes precautions and safe practice for averting these hazards.

Ventilation. All mixing, handling and use of etchants should be done in a well-ventilated area, preferably in an exhaust hood, to prevent exposure to or inhalation of toxic and corrosive fumes. Use of an exhaust hood is mandatory whenever large quantities of chemicals are handled or large areas of metal are etched (as in macroetching), and when carrying out lengthy electropolishing operations or electropolishing large areas.

Protection of Personnel. When chemicals and etchants are being poured, mixed or handled, and when etchants are being used, suitable protective equipment and clothing (glasses, face shield, gloves, apron, and other items, as appropriate) should always be worn, to prevent contact of chemicals with the eyes, skin or clothing.

If chemicals contact the skin, they should be washed off promptly with water and soap. Medical attention should be obtained as soon as possible for chemical burns, especially if at cuts or abrasions in the skin. If chemicals contact the eyes, the eyes should be flushed *at once* with large quantities of water, and medical attention should be obtained *without delay*. A face-and-eye fountain should be available for use wherever chemicals or etchants are stored or handled. Wherever quantities large enough to be hazardous are stored or handled, a safety shower is needed also. This washing equipment *should be readily available, and should be tested at scheduled intervals* to ensure dependable performance in an emergency.

Hydrofluoric and fluosilicic acids can cause painful and serious ulcers on contacting the skin, unless washed off immediately. Also especially harmful to the skin are concentrated HNO_3, H_2SO_4, CrO_3, H_2O_2 (30% or 50%), NaOH, KOH, Br_2 and anhydrous $AlCl_3$. Inhalation of vapors or mist from these chemicals or etchants containing them can also cause irritation or serious damage to the respiratory system.

Container Material and Design. In preparing, storing and handling etchants, use containers and equipment made of materials suitable for the chemicals used. Glass is resistant to nearly all chemicals. Polyethylene, polypropylene and similarly inert plastics are resistant to hydrofluoric, fluosilicic and fluoboric acids, as well as to solutions containing salts of these acids. These inert plastics are also recommended for prolonged storage of strongly alkaline solutions and strong solutions of phosphoric acid, both of which attack glass (especially, ordinary grades of glass).

Certain mixtures of chemicals can generate gaseous reaction products over a period of time or if inadvertently exposed to heat, and can build up dangerous pressures if stored in tightly sealed containers.

The use of vented or pressure-relief types of stoppers instead of tightly sealed screw caps or conventional stoppers on bottles of etchants that

Table 4. Descriptions of miscellaneous chemicals used in metallographic etchants

aluminum chloride, anhydrous. Solid; $AlCl_3$; reacts violently with water, evolving HCl gas; use of hydrated form, $AlCl_3 \cdot 6H_2O$, is preferred.

ammonium molybdate. Crystals; also called ammonium paramolybdate or heptamolybdate; $(NH_4)_6Mo_7O_{24} \cdot 4H_2O$; can be used interchangeably with "molybdic acid, 85%."

benzalkonium chloride. Crystals; essentially alkyl-dimethyl-benzyl-ammonium chloride. May not be readily available in this form; see *zephiran chloride*.

1-butanol. See *n-butyl alcohol*.

2-butoxyethanol. See *butyl cellosolve*.

n-butyl alcohol. Liquid; normal butyl alcohol; also called butyl alcohol and 1-butanol.

butyl carbitol. Liquid; diethylene glycol monobutyl ether.

butyl cellosolve. Liquid; ethylene glycol monobutyl ether; also called 2-butoxyethanol.

carbitol. Liquid; diethylene glycol monoethyl ether.

cellosolve. Liquid; ethylene glycol monoethyl ether.

chromic acid. Dark-red crystals or flakes; CrO_3; also called chromic anhydride, chromic acid anhydride, and chromium trioxide. (See *chromic oxide*, Cr_2O_3.)

chromic anhydride. See *chromic acid*.

chromic oxide. Fine green powder; Cr_2O_3; a polishing abrasive. Do not confuse with chromic acid (CrO_3), which is a strong acid and a component of many etchants.

cupric ammonium chloride. Crystals; a double salt, $CuCl_2 \cdot 2NH_4Cl \cdot 2H_2O$. If not available, substitute 0.6 g $CuCl_2 \cdot 2H_2O$ plus 0.4 g NH_4Cl for each gram of the double salt.

diethylene glycol. Syrupy liquid; also called 2.2'-oxydiethanol and dihydroxydiethyl ether; $(HOCH_2CH_2)_2O$. More viscous than ethylene glycol; otherwise similar in behavior.

diethylene glycol monobutyl ether. See *butyl carbitol*.

diethylene glycol monoethyl ether. See *carbitol*.

diethyl ether. See *ether*.

ether. Liquid; also called ethyl ether and diethyl ether; very low flash point, highly explosive; boiling point is 34.4 °C (94 °F).

ethylene glycol. Syrupy liquid; also called 1,2-ethanediol and dihydroxyethane; $(CH_2)_2/(OH)_2$. Less viscous than diethylene glycol; otherwise similar in behavior.

ethylene glycol monobutyl ether. Liquid; also called 2-butoxyethanol or butyl cellosolve.

ethylene glycol monoethyl ether. See *cellosolve*.

ethyl ether. See *ether*.

ferric nitrate. Crystals; $Fe(NO_3)_3 \cdot 9H_2O$. There is no anhydrous form of this salt.

fluoboric acid, 48%. Liquid; HBF_4; if not readily available in small quantities, substitute 10.3 mL HF (48%) plus 4.4 g H_3BO_3, for each 10 mL of 48% fluoboric acid specified.

glycerol. Syrupy liquid; also called glycerin or glycerine; $C_3H_5(OH)_3$; contains up to 5% (by weight) water.

molybdic acid, 85%. Crystals or powder containing the equivalent of 85% MoO_3. This misnamed chemical consists mostly of ammonium molybdate (or paramolybdate), which is $(NH_4)_6Mo_7O_{24} \cdot 4H_2O$. The two chemicals can be used interchangeably. See *ammonium molybdate*.

muriatic acid. Liquid; technical grade HCl (see Table 1).

picric acid. Crystals; 2,4,6-trinitrophenol; crystals of laboratory chemical contain 10 to 15% water; explosive; its crystalline metallic salts are even more explosive. Do not use grades that do not have the 10 to 15% water content.

pyrophosphoric acid. Crystals or viscous liquid; $H_4P_2O_7$, anhydrous; hydrolyzes to phosphoric acid (H_3PO_4) slowly in cold water and rapidly in hot water.

zephiran chloride. Aqueous solution; a proprietary material produced in grades containing about 12% and 17% (by weight) benzalkonium chloride (alkyl-dimethyl-benzyl-ammonium chloride) as the active constituent, plus some ammonium acetate; also called sephiran chloride. Available from pharmacies or pharmaceutical distributors. See *benzalkonium chloride*.

are prepared in quantity and stored is a worthwhile safety precaution.

Heat Evolution in Preparing Etchants. Exercise caution and follow accepted laboratory procedures when mixing chemicals. In general, heat is evolved, sometimes in large amounts, when strong acids (particularly H_2SO_4), alkalis (NaOH and KOH), anhydrous $AlCl_3$ or their concentrated solutions are added to water, alcohols or solutions of other chemicals, and when combining acidic with alkaline substances or solutions.

Always add the acid, alkali or anhydrous $AlCl_3$ to the water, alcohol or solution. Introduce these chemicals slowly, while stirring continuously to avoid local overheating. Incomplete mixing can permit layering, with danger of a delayed violent reaction. Special attention and special cooling procedures may be needed when large quantities of etchants are prepared and large areas of metal are etched, as in some macroetching; and when high currents are used in electropolishing.

Mixing of Oxidizing Agents With Reducing Agents. Exercise special care in mixing oxidizing agents (such as HNO_3, H_2SO_4, $HClO_4$, CrO_3, salts of these acids, persulfates, Br_2 and H_2O_2) with reducing agents (such as alcohols and other organic solvents, acetic acid, acetic anhydride and most organic compounds); failure to follow accepted safe procedures can result in violent or explosive reactions. Acetic anhydride cannot be used safely in electropolishing solutions except in limited ranges of composition and water content, and its use is not recommended.

Care With Cyanides. The use of etchants that contain cyanides presents special toxicity hazards, because poisoning can result from inhaling hard-to-detect small amounts of HCN gas evolved from acidic solutions, from ingesting small amounts of cyanides, and from absorbing cyanides through the skin or exposed body tissues. Careful handling and the use of an effective exhaust hood are especially important. Used cyanide-containing solutions should be made slightly alkaline with ammonia and poured into a chemically resistant waste-disposal drain, and the drain flushed thoroughly with a copious amount of water.

Disposal of Etchants. Spent etchant solutions should be discarded individually promptly after use. They should be poured slowly into a chemically resistant waste-disposal drain in an exhaust hood, while at the same time running a substantial flow of tap water down the drain, after which the drain should be flushed thoroughly with a copious amount of water. Strongly acidic, strongly alkaline, corrosive or toxic solutions should be handled with extra care when being disposed of, because of the hazards described above, under "Protection of Personnel."

Special attention should be given to the safe discarding of used solutions that contain substantial amounts of volatile solvents, so as to avoid the creation of toxicity, fire or explosion hazards from vapors of the solvents.

Measurement of Microstructure

By George F. Vander Voort, Carpenter Technology Corp.

IN RECENT YEARS considerable success has been achieved in developing procedures for quantifying microstructural parameters and using such data to develop structure/property relation-

Table 5. Standard notation recommended by International Society for Stereology(a)

Symbol	Units	Description	Common name
P	...	Number of point elements or test points	...
P_P	...	Point fraction—i.e., number of point elements per total number of test points	Point count
L	mm	Length of linear elements or test-line length	...
P_L	mm^{-1}	Number of point intersections per unit length of test line	...
L_L	mm/mm	Sum of linear intercept lengths divided by total test-line length	Lineal fraction
A	mm^2	Planar area of intercepted features or test area	...
S	mm^2	Surface area or interface area, generally reserved for curved surfaces	...
V	mm^3	Volume of three-dimensional structural elements or test volume	...
A_A	mm^2/mm^2	Sum of areas of intercepted features divided by total test area	Areal fraction
S_V	mm^2/mm^3	Surface or interface area divided by total test volume—i.e., surface-to-volume ratio	...
V_V	mm^3/mm^3	Sum of volumes of structural features divided by total test volume	Volume fraction
N	...	Number of features	...
N_L	mm^{-1}	Number of interceptions of features divided by total test-line length	Lineal density
P_A	mm^{-2}	Number of point features divided by total test area	...
L_A	mm/mm^2	Sum of lengths of linear features divided by total test area	Perimeter (total)
N_A	mm^{-2}	Number of interceptions of features divided by total test area	Areal density
P_V	mm^{-3}	Number of points per test volume	...
L_V	mm/mm^3	Length of features per test volume	...
N_V	mm^{-3}	Number of features per test volume	Volumetric density
$\bar{L}$	mm	Mean linear intercept distance, L_L/N_L	...
$\bar{A}$	mm^2	Mean areal intercept, A_A/N_A	...
$\bar{S}$	mm^2	Mean particle surface area, S_V/N_V	...
$\bar{V}$	mm^3	Mean particle volume, V_V/N_V	...

(a) Note: Fractional parameters are expressed per unit length, area or volume. Source: *Quantitative Stereology*, by Ervin E. Underwood, Addison-Wesley, Reading, MA, 1970.

ships. Chart methods for rating microstructures have been used for many years to evaluate microstructures, chiefly for conformance to specifications. At this time, true quantitative procedures have not replaced chart methods for such purposes but they have gained wide usage in quality control and research studies. Examples of the applications of quantitative metallography have been reviewed by Underwood (Ref 1).

Basically, two types of measurements of microstructures are made. The first group includes measurements of depths—i.e., depth of decarburization, depth of surface hardening, or coating thicknesses. These measurements are made at a specific location (the surface) and are subject to considerable variation. To obtain reproducible data, these surface conditions must be measured at a number of positions on a given specimen, and on several specimens if the material being sampled is rather large.

The second group of measurements belong to the field referred to as stereology. This is the body of measurements that describe the relationship between measurements made on the two-dimensional plane of polish and the three-dimensional microstructural features sampled. To facilitate communications, the International Society for Stereology (ISS) has proposed a standard system of notation, as shown in Table 5 (Ref 2).

These measurements are made manually with the aid of templates outlining a fixed field area, systems of straight or curved lines of known length, or a number of systematically spaced points. The simple counting measurements, P_P, P_L, N_L, P_A and N_A, are most important and are easily made. These measurements are useful by themselves and can be used to derive other useful relationships.

VOLUME FRACTION

One of the simplest and most useful measurements is the point count (described in ASTM E562) used to estimate volume fractions of mi-

crostructural constituents. While other procedures can be employed, the point count is most efficient—i.e., it gives the best accuracy with minimum effort (Ref 3). To perform this test, a clear plastic grid with a number of systematically spaced points (usually crosses are employed, where the "point" is the intersection of the arms), typically 9, 16, 25, 49, 64 and 100, is placed on a micrograph, on a projection screen, or inserted as an eyepiece reticle. The number of points lying on the phase or constituent of interest is counted and divided by the total number of grid points. Points lying on a boundary are counted as half-points. This procedure is repeated on a number of fields selected without bias—i.e., without looking at the screen.

The point fraction, P_P, is given by:

$$P_P = \frac{P_\alpha}{P_T} \qquad \text{(Eq 1)}$$

where P_α is the number of grid points lying inside the feature of interest, α, plus one-half the number of grid points lying on particle boundaries and P_T is the total number of grid points. Studies have shown that the point fraction is equal to the area fraction, A_A, and volume fraction, V_V, of the second-phase particles:

$$P_P = A_A = V_V \qquad \text{(Eq 2)}$$

The volume fraction can also be estimated by dividing the total length of linear elements of a test pattern lying within the phase by the total length of the test pattern. The lineal fraction, L_L, is also equal to the parameters in Eq 2. Point counting is always performed on the minor phases—i.e., where $V_V \leq 0.5$. The amount of the major (matrix) phase can be determined by difference.

The fields measured should be selected at locations over the entire polished surface—i.e., not confined to a small portion of the sample surface. The field measurements should be averaged, and the standard deviation can be used to

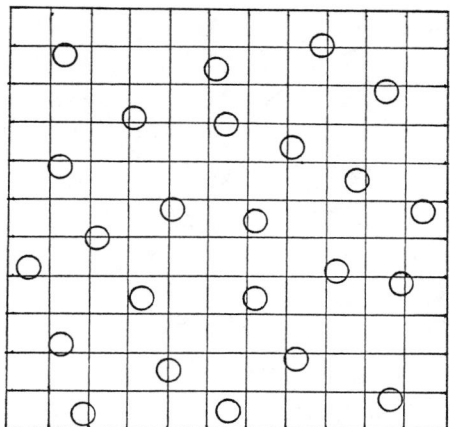

Test area is 12 100 mm². Test lines, ten horizontal and ten vertical, are 110 mm long.

Fig. 27. Synthetic microstructure of uniformly shaped, identical spherical particles in a matrix phase

assess the relative accuracy of the measurement, as described in ASTM E562.

In general, the number of points on the grid should be increased as the volume fraction of the feature of interest decreases. One study (Ref 3) suggested that the optimum number of points is $3/V_V$. Hence, for volume fractions of 0.5 (50%) and 0.01 (1%), the optimum numbers of grid points are 6 and 300, respectively. If the structure is heterogeneous, measurement accuracy is improved by using a low-point-density grid and increasing the number of fields measured.

To illustrate the point-counting procedure, Fig. 27 shows a synthetic microstructure consisting of 24 6-mm-diam circular particles within a field area of 12 100 mm². The total area of the circular particles is 678.6 mm², which is an area fraction of 0.056 (i.e., 5.6%). A square grid pattern has been drawn over this field, producing 100 intersection points. Four of these intersections are completely within the particles and four lie on the particle interface. The number of "hits" is, therefore, $4 + \frac{1}{2}(4)$, or 6. Thus, P_P is 6/100 or 0.06 (i.e., 6%), which agrees very closely with the theoretically calculated area fraction. The area fraction, A_A, is equal to the volume fraction, V_V, as long as the sectioning plane intersects the structural features at random.

NUMBER PER UNIT AREA

The count of the number of particles within a given measurement area, N_A, is a useful microstructural parameter and is used in other calculations. Referring again to Fig. 27, there are 24 particles in the measurement area (12 100 mm²). Hence, the number of particles per unit area, N_A, is 0.001 98 mm⁻². The average area of the particles can be calculated by dividing the volume fraction, V_V, by N_A:

$$\bar{A} = \frac{V_V}{N_A} \qquad \text{(Eq 3)}$$

This yields an average area, $\bar{A}$, of 28.23 mm², which agrees well with the calculated area of a 6-mm-diam particle of 28.27 mm².

The above example illustrates the calculation of the average area of particles in a two-phase

microstructure. For the case of a single-phase structure, the reciprocal of the count of the number of grains per unit area gives a simple measurement of the average grain size which can be used to calculate the ASTM grain size, as described later in this article.

P_L AND N_L

Counting of the number of intersections of a line of known length with particle or grain features, P_L, or the number of interceptions of particles or grains by a line of known length, N_L, provides two very useful microstructural parameters. For space-filling grain structures (single phase), $P_L = N_L$, while for two-phase structures, $P_L = 2N_L$ (this may differ by one count in actual cases).

Grain-Structure Measurements. For single-phase grain structures, it is usually easier to count the grain-boundary intersections with a line of known length. This is the basis of the Heyn intercept grain-size procedure described in ASTM E112. For most work, a circular test grid composed of three concentric circles with a total line length of 500 mm is preferred. Grain size is defined by the mean lineal intercept length, $\bar{L}_3$:

$$\bar{L}_3 = \frac{1}{P_L} = \frac{1}{N_L} \qquad \text{(Eq 4)}$$

This equation must be modified, as described later, for two-phase structures. $\bar{L}_3$ can be used to calculate the ASTM grain size.

P_L measurements can be utilized to define the surface area per unit volume, S_V, and the length per unit area, L_A, of grain boundaries:

$$S_V = 2P_L \qquad \text{(Eq 5)}$$

and

$$L_A = \frac{\pi}{2} P_L \qquad \text{(Eq 6)}$$

For single-phase structures P_L and N_L are equal, and either measurement can be used. For two-phase structures, it is best to measure P_L to determine the phase-boundary surface area per unit volume, or phase-boundary length per unit area.

Oriented Structures. P_L measurements are utilized to assess the degree of orientation of lines or surfaces. Certain microstructures exhibit a high degree of preferred directionality on the plane of polish or within the sample volume. A structure is completely oriented if all of its elements are parallel. Partially oriented systems are those with features having both random and oriented elements. To assess the degree of orientation, the plane of polish must be either longitudinal, planar or transverse with respect to the deformation axis.

Several approaches can be utilized to assess the degree of orientation of a microstructure. For single-phase grain structures, a simple procedure is to make P_L measurements parallel and perpendicular to the deformation axis on a longitudinally oriented specimen. The degree of grain orientation is the ratio of perpendicular to parallel P_L values—i.e., $P_{L\perp}/P_{L\parallel}$. Another very useful procedure is to calculate the degree of orientation, Ω, using these P_L values:

$$\Omega = \frac{P_{L\perp} - P_{L\parallel}}{P_{L\perp} + 0.571 P_{L\parallel}} \qquad \text{(Eq 7)}$$

To illustrate these measurements, a section of low-carbon steel sheet was cold rolled to reductions

Table 6. Degrees of grain orientation for four samples of low-carbon steel sheet

Sample	$P_{L\perp}$(a)	$P_{L\parallel}$(a)	$P_{L\perp}/P_{L\parallel}$	Ω, %
As-received	114.06	98.86	1.15	8.9
Cold rolled:				
12% reduction. .	126.04	75.97	1.66	29.6
30% reduction. .	167.71	60.6	2.77	52.9
70% reduction. .	349.4	34.58	10.1	85.3
(a) Number of grain-boundary intersections per millimetre.				

in thickness of 12, 30 and 70%. $P_{L\perp}$ and $P_{L\parallel}$ measurements were made using a grid with parallel straight test lines on a longitudinal section from each of four samples (one sample for each of the three reductions, plus one sample of as-received material). The results are given in Table 6. As shown, cold working produces an increased orientation of the grains in the longitudinal direction.

Spacings. N_L measurements are utilized in the study of the spacing of two-phase structures. One of the commonest spacing measurements is that of lamellar pearlite or other eutectoid or eutectic structures (Ref 4). The true interlamellar spacing, σ_t, is difficult to measure, but the mean random spacing, σ_r, is readily assessible and is directly related to the mean true spacing:

$$\sigma_t = \sigma_r/2 \qquad \text{(Eq 8)}$$

The mean random spacing is determined by placing a test grid consisting of one or more concentric circles on the pearlite lamellae in an unbiased manner. The number of interceptions of the carbide with the test line(s) is counted and divided by the true length of the test line, to obtain N_L. The reciprocal of N_L is the mean random spacing:

$$\sigma_r = \frac{1}{N_L} \qquad \text{(Eq 9)}$$

The mean true spacing, σ_t, is $\frac{1}{2} \sigma_r$. To make accurate measurements, the lamellae must be clearly resolved; hence, use of transmission electron microscope replicas is quite common.

N_L measurements are also used to measure interparticle spacings in two-phase alloys. The mean center-to-center spacing of particles, σ, is the reciprocal of N_L. For the second-phase particles in the idealized two-phase structure shown in Fig. 27, a count of the number of particles intercepted by the horizontal and vertical test lines yields 31 interceptions. The total line length is 2200 mm; hence, $N_L = 0.0141$ mm⁻¹ and $\sigma = 70.9$ mm.

The mean edge-to-edge distance between such particles, known as the mean free path, λ, is determined in like manner but requires knowledge of the volume fraction of the particles. The mean free path is calculated from:

$$\lambda = \frac{1 - V_V}{N_L} \qquad \text{(Eq 10)}$$

For the structure illustrated in Fig. 27, the volume fraction of the particles was estimated as 0.056. Hence, $\lambda = 66.95$ mm.

The mean lineal intercept distance, $\bar{L}_{3\alpha}$, for these particles is determined by:

$$\bar{L}_{3\alpha} = \sigma - \lambda \qquad \text{(Eq 11)}$$

For this example, $\bar{L}_{3\alpha} = 3.95$ mm. This value is smaller than the caliper diameter of the particles because the test lines intercept the particles at random, not only at the maximum dimension.

GRAIN SIZE

One of the commonest quantitative microstructural measurements is that of the grain size of metals and alloys. Numerous procedures have been developed to estimate grain size; these procedures are summarized in detail in ASTM E112 and illustrated in Ref 5. Several types of grain size are measured: ferrite grain size, austenite grain size, and prior-austenite grain size. Each type presents particular problems associated with revealing these boundaries so that an accurate rating can be obtained (Ref 5). To complicate matters, a variety of parameters are utilized to define grain size:

- Average grain diameter, $\bar{d}$
- Average grain area, $\bar{A}$
- Number of grains per unit area, N_A
- Average intercept length, $\bar{L}_3$
- Number of grains intercepted by a line of fixed length
- Number of grains per unit volume, N_V
- Average grain volume, $\bar{V}$

These parameters can be related to ASTM grain size, G.

The ASTM grain-size scale was established using the English system of units, but no difficulty is introduced by metric measurements. The ASTM grain-size equation is:

$$n = 2^{G-1} \qquad \text{(Eq 12)}$$

where n is the number of grains per square inch at $100\times$. Multiplication of n by 15.5 gives the number of grains per square millimetre at $1\times$, N_A.

Planimetric Method. The oldest procedure for measuring the grain size of metals is the planimetric method. A circle of known size (generally 79.8-mm diameter, 5000-mm^2 area) is drawn on a photomicrograph or used as a template on a projection screen. The number of grains completely within the circle, n_1, and the number of grains intersecting the circle, n_2, are counted. For accurate counts, the grains must be marked off as they are counted, which makes this method slow. The number of grains per square millimetre at $1\times$, N_A, is determined by:

$$N_A = f(n_1 + n_2/2) \qquad \text{(Eq 13)}$$

where f is the magnification squared divided by 5000 (the circle area). The average grain area, $\bar{A}$, in square millimetres, is:

$$\bar{A} = \frac{1}{N_A} \qquad \text{(Eq 14)}$$

and the average grain diameter, $\bar{d}$, in millimetres, is:

$$\bar{d} = (\bar{A})^{1/2} = \frac{1}{N_A^{1/2}} \qquad \text{(Eq 15)}$$

The ASTM grain size, G, can be found by using the tables in ASTM E112 or by the following equation:

$$G = [3.322 (\log N_A)] - 2.95 \qquad \text{(Eq 16)}$$

Figure 28 illustrates the planimetric method.

Intercept Method. The intercept method is faster than the planimetric method because the micrograph or template does not require marking to obtain an accurate count. ASTM E112 recommends use of a template consisting of three concentric circles with a total line length of 500 mm

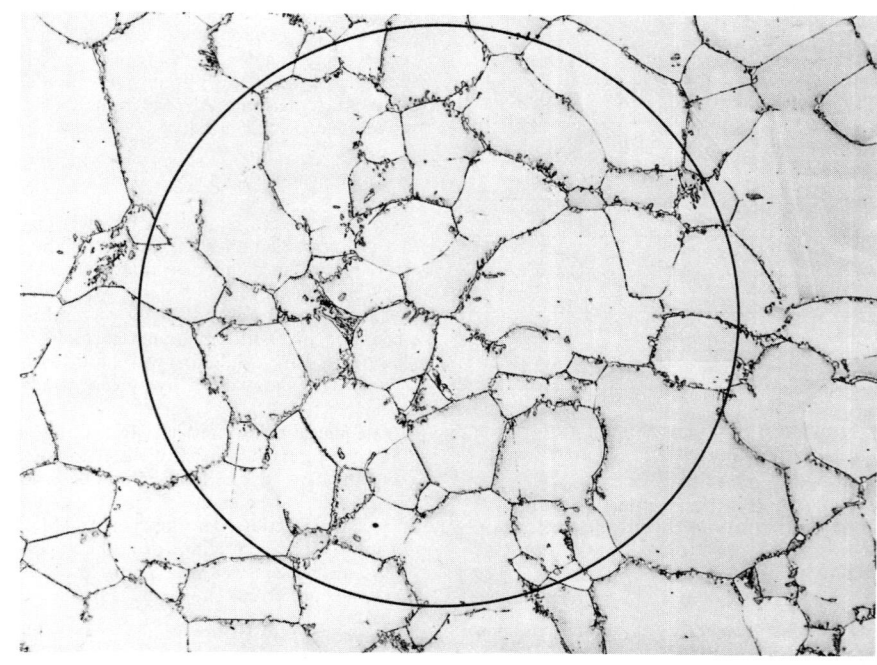

Micrograph, at 100×, of an austenitic manganese steel that was solution annealed at 1040 °C (1900 °F) and aged at 620 °C (1150 °F) to decorate the grain boundaries with fine pearlite. The circle has a diameter of 79.8 mm, for an area of 5000 mm². A count is made to determine the number of grains completely within the circle (44) and the number intersecting the circle (25). The Jeffries multiplier for 100× is 2. The number of grains per square millimetre at 1× is (2)(44 + 25/2) = 113. The ASTM grain size, G, is (3.322 log 113) − 2.95 = 3.87 (round to 3.9 or 4).

Fig. 28. Example of grain-size measurement by the planimetric (Jeffries) method

(template available from ASTM). The template is placed over the grain structure without bias, and the number of grain-boundary intersections, P, or the number of grains intercepted, N, is counted. Dividing P or N by the true line length gives P_L or N_L, which are identical (N or P can differ slightly due to tangent hits) for a single-phase grain structure. It is usually easier to count grain-boundary intersections for single-phase structures. If a grain boundary is tangent to the line, it is counted as $1/2$ of an intersection. If a triple-point line junction is intersected it is counted as $1 1/2$ or 2. The latter is preferred because the small diameter of the inner circle introduces a slight bias to the measurement which is offset by weighing triple-line intersections as 2 hits.

The mean lineal intercept length, $\bar{L}_3$, determined as shown in Eq 4, is a measure of ASTM grain size. $\bar{L}_3$ is smaller than $\bar{d}$ because the test lines do not intersect each grain at its maximum breath. The ASTM grain size, G, can be determined by use of the tables in ASTM E112 or can be calculated from:

$$G = [-6.6457 (\log \bar{L}_3)] - 3.298 \qquad \text{(Eq 17)}$$

where $\bar{L}_3$ is in millimetres. Figure 29 illustrates the intercept method for a single-phase alloy.

Nonequiaxed Grains. Nonequiaxed grain structures require measurements on the three principal planes—i.e., the longitudinal, planar and transverse planes. For such structures, the intercept method is preferred, but the test grid should consist of a number of straight, parallel test lines of known length rather than circles. Because the ends of the straight lines generally end within grains, these interceptions are counted as half-hits. Three mutually perpendicular orientations are evaluated using grain-interception counts:

1. N_{L_l}– parallel to the grain elongation, longitudinal plane
2. N_{L_t}– perpendicular to the grain elongation (through-thickness direction), transverse plane
3. N_{L_p}– perpendicular to the grain elongation (across width), planar surface

These three values are totaled and divided by three to obtain the average $\bar{N}_L$ value, and G is determined by reference to the tables in ASTM E112 or by Eq 17.

Two-Phase Grain Structures. The grain size of a particular phase in a two-phase structure requires determination of the volume fraction of the phase of interest—e.g., by point counting. The minor phase (second phase) is point-counted and the volume fraction of the major phase (matrix phase) is determined by difference.

Next, a circular test grid is applied to the microstructure without bias and the number of grains of the phase of interest intercepted by the test line, N_α, is counted. The mean lineal intercept length of the alpha grains, $\bar{L}_{3\alpha}$ is determined by:

$$\bar{L}_{3\alpha} = \frac{(V_V)(L/M)}{N_\alpha} \qquad \text{(Eq 18)}$$

where L is the line length and M is the magnification. The ASTM grain size can be determined from the tables in ASTM E112 or by use of Eq 17. The method is illustrated in Fig. 30.

INCLUSION CONTENT

Assessment of inclusion types and contents is commonly performed on high-quality steels. Production evaluations utilize comparison-chart methods such as those described in ASTM E45, SAE J422a, or the German standard SEP 1570.

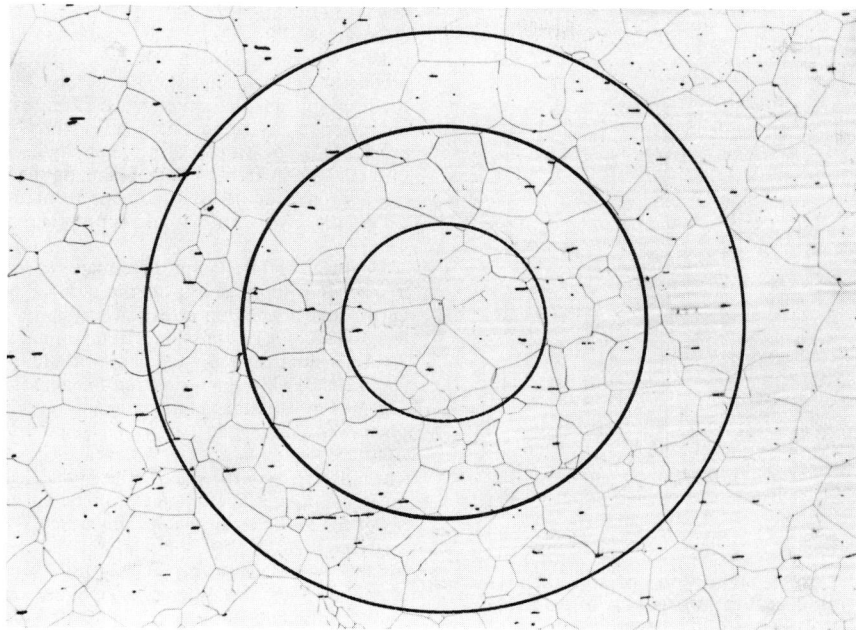

The three concentric circles have diameters of 79.5, 47.8 and 31.8 mm, for a total line length of 500 mm. The magnification of this micrograph is 500×, and hence the true line length is 1 mm. A count of the number of grain-boundary intersections (several are not well delineated by the nital etch—a common problem in such ratings, especially if automated devices are used) revealed 60 grain-boundary intersections and 7 triple line-junction intersections. Hence, $P = 7(1.5) + 60 = 70.5$, and $P_L = 70.5/1$ mm = 70.5/mm. Therefore, $L_3 = 1/P_L = 0.0142$ mm. The ASTM grain-size number can be calculated using Eq 17: $G = (-6.6457 \log 0.0142) - 3.298 = 8.98$ (round to 9).

Fig. 29. Example of grain-size measurement using the Heyn intercept method

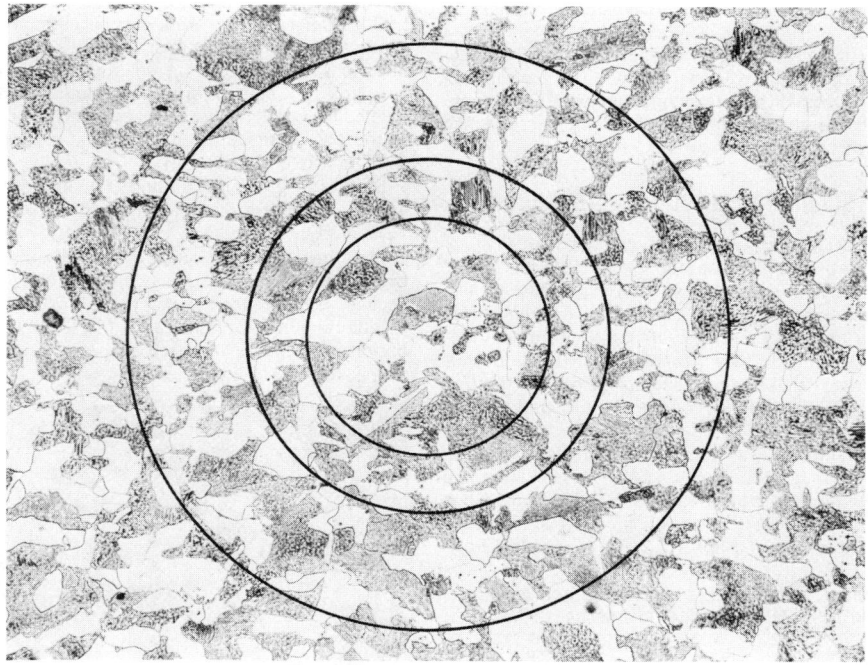

Micrograph, at 500×, of a specimen of alloy Ti-6Al-2Sn-4Zr-2Mo that was forged at 955 °C (1750 °F) in the alpha-beta region and annealed at 970 °C (1775 °F) in the alpha-beta region, producing primary alpha (white) and an alpha-beta eutectoid. The specimen was etched with Kroll's reagent. Point counting was used to determine the amount of equiaxed alpha (48.5%). The three concentric circles have a total circumference of 500 mm. A count is made of the number of equiaxed alpha grains intercepted by the three circles (76). The mean lineal intercept, $\bar{L}_{3_\alpha}$, is calculated as follows: $\bar{L}_{3_\alpha} = (0.485)(500/500)/76 = 0.006382$ mm. The ASTM grain size, G, is calculated using Eq 17: $G = (-6.646 \log 0.006382) - 3.298 = 11.29$ (round to 11.3 or 11.5).

Fig. 30. Example of the determination of grain size in a two-phase alloy using point counting and intercept counting

In these chart methods, the inclusion pictures are separated by type and graded by severity (amount). Either qualitative procedures (worst rating of each type observed) or quantitative procedures (all fields in a given area measured) are employed. Only the Japanese standard JIS-G-0555 uses actual volume-fraction measurements for rating of inclusion content.

Manual measurement of the volume fraction of inclusions requires substantial effort to obtain acceptable measurement accuracy due to the rather low volume fractions usually encountered (Ref 6). Consequently, extensive use of image analyzers has been made to overcome this problem. Image analyzers separate the oxide and sulfide inclusions on the basis of their gray-level differences. By using automated stage movement and autofocusing, enough field measurements can be made in a relatively short time to obtain good statistical accuracy. Image analysis is also employed to measure the length of inclusions or to determine stringer lengths.

MEASUREMENT STATISTICS

In performing stereological measurements, it is necessary to make these measurements on a number of fields and average the results. Measurements on a single field may not be representative of the bulk conditions, because few (if any) materials are sufficiently homogeneous. Calculation of the standard deviation of the field measurements gives a good indication of measurement variability. Calculation of the standard deviation can be done quite simply with an inexpensive pocket calculator.

A further refinement of statistical analysis is calculation of the 95% confidence limit based on the standard deviation, s, of the field measurements. The 95% confidence limit is calculated from:

$$95\% \text{ CL} = \frac{ts}{(N - D)^{1/2}} \qquad \text{(Eq 19)}$$

where t is the student t value that varies with N, the number of measurements. Many users standardize on a single value of t, 2, for calculations irrespective of N. The measurement value is expressed as the average, $\bar{X}$, $\pm$ the 95% CL value. This means that if the test were conducted 100 times, the average values would be between plus and minus the average, $\bar{X}$, in 95 of the measurements. Next, one can calculate the relative accuracy, % RA, of the measurement by:

$$\% \text{ RA} = \frac{95\% \text{ CL}}{\bar{X}} \times 100 \qquad \text{(Eq 20)}$$

Usually, a 10% relative accuracy is considered to be adequate. DeHoff (Ref 7) has developed a simple formula to determine how many fields, N, must be measured to obtain a specific desired degree of relative accuracy at the 95% confidence limit:

$$N = \left[\frac{200}{(\% \text{ RA})} - \frac{s}{\bar{X}} \right]^2 \qquad \text{(Eq 21)}$$

IMAGE ANALYSIS

The measurements described in this brief review, and other measurements not discussed, can be made by use of automatic image analyzers. These devices rely primarily on the gray level of the image on the television monitor to detect the desired features. In some instances, complex image editing can be utilized to aid separation. Some structures, however, cannot be separated completely, and for these structures the semiautomatic digital tracing devices can be utilized to improve measurement speed. These devices are discussed elsewhere in this text.

SUMMARY

This article has reviewed many of the simple stereological counting measurements and simple relationships based on these parameters. More complex measurements are discussed at length in Ref 1. The measurements described are easy to learn and use. Their application enables the metallographer to discuss microstructures in a more quantitative manner and reveals relationships between the structure and properties of the material.

REFERENCES

1. Applications of Quantitative Metallography, by E. E. Underwood: in *Metals Handbook*, 8th Ed., Vol 8, American Society for Metals, Metals Park, OH, 1973, p 37-47
2. *Quantitative Stereology,* by E. E. Underwood: Addison-Wesley, Reading, MA, 1970
3. An Evaluation of Procedures in Quantitative Metallography for Volume-Fraction Analysis, by J. E. Hilliard and J. W. Cahn: *Trans. AIME*, Vol 221, Apr 1961, p 344-352
4. Measurement of the Interlamellar Spacing of Pearlite, by G. F. Vander Voort and A. Roosz: *Metallography*, Vol 17, Feb 1984, p 1-17
5. Grain Size Measurement, by G. F. Vander Voort: in *Practical Applications of Quantitative Metallography*, STP 839, American Society for Testing and Materials, Philadelphia, 1984, p 85-131
6. Inclusion Measurement, by G. F. Vander Voort: in *Metallography as a Quality Control Tool*, Plenum Press, New York, 1980, p 1-88
7. Quantitative Metallography, by R. T. DeHoff: in *Techniques of Metals Research*, Vol II, Part 1, Interscience, New York, 1968, p 221-253

Application of Electron Microscopes in Metallurgy

By Gary M. Michal, Case Western Reserve University; Barbra Gabriel, Packer Engineering Associates, Inc.; and Edward Kenik, Martin Marietta Energy Systems, Inc.

THE USE of electron microscopes can greatly extend the metallurgist's ability to define the structure of metals compared with more classical techniques of optical microscopy and x-ray analysis. Electron microscopes offer much better spacial resolution due to the use of very-short-wavelength electron radiation (<0.1 Å) and electron beams that can be focused to very small spot sizes (<100 Å).

This enhanced spacial resolution in electron microscopes can be used not only to define the size, shape and distribution of very small microstructural features, but also to define the chemical composition and detailed crystallographic nature of the features.

A variety of events occur when a beam of electrons impinges on a specimen. Several types of electron, x-ray and optical signals may be created. These interactions are illustrated in Fig. 31. The type of signal analyzed determines the general type of electron microscope being em-

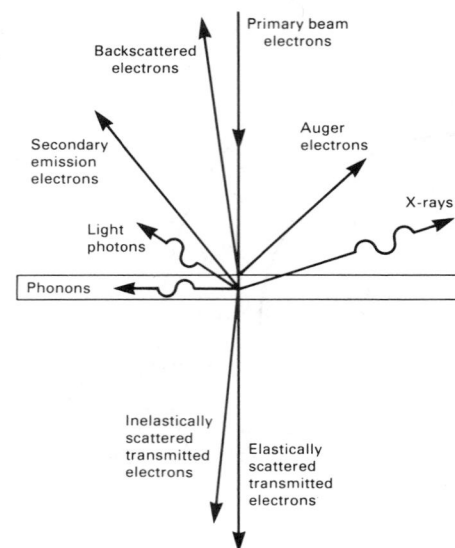

Fig. 31. Nine types of signals created by the interaction of an electron beam with a specimen

ployed. Modern scanning electron microscopes (SEM) generally examine thick specimens and analyze the backscattered and secondary emission electron and x-ray signals. Transmission electron microscopes (TEM) examine thin specimens (<0.5 μm) and analyze primarily inelastically and elastically scattered transmitted electrons and x-ray signals. More recently, many TEM's have been equipped to detect secondary emission and backscattered electron signals as well.

The following generally outlines the capabilities of these two general types of electron microscopes for microstructural analysis. For a more detailed description of these microscopes the reader should refer to Ref 1.

SCANNING ELECTRON MICROSCOPES

The scanning electron microscope (SEM) provides both high resolution and high depth of field for analysis of surface structure (Ref 2 to 8). The SEM improves on the capabilities of both binocular analysis (for depth of field) and optical metallograph analysis (for resolution). Table 7 illustrates these improvements in both resolution and depth of field compared with optical systems.

In the SEM, a beam of electrons is scanned in an X-Y fashion over a region of a conductive specimen. For each location of the beam some of it is reflected back as elastically backscattered electrons or first absorbed and then released as

Table 7. Comparison of resolution and depth of field for SEM and optical microscopes

Magnifi-cation	Resolution		Depth of field	
	SEM	Optical	SEM	Optical
20×	5 μm	5 μm	1 mm	5 μm
100×	1 μm	1 μm	200 μm	2 μm
200×	0.5 μm	0.5 μm	100 μm	0.7 μm
1 000×	0.1 μm	0.1 μm(a)	20 μm	0.2 μm
2 000×	500 Å	...	10 μm	...
10 000×	100 Å	...	2 μm	...
20 000×	50 Å	...	1 μm	...

(a) Best possible optical resolution.

lower-energy inelastically scattered secondary electrons (Ref 9 and 10). These interactions also induce other radiative transitions, such as the production of characteristic x-rays and light in the form of cathodoluminescence (Ref 11), and nonradiative transitions such as phonon (heat) production. The variation in the strength of any of these signals from various points on the specimen surface can be used to create a contrast on a CRT display tied in with an appropriate detector.

Contrast can arise from three major sources in a specimen: topography, atomic-number differences, and orientation effects in crystalline materials. Secondary-emission electron imaging is most sensitive to topographical variations in a specimen. Backscattered-electron imaging is most sensitive to variations in the orientation of crystal planes in a specimen. X-ray emission is most sensitive to atomic-number differences in a specimen (Ref 12 to 16). Thus, the signal used to create an image should be determined on the basis of the variation in the specimen that is to be analyzed.

Specimen preparation for SEM is usually rather simple. Either metallographic mounts or unmounted fracture surfaces can be viewed directly as long as they are grounded. The major criteria for SEM specimens are: (a) the specimen surface must be clean (fracture-surface cleaning is discussed in Ref 17 to 22); (b) the specimen must be conductive (Ref 23 to 25); and (c) the specimen must either be small enough to fit in the specimen chamber (about 1 in. on a side) or be capable of being sectioned.

Imaging Capabilities. In metallography, SEM is used to characterize microstructures beyond the resolution limit of the light microscope (Ref 26 to 30). Deep etching often can be employed to create topographical contrast in phases and bring inclusions into relief for the determination of their morphology. This information can be supplemented by x-ray analyses of the elements present in the phases to define the conditions for their formation (Ref 31 and 32).

SEM is ideal for the study of fracture surfaces and for definition of modes of failure (see Fig. 32 and 33). Unique features exhibited by different fracture modes are readily resolved and identified (Ref 4, and 33 to 36). Artifacts are avoided because the specimen is examined directly, and a wide range of magnifications (20 to 20 000×) permits correlation of macroscopic with microscopic features.

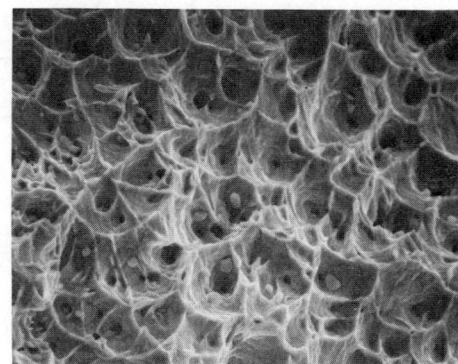

Magnification, 750×; shown here at 60%.

Fig. 32. Scanning electron micrograph showing dimpled rupture in a copper tensile-test specimen

Magnification, 5000×; shown here at 60%.

Fig. 33. Scanning electron micrograph showing fatigue striations in an aluminum alloy 2014-T6 forging

Unpolished specimen. Magnification (both), 240×; shown here at 140%.

Fig. 34. SEM backscattered-electron image (top), and thermal-wave image (bottom), of same region in type 309S stainless steel

A nonradiative transition induced during electron irradiation is phonon (heat) production. A recently introduced technique used in conjunction with SEM is thermal-wave microscopy, which serves to provide subsurface images resulting from changes in thermal conductivity (Ref 37 and 38). Figure 34 compares SEM and thermal-wave images of a specimen of type 309S stainless steel. In this case, the grain structure would be visible only after more stringent metallographic preparation. Clearly, this development further increases the applications of the SEM.

Analytical Capabilities. The SEM is capable not only of creating differences in contrast between areas of a specimen with different chemical compositions, but also of determining the chemical composition of various areas. Specimens irradiated with electrons emit characteristic x-rays of unique energy/wavelength that correspond to particular elemental compositions. These x-ray outputs can be converted to chemical composition for areas as small as about 1 to 2 μm in bulk SEM specimens (Ref 16).

The two basic types of x-ray microanalysis used in electron microscopy are energy- and wavelength-dispersive spectroscopy, which discriminate, respectively, among the energies and wavelengths of characteristic x-rays both qualitatively and quantitatively (Ref 12, 13, 15 and 22). Although wavelength-dispersive spectroscopy has been traditionally associated with microprobes, it has become popular in EM for its high-resolution ability to sort individual x-ray wavelengths with a crystal spectrometer (Ref 39). The energy-dispersive spectrometer uses a lithium-drifted silicon diode to detect, simultaneously, all x-rays emitted by the specimen, although only those x-rays emitted by elements including and heavier than sodium are analyzed (Ref 40 to 43).

Current developments in energy-dispersive x-ray microanalysis include windowless detectors for analysis of elements as light as carbon (Ref 44 to 46), and a complimentary microanalysis technique that combines x-ray data with the backscattered-electron signal to analyze specimens according to compound formulas rather than only elemental composition (Ref 47).

Diffraction Capabilities. If an SEM specimen contains relatively perfect crystals of a size greater than about 5 μm, diffraction information about the orientation of the crystals can be obtained. The techniques that can be used include selected-area electron channeling (Fig. 35) and micro Kossel x-ray diffraction. These techniques provide crystallographic information from selected areas of bulk specimens, and represent an intermediate between x-ray diffraction and transmission electron diffraction of thin foils (Ref 48 to 52).

TRANSMISSION ELECTRON MICROSCOPES

There are now two major types of nominally 100-to-200-kV transmission electron microscopes: conventional (CTEM, or simply TEM) and scanning (STEM). Many new instruments can operate in both modes. The CTEM operates in a fashion analogous to a transmitted light optical microscope. The specimen must be thin enough (<0.5 μm) so that a reasonable number of electrons pass through it. The percentage of the beam that is directly transmitted or diffracted at various angles differs from point to point over the total area irradiated. This gives rise to the major source of contrast. Placement of an objective aperture determines whether an image will be created by the transmitted beam (bright field) or a diffracted beam (dark field). The theory and operation of the CTEM is well documented, and the reader is referred to Ref 52 to 58. The STEM operates in a mode that combines scanning of a small beam spot over a specimen with detection of either the transmitted or diffracted beams from each spot location. Detectors located below the specimen determine the strength of the signals from these beams. These signals are used to create an image on a CRT in a manner equivalent to the SEM.

Specimen Preparation. The geometrical requirements of a specimen to be examined in transmission are rather exacting. In over-all dimensions the specimen must be smaller than a disk 3 mm in diameter and about 0.25 mm thick. These requirements necessitate several initial sectioning operations for most bulk materials. The portion of the disk that will actually be examined must be thinned to less than 0.5 μm. The methods used for thinning of metallic specimens include preparation of thin foils and evaporated thin films for *in situ* examination of metals, while surface microstructure may be examined via replication techniques.

Creation of a thin-foil region is usually the most difficult step in the specimen-preparation process. Electropolishing is a traditional thinning method that has evolved from an art into a more exact science, although skill is a major factor in success (Ref 58 to 63). The major advantage of electropolishing is that thickness is reduced without mechanical deformation or a major loss of structure, and thus the foil is representative of the bulk material. Thin areas as large as 50 by 5 μm can be prepared by this technique.

In comparison, thin films only a few hundred angstroms thick may be evaporated or sputtered from a pure-metal or alloyed target. In thermal evaporation, also known as resistance heating, thin films are prepared under high-vacuum conditions by passing a current through the target metal until vaporization occurs. The vaporized atoms recondense as a thin film, the thickness being dependent on characteristics of the evaporant and on operating parameters (Ref 4, 64, 65 and 66 discuss theoretical and practical aspects of thermal evaporation). Resistance heating is employed in several areas of specimen preparation, including carbon replication and enhancement of surface electron conductivity for SEM.

Although refractory metals cannot be evaporated by resistance heating, electron-gun evaporation produces extremely fine-grain thin films of these metals (Ref 24 and 67). Another method of producing thin films of either refractory or nonrefractory metals is sputtering, or the erosion of atoms from a metal target by an energetic plasma (Ref 23, 25, and 68 to 70).

TEM specimens may also be prepared by one of several replication techniques. Replicas are high-fidelity surface reproductions, and thus they serve a purpose different from that of thin foils or films. Replicas are used when high-resolution surface studies are required, or when another preparation method will destroy the surface of interest, such as a fracture surface (Ref 35, and 71 to 73). The extraction replica technique involves removal of inclusions or precipitates from a surface, which are subsequently imaged and analyzed by electron diffraction and various analytic techniques. These techniques were most popular prior to the advent of the SEM (which directly analyzes surfaces), although several situations exist where replication is necessary. For example, replicas are ideal chronological records of the progression of surface defects under experimental conditions such as those of gear-wear studies (Ref 74), and for examining specimens too large for SEM.

Single-stage replicas are negative impressions of a surface. Carbon or oxide replicas of polished and etched specimens faithfully duplicate metallographic structure which is beyond the resolution limit of light or even SEM analysis (Ref 58).

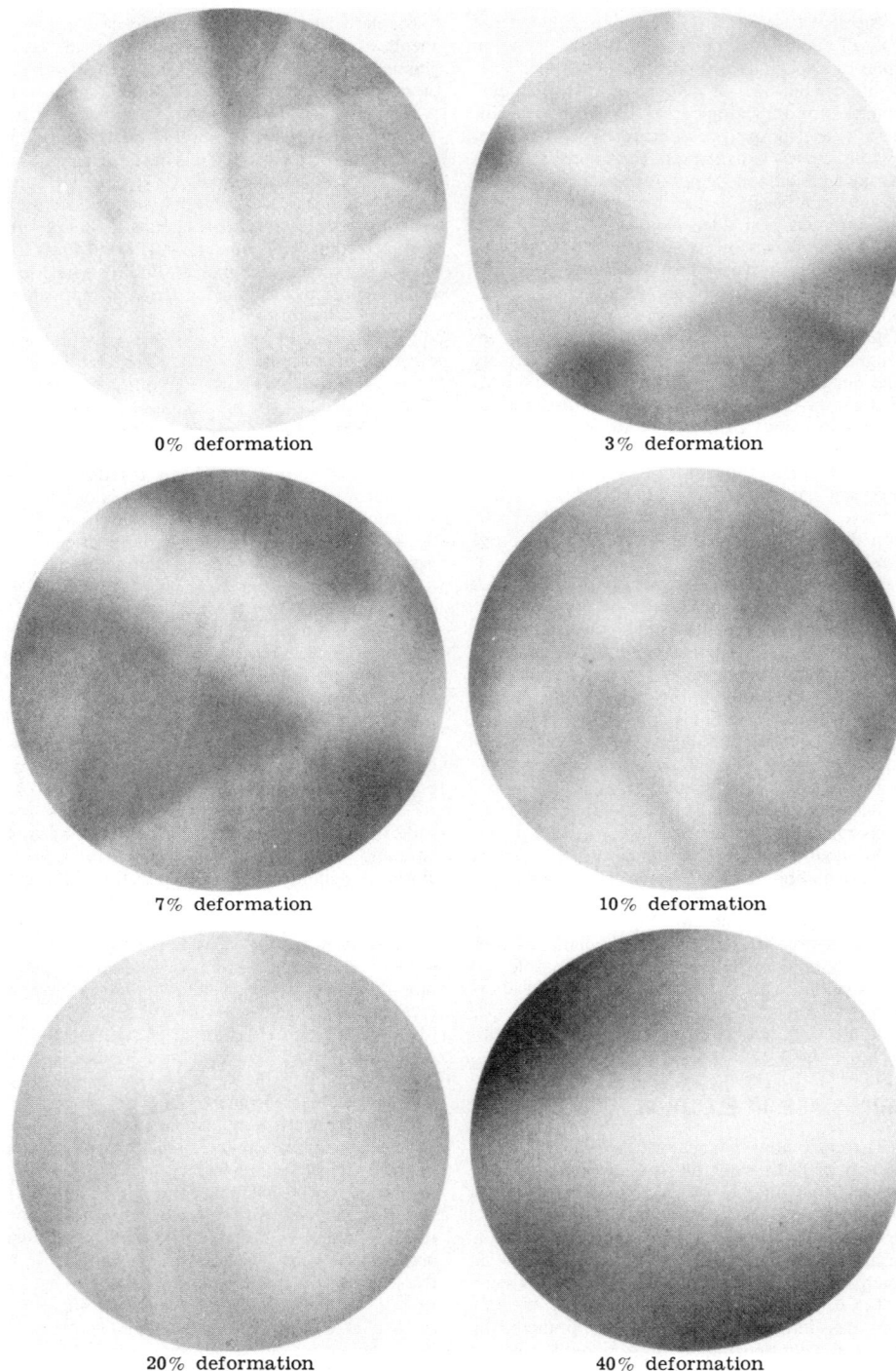

0% deformation

3% deformation

7% deformation

10% deformation

20% deformation

40% deformation

Specimens were taken from the gage lengths of tensile bars that had been elongated at room temperature to produce the indicated percentages of deformation.

Fig. 35. Selected-area electron-channeling patterns obtained in a scanning electron microscope in deformation studies of type 316 stainless steel

Carbon replicas are prepared under vacuum by thermal evaporation using carbon rods (Ref 72, 75 and 76 discuss use of a carbon yarn in place of rods). The desired film thickness is variable from less than 100 Å (for very high resolution but very fragile replicas) to several hundred angstroms (for moderate resolution and mechanically stable replicas). The carbon replica is removed from the specimen by electropolishing,

mounted on a specimen-support grid, and shadowed with a heavy metal, such as chromium or gold, to increase electron contrast (Ref 53, 77 and 78). Single-stage carbon replicas, when properly prepared, provide the highest resolution of any replicating medium (Ref 71).

Extraction replicas encapsulate inclusions or precipitates, and are used when electron diffraction or high-resolution analytic analysis of par-

ticles is required (Ref 58 and 79). A polished specimen is coated with a carbon film, then etched through the carbon film to release the inclusions or precipitates. The liberated particles are captured by the replica.

Aluminum and its alloys form an anodic oxide film which can be removed from a specimen surface and examined (Ref 58 and 80). Reference 73 discusses other alloys appropriate for oxide replication.

Two-stage plastic-carbon replicas are prepared when the surface of interest must remain in place. This nondestructive technique is especially useful in fractography (Ref 4, 35 and 81). Basically, an initial impression of the surface is made with cellulose acetate tape softened with acetone. The tape is held in place over the area of interest, then dried. The cellulose acetate negative replica is carefully stripped and used as the template for the carbon, or positive, replica prepared as described above. Two-stage replicas are prone to artifact formation and must be carefully evaluated for valid interpretation (Ref 71 comprehensively summarizes the interpretation of replicas).

Imaging Capabilities. Although considerable effort is usually required to prepare a specimen for analysis in transmission, structural details smaller than 10 Å can be easily resolved using CTEM once the specimen has been prepared. Ultimate resolution in modern TEM's, such as the Philips Em 420 and JEOL JEM 200CX, approaches 2 Å under ideal conditions. Because the contrast in CTEM comes from primarily crystallographic differences in a specimen, this technique is especially sensitive to analysis of defects such as dislocations in metals (Fig. 36) and to phase identification associated with transformation in alloys (Fig. 37 to 39).

For most thin-foil applications the resolving power of CTEM will surpass that of STEM. The resolution limit of STEM is set by the size of the electron beam. Although very small beams can be made to impinge on a specimen (Ref 82 to 85), the specimen itself will typically cause beam broadening during transmission. This effect is illustrated in Fig. 40. Thus, the smallest effective spot size and best resolution during STEM operation are about 25 Å, unless very thin specimens — i.e., evaporated films — are being examined.

Analytical Capabilities. By working in transmission, elemental analysis can be determined from much smaller areas than can be determined in an SEM. This has to do with broadening of the electron beam by the specimen, as illustrated in Fig. 40. Even if a 100-Å beam were used in an SEM, the specimen would broaden the beam such that x-rays would be created from a region less than 1 μm across in the specimen. This problem is greatly reduced when thin-foil specimens are used. Areas as small as a few hundred angstroms can be analyzed with modern CTEM or STEM instruments (Ref 22 and 86). When used for elemental analyses, these microscopes are commonly referred to as analytic electron microscopes (AEM). The advantage of STEM operation is that it permits line-scan and dot-map analyses equivalent to those performed in SEM's and microprobe analyzers.

In an AEM, an energy-dispersive x-ray spectrometer (EDXS) is used for analysis of elements as light as sodium. Elemental analyses can also be performed on the inelastically scattered transmitted electrons in CTEM and STEM instruments. This technique is called electron energy-

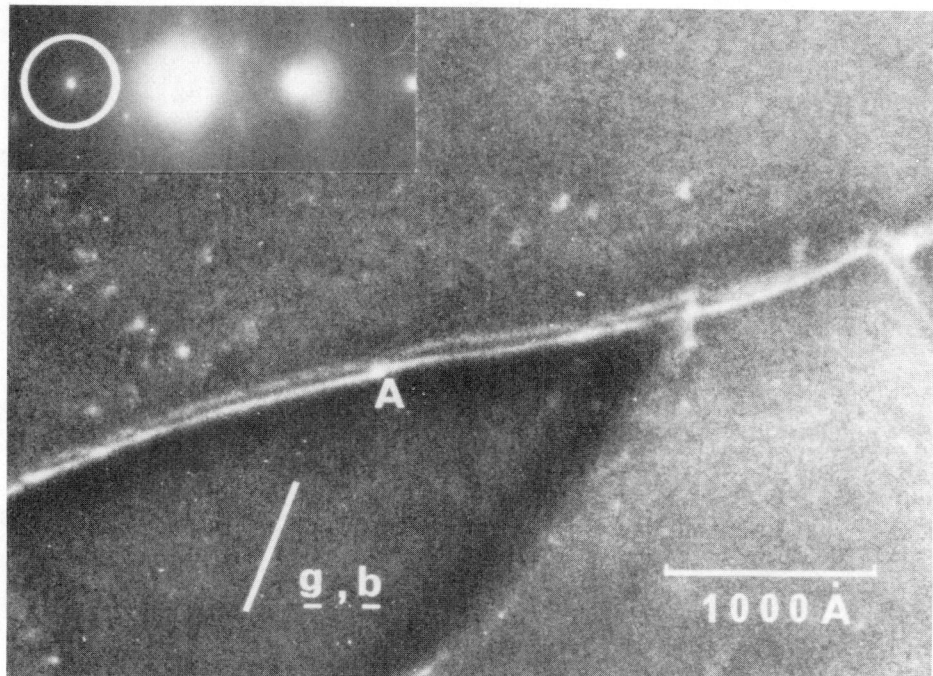

Fig. 36. Weak-beam 220 dark-field TEM image (magnification, 275 000×) of a dissociated dislocation in a thin-foil specimen of silver, showing partial dislocations pinned at A (Ref 93)

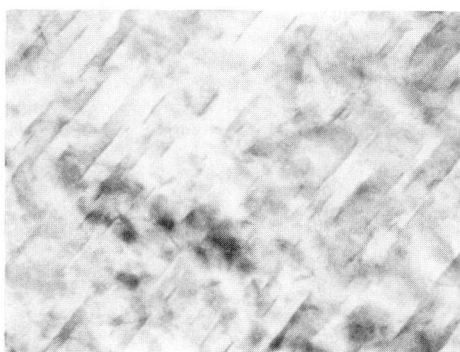

Nitrided for 4 h at 575 °C (1065 °F) in an atmosphere of 8% ammonia and 92% hydrogen. Zones have formed that are believed to correspond to $(FeMo)_{16}N_2$. Thin-foil specimen. Magnification not reported.

Fig. 37. Transmission electron micrograph showing structure of nitrided Fe-5Mo alloy

Copper was plastically deformed to 99% deformation by rolling at 20 °C (68 °F). Thin-foil specimen. Magnification, 10 000×.

Fig. 38. Transmission electron micrograph showing recrystallized grain in plastically deformed OFHC copper (Ref 94)

Thin-foil specimen. Magnification, 12 000×.

Fig. 39. Transmission electron micrograph showing domains of tetragonality in 70Cu-30Mn alloy (Ref 95)

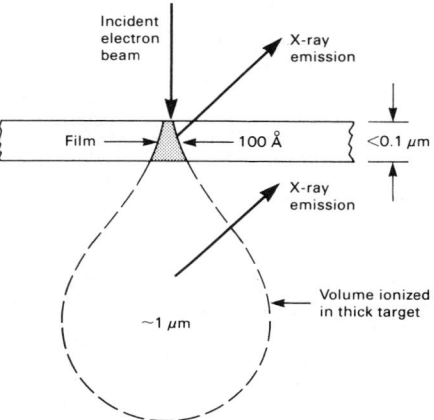

Fig. 40. Schematic illustration of the amount of beam broadening that can result from use of a thin-foil specimen compared with a bulk (thick) specimen

loss spectroscopy (EELS) (Ref 87). EELS is complementary to EDXS in that it provides high sensitivity to light elements ($z < 11$). The energy distribution of the transmitted electrons is measured by a magnetic, electrostatic or combined magnetic-electrostatic analyzer. Either plasmon or inner-shell excitation losses can be used for elemental analysis. An additional benefit of EELS is that the degradation of spacial resolution associated with beam broadening by the specimen should be less than for EDXS. This technique has the drawback of requiring the examined area to be especially thin (<500 Å): otherwise, the sensitivity to light elements is poor.

Diffraction Capabilities. The CTEM has a very powerful capability of discerning subtle diffraction effects from very small regions of a speci-

men. Using conventional selected-area diffraction (SAD), diffraction patterns can be obtained from areas as small as 0.5 μm across in thin foils, thin films or extraction replicas (Ref 55, 56, 88 and 89).

The small incident electron probes and high collection angles of current AEM's provide new and powerful electron diffraction techniques. Convergent-beam electron diffraction (CBED) is one such technique where the effective electron-probe size defines the size of the area producing the diffraction pattern. Diffraction patterns can routinely be obtained from areas less than a few hundred angstroms across. CBED can provide information about specimen thickness, accurate lattice-parameter determinations, and crystallographic space groups (Ref 90 to 92).

REFERENCES

1. D. K. Bowen and C. R. Hall, *Microscopy of Materials*, John Wiley & Sons, New York, 1975
2. T. E. Everhart and T. L. Hayes, *Sci. Amer.*, Vol 225, 1972, p 54
3. B. L. Gabriel, *Biological Electron Microscopy*, Van Nostrand Reinhold, New York, 1982
4. B. L. Gabriel, *SEM: A User's Guide to Materials Science*, ASM, Metals Park, OH, 1982
5. S. Komoto and J. C. Russ, *Amer. Sci.*, Vol 57, No. 1, 1969, p 112
6. G. F. Pfefferkorn *et al.*, *SEM, Inc. Proc.*, Vol 1, 1978, p 1
7. J. C. Russ, in *Applications of Modern Metallographic Techniques*, ASTM, Philadelphia, 1970, p 214
8. O. C. Wells, *Scanning Electron Microscopy*, McGraw-Hill, New York, 1974
9. D. E. Newbury, *IITRI/SEM Proc.*, Vol 1, 1977, p 553
10. V. N. E. Robinson and E. P. George, *SEM, Inc. Proc.*, Vol 1, 1978, p 859
11. D. B. Holt and S. Datta, *SEM, Inc. Proc.*, Vol 1, 1980, p 259
12. L. S. Birks, *Electron Probe Microanalysis*, 2nd Ed., Wiley Interscience, New York, 1971
13. J. D. Geller, *IITRI/SEM Proc.*, Vol 1, 1977, p 281
14. J. I. Goldstein *et al.*, *IITRI/SEM Proc.*, Vol 1, 1977, p 315
15. J. I. Goldstein *et al.*, *Scanning Electron Mi-*

croscopy and X-ray Microanalysis, Plenum Press, New York, 1981

16. D. E. Newbury, *SEM, Inc.,* Vol 2, 1979, p 1
17. E. P. Dahlberg, *IITRI/SEM Proc.,* 1974, p 911
18. E. P. Dahlberg, *ibid.,* Vol 1, 1976, p 715
19. I. Kayafas, *Corrosion,* Vol 36, No. 8, 1980, p 443
20. R. Lohberg *et al., SEM, Inc. Proc.,* Vol 4, 1983, p 1571
21. P. M. Yuzawich and C. W. Hughes, *Pract. Metallog.,* Vol 15, 1978, p 184
22. N. J. Zaluzec, in *Introduction to Analytical Electron Microscopy,* Plenum Press, New York, 1979, p 121
23. P. Echlin, *SEM, Inc. Proc.,* Vol 1, 1981, p 79
24. P. Echlin and G. Kaye, *ibid.,* Vol 2, 1979, p 21
25. P. Echlin *et al., ibid.,* Vol 1, 1980, p 163
26. D. L. Davidson, *SEM, Inc. Proc.,* Vol 1, 1981, p 403
27. M. G. Hall, *SEM, Inc. Proc.,* Vol 1, 1981, p 409
28. O. Johari *et al., IITRI/SEM Proc.,* 1969, p 279
29. V. Thien *et al., SEM, Inc. Proc.,* Vol 2, 1983, p 577
30. R. C. Voigt and C. R. Loper, Jr., *ibid.,* Vol 3, 1983, p 1067
31. J. Howe, *Metallography,* Vol 16, 1983, p 275
32. M. Pohl *et al., SEM, Inc. Proc.,* Vol 4, 1983, p 1563
33. S. Bhattacharyya *et al., IITRI Fracture Handbook,* IITRI, Chicago, 1979
34. J. L. McCall, *Fracture Analysis by Scanning Electron Microscopy,* Battelle, Columbus, OH, 1972
35. A. Phillips *et al., Electron Fractography Handbook,* Battelle, Columbus, OH, 1976
36. G. F. Pittinato *et al., SEM/TEM Fractography Handbook, ibid.,* 1975
37. D. G. Davies, *ibid.,* Vol 3, 1983, p 1163
38. A. Rosencwaig, *Science,* Vol 218, 1982, p 223
39. I. Pozsgai and A. Barna, *ibid.,* Vol 2, 1983, p 585
40. N. C. Barbi, *Electron Probe Microanalysis Using Energy Dispersive X-ray Spectroscopy,* PGT, Inc., Princeton, NJ, 1981
41. K. F. J. Heinrich *et al., Energy Dispersive X-ray Spectrometry,* U.S. Govt. Printing Office, Washington, 1981
42. K. F. J. Heinrich, *Electron Beam X-ray Microanalysis,* Van Nostrand Reinhold, New York, 1981
43. R. Woldseth, *X-ray Energy Spectrometry,* Kevex Corp., Burlingame, CA, 1973
44. N. C. Barbi and J. C. Russ, *IITRI/SEM Proc.,* 1975, p 85
45. J. C. Russ, *IITRI/SEM Proc.,* Vol 1, 1977, p 289

46. J. C. Russ and A. O. Sandborg, in *Energy Dispersive X-ray Spectrometry,* U.S. Government Printing Office, Washington, 1981, p 297
47. V. N. E. Robinson *et al., ibid.,* to be published in 1984
48. D. L. Davidson, *Res. & Dev.,* Vol 25, 1974, p 34
49. D. J. Dingley, *ibid.,* Vol 1, 1978, p 869
50. D. J. Dingley, *ibid.,* Vol 4, 1981, p 273
51. C. J. Harland *et al., J. Phys. E.,* Vol 14, 1981, p 175
52. J. A. Venables and J. C. Harland, *Phil. Mag.,* Vol 27, 1973, p 1193
53. B. L. Gabriel, *Biological Electron Microscopy,* Van Nostrand Reinhold, New York, 1982
54. R. D. Heidenreich, *Fundamentals of Transmission Electron Microscopy,* Interscience, New York, 1964
55. P. B. Hirsch *et al., Electron Microscopy of Thin Crystals,* Krieger Publishing Co., New York, 1977
56. L. E. Murr, *Electron and Ion Microscopy and Microanalysis,* Marcel Dekker, Inc., New York, 1982
57. J. W. Edington, *Practical Electron Microscopy in Materials Science,* Van Nostrand Reinhold, New York, 1976
58. G. Thomas and J. Goringe, *Transmission Electron Microscopy of Materials,* John Wiley & Sons, New York, 1982
59. P. J. Goodhew, in *Practical Methods in Electron Microscopy,* Vol 1, Part 1, Elsevier North Holland, New York, 1973
60. A. Howie, in *Techniques for Electron Microscopy,* 2nd Ed., Blackwell, Oxford, 1967, p 433
61. A. G. Jackson *et al., Characterization of Microstructures,* Systems Research Labs, Inc., Dayton, OH, 1977
62. B. J. Kestel, *Polishing Methods for Metallic and Ceramic Transmission Electron Microscope Specimens,* Argonne National Lab., Argonne, IL, 1981
63. I. S. Brammer, *ibid.,* p 356
64. G. A. Bassett, in *Techniques for Electron Microscopy,* 2nd Ed., Blackwell, Oxford, 1967, p 411
65. L. Holland, *Vacuum Deposition of Thin Films,* Chapman & Hall, Ltd., London, 1970
66. C. C. Shiflett, in *Thin Film Technology,* Van Nostrand Reinhold, New York, 1968, p 113
67. H. F. Hagler *et al., J. Micros.,* Vol 110, 1977, p 149
68. A. N. Broers and E. Spiller, *SEM, Inc. Proc.,* Vol 1, 1980, p 201
69. A. Renou and M. Gillet, *J. Cryst. Growth,* Vol 44, 1978, p 190
70. G. K. Wehner and G. S. Anderson, in *Handbook of Thin Film Technology,* McGraw-Hill, New York, 1970, C-3

71. C. D. Beachem, in *Fracture,* Vol 1, Academic Press, New York, 1969, p 243
72. D. E. Bradley, in *Techniques for Electron Microscopy,* 2nd Ed., Blackwell, Oxford, 1967, p 96
73. W. E. Warke *et al.,* in *Electron Fractography,* ASTM, Philadelphia, 1968, p 212
74. S. Anderson, *Wear,* Vol 29, 1974, p 271
75. D. E. Bradley, *J. Appl. Phys.,* Vol 27, 1956, p 1399
76. K. R. Peters, *J. Micros.,* Vol 133, 1984, p 17
77. R. C. Williams and R. W. G. Wyckoff, *J. Appl. Phys.,* Vol 17, 1946, p 23
78. J. H. M. Willison and A. J. Rowe, in *Practical Methods in Electron Microscopy,* Vol 8, Elsevier North Holland, New York, 1980
79. R. M. Fisher, *J. Appl. Phys.,* Vol 24, 1953, p 113
80. C. D. Beachem and R. M. N. Pelloux, in *Fracture Toughness Testing and Its Applications,* ASTM, Philadelphia, 1964, p 210
81. I. LeMay, in *Microstructural Analysis: Tools and Techniques,* Plenum Press, New York, 1973, p 153
82. A. V. Crewe, *Science,* Vol 221, 1983, p 325
83. A. V. Crewe *et al., Rev. Sci. Instr.,* Vol 39, 1968, p 576
84. A. V. Crewe *et al., Science,* Vol 168, 1970, p 1338
85. J. P. Langmore, in *Principles and Techniques of Electron Microscopy,* Vol 9, Van Nostrand Reinhold, New York, 1978, p 1
86. J. I. Goldstein, in *Introduction to Analytical Electron Microscopy,* Plenum Press, New York, 1979, p 83
87. D. C. Joy, in *Introduction to Analytical Electron Microscopy,* Plenum Press, New York, 1979, p 223
88. R. H. Alderson and J. S. Halliday, in *Techniques for Electron Microscopy,* 2nd Ed., Blackwell, Oxford, 1967, p 478
89. B. E. P. Beeston *et al.,* in *Practical Methods in Electron Microscopy,* Vol 1, Part 2, Elsevier North Holland, New York, 1973
90. B. F. Buxton *et al., Phil. Trans.,* Vol 281, 1976, p 171
91. J. W. Steeds, in *Introduction to Analytical Electron Microscopy,* Plenum Press, New York, 1979, p 387
92. L. P. Stoter, *J. Mat. Sci.,* Vol 16, 1981, p 1356
93. P. B. Hirsch, in *Electron Microscopy and the Structure of Materials,* University of California Press, 1972, p 1-20
94. J. H. Cairns, J. Clough, M. A. P. Dewey and J. Nutting, *J. Inst. Metals,* Vol 99, 1971, p 93-97
95. E. P. Butler and P. M. Kelly, *Trans. Met. Soc. AIME,* Vol 242, 1968, p 2099-2106

Metallographic Technique for Ferrous Metals

Technique for Carbon and Alloy Steels

PROCEDURES described in the article on "Metallographic Methods" (see first page of this section) regarding sectioning and polishing of specimens are, for the most part, applicable to carbon and alloy steels. However, some steps in the preparation of specimens of carbon and alloy steels require special consideration, and the special procedures needed are discussed in this article.

SECTIONING

Proper sectioning of the surface to be examined is a very important step in preparing steel specimens, because in this step the metallurgical structure of the specimen can easily be altered. Care in sectioning is necessary for all steels, but especially for fully hardened (untempered) steels. Precautions against overheating and excessive mechanical deformation of the steel during sectioning are mandatory.

The first step in preventing damage to the metallurgical structure is to minimize the amount of sectioning that is done. Regardless of whether the test pieces are separate test pieces or are coupons attached to forgings or castings, they should be designed so as to require a minimum amount of sectioning.

Abrasive-wheel cutting is the method most used for sectioning of specimens. It is the required method for sectioning hardened steels, and it is a convenient method for sectioning all steels. When the specimen must be removed from a large mass of material, sectioning by sawing (provided that the steel is not too hard for sawing) usually precedes cutting with an abrasive wheel.

Rubber-bonded aluminum oxide wheels 0.75 to 1.0 mm (0.030 to 0.040 in.) thick are recommended. Some advantage can be gained, in maximizing cutting speed and in minimizing damage to the specimen, by selecting a grade of cutoff wheel related to the hardness of the steel being cut. For cutting hardened steels, a relatively soft wheel is preferred, such as grade H.

Heating can be minimized by directing an ample flow of water or water-soluble coolant over the surface being cut. Underwater cutting is often used, but a specimen can be burned under water if it is cut too fast. Slow cutting of submerged specimens with a forced stream of coolant is the safest procedure.

Separate Test Pieces. Because a major use of metallography is for process control (principally, control of heat treating processes), separate test

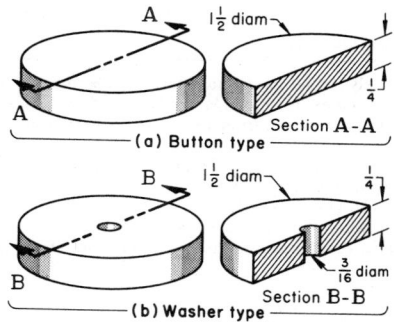

Fig. 1. Two types of test pieces used for metallographic control of heat treating processes

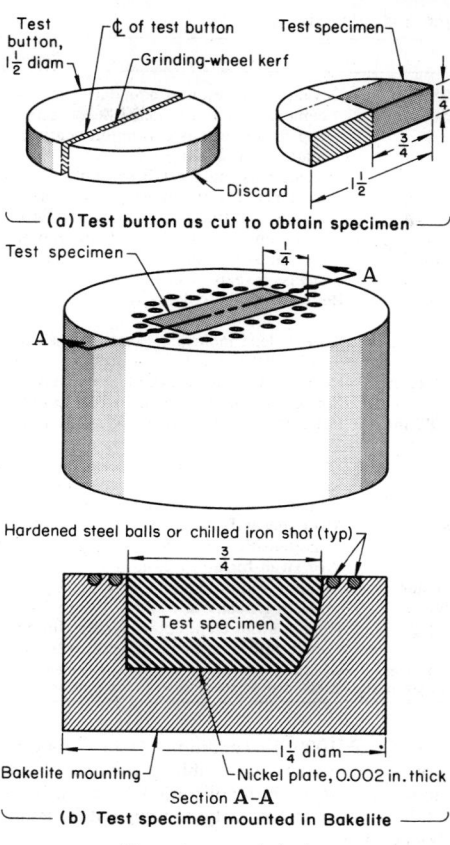

Fig. 2. Methods of (a) cutting a test specimen from a test button and (b) mounting the test specimen to retain flatness for metallographic examination

pieces are widely used. They must have the same composition and thermal history as the steel for which the tests are run.

A single test piece often suffices as a tensile-test bar and as a metallographic specimen. The size and design of the test piece depend largely on which tests are planned. When the test piece is to be used only for metallography, it is usually designed so that it can be cut into two or more smaller specimens.

Two types of test pieces used for control of heat treating processes are shown in Fig. 1. The solid, button-type test piece shown in Fig. 1(a) is easy to prepare and is usually adequate. The washer-type test piece shown in Fig. 1(b) has two advantages over the solid, button type. First, the washer-type piece can be more easily hung or otherwise secured in various locations of a heat treating furnace. Second, when it is sectioned, inside edges, which are less susceptible to edge rounding during grinding and polishing, are available for examination.

To produce metallographic specimens, the test pieces are cut slightly off-center so that a full half is available, and further specimen cutting proceeds as shown in Fig. 2.

MOUNTING

A thermosetting phenolic such as Bakelite is commonly used for mounting steel specimens. Bakelite offers the advantages of quick mounting and relative hardness; the latter helps in maintaining flatness of the specimen during grinding and polishing. A disadvantage of Bakelite is that its thermosetting curing temperature (135 to 150 °C, or 275 to 300 °F) is one to which many steel specimens must not be heated because it will affect their microstructure. Thus, close temperature control of the process is often essential when steel specimens are mounted in Bakelite, and some steels (for example, untempered tool steels) cannot be mounted in Bakelite. Another disadvantage is that fragile specimens may be damaged by the pressure required during preparation of the mount. Bakelite, by itself, does not provide adequate support for edge retention. Powdered epoxy, which possesses many of the same characteristics as Bakelite, is an improved thermosetting mounting product. The required time, temperature and pressure of mounting are similar to those for Bakelite, but the hot flow characteristics are much better, producing a superior mount with better edge retention.

Plastics that set at room temperature, such as epoxy resins, are preferred by many metallog-

raphers. Because use of these mounting materials involves casting, many molds can be poured at one time. Room-temperature plastics usually require several hours (or overnight) to set; some formulations, however, set in 1 to 2 h, depending on the amount of hardener used. Such mounts are usually fairly soft and thus require the same mounting techniques for edge retention as those used for thermoplastic materials. Care must be taken in mounting large specimens with castable epoxy. In some cases, high temperatures can be generated, resulting in cracking of the mount.

Mounting Techniques. The specific technique used for mounting carbon and alloy steel specimens depends on the magnification to be used, requirements for edge preservation, and mounting material. When examination will be made at no more than 100 diameters, when edges are of no particular interest, and when the maximum temperature incurred in mounting is not a problem, no special measures are necessary and all of the mounting materials discussed are satisfactory. In contrast, when a magnification of more than 100 diameters will be used and when edge preservation is of prime concern (especially when one of the softer mounting materials must be used), mounting must be done with great care. Special techniques are required to obtain the degree of flatness required for edge preservation or examination under high magnification.

Plating the test piece with nickel is common practice. The nickel plate protects and preserves the edges during final cutting and helps maintain

flatness during grinding and polishing. A plate thickness of about 0.05 mm (0.002 in.) is usually sufficient.

Another practice that is effective in obtaining and maintaining maximum flatness is shown in Fig. 2. In this practice, a test button such as that illustrated in Fig. 1 is sectioned by first cutting slightly off-center and then cutting off a 19-by-6.4-mm (^{3}/$_4$-by-^{1}/$_4$-in.) specimen, as shown at the right in Fig. 2(a). The specimen is nickel plated and then mounted in Bakelite ready for grinding and polishing (see section A-A in Fig. 2b for specimen placement). The mount also features the use of hard metal pieces surrounding the specimen (Fig. 2b). Often, these are small hardened steel balls, but chilled iron shot costs less and is equally satisfactory for most purposes. An acceptable alternative is to surround the specimen with a pipe section. For the specimen shown in Fig. 2, the inside diameter of the pipe section would have to be about 22 mm (^{7}/$_8$ in.). These techniques are less effective than plating.

GRINDING AND POLISHING

Steel specimens require coarse and fine grinding. Coarse grinding is done with 80, 100 or 120-mesh abrasive. Fine grinding is usually done with three or four papers of progressively finer grit size (usually, No. 240, 320, 400 and 600 mesh).

Mechanical Polishing. Steel specimens should be polished in two steps—rough polishing and fine polishing.

For rough polishing, a napless cloth impregnated with diamond paste is preferred. Nylon, silk, thin cotton, canvas and cotton drill cloth are suitable. Polishing with 6-μm diamond paste usually is sufficient, but some metallographers prefer to polish with 9-μm followed by 3-μm paste or to use 6- and 1-μm diamond.

Powdered alpha alumina (Al_2O_3) in the same range of sizes as diamond can be substituted for diamond paste, but diamond is less likely to cause distortion of the specimen surface, to round off hard phases or particles, or to pull out nonmetallic inclusions.

Rough polishing is most often done on a cloth-covered disk rotating at 150 to 300 rpm. The specimen is held by hand and moved in the direction opposite to the direction of wheel rotation. Alternatively, automatic devices can be used for grinding and polishing.

Fine polishing is done on a napped cloth with a slurry of gamma alumina and water. Almost any soft cloth can be used, but the most satisfactory is one with synthetic fibers on a cotton backing. Alumina in a 0.05-μm size is used most often, although some metallographers prefer to fine polish in two steps—first with 0.3-μm and then with 0.05-μm alumina. Wheel speed and specimen rotation for fine polishing are the same as for rough polishing. Time required for rough and fine polishing varies with hardness and size of the specimen, but the average time is 2 to 4 min, using the technique described above.

Electrolytic polishing has only limited application to carbon and alloy steels, because it removes inclusions, severely rounds edges, and results in preferential attack in multiphase systems. Electropolishing has the advantage of speed; a high-quality polish that will reveal ferritic grain size can be obtained in about 2 min. In most instances, it is not necessary to mount the specimen for electrolytic polishing, although a conductive mounting can be used.

Table 1. Etchants for microscopic examination of carbon and alloy steels

No.	Etchant	Purpose, or characteristic revealed
1	Nital: 1 to 5 mL HNO$_3$ in 99 to 95 mL ethanol (95%) or methanol (95%)	Develops ferrite grain boundaries in low-carbon steels; produces contrast between pearlite and a cementite or ferrite network; develops ferrite boundaries in structures consisting of martensite and ferrite; etches chromium-bearing low-alloy steels resistant to action of picral. Preferred for martensitic structures.
2	Picral: 4 g picric acid in 100 mL ethanol (95%) or methanol (95%)	Reveals maximum detail in pearlite, and bainite; reveals undissolved carbide particles in martensite; differentiates ferrite, martensite and massive carbide by coloration; differentiates bainite and fine pearlite; reveals carbide particles in grain boundaries of low-carbon steel. Addition of about 0.5 to 1 mL zephiran chloride wetting agent increases speed of attack.
3	Vilella's reagent: 5 mL HCl, 1 g picric acid, 100 mL ethanol (95%) or methanol (95%)	For contrast etching(a); reveals prior austenite grains in tempered martensite, and in austempered steels; reveals pearlite colonies.
4	1 to 1.5 mL HCl (conc), 2 to 3 g picric acid, 100 mL ethanol (95%)	Reveals pearlite colonies(b).
5	30 g K$_2$Cr$_2$O$_7$ in 225 mL hot distilled water; and 30 mL acetic acid (glacial)	Reveals lead inclusions, causing them to appear yellow or gold when specimen is examined under polarized light(c).
6	16 g CrO$_3$ in 145 mL distilled water; add 80 g NaOH(d)	Reveals intergranular oxidation; used for medium-carbon alloy steels that contain nickel(e).
7	10 g potassium metabisulfite, 100 mL water	For resolution of hardened structures. Should be preceded by an etch in nital or picral.
8	Howarth's reagent: 10 mL H$_2$SO$_4$, 10 mL HNO$_3$, 80 mL water	For detection of overheating and burning, and for examination of steel forgings.
9	8 g sodium metabisulfite in 100 mL water	Produces good contrast in as-quenched martensitic structures.
10	1 g KCN in 100 mL water, mixed with 0.25 g diphenylthiocarbazone in 10 mL chloroform	Reveals lead inclusions by coloring them red; coloration is most visible when specimens are viewed under polarized light.
11	Saturated aqueous picric acid plus 1 g/100 mL sodium tridecylbenzene sulfonate	Most successful etch for revealing prior austenite grain boundaries in medium- or high-carbon martensite steels. Steels should be untempered, or tempered below 540 °C (1000 °F). Immerse or swab for up to 20 min.
12	2 g picric acid, 25 g sodium hydroxide, 100 mL water	Alkaline sodium picrate. Use boiling for 30 s or more to darken cementite. Can be used electrolytically at 6 V dc, 0.5 to 24/in.^{2}, for 30 to 120 s.
13	50 mL cold saturated aqueous sodium thiosulfate, 1 g potassium metabisulfite	Klemm's tint etch; colors ferrite. Immerse for 40 to 100 s until surface is colored. A light pre-etch with nital or picral improves sharpness.

(a) Specimen should be tempered for 20 to 30 min at 315 °C (600 °F). (b) Immerse specimen for 5 to 10 s in solution at room temperature. (c) Etch for 10 to 20 s in solution at room temperature, rinse in hot water and dry. (d) Sodium hydroxide (NaOH) must be added slowly, with constant stirring. (e) Immerse specimen in boiling solution for 10 to 30 min, rinse in hot water, dry in air blast.

ETCHING

Compositions of etchants used for microscopic examination of carbon and alloy steels are given in Table 1. Table 1 also lists the major characteristics of microstructures that each etchant reveals. Nital is by far the most widely used etchant for microetching of carbon and alloy steels.

Nital vs Picral. Although nital and picral are so similar in etching action that they are often regarded as being interchangeable, they differ in several important respects. For some structures there is little if any difference in etching action between nital and picral; for others, the difference is significant.

For etching a plain carbon steel that has been annealed to produce coarse pearlite, either nital or picral can be used for examination at magnifications between 500 and 1000 diameters. However, if a specimen has been heat treated to produce a moderately fine pearlite, the use of nital or picral will make a significant difference. In general, picral is preferred for nonmartensitic structures while nital is preferred for martensitic structures.

Special-Purpose Etchants. Addition of hydrochloric acid to picral (Vilella's reagent in Table 1) brings about radical changes in etching behavior. For medium-carbon alloy steels quenched and tempered between 315 and 480 °C (600 and 900 °F), Vilella's reagent may reveal the prior austenite grain structure by producing grain contrast. The color differences between grains in a specimen result from deeper attack of some grains than of others (contrast etching). In general, large austenite grains are more sharply and clearly resolved than small ones.

The most successful etchant for revealing prior austenite grain boundaries is saturated aqueous picric acid plus about 1 g/100 mL of a wetting agent (sodium tridecylbenzene sulfonate is the most popular wetting agent). This procedure works best on medium-carbon, high-carbon and alloy steels in the as-quenched or tempered (up to about

540 °C, or 1000 °F) condition. Etching may be done by immersion or swabbing for times up to about 20 min. Light repolishing to remove some of the structure etching helps to bring out the grain-boundary attack.

An etchant consisting of 8 g of sodium metabisulfite in 100 mL of water develops good contrast in as-quenched martensitic structures. Several other special-purpose etchants are listed in Table 1.

PREPARATION OF REPLICAS OF STEEL SPECIMENS FOR ELECTRON MICROSCOPY

Specimen preparation before replication, rather than the electron microscope or the replica material, limits the resolution, or ability to reproduce surface detail accurately.

A material used for making replicas should be structureless and of low atomic number, so that it does not scatter the electron beam. It must be strong enough to bridge the grid openings. It should have thermal conductivity, stability and strength adequate to allow it to withstand high-intensity electron beams without acquiring intense electrostatic charges, tearing or polymerizing. Replica making is generally divided into single-stage and two-stage techniques.

Single-Stage Replicas

Single-stage replicas are made by placing a few drops of a plastic solution of collodion or a polyvinyl formal resin on a prepared surface and allowing it to dry. The replica is then dry stripped by placing a piece of plastic tape over the replica and then lifting to remove both replica and tape. This replicating method is little used, because a single-stage replica is limited in resolution by its internal structure, is difficult to strip, and is easily torn.

Two-Stage Carbon Replicas

Preparation of two-stage carbon replicas entails the following operations, which are described below: (a) production of a plastic, first-stage replica; (b) shadowing the plastic replica with chromium, to obtain contrast; (c) evaporation and deposition of carbon, to obtain the second stage of the replica; and (d) washing the two-stage replica to remove the plastic replica but retain both the carbon replica and the chromium from shadowing.

Production of the first-stage replica is begun by cutting a piece of acetylcellulose sheet-plastic film 0.034 mm (0.0013 in.) thick and slightly larger than the surface area to be replicated. The mounted and etched steel specimen is flooded with acetone, and any air bubbles trapped in the junction between the specimen and the mount are allowed to escape. The small piece of sheet plastic is picked up by tweezers on a curled side (small pieces of sheet plastic will curl) and then softened by immersion in acetone, or softened on one side with acetone applied from an eyedropper or a squeeze bottle. Just before all of the curl disappears, one end of the softened sheet is placed on the mount and the other end is lowered over the area to be replicated so as to push away the excess acetone on the specimen and prevent air entrapment. Any excess acetone remaining is drained by turning the mount on its side and placing it on a paper towel.

After the replica is dry, it is stripped from the specimen by lifting one corner at a time with tweezers until it pops off. Any adhesion at this time indicates that the replica is not dry, and removal might cause distortion or artifacts. The stripped replica is trimmed and placed replication side up on a glass microscope slide that has been covered with lined double-coated tape. The edges of the replica are taped down to the slide.

Shadowing. Contrast is obtained by shadowing the plastic replicas. Shadow length depends on surface topography and shadow angle. Metals having high atomic weight, such as chromium, gold, platinum and palladium, produce good results. Of these, chromium is most commonly used for shadowing.

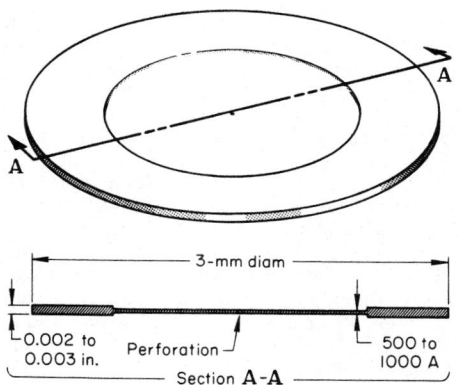

The thinned area (500 to 1000 angstroms thick) surrounding the perforation is the portion used for examination in the electron microscope.

Fig. 3. Thin-foil steel specimen after final thinning

A small pellet of chromium (about the size of an ordinary BB) is placed in a tungsten-wire basket in specially designed evaporation equipment. The basket is positioned 5.7 cm (2.2 in.) below a replica holder, and at a 45° angle to it. The replica is mounted in the holder, and the vacuum chamber is pumped down to 1 torr minimum. At 50 to 55 volts at 2 amp, the chromium pellet is resistance heated to its vaporization temperature, where it is held for a few seconds.

Evaporation of Carbon. The second stage of the replica entails the deposition of carbon, evaporated from an arc, on the shadowed plastic replica. To accomplish this, the plastic replica remains under vacuum as described above. The specimen is removed at regular intervals, washed in water and then in methanol, and dried in hot air. After each interval the specimen thickness is checked with micrometers. A few cycles as described above will indicate the thinning rate, which may vary for different steel compositions.

Final thickness from preliminary thinning should be 0.05 to 0.08 mm (0.002 to 0.003 in.). The final specimen is obtained by punching out a circle 3 mm (1/8 in.) in diameter, using a circular specimen punch specially designed for this purpose. Common practice is to punch out several circles (up to 5 from a 10-mm or 0.4-in. square), because more than one may be needed before final preparation and examination are completed.

Final (electrolytic) thinning is usually done in an apparatus specially designed for the purpose. An electrolyte commonly used for final thinning is composed of 515 mL of glacial acetic acid, 100 g of sodium chromate, and 50 g of chromic acid. It is important to add the sodium chromate very slowly to the acetic acid while stirring constantly; otherwise, difficulty will be encountered in dissolving the sodium chromate.

The circular specimen, with its edges clamped in a holder, is placed in the thinning apparatus and thinned until a perforation develops in the specimen. To determine the proper current for this operation, it is necessary to experiment by trying different settings and prematurely removing the specimen. Pitting of the specimen surface indicates too high a current; etching of the surface indicates too low a current.

The final specimen should have a cross section like that shown in Fig. 3. The tiny perforation in the center is surrounded by a reasonably large thinned area having a thickness of 500 to 1000 angstroms; the amount of suitably thinned area can vary considerably. The edges of the 3-mm-diam circle will retain their thickness of 0.05 to 0.08 mm (0.002 to 0.003 in.) because they are clamped in the specimen holder and therefore protected during thinning.

Washing. Immediately after being removed from the electrolyte, the specimen and specimen holder should be rinsed twice in glacial acetic acid, then rinsed twice in methanol. In the second methanol rinse, the specimen is removed from the holder with tweezers, then rinsed three more times in methanol. The specimen is placed briefly on filter paper to dry and then inserted immediately into the electron microscope.

MACROETCHING OF CARBON AND ALLOY STEELS

Macroetching of carbon and alloy steels is an inspection procedure for revealing certain aspects of the quality and structure of a steel by subjecting it to the corrosive action of an etchant and examining it visually or at low magnification.

Macroetching is widely used for inspection of bars, billets, forgings, castings and other steel products. Specimens usually are selected to represent a given batch or lot of metal and are tested by etching in acid until the structural characteristics or conditions are revealed.

Equipment for Macroetching

The principal equipment for macroetching comprises a container for the etchant, provision for heating the etchant (if necessary), a ventilating system to contain and carry away corrosive fumes, and some means for washing and drying the specimen.

Etchants

Table 2 lists the etchants that are most commonly used for macroetching of carbon and alloy steels.

A 50% solution composed of one part hydrochloric acid and one part water (etchant 1 in Table 2) is the most widely used etchant and is capable of revealing a number of characteristics in carbon and alloy steels (see last column in Table 2).

Etching Procedure

The usual procedure in macroetching is to put the prepared specimens directly into the etching solution, with the surfaces to be examined either face up or vertical to permit the gas generated to escape freely. Nonuniform etching will result if

Table 2. Etchants and recommendations for macroetching of carbon and alloy steels

No.	Etchant Composition (parts listed are by volume)(a)	Etching time(b)	Surface required(c)	Purpose, or characteristic revealed
Etchants for use at 71 to 82 °C (160 to 180 °F)(b)				
1	1 part HCl, 1 part water	15 to 60 min	A or B	Segregation, porosity, hardness penetration, cracks, inclusions, dendrites, flow lines, soft spots, structure, weld examination
2	Concentrated HCl	15 to 60 min	A or B	Same as for etchant 1
3	2 parts H_2SO_4, 1 part HCl, 3 parts water	30 to 60 min	A	Same as for etchant 1
4	50 parts HCl, 7 parts H_2SO_4, 18 parts water	30 to 60 min	A	Same as for etchant 1
5	10 to 40 parts HNO_3, 4 to 10 parts HF (48%), 50 to 87 parts water	Until desired etch is obtained	B or C	Same as for etchant 1. (Ratio of HNO_3 to HF can vary as shown at left.)
6	38 parts HCl, 12 parts H_2SO_4, 50 parts water	30 to 60 min	B or C	Same as for etchant 1
7	10 parts H_2SO_4, 90 parts water	15 to 60 min	A	Sulfide and oxide inclusions
Etchants for use at room temperature				
8	2 to 25% HNO_3 in water or ethanol	5 to 30 min	B or C	Carburization and decarburization, hardness penetration, cracks, segregation, weld examination
9	2.5 g $CuCl_2 \cdot 2H_2O$, 20 g $MgCl_2 \cdot 6H_2O$, 10 mL HCl, 500 mL ethanol	Until coppery sheen appears	B or C	Phosphorus-rich areas, banding
10	50 g $(NH_4)_2S_2O_8$, 500 ml water	Swab until desired etch is obtained	C	Grain size, weld examination
11	40 g $FeCl_3$, 3 g $CuCl_2$, 40 mL HCl, 500 mL water	15 to 30 s	B or C	Dendritic structure of cast steel. (Precede use of this etchant with etch in 10% nital for 10 to 20 s.)
12	30 g $FeCl_3$, 1 g $CuCl_2$, 0.5 g $SnCl_2$, 50 mL HCl, 500 mL ethanol, 500 mL water	30 s to 2 min	C	Dendritic structure of cast steel. (Overetching deposits excessive copper, which may obscure details of structure.)
13	4 g picric acid in 100 mL methanol	3 to 5 h	C	Carbon segregation

(a) All acids listed are of concentrated strength; commercial grades ordinarily can be used instead of laboratory or reagent grades. Water or alcohol should never be poured into an acid; rather, the acid should always be poured and gradually stirred into the other liquid. (b) See text for discussion of variations in time and temperature. (c) A = saw-cut or machined surface; B = average ground surface; C = polished surface.

the specimens overlap or are racked too close together.

Heated Solutions. When heated etchants are used, best practice is to have the etching solution at the pre-established temperature before the specimens are immersed.

Temperature of Acid. For the seven etchants in Table 2 used above room temperature, an etching temperature of 71 to 82 °C (160 to 180 °F) is recommended. At temperatures on the lower side of this range, etching reaction is sufficiently vigorous to provide effective etching and yet evaporation of solution is not excessive. However, in production-control etching, temperatures on the higher side of the range are preferred, because they decrease time of etching.

Time of etching depends on the type of steel, surface condition and physical condition of the specimen. For best reproducibility of results with a large number of specimens, the time should be measured and should not vary from batch to batch. If etching time is too short, not all of the desired information will be revealed.

Specimen Preservation. After a specimen has been properly etched, it should be removed from the etchant, rinsed thoroughly under running water, scrubbed with a stiff fiber brush to remove the deposit of "smut" from the surface, rinsed again, and dried by a warm-air blast or by blotting with paper or cloth towels. A cold-water rinse usually will result in a better-appearing etched surface than that obtained when a hot-water rinse is used. Immersion in the etchant for a few minutes after scrubbing results in better definition, especially of flow lines.

As a temporary means of avoiding rust, the etched specimen may be rinsed in water, dipped in a dilute alkaline solution (such as aqueous ammonium hydroxide) to neutralize the remaining traces of acid, and washed in hot water. The residual acid can also be neutralized by dipping the specimen in a dilute solution of potassium carbonate or sodium carbonate; in addition to neutralizing the acid, these solutions also leave a film that prevents rust after drying.

TECHNIQUE FOR LOW-CARBON STEEL SHEET

Complete examination of a sample of low-carbon steel sheet requires at least three specimens: (*a*) a longitudinal cross section (one running parallel to the rolling direction), (*b*) a transverse cross section (one running perpendicular to the rolling direction), and (*c*) a surface specimen taken parallel to the rolling plane and viewed perpendicular to the rolling direction. If segregation is suspected, additional specimens from various locations must be examined; this may require sections taken at various angles to the rolling direction.

Mounting

Mounting techniques for steel sheet specimens are basically the same as those described in the article "Metallographic Methods." A thermosetting mounting material such as Bakelite is commonly used, although compression-mounting epoxy mounts provide much better edge retention. When the microstructure is likely to be affected by the curing temperature of Bakelite (150 °C, or 300 °F), or when steel sheet specimens are so thin that they may be crushed or distorted by the mounting pressure, cold mounting is used.

Clamp mounting has proved quite successful for

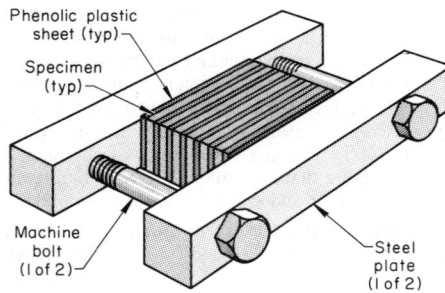

Fig. 4. Typical arrangement for clamp mounting of longitudinal or transverse sections of steel sheet during grinding and polishing

holding steel sheet specimens during preparation. Sometimes this technique is used where no mounting press is available, but more often it is used where many specimens are being prepared or where speed of preparation is important.

In clamp mounting, several specimens can be clamped in sandwich form between two plates (Fig. 4). Bleeding problems are eliminated if the clamp is tightened very securely. Some users insert thin spacers, generally of plastic, between the specimens before tightening. Edge retention with clamps is excellent.

TECHNIQUE FOR COATED STEEL SHEET

Most techniques for preparing metallographic specimens of coated steel sheet are the same as those used for preparing specimens of uncoated steel sheet.

Galvanized Steel Sheet

Care in cutting galvanized steel sheet is necessary in order to avoid delamination of cut edges at the interface of the zinc coating and the steel backing. Cutting should be done with an abrasive wheel, rather than by shearing.

Aluminum-Coated Steel Sheet

Preparation of hot dip aluminum-coated steel for metallographic study is similar to preparation of galvanized steel; both types of coating are relatively soft. Staining is not a problem with aluminum-coated steel sheet, and so water can be used as a suspending fluid and lubricant.

Stainless-Clad and Chromized Steel Sheet

Stainless-clad and chromized steel sheets, although manufactured by different processes, consist of a relatively soft low-carbon steel core sandwiched between stainless steel or chromium-rich surface layers. The surface layers are harder than the low-carbon steel base metal.

Porcelain Enameled Steel Sheet

Because of the brittleness of porcelain enamel coatings, cutting a specimen from porcelain enameled steel sheet requires the use of procedures that will minimize or prevent chipping of the friable coating. Two procedures have been used successfully, one employing a thin alumina cutoff wheel and the other a jeweler's coping saw. Low-speed diamond saws are also quite effective. When the cutoff wheel is used, the cut should be made about 3 mm (1/8 in.) away from the area to be examined. The specimen is then mounted and ground back about 3 mm to remove areas damaged in cutting. Use of a jeweler's coping

saw or a low-speed diamond saw permits cutting directly into the area to be examined.

Mounting. Specimens of porcelain enameled steel sheet may present a problem in mounting, because they cannot be bent into L or Z shapes without damaging the coating. There are, however, several techniques that may be used.

Clamp mounting is applicable to porcelain enameled specimens, and is preferred in some laboratories. Another method for mounting several specimens simultaneously is to clip them together, or to glue them together with an epoxy resin. The specimens may be interleaved with strips of soft metal such as copper or lead. Either the clipping or the gluing technique will provide a section thickness sufficient to be self-supporting when placed upright in a mold.

TECHNIQUE FOR STEEL PLATE, TUBE AND PIPE

Steel tube and pipe do not present any special problems in specimen extraction; steel plate does. Because plate may be quite thick (up to 12 in. or more), it often is necessary to examine specimens from three or more locations in order to observe true conditions.

Extraction of specimens from a thick plate is likely to require a considerable amount of cutting. Various machining methods, including sawing and hollow boring, are used to obtain the initial test pieces. During machining, care must be taken to prevent the specimen from becoming overheated, which might change the structure of the steel.

Examination of Welded Joints

Steel plate, tube and pipe are often welded, and therefore techniques for metallographic examination of welded joints are needed. Welds are usually examined at a low magnification so that a large area can be viewed. Specimens from weldments are sectioned, mounted, ground and polished in the same manner as other metallographic specimens.

TECHNIQUE FOR CARBON AND ALLOY STEELS OF MEDIUM CARBON CONTENT

This group of steels is comprised of wrought steels that have carbon contents ranging from about 0.25 to 0.50% and, for the alloy grades, total alloy contents up to about 7% (but mostly less than 3%).

The methods of preparing metallographic specimens discussed so far in the present article, and in the article on "Metallographic Methods" beginning on the first page of this section, are generally applicable to carbon and alloy steels of medium carbon content.

Many large forgings, because they are too costly to destroy or too difficult to section, are designed with an allowance of metal on the end from which test coupons can be taken — usually by sawing or hollow boring. Small forgings (up to about 5 lb) may be sectioned instead of being designed with test coupons. Test coupons are cut into specimens, usually by sawing or abrasive-wheel cutting. Often, specially designed test specimens are used.

TECHNIQUE FOR CARBON AND ALLOY STEELS OF HIGH CARBON CONTENT

In general, high-carbon steels are more easily damaged by abrasive cutting than low-carbon steels. Often, a high-carbon steel specimen should not be heated to more than 150 °C (300 °F) during cutting, and sometimes (for example, when the steel is to be examined in the as-quenched condition) the specimen should not be heated to more than about 66 °C (150 °F). Even slight overheating will temper the martensite and change its appearance, and could result in misinterpretation of the structure of the steel.

Even when cutting is done with a coolant, partial burning of the cut surface can occur. Etching of the cut specimen in nital will reveal this condition and indicate the corrective measures that must be taken to bring out the original structure of the specimen, such as removal of more stock from the surface to be examined.

Technique for Case Hardening Steel

TECHNIQUES described in the first part of this article are oriented primarily toward preparation of specimens from carburized and carbonitrided steels, although they are also applicable to preparation of specimens from cyanided steels. For preparation of specimens from nitrided steels, special procedures are required; these are described in the second part of this article.

CARBURIZED STEEL AND CARBONITRIDED STEEL

Most steels used for carburizing and carbonitriding have an initial carbon content ranging from 0.10 to 0.30%, and they may or may not contain one or more alloying elements. Sometimes, to satisfy special requirements, medium-carbon steels are carburized.

Because the carbon content of a carburized steel may range from that of the uncarburized core (0.10 to 0.30% C) to as much as 1.20% in the case, specimens are difficult to extract and prepare. Even for a steel that has been carburized but not hardened, hardness varies considerably from surface to center of a cross section, and when the carburized steel has been hardened, the hardness difference between the core and the outermost portion of the case may exceed 40 points on the HRC scale.

Sectioning. Small separate test pieces designed so as to minimize the amount of sectioning required are often used.

If the production parts are very small, one or more of them are sectioned for metallographic study.

When the specimen must be removed from a large mass of material, the same meticulous care must be used.

Mounting. The mounting techniques and precautions used for specimens of high-carbon steels are applicable to specimens of carburized steels. Overheating of specimens must be avoided to prevent changes in structure. The use of cold mounting, and of wet grinding, is required for examination for retained austenite.

Preservation of the edges of carburized cases is of the utmost importance. For edge preservation, any of three recommended methods can be used: (*a*) use of hard mounting materials or iron-filled epoxy mounting materials; (*b*) surrounding the specimen in the mount with bits of ceramic or chilled iron shot, or (*c*) nickel plating of the specimen.

Grinding and polishing techniques are the same as those described for specimens of other steels.

It is highly desirable to minimize the amount of polishing. Because of the wide differences in hardness within a single specimen, some constituents will polish in relief, and flatness of the specimen will be destroyed. Frequently, specimens that have been overpolished must be completely reworked to regain flatness.

Etching. Nital is a general-purpose etchant for routine metallographic study. It is well suited for revealing case depth, retained austenite, carbide networks, and surface decarburization.

Picral is equivalent to nital for revealing the characteristics noted above. Alkaline sodium picrate is recommended for revealing excess cementite. In addition, picral (usually 2 to 4%), or 4% picral to which 0.01% hydrochloric acid has been added, will produce sharper definition of nonmartensitic structures than will nital.

NITRIDED STEELS

Preparation of nitrided steel specimens to reveal the true microstructure requires more exacting techniques than are normally required. In the first place, the extreme hardness differential between the case and the core (case hardness is often 1000 HK or more) makes for problems. Also, the nitride concentration at the surface (white layer), which often is the primary subject of metallographic examination, is invariably partly or entirely destroyed by ordinary preparation techniques.

Test Pieces. The principal use of metallography for nitrided steels is quality control of production nitriding. Often, the workpieces being nitrided are too valuable to permit destructive testing, and so separate test pieces are prepared of the same steel with the same preliminary heat treatment as the workpieces.

Test pieces must be made with great care. Bars of convenient length and 41 mm (1⅝ in.) in diameter are quenched and tempered, unless they are already in the heat treated condition. They are then turned and ground to a diameter of 38 mm (1½ in.) so as to remove any decarburization or carburization that may have occurred in the preliminary heat treatment. The ground surface should be no rougher than 0.13 μm (5 μin.). Next, the turned and ground bars are cut into slices about 6.4 mm (¼ in.) thick. This is preferably done in a lathe, but an abrasive cutoff machine can be used if coolant is supplied copiously to prevent burning. A hole can be drilled in the completed test piece, permitting it to be hung in the furnace to prevent its being lost during nitriding.

Plating of Test Pieces. Plating of nitrided test pieces prior to sectioning is essential for protection of the edges and preservation of the white layer. Plating must be done without blasting or severe etching. Nickel plating has proved best. The thickness of the deposit is not critical; 0.05 mm (0.002 in.) is usually sufficient, but a greater thickness is not harmful. Chromium plating is unsatisfactory because the plate does not adhere well to nitrided surfaces without an unacceptable amount of surface preparation. Copper, zinc and cadmium plates are too soft.

Mounting of Specimens. Bakelite is usually preferred for mounting specimens of nitrided steel. The high mounting temperature of Bakelite, which is critical for some steels, does not affect nitriding steels. Placing chilled iron shot or hardened steel balls around the specimen is recommended

for preserving flatness of the specimen during grinding and polishing.

Grinding and polishing of nitrided steel specimens is the same as grinding and polishing of other steel specimens.

Etching. Best practice is to etch the polished specimen, immediately repolish very lightly (just enough to remove evidence of etching), and then etch again. Some metallographers continue this procedure for two or more cycles to ensure that all disturbed metal is removed and that the true structure is revealed. This practice is not needed if proper care is taken during grinding and polishing.

For routine examination, nitrided steel specimens are most often etched in nital—2% nital for microscopic examination and up to 5% nital for case-depth measurements at low magnification. Nital reveals the white layer, the depth of case, and the structure of nitrided alloy steels such as AMS 6470, 4140 and H13.

Technique for Steel Castings

SPECIMENS from carbon and alloy steel castings are prepared by the same techniques as those used for preparing specimens of wrought carbon and alloy steels of the same composition.

EXTRACTION OF SPECIMENS

In a single steel casting, section thickness can vary from a fraction of an inch to several inches. These differences in section thickness will result in different cooling rates, and thus in different microstructures, within the casting. This is true both for as-cast and for heat treated castings. Therefore, for complete examination of a casting, several specimens may have to be extracted.

Steel castings, both as-cast and heat treated, are usually soft enough to permit sawing or hollow boring for initial extraction of test pieces, after which the oversize pieces are reduced to specimen size by sawing or abrasive-wheel cutting. If the casting is hard, abrasive-wheel cutting is used for all operations. Precautions must be taken to avoid overheating during cutting.

MOUNTING

Bakelite is often used for mounting specimens. The microstructures of most steel castings are not affected by the high thermosetting temperature of Bakelite.

GRINDING AND POLISHING FOR PRESERVATION OF INCLUSIONS

Frequently, steel castings are examined for the presence and identification of inclusions; thus, methods of inclusion preservation are essential. Initial specimen flatness frequently is obtained by use of a belt grinder for the coarsest grit size— e.g., 120-grit SiC. Next, the specimen is ground sequentially on water-cooled SiC paper using rotating wheels in the sequence 240, 320, 400 and 600 grit. The specimen is then rough polished on a nylon or canvas polishing cloth charged with 3-to-9-μm diamond paste. This often is followed by polishing with 1-μm diamond paste on a low or medium nap cloth. For either diamond polishing step, the proper degree of lubrication and charging is required to promote cutting and to maximize inclusion retention. Final polishing is done on a medium-nap synthetic cloth charged with a slurry of water containing 0.05-μm alumina. Excessive polishing must be avoided, to prevent extraction of inclusions. Examination for inclusions is done prior to etching.

ETCHING

Nital is the etchant most often used for specimens from steel castings.

Picral is sometimes used for etching specimens from steel castings, especially those from castings with carbon contents of more than 0.30%. Often, carbide structures are resolved better with picral than with nital.

Electrolytic etching is only rarely employed for specimens from carbon or low-alloy steel castings.

Technique for Cast Irons

THE PRESENCE of graphite in gray, malleable and ductile cast irons prevents the use of many of the metallographic preparation techniques normally used for other metals. The soft and friable nature of the graphite phase, in combination with the relatively hard matrix, presents difficulties during preparation not normally experienced with other ferrous metals. Techniques that will promote a scratch-free and distortion-free surface on the matrix may not allow retention and subsequent polishing of the graphite phase. Similarly, techniques for polishing soft and friable phases such as graphite do not easily lend themselves to successful preparation of the harder matrix. The peculiar friability of graphite in cast iron results in the phase being torn out wholly, or in part, leaving cavities that are either opened up or burnished over, depending on the preparation procedure adopted. As a result, the shape and size of the graphite may not be truly represented and an accurate assessment of the microstructure cannot be made. To obtain a scratch-free and distortion-free matrix containing well-preserved and polished graphite, special preparation procedures specific to cast irons containing graphite have to be adopted. The preparation of white cast iron microspecimens does not have the limitations imposed by the presence of graphite, and conventional techniques may be used.

SECTIONING

Specimens can be obtained from iron castings using conventional cutting techniques such as hand hacksawing, mechanical hacksawing or, for hard materials, abrasive cutoff. Sawing produces a rough-cut face and deforms the base metal to a greater extent than does abrasive cutting. When specimens are obtained by sawing, a lubricant should be used, and grinding must commence with coarser-grit SiC papers than would be needed if the specimens were obtained by abrasive cutting. In cutting of hard materials, such as white irons, it is necessary to use a cutoff machine employing thin, bonded, silicon carbide or aluminum oxide disks, and overheating must be avoided. Cutting under water or with copious water flow is recommended to minimize heat effects.

MOUNTING

Specimens that are difficult to handle should be mounted, using the hot compression mounting technique. In many laboratories, it may be convenient to mount all specimens, and mounting is often essential if polishing is to be done on automatic equipment. A suitable thermosetting phenolic plastic should be used as the mounting medium, because this provides sufficient adhesion and edge support to the specimen, and minimizes the rounding of edges that results in loss of sharpness during examination.

ROUGH GRINDING

After sectioning and mounting, the specimen should be rough ground to remove distorted surface layers and to give an essentially flat surface. Coarse and fine grinding wheels, mounted on a double-end grinding machine, can be used for this purpose. (Belt grinders can also be used, but serious rounding of the edges of the specimen is likely to result; for this reason, use of belt grinders is not recommended.)

In grinding, the specimen should not be held in one position but should be moved backward and forward, from the edges toward the center, on the side of the wheel. Wet grinding is preferred for best retention of graphite, minimization of distortion and improved cutting.

ABRASION (FINE GRINDING)

After being rough ground, the specimen is abraded (fine ground) to a condition suitable for polishing. The marks from rough grinding are replaced by successively finer and finer scratches from a series of waterproof silicon carbide abrasive papers. The correct sequence for optimum results is 120, 280, 400 and 600-grit papers. Water is used as the lubricant and to remove abrasion debris from the paper. Many metallographers use a second 600-grit grinding operation without water (i.e., dry) to maximize graphite retention.

When the manual method is used with stationary grinding papers, the specimen is held with one or both hands and rubbed unidirectionally against the coarsest grade of abrasive paper. Heavy pressure should be applied on the forward stroke with relaxation (but not complete removal of pressure) on the return. This procedure is repeated until the scratches are parallel and in one direction. Further processing on subsequent papers is carried out with the specimen rotated at right angles to the previous set of scratches.

If grinding is performed on rotating wheels the specimen is held against the paper, being careful not to rock the specimen. It is moved from edge to center as the wheel rotates generally at 150 to 300 rpm. After grinding with each paper, the specimen is reoriented 45 to 90° from the prior direction and ground with the next paper. When automatic devices are used, a multidirectional scratch pattern is produced, and thus specimen reorientation is not required. Automatic devices provide best flatness and graphite retention.

Abrasion on the final paper is the most important stage in the preparation procedure, because it is here that the graphite phase may be damaged. For most gray irons containing fine graphite structures, white irons, and alloy irons, careful abrasion on 600-grit paper is normally sufficient before polishing. A two-stage 600-grit

Note enlarged cavities where the graphite phase has been torn out. Not polished, not etched; magnification, 100×.

Fig. 5. Gray iron, containing coarse graphite flakes, that was incorrectly prepared at the abrasion (final grinding) stage

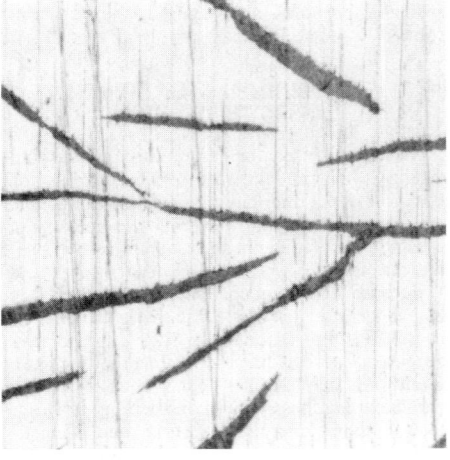

Some burnishing of graphite flakes occurred, and their apparent width was reduced. Not polished, not etched; magnification, 100×.

Fig. 6. Same field as in Fig. 5 after correct abrasion, showing well-preserved graphite flakes

grinding procedure generally is recommended—the first stage with water lubrication, and the second stage dry.

Figure 5 shows a coarse-flake gray iron that has been abraded on a poor-quality abrasive paper. The graphite flakes have been torn out and the cavities enlarged. This effect can also be due to other factors, such as grinding with worn or loaded papers, or grinding in the same direction with each paper grade.

Figure 6 shows essentially the same field as that in Fig. 5, but after the specimen has undergone correct abrasion. The graphite flakes are well preserved, and the matrix is uniformly scratched.

Figures 7 and 8 show, at high magnification, a single graphite spherulite in a ductile cast iron in the incorrectly and correctly abraded states.

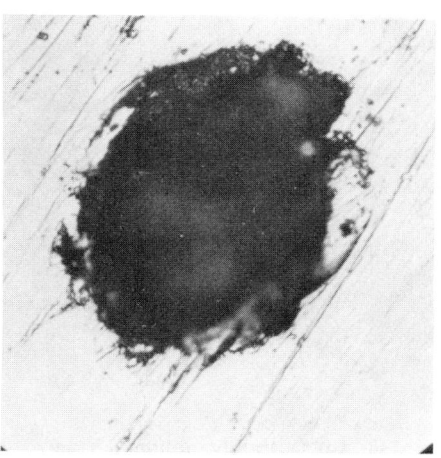

The graphite phase has been torn away. Not polished, not etched; magnification, 800×.

Fig. 7. Site of graphite spherulite in ductile iron, showing incorrect abrasion (final grinding)

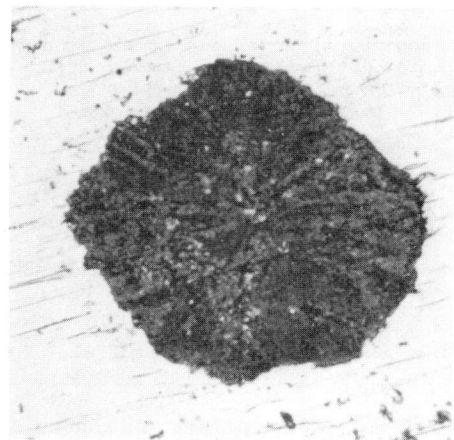

Note that the graphite spherulite has been well preserved. As polished (not etched); magnification, 800×.

Fig. 8. Same field as in Fig. 7, but after correct abrasion (final grinding)

An incorrectly abraded ductile iron can also be seen in Fig. 9, which shows burnished cavities where the graphite phase has been torn out. The same ductile iron as in Fig. 9, but after correct abrasion, is shown in Fig. 10, and the same ductile iron as in Fig. 7 and 8, after final polishing, is shown in Fig. 11.

POLISHING

For cast iron, polishing with diamond-impregnated pastes has been found to be the most satisfactory. Polishing in two stages is recommended. Coarse polishing frequently is performed with 6-μm diamond, and fine polishing with 1-μm diamond. In some instances, finer diamond pastes can be employed. The paste is supplied in syringe applicators from which it is squeezed in the form of a thin ribbon. The action of diamond-impregnated carrier pastes produces rapid polishing with minimum deformation.

Note burnished cavities where the graphite phase has been torn out. Not polished, not etched; magnification, 300×.

Fig. 9. Ductile iron that was incorrectly prepared at the abrasion (final grinding) stage

Note the well-preserved coarse graphite spherulites. Not polished, not etched; magnification, 300×.

Fig. 10. Same field as in Fig. 9, but after correct abrasion (final grinding)

Polishing is an important phase of the preparation sequence, because properly ground specimens can be damaged by poor polishing technique. Use of inadequate lubricant, a worn cloth, an inadequately charged cloth, or excessive pressure or wheel speed can degrade graphite retention, enlarge the apparent size of the graphite, produce scratched surfaces and reduce interface sharpness.

For gray irons, in order to polish the graphite and remove distorted and smeared metal, alternate polishing and etching may be practiced.

The type of polishing cloth used with the diamond paste has a significant effect on retention of graphite. Polishing cloths with a high nap or pile are extremely effective in removing scratches

Etched in 4% picral; magnification, 800×. Plane polarized light has been used to enhance detail.

Fig. 11. Same field as in Fig. 7 and 8, but after final polishing on 0.25-μm diamond paste

from previous operations, but they have an adverse effect on retention of graphite. Napless or short-pile cloths are more suitable for retaining graphite but suffer from the limitation that they are less efficient in removing abrasion scratches, and sometimes cause further scratching of the specimen surface. The use of a synthetic velvet cloth is a reasonable compromise between these two extremes; it gives excellent polishing rates, removes abrasion scratches, and does not dislodge the graphite phase. With two-stage polishing, a napless cloth, such as canvas, is used for the rough polishing stage while a low- or medium-nap cloth is used for the final diamond polishing stage. A prerequisite for polishing of coarse graphite structures, and particularly ferritic blackheart malleable irons and ductile irons, is that the cloth be "run in" on a prepared specimen. After about 3 or 4 min of prepolishing, the synthetic velvet cloth acts much like a napless cloth and is suitable for all types of coarse-graphite irons.

After polishing, the lubricant may be removed from the specimen surface by swabbing with an alcohol from a squeeze-type wash bottle. The specimen is held in the fingers and lightly stroked with the thumb until the lubricant is removed. If cavities are present, ultrasonic cleaning may be needed. The specimen is dried by flooding it with high-purity alcohol and then holding it under the hot air blast from a specimen dryer.

Figure 12 shows the same field as in Fig. 6 after a single polish on 1-μm diamond paste. Partial dislodgement of the graphite flakes is evident.

An incorrectly polished gray iron specimen is shown in Fig. 13. The graphite flakes are torn out and damaged because the specimen was polished on an unprepared synthetic velvet cloth that had been freshly loaded with 0.25-μm diamond paste. For comparison, the same gray iron is shown in Fig. 14 after it had been lightly etched in picral and correctly polished on a synthetic velvet cloth that had been suitably prepared.

ETCHING

If an assessment of graphite shape, size and distribution is required, this is best done when

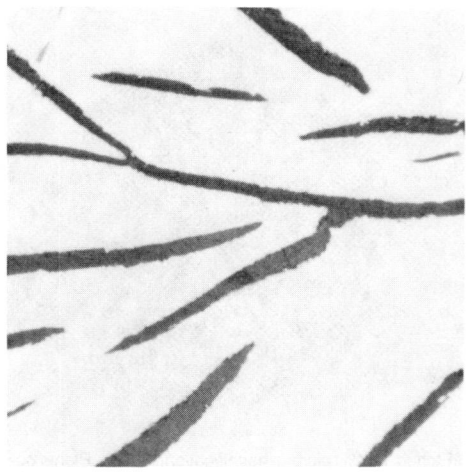

Note fine scratches on the matrix, and that the graphite flakes are poorly polished. As first polished; magnification, 100×.

Fig. 12. Same field as in Fig. 6 after a single etch in 4% picral and a 2-min polish with 1-μm diamond paste on synthetic velvet cloth

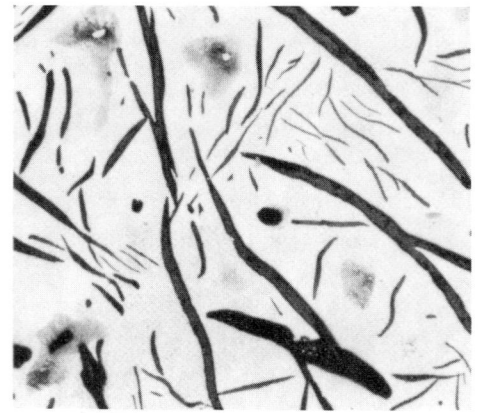

Graphite phase is poorly preserved; shallow flakes have been torn out. Etched in 4% picral; magnification, 60×.

Fig. 13. Incorrectly polished gray iron, showing a ferritic matrix with small areas of pearlite. See Fig. 14.

Etched in 4% picral; magnification, 60×.

Fig. 14. Correctly polished gray iron, showing well-preserved graphite flakes in a predominantly ferritic matrix containing small areas of pearlite. Compare with Fig. 13.

the specimen is in the as-polished (not etched) condition. Examination for porosity or the presence of nonmetallic inclusions may also be more easily achieved when the specimen is polished but not etched. However, observation of other important structural details can be made only after the specimen has been suitably etched. This is normally done by holding the specimen in a pair of tongs and immersing it, with intermittent agitation, in the desired reagent for a suitable period of time.

The production and interpretation of correctly etched cast iron structures is often dependent on practice and experience, because consideration usually has to be given to both the characteristics of the reagents used and the nature of the investigation. Table 3 gives some of the more common etching reagents employed for cast iron, together with techniques for use and their principal applications. Most cast irons can be satisfactorily etched in either 4% picral or 2% nital.

Picral is the best general-purpose etchant for all predominantly pearlitic gray, malleable and ductile cast irons. It gives a clean, uniform etch to both coarse and fine pearlitic structures. Pearlite should be etched until it is a medium-to-light brown in color, so that graphite present in the

Table 4. Etchants for macroscopic examination of cast irons

Etchant	Composition	Etching technique	Application
Stead's reagent	10 g cupric chloride, 40 g magnesium chloride, 20 mL hydrochloric acid, 1000 mL ethanol(a)	Immersion for up to 3 h	Used to reveal the eutectic cell number in gray cast irons
Rapid cell-etching reagent	10 g cupric chloride, 50 mL water, 100 mL hydrochloric acid	Dip etch for about 60 s	As above, but results are less distinct
Ammonium persulfate	10 g ammonium persulfate, 100 mL water, few drops H_2SO_4 (conc)(b)	Immersion and swabbing	Reveals carbide and phosphide distribution
Nital	5 or 10% nitric acid, 95 or 90% ethanol	Dip etch for up to 3 min	Used to reveal macrostructure in white irons
4% picral	4% picric acid, 96% ethanol	Dip etch for up to 3 min	Used to reveal macrostructure in white irons

(a) Dissolve cupric chloride in a minimum quantity of hot water (10-15 mL); add magnesium chloride and dissolve; add ethanol, then hydrochloric acid. (b) Add H_2SO_4 just before use.

Table 3. Etchants for microscopic examination of cast irons

Etchant	Composition	Etching technique	Applications
Picral	4% picric acid, 96% ethanol	Dip etch for 2-10 s	General-purpose etching of all pearlitic gray, malleable and ductile cast irons. Best etchant for pearlite. Etches some austenitic cast irons, Ni-Hard and acicular irons.
Nital, 5%	5% nitric acid, 95% ethanol	(1) Dip etch for 2-10 s	(1) General-purpose etching of all ferritic gray, malleable and ductile cast irons. Etches grain boundaries. Etches some austenitic irons and irons containing martensite.
		(2) Electrolytic etch(a)	(2) High-chromium irons
Nital, 2%	2% nitric acid, 98% ethanol	Dip etch for 2-10 s	Observation of ferritic grain boundaries at high magnification
Ferric chloride	10 g ferric chloride, 100 mL water	Dip etch for 3-20 s	Austenitic cast irons
Mixed acid in glycerol	10 mL HNO_3, 20 mL HF, 40 mL glycerol	Dip etch for 10-40 s	High-silicon irons (14 to 16% Si)
Vilella's reagent	1 vol HNO_3, 2 vol HCl, 3 vol glycerol	Dip etch for up to 20 s	High-chromium irons
Potassium ferricyanide	10% alkaline aqueous solution of potassium ferricyanide	Dip etch for 5-30 s, in etchant at 50 °C (122 °F).	High-chromium irons
Murakami's reagent	10 g KOH, 10 g $K_3Fe(CN)_6$	(1) Dip etch for 2-3 min	(1) 30% chromium irons
		(2) Dip etch for 10-30 s	(2) High-phosphorus irons, to distinguish between iron phosphide and iron carbide
Alkaline sodium picrate	2 g picric acid, 25 g NaOH, 100 mL water; warm to dissolve	(1) Dip etch for 10 s to 2 min at boiling point	(1) Blackens cementite
		(2) Electrolytic etch(b)	(2) Blackens cementite

(a) Specimen is anode; platinum cathode. Current density, 0.13 to 0.31 A/cm^2 (0.5 to 2.0 $A/in.^2$) for up to 2 min. (b) Specimen is anode; stainless steel cathode. Current density, 0.13 to 0.31 A/cm^2 (0.5 to 2.0 $A/in.^2$) for up to 2 min in cold solution.

structure can be clearly seen. Picral does not damage graphite but gives an added sharpness to the outline of the flakes or nodules.

MACROEXAMINATION

Macroetching of cast iron is used to reveal such features as chill formation and segregation. Table 4 gives the more commonly used macroetchants for cast iron. Usually, picral, nital and ammonium persulfate are the most suitable reagents because they do not cause extensive staining. The specimens should be immersed in the solution and swabbed with cotton wool. After removal from the etchant they are flooded with an alcohol, rinsed with acetone, and dried with hot air, as described previously.

Eutectic cells in gray irons are generally revealed by etching in Stead's reagent. The reagent selectively darkens low-phosphorus regions in the iron, leaving high-phosphorus regions (normally segregated at cell boundaries) unattacked and light. The specimens can be etched by immersion in a stock solution for periods of up to 3 h. The time of immersion is found by trial but, in general, irons containing coarse cell structures take longer than irons containing fine cell structures.

Technique for Tool Materials

METALLOGRAPHIC techniques for tool steels, cast cobalt alloy tool materials, cemented carbides, and ceramic tool materials are described here.

TOOL STEELS

The procedures used for specimen preparation and macroscopic and microscopic examination of tool steels are essentially the same as those used for carbon and alloy steels.

Macroetching. A solution containing one part hydrochloric acid and one part water at 71 to 82 °C (160 to 180 °F) is the etchant most often used for revealing segregation, porosity, cracks, inclusions and flow lines. This etchant also can be used for revealing hardness penetration and soft spots.

A solution of nitric acid (2 to 25% in water or 2 to 5% in ethanol) is often favored for showing depth of hardened zone, carburized case depth, and decarburized area; it is used to a lesser extent for revealing defects.

Preparation of Specimens for Microscopy. The test piece is obtained from the larger mass by breaking, hacksawing, or abrasive cutoff. Use of abrasive cutoff for sectioning tool steels requires extreme care, to prevent overheating. In sectioning of pieces that have been hardened but not tempered, it is inadvisable to permit them to exceed a temperature of about 66 °C (150 °F). Steels that contain large amounts of massive carbide particles, such as D7 and T15, are very difficult to cut, even when in the annealed condition, and require the ultimate in care when abrasive cutoff is used. A soft grade of cutoff wheel, a copious supply of coolant, and slow cutting speed are mandatory.

Whenever possible, it is advisable to remove the specimen in such a manner that the cut surface is not the surface to be examined.

Table 5. Special-purpose etchants used for etching of tool steels

Feature	Etchant	Comments
Chromium and complex (M_6C) carbides	4 g NaOH, 10 g $KMnO_4$, 85 mL water	Immerse specimen in boiling solution
Vanadium (MC) carbide	1% chromic acid	Electrolytically etch at 2 V
Grain size	10 g picric acid, 5 drops HCl, 100 mL ethanol	Immerse specimen at room temperature
Ferrite	Kalling's reagent	Immerse or swab specimen

Mounting techniques are the same for tool steels as for other steels. Bakelite or epoxies are most often used as the mounting materials. Bakelite is a satisfactory material except for mounting hardened but untempered specimens, which should not be subjected to the thermosetting temperatures required for Bakelite. A mounting material, such as an epoxy, that hardens at room temperature, or at no more than about 66 °C (150 °F), is preferred for most hardened but untempered tool steels.

Grinding and polishing techniques are basically the same for tool steel specimens as for specimens of carbon and alloy steels. It is important to keep polishing time as short as possible, especially for steels such as D7 and T15. Excessive polishing produces a relief effect, because of the carbide particles, and may pull out or drag inclusions.

Microetching With Nital. For routine metallographic examination, tool steel specimens are most often etched in 2 to 5% nital (2 to 5 mL of nitric acid to ethanol or methanol to make 100 mL of etchant). Nital is a general-purpose etchant and can be used on all tool steels in all conditions—annealed, quenched, and quenched-and-tempered.

Because of the many variables which affect etching time, it is impractical to prescribe specific etching times. Common practice is to immerse the specimen in the etchant with the polished side up, agitate it gently, and observe it carefully. The surface of the specimen first changes from a polished appearance to light brown. If etching proceeds too far, the specimen turns dark gray. Overetching causes a loss of detail in the matrix and shows the carbides in relief. Overetching may result in removal of the carbide particles. Very often, alternate etching and microscopic observation are essential to obtain the ideal etch for the magnification to be used.

Etching of quenched but untempered steels for grain-size observation requires 4 or 5 min in 2% nital. For this purpose, etching by specific timing is often practiced.

Microetching With Picral. Picral, a 4% solution of picric acid in ethanol, is applicable for etching all grades of tool steel in all conditions. Picral may be used when the quality required for the micrograph, in terms of structural detail, is higher than that required for routine work. Also, various carbide structures are shown better if the specimen is etched with picral rather than with nital.

Techniques for etching with picral are essentially the same as those for etching with nital.

Special-Purpose Etchants. Several special-purpose etchants are frequently used for etching of tool steels. They are used for greater distinction of specific microconstituents (see Table 5).

CAST COBALT ALLOY TOOL MATERIALS

The cast cobalt alloy tool materials are a unique family of proprietary alloys that are hard in the as-cast condition and are used without further heat treatment. Most of them contain 38 to 50% Co, 25 to 32% Cr, 4 to 25% W, and 1 to 4% C. Some of these alloys contain up to 4% V, and others have up to about 5.5% Nb and 3.5% Fe.

Microstructure. The most noticeable constituent is large carbides that appear in elongated or acicular form and have been identified as chromium carbide Cr_7C_3, in which some of the chromium may be replaced by cobalt or tungsten. M_6C and $M_{23}C_6$ carbides have been identified in some alloys. The matrix usually consists of various binary and ternary eutectics containing all constituents of the alloy.

Specimen Preparation. Sectioning is done by abrasive cutoff. The greatest of care must be used to prevent damage from overheating.

If Bakelite is used as the mounting material, the heat involved in curing is not sufficient to alter the structure of the alloys.

Grinding and polishing are done by the same procedures as those used for grinding and polishing carbon and alloy steels.

Etching. Etching procedures and etchants for cast cobalt alloy tool materials differ among various laboratories. The two procedures that follow are the most commonly used:

1. Activate the surface of the polished specimen with a short (about 2-s) electrolytic etch in 2% chromic acid, then immerse the specimen for 10 s in an alkaline permanganate solution. (The alkaline permanganate solution is prepared immediately before use by mixing equal parts of a saturated solution of potassium permanganate and an 8% solution of sodium hydroxide. It should be discarded immediately after use.) After etching, wash the specimen in running water and then in ethanol or methanol, and dry it in an air blast. The etched specimen must not be rubbed.
2. Etch the specimen for about a minute (the exact time is determined by trial and error) in a solution of six parts of concentrated hydrochloric acid and one part of 30% hydrogen peroxide. Etching must be done under a hood.

CEMENTED CARBIDES

Both low and high magnifications are used for examining structures in cemented carbides.

Examination at Low Magnifications. Macroscopic examination of cemented carbides is usually practiced with a low-power microscope at 20 or 30 diameters for detecting pits, pressing flaws, contamination, segregation, free carbon, and carbon deficiency (eta phase).

Free carbon is characterized on an as-sintered surface or a fracture surface by clustered dark spots. Often, a specimen with excessive free carbon has an as-sintered surface that feels slippery to the touch.

Carbon deficiency is manifested by the appearance of shiny stringers, dots and clusters which turn dark when etched with Murakami's reagent.

Because defects such as pits and pressing flaws lower the strength of the material, fracture will initiate in them and propagate through them. For this reason, they are easy to identify on a fracture surface.

Preparation of Specimens for Microscopy. The specimen is extracted from a larger mass by breaking or by abrasive-wheel cutting. Cutting with a low-speed diamond saw is widely practiced because it produces little damage and yields a surface requiring a minimum of grinding. Because cemented carbides are generally difficult to polish (due to the structure and hardness), it is recommended that the size of the specimen be kept small. Specimens having a face area no larger than 160 mm² (¼ in.²) are preferred.

The mounting materials and techniques that are applicable to steels are applicable also to cemented carbides. Bakelite is the mounting material most often used.

Because cemented carbides are very hard, hand grinding and polishing is quite tedious and use of automatic devices is recommended. If the specimen has been cut with a low-speed diamond saw, a high-quality surface is obtained which requires a minimum of grinding. Although SiC paper can be used, metal- or resin-bonded diamond disks are preferred. These are available with nominal diamond sizes of 45, 30 and 15 μm. If the surface to be polished has been sectioned by fracturing, start with the coarsest diamond size. If the surface was cut with a low-speed diamond saw, grinding can commence with the 30-μm diamond disk. Grinding is done with moderately high pressure, copious water cooling, and a speed of 100 to 200 rpm.

Polishing generally is conducted in several stages using diamond paste on napless or low-nap cloths. Nylon is frequently used. Diamond charging is heavier than for polishing of steels. The cloth should be moistened with extender lubricant and not permitted to dry out. Polishing with 6- and 1-μm diamond paste is recommended. Again, the pressure is somewhat higher than that used for steels, and automatic devices are preferred. For most work, final polishing with 1-μm diamond is adequate. A short 30-s final hand polish with colloidal silica may be desirable for best results. This is performed using a medium-nap synthetic cloth attached to a glass plate (i.e., stationary) using a figure-eight motion. This step will reveal the interfaces between the cobalt finder and the carbides with a minimum of relief.

Etching. The most commonly used etchant for cemented carbides is alkaline potassium ferricyanide, known as Murakami's reagent (10 g potassium ferricyanide, 10 g potassium hydroxide or 7 g sodium hydroxide, and 100 mL water. A short etch (2 to 10 s) colors only eta phase. Longer etching times attack the eta phase but reveal the phase boundaries. Etching is performed at room temperature.

To preferentially color the cobalt binder phase, etch for 1 to 5 min at room temperature in hydrochloric acid saturated with ferric chloride (Chaporova's reagent). The cobalt binder phase also may be revealed preferentially by heat tinting the polished specimen (unmounted) at 315 °C (600 °F) for 5 min in a laboratory furnace (Ref 1). This colors the cobalt rust brown. Higher temperatures will color the carbide grains. Considerable use has also been made of the Pepperhoff vacuum-deposition interference-film technique (Ref 2).

Etch Polishing. The surface of the specimen is first prepared by mechanically polishing with 6-μm diamond and is then subjected to vibratory polishing. The vibratory polisher is charged with an aqueous slurry of 0.3-μm alpha alumina and an etchant composed of 20 mL of a 5% solution of sodium hydroxide, 30 mL of a 5% solution of potassium ferricyanide, and 5 to 10 drops of 30% hydrogen peroxide. From 50 to 100 mL of this etchant is sufficient to charge a vibratory polisher having a 200-mm (8-in.) diam bowl. An etch-polishing time of 3 to 4 h is typical.

Apparent porosity is a term construed to include all microstructural features observed at a magnification of 200 diameters on properly prepared, unetched surfaces of cemented carbides. Porosity rating is made by comparing the observed field with a porosity chart such as that provided by ASTM B276. Free carbon can be confirmed as C-type apparent porosity at 200 diameters. Segregation, contamination and binder laking can be observed also.

Microstructure. Examination of the microconstituents of cemented carbides is best done at magnification of 1000 to 1500 diameters after etching by immersion in Murakami's reagent at room temperature. Tungsten carbide is revealed as gray angular crystals; cobalt, unattacked by the etch, appears as a white or light-cream matrix. If tantalum carbide, titanium carbide or solid-solution phases are present, they will appear brownish or gold in color, and will stand above the polished surface when etched. The different phases are distinguished by slight differences in color. For careful examination of cemented carbide materials containing these phases, the etched surface should be gently wiped with a soft tissue; a drop of immersion oil (oil used with oil-immersion objectives) placed on the etched surface before wiping will preserve the integrity of the etch during wiping.

Because each producer manufactures cemented carbide materials for performance under given conditions according to his own specifications as to microstructure, considerable differences will be observed in the microstructures of materials from different producers. Thus, the value of a microscopic examination, beyond evaluation of apparent porosity and examination for extraneous phases, will depend on the correlation developed between a specific microstructure and performance.

To estimate the grain size of cemented carbide materials, the microstructure is compared with some set of standard charts or micrographs. A visual-comparison procedure for evaluating the apparent grain size of cemented tungsten carbide materials is described in ASTM B390.

CERAMIC TOOL MATERIALS

The steps used in preparing and polishing specimens from ceramic (oxide-base) tool materials for microscopic examination are essentially the same as those for preparing specimens from cemented carbides. Because of the absence of a ductile second phase in most ceramic tool materials, extreme care is required during polishing to prevent pullout of hard, brittle particles. Specimens to be etched are not mounted in plastic, because of the high temperatures required for etching.

The polished, unetched specimen is viewed at magnifications of 200 and 750 diameters to determine porosity, flaws and segregation. Because complete freedom from pullouts is almost impossible to attain, density measurement is used as the primary means of determining porosity,

with microscopic examination employed for confirmation.

If etching is desired, the polished, unmounted specimen is immersed in 85% H_3PO_4 at 180 to 250 °C (360 to 480 °F) for 8 to 10 min. The etchant preferentially attacks second phases and grain boundaries, and this must be taken into account in an interpretation of the etched microstructure.

REFERENCES

1. Color Shows Up the Unknown in Metallography, by J. H. Powers and W. J. Loach: *Steel*, Vol 133, Oct 15, 1953, p 93-96
2. Revealing of Hard Metal Structures by Interference Vapor-Deposition, by W. Peter, E. Kohlhaas and O. Jung: *Prakt Metallographie*, Vol 4, June 1967, p 284-290

Technique for Ferrous Powder Metallurgy Alloys

THE MAJOR difference between parts made from metal powder and those made from wrought metal is in the amount of porosity (voids). Powder metallurgy parts are seldom compacted to full density, with the result that there are invariably some pores, the number and size of which vary with density.

The presence of pores is likely to cause difficulty in the preparation of metallographic specimens. First, there is the possibility that grinding debris, abrasive or particles of other metals will become embedded in the pores during cutting, grinding and polishing. Also, moisture may become entrapped in the pores and may bleed out during etching, resulting in staining. Another possibility is the entrapment of etchant, which may bleed out and stain the surface.

SPECIMEN SELECTION AND SECTIONING

Selection of representative specimens of powder metallurgy alloys requires careful consideration, primarily because porosity is seldom uniform. Density (and therefore porosity) may vary from surface to center of a relatively thick part, thus necessitating examination of the entire cross section. Also, when a powder metallurgy part is subjected to a furnace atmosphere, as in sintering or heat treatment, microstructural characteristics may be different within a given cross section.

WASHING

Following sectioning, specimens should be washed to remove foreign material that may have infiltrated the pores. Two devices used for removing contaminants from pores are an extractor-condenser and an ultrasonic cleaner.

Extractor-Condenser. The use of an extractor-condenser is the more efficient and the less expensive of the two methods of washing. The extractor-condenser consists of a flask, a siphon cup, and a condensing-coil unit that fits on the top of the flask (see Fig. 15). A solvent, such as acetone, is placed in the flask, and the specimens to be washed are placed in the siphon cup. A cold-water line is connected to the condensing coil. The flask is heated to the boiling temperature of the solvent. The solvent evaporates and, when the vapor comes in contact with the cold con-

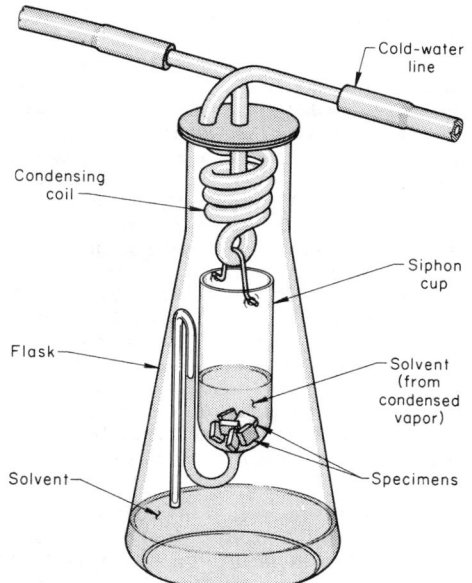

See text for description of operation.

Fig. 15. Extractor-condenser used for washing of P/M specimens to remove contaminants from pores

densing coil, it condenses and drips into the siphon cup. When the siphon cup becomes filled to a predetermined level, it empties, returning the solvent to the flask. Recycling allows a continuous flow of clear solvent over the specimen.

Ultrasonic Cleaner. The ultrasonic cleaner used for washing powder metallurgy specimens consists of a power supply and a small tank, which holds a solvent bath.

IMPREGNATION

After the specimen has been washed, some means of preventing abrasives, water and etchant from entering the pores should be provided. For specimens that are to be ground and polished without mounting or that are to be mounted using heat and pressure (as required for Bakelite), wax impregnation is recommended. Specimens are impregnated by soaking them in a molten synthetic wax at 175 °C (350 °F) for 2 to 4 h. After cooling and removal of the surface wax, the specimens are ready for mounting, or for grinding and polishing without mounting.

Vacuum impregnation of epoxy is also commonly performed. Impregnation is done after grinding. When the epoxy is dry, the surface is lightly reground with the finest grit used before polishing.

MOUNTING

Mounting of specimens is always preferable, and it is required for very small specimens and for edge preservation. The materials and techniques used for mounting powder metallurgy specimens are similar to those for mounting specimens of other metals.

GRINDING

Procedures for rough grinding of powder metallurgy specimens are generally the same as for specimens of other metals.

POLISHING

Rough polishing is best accomplished on a napless cloth, such as nylon, charged with abrasive—diamond paste, aluminum oxide or magnesium oxide. Abrasive-particle size is usually 1 to 0.3 μm. Specimens should be rough polished only as much as is required to remove marks from the last operation. Excessive relief introduced during polishing will enlarge the porosity and give a false impression of its size and amount.

Fine polishing of powder metallurgy alloys is generally the same as for other metals.

ETCHING

Ferrous powder metallurgy alloys are almost always etched by immersion in 2% nital, although picral is sometimes used and is preferred by some metallographers for revealing certain structures. Staining problems usually are easier to control after etching with nital than with picral.

Technique for Stainless Steel Casting Alloys

STAINLESS STEEL casting alloys referred to in this article are widely used to resist corrosion by aqueous solutions at or near room temperature, and by hot gases and high-boiling-point liquids at elevated temperatures up to 650 °C (1200 °F). All these alloys, which are identified by ACI designations, contain more than 11% chromium and various amounts of nickel up to 30%.

Microstructures of stainless steel casting alloys generally are similar to those of their wrought counterparts except that the segregation inherent in the casting process is not altered. Hence, the microstructures usually are coarser and more diverse, and porosity or shrinkage cavities may be observed.

Grinding and Polishing. The techniques and equipment used for mounting, grinding and polishing specimens of stainless steel casting alloys for metallographic examination are essentially the same as those used for wrought stainless steels. Coarse and fine grinding require the successive use of increasingly finer abrasive papers. Coarse grinding is generally done on SiC papers with grit sizes of 80, 100 and 120 mesh, and is followed by fine grinding on papers with grit sizes of 240, 320, 400 and 600 mesh. Rough polishing entails the successive use of rotating disks covered with napless cloth impregnated with pastes of diamond dust, or slurries of alumina, with particle sizes of 9, 6, 3 and 1 μm. Rotating wheels covered with medium-nap cloths and slurries of alumina of 0.3 and 0.05-μm particle size are used for fine polishing.

Throughout grinding and polishing, care should be exercised to minimize the occurrence of disturbed metal.

Etching. Among the etchants used to delineate the microstructure of stainless steel casting alloys are those for which compositions are given in the upper part of Table 6. As shown in the lower part of Table 6, the alloys are etched to reveal general microstructure or to emphasize a microconstituent of major interest, such as ferrite or carbide. The presence of ferrite in the CF-type alloys is of practical significance, because small amounts of ferrite improve resistance to intergranular penetration.

Table 6. Compositions and applications of etchants for stainless steel casting alloys

Etchant No. and name	Composition
1 Oxalic acid (electrolytic, 6 V)	10 g oxalic acid, 100 mL water
2 Vilella's reagent	5 mL HCl, 1 g picric acid, 100 mL ethanol (95%) or methanol (95%)
3 Kalling's reagent 2	100 mL HCl, 5 g CuCl₂, 100 mL ethanol (95%)
4 Murakami's reagent (unheated)	1 to 4 g K₃Fe(CN)₆, 10 g KOH (or 7 g NaOH), 100 mL water
5 Murakami's reagent (boiling)	Same composition as etchant 4, above, but heated to boiling temperature for use
6 Chromic acid (electrolytic, 6 V)	10 g CrO₃, 100 mL water
7 10N potassium hydroxide (electrolytic, 6 V)	560 g KOH diluted with distilled water to a volume of 1000 mL
8 HCl, HNO₃, acetic acid	15 mL HCl, 10 mL HNO₃, 10 mL acetic acid
9 Acid ferric chloride	Saturated solution of FeCl₃·6H₂O in HCl (conc); add few drops HNO₃
10 Glyceregia	10 mL HNO₃, 20 to 50 mL HCl, 30 mL glycerol
11 Sodium cyanide (electrolytic, 6 V)	10 g NaCN, 90 mL water

Alloy	Normal heat treatment	General microstructure	Ferrite	Carbide	Sigma phase
Application of etchants above to examination of specific alloys					
CA-6NM	Hardened and tempered(a)	2	3	4	. . .
CA-15	Hardened and tempered(a)	2 or 9	3	4	. . .
CD-4MCu	Annealed(b)	1, 2, or 6	. . .	. . .	2 then 7
CE-30	As cast	2	3	4	2 then 7; or 11
CF-3	Annealed(c)	7	3	4	2 then 7
CF-3M	Annealed(c)	8	3	4	2 then 7; or 11
CF-8	Annealed(d)	7 or 10	3	4	2 then 7
CF-8C	Annealed(d)	1 or 6	3	4	2 then 7; or 11
CF-8M	Annealed(d)	9	3	4	2 then 7
CF-20	Annealed(d)	1	3	4	. . .
CG-8M	Annealed(d)	1	3	4	5; or 7 then 11
CN-7M	Annealed(e)	1 or 6	. . .	. . .	. . .

(a) Heat to 955 °C (1750 °F) min, air cool and temper at 595 °C (1100 °F) min. (b) Heat to 1120 °C (2050 °F) min, furnace cool to 1040 °C (1900 °F), quench in water or oil. (c) Heat to 1040 °C (1900 °F) min, rapid cool. (d) Heat to 1040 °C (1900 °F) min, water quench. (e) Heat to 1120 °C (2050 °F) min, water quench.

Technique for Fe-Cr-Ni Heat-Resistant Casting Alloys

THE ALLOYS discussed in this article are iron-base and nickel-base heat-resistant casting alloys containing enough chromium to confer oxidation resistance.

Microstructures

Alloy HA and sometimes, depending on specific composition, alloy HB are hardened by the transformation of austenite during quenching. Thus, austenite transformation products such as bainite and martensite may appear in their microstructures. Also, depending on specific composition, the HB and HC alloys may have a matrix of ferrite or austenite, or a mixture of ferrite and austenite; the austenite may be unstable and transform, at least partly, on cooling to room temperature.

Except for the nickel-free alloy HA and the low-nickel alloys HB and HC, the matrix of the remainder of the alloys in the series is austenite. Other microconstituents usually observed are ferrite, several types of chromium carbide, sigma phase, inclusions and oxides. The HD and HE alloys are intended to be partly ferritic; the presence of ferrite results in lower creep resistance and higher ductility at temperatures of 870 to 1150 °C (1600 to 2100 °F) than for the wholly austenitic alloys. Alloys HD and HE are also more susceptible to the development of the undesirable

sigma phase if exposed to temperatures in the range of 650 to 870 °C (1200 to 1600 °F). The HF, HH, HI, HK and HL alloys are borderline; they can be partly ferritic and partly austenitic or wholly austenitic, depending on composition. As usually supplied, the HI and HK alloys seldom show ferrite. The HN, HP, HT, HU, HW and HX alloys are wholly austenitic; neither ferrite nor sigma phase is likely to be present.

Mounting, Grinding and Polishing

The techniques and equipment used for mounting, grinding and polishing specimens of Fe-Cr-Ni heat-resistant casting alloys for metallographic examination are essentially the same as those used for wrought and cast stainless steels (see previous article).

Etching

The etchants used for etching the Fe-Cr-Ni alloys are given in Table 7. They fall into three categories: (*a*) delineating etchants (usually acid), which provide contrast and reveal general structure; (*b*) staining or film-forming etchants (alkaline); and (*c*) solutions for electrolytic etching.

Delineating Etchants. Glyceregia is the most widely used delineating etchant and is also used for etch-and-polish cycles. Marble's reagent and aqua regia are sometimes used, but to a far lesser extent than glyceregia.

Hydrochloric acid (50%) is sometimes used for outlining ferrite. It may or may not be followed by etching with a staining etchant.

Staining etchants form films of reaction products on the surface of the specimen. The films have

a color, the hue of which depends in part on film thickness, which is controlled by etching time, temperature and the etchant used. The etchants are generally aqueous solutions of potassium or sodium hydroxide with an oxidizing agent added. Picrates, potassium permanganate, hydrogen peroxide, and ferricyanides are used as oxidizing agents.

Murakami's reagent, which contains potassium hydroxide with potassium ferricyanide as the oxidizing agent, is a versatile staining etchant. By staining in different tints, it permits differentiation of several types of carbide and sigma phase. Murakami's reagent is used cold, warm or boiling to obtain a variety of effects, but it must be used with discrimination. Because the response of the reagent indicates sensitivity to the composition of the phase being stained, a given constituent does not respond identically when it appears in alloys of different composition.

Electrolytic etching, when controlled by an electronic timer, has the advantages of precision and reproducibility. Normally, the specimen to be etched is made the anode; the cathode usually is stainless steel. The current can be supplied by one or more dry-cell batteries or a rectified power supply. Current density ranges from less than 0.15 to more than 2.0 A/cm^2 (less than 1 to more than 13 A/in.^{2}) of specimen surface.

Unmounted specimens are held with stainless steel tongs. If the specimen is mounted in a nonconducting material, the electrical connection can be conveniently made by means of a brass machine screw that contacts the underside of the specimen through a tapped hole.

Table 7. Etchants for microscopic examination of Fe-Cr-Ni heat-resistant casting alloys

Common name	Composition	Remarks on use
Etchants for delineating general structure		
Aqua regia	20 mL HNO$_3$, 60 mL HCl	Immerse specimen
Glyceregia	10 mL HNO$_3$, 20-50 mL HCl, 30 mL glycerol	Immerse specimen; use a hood
Hydrochloric acid (50%)	50 mL HCl, 50 mL water	Outlines ferrite; immerse specimen
Marble's reagent	10 g CuSO$_4$, 50 mL HCl, 50 mL water	Immerse specimen
Vilella's reagent	1 g picric acid, 5 mL HCl, 100 mL ethanol	Immerse specimen
Etchants for staining or film-forming		
Alkaline hydrogen peroxide	25 mL NH$_4$OH, 50 mL H$_2$O$_2$ (3%), 25 mL water	Ordinarily used after a delineating etchant; immerse specimen
Alkaline potassium ferricyanide	10 g K$_3$Fe(CN)$_6$, 10 g NaOH, 100 mL water	Same as above
Alkaline potassium permanganate	4 g NaOH, 10 g KMnO$_4$, 85 mL water	Same as above
Alkaline sodium picrate	2 g picric acid, 25 g NaOH, 100 mL water	Same as above
Emmanuel's reagent	30 g K$_3$Fe(CN)$_6$, 30 g KOH, 60 mL water	Attacks sigma phase with little or no effect on carbide particles; immerse specimen
Murakami's reagent	10 g K$_3$Fe(CN)$_6$, 10 g KOH, 100 mL water	Stains carbide particles without staining sigma phase(a); immerse specimen
Solutions for electrolytic etching		
Ammonium hydroxide	Concentrated NH$_4$OH	Final electrolytic etch after etching in Vilella's reagent and in 10N KOH (electrolytic)
Cadmium acetate	10 g cadmium acetate, 100 mL water	Attacks (Cr,Fe)$_{23}$C$_6$ carbide particles
Chromic acid	2-10 g Cr$_2$O$_3$, 100 mL water	Outlines carbide particles; extracts sigma phase
Lead acetate (2N)	38 g Pb(C$_2$H$_3$O$_2$)$_2$·3H$_2$O, distilled water to make 100 mL	Stains austenite, then sigma phase, then carbide particles; 1.5 V for 30 s
Oxalic acid	10 g oxalic acid, 100 mL water	Outlines carbide and sigma; 6 V, 1 to 5 s
Potassium hydroxide (1N)	5.6 g KOH, 100 mL water	Blackens sigma phase without outlining other phases; 1.5 V for 1 s
Potassium hydroxide (10N)	56 g KOH, 100 mL water	Intermediate etch between Vilella's and ammonium hydroxide (electrolytic)
Sodium cyanide	10 g NaCN, 100 mL water	Used after glyceregia; outlines carbide particles, stains sigma phase; use at 0.15 A/cm^2 (1 A/in.^{2}) for 1 to 5 s, under hood

(a) Sometimes sigma phase is stained. Behavior must be established on a given composition.

Typical Microstructures of Iron-Base Alloys

By George F. Vander Voort, Carpenter Technology Corp.

IN THIS ARTICLE, the more common microstructures observed in iron-base alloys (carbon and alloy steels, cast irons, tool steels, and stainless steels) are described and illustrated. These microstructures occur as a result of variations in chemical analysis and processing. A more comprehensive coverage of the microstructures of ferrous alloys is presented in Vol 7 and 8 of the 8th Edition of Metals Handbook.

A wide range of constituents (a phase, or combination of two phases, in a specific configuration) are encountered in iron-base alloys. Single-phase constituents in iron-base alloys include: austenite, ferrite, delta ferrite, cementite, various alloy carbides, graphite, martensite, and a variety of intermetallic phases, nitrides and nonmetallic inclusions. Two-phase constituents include: tempered martensite, pearlite and bainite. Nonmetallic inclusions that consist of two or more phases can be present in steels. The nomenclature used to describe the constituents observed in steels has evolved over many years (Ref 1). The literature documents many names for the same constituents, which leads to some confusion.

If only matrix phases are considered—i.e., if very small amounts of inclusions or nitrides are ignored—the only single-phase ferrous metals or alloys are those consisting solely of ferrite, austenite or martensite. Because as-quenched martensite is rather brittle, tempering is almost always performed before commercial utilization.

COMMON CONSTITUENTS IN FERROUS ALLOYS

Ferrite. Fully ferritic steels are only obtained when the carbon content is quite low. Examples include electrolytic iron, certain sheet steels, Fe-Si and Fe-Ni electrical alloys and ferritic stainless steel. The most obvious microstructural features in such metals are the ferrite grain boundaries, as shown in Fig. 1. Ferrite is a very soft, low-strength phase. If the ferrite grain size (measurement of grain size is discussed in the article "Measurement of Microstructure") is quite fine, good ductility and formability are obtained. Ferritic stainless steels, due to their alloy content, are not nearly as ductile as ferritic sheet steel. Because ferrite has a body-centered-cubic (bcc) crystal structure, ferritic metals and alloys exhibit a transition from ductile to brittle behavior as the temperature decreases or as the strain rate increases.

Austenite. Obtainment of fully austenitic steels requires careful balancing of chemical composition—i.e., rather large amounts of the austenite-stabilizing elements (carbon, nitrogen, nickel and manganese) must be present compared with those elements that stabilize ferrite. Examples of fully austenitic ferrous alloys are austenitic stainless steels and austenitic manganese steel. Again, the most visible microstructural features of these single-phase alloys are the austenite grain boundaries. However, unlike ferritic structures, these alloys will also contain annealing twin boundaries in the wrought, solution-annealed condition, as shown in Fig. 2. Austenite is also a rather soft, low-strength phase; however, cold working produces substantial strengthening and, if extensive, will produce strain-induced martensite. Due to their face-centered-cubic (fcc) crystal structures, austenitic alloys remain ductile irrespective of temperature or strain rate unless detrimental phase changes occur.

Cementite, or iron carbide, contains 6.67% carbon (by weight), corresponding to the formula Fe_3C. In carbon and alloy steels, some of the carbide-forming elements (e.g., Mn and Cr) will re-place some of the iron in cementite. Hence, the formula for cementite is often referred to as M_3C, where M stands for the carbide-forming elements present. Pure Fe_3C is quite hard (about 800 HV) and brittle. Substitution of other elements (e.g., Cr) for some of the iron in cementite will increase the hardness appreciably. Because of the brittleness of cementite, only limited amounts are present in steels—i.e., no alloy consisting solely of cementite is produced commercially. The highest cementite contents are observed in white cast irons (Fig. 3), which are used in applications where high wear resistance is desired, and the brittle nature of such alloys is not a problem.

Graphite is a very soft, brittle phase produced in certain steels and cast irons by adjustment of chemical composition. Because cementite is not a true equilibrium phase under conditions of long-time elevated-temperature exposure, the cement-

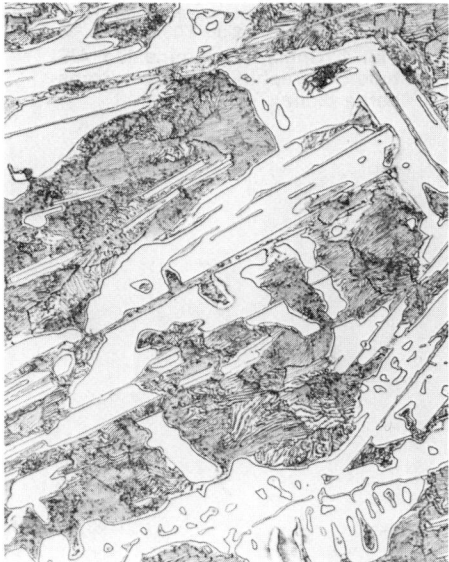

Etched with 4% picral. Magnification, 500×.

Fig. 3. White cast iron containing massive cementite (white) and pearlite (dark)

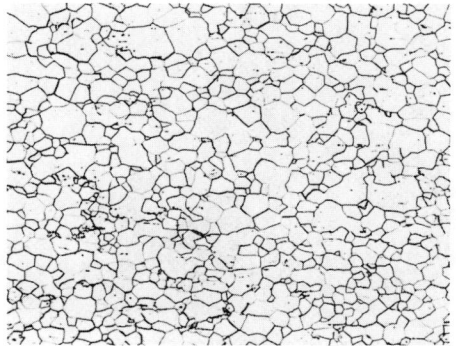

Fig. 1. Low-carbon ferritic sheet steel etched with 2% nital to reveal ferrite grain boundaries. Magnification, 100×.

(Left) AISI type 316 stainless steel. Etched with $HCl/HNO_3/H_2O$ (equal parts). Magnification, 100×. (Center) Hadfield's manganese steel. Etched with 2% nital (3 s) and 20% $Na_2S_2O_5$ (20 s). Magnification, 100×. (Right) Fe-48%Ni alloy (annealed in H_2). Etched with $FeCl_3/HCl/H_2O$. Magnification, 1×.

Fig. 2. Examples of fully austenitic iron-base alloys in the solution-annealed condition

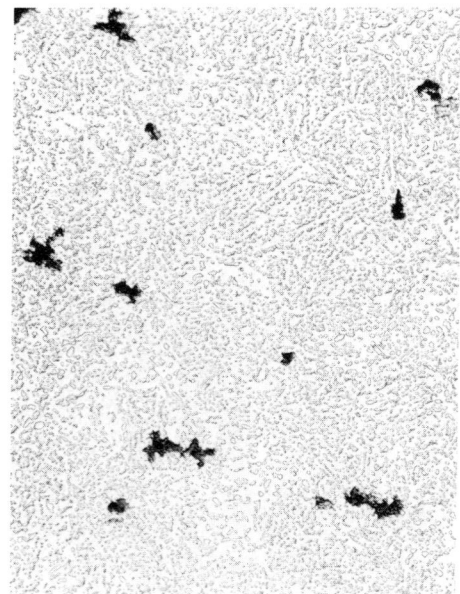

The irregular black particles are graphite. The matrix is ferrite containing spheroidized cementite. Etched with 4% picral. Magnification, 500×.

Fig. 4. Spheroidize annealed AISI type O6 graphitic tool steel (transverse plane)

ite in certain carbon and alloy steels will decompose. Several tool steels are deliberately designed to produce a small amount of graphite in the structure to enhance machinability, as shown in Fig. 4. The graphite in cast irons is produced in a variety of forms — e.g., flake, compacted and spheroidal. Gray cast iron containing flake graphite is quite brittle; however, if the graphite is spheroidal, as in ductile iron, good ductility is achieved. Figure 5 illustrates flake and spheroidal graphite shapes. As with cementite, graphite is only present as a minor phase, even in cast irons.

Martensite is not an equilibrium phase in steels. Formation of martensite is dependent on chemical composition and cooling rate from the high-temperature austenite region. Unlike other austenite transformation products, martensite forms instantaneously once the specimen is cooled below a specific temperature, the martensite start or M_s temperature, which is a function of the carbon and alloy content of the parent austenite phase. The transformation is completed when the specimen reaches a lower temperature, the martensite finish or M_f temperature. The hardness of martensite is governed primarily by carbon content but is also influenced slightly by alloy content. The ability to form martensite in a steel as a function of section size and quench rate depends on the hardenability of the steel. Hardenability is increased by increasing carbon and alloy contents and by enlargement of grain size (coarsen-

ing of grain size to improve hardenability is rarely done with wrought steels, because most mechanical properties are impaired).

The nature of martensite is influenced markedly by carbon content. Basically, two types of martensite can be formed in steels. At low carbon contents, lath martensite is formed. The laths are present in a packet arrangement where the individual laths within the packet have essentially the same orientation. At high carbon contents, plate martensite is formed. The plates form as individual lenticular crystals in a wide range of sizes. Many terms have been used to denote these two types of martensite, but lath and plate are the preferred terms. At intermediate carbon contents, mixtures of lath and plate martensite are obtained. Figure 6 illustrates the appearance of lath and plate martensite; the shape of the latter is best observed when only a small amount of martensite has been formed, as illustrated.

Because of the important influence of grain size on the properties of martensitic steels, much effort has been expended on developing the grain size of such steels. However, unlike ferritic and austenitic alloys, the critical grain size for martensitic steels is that of the parent austenite phase — i.e., the prior austenite grain size. Delineation of the prior austenite grain boundaries in martensitic steels by use of selective etchants (Ref 2) is difficult but can often be achieved as illustrated in Fig. 7. In general, the low-carbon martensitic steels are more difficult to etch in this

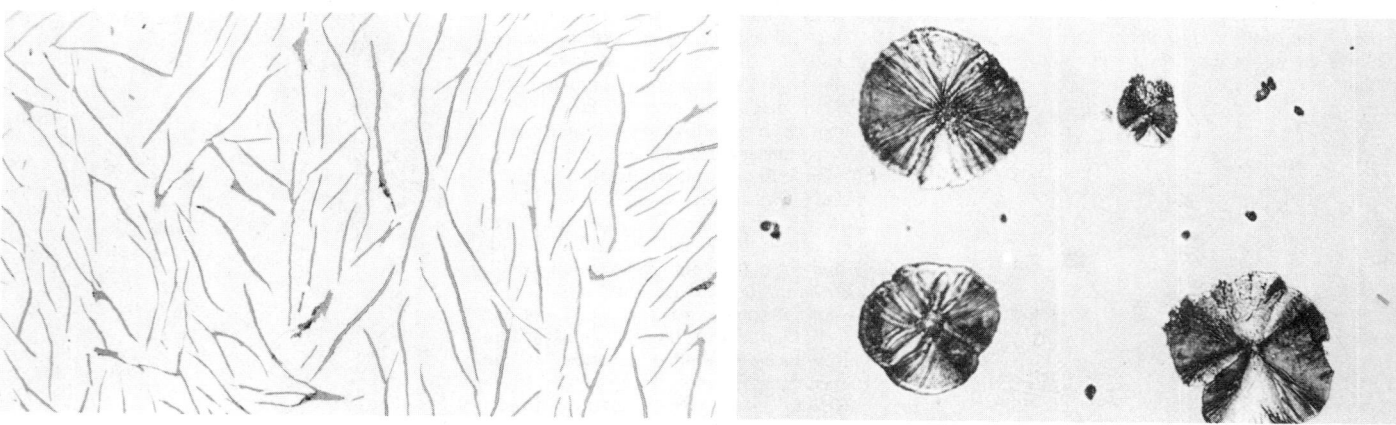

(Left) Flake graphite in gray iron. Magnification, 100×. (Right) Spheroidal graphite in ductile iron (viewed in polarized light). Magnification, 500×.

Fig. 5. Examples of two different graphite shapes in cast irons

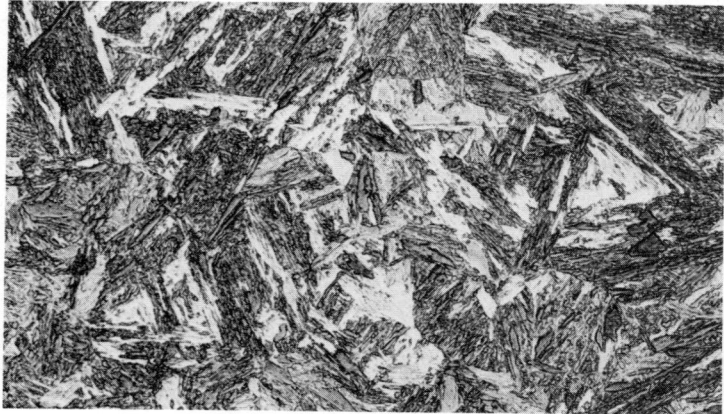
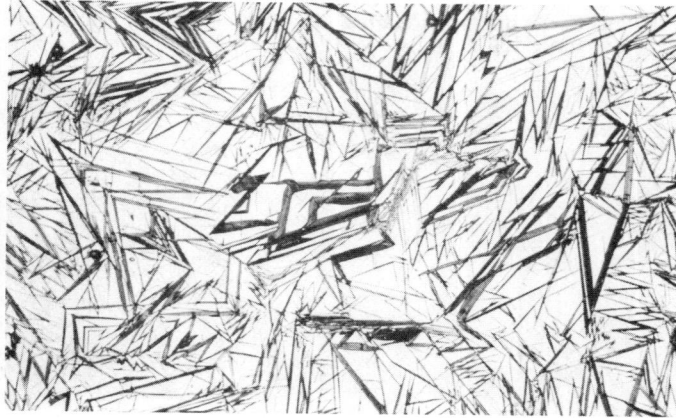

(Left) Lath martensite in a heat treated low-carbon alloy steel. Magnification, 500×. (Right) Plate martensite (retained-austenite matrix) in an Fe-1.4%C alloy. Magnification, 200×. Both were heat treated at higher-than-normal temperature to reveal the martensite more clearly. Both etched with 2% nital.

Fig. 6. Examples of lath and plate martensite.

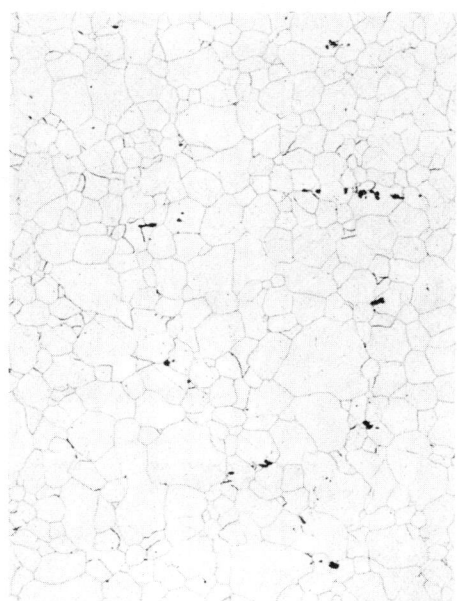

Etched with saturated aqueous picric acid plus a wetting agent. Magnification, 100×.

Fig. 7. Prior austenite grain boundaries revealed by special etching of a heat treated medium-carbon alloy steel

manner than medium- and high-carbon steels. In the case of lath martensite, the packet size has also been found to be an important microstructural measurement.

As the carbon content of martensite increases, its hardness and strength also increase; however, it also becomes more brittle. Martensite has a body-centered-tetragonal (bct) structure. The degree of tetragonality increases with carbon content. Tempering decreases the strength of martensite but increases its toughness. However, tempering of alloy steels within certain temperature ranges can reduce toughness due to embrittlement problems (temper martensite embrittlement or temper embrittlement). However, the tempering process, along with composition selections, can permit achievement of a very wide range of useful strengths and toughness, a factor that makes tempered martensite the most commercially important constituent in steels.

Tempering is a complex process from a microstructural viewpoint. In commercial practice, tempering is conducted at temperatures from about 125 °C (255 °F) to values slightly below the lower critical temperature, depending on the alloy in question and the intended application. Tempering of carbon steels is simpler than tempering of alloy steels. In the case of carbon steels, four stages have been identified.

The first stage covers the range up to about 250 °C (480 °F), where the tetragonality of the bct lattice is lost and a transition carbide, Fe$_{2.4}$C, precipitates. This carbide is extremely small and cannot be resolved by light microscopy. In the second stage, generally in the range of about 200 to 300 °C (390 to 570 °F), retained austenite (when present) decomposes. In order for retained austenite to be resolvable by light microscopy, it must be present in amounts of at least 10 to 20% — a condition obtained only in high-carbon steels. The third stage occurs in the range of about 200 to

350 °C (390 to 660 °F) and is characterized by formation of cementite. The fourth stage occurs between about 300 °C (570 °F) and the lower critical temperature and concerns the growth of cementite to sizes visible in a light microscope and production of equiaxed ferrite by recrystallization. These changes are illustrated for a low-carbon alloy steel in Fig. 8 and for a high-carbon alloy steel in Fig. 9. Neither of these grades contains enough strong carbide formers to produce more complex alloy carbides. When substantial quantities of elements such as chromium, molybdenum, vanadium and tungsten are present, alloy carbides are formed in the region of about 500 to 600 °C (930 to 1110 °F) along with a modest increase in hardness, termed "secondary hardening."

Pearlite is a mixture of ferrite and cementite where the two phases are formed from austenite in an alternating lamellar pattern. Formation of pearlite requires relatively slow cooling from the austenite region. Pearlite forms at temperatures close to the lower critical temperature of the steel in question and may be formed either isothermally or by continuous cooling. As the hardenability of the steel decreases, the cooling rate can be increased without forming other constituents. As the isothermal reaction temperature decreases or the cooling rate increases, the interlamellar spacing decreases. The strength and toughness of pearlitic steels are increased as the interlamellar spacing decreases.

For rather slow cooling rates, the relative amounts of ferrite and pearlite in carbon steels can be estimated using the iron-carbon equilibrium diagram. The fact that the maximum solubility of carbon in ferrite is nearly zero, and the fact that a fully pearlitic microstructure is obtained when a steel containing 0.8% C is slowly cooled from the austenite region, can be used to estimate the volume fractions of ferrite and pearlite.

To illustrate the influence of carbon content on the amounts of ferrite and pearlite in Fe-C alloys, Fig. 10 shows the microstructures of Fe-C alloys containing 0.2, 0.4, 0.6, 0.8 and 1.0% C. The amount of pearlite in the Fe-0.2%C alloy is simply 0.2/0.8 or 25%, and the remainder of the structure (75%) is ferrite. In like manner, the amount of pearlite in the 0.4%C alloy is 50%, and the amount in the 0.6%C alloy is 75%.

In these alloys, the ferrite forms prior to the eutectoid reaction (which produces the pearlite) and is termed "proeutectoid ferrite." Below about 0.4% C, the proeutectoid ferrite forms as equiaxed patches and is the continuous phase. Above about 0.4% C, the proeutectoid ferrite generally exists as isolated, equiaxed patches or as a grain-boundary layer.

Carbon steels are referred to as hypoeutectoid, eutectoid or hypereutectoid when their carbon contents are below 0.8%, about 0.8% or above 0.8%, respectively. In the case of hypereutectoid steels, such as the Fe-1%C alloy shown in Fig. 10, excess cementite above the amount required to form pearlite will precipitate in the austenite grain boundaries prior to the eutectoid reaction. This excess cementite is referred to as "proeutectoid cementite." A grain-boundary cementite network renders such steels quite brittle.

The strength and hardness of ferrite-pearlite steels increase with increasing pearlite content and are further increased by reductions in the interlamellar spacing. Pure ferrite (no carbon) has a hardness of about 70 HV, whereas very fine

pearlite in a eutectoid carbon steel has a hardness of nearly 400 HV. Fine pearlite is the most desirable structure for wire drawing, where extremely high strengths can be obtained. Figure 11 shows the microstructures of cold drawn low-carbon and high-carbon wire (longitudinal views).

Carbon steels are widely utilized in the hot rolled condition. The austenite grain size of the steel as it enters the final rolling pass establishes the relative sizes of the ferrite and pearlite produced during subsequent air cooling, whereas the cooling rate influences the fineness of the pearlite, the morphology of the proeutectoid ferrite and the amounts of the various constituents. The influence of austenite grain size on the sizes, shapes and amounts of ferrite and pearlite is illustrated in Fig. 12, where an AISI 1040 carbon steel has been air cooled from 900, 980 and 1095 °C (1650, 1800 and 2000 °F).

To obtain maximum ductility and formability, carbon steels may be subjected to a special annealing treatment that spheroidizes the cementite. This treatment is commonly used on tool steels to improve their machinability. To best observe the spheroids of cementite, specimens of steels so treated should be etched with 4% picral, which reveals the ferrite/carbide interfaces but not the ferrite grain boundaries. Figure 13 illustrates this problem of structure visibility using a spheroidized low-carbon steel. Etching with 2% nital obscures the spheroidized cementite. As the carbon content increases, the volume fraction of cementite increases. Figure 14 shows the microstructure of spheroidized AISI W2 (1.1% C) tool steel.

Bainite, an austenite transformation product, is a lathlike aggregate of ferrite and cementite that forms under conditions intermediate to those that result in formation of pearlite and martensite. Bainite is commonly classified as either upper bainite or lower bainite. Upper bainite forms either isothermally or during continuous cooling at temperatures just below those that produce pearlite. Lower bainite forms at still lower temperatures, down to the M_s temperature, or slightly below in certain cases.

Formation of upper bainite begins by growth of long ferrite laths that are devoid of carbon. Because the carbon content of the ferrite laths is very low, the austenite at the lath boundaries is enriched in carbon. The shape of the cementite formed at the lath boundaries varies with carbon content. In low-carbon steels, the cementite will precipitate as discontinuous stringers and isolated particles, whereas at higher carbon contents the stringers are more continuous. In some instances, carbide is not precipitated but is retained as austenite or transforms to plate martensite. Figure 15 illustrates the microstructure of upper bainite.

Lower bainite has a more acicular appearance compared with that of upper bainite. The ferrite plates are broader than those in upper bainite and are more similar in appearance to plate martensite. As with upper bainite, the appearance of lower bainite varies with carbon content. Lower bainite is characterized by formation of rodlike cementite or Fe$_{2.4}$C (depending on the transformation temperature or steel composition) within the ferrite plates. Figure 16 illustrates the microstructure of lower bainite.

Nonmetallic Inclusions. Inclusions in steel are either indigenous or exogenous in origin. Indigenous inclusions form as a natural result of the decrease in solubility of oxygen or sulfur that occurs as

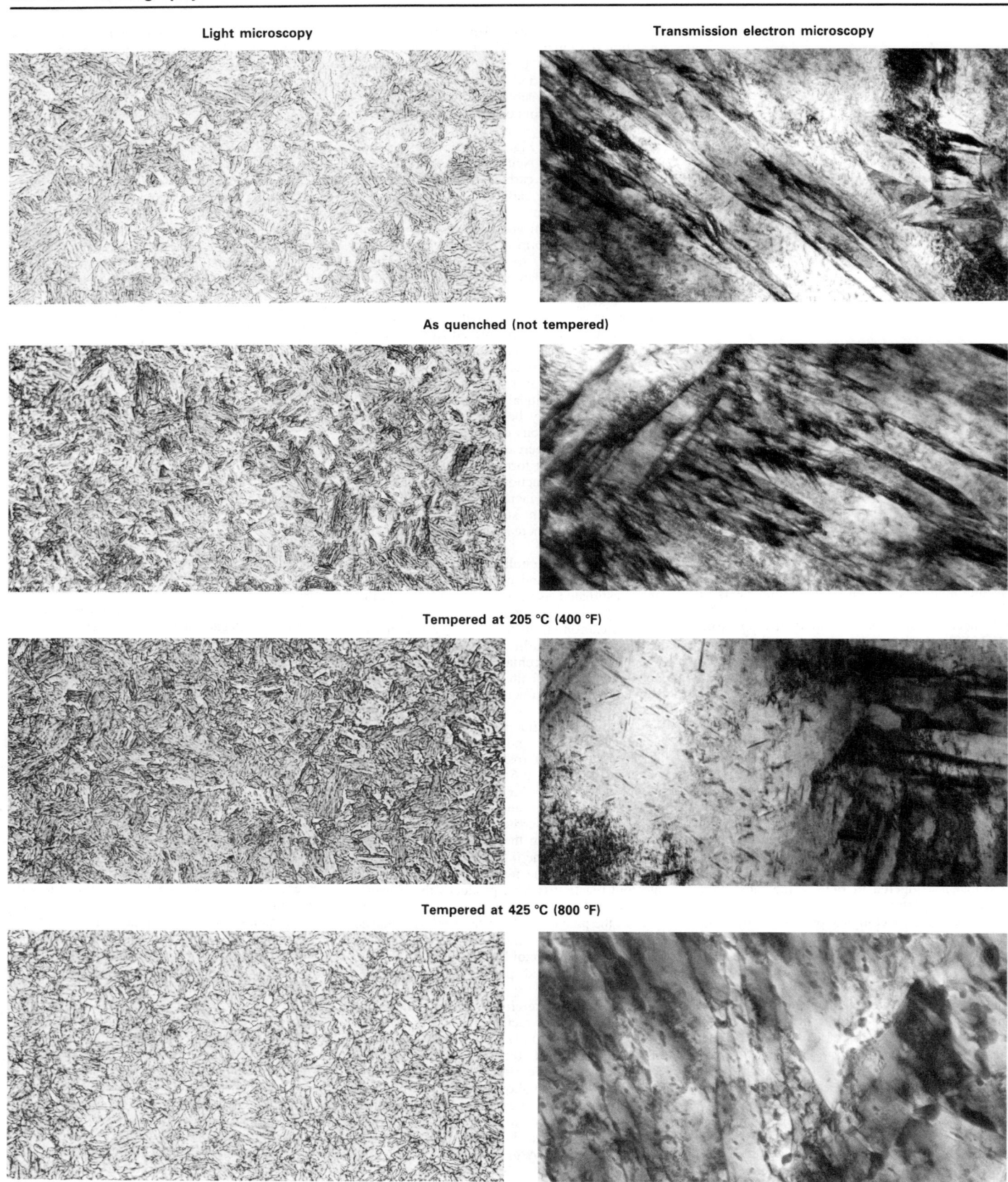

Fig. 8. Light-microscopy and transmission-electron-microscopy (thin-foil) views of AISI 8620 alloy steel after tempering at various temperatures. All specimens were water quenched from 900 °C (1650 °F) prior to tempering. Light microscopy: 2% nital, 500×. TEM: 25 000×.

Light microscopy **Transmission electron microscopy**

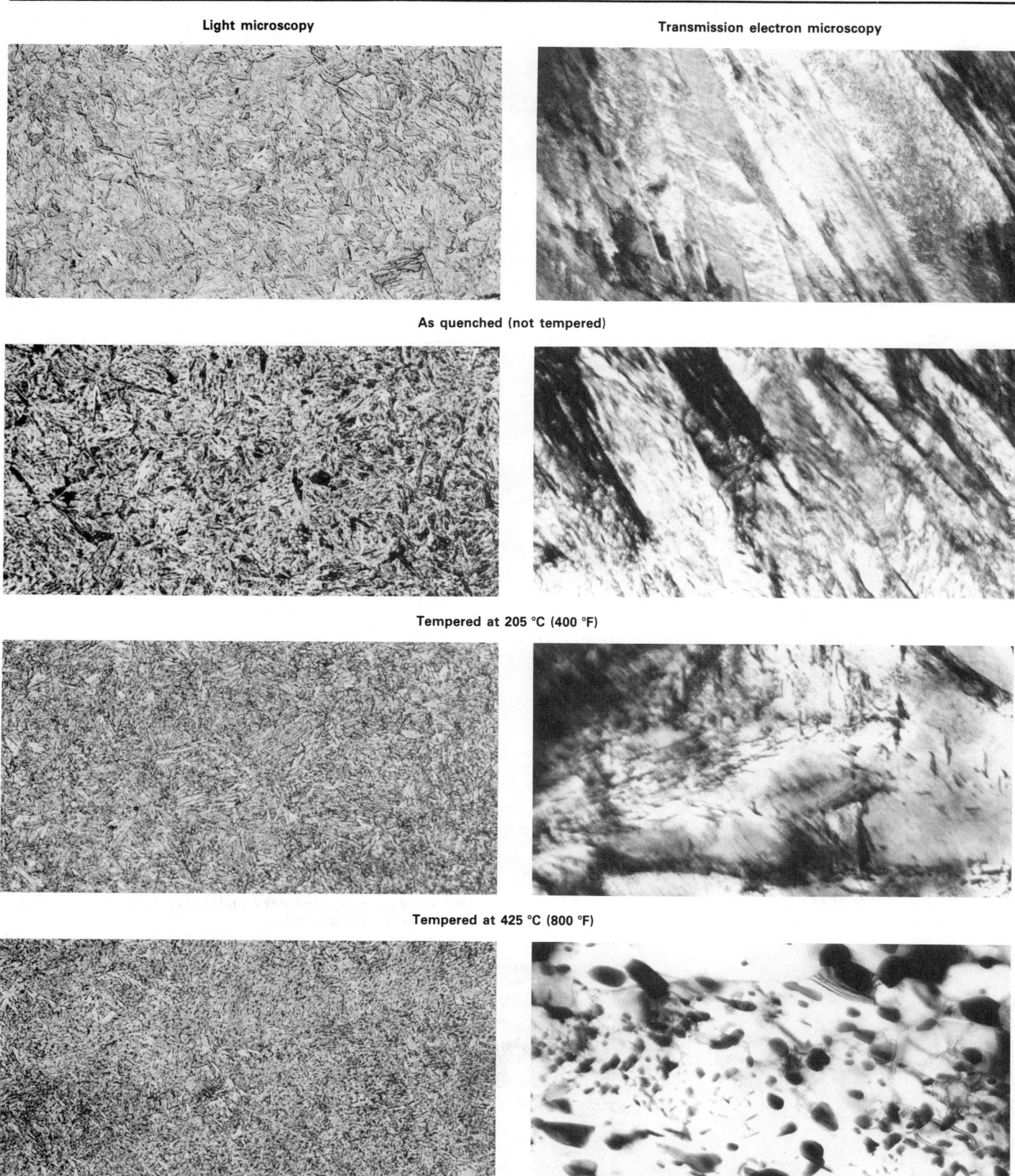

As quenched (not tempered)

Tempered at 205 °C (400 °F)

Tempered at 425 °C (800 °F)

Tempered at 650 °C (1200 °F)

Fig. 9. Light-microscopy and transmission-electron-microscopy (thin-foil) views of AISI 5160 alloy steel after tempering at various temperatures. All specimens were oil quenched from 830 °C (1525 °F) prior to tempering. Light microscopy: 2% nital, 500×. TEM: 25 000×.

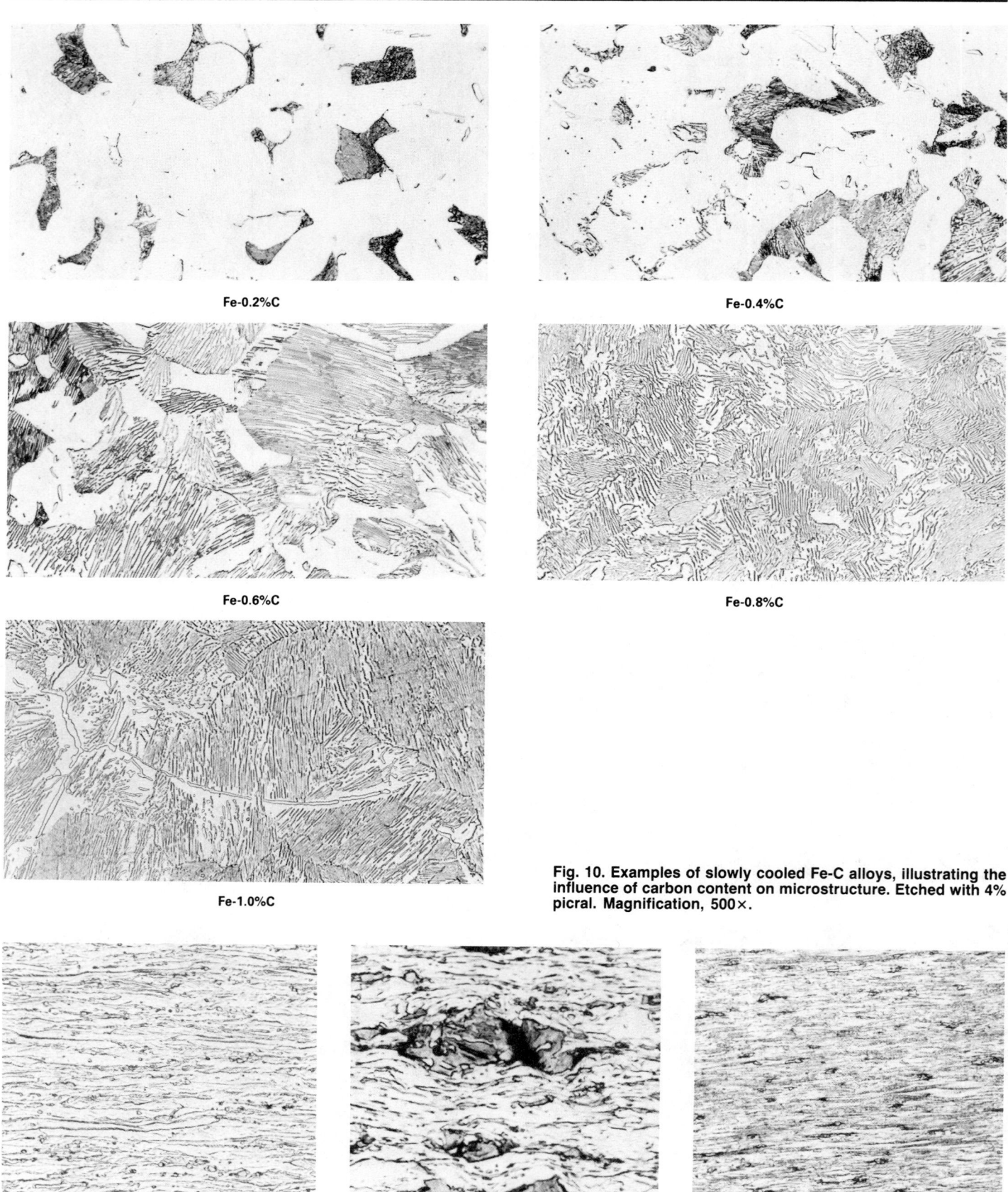

Fe-0.2%C

Fe-0.4%C

Fe-0.6%C

Fe-0.8%C

Fe-1.0%C

Fig. 10. Examples of slowly cooled Fe-C alloys, illustrating the influence of carbon content on microstructure. Etched with 4% picral. Magnification, 500×.

(Left) AISI 1008. Magnification, 500×. (Center) AISI 1020. Magnification, 1000×. (Right) AISI 1080. Magnification, 500×.
Fig. 11. Examples of cold worked carbon steel microstructures (longitudinal planes; etched with 2% nital)

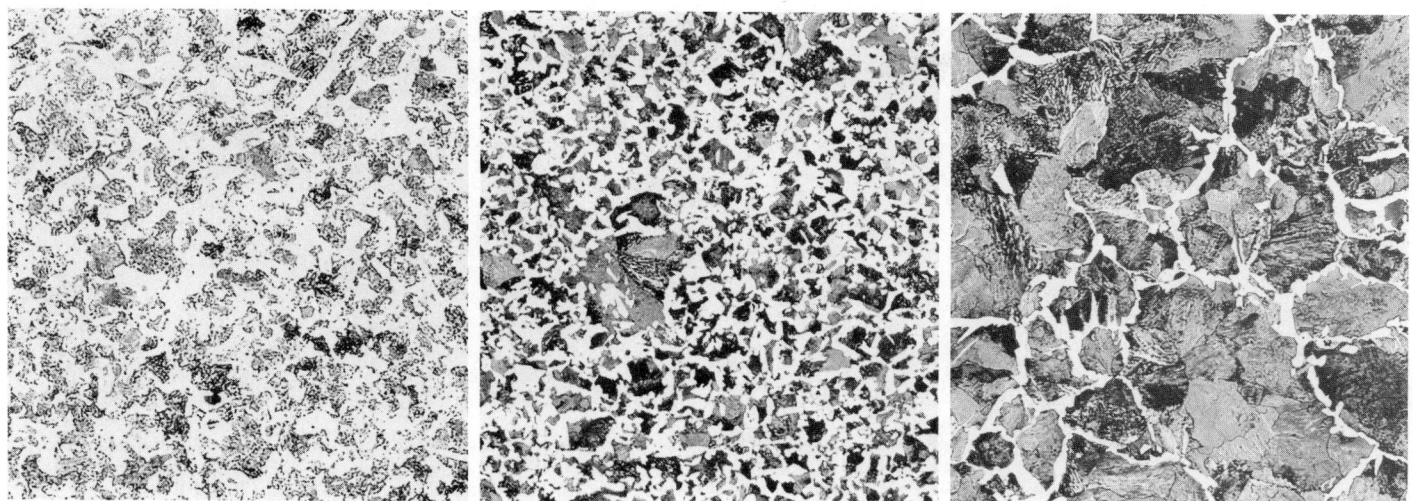

(Left) Austenitized at 900 °C (1650 °F), air cooled. (Center) Austenitized at 980 °C (1800 °F), air cooled. (Right) Austenitized at 1095 °C (2000 °F), air cooled. All etched with 4% picral. Magnification, 200×.

Fig. 12. Illustration of influence of austenitization temperature on amount, size and fineness of the transformation products, ferrite and pearlite, of normalized AISI 1040 carbon steel

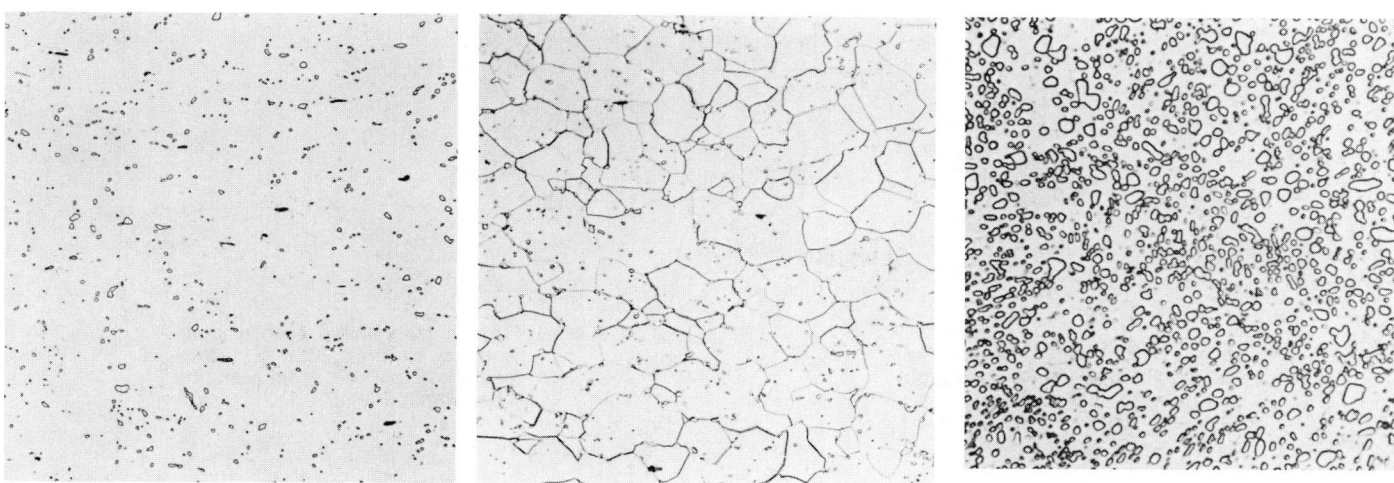

(Left) Etched with 4% picral. (Right) Etched with 2% nital. Magnification (both), 500×.

Fig. 13. Influence of etchant on the ability to observe spheroidal cementite in AISI 1008 sheet steel

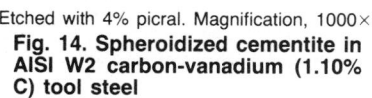

Etched with 4% picral. Magnification, 1000×.

Fig. 14. Spheroidized cementite in AISI W2 carbon-vanadium (1.10% C) tool steel

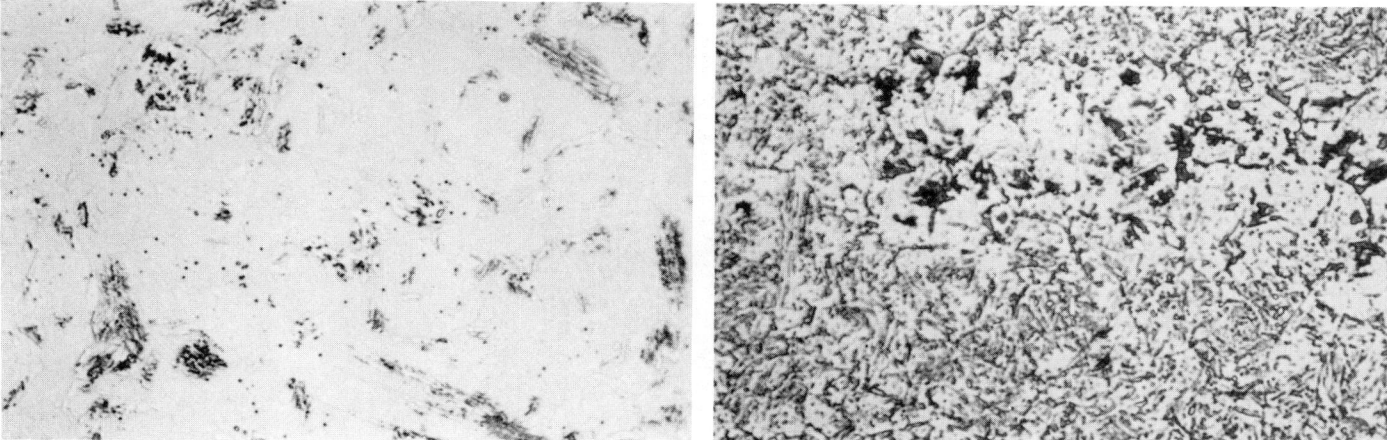

(Left) 845 °C (1550 °F); 480 °C (900 °F) for 30 s; water quenched. (Right) 845 °C; 480 °C for 2 h; water quenched. Both etched with 4% picral. Magnification, 1000×.

Fig. 15. Microstructures of upper bainite formed in AISI 4142 alloy steel isothermally transformed at 480 °C (900 °F)

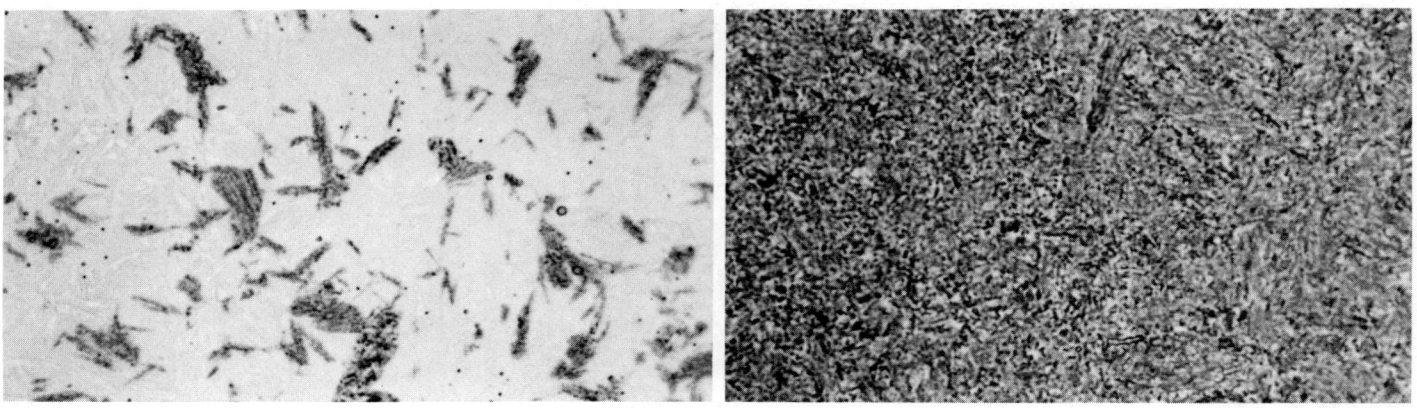

(Left) 845 °C (1550 °F); 370 °C (700 °F) for 60 s; water quenched. (Right) 845 °C; 370 °C for 30 min; water quenched. Both etched with 4% picral. Magnification, 1000×.

Fig. 16. Microstructures of lower bainite formed isothermally in AISI 4142 alloy steel at 370 °C (700 °F)

Etched with 2% nital. Magnification, 50×.

Fig. 17. Microstructure of wrought iron (longitudinal plane)

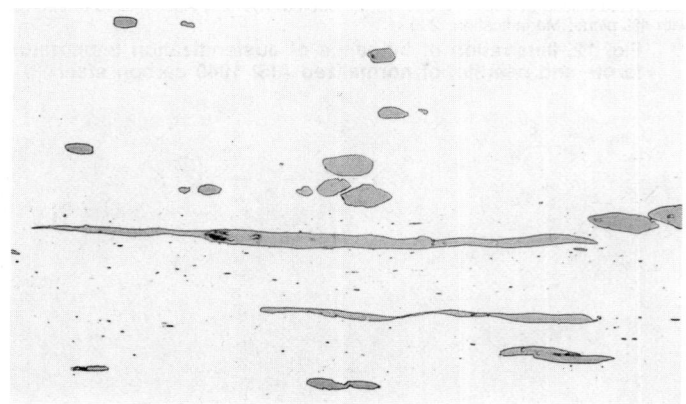

Fig. 18. Manganese sulfide inclusions in wrought AISI 1214 free-machining carbon steel (longitudinal plane; 200×)

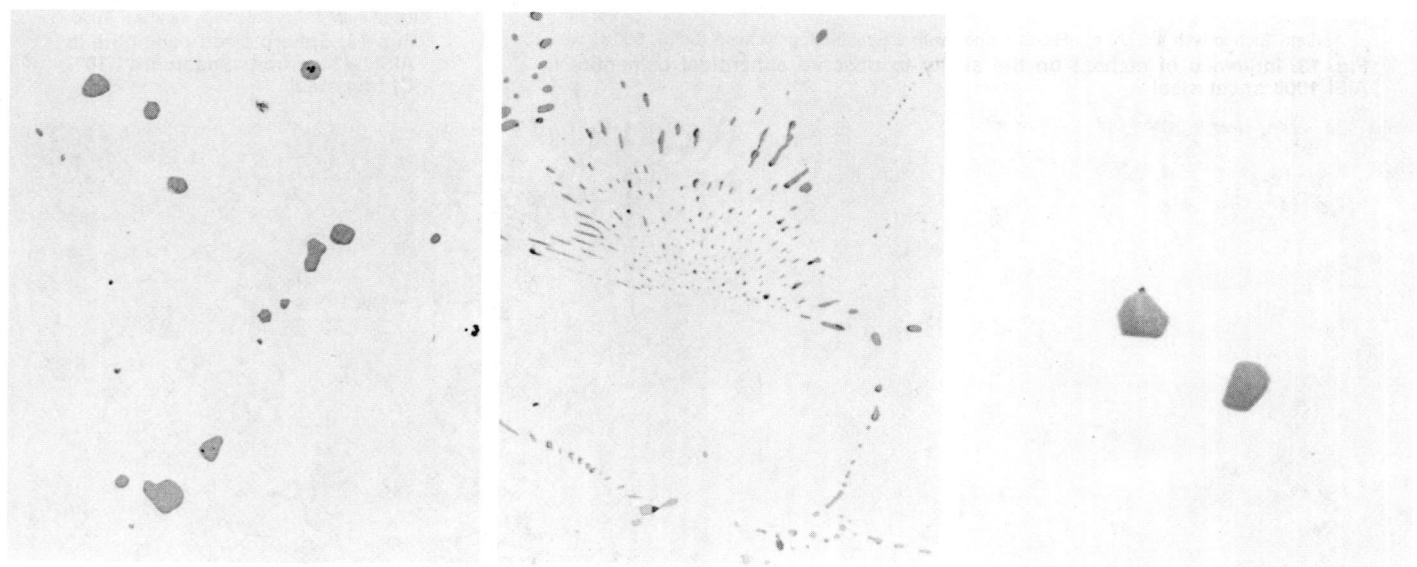

(Left) Type I. Magnification, 250×. (Center) Type II. Magnification, 250×. (Right) Type III. Magnification, 1600×.

Fig. 19. Examples of the three most common forms of manganese sulfides viewed in the as-cast condition

the metal freezes. Exogenous inclusions are introduced from external sources (e.g., slag or refractories) that enter the steel and become trapped during solidification. In most instances, these included phases are undesirable.

There are, however, two basic types of ferrous alloys that are deliberately made with additions of either exogenous or indigenous inclusions. Wrought iron (Fig. 17) is made by mixing together slag and relatively pure iron. In the hot worked condition, the slag becomes elongated, thus promoting considerable anisotropy. The toughness of wrought iron is very good if the crack is perpendicular to the fiber axis but very poor if it is parallel to the fiber axis. Free-machining steels are a classic example of the deliberate addition of indigenous inclusions, most commonly manganese sulfide, to enhance machinability. Figure 18 illustrates the appearance of MnS in a free-machining steel. Some free-machining steels are also given deliberate additions of either lead or bismuth. Both of these elements are virtually insoluble in solid steel and exist as small elemental globules sometimes associated with the sulfides.

Manganese sulfides precipitate in different forms depending primarily on the oxygen content of the steel. Very high oxygen contents (low carbon contents; no strong deoxidizers) promote formation of large, globular isolated manganese sulfides referred to as type I sulfides. Type I sulfides generally are desirable in free-machining steels. Lower oxygen contents and moderate deoxidation, particularly by small amounts of aluminum, promote precipitation of sulfides in a fanlike or chainlike pattern at the grain boundaries; these sulfides are referred to as type II sulfides. Type II sulfides are most detrimental to toughness and ductility. Very low oxygen contents and strong deoxidation by aluminum promote formation of small, angular, dispersed sulfides, referred to as type III sulfides. These are

less harmful to properties than are type II sulfides. Figure 19 illustrates these three classic sulfide forms. These forms are most easily observed in the as-cast condition. Other more complex types of sulfides can be produced due to the addition of other elements that influence sulfide composition and the precipitation reaction (e.g., rare earth elements).

EXAMPLES OF COMPLEX MIXTURES OF CONSTITUENTS

The micrographs previously presented have illustrated many of the more common complex mixtures of microstructures in ferrous alloys. There are a few others, however, that deserve illustration. Figure 20, for example, shows the microstructure of a dual-phase sheet steel that was austenitized in the two-phase region and then slowly cooled. The white matrix phase is ferrite, the small outlined white particles are retained austenite, and the small dark patches are autotempered martensite (and a minor amount of retained austenite). A large nitride particle is also shown at the bottom of the field.

The higher-alloy, high-carbon tool steels contain undissolved alloy carbides in a matrix of tempered martensite after commercial heat treatment. When properly heat treated, many of these alloys also contain small amounts of retained austenite, although such amounts are usually too low to be resolved by light microscopy unless an excessively high austenitizing temperature was used by error. Figure 21 shows two such alloys: AISI D2 and M2 tool steels. Both were austenitized at the recommended temperatures and then

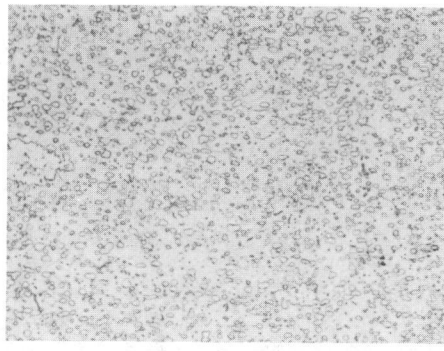
(a)

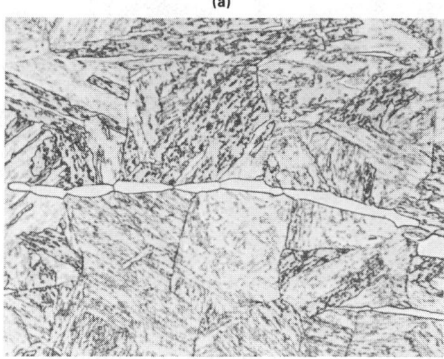
(b)

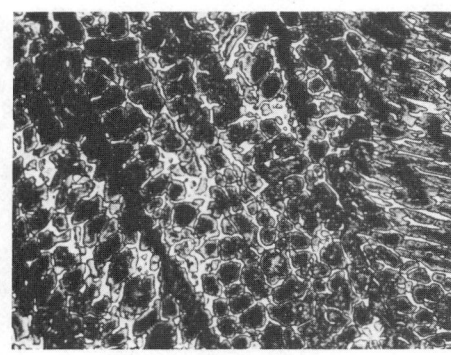
(c)

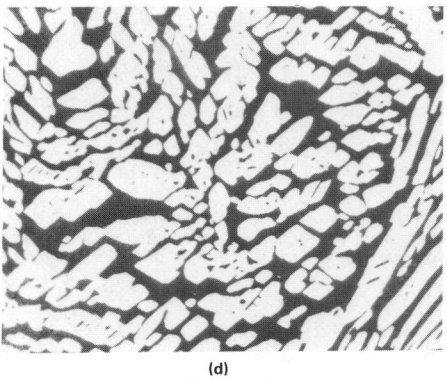
(d)

(a) AISI 440C, annealed. Structure: ferrite and carbides. Vilella's reagent; 500×. (b) 17-4 PH, solution annealed and aged. Structure: martensite and delta ferrite (white). Fry's reagent; 500×. (c) Austenitic AISI 304, welded. Structure: delta ferrite and austenite (dark). Heat tinted; 500×. (d) Duplex alloy, as cast and annealed. Structure: austenite and ferrite (dark). Tint etched; 200×.

Fig. 22. Examples of two-phase stainless steels with various microstructural combinations

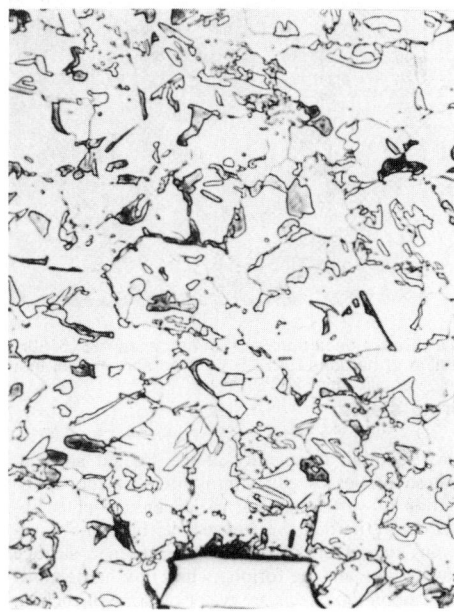

Etched with 20% Na₂S₂O₅ in H₂O. Magnification, 1000×.

Fig. 20. Microstructure of heat treated dual-phase low-carbon sheet steel

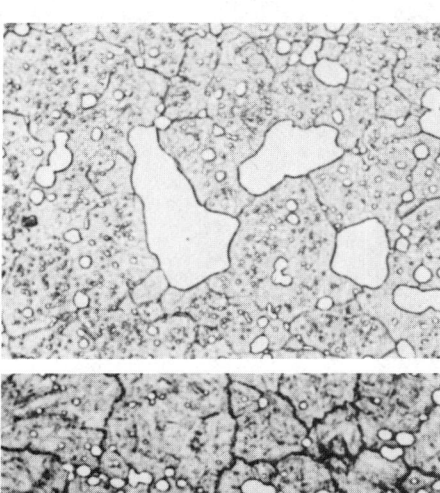

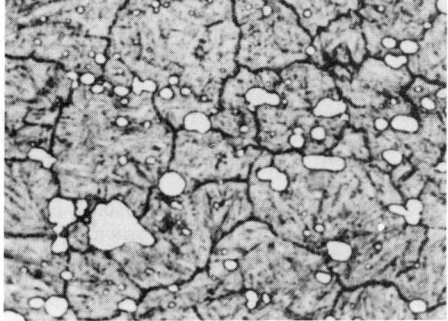

(Top) AISI D2 tool steel. Etched with 4% picral plus HCl. (Bottom) AISI M2 tool steel. Etched with 10% nital. Magnification (both), 1000×.

Fig. 21. Microstructures of two common high-carbon, high-alloy tool steels in the heat treated condition

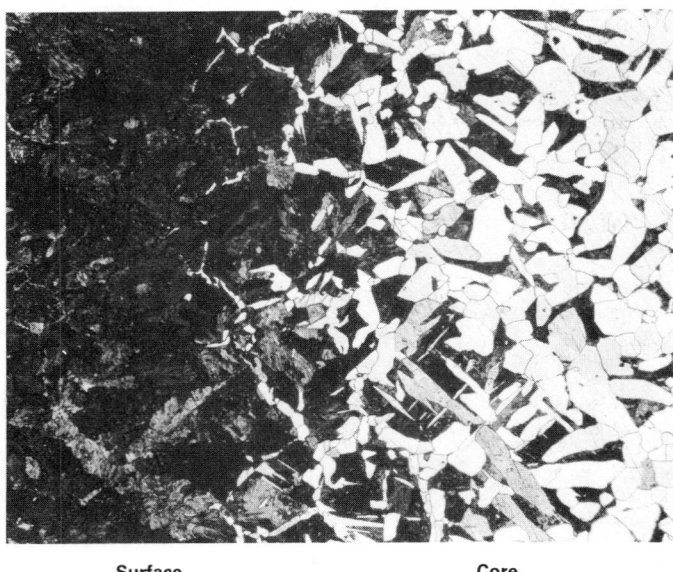

Surface   Core

Fig. 23. Microstructures at surface and core of carburized AISI 1020 carbon steel slowly cooled from the carburizing temperature. 2% nital; 200×.

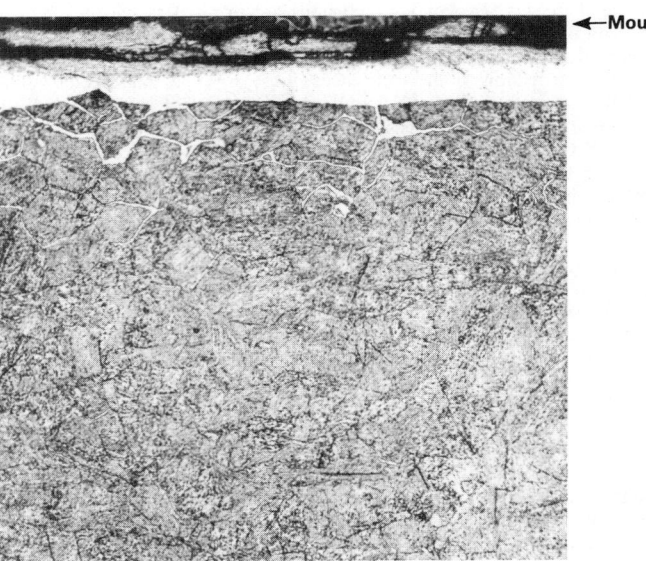

←Mount

Fig. 24. Microstructure at the surface of heavily nitrided AISI 4150 alloy steel that was heat treated prior to nitriding. Note heavy iron nitride "white" layer and grain-boundary penetration at the surface. 4% picral; 200×.

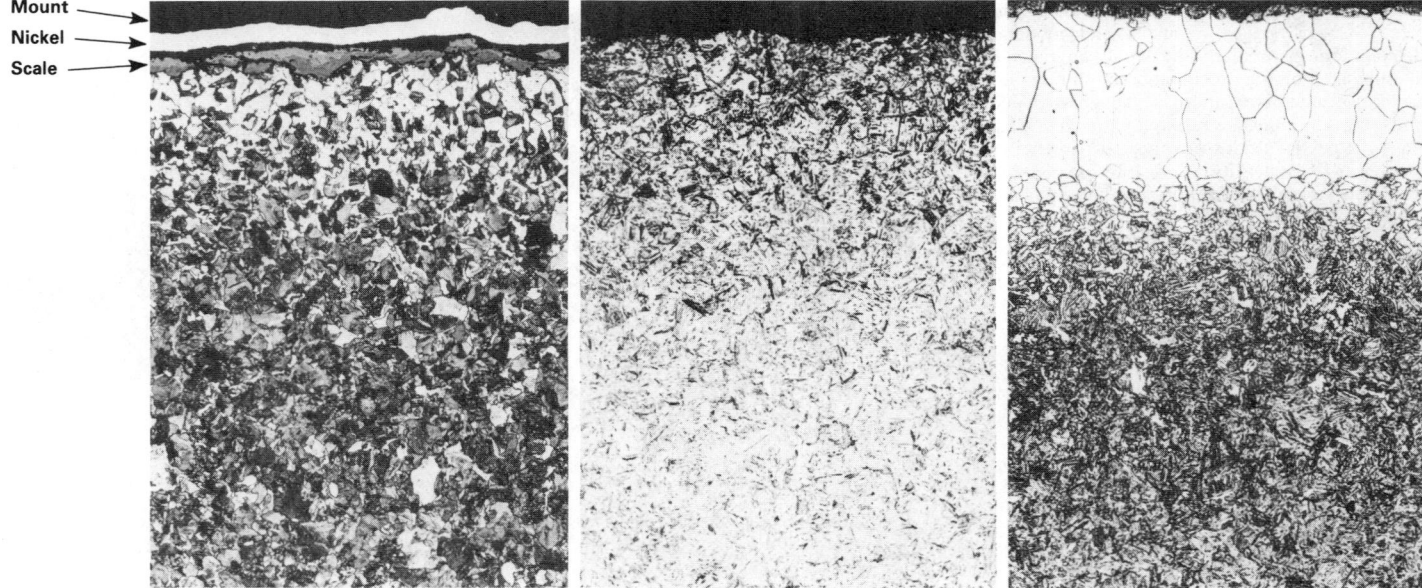

Mount
Nickel
Scale

(Left) Partial decarburization of as-rolled AISI 9260 Mod spring steel that was nickel plated for edge protection. (Center) Partial decarburization of hardened AISI 5160H spring steel whose surface was turned prior to hardening (no protective atmosphere). (Right) Complete decarburization of hardened AISI 5160 Mod spring steel that was given no surface preparation prior to hardening (no protective atmosphere). All specimens etched with 2% nital. Magnification, 200×.

Fig. 25. Examples of different degrees of decarburization in spring steels

tempered. The white outlined particles are the alloy carbides, also called primary carbides, and the matrix is tempered martensite plus a small amount of retained austenite not visible by light microscopy. Because these specimens were tempered at relatively low temperatures, the prior austenite grain boundaries are visible after etching.

Stainless steels can also exhibit a wide range of complex structures which can be even more complex after long-time, high-temperature exposure. Figure 22 illustrates some of the various combinations possible in stainless steels. AISI type 440C martensitic stainless steel contains undissolved alloy carbides in a high-carbon martensitic matrix (some retained austenite is also present but cannot be resolved by light microscopy). The AISI type 304 austenitic weld metal contains delta ferrite, which prevents hot cracking. The AISI type 17-4 PH (precipitation-hardening) stainless steel contains delta ferrite stringers in a low-carbon martensitic matrix. The duplex stainless steel (as cast and annealed) contains austenite in a ferritic matrix.

Various surface treatments can also be applied to ferrous alloys to produce considerable microstructural complexity at the treated surface. These processes can simply involve the application of sufficient heat to austenitize and subsequently harden the surface, as in flame hardening or induction hardening. Hardened surface layers may also be formed when low-hardenability carbon tool steels are water or brine quenched. The net result of these treatments is a martensitic surface layer and an unhardened core with a transition layer between them. The transition-layer microstructure may be more complex for flame- or induction-hardened specimens than for quenched carbon tool steels.

Other processes are also utilized that change the composition of the surface layer so as to develop unique combinations of properties—i.e., a strong and highly wear-resistant surface on a more-ductile, lower-strength core. The most common treatments used to produce these effects are carburizing, nitriding and carbonitriding. These practices are usually applied to low- or medium-carbon steels. Figure 23 shows the carburized surface of a low-carbon steel. At the extreme surface, the carbon content is above the eutectoid carbon content and a grain-boundary cementite layer is observed. Beneath this zone, the structure is fully pearlitic. At still greater depths, the amount of pearlite decreases until the matrix is reached. This specimen was cooled slowly after carburizing. Before such a steel would be placed in service, it would be heat treated to form martensite in the carburized case.

Nitriding is commonly performed on medium-carbon alloy steels that have been austenitized, quenched, and tempered between about 540 and 620 °C (1000 and 1150 °F). Nitriding is generally performed at temperatures slightly below this range for rather long times compared with those used for carburizing. Diffusion of nitrogen into the surface layer produces iron nitrides, which promote high surface hardness and beneficial residual compressive stresses. Figure 24 shows a nitrided specimen of AISI type 4150 steel on which has formed a rather thick surface layer of undesirable, brittle iron nitride.

Decarburization due to improper control of furnace atmosphere also produces a variation in surface microstructure. The micrograph at left in Fig. 25 shows a partially decarburized surface layer on hot rolled AISI type 9260 Mod spring steel. The white layer at the surface is nickel plating (applied for edge protection). The gray layer beneath the nickel plating is scale. The decarburized surface contains a range of ferrite and pearlite, and the matrix is nearly all pearlite. The other two micrographs in Fig. 25 show decarburization on hardened AISI 5160H and 5160 Mod spring steels. The view at center shows a very minor loss of carbon on a bar whose surface was turned prior to heat treatment. No free ferrite is present. The structure at the surface is a mixture of bainite and martensite. The micrograph at right in Fig. 25 shows a thick layer of free ferrite (total loss of carbon) with a columnar grain structure after heat treatment. Note the narrow zone beneath the free ferrite, where the structure changes gradually to all martensite.

SUMMARY

The examples presented in this article illustrate many of the microstructures commonly encountered in ferrous alloys. The natures of the structures present depend on the composition of the alloy and the manner of processing. To reveal these constituents clearly, the specimens must be properly prepared and etched with the reagent most appropriate for the constituents present.

REFERENCES

1. Nomenclature in the Metallography of Steel, by N. F. Kennon: *J. Australian Inst. Metals,* Vol 18, No. 2, June 1973, p 57-73
2. *Metallography: Principles and Practice,* by G. F. Vander Voort: McGraw-Hill, New York, 1984

Metallographic Technique for Nonferrous Metals and Special-Purpose Alloys

Technique for Refractory Metals and Alloys

MARKED DIFFERENCES in ductility, abrasion resistance, response to work hardening, and other pertinent properties among the refractory metals (tungsten, molybdenum, niobium and tantalum) and their alloys necessitate the use of a variety of techniques for preparation of metallographic specimens. A particular product form, such as wire, may also require special preparation techniques.

TUNGSTEN

Sectioning of tungsten and tungsten alloys should be performed using a resinoid-bonded, 70-grit aluminum oxide abrasive wheel, such as A-70-T-B. Wire can be cut with wet or dry abrasive wheels, preferably after a heavy nickel plate has been applied for protection. Nickel plating also aids in edge retention and in keeping the wire sections flat. Because tungsten wire can delaminate, it should never be cut with mechanical wire cutters.

Mounting. Most specimens of tungsten or tungsten alloys are mounted in the same manner as specimens of other metals.

Electrolytic Polishing. Aqueous electrolytes containing 1 to 10% sodium hydroxide or potassium hydroxide can be used successfully for electrolytic polishing of tungsten and tungsten alloys. Usually, a 2 to 4% sodium hydroxide aqueous solution is used, with a nickel cathode. Better results may be obtained if half the volume of water in the solution is replaced by glycerol and the sodium hydroxide content is increased to 5%.

Etching. Murakami's reagent, conventional or modified, is most often used for etching tungsten and tungsten alloys, although other etchants are sometimes used (see Table 1). In addition to the etchants shown in Table 1, electrolytic etching in a 4% sodium hydroxide aqueous solution, using 1 to 5 A ac or 0.5 to 1 A dc, has been successfully used in some laboratories for improving grain-boundary contrast.

MOLYBDENUM

Because molybdenum is relatively soft, scratches and distorted metal developed in sectioning, grinding and polishing are difficult to eliminate, and therefore electrolytic polishing is commonly used for molybdenum.

Electrolytic Polishing. An excellent polish, particularly with difficult-to-polish alloys, such as TZC,

Table 1. Etchants for metallographic specimens of refractory metals

Etchant name or ASTM number (E407)	Composition
Etchants for tungsten and molybdenum and their alloys	
Murakami's reagent (etchant 98c)	10 g $K_3Fe(CN)_6$, 10 g KOH or NaOH, 100 mL water
Murakami's reagent (mod)	15 g $K_3Fe(CN)_6$, 2 g NaOH, 100 mL water
Etchant 131 (electrolytic)	5 mL H_2SO_4, 1 mL HF, 100 mL methanol (95%)
Etchant 132(a)	5 mL HF, 10 mL HNO_3, 30 mL lactic acid
Etchant 209	15 mL HNO_3, 3 mL HF, 80 mL water
Additional etchants for molybdenum and molybdenum alloys	
Etchant 129	10 mL HF, 30 mL HNO_3, 60 mL lactic acid
Etchant 130 (electrolytic)	25 mL HCl, 10 mL H_2SO_4, 75 mL methanol
Etchants for niobium and tantalum and their alloys	
Etchant 66	30 mL HF, 15 mL HNO_3, 30 mL HCl
Etchant 158	10 mL HF, 10 mL HNO_3, 20 mL glycerol
Etchant 159	5 mL HF, 20 mL HNO_3, 50 mL acetic acid
Etchant 161	25 mL HNO_3, 5 mL HF, 50 mL water
Etchant 163	30 mL H_2SO_4, 30 mL HF, 3 to 5 drops H_2O_2 (30%), 30 mL water
Etchant 164	50 mL HNO_3, 30 g ammonium bifluoride, 20 mL water
Additional etchants for niobium and niobium alloys	
Etchant 160	20 mL HF, 15 mL H_2SO_4, 5 mL HNO_3, 50 mL water
Etchant 162B	30 mL lactic acid, 10 mL HNO_3, 10 mL HF
HNO_3-HF-water	20 mL HNO_3, 10 mL HF, 70 mL water
HCl-H_2SO_4-HNO_3-water	15 mL HCl, 15 mL H_2SO_4, 8 mL HNO_3, 62 mL water
Additional etchants for tantalum and tantalum alloys	
Etchant 177	10 g NaOH, 100 mL water
Etchant 178	20 mL HF, 20 mL HNO_3, 60 mL lactic acid
Etchant 179B (electrolytic)	10 mL HF, 90 mL H_2SO_4

(a) Procedure: Swab with heavy pressure for 5 to 10 s, water rinse, alcohol rinse, dry, etch with Murakami's reagent (etchant 98c).

can be obtained using an electrolyte consisting of 12.5% sulfuric acid, remainder ethanol or methanol. Nickel is used as the cathode. Voltages between 8 and 50 V dc have been used successfully.

Etching. A modified Murakami's reagent is recommended for etching molybdenum. It provides good grain-boundary contrast and yet minimizes etch pitting. A typical mixture contains 15 g of potassium ferricyanide, 2 g of sodium hydroxide and 100 mL of water.

NIOBIUM AND TANTALUM

Specimens of niobium and tantalum, and their alloys, are prepared by techniques similar to those used for preparing stainless steel specimens.

Polishing. A typical method of rough polishing niobium and tantalum employs a wax wheel and 15-μm levigated alumina. Intermediate polishing is done on a microcloth-covered wheel using 1-μm alumina, and final polishing is also done on a microcloth-covered wheel using 0.3-μm alumina.

Etchants used for metallographic specimens of niobium and tantalum, and their alloys, are listed in Table 1.

Technique for Magnetic Materials

METALLOGRAPHIC technique differs widely for the several groups of magnetic materials, which include silicon steels (also called electrical steels), iron-nickel alloys, iron-cobalt alloys, ferrites and garnets, and a variety of metals used for permanent magnets.

SILICON STEELS

Techniques for cutting, mounting, grinding and polishing are generally the same for specimens of silicon steels as for specimens of other low-carbon sheet steels.

Etching. For some purposes, such as revealing precipitated iron carbide in high-silicon steels, specimens are examined in the as-polished condition. Nital (3 to 10% nitric acid in methanol) is used extensively for etching. Picral (4% picric acid in ethanol) is sometimes used instead of nital, especially for revealing grain-boundary contaminants.

Grain orientation is often of particular interest in metallographic studies of silicon (electrical) steels. Pitting etchants, which selectively expose certain crystallographic faces, often are used to determine orientation with an optical goniometer. Results are not always consistent; details of preparation, including methods of rinsing and drying and other steps in the procedure, often must be determined individually for best results, and variations of the etchant are sometimes necessary. Pitting etchants that have been used successfully are listed in Table 2.

Special techniques for observing magnetic domains include use of Bitter patterns, use of the magneto-optic Kerr effect, and the Lorenz technique.

IRON-NICKEL AND IRON-COBALT ALLOYS

Techniques used for preparing metallographic specimens of iron-nickel and iron-cobalt ferromagnetic alloys generally are similar to those used for preparing specimens of other soft metals.

Macroetching. Macroexaminations of iron-nickel alloys frequently are made on specimens that have been subjected to thermal etching. In thermal etching, a polished specimen is placed in a furnace, usually with a charge of material for annealing, and is heated and cooled with the charge; annealing temperature is usually 1175 °C (2150 °F).

Mounting. When the surface of strip is to be examined, mounting is no problem. Bakelite is the mounting material most often used, although other materials can be used.

Mechanical grinding and polishing of iron-nickel and iron-cobalt alloys are basically the same as for strip or sheet steel specimens. Common practice is to grind dry on 120-grit silicon carbide and then wet on 400-grit and 600-grit silicon carbide. Wet grinding minimizes embedding of abrasive particles in the soft-metal specimen.

Rough polishing is done with 3-μm diamond paste on a napless cloth-covered wheel. Final polishing is preferably done with gamma alumina on a napped cloth.

Electropolishing is sometimes used for iron-nickel and iron-cobalt alloys. Common practice is to mechanically finish the specimen through 600-grit silicon carbide and then to electropolish. Two widely used electrolytes and typical electropolishing conditions are given in Table 3.

Microetching (Chemical). Table 4 lists several etchants that may be used for iron-nickel or iron-cobalt alloys. As shown in Table 5, some of the etchants in Table 4 are suitable for both groups of alloys; selection of etchant is often arbitrary. The first etchant listed in Table 4 (HCl, CuCl₂, FeCl₃, HNO₃, methanol and water) is most often used. Some metallographers prefer to obtain an electrolytic etch by reducing the current following electropolishing with one of the two solutions

Table 3. Electrolytes and conditions for electropolishing of iron-nickel and iron-cobalt magnetic alloys

Alloy	Electrolyte	Conditions for use
Fe-Ni only	135 mL acetic acid (glacial), 25 g CrO₃, 7 mL water	80 V, 0.8 to 1.6 A/cm², 5 to 30 s at 7 °C (45 °F) max
Fe-Ni or Fe-Co	100 mL acetic acid (glacial), 10 mL perchloric acid	45 V, 0.2 A/cm², 3 to 4 min at 24 °C (75 °F)

Table 4. Etchants for microscopic examination of iron-nickel and iron-cobalt magnetic alloys
See Table 5 for applicability

Etchant	Composition	Conditions for use(a)
Chemical etching		
1	100 mL HCl, 2 g CuCl₂, 7 g FeCl₃, 5 mL HNO₃, 200 mL methanol, 100 mL water	Immerse or swab for 10 to 15 s
2(b) ..	15 mL HCl, 5 g FeCl₃ (anhydrous), 60 mL ethanol	Immerse for 5 to 10 s
3	3 mL HCl, 1 mL HNO₃ saturated with CuCl₃	Swab for 2 to 3 s
4	15 mL HCl, 5 mL HNO₃, 10 mL glycerol	Swab for 10 to 15 s
5	Ammonium persulfate (saturated aqueous solution)	Immerse for 20 to 30 s
6	2 to 10% nital (HNO₃ in ethanol or methanol)	Immerse for 5 to 10 s
7	50 mL HCl, 10 g CuSO₄, 50 mL water (Marble's reagent)	Immerse or swab for 5 s
Electrolytic etching		
8	5 to 10 mL HCl, 100 mL water	2-5 s at 250-500 mA/cm²
9	2 g CrO₃, 100 mL water	2-5 s at 100-200 mA/cm²
10	3% sulfuric acid	5 to 10 s

(a) All etchants are used at room temperature. (b) Recommended for electron metallography.

Table 2. Pitting etchants for determination of grain orientation in silicon steels by optical microscopy

Etchant	Composition (parts are by volume)	Conditions for use	Purpose
1	1 part HF, 1 part HNO₃, 4 parts water	Immerse for 10 s.	Exposes {100} crystallographic faces in (110)[001] (cube-on-edge) oriented 3.25% Si steel.
2	2 parts HF, 1 part HNO₃, 3 parts methanol, 4 parts glycerol	Swab for 1 min.	Same as for etchant 1.
3	A: 6 mL H₂O₂ (30%), 0.1 mL HCl, 100 mL water B: 40 mL FeCl₃·6H₂O, 40 mL ethanol, 20 mL water	Immerse in A for 10 s, rinse and dry; then immerse in B for 3 s, rinse and dry.	Develops etch pits in (110)[001] (cube-on-edge) oriented 3.25% silicon steel.
4	100 g ferric sulfate, 100 mL H₂SO₄, 1000 mL water	Immerse for 15 s in solution heated to 80 to 90 °C (175 to 195 °F).	Develops etch pits in (100)[001] (cube-on-face) oriented 3.25% silicon steel.
5	A: 5 mL HF, 95 mL methanol B: 100 mL H₂O₂ (3%), 100 mL water, 2 drops HCl C: 5 mL HCl, 95 mL methanol	Polish, etch heavily in nital. Repolish, etch in nital to reveal grain boundaries. Immerse 10 s in A, rinse, dry; 2 s in B, rinse, dry; 30 s in C, rinse, dry.	Exposes {100} crystallographic faces in primary recrystallized 3.25% silicon steel and nonoriented silicon steels.
6	600 g FeCl₃, 10 g ammonium bisulfate, 600 mL HCl, 150 mL HNO₃, 1650 mL water	Immerse for 1 min in solution heated to 50 to 60 °C (120 to 140 °F).	Exposes {111} crystallographic faces in secondary recrystallized 50Ni-50Fe.

Table 5. Etchants in Table 4 recommended for microscopic examination of iron-nickel and iron-cobalt magnetic alloys

Etchant	Characteristic revealed
Iron-nickel alloys	
1, 2, 4, 5(a), 6	Grain size, structure
3(b), 7	Grain size
Iron-cobalt alloys	
1, 5, 6, 8, 9, 10	Grain size, structure

(a) For etching 50Fe-50Ni. (b) For etching high-nickel alloys such as Moly Permalloy.

for which compositions are given in Table 3.

The pros and cons of electrolytic etching for iron-nickel and iron-cobalt alloys are much the same as for other metals. Chemical etching is preferred in most laboratories.

FERRITES AND GARNETS

Because ferrites and garnets are electrical insulators, they cannot be electropolished or etched electrolytically. In addition, these materials are chemically inert, and thus are not readily attacked by most etchants.

Preparation of specimens up to the point of etching is essentially the same as for metals. Magnetically soft and magnetically hard ferrites, including nickel and barium ferrites, are very often examined in the as-polished condition. This technique reveals voids and second phases. Grain size can be estimated (*a*) from fractographs of broken surfaces (ferrites are brittle and fracture along grain boundaries); and (*b*) by etching polished surfaces with strong reagents, which results in attack on grain boundaries.

Most etchants for nickel ferrites contain hydrofluoric acid, combined with nitric, sulfuric or orthophosphoric acid, and are usually heated to 60 to 90 °C (140 to 195 °F). Etching time may range from a few seconds to 30 min.

ALNICO PERMANENT MAGNET ALLOYS

All Alnico alloys are hard (40 to 60 HRC) and brittle, so that specimens must be extracted by breaking or abrasive-wheel cutting, or both. When abrasive cutoff wheels are used, extreme care is necessary to avoid cracking and burning of specimens; slow feeds and an adequate flow of coolant must be used. Best practice is to remove at least 1.6 mm ($^1/_{16}$ in.) from the cut face or to avoid using the cut face for examination. Because of the high hardness, grinding and polishing are difficult and slow; consequently, it is advisable to use relatively small specimens (polished face no greater than 160 mm^2, or $^1/_4$ in.^{2}).

Mounting prior to grinding and polishing is preferred for small specimens, although it is not essential for large specimens. Bakelite is most often used as the mounting material, but other materials may be used.

Grinding and Polishing. Grinding is normally done by using silicon carbide of progressively finer grit sizes through 600 mesh. Rougher grinding may be done dry, but preferred practice is to use kerosine or light spindle oil for grinding with abrasive finer than 400 mesh.

A polishing procedure satisfactory for routine examination of Alnico alloys consists of wet polishing the specimen (previously ground to the finish obtained with 600-mesh abrasive) on a conventional rotating wheel with a billiard-cloth

cover. Levigated alumina 0.05-μm size or chromic oxide of 0.5-μm size can be used as the polishing abrasive.

Etching. Specimens of Alnico alloys sometimes are examined without etching, principally for voids such as porosity and microshrinkage. Also, unetched specimens are sometimes used for electron microscopy.

Alnico alloys resist attack by many chemicals, and thus a relatively severe etchant is required. Marble's reagent is most often used.

Procedures for other permanent magnet materials, such as Cunife and rare-earth–cobalt alloys, may resemble those discussed above. In some instances, etch-polishing or electrolytic techniques are necessary. Again, nital is a commonly used etchant.

Technique for Electrical Contact Materials

PROCEDURES for preparing specimens of electrical contact materials are basically the same as those for preparing specimens of most other metals. When electrical contacts are small, which they often are, sectioning is more difficult. Electrical contact materials have many compositions. Some are made of metals that are quite soft, some are composites containing metals that vary widely in hardness, and some are composed partly or completely of noble metals, which are difficult to etch. Consequently, some problems are likely to be encountered in the preparation of specimens, and some techniques of specimen preparation will differ from those of other materials.

EXAMINATION WITHOUT ETCHING

Microscopic examination of electrical contact materials is often done without etching the specimen—more often than for most other metals. A scanning electron microscope is commonly used to examine unetched contacts, especially for their surface character. In many contact materials, the constituents vary so widely in hardness that one metal will polish in relief, thus revealing distribution of one metal in the other Tungsten compacts infiltrated with silver, copper-tungsten mixtures, silver-graphite and silver-nickel combinations are often examined without being etched.

CHEMICAL ETCHING

The solutions used for etching specimens of electrical contact materials are listed in Table 6. Often, a specific metal or combination of metals can be etched successfully with two or more different etchants. Also, a specific etchant often is used for two or more different materials. The choice of etchant depends on the material or combination of materials being etched; whereas several etchants could be used, the final choice is often arbitrary, or may depend on conditions in the specific laboratory.

Table 7 suggests the preferred etchants for the most common electrical contact materials. Etchants composed of ammonium hydroxide and hydrogen peroxide, with and without added water, are frequently used to etch several different copper-base and most silver-base alloys (see etchants 1 and 3 in Table 6).

An etchant composed of potassium ferricyanide, sodium hydroxide and water (Murakami's reagent or a modification thereof) is used extensively for etching.

Table 6. Etchants and etching procedures for electrical contact materials

No.	Etchant Composition	Procedure for use
1	20 mL NH$_4$OH, 10-20 mL H$_2$O$_2$ (30%), 10-20 mL water	Swab at room temperature, 3-10 s. Use fresh. More water, less H$_2$O$_2$ for copper alloys; vice versa for silver alloys.
2	2 g K$_2$Cr$_2$O$_7$, 1.5 g NaCl, 8 mL H$_2$SO$_4$ (conc), 100 mL water	Swab at room temperature, 5-10 s. Good for etching hard-to-etch copper alloys.
3	50 mL NH$_4$OH, 10-30 mL H$_2$O$_2$ (30%)	Swab at room temperature for 3-10 s. Use fresh.
4	10 g FeCl$_3$, 90 mL water	Swab or immerse.
5	A: 100 mL saturated aqueous solution of K$_2$Cr$_2$O$_7$, 2 mL saturated aqueous solution of NaCl, 10 mL H$_2$SO$_4$ B: 1 part solution A, 10 parts water C: 98 mL water, 3 g CrO$_3$, 2 mL H$_2$SO$_4$	Use solution A, then solution B, then solution C. Swab at room temperature for 15-20 s with each solution. Rinse in water between solutions.
6	20 g CrO$_3$, 4.5 g NH$_4$Cl, 18 mL HNO$_3$ (conc), 15 mL H$_2$SO$_4$ (conc), water to make $^1/_2$ litre (Waterbury reagent)	Dilute 2:1 with water at time of use. Swab at room temperature for 3-10 s.
7	A: 25 mL HNO$_3$, 1 g K$_2$Cr$_2$O$_7$, 100 mL water B: 40 g CrO$_3$, 3 g Na$_2$SO$_4$, 200 mL water	Mix equal parts of A and B. Swab at room temperature for 5-10 s.
8	20 mL HNO$_3$ (conc), 20 mL acetic acid (glacial), 20 mL glycerol	Swab at 38-42 °C (100-108 °F) for 3-10 s.
9	0.2% CrO$_3$ and 0.2% H$_2$SO$_4$, in water	Swab for 1 min.
10	A: 200 mL HNO$_3$ (50%), 2 g K$_2$Cr$_2$O$_7$ B: 20 g CrO$_3$, 1.5 g Na$_2$SO$_4$, 100 mL water	Mix 1 part A with 20 parts B at time of use. Swab at room temperature, 3-15 s.
11	10 mL K$_3$Fe(CN)$_6$ (30%), 10 mL NaOH (10%) (Murakami's reagent)	Swab at room temperature for 5-15 s. Use at half strength for more control.
12	20 mL HNO$_3$ (conc), 20 mL acetic acid (glacial)	Immerse at room temperature, 10-20 s.
13	20 mL KCN (10%), 20 mL (NH$_4$)$_2$S$_2$O$_8$ (10%)	Use in a hood. Immerse at room temperature for 10-30 s.
14	A: 5% nital B: 5% FeCl$_3$ in methanol	Immerse specimen alternately in A and B.
15	10 mL HNO$_3$, 20 mL HCl, 10 mL glycerol	Swab at room temperature for 3-10 s.
16	30 mL HCl, 10 mL water	Electrolytic. Up to 5 V dc; 1.5 A/cm^2. Room temperature, 1-3 min.
17	A: 50 g KI, 30 g I, 50 mL water B: NH$_4$OH	Dilute A with water 1:1. Swab for 2 s. Then swab with B for 3 s, to remove stain.

Table 7. Suggested etchants for specific electrical contact materials

Material	Etchant No. in Table 6(a)
Copper-graphite	4
Chromium-copper	1
Cadmium-copper	1
Copper-cobalt-beryllium	1, 2
Copper-cobalt-silicon	1, 2
Copper-tungsten	1, 11
Silver (99.9% Ag)	3, 4, 5
Silver-copper	6, 7
Silver-copper-cadmium	7
Silver-cadmium	6 then 3 then 8
Silver-cadmium brazed to brass	6 then 3 then 8
Silver-graphite	6 then 3 then 8
Silver-nickel	9, 10
Silver-magnesium-nickel	9
Silver-molybdenum	11
Silver-tungsten	3, 11
Silver-tungsten carbide	11
Tungsten	11
Tungsten-copper	11
Tungsten-nickel	11
Molybdenum	11
Molybdenum-silver	11
Gold-silver-platinum	13
Gold-plated nickel-iron	11
Gold-silver clad palladium	12
Platinum-ruthenium	16
Platinum-iridium	16
Palladium welded to nickel silver	12
Palladium-ruthenium	13
Palladium-copper	13
Palladium-silver	13
Palladium-platinum-gold-silver-copper-zinc	16
Rhodium and gold-plated nickel-iron	14, 15
Hard gold plating on soft gold	17

(a) See Table 6 for compositions of etchants and procedures for using them.

Gold, palladium, and noble-metal alloys can be etched with a solution of potassium cyanide and ammonium persulfate.

Special procedures may be necessary, such as using different etchants consecutively to delineate different layers, or using electrolytic polishing and/or etching.

Technique for Sleeve Bearing Materials

SLEEVE BEARINGS, which usually have one or more layers of fairly soft metal bonded to a relatively thick steel backing, require metallographic-preparation techniques somewhat different from those for other metals and alloys.

SECTIONING

Metallographic specimens to be mounted should not be so large as to require excessive polishing time. A suggested maximum size is 19 mm ($^3/_4$ in.) long by 6.4 mm ($^1/_4$ in.) wide.

In sectioning a piece of bearing material to produce a specimen, it is advisable to avoid a large excess of steel backing compared with the soft-metal overlay. A thick steel layer increases the likelihood that relief will be produced in polishing because the soft-metal overlay is polished away.

Cutting is done with a power saw, a handsaw or an abrasive cutoff wheel. The direction of cutting should be from the soft metal into the steel, or with the soft metal–steel interface parallel with the direction of cutting. The direction of cutting should never be from the steel into the soft metal. Severely distorted soft metal should be removed by hand filing with a sharp mill file, again by cutting from the soft metal into the steel or with the interface parallel to the direction of cutting.

MOUNTING

Specimens from sleeve bearing materials are mounted in basically the same way as specimens of other metals.

When several small specimens are mounted in one block, they should be positioned so that the soft-metal layer of each faces in the same direction.

Bakelite and Lucite are often used as mounting materials, but for soft-metal specimens to be examined for internal voids, they are unsuitable; the time-temperature-pressure combination required is likely to result in collapse of the voids. For these specimens, self-curing acrylics and epoxy liquids that harden at room temperature and atmospheric pressure must be used. To eliminate some of the entrapped air and to obtain better penetration of the liquid mounting material, a brief vacuum treatment is frequently employed immediately after pouring the mounting liquid over the specimens.

GRINDING

Grinding with abrasive paper is done on a flat plate or on a low-speed disk grinder (550 rpm is suggested). A suitable series of abrasives is: 180-grit aluminum oxide; 320-grit aluminum oxide; then 0, 00, and 000 (500- to 600-grit) abrasive. Grinding action should proceed from the soft metal into the steel, or with the soft metal–steel interface parallel with the direction of grinding. Pickup of loosened abrasive particles and debris by the soft metal can be minimized by frequent cleaning of the abrasive paper with a suction cleaner.

POLISHING

Rough polishing is done on medium-speed (500-to-600-rpm) disks with billiard-cloth or broad-cloth coverings. Polishing abrasive, usually 0.3-μm high-purity alumina in a water suspension, is applied from a shaker bottle. A means of supplying a constant water drip at about 5-s intervals is helpful in keeping the wheel wet.

Finish polishing is done on a low-speed wheel (operating at less than 300 rpm), using 0.1-μm alumina on a napped cloth. During the final seconds of polishing, the wheel is flooded with water so that almost all of the abrasive is removed. The specimen is then washed in warm running water and dried in an air blast. Occasionally, disk polishing is followed by final polishing on the pad using magnesium oxide abrasive. This practice results in a virtually scratch-free surface and is particularly suitable for specimens of aluminum alloys.

Techniques that are standard for polishing most metals are not always good for polishing specimens that have large differences in hardness, such as specimens of babbitt overlays on steel. For sleeve bearing materials, polishing is done with

Table 8. Etchants for microscopic examination of sleeve bearing materials

Etchant	Some applications
NH$_4$OH, H$_2$O$_2$(a)	Commercial bronze liner
	Copper-lead alloy liner
	Copper-lead-tin alloy liner
	High-leaded tin bronze liner
	Leaded tin bronze liner
	Lead-tin-copper overlay on copper-lead alloy liner
	Nickel bronze infiltrated with lead-base babbitt
	Nickel-tin bronze infiltrated with lead-base babbitt
	Silver electroplate on steel
	Silver-lead alloy electroplate on steel
	Tin-base babbitt overlay on copper-lead-tin alloy liner
	Tin bronze infiltrated with lead-base babbitt
	Tin bronze infiltrated with Teflon
	Trimetal bearing: lead-tin-copper electroplated overlay, brass electroplated barrier, copper-lead alloy
0.5% HF	Aluminum alloy clad to steel
	Aluminum-silicon alloy clad to steel
	High-tin aluminum alloy clad with unalloyed aluminum
	Lead-tin-copper overlay on aluminum alloy liner
	Low-tin aluminum alloy clad to steel
	Trimetal bearing: lead-tin-copper electroplated overlay, copper electroplated barrier, aluminum-silicon-cadmium alloy
5% nital	High-tin aluminum alloy clad to nickel-plated steel
	Lead-base babbitt liner
	Tin-base babbitt liner
	Steel backing of any bearing alloy
Keller's reagent	Lead-tin-copper overlay on aluminum-cadmium alloy
Ferric chloride(b)	Cadmium alloy liner

(a) Equal parts of conc NH$_4$OH and water with 2 to 4 drops of H$_2$O$_2$ (30%) per 10 mL of solution. (b) 10 g FeCl$_3$, 90 mL ethanol.

an in-and-out radial, not a circular, movement of the specimen. The specimen is held so that the abrasive action is across the steel, into the soft metal (unlike grinding). This technique helps to hold the bond line level and to minimize the formation of a ridge at the interface between the steel and the soft metal.

If relief is produced because too much of the soft metal is polished away, a dark line will be visible at the interface between the steel and the soft metal when the specimen is viewed at low magnification. This line could easily be interpreted as a void or discontinuity at the bond line. Examination at high magnification, such as 500 diameters, will show the true nature of the bond line. To minimize preferential erosion of the soft metal, specimens in a mount should be few and small, polishing time should be restricted to an absolute minimum, and the polishing technique described above should be used. Minimizing the thickness of the steel backing layer will also help in the avoidance of polishing relief.

A small amount of relief is usually no problem during visual examination. However, in making micrographs, part of the specimen may be out of focus and it may be necessary to shim up the edge of the mount on the steel side with thin strips of paper or metal foil in order to compensate for the small tilt in the polished surface. Depending on the type of microscope used, the same result may be achieved easier by shimming the central support plate of the microscope stage.

ETCHING

Specimens of sleeve bearing materials are usually etched for metallographic examination. Etching is usually done by immersion. Only a few etchants are used. Five of these etchants, and some of the sleeve bearing materials to which they are best adapted, are listed in Table 8.

Technique for —— Aluminum Alloys——

ALUMINUM ALLOYS encompass a wide range of chemical composition and, as a result, a wide range of hardness; consequently, there is considerable variation in the techniques required for metallographic preparation and examination. The softer alloys generally are more difficult to prepare by mechanical polishing, because deformation caused by cutting and grinding extends to a greater depth, embedding of abrasive particles in the metal during polishing is more likely, and relief between the matrix and second-phase particles, which are much harder than the matrix, will develop more readily during polishing.

METALLOGRAPHY ON THE MACROSCOPIC SCALE

The same principles of preparation for macroscopic examination apply to aluminum alloys as apply to metals in general.

Chemical Preparation. Removal of cutting oils and other greasy contaminants from aluminum surfaces before etching is helpful but not always necessary. The caustic etch (etchant 1 in Table 9) is, in itself, an excellent degreaser. The acidic etchants are more likely than the caustic etch to act in an uneven manner if the surface is not precleaned.

Table 9 lists a selection of etchants and etching methods that will accomplish all necessary tasks of preparing specimens for macroexamination.

Table 10 indicates the etchants in Table 9 that are most applicable to various classes of alloys. Where Table 10 presents a choice between caustic and mixed-acid etching, the selection should be based on the primary purpose of the examination.

METALLOGRAPHY ON THE MICROSCOPIC SCALE

The determination of the best procedure for microscopic examination is governed by the same considerations as in macroscopic examination, but the area that can conveniently be examined is usually more limited.

Mounting in a plastic medium to form a cylindrical piece is the generally accepted procedure, unless the specimen is of such dimensions that it can easily be hand held for subsequent grinding and polishing.

Special problems relating to choice of mounting method or material may be caused by the inclusion of alloys of greatly dissimilar hardness in the same mount, by the need for preservation of flatness to the very edge, by the need to mount thin sheet specimens that must be polished in a plane perpendicular to the rolled surface, and by the need for connecting electrical leads to one or more specimens for subsequent electropolishing or electrolytic etching.

Grinding and Rough Polishing. Because aluminum and its alloys are relatively soft compared with many other metals, most abrasives are satisfactory for grinding, although SiC papers are most commonly used. Grinding should progress through the standard sequence of grit sizes—i.e., 180, 240, 320, 400 and 600 grit. For control of relief, it is possible to use grinding paper with still finer grits. Rough polishing follows, using a napless or low-nap cloth for relief control or a medium-nap cloth if relief is not a problem. The most commonly used abrasive for rough polishing is 6-μm diamond paste, with an extender.

Final Polishing. The most commonly used final-polishing medium for aluminum alloys is a pure heavy grade of magnesium oxide powder, used on a medium-nap or short-nap cloth of uniform texture. The abrasive is rendered pastelike by applying about a teaspoonful near the center of the cloth and moistening it with distilled or deionized water, working it to a proper consistency with the finger. A low wheel speed, such as 300 rpm, is required for final polishing. High hand pressure and frequent rotation of the specimen are used for the first few minutes, and only enough water is added to avoid dryness and pulling of the specimen by the cloth. Gradually, pressure is relaxed and more water is added to wash away excess abrasive. Toward the end of the polish, copious quantities of water can be used to remove all abrasive, and the polishing cloth, in effect, serves to wipe the specimen clean. Residual abrasive may be removed by lightly applying a clean, wet cotton swab.

Etching for Microscopic Examination. Microscopic examination of the polished specimen before etching is often advisable, because etching can obscure as well as reveal important details, such as incipient melting, fine cracks, and nonmetallic inclusions.

Table 11 gives a selection of etchants that encompass all the ordinary purposes of microscopic examination of commercial aluminum alloys. Table 12 describes these purposes and suggests etchants that are most likely to be successful on various classes of alloys. Often, it may be possible to apply a second etch directly over the first, without repolishing, as dictated by experience. A general rule-of-thumb is that the etchants that reveal grain structure are the most aggressive and should be applied last. When use of more than one etchant is anticipated and when these etchants cannot be used in tandem, valuable repolishing time can be saved by immersing just a portion of the polished specimen area, saving the remainder for another etchant.

Table 10. Applicability of etchants in Table 9 to macroexamination of aluminum alloys

Alloy	Etchant
High-purity aluminum	4 or 5
Commercial-purity aluminum:	
1xxx series	4, 2 or 1
All high-copper alloys:	
2xxx series and casting alloys	1
Al-Mn alloys:	
3xxx series	4, 2 or 1
Al-Si alloys:	
4xxx series and casting alloys(a)	4, 2 or 3
Al-Mg alloys:	
5xxx series and casting alloys	4, 2 or 1
Al-Mg-Si alloys:	
6xxx series and casting alloys	4, 2 or 1
Al-Cu-Mg-Zn alloys:	
7xxx series and casting alloys	1

(a) Also, welds and brazed joints made with the use of these alloys as filler metals.

Table 9. Etchants for use in macroscopic examination of aluminum alloys
See Table 10 for applicability to specific alloys.

Etchant	Composition (parts are by volume)	Procedure for use
1 (caustic etch)	10 g NaOH to each 90 mL water	Immerse specimen 5 to 15 min in solution heated to 60 to 70 °C (140 to 160 °F)(a), rinse in water, dip in 50% HNO₃ solution to desmut, rinse in water, dry.
2 (Tucker's reagent)	45 parts HCl (conc), 15 parts HNO₃ (conc), 15 parts HF (48%), 25 parts water	Mix fresh before using. Immerse or swab specimen for 10 to 15 s, rinse in warm water, dry, and examine for desired effect. Repeat as much as necessary until desired effect is obtained.
3	1 part HF (48%), 9 parts water	Requires fairly smooth surface. Immerse until desired effect is obtained, hot-water rinse, dry.
4 (Poulton's reagent)	12 parts HCl (conc), 6 parts HNO₃ (conc), 1 part HF (48%), 1 part water	May be premixed and stored(b) for long periods. Etch by brief immersion or by swabbing. Rinse in cool water, and do not allow either the etchant or the specimen to heat up during etching.
5	50 parts HCl (conc), 15 parts HNO₃ (conc), 3 parts HF (48%), 5 parts FeCl₃ soln (conc)	Mix fresh before use. Cool solution to 10 to 15 °C (50 to 60 °F) with jacket of cold water. Immerse a few seconds, rinse in cold water; repeat until desired effect is obtained.

(a) This etchant may be used without being heated, but the etching action will be slower. (b) Solution should be stored in a vented container, preferably under a fume hood, to prevent buildup of gas pressure. The container should be made of polyethylene or be lined with wax.

Table 11. Etchants for use in microscopic examination of aluminum alloys
See Table 12 for applicability to specific alloys.

Etchant	Composition	Procedure for use
1 (hydrofluoric acid etch)	1 mL HF (48%), 200 mL water	Swab for 15 s, or immerse for 30 to 45 s.
2	1 g NaOH, 100 mL water	Swab for 5 to 10 s.
3A (Keller's reagent)	2 mL HF (48%), 3 mL HCl (conc), 5 mL HNO$_3$ (conc), 190 mL water	Immerse for 8 to 15 s, wash in stream of warm water, blow dry. Do not remove etching products from surface.
3B (dilute Keller's reagent)	20 mL etchant 3A, 80 mL water	Mix fresh before using. Immerse specimen for 5 to 10 s.
4 (modified Keller's reagent)	2 mL HF (48%), 3 mL HCl (conc), 20 mL HNO$_3$ (conc), 175 mL water	Immerse for 10 to 60 s, wash in stream of warm water, blow dry. Do not remove etching products from surface.
5 (Barker's reagent)	4 to 5 mL HBF$_4$ (48%), 200 mL water	Electrolytic: use Al, Pb or stainless steel for cathode; specimen is anode. Anodize 40 to 80 s at about 0.2 A/cm^2 (about 20 V dc). Check results on microscope with crossed polarizers.
6	25 mL HNO$_3$ (conc), 75 mL water	Immerse in solution at 70 °C (160 °F) for 45 to 60 s.
7	20 mL H$_2$SO$_4$ (conc), 80 mL water	Immerse at 70 °C (160 °F) for 30 s; rinse in cold water.
8	10 mL H$_3$PO$_4$ (85%), 90 mL water	Immerse at 50 °C (120 °F) 1 min, or 3 to 5 min (see Table 12).
9	5 mL HF (48%), 10 mL H$_2$SO$_4$, 85 mL water	Immerse for 30 s.
10	4 g KMnO$_4$, 2 g Na$_2$CO$_3$, 94 mL water, few drops wetting agent	Specimen surface must be well polished, and be precleaned in 20% H$_3$PO$_4$ at 95 °C (205 °F) for uniform wettability. After precleaning, rinse in cold water and immediately immerse in etchant for 30 s.
11	2 g NaOH, 5 g NaF, 93 mL water	Immerse for 2 to 3 min.
12	50 mL Poulton's reagent (etchant 4 in Table 9), 25 mL HNO$_3$ (conc), 40 mL of solution of 3 g chromic acid per 10 mL of water	Put a few drops on as-rolled or as-extruded surface for 1 to 4 min, rinse, and swab to desmut. Examine on microscope with crossed polarizers to show grains. Repeat etching, if necessary. For some 5xxx alloys, increase amount of HNO$_3$ in solution to 50 mL.
13	8 mL HNO$_3$ (conc), 2 mL HCl (conc), 45 mL water, 45 mL methanol	Immerse for 10 s.
14	5 mL acetic acid (glacial), 1 mL HNO$_3$ (conc), 94 mL water	Immerse for 20 to 30 min.

Table 12. Applicability of etchants in Table 11 to microscopic examination of aluminum alloys

Alloy	Etchant	Evidence revealed
Examination for grain size and shape		
1xxx, 3xxx, 5xxx, 6xxx series; most casting alloys	5 or 12	Grain contrast when using crossed polarizers, with or without sensitive tint
2xxx, 7xxx series; Al-Cu or Al-Zn casting alloys	3A or 11	Grain contrast or grain-boundary lines
5xxx series alloys with more than 3% Mg	8 (3-5 min)	Precipitation in grain boundaries
Examination for cold working		
1xxx, 3xxx, 5xxx, 6xxx series alloys	5 or 12	Deformation bands or markings that cause streaked effect when using crossed polarizers
2xxx, 7xxx series alloys	3A or 11	Deformation bands or markings that accompany relatively great amounts of cold working
5xxx series alloys with more than 3% Mg	8 (3 to 5 min)	Precipitation in bands of slip
Examination for incomplete recrystallization		
1xxx, 3xxx, 5xxx, 6xxx series alloys	5 or 12	Even-toned, well-outlined grains that are recrystallized, otherwise streaked, or banded
2xxx series alloys, hot worked and heat treated	3A or 11	Unrecrystallized grains made up of multiple, very fine subgrains
6xxx series alloys, hot worked and heat treated	9	Unrecrystallized grains made up of multiple, very fine subgrains
7xxx series alloys, hot worked and heat treated	8 (3 to 5 min) or 14	Unrecrystallized grains made up of multiple, very fine subgrains
Examination for preferred orientation		
1xxx, 3xxx, 5xxx, 6xxx series alloys	5 or 12	Predominance of certain gray tones when crossed polarizers are used, lack of randomness
2xxx series alloys in T4 temper	3A or 11	Lack of randomness in grain contrast
Examination for overheating (partial melting)		
2xxx series alloys	8 (1 min)	Rosettes and grain-boundary eutectic
6xxx series alloys	2	Grain-boundary eutectic formations
7xxx series alloys	3B	Rosettes and grain-boundary eutectic formations
Examination for general constituent size and distribution		
All wrought alloys and casting alloys	1, 8 (1 min) or any etchant that does not pit solid-solution matrix	Coarse insoluble particles and fine precipitate particles. Longer etching time exaggerates size of fine particles.
Examination for distinction between solution heat treated (T4) and artificially aged (T6) tempers		
2xxx series alloys	3A or 11	Loss of grain contrast, general darkening, in T6 compared with T4
6061	9	Clear outlining of grain boundaries in T6; faint outlining in T4
7075, recrystallized	4	More grain contrast, sharper grain-boundary outlining, in T4
Examination for over-aging or poor quenching of solution heat treated alloy		
2017 and 2024, in T4 temper	6	Faint dark precipitate at grain boundaries
Examination for cladding thickness		
Alclad 2014, 2024, 7075	3A or 11	Boundary between high grain contrast or outlining of alloy core and lighter-etching cladding
Brazing sheet	1 (swab) or 13	Boundary of high-silicon cladding alloy
Other clad alloys	1 (immerse), 2, 3A, 5 or 11	Any differences in structure that demarcate one layer from another
Examination for solid-solution coring or segregation, and diffusion effects		
3xxx, 5xxx series; Al-Mg casting alloys	10	Interference colors due to differences in thickness of tarnish films laid down on the surface
2xxx series alloys and others with more than 1% Cu	3A or 11	Brownish-colored films due to redeposition of copper

Technique for Copper and Copper Alloys

PREPARATION of specimens of wrought and cast coppers and copper alloys for macroexamination and microexamination is subject to the same general rules and recommendations as those that apply to preparation of specimens of other metals.

TECHNIQUES FOR MACROEXAMINATION

Specimens for macroscopic examination are extracted from larger masses using common cutting tools. The tools must be kept sharp to minimize cold working of the specimen.

Surface Preparation. Surfaces suitable for macroetching usually can be obtained in two machining operations. In the first operation, a heavy cut is taken to remove the metal that was cold worked during sectioning; in the second operation, a light cut is taken using a V-shape tool, or the specimen is ground, to remove the remaining effects of cold work.

Etching. Deep etching removes the effects of cold work but produces a rough surface, and so it is common practice to deep etch the machined or rough-ground surface, regrind it lightly, and then give it a light etch.

Selection of an etchant for a macrospecimen depends mainly on the alloy to be etched and the features to be examined. Frequently, the capabilities of two or more etchants overlap (or are the same) so that selection of a specific etchant is arbitrary. Table 13 gives compositions of the more commonly used macroetchants, along with etching procedures, purposes of the etchants or characteristics revealed, and alloys for which they are ordinarily used.

TECHNIQUES FOR MICROEXAMINATION

Specimens of copper and copper alloys for microscopic examination are extracted from larger masses by sawing, shearing, filing, hollow boring and abrasive-wheel cutting.

Mounting. In general, the practice used for mounting copper and copper alloy specimens is the same as for mounting specimens of other metals.

Grinding. Wet grinding is preferred for all coppers and copper alloys. Common practice is first to rough grind the specimen surface so as to remove metal that has been cold worked, then finish grind to obtain a suitable surface. Finish grinding is done on flat wheels using silicon carbide papers of progressively finer grit sizes— usually, 240, 320, 400 and 600 mesh.

Rough Polishing. Most coppers and copper alloys are relatively soft, and thus require a polishing medium that provides maximum cutting action

Table 13. Etchants for macroscopic examination of coppers and copper alloys
Procedure for use: immerse at room temperature, rinse in warm water, dry.

Composition of etchant	Copper or copper alloy	Purpose, or characteristic revealed
50 mL HNO_3, 0.5 g $AgNO_3$, 50 mL water	All coppers and copper alloys	Produces a brilliant, deep etch
10 mL HNO_3, 90 mL water	Coppers and all brasses	Grains; cracks and other defects
50 mL HNO_3, 50 mL water(a)	Coppers, all brasses, aluminum bronze(b)	Same as above; reveals grain contrast
30 mL HCl, 10 mL $FeCl_3$, 120 mL water or methanol	Coppers and all brasses	Same as etchant above(c)
20 mL acetic acid, 10 mL 5% CrO_3, 5 mL 10% $FeCl_3$, 100 mL water(d)	All brasses	Produces a brilliant, deep etch
2 g $K_2Cr_2O_7$, 4 mL saturated solution of NaCl, 8 mL H_2SO_4, 100 mL water(e)	Coppers, high-Cu alloys, phosphor bronze	Grain boundaries, oxide inclusions
40 g CrO_3, 7.5 g NH_4Cl, 50 mL HNO_3, 100 mL water	Silicon brass, silicon bronze	General macrostructure

(a) Solution should be agitated during etching, to prevent pitting of some alloys. (b) Aluminum bronzes may form smut, which can be removed by brief immersion in concentrated HNO_3. (c) Excellent for grain contrast. (d) Amount of water can be varied as desired. (e) Immerse specimen for 15 to 30 min, then swab with fresh solution.

Table 14. Etchants and procedures for microetching of coppers and copper alloys

Etchant composition(a)	Procedure	Copper or copper alloy
20 mL NH_4OH, 0 to 20 mL water, 8 to 20 mL H_2O_2 (3%)	Immersion or swabbing for 1 min. (H_2O_2 content varies with copper content of alloy to be etched; use fresh H_2O_2 for best results.)(b)	Coppers and copper alloys. Film on etched aluminum bronze can be removed with weak Grard's solution.
1 g $Fe(NO_3)_3$, 100 mL water	Add dropwise to the cloth during polishing with an abrasive, such as alumina.	Attack polishing of coppers and copper alloys
25 mL NH_4OH, 25 mL water, 50 mL $(NH_4)_2S_2O_8$ (2.5%)	Add dropwise to the cloth during polishing with an abrasive, such as alumina.	Attack polishing of coppers and some copper alloys
2 g $K_2Cr_2O_7$, 8 mL H_2SO_4, 4 mL NaCl (saturated solution), 100 mL water	Immersion. (NaCl replaceable by 1 drop HCl per 25 mL soln; add just before using.) Follow with $FeCl_3$ or other contrast etch.	Coppers; copper alloys of beryllium, manganese and silicon; nickel silver; bronzes; chromium copper
CrO_3 (saturated aqueous solution)	Immersion or swabbing	Coppers, brasses, bronzes, nickel silver
50 mL CrO_3 (10 to 15%), 1 to 2 drops HCl	Immersion (add HCl at time of use)	Same as above. Color by electrolytic etching or with $FeCl_3$ etchants.
10 g $(NH_4)_2S_2O_8$, 90 mL water	Immersion (use either cold or boiling)	Coppers, brasses, bronzes, nickel silver, aluminum bronze
10% aqueous solution of copper ammonium chloride plus ammonium hydroxide to neutrality or alkalinity	Immersion. Wash specimen thoroughly.	Coppers, brasses, nickel silver; darkening large areas of beta in alpha-beta brass
$FeCl_3$, g — HCl, mL — Water, mL: 5 — 50 — 100; 20 — 5 — 100(c)(d); 25 — 25 — 100; 1 — 20 — 100; 8 — 25 — 100; 5 — 10 — 100(e)(f)	Immersion or swabbing. Etch lightly or by successive light etches to required results	Coppers, brasses, bronzes, aluminum bronze; darkens beta in brass; gives contrast following dichromate and other etches
5 g $FeCl_3$, 100 mL ethanol, 5 to 30 mL HCl	Immersion or swabbing for 1 s to several minutes.	Coppers and copper alloys
Nitric acid (various concentrations)	Immersion or swabbing. $AgNO_3$ (0.15 to 0.3%) added to 1:1 solution gives a brilliant, deep etch.	Coppers and copper alloys
Ammonium hydroxide (dilute solutions)	Add dropwise to the cloth during polishing with an abrasive.	Attack polishing of brasses and bronzes
50 mL HNO_3, 20 g CrO_3, 30 mL water	Immersion	Aluminum bronze, free-cutting brass. Film from polishing can be removed with 10% HF.
5 mL HNO_3, 20 g CrO_3, 75 mL water	Immersion	Same as above
50 mL cold saturated solution of sodium thiosulfate, 1 g potassium metabisulfite	Lightly pre-etch specimen with 10% aqueous ammonium persulfate. Immerse in solution until surface is colored. Gently agitate.	Tint etch for many copper and copper alloys. Produces brilliant colors. Colors beta more rapidly than alpha. Can be stored and used over.
5 g ferric nitrate, 25 mL HCl, 70 mL water	Immerse in solution for 10 to 30 s.	Excellent general-purpose etchant for coppers and copper alloys. Good grain-boundary etchant.
3 g ammonium persulfate, 1 mL ammonium hydroxide, 100 mL water	Swab for 5 to 20 s.	Excellent etchant for coppers and copper alloys, particularly beryllium copper. Reveals grain boundaries in solution-annealed specimens.

(a) The use of concentrated etchants is intended unless otherwise specified. (b) This etchant may be alternated with $FeCl_3$. (c) Grard's No. 1 etchant. (d) Plus 1 g CrO_2. (e) Grard's No. 2 etchant. (f) Plus 1 g $CuCl_2$ and 0.05 g $SnCl_2$.

with minimum rubbing. Rough polishing should be done on diamond-impregnated nylon cloth. Duck, canvas, wool broadcloth, and cotton (listed in order of decreasing preference) are also used for polishing.

Although diamond paste is more expensive than other coarse abrasives, it is preferred by metallographers because it provides faster and more effective polishing. Diamond paste of 3- to 9-μm size is the preferred abrasive for rough polishing on any of the cloths mentioned above. A wheel speed of about 300 to 500 rpm is generally recommended.

Finish Polishing. Generally, napped cloths are preferred for finish polishing. Abrasive is usually 0.3-μm alpha alumina or 0.05-μm gamma alumina; both abrasives are used with water as a vehicle. Other abrasives that have proved satisfactory for finish polishing are magnesium oxide in distilled water, and colloidal silica. A wheel speed of 150 to 300 rpm is recommended.

There are numerous opinions on how the specimen should be rotated during polishing. In hand polishing, the specimen generally is rotated in a direction opposite the direction of wheel rotation while moving from edge to center.

After polishing, the specimen is rinsed in water and dried with tissue or warm air. Automatic polishing (usually vibratory) has proved highly successful for polishing copper alloys. In automatic polishing, human variables are greatly minimized. Attack polishing (combined polishing and etching), with ferric nitrate or with ammonium hydroxide — ammonium persulfate (see Table 14) can be more safely done with automatic equipment than by hand.

Chemical Etching. Table 14 lists chemical etchants that are used for microetching coppers and copper alloys, and includes etching procedures and the alloys to which each etchant is commonly applied.

The ammonium hydroxide – hydrogen peroxide – water solution (listed first in Table 14) is by far the most widely used etchant. It is probably the best for routine work and is applicable to many coppers and copper alloys.

Technique for Lead and Lead Alloys

PREPARATION of lead and lead alloy specimens for metallographic examination presents problems that are not common to preparation of most other metals, because lead and lead alloys are so soft that they deform even under low pressure. Consequently, considerable surface flow or distortion can occur, and abrasives can become embedded during grinding and polishing and, if not removed can obscure the true structure of (or produce a pseudostructure in) the specimen being examined. Proper preparation entails grinding with 400-grit SiC abrasive, rough polishing with 9-μm diamond paste on a napless cloth and 0.3-μm Al₂O₃ slurry on a medium-nap cloth, and final polishing with colloidal silica on a medium-nap cloth using very light pressure, until the surface is bright and scratch-free. In addition, because of the low melting points of lead and lead alloys, deformation of the specimen in metallographic preparation can be sufficient to cause heating and recrystallization. Therefore, in preparation of lead and lead alloy specimens, it is mandatory to employ a technique that will result in minimum surface distortion and heating. Use of a cold mounting procedure will enable the metallographer to avoid microstructural and crystallographic changes.

ETCHING

Table 15 gives compositions of the etchants most often used for macroetching and microetching of lead and lead alloy specimens. Etching procedures and specific uses for the various etchants are also included in Table 15. All but two of these etchants are used by immersing or swabbing the specimen; only etchants 9 and 10 are used for electrolytic etching.

Primarily because lead and lead alloys smear so easily, etch-polishing alternated with examination under the microscope is required. Some-

Table 15. Recommended etchants and procedures for macroscopic and microscopic examination of lead and lead alloys

Etchant No.	Composition (parts are by volume)	Procedure	Use
1	1 part acetic acid (glacial) 1 part nitric acid (conc) 4 parts glycerol	Use freshly prepared solution at 80 °C (175 °F); discard after use. For macroetching: etch several minutes, rinse in water. For microetching: etch several seconds. For best results, alternate etching with polishing.	Macroetching of lead; development of microstructures and grain boundaries in lead, and in lead-calcium, lead-antimony and lead-tin (low-Sn) alloys
2	100 parts acetic acid (glacial) 10 parts hydrogen peroxide (30%)	Etch for 10 to 30 min, depending on the depth of the disturbed layer. Dry and clean with concentrated nitric acid if required.	Microetching of lead-antimony alloys containing up to 2% antimony
3	3 parts acetic acid (glacial) 1 part hydrogen peroxide (30%)	Etch by immersing specimen in solution for 6 to 15 s. Dry with alcohol.	Microetching of lead, lead-calcium alloys, and lead-antimony alloys containing more than 2% antimony. Also removes disturbed metal.
4	Solution A: 15 g ammonium molybdate 100 mL distilled water Solution B: 6 parts nitric acid (conc) 4 parts distilled water	Mix equal quantities of solutions A and B. Etch by alternately swabbing specimen and washing in running water.	Macroetching of lead. A very rapid etchant; well suited for removing thick layers of disturbed metal from specimens.
5	3 parts acetic acid (glacial) 4 parts nitric acid (conc) 16 parts distilled water	Use freshly prepared solution at 40 to 42 °C (104 to 108 °F). Immerse specimen for 4 to 30 min until disturbed layer is removed. Clean with cotton in running water.	Microetching of unalloyed lead, and lead-tin alloys containing up to 3% tin
6	2 parts acetic acid (glacial) 2 parts nitric acid (conc) 2 parts hydrogen peroxide (30%) 5 parts distilled water	Etch for 2 to 10 s by swabbing. Rinse specimen in running water and dry with alcohol.	Macroetching of unalloyed lead, and of lead-bismuth, lead-tellurium and lead-nickel alloys
7	1 part nitric acid (conc) 1 part distilled water	Etch for 5 to 10 min by immersion. If thick layer of disturbed metal is to be removed, solution can be heated to boiling. Rinse in running water, rinse in alcohol and dry.	Developing macrostructure of welds and laminations in lead products
8	Solution A: 10% aqueous solution of ammonium persulfate Solution B: 30% aqueous solution of tartaric acid	Mix 5 mL of solution A with 2 mL of solution B. Swab specimen for 5 to 10 s. Rinse in running water.	Microetching to distinguish cuboidal SbSn phase from Sb-rich phases in Pb-Sb-Sn alloys such as bearing alloys or type metals. Solution A blackens SbSn phase; solution B etches Sb-rich phases.
9	6 parts perchloric acid (70%) 4 parts water	Immerse specimen (cathode) in electrolyte; anode is platinum spiral. Etch 45 to 90 s at 6 V, 4 A, from a rectifier.	Electrolytic etching of lead-antimony alloys containing more than 2% antimony
10	1 part hydrochloric acid (conc) 9 parts water	Same as for etchant 9	Same as for etchant 9
11	10 g ammonium molybdate 10 g citric acid 100 mL water	Etch for 15 to 30 s by immersion, rinse in warm water and alcohol and blow dry.	Delicate etch for lead and lead alloys. Ideal for revealing phases for high-magnification examination, microstructures and grain boundaries in lead and in lead-calcium, lead-antimony and lead-tin (low-Sn) alloys

times, several cycles of polishing, etching and examination are necessary; thus, a great deal of skill and patience is required. In addition to smearing, lead-antimony alloys are susceptible to tarnishing (or blackening) during etching. One or two light turns on the final polishing wheel, followed by immersion of the specimen in water, rinsing in ethanol or methanol, and drying in warm air, is often sufficient to remove this tarnish and reveal the true structure. Another technique involves use of a camel hair brush and liquid hand soap to remove the black etching product followed by rinsing in a strong stream of running water and drying with alcohol and a warm air blast. Discoloration resulting from overetching or oxidation, or both, can sometimes be removed by lightly swabbing the specimen with EDTA (a solution of ethylenediamine tetraacetic acid disodium salt).

Finally, lead and lead alloys begin to oxidize immediately after metallographic preparation and should be examined as soon as possible. Although EDTA sometimes removes the discoloration resulting from oxidation, the best solution is thorough drying of the specimen after etching. EDTA will not solve the problem of overetching.

SELECTED REFERENCES

Improved Metallographic Preparation of Lead and Lead Alloys, by R. M. Stepian and G. A. Blann: *Metallography*, Vol 12, No. 3, 1979, p 195-214
Metallographic Technique for Lead-Acid Battery Plates, by M. D. Allen: *Microstructural Science*, Vol 6, Elsevier, New York, 1978, p 31-45

Technique for Magnesium Alloys

SPECIMENS of magnesium alloys for metallographic examination should be chosen with care so as to be representative. For example, both longitudinal and transverse specimens of sheet should be examined, edge and center sections of large ingots should be considered, and the mouth of a casting crack should be studied as the point of origin of the crack.

SPECIMEN PREPARATION

Specimens are removed from the metal mass by band sawing or hacksawing. Care should be exercised to prevent cold working of the metal, because cold working can alter the microstructure and complicate interpretation of constituents.

Specimens that have been severely cold worked by rough sawing, squeezing in a vise, or heavy stamping are likely to be mechanically twinned close to the worked surface. The cold worked surface can be easily removed by extending the time of each grinding and polishing operation up to twice the time required for removing scratches from the preceding stage. Removal of metal to a depth of approximately 1 mm (0.040 in.) during preparation of the specimen is recommended.

Grinding. Dry and wet grinding are done on abrasive belts or rotating disks, or on abrasive papers by hand.

Mechanical polishing is done in two stages: rough and finish. Rough polishing removes the major part of the disturbed metal remaining after the final grinding step. Finish polishing removes the superficial scratches that remain after rough polishing.

Wheels used for both polishing stages are covered with medium-nap cloths. A suspension of 600-grit aluminum oxide powder in distilled water (35 g of aluminum oxide per 500 mL of water) is used on the wheel for rough polishing; for best results, the cloth is maintained just moist enough to prevent seizure of the specimen. The specimen is rotated counter to wheel direction. Polishing time should be twice as long as is needed to remove the scratches from the final grinding operation. A moderate amount of pressure should be used to minimize relief polishing caused by

the difference in rate of stock removal between the harder and softer microconstituents.

For finish polishing, a suspension of alpha alumina in distilled water (10 g of alumina per 500 mL of water) is used on the wheel. Sometimes, to facilitate polishing, 15 mL of filtered or other soft soap is added to the abrasive solution.

Macroetching. The macroetchant most often used for showing discontinuities in castings and flow lines in forgings is an aqueous solution containing 5 to 20% acetic acid. The etchant is swabbed on the prepared surface for 10 s to 3 min and then washed away in running water.

To show the grain structure of impact extruded or forged parts that have a homogeneous recrystallized structure and a minimum of alloy gradients, and for castings, one of the acetic-picral etchants listed in Table 16 as etchants 6, 7 and 8 should be used. The part is immersed in the etchant for 10 s to 1 min, transferred to a pan of ethanol to rinse the etched surface uniformly, rinsed in flowing ethanol, and then dried in a blast of air.

MICROEXAMINATION

The etchants used and etching times depend on the physical condition and temper of the specimen. Time may vary from 5 to 10 s for a specimen of an alloy in the as-cast or aged condition to 30 s for one in the solution heat treated condition.

Etchants used for specimens of magnesium alloys are listed in Table 16, along with their compositions, etching procedures, characteristics and uses.

For sand, permanent mold, and die cast alloys in the as-cast condition and for virtually all of the alloys in the aged condition, the glycol etchant is used; it is especially good for the magnesium–rare earth alloys and magnesium-thorium alloys. Acetic-picral etchants are used to selectively stain the grains, especially for viewing with sensitive-tint or polarized light.

Table 16. Selected etchants for macroscopic and microscopic examination of magnesium alloys

Etchant No.	Composition	Etching procedure	Characteristics and use
1	*Nital:* 1 to 5 mL HNO$_3$ (conc), 100 mL ethanol (95%) or methanol (95%)	Swab or immerse specimen for a few seconds to 1 min. Wash in water then alcohol and dry.	Shows general structure.
2	*Glycol:* 1 mL HNO$_3$ (conc), 24 mL water, 75 mL ethylene glycol	Immerse specimen face up and swab with cotton for 3 to 5 s for as-cast or aged metal, and up to 1 min for heat treated metal. Wash in water, then alcohol and dry.	Shows general structure. Reveals constituents in Mg–rare earth and Mg-Th alloys.
3	*Acetic glycol:* 20 mL acetic acid, 1 mL HNO$_3$ (conc), 60 mL ethylene glycol, 20 mL water	Immerse specimen face up with gentle agitation for 1 to 3 s for as-cast or aged metal, and for 10 s for heat treated metal. Wash in water, then alcohol and dry.	Shows general structure and grain boundaries in heat treated castings. Shows grain boundaries in Mg–rare earth and Mg-Th alloys.
4	10 mL HF (48%), 90 mL water	Immerse specimen face up for 1 to 2 s. Wash in water, then alcohol and dry.	Darkens Mg$_{17}$Al$_{12}$ phase and leaves Mg$_{32}$(Al,Zn)$_{49}$ phase unetched and white.
5	*Phospho-picral:* 0.7 mL H$_3$PO$_4$, 4 to 6 g picric acid, 100 mL ethanol (95%)	Immerse specimen face up for about 10 to 20 s, or until polished surface is darkened. Wash in alcohol and dry.	For estimating the amount of massive phase. Stains matrix and leaves phase white. Staining improves as magnesium-ion content increases with use.
6	*Acetic-picral:* 5 mL acetic acid, 6 g picric acid, 10 mL water, 100 mL ethanol (95%)	Immerse specimen face up with gentle agitation until face turns brown. Wash in a stream of alcohol and dry with a blast of air.	A universal etchant. Defines grain boundaries in most alloys and tempers by etch rate and color of stain. Reveals cold work and twinning readily.
7	*Acetic-picral:* 20 mL acetic acid, 3 g picric acid, 20 mL water, 50 mL ethanol (95%)	Same as for etchant 6, above, but etch for at least 15 s to develop a heavy film.	Orientation of crackled film is parallel to trace of basal plane. Film crackles in high-alloy areas. Distinguishes between fusion voids surrounded by normal level of alloy and microshrinkage with low alloy content.
8	*Acetic-picral:* 10 mL acetic acid, 4.2 g picric acid, 10 mL water, 70 mL ethanol (95%)	Same as for etchant 6, above	Reveals grain boundaries more readily than etchant 6, above, especially in dilute alloys.
9	0.6 g picric acid, 10 mL ethanol (95%), 90 mL water	Immerse specimen face up for 15 to 30 s. Wash in alcohol and dry.	Used after HF etchant to darken matrix to give better contrast between matrix and white ternary phase.
10	2 mL HF (48%), 2 mL HNO$_3$ (conc), 96 mL water	Immerse specimen face up with gentle agitation. Do not swab.	Grain structure and coring in Mg-Zn-Zr alloys.

Technique for Nickel and Nickel Alloys

By George F. Vander Voort, Carpenter Technology Corp.

THIS ARTICLE concerns the preparation of metallographic specimens of two classes of nickel alloys: (*a*) alloys containing 96% or more nickel (Nickel 200, Nickel 270 and Duranickel 301); and (*b*) nickel-copper alloys (Monel 400, Monel R-405 and Monel K-500). Micrographs of these alloys are shown on pages 314 and 315 in Volume 7 of the 8th Edition of Metals Handbook. Techniques for nickel-base magnetic alloys and wrought heat-resisting alloys are given elsewhere in this volume.

The procedures and materials used in sectioning, mounting, grinding and polishing of specimens are essentially the same for all of the nickel alloys. In preparing specimens for metallographic examination, it is important to prevent cold working of the surface.

PREPARATION FOR MACROSCOPIC EXAMINATION

The preparation steps and macroetching procedures for these alloys are essentially the same as those for steels. Resolution of detail is improved as the quality of the ground surface increases.

Macroetching of nickel alloys is performed by immersion or swabbing using a cold solution of nitric acid and acetic acid (equal parts by volume) for times less than about 30 s. For nickel-copper alloys, fresh concentrated nitric acid is preferred. Times up to 5 min may be needed. Stale nitric acid may produce staining. Shorter times are recommended to observe sulfur embrittlement in Monel welds whereas longer times reveal macrostructural features.

PREPARATION FOR MICROSCOPIC EXAMINATION

Sectioning, mounting, grinding and polishing steps for nickel and nickel-copper alloys are quite similar to those for austenitic stainless steels. The chief danger is introduction of cold work to the structure. As with austenitic stainless steels, removal of fine scratches requires careful final polishing.

SiC papers are recommended for grinding, using the sequence 120, 240, 320, 400 and 600 grit with water cooling. Rough polishing with 6-μm diamond paste on a medium-nap cloth followed by 1-μm diamond on the same type of cloth will produce excellent results. Final polishing with a 0.05-μm alumina slurry or colloidal silica on a medium-nap cloth produces excellent results. Vibratory polishing with either of these abrasives may be required to obtain optimum results.

An alternative procedure is to electropolish specimens after mechanical polishing to remove fine scratches and deformation. Recommended electrolytes and current densities for electropolishing these alloys are given in Table 17. A platinum cathode should be used, and the electrolyte should be water cooled and continuously stirred.

Etching. The solutions and conditions for etching the various nickel alloys for microscopic examination are described in Table 18. The acids used should be concentrated; when water is indicated, use distilled water only.

Technique for Wrought Heat-Resisting Alloys

WROUGHT heat-resisting alloys are of three major classes: iron-nickel-chromium alloys, nickel-base alloys and cobalt-base alloys. Microstructures of these alloys are shown on pages 158 to 176 in Volume 7 of the 8th Edition of Metals Handbook.

Although the three classes of alloys differ significantly in composition, microstructure and elevated-temperature properties, the techniques and equipment used for preparation of metallographic specimens are essentially the same for all three classes. Further, the techniques used for these wrought alloys are basically the same for their cast counterparts and are similar to those used for other metals.

PREPARATION OF SPECIMENS FOR OPTICAL MICROSCOPY

Sectioning and Grinding. The initial preparation stages of sectioning and mounting (if required) are virtually identical to those employed for steels. Likewise, coarse and fine grinding procedures are also similar, using water-cooled SiC papers of 120, 240, 320, 400 and 600 grit. If grinding is done manually, the direction of grinding must be rotated 45 to 90° between each step.

Polishing. Rough polishing is generally performed with 3-, 6- or 9-μm diamond paste on a napless or low-nap cloth, such as nylon or silk, to minimize relief at the hard constituents. This may be followed by a similar step using 0.5- or 1-μm diamond paste on a napless or low-nap cloth. If these steps are done manually, the specimen should be rotated against the cloth in the direction opposite to wheel rotation until scratches from the previous step are removed.

Table 17. Electrolytes and current densities for electropolishing of nickel and nickel-copper alloys

Composition of electrolyte	Applicable alloys	Current density A/cm²	A/in.²
37 mL H₃PO₄ (conc), 56 mL glycerol, 7 mL water	Nickel 200 .... 1.4 to 1.6		9 to 10
	Nickel 270 .... 1.6 to 1.9		10 to 12
	Duranickel 301 .... 1.2 to 1.6		8 to 10
	Monel 400 .... 0.9 to 1.1		6 to 7
33 mL HNO₃ (conc), 66 mL methanol	Monel 400, R-405, K-500 .... 1.6 to 2.3		10 to 15

Table 18. Etchants for microscopic examination of nickel and nickel-copper alloys for grain boundaries and general structure

Composition of etchant	Conditions for use
Etchants for Nickel 200 and 270; Duranickel 301; and Monel 400, R-405 and K-500	
1 part 10% aqueous solution of sodium cyanide, 1 part 10% aqueous solution of ammonium persulfate; mix solutions when ready to use.	Immerse or swab specimen for 5 to 90 s. (*Caution:* use fume hood; solutions release toxic fumes when mixed.)
1 part nitric acid (conc), 1 part acetic acid (glacial). Use fresh solution.	For revealing grain boundaries. Immerse or swab specimen for 5 to 20 s.
Alternative etchant for Monel K-500	
Glyceregia: 10 mL nitric acid (conc), 20 mL hydrochloric acid (conc), 30 to 40 mL glycerol	Etch by immersing or swabbing the specimen for 30 s to 5 min.

Table 19. Microetching procedures for wrought iron-nickel-chromium heat-resisting alloys

Etchant No.(a)	Etching method	Etching time, s	Cell voltage	Purpose, or characteristics revealed
Alloy A-286 (AISI 660)				
1	Swab	3 to 20	...	General structure
2	Swab	5 to 60	...	General structure; may stain or pit
3	Immerse	10 to 60	...	General structure
Incoloy 800				
4	Electrolytic	15 to 30	5 to 10	General structure; grain boundaries
5	Electrolytic	10 to 20	5 to 10	Grain boundaries; carbide particles
6	Electrolytic	10 to 15	20	Carbide particles
7	Swab	15 to 30	...	Grain boundaries(b); carbide. No staining.
8	Electrolytic	10 to 30	10	Preferential attack at grain boundaries
Incoloy 825				
4	Electrolytic	15 to 30	5 to 10	General structure; grain boundaries
5	Electrolytic	10 to 20	5 to 10	Grain boundaries; carbide particles
7	Swab	15 to 30	...	Grain boundaries; carbide. No staining.
9	Swab or immerse	(c)	...	General structure
RA 330				
8	Electrolytic	5 to 10	5	For etch pitting
10	Electrolytic	2 to 10	3	General structure; precipitates
(a) See Table 22 for compositions of etchants. (b) Grain boundaries are faint if free of carbide particles. (c) Etching time varies from a few seconds to 12 minutes.				

Final polishing can be performed using a variety of abrasives and cloths (napless to medium nap). If two diamond polishing steps were used, final polishing may be conducted using abrasives such as 0.05-μm alumina (water slurry) or colloidal silica. If only one diamond polishing step was used—e.g., 6-μm diamond—two different alumina abrasive steps, such as 0.3 and 0.05 μm, may be desired. Polishing pressure should be light to moderate depending on the alloy being polished.

In recent years, considerable use has been made of automatic polishing devices for the grinding and polishing steps. The more fully automated devices permit precise duplication of the polishing cycle, which improves reproducibility of polishing results and generally produces better control of relief than can be achieved with manual polishing.

Vibratory polishing has often been used for final polishing, with both fine diamond and fine abrasive slurries. Although polishing times are substantially longer, excellent results can be obtained. Typically, times up to about 30 min are required with each abrasive used in this manner.

Etching procedures for specific wrought heat-resisting alloys of the three major classes are listed in Tables 19 to 21. Compositions of the etchants used are given in Table 22.

Table 21. Microetching procedures for wrought cobalt-base heat-resisting alloys

Etchant No.(a)	Etching method	Etching time, s	Purpose
Haynes 25 (AISI 670), Haynes 188			
26	Electrolytic, 6 V	2 to 5	(b)
Stellite 6B			
26	Electrolytic, 6 V	2 to 5	(b)
27:			
Stage 1 . . .	Electrolytic, 6 V	2 to 5	(b)
Stage 2 . . .	Immersion	5 to 10	(c)

(a) See Table 22 for compositions of etchants. (b) General structure. (c) Carbide particles.

ELECTRON MICROSCOPY

Because these alloys contain a number of fine precipitates that are usually too small to be observed with a light optical microscope, many specimens are also examined using a scanning electron microscope (SEM), a transmission electron microscope (TEM) or a scanning transmission electron microscope (STEM).

The initial specimen-preparation stages for such observation work are often quite similar, using these instruments except for the etching step. Specimens can be directly observed in the SEM, or replicas or thin foils may be examined in the TEM or STEM. Procedures for preparing replicas or thin foils are described in many standard textbooks and in Volume 8 of the 8th Edition of Metals Handbook, "Metallography, Structures and Phase Diagrams," (pages 100 to 104).

Technique for Tin and Tin Alloys

TIN AND TIN ALLOYS are extremely soft and have low recrystallization temperatures, and thus preparation of tin and tin alloy specimens for metallographic examination presents special problems that are not often encountered in preparing specimens of other metals. However, many of the steps in preparation are similar or identical to those for other metals.

MOUNTING

Mounting of tin and tin alloy specimens should be done at room temperature, because elevated temperatures can cause structural changes, such as precipitation of a second phase from the supersaturated solid solution. Compression mounting compounds are therefore unsuitable.

Table 20. Microetching procedures for wrought nickel-base heat-resisting alloys

Etchant No.(a)	Etching method	Etching time, s	Cell voltage	Purpose, or characteristics revealed
Hastelloy C				
11	Electrolytic	2 to 10	3	General structure
Hastelloy W				
12	Immerse	. . .	. . .	General structure
Hastelloy X (AISI 680)				
13	Electrolytic	2 to 10	6	General structure. Remove stains with HNO_3.
Inconel 600 and 601				
4	Electrolytic	15 to 20	5 to 10	General structure. No pitting.
5	Electrolytic	15 to 20	5 to 10	General structure; grain boundaries; carbide
6	Electrolytic	15 to 20	5 to 10	General structure; excellent for revealing carbide particles
7	Swab or immerse	(b)	. . .	Grain-boundary contrast fair; carbide
14	Immerse	(c)(d)	. . .	General structure; carbide particles
Inconel 625				
8	Electrolytic	1 to 2	50	Grain-boundary films. Results vary with thermal history of specimen.
15	Electrolytic	10 to 20	5 to 10	Grain boundaries. No staining. Results vary with thermal history of specimen.
16	Electrolytic	15 to 20	5 to 10	General structure; grain boundaries
17	Electrolytic	15 to 20	5 to 10	General structure
18	Electrolytic	8 to 20	2 to 10	Outlines phases. May cause pitting. Poor results on cold worked metal.
Inconel 706 and Alloy 718				
8	Electrolytic	1 to 2	50	Grain-boundary films. Shows grain boundaries in relief.
15	Electrolytic	10 to 20	5 to 10	Good for general structure and phase outline; grain boundaries
16	Electrolytic	15 to 20	5 to 10	Good for general structure, phase outline, and matrix segregation for most heat treated conditions; grain boundaries
17	Electrolytic	15 to 20	5 to 10	General structure; precipitate phases in fully heat treated material
19	Swab or immerse	(b)	. . .	General structure; microsegregation
20	Immerse	(e)	. . .	Carbide particles. Chromium carbide particles darken faster than nitrides and Laves phase.
Inconel X-750 (AISI 688)				
4	Electrolytic	15 to 20	5 to 10	General structure. No pitting.
5	Electrolytic	15 to 20	5 to 10	Grain boundaries; carbide. No pitting.
15	Electrolytic	10 to 20	5 to 10	Good for revealing grain boundaries and carbide particles
21	Swab	2 to 10	. . .	Excellent for showing details of overaged gamma prime
22	Swab	5 to 60	. . .	General structure; microsegregation
U-700 (AISI 687)				
23	Swab or immerse	10 to 20	. . .	Good for contrast(f)
24	Swab or immerse	(c)	. . .	General structure; grain boundaries. No staining.
28	Electrolytic	5 to 20	5 to 10	General structure; grain boundaries
Waspaloy (AISI 685)				
1	Swab	3 to 20	. . .	General structure
24	Swab or immerse	(c)	. . .	General structure. No staining.
25	Swab or immerse	5 to 30	. . .	General structure

(a) See Table 22 for compositions of etchants. (b) $^1/_2$ to 5 minutes. (c) 1 to 5 minutes. (d) Heat specimen to reduce etching time. (e) 5 to 10 minutes. (f) Use well-prepared specimen.

Table 22. Etchants for microscopic examination of wrought heat-resisting alloys
See Tables 19, 20 and 21 for applicability to specific alloys.

Etchant No. and name	Composition(a)	Remarks on preparation and use
1 HCl, HNO₃, acetic acid	15 mL HCl, 10 mL HNO₃, 10 mL acetic acid	...
2 Chrome regia	2 g CrO₃, 20 mL HCl, 80 mL water	CrO₃ may be increased, but staining may result.
3 Ferric chloride–hydrochloric	5 g FeCl₃, 15 mL HCl, 100 mL methanol	...
4 Nital	5 mL HNO₃, 95 mL methanol	Use colorless acid and absolute methanol.
5 Oxalic acid	10 g oxalic acid, 100 mL water	Can be stored.
6 Phosphoric acid	80 mL H₃PO₄, 20 mL water	Change to a 1-to-1 solution for specific results.
7 Glyceregia	10 mL HNO₃, 20 mL HCl, 40 mL glycerol	Must be freshly prepared.
8 Hydrochloric-methanol	10 mL HCl, 90 mL methanol	Water can be substituted for methanol to show segregation.
9 Vilella's reagent	5 mL HCl, 1 g picric acid, 100 mL methanol	A few drops of 3% H₂O₂ will speed etching reaction.
10 HCl-H₂O	5 mL HCl, 95 mL water	...
11 Chromic acid	2 to 10 g CrO₃, 100 mL water	
12 Hydrochloric-chromic	80 mL HCl, 20 mL 50% chromic acid	Use fresh solution.
13 Oxalic acid	10 g oxalic acid, 90 mL water	
14 Nitric-hydrofluoric	20 mL HNO₃, 3 mL HF	Use colorless acids. Remove thoroughly by water rinse.
15 Chromic-acetic	25 g CrO₃, 7 mL water, 130 mL acetic acid	Can be stored for up to one month.
16 Chromic acid	5 g CrO₃, 100 mL water	...
17 47-41-12	47 mL H₂SO₄, 41 mL HNO₃, 12 mL H₃PO₄	Add H₂SO₄ last, and slowly. Produces noxious fumes and is highly corrosive. When correctly etched, the specimen surface is colored brown.
18 Hydrochloric-acetic	10 mL acetic acid, 3 drops HCl, 90 mL water	...
19 Inverted glyceregia	50 mL HCl, 10 mL glycerol, 10 mL HNO₃	
20 Murakami's reagent	10 g KOH or NaOH, 10 g K₃Fe(CN)₆, 100 mL water	Dissolve KOH (or NaOH) and K₃Fe(CN)₆ in boiling water; etch specimen in boiling solution. Prepare fresh for use.
21 Nitric-hydrofluoric	50 mL HNO₃, 50 drops HF	Use colorless acids.
22 Hydrochloric-hydrofluoric-nitric	80 mL HCl, 13 mL HF, 7 mL HNO₃	...
23 Marble's reagent	4 g CuSO₄·5H₂O, 20 mL HCl, 20 mL water	Dissolve CuSO₄ in water and add HCl.
24 Kalling's reagent	2 g CuCl₂, 40 mL HCl, 80 mL methanol	Can be stored.
25 92-5-3	92 mL HCl, 5 mL H₂SO₄, 3 mL HNO₃	Must be freshly prepared.
26 Hydrochloric–hydrogen peroxide	97 mL HCl, 3 mL 3% H₂O₂	Must be freshly prepared.
27 Grosbeck's reagent (two-stage)	Stage 1: 2 to 10% CrO₃ in water; stage 2: equal parts 20% KMnO₃, 8% NaOH	Mix second stage immediately before use.
28 HCl-ethanol-H₂O₂	35 mL HCl, 65 mL ethanol (95%), 7 drops H₂O₂ (30%)	Must be freshly prepared.

(a) Use concentrated acids, unless indicated otherwise. Use distilled water to avoid staining.

Table 23. Etchants for use in microscopic examination of tin and tin alloys

Etchant composition	Uses
5 mL HCl, 2 g FeCl₃, 30 mL water, 60 mL absolute alcohol	General use for tin and tin alloys
2 mL HCl, 98 mL methanol (95%) or ethanol (95%)	Grain-boundary etch for pure tin
10 mL HNO₃, 10 mL acetic acid, 80 mL glycerol	Darkens the lead in the eutectic of tin-rich tin-lead alloys
5% silver nitrate in water	Darkens primary and eutectic lead in lead-rich tin-lead alloys
2% nital	Recommended for etching tin-antimony alloys; darkens tin-rich matrix, leaving intermetallic compounds unattacked. Often used for etching specimens of babbitted bearings.
Picral	For etching tin-coated steel and tin-coated cast iron
1 drop HNO₃ (conc), 2 drops HF, 25 mL glycerol; then picral	For etching tin-coated steel
Dilute ammonium hydroxide with a few drops of 30% hydrogen peroxide	For etching tin-coated copper and copper alloys

GRINDING AND POLISHING

Distortion of the surface layers of tin and tin alloys may cause recrystallization, which will mask the true structure. For tin alloys in a metastable state, distortion of the surface regions as a result of working may cause structural changes, such as precipitation from a supersaturated solid solution.

To avoid working of the surface regions, extreme care must be taken during turning, filing, grinding and polishing, because of the susceptibility to distortion during those operations.

Grinding. After the specimen is flattened with a file, or by careful turning in a lathe, it is wet ground on silicon carbide papers of progressively finer grit sizes (180, 240, 320, 400 and 600 mesh). The papers are kept wet by a continuous stream of lubricant, usually water, that washes away the particles of metal as they are cut from the surface

of the specimen. This prevents clogging of the papers, which would lead to surface flowing of the metal instead of cutting. For alloys containing zinc-rich and aluminum-rich phases, which may be stained by water, kerosine is used as the lubricant.

Polishing. A light positive pressure is preferred during all stages of polishing, and each stage of polishing should be continued for at least twice as long as it takes to remove the scratches from the previous stage. Polishing is usually carried out using different grades of diamond suspended in an oil-base vehicle. Every effort must be made to ensure that the polishing wheels and the pads used for diamond pastes are kept free from dust, grit and the diamond particles from the previous polishing stage. It is essential that coarse grinding and fine polishing be done at different locations; the environment where fine polishing is conducted must be free from corrosive fumes.

Scratches from the 600-mesh silicon carbide papers are removed by polishing for several minutes on a wheel covered with a short-nap or napless cloth impregnated with 6-μm diamond paste. Next, the specimen is polished on a wheel covered with a short-nap cloth impregnated with 1-μm diamond paste.

ETCHING

Although the method of polishing is the same for all types of tin and tin alloys, different etchants are used for etching the various alloys (see Table 23).

Technique for Titanium and Titanium Alloys

PREPARATION of metallographic specimens of unalloyed titanium, and of alpha, alpha-beta and beta titanium alloys, is described in this article. Pages 322 to 334 in Volume 7 of the 8th Edition of Metals Handbook contain 127 micrographs of titanium alloys.

Sectioning of titanium and titanium alloys follows conventional procedures, but deformation and overheating must be avoided. Deformation can result in mechanical twinning and strain-induced transformation products; overheating, in changes in structure. Abrasive cutting with silicon carbide wheels is satisfactory if adequate coolant is used. Use of bandsaws is common for these alloys. Sectioning using an oil-cooled low-speed diamond saw is recommended for small specimens, to avoid overheating.

Table 24. Etchants for microscopic examination of titanium and titanium alloys

Specimen metal	Composition of etchant	Purpose
Unalloyed titanium	1 to 3 mL HF, 10 mL HNO₃, 30 mL lactic acid	Reveals hydrides
	1 mL HF, 30 mL HNO₃, 30 mL lactic acid	Reveals hydrides
Most titanium alloys	Kroll's reagent: 1 to 3 mL HF, 2 to 6 mL HNO₃, water to 1000 mL	General-purpose etch
	10 mL HF, 5 mL HNO₃, 85 mL water	General-purpose etch
	1 mL HF, 2 mL HNO₃, 50 mL H₂O₂, 47 mL water	Removes stain
	10 mL HF, 10 mL HNO₃, 30 mL lactic acid	Chemical polish and etch
	2 mL HF, 98 mL water	Reveals alpha case
	1 to 2 mL HF, 4 to 5 mL H₂O₂, water to 1000 mL	Nonstaining etch
Near-alpha titanium alloys	2 mL HF, 98 mL water; then 1 mL HF, 2 mL HNO₃, 97 mL water	General-purpose etch(a)
Alpha-beta titanium alloys	10 mL KOH (40%), 5 mL H₂O₂, 20 mL water	Stains alpha, transformed beta
Ti-Al-Zr and Ti-Si alloys	18.5 g benzalkonium chloride, 33 mL ethanol, 40 mL glycerol, 25 mL HF	General-purpose etch
Ti-3Al-8V-6Cr-4Mo-4Zr	30 mL H₂O₂, 3 drops HF	General-purpose etch
Ti-8Mn; aged Ti-13V-11Cr-3Al	2 mL HF, 4 mL HNO₃, 94 mL water	General-purpose etch
Ti-Si alloys	2 drops HF, 1 drop HNO₃, 3 mL HCl, 25 mL glycerol	General-purpose etch

(a) First etchant stains alpha phase; second etchant removes stain.

MOUNTING

Titanium and its alloys are mounted in the common materials, such as Bakelite, methyl methacrylate and diallyl phthalate; when edge preservation is important, compression-mounting epoxy is recommended. Nickel plating of specimens before mounting also assists in edge preservation. In general, the temperatures encountered in using Bakelite, methyl methacrylate or diallyl phthalate do not cause problems. However, when metallographic examination involves the hydride phase (TiH), it is best to leave the specimen unmounted or to mount it in a room-temperature-setting epoxy resin because of the increased solubility of hydrogen in titanium at the higher temperatures. Sometimes, mounting in a thermosetting material can cause solution of existing hydride and, on cooling, precipitation of the hydride in an altered form, usually a fine dispersion. Adding 50 vol % of silica powder (between 200 and 325 mesh) to the epoxy resin or nickel plating with help in edge preservation.

GRINDING

The procedure for grinding titanium specimens is similar to that for grinding steel specimens. The specimens are ground on successive grades of silicon carbide paper, starting with 180-grit and proceeding to 240, 320, 400 and 600-grit papers, using water to keep the specimens cool and flush away loose particles of metal and abrasive. It is possible to start grinding on a paper with a grit size as coarse as 80 mesh (if the cut surface is rough), provided that pressure is light so as to minimize cold working.

Hand, disk and belt grinding are used. To avoid embedding abrasive particles in the ground surface, the papers may be dressed with solid stick wax. Wax is also used in dry final grinding, when this grinding method is preferred.

POLISHING

Manual, automatic, vibratory and electrolytic polishing are used for specimens of titanium and titanium alloys.

Manual polishing is performed in three stages — rough, intermediate and final polishing. Rough polishing is performed on a high-speed polishing wheel covered with a lintless rayon or silk cloth, using medium pressure and a slurry consisting of 35 mL of water, 15 g of 1- or 3-μm alpha (levigated) alumina, and 5 mL of 20% chromic acid (optional). An alternative procedure for rough polishing is to use a nylon cloth and 6-μm diamond paste. Intermediate polishing is done with the same equipment and procedure, except that 0.3-μm alumina is used in the slurry. Final polishing is performed on a low-speed wheel covered with microcloth, using a slurry containing 15 g of 0.05-μm alumina and 35 mL of water, and with light pressure on the specimen.

At the start of intermediate polishing, rather heavy pressure is applied. As polishing progresses, pressure is decreased. Several cycles of polishing and etching are needed to remove flowed metal. Etching should be light to avoid pitting. The specimen is unetched when final polishing starts. Several cycles of polishing and etching may also be necessary during final polishing.

For rough polishing of unalloyed titanium, which is considerably softer than the alloys, use of a wheel covered with 6.4 to 12.7 mm (¹/₄ to ¹/₂ in.) of paraffin and a slurry of 3-μm levigated alumina is recommended. For intermediate polishing, the same wheel may be used with a slurry of 0.3-μm alumina. Final polishing may be performed on a wheel covered with a short-nap cloth and using a slurry of 0.05-μm alumina.

Automatic polishing has an advantage in that the pressure exerted on the specimen can be controlled. Specimens are ground through 180, 240, 320, 400 and 600-grit silicon carbide paper using water or kerosine as the coolant. Final polishing is done using microcloth and 0.05-μm gamma alumina for about 10 min. Several specimens can be prepared simultaneously.

Vibratory polishing, although a slower process than electrolytic or mechanical polishing, produces good results. Vibratory polishing is a two-stage operation that follows grinding on abrasive papers. Preliminary polishing is performed on a canvas cloth for 2 to 4 h, using a slurry of 0.5-μm alumina, or on a silk cloth for 30 to 45 min, using a slurry of 0.3-μm alumina. Final polishing is performed on a short-nap cloth for 15 min to 4 h, using a slurry of 0.05-μm alumina. Several short polishing and etching cycles may be used to remove disturbed metal.

Electropolishing is considerably faster than mechanical polishing. An electrolyte recommended for electropolishing contains 600 mL of methanol, 360 mL of ethylene glycol, and 60 mL of perchloric acid. Polishing time is 15 to 25 s at a current density of 1 to 1.5 A/cm², depending on specimen size and polishing area. The electrolyte given above, with a low concentration of perchloric acid, is nonexplosive and can be stored for several weeks. However, care should be exercised in handling concentrated perchloric acid, because it can react explosively with organic materials.

ETCHING

At least three different etchants are used for macroetching of titanium and titanium alloys. Two of these are general-purpose etchants; the choice between the two is usually arbitrary. One of the general-purpose etchants is composed of equal parts, by volume, of hydrofluoric acid, nitric acid and water; the other is composed of 5 mL of hydrofluoric acid, 35 mL of nitric acid and 60 mL of water. An etchant composed of 50 mL of concentrated hydrochloric acid and 50 mL of water is often preferred for macroetching of alpha-beta titanium alloys.

Microetching is accomplished by swabbing or immersion. Because etching by immersion can promote reprecipitation of dissolved alloying elements such as tin back onto the specimen surface, it is recommended that both immersion and swabbing techniques be initially evaluated to identify the presence of such etching artifacts. Swabbing usually produces a brighter, clearer etching response than immersion. Etching times are usually short, ranging from 3 to 10 s. The specimen should be examined after light etching, to avoid overetching.

Nearly all etchants for titanium and titanium alloys contain hydrofluoric acid and an oxidizing agent, such as nitric acid. Table 24 gives the compositions and purposes of etchants suitable for use on unalloyed titanium and on titanium alloys. Kroll's reagent is the etchant most widely used for commercial titanium alloys.

Technique for Zinc and Zinc Alloys

METALLOGRAPHIC techniques that are special to the preparation of specimens of zinc and zinc alloys are discussed in this article.

SECTIONING AND MOUNTING

The initial sample may be removed from a larger mass of material by sawing, breaking or shearing — much the same as for other metals.

Mounting. Specimens of rolled zinc and zinc alloys are usually held by clamping; they are not mounted in plastic. Several specimens are mounted together in a screw clamp, with thin spacers of soft zinc between specimens and heavy strips of zinc between the clamp plates and the outermost specimens. The assembly is tightly clamped to prevent seepage of etchants between specimens. The zinc spacers are of known structure and serve as convenient standards of comparison for determining if the specimens have been correctly prepared.

Most specimens other than those of rolled metal are mounted in conventional plastic molding materials (see the description of mounting materials in the article entitled "Metallographic Methods").

GRINDING AND POLISHING

In grinding and polishing of cast zinc, distortion can occur to a depth 20 to 100 times as great as the deepest scratch. Therefore, in each stage of grinding and polishing, considerably more metal should be removed than the amount required for eliminating the scratches that remain from the previous stage. It is easier to prepare a distortion-free surface on specimens of fine-grain zinc than on specimens of coarse-grain, soft zinc.

Grinding (wet or dry) on a belt grinding machine using silicon carbide abrasives of grit sizes 60, 120 and 240 mesh is suitable for zinc and zinc alloys. Local heating from grinding must be minimized as much as possible by water cooling. Otherwise, heat can cause structural changes to a depth too great to be removable by polishing.

Grinding is done with silicon carbide papers, using successively finer grit sizes of 280, 320, 400 and 600 mesh. These papers are less susceptible to loading than emery papers. Grinding can be done by hand on papers supported on a flat surface, or by holding the specimen against the same grades of paper on a grinding wheel. Overheating of the specimen during wheel grinding should be minimized; this can be done by using a low wheel speed (250 rpm max), by applying water to the silicon carbide papers during grinding, and by grinding in intervals of a few seconds and allowing the specimen to cool between intervals before grinding is resumed.

Rough polishing may be performed using 6-μm diamond paste charged onto napless, low-nap or medium-nap cloths. Final polishing can be done using alumina or magnesium oxide slurries on medium-nap cloths. Colloidal silica is excellent for final polishing of zinc alloys but may cause pitting of pure zinc.

MACROETCHING

Use of concentrated hydrochloric acid at room temperature, followed by rinsing and wiping off the resulting black deposit, produces good grain contrast on copper-free zinc and zinc alloys. Etchant 1 in Table 25 may be used for zinc containing 1% Cu or less. With this etchant, grain contrast is not well defined. An etchant equal to hydrochloric acid for producing grain contrast has not been found for zinc alloys containing copper.

MICROETCHING

The most useful etchants for microscopic examination of zinc and zinc alloys are aqueous solutions of chromic acid (CrO_3, or chromic anhydride) to which sodium sulfate (Na_2SO_4) has been added. Grades of chromic acid used for chromium plating are satisfactory. Table 25 gives compositions of etchants commonly used.

Table 25. Etchants for zinc and zinc alloys

Etchant	Composition
1(a)	200 g CrO_3, 15 g Na_2SO_4, 1000 mL water
2(b)	50 g CrO_3, 4 g Na_2SO_4, 1000 mL water
3	200 g CrO_3, 1000 mL water

(a) For rolled zinc-copper alloys, the Na_2SO_4 content can be reduced to 7.5 g. If desired, a smoothly etched surface can be obtained by increasing the Na_2SO_4 to 30 g. (b) This etchant can be made by mixing one part (by volume) of etchant 1 and three parts of water.

Table 26. Etchants and etching times for zinc and zinc die-casting alloys

Specimen metal	Etchant (from Table 25)	Time, s, for examination at: 250×	1000×
Cast or rolled zinc	1	5	1
Alloy AC41A or AG40A	2	1	1

Etching should follow soon after final polishing. The specimen should be cleaned in alcohol and then in running water, and etched while still wet. To avoid staining, the use of etchant 1 or 2 in Table 25 should be followed immediately by a rinse in etchant 3. The specimen is then thoroughly washed in running water, dipped in alcohol, and dried with a stream of warm, clean air.

Table 26 gives recommendations for etchants and etching times for zinc and zinc die-casting alloys. The etching time may be longer or shorter for specific etching conditions (a minor difference in solution temperature may affect etching time). Also, as indicated in Table 26 for cast or rolled zinc, etching time is often decreased as the magnification to be used is increased.

36 FRACTOGRAPHY

Edited by Wilson G. Dobson, Binary Engineering Associates, Inc.
Review Committee: R. M. N. Pelloux, Massachusetts Institute of Technology; R. Biederman, Worcester Polytechnic Institute; R. R. Boyer, Boeing Commercial Airplane Co.; and R. G. Ballinger, Massachusetts Institute of Technology

This section was condensed from Metals Handbook, Eighth Edition, Volume 9, Fractography and Atlas of Fractographs, pages 1 to 124. For more detailed information on the subjects covered herein, the reader is referred to the larger work.

Principles and Procedures

THE TERM "fractography" was coined in 1944 to describe the science of studying fracture surfaces, although fractures and fracture surfaces had been examined and studied for centuries as part of the science of metallurgy. The two chief metallurgical benefits of fractography are: (*a*) it is a useful tool in failure analysis, and (*b*) it provides a means of correlating the influence of microstructure on the fracture mode of the material.

Study of the characteristics of fracture surfaces by optical (light-microscope) fractography is widely practiced, particularly when a low magnification of the fracture surface is adequate. This study can be accomplished with relatively simple equipment by personnel with moderate training.

Preparation and Preservation of Fracture Specimens

WHEN A FRACTURE OCCURS and there is even a slight chance that it will be subjected to laboratory examination, several important steps must be followed if fracture-surface information is to be preserved.

Mechanical damage may arise from several sources, including simply permitting the fracture surface to strike other objects. This can occur during the fracture event (as is often observed in fatigue failures) or in removing a fractured piece from its surroundings or transporting it from one place to another. There is also a tendency for people to put the mating halves of the fracture together; this should be avoided, particularly in view of the microscopic features examined in microfractography. Subtle fracture-surface details are easily damaged.

Chemical (corrosion) damage to a fracture can be prevented in a number of ways, and each laboratory has its preferred method. Because the

identification of foreign material present on a fracture surface may be important in the over-all interpretation of the cause of fracture, many laboratories prefer not to use corrosion-preventive coatings on fractures. If analysis of foreign material on the fracture surface will not be required (for example, in analysis of laboratory-produced fractures), a satisfactory method of preserving the fracture and preventing it from corroding is to apply a coating. When in the field, the fracture can be coated with a good grade of fresh axle grease.

CLEANING OF METAL FRACTURES

Numerous cleaning procedures have been developed, each suited for a particular metal or the preference of a particular laboratory. Most fractured parts that have undergone extensive exposure to corrosive environmental conditions will typically require some cleaning. The cleaning techniques commonly employed include:

- A dry air blast
- Treating with inorganic solvents
- Treating with mild acid or alkaline solutions
- Ultrasonic cleaning with water-base detergents or solvents
- Applying and stripping of plastic replicas.

Because identification of foreign material present on a fracture may be important, many laboratories examine the original fracture surface before and after cleaning. Wire brushes, grinding wheels, rust removing pastes or similar methods should never be used for cleaning fracture surfaces.

SECTIONING OF FRACTURES

It is frequently necessary to remove the portion containing a fracture from the total part (because the total part is to be repaired) or to reduce the

size of the specimen to be examined to a size that is convenient to handle.

For large parts, a common method of specimen removal is flame cutting. The cutting must be done at a sufficient distance from the fracture so that the microstructure of the metal underlying the fracture surface is not altered by the heat of the flame and so that none of the molten metal from flame cutting is deposited on the fracture surface.

Saw cutting and cutoff-wheel cutting can be used for parts in a wide range of sizes. Dry cutting is preferable, because coolants may corrode the fracture or may wash away foreign matter from the fracture. A coolant may be required, however, if a dry cut cannot be made at a sufficient distance from the fracture to avoid heat damage.

Often the fracture can be sectioned by hand with a hacksaw. Temperature and corrosion can be kept to a minimum using this method.

OPENING OF SECONDARY CRACKS

When a primary fracture has been damaged or corroded to a degree that prevents it from providing sufficient information, it is desirable to open any secondary cracks to expose their fracture surfaces for examination and study. These cracks may provide more information than the primary fracture. This practice must be used with caution, however, inasmuch as the mode of propagation of a secondary crack need not be the same as that of the primary fracture. A secondary crack may initiate by stress corrosion and propagate by fatigue, or vice versa. In this instance, analysis of the crack could be misleading as to the initial cause of the failure.

To avoid cutting, a crack may be forced open, forming a "lab fracture" region. For steels, this crack-opening practice may be aided by immersing the specimen in liquid nitrogen until it is uniformly cold. This makes the specimen brittle, and

36•1

thus less energy is required to complete the fracture.

EFFECT OF NONDESTRUCTIVE INSPECTION

Many of the so-called nondestructive inspection methods are not entirely nondestructive. The liquid penetrants used for crack detection may corrode fractures in some metals, and they certainly will deposit foreign compounds on the fracture surfaces—both of which could lead to misinterpretation of the nature of the fracture.

Even magnetic-particle inspection, which is often used to locate cracks in ferrous parts, may affect later examination. For example, arcing that may occur across tight cracks can affect their fracture surfaces.

Photography of Fractured Parts and Fracture Surfaces

PHOTOGRAPHY of a fracture surface should be preceded by a detailed and thorough examination of the specimen in the as-received condition to determine which features are most important, which aspects are extraneous (such as dirt or postfracture mechanical damage), and whether special treatment of the surface will be required. This scrutiny should begin with unassisted visual examination, continue with inspection with a hand-held magnifier (and with a penlight if helpful), and perhaps include study with a low-power stereomicroscope. Observations made in these surveys should be recorded for future review and determination of the probable causes of fracture. A list should be made of the features deserving photography, with the magnification that will probably be required.

The next step should be general photography of the entire fractured part and of the broken pieces, to record their size and condition and to show how the fracture is related to the components of the part.

SETUPS FOR PHOTOGRAPHY OF FRACTURED PARTS

If possible, it is preferable to bring the parts or assemblies into the laboratory for photographing, where factors such as lighting can be carefully controlled. For larger parts, background paper, which is available in white or in colors and in rolls of various widths, can be used to provide an uncluttered background.

A workable setup for photographing specimens and small parts is with a view camera mounted vertically on an enlarger (or equivalent) stand. The longest-focal-length lens that will conveniently give the desired magnification to fill the image area should be chosen. Simple basic setups that should be satisfactory for photographing specimens and most small parts are shown in Fig. 1.

CAMERAS

The camera selected for photographing fractured parts and fracture surfaces should be the one that will do the best job for the major part of the work of the laboratory. There is no one ideal camera.

Microscope Systems. Low-magnification stereo and monocular microscopes with attached cameras are

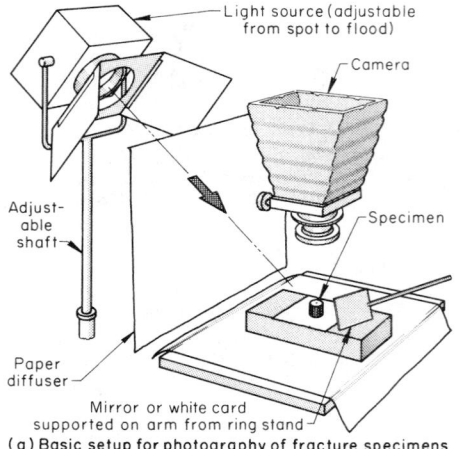

(a) Basic setup for photography of fracture specimens

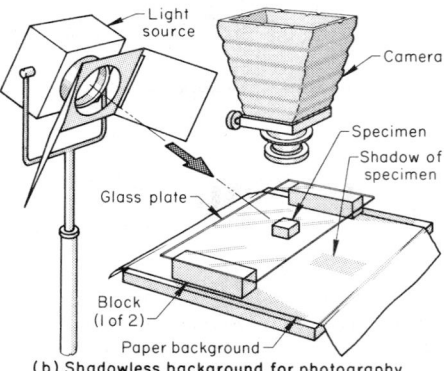

(b) Shadowless background for photography of fracture specimens

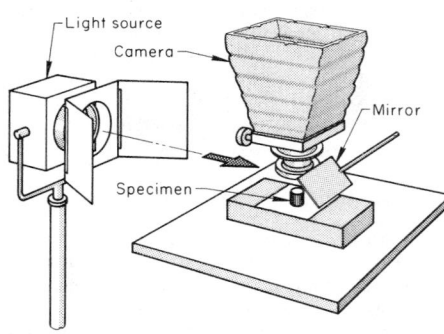

(c) Horizontal illumination for close-up photography of fracture specimens

(a) General arrangement of camera, light source with diffuser, and specimen. Size and angle of the beam of light should be adjusted to give the best display of texture. A reflecting mirror or white card can be used to fill in shadows.
(b) Placing the specimen on a raised glass plate will throw the shadow beyond the view range of the camera and provide a shadowless background.
(c) If the magnification desired requires the close approach of the camera to the specimen, which would throw a shadow on the fracture surface, a horizontal light beam reflected by a small mirror positioned near the camera lens will provide the proper angle of illumination.

Fig. 1. Basic setups and lighting employed in photography of fracture specimens and small parts

suitable for photography of fracture surfaces. In general, these types of microscopes will cover a much smaller field than is obtainable with the single photographic objective lens. Some systems permit the eyepiece on the microscope to

be removed to increase the size of the field.

Whereas camera systems have practical magnification limits of 15 to 20 diameters, much higher magnifications can be obtained with a microscope, the magnification being limited primarily by the diminishing depth of field at higher powers.

LIGHTING TECHNIQUES

For indoor photography, the usual technique for illumination of a part is to place one light to one side of and above the camera to light the part at a 45° angle, and a second light to the other side, at about camera level, as a fill light. At times, a third light is placed slightly behind the part for backlighting.

Lighting is very critical in photography. The specimen should be placed on a background which will not detract from the resolution of the fracture surface. For basic lighting, one spotlight is suggested. The light is then raised or lowered, and the beam adjusted from flood to spot, to find the quality of lighting desired. Generally, the flatter the fracture surface, the lower the light should be. The effects of the angle of lighting are illustrated in Fig. 2. The angle of the incident light is much more important than its intensity.

A spotlight beam without a diffuser, or a bare bulb with a small filament, will provide high-contrast lighting. If desired, the light can be diffused by placing a piece of tracing paper between the light source and the specimen.

FILM

The black-and-white film used for over-all photography and photomacrography should be a moderately fast, fine-grain film with a long scale of gradation and good exposure latitude. Selection of a particular film is not as important as deciding on one film and using it consistently to become completely familiar with its response and characteristics. An advantage of using an ASA (ANSI) 400-speed cut film such as Tri-X is that type 52 ASA (ANSI) 400 Polaroid film can be used to check exposures and lighting and no readjustments are needed in final photography on cut film. Other self-processing films available for 4-by-5-in. cameras are: Polaroid type 51 high-contrast film (ASA or ANSI 125 with floods, 200 with flash), type 55 P-N (positive-negative) ASA (ANSI) 50 film, type 57 ASA (ANSI) 3000 film, and type 58 ASA (ANSI) 75 color film.

When both monochromatic and color film are used, there is an advantage in selecting types that provide equivalent film speeds, thereby avoiding the need to change exposure when changing film type.

STEREO VIEWS

Stereo-pair photographs of fracture surfaces provide a means of viewing the fracture contours in simulated three dimensions. The basic technique for preparing such photographs entails taking two pictures of a subject area, the second from an angle slightly different from the first. The photographs are then examined under a visual condition in which, for example, the viewer's left eye focuses on the first picture and his right eye focuses on the second. Stereo viewers are available, and in most cases are necessary, to ease the task of viewing stereo pairs. The effect is to convince the brain that the eyes are indeed seeing a three-dimensional scene. If the angular displace-

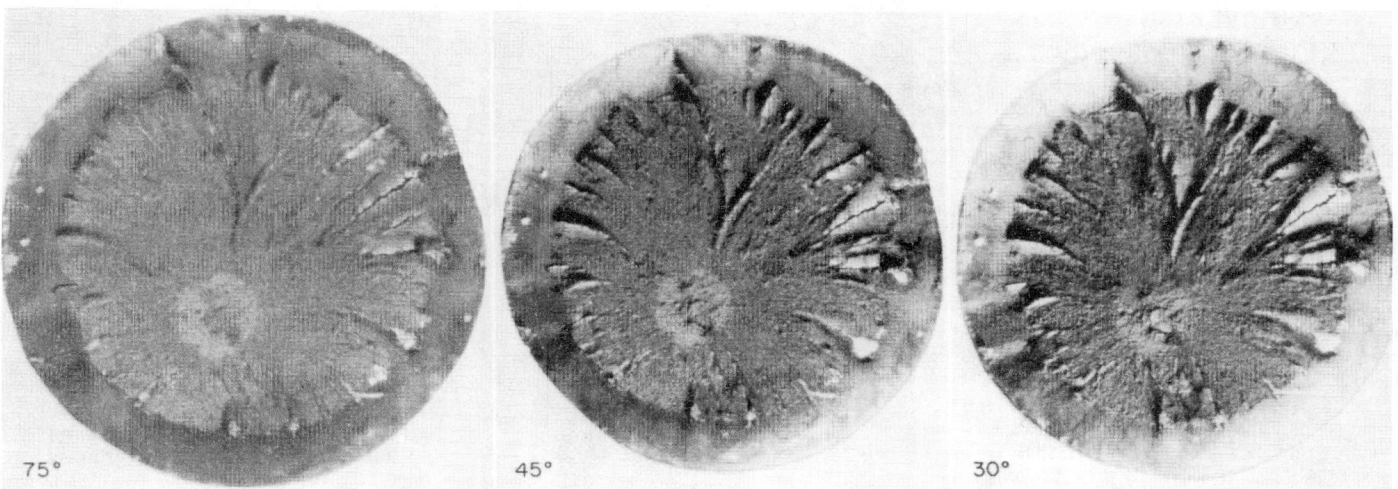

Fig. 2. Light fractographs (at 7×) showing effect of angle of incident light in delineating texture of a fracture surface of an unnotched laboratory tensile-test specimen of 4340 steel. The microstructure was tempered martensite. As the angle of illumination approaches 90°, the surface becomes featureless. The precise optimum angle of obliquity depends on the nature of the fracture markings.

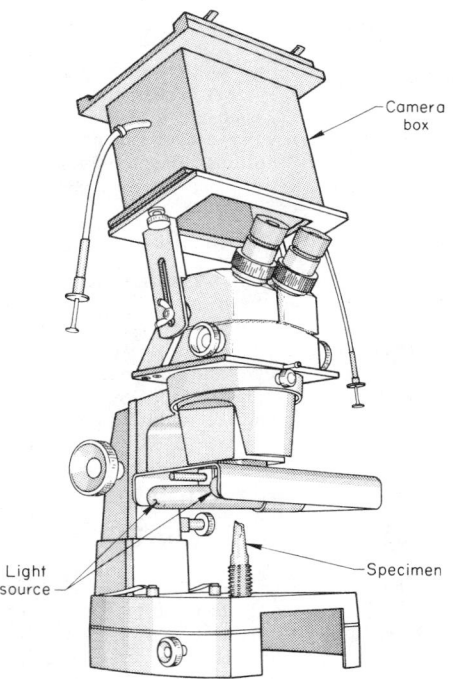

Fig. 3. Optical stereomicroscope with camera box partly removed for direct viewing. The light source is two 40-watt incandescent tubes above the specimen.

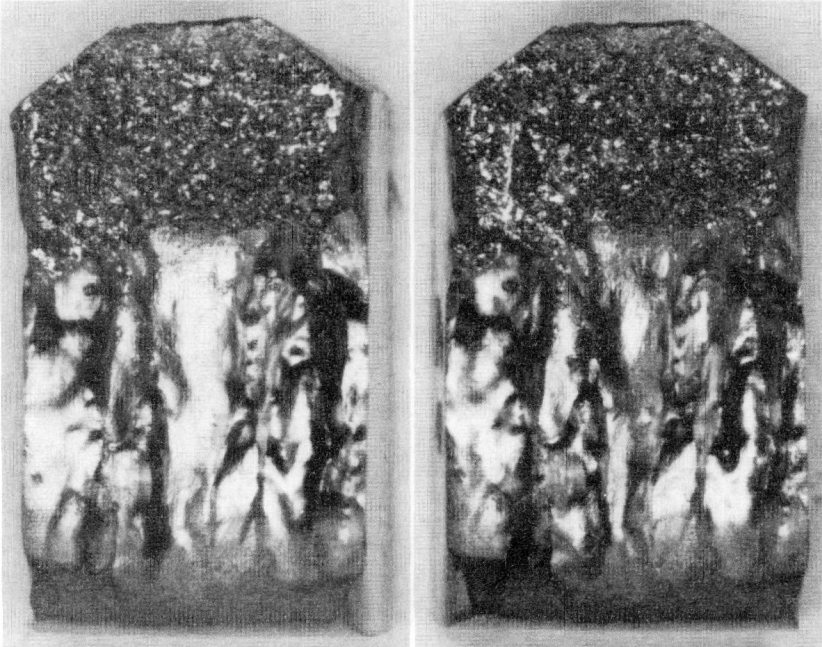

Light fractograph (stereo pair)   3×

The smooth columnar contours in the lower portions of the photographs were the result of cracking along prior austenite grain boundaries during straightening. The upper portions of the photographs also show intergranular fracture, but along ferrite grain junctions, produced by impact in the laboratory to expose the original crack surface.

Fig. 4. Stereo view of a fracture surface of cast experimental low-carbon steel

ment between the two pictures is appropriate (an included angle of 12° or 14° is desirable), the illusion is very vivid.

Stereo images by light microscopy have been used only to a limited extent, because of the restricted depth of field. They provide a useful means of studying fractures at magnifications generally not greater than 200×.

Stereo-pair photographs can be taken using a single-lens camera, if it is provided with a mount that will pivot about a horizontal axis through the subject. After the subject is properly aligned be-

neath the camera with the mount vertical, the camera is swung to an angle of 6° to 7° from vertical and an exposure is made. The second exposure is made with the camera at the same angle as for the first exposure, but swung to the other side of the vertical, or zero, position.

With the stereomicroscope shown in Fig. 3, it is possible to observe the voids that are a part of ductile dimpled fracture, and, in brittle fractures, cleavage facets and some of the river marks can be discerned. It is most useful for preliminary examination of fracture surfaces, leaving final

documentation of fine details for electron microscopy.

An example of a light-microscope stereo pair is shown in Fig. 4. The subject is a fracture surface in a low-carbon steel casting that cracked along prior austenite grain boundaries during a straightening operation. The bold relief contours of the columnar fracture are evident when viewed in stereo.

If a tool such as a parallax bar is used, quantitative measurements of topographic depths and elevations can be obtained.

Fractographic Features Revealed by Light Microscopy

TENSILE-FRACTURE SURFACE MARKS IN UNNOTCHED SPECIMENS

The macroscopic appearance of a fracture surface has often been used to appraise the degree of ductility and of toughness present in a metal. According to the concepts of fracture mechanics, toughness is the critical material property associated with overload or rapid fracture. The fracture surface contains vestigial marks that indicate the amounts of high-energy (ductile, or tough) and low-energy (brittle) crack extension that produced the fracture.

Tensile-fracture surface marks have been classified into three zones by configuration— namely, (*a*) the fibrous zone, (*b*) the radial zone, and (*c*) the shear-lip zone; this shear-lip zone is the highest-energy portion of the fracture. The relative amount of shear lip provides an indicator of the toughness of the material. The three zones are illustrated schematically in Fig. 5.

Fractures consisting solely of one zone occur only under conditions of extreme ductility or brittleness.

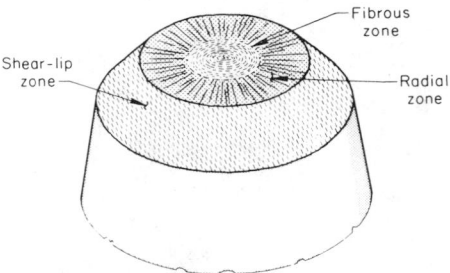

Fig. 5. Schematic representation of zones within a typical tensile fracture of an unnotched, cylindrical specimen. The surfaces of the fibrous and radial zones are usually normal to the tensile axis, as shown. The shear-lip surface is always at about 45° to the tensile axis.

FATIGUE MARKS

The formation of cracks under conditions of repeated or cyclic stress has been denoted as fatigue cracking. Zones of crack propagation on fatigue fractures exhibit several types of surface marks, such as beach marks, striations and ratchet marks.

Beach Marks. The term "beach mark" describes the macroscopic features present on the fracture surface as illustrated in Fig. 6. Beach marks indicate a local region of variation in crack-growth rate. Such beach marks may, but do not always, indicate fatigue as the mode of cracking. Stress-corrosion fractures may also show beach marks.

Striations. The term "striation" refers to a "line" on the fracture surface indicating the position of a crack front after an increment of crack propagation has occurred. Each increment of propagation is due to a cycle of stress, i.e., a cyclic load. The distance between striations indicates the advance of the crack front during each succeeding cycle. Fatigue striations may be mac-

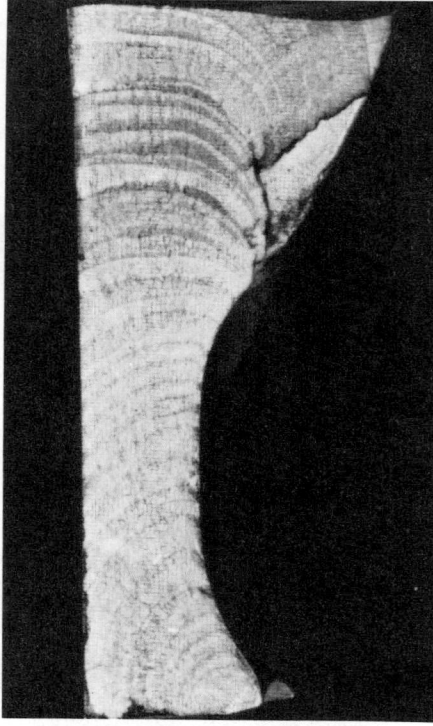

Light fractograph   **Actual size**

The light-colored, reflective bands are zones of slow crack propagation. At high magnifications, thousands of fatigue striations can be resolved within each band. The dull, fibrous bands are zones of crack propagation by microvoid coalescence.

Fig. 6. Beach marks on a fatigue fracture in an aluminum alloy 7075-T73 forging

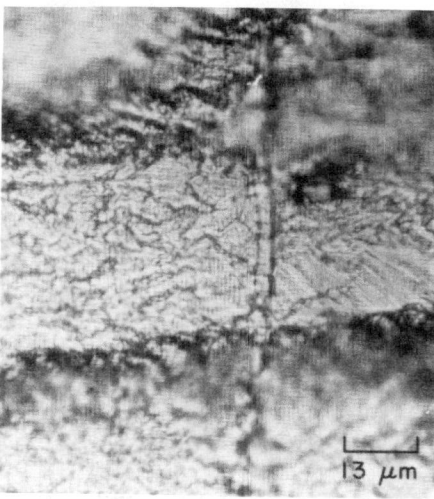

Fig. 7. Fatigue striations in aluminum alloy 7075-T651 plate. Note the fine, closely spaced, vertical striations at the left of the vertical step.

roscopic, but they are most often microscopic and may often be found within the fine structure between individual beach marks (Fig. 7).

Ratchet marks are macroscopic features that may be seen on fatigue fractures in shafts and flat leaf

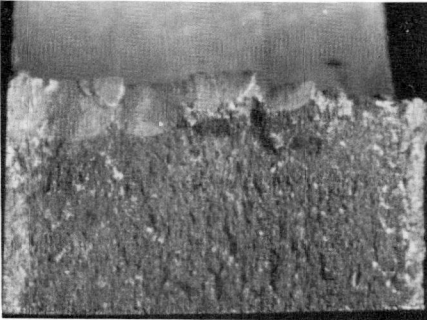

Light fractograph   **7¹/₂×**

Fig. 8. Ratchet marks in aluminum alloy 7075-T6 plate. The stepped appearance of these marks is characteristic whenever fatigue cracks emanate from several sites and later join to form one primary crack front.

springs, and they may also occur in ductile fractures in overtorqued fasteners. In fatigue fractures, ratchet marks are the result of multiple fatigue-crack origins, each producing a separate fatigue-crack zone. As two approaching cracks meet, a small step is formed. The small steps are the ratchet marks.

Although ratchet marks are most apparent on the peripheries of fractures in shafts, the stepped appearance is characteristic whenever fatigue cracks emanate from several origins and subsequently meet to form one principal crack front (see Fig. 8).

DISCONTINUITIES

Fractures originate from a broad variety of discontinuities within the metal structure, such as laps and seams from primary metal forming, shrinkage and gas cavities in cast structures, hot tears, inclusions, segregation of impurities, and imperfections in welds. Many of these features are familiar and do not need illustration.

Interpretation of Optical Fractographs

A RUDIMENTARY KNOWLEDGE of how to "read" a fracture surface must be gained so that meaningful fractographs can be taken from which to describe the fracture process. Fractography can provide information about the conditions of stress, the effects of temperature and chemical environment, and the origin of the fracture and how the crack progressed to final rupture.

STATES OF STRESS

Information regarding the stresses that caused a fracture can be learned from a casual examination of the fractured part.

In many types of fracture, the general plane of fracture is perpendicular to the maximum principal tensile stress. These types, called group I fractures in the following discussion, include cleavage and other brittle fractures, ductile fractures (also called microvoid coalescence and dimpled rupture) under plane-strain conditions (in thick sections), fatigue fractures (at least in the

Photograph   **About 2×**

(a) Photograph of the shaft showing the deformation of the splines in the region of fracture, which would not occur if the fracture were caused by fatigue. The shaft, 28 mm (1³/₃₂ in.) in diameter, was made of 6118 steel and had a hardness of 23 HRC. Being made of a ductile metal, it was twisted in pure torsion with a single overload, yielding the fracture on the transverse shear plane shown in (b).

Light fractograph   **About 2×**

(b) Fracture surface of the shaft, showing the rotary deformation characteristic of a single-torsion-overload fracture in a ductile metal. If there is a combined bending component, the region of final, fast fracture will be offset from the center of the section. This type of fracture should not be confused with one resulting from rotating-bending fatigue, which does not have the gross distortion seen here.

Fig. 9. Splined shaft of 6118 steel that fractured from a single torsional overload

Light fractograph   **About 0.95×**

Fig. 11. A torsional-fatigue fracture in an induction-hardened 1037 steel shaft 25 mm (1 in.) in diameter that finally fractured in longitudinal shear. No clear point of origin is visible because the surfaces rubbed as the crack propagated.

intermediate stages), and stress-corrosion cracks.

Other types of fracture propagate along planes of maximum shear stress. These types, called group II fractures here, include ductile fractures under plane stress (that is, in thin sections or near free surfaces), shear fractures, and the very early stages of fatigue fractures in pure or relatively impurity-free metals. In a ductile material, the shear stresses cause considerable deformation prior to fracture, although the deformation is not always obvious because the shape of the part is not changed except for flow on the surface. Figure 9(a) is a photograph showing deformation in a fractured shaft. Torsional single-overload fractures (group II) of a ductile material usually occur on the transverse shear plane, straight across the cylinder, and exhibit a telltale swirled appearance (Fig. 9b). The final-fracture area will be at the center of the bar. A brittle material in pure torsion will fracture in a plane perpendicular to the tensile-stress component, which is 45° to the specimen axis (group I fracture). A spiral-type fracture is one characteristic of this type of loading and material, and can be demonstrated by twisting a piece of chalk to fracture (see Fig. 10). The elastic-stress distribution in pure torsion is maximum at the surface and zero at the center. Thus, fracture normally originates at the highest-stressed region (the surface) in pure torsion. Longitudinal torsional fractures are sometimes observed (for example, Fig. 11), because longitudinal planes have the same magnitude of shear stress as transverse planes and longitudinal planes usually have lower toughness, due to the shape and distribution of inclusions.

CRACK ORIGINS

An interest in the exact location of the point of origin of a fracture derives from the importance of determining what initiated the fracture. The initial examination of a fracture is concerned with the recognition of all features that may point to the crack origin.

Gross Aspects of Fractures. Some indications of crack-propagation direction can be seen by examination of the gross aspects of a broken part. They relate to the order in which events oc-

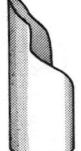

Fig. 10. Torsional brittle fracture of chalk. Fracture follows the 45° direction of maximum tensile stress.

curred, sometimes called "fracture sequencing." The fragments of a fractured structure can be reassembled in approximate juxtaposition, without allowing the fracture surfaces to touch, and then the telltale indications should be sought.

First, a fast-running crack in sheet or plate will frequently branch as it propagates but will almost never join another crack to continue as a single crack. Second, if a running crack joins a pre-existing fracture, it will usually meet it at approximately a 90° angle, not at a shallow angle. Third, it is almost impossible for an intersecting crack to cross and propagate beyond a pre-existing fracture. These considerations lead to the following useful guidelines concerning crack origins:

- The direction to the crack origin is always opposite to that of crack branching, as shown in Fig. 12.
- If a crack meets another at about 90°, it occurred later and the origin should not be sought in it but in the earlier crack. This is known as the T-junction method of crack-origin location (see Fig. 13).

The initial section of fracture (containing the crack origin) transfers its original load to adjoining sections, in all probability overstressing them. If these sections do not contain imperfections, succeeding fractures (assuming a normally ductile material) will be preceded by a certain amount of plastic deformation.

Fibrous marks, tear ridges and beach marks can also indicate the location of a crack origin. Chev-

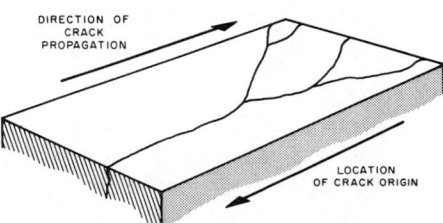

Fig. 12. Schematic representation of the information conveyed by crack branching with regard to the location of the crack origin

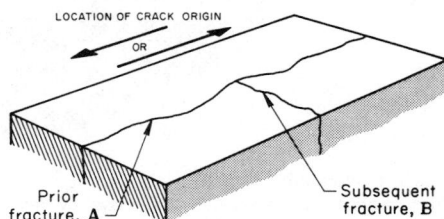

Fig. 13. Schematic representation of the T-junction method of determining which fracture surface to search to locate the crack origin. Because B does not cross A but meets it at about 90°, B occurred later and cannot contain the crack origin.

ron markings also can be used. Where the curvature of such marks is slight, the origin is generally on the concave side of the crack-front curve.

In general, the region of crack initiation will be flat and will lack any free-surface shear-lip zone. The shear-lip zone appears only at some

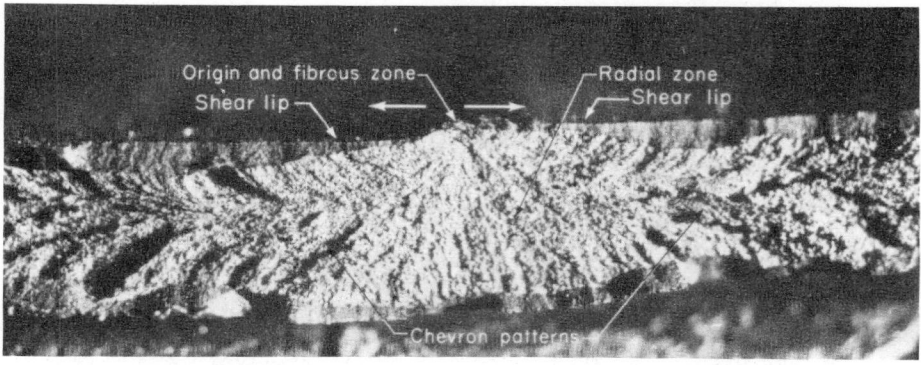

Light fractograph   6×

The shear lip increases in width with increasing distance from the origin. The radial marks below the origin and the chevron patterns to the left and right also indicate the directions of fracture.

Fig. 14. Fracture in a welded pressure vessel of 4340 steel displaying a flat origin at the top with a shear lip beginning on either side of it

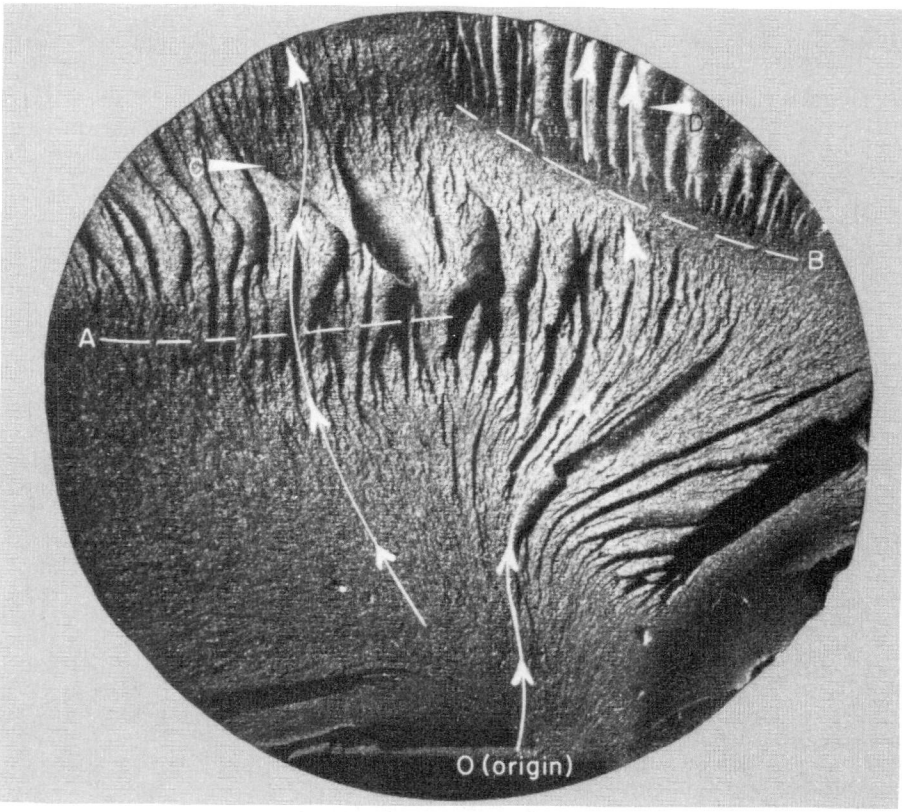

Light fractograph   3×

Fracture origin can be found in three ways: by tracing the radial marks in the lower portion of the fracture to their point of convergence (the arrows on the curved lines indicate the direction of crack propagation); by drawing normals to the crack-arrest fronts labeled A and B; and by projecting the tangents to the final radial marks at C and D toward the bottom. The crack came to a full stop at B with the first hammer blow and resumed motion at the second hammer blow.

Fig. 15. Locating the origin in an impact fracture, produced by two hammer blows, in a notched bar of 12% Cr steel

distance from the origin and becomes larger as the distance increases (see Fig. 14).

Location of Origins in Impact-Overload Fractures. Figure 15 shows the fracture of a 12% Cr steel bar that was notched and then struck with a hammer. Two blows were necessary to complete the fracture. The fracture marks are radial. They may be traced downward to a common intersection. Also

present are crack arrests, one at A and a second at B, where fracture progress came to a complete halt before the second blow was struck. The arrest marks are parallel to the crack front, and lines drawn normal to them should intersect at or near the origin. The contours of the final fracture marks, at C and D, also point to the general location of the beginning of fracture.

FRACTURE PROGRESS

Many types of fractures, including most service fractures, occur by a sequence involving crack initiation, subcritical crack propagation (due to ductile crack extension, fatigue, corrosion fatigue, stress-corrosion cracking or hydrogen embrittlement), and fast fracture, which occurs when the remaining cross section can no longer support the applied load. The fracture processes leave telltale marks on the fracture surfaces, which enable a trained investigator to locate the initiation sites, to discern the propagation direction and crack-front shape, and to distinguish the fast-fracture zone. This information can lead to an understanding of the stress levels and conditions leading to fracture.

FRACTURE CHANGES DURING CRACK PROPAGATION

Several influences may affect the growth of a crack, causing it to progress thereafter by a mechanism of fracture different from that in effect when cracking started. These influences include: (*a*) local differences in microstructure; (*b*) changes in stress-intensity factor, K; (*c*) changes in chemical or thermal environment; (*d*) differences in stress state.

Changes Caused by Local Differences in Structure. Microstructure exerts a pronounced influence on local fracture appearance. The presence of two or more types of microstructure may result in different fracture mechanisms being involved, and a different fracture appearance. A simple example is a fracture in a chilled white iron part. Fracture is by cleavage through the chilled zone and is fibrous in the pearlitic zone.

Light fractograph   5.7×

Small corrosion pits formed beneath the layer of chromium plate, as at A, and generated stress-corrosion cracks B. Growth of these cracks altered the stress intensity at the crack tips, leading to propagation of fatigue cracks C. Final, fast fracture D occurred when the critical crack-tip stress-intensity value was reached.

Fig. 16. Changes in fracture mechanism and appearance that were caused by changes in chemical and stress environment for a chromium-plated aluminum alloy 7079-T6 forging

Another structure difference is that of case and core in carburized, flame-hardened and induction-hardened parts. The difference in properties between such structures can cause a crack to proceed by quite different fracture mechanisms in adjacent regions.

Changes Caused by Altered Environments. A fracture-mechanism change as a result of different chemical and stress environmental conditions is shown in Fig. 16. A corrodent generated small pits below a layer of chromium plate and provided the environment for the growth of stress-corrosion cracks, which originated at the pits. The stress may have been residual or applied. As the stress-corrosion cracks grew, the stress intensity at the crack tip increased for the applied cyclic loads. At some critical level of environment and cyclic-stress intensity, the fracture mechanism changed to one of fatigue. The fatigue cracks propagated until the critical crack-tip stress-intensity values were reached, and then unstable fracture occurred in an essentially ductile manner.

The Scanning Electron Microscope and Its Application to Fractography

PRINCIPAL FEATURES OF SCANNING ELECTRON MICROSCOPES

A scanning electron microscope (SEM) is a combination of electron-optical, vacuum, and electronic-control devices for impinging a beam of electrons on a pinpointed spot on the surface of a target specimen and collecting and displaying the secondary electrons given off from that target. The SEM was developed initially to obtain information about surface topography and was thought of as complementing the capabilities of the light microscope and the transmission electron microscope. However, the development of an assortment of related capabilities has converted the SEM into a tool that can be used for total material characterization.

The SEM is capable of magnifications from about 5 to 240 000 diameters, but its useful upper limit in fractography is perhaps 30 000 diameters. It has a resolution limit as low as 5 nm (0.2 μin.) depending on the equipment used and on the special surface-preparation techniques employed. The depth of field is about 300 times that of the light microscope. This corresponds to a useful depth of field of over 1 mm (0.04 in.) at 100 diameters and of about 10 μm (400 μin.) at 10 000 diameters. Specimens can be tilted as much as 45° in either direction and still be in focus. The working distance from the objective aperture to the specimen is typically 10 to 25 mm (0.4 to 1.0 in.).

A typical arrangement of the components of a scanning electron microscope for use in fractography is shown in Fig. 17. An electron gun at the top of a column emits electrons from a heated filament and accelerates them in a constant stream down the column. The specimen being examined is near the lower end of the column. The electrons pass through two or more electromagnetic lenses that focus the stream of electrons into the shape of a small beam so that the impact spot on the specimen is of minimum size. Instrument resolution is a function of minimum spot size.

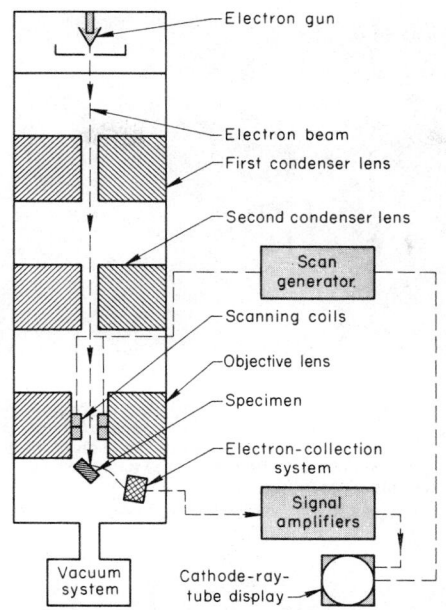

Fig. 17. Essential components of a scanning electron microscope

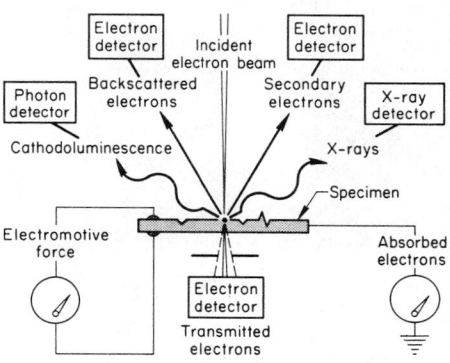

Fig. 18. Useful signals generated by an electron beam

MODES OF OPERATION OF SCANNING ELECTRON MICROSCOPES

Figure 18 shows the useful signals generated by the various interactions of the incident electron beam with the specimen, and the devices used to detect or collect the signals. Although both secondary and backscattered electrons are used for fractography, the secondary electrons are almost always the preferred signal because they offer better resolution, produce an abundant signal, and permit viewing of areas of the specimen that are not in a direct line of sight with the collector. There are occasions, however, where it is necessary to sacrifice resolution in order to improve image contrast, especially with smooth specimens and at low magnifications. This is accomplished by using the backscattered-electron mode of operation. Backscattered electrons possess high energy, travel at high velocity, and travel from the specimen to the detector in straight-line paths that produce shadow effects and, consequently, high image contrast. The backscattered-electron mode of operation is particularly useful for stereography.

A considerable loss of resolution is inherent in

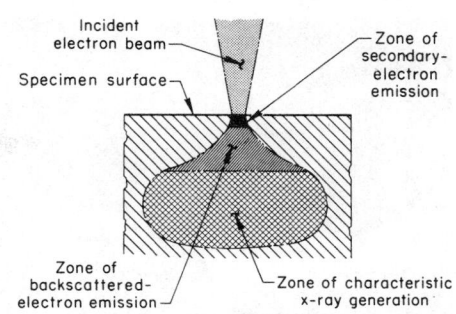

Fig. 19 Pattern of zones in a specimen that are sources of signals generated by an incident electron beam

the backscattered-electron mode of operation. Backscattered electrons are electrons with high energy (equivalent to that provided by the accelerating voltage, which is usually in the range of 20 to 25 kV). The electrons escape from as deep as 30 nm (1.2 μin.) in the specimen. Because of scatter within the specimen, the resultant source is much larger in diameter than the incident-beam diameter (see Fig. 19).

SPECIMEN PREPARATION

There are three main requirements of a specimen for examination by scanning electron microscopy. The specimen must be small enough to fit into the SEM chamber. This frequently requires sectioning of the fractured part. The specimen must be clean to prevent contamination of the microscope, and application of a conductive coating may be necessary. When nonmetallics, such as rust, are present on the fracture surface, a buildup of electrical charge will occur. This severely hampers surface examination. Conductive coatings satisfactorily eliminate this problem so that a thorough fracture-surface analysis can take place.

STEREOGRAPHY

Because of its great depth of field, a scanning electron microscope produces a realistic, three-dimensional image and makes qualitative interpretation of fracture topography relatively easy. If it is desirable to increase the three-dimensional effect, stereo pairs of photographs may be taken of the same area on the specimen.

Interpretation of Scanning-Electron-Microscope Fractographs

AT LOW MAGNIFICATIONS, the features in scanning-electron-microscope (SEM) fractographs strongly resemble the aspects of the fracture apparent to the naked eye; but at high magnifications, more detail is visible which needs to be categorized and interpreted if the fractograph is to be related to the micromechanisms of fracture that were active.

It is important to realize that microscopic features of fractures ordinarily differ widely within a small area. The principal categories of fracture features are as follows:

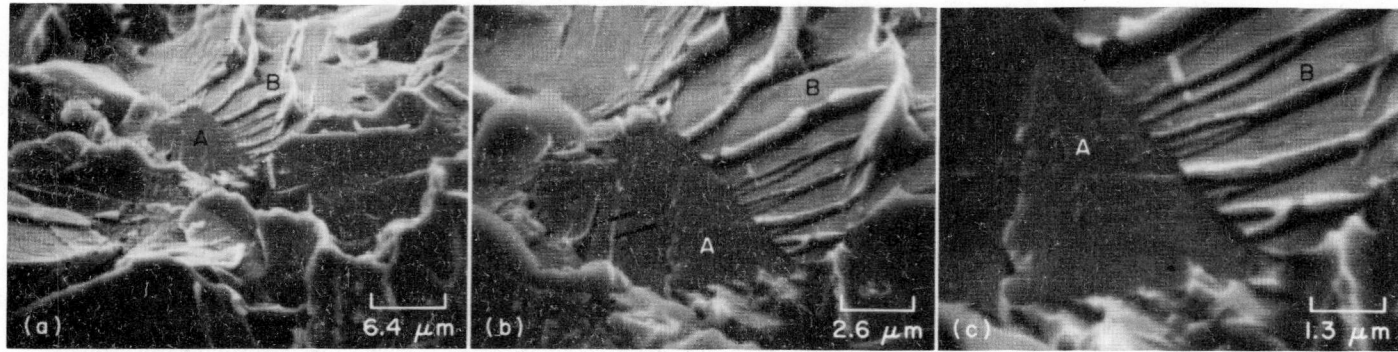

SEM fractographs

1560× (a); 3900× (b); 7800× (c)

The specimen was tilted in the scanning electron microscope at an angle of 40° to the electron beam. The cleavage planes followed by the crack show various alignments, as influenced by the orientations of the individual grains. Grain A, at the center in fractograph (a), shows two sets of tongues (see arrowheads in fractograph b) as the result of local cleavage along the {112} planes of microtwins created by plastic deformation at the tip of the main crack on {100} planes. Grain B and many other facets show the cleavage steps of river patterns. The junctions of the steps point in the direction of crack propagation from grain A through grain B, at an angle of about 22° to the horizontal plane. The details of these forks are clear in fractograph (c).

Fig. 20. Cleavage fracture in a notched impact specimen of hot rolled 1040 steel broken at −196 °C (−321 °F), shown at three magnifications

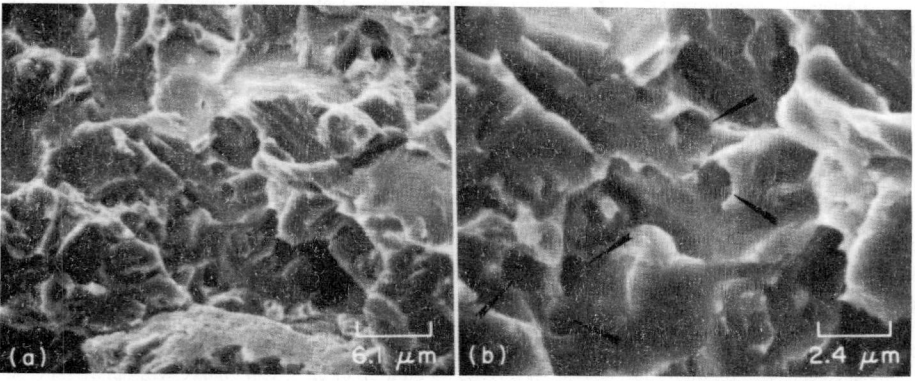

SEM fractographs

1650× (a); 4140× (b)

The small cleavage facets in martensite platelets contain river patterns and are separated by tear ridges. Shallow dimples, marked by arrowheads, are also visible. Direction of crack propagation is from bottom to top in each fractograph. The specimen was heat treated for 1 h at 843 °C (1550 °F), oil quenched, and tempered for 1 h at 427 °C (800 °F). Fracture was by Charpy impact at −196 °C (−321 °F).

Fig. 21. Quasicleavage in the surface of an impact fracture in a specimen of 4340 steel. The same area is shown in both fractographs, but at different magnifications.

- Cleavage features (tongues, microtwins, and location of cleavage-crack origins)
- Quasicleavage features
- Dimples from microvoid coalescence
- Tear ridges
- Fatigue striations
- Separated-grain facets—i.e., intergranular fracture
- Mixed fracture features, including binary combinations of cleavage features, dimples, tears, fatigue striations and intergranular-fracture features
- Features of fractures resulting from chemical and thermal environments.

TRANSGRANULAR CLEAVAGE FEATURES

In cleavage fracture, the fracture path follows a transgranular plane that is usually a well-defined crystallographic plane. This plane of fracture is one of the {100} planes in most body-centered-cubic metals. Cleavage fracture is produced, usually at low temperature, under a condition of high triaxial stress—that is, at the root of a notch—or at a high deformation rate, as, for example, by impact loading, or as a result of environmental factors.

Figure 20 provides three views, at increasing magnification, of an area in an impact fracture exhibiting features that are typical of cleavage. It is apparent that the fracture plane changes orientation from grain to grain. As a result, the average grain size can be measured on the fractograph and related to grain-size measurements on a metallographic section. The change of orientation from grain to grain leads to a branching of the crack along different planes and to a very chaotic over-all appearance of the fracture surface. At higher magnification, many features typical of cleavage can be identified. In Fig. 20(b), the evidence of change in orientation between grain A and grain B is particularly clear because of the river patterns that begin in grain B at the interface. The river patterns, which represent steps between different local cleavage facets of the same general cleavage plane, are well defined.

QUASICLEAVAGE FEATURES

In steels that have been quenched to form martensite and then tempered to precipitate a fine network of carbide particles, the size and ori-

entation of the available cleavage planes within a grain of prior austenite may be poorly defined. True cleavage planes have been replaced by smaller, ill-defined cleavage facets, which usually are initiated at carbide particles or large inclusions. The small cleavage facets have been referred to as quasicleavage planes because, although they look like cleavage planes with river patterns radiating from the initiation sites, until recently they have not been clearly identified as crystallographic planes. Quasicleavage features tend to be more rounded, indicating a somewhat higher energy absorption than that of true cleavage.

Quasicleavage facets on a fracture surface of a quenched-and-tempered 4340 steel specimen broken by impact at −196 °C (−321 °F) are shown in Fig. 21. The poorly defined cleavage facets are connected by tear ridges and shallow dimples.

Quasicleavage, or cleavage in complex microstructures, is more difficult to identify than the cleavage found in low-carbon steel made up of ferrite and pearlite. When identification is uncertain, it is essential to relate the fracture features to the microstructure, including the prior austenite grain size, the martensite plate size, and the distribution, size, spacing and volume fraction of fine carbide particles precipitated during tempering.

DIMPLES FORMED BY MICROVOID COALESCENCE

If the temperature of fracture is raised from very low (for example, liquid-nitrogen temperature) to higher levels, the fracture mechanism changes from brittle or cleavage fracture to ductile fracture by microvoid initiation, growth and coalescence. At low magnification, the fracture surface may exhibit both fibrous and cleavage regions. Fibrous regions are usually observed (*a*) near the free surface in a shear-lip zone, or (*b*) at the origin of fracture at the center of a smooth (unnotched) tensile specimen. Cleavage regions are typically observed in the surfaces of flat, plane-strain fractures and in regions of high crack velocity. However, at high magnification, the fibrous-fracture region, and also the cleavage region, may show successive fracture mechanisms—that is, the fracture features may include both cleavage facets and dimples. Both dimples

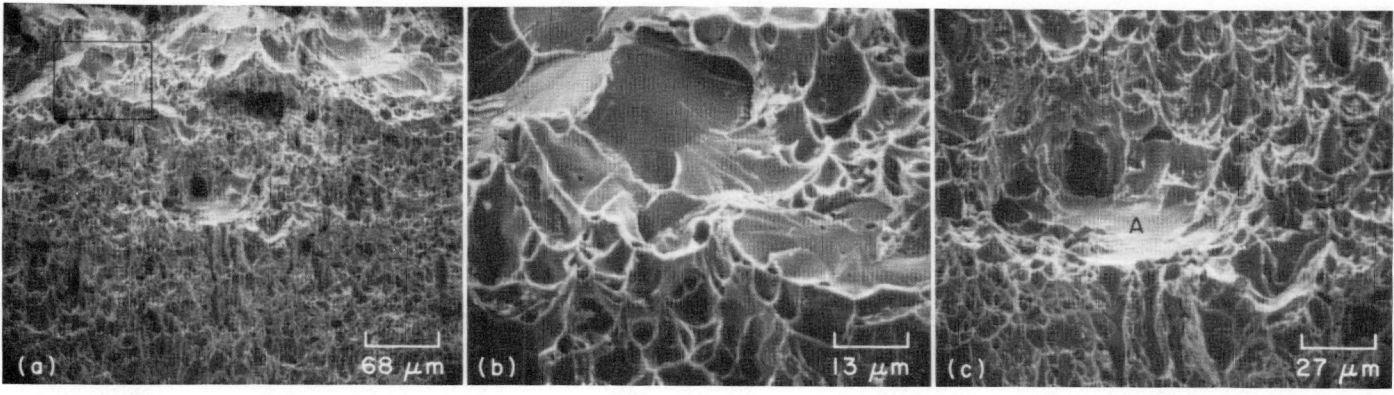

SEM fractographs   147× (a); 750× (b); 370× (c)

The machined notch of the specimen was below the region shown in (a). The over-all direction of crack propagation was upward. Although equiaxed dimples predominate, certain grain orientations near the top of (a) were unfavorable for ductile fracture by microvoid coalescence and local cleavage occurred, as shown in detail in (b), which is a higher-magnification view of the outlined area in (a). Fractograph (c), a higher-magnification view of the region at the center of (a), shows a deep dimple, which initiated the local ductile fracture immediately surrounding it. The smooth surface at A shows no river patterns and should not be identified as a cleavage facet; it could be a grain-boundary surface, or perhaps a region of stretching.

Fig. 22. Dimples and cleavage facets exhibited in three aspects of a Charpy impact fracture at room temperature in a specimen of hot rolled 1040 steel, tilted in the scanning electron microscope at an angle of 30° to the electron beam

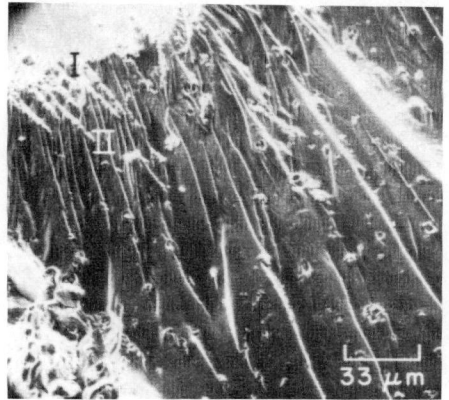

SEM fractograph   300×

The transition from stage I (upper left) to stage II is well defined. The presence of (Fe,Si)-rich inclusions did not affect the fracture path markedly. The inclusions, which were fractured, range from 5 to 25 μm (200 to 1000 μin.) in diameter. The stage II area shows a large number of approximately parallel fatigue patches containing very fine fatigue striations that are not resolved at this magnification.

Fig. 23. Transition from stage I to stage II of a fatigue fracture in a coarse-grain specimen of aluminum alloy 2024-T3

and cleavage facets are visible in the Charpy impact fracture shown in Fig. 22. An over-all view of the region of transition between ductile and cleavage fracture is shown in Fig. 22(a). The upper portion of the fracture surface in this view is a result of cleavage, and the lower portion is a result of fracture by microvoid coalescence. The cleavage facets in Fig. 22(a) and (b) are at a considerable distance from the notch, which is below the region shown in the fractograph. In the intermediate-magnification view of the region of microvoid coalescence (Fig. 22c), facet A cannot be positively identified as a cleavage-fracture feature; it could be the result of a fracture along a grain boundary, or even possibly a microvoid surface that has stretched. Around this facet is a region of ductile fracture, which originated at the interface between the matrix and a carbide particle (in the deep dimple).

TEARING

Tearing designates a mechanism of local fracture that is often found at a discontinuity in the crack advance by another fracture mechanism. It occurs when small regions or ligaments fracture by plastic flow or necking. Tearing is frequently observed when small unbroken areas remain behind the main crack front. The occurrence of tearing is accompanied by the formation of tear ridges, which are typically sharp and thus produce bright contrast in the SEM image. Tearing may also produce flat-topped, featureless areas having some of the characteristics of local glide-plane decohesion, similar to facet A in Fig. 22(c).

FATIGUE STRIATIONS

The advantages offered by the use of the SEM include:

- Easier identification and evaluation of the origin of the fatigue fracture, whether near a free surface, at an edge, or at the bottom of a notch or a groove
- Better differentiation between stages I and II of fatigue-fracture progress by viewing the over-all fracture at low magnification not obtainable with a TEM
- Estimates of crack-growth rates, which are used in fracture-mechanics evaluation of loads or for estimation of total number of cycles to failure
- Simpler quantitative analysis of fracture surfaces to determine which portions of fracture surfaces resulted from microvoid coalescence, from intergranular separation and from cleavage fracture.

The main disadvantage of scanning electron microscopy for the investigation of fatigue fractures is that fatigue striations are not as sharply defined as with transmission electron microscopy. This lack of resolution occurs because a striation represents only a small surface displacement, which often fades out in the electron image. By shadowing the fracture surface with a gold-palladium film 10 nm (0.4 μin.) thick, it is possible to enhance striation contrast markedly. Striation spacings as small as 25 nm (1.0 μin.) have been measured in aluminum alloys with the SEM. With scanning electron microscopes that have resolutions better than 10 nm, spacings as small as 10 nm can be resolved.

It is desirable to reveal as much as possible of

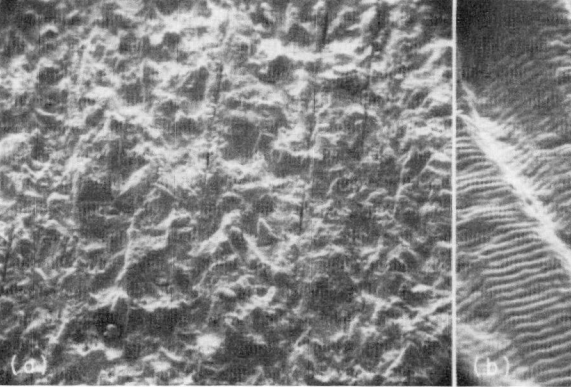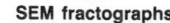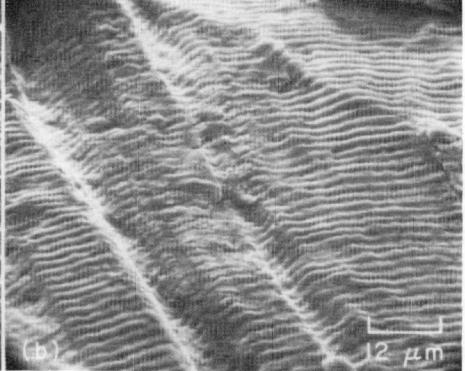

SEM fractographs   39× (a); 810× (b)

Fig. 24. Fatigue fracture in type 304 stainless steel tested at room temperature. The vertical secondary cracks in (a) are grain-boundary separations. The well-defined striations in (b) resulted from the planar slip characteristic of stainless steels.

 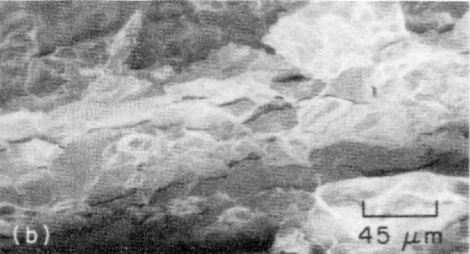 

SEM fractographs   700× (a); 220× (b); 560× (c)

(a) Sintered tungsten rod drawn to 1.5-mm (0.060-in.) diam, recrystallized for 100 h at 10^{-6} torr and 2600 °C (4712 °F), and fractured in tension. (b) Iridium sheet annealed for 50 h in purified helium at 1700 °C (3092 °F), and broken by bending. (c) Tungsten−3 wt % rhenium alloy that was prepared in the same manner as the sintered tungsten rod in fractograph (a). Microvoids ("bubbles") at grain boundaries resulted from segregation of potassium (an impurity).

Fig. 25. Intergranular brittle fractures in tungsten, iridium, and a tungsten−3 wt % rhenium alloy

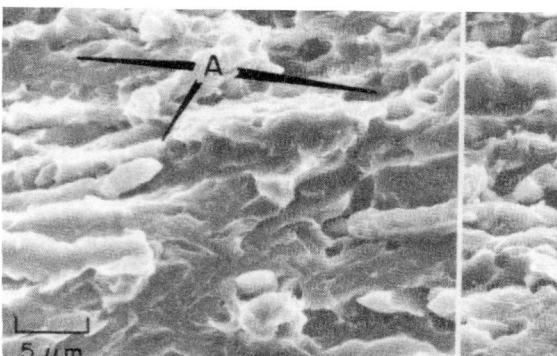  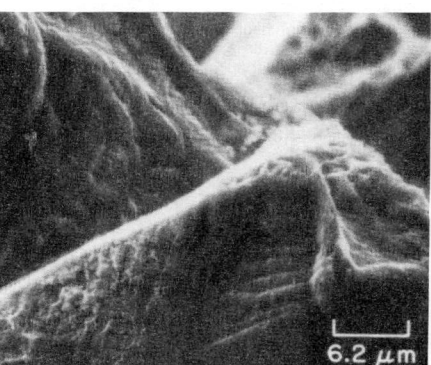

SEM fractographs   Both at 2000×

Fig. 26. Intermingled cleavage facets and dimples in two views of a stress-corrosion fracture in a step-cooled two-phase Ti-6Al-2Sn-4Zr-6Mo alloy exposed to a $3^1/_2$% NaCl aqueous solution. Cleavage facets formed in the alpha phase, and poorly developed dimples, such as at the sites marked A, formed in the beta phase.

SEM fractograph   1620×

Fig. 27. Intergranular fracture in copper alloy CDA No. 715 (copper nickel, 30%) that became embrittled by grain-boundary oxidation during extended exposure to high-temperature steam in a heat exchanger. Crack penetration (which was cyclic, as intergranular layers of oxide formed, broke and reformed) produced fine striations that could be mistaken for fatigue striations.

the surface details within all secondary cracks associated with fatigue-crack branching. This can best be achieved with the specimen oriented with the crack-propagation direction pointing toward the secondary-electron collector to gain maximum penetration of the primary electron beam into the depth of the secondary cracks.

The SEM fractographs in Fig. 23 and 24 show some of the characteristic features of fatigue-fracture surfaces.

INTERGRANULAR FRACTURE

Intergranular fracture is simply described as grain-boundary separation. It can occur by catastrophic brittle separation, or by separation plus microvoid coalescence on the interfaces of grains. Such fractures are regarded as the result of a severe reduction in grain-boundary energy by a Gibbsian (thermodynamic) adsorption mechanism. In its simplest form, segregation of metallic or gas-metal impurities can alter the grain-boundary free energy. Furthermore, grain-boundary energy can vary over a temperature range, leading to brittle-ductile fracture transitions and thermally induced brittle or ductile fractures. Variations in segregation at grain boundaries can also lead to mixed mechanisms of fracture, characterized by the appearance of dimpled areas in an intergranular fracture. Intergranular brittle fracture in the absence of an aggressive environment may be the result of segregation of a thermally activated impurity, which

allows the grains to separate along smooth interfacial planes. Typical examples of fractures involving grain-boundary segregation are shown for tungsten, iridium and a tungsten-rhenium alloy in Fig. 25.

FEATURES INDICATIVE OF MIXED MECHANISMS OF FRACTURE

A fracture that occurs by operation of two or more intermingled mechanisms of fracture is generally labeled a "mixed-mode" fracture. This is not to be confused with the successive operation of different fracture mechanisms, which can be analyzed sequentially and therefore require no special discussion. The occurrence of fracture by mixed mechanisms often indicates (*a*) that the usual factors that determine the operative mechanism, such as state of stress, loading history, microstructure and environment, favor both mechanisms; and (*b*) that the local fracture mechanism is determined by a combination of deviations in these factors and the influence of secondary variables, such as local grain orientation.

The reasons for mixed mechanisms of fracture are as diverse as the fractures in which they have been observed. However, it is useful to identify the individual fracture mechanisms that contribute to such mixtures and to establish the circumstances that can lead to their individual occurrence. These circumstances establish limits. The occurrence of mixed mechanisms of fracture usu-

ally indicates that interacting influences have caused the fracture to depart from either limiting mechanism.

Mixed fracture mechanisms can result from quite different causes, and caution must be exercised in inferring the cause of fracture from the fracture features alone.

FEATURES OF FRACTURES RESULTING FROM CHEMICAL AND THERMAL ENVIRONMENTS

For a specific metal or alloy, certain fracture mechanisms, such as cleavage, microvoid coalescence, fatigue, and intergranular separation, are often associated with particular environmental and stress states. The accumulation of experience with such fractures arising from known states of environment and stress makes it possible to ascertain, with the aid of fractography, the causes of fractures occurring under unknown service conditions. Some instances in which environment caused a specific fracture response that can be characterized through electron-microscope fractography are illustrated in Fig. 26 and 27.

Use of Fractography for Failure Analysis

Aspects of Failure That Can Be Resolved by Fractography

EXAMINATION of a fracture begins with visual scrutiny, which establishes (*a*) whether there is gross evidence of mechanical abuse, (*b*) whether there are indications of excessive corrosion, (*c*) whether the part is deformed, (*d*) whether the origin of the crack can be readily identified, (*e*) whether the direction of crack propagation can be easily recognized and (*f*) whether there are obvious secondary fractures. Often it is helpful to have an undamaged part of the same design as the fractured part available during this portion of the examination. The findings of this scrutiny will permit many conclusions to be drawn concerning the service conditions existing at and prior to the time of fracture. These findings can then be extended by an examination of the fracture surface at low magnification with a stereomicroscope and then at high magnification by scanning electron microscopy, transmission electron microscopy, metallography, or some combination of these examination techniques. A survey at low magnification is important for identification of those areas that need further inspection at high magnification. The salient features are recorded in fractographs of appropriate magnification for report purposes, and for future reference should subsequent handling or sectioning destroy evidence needed for failure analysis.

Care must be taken in handling fractured parts to be studied and in preparing fracture specimens. Evidence that may be useful in establishing the origin and mechanism of fracture can easily be altered or destroyed by carelessness in collecting fractured parts or performing failure analysis.

Crack Origin and Crack-Propagation Path. Establishing the origin of a fracture is essential in failure analysis, and the location of the origin may have a large influence in determining which measures should be taken to prevent a repetition of the fracture. For example, if the crack origin is at the surface, surface alteration by such means as case hardening, shot peening, cold working of holes and corrosion protection may eliminate recurrence of the problem. If the origin is subsurface, higher over-all strength, a greater cross section or fewer internal discontinuities may be needed. The features that help identify the crack origin, such as concentric fibrous marks, radial marks and beach marks, and the fracture-surface characteristics that show the direction of crack propagation (and conversely, the direction toward the origin), such as chevron marks, crack branching and river patterns, have been treated in other sections.

Type of Loading. The design of a part anticipates a specific mode of loading in service. Whether or not loading was as expected is of prime concern in studying a fracture. Much can be learned regarding the type and magnitude of applied stress from fractographic inspection. Obviously, if the type of stress that caused fracture is radically different from the type that was anticipated, a change of treatment in service or a new design may be required.

Fatigue fractures result from progressive growth of one or more fatigue cracks generated by repetitive cyclic loading. The cracks may start at any one of several surface or subsurface discontinuities. Usually, the repetitive nature of the load cycles is reflected in a stepwise growth of the fatigue crack, producing the parallel microscopic marks known as fatigue striations. These can be recognized by following the guidelines described in the article on interpretation of SEM fractographs.

Evidence of a wear pattern in the form of an arc on the inner surface of one of the fingers of the clevis shown in Fig. 1 suggested that the shank had been subjected to bending-moment forces oriented favorably for crack initiation at the observed origin. Investigation indicated that the wear pattern and the cyclic bending loads had resulted from malfunction of a bearing in the linkage component attached to the clevis. This malfunction was attributed to faulty assembly procedures during installation of the bearing.

The absence of fatigue striations should not be interpreted as definite proof that the part did not undergo cyclic loading, because striations are sometimes masked or erased by corrosion. Some materials (high-strength steels, for example) form very fine striations which are easily obscured or destroyed by rubbing and numerous environmental factors. How distinct striations are depends on the environment to which the fracture surface has been exposed, and not on their formation. If the evidence shows that a fracture occurred by fatigue when none was anticipated, the service application was evidently in error and should be corrected. To accomplish this, either the peak stress should be reduced or the material should be improved. In steels, for example, at best the fatigue limit is about half the yield strength, and may be much lower if the surface is decarburized or has small discontinuities.

The orientation of the principal applied stress cannot always be explicitly deduced from an inspection of the fracture surface, but certain approximations can be made. If the dimples are equiaxed, the fracture probably resulted from simple tension and the stress axis probably was normal to the approximate plane of the fracture.

If the dimples are elongated, the fracture was produced either by shear or by tearing. In a shear fracture, the magnitude of applied stress can be qualitatively inferred from fractographic examination. Semiquantitative assessments of load levels may be obtained from striation-spacing measurements, or, macroscopically, from the critical crack length before the onset of rapid fracture. One appraisal depends on noting whether or not the part has undergone marked distortion or obvious plastic flow. Appreciable deformation results only from a stress that is high relative to the yield strength of the material. Other clues are offered by the relative proportions of various fracture features.

Material of the Part. Many questions concerning the relation of the material of a part to fracture can be answered through fractography; others require additional techniques, such as mechanical testing, metallography, chemical analysis, electron microprobe analysis, x-ray diffraction analysis, imaging by cathodoluminescence, and Auger spectroscopy. When these investigations are performed by auxiliary equipment on either a scanning or a transmission electron microscope, they may be considered to be fractographic procedures.

Brittleness or ductility may be revealed by fractography. In general, a ductile alloy will fracture by microvoid coalescence and a brittle alloy will fracture by transgranular cleavage or by intergranular separation. This is an oversimplification (most fractures display a mixture of features), but it is the basic premise.

Metallographic examination of sections normal to the fracture surface will disclose the nature of the microstructure and whether the heat treatment was satisfactory. If subsequent overheating occurred, a micrograph will reveal the effect it had. If the grain boundaries contain brittle films contributing to intergranular fracture, a micrograph may show them.

High-magnification fractographs will show whether or not a metal was "clean" — that is, whether there were few or many inclusions. Micrographs will also reveal inclusions, but frac-

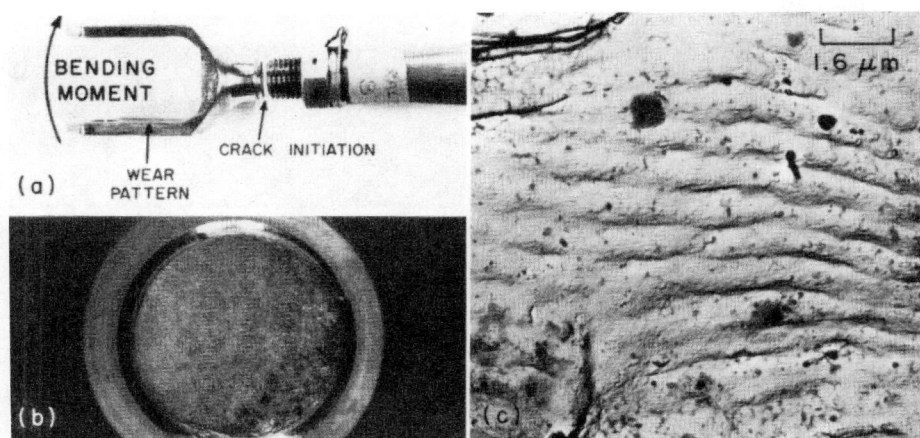

(a) Photograph, at 0.4×, of fractured clevis, showing orientation of bending moment, wear-pattern location, and crack-initiation site in shank. (b) Light fractograph, at 2.8×, showing fracture surface of clevis. (c) TEM fractograph (p-c replica), at 6400×, showing fatigue striations on fracture surface.

Fig. 1. Fracture of a titanium alloy Ti-6Al-4V clevis from a control system for a rotary-wing aircraft

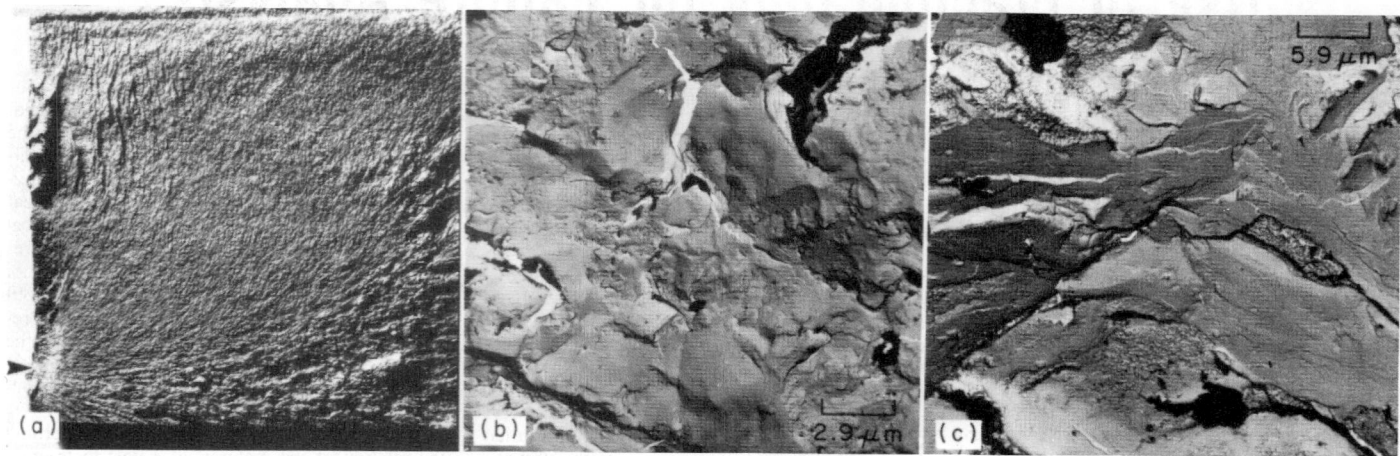

(a) Light fractograph, at 0.4×, showing crack origin (arrow) in weld deposit. (b) TEM fractograph (p-c replica), at 3400×, showing intergranular fracture in weld deposit. (c) TEM fractograph (p-c replica), at 1700×, showing cleavage fracture in an area outside weld deposit.

Fig. 2. Fracture in a welded 4340 steel axle. See also Fig. 3.

tographs will give more information regarding their shapes, because the fracture process usually separates most of the inclusion particle from its original pocket, affording a nearly three-dimensional view.

If segregation of elements or alloy phases occurred in an alloy, fractography may sometimes detect the pattern of segregation as a departure

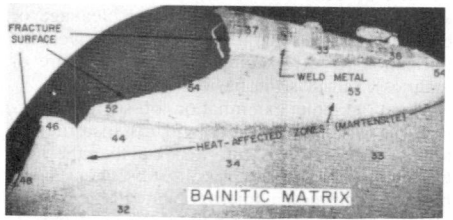

Macrograph (etchant: 2% nital)   12×

Fig. 3. Metallographic section of the 4340 steel axle of Fig. 2 in the region of crack origin, showing the weld metal, the heat-affected zone adjacent to the weld metal, and the Rockwell C hardness at various locations

from the normal arrangement of fracture features. The degree of success in identifying segregation will depend somewhat on how gross the pattern of segregation is. If the pattern is extremely fine, its discovery may require the use of metallography and electron microprobe analysis.

Study of the fracture surface will reveal whether discontinuities from processing led to the initiation of fracture. Forging laps and seams are readily recognized because the surfaces of these imperfections bear none of the characteristic features of fracture. Shrinkage porosity in castings displays dendrite contours within the shrinkage cavities that are unique in shape and that are not to be confused with fracture.

Welds are also potential sources of cracks; the cracks can occur either within or at the surface of the weld metal, or can form in the heat-affected zone of the base metal. If they were formed at ambient temperature, they are likely to resemble the final fracture in surface appearance. If they occurred before the part cooled from the welding temperature, oxide films should be visible on the surfaces of those cracks that were open to the external surface. Cracks that were entirely internal will have escaped oxidation. Fracto-

graphic examination will locate all open, oxidized cracks, but more positive identification of welding cracks is provided by metallographic examination of transverse sections through the weld and the heat-affected zone for the purpose of studying the relationship between the cracks and the weld, as in the following example.

Fracture started in a weld deposit (Fig. 2a) on a 4340 steel axle and propagated through the 150-by-150-mm (6-by-6-in.) cross section of the axle. Examination of TEM replicas from within the weld showed the fracture to be intergranular (Fig. 2b). The high hardness of the heat-affected zone (Fig. 3) indicated martensite formed by the mass quenching of the heavy section, possibly generating small quench cracks that served as fracture initiators. The fracture through the axle was brittle, as confirmed by fractographic evidence of cleavage (Fig. 2c) in all areas outside the weld deposit. Ambient temperature at fracture was about 10 °C (50 °F).

Shape and Surface of the Part. Overloading can produce noticeable distortion of a part. The initial visual scrutiny of a part, both with the unaided eye and with a low-power stereomicroscope, should include a critical examination for evidence of distortion. However, this will pro-

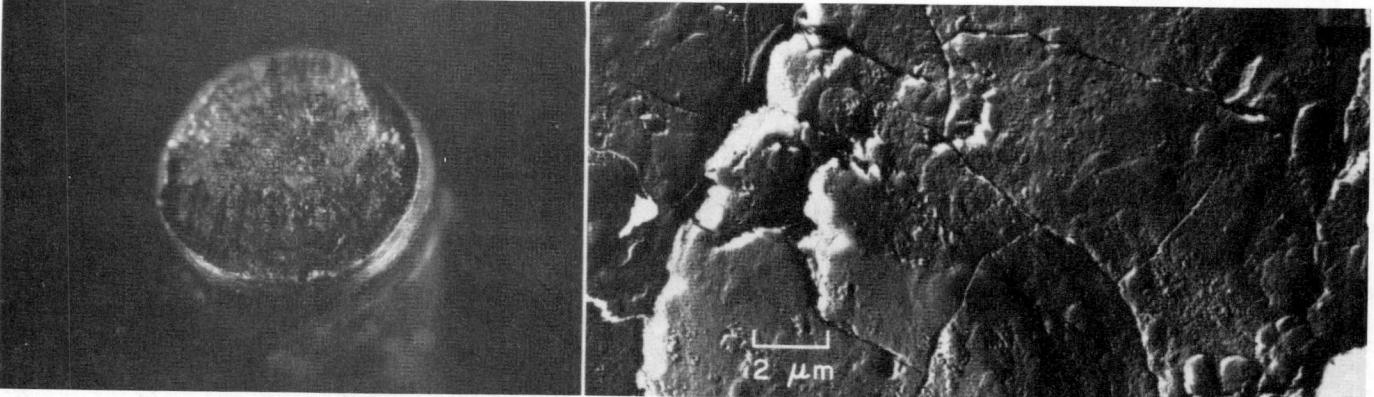

(Left) Light fractograph, at 6×, showing the surface of a fracture resulting from stress-corrosion cracking in a threaded bolt of H11 tool steel (tensile strength: 2041 MPa, or 296 ksi) stressed to 75% of tensile strength while immersed in a solution of 3.5% NaCl in water. This fracture, which originated around more than 50% of the circumference of the specimen (portion at bottom and at right), yielded a fibrous surface. (Right) TEM fractograph (p-c replica), at 5000×, of a region in the stress-corrosion portion of the fracture surface shown at left. The features revealed are separated-grain facets containing, in this area at least, comparatively little evidence of corrosion products. There are many secondary cracks within the grains and a few large secondary cracks between grains.

Fig. 4. Fracture caused by stress-corrosion cracking in a threaded bolt of H11 tool steel

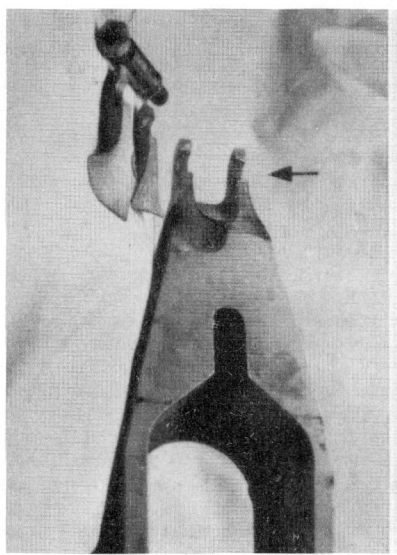
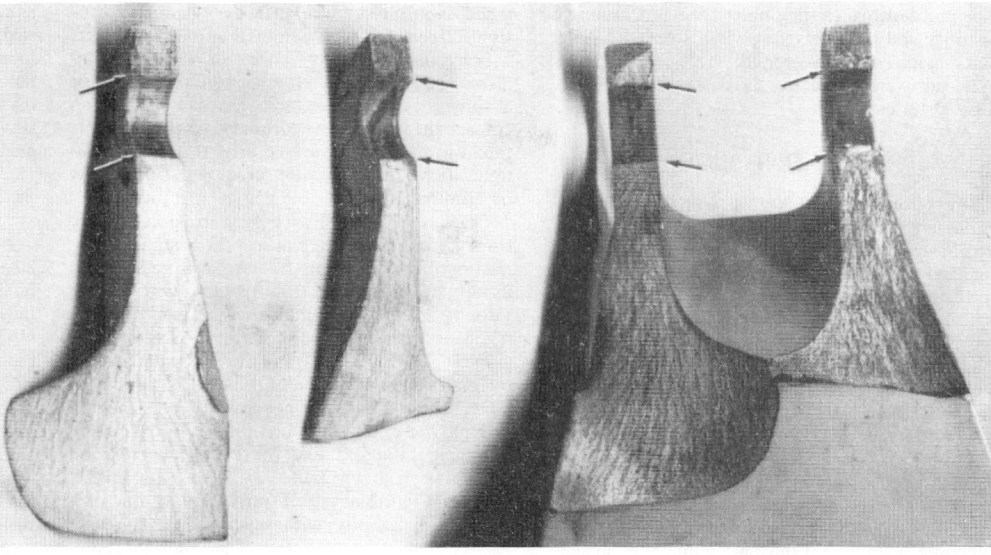

Photograph   1/6×

(a) Portion of a broken aircraft landing-gear actuator beam of aluminum alloy 7079-T6, showing stress-corrosion fractures (at arrow). See also (b) to (f).

Light fractograph   5/8×

(b) Higher-magnification view of the surfaces of the stress-corrosion fractures shown in (a). These fractures originated at the interior edges of boltholes in clevis-attachment lugs, at the locations designated by the arrows. An appreciable amount of evidence of both corrosion and secondary cracking was found at these origins. Visible are chevron marks, which also confirm these locations as the origins of fracture.

Light fractograph   4×

(c) Fracture surface outboard of the bolthole in the clevis-attachment lug at right in (b). Entire surface is typical of stress-corrosion cracking. Origin is at arrow (lower right corner of surface).

Light fractograph   4×

(d) Fracture surface inboard of the bolthole in the clevis-attachment lug in (c). The primary stress-corrosion crack, which originated at arrow, is within the area enclosed by the dashed line (lower left). The remainder of this surface shows chevron marks typical of final, fast fracture.

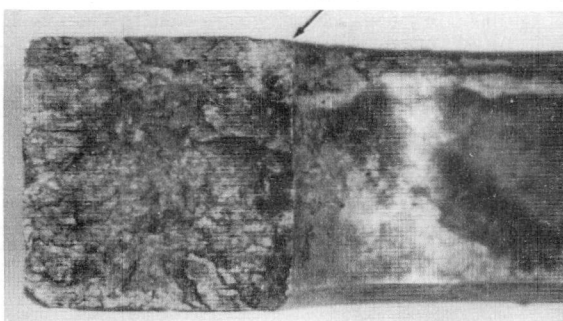
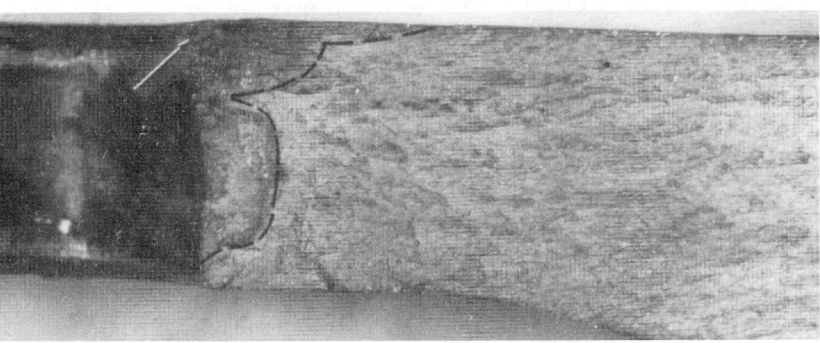

Light fractograph   4×

(e) Fracture surface outboard of the bolthole in the lug at left in (b). Here, as in (c), the entire surface is typical of stress-corrosion cracking. Origin is at arrow (upper right corner of surface).

Light fractograph   4×

(f) Fracture surface inboard of the bolthole in the clevis-attachment lug in (e). The primary stress-corrosion crack, which originated at the arrow, and which is enclosed by the dashed line, is larger than that in (d). This actuator beam had hardness normal for 7079-T6: 152-157 dph (10-kg load).

Fig. 5. Aluminum alloy 7079-T6 aircraft landing-gear actuator beam fractured because of stress-corrosion cracking

vide proof of overloading only if the part has high ductility and relatively low yield strength. With parts made of high-strength steels, proof that overloads were well above the design limit is more difficult to establish.

CHEMICAL ENVIRONMENTS

Stress-corrosion cracking is a mechanical-environmental failure process in which sustained tensile stress and chemical attack combine to initiate and propagate fracture in a metal part. Stress-corrosion cracking is produced by the synergistic action of sustained tensile stress and a specific corrosive environment, causing failure in less time than would the separate effects of the stress and the corrosive environment if simply added together.

Failure by stress-corrosion cracking frequently is caused by simultaneous exposure to a seemingly mild chemical environment and to a tensile stress well below the yield strength of the metal (Fig. 4). Under such conditions, fine cracks can penetrate deeply into the part while the surface exhibits only faint signs of corrosion. Hence, there may be no macroscopic indications of an impending failure.

MACROSCOPIC EXAMINATION

Macroscopically, fractures produced by stress-corrosion cracking always appear brittle, exhibiting little or no ductility even in very tough materials; in this way they resemble corrosion-fatigue fractures. Many transgranular stress-corrosion cracks characteristically change fracture planes as they propagate, producing flat facets. Both these flat facets and the grain facets of intergranular stress-corrosion cracks ordinarily are observable at low magnification.

The fracture surfaces usually contain easily identifiable regions of crack initiation, slow crack propagation and final rupture (see Fig. 5). Final rupture usually occurs by tensile overload. Thus, the area of final fracture often shows some evidence of ductility such as a shear lip or a herringbone pattern emanating from the zone of slow cracking.

The area of slow crack growth often contains corrosion products or is stained or otherwise discolored with respect to the area of final, fast fracture. However, the presence of staining or corrosion products on the fracture surface is by no means positive proof of stress-corrosion cracking. Some stress-corrosion-cracking fractures are not stained or discolored, especially in materials with good corrosion resistance; in addition, many fractures become corroded before inspection can be accomplished.

The zones of slow propagation of stress-corrosion cracks usually are much rougher in appearance than those of corrosion-fatigue cracks and do not contain fatigue beach marks or macroscopic evidence of cold work. However, if either the stress component is removed or the environment becomes inactive, crack propagation will discontinue until these factors again become operative. This sometimes results in macroscopic markings similar in appearance to the beach marks characteristic of fatigue fractures.

Both types of fracture can, and often do, initiate at corrosion pits or other stress raisers on the surface of the part. However, if the environment is sufficiently corrosive and if tensile stresses are present, stress-corrosion cracking can be initiated even on smooth surfaces, whereas corrosion fatigue always is initiated at a stress raiser.

Fractures caused by either of these mechanisms always initiate at the surface of the part or at some location where the aggressive species can contact the metal. By corollary, if the crack is found to have a subsurface origin, corrosion fatigue and stress-corrosion cracking usually can be eliminated as possible causes of failure.

Cracks produced by stress corrosion and by some hydrogen-damage processes generally have macroscopically rough fracture features, which usually are discolored from reaction with the environment. These mechanisms are easily recognized as macroscopically different from corrosion fatigue, but identifying the exact mechanism by macroexamination alone is not always possible. These cracks do not always have distinct origins; there may be much pitting in the region of crack initiation, and the crack may exhibit branching in this region.

Careful correlation of microscopic fracture-surface topography with macroscopic fracture-surface features is essential. The crack-initiation region and the directions of crack growth must be identified accurately so that information concerning the sequence of events and the micro-mechanisms of fracture, as observed by electron microscopy, can be correlated with the circumstances of crack initiation and the mechanism of crack propagation. Frequently, several different fracture micromechanisms are observed on a single fracture surface. Accordingly, correct identification of the initiating fracture mechanism and of any changes in micromechanism during fracture propagation is of vital importance in arriving at a correct understanding of the failure.

It is difficult to distinguish between service failures by stress-corrosion cracking and by hydrogen damage solely from microfractographic evidence. Fractures of both types mainly follow intergranular paths, although stress-corrosion cracking sometimes is transgranular, and the surfaces of both types of fracture may be substantially corroded. Also, hydrogen evolution during stress-corrosion cracking may be a factor in the cracking process, so that the combination of hydrogen embrittlement and corrosion at the crack tip may be instrumental in producing grain-boundary separation.

VISUAL EXAMINATION AND CLEANING

Metal samples are first examined visually and with the aid of a hand magnifier or other suitable viewing aid. At this stage, features such as the extent of damage, general appearance of the damage zone, and the color, texture and quantity of surface residues are of primary interest. If substantial amounts of foreign matter are visible, cleaning is necessary before further examination. The residues can be removed in some areas, leaving portions of the failure region in the as-received condition to preserve evidence. When only small amounts of foreign matter are present, it is sometimes preferable to defer cleaning so that the surface can be examined with a stereomicroscope before and after cleaning, or to defer cleaning until necessary for surface examination at higher magnifications or for the preparation of metallographic specimens.

Washing with water or solvent, with or without the aid of an ultrasonic bath, is usually adequate to remove soft residues that obscure the view. Inhibited pickling solutions will remove adherent rust or scale. Usually, it is advisable to save the cleaning solutions for later analysis and identification of the substance removed. Alternatively, plastic replicas can be used for cleaning; besides cleaning, the replicas retain and preserve surface contaminants, thus making them available for analysis.

MICROSCOPIC EXAMINATION

Examination by both light microscopy and electron microscopy can be used to observe minute features on corroded surfaces, to evaluate the microstructures of the metallic parts, and to observe the manner in which, and extent to which, the metal was attacked by the corrodent.

Corroded Surfaces. Viewing the cleaned surface with a stereomicroscope clearly shows gross topographic features such as pitting, cracking and surface patterns that can provide information about the failure mechanism — whether corrosion was the sole phenomenon involved; the type of corrosion; and whether other mechanisms, such as wear and fracture, also were operative.

If the features cannot be observed clearly using a stereomicroscope, a scanning electron microscope often can resolve the features, especially on very rough surfaces.

Microstructure. Microscopic examination of polished or polished-and-etched sections can reveal not only microstructural features and additional damage such as cracking, but also the manner in which the corrodent has attacked the metal (such as grain-boundary attack or selective leaching).

It is desirable to retain the corrosion products if they possess sufficient coherence and hardness to be polished. One method of keeping the surface material in place is to impregnate the sample with a casting-type resin, which is allowed to harden before specimens are cut. Polishing on napless cloths with diamond abrasives is recommended to secure maximum quality of edge retention.

Defects Leading to Fracture That Are Revealed by Fractography

DISCONTINUITIES such as laps, seams, cold shuts, previous cracks, porosity, inclusions, segregation, and unfavorable grain flow in forgings often serve as nuclei for fatigue fractures or stress-corrosion fractures because they increase both local stresses and reactions to detrimental environments. Large discontinuities may reduce the strength of a part to such an extent that it will fracture under a single application of load. However, a discontinuity should not be singled out as the sole cause of fracture without considering other possible causes or contributing factors. Thorough failure analysis may show that the fracture would have occurred even if the discontinuity had not been present.

LAPS, SEAMS AND COLD SHUTS

An observer who has some familiarity with the characteristics of various types of fractures in the material being examined can usually find indications of a discontinuity if one was present at the fracture origin. A flat area that, when viewed

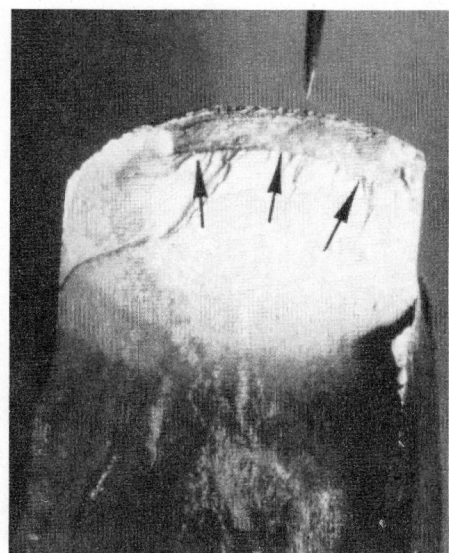

Light fractograph 4.2×

The seam was in the area between the fracture edge indicated by the pointer and the line indicated by the arrows. Hardness of the steel was 255 HB.

Fig. 6. Fatigue fracture that propagated from a seam in the counterbalance cheek of a forged 1046 (mod) steel crankshaft

without magnification, appears black or dull gray and does not exhibit the normal characteristics of fracture indicates the presence of a lap, a seam or a cold shut. (Such an area may appear to have resulted from the peeling apart of two metal surfaces that were in intimate contact, but not strongly bonded together.) The surface of a seam that resulted in a fatigue fracture of a 1046 (mod) steel crankshaft is shown in Fig. 6. A lap, a seam or a cold shut is fairly easy to identify under a low-power stereomicroscope also, because the area of any of these discontinuities is distinctly different in texture and color from the remainder of the fracture surface.

CRACKS

The cause and size of a pre-existing crack are of primary importance in fracture mechanics, as well as in failure analysis, because of their relationship to the critical crack length for unstable crack growth. Figure 7 shows a fracture in a highly stressed 4340 steel part. A narrow zone of corroded intergranular fracture at the surface of the part is adjoined by a zone of uncorroded intergranular fracture, which in turn is adjoined by a dimpled region. The part had been reworked to remove general corrosion products shortly before fracture. It was concluded that the rework failed to remove about 0.1 mm (0.004 in.) of a pre-existing stress-corrosion crack, which continued to grow after the part was returned to service.

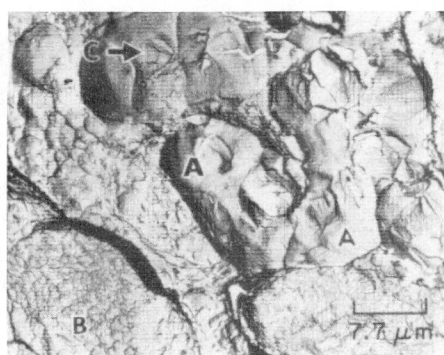

TEM fractograph 1300×

Fig. 9. Fracture surface of aluminum alloy 2124-T851, showing large dimples (at A's) that initiated at large inclusions (arrow at C). Cracking was produced by relatively little strain. During final, fast fracture, many small dimples (such as at B) formed at fine particles.

In Fig. 8, the light fractograph shows a pre-existing cleavage crack that grew from several origins and served as the nucleus for an overload fracture in a 300M steel part that had been heat treated to a tensile strength of 1930 to 2070 MPa (280 to 300 ksi); the transmission-electron-microscope fractograph in Fig. 8 shows the pres-

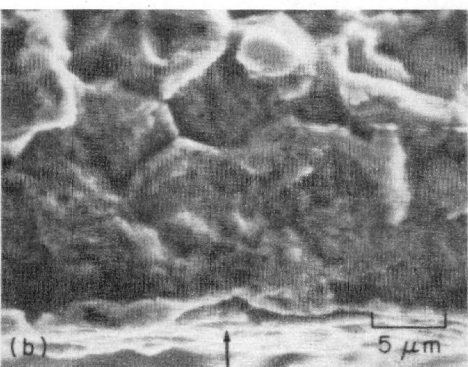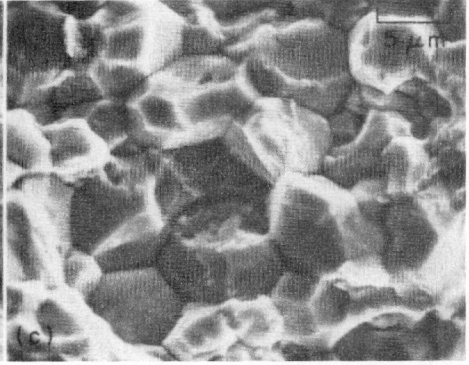

SEM fractographs 80× (a); 2000× (b and c)

(a) and (b) Remains of an old crack along the edge of the surface of the part (arrows); note dark zone in (a) and extensively corroded separated-grain facets in (b). (c) Clean intergranular portion of crack surface, formed at the time of final fracture.

Fig. 7. Fracture caused by a portion of an old intergranular stress-corrosion crack that was not removed in reworking. Part was made of 4340 steel, heat treated to a tensile strength of 1790 to 1930 MPa (260 to 280 ksi).

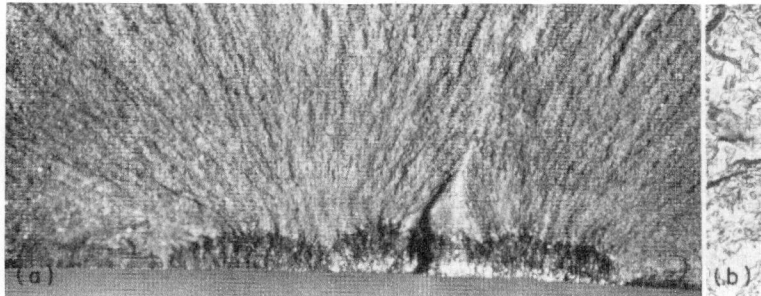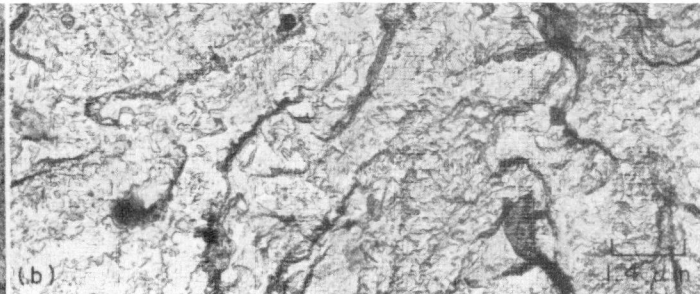

Light fractograph (a); TEM fractograph (b) About 5× (a); 7000× (b)

Fig. 8. Pre-existing cleavage crack (in light fractograph) that grew from several origins and served as the nucleus for an overload fracture in a 300M steel part that had been heat treated to a tensile strength of 1930 to 2070 MPa (280 to 300 ksi). TEM fractograph shows corrosion products on the fracture surface.

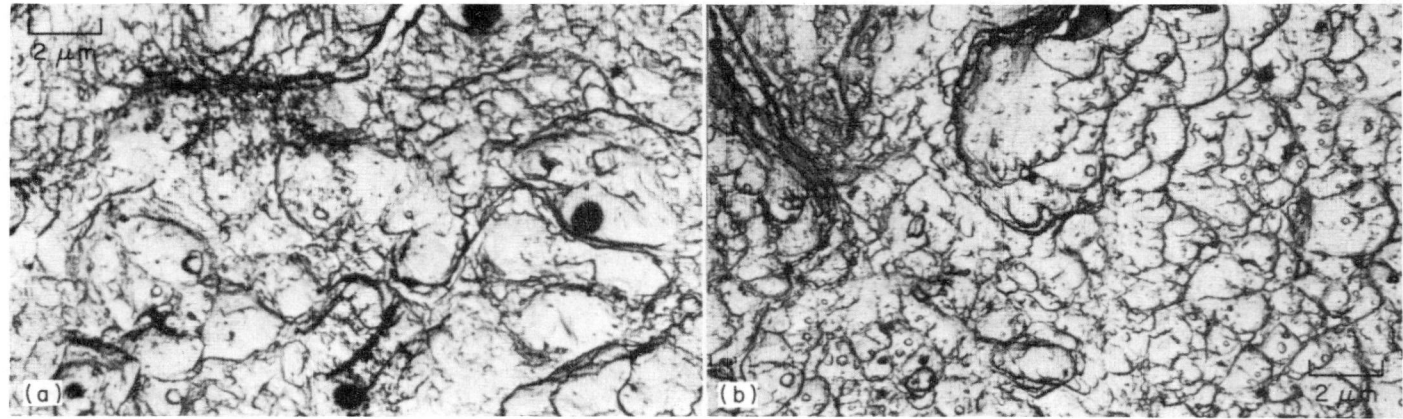

TEM fractographs   **Both at 5000×**

(a) Specimen from heat containing 0.049% sulfur. Large, shallow dimples were nucleated by relatively coarse sulfide particles. (b) Specimen from heat containing 0.008% sulfur. Many small dimples were nucleated by small carbide particles, but there were relatively few large dimples derived from sulfide particles.

Fig. 10. Fracture surfaces of two plane-strain fracture-toughness specimens from experimental heats of martensitic 0.45% C nickel-chromium-molybdenum steel containing different amounts of sulfur. Both specimens were tempered at 427 °C (800 °F).

ence of corrosion products on the fracture surface.

INCLUSIONS

Discontinuities in the form of inclusions, such as oxides, sulfides and silicates, can initiate fatigue fractures in parts subjected to cyclic loading. In addition, such inclusions have been identified as initiation sites of ductile fractures in aluminum alloys and steels by several investigators. At relatively low strains, microvoids form at inclusions, either by fracture of the inclusion or by decohesion of the matrix-inclusion interface. Figure 9 shows a fracture surface of aluminum alloy 2124-T851 containing large inclusions (one of which is shown in Fig. 9) that initially fractured to form microvoids. With additional straining, the microvoids grew to the size represented by the large dimples. During final fracture, many small dimples formed at fine particles.

Figure 10 shows typical fracture-surface characteristics produced by plane-strain fracture-toughness tests of two experimental heats of quenched-and-tempered 0.45% C nickel-chromium-molybdenum steel with extremes of sulfur content. Because high-purity materials and carefully controlled melting procedures were used, both heats were remarkably free of nonmetallic inclusions, except for sulfide particles. The fracture surfaces of specimens from both heats of steel consisted predominantly of equiaxed dimples. The fracture surface of the 0.049% S steel (see Fig. 10a) consists of many large, shallow dimples that appear to have been nucleated during the fracture process by numerous, relatively coarse sulfide particles. In contrast, the fracture surface of the 0.008% S steel (see Fig. 10b) shows relatively few large dimples (and associated sulfide particles) and many small dimples, which evidently were nucleated by small carbide particles.

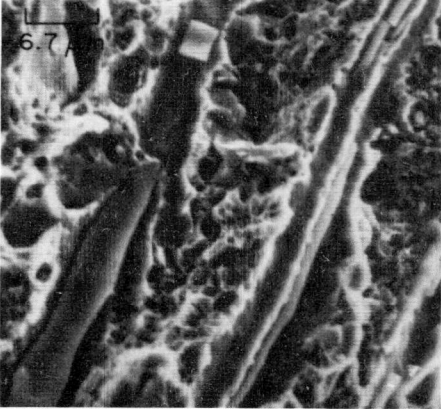

SEM fractograph   **1500×**

Stringers are visible as parallel features inclined about 30° to right of vertical. They were not identified as to composition, but may be accidentally entrapped slag that was elongated in the major direction of flow during forging.

Fig. 11. Stringers on the surface of a fracture that occurred during straightening of a 4340 steel forging that had been heat treated to a tensile strength of 1380 to 1520 MPa (200 to 220 ksi)

Stringers are elongated nonmetallic inclusions, or metallic or nonmetallic constituents, oriented in the direction of working. Nonmetallic stringers usually form from deoxidation products or slag, but may also result from the intentional addition of elements such as sulfur to enhance machinability. Figure 11 shows unidentified stringers on the fracture surface of a 4340 steel forging. Overload cracking occurred during straightening after the forging had been heat treated to a tensile strength of 1380 to 1520 MPa (200 to 220 ksi).

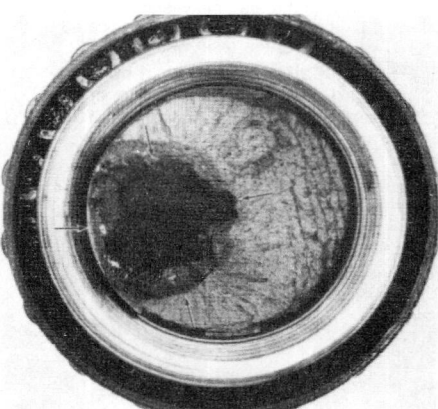

Light fractograph   **0.8×**

Fig. 12. Fracture through a large shrinkage cavity (arrows) in a spindle shaft made of cast 0.20% C steel. A dendritic structure is present on the cavity surface.

POROSITY

Porosity is the name applied to a condition of fine holes or pores in a metal. It is most common in castings and welds, but residual porosity from the cast ingot sometimes still persists in forgings. In fractures that occur through regions of excessive porosity, numerous small depressions or voids (sometimes appearing as round bottom pits) or areas with a dendritic appearance can be observed. At low magnification, fractures through regions of excessive porosity may appear "dirty" or "sooty" because of the large number of small voids, which look like black spots. A large shrinkage cavity in a fracture through a cast 0.20% C steel spindle shaft is shown in Fig. 12. A dendritic structure is visible on the cavity surface.

37 QUALITY CONTROL

This section was condensed from Metals Handbook, Eighth Edition, Volume 11, Nondestructive Inspection and Quality Control, pages 399 to 424. For more detailed information on the topics covered herein, the reader is referred to the larger work. For additional information on Quality Control within this volume, the reader should consult the index.

Quality Concepts and Statistical Methods

By R. A. Abbott, General Electric Company

QUALITY CONTROL is an accepted decision-making tool of scientific management. Like many other management techniques, quality control began as a practical manufacturing technique. Today, quality control is used by many management levels, in both manufacturing and nonmanufacturing departments, and influences a wide variety of business decisions.

Sound business decisions are based on observing and measuring effects, evaluating relative importance of observed effects, determining causes of observed effects, and determining which causes are controllable and which are not. Many sound business decisions are made by business owners, supervisors, foremen, and individual production workers through the informal application of business principles based on experience. However, informal techniques are not always satisfactory, because experience teaches different people different things, and because many complex problems do not yield easily to informal techniques. Quality control provides a formal method for analyzing and solving business problems. Statistical quality-control techniques can derive answers that are not otherwise possible, and can estimate the probable outcomes of events that are associated with alternative decisions.

The usefulness of quality control as a decision-making tool is based on recognition of variability as a major factor in the production of goods and the delivery of services. All products and services have variability because it is impossible for all units of a product or service to be made exactly alike. When a product is made with highly precise manufacturing equipment, variability may be difficult to measure, but it is always present. Sometimes, variability is blatantly obvious, as evidenced by excessive scrap, rework, returned goods, or service calls. There usually is an unacceptably high cost associated with obvious variability.

The purpose of a quality-control program is to reduce or maintain variability at an economic minimum consistent with management objectives. In this context, variability is an effect to be observed and measured. Part of the variability is beyond control; it is inherent in the process and materials. This portion of variability can be changed only by changing the basic process or material. The remainder of the variability is due to assignable causes — causes that can be reduced or eliminated if they can be identified. Quality control uses statistical methods to separate variability due to controllable causes from that due to uncontrollable causes. In this respect, quality control is a marriage of the techniques of manufacturing engineering with those of statistical mathematics.

QUALITY CONCEPTS

As broadly defined by the American Society for Quality Control (ASQC), quality is the totality of features and characteristics of a product or service that bear on its ability to satisfy a given need. This definition is based on the "fitness for use" concept of product quality, which is discussed later in this section.

If quality is defined as fitness for use, then quality control is the broad task of maintaining a state of product and service quality that assures a customer of a useful product and a manufacturer of a viable business enterprise. This has particular meaning for metalworking, metal-finishing, and metal-treating companies. Too often, an existing specification, process description or control parameter is taken for granted without asking whether the process or specification will really result in customer satisfaction. This is particularly true when choosing a nondestructive test to evaluate a particular quality characteristic.

Quality assurance is basically a management technique that uses independent audits to evaluate the contribution to product quality of all the business functions within a company. Some companies have adopted a multidisciplinary concept that combines control of the quality aspects of design, control of the as-manufactured quality of the product, quality assurance, and a customer-oriented evaluation of availability of a product into a single product-assurance function. This concept is sometimes identified as a "product integrity function." Regardless of the label, each of these functions contributes to the broad task of delivering a product or service that fulfills the needs of the customer. It is more important to develop an understanding of the common work elements involved in controlling quality than to understand the labels that have been applied.

Fitness for Use. Among nonprofessionals, quality is often viewed subjectively by intuitively assessing desirable features. For instance, a luxury car is often considered to be of higher quality than a compact car. Alternatively, quality can be viewed as the degree of conformance to specifications. In this view, a compact car and a luxury car can be of equal quality if each meets overall design and performance specifications and if each operates as intended by the original design concept. Both of these concepts of quality, however, are lacking in one important aspect — the consumer's view of product usefulness.

The fitness-for-use concept of quality can be broken down into four major divisions and 12 subdivisions (Ref 1):

- **Quality of design:**
  Market research
  Design concept
  Specification
- **Quality of conformance:**
  Manufacturing technology
  Manpower
  Management
- **Availability:**
  Reliability
  Maintainability
  Logistic support
- **Field service:**
  Promptness
  Competence
  Integrity

Several of the subdivisions encompass the consumer's view, especially market research under quality of design and the subdivisions under both availability and field service. In market research, the needs of potential consumers are evaluated, often with considerable attention to technical details of desirable features of the product or service. Availability encompasses reliability, which is the probability that the product will perform its function without failure for a specific service life under a specific set of conditions;

maintainability, which is the ease of keeping the product in good running order; and logistic support, which includes the cost and ease of obtaining supplies necessary to the operation and preventive maintenance of the product. Field service encompasses all those services provided by the manufacturer after the product has been delivered to the customer, including equipment installation, operator training, repair service, warranty service and adjustment of claims. The promptness, competence and integrity of each field service that is provided are factors in determining the ultimate value of the product to the customer, and therefore contribute to over-all product quality in terms of satisfying the needs of the customer. Field service also provides a major source of input to the quality-control system by generating feedback through direct customer contact. This closes the loop of quality assurance by initiating corrective action on products that are not fit for use.

Quality control is a term that has been used in both a narrow sense and a broad sense. In the narrow sense, quality control means quality of conformance—that is, control of quality by inspection and process control on the manufacturing floor. This is the usual layman's concept of quality control. Among those who have made quality control their profession, it is recognized that quality control has a much broader scope, which involves quality of conformance *plus* control of the quality of design and evaluation of the degree of customer satisfaction. Because of the confusion generated by using the term "quality control" to mean both the narrow concept and the broad concept, the term "total quality control" is often applied to the broad concept.

STATISTICAL METHODS

Quality cannot be controlled unless there is a means for quantitatively measuring the level of quality and for distinguishing controllable variables from uncontrollable variables. The task of evaluating these factors is accomplished largely by the application of statistical methods.

Development of Statistical Quality-Control Techniques. Variability in a manufactured product was recognized by Venetian shipbuilders of the 12th and 13th centuries, who were concerned with building many ships as much alike as possible for greater ease and speed of construction. Similarly, Eli Whitney in the early 19th century developed the first sense of variability in mass production of interchangeable parts in the manufacture of firearms for the United States Army. However, it was largely the work of Dr. Walter A. Shewhart and colleagues at the Bell Telephone Laboratories in the 1920's and 1930's that proved the value of statistical concepts. Dr. Shewhart developed a technique of using control charts as a means of distinguishing between the normal variation introduced by chance causes and the excessive variation introduced by assignable causes.

The original Shewhart control chart was in two parts: one for observing the average level about which it was desired to control results, called the X-bar chart; and one for evaluating the standard deviation, or statistical measure of observed variability. The statistical concept of standard deviation as a measure of the dispersion of individual measurements is not readily grasped nor easily calculated. Fortunately, a chart of ranges (differences between the highest and lowest mea-

sured values in each sample) can be used instead of a chart of standard deviations to evaluate variability. Numerical factors for establishing control limits for both average and range were developed by Leonard H. C. Tippett and were first published in Great Britain in 1935. Although the use of range charts and associated numerical factors simplified calculations required for control charts, acceptance of this method did not become widespread until World War II.

The great production demands of World War II exposed the inadequacy of prewar manufacturing procedures. Quality control in industry was highly variable, and it was United States military agencies that mandated a program of statistical sampling. Through the combined efforts of industry and government, statistical quality control gained acceptance as an organizational function. Military procurement agencies provided additional support to the quality discipline by requiring quality-control clauses to be included in all purchase contracts. One of the most significant and long-lasting developments was the publication and use of MIL-STD-105 for acceptance sampling by attributes. This technique was the means by which unsatisfactory lots were separated from acceptable lots at minimum cost and minimum known risk to both producer and user.

Since 1946, three standards on definitions and terminology, three on control charts and one on a total quality-system specification have been developed and revised by ASQC. Since 1975, ASQC has developed, updated and internationally coordinated versions of MIL-STD-105D (the military fourth revision) and MIL-STD-414, a sampling plan for acceptance of lots through measurement of variable characteristics. In addition, five manuals on the practice of various statistical aspects of quality control have been published by ASTM. All of the above standards have been through the consensus process of the American National Standards Institute and are approved American National Standards (see Table 1). Fundamentals of control charts and acceptance sampling, which are the most widely used statistical methods for evaluating variability, are discussed subsequently.

Definitions. The terms listed below are widely used in statistical quality control and have special meanings in that field.

characteristic. A property that can serve to differentiate between units of product. The differentiation may be qualitative (*method of attributes*) or quantitative (*method of variables*).

distribution. Position, arrangement or frequency of occurrence of individual attributes or variables within a *sample* or *population*.

lot. A group of individual items that are produced or sold as a unit.

method of attributes. Measurement of quality that consists of noting the presence (or absence) of some characteristic in each of the units in the group under consideration and counting the number of units that do (or do not) possess the characteristic. Alternatively, the total number of *nonconformances* in a group of units can be counted.

method of variables. Measurement of quality that consists of measuring (against some kind of scale) the magnitude of a quality characteristic for each of the units in the group under consideration.

nonconformance. Failure to meet any specified quality requirement.

population. The totality of individual items of a single design that are available during a given period of interest, and from which *samples* can be taken for statistical analysis.

random sampling. A process of withdrawing a *sample* from a *lot* or *population* in which each unit in the lot or population has an equal chance of being included in the sample.

range. The difference between the largest value and the smallest value in a given set of observations.

sample. A finite part of a statistical *population* whose properties are studied to gain information about the population as a whole.

Variability is evaluated as the dispersion, or spread, of a measured attribute about the most common value of that attribute. Both the limits of the spread and the frequency distribution of measured values within the spread must be defined in order to evaluate variability accurately. Frequency distributions can be of many forms. Most industrial processes possess measurable characteristics that exhibit a normal (Gaussian) distribution, where the most common value is centered within the spread and where the frequency of occurrence of any given value diminishes the farther that value is from the most common value. Sometimes, variability is more accurately described by a distribution in which the most common value is not centered in the spread (skewed distribution) or by a distribution that contains two or more peak values of frequency of occurrence. However, as discussed in the next section of this article, the variability of

Table 1. Approved American National Standards covering quality-assurance subjects

Title	ANS number
Guide for Quality Control	ASQC B1
Control Chart Method of Analyzing Data	ASQC B2
Control Chart Method of Controlling Quality During Production	ASQC B3
Sampling Procedures and Tables for Inspection by Attributes	ASQC Z1.4 (a)
Definitions, Symbols, Formulas and Tables for Control Charts	ASQC A1 (b)
Definitions and Symbols for Acceptance Sampling by Attributes	ASQC A2 (c)
Glossary of General Terms Used in Quality Control	ASQC A3 (d)
General Requirements for a Quality Program	ASQC C1
Sampling Procedures and Tables for Inspection by Variables for Percent Defective	ASQC Z1.9 (e)
Practice for Probability Sampling of Materials	ASTM E105-61
Practice for Choice of Sample Size to Estimate the Average Quality of a Lot or Process	ASTM E122-72
Practice for Acceptance of Evidence Based on the Results of Probability Sampling	ASTM E141-69
Practice for Use of the Terms "Precision" and "Accuracy" as Applied to Measurement of a Property of a Material	ASTM E177-71
Practice for Dealing with Outlying Observations	ASTM E178-80
Generic Guidelines for Quality Systems	ASQC Z1.15

(a) Adapted from MIL-STD-105D. (b) Formerly Z1.5. (c) Formerly Z1.6. (d) Formerly Z1.7. (e) Adapted from MIL-STD-414.

average values of measurable characteristics of samples withdrawn from such a population will have a normal distribution if the sample size is large enough. This property of averages is very useful for controlling an industrial process within specification limits.

Theoretically, a normal distribution occurs when variability is the result of many independent random causes. Although these conditions are not strictly fulfilled for many industrial processes, inherent variability usually can be approximated by a normal distribution. For instance, if bar stock is automatically fed into a machine that cuts the stock into 2-in. lengths, the frequency distribution of length measurements on a lot consisting of 500 consecutively cut pieces might be of the form shown in Fig. 1. The peak or central value, called the process average, measures 2.00 in. The spread of individual values is from 1.91 to 2.09 in.

The spread can be expressed mathematically as the root-mean-square (rms) value of deviations from the process average. The rms deviation is called the standard deviation, and usually is designated by the Greek letter sigma (σ); for the data plotted in Fig. 1, the standard deviation is 0.03 in. It is inherent in all normal (Gaussian) distributions that 99.73% of the individual values fall within plus or minus three standard deviations ($\pm 3 \sigma$) of the process average. Hence, the length of no more than one of the 500 pieces would be expected to fall outside the 1.91-to-2.09-in. spread — and, in fact, all 500 were within ± 3 σ of the average.

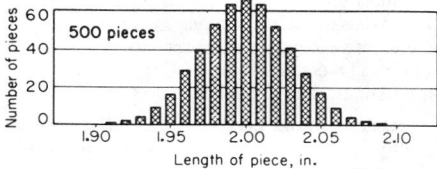

Frequency distribution is for length measurements of 500 pieces of bar stock that were cut to 2-in. lengths in an automatic machine.

Fig. 1. Histogram showing a normal (Gaussian) distribution

Control charts of the type developed by W. A Shewhart are prepared by withdrawing samples from a population, measuring the same characteristic of every unit in each sample, and plotting the average and range of that characteristic in each sample against a sample number or other means of identifying the sample.

The Shewhart control chart relies on two fundamental principles — the central-limit theorem and the relation between chart sensitivity and sample size. In simple terms, the central-limit theorem shows that, regardless of the shape of the frequency distribution of individual measurements, the distribution of averages for samples that each contain n units will tend toward a normal distribution as n increases. For most industrial processes, the central-limit theorem holds for all sample sizes of five or more units, which allows statistical generalization to be applied to a wide variety of industrial processes. (Sample sizes of five or ten units are usually the most convenient to use.)

The second fundamental feature of the Shewhart control chart is that the sensitivity of the chart to small fluctuations in the production process increases as the sample size (n) increases. Specif-

Table 2. Factors for X-bar, R and sigma control charts — trial control limits(a)

Number of observations in subgroup	Factors for $\bar{X}$ chart — From $\bar{R}$ (A_2)	Factors for $\bar{X}$ chart — From $\bar{\sigma}$ (A_1)	Factors for R chart — Lower (D_3)	Factors for R chart — Upper (D_4)	Factors for σ chart — Lower (B_3)	Factors for σ chart — Upper (B_4)
2	1.880	3.760	0	3.267	0	3.267
3	1.023	2.394	0	2.575	0	2.568
4	0.729	1.880	0	2.282	0	2.266
5	*0.577*	1.596	*0*	*2.115*	0	2.089
6	0.483	1.410	0	2.004	0.030	1.970
7	0.419	1.277	0.076	1.924	0.118	1.882
8	0.373	1.175	0.136	1.864	0.185	1.815
9	0.337	1.094	0.184	1.816	0.239	1.761
10	0.308	1.028	0.223	1.777	0.284	1.716
11	0.285	0.973	0.256	1.744	0.321	1.679
12	0.266	0.925	0.284	1.716	0.354	1.646
13	0.249	0.884	0.308	1.692	0.382	1.618
14	0.235	0.848	0.329	1.671	0.406	1.594
15	0.223	0.816	0.348	1.652	0.428	1.572
16	0.212	0.788	...	...	0.448	1.552
17	0.203	0.762	...	...	0.466	1.534
18	0.194	0.738	...	...	0.482	1.518
19	0.187	0.717	...	...	0.497	1.503
20	0.180	0.697	...	...	0.510	1.490
21	0.173	0.679	...	...	0.523	1.477
22	0.167	0.662	...	...	0.534	1.466
23	0.162	0.647	...	...	0.545	1.455
24	0.157	0.632	...	...	0.555	1.445
25	0.153	0.619	...	...	0.565	1.435
Over 25	(b)	(b)	...	...	(b)	(b)

Equations for computing control limits using tabulated factors

Upper control limit for $\bar{X} = \bar{\bar{X}} + A_2\bar{R}$
Lower control limit for $\bar{X} = \bar{\bar{X}} - A_2\bar{R}$
or
Upper control limit for $\bar{X} = \bar{\bar{X}} + A_1\bar{\sigma}$
Lower control limit for $\bar{X} = \bar{\bar{X}} - A_1\bar{\sigma}$

Upper control limit for $R = D_4\bar{R}$
Lower control limit for $R = D_3\bar{R}$

Upper control limit for $\sigma = B_4\bar{\sigma}$
Lower control limit for $\sigma = B_3\bar{\sigma}$

(a) The factors listed here are from "ASTM Manual on Quality Control of Materials," STP 15-C, American Society for Testing and Materials, Philadelphia, 1951. All factors in this table are based on a normal distribution. (b) Values of these constants may be determined for larger sample sizes from the formulas given in the Glossary of Symbols, Appendix 1 of Ref 1.

ically, control-chart sensitivity varies with the square root of the sample size. For instance, compared with a plot of individual measurements, a plot of sample averages is about twice as sensitive to a process change when the sample contains five units, about three times as sensitive for a sample size of ten units and about four times as sensitive for a sample size of 16. The width of the scatter band between control limits on the chart is inversely proportional to the square root of sample size.

By computing the range for each sample, variability of the population can be estimated. Then, using appropriate factors developed by Tippett (see Table 2), the average of these ranges can be converted into "control limits" for judging overall variability, and for monitoring and controlling future production. For instance, using the same example of cutting bar stock into 2-in. lengths that was used for constructing Fig. 1, if samples of five units each were withdrawn at five intervals during the process, length measurements of the samples might typically be as listed in the table that accompanies Fig. 2. Also listed in this table are the average value ($\bar{X}$) and range (R) for each sample, and the process average ($\bar{\bar{X}}$) and the average range ($\bar{R}$) for all five samples.

The control chart is nothing more than a plot of average and range for each sample on a time-based scale (see Fig. 2). The process average ($\bar{\bar{X}}$) and the average range ($\bar{R}$) are drawn as solid horizontal lines in Fig. 2(a) and (b), respectively. The upper and lower control limits for $\bar{X}$ and R that are shown in Fig. 2 are computed by applying the italicized factors (*0.577, 0* and *2.115*) in the line for five observations in Table 2 to the appropriate equations in the list of equations in Table 2; these control limits are drawn in Fig. 2

as dashed lines above and below the process average and range average. The upper and lower control limits define the ± 3 σ spread for the sample average and range — that is, the spread within which 99.73% of all sample averages or sample ranges are expected to fall, assuming that there is no change in the process average or variability.

If the level of control existing during collection of the samples used to compute the control limits is expected to remain the same, the control chart becomes a guidepost for future decisions. The probability that an individual sample average or range will be above or below the control limits is less than 13 in 10 000 (splitting the probability evenly between upper and lower "tails"). Consequently, if the sample average or range falls outside the control limits, it is a signal to immediately look for assignable causes in order to effect a change that will bring the process back under control. For instance, if the process of cutting bar stock into 2-in. lengths were continued indefinitely with five-piece samples taken and evaluated at regular intervals, the control chart might be as shown in Fig. 3. Note that the last nine samples plotted in Fig. 3(a) show a trend toward the upper control limit, and that the last six samples represent a 1.5-σ shift in the process average. This is a signal to look for an assignable cause for the deviation, such as a need to replace the cutoff tool or to readjust the length stop on the machine.

There are other types of control charts besides $\bar{X}$ charts and R charts. Factors for computing control limits for X charts and σ charts, factors for computing control limits for $\bar{X}$ charts or X charts when σ rather than R is known, and complete instructions for constructing and using all

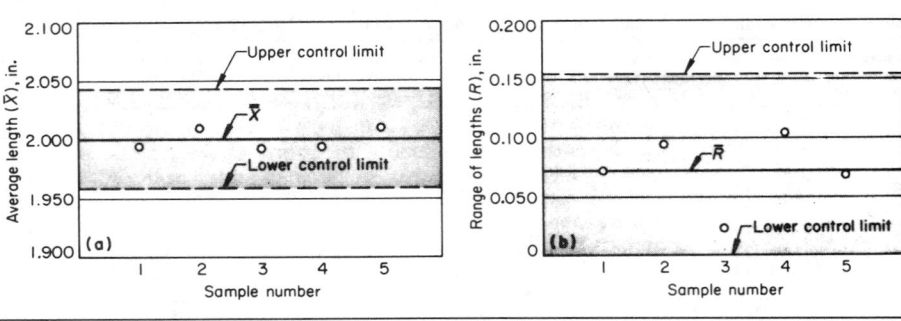

Unit in sample	Length, in.				
	Sample 1	Sample 2	Sample 3	Sample 4	Sample 5
A	2.016	2.025	2.002(a)	1.973	2.033
B	2.027(a)	1.963(b)	1.988	2.046(a)	2.003
C	1.994	2.015	1.999	1.941(b)	2.037(a)
D	1.954(b)	2.059(a)	1.996	2.001	1.968(b)
E	1.985	1.990	1.978(b)	2.009	2.013
Average	1.995	2.010	1.993	1.994	2.011
Range	0.073	0.096	0.024	0.105	0.069
Process average					2.001
Average range					0.073

(a) Largest measurement in sample. (b) Smallest measurement in sample.

Fig. 2. Control charts for (a) sample average and (b) range for the population of 2-in.-long pieces of bar stock whose frequency distribution is given in Fig. 1. See text.

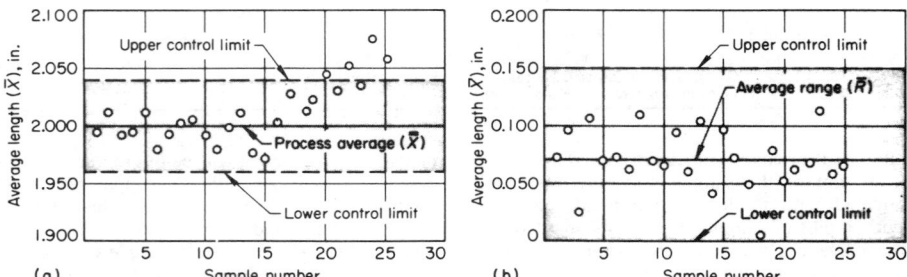

Fig. 3. Control charts for (a) sample average and (b) range for a continuation of the process of cutting bar stock into 2-in.-long pieces illustrated in Fig. 2. Samples 16 to 25 show a trend toward the upper control limit, and samples 20 to 25 represent a 1.5-σ shift in the process average with no change in process variability.

of these charts can be found in ASQC standard B2 (ANS Z1.2).

ACCEPTANCE SAMPLING

Another tool of extremely wide use and value is the practice of lot acceptance by inspecting samples. In many industries, it is extremely useful to be able to make a judgment for purposes of accepting, rejecting or reworking a production lot by inspecting only a few pieces from that lot. This approach to inspection is especially useful for high-rate production, where the cost of inspecting every piece in the lot is often prohibitive. The principles of statistical sampling were developed in the 1920's and 1930's by Dodge and Shewhart and their many colleagues. Two of the most useful tables for performing statistical sampling are the Dodge-Romig tables and MIL-STD-105, which were developed before the end of World War II.

Operating-Characteristic Curves. The underlying principles of sampling must be appreciated to understand the two types of tables mentioned above. Sampling plans depend on two major factors for

their accuracy. One is randomness of the sample, which determines whether a given sample is a valid group by which to judge the lot. The other is the actual number of units in the sample (not the percentage of the lot that is contained in the sample). An acceptance sampling plan is best described in graphical terms on an operating-characteristic (OC) curve. An OC curve is a plot of the actual number of nonconforming units in a lot (expressed as a percentage) against the probability that the lot will be accepted when sampled according to the plan. The shape of an OC curve is determined primarily by sample size, n, and acceptance number, c, although there is a small effect of lot size, N. (Acceptance number, c, is the largest number of nonconforming units, or nonconformances, that may be found in the sample without causing rejection of the lot.)

Figures 4, 5 and 6 present OC curves for three different families of sampling plans. The family of four plans shown in Fig. 4 is based on a constant-percentage sample size (n equals 10% of N) and a constant acceptance number (c equals zero). The OC curves for such constant-percentage families of plans are widely separated and

indicate that, characteristically, large lots are far less likely to be accepted than small lots. For instance, in the family of plans shown in Fig. 4, lots of 200 units that contain four nonconforming units each (2% nonconformance) have about a 67% probability of being accepted, whereas lots of 50 units that contain one nonconforming unit each (also 2% nonconformance) have about a 90% probability of being accepted. Even lots of 50 units that each contain six nonconforming units (12% nonconformance) have about a 50% probability of being accepted. Clearly, with constant-percentage sampling, both the high probability of acceptance for small lots that contain several units of unacceptable quality and the low probability of acceptance for large lots would be intolerable to most manufacturers.

The ideal plan would be one whose OC curve was Z-shaped, so that all lots submitted according to the plan would be accepted if they contained a predetermined allowable percentage of nonconforming units or less, and any lot that contained more than the allowable percentage of nonconforming units would be rejected. Theoretically, such a plan could be devised, but it would require inspection of the entire lot, and would assume perfect inspection. The first requisite is usually not economically possible, and is in direct opposition to the purpose of sampling. Perfect inspection is often mistakenly assumed by users of sampling plans. Because of an inherent probability of error in any inspection process, perfect inspection requires 300% to 400% inspection even when the most simple inspection task is performed at a leisurely pace. The plans that most closely approach the ideal Z shape are those whose absolute sample sizes are fairly large and whose acceptance numbers are not zero. Figures 5 and 6 contain several OC curves that closely approximate the ideal shape.

Acceptable Quality Level. On the OC curve, the quality level at which there is a 95% probability of acceptance is commonly referred to as the acceptable quality level, or AQL. The AQL is defined as either the maximum percentage of nonconforming units or the maximum number of nonconformances that, for purposes of acceptance sampling, can be considered acceptable as a process average. The numerical value of AQL that is selected for a given inspection procedure will mainly depend on quality objectives set by the customer or the manufacturer. However, in selecting the AQL, it is impractical to choose a value that cannot be achieved with the existing process variability.

The acceptance number for lot inspection represents the maximum number of nonconformances that may be found in a sample without rejecting the entire lot and setting the lot aside for sorting and rework. When AQL is defined as percentage of nonconformance, each unit that does not meet specifications for one or more attributes is counted only once. However, when AQL is specified as nonconformances per hundred units, each occurrence of a nonconforming characteristic is counted; for instance, if a single unit has three characteristics that do not meet specifications, three nonconformances are counted.

The acceptance-sampling standard ANSI/ASQC Z1.4-19, the coordinated version of MIL-STD-105D, employs a family of sampling plans whose OC curves all have approximately 95% probability of acceptance at the designated AQL (see Fig. 5). The plans of Z1.4, therefore, are known as AQL plans and are designed to accept, on the

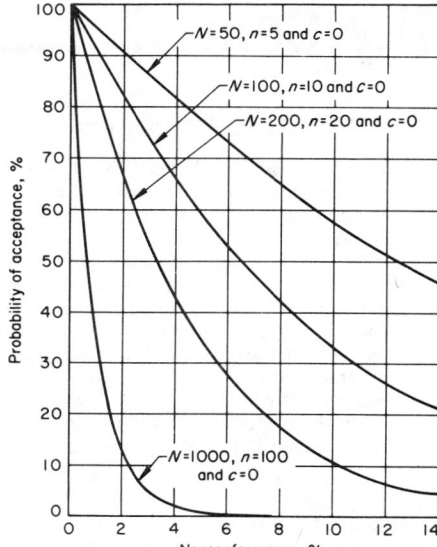

Fig. 4. Operating-characteristic curves for a family of four sampling plans based on a sample size of 10% and an acceptance number of zero (Ref 2)

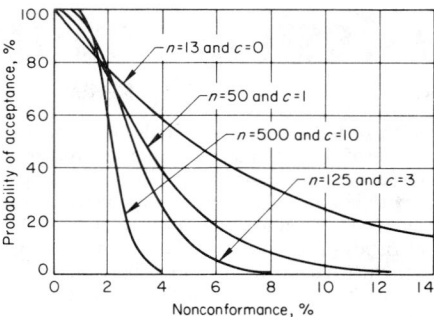

Fig. 5. Operating-characteristic curves for a family of four sampling plans from MIL-STD-105D, all with an acceptable quality level (AQL) of 1% (Ref 2)

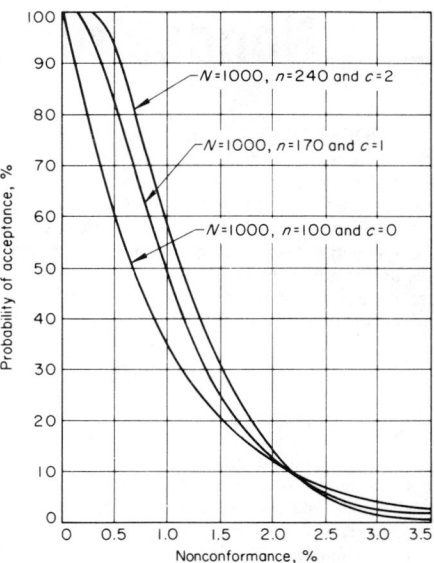

Fig. 6. Operating-characteristic curves for a family of three "lot tolerance percent defective" (LTPD) sampling plans, all with 10% probability of acceptance for a 2.2%-nonconforming lot (Ref 2)

average, most of the lots containing a percentage of nonconforming units equal to or less than the designated AQL.

Producers' Risk and Consumers' Risk. Two points on an OC curve have become associated with the so-called producers' risk and consumers' risk. The producers' risk is generally associated with a high probability of acceptance, usually 95%. The manufacturer can expect that lots whose percentage of nonconformance corresponds to a point on the curve representing 95% probability of acceptance will, on the average, be accepted 95% of the time. The other 5% represents a manufacturer's risk that lots whose actual percentage of nonconformance is equal to or less than the AQL will be rejected.

The point on an OC curve corresponding to a 10% probability of acceptance was designated by Dodge and Romig as the "lot tolerance percent defective" (LTPD), and is associated with consumers' risk. In families of LTPD plans, lots that contain a percentage of nonconforming units equal to the LTPD quality level have a probability of acceptance of about 10%. Three sampling plans are shown in the OC curve of Fig. 6 for an LTPD of 2.2% nonconformance. Note that the AQL or 95% point on these three plans differs quite markedly even though the consumer can expect that lots containing 2.2% nonconforming units will be accepted 10% of the time or less.

Besides AQL and LTPD plans, a set of sampling plans for reinspection of sorted or reworked

lots that were rejected during original inspection (called "average outgoing quality level," or AOQL, plans) and a set of sampling plans to be used for continuous-flow production (called "continuous-sampling plans," or CSP) also are available as industry standards. Families of AOQL or CSP plans can be chosen to give uniform producers' risk or consumers' risk. All the published industry-standard plans mentioned above contain detailed procedural instructions and are meant to be used on a continuing basis for successive lots. The statistics used to develop the plans are based on continual use, rather than on individual single-lot acceptance probabilities.

Product-Quality Auditing. Although a statistical sampling plan in which the consumers' risk is limited is usually chosen, the actual probability that the consumer will receive lots having a greater percentage of nonconformance than the LTPD for that plan depends on the quality of the inspection process. The usual method of evaluating the effectiveness of inspection and quality control is to perform an audit by randomly sampling and reinspecting goods that have been released for shipment. Reinspection often is done using an AOQL plan. Rejection of any lot of goods that is ready for shipment usually is taken to mean that control of the inspection process itself needs to be tightened, at least until it can be determined that there has been no change in the inspection or sampling procedures. In critical cases, it may be necessary to determine the reliability of the inspection process to be sure that no change has occurred.

Quality-system auditing is a much broader form of auditing that constitutes the major portion of the function of quality assurance. This form of auditing, which may be required under terms of a sales contract, may include product-quality auditing described above but usually concentrates on conformance to established systems and procedures. Quality-system auditing employs independent evaluation of a sample of the operating procedures and records of each department that

is responsible for any phase of the design, manufacture or quality control of the company's products. Even without contractual requirement, some companies have found it advantageous to use quality-system auditing as a means of maintaining management control over internal operations. Improved product reliability, better product design and reduced liability losses often can be achieved by rigorous periodic auditing. Standards such as ASQC C1 (ANS Z1.8), MIL-Q-9858A and NASA NHB 5300.4(1A) cover numerous aspects of the quality-assurance function, including quality management, design information, procurement, material control, manufacture, acceptance, measurement, quality information and quality-program audits.

REFERENCES

1. *Quality Control Handbook,* 3rd Ed., edited by J. M. Juran: McGraw-Hill, New York, 1974
2. *Statistical Quality Control,* 4th Ed., by E. L. Grant and R. S. Leavenworth: McGraw-Hill, New York, 1972

SELECTED REFERENCES

W. A. Shewhart, in *Statistical Method from the Viewpoint of Quality Control,* edited by E. W. Deming: The Graduate School, U.S. Department of Agriculture, Washington, 1939

Technological Applications of Statistics, by L. H. C. Tippett: John Wiley, New York, 1950

Management Considerations in
Nondestructive Testing and Quality Control

NONDESTRUCTIVE TESTING is a powerful tool for reducing costs, improving product quality, and maintaining quality levels. In some instances, the competitive position of a manufacturer may depend on intelligent use of nondestructive tests. On the other hand, ineffective use of nondestructive tests can be disastrous, particularly in industries whose products are judged by their safety features.

COSTS AND BENEFITS

A vital responsibility of management is to evaluate all costs and benefits, including incidental sales and advertising costs, exposure to product-liability or consumer-action lawsuits, and promotional and competitive advantages accrued or lost because of management decisions. Generally, casual or occasional observations alone are not enough to estimate the costs or benefits of nondestructive testing and inspection. All factors must be carefully analyzed. If there is a possibility that production costs are being wasted on improper or unnecessary nondestructive tests, consideration by management is indicated. As an example, if a costly nondestructive test (such as film radiography) is being used where a much less expensive test (such as fluoroscopy) could be employed with equal reliability, the latter should be used unless prohibited by industry standards or customer specifications.

Control of raw materials, in coordination with design, is the first stage of manufacturing where nondestructive tests can be used to reduce costs. Designers, materials engineers and nondestructive-testing engineers can study possible areas in which savings of materials could be made if the quality of the raw material were controlled adequately at a reasonable cost. Then, the designer can use less material in each component because there is less need to allow for variations in characteristics of the raw material. As an extra benefit, such reductions can often be coupled to a design that has been improved through stress analysis.

Production monitoring is the next stage of manufacturing where nondestructive tests can yield substantial benefits. Parts are inspected for processing flaws after each basic operation that imparts a critical property to the final product. In most instances, the optimum point at which to check for processing flaws is immediately following the processing step in which they could originate.

Process Control. Nondestructive-testing devices can serve as basic feedback components of process-control systems, because all such tests are based on measurements that do not damage the parts being inspected.

Equipment Maintenance. Production equipment is often an integral part of a continuous production line. Here, equipment failures can cost far more in lost production time than the repair cost of any specific machine that has failed.

Monitoring the Labor Force. Because the cost of skilled labor is a major factor in many production and assembly processes, control of quality requires continuing supervision of output. In the welding of pipelines, welds are radiographed primarily to monitor the technique used by skilled pipeline welders, rather than to predict serviceability of the welds.

ORGANIZING FOR QUALITY

A major responsibility of management is to organize the functions of a company and each of its operating units to achieve certain defined objectives. Although the discussion in this article focuses on the organization needed to achieve consistently high quality in manufactured goods, it should be recognized that every company has additional objectives. Inevitably, some of these objectives—particularly those of achieving near-capacity production rates and maximizing profits—will conflict with the objective of achieving consistently high quality. Resolving such conflicts is the responsibility of top management.

Sales provides a definition or interpretation of the quality needs of the customer.

Design creates product drawings and specifications that accurately reflect both the needs of the customer and the manufacturing capability of the producer.

Manufacturing engineering specifies processes, tools and machinery capable of producing goods that conform to product drawings and specifications.

Purchasing ensures that manufacturing supplies and outside services do not introduce unwanted or uncontrollable defects.

Production maintains control of the manufacturing cycle.

Quality control defines internal standards of acceptability, measures the actual level of quality, assesses the degree of customer satisfaction and advises management when changes are needed.

Shipping ensures that finished goods are packaged and transported to the customer without degradation of quality.

Quality-Control Applications
of Nondestructive Inspection

QUALITY CONTROL of manufactured goods is accomplished by measuring dimensions, properties or other characteristics, comparing the measurements with predetermined standards, and varying the manufacturing process as necessary to control those characteristics. Often, direct measurement of characteristics can be accomplished only by destroying the parts. Obviously, a product that has been destroyed cannot be sold. The commercial impact of this fact is twofold—costs were incurred to make the product, yet no profit can be made from its sale. However, if the same information can be obtained without destroying the part, even if only as an indirect measurement, then the part can be sold for a profit after it has been tested. The commercial incentive to test nondestructively is large when small quantities and large profit margins are involved, and is crucial with one-of-a-kind products.

Various methods have been developed for accurately and reliably measuring characteristics of parts without affecting their commercial value. Many of these are indirect methods, but they have gained acceptance as tools that can aid both management and production personnel in reducing costs and improving product quality. Also, use of nondestructive inspection has become necessary as a means of meeting certain legal and contractual requirements affecting the production and sale of a wide variety of manufactured products.

QUALITY OF INSPECTION

Successful application of nondestructive methods to the inspection of manufactured goods requires that (a) the test system and procedure be suited to both inspection objectives and types of flaws to be detected, (b) the operator have sufficient training and experience, and (c) the standard for acceptance appropriately define undesirable characteristics of a nonconforming part. If any of these prerequisites is not met, there is a potential for error in meeting quality objectives. For instance, with inappropriate equipment or with a poorly trained operator, gross errors are possible in detecting and characterizing flaws. This is of particular concern if it means chronic failure to detect flaws that seriously impair service performance. With inadequate standards, flaws having little or no bearing on product performance may be deemed serious, or significant flaws may be deemed unimportant.

It is necessary that the types of flaws that can be induced by each manufacturing operation be understood. Only then is it practical to define the nondestructive inspection that should be used. For instance, if a forging is inspected for internal forging cracks by radiography, it is important to determine the direction of grain flow (and hence the most probable direction of cracking), because any cracks that are not aligned with the radiation beam usually will not be detected. Even when the direction of grain flow is known, it may be difficult to orient the radiation beam properly, but it is usually easy and effective to inspect the part ultrasonically.

As used in nondestructive testing and quality control, the term "flaw" means a detectable lack of continuity or a detectable imperfection in a physical or dimensional attribute of a part. The fact that a part contains one or more flaws does not necessarily imply that the part is nonconforming to specification or is unfit for use. Similarly, the term "nonconforming" means only that a part is deficient in one or more specified characteristics. It should not be automatically assumed that a nonconforming part is unfit for use. In many instances, a nonconforming part is entirely capable of performing its intended function, even in its nonconforming condition. In other instances, a nonconforming part can be reworked to make it conform to specifications. Of course, sometimes a nonconforming part can neither be used nor reworked, and must be scrapped.

Human Factors. Education of all levels of personnel engaged in nondestructive inspection, including formal training and certification in accordance with government, technical-society or industry standards, is probably the greatest single factor affecting the quality of nondestructive inspection. All methods of nondestructive inspection are highly dependent on operators for obtaining and interpreting data. Inadequate education of personnel jeopardizes the reliability of inspection. This applies even to automated inspection, which is controlled by the accept-reject criteria programed into the process. Automatic data-analysis techniques must be established, proven and monitored by competent nondestructive-inspection personnel. In general, inspection should be performed by personnel who are trained to the equivalency of ASNT TC 1A, Level I. Supervisory personnel should have skill equivalent to ASNT TC 1A, Level II.

The effects of human factors on the nondestructive-inspection process also must be considered. It has been found through independent statistical studies that different people have widely differing abilities to find all the flaws in a part, even when the same nondestructive process and specific inspection procedure are used. This variability usually is more pronounced with small flaws. There is also a pronounced variation in the effects of factors such as heat, lighting, ventilation, fatigue and attitude on the performance of properly trained and qualified operators. As a result of these studies, confidence curves have been established showing the probability of detection versus flaw size for each of the major nondestructive-inspection processes.

Human factors should always be considered by the design engineer when setting maximum allowable flaw sizes, and by quality engineers when setting accept-reject criteria.

Acceptance Limits. The setting of accept-reject criteria is important to the quality of nondestructive inspection. Limits that are too strict unnecessarily increase both manufacturing and inspection costs, and often require special manufacturing techniques to meet the strict acceptability limits. Acceptance limits are usually indicated on the design drawing or specification; often, however, these limits have been selected arbitrarily. It is a function of quality engineering to review acceptance criteria, ascertain that they are appropriate and can be met in production, and then approve them. It is often necessary, after production experience and reports of service usage have been obtained, to review acceptance limits to see if changes are needed. An acceptance limit that is too strict increases cost, but one that is too lax can contribute to failure to meet service requirements.

Inspection standards should be established so that decisions to accept, rework or scrap parts are based on the probable effect that a given flaw will have on service life or product safety. Once such standards are established, nondestructive inspection can characterize flaws in terms of a real effect rather than on an arbitrary basis that may impose useless or redundant quality requirements.

In general, the principles below should be followed when choosing the point of inspection:

* Inspect raw material for flaws that may have been missed by the supplier's inspection and can interfere with manufacturing operations or will reduce performance of the finished part.
* Perform intermediate inspection following each operation or series of operations that have a significant probability of introducing serious flaws.
* Perform intermediate inspection when part shape affords easiest access to the region to be examined.
* Limit the extent of nondestructive inspection to detection of flaws having a size, type and location that will significantly affect later manufacturing operations or service performance.
* Use different inspection methods to detect different types of flaws, particularly when no single method yields an optimal balance between inspection costs and sensitivity to the various types of flaws that may be present.
* Perform final nondestructive inspection only to detect those flaws that could have been introduced after the last previous intermediate inspection or to serve as a check (audit) of intermediate inspection.

Characteristically, nondestructive tests are easiest to perform and most effective when applied to incoming stock or at intermediate points in the manufacturing process rather than at final inspection.

HARDNESS TESTING

Various methods of measuring hardness are discussed in the article on Hardness Testing in this volume. Hardness measurements are used mainly as a rapid, reliable and economical method of estimating material properties, especially tensile strength. For a few types of parts, such as ball or roller bearings and automotive engine valves, indentation hardness testing is considered a destructive test, but on most other types of parts the small impression produced during testing is not objectionable.

LIQUID-PENETRANT INSPECTION

Liquid-penetrant inspection is an aided visual inspection used primarily for detection of cracks or cracklike discontinuities that are open to the surface of a part. The process is relatively simple, can be used on both ferrous and nonferrous metals, and can be adapted for inspection of either single parts or mass-production quantities. The process of applying penetrant, removing excess penetrant and applying developer can be automated, but viewing of processed parts and interpreting indications are almost exclusively human functions.

MAGNETIC-PARTICLE INSPECTION

Magnetic-particle inspection is used on ferromagnetic materials and is most effective on alloys that have high magnetic permeability. When magnetic-particle inspection is used, care should be taken in selecting the point of inspection relative to manufacturing steps. Magnetic-particle inspection normally should be done after all processing has been completed, except that magnetic-particle inspection should be done before and after chromium plating, before electroless and vacuum plating, after welding, after each heat treating operation, before painting, and before shot peening.

EDDY-CURRENT INSPECTION

Eddy-current instruments operate by measuring impedance changes in a test coil. When an eddy-current instrument has the electrical circuit, the test-probe coil, and the part to be inspected in a balanced condition, any change in the test part (such as electrical conductivity, magnetic permeability, or dimension) will affect the impedance (combined effect of resistance and reactance) and can be monitored on a meter or a cathode-ray tube. Inspection results are mainly used to make "go, no-go" decisions.

Flaws in metal can cause a change in the flow pattern of the eddy currents, which, in turn, changes the output reading. Although under ideal conditions certain flaws can readily be detected by other methods, eddy-current inspection should be considered when (a) limited access to an installed part may prevent use of radiography, (b) large part size may prevent use of magnetic-particle inspection, (c) unfavorable orientation of the area of interest may prevent use of magnetic-rubber inspection, (d) coatings on nonferrous parts may prevent use of liquid-penetrant inspection.

RADIOGRAPHIC INSPECTION

Industrial radiography is used mainly for nondestructive examination of the interior of opaque parts and assemblies. With radiography, visual examination is extended from detecting surface flaws to detecting those below the surface. Both x-ray tubes and gamma-ray sources emit high-energy electromagnetic radiation capable of penetrating relatively thick sections. Special films, scintillating screens or radiation-sensitive meters are used to detect and record differences in the intensity of radiation emerging from the side of a test piece opposite the radiation source. Flaws within materials are usually recorded as images on a film or a screen, and the images are interpreted visually.

Many factors influence detectability of flaws. Some of these factors are: alloy content and distribution; test-piece thickness and shape; location, size, shape and orientation of flaws; type, intensity, energy and direction of radiation; type and sensitivity of the recording medium; and viewing conditions.

ULTRASONIC INSPECTION

Ultrasonic inspection can be used to detect both large and small discontinuities, located either at the surface or deep within a part. The part can be made of a ferrous or nonferrous metal or of a nonmetal. Testing can be done by manual scanning or can be fully automated, with either visual interpretation or permanent recording of results. Besides flaw detection, ultrasonics can be used to measure thickness of material between two parallel surfaces (often with greater accuracy than can be achieved by mechanical measurement) or to assess material properties (such as the measurement of graphite-flake size in cast iron, which is directly related to strength). Ultrasonic inspection can be performed on either flat or curved surfaces.

Quality Assurance

QUALITY ASSURANCE is the term most often used to describe a formal system within a company for managing quality-control activities; in effect, it is "quality control of the quality-control function." A quality-assurance system formally defines and controls all company activities that affect customer satisfaction with the quality of goods or services.

Inspection alone does not assure quality or, more precisely, does not assure conformance to requirements of the contract or customer order. This applies not only to complex products such as satellites or nuclear submarines, but also to simple products such as nails or pipe fittings. Regardless of the product or service involved, the essentials of an effective quality-assurance system include:

- Independence of the quality-assurance department from the design and production departments
- Standards of quality that reflect both the needs of the customer and characteristics of the manufacturing process
- Written procedures covering all phases of design, production, inspection, installation and service, with a program for continuous review and update of these procedures
- Control of the flow of documents such as order entry, order changes, specifications, drawings, route slips, inspection tickets and shipping papers
- Methods for maintenance of part identity, which must establish traceability throughout the process
- Methods for timely detection and segregation of nonconforming material, which also must include programs for corrective action
- Schedules for periodic calibration of inspection equipment
- Schedules for retaining important records
- Programs for training and qualification of key production and inspection personnel
- Systems for control of specifications incorporated into purchase orders, for control of the quality of purchased goods, and for appropriate inspection of purchased goods
- Systems for control of manufacturing, assembly and packaging processes, including inspection at key points in the process flow
- A system for periodic auditing of any or all of the above by persons having no direct responsibility in the area being audited.

INDEPENDENCE OF QUALITY-ASSURANCE DEPARTMENT

Responsibility for the development, operation and monitoring of an effective quality-assurance program in a plant usually rests with the quality-assurance manager. Companies having several plants may have a corporate quality-assurance department that reviews and coordinates the system for the entire organization. To be effective, this department should be an independent staff department that reports directly to an upper-level manager such as a general superintendent, vice president or president. The quality-assurance department should have the freedom to devise and recommend specific systems and procedures, and to require corrective action at their discretion.

ESTABLISHMENT OF QUALITY STANDARDS

In most plants, it is the responsibility of the quality-assurance manager to interpret national codes and standards or terms of the purchase order, and from these to devise process rules uniquely suited to the specific products and manufacturing methods used in that particular plant. The set of process rules thus devised may be known by various names; in this article, it will be called an "operating-practice description." There may be thousands of operating-practice descriptions in plant files, each varying from the others as dictated by code or customer requirements, limits on chemical composition or mechanical properties, or other special characteristics. Large plants may have computerized storage systems permitting immediate retrieval of part or all of the operating practice descriptions at key locations throughout the plant.

CONTROL OF DOCUMENT FLOW

The original purchase order, which is often less than one page in length, may generate hundreds of other working papers before the ordered material or part is shipped. All paperwork must be accurate, and must reach each work station on time. In some industries, where there may be an average of two or more specification or drawing changes per order, an effective system of material tracking that is separate and distinct from material identification is necessary.

MAINTAINING IDENTITY AND TRACEABILITY OF MATERIALS

In high-speed manufacturing operations, particularly those involving hot work, identity markings on the raw material, such as paint marks, stencils or stamps, are usually destroyed during processing. In such instances, procedures must be devised for maintaining identity—not by marking alone but also by location and count. These procedures sometimes must provide for traceability of individual units of product by a method suitable for the product and process, and must include any additional identity that the customer may require. Ultimately, both producer and customer must be confident that the goods actually delivered are described accurately in the shipping papers, test reports and certificates of compliance. This confidence is of great importance in certain applications in the aerospace and nuclear industries.

NONCONFORMING MATERIAL AND CORRECTIVE ACTION

A system for detection and segregation of nonconforming material requires (a) written inspection instructions that can be clearly understood; (b) identified, segregated holding areas for parts that have been rejected; and (c) a structured group (sometimes called a materials review board) to evaluate rejected material, make final judgment on its fitness for use, decide what is to be done with nonconforming material, and prescribe corrective action for the cause of rejection.

In many instances, rejected parts are only slightly out of tolerance and their usefulness is not impaired. Even so, all decisions of a materials review board to accept nonconforming material must be unanimous. (In the absence of unanimity, the problem may be referred to top management for a decision based on over-all business judgment.) In some companies, the authority of the materials review board is limited to merely deciding whether or not nonconforming material is fit for use. However, in many companies the board also determines what is to be done with nonconforming lots—whether they are to be shipped "as is," sorted, repaired or scrapped—and fixes the accountability for incurred losses. When corrective action is recommended by a materials review board, it is usually "system oriented"—that is, intended to prevent recurrence of the nonconformity by avoiding its cause. In instances where a lot has been rejected because the acceptance number for a sampling plan has been exceeded, decisions concerning disposition of the lot often are made on the basis of costs; the solution that results in the least total loss to both producer and customer is adopted. Sometimes, material that is slightly out of tolerance and therefore not fit for use by one customer may meet the specifications of another customer.

PERIODIC AUDITING

Periodic auditing of system performance against written standards is needed to detect corner-cutting, noncompliance and intentional violations of established quality procedures. To be as unbiased as possible, such audits should be performed by persons not having responsibility in the area being audited. In companies having multiple plants, each individual plant may conduct its own internal audits, but in addition should be subject to audit by corporate-staff personnel. The most important activities of a corporate staff, aside from auditing, are review of the quality system with the highest level of plant management, and follow-up to approve corrective action for any discrepancies found during an audit.

Periodic review of the quality-assurance system and reaffirmation of quality objectives by top management should be part of company policy. This will, in part, ensure long-range viability of the business enterprise.

INDEX

Compiled by Marjorie R. Hyslop
Consultant – Information Science and Technology

Although this Handbook is organized in four principal parts — General Information, Properties and Selection, Processing, and Testing and Inspection — the alphabetical demands of a total index made it necessary to intermix all the concepts involved in a single listing. In so doing, no details of processing, properties and inspection have been neglected, but the fact that this is a "metals" handbook has been held paramount.

In the indexing of individual metals and alloys no provision was made for a separate section on "numbered alloys" that has been a feature of many previous handbook indexes. There are, however, alphanumeric index entries — e.g., A286 superalloy, B-66 niobium alloy, Mx and Mxx silicon steels (where the x's represent numerals). Common trade names are also indexed.

The principal industrial metallic materials traditionally are irons and steels, aluminum alloys, copper alloys, and a number of others. Each has "grown up" under an inherent classification system peculiar to its own technology. Therefore, each required a different approach in indexing the "members of the family."

For a number of these families explanatory notes have been provided at the beginning of the entries for that material. Such notes will be found following the entries for Aluminum, Copper, Stainless steels, Steel, Titanium and Tool steel. Each of these families of alloys has a unique system of classification and alloy designation that dictates an individually tailored indexing approach. Some are indexed by serial number classification, some by name, some by composition. The explanatory notes are intended to help the reader find information about particular alloys or types within a manageable range of compositions or classes.

In the interests of making the index as compact yet informative as possible, some liberties have been taken with certain terminology. For example, the term "mechanical properties" has been used very sparingly as a sub-head, particularly in indexing tabular matter, because it is so broad. Instead, "tensile properties" is indexed at second-order level, but includes not only tensile strength, elongation and reduction of area (ductility) but also yield and compressive strengths — which might offend a language purist. However, other strengths and mechanical properties, such as shear strength, fatigue properties, creep properties and fracture toughness, are indexed separately at second-order level.

On the premise that an index should be more than a simple alphabetical list of keywords, an effort has been made to bring together all aspects of a topic under a single main heading, with cross references to detail under other main headings. This has sometimes resulted in a proliferation of "see" and "see also" references. The convenience to readers of finding all the information they seek on a particular topic, no matter what words they use in their search, should outweigh any inconvenience caused by cross-reference page-turning.

The letters (F) and (T) signify that information on the topic will be found in a figure or a table, respectively.

Appendix: MetalSelector

Programmed by Martin E. Heller, Physical Sciences Inc.

Technical direction provided by the Metals Handbook Desk Edition editorial staff: Howard E. Boyer, Editor; Timothy L. Gall, Project Director

IBM PC, IBM PC /XT, and IBM PC /AT are trademarks of International Business Machines, Inc. Lotus and
1-2-3 are trademarks of Lotus Development Corporation. Multiplan is a trademark of Microsoft Corporation.

THE PERSONAL COMPUTER is quickly becoming an indispensable tool of the materials engineer. This appendix explains a PC software program named MetalSelector that was developed by ASM as a supplement to this Desk Edition. It is available through the ASM Publications Department. MetalSelector provides the following functions: *metals selection, graphics, unit conversions, report writing,* and *communications.* Each function will be described in greater detail in the material that follows. Instructional documentation explaining the actual operation of the software is provided with the MetalSelector package itself and will not be covered here.

Hardware Requirements. The program runs on IBM-PC, XT, AT, or other 100% IBM compatible personal computers with 256K min. RAM and 2 double-sided disk drives. It can also be installed on a hard disk. A graphics adapter is required to run the graphics program and a modem is required for the communications program. Plans are also under way to make MetalSelector available for personal computers other than the IBM PC.

The MetalSelector Diskettes. The MetalSelector package contains two diskettes. One is labeled "System Disk," and the other is labeled "Database Disk." The System Disk contains the MetalSelector program itself; the Database Disk contains the actual data used by the MetalSelector program.

The System Disk is copy-protected. That means that if you attempt to copy the entire diskette at once (with the DISKCOPY program, for example), you can permanently damage the diskette and make the program unusable.

On the other hand, you can copy the Database Disk freely. In fact, you should make a backup copy of the Database Disk before you ever use the program, and again every time you make a change to the database. (The program itself will remind you in case you forget.)

Even though the System Disk cannot be copied, you are protected. If you somehow damage the System Disk so that the program does not work, as a registered user you can mail the diskette to the American Society for Metals, together with the cost of postage and handling, and the Society will then send you a replacement copy of the program by an express mail service.

METALS SELECTION

Select Metals. The metals selection program contains a database of 500 alloys. To search the database for alloys meeting your criteria, choose the "Select Metals" option on the Main Menu. This routine allows you to:

1. Select only the alloys you want to see
2. Sort selected alloys by one of several criteria
3. Display selected alloys on the screen, print them on a printer, or write them into a file on disk.

Throughout the Alloy Selection routine, there are menus of selections arranged in columns. The same keys are used to move the cursor around the screen as in single-column menus. There are also pages of Help information about how to use the program.

Sifting Alloys. The Sifting portion of the Alloy Selection routine permits you to narrow down the alloys currently selected from the total number of alloys in the database to only those records that you actually need.

Four different menu screens are presented in order, each containing a category of items to choose from:

1. Sifting by alloy class
2. Sifting by properties
3. Sifting by forms
4. Sifting by additional factors (such as UNS number, strength, hardness, or processing cost, for instance).

At the beginning of the Alloy Selection routine, every alloy record in the database is selected. You may indicate as many options as you want from any of the four screens to define exactly the records you want to see. A number in the upper right-hand corner of the screen indicates the number of currently selected records and changes as you make your choices. If, however, the number of currently selected alloys becomes zero, you can reselect the entire database and begin selection over again.

The first Sifting screen, for sifting by major alloy class, appears automatically when you choose the "Select Metals" option from the Main Menu:

Accept classes selected   Selected: 500

Sift by Alloy Class

CAST IRON	CARBON STEEL	ALLOY STEEL	STAINLESS STEEL
TOOL MATERIALS	ALUMINUM	COPPER	MAGNESIUM
TITANIUM	NICKEL	COBALT	ZINC
TIN	LEAD	GOLD	SILVER
PLATINUM	LOW MELTING	REACT&REFRACT	RARE EARTH
MISC.	NON-METALLIC	HELP	QUIT
REDO	NEXT		

As you move the cursor from one menu item to another, a message appears in the upper left-hand corner of the screen, explaining the current option more fully.

When you choose the NEXT option on the Alloy Class screen, the second Sifting screen, for sifting by major alloy property classes, appears:

Accept properties selected   Selected: 500

Sift by Properties

CORROSION RESISTANT	STRESS CORROSION RESISTANT
FATIGUE RESISTANT	WEAR RESISTANT
HEAT RESISTANT	HIGH STRENGTH
HIGH TOUGHNESS	MAGNETIC MATERIALS
ELECTRICAL MATERIALS	LOW EXPANSION MATERIALS
HELP	QUIT
REDO	NEXT

When you choose the NEXT option on the Alloy Properties screen, the third Sifting screen, for sifting by major alloy product forms, appears:

Accept forms selected   Selected: 500
                        Sift by Forms
CAST   WROUGHT   FORGING   EXTRUSION
SHAPES   SHEET   STRIP   PLATE
BAR   WIRE   TUBE   FILLER
P/M   HELP   QUIT   REDO
NEXT

When you choose the NEXT option on the Alloy Forms screen, the final Sifting screen, for sifting by DES number and alloy properties, appears:

Accept Sift and go to Sort   Selected: 500
                        Further Sifting
DES   TENSILE   YIELD   ELONG
HARDNESS   USER1   FORM   WELD
MACHIN   HARDEN   AVAIL   PROCCOST
HELP   QUIT   REDO   NEXT

When you choose the NEXT option on the Further Sifting screen, the program displays the Alloy Display screen containing only the currently selected alloys.

Displaying Alloys. Below is an example of the Alloy Display screen. This page is used to look through the alloys you have selected. Only the first page is displayed. The cursor starts out on the first alloy in the list.

DES	T	Ten-sile	Yield	Elong	Hard-ness	Form	Weld	Mach	Hrdn	Avail	ProcCost	User
G10080		45	33	40.0	52.0 RB	A	A	C	F	A	A	0
G10180		56	40	37.0	95.0 RB	B	A	B	F	A	B	0
G10200		57	43	36.5	111.0 HB	B	A	B	F	A	C	0
G10250		60	40	35.0	124.0 HB	C	B	B	F	B	C	0
G10300	A	67	50	31.0	126.0 HB	D	C	B	E	B	C	0
G10300	B	116	90	19.0	401.0 HB	X	D	C	E	B	B	0
G10350		70	51	30.0	143.0 HB	C	C	B	F	B	C	0
G10400	A	75	51	30.0	149.0 HB	D	C	B	E	B	C	0
G10400	B	113	86	20.0	255.0 HB	X	D	C	E	C	B	0
G10450		86	52	27.5	171.0 HB	E	C	C	F	B	C	0
G10500	A	92	53	23.5	187.0 HB	F	D	C	E	C	D	0
G10500	B	142	105	14.0	321.0 HB	X	E	D	E	D	C	0
G10550		96	56	22.0	190.0 HB	F	E	D	E	D	D	0
G10600	A	91	54	22.5	179.0 HB	X	F	C	E	C	D	0
G10600	B	160	113	13.0	321.0 HB	X	X	E	E	E	C	0
G10700		97	58	21.0	190.0 HB	X	F	C	E	C	D	0
G10800	A	89	55	24.5	174.0 HB	F	F	C	E	D	D	0
G10800	B	189	142	12.0	388.0 HB	X	F	E	E	E	C	0
G10950	A	95	55	13.0	192.0 HB	F	F	C	E	C	D	0
G10950	B	183	118	10.0	375.0 HB	X	F	E	E	D	C	0
G11100		50	35	35.0	95.0 RB	F	F	A	F	B	A	0
G11180		65	41	34.5	131.0 HB	X	X	A	E	B	A	0

F1 = Help   F2 = SORT   F3 = SIFT   ←J = SELECT   Esc = QUIT   ↑↓ Home End PgUp PgDn

If you move the cursor to an alloy that you want to examine more carefully and press the carriage return, a screen like the following appears, called the Alloy Data screen:

                        Alloy Data
DES G10080   Common 1008   User   JIS
BS 015A03   DIN 1.0312   AFNOR FD4   Class CARBON STEEL
Temper/condition ANNEALED
     Tensile strength 45   KSI   Hardness   52   RB
     Yield strength   33   KSI   User rating 0
     Elongation   40   %   Reference   4 .7
Formability A Weldability A Machina'ty C Hardena'ty F Availa'ty A Proc. cost A
Available as WROUGHT   FORGING   EXTRUSION SHEET   STRIP   BAR

Props
Comments: AISI grade 1008 is usually produced as rimmed, capped, semikilled, and fully
          killed steels. Rimmed steel of this grade has exceptionally good cold formability.
          Because of low tensile strengths, this grade is used where strength is of minor
          importance. Aluminum killed steels are very soft, mild steels of low strength
          and high ductility, which are easy to forge or to form and draw cold. It is not
          heat treatable by quenching and tempering, but can be hardened or strengthened
          by cold working and be restored to initial soft condition by annealing. Weldability
          (spot, projection, butt, and fusion) and brazeability are excellent. Used for ex-
          truded, cold headed, cold upset, and cold pressed

F1 = HELP F2 = PRINT F3 = DISK ←J = SELECT Esc = QUIT Home PgUp PgDn (comments)

On the Alloy Data screen, you can scroll the comments section up and down so that you can see long comments in total. Use the PgDn key to move forward, the PgUp key to move backward, and the Home key to move to the beginning. The reference directs you to the section and page in the Desk Edition where you can find more information on this particular alloy.

Sorting Alloys. Below is a sample of the Sorting screen:

Return to alloy display   Selected: 46
                        Sorting Orders
DES   TENSILE   YIELD   ELONG
HARDNESS   USER1   FORM   WELD
MACHIN   HARDEN   AVAIL   PROCCOST
HELP   QUIT   REDO   RETURN

This screen allows you to sort the alloys on the display screen by any of the criteria listed. For example, you can move the cursor to YIELD and the computer will sort the alloys by decreasing yield strength.

Change the Database. It is likely that you will want to stock the database with alloys that you frequently use, thereby conforming it to your own particular needs. This program allows you to change the database by adding or deleting alloy records. You can also change any of the entries in existing records.

GRAPHICS

The Graphics program in the MetalSelector will allow you to create graphs. You can plot points, join points with lines and fit curves. You can adjust the X and Y scales, add tick marks, grid lines, and symbols, label the axis, and use up to five lines for titles. You can graph the data in the metals database or accept data from spreadsheets like Lotus 1-2-3 or Multiplan. Graphs can be printed or saved in a file for further reference.

Sample abstract

771587 81-510668
Trends in Casting Technology.
Chandler, H E ; Baxter Jr, D F
Met. Prog.   , Jan. 1981,   119, (1),   96-100
Language: ENGLISH
Document Type: ARTICLE
New die-casting applications for Al in the U.S. auto industry will include: engine blocks, cylinder heads, intake manifolds and pistons; the use of cold chamber cast Mg for transmission cases is being considered. Improvements in casting technology include: use of hot isostatic pressing to cast complex Ti components for the aerospace industry and to improve performance of Al castings; investment casting of stainless steels and Ti; spin casting of iron rolls and Al. Improvements in quality will result from automation, use of the AOD (Ar   O decarburization) process, metal stream inoculation, in-mold measurement of gas pressure and ceramic foam filter for liquid metal filtration. --M.G.S.
  Descriptors: Ferrous alloys, Casting; Nonferrous alloys, Casting; Automotive components, Casting; Foundry practice
  Section Heading: 51 (FOUNDRY) Journal Announcement: 8107

Fig. 1. Metadex abstract

UNIT CONVERSIONS

This program converts between English and Metric units of measurement. Many common conversion factors are included. The program also allows you to enter conversion factors that are not provided.

REPORT WRITING

While selecting metals or creating graphs, you may wish to jot down some notes about the material with which you are working. Although not intended for heavy word processing, the report-writing program will allow you to capture your thoughts without having to move out of the MetalSelector itself. It will also allow you to print your report in a format suitable for presentation or save it for future reference.

COMMUNICATIONS

This program will help you communicate with Metadex. To utilize the program, your PC must be equipped with a modem.

Metadex. The storage of technical information in computer data banks is increasing. On-line databases currently exist that contain information on everything from airline flight schedules to restaurant reviews. In this practice, personal computers equipped with modems and linked to a central database by telephone are used to search the stored information and retrieve relevant data. The databases are accessible by anyone with the necessary hardware.

The American Society for Metals maintains Metadex, a database on metals and materials. The database covers the international metals literature from January 1966 to the present. Each month ASM editors abstract documents for the Metadex database. These documents include:

- Seminar proceedings
- Journals
- Magazines
- Books
- Technical reports
- Patents
- Dissertations
- Translations

The document information is compiled into an abstract, similar to the example in Figure 1, which is entered into the database file. The Metadex database consists of more than 550,000 abstracts, with approximately 3500 new abstracts added monthly.

Retrieving Information. Subject headings, standard indexing terms (descriptors) and key words are used to search for abstracts in the database. For example, the computer can be used to select every abstract which:

- Is categorized in subject heading 51 (Foundry) *and*
- Contains the standard indexing term "casting" in the descriptor field *and*
- Contains the phrase "improvements in quality" somewhere in the abstract text.

These three "commands," when used together, would retrieve the abstract shown in Figure 1 and display it on the computer terminal. It is easy to see the benefits from being able to go "on-line."

Unfortunately, for those who have no experience with searching on-line databases, the "log-on" procedures and the difficult to use and remember search commands can become an insurmountable barrier. Especially when one considers that once on-line the "connect charges" can be expensive. The communications program in MetalSelector is designed to take the difficulty and some of the expense out of this procedure. It is designed to help you search Metadex with your personal computer.

The program asks you questions that will help you formulate a search strategy before you even pick up the phone, thereby limiting on-line time to the execution of the search and not to its formulation. When you are ready to go on-line you simply dial the phone and, when you hear your connect signal, press the return key. Your PC automatically runs your search and reports back the results.

As more and more data are stored in databases, rather than in books and reports, engineers will increasingly be going on-line for the technical information they need. The communication package in MetalSelector is an important first step in understanding this research tool.

For additional information on Metadex and other ASM databases, please contact Metals Information, American Society for Metals, Metals Park, OH 44073.